SEMICONDUCTOR CROSS REFERENCE BOOK

Fourth Edition

From the Engineers of Howard W. Sams & Company

PROMPT®

PUBLICATIONS

A Division of
Howard W. Sams & Company
Indianapolis, IN
A Bell Atlantic Company

FOURTH EDITION, 1996

PROMPT® Publications is an imprint of Howard W. Sams & Company, a Bell Atlantic Company, 2647 Waterfront Parkway, E. Dr., Suite 100, Indianapolis, IN 46214-2041.

International Standard Book Number: 0-7906-1080-9

Editor: Candace M. Hall
Assistant Editors: Barry Buchanan, Natalie Harris, Karen Mittelstadt
Compilation: Barry Buchanan, Randy Lam, Greg Lowell
Editing: Barry Buchanan, the Howard W. Sams PHOTOFACT® Department
Typesetting: Matt Floyd, Jodi Koehl, Leah Marckel
Layout: Karen Boger, C.A. Franklin, Suzanne Lincoln, Scott Stadler, Kelli Ternet, Phil Velikan
Cover Design: Christy Pierce

Printed in the United States of America

9 8 7 6 5 4 3 2 1

TABLE OF CONTENTS

How To Use This Book

To find a suitable replacmeent for a semiconductor, follow these simple steps:

1 **Locate the part number stamped on the part** you wish to replace, or find the part number in a parts list.

2 **Look up the old part number in SEction 1** of this cross-reference. To the right of the number will be a replacement code/block number.

3 **Look up the replacement code in Section 2** (pages identified with black bars). To find right, find part numbers of suitable replacements from NTE, ECG, Radio Shack, and Thompson Consumer Electronics.

Other Important Information

The engineering staff of Howard W. Sams & Company has assembled this cross-reference guide to help you find replacements or substitutions for more than 490,000 semicondcutors. It has been compiled from manufacturer's data and from the analysis of consumer electronics devices for PHOTOFACT® service data, which has been relied upon since 1946 by service technicians worldwide. Here are some important notes to help you use the two sections of this book effectively.

Section 1—Original Device Types

This section lists device types in alphanumeric order by manufacturer's part number, type number, or other identification, including numbers from the U.S., Europe, and the Far East. At the right of the part number is a replacement code that you will use to look up replacements in section 2.

We have included part numbers for which alternates have not yet been identified from NTE, ECG, Radio Shack, or TCE. For these items, the abbreviation "OEM" appears at the right of the part number. Contact the original equipment manufacturer for appropriate replacements for these parts. For your convenienve, a list of manufacturers' addresses is provided.

Occasionally you will find the same part number listed twice. This is because different manufacturers have used the same part number for different devices. In these cases, we provide identifying notes (device type, manufacturer, or color) in parantheses, following the part number. For example, AR-31 (DIO) is the part number for a diode, and AR-31 (XSTR) is the part number for a transistor.

Section 2—Replacements

This section provides substitutions and replacements for the semiconductors listed in section 1. The data for this cross-reference was developed from the application parameters found during our PHOTOFACT® analysis. Using manufacturer's specifications and their suggested replacements, we have compiled the most complete replacement guide available.

Some of the replacement numbers are followed by the letter N and another number. These notations provide additional information about the use of the recommended replacement. The key to these notes is as follows:

N1 Two required, connect in series—anode to anode.

N2 Two required, connect in series—cathode to cathode.

N3 Two required, connect in series—anode to cathode.

N4 Caution: Replacement device threads may differ from the original.

N5 Four required, connect as in original circuit.

N6 Five required, connect as in original circuit.

N7 Six required, connect as in original circuit.

N8 Seven required, connect as in original circuit.

N9 Eight required, connect as in original circuit.

* Lead configuration may vary from original.

% Use insulating hardware supplied with replacement.

Some manufacturers make modules of discrete components (transistor, resistors, and capacitors) dipped into potting compound. The modules may be completely encapsulated or may have one or more discrete components exposed. The ICs listed as substitutes for these modules are electrically suitable replacments but may not have the same pin-basing arrangement.

Manufacturers and Importers

Do you have a question about information provided by a manufacturer or are you interested in finding service data directly from a manufacturer? Are you looking for parts for foreign consumer electronics, but don't know how to find a company that imports the materials you need? As an added resource, the following pages contain the contact information you need to fill your requirements.

A

A.C.T. ELECTRONICS★†
2345 E. ANAHEIM STREET
LONG BEACH, CA 90804
310-433-0475

AC-DELCO GENERAL MOTORS CORP.★
SERVICE PARTS OPERATIONS
NEW CENTER ONE BUILDING
3031 W. GRAND BOULEVARD
PO BOX 48232
DETROIT, MI 48202
313-974-0190

ADCOM
11 ELKINS ROAD
EAST BRUNSWICK, NJ 08816
908-390-1130
908-390-5657 FAX

AIWA AMERICA, INC.★†
800 CORPORATE DRIVE
MAHWAH, NJ 07430
201-512-3600

AKAI AMERICA LTD.
SEE MITSUBISHI
800-553-7278

ALARON, INC.★†
1026 DORIS ROAD
AUBURN HILLS, MI 48326
810-340-7500

ALPINE ELECTRONICS OF AMERICA, INC.★†
19145 GRAMERCY PLACE
ATTN: PARTS DEPARTMENT
TORRANCE, CA 90501
310-326-8000
310-533-0369 FAX

ALTEC LANSING INTERNATIONAL
CUSTOMER SERVICE/REPAIR
10500 W. RENO AVENUE
OKLAHOMA CITY, OK 72317
405-324-5311

AM-COMM SYSTEMS, INC.
2398 PINE STREET
SAN FRANCISCO, CA 94112
415-239-4188

AMERICAN ELECTRONICS, INC.★†
PO BOX 301
GREENWOOD, IN 46142
317-888-7265

AMERICAN MOTORS CORP.
1200 CHRYSLER DRIVE
DETROIT, MI 48288-1118

AMPEX CORP★†
401 BROADWAY
REDWOOD CITY, CA 94063-3199
415-367-2011

AMPRO CORPORATION
525 JOHN RODES BOULEVARD
MELBOURNE, FL 32934
800-327-3644

ANDREA ELECTRONICS CORP.★†
11-40 45TH ROAD
LONG ISLAND CITY, NY 11101
718-729-8500

AOC INTERNATIONAL (USA) LTD.★†
311 SINCLAIR FRONTAGE ROAD
MILPITAS, CA 95035
800-343-5777

APELCO CO.
676 ISLAND POND ROAD
MANCHESTER, NH 03109
603-647-7530

APPLE COMPUTER
1 INFINITE LOOP
CUPERTINO, CA 95014
408-996-1010

ARTHUR FULMER, INC.
122 GAYOSO
MEMPHIS, TN 38103
901-525-5711

ARVIN INDUSTRIES, INC.
1531 13TH STREET
COLUMBUS, IN 47201
812-379-3000

ASSOCIATED MERCHANDISING CORP.
1440 BROADWAY
NEW YORK, NY 10018
212-596-4000

ASTATIC/CTI★†
PO BOX 120
HARBOR & JACKSON STREETS
CONNEAUT, OH 44030-0120
216-593-1111

AT&T
CUSTREETOMER INFORMATION
DOCUMENTS ONLY
PO BOX 19901
INDIANAPOLIS, IN 46219
800-432-6600

ATARI CORP.
1196 BORREGAS AVENUE
SUNNYVALE, CA 94089
408-745-2000

AUDIO-TECHNICA US, INC.
ATTN: SERVICE DEPARTMENT
1221 COMMERCE DRIVE
STOW, OH 44224
216-686-2600

AUDIOVOX CORP.
150 MARCUS BOULEVARD
HAUPPAUGE, NY 11788
516-231-7750

AUTORADIO U.S.A.★†
613-19 S. 24TH STREET
PHILADELPHIA, PA 19146
215-545-2434

B

BARCUS-BERRY ELECT, INC.
5381 PRODUCTION DRIVE
HUNTINGTON BEACH, CA 92649
714-897-6766

BECKER OF NORTH AMERICA, INC.★†
16 PARK WAY
UPPER SADDLE RIVER, NJ 07458
201-327-9061
201-327-2084 FAX

BELL & HOWELL
A DIVISION OF E.I.K.I. INTERNATIONAL, INC.
800-247-3724

BLONDER-TONGUE LABS, INC.★†
ONE JAKE BROWN ROAD
OLD BRIDGE, NJ 08857
908-679-4000

BOGEN COMMUNICATIONS, INC.★†
50 SPRING STREET
PO BOX 575
RAMSEY, NJ 07446
201-934-8500

BORG-WARNER AUTOMOTIVE CORP.
700 S. 25TH AVENUE
BELLWOOD, IL 60104
708-547-2600

BRUNSWICK CORPORATION
525 W. LAKETON AVENUE
MUSKEGON, MI 49443
616-725-3300

C

C.ITOH DIGITAL ELECTRONICS, INC.★†
2701 DOW AVENUE
TUSTIN, CA 92680
714-573-2942
714-757-4488 FAX

CAB SERVICE & PARTS CORP.
1157 W. MONROE
CHICAGO, IL 60607
312-421-1122

CANON U.S.A., INC. (EAST)
100 JAMESBURG ROAD
PO BOX 1000
JAMESBURG, NJ 08831-1000
908-521-7230
800-258-0900 FAX

CANTON ELECTRONICS CORP., INC.
915 WASHINGTON AVENUE SO.
MINNEAPOLIS, MN 55415-1245
612-333-1150

CASIO, INC.
570 MT. PLEASANT AVENUE
DOVER, NJ 07801
201-361-5400

CHANNEL MASTER★†
DIVISION OF AVNET, INC.
1315 INDUSTRIAL PARK DRIVE
PO BOX 1416
SMITHFIELD, NC 27577
919-934-9711

★ **SERVICE DATA AVAILABLE ON REQUEST**

CLARION CORP. OF AMERICA★†
661 W. REDONDO BEACH BOULEVARD
GARDENA, CA 90247-4201
310-327-9100
310-327-9377 FAX

COBRA ELECTRONICS CORP.
6500 W. CORTLAND STREET
CHICAGO, IL 60635
312-889-8870

COMMAND ELECTRONICS, INC.
11811 SHAVER ROAD
SCHOOLCRAFT, MI 49087

COMPAQ COMPUTER CORP.
PO BOX 692000
HOUSTON, TX 77269-2000
713-370-0670

CORNELL-DUBILIER/
SANGAMO COMPONENT DIV.
INSIDE SALES
1605 RODNEY FRENCH BOULEVARD
NEW BEDFORD, MA 02744
508-996-8561

CRAIG CONSUMER ELECT. CORP.★
13845 ARTESIA BOULEVARD
CERRITOS, CA 90703-9000
310-926-9944
310-926-9269 FAX

CROWN INTERNATIONAL, INC.★†
1718 W. MISHAWAKA ROAD
ELKHART, IN 46517
219-294-8000

CUBIC COMMUNICATIONS, INC.
PO BOX 85587
SAN DIEGO, CA 92186-5587
619-277-6780

CURTIS MATHES CORP.
1111 COMMERCIAL DRIVE
PO BOX 112
ATHENS, TX 75751
903-675-6886

D

DAEWOO ELECTRONICS OF AMERICA★†
120 CHUBB AVENUE
LYNHURST, NJ 07071
201-935-8700
201-935-6491 FAX

DELL COMPUTER CORP.
2214 W. BRAKER
AUSTIN, TX 78748
800-624-9897
800-879-3355 SALES

DENON AMERICA, INC.★†
222 NEW ROAD
PARSIPPANY, NJ 07054-5370
201-882-7490
201-575-1213 FAX

DYNASCAN CORP.★†
COBRA ELECTRONICS CORP.
6500 W. CORTLAND STREET
CHICAGO, IL 60635
312-889-8870

E

EASTMAN KODAK COMPANY
800 LEE ROAD
ROCHESTER, NY 14650
716-724-7278

EICO ELECTRONIC INSTRUMENT CO., INC.★†
363 MERRICK ROAD
LYNBROOK, NY 11563
212-219-8172

ELECTRO BRAND, INC.★†
5410 W. ROOSEVELT ROAD
CHICAGO, IL 60650
312-261-5000

ELECTRO-VOICE, INC.★†
600 CECIL STREET
BUCHANAN, MI 49107
616-695-6831

ELECTRONIC ENGINEERS, INC.★†
2522 W. ARMITAGE AVENUE
CHICAGO, IL 60647
312-227-2600

ELECTRONIC SYSTEMS PRODUCTS
SEE AMPRO CORP.
800-327-3644

EMERSON RADIO CORP.★†
NATIONAL PARTS DISTRIBUTION
9 ENTIN ROAD
PARSIPPANY, NJ 07054-0432
201-952-0077

F

FANON COURIER, INC.★†
14811 MYFORD ROAD
TUSTIN, CA 92680-7253
714-669-9890
714-669-1081 FAX
85-2263 TELEX

FARADAY, INC.
805 S. MAUMEE STREET
TECUMSEH, MI 49286
517-423-2111

FIRESTIK ANTENNA CO.
A DIV. OF PAL INTERNATIONAL CORP.
2614 E. ADAMS
PHOENIX, AZ 85034
602-273-7151
800-528-8113

FISHER
See SFS Corp.
800-421-5013

FUJITSU TEN CORP. OF AMERICA
NATIONAL SERVICE HEADQUARTERS
19600 S. VERMONT AVENUE
TORRANCE, CA 90502
800-423-8161

FUNAI USA CORPORATION
PARTS DEPARTMENT
100 N STREET
TETERBORO, NJ 07608
800-242-7158
201-288-2666

G

GENERAL TIME CORP.★†
WESTCLOX/SETH THOMAS
520 GUTHRIDGE COURT
NORCROSS, GA 30092
404-447-5300

GOLD STAR ELECTRONICS INTERNATIONAL★†
SERVICE DIVISION
201 JAMES RECORD ROAD
HUNTSVILLE, AL 35824-0166
800-562-0244
800-442-2561 FAX

GRAN PRIX ELECTRONICS★†
108 MADISON STREET
ST. LOUIS, MO 63102
314-621-3314

H

HARMAN KARDON, INC.★†
JBL, EPI, CONCORD
80 CROSSWAYS PARK WEST
WOODBURY, NY 11797
516-496-3400

HITACHI SALES CORP. OF AMERICA★†
905 E. SOUTH STREET
ANAHEIM, CA 92805
800-448-2244
404-279-5665

I

IBM CORP.
1000 N.W. 51ST STREET
BOCA RATON, FL 33432
407-443-2000

INTERNATIONAL CRYSTAL MFG. CO., INC.
10 N. LEE
PO BOX 26330
OKLAHOMA CITY, OK 73126
405-236-3741

ITT CORPORATION★†
1330 AVENUE OF THE AMERICAS
NEW YORK, NY 10019
212-258-1000
212-752-6000

J

JCPENNEY CO., INC.★†
NATIONAL PARTS CENTER
6840 BARTON ROAD
MORROW, GA 30260
800-933-7115
404-961-8408 FAX

JENSEN
A DIVISION OF INTERNATIONAL JENSEN, INC.
SUITE 400
25 TRI-STATE INTERNATIONAL OFFICE CENTER
LINCOLNSHIRE, IL 60069
800-323-4815
708-317-3836 FAX

E.F. JOHNSON CO.★†
PO BOX 1249
299 JOHNSON AVENUE
WASECA, MN 56093
507-835-6222

JVC (MIDWEST REGION)
705 ENTERPRISE STREET
AURORA, IL 60504-8149
708-851-7855
708-851-0036 FAX

JVC (WEST COAST REGION)
5665 CORPORATE AVENUE
CYPRESS, CA 90630
714-527-7500

JVC★†
DIV. OF US JVC CORP.
107 LITTLE FALLS ROAD
FAIRFIELD, NJ 07004
201-808-2100
800-882-2345
800-446-0085 FAX

K

KENWOOD
PO BOX 22745
LONG BEACH, CA 90801-5745
800-367-7514
310-609-2127 FAX

KRACO ENTERPRISES, INC.
505 E. EUCLID AVENUE
COMPTON, CA 90224
800-421-1910

S.S. KRESGE CO.
KMART
3100 W. BIG BEAVER
TROY, MI 48084
800-635-6278

KTV, INC.★†
SERVICE DEPARTMENT
205 MOONACHIE ROAD
MOONACHIE, NJ 07074
201-440-9090

† **PARTS AVAILABLE ON REQUEST**

MANUFACTURERS AND IMPORTERS

KYOCERA ELECTRONICS, INC.
ATTN: PARTS DEPARTMENT
100 RANDOLPH ROAD
SOMERSET, NJ 08875-6727
908-560-3400
908-560-0350 FAX-PARTS

L

LLOYD'S ELECTRONICS, INC.
SEE COBRA
312-889-8870

M

MAGNADYNE CORP.
1111 W. VICTORIA STREET
COMPTON, CA 90220
310-639-2200

MARANTZ AMERICA, INC.★†
A DIV. OF BANG & OLUFEN OF AMERICA, INC.
440 MEDINAH ROAD
ROSELLE, IL 60172
708-307-3100
708-307-2687 FAX

MATSUSHITA SERVICES CO.†
PARTS DIVISION
50 MEADOWLAND PARKWAY
SECAUCUS, NJ 07094
800-447-4700

McINTOSH LABORATORY, INC.★†
2 CHAMBERS STREET
BINGHAMTON, NY 13903
607-723-3512

MGA-MITSUBISHI
800-553-7278 PARTS
800-552-8324 TECH SUPPORT

MIDLAND INTERNATIONAL CORP.★†
CUSTOMER SERVICE (PARTS DEPT.)
1690 NORTH TOPPING AVENUE
KANSAS CITY, MO 64120
800-643-5263 1-800-MIDLAND

MITSUBISHI ELECTRIC SALES AMERICA, INC.★†
PRODUCT SUPPORT GROUP
5757 PLAZA DRIVE
PO BOX 6007
CYPRESS, CA 90630-0007
800-553-7278
800-825-6655 FAX

MONTGOMERY WARD
MONTGOMERY WARD PLAZA
CHICAGO, IL 60671
800-323-1965

N

NAD (USA), INC.
89 DOUG BROWN WAY
HOLLISTON, MA 01746
508-429-2525

NATM BUYING CORP.★†
110 PENN PLAZA
NEW YORK, NY 10119
212-239-7222

NEC TECHNOLOGIES, INC.★†
ATTN: PARTS ORDER DEPT.
1255 MICHAEL DRIVE
WOOD DALE, IL 60191-1094
708-860-0335
800-366-3632 PARTS
800-356-2415 FAX

NICHOLS ELECTRONICS, INC.★†
435 SCHOOL STREET
TILTON, NH 03276-7701
603-286-4421

NORTRONICS CO. INC.†
1000 SUPERIOR
WAYZATA, MN 55391
612-476-7356
612-476-7316 FAX

0

OKIDATA
800-654-3282

ONKYO U.S.A. CORP.★†
ATTN: PARTS DEPARTMENT
200 WILLIAMS DRIVE
RAMSEY, NJ 07446
201-825-7950
201-934-1845 FAX

ORTOFON, INC.
65 EAST BETHPAGE ROAD
PLAINVILLE, NY 11803
516-454-6570
516-454-6515 FAX

P

PANASONIC
50 MEADOWLANDS PARKWAY
SECAUCUS, NJ 07094
800-545-2672

PERMA POWER ELECTRONICS, INC.★†
4001-11 N. AMERICAN STREET
PHILADELPHIA, PA 19140
800-323-4255

PHILIPS CONSUMER ELECTRONICS CO.★†
PRODUCT SERVICES PARTS ORGANIZATION
112 POLK STREET
PO BOX 967
GREENVILLE, TN 37744
800-851-8885
800-535-3715 FAX

PHILIPS ECG
907 SNAPP FERRY ROAD
PLANT #2
GREENVILLE, TN 37743
800-526-9354
800-346-6621 FAX

PILOT AUDIO/VIDEO SYSTEMS
SEE CURTIS MATHES
903-675-6886

PIONEER ELECTRONICS SERVICE, INC.★†
1925 EAST DOMINGUEZ STREET
LONG BEACH, CA 90810
800-457-2881
310-952-2247 FAX

PROSCAN
SEE THOMPSON CONSUMER ELECTRONICS
800-336-1900

PROTON†
PROTON PARTS DEPARTMENT
13855 STRUIKMAN ROAD
CERRITOS, CA 90703
310-404-2222

Q

QUASAR
50 MEADOWLANDS PARKWAY
SECAUCUS, NJ 07094
800-545-2672

R

RADIO SHACK
NATIONAL PARTS DEPARTMENT
900 E. NORTHSIDE DRIVE
FT. WORTH, TX 76102
800-442-2425
817-870-5737 FAX

RANGAIRE CORP.★
PO BOX 177
CLEBURNE, TX 76033
817-556-6500

RAY JEFFERSON★†
4200 MITCHELL STREET
PHILADELPHIA, PA 19128
215-487-2800

READERS DIGEST
READERS DIGEST ROAD
PLEASANTVILLE, NY 10570
800-234-9000

REGENCY ELECTRONICS, INC.★†
SEE RELM COMMUNICATIONS, INC.
317-545-4281

RELM COMMUNICATIONS, INC.
7707 RECORDS STREET
INDIANAPOLIS, IN 46226
317-545-4281

RF PARTS CO.★
435 S. PACIFIC STREET
SAN MARCOS, CA 92069
619-744-0700
619-744-1943 FAX

ROBERT BOSCH CORP.★†
2800 S. 25TH AVENUE
BROADVIEW, IL 60153
708-865-5200

ROTEL OF AMERICA
A DIVISION OF EQUITY INTERNATIONAL, INC.
54 CONCORD STREET
PO BOX 8
NORTH READING, MA 01864-0008
800-543-0471

ROYAL CONSUMER BUSINESS PRODUCTS★†
C/O OLIVETTI USA
765 US HWY 202
BRIDGEWATER, NJ 08807
800-243-3234
800-222-2310

S

SAAB-SCANIA OF AMERICA, INC.
ATTN: PARTS
4405A SAAB DRIVE
PO BOX 9000
NORCROSS, GA 30091
404-279-0100
800-955-9007

SAMPO CORP. OF AMERICA★†
PARTS & SERVICE
5550 PEACHTREE INDUSTRIAL BOULEVARD
NORCROSS, GA 30071
404-449-6220
404-447-1109 FAX

SAMSUNG ELECTRONICS AMERICA, INC.★†
PARTS DEPARTMENT
ONE SAMSUNG PLACE
LEDGEWOOD, NJ 07852
800-634-8276
201-691-6200
800-248-0498 FAX

SANSUI ELECTRONICS CORP.★†
17150 S. MARGAY AVENUE
PO BOX 4687
CARSON, CA 90746
310-604-7300
310-604-1664 FAX

SANYO
SEE SFS CORP. LISTING
800-421-5013

H.H. SCOTT, INC.
SEE EMERSON RADIO CORP.
201-952-0077

SFS CORP. (WESTERN REGION)★†
PARTS ORDERING DEPT.
1200 ARTESIA BOULEVARD.
COMPTON, CA 90224-9038
310-605-6741
310-605-6744 FAX

SFS CORP. (EASTERN REGION)
PARTS ORDERING DEPT.
210 RISER ROAD
LITTLE FERRY, NJ 07643
201-641-3000

★ SERVICE DATA AVAILABLE ON REQUEST

SFS CORP. (MIDWEST REGION)
PARTS ORDERING DEPT.
900 N. ARLINGTON HEIGHTS
HITASCA, IL 60143
708-775-1414

SFS CORP. (SOUTHERN REGION)
PARTS ORDERING DEPT.
1790 CORPORATE DRIVE #340
NORCROSS, GA 30093
404-925-8900
404-925-9308 FAX

SHAKESPEARE ELECTRONICS &
FIBERGLASS DIV.
PO BOX 733
NEWBERRY, SC 29108
803-276-5504

SHARP ELECTRONICS CORP.
PO BOX EE
SHARP PLAZA
MAHWAH, NJ 07430-2135
800-237-4277
201-529-9284 FAX

SHERWOOD/INKEL CORP.★†
14830 ALONDRA BOULEVARD
LA MIRANDA, CA 90638
714-521-6100

SHINTOM WEST CORP. OF AMERICA
20435 S. WESTERN AVENUE
TORRANCE, CA 90501
310-328-7200

SHURE BROTHERS, INC.★†
ATTN: SERVICE DEPARTMENT
222 HARTREY AVENUE
EVANSTON, IL 60202-3696
708-866-2553 SERVICE MANUALS
708-866-5732 PARTS
800-488-3297 FAX

SIGNET
25 ESNA PARK DRIVE
MARKHAM, ON, CANADA L3R1C9
905-474-9129

SMITH CORONA NATL. SERVICE STORES★†
ROUTE 13
PO BOX 2020
CORTLAND, NY 13045
607-753-6011
800-523-2881 FAX

SONAR RADIO CORP.★†
3000 STIRLING ROAD
HOLLYWOOD, FL 33021
305-981-8800

SONY SERVICE CO.★†
PARTS DIVISION
8281 N.W. 107TH TERRACE
KANSAS CITY, MO 64153
816-891-7550
916-891-2580 FAX

SOUNDCRAFTSMEN
MTX CORP.
4545 E. BASELINE ROAD
PHOENIX, AZ 85044
602-438-4545

SOUNDESIGN CORP. (SDI)★†
800 FEDERAL BOULEVARD
CARTERET, NJ 07008
800-888-4491
908-855-0224 FAX

SPARKOMATIC CORPORATION†
ATTN: PARTS DEPT.
PO BOX 277
MILFORD, PA 18337
800-338-5085
717-296-1230

SPIEGEL CUSTOMER SERVICE
PO BOX 6105
RAPID CITY, SD 57709-6105
605-348-8100

STANDARD COMMUNICATIONS CORP.★†
PO BOX 92151
LOS ANGELES, CA 90009-2151
310-532-5300

STUDER REVOX CANADA LTD.
TORONTO, CANADA
416-510-1294

STUDOR-EDITECH CORP.
1865 AIR LINE DRIVE
SUITE 12
NASHVILLE, TN 37210
615-391-3399
615-391-5974 FAX

SYMPHONIC
100 NORTH STREET
FUNAI USA CORP.
TETERBORO, NJ 07608
800-242-7158
201-288-2666

T

TANDY NATIONAL PARTS★†
A DIVISION OF TANDY ELECTRONICS
900 E. NORTHSIDE DRIVE
FORT WORTH, TX 76102
800-442-2425
817-870-5737 FAX

TATUNG CO. OF AMERICA, INC.★†
2850 EL PRESIDIO STREET
LONG BEACH, CA 90810
310-637-2105
213-979-7055

TCE PUBLICATIONS★
10003 BUNSEN WAY
LOUISVILLE, KY 40299
502-491-8110

TEAC CORP. OF AMERICA★†
7733 TELEGRAPH ROAD
PO BOX 750
MONTEBELLO, CA 90640
213-726-0303
800-366-8868 FAX

TECHNICS
50 MEADOWLANDS PARKWAY
SECAUCUS, NJ 07094
800-545-2672

TEKNIKA ELECTRONICS CORP.★†
DIVISION OF FUJITSU LTD.
ATTN: PARTS DEPT.
353 ROUTE 46 WEST
FAIRFIELD, NJ 07004
201-575-0380
800-835-6452

TELEDYNE ACOUSTIC RES
ATTN: PARTS DEPT.
535 GETTY CT. BLDG. A
BENECIA, CA 94510

TELEX COMMUNICATIONS, INC.★†
9600 ALDRICH AVENUE S.
MINNEAPOLIS, MN 55420
612-884-4051 TECH. CUST. SERV.

TESI
100 RISER ROAD
LITTLE FERRY, NJ 07643
201-641-3680

THOMSON CONSUMER ELECTRONICS
10003 BUNSEN WAY
LOUISVILLE, KY 40299
502-491-8110

TIMEX★†
A DIVISION OF SDI TECHNOLOGIES
800 FEDERAL BOULEVARD
CARTERET, NJ 07008-1002
800-695-4850
908-855-0224

TMK
SEE TESI
201-641-3680

TOMEN
SEE TESI
201-641-3680

TORO WHEEL HORSE CO.
8111 LYNDALE AVENUE SOUTH
BLOOMINGTON, MN 55420
612-887-8260

TOSHIBA AMERICA
CONSUMER PRODUCTS, INC.★†
NATIONAL PARTS CENTER
1420 TOSHIBA DRIVE
LEBANON, TN 37087
800-341-9785
615-444-7481 FAX

TOYOMENKA (AMERICA), INC.
SEE TESI
201-641-3680

U

UNIDEN CORP. OF AMERICA★†
8707 NORTH BY NORTHEAST BOULEVARD
INDIANAPOLIS, IN 46250
317-842-1036

UNISONIC PRODUCTS CORP.
FACTORY PARTS CTR.
16 WEST 25TH STREET
NEW YORK, NY 10010
212-255-5400

UTAH-AMERICAN CORP.
DIVISION OXFORD SPEAKER CO.
4237 W. 42ND PLACE
CHICAGO, IL 60632
312-927-3715

V

VOLKSWAGEN OF AMERICA, INC.
3800 HAMLIN ROAD
AUBURN HILLS, MI 48326
810-340-5000
800-822-8987 VOLKSWAGEN
800-822-2834 AUDI

W

WELLS-GARDNER ELECTRONICS CORP.★†
SERVICE PARTS DEPT.
2701 N. KILDARE AVENUE
CHICAGO, IL 60639
312-252-8220

WINEGARD CO.★†
3000 KIRKWOOD
BURLINGTON, IA 52601-1007
319-753-0121

Y

YAMAHA ELECTRONICS CORP. USA
ATTN: PARTS OFFICE
6660 ORANGETHORPE AVENUE
BUENA PARK, CA 90620
714-522-9888
800-634-0355 FAX

YORX ELECTRONICS CORP.★†
ATTN: PARTS DEPT.
211 2ND STREET
ST. LOUIS, MO 63102
314-621-1555

Z

ZENITH ELECTRONICS CORP.
1900 N. AUSTIN AVENUE
CHICAGO, IL 60639
312-745-2000

ZENITH DATA SYSTEMS.★
2150 E. LAKE COOK ROAD
BUFFALO GROVE, IL 60089
708-808-5000

† PARTS AVAILABLE ON REQUEST

SECTION 1

Original Device Types

and Replacment Codes

DEVICE TYPE	REPL CODE
0000000FR1	0015
0000000FR2	0023
0000000FRI	0015
0000000MV4	0133
0000000S15	0102
0000001N60	0019
0000001S85	0720
00000010D1	0023
000000DG13	0924
000000DS17	0015
000000DS18	0015
000000DS-38	0015
000000DS38	0015
000000SD1A	0102
000000SD1AB	0015
000000SD1AB(BOOST)	0102
000000SD1AB/(BOOST)	0102
000000SD1Y	0023
000000SD46	0143
000000SV31	0164
000000SZT8	0649
000001S188	0143
000001S334	0012
000001S553	0760
000001S990	0133
000002SB22	0164
0000010DC1	0015
0000015330A	0041
00000DS18	0023
00000DS-38	0071
00000DS131	0015
00000DS410	0133
00000FR202	2079
00000MZ206	0157
00000SIB01	0015
00000WZ090	0057
0000-00011-053	0849
0000-0071	0681
0000-04	0079
0000-0141	0006
0000-0150	0006
0000-0300	0006
00001S155	0133
00001S188	0143
00001S330A	0041
00001S1210	0139
00001S1555	0133
00001S1849	0015
00001S2076	0133
00001S4460	0143
00002SA10	0050
00002SA101	0136
00002SA202	0633
00002SA550	0037
00002SB22	0004
00002SB185	0004
00002SB185-0	0004
00002SB186	0004
00002SB186-0	0004
00002SB187	0004
00002SB303	0004
00002SB303-0	0004
00002SB405	0164
00002SB405R	0164
00002SB435	0925
00002SB460	0127
00002SB474	1010
00002SB481	0822
00002SB492	0847
00002SB968	0079
00002SC373	0076
00002SC460	0151
00002SC460A	0144
00002SC460B	0144
00002SC460C	0144
00002SC461	0151
00002SC492	0847
00002SC535	0127
00002SC536	0537
00002SC537	0191
00002SC606	0144
00002SC609	1411
00002SC644	0111
00002SC668	0127
00002SC735	0191
00002SC772	1144
00002SC828	1220
00002SC829	0127
00002SC838	0191
00002SC858	0111
00002SC870	0016
00002SC870A	0016
00002SC870B	0016
00002SC870C	0016
00002SC929	1066
00002SC930	2209
00002SC945	0076
00002SC968	0198
00002SC1023	0079
00002SC1026	0007
00002SC1032	0007
00002SC1061	0423
00002SD235	0042
00002SD261	0595
0000101	0136
0000102	0136
0000103	0004
0000104	0004
00001201	0579
000071090	0222
000071120	1267
000071130	1051
000071131	1051
000071150	0037
000071151	0037
000072020	0015
000072050	0015
000072090	0143
000072150	0012
000072160	0143
000072180	0244
000072190	0012
000073070	0111
000073080	0111
000073090	0016
000073100	0016
000073110	0086
000073120	0016
000073130	0016
000073140	0224
000073230	0016
000073231	0016
000073280	0560
000073290	0016
000073300	0595
000073301	0595
000073302	0595
000073303	0595
000073310	0016
000073320	0560
000073332	0016
000073333	0016
000073351	0111
000073361	0111
000073370	0079
000073380	0560
000073381	0219
000073390	0016
000073391	0016
000074010	1215
000074020	7141
000074030	1478
0000DS410R	0133
0000FR202	0015
0000LA3301	1215
0000LD3000	2404
0000RS1542	0015
0000SD-1AUF	0133
000-04	0016
0001S188	0143
0001S188AM	0143
0001S188FM	0143
0002SB187	0004
0002SB303-0	0004
0002SB405R	0164
0002SB1860	0211
0002SB3030	0004
0002SC373	0016
0002SC373W	0076
0002SC458B	0076
0002SC458C	0076
0002SC537F	0155
0002SC644Q	0111
0002SC644S,R,Q	0111
0002SC710B	0286
0002SC710C	0286
0002SC772C	0127
0002SC828	1220
0002SC828H	1220
0002SC828Q	1220
0002SC930D	0111
0002SC930E	0111
0002SC968P	2929
0002SC1023	0127
0002SC1026	0127
0002SC1026A	0127
0002SC1026B	0127
0002SC1026C	0127
0002SC1032	0127
0002SC1032A	0127
0002SC1032B	0127
0002SC1032C	0127
0002SC1061	0042
0002SC1061A	0042
0002SC1061B	0042
0002SC1061C	0042
0002SC1162	0563
0003-009700	0137
000157GN	0143
000188-4	0178
000188GN	0143
000192-3	0178
000192GN	4511
000546-1	0016
000704	0079
0001849	0015
0001849R	0015
0004203	0222
00023645	0079
000583954	0319
0002500040	3367
00031011049	0079
00031013045	0079
00031014007	0023
00031014021	0143
00031014022	0015
000HA1306U	2262
000LA1201B	0579
000LA1306U	2262
000LD1020A	0021
000WG1010	0124
00	0111
00-06-006	OEM
001	1669
001-0000-00	0143
001-00	0016
001-002	5367
001-003	5367
001-004	5367
001-007-00	0015
001-0010-00	0143
001-0020-00	0123
001-0020-0	0143
001-0020-02	1018
001-0022-00	0143
001-0036	0443
001-0072-00	0015
001-0072-00(MUNTZ)	0196
001-0077-00	0023
001-0081	0143
001-0081-00	0229
001-0082-00	0636
001-0085-00	0030
001-0091	2872
001-0095-00	0133
001-0095-02	0124
001-0099-00	0649
001-0099-01	0062
001-0099-02	0077
001-007100	OEM
001-02	0037
001-03	0037
001-04	0037
001-0101-01	1083
001-0112-00	0124
001-0125-00	0133
001-0127-00	0064
001-0130-00	0628
001-0130-01	0628
001-0140-00	OEM
001-0151-00	0133
001-0151-01	0133
001-0152-00	0244
001-0153-00	0015
001-0156-00	OEM
001-0160-00	0030
001-0161-00	0244
001-0163-02	0041
001-0163-04	0253
001-0163-10	0077
001-0163-13	0165
001-0163-15	0012
001-0176-00	0030
001-01101-0	0600
001-01202-1	0280
001-01203-1	0280
001-01204	0440
001-01204-0	0085
001-01205-0	0085
001-01205-1	0085
001-01206-0	0211
001-01250	0160
001-01251	0160
001-01252	0160
001-01253	0160
001-01254	0160
001-01255	0160
001-01256	0160
001-01257	0160
001-01258	0160
001-01259	0160
001-01501-0	0143
001-02101-0	0016
001-02101-1	0016
001-02102-0	0016
001-02103-0	0016
001-02104-0	0016
001-02105-0	0016
001-02106-0	0016
001-02107-0	0016
001-02108-0	0016
001-02109-0	0016
001-02110-0	0016
001-02111-0	0079
001-02111-1	0016
001-02113-2	0016
001-02113-3	0016
001-02113-4	0016
001-02113-5	0016
001-02114-0	0168
001-02115-0	0142
001-02116-0	0142
001-02117-2	0791
001-02119-0	0086
001-02121-0	0016
001-02201-0	0037
001-02303-0	0162
001-02303-3	0636
001-02303-4	0002
001-02405-0	0015
001-02405-1	0015
001-02405-2	0015
001-02406-0	0087
001-02406-1	0087
001-02601-0	0015
001-02603-0	0015
001-02701-1	0325
001-02702-0	0325
001-02703-0	0325
001-011010	0600
001-012010	0211
001-012011	0211
001-012020	0280
001-012021	0280
001-012030	0280
001-012031	0280
001-012040	0085
001-012050	0604
001-012051	0085
001-012052	0085
001-012053	0085
001-012060	0211
001-015010	0143
001-015011	0143
001-021010	0155
001-021011	0155
001-021020	0155
001-021030	0155
001-021040	0155
001-021050	0155
001-021060	0155
001-021070	0111
001-021080	0155
001-021090	0155
001-021100	0861
001-021110	0086
001-021111	0086
001-021130	0155
001-021131	0155
001-021132	0155
001-021133	0155
001-021134	0155
001-021135	0155
001-021136	0111
001-021140	0187
001-021150	0142
001-021160	0142
001-021161	0142
001-021163	0142
001-021172	0037
001-021180	0103
001-021190	0086
001-021200	0103
001-021210	0016
001-021218	0016
001-021230	0218
001-021231	0218
001-021232	0218
001-021270-1	0103
001-021280	0130
001-021290	0086
001-022010	0037
001-022020	0006
001-022030	0218
001-022032	1242
001-022050	0316
001-023020	0221
001-023021	1736
001-023027	0441
001-023030	0162
001-023031	0565
001-023033	0636
001-023034	0002
001-023035	0170
001-023036	0456
001-023037	0451
001-023038	0420
001-023041	0247
001-023042	0480
001-024010	0015
001-024020	0015
001-024030	0015
001-024040	1124
001-024050	0015
001-024051	0015
001-024052	0015
001-024060	0953
001-024061	0953
001-024070	3189
001-024080	0155
001-026010	0133
001-026030	0133
001-026050	0316
001-026060	0124
001-027010	2987
001-027020	2987
001-027030	0325
001-044272-002	0155
001-044273-002	0086
001-044275-002	0178
001-044277-002	0015
001-044672-001	0004
001-044673-001	0004
001-044674-001	0155
001-044676-001	0133
001-044677-001	0111
001-1	0400
001-157-00	OEM
001-163-10	OEM
001-21011	0016
001-223027	0441
001-223034	0002
001-226030	0015
002	0829
002-00840	0211
002-00900	0050
002-005100	0004
002-006300	0211
002-006500	0198
002-006600	0211
002-006800	0004
002-006900	0211
002-007000	0085
002-007100	0136
002-007200	0050
002-007300	0004
002-007400	0050
002-008100	0085
002-008200	OEM
002-008300	0111
002-008400	0164
002-008800	0160
002-009000	0050
002-009100	0142
002-009500	0016
002-009500H	OEM
002-009501	0004
002-009502	0016
002-009502-12	0198
002-009600	0144
002-009601	0144
002-009601-12	0007
002-009700	0085
002-009701	0085
002-009800	0037
002-009800-12	0126
002-009800A	0006
002-009900	0016
002-03	0016
002-010100	0160
002-010300	0037
002-010300-6	0126
002-010300A	0037
002-010400	0016
002-010500	0037
002-010500A	0037
002-010600	0086
002-010700	0126
002-010800	0016
002-010900	0037
002-010900A	0037
002-011000	0004
002-011100	OEM
002-011400	0007
002-011500	0007
002-011600	0136
002-011700	0208
002-011800	0847
002-011900	0004
002-012000	0198
002-012100	0460
002-012200	0161
002-012300	0925
002-012400	0042
002-012500	0086
002-012600	0126
002-012700	0160
002-012700-12	0975
002-012800	0037
002-012800A	0037
002-9501	0111
002-9502	0111
002-9502-12	0111
002-9601	0144
002-9601-12	0144
002-9700	0160
002-9800-12	0126
002-9800-A	0037
002-11700	0208
002-11800	0004
002-11900	0004
002-12000	0016
002-SANYO	OEM
002D235RY	0042
002SA203AA	0136
002SB185AA	0004
002SB435RY	0925
002SC203	0595
002SC203A	0595
002SC203AA	0595
002SC668D	0127
002SC735-OY	0079
002SC7350Y	0191
002SC1061	0560
002SC1209C	0076
002SC7350Y	0155
002SD235RY	0228
003	0123
003-00	0224
003-00200	0123
003-002200	0196
003-004200	0123
003-005400	0143
003-006700	0143
003-007500	0123
003-009000	0143
003-009100	0002
003-009200	0143
003-009400	0015
003-009600	0143
003-009700	0137
003-009900	0015
003-0	0901
003-01	0224
003-02000	0123
003-010000	0137
003PV-T018	OEM
004	3958
004(TOYOTA)	3958
004-00	0016
004-001	0604
004-0027-00	0015
004-00900	0133
004-002000	0015
004-002700	0015
004-002800	0015
004-002900	0196
004-003000	0015
004-003100	1303
004-003200	0774
004-003300	0015
004-003400	0015
004-003500	0102
004-003600	0015
004-003700	0474
004-003900	0015
004-004000	0071
004-004100	0015
004-009200	0143
004-03100	1303
004-03200	0774
004-03300	0015
004-03500	0015
004-03600	0015
004-03700	0015
004-1	0382
004-1A	0126
004-2	0042
004-175-9-001	0139
004-8000	0160
005-02	0016
005-010300	0190
005-010400	1501
005-010800	1501
005-010900	1501
005-011000	1501
005-011200	1501
005-012200	2982
005-013500	0409
005-013600	2804
005-1	0042
006-0000137A	0103
006-0000146	0232
006-0000147	0512
006-0000151	1426
006-0000155A	0004
006-0000162	0942
006-000138B	0130
006-0001003	0233
006-0001011	0233
006-0002034	3624
006-0002088	0233
006-0004443	0325
006-0004779	1350
006-0005545	1929
006-0006903	0439
006-02	0037
006B2M	0701
007-00	0285
007-009-00	2617
007-0051-00	0086
007-0051-01	0086
007-0072-00	0130
007-0074	0178
007-0112-00	0423
007-0112-04	0423
007-0150-00	2501
007-0189-00	0212
007-0197-00	2017
007-0214-00	0325
007-0214-01	0325
007-023200	0338
007-117-02	0212
007-214-00	0325
007-25853-01RB	4223
007-25853-02	4223
007-73203-01	0004
007-73401-02	0050
007-1671101	1696
007-1681301	1929
007-1695001	0232
007-1695101	0697
007-1695301	0361
007-1695701	1024
007-1695901	0512
007-1696001	0374
007-1696101	2243
007-1696201	0314
007-1696301	3520
007-1696701	0527
007-1696801	0619
007-1696901	1206
007-1697701	1829
007-1697801	1038
007-1698301	1921
007-1698401	1926
007-1698901	0743
007-1699201	0527
007-1699301	0467
007-1699401	0335
007-1699801	1313
007-7450301C	0103
009-00	0065
0010-0150	0718
0010-0180	1648
0010-0190	0460
0018	0079
0018-162-60002	OEM
0018-162-60203	OEM
0018-162-70005	OEM
0018-165-60005	OEM
0018-165-60304	OEM
0019-003485	0178
0019-003485A	0006
0019-1313	2623
0020-0191	0076
0020-0221	0390
0020-0250	0191
0020-0321	1066
0020-0322	1066
0020-0330	0076
0020-0332	0076
0020-0351	1513
0020-0381	1513
0020-0442	0076
0020-0521	0224
0020-0531	0127
0020-0540	1327
0020-0541	1327
0020-0630	0842
0020-0643	0276
0020-0650	0728
0020-0710	0261
0020-0711	0261
0020-0720	0219
0020-0761	0168
0020-0800	0955
0020-0801	0955
0030-007-0	OEM
0030-0041	1790
0030-0091	1284
0030-0150	0728
0030-0161	1284
0030-0370	0055
0036-001	0037
0036-181-32215	OEM
0050-0011	0019
0050-0014	0143
0050-0021	0123
0050-0023	0123
0050-0032	0123
0050-0033	0123
0050-0050	1925
0050-0070	0124
0050-0130	0133
0050-0250	0124
0050-0300	0124
0050-0301	0124
0050-0302	0124
0051-0010	0023
0051-0070	0023
0051-0100	0023
0051-0110	0023
0051-0130	0023
0051-0160	0015
0051-0170	0102
0051-0290	1097
0051-0291	1802
0051-0340	0023
0051-0400	0095
0052-0060	0720
0053-015-71100	0064
0053-015-71330	0166
0053-015-71629	0471
0053-015-71829	0165
0053-015-71919	0057
0053-017-11002	0609
0053-018-11001	0133
0053-018-11002	0124
0053-024-71001	0917
0053-024-91439	OEM
0053-024-91519	0589
0053-024-91629	0295
0053-026-01002	0031
0053-026-11001	0023
0053-026-21001	0071
0053-026-91002	OEM
0053-027-40002	OEM
0053-029-70001	OEM
0053-029-91002	1090
0053-030-51003	0031
0053-030-61001	0953
0053-031-01001	0031
0053-032-11100	OEM
0053-032-91439	0275
0053-032-91629	0295
0053-033-71120	0052
0053-033-71330	0166
0053-033-71479	0140
0053-033-71629	2522
0053-033-71919	0057
0053-901-30479	0140
0053-901-50479	0140
0053-901-50519	0041
0053-912-00520	0015
0053-915-00270	0124
0053-915-00330	OEM
0053-915-00610	0031
0053-915-00800	OEM
0053-919-00100	0057
0054-0010	0376
0054-0020	0140
0054-0050	0253

DEVICE TYPE	REPL CODE
0054-0060	0077
0054-0091	0057
0054-0110	0181
0054-0115	0181
0054-0152	0195
0054-0191	0041
0054-0240	1005
0054-0250	0124
0054-0611	0455
0061-015-80003	0037
0061-022-39001	OEM
0061-022-40001	0079
0061-022-49001	0111
0061-043-49001	0079
0061-043-49002	0037
0061-043-49003	0037
0061-043-59001	0079
0061-043-59002	0079
0061-044-00019	OEM
0061-044-00119	OEM
0061-044-00120	0076
0061-044-00123	0148
0061-044-30001	3799
0061-044-39001	OEM
0061-049-99001	0079
0061-055-49001	0261
0061-055-50002	1541
0061-056-30001	OEM
0061-058-50001	0168
0061-058-70001	0556
0061-232-70001	4601
0061-238-00001	0101
0061-247-90001	0624
0061-247-90004	OEM
0061-253-54078	OEM
0061-257-40001	2957
0061-258-60003	3955
0061-258-60004	2364
0061-261-90001	OEM
0061-262-50001	0732
0061-266-30004	OEM
0061-900-01002	0261
0061-900-04412	2691
0061-900-05200	4007
0061-900-05600	0866
0061-901-00660	0148
0061-901-01401	OEM
0061-901-02160	4125
0061-920-01330	4422
0061-920-20620	OEM
0061-920-60090	0976
0061-969-90120	OEM
0061-969-90130	1292
0061-969-90140	OEM
0062-0061	OEM
0070-0120	4652
0070-0210	2896
0070-0500	2194
0071-0030	2633
0073-0020	1329
0073-0030	2926
0073-0180	0069
0079-0060	4635
0079-0150	5193
0079-0160	3609
0079-0170	2019
0079-0220	5229
0079-0500	0912
0079-0510	0332
0079-0780	0490
0079-1313	2623
0081-0050	0120
0086-0173	OEM
0098-0520	OEM
0098-0530	OEM
0098-0550	OEM
0098-0570	OEM
0098-0580	OEM
0098-0620	OEM
0098-0640	OEM
0098-0660	OEM
0098-0670	OEM
0098-0680	OEM
0098-0690	OEM
0098-0710	2048
0099-1030	0023
0099-1040	0124
00192-6	0178
00192GN	4511
00508	OEM
00509	OEM
00764-16	0178
00764GN	4511
00765A	0178
00765GN	4511
00766-18	0178
00766GN	4511
00963	OEM
001422	0079
002112A010	OEM
002112P010	OEM
002132Q010	OEM
002132Q030	OEM
003002	0143
003007	0015
003008	0133
003015	0015
003016	0019
003017	0133

DEVICE TYPE	REPL CODE
003023	0023
003102	0012
003111	0057
003113	0162
003114	1005
003307	0124
003449	0224
003460	0111
003461	0111
003501	0579
003515	0520
003516	0416
003519	2019
003522	0433
003526	2534
003536	1816
003542	0335
004567	0015
004746	0037
004747	3538
004763	0133
004781	0854
004792	0127
004795	2617
004887	0137
0022481	0222
0023645	0127
0023828	0016
0023829	0127
0030058	OEM
0044028-014	0016
0044028-14	0016
0048000	0004
0099201-325	0057
0099202-128	0039
0099203-005	0015
0099203-007	0015
00148907	OEM
00200118	0080
002007300	OEM
008010028	0124
0021120030	OEM
0021120150	OEM
0021121050	OEM
0021121090	OEM
0021150030	OEM
0021151050	1778
0021320010	OEM
0021321070	OEM
0021321080	OEM
0021321110	OEM
0021520130	OEM
0021521100	OEM
0021721010	OEM
0031013026	0079
0031013029	0144
0031013045	0079
0031013049	0079
0031013050	0037
0031013052	0037
0031013056	0224
0031013064	1051
0031013065	0791
0031014003	0023
0031014007	0023
0031014008	0023
0031014009	0077
0031014021	0143
0031014022	0015
0031014024	0023
0031014029	0124
0031014030	0023
0031015012	0167
0042122001	4327
0063050118	0015
0072585411A	4223
00HA1306PU	2262
00S08	1048
00S09	OEM
0-00(AMPEX)	0126
01-031-018	OEM
01-03945	0111
01-010473	0925
01-010495	0006
01-010562	0006
01-010564	0203
01-010628	0006
01-010673	0148
01-010719	0006
01-010733	0006
01-010844	0693
01-020562	0532
01-020566	0681
01-030372	0076
01-030373	0111
01-030380	0224
01-030388	0330
01-030394	0155
01-030454	0076
01-030458	0076
01-030460	0151
01-030495	1591
01-030509	0155
01-030536	0537
01-030643	0313
01-030668	0127
01-030682	0047
01-030710	0368
01-030711	0380

DEVICE TYPE	REPL CODE
01-030719	0006
01-030732	0111
01-030733	0006
01-030734	0076
01-030735	0191
01-030763	0127
01-030784	1144
01-030785	1144
01-030828	1220
01-030829	0151
01-030900	1221
01-030930	2209
01-030945	0076
01-030983	0066
01-031018	2050
01-031047	0113
01-031096	0390
01-031166	0733
01-031173	0042
01-031175	0191
01-031213	0191
01-031239	1908
01-031293	1066
01-031306	0839
01-031317	0155
01-031318	0155
01-031327	1221
01-031359	0151
01-031360	1663
01-031364	0076
01-031446	0643
01-031507	0955
01-031514	1085
01-031520	0168
01-031550	0276
01-031664	0629
01-031674	1066
01-031675	0076
01-031678	0936
01-031685	0286
01-031686	0224
01-031687	0224
01-031688	0016
01-031756	0643
01-031760	1946
01-031815	0076
01-031846	0786
01-031855	0224
01-031906	0127
01-031909	0936
01-031921	0261
01-031923	0224
01-031957	2710
01-031964	2495
01-031974	0836
01-032028	1591
01-032029	0936
01-032057	0224
01-032076	0079
01-032078	0936
01-032092	0936
01-032314	1591
01-040201	0867
01-040234	0042
01-040243	0228
01-040299	0223
01-040313	0423
01-040389	0042
01-040471	0018
01-040476	0423
01-040724	0392
01-070019	3072
01-070030	0325
01-080019	3072
01-080045	0414
01-080059	0212
01-0318145	0111
01-201-0	0079
01-9011-5/2221-3	0079
01-9013-7/2221-3	0079
01-9014-2/2221-3	0079
01-9016-4/2221-3	0079
01-9018-6/2221-3	0079
01-30828	0224
01-30829	0079
01-57291	0042
01-117005	0144
01-117006	0144
01-119185-01	0379
01-119185-02	0379
01-119185-03	0379
01-121365	0167
01-349418	0086
01-349423	0016
01-349426	0224
01-349634	0079
01-349681	0037
01-472814	0155
01-571591	0037
01-571751	0037
01-571794	0007
01-571804	0330
01-571811	0079
01-571821	0079
01-571831	0233
01-571921	0142
01-571941	0016
01-572088	0285
01-572588	0037

DEVICE TYPE	REPL CODE
01-572631	0313
01-572774	0925
01-572784	0042
01-572791	0042
01-572811	0006
01-572814	0155
01-572831	0233
01-572861	0042
01-680415	0715
01-680815	0079
01-690261	0595
01-690471	0018
01-690564	0532
01-690643	1242
01-690733	0037
01-690945	0076
01-691187	0224
01-691674	0144
01-700542	0037
01K-3.6E	0188
01K-3.8E	0456
01K-4.6E	0451
01K-5.0E	0162
01K-5.2E	0162
01K-5.4E	0157
01K5.4E	0162
01K-5.8E	0157
01K-6.5E	0636
01K6.5E	0636
01M1371008B	OEM
01S60	OEM
01S67	OEM
01ST-2SC945P	0076
01ST-MPS9700D	0079
02-004558	0360
02-07150-01	OEM
02-010103	1540
02-010241	4038
02-010245	3269
02-010331	2602
02-010612	4042
02-091128	0167
02-091366	1128
02-2SC458LGC	0076
02-1001-1/2221-3	0143
02-1006-2/2221-3	0133
02-1006-22221-3	0133
02-3002-2/2221-3	0015
02-3002-22221-3	0015
02-5004	2018
02-5004-6	2017
02-5184-11	0147
02-33379-6	0086
02-33379-7	0400
02-103898	2728
02-121201	0579
02-121365	0167
02-124400	3318
02-124422	1050
02-151204	0872
02-161204	0872
02-161521	5340
02-165143	0167
02-173710	1816
02-173756	4648
02-217660	0785
02-235104	2616
02-252222	4652
02-257130	2019
02-257205	1050
02-257222	4652
02-257310	1540
02-263001	1540
02-269106	2828
02-280007	4645
02-300023	0528
02-300574	1329
02-300577	2261
02-301181	0754
02-301182	0756
02-301380	1055
02-310023	0528
02-311380	1055
02-341125	0167
02-341128	0167
02-341358	0167
02-343065	0167
02-360002	1714
02-360003	3559
02-373001	1540
02-392816	2328
02-403065	0167
02-435624	4639
02-437205	1050
02-437310	1540
02-455804	0360
02-507120	4850
02-507130	OEM
02-537666	0167
02-540800	4147
02-561171	1337
02-561201	0802
02-561251	0848
02-561410	0352
02-561492	OEM
02-561551	OEM
02-561651	OEM
02-781050	1292
02-781060	2301

DEVICE TYPE	REPL CODE
02-ST-2SD1178C	0563
02A20A	OEM
02BZ2R7	0760
02BZ2.2	1276
02BZ2.7	0760
02BZ3.3	0292
02BZ3.9	0456
02BZ4.7	0451
02CZ2.0	OEM
02CZ2.2	OEM
02CZ2.4	OEM
02CZ2.7	OEM
02CZ3.0	OEM
02CZ3.3	OEM
02CZ3.6	OEM
02CZ3.9	OEM
02CZ4.3	OEM
02CZ4.7	OEM
02CZ5.1	OEM
02CZ5.6	OEM
02CZ6.2	OEM
02CZ6.8	OEM
02CZ7.5	OEM
02CZ8.2	OEM
02CZ9.1	OEM
02CZ10	OEM
02CZ11	OEM
02CZ12	OEM
02CZ13	OEM
02CZ15	OEM
02CZ16	OEM
02CZ18	OEM
02CZ20	OEM
02CZ22	OEM
02CZ24	OEM
02CZ27	OEM
02CZ30	OEM
02CZ33	OEM
02CZ36	OEM
02CZ39	OEM
02DZ3.3	OEM
02P1B	0037
02RECT-UG-1004	0106
02SI-TDB2033	OEM
02ST-2SC1815	0076
02ST-2SC1815Y	0076
02ST-2SC1875	2660
02ST-2SD1178C	0563
02Z5.6	0157
02Z5.6A	0253
02Z5.6Y	0253
02Z-6.2A	0636
02Z6.2A	0471
02Z6.2W	0471
02Z6.2W(FA-1)	0471
02Z6.2WFA-1	0471
02Z6.8A	0025
02Z7.5	0649
02Z-7.5A	0649
02Z7.5A	0077
02Z8.2	0244
02Z8.2A	0165
02Z9.1	0012
02Z-9.1A	0057
02Z9.1A	0057
02Z-10A	0170
02Z10A	0064
02Z10A-U	0170
02Z11A	0317
02Z12A	0052
02Z12GR	0052
02Z13A	0053
02Z15	0002
02Z15A	0445
02Z16A	0420
02Z18	0531
02Z18A	0495
02Z20A	0700
02Z22A	0705
02Z24	0402
02Z24A	0494
02Z62A	0636
02Z82A	0628
03-0018-0	0015
03-0020-0	0211
03-0023-0	0211
03-0247	OEM
03-0248	OEM
03-156B	0004
03-160	0143
03-460C	0144
03-461B	0144
03-535A	0144
03-1585/G	0016
03-3016	0015
03-931051	0143
03-931601	0015
03-931609	0015
03-931641	0604
03-931642	0133
03-931645	0133
03-931771	0143
03-931971	0015
03-933935	0350
03-933943	0137
03-936011	0102
03A02	0018
03A03	0016
03A04	0006

DEVICE TYPE	REPL CODE
03A05	0016
03A06	0532
03A09	0325
03A10	0212
03A10	0212
03A11	0079
03A12	0079
03K-3.5E	0188
03K-4.2E	0456
03P05M	0344
03P1M	0901
03P2M	0058
03P3M	0407
03P4M	0407
03P4MG	OEM
03P4MGC	OEM
03P5MG	OEM
03P5MGC	OEM
03P6MGC	OEM
03SI-MC1352P	0395
03SZ-RD12EB	1005
03SZ-RD12EB(AO)	1005
03SZ-RD12EB(AU)	0052
04-00072-01	0208
04-000655-1	0124
04-00156-03	0164
04-00460-03	0016
04-00461-02	0127
04-00535-02	0127
04-00535-06	0144
04-01585-06	0016
04-01585-07	0016
04-01585-08	0127
04-02090-02	0079
04-07150-01	0525
04-2	0554
04-4600-02	0224
04-11620-01	0563
04-15850-06	0016
04-38190-01	0325
04-46000-02	0224
04-67000-01	0037
04AZ2.0	OEM
04AZ2.0-Z	OEM
04AZ2.0Z	OEM
04AZ2.2	OEM
04AZ2.2-Z	OEM
04AZ2.2Z	OEM
04AZ2.4	1276
04AZ2.7	0760
04AZ2.7X	OEM
04AZ2.7Y	OEM
04AZ2.7Z	0760
04AZ3.0	0118
04AZ3.3	0299
04AZ3.6	0376
04AZ3.9	0036
04AZ4.3	0275
04AZ4.3-Y	0275
04AZ4.3X	0275
04AZ4.3Y	0275
04AZ4.3Z	0275
04AZ4.7	0140
04AZ4.7-X	0589
04AZ4.7-Y	0140
04AZ4.7X	0589
04AZ4.7Y	0140
04AZ4.7Z	OEM
04AZ5.1	0041
04AZ5.1-X	0589
04AZ5.1X	0041
04AZ5.1Y	0589
04AZ5.1ZTPA7	OEM
04AZ5.6	0253
04AZ5.6-X	0883
04AZ5.6X	0041
04AZ5.6Y	0253
04AZ5.6Z	0883
04AZ6.2	0471
04AZ6.2-Y	0295
04AZ6.2X	0295
04AZ6.2Y	0471
04AZ6.2Z	0295
04AZ6.8	0062
04AZ6.8-Y	0062
04AZ6.8-Z	0062
04AZ6.8X	0062
04AZ6.8Y	0062
04AZ6.8Z	0062
04AZ7.5	0077
04AZ7.5-X	0077
04AZ7.5X	0062
04AZ7.5Y	OEM
04AZ7.5Z	0077
04AZ8.2	0165
04AZ8.2X	0077
04AZ8.2Z	0165
04AZ9.1	0057
04AZ9.1X	0057
04AZ9.1Y	0057
04AZ9.1Z	0057
04AZ9.1ZTPA7	OEM
04AZ10	0064
04AZ10-X	0248
04AZ10X	0057
04AZ10Y	OEM
04AZ10Z	OEM
04AZ11	0181
04AZ12	0052

DEVICE TYPE	REPL CODE
04AZ12-X	1005
04AZ12-Y	1005
04AZ12X	1005
04AZ12Y	1005
04AZ12Z	1005
04AZ13	0053
04AZ13X	OEM
04AZ15	0686
04AZ15-Z	0686
04AZ15X	4526
04AZ15Y	4526
04AZ15Z	0686
04AZ16	0445
04AZ16ZTPA7	OEM
04AZ18	0375
04AZ20	0700
04AZ20-Y	0671
04AZ20X	0671
04AZ20Y	0671
04AZ20Z	0671
04AZ22	0705
04AZ24	0494
04AZ27	0455
04AZ30	0195
04AZ30Y	OEM
04AZ33	0166
04AZ33-Y	OEM
04AZ33Y	OEM
04AZ36	0010
04AZ36X	OEM
04AZ39	0032
04AZ39Z	OEM
04BZ4.7	0140
04BZ5.1	0041
04BZ5.6	0253
04BZ6.2	0471
04BZ6.8	0062
04BZ7.5	0077
04BZ8.2	0165
04BZ9.1	0057
04BZ10	0064
04BZ11	0181
04BZ12	0052
04BZ13	0053
04BZ15	0686
04BZ16	0445
04BZ18	0375
04BZ20	0700
04BZ22	0705
04BZ24	0494
04DZ2.0	OEM
04DZ2.2	OEM
04DZ2.4	1276
04DZ2.7	0760
04DZ3.0	0118
04DZ3.3	0299
04DZ3.6	0376
04DZ3.9	0036
04DZ4.3	0275
04DZ4.7	0140
04DZ5.1	0041
04HFL40S05	OEM
04HFL80S05	OEM
04S05	2050
04S06	0936
04S4	6413
04S30	0037
04S48	0079
04S51	0715
04ST-2SC945Q(AU)E	0076
04SZ-UZ-6.2B	0471
04Z2.0	OEM
04Z2.2	OEM
04Z2.4	OEM
04Z2.7	OEM
04Z3.0	OEM
04Z3.3	OEM
04Z3.6	OEM
04Z3.9	OEM
04Z4.3	OEM
04Z4.7	OEM
04Z5.1	OEM
04Z5.6	OEM
04Z6.2	OEM
04Z6.8	OEM
04Z7.5	OEM
04Z8.2	OEM
04Z9.1	OEM
04Z10	OEM
04Z11	OEM
04Z12	OEM
04Z13	OEM
04Z15	OEM
04Z16	OEM
04Z18	OEM
04Z20	OEM
04Z22	OEM
04Z24	OEM
05-00000-00	0143
05-00060-00	0143
05-00060-01	0019
05-00085-00	0030
05-000104	0023
05-0085-00	OEM
05-00160-01	0143
05-00608-01	OEM
05-001100	0015
05-002139	0005
05-02160-01	0133

If replacement code is OEM, contact original manufacturer for replacement.

Original Device Types

DEVICE TYPE	REPL CODE
05-02638-01	0030
05-03016-01	0015
05-04001-02	0023
05-04800-02	0244
05-040004	0102
05-040006	0102
05-060110	0317
05-18155	0133
05-69669-01	0025
05-95017	0495
05-110046	0102
05-110106	0091
05-110107	0062
05-110108	0062
05-110442	0133
05-111011	0317
05-112404	0023
05-112406	0102
05-123103	0023
05-141025	0015
05-150018	0023
05-150046	0120
05-170034	0123
05-170060	0019
05-174002	0015
05-174004	0795
05-180034	0143
05-180053	0133
05-180188	0143
05-180553	0760
05-180953	0124
05-181555	0133
05-181658	0030
05-181885	0023
05-182076	0124
05-182236	0005
05-182688	0030
05-190060	0143
05-190061	0023
05-200310	0628
05-200410	0030
05-320301	0133
05-328518	1925
05-330150	0133
05-330161	0133
05-331091	0057
05-370151	0288
05-429602	0015
05-472209	0030
05-480086	0322
05-480204	0456
05-480205	0162
05-480209	0628
05-480306	0301
05-490090	0019
05-490091	0019
05-490095	0143
05-540001	0023
05-540082	0165
05-540091	0012
05-540094	0012
05-540112	0023
05-610046	0143
05-740012	0102
05-750010	0015
05-780251	0030
05-800015	0182
05-860002	0023
05-860091	0012
05-900006	0023
05-931601	0015
05-931609	0023
05-931642	0133
05-931645	0133
05-931771	0143
05-931971	0015
05-932510	0143
05-933935	0350
05-933943	0137
05-933945	0402
05-933948	1649
05-933949	0299
05-933950	0420
05-935201	0023
05-936010	0628
05-936470	0133
05-950177	1649
05-990094	0012
05-2000310	0628
05A	0140
05A01	0133
05A03	0143
05A05	0133
05A06	0124
05A07	0015
05A08	0030
05AZ2.0	OEM
05AZ2.2	OEM
05AZ2.4	1276
05AZ2.7	0760
05AZ3.0	0118
05AZ3.3	0299
05AZ3.6	0376
05AZ3.9	0036
05AZ4.3	0275
05AZ4.7	0140
05AZ5.1	0041
05AZ5.6	0253
05AZ5.6Y	0883
05AZ6.2	0471
05AZ6.8	0062
05AZ7.5	0077
05AZ8.2	0165
05AZ9.1	0057
05AZ10	0064
05AZ11	0181
05AZ12	0052
05AZ13	0053
05AZ15	0686
05AZ16	0445
05AZ18	0375
05AZ20	0700
05AZ22	0705
05AZ24	0494
05AZ27	0455
05AZ30	0195
05AZ33	0166
05AZ36	0010
05AZ39	0032
05AZ43	0054
05AZ47	0068
05AZ51	0092
05AZ56	0125
05AZ62	0152
05AZ68	0173
05AZ75	0094
05AZ75-R	0094
05AZ82	0049
05AZ91	0156
05AZ100	0189
05B	0140
05B2B	0167
05B2D	0167
05B2M	0167
05B486	0007
05C	0041
05D	0041
05M1	0041
05RECT-SI-56	OEM
05RECT-SI-156	0795
05RM80	OEM
05RM100	OEM
05RM120	OEM
05RM150	OEM
05RM200	OEM
05RM250	OEM
05SM05	OEM
05SM1	1335
05ST-2SA844C	0693
05ST-2SC2482	0261
05V-50	0080
05V50	0015
05WM05	1335
05Z2.0	OEM
05Z2.0-X	OEM
05Z2.0X	OEM
05Z2.2	OEM
05Z2.4	1276
05Z2.4X	1278
05Z2.4Z	5971
05Z2.7	0760
05Z3.0	0118
05Z3.0-Y	0118
05Z3.0X	4320
05Z3.0Y	0299
05Z3.0Z	6037
05Z3.3	0299
05Z3.3XLC-5	0299
05Z3.6	0376
05Z3.6X	6042
05Z3.6Y	6042
05Z3.6Z	OEM
05Z3.9	0036
05Z3.9X	0036
05Z3.9Y	0036
05Z3.9Z	5235
05Z4-3Y	0275
05Z4.3	0275
05Z4.3-X	0275
05Z4.3-Y	0275
05Z4.3X	0275
05Z4.3Y	0275
05Z4.3Z	6125
05Z4.7	0140
05Z4.7-X	0140
05Z4.7-Z	0140
05Z4.7X	0140
05Z4.7Y	6125
05Z4.7Z	6125
05Z5-1X	0041
05Z5-1Y	0041
05Z5.1	0041
05Z5.1-X	0041
05Z5.1L	0041
05Z5.1U	0041
05Z5.1X	0041
05Z5.1Y	0041
05Z5.1Z	0041
05Z5.4Z	OEM
05Z5.6	0253
05Z5.6-X	0253
05Z5.6U	6208
05Z5.6X	0041
05Z5.6Y	0253
05Z5.6Z	0253
05Z6.2	0471
05Z6.2-Z	0471
05Z-6.2L	0471
05Z6.2L	0091
05Z6.2U	6258
05Z6.2W	0471
05Z6.2WFA-1	0471
05Z6.2X	6260
05Z6.2Y	0471
05Z6.2Z	0471
05Z6.8	0062
05Z6.8-X	0062
05Z6.8-Y	0062
05Z6.8U	OEM
05Z6.8X	0062
05Z-6.8Y	0062
05Z6.8Y	0062
05Z6.8Z	0062
05Z7.5	0077
05Z7.5-UNI	0649
05Z7.5-Y	0077
05Z7.5X	6377
05Z7.5Y	0077
05Z7.5Z	6377
05Z8.2	0165
05Z8.2-U	0244
05Z8.2X	0244
05Z8.2Y	4309
05Z8.2Z	0057
05Z9.1	0057
05Z9.1-Z	0057
05Z9.1L	0012
05Z9.1U	0012
05Z9.1X	6564
05Z9.1Z	0057
05Z-10	6813
05Z10	0006
05Z10L	0170
05Z10X	6243
05Z10Y	6243
05Z10Z	6243
05Z11	0181
05Z11-Z	0181
05Z11Y	0181
05Z11Z	0181
05Z12	0052
05Z12-X	0052
05Z12-Y	0052
05Z12-Z	0365
05Z12X	1005
05Z12Y	0137
05Z12Z	0365
05Z13	0053
05Z13X	0053
05Z13Y	0053
05Z13Z	0365
05Z15	0686
05Z15-Y	0686
05Z15X	0879
05Z15Y	0686
05Z15Z	0686
05Z16	0445
05Z16X	0445
05Z16Y	1229
05Z18	0375
05Z18L	0495
05Z18X	4848
05Z18Y	4848
05Z18Z	4848
05Z20	0700
05Z20-X	0700
05Z20-Y	0700
05Z20X	0671
05Z20Y	0700
05Z22	0705
05Z22L	0705
05Z-22X	0705
05Z22X	0705
05Z22Z	4942
05Z24	0494
05Z24I	0494
05Z24K	OEM
05Z24L	0494
05Z24X	2467
05Z24Y	0494
05Z24Z	2467
05Z27	0455
05Z30	0195
05Z30Y	2436
05Z33	0166
05Z33-Y	0166
05Z33Y	0166
05Z36	0010
05Z36X	0010
05Z36Y	0010
05Z36Z	0010
05Z39	0032
05Z43	0054
05Z47	0068
05Z51	0092
05Z56	0125
05Z62	0152
05Z68	0173
05Z72	1189
05Z72(TP)	1189
05Z75	0094
05Z82	0049
05Z91	0156
05Z100	0189
06A	0253
06B1AH	0701
06B1B	0701
06B1D	0701
06B1M	0701
06C	0091
06D	0091
06M	0091
06M6	0091
06SI-SAB3034	OEM
06SP-LT-3233G	OEM
06ST-2SC1906	0127
06X90514A75	0058
07-07159	3046
07-07459	0007
07-72	0111
07-1077-02	0004
07-1777-03	0363
07-5085-36	0030
07-5134-14	0143
07-5160-15	0123
07-5331-86	0636
07-22816-90	0124
07-28777-40	4034
07-28777-41	0167
07-28777-42	2960
07-28777-45	4067
07-28777-49	4079
07-28777-56	4746
07-28777-70	0203
07-28777-71	0006
07-28777-72	0525
07-28777-76	2223
07-28777-83	0786
07-28777-84	0582
07-28777-87	0563
07-28777-92	0133
07-28777-95	0120
07-28777-96	0917
07-28777-100	0471
07-28778-70	4744
07-28778-72	4758
07-28778-73	4752
07-28778-74	4747
07-28778-75	4751
07-28778-76	5071
07-28778-77	4759
07-28778-78	4162
07-28778-79	0563
07-28778-80	0155
07-28778-82	4139
07-28778-84	0532
07-28778-85	0325
07-28778-86	0133
07-28815-75	0359
07-28815-76	0825
07-28815-77	0825
07-28815-82	2223
07-28815-83	2223
07-28815-84	0582
07-28815-85	0582
07-28815-86	0155
07-28815-89	0582
07-28815-92	0023
07-28815-97	0253
07-28815-100	3862
07-28815-101	2259
07-28815-103	0925
07-28815-108	0825
07-28815-116	0124
07-28815-118	0199
07-28815-119	1015
07-28816-70	1580
07-28816-71	1582
07-28816-72	1585
07-28816-73	1524
07-28816-74	0681
07-28816-75	0286
07-28816-76	0019
07-28816-77	0023
07-28816-78	0036
07-28816-81	1329
07-28816-82	OEM
07-28816-83	1599
07-28816-84	1143
07-28816-85	0398
07-28816-86	0119
07-28816-87	1603
07-28816-88	1220
07-28816-89	0124
07-28816-90	0124
07-28816-91	0143
07-28816-92	0100
07-28816-93	0039
07-28816-94	1613
07B	0062
07B2B	0664
07B2Z	2755
07B3B	0664
07B3M	0664
07B3Z	0664
07B27	0664
07D	0077
07M	0062
07M6	0062
07SD-1S1555	0133
07SD-1S1887	0023
07SD-1S1887(AU)	0023
07ST-2SC383W	0076
07SZ-UZP-11B	0317
08-0040	0015
08-0821	0015
08-08112	0019
08-302152	0224
08A	0077
08A159-001	0139
08A159-006	OEM
08A159-007	0133
08A159-009	0139
08A165-001	0133
08D	0087
08P-2-12	0004
08SD-1K34A	2232
08SI-UPD4556BC	3431
08ST-2SC945Q(AU)E	0076
08ST-2SC945R(AU)E	0076
08ST-JE9015C	0037
08V-50	0080
08V-100	0609
08V-200	0795
09-004	0654
09-011	0443
09-05032	1211
09-030214	0198
09-033006	0208
09-309-070	0111
09-309-071	0042
09-30012	0050
09-30036	0126
09-30126	0004
09-30205	0630
09-30313	0004
09-30514	0325
09-30636	0123
09-30924	OEM
09-30970	0111
09-36154	0124
09-207039	0139
09-300002	0136
09-300005	0164
09-300006	0633
09-300007	0136
09-300011	0136
09-300012	0136
09-300015	0136
09-300016	0136
09-300017	0004
09-300021	0050
09-300024	0004
09-300026	0203
09-300027	0136
09-300028	0050
09-300029	0136
09-300036	0203
09-300037	0006
09-300037A	0160
09-300043	0126
09-300059	0006
09-300061	0037
09-300062	0037
09-300063	0006
09-300064	0006
09-300067	1430
09-300068	0681
09-300070	0006
09-300071	0532
09-300072	1242
09-300073	0681
09-300074	0037
09-300076	0532
09-300077	0037
09-300078	0136
09-300079	0136
09-300080	0006
09-300081	0006
09-300084	0050
09-300090	0925
09-300307	0037
09-301001	0164
09-301002	0004
09-301002-6	0004
09-301003	0004
09-301004	0004
09-301005	0004
09-301006	0004
09-301007	0004
09-301008	0164
09-301008-18	0004
09-301009	0004
09-301010	0085
09-301012	0164
09-301014	0164
09-301015	0847
09-301016	0004
09-301019	0164
09-301020	0004
09-301022	1062
09-301023	0004
09-301024	0222
09-301025	0004
09-301025-6	0004
09-301026	0004
09-301027	0164
09-301030	1010
09-301031	0847
09-301032	0004
09-301034	0222
09-301036	0004
09-301039	0007
09-301048	0004
09-301050	0532
09-301052	0160
09-301054	0004
09-301056	0004
09-301066	0164
09-301071	0975
09-301072	0004
09-301073	0975
09-301074	0004
09-301075	0222
09-301077	0925
09-301079	0975
09-301081	1911
09-302002	0286
09-302003	0155
09-302004	0151
09-302005	0144
09-302006	0127
09-302007	0191
09-302009	0127
09-302010	1144
09-302012	0368
09-302014	0127
09-302015	1400
09-302016	0842
09-302017	0144
09-302019	0716
09-302020	0076
09-302026	OEM
09-302028	0007
09-302030	0208
09-302032	0127
09-302033	0151
09-302034	0076
09-302036	1066
09-302037	2209
09-302038	1220
09-302039	0155
09-302040	0191
09-302041	0079
09-302042	OEM
09-302044	0127
09-302045	0016
09-302045-12	0016
09-302046	2046
09-302050	1411
09-302051	1908
09-302053	0537
09-302054	0191
09-302055	1593
09-302056	1411
09-302058	0286
09-302060	0475
09-302061	0549
09-302062	0155
09-302063	0475
09-302068	1400
09-302072	0475
09-302073	0127
09-302074	0788
09-302074(SHARP)	0111
09-302075	0079
09-302078	0788
09-302079	0127
09-302080	2046
09-302081	1400
09-302082	0630
09-302083	0042
09-302085	0111
09-302086	0537
09-302090	0076
09-302090(DIODE)	0133
09-302092	0191
09-302093	1221
09-302095	1144
09-302097	0552
09-302099	0233
09-302101	0286
09-302102	2050
09-302103	0127
09-302106	0151
09-302107	0111
09-302113	0560
09-302117	4208
09-302118	0076
09-302119	0086
09-302121	0219
09-302122	0620
09-302123	0219
09-302124	0076
09-302125	0076
09-302126	0390
09-302127	0111
09-302128	0047
09-302129	0127
09-302130	0861
09-302131	0016
09-302132	0042
09-302135	2050
09-302136	0786
09-302138	1144
09-302139	0111
09-302140	0155
09-302141	0144
09-302142	0127
09-302143	0127
09-302144	0127
09-302145	0076
09-302146	0142
09-302148	0216
09-302149	1154
09-302150	0698
09-302151	0127
09-302152	0127
09-302153	0016
09-302155	0018
09-302156	0142
09-302157	0003
09-302158	0642
09-302159	0074
09-302160	0142
09-302161	0791
09-302162	0127
09-302164	0042
09-302165	0076
09-302166	0622
09-302169	1908
09-302170	0042
09-302171	0086
09-302172	0016
09-302173	0155
09-302174	0007
09-302175	0191
09-302176	2437
09-302177	0642
09-302185	0142
09-302186	0168
09-302187	0065
09-302188	1027
09-302189	0016
09-302190	0475
09-302191	0475
09-302192	0936
09-302193	0836
09-302194	0111
09-302199	0007
09-302201	0007
09-302202	1978
09-302204	0016
09-302206	0286
09-302207	0155
09-302212	0076
09-302215	0224
09-302216	0224
09-302218	0178
09-302219	1173
09-302220	0127
09-302222	0155
09-302224	1066
09-302225	0076
09-302226	0218
09-302227	0155
09-302236	0042
09-302237	0955
09-302238	1150
09-302240	0007
09-302241	0224
09-302242	0224
09-302243	0066
09-302244	0191
09-302245	1150
09-302246	2660
09-302325	0636
09-303005	0178
09-303006	0208
09-303012	0208
09-303013	0208
09-303018	0228
09-303019	0218
09-303020	0423
09-303021	0042
09-303022	0560
09-303023	0208
09-303025	1419
09-303028	0142
09-303029	0168
09-303030	0004
09-303031	0423
09-303032	1607
09-303033	0042
09-303034	0392
09-303036	0018
09-303042	0224
09-303058	0198
09-304011	0050
09-304012	0037
09-304017	1392
09-304019	0127
09-304042	0224
09-304043	0224
09-304044	0016
09-304045	0016
09-304046	0285
09-304047	0037
09-304048	0079
09-304049	0037
09-304050	0037
09-304051	0037
09-304052	0338
09-304055	0561
09-304056	0065
09-304057	0003
09-304058	0016
09-304264	OEM
09-305006	0007
09-305007	0127
09-305011	0007
09-305014	0688
09-305021	3046

If replacement code is OEM, contact original manufacturer for replacement.

DEVICE TYPE	REPL CODE
09-305023	0325
09-305024	0037
09-305031	0325
09-305032	0325
09-305033	0127
09-305034	0016
09-305035	OEM
09-305036	0127
09-305040	0353
09-305041	0127
09-305048	0111
09-305049	2046
09-305050	0224
09-305051	2209
09-305052	0111
09-305058	0693
09-305062	0016
09-305063	0076
09-305068	0016
09-305069	0144
09-305072	0144
09-305074	0144
09-305077	0016
09-305091	1908
09-305092	0733
09-305093	0155
09-305094	1144
09-305095	0042
09-305124	1978
09-305126	0111
09-305131	0004
09-305132	1154
09-305133	0325
09-305134	0006
09-305135	0325
09-305136	0839
09-305137	0836
09-305138	0103
09-305139	0076
09-305140	0042
09-305148	0079
09-305149	0037
09-305150	0715
09-305151	0342
09-305152	0079
09-306002	0143
09-306008	0030
09-306009	0143
09-306010	0143
09-306012	0143
09-306014	0030
09-306015	0030
09-306018	0030
09-306019	0123
09-306020	0123
09-306021	0030
09-306023	0005
09-306024	0002
09-306024(ZENER)	0002
09-306028	6323
09-306030	0102
09-306031	0920
09-306033	0015
09-306034	0015
09-306036	0019
09-306037	0143
09-306039	0030
09-306040	0143
09-306042	0015
09-306042(DIO)	2637
09-306042(ZENER)	0244
09-306047	0143
09-306049	0143
09-306050	0023
09-306051	0143
09-306052	0062
09-306053	0087
09-306054	0015
09-306055	0636
09-306057	0123
09-306058	0143
09-306059	0015
09-306060	0139
09-306061	0143
09-306062	0133
09-306063	0015
09-306073	0301
09-306077	0917
09-306083	0087
09-306084	OEM
09-306088	0015
09-306089	0917
09-306093	0143
09-306100	0023
09-306101	0196
09-306102	0071
09-306103	0015
09-306104	0133
09-306106	0012
09-306107	0143
09-306108	0143
09-306109	0301
09-306110	0139
09-306110(ZENER)	0002
09-306111	0133
09-306112	1149
09-306113	0120
09-306114	0139
09-306115	0015
09-306119	0015
09-306124	0012
09-306125	0015
09-306126	0030
09-306127	0365
09-306129	0139
09-306134	0133
09-306135	0133
09-306138	0015
09-306141	0102
09-306144	0102
09-306145	0133
09-306148	0143
09-306149	0023
09-306151	0124
09-306154	0133
09-306157	0015
09-306158	0012
09-306159	0133
09-306160	0015
09-306161	0133
09-306162	0015
09-306163	0133
09-306165	0012
09-306167	0030
09-306168	1149
09-306169	0015
09-306170	0133
09-306171	0133
09-306172	0015
09-306173	0053
09-306176	0015
09-306177	0015
09-306178	0023
09-306179	0137
09-306180	0012
09-306181	0649
09-306183	0091
09-306191	0041
09-306192	0015
09-306193	0196
09-306194	0012
09-306195	0133
09-306196	0244
09-306197	0649
09-306198	0133
09-306199	0139
09-306200	0139
09-306201	0030
09-306202	0133
09-306204	0133
09-306205	0015
09-306206	0133
09-306208	0301
09-306209	0917
09-306210	0917
09-306211	0133
09-306211(RECT)	0133
09-306212	0077
09-306212(RECT)	0015
09-306213	0015
09-306214	0030
09-306215	0025
09-306216	0917
09-306219	0143
09-306220	0002
09-306221	0133
09-306222	0133
09-306223	0139
09-306224	0139
09-306225	0071
09-306226	0071
09-306227	0015
09-306228	0064
09-306229	0019
09-306230	0030
09-306231	0133
09-306233	0139
09-306235	0170
09-306236	0133
09-306237	0102
09-306238	0165
09-306239	0012
09-306240	0720
09-306241	0012
09-306242	0077
09-306244	0133
09-306245	0023
09-306247	0012
09-306248	0133
09-306249	0015
09-306250	0015
09-306251	0030
09-306253	0030
09-306254	0015
09-306255	0015
09-306256	0139
09-306257	0071
09-306258	0182
09-306259	0102
09-306260	0102
09-306261	0917
09-306262	0917
09-306263	0015
09-306264	0015
09-306265	0030
09-306266	0133
09-306270	0143
09-306274	0015
09-306276	0133
09-306277	0244
09-306278	0100
09-306283	0133
09-306285	0015
09-306286	0062
09-306287	0012
09-306288	0163
09-306289	0533
09-306290	0143
09-306300	0015
09-306301	0924
09-306302	0015
09-306303	0133
09-306304	0087
09-306305	0649
09-306306	0120
09-306309	0133
09-306310	0757
09-306311	0071
09-306312	0015
09-306313	0133
09-306314	0244
09-306315	0080
09-306323	0015
09-306324	0420
09-306325	0025
09-306326	0124
09-306327	0012
09-306329	0030
09-306330	0143
09-306332	0057
09-306333	0015
09-306334	0143
09-306335	0123
09-306336	0019
09-306349	0143
09-306350	0015
09-306351	0012
09-306352	0030
09-306353	0139
09-306354	0244
09-306355	1649
09-306356	0165
09-306359	0005
09-306365	0015
09-306366	0023
09-306367	0002
09-306368	0133
09-306369	0005
09-306370	0143
09-306371	5116
09-306372	0120
09-306373	0133
09-306374	0030
09-306375	0012
09-306376	0015
09-306377	0091
09-306378	0244
09-306379	0087
09-306380	0456
09-306381	0244
09-306382	0012
09-306384	0015
09-306389	0015
09-306390	0133
09-306391	0137
09-306392	0102
09-306394	0015
09-306400	OEM
09-306401	0012
09-306417	0023
09-306418	0071
09-306419	0025
09-306420	0378
09-306421	0071
09-306422	0015
09-306423	0023
09-306424	0348
09-306425	0348
09-306426	0133
09-306427	0015
09-306428	0317
09-306429	0317
09-306430	0288
09-306431	0023
09-306432	0015
09-306433	0023
09-307037	0139
09-307039	0133
09-307043	0139
09-307045	0139
09-307055	0133
09-307056	0120
09-307080	0133
09-307081	0120
09-307082	0649
09-307083	0133
09-307084	0139
09-307085	0139
09-307088	0025
09-307089	0133
09-307093	0005
09-307094	OEM
09-308002	0823
09-308003	0823
09-308004	0632
09-308005	OEM
09-308007	0823
09-308008	0579
09-308009	0528
09-308010	2617
09-308011	2576
09-308012	OEM
09-308013	0632
09-308015	OEM
09-308017	2624
09-308019	0632
09-308021	0141
09-308022	0232
09-308024	2155
09-308025	0946
09-308026	0964
09-308027	2882
09-308028	0930
09-308029	2887
09-308030	0638
09-308031	0673
09-308033	1750
09-308034	1215
09-308041	0911
09-308043	1994
09-308044	2031
09-308045	0354
09-308046	0354
09-308047	0666
09-308048	0666
09-308049	2509
09-308050	2605
09-308051	2158
09-308052	2261
09-308053	0470
09-308054	2393
09-308055	OEM
09-308056	OEM
09-308059	0911
09-308061	3263
09-308062	0579
09-308063	0579
09-308064	1215
09-308065	2406
09-308066	2630
09-308067	3670
09-308069	2882
09-308070	1393
09-308071	1478
09-308072	0127
09-308074	2112
09-308076	0433
09-308077	0021
09-308079	0352
09-308080	1050
09-308083	0653
09-308084	4420
09-308086	3366
09-308089	0878
09-308090	0395
09-308094	4634
09-308095	1018
09-308096	2316
09-308098	0167
09-308099	1337
09-308100	0167
09-308102	0653
09-308103	5746
09-308208	0636
09-309006	0224
09-309007	0144
09-309012	0016
09-309013	0144
09-309015	0144
09-309023	0016
09-309024	0144
09-309027	0144
09-309028	0144
09-309029	0400
09-309030	0791
09-309031	0595
09-309032	0144
09-309038	0037
09-309040	OEM
09-309042	0037
09-309049	0016
09-309050	0016
09-309059	0111
09-309060	0016
09-309061	0155
09-309062	0698
09-309069	0127
09-309070	0111
09-309071	0219
09-309072	0937
09-309073	0144
09-309075	0208
09-309076	0016
09-309171	0133
09-310171	0975
09-395148	OEM
09-3060111	0133
09-ST-BF459	0276
09A001-00	1030
09A02	0785
09A04	1692
09A05	1661
09A07	2257
09A08	3262
09A10	4401
09C5	0057
09D	0057
09ST-2SA755C	0925
09ST-BF459	0276
010-694	0285
010PIN-T05	OEM
010PV-T05	OEM
010UV-PV-T05	OEM
011-00	0338
011H01	0280
012-00	0561
012-1020	OEM
012-1020-005	0133
012-1021-001	0133
012-1022-002	0484
012-1023-007	0157
012-1024-001	0133
012-1025-002	0087
012E	0016
012H01	0280
013-005002-06	0723
013-005005	1876
013-005005-6	1876
013-005007	1876
013-005007-6	1876
013-005009-6	0623
013-0144-6438	0015
013-1	4337
013-3	6255
013-339	0015
014-611	0037
014-652	0037
014-698	0016
014-784	0016
014-862	0111
015-002	0015
015-006	0015
015-1	1137
016	0594
016B12	0016
016B810	0016
016B812	0016
017E824	0016
018-00001	0037
018-00002	0037
018-00003	0016
018-00004	1267
018-00005	0560
018-00006	0015
018-00007	0015
018-00008	0015
018-00009	0015
018-0008	0015
018-0009	0015
019-00009	0016
019-00010	0016
019-001918	0143
019-001980	0143
019-002935	0015
019-002964	0133
019-003315	0050
019-003317	0038
019-003318	0038
019-003319	0038
019-003324	0280
019-003342	0280
019-003343	0280
019-003349	0086
019-003411	0170
019-003415	0164
019-003416	0164
019-003420	0015
019-003485	0178
019-003691	0698
019-003692	1400
019-003777	0136
019-003778	0050
019-003870-013	0015
019-003870-020	0015
019-003929	0144
019-003932	0016
019-003934	0016
019-005006	0016
019-005010	0037
019-005043	0143
019-005045	0965
019-102311	2232
019-301980	0143
020-00011	0123
020-00012	0123
020-00023	0136
020-00024	0127
020-00025	0127
020-00026	0144
020-00027	0144
020-00028	0127
020-00030	0143
020-00031	0825
020-1110-004	0150
020-1110-004(SCOTT)	0037
020-1110-004C	0037
020-1110-005	0693
020-1110-005(SCOTT)	0037
020-1110-005C	0006
020-1110-006	0050
020-1110-006(SCOTT)	0050
020-1110-007(SCOTT)	0126
020-1110-008	0086
020-1110-008(SCOTT)	0086
020-1110-009	0126
020-1110-009(SCOTT)	0126
020-1110-010	0111
020-1110-010(SCOTT)	0016
020-1110-011	0079
020-1110-011(SCOTT)	0016
020-1110-012	0111
020-1110-012(SCOTT)	0016
020-1110-013	0111
020-1110-013(SCOTT)	0016
020-1110-014	0126
020-1110-014(SCOTT)	0233
020-1110-015(SCOTT)	0126
020-1110-016	0325
020-1110-016(SCOTT)	0325
020-1110-017	0079
020-1110-017(SCOTT)	0016
020-1110-018	0325
020-1110-018(SCOTT)	0325
020-1110-021	0325
020-1110-022	0325
020-1110-025	0164
020-1110-025(SCOTT)	0004
020-1110-027	0693
020-1110-038	0861
020-1111-002	0103
020-1111-002(SCOTT)	0103
020-1111-003	0074
020-1111-003(SCOTT)	0103
020-1111-004	0925
020-1111-005	0042
020-1111-006(SCOTT)	0604
020-1111-007	0103
020-1111-007(SCOTT)	0103
020-1111-008	0103
020-1111-008(SCOTT)	0103
020-1111-017	0126
020-1111-018	0086
020-1112-001	0016
020-1112-002	0325
020-1112-004	0127
020-1112-004(SCOTT)	0016
020-1112-005	0325
020-1112-006	0325
020-1112-007	0325
020-1112-008	0325
020-1112-009	0325
020-1114-004	OEM
020-1114-006	0654
020-1114-007	0632
020-1114-008	0632
020-1114-009	0443
020-1114-010	OEM
020-1114-016	1696
022-006500	0198
022-009600	0079
022-2823-001	0564
022-2823-002	0012
022-2823-003	0019
022-2823-004	0133
022-2823-005	0133
022-2823-006	0123
022-2823-007	0123
022-2823-008	0143
022-2823-009	0015
022-2823-010	0133
022-2823-011	0015
022-2823-501	0720
022-2823-503	0030
022-2844-001	4401
022-2844-002	7060
022-2844-501	2301
022-2844-502	OEM
022-2844-701	2301
022-2844-702	1899
022-2844-703	4401
022-2876-001	0325
022-2876-002	0208
022-2876-003	0076
022-2876-004	0037
022-2876-005	0325
022-2876-006	0325
022-2876-007	0368
022-2876-008	1031
022-2876-009	0688
022-2876-010	0161
022-2876-011	0839
022-2876-012	0836
022-2876-704	0936
022-3504-060	0211
022-3505-910	0211
022-3511-770	0050
022-3511-780	0050
022-3511-790	0050
022-3516-380	0136
022-3640-050	0160
022-3640-080	0144
022-3640-081	0698
022-3640-082	0698
022-3640-253	0164
022-3901-001	0143
022-3905-001	0015
022-5311-770	0136
022-5311-780	0136
022-5511-790	0136
022-3511780	0050
022-3511790	0050
022D	0015
022.3504-040	0211
024	0905
025-009600	0127
025-1	2743
025-100003	0144
025-100004	0144
025-100005	OEM
025-100006	OEM
025-100008	0917
025-100009	0144
025-100010	OEM
025-100011	0030
025-100012	0007
025-100013	0144
025-100014	0144
025-100015	0086
025-100016	0015
025-100017	0111
025-100018	0079
025-100019	OEM
025-100024	0015
025-100026	0127
025-100027	0143
025-100028	0015
025-100029	0015
025-100030	0016
025-100031	0164
025-100035	0015
025-100036	0007
025-100036-12	0007
025-100037	0007
025-100037-12	0007
025-100038	0007
025-100038-12	0007
025-100039	0030
025-100040	0016
025B	0016
025B1C	2743
025B-YEL	0016
026-1000-20	0085
026-100003	0160
026-100004	0265
026-100005	0211
026-100012	0211
026-100013	0086
026-100017	0198
026-100018	0211
026-100020	0265
026-100026	0016
026-100028	0160
027-000296	0015
027-000306	0015
027-000312	0015
027-300226	0196
028	0142
028-300-226	0123
030	0087
030-007-0	0693
030-034-0	0208
031	0087
031-002-0	2046
031-025-0	0368
031-098-0	0936
031A	0016
032	0087
033-00234	1027
033-005400	OEM
033-014-0	0414
033-3	0086
033-4	OEM
033A	0111
034-001-0	0123
034-018-0	0139
034-032-0	0124
035	0087
035-013-0	0005
037	0016
037-250	0402
038	0086
039	0126
040-014-0	OEM
041	0016
041A	0212
042	0016
044-0603-210	0259
045	0353
045-1	0127
045-1(SYLVANIA)	0144
045-2	0127
045-2(SYLVANIA)	0144
046-0134	0133
046-0909	0133
046-07037	0748
046-1	0560
046-1(SYLVANIA)	0560
046-40209	0090
047	0353
047-1	1267
047-1(SYLVANIA)	1267
048	0849
050-0011-00	0015
050A1GOF	1704
050A1OOF	1704
050A2GOF	OEM
050A2OOF	OEM
050A3GOF	OEM
050A3OOF	OEM
050A4GOF	OEM
050A4OOF	OEM
050AAGOF	1704
050AAOOF	1704

If replacement code is OEM, contact original manufacturer for replacement.

DEVICE TYPE	REPL CODE
050BBGOF	1704
050BBOOF	1704
050PIN-RM	OEM
050PIN-T08	OEM
050PV-RM	OEM
051-0003	0133
051-0006	0015
051-003	0133
051-0010-00	4557
051-0011-00	1994
051-0011-00-04	1994
051-0011-00-05	3670
051-0011-04	2926
051-0012-00	0443
051-0012-11	0443
051-0016-00	0015
051-0017-00	0701
051-0020	0133
051-0020-00	1018
051-0020-02	1018
051-0021-00	0785
051-0022-00	0678
051-0035-01	1018
051-0035-02	1018
051-0035-03	1018
051-0035-04	1018
051-0035-4	OEM
051-0036-00	4093
051-0036-01	4093
051-0036-02	4093
051-0036-03	4093
051-0036-0102	4093
051-0038-00	2568
051-0039-00	1478
051-0046	0016
051-0047	0016
051-0049	0127
051-0050-00	0520
051-0050-01	0520
051-0055-02	1050
051-0055-03	1050
051-0055-0203	1050
051-0062	0050
051-0063	0050
051-0066-00	4089
051-0068-02	0579
051-0079	0136
051-0086-00	0706
051-0087-00	3815
051-0088-00	0416
051-0088-00101	0416
051-0099-00	OEM
051-0100-00	1714
051-0107	0037
051-0107B	0006
051-0151	0178
051-0155	0016
051-0156	0698
051-0157	0698
052	0400
052A	0400
052ALRLF	OEM
053-602	OEM
054-50	OEM
055	0016
055C	0324
056	0037
057	0016
057B474H	0007
058	0015
058-001138	1143
059	0142
062	0016
063N	OEM
065	0178
065-001	0325
065-002	0325
065-004	0016
065-006	0111
065-007	0219
065-008	0595
065-011	OEM
065-012	0133
065-013	0143
065-014	0123
065-015	0015
065-016	0087
065-1	0050
065-1-12	0050
065-1-12-7	0050
065-2	0208
065-2-12	0208
065-4	0600
065-4-12	0600
065-4-12-7	0600
065-14	0143
065-210	0836
065A	0050
065A-12	0050
065A-12-7	0050
065B	0136
065B-12	0136
065B-12-7	0136
065C	0160
065C-12	0160
065C-12-7	0160
066	0086
066-1	0004
066-1-12	0004
066-1-12-7	0004
066-2	0004
066-2-12	0004
066-2-12-7	0004
066-3	0004
066-3-12	0004
066-3-12-7	0004
066C	0436
069	0016
070	0103
070-001	0050
070-003	0123
070-004	0071
070-005	0071
070-006	0071
070-007	0071
070-008	0071
070-009	0071
070-010	0071
070-011	0170
070-013	0071
070-014	0071
070-015	0071
070-016	0071
070-017	0071
070-019	0015
070-020	0280
070-021	0002
070-022	0133
070-024	0170
070-027	0922
070-028	0071
070-030	0071
070-031	0087
070-032	0071
070-033	0071
070-035	0087
070-036	0922
070-041	0087
070-043	0922
070-047	0133
072-1	0133
074-1	0219
074-3	0280
074-3-12-7	0280
075-045037	1313
075-046270	1426
075B	OEM
077A	0440
077A-12	0440
077A-12-7	0440
077C	0160
077C-12	0160
077C-12-7	0160
078-0016	0015
078-1696	0015
078-2400	0015
078-5001	0015
083	OEM
083-1	0037
083-2	OEM
083-6	0037
084A	0160
084A-12	0160
084A-12-7	0160
085	0103
088C	0280
088C-12	0280
088C-12-7	0280
089-4	0187
089-4(SYL)	1708
089-4(SYLVANIA)	0276
089-214	0144
089-215	0144
089-216	0144
089-220	0136
089-222	0004
089-223	0016
089-226	0016
089-231	0164
089-233	0038
089-235	0030
089-236	0143
089-241	0133
089-248	0143
089-252	0229
089-293	0143
090A64-1	0143
092-1	0440
092-1-12	0440
092-1-12-7	0440
093	OEM
094-007	0133
094-010	0023
094-011	0023
094-012	0023
094-013	0160
094-014	0019
094K	OEM
095C	3948
098-1	0015
098GI219	0071
099-1	0042
099-1(SYL)	0042
0100	0015
0100(IC)	OEM
0101-0034	0136
0101-0222	0136
0101-0439	0791
0101-0448	1267
0101-0448A	1267
0101-0465	0925
0101-0466	1267
0101-0491	0016
0101-0531	0144
0101-0540	0016
0101-439	0126
0102	0015
0103	0325
0103-0014	0016
0103-0014(R,S)	0086
0103-0014R	0086
0103-0014S	0086
0103-0014T	0086
0103-0014U	0086
0103-0051	0086
0103-0060	0144
0103-0060(B)	0079
0103-0060-6438	0144
0103-0060A	0079
0103-0060B	0144
0103-0088	0111
0103-0088/4460	0111
0103-0088H	0016
0103-0088R	0016
0103-0088S	0016
0103-00608	0079
0103-052	0007
0103-0191	0144
0103-0389	0144
0103-0419	0560
0103-0419A	0560
0103-0473	0016
0103-0491	0076
0103-0491-6438	0198
0103-0503	0155
0103-0503S	0086
0103-0504	0016
0103-0521	0144
0103-0521(B)	0079
0103-0521B	0144
0103-0531	0076
0103-0531-4460	0016
0103-0531-6438	0016
0103-0540	0016
0103-0568	2381
0103-0607	0086
0103-0616	0086
0103-389	0144
0103-512M	0560
0104-0013	0086
0104-0013-6438	0086
0105-0012	0325
0110	0015
0110-0011	0015
0110-0141	0015
0110-0141-6438	0015
0110-0209	0015
0110-0209-6438	0015
0111	0133
0112	0015
0112-0019	0143
0112-0028	0143
0112-0028/5462	0143
0112-0028-6438	0019
0112-0037	0143
0112-0037-6438	0143
0112-0046	0143
0112-0073	0143
0112-0082	0143
0113-0027	0030
0113-0036	0030
0113-0036/4460	0143
0113-0036-6438	1217
0114	0133
0114-0017	0012
0114-0026	OEM
0114-0090	0012
0114-0090-6438	0012
0114-0260	0062
0117-02	0353
0121	0016
0122-0028-6438	0143
0124	0161
0124(KNIGHT)	0037
0124A	0037
0125	0016
0126	0016
0126(WARDS)	0382
0127	0016
0128	0016
0130	0086
0131	0016
0131-000101	0004
0131-000102	0004
0131-000192	0160
0131-000335	0037
0131-000336	0160
0131-000337	0160
0131-000418	0050
0131-000419	0050
0131-000473	0016
0131-000498	0050
0131-000561	0086
0131-000562	0160
0131-000563	0004
0131-000704	0016
0131-000802	0050
0131-000859	0050
0131-000862	0050
0131-000863	0050
0131-0026	0015
0131-0044	0015
0131-0053	0015
0131-001050	0164
0131-001056	0164
0131-001182	0050
0131-001314	0050
0131-001328	0037
0131-001329	0037
0131-001332	0050
0131-001417	0016
0131-001418	0016
0131-001419	0004
0131-001420	0037
0131-001421	0016
0131-001422	0016
0131-001423	0016
0131-001424	0016
0131-001425	0160
0131-001426	0004
0131-001427	0126
0131-001428	0126
0131-001429	0086
0131-001430	0086
0131-001433	0050
0131-001434	0050
0131-001435	0050
0131-001436	0050
0131-001438	0150
0131-001439	0037
0131-001464	0016
0131-001597	0103
0131-001697	0050
0131-001864	0016
0131-002049	0925
0131-002068	0103
0131-002656	0004
0131-003029	0050
0131-004560	0178
0131-005807	0086
0131-026	0015
0131-053	0015
0131-0144	OEM
0131-0144-6438	0139
0131-5352	0423
0132	0103
0133	0037
0133-0024	OEM
0134	0016
0135-1	0953
0142	0144
0148	4403
0154	1413
0200	1250
0201	1250
0210	1250
0211	1250
0212A	OEM
0220	1250
0221	1250
0234	0015
0243	1000
0244	0015
0245	2081
0248	0087
0276.8A	0025
0300	0015
0301	0015
0302	0015
0304	0015
0307	0071
0310	0015
0311	0015
0312	0015
0314	0015
0317	0071
0320	0015
0321	0015
0322	0015
0324	0015
0327	0071
0361S	OEM
0366L	OEM
0400	0015
0401	0015
0402	0015
0404	OEM
0404-1	OEM
0404-2	OEM
0410	0015
0411	0015
0412	0015
0413	OEM
0414	OEM
0415	0423
0450	0015
0460	0015
0461	OEM
0462	OEM
0463	OEM
0491	OEM
0492	OEM
0493	OEM
0500	0133
0516-3101-250	0475
0516-3101-406	0475
0557-010	0015
0575-005	0143
0602	0109
0604	0122
0606	0145
0615	0681
0648R	0025
0684R	0025
0700	0015
0701	0015
0702	0015
0703	0144
0704	0015
0707	0071
0710	0015
0727-50	0015
0737R	0025
0770	0460
0772	0161
0773	0460
0783Z	0678
0801	OEM
0802	OEM
0803	OEM
0831	0693
0847T0R402	3914
0847TR2501	1807
0912	0342
0912-7	OEM
0912-25	OEM
0912-45	OEM
0912-125A	OEM
0912-125B	OEM
0912P200	OEM
0912P250	OEM
0912P250A	OEM
0912P250B	OEM
0912P400A	OEM
0912P400B	OEM
0912P600A	OEM
0912P600B	OEM
01103-0036	0628
01122-0073	0143
01339	0015
02164	OEM
0220100F	0712
0220100L	1650
0220100N	1848
0220200F	0469
0220200L	1650
0220200N	1848
0220300F	0722
0220300L	2647
0220300N	1855
0220400F	0722
0220400L	2647
0220400N	1855
0220500F	0725
0220500L	2649
0220500N	3215
0230100F	0712
0230100N	3939
0230200F	0469
0230200N	1848
0230300F	0722
0230300N	1855
0230400F	0722
0230400N	1855
0230500F	0725
0230500N	3215
0230600F	0725
0230600N	3215
02375-A	0178
02724A	0402
03002	0130
03008-1	0016
03500	0143
03507FWF	OEM
03507FXF	OEM
03507GUF	OEM
03507GWF	OEM
03507GXF	OEM
03507HWF	OEM
03507HXF	OEM
03507IWF	OEM
03507IXF	OEM
03508FWF	OEM
03508FXF	OEM
03508GUF	OEM
03508GXF	OEM
03508HWF	OEM
03508HXF	OEM
03508IWF	OEM
03508IXF	OEM
03509FWF	OEM
03509FXF	OEM
03509GUF	OEM
03509GWF	OEM
03509GXF	OEM
03509HWF	OEM
03509HXF	OEM
03509IWS	OEM
03509IXF	OEM
03510FWF	OEM
03510FXF	OEM
03510GUF	OEM
03510GWF	OEM
03510GXF	OEM
03510HWF	OEM
03510HXF	OEM
03510IWF	OEM
03510IXF	OEM
03511FWF	OEM
03511FXF	OEM
03511GUF	OEM
03511GWF	OEM
03511GXF	OEM
03511HWF	OEM
03511HXF	OEM
03511IWF	OEM
03511IXF	OEM
03512FWF	OEM
03512FXF	OEM
03512GUF	OEM
03512GWF	OEM
03512GXF	OEM
03512HWF	OEM
03512HXF	OEM
03512IWF	OEM
03512IXF	OEM
03571	0143
03901FPF	OEM
03901FRF	OEM
03901FUF	OEM
03901GPF	OEM
03901GRF	OEM
03901GUF	OEM
03901HPF	OEM
03901HRF	OEM
03901HUF	OEM
03902FPF	OEM
03902FRF	OEM
03902FUF	OEM
03902GPF	OEM
03902GRF	OEM
03902GUF	OEM
03902HPF	OEM
03902HRF	OEM
03902HUF	OEM
03903FPF	OEM
03903FRF	OEM
03903FUF	OEM
03903GPF	OEM
03903GRF	OEM
03903GUF	OEM
03903HPF	OEM
03903HRF	OEM
03903HUF	OEM
03904FPF	OEM
03904FRF	OEM
03904FUF	OEM
03904GPF	OEM
03904GRF	OEM
03904GUF	OEM
03904HPF	OEM
03904HRF	OEM
03904HUF	OEM
03905FPF	OEM
03905FRF	OEM
03905FUF	OEM
03905GPF	OEM
03905GRF	OEM
03905GUF	OEM
03905HPF	OEM
03905HRF	OEM
03905HUF	OEM
03906FPF	OEM
03906FRF	OEM
03906FUF	OEM
03906GPF	OEM
03906GRF	OEM
03906GUF	OEM
03906HPF	OEM
03906HRF	OEM
03906HUF	OEM
04049B	0015
04170	0277
04306A	OEM
04770	0143
04970	0071
05001GOF	OEM
0500100F	1704
05002GOF	1704
0500200F	1704
05003GOF	4031
0500300F	4031
05004GOF	4031
0500400F	4031
05005GOF	4031
0500500F	4031
05007GOF	0679
0500700F	0679
05008GOF	0679
0500800F	0679
05009GOF	0679
0500900F	0679
05010GOF	0679
0501000F	0679
0501100F	0679
05012GOF	0679
0501200F	0679
05070	0924
05201GOA	0608
05201GOD	0641
0520100A	0608
0520100D	OEM
05202GOA	0608
05202GOD	0641
0520200A	0608
0520200D	OEM
05203GOA	0610
05203GOD	0217
0520300A	0610
0520300D	OEM
05204GOA	0610
05204GOD	0217
0520400A	0610
0520400D	OEM
05205GOA	0610
05205GOD	0217
0520500A	0610
0520500D	OEM
05206-0	0111
05206GOA	0610
05206GOD	0217
0520600A	0610
0520600D	0610
05253	0160
05470	0649
07101GOA	0608
07101GOD	0641
0710100A	OEM
0710100D	OEM
07102GOA	0608
07102GOD	0641
0710200D	OEM
0710200Z	OEM
07103GOA	0610
07103GOD	0217
0710300A	OEM
0710300D	OEM
07104GOA	0610
07104GOD	0217
0710400A	OEM
0710400D	OEM
07105GOA	0610
07105GOD	0217
0710500	OEM
07106GOA	0610
07106GOD	0217
0710600A	OEM
0710600D	OEM
07607GUA	OEM
07607GUD	OEM
07607GWA	OEM
07607GWD	OEM
07607GXA	OEM
07607GXD	OEM
07607HUA	OEM
07607HUD	OEM
07607HWA	OEM
07607HWD	OEM
07607HXA	OEM
07607HXD	OEM
07607IUA	OEM
07607IUD	OEM
07607IWA	OEM
07607IWD	OEM
07607IXA	OEM
07607IXD	OEM
07608GUA	OEM
07608GUD	OEM
07608GWA	OEM
07608GWD	OEM
07608GXA	OEM
07608GXD	OEM
07608HUA	OEM
07608HUD	OEM
07608HWA	OEM
07608HWD	OEM
07608HXA	OEM
07608HXD	OEM
07608IUA	OEM
07608IUD	OEM
07608IWA	OEM
07608IWD	OEM
07608IXA	OEM
07608IXD	OEM
07609GUA	OEM
07609GUD	OEM
07609GWA	OEM
07609GWD	OEM
07609GXA	OEM
07609GXD	OEM
07609HUA	OEM
07609HUD	OEM
07609HWA	OEM
07609HWD	OEM
07609HXD	OEM
07609IUA	OEM
07609IUD	OEM
07609IWA	OEM
07609IWD	OEM
07609IXA	OEM
07609IXD	OEM
07609NXA	OEM
07610GUA	OEM
07610GUD	OEM
07610GWA	OEM
07610GWD	OEM
07610GXA	OEM
07610GXD	OEM
07610HUA	OEM
07610HUD	OEM
07610HWA	OEM
07610HWD	OEM

If replacement code is OEM, contact original manufacturer for replacement.

DEVICE TYPE	REPL CODE
07610HXA	OEM
07610HXD	OEM
07610IUA	OEM
07610IUD	OEM
07610IWA	OEM
07610IWD	OEM
07610IXA	OEM
07610IXD	OEM
07611GUA	OEM
07611GUD	OEM
07611GWA	OEM
07611GWD	OEM
07611GXA	OEM
07611GXD	OEM
07611HUA	OEM
07611HUD	OEM
07611HWA	OEM
07611HWD	OEM
07611HXA	OEM
07611HXD	OEM
07611IUA	OEM
07611IUD	OEM
07611IWA	OEM
07611IWD	OEM
07611IXA	OEM
07611IXD	OEM
07612GUA	OEM
07612GUD	OEM
07612GWA	OEM
07612GWD	OEM
07612GXA	OEM
07612GXD	OEM
07612HUD	OEM
07612HWA	OEM
07612HWD	OEM
07612HXA	OEM
07612HXD	OEM
07612IUA	OEM
07612IUD	OEM
07612IWA	OEM
07612IWD	OEM
07612IXA	OEM
07612IXD	OEM
07621HUA	OEM
07901GRA	OEM
07901GRD	OEM
07901GUA	OEM
07901GUD	OEM
07901GWA	OEM
07901HRA	OEM
07901HUA	OEM
07901HUD	OEM
07901HWA	OEM
07901HWD	OEM
07901IRA	OEM
07901IRD	OEM
07901IUA	OEM
07901IUD	OEM
07902GRA	OEM
07902GRD	OEM
07902GUA	OEM
07902GUD	OEM
07902GWA	OEM
07902GWD	OEM
07902HRA	OEM
07902HRD	OEM
07902HUA	OEM
07902HUD	OEM
07902HWA	OEM
07902HWD	OEM
07902IRA	OEM
07902IRD	OEM
07902IUA	OEM
07902IUD	OEM
07903GRA	OEM
07903GRD	OEM
07903GUA	OEM
07903GUD	OEM
07903GWA	OEM
07903GWD	OEM
07903HRA	OEM
07903HRD	OEM
07903HUA	OEM
07903HUD	OEM
07903HWA	OEM
07903HWD	OEM
07903IRA	OEM
07903IRD	OEM
07903IUA	OEM
07903IUD	OEM
07904GRA	OEM
07904GRD	OEM
07904GUA	OEM
07904GUD	OEM
07904GWA	OEM
07904GWD	OEM
07904HRA	OEM
07904HRD	OEM
07904HUA	OEM
07904HUD	OEM
07904HWA	OEM
07904HWD	OEM
07904IRA	OEM
07904IRD	OEM
07904IUA	OEM
07904IUD	OEM
07905GRA	OEM
07905GRD	OEM
07905GUA	OEM
07905GUD	OEM
07905GWA	OEM
07905GWD	OEM
07905HRA	OEM
07905HRD	OEM
07905HUA	OEM
07905HUD	OEM
07905HWA	OEM
07905HWD	OEM
07905IRA	OEM
07905IRD	OEM
07905IUA	OEM
07905IUD	OEM
07906GRA	OEM
07906GRD	OEM
07906GUA	OEM
07906GUD	OEM
07906GWA	OEM
07906GWD	OEM
07906HRA	OEM
07906HRD	OEM
07906HUA	OEM
07906HUD	OEM
07906HWA	OEM
07906HWD	OEM
07906IRA	OEM
07906IRD	OEM
07906IUA	OEM
07906IUD	OEM
08001G0A	0608
08001G0D	0641
0800100A	OEM
0800100D	OEM
08002G0A	0608
08002G0D	0641
0800200A	OEM
0800200D	OEM
08003G0A	0610
08003G0D	0641
0800300A	OEM
0800300D	OEM
08004G0A	0610
08004G0D	0217
0800400A	OEM
0800400D	OEM
08005G0A	0610
08005G0D	0217
0800500A	OEM
0800500D	OEM
08006G0A	0610
08006G0D	0217
0800600A	OEM
0800600D	OEM
08050	0015
09501	0111
09502	0111
09502-8	0016
09800	0037
09800-12	0126
09800C	0006
010562	0006
010694	1708
011103	0196
011119	0123
011406	OEM
013339	0229
013339(RECTIFIER)	0071
013684	0122
014382	0160
014558	0233
015040/7	0361
018069	0126
018077	0086
020156	0280
021154	0015
022939	0187
023606	0325
023754	0178
023762	0103
025026	0015
025056	0015
025072	0071
026237	0079
030010	0079
030010-1	0079
030011	0086
030011-1	0086
030011-2	0086
030387	0940
030512-1	0079
030512-2	0079
030515	0079
030515-4	0079
030527	0079
030531-1	0440
030536	0079
030536-1	0079
030537	0079
030537-1	0079
030537-2	0079
030538	0079
030539-1	0103
030542	0079
030542-1	0079
030543	0079
030543-1	0079
030543-2	0079
030548	0079
030828	0111
030930	0224
031033	0143
031034	0015
031040	0143
031450	0242
033370	0920
033421	0920
033549	0920
033563	0007
033571	0126
033589	0037
036001	0632
037077	0178
037085	0103
037085BL	0130
037085BR	0130
037085GR	0130
037085OR	0130
037085R	0130
037085Y	0130
040001	0127
040048	1010
041001	0006
041200-30110	0133
041616	0133
043001	0144
053328	0050
053480	0007
055210	0229
055210H	0015
055228	0071
055228H	0071
057001	0143
057001H	0143
057005	0123
057005H	0143
057019H	0143
057174	0744
057500	0143
057524	0030
059395	0143
061366	0079
062971	1413
071964-001	0103
075005	0143
075085-36	0030
080001	0136
080003	0004
080004	0004
080006	0144
080021	0144
080022	0144
080023	0144
080026	0050
080027	0050
080028	0050
080040	0015
080041	0144
080042	0144
080043	0004
080047	0004
080048	0160
080050	0015
080052	0004
080059	0144
080060	0144
080061	0050
080071	0050
080072	0136
080073	0004
080114	0050
080206	0136
080224	0136
080225	0136
080228	0136
080236	0136
080244	0050
080245	0050
080253	0136
080258	0050
080266	0050
080267	0050
080269	0050
080274	0136
080275	0136
080276	0136
080277	0136
081001	0004
081018	0004
081019	0004
081026	0164
081027	0164
081029	0164
081038	0004
081042	0822
081046	0004
081047	0004
081048	0004
081049	0004
081050	0004
081056	0004
081059	0164
082006	0079
082019	0079
082020	0330
082022	0698
082025	1411
082028	0155
082029	1411
082033	0698
084001	0037
084001C	0160
085002	0143
085003	0133
085004	0143
085005	0123
085006	0143
085016	0143
085026	0123
087003	0004
094013	0604
0112945	0277
0112945(ELGIN)	0277
0170301	0133
0201201	0579
0207046	1393
0207120	1018
0207205	1050
0220600L	2649
0230100N	1848
0260540	0012
0300005	0004
0300006	0208
0300008B	0004
0303303	OEM
0310002	0847
0320031	0079
0320051	0233
0320064	0155
0330302	0137
0404011-001	1835
0404012-001	1844
0444028-010	0016
0444028-014	0016
0510006	0015
0510079	0050
0510079H	0050
0517022	0143
0517132	0133
0517133	0133
0517261B	0030
0517550	0015
0517550-3	0015
0517750	0015
0517750H	0015
0517826	0143
0517828	0143
0517829	0143
0518926	0133
0525001	0030
0525002	0143
0525002H	0143
0526224	0143
0526232	0133
0531014	0004
0531015	0004
0531016	0004
0535001	0143
0535005	0143
0537125	0004
0537640	0004
0537640(DIO)	0015
0537820	0143
0539860	0030
0551029	0277
0551029H	0277
0552005	0102
0552005H	0229
0552006	0015
0552006H	0015
0552007	0102
0552007H	0229
0552010	0015
0552010H	0015
0563012H	0004
0570519	0143
0572480	0076
0573001	0004
0573001-14	0004
0573001H	0004
0573002	0136
0573003	0004
0573003H	0004
0573004	0004
0573005	0004
0573005-14	0004
0573005H	0004
0573011	0004
0573012	0004
0573012H	0004
0573018	0004
0573018H	0004
0573022	0004
0573022H	0004
0573023	0004
0573023A	0004
0573023H	0004
0573024	0004
0573024-14	0004
0573025	0004
0573030	0222
0573030-14	0222
0573031	0822
0573034	0004
0573036	0004
0573036H	0004
0573037	0208
0573037H	0208
0573040	0160
0573055	0164
0573056	0004
0573066	0016
0573103	0211
0573114	0004
0573114H	0004
0573117	0004
0573117-14	0004
0573119	0004
0573124H	0136
0573125	0004
0573131	0004
0573139	0208
0573142	0004
0573142H	0004
0573152	0004
0573153	0004
0573153H	0004
0573166	0085
0573184	0164
0573185	0847
0573187	0004
0573199	0975
0573199H	0975
0573200	0004
0573202	0016
0573204	0164
0573205	0160
0573212	0222
0573212H	0975
0573328	0004
0573330	0136
0573335	0050
0573366	0050
0573398	0136
0573415	0142
0573418	0016
0573422	0004
0573422H	0004
0573427	0136
0573428	0050
0573429	0004
0573430	0016
0573468	0016
0573468(HITACHI)	0127
0573469	0016
0573469H	0016
0573471	0136
0573474	0047
0573474H	0007
0573475	0007
0573479	0155
0573479H	0016
0573480	0076
0573480H	0086
0573481	0076
0573481H	0016
0573485	0127
0573486	0127
0573486H	0330
0573487	0151
0573487H	0330
0573490	0016
0573491	0076
0573491H	0155
0573492	0076
0573494	0144
0573495	0144
0573497	OEM
0573501	0233
0573506	0144
0573506H	0144
0573507	0127
0573507H	0144
0573508	0127
0573509	0224
0573509H	0127
0573510	0127
0573510H	0127
0573511	0144
0573511H	0127
0573515	0142
0573515H	0142
0573517	0955
0573518	0136
0573519	0233
0573523	0016
0573525	0178
0573526	2437
0573527	0086
0573529	0016
0573532	0016
0573541	1411
0573542	0126
0573556	0016
0573557	0086
0573559	0126
0573560	0126
0573562	0103
0573570	0144
0573607	0127
0573742	0004
0573981	0016
0575001	0123
0575001H	0143
0575002	0143
0575002AH	0143
0575002H	0143
0575004	0143
0575005	0123
0575005H	0143
0575007	0143
0575010	0030
0575010H	0030
0575019	0123
0575019H	0123
0575024	0030
0575024(8)	0030
0575024H	0030
0575027	0030
0575047	0102
0575047H	0102
0575048	0102
0575048H	0924
0575049	0102
0575049H	0102
0575050	0071
0575051	0015
0575054	0102
0575066	0102
0575067	0143
0575099	0143
0576001	0004
0576054	0015
0576054(BIAS)	0015
0576054(SW)	0133
0577001	0143
0610007	OEM
0610008	OEM
0610083	0006
0610084	0148
0612016/1	0261
0613761A02	1993
0613785A03	OEM
0620001	OEM
0620017	0143
0621063	0178
0622070	0023
0630049	0912
0690403A47	1993
0690524A04	OEM
0800040	OEM
0899024	0323
0900296	0823
01310131-0035	0015
02600043	0143
04440028-002	0016
04440028-007	0016
04450023-007	0130
04450300-1	0147
04450303-001	0744
04450303-002	0744
04450304	0744
05320023	0016
05320074	0016
05734815	0076
05750011	0123
06100007	1308
06100008	1308
06100015	0148
06100016	0148
06100030	0006
06100033	0925
06100034	0006
06100035	0006
06100047	0006
06100053	0693
06100058	0648
06100061	0693
06100062	0006
06100063	1911
06100076	1911
06100081	0006
06100082	0006
06100083	0006
06100084	0148
06100085	0006
06100086	0532
06100089	1648
06100090	0148
06100094	0006
06100095	0006
06100096	0006
06100131	2129
06100132	OEM
06100137	0532
06110032	1648
06120001	0716
06120003	0042
06120005	0076
06120006	0076
06120008	0151
06120009	0127
06120012	2437
06120015	0127
06120018	0042
06120021	0042
06120025	0842
06120026	0286
06120028	0155
06120029	0079
06120030	0042
06120041	0643
06120044	2437
06120053	0276
06120063	0191
06120073	0224
06120083	1085
06120085	0224
06120087	0286
06120088	0168
06120091	0168
06120096	0111
06120097	0168
06120098	0168
06120100	1496
06120109	0168
06120110	0168
06120126	0261
06120128	0261
06120129	0261
06120130	0130
06120134	4155
06120141	0955
06120144	0276
06120145	0392
06120148	1284
06120149	0643
06120150	0643
06120151	0951
06120152	0261
06120153	0313
06120155	0276
06120156	0276
06120161	0066
06120162	0261
06120168	0286
06120169	0286
06120170	0286
06120171	0286
06120172	0286
06120173	0155
06120174	0155
06120175	0155
06120181	0219
06120182	0219
06120186	0168
06120187	0168
06120200	0276
06120201	1284
06120202	1284
06120204	0076
06120207	1284
06120208	2660
06120212	1386
06120217	0261
06120218	0066
06120220	0643
06120221	0643
06120224	0079
06120225	0261
06120227	0079
06120228	0079
06120229	0079
06120230	0079
06120231	0037
06120232	0037
06120233	0079
06120235	0842
06120236	0261
06120238	0261
06120239	0286
06120240	0286
06120253	0148
06120255	0261
06120365	0866
06120619	0286
06130014	1284
06130017	1284
06130018	1284
06130019	0313
06130033	0461
06130034	0042
06130036	1165
06130037	0065
06130040	1978
06130042	0055
06130046	0392
06130047	0392
06130052	0392
06130053	0392
06130063	0718
06130086	3494
06130118	1541
06130134	0556
06150012	0261
06150031	0079
06160006	0055
06160008	0127
06170014	0079
06170028	1541
06170030	1541
06179462	OEM
06179495	0261
06179501	0342
06179502	0261
06200001	0123
06200002	0019
06200003	0123
06200005	0071
06200009	0015
06200010	0023
06200012	0025
06200013	0133
06200014	0023
06200017	0143

If replacement code is OEM, contact original manufacturer for replacement.

DEVICE TYPE	REPL CODE
06200027	0023
06200030	0023
06200032	0317
06200037	0965
06200038	0182
06200039	0023
06200041	0143
06200044	0133
06200045	0143
06200046	0120
06200047	0139
06200048	0133
06200050	0102
06200051	0023
06200052	0015
06200055	0636
06200056	0100
06200057	0025
06200058	0137
06200059	0137
06200060	0296
06200061	6557
06200062	0023
06200063	0795
06200064	0124
06200065	0071
06200067	0071
06200069	0031
06200071	0795
06200072	3077
06200073	0133
06200076	0288
06200083	0091
06200084	0181
06200085	0057
06200088	0023
06200089	0120
06200090	0564
06200091	1268
06200095	2232
06200096	0143
06200100	0023
06200102	0023
06200103	0023
06200104	0023
06200105	0023
06200106	0102
06200107	0071
06200108	0471
06200109	0071
06200112	0471
06200113	0057
06200114	0181
06200115	0133
06200116	0133
06200117	0133
06200118	0080
06200119	0795
06200120	0348
06200121	6848
06200122	0023
06200123	0023
06200124	0015
06200127	0137
06200129	0471
06200130	0133
06200149	0248
06200150	0124
06200159	0023
06200164	0253
06200167	0124
06200168	0137
06200170	0071
06200177	0181
06200179	0133
06200180	1005
06200181	0023
06200182	0023
06200183	0023
06200186	0143
06200189	0023
06200194	0317
06200202	0023
06200203	0023
06200206	0181
06200207	0023
06200211	0023
06200212	0023
06200223	0181
06200226	0124
06200228	0181
06200229	0023
06200230	0023
06200231	0023
06200232	0023
06200233	0181
06200255	0023
06200263	0031
06200272	0041
06200283	0023
06200284	0023
06200285	0023
06200286	0023
06200287	0102
06200288	0023
06200295	0248
06200323	0023
06210010	OEM
06210011	2628
06210017	OEM
06210018	OEM
06210018(DS2(LED))	OEM
06210020	OEM
06220017	0023
06220021	0023
06220022	0953
06220069	0080
06220070	0609
06220071	0795
06220120	0057
06220127	0137
06220132	0041
06220137	OEM
06220167	0124
06220168	0376
06220227	0041
06220228	0253
06220233	0057
06220235	0181
06220241	OEM
06220242	0195
06300001	3273
06300003	3269
06300004	1055
06300005	0528
06300009	0167
06300028	OEM
06300029	OEM
06300030	3282
06300031	1429
06300034	0490
06300039	0624
06300049	0912
06300050	2054
06300051	2054
06300052	4134
06300056	2090
06300058	3502
06300066	2533
06300069	0167
06300071	0683
06300072	2090
06300080	1201
06300089	0624
06300090	1201
06300095	2026
06300104	OEM
06300105	0631
06300106	OEM
06300107	4138
06300108	OEM
06300111	2103
06300113	2116
06300115	2122
06300116	2103
06300117	OEM
06300120	0167
06300121	2666
06300144	0624
06300154	4146
06300160	3772
06300163	OEM
06300191	0035
06300192	0035
06300219	0334
06300248	1595
06300253	OEM
06300311	1899
06300315	1014
06300316	0413
06300317	1551
06300318	OEM
06300320	OEM
06300323	4189
06300330	OEM
06300331	OEM
06300332	OEM
06300334	3254
06300335	0541
06300336	4194
06300337	0315
06300342	0624
06300357	0539
06300359	0557
06300360	0732
06300365	2122
06300373	OEM
06300394	0631
06300395	OEM
06300403	3953
06300411	OEM
06300440	0315
06300441	OEM
06300450	OEM
06300456	3083
06300457	2870
06300488	3635
06300491	2957
06300508	1945
06300514	0523
06300516	OEM
06300517	OEM
06300519	0462
06300536	OEM
06300537	OEM
06300546	1201
06300554	0024
06300564	OEM
06300565	0770
06300574	0539
06300575	OEM
06300615	OEM
06300616	0557
06300621	3635
06300629	2364
06300634	OEM
06300656	1239
06300661	OEM
06300662	OEM
06300688	1641
06300714	OEM
06300724	OEM
06300733	OEM
06300736	OEM
06300737	OEM
06300738	0101
06300739	OEM
06300740	OEM
06300741	OEM
06300742	OEM
06300743	OEM
06300744	OEM
06300745	OEM
06300746	OEM
06300761	3391
06300763	3515
06300764	OEM
06300765	OEM
06300766	3253
06300767	1028
06300769	OEM
06300778	OEM
06300779	OEM
06363040	OEM
06363045	OEM
06400001	0058
06400002	0239
06400537	OEM
06500004	0120
07009502	1348
07011751	3625
07011752	0148
07011757	0148
07012611	1199
07107071	0854
07107951	2897
07108251	0854
07108841	0602
07108861	2237
07108951	2897
07109741	0602
07110121	OEM
07110991	0602
07111351	4125
07111681	0460
07113591	OEM
07220031	0286
07227851	0249
07227853	0151
07233301	2954
07233312	0018
07307741	2030
07308631	1386
07308821	0161
07309861	2970
07311951	1212
07311961	2235
07312371	1165
07312731	0928
07312771	1212
07313081	3136
07313761	3703
07313951	2366
07315931	OEM
07316371	1061
07317251	0161
07320101	OEM
07506562	0576
07602002	0832
07650003	OEM
07650004	OEM
07650013	OEM
07650016	3967
07650017	OEM
07650020	OEM
07650021	OEM
07650023	OEM
07650024	OEM
07650027	OEM
07650029	OEM
07650031	OEM
08000001	0133
08000015	0124
08000016	0087
08000017	0922
08000018	0133
08000022	0071
08000028	0124
08000039	0133
08000040	1090
08000044	0080
08000045	1090
08000048	1090
08020007	0636
08020020	0531
08020022	1606
08020025	0023
08020028	0053
08020029	0471
08020030	0140
08020033	0062
08020036	0760
08020037	0274
08020040	0376
08020052	0039
08020054	0441
08020056	0064
08020059	OEM
08020066	OEM
08020067	0275
08020073	0253
08020075	0140
08020077	0165
08030001	2542
08030002	0023
08030003	2234
08030005	2542
08200040	0872
08200043	0624
08200045	1828
08200050	1836
08200064	1836
08200068	OEM
08200069	1292
08200070	2921
08200071	OEM
08200073	OEM
08200074	OEM
08200077	2921
08200086	OEM
08200091	0631
08200093	2539
08200095	OEM
08200096	OEM
08200101	OEM
08200103	OEM
08200105	OEM
08200109	OEM
08200111	OEM
08200112	OEM
08200114	OEM
08200115	OEM
08202003	OEM
08210001	1527
08210002	1595
08210004	1841
08210006	1206
08210008	0973
08210017	1608
08210022	1888
08210023	1698
08210029	0427
08210030	0209
08210032	1633
08210033	0243
08210038	0709
08210039	1161
08210040	1744
08210043	5045
08210046	0153
08210047	0508
08210048	0511
08210050	0775
08210052	0713
08210053	0088
08210058	1667
08210059	0979
08210063	1775
08210064	0819
08210068	0943
08210069	0458
08210070	0291
08210071	1646
08210072	1559
08210103	OEM
08220014	2697
08220027	1898
08220028	OEM
08220029	OEM
08220030	0812
08220031	1898
08220032	OEM
08220033	0812
08220036	OEM
08220037	OEM
08220038	OEM
08220039	OEM
08220041	OEM
08220042	OEM
08220043	OEM
08220047	1638
08220048	OEM
08220070	OEM
08220072	1559
08220073	OEM
08220082	OEM
08220086	OEM
08220090	0812
08220092	0176
08220096	OEM
08220099	OEM
08220101	OEM
08220102	OEM
08220102(HD63A03YF)	OEM
08220103	OEM
08220106	OEM
08220108	OEM
08220116	OEM
08221002	OEM
08221007	OEM
08230028	OEM
08230040	OEM
08240001	OEM
08240002	OEM
08240013	OEM
08250001	OEM
08300027	OEM
08300028	OEM
08300032	OEM
08300055	OEM
08300058	OEM
08990202	3023
08990207	0323
08990208	0277
08990210	0323
08990211	0473
08990214	0323
09301027	0164
09302090	0076
09306244	0133
09391000A	2890
061300053	0392
0847000202	OEM
0A10	0965
0A70	0143
0A90	0019
0A90A-G	0019
0A90A-M	0019
0A90AM	0019
0A90FM	0123
0A90G	0019
0A90LF	0143
0A90M	0143
0A90MLF	0143
0A90Z	0143
0A90ZA	0143
0A91	0019
0A95	0123
0A99	0143
0A126/10	0181
0A126/12	0137
0AZ209	0162
0AZ243	0636
0AZ244	0062
0AZ246	0165
0AZ269	0162
0C3K	0211
0C-16	0160
0C16	0085
0C19	0160
0C20	0160
0C-22	0160
0C22	0085
0C-23	0160
0C23	0085
0C-24	0160
0C24	0085
0C-25	0160
0C25	0085
0C-26	0160
0C26	0085
0C-30A	0160
0C32	0211
0C-35	0160
0C35	0085
0C-36	0160
0C36	0085
0C41	0211
0C-44	0280
0C-45	0023
0C-46	0280
0C-47	0280
0C56	0211
0C74N	0164
0C83	0164
0C84	0164
0C-130	0280
0C169	0050
0C-304	0847
0C308	0164
0C463K	0861
0DD2000AF	0023
0DD20000AE	0564
0DD20000AF	0023
0DD100009AC	0023
0DD100009AE	0023
0DD100009BA	0023
0DD120000BB	1127
0DD150009CA	0031
0DD200009AC	0031
0DD200009AE	0564
0DD200009AF	0023
0DD200009AH	0023
0DD247109AA	0124
0DD300000EJ	0023
0DD300009AB	0283
0DD400000AE	0031
0DD400000AF	0102
0DD400000AG	0031
0DD400209AA	0080
0DD400409AA	4211
0DD400509BA	6022
0DD414809ED	0133
0DD606000AA	1045
0DL206000AA	OEM
0DL206000AC	OEM
0DZ10009BA	0057
0DZ110009AA	0181
0DZ112009AD	OEM
0DZ120009AA	0052
0DZ120009AC	OEM
0DZ240009BB	6014
0DZ300009BA	0195
0DZ360000BA	0376
0DZ360009AB	0376
0DZ510009AB	0041
0DZ512000AA	4111
0DZ520009AA	OEM
0DZ560009AA	0253
0DZ620000BA	0636
0DZ620009AA	5811
0DZ680009AA	0062
0DZ750009AA	0077
0DZ910009BA	0012
0DZ911000AC	0057
0F66	0102
0F160	0015
0F162	0133
0F173	0143
0F612	OEM
0F643	OEM
0G02	0105
0IGL301300A	2364
0IGS348000A	OEM
0IGS381200A	OEM
0IGS382000A	OEM
0IGS406600A	0101
0IGS455800A	OEM
0IGS781200A	4413
0IGS781500A	1321
0IHI115660A	OEM
0IHY934600A	OEM
0IKE703300A	OEM
0IKE780120A	0334
0IM0800809A	OEM
0IM1371008B	OEM
0IMA300800A	OEM
0IMA310100A	OEM
0IMA530200A	OEM
0IMI371008B	OEM
0IMI514940A	OEM
0IMI820810A	OEM
0IMI1514940A	OEM
0IM0800809A	OEM
0INE187000A	OEM
0INE187100A	OEM
0INNS745950A	OEM
0INS199200A	OEM
0INS745950A	OEM
0INS934600B	OEM
0INS1999200A	OEM
0IO111000CA	OEM
0IPH842500A	OEM
0IS0111000A	OEM
0IS0112400A	OEM
0IS0163701A	OEM
0ISA426100A	3253
0ISA701600A	1028
0ISA722200A	OEM
0ISA755500A	OEM
0ISA757500A	OEM
0ISA783000A	0732
0ISA783100A	2957
0ISA783300A	OEM
0ISA840819A	OEM
0ISG382000A	OEM
0ISK120000A	OEM
0ISK500410A	3344
0ISM460500A	OEM
0ISM4605500A	OEM
0IS0111000A	OEM
0IS0112400A	OEM
0ISS330000A	1329
0IT063100A	0315
0IT0631000A	OEM
0IT0762900A	OEM
0IT0820000A	OEM
0IT0868000A	3515
0IT0882500B	OEM
0IT0890802D	OEM
0IT0631000A	0315
0IT0820000A	OEM
0IX1240200B	OEM
0LD206000AC	OEM
0N008341C-2	0004
0N008351	0280
0N049874	3897
0N064071-1	0004
0N088598	2689
0N097164	0142
0N271	0079
0N1105	OEM
0N1112	OEM
0N2170-R	OEM
0N2170-RLF	OEM
0N2170-RS	OEM
0N3161-QR	OEM
0N101880	2687
0N101881	2537
0N161646C	0004
0N188320-2	0478
0N198382-2	1143
0PP28	OEM
0S40	OEM
0SD-0033	0649
0SD0033	0649
0T3	OEM
0TF325000AA	OEM
0TR100100AA	OEM
0TR101509AA	0006
0TR105009AB	OEM
0TR106009AC	OEM
0TR126609AA	4519
0TR127009AA	0006
0TR127109AA	0477
0TR135100AB	0406
0TR136600AB	OEM
0TR155400AA	1541
0TR162700AC	1386
0TR162709AA	1386
0TR162709AB	1386
0TR165100AA	0556
0TR181509AB	0286
0TR187900AA	2129
0TR188600AA	3494
0TR195909AA	0286
0TR206800BA	0643
0TR206800BB	0643
0TR222909AB	0261
0TR223009AB	0261
0TR223500AA	OEM
0TR223800AA	0392
0TR248200AA	0066
0TR248209AA	0261
0TR319709AB	OEM
0TR319809AA	5954
0TR320209AA	0383
0TR320309AA	OEM
0TR320609AB	0261
0TR322700AC	0324
0TR322709AA	OEM
0TR322709AB	OEM
0TR322900AA	7024
0TR387500AA	0724
0TR388000AC	0842
0TR388009AA	0383
0TR388100AA	OEM
0TR508000EA	3360
0TR562009AA	0006
0TR595000AB	0718
0TR880000AB	4518
0TR949000AB	1522
0TR966000AA	OEM
0TR968000AA	1648
0V02	0102
0Z27T5	0441
0Z27T10	0441
0.4M9.1SZ5	OEM
0.5Z5.1U	0041
.2KT12.8A	0365
.4M3.3AZ5	0292
.4M3.3AZ10	0292
.4M3.6AZ5	0188
.4M3.6AZ10	0188
.4M3.9AZ5	0456
.4M3.9AZ10	0456
.4M4.3AZ5	0533
.4M4.3AZ10	0533
.4M4.7AZ5	0451
.4M4.7AZ10	0451
.4M10AZ5	0064
.4M10AZ10	0064
.4M13Z5	0365
.4M16Z5	0420
.4M18Z5	0495
.4M30Z5	0726
.4T5.6	0157
.4T5.6A	0157
.4T5.6B	0157
.4T6.8	0025
.4T6.8B	0025
.4T7.5	0649
.4T7.5A	0649
.4T7.5B	0649
.4T8.2	0244
.4T9.1	0012
.4T9.1A	0012
.4T9.1B	0012
.4T10	0170
.4T10A	0170
.4T10B	0170
.4T12A	0137
.4T12B	0137
.4T15	0002
.4T15A	0002
.4T15B	0002
.4T18	0495
.4T18A	0495
.4T18B	0495
.4T22	0565
.4T22A	0565
.4T22B	0565
.4T22V	0565
.4T27	0441
.4T27A	0441
.4T27B	0441
.4T33	0039
.4T33A	0039
.4T33B	0039
.4T39	0350
.4T39A	0350
.4T39B	0350
.4T47	0999

If replacement code is OEM, contact original manufacturer for replacement.

DEVICE TYPE	REPL CODE
.4T47A	0999
.4T47B	0999
.4T56	0869
.4T56A	0869
.4T56B	0869
.4T68	3204
.4T68A	3204
.4T68B	3204
.4T82	0331
.4T82A	0331
.4T82B	0331
.4T100	0098
.4T100A	0098
.4T100B	0098
.4T110B	0098
.4T120	0186
.4T120A	0186
.4T120B	0186
.4T150	0028
.4T150A	0028
.4T150B	0028
.4T180	0367
.4T180A	0367
.4T180B	0367
.4T688	1268
.4Z14D	0879
.4Z14D5	0879
.4Z14D10	0879
.4Z17D	0210
.4Z17D5	0210
.4Z17D10	0210
.4Z19D	0671
.4Z19D5	0671
.4Z19D10	0671
.4Z25D	0714
.4Z25D5	0714
.4Z25D10	0714
.4Z45D	0054
.4Z45D5	0054
.4Z45D10	0054
.4Z50D	0092
.4Z50D5	0092
.4Z50D10	0092
.4Z52D	0092
.4Z52D5	0092
.4Z52D10	0092
.4Z105D	0189
.4Z105D5	0189
.4Z105D10	0189
.4Z140D	0252
.4Z140D5	0252
.4Z140D10	0252
.4Z175D	0394
.4Z175D5	0394
.4Z175D10	0394
.5E05	0015
.5E1	0015
.5E2	0015
.5E3	0015
.5E4	0015
.5E5	0015
.5E6	0015
.5E7	0071
.5E8	0071
.5E10	0071
.5E12	0102
.5J05	0015
.5J1	0015
.5J2	0015
.5J3	0015
.5J4	0015
.5J5	0015
.5J6	0015
.5J7	0071
.5J8	0071
.5J10	0071
.5J12	0102
.5L1-2	OEM
.5LS1-3	OEM
.5LS1-3S	OEM
.5LS1.2-3.6	OEM
.5LS1.2-3.6S	OEM
.5LS1.5-3	OEM
.5M3.3ZS5	0292
.5M3.3ZS10	0292
.5M3.6ZS5	0188
.5M3.6ZS10	0188
.5M3.9ZS5	0456
.5M3.9ZS10	0456
.5M4.3ZS5	0533
.5M4.3ZS10	0533
.5M4.7ZS5	0451
.5M4.7ZS10	0451
.5M5.1ZS3	0041
.5M8.7ZS5	1083
.5M13ZS5	0365
.5M16ZS5	0420
.5M18ZS5	0495
.5M25Z55	1606
.5M28Z55	1674
.5M30ZS5	0726
.5PL.5-1.5	OEM
.5PL.5-1.5S	OEM
.5PL.8-2.4	OEM
.5PL.8-2.4S	OEM
.5P.2-.4	OEM
.5P.2-.4S	OEM
.5P.5-1	OEM
.5P.5-1.5	OEM
.5P.25-.5	OEM
.5P.25-.5S	OEM
.7E05	0015
.7E1	0015
.7E2	0015
.7E3	0015
.7E4	0015
.7E5	0015
.7E6	0015
.7E7	0071
.7E8	0071
.7E10	0071
.7E12	0102
.7J05	0015
.7J1	0015
.7J2	0015
.7J3	0015
.7J5	0015
.7J6	0015
.7J7	0071
.7J8	0071
.7J10	0071
.7J12	0102
.7J2100	0098
.7J2120	0186
.7JM13A	0365
.7JM13B	0365
.7JM13C	0365
.7JM13D	0365
.7JZ6.8	0025
.7JZ7.5	0649
.7JZ9.1	0012
.7JZ10	0170
.7JZ12	0137
.7JZ15	0002
.7JZ18	0495
.7JZ22	0565
.7JZ27	0441
.7JZ33	0039
.7JZ39	0350
.7JZ45	0999
.7JZ47	0999
.7JZ68	1268
.7JZ82	0331
.7JZ100	0098
.7JZ105	0098
.7JZ110	0405
.7JZ120	0186
.7JZ130	0213
.7JZ140	0245
.7JZ150	0028
.7JZ160	0255
.7JZ175	0877
.7JZ180	0367
.7JZ200	2858
.7JZ220	2351
.7Z6.8	0025
.7Z6.8A	0025
.7Z6.8B	0025
.7Z6.8C	0025
.7Z6.8D	0025
.7Z7.5	0649
.7Z7.5A	0649
.7Z7.5B	0649
.7Z7.5C	0649
.7Z7.5D	0649
.7Z8.2	0244
.7Z8.2A	0244
.7Z8.2B	0244
.7Z8.2C	0244
.7Z8.2D	0244
.7Z9.1	0012
.7Z9.1A	0012
.7Z9.1B	0012
.7Z9.1C	0012
.7Z9.1D	0012
.7Z10	0170
.7Z10A	0170
.7Z10B	0170
.7Z10C	0170
.7Z10D	0170
.7Z11	0317
.7Z11A	0317
.7Z11B	0317
.7Z11C	0317
.7Z11D	0317
.7Z12	0137
.7Z12A	0137
.7Z12B	0137
.7Z12C	0137
.7Z12D	0137
.7Z13	0365
.7Z13A	0365
.7Z13B	0365
.7Z13C	0365
.7Z13D	0365
.7Z14	0100
.7Z14A	0002
.7Z14B	0100
.7Z14C	0002
.7Z14D	0100
.7Z15	0002
.7Z15A	0002
.7Z15B	0137
.7Z15C	0002
.7Z15D	0002
.7Z16	0420
.7Z16A	0420
.7Z16B	0420
.7Z16C	0420
.7Z16D	0420
.7Z17	1649
.7Z17A	1649
.7Z17B	1649
.7Z17C	1649
.7Z17D	1649
.7Z18	0495
.7Z18A	0495
.7Z18B	0495
.7Z18C	0495
.7Z18D	0495
.7Z19	0949
.7Z19A	0949
.7Z19B	0949
.7Z19C	0949
.7Z19D	0949
.7Z20	0531
.7Z20A	0531
.7Z20B	0531
.7Z20C	0531
.7Z20D	0531
.7Z22	0565
.7Z22A	0565
.7Z22B	0565
.7Z22C	0565
.7Z22D	0565
.7Z24	0402
.7Z24A	0402
.7Z24B	0402
.7Z24C	0402
.7Z24D	0402
.7Z25	1606
.7Z25A	1606
.7Z25B	1606
.7Z25C	1606
.7Z25D	1606
.7Z27	0441
.7Z27A	0441
.7Z27B	0441
.7Z27C	0441
.7Z27D	0441
.7Z30	0726
.7Z30A	0726
.7Z30B	0726
.7Z30C	0726
.7Z30D	0726
.7Z33	0039
.7Z33A	0039
.7Z33B	0039
.7Z33C	0039
.7Z33D	0039
.7Z36	0820
.7Z36A	0820
.7Z36B	0820
.7Z36C	0820
.7Z36D	0820
.7Z39	0350
.7Z39A	0350
.7Z39B	0350
.7Z39C	0350
.7Z39D	0350
.7Z43	0931
.7Z43A	0931
.7Z43B	0931
.7Z43C	0931
.7Z43D	0931
.7Z45	0931
.7Z45A	0999
.7Z45B	0999
.7Z45C	0999
.7Z45D	0999
.7Z47	0999
.7Z47A	0999
.7Z47B	0999
.7Z47C	0999
.7Z47D	0999
.7Z50	0502
.7Z50A	0502
.7Z50B	0502
.7Z50C	0502
.7Z50D	0502
.7Z51	0502
.7Z51A	0502
.7Z51B	0502
.7Z51C	0502
.7Z51D	0502
.7Z52	0502
.7Z52A	0502
.7Z52B	0502
.7Z52C	0502
.7Z52D	0502
.7Z56	0869
.7Z56A	1834
.7Z56B	0869
.7Z56C	1834
.7Z56D	1834
.7Z62	0783
.7Z62A	0783
.7Z62B	0783
.7Z62C	0783
.7Z62D	0783
.7Z68	2157
.7Z68A	1268
.7Z68B	1268
.7Z68C	1268
.7Z68D	1268
.7Z75	1189
.7Z75A	1189
.7Z75B	1189
.7Z75C	1189
.7Z75D	1189
.7Z82	0331
.7Z82A	0331
.7Z82B	0331
.7Z82C	0331
.7Z82D	0331
.7Z91	1311
.7Z91A	1311
.7Z91B	1311
.7Z91C	1311
.7Z91D	1311
.7Z100	0098
.7Z100A	0098
.7Z100B	0098
.7Z100C	0098
.7Z100D	0098
.7Z105	0098
.7Z105A	0098
.7Z105B	0098
.7Z105C	0098
.7Z105D	OEM
.7Z110	0149
.7Z110A	0149
.7Z110B	0149
.7Z110C	0149
.7Z110D	0149
.7Z120	0186
.7Z120A	0186
.7Z120B	0186
.7Z120C	0186
.7Z120D	0186
.7Z130	0213
.7Z130A	0213
.7Z130B	0213
.7Z130C	0213
.7Z130D	0213
.7Z140	0245
.7Z140A	0245
.7Z140B	0245
.7Z140C	0245
.7Z140D	0245
.7Z150	0028
.7Z150A	0028
.7Z150B	0028
.7Z150C	0028
.7Z150D	0028
.7Z160	0255
.7Z160A	0255
.7Z160B	0255
.7Z160C	0255
.7Z160D	0255
.7Z175	0877
.7Z175A	0877
.7Z175B	0877
.7Z175C	0877
.7Z175D	OEM
.7Z180	0367
.7Z180A	0367
.7Z180B	0367
.7Z180C	0367
.7Z180D	0367
.7Z200	0421
.7Z200A	0421
.7Z200B	0421
.7Z200C	0421
.7Z200D	0421
.7Z220	OEM
.7Z220A	2351
.7Z220B	2351
.7Z220C	2351
.7Z220D	2351
.7ZM6.8	0025
.7ZM6.8A	0025
.7ZM6.8B	0025
.7ZM6.8C	0025
.7ZM6.8D	0025
.7ZM7.5	0649
.7ZM7.5A	0649
.7ZM7.5B	0649
.7ZM7.5C	0649
.7ZM7.5D	0649
.7ZM8.2	0244
.7ZM8.2A	0244
.7ZM8.2B	0244
.7ZM8.2C	0244
.7ZM8.2D	0244
.7ZM9.1	0012
.7ZM9.1A	0012
.7ZM9.1B	0012
.7ZM9.1C	0012
.7ZM9.1D	0012
.7ZM10	0170
.7ZM10A	0170
.7ZM10B	0170
.7ZM10C	0170
.7ZM10D	0170
.7ZM11	0317
.7ZM11A	0317
.7ZM11B	0317
.7ZM11C	0317
.7ZM11D	0317
.7ZM12	0137
.7ZM12A	0137
.7ZM12B	0137
.7ZM12C	0137
.7ZM12D	0137
.7ZM13	0365
.7ZM13A	0365
.7ZM13B	0365
.7ZM13C	0365
.7ZM13D	0365
.7ZM14	0100
.7ZM14A	0100
.7ZM14B	0100
.7ZM14C	0100
.7ZM14D	0100
.7ZM15	0002
.7ZM15A	0002
.7ZM15B	0002
.7ZM15C	0002
.7ZM15D	0002
.7ZM16	0420
.7ZM16A	0420
.7ZM16B	0420
.7ZM16C	0420
.7ZM16D	0420
.7ZM17	1649
.7ZM17A	1649
.7ZM17B	1649
.7ZM17C	1649
.7ZM17D	1649
.7ZM18	0495
.7ZM18A	0495
.7ZM18B	0495
.7ZM18C	0495
.7ZM18D	0495
.7ZM19	0949
.7ZM19A	0949
.7ZM19B	0949
.7ZM19C	0949
.7ZM19D	0949
.7ZM20	0531
.7ZM20A	0531
.7ZM20B	0531
.7ZM20C	0531
.7ZM20D	0531
.7ZM22	0565
.7ZM22A	0565
.7ZM22B	0565
.7ZM22C	0565
.7ZM22D	0565
.7ZM24	0402
.7ZM24A	0402
.7ZM24B	0402
.7ZM24C	0402
.7ZM24D	0402
.7ZM25	1606
.7ZM25A	1606
.7ZM25B	1606
.7ZM25C	1606
.7ZM25D	1606
.7ZM27	0441
.7ZM27A	0441
.7ZM27B	0441
.7ZM27C	0441
.7ZM27D	0441
.7ZM30	0726
.7ZM30A	0726
.7ZM30B	0726
.7ZM30C	0726
.7ZM30D	0726
.7ZM33	0039
.7ZM33A	0039
.7ZM33B	0039
.7ZM33C	0039
.7ZM33D	0039
.7ZM36	0820
.7ZM36A	0820
.7ZM36B	0820
.7ZM36C	0820
.7ZM36D	0820
.7ZM39	0350
.7ZM39A	0350
.7ZM39B	0350
.7ZM39C	0350
.7ZM39D	0350
.7ZM43	0931
.7ZM43A	0931
.7ZM43B	0931
.7ZM43C	0931
.7ZM43D	0931
.7ZM45	0931
.7ZM45A	0931
.7ZM45B	0931
.7ZM45C	0931
.7ZM45D	OEM
.7ZM47	0999
.7ZM47A	0999
.7ZM47B	0999
.7ZM47C	0999
.7ZM47D	0999
.7ZM50	0502
.7ZM50A	0502
.7ZM50B	0502
.7ZM50C	0502
.7ZM50D	0502
.7ZM51	0502
.7ZM51A	0502
.7ZM51B	0502
.7ZM51C	0502
.7ZM51D	0502
.7ZM52	0502
.7ZM52A	0502
.7ZM52B	0502
.7ZM52C	0502
.7ZM52D	0502
.7ZM56	0869
.7ZM56A	0869
.7ZM56B	0869
.7ZM56C	0869
.7ZM56D	0869
.7ZM62	0783
.7ZM62A	0783
.7ZM62B	0783
.7ZM62C	0783
.7ZM62D	0783
.7ZM68	2157
.7ZM68A	1268
.7ZM68B	1268
.7ZM68C	1268
.7ZM68D	1268
.7ZM75	1189
.7ZM75A	1189
.7ZM75B	1189
.7ZM75C	1189
.7ZM75D	1189
.7ZM82	0331
.7ZM82A	0331
.7ZM82B	0331
.7ZM82C	0331
.7ZM82D	0331
.7ZM91	1311
.7ZM91A	1311
.7ZM91B	1311
.7ZM91C	1311
.7ZM91D	1311
.7ZM100	0098
.7ZM100A	0098
.7ZM100B	0098
.7ZM100C	0098
.7ZM100D	0098
.7ZM105	0098
.7ZM105A	0098
.7ZM105B	0098
.7ZM105C	0098
.7ZM105D	OEM
.7ZM110	0149
.7ZM110A	0149
.7ZM110B	0149
.7ZM110C	0149
.7ZM110D	0149
.7ZM120	0186
.7ZM120A	0186
.7ZM120B	0186
.7ZM120C	0186
.7ZM120D	0186
.7ZM130	0213
.7ZM130A	0213
.7ZM130B	0213
.7ZM130C	0213
.7ZM130D	0213
.7ZM140	0245
.7ZM140A	0245
.7ZM140B	0245
.7ZM140C	0245
.7ZM140D	0245
.7ZM150	0028
.7ZM150A	0028
.7ZM150B	0028
.7ZM150C	0028
.7ZM150D	0028
.7ZM160	0255
.7ZM160A	0255
.7ZM160B	0255
.7ZM160C	0255
.7ZM160D	0255
.7ZM175	0877
.7ZM175A	0877
.7ZM175B	0877
.7ZM175C	0877
.7ZM175D	OEM
.7ZM180	0367
.7ZM180A	0367
.7ZM180B	0367
.7ZM180C	0367
.7ZM180D	0367
.7ZM200	0421
.7ZM200A	0421
.7ZM200B	0421
.7ZM200C	0421
.7ZM200D	0421
.7ZM220	OEM
.7ZM220A	2351
.7ZM220B	2351
.7ZM220C	2351
.7ZM220D	OEM
.25N6.8	0062
.25N7.5	0077
.25N8.2	0165
.25N9.1	0057
.25N10	0064
.25N11	0181
.25N12	0052
.25N13	0053
.25N14	0879
.25N15	0686
.25N16	0445
.25N17	0210
.25N18	0375
.25N19	0671
.25N20	0700
.25N22	0705
.25N24	0494
.25N25	0714
.25N27	0455
.25N30	0195
.25N33	0166
.25N36	0010
.25N39	0032
.25N43	0054
.25N45	0068
.25N47	0068
.25N50	0068
.25N52	0068
.25N56	0125
.25N62	0152
.25N68	0173
.25N75	0094
.25N82	0049
.25N91	0156
.25N100	0189
.25N105	0099
.25N110	0099
.25N120	0089
.25N130	0287
.25N140	0252
.25N150	0340
.25N175	0425
.25N200	1473
.25NN15	0686
.25P25-.75S	OEM
.25P25A-.75	OEM
.25PL4-1.2	OEM
.25PL4-1.2S	OEM
.25P.35-1	OEM
.25P.35-1S	OEM
.25T5.6	0157
.25T5.6A	0157
.25T5.6B	0157
.25T6.8	0025
.25T6.8A	0025
.25T7.1	0025
.25T7.1A	0025
.25T7.5	0649
.25T7.5A	0649
.25T7.5B	0649
.25T8.2	0244
.25T8.8A	1083
.25T9.1	0012
.25T9.1A	0012
.25T9.1B	0012
.25T10	0170
.25T10A	0170
.25T10B	0170
.25T10.5	0170
.25T10.5A	0170
.25T12	0137
.25T12A	0137
.25T12B	0137
.25T15	0002
.25T15A	0002
.25T15B	0002
.25T18	0495
.25T18A	0495
.25T22	0565
.25T22A	0565
.25T24.	0402
.25T27	0441
.25T27A	0441
.25T27B	0441
.25T33	0039
.25T33A	0039
.25T33B	0039
.25T39	0350
.25T39A	0350
.25T47	0999
.25T47A	0999
.25T56	0869
.25T56A	0869
.25T68	3204
.25T68A	3204
.25T82	0331
.25T82A	0331
.25T82B	0331
.25T87A	3026
.25T100	0098
.25T100A	0098
.25T110A	0405
.25T120	0186
.25T120A	0186
.25T140	0245
.25T150	0028
.25T150A	0028
.25T180	0367
.25T180A	0367
.72M6.8A	0025
.72M6.8B	0025
.72M6.8C	0025
.72M6.8D	0025
.72M16A	0420
.72M20B	0531
.72M20C	0531
.72M20D	0531
.72M36C	0820
.72M36D	0820
.75N5	0002
.75N5.1	0162
.75N5.6	0157
.75N6.2	0636
.75N7.5	0649
.75N12	0137
.75N27	0441
.75N33	0039
.75N82	0331

DEVICE TYPE	REPL CODE
.75N110	0149
.728.2A	0244
.728.2B	0244
.728.2C	0244
.MZ18D	0495
.MZ24C	0402
.MZ110B	0149
.MZM14A	0100
1/47M.1AZ5	0162
1/4A6.8	0062
1/4A6.8A	0062
1/4A6.8B	0062
1/4A7.5	0077
1/4A7.5A	0077
1/4A7.5B	0077
1/4A8.2	0165
1/4A8.2A	0165
1/4A8.2B	0165
1/4A9.1	0057
1/4A9.1A	0057
1/4A9.1B	0057
1/4A10	0170
1/4A10A	0170
1/4A10B	0170
1/4A11	0181
1/4A11A	0181
1/4A11B	0181
1/4A12	0052
1/4A12A	0052
1/4A12B	0052
1/4A13	0053
1/4A13A	0053
1/4A13B	0053
1/4A14	0100
1/4A14A	0100
1/4A14B	0100
1/4A15	0686
1/4A15A	0686
1/4A15B	0686
1/4A16	0445
1/4A16A	0445
1/4A16B	0445
1/4A17	0210
1/4A17A	0210
1/4A17B	0210
1/4A18	0375
1/4A18A	0375
1/4A18B	0375
1/4A19	0671
1/4A19A	0671
1/4A19B	0671
1/4A20	0700
1/4A20A	0700
1/4A20B	0700
1/4A22	0705
1/4A22A	0705
1/4A22B	0705
1/4A24	0494
1/4A24A	0494
1/4A24B	0494
1/4A25	0714
1/4A25A	0714
1/4A25B	0714
1/4A27	0441
1/4A27A	0441
1/4A27B	0441
1/4A30	0195
1/4A30A	0195
1/4A30B	0195
1/4A33	0039
1/4A33A	0039
1/4A33B	0039
1/4A36	0010
1/4A36A	0010
1/4A36B	0010
1/4A39	0032
1/4A39A	0032
1/4A39B	0032
1/4A43	0054
1/4A43A	0054
1/4A43B	0054
1/4A45	0054
1/4A45A	0054
1/4A45B	0054
1/4A47	0068
1/4A47A	0068
1/4A47B	0068
1/4A50	0092
1/4A50A	0092
1/4A50B	0092
1/4A52	0092
1/4A52A	0092
1/4A52B	0092
1/4A56	0125
1/4A56A	0125
1/4A56B	0125
1/4A62	0783
1/4A62A	0783
1/4A62B	0783
1/4A68	0173
1/4A68A	0173
1/4A68B	0173
1/4A75	0094
1/4A75A	0094
1/4A75B	0094
1/4A82	0331
1/4A82A	0331
1/4A82B	0331
1/4A91	0156

DEVICE TYPE	REPL CODE
1/4A91A	0156
1/4A91B	0156
1/4A100	0189
1/4A100A	0189
1/4A100B	0189
1/4A105	0189
1/4A105A	0189
1/4A105B	0394
1/4A110	0149
1/4A110A	0149
1/4A110B	0149
1/4A120	0089
1/4A120A	0089
1/4A120B	0089
1/4A130	0287
1/4A130A	0287
1/4A130B	0287
1/4A140	0252
1/4A140A	0252
1/4A140B	0252
1/4A150	0340
1/4A150A	0340
1/4A150B	0340
1/4A175	0394
1/4A175A	0394
1/4A175B	0394
1/4A200	1473
1/4A200A	1473
1/4A200B	1473
1/4AZ2.2D	OEM
1/4AZ2.2D5	OEM
1/4AZ2.2D10	OEM
1/4AZ2.4D5	1276
1/4AZ2.4D10	1276
1/4AZ2.7D	0760
1/4AZ2.7D5	0760
1/4AZ2.7D10	0760
1/4AZ3.0D	0118
1/4AZ3.0D5	0118
1/4AZ3.0D10	0118
1/4AZ3.3D	0299
1/4AZ3.3D5	0299
1/4AZ3.3D10	0299
1/4AZ3.6D	0376
1/4AZ3.6D5	0188
1/4AZ3.6D10	0188
1/4AZ3.9D	0036
1/4AZ3.9D5	0036
1/4AZ3.9D10	0036
1/4AZ4.3D	0275
1/4AZ4.3D5	0275
1/4AZ4.3D10	0275
1/4AZ4.7D	0140
1/4AZ4.7D5	0140
1/4AZ4.7D10	0140
1/4AZ5.1D	0162
1/4AZ5.1D5	0162
1/4AZ5.1D10	0162
1/4AZ5.6D	0157
1/4AZ5.6D5	0157
1/4AZ5.6D10	0157
1/4AZ6.2D	0471
1/4AZ6.2D5	0471
1/4AZ6.2D10	0471
1/4AZ6.8D	0062
1/4AZ6.8D5	0062
1/4AZ6.8D10	0062
1/4LZ2.2D	OEM
1/4LZ2.2D5	OEM
1/4LZ2.2D10	OEM
1/4LZ2.4D	1276
1/4LZ2.4D5	1276
1/4LZ2.4D10	1276
1/4LZ2.7D	0760
1/4LZ2.7D5	0760
1/4LZ2.7D10	0760
1/4LZ3.0D	0118
1/4LZ3.0D5	0118
1/4LZ3.0D10	0118
1/4LZ3.3D	0299
1/4LZ3.3D5	0299
1/4LZ3.3D10	0299
1/4LZ3.6D	0211
1/4LZ3.6D5	0211
1/4LZ3.6D10	0376
1/4LZ3.9D	0036
1/4LZ3.9D5	0036
1/4LZ3.9D10	0036
1/4LZ4.3D	0275
1/4LZ4.3D5	0275
1/4LZ4.3D10	0275
1/4LZ4.7D	0140
1/4LZ4.7D5	0140
1/4LZ4.7D10	0140
1/4LZ5.1D	0041
1/4LZ5.1D5	0041
1/4LZ5.1D10	0041
1/4LZ5.6D	0157
1/4LZ5.6D5	0157
1/4LZ5.6D10	0157
1/4LZ6.2D	0471
1/4LZ6.2D5	0471
1/4LZ6.2D10	0471
1/4LZ6.8D	0062
1/4LZ6.8D5	0062
1/4LZ6.8D10	0062
1/4M2.4AZ	1276
1/4M2.4AZ5	1276
1/4M2.4AZ10	1276

DEVICE TYPE	REPL CODE
1/4M2.4AZ10.5	OEM
1/4M2.7AZ	0760
1/4M2.7AZ5	0760
1/4M2.7AZ10	0760
1/4M2.7AZ10.5	OEM
1/4M3.0AZ	0118
1/4M3.0AZ5	0118
1/4M3.0AZ10	0118
1/4M3.3AZ	0299
1/4M3.3AZ5	0299
1/4M3.3AZ10	0299
1/4M3.3AZ10.5	OEM
1/4M3.6AZ	0376
1/4M3.6AZ5	0188
1/4M3.6AZ10	0188
1/4M3.6AZ10.5	OEM
1/4M3.9AZ	0036
1/4M3.9AZ5	0036
1/4M3.9AZ10	0036
1/4M4.3AZ	0275
1/4M4.3AZ5	0275
1/4M4.3AZ10	0275
1/4M4.3AZ10.5	OEM
1/4M4.7AZ	0140
1/4M4.7AZ5	0140
1/4M4.7AZ10	0140
1/4M5.1	0041
1/4M5.1AZ	0041
1/4M5.1AZ5	0162
1/4M5.1AZ10	0041
1/4M5.1AZ25	0162
1/4M5.6AZ	0157
1/4M5.6AZ5	0157
1/4M5.6AZ10	0253
1/4M5.6AZ10.5	OEM
1/4M5.6AZ25	0157
1/4M6.2AZ	0471
1/4M6.2AZ5	0471
1/4M6.2AZ10	0471
1/4M6.2AZ10.5	OEM
1/4M6.8AZ	0062
1/4M6.8AZ5	0062
1/4M6.8AZ10	0062
1/4M6.8Z	0062
1/4M6.8Z5	0062
1/4M6.8Z10	0062
1/4M6.8Z10.5	OEM
1/4M7.5Z	0077
1/4M7.5Z5	0077
1/4M7.5Z10	0077
1/4M8/27	0165
1/4M8.2Z	0165
1/4M8.2Z5	0165
1/4M8.2Z10	0165
1/4M9.1Z	0057
1/4M9.1Z5	0057
1/4M9.1Z10	0057
1/4M9.1Z10.5	OEM
1/4M10Z	0170
1/4M10Z5	0064
1/4M10Z10	0170
1/4M10Z10.5	OEM
1/4M11Z5	0181
1/4M11Z10	0181
1/4M12Z	0052
1/4M12Z5	0052
1/4M12Z10	0052
1/4M12Z10.5	OEM
1/4M13Z	OEM
1/4M13Z5	0365
1/4M13Z10	0053
1/4M14	0100
1/4M14Z	0100
1/4M14Z5	0100
1/4M14Z10	0064
1/4M15Z	0002
1/4M15Z5	0686
1/4M15Z10	0686
1/4M16Z5	0420
1/4M16Z10	0445
1/4M17Z10	0210
1/4M18Z5	0495
1/4M18Z10	0375
1/4M19Z10	0671
1/4M20Z10	0700
1/4M22Z	0705
1/4M22Z10	0705
1/4M24Z	0494
1/4M24Z5	0494
1/4M24Z10	0494
1/4M24Z10.5	OEM
1/4M25Z5	0714
1/4M25Z10	0714
1/4M27Z	0441
1/4M27Z5	0441
1/4M27Z10	0455
1/4M30Z10	0195
1/4M33Z	0039
1/4M33Z5	0039
1/4M33Z10	0166
1/4M36Z10	0010
1/4M39Z10	0032
1/4M43Z10	0054
1/4M45Z10	0068
1/4M47Z10	0068
1/4M50Z10	0092
1/4M51Z	0092
1/4M51Z5	0092
1/4M51Z10	0092

DEVICE TYPE	REPL CODE
1/4M52Z10	0092
1/4M56Z10	0125
1/4M62Z	0783
1/4M62Z5	0783
1/4M62Z10	0152
1/4M68Z10	0173
1/4M68.Z10	0025
1/4M75Z10	0094
1/4M82Z	0331
1/4M82Z5	0331
1/4M82Z10	0049
1/4M91Z10	0156
1/4M100Z10	0189
1/4M105Z10	0099
1/4M110Z	0149
1/4M110Z5	0149
1/4M110Z10	0149
1/4M120Z10	0089
1/4M150Z	0340
1/4M150Z5	0340
1/4M150Z10	0340
1/4M160Z	0370
1/4M160Z5	0370
1/4M160Z10	0370
1/4M175Z	0394
1/4M175Z5	0394
1/4M175Z10	0394
1/4M180	0425
1/4M180Z	0425
1/4M180Z5	0425
1/4M180Z10	0425
1/4M200Z	1473
1/4M200Z5	1473
1/4M200Z10	1473
1/4M227	0565
1/4M3025	0726
1/4Z5.6T5	0253
1/4Z6.8D	0062
1/4Z6.8D5	0062
1/4Z6.8D10	0062
1/4Z7.5D	0077
1/4Z7.5D5	0077
1/4Z7.5D10	0077
1/4Z8.2D	0165
1/4Z8.2D5	0165
1/4Z8.2D10	0165
1/4Z9.1D	0057
1/4Z9.1D5	0140
1/4Z9.1D10	0057
1/4Z10D	0064
1/4Z10D5	0064
1/4Z10D10	0064
1/4Z11D	0181
1/4Z11D5	0181
1/4Z11D10	0181
1/4Z12D	0052
1/4Z12D5	0052
1/4Z12D10	0052
1/4Z12T5	0137
1/4Z13D	0053
1/4Z13D5	0053
1/4Z13D10	0053
1/4Z14D	0100
1/4Z14D5	0100
1/4Z14D10	0100
1/4Z15D	0686
1/4Z15D5	0686
1/4Z15D10	0686
1/4Z15T5	0686
1/4Z16D	0445
1/4Z16D5	0445
1/4Z16D10	0445
1/4Z17D	0210
1/4Z17D5	0210
1/4Z17D10	0210
1/4Z18D	0375
1/4Z18D5	0375
1/4Z18D10	0375
1/4Z19D	0671
1/4Z19D5	0671
1/4Z19D10	0671
1/4Z20D	0700
1/4Z20D5	0700
1/4Z20D10	0700
1/4Z22D	0705
1/4Z22D5	0705
1/4Z22D10	0705
1/4Z24D	0494
1/4Z24D5	0494
1/4Z24D10	0494
1/4Z25D	0714
1/4Z25D5	0714
1/4Z25D10	0714
1/4Z27D	0441
1/4Z27D5	0441
1/4Z27D10	0441
1/4Z27T5	0455
1/4Z30D	0195
1/4Z30D5	0195
1/4Z30D10	0195
1/4Z33D	0039
1/4Z33D5	0039
1/4Z33D10	0039
1/4Z33T5	0166
1/4Z36D	0010
1/4Z36D5	0010
1/4Z36D10	0010
1/4Z39D	0032
1/4Z39D5	0032

DEVICE TYPE	REPL CODE
1/4Z39D10	0032
1/4Z43D	0054
1/4Z43D5	0054
1/4Z43D10	0054
1/4Z45D	0054
1/4Z45D5	0054
1/4Z45D10	0054
1/4Z47D	0068
1/4Z47D5	0068
1/4Z47D10	0068
1/4Z50D	0092
1/4Z50D5	0092
1/4Z50D10	0092
1/4Z51D	0092
1/4Z51D5	0092
1/4Z51D10	0092
1/4Z52D	0092
1/4Z52D5	0092
1/4Z52D10	0092
1/4Z56D	0125
1/4Z56D5	0125
1/4Z56D10	0125
1/4Z62D	0783
1/4Z62D5	0783
1/4Z62D10	0783
1/4Z68D	0173
1/4Z68D5	0173
1/4Z68D10	0173
1/4Z75D	0094
1/4Z75D5	0094
1/4Z75D10	0094
1/4Z82D	0331
1/4Z82D5	0049
1/4Z82D10	0331
1/4Z91D	0156
1/4Z91D5	0156
1/4Z91D10	0156
1/4Z100D	0189
1/4Z100D5	0189
1/4Z100D10	0189
1/4Z105D	0189
1/4Z105D5	0189
1/4Z105D10	0189
1/4Z110D	0149
1/4Z110D5	0149
1/4Z110D10	0149
1/4Z120D	0089
1/4Z120D5	0089
1/4Z120D10	0089
1/4Z130D	0287
1/4Z130D5	0287
1/4Z130D10	0287
1/4Z140D	0252
1/4Z140D5	0252
1/4Z140D10	0252
1/4Z150D	0340
1/4Z150D5	0340
1/4Z150D10	0340
1/4Z160D	0370
1/4Z160D5	0370
1/4Z160D10	0370
1/4Z175D	0394
1/4Z175D5	0394
1/4Z175D10	0394
1/4Z180D	0425
1/4Z180D5	0425
1/4Z180D10	0425
1/4Z200D	1473
1/4Z200D5	1473
1/4Z200D10	1473
1/276.5T5	0649
1/2Z2TT5	0441
1/2Z3.3T5	0299
1/2Z3.5T5	OEM
1/2Z3.6T5	0376
1/2Z3.9T5	0036
1/2Z4.3T5	0275
1/2Z4.7T5	0140
1/2Z5.1T5	0041
1/2Z5.6T5	0253
1/2Z6.2T5	0471
1/2Z6.5T5	0157
1/2Z6.8T5	0062
1/2Z7.5T5	0077
1/2Z8.2T5	0165
1/2Z9.1T5	0057
1/2Z10T5	0064
1/2Z11T5	0181
1/2Z12T5	0052
1/2Z13T5	0053
1/2Z15T5	0686
1/2Z16T5	0445
1/2Z18T5	0375
1/2Z20T5	0700
1/2Z22T5	0705
1/2Z24T5	0494
1/2Z27T5	0455
1/2Z30T5	0195
1/2Z33T5	0170
1/2Z36T5	0010
1/2Z39T5	0032
1/2Z43T5	0054
1/2Z47T5	0068
1/2Z51T5	0068
1/2Z56T5	0125
1/2Z62T5	0152
1/2Z68T5	0173
1/2Z75T5	0094
1/2Z82T5	0049

DEVICE TYPE	REPL CODE
1/2Z91T5	0156
1/2Z100T5	0189
1/2Z105T5	0170
1/2Z276.5T5	0012
3/4D25D5	OEM
3/4M14Z	0100
3/4M14Z5	0100
3/4M14Z10	0100
3/4M17Z	1649
3/4M17Z5	1649
3/4M17Z10	1649
3/4M19Z	0949
3/4M19Z5	0949
3/4M19Z10	0949
3/4M25Z	1606
3/4M25Z5	1606
3/4M25Z10	1606
3/4M140Z	0245
3/4M140Z5	0245
3/4M140Z10	0245
3/4T12D5	0137
3/4Z12D5	0137
3/4Z12D10	0137
3/4Z14D	0100
3/4Z14D5	0100
3/4Z14D10	0100
3/4Z17D	1649
3/4Z17D5	1649
3/4Z17D10	1649
3/4Z19D	0949
3/4Z19D5	0949
3/4Z19D10	0949
3/4Z25D	1606
3/4Z25D10	1606
3/4Z45D	OEM
3/4Z45D5	OEM
3/4Z45D10	OEM
3/4Z50D	OEM
3/4Z50D5	OEM
3/4Z50D10	OEM
3/4Z52D	OEM
3/4Z52D5	OEM
3/4Z52D10	OEM
3/4Z105D	OEM
3/4Z105D5	OEM
3/4Z105D10	OEM
3/4Z140D	0245
3/4Z140D5	0245
3/4Z140D10	0245
3/4Z175D	OEM
3/4Z175D5	OEM
3/4Z175D10	OEM
1/042/2207	0079
1-00004-064	0124
1-00006-003	0079
1-00006-007	0143
1-00006-281	0124
1-000-099-00	1208
1-0002-001	0124
1-00011-043	0012
1-00030-010	0229
1-00032-027	0208
1-00038-008	0124
1-001	OEM
1-001/2207	0133
1-001-003-15	0016
1-014/2207	0133
1-034/2207	0037
1-035/2207	0086
1-037/2207	0143
1-041/2207	0144
1-042/2207	0016
1-043/2207	0037
1-044/2207	0016
1-20-001-890	0015
1-20-001890	0015
1-21-73	0280
1-21-74	0280
1-21-75	0280
1-21-76	0280
1-21-78	0280
1-21-83	0280
1-21-91	0280
1-21-92	0280
1-21-93	0280
1-21-95	0004
1-21-96	0004
1-21-100	0280
1-21-102	0280
1-21-103	0280
1-21-104	0280
1-21-105	0280
1-21-106	0004
1-21-107	0004
1-21-120	0004
1-21-128	0280
1-21-135	0050
1-21-137	0050
1-21-138	0136
1-21-139	0050
1-21-148	0004
1-21-150	0050
1-21-157	0050
1-21-161	0280
1-21-162	0280
1-21-164	0004
1-21-179	0280
1-21-180	0280
1-21-184	0004

DEVICE TYPE	REPL CODE
1-21-186	0280
1-21-189	0136
1-21-190	0050
1-21-191	0004
1-21-192	0004
1-21-225	0004
1-21-226	0004
1-21-227	0004
1-21-228	0050
1-21-229	0050
1-21-230	0050
1-21-231	0050
1-21-232	0004
1-21-233	0050
1-21-234	0280
1-21-235	0280
1-21-236	0280
1-21-240	0280
1-21-241	0280
1-21-242	0136
1-21-243	0136
1-21-244	0136
1-21-246	0004
1-21-254	0280
1-21-256	0050
1-21-257	0136
1-21-258	0050
1-21-259	0050
1-21-260	0050
1-21-266	0004
1-21-267	0004
1-21-270	0160
1-21-271	0160
1-21-272	0004
1-21-273	0280
1-21-274	0004
1-21-275	0280
1-21-276	0016
1-21-277	0016
1-21-278	0016
1-21-279	0016
1-21-289	0280
1-63-5209	OEM
1-101	0015
1-231-416-00	OEM
1-231-443	OEM
1-231-443-00	OEM
1-231-462-00	OEM
1-231-462-11	OEM
1-231-705-00	OEM
1-232-680-11	0471
1-235-031-00	OEM
1-235-034-00	OEM
1-235-058-00	OEM
1-235-120-00	OEM
1-235-217-00	OEM
1-235-217-11	OEM
1-235-223-11	OEM
1-235-444-11	1421
1-235-783-11	OEM
1-235-783-21	OEM
1-235-784-11	OEM
1-235-784-12	OEM
1-235-963-11	OEM
1-235-971-12	0520
1-424-115-11	3574
1-425-636	0143
1-453-032-13	OEM
1-453-032-16	OEM
1-453-041-12	OEM
1-453-043-00	OEM
1-453-050-00	OEM
1-453-050-31	OEM
1-453-063-00	OEM
1-453-067-13	OEM
1-453-071-00	OEM
1-453-082-00	OEM
1-464-756-21	OEM
1-506-319	0087
1-530-012-11	0015
1-531-016	0015
1-531-024	0289
1-531-027	0015
1-531-028	0288
1-531-028-21	0288
1-531-036-11	0774
1-531-052	0201
1-531-055	0201
1-531-105	0015
1-531-105-11	0015
1-531-105-13	0015
1-531-106	0015
1-531-106-13	0015
1-531-106-17	0015
1-534-105-13	0015
1-534-106-13	0015
1-590-1261	1024
1-800-662-11	1479
1-800-945-00	OEM
1-801-003	0016
1-801-003-12	0144
1-801-003-13	0144
1-801-003-14	0144
1-801-003-15	0144
1-801-004	0016
1-801-004-17	0016
1-801-005	0164
1-801-005-23	0004
1-801-006	0164

DEVICE TYPE	REPL CODE
1-801-006-12	0004
1-801-006-14	0004
1-801-009-11	2353
1-801-010	OEM
1-801-099-11	4233
1-801-301-13	0142
1-801-301-14	0142
1-801-301-15	0142
1-801-304-15	0127
1-801-305-13	0144
1-801-306	0144
1-801-306-13	0144
1-801-306-14	0144
1-801-306-15	0144
1-801-308	0004
1-801-308-24	0004
1-801-309	0208
1-801-310	0164
1-801-314	0016
1-801-314-15	0016
1-801-314-16	0016
1-801-951	OEM
1-805-104	OEM
1-805-104-11	OEM
1-805-105-11	1634
1-806-214-11	OEM
1-806-476-00	1414
1-806-549-41	0017
1-807-643-11	OEM
1-807-745-11	OEM
1-808-592-11	0031
1-808-919-11	OEM
1-808-948-11	OEM
1-809-011-11	OEM
1-809-120-21	OEM
1-809-365-11	OEM
1-809-366-11	OEM
1-809-401-11	OEM
1-809-401-31	OEM
1-810-039-11	OEM
1-810-050-11	OEM
1-810-052-21	OEM
1-6732	0423
1-8259	0015
1-12689	0123
1-16549	0133
1-18341	0030
1-20363	0057
1-20398	0064
1-40159E01	OEM
1-79990A	0086
1-464607-3	0409
1-464607-5	0409
1-464860-1	0409
1-464865-1	0409
1-464984-1	0409
1-466860-1	0409
1-52221011	0004
1-522210111	0004
1-522210131	0050
1-522210300	0050
1-522210921	0050
1-522211021	0050
1-522211200	0004
1-522211328	0004
1-522211921	0050
1-522214400	0050
1-522214411	0050
1-522214435	0050
1-522214821	0050
1-522214831	0050
1-522216500	0004
1-522216600	0050
1-522217400	0050
1-522223720	0016
1-6147191229	0016
1-6171191368	0079
1-6207190405	0004
1-6501190016	0015
1-B527-062	0196
1-TD-2684	0004
1-TR-016	2295
1+12/1	0015
1A0013	0622
1A0020	0475
1A0021	0155
1A0022	0155
1A0024	0111
1A0025	0079
1A0027	0178
1A0029	0079
1A0032	0079
1A0033	0155
1A0034	0016
1A0035	0016
1A0037	0155
1A0038	0142
1A0043	0016
1A0044	0127
1A0045	0155
1A0046	0236
1A0048	0178
1A0051	0155
1A0055	0004
1A0056	0004
1A0058	0042
1A0059	0236
1A0063	0016
1A0066	0086

DEVICE TYPE	REPL CODE
1A0067	0155
1A0070	0079
1A0076	0079
1A0077	0079
1A0078	0079
1A0079	0016
1A0080	0016
1A0081	0016
1A0083	0155
1A0084	0155
1A01	0086
1A02	0126
1A03	0086
1A04	4587
1A05	0126
1A06	0086
1A07	0086
1A08	0126
1A09	0187
1A1/18	OEM
1A1/52	OEM
1A2/18	OEM
1A2/52	OEM
1A3	OEM
1A3-F	OEM
1A3F	OEM
1A5/18	OEM
1A5/52	OEM
1A6.8M	0025
1A6.8MA	0025
1A6.8MB	0025
1A7.5M	0649
1A7.5MA	0649
1A7.5MB	0649
1A8-1A82	0050
1A8.2M	0244
1A8.2MA	0244
1A8.2MB	0244
1A9.1M	0012
1A9.1MA	0012
1A9.1MB	0012
1A10M	0170
1A10MA	0170
1A10MB	0170
1A11	0187
1A11M	0317
1A11MA	0317
1A11MB	0317
1A12	OEM
1A12/18	OEM
1A12M	0137
1A12MA	0137
1A12MB	0137
1A13	0187
1A13/18	OEM
1A13M	0365
1A13MA	0365
1A13MB	0365
1A14	OEM
1A14/18	OEM
1A15	0187
1A15/18	OEM
1A15M	0002
1A15MA	0002
1A15MB	0002
1A16	0338
1A16/18	OEM
1A16M	0420
1A16MA	0420
1A16MB	0420
1A17	0086
1A18	0086
1A18M	0495
1A18MA	0495
1A18MB	0495
1A19	0126
1A20M	0531
1A20MA	0531
1A20MB	0531
1A22	OEM
1A22/18	OEM
1A22M	0565
1A22MA	0565
1A22MB	0565
1A23	OEM
1A23/18	OEM
1A24	OEM
1A24/18	OEM
1A24M	0402
1A24MA	0402
1A24MB	0402
1A25	OEM
1A25/18	OEM
1A26	OEM
1A26/18	OEM
1A27M	0441
1A27MA	0441
1A27MB	0441
1A30M	0726
1A30MA	0726
1A30MB	0726
1A33M	0039
1A33MA	0039
1A33MB	0039
1A34(R)	0595
1A36M	0820
1A36MA	0820
1A36MB	0820
1A37	OEM

DEVICE TYPE	REPL CODE
1A38	0595
1A38(R)	0595
1A38R	0595
1A39M	0350
1A39MA	0350
1A39MB	0350
1A40	OEM
1A43M	0931
1A43MA	0931
1A43MB	0931
1A47M	0999
1A47MA	0999
1A47MB	0999
1A48B	OEM
1A48C	OEM
1A48FB	OEM
1A48FC	OEM
1A51M	0502
1A51MA	0502
1A51MB	0502
1A52	1276
1A52/18	OEM
1A53	0760
1A53/18	OEM
1A54	0118
1A54/18	OEM
1A55	0299
1A55/18	OEM
1A56	0376
1A56/18	OEM
1A56M	1834
1A56MA	1834
1A56MB	0869
1A57	0036
1A58	0275
1A59	0140
1A60	0041
1A61	0782
1A62	0471
1A62M	0783
1A62MA	0783
1A62MB	0783
1A68M	2157
1A68MA	2157
1A68MB	2157
1A75M	1189
1A75MA	1189
1A75MB	1189
1A82M	0331
1A82MA	0331
1A82MB	0331
1A83	OEM
1A91M	1311
1A91MA	1311
1A91MB	1311
1A100M	0098
1A100MA	0098
1A100MB	0098
1A104	OEM
1A110M	0149
1A110MA	0149
1A110MB	0149
1A120M	0186
1A120MA	0186
1A120MB	0186
1A124	OEM
1A129	OEM
1A130	OEM
1A130M	0213
1A130MA	0213
1A130MB	0213
1A150M	0028
1A150MA	0028
1A150MB	0028
1A160M	0255
1A160MA	0255
1A160MB	0255
1A180M	0367
1A180MA	0367
1A180MB	0367
1A200M	0421
1A200MA	0421
1A200MB	0421
1A348(R)	0595
1A4102	3190
1A10425	0015
1A10952	0102
1A11184	0015
1A11306	0143
1A11671	0015
1A12214	0015
1A12407	0015
1A12687	0030
1A12688	0012
1A12689	0143
1A12690	0015
1A13219	0015
1A13719	0015
1A13720	0015
1A14384	0143
1A15790	0015
1A16549	0133
1A16550	0015
1A16551	0133
1A17273	OEM
1A66102-02C	OEM
1A66102-03C	OEM
1A66400-05B	OEM
1AC12	0137

DEVICE TYPE	REPL CODE
1AC12A	0137
1AC12B	0137
1AC15	0002
1AC15A	0002
1AC15B	0002
1AC27	0441
1AC27A	0441
1AC27B	0441
1AC33	0039
1AC33A	0039
1AC33B	0039
1AC62	0783
1AC62A	0783
1AC62B	0783
1AC82	0331
1AC82A	0331
1AC82B	0331
1AC110	0149
1AC110A	0149
1AC110B	0149
1AMH2	0348
1AR	OEM
1AS027	0087
1AS029	0087
1AU3	1306
1AU601	OEM
1AU601A	OEM
1AU601B	OEM
1AV20	OEM
1AV30	OEM
1AV35	OEM
1AV40	OEM
1AV48	OEM
1AV50	OEM
1AV51	OEM
1AV56	OEM
1AV57	OEM
1AV58	OEM
1AV59	OEM
1AV60	OEM
1AV61	OEM
1AV62	OEM
1AV65	OEM
1AV67	OEM
1AV68	OEM
1AV69	OEM
1AV80	OEM
1AV81	OEM
1AV82	OEM
1AV83	OEM
1AV84	OEM
1AV85	OEM
1AV86	OEM
1AV87	OEM
1AV88	OEM
1AV89	OEM
1AV95	OEM
1AV96	OEM
1AV97	OEM
1AV97A	OEM
1AV98	OEM
1AV98A	OEM
1AV99	OEM
1AV99A	OEM
1AV100	OEM
1AV120	OEM
1AV130	OEM
1AZ6.8	0025
1AZ7.5	0649
1AZ8.2	0244
1AZ9.1	0012
1AZ10	0170
1AZ11	0317
1AZ12	0137
1AZ13	0365
1AZ15	0002
1AZ16	0420
1AZ18	0495
1AZ20	0531
1AZ22	0565
1AZ24	0274
1AZ27	0441
1AZ30	0726
1AZ33	0039
1AZ36	0820
1AZ47	0999
1AZ51	0502
1AZ200	0421
1AZ270	OEM
1AZ300	OEM
1AZ330	OEM
1B01	0620
1B04	0074
1B05	0277
1B05J05	0277
1B05J20	0015
1B05J40	0015
1B06	0177
1B07	2427
1B08	2427
1B08T05	0277
1B08T10	0277
1B08T20	0289
1B09	0363
1B1	0277
1B1/18	OEM
1B1/52	OEM
1B1.1	0106
1B2	0289

DEVICE TYPE	REPL CODE
1B2/18	OEM
1B2/52	OEM
1B2C1	0196
1B2C1-LC2	OEM
1B2Z1	0015
1B2Z1-LC2	OEM
1B2Z1AY	0015
1B3	1306
1B4	0296
1B4B1	0106
1B4B41	0277
1B4B42	0241
1B4.1	0106
1B5/18	OEM
1B5/52	OEM
1B6	0302
1B8	0250
1B10	0250
1B10J20	0015
1B12	OEM
1B12/18	OEM
1B13	OEM
1B13/18	OEM
1B14	OEM
1B14/18	OEM
1B15	OEM
1B15/18	OEM
1B15CL05	OEM
1B15CL10	OEM
1B15CL20	OEM
1B15CL30	OEM
1B15CL40	OEM
1B16	OEM
1B16/18	OEM
1B22	OEM
1B22/18	OEM
1B23	OEM
1B23/18	OEM
1B24	OEM
1B24/18	OEM
1B24A	OEM
1B24A/MXT14	OEM
1B25	OEM
1B25/18	OEM
1B26	OEM
1B26/18	OEM
1B27	OEM
1B35A	OEM
1B35A/MXA14	OEM
1B37A	OEM
1B37A/MXA19	OEM
1B38	OEM
1B38/MST26	OEM
1B40CL05	OEM
1B40CL10	OEM
1B40CL20	OEM
1B40CL30	OEM
1B40CL40	OEM
1B42IG1	0901
1B44	OEM
1B44/MSA12	OEM
1B52	OEM
1B52/18	OEM
1B52/MSA21	OEM
1B53	OEM
1B53/18	OEM
1B53/MSA15	OEM
1B54	OEM
1B54/18	OEM
1B54/MST27	OEM
1B55	OEM
1B55/18	OEM
1B55/MST18	OEM
1B56	OEM
1B56/18	OEM
1B56/MSA11	OEM
1B58A	OEM
1B58A/MST12	OEM
1B63A	OEM
1B63A/MXT15	OEM
1B63B	OEM
1B759	0137
1B1055	OEM
1B12421	0122
1BH62	0080
1BZ61	0015
1BZ61-A	0015
1BZ61-B	0015
1BZ61-C	0015
1BZ61-D	0015
1BZ61-K	0015
1BZ61-L	0015
1BZ61-OR	0015
1BZ61-Y	0015
1C0009	0015
1C0017	0133
1C0020	0649
1C0025	0015
1C0026	0015
1C0029	0143
1C0031	0015
1C0038	0636
1C03	0236
1C04	0854
1C05	0015
1C06	0854
1C07	0561
1C08	1199
1C09	0042

DEVICE TYPE	REPL CODE
1C2	0015
1C4	0015
1C6	4213
1C8	0789
1C9.1Z	0012
1C9.1ZA	0012
1C10	0071
1C10Z	0170
1C10ZA	0170
1C11	0854
1C11Z	0317
1C11ZA	0317
1C12	2830
1C12Z	0137
1C12ZA	0137
1C13	0354
1C13Z	0365
1C13ZA	0365
1C14	0423
1C15	0167
1C15A	0002
1C15Z	0002
1C15ZA	0002
1C16	2237
1C16Z	0420
1C16ZA	0420
1C17	0785
1C18	1808
1C18Z	0495
1C18ZA	0495
1C-19	0372
1C20	1077
1C20Z	0531
1C20ZA	0531
1C21	7140
1C22Z	0565
1C22ZA	0565
1C-24	0579
1C24Z	0402
1C24ZA	0402
1C26	0167
1C27Z	0441
1C27ZA	0441
1C-30	1611
1C30Z	0726
1C30ZA	0726
1C33Z	0039
1C33ZA	0039
1C36Z	0820
1C36ZA	0820
1C39Z	0350
1C39ZA	0350
1C43Z	0931
1C43ZA	0931
1C47Z	0999
1C47ZA	0999
1C51Z	0502
1C51ZA	0502
1C56Z	0869
1C56ZA	0869
1C62Z	0783
1C62ZA	0783
1C68Z	2157
1C68ZA	2157
1C75Z	1189
1C75ZA	1189
1C82Z	0331
1C82ZA	0331
1C91	0632
1C91Z	1311
1C91ZA	1311
1C100Z	0098
1C100ZA	0098
1C-101	0400
1C110Z	0149
1C110ZA	0149
1C-112	0855
1C120Z	0186
1C120ZA	0186
1C-126	4582
1C129M	OEM
1C130Z	0213
1C130ZA	0213
1C-142	0167
1C150Z	0028
1C150ZA	0028
1C160Z	0255
1C160ZA	0255
1C180Z	0367
1C180ZA	0367
1C200Z	0421
1C200ZA	0421
1C202	1808
1C204	3219
1C288	0372
1C-289	1753
1C311	2743
1C319	2714
1C-502	0753
1C504	0664
1C505	0678
1C-507	0579
1C507	2393
1C508	0666
1C509	0352
1C510	0354
1C511	0654
1C513	0701
1C-602	2557

DEVICE TYPE	REPL CODE
1C821	OEM
1C821A	OEM
1C823	OEM
1C823A	OEM
1C825	OEM
1C825A	OEM
1C2992	OEM
1C2993	OEM
1C2995	OEM
1C3003	OEM
1C3004	OEM
1C3005	OEM
1C3006	OEM
1C3007	OEM
1C3008	OEM
1C3009	OEM
1C3010	OEM
1C3011	OEM
1C3012	OEM
1C3014	OEM
1C3015	OEM
1C3041	OEM
1C3993	OEM
1C4728	OEM
1C4729	OEM
1C4730	OEM
1C4731	OEM
1C4732	OEM
1C4733	OEM
1C4734	OEM
1C4735	OEM
1C4736	OEM
1C4737	OEM
1C4738	OEM
1C4739	OEM
1C4740	OEM
1C4741	OEM
1C4742	OEM
1C4743	OEM
1C4744	OEM
1C4745	OEM
1C4746	OEM
1C4747	OEM
1C4748	OEM
1C4749	OEM
1C4750	OEM
1C4751	OEM
1C4752	OEM
1C4753	OEM
1C4754	OEM
1C4755	OEM
1C4756	OEM
1C4757	OEM
1C4758	OEM
1C4759	OEM
1C4760	OEM
1C4761	OEM
1C4762	OEM
1C4763	OEM
1C4764	OEM
1C5247	OEM
1C5283	OEM
1C5284	OEM
1C5285	OEM
1C5286	OEM
1C5287	OEM
1C5288	OEM
1C5289	OEM
1C5290	OEM
1C5291	OEM
1C5292	OEM
1C5293	OEM
1C5294	OEM
1C5295	OEM
1C5296	OEM
1C5297	OEM
1C5298	OEM
1C5299	OEM
1C5300	OEM
1C5301	OEM
1C5302	OEM
1C5303	OEM
1C5304	OEM
1C5305	OEM
1C5306	OEM
1C5307	OEM
1C5308	OEM
1C5309	OEM
1C5310	OEM
1C5311	OEM
1C5312	OEM
1C5313	OEM
1C5314	OEM
1C5333	OEM
1C5334	OEM
1C5335	OEM
1C5336	OEM
1C5337	OEM
1C5338	OEM
1C5339	OEM
1C5340	OEM
1C5341	OEM
1C5342	OEM
1C5343	OEM
1C5344	OEM
1C5345	OEM
1C5346	OEM
1C5347	OEM
1C5348	OEM

DEVICE TYPE	REPL CODE
1C5349	OEM
1C5350	OEM
1C5351	OEM
1C5352	OEM
1C5353	OEM
1C5354	OEM
1C5355	OEM
1C5356	OEM
1C5357	OEM
1C5358	OEM
1C5359	OEM
1C5360	OEM
1C5361	OEM
1C5362	OEM
1C5363	OEM
1C5364	OEM
1C5365	OEM
1C5367	OEM
1C5368	OEM
1C5369	OEM
1C5370	OEM
1C5371	OEM
1C5372	OEM
1C5373	OEM
1C5374	OEM
1C5375	OEM
1C5376	OEM
1C5377	OEM
1C5378	OEM
1C5379	OEM
1C5380	OEM
1C5381	OEM
1C5382	OEM
1C5383	OEM
1C5384	OEM
1C5385	OEM
1C5386	OEM
1C5387	OEM
1C5388	OEM
1C6267	OEM
1C6267A	OEM
1C6268	OEM
1C6268A	OEM
1C6269	OEM
1C6269A	OEM
1C6270	OEM
1C6270A	OEM
1C6271	OEM
1C6271A	OEM
1C6272	OEM
1C6272A	OEM
1C6273	OEM
1C6273A	OEM
1C6274	OEM
1C6274A	OEM
1C6275	OEM
1C6275A	OEM
1C6276	OEM
1C6276A	OEM
1C6277	OEM
1C6277A	OEM
1C6278	OEM
1C6278A	OEM
1C6279	OEM
1C6279A	OEM
1C6280	OEM
1C6280A	OEM
1C6281	OEM
1C6281A	OEM
1C6282	OEM
1C6282A	OEM
1C6283	OEM
1C6283A	OEM
1C6284	OEM
1C6284A	OEM
1C6285	OEM
1C6285A	OEM
1C6286	OEM
1C6286A	OEM
1C6287	OEM
1C6287A	OEM
1C6288	OEM
1C6288A	OEM
1C6289	OEM
1C6289A	OEM
1C6290	OEM
1C6290A	OEM
1C6291	OEM
1C6291A	OEM
1C6292	OEM
1C6292A	OEM
1C6293	OEM
1C6293A	OEM
1C6294	OEM
1C6294A	OEM
1C6295	OEM
1C6295A	OEM
1C6296	OEM
1C6296A	OEM
1C6297	OEM
1C6297A	OEM
1C6298	OEM
1C6298A	OEM
1C6299	OEM
1C6299A	OEM
1C6300	OEM
1C6300A	OEM
1C6301	OEM
1C6301A	OEM
1C6302	OEM
1C6302A	OEM
1C6303	OEM
1C6303A	OEM
1C25681	2546
1C73045	0042
1C744005	0007
1C744006	0018
1C744009	0836
1CL-101-TY	0093
1CL-101A-TY	0093
1CL-201-TY	0093
1CL-201A-TY	0093
1CL-301A-PA	0093
1CL-301A-TY	0093
1CL-748-TY	0093
1CP-N5	1807
1D08	0002
1D059	1267
1D2	0015
1D2C1	0199
1D2C1LC2	0199
1D2Z1	0015
1D3.9	0456
1D3.9A	0456
1D3.9B	0456
1D4B1	0106
1D4B41	0289
1D4B42	1875
1D4.7	0451
1D4.7A	0451
1D4.7B	0451
1D5.1A	0162
1D5.1B	0442
1D5.6	0157
1D5.6A	0157
1D5.6B	0157
1D6.2A	0636
1D6.2B	0636
1D6.2SA	0636
1D6.2SB	0636
1D8	0071
1D10	0071
1D12	OEM
1D20M	OEM
1D27Z5	0441
1D40M	OEM
1D60M	OEM
1D80M	OEM
1D80MS	OEM
1D100	1137
1D100M	OEM
1D100MS	OEM
1D101	0344
1D102	0901
1D103	2342
1D104	0058
1D150M	OEM
1D150MS	OEM
1D200M	OEM
1D200MA	OEM
1D261	0015
1D300MA	OEM
1D600M	OEM
1DA1Z8.2A	0244
1DA1Z8.2B	0244
1DA1Z8.2C	0244
1DA1Z8.20	0244
1DA1Z18B	0495
1DAZ3401	0244
1DAZ3401A	0244
1DAZ3401B	0244
1DAZ3401C	0244
1DAZ34010	0244
1DAZ3405B	0495
1DB050	1875
1DB100	1875
1DB200	1875
1DB400	1875
1DB600	1875
1DC1	0015
1DF251	OEM
1DG2	0143
1DH62	0609
1DMB10	0241
1DMB20	0241
1DMB40	4581
1DMB60	1875
1D-S	OEM
1DZ12	0137
1DZ61	0015
1DZ61-A	0015
1DZ61-B	0015
1DZ61-C	0015
1DZ61-D	0015
1DZ61-K	0015
1DZ61-L	0015
1DZ61-OR	0015
1DZ61-Y	0015
1DZ120	0186
1E02	0015
1E05	0015
1E-1	0086
1E1	0015
1E2	0015
1E3	0015
1E4	0015
1E5	0015
1E6	0015
1E6.8Z5	0025
1E6.8Z10	0025
1E7	0071
1E7.5Z	0649
1E7.5Z5	0649
1E7.5Z10	0649
1E8	0071
1E8.2Z	0244
1E8.2Z5	0244
1E8.2Z10	0244
1E-9A	0050
1E9.1D	0012
1E9.1D5	0012
1E9.1D10	0012
1E9.1Z	0012
1E9.1Z5	0012
1E9.1Z10	0012
1E10	0071
1E10Z	0170
1E10Z5	0170
1E10Z10	0170
1E11Z	0794
1E11Z5	0317
1E11Z10	0794
1E12	OEM
1E12Z	0137
1E12Z5	0137
1E12Z10	0137
1E13Z	0365
1E13Z5	0365
1E13Z10	0365
1E14	OEM
1E14Z	0100
1E14Z5	0100
1E14Z10	0100
1E15Z	0002
1E15Z5	0002
1E15Z10	0002
1E16	OEM
1E16Z	0420
1E16Z5	0420
1E16Z10	0420
1E17Z	1649
1E17Z5	1649
1E17Z10	1649
1E18	OEM
1E18Z	0495
1E18Z5	0495
1E18Z10	0495
1E19Z	0949
1E19Z5	0949
1E19Z10	0949
1E20	OEM
1E20Z	0531
1E20Z5	0531
1E20Z10	0531
1E22Z	0565
1E22Z5	0565
1E22Z10	0565
1E24Z	0402
1E24Z5	0402
1E24Z10	0402
1E25	OEM
1E25Z	1606
1E25Z5	1606
1E25Z10	1606
1E27Z	0441
1E27Z5	0441
1E27Z10	0441
1E30Z	0726
1E30Z5	0726
1E30Z10	0726
1E33Z	0039
1E33Z5	0039
1E33Z10	0039
1E36Z	0820
1E36Z5	0820
1E36Z10	0820
1E39Z	0350
1E39Z5	0350
1E39Z10	0350
1E43Z	0931
1E43Z5	0931
1E43Z10	0931
1E45Z	0999
1E45Z5	0999
1E45Z10	0999
1E47Z	0999
1E47Z5	0999
1E47Z10	0999
1E50A	OEM
1E50Z	0502
1E50Z5	0502
1E50Z10	0502
1E51Z	0502
1E51Z5	0502
1E51Z10	0502
1E52Z	0502
1E52Z5	0502
1E52Z10	0502
1E56Z	0869
1E56Z5	0869
1E56Z10	0869
1E62Z	0783
1E62Z5	0783
1E62Z10	0783
1E68Z	2157
1E68Z5	2157
1E68Z10	2157
1E75Z	1189
1E75Z5	1189
1E75Z10	1189
1E82Z	0331
1E82Z5	0331
1E82Z10	0331
1E91Z	1311
1E91Z5	1311
1E91Z10	1311
1E100Z	0098
1E100Z5	0098
1E100Z10	0098
1E110Z	0149
1E110Z5	0149
1E110Z10	0149
1E120Z	0186
1E120Z5	0186
1E120Z10	0186
1E130Z	0213
1E130Z5	0213
1E130Z10	0213
1E150Z	0028
1E150Z5	0028
1E150Z10	0028
1E160Z	0255
1E160Z5	0255
1E160Z10	0255
1E180Z	0367
1E180Z5	0367
1E180Z10	0367
1E200Z	0421
1E200Z5	0421
1E200Z10	0421
1E535A	0224
1E535B	0127
1E703E	0632
1EA10A	0229
1EA20A	0229
1EA30A	0229
1EA40A	0229
1EA50A	0229
1EA60A	0229
1EA70A	4257
1EA80A	4257
1EA100A	0071
1EA120A	OEM
1EB10A	0090
1EB20A	0097
1EB30A	0105
1EB40A	0109
1EB50A	0204
1EB60A	0122
1EB70A	0131
1EB80A	0131
1EB100A	0145
1EK200	OEM
1EK220	OEM
1EK250	OEM
1EK300	OEM
1ER100A	0145
1ET02	0015
1ET05	0015
1ET1	0015
1ET2	0015
1ET3	0015
1ET4	0015
1ET5	0015
1ET6	0015
1ET7	0071
1ET8	0071
1ET10	0071
1EZ3.3D	0292
1EZ3.3D5	0292
1EZ3.3D10	0292
1EZ3.3T10	0292
1EZ3.6	0188
1EZ3.6D	0188
1EZ3.6D5	0188
1EZ3.6D10	0188
1EZ3.6T10	0188
1EZ3.9D	0456
1EZ3.9D5	0456
1EZ3.9D10	0456
1EZ3.9T10	0456
1EZ4.3D	0533
1EZ4.3D5	0533
1EZ4.3D10	0533
1EZ4.3T10	0533
1EZ4.7D	0451
1EZ4.7D5	0451
1EZ4.7D10	0451
1EZ4.7T10	0451
1EZ-5Z10	0649
1EZ5.1	0162
1EZ5.1D	0162
1EZ5.1D5	0162
1EZ5.1D10	0162
1EZ5.1T10	0162
1EZ5.6	0157
1EZ5.6D	0157
1EZ5.6D5	0157
1EZ5.6D10	0157
1EZ5.6T10	0157
1EZ6.2D	0636
1EZ6.2D5	0636
1EZ6.2D10	0636
1EZ6.2T10	0636
1EZ6.8D	0025
1EZ6.8D5	0025
1EZ6.8D10	0025
1EZ6.8T10	0025
1EZ7.5D	0649
1EZ7.5D5	0649
1EZ7.5D10	0649
1EZ7.5T10	0649
1EZ7.5Z	0649
1EZ8.2D	0244
1EZ8.2D5	0244
1EZ8.2D10	0244
1EZ8.2T10	0244
1EZ9.1	0012
1EZ9.1D	0012
1EZ9.1D5	0012
1EZ9.1D10	0012
1EZ9.1T10	0012
1EZ10D	0170
1EZ10D5	0170
1EZ10D10	0170
1EZ10T10	0170
1EZ11D	0317
1EZ11D5	0317
1EZ11D10	0317
1EZ11T10	0317
1EZ12	0137
1EZ12D	0137
1EZ12D5	0137
1EZ12D10	0137
1EZ12T10	0137
1EZ13D	0365
1EZ13D5	0365
1EZ13D10	0365
1EZ13T10	0365
1EZ14D	0100
1EZ14D5	0100
1EZ14D10	0100
1EZ15	0002
1EZ15D	0002
1EZ15D5	0002
1EZ15D10	0002
1EZ15T10	0002
1EZ16D	0420
1EZ16D5	0420
1EZ16D10	0420
1EZ16T10	0420
1EZ17D	1649
1EZ17D5	1649
1EZ17D10	1649
1EZ18D	0495
1EZ18D5	0495
1EZ18D10	0495
1EZ18T10	0495
1EZ19D	0949
1EZ19D5	0949
1EZ19D10	0949
1EZ20D	0531
1EZ20D5	0531
1EZ20D10	0531
1EZ20T10	0531
1EZ22D	0565
1EZ22D5	0565
1EZ22D10	0565
1EZ22T10	0565
1EZ24D	0402
1EZ24D5	0402
1EZ24D10	0402
1EZ24T10	0402
1EZ27	0441
1EZ27D	0441
1EZ27D5	0441
1EZ27D10	0441
1EZ27T10	0441
1EZ30D	0726
1EZ30D5	0726
1EZ30D10	0726
1EZ30T10	0726
1EZ33D	0039
1EZ33D5	0039
1EZ33D10	0039
1EZ36D	0820
1EZ36D5	0820
1EZ36D10	0820
1EZ39D	0350
1EZ39D5	0350
1EZ39D10	0350
1EZ43D	0931
1EZ43D5	0931
1EZ43D10	0931
1EZ47D	0999
1EZ47D5	0999
1EZ47D10	0999
1EZ50D	OEM
1EZ50D5	OEM
1EZ50D10	OEM
1EZ51D	0502
1EZ51D5	0502
1EZ51D10	0502
1EZ52D	OEM
1EZ52D5	OEM
1EZ52D10	OEM
1EZ56D	0869
1EZ56D5	0869
1EZ56D10	0869
1EZ62D	0783
1EZ62D5	0783
1EZ62D10	0783
1EZ68D	2157
1EZ68D5	2157
1EZ68D10	2157
1EZ75D	1189
1EZ75D5	1189
1EZ75D10	1189
1EZ82D	0331
1EZ82D5	0331
1EZ82D10	0331
1EZ91D	1311
1EZ91D5	1311
1EZ91D10	1311
1EZ100D	0098
1EZ100D5	0098
1EZ100D10	0098
1EZ110D	0149
1EZ110D5	0149
1EZ110D10	0149
1EZ120D	0186
1EZ120D5	0186
1EZ120D10	0186
1EZ130D	0213
1EZ130D5	0213
1EZ130D10	0213
1EZ140D	0245
1EZ140D5	0245
1EZ140D10	0245
1EZ150D	0028
1EZ150D5	0028
1EZ150D10	0028
1EZ160D	0255
1EZ160D5	0255
1EZ160D10	0255
1EZ170D	0877
1EZ170D5	0877
1EZ170D10	0877
1EZ180D	0367
1EZ180D5	0367
1EZ180D10	0367
1EZ190D	2858
1EZ190D5	2858
1EZ190D10	2858
1EZ200D	0421
1EZ200D5	0421
1EZ200D10	0421
1EZ1905	2858
1E.5	OEM
1F05	0015
1F1	1090
1F2	0015
1F3	1090
1F4	OEM
1F6	OEM
1F8	0071
1F10	0071
1F12	OEM
1F14	OEM
1F14A	0015
1F16	OEM
1F18	OEM
1F20	OEM
1F101	OEM
1F104	OEM
1F105	2518
1F106	2518
1F112	OEM
1F911	OEM
1F911N	OEM
1F916	OEM
1F918H	OEM
1F935	OEM
1F941M	OEM
1F941N	OEM
1F942N	OEM
1F943M	OEM
1FB100R	OEM
1FJ0	OEM
1FJ0(F)	OEM
1FJ1	OEM
1FJ1(F)	OEM
1FJ2	OEM
1FJ2(F)	OEM
1FJ3	OEM
1FJ3(F)	OEM
1FM2	0015
1FP70020H	2571
1G	OEM
1G01	0143
1G1	OEM
1G2	0071
1G2C1	0196
1G2Z	0015
1G2Z1	1097
1G3	1306
1G3(RECT)	0071
1G3.3	0292
1G3.3A	0292
1G3.6	0188
1G3.6A	0188
1G3.9	0456
1G3.9A	0456
1G4	0071
1G4B1	0106
1G4B41	0296
1G4B42	1875
1G4.3	0533
1G4.3A	0533
1G4.7	0451
1G4.7A	0451
1G5	OEM
1G5.1	0162
1G5.1A	0162
1G5.6	0157
1G5.6A	0157
1G6	0071
1G6.2	0636
1G6.2A	0636
1G6.8	0025
1G6.8A	0025
1G7.5	0649
1G7.5A	0649
1G8	0071
1G8.2	0244
1G8.2A	0244
1G9.1	0012
1G9.1A	0012
1G10	0170
1G10A	0170
1G11	0317
1G11A	0317
1G12	0137
1G12A	0137
1G13	0365
1G13A	0365
1G15	0100
1G15A	0002
1G16	0420
1G16A	0420
1G18	0495
1G18A	0495
1G20	0531
1G20A	0531
1G22	0565
1G22A	0565
1G24	0402
1G24A	0402
1G25	0143
1G27	0441
1G27A	0441
1G30	0726
1G30A	0726
1G33	0039
1G33A	0039
1G36	0820
1G36A	0820
1G39	0350
1G39A	0350
1G43	0931
1G43A	0931
1G47	0999
1G47A	0999
1G51	0502
1G51A	0502
1G56	0869
1G56A	0869
1G62	0783
1G62A	0783
1G68	2157
1G68A	2157
1G75	1189
1G75A	1189
1G82	0331
1G82A	0331
1G86	0143
1G91	0156
1G91A	1311
1G100	0440
1G100(ZENER)	0098
1G100A	0098
1G110	0149
1G110A	0149
1G120	0186
1G120A	0186
1G130	0213
1G130A	0213
1G150	0028
1G150A	0028
1G160	0255
1G160A	0255
1G180	0367
1G180A	0367
1G200	0421
1G200A	0421
1GD2	0143
1GD4	0143
1GD5	0030
1GD5X	0143
1GD6	0143
1GD10	0143
1GH11	0182
1GH11F-FA	0102
1GH62	0023
1GWJ43	0735
1GZ61	0015
1GZ61-A	0015
1GZ61-B	0015
1GZ61-C	0015
1GZ61-D	0015
1GZ61-K	0015
1GZ61-L	0015
1GZ61-OR	0015
1GZ61-Y	0015
1GZZ1	0015
1H3	0182
1H20M	OEM
1H40M	OEM
1H40MS	OEM
1H60H	OEM
1H60M	OEM
1H60S	OEM
1H80M	OEM

If replacement code is OEM, contact original manufacturer for replacement.

DEVICE TYPE	REPL CODE
1H80MS	OEM
1H100M	OEM
1H120M	OEM
1H120MS	OEM
1H140M	OEM
1H160M	OEM
1H160MS	OEM
1H180M	OEM
1H200M	OEM
1H200MS	OEM
1H250M	OEM
1H300M	OEM
1H400M	OEM
1HC-10F	0015
1HC-10R	0015
1HC-15F	0015
1HC-15R	0015
1HC-20F	0015
1HC-20R	0015
1HC-25F	0015
1HC-25R	0015
1HC-30F	0015
1HC-30R	0015
1HC-40F	0015
1HC-40R	0015
1HC-50F	0015
1HC-50R	0015
1HC-60F	0015
1HC-60R	0015
1HC-80F	0071
1HC-80R	0071
1HC-100F	0071
1HC-100R	0071
1HC-120F	0102
1HC-120R	0102
1HY40	0015
1HY50	0015
1HY80	0071
1HY100	0071
1J1	0016
1J2C1	0199
1J2Z1	1015
1J3	1306
1J4B1	2010
1J4B41	0302
1J4B42	1875
1JA16	0160
1JH11	OEM
1JH11F	0015
1JH62	0023
1JK30	OEM
1JK60	OEM
1JK90	OEM
1JK100	OEM
1JK120	OEM
1JK150	OEM
1JK160	OEM
1JK180	OEM
1JL50	0015
1JL51	0015
1JL52	0015
1JL53	0015
1JL56	0015
1JL57	0015
1JL58	0071
1JL59	0071
1JL63	0071
1JL64	0071
1JLA10	OEM
1JM11F	OEM
1JUS.540A	OEM
1JZ61	0023
1JZ61-A	0015
1JZ61-B	0015
1JZ61-C	0015
1JZ61-D	0015
1JZ61-K	0015
1JZ61-L	0015
1JZ61-OR	0015
1JZ61-Y	0015
1K	OEM
1K01	OEM
1K2	0242
1K3	1306
1K-3.5E	0188
1K34A	2232
1K60	0143
1K60A	0143
1K60P	0143
1K60R	0143
1K75CH	OEM
1K75CK	OEM
1K75CK8199	OEM
1K75CLA	OEM
1K75CS	OEM
1K90	0143
1K110	0143
1K125CA	OEM
1K125CB	OEM
1K125CB-B	OEM
1K125CD	OEM
1K188	0143
1K188F1	OEM
1K188FM	0143
1K188FM-1	0143
1K188FM1	OEM
1K261	0143
1KAB100E	OEM
1KF20-04	0143
1KK-HA1196	4799
1KK-HA11225	2252
1L1-2R	OEM
1LA7837	4359
1LE11	0087
1LE11-A	0071
1LE11-B	0071
1LE11-C	0071
1LE11-D	0071
1LE11-K	0071
1LE11-L	0071
1LE11-OR	0071
1LE11-Y	0071
1LE11A	0071
1LE11B	0071
1LE11C	0071
1LE11D	0071
1LE11K	0071
1LE11L	0071
1LE11R	0071
1LE11Y	0071
1LF11	0087
1LH62	0072
1LM340T12	0334
1LZ61	0071
1LZ61-A	0071
1LZ61-B	0071
1LZ61-C	0071
1LZ61-D	0071
1LZ61-K	0071
1LZ61-L	0071
1LZ61-OR	0071
1LZ61-Y	0071
1M3.3AZ3	0299
1M3.3AZ10	0292
1M3.6AZ3	0188
1M3.6AZ5	0188
1M3.6AZ10	0188
1M3.6ZS10	0188
1M3.9AZ3	OEM
1M3.9AZ10	0456
1M3.9ZS5	0456
1M3.9ZS10	0456
1M4.3AZ3	OEM
1M4.3AZ10	0533
1M4.7AZ3	OEM
1M4.7AZ10	0451
1M4.7ZS5	0451
1M4.7ZS10	0451
1M5.1AZ3	0162
1M5.1AZ10	0162
1M5.1ZS5	0442
1M5.1ZS10	0162
1M5.6AZ3	OEM
1M5.6AZ10	0157
1M5.6ZS5	0157
1M5.6ZS10	0157
1M6.2AZ3	0636
1M6.2AZ10	0636
1M6.2ZS10	0636
1M6.8AZ3	OEM
1M6.8AZ10	0025
1M6.8Z	0025
1M6.8Z3	OEM
1M6.8Z5	0025
1M6.8Z10	0025
1M6.8ZS5	0025
1M6.8ZS10	0025
1M7.5AZ3	OEM
1M7.5AZ10	0649
1M7.5Z	0649
1M7.5Z3	OEM
1M7.5Z5	0649
1M7.5Z10	0649
1M7.5ZS5	0649
1M7.5ZS10	0649
1M8.2Z	0244
1M8.2Z3	3100
1M8.2Z5	0244
1M8.2Z10	0244
1M8.2ZS5	0244
1M8.2ZS10	0244
1M9.1Z	0012
1M9.1Z3	OEM
1M9.1Z5	0012
1M9.1Z10	0012
1M9.1ZS5	0012
1M9.1ZS10	0012
1M10Z	0170
1M10Z3	OEM
1M10Z5	0170
1M10Z10	0170
1M10ZS5	0170
1M10ZS10	0170
1M11Z	0317
1M11Z3	OEM
1M11Z5	0317
1M11Z10	0317
1M11ZS5	0317
1M11ZS10	0317
1M12Z	0137
1M12Z3	0137
1M12Z5	0137
1M12Z10	0137
1M12ZS5	0137
1M12ZS10	0137
1M13Z3	OEM
1M13Z5	0365
1M13Z10	0365
1M13ZS10	0365
1M140Z	0245
1M14Z	0100
1M14Z5	0100
1M14Z10	0100
1M15/1B	OEM
1M15Z	0002
1M15Z3	0002
1M15Z5	0002
1M15Z10	0002
1M15ZS5	0002
1M15ZS10	0002
1M16Z	0420
1M16Z3	OEM
1M16Z5	0420
1M16Z10	0420
1M16ZS5	0420
1M16ZS10	0420
1M17Z	1649
1M17Z5	1649
1M17Z10	1649
1M18Z	0495
1M18Z3	OEM
1M18Z5	0495
1M18Z10	0495
1M18ZS5	0495
1M18ZS10	0495
1M19Z	0949
1M19Z5	0949
1M19Z10	0949
1M20Z	0531
1M20Z3	0531
1M20Z5	0531
1M20Z10	0531
1M20ZS5	0186
1M20ZS10	0531
1M22Z	0565
1M22Z3	OEM
1M22Z5	0565
1M22Z10	0565
1M22ZS5	0565
1M22ZS10	0565
1M24Z	0402
1M24Z3	OEM
1M24Z5	0402
1M24Z10	0402
1M24ZS5	0402
1M24ZS10	0402
1M25Z	1606
1M25Z5	1606
1M25Z10	1606
1M27Z	0441
1M27Z3	OEM
1M27Z5	0441
1M27Z10	0441
1M27ZS5	0441
1M27ZS10	0441
1M30Z	0726
1M30Z3	OEM
1M30Z5	0726
1M30Z10	0726
1M33Z	0039
1M33Z3	OEM
1M33Z5	0039
1M33Z10	0039
1M33ZS5	0039
1M33ZS10	0039
1M36Z	0820
1M36Z3	OEM
1M36Z5	0820
1M36Z10	0820
1M39Z	0350
1M39Z3	OEM
1M39Z5	0350
1M39Z10	0350
1M39ZS5	0350
1M39ZS10	0350
1M43Z	0931
1M43Z3	OEM
1M43Z5	0931
1M43Z10	0931
1M47Z	0999
1M47Z3	OEM
1M47Z5	0999
1M47Z10	0999
1M47ZS5	0999
1M47ZS10	0999
1M51Z	0502
1M51Z3	OEM
1M51Z5	0502
1M51Z10	0502
1M56Z	0869
1M56Z3	OEM
1M56Z5	0869
1M56Z10	0869
1M56ZS5	0869
1M56ZS10	0869
1M62Z	0783
1M62Z5	0783
1M62Z10	0783
1M62ZS5	0783
1M62ZS10	0783
1M68Z	1268
1M68Z3	1268
1M68Z5	1268
1M68Z10	1268
1M68ZS5	1268
1M68ZS10	1268
1M70A	OEM
1M75Z	1189
1M75Z3	1189
1M75Z5	1189
1M75Z10	1189
1M82Z	0331
1M82Z3	OEM
1M82Z5	0331
1M82Z10	0331
1M82ZS5	0331
1M82ZS10	0331
1M87ZS5	3026
1M91Z	1311
1M91Z3	1311
1M91Z5	1311
1M91Z10	1311
1M99	OEM
1M100Z	0098
1M100Z3	OEM
1M100Z5	0098
1M100Z10	0098
1M100ZS5	0098
1M100ZS10	0098
1M110Z	0149
1M110Z3	0149
1M110Z5	0149
1M110Z10	0149
1M110ZS5	0149
1M110ZS5RL	OEM
1M110ZS10	0149
1M110ZS10RL	OEM
1M110ZS10.5	OEM
1M120Z	0186
1M120Z3	0186
1M120Z5	0186
1M120Z10	0186
1M120ZS5	0186
1M120ZS5RL	OEM
1M120ZS10	0186
1M120ZS10RL	OEM
1M120ZS10.5	OEM
1M130Z	0213
1M130Z3	0213
1M130Z5	0213
1M130Z10	0213
1M130ZS5	0213
1M130ZS5RL	OEM
1M130ZS10	0213
1M130ZS10RL	OEM
1M130ZS10.5	OEM
1M130Z.5	OEM
1M140Z	0245
1M140Z5	0245
1M140Z10	0245
1M140ZS5	0245
1M140ZS5RL	OEM
1M140ZS10	0245
1M140ZS10RL	OEM
1M150Z	0028
1M150Z3	0028
1M150Z5	0028
1M150Z10	0028
1M150ZS5	0028
1M150ZS5RL	OEM
1M150ZS10	0028
1M150ZS10RL	OEM
1M150Z.5	OEM
1M151	OEM
1M160Z	0255
1M160Z3	0255
1M160Z5	0255
1M160Z10	0255
1M160ZS5	0255
1M160ZS5RL	OEM
1M160ZS10	0255
1M160ZS10RL	OEM
1M160Z.5	OEM
1M170ZS5	0877
1M175ZS5	0367
1M175ZS5RL	OEM
1M175ZS10	0367
1M175ZS10RL	OEM
1M180Z	0367
1M180Z3	0367
1M180Z5	0367
1M180Z10	0367
1M180ZS5	0367
1M180ZS5RL	OEM
1M180ZS10	0367
1M180ZS10RL	OEM
1M180Z.5	OEM
1M200Z	0421
1M200Z3	0421
1M200Z5	0421
1M200Z10	0421
1M200ZS5	0421
1M200ZS5RL	OEM
1M200ZS10	0421
1M200ZS10RL	OEM
1M200Z.5	OEM
1M383T	1048
1M392S10	0350
1M430B	OEM
1M1140Z5	0245
1M2086	0015
1M4796D	OEM
1M8513A	0015
1M39000A30	OEM
1MA4	0015
1MC110Z10	OEM
1MC120Z10	OEM
1MC130Z10	OEM
1MC150Z10	OEM
1MC160Z10	OEM
1MC180Z10	OEM
1MC200Z10	OEM
1MC1358P	0167
1MJJ3.6	0188
1MJJ3.6A	0188
1MJJ6.2	0636
1MJJ6.2A	0636
1MJJ9.1	0774
1MJJ9.1A	0774
1MJJ12	0137
1MJJ12A	0137
1MZ9.1T5	0012
1MZ9.1T10	0012
1MZ9.1T20	0012
1MZS22ZS10	0565
1N0188A	OEM
1N1055	OEM
1N2	1306
1N4020A	OEM
1N4084A	OEM
1N6E11	OEM
1N9	OEM
1N21	1113
1N21A	0606
1N21B	0606
1N21BM	0606
1N21BMR	0606
1N21BR	0606
1N21C	0606
1N21CM	0606
1N21CMR	0606
1N21CR	0606
1N21D	0606
1N21DM	0606
1N21DMR	0606
1N21DR	0606
1N21E	0606
1N21EM	0606
1N21EMR	0606
1N21ER	0606
1N21F	0606
1N21FM	0606
1N21FMR	0606
1N21FR	0606
1N21G	0606
1N21GM	0606
1N21GMR	0606
1N21GR	0606
1N21WD	0606
1N21WDM	0606
1N21WE	0606
1N21WEM	0606
1N21WF	0606
1N21WFM	0606
1N21WG	0606
1N21WGM	0606
1N-22	0143
1N22	1114
1N22WA	OEM
1N22WB	OEM
1N23	1115
1N23A	0045
1N23B	0606
1N23BM	0606
1N23BMR	0606
1N23BR	0606
1N23C	1115
1N23CM	1115
1N23CMR	1115
1N23CR	1115
1N23D	0045
1N23DM	0045
1N23DMR	0045
1N23DR	0045
1N23E	0045
1N23EM	0045
1N23EMR	0045
1N23ER	0045
1N23F	0606
1N23FM	0606
1N23FMR	0606
1N23FR	0606
1N23G	0606
1N23GM	0606
1N23GMR	0606
1N23GR	0606
1N23H	0917
1N23WD	0606
1N23WDM	0606
1N23WE	0045
1N23WEM	0045
1N23WF	0606
1N23WFM	0606
1N23WG	0606
1N23WGM	0606
1N25	0162
1N25A	0133
1N25B	OEM
1N25WA	OEM
1N26	OEM
1N26A	OEM
1N26B	OEM
1N26C	OEM
1N27C	OEM
1N28G	0917
1N28J	OEM
1N31	OEM
1N31A	0143
1N32	OEM
1N32A	OEM
1N34	0143
1N34A	0123
1N34A(AU)	0143
1N34A(H)	0143
1N34A(K)	0143
1N34AM	0143
1N34AS	0143
1N34A-T	0143
1N34ATF1	0143
1N34A-Z	0143
1N34G	0143
1N34GA	0143
1N34M	0123
1N34N	0143
1N34Z	0143
1N35	0143
1N35(H)	OEM
1N36	0143
1N38	0143
1N38A	0143
1N38A(H)	0143
1N38B	0143
1N39	OEM
1N39A	0609
1N39B	OEM
1N40	0143
1N41	0143
1N42	0143
1N43	0143
1N44	0143
1N44A1	OEM
1N45	0143
1N46	0143
1N46A	0143
1N47	0143
1N48	0143
1N48A	0143
1N48N	0143
1N49	0143
1N50	0143
1N51	0143
1N52	0143
1N52A	0143
1N52A1	OEM
1N53	OEM
1N53A	OEM
1N53B	OEM
1N53C	OEM
1N53D	OEM
1N54	0143
1N54A	0143
1N54G	0143
1N54GA	0143
1N55	OEM
1N55A	0143
1N55B	OEM
1N56	0143
1N56A	0143
1N57	0143
1N57A	0143
1N58	0143
1N58A	0143
1N59	OEM
1N60	0019
1N60(GA)	0019
1N60(GB)	0019
1N60(H)	0019
1N60/7825B	0019
1N60-1	0019
1N60-5	0123
1N60-FA1	0019
1N60-FD1	0019
1N60-FM	0019
1N60-M3	0019
1N60-S	0019
1N60-T	0019
1N60-TF1	0019
1N60-Z	0019
1N60A	0019
1N60AM	0019
1N60AM2000-302	OEM
1N60B	0019
1N60C	0019
1N60D	0019
1N60F	0019
1N60FA1	0019
1N60FD1	0019
1N60FM	0019
1N60FM-1	0019
1N60FM1	0019
1N60FMX	0019
1N60G	0019
1N60GA	0019
1N60GB	0019
1N60M	0019
1N60M3	0019
1N60MP	0123
1N60N	0019
1N60P	0123
1N60R	0019
1N60S	0019
1N60SD60	0019
1N60T	0019
1N60TF1	0019
1N60TPGL	0019
1N60TV	0019
1N60TVG1	1114
1N60TVGL	0019
1N60TV-TDGL	1114
1N60TV-TOGL	0019
1N60TVTP	0019
1N60TV-TPGL	0019
1N60TVTPGL	0019
1N60Z	0019
1N61	OEM
1N61A	0143
1N61A1	OEM
1N62	0143
1N63	0143
1N63A	0143
1N64	0143
1N64A	0143
1N64B	0143
1N64G	0143
1N64GA	0143
1N64N	OEM
1N64P	0143
1N64S	0124
1N65	0143
1N65A	0143
1N66	0143
1N66A	0143
1N67	0143
1N67A	0143
1N67D	0143
1N68	0143
1N68A	0143
1N69	0143
1N69A	0143
1N70	0143
1N70A	0143
1N71	0143
1N72	0143
1N72G	0143
1N73	0143
1N74	0143
1N75	0143
1N75A	0143
1N76	0143
1N76A	0143
1N76C	0143
1N76G	0143
1N78	OEM
1N78A	OEM
1N78B	OEM
1N78C	OEM
1N78D	OEM
1N78E	OEM
1N78F	OEM
1N78G	OEM
1N81	0143
1N81A	0143
1N82	0917
1N82A	0917
1N82AD	0917
1N82AG	0917
1N82G	0917
1N83	OEM
1N84	0143
1N86	0143
1N86AG	0143
1N87	0143
1N87A	0143
1N87AN	OEM
1N87G	0143
1N87GA	0143
1N87S	0143
1N87T	0143
1N88	0123
1N89	0143
1N90	0143
1N90G	0143
1N90GA	0143
1N91	0143
1N92	0015
1N93	0015
1N93A	0015
1N95	0143
1N96	0143
1N96A	0143
1N96B	OEM
1N97	0143
1N97A	0143
1N98	0143
1N98A	0143
1N99	0143
1N99A	0143
1N100	0143
1N100A	0143
1N102	OEM
1N103	0143
1N104	0143
1N105	0143
1N107	0143
1N108	0143
1N109	0143
1N110	0917
1N111	0143
1N112	0143
1N113	0143
1N114	0143
1N115	0143
1N116	0143

Original Device Types

DEVICE TYPE	REPL CODE
1N116A	0143
1N117	0143
1N117A	0143
1N118	0143
1N118A	0143
1N119	0143
1N119A	0143
1N120	0143
1N120A	0143
1N122B	0229
1N124	0917
1N124A	0917
1N125	0143
1N126	0143
1N126A	0143
1N127	0143
1N127A	0143
1N128	0143
1N128A	0143
1N132	0143
1N133	0143
1N134	0143
1N135	0133
1N136	0143
1N137	0143
1N137A	0133
1N137B	0133
1N138	0133
1N138A	0133
1N138B	0133
1N139	0143
1N140	0143
1N141	0143
1N142	0143
1N143	0143
1N144	0143
1N145	0143
1N147	0917
1N147A	0917
1N148	0143
1N149	OEM
1N150	OEM
1N151	0015
1N152	0015
1N153	0015
1N156	OEM
1N158	0015
1N160	OEM
1N169	0015
1N172	0917
1N173	0917
1N175	OEM
1N191	0143
1N192	0143
1N194	0133
1N194A	0133
1N194B	0133
1N195	0133
1N196	0133
1N197B	OEM
1N198	0143
1N198A	0143
1N198B	0143
1N198N	0143
1N200	0133
1N200AR	0133
1N201	0133
1N202	0133
1N203	0133
1N204	0133
1N205	0133
1N206	0133
1N207	0133
1N208	0133
1N209	0133
1N210	0133
1N211	0133
1N212	0133
1N213	0133
1N214	0133
1N215	0133
1N216	0133
1N217	0133
1N218	0133
1N219	0795
1N220	0795
1N221	0795
1N222	0795
1N225	0322
1N225A	0322
1N226	0064
1N226A	0064
1N227	0053
1N227A	0053
1N228	0445
1N228A	0445
1N229	0671
1N229A	0671
1N230	0494
1N230A	0494
1N231	0257
1N231A	0257
1N232	0010
1N232A	0010
1N233	0054
1N233A	0054
1N234	0068
1N234A	0068
1N235	2317

DEVICE TYPE	REPL CODE
1N235A	2317
1N236	OEM
1N237	0104
1N237A	0104
1N238	OEM
1N239	0287
1N248	3189
1N248A	3189
1N248AR	1630
1N248B	3189
1N248BR	1630
1N248C	1124
1N248CR	1630
1N248R	1630
1N248RA	1630
1N248RB	1630
1N248RC	1630
1N249	1124
1N249A	2901
1N249AR	0254
1N249B	2901
1N249BR	0254
1N249C	2901
1N249CR	0254
1N249R	1107
1N249RA	0254
1N249RB	0254
1N249RC	0254
1N250	1250
1N250A	1124
1N250AR	1107
1N250B	1124
1N250BR	1107
1N250C	0805
1N250CR	1107
1N250R	1577
1N250RA	1107
1N250RB	1107
1N250RC	0258
1N251	0133
1N251A	0133
1N252	0133
1N252A	0133
1N253	0580
1N253R	2291
1N254	0580
1N254R	OEM
1N255	1000
1N255R	0476
1N256	2081
1N256R	0476
1N258	OEM
1N263	0917
1N265	0143
1N266	0143
1N267	0143
1N268	0143
1N270	0143
1N273	0143
1N276	0023
1N277	0143
1N278	0143
1N279	0143
1N281	0143
1N282	0143
1N283	0143
1N285	0143
1N286	OEM
1N286A	OEM
1N287	0143
1N288	0143
1N289	0143
1N290	0143
1N291	0143
1N292	0143
1N294	0143
1N294A	0143
1N295	0143
1N295A	0143
1N295S	0143
1N295X	0143
1N296	0143
1N297	0143
1N297A	0143
1N298	0143
1N298A	0143
1N299	0917
1N300	0133
1N300A	0133
1N300B	0133
1N301	0133
1N301A	0133
1N301B	0133
1N302	0133
1N302A	0133
1N302B	0133
1N303	0133
1N303A	0133
1N303B	0133
1N304	0143
1N305	0143
1N306	0143
1N307	0143
1N308	0143
1N309	0143
1N310	0143
1N312	0143
1N313	0143
1N314	0143

DEVICE TYPE	REPL CODE
1N315	0015
1N315A	0015
1N316	0015
1N316A	0015
1N317	0015
1N317A	0015
1N318	0015
1N318A	0015
1N319	0015
1N319A	0015
1N320	0015
1N320A	0015
1N321	0071
1N321A	0071
1N322	0071
1N322A	0071
1N323	0015
1N323A	0015
1N324	0015
1N324A	0015
1N325	0015
1N325A	0015
1N326	0015
1N326A	0015
1N326B	OEM
1N327	0015
1N327A	0015
1N328	0071
1N328A	0071
1N329	0071
1N329A	0071
1N330	0133
1N331	0133
1N332	1000
1N332R	1012
1N333	1000
1N333R	1012
1N334	2060
1N334R	4523
1N335	1000
1N335R	4523
1N336	0580
1N336R	0947
1N337	0580
1N337R	0947
1N338	0580
1N338R	5517
1N339	0580
1N339R	5517
1N340	0580
1N340R	5517
1N341	1000
1N341R	1012
1N342	1000
1N342R	1012
1N343	1000
1N343R	4523
1N344	1000
1N344R	4523
1N345	0580
1N345R	0947
1N346	0580
1N346R	0947
1N347	0580
1N347R	5517
1N348	0580
1N348R	5517
1N349	0580
1N349R	5517
1N350	0133
1N351	0133
1N352	0133
1N353	0133
1N354	0795
1N355	0143
1N358	OEM
1N358A	OEM
1N359	0015
1N359A	0015
1N360	0015
1N360A	0015
1N361	0015
1N361A	0015
1N362	0015
1N362A	0015
1N363	0015
1N363A	0015
1N364	0071
1N364A	0071
1N365	0071
1N365A	0071
1N367	0143
1N367B	0143
1N368	0015
1N369	OEM
1N369A	OEM
1N370	OEM
1N371	1276
1N372	1713
1N373	0188
1N374	0275
1N375	0275
1N376	0162
1N377	0301
1N378	0062
1N379	0133
1N380	0133
1N381	0133
1N382	0133

DEVICE TYPE	REPL CODE
1N383	0133
1N384	0133
1N385	0133
1N386	0133
1N387	0133
1N388	0133
1N389	0133
1N390	0133
1N391	0133
1N392	0133
1N393	0133
1N394	0133
1N400	0015
1N400B	0015
1N411B	0599
1N411RB	1347
1N412B	0599
1N412RB	1347
1N413B	0599
1N413RB	1347
1N415B	1115
1N415C	1115
1N415D	0045
1N415E	0045
1N415F	OEM
1N415G	OEM
1N415H	OEM
1N416B	0917
1N416C	1113
1N416D	0606
1N416E	0606
1N416F	OEM
1N416G	OEM
1N417	0143
1N417SA	0820
1N418	0143
1N419	0143
1N422A	OEM
1N429	0471
1N430	0165
1N430A	0165
1N430B	0165
1N431	0133
1N432	0133
1N432A	0133
1N432B	0133
1N433	0133
1N433A	0133
1N433B	0133
1N434	0133
1N434A	0133
1N434B	0133
1N435	0143
1N440	0015
1N440B	0015
1N441	0015
1N441A	0531
1N441B	0015
1N442	0015
1N442B	0015
1N443	0015
1N443B	0015
1N444	0015
1N444B	0015
1N445	0015
1N445B	0015
1N446	0143
1N447	0143
1N447A	OEM
1N448	0143
1N449	0143
1N450	0143
1N451	0143
1N452	0143
1N453	0143
1N454	0143
1N455	0143
1N456	0133
1N456A	0133
1N456AM	OEM
1N457	0133
1N457A	0133
1N457AM	0080
1N457M	0133
1N458	0133
1N458A	0133
1N458AM	0133
1N458M	0133
1N459	0133
1N459A	0133
1N459M	0133
1N460	0133
1N460A	0133
1N460B	0133
1N461	0133
1N461A	0133
1N461AM	OEM
1N461M	OEM
1N462	0133
1N462A	0080
1N462AM	0080
1N462M	0133
1N463	0133
1N463A	0133
1N464	0133
1N464A	0133
1N465	0760
1N465A	0760
1N465B	OEM

DEVICE TYPE	REPL CODE
1N466	0376
1N466A	0376
1N466B	OEM
1N467	0036
1N467A	0036
1N467B	OEM
1N468	0140
1N468A	0140
1N468B	0589
1N469	0253
1N469A	0062
1N469B	0883
1N470	0077
1N470A	0062
1N470B	0295
1N471	0376
1N471A	0376
1N471B	OEM
1N472	0275
1N472A	0275
1N472B	0275
1N473	0041
1N473A	0041
1N473B	0162
1N474	0091
1N474A	0091
1N474B	0091
1N475	0062
1N475A	0062
1N475B	0062
1N476	0143
1N477	0143
1N478	0143
1N479	0143
1N480	0143
1N480B	0015
1N481A	OEM
1N482	0133
1N482A	0133
1N482AM	0133
1N482B	0133
1N482BM	0133
1N482C	0133
1N482M	0133
1N482TH	0015
1N483	0133
1N483A	0133
1N483AM	0133
1N483B	0133
1N483BM	0133
1N483C	0133
1N483M	0080
1N483TH	0015
1N484	0133
1N484A	0133
1N484B	0133
1N484C	0133
1N484TH	0015
1N485	0133
1N485A	0133
1N485B	0133
1N485C	0133
1N485TH	0015
1N486	0133
1N486A	0133
1N486B	0133
1N486TH	0015
1N487	0102
1N487A	0102
1N487B	0102
1N487TH	0015
1N488	0102
1N488A	0102
1N488B	0102
1N488TH	0015
1N490	0143
1N492	0794
1N493	0794
1N497	0143
1N498	0143
1N499	0143
1N500	0143
1N501	0143
1N502	OEM
1N503	0015
1N504	0015
1N505	0015
1N506	0015
1N507	0015
1N508	0015
1N509	0071
1N510	0071
1N511	0015
1N512	0015
1N513	0015
1N514	0015
1N515	0015
1N516	0015
1N517	0071
1N518	0071
1N519	0015
1N520	0015
1N521	0015
1N522	0015
1N523	0015
1N524	0015
1N525	0071
1N526	0071
1N527	0143

DEVICE TYPE	REPL CODE
1N530	0015
1N531	0015
1N532	0015
1N533	0015
1N534	0015
1N535	0015
1N536	0015
1N536ARL	OEM
1N537	0015
1N538	0015
1N539	0015
1N540	0015
1N541	0123
1N542	0143
1N542A	0143
1N542MP	0143
1N543	0102
1N543A	0102
1N547	0015
1N547A	OEM
1N548	0071
1N549	0102
1N550	4140
1N550R	5517
1N551	0580
1N551R	0947
1N552	2060
1N552R	4523
1N553	1000
1N553R	1012
1N554	2076
1N554R	5567
1N555	2081
1N555R	1073
1N560	0071
1N561	0071
1N562	2088
1N562R	1138
1N563	0612
1N563R	1188
1N564	0143
1N566	OEM
1N567	OEM
1N568	0143
1N569	0143
1N570	2088
1N571	0143
1N580A	OEM
1N588	0017
1N589	0017
1N590	OEM
1N591	OEM
1N596	0015
1N597	0071
1N598	0071
1N599	0015
1N599A	0015
1N600	0015
1N600A	0015
1N600KB	0133
1N601	0015
1N601A	0015
1N601S34	0143
1N602	0015
1N602A	0015
1N603	0015
1N603A	0015
1N604	0015
1N604A	0015
1N605	0015
1N605A	0015
1N606	0015
1N606A	0015
1N607	0708
1N607A	0708
1N607R	2291
1N608	0580
1N608A	0580
1N608R	2291
1N609	0580
1N609A	0580
1N609R	2291
1N610	0580
1N610A	0580
1N610R	2291
1N611	0947
1N611A	0947
1N611R	0476
1N612	1000
1N612A	1000
1N612R	0476
1N613	2076
1N613A	2076
1N613R	0476
1N614	2081
1N614A	2081
1N614R	0476
1N616	0143
1N617	0143
1N618	0143
1N619	0133
1N619M	OEM
1N622	0133
1N625	0133
1N625A	0133
1N625M	0133
1N626	0133
1N626A	0133
1N626M	0133

DEVICE TYPE	REPL CODE
1N627	0133
1N627A	0133
1N628	0133
1N628A	0133
1N629	0133
1N629A	0133
1N630	OEM
1N630A	OEM
1N631	0143
1N632	0143
1N633	0133
1N634	OEM
1N635	OEM
1N636	0143
1N636A	0143
1N643	0133
1N643A	0133
1N645	0015
1N645-1	0015
1N645A	0795
1N645B	0795
1N645J	0795
1N645TH	0015
1N646	0015
1N646TH	0015
1N647	0015
1N647-1	0015
1N647TH	0015
1N648	0015
1N648TH	0015
1N649	0015
1N649-1	0015
1N649TH	0015
1N658	0133
1N658A	0133
1N658M	0133
1N659	0133
1N659/A	0133
1N659A	0133
1N659AM	OEM
1N659M	OEM
1N660	0133
1N660A	0133
1N661	0133
1N661A	0133
1N662	0133
1N662A	0133
1N663	0133
1N663A	0133
1N663M	0133
1N664	0165
1N665	0052
1N666	0686
1N667	0375
1N668	0705
1N669	0441
1N670	0173
1N671	0189
1N672	0340
1N673	0015
1N674	0140
1N675	0471
1N676	0015
1N677	0015
1N678	0015
1N679	0015
1N681	0015
1N682	0015
1N683	0015
1N684	0015
1N685	0015
1N686	0015
1N687	0015
1N689	0015
1N690	0133
1N691	0133
1N692	0133
1N693	0133
1N695	0143
1N695A	0143
1N696	0133
1N697	0133
1N698	0143
1N699	0143
1N701	0064
1N702	0760
1N702A	0760
1N702A0	OEM
1N702A1	OEM
1N702A3	OEM
1N702A4	OEM
1N702A5	OEM
1N702A7	OEM
1N702A8	OEM
1N702A9	OEM
1N703	0376
1N703A	0376
1N703A0	OEM
1N703A1	OEM
1N703A3	OEM
1N703A4	OEM
1N703A5	OEM
1N703A6	OEM
1N704	0036
1N704A	0036
1N704A1	OEM
1N704A2	OEM
1N704A3	OEM
1N704A4	OEM

If replacement code is OEM, contact original manufacturer for replacement.

DEVICE TYPE	REPL CODE
1N704A5	OEM
1N705	0140
1N705A	0140
1N705A0	OEM
1N705A1	OEM
1N705A2	OEM
1N705A4	OEM
1N705A5	OEM
1N705A6	OEM
1N705A7	OEM
1N705A8	OEM
1N706	0091
1N706A	0091
1N706A1	OEM
1N706A4	OEM
1N706A6	OEM
1N706A8	OEM
1N706A9	OEM
1N707	0062
1N707A	0062
1N707A0	OEM
1N707A2	OEM
1N707A3	OEM
1N707A6	OEM
1N708	0253
1N708A	0253
1N708B	0253
1N709	0471
1N709A	0471
1N709B	0471
1N710	0062
1N710A	0062
1N710B	0062
1N711	0077
1N711A	0077
1N711B	0099
1N712	0165
1N712A	0165
1N712B	0165
1N713	0057
1N713A	0057
1N713B	0057
1N714	0064
1N714A	0064
1N714B	0064
1N715	0181
1N715A	0181
1N715B	0181
1N716	0052
1N716A	0052
1N716B	0137
1N717	0053
1N717A	0053
1N717B	0053
1N718	0686
1N718A	0686
1N718B	0002
1N719	0445
1N719A	0445
1N719B	0445
1N720	0375
1N720A	0375
1N720B	0375
1N721	0700
1N721A	0700
1N721B	0700
1N722	0705
1N722A	0705
1N722B	0705
1N723	0494
1N723A	0494
1N723B	0494
1N724	0455
1N724A	0455
1N724B	0441
1N725	0195
1N725A	0195
1N725B	0195
1N726	0166
1N726A	0166
1N726B	0166
1N727	0010
1N727A	0010
1N727B	0820
1N728	0032
1N728A	0032
1N728B	0032
1N729	0054
1N729A	0054
1N729B	0054
1N729RL	OEM
1N730	0068
1N730A	0068
1N730B	0068
1N731	0092
1N731A	0092
1N731B	0092
1N732	0125
1N732A	0125
1N732B	0125
1N733	0783
1N733A	0152
1N733B	0152
1N734	0173
1N734A	0173
1N734B	0173
1N735	0094
1N735A	0094
1N735B	0094
1N736	0331
1N736A	0049
1N736B	0049
1N737	0156
1N737A	0156
1N737B	0156
1N738	0189
1N738A	0189
1N738B	0189
1N739	0149
1N739A	0099
1N739B	0099
1N740	0089
1N740A	0089
1N740B	0089
1N741	0287
1N741A	0287
1N741B	0340
1N742	0340
1N742A	0340
1N742B	0340
1N743	0370
1N743A	0370
1N743B	0370
1N744	0425
1N744A	0425
1N744B	0425
1N745	1473
1N745A	1473
1N745B	1473
1N745N	OEM
1N746	0299
1N746A	0299
1N746RL	OEM
1N747	0376
1N747A	0376
1N747B	0188
1N747RL	OEM
1N748	0036
1N748A	0036
1N748RL	OEM
1N749	0275
1N749A	0275
1N749ARL	OEM
1N749RL	OEM
1N750	0140
1N750A	0140
1N750ARL	OEM
1N750RL	OEM
1N751	0041
1N751A	0041
1N751ARL	OEM
1N751RL	OEM
1N752	0253
1N752A	0253
1N752ARL	OEM
1N752RL	OEM
1N753	0471
1N753A	0471
1N753ARL	OEM
1N753D	OEM
1N753RL	OEM
1N754	0062
1N754A	0062
1N754ARL	OEM
1N754RL	OEM
1N755	0077
1N755A	0077
1N755ARL	OEM
1N755RL	OEM
1N756	0165
1N756A	0165
1N756ARL	OEM
1N756RL	OEM
1N757	0057
1N757A	0057
1N757ARL	OEM
1N757RL	OEM
1N758	0064
1N758A	0064
1N758ARL	OEM
1N758RL	OEM
1N759	0052
1N759A	0052
1N759ARL	OEM
1N759RL	OEM
1N760	0133
1N761	0140
1N761-1	0140
1N761-2	0041
1N761A	0162
1N762	0091
1N762-1	0253
1N762-2	0091
1N762A	0157
1N763	0077
1N763-1	0062
1N763-2	0062
1N763-3	0077
1N763A	0077
1N764	0322
1N764-1	0165
1N764-2	0322
1N764-3	0057
1N764-4	0057
1N764A	0322
1N765	0181
1N765-1	0064
1N765-2	0181
1N765A	0064
1N766	0053
1N766-1	0052
1N766-2	0053
1N766-3	0100
1N766A	0365
1N767	0445
1N767-1	0686
1N767-2	0445
1N767-3	0210
1N767-22	0445
1N767A	0445
1N768	0671
1N768-1	0375
1N768-2	0671
1N768-3	0700
1N768A	0671
1N769	0494
1N769-1	0705
1N769-2	0494
1N769-3	0455
1N769-4	0257
1N769A	0494
1N770	0143
1N771	0143
1N771A	0143
1N771B	0143
1N772	0133
1N772A	0143
1N773	0143
1N773A	0143
1N774	0143
1N774A	0143
1N775	0143
1N776	0143
1N777	0143
1N778	0133
1N779	0133
1N781	0143
1N781A	0143
1N788	0133
1N789	0133
1N789M	0133
1N790	0133
1N790M	0133
1N791	0133
1N791M	0133
1N792	0133
1N792M	0133
1N793	0133
1N793M	0133
1N794	0133
1N794M	OEM
1N795	0133
1N795M	OEM
1N796	0133
1N796M	OEM
1N797	0133
1N798	0133
1N799	0133
1N800	0133
1N801	0133
1N802	0133
1N803	0133
1N804	0133
1N805	0143
1N806	0133
1N807	0133
1N808	0133
1N809	0133
1N810	0133
1N811	0133
1N811M	0133
1N812	0133
1N812M	0133
1N813	0133
1N813M	0133
1N814	0133
1N814M	0133
1N815	0133
1N815M	0133
1N816	0133
1N817	0133
1N818	0133
1N818M	0133
1N819	0015
1N821	0471
1N821-1	OEM
1N821A	0471
1N821A1	OEM
1N821ARL	OEM
1N821RL	OEM
1N822	0471
1N822A	0471
1N822AG	0471
1N823	0471
1N823-1	OEM
1N823A	0471
1N823A1	OEM
1N823ARL	OEM
1N823RL	OEM
1N824	0471
1N824A	0471
1N825	0471
1N825-1	OEM
1N825A	0471
1N825A1	OEM
1N825ARL	OEM
1N825RL	OEM
1N826	0062
1N826A	0062
1N827	0471
1N827-1	OEM
1N827A	0471
1N827A1	OEM
1N827ARL	OEM
1N827RL	OEM
1N828	0062
1N828A	0062
1N829	0471
1N829-1	OEM
1N829A	0471
1N829A1	OEM
1N829ARL	OEM
1N829RL	OEM
1N830	OEM
1N830A	OEM
1N831	OEM
1N831A	OEM
1N831B	OEM
1N831C	OEM
1N832	OEM
1N832A	OEM
1N832B	OEM
1N832C	OEM
1N832D	OEM
1N833	OEM
1N833A	OEM
1N835	0143
1N837	0133
1N837A	0133
1N838	0133
1N839	0133
1N840	0133
1N840M	0133
1N841	0133
1N842	0133
1N843	0133
1N844	0133
1N845	0133
1N846	0015
1N847	0015
1N848	0015
1N849	0015
1N850	0015
1N851	0015
1N852	0015
1N853	0071
1N854	0071
1N855	0071
1N856	0071
1N857	0015
1N858	0015
1N859	0015
1N860	0015
1N861	0015
1N862	0015
1N863	0015
1N864	0071
1N865	0071
1N866	0071
1N867	0071
1N868	0015
1N869	0015
1N870	0015
1N871	0015
1N872	0015
1N873	0015
1N874	0015
1N875	0071
1N876	0071
1N877	0071
1N878	0071
1N879	0015
1N880	0015
1N881	0015
1N881B	0015
1N882	0015
1N883	0015
1N884	0015
1N885	0015
1N886	0071
1N887	0071
1N888	0071
1N889	0071
1N890	0133
1N890M	0133
1N891	0133
1N891M	OEM
1N892	0133
1N893	0102
1N894	OEM
1N897	0133
1N898	0133
1N899	0133
1N900	0133
1N901	0133
1N902	0133
1N903	0133
1N903/A	0133
1N903A	0133
1N903AM	0133
1N903M	0133
1N904	0133
1N904/A	0133
1N904A	0133
1N904AM	0133
1N904M	0133
1N905	0133
1N905/A	0133
1N905A	0133
1N905AM	0133
1N905M	0133
1N906	0133
1N906/A	0133
1N906A	0133
1N906AM	0133
1N906M	0133
1N907	0133
1N907/A	0133
1N907A	0133
1N907AM	0133
1N907M	0133
1N908	0133
1N908/A	0133
1N908A	0133
1N908AM	0133
1N908M	0133
1N909	0143
1N910	0143
1N911	0143
1N912	OEM
1N912A	OEM
1N913	OEM
1N913A	OEM
1N914	0124
1N914/A/B	0124
1N914A	0124
1N914B	0124
1N914F	0133
1N914M	0124
1N915	0133
1N916	0124
1N916A	0133
1N916B	0133
1N917	0133
1N918	OEM
1N919	0133
1N920	0133
1N921	0133
1N922	0133
1N923	0133
1N924	0133
1N925	0133
1N926	0133
1N927	0133
1N928	0133
1N929	0133
1N930	0133
1N931	0133
1N932	0133
1N933	0133
1N934	0133
1N935	0057
1N935-1	OEM
1N935A1	OEM
1N935ARL	OEM
1N935B1	OEM
1N935BRL	OEM
1N935RL	OEM
1N936	0057
1N936A	0057
1N936ARL	OEM
1N936B	0057
1N936BRL	OEM
1N936RL	OEM
1N937	0057
1N937-1	OEM
1N937A	0057
1N937A1	OEM
1N937ARL	OEM
1N937B	0057
1N937B1	OEM
1N937BRL	OEM
1N937RL	OEM
1N938	0057
1N938-1	OEM
1N938A	0057
1N938A1	OEM
1N938B	0057
1N938B1	OEM
1N939	0057
1N939A	0057
1N939ARL	OEM
1N939B	0057
1N939BRL	OEM
1N939RL	OEM
1N940	0057
1N940A	0057
1N940B	0057
1N941	0794
1N941-1	OEM
1N941A	0794
1N941A1	OEM
1N941ARL	OEM
1N941B	0794
1N941B1	OEM
1N941RL	OEM
1N942	0794
1N942A	0137
1N942B	0137
1N942BRL	OEM
1N942RL	OEM
1N943	0137
1N943-1	OEM
1N943A	0137
1N943A1	OEM
1N943ARL	OEM
1N943B	0137
1N943B1	OEM
1N943BRL	OEM
1N943RL	OEM
1N944	0137
1N944-1	OEM
1N944A	0137
1N944A1	OEM
1N944ARL	OEM
1N944B	0137
1N944B1	OEM
1N944BRL	OEM
1N944RL	OEM
1N945	0794
1N945-1	OEM
1N945A	0794
1N945A1	OEM
1N945B	0794
1N945B1	OEM
1N946	0794
1N946A	0794
1N946B	0794
1N947	0015
1N948	0133
1N949	0143
1N950	OEM
1N950A	OEM
1N950B	OEM
1N951	0064
1N951A	0064
1N951B	OEM
1N952	OEM
1N952A	OEM
1N952B	OEM
1N953	OEM
1N953A	OEM
1N953B	OEM
1N954	0030
1N954A	OEM
1N954B	OEM
1N955	OEM
1N955A	OEM
1N955B	OEM
1N956	OEM
1N956A	OEM
1N956B	OEM
1N957	0062
1N957A	0062
1N957ARL	OEM
1N957B	0062
1N957BRL	OEM
1N958	0077
1N958A	0077
1N958ARL	OEM
1N958B	0077
1N958BRL	OEM
1N959	0165
1N959A	0165
1N959ARL	OEM
1N959B	0165
1N959BRL	OEM
1N960	0057
1N960A	0057
1N960ARL	OEM
1N960B	0057
1N960BRL	OEM
1N961	0170
1N961A	0064
1N961ARL	OEM
1N961B	0064
1N961BRL	OEM
1N962	0181
1N962A	0181
1N962ARL	OEM
1N962B	0181
1N962BRL	OEM
1N963	0052
1N963A	0052
1N963ARL	OEM
1N963B	0052
1N963BRL	OEM
1N964	0053
1N964A	0053
1N964ARL	OEM
1N964B	0053
1N964B-F	0365
1N964BRL	OEM
1N965	0686
1N965A	0686
1N965ARL	OEM
1N965B	0686
1N965BRL	OEM
1N966	0445
1N966A	0445
1N966ARL	OEM
1N966B	0445
1N966B-F	0445
1N966BRL	OEM
1N967	0375
1N967A	0375
1N967ARL	OEM
1N967B	0375
1N967B-F	0495
1N967BRL	OEM
1N968	0700
1N968A	0700
1N968ARL	OEM
1N968B	0700
1N968B-F	
1N968BRL	
1N969	
1N969A	
1N969ARL	
1N969B	
1N969BRL	
1N970	0
1N970A	0
1N970ARL	
1N970B	0
1N970BRL	0
1N971	04
1N971A	04
1N971ARL	O
1N971B	04
1N971BRL	OE
1N972	01
1N972A	01
1N972ARL	OE
1N972B	019
1N972B-F	072
1N972BRL	OE
1N973	016
1N973A	016
1N973ARL	OE
1N973B	016
1N973BRL	OEM
1N974	0010
1N974A	0010
1N974ARL	OEM
1N974B	0010
1N974BRL	OEM
1N975	0032
1N975A	0032
1N975ARL	OEM
1N975B	0032
1N975BRL	OEM
1N976	0054
1N976A	0054
1N976ARL	OEM
1N976B	0054
1N976BRL	OEM
1N977	0068
1N977A	0068
1N977ARL	OEM
1N977B	0068
1N977BRL	OEM
1N978	0092
1N978A	0092
1N978ARL	OEM
1N978B	0092
1N978BRL	OEM
1N979	0125
1N979A	0125
1N979ARL	OEM
1N979B	0125
1N979BRL	OEM
1N980	0152
1N980A	0152
1N980ARL	OEM
1N980B	0152
1N980BRL	OEM
1N981	0173
1N981A	0173
1N981ARL	OEM
1N981B	0173
1N981BRL	OEM
1N982	0094
1N982A	0094
1N982ARL	OEM
1N982B	0094
1N982BRL	OEM
1N983	0049
1N983A	0049
1N983ARL	OEM
1N983B	0049
1N983BRL	OEM
1N984	0156
1N984A	0156
1N984ARL	OEM
1N984B	0156
1N984BRL	OEM
1N985	0189
1N985A	0189
1N985ARL	OEM
1N985B	0189
1N985BRL	OEM
1N986	0099
1N986A	0099
1N986ARL	OEM
1N986B	0099
1N987	0089
1N987A	0089
1N987ARL	OEM
1N987B	0089
1N987BRL	OEM
1N988	0287
1N988A	0287
1N988ARL	OEM
1N988B	0287
1N988BRL	OEM
1N989	0340
1N989A	0340
1N989ARL	OEM
1N989B	0340
1N989BRL	OEM
1N990	0370
1N990A	0370

DEVICE TYPE	REPL CODE	DEVICE TYPE	REPL CODE	DEVICE TYPE	REPL CODE	DEVICE TYPE	REPL CODE	DEVICE TYPE	REPL CODE	DEVICE TYPE	REPL CODE	DEVICE TYPE	REPL CODE
	OEM	1N1100	0015	1N1187	0805	1N1203RC	OEM	1N1259(WEST)	1195	1N1345B	0993	1N1360R	1756
	0370	1N1101	0015	1N1187A	1126	1N1204	0109	1N1260	0071	1N1345BR	0993	1N1360RA	1756
	OEM	1N1102	0015	1N1187AR	1111	1N1204A	0109	1N1260(WEST)	0319	1N1345C	OEM	1N1360RB	1756
	0425	1N1103	0015	1N1187R	1111	1N1204AR	0538	1N1261	0071	1N1345R	0993	1N1361	1761
	0425	1N1104	0015	1N1187RA	1111	1N1204B	0109	1N1262	OEM	1N1345RA	0993	1N1361A	1761
	OEM	1N1104R	0015	1N1188	0805	1N1204BR	0538	1N1281	OEM	1N1346	0197	1N1361B	1761
	0425	1N1105	0015	1N1188A	0805	1N1204C	1000	1N1282	OEM	1N1346A	0197	1N1361C	1782
	OEM	1N1108	0071	1N1188AR	0258	1N1204R	0538	1N1283	OEM	1N1346AR	0997	1N1361CA	OEM
	1473	1N1109	OEM	1N1188R	0258	1N1204RA	0538	1N1284	OEM	1N1346B	0197	1N1361R	1782
	1473	1N1110	OEM	1N1188RA	0258	1N1204RB	0476	1N1285	OEM	1N1346BR	0997	1N1361RA	1782
	OEM	1N1111	OEM	1N1189	1195	1N1204RC	1012	1N1286	OEM	1N1346C	OEM	1N1361RB	1782
	1473	1N1112	OEM	1N1189A	1195	1N1205	0116	1N1287	OEM	1N1346R	0997	1N1361RC	1782
	OEM	1N1113	OEM	1N1189AR	1644	1N1205A	0116	1N1291	OEM	1N1346RA	0997	1N1362	1772
	0133	1N1115	0580	1N1189R	1644	1N1205AR	0801	1N1292	OEM	1N1346RB	0997	1N1362A	1772
	0143	1N1115R	2291	1N1189RA	1644	1N1205B	0116	1N1293	OEM	1N1347	0200	1N1362B	1772
	0143	1N1116	0580	1N1190	0319	1N1205BR	0801	1N1294	OEM	1N1347A	0200	1N1362C	OEM
	0143	1N1116R	2291	1N1190A	0319	1N1205C	1637	1N1295	OEM	1N1347AR	1001	1N1362CA	OEM
	0143	1N1117	2060	1N1190AR	0267	1N1205R	0801	1N1296	OEM	1N1347B	1001	1N1362R	1794
	0133	1N1117R	0476	1N1190R	0267	1N1205RA	0801	1N1297	OEM	1N1347BR	1001	1N1362RA	1794
	0015	1N1118	1000	1N1190RA	0267	1N1205RB	0476	1N1301	3189	1N1347C	OEM	1N1362RB	1794
	0133	1N1118R	0476	1N1191	3189	1N1205RC	5578	1N1301R	2813	1N1347R	1001	1N1363	1788
	0015	1N1119	2076	1N1191A	3189	1N1206	0122	1N1302	2901	1N1347RA	1001	1N1363A	1788
	0015	1N1119R	0476	1N1191AR	1630	1N1206A	0122	1N1302R	2813	1N1347RB	1001	1N1363B	1788
	0015	1N1120	2081	1N1191R	1630	1N1206AR	0816	1N1303	2813	1N1348	0204	1N1363C	OEM
	0015	1N1120R	0476	1N1191RA	1630	1N1206B	0122	1N1303R	2813	1N1348A	0204	1N1363CA	OEM
	0015	1N1122A	0015	1N1192	2901	1N1206BR	0816	1N1304	1124	1N1348AR	0515	1N1363R	1799
	0015	1N1124	0580	1N1192A	2901	1N1206C	2076	1N1304R	2813	1N1348B	0204	1N1363RA	1799
	0015	1N1124A	0989	1N1192AR	0254	1N1206R	0816	1N1305	2813	1N1348BR	0515	1N1363RB	1799
	0708	1N1124AR	0990	1N1192R	0254	1N1206RA	0816	1N1305R	2813	1N1348C	OEM	1N1364	1796
	5517	1N1124R	0947	1N1192RA	0254	1N1206RB	0476	1N1306	1126	1N1348R	0515	1N1364A	1796
	0947	1N1124RA	0476	1N1193	1124	1N1206RC	5567	1N1306R	2813	1N1348RA	0515	1N1364B	1796
7	0947	1N1125	2060	1N1193A	1124	1N1216R	OEM	1N1313	0322	1N1348RB	0515	1N1364C	OEM
8	4523	1N1125A	3743	1N1193AR	1107	1N1217	0023	1N1313A	0322	1N1351	0510	1N1364CA	OEM
39	1012	1N1125AR	0993	1N1193R	1107	1N1217A	0071	1N1314	0181	1N1351A	0510	1N1364R	1809
40	0933	1N1125R	4523	1N1193RA	1107	1N1217B	0071	1N1314A	0181	1N1351B	0510	1N1364RA	1809
041	5517	1N1125RA	0476	1N1194	1124	1N1218	0071	1N1315	0053	1N1351C	0510	1N1364RB	1809
042	0947	1N1126	1000	1N1194A	1124	1N1218A	0071	1N1315A	0053	1N1351CA	OEM	1N1365	1804
1043	0947	1N1126A	0197	1N1194AR	1107	1N1218B	0071	1N1316	0445	1N1351R	0992	1N1365A	1804
1044	4523	1N1126AR	0997	1N1194R	1107	1N1219	0071	1N1316A	0445	1N1351RA	0992	1N1365B	1804
N1045	1012	1N1126R	1012	1N1194RA	1107	1N1219A	0071	1N1317	0671	1N1351RB	0992	1N1365C	OEM
N1046	0933	1N1126RA	0476	1N1195	1126	1N1219B	0071	1N1317A	0671	1N1352	0691	1N1365CA	OEM
N1047	5517	1N1127	2076	1N1195A	1126	1N1220	0242	1N1318	0494	1N1352A	0691	1N1365R	1817
1N1048	0947	1N1127A	0200	1N1195AR	1111	1N1220A	0242	1N1318A	0494	1N1352B	0691	1N1365RA	1817
1N1049	0947	1N1127AR	1001	1N1195R	1111	1N1220B	0242	1N1319	0257	1N1352C	0691	1N1365RB	1817
1N1050	4523	1N1127R	5567	1N1195RA	0258	1N1221	1747	1N1319A	0257	1N1352CA	OEM	1N1366	1194
1N1051	1012	1N1127RA	0476	1N1196	0805	1N1221A	1747	1N1320	0166	1N1352R	0995	1N1366A	1194
1N1052	0015	1N1128	2081	1N1196A	0805	1N1221B	1747	1N1320A	0166	1N1352RA	0995	1N1366B	1194
1N1053	0015	1N1128A	0204	1N1196AR	0258	1N1222	0540	1N1321	0032	1N1352RB	0995	1N1366C	OEM
1N1054	0071	1N1128AR	0515	1N1196R	0258	1N1222A	0540	1N1321A	0032	1N1353	0870	1N1366CA	OEM
1N1055	0071	1N1128R	1073	1N1196RA	0258	1N1222B	0540	1N1322	0068	1N1353A	0870	1N1366R	1826
1N1056	0087	1N1128RA	0449	1N1197	1195	1N1223	1771	1N1322A	0068	1N1353B	0870	1N1366RA	1826
1N1057	0087	1N1130	OEM	1N1197A	1195	1N1223A	1771	1N1323	2317	1N1353C	0870	1N1366RB	1826
1N1058	0985	1N1131	OEM	1N1197AR	1644	1N1223B	1771	1N1323A	2317	1N1353CA	OEM	1N1367	0022
1N1059	0910	1N1132	OEM	1N1197R	1644	1N1224	0965	1N1324	0173	1N1353R	1263	1N1367A	0022
1N1060	0990	1N1133	OEM	1N1197RA	0267	1N1224A	0965	1N1324A	2157	1N1353RA	1263	1N1367B	0022
1N1061	0990	1N1134	OEM	1N1198	0319	1N1224B	0965	1N1325	0104	1N1353RB	1263	1N1367C	OEM
1N1062	0993	1N1135	OEM	1N1198A	0319	1N1225	0071	1N1325A	0104	1N1354	1020	1N1367CA	OEM
1N1063	0997	1N1136	OEM	1N1198AR	0267	1N1225A	0071	1N1326	0099	1N1354A	1020	1N1367R	1853
1N1064	0985	1N1137	OEM	1N1198R	0267	1N1225B	0087	1N1326A	0149	1N1354B	1020	1N1367RA	1853
1N1064R	2291	1N1138	OEM	1N1198RA	0267	1N1226	0071	1N1327	0287	1N1354C	1020	1N1367RB	1853
1N1065	0910	1N1139	OEM	1N1199	0084	1N1226A	0071	1N1327A	0213	1N1354CA	OEM	1N1368	0132
1N1065R	2291	1N1140	OEM	1N1199A	0084	1N1226B	0087	1N1329	OEM	1N1354R	1249	1N1368A	0132
1N1066	0989	1N1141	OEM	1N1199AR	0534	1N1227	0708	1N1330	OEM	1N1354RA	1249	1N1368B	0132
1N1066R	2291	1N1142	OEM	1N1199B	0084	1N1227A	0708	1N1331	OEM	1N1354RB	1249	1N1368C	OEM
1N1067	0989	1N1143	OEM	1N1199BR	0534	1N1227B	0708	1N1332	OEM	1N1355	1274	1N1368CA	OEM
1N1067R	2291	1N1143A	OEM	1N1199C	0084	1N1228	0580	1N1333	OEM	1N1355A	1274	1N1368R	1866
1N1068	3743	1N1144	OEM	1N1199R	0534	1N1228A	0580	1N1334	OEM	1N1355B	1274	1N1368RA	1866
1N1068R	0476	1N1145	OEM	1N1199RA	0534	1N1228B	0580	1N1335	OEM	1N1355C	1274	1N1368RB	1866
1N1069	0997	1N1146	OEM	1N1199RB	2291	1N1229	0580	1N1336	OEM	1N1355CA	OEM	1N1369	0207
1N1069R	0476	1N1147	OEM	1N1199RC	0084	1N1229A	0580	1N1337-5	0015	1N1355R	1639	1N1369A	0207
1N1070	0985	1N1148	OEM	1N1200	0090	1N1229B	0580	1N1341	0970	1N1355RA	1639	1N1369B	0207
1N1071	0910	1N1149	OEM	1N1200A	0090	1N1230	0580	1N1341A	0970	1N1355RB	1639	1N1369C	OEM
1N1072	0990	1N1150	OEM	1N1200AR	0748	1N1230A	0580	1N1341AR	0985	1N1356	1402	1N1369CA	OEM
1N1073	0990	1N1150A	OEM	1N1200B	0090	1N1230B	0580	1N1341B	0970	1N1356A	1402	1N1369R	1884
1N1074	0993	1N1157	OEM	1N1200BR	0748	1N1231	2060	1N1341BR	0985	1N1356B	1402	1N1369RA	1884
1N1075	0997	1N1158	OEM	1N1200C	0580	1N1231A	2060	1N1341C	OEM	1N1356C	1402	1N1369RB	1884
1N1076	5005	1N1159	OEM	1N1200R	0748	1N1231B	2060	1N1341R	0985	1N1356CA	OEM	1N1370	0263
1N1077	1240	1N1160	OEM	1N1200RA	0765	1N1232	1000	1N1341RA	0985	1N1356R	1703	1N1370A	0263
1N1078	1250	1N1161	OEM	1N1200RB	2291	1N1232A	1000	1N1341RB	0985	1N1356RA	1703	1N1370B	0263
1N1079	1241	1N1162	OEM	1N1200RC	5517	1N1232B	1000	1N1342	3733	1N1356RB	1703	1N1370C	OEM
1N1080	3275	1N1163	OEM	1N1201	0097	1N1233	2076	1N1342A	3733	1N1357	1078	1N1370CA	OEM
1N1081	0015	1N1164	OEM	1N1201A	0097	1N1233A	2076	1N1342AR	0910	1N1357A	1078	1N1370R	1895
1N1081A	0015	1N1165	OEM	1N1201AR	0765	1N1233B	2076	1N1342B	3733	1N1357B	1078	1N1370RA	1895
1N1082	0015	1N1166	OEM	1N1201B	0097	1N1234	2081	1N1342BR	0910	1N1357C	OEM	1N1370RB	1895
1N1082A	0015	1N1167	OEM	1N1201BR	0765	1N1234A	2081	1N1342C	OEM	1N1357CA	OEM	1N1371	0310
1N1083	0015	1N1168	OEM	1N1201C	OEM	1N1234B	2081	1N1342R	0910	1N1357R	1716	1N1371A	0310
1N1083A	0015	1N1169	0015	1N1201R	0765	1N1235	2088	1N1342RA	0910	1N1357RA	1716	1N1371B	0310
1N1083C	OEM	1N1169A	0015	1N1201RA	0765	1N1235A	2088	1N1342RB	0910	1N1357RB	1716	1N1371C	OEM
1N1084	0015	1N1182	0319	1N1201RB	2291	1N1235B	2088	1N1343	0989	1N1358	1717	1N1371CA	OEM
1N1084A	0015	1N1183	3189	1N1201RC	OEM	1N1236	2088	1N1343A	0989	1N1358A	1717	1N1371R	1902
1N1085	OEM	1N1183A	3189	1N1202	0097	1N1236A	2088	1N1343AR	0990	1N1358B	1717	1N1371RA	1902
1N1085A	OEM	1N1183AR	1630	1N1202A	0097	1N1236B	2088	1N1343B	0989	1N1358C	OEM	1N1371RB	1902
1N1086	OEM	1N1183R	1630	1N1202AR	0765	1N1237	OEM	1N1343BR	0990	1N1358CA	OEM	1N1372	0329
1N1086A	OEM	1N1183RA	1630	1N1202B	0097	1N1238	OEM	1N1343C	OEM	1N1358R	1731	1N1372A	0329
1N1087	OEM	1N1184	2901	1N1202BR	0765	1N1239	OEM	1N1343R	0990	1N1358RA	1731	1N1372B	0329
1N1087A	OEM	1N1184A	2901	1N1202C	0580	1N1251	0015	1N1343RA	0990	1N1358RB	1731	1N1372C	OEM
1N1088	OEM	1N1184AR	0254	1N1202R	0765	1N1252	0015	1N1343RB	2291	1N1359	1723	1N1372CA	OEM
1N1088A	OEM	1N1184R	0254	1N1202RA	0765	1N1253	0015	1N1344	0989	1N1359A	1723	1N1372R	0736
1N1089	OEM	1N1184RA	1107	1N1202RB	2291	1N1254	0015	1N1344A	0989	1N1359B	1723	1N1372RA	0736
1N1089A	OEM	1N1185	1124	1N1202RC	0947	1N1254(WEST)	3189	1N1344AR	0990	1N1359C	OEM	1N1372RB	0736
1N1090	OEM	1N1185A	1124	1N1203	0105	1N1255	0015	1N1344B	0989	1N1359CA	OEM	1N1373	0356
1N1090A	OEM	1N1185AR	1107	1N1203A	0105	1N1255(WEST)	2901	1N1344BR	0990	1N1359R	0727	1N1373A	0356
1N1091	OEM	1N1185R	1107	1N1203AR	0777	1N1255A	0015	1N1344C	OEM	1N1359RA	0727	1N1373B	0356
1N1091A	OEM	1N1185RA	1107	1N1203B	0105	1N1256	0015	1N1344R	0990	1N1359RB	0727	1N1373C	OEM
1N1092	OEM	1N1186	1124	1N1203BR	0777	1N1256(WEST)	1124	1N1344RA	0990	1N1360	1736	1N1373CA	OEM
1N1092A	OEM	1N1186A	1124	1N1203C	OEM	1N1257	0015	1N1344RB	0990	1N1360A	1736	1N1373R	1909
1N1093	0143	1N1186AR	1107	1N1203R	0777	1N1257(WEST)	1126	1N1345	3743	1N1360B	1736	1N1373RA	1909
1N1095	0015	1N1186R	1107	1N1203RA	0777	1N1258	0071	1N1345A	3743	1N1360C	OEM	1N1373RB	1909
1N1096	0015	1N1186RA	1107	1N1203RB	0476	1N1259	0071	1N1345AR	0993	1N1360CA	OEM	1N1374	0412

Original Device Types

DEVICE TYPE	REPL CODE
1N1374A	0412
1N1374B	0412
1N1374C	OEM
1N1374CA	OEM
1N1374R	1914
1N1374RA	1914
1N1374RB	1914
1N1375	0438
1N1375A	0438
1N1375B	0438
1N1375C	OEM
1N1375CA	OEM
1N1375R	1163
1N1375RA	1163
1N1375RB	1163
1N1376	OEM
1N1377	OEM
1N1378	OEM
1N1379	OEM
1N1380	OEM
1N1381	OEM
1N1382	OEM
1N1391	0143
1N1396	1560
1N1396R	1347
1N1397	0599
1N1397R	1347
1N1398	0599
1N1398R	1347
1N1399	2657
1N1399R	1347
1N1400	0599
1N1400R	1347
1N1401	2850
1N1401R	1905
1N1402	1986
1N1402R	1905
1N1403	1986
1N1403R	1905
1N1406	0015
1N1407	0071
1N1408	0071
1N1409	0102
1N1410	0182
1N1411	0348
1N1412	0348
1N1413	OEM
1N1415	0015
1N1416	1490
1N1416R	1622
1N1417	0870
1N1417R	1263
1N1418	1274
1N1418R	1639
1N1419	1078
1N1419R	1716
1N1420	1723
1N1420R	0727
1N1421	1761
1N1421R	1782
1N1422	0310
1N1422R	1902
1N1423	0438
1N1423R	1163
1N1424	0063
1N1424R	1961
1N1425	0244
1N1426	0137
1N1427	0002
1N1428	0495
1N1429	0565
1N1430	0441
1N1431	1268
1N1432	0098
1N1433	0028
1N1434	3189
1N1434R	0319
1N1435	2901
1N1435R	0319
1N1436	1124
1N1436R	0319
1N1437	0805
1N1437R	0319
1N1438	0319
1N1438R	0319
1N1439	0015
1N1440	0015
1N1441	0015
1N1442	0015
1N1443	0071
1N1443A	0071
1N1443B	0071
1N1444	0612
1N1444A	0612
1N1444B	0612
1N1446	0580
1N1447	0580
1N1448	2060
1N1449	1000
1N1450	0580
1N1451	0580
1N1452	2060
1N1453	1000
1N1454	OEM
1N1455	OEM
1N1456	OEM
1N1457	OEM
1N1458	OEM
1N1459	OEM

DEVICE TYPE	REPL CODE
1N1460	OEM
1N1461	OEM
1N1462	OEM
1N1463	OEM
1N1464	OEM
1N1465	OEM
1N1466	OEM
1N1467	OEM
1N1468	OEM
1N1469	OEM
1N1470	OEM
1N1471	OEM
1N1472	OEM
1N1473	OEM
1N1474	OEM
1N1475	OEM
1N1476	OEM
1N1477	OEM
1N1478	OEM
1N1479	OEM
1N1480	OEM
1N1481	OEM
1N1482	1438
1N1482R	2113
1N1483	2221
1N1483R	0696
1N1484	0451
1N1485	0636
1N1486	0015
1N1487	0015
1N1488	0015
1N1489	0015
1N1490	0015
1N1491	0015
1N1492	0015
1N1507	0456
1N1507A	0456
1N1508A	0451
1N1509	0157
1N1509A	0157
1N1510	0025
1N1510A	0025
1N1511	0244
1N1511A	0244
1N1512	0170
1N1512A	0170
1N1513	0137
1N1513A	0137
1N1514	0002
1N1514A	0002
1N1515	0495
1N1515A	0495
1N1516	0565
1N1516A	0565
1N1517	0441
1N1517A	0441
1N1518	0456
1N1518A	0456
1N1519	0451
1N1519A	0451
1N1520	0157
1N1520A	0157
1N1521	0025
1N1521A	0025
1N1522	0244
1N1522A	0244
1N1523	0170
1N1523A	0170
1N1524	0137
1N1524A	0137
1N1524R	0816
1N1525	0002
1N1525A	0002
1N1526	0495
1N1526A	0495
1N1527	0565
1N1527A	0565
1N1528	0441
1N1528A	0441
1N1530	1083
1N1530A	1083
1N1531A	0441
1N1537	0708
1N1537R	0933
1N1538	0580
1N1538R	0947
1N1539	0580
1N1539R	0947
1N1540	0580
1N1540R	0947
1N1541	0947
1N1541R	4523
1N1542	1000
1N1542R	1012
1N1543	2076
1N1543R	5567
1N1544	2081
1N1544A	0879
1N1544R	1073
1N1551	0580
1N1551R	2291
1N1552	0580
1N1552R	2291
1N1553	2060
1N1553R	0476
1N1554	1000
1N1554R	0476
1N1555	2076
1N1555A	0257

DEVICE TYPE	REPL CODE
1N1555R	0476
1N1556	0953
1N1557	0242
1N1558	1747
1N1559	0540
1N1560	1771
1N1561	0143
1N1562	0143
1N1563	0953
1N1563A	0015
1N1564	0242
1N1564A	0015
1N1565	1747
1N1565A	0015
1N1566	0540
1N1566A	0015
1N1567	1771
1N1567A	0015
1N1568	0965
1N1568A	0015
1N1575	0094
1N1576	OEM
1N1577	OEM
1N1578	OEM
1N1581	0708
1N1581R	0933
1N1582	0580
1N1582R	5517
1N1583	0580
1N1583R	0947
1N1584	2060
1N1584R	4523
1N1585	1000
1N1585R	1012
1N1586	2076
1N1586R	5567
1N1587	2081
1N1587R	1073
1N1588	0547
1N1588A	0547
1N1588AR	2035
1N1588B	0547
1N1588R	2035
1N1588RA	2035
1N1588RB	2035
1N1589	2113
1N1589A	2113
1N1589AR	1438
1N1589B	2113
1N1589R	1438
1N1589RA	2401
1N1589RB	2401
1N1590	1901
1N1590A	1901
1N1590AR	1445
1N1590B	1901
1N1590R	1445
1N1590RA	1445
1N1590RB	1445
1N1591	1601
1N1591A	1601
1N1591AR	1458
1N1591B	1601
1N1591R	1458
1N1591RA	1458
1N1591RB	1458
1N1592	1622
1N1592A	1622
1N1592AR	1490
1N1592B	1622
1N1592R	1490
1N1592RA	1490
1N1592RB	1490
1N1593	0992
1N1593A	0992
1N1593AR	0510
1N1593B	0992
1N1593R	0510
1N1593RA	0510
1N1593RB	0510
1N1594	1263
1N1594A	1263
1N1594AR	0870
1N1594B	1263
1N1594R	0870
1N1594RA	0870
1N1594RB	0870
1N1595	1639
1N1595A	1639
1N1595AR	1274
1N1595B	1639
1N1595R	1274
1N1595RA	1274
1N1595RB	1274
1N1596	1716
1N1596A	1716
1N1596AR	1078
1N1596B	1716
1N1596R	1078
1N1596RA	1078
1N1596RB	1078
1N1597	0727
1N1597A	0727
1N1597AR	1723
1N1597B	0727
1N1597R	1723
1N1597RA	1723
1N1597RB	1723
1N1598	1782

DEVICE TYPE	REPL CODE
1N1598A	1782
1N1598AR	1761
1N1598B	1782
1N1598R	1761
1N1598RA	1761
1N1598RB	1761
1N1599	0547
1N1599A	0547
1N1599AR	2035
1N1599R	2035
1N1599RA	2035
1N1600	2113
1N1600A	0547
1N1600AR	1438
1N1600R	1438
1N1600RA	1438
1N1601	1901
1N1601A	1901
1N1601AR	1445
1N1601R	1445
1N1601RA	1445
1N1602	1601
1N1602A	1601
1N1602AR	1458
1N1602R	1458
1N1602RA	1458
1N1603	1622
1N1603A	1622
1N1603AR	1490
1N1603R	1490
1N1603RA	1490
1N1604	0992
1N1604A	0992
1N1604AR	0510
1N1604R	0510
1N1604RA	0510
1N1605	1263
1N1605A	1263
1N1605AR	0870
1N1605R	0870
1N1605RA	0870
1N1606	1639
1N1606A	1639
1N1606AR	1274
1N1606R	1274
1N1606RA	1274
1N1607	1716
1N1607A	1716
1N1607AR	1078
1N1607R	1078
1N1607RA	1078
1N1608	0727
1N1608A	0727
1N1608AR	1723
1N1608R	1723
1N1608RA	1723
1N1609	1782
1N1609A	1782
1N1609AR	1761
1N1609R	1761
1N1609RA	1761
1N1610	OEM
1N1611	OEM
1N1611A	OEM
1N1611B	OEM
1N1612	0970
1N1612(REV.POL.)	0985
1N1612A	0970
1N1612AR	0985
1N1612R	0985
1N1612RA	2291
1N1613	3733
1N1613(REV.POL.)	0910
1N1613A	3733
1N1613AR	0910
1N1613R	0910
1N1613RA	2291
1N1614	0989
1N1614(REV.POL.)	0990
1N1614A	0989
1N1614AR	0990
1N1614B	0990
1N1614R	0990
1N1614RA	2291
1N1615	0197
1N1615A	0197
1N1615AR	0997
1N1615R	0997
1N1615RA	0476
1N1616	0204
1N1616(REV.POL.)	0515
1N1616A	0204
1N1616AR	0515
1N1616R	0515
1N1616RA	0476
1N1617	0953
1N1617AR	0953
1N1618	0015
1N1619	0015
1N1620	0015
1N1621	0090
1N1622	0097
1N1623	0105
1N1624	0109
1N1625	OEM
1N1625A	OEM
1N1626	OEM
1N1626A	OEM
1N1627	OEM

DEVICE TYPE	REPL CODE
1N1628	OEM
1N1629	OEM
1N1630	0133
1N1631	OEM
1N1632	OEM
1N1633	OEM
1N1634	OEM
1N1635	OEM
1N1636	OEM
1N1637	OEM
1N1638	0133
1N1639	OEM
1N1640	OEM
1N1641	OEM
1N1642	OEM
1N1644	0110
1N1645	0953
1N1646	0242
1N1647	0242
1N1648	1747
1N1649	1747
1N1650	0540
1N1651	0540
1N1652	1771
1N1653	0965
1N1660	1023
1N1661	1023
1N1662	1023
1N1663	1023
1N1664	1023
1N1665	1036
1N1666	1036
1N1670	1023
1N1671	1023
1N1672	1023
1N1673	1023
1N1674	1023
1N1675	1036
1N1676	1036
1N1680	0599
1N1680R	1347
1N1681	0599
1N1681R	1347
1N1682	0599
1N1682R	1347
1N1683	0599
1N1683R	1905
1N1684	1986
1N1684R	1905
1N1685	1986
1N1685R	1905
1N1686	1986
1N1686R	1905
1N1687	1986
1N1687R	1905
1N1688	0657
1N1688R	0193
1N1689	0657
1N1689R	0193
1N1690	0657
1N1690R	0193
1N1691	0657
1N1691R	0193
1N1692	0015
1N1693	0015
1N1694	0015
1N1695	0015
1N1696	0015
1N1697	0015
1N1698	OEM
1N1699	OEM
1N1700	OEM
1N1701	0015
1N1702	0015
1N1703	0015
1N1704	0015
1N1705	0015
1N1706	0015
1N1707	0015
1N1708	0015
1N1709	0015
1N1710	0015
1N1711	0015
1N1712	0015
1N1713	0137
1N1730	0071
1N1730A	0071
1N1731	0348
1N1731A	0348
1N1732	0348
1N1732A	0348
1N1733	3889
1N1733A	3889
1N1734	3889
1N1734A	3889
1N1735	0471
1N1736	0052
1N1736A	0052
1N1737	0949
1N1737A	0949
1N1738	1606
1N1738A	1606
1N1739	0726
1N1739A	0726
1N1740	0320
1N1740A	0320
1N1741	0337
1N1741A	0337
1N1742	0027

DEVICE TYPE	REPL CODE
1N1742A	0027
1N1743	0510
1N1743R	0992
1N1744	0170
1N1745	OEM
1N1746	OEM
1N1747	OEM
1N1748	OEM
1N1749	OEM
1N1750	OEM
1N1751	OEM
1N1752	OEM
1N1753	OEM
1N1754	OEM
1N1755	OEM
1N1756	OEM
1N1757	OEM
1N1758	OEM
1N1759	OEM
1N1760	OEM
1N1761	OEM
1N1762	OEM
1N1763	0023
1N1763A	0540
1N1764	0015
1N1764A	1771
1N1765	0157
1N1765A	0157
1N1766	0636
1N1766A	0636
1N1767	0025
1N1767A	0025
1N1768	0649
1N1768A	0649
1N1769	0244
1N1769A	0244
1N1770	0012
1N1770A	0012
1N1771	0170
1N1771A	0170
1N1772	0317
1N1772A	0317
1N1773	0137
1N1773A	0137
1N1774	0365
1N1774A	0365
1N1775	0002
1N1775A	0002
1N1776	0420
1N1776A	0420
1N1777	0495
1N1777A	0495
1N1778	0531
1N1778A	0531
1N1779	0301
1N1779A	0565
1N1780	0402
1N1780A	0531
1N1781	0441
1N1781A	0441
1N1782	0726
1N1782A	0726
1N1783	0365
1N1783A	0039
1N1784	0820
1N1784A	0820
1N1785	0350
1N1785A	0350
1N1786	0931
1N1786A	0931
1N1787	0999
1N1787A	0999
1N1788	0502
1N1788A	0502
1N1789	0869
1N1789A	0869
1N1790	0783
1N1790A	0783
1N1791	1268
1N1791A	1268
1N1792	1189
1N1792A	1189
1N1793	0331
1N1793A	0331
1N1794	1311
1N1794A	1311
1N1795	0098
1N1795A	0098
1N1796	0149
1N1796A	0149
1N1797	0186
1N1797A	0186
1N1798	0213
1N1798A	0213
1N1799	0028
1N1799A	0028
1N1800	0255
1N1800A	0255
1N1801	0367
1N1801A	0367
1N1802	0421
1N1802A	0421
1N1803	1445
1N1803A	1445
1N1803AR	1901
1N1803B	1445
1N1803BR	1901
1N1803C	1445
1N1803CA	OEM

DEVICE TYPE	REPL CODE
1N1803R	1901
1N1803RA	1901
1N1804	2221
1N1804A	2221
1N1804B	2221
1N1804C	2221
1N1804R	0696
1N1804RA	0696
1N1805	1458
1N1805A	1458
1N1805AR	1601
1N1805B	1458
1N1805BR	1458
1N1805C	1458
1N1805CA	OEM
1N1805R	1601
1N1805RA	1601
1N1805RB	1601
1N1806	0221
1N1806A	0221
1N1806B	0213
1N1806C	0221
1N1806CA	OEM
1N1806R	1616
1N1806RA	1616
1N1807	1490
1N1807A	1490
1N1807AR	1622
1N1807B	1490
1N1807BR	1622
1N1807C	1490
1N1807CA	OEM
1N1807R	1622
1N1807RA	1622
1N1808	1618
1N1808A	1618
1N1808B	1618
1N1808C	1618
1N1808CA	OEM
1N1808R	0627
1N1808RA	0627
1N1809	0488
1N1809A	0488
1N1809B	0488
1N1809C	0488
1N1809CA	OEM
1N1809R	1933
1N1809RA	1933
1N1810	0509
1N1810A	0509
1N1810AR	1941
1N1810B	0509
1N1810BR	1941
1N1810C	0509
1N1810CA	OEM
1N1810R	1941
1N1810RA	1941
1N1811	0524
1N1811A	0524
1N1811B	0524
1N1811C	0524
1N1811CA	OEM
1N1811R	1947
1N1811RA	1947
1N1812	0063
1N1812A	0063
1N1812AR	1961
1N1812B	0063
1N1812BR	1961
1N1812C	0063
1N1812CA	OEM
1N1812R	1961
1N1812RA	1961
1N1813	0401
1N1813A	0401
1N1813B	0401
1N1813C	0401
1N1813CA	OEM
1N1813R	0357
1N1813RA	0357
1N1814	0634
1N1814A	0634
1N1814AR	0776
1N1814B	0634
1N1814BR	0776
1N1814C	0634
1N1814CA	OEM
1N1814R	0776
1N1814RA	0776
1N1815	0668
1N1815A	0668
1N1815B	0668
1N1815C	0668
1N1815CA	OEM
1N1815R	1071
1N1815RA	1071
1N1816	1020
1N1816A	1020
1N1816C	1020
1N1816R	1249
1N1816RA	1249
1N1817	1274
1N1817A	1274
1N1817AR	1639
1N1817B	1274
1N1817BR	1639
1N1817C	1274
1N1817R	1639
1N1817RA	1639

DEVICE TYPE	REPL CODE
1N1818	1402
1N1818A	1402
1N1818C	1402
1N1818R	1703
1N1818RA	1703
1N1819	1078
1N1819A	1078
1N1819AR	1716
1N1819B	1078
1N1819BR	1716
1N1819C	1078
1N1819R	1716
1N1819RA	1716
1N1820	1717
1N1820A	1717
1N1820C	1717
1N1820R	1731
1N1820RA	1731
1N1821	1723
1N1821A	1723
1N1821AR	0727
1N1821B	1723
1N1821BR	0727
1N1821C	1723
1N1821R	0727
1N1821RA	0727
1N1822	1736
1N1822A	1736
1N1822C	1736
1N1822R	1756
1N1822RA	1756
1N1823	1761
1N1823A	1761
1N1823AR	1782
1N1823B	1761
1N1823BR	1782
1N1823C	1761
1N1823R	1782
1N1823RA	1782
1N1824	1772
1N1824A	1772
1N1824C	1772
1N1824CA	OEM
1N1824R	1794
1N1824RA	1794
1N1825	1788
1N1825A	1788
1N1825AR	1799
1N1825B	1788
1N1825BR	1799
1N1825C	1788
1N1825R	1799
1N1825RA	1799
1N1826	1796
1N1826A	1796
1N1826C	1796
1N1826R	1809
1N1826RA	1809
1N1827	1804
1N1827A	1804
1N1827AR	1817
1N1827B	1804
1N1827BR	1817
1N1827C	1804
1N1827R	1817
1N1827RA	1817
1N1828	1194
1N1828A	1194
1N1828C	1194
1N1828R	1826
1N1828RA	1826
1N1829	0022
1N1829A	0022
1N1829AR	1853
1N1829B	0022
1N1829BR	1853
1N1829C	0022
1N1829R	1853
1N1829RA	1853
1N1830	0132
1N1830A	0132
1N1830C	0132
1N1830R	1866
1N1830RA	1866
1N1831	0207
1N1831A	0207
1N1831AR	1884
1N1831B	0207
1N1831BR	1884
1N1831C	0207
1N1831R	1884
1N1831RA	1884
1N1832	0263
1N1832A	0263
1N1832C	0263
1N1832R	1895
1N1832RA	1895
1N1833	0310
1N1833A	0310
1N1833AR	1902
1N1833B	0310
1N1833BR	1902
1N1833C	0310
1N1833R	1902
1N1833RA	1902
1N1834	0329
1N1834A	0329
1N1834C	0329
1N1834R	0736
1N1834RA	0736
1N1835	0356
1N1835A	0356
1N1835AR	1909
1N1835B	0356
1N1835BR	1909
1N1835C	0356
1N1835R	1909
1N1835RA	1909
1N1836	0412
1N1836A	0412
1N1836C	0412
1N1836R	1914
1N1836RA	1914
1N1838	OEM
1N1839	0133
1N1840	0133
1N1841	0133
1N1842	0133
1N1843	0133
1N1844	0133
1N1845	0133
1N1846	0133
1N1847	0133
1N1848	0133
1N1849	0795
1N1850	0795
1N1875	0244
1N1875A	0244
1N1875B	0244
1N1876	0170
1N1876A	0170
1N1876B	OEM
1N1877	0137
1N1877A	0137
1N1877B	OEM
1N1878	0002
1N1878A	0002
1N1878B	OEM
1N1879	0495
1N1879A	0495
1N1879B	OEM
1N1880	0565
1N1880A	0565
1N1880B	OEM
1N1881	0441
1N1881A	0441
1N1881B	OEM
1N1882	0039
1N1882A	0039
1N1882B	OEM
1N1883	0350
1N1883A	0350
1N1883B	OEM
1N1884	0999
1N1884A	0999
1N1884B	OEM
1N1885	0869
1N1885A	0869
1N1885B	OEM
1N1886	1268
1N1886A	1268
1N1886B	OEM
1N1887	0331
1N1887A	0331
1N1887B	OEM
1N1888	0098
1N1888A	0098
1N1888B	OEM
1N1889	0186
1N1889A	0509
1N1889B	OEM
1N1890	0028
1N1890A	0028
1N1890B	OEM
1N1891	1622
1N1891A	1622
1N1891R	1490
1N1891RA	1490
1N1892	0992
1N1892A	0992
1N1892R	0510
1N1892RA	0510
1N1893	1263
1N1893A	1263
1N1893R	0870
1N1893RA	0870
1N1894	1639
1N1894A	1639
1N1894R	1274
1N1894RA	1274
1N1895	1716
1N1895A	1716
1N1895R	1078
1N1895RA	1078
1N1896	0727
1N1896A	0727
1N1896R	1723
1N1896RA	1723
1N1897	1782
1N1897A	1782
1N1897R	1761
1N1897RA	1761
1N1898	1799
1N1898A	1799
1N1898R	1788
1N1898RA	1788
1N1899	1817
1N1899A	1817
1N1899R	1804
1N1899RA	1804
1N1900	1853
1N1900A	1853
1N1900R	0022
1N1900RA	0022
1N1901	1884
1N1901A	1884
1N1901R	0207
1N1901RA	0207
1N1902	1902
1N1902A	1902
1N1902R	0310
1N1902RA	0310
1N1903	1909
1N1903A	1909
1N1903R	0356
1N1903RA	0356
1N1904	1163
1N1904A	1163
1N1904R	0438
1N1904RA	0438
1N1905	0509
1N1906	0063
1N1907	0015
1N1908	0015
1N1909	0015
1N1910	1747
1N1911	0015
1N1912	0015
1N1913	0015
1N1914	0071
1N1915	0071
1N1916	0071
1N1917	0970
1N1918	OEM
1N1919	OEM
1N1920	OEM
1N1921	OEM
1N1922	OEM
1N1923	OEM
1N1924	OEM
1N1925	OEM
1N1926	OEM
1N1927	0036
1N1927A	0036
1N1927B	OEM
1N1928	0140
1N1928A	0140
1N1928B	OEM
1N1929	0253
1N1929A	0253
1N1929B	0253
1N1930	0062
1N1930A	0062
1N1930B	OEM
1N1931	0165
1N1931A	0165
1N1931B	0244
1N1932	0064
1N1932A	0170
1N1932B	0064
1N1933	0052
1N1933A	0052
1N1933B	1005
1N1934	0686
1N1934A	0002
1N1934B	0686
1N1935	0375
1N1935A	0375
1N1935B	OEM
1N1936	0705
1N1936A	0705
1N1936B	OEM
1N1937	0455
1N1937A	0441
1N1937B	0455
1N1938	0166
1N1938A	0039
1N1938B	0166
1N1939	0032
1N1939A	0032
1N1939B	OEM
1N1940	0068
1N1940A	0068
1N1940B	OEM
1N1941	0125
1N1941A	0125
1N1941B	OEM
1N1942	0125
1N1942A	0173
1N1942B	OEM
1N1943	0049
1N1943A	0331
1N1943B	0049
1N1944	0189
1N1944A	0189
1N1944B	OEM
1N1945	0089
1N1945A	0186
1N1945B	OEM
1N1946	0340
1N1946A	0340
1N1946B	OEM
1N1947	0425
1N1947A	0425
1N1947B	OEM
1N1948	OEM
1N1948A	OEM
1N1948B	OEM
1N1949	OEM
1N1949A	OEM
1N1949B	OEM
1N1950	OEM
1N1950A	OEM
1N1950B	OEM
1N1951	OEM
1N1951A	OEM
1N1951B	OEM
1N1952	OEM
1N1952A	OEM
1N1952B	OEM
1N1953	OEM
1N1953A	OEM
1N1953B	OEM
1N1954	0036
1N1954A	0036
1N1954B	0036
1N1955	0140
1N1955A	0140
1N1955B	OEM
1N1956	0253
1N1956A	0253
1N1956B	OEM
1N1957	0062
1N1957A	0062
1N1957B	OEM
1N1958	0165
1N1958A	0165
1N1958B	OEM
1N1959	0064
1N1959A	0064
1N1959B	OEM
1N1960	0052
1N1960A	0052
1N1960B	0052
1N1961	0879
1N1961A	0879
1N1961B	OEM
1N1962	0375
1N1962A	0375
1N1962B	OEM
1N1963	0705
1N1963A	0705
1N1963B	OEM
1N1964	0455
1N1964A	0441
1N1964B	0455
1N1965	0166
1N1965A	0039
1N1965B	0166
1N1966	0032
1N1967	0068
1N1968	0125
1N1969	0173
1N1970	0049
1N1971	0189
1N1972	0089
1N1973	0340
1N1973/T59277	OEM
1N1974	0425
1N1975	2351
1N1981	0036
1N1981A	0036
1N1981B	0036
1N1982	0140
1N1982A	0140
1N1982B	OEM
1N1983	0253
1N1983A	0253
1N1983B	0253
1N1984	0062
1N1984A	0062
1N1984B	OEM
1N1985	0165
1N1985A	0165
1N1985B	0165
1N1986	0064
1N1986A	0170
1N1986B	0064
1N1987	0052
1N1987A	0052
1N1987B	0052
1N1988	0686
1N1988A	0002
1N1988B	0686
1N1989	0375
1N1989A	0375
1N1989B	0375
1N1990	0705
1N1990A	0705
1N1990B	OEM
1N1991	0455
1N1991A	0441
1N1991B	0441
1N1992	0166
1N1992A	0039
1N1992B	0039
1N1993	0032
1N1993A	0032
1N1993B	OEM
1N1994	0068
1N1994A	0068
1N1994B	OEM
1N1995	0125
1N1995A	0125
1N1995B	OEM
1N1996	0173
1N1996A	0173
1N1996B	OEM
1N1997	0049
1N1997A	0331
1N1997B	4510
1N1998	0189
1N1998A	0189
1N1998B	OEM
1N1999	0089
1N1999A	0089
1N1999B	OEM
1N2000	0340
1N2000A	0340
1N2000B	OEM
1N2001	0425
1N2001A	0425
1N2001B	0425
1N2002	OEM
1N2002A	OEM
1N2002B	OEM
1N2003	6646
1N2003A	OEM
1N2003B	OEM
1N2004	OEM
1N2004A	OEM
1N2004B	OEM
1N2005	OEM
1N2005A	OEM
1N2005B	OEM
1N2006	OEM
1N2006A	OEM
1N2006B	OEM
1N2007	OEM
1N2007A	OEM
1N2007B	OEM
1N2008	0438
1N2008A	0438
1N2008AR	1163
1N2008C	0438
1N2008R	1163
1N2008RA	1163
1N2009	0488
1N2009A	0488
1N2009C	0488
1N2009R	1933
1N2009RA	1933
1N2010	0509
1N2010A	0509
1N2010AR	1941
1N2010C	0509
1N2010R	1941
1N2010RA	1941
1N2011	0524
1N2011A	0524
1N2011C	0524
1N2011R	1947
1N2011RA	1947
1N2012	0063
1N2012A	0063
1N2012AR	1961
1N2012C	0063
1N2012R	1961
1N2012RA	1961
1N2013	0110
1N2014	0953
1N2015	0242
1N2016	0242
1N2017	1747
1N2018	1747
1N2019	0540
1N2020	0540
1N2021	1250
1N2021R	0267
1N2022	5587
1N2022R	0267
1N2023	5587
1N2023R	0267
1N2024	1581
1N2024R	0267
1N2025	1581
1N2025R	0267
1N2026	0110
1N2026R	2291
1N2027	0580
1N2027R	2291
1N2028	2060
1N2028R	0476
1N2029	1000
1N2029R	0476
1N2030	2076
1N2031	2081
1N2032	0451
1N2032-1	0451
1N2032-2	0162
1N2032A	0451
1N2033	0301
1N2033-1	0157
1N2033-2	0301
1N2033A	0157
1N2034	0025
1N2034-1	0025
1N2034-2	0025
1N2034-3	0649
1N2034A	0025
1N2035	1083
1N2035-1	0244
1N2035-2	1083
1N2035-3	0012
1N2035-4	0012
1N2035A	0244
1N2036	0170
1N2036-1	0170
1N2036-2	0317
1N2036A	0170
1N2037	0365
1N2037-1	0137
1N2037-2	0365
1N2037-3	0100
1N2037A	0137
1N2038	0420
1N2038-1	0002
1N2038-2	0420
1N2038-3	1649
1N2038A	0420
1N2039	0949
1N2039-1	0495
1N2039-2	0949
1N2039-3	0531
1N2039A	0949
1N2040	0402
1N2040-1	0565
1N2040-2	0402
1N2040-3	1606
1N2040A	0402
1N2041	2113
1N2041-1	2113
1N2041-2	2113
1N2041A	2113
1N2041B	2410
1N2041C	2410
1N2041RA	1438
1N2041RB	2407
1N2042	1901
1N2042-1	1901
1N2042-2	2416
1N2042A	1901
1N2042B	0696
1N2042C	2416
1N2042RA	1445
1N2042RB	2221
1N2043	1601
1N2043-1	1601
1N2043-2	1601
1N2043-3	1616
1N2043A	1601
1N2043B	1601
1N2043C	1616
1N2043RA	1458
1N2043RB	1458
1N2043RC	0221
1N2044	2424
1N2044-1	1622
1N2044-2	2424
1N2044-3	0627
1N2044-4	0627
1N2044A	1622
1N2044B	1622
1N2044C	0627
1N2044D	0627
1N2044RA	1490
1N2044RB	1490
1N2044RC	1618
1N2044RD	1618
1N2045	0992
1N2045-1	0995
1N2045-2	0995
1N2045A	0992
1N2045B	0995
1N2045C	0995
1N2045RA	0510
1N2045RB	0691
1N2046	1249
1N2046-1	1249
1N2046-2	1249
1N2046-3	1636
1N2046A	1263
1N2046B	1249
1N2046C	1636
1N2046R	1020
1N2046RA	0870
1N2046RB	1020
1N2046RC	1153
1N2047	1703
1N2047-1	1703
1N2047-2	1703
1N2047-3	1640
1N2047A	1639
1N2047B	1703
1N2047C	1640
1N2047RA	1274
1N2047RB	1402
1N2047RC	1532
1N2048	1720
1N2048-1	1720
1N2048-2	1720
1N2048-3	1731
1N2048A	1716
1N2048B	1720
1N2048C	1731
1N2048RA	1078
1N2048RB	1711
1N2048RC	1717
1N2049	1756
1N2049-1	1756
1N2049-2	1756
1N2049-3	1782
1N2049-4	2453
1N2049A	0727
1N2049B	1756
1N2049C	1768
1N2049D	1782
1N2049RA	1723
1N2049RB	1736
1N2049RC	1748
1N2049RD	1761
1N2050AR	1107
1N2050BR	1107
1N2054	1023
1N2054-68	1023
1N2054R	0501
1N2055	1023
1N2055R	0501
1N2056	1023
1N2056R	0501
1N2057	1023
1N2057R	0501
1N2058	1023
1N2058R	0501
1N2059	1023
1N2059R	0501
1N2060	1023
1N2060R	1777
1N2061	1023
1N2061R	1777
1N2062	1036
1N2062R	1777
1N2063	1036
1N2063R	1777
1N2064	1036
1N2064R	1777
1N2065	1046
1N2065R	1789
1N2066	1046
1N2066R	1789
1N2067	1046
1N2067R	1789
1N2068	1046
1N2068R	1789
1N2069	0015
1N2069A	0015
1N2070	0023
1N2070A	0015
1N2071	0015
1N2071A	0015
1N2072	0015
1N2073	0015
1N2074	0015
1N2075	0015
1N2076	0015
1N2077	0015
1N2078	0015
1N2079	0015
1N2080	0015
1N2081	0015
1N2082	0015
1N2083	0015
1N2084	0015
1N2085	0015
1N2086	0015
1N2088	0015
1N2089	0015
1N2090	0015
1N2091	0015
1N2092	0015
1N2093	0015
1N2094	0015
1N2095	0015
1N2096	0015
1N2102	OEM
1N2103	0015
1N2104	0015
1N2105	0015
1N2106	0015
1N2107	0015
1N2108	0015
1N2109	0708
1N2110	OEM
1N2115	0015
1N2116	0015
1N2117	0015
1N2127	OEM
1N2127A	OEM
1N2128	3762
1N2128A	3762
1N2128AR	2665
1N2128R	2665
1N2128RA	1783
1N2129	2653
1N2129A	2653
1N2129AR	2695
1N2129R	2695
1N2129RA	1783
1N2130	2657
1N2130A	2657
1N2130AR	2768
1N2130R	2768
1N2130RA	1783
1N2131	2657
1N2131A	2657
1N2131AR	2768
1N2131R	2768
1N2131RA	1783
1N2132	2664
1N2132A	2664
1N2132AR	2855
1N2132R	2855
1N2132RA	1783

If replacement code is OEM, contact original manufacturer for replacement.

DEVICE TYPE	REPL CODE
1N2133	2664
1N2133A	2664
1N2133AR	2855
1N2133R	2855
1N2133RA	1783
1N2134	2006
1N2134A	2006
1N2134AR	2907
1N2134R	2907
1N2134RA	1783
1N2135	2006
1N2135A	2006
1N2135AR	2907
1N2135R	2907
1N2135RA	1783
1N2136	2677
1N2136A	2677
1N2136AR	2974
1N2136R	2974
1N2136RA	1783
1N2137	2677
1N2137A	2677
1N2137AR	2974
1N2137R	2974
1N2137RA	1783
1N2138	2682
1N2138A	2682
1N2138AR	3051
1N2138R	3051
1N2138RA	1783
1N2139	2682
1N2139A	2682
1N2139AR	3051
1N2139R	3051
1N2147	0970
1N2147A	0970
1N2148	3733
1N2148A	3733
1N2149	0989
1N2149A	0989
1N2150	3743
1N2150A	3743
1N2151	0200
1N2151A	0197
1N2152	5514
1N2152A	0200
1N2153	0204
1N2153A	0204
1N2154	3189
1N2154R	1630
1N2155	2901
1N2155R	0254
1N2156	1124
1N2156R	1107
1N2157	1126
1N2157R	1111
1N2158	0805
1N2158R	0258
1N2159	1195
1N2159R	1644
1N2160	0319
1N2160R	0267
1N2163	0012
1N2163A	0012
1N2164	0012
1N2164A	0012
1N2165	0012
1N2165A	0012
1N2166	0012
1N2166A	0012
1N2167	0012
1N2167A	0012
1N2168	0012
1N2168A	0012
1N2169	0012
1N2169A	0012
1N2170	0012
1N2170A	0012
1N2171	0012
1N2171A	0012
1N2172	0599
1N2172R	1347
1N2173	0599
1N2173R	1347
1N2174	0599
1N2174R	1347
1N2181	0015
1N2214	0157
1N2216	0708
1N2216R	2291
1N2217	0708
1N2218	2076
1N2218R	2291
1N2219	2076
1N2220	2081
1N2220R	2291
1N2221	2081
1N2222	2088
1N2222A	2088
1N2222R	0449
1N2223	2088
1N2223A	2088
1N2224	0612
1N2224A	0612
1N2224R	0449
1N2225	0612
1N2225A	0612
1N2226	0817
1N2226A	0817

DEVICE TYPE	REPL CODE
1N2227	OEM
1N2227A	OEM
1N2228	0970
1N2228A	0970
1N2228R	2900
1N2228RA	2900
1N2229	0970
1N2229A	0970
1N2230	0989
1N2230A	0989
1N2231	0989
1N2231A	0989
1N2232	3743
1N2232A	3743
1N2233	3743
1N2233A	3743
1N2234	0197
1N2234A	0197
1N2235	0197
1N2235A	0197
1N2236	0200
1N2236A	0200
1N2237	0200
1N2237A	0200
1N2238	0204
1N2238A	0204
1N2239	0204
1N2239A	0204
1N2240	0206
1N2240A	0206
1N2241	0206
1N2241A	0206
1N2242	0586
1N2242A	0586
1N2243	0586
1N2243A	0586
1N2244	OEM
1N2244A	OEM
1N2245	OEM
1N2245A	OEM
1N2246	0084
1N2246A	0084
1N2246R	2291
1N2246RA	2291
1N2247	0084
1N2247A	0084
1N2248	0090
1N2248A	0090
1N2248R	2291
1N2248RA	2291
1N2249	0090
1N2249A	0090
1N2250	0097
1N2250A	0097
1N2250R	2291
1N2250RA	2291
1N2251	0097
1N2251A	0097
1N2252	0105
1N2252A	0105
1N2252R	0476
1N2252RA	0476
1N2253	0105
1N2253A	0105
1N2254	0109
1N2254A	0109
1N2254R	0476
1N2254RA	0476
1N2255	0109
1N2255A	0109
1N2256	0116
1N2256A	0116
1N2256R	0476
1N2256RA	0476
1N2257	0116
1N2257A	0116
1N2258	0122
1N2258A	0122
1N2258R	0476
1N2258RA	0476
1N2259	0122
1N2259A	0122
1N2260	0131
1N2260A	0131
1N2260R	0449
1N2260RA	0449
1N2261	0131
1N2261A	0131
1N2262	0145
1N2262A	0145
1N2262R	0449
1N2262RA	0449
1N2263	0145
1N2263A	0145
1N2264	OEM
1N2264A	OEM
1N2265	OEM
1N2265A	OEM
1N2266	0708
1N2266R	2291
1N2267	0708
1N2268	2076
1N2268R	0476
1N2269	2076
1N2270	2081
1N2270R	0476
1N2271	2081
1N2272	1600
1N2272R	0267

DEVICE TYPE	REPL CODE
1N2273	0590
1N2273R	0267
1N2274	1250
1N2274R	0267
1N2275	5587
1N2275R	0267
1N2276	1581
1N2276R	0267
1N2277	5589
1N2277R	0267
1N2278	1586
1N2278R	0267
1N2279	1132
1N2279R	0281
1N2280	0044
1N2281	OEM
1N2282	1126
1N2282R	0267
1N2283	0805
1N2283R	0267
1N2284	1195
1N2284R	0267
1N2285	0319
1N2285R	0267
1N2286	1132
1N2286R	0281
1N2287	0044
1N2287R	0281
1N2288	OEM
1N2289	0708
1N2289A	0708
1N2290	0708
1N2290A	0708
1N2291	0580
1N2291A	2060
1N2292	2060
1N2292A	2060
1N2293	1000
1N2293A	1000
1N2323	0015
1N2326	3683
1N2327	0102
1N2328	OEM
1N2348	0708
1N2348R	5038
1N2349	0580
1N2349R	2291
1N2350	0580
1N2350R	2291
1N2357	0102
1N2358	0348
1N2359	0348
1N2360	0348
1N2361	0348
1N2362	OEM
1N2362A	OEM
1N2362B	OEM
1N2363	OEM
1N2363A	OEM
1N2363B	OEM
1N2364	OEM
1N2364A	OEM
1N2364B	OEM
1N2365	OEM
1N2365A	OEM
1N2365B	OEM
1N2366	OEM
1N2366A	OEM
1N2366B	OEM
1N2367	OEM
1N2367A	OEM
1N2367B	OEM
1N2368	OEM
1N2368A	OEM
1N2368B	OEM
1N2369	OEM
1N2369A	OEM
1N2369B	OEM
1N2370	OEM
1N2370A	OEM
1N2370B	OEM
1N2371	OEM
1N2371A	OEM
1N2371B	OEM
1N2372	0612
1N2373	0965
1N2374	0817
1N2375	0182
1N2376	0348
1N2377	OEM
1N2378	OEM
1N2379	OEM
1N2380	OEM
1N2381	1791
1N2382	3889
1N2382A	3889
1N2383	3889
1N2383A	3889
1N2384	OEM
1N2384A	OEM
1N2385	OEM
1N2385A	OEM
1N2387	0726
1N2389	OEM
1N2390	0110
1N2391	0953
1N2392	0242
1N2393	1747
1N2394	0540

DEVICE TYPE	REPL CODE
1N2395	1771
1N2396	0965
1N2397	0817
1N2398	0817
1N2399	0110
1N2400	0953
1N2401	0242
1N2402	1747
1N2403	0540
1N2404	1771
1N2405	0965
1N2406	0817
1N2407	0817
1N2408	0110
1N2409	0953
1N2410	0242
1N2411	1747
1N2412	0540
1N2413	1771
1N2414	0965
1N2415	0817
1N2416	0817
1N2417	0071
1N2418	0071
1N2419	0071
1N2420	0071
1N2421	0071
1N2422	0071
1N2423	0071
1N2424	0071
1N2425	0071
1N2426	0599
1N2426R	1347
1N2427	0599
1N2427R	1347
1N2428	0599
1N2428R	1347
1N2429	0599
1N2429R	1347
1N2430	0599
1N2430R	1347
1N2431	0599
1N2431R	1347
1N2432	0599
1N2432R	1777
1N2433	0599
1N2433R	1777
1N2434	1986
1N2434R	1777
1N2435	1986
1N2435R	1777
1N2436	0599
1N2436R	1347
1N2437	0599
1N2437R	1347
1N2438	0599
1N2438R	1347
1N2439	0599
1N2439R	1347
1N2440	0599
1N2440R	1347
1N2441	0599
1N2441R	1347
1N2442	1986
1N2442R	1905
1N2443	0599
1N2443R	1905
1N2444	1986
1N2444R	1905
1N2445	1986
1N2445R	1905
1N2446	3189
1N2446R	1630
1N2447	2901
1N2447R	0254
1N2448	1124
1N2448R	1107
1N2449	1124
1N2449R	1107
1N2450	1126
1N2450R	1111
1N2451	1126
1N2451R	1111
1N2452	0805
1N2452R	0258
1N2453	0805
1N2453R	0258
1N2454	1195
1N2454R	1644
1N2455	0319
1N2455R	0267
1N2456	1132
1N2456R	1119
1N2457	1132
1N2457R	1119
1N2458	3189
1N2458R	1630
1N2459	2901
1N2459R	0254
1N2460	1124
1N2460R	1107
1N2461	1124
1N2461R	1107
1N2462	1126
1N2462R	1111
1N2463	1126
1N2463R	1111
1N2464	0805
1N2464R	0258

DEVICE TYPE	REPL CODE
1N2465	0805
1N2465R	0258
1N2466	1195
1N2466R	1644
1N2467	0319
1N2467R	0267
1N2468	1132
1N2468R	1119
1N2469	1132
1N2469R	1119
1N2482	0015
1N2483	0015
1N2484	0015
1N2485	0015
1N2486	0015
1N2487	0015
1N2488	0015
1N2489	0015
1N2490	OEM
1N2491	0970
1N2491R	2291
1N2492	3733
1N2492R	2291
1N2493	0989
1N2493R	2291
1N2494	3743
1N2494R	0476
1N2495	0197
1N2495R	0476
1N2496	0200
1N2496R	0476
1N2497	0204
1N2497R	0476
1N2498	0510
1N2498A	0510
1N2498AR	0992
1N2498C	0510
1N2498R	0992
1N2498RA	0992
1N2498RC	0992
1N2499	0691
1N2499A	0691
1N2499C	0691
1N2499R	0995
1N2499RA	0995
1N2499RC	0995
1N2500	0870
1N2500A	0870
1N2500AR	1263
1N2500B	0870
1N2500C	0870
1N2500CA	0870
1N2500R	1263
1N2500RA	1263
1N2500RB	1263
1N2500RC	1263
1N2500RCA	1263
1N2501	0817
1N2502	0922
1N2503	0102
1N2504	0348
1N2505	0817
1N2506	0922
1N2507	0102
1N2508	0348
1N2509	OEM
1N2510	OEM
1N2512	0989
1N2512R	2291
1N2513	0989
1N2513R	2291
1N2514	1112
1N2514R	0476
1N2515	0197
1N2515R	0476
1N2516	0200
1N2516R	0476
1N2517	0204
1N2517R	0476
1N2518	3733
1N2519	0989
1N2520	2060
1N2521	1000
1N2522	0200
1N2523	2081
1N2524	0708
1N2525	0580
1N2526	0580
1N2527	2060
1N2528	1000
1N2529	2076
1N2530	2081
1N2531	2088
1N2532	2088
1N2533	0612
1N2534	0612
1N2535	0708
1N2536	0580
1N2537	0580
1N2538	2060
1N2539	1000
1N2540	2076
1N2541	2081
1N2542	2088
1N2543	2088
1N2544	0612
1N2545	0612
1N2546	0708
1N2547	0580

DEVICE TYPE	REPL CODE
1N2548	0580
1N2549	2060
1N2550	1000
1N2551	2076
1N2552	2081
1N2553	2088
1N2554	2088
1N2555	0612
1N2556	0612
1N2557	0206
1N2558	0206
1N2559	0586
1N2560	0586
1N2561	0206
1N2562	0206
1N2563	0586
1N2564	0586
1N2565	0970
1N2566	3733
1N2567	0989
1N2568	3743
1N2569	0197
1N2570	0200
1N2571	0204
1N2572	0206
1N2573	0206
1N2574	0586
1N2575	0586
1N2576	0084
1N2577	0090
1N2578	0097
1N2579	0105
1N2580	0109
1N2581	0116
1N2582	0122
1N2583	0131
1N2584	0131
1N2585	0145
1N2586	0145
1N2587	0084
1N2588	0090
1N2589	0097
1N2590	0105
1N2591	0109
1N2592	0116
1N2593	0122
1N2594	0131
1N2595	0131
1N2596	0145
1N2597	0145
1N2598	0084
1N2598R	1112
1N2599	0090
1N2600	0097
1N2601	0105
1N2602	0109
1N2603	0116
1N2604	0122
1N2605	0131
1N2606	0131
1N2607	0145
1N2608	0145
1N2609	0015
1N2610	0015
1N2611	0015
1N2612	0015
1N2613	0540
1N2614	0015
1N2615	0015
1N2616	0071
1N2617	0071
1N2618	0102
1N2619	0348
1N2620	0012
1N2620A	0012
1N2620B	0012
1N2621	0012
1N2621A	0012
1N2621B	0012
1N2622	0012
1N2622A	0012
1N2622B	0012
1N2623	0012
1N2623A	0012
1N2623B	0012
1N2624	0012
1N2624A	0012
1N2624B	0012
1N2625	OEM
1N2626	OEM
1N2630	OEM
1N2631	3331
1N2632	OEM
1N2633	OEM
1N2634	4405
1N2635	OEM
1N2636	OEM
1N2637	OEM
1N2638	0015
1N2641	OEM
1N2644	OEM
1N2647	OEM
1N2650	0015
1N2653	0071
1N2656	OEM
1N2659	OEM
1N2662	OEM
1N2664	OEM
1N2666	OEM

DEVICE TYPE	REPL CODE
1N2667	OEM
1N2668	OEM
1N2669	OEM
1N2673	OEM
1N2677	OEM
1N2681	OEM
1N2685	OEM
1N2685A	OEM
1N2687	OEM
1N2689	OEM
1N2689A	OEM
1N2690	OEM
1N2691	OEM
1N2692	OEM
1N2694	OEM
1N2696	OEM
1N2698	OEM
1N2700	OEM
1N2701	OEM
1N2702	0580
1N2705	OEM
1N2708	OEM
1N2711	OEM
1N2714	OEM
1N2717	OEM
1N2720	OEM
1N2722	OEM
1N2723	OEM
1N2724	OEM
1N2725	OEM
1N2728	OEM
1N2731	OEM
1N2734	OEM
1N2737	OEM
1N2738	OEM
1N2739	OEM
1N2740	OEM
1N2742	OEM
1N2744	OEM
1N2746	OEM
1N2748	OEM
1N2749	OEM
1N2750	0580
1N2753	OEM
1N2756	OEM
1N2759	OEM
1N2762	OEM
1N2763	OEM
1N2764	OEM
1N2765	0062
1N2765A	0062
1N2766	0879
1N2766A	0879
1N2767	0531
1N2767A	0531
1N2768	0441
1N2768A	0441
1N2769	OEM
1N2769A	OEM
1N2770	OEM
1N2770A	OEM
1N2771	OEM
1N2772	0817
1N2773	0817
1N2774	0922
1N2775	0922
1N2776	0102
1N2777	0102
1N2778	0102
1N2779	0102
1N2780	OEM
1N2781	OEM
1N2783	0386
1N2784	0871
1N2785	0853
1N2786	1250
1N2786R	1577
1N2787	1581
1N2787R	3281
1N2788	2657
1N2788R	2768
1N2789	2006
1N2789R	2907
1N2790	1083
1N2791	OEM
1N2792	OEM
1N2792A	OEM
1N2792B	OEM
1N2793	2002
1N2793R	0267
1N2794	0590
1N2794R	0267
1N2795	1250
1N2795R	0267
1N2796	1250
1N2796R	0267
1N2797	5587
1N2797R	0267
1N2798	5587
1N2798R	0267
1N2799	1581
1N2799R	0267
1N2800	1581
1N2800A	0267
1N2801	0143
1N2802	OEM
1N2804	OEM
1N2804A	OEM
1N2804B	OEM

If replacement code is OEM, contact original manufacturer for replacement.

DEVICE TYPE	REPL CODE
1N2804R	OEM
1N2804RA	OEM
1N2804RB	OEM
1N2805	OEM
1N2805A	OEM
1N2805B	OEM
1N2805R	OEM
1N2805RA	OEM
1N2805RB	OEM
1N2806	OEM
1N2806A	OEM
1N2806B	OEM
1N2806R	OEM
1N2806RA	OEM
1N2806RB	OEM
1N2807	OEM
1N2807A	OEM
1N2807B	OEM
1N2807R	OEM
1N2807RA	OEM
1N2807RB	OEM
1N2808	OEM
1N2808A	OEM
1N2808B	OEM
1N2808R	OEM
1N2808RA	OEM
1N2808RB	OEM
1N2809	OEM
1N2809A	OEM
1N2809B	OEM
1N2809R	OEM
1N2809RA	OEM
1N2809RB	OEM
1N2810	OEM
1N2810A	OEM
1N2810B	OEM
1N2810R	OEM
1N2810RA	OEM
1N2810RB	OEM
1N2811	OEM
1N2811A	OEM
1N2811B	OEM
1N2811R	OEM
1N2811RA	OEM
1N2811RB	OEM
1N2812	OEM
1N2812A	OEM
1N2812B	OEM
1N2812R	OEM
1N2812RA	OEM
1N2812RB	OEM
1N2813	OEM
1N2813A	OEM
1N2813B	OEM
1N2813R	OEM
1N2813RA	OEM
1N2813RB	OEM
1N2814	OEM
1N2814A	OEM
1N2814B	OEM
1N2814R	OEM
1N2814RA	OEM
1N2814RB	OEM
1N2815	OEM
1N2815A	OEM
1N2815B	OEM
1N2815R	OEM
1N2815RA	OEM
1N2815RB	OEM
1N2816	OEM
1N2816A	OEM
1N2816B	OEM
1N2816R	OEM
1N2816RA	OEM
1N2816RB	OEM
1N2817	OEM
1N2817A	OEM
1N2817B	OEM
1N2817R	OEM
1N2817RA	OEM
1N2817RB	OEM
1N2818	OEM
1N2818A	OEM
1N2818B	OEM
1N2818R	OEM
1N2818RA	OEM
1N2818RB	OEM
1N2819	OEM
1N2819A	OEM
1N2819B	OEM
1N2819R	OEM
1N2819RA	OEM
1N2819RB	OEM
1N2820	OEM
1N2820A	OEM
1N2820B	OEM
1N2820R	OEM
1N2820RA	OEM
1N2820RB	OEM
1N2821	OEM
1N2821A	OEM
1N2821B	OEM
1N2821R	OEM
1N2821RA	OEM
1N2821RB	OEM
1N2822	OEM
1N2822A	OEM
1N2822B	OEM
1N2822R	OEM
1N2822RA	OEM
1N2822RB	OEM
1N2823	OEM
1N2823A	OEM
1N2823B	OEM
1N2823R	OEM
1N2823RA	OEM
1N2823RB	OEM
1N2824	OEM
1N2824A	OEM
1N2824B	OEM
1N2824R	OEM
1N2824RA	OEM
1N2824RB	OEM
1N2825	OEM
1N2825A	OEM
1N2825B	OEM
1N2825R	OEM
1N2825RA	OEM
1N2825RB	OEM
1N2826	OEM
1N2826A	OEM
1N2826B	OEM
1N2826R	OEM
1N2826RA	OEM
1N2826RB	OEM
1N2827	OEM
1N2827A	OEM
1N2827B	OEM
1N2827R	OEM
1N2827RA	OEM
1N2827RB	OEM
1N2828	OEM
1N2828A	OEM
1N2828B	OEM
1N2828R	OEM
1N2828RA	OEM
1N2828RB	OEM
1N2829	OEM
1N2829A	OEM
1N2829B	OEM
1N2829R	OEM
1N2829RA	OEM
1N2829RB	OEM
1N2830	OEM
1N2830A	OEM
1N2830B	OEM
1N2830R	OEM
1N2830RA	OEM
1N2830RB	OEM
1N2831	OEM
1N2831A	OEM
1N2831B	OEM
1N2831R	OEM
1N2831RA	OEM
1N2831RB	OEM
1N2832	OEM
1N2832A	OEM
1N2832B	OEM
1N2832R	OEM
1N2832RA	OEM
1N2832RB	OEM
1N2833	OEM
1N2833A	OEM
1N2833B	OEM
1N2833R	OEM
1N2833RA	OEM
1N2833RB	OEM
1N2834	OEM
1N2834A	OEM
1N2834B	OEM
1N2834R	OEM
1N2834RA	OEM
1N2834RB	OEM
1N2835	OEM
1N2835A	OEM
1N2835B	OEM
1N2835R	OEM
1N2835RA	OEM
1N2835RB	OEM
1N2836	OEM
1N2836A	OEM
1N2836B	OEM
1N2836R	OEM
1N2836RA	OEM
1N2836RB	OEM
1N2837	OEM
1N2837A	OEM
1N2837B	OEM
1N2837R	OEM
1N2837RA	OEM
1N2837RB	OEM
1N2838	OEM
1N2838A	OEM
1N2838B	OEM
1N2838R	OEM
1N2838RA	OEM
1N2838RB	OEM
1N2839	OEM
1N2839A	OEM
1N2839B	OEM
1N2839R	OEM
1N2839RA	OEM
1N2839RB	OEM
1N2840	OEM
1N2840A	OEM
1N2840B	OEM
1N2840R	OEM
1N2840RA	OEM
1N2840RB	OEM
1N2841	OEM
1N2841A	OEM
1N2841B	OEM
1N2841R	OEM
1N2841RA	OEM
1N2841RB	OEM
1N2842	OEM
1N2842A	OEM
1N2842B	OEM
1N2842R	OEM
1N2842RA	OEM
1N2842RB	OEM
1N2843	OEM
1N2843A	OEM
1N2843B	OEM
1N2843R	OEM
1N2843RA	OEM
1N2843RB	OEM
1N2844	OEM
1N2844A	OEM
1N2844B	OEM
1N2844R	OEM
1N2844RA	OEM
1N2844RB	OEM
1N2845	OEM
1N2845A	OEM
1N2845B	OEM
1N2845R	OEM
1N2845RA	OEM
1N2845RB	OEM
1N2846	OEM
1N2846A	OEM
1N2846B	OEM
1N2846R	OEM
1N2846RA	OEM
1N2846RB	OEM
1N2847	0580
1N2848	0580
1N2849	2060
1N2850	1000
1N2851	2076
1N2852	2081
1N2858	0015
1N2858A	0110
1N2859	0953
1N2859A	0953
1N2860	0023
1N2860A	0242
1N2861	0015
1N2861A	1747
1N2862	0015
1N2862A	0540
1N2863	0015
1N2863A	1771
1N2864	0015
1N2864A	0965
1N2865	0817
1N2866	0922
1N2867	0817
1N2868	0922
1N2878	0071
1N2879	0071
1N2880	0071
1N2881	0071
1N2882	0071
1N2883	0071
1N2884	0102
1N2885	0102
1N2886	0182
1N2887	0182
1N2887R	0267
1N2888	0348
1N2889	0348
1N2890	0348
1N2891	0348
1N2892	OEM
1N2893	OEM
1N2894	OEM
1N2895	OEM
1N2896	OEM
1N2897	OEM
1N2898	OEM
1N2899	OEM
1N2900	OEM
1N2901	OEM
1N2902	OEM
1N2903	OEM
1N2904	OEM
1N2905	OEM
1N2906	OEM
1N2907	OEM
1N2908	OEM
1N2909	OEM
1N2910	OEM
1N2911	OEM
1N2912	OEM
1N2913	OEM
1N2914	OEM
1N2915	OEM
1N2916	OEM
1N2917	OEM
1N2918	OEM
1N2919	OEM
1N2920	OEM
1N2921	OEM
1N2922	OEM
1N2923	OEM
1N2924	OEM
1N2925	OEM
1N2926	OEM
1N2926A	OEM
1N2929A	OEM
1N2930A	OEM
1N2937	OEM
1N2938	OEM
1N2939	OEM
1N2939A	OEM
1N2940	OEM
1N2940A	OEM
1N2941	OEM
1N2941A	OEM
1N2942	0958
1N2942R	1225
1N2943	0994
1N2943R	1237
1N2944	1009
1N2944R	1252
1N2945	1019
1N2945R	1269
1N2946	0889
1N2946R	1277
1N2947	1049
1N2947R	1294
1N2948	0932
1N2948R	1302
1N2949	1079
1N2949R	1310
1N2950	1106
1N2950R	1324
1N2951	1135
1N2951R	1333
1N2952	1152
1N2952R	1344
1N2953	1164
1N2953R	1356
1N2954	1184
1N2954R	1371
1N2955	1210
1N2955R	1387
1N2956	1223
1N2956R	1406
1N2957	1232
1N2957R	1415
1N2958	1246
1N2958R	1428
1N2959	1265
1N2959R	1440
1N2960	1290
1N2960R	1447
1N2961	1307
1N2961R	1461
1N2962	1331
1N2962R	1474
1N2963	1353
1N2963R	0613
1N2964	1365
1N2964R	1510
1N2965	1384
1N2965R	1523
1N2966	1401
1N2966R	1537
1N2967	1412
1N2967R	1550
1N2968	1423
1N2968R	1575
1N2969	1412
1N2969A	OEM
1N2970	1458
1N2970A	1458
1N2970AR	1601
1N2970B	1458
1N2970BR	1601
1N2970R	1601
1N2970RA	1601
1N2970RB	1601
1N2971	0221
1N2971A	0221
1N2971B	0221
1N2971R	1616
1N2971RA	1616
1N2971RB	1616
1N2972	1490
1N2972A	1490
1N2972AR	1622
1N2972B	1490
1N2972BR	1622
1N2972R	1622
1N2972RA	1622
1N2972RB	1622
1N2973	1618
1N2973A	1618
1N2973B	1618
1N2973R	0627
1N2973RA	0627
1N2973RB	0627
1N2974	0510
1N2974A	0510
1N2974AR	0992
1N2974B	0510
1N2974BR	0992
1N2974R	0992
1N2974RA	0992
1N2974RB	0992
1N2975	0691
1N2975A	0691
1N2975B	0691
1N2975R	0995
1N2975RA	0995
1N2975RB	0995
1N2976	0870
1N2976A	0870
1N2976AR	1263
1N2976B	0870
1N2976BR	1263
1N2976R	1263
1N2976RA	1263
1N2976RB	1263
1N2976RC	1263
1N2977	1020
1N2977A	1020
1N2977B	1020
1N2977R	1249
1N2977RA	1249
1N2977RB	1249
1N2978	1153
1N2978A	1153
1N2978B	1153
1N2978R	1636
1N2978RA	1636
1N2978RB	1636
1N2979	1274
1N2979A	1274
1N2979AR	1639
1N2979B	1274
1N2979BR	1639
1N2979R	1639
1N2979RA	1639
1N2979RB	1639
1N2980	1402
1N2980A	1402
1N2980B	1402
1N2980R	1703
1N2980RA	1703
1N2980RB	1703
1N2981	1532
1N2981A	1532
1N2981B	1532
1N2981R	1640
1N2981RA	1640
1N2981RB	1640
1N2982	1078
1N2982A	1078
1N2982AR	1716
1N2982B	1078
1N2982BR	1716
1N2982R	1716
1N2982RA	1716
1N2982RB	1716
1N2983	1711
1N2983A	1711
1N2983B	1711
1N2983R	1720
1N2983RA	1720
1N2983RB	1720
1N2984	1717
1N2984A	1717
1N2984B	1717
1N2984R	1731
1N2984RA	1731
1N2984RB	1731
1N2985	1723
1N2985A	1723
1N2985AR	0727
1N2985B	1723
1N2985BR	0727
1N2985R	0727
1N2985RA	0727
1N2985RB	0727
1N2986	1736
1N2986A	1736
1N2986AN	1756
1N2986B	1736
1N2986R	1756
1N2986RA	1756
1N2986RB	1756
1N2987	1748
1N2987A	1748
1N2987B	1748
1N2987R	1768
1N2987RA	1768
1N2987RB	1768
1N2988	1761
1N2988A	1761
1N2988AR	1782
1N2988B	1761
1N2988BR	1782
1N2988R	1782
1N2988RA	1782
1N2988RB	1782
1N2989	1772
1N2989A	1772
1N2989B	1772
1N2989R	1794
1N2989RA	1794
1N2989RB	1794
1N2990	1788
1N2990A	1788
1N2990AR	1799
1N2990B	1788
1N2990BR	1799
1N2990R	1799
1N2990RA	1799
1N2990RB	1799
1N2991	1796
1N2991A	1796
1N2991B	1796
1N2991R	1809
1N2991RA	1809
1N2991RB	1809
1N2992	1804
1N2992A	1804
1N2992AR	1817
1N2992B	1804
1N2992BR	1817
1N2992R	1817
1N2992RA	1817
1N2992RB	1817
1N2993	1194
1N2993A	1194
1N2993B	1194
1N2993R	1826
1N2993RA	1826
1N2993RB	1826
1N2994	1821
1N2994A	1821
1N2994B	1821
1N2994R	1840
1N2994RA	1840
1N2994RB	1840
1N2995	0022
1N2995A	0022
1N2995AR	1853
1N2995B	0022
1N2995BR	1853
1N2995R	1853
1N2995RA	1853
1N2995RB	1853
1N2996	0070
1N2996A	0070
1N2996B	0070
1N2996R	1861
1N2996RA	1861
1N2996RB	1861
1N2997	0132
1N2997A	0132
1N2997B	0132
1N2997R	1866
1N2997RA	1866
1N2997RB	1866
1N2998	0172
1N2998A	0172
1N2998B	0172
1N2998R	1874
1N2998RA	1874
1N2998RB	1874
1N2999	0207
1N2999A	0207
1N2999AR	1884
1N2999B	0207
1N2999BR	1884
1N2999R	1884
1N2999RA	1884
1N2999RB	1884
1N3000	0263
1N3000A	0263
1N3000B	0263
1N3000R	1895
1N3000RA	1895
1N3000RB	1895
1N3001	0310
1N3001A	0310
1N3001AR	1902
1N3001B	0310
1N3001BR	1902
1N3001R	1902
1N3001RA	1902
1N3001RB	1902
1N3002	0329
1N3002A	0329
1N3002B	0329
1N3002R	0736
1N3002RA	0736
1N3002RB	0736
1N3003	0356
1N3003A	0356
1N3003AR	1909
1N3003B	0356
1N3003BR	1909
1N3003R	1909
1N3003RA	1909
1N3003RB	1909
1N3004	0412
1N3004A	0412
1N3004B	0412
1N3004R	1914
1N3004RA	1914
1N3004RB	1914
1N3005	0438
1N3005A	0438
1N3005AR	1163
1N3005B	0438
1N3005BR	1163
1N3005R	1163
1N3005RA	1163
1N3005RB	1163
1N3006	0464
1N3006A	0464
1N3006B	0464
1N3006R	1924
1N3006RA	1924
1N3006RB	1924
1N3007	0488
1N3007A	0488
1N3007B	0488
1N3007R	1933
1N3007RA	1933
1N3007RB	1933
1N3008	0509
1N3008A	0509
1N3008AR	1941
1N3008B	0509
1N3008BR	1941
1N3008R	1941
1N3008RA	1941
1N3008RB	1941
1N3009	0524
1N3009A	0524
1N3009B	0524
1N3009R	1947
1N3009RA	1947
1N3009RB	1947
1N3010	0542
1N3010A	0542
1N3010B	0542
1N3010R	1953
1N3010RA	1953
1N3010RB	1953
1N3011	0063
1N3011A	0063
1N3011AR	1961
1N3011B	0063
1N3011BR	1961
1N3011R	1961
1N3011RA	1961
1N3011RB	1961
1N3012	0401
1N3012A	0401
1N3012B	0401
1N3012R	0357
1N3012RA	0357
1N3012RB	0357
1N3013	0616
1N3013A	0616
1N3013B	0616
1N3013R	0670
1N3013RA	0670
1N3013RB	0670
1N3014	0634
1N3014A	0634
1N3014AR	0776
1N3014B	0634
1N3014BR	0776
1N3014R	0776
1N3014RA	0776
1N3014RB	0776
1N3015	0668
1N3015A	0668
1N3015B	0668
1N3015R	1071
1N3015RA	1071
1N3015RB	1071
1N3016	0025
1N3016A	0025
1N3016B	0025
1N3017	0649
1N3017A	0649
1N3017B	0649
1N3018	0244
1N3018A	0244
1N3018B	0244
1N3018BA	0244
1N3019	0012
1N3019A	0012
1N3019B	0012
1N3020	0170
1N3020A	0170
1N3020B	0170
1N3021	0317
1N3021A	0317
1N3021B	0317
1N3022	0137
1N3022A	0137
1N3022B	0137
1N3023	0365
1N3023A	0365
1N3023B	0365
1N3024	0002
1N3024A	0002
1N3024B	0002
1N3025	0420
1N3025A	0420
1N3025B	0420
1N3026	0495
1N3026A	0495
1N3026B	0495
1N3027	0531
1N3027A	0531
1N3027B	0531
1N3028	0565
1N3028A	0565
1N3028B	0565
1N3029	0402
1N3029A	0402
1N3029B	0402
1N3030	0441
1N3030A	0441
1N3030B	0441
1N3031	0726
1N3031A	0726
1N3031B	0726
1N3032	0039
1N3032A	0039
1N3032B	0039
1N3033	0820

If replacement code is OEM, contact original manufacturer for replacement.

DEVICE TYPE	REPL CODE	DEVICE TYPE	REPL CODE	DEVICE TYPE	REPL CODE	DEVICE TYPE	REPL CODE	DEVICE TYPE	REPL CODE	DEVICE TYPE	REPL CODE	DEVICE TYPE	REPL CODE
1N3033A	0820	1N3098	0186	1N3177	OEM	1N3272	1046	1N3307RB	1225	1N3324	1210	1N3340RB	1575
1N3033B	0820	1N3098A	0186	1N3179	0609	1N3272R	1789	1N3308	0994	1N3324A	1210	1N3341	1436
1N3034	0350	1N3099	0028	1N3180	0609	1N3273	1046	1N3308A	0994	1N3324B	1210	1N3341A	1436
1N3034A	0350	1N3099A	0028	1N3181	0649	1N3273R	1789	1N3308B	0994	1N3324R	1387	1N3341B	1436
1N3034B	0350	1N3100	0367	1N3182	0030	1N3274	3928	1N3308R	1237	1N3324RA	1387	1N3341R	2802
1N3035	0931	1N3100A	0367	1N3183	0015	1N3274R	3531	1N3308RA	1237	1N3324RB	1387	1N3341RA	2802
1N3035A	0931	1N3102	1941	1N3184	0015	1N3275	2832	1N3308RB	1237	1N3325	1223	1N3341RB	2802
1N3035B	0931	1N3102A	1941	1N3185	0071	1N3275R	3167	1N3309	1009	1N3325A	1223	1N3342	1444
1N3036	0999	1N3102R	0509	1N3186	0071	1N3276	OEM	1N3309A	1009	1N3325AR	1406	1N3342A	1444
1N3036A	0999	1N3102RA	0509	1N3187	OEM	1N3276R	OEM	1N3309AR	1252	1N3325B	1223	1N3342B	1444
1N3036B	0999	1N3103	1961	1N3188	OEM	1N3277	0015	1N3309B	1009	1N3325BR	1406	1N3342R	2803
1N3037	0502	1N3103A	1961	1N3189	0015	1N3278	0015	1N3309BR	1252	1N3325R	1406	1N3342RA	2803
1N3037A	0502	1N3103R	0063	1N3190	0015	1N3279	0015	1N3309R	1252	1N3325RA	1406	1N3342RB	2803
1N3037B	0502	1N3103RA	0063	1N3191	0015	1N3280	0071	1N3309RA	1252	1N3325RB	1406	1N3343	1457
1N3038	0869	1N3104	0776	1N3193	0015	1N3281	0071	1N3309RB	1252	1N3326	1232	1N3343A	1457
1N3038A	0869	1N3104A	0776	1N3194	0023	1N3282	0071	1N3310	1019	1N3326A	1232	1N3343B	1457
1N3038B	0869	1N3104R	0634	1N3195	0023	1N3283	0348	1N3310A	1019	1N3326B	1232	1N3343R	1507
1N3039	0783	1N3104RA	0634	1N3196	0023	1N3284	OEM	1N3310B	1019	1N3326R	1415	1N3343RA	1507
1N3039A	0783	1N3105	OEM	1N3197	0133	1N3285	OEM	1N3310R	1269	1N3326RA	1415	1N3343RB	1507
1N3039B	0783	1N3105A	OEM	1N3198	OEM	1N3286	OEM	1N3310RA	1269	1N3326RB	1415	1N3344	1470
1N3040	1268	1N3105R	OEM	1N3199	0165	1N3287	0143	1N3310RB	1269	1N3327	1246	1N3344A	1470
1N3040A	1268	1N3105RA	OEM	1N3200	OEM	1N3287N	0143	1N3311	0889	1N3327A	1246	1N3344B	1470
1N3040B	1268	1N3106	0965	1N3201	OEM	1N3287W	0143	1N3311A	0889	1N3327AR	1428	1N3344R	2805
1N3041	1189	1N3107	0922	1N3202	OEM	1N3288	0599	1N3311AR	1277	1N3327B	1246	1N3344RA	2805
1N3041A	1189	1N3108	0012	1N3203	0015	1N3288A	0599	1N3311B	0889	1N3327BR	1428	1N3344RB	2805
1N3041B	1189	1N3109	0612	1N3204	0143	1N3288AR	1347	1N3311BR	1277	1N3327R	1428	1N3345	1484
1N3042	0331	1N3110	0143	1N3205	OEM	1N3288R	1347	1N3311R	1277	1N3327RA	1428	1N3345A	1484
1N3042A	0331	1N3111	0599	1N3206	0133	1N3288RA	1347	1N3311RA	1277	1N3327RB	1428	1N3345B	1484
1N3042B	0331	1N3111R	1347	1N3207	0133	1N3289	0599	1N3311RB	1277	1N3328	1265	1N3345R	2807
1N3043	1311	1N3112	0086	1N3208	2002	1N3289A	0599	1N3312	1049	1N3328A	1265	1N3345RA	2807
1N3043A	1311	1N3118	OEM	1N3208R	2003	1N3289AR	1347	1N3312A	1049	1N3328B	1265	1N3345RB	2807
1N3043B	1311	1N3121	0143	1N3209	0590	1N3289R	1347	1N3312B	1049	1N3328R	1440	1N3346	1505
1N3044	0098	1N3122	0143	1N3209R	3265	1N3289RA	1347	1N3312R	1294	1N3328RA	1440	1N3346A	1505
1N3044A	0098	1N3123	0133	1N3210	1250	1N3290	0599	1N3312RA	1294	1N3328RB	1440	1N3346B	1505
1N3044B	0098	1N3124	0133	1N3210R	1577	1N3290A	0599	1N3312RB	1294	1N3329	1273	1N3346R	2810
1N3045	0149	1N3125	0143	1N3211	5587	1N3290AR	1347	1N3313	1058	1N3329A	1273	1N3346RA	2810
1N3045A	0149	1N3128	OEM	1N3211R	3275	1N3290R	1347	1N3313A	1058	1N3329B	1273	1N3346RB	2810
1N3045B	0149	1N3129	OEM	1N3212	1581	1N3290RA	1347	1N3313B	1058	1N3329R	2772	1N3347	1521
1N3046	0186	1N3130	6761	1N3212R	3281	1N3291	0599	1N3313R	2752	1N3329RA	2772	1N3347A	1521
1N3046A	0186	1N3138	OEM	1N3213	5589	1N3291A	0599	1N3313RA	2752	1N3329RB	2772	1N3347B	1521
1N3046B	0186	1N3139	1560	1N3213R	3287	1N3291AR	1347	1N3313RB	2752	1N3330	1290	1N3347R	2815
1N3047	0213	1N3140	1560	1N3214	1586	1N3291R	1347	1N3314	0932	1N3330A	1290	1N3347RA	2815
1N3047A	0213	1N3141	OEM	1N3214R	3294	1N3291RA	1347	1N3314A	0932	1N3330B	1290	1N3347RB	2815
1N3047B	0213	1N3142	2657	1N3215	OEM	1N3292	1986	1N3314AR	1302	1N3330R	1447	1N3348	1531
1N3048	0028	1N3143	OEM	1N3217	OEM	1N3292A	1986	1N3314B	0932	1N3330RA	1447	1N3348A	1531
1N3048A	0028	1N3146	0143	1N3218	OEM	1N3292AR	1905	1N3314BR	1302	1N3330RB	1447	1N3348B	1531
1N3048B	0028	1N3147	0133	1N3219	OEM	1N3292B	1986	1N3314R	1302	1N3331	1299	1N3348R	2818
1N3049	0255	1N3148	0322	1N3220	OEM	1N3292BR	1905	1N3314RA	1302	1N3331A	1299	1N3348RA	2818
1N3049A	0255	1N3149	OEM	1N3221	0071	1N3292R	1905	1N3314RB	1302	1N3331B	1299	1N3348RB	2818
1N3049B	0255	1N3149A	OEM	1N3222	OEM	1N3292RA	1905	1N3315	1079	1N3331R	2776	1N3349	1548
1N3050	0367	1N3150	OEM	1N3223	0133	1N3292RB	1905	1N3315A	1079	1N3331RA	2776	1N3349A	1548
1N3050A	0367	1N3151	OEM	1N3224	OEM	1N3293	1986	1N3315B	1079	1N3331RB	2776	1N3349B	1548
1N3050B	0367	1N3154	0244	1N3225	OEM	1N3293A	1986	1N3315R	1310	1N3332	1307	1N3349R	2820
1N3051	0421	1N3154-1	OEM	1N3227	0953	1N3293AR	1905	1N3315RA	1310	1N3332A	1307	1N3349RA	2820
1N3051A	0421	1N3154A	0244	1N3228	0242	1N3293R	1905	1N3315RB	1310	1N3332B	1307	1N3349RB	2820
1N3051B	0421	1N3154A1	OEM	1N3229	1771	1N3293RA	1905	1N3316	1096	1N3332R	1461	1N3350	1587
1N3052	5731	1N3155	0244	1N3230	0965	1N3294	0657	1N3316A	1096	1N3332RA	1461	1N3350A	1587
1N3053	5731	1N3155A	0244	1N3231	0817	1N3294A	0657	1N3316B	1096	1N3332RB	1461	1N3350B	1587
1N3054	5731	1N3155A-1	OEM	1N3232	0922	1N3294AR	0193	1N3316R	2756	1N3333	1322	1N3350R	2824
1N3055	5731	1N3155A1	OEM	1N3233	0102	1N3294R	0193	1N3316RA	2756	1N3333A	1322	1N3350RA	2824
1N3056	5731	1N3156	0244	1N3234	6603	1N3294RA	0193	1N3316RB	2756	1N3333B	1322	1N3350RB	2824
1N3057	5731	1N3156-1	OEM	1N3235	0348	1N3295	0657	1N3317	1106	1N3333R	2779	1N3353	OEM
1N3058	5731	1N3156A	0244	1N3236	0348	1N3295A	0657	1N3317A	1106	1N3333RA	2779	1N3354	OEM
1N3059	5731	1N3156A1	OEM	1N3237	0110	1N3295AR	0193	1N3317AR	1324	1N3333RB	2779	1N3355	OEM
1N3060	5731	1N3157	0244	1N3238	0953	1N3295R	0193	1N3317B	1106	1N3334	1331	1N3356	OEM
1N3061	5731	1N3157-1	OEM	1N3239	0242	1N3295RA	0193	1N3317BR	1324	1N3334A	1331	1N3357	OEM
1N3062	0124	1N3157A	0244	1N3240	0540	1N3296	3309	1N3317R	1324	1N3334AR	1474	1N3358	OEM
1N3063	0133	1N3157A1	OEM	1N3241	0965	1N3296A	3309	1N3317RA	1324	1N3334B	1331	1N3359	OEM
1N3064	0133	1N3159	OEM	1N3242	0817	1N3296AR	0202	1N3317RB	1324	1N3334BR	1474	1N3360	OEM
1N3065	0133	1N3160	0015	1N3243	0922	1N3296R	0202	1N3318	1123	1N3334R	1474	1N3361	OEM
1N3066	0133	1N3161	1023	1N3244	0102	1N3296RA	0202	1N3318A	1123	1N3334RA	1474	1N3362	OEM
1N3066M	OEM	1N3161R	0501	1N3245	6603	1N3297	OEM	1N3318B	1123	1N3334RB	1474	1N3363	OEM
1N3067	0133	1N3161RC	1782	1N3246	0110	1N3297A	3309	1N3318R	2758	1N3335	1353	1N3364	OEM
1N3068	0133	1N3162	1023	1N3247	0953	1N3297AR	0202	1N3318RA	2758	1N3335A	1353	1N3365	OEM
1N3069	0133	1N3162R	0501	1N3248	0242	1N3297RA	3309	1N3318RB	2758	1N3335B	1353	1N3366	OEM
1N3069M	0133	1N3163	1023	1N3249	0540	1N3298	0023	1N3319	1135	1N3335R	0613	1N3367	OEM
1N3070	0133	1N3163R	0501	1N3250	0965	1N3298A	OEM	1N3319A	1135	1N3335RA	0613	1N3368	OEM
1N3071	0133	1N3164	1023	1N3251	0817	1N3299	OEM	1N3319B	1135	1N3335RB	0613	1N3369	OEM
1N3072	0110	1N3164R	0501	1N3252	0922	1N3300	3648	1N3319R	1333	1N3336	1365	1N3370	OEM
1N3073	0953	1N3165	1023	1N3253	0015	1N3300A	3648	1N3319RA	1333	1N3336A	1365	1N3371	OEM
1N3074	0242	1N3165R	0501	1N3254	0015	1N3301	3331	1N3319RB	1333	1N3336B	1365	1N3372	OEM
1N3075	0242	1N3166	1023	1N3255	0015	1N3301A	3331	1N3320	1152	1N3336R	1510	1N3373	OEM
1N3076	1747	1N3166R	0501	1N3256	0071	1N3302	2730	1N3320A	1152	1N3336RA	1510	1N3374	OEM
1N3077	1747	1N3167	1023	1N3257	0133	1N3302A	2730	1N3320AR	1344	1N3336RB	1510	1N3375	OEM
1N3078	0540	1N3167R	1777	1N3258	0133	1N3303	OEM	1N3320B	1152	1N3337	1384	1N3376	OEM
1N3079	0540	1N3168	1023	1N3260	1023	1N3303A	2730	1N3320BR	1344	1N3337A	1384	1N3377	OEM
1N3080	0817	1N3168R	1777	1N3260R	0501	1N3304	OEM	1N3320R	1344	1N3337B	1384	1N3378	OEM
1N3081	0031	1N3169	1036	1N3261	1023	1N3304A	OEM	1N3320RA	1344	1N3337R	1523	1N3379	OEM
1N3082	0242	1N3169R	1777	1N3261R	0501	1N3305	0926	1N3320RB	1344	1N3337RA	1523	1N3380	OEM
1N3083	0540	1N3170	1036	1N3262	1023	1N3305A	0926	1N3321	1164	1N3337RB	1523	1N3381	1335
1N3084	0965	1N3170R	1777	1N3262R	0501	1N3305AR	2740	1N3321A	1164	1N3338	1401	1N3382	0133
1N3085	0599	1N3171	1046	1N3263	1023	1N3305B	0926	1N3321B	1164	1N3338A	1401	1N3383	1335
1N3085R	1347	1N3171A	1046	1N3263R	0501	1N3305BR	2740	1N3321R	1356	1N3338B	1401	1N3384	0080
1N3086	0599	1N3171R	1789	1N3264	1023	1N3305R	2740	1N3321RA	1356	1N3338R	1537	1N3385	0080
1N3086R	1347	1N3171RA	1789	1N3264R	0501	1N3305RA	2740	1N3321RB	1356	1N3338RA	1537	1N3386	0609
1N3087	0599	1N3172	1046	1N3265	1023	1N3305RB	2740	1N3322	1174	1N3338RB	1537	1N3387	0609
1N3087R	1347	1N3172A	1046	1N3265R	0501	1N3306	0944	1N3322A	1174	1N3339	1412	1N3388	0795
1N3088	1986	1N3172R	1789	1N3266	1023	1N3306A	0944	1N3322B	1174	1N3339A	1412	1N3389	0795
1N3088R	1905	1N3172RA	1789	1N3266R	0501	1N3306B	0944	1N3322R	2760	1N3339B	1412	1N3390	0795
1N3089	1986	1N3173	1046	1N3267	1023	1N3306R	2744	1N3322RA	2760	1N3339R	1550	1N3391	OEM
1N3089R	1905	1N3173A	1046	1N3267R	0501	1N3306RA	2744	1N3322RB	2760	1N3339RA	1550	1N3392	OEM
1N3090	1986	1N3173R	1789	1N3268	1036	1N3306RB	2744	1N3323	1184	1N3339RB	1550	1N3393	OEM
1N3090R	1905	1N3173RA	1789	1N3268R	1777	1N3307	0958	1N3323A	1184	1N3340	1423	1N3394	OEM
1N3091	0657	1N3174	1046	1N3269	1036	1N3307A	0958	1N3323AR	1371	1N3340A	1423	1N3395	OEM
1N3091R	0193	1N3174A	1046	1N3269R	0805	1N3307AR	1225	1N3323B	1184	1N3340AR	1575	1N3396	0299
1N3092	0657	1N3174R	1789	1N3270	1046	1N3307B	0958	1N3323BR	1371	1N3340B	1423	1N3397	OEM
1N3092R	0193	1N3174RA	1789	1N3270R	1789	1N3307BR	1225	1N3323R	1371	1N3340BR	1575	1N3398	OEM
1N3094	0015	1N3175	3928	1N3271	1046	1N3307R	1225	1N3323RA	1371	1N3340R	1575	1N3399	0157
1N3097	OEM	1N3176	OEM	1N3271R	1789	1N3307RA	1225	1N3323RB	1371	1N3340RA	1575	1N3400	0062

DEVICE TYPE	REPL CODE
1N3401	0165
1N3401-0	0244
1N3401O	0244
1N3402	0064
1N3403	0052
1N3404	0686
1N3405	0375
1N3406	0705
1N3407	0455
1N3408	0166
1N3409	0032
1N3410	0068
1N3411	0471
1N3412	0062
1N3413	0077
1N3414	0165
1N3415	0064
1N3416	0052
1N3417	0686
1N3418	0375
1N3419	0705
1N3420	0455
1N3421	0195
1N3422	0166
1N3423	0032
1N3424	0068
1N3425	0125
1N3426	0173
1N3427	0049
1N3428	0189
1N3429	0089
1N3430	0340
1N3431	0425
1N3432	OEM
1N3433	1490
1N3433A	1490
1N3433C	OEM
1N3433R	1622
1N3433RA	1622
1N3434	0510
1N3434A	0510
1N3434C	OEM
1N3434R	0992
1N3434RA	0992
1N3435	0870
1N3435A	0870
1N3435C	OEM
1N3435R	1263
1N3435RA	1263
1N3436	1274
1N3436A	1274
1N3436C	OEM
1N3436R	1639
1N3436RA	1639
1N3437	1078
1N3437A	1078
1N3437C	OEM
1N3437R	1716
1N3437RA	1716
1N3438	1723
1N3438A	1723
1N3438C	OEM
1N3438R	0727
1N3438RA	0727
1N3439	1761
1N3439A	1761
1N3439C	OEM
1N3439R	1782
1N3439RA	1782
1N3440	1788
1N3440A	1788
1N3440C	OEM
1N3440R	1799
1N3440RA	1799
1N3441	1804
1N3441A	1804
1N3441C	OEM
1N3441R	1817
1N3441RA	1817
1N3442	0022
1N3442A	0022
1N3442C	OEM
1N3442R	1853
1N3442RA	1853
1N3443	0696
1N3443A	0696
1N3443R	2221
1N3443RA	2221
1N3444	1601
1N3444A	1601
1N3444R	1458
1N3444RA	1458
1N3445	1616
1N3445A	1616
1N3445R	0221
1N3445RA	0221
1N3446	0992
1N3446A	0992
1N3446R	0510
1N3446RA	0510
1N3447	1263
1N3447A	1263
1N3447R	0870
1N3447RA	0870
1N3448	1639
1N3448A	1639
1N3448R	1274
1N3448RA	1274
1N3449	1716
1N3449A	1716
1N3449R	1078
1N3449RA	1078
1N3450	0727
1N3450A	0727
1N3450R	1723
1N3450RA	1723
1N3451	1782
1N3451A	1782
1N3451R	1761
1N3451RA	1761
1N3452	1794
1N3452A	1794
1N3452R	1772
1N3452RA	1772
1N3453	1799
1N3453A	1799
1N3453R	1788
1N3453RA	1788
1N3454	1817
1N3454A	1817
1N3454R	1804
1N3454RA	1804
1N3455	1853
1N3455A	1853
1N3455R	0022
1N3455RA	0022
1N3456	1884
1N3456A	1884
1N3456R	0207
1N3456RA	0207
1N3457	1902
1N3457A	1902
1N3457R	0310
1N3457RA	0310
1N3458	1909
1N3458A	1909
1N3458R	0356
1N3458RA	0356
1N3459	1163
1N3459A	1163
1N3459R	0438
1N3459RA	0438
1N3460	1941
1N3460A	1941
1N3460R	0509
1N3460RA	0509
1N3461	1961
1N3461A	1961
1N3461R	0063
1N3461RA	0063
1N3462	0776
1N3462A	0776
1N3462R	0634
1N3462RA	0634
1N3463	OEM
1N3463A	OEM
1N3463R	OEM
1N3463RA	OEM
1N3464	OEM
1N3465	0143
1N3466	0143
1N3467	0143
1N3468	0143
1N3469	0143
1N3470	0143
1N3471	0133
1N3473	0015
1N3474	0015
1N3475	0015
1N3476	0071
1N3477	OEM
1N3477A	OEM
1N3481	OEM
1N3482	0143
1N3483	0143
1N3484	0143
1N3485	OEM
1N3486	0953
1N3487	0102
1N3488	OEM
1N3489	OEM
1N3489A	OEM
1N3490	OEM
1N3491	5026
1N3491R	2559
1N3492	2615
1N3492R	2566
1N3493	2615
1N3493R	2566
1N3494	5026
1N3494R	2559
1N3495	5026
1N3495R	2559
1N3496	0471
1N3497	0471
1N3498	0471
1N3499	0471
1N3500	0471
1N3501	OEM
1N3502	OEM
1N3503	OEM
1N3503A	OEM
1N3504	OEM
1N3504A	OEM
1N3506	0299
1N3507	0376
1N3508	0036
1N3509	0275
1N3510	0140
1N3511	0162
1N3512	0253
1N3513	0471
1N3514	0062
1N3514R	0267
1N3515	0077
1N3516	0165
1N3517	0057
1N3518	0064
1N3519	0181
1N3520	0052
1N3521	0053
1N3522	0686
1N3523	0445
1N3524	0375
1N3525	0700
1N3526	0705
1N3527	0494
1N3528	0455
1N3529	0195
1N3530	0166
1N3531	0820
1N3532	0032
1N3533	0054
1N3534	0068
1N3537	0137
1N3539	OEM
1N3539A	OEM
1N3540	OEM
1N3540A	OEM
1N3541	OEM
1N3541A	OEM
1N3541B	0122
1N3542	OEM
1N3542A	OEM
1N3542B	0122
1N3543	OEM
1N3543A	OEM
1N3544	0015
1N3544B	0122
1N3545	0015
1N3545B	0965
1N3546	0015
1N3546B	0122
1N3547	0015
1N3547B	0122
1N3548	0015
1N3548B	0122
1N3549	0015
1N3550	0133
1N3551	OEM
1N3552	OEM
1N3553	0471
1N3554	0720
1N3555	OEM
1N3556	OEM
1N3557	OEM
1N3558	OEM
1N3559	0143
1N3560	OEM
1N3561	OEM
1N3562	OEM
1N3563	0922
1N3564	0143
1N3565	0708
1N3566	OEM
1N3567	OEM
1N3568	OEM
1N3569	3733
1N3569R	2291
1N3570	0989
1N3570R	2291
1N3571	3743
1N3571R	0476
1N3572	0197
1N3572R	0476
1N3573	0200
1N3573R	0476
1N3574	0204
1N3574R	0476
1N3575	0133
1N3576	0133
1N3577	0133
1N3578	0795
1N3579	0795
1N3580	0794
1N3580A	0794
1N3580B	0794
1N3581	0794
1N3581A	0794
1N3581B	0794
1N3582	0794
1N3582A	0794
1N3582B	0794
1N3583	0137
1N3583A	0137
1N3583B	0137
1N3584	0794
1N3584A	0794
1N3584B	0794
1N3585	OEM
1N3586	OEM
1N3587	OEM
1N3588	OEM
1N3589	OEM
1N3590	OEM
1N3591	OEM
1N3592	0143
1N3593	0133
1N3594	0133
1N3595	0133
1N3596	0133
1N3597	OEM
1N3598	0133
1N3599	0133
1N3600	0124
1N3601	0133
1N3602	0133
1N3603	0133
1N3604	0133
1N3605	0133
1N3606	0133
1N3607	0133
1N3608	0133
1N3609	0133
1N3610	2900
1N3611	0609
1N3611GP	0609
1N3612	0795
1N3612GP	0795
1N3613	0015
1N3613GP	0015
1N3614	0072
1N3614GP	0072
1N3615	5514
1N3615R	1238
1N3616	5582
1N3616R	1240
1N3617	2900
1N3617R	1241
1N3618	2900
1N3618R	1241
1N3619	3751
1N3619R	2365
1N3620	3751
1N3620R	2365
1N3621	3757
1N3621R	1253
1N3622	1609
1N3622R	2376
1N3623	3768
1N3623R	1264
1N3624	3011
1N3624R	0449
1N3625	0133
1N3626	0133
1N3627	OEM
1N3628	OEM
1N3629	0015
1N3630	0015
1N3631	0015
1N3632	0015
1N3633	0015
1N3634	0015
1N3635	0071
1N3636	0071
1N3637	0071
1N3638	0071
1N3639	0015
1N3640	0015
1N3641	0015
1N3642	0071
1N3643	0373
1N3644	0348
1N3645	0348
1N3646	OEM
1N3647	OEM
1N3648	OEM
1N3649	0206
1N3649(REV.POL.)	1008
1N3649R	1008
1N3650	0586
1N3650R	0449
1N3653	0133
1N3654	0133
1N3655	OEM
1N3655A	0606
1N3655B	OEM
1N3656	0242
1N3657	0540
1N3658	0965
1N3659	5026
1N3659R	2559
1N3660	5026
1N3660R	2559
1N3661	5026
1N3661R	2559
1N3662	5026
1N3662R	2559
1N3663	5026
1N3663R	2559
1N3664	2615
1N3665	2615
1N3665R	OEM
1N3666	0133
1N3667	OEM
1N3668	0133
1N3669	0015
1N3670	0131
1N3670A	0131
1N3670AR	0545
1N3670R	0545
1N3670RA	0545
1N3671	0131
1N3671A	0131
1N3671AR	0545
1N3671R	0545
1N3671RA	0545
1N3672	0145
1N3672A	0145
1N3672AR	0550
1N3672R	0550
1N3672RA	0550
1N3673	0145
1N3673A	0145
1N3673AR	0550
1N3673R	0550
1N3673RA	0550
1N3675	0025
1N3675A	0025
1N3675B	0025
1N3676	0649
1N3676A	0649
1N3676B	0649
1N3677	0244
1N3677A	0244
1N3677B	0244
1N3678	0012
1N3678A	0012
1N3678B	0012
1N3679	0170
1N3679A	0170
1N3679B	0170
1N3680	0317
1N3680A	0317
1N3680B	0317
1N3681	0137
1N3681A	0137
1N3681B	0137
1N3682	0365
1N3682A	0365
1N3682B	0365
1N3683	0002
1N3683A	0002
1N3683B	0002
1N3684	0420
1N3684A	0420
1N3684B	0420
1N3685	0495
1N3685A	0495
1N3685B	0495
1N3686	0531
1N3686A	0531
1N3686B	0531
1N3687	0565
1N3687A	0565
1N3687B	0565
1N3688	0402
1N3688A	0402
1N3688B	0402
1N3689	0441
1N3689A	0441
1N3689B	0441
1N3690	0726
1N3690A	0726
1N3690B	0726
1N3691	0039
1N3691A	0039
1N3691B	0039
1N3692	0820
1N3692A	0820
1N3692B	0820
1N3693	0350
1N3693A	0350
1N3693B	0350
1N3694	0931
1N3694A	0931
1N3694B	0931
1N3695	0999
1N3695A	0999
1N3695B	0999
1N3696	0502
1N3696A	0502
1N3696B	0502
1N3697	0869
1N3697A	0869
1N3697B	0869
1N3698	0783
1N3698A	0783
1N3698B	0783
1N3699	1268
1N3699A	1268
1N3699B	1268
1N3700	1189
1N3700A	1189
1N3700B	1189
1N3701	0331
1N3701A	0331
1N3701B	0331
1N3702	1311
1N3702A	1311
1N3702B	1311
1N3703	0098
1N3703A	0098
1N3703B	0098
1N3704	0149
1N3704A	0149
1N3704B	0149
1N3705	0186
1N3705A	0186
1N3705B	0186
1N3706	0213
1N3706A	0213
1N3706B	0213
1N3707	0028
1N3707A	0028
1N3707B	0028
1N3708	0255
1N3708A	0255
1N3708B	0255
1N3709	0367
1N3709A	0367
1N3709B	0367
1N3710	0421
1N3710A	0421
1N3710B	0421
1N3711	OEM
1N3712	OEM
1N3713	OEM
1N3714	OEM
1N3715	OEM
1N3716	OEM
1N3717	OEM
1N3718	OEM
1N3719	OEM
1N3720	OEM
1N3721	OEM
1N3722	0133
1N3723	0922
1N3724	0102
1N3725	0102
1N3726	0348
1N3727	0348
1N3728	0795
1N3729	0102
1N3730	OEM
1N3731	0023
1N3732	0162
1N3733	OEM
1N3734	OEM
1N3735	1023
1N3735R	0501
1N3736	1023
1N3736A	0301
1N3736R	0501
1N3737	1023
1N3737R	0501
1N3738	1023
1N3738R	0501
1N3739	1036
1N3739R	1777
1N3740	1036
1N3740R	1777
1N3741	1046
1N3741R	1789
1N3742	1046
1N3742R	1789
1N3743	3928
1N3743R	2792
1N3744	2832
1N3744R	3167
1N3745	1115
1N3746	OEM
1N3747	OEM
1N3747W	OEM
1N3748	0242
1N3749	0540
1N3750	0965
1N3751	0817
1N3752	0922
1N3753	0143
1N3754	0015
1N3755	0242
1N3756	0023
1N3757	0242
1N3758	0540
1N3759	0965
1N3760	0817
1N3761	0922
1N3762	OEM
1N3763	0262
1N3764	OEM
1N3765	1132
1N3765R	1119
1N3766	1132
1N3766R	1119
1N3767	0044
1N3767R	0281
1N3768	0044
1N3768R	0281
1N3769	0143
1N3770	OEM
1N3771	OEM
1N3772	OEM
1N3773	0143
1N3774	OEM
1N3775	OEM
1N3776	0170
1N3777	1132
1N3778	OEM
1N3779	0062
1N3780	0062
1N3781	0062
1N3782	0794
1N3782A	0794
1N3782B	0794
1N3783	0062
1N3784	0794
1N3784A	0794
1N3784B	0794
1N3785	0205
1N3785A	0205
1N3785B	0205
1N3786	0480
1N3786A	0480
1N3786B	0480
1N3787	0504
1N3787A	0504
1N3787B	0504
1N3788	0684
1N3788A	0684
1N3788B	0684
1N3789	0225
1N3789A	0225
1N3789B	0225
1N3790	0230
1N3790A	0230
1N3790B	0230
1N3791	0234
1N3791A	0234
1N3791B	0234
1N3792	0237
1N3792A	0237
1N3792B	0237
1N3793	0247
1N3793A	0247
1N3793B	0247
1N3794	0251
1N3794A	0251
1N3794B	0251
1N3795	0256
1N3795A	0256
1N3795B	0256
1N3796	0262
1N3796A	0262
1N3796B	0262
1N3797	0269
1N3797A	0269
1N3797B	0269
1N3798	0274
1N3798A	0274
1N3798B	0274
1N3799	0294
1N3799A	0294
1N3799B	0294
1N3800	0309
1N3800A	0309
1N3800B	0309
1N3801	0318
1N3801A	0318
1N3801B	0318
1N3802	0320
1N3802A	0320
1N3802B	0320
1N3803	0326
1N3803A	0326
1N3803B	0326
1N3804	0337
1N3804A	0337
1N3804B	0337
1N3805	0347
1N3805A	0347
1N3805B	0347
1N3806	0027
1N3806A	0027
1N3806B	0027
1N3807	0266
1N3807A	0266
1N3807B	0266
1N3808	0386
1N3808A	0386
1N3808B	0386
1N3809	0405
1N3809A	0405
1N3809B	0405
1N3810	0426
1N3810A	0426
1N3810B	0426
1N3811	0444
1N3811A	0444
1N3811B	0444
1N3812	0238
1N3812A	0238
1N3812B	0238
1N3813	1180
1N3813A	1180
1N3813B	1180
1N3814	1190
1N3814A	1190
1N3814B	1190
1N3815	1207
1N3815A	1207
1N3815B	1207
1N3816	1218
1N3816A	1218
1N3816B	1218
1N3817	0647
1N3817A	0647
1N3817B	0647
1N3818	1255
1N3818A	1255
1N3818B	1255
1N3819	1279
1N3819A	1279
1N3819B	1279
1N3820	0605
1N3820A	0605
1N3820B	0605
1N3821	0292
1N3821A	0292
1N3822	0188
1N3822A	0188
1N3823	0456
1N3823A	0456

If replacement code is OEM, contact original manufacturer for replacement.

DEVICE TYPE	REPL CODE
1N3824	0533
1N3824A	0533
1N3825	0451
1N3825A	0451
1N3826	0162
1N3826A	0162
1N3827	0157
1N3827A	0157
1N3828	0636
1N3828A	0636
1N3829	0025
1N3829A	0025
1N3829B	OEM
1N3830	0649
1N3830A	0649
1N3831	OEM
1N3832	OEM
1N3833	OEM
1N3834	OEM
1N3835	OEM
1N3836	OEM
1N3837	OEM
1N3838	OEM
1N3839	OEM
1N3840	OEM
1N3841	OEM
1N3842	OEM
1N3843	OEM
1N3844	OEM
1N3845	OEM
1N3846	OEM
1N3847	OEM
1N3848	OEM
1N3849	OEM
1N3850	OEM
1N3851	OEM
1N3852	OEM
1N3853	OEM
1N3854	OEM
1N3855A	0012
1N3856	OEM
1N3857	OEM
1N3858	OEM
1N3859	OEM
1N3860	OEM
1N3862	OEM
1N3864	0133
1N3865	1335
1N3866	0242
1N3867	0540
1N3868	0965
1N3869	0922
1N3870	0182
1N3871	OEM
1N3872	0133
1N3873	0133
1N3874	0920
1N3875	0015
1N3876	OEM
1N3877	OEM
1N3878	OEM
1N3879	1567
1N3879A	1567
1N3879R	1547
1N3879RA	1547
1N3880	1567
1N3880A	1567
1N3880R	1547
1N3880RA	1547
1N3881	1567
1N3881A	1567
1N3881R	1547
1N3881RA	1547
1N3882	2153
1N3882A	2153
1N3882R	3139
1N3882RA	3139
1N3883	2153
1N3883A	2153
1N3883R	3139
1N3883RA	3139
1N3884	OEM
1N3885	OEM
1N3886	OEM
1N3887	OEM
1N3888	OEM
1N3889	1567
1N3889A	1276
1N3889R	1547
1N3889RA	1547
1N3890	1567
1N3890A	1567
1N3890R	1547
1N3890RA	1547
1N3891	1567
1N3891A	1567
1N3891R	1547
1N3891RA	1547
1N3892	2153
1N3892A	2153
1N3892R	3139
1N3892RA	3139
1N3893	2153
1N3893A	2153
1N3893R	3139
1N3893RA	3139
1N3894	0795
1N3895	0795
1N3896	OEM

DEVICE TYPE	REPL CODE
1N3897	OEM
1N3898	OEM
1N3899	1530
1N3899R	1520
1N3900	1530
1N3900R	1520
1N3901	1530
1N3901R	1520
1N3902	0029
1N3902R	1847
1N3903	0029
1N3903R	1847
1N3904	OEM
1N3905	OEM
1N3906	OEM
1N3907	OEM
1N3908	OEM
1N3909	1530
1N3909R	1520
1N3910	1530
1N3910R	1520
1N3911	1530
1N3911R	1520
1N3912	0029
1N3912R	1847
1N3913	0029
1N3913R	1847
1N3914	0920
1N3915	0102
1N3916	OEM
1N3917	OEM
1N3918	OEM
1N3919	0586
1N3920	OEM
1N3921	OEM
1N3922	OEM
1N3923	OEM
1N3924	0145
1N3925	OEM
1N3926	OEM
1N3927	OEM
1N3928	OEM
1N3929	0071
1N3930	OEM
1N3931	OEM
1N3932	OEM
1N3933	OEM
1N3934	OEM
1N3935	OEM
1N3936	OEM
1N3938	0242
1N3939	0540
1N3940	0965
1N3941	0817
1N3942	0922
1N3943	OEM
1N3944	OEM
1N3945	OEM
1N3946	OEM
1N3947	0636
1N3949	1717
1N3949A	1717
1N3950	0262
1N3951	2399
1N3952	0242
1N3953	0133
1N3954	0133
1N3955	OEM
1N3956	0133
1N3957	0071
1N3957GP	0071
1N3958	0097
1N3958C	OEM
1N3959	0097
1N3959C	OEM
1N3960	0097
1N3960C	OEM
1N3961	0097
1N3961C	OEM
1N3962	OEM
1N3962C	OEM
1N3963	OEM
1N3963C	OEM
1N3964	0871
1N3965	0853
1N3966	1609
1N3966R	1205
1N3967	1610
1N3968	2657
1N3968R	2768
1N3969	2006
1N3969R	2907
1N3970	2682
1N3970R	3051
1N3971	3900
1N3971R	3117
1N3972	0599
1N3973	0599
1N3974	1986
1N3975	0657
1N3976	1023
1N3977	1023
1N3978	1250
1N3979	OEM
1N3980	OEM
1N3981	0242
1N3982	0540
1N3983	0965
1N3984	1901

DEVICE TYPE	REPL CODE
1N3984A	1901
1N3985	2416
1N3985A	2416
1N3986	0696
1N3986A	0696
1N3987	0206
1N3987R	1008
1N3988	0206
1N3988R	1008
1N3989	0586
1N3989R	0948
1N3990	0586
1N3990R	0550
1N3991	0143
1N3992	OEM
1N3993	0547
1N3993A	0547
1N3993B	0547
1N3993R	2035
1N3993RA	2035
1N3993RB	2035
1N3994	2403
1N3994A	2403
1N3994B	2403
1N3994R	2401
1N3994RA	2401
1N3994RB	2401
1N3995	2113
1N3995A	2113
1N3995B	2113
1N3995R	1438
1N3995RA	1438
1N3995RB	1438
1N3996	2410
1N3996A	2410
1N3996B	2410
1N3996R	2407
1N3996RA	2407
1N3996RB	2407
1N3997	1901
1N3997A	1901
1N3997B	1901
1N3997R	1445
1N3997RA	1445
1N3997RB	1445
1N3998	0696
1N3998A	0696
1N3998R	2221
1N3998RA	2221
1N3998RB	2221
1N3999	1601
1N3999A	1601
1N3999R	1458
1N3999RA	1458
1N4000	1616
1N4000A	1616
1N4000R	0221
1N4000RA	0221
1N4001	5473
1N4001A	1335
1N4001F	OEM
1N4001G	1335
1N4001GP	1335
1N4001RL	OEM
1N4002	0080
1N4002A	0080
1N4002F	OEM
1N4002G	0080
1N4002GP	6557
1N4002ID	0080
1N4002IO	0080
1N4002L	0080
1N4002RL	OEM
1N4002TA	6557
1N4003	5733
1N4003-T3	0023
1N4003-Z	0609
1N4003A	0609
1N4003F	OEM
1N4003G	0609
1N4003GP	0023
1N4003K	OEM
1N4003PX	0609
1N4003RL	0609
1N4003TA	0609
1N4004	0795
1N4004A	0795
1N4004F	OEM
1N4004G	0795
1N4004GP	0795
1N4004RL	OEM
1N4004TA	0795
1N4004TP	OEM
1N4005	0015
1N4005BP	0015
1N4005F	OEM
1N4005G	0015
1N4005GP	0015
1N4005RL	OEM
1N4006	0072
1N4006F	OEM
1N4006G	0071
1N4006GP	0072
1N4006RL	OEM
1N4007	0023
1N4007F	OEM
1N4007G	0071
1N4007GP	0071
1N4007RL	OEM

DEVICE TYPE	REPL CODE
1N4008	OEM
1N4009	0133
1N4010	0471
1N4011	0071
1N4012	0131
1N4012R	0545
1N4013	0131
1N4013R	0545
1N4014	0145
1N4014R	0550
1N4015	0145
1N4015R	0550
1N4016	1490
1N4016A	1490
1N4016B	1490
1N4017	1618
1N4017A	1618
1N4017B	1618
1N4018	0510
1N4018A	0510
1N4018B	0510
1N4019	0691
1N4019A	0691
1N4019B	0691
1N4020	0870
1N4020A	0870
1N4020B	0870
1N4021	1020
1N4021A	1020
1N4021B	1020
1N4022	1274
1N4022A	1274
1N4022B	1274
1N4023	1402
1N4023A	1402
1N4023B	1402
1N4024	1078
1N4024A	1078
1N4024B	1078
1N4025	1717
1N4025A	1717
1N4025B	1717
1N4026	1723
1N4026A	1723
1N4026B	1723
1N4027	1736
1N4027A	1736
1N4027B	1736
1N4028	1761
1N4028A	1761
1N4028B	1761
1N4029	1772
1N4029A	1772
1N4029B	0889
1N4030	1788
1N4030A	1788
1N4030B	1788
1N4031	1796
1N4031A	1796
1N4031B	1796
1N4032	1804
1N4032A	1804
1N4032B	1804
1N4033	1194
1N4033A	1194
1N4033B	1194
1N4034	0022
1N4034A	0022
1N4034B	0022
1N4035	0132
1N4035A	0132
1N4035B	0132
1N4036	0207
1N4036A	0207
1N4036B	0207
1N4037	0263
1N4037A	0263
1N4037B	0263
1N4038	0310
1N4038A	0310
1N4038B	0310
1N4039	0329
1N4039A	0329
1N4039B	0329
1N4040	0356
1N4040A	0356
1N4040B	0356
1N4041	0412
1N4041A	0412
1N4041B	0412
1N4042	0438
1N4042A	0438
1N4042B	0438
1N4043	0133
1N4044	1023
1N4044R	0501
1N4045	1023
1N4045R	0501
1N4046	1023
1N4046R	0501
1N4047	1023
1N4047R	0501
1N4048	1023
1N4048R	0501
1N4049	1023
1N4049R	0501
1N4050	1023
1N4050R	0501
1N4051	1036

DEVICE TYPE	REPL CODE
1N4051R	1777
1N4052	1036
1N4052R	1777
1N4053	1046
1N4053R	1789
1N4054	1046
1N4054R	1789
1N4055	1046
1N4055R	1789
1N4056	1046
1N4056R	1789
1N4057	0234
1N4057A	0234
1N4058	0247
1N4058A	0247
1N4059	1178
1N4059A	1178
1N4060	2395
1N4060A	2395
1N4061	0262
1N4061A	0262
1N4062	0269
1N4062A	0269
1N4063	0294
1N4063A	0294
1N4064	0309
1N4064A	0309
1N4065	0318
1N4065A	0318
1N4066	0320
1N4066A	0320
1N4067	0337
1N4067A	0337
1N4068	0347
1N4068A	0347
1N4069	0027
1N4069A	0027
1N4070	0266
1N4070A	0266
1N4071	0386
1N4071A	0386
1N4072	0405
1N4072A	0405
1N4073	0426
1N4073A	0426
1N4074	0444
1N4074A	0444
1N4075	3028
1N4075A	3028
1N4076	0238
1N4076A	0238
1N4077	1180
1N4077A	1180
1N4078	1190
1N4078A	1190
1N4079	1190
1N4079A	1190
1N4080	1207
1N4080A	1207
1N4081	1218
1N4081A	1218
1N4082	1881
1N4082A	1881
1N4083	0647
1N4083A	0647
1N4084	1279
1N4084A	1279
1N4085	0605
1N4085A	0605
1N4086	2513
1N4087	0133
1N4088	0143
1N4089	0540
1N4090	OEM
1N4091	OEM
1N4092	0133
1N4093	0913
1N4094	OEM
1N4095	0041
1N4096	0238
1N4097	1180
1N4098	0647
1N4099	0062
1N4099-1	OEM
1N4099RL	OEM
1N4100	0077
1N4100-1	OEM
1N4100RL	OEM
1N4101	0165
1N4101-1	OEM
1N4101RL	OEM
1N4102	0322
1N4102-1	OEM
1N4102RL	OEM
1N4103	0057
1N4103-1	OEM
1N4103A	0012
1N4103RL	OEM
1N4104	0064
1N4104-1	OEM
1N4104RL	OEM
1N4105	0181
1N4105-1	OEM
1N4106	0052
1N4106-1	OEM
1N4106A	0137
1N4106RL	OEM
1N4107	0053
1N4107-1	OEM

DEVICE TYPE	REPL CODE
1N4107RL	OEM
1N4108	0879
1N4108-1	OEM
1N4108A	0100
1N4108B	0100
1N4108RL	OEM
1N4109	0686
1N4109-1	OEM
1N4109RL	OEM
1N4110	0445
1N4110-1	OEM
1N4110RL	OEM
1N4111	0210
1N4111RL	OEM
1N4112	0375
1N4112-1	OEM
1N4112RL	OEM
1N4113	0671
1N4113-1	OEM
1N4113RL	OEM
1N4114	0700
1N4114-1	OEM
1N4114RL	OEM
1N4115	0705
1N4115JAN	OEM
1N4115JTX	OEM
1N4115JTXV	OEM
1N4115RL	OEM
1N4116	0494
1N4116-1	OEM
1N4116RL	OEM
1N4117	0714
1N4117-1	OEM
1N4117RL	OEM
1N4118	0455
1N4118-1	OEM
1N4118RL	OEM
1N4119	0257
1N4119-1	OEM
1N4119RL	OEM
1N4120	0195
1N4120-1	OEM
1N4120RL	OEM
1N4121	0166
1N4121A	0039
1N4121B	0039
1N4121RL	OEM
1N4122	0010
1N4122RL	OEM
1N4123	0032
1N4123RL	OEM
1N4124	0054
1N4124RL	OEM
1N4125	0068
1N4125RL	OEM
1N4126	0068
1N4126RL	OEM
1N4127	0125
1N4127A	0869
1N4127B	0869
1N4127RL	OEM
1N4128	2317
1N4128RL	OEM
1N4129	0152
1N4129A	0783
1N4129B	0783
1N4129RL	OEM
1N4130	0173
1N4130RL	OEM
1N4131	0094
1N4131RL	OEM
1N4132	0049
1N4132A	0331
1N4132B	0331
1N4132RL	OEM
1N4133	0104
1N4133RL	OEM
1N4134	0156
1N4134RL	OEM
1N4135	0189
1N4135RL	OEM
1N4136	2840
1N4136R	1783
1N4137	2850
1N4137R	1783
1N4138	2871
1N4138R	1783
1N4139	0110
1N4140	0953
1N4141	0242
1N4142	0540
1N4143	0965
1N4144	0817
1N4145	0922
1N4146	0102
1N4147	0133
1N4148	0124
1N4148(AU)	0124
1N4148B	OEM
1N4148F	0124
1N4148M	0124
1N4148RE	OEM
1N4148T	0124
1N4148TA	0124
1N4148TD	0124
1N4148TE	0124
1N4148V	0028
1N4149	0124
1N4150	0124

DEVICE TYPE	REPL CODE
1N4150V	0133
1N4151	0133
1N4152	0133
1N4153	0124
1N4154	0124
1N4155	0102
1N4156	0133
1N4157	0133
1N4158	0025
1N4158A	0025
1N4158B	0025
1N4159	0649
1N4159A	0649
1N4159B	0649
1N4160	0244
1N4160A	0244
1N4160B	0244
1N4161	0012
1N4161A	0012
1N4161B	0012
1N4162	0170
1N4162A	0170
1N4162B	0170
1N4163	0317
1N4163A	0317
1N4163B	0317
1N4164	0137
1N4164A	0137
1N4164B	0137
1N4165	0365
1N4165A	0365
1N4165B	0365
1N4166	0002
1N4166A	0002
1N4166B	0002
1N4167	0420
1N4167A	0420
1N4167B	0420
1N4168	0495
1N4168A	0495
1N4168B	0495
1N4169	0531
1N4169A	0531
1N4169B	0531
1N4170	0565
1N4170A	0565
1N4170B	0565
1N4171	0402
1N4171A	0402
1N4171B	0402
1N4172	0441
1N4172A	0441
1N4172B	0441
1N4173	0726
1N4173A	0726
1N4173B	0726
1N4174	0039
1N4174A	0039
1N4174B	0039
1N4175	0820
1N4175A	0820
1N4175B	0820
1N4176	0350
1N4176A	0350
1N4176B	0350
1N4177	0931
1N4177A	0931
1N4177B	0931
1N4178	0999
1N4178A	0999
1N4178B	0999
1N4179	0502
1N4179A	0502
1N4179B	0502
1N4180	0869
1N4180A	0869
1N4180B	0869
1N4181	0783
1N4181A	0783
1N4181B	0783
1N4182	1268
1N4182A	1268
1N4182B	1268
1N4183	1189
1N4183A	1189
1N4183B	1189
1N4184	0331
1N4184A	0331
1N4184B	0331
1N4185	1311
1N4185A	1311
1N4185B	1311
1N4186	0098
1N4186A	0098
1N4186B	0098
1N4187	0149
1N4187A	0149
1N4187B	0149
1N4188	0186
1N4188A	0186
1N4188B	0186
1N4189	0213
1N4189A	0213
1N4189B	0213
1N4190	0028
1N4190A	0028
1N4190B	0028
1N4191	0255
1N4191A	0255

If replacement code is OEM, contact original manufacturer for replacement.

Original Device Types

DEVICE TYPE	REPL CODE
1N4191B	0255
1N4192	0367
1N4192A	0367
1N4192B	0367
1N4193	0421
1N4193A	0421
1N4193B	0421
1N4194	1458
1N4194A	1458
1N4194B	OEM
1N4195	0221
1N4195A	0221
1N4195B	OEM
1N4196	1490
1N4196A	1490
1N4196B	OEM
1N4197	1618
1N4197A	1618
1N4197B	OEM
1N4198	0510
1N4198A	0510
1N4198B	OEM
1N4199	0691
1N4199A	0691
1N4199B	OEM
1N4200	0870
1N4200A	0870
1N4200B	OEM
1N4201	1020
1N4201A	1020
1N4201B	OEM
1N4202	1153
1N4202A	1153
1N4202B	OEM
1N4203	1274
1N4203A	1274
1N4203B	OEM
1N4204	1402
1N4204A	1402
1N4204B	OEM
1N4205	1532
1N4205A	1532
1N4205B	OEM
1N4206	1078
1N4206A	1078
1N4206B	OEM
1N4207	1711
1N4207A	1711
1N4207B	OEM
1N4208	1717
1N4208A	1717
1N4208B	OEM
1N4209	1723
1N4209A	1723
1N4209B	OEM
1N4210	1736
1N4210A	1736
1N4210B	OEM
1N4211	1748
1N4211A	1748
1N4211B	OEM
1N4212	1761
1N4212A	1761
1N4212B	OEM
1N4213	1772
1N4213A	1772
1N4213B	OEM
1N4214	1788
1N4214A	1788
1N4214B	OEM
1N4215	1796
1N4215A	1796
1N4215B	OEM
1N4216	1804
1N4216A	1804
1N4216B	OEM
1N4217	1194
1N4217A	1194
1N4217B	OEM
1N4218	1821
1N4218A	1821
1N4218B	OEM
1N4219	0022
1N4219A	0022
1N4219B	OEM
1N4220	0070
1N4220A	0070
1N4220B	OEM
1N4221	0132
1N4221A	0132
1N4221B	OEM
1N4222	0172
1N4222A	0172
1N4222B	OEM
1N4223	0207
1N4223A	0207
1N4223B	OEM
1N4224	0263
1N4224A	0263
1N4224B	OEM
1N4225	0310
1N4225A	0310
1N4225B	OEM
1N4226	0329
1N4226A	0329
1N4226B	OEM
1N4227	0356
1N4227A	0356
1N4227B	OEM
1N4228	0412
1N4228A	0412
1N4228B	OEM
1N4229	0438
1N4229A	0438
1N4229B	OEM
1N4230	0464
1N4230A	0464
1N4230B	OEM
1N4231	0488
1N4231A	0488
1N4231B	OEM
1N4232	0509
1N4232A	0509
1N4232B	OEM
1N4233	0524
1N4233A	0524
1N4233B	OEM
1N4234	0542
1N4234A	0542
1N4234B	OEM
1N4235	0063
1N4235A	0063
1N4235B	OEM
1N4236	0401
1N4236A	0401
1N4236B	OEM
1N4237	0616
1N4237A	0616
1N4237B	OEM
1N4238	0634
1N4238A	0634
1N4238B	OEM
1N4239	0668
1N4239A	0668
1N4239B	OEM
1N4240	2407
1N4241	2415
1N4242	0133
1N4243	0133
1N4244	0124
1N4245	0609
1N4245GP	0609
1N4246	0015
1N4246GP	0795
1N4247	0015
1N4247GP	0015
1N4248	0071
1N4248GP	0072
1N4249	0071
1N4249GP	0071
1N4250	0071
1N4251	0071
1N4252	0102
1N4253	6603
1N4254	0017
1N4255	0348
1N4256	OEM
1N4257	OEM
1N4258	1458
1N4258A	1458
1N4258B	OEM
1N4259	0221
1N4259A	0221
1N4259B	OEM
1N4260	1490
1N4260A	1490
1N4260B	OEM
1N4261	1618
1N4261A	1618
1N4261B	OEM
1N4262	0510
1N4262A	0510
1N4262B	OEM
1N4263	0691
1N4263A	0691
1N4263B	0691
1N4264	0870
1N4264A	0870
1N4264B	OEM
1N4265	1020
1N4265A	1020
1N4265B	OEM
1N4266	1274
1N4266A	1274
1N4266B	OEM
1N4267	1402
1N4267A	1402
1N4267B	OEM
1N4268	1078
1N4268A	1078
1N4268B	OEM
1N4269	1717
1N4269A	1717
1N4269B	OEM
1N4270	1723
1N4270A	1723
1N4270B	OEM
1N4271	1736
1N4271A	1736
1N4271B	OEM
1N4272	1761
1N4272A	1761
1N4272B	OEM
1N4273	1772
1N4273A	1772
1N4273B	OEM
1N4274	1788
1N4274A	1788
1N4274B	OEM
1N4275	1796
1N4275A	1796
1N4275B	OEM
1N4276	1804
1N4276A	1804
1N4276B	OEM
1N4277	1194
1N4277A	1194
1N4277B	OEM
1N4278	0022
1N4278A	0022
1N4278B	OEM
1N4279	0132
1N4279A	0132
1N4279B	OEM
1N4280	0207
1N4280A	0207
1N4280B	OEM
1N4281	0263
1N4281A	0263
1N4281B	OEM
1N4282	0310
1N4282A	0310
1N4282B	OEM
1N4283	0329
1N4283A	0329
1N4283B	OEM
1N4284	0356
1N4284A	0356
1N4284B	OEM
1N4285	0412
1N4285A	0412
1N4285B	OEM
1N4286	0438
1N4286A	0438
1N4286B	OEM
1N4287	0488
1N4287A	0488
1N4287B	OEM
1N4288	0509
1N4288A	0509
1N4288B	OEM
1N4289	0524
1N4289A	0524
1N4289B	OEM
1N4290	0063
1N4290A	0063
1N4290B	OEM
1N4291	0401
1N4291A	0401
1N4291B	OEM
1N4292	0634
1N4292A	0634
1N4292B	OEM
1N4293	0668
1N4293A	0668
1N4293B	0668
1N4294	OEM
1N4295	0170
1N4295A	0170
1N4296	0170
1N4296A	0170
1N4297	1083
1N4297A	1083
1N4297B	OEM
1N4298	1083
1N4298A	1083
1N4298B	OEM
1N4299	0317
1N4299A	0317
1N4299B	0794
1N4300	0317
1N4300A	0317
1N4300B	0794
1N4301	0317
1N4301A	0317
1N4301B	OEM
1N4302	1083
1N4302A	1083
1N4302B	OEM
1N4303	0317
1N4303A	0317
1N4303B	0794
1N4304	0317
1N4304A	0317
1N4304B	0794
1N4305	0124
1N4305F	0124
1N4306	0133
1N4307	0133
1N4308	0133
1N4309	0133
1N4310	OEM
1N4311	0133
1N4312	OEM
1N4313	OEM
1N4314	OEM
1N4315	0133
1N4316	OEM
1N4317	OEM
1N4318	0133
1N4319	OEM
1N4320	OEM
1N4321	0502
1N4322	0133
1N4323	0025
1N4323A	0025
1N4323B	0025
1N4324	0649
1N4324A	0649
1N4324B	0649
1N4325	0244
1N4325A	0244
1N4325B	0244
1N4326	0012
1N4326A	0012
1N4326B	0012
1N4327	0170
1N4327A	0170
1N4327B	0170
1N4328	0317
1N4328A	0317
1N4328B	0317
1N4329	0137
1N4329A	0137
1N4329B	0137
1N4330	0365
1N4330A	0365
1N4330B	0365
1N4331	0002
1N4331A	0002
1N4331B	0002
1N4332	0420
1N4332A	0420
1N4332B	0420
1N4333	0495
1N4333A	0495
1N4333B	0495
1N4334	0531
1N4334A	0531
1N4334B	0531
1N4335	0565
1N4335A	0565
1N4335B	0565
1N4336	0402
1N4336A	0402
1N4336B	0402
1N4337	0441
1N4337A	0441
1N4337B	0039
1N4338	0726
1N4338A	0726
1N4338B	0726
1N4339	0039
1N4339A	0039
1N4339B	0039
1N4340	0820
1N4340A	0820
1N4340B	0820
1N4341	0350
1N4341A	0350
1N4341B	0350
1N4342	0931
1N4342A	0931
1N4342B	0931
1N4343	0999
1N4343A	0999
1N4343B	0999
1N4344	0502
1N4344A	0502
1N4344B	0502
1N4345	0869
1N4345A	0869
1N4345B	0869
1N4346	0783
1N4346A	0783
1N4346B	0783
1N4347	1268
1N4347A	1268
1N4347B	1268
1N4348	1189
1N4348A	1189
1N4348B	1189
1N4349	0331
1N4349A	0331
1N4349B	0331
1N4350	1311
1N4350A	1311
1N4350B	1311
1N4351	0098
1N4351A	0098
1N4351B	0098
1N4352	0149
1N4352A	0149
1N4352B	0149
1N4353	0186
1N4353A	0186
1N4353B	0186
1N4354	0213
1N4354A	0213
1N4354B	0213
1N4355	0028
1N4355A	0028
1N4355B	0028
1N4356	0255
1N4356A	0255
1N4356B	0255
1N4357	0367
1N4357A	0367
1N4357B	0367
1N4358	0421
1N4358A	0421
1N4358B	0421
1N4359	OEM
1N4360	1276
1N4361	0071
1N4362	OEM
1N4363	0133
1N4364	0015
1N4365	0277
1N4366	0277
1N4367	0015
1N4368	0015
1N4369	0015
1N4370	1276
1N4370A	1276
1N4370ARL	OEM
1N4370RL	OEM
1N4371	0760
1N4371A	0760
1N4371ARL	OEM
1N4371RL	OEM
1N4372	0118
1N4372A	0118
1N4372ARL	OEM
1N4372RL	OEM
1N4374	OEM
1N4375	0133
1N4376	0133
1N4377	OEM
1N4378	OEM
1N4379	OEM
1N4380	OEM
1N4381	OEM
1N4382	0133
1N4383	0015
1N4383GP	0609
1N4384	0015
1N4384GP	0795
1N4385	0015
1N4385GP	0015
1N4386	OEM
1N4387	OEM
1N4388	OEM
1N4389	0133
1N4390	OEM
1N4391	OEM
1N4392	0133
1N4393	OEM
1N4393A	OEM
1N4393B	OEM
1N4394	OEM
1N4394A	OEM
1N4394B	OEM
1N4395	0133
1N4395A	0133
1N4395B	OEM
1N4396	OEM
1N4396A	OEM
1N4396B	OEM
1N4397	OEM
1N4397A	OEM
1N4397B	OEM
1N4398	OEM
1N4398A	OEM
1N4398B	OEM
1N4399	OEM
1N4399A	OEM
1N4399B	OEM
1N4400	0025
1N4400A	0025
1N4400B	0025
1N4401	0649
1N4401A	0649
1N4401B	0649
1N4402	0244
1N4402A	0244
1N4402B	0244
1N4403	0012
1N4403A	0012
1N4403B	0012
1N4404	0170
1N4404A	0170
1N4404B	0170
1N4405	0317
1N4405A	0317
1N4405B	0317
1N4406	0137
1N4406A	0137
1N4406B	0137
1N4407	0365
1N4407A	0365
1N4407B	0365
1N4408	0002
1N4408A	0002
1N4408B	0002
1N4409	0420
1N4409A	0420
1N4409B	0420
1N4410	0495
1N4410A	0495
1N4410B	0495
1N4411	0531
1N4411A	0531
1N4411B	0531
1N4412	0565
1N4412A	0565
1N4412B	0565
1N4413	0402
1N4413A	0402
1N4413B	0402
1N4414	0441
1N4414A	0441
1N4414B	0441
1N4415	0726
1N4415A	0726
1N4415B	0726
1N4416	0039
1N4416A	0039
1N4416B	0039
1N4417	0820
1N4417A	0820
1N4417B	0820
1N4418	0350
1N4418A	0350
1N4418B	0350
1N4419	0931
1N4419A	0931
1N4419B	0931
1N4420	0999
1N4420A	0999
1N4420B	0999
1N4421	0502
1N4421A	0502
1N4421B	0502
1N4422	0869
1N4422A	0869
1N4422B	0869
1N4423	0783
1N4423A	0783
1N4423B	0783
1N4424	1268
1N4424A	1268
1N4424B	1268
1N4425	1189
1N4425A	1189
1N4425B	1189
1N4426	0331
1N4426A	0331
1N4426B	0331
1N4427	1311
1N4427A	1311
1N4427B	1311
1N4428	0098
1N4428A	0098
1N4428B	0098
1N4429	0149
1N4429A	0149
1N4429B	0149
1N4430	0186
1N4430A	0186
1N4430B	0186
1N4431	0213
1N4431A	0213
1N4431B	0213
1N4432	0028
1N4432A	0028
1N4432B	0028
1N4433	0255
1N4433A	0255
1N4434	0367
1N4434A	0367
1N4434B	0367
1N4435	0421
1N4435A	0421
1N4435B	0421
1N4436	OEM
1N4436F	OEM
1N4436S	OEM
1N4436T	OEM
1N4437	OEM
1N4437F	OEM
1N4437S	OEM
1N4437T	OEM
1N4438	OEM
1N4438F	OEM
1N4438S	OEM
1N4438T	OEM
1N4440	OEM
1N4441	OEM
1N4444	0133
1N4445	OEM
1N4445A	OEM
1N4446	0124
1N4447	0124
1N4448	0124
1N4449	0124
1N4450	0133
1N4451	0133
1N4452	OEM
1N4453	0133
1N4453A	OEM
1N4454	0124
1N4455	0133
1N4456	OEM
1N4457	OEM
1N4458	0206
1N4458R	1008
1N4459	0586
1N4459R	0948
1N4460	0185
1N4461	0205
1N4462	0480
1N4463	0504
1N4464	0684
1N4465	0225
1N4466	0230
1N4467	0234
1N4468	0237
1N4469	0247
1N4470	0251
1N4471	0256
1N4472	0262
1N4472B	OEM
1N4473	0269
1N4474	0274
1N4475	0294
1N4476	0309
1N4477	0318
1N4478	0320
1N4479	0326
1N4480	0337
1N4481	0347
1N4482	0027
1N4483	0266
1N4484	0386
1N4485	0405
1N4486	0426
1N4487	0444
1N4488	0238
1N4489	1180
1N4490	1190
1N4491	1207
1N4492	1218
1N4493	0647
1N4494	1255
1N4495	1279
1N4496	0605
1N4497	OEM
1N4498	OEM
1N4499	0636
1N4500	0965
1N4501	0062
1N4502	0143
1N4503	0318
1N4504	0605
1N4505	OEM
1N4506	0097
1N4507	0109
1N4508	0122
1N4509	0131
1N4510	0145
1N4510R	OEM
1N4511	OEM
1N4511R	OEM
1N4512	1335
1N4513	OEM
1N4514	0817
1N4517	0242
1N4523	0143
1N4524	OEM
1N4525	1124
1N4526	0805
1N4527	0319
1N4528	1132
1N4528B	OEM
1N4529	0044
1N4529R	OEM
1N4530	OEM
1N4530R	OEM
1N4531	0133
1N4532	0133
1N4533	0133
1N4534	0133
1N4535	0292
1N4536	0133
1N4537	OEM
1N4538	OEM
1N4539	OEM
1N4540	OEM
1N4541	0795
1N4542	0795
1N4543	OEM
1N4544	OEM
1N4545	0373
1N4546	OEM
1N4547	0133
1N4548	0133
1N4549	0815
1N4549A	0815
1N4549B	0815
1N4549R	2721
1N4549RA	2721
1N4549RB	2721
1N4550	0827
1N4550A	0827
1N4550B	0827
1N4550R	2724
1N4550RA	2724
1N4550RB	2724
1N4551	0846
1N4551A	0846
1N4551B	0846
1N4551R	2726
1N4551RA	2726
1N4551RB	2726
1N4552	0868
1N4552A	0868
1N4552B	0868
1N4552R	2729
1N4552RA	2729
1N4552RB	2729
1N4553	0885
1N4553A	0885
1N4553B	0885
1N4553R	2732
1N4553RA	2732
1N4553RB	2732
1N4554	0914
1N4554A	0914
1N4554B	0914
1N4554R	2738
1N4554RA	2738
1N4554RB	2738

If replacement code is OEM, contact original manufacturer for replacement.

Original Device Types

DEVICE TYPE	REPL CODE	DEVICE TYPE	REPL CODE	DEVICE TYPE	REPL CODE	DEVICE TYPE	REPL CODE	DEVICE TYPE	REPL CODE	DEVICE TYPE	REPL CODE	DEVICE TYPE	REPL CODE
1N4555	0926	1N4574A	0636	1N4616	OEM	1N4690	0253	1N4734	0157	1N4758RL	OEM	1N4789D	OEM
1N4555A	0926	1N4574A1	OEM	1N4616RL	OEM	1N4690RL	OEM	1N4734A	0157	1N4759	0783	1N4790	OEM
1N4555AR	2740	1N4574B	OEM	1N4617	1276	1N4691	0471	1N4734ARL	OEM	1N4759A	0783	1N4790A	OEM
1N4555B	0926	1N4575	0636	1N4617RL	OEM	1N4691RL	OEM	1N4734RL	OEM	1N4759ARL	OEM	1N4790B	OEM
1N4555BR	2740	1N4575-1	OEM	1N4618	0760	1N4692	0062	1N4735	0636	1N4759RL	OEM	1N4790C	OEM
1N4555R	2740	1N4575A	0636	1N4618RL	OEM	1N4692RL	OEM	1N4735A	0636	1N4760	1268	1N4790D	OEM
1N4555RA	2740	1N4575A1	OEM	1N4619	0118	1N4693	0077	1N4735ARL	OEM	1N4760A	1268	1N4791	OEM
1N4555RB	2740	1N4575B	OEM	1N4619RL	OEM	1N4693RL	OEM	1N4735RL	OEM	1N4760AE	1268	1N4791A	OEM
1N4556	0944	1N4576	0636	1N4620	0299	1N4694	0165	1N4736	0025	1N4760AG	1268	1N4791B	OEM
1N4556A	0944	1N4576-1	OEM	1N4620RL	OEM	1N4694RL	OEM	1N4736A	0025	1N4760ARL	OEM	1N4791C	OEM
1N4556B	0944	1N4576A	0636	1N4621	0376	1N4695	0322	1N4736ARL	OEM	1N4760RL	OEM	1N4791D	OEM
1N4556R	2744	1N4576A1	OEM	1N4621RL	OEM	1N4695RL	OEM	1N4736RL	OEM	1N4761	1189	1N4792	0628
1N4556RA	2744	1N4576B	OEM	1N4622	0036	1N4696	0057	1N4737	0649	1N4761A	1189	1N4792A	0628
1N4556RB	2744	1N4577	0636	1N4622RL	OEM	1N4696RL	OEM	1N4737A	0649	1N4761ARL	OEM	1N4792B	OEM
1N4557	OEM	1N4577-1	OEM	1N4623	0275	1N4697	0064	1N4737ARL	OEM	1N4761RL	OEM	1N4792C	OEM
1N4557A	OEM	1N4577A	0636	1N4623RL	OEM	1N4697RL	OEM	1N4737RL	OEM	1N4762	0331	1N4792D	OEM
1N4557B	OEM	1N4577A1	OEM	1N4624	0140	1N4698	0181	1N4738	0244	1N4762A	0331	1N4793	OEM
1N4557R	OEM	1N4577B	OEM	1N4624RL	OEM	1N4698RL	OEM	1N4738A	0244	1N4762ARL	OEM	1N4793A	OEM
1N4557RA	OEM	1N4578	0636	1N4625	0041	1N4699	0052	1N4738ARL	OEM	1N4762RL	OEM	1N4793B	OEM
1N4557RB	OEM	1N4578-1	OEM	1N4625-1	OEM	1N4699RL	OEM	1N4738RL	OEM	1N4763	1311	1N4793C	OEM
1N4558	OEM	1N4578A	0636	1N4625RL	OEM	1N4700	0053	1N4739	0012	1N4763A	1311	1N4793D	OEM
1N4558A	OEM	1N4578A1	OEM	1N4626	0253	1N4700RL	OEM	1N4739A	0012	1N4763AE	1311	1N4794	0030
1N4558B	OEM	1N4578B	OEM	1N4626-1	OEM	1N4701	0879	1N4739ARL	OEM	1N4763AG	1311	1N4794A	0030
1N4558R	OEM	1N4579	0636	1N4626RL	OEM	1N4701RL	OEM	1N4739RL	OEM	1N4763ARL	OEM	1N4794B	OEM
1N4558RA	OEM	1N4579-1	OEM	1N4627	0471	1N4702	0686	1N4740	0170	1N4763RL	OEM	1N4794C	OEM
1N4558RB	OEM	1N4579A	0636	1N4627-1	OEM	1N4702RL	OEM	1N4740A	0170	1N4764	0098	1N4794D	OEM
1N4559	OEM	1N4579A1	OEM	1N4627RL	OEM	1N4703	0445	1N4740ARL	OEM	1N4764A	0098	1N4795	OEM
1N4559A	OEM	1N4579B	OEM	1N4628	0025	1N4703A	OEM	1N4740RL	OEM	1N4764ARL	OEM	1N4795A	OEM
1N4559B	OEM	1N4580	0636	1N4629	0649	1N4703B	OEM	1N4741	0317	1N4764RL	OEM	1N4795B	OEM
1N4559R	OEM	1N4580-1	OEM	1N4630	0244	1N4703C	OEM	1N4741A	0317	1N4765	0057	1N4795C	OEM
1N4559RA	OEM	1N4580A	0636	1N4631	0012	1N4703D	OEM	1N4741ARL	OEM	1N4765A	0057	1N4795D	OEM
1N4559RB	OEM	1N4580A1	OEM	1N4632	0170	1N4703RL	OEM	1N4741RL	OEM	1N4765B	0057	1N4796	OEM
1N4560	OEM	1N4580B	OEM	1N4632A	OEM	1N4704	0210	1N4742	0137	1N4766	0012	1N4796A	OEM
1N4560A	OEM	1N4581	0636	1N4633	0317	1N4704RL	OEM	1N4742A	0137	1N4766A	0012	1N4796B	OEM
1N4560B	OEM	1N4581-1	OEM	1N4633A	OEM	1N4705	0375	1N4742ARL	OEM	1N4766B	0057	1N4796C	OEM
1N4560R	OEM	1N4581A	0636	1N4634	0137	1N4705RL	OEM	1N4742RL	OEM	1N4767	0012	1N4796D	OEM
1N4560RA	OEM	1N4581A1	OEM	1N4634A	OEM	1N4706	0671	1N4743	0365	1N4767A	0012	1N4797	OEM
1N4560RB	OEM	1N4581B	OEM	1N4635	0365	1N4706RL	OEM	1N4743A	0365	1N4767B	0057	1N4797A	OEM
1N4561	OEM	1N4582	0636	1N4635A	OEM	1N4707	0700	1N4743ARL	OEM	1N4768	0012	1N4797B	OEM
1N4561A	OEM	1N4582-1	OEM	1N4636	0002	1N4707RL	OEM	1N4743RL	OEM	1N4768A	0012	1N4797C	OEM
1N4561B	OEM	1N4582A	0636	1N4636A	OEM	1N4708	0705	1N4744	0002	1N4768B	0057	1N4797D	OEM
1N4561R	OEM	1N4582A1	OEM	1N4637	0420	1N4708RL	OEM	1N4744A	0002	1N4769	0012	1N4798	OEM
1N4561RA	OEM	1N4582B	OEM	1N4637A	OEM	1N4709	0494	1N4744ARL	OEM	1N4769A	0012	1N4798A	OEM
1N4561RB	OEM	1N4583	0794	1N4638	0495	1N4709RL	OEM	1N4744RL	OEM	1N4769B	0057	1N4798B	OEM
1N4562	OEM	1N4583-1	OEM	1N4638A	OEM	1N4710	0714	1N4745	0420	1N4770	0057	1N4798C	OEM
1N4562A	OEM	1N4583A	0794	1N4639	0531	1N4710RL	OEM	1N4745(A)	0420	1N4770A	0057	1N4798D	OEM
1N4562B	OEM	1N4583A1	OEM	1N4639A	OEM	1N4711	0455	1N4745A	0420	1N4770B	0057	1N4799	OEM
1N4562R	OEM	1N4583B	0636	1N4640	0565	1N4711RL	OEM	1N4745ARL	OEM	1N4771	0057	1N4799A	OEM
1N4562RA	OEM	1N4584	0636	1N4641	0402	1N4712	0257	1N4745RL	OEM	1N4771A	0057	1N4799B	OEM
1N4562RB	OEM	1N4584-1	OEM	1N4642	0441	1N4712RL	OEM	1N4746	0495	1N4771B	0057	1N4799C	OEM
1N4563	OEM	1N4584A	0636	1N4643	0726	1N4713	0195	1N4746A	0495	1N4772	0057	1N4799D	OEM
1N4563A	OEM	1N4584A1	OEM	1N4644	0039	1N4713RL	OEM	1N4746ARL	OEM	1N4772A	0057	1N4800	OEM
1N4563B	OEM	1N4584B	OEM	1N4645	0820	1N4714	0166	1N4746RL	OEM	1N4772B	0057	1N4800A	OEM
1N4563R	OEM	1N4585	0817	1N4646	0350	1N4714RL	OEM	1N4747	0531	1N4773	0057	1N4800B	OEM
1N4563RA	OEM	1N4585GP	0072	1N4647	0931	1N4715	0010	1N4747A	0531	1N4773A	0057	1N4800C	OEM
1N4563RB	OEM	1N4586	0817	1N4648	0999	1N4715RL	OEM	1N4747ARL	OEM	1N4773B	0057	1N4800D	OEM
1N4564	OEM	1N4586GP	0071	1N4649	0292	1N4716	0032	1N4747RL	OEM	1N4774	0057	1N4801	0554
1N4564A	OEM	1N4587	0599	1N4650	0188	1N4716RL	OEM	1N4748	0565	1N4774A	0057	1N4801A	0554
1N4564B	OEM	1N4587R	1347	1N4651	0456	1N4717	0054	1N4748A	0565	1N4774B	0057	1N4801B	OEM
1N4564R	OEM	1N4588	0599	1N4652	0533	1N4717RL	OEM	1N4748ARL	OEM	1N4775	1083	1N4801C	OEM
1N4564RA	OEM	1N4588R	1347	1N4653	0451	1N4718	0920	1N4748RL	OEM	1N4775A	1083	1N4801D	OEM
1N4564RB	OEM	1N4589	0599	1N4654	0162	1N4719	0110	1N4749	0402	1N4775B	OEM	1N4802	OEM
1N4565	0636	1N4589R	1347	1N4655	0157	1N4719E	OEM	1N4749(A)	0402	1N4776	1083	1N4802A	OEM
1N4565-1	OEM	1N4590	0599	1N4656	0636	1N4719R	0087	1N4749A	0402	1N4776A	1083	1N4802B	OEM
1N4565A	0636	1N4590R	1905	1N4657	0025	1N4720	0953	1N4749ARL	OEM	1N4776B	OEM	1N4802C	OEM
1N4565A1	OEM	1N4591	1986	1N4658	0649	1N4720E	OEM	1N4749RL	OEM	1N4777	1083	1N4802D	OEM
1N4565B	OEM	1N4591R	1905	1N4659	0244	1N4720R	0087	1N4750	0441	1N4777A	1083	1N4803	OEM
1N4566	0636	1N4592	1986	1N4660	0012	1N4721	0242	1N4750A	0441	1N4777B	OEM	1N4803A	OEM
1N4566-1	OEM	1N4592R	1905	1N4661	0170	1N4721E	OEM	1N4750ARL	OEM	1N4778	1083	1N4803B	OEM
1N4566A	0636	1N4593	0657	1N4662	0317	1N4721R	0242	1N4750RL	OEM	1N4778A	1083	1N4803C	OEM
1N4566A1	OEM	1N4593R	0193	1N4663	0137	1N4722	0540	1N4751	0726	1N4778B	OEM	1N4803D	OEM
1N4566B	OEM	1N4594	0657	1N4664	0365	1N4722E	OEM	1N4751A	0726	1N4779	1083	1N4804	OEM
1N4567	0636	1N4594R	0193	1N4665	0002	1N4722R	0087	1N4751ARL	OEM	1N4779A	1083	1N4804A	OEM
1N4567-1	OEM	1N4595	3309	1N4666	0420	1N4723	0965	1N4751RL	OEM	1N4779B	OEM	1N4804B	OEM
1N4567A	0636	1N4595R	0202	1N4667	0495	1N4723E	OEM	1N4752	0039	1N4780	1083	1N4804C	OEM
1N4567A1	OEM	1N4596	3309	1N4668	0531	1N4723R	0087	1N4752A	0039	1N4780A	1083	1N4804D	OEM
1N4567B	OEM	1N4596R	0202	1N4669	0565	1N4724	0817	1N4752ARL	OEM	1N4780B	OEM	1N4805	OEM
1N4568	0636	1N4597	OEM	1N4670	0402	1N4724E	OEM	1N4752RL	OEM	1N4781	1083	1N4805A	OEM
1N4568-1	OEM	1N4598	OEM	1N4671	0441	1N4724R	0087	1N4753	0820	1N4781A	1083	1N4805B	OEM
1N4568A	0636	1N4599	OEM	1N4672	0726	1N4725	0922	1N4753A	0820	1N4781B	OEM	1N4805C	OEM
1N4568A1	OEM	1N4600	OEM	1N4673	0039	1N4725E	OEM	1N4753ARL	OEM	1N4782	1083	1N4805D	OEM
1N4568B	OEM	1N4601	OEM	1N4674	0820	1N4725R	0087	1N4753RL	OEM	1N4782A	1083	1N4806	OEM
1N4569	0636	1N4602	OEM	1N4675	0350	1N4726	0133	1N4754	0350	1N4782B	OEM	1N4806A	OEM
1N4569-1	OEM	1N4603	OEM	1N4676	0931	1N4727	0133	1N4754A	0350	1N4783	1083	1N4806B	OEM
1N4569A	0636	1N4604	OEM	1N4677	0999	1N4728	0292	1N4754ARL	OEM	1N4783A	1083	1N4806C	OEM
1N4569A1	OEM	1N4605	OEM	1N4678	OEM	1N4728A	0292	1N4754RL	OEM	1N4783B	OEM	1N4806D	OEM
1N4569B	OEM	1N4606	0133	1N4678RL	OEM	1N4728AG	0292	1N4755	0931	1N4784	1083	1N4807	OEM
1N4570	0636	1N4606R	0124	1N4679	OEM	1N4728ARL	OEM	1N4755A	0931	1N4784A	1083	1N4807A	OEM
1N4570-1	OEM	1N4607	0133	1N4679RL	OEM	1N4728RL	OEM	1N4755AE	0931	1N4784B	OEM	1N4807B	OEM
1N4570A	0636	1N4608	0133	1N4680	0292	1N4729	0188	1N4755AG	0931	1N4785	OEM	1N4807C	OEM
1N4570A1	OEM	1N4609	0628	1N4680RL	OEM	1N4729A	0188	1N4755ARL	OEM	1N4786	0554	1N4807D	OEM
1N4570B	OEM	1N4609A	0628	1N4681	1276	1N4729ARL	OEM	1N4755RL	OEM	1N4786A	0554	1N4808	OEM
1N4571	0636	1N4610	0133	1N4681RL	OEM	1N4730	0456	1N4756	0999	1N4786B	OEM	1N4808A	OEM
1N4571-1	OEM	1N4611	0025	1N4682	0760	1N4730A	0456	1N4756A	0999	1N4786C	OEM	1N4808B	OEM
1N4571A	0636	1N4611A	OEM	1N4682RL	OEM	1N4730AG	0456	1N4756AE	0999	1N4786D	OEM	1N4808C	OEM
1N4571A1	OEM	1N4611B	OEM	1N4683	0118	1N4730ARL	OEM	1N4756AG	0999	1N4787	OEM	1N4808D	OEM
1N4571B	OEM	1N4611C	OEM	1N4683RL	OEM	1N4730RL	OEM	1N4756ARL	OEM	1N4787A	OEM	1N4809	OEM
1N4572	0636	1N4612	0025	1N4684	0299	1N4731	0760	1N4756RL	OEM	1N4787B	OEM	1N4809A	OEM
1N4572-1	OEM	1N4612A	0025	1N4684RL	OEM	1N4731A	0533	1N4757	0502	1N4787C	OEM	1N4809B	OEM
1N4572A	0636	1N4612B	0025	1N4685	0376	1N4731ARL	OEM	1N4757A	0502	1N4787D	OEM	1N4809C	OEM
1N4572A1	OEM	1N4612C	0025	1N4685RL	OEM	1N4731RL	OEM	1N4757AE	0502	1N4788	0005	1N4809D	OEM
1N4572B	OEM	1N4613	0025	1N4686	0036	1N4732	0451	1N4757AG	0502	1N4788A	0005	1N4810	OEM
1N4573	0636	1N4613A	OEM	1N4686RL	OEM	1N4732A	0451	1N4757ARL	OEM	1N4788B	OEM	1N4810A	OEM
1N4573-1	OEM	1N4613B	OEM	1N4687	0275	1N4732ARL	OEM	1N4757RL	OEM	1N4788C	OEM	1N4810B	OEM
1N4573A	0636	1N4613C	OEM	1N4687RL	OEM	1N4732RL	OEM	1N4758	0869	1N4788D	OEM	1N4810C	OEM
1N4573A1	OEM	1N4614	OEM	1N4688	0140	1N4733	0162	1N4758A	0869	1N4789	0720	1N4810D	OEM
1N4573B	OEM	1N4614RL	OEM	1N4688RL	OEM	1N4733A	0442	1N4758AG	0869	1N4789A	0720	1N4811	OEM
1N4574	0636	1N4615	OEM	1N4689	0041	1N4733ARL	OEM	1N4758ARL	OEM	1N4789B	OEM	1N4811A	OEM
1N4574-1	OEM	1N4615RL	OEM	1N4689RL	OEM	1N4733RL	OEM	1N4758E	0869	1N4789C	OEM	1N4811B	OEM

If replacement code is OEM, contact original manufacturer for replacement.

DEVICE TYPE	REPL CODE
1N4811C	OEM
1N4811D	OEM
1N4812	OEM
1N4812A	OEM
1N4812B	OEM
1N4812C	OEM
1N4812D	OEM
1N4813	OEM
1N4813A	OEM
1N4813B	0684
1N4813C	OEM
1N4813D	OEM
1N4814	OEM
1N4814A	OEM
1N4814B	OEM
1N4814C	OEM
1N4814D	OEM
1N4815	OEM
1N4815A	OEM
1N4815B	OEM
1N4815C	OEM
1N4815D	OEM
1N4816	0110
1N4817	0071
1N4818	0071
1N4819	1747
1N4820	0540
1N4821	1771
1N4822	0965
1N4823	0015
1N4824	0015
1N4825	0015
1N4826	0102
1N4827	0133
1N4828	0133
1N4829	0133
1N4830	0133
1N4831	0684
1N4831A	0684
1N4831B	0684
1N4832	0225
1N4832A	0225
1N4832B	0225
1N4833	0230
1N4833A	0230
1N4833B	0230
1N4834	0234
1N4834A	0234
1N4834B	0234
1N4835	0237
1N4835A	0237
1N4835B	0237
1N4836	0247
1N4836A	0247
1N4836B	0247
1N4837	0251
1N4837A	0251
1N4837B	0251
1N4838	0256
1N4838A	0256
1N4838B	0256
1N4839	0262
1N4839A	0262
1N4839B	0262
1N4840	0269
1N4840A	0269
1N4840B	0269
1N4841	0274
1N4841A	0274
1N4841B	0274
1N4842	0294
1N4842A	0294
1N4842B	0294
1N4843	0309
1N4843A	0309
1N4843B	0309
1N4844	0318
1N4844A	0318
1N4844B	0318
1N4845	0320
1N4845A	0320
1N4845B	0320
1N4846	0326
1N4846A	0326
1N4846B	0326
1N4847	0337
1N4847A	0337
1N4847B	0337
1N4848	0347
1N4848A	0347
1N4848B	0347
1N4849	0027
1N4849A	0027
1N4849B	0027
1N4850	0266
1N4850A	0266
1N4850B	0266
1N4851	0386
1N4851A	0386
1N4851B	0386
1N4852	0405
1N4852A	0405
1N4852B	0405
1N4853	0426
1N4853A	0426
1N4853B	0426
1N4854	0444
1N4854A	0444
1N4854B	0444

DEVICE TYPE	REPL CODE
1N4855	0238
1N4855A	0238
1N4855B	0238
1N4856	1180
1N4856A	1180
1N4856B	1180
1N4857	1190
1N4857A	1190
1N4857B	1190
1N4858	1207
1N4858A	0186
1N4858B	1207
1N4859	0213
1N4859A	0213
1N4859B	1218
1N4860	0028
1N4860A	0028
1N4860B	0647
1N4861	0133
1N4862	0133
1N4863	0133
1N4864	0133
1N4865	OEM
1N4866	OEM
1N4867	OEM
1N4868	OEM
1N4869	OEM
1N4870	OEM
1N4871	OEM
1N4872	OEM
1N4873	OEM
1N4874	OEM
1N4875	OEM
1N4876	OEM
1N4877	OEM
1N4878	0599
1N4879	1023
1N4880	1023
1N4881	0262
1N4882	0320
1N4883	0234
1N4884	0262
1N4885	OEM
1N4886	OEM
1N4887	OEM
1N4888	OEM
1N4889	0386
1N4890	OEM
1N4890A	OEM
1N4891	OEM
1N4891A	OEM
1N4892	OEM
1N4892A	OEM
1N4893	OEM
1N4893A	OEM
1N4894	OEM
1N4894A	OEM
1N4895	OEM
1N4895A	OEM
1N4896	0053
1N4896A	0053
1N4897	0053
1N4897A	0053
1N4898	0053
1N4898A	0053
1N4899	0053
1N4899A	0053
1N4900	0053
1N4900A	0053
1N4901	0053
1N4901A	0053
1N4902	0053
1N4902A	0053
1N4903	0053
1N4903A	0053
1N4904	0053
1N4904A	0053
1N4905	0053
1N4905A	0053
1N4906	0053
1N4906A	0053
1N4907	0053
1N4907A	0053
1N4908	0053
1N4908A	0053
1N4909	0053
1N4909A	0053
1N4910	0053
1N4910A	0053
1N4911	0053
1N4911A	0053
1N4912	0053
1N4912A	0053
1N4913	0053
1N4913A	0053
1N4914	0053
1N4914A	0053
1N4915	0053
1N4915A	0053
1N4916	0949
1N4916A	0949
1N4917	0949
1N4917A	0949
1N4918	0949
1N4918A	0949
1N4919	0949
1N4919A	0949
1N4920	0949
1N4920A	0949

DEVICE TYPE	REPL CODE
1N4921	0949
1N4921A	0949
1N4922	0949
1N4922A	0949
1N4923	0949
1N4923A	0949
1N4924	0949
1N4924A	0949
1N4925	0949
1N4925A	0949
1N4926	0949
1N4926A	0949
1N4927	0949
1N4927A	0949
1N4928	0949
1N4928A	0949
1N4929	0949
1N4929A	0949
1N4930	0949
1N4930A	0949
1N4931	0949
1N4931A	0949
1N4932	0949
1N4932A	0949
1N4933	0023
1N4933G	OEM
1N4933GP	0023
1N4933RL	OEM
1N4934	0023
1N4934G	OEM
1N4934GP	0023
1N4934RL	OEM
1N4935	0023
1N4935G	OEM
1N4935GP	0023
1N4935RL	OEM
1N4936	0023
1N4936G	OEM
1N4936GP	0023
1N4936RL	OEM
1N4937	0023
1N4937G	OEM
1N4937GP	0023
1N4937RL	OEM
1N4938	0133
1N4939	OEM
1N4940	OEM
1N4941	OEM
1N4942	0023
1N4942G	OEM
1N4942GP	0023
1N4943	0087
1N4944	0023
1N4944G	OEM
1N4944GP	0023
1N4945	0102
1N4946	0023
1N4946G	OEM
1N4946GP	0023
1N4947	0017
1N4947G	OEM
1N4947GP	0017
1N4948	0017
1N4948G	OEM
1N4948GP	0017
1N4949	0133
1N4950	0133
1N4951	0133
1N4952	0133
1N4953	OEM
1N4954	0205
1N4954A	OEM
1N4954B	OEM
1N4955	0480
1N4955A	OEM
1N4955B	OEM
1N4956	0504
1N4956A	OEM
1N4956B	OEM
1N4957	0684
1N4957A	OEM
1N4957B	OEM
1N4958	0225
1N4958A	OEM
1N4958B	OEM
1N4959	0230
1N4959A	1987
1N4959B	OEM
1N4960	0234
1N4960A	OEM
1N4960B	OEM
1N4960D	0234
1N4961	0237
1N4961A	OEM
1N4961B	OEM
1N4962	0247
1N4962A	OEM
1N4962B	OEM
1N4963	0251
1N4963A	OEM
1N4963B	OEM
1N4964	0256
1N4964A	OEM
1N4964B	OEM
1N4965	0262
1N4965A	OEM
1N4965B	OEM
1N4966	0269
1N4966A	OEM

DEVICE TYPE	REPL CODE
1N4966B	OEM
1N4967	0274
1N4967A	OEM
1N4967B	OEM
1N4968	0294
1N4968A	OEM
1N4968B	OEM
1N4969	0309
1N4969A	OEM
1N4969B	OEM
1N4970	0318
1N4970A	OEM
1N4970B	OEM
1N4971	0320
1N4971A	OEM
1N4971B	OEM
1N4972	0326
1N4972A	OEM
1N4972B	OEM
1N4973	0337
1N4973A	OEM
1N4973B	OEM
1N4974	0347
1N4974A	OEM
1N4974B	OEM
1N4975	0027
1N4975A	OEM
1N4975B	OEM
1N4976	0266
1N4976A	OEM
1N4976B	OEM
1N4977	0386
1N4977A	OEM
1N4977B	OEM
1N4978	0405
1N4978A	OEM
1N4978B	OEM
1N4979	0426
1N4979A	OEM
1N4979B	OEM
1N4980	0444
1N4980A	OEM
1N4980B	OEM
1N4981	0238
1N4981A	OEM
1N4981B	OEM
1N4982	1180
1N4982A	OEM
1N4982B	OEM
1N4983	1190
1N4983A	OEM
1N4983B	OEM
1N4984	1207
1N4984A	OEM
1N4984B	OEM
1N4985	1218
1N4985A	OEM
1N4985B	OEM
1N4986	0647
1N4986A	OEM
1N4986B	OEM
1N4987	1255
1N4987A	OEM
1N4987B	OEM
1N4988	1279
1N4988A	OEM
1N4988B	OEM
1N4989	0605
1N4989A	OEM
1N4989B	OEM
1N4990	OEM
1N4990A	OEM
1N4990B	OEM
1N4991	OEM
1N4991A	OEM
1N4991B	OEM
1N4991D	OEM
1N4992	OEM
1N4992A	OEM
1N4992B	OEM
1N4993	OEM
1N4993A	OEM
1N4993B	OEM
1N4994	OEM
1N4994A	OEM
1N4994B	OEM
1N4995	OEM
1N4995A	OEM
1N4995B	OEM
1N4996	OEM
1N4996A	OEM
1N4996B	OEM
1N4997	0110
1N4997R	0087
1N4998	0953
1N4998R	0087
1N4999	0242
1N4999R	0087
1N5000	0540
1N5000R	0087
1N5001	0965
1N5001R	0087
1N5002	0817
1N5002R	0087
1N5003	0922
1N5003R	0087
1N5004	0953
1N5005	0242
1N5006	0540

DEVICE TYPE	REPL CODE
1N5007	0965
1N5008	0782
1N5008A	0782
1N5009	0796
1N5009A	0796
1N5010	0806
1N5010A	0806
1N5011	0821
1N5011A	0821
1N5012	0833
1N5012A	0833
1N5013	0442
1N5013A	0442
1N5014	0876
1N5014A	0876
1N5015	0185
1N5015A	0185
1N5016	0205
1N5016A	0205
1N5017	0480
1N5017A	0480
1N5018	0504
1N5018A	0504
1N5019	0684
1N5019A	0684
1N5020	0225
1N5020A	0225
1N5021	0230
1N5021A	0230
1N5022	0234
1N5022A	0234
1N5023	0237
1N5023A	0237
1N5024	1397
1N5024A	1397
1N5025	0247
1N5025A	0247
1N5026	0251
1N5026A	0251
1N5027	1178
1N5027A	1178
1N5028	0256
1N5028A	0256
1N5029	2395
1N5029A	2395
1N5030	0262
1N5030A	0262
1N5031	0269
1N5031A	0269
1N5032	0274
1N5032A	0274
1N5033	2399
1N5033A	2399
1N5034	0294
1N5034A	0294
1N5035	0309
1N5035A	0309
1N5036	0318
1N5036A	0318
1N5037	0320
1N5037A	0320
1N5038	0326
1N5038A	0326
1N5039	0337
1N5039A	0337
1N5040	0347
1N5040A	0347
1N5041	0347
1N5041A	0347
1N5042	0027
1N5042A	0027
1N5043	0027
1N5043A	0027
1N5044	0027
1N5044A	0027
1N5045	0266
1N5045A	0266
1N5046	0386
1N5046A	0386
1N5047	0405
1N5047A	0405
1N5048	0426
1N5048A	0426
1N5049	0444
1N5049A	0444
1N5050	0238
1N5050A	0238
1N5051	1180
1N5051A	1180
1N5052	0817
1N5053	0817
1N5054	0071
1N5055	0953
1N5056	0242
1N5057	1747
1N5058	0540
1N5059	0102
1N5059GP	0609
1N5060	0102
1N5060GP	0795
1N5061	0023
1N5061GP	0015
1N5062	0071
1N5062GP	0072
1N5063	0205
1N5064	0480
1N5065	0504
1N5066	0684
1N5067	0225

DEVICE TYPE	REPL CODE
1N5068	0230
1N5069	0237
1N5070	1397
1N5071	0247
1N5072	0251
1N5073	0256
1N5074	0269
1N5075	0274
1N5076	0294
1N5077	0309
1N5078	0318
1N5079	0320
1N5080	0326
1N5081	0326
1N5082	0337
1N5083	0347
1N5084	0347
1N5085	0027
1N5086	0027
1N5087	0266
1N5088	2856
1N5089	0386
1N5090	0405
1N5091	OEM
1N5092	0426
1N5093	0444
1N5094	0444
1N5095	0238
1N5096	1190
1N5097	1207
1N5098	1218
1N5099	1881
1N5100	1255
1N5101	2102
1N5102	1279
1N5103	2225
1N5104	0605
1N5105	OEM
1N5106	OEM
1N5107	OEM
1N5108	OEM
1N5109	OEM
1N5110	OEM
1N5111	OEM
1N5112	OEM
1N5113	OEM
1N5114	OEM
1N5115	OEM
1N5116	OEM
1N5117	OEM
1N5118	1397
1N5119	0326
1N5120	0347
1N5121	0027
1N5122	2856
1N5123	0405
1N5124	0444
1N5125	0238
1N5126	1881
1N5127	2102
1N5128	2225
1N5129	6880
1N5130	OEM
1N5131	6881
1N5132	6882
1N5133	6883
1N5134	6884
1N5135	OEM
1N5136	OEM
1N5136A	OEM
1N5137	OEM
1N5137A	OEM
1N5138	OEM
1N5138A	OEM
1N5139	0554
1N5139A	0554
1N5140	0005
1N5140A	0005
1N5141	0720
1N5141A	0720
1N5142	0720
1N5142A	0720
1N5143	0628
1N5143A	0628
1N5144	0628
1N5144A	0628
1N5145	0628
1N5145A	0628
1N5146	0030
1N5146A	0030
1N5147	0030
1N5147A	0030
1N5148	0030
1N5148A	0030
1N5149	OEM
1N5150	OEM
1N5150A	OEM
1N5151	OEM
1N5152	OEM
1N5152A	OEM
1N5153	OEM
1N5153A	OEM
1N5154	OEM
1N5154B	0441
1N5155	OEM
1N5155A	OEM
1N5156	OEM
1N5157	OEM
1N5158	OEM

DEVICE TYPE	REPL CODE
1N5159	OEM
1N5160	OEM
1N5161	OEM
1N5162	OEM
1N5163	OEM
1N5164	OEM
1N5165	OEM
1N5166	OEM
1N5167	OEM
1N5168	OEM
1N5169	OEM
1N5170	0087
1N5171	0110
1N5172	0087
1N5173	1747
1N5174	0540
1N5175	1771
1N5176	0965
1N5177	0087
1N5178	0087
1N5179	0133
1N5180	1282
1N5181	OEM
1N5182	OEM
1N5183	OEM
1N5184	OEM
1N5185	0965
1N5185GP	0965
1N5186	0965
1N5186GP	0965
1N5187	0965
1N5187GP	0965
1N5188	0102
1N5188A	0965
1N5188GP	0965
1N5189	0102
1N5189GP	0965
1N5190	0102
1N5190GP	0087
1N5194	0133
1N5195	0133
1N5196	0133
1N5197	0110
1N5198	0229
1N5199	0242
1N5200	0540
1N5201	0965
1N5206	0965
1N5207	1293
1N5208	0133
1N5209	0133
1N5210	0133
1N5211	0015
1N5212	0015
1N5213	0015
1N5214	0071
1N5215	0015
1N5216	0015
1N5217	0015
1N5218	0071
1N5219	0133
1N5220	0133
1N5221	1276
1N5221A	1276
1N5221ARL	OEM
1N5221B	1276
1N5221BRL	OEM
1N5222	2874
1N5222A	2874
1N5222ARL	OEM
1N5222B	1276
1N5222BRL	OEM
1N5222RL	OEM
1N5223	0760
1N5223A	0760
1N5223ARL	OEM
1N5223B	0760
1N5223BRL	OEM
1N5223RL	OEM
1N5224	0677
1N5224A	0677
1N5224ARL	OEM
1N5224B	0677
1N5224BRL	OEM
1N5224RL	OEM
1N5225	0118
1N5225A	0118
1N5225ARL	OEM
1N5225B	0118
1N5225BRL	OEM
1N5225RL	OEM
1N5226	0299
1N5226A	0299
1N5226ARL	OEM
1N5226B	0299
1N5226BRL	OEM
1N5226C	0299
1N5226D	OEM
1N5226RL	OEM
1N5227	0376
1N5227A	0376
1N5227ARL	OEM
1N5227B	0188
1N5227BRL	OEM
1N5227C	OEM
1N5227D	OEM
1N5227RL	OEM
1N5228	0036
1N5228A	0036

If replacement code is OEM, contact original manufacturer for replacement.

DEVICE TYPE	REPL CODE
1N5228ARL	OEM
1N5228B	0036
1N5228BRL	OEM
1N5228C	OEM
1N5228D	OEM
1N5228RL	OEM
1N5229	0275
1N5229A	0275
1N5229ARL	OEM
1N5229B	0533
1N5229BRL	OEM
1N5229C	OEM
1N5229D	OEM
1N5229RL	OEM
1N5230	0140
1N5230A	0140
1N5230ARL	OEM
1N5230B	0140
1N5230BRL	OEM
1N5230C	OEM
1N5230D	OEM
1N5230RL	OEM
1N5231	0041
1N5231A	0041
1N5231ARL	OEM
1N5231B	0041
1N5231BRL	OEM
1N5231C	OEM
1N5231D	OEM
1N5232	0253
1N5232A	0253
1N5232ARL	OEM
1N5232B	0253
1N5232BRL	OEM
1N5232C	OEM
1N5232D	OEM
1N5232RL	OEM
1N5233	0091
1N5233A	0091
1N5233ARL	OEM
1N5233B	0091
1N5233BRL	OEM
1N5233C	OEM
1N5233D	OEM
1N5233RL	OEM
1N5234	0471
1N5234A	0471
1N5234ARL	OEM
1N5234B	0471
1N5234BRL	OEM
1N5234C	OEM
1N5234D	OEM
1N5234RL	OEM
1N5235	0062
1N5235A	0062
1N5235ARL	OEM
1N5235B	0062
1N5235BRL	OEM
1N5235C	OEM
1N5235D	OEM
1N5235RL	OEM
1N5236	0077
1N5236A	0077
1N5236ARL	OEM
1N5236B	0077
1N5236BRL	OEM
1N5236C	OEM
1N5236D	OEM
1N5236RL	OEM
1N5237	0165
1N5237A	0165
1N5237ARL	OEM
1N5237B	0165
1N5237BRL	OEM
1N5237C	OEM
1N5237D	OEM
1N5237RL	OEM
1N5238	0322
1N5238-BRL	1083
1N5238A	0322
1N5238ARL	OEM
1N5238B	0322
1N5238BRL	OEM
1N5238C	OEM
1N5238D	OEM
1N5238RL	OEM
1N5239	0057
1N5239A	0057
1N5239ARL	OEM
1N5239B	0057
1N5239BRL	OEM
1N5239C	OEM
1N5239D	OEM
1N5239RL	OEM
1N5240	0064
1N5240A	0064
1N5240ARL	OEM
1N5240B	0064
1N5240BRL	OEM
1N5240C	OEM
1N5240D	OEM
1N5240RL	OEM
1N5241	0181
1N5241A	0181
1N5241ARL	OEM
1N5241B	0181
1N5241BRL	OEM
1N5241C	OEM
1N5241D	OEM
1N5241RL	OEM
1N5242	0052
1N5242A	0052
1N5242ARL	OEM
1N5242B	0052
1N5242BRL	OEM
1N5242C	OEM
1N5242D	OEM
1N5242RL	OEM
1N5243	0053
1N5243-BRL	0053
1N5243A	0053
1N5243ARL	OEM
1N5243B	0053
1N5243BRL	OEM
1N5243C	OEM
1N5243D	OEM
1N5243RL	OEM
1N5244	0879
1N5244A	0879
1N5244ARL	OEM
1N5244B	0879
1N5244BRL	OEM
1N5244C	OEM
1N5244D	OEM
1N5244RL	OEM
1N5245	0686
1N5245A	0686
1N5245ARL	OEM
1N5245B	0686
1N5245BRL	OEM
1N5245C	OEM
1N5245D	OEM
1N5245RL	OEM
1N5246	0445
1N5246A	0445
1N5246ARL	OEM
1N5246B	0445
1N5246BRL	OEM
1N5246C	OEM
1N5246D	OEM
1N5246RL	OEM
1N5247	0210
1N5247A	0210
1N5247ARL	OEM
1N5247B	0210
1N5247BRL	OEM
1N5247C	OEM
1N5247D	OEM
1N5247RL	OEM
1N5248	0375
1N5248A	0375
1N5248ARL	OEM
1N5248B	0375
1N5248BRL	OEM
1N5248C	OEM
1N5248D	OEM
1N5248RL	OEM
1N5249	0671
1N5249A	0671
1N5249ARL	OEM
1N5249B	0671
1N5249BRL	OEM
1N5249C	OEM
1N5249D	OEM
1N5249RL	OEM
1N5250	0700
1N5250A	0700
1N5250ARL	OEM
1N5250B	0700
1N5250BRL	OEM
1N5250C	OEM
1N5250D	OEM
1N5250RL	OEM
1N5251	0705
1N5251A	0705
1N5251ARL	OEM
1N5251B	0705
1N5251BRL	OEM
1N5251C	OEM
1N5251D	OEM
1N5251RL	OEM
1N5252	0494
1N5252A	0494
1N5252ARL	OEM
1N5252B	0494
1N5252BRL	OEM
1N5252C	OEM
1N5252D	OEM
1N5252RL	OEM
1N5253	0714
1N5253A	0714
1N5253ARL	OEM
1N5253B	0714
1N5253BRL	OEM
1N5253C	OEM
1N5253D	OEM
1N5253RL	OEM
1N5254	0455
1N5254A	0455
1N5254ARL	OEM
1N5254B	0455
1N5254BRL	OEM
1N5254C	OEM
1N5254D	OEM
1N5254RL	OEM
1N5255	0257
1N5255A	0257
1N5255ARL	OEM
1N5255B	0257
1N5255BRL	OEM
1N5255C	OEM
1N5255D	OEM
1N5255RL	OEM
1N5256	0195
1N5256A	0195
1N5256ARL	OEM
1N5256B	0195
1N5256BRL	OEM
1N5256C	OEM
1N5256D	OEM
1N5256RL	OEM
1N5257	0166
1N5257A	0166
1N5257ARL	OEM
1N5257B	0166
1N5257BRL	OEM
1N5257C	OEM
1N5257D	OEM
1N5257RL	OEM
1N5258	0010
1N5258A	0010
1N5258ARL	OEM
1N5258B	0010
1N5258BRL	OEM
1N5258C	OEM
1N5258D	OEM
1N5258RL	OEM
1N5259	0032
1N5259A	0032
1N5259ARL	OEM
1N5259B	0032
1N5259BRL	OEM
1N5259C	OEM
1N5259D	OEM
1N5259RL	OEM
1N5260	0054
1N5260A	0054
1N5260ARL	OEM
1N5260B	0054
1N5260BRL	OEM
1N5260C	OEM
1N5260D	OEM
1N5260RL	OEM
1N5261	0068
1N5261A	0068
1N5261ARL	OEM
1N5261B	0068
1N5261BRL	OEM
1N5261C	OEM
1N5261D	OEM
1N5261RL	OEM
1N5262	0092
1N5262A	0092
1N5262ARL	OEM
1N5262B	0092
1N5262BRL	OEM
1N5262C	OEM
1N5262D	OEM
1N5262RL	OEM
1N5263	0125
1N5263A	0125
1N5263ARL	OEM
1N5263B	0125
1N5263BRL	OEM
1N5263C	OEM
1N5263D	OEM
1N5263RL	OEM
1N5264	2317
1N5264A	2317
1N5264ARL	OEM
1N5264B	2317
1N5264BRL	OEM
1N5264C	OEM
1N5264D	OEM
1N5264RL	OEM
1N5265	0152
1N5265A	0152
1N5265ARL	OEM
1N5265B	0152
1N5265BRL	OEM
1N5265C	OEM
1N5265D	OEM
1N5265RL	OEM
1N5266	0173
1N5266A	0173
1N5266ARL	OEM
1N5266B	0173
1N5266BRL	OEM
1N5266C	OEM
1N5266D	OEM
1N5266RL	OEM
1N5267	0094
1N5267A	0094
1N5267ARL	OEM
1N5267B	0094
1N5267BRL	OEM
1N5267RL	OEM
1N5268	0049
1N5268A	0049
1N5268ARL	OEM
1N5268B	0049
1N5268BRL	OEM
1N5268RL	OEM
1N5269	0104
1N5269A	0104
1N5269ARL	OEM
1N5269B	0104
1N5269BRL	OEM
1N5269RL	OEM
1N5270	0156
1N5270A	0156
1N5270ARL	OEM
1N5270B	0156
1N5270BRL	OEM
1N5270RL	OEM
1N5271	0189
1N5271A	0189
1N5271ARL	OEM
1N5271B	0189
1N5271BRL	OEM
1N5271RL	OEM
1N5272	0099
1N5272A	0099
1N5272ARL	OEM
1N5272B	0099
1N5272BRL	OEM
1N5272RL	OEM
1N5273	0089
1N5273A	0089
1N5273ARL	OEM
1N5273B	0089
1N5273BE	0089
1N5273BRL	OEM
1N5273RL	OEM
1N5274	0287
1N5274A	0287
1N5274ARL	OEM
1N5274B	0287
1N5274BRL	OEM
1N5274RL	OEM
1N5275	0252
1N5275A	0252
1N5275ARL	OEM
1N5275B	0252
1N5275BRL	OEM
1N5275RL	OEM
1N5276	0340
1N5276A	0340
1N5276ARL	OEM
1N5276B	0028
1N5276BRL	OEM
1N5276RL	OEM
1N5277	0370
1N5277A	0370
1N5277ARL	OEM
1N5277B	0255
1N5277BRL	OEM
1N5277RL	OEM
1N5278	0394
1N5278A	0394
1N5278ARL	OEM
1N5278B	0877
1N5278BRL	OEM
1N5278RL	OEM
1N5279	0425
1N5279A	0425
1N5279ARL	OEM
1N5279B	0367
1N5279BRL	OEM
1N5279RL	OEM
1N5280	0453
1N5280A	0453
1N5280ARL	OEM
1N5280B	2858
1N5280BRL	OEM
1N5280RL	OEM
1N5281	1473
1N5281A	1473
1N5281ARL	OEM
1N5281B	0421
1N5281BRL	OEM
1N5281RL	OEM
1N5282	0124
1N5283	OEM
1N5284	OEM
1N5285	OEM
1N5286	OEM
1N5287	OEM
1N5288	OEM
1N5289	OEM
1N5290	OEM
1N5291	OEM
1N5292	OEM
1N5293	OEM
1N5294	OEM
1N5295	OEM
1N5296	OEM
1N5297	OEM
1N5298	OEM
1N5299	OEM
1N5300	OEM
1N5301	OEM
1N5302	OEM
1N5303	OEM
1N5304	OEM
1N5305	OEM
1N5306	OEM
1N5307	OEM
1N5308	OEM
1N5309	OEM
1N5310	OEM
1N5311	OEM
1N5312	OEM
1N5313	OEM
1N5314	OEM
1N5315	0133
1N5316	0133
1N5317	0133
1N5318	0133
1N5319	0133
1N5320	0133
1N5321	OEM
1N5322	OEM
1N5323	OEM
1N5324	0471
1N5324A	0471
1N5324B	0253
1N5326	1282
1N5329	OEM
1N5330	0442
1N5331	OEM
1N5331R	OEM
1N5332	OEM
1N5332R	OEM
1N5333	0782
1N5333A	0782
1N5333ARL	OEM
1N5333B	0782
1N5333BE	0782
1N5333BRL	OEM
1N5334	0796
1N5334A	0796
1N5334ARL	OEM
1N5334B	0796
1N5334BE	0796
1N5334BRL	OEM
1N5335	0806
1N5335A	0806
1N5335ARL	OEM
1N5335B	0806
1N5335BE	0806
1N5335BRL	OEM
1N5336	0821
1N5336A	0821
1N5336ARL	OEM
1N5336B	0821
1N5336BE	0821
1N5336BRL	OEM
1N5337	0833
1N5337A	0833
1N5337ARL	OEM
1N5337B	0833
1N5337BE	0833
1N5337BRL	OEM
1N5338	0442
1N5338A	0442
1N5338ARL	OEM
1N5338B	0442
1N5338BE	0442
1N5338BRL	OEM
1N5339	0876
1N5339A	0876
1N5339ARL	OEM
1N5339B	0876
1N5339BE	0876
1N5339BRL	OEM
1N5340	3127
1N5340A	3127
1N5340ARL	OEM
1N5340B	3127
1N5340BE	3127
1N5340BRL	OEM
1N5341	0185
1N5341A	0185
1N5341ARL	OEM
1N5341B	0185
1N5341BE	0185
1N5341BRL	OEM
1N5342	0205
1N5342A	0205
1N5342ARL	OEM
1N5342B	0205
1N5342BE	0205
1N5342BRL	OEM
1N5343	0480
1N5343A	0480
1N5343ARL	OEM
1N5343B	0480
1N5343BE	0480
1N5343BRL	OEM
1N5344	0504
1N5344A	0504
1N5344ARL	OEM
1N5344B	0504
1N5344BE	0504
1N5344BRL	OEM
1N5345	3317
1N5345A	3317
1N5345ARL	OEM
1N5345B	3317
1N5345BE	3317
1N5345BRL	OEM
1N5346	0684
1N5346A	0684
1N5346ARL	OEM
1N5346B	0684
1N5346BE	0684
1N5346BRL	OEM
1N5347	0225
1N5347A	0225
1N5347ARL	OEM
1N5347B	0225
1N5347BE	0225
1N5347BRL	OEM
1N5348	0230
1N5348A	0230
1N5348ARL	OEM
1N5348B	0170
1N5348BE	0230
1N5348BRL	OEM
1N5349	0234
1N5349A	0234
1N5349ARL	OEM
1N5349B	0234
1N5349BE	0234
1N5349BRL	OEM
1N5350	0237
1N5350A	0237
1N5350ARL	OEM
1N5350B	0237
1N5350BE	0237
1N5350BRL	OEM
1N5351	1397
1N5351A	1397
1N5351ARL	OEM
1N5351B	1397
1N5351BE	1397
1N5351BRL	OEM
1N5352	0247
1N5352A	0247
1N5352ARL	OEM
1N5352B	0251
1N5352BE	0247
1N5352BRL	OEM
1N5353	0251
1N5353A	0251
1N5353ARL	OEM
1N5353B	0251
1N5353BE	0251
1N5353BRL	OEM
1N5354	1178
1N5354A	1178
1N5354ARL	OEM
1N5354B	1178
1N5354BE	1178
1N5354BRL	OEM
1N5355	0256
1N5355A	0256
1N5355ARL	OEM
1N5355B	0256
1N5355BE	0256
1N5355BRL	OEM
1N5356	2395
1N5356A	2395
1N5356ARL	OEM
1N5356B	3028
1N5356BE	2395
1N5356BRL	OEM
1N5357	0262
1N5357A	0262
1N5357ARL	OEM
1N5357B	0262
1N5357BE	0262
1N5357BRL	OEM
1N5358	0269
1N5358A	0269
1N5358ARL	OEM
1N5358B	0269
1N5358BE	0269
1N5358BRL	OEM
1N5359	0274
1N5359A	0274
1N5359ARL	OEM
1N5359B	0274
1N5359BE	0274
1N5359BRL	OEM
1N5360	2399
1N5360A	2399
1N5360ARL	OEM
1N5360B	2399
1N5360BE	2399
1N5360BRL	OEM
1N5361	0294
1N5361A	0294
1N5361B	0294
1N5361BE	0294
1N5361BRL	OEM
1N5362	1177
1N5362A	1177
1N5362ARL	OEM
1N5362B	1177
1N5362BE	1177
1N5362BRL	OEM
1N5363	0309
1N5363A	0309
1N5363ARL	OEM
1N5363B	0309
1N5363BE	0309
1N5363BRL	OEM
1N5364	0318
1N5364A	0318
1N5364ARL	OEM
1N5364B	0318
1N5364BE	0318
1N5364BRL	OEM
1N5365	0320
1N5365A	0320
1N5365ARL	OEM
1N5365B	0320
1N5365BE	0320
1N5365BRL	OEM
1N5366	0326
1N5366A	0326
1N5366ARL	OEM
1N5366B	0326
1N5366BE	0326
1N5366BRL	OEM
1N5367	0337
1N5367A	0337
1N5367ARL	OEM
1N5367B	0337
1N5367BE	0337
1N5367BRL	OEM
1N5368	0347
1N5368A	0347
1N5368ARL	OEM
1N5368B	0347
1N5368BE	0347
1N5368BRL	OEM
1N5369	0027
1N5369A	0027
1N5369ARL	OEM
1N5369B	0027
1N5369BE	0027
1N5369BRL	OEM
1N5370	0266
1N5370A	0266
1N5370ARL	OEM
1N5370B	0266
1N5370BE	0266
1N5370BRL	OEM
1N5371	2856
1N5371A	2856
1N5371ARL	OEM
1N5371B	2856
1N5371BE	2856
1N5371BRL	OEM
1N5372	0386
1N5372A	0386
1N5372ARL	OEM
1N5372B	0386
1N5372BRL	OEM
1N5373	0405
1N5373A	0405
1N5373ARL	OEM
1N5373B	0405
1N5373BE	0405
1N5373BRL	OEM
1N5374	0426
1N5374A	0426
1N5374ARL	OEM
1N5374B	0426
1N5374BE	0405
1N5374BRL	OEM
1N5375	0444
1N5375A	0444
1N5375ARL	OEM
1N5375B	0444
1N5375BE	0444
1N5375BRL	OEM
1N5376	3028
1N5376A	3028
1N5376ARL	OEM
1N5376B	3028
1N5376BE	3028
1N5376BRL	OEM
1N5377	0238
1N5377A	0238
1N5377ARL	OEM
1N5377B	0238
1N5377BE	0238
1N5377BRL	OEM
1N5378	1180
1N5378A	1180
1N5378ARL	OEM
1N5378B	1180
1N5378BE	1180
1N5378BRL	OEM
1N5379	1190
1N5379A	1190
1N5379ARL	OEM
1N5379B	1190
1N5379BE	1190
1N5379BRL	OEM
1N5380	1207
1N5380A	1207
1N5380ARL	OEM
1N5380B	1207
1N5380BE	1207
1N5380BRL	OEM
1N5381	1218
1N5381A	1218
1N5381ARL	OEM
1N5381B	1218
1N5381BE	1218
1N5381BRL	OEM
1N5382	1881
1N5382A	1881
1N5382ARL	OEM
1N5382B	1881
1N5382BE	1881
1N5382BRL	OEM
1N5383	0647
1N5383A	0647
1N5383ARL	OEM
1N5383B	0647
1N5383BE	0647
1N5383BRL	OEM
1N5384	1255
1N5384A	1255
1N5384ARL	OEM
1N5384B	1255
1N5384BE	1255

If replacement code is OEM, contact original manufacturer for replacement.

DEVICE TYPE	REPL CODE	DEVICE TYPE	REPL CODE	DEVICE TYPE	REPL CODE	DEVICE TYPE	REPL CODE	DEVICE TYPE	REPL CODE	DEVICE TYPE	REPL CODE	DEVICE TYPE	REPL CODE
1N5384BRL	OEM	1N5430	OEM	1N5459	OEM	1N5520	0036	1N5539A	0671	1N5576B	0820	1N5630P1	OEM
1N5385	2102	1N5431	OEM	1N5459A	OEM	1N5520A	0036	1N5539B	0671	1N5577	0350	1N5631	OEM
1N5385A	2102	1N5432	OEM	1N5459B	OEM	1N5520B	0036	1N5539B1	0671	1N5577A	0350	1N5631A	OEM
1N5385ARL	OEM	1N5433	0102	1N5459C	OEM	1N5520C	0036	1N5539C	OEM	1N5577B	0820	1N5631AP1	OEM
1N5385B	2102	1N5434	0102	1N5459D	OEM	1N5520D	0036	1N5539D	OEM	1N5578	0931	1N5631P1	OEM
1N5385BRL	OEM	1N5435	0131	1N5460	OEM	1N5521	0275	1N5540	0700	1N5578A	0931	1N5632	OEM
1N5386	1279	1N5436	OEM	1N5460A	OEM	1N5521A	0275	1N5540A	0700	1N5578B	0931	1N5632A	OEM
1N5386A	1279	1N5437	OEM	1N5460B	OEM	1N5521B	0275	1N5540B	0700	1N5579	0999	1N5632AP1	OEM
1N5386ARL	OEM	1N5438	OEM	1N5460C	OEM	1N5521C	OEM	1N5540B1	0700	1N5579A	0999	1N5632P1	OEM
1N5386B	1279	1N5439	OEM	1N5460D	OEM	1N5521D	OEM	1N5540C	OEM	1N5579B	0999	1N5633	OEM
1N5386BE	1279	1N5439A	OEM	1N5461	0554	1N5522	0140	1N5540D	OEM	1N5580	0502	1N5633A	OEM
1N5386BRL	OEM	1N5439B	OEM	1N5461A	0554	1N5522A	0140	1N5541	0705	1N5580A	0502	1N5633AP1	OEM
1N5387	2225	1N5439C	OEM	1N5461B	0554	1N5522B	0140	1N5541A	0705	1N5580B	0502	1N5633P1	OEM
1N5387A	2225	1N5439D	OEM	1N5461C	0554	1N5522C	OEM	1N5541B	0705	1N5581	0869	1N5634	OEM
1N5387ARL	OEM	1N5440	OEM	1N5461D	OEM	1N5522D	OEM	1N5541B1	0705	1N5581A	0869	1N5634A	OEM
1N5387B	2225	1N5440A	OEM	1N5462	OEM	1N5523	0041	1N5541C	OEM	1N5581B	0869	1N5634AP1	OEM
1N5387BRL	OEM	1N5440B	OEM	1N5462A	OEM	1N5523A	0041	1N5541D	OEM	1N5582	0783	1N5634P1	OEM
1N5388	0605	1N5440C	OEM	1N5462B	OEM	1N5523B	0041	1N5542	0494	1N5582A	0783	1N5635	OEM
1N5388A	0605	1N5440D	OEM	1N5462C	OEM	1N5523C	0471	1N5542A	0494	1N5582B	0783	1N5635A	OEM
1N5388ARL	OEM	1N5441	0554	1N5462D	OEM	1N5523D	0589	1N5542B	0494	1N5583	1268	1N5635AP1	OEM
1N5388B	0605	1N5441A	0554	1N5463	0005	1N5524	0253	1N5542B1	0494	1N5583A	1268	1N5635P1	OEM
1N5388BE	0605	1N5441B	0554	1N5463A	0005	1N5524A	0253	1N5542C	OEM	1N5583B	1268	1N5636	3172
1N5388BRL	OEM	1N5441C	0554	1N5463B	0005	1N5524B	0253	1N5542D	OEM	1N5584	1189	1N5636A	3172
1N5389	OEM	1N5441D	OEM	1N5463C	0005	1N5524C	0883	1N5543	0714	1N5584A	1189	1N5636AP1	OEM
1N5390	0917	1N5442	OEM	1N5463D	OEM	1N5524D	0883	1N5543A	0714	1N5584B	1189	1N5636P1	OEM
1N5391	0110	1N5442A	0554	1N5464	0720	1N5525	0471	1N5543B	0714	1N5585	0331	1N5637	OEM
1N5391G	0071	1N5442B	0554	1N5464A	0720	1N5525A	0471	1N5543B1	0714	1N5585A	0331	1N5637A	OEM
1N5391GP	0071	1N5442C	0554	1N5464B	0720	1N5525B	0471	1N5543C	OEM	1N5585B	0331	1N5637AP1	OEM
1N5391RL	OEM	1N5442D	OEM	1N5464C	0720	1N5525C	OEM	1N5543D	OEM	1N5586	1311	1N5637P1	OEM
1N5392	0071	1N5443	0005	1N5464D	OEM	1N5525D	OEM	1N5544	0257	1N5586A	1311	1N5638	OEM
1N5392(GP)	0087	1N5443A	0005	1N5465	OEM	1N5526	0062	1N5544A	0257	1N5586B	1311	1N5638A	OEM
1N5392G	0071	1N5443B	0005	1N5465A	OEM	1N5526A	0062	1N5544B	0257	1N5587	0098	1N5638AP1	OEM
1N5392G2	0953	1N5443C	0005	1N5465B	OEM	1N5526B	0062	1N5544B1	0257	1N5587A	0098	1N5638P1	OEM
1N5392GP	0071	1N5443D	OEM	1N5465C	OEM	1N5526B1	0062	1N5544C	OEM	1N5587B	0098	1N5639	3191
1N5392RL	OEM	1N5444	0720	1N5465D	OEM	1N5526C	OEM	1N5544D	OEM	1N5588	0149	1N5639A	3191
1N5393	0242	1N5444A	0720	1N5466	OEM	1N5526D	OEM	1N5545	0195	1N5588A	0149	1N5639AP1	OEM
1N5393G	0071	1N5444B	0720	1N5466A	OEM	1N5527	0077	1N5545A	0195	1N5588B	0149	1N5639P1	OEM
1N5393GP	0071	1N5444C	0720	1N5466B	OEM	1N5527A	0077	1N5545B	0195	1N5589	0186	1N5640	3201
1N5393L	OEM	1N5444D	OEM	1N5466C	OEM	1N5527B	0077	1N5545B1	0195	1N5589A	0186	1N5640A	3201
1N5394	1747	1N5445	OEM	1N5466D	OEM	1N5527B1	0077	1N5545C	OEM	1N5589B	0186	1N5640AP1	OEM
1N5394G	0071	1N5445A	0720	1N5467	OEM	1N5527C	OEM	1N5545D	OEM	1N5590	0213	1N5640P1	OEM
1N5394GP	0071	1N5445B	0720	1N5467A	OEM	1N5527D	OEM	1N5546	0166	1N5590A	0213	1N5641	OEM
1N5395	0540	1N5445C	0720	1N5467B	OEM	1N5528	0165	1N5546A	0166	1N5590B	0213	1N5641A	OEM
1N5395G	0071	1N5445D	OEM	1N5467C	OEM	1N5528A	0165	1N5546B	0166	1N5591	0028	1N5641AP1	OEM
1N5395GP	0071	1N5446	OEM	1N5467D	OEM	1N5528B	0165	1N5546B1	0166	1N5591A	0028	1N5641P1	OEM
1N5395RL	OEM	1N5446A	0628	1N5468	0628	1N5528B1	0165	1N5546C	OEM	1N5591B	0028	1N5642	OEM
1N5396	1771	1N5446B	0628	1N5468A	0628	1N5528C	4309	1N5546D	OEM	1N5592	0255	1N5642A	OEM
1N5396G	0071	1N5446C	0628	1N5468B	0628	1N5528D	4309	1N5550	0087	1N5592A	0255	1N5642AP1	OEM
1N5396GP	0071	1N5446D	OEM	1N5468C	0628	1N5529	0057	1N5551	0087	1N5592B	0255	1N5642P1	OEM
1N5396RB	2407	1N5447	0628	1N5468D	OEM	1N5529A	0057	1N5552	0087	1N5593	0367	1N5643	1915
1N5397	0965	1N5447A	0628	1N5469	OEM	1N5529B	0057	1N5553	0102	1N5593A	0367	1N5643A	1915
1N5397G	0071	1N5447B	0628	1N5469A	OEM	1N5529B1	0057	1N5554	0102	1N5593B	0367	1N5643AP1	OEM
1N5397GP	0071	1N5447C	0628	1N5469B	OEM	1N5529C	OEM	1N5555	OEM	1N5594	0421	1N5643P1	OEM
1N5397RL	OEM	1N5447D	OEM	1N5469C	OEM	1N5529D	OEM	1N5556	OEM	1N5594A	0421	1N5644	OEM
1N5398	0817	1N5448	0628	1N5469D	OEM	1N5530	0064	1N5557	OEM	1N5594B	0421	1N5644A	OEM
1N5398G	0071	1N5448A	0628	1N5470	0030	1N5530A	0064	1N5558	OEM	1N5595	OEM	1N5644AP1	OEM
1N5398GP	0071	1N5448B	0628	1N5470A	0030	1N5530B	0064	1N5559	0025	1N5596	OEM	1N5644P1	OEM
1N5398RL	OEM	1N5448C	0628	1N5470B	0030	1N5530B1	0064	1N5559A	0025	1N5597	OEM	1N5645	OEM
1N5399	0087	1N5448D	OEM	1N5470C	0030	1N5530C	0248	1N5559B	0025	1N5598	OEM	1N5645A	OEM
1N5399G	0071	1N5449	OEM	1N5470D	OEM	1N5530D	0248	1N5560	0649	1N5599	OEM	1N5645AP1	OEM
1N5399GP	0071	1N5449A	0628	1N5471	OEM	1N5531	0181	1N5560A	0649	1N5600	OEM	1N5645P1	OEM
1N5399RL	OEM	1N5449B	0628	1N5471A	OEM	1N5531A	0181	1N5560B	0649	1N5601	OEM	1N5646	OEM
1N5400	0110	1N5449C	0628	1N5471B	OEM	1N5531B	0181	1N5561	0244	1N5602	OEM	1N5646A	OEM
1N5400G	0110	1N5449D	OEM	1N5471C	OEM	1N5531B1	0181	1N5561A	0244	1N5603	OEM	1N5646AP1	OEM
1N5400RL	OEM	1N5450	0030	1N5471D	OEM	1N5531C	OEM	1N5561B	0244	1N5604	OEM	1N5646P1	OEM
1N5401	0953	1N5450A	0030	1N5472	OEM	1N5531D	OEM	1N5562	0012	1N5605	0133	1N5647	OEM
1N5401G	0953	1N5450B	0030	1N5472A	OEM	1N5532	0052	1N5562A	0012	1N5606	0133	1N5647A	OEM
1N5401RL	OEM	1N5450C	0030	1N5472B	OEM	1N5532A	0052	1N5562B	0649	1N5607	0133	1N5647AP1	OEM
1N5402	0242	1N5450D	OEM	1N5472C	OEM	1N5532B	0052	1N5563	0170	1N5608	0133	1N5647P1	OEM
1N5402-M21	OEM	1N5451	OEM	1N5472D	OEM	1N5532B1	0052	1N5563A	0170	1N5609	0133	1N5648	OEM
1N5402G	0242	1N5451A	OEM	1N5473	OEM	1N5532C	1005	1N5563B	0012	1N5610	0133	1N5648A	OEM
1N5402RL	OEM	1N5451B	OEM	1N5473A	OEM	1N5532D	1005	1N5564	0317	1N5611	OEM	1N5648AP1	OEM
1N5403	1747	1N5451C	OEM	1N5473B	OEM	1N5533	0053	1N5564A	0317	1N5614	0023	1N5648P1	OEM
1N5403G	OEM	1N5451D	OEM	1N5473C	OEM	1N5533A	0053	1N5564B	0317	1N5614GP	0023	1N5649	OEM
1N5404	0540	1N5452	OEM	1N5473D	OEM	1N5533B	0053	1N5565	0137	1N5615	0023	1N5649A	OEM
1N5404G	0540	1N5452A	OEM	1N5474	OEM	1N5533B1	0053	1N5565A	0137	1N5615GP	0023	1N5649AP1	OEM
1N5404RL	OEM	1N5452B	OEM	1N5474A	OEM	1N5533C	OEM	1N5565B	0137	1N5616	0023	1N5649P1	OEM
1N5405	1771	1N5452C	OEM	1N5474B	OEM	1N5533D	OEM	1N5566	0365	1N5616GP	0023	1N5650	0568
1N5405G	OEM	1N5452D	OEM	1N5474C	OEM	1N5534	0879	1N5566A	0365	1N5617	0023	1N5650A	0568
1N5406	0965	1N5453	OEM	1N5474D	OEM	1N5534A	0879	1N5566B	0365	1N5617GP	0023	1N5650AP1	OEM
1N5406G	0965	1N5453A	OEM	1N5475	OEM	1N5534B	0879	1N5567	0002	1N5618	0017	1N5650P1	OEM
1N5406RL	OEM	1N5453B	OEM	1N5475A	OEM	1N5534B1	0879	1N5567A	0002	1N5618GP	0017	1N5651	OEM
1N5407	0817	1N5453C	OEM	1N5475B	OEM	1N5534C	OEM	1N5567B	0002	1N5619	0017	1N5651A	OEM
1N5407G	0817	1N5453D	OEM	1N5475C	OEM	1N5534D	OEM	1N5568	0420	1N5619GP	0017	1N5651AP1	OEM
1N5408	0922	1N5454	OEM	1N5475D	OEM	1N5535	0686	1N5568A	0420	1N5620	0017	1N5651P1	OEM
1N5408G	0922	1N5454A	OEM	1N5476	OEM	1N5535A	0686	1N5568B	0420	1N5620GP	0017	1N5652	OEM
1N5409	1126	1N5454B	OEM	1N5476A	OEM	1N5535B	0686	1N5569	0495	1N5621	0017	1N5652A	OEM
1N5409R	1111	1N5454C	OEM	1N5476B	OEM	1N5535B1	0686	1N5569A	0495	1N5621GP	0017	1N5652AP1	OEM
1N5410	0105	1N5454D	OEM	1N5476C	OEM	1N5535C	OEM	1N5569B	0495	1N5622	0017	1N5652P1	OEM
1N5410R	0777	1N5455	OEM	1N5476D	OEM	1N5535D	OEM	1N5570	0531	1N5622GP	0017	1N5653	0831
1N5411	2730	1N5455A	OEM	1N5477	OEM	1N5536	0445	1N5570A	0531	1N5623	0017	1N5653A	0831
1N5412	0229	1N5455B	OEM	1N5478	OEM	1N5536A	0445	1N5570B	0531	1N5623GP	0017	1N5653AP1	OEM
1N5413	0133	1N5455C	OEM	1N5479	OEM	1N5536B	0445	1N5571	0565	1N5624	0242	1N5653P1	OEM
1N5414	0133	1N5455D	OEM	1N5480	OEM	1N5536B1	0445	1N5571A	0565	1N5624GP	0242	1N5654	OEM
1N5415	0920	1N5456	OEM	1N5481	OEM	1N5536C	OEM	1N5571B	0565	1N5625	0540	1N5654A	OEM
1N5416	0031	1N5456A	OEM	1N5482	OEM	1N5536D	OEM	1N5572	0402	1N5625GP	0540	1N5654AP1	OEM
1N5417	0031	1N5456B	OEM	1N5483	OEM	1N5537	0210	1N5572A	0402	1N5626	0965	1N5654P1	OEM
1N5418	0102	1N5456C	OEM	1N5484	OEM	1N5537A	0210	1N5572B	0402	1N5626G	OEM	1N5655	OEM
1N5419	0102	1N5456D	OEM	1N5485	OEM	1N5537B	0210	1N5573	0441	1N5626GP	0965	1N5655A	OEM
1N5420	0087	1N5457	OEM	1N5518	0299	1N5537B1	0210	1N5573A	0441	1N5627	0817	1N5655AP1	OEM
1N5421	OEM	1N5457A	OEM	1N5518A	0299	1N5537C	OEM	1N5573B	0441	1N5627GP	0817	1N5655P1	OEM
1N5422	OEM	1N5457B	OEM	1N5518B	0299	1N5537D	OEM	1N5574	0726	1N5628	OEM	1N5656	OEM
1N5423	OEM	1N5457C	OEM	1N5518C	OEM	1N5538	0375	1N5574A	0726	1N5629	3114	1N5656A	OEM
1N5424	OEM	1N5457D	OEM	1N5518D	OEM	1N5538A	0375	1N5574B	0726	1N5629A	3114	1N5656AP1	OEM
1N5425	OEM	1N5458	OEM	1N5519	0376	1N5538B	0375	1N5575	0039	1N5629AP1	OEM	1N5656P1	OEM
1N5426	0133	1N5458A	OEM	1N5519A	0376	1N5538B1	0375	1N5575A	0039	1N5629P1	OEM	1N5657	OEM
1N5427	OEM	1N5458B	OEM	1N5519B	0376	1N5538C	OEM	1N5575B	0039	1N5630	OEM	1N5657A	OEM
1N5428	OEM	1N5458C	OEM	1N5519C	OEM	1N5538D	OEM	1N5576	0820	1N5630A	OEM	1N5657AP1	OEM
1N5429	OEM	1N5458D	OEM	1N5519D	OEM	1N5539	0671	1N5576A	0820	1N5630AP1	OEM	1N5657P1	OEM

Original Device Types

DEVICE TYPE	REPL CODE
1N5658	OEM
1N5658A	OEM
1N5658AP1	OEM
1N5658P1	OEM
1N5659	OEM
1N5659A	OEM
1N5659AP1	OEM
1N5659P1	OEM
1N5660	OEM
1N5660A	OEM
1N5660AP1	OEM
1N5660P1	OEM
1N5661	OEM
1N5661A	OEM
1N5661AP1	OEM
1N5661P1	OEM
1N5662	OEM
1N5662A	OEM
1N5662AP1	OEM
1N5662P1	OEM
1N5663	OEM
1N5663A	OEM
1N5663AP1	OEM
1N5663P1	OEM
1N5664	OEM
1N5664A	OEM
1N5664AP1	OEM
1N5664P1	OEM
1N5665	1405
1N5665A	1405
1N5665AP1	OEM
1N5665P1	OEM
1N5670A	2874
1N5671A	0677
1N5676A	OEM
1N5678A	OEM
1N5679	0015
1N5680	0015
1N5681	0554
1N5681A	0554
1N5681B	OEM
1N5682	OEM
1N5682A	OEM
1N5682B	OEM
1N5683	0005
1N5683A	0005
1N5683B	OEM
1N5684	0720
1N5684A	0720
1N5684B	OEM
1N5685	OEM
1N5685A	OEM
1N5685B	OEM
1N5686	OEM
1N5686A	OEM
1N5686B	OEM
1N5687	0628
1N5687A	0628
1N5687B	OEM
1N5688	OEM
1N5688A	OEM
1N5688B	OEM
1N5689	0030
1N5689A	0030
1N5689B	OEM
1N5690	OEM
1N5690A	OEM
1N5690B	OEM
1N5691	OEM
1N5691A	OEM
1N5691B	OEM
1N5692	OEM
1N5692A	OEM
1N5692B	OEM
1N5693	OEM
1N5693A	OEM
1N5693B	OEM
1N5694	OEM
1N5694A	OEM
1N5694B	OEM
1N5695	OEM
1N5695A	OEM
1N5695B	OEM
1N5696	0554
1N5696A	0554
1N5696B	OEM
1N5697	OEM
1N5697A	OEM
1N5697B	OEM
1N5698	OEM
1N5698A	OEM
1N5698B	OEM
1N5699	0720
1N5699A	0720
1N5699B	OEM
1N5700	OEM
1N5700A	OEM
1N5700B	OEM
1N5701	OEM
1N5701A	OEM
1N5701B	OEM
1N5702	0628
1N5702A	0628
1N5702B	OEM
1N5703	OEM
1N5703A	OEM
1N5703B	OEM
1N5704	OEM
1N5704A	OEM
1N5704B	OEM
1N5705	OEM
1N5705A	OEM
1N5705B	OEM
1N5706	OEM
1N5706A	OEM
1N5706B	OEM
1N5707	OEM
1N5707A	OEM
1N5707B	OEM
1N5708	OEM
1N5708A	OEM
1N5708B	OEM
1N5709	OEM
1N5709A	OEM
1N5709B	OEM
1N5710	OEM
1N5710A	OEM
1N5710B	OEM
1N5711	0133
1N5712	6937
1N5713	0133
1N5714	OEM
1N5715	OEM
1N5716	OEM
1N5717	OEM
1N5718	OEM
1N5719	0133
1N5720	0124
1N5721	0124
1N5722	OEM
1N5723	OEM
1N5724	OEM
1N5725	OEM
1N5726	0124
1N5726A	OEM
1N5727	0124
1N5728	0140
1N5728B	0140
1N5728C	OEM
1N5728D	OEM
1N5729	0041
1N5729B	0041
1N5729C	0589
1N5729D	0589
1N5730	0253
1N5730B	0253
1N5730C	OEM
1N5730D	OEM
1N5731	0471
1N5731B	0471
1N5731C	OEM
1N5731D	OEM
1N5732B	0062
1N5732C	OEM
1N5732D	OEM
1N5733	0077
1N5733B	0077
1N5733C	OEM
1N5733D	OEM
1N5734	0165
1N5734A	0165
1N5734B	0244
1N5734C	0165
1N5734D	0165
1N5735B	0057
1N5735C	OEM
1N5735D	OEM
1N5736	0064
1N5736B	0052
1N5736C	OEM
1N5736D	OEM
1N5737B	0181
1N5737C	OEM
1N5737D	OEM
1N5738	0052
1N5738B	0052
1N5738C	OEM
1N5738D	OEM
1N5739A	0053
1N5739B	0053
1N5739C	OEM
1N5739D	OEM
1N5740	0686
1N5740B	0686
1N5740C	OEM
1N5740D	OEM
1N5741	0445
1N5741B	0420
1N5741C	OEM
1N5741D	OEM
1N5742	0375
1N5742B	0495
1N5742C	OEM
1N5742D	OEM
1N5743B	0700
1N5743C	OEM
1N5743D	OEM
1N5744	0705
1N5744B	0705
1N5744C	OEM
1N5744D	OEM
1N5745B	0494
1N5745C	OEM
1N5745D	OEM
1N5746	0455
1N5746B	0455
1N5746C	0455
1N5746D	0455
1N5747B	0726
1N5747C	OEM
1N5747D	OEM
1N5748B	0166
1N5748C	OEM
1N5748D	OEM
1N5749	0010
1N5749B	0010
1N5749C	OEM
1N5749D	OEM
1N5750	0032
1N5750B	0032
1N5750C	OEM
1N5750D	OEM
1N5751	0054
1N5751B	0054
1N5751C	OEM
1N5751D	OEM
1N5752	0068
1N5752B	0068
1N5752C	OEM
1N5752D	OEM
1N5753	0092
1N5753B	0092
1N5753C	OEM
1N5753D	OEM
1N5754B	0125
1N5754C	OEM
1N5754D	OEM
1N5755B	0152
1N5755C	OEM
1N5755D	OEM
1N5756B	0173
1N5756C	OEM
1N5756D	OEM
1N5757B	0094
1N5757C	OEM
1N5757D	OEM
1N5758	3648
1N5758A	3648
1N5759	3331
1N5759A	3331
1N5760	3331
1N5760A	2730
1N5761	2730
1N5761A	3331
1N5762	OEM
1N5762A	OEM
1N5763	OEM
1N5764	OEM
1N5765	OEM
1N5767	0133
1N5768	OEM
1N5769	OEM
1N5770	OEM
1N5771	OEM
1N5772	OEM
1N5773	OEM
1N5774	OEM
1N5775	OEM
1N5779	OEM
1N5780	OEM
1N5781	OEM
1N5782	OEM
1N5783	OEM
1N5784	OEM
1N5785	OEM
1N5786	OEM
1N5787	OEM
1N5788	OEM
1N5789	OEM
1N5790	OEM
1N5791	OEM
1N5792	OEM
1N5793	OEM
1N5794	OEM
1N5795	OEM
1N5796	OEM
1N5797	OEM
1N5798	OEM
1N5799	OEM
1N5800	OEM
1N5802	0124
1N5803	0124
1N5804	0133
1N5805	0133
1N5806	0133
1N5807	0124
1N5808	0124
1N5809	0133
1N5810	0133
1N5811	0133
1N5812	OEM
1N5813	OEM
1N5814	OEM
1N5815	OEM
1N5816	OEM
1N5817	1335
1N5817RL	OEM
1N5818	1335
1N5818RL	OEM
1N5819	0735
1N5819RL	OEM
1N5820	0087
1N5820RL	OEM
1N5821	2542
1N5821RL	OEM
1N5822	2542
1N5822RL	OEM
1N5823	OEM
1N5824	OEM
1N5825	OEM
1N5826	0615
1N5827	0615
1N5828	0615
1N5829	0615
1N5830	6948
1N5831	6948
1N5832	1545
1N5833	1545
1N5834	1545
1N5835	0068
1N5836	0068
1N5837	1276
1N5837A	1276
1N5837B	1276
1N5837C	OEM
1N5837D	OEM
1N5838	2874
1N5838A	2874
1N5838B	2874
1N5838C	OEM
1N5838D	OEM
1N5839	0760
1N5839A	0760
1N5839B	0760
1N5839C	OEM
1N5839D	OEM
1N5840	0677
1N5840A	0677
1N5840B	0677
1N5840C	OEM
1N5840D	OEM
1N5841	0118
1N5841A	0118
1N5841B	0118
1N5841C	OEM
1N5841D	OEM
1N5842	0299
1N5842A	0299
1N5842B	0299
1N5842C	OEM
1N5842D	OEM
1N5843	0376
1N5843A	0376
1N5843B	0376
1N5843C	OEM
1N5843D	OEM
1N5844	0036
1N5844A	0036
1N5844B	0036
1N5844C	OEM
1N5844D	OEM
1N5845	0275
1N5845A	0275
1N5845B	0533
1N5845C	OEM
1N5845D	OEM
1N5846	0140
1N5846A	0140
1N5846B	0140
1N5846C	OEM
1N5846D	OEM
1N5847	0041
1N5847A	0041
1N5847B	0041
1N5847C	OEM
1N5847D	OEM
1N5848	0253
1N5848A	0253
1N5848B	0253
1N5848C	OEM
1N5848D	OEM
1N5849	0091
1N5849A	0091
1N5849B	0091
1N5849C	OEM
1N5849D	OEM
1N5850	0471
1N5850A	0471
1N5850B	0471
1N5850C	OEM
1N5850D	OEM
1N5851	0062
1N5851A	0062
1N5851B	0025
1N5851C	OEM
1N5851D	OEM
1N5852	0649
1N5852A	0649
1N5852B	0077
1N5852C	OEM
1N5852D	OEM
1N5853	0165
1N5853A	0165
1N5853B	0165
1N5853C	OEM
1N5853D	OEM
1N5854	0322
1N5854A	0322
1N5854B	0322
1N5854C	OEM
1N5854D	OEM
1N5855	0057
1N5855A	0057
1N5855B	0057
1N5855C	OEM
1N5855D	OEM
1N5856	0064
1N5856A	0064
1N5856B	0064
1N5856C	OEM
1N5856D	OEM
1N5857	0181
1N5857A	0181
1N5857B	0181
1N5857C	OEM
1N5857D	OEM
1N5858	0052
1N5858A	0052
1N5858B	0052
1N5858C	OEM
1N5858D	OEM
1N5859	0053
1N5859A	0053
1N5859B	0053
1N5859C	OEM
1N5859D	OEM
1N5860	0879
1N5860A	0879
1N5860B	0879
1N5860C	OEM
1N5860D	OEM
1N5861	0686
1N5861A	0686
1N5861B	0686
1N5861C	OEM
1N5861D	OEM
1N5862	0445
1N5862A	0445
1N5862B	0445
1N5862C	OEM
1N5862D	OEM
1N5863	0210
1N5863A	0210
1N5863B	0210
1N5863C	OEM
1N5863D	OEM
1N5864	0375
1N5864A	0375
1N5864B	0375
1N5864C	OEM
1N5864D	OEM
1N5865	0671
1N5865A	0671
1N5865B	0671
1N5865C	OEM
1N5865D	OEM
1N5866	0700
1N5866A	0700
1N5866B	0700
1N5866C	OEM
1N5866D	OEM
1N5867	0705
1N5867A	0705
1N5867B	0705
1N5867C	OEM
1N5867D	OEM
1N5868	0494
1N5868A	0494
1N5868B	0494
1N5868C	OEM
1N5868D	OEM
1N5869	0714
1N5869A	0714
1N5869B	0714
1N5869C	OEM
1N5869D	OEM
1N5870	0455
1N5870A	0455
1N5870B	0455
1N5870C	OEM
1N5870D	OEM
1N5871	0257
1N5871A	0257
1N5871B	0257
1N5871C	OEM
1N5871D	OEM
1N5872	0195
1N5872A	0195
1N5872B	0195
1N5872C	OEM
1N5872D	OEM
1N5873	0166
1N5873A	0166
1N5873B	0166
1N5873C	OEM
1N5873D	OEM
1N5874	0010
1N5874A	0010
1N5874B	0010
1N5874C	OEM
1N5874D	OEM
1N5875	0032
1N5875A	0032
1N5875B	0032
1N5875C	OEM
1N5875D	OEM
1N5876	0054
1N5876A	0054
1N5876B	0054
1N5876C	OEM
1N5876D	OEM
1N5877	0068
1N5877A	0068
1N5877B	0068
1N5877C	OEM
1N5877D	OEM
1N5878	0092
1N5878A	0092
1N5878B	0092
1N5878C	OEM
1N5878D	OEM
1N5879	0125
1N5879A	0125
1N5879B	0125
1N5879C	OEM
1N5879D	OEM
1N5880	2317
1N5880A	2317
1N5880B	7163
1N5880C	OEM
1N5880D	OEM
1N5881	0152
1N5881A	0152
1N5881B	0152
1N5881C	OEM
1N5881D	OEM
1N5882	0173
1N5882A	0173
1N5882B	0173
1N5882C	OEM
1N5882D	OEM
1N5883	0094
1N5883A	0094
1N5883B	0094
1N5883C	OEM
1N5883D	OEM
1N5884	0049
1N5884A	0049
1N5884B	0049
1N5884C	OEM
1N5884D	OEM
1N5885	0104
1N5885A	0104
1N5885B	0104
1N5885C	OEM
1N5885D	OEM
1N5886	0156
1N5886A	0156
1N5886B	0156
1N5886C	OEM
1N5886D	OEM
1N5887	0189
1N5887A	0189
1N5887B	0189
1N5887C	OEM
1N5887D	OEM
1N5888	0099
1N5888A	0099
1N5888B	0099
1N5888C	OEM
1N5888D	OEM
1N5889	0089
1N5889A	0244
1N5889B	0089
1N5889C	OEM
1N5889D	OEM
1N5890	0213
1N5890A	0213
1N5890B	0287
1N5890C	OEM
1N5890D	OEM
1N5891	0245
1N5891A	0245
1N5891B	0252
1N5891C	OEM
1N5891D	OEM
1N5892	0340
1N5892A	0340
1N5892B	0340
1N5892C	OEM
1N5892D	OEM
1N5893	0370
1N5893A	0370
1N5893B	0370
1N5893C	OEM
1N5893D	OEM
1N5894	0394
1N5894A	0394
1N5894B	0394
1N5894C	OEM
1N5894D	OEM
1N5895	0425
1N5895A	0425
1N5895B	0425
1N5895C	OEM
1N5895D	OEM
1N5896	0453
1N5896A	0453
1N5896B	0453
1N5896C	OEM
1N5896D	OEM
1N5897	0453
1N5897A	0453
1N5897B	0453
1N5897C	OEM
1N5897D	OEM
1N5898	0110
1N5899	0953
1N5900	0242
1N5901	0540
1N5902	0965
1N5903	0817
1N5904	0922
1N5905	OEM
1N5906	OEM
1N5907	2396
1N5908	2396
1N5908C	OEM
1N5913	0782
1N5913A	0782
1N5913ARL	OEM
1N5913B	0782
1N5913BRL	OEM
1N5913C	OEM
1N5913D	OEM
1N5914	0796
1N5914A	0796
1N5914ARL	OEM
1N5914B	0796
1N5914BRL	OEM
1N5914C	OEM
1N5914D	0589
1N5915	0806
1N5915A	0806
1N5915ARL	OEM
1N5915B	0806
1N5915BRL	OEM
1N5915C	OEM
1N5915D	OEM
1N5916	0821
1N5916A	0821
1N5916ARL	OEM
1N5916B	0821
1N5916BRL	OEM
1N5916C	OEM
1N5916D	OEM
1N5917	0833
1N5917A	0833
1N5917ARL	OEM
1N5917B	0833
1N5917BRL	OEM
1N5917C	OEM
1N5917D	OEM
1N5918	0442
1N5918A	0442
1N5918ARL	OEM
1N5918B	0442
1N5918BRL	OEM
1N5918C	OEM
1N5918D	0589
1N5919	0876
1N5919A	0876
1N5919ARL	OEM
1N5919B	0876
1N5919BRL	OEM
1N5919C	OEM
1N5919D	0883
1N5920	0185
1N5920A	0185
1N5920ARL	OEM
1N5920B	0185
1N5920BRL	OEM
1N5920C	OEM
1N5920D	0295
1N5920E	OEM
1N5921	0205
1N5921A	0205
1N5921ARL	OEM
1N5921B	0205
1N5921BRL	OEM
1N5921C	OEM
1N5921D	OEM
1N5922	0480
1N5922A	0480
1N5922ARL	OEM
1N5922B	0480
1N5922BRL	OEM
1N5922C	OEM
1N5922D	OEM
1N5923	0504
1N5923A	0504
1N5923ARL	OEM
1N5923B	0504
1N5923BRL	OEM
1N5923C	OEM
1N5923D	OEM
1N5924	0684
1N5924A	0684
1N5924ARL	OEM
1N5924B	0684
1N5924BRL	OEM
1N5924C	OEM
1N5924D	OEM
1N5925	0225
1N5925A	0225
1N5925ARL	OEM
1N5925B	0225
1N5925BRL	OEM
1N5925C	OEM
1N5925D	0248
1N5926	0230
1N5926A	0230
1N5926ARL	OEM
1N5926B	0230
1N5926BRL	OEM
1N5926C	OEM
1N5926D	OEM
1N5927	0234
1N5927A	0234
1N5927ARL	OEM
1N5927B	0234
1N5927BRL	OEM
1N5927C	OEM

If replacement code is OEM, contact original manufacturer for replacement.

DEVICE TYPE	REPL CODE
1N5927D	1005
1N5928	0237
1N5928A	0237
1N5928ARL	OEM
1N5928B	0237
1N5928BRL	OEM
1N5928C	OEM
1N5928D	OEM
1N5929	0247
1N5929A	0247
1N5929ARL	OEM
1N5929B	0247
1N5929BRL	OEM
1N5929C	OEM
1N5929D	OEM
1N5930	0251
1N5930A	0251
1N5930ARL	OEM
1N5930B	0251
1N5930BRL	OEM
1N5930C	OEM
1N5930D	OEM
1N5931	0256
1N5931A	0256
1N5931ARL	OEM
1N5931B	0256
1N5931BRL	OEM
1N5931C	OEM
1N5931D	OEM
1N5932	0262
1N5932A	0262
1N5932ARL	OEM
1N5932B	0262
1N5932BRL	OEM
1N5932C	OEM
1N5932D	OEM
1N5933	0269
1N5933A	0269
1N5933ARL	OEM
1N5933B	0269
1N5933BRL	OEM
1N5933C	OEM
1N5933D	OEM
1N5934	0274
1N5934A	0274
1N5934ARL	OEM
1N5934B	0274
1N5934BRL	OEM
1N5934C	OEM
1N5934D	OEM
1N5935	0294
1N5935A	0294
1N5935ARL	OEM
1N5935B	0294
1N5935BRL	OEM
1N5935C	OEM
1N5935D	OEM
1N5936	0309
1N5936A	0309
1N5936ARL	OEM
1N5936B	0309
1N5936BRL	OEM
1N5936C	OEM
1N5936D	OEM
1N5937	0318
1N5937A	0318
1N5937ARL	OEM
1N5937B	0318
1N5937BRL	OEM
1N5937C	OEM
1N5937D	OEM
1N5938	0320
1N5938A	0320
1N5938ARL	OEM
1N5938B	0320
1N5938BRL	OEM
1N5938C	OEM
1N5938D	OEM
1N5939	0326
1N5939A	0326
1N5939ARL	OEM
1N5939B	0326
1N5939BRL	OEM
1N5939C	OEM
1N5939D	OEM
1N5940	0337
1N5940A	0337
1N5940ARL	OEM
1N5940B	0337
1N5940BRL	OEM
1N5940C	OEM
1N5940D	OEM
1N5941	0347
1N5941A	0347
1N5941ARL	OEM
1N5941B	0347
1N5941BRL	OEM
1N5941C	OEM
1N5941D	OEM
1N5942	0027
1N5942A	0027
1N5942ARL	OEM
1N5942B	0027
1N5942BRL	OEM
1N5942C	OEM
1N5942D	OEM
1N5943	0266
1N5943A	0266
1N5943ARL	OEM
1N5943B	0266
1N5943BRL	OEM
1N5943C	OEM
1N5943D	OEM
1N5944	0386
1N5944A	0386
1N5944ARL	OEM
1N5944B	0386
1N5944BRL	OEM
1N5944C	OEM
1N5944D	OEM
1N5945	0405
1N5945A	0405
1N5945ARL	OEM
1N5945B	0405
1N5945BRL	OEM
1N5945C	OEM
1N5945D	OEM
1N5946	0426
1N5946A	0426
1N5946ARL	OEM
1N5946B	0426
1N5946BRL	OEM
1N5946C	OEM
1N5946D	OEM
1N5947	0444
1N5947A	0444
1N5947ARL	OEM
1N5947B	0444
1N5947BRL	OEM
1N5947C	OEM
1N5947D	OEM
1N5948	0238
1N5948A	0238
1N5948ARL	OEM
1N5948B	0238
1N5948BRL	OEM
1N5948C	OEM
1N5948D	OEM
1N5949	1180
1N5949A	1180
1N5949ARL	OEM
1N5949B	1180
1N5949BRL	OEM
1N5949C	OEM
1N5949D	OEM
1N5949GP	1180
1N5950	1190
1N5950A	1190
1N5950ARL	OEM
1N5950B	1190
1N5950BRL	OEM
1N5950C	OEM
1N5950D	OEM
1N5950GP	1190
1N5951	1207
1N5951A	1207
1N5951ARL	OEM
1N5951B	1207
1N5951BRL	OEM
1N5951C	OEM
1N5951D	OEM
1N5951GP	1207
1N5952	1218
1N5952A	1218
1N5952ARL	OEM
1N5952B	1218
1N5952BRL	OEM
1N5952C	OEM
1N5952D	OEM
1N5952GP	1218
1N5953	0647
1N5953A	0647
1N5953ARL	OEM
1N5953B	0647
1N5953BRL	OEM
1N5953C	OEM
1N5953D	OEM
1N5953GP	0647
1N5954	1255
1N5954A	1255
1N5954ARL	OEM
1N5954B	1255
1N5954BRL	OEM
1N5954C	OEM
1N5954D	OEM
1N5954GP	1255
1N5955	1279
1N5955A	1279
1N5955ARL	OEM
1N5955B	1279
1N5955BRL	OEM
1N5955C	OEM
1N5955D	OEM
1N5955GP	1279
1N5956	0605
1N5956A	0605
1N5956ARL	OEM
1N5956B	0605
1N5956BRL	OEM
1N5956C	OEM
1N5956D	OEM
1N5956GP	0605
1N5957	OEM
1N5968	0005
1N5968A	0005
1N5969	0185
1N5985	1276
1N5985A	1276
1N5985ARL	OEM
1N5985B	1276
1N5985BRL	OEM
1N5985C	OEM
1N5985D	OEM
1N5986	0760
1N5986A	0760
1N5986ARL	OEM
1N5986B	1312
1N5986BRL	OEM
1N5986C	OEM
1N5986D	OEM
1N5987	0118
1N5987A	0118
1N5987ARL	OEM
1N5987B	1713
1N5987BRL	OEM
1N5987C	OEM
1N5987D	OEM
1N5988	0299
1N5988A	0299
1N5988ARL	OEM
1N5988B	0292
1N5988BRL	OEM
1N5988C	OEM
1N5988D	OEM
1N5989	0376
1N5989A	0376
1N5989ARL	OEM
1N5989B	0376
1N5989BRL	OEM
1N5989C	OEM
1N5989D	OEM
1N5990	0036
1N5990A	0036
1N5990ARL	OEM
1N5990B	0036
1N5990BRL	OEM
1N5990C	OEM
1N5990D	OEM
1N5991	0275
1N5991A	0275
1N5991ARL	OEM
1N5991B	0275
1N5991BRL	OEM
1N5991C	OEM
1N5991D	OEM
1N5992	0140
1N5992A	0140
1N5992ARL	OEM
1N5992B	0140
1N5992BRL	OEM
1N5992C	OEM
1N5992D	OEM
1N5993	0041
1N5993A	0041
1N5993ARL	OEM
1N5993B	0041
1N5993BRL	OEM
1N5993C	0589
1N5993D	0589
1N5994	0253
1N5994A	0253
1N5994ARL	OEM
1N5994B	0253
1N5994BRL	OEM
1N5994C	0883
1N5994D	0883
1N5995	0471
1N5995A	0471
1N5995ARL	OEM
1N5995B	0471
1N5995BRL	OEM
1N5995C	0295
1N5995D	0295
1N5996	0062
1N5996A	0062
1N5996ARL	OEM
1N5996B	0062
1N5996BRL	OEM
1N5996C	OEM
1N5996D	OEM
1N5997	0077
1N5997A	0077
1N5997ARL	OEM
1N5997B	0077
1N5997BRL	OEM
1N5997C	OEM
1N5997D	OEM
1N5998	0165
1N5998A	0165
1N5998ARL	OEM
1N5998B	0165
1N5998BRL	OEM
1N5998C	OEM
1N5998D	OEM
1N5999	0057
1N5999A	0057
1N5999ARL	OEM
1N5999B	0057
1N5999BRL	OEM
1N5999C	OEM
1N5999D	OEM
1N6000	0064
1N6000A	0064
1N6000ARL	OEM
1N6000B	0064
1N6000BRL	OEM
1N6000C	0248
1N6000D	0248
1N6001	0181
1N6001A	0181
1N6001ARL	OEM
1N6001B	0181
1N6001BRL	OEM
1N6001C	OEM
1N6001D	OEM
1N6002	0052
1N6002A	0052
1N6002ARL	OEM
1N6002B	0052
1N6002BRL	OEM
1N6002C	1005
1N6002D	1005
1N6003	0053
1N6003A	0053
1N6003ARL	OEM
1N6003B	0365
1N6003BRL	OEM
1N6003C	0053
1N6003D	OEM
1N6004	0686
1N6004A	0686
1N6004ARL	OEM
1N6004B	0686
1N6004BRL	OEM
1N6004C	OEM
1N6004D	OEM
1N6005	0445
1N6005A	0445
1N6005ARL	OEM
1N6005B	0445
1N6005BRL	OEM
1N6005C	OEM
1N6005D	OEM
1N6006	0375
1N6006A	0375
1N6006ARL	OEM
1N6006B	0495
1N6006BRL	OEM
1N6006C	OEM
1N6006D	OEM
1N6007	0700
1N6007A	0700
1N6007ARL	OEM
1N6007B	0700
1N6007BRL	OEM
1N6007C	OEM
1N6007D	OEM
1N6008	0705
1N6008A	0705
1N6008ARL	OEM
1N6008B	0705
1N6008BRL	OEM
1N6008C	OEM
1N6008D	OEM
1N6009	0494
1N6009A	0494
1N6009ARL	OEM
1N6009B	0494
1N6009BRL	OEM
1N6009C	OEM
1N6009D	OEM
1N6010	0455
1N6010A	0455
1N6010ARL	OEM
1N6010B	0455
1N6010BRL	OEM
1N6010C	OEM
1N6010D	OEM
1N6011	0257
1N6011A	0257
1N6011ARL	OEM
1N6011B	0257
1N6011C	OEM
1N6011D	OEM
1N6012	0166
1N6012A	0166
1N6012ARL	OEM
1N6012B	0166
1N6012BRL	OEM
1N6012C	OEM
1N6012D	OEM
1N6013	0010
1N6013A	0010
1N6013ARL	OEM
1N6013B	0010
1N6013BRL	OEM
1N6013C	OEM
1N6013D	OEM
1N6014	0032
1N6014A	0032
1N6014ARL	OEM
1N6014B	0032
1N6014BRL	OEM
1N6014C	OEM
1N6014D	OEM
1N6015	0054
1N6015A	0054
1N6015ARL	OEM
1N6015B	0054
1N6015BRL	OEM
1N6015C	OEM
1N6015D	OEM
1N6016	0068
1N6016A	0068
1N6016ARL	OEM
1N6016B	0068
1N6016BRL	OEM
1N6016C	OEM
1N6016D	OEM
1N6017	0092
1N6017A	0092
1N6017ARL	OEM
1N6017B	0092
1N6017BRL	OEM
1N6017C	OEM
1N6017D	OEM
1N6018	0125
1N6018A	0125
1N6018B	0125
1N6018BRL	OEM
1N6018C	OEM
1N6018D	OEM
1N6019	0152
1N6019A	0152
1N6019ARL	OEM
1N6019B	0152
1N6019BRL	OEM
1N6019C	OEM
1N6019D	OEM
1N6020	0173
1N6020A	0173
1N6020ARL	OEM
1N6020B	0173
1N6020BRL	OEM
1N6020C	OEM
1N6020D	OEM
1N6021	0094
1N6021A	0094
1N6021B	0094
1N6021BRL	OEM
1N6021C	OEM
1N6021D	OEM
1N6022	0049
1N6022A	0049
1N6022ARL	OEM
1N6022B	0049
1N6022BRL	OEM
1N6022C	OEM
1N6022D	OEM
1N6023	0156
1N6023A	0156
1N6023B	0156
1N6023BRL	OEM
1N6023C	OEM
1N6023D	OEM
1N6024	0189
1N6024A	0189
1N6024B	0099
1N6024C	OEM
1N6024D	OEM
1N6025	0099
1N6025A	0099
1N6025B	0099
1N6025C	OEM
1N6025D	OEM
1N6026	0089
1N6026A	0089
1N6026B	0089
1N6026C	OEM
1N6026D	OEM
1N6027	0287
1N6027A	0287
1N6027B	0287
1N6027C	OEM
1N6027D	OEM
1N6028	0340
1N6028A	0340
1N6028B	0340
1N6028C	OEM
1N6028D	OEM
1N6029	0370
1N6029A	0370
1N6029B	0370
1N6029C	OEM
1N6029D	OEM
1N6030	0425
1N6030A	0425
1N6030B	0425
1N6030C	OEM
1N6030D	OEM
1N6031	1473
1N6031A	1473
1N6031B	0421
1N6031C	OEM
1N6031D	OEM
1N6036	OEM
1N6036A	2133
1N6037	OEM
1N6037A	OEM
1N6038	OEM
1N6038A	OEM
1N6039	OEM
1N6039A	OEM
1N6040	OEM
1N6040A	OEM
1N6041	OEM
1N6041A	OEM
1N6042	1169
1N6042A	1169
1N6043	OEM
1N6043A	OEM
1N6044	OEM
1N6044A	OEM
1N6045	1767
1N6045A	1767
1N6046	1932
1N6046A	1932
1N6047	OEM
1N6047A	OEM
1N6048	OEM
1N6048A	OEM
1N6049	1952
1N6049A	1952
1N6050	OEM
1N6050A	OEM
1N6051	OEM
1N6051A	OEM
1N6052	OEM
1N6052A	OEM
1N6053	OEM
1N6053A	OEM
1N6054	OEM
1N6054A	OEM
1N6055	OEM
1N6055A	OEM
1N6056	1972
1N6056A	1972
1N6057	OEM
1N6057A	OEM
1N6058	OEM
1N6058A	OEM
1N6059	1987
1N6059A	1987
1N6060	OEM
1N6060A	OEM
1N6061	OEM
1N6061A	OEM
1N6062	OEM
1N6062A	OEM
1N6063	OEM
1N6063A	OEM
1N6064	OEM
1N6064A	OEM
1N6065	OEM
1N6065A	OEM
1N6066	OEM
1N6066A	OEM
1N6067	OEM
1N6067A	OEM
1N6068	OEM
1N6068A	OEM
1N6069	OEM
1N6069A	OEM
1N6070	OEM
1N6070A	OEM
1N6071	1408
1N6071A	1408
1N6072	OEM
1N6072A	OEM
1N6073	0546
1N6074	2523
1N6075	3077
1N6076	OEM
1N6077	OEM
1N6078	OEM
1N6079	OEM
1N6080	OEM
1N6081	OEM
1N6082	0275
1N6082A	0275
1N6082B	0275
1N6082C	OEM
1N6082D	OEM
1N6083	0140
1N6083A	0140
1N6083B	0140
1N6083C	0589
1N6083D	0589
1N6084	0041
1N6084A	0041
1N6084B	0041
1N6084C	0883
1N6084D	0883
1N6085	0253
1N6085A	0253
1N6085B	0253
1N6085C	OEM
1N6085D	OEM
1N6086	0471
1N6086A	0471
1N6086B	0471
1N6086C	OEM
1N6086D	OEM
1N6087	0062
1N6087A	0062
1N6087B	0062
1N6087C	OEM
1N6087D	OEM
1N6088	0077
1N6088A	0077
1N6088B	0077
1N6088C	OEM
1N6088D	OEM
1N6089	0165
1N6089A	0165
1N6089B	0165
1N6089C	OEM
1N6089D	OEM
1N6090	0057
1N6090A	0057
1N6090B	0057
1N6090C	OEM
1N6090D	OEM
1N6091	0064
1N6091A	0064
1N6091B	0064
1N6091C	OEM
1N6091D	OEM
1N6092	OEM
1N6093	OEM
1N6094	OEM
1N6095	0615
1N6096	0615
1N6097	1545
1N6098	1545
1N6099	0133
1N6099A	OEM
1N6102	OEM
1N6102A	OEM
1N6103	OEM
1N6103A	OEM
1N6104	OEM
1N6104A	OEM
1N6105	OEM
1N6105A	OEM
1N6106	OEM
1N6106A	OEM
1N6107	OEM
1N6107A	OEM
1N6108	OEM
1N6108A	OEM
1N6109	OEM
1N6109A	OEM
1N6110	OEM
1N6110A	OEM
1N6111	OEM
1N6111A	OEM
1N6112	OEM
1N6112A	OEM
1N6113	OEM
1N6113A	OEM
1N6114	OEM
1N6114A	OEM
1N6115	OEM
1N6115A	OEM
1N6116	OEM
1N6116A	OEM
1N6117	OEM
1N6117A	OEM
1N6118	OEM
1N6118A	OEM
1N6119	OEM
1N6119A	OEM
1N6120	OEM
1N6120A	OEM
1N6121	OEM
1N6121A	OEM
1N6122	OEM
1N6122A	OEM
1N6123	OEM
1N6123A	OEM
1N6124	OEM
1N6124A	OEM
1N6125	OEM
1N6125A	OEM
1N6126	OEM
1N6126A	OEM
1N6127	OEM
1N6127A	OEM
1N6128	OEM
1N6128A	OEM
1N6129	OEM
1N6129A	OEM
1N6130	OEM
1N6130A	OEM
1N6131	OEM
1N6131A	OEM
1N6132	OEM
1N6132A	OEM
1N6133	OEM
1N6133A	OEM
1N6134	OEM
1N6134A	OEM
1N6135	OEM
1N6135A	OEM
1N6136	OEM
1N6136A	OEM
1N6137	OEM
1N6137A	OEM
1N6138	OEM
1N6138A	OEM
1N6139	OEM
1N6139A	OEM
1N6140	OEM
1N6140A	OEM
1N6141	OEM
1N6141A	OEM
1N6142	OEM
1N6142A	OEM
1N6143	OEM
1N6143A	OEM
1N6144	OEM
1N6144A	OEM
1N6145	OEM
1N6145A	OEM
1N6146	OEM
1N6146A	OEM
1N6147	OEM
1N6147A	OEM
1N6148	OEM
1N6148A	OEM
1N6149	OEM
1N6149A	OEM

If replacement code is OEM, contact original manufacturer for replacement.

DEVICE TYPE	REPL CODE
1N6150	OEM
1N6150A	OEM
1N6151	OEM
1N6151A	OEM
1N6152	OEM
1N6152A	OEM
1N6153	OEM
1N6153A	OEM
1N6154	OEM
1N6154A	OEM
1N6155	OEM
1N6155A	OEM
1N6156	OEM
1N6156A	OEM
1N6157	OEM
1N6157A	OEM
1N6158	OEM
1N6158A	OEM
1N6159	OEM
1N6159A	OEM
1N6160	OEM
1N6160A	OEM
1N6161	OEM
1N6161A	OEM
1N6162	OEM
1N6162A	OEM
1N6163	OEM
1N6163A	OEM
1N6164	OEM
1N6164A	OEM
1N6165	OEM
1N6165A	OEM
1N6166	OEM
1N6166A	OEM
1N6167	OEM
1N6167A	OEM
1N6168	OEM
1N6168A	OEM
1N6169	OEM
1N6169A	OEM
1N6170	OEM
1N6170A	OEM
1N6171	OEM
1N6171A	OEM
1N6172	OEM
1N6172A	OEM
1N6173	OEM
1N6173A	OEM
1N6176A	OEM
1N6262	OEM
1N6263	3891
1N6264	OEM
1N6265	OEM
1N6266	OEM
1N6267	3114
1N6267A	3114
1N6267C	2133
1N6267CA	2133
1N6268	OEM
1N6268A	OEM
1N6268C	OEM
1N6268CA	OEM
1N6269	OEM
1N6269A	OEM
1N6269C	OEM
1N6269CA	OEM
1N6270	OEM
1N6270A	OEM
1N6270C	OEM
1N6270CA	OEM
1N6271	OEM
1N6271A	OEM
1N6271CA	OEM
1N6272	OEM
1N6272A	OEM
1N6272C	OEM
1N6272CA	OEM
1N6273	OEM
1N6273A	OEM
1N6273C	OEM
1N6273CA	OEM
1N6274	3172
1N6274A	3172
1N6274C	1169
1N6274CA	1169
1N6275	OEM
1N6275A	OEM
1N6275C	OEM
1N6275CA	OEM
1N6276	OEM
1N6276A	OEM
1N6276C	OEM
1N6276CA	OEM
1N6277	3191
1N6277A	3191
1N6277C	1767
1N6277CA	1767
1N6278	3201
1N6278A	3201
1N6278C	1932
1N6278CA	1932
1N6279	OEM
1N6279A	OEM
1N6279C	OEM
1N6279CA	OEM
1N6280	OEM
1N6280A	OEM
1N6280C	OEM
1N6280CA	OEM

DEVICE TYPE	REPL CODE
1N6281	1915
1N6281A	1915
1N6281C	1952
1N6281CA	1952
1N6282	OEM
1N6282A	OEM
1N6282C	OEM
1N6282CA	OEM
1N6283	OEM
1N6283A	OEM
1N6283C	OEM
1N6283CA	OEM
1N6284	OEM
1N6284A	OEM
1N6284C	OEM
1N6284CA	OEM
1N6285	OEM
1N6285A	OEM
1N6285C	OEM
1N6285CA	OEM
1N6286	OEM
1N6286A	OEM
1N6286C	OEM
1N6286CA	OEM
1N6287	OEM
1N6287A	OEM
1N6287C	OEM
1N6287CA	OEM
1N6288	0568
1N6288A	0568
1N6288C	1972
1N6288CA	1972
1N6289	OEM
1N6289A	OEM
1N6289C	OEM
1N6289CA	OEM
1N6290	OEM
1N6290A	OEM
1N6290C	OEM
1N6290CA	OEM
1N6291	0831
1N6291A	0831
1N6291C	1987
1N6291CA	1987
1N6292	OEM
1N6292A	OEM
1N6292C	OEM
1N6292CA	OEM
1N6293	OEM
1N6293A	OEM
1N6293C	OEM
1N6293CA	OEM
1N6294	OEM
1N6294A	OEM
1N6294C	OEM
1N6294CA	OEM
1N6295	OEM
1N6295A	OEM
1N6295C	OEM
1N6295CA	OEM
1N6296	OEM
1N6296A	OEM
1N6296C	OEM
1N6296CA	OEM
1N6297	OEM
1N6297A	OEM
1N6297C	OEM
1N6297CA	OEM
1N6298	OEM
1N6298A	OEM
1N6298C	OEM
1N6298CA	OEM
1N6299	OEM
1N6299A	OEM
1N6299C	OEM
1N6299CA	OEM
1N6300	OEM
1N6300A	OEM
1N6300C	OEM
1N6300CA	OEM
1N6301	OEM
1N6301A	OEM
1N6301C	OEM
1N6301CA	OEM
1N6302	OEM
1N6302A	OEM
1N6302C	OEM
1N6302CA	OEM
1N6303	1405
1N6303A	1405
1N6303C	1408
1N6303CA	1408
1N6304	OEM
1N6305	OEM
1N6306	OEM
1N6309	1276
1N6310	0760
1N6311	0118
1N6312	0299
1N6313	0376
1N6314	0036
1N6315	0275
1N6316	0140
1N6317	0041
1N6318	0253
1N6319	0471
1N6320	0062
1N6321	0077
1N6322	0165

DEVICE TYPE	REPL CODE
1N6323	0057
1N6324	0064
1N6325	0181
1N6326	0052
1N6327	0053
1N6328	0879
1N6329	0445
1N6330	0375
1N6331	0700
1N6332	0705
1N6333	0494
1N6334	0455
1N6335	0195
1N6336	0166
1N6337	0010
1N6338	0032
1N6339	0054
1N6340	0068
1N6341	0092
1N6342	0125
1N6343	2317
1N6344	0173
1N6345	0094
1N6346	0049
1N6347	0156
1N6348	0189
1N6349	0099
1N6350	0089
1N6351	0287
1N6352	0340
1N6353	0370
1N6354	0425
1N6355	1473
1N6356	OEM
1N6357	OEM
1N6358	OEM
1N6359	OEM
1N6360	OEM
1N6361	OEM
1N6362	OEM
1N6363	OEM
1N6364	OEM
1N6365	OEM
1N6366	OEM
1N6367	OEM
1N6368	OEM
1N6369	OEM
1N6370	OEM
1N6371	OEM
1N6371C	OEM
1N6372	OEM
1N6373	OEM
1N6373C	OEM
1N6374	OEM
1N6374C	OEM
1N6375	OEM
1N6375C	OEM
1N6376	OEM
1N6376C	OEM
1N6377	OEM
1N6377C	OEM
1N6378	OEM
1N6378C	OEM
1N6379	OEM
1N6379C	OEM
1N6380	OEM
1N6380C	OEM
1N6381	OEM
1N6381C	OEM
1N6382	OEM
1N6382C	OEM
1N6383	OEM
1N6383C	OEM
1N6384	OEM
1N6384C	OEM
1N6385	OEM
1N6385C	OEM
1N6386	OEM
1N6386C	OEM
1N6387	OEM
1N6387C	OEM
1N6388	OEM
1N6388C	OEM
1N6389	OEM
1N6389C	OEM
1N6391	OEM
1N6392	OEM
1N6402	OEM
1N6402A	OEM
1N6403	OEM
1N6403A	OEM
1N6404	OEM
1N6404A	OEM
1N6405	OEM
1N6405A	OEM
1N6406	OEM
1N6406A	OEM
1N6407	OEM
1N6407A	OEM
1N6408	OEM
1N6408A	OEM
1N6409	OEM
1N6409A	OEM
1N6410	OEM
1N6410A	OEM
1N6411	OEM
1N6411A	OEM
1N6412	OEM
1N6412A	OEM

DEVICE TYPE	REPL CODE
1N6413	OEM
1N6413A	OEM
1N6414	OEM
1N6414A	OEM
1N6415	OEM
1N6415A	OEM
1N6416	OEM
1N6416A	OEM
1N6417	OEM
1N6417A	OEM
1N6418	OEM
1N6418A	OEM
1N6419	OEM
1N6419A	OEM
1N6420	OEM
1N6420A	OEM
1N6421	OEM
1N6421A	OEM
1N6422	OEM
1N6422A	OEM
1N6423	OEM
1N6423A	OEM
1N6424	OEM
1N6424A	OEM
1N6425	OEM
1N6425A	OEM
1N6426	OEM
1N6426A	OEM
1N6427	OEM
1N6427A	OEM
1N6428	OEM
1N6428A	OEM
1N6429	OEM
1N6429A	OEM
1N6430	OEM
1N6430A	OEM
1N6431	OEM
1N6431A	OEM
1N6432	OEM
1N6432A	OEM
1N6433	OEM
1N6433A	OEM
1N6434	OEM
1N6434A	OEM
1N6435	OEM
1N6435A	OEM
1N6436	OEM
1N6436A	OEM
1N6437	OEM
1N6437A	OEM
1N6438	OEM
1N6438A	OEM
1N6439	OEM
1N6439A	OEM
1N6440	OEM
1N6440A	OEM
1N6441	OEM
1N6441A	OEM
1N6442	OEM
1N6442A	OEM
1N6443	OEM
1N6443A	OEM
1N6444	OEM
1N6444A	OEM
1N6445	OEM
1N6445A	OEM
1N6446	OEM
1N6446A	OEM
1N6447	OEM
1N6447A	OEM
1N6448	OEM
1N6448A	OEM
1N6449	OEM
1N6449A	OEM
1N6450	OEM
1N6450A	OEM
1N6451	OEM
1N6451A	OEM
1N6452	OEM
1N6452A	OEM
1N6453	OEM
1N6453A	OEM
1N6454	OEM
1N6454A	OEM
1N6455	OEM
1N6455A	OEM
1N6456	OEM
1N6456A	OEM
1N6478	OEM
1N6479	OEM
1N6480	OEM
1N6481	OEM
1N6482	OEM
1N6483	OEM
1N6484	OEM
1N9221RA	OEM
1N19161	OEM
1N29864A	OEM
1N33974	0012
1N36001	0124
1N40021D	0023
1N43182	0030
1N47551	OEM
1N63402	OEM
1N70980	0133
1N601000-17	0143
1NA4	0143
1NA4G	0143
1NAG1	0143

DEVICE TYPE	REPL CODE
1NC61684	0015
1NE11	0087
1NE11-OR	0071
1NE11-A	0071
1NE11-B	0071
1NE11-C	0071
1NE11-D	0071
1NE11-K	0071
1NE11-L	0071
1NE11-Y	0071
1NE11A	0071
1NE11B	0071
1NE11C	0071
1NE11D	0071
1NE11K	0071
1NE11L	0071
1NE11R	0071
1NE11Y	0071
1NF11	0087
1NH61	0071
1NH62	0017
1NJ27	0071
1NJ33233	0143
1NJ60284	0143
1NJ61224	0143
1NJ61225	0157
1NJ61433	0133
1NJ61675	0143
1NJ61676	0015
1NJ61677	0133
1NJ61725	0133
1NJ70972	0196
1NJ70973	0143
1NJ70976	0102
1NJ70980	0133
1NJ70989	0143
1NJ71126	0920
1NJ71185	0143
1NJ71186	0920
1NJ71223	0196
1NJ71224	0133
1NK862B	0420
1NL749A	OEM
1NL750A	OEM
1NL751A	OEM
1NL752A	OEM
1NL753A	OEM
1NL754A	OEM
1NL755A	OEM
1NL756A	OEM
1NL757A	OEM
1NL914	OEM
1NL914A	OEM
1NL914B	OEM
1NL915	OEM
1NL4148	OEM
1NL4153	OEM
1NLD914	OEM
1NLD914A	OEM
1NLD914B	OEM
1NLD915	OEM
1NLD4148	OEM
1NLD4153	OEM
1NT175	OEM
1NZ61	0071
1NZ61-A	0071
1NZ61-B	0071
1NZ61-C	0071
1NZ61-D	0071
1NZ61-K	0071
1NZ61-L	0071
1NZ61-OR	0071
1NZ61-Y	0071
10M30	0805
1P	OEM
1P5-1	OEM
1P20-0006	0086
1P20-0023	0102
1P20-0029	0144
1P20-0032	0155
1P20-0034	0144
1P20-0036	0423
1P20-0037	0144
1P20-0041	0079
1P20-0046	0006
1P20-0060	0143
1P20-0061	0124
1P20-0083	0042
1P20-0086	0162
1P20-0103	1591
1P20-0110	0144
1P20-0120	0015
1P20-0122	0079
1P20-0155	0836
1P20-0157	0212
1P20-0159	0037
1P20-0167	0178
1P20-0172	1984
1P20-0213	0006
1P20-0214	0198
1P35-1	OEM
1P60F	0019
1P100	1137
1P101	0344
1P102	0901
1P103	2342
1P541	0143
1P542	0143
1P643	0023

DEVICE TYPE	REPL CODE
1P644	0023
1P645	0023
1P646	0023
1P647	0023
1P648	0271
1P649	0023
1PT	OEM
1PX	OEM
1P.5-1	OEM
1P.5-1R	OEM
1P.5-1S	OEM
1P.25-5	OEM
1P.25-5S	OEM
1P.375-75	OEM
1P.375-75S	OEM
1QE11	0102
1QZ61	0102
1R0	0133
1R01	0015
1R0E	0071
1R0F	0071
1R0H	0015
1R1D	0015
1R1K	0015
1R2	0133
1R2A	0015
1R2D	0015
1R2E	0015
1R2E01	OEM
1R2E02	OEM
1R2E16	OEM
1R3A	0133
1R3D	0015
1R3G	0015
1R3J	0015
1R3.6	0188
1R4JA	OEM
1R5B	0015
1R5BF-A	OEM
1R5BZ61	0087
1R5BZ61-A	0071
1R5BZ61-B	0071
1R5BZ61-C	0071
1R5BZ61-D	0071
1R5BZ61-K	0071
1R5BZ61-L	0071
1R5BZ61-OR	0071
1R5BZ61-Y	0071
1R5DZ61	0242
1R5DZ61-A	0071
1R5DZ61-B	0071
1R5DZ61-C	0071
1R5DZ61-D	0071
1R5DZ61-K	0365
1R5DZ61-L	0071
1R5DZ61-OR	0071
1R5DZ61-Y	0071
1R5G	0564
1R5GU41	3087
1R5GZ41	0031
1R5GZ61	0071
1R5GZ61(FA-1)	0071
1R5GZ61-A	0071
1R5GZ61-B	0071
1R5GZ61-C	0071
1R5GZ61-D	0071
1R5GZ61-K	0071
1R5GZ61-L	0071
1R5GZ61-OR	0071
1R5GZ61-Y	0071
1R5GZ61FA-1	0071
1R5GZ61FA1	0071
1R5H	0015
1R5JZ61	0965
1R5JZ61-A	0071
1R5JZ61-B	0071
1R5JZ61-C	0071
1R5JZ61-D	0071
1R5JZ61-K	0071
1R5JZ61-L	0071
1R5JZ61-OR	0071
1R5JZ61-V	0071
1R5TH61	0017
1R5.1	0162
1R6.8	OEM
1R6.8A	OEM
1R6.8B	OEM
1R7.5	0649
1R7.5A	0649
1R7.5B	OEM
1R9	0015
1R9D	0015
1R9H	0015
1R9I	0015
1R9J	0015
1R9L	0015
1R9P	0015
1R9U	0015
1R9.1	0222
1R9.1A	0012
1R9.1B	0012
1R10D3K	0133
1R10D3L	0102
1R11	0317
1R11A	0317
1R11B	0317
1R12	0137
1R12A	0137
1R12B	0137

DEVICE TYPE	REPL CODE
1R13	0365
1R13A	0365
1R13B	0365
1R15	0002
1R15A	0002
1R15B	0002
1R18	0495
1R18A	0495
1R18B	0495
1R20	0531
1R22	0565
1R22A	0565
1R22B	0565
1R27	0441
1R27A	0441
1R27B	0441
1R31	0102
1R33	0039
1R33A	0039
1R33B	0039
1R39A	0350
1R39B	0350
1R47	0999
1R47A	0999
1R47B	0999
1R56	0869
1R56A	0869
1R56B	0869
1R61	OEM
1R62	0783
1R62A	0783
1R62B	0783
1R68	1268
1R68A	1268
1R68B	1268
1R81	0015
1R82	0331
1R82A	0331
1R82B	0331
1R90	0015
1R91	1311
1R96	0071
1R100	0098
1R100A	0098
1R100B	0098
1R106B1	0940
1R106B1-C	0940
1R106B2	0940
1R106B3	0940
1R106B4	0940
1R106B41	0940
1R106D1	0095
1R106D2	0095
1R106D3	0095
1R106D4	0095
1R106D41	0095
1R110	0149
1R110A	0149
1R110B	0149
1R120	0186
1R120A	0186
1R120B	0186
1R140	0245
1R140A	0245
1R140B	0245
1R150	0028
1R150A	0028
1R150B	0028
1R161R16B	0420
1R180	0367
1R180A	0367
1R180B	0367
1R914	0124
1R1600	0240
1R1601	0240
1R1602	0240
1R1771	0240
1R1772	0240
1R1774	0240
1RBB2T	OEM
1RC5	1259
1RC10	0447
1RC20	0940
1RC30	1222
1RC40	0095
1RC60	OEM
1RE02	OEM
1RE10	0015
1RE20	OEM
1RE40	OEM
1RE60	OEM
1RE80	OEM
1RE100	OEM
1RE120	OEM
1RM40	OEM
1RM60	OEM
1RM80	OEM
1RM100	OEM
1RM120	OEM
1RM150	OEM
1RM200	OEM
1RM250	OEM
1ROE	0133
1ROF	0015
1ROH	0015
1RRGZ61	OEM
1RS	OEM
1RS35-200AT	OEM
1R-TR24	0144

DEVICE TYPE	REPL CODE
1R-TR25	0144
1RTR51	0111
1R-TR57	2096
1RTR75	0338
1R-TR77	0854
1R-TR87	0218
1S005	0015
1S016	0015
1S020	0953
1S021	0242
1S023	0540
1S025	0965
1S027	0817
1S030	0110
1S031	0953
1S032	0242
1S034	0540
1S036	0965
1S038	0817
1S054	0015
1S058	0071
1S-0142	1443
1S0S2-2.65	OEM
1S1Z09	0015
1S11	0143
1S11A	0157
1S12	0143
1S13	0143
1S14	0143
1S15	0143
1S16	0143
1S17	0143
1S17D1	0143
1S18	0143
1S19	0143
1S20	0143
1S21	OEM
1S22	OEM
1S23	OEM
1S24	OEM
1S25	OEM
1S28	0505
1S32	0143
1S33	0143
1S34	0143
1S34(TP)	0143
1S34(TV)	0143
1S34A	0143
1S34S	0143
1S35	0143
1S36	0015
1S37	0301
1S38	0133
1S39	0071
1S40	0953
1S41	0760
1S42	1747
1S43	0540
1S44	0023
1S45	0965
1S46	0817
1S47	0015
1S48	0720
1S48A	0628
1S49	0030
1S50	0123
1S51	0451
1S52	0157
1S53	0025
1S54	0649
1S55	0012
1S56	0012
1S57	OEM
1S58	0143
1S60	0123
1S60(RECT)	0015
1S60P	0143
1S61	0015
1S62	0015
1S63	0015
1S64	0015
1S65	0015
1S66	0015
1S71	0015
1S72	0143
1S73	0143
1S73A	0143
1S74	0143
1S75	0143
1S77	0143
1S77(H)	0143
1S77H	0143
1S78	0143
1S78(H)	0143
1S78H	0143
1S79	0143
1S79(H)	0143
1S79H	0143
1S80	0143
1S81	0015
1S82	0123
1S83	0015
1S83(H)	0015
1S84	0133
1S84H	0015
1S85	0720
1S85(Y)	0030
1S85(YL)	OEM
1S85/0575024	0030
1S85B0	0030
1S85H	0720
1S85L	0030
1S85M	0030
1S85SW	0030
1S85V	0015
1S85W	0030
1S85W7	0030
1S85WR	0030
1S85WT	0030
1S85Y	0030
1S85YL	0030
1S86	0760
1S87	0143
1S88	0143
1S89	0133
1S90	0015
1S90R	0133
1S91	0015
1S91R	0015
1S92	0015
1S92R	0015
1S93	0015
1S93/SGJ	0015
1S93R	0015
1S94	0015
1S94R	0015
1S95	0015
1S95R	0015
1S96	0071
1S96R	0071
1S97	0071
1S97R	0071
1S98	0071
1S98R	0071
1S99	0071
1S99A	0015
1S99R	0015
1S100	0015
1S100R	0015
1S101	0015
1S101R	0015
1S102	0015
1S102R	0015
1S103	0015
1S103R	0015
1S104	0015
1S104R	0015
1S105	0015
1S105R	0015
1S106	0015
1S106A	5525
1S106R	0015
1S107	0071
1S107R	0071
1S108	0071
1S108R	0071
1S109	0071
1S109R	0071
1S110	0015
1S110A	5525
1S111	0015
1S112	0015
1S113	0015
1S113A	0015
1S114	0015
1S115	0015
1S116	0015
1S117	0071
1S118	0015
1S119	0015
1S120	0015
1S121	0015
1S121(RECT)	0015
1S122	0015
1S123	0015
1S124	0015
1S125	0015
1S126	0015
1S127	0143
1S128	0143
1S128FM	0123
1S129	0143
1S130	0133
1S131	0133
1S132	0133
1S133	1083
1S134	0140
1S135	0023
1S135E	OEM
1S136	0062
1S136(RECT)	0015
1S136Q	0795
1S137	0077
1S138	0322
1S138(RECT)	0071
1S138(ZENER)	0322
1S139	0057
1S140	0181
1S141	0052
1S142	0686
1S143	0445
1S144	0133
1S145	0005
1S146	4140
1S147	0015
1S148	0015
1S149	0015
1S150	0015
1S151	0097
1S152	0097
1S153	0805
1S154	0805
1S155	0133
1S155-1	0133
1S156	2088
1S157	4140
1S158	0580
1S159	2060
1S160	0084
1S161	0090
1S162	0097
1S163	0105
1S164	0109
1S165	0116
1S166	0122
1S167	0871
1S168	0871
1S169	0853
1S170	1600
1S171	5282
1S172	0871
1S173	5283
1S174	0853
1S175	5286
1S176	1609
1S177	0871
1S178	0871
1S179	0853
1S180	0133
1S180/5GB	0015
1S180-M	0015
1S180B	0133
1S181	0023
1S181-M	0015
1S181FA	0133
1S182	0133
1S182-M	0015
1S183	0015
1S183-M	0015
1S184	0913
1S185	0143
1S186	0143
1S186(FM)	0143
1S186FM	0123
1S186G	0015
1S187	0143
1S187(S)	0143
1S187S	3316
1S-188	0019
1S188	0143
1S188(AM)	0143
1S188(FM)	1114
1S188-FM	2232
1S188-TV-M	0143
1S188A	0143
1S188AM	0143
1S188AM-T	OEM
1S188AR	0143
1S188F	0143
1S188F1	OEM
1S188FM	0143
1S188FM-1	0143
1S188FM1	0143
1S188FM1A	0143
1S188FM-2	0143
1S188FM2	0143
1S188FMA	0143
1S188FMI	0123
1S188G	0143
1S188MPX	0123
1S188P	0123
1S188S	0143
1S188TV	0143
1S188TV(MP)	0123
1S189	0143
1S189FM	0123
1S189FM(MP)	0123
1S190	0041
1S191	0091
1S192	0062
1S193	0244
1S194	0057
1S195	0064
1S196	0181
1S197	0052
1S198	0100
1S199	0495
1S200	0143
1S204	0015
1S205	0015
1S206	0015
1S207	0015
1S208	0015
1S208/2SJ2A	0229
1S209	0015
1S209/2SJ4A	0229
1S210	0071
1S210/2SJ6A	0071
1S211	0071
1S211/2SJ8A	0071
1S211A	0071
1S212	0322
1S213	0322
1S213C	0030
1S214	0322
1S215	0170
1S216	0170
1S217	0012
1S218	0133
1S219	OEM
1S220	0533
1S221	0157
1S222	0025
1S223	0649
1S224	1083
1S225	0170
1S226	0317
1S227	0137
1S228	0365
1S229	0100
1S230	0002
1S231	0420
1S232	1649
1S233	0495
1S234	0949
1S235	0531
1S236	0565
1S237	0402
1S238	1606
1S239	0441
1S240	0726
1S241	0039
1S242	0820
1S243	0350
1S244	0931
1S245	0999
1S246	0999
1S247	0133
1S248	0502
1S249	0869
1S250	0783
1S251	1268
1S251B	0030
1S252	1189
1S253	0331
1S254	1311
1S255	0098
1S256	0149
1S257	0149
1S258	0186
1S259	0213
1S260	0245
1S261	0028
1S262	2113
1S263	1901
1S264	1601
1S265	1616
1S266	2424
1S267	0992
1S268	0691
1S268A	0995
1S268AA	0995
1S268AA,T	0015
1S268T	0015
1S269	1263
1S270	1249
1S271	1636
1S272	1639
1S273	1703
1S274	1640
1S275	1716
1S276	1720
1S277	1731
1S278	0727
1S279	1756
1S279WT	OEM
1S280	1768
1S281	1782
1S282	1794
1S283	1799
1S284	1809
1S285	1817
1S286	1826
1S287	1853
1S288	1853
1S289	1861
1S290	1874
1S291	1884
1S292	1895
1S293	1902
1S294	0736
1S295	1909
1S296	1914
1S297	1163
1S298	1924
1S299	1933
1S300	1941
1S301	0524
1S301GR	1947
1S301.5	0002
1S302	1953
1S303	1961
1S304	0322
1S305	0322
1S306	0124
1S306M	0124
1S307	0133
1S308	OEM
1S309	0015
1S310	0015
1S310H	0015
1S311	0229
1S311H	0015
1S311N	0015
1S312	0015
1S312A	0015
1S312H	0015
1S313	0015
1S313H	0015
1S314	0023
1S314H	0015
1S314N	0015
1S315	0023
1S315FR1MD	0071
1S315H	0015
1S316	0071
1S317	0071
1S318	0143
1S319	0143
1S320	0162
1S322	0133
1S323	0143
1S324	0124
1S325	0124
1S326	0124
1S327	0015
1S328	0015
1S329	0071
1S330	0041
1S330A	0041
1S-331	0636
1S331	0091
1S331A	0091
1S331AZ	0091
1S332	0649
1S332M	0025
1S333	0244
1S333Y	0244
1S334	0012
1S334M	0012
1S334N	0012
1S335	0064
1S336	0181
1S337	0052
1S337-Y	0137
1S337A	0137
1S337E	0137
1S337Y	0137
1S338	0100
1S338Q	0365
1S338U	0100
1S339	0375
1S340	0442
1S341	3127
1S341W1	0030
1S341W2	0030
1S341W3	0030
1S342	0205
1S343	0504
1S344	0684
1S345	0225
1S346	0230
1S347	0301
1S348	1397
1S349	0256
1S351	0005
1S351-S	0005
1S351B	0005
1S351M	0005
1S351R	0005
1S351S	0005
1S351W	0005
1S351W1	0005
1S351W2	0005
1S351W3	0005
1S352	0030
1S352A	0030
1S352M	0030
1S352N	0720
1S352R	0030
1S353	0628
1S353M	0030
1S354	0143
1S355	0143
1S356	0143
1S357	0143
1S358	0015
1S358A	0015
1S358S	0015
1S359	0015
1S360	OEM
1S361	OEM
1S362	OEM
1S362.3	OEM
1S363	OEM
1S367	0015
1S368	0015
1S369	0015
1S370	0015
1S371	0015
1S372	0015
1S373	0015
1S374	0015
1S375	0015
1S376	0015
1S377	0015
1S378	0015
1S379	0015
1S380	0015
1S381	0015
1S382	0015
1S383	0871
1S384	0871
1S385	0853
1S388	0871
1S389	0871
1S390	0015
1S393	0015
1S394	0015
1S395	0015
1S396	0015
1S397	0071
1S398	0071
1S399	0015
1S400	0015
1S401	0071
1S402	0071
1S403	0102
1S404	0015
1S405	0071
1S406	0071
1S407	0102
1S408	OEM
1S409	OEM
1S410	0580
1S411	0580
1S412	OEM
1S413	1000
1S414	OEM
1S415	2081
1S416	OEM
1S417	2088
1S418	0871
1S419	0612
1S420	0090
1S421	0097
1S422	0853
1S423	0109
1S425	0122
1S426	0123
1S426G	0143
1S426GFM	0143
1S427	0879
1S428	0143
1S430	0110
1S431	0953
1S431R	OEM
1S432	0242
1S432R	OEM
1S433	OEM
1S433R	OEM
1S434	0540
1S434R	OEM
1S435	OEM
1S435R	OEM
1S436	0965
1S436R	OEM
1S437	OEM
1S438	0817
1S438R	OEM
1S439	OEM
1S440	OEM
1S440R	OEM
1S441	0143
1S441(RECT)	0708
1S442	0143
1S442(RECT)	0708
1S443	0708
1S443(RECT)	0708
1S444	0133
1S444(RECT)	0580
1S445	OEM
1S445(RECT)	0580
1S446	0143
1S446(RECT)	0580
1S446D	0143
1S447	0143
1S448	0143
1S449	0143
1S450	OEM
1S451	0143
1S452	0143
1S453	0143
1S454	0143
1S455	0143
1S456	0124
1S457	0124
1S458	0023
1S459	0023
1S460	0124
1S461	0124
1S462	0124
1S463	0242
1S464	0133
1S465	0124
1S466	0163
1S467	OEM
1S467R	0143
1S468	0124
1S470	0091
1S471	0062
1S472	0057
1S473	0181
1S474	0879
1S475	0375
1S476	0705
1S477	0257
1S478	0010
1S479	0068
1S480	0301
1S481	0025
1S482	0012
1S483	0317
1S484	0100
1S485	0495
1S486	0565
1S487	1674
1S488	0820
1S489	0999
1S490	2416
1S491	0205
1S492	0627
1S493	0230
1S494	1153
1S495	1716
1S496	1723
1S497	2453
1S498	1809
1S499	0347
1S500	0124
1S501	0124
1S501M	0124
1S502	OEM
1S503	OEM
1S504	OEM
1S505	OEM
1S506	OEM
1S507	OEM
1S508	OEM
1S509	OEM
1S510	0952
1S511	OEM
1S512	OEM
1S513	OEM
1S514	OEM
1S515	OEM
1S516	OEM
1S517	OEM
1S518	OEM
1S519	OEM
1S520	0952
1S521	OEM
1S522	OEM
1S523	OEM
1S524	OEM
1S525	OEM
1S526	OEM
1S527	OEM
1S528	OEM
1S529	OEM
1S530	OEM
1S531	OEM
1S532	OEM
1S533	OEM
1S534	OEM
1S535	0554
1S535-1	0030
1S536	OEM
1S537	OEM
1S538	OEM
1S539	OEM
1S540	OEM
1S541	OEM
1S542	0143
1S543	0917
1S544	0131
1S545	0145
1S546	1610
1S547	1614
1S548	0124
1S550	0165
1S551	0165
1S552	0165
1S553	0760
1S553T	0760
1S554	0628
1S554T	0628
1S555	0628
1S556	0030
1S557	0023
1S557H	0071
1S558	0023
1S559	0015
1S560	0133
1S560(H)	OEM
1S560H	0133
1S561	OEM
1S561A	OEM
1S562	OEM
1S562A	OEM
1S563	OEM
1S563A	OEM
1S564	OEM
1S564A	OEM
1S565	OEM
1S565A	OEM
1S566	OEM
1S566A	OEM
1S567	OEM
1S567A	OEM
1S568	0133
1S568A	OEM
1S569	OEM
1S569A	OEM
1S582	OEM
1S588	0015
1S589	0143
1S593	0015
1S597	0002
1S600	OEM
1S601	OEM
1S602	OEM
1S603	OEM

DEVICE TYPE	REPL CODE
1S604	OEM
1S610	0952
1S611	0760
1S612	0376
1S613	0533
1S614	0533
1S615	0253
1S616	0649
1S617	0649
1S618	0170
1S619	0365
1S620	0420
1S621	0671
1S622	0402
1S623	1674
1S624	0010
1S630	OEM
1S631	OEM
1S632	0376
1S633	0451
1S634	0162
1S635	0301
1S636	0025
1S661	0015
1S662	0071
1S663	0071
1S664	0102
1S665	OEM
1S666	OEM
1S667	OEM
1S668	OEM
1S669	0853
1S673	0853
1S685	0015
1S686	0071
1S687	0071
1S688	0102
1S689	0924
1S689A	0924
1S689BK	0924
1S690	0456
1S691	0451
1S692	0157
1S693	0133
1S694	0244
1S695	0170
1S696	0137
1S697	0002
1S698	0495
1S699	0565
1S700	0441
1S701	2035
1S702	1438
1S703	1445
1S704	1458
1S704A	0275
1S704B	0275
1S705	1490
1S706	0510
1S707	0870
1S708	1274
1S709	1078
1S710	1723
1S711	1761
1S712	2035
1S713	1438
1S714	1445
1S715	1458
1S716	1490
1S717	0510
1S718	0870
1S719	1274
1S720	1078
1S721	1723
1S722	1761
1S731	0124
1S732	OEM
1S733	OEM
1S734	OEM
1S735	OEM
1S735R	OEM
1S736	OEM
1S736R	OEM
1S737	OEM
1S738	OEM
1S738R	OEM
1S739	OEM
1S740R	OEM
1S741	OEM
1S742	OEM
1S742R	OEM
1S743	OEM
1S744	0143
1S745	0143
1S746	0143
1S747	0143
1S748	0143
1S749	OEM
1S750	0917
1S750(2)	0917
1S750S	0917
1S751	0924
1S752	0760
1S752(H)	0760
1S753	0376
1S753(H)	0299
1S753H	0292
1S754	0275
1S754(H)	0275
1S754H	0456
1S755	0140
1S755(H)	0041
1S755H	0451
1S756	0091
1S756(H)	0253
1S757	0062
1S757(H)	0077
1S757A	0012
1S758	0322
1S758(H)	0322
1S758GS	0012
1S758H	1083
1S759	0064
1S759(H)	0181
1S759H	0170
1S760	0053
1S760(H)	0053
1S760H	0365
1S761	0445
1S761(H)	0445
1S762	0671
1S762(H)	0671
1S763	0494
1S763(H)	0494
1S764	0455
1S764(H)	0257
1S765	0039
1S765(H)	0010
1S765H	0039
1S766	0005
1S767	0005
1S767A	0005
1S768	OEM
1S769	0005
1S770	OEM
1S770R	OEM
1S773	0124
1S773A	0124
1S774	0124
1S774A	0124
1S775	0124
1S775FA-1	0023
1S776	0124
1S776A	0124
1S777	0015
1S778	0871
1S779	OEM
1S780	OEM
1S781	0853
1S782	OEM
1S783	OEM
1S784	OEM
1S785	OEM
1S786	OEM
1S787	OEM
1S788	0143
1S789	OEM
1S790	0853
1S791	OEM
1S792	OEM
1S793	OEM
1S794	OEM
1S796	OEM
1S797	OEM
1S798	OEM
1S799	OEM
1S800	OEM
1S801	0124
1S802	0124
1S803	0124
1S804	0124
1S805	0124
1S806	0124
1S807	0124
1S808	0124
1S811	0015
1S812	0015
1S813	0071
1S814	0071
1S815	0102
1S816	0015
1S817	0015
1S818	0071
1S819	0071
1S820	0102
1S821	0853
1S822	0853
1S823	OEM
1S824	OEM
1S825	OEM
1S826	0853
1S827	0853
1S831	0853
1S832	0853
1S836	0853
1S837	0853
1S841	0015
1S842	0015
1S843	0015
1S844	0015
1S844N	0015
1S845	0015
1S846	0015
1S846N	0015
1S847	0071
1S848	0071
1S848N	0071
1S849	0015
1S849,R	0015
1S849R	0015
1S850	0071
1S850N	0071
1S851	0143
1S852	0102
1S853	0030
1S854	0540
1S854L	0030
1S855	0965
1S856	0965
1S857	0817
1S858	0817
1S859	0922
1S860	0922
1S871	0015
1S871A	0015
1S871B	0071
1S871C	0071
1S871D	0102
1S871E	0102
1S872	0015
1S872A	0015
1S872B	0071
1S872C	0071
1S872D	0102
1S873	0853
1S874	0853
1S881	OEM
1S881A	0015
1S881B	0071
1S881C	0071
1S881D	0102
1S881E	0102
1S882	0015
1S882A	0015
1S882B	0071
1S882C	0071
1S882D	0102
1S882E	0102
1S885	0015
1S887	0015
1S891	0005
1S892	0720
1S893	0628
1S894	0201
1S895	0201
1S896	0201
1S897	0201
1S920	0133
1S921	0133
1S921(RECT)	0242
1S922	0540
1S922(DIODE)	0133
1S922(RECT)	0540
1S923	0133
1S923(RECT)	0965
1S923A	OEM
1S924	0817
1S925	0922
1S926	0097
1S927	0109
1S928	0122
1S929	0131
1S930	0145
1S931	0599
1S932	1986
1S933	0657
1S934	0657
1S935	0657
1S936	1124
1S937	OEM
1S938	OEM
1S939	0657
1S940	0657
1S941	0133
1S942	0133
1S943	0015
1S946	0015
1S947	0015
1S948	0015
1S949	0071
1S950	0071
1S951	0124
1S952	0124
1S953	0133
1S954	0133
1S954(2)	0124
1S955	0124
1S956	0242
1S957	0540
1S958	0965
1S959	0817
1S960	0922
1S961	0124
1S962	0853
1S963	0229
1S967	0853
1S977	0124
1S978	0124
1S981	0124
1S982	0124
1S983	0133
1S984	0124
1S985	0124
1S986	0124
1S987	0124
1S988	0124
1S989	1720
1S990	0133
1S990-A	OEM
1S990-AM	0133
1S990A	0133
1S990AM	0143
1S990S	0002
1S991	OEM
1S993	0118
1S994	0456
1S994A	0133
1S994S	0451
1S994Y	0649
1S995	0005
1S1000	0917
1S1000A	0917
1S1001	0133
1S1004	0015
1S1005	OEM
1S1006	0143
1S1007	0143
1S1007S	0143
1S1008	0143
1S1009	0143
1S1010	0143
1S1011	OEM
1S1012	0015
1S1013	0015
1S1014	0071
1S1015	0071
1S1016	0102
1S1017	0102
1S1021	0015
1S1022	0015
1S1023	0071
1S1024	0071
1S1025	0102
1S1026	0102
1S1027	0015
1S1028	0015
1S1029	0071
1S1030	0071
1S1031	0102
1S1032	0102
1S1033	1023
1S1034	1023
1S1035	1023
1S1036	1036
1S1037	1046
1S1038	1046
1S1039	0540
1S1040	0965
1S1041	0817
1S1042	0922
1S1043	0015
1S1044	0015
1S1045	0071
1S1046	0071
1S1047	0102
1S1048	0102
1S1052	0133
1S-1053	0124
1S1053	0133
1S1061	0015
1S1062	0015
1S1063	0015
1S1064	0015
1S1065	0071
1S1066	0071
1S1067	0102
1S1068	0102
1S1071	0087
1S1072	0242
1S1073	0540
1S1074	0965
1S1075	0817
1S1076	0087
1S1076TF1	0124
1S1081	0871
1S1094	0077
1S1095	0057
1S1096	0015
1S1097	0365
1S1098	0445
1S1099	0671
1S1100	0494
1S1101	0257
1S1102	0820
1S1103	OEM
1S1104	OEM
1S1105	OEM
1S1106	OEM
1S1107	OEM
1S1108	OEM
1S1109	OEM
1S1110	OEM
1S1111	OEM
1S1112	OEM
1S1113	OEM
1S1114	0649
1S1114A	0649
1S1115	OEM
1S1116	OEM
1S1117	OEM
1S1118	OEM
1S1119	OEM
1S1120	OEM
1S1121	0277
1S1122	0289
1S1123	0296
1S1124	0296
1S1125	0302
1S1126	0250
1S1127	0250
1S1128	0323
1S1129	1414
1S1130	0473
1S1131	0473
1S1132	0446
1S1135	0323
1S1136	1414
1S1137	0473
1S1138	0473
1S1139	0446
1S1144	OEM
1S1145	OEM
1S1146	0015
1S1147	0124
1S1148	0124
1S1149	0124
1S1154	0077
1S1155	0165
1S1156	0064
1S1160	0376
1S1161	0140
1S1162	0091
1S1163	0077
1S1164	0165
1S1165	0064
1S1166	0053
1S1167	0445
1S1168	0671
1S1169	0494
1S1170	0257
1S1171	0010
1S1172	0451
1S1173	0301
1S1174	0025
1S1175	1083
1S1176	0317
1S1177	0002
1S1178	0420
1S1179	0949
1S1180	0402
1S1181	1674
1S1182	0820
1S1183	0931
1S1184	0502
1S1185	1438
1S1186	1445
1S1187	1458
1S1188	1490
1S1189	0510
1S1190	0870
1S1191	1532
1S1192	1078
1S1193	1723
1S1194	1761
1S1195	1804
1S1196	0022
1S1197	0207
1S1198	OEM
1S1198A	OEM
1S1199	OEM
1S1199A	OEM
1S1200	OEM
1S1200A	OEM
1S1201	0471
1S1201(H)	0471
1S1202	0062
1S1202(H)	0062
1S1203	0165
1S1204	0139
1S1207	0133
1S1208	0124
1S1209	0139
1S1210	0139
1S1211	0139
1S1212	0139
1S1212A	0139
1S1213	0133
1S1214	0133
1S1215	0133
1S1216	0133
1S1217	0133
1S1218	0133
1S1219	0133
1S1219H	0124
1S1220	0133
1S1221	0015
1S1222	0015
1S1223	0071
1S1224	0015
1S1225	0071
1S1225A	0071
1S1226	0102
1S1230	0015
1S1230H	0015
1S1231	0015
1S1231H	0015
1S1232	0015
1S1232H	0015
1S1232N	0015
1S1233	0071
1S1233H	0071
1S1233N	0071
1S1234	0071
1S1234H	0071
1S1234N	0071
1S1235	0331
1S1236	0965
1S1237	0023
1S1238	0102
1S1239	0143
1S1244	0143
1S1245	0143
1S1246	0143
1S1255A	0071
1S1260	0871
1S1260R	1635
1S1261	0871
1S1261R	1635
1S1262	0853
1S1262R	1251
1S1263	0853
1S1263R	1251
1S1264	1609
1S1264R	1205
1S1265	1609
1S1265R	1205
1S1266	1610
1S1266R	2137
1S1267	1610
1S1267R	2137
1S1268	1614
1S1268R	2251
1S1269	1614
1S1269R	2251
1S1270	0871
1S1270R	1635
1S1271	0871
1S1271R	1635
1S1272	0853
1S1272R	1251
1S1273	0853
1S1280	0871
1S1281	0853
1S1282	0853
1S1288	5282
1S1288R	OEM
1S1289	0871
1S1290	0853
1S1295	0124
1S1296	0124
1S1297	0277
1S1298	0015
1S1299	0015
1S1300	OEM
1S1301	0124
1S1302	0124
1S1303	0124
1S1304	0133
1S1305	0133
1S1308	0365
1S1314	1335
1S1315	0005
1S1316	0124
1S1317	0124
1S1318	0023
1S1319	0023
1S1320	0023
1S1321	1282
1S1322	1282
1S1323	1287
1S1324	1287
1S1325	1293
1S1326	1293
1S1327	1296
1S1328	1296
1S1329	1296
1S1329B	0030
1S1330	1296
1S1331	OEM
1S1332	OEM
1S1333	0023
1S1333(RECT)	OEM
1S1333R	OEM
1S1334	OEM
1S1334R	OEM
1S1335	OEM
1S1335R	OEM
1S1336	OEM
1S1336R	OEM
1S1337	OEM
1S1337R	OEM
1S1338	OEM
1S1338R	OEM
1S1339	OEM
1S1339R	OEM
1S1340	OEM
1S1341	0015
1S1342	0015
1S1343	0015
1S1344	0015
1S1345	0071
1S1346	0015
1S1347	0071
1S1348	0071
1S1349	0102
1S1350	1600
1S1351	4140
1S1352	0580
1S1353	2060
1S1354	1000
1S1355	2081
1S1356	2088
1S1357	0612
1S1358	OEM
1S1359	0015
1S1359A	0015
1S1360	0015
1S1360A	0015
1S1361	0015
1S1361A	0015
1S1362	0071
1S1362A	0071
1S1363	0071
1S1363A	0071
1S1364	0102
1S1364A	0102
1S1365	0015
1S1365A	0015
1S1366	0015
1S1366A	0015
1S1367	0015
1S1367A	0015
1S1368	0071
1S1368A	0071
1S1369	0071
1S1369A	0071
1S1370	0102
1S1370A	0102
1S1371	2874
1S1371A	2874
1S1372	0376
1S1372A	0376
1S1373	0533
1S1373A	0533
1S1374	0157
1S1374A	0157
1S1375	0636
1S1375A	0636
1S1376	0025
1S1376A	0025
1S1377	1083
1S1377A	1083
1S1378	0170
1S1378A	0064
1S1379	0365
1S1379A	0365
1S1380	0420
1S1380A	0420
1S1381	0700
1S1381A	0700
1S1382	1606
1S1382A	1606
1S1383	0726
1S1383A	0726
1S1384	0820
1S1384A	0820
1S1385	0054
1S1385A	0054
1S1386	0760
1S1386A	0760
1S1387	0376
1S1387A	0376
1S1388	0533
1S1388A	0533
1S1389	0157
1S1389A	0157
1S1390	0636
1S1390A	0636
1S1391	0649
1S1391A	0649
1S1392	1083
1S1392A	1083
1S1393	0170
1S1393A	0170
1S1394	0365
1S1394A	0365
1S1395	0420
1S1395A	0420
1S1396	0700
1S1396A	0700
1S1397	1606
1S1397A	1606
1S1398	0726
1S1398A	0726
1S1399	0820
1S1399A	0820
1S1400	0054
1S1400A	0054
1S1401	0502
1S1401A	0502
1S1402	1312
1S1402R	1312
1S1403	0188
1S1403R	0188
1S1404	0451
1S1404R	0451
1S1405	0157
1S1405R	0157
1S1406	0636
1S1406R	0636
1S1407	0649
1S1407R	0649
1S1408	1083
1S1408R	1083
1S1409	0170
1S1409R	0170
1S1410	0100
1S1410R	0100
1S1411	0420
1S1411R	0420
1S1412	0531
1S1412R	0531
1S1413	1606

DEVICE TYPE	REPL CODE
1S1413R	1606
1S1414	0726
1S1414R	0726
1S1415	0820
1S1415R	0820
1S1417	1119
1S1418	0281
1S1419	OEM
1S1420	OEM
1S1420(H)	0133
1S1420H	0133
1S1421	0931
1S1421R	0931
1S1422	1834
1S1422R	1834
1S1424	2035
1S1424R	0547
1S1425	1438
1S1425R	2113
1S1426	1445
1S1426R	1901
1S1427	1458
1S1427R	1601
1S1429	1490
1S1429R	1622
1S1430	0510
1S1430R	0992
1S1439	1438
1S1439R	2113
1S1440	1445
1S1440R	1901
1S1441	1458
1S1441R	1601
1S1443	1490
1S1443R	1622
1S1444	0510
1S1444R	0992
1S1445	0870
1S1445R	1263
1S1454	0853
1S1461	OEM
1S1462	OEM
1S1463	OEM
1S1464	0917
1S1465	0924
1S1471	0071
1S1472	0229
1S1473	0124
1S1474	0242
1S1475	0540
1S1476	0965
1S1477	0817
1S1478	0922
1S1480	0871
1S1481	0853
1S1486	0015
1S1487	0015
1S1488	0242
1S1489	0540
1S1490	0965
1S1491	0817
1S1492	0922
1S1500	0554
1S1501	0554
1S1502	0005
1S1502C	0005
1S1503	0760
1S1514	0133
1S1515	0133
1S1516	0133
1S1517	0182
1S1517A	0102
1S1532	0133
1S1533	0030
1S1540	OEM
1S1544	0163
1S1544A	0163
1S1545	0163
1S1546	0015
1S1547	0015
1S1548	0015
1S1549	0917
1S1551	OEM
1S1552	0124
1S1553	0133
1S1553(TV)	0133
1S1553V	0133
1S1553V(TV)	0133
1S1554	0133
1S1554(TV)	0133
1S1554V	0133
1S1554V(TV)	0133
1S1555	0133
1S1555(TV)	0133
1S1555-1	0133
1S1555-5	0102
1S1555-K	0133
1S1555-S	0133
1S-1555-Z	0133
1S1555FA-1	0133
1S1555FA1	0133
1S1555TP	0133
1S1555TP3	0133
1S1555TV	4179
1S-1555V	0133
1S1555V	0133
1S1555V(TV)	0133
1S1555Z	0133
1S1556	OEM

DEVICE TYPE	REPL CODE
1S1557	OEM
1S1558	0030
1S1560	OEM
1S1560R	OEM
1S1561	OEM
1S1562	OEM
1S1563	OEM
1S1564	OEM
1S1571	OEM
1S1572	OEM
1S1573	OEM
1S1574	OEM
1S1575	0109
1S1576	0122
1S1577	0131
1S1578	0145
1S1579	0133
1S1580	0133
1S1581	OEM
1S1582	0554
1S1583	OEM
1S1585	0124
1S1585(TPE3)	0124
1S1586	0124
1S1587	0124
1S1588	0133
1S1588-FM	0143
1S1588-T	OEM
1S1588T	OEM
1S1588V	0124
1S1589	0133
1S1591	0102
1S1592	0102
1S1593	0102
1S1594	0102
1S1595	0102
1S1596	0102
1S1597	0015
1S1598	0015
1S1599	0015
1S1600	0023
1S1601	0023
1S1602	0023
1S1604	0015
1S1605	0071
1S1606	0871
1S1607	0853
1S1608	0071
1S1609	0102
1S1610	0102
1S1611	0201
1S1612	0201
1S1613	0378
1S1614	2833
1S1615	2473
1S1616	OEM
1S1617	OEM
1S1618	OEM
1S1619	OEM
1S1621	0133
1S1621-O	0133
1S1621-R	0133
1S1621-Y	0133
1S1622	0015
1S1623	0015
1S1624	0015
1S1625	0015
1S1626	0206
1S1627	0586
1S1628	OEM
1S1629	2088
1S1630	0612
1S1631	OEM
1S1632	OEM
1S1640	0871
1S1641	0871
1S1642	OEM
1S1642R	OEM
1S1643	0599
1S1643R	1347
1S1644	0599
1S1644R	1347
1S1645	OEM
1S1646	OEM
1S1647	OEM
1S1648	0030
1S1650	0133
1S1651	0133
1S1652	1250
1S1652R	0765
1S1653	5587
1S1653R	0777
1S1654	1124
1S1654R	1107
1S1655	1126
1S1655R	1111
1S1656	OEM
1S1657	0924
1S1658	0030
1S1658(FA-3)	OEM
1S1658FA-2	0030
1S1658FA2	0030
1S1658FA-3	0030
1S1659	OEM
1S1660	0580
1S1660R	0947
1S1661	2060
1S1661R	4523
1S1662	0989

DEVICE TYPE	REPL CODE
1S1662R	0990
1S1663	3743
1S1663R	0993
1S1664	0015
1S1665	0229
1S1666	0015
1S1667	0015
1S1668	0015
1S1668F	0015
1S1669	0071
1S1670	0277
017031	0289
1S1671	0289
1S1672	0296
1S1673	0302
1S1674	0250
1S1675	0250
1S1676	OEM
1S1677	OEM
1S1678	0015
1S1679	0015
1S1680	0015
1S1681	0015
1S1682	0071
1S1683	0071
1S1684	0015
1S1685	0015
1S1686	0015
1S1687	0015
1S1688	0071
1S1689	0071
1S1690	0143
1S1690A	0015
1S1691	0015
1S1691A	0015
1S1692	0229
1S1692A	0015
1S1693	0015
1S1693A	0015
1S1694	0229
1S1694A	0015
1S1695	0071
1S1695A	0015
1S1696	0015
1S1696A	0015
1S1697	0015
1S1698	0015
1S1699	0167
1S1700	0167
1S1701	5594
1S1702	0102
1S1703	0102
1S1704	0102
1S1704A	0102
1S1705	0102
1S1705A	0102
1S1706	0102
1S1707	0102
1S1707A	0102
1S1708	0102
1S1709	0102
1S1710	0124
1S1711	0124
1S1712	0124
1S1712A	0124
1S1713	0124
1S1714	0124
1S1715	0636
1S1716	0077
1S1717	1083
1S1717L	0012
1S1718	0012
1S1719	OEM
1S1720	0087
1S1720R	0087
1S1721	0917
1S1722	0917
1S1723	0015
1S1724	0015
1S1724R	0015
1S1725	0015
1S1725R	0015
1S1726	0015
1S1726R	0015
1S1727	0015
1S1727R	0015
1S1728	0012
1S1728R	0015
1S1729	0015
1S1729R	0015
1S1730	0015
1S1730R	0015
1S1731	0071
1S1731R	0071
1S1732	0071
1S1732R	0071
1S1733	0102
1S1733R	0102
1S1734	0162
1S1735	0636
1S1736	0649
1S1737	1083
1S1738	0317
1S1739	0137
1S1740	0002
1S1741	1649
1S1742	0531
1S1743	1606
1S1744	0726

DEVICE TYPE	REPL CODE
1S1745	0820
1S1746	0931
1S1747	1834
1S1748	0783
1S1749	1189
1S1750	0331
1S1751	1311
1S1752	0098
1S1753	0149
1S1754	0186
1S1755	0213
1S1756	0245
1S1757	0628
1S1758	0628
1S1759	0628
1S1760	OEM
1S1761	OEM
1S1762	OEM
1S1763	OEM
1S1764	OEM
1S1765	OEM
1S1766	0162
1S1767	0301
1S1768	0649
1S1769	1083
1S1770	0317
1S1771	0365
1S1772	0002
1S1773	1649
1S1774	0700
1S1775	0402
1S1776	0726
1S1777	0039
1S1781	1083
1S1782	1083
1S1783	1083
1S1784	0783
1S1785	2157
1S1786	0331
1S1787	0098
1S1788	0186
1S1789	0213
1S1790	0028
1S1791	0367
1S1792	0421
1S1796	0124
1S1798	0097
1S1798R	0765
1S1799	0097
1S1799R	0765
1S1800	OEM
1S1800R	OEM
1S1801	0109
1S1801-02	0124
1S1801R	2966
1S1802	0122
1S1802R	0122
1S1803	0131
1S1803R	0545
1S1804	0145
1S1804R	0550
1S1805	0005
1S1805A	0005
1S1806	0005
1S1806A	0005
1S1807	0917
1S1820	0917
1S1821	0072
1S1821R	0072
1S1822	OEM
1S1822R	OEM
1S1823	OEM
1S1823R	OEM
1S1824	OEM
1S1824R	OEM
1S1825	0133
1S1825R	OEM
1S1829	0071
1S1829A	0071
1S1829B	0071
1S1829C	0071
1S1829D	0071
1S1829K	0071
1S1829L	0071
1S1829Y	0071
1S1830	0071
1S1830A	0071
1S1830B	0071
1S1830C	0071
1S1830D	0071
1S1830K	0071
1S1830L	0071
1S1830Y	0071
1S1832	0182
1S1833	0023
1S1834	0023
1S1835	0023
1S1836	0288
1S1837	0378
1S1838	OEM
1S1841	0805
1S1842	0319
1S1843	1132
1S1844	0044
1S1845	0015
1S1845A	0015
1S1845B	0015
1S1845C	0071
1S1845D	0071

DEVICE TYPE	REPL CODE
1S1845E	0102
1S1846	0015
1S1846A	0015
1S1846B	0015
1S1846C	0071
1S1846D	0071
1S1846E	0102
1S1847A	0853
1S1848	0015
1S1848A	0853
1S1849	0015
1S1849R	0023
1S1850	0023
1S1850R	0023
1S1851	0015
1S1851R	0015
1S1855	0124
1S1856	OEM
1S1857	OEM
1S1858	OEM
1S1860	0917
1S1876	OEM
1S1881AM	0015
1S1882	0102
1S1885	0023
1S1885(FA)	0023
1S1885-3	0015
1S1885A	0609
1S1885FA	0609
1S1886	0023
1S1886A	0609
1S1887	0023
1S1887(AU)	0795
1S1887(FA)	0795
1S1887A	0023
1S1887A(FA)	OEM
1S1887A(FA-1)	0015
1S1887AFA	0015
1S1887AFA-1	0795
1S1887ALC	0795
1S1887A-LC7	0023
1S1887ALC-7	0795
1S1887A-T3	0023
1S1887A-Z	0795
1S1887FA	0015
1S1887FA-1	0015
1S1888	5473
1S1888H	OEM
1S1888J	OEM
1S1888V	0015
1S1890	0965
1S1891	0023
1S1892	0087
1S1893	0554
1S1894A	OEM
1S1894B	OEM
1S1894C	OEM
1S1895	0720
1S1895A	OEM
1S1895B	OEM
1S1895C	OEM
1S1895D	OEM
1S1897	0965
1S1898	0817
1S1899	0922
1S1906	0015
1S1910	0102
1S1911	0102
1S1912	0201
1S1914	0953
1S1915	0242
1S1916	0540
1S1917	0965
1S1918	0817
1S1919	0922
1S1920	0182
1S1921A	0023
1S1921B	0023
1S1921C	0023
1S1921D	0023
1S1921E	0023
1S1921F	0023
1S1922	0554
1S1922DS	OEM
1S1922G	0917
1S1923	0005
1S1923A	0554
1S1924	0720
1S1925	0917
1S1926	0917
1S1926K	0917
1S1926P	0917
1S1928	0917
1S1934	0871
1S1934R	1635
1S1935	0871
1S1935R	1635
1S1936	0853
1S1936R	1251
1S1937	1609
1S1937R	1205
1S1938	1610
1S1938R	2137
1S1939	1614
1S1939R	2251
1S1941	0023
1S1941B	0133
1S1941L	1335
1S1942	0023

DEVICE TYPE	REPL CODE
1S1943	0015
1S1944	0023
1S1944H	OEM
1S1944J	OEM
1S1945	OEM
1S1946	OEM
1S1947	OEM
1S1948	0015
1S1949	0015
1S1950	0015
1S1951	0039
1S1953	1713
1S1954	OEM
1S1955	OEM
1S1956	0301
1S1957	0025
1S1958	0244
1S1959	0012
1S1960	0170
1S1961	0317
1S1962	0137
1S1963	0100
1S1964	0002
1S1965	0420
1S1966	0495
1S1967	0531
1S1968	0565
1S1969	0402
1S1970	0441
1S1971	0726
1S1972	0124
1S1972-M	0124
1S1973	0124
1S1973-M	0124
1S1991	0242
1S1992	2518
1S1993	2518
1S1994	2518
1S1995	2518
1S1996	0871
1S1997	0853
1S2000	0853
1S2003	OEM
1S2003R	OEM
1S2004	OEM
1S2004R	OEM
1S2005	0853
1S2019	0071
1S2030A	0118
1S2031	0036
1S2033	0299
1S2033A	0299
1S2034	OEM
1S2034R	OEM
1S2035	OEM
1S2035R	OEM
1S2036	0376
1S2036A	0376
1S2036R	OEM
1S2037	OEM
1S2037R	OEM
1S2039	0456
1S2039A	0036
1S2043	0275
1S2043A	0275
1S2045(STUD)	0547
1S2046(STUD)	2113
1S2047	0140
1S2047(STUD)	1901
1S2047A	0140
1S2049(STUD)	1622
1S2051	0041
1S2051(STUD)	0992
1S2051A	0041
1S2051B	0162
1S2053(STUD)	1263
1S2055(STUD)	1274
1S2056	0253
1S2056A	0253
1S2056B	0253
1S2057(STUD)	1716
1S2059(STUD)	0727
1S2061(STUD)	1782
1S2062	0471
1S2062A	0471
1S2063	0071
1S2064	0922
1S2067	0322
1S2068	0322
1S2068A	0062
1S2069	0322
1S2070	0124
1S2071	0124
1S2071A	0124
1S2072	0124
1S2073-A	OEM
1S2074	0124
1S2074H	0133
1S2075	0133
1S2075A	0077
1S2075K	0133
1S2076	0133
1S2076-27	0124
1S2076-TF1	0124
1S2076-TFI	0124
1S2076A	0124
1S2076A/-07-27	0124
1S2076A-07	0124
1S2076A-07-27	0124

DEVICE TYPE	REPL CODE
1S2076MC	0124
1S2076P	0124
1S2076S4	OEM
1S2076T	0124
1S2076TF1	0124
1S2076TFI	0396
1S2080	0023
1S2081	0015
1S2082	0165
1S2082A	0165
1S2083	OEM
1S2083R	OEM
1S2085	0720
1S2085A	0720
1S2086	0720
1S2086A	0720
1S2087	0720
1S2087A	0720
1S2088	0720
1S2088A	0720
1S2089	0005
1S2090	0554
1S2091	0133
1S2091-BK	0133
1S2091-BL	0133
1S2091-W	0133
1S2091A	0057
1S2091BK	0133
1S2091BL	0133
1S2091W	0133
1S2092	0133
1S2093	2730
1S2094	0005
1S2095	0124
1S2095A	0124
1S2097	0133
1S2098	0133
1S2099	0124
1S2100	0064
1S2100A	0064
1S2107	0456
1S2107A	0456
1S2108	0533
1S2108A	0533
1S2109	0140
1S2109A	0451
1S2110	0181
1S2110A	0181
1S2111	0157
1S2111A	0157
1S2112	0091
1S2112A	0471
1S2112A(FA-1)	OEM
1S2112AFA-1	0636
1S2112FA-1	OEM
1S2113	0025
1S2113A	0025
1S2114	0649
1S2114A	0649
1S2115	0244
1S2115A	0244
1S2116	0012
1S2116A	0012
1S2117	5653
1S2117A	0170
1S2118	0317
1S2118A	0317
1S2119	0052
1S2119A	0052
1S2120	0052
1S2120A	0052
1S2121	0002
1S2121A	0002
1S2122	1649
1S2122A	0420
1S2123	0495
1S2123A	0495
1S2124	0015
1S2125	0091
1S2126	0322
1S2127	0052
1S2128	0301
1S2129	1083
1S2130	0137
1S2130A	0053
1S2131	0949
1S2131C	0030
1S2132	0098
1S2133	0213
1S2134	0124
1S2135	0124
1S2136	0091
1S2137	1083
1S2138	0137
1S2139	0005
1S2139-A	0554
1S2139-B	0005
1S2139-C	0720
1S2139A	0554
1S2139B	0005
1S2139C	0720
1S2139D	0030
1S2139P	0005
1S2140	0015
1S2142	0005
1S2143	0005
1S2144	0133
1S2144A	0133
1S-2144Z	0133

If replacement code is OEM, contact original manufacturer for replacement.

DEVICE TYPE	REPL CODE
1S2144Z	0133
1S2145	0124
1S2147	0554
1S2147A	0720
1S2147B	0720
1S2147C	0720
1S2147D	OEM
1S2148	OEM
1S2149	OEM
1S2150	0686
1S2150A	0686
1S2160	4938
1S2160A	0445
1S2175	OEM
1S2176	OEM
1S2180	0375
1S2180A	0375
1S2181	OEM
1S2186	0133
1S2186GR	4179
1S2187	0917
1S2188	0005
1S2189	0030
1S2190	0165
1S2191	0165
1S2192	0165
1S2193	0165
1S2194	0091
1S2195	0322
1S2196	0052
1S2197	0030
1S2197A	0030
1S2198	0917
1S2198HC	0917
1S2199	OEM
1S2200	0700
1S2200A	0700
1S2201	OEM
1S2203	OEM
1S2204	0124
1S2206	0030
1S-2207	0554
1S2207	0030
1S2207-9	0005
1S2208	1029
1S2208B	5633
1S2209	1029
1S2209B	OEM
1S2210	2518
1S2211	2518
1S2212	0124
1S2220	0705
1S2220A	0705
1S2222	2518
1S2226	0015
1S2227	0015
1S2228	0015
1S2229	0015
1S2230	0015
1S2231	0071
1S2233	0965
1S2234	0817
1S2235	0922
1S2236	0005
1S2237	OEM
1S2238	0071
1S2239	0071
1S2239C	0720
1S2240	0071
1S2240A	0494
1S2241	0071
1S2242	0071
1S2243	0071
1S2244	0023
1S2245	0023
1S2246	0023
1S2247	OEM
1S2248	OEM
1S2249	OEM
1S2250	OEM
1S2251	OEM
1S2252	OEM
1S2253	OEM
1S2254	OEM
1S2255	OEM
1S2256	OEM
1S2257	OEM
1S2258	OEM
1S2259	OEM
1S2260	OEM
1S2261	OEM
1S2262	OEM
1S2263	OEM
1S2264	OEM
1S2265	OEM
1S2266	OEM
1S2267	0720
1S2268	0720
1S2269	0015
1S2270	0455
1S2270A	0455
1S2271	0015
1S2272	0015
1S2273	0071
1S2274	0071
1S2275FA-1	0015
1S2276	0133
1S2277	0023
1S2278	0023
1S2279	0023
1S2280	0017
1S2281	0017
1S2283(STUD)	0547
1S2284(STUD)	2113
1S2285(STUD)	1901
1S2286	1458
1S2286(STUD)	1601
1S2286A	1458
1S2287	1490
1S2287(STUD)	1622
1S2287A	1490
1S2289	0510
1S2289(STUD)	0992
1S2289A	0510
1S2291	0870
1S2291(STUD)	1263
1S2291A	0870
1S2293	1274
1S2293(STUD)	1639
1S2293A	1274
1S2295	1078
1S2295(STUD)	1716
1S2295A	1078
1S2297	1723
1S2297(STUD)	0727
1S2297A	1723
1S2299	1723
1S2299(STUD)	1782
1S2299A	1761
1S2300	1772
1S2300A	1772
1S2302	0292
1S2303	0456
1S2304	0162
1S2305	OEM
1S2306	0102
1S2307	0102
1S2308	0102
1S2309	0102
1S2310	0102
1S2311	0071
1S2312	0071
1S2313	0015
1S2314	0071
1S2315	0071
1S2316	0023
1S2317	0023
1S2318	0017
1S2319	0017
1S2323	0017
1S2324	0017
1S2325	0017
1S2326	0017
1S2329	OEM
1S2330	0166
1S2330A	0166
1S2331	OEM
1S2332	OEM
1S2333	OEM
1S2334	OEM
1S2335	OEM
1S2336	0005
1S2337	OEM
1S2339	0554
1S2339-C	0554
1S2339C	0005
1S2339K	0030
1S2340	0005
1S2342	OEM
1S2343	OEM
1S2344	OEM
1S2345	OEM
1S2346	OEM
1S2347	OEM
1S2348	0124
1S2348H	OEM
1S2349	OEM
1S2350	0540
1S2351	0015
1S2352	0015
1S2353	0071
1S2354	0071
1S2355	0102
1S2356	0015
1S2357	0015
1S2358	0015
1S2359	0015
1S2361	0015
1S2362	0015
1S2363	0015
1S2364	0071
1S2365	0071
1S2366	0102
1S2367	0023
1S2368	0102
1S2369	0102
1S2370	0017
1S2371	0277
1S2371A	0277
1S2372	0015
1S2372A	0289
1S2373	0015
1S2373A	0296
1S2374	0015
1S2374A	0302
1S2375	0015
1S2375A	0289
1S2376	0015
1S2376A	0296
1S2377	0087
1S2378	0580
1S2378R	0947
1S2379	0540
1S2380	0965
1S2387	OEM
1S2388	OEM
1S2389	OEM
1S2390	0015
1S2391	0015
1S2392	0015
1S2393	0015
1S2394	0015
1S2395	0015
1S2396	0015
1S2397	0071
1S2398	0071
1S2399	0071
1S2400	0071
1S2401	0015
1S2402	0015
1S2403	0071
1S2404	0015
1S2405	0071
1S2406	0071
1S2407	0102
1S2408	0110
1S2409	0087
1S2410	0242
1S2411	1747
1S2412	0540
1S2413	1771
1S2414	0965
1S2415	0110
1S2416	0953
1S2417	0242
1S2418	1747
1S2419	0540
1S2420	1771
1S2421	0965
1S2422	5026
1S2422R	2559
1S2423	5026
1S2423R	2559
1S2424	5026
1S2424R	2559
1S2425	5026
1S2425R	2559
1S2426	5026
1S2426R	2559
1S2427	2615
1S2427R	2566
1S2428	OEM
1S2428R	OEM
1S2429	OEM
1S2429R	OEM
1S2430	OEM
1S2430R	OEM
1S2431	OEM
1S2431R	OEM
1S2432	OEM
1S2432R	OEM
1S2433	OEM
1S2433R	OEM
1S2434	2002
1S2434R	2003
1S2435	1250
1S2435R	1577
1S2436	1250
1S2436R	1577
1S2437	1581
1S2437R	3281
1S2438	1586
1S2438R	3294
1S2439	1586
1S2439R	3294
1S2440	OEM
1S2440R	OEM
1S2441	OEM
1S2441R	OEM
1S2442	OEM
1S2442R	OEM
1S2443	OEM
1S2443R	OEM
1S2444	OEM
1S2444R	OEM
1S2445	OEM
1S2445R	OEM
1S2446	3189
1S2446R	1630
1S2447	2901
1S2447R	0254
1S2448	1124
1S2448R	1107
1S2449	1126
1S2449R	1111
1S2450	0805
1S2450R	0258
1S2451	1195
1S2451R	1644
1S2452	0636
1S2453	0636
1S2454	0636
1S2455	0564
1S2456	0242
1S2457	0540
1S2458	1217
1S2459	1771
1S2460	0133
1S2461	0133
1S2461V-Y	0124
1S2462	0023
1S2463	0023
1S2464	OEM
1S2469	0378
1S2470	0378
1S2471	0124
1S2471A	0124
1S2471B	0124
1S2471C	OEM
1S2471TA	0133
1S-2472	0133
1S2472	0133
1S2472T-77	0133
1S2473	0133
1S2473*7A	3898
1S2473*7K	3898
1S2473-2	0124
1S2473-K	0124
1S2473-T72	0124
1S2473C	OEM
1S2473H	0133
1S2473HC	0133
1S2473H-R	0133
1S2473H-Y	0124
1S2473I	0124
1S2473K	0124
1S2473T	0133
1S2473T72	0124
1S2473T-83	OEM
1S2473TA	0133
1S2473TV	OEM
1S2473UJ	0133
1S2473V	0124
1S2473VE	0133
1S2473VH	OEM
1S2479	0143
1S2480	1312
1S2480R	1312
1S2481	1713
1S2481R	1713
1S2482	0292
1S2482R	0292
1S2483	0188
1S2483R	0188
1S2484	0456
1S2484R	0456
1S2485	0533
1S2485R	0533
1S2486	0451
1S2486R	0451
1S2487	0162
1S2488	0157
1S2489	0636
1S2490	0025
1S2491	0649
1S2492	0244
1S2493	0012
1S2494	0170
1S2495	0317
1S2496	0137
1S2497	0365
1S2498	0100
1S2499	0420
1S2500	0495
1S2501	0531
1S2502	0565
1S2503	0402
1S2504	0441
1S2505	0726
1S2506	0039
1S2507	0820
1S2508	0350
1S2509	0931
1S2510	0999
1S2511	0502
1S2512	1834
1S2513	0783
1S2514	5718
1S2515	1189
1S2516	0331
1S2517	1311
1S2518	0098
1S2519	0149
1S2520	0186
1S2521	0213
1S2522	0028
1S2523	0255
1S2524	0367
1S2525	0421
1S2536	1276
1S2536R	1276
1S2537	0760
1S2537R	0760
1S2538	0118
1S2538R	0118
1S2539	0292
1S2539R	0376
1S2540	0188
1S2540R	0188
1S2541	0456
1S2541R	0456
1S2542	0533
1S2542R	0533
1S2543	0451
1S2543R	0451
1S2544	0157
1S2545	0157
1S2546	0301
1S2547	0636
1S2548	0649
1S2549	0165
1S2550	0012
1S2551	0170
1S2552	0317
1S2553	0137
1S2554	0365
1S2555	0100
1S2556	0420
1S2557	0495
1S2558	0700
1S2559	0565
1S2560	0402
1S2561	0441
1S2562	1674
1S2563	0039
1S2564	0820
1S2565	0533
1S2566	0054
1S2567	0068
1S2568	0092
1S2569	1834
1S2570	0783
1S2571	2157
1S2572	0094
1S2573	0331
1S2574	0156
1S2575	0189
1S2588	0124
1S2589	0015
1S2590	0015
1S2591	0023
1S2592	0017
1S2593	0017
1S2594	0017
1S2595	0017
1S2596	0023
1S2597	0133
1S2598	0965
1S2602	0015
1S2603	0015
1S2604	0015
1S2605	0015
1S2606	0015
1S2607	0071
1S2608	0071
1S2609	0071
1S2610	0071
1S2611	0017
1S2612	0017
1S2613	0017
1S2614	0017
1S2615	OEM
1S2616	OEM
1S2617	OEM
1S2618	0071
1S2619	0201
1S2620	0378
1S2628	0871
1S2629	0871
1S2630	0871
1S2631	0871
1S2632	0871
1S2633	0871
1S2638	0005
1S2638VVC	0030
1S2666	OEM
1S2667	OEM
1S2668	OEM
1S2669	OEM
1S2670	OEM
1S2671	OEM
1S2672	0102
1S2673	0102
1S2674	0102
1S2675	0023
1S2676	0023
1S2677	0017
1S2678	0017
1S2679	0017
1S2682D	OEM
1S2685	OEM
1S2686	0005
1S2686D	0005
1S2687	0720
1S2687-1	0720
1S2687A	0720
1S2687AA	0720
1S2687D	0720
1S2688	0030
1S2688A	OEM
1S2688B	0030
1S2688C	0030
1S2688E	0030
1S2688EA	0030
1S2688EB	0030
1S2688T	OEM
1S2689	0030
1S2689B	0030
1S2689K	0030
1S2692	2518
1S2692A	2518
1S2692AB	0124
1S2692B	6191
1S2701	OEM
1S2701R	OEM
1S2702	OEM
1S2702R	OEM
1S2703	OEM
1S2703R	OEM
1S2711	0182
1S2712	OEM
1S2712F-FA	0102
1S2713	OEM
1S2714	0871
1S2715	0871
1S2716	0853
1S2717	0853
1S2724	OEM
1S2725	OEM
1S2726	OEM
1S2745	1217
1S2746	1217
1S2747	0922
1S2756	0023
1S2757	0015
1S2758	0922
1S2759	0922
1S2760	0110
1S2761	0953
1S2762	0242
1S2763	0540
1S2764	0540
1S2765	0965
1S2766	0965
1S2766A	1217
1S2767	0817
1S2768	0817
1S2768A	0087
1S2769	0062
1S2770	0062
1S2771	0062
1S2772	0062
1S2773	0062
1S2774	0062
1S2775	0023
1S2775(FA-1)	0023
1S2775FA-1	0499
1S2775FA1	0015
1S2775TPA2	0499
1S2776	0087
1S2777	0015
1S2778	0965
1S2779	0540
1S2780	0965
1S2781	0817
1S2782	0922
1S2787	0124
1S2788	0124
1S2788B	0133
1S2789	0720
1S2789W	0005
1S2790	0628
1S2790W	0628
1S2790WT	0628
1S2790Y	OEM
1S2790YL	0030
1S2791	0917
1S2792	0871
1S2793	0853
1S2794	0853
1S2795	0853
1S2827A	0242
1S2828A	0540
1S2830A	0817
1S2831A	0922
1S2832	0015
1S2833	0037
1S2834	0102
1S2835	0903
1S2835-T1	0903
1S2835CHIP	OEM
1S2836	0133
1S2837	0907
1S2837CHIP	OEM
1S2838	0907
1S2867T	0720
1S2880	OEM
1S2885	0499
1S2888	OEM
1S2939	OEM
1S2970WT	OEM
1S2976	OEM
1S3006	0025
1S3007	0649
1S3008	0244
1S3009	0012
1S3010	0170
1S3011	0317
1S3012	0137
1S3013	0365
1S3016	0002
1S3016(RECT)	0015
1S3016(ZENER)	0420
1S3016R	0015
1S3018	0495
1S3022	0565
1S3027	0441
1S3033	0039
1S3039	0350
1S3047	0999
1S3056	0869
1S3068	1268
1S3082	0331
1S3091A	OEM
1S3100	0098
1S3100A	OEM
1S3110	0149
1S3120	0186
1S3150	0028
1S3180	0367
1S3195	0015
1S3302	OEM
1S3305	0041
1S3332	0025
1S3585	0143
1S4006	0205
1S4006A	0205
1S4007	0480
1S4007A	0480
1S4008	0504
1S4008A	0504
1S4009	0684
1S4009A	0684
1S4010	0225
1S4010A	0225
1S4011	0230
1S4011A	0230
1S4012	0234
1S4012A	0234
1S4013	0237
1S4013A	0237
1S4015	0247
1S4015A	0247
1S4016	0251
1S4016A	0251
1S4018	0256
1S4018A	0256
1S4020	0262
1S4020A	0262
1S4022	0269
1S4022A	0269
1S4024	0274
1S4024A	0274
1S4027	0294
1S4027A	0294
1S4030	0309
1S4030A	0309
1S4033	0318
1S4033A	0318
1S4036	0320
1S4036A	0320
1S4039	0326
1S4039A	0326
1S4043	0337
1S4043A	0337
1S4047	0347
1S4047A	0347
1S4050	OEM
1S4051	0027
1S4051A	0027
1S4056	0266
1S4056A	0266
1S4062	0386
1S4062A	0386
1S4068	0405
1S4068A	0405
1S4075	0426
1S4075A	0426
1S4082	0444
1S4082A	0444
1S4091	0238
1S4091A	0238
1S4100	1180
1S4100A	1180
1S4110	1190
1S4110A	1190
1S4120	1207
1S4120A	1207
1S4130	1218
1S4130A	1218
1S4148	OEM
1S4150	0647
1S4150A	0647
1S4160	1255
1S4160A	1255
1S4180	1279
1S4180A	1279
1S4200	0605
1S4200A	0605
1S4266	0143
1S4266FM	0143
1S4735	0041
1S5006	1458
1S5006A	1458
1S5007	0221
1S5007A	0221
1S5008	1490
1S5008A	1490
1S5009	1618
1S5009A	1618
1S5010	0510
1S5010A	0510
1S5011	0691
1S5011A	0691
1S5012	0870
1S5012A	0870
1S5013	1020
1S5013A	1020
1S5015	1274
1S5015A	1274
1S5015C	OEM
1S5016	1402
1S5016A	1402

If replacement code is OEM, contact original manufacturer for replacement.

DEVICE TYPE	REPL CODE
1S5016C	OEM
1S5018	1078
1S5018A	1078
1S5018C	OEM
1S5020	1717
1S5020A	1717
1S5020C	OEM
1S5022	1723
1S5022A	1723
1S5022C	OEM
1S5024	1736
1S5024A	1736
1S5024C	OEM
1S5027	1761
1S5027A	1761
1S5027C	OEM
1S5030	1772
1S5030A	1772
1S5030C	OEM
1S5033	1788
1S5033A	1788
1S5033C	OEM
1S5036	1796
1S5036A	1796
1S5036C	OEM
1S5039	1804
1S5039A	1804
1S5039C	OEM
1S5043	1194
1S5043A	1194
1S5043C	OEM
1S5047	0022
1S5047A	0022
1S5047C	OEM
1S5051	0132
1S5051A	0132
1S5051C	OEM
1S5056	0207
1S5056A	0207
1S5056C	OEM
1S5062	0263
1S5062A	0263
1S5062C	OEM
1S5068	0310
1S5068A	0310
1S5068C	OEM
1S5075	0329
1S5075A	0329
1S5075C	OEM
1S5082	0356
1S5082A	0356
1S5082C	OEM
1S5091	0412
1S5091A	0412
1S5091C	OEM
1S5100	0438
1S5100A	0438
1S5100C	OEM
1S5110	0488
1S5110A	0488
1S5110C	OEM
1S5120	0509
1S5120A	0509
1S5120C	OEM
1S5130	0524
1S5130A	0524
1S5130C	OEM
1S5150	0063
1S5150A	0063
1S5150C	OEM
1S5277B	0080
1S5295G	0023
1S5295J	0023
1S5351	0030
1S5454	0143
1S5566B	OEM
1S6006	1458
1S6006A	1458
1S6007	0221
1S6007A	0221
1S6008	1490
1S6008A	1490
1S6009	1618
1S6009A	1618
1S6010	0510
1S6010A	0510
1S6011	0691
1S6011A	0691
1S6012	0870
1S6012A	0870
1S6013	1020
1S6013A	1020
1S6015	1274
1S6015A	1274
1S6016	1402
1S6016A	1402
1S6018	1078
1S6018A	1078
1S6020	1717
1S6020A	1717
1S6022	1723
1S6022A	1723
1S6024	1736
1S6024A	1736
1S6027	1761
1S6027A	1761
1S6030	1772
1S6030A	1772
1S6033	1788
1S6033A	1788
1S6036	1796
1S6036A	1796
1S6039	1804
1S6039A	1804
1S6043	1194
1S6043A	1194
1S6047	0022
1S6047A	0022
1S6051	0132
1S6051A	0132
1S6056	0207
1S6056A	0207
1S6062	0263
1S6062A	0263
1S6068	0310
1S6068A	0310
1S6075	0329
1S6075A	0329
1S6082	0356
1S6082A	0356
1S6091	0412
1S6091A	0412
1S6100	0438
1S6100A	0438
1S6110	0488
1S6110A	0488
1S6120	0509
1S6120A	0509
1S6130	0524
1S6130A	0524
1S6150	0063
1S6150A	0063
1S6160	0401
1S6160A	0401
1S6180	0634
1S6180A	0634
1S6200	0668
1S6200A	0668
1S6288B	0720
1S6289	0720
1S7026A	OEM
1S7026B	OEM
1S7030A	0118
1S7033	0299
1S7033A	0299
1S7033B	0299
1S7036	0376
1S7036A	0376
1S7036B	0376
1S7039	0036
1S7039A	0036
1S7039B	0036
1S7043	0275
1S7043A	0275
1S7043B	0275
1S7047	0140
1S7047A	0140
1S7047B	0140
1S7051	0041
1S7051A	0041
1S7051B	0041
1S7056	0253
1S7056A	0253
1S7056B	0253
1S7062	0471
1S7062A	0471
1S7062B	0471
1S7068	0062
1S7068A	0062
1S7068B	0062
1S7075	0077
1S7075A	0077
1S7075B	0077
1S7082	0165
1S7082A	0165
1S7082B	0165
1S7091	0057
1S7091A	0057
1S7091B	0057
1S7100	0064
1S7100A	0064
1S7100B	0064
1S7110	0181
1S7110A	0181
1S7110B	0181
1S7120	0052
1S7120A	0052
1S7120B	0052
1S7130	0053
1S7130A	0053
1S7130B	0053
1S7150	0686
1S7150A	0686
1S7150B	0686
1S7160A	0445
1S7209	0139
1S7230A	OEM
1S7581S759	OEM
1S9413	0015
1S9908	0143
1S15188	0124
1S16009	1618
1S16009A	1618
1S18813	OEM
1S18853	OEM
1S19413	0015
1S19413T	0015
1S21447	0124
1S22079	0720
1S23396	OEM
1S26870	0720
1SB01-02	0015
1SB34	OEM
1SB50S	0030
1SB69-06	0720
1SC301A	OEM
1SC301B	OEM
1SC301C	OEM
1SC301D	OEM
1SC301E	OEM
1SC683	0007
1SC929	0007
1SC1566	0276
1SC2367	0133
1SD-2	0015
1SD2	0015
1SD1212	0015
1SF1	1567
1SF2	1567
1SF4	2153
1SG11	OEM
1SG12	OEM
1SG13	OEM
1SG14	OEM
1SG15	OEM
1SG16	OEM
1SG17	OEM
1SG18	OEM
1SG19	OEM
1SG20	OEM
1SG21	OEM
1SG22	OEM
1SG23	OEM
1SG24	OEM
1SG25	OEM
1SG26	OEM
1SG27	OEM
1SG28	OEM
1SG29	OEM
1SG30	OEM
1SIZ09	0015
1SM1	0080
1SM3	0031
1SM6	OEM
1SM15	OEM
1SM25	OEM
1SM40	OEM
1SM55	OEM
1SM70	OEM
1SM100	OEM
1SM150	OEM
1SM-150A	0102
1SM200	OEM
1S-M2511-8010S035B	0124
1SN1835	0102
1SO30	0015
1SO31	0015
1SO32	0015
1SO34	0015
1SO36	0015
1SO38	0071
1SO58	0071
1SR1FM4	0015
1SR1K	0015
1SR11-100	0087
1SR11-200	0015
1SR11-400	0087
1SR11-600	0015
1SR11-800	0072
1SR11-1000	0071
1SR11-1200	0102
1SR11-1500	0102
1SR12-50	0087
1SR12-100	0087
1SR12-200	0015
1SR12-400	0087
1SR12-600	0015
1SR12-800	0037
1SR12-1000	0071
1SR13-50	0087
1SR13-100	0087
1SR13-200	0015
1SR13-400	0087
1SR13-600	0015
1SR13-800	0037
1SR13-1000	0071
1SR14-50	0087
1SR14-100	0087
1SR14-200	0015
1SR14-400	0087
1SR15-50	0110
1SR15-100	0953
1SR15-200	0242
1SR15-400	0540
1SR15-600	0965
1SR15-800	0817
1SR15-1000	0922
1SR16-50	0110
1SR16-100	0953
1SR16-200	0242
1SR16-400	0540
1SR16-600	0965
1SR16-800	0817
1SR16-1000	0922
1SR17-50	0110
1SR17-100	0953
1SR17-200	0242
1SR17-400	0540
1SR17-600	0965
1SR17-800	0817
1SR17-1000	0922
1SR19-100	OEM
1SR19-250	OEM
1SR19-400	OEM
1SR20-100	OEM
1SR20-250	OEM
1SR20-400	OEM
1SR29-200	OEM
1SR29-400	OEM
1SR29-600	OEM
1SR29-800	OEM
1SR29-1000	OEM
1SR30-200	0015
1SR30-400	0087
1SR30-600	0015
1SR31-800	0037
1SR31-1000	0071
1SR32-1200	0102
1SR32-1400	0102
1SR32-1500	0102
1SR34	0023
1SR34-200	4240
1SR35	0023
1SR35-50	1335
1SR35-50A	1335
1SR-35-100	0015
1SR35-100	0080
1SR35-100-F	0015
1SR35-100-HT	0080
1SR35-100-T3	0080
1SR35-100-Z	0080
1SR35-100A	0023
1SR35-100A-T2	0023
1SR35-100A-Y	0080
1SR35-200	0023
1SR35-200A	0609
1SR-35-400	0795
1SR35-400	0795
1SR35-400A	0795
1SR36	0102
1SR37	0871
1SR55-50	0087
1SR55-100	0087
1SR55-200	0015
1SR56-50	0102
1SR56-100	0102
1SR56-200	0102
1SR57-200	OEM
1SR57-300	OEM
1SR58-200	OEM
1SR58-300	OEM
1SR62-2000	OEM
1SR62-2500	OEM
1SR62-2800	OEM
1SR62-3000	OEM
1SR63-2000	OEM
1SR63-2500	OEM
1SR64-200	OEM
1SR64-400	OEM
1SR64-600	OEM
1SR67-200	1414
1SR68	0071
1SR69	0015
1SR70-100	0277
1SR70-200	0289
1SR70-400	0296
1SR70-800	0250
1SR70-1000	0250
1SR71-30	1600
1SR71-40	1600
1SR74-200	0871
1SR74-300	0853
1SR74-400	0853
1SR74-600	OEM
1SR74-800	OEM
1SR74-1000	OEM
1SR74R-200	OEM
1SR74R-300	OEM
1SR74R-400	OEM
1SR74R-600	OEM
1SR74R-800	OEM
1SR74R-1000	OEM
1SR75-200	0871
1SR75-300	0853
1SR75-400	0853
1SR75-600	OEM
1SR75-800	OEM
1SR75-1000	OEM
1SR75R-200	OEM
1SR75R-300	OEM
1SR75R-400	OEM
1SR75R-600	OEM
1SR75R-800	OEM
1SR75R-1000	OEM
1SR77-200	0015
1SR77-400	0087
1SR77-600	0015
1SR77-800	0072
1SR77-1000	0071
1SR78-200	0015
1SR78-400	0087
1SR79-800	0037
1SR79-1000	0071
1SR80-100	0871
1SR80-200	0871
1SR80-400	0853
1SR80-600	OEM
1SR80R-100	OEM
1SR80R-200	OEM
1SR80R-400	OEM
1SR80R-600	OEM
1SR81-800	OEM
1SR81R-800	OEM
1SR82-100	0871
1SR82-200	0871
1SR82-400	0853
1SR82-600	OEM
1SR82-800	OEM
1SR82-1000	OEM
1SR82R-100	OEM
1SR82R-200	OEM
1SR82R-600	OEM
1SR83-800	OEM
1SR83R-800	OEM
1SR84-100	OEM
1SR84-200	OEM
1SR84-400	OEM
1SR90-200	OEM
1SR90-400	OEM
1SR90-600	OEM
1SR90-800	OEM
1SR90R-200	OEM
1SR90R-400	OEM
1SR90R-600	OEM
1SR90R-800	OEM
1SR91-100	0087
1SR91-200	0087
1SR91-400	0087
1SR95-1600	0348
1SR95A-1600	0348
1SR96-200	0609
1SR96-400	0795
1SR96-600	0015
1SR97-800	0072
1SR97-1000	0071
1SR98-100	0031
1SR98-200	0031
1SR99-100	0080
1SR99-200	0609
1SR99-400	0795
1SR99A-100	0080
1SR99A-200	0609
1SR99A-400	0795
1SR100-200	0609
1SR100-400	0795
1SR100-600	0015
1SR100-800	0072
1SR100-1200	0348
1SR100-1500	0348
1SR101-800	0072
1SR101-1200	0348
1SR101-1500	0348
1SR102-2000	0348
1SR106-200	OEM
1SR106-400	OEM
1SR106-600	OEM
1SR106-800	OEM
1SR106R-200	OEM
1SR106R-400	OEM
1SR106R-600	OEM
1SR106R-800	OEM
1SR107-50	OEM
1SR107-100	OEM
1SR107-200	OEM
1SR107R-50	OEM
1SR107R-100	OEM
1SR107R-200	OEM
1SR108-50	OEM
1SR108-100	OEM
1SR108-200	OEM
1SR108R-50	OEM
1SR108R-100	OEM
1SR108R-200	OEM
1SR109-50	OEM
1SR109-100	OEM
1SR109-200	OEM
1SR109R-50	OEM
1SR109R-100	OEM
1SR109R-200	OEM
1SR112-800	0072
1SR112-1000	0071
1SR112-1300	0348
1SR112-1500	0348
1SR117-100	0087
1SR117-200	0087
1SR117-400	0087
1SR117-600	0087
1SR117-800	0087
1SR119-100	0080
1SR119-200	0609
1SR119-400	0795
1SR124-50	1335
1SR124-50A	1335
1SR124-100	0023
1SR124-100A	0023
1SR124-200	0023
1SR124-200A	0023
1SR124-400	0023
1SR124-400-Z	0023
1SR124-400A	0023
1SR124-400V-Z	0023
1SR133-200	OEM
1SR133-200R	OEM
1SR133-400	OEM
1SR133-400R	OEM
1SR133-600	OEM
1SR133-600R	OEM
1SR135	0182
1SR135-100	OEM
1SR135-100-Z	OEM
1SR136	OEM
1SR137	OEM
1SR138	0124
1SR139	0031
1SR139-100	0424
1SR139-100AT	OEM
1SR139-200	0424
1SR139-400	0424
1SR140	0031
1SR141	0031
1SR142	0031
1SR143	OEM
1SR35200TB	OEM
1SRBA20Z	0289
1SS11	OEM
1SS12	OEM
1SS13	OEM
1SS14	OEM
1SS15	OEM
1SS-16	0917
1SS16	0917
1SS19	0124
1SS23	0143
1SS24	0913
1SS25	0913
1SS26	0913
1SS27	0143
1SS28	0913
1SS29	0913
1SS30	0124
1SS31	0124
1SS40	OEM
1SS41	0124
1SS43	0917
1SS48	OEM
1SS49	0124
1SS50	0124
1SS51	0124
1SS52	2518
1SS-53	0133
1SS53	0133
1SS54	0133
1SS54-T4	0124
1SS55	0023
1SS68	0124
1SS69	OEM
1SS70	OEM
1SS78	0917
1SS81	0133
1SS81-F1	OEM
1SS81-R	0133
1SS81-T2	0133
1SS81-T5	0133
1SS81-Y	0133
1SS-82	0023
1SS82	0023
1SS82-Y	0023
1SS83	0133
1SS84	0133
1SS85	0163
1SS86	0917
1SS87	0917
1SS88	0339
1SS90	0917
1SS91	OEM
1SS91-G	0163
1SS91-S	0163
1SS92	0124
1SS93	0124
1SS94	0124
1SS95	0133
1SS96	OEM
1SS97	0339
1SS98	0917
1SS99	0917
1SS101	3891
1SS102	0133
1SS103	2518
1SS104	0133
1SS104(FA)	0133
1SS104FA	0133
1SS105	OEM
1SS105A	OEM
1SS106	3173
1SS107	0339
1SS108	3891
1SS109	0124
1SS110	0015
1SS110FQ	0124
1SS110T	0124
1SS110Y	0124
1SS111	0163
1SS112	0163
1SS113	0396
1SS114	OEM
1SS115	OEM
1SS116	0917
1SS117	0124
1SS118	0124
1SS118-S3	OEM
1SS119	0124
1SS119/VHD1SS119//-1	OEM
1SS119-041T	0124
1SS119-14	0124
1SS119-14FQ	0124
1SS119-25	4468
1SS119-T	0124
1SS119TD	0124
1SS119TG	0124
1SS120	0124
1SS120-7	OEM
1SS121	0124
1SS122	0023
1SS122H	OEM
1SS123	0124
1SS124	OEM
1SS125	OEM
1SS130	0133
1SS130T	0133
1SS131	0124
1SS131T-77	0124
1SS131T-177	OEM
1SS132	0124
1SS132T	0124
1SS132T-77	0124
1SS133	0133
1SS133-T2	0133
1SS133-Y	0124
1SS133HV	0133
1SS133T	0133
1SS133T-72	OEM
1SS133VT	OEM
1SS134	0124
1SS135	0124
1SS136	0124
1SS137	0124
1SS138	0124
1SS139	0133
1SS140	0133
1SS141	0133
1SS142	0023
1SS143	0023
1SS144	0023
1SS145	0023
1SS146	0023
1SS146-T2	0023
1SS146-Y	0023
1SS147	0023
1SS148	0133
1SS149	0023
1SS149H	OEM
1SS150	0124
1SS150(A)	0133
1SS151	0917
1SS151(A)	OEM
1SS152	0124
1SS152(A)	0133
1SS153	3896
1SS154	OEM
1SS155	0396
1SS156	OEM
1SS157	OEM
1SS158	OEM
1SS162	OEM
1SS163	0124
1SS164	0124
1SS165	0163
1SS165(A)	OEM
1SS166	0163
1SS166(A)	OEM
1SS166-03TE	OEM
1SS168	0163
1SS168(A)	0133
1SS169	0124
1SS170	0124
1SS171	0124
1SS172	0023
1SS173	0163
1SS173A	0124
1SS174	0917
1SS175	0133
1SS175A	0133
1SS176	0133
1SS176(FA)	0124
1SS176-TPA7	OEM
1SS176FA	0124
1SS176TPA7	OEM
1SS177	0124
1SS178	0124
1SS178(FA)	0124
1SS178-Y	0124
1SS178FA	0124
1SS179	OEM
1SS181	0903
1SS182	OEM
1SS184	0907
1SS184-TE85L	0907
1SS184TE85L	0907
1SS184TE85R	OEM
1SS187	0124
1SS188-03TD	OEM
1SS190	0124
1SS193	0124
1SS196	0124
1SS197	0163
1SS198	0339
1SS199	OEM
1SS200	6213
1SS201	0124
1SS202	0124
1SS202-1	OEM
1SS203	0124
1SS204	0124

DEVICE TYPE	REPL CODE
1SS205	0133
1SS206	0124
1SS207	0133
1SS208	0124
1SS209	0124
1SS210	0124
1SS211	0124
1SS212	0124
1SS213	0124
1SS214	0124
1SS215	0124
1SS216	0163
1SS217	0133
1SS218	0124
1SS219	0133
1SS220	0124
1SS221	0124
1SS222	0124
1SS223	0124
1SS226	OEM
1SS227	1091
1SS228	OEM
1SS229	0694
1SS230	0694
1SS231	2203
1SS232	2203
1SS233	0694
1SS233F	3192
1SS234	0694
1SS235	2203
1SS236	2203
1SS237	0339
1SS238	OEM
1SS239	0339
1SS240	0124
1SS241	0163
1SS242	0917
1SS243	0133
1SS244	1090
1SS244-T2	1090
1SS244-T77	OEM
1SS245	1090
1SS246	OEM
1SS247	OEM
1SS248	0396
1SS248TP3	5358
1SS249	OEM
1SS250	0133
1SS251	0023
1SS252	0133
1SS253	0124
1SS254	0124
1SS267	0023
1SS268	0907
1SS269	0903
1SS270	0133
1SS270A	0124
1SS270TA	0133
1SS270TD	OEM
1SS271	OEM
1SS272	0124
1SS273	0124
1SS274	0917
1SS275	0917
1SS276	0917
1SS277	0163
1SS278	0163
1SS279	2518
1SS280	2518
1SS281	0339
1SS282	0339
1SS283	0339
1SS284	0339
1SS285	3891
1SS290	0133
1SS291	0133
1SS292	0133
1SS292T-72	OEM
1SS312	OEM
1SS355	0124
1SS585	OEM
1SS916	OEM
1SS1555	0396
1SS53125	OEM
1ST11	OEM
1ST11-UR	OEM
1ST12	OEM
1ST12-UR	OEM
1ST13	OEM
1ST13-UR	OEM
1ST14	OEM
1ST14-UR	OEM
1ST15	OEM
1ST15-UR	OEM
1ST16	OEM
1ST16-UR	OEM
1ST17	OEM
1ST17-UR	OEM
1ST18	OEM
1ST18-UR	OEM
1ST20	OEM
1ST21	OEM
1ST22	OEM
1ST37	OEM
1SV11	OEM
1SV25	OEM
1SV25/S3015A	OEM
1SV26	OEM
1SV27	OEM
1SV28	OEM
1SV29	OEM
1SV30	OEM
1SV31	OEM
1SV32	OEM
1SV33	OEM
1SV34	0124
1SV35	0124
1SV36	OEM
1SV37	OEM
1SV45	OEM
1SV46	OEM
1SV47	OEM
1SV48	OEM
1SV48-1	OEM
1SV48-2	OEM
1SV49	OEM
1SV50	0030
1SV50(1)	OEM
1SV50-1	OEM
1SV50S	0030
1SV52	OEM
1SV53	0030
1SV53-A	OEM
1SV53-B	OEM
1SV53A	OEM
1SV54	OEM
1SV54-G	OEM
1SV55	0030
1SV56	OEM
1SV57	OEM
1SV58	1029
1SV59	1029
1SV65	OEM
1SV66	OEM
1SV68	0030
1SV68-01	OEM
1SV69	1029
1SV69-06	0720
1SV70	1029
1SV70-06	0720
1SV71	OEM
1SV72	OEM
1SV73	OEM
1SV74	OEM
1SV75	OEM
1SV77	0005
1SV80	OEM
1SV84	1029
1SV87	OEM
1SV88	OEM
1SV89	0720
1SV89A	0720
1SV89B01	0720
1SV89BFG	0720
1SV90	OEM
1SV90A	3688
1SV91	3688
1SV92	OEM
1SV93	OEM
1SV94	OEM
1SV95	OEM
1SV96	OEM
1SV97	0628
1SV98	OEM
1SV100	OEM
1SV101	OEM
1SV102	OEM
1SV103	OEM
1SV108	OEM
1SV109	3687
1SV110	OEM
1SV110-07	OEM
1SV111	1029
1SV111-06	1029
1SV111-14	OEM
1SV112	OEM
1SV113	0628
1SV114	OEM
1SV114(A)	OEM
1SV115	OEM
1SV116	OEM
1SV117	OEM
1SV118	OEM
1SV121	OEM
1SV122	OEM
1SV123	0720
1SV124	0628
1SV125	OEM
1SV125(A)	OEM
1SV126	OEM
1SV128	OEM
1SV132	OEM
1SV132(A)	OEM
1SV133	0628
1SV134	OEM
1SV135	OEM
1SV136	OEM
1SV136(A)	OEM
1SV141	OEM
1SV141(A)	OEM
1SV142	OEM
1SV143	OEM
1SV144	OEM
1SV145	OEM
1SV146	OEM
1SV147	OEM
1SV148	OEM
1SV149	OEM
1SV151S	OEM
1SV153	OEM
1SV154	OEM
1SV154FQ	OEM
1SV156	OEM
1SV157	OEM
1SV158	OEM
1SV160	OEM
1SV161	OEM
1SV161S	OEM
1SV162	OEM
1SV164	OEM
1SV165	OEM
1SV166	OEM
1SV166(1)	OEM
1SV166S	OEM
1SV167	OEM
1SV168	OEM
1SV169	OEM
1SV170	OEM
1SV172	OEM
1SV173	OEM
1SV174	OEM
1SV175	OEM
1SV176	OEM
1SV177	OEM
1SV178	OEM
1SV179	OEM
1SV180	OEM
1SV181	OEM
1SV182	OEM
1SV186	OEM
1SV187	OEM
1SV188	OEM
1SV189	OEM
1SV190	OEM
1SV191	OEM
1SV192	OEM
1SV193	OEM
1SV194	OEM
1SV195	OEM
1SV196	OEM
1SV197	OEM
1SV198	OEM
1SV201	OEM
1SV203	OEM
1SV206	OEM
1SV207	OEM
1SV209	OEM
1SV1615	OEM
1SX170	0023
1SX171	0023
1SX172	0023
1SX173	0023
1SX174	0023
1SX175	0023
1SZ11	0210
1SZ12	0671
1SZ13	0010
1SZ14	OEM
1SZ21	1649
1SZ22	0375
1SZ23	0010
1SZ24	OEM
1SZ25	0165
1SZ26	0471
1SZ27	0165
1SZ28	0445
1SZ29	0714
1SZ30	0064
1SZ39	0636
1SZ40-11	0317
1SZ40-12	0137
1SZ40-13	0365
1SZ40-15	0002
1SZ40-16	0420
1SZ40-18	0495
1SZ40-20	0531
1SZ40-22	0565
1SZ40-24	0402
1SZ40-27	0441
1SZ40-30	0726
1SZ40-33	0039
1SZ41	OEM
1SZ42	OEM
1SZ43	0157
1SZ45	0295
1SZ45A	0295
1SZ46	0295
1SZ46A	0295
1SZ47	0295
1SZ47A	0295
1SZ48	0295
1SZ48A	0295
1SZ49	0295
1SZ49A	0295
1SZ50	0471
1SZ51	0471
1SZ52	0471
1SZ53	0471
1SZ54	0471
1SZ55-C2V2	OEM
1SZ55-C2V4	1276
1SZ55-C2V7	0760
1SZ55-C2V0	OEM
1SZ55-C3V0	0118
1SZ55-C3V3	0299
1SZ55-C3V6	0376
1SZ55-C3V9	0036
1SZ55-C4V3	0275
1SZ55-C4V7	0140
1SZ55-C5V1	0041
1SZ55-C5V6	0253
1SZ55-C6V2	0471
1SZ55-C6V8	0062
1SZ55-C7V5	0077
1SZ55-C8V2	0165
1SZ55-C9V1	0057
1SZ55-C10	0064
1SZ55-C11	0181
1SZ55-C12	0052
1SZ55-C13	0053
1SZ55-C15	0686
1SZ55-C16	0445
1SZ55-C18	0375
1SZ55-C20	0700
1SZ55-C22	0705
1SZ55-C24	0494
1SZ55-C27	0455
1SZ55-C30	0195
1SZ55-C33	0166
1SZ55-C36	0010
1SZ55-C39	0032
1SZ56-07	0185
1SZ56-08	0480
1SZ56-09	3317
1SZ56-10	0684
1SZ56-11	0225
1SZ56-12	0230
1SZ56-13	0234
1SZ56-15	1397
1SZ56-16	0247
1SZ56-18	1178
1SZ56-20	2395
1SZ56-22	0262
1SZ56-24	0269
1SZ56-27	2399
1SZ56-30	1177
1SZ56-33	0309
1SZ57	0295
1SZ58	0295
1SZ59	0295
1SZ60	0295
1SZ61	0471
1SZ62	0165
1SZ63	0052
1SZ64	0456
1SZ65	OEM
1SZ66	0166
1SZ67	0062
1SZ68	OEM
1SZ69	OEM
1SZ70	OEM
1SZ71	OEM
1SZ72	OEM
1T06	0555
1T08	0555
1T010	0772
1T015	0772
1T1	OEM
1T1TA14	0100
1T2C1	0015
1T2Z1	0015
1T3	OEM
1T5.6	0157
1T5.6B	0157
1T6	OEM
1T6.2	0636
1T6.2A	0636
1T6.2B	0636
1T7.5	0649
1T7.5A	0649
1T7.5B	0649
1T9.1	0012
1T9.1A	0012
1T9.1B	0012
1T10	0170
1T10B	0170
1T11	0317
1T11A	0317
1T11B	0317
1T11.5A	0794
1T12	0137
1T12A	0137
1T12B	0137
1T13	0917
1T13A	0917
1T13B	0917
1T14	0100
1T14A	0100
1T14B	0100
1T15	0002
1T15B	0002
1T16	0555
1T16A	0420
1T16B	0420
1T18	0555
1T19-15B	0002
1T20	0531
1T20A	0531
1T20B	0531
1T22	0143
1T22(A)	0143
1T22A	0143
1T22AJ	0143
1T22AM	0143
1T22B	0143
1T22G	0143
1T22M	0143
1T23	0143
1T23A	0143
1T23B	0143
1T23G	0143
1T23J	0143
1T23M	0143
1T23S	0123
1T24	0402
1T24A	0402
1T24B	0402
1T25	0720
1T25-4	OEM
1T26	0143
1T26-2	0143
1T26TRIAC	0772
1T27	0441
1T27B	0441
1T28	0555
1T30	0726
1T30A	0726
1T30B	0726
1T31	OEM
1T32	OEM
1T32-2	OEM
1T33	0039
1T33A	0039
1T33B	0039
1T33C	OEM
1T33CT	OEM
1T36	0555
1T36A	0820
1T36B	0820
1T38	0744
1T39	0350
1T39A	0350
1T39B	0350
1T40	0133
1T43	0931
1T43A	0931
1T43B	0931
1T46	0555
1T47	0999
1T47A	0999
1T47B	0999
1T48	0555
1T51	0502
1T51A	0502
1T51B	0502
1T56	0869
1T56A	0869
1T56B	0869
1T58	0555
1T62	0783
1T62A	0783
1T62B	0783
1T66	0555
1T68	0555
1T68A	2157
1T68B	2157
1T75	1189
1T75A	1189
1T75B	1189
1T82	0331
1T82A	0331
1T82B	0331
1T91	1311
1T91A	1311
1T91B	1311
1T100	0098
1T100A	0098
1T100B	0098
1T110	0555
1T110A	0149
1T110B	0149
1T120	0186
1T120A	0186
1T120B	0186
1T130	0213
1T130A	0213
1T130B	0213
1T150	0028
1T150A	0028
1T150B	0028
1T160	0255
1T160A	0255
1T160B	0255
1T180	0367
1T180A	0367
1T180B	0367
1T188	0143
1T200	0421
1T200A	0421
1T200B	0421
1T205A	0050
1T210	0555
1T213	0143
1T215	0772
1T220	0325
1T229-2	0015
1T231	0143
1T236	0143
1T240	0143
1T240A	0143
1T243	0133
1T243M	0012
1T261	0019
1T261T	0143
1T262	0143
1T263	0502
1T264	0502
1T310	0555
1T315	0744
1T377	OEM
1T378	0015
1T410	0555
1T415	0744
1T495	0164
1T501	0015
1T502	0015
1T503	0015
1T504	0015
1T505	0015
1T506	0015
1T507	0071
1T508	0071
1T509	0071
1T510	0071
1T515	0617
1T610	0555
1T615	0617
1T929	0079
1T930	0079
1T1101	OEM
1T1102	OEM
1T1103	OEM
1T1104	OEM
1T1110	OEM
1T1331	0123
1T1401	OEM
1T1402	OEM
1T2011	0015
1T2012	0015
1T2013	0015
1T2014	0015
1T2015	0015
1T2016	0015
1T20105	0015
1TA5.6	0157
1TA5.6A	0157
1TA5.6B	0157
1TA6.2	0636
1TA6.2A	0636
1TA6.2B	0636
1TA6.8	0025
1TA7.5	0649
1TA7.5A	0649
1TA8.2	0244
1TA9.1	0012
1TA9.1A	0012
1TA9.1B	0012
1TA10	0170
1TA10A	0170
1TA11	0317
1TA12	0137
1TA12A	0137
1TA13	0365
1TA13A	0365
1TA13B	0365
1TA14	0100
1TA14A	0100
1TA14B	0100
1TA15	0002
1TA15A	0002
1TA16	0420
1TA18	0495
1TA20	0531
1TA22	0565
1TA24	0402
1TA25A	0402
1TA27	0441
1TA27A	0441
1TA30	0726
1TA33	0039
1TA33A	0039
1TA36	0820
1TA39	0350
1TA43	0931
1TA47	0999
1TA51	0502
1TA56	0869
1TA62	0783
1TA68	1268
1TA75	1189
1TA82	0331
1TA82A	0331
1TA91	1311
1TA100	0098
1TA110	0149
1TA120	0186
1TA130	0213
1TA150	0028
1TA160	0255
1TA180	0367
1TA200	0421
1TA7607AP	0912
1TA7608CP5	2054
1TA7609P-2	4134
1TB06	0015
1TH60	OEM
1TH61	0071
1TH62	0017
1TMP4N3119	OEM
1TS05	0015
1TT210	OEM
1TT-410	0030
1TT410	0030
1TT921	0102
1TT1168	0030
1TT7215	0102
1TV24237	0474
1TW85B	0030
1TWC8H	1083
1TWC8L	1083
1TWC8M	1083
1TZ61	0017
1U585F	0016
1UPC1363C	0683
1UPD1937C	2090
1UPD1986C	4788
1V2	1449
1V3074A20	0015
1V3074A21	0015
1V9002	0143
1V68611A47	0037
1V68611A47A	0037
1VA10	0015
1VA20	OEM
1VA30	OEM
1VA40	OEM
1VA50	OEM
1VA60	OEM
1VA70	OEM
1VA80	OEM
1VA100	OEM
1VA120	OEM
1VR10A	OEM
1VR10B	OEM
1W01-08J	0317
1W20	OEM
1W20A	OEM
1W50	OEM
1W80	OEM
1W1706	0325
1W9640	0037
1W9640A	0037
1W9728	0037
1W9728A	0037
1W9782	0037
1W9782A	0037
1W9787	0016
1W9810	0037
1W9810A	0037
1W9810S	0037
1W9810SA	0037
1W11700	0037
1W11700A	0037
1W11702	0037
1W11702A	0037
1W11706	0325
1W11708-1	0016
1W11708-2	0016
1W11711	0037
1W11711A	0006
1WB-5A	0162
1WB-6A	0157
1WB-7A	0025
1WB-9D	1083
1WB-11D	0317
1WB-13D	0365
1WB-15D	0002
1WB-17D	1649
1WB-20D	0531
1WB-25D	0402
1WB-30D	0726
1WB-35D	0820
1WB-45D	0931
1WB-55D	1834
1WB-65D	0783
1WB-75D	1189
1WB-85D	0331
1WB-95D	1311
1WB-100D	0098
1WB-110D	0149
1WB-120D	0186
1WB-130D	0213
1WB-140D	0245
1WM6	0015
1WS1	0087
1WS2	0087
1WS4	0540
1WS6	0965
1WS8	0087
1WS10	0087
1X8055	0133
1X9179	0133
1X9805	0133
1X9809	0133
1XG1	OEM
1XG2	OEM
1XV1	OEM
1XV2	OEM
1XV3	OEM
1Z3.3	0292
1Z3.3A	0292
1Z3.3B	0292
1Z3.3T10.5	0292
1Z3.6	0188
1Z3.6A	0188
1Z3.6B	0188
1Z3.6T5	0188
1Z3.6T10	0188
1Z3.6TS	0188
1Z3.9	0456
1Z3.9A	0456
1Z3.9B	0456
1Z4.3	0533

DEVICE TYPE	REPL CODE
1Z4.3A	0533
1Z4.3B	0533
1Z4.7	0451
1Z4.7A	0451
1Z4.7B	0451
1Z5	OEM
1Z5.1	0162
1Z5.1A	0162
1Z5.1B	0162
1Z5.1T5	0162
1Z5.6	0157
1Z5.6A	0157
1Z5.6B	0157
1Z6.2	0636
1Z6.2A	0636
1Z6.2B	0636
1Z6.2T5	0636
1Z6.2T20	0636
1Z6.8	0025
1Z6.8A	0025
1Z6.8B	0025
1Z6.8C	OEM
1Z6.8D	0025
1Z6.8D5	0025
1Z6.8D10	0025
1Z7.5	0649
1Z7.5A	0649
1Z7.5B	0649
1Z7.5C	OEM
1Z7.5D	0649
1Z7.5D5	0649
1Z7.5D10	0649
1Z8.2	0244
1Z8.2A	0244
1Z8.2B	0244
1Z8.2C	OEM
1Z8.2D	0244
1Z8.2D5	0244
1Z8.2D10	0244
1Z8.3	0244
1Z9.1	0012
1Z9.1A	0012
1Z9.1B	0012
1Z9.1C	0012
1Z9.1D	0012
1Z9.1D5	0012
1Z9.1D10	0012
1Z9.1D15	0012
1Z9.1T5	0012
1Z9.1T10	0012
1Z9.1T20	0012
1Z10	0170
1Z10A	0170
1Z10B	0170
1Z10C	OEM
1Z10D	0170
1Z10D5	0170
1Z10D10	0170
1Z10T10	0170
1Z10T20	0170
1Z11	0317
1Z11A	0317
1Z11B	0317
1Z11C	OEM
1Z11D	0317
1Z11D5	0317
1Z11D10	0317
1Z11T5	0317
1Z11T10	0317
1Z11T20	0317
1Z-12	0137
1Z12	0137
1Z12A	0137
1Z12B	0137
1Z12C	0137
1Z12D	0137
1Z12D5	0137
1Z12D10	0137
1Z12T5	0137
1Z12T10	0137
1Z12T20	0137
1Z13	0365
1Z13A	0365
1Z13B	0365
1Z13C	0365
1Z13D	0365
1Z13D5	0365
1Z13D10	0365
1Z13T5	0365
1Z13T10	0365
1Z13T20	0365
1Z14	0100
1Z14A	0100
1Z14B	0100
1Z14C	0100
1Z14D	0100
1Z14D5	0100
1Z14D10	0100
1Z15	0002
1Z15A	0002
1Z15B	0002
1Z15C	0002
1Z15D	0002
1Z15D5	0002
1Z15D10	0002
1Z15D20	0002
1Z15T5	0002
1Z15T10	0002
1Z15T20	0002
1Z16	0420
1Z16A	0420
1Z16B	0420
1Z16C	OEM
1Z16D	0420
1Z16D5	0420
1Z16D10	0420
1Z17	1649
1Z17A	1649
1Z17B	1649
1Z17C	OEM
1Z17D	1649
1Z17D5	1649
1Z17D10	1649
1Z18	0495
1Z18A	0495
1Z18B	0495
1Z18C	OEM
1Z18D	0495
1Z18D5	0495
1Z18D10	0495
1Z18T5	0495
1Z18T10	0495
1Z18T20	0495
1Z19	0949
1Z19A	0949
1Z19B	0949
1Z19C	OEM
1Z19D	0949
1Z19D5	0949
1Z19D10	0949
1Z20	0531
1Z20A	0531
1Z20B	0531
1Z20C	OEM
1Z20D	0531
1Z20D5	0531
1Z20D10	0531
1Z22	0565
1Z22A	0565
1Z22B	0565
1Z22C	OEM
1Z22D	0565
1Z22D5	0565
1Z22D10	0565
1Z22T5	0565
1Z22T10	0565
1Z22T20	0565
1Z24	0274
1Z24A	0402
1Z24B	0402
1Z24C	OEM
1Z24D	0402
1Z24D5	0402
1Z24D10	0402
1Z25	1606
1Z25A	1606
1Z25B	1606
1Z25C	OEM
1Z25D	1606
1Z25D5	1606
1Z25D10	1606
1Z26A	2575
1Z27	0441
1Z27A	0441
1Z27B	0441
1Z27C	0441
1Z27D	0441
1Z27D5	0441
1Z27D10	0441
1Z27T5	0441
1Z27T10	0441
1Z27T20	0441
1Z30	0726
1Z30A	0726
1Z30B	0726
1Z30C	OEM
1Z30D	0726
1Z30D5	0726
1Z30D10	0726
1Z33	0039
1Z33-A	0039
1Z33A	0039
1Z33B	0039
1Z33C	0039
1Z33D	0039
1Z33D5	0039
1Z33D10	0039
1Z33T5	0039
1Z33T10	0039
1Z33T20	0039
1Z36	0820
1Z36B	0820
1Z36C	OEM
1Z36D	0820
1Z36D5	0820
1Z36D10	0820
1Z39	0350
1Z39A	0350
1Z39B	0350
1Z39C	OEM
1Z39D	0350
1Z39D5	0350
1Z39D10	0350
1Z43	0931
1Z43A	0931
1Z43B	0931
1Z43C	OEM
1Z43D	0931
1Z43D5	0931
1Z43D10	0931
1Z45	0999
1Z45A	0999
1Z45B	0999
1Z45C	OEM
1Z45D	0999
1Z45D5	0999
1Z45D10	0999
1Z47	0999
1Z47A	0999
1Z47B	0999
1Z47C	OEM
1Z47D	0999
1Z47D5	0999
1Z47D10	0999
1Z50	0502
1Z50A	0502
1Z50B	0502
1Z50C	OEM
1Z50D	0502
1Z50D5	0502
1Z50D10	0502
1Z51	0502
1Z51A	0502
1Z51B	0502
1Z51C	OEM
1Z51D	0502
1Z51D5	0502
1Z51D10	0502
1Z52	0502
1Z52A	0502
1Z52B	0502
1Z52C	OEM
1Z52D	0502
1Z52D5	0502
1Z52D10	0502
1Z56	0869
1Z56A	0869
1Z56B	0869
1Z56C	1834
1Z56D	0869
1Z56D5	0869
1Z56D10	0869
1Z62	0783
1Z62A	0783
1Z62B	0783
1Z62C	0783
1Z62D	0783
1Z62D5	0783
1Z62D10	0783
1Z68	1268
1Z68A	1268
1Z68B	2157
1Z68C	OEM
1Z68D	2157
1Z68D5	2157
1Z68D10	2157
1Z72	1189
1Z75	1189
1Z75A	1189
1Z75B	1189
1Z75C	OEM
1Z75D	1189
1Z75D5	1189
1Z75D10	1189
1Z75FA	1189
1Z82	0012
1Z82A	0331
1Z82B	0331
1Z82C	0331
1Z82D	0331
1Z82D5	0331
1Z82D10	0331
1Z91	1311
1Z91A	1311
1Z91B	1311
1Z91C	OEM
1Z91D	1311
1Z91D5	1311
1Z91D10	1311
1Z91.1D10	0012
1Z91.D10	0012
1Z100	0098
1Z100A	0098
1Z100B	0098
1Z100C	OEM
1Z100D	0098
1Z100D5	0098
1Z100D10	0098
1Z105	0149
1Z105A	0149
1Z105B	0149
1Z105C	OEM
1Z105D	0149
1Z105D5	0149
1Z105D10	0149
1Z110	0149
1Z110A	0149
1Z110B	0149
1Z110C	0149
1Z110D	0149
1Z110D5	0149
1Z110D10	0149
1Z120	0186
1Z120A	0186
1Z120B	0186
1Z120C	OEM
1Z120D	0186
1Z120D5	0186
1Z120D10	0186
1Z130	0213
1Z130A	0213
1Z130B	0213
1Z130C	OEM
1Z130D	0213
1Z130D5	0213
1Z130D10	0213
1Z140	0245
1Z140A	0245
1Z140B	0245
1Z140C	OEM
1Z140D	0245
1Z140D5	0245
1Z140D10	0245
1Z150	0028
1Z150A	0028
1Z150B	0028
1Z150C	OEM
1Z150D5	0028
1Z150D10	0028
1Z160	0255
1Z160A	0255
1Z160B	0255
1Z160C	OEM
1Z160D	0255
1Z160D5	0255
1Z160D10	0255
1Z175	0367
1Z175A	0367
1Z175B	0367
1Z175C	OEM
1Z175D	0367
1Z175D5	0367
1Z175D10	0367
1Z180	0367
1Z180A	0367
1Z180B	0367
1Z180C	OEM
1Z180D	0367
1Z180D5	0367
1Z180D10	0367
1Z200	0421
1Z200A	0421
1Z200B	0421
1Z200C	OEM
1Z200D	0421
1Z200D5	0421
1Z200D10	0421
1Z261-02	OEM
1Z330	OEM
1Z390	OEM
1ZB10	0170
1ZB10B	0170
1ZB12	0137
1ZB12B	0137
1ZB15	0002
1ZB15B	0002
1ZB22	0565
1ZB22B	0565
1ZB27	0441
1ZB27B	0441
1ZB33	0039
1ZB33B	0039
1ZB39	0350
1ZB39B	0350
1ZB47	0999
1ZB47B	0999
1ZB56	0869
1ZB56B	0869
1ZB68	2157
1ZB68B	2157
1ZB82	0331
1ZB82B	0331
1ZB100	0098
1ZB100B	0098
1ZC1LT10	0420
1ZC3.6	0188
1ZC3.6T5	0376
1ZC3.6T10	0376
1ZC5.1	0162
1ZC5.6	0157
1ZC5.6T10.5	0157
1ZC6.2	0636
1ZC9.1	0012
1ZC9.1T5	0012
1ZC9.1T10	0012
1ZC10T5	0170
1ZC12	0137
1ZC12T5	0137
1ZC12T10	0137
1ZC12T20	0137
1ZC15	0002
1ZC15T5	0002
1ZC15T10	0002
1ZC18T5	0495
1ZC18T10	0495
1ZC22T5	0565
1ZC22T10	0565
1ZC27	0441
1ZC27T5	0441
1ZC27T10	0441
1ZCT10	0170
1ZD6	0649
1ZD8-2	0244
1ZD8-5	0244
1ZD8.2	0244
1ZD8.2V	0244
1ZD12	0137
1ZD82S	0244
1ZD120	0186
1ZD825	0244
1ZD892	0244
1ZF5.1T5	0162
1ZF5.1T10	0162
1ZF5.1T20	0162
1ZF5.6T5	0157
1ZF5.6T10	0157
1ZF5.6T20	0157
1ZF9.1T5	0012
1ZF9.1T10	0012
1ZF9.1T20	0012
1ZF12T5	0137
1ZF12T10	0137
1ZF12T20	0137
1ZF15T5	0002
1ZF15T10	0002
1ZF15T20	0002
1ZF27T5	0441
1ZF27T10	0441
1ZF27T20	0441
1ZF33T5	0039
1ZF33T10	0039
1ZF33T20	0039
1ZG10	OEM
1ZG10B	OEM
1ZG15	OEM
1ZG15B	OEM
1ZG18	OEM
1ZG18B	OEM
1ZG22	OEM
1ZG22B	OEM
1ZG27	OEM
1ZG27B	OEM
1ZG33	OEM
1ZG33B	OEM
1ZG39	OEM
1ZG39B	OEM
1ZG47	OEM
1ZG47B	OEM
1ZG56	OEM
1ZG56B	OEM
1ZG68	OEM
1ZG68B	OEM
1ZG82	OEM
1ZG82B	OEM
1ZG100	OEM
1ZG100B	OEM
1ZK10	OEM
1ZK10B	OEM
1ZK12	OEM
1ZK12B	OEM
1ZK15	OEM
1ZK15B	OEM
1ZK18	OEM
1ZK18B	OEM
1ZK22	OEM
1ZK22B	OEM
1ZK27	OEM
1ZK27B	OEM
1ZK33	OEM
1ZK33B	OEM
1ZK39	OEM
1ZK39B	OEM
1ZK47	OEM
1ZK47B	OEM
1ZK56	OEM
1ZK56B	OEM
1ZK68	OEM
1ZK68B	OEM
1ZK82	OEM
1ZK82B	OEM
1ZK100	OEM
1ZK100B	OEM
1ZM3.3T5	0292
1ZM3.3T10	0292
1ZM3.3T20	0292
1ZM3.6T5	0188
1ZM3.6T10	0188
1ZM3.6T20	0188
1ZM3.9T5	0456
1ZM3.9T10	0456
1ZM3.9T20	0456
1ZM4.3T5	0533
1ZM4.3T10	0533
1ZM4.3T20	0533
1ZM4.7T5	0451
1ZM4.7T10	0451
1ZM4.7T20	0451
1ZM5.1T5	0162
1ZM5.1T10	0162
1ZM5.1T20	0162
1ZM5.6T5	0157
1ZM5.6T10	0157
1ZM5.6T20	0157
1ZM6.2T5	0636
1ZM6.2T10	0636
1ZM6.2T20	0636
1ZM6.8T5	0025
1ZM6.8T10	0025
1ZM6.8T20	0025
1ZM7.5T5	0649
1ZM7.5T10	0649
1ZM7.5T20	0649
1ZM8.2T5	0244
1ZM8.2T10	0244
1ZM8.2T20	0244
1ZM9.1T5	0012
1ZM9.1T10	0012
1ZM9.1T20	0012
1ZM10T5	0170
1ZM10T10	0170
1ZM10T20	0170
1ZM11T5	0317
1ZM11T10	0317
1ZM11T20	0317
1ZM12T5	0137
1ZM12T10	0137
1ZM12T20	0137
1ZM13T5	0365
1ZM13T10	0365
1ZM13T20	0365
1ZM15T5	0002
1ZM15T10	0002
1ZM15T20	0002
1ZM16T5	0420
1ZM16T10	0420
1ZM16T20	0420
1ZM18T5	0495
1ZM18T10	0495
1ZM18T20	0495
1ZM20T5	0531
1ZM20T10	0531
1ZM20T20	0531
1ZM22T5	0565
1ZM22T10	0565
1ZM24T5	0402
1ZM24T10	0402
1ZM27	0441
1ZM27T5	0441
1ZM27T10	0441
1ZM30	0726
1ZM30T5	0726
1ZM30T10	0726
1ZM33T5	0039
1ZM33T10	0039
1ZM36T5	0820
1ZM36T10	0820
1ZM39T5	0350
1ZM39T10	0350
1ZM43T5	0931
1ZM43T10	0931
1ZM47	0999
1ZM47T5	0999
1ZM47T10	0999
1ZM50	0502
1ZM51T5	0502
1ZM51T10	0502
1ZM56T5	0869
1ZM56T10	0869
1ZM62T5	0783
1ZM62T10	0783
1ZM68T5	1268
1ZM68T10	2157
1ZM75T5	1189
1ZM75T10	1189
1ZM82T5	0331
1ZM82T10	0331
1ZM91T5	1311
1ZM91T10	1311
1ZM100	0098
1ZM100T5	0098
1ZM100T10	0098
1ZM110T5	0149
1ZM110T10	0149
1ZM120T5	0186
1ZM120T10	0186
1ZM130T5	0213
1ZM130T10	0213
1ZM150T5	0028
1ZM150T10	0028
1ZM160T5	0255
1ZM160T10	0255
1ZM180	0367
1ZM180T5	0367
1ZM180T10	0367
1ZM200T5	0421
1ZM200T10	0421
1ZM330	OEM
1ZM390	OEM
1ZPD3.9	0036
1ZS13A	0365
1ZS16A	0420
1ZS18A	0495
1ZT10	0170
1ZT10B	0170
1ZT12	0137
1ZT12B	0137
1ZT15	0002
1ZT15B	0002
1ZT18	0495
1ZT18B	0495
1ZT22	0565
1ZT22B	0565
1ZT27	0441
1ZT27B	0441
1ZT33	0039
1ZT33B	0039
1ZT39	0350
1ZT39B	0350
1ZT47	0999
1ZT47B	0999
1ZT56	0869
1ZT56B	0869
1ZT68	2157
1ZT68B	2157
1ZT82	0331
1ZT82B	0331
1ZT100	0098
1ZT100B	0098
1ZT110A	0149
1ZT120A	0186
1ZT130A	0213
1ZT150A	0028
1ZT160A	0255
1ZT180A	0367
1ZT200A	0421
1.000.111-00	0037
1.4Z45D10	OEM
1.5B05	0087
1.5B1	2523
1.5B2	2523
1.5B4	2527
1.5B6	2527
1.5B8	2528
1.5B10	6451
1.5B12	0182
1.5B14	0182
1.5B16	0348
1.5B18	0348
1.5B20	0348
1.5BZ100D	1180
1.5BZ110B	1190
1.5BZ120	1207
1.5BZ120C	1207
1.5BZ130D	1218
1.5BZ140B	1881
1.5BZ150	0647
1.5BZ150C	0647
1.5C05	0087
1.5C1	2523
1.5C2	2523
1.5C4	2527
1.5C6	2527
1.5C8	2528
1.5C10	6451
1.5C12	0182
1.5C14	0182
1.5C16	0348
1.5C18	0348
1.5C20	0348
1.5DKZ6.8	0205
1.5DKZ6.8A	0205
1.5DKZ6.8B	0205
1.5DKZ7.5	0480
1.5DKZ7.5A	0480
1.5DKZ7.5B	0480
1.5DKZ8.2	0504
1.5DKZ8.2A	0504
1.5DKZ8.2B	0504
1.5DKZ9.1	0684
1.5DKZ9.1A	0684
1.5DKZ9.1B	0684
1.5DKZ10	0225
1.5DKZ10A	0225
1.5DKZ10B	0225
1.5DKZ11	0230
1.5DKZ11A	0230
1.5DKZ11B	0230
1.5DKZ12	0234
1.5DKZ12A	0234
1.5DKZ12B	0234
1.5DKZ13	0237
1.5DKZ13A	0237
1.5DKZ13B	0237
1.5DKZ15	0247
1.5DKZ15A	0247
1.5DKZ15B	0247
1.5DKZ16	0251
1.5DKZ16A	0251
1.5DKZ16B	0251
1.5DKZ18	0256
1.5DKZ18A	0256
1.5DKZ18B	0256
1.5DKZ20	0262
1.5DKZ20A	0262
1.5DKZ20B	0262
1.5DKZ22	0269
1.5DKZ22A	0269
1.5DKZ22B	0269
1.5DKZ24	0274
1.5DKZ24A	0274
1.5DKZ24B	0274
1.5DKZ27	0294
1.5DKZ27A	0294
1.5DKZ27B	0294
1.5DKZ30	0309
1.5DKZ30A	0309
1.5DKZ30B	0309
1.5DKZ33	0318
1.5DKZ33A	0318
1.5DKZ33B	0318
1.5DKZ36	0320
1.5DKZ36A	0320
1.5DKZ36B	0320
1.5DKZ39	0326
1.5DKZ39A	0326
1.5DKZ39B	0326
1.5DKZ43	0337
1.5DKZ43A	0337
1.5DKZ43B	0337
1.5DKZ47	0347
1.5DKZ47A	0347
1.5DKZ47B	0347
1.5DKZ51	0027

If replacement code is OEM, contact original manufacturer for replacement.

DEVICE TYPE	REPL CODE
1.5DKZ51A	0027
1.5DKZ51B	0027
1.5DKZ56	0266
1.5DKZ56A	0266
1.5DKZ56B	0266
1.5DKZ62	0386
1.5DKZ62A	0386
1.5DKZ62B	0386
1.5DKZ68	0405
1.5DKZ68A	0405
1.5DKZ68B	0405
1.5DKZ75	0426
1.5DKZ75A	0426
1.5DKZ75B	0426
1.5E05	0110
1.5E1	0953
1.5E2	0242
1.5E3	0087
1.5E4	0540
1.5E5	1771
1.5E6	0965
1.5E7	0087
1.5E8	0087
1.5E10	0087
1.5E12	0102
1.5E14	0182
1.5E16	0348
1.5E18	0348
1.5E20	0348
1.5E25	OEM
1.5E50A	OEM
1.5E100	OEM
1.5E140	1881
1.5F05	OEM
1.5F1	2523
1.5F2	2523
1.5F4	2527
1.5F6	2527
1.5F8	2528
1.5F10	6451
1.5F12	0182
1.5F14	0182
1.5F16	0348
1.5F18	0348
1.5F20	0348
1.5J05	0110
1.5J1	0953
1.5J2	0242
1.5J3	1747
1.5J4	0540
1.5J5	1771
1.5J6	0965
1.5J7	0087
1.5J8	0087
1.5J10	0087
1.5J12	0102
1.5JA19B	2395
1.5JZ6.8	0205
1.5JZ6.8A	0205
1.5JZ6.8B	0205
1.5JZ6.8C	0205
1.5JZ6.8D	0205
1.5JZ7.5	0480
1.5JZ7.5A	0480
1.5JZ7.5B	0480
1.5JZ7.5C	0480
1.5JZ7.5D	0480
1.5JZ8.2	0504
1.5JZ8.2A	0504
1.5JZ8.2B	0504
1.5JZ8.2C	0504
1.5JZ8.2D	0504
1.5JZ9.1	0684
1.5JZ9.1A	0684
1.5JZ9.1B	0684
1.5JZ9.1C	0684
1.5JZ9.1D	0684
1.5JZ10	0225
1.5JZ10A	0225
1.5JZ10B	0225
1.5JZ10C	0225
1.5JZ10D	0225
1.5JZ11	0230
1.5JZ11A	0230
1.5JZ11B	0230
1.5JZ11C	0230
1.5JZ11D	0230
1.5JZ12	0234
1.5JZ12A	0234
1.5JZ12B	0234
1.5JZ12C	0234
1.5JZ12D	0234
1.5JZ13	0237
1.5JZ13A	0237
1.5JZ13B	0237
1.5JZ13C	0237
1.5JZ13D	0237
1.5JZ14	1397
1.5JZ14A	1397
1.5JZ14B	1397
1.5JZ14C	1397
1.5JZ14D	1397
1.5JZ15	0247
1.5JZ15A	0247
1.5JZ15B	0247
1.5JZ15C	0247
1.5JZ15D	0247
1.5JZ16	0251
1.5JZ16A	0251
1.5JZ16B	0251
1.5JZ16C	0251
1.5JZ16D	0251
1.5JZ17	1178
1.5JZ17A	1178
1.5JZ17B	1178
1.5JZ17C	1178
1.5JZ17D	1178
1.5JZ18	0256
1.5JZ18A	0256
1.5JZ18B	0256
1.5JZ18C	0256
1.5JZ18D	0256
1.5JZ19	2395
1.5JZ19A	2395
1.5JZ19B	2395
1.5JZ19C	2395
1.5JZ19D	2395
1.5JZ20	0262
1.5JZ20A	0262
1.5JZ20B	0262
1.5JZ20C	0262
1.5JZ20D	0262
1.5JZ22	0269
1.5JZ22A	0269
1.5JZ22B	0269
1.5JZ22C	0269
1.5JZ22D	0269
1.5JZ24	0274
1.5JZ24A	0274
1.5JZ24B	0274
1.5JZ24C	0274
1.5JZ24D	0274
1.5JZ25	2399
1.5JZ25A	2399
1.5JZ25B	2399
1.5JZ25C	1178
1.5JZ25D	2399
1.5JZ27	0294
1.5JZ27A	0294
1.5JZ27B	0294
1.5JZ27C	0294
1.5JZ27D	0294
1.5JZ30	0309
1.5JZ30A	0309
1.5JZ30B	0309
1.5JZ30C	0309
1.5JZ30D	0309
1.5JZ33	0318
1.5JZ33A	0318
1.5JZ33B	0318
1.5JZ33C	0318
1.5JZ33D	0318
1.5JZ36	0320
1.5JZ36A	0320
1.5JZ36B	0320
1.5JZ36C	0320
1.5JZ36D	0320
1.5JZ39	0326
1.5JZ39A	0326
1.5JZ39B	0326
1.5JZ39C	0326
1.5JZ39D	0326
1.5JZ43	0337
1.5JZ43A	0337
1.5JZ43B	0337
1.5JZ43C	0337
1.5JZ43D	0337
1.5JZ45	0337
1.5JZ45A	0337
1.5JZ45B	0337
1.5JZ45C	0337
1.5JZ45D	OEM
1.5JZ47	0347
1.5JZ47A	0347
1.5JZ47B	0347
1.5JZ47C	0347
1.5JZ47D	0347
1.5JZ50	0027
1.5JZ50A	0027
1.5JZ50B	0027
1.5JZ50C	0027
1.5JZ50D	OEM
1.5JZ51	0027
1.5JZ51A	0027
1.5JZ51B	0027
1.5JZ51C	0027
1.5JZ51D	0027
1.5JZ52	0027
1.5JZ52A	0027
1.5JZ52B	0027
1.5JZ52C	0027
1.5JZ52D	OEM
1.5JZ56	0266
1.5JZ56A	0266
1.5JZ56B	0266
1.5JZ56C	0266
1.5JZ56D	0266
1.5JZ62	0386
1.5JZ62A	0386
1.5JZ62B	0386
1.5JZ62C	0386
1.5JZ62D	0386
1.5JZ68	0405
1.5JZ68A	0405
1.5JZ68B	0405
1.5JZ68C	0405
1.5JZ68D	OEM
1.5JZ75	0426
1.5JZ75A	0426
1.5JZ75B	0426
1.5JZ75C	0426
1.5JZ75D	0426
1.5JZ82	0444
1.5JZ82A	0444
1.5JZ82B	0444
1.5JZ82C	0444
1.5JZ82D	0444
1.5JZ91	0238
1.5JZ91A	0238
1.5JZ91B	0238
1.5JZ91C	0238
1.5JZ91D	0238
1.5JZ100	1180
1.5JZ100A	1180
1.5JZ100B	1180
1.5JZ100C	1180
1.5JZ100D	OEM
1.5JZ105	1180
1.5JZ105A	1180
1.5JZ105B	1180
1.5JZ105C	1180
1.5JZ105D	OEM
1.5JZ110	1190
1.5JZ110A	1190
1.5JZ110B	1190
1.5JZ110C	1190
1.5JZ110D	1190
1.5JZ120	1207
1.5JZ120A	1207
1.5JZ120B	1207
1.5JZ120C	1207
1.5JZ120D	1207
1.5JZ130	1218
1.5JZ130A	1218
1.5JZ130B	1218
1.5JZ130C	1218
1.5JZ130D	1218
1.5JZ140	1881
1.5JZ140A	1881
1.5JZ140B	1881
1.5JZ140C	1881
1.5JZ140D	1881
1.5JZ150	0647
1.5JZ150A	0647
1.5JZ150B	0647
1.5JZ150C	0647
1.5JZ150D	0647
1.5JZ160	1255
1.5JZ160A	1255
1.5JZ160B	1255
1.5JZ160C	1255
1.5JZ160D	1255
1.5JZ175	2102
1.5JZ175A	2102
1.5JZ175B	2102
1.5JZ175C	2102
1.5JZ175D	OEM
1.5JZ180	1279
1.5JZ180A	1279
1.5JZ180B	1279
1.5JZ180C	1279
1.5JZ180D	OEM
1.5JZ200	0605
1.5JZ200A	0605
1.5JZ200B	0605
1.5JZ200C	0605
1.5JZ200D	0605
1.5JZ220	OEM
1.5JZ220A	OEM
1.5JZ220B	OEM
1.5JZ220C	OEM
1.5JZ220D	OEM
1.5K6.8	3114
1.5K6.8A	3114
1.5K6.8R	3114
1.5K6.8RA	3114
1.5K7.5	OEM
1.5K7.5A	OEM
1.5K7.5R	OEM
1.5K7.5RA	OEM
1.5K8.2	OEM
1.5K8.2A	OEM
1.5K8.2R	OEM
1.5K8.2RA	OEM
1.5K9.1	OEM
1.5K9.1A	OEM
1.5K9.1R	OEM
1.5K9.1RA	OEM
1.5K10	OEM
1.5K10A	OEM
1.5K10R	OEM
1.5K10RA	OEM
1.5K11	OEM
1.5K11A	OEM
1.5K11R	OEM
1.5K11RA	OEM
1.5K12	OEM
1.5K12A	OEM
1.5K12R	OEM
1.5K12RA	OEM
1.5K13	3172
1.5K13A	3172
1.5K13R	3172
1.5K13RA	3172
1.5K15	OEM
1.5K15A	OEM
1.5K15R	OEM
1.5K15RA	OEM
1.5K16	OEM
1.5K16A	OEM
1.5K16R	OEM
1.5K16RA	OEM
1.5K18	3191
1.5K18A	3191
1.5K18R	3191
1.5K18RA	3191
1.5K20	3201
1.5K20A	3201
1.5K20R	3201
1.5K20RA	3201
1.5K22	OEM
1.5K22A	OEM
1.5K22R	OEM
1.5K22RA	OEM
1.5K24	OEM
1.5K24A	OEM
1.5K24R	OEM
1.5K24RA	OEM
1.5K27	1915
1.5K27A	1915
1.5K27R	1915
1.5K27RA	1915
1.5K30	OEM
1.5K30A	OEM
1.5K30R	OEM
1.5K30RA	OEM
1.5K33	OEM
1.5K33A	OEM
1.5K33R	OEM
1.5K33RA	OEM
1.5K36	OEM
1.5K36A	OEM
1.5K36R	OEM
1.5K36RA	OEM
1.5K39	OEM
1.5K39A	OEM
1.5K39R	OEM
1.5K39RA	OEM
1.5K43	OEM
1.5K43A	OEM
1.5K43R	OEM
1.5K43RA	OEM
1.5K47	OEM
1.5K47A	OEM
1.5K47R	OEM
1.5K47RA	OEM
1.5K51	0568
1.5K51A	0568
1.5K51R	0568
1.5K51RA	0568
1.5K56	OEM
1.5K56A	OEM
1.5K56R	OEM
1.5K56RA	OEM
1.5K62	OEM
1.5K62A	OEM
1.5K62R	OEM
1.5K62RA	OEM
1.5K68	0831
1.5K68A	0831
1.5K68R	0831
1.5K68RA	0831
1.5K75	OEM
1.5K75(1N5654)	OEM
1.5K75A	OEM
1.5K75R	OEM
1.5K75RA	OEM
1.5K82	OEM
1.5K82A	OEM
1.5K82R	OEM
1.5K82RA	OEM
1.5K91	OEM
1.5K91A	OEM
1.5K91R	OEM
1.5K91RA	OEM
1.5K100	OEM
1.5K100A	OEM
1.5K100R	OEM
1.5K100RA	OEM
1.5K110	OEM
1.5K110A	OEM
1.5K110R	OEM
1.5K110RA	OEM
1.5K120	OEM
1.5K120A	OEM
1.5K120R	OEM
1.5K120RA	OEM
1.5K130	OEM
1.5K130A	OEM
1.5K130R	OEM
1.5K130RA	OEM
1.5K150	OEM
1.5K150A	OEM
1.5K150R	OEM
1.5K150RA	OEM
1.5K160	OEM
1.5K160A	OEM
1.5K160R	OEM
1.5K160RA	OEM
1.5K170	OEM
1.5K170A	OEM
1.5K170R	OEM
1.5K170RA	OEM
1.5K180	OEM
1.5K180A	OEM
1.5K180R	OEM
1.5K180RA	OEM
1.5K200	1405
1.5K200A	1405
1.5K200R	1405
1.5K200RA	1405
1.5KC6.8	OEM
1.5KC6.8A	3114
1.5KC7.5	OEM
1.5KC7.5A	OEM
1.5KC7.5C	OEM
1.5KC7.5CA	OEM
1.5KC8.2	OEM
1.5KC8.2A	OEM
1.5KC8.2C	OEM
1.5KC8.2CA	OEM
1.5KC9.1	OEM
1.5KC9.1A	OEM
1.5KC9.1C	OEM
1.5KC9.1CA	OEM
1.5KC10	OEM
1.5KC10A	OEM
1.5KC10C	OEM
1.5KC10CA	OEM
1.5KC11	OEM
1.5KC11A	OEM
1.5KC11C	OEM
1.5KC11CA	OEM
1.5KC12	OEM
1.5KC12A	OEM
1.5KC13	OEM
1.5KC13A	3172
1.5KC15	OEM
1.5KC15A	OEM
1.5KC16	OEM
1.5KC16A	OEM
1.5KC18	OEM
1.5KC18A	3191
1.5KC20	OEM
1.5KC20A	3201
1.5KC22	OEM
1.5KC22A	OEM
1.5KC24	OEM
1.5KC24A	OEM
1.5KC27	OEM
1.5KC27A	1915
1.5KC30	OEM
1.5KC30A	OEM
1.5KC33	OEM
1.5KC33A	OEM
1.5KC36	OEM
1.5KC36A	OEM
1.5KC39	OEM
1.5KC39A	OEM
1.5KC43	OEM
1.5KC43A	OEM
1.5KC47	OEM
1.5KC47A	OEM
1.5KC51	OEM
1.5KC51A	0568
1.5KC56	OEM
1.5KC56A	OEM
1.5KC62	OEM
1.5KC62A	OEM
1.5KC68	OEM
1.5KC68A	0831
1.5KC75	OEM
1.5KC75A	OEM
1.5KC82	OEM
1.5KC82A	OEM
1.5KC91	OEM
1.5KC91A	OEM
1.5KC100	OEM
1.5KC100A	OEM
1.5KC110	OEM
1.5KC110A	OEM
1.5KE6V8A	OEM
1.5KE6V8CA	OEM
1.5KE6V8CP	OEM
1.5KE6V8P	OEM
1.5KE6.8	3114
1.5KE6.8A	3114
1.5KE6.8C	2133
1.5KE6.8CA	2133
1.5KE7V5A	OEM
1.5KE7V5CA	OEM
1.5KE7V5CP	OEM
1.5KE7V5P	OEM
1.5KE7.5	OEM
1.5KE7.5A	OEM
1.5KE7.5C	OEM
1.5KE7.5CA	OEM
1.5KE8V2A	OEM
1.5KE8V2CA	OEM
1.5KE8V2CP	OEM
1.5KE8V2P	OEM
1.5KE8.2	OEM
1.5KE8.2A	OEM
1.5KE8.2C	OEM
1.5KE8.2CA	OEM
1.5KE9V1A	OEM
1.5KE9V1CA	OEM
1.5KE9V1CP	OEM
1.5KE9V1P	OEM
1.5KE9.1	OEM
1.5KE9.1A	OEM
1.5KE9.1C	OEM
1.5KE9.1CA	OEM
1.5KE10	OEM
1.5KE10A	OEM
1.5KE10C	OEM
1.5KE10CA	OEM
1.5KE10CP	OEM
1.5KE10P	OEM
1.5KE11	OEM
1.5KE11A	OEM
1.5KE11C	OEM
1.5KE11CA	OEM
1.5KE11CP	OEM
1.5KE11P	OEM
1.5KE12	OEM
1.5KE12A	OEM
1.5KE12C	OEM
1.5KE12CA	OEM
1.5KE12CP	OEM
1.5KE12P	OEM
1.5KE13	3172
1.5KE13A	3172
1.5KE13C	1169
1.5KE13CA	1169
1.5KE13CP	OEM
1.5KE13P	OEM
1.5KE15	OEM
1.5KE15A	OEM
1.5KE15C	OEM
1.5KE15CA	OEM
1.5KE15CP	OEM
1.5KE15P	OEM
1.5KE16	OEM
1.5KE16A	OEM
1.5KE16C	OEM
1.5KE16CA	OEM
1.5KE16CP	OEM
1.5KE16P	OEM
1.5KE18	3191
1.5KE18A	3191
1.5KE18C	1767
1.5KE18CA	1767
1.5KE18CP	OEM
1.5KE18P	OEM
1.5KE20	3201
1.5KE20A	3201
1.5KE20C	1932
1.5KE20CA	1932
1.5KE20CP	OEM
1.5KE20P	OEM
1.5KE22	OEM
1.5KE22A	OEM
1.5KE22C	OEM
1.5KE22CA	OEM
1.5KE22CP	OEM
1.5KE22P	OEM
1.5KE24	OEM
1.5KE24A	OEM
1.5KE24C	OEM
1.5KE24CA	OEM
1.5KE24CP	OEM
1.5KE24P	OEM
1.5KE27	1915
1.5KE27A	1915
1.5KE27C	1952
1.5KE27CA	1952
1.5KE27CP	OEM
1.5KE27P	OEM
1.5KE30	OEM
1.5KE30A	OEM
1.5KE30C	OEM
1.5KE30CA	OEM
1.5KE30CP	OEM
1.5KE30P	OEM
1.5KE33	OEM
1.5KE33A	OEM
1.5KE33C	OEM
1.5KE33CA	OEM
1.5KE33CP	OEM
1.5KE33P	OEM
1.5KE36	OEM
1.5KE36A	OEM
1.5KE36C	OEM
1.5KE36CA	OEM
1.5KE36CP	OEM
1.5KE36P	OEM
1.5KE39	OEM
1.5KE39A	OEM
1.5KE39C	OEM
1.5KE39CA	OEM
1.5KE39CP	OEM
1.5KE39P	OEM
1.5KE43	OEM
1.5KE43A	OEM
1.5KE43C	OEM
1.5KE43CA	OEM
1.5KE43CP	OEM
1.5KE43P	OEM
1.5KE47	OEM
1.5KE47A	OEM
1.5KE47C	OEM
1.5KE47CA	OEM
1.5KE47CP	OEM
1.5KE47P	OEM
1.5KE51	0568
1.5KE51A	0568
1.5KE51C	1972
1.5KE51CA	1972
1.5KE51CP	OEM
1.5KE51P	OEM
1.5KE56	OEM
1.5KE56A	OEM
1.5KE56C	OEM
1.5KE56CA	OEM
1.5KE56CP	OEM
1.5KE56P	OEM
1.5KE62	OEM
1.5KE62A	OEM
1.5KE62C	OEM
1.5KE62CA	OEM
1.5KE62CP	OEM
1.5KE62P	OEM
1.5KE68	0831
1.5KE68A	0831
1.5KE68C	1987
1.5KE68CA	1987
1.5KE68CP	OEM
1.5KE68P	OEM
1.5KE75	OEM
1.5KE75A	OEM
1.5KE75C	OEM
1.5KE75CA	OEM
1.5KE75CP	OEM
1.5KE75P	OEM
1.5KE82	OEM
1.5KE82A	OEM
1.5KE82C	OEM
1.5KE82CA	OEM
1.5KE82CP	OEM
1.5KE82P	OEM
1.5KE91	OEM
1.5KE91A	OEM
1.5KE91C	OEM
1.5KE91CA	OEM
1.5KE91CP	OEM
1.5KE91P	OEM
1.5KE100	OEM
1.5KE100A	OEM
1.5KE100C	OEM
1.5KE100CA	OEM
1.5KE100CP	OEM
1.5KE100P	OEM
1.5KE110	OEM
1.5KE110A	OEM
1.5KE110C	OEM
1.5KE110CA	OEM
1.5KE110CP	OEM
1.5KE110P	OEM
1.5KE120	OEM
1.5KE120A	OEM
1.5KE120C	OEM
1.5KE120CA	OEM
1.5KE120CP	OEM
1.5KE120P	OEM
1.5KE130	OEM
1.5KE130A	OEM
1.5KE130C	OEM
1.5KE130CA	OEM
1.5KE130CP	OEM
1.5KE130P	OEM
1.5KE150	OEM
1.5KE150A	OEM
1.5KE150C	OEM
1.5KE150CA	OEM
1.5KE160	OEM
1.5KE160A	OEM
1.5KE160C	OEM
1.5KE160CA	OEM
1.5KE170	OEM
1.5KE170A	OEM
1.5KE170C	OEM
1.5KE170CA	OEM
1.5KE180	OEM
1.5KE180A	OEM
1.5KE180C	OEM
1.5KE180CA	OEM
1.5KE180CP	OEM
1.5KE200	1405
1.5KE200A	1405
1.5KE200C	1408
1.5KE200CA	1408
1.5KE200CP	OEM
1.5KE200P	OEM
1.5KE220	OEM
1.5KE220A	OEM
1.5KE220C	OEM
1.5KE220CA	OEM
1.5KE220CP	OEM
1.5KE220P	OEM
1.5KE250	OEM
1.5KE250A	OEM
1.5KE250C	OEM
1.5KE250CA	OEM
1.5KE250CP	OEM
1.5KE250P	OEM
1.5KE280	OEM
1.5KE280A	OEM
1.5KE280CA	OEM
1.5KE280CP	OEM
1.5KE280P	OEM
1.5KE300	OEM
1.5KE300A	OEM
1.5KE300CA	OEM
1.5KE300CP	OEM
1.5KE300P	OEM
1.5KE320	OEM
1.5KE320A	OEM
1.5KE320CA	OEM
1.5KE320CP	OEM
1.5KE320P	OEM
1.5KE350	OEM
1.5KE350A	OEM

DEVICE TYPE	REPL CODE
1.5KE350CA	OEM
1.5KE350CP	OEM
1.5KE350P	OEM
1.5KE400	OEM
1.5KE400A	OEM
1.5KE400CA	OEM
1.5KE400CP	OEM
1.5KE400P	OEM
1.5KE440	OEM
1.5KE440A	OEM
1.5KE440CA	OEM
1.5KE440CP	OEM
1.5KE440P	OEM
1.5L56A	OEM
1.5M14Z	1397
1.5M14Z5	1397
1.5M14Z10	1397
1.5M17Z	1178
1.5M17Z5	1178
1.5M17Z10	1178
1.5M19Z	2395
1.5M19Z5	2395
1.5M19Z10	2395
1.5M25Z	2399
1.5M25Z5	2399
1.5M25Z10	2399
1.5M140Z	1881
1.5M140Z5	1881
1.5M140Z10	1881
1.5R7.5	0480
1.5R7.5A	0480
1.5R7.5B	0480
1.5R8.2	0504
1.5R8.2A	0504
1.5R8.2B	0504
1.5R9.1	0684
1.5R9.1A	0684
1.5R9.1B	0684
1.5R10	0225
1.5R10A	0225
1.5R10B	0225
1.5R11	0230
1.5R11A	0230
1.5R11B	0230
1.5R12	0234
1.5R12A	0234
1.5R12B	0234
1.5R13	0237
1.5R13A	0237
1.5R13B	0237
1.5R15	0247
1.5R15A	0247
1.5R15B	0247
1.5R16	0251
1.5R16A	0251
1.5R16B	0251
1.5R18	0256
1.5R18A	0256
1.5R18B	0256
1.5R20	0262
1.5R20A	0262
1.5R20B	0262
1.5R22	0269
1.5R22A	0269
1.5R22B	0269
1.5R24	0274
1.5R24A	0274
1.5R24B	0274
1.5R27	0294
1.5R27A	0294
1.5R27B	0294
1.5R30	0309
1.5R30A	0309
1.5R30B	0309
1.5R31	0326
1.5R33	0318
1.5R33A	0318
1.5R33B	0318
1.5R36	0320
1.5R36A	0320
1.5R36B	0320
1.5R39A	0326
1.5R39B	0326
1.5R43	0337
1.5R43A	0337
1.5R43B	0337
1.5R47	0347
1.5R47A	0347
1.5R47B	0347
1.5R51	0027
1.5R51A	0027
1.5R51B	0027
1.5R56	0266
1.5R56A	0266
1.5R56B	0266
1.5R62	0386
1.5R62A	0386
1.5R62B	0386
1.5R68	0405
1.5R68A	0405
1.5R68B	0405
1.5R91	0238
1.5R91A	0238
1.5R91B	0238
1.5R100	1180
1.5R100A	1180
1.5R100B	1180
1.5R110	1190
1.5R110A	1190
1.5R110B	1190
1.5R120	1207
1.5R120A	1207
1.5R120B	1207
1.5R130	1218
1.5R130A	1218
1.5R130B	1218
1.5R140	1881
1.5R140A	1881
1.5R140B	1881
1.5R150	0647
1.5R150A	0647
1.5R150B	0647
1.5R160	1255
1.5R160A	1255
1.5R160B	1255
1.5R180	1279
1.5R180A	1279
1.5R180B	1279
1.5R200	0605
1.5R200A	0605
1.5R200B	0605
1.5S220	OEM
1.5S220A	OEM
1.5S250	OEM
1.5S250A	OEM
1.5S300	OEM
1.5S300A	OEM
1.5S350	OEM
1.5S350A	OEM
1.5S400	OEM
1.5S400A	OEM
1.5SE6.8	3114
1.5SE6.8A	3114
1.5SE7.5	OEM
1.5SE7.5A	OEM
1.5SE8.2	OEM
1.5SE8.2A	OEM
1.5SE9.1	OEM
1.5SE9.1A	OEM
1.5SE10	OEM
1.5SE10A	OEM
1.5SE11	OEM
1.5SE11A	OEM
1.5SE12	OEM
1.5SE12A	OEM
1.5SE13	3172
1.5SE13A	3172
1.5SE15	OEM
1.5SE15A	OEM
1.5SE16	OEM
1.5SE16A	OEM
1.5SE18	3191
1.5SE18A	3191
1.5SE20	3201
1.5SE20A	3201
1.5SE22	OEM
1.5SE22A	OEM
1.5SE24	OEM
1.5SE24A	OEM
1.5SE27	1915
1.5SE27A	1915
1.5SE30	OEM
1.5SE30A	OEM
1.5SE33	OEM
1.5SE33A	OEM
1.5SE36	OEM
1.5SE36A	OEM
1.5SE39	OEM
1.5SE39A	OEM
1.5SE43	OEM
1.5SE43A	OEM
1.5SE47	OEM
1.5SE47A	OEM
1.5SE51	0568
1.5SE51A	0568
1.5SE56	OEM
1.5SE56A	OEM
1.5SE62	OEM
1.5SE62A	OEM
1.5SE68	0831
1.5SE68A	0831
1.5SE75	OEM
1.5SE75A	OEM
1.5SE82	OEM
1.5SE82A	OEM
1.5SE91	OEM
1.5SE91A	OEM
1.5SE100	OEM
1.5SE100A	OEM
1.5SE110	OEM
1.5SE110A	OEM
1.5SE120	OEM
1.5SE120A	OEM
1.5SE130	OEM
1.5SE130A	OEM
1.5SE150	OEM
1.5SE150A	OEM
1.5SE160	OEM
1.5SE160A	OEM
1.5SE170	OEM
1.5SE170A	OEM
1.5SE180	OEM
1.5SE180A	OEM
1.5SE200	1405
1.5SE200A	1405
1.5SE220	OEM
1.5SE220A	OEM
1.5SE250	OEM
1.5SE250A	OEM
1.5SE300	OEM
1.5SE300A	OEM
1.5SE350	OEM
1.5SE350A	OEM
1.5SE400	OEM
1.5SE400A	OEM
1.5X52D5	OEM
1.5X180A	OEM
1.5Z6.8	0205
1.5Z6.8A	0205
1.5Z6.8B	0205
1.5Z6.8C	0205
1.5Z6.8D	0205
1.5Z7.5	0480
1.5Z7.5A	0480
1.5Z7.5B	0480
1.5Z7.5C	0480
1.5Z7.5D	0480
1.5Z8.2	0504
1.5Z8.2A	0504
1.5Z8.2B	0504
1.5Z8.2C	0504
1.5Z8.2D	0504
1.5Z9.1	0684
1.5Z9.1A	0684
1.5Z9.1B	0684
1.5Z9.1C	0684
1.5Z9.1D	0684
1.5Z10	0225
1.5Z10A	0225
1.5Z10B	0225
1.5Z10C	0225
1.5Z10D	0225
1.5Z11	0230
1.5Z11A	0230
1.5Z11B	0230
1.5Z11C	0230
1.5Z11D	0230
1.5Z12	0234
1.5Z12A	0234
1.5Z12B	0234
1.5Z12C	0234
1.5Z12D	0234
1.5Z13	0237
1.5Z13A	0237
1.5Z13B	0237
1.5Z13C	0237
1.5Z13D	0237
1.5Z14	1397
1.5Z14A	1397
1.5Z14B	1397
1.5Z14C	1397
1.5Z14D	1397
1.5Z15	0247
1.5Z15A	0247
1.5Z15B	0247
1.5Z15C	0247
1.5Z15D	0247
1.5Z16	0251
1.5Z16A	0251
1.5Z16B	0251
1.5Z16C	0251
1.5Z16D	0251
1.5Z17	1178
1.5Z17A	1178
1.5Z17B	1178
1.5Z17C	1178
1.5Z17D	1178
1.5Z17D5	1178
1.5Z17D10	1178
1.5Z18	0256
1.5Z18A	0256
1.5Z18B	0256
1.5Z18C	0256
1.5Z18D	0256
1.5Z19	2395
1.5Z19A	2395
1.5Z19B	2395
1.5Z19C	2395
1.5Z19D	2395
1.5Z19D5	2395
1.5Z19D10	2395
1.5Z20	0262
1.5Z20A	0262
1.5Z20B	0262
1.5Z20C	0262
1.5Z20D	0262
1.5Z22	0269
1.5Z22A	0269
1.5Z22B	0269
1.5Z22C	0269
1.5Z22D	0269
1.5Z24	0274
1.5Z24A	0274
1.5Z24B	0274
1.5Z24C	0274
1.5Z24D	0274
1.5Z25	2399
1.5Z25A	2399
1.5Z25B	2399
1.5Z25C	2399
1.5Z25D	2399
1.5Z25D5	2399
1.5Z25D10	2399
1.5Z27	0294
1.5Z27A	0294
1.5Z27B	0294
1.5Z27C	0294
1.5Z27D	0294
1.5Z30	0309
1.5Z30A	0309
1.5Z30B	0309
1.5Z30C	0309
1.5Z30D	0309
1.5Z31C	0326
1.5Z33	0318
1.5Z33A	0318
1.5Z33B	0318
1.5Z33C	0318
1.5Z33D	0318
1.5Z36	0320
1.5Z36A	0320
1.5Z36B	0320
1.5Z36C	0320
1.5Z36D	0320
1.5Z39	0326
1.5Z39A	0326
1.5Z39B	0326
1.5Z39C	0326
1.5Z39D	0326
1.5Z43	0337
1.5Z43A	0337
1.5Z43B	0337
1.5Z43C	0337
1.5Z43D	0337
1.5Z45	0337
1.5Z45A	0337
1.5Z45B	0337
1.5Z45C	0337
1.5Z45D	0337
1.5Z45D5	0337
1.5Z47	0347
1.5Z47A	0347
1.5Z47B	0347
1.5Z47C	0347
1.5Z47D	0347
1.5Z50	0027
1.5Z50A	0027
1.5Z50B	0027
1.5Z50C	0027
1.5Z50D	0027
1.5Z50D5	0027
1.5Z50D10	0027
1.5Z51	0027
1.5Z51A	0027
1.5Z51B	0027
1.5Z51C	0027
1.5Z51D	0027
1.5Z52	0027
1.5Z52A	0027
1.5Z52B	0027
1.5Z52C	0027
1.5Z52D	0027
1.5Z52D10	0027
1.5Z56	0266
1.5Z56A	0266
1.5Z56B	0266
1.5Z56C	0266
1.5Z56D	0266
1.5Z62	0386
1.5Z62A	0386
1.5Z62B	0386
1.5Z62C	0386
1.5Z62D	0386
1.5Z68	0405
1.5Z68A	0405
1.5Z68B	0405
1.5Z68C	0405
1.5Z68D	0405
1.5Z75	0426
1.5Z75A	0426
1.5Z75B	0426
1.5Z75C	0426
1.5Z75D	0426
1.5Z82	0444
1.5Z82A	0444
1.5Z82B	0444
1.5Z82C	0444
1.5Z82D	0444
1.5Z91	0238
1.5Z91A	0238
1.5Z91B	0238
1.5Z91C	0238
1.5Z91D	0238
1.5Z100	1180
1.5Z100A	1180
1.5Z100B	1180
1.5Z100C	1180
1.5Z100D	1180
1.5Z105	1180
1.5Z105A	1180
1.5Z105B	1180
1.5Z105C	1180
1.5Z105D	1180
1.5Z105D5	0247
1.5Z105D10	1180
1.5Z110	1190
1.5Z110A	1190
1.5Z110B	1190
1.5Z110C	1190
1.5Z110D	1190
1.5Z120	1207
1.5Z120A	1207
1.5Z120B	1207
1.5Z120C	1207
1.5Z120D	1207
1.5Z130	1218
1.5Z130A	1218
1.5Z130B	1218
1.5Z130C	1218
1.5Z130D	1218
1.5Z140	1881
1.5Z140A	1881
1.5Z140B	1881
1.5Z140C	1881
1.5Z140D	1881
1.5Z140D5	1881
1.5Z140D10	1881
1.5Z150	0647
1.5Z150A	0647
1.5Z150B	0647
1.5Z150C	0647
1.5Z150D	0647
1.5Z160	1255
1.5Z160A	1255
1.5Z160B	1255
1.5Z160C	1255
1.5Z160D	1255
1.5Z175	2102
1.5Z175A	2102
1.5Z175B	2102
1.5Z175C	2102
1.5Z175D	2102
1.5Z175D5	2102
1.5Z175D10	2102
1.5Z180	1279
1.5Z180A	1279
1.5Z180B	1279
1.5Z180C	1279
1.5Z180D	1279
1.5Z200	0605
1.5Z200A	0605
1.5Z200B	0605
1.5Z200C	0605
1.5Z200D	0605
1.5Z220	OEM
1.5Z220A	OEM
1.5Z220B	OEM
1.5Z220C	OEM
1.5Z220D	OEM
1.25AGD	OEM
1.ZZ9.1	OEM
2(3395	0079
2-00020-327F	0319
2-00037-955	0160
2-00045-345	0079
2-00045-424	0079
2-0A70	OEM
2-0A90	0123
2-0A90-H	0143
2-1K60	0143
2-1K261	0143
2-5SR3	0079
2-80	0917
2-131	2983
2-136	2983
2-140	2983
2-140-01	2983
2-312	OEM
2-412	OEM
2-5004-6	2017
2-5004-14	2018
2-5184-11	0147
2-D2K	0015
2-D2L	0015
2-D2R	0015
2-0A90	0143
2-0A90-M	0143
2-0A90A	0143
2-0A90A-H	0143
2-0A90A-M	0143
2-0A99	0143
2-0A99A	0143
2A	0136
2A1/18	OEM
2A1/52	OEM
2A2/18	OEM
2A2/52	OEM
2A5/18	OEM
2A5/52	OEM
2A12	0126
2A12/18	OEM
2A12/52	OEM
2A13	OEM
2A13/18	OEM
2A13/52	OEM
2A14	OEM
2A14/18	OEM
2A14/52	OEM
2A15	OEM
2A15/18	OEM
2A15/52	OEM
2A16	OEM
2A16/18	OEM
2A16/52	OEM
2A21	0041
2A21(ZENER)	0041
2A22	0041
2A22(ZENER)	0041
2A22/18	OEM
2A23	OEM
2A23/18	OEM
2A24	OEM
2A24/18	OEM
2A25	0188
2A25/18	OEM
2A26	OEM
2A26/18	OEM
2A28(ZENER)	0118
2A30	0087
2A43	0677
2A46	0456
2A50	0124
2A52	OEM
2A52/18	OEM
2A53	OEM
2A53/18	OEM
2A54	OEM
2A54/18	OEM
2A55	OEM
2A55/18	OEM
2A56	OEM
2A56/18	OEM
2A64	0166
2A100	0087
2A112	OEM
2A113	OEM
2A119	0143
2A200	0087
2A300	0071
2A400	0071
2A500	0071
2A600	0071
2A800	0071
2A1000	0071
2A1200	0102
2AA112	0019
2AA113	0143
2AA116	0143
2AA117	OEM
2AA118	OEM
2AA119	0123
2AA119(MP)	0123
2AC127	OEM
2AC128	0136
2AC128-01	OEM
2AC132	0004
2AC187	0847
2AC188	0164
2AC188-01	OEM
2AD140	0160
2AD149	0265
2AD161	OEM
2AF1	0071
2AF1LH	OEM
2AF1LV	OEM
2AF2	0071
2AF2LH	OEM
2AF2LV	OEM
2AF4	0071
2AF4LH	OEM
2AF4LV	OEM
2AG	0007
2AH	0007
2AM-1	0940
2AS160	OEM
2AS429	OEM
2ASCP30	OEM
2ASCP40	OEM
2ASCP45	OEM
2ASCP120	OEM
2ASLD1	OEM
2ASLD2	OEM
2ASLD3	OEM
2ASLD4	OEM
2ASLD5	OEM
2ASLD6	OEM
2ASLD7	OEM
2ASLD8	OEM
2ASLD9	OEM
2ASLD10	OEM
2ASLD10TRA	OEM
2ASLD10TRB	OEM
2ASLD15	OEM
2ASLD15TR	OEM
2ASLD20	OEM
2ASLD25	OEM
2ASLD30	OEM
2ASLD35	OEM
2ASLD40	OEM
2ASLD40TR	OEM
2ASLD45	OEM
2ASLD50	OEM
2ASLD75	OEM
2ASLD100	OEM
2ASLD120	OEM
2ASPC1	OEM
2ASPC2	OEM
2ASPC3	OEM
2ASPC4	OEM
2ASPC5	OEM
2ASPC6	OEM
2ASPC7	OEM
2ASPC8	OEM
2ASPC9	OEM
2ASPC10	OEM
2ASPC15	OEM
2ASPC20	OEM
2ASPC25	OEM
2ASPC30	OEM
2ASPC35	OEM
2ASPC40	OEM
2ASPC45	OEM
2ASPC50	OEM
2ASPC75	OEM
2ASPC100	OEM
2ASPC120	OEM
2ASPF16	OEM
2ASPF20	OEM
2ASPF30	OEM
2ASPF40	OEM
2ASPF50	OEM
2ASPF60	OEM
2ASPF70	OEM
2ASPF80	OEM
2ASPF90	OEM
2ASPF100	OEM
2ASPF150	OEM
2ASPF200	OEM
2ASPF250	OEM
2ASPF300	OEM
2ASPF350	OEM
2ASPF400	OEM
2ASPF450	OEM
2ASPF500	OEM
2ASPF750	OEM
2ASPF1000	OEM
2ASPF1200	OEM
2ASPS16	OEM
2ASPS20	OEM
2ASPS30	OEM
2ASPS40	OEM
2ASPS50	OEM
2ASPS60	OEM
2ASPS70	OEM
2ASPS80	OEM
2ASPS90	OEM
2ASPS100	OEM
2ASPS150	OEM
2ASPS200	OEM
2ASPS250	OEM
2ASPS300	OEM
2ASPS350	OEM
2ASPS400	OEM
2ASPS450	OEM
2ASPS500	OEM
2ASPS750	OEM
2ASPS1000	OEM
2ASPS1200	OEM
2ASZ15	OEM
2AT128	OEM
2AT329	OEM
2AT331	OEM
2ATPF16	OEM
2ATPF20	OEM
2ATPF30	OEM
2ATPF40	OEM
2ATPF50	OEM
2ATPF60	OEM
2ATPF70	OEM
2ATPF80	OEM
2ATPF90	OEM
2ATPF100	OEM
2ATPF150	OEM
2ATPF200	OEM
2ATPF250	OEM
2ATPF300	OEM
2ATPF350	OEM
2ATPF400	OEM
2ATPF450	OEM
2ATPF500	OEM
2ATPF750	OEM
2ATPF1000	OEM
2ATPF1200	OEM
2ATPH16	OEM
2ATPH20	OEM
2ATPH30	OEM
2ATPH40	OEM
2ATPH50	OEM
2ATPH60	OEM
2ATPH70	OEM
2ATPH80	OEM
2ATPH90	OEM
2ATPH100	OEM
2ATPH150	OEM
2ATPH200	OEM
2ATPH250	OEM
2ATPH300	OEM
2ATPH350	OEM
2ATPH400	OEM
2ATPH450	OEM
2ATPH500	OEM
2ATPH750	OEM
2ATPH1000	OEM
2ATPH1200	OEM
2AV2	1449
2AV30	OEM
2AV41	OEM
2B	0280
2B1/18	OEM
2B1/52	OEM
2B-2	0102
2B2/18	OEM
2B2/52	OEM
2B-2D10DE	0102
2B-2DIODE	0102
2B2DM	0277
2B2M	0277
2B2MS	0277
2B3.L	0323
2B4B41	0323
2B4DM	0289
2B4M	0289
2B4MS	0289

If replacement code is OEM, contact original manufacturer for replacement.

DEVICE TYPE	REPL CODE
2B4.E	1414
2B5/18	OEM
2B5/52	OEM
2B6DM	0296
2B6M	0296
2B6MS	0296
2B8	OEM
2B8DM	0296
2B8M	0296
2B8MS	0296
2B10DM	0302
2B10M	0302
2B10MS	0302
2B12	OEM
2B12/18	OEM
2B12DM	0302
2B12DS	0302
2B12M	0250
2B12MS	0250
2B13	OEM
2B13/18	OEM
2B14	OEM
2B14/18	OEM
2B15	OEM
2B15/18	OEM
2B16	OEM
2B16/18	OEM
2B16DM	0250
2B16M	0250
2B20A	OEM
2B20B	OEM
2B20DM	0250
2B20M	0250
2B22	OEM
2B22/18	OEM
2B22J	OEM
2B22K	OEM
2B22L	OEM
2B23	OEM
2B23/18	OEM
2B24	OEM
2B24/18	OEM
2B25	OEM
2B25/18	OEM
2B26	OEM
2B26/18	OEM
2B35J	OEM
2B35K	OEM
2B50B	OEM
2B50C	OEM
2B50FB	OEM
2B50FC	OEM
2B52	OEM
2B52/18	OEM
2B52A-1J01	OEM
2B52A-1J05	OEM
2B52A-1K01	OEM
2B52A-1K04	OEM
2B52A-1K06	OEM
2B52A-1T01	OEM
2B53	OEM
2B53/18	OEM
2B53A-1J01	OEM
2B53A-1J03	OEM
2B53A-1J05	OEM
2B53A-1K01	OEM
2B53A-1K04	OEM
2B53A-1K06	OEM
2B53A-1T01	OEM
2B53A-1T02	OEM
2B54	OEM
2B54/18	OEM
2B55	OEM
2B55/18	OEM
2B56	OEM
2B56/18	OEM
2B56A	OEM
2B57A	OEM
2B58A-1-1-01	OEM
2B58A-1-1-02	OEM
2B58A-1-1-03	OEM
2B58A-1-1-04	OEM
2B59A-0-1-01	OEM
2B59A-0-1-02	OEM
2B59A-0-1-03	OEM
2B59A-0-1-04	OEM
2B59A-02-01	OEM
2B59A-02-02	OEM
2B59A-02-03	OEM
2B59A-02-04	OEM
2B1202	0440
2BA2	1449
2BC119	OEM
2BC138	OEM
2BC139	OEM
2BC142	OEM
2BC143	OEM
2BC144	OEM
2BC221	OEM
2BC222	OEM
2BC286	OEM
2BC288	OEM
2BD117	OEM
2BD181	1514
2BD182	OEM
2BD183	1514
2BDY20	OEM
2BDY38	OEM
2B-E	OEM

DEVICE TYPE	REPL CODE
2BG2C41	0199
2BG2Z41	1015
2BZ61	OEM
2C	0280
2C05	0087
2C05/18	OEM
2C05/52	OEM
2C054	OEM
2C054/18	OEM
2C055	OEM
2C055/18	OEM
2C056	OEM
2C056/18	OEM
2C057	OEM
2C057/18	OEM
2C1	OEM
2C1/18	OEM
2C1/52	OEM
2C2	OEM
2C2/18	OEM
2C2/52	OEM
2C2DM	0199
2C2M	0199
2C4-6S	OEM
2C4DM	0199
2C4M	0199
2C6	0087
2C6-7S	OEM
2C6DM	0199
2C6M	0199
2C8	0087
2C8DM	0199
2C8M	0199
2C10	0087
2C10DM	0199
2C10M	0199
2C12	OEM
2C12DM	0199
2C12M	0199
2C13B	0237
2C14	OEM
2C14/18	OEM
2C15	OEM
2C15/18	OEM
2C16	OEM
2C16/18	OEM
2C16DM	0199
2C16M	0199
2C17	OEM
2C17/18	OEM
2C18	OEM
2C20	OEM
2C20DM	0199
2C20M	0199
2C24	OEM
2C24/18	OEM
2C25	OEM
2C25/18	OEM
2C26	OEM
2C26/18	OEM
2C27	OEM
2C27/18	OEM
2C34	OEM
2C91G	0600
2C111	OEM
2C415	OEM
2C425	OEM
2C444	OEM
2C918	OEM
2C2102	OEM
2C2222A	OEM
2C2369A	OEM
2C2484	OEM
2C2546	OEM
2C2857	OEM
2C2894	OEM
2C2907A	OEM
2C2946	OEM
2C3303	OEM
2C3308	OEM
2C3439	OEM
2C3468	OEM
2C3501	OEM
2C3506	OEM
2C3511	OEM
2C3553	OEM
2C3635	OEM
2C3724	OEM
2C3762	OEM
2C3767	OEM
2C3796	OEM
2C3799	OEM
2C3866	OEM
2C3868	OEM
2C3904	OEM
2C3906	OEM
2C3960	OEM
2C4091	OEM
2C4117	OEM
2C4220	OEM
2C4261	OEM
2C4352	OEM
2C4405	OEM
2C4416	OEM
2C4427	OEM
2C4900	OEM
2C4920	OEM
2C4923	OEM
2C4957	OEM

DEVICE TYPE	REPL CODE
2C5031	OEM
2C5038	OEM
2C5108	OEM
2C5109	OEM
2C5160	OEM
2C5179	OEM
2C5192	OEM
2C5195	OEM
2C5208	OEM
2C5339	OEM
2C5345	OEM
2C5401	OEM
2C5416	OEM
2C5462	OEM
2C5551	OEM
2C5583	OEM
2C5631	OEM
2C5641	OEM
2C5642	OEM
2C5680	OEM
2C5682	OEM
2C5760	OEM
2C5836	OEM
2C5841	OEM
2CN3	1306
2CS537FC	0111
2CY30	0861
2CY31	0861
2CY32	0861
2CY33	0861
2CY34	0861
2CY38	0037
2CY39	0037
2CZ6.8	OEM
2CZ6.8A	OEM
2CZ6.8B	OEM
2CZ7.5	OEM
2CZ7.5A	OEM
2CZ7.5B	OEM
2CZ8.2	OEM
2CZ8.2A	OEM
2CZ8.2B	OEM
2CZ9.1	OEM
2CZ9.1A	OEM
2CZ9.1B	OEM
2CZ10	OEM
2CZ10A	OEM
2CZ10B	OEM
2CZ11	OEM
2CZ11A	OEM
2CZ11B	OEM
2CZ12	OEM
2CZ12A	OEM
2CZ12B	OEM
2CZ13	OEM
2CZ13A	OEM
2CZ13B	OEM
2CZ15	OEM
2CZ15A	OEM
2CZ15B	OEM
2CZ16	OEM
2CZ16A	OEM
2CZ16B	OEM
2CZ18	OEM
2CZ18A	OEM
2CZ18B	OEM
2CZ20	OEM
2CZ20A	OEM
2CZ20B	OEM
2CZ22	OEM
2CZ22A	OEM
2CZ22B	OEM
2CZ24	OEM
2CZ24A	OEM
2CZ24B	OEM
2CZ27	OEM
2CZ27A	OEM
2CZ27B	OEM
2CZ30	OEM
2CZ30A	OEM
2CZ30B	OEM
2CZ33	OEM
2CZ33A	OEM
2CZ33B	OEM
2CZ36	OEM
2CZ36A	OEM
2CZ36B	OEM
2CZ39	OEM
2CZ39A	OEM
2CZ39B	OEM
2CZ43	OEM
2CZ43A	OEM
2CZ43B	OEM
2CZ47	OEM
2CZ47A	OEM
2CZ47B	OEM
2CZ51	OEM
2CZ51A	OEM
2CZ51B	OEM
2CZ56	OEM
2CZ56A	OEM
2CZ56B	OEM
2CZ62	OEM
2CZ62A	OEM
2CZ62B	OEM
2CZ68	OEM
2CZ68A	OEM
2CZ68B	OEM
2CZ75	OEM

DEVICE TYPE	REPL CODE
2CZ75A	OEM
2CZ75B	OEM
2CZ82	OEM
2CZ82A	OEM
2CZ82B	OEM
2CZ91	OEM
2CZ91A	OEM
2CZ91B	OEM
2CZ100	OEM
2CZ100A	OEM
2CZ100B	OEM
2CZ110	OEM
2CZ110A	OEM
2CZ110B	OEM
2CZ120	OEM
2CZ120A	OEM
2CZ120B	OEM
2CZ130	OEM
2CZ130A	OEM
2CZ130B	OEM
2CZ150	OEM
2CZ150A	OEM
2CZ150B	OEM
2CZ160	OEM
2CZ160A	OEM
2CZ160B	OEM
2CZ170	OEM
2CZ170A	OEM
2CZ170B	OEM
2CZ180	OEM
2CZ180A	OEM
2CZ180B	OEM
2CZ200	OEM
2CZ200A	OEM
2CZ200B	OEM
2D	0280
2D001	0160
2D002	0079
2D002-41	0079
2D002-168	0016
2D002-169	0016
2D002-170	0079
2D002-171	0285
2D002-175	0079
2D004	0160
2D004-9	0160
2D010	0103
2D013-13	0211
2D013-54	0211
2D013-109	0211
2D013-160	0211
2D016	0211
2D016-45	0211
2D016-54	0211
2D017-165	0037
2D017-166	0037
2D017-167	0037
2D017-169	0037
2D020-173	0861
2D020-174	0861
2D021	0280
2D021-8	0280
2D021-11	0280
2D021-56	0280
2D022	1669
2D022-211	1669
2D023	0211
2D026	0079
2D026-274	0079
2D027	0037
2D030	1893
2D031	2950
2D033	0155
2D036	0004
2D038	0475
2D039	0004
2D040	0187
2D1	OEM
2D1PDA	OEM
2D4B41	OEM
2D7A	0098
2D7AM	0098
2D41	1137
2D600M	OEM
2DB2A	0289
2DB4A	0296
2DB6A	0302
2DF12	OEM
2DG2	0143
2DG2C41	0199
2DG2Z41	1015
2DL8	OEM
2DL10	OEM
2DL-15	0102
2DL15	0017
2DL15A	0017
2DL18A	0348
2DL20	OEM
2D021-56	0164
2D023	0164
2D033	0079
2DS24YD	0142
2DS500	OEM
2DS503	OEM
2DS506	OEM
2E	0136
2E05	0015
2E1	0015
2E2	0087

DEVICE TYPE	REPL CODE
2E2DM	1015
2E4	0015
2E4DM	1015
2E6	0087
2E6DM	1015
2E8	0087
2E8DM	1015
2E10	0087
2E10DM	1015
2E12	OEM
2E12DM	1015
2E14	OEM
2E16	OEM
2E16DM	1015
2E18	OEM
2E20	OEM
2E20DM	1015
2E40A	OEM
2EM03	0071
2EM05	0071
2EM015	0087
2EM1	OEM
2EM2	OEM
2EM3	OEM
2EM4	OEM
2EM5	OEM
2EM6	0087
2EM8	0087
2EM10	0087
2EM12	OEM
2EZ3.6D	0796
2EZ3.6D5	0796
2EZ3.6D10	0796
2EZ3.9D	0806
2EZ3.9D5	0806
2EZ3.9D10	0806
2EZ4.3D	0821
2EZ4.3D5	0821
2EZ4.3D10	0821
2EZ4.7D	0833
2EZ4.7D5	0833
2EZ4.7D10	0833
2EZ5.1D	0442
2EZ5.1D5	0442
2EZ5.1D10	0442
2EZ5.6D	0876
2EZ5.6D5	0876
2EZ5.6D10	0876
2EZ6.2D	0185
2EZ6.2D5	0185
2EZ6.2D10	0185
2EZ6.3D10	OEM
2EZ6.8D	0205
2EZ6.8D5	0205
2EZ6.8D10	0205
2EZ7.5D	0480
2EZ7.5D5	0480
2EZ7.5D10	0480
2EZ8.2D	0504
2EZ8.2D5	0504
2EZ8.2D10	0504
2EZ9.1D	0684
2EZ9.1D5	0684
2EZ9.1D10	0684
2EZ10D	0225
2EZ10D5	0225
2EZ10D10	0225
2EZ11D	0230
2EZ11D5	0230
2EZ11D10	0230
2EZ12D	0234
2EZ12D5	0234
2EZ12D10	0234
2EZ13D	0237
2EZ13D5	0237
2EZ13D10	0237
2EZ14D	1397
2EZ14D5	1397
2EZ14D10	1397
2EZ15D	0247
2EZ15D5	0247
2EZ15D10	0247
2EZ16D	0251
2EZ16D5	0251
2EZ16D10	0251
2EZ17D	1178
2EZ17D5	1178
2EZ17D10	1178
2EZ18D	0256
2EZ18D5	0256
2EZ18D10	0256
2EZ19D	2395
2EZ19D5	2395
2EZ19D10	2395
2EZ20D	0262
2EZ20D5	0262
2EZ20D10	0262
2EZ22D	0269
2EZ22D5	0269
2EZ22D10	0269
2EZ24C	0274
2EZ24D	0274
2EZ24D5	0274
2EZ24D10	0274
2EZ27D	0294
2EZ27D5	0294
2EZ27D10	0294
2EZ30D	0309
2EZ30D5	0309

DEVICE TYPE	REPL CODE
2EZ30D10	0309
2EZ33D	0318
2EZ33D5	0318
2EZ33D10	0318
2EZ36D	0320
2EZ36D5	0320
2EZ36D10	0320
2EZ39D	0326
2EZ39D5	0326
2EZ39D10	0326
2EZ43D	0337
2EZ43D5	0337
2EZ43D10	0337
2EZ47D	0347
2EZ47D5	0347
2EZ47D10	0347
2EZ51D	0027
2EZ51D5	0027
2EZ51D10	0027
2EZ56D	0266
2EZ56D5	0266
2EZ56D10	0266
2EZ62D	0386
2EZ62D5	0386
2EZ62D10	0386
2EZ68D	0405
2EZ68D5	0405
2EZ68D10	0405
2EZ75D	0426
2EZ75D5	0504
2EZ75D10	0426
2EZ82D	0444
2EZ82D5	0444
2EZ82D10	0444
2EZ91D	0238
2EZ91D5	0238
2EZ91D10	0238
2EZ100D	1180
2EZ100D5	1180
2EZ100D10	1180
2EZ110D	1190
2EZ110D5	1190
2EZ110D10	1190
2EZ120D	1207
2EZ120D5	1207
2EZ120D10	1207
2EZ130D	1218
2EZ130D5	1218
2EZ130D10	1218
2EZ140D	1881
2EZ140D5	1881
2EZ140D10	1881
2EZ150D	0647
2EZ150D5	0647
2EZ150D10	0647
2EZ160D	1255
2EZ160D5	1255
2EZ160D10	1255
2EZ170D	2102
2EZ170D5	2102
2EZ170D10	2102
2EZ180D	1279
2EZ180D5	1279
2EZ180D10	1279
2EZ190D	2225
2EZ190D5	2225
2EZ190D10	2225
2EZ200D	0605
2EZ200D5	0605
2EZ200D10	0605
2E.5	OEM
2F	0050
2F05	0110
2F1	OEM
2F2	OEM
2F2CN1	0087
2F2CN5	0087
2F3	OEM
2F4	0015
2F5CN1	0071
2F6	0087
2F8	0087
2F10	0087
2F12	OEM
2F14	OEM
2F16	OEM
2F18	OEM
2F20	OEM
2FB050	0277
2FB050R	0277
2FB100	0277
2FB100R	0277
2FB200	0289
2FB200R	0289
2FB400	0296
2FB400R	0296
2FB600	0302
2FB600R	0302
2FB800	0250
2FB800R	0250
2FB1000	0250
2FB1000R	0250
2G	0136
2G022	0211
2G1	0071
2G2	0071
2G4	0071
2G4B41	0473
2G6	0071

DEVICE TYPE	REPL CODE
2G8	0071
2G10	0071
2G13	0015
2G101	0136
2G102	0050
2G103	0050
2G104	0050
2G106	0050
2G108	0211
2G109	0050
2G110	0050
2G138	0280
2G139	0280
2G140	0280
2G141	OEM
2G201	0050
2G202	0050
2G210	0160
2G220	0160
2G221	2717
2G222	0160
2G223	0085
2G224	0160
2G225	0160
2G226	0604
2G227	0604
2G228	0604
2G229	OEM
2G230	OEM
2G231	OEM
2G240	0160
2G270	0211
2G271	0211
2G301	0050
2G302	0211
2G303	0004
2G304	0004
2G306	0211
2G308	0004
2G309	0004
2G319	0211
2G320	0211
2G321	0211
2G322	0211
2G323	0211
2G324	0211
2G339	0038
2G339A	0600
2G344	0004
2G345	0004
2G371	0004
2G371A	0004
2G374	0004
2G374A	0004
2G376	0004
2G377	0004
2G381	0004
2G381A	0004
2G382	0004
2G383	0280
2G384	0211
2G385	0211
2G386	0211
2G387	0211
2G394	0280
2G395	0211
2G396	0280
2G397	0280
2G398	OEM
2G401	0136
2G402	0136
2G403	0050
2G404	0050
2G413	0050
2G414	0136
2G415	0136
2G416	0136
2G417	0136
2G508	0211
2G509	0211
2G524	0280
2G525	0280
2G526	0280
2G527	0280
2G577	0280
2G601	0280
2G602	0280
2G603	0280
2G604	0280
2G605	0280
2G805	0015
2G1024	0280
2G1025	0280
2G1026	0280
2G1027	0211
2GA	0015
2GG2C41	0199
2GG2Z41	1015
2GL05	0280
2GU026	0280
2H1254	0037
2H1255	0037
2H1256	0150
2H1257	0150
2H1258	0037
2H1259	0037
2HR3J	0071
2HR3M	0071
2HR3N	OEM

DEVICE TYPE	REPL CODE
2HR3P	OEM
2HT-2L	0774
2HT-6	0757
2HT-9	OEM
2HT-11C	0288
2HT-11L	0288
2HT-13C	0288
2HT-13L	0288
2HT-16	0201
2HT-18C	0757
2HT-20C	0757
2ISD235-R	0042
2J4B41	0446
2J30	OEM
2J31	OEM
2J32	OEM
2J33	OEM
2J34	OEM
2J42	OEM
2J42A	OEM
2J42A-6027	OEM
2J42H	OEM
2J50	OEM
2J51	OEM
2J51A	OEM
2J53	OEM
2J55	OEM
2J70A	OEM
2J70B	OEM
2J72	0050
2J73	0050
2J8138	0015
2JG2C41	0199
2JG2Z41	1015
2JUS1.08A	OEM
2JUS1.08B	OEM
2JUS1.23A	OEM
2JUS1.23B	OEM
2JUS1.30A	OEM
2JUS1.30B	OEM
2JUS1.46A	OEM
2JUS1.46B	OEM
2JUS1.52A	OEM
2JUS1.52B	OEM
2K02	0071
2K19H	0325
2K22	OEM
2K25	OEM
2K25A	OEM
2K26	OEM
2K28	OEM
2K28A	OEM
2K29	OEM
2K45	OEM
2K48	0050
2K56	OEM
2KBP005	0277
2KBP01	0277
2KBP02	0289
2KBP04	0296
2KBP06	0302
2KBP08	0250
2KBP10	0250
2KBP20	OEM
2KZ6.8	OEM
2KZ6.8A	OEM
2KZ6.8B	OEM
2KZ7.5	OEM
2KZ7.5A	OEM
2KZ7.5B	OEM
2KZ8.2	OEM
2KZ8.2A	OEM
2KZ8.2B	OEM
2KZ9	OEM
2KZ9A	OEM
2KZ9B	OEM
2KZ9.1	OEM
2KZ9.1A	OEM
2KZ9.1B	OEM
2KZ10	OEM
2KZ10A	OEM
2KZ10B	OEM
2KZ11	OEM
2KZ11A	OEM
2KZ11B	OEM
2KZ12	OEM
2KZ12A	OEM
2KZ12B	OEM
2KZ13	OEM
2KZ13A	OEM
2KZ13B	OEM
2KZ15	OEM
2KZ15A	OEM
2KZ15B	OEM
2KZ16	OEM
2KZ16A	OEM
2KZ16B	OEM
2KZ18	OEM
2KZ18A	OEM
2KZ18B	OEM
2KZ20	OEM
2KZ20A	OEM
2KZ20B	OEM
2KZ22	OEM
2KZ22A	OEM
2KZ22B	OEM
2KZ24	OEM
2KZ24A	OEM
2KZ24B	OEM

DEVICE TYPE	REPL CODE
2KZ27	OEM
2KZ27A	OEM
2KZ27B	OEM
2KZ30	OEM
2KZ30A	OEM
2KZ30B	OEM
2KZ33	OEM
2KZ33A	OEM
2KZ33B	OEM
2KZ36	OEM
2KZ36A	OEM
2KZ36B	OEM
2KZ39	OEM
2KZ39A	OEM
2KZ39B	OEM
2KZ43	OEM
2KZ43A	OEM
2KZ43B	OEM
2KZ47	OEM
2KZ47A	OEM
2KZ47B	OEM
2KZ51	OEM
2KZ51A	OEM
2KZ51B	OEM
2KZ56	OEM
2KZ56A	OEM
2KZ56B	OEM
2KZ62	OEM
2KZ62A	OEM
2KZ62B	OEM
2KZ68	OEM
2KZ68A	OEM
2KZ68B	OEM
2KZ75	OEM
2KZ75A	OEM
2KZ75B	OEM
2KZ82	OEM
2KZ82A	OEM
2KZ82B	OEM
2KZ90	OEM
2KZ90A	OEM
2KZ90B	OEM
2KZ91	OEM
2KZ91A	OEM
2KZ91B	OEM
2KZ100	OEM
2KZ100A	OEM
2KZ100B	OEM
2KZ110	OEM
2KZ110A	OEM
2KZ110B	OEM
2KZ120	OEM
2KZ120A	OEM
2KZ120B	OEM
2KZ130	OEM
2KZ130A	OEM
2KZ130B	OEM
2KZ150	OEM
2KZ150A	OEM
2KZ150B	OEM
2KZ160	OEM
2KZ160A	OEM
2KZ160B	OEM
2KZ170	OEM
2KZ170A	OEM
2KZ170B	OEM
2KZ180	OEM
2KZ180A	OEM
2KZ180B	OEM
2KZ200	OEM
2KZ200A	OEM
2KZ200B	OEM
2L02	OEM
2L08	OEM
2L1-2	OEM
2L1-2S	OEM
2L15	0348
2L15A	OEM
2L15B	OEM
2L15C	OEM
2L20	OEM
2L25	OEM
2L30	OEM
2L35	OEM
2L40	OEM
2L45	OEM
2L50	OEM
2L55	OEM
2L60	OEM
2L65	OEM
2L70	OEM
2L75	OEM
2L80	OEM
2L85	OEM
2L90	OEM
2L95	OEM
2L100	OEM
2L110	OEM
2L120	OEM
2L130	OEM
2L140	OEM
2L150	OEM
2L160	OEM
2L170	OEM
2L180	OEM
2L190	OEM
2L200	OEM
2L210	OEM
2L220	OEM

DEVICE TYPE	REPL CODE
2L230	OEM
2L240	OEM
2L250	OEM
2L260	OEM
2L270	OEM
2L280	OEM
2L290	OEM
2L300	OEM
2LS1.5-3	OEM
2M13	OEM
2M53	OEM
2M53G	OEM
2M54	OEM
2M57A	OEM
2M60A	OEM
2M66	OEM
2M68	OEM
2M76	0142
2M78	0600
2M105	OEM
2M107A	OEM
2M109	OEM
2M110	OEM
2M115	OEM
2M115G	OEM
2M120	OEM
2M121	OEM
2M125	OEM
2M125H	OEM
2M125J	OEM
2M130	OEM
2M131	OEM
2M132	OEM
2M132A	OEM
2M140	OEM
2M140A	OEM
2M142	OEM
2M142G	OEM
2M142H	OEM
2M142J	OEM
2M151A	OEM
2M152	OEM
2M152G	OEM
2M152H	OEM
2M152J	OEM
2M158	OEM
2M158H	OEM
2M158J	OEM
2M162	OEM
2M162G	OEM
2M162H	OEM
2M162J	OEM
2M164	OEM
2M172	OEM
2M172G	OEM
2M172H	OEM
2M172J	OEM
2M182	OEM
2M214	1083
2M1127	0211
2M1303	0164
2M1305	0050
2M1849	0725
2M6219	0144
2M6504	OEM
2MA	1303
2MA509	0436
2MC	0136
2MC509	0018
2MN6374	0144
2MW665	0143
2N0681	OEM
2N0682	OEM
2N0682J	OEM
2N0682TX	OEM
2N0683	OEM
2N0683J	OEM
2N0683TX	OEM
2N0684	OEM
2N0685	OEM
2N0685J	OEM
2N0685TX	OEM
2N0686	OEM
2N0686J	OEM
2N0686TX	OEM
2N0687	OEM
2N0687J	OEM
2N0687TX	OEM
2N0688	OEM
2N0688J	OEM
2N0688TX	OEM
2N0689	OEM
2N0690	OEM
2N0690J	OEM
2N0690TX	OEM
2N0697	OEM
2N0699	OEM
2N2G374	0211
2N2J324	0086
2N2J374	0595
2N404A	OEM
2N15	0211
2N21	0164
2N21A	OEM
2N22	0164
2N23	0211
2N24	0211
2N25	0211
2N26	0211

DEVICE TYPE	REPL CODE
2N27	0280
2N28	0600
2N29	0600
2N30	0280
2N31	0280
2N32	0211
2N32A	0211
2N33	0136
2N34	0004
2N34/5	OEM
2N34A	0004
2N35	0208
2N36	0004
2N37	0136
2N38	0136
2N38A	0004
2N39	0211
2N40	0211
2N41	0211
2N42	0211
2N43	0211
2N43A	0211
2N44	0211
2N44A	0004
2N45	0004
2N45A	0004
2N46	0211
2N47	0211
2N48	0211
2N48A	0211
2N49	0211
2N50	0211
2N51	0211
2N52	0211
2N53	0211
2N54	0211
2N55	0211
2N56	0211
2N57	0085
2N58	0211
2N59	0280
2N59A	0211
2N59B	0211
2N59C	0211
2N59D	0211
2N60	0280
2N60A	0211
2N60B	0211
2N60C	0211
2N60R	0211
2N61	0280
2N61A	0211
2N61B	0211
2N61C	0211
2N62	0004
2N63	0004
2N64	0004
2N65	0211
2N66	0085
2N67	0085
2N68	0085
2N70	OEM
2N71	0211
2N72	0136
2N73	0211
2N74	0211
2N75	0136
2N76	0211
2N77	0004
2N78	0600
2N78A	0600
2N79	0211
2N80	0211
2N81	0211
2N82	0211
2N82AG	0917
2N83	0280
2N84	0050
2N85	0136
2N86	0136
2N87	0136
2N88	0136
2N89	0136
2N90	0136
2N94	0038
2N94A	0038
2N95	2181
2N96	0211
2N97	0038
2N97A	0600
2N98	0600
2N98A	0600
2N99	0600
2N100	0600
2N102	0600
2N103	0600
2N104	0004
2N105	0004
2N106	0004
2N107	0280
2N108	0004
2N109	0086
2N109/5	0211
2N109BLU	0004
2N109GRN	0004
2N109M1	0004
2N109M2	0004
2N109WHT	0004
2N109YEL	0004

DEVICE TYPE	REPL CODE
2N110	OEM
2N111	0211
2N111A	0211
2N111B	0136
2N111M1	0136
2N111M2	0136
2N112	0136
2N112A	0211
2N112M1	0136
2N113	0280
2N114	0280
2N115	0160
2N117	0079
2N117(T1)	0861
2N118	0079
2N118(T1)	0861
2N118A	0079
2N118A(T1)	0861
2N119	0079
2N119(T1)	0861
2N120	0079
2N120(T1)	0861
2N122	0168
2N123	0280
2N123(GE)	0211
2N123/5	0050
2N123A	0136
2N123A5	0050
2N124	0600
2N124(T1)	0600
2N125	0600
2N125(T1)	0600
2N126	0600
2N126(T1)	0600
2N127	0600
2N128	0136
2N129	0136
2N130	0211
2N130A	0211
2N130B	0038
2N131	0211
2N131A	0211
2N132	0211
2N132A	0004
2N133	0211
2N133A	0004
2N135	0280
2N136	0280
2N137	0280
2N138	0136
2N138A	0004
2N138B	0004
2N139	0136
2N140	0136
2N140M1	0136
2N140M2	0136
2N141	0160
2N143	0160
2N145	0600
2N146	0600
2N147	0600
2N148	0208
2N148A	0208
2N148B	0208
2N148C	0208
2N148C/D	0208
2N148D	0208
2N149	0208
2N149A	0208
2N149B	0164
2N150	0600
2N150A	0600
2N151	0211
2N155	0160
2N156	0160
2N157	0160
2N157A	0160
2N158	0160
2N158A	0160
2N159	0211
2N160	0079
2N160A	0079
2N161	0079
2N161A	0079
2N162	0079
2N162A	0127
2N163	0079
2N163A	0079
2N164	0600
2N164A	0600
2N165	0600
2N166	0600
2N167	0600
2N167A	0600
2N168	0600
2N168A	0600
2N169	0600
2N169A	0600
2N170	0600
2N171	0038
2N172	0600
2N173	0040
2N174	0040
2N174A	0040
2N174RED	0040
2N175	0004
2N176	0160
2N176(WHT/BRN)	0160
2N176(WHT/GRN	0160

DEVICE TYPE	REPL CODE
2N176(WHT/GRN)G	0160
2N176(WHT/RED)	0160
2N176-1	0085
2N176-1BLU	0160
2N176-1WHT	0160
2N176-1YEL	0160
2N176-3PUR	0160
2N176-4PUR	0160
2N176-5WHT	0160
2N176-6WHT	0160
2N176A	0085
2N176BLK	0160
2N176BLU	0160
2N176G	0085
2N176GRN	0160
2N176PUR	0160
2N176RED	0160
2N176W	0085
2N176WHT	0160
2N176YEL	0160
2N178	0085
2N178A	OEM
2N179	0085
2N180	0004
2N181	0004
2N182	0208
2N183	0600
2N184	0600
2N185	0004
2N185BLU	0004
2N186	0004
2N186A	0004
2N187	0004
2N187A	0211
2N188	0211
2N188A	0211
2N189	0211
2N190	0211
2N191	0211
2N192	0211
2N193	0038
2N194	0038
2N194A	0038
2N195	0211
2N196	0211
2N197	0211
2N198	0211
2N199	0211
2N200	0211
2N204	0211
2N205	0211
2N206	0211
2N207	0211
2N207A	0211
2N207B	0211
2N207BLU	0211
2N211	0038
2N211M2	0050
2N212	0038
2N213	0208
2N213A	0208
2N214	0208
2N214A	0038
2N214MP	OEM
2N215	0004
2N216	0038
2N217	0164
2N217A	0004
2N217EQ	OEM
2N217RED	0004
2N217WHT	0004
2N217YEL	0004
2N218	0136
2N219	0050
2N220	0004
2N222	0004
2N222A	0004
2N223	0004
2N224	0004
2N225	0004
2N226	0004
2N227	0004
2N228	0208
2N228A	0164
2N229	0038
2N230	0085
2N231	0136
2N231-YEL-RED	0136
2N231BLU	0136
2N231RED	0136
2N231YEL	0136
2N232	0136
2N233	0038
2N233A	0038
2N234	0160
2N234A	0160
2N235	0160
2N235A	0160
2N235B	0160
2N236	0160
2N236A	0160
2N236B	0160
2N237	0211
2N238	0004
2N238-ORN	0004
2N238D	0004
2N238E	0004
2N238F	0004
2N240	0136

DEVICE TYPE	REPL CODE
2N241	0004
2N241A	0004
2N242	0160
2N243	0018
2N244	0018
2N245	0595
2N246	0595
2N247	0136
2N247/33	0136
2N248	0136
2N249	0004
2N250	0085
2N250A	0085
2N251	0160
2N251A	0160
2N252	0136
2N253	0600
2N254	0600
2N255	0085
2N255A	0085
2N256	0085
2N256A	0085
2N257	0160
2N257-B	0085
2N257-BLK	0975
2N257-GRN	0975
2N257-W	0085
2N257-WHT	0604
2N257A	0085
2N257B	0085
2N257G	0085
2N257GRN	0085
2N257W	0085
2N260	OEM
2N260A	OEM
2N261	OEM
2N262	0280
2N262A	OEM
2N263	0079
2N264	0079
2N265	0211
2N266	0211
2N267	0050
2N268	0604
2N268A	0604
2N269	0004
2N270	0004
2N270A	0211
2N270SM	0280
2N271	0280
2N271A	0280
2N272	0211
2N273	0280
2N274	0050
2N274BLU	0050
2N274WHT	0050
2N275	0136
2N275W	0085
2N276	0050
2N277	0040
2N278	0040
2N279	0004
2N280	0004
2N281	0004
2N282	0211
2N283	0004
2N284	0004
2N284A	0004
2N285	0085
2N285A	0085
2N285B	0085
2N286	0050
2N289	0050
2N290	0440
2N291	0211
2N292	0600
2N292A	0600
2N293	0600
2N296	0085
2N297	0160
2N297A	0160
2N299	0136
2N300	0050
2N301	0160
2N301A	0160
2N301B	0160
2N301G	0136
2N301W	0160
2N302	0211
2N303	0211
2N306	0208
2N306A	0208
2N307	0604
2N307A	0604
2N307B	0604
2N308	0280
2N309	0280
2N310	0211
2N311	0280
2N312	0600
2N313	0600
2N314	0600
2N315	0280
2N315A	0280
2N315B	0280
2N316	0280
2N316A	0211
2N317	0280
2N317A	0280

DEVICE TYPE	REPL CODE
2N318	0136
2N319	0211
2N320	0004
2N321	0211
2N322	0211
2N323	0211
2N324	0211
2N325	0085
2N327	0280
2N327A	0280
2N327B	0126
2N328	0126
2N328A	0126
2N328B	0126
2N329	0126
2N329A	0126
2N329B	0126
2N330	0126
2N330A	0126
2N331	0050
2N331M	0211
2N332	0198
2N332A	0198
2N333	0198
2N333A	0198
2N334	0198
2N334A	0198
2N334B	0198
2N335	0198
2N335A	0198
2N335B	0198
2N336	0198
2N336A	0198
2N337	0198
2N337A	0198
2N338	0198
2N338A	0198
2N339	0086
2N339A	0086
2N3390-U29	0079
2N340	0086
2N340A	0086
2N341	0233
2N341A	0233
2N342	0086
2N342A	0086
2N342B	0086
2N343	0086
2N343A	0086
2N343B	0595
2N344	0136
2N345	0136
2N346	0136
2N347	OEM
2N348	OEM
2N349	OEM
2N350	0085
2N350A	0085
2N351	0085
2N351A	0085
2N352	0160
2N353	0160
2N355	0861
2N356	0600
2N356A	0038
2N357	0600
2N357A	0038
2N358	0600
2N358A	0038
2N359	0211
2N360	0211
2N361	0211
2N362	0211
2N363	0211
2N364	0038
2N365	0038
2N366	0038
2N367	0211
2N368	0211
2N369	0211
2N370	0136
2N370/33	0136
2N370A	0136
2N371	0136
2N371/33	0136
2N372	0136
2N372/33	0136
2N373	0136
2N374	0136
2N375	0085
2N376	0085
2N376A	0085
2N377	0600
2N377A	0600
2N378	0085
2N379	0604
2N380	0085
2N381	0211
2N382	0211
2N383	0211
2N384	0050
2N384/33	0050
2N385	0600
2N385/46	0079
2N385A	0600
2N386	0085
2N387	0085
2N388	0600
2N388A	0600
2N389/I	OEM
2N389A/I	OEM
2N391	OEM
2N392	0085
2N393	0136
2N394	0280
2N394A	0280
2N395	0280
2N396	0280
2N396A	0280
2N397	0280
2N399	0085
2N400	0085
2N401	0085
2N401A	0085
2N402	0211
2N403	0280
2N404	0004
2N404A	0211
2N405	0004
2N406	0004
2N406BLU	0004
2N406BRN	0004
2N406GRN	0004
2N406GRN-YEL	0004
2N406ORN	0004
2N406RED	0004
2N407	0136
2N407BLK	0004
2N407GRN	0004
2N407J	0004
2N407RED	0004
2N407WHT	0004
2N407YEL	0004
2N408	0004
2N408J	0004
2N408WHT	0004
2N409	0136
2N410	0136
2N411	0136
2N412	0136
2N413	0280
2N413A	0280
2N414	0280
2N414A	0280
2N414B	0211
2N414C	0211
2N415	0280
2N415A	0280
2N416	0280
2N417	0280
2N418	0604
2N419	0085
2N420	0160
2N420A	0160
2N421	OEM
2N422	0211
2N422/A	0325
2N422A	0211
2N424/I	OEM
2N424A/I	OEM
2N425	0211
2N425A	0150
2N426	0280
2N427	0280
2N427A	0004
2N428	0280
2N428A	0280
2N431	0198
2N432	0198
2N433	0198
2N435	0164
2N438	0038
2N438A	0038
2N439	0038
2N439A	0280
2N440	0600
2N440A	0038
2N441	0040
2N441BLU	0437
2N442	0040
2N443	0040
2N444	0600
2N444A	0038
2N445	0600
2N445A	0038
2N446	0600
2N446A	0038
2N447	0600
2N447A	0038
2N447B	0038
2N448	0600
2N449	0600
2N450	0004
2N451	OEM
2N452	OEM
2N453	OEM
2N454	OEM
2N456	0085
2N456A	0085
2N456B	0085
2N457	0085
2N457A	0085
2N457B	0085
2N458	0085
2N458A	0085
2N458B	0160
2N459	0604
2N459A	0604
2N460	0211
2N461	0211
2N462	0211
2N464	0211
2N465	0211
2N466	0211
2N467	0211
2N468	0211
2N469	0164
2N470	0198
2N471	0198
2N471A	0198
2N472	0016
2N472A	0016
2N473	0198
2N474	0198
2N474A	0198
2N475	0016
2N475A	0016
2N476	0198
2N476A	0127
2N477	0198
2N477A	0861
2N478	0198
2N478A	0861
2N479	0198
2N479A	0198
2N480	0016
2N480A	0016
2N480B	0086
2N481	0280
2N482	0280
2N483	0280
2N483-6M	0211
2N483-B	0280
2N483B	0211
2N484	0280
2N485	0280
2N486	0280
2N486B	0280
2N487	0280
2N489	OEM
2N489A	OEM
2N489B	OEM
2N490	OEM
2N490A	OEM
2N490B	OEM
2N490C	OEM
2N491	1669
2N491A	1669
2N491B	1669
2N492	OEM
2N492A	OEM
2N492B	OEM
2N492C	OEM
2N493	OEM
2N493A	OEM
2N493B	OEM
2N494	OEM
2N494A	OEM
2N494B	OEM
2N494C	OEM
2N495	0150
2N495/18	0150
2N496	0150
2N496/18	OEM
2N497	0086
2N497A	0086
2N498	0086
2N498A	0168
2N499	0136
2N499A	0136
2N500	0136
2N500BLU	0136
2N500RED	0136
2N500WHT	0136
2N501	0136
2N501/18	0136
2N501A	0136
2N502	0136
2N502A	0136
2N502B	0136
2N502BRN	0050
2N502GRN	0050
2N502RED	0050
2N503	0136
2N504	0136
2N505	0280
2N506	0136
2N507	0208
2N508	0211
2N508A	0211
2N509	0050
2N511	0085
2N511A	0085
2N511B	0085
2N512	0085
2N512A	0085
2N512B	0085
2N513	0085
2N513A	0085
2N513B	0085
2N514	0604
2N514A	0604
2N514B	0160
2N515	0038
2N516	0038
2N517	0038
2N518	0280
2N519	0280
2N519A	0280
2N520	0280
2N520A	0280
2N521	0280
2N521A	0280
2N522	0280
2N522A	0280
2N523	0280
2N523A	0280
2N524	0211
2N524A	0211
2N525	0211
2N525A	0211
2N526	0211
2N526A	0211
2N527	0211
2N527A	0211
2N529	0280
2N529/N	0600
2N529/P	0050
2N530	0280
2N530/N	0600
2N530/P	0050
2N531	0280
2N531/N	0600
2N531/P	0847
2N532	0280
2N532/N	0600
2N532/P	0050
2N533	0280
2N533/N	0600
2N533/P	0050
2N534	0004
2N535	0004
2N535A	0004
2N535B	0004
2N536	0004
2N537	0050
2N538	0160
2N538A	0160
2N539	0085
2N539A	0085
2N540	0085
2N540A	0085
2N541	0198
2N541/A	OEM
2N541A	0007
2N542	0086
2N542A	0198
2N543	0016
2N543A	0016
2N544	0136
2N544/33	0136
2N545	0086
2N546	0016
2N547	0086
2N548	0016
2N549	0086
2N550	0016
2N551	0016
2N552	0198
2N553	0160
2N554	0160
2N555	0160
2N556	0600
2N557	0600
2N558	0600
2N559	0050
2N560	0086
2N561	0604
2N563	0004
2N564	0211
2N565	0004
2N566	0211
2N567	0004
2N568	0211
2N569	0004
2N570	0211
2N571	0004
2N572	0280
2N573	0211
2N573BLU	0211
2N573BRN	0211
2N573ORN	0211
2N573RED	0211
2N574	0440
2N574A	0440
2N575	0040
2N575A	0040
2N576	0600
2N576A	0600
2N577	OEM
2N578	0280
2N579	0280
2N580	0280
2N581	0280
2N582	0280
2N583	0280
2N584	0004
2N585	0600
2N586	0280
2N587	0600
2N588	0136
2N588A	0136
2N589	0160
2N591	0164
2N591/5	0004
2N591A	0004
2N592	0280
2N593	0280
2N594	0600
2N595	0600
2N596	0600
2N597	0280
2N598	0280
2N599	0280
2N600	0280
2N601	0050
2N602	0136
2N602A	0136
2N603	0136
2N603A	0136
2N604	0136
2N604A	0136
2N605	0136
2N606	0136
2N607	0136
2N608	0136
2N609	0211
2N610	0211
2N611	0211
2N612	0211
2N613	0211
2N614	0280
2N615	0280
2N616	0280
2N617	0280
2N618	0085
2N619	0016
2N620	0016
2N621	0016
2N622	0016
2N623	0136
2N624	0136
2N625	0038
2N626	0150
2N627	0604
2N628	0604
2N628A	OEM
2N629	0604
2N630	0604
2N631	0211
2N632	0211
2N633	0211
2N633B	0211
2N634	0600
2N634A	0600
2N635	0600
2N635A	0600
2N636	0600
2N636A	0600
2N637	0085
2N637A	0085
2N637B	0085
2N638	0085
2N638A	0085
2N638B	0085
2N639	0085
2N639A	0085
2N639B	0085
2N640	0136
2N641	0136
2N641REDM/F	0136
2N642	0136
2N643	0136
2N644	0136
2N645	0136
2N646	0208
2N647	0208
2N647/22	0208
2N648	0038
2N649	0208
2N649/5	0208
2N649/22	0038
2N650	0211
2N650A	0211
2N651	0211
2N651A	0211
2N651BLA	0211
2N652	0211
2N652A	0211
2N653	0211
2N654	0211
2N654(TAD)	0164
2N655	0211
2N655(TAD)	0211
2N655A	0086
2N655GRN	0211
2N655RED	0211
2N656	0086
2N656A	0086
2N656S	0086
2N657	0086
2N657A	0168
2N657S	0086
2N658	0126
2N659	0126
2N660	0847
2N661	0847
2N662	0085
2N663	0160
2N665	0160
2N669	0085
2N669TX	OEM
2N670	0164
2N671	0164
2N672	0847
2N673	0164
2N674	0280
2N675	0164
2N676	OEM
2N677	0604
2N677A	0604
2N677B	0604
2N677C	0604
2N678	0604
2N678A	0604
2N678B	0604
2N678C	0604
2N679	0600
2N680	0164
2N681	3307
2N681A	2188
2N682	2188
2N682A	2188
2N683	0567
2N683A	0567
2N683AS	OEM
2N683S	OEM
2N684	3258
2N684A	3258
2N685	0762
2N685A	0762
2N685AS	OEM
2N685S	OEM
2N686	3268
2N686A	3268
2N687	3271
2N687-300R	OEM
2N687A	3271
2N687AS	OEM
2N687S	OEM
2N688	0740
2N688A	0740
2N688AS	OEM
2N688S	OEM
2N689	3291
2N689A	3291
2N689AS	OEM
2N689S	OEM
2N690	0764
2N690A	0764
2N690AS	OEM
2N690S	OEM
2N691	0766
2N691A	0766
2N691AS	OEM
2N691S	OEM
2N692	0766
2N692A	0766
2N692AS	OEM
2N692S	OEM
2N694	0136
2N695	0136
2N696	0198
2N696A	0086
2N696S	0086
2N697	0086
2N697A	0086
2N698	0187
2N699	0086
2N699A	0086
2N699A,B	0086
2N699B	0086
2N700	0050
2N700/18	0050
2N700A	0050
2N700A/18	0050
2N701	0198
2N702	0016
2N703	0016
2N705	0136
2N705A	0136
2N706	0016
2N706/46	0016
2N706/51	0016
2N706/KVT	OEM
2N706/TNT	0016
2N706/TPT	OEM
2N706A	0016
2N706A/46	0016
2N706A/51	0016
2N706A/TNT	0016
2N706A/TPT	OEM
2N706B	0016
2N706B/46	0016
2N706B/51	0016
2N706C	0016
2N706C/46	OEM
2N706C/51	OEM
2N706M-JAN	0016
2N707	0493
2N707A	0016
2N708	0016
2N708/46	0016
2N708/51	0016
2N708/KVT	OEM
2N708/TNT	0016
2N708/TPT	OEM
2N708A	0016
2N709	0144
2N709/46	0418
2N709/51	0418
2N709/52	0144
2N709/KVT	OEM
2N709/TNT	0418
2N709/TPT	OEM
2N709A	0144
2N709A/46	OEM
2N709A/51	0418
2N709A46	0418
2N710	0136
2N710A	0136
2N711	0136
2N711/46	0144
2N711A	0136
2N711B	0136
2N715	0016
2N716	0079
2N717	0016
2N717A	0144
2N718	0016
2N718A	0016
2N719	0861
2N719A	0861
2N720	0861
2N720A	0861
2N721	0037
2N721A	0037
2N721AB	0006
2N722	0037
2N722A	0037
2N722AB	0037
2N723	0150
2N725	0136
2N726	0150
2N727	0150
2N728	0198
2N729	0016
2N730	0086
2N731	0086
2N734	0086
2N734A	0086
2N735	0086
2N735A	0086
2N736	0086
2N736A	0086
2N736B	0086
2N738	0233
2N738A	0715
2N738B	OEM
2N739	0086
2N739A	0086
2N740	0086
2N740A	0086
2N741	0136
2N741A	0050
2N742	0016
2N742A	0016
2N743	0016
2N743/46	0595
2N743/51	0595
2N743A	0493
2N744	0016
2N744/46	0595
2N744/51	0595
2N744A	0493
2N745	0016
2N746	0016
2N747	0016
2N748	0016
2N749	0016
2N750	0861
2N751	0861
2N752	0086
2N753	0016
2N753/46	0079
2N753/51	0595
2N754	0016
2N755	0086
2N756	0016
2N756A	0016
2N757	0016
2N757A	0016
2N757B	0155
2N758	0016
2N758A	0016
2N758B	0016
2N759	0016
2N759A	0016
2N759B	0016
2N760	0016
2N760A	0016
2N760B	0016
2N761	0016
2N762	0016
2N764	OEM
2N765	OEM
2N766	OEM
2N767	OEM
2N768	0136
2N769	0136
2N770	0198
2N771	0198
2N772	0198
2N773	0198
2N774	0198
2N775	0198
2N776	0198
2N777	0198
2N778	0198
2N779	0136
2N779A	0136
2N779B	0050
2N780	0016

If replacement code is OEM, contact original manufacturer for replacement.

DEVICE TYPE	REPL CODE
2N781	0136
2N782	0136
2N783	0016
2N784	0016
2N784/51	0198
2N784A	0016
2N784A/46	0016
2N784A/51	0016
2N784A46	0079
2N789	0198
2N790	0198
2N791	0198
2N792	0198
2N793	0198
2N794	0136
2N795	0136
2N796	0136
2N797	0038
2N799	0211
2N800	0004
2N801	0280
2N802	0280
2N803	0164
2N804	0164
2N805	0164
2N806	0164
2N807	0004
2N808	0004
2N809	0280
2N810	0280
2N811	0280
2N812	0280
2N813	0004
2N814	0004
2N815	0004
2N816	0004
2N817	0164
2N818	0164
2N819	0004
2N820	0004
2N821	0600
2N822	0600
2N823	0600
2N824	0600
2N825	0211
2N826	0211
2N827	0136
2N828	0136
2N828A	0136
2N829	0136
2N834	0016
2N834/46	0016
2N834/51	0016
2N834A	0016
2N835	0016
2N835/46	0016
2N835/51	0016
2N837	0136
2N838	0136
2N839	0016
2N840	0016
2N841	0016
2N841/46	0016
2N841/51	OEM
2N841/KVT	0016
2N841/TNT	OEM
2N841/TPT	OEM
2N842	0016
2N843	0016
2N844	0016
2N845	0086
2N846	0050
2N846A	0136
2N846B	0050
2N847	OEM
2N848	OEM
2N849	0144
2N850	0144
2N851	0144
2N852	0144
2N858	0037
2N858C	0037
2N859	0037
2N859D	0037
2N860	0037
2N860A	0037
2N861	0037
2N861A	0037
2N862	0037
2N862A	0037
2N863	0037
2N863A	0037
2N864	0037
2N864A	0150
2N865	0150
2N865A	0150
2N866	0016
2N867	0016
2N869	0150
2N869A	0150
2N870	0086
2N871	0086
2N876	1137
2N877	1137
2N878	0344
2N879	0901
2N880	2342
2N881	0058
2N882	OEM

DEVICE TYPE	REPL CODE
2N883	OEM
2N884	OEM
2N884A	OEM
2N885	1137
2N885A	1137
2N886	0344
2N886A	OEM
2N887	OEM
2N887A	OEM
2N888	OEM
2N888A	2342
2N889	OEM
2N889A	OEM
2N890	OEM
2N891	OEM
2N892	OEM
2N893	OEM
2N894	OEM
2N895	OEM
2N896	OEM
2N897	OEM
2N898	OEM
2N899	OEM
2N900	OEM
2N901	OEM
2N902	0007
2N903	0007
2N904	0007
2N904AS	0861
2N905	0007
2N906	0079
2N907	0595
2N908	0086
2N909	0016
2N910	0086
2N911	0086
2N912	0086
2N913	0016
2N914	0016
2N914/46	0016
2N914/51	0144
2N914A	0016
2N915	0016
2N915A	0016
2N916	0016
2N916A	0016
2N916B	0016
2N917	0007
2N917/46	0224
2N917/51	0127
2N917A	0007
2N918	0144
2N918/46	0144
2N918/51	0144
2N919	0016
2N920	0016
2N921	0016
2N922	0016
2N923	0037
2N923A	0037
2N924	0037
2N924A	0037
2N925	0037
2N925A	0037
2N926	0037
2N926A	0037
2N927	0037
2N927A	0037
2N928	0037
2N928A	0037
2N929	0016
2N929/46	0016
2N929/51	OEM
2N929A	0016
2N930	0016
2N930/5P5	OEM
2N930/46	0016
2N930/51	OEM
2N930/KVT	0016
2N930/TNT	0016
2N930A	0016
2N930A/46	0016
2N930A/51	OEM
2N930B	0016
2N933	0136
2N934	0136
2N935	0037
2N935A	0037
2N936	0037
2N936A	0037
2N937	0037
2N937A	0037
2N938	0037
2N938A	0037
2N939	0037
2N939A	0037
2N940	0037
2N940A	0037
2N941	0037
2N941A	0037
2N942	0037
2N942A	0342
2N943	0037
2N943A	0037
2N944	0037
2N944A	0037
2N945	0037
2N945A	0037
2N946	0037

DEVICE TYPE	REPL CODE
2N946A	0037
2N947	0016
2N948	OEM
2N949	OEM
2N950	OEM
2N951	0079
2N955	0600
2N955A	0600
2N956	0016
2N957	0016
2N958	0079
2N959	0079
2N960	0136
2N960/46	0050
2N961	0136
2N961/46	0136
2N962	0136
2N962/46	0136
2N963	0136
2N964	0136
2N964/46	0050
2N964A	0136
2N965	0136
2N966	0136
2N967	0136
2N968	0050
2N969	0136
2N970	0136
2N971	0136
2N972	0136
2N973	0136
2N974	0136
2N975	0136
2N976	0050
2N977	0136
2N978	0150
2N979	0136
2N980	0136
2N981	0086
2N982	0136
2N983	0136
2N984	0136
2N985	0136
2N986	0136
2N987	0050
2N988	0016
2N989	0016
2N990	0136
2N991	0136
2N992	0136
2N993	0136
2N994	0136
2N995	0150
2N995A	0150
2N996	0150
2N997	0400
2N998	0400
2N999	0400
2N1000	0600
2N1003	0136
2N1004	0136
2N1005	0144
2N1006	0198
2N1007	0085
2N1008	0211
2N1008A	0211
2N1008B	0211
2N1009	0211
2N1010	0208
2N1011	0604
2N1012	0600
2N1013	OEM
2N1014	0604
2N1016B/M	OEM
2N1016C/M	OEM
2N1017	0050
2N1018	0050
2N1019	OEM
2N1020	0160
2N1021	0604
2N1021A	0604
2N1022	0604
2N1022A	0604
2N1023	0050
2N1024	0150
2N1025	0037
2N1025A	0037
2N1025M	0037
2N1026	0037
2N1026A	0037
2N1026AC	0037
2N1027	0037
2N1027A	0037
2N1028	0037
2N1028A	0037
2N1029	0604
2N1029A	0604
2N1029B	0604
2N1029C	0604
2N1030	0604
2N1030A	0604
2N1030B	0604
2N1030C	0604
2N1031	0604
2N1031A	0604
2N1031B	0604
2N1031C	0604
2N1032	0604
2N1032A	0604

DEVICE TYPE	REPL CODE
2N1032B	0604
2N1032C	0604
2N1033	0160
2N1034	0126
2N1035	0126
2N1036	0126
2N1037	0126
2N1038	0847
2N1039	0160
2N1040	0085
2N1041	0160
2N1042	0136
2N1043	0136
2N1044	0211
2N1045	0211
2N1046	0604
2N1046A	0975
2N1046B	0160
2N1047	0042
2N1048	0042
2N1049	0042
2N1050	0168
2N1051	0086
2N1052	0233
2N1053	0233
2N1054	0233
2N1055	0086
2N1056	0211
2N1057	0211
2N1058	0600
2N1059	0208
2N1059-1	0038
2N1060	0144
2N1065	0050
2N1066	0050
2N1067	0560
2N1068	0560
2N1069	0103
2N1070	0103
2N1072	0103
2N1073	0604
2N1073A	0604
2N1073B	0975
2N1077	0016
2N1078	0208
2N1081	0086
2N1082	0198
2N1084	0861
2N1085	0144
2N1086	0600
2N1086A	0600
2N1087	0600
2N1090	0600
2N1091	0600
2N1092	0086
2N1093	0211
2N1094	0050
2N1095	0038
2N1096	0038
2N1097	0211
2N1098	0211
2N1099	0440
2N1100	0440
2N1101	0208
2N1102	0208
2N1102/5	0600
2N1103	0016
2N1104	0198
2N1105	0861
2N1106	0630
2N1107	0136
2N1108	0136
2N1108RED	0136
2N1109	0136
2N1110	0136
2N1111	0136
2N1111A	0136
2N1111B	0136
2N1111RED	0136
2N1112	0600
2N1114	0600
2N1115	0004
2N1115A	0050
2N1116	0086
2N1117	0086
2N1118	0037
2N1118A	0037
2N1118AB	0037
2N1119	0037
2N1119A	0037
2N1120	0604
2N1121	0600
2N1122	0050
2N1122A	0050
2N1123	0164
2N1124	0211
2N1125	0211
2N1126	0211
2N1127	0164
2N1128	0847
2N1129	0164
2N1130	0004
2N1131	0037
2N1131/46	0037
2N1131/51	0037
2N1131A	0037
2N1131A/51	0037
2N1131AD	0037
2N1131AS	0037

DEVICE TYPE	REPL CODE
2N1131S	0037
2N1132	0037
2N1132/46	0037
2N1132/51	0037
2N1132/KVT	OEM
2N1132/TNT	0037
2N1132/TPT	OEM
2N1132A	0037
2N1132A/46	0037
2N1132A/51	0037
2N1132A46	0037
2N1132AC	0037
2N1132B	0037
2N1132B/46	0037
2N1132B/51	0037
2N1132B46	0037
2N1135	0037
2N1135A	0037
2N1136	0085
2N1136A	0085
2N1136B	0085
2N1136C	0085
2N1137	0604
2N1137A	0604
2N1137B	0604
2N1138	0604
2N1138A	0604
2N1138B	0604
2N1139	0198
2N1140	0016
2N1141	0050
2N1141A	0050
2N1142	0050
2N1142A	0050
2N1143	0050
2N1143A	0050
2N1144	0211
2N1145	0211
2N1146	0604
2N1146A	0604
2N1146B	0604
2N1146C	0604
2N1147	0604
2N1147A	0604
2N1147B	0604
2N1147C	0604
2N1149	0079
2N1150	0079
2N1150/904	0079
2N1151	0079
2N1151/904A	0079
2N1152	0079
2N1153	0079
2N1153/910	0079
2N1154	0018
2N1154/951	0018
2N1155	0018
2N1155/952	0018
2N1156	0086
2N1156/953	0086
2N1157	0040
2N1157A	0040
2N1158	0136
2N1158A	0136
2N1159	0604
2N1160	0604
2N1162	0604
2N1162A	0604
2N1163	0604
2N1163A	0604
2N1164	0604
2N1164A	0604
2N1165	0085
2N1165A	0604
2N1166	0604
2N1166A	0604
2N1167	0975
2N1167A	0975
2N1168	0160
2N1169	0038
2N1170	0038
2N1171	0280
2N1172	0085
2N1173	0208
2N1173W	0208
2N1174	0211
2N1174W	0847
2N1175	0211
2N1175A	0211
2N1176	0280
2N1176A	0280
2N1176B	0280
2N1177	0050
2N1178	0050
2N1179	0050
2N1180	0136
2N1182	0604
2N1183	0847
2N1183A	0847
2N1183B	0847
2N1184	0847
2N1184A	0847
2N1184B	0847
2N1185	0280
2N1186	0280
2N1187	0280
2N1188	0280
2N1189	0211
2N1190	0211

DEVICE TYPE	REPL CODE
2N1191	0211
2N1192	0211
2N1193	0211
2N1194	0211
2N1195	0050
2N1196	0037
2N1197	0126
2N1198	0600
2N1199	0198
2N1199A	0198
2N1200	0198
2N1201	0198
2N1202	0440
2N1203	0440
2N1204	0136
2N1204A	0136
2N1205	0198
2N1206	0595
2N1207	0233
2N1208	2662
2N1208/I	OEM
2N1209	2662
2N1209/I	OEM
2N1210/I	OEM
2N1211/I	OEM
2N1212	2662
2N1212/I	OEM
2N1213	0136
2N1214	0136
2N1215	0136
2N1216	0136
2N1217	0600
2N1218	0604
2N1219	0037
2N1220	0037
2N1221	0037
2N1222	0037
2N1223	0037
2N1224	0136
2N1225	0136
2N1226	0136
2N1227	0085
2N1227-3	0085
2N1227-4	0085
2N1227-4A	0085
2N1227-4R	0085
2N1227A	0085
2N1228	0126
2N1229	0126
2N1230	0126
2N1231	0126
2N1232	0126
2N1232A	OEM
2N1233	0126
2N1234	0126
2N1238	0150
2N1239	0150
2N1240	0150
2N1241	0150
2N1242	0791
2N1242A	OEM
2N1243	0791
2N1245	0085
2N1246	0085
2N1247	0198
2N1248	0198
2N1249	0198
2N1250/I	OEM
2N1251	0208
2N1252	0198
2N1252A	0198
2N1253	0198
2N1253A	0198
2N1254	0126
2N1255	0126
2N1256	0037
2N1257	0037
2N1258	0126
2N1259	0126
2N1261	0440
2N1261A	OEM
2N1262	0440
2N1262A	OEM
2N1263	0440
2N1263A	OEM
2N1264	0280
2N1264/13	OEM
2N1265	0211
2N1265/5	0280
2N1265A	0211
2N1266	0136
2N1267	0198
2N1268	0198
2N1269	0198
2N1270	0198
2N1271	0198
2N1272	0198
2N1273	0004
2N1273BLU	0211
2N1273GRN	0211
2N1273ORN	0211
2N1273RED	0211
2N1273YEL	0211
2N1274	0004
2N1274BLU	0211
2N1274BRN	0211
2N1274GRN	0211
2N1274ORN	0211
2N1274PUR	0211

DEVICE TYPE	REPL CODE
2N1274RED	0211
2N1274V10	0847
2N1274VIO	0847
2N1275	0126
2N1276	0016
2N1277	0016
2N1278	0198
2N1279	0016
2N1280	0280
2N1281	0280
2N1282	0280
2N1284	0280
2N1285	0050
2N1287	0211
2N1287A	0211
2N1288	0600
2N1289	0600
2N1290	0280
2N1291	0085
2N1292	2713
2N1293	0160
2N1294	0604
2N1295	0604
2N1296	2717
2N1297	0604
2N1298	0208
2N1299	0600
2N1300	0136
2N1301	0050
2N1302	0208
2N1303	0211
2N1304	0600
2N1305	0211
2N1306	0600
2N1307	0280
2N1308	0600
2N1309	0280
2N1309A	0280
2N1310	0600
2N1311	0600
2N1312	0038
2N1313	0136
2N1314	0085
2N1314R	0085
2N1315	0847
2N1316	0280
2N1317	0280
2N1318	0280
2N1319	0280
2N1330	0155
2N1335	0233
2N1336	0233
2N1337	0233
2N1338	0086
2N1339	0233
2N1340	0233
2N1341	0233
2N1342	0233
2N1342A#	0233
2N1343	0280
2N1344	0280
2N1345	0280
2N1346	0280
2N1347	0280
2N1348	0211
2N1349	0280
2N1350	0280
2N1351	0280
2N1352	0211
2N1353	0211
2N1354	0280
2N1355	0280
2N1356	0280
2N1357	0280
2N1358	0040
2N1358A	0040
2N1358M	0040
2N1359	0160
2N1360	0604
2N1361	0280
2N1361A	0280
2N1362	0604
2N1363	0604
2N1364	0160
2N1365	0160
2N1366	0600
2N1367	0600
2N1370	0211
2N1371	0211
2N1372	0211
2N1373	0211
2N1374	0211
2N1375	0211
2N1376	0211
2N1377	0211
2N1378	0211
2N1379	0211
2N1380	0211
2N1381	0211
2N1382	0211
2N1383	0211
2N1384	0136
2N1385	0050
2N1386	0198
2N1387	0198
2N1388	0016
2N1389	0016
2N1390	0198
2N1391	0038

If replacement code is OEM, contact original manufacturer for replacement.

DEVICE TYPE	REPL CODE
2N1392	0211
2N1393	0280
2N1394	0050
2N1395	0280
2N1396	0136
2N1397	0136
2N1398	0136
2N1399	0136
2N1400	0136
2N1401	0136
2N1401A	0136
2N1402	0136
2N1403	0050
2N1404	0211
2N1404A	0280
2N1405	0050
2N1406	0050
2N1407	0050
2N1408	0050
2N1409	0016
2N1409A	0198
2N1410	0086
2N1410A	0086
2N1411	0050
2N1412	0437
2N1412A	0437
2N1413	0211
2N1414	0211
2N1415	0211
2N1416	0004
2N1417	0198
2N1418	0198
2N1419	0604
2N1420	0086
2N1420A	0086
2N1422	0103
2N1423	0103
2N1425	0136
2N1426	0136
2N1427	0136
2N1428	0037
2N1429	0126
2N1430	0160
2N1431	0038
2N1432	0211
2N1433	0440
2N1434	0440
2N1435	0440
2N1436	0050
2N1437	0160
2N1438	0160
2N1439	0037
2N1440	0037
2N1441	0037
2N1442	0037
2N1443	0037
2N1444	0086
2N1445	0187
2N1446	0211
2N1447	0211
2N1448	0211
2N1449	0211
2N1450	0136
2N1451	0211
2N1452	0211
2N1468	OEM
2N1469	0280
2N1470	0280
2N1471	0280
2N1472	0086
2N1473	0600
2N1474	0037
2N1474A	0037
2N1475	0037
2N1476	0037
2N1477	0037
2N1478	0211
2N1479	1480
2N1479J	OEM
2N1480	1480
2N1480J	OEM
2N1481	1480
2N1481J	OEM
2N1482	1480
2N1482J	OEM
2N1483	0042
2N1483J	OEM
2N1483TX	OEM
2N1484	0042
2N1484TX	OEM
2N1485	0042
2N1485J	OEM
2N1485TX	OEM
2N1486	0042
2N1486J	OEM
2N1486TX	OEM
2N1487	0177
2N1487J	OEM
2N1488	0177
2N1488J	OEM
2N1489	0177
2N1489J	OEM
2N1490	0363
2N1490J	OEM
2N1491	0086
2N1492	0086
2N1493	0233
2N1494	0050
2N1495	0050

DEVICE TYPE	REPL CODE
2N1499	0136
2N1499A	0136
2N1499B	0136
2N1500	0136
2N1500/18	0050
2N1501	0085
2N1502	0085
2N1504/10	OEM
2N1505	0016
2N1506	0016
2N1506A	0016
2N1507	0086
2N1508	0086
2N1509	0086
2N1510	0600
2N1515	0136
2N1516	0136
2N1517	0136
2N1517A	0050
2N1518	0040
2N1519	0040
2N1520	0040
2N1521	0040
2N1522	0437
2N1523	0440
2N1524	0050
2N1524/33	0050
2N1524-1	0050
2N1524-2	0050
2N1525	0050
2N1526	0050
2N1526/33	0050
2N1527	0136
2N1528	0198
2N1529	0604
2N1529A	0604
2N1530	0604
2N1530A	0160
2N1531	0604
2N1531A	0604
2N1532	0604
2N1532A	0604
2N1533	0160
2N1534	0604
2N1534A	0604
2N1535	0604
2N1535A	0160
2N1536	0604
2N1536A	0160
2N1537	0604
2N1537A	0604
2N1538	0160
2N1538A	0160
2N1539	0604
2N1539A	0604
2N1540	0604
2N1540A	0604
2N1541	0604
2N1541A	0604
2N1542	0604
2N1542A	0604
2N1543	0604
2N1543A	0604
2N1544	0604
2N1544A	0604
2N1545	0604
2N1545A	0604
2N1546	0604
2N1546A	0604
2N1547	0604
2N1547A	0604
2N1548	0604
2N1548A	0604
2N1549	0604
2N1549A	0604
2N1550	0604
2N1550A	0604
2N1551	0604
2N1551A	0604
2N1552	0604
2N1552A	0604
2N1553	0604
2N1553A	0604
2N1554	0604
2N1554A	0604
2N1555	0604
2N1555A	0604
2N1556	0604
2N1556A	0604
2N1557	0604
2N1557A	0604
2N1558	0604
2N1558A	0604
2N1559	0604
2N1559A	0604
2N1560	0604
2N1560A	0604
2N1561	0211
2N1562	0211
2N1564	0086
2N1565	0086
2N1566	0086
2N1566A	0086
2N1567	0440
2N1568	0440
2N1569	0440
2N1570	0280
2N1572	0233
2N1573	0233

DEVICE TYPE	REPL CODE
2N1574	0233
2N1581	0280
2N1583	0280
2N1584	0280
2N1585	0600
2N1586	0127
2N1587	0079
2N1588	0079
2N1589	0127
2N1590	0079
2N1591	0079
2N1592	0127
2N1593	0079
2N1594	0079
2N1595	0179
2N1595A	0179
2N1596	0940
2N1596A	0179
2N1597	0940
2N1597A	0179
2N1598	0346
2N1598A	0346
2N1599	0346
2N1599A	0346
2N1600	1110
2N1601	2932
2N1602	0963
2N1603	2933
2N1604	2936
2N1605	0600
2N1605-A	0038
2N1605A	0600
2N1606	0126
2N1607	0037
2N1608	0037
2N1609	0085
2N1610	0085
2N1611	0160
2N1612	0160
2N1613	0076
2N1613/46	0016
2N1613/51	OEM
2N1613/KVT	OEM
2N1613/TNT	OEM
2N1613/TPT	OEM
2N1613A	0016
2N1613B	0016
2N1613S	0086
2N1614	0164
2N1615	0086
2N1616	2662
2N1616/I	OEM
2N1616A	2662
2N1616A/I	OEM
2N1617	2662
2N1617/I	OEM
2N1617A	2662
2N1617A/I	OEM
2N1618	2662
2N1618/I	OEM
2N1618A	2662
2N1618A/I	OEM
2N1619	OEM
2N1620/I	OEM
2N1622	0600
2N1623	0126
2N1624	0600
2N1625	0050
2N1631	0136
2N1632	0136
2N1633	0136
2N1634	0136
2N1635	0136
2N1636	0136
2N1637	0050
2N1637/33	0050
2N1638	0136
2N1638/33	0136
2N1639	0136
2N1639/33	0050
2N1640	0150
2N1641	0150
2N1642	0150
2N1643	0037
2N1644	0086
2N1644A	0086
2N1645	OEM
2N1646	0050
2N1647	2964
2N1648	2964
2N1649	2964
2N1650	2964
2N1651	0604
2N1652	0604
2N1653	0160
2N1654	0126
2N1655	0342
2N1656	0342
2N1657	OEM
2N1662	0136
2N1663	0198
2N1664	0280
2N1665	0050
2N1666	0604
2N1667	0160
2N1668	0085
2N1669	0085
2N1670	0050
2N1671	1669

DEVICE TYPE	REPL CODE
2N1671A	1669
2N1671B	1669
2N1671BP	OEM
2N1671C	1669
2N1671CX	OEM
2N1672	0208
2N1672A	0208
2N1673	0050
2N1674	0016
2N1676	0037
2N1677	0037
2N1678	0050
2N1679	0630
2N1680	0595
2N1681	0211
2N1682	0198
2N1683	0136
2N1684	0280
2N1685	0600
2N1686	OEM
2N1687	OEM
2N1688	OEM
2N1689	OEM
2N1691	0168
2N1694	0600
2N1699	0136
2N1700	0086
2N1701	0042
2N1702	0363
2N1703	0103
2N1704	0016
2N1705	0211
2N1706	0211
2N1707	0211
2N1708	0016
2N1708A	0016
2N1709	0042
2N1710	0560
2N1711	0086
2N1711/46	0016
2N1711/51	OEM
2N1711/KVT	OEM
2N1711/TNT	0861
2N1711/TPT	OEM
2N1711A	0086
2N1711B	0086
2N1711S	0086
2N1713	0136
2N1714	0086
2N1715	0187
2N1716	0086
2N1717	0187
2N1718	0042
2N1719	OEM
2N1720	0042
2N1721	OEM
2N1722	0561
2N1722/I	OEM
2N1722A/I	OEM
2N1723	0060
2N1724	2662
2N1724/I	OEM
2N1724A	2662
2N1724A/I	OEM
2N1725	2662
2N1726	0136
2N1727	0136
2N1728	0136
2N1729	0280
2N1730	0600
2N1731	0280
2N1732	0600
2N1741	OEM
2N1742	0136
2N1743	0136
2N1744	0136
2N1745	0136
2N1746	0136
2N1747	0136
2N1748	0136
2N1748A	0136
2N1749	0050
2N1750	0050
2N1751	0604
2N1752	0136
2N1753	0004
2N1754	0136
2N1755	0085
2N1756	0085
2N1757	0085
2N1758	0085
2N1759	0085
2N1760	0085
2N1761	0085
2N1762	0085
2N1763	0198
2N1764	0086
2N1765	OEM
2N1768	0042
2N1769	0042
2N1770	1103
2N1770A	3419
2N1771	1110
2N1771A	1103
2N1772	2932
2N1772A	2491
2N1773	0240
2N1773A	0240
2N1774	0963

DEVICE TYPE	REPL CODE
2N1774A	0240
2N1775	0676
2N1775A	2659
2N1776	3053
2N1776A	2659
2N1776B	2659
2N1777	2936
2N1777A	0676
2N1778	3055
2N1778A	2809
2N1779	0600
2N1780	0600
2N1781	0600
2N1782	0050
2N1783	0600
2N1784	0050
2N1785	0136
2N1786	0136
2N1787	0136
2N1788	0050
2N1789	0050
2N1790	0050
2N1792	0641
2N1793	0641
2N1794	0641
2N1795	0641
2N1796	0217
2N1797	0217
2N1798	0217
2N1799	0217
2N1800	0217
2N1801	0658
2N1802	0658
2N1803	0658
2N1804	0658
2N1805	0610
2N1806	0610
2N1807	0468
2N1808	0600
2N1837	0086
2N1837A	0086
2N1837B	0086
2N1838	0086
2N1839	0086
2N1839A	0086
2N1840	0016
2N1841	OEM
2N1842	3102
2N1842A	3102
2N1842B	0722
2N1843	2519
2N1843A	2519
2N1843B	2519
2N1844	0741
2N1844A	0741
2N1844B	0741
2N1845	3105
2N1845A	3105
2N1845B	0745
2N1846	0745
2N1846A	0745
2N1846B	0745
2N1847	3109
2N1847A	3109
2N1847B	3109
2N1848	2917
2N1848A	2917
2N1848B	2917
2N1849	0747
2N1849A	0747
2N1849B	0747
2N1850	0752
2N1850A	3243
2N1850B	3243
2N1853	0136
2N1853/18	0050
2N1854	0136
2N1858	0038
2N1864	0136
2N1865	0136
2N1866	0050
2N1867	0050
2N1868	0136
2N1869	OEM
2N1869A	OEM
2N1870	OEM
2N1870A	OEM
2N1871	OEM
2N1871A	OEM
2N1872	0050
2N1872A	OEM
2N1873	0050
2N1873A	OEM
2N1874	0050
2N1874A	OEM
2N1875	0050
2N1875A	OEM
2N1876	OEM
2N1876A	OEM
2N1877	OEM
2N1877A	OEM
2N1878	OEM
2N1878A	OEM
2N1879	OEM
2N1879A	OEM
2N1880	OEM
2N1880A	OEM
2N1881	OEM
2N1882	OEM

DEVICE TYPE	REPL CODE
2N1883	OEM
2N1884	OEM
2N1885	OEM
2N1886	0042
2N1889	0086
2N1890	0086
2N1890S	0086
2N1891	0600
2N1892	0211
2N1893	0233
2N1893/46	0233
2N1893/51	OEM
2N1893/KVT	OEM
2N1893/T59235A	OEM
2N1893/TNT	OEM
2N1893/TPT	OEM
2N1893A	0233
2N1893S	0086
2N1905	0975
2N1906	0604
2N1907	0604
2N1907A	0604
2N1908	0604
2N1908A	0604
2N1909	0608
2N1910	0608
2N1911	0608
2N1912	0608
2N1913	0608
2N1914	0610
2N1914A	OEM
2N1914B	OEM
2N1915	0610
2N1916	0610
2N1916A	OEM
2N1917	0342
2N1918	0342
2N1919	0037
2N1920	0037
2N1921	0037
2N1922	0126
2N1923	0086
2N1924	0004
2N1925	0004
2N1926	0004
2N1929	1137
2N1930	OEM
2N1931	0240
2N1932	0240
2N1933	0240
2N1934	OEM
2N1935	OEM
2N1940	0280
2N1941	0086
2N1942	0280
2N1943	0086
2N1944	0016
2N1945	0016
2N1946	0016
2N1947	0016
2N1948	0016
2N1949	0016
2N1950	0016
2N1951	0016
2N1952	0016
2N1953	0086
2N1954	0004
2N1955	0004
2N1956	0004
2N1957	0004
2N1958	0086
2N1958/18	0086
2N1958A	0086
2N1958A/51	0086
2N1959	0086
2N1959/18	0086
2N1959A	0086
2N1959A/51	0086
2N1960	0050
2N1960/46	0050
2N1961	0050
2N1961/46	0211
2N1962	0016
2N1962/46	0016
2N1963	0016
2N1963/46	0016
2N1964	0016
2N1964/46	0016
2N1965	0016
2N1965/46	0016
2N1969	0280
2N1970	0040
2N1971	0160
2N1972	0086
2N1973	0086
2N1974	0086
2N1975	0016
2N1978	OEM
2N1980	0040
2N1981	0040
2N1982	0040
2N1983	0086
2N1983/T59276	OEM
2N1984	0086
2N1985	0086
2N1986	0086
2N1987	0086
2N1988	0086
2N1989	0086

DEVICE TYPE	REPL CODE
2N1990	0086
2N1990/46	0715
2N1990R	0715
2N1990S	0715
2N1990W	0715
2N1991	0037
2N1991S	0150
2N1992	0016
2N1993	0038
2N1994	0600
2N1995	0600
2N1996	0600
2N1997	0280
2N1998	0280
2N1999	0050
2N2000	0004
2N2001	0211
2N2002	0037
2N2003	0037
2N2004	0037
2N2005	0037
2N2006	0037
2N2007	0037
2N2008	0233
2N2009	0050
2N2010	OEM
2N2011	OEM
2N2012	OEM
2N2013	OEM
2N2014	OEM
2N2017	0086
2N2018	3014
2N2019	3014
2N2020	3014
2N2021	3014
2N2022	0050
2N2023	0608
2N2024	0608
2N2025	0608
2N2026	0608
2N2027	0608
2N2028	0610
2N2029	0610
2N2030	0610
2N2031	OEM
2N2032	0144
2N2032/I	OEM
2N2033	0622
2N2033/S	OEM
2N2034	0622
2N2034/S	OEM
2N2035	0042
2N2036	2314
2N2038	0086
2N2039	0086
2N2040	0086
2N2041	0086
2N2044	OEM
2N2045	OEM
2N2046	OEM
2N2047	OEM
2N2048	0136
2N2048A	0136
2N2049	0086
2N2059	0136
2N2060	2045
2N2060A	2045
2N2060B	OEM
2N2061	0085
2N2061A	0085
2N2062	0085
2N2062A	0085
2N2063	0085
2N2063A	0085
2N2064	0085
2N2064A	0085
2N2065	0085
2N2065A	0085
2N2066	0085
2N2066A	0085
2N2067	0085
2N2067-O	OEM
2N2067B	0085
2N2067G	0085
2N2067W	0085
2N2068-O	OEM
2N2069	0604
2N2070	0604
2N2071	0160
2N2072	0160
2N2075	0040
2N2075A	0040
2N2076	0040
2N2076A	0040
2N2077	0040
2N2077A	0040
2N2078	0040
2N2078A	0040
2N2079	0040
2N2079A	0040
2N2080	0040
2N2080A	0040
2N2081	0040
2N2081A	0040
2N2082	0040
2N2082A	0440
2N2083	3216
2N2084	0050
2N2085	0600

DEVICE TYPE	REPL CODE
2N2086	0086
2N2087	0079
2N2089	0136
2N2090	0136
2N2091	0136
2N2092	0136
2N2093	0136
2N2094	0086
2N2094A	0086
2N2095	0164
2N2095A	0164
2N2096	0164
2N2096A	0016
2N2097	0050
2N2097A	0016
2N2098	0050
2N2099	0136
2N2100	0050
2N2100A	0164
2N2102	0086
2N2102A	0086
2N2102S	0086
2N2104	0037
2N2105	0037
2N2106	0086
2N2107	0086
2N2108	0086
2N2121	0037
2N2137	0085
2N2137A	0085
2N2138	0085
2N2138A	0085
2N2139	0085
2N2139A	0085
2N2140	0085
2N2140A	0085
2N2141	0085
2N2141A	0085
2N2142	0085
2N2142A	0085
2N2143	0085
2N2143A	0085
2N2144	0085
2N2144A	0085
2N2145	0085
2N2145A	0085
2N2146	0085
2N2146A	0085
2N2147	0160
2N2148	0160
2N2149	OEM
2N2150	2964
2N2151	2964
2N2152	0040
2N2152A	0040
2N2153	0040
2N2153A	0040
2N2154	0040
2N2154A	0040
2N2155	0040
2N2155A	0040
2N2156	0040
2N2156A	0040
2N2157	0040
2N2157A	0040
2N2158	0040
2N2158A	0040
2N2159	0040
2N2159A	0040
2N2160	1669
2N2161	0086
2N2162	0037
2N2163	0037
2N2164	0037
2N2165	0037
2N2166	0037
2N2167	0037
2N2168	0136
2N2169	0136
2N2170	0136
2N2171	0280
2N2172	0280
2N2173	0847
2N2174	OEM
2N2175	0150
2N2176	0150
2N2177	0150
2N2178	0037
2N2180	0136
2N2181	0037
2N2182	0037
2N2183	0037
2N2184	0037
2N2185	0037
2N2186	0037
2N2187	0037
2N2188	0136
2N2189	0136
2N2190	0050
2N2191	0050
2N2192	0086
2N2192A	0086
2N2192B	0086
2N2193	0086
2N2193A	0086
2N2193AS	0086
2N2193B	0086
2N2194	0086
2N2194A	0086

DEVICE TYPE	REPL CODE
2N2194B	0086
2N2195	0086
2N2195A	0086
2N2195B	0086
2N2196	0630
2N2197	0144
2N2198	0086
2N2199	0136
2N2200	0136
2N2204	0142
2N2205	0016
2N2206	0016
2N2207	0136
2N2208	0136
2N2209	0211
2N2210	0440
2N2212	0160
2N2213	OEM
2N2214	0198
2N2216	0285
2N2217	0086
2N2217/51	0079
2N2217A	OEM
2N2218	0086
2N2218/51	0079
2N2218/TNT	0079
2N2218/TPT	OEM
2N2218A	0086
2N2218AS	0016
2N2218S	0016
2N2219	0086
2N2219/51	0079
2N2219/TNT	0079
2N2219/TPT	OEM
2N2219A	0086
2N2219AS	0016
2N2219S	0016
2N2220	0016
2N2220A	0016
2N2221	0016
2N2221A	0016
2N2222	0016
2N2222A	0016
2N2223	2045
2N2223A	0715
2N2224	0016
2N2225	0136
2N2226	OEM
2N2230S	0198
2N2233A	2045
2N2234	0198
2N2235	0198
2N2236	0016
2N2237	0016
2N2238	0050
2N2239	3278
2N2240	0198
2N2241	0198
2N2242	0016
2N2243	0086
2N2243A	0086
2N2244	0016
2N2245	0016
2N2246	0016
2N2247	0016
2N2248	0016
2N2249	0016
2N2250	0016
2N2251	0016
2N2252	0016
2N2253	0016
2N2254	0016
2N2255	0016
2N2256	0016
2N2257	0016
2N2258	0136
2N2259	0136
2N2266	0440
2N2267	0440
2N2268	0440
2N2269	0440
2N2270	0086
2N2270S	0086
2N2271	0211
2N2272	0016
2N2273	0136
2N2274	0037
2N2275	0037
2N2276	0037
2N2277	0037
2N2278	0037
2N2279	0037
2N2279/51	0595
2N2280	0037
2N2281	0037
2N2282	0085
2N2285	0604
2N2286	0604
2N2287	0160
2N2288	0604
2N2289	0604
2N2290	0604
2N2291	0604
2N2292	0604
2N2293	0160
2N2294	0160
2N2295	0160
2N2296	0160
2N2297	0086

DEVICE TYPE	REPL CODE
2N2297/51	0079
2N2297S	0086
2N2303	0037
2N2303/46	0037
2N2303/51	OEM
2N2303/KVT	OEM
2N2303/TNT	0037
2N2303/TPT	OEM
2N2304	0086
2N2305	0103
2N2307	OEM
2N2309	0198
2N2310	0715
2N2311	0715
2N2312	0715
2N2313	0715
2N2314	0016
2N2315	0016
2N2316	0715
2N2317	0086
2N2318	0016
2N2319	0016
2N2320	0016
2N2322	0940
2N2322A	OEM
2N2323	0940
2N2323A	OEM
2N2324	0940
2N2324A	OEM
2N2325	0940
2N2325A	OEM
2N2326	0940
2N2326A	OEM
2N2327	1222
2N2327A	OEM
2N2328	0346
2N2328A	OEM
2N2329	3324
2N2329A	OEM
2N2330	0198
2N2330S	0018
2N2331	0016
2N2331S	0198
2N2332	0037
2N2333	0037
2N2334	0037
2N2335	0037
2N2336	0037
2N2337	0037
2N2339	0042
2N2340	1411
2N2341	1411
2N2342	0325
2N2343	0325
2N2344	OEM
2N2345	OEM
2N2346	OEM
2N2347	OEM
2N2348	OEM
2N2349	0016
2N2350	0086
2N2350A	0086
2N2351	0086
2N2351A	0086
2N2352	0086
2N2352A	0086
2N2353	0086
2N2353A	0086
2N2354	0038
2N2356	OEM
2N2356A	0715
2N2357	0160
2N2358	0160
2N2360	0050
2N2361	0050
2N2362	0050
2N2363	0136
2N2364	0187
2N2364A	0187
2N2368	0016
2N2368/51	0595
2N2369	0016
2N2369/46	0016
2N2369/51	0595
2N2369/KVT	OEM
2N2369/TNT	OEM
2N2369/TPT	OEM
2N2369A	0493
2N2370	0037
2N2371	0037
2N2372	0037
2N2373	0037
2N2374	0211
2N2375	0211
2N2376	0211
2N2377	0037
2N2378	0037
2N2379	0440
2N2380	0086
2N2380A	0086
2N2381	0086
2N2382	0050
2N2383	0103
2N2384	0103
2N2385	0222
2N2386	1141
2N2386A	1141
2N2387	0079
2N2388	0079

DEVICE TYPE	REPL CODE
2N2389	0198
2N2390	0198
2N2391	0861
2N2392	0861
2N2393	0037
2N2394	0037
2N2395	0150
2N2396	0198
2N2397	0198
2N2398	0050
2N2399	0050
2N2400	0136
2N2401	0136
2N2402	0136
2N2403	0595
2N2404	0595
2N2405	0644
2N2405L	0644
2N2405S	0086
2N2406	0847
2N2410	0086
2N2410/51	0018
2N2410A	OEM
2N2411	0037
2N2412	0037
2N2413	0016
2N2414	0324
2N2415	0050
2N2416	0050
2N2417	0016
2N2417A	OEM
2N2417B	OEM
2N2418	OEM
2N2418A	OEM
2N2418B	OEM
2N2419	OEM
2N2419A	OEM
2N2419B	OEM
2N2420	OEM
2N2420B	OEM
2N2421	0164
2N2421A	OEM
2N2421B	OEM
2N2422	OEM
2N2422A	OEM
2N2422B	OEM
2N2423	0604
2N2424	0037
2N2425	0037
2N2426	0600
2N2427	0016
2N2428	0004
2N2429	0004
2N2429/353	0211
2N2430	0208
2N2431	0164
2N2431B	0211
2N2432	0016
2N2432A	0111
2N2433	0086
2N2434	0086
2N2435	0715
2N2436	0715
2N2437	0715
2N2438	0715
2N2439	0715
2N2440	0233
2N2443	0233
2N2446	0604
2N2447	0004
2N2448	0004
2N2449	0004
2N2450	0004
2N2451	0136
2N2453	2045
2N2454	2936
2N2455	0050
2N2456	0050
2N2457	OEM
2N2458	OEM
2N2459	0715
2N2460	0715
2N2461	0715
2N2462	0715
2N2463	0086
2N2464	0086
2N2465	0086
2N2466	0086
2N2468	0211
2N2469	0211
2N2472	0079
2N2473	0079
2N2474	0037
2N2475	0144
2N2475/46	0224
2N2475/51	0224
2N2476	0016
2N2477	0016
2N2478	0086
2N2479	0086
2N2480	2045
2N2480A	2045
2N2481	0016
2N2482	0038
2N2483	0016
2N2484	0127
2N2484A	0595
2N2485	OEM
2N2486	OEM

DEVICE TYPE	REPL CODE
2N2487	0136
2N2488	0136
2N2489	0136
2N2489A	0086
2N2490	0040
2N2491	0040
2N2492	0040
2N2493	0040
2N2494	0136
2N2495	0050
2N2496	0050
2N2497	2987
2N2498	1141
2N2499	OEM
2N2500	OEM
2N2501	0016
2N2503	OEM
2N2504	OEM
2N2505	OEM
2N2506	OEM
2N2507	OEM
2N2508	OEM
2N2509	0233
2N2510	0086
2N2511	0086
2N2512	0050
2N2514	0016
2N2515	0016
2N2516	0086
2N2517	0715
2N2518	0715
2N2519	0715
2N2520	0715
2N2521	0715
2N2522	0715
2N2523	0016
2N2524	0016
2N2526	0604
2N2527	0975
2N2528	0975
2N2529	0016
2N2530	0016
2N2531	0016
2N2532	0016
2N2533	0016
2N2534	0016
2N2537	0086
2N2538	0086
2N2539	0016
2N2540	0016
2N2541	0847
2N2542	0738
2N2543	0738
2N2544	0738
2N2545	0738
2N2546	0738
2N2547	0738
2N2548	0738
2N2549	0738
2N2550	OEM
2N2551	0342
2N2564	0211
2N2564/5	0847
2N2565	0211
2N2568	OEM
2N2569	0016
2N2570	0016
2N2571	0016
2N2572	0016
2N2573	1825
2N2574	1825
2N2575	1825
2N2576	1825
2N2577	1825
2N2578	1825
2N2579	1825
2N2586	0016
2N2587	0050
2N2588	0136
2N2590	0126
2N2591	0126
2N2592	0126
2N2593	0126
2N2594	0086
2N2594/TNT	OEM
2N2594/TPT	OEM
2N2595	0037
2N2596	0037
2N2597	0037
2N2598	0342
2N2599	0342
2N2599A	0342
2N2600	0342
2N2600A	0037
2N2601	0037
2N2602	0037
2N2603	0037
2N2604	0037
2N2605	0037
2N2605A	0037
2N2606	OEM
2N2607	OEM
2N2608	OEM
2N2609	1141
2N2610	0127
2N2611	0264
2N2612	0604
2N2613	0004
2N2614	0004

DEVICE TYPE	REPL CODE
2N2615	0144
2N2616	0144
2N2617	0079
2N2618	0233
2N2618/46	0079
2N2619	3439
2N2620	OEM
2N2621	0136
2N2622	0136
2N2623	0136
2N2624	0136
2N2625	0136
2N2626	0136
2N2627	0136
2N2628	0136
2N2629	0136
2N2630	0136
2N2631	2873
2N2632	2873
2N2633	2873
2N2635	0136
2N2636	0604
2N2637	0604
2N2638	0604
2N2639	2045
2N2640	2045
2N2641	2045
2N2642	2045
2N2643	2045
2N2644	2045
2N2645	0016
2N2646	2136
2N2646P	OEM
2N2647	1175
2N2648	0280
2N2649	OEM
2N2650	OEM
2N2650A	OEM
2N2651	0016
2N2652	0861
2N2652A	0861
2N2653	2936
2N2654	0050
2N2655	0144
2N2656	0016
2N2657	0042
2N2671	0050
2N2672	0847
2N2672A	0050
2N2672BLK	0050
2N2672GRN	0050
2N2673	0016
2N2674	0016
2N2675	0016
2N2676	0016
2N2677	0016
2N2678	0016
2N2679	1137
2N2679A	OEM
2N2680	0344
2N2680A	0344
2N2681	OEM
2N2681A	OEM
2N2682	OEM
2N2682A	OEM
2N2683	OEM
2N2683A	OEM
2N2684	OEM
2N2684A	OEM
2N2685	OEM
2N2685A	OEM
2N2686	OEM
2N2686A	OEM
2N2687	1137
2N2688	0344
2N2689	0901
2N2690	0058
2N2691	0604
2N2691A	0604
2N2692	0016
2N2693	0016
2N2694	0111
2N2695	0037
2N2696	0037
2N2699	0600
2N2706	0164
2N2706MP	OEM
2N2707	0164
2N2708	0259
2N2709	0037
2N2710	0493
2N2711	0155
2N2712	0076
2N2712BLUE	0144
2N2712ORANGE	0144
2N2713	0155
2N2714	0155
2N2715	0127
2N2716	0127
2N2717	0136
2N2718	0136
2N2719	0016
2N2720	2045
2N2721	2045
2N2722	2045
2N2723	2258
2N2724	2258
2N2725	0400
2N2726	0233

DEVICE TYPE	REPL CODE
2N2727	0285
2N2728	0440
2N2729	0493
2N2730	0440
2N2731	0440
2N2732	0440
2N2757	3489
2N2758	3489
2N2759	3489
2N2763	3489
2N2764	3489
2N2765	3489
2N2769	3489
2N2770	3489
2N2771	3489
2N2775	3489
2N2776	3489
2N2777	3489
2N2783	0136
2N2784	0144
2N2784/46	0418
2N2784/52	0144
2N2784/KVT	OEM
2N2784/TNT	0079
2N2784/TPT	OEM
2N2786	0050
2N2786A	0050
2N2787	0086
2N2788	0086
2N2789	0086
2N2789S	0086
2N2790	0016
2N2791	0016
2N2792	0016
2N2793	0440
2N2794	OEM
2N2795	0136
2N2796	0136
2N2797	0050
2N2798	0050
2N2799	0050
2N2800	0126
2N2800/46	0037
2N2800/51	0150
2N2800S	0126
2N2801	0037
2N2801/46	0126
2N2801/51	0150
2N2802	2468
2N2803	2468
2N2804	2468
2N2805	2468
2N2806	2468
2N2807	2468
2N2808	0079
2N2808A	0079
2N2809	0007
2N2809A	0079
2N2810	0007
2N2810A	0079
2N2811	2662
2N2812	2662
2N2813	2662
2N2814	2662
2N2815	3512
2N2816	3512
2N2819	3489
2N2820	3489
2N2821	3489
2N2822	3489
2N2823	3489
2N2824	3489
2N2825	3489
2N2828	0042
2N2829	0042
2N2831	0016
2N2832	0604
2N2835	0222
2N2836	0160
2N2837	0037
2N2838	0037
2N2840	OEM
2N2841	OEM
2N2842	OEM
2N2843	OEM
2N2844	OEM
2N2845	0016
2N2846	0016
2N2847	0016
2N2848	0016
2N2849	0622
2N2849-1	0622
2N2849-2	2873
2N2850	0622
2N2850-1	0622
2N2850-2	2873
2N2851	0622
2N2851-1	0622
2N2851-2	2873
2N2852	0622
2N2852-1	0622
2N2852-2	2873
2N2853	0622
2N2853-1	0622
2N2853-2	0622
2N2854	0622
2N2854-1	0622
2N2854-2	0622
2N2855	0622

If replacement code is OEM, contact original manufacturer for replacement.

DEVICE TYPE	REPL CODE
2N2855-1	0622
2N2855-2	0622
2N2855-3	0622
2N2856	0622
2N2856-1	0622
2N2856-2	0622
2N2857	0259
2N2857J	OEM
2N2860	0050
2N2861	0037
2N2862	0037
2N2863	0086
2N2864	0086
2N2865	0007
2N2868	0086
2N2869	0604
2N2869/2N301	0604
2N2870	0160
2N2870/2N301A	0975
2N2873	0050
2N2874	0630
2N2875	0925
2N2876	0836
2N2877	2964
2N2878	2964
2N2879	2964
2N2880	2964
2N2882	0892
2N2883	0016
2N2884	0016
2N2885	0111
2N2886	0018
2N2888	0762
2N2889	3268
2N2890	0086
2N2891	0086
2N2892	2873
2N2893	2873
2N2894	0037
2N2894A	0037
2N2895	0086
2N2896	0644
2N2897	0086
2N2898	0187
2N2899	0187
2N2900	0086
2N2901	0127
2N2903	2045
2N2903A	2045
2N2904	0148
2N2904/TNT	0037
2N2904/TPT	OEM
2N2904A	0126
2N2904AS	0037
2N2904S	0037
2N2905	0126
2N2905A	0126
2N2905AS	0037
2N2905S	0037
2N2906	0037
2N2906A	0037
2N2907	0037
2N2907A	5144
2N2909	0086
2N2910	0079
2N2913	2045
2N2914	0079
2N2915	2045
2N2915A	2045
2N2916	0016
2N2916A	0079
2N2917	2045
2N2918	0079
2N2919	2045
2N2919A	2045
2N2920	0016
2N2920A	0079
2N2921	0155
2N2922	0155
2N2923	0155
2N2924	0155
2N2924-8	OEM
2N2924B	OEM
2N2925	0155
2N2926	0076
2N2926-6	0016
2N2926-B	0007
2N2926-BRN	0111
2N2926-G	0079
2N2926-GRN	0111
2N2926-O	0007
2N2926-ORG	0111
2N2926-R	0007
2N2926-RED	0111
2N2926-Y	0079
2N2926-YEL	0111
2N2926G	0016
2N2926GRN	0016
2N29260	0595
2N29260RN	0016
2N2926R	0595
2N2926Y	0079
2N2927	0037
2N2927/46	0037
2N2927/51	0150
2N2928	0050
2N2929	0050
2N2930	0211
2N2931	0111
2N2932	0111
2N2933	0111
2N2934	0111
2N2935	0111
2N2936	0079
2N2937	0079
2N2938	0016
2N2939	0086
2N2940	0187
2N2941	0187
2N2942	0050
2N2943	0050
2N2944	0037
2N2944A	0037
2N2945	0037
2N2945A	0037
2N2946	0037
2N2946A	0037
2N2947	0042
2N2948	0103
2N2949	0698
2N2950	1591
2N2951	0016
2N2951S	0086
2N2952	0016
2N2953	0004
2N2954	0079
2N2955	0136
2N2956	0136
2N2957	0136
2N2958	0016
2N2959	0016
2N2959S	0016
2N2960	0016
2N2961	0016
2N2966	0280
2N2967	0007
2N2968	0150
2N2968A	0325
2N2969	0150
2N2970	0150
2N2971	0150
2N2972	2045
2N2973	0079
2N2974	2045
2N2975	0079
2N2976	2045
2N2977	0079
2N2978	2045
2N2979	0079
2N2980	0086
2N2987	0168
2N2988	0168
2N2989	0168
2N2990	0168
2N2991	0136
2N2992	0136
2N2993	0136
2N2994	0136
2N2995	0630
2N2996	0050
2N2997	0050
2N2998	0050
2N2999	0050
2N3001	OEM
2N3002	OEM
2N3003	OEM
2N3004	OEM
2N3005	1137
2N3006	0344
2N3007	0901
2N3008	0058
2N3009	0016
2N3010	0144
2N3011	0016
2N3012	0037
2N3013	0016
2N3014	0016
2N3015	0016
2N3016	0566
2N3017	0566
2N3019	0086
2N3019S	0187
2N3020	0086
2N3020S	0086
2N3021	0925
2N3022	0925
2N3023	0925
2N3024	0925
2N3025	0925
2N3026	0925
2N3027	1137
2N3028	0344
2N3029	0901
2N3030	1137
2N3031	0344
2N3032	0901
2N3033	0861
2N3034	0861
2N3035	0144
2N3036	0086
2N3037	0715
2N3038	0086
2N3039	0126
2N3040	0126
2N3043	0079
2N3044	0079
2N3045	0079
2N3046	0079
2N3047	0079
2N3048	0079
2N3049	0150
2N3050	0150
2N3051	0150
2N3052	0595
2N3053	0076
2N3053/4053	0630
2N3053/40053	0595
2N3053A	0086
2N3053S	0630
2N3054	0429
2N3054A	0178
2N3055	0103
2N3055-1	0103
2N3055-2	0103
2N3055-3	0130
2N3055-4	0103
2N3055-5	0103
2N3055-6	0130
2N3055-7	0130
2N3055-8	0130
2N3055-9	0103
2N3055-10	0103
2N3055A	0103
2N3055AH	0103
2N3055C	OEM
2N3055H	0103
2N3055HOM	1833
2N3055J	OEM
2N3055MP	0103
2N3055S	0103
2N3055SD	0103
2N3055TX	OEM
2N3055UB	0103
2N3056	0086
2N3056A	0187
2N3057	0715
2N3057A	0187
2N3058	0037
2N3059	0037
2N3059AS	0086
2N3060	0037
2N3061	0037
2N3062	0037
2N3063	0342
2N3064	0342
2N3065	0342
2N3066	0325
2N3066A	OEM
2N3067	0325
2N3067A	OEM
2N3068	0325
2N3068A	OEM
2N3069	0325
2N3069A	OEM
2N3070	0325
2N3070A	OEM
2N3071	0325
2N3072	0037
2N3073	0037
2N3074	0050
2N3075	0280
2N3077	0086
2N3078	0086
2N3081	0037
2N3081/46	0791
2N3081/51	OEM
2N3082	0127
2N3083	0127
2N3084	0325
2N3085	0325
2N3086	0325
2N3087	0325
2N3088	0325
2N3088A	0325
2N3089	0325
2N3089A	0325
2N3091	0610
2N3091-96	OEM
2N3092	0468
2N3093	0468
2N3094	0468
2N3095	0468
2N3096	0468
2N3097	0468
2N3098	OEM
2N3099	OEM
2N3100	OEM
2N3101	OEM
2N3102	OEM
2N3103	OEM
2N3104	OEM
2N3105	OEM
2N3106	OEM
2N3107	0086
2N3108	0086
2N3109	0086
2N3110	0086
2N3110S	0086
2N3112	OEM
2N3113	OEM
2N3114	0233
2N3114S	0338
2N3115	0016
2N3116	0016
2N3117	0111
2N3118	0086
2N3119	0285
2N3120	0126
2N3121	0037
2N3122	0086
2N3123	0016
2N3124	0604
2N3125	0604
2N3126	0160
2N3127	0050
2N3128	0079
2N3129	0079
2N3130	0079
2N3131	0007
2N3132	0604
2N3133	0126
2N3133S	0233
2N3134	0126
2N3135	0037
2N3136	0037
2N3137	0493
2N3138	0042
2N3140	0042
2N3142	0042
2N3144	0042
2N3146	0160
2N3147	0160
2N3148	0050
2N3153	0050
2N3162	OEM
2N3171	1681
2N3172	1681
2N3173	1681
2N3174	1681
2N3183	1681
2N3184	1681
2N3185	1681
2N3186	1681
2N3195	1681
2N3196	1681
2N3197	1681
2N3198	1681
2N3199	0925
2N3200	0925
2N3202	0126
2N3203	0126
2N3205	0925
2N3206	0925
2N3208	0126
2N3209	0150
2N3210	0016
2N3211	0016
2N3212	2713
2N3213	0160
2N3214	2713
2N3215	2713
2N3216	0280
2N3217	0037
2N3218	0037
2N3219	0037
2N3224	0439
2N3225	0439
2N3226	0103
2N3227	0493
2N3228	3696
2N3229	0042
2N3230	OEM
2N3231	OEM
2N3232	0103
2N3233	0103
2N3234	0103
2N3235	0103
2N3236	3701
2N3237	0130
2N3238	0130
2N3239	0130
2N3241	0016
2N3241A	0016
2N3242	0016
2N3242A	0016
2N3244	0126
2N3244S	0126
2N3245	0126
2N3245S	0126
2N3246	0016
2N3247	0016
2N3248	0037
2N3249	0037
2N3250	0037
2N3250A	0037
2N3251	0037
2N3251A	0037
2N3252	0595
2N3252S	0630
2N3253	0836
2N3253S	0630
2N3254	0630
2N3255	OEM
2N3256	OEM
2N3257	OEM
2N3258	OEM
2N3259	0344
2N3261	0016
2N3262	0622
2N3265	3512
2N3266	3512
2N3267	0050
2N3268	0016
2N3269	OEM
2N3270	OEM
2N3271	OEM
2N3272	OEM
2N3273	OEM
2N3274	OEM
2N3275	OEM
2N3276	OEM
2N3277	OEM
2N3278	OEM
2N3279	0050
2N3280	0050
2N3281	0050
2N3282	0050
2N3283	0050
2N3284	0050
2N3285	0050
2N3286	0050
2N3287	0007
2N3288	0007
2N3289	0007
2N3290	0007
2N3291	0007
2N3292	0007
2N3293	0007
2N3294	0007
2N3295	0595
2N3296	0560
2N3297	0103
2N3298	0079
2N3299	0016
2N3299S	0493
2N3300	0016
2N3300S	0493
2N3301	0016
2N3302	0016
2N3303	2700
2N3304	0150
2N3305	0037
2N3306	0037
2N3307	0037
2N3308	0037
2N3309	0936
2N3309A	OEM
2N3310	0016
2N3311	0040
2N3312	0040
2N3313	0040
2N3314	0040
2N3315	0040
2N3316	0440
2N3317	0037
2N3318	0037
2N3319	0037
2N3320	0136
2N3321	0136
2N3322	0136
2N3323	0136
2N3324	0136
2N3325	0136
2N3326	0086
2N3328	0050
2N3329	1141
2N3330	1141
2N3331	1141
2N3332	1141
2N3333	OEM
2N3334	OEM
2N3335	OEM
2N3336	OEM
2N3337	0007
2N3338	0007
2N3339	0007
2N3340	0016
2N3341	0037
2N3342	0150
2N3343	0126
2N3344	0126
2N3345	0126
2N3346	0037
2N3347	2468
2N3348	2468
2N3349	2468
2N3350	2468
2N3351	2468
2N3352	2468
2N3353	OEM
2N3354	OEM
2N3355	OEM
2N3356	OEM
2N3357	OEM
2N3358	OEM
2N3359	OEM
2N3360	OEM
2N3361	OEM
2N3362	OEM
2N3363	OEM
2N3364	OEM
2N3365	0325
2N3366	0325
2N3367	0325
2N3368	0325
2N3369	0325
2N3370	0325
2N3371	0050
2N3374	0595
2N3375	0560
2N3375J	OEM
2N3375TX	OEM
2N3376	1141
2N3377	1141
2N3378	1141
2N3379	1141
2N3380	OEM
2N3381	OEM
2N3382	OEM
2N3383	OEM
2N3384	OEM
2N3385	OEM
2N3386	OEM
2N3387	OEM
2N3388	0233
2N3389	0233
2N3390	0111
2N3390-U29	0111
2N3391	0111
2N3391-U29	0111
2N3391A	0111
2N3392	0155
2N3392-U29	0111
2N3393	0155
2N3393-U29	0111
2N3394	0155
2N3394-U29	0111
2N3395	0155
2N3395-W	0079
2N3395-WHT	0111
2N3395-Y	0079
2N3395-YEL	0111
2N3396	0155
2N3396-O	0086
2N3396-ORG	0111
2N3396-W	0079
2N3396-WHT	0111
2N3396-Y	0079
2N3396-YEL	0111
2N3397	0155
2N3397-O	0086
2N3397-ORG	0111
2N3397-R	0086
2N3397-RED	0111
2N3397-W	0079
2N3397-WHT	0111
2N3397-Y	0079
2N3397-YEL	0111
2N3398	0111
2N3398-B	0079
2N3398-BLU	0111
2N3398-O	0086
2N3398-ORG	0111
2N3398-R	0086
2N3398-RED	0111
2N3398-W	0079
2N3398-WHT	0111
2N3398-YEL	0224
2N3399	0050
2N3400	0136
2N3401	0037
2N3402	0286
2N3403	0286
2N3404	0286
2N3405	0286
2N3406	OEM
2N3407	0007
2N3408	OEM
2N3409	0079
2N3410	0079
2N3411	0079
2N3412	0136
2N3413	0342
2N3414	0155
2N3415	0016
2N3416	0155
2N3417	0218
2N3418	0560
2N3420	0560
2N3423	0079
2N3424	0079
2N3425	0224
2N3426	0698
2N3427	0211
2N3428	0211
2N3435	0595
2N3436	0325
2N3437	0325
2N3438	0325
2N3439	0187
2N3439J	OEM
2N3439TX	OEM
2N3440	0187
2N3440J	OEM
2N3440S	0187
2N3440TX	OEM
2N3441	1027
2N3441J	OEM
2N3441TX	OEM
2N3442	0543
2N3442J	OEM
2N3442TX	OEM
2N3443	0136
2N3444	0630
2N3444S	0630
2N3445	0103
2N3446	1966
2N3447	2013
2N3448	0103
2N3449	0136
2N3450	0187
2N3451	0037
2N3452	0325
2N3453	0325
2N3454	0325
2N3455	0325
2N3456	0325
2N3457	0325
2N3458	0325
2N3459	0325
2N3460	0325
2N3461	0164
2N3462	0016
2N3463	0016
2N3464	0037
2N3465	0325
2N3466	0325
2N3467	0126
2N3467S	0126
2N3468	0126
2N3468S	0126
2N3469	0086
2N3478	0259
2N3479	OEM
2N3480	OEM
2N3481	OEM
2N3482	OEM
2N3483	0178
2N3484	OEM
2N3485	0037
2N3485A	0037
2N3486	0037
2N3486A	0037
2N3487	0561
2N3488	0561
2N3490	0561
2N3491	0561
2N3493	0007
2N3494	0126
2N3494S	0037
2N3495	0342
2N3495S	0342
2N3496	0037
2N3497	0342
2N3498	0086
2N3498S	0338
2N3499	0086
2N3499S	0338
2N3500	0168
2N3500S	0338
2N3501	0086
2N3501S	0338
2N3502	0126
2N3503	0126
2N3504	0037
2N3505	0037
2N3506	0622
2N3506S	0630
2N3507	0622
2N3507S	0630
2N3508	0016
2N3509	0016
2N3510	0016
2N3511	0016
2N3512	0016
2N3513	OEM
2N3514	OEM
2N3515	OEM
2N3516	OEM
2N3517	OEM
2N3518	OEM
2N3519	OEM
2N3520	OEM
2N3521	OEM
2N3522	OEM
2N3523	OEM
2N3524	OEM
2N3525	3383
2N3525A	0560
2N3526	0233
2N3527	0037
2N3528	2394
2N3529	0710
2N3530	OEM
2N3531	OEM
2N3532	OEM
2N3533	OEM
2N3533J	OEM
2N3533TX	OEM
2N3534	OEM
2N3535	OEM
2N3536	OEM
2N3537	OEM
2N3538	OEM
2N3539	OEM
2N3540	OEM
2N3541	OEM
2N3542	0079
2N3544	0144
2N3545	0037
2N3546	0037
2N3547	0037
2N3548	0037
2N3549	0037
2N3550	0037
2N3553	1598
2N3554	0698
2N3555	OEM
2N3556	OEM
2N3557	OEM
2N3558	OEM
2N3559	OEM
2N3560	OEM

DEVICE TYPE	REPL CODE
2N3561	OEM
2N3562	0144
2N3563	0144
2N3563-1	0144
2N3564	0016
2N3565	0016
2N3566	0076
2N3567	0076
2N3568	0086
2N3569	0076
2N3570	0144
2N3571	0259
2N3572	0259
2N3573	OEM
2N3574	OEM
2N3575	OEM
2N3576	0150
2N3577	OEM
2N3578	OEM
2N3579	0037
2N3580	0037
2N3581	0037
2N3582	0037
2N3583	1027
2N3584	0429
2N3584J	OEM
2N3584TX	OEM
2N3585	1027
2N3585J	OEM
2N3585TX	OEM
2N3586	OEM
2N3587	2045
2N3588	0050
2N3589	0168
2N3590	0168
2N3591	0168
2N3592	0168
2N3593	0168
2N3594	0168
2N3597	3512
2N3598	3512
2N3599	3512
2N3600	0007
2N3601	OEM
2N3602	OEM
2N3603	OEM
2N3604	OEM
2N3605	0155
2N3605A	0155
2N3606	0155
2N3606A	0155
2N3607	0155
2N3608	0325
2N3609	OEM
2N3610	OEM
2N3611	0604
2N3612	0604
2N3613	0604
2N3614	0604
2N3615	0604
2N3616	0604
2N3617	0604
2N3618	0604
2N3619	0560
2N3620	0560
2N3621	0939
2N3622	2662
2N3623	0560
2N3624	0560
2N3625	0939
2N3626	0042
2N3627	0042
2N3628	0560
2N3629	0939
2N3630	0042
2N3631	3628
2N3632	0042
2N3633	0418
2N3633/46	0127
2N3633/51	OEM
2N3633/52	0595
2N3633/KVT	0861
2N3633/TNT	0127
2N3633/TPT	OEM
2N3637S	1825
2N3638	0006
2N3638A	0037
2N3639	0037
2N3640	0037
2N3641	0016
2N3642	0016
2N3643	0016
2N3644	0006
2N3645	0006
2N3646	0016
2N3647	0016
2N3648	0016
2N3649	0804
2N3650	0804
2N3651	0804
2N3652	0804
2N3653	0804
2N3654	0804
2N3655	0804
2N3656	0804
2N3657	0804
2N3658	0804
2N3659	0168
2N3660	0892
2N3661	0892
2N3662	0127
2N3663	0127
2N3665	0086
2N3666	0086
2N3667	0103
2N3668	1825
2N3669	3857
2N3670	1825
2N3671	0126
2N3672	0037
2N3673	0037
2N3675	0042
2N3677	0126
2N3678	0086
2N3679	OEM
2N3681	0007
2N3682	0144
2N3683	0007
2N3684	3133
2N3684A	0325
2N3685	0325
2N3685A	0325
2N3686	0325
2N3686A	0325
2N3687	0325
2N3687A	0493
2N3688	0224
2N3689	0016
2N3690	0016
2N3691	0016
2N3692	0016
2N3693	0016
2N3694	0076
2N3695	0144
2N3696	OEM
2N3697	3622
2N3698	OEM
2N3700	0233
2N3701	0233
2N3702	0006
2N3703	0037
2N3704	0076
2N3705	0076
2N3706	0079
2N3707	0155
2N3708	1221
2N3708-BLU	0111
2N3708-BRN	0111
2N3708-G	0079
2N3708-GRN	0111
2N3708-O	0086
2N3708-ORG	0111
2N3708-R	0086
2N3708-RED	0111
2N3708-V	0079
2N3708-VIO	0111
2N3708-Y	0086
2N3708-YEL	0111
2N3709	0127
2N3710	1221
2N3711	0155
2N3712	0233
2N3712S	0233
2N3713	3873
2N3713HS	0543
2N3714	3873
2N3714HS	0543
2N3715	3873
2N3716	3873
2N3716HS	1966
2N3717	OEM
2N3719	0126
2N3720	0126
2N3721	0155
2N3722	0086
2N3723	0086
2N3724	0493
2N3724A	0936
2N3725	0016
2N3725A	0423
2N3726	2468
2N3727	2468
2N3728	OEM
2N3729	OEM
2N3730	0975
2N3731	0975
2N3732	0975
2N3733	0836
2N3734	0936
2N3734S	0630
2N3735	0630
2N3735S	0630
2N3736	0079
2N3737	0630
2N3738	1027
2N3739	1027
2N3740	0905
2N3740A	0925
2N3740Q	OEM
2N3741	0681
2N3741A	0681
2N3742	0233
2N3742S	0233
2N3743	0233
2N3743S	0439
2N3744	0042
2N3745	0042
2N3747	0042
2N3748	0042
2N3753	2449
2N3754	0804
2N3755	OEM
2N3756	OEM
2N3757	OEM
2N3758	OEM
2N3759	OEM
2N3760	OEM
2N3761	OEM
2N3762	0126
2N3762A	0630
2N3762S	0630
2N3763	0126
2N3763S	0630
2N3764	0126
2N3764A	0630
2N3765	0532
2N3766	0429
2N3767	0429
2N3770	0050
2N3771	0130
2N3771J	OEM
2N3771TX	OEM
2N3772	0130
2N3772C	5640
2N3772J	OEM
2N3772TX	OEM
2N3773	0543
2N3774	0126
2N3775	0126
2N3776	0126
2N3777	0892
2N3778	0126
2N3779	0126
2N3780	0126
2N3781	1984
2N3782	0126
2N3783	0050
2N3784	0050
2N3785	0050
2N3788	0642
2N3789	0491
2N3790	0491
2N3791	0491
2N3792	3909
2N3793	0079
2N3794	0079
2N3795	0892
2N3796	3628
2N3797	3628
2N3798	0037
2N3798A	0037
2N3799	0037
2N3799A	0037
2N3800	2468
2N3801	OEM
2N3802	OEM
2N3803	OEM
2N3804	OEM
2N3804A	OEM
2N3805	OEM
2N3805A	OEM
2N3806	2468
2N3807	0016
2N3808	2468
2N3809	0016
2N3810	2468
2N3810A	2468
2N3811	0016
2N3811A	0016
2N3812	0037
2N3813	0016
2N3814	0037
2N3815	0016
2N3816	0037
2N3816A	0037
2N3817	0016
2N3817A	0016
2N3818	0042
2N3819	0325
2N3820	2987
2N3821	0325
2N3822	3133
2N3823	0849
2N3824	3133
2N3825	0155
2N3826	0155
2N3827	0155
2N3828	0079
2N3829	0037
2N3830	0086
2N3831	0622
2N3832	0007
2N3833	0418
2N3834	0418
2N3835	0418
2N3836	OEM
2N3837	OEM
2N3838	3919
2N3839	0259
2N3840	0037
2N3841	0342
2N3842	0037
2N3843	0155
2N3843A	0155
2N3844	0155
2N3844A	0155
2N3845	0155
2N3845A	0155
2N3846	0127
2N3850	2873
2N3851	2873
2N3852	2873
2N3853	2873
2N3854	0155
2N3854A	0155
2N3855	0155
2N3855A	0127
2N3856	0155
2N3856A	0155
2N3857	0037
2N3858	0155
2N3858A	0155
2N3859	0155
2N3859A	0155
2N3860	0155
2N3861	3280
2N3862	0493
2N3863	0103
2N3864	0103
2N3865	0074
2N3866	0493
2N3866A	0493
2N3866AJ	OEM
2N3866ATX	OEM
2N3866J	OEM
2N3866TX	OEM
2N3867	0126
2N3868	0126
2N3869	0493
2N3870	1650
2N3871	1650
2N3872	2647
2N3873	2649
2N3877	0086
2N3877A	0155
2N3878	0429
2N3879	0178
2N3879J	3362
2N3879TX	OEM
2N3880	0259
2N3881	0086
2N3882	OEM
2N3883	0136
2N3884	1084
2N3885	1084
2N3886	1084
2N3887	1086
2N3888	1086
2N3889	1086
2N3890	1086
2N3891	1102
2N3892	1102
2N3893	1102
2N3894	1102
2N3895	1102
2N3896	0567
2N3897	0469
2N3898	0722
2N3899	0725
2N3900	0155
2N3900A	0155
2N3901	0111
2N3902	0363
2N3903	0076
2N3904	0076
2N3905	0037
2N3906	0006
2N3907	2045
2N3908	2045
2N3909	OEM
2N3909A	OEM
2N3910	0037
2N3911	0037
2N3912	0037
2N3913	0037
2N3914	0037
2N3915	0150
2N3916	0168
2N3917	0103
2N3918	0103
2N3919	0620
2N3920	0620
2N3921	OEM
2N3922	OEM
2N3923	0233
2N3924	0198
2N3925	0560
2N3926	3321
2N3927	3322
2N3930	0037
2N3931	0037
2N3932	0007
2N3933	0007
2N3934	OEM
2N3935	OEM
2N3936	0240
2N3937	0240
2N3938	0676
2N3939	0676
2N3940	0336
2N3941	OEM
2N3942	OEM
2N3943	OEM
2N3944	OEM
2N3945	0936
2N3946	0016
2N3947	0016
2N3948	0493
2N3950	0836
2N3953	0259
2N3954	OEM
2N3954A	OEM
2N3955	OEM
2N3955A	OEM
2N3956	OEM
2N3957	2945
2N3958	OEM
2N3959	0418
2N3960	0418
2N3961	0560
2N3962	0037
2N3963	0037
2N3964	0037
2N3965	0037
2N3966	0325
2N3967	0325
2N3967A	0325
2N3968	0325
2N3968A	0325
2N3969	0325
2N3969A	0325
2N3970	1155
2N3971	1155
2N3972	OEM
2N3973	0018
2N3974	0018
2N3975	0018
2N3976	0018
2N3977	0037
2N3978	0037
2N3979	0037
2N3980	1893
2N3981	0595
2N3982	0595
2N3983	0127
2N3984	0127
2N3985	0127
2N3986	0468
2N3987	0468
2N3988	0468
2N3989	0468
2N3990	0468
2N3991	0468
2N3992	0468
2N3993	OEM
2N3993A	OEM
2N3994	0155
2N3994A	OEM
2N3995	0050
2N3996	2964
2N3997	2964
2N3998	2873
2N3999	2873
2N4000	0168
2N4001	0622
2N4002	3489
2N4003	3489
2N4006	0037
2N4007	0037
2N4008	0037
2N4009	OEM
2N4010	OEM
2N4011	OEM
2N4012	0560
2N4013	0016
2N4014	0016
2N4015	2468
2N4016	0037
2N4018	2468
2N4020	OEM
2N4021	2468
2N4022	OEM
2N4023	OEM
2N4024	2468
2N4025	OEM
2N4026	0126
2N4027	0126
2N4028	0126
2N4029	0126
2N4030	0126
2N4031	0126
2N4032	0126
2N4033	0126
2N4034	0150
2N4035	0150
2N4036	0126
2N4037	0126
2N4038	0983
2N4039	0983
2N4040	0560
2N4041	0560
2N4042	OEM
2N4043	OEM
2N4044	0079
2N4045	0079
2N4046	0016
2N4047	0086
2N4048	4012
2N4049	4012
2N4050	4012
2N4051	0437
2N4052	4012
2N4053	0437
2N4054	0276
2N4055	0276
2N4056	0276
2N4057	0276
2N4058	0150
2N4059	0150
2N4060	0150
2N4061	0150
2N4062	0150
2N4063	4021
2N4064	4021
2N4065	0844
2N4066	OEM
2N4067	OEM
2N4068	0233
2N4069	0168
2N4070	3554
2N4071	1966
2N4071MP	0177
2N4072	0016
2N4073	0493
2N4074	0016
2N4075	4029
2N4076	4029
2N4077	2763
2N4078	0222
2N4079	3545
2N4080	4032
2N4081	0007
2N4082	OEM
2N4083	OEM
2N4084	OEM
2N4085	OEM
2N4086	0111
2N4087	0111
2N4087A	0111
2N4088	OEM
2N4089	OEM
2N4090	OEM
2N4091	1155
2N4091A	OEM
2N4092	1155
2N4092A	OEM
2N4093	3563
2N4093A	OEM
2N4094	OEM
2N4095	OEM
2N4096	OEM
2N4097	OEM
2N4098	OEM
2N4099	OEM
2N4100	0079
2N4101	3624
2N4102	OEM
2N4103	4040
2N4104	0079
2N4105	0038
2N4106	0164
2N4107	0164
2N4108	OEM
2N4109	OEM
2N4110	OEM
2N4111	0620
2N4112	0103
2N4113	0620
2N4114	0103
2N4117	0325
2N4117A	0325
2N4118	0325
2N4118A	3384
2N4119	0325
2N4119A	0325
2N4120	0844
2N4121	0037
2N4122	0150
2N4123	0079
2N4124	0076
2N4125	0037
2N4126	0037
2N4127	1233
2N4128	1977
2N4130	0103
2N4134	0007
2N4135	0007
2N4136	0208
2N4137	0493
2N4138	0016
2N4139	OEM
2N4140	0016
2N4141	0016
2N4142	0037
2N4143	0037
2N4144	1137
2N4145	1137
2N4146	0344
2N4147	0901
2N4148	0058
2N4149	0058
2N4150	0622
2N4151	1487
2N4152	1487
2N4153	1487
2N4154	1487
2N4155	1487
2N4156	1487
2N4157	OEM
2N4158	OEM
2N4159	1487
2N4160	1487
2N4161	1487
2N4162	1487
2N4163	1487
2N4164	1487
2N4165	OEM
2N4166	OEM
2N4167	3419
2N4168	1103
2N4169	2491
2N4170	0240
2N4171	2659
2N4172	0676
2N4173	2809
2N4174	0336
2N4175	1103
2N4176	3419
2N4177	2491
2N4178	0240
2N4179	2659
2N4180	0676
2N4181	0336
2N4182	0336
2N4183	1487
2N4184	1487
2N4185	1487
2N4186	1487
2N4187	1487
2N4188	1487
2N4189	OEM
2N4190	OEM
2N4191	1487
2N4192	1487
2N4193	1487
2N4194	1487
2N4195	1487
2N4196	1487
2N4197	OEM
2N4198	OEM
2N4199	OEM
2N4200	OEM
2N4201	OEM
2N4202	OEM
2N4203	OEM
2N4204	OEM
2N4205	OEM
2N4207	0037
2N4208	0037
2N4209	0037
2N4210	3512
2N4211	3512
2N4212	OEM
2N4213	OEM
2N4214	OEM
2N4215	OEM
2N4216	OEM
2N4217	OEM
2N4218	0407
2N4219	OEM
2N4220	3622
2N4220A	0325
2N4221	3622
2N4221A	3622
2N4222	0325
2N4222A	0325
2N4223	0325
2N4224	0325
2N4225	0560
2N4226	0560
2N4227	0016
2N4228	0037
2N4231	0178
2N4231A	0161
2N4232	0178
2N4232A	0423
2N4233	0178
2N4233A	0423
2N4234	0126
2N4235	0126
2N4236	0126
2N4237	0086
2N4238	0086
2N4239	0086
2N4240	1027
2N4241	0160
2N4241MP	OEM
2N4242	0604
2N4243	0604
2N4244	0085
2N4245	0604
2N4246	0604
2N4247	0085
2N4248	0006
2N4249	0037
2N4250	0037
2N4250A	0037
2N4251	0418
2N4252	0007
2N4253	0007
2N4254	0127
2N4255	0127
2N4256	0155
2N4257	0037
2N4257A	0037
2N4258	0037
2N4258A	0037
2N4259	0007
2N4260	1591
2N4261	3607
2N4261JAN	OEM
2N4261JTX	OEM
2N4262	OEM

If replacement code is OEM, contact original manufacturer for replacement.

DEVICE TYPE	REPL CODE
2N4263	OEM
2N4264	0079
2N4265	0079
2N4267	0844
2N4269	0233
2N4270	0233
2N4271	0168
2N4273	0429
2N4274	0016
2N4275	0016
2N4276	0085
2N4277	0085
2N4278	0085
2N4279	0085
2N4280	4099
2N4281	0085
2N4282	0085
2N4283	0085
2N4284	0037
2N4285	0037
2N4286	0155
2N4287	0155
2N4288	0037
2N4289	0037
2N4290	0037
2N4291	0037
2N4292	0144
2N4293	0144
2N4294	0079
2N4295	0007
2N4296	1027
2N4297	1027
2N4298	0142
2N4299	0142
2N4301	2662
2N4302	2950
2N4303	2950
2N4304	2950
2N4307	0042
2N4308	0042
2N4311	0042
2N4312	0042
2N4313	0037
2N4314	0126
2N4315	OEM
2N4316	0505
2N4317	0505
2N4318	0505
2N4319	OEM
2N4320	OEM
2N4321	OEM
2N4322	OEM
2N4323	OEM
2N4324	OEM
2N4325	OEM
2N4326	OEM
2N4327	OEM
2N4328	OEM
2N4329	OEM
2N4330	OEM
2N4331	OEM
2N4332	OEM
2N4333	OEM
2N4334	OEM
2N4335	OEM
2N4336	OEM
2N4337	OEM
2N4338	0325
2N4339	0325
2N4340	3133
2N4341	0325
2N4342	OEM
2N4343	OEM
2N4346	0975
2N4347	0177
2N4348	0177
2N4349	0936
2N4350	0560
2N4351	0983
2N4352	0844
2N4353	0844
2N4354	0037
2N4355	0037
2N4356	0037
2N4357	0342
2N4359	0037
2N4360	2987
2N4361	0468
2N4362	0468
2N4363	0468
2N4364	0468
2N4365	0468
2N4366	0468
2N4367	0468
2N4368	OEM
2N4369	OEM
2N4370	OEM
2N4371	0658
2N4372	0658
2N4373	0658
2N4374	0658
2N4375	0658
2N4376	0658
2N4377	0658
2N4378	OEM
2N4379	OEM
2N4380	OEM
2N4381	1141
2N4382	OEM
2N4383	0086
2N4384	0018
2N4385	0086
2N4386	0018
2N4387	0925
2N4388	0925
2N4389	0037
2N4390	0861
2N4391	1155
2N4392	1155
2N4393	OEM
2N4395	0103
2N4396	0103
2N4397	0007
2N4398	1681
2N4399	1681
2N4400	0076
2N4401	0076
2N4402	0037
2N4403	0037
2N4404	0126
2N4405	0126
2N4406	0126
2N4407	0126
2N4409	0861
2N4410	0076
2N4411	0037
2N4412	0126
2N4412A	0126
2N4413	0037
2N4413A	0037
2N4414	0126
2N4414A	0126
2N4415	0037
2N4415A	0037
2N4416	2888
2N4416A	2888
2N4417	2888
2N4418	0016
2N4419	0016
2N4420	0079
2N4421	0079
2N4422	0079
2N4423	0150
2N4424	0155
2N4425	0218
2N4427	2041
2N4427M	2041
2N4428	0560
2N4429	0560
2N4430	0560
2N4432	0016
2N4432A	0016
2N4433	0007
2N4434	0007
2N4435	0007
2N4436	0016
2N4437	0016
2N4440	0560
2N4440J	OEM
2N4440TX	OEM
2N4441	0429
2N4442	2270
2N4443	3402
2N4444	3404
2N4445	OEM
2N4446	OEM
2N4447	OEM
2N4448	OEM
2N4449	0493
2N4450	0016
2N4451	0150
2N4452	0037
2N4453	0150
2N4625	0079
2N4836M	0847
2N4848A	OEM
2N4851	2136
2N4852	1175
2N4853	1175
2N4854	0861
2N4855	0861
2N4856	1155
2N4856A	1155
2N4857	1155
2N4857A	1155
2N4858	1155
2N4858A	1155
2N4859	1155
2N4859A	1155
2N4860	1155
2N4860A	1155
2N4861	1155
2N4861A	1155
2N4864	0178
2N4867	0325
2N4867A	0325
2N4868	0325
2N4868A	0325
2N4869	3133
2N4869A	3133
2N4870	1893
2N4871	1893
2N4872	0150
2N4873	0493
2N4874	0493
2N4875	0493
2N4876	0493
2N4877	0622
2N4878	0079
2N4879	0079
2N4880	0079
2N4881	OEM
2N4882	OEM
2N4884	OEM
2N4885	OEM
2N4886	OEM
2N4888	0342
2N4889	0037
2N4890	0126
2N4890S	0126
2N4891	1669
2N4892	OEM
2N4893	2136
2N4894	OEM
2N4895	0622
2N4896	0622
2N4897	0622
2N4898	0905
2N4899	0905
2N4900	0905
2N4901	0491
2N4902	0491
2N4903	0491
2N4904	2013
2N4905	2013
2N4906	2013
2N4907	0491
2N4908	0491
2N4909	0491
2N4910	0429
2N4911	0429
2N4912	0429
2N4913	0177
2N4914	0177
2N4915	0177
2N4916	0006
2N4917	0037
2N4918	0460
2N4919	0460
2N4920	0460
2N4921	0161
2N4922	0161
2N4923	0161
2N4924	4225
2N4924S	0264
2N4925	0233
2N4926	0233
2N4927	0233
2N4928	0126
2N4928S	0382
2N4929	0439
2N4930	0439
2N4930S	0439
2N4931	0439
2N4931S	0439
2N4932	0622
2N4934	0259
2N4935	0493
2N4936	0493
2N4937	0150
2N4938	0150
2N4939	0150
2N4940	0150
2N4941	0150
2N4942	0150
2N4943	0086
2N4944	0086
2N4945	0861
2N4946	0086
2N4947	OEM
2N4948	1175
2N4949	1669
2N4950	0079
2N4951	0079
2N4952	0079
2N4953	0079
2N4954	0155
2N4955	OEM
2N4956	OEM
2N4957	3607
2N4958	3607
2N4959	3607
2N4960	0086
2N4961	0086
2N4962	0086
2N4963	0086
2N4964	0086
2N4965	0037
2N4966	0016
2N4967	0111
2N4968	0016
2N4969	0016
2N4970	0079
2N4971	0037
2N4971A	0791
2N4972	0037
2N4972A	0791
2N4973	OEM
2N4976	0560
2N4977	OEM
2N4978	OEM
2N4979	OEM
2N4980	0037
2N4981	0037
2N4982	0037
2N4982A	0791
2N4983	3407
2N4984	OEM
2N4985	OEM
2N4986	OEM
2N4987	3407
2N4988	OEM
2N4989	OEM
2N4990	3407
2N4991	4233
2N4992	4233
2N4993	4233
2N4994	0079
2N4995	0079
2N4996	0144
2N4997	0076
2N4998	4029
2N5000	4029
2N5006	0939
2N5008	0939
2N5010	0168
2N5018	OEM
2N5019	OEM
2N5020	OEM
2N5021	OEM
2N5022	0126
2N5023	0126
2N5024	0259
2N5027	0079
2N5028	0079
2N5029	0079
2N5030	0079
2N5031	0007
2N5032	0127
2N5033	2987
2N5034	0103
2N5035	0103
2N5036	0103
2N5037	0103
2N5038	2485
2N5038-1	2485
2N5038J	OEM
2N5038TX	OEM
2N5039	2485
2N5039-1	2485
2N5039J	OEM
2N5039TX	OEM
2N5040	0126
2N5041	0126
2N5042	0126
2N5045	0325
2N5046	0325
2N5047	0325
2N5048	2662
2N5049	2662
2N5050	0429
2N5051	0429
2N5052	0007
2N5053	0259
2N5054	0259
2N5055	0791
2N5058	0233
2N5058S	0144
2N5059	0233
2N5059S	0144
2N5060	1137
2N5061	0344
2N5062	0901
2N5063	2342
2N5064	0058
2N5066	0079
2N5067	0177
2N5068	0177
2N5069	0177
2N5070	0936
2N5071J	OEM
2N5071TX	OEM
2N5073	0233
2N5074	4252
2N5075	4252
2N5076	4252
2N5077	4252
2N5078	2888
2N5078A	0325
2N5079	0560
2N5080	0560
2N5081	0016
2N5082	0016
2N5086	0037
2N5086A	0791
2N5087	0037
2N5087A	0791
2N5088	1221
2N5089	1221
2N5090	0622
2N5092	0168
2N5095	0168
2N5097	0103
2N5101	3280
2N5102	0130
2N5103	0079
2N5103A	0325
2N5104	OEM
2N5104A	0325
2N5105	OEM
2N5105A	0325
2N5106	0018
2N5107	0016
2N5108	0418
2N5108A	0418
2N5109	0418
2N5109J	OEM
2N5109TX	OEM
2N5110	0126
2N5111	0126
2N5112	0925
2N5114	1155
2N5115	OEM
2N5116	OEM
2N5117	OEM
2N5118	OEM
2N5119	OEM
2N5120	0086
2N5121	OEM
2N5122	OEM
2N5123	OEM
2N5124	OEM
2N5125	OEM
2N5126	0007
2N5127	0016
2N5128	0016
2N5129	0016
2N5130	0007
2N5131	0079
2N5132	0007
2N5133	0007
2N5134	0016
2N5135	0198
2N5136	0198
2N5137	0016
2N5138	0037
2N5138A	0791
2N5139	0037
2N5139A	0791
2N5140	0150
2N5141	0150
2N5142	0037
2N5142A	0791
2N5143	0037
2N5143A	0791
2N5144	0079
2N5145	0016
2N5146	OEM
2N5148	0622
2N5150	0622
2N5155	0975
2N5156	0604
2N5157	0363
2N5158	OEM
2N5159	OEM
2N5160	0126
2N5161	0925
2N5163	0325
2N5163A	0325
2N5164	1651
2N5165	1651
2N5166	1584
2N5167	1665
2N5168	2188
2N5169	0762
2N5170	0722
2N5171	0764
2N5172	0155
2N5174	0155
2N5175	0861
2N5176	0233
2N5179	0259
2N5180	0007
2N5181	0007
2N5182	0007
2N5183	0198
2N5184	0233
2N5185	0233
2N5186	0016
2N5187	0016
2N5188	0086
2N5189	4271
2N5190	0236
2N5191	0161
2N5192	0161
2N5193	0681
2N5194	0460
2N5195	0460
2N5196	2945
2N5197	2945
2N5198	2945
2N5199	2945
2N5200	0418
2N5201	0715
2N5202	0429
2N5204	0725
2N5205	0766
2N5206	0679
2N5207	0679
2N5208	0150
2N5209	0079
2N5210	0079
2N5211	0086
2N5213	0423
2N5218	2662
2N5219	0079
2N5220	0079
2N5221	0006
2N5222	0224
2N5223	0079
2N5224	0079
2N5225	0076
2N5226	0037
2N5226A	0791
2N5227	0037
2N5227A	0791
2N5228	0150
2N5229	0037
2N5230	0007
2N5231	0037
2N5232	0111
2N5232A	0111
2N5233	0324
2N5234	0324
2N5235	0324
2N5236	0493
2N5238	OEM
2N5239	0270
2N5240	0270
2N5241	0363
2N5242	0126
2N5243	0126
2N5245	0325
2N5245A	0325
2N5246	0325
2N5246A	0325
2N5247	0325
2N5247A	0325
2N5248	0325
2N5248A	0325
2N5249	0111
2N5249A	0111
2N5252	0168
2N5253	0168
2N5254	OEM
2N5255	OEM
2N5256	OEM
2N5257	OEM
2N5258	0325
2N5259	OEM
2N5260	OEM
2N5261	OEM
2N5262	4271
2N5264	1966
2N5265	OEM
2N5266	OEM
2N5267	0325
2N5268	0325
2N5269	0325
2N5270	0325
2N5271	OEM
2N5272	0493
2N5273	3150
2N5274	3152
2N5275	2018
2N5276	OEM
2N5277	0325
2N5278	4282
2N5278A	0325
2N5279	0168
2N5281	0439
2N5282	0439
2N5288	0939
2N5289	0939
2N5291	0439
2N5292	0493
2N5293	0042
2N5294	0423
2N5295	0042
2N5296	0236
2N5297	0423
2N5298	0042
2N5300	0086
2N5301	0130
2N5302	0130
2N5302J	OEM
2N5302TX	OEM
2N5303	0130
2N5303J	OEM
2N5303TX	OEM
2N5304	0144
2N5305	0400
2N5306	0400
2N5306A	0400
2N5307	0400
2N5308	0400
2N5308A	0400
2N5309	0155
2N5310	0155
2N5311	0155
2N5313	2662
2N5315	2662
2N5317	0939
2N5319	0939
2N5320	3329
2N5320HS	0595
2N5321	0622
2N5322	0126
2N5323	0126
2N5324	0975
2N5325	0975
2N5328	0940
2N5331	3489
2N5334	0042
2N5335	0622
2N5336	0622
2N5337	0622
2N5338	0622
2N5339	0622
2N5344	0168
2N5345	0168
2N5346	4029
2N5347	4029
2N5348	4029
2N5349	2873
2N5352	0150
2N5354	0006
2N5354A	0791
2N5355	0006
2N5355A	0791
2N5356	0037
2N5356A	0791
2N5358	0325
2N5359	0325
2N5360	0325
2N5360A	0325
2N5361	0325
2N5362	0325
2N5363	0325
2N5364	0325
2N5365	0037
2N5365A	0791
2N5366	0037
2N5366A	0791
2N5367	0037
2N5367A	0791
2N5368	0079
2N5369	0079
2N5370	0079
2N5371	0079
2N5372	0037
2N5372A	0791
2N5373	0037
2N5373A	0791
2N5374	0037
2N5374A	0791
2N5375	0037
2N5375A	0791
2N5376	1221
2N5377	0079
2N5378	0037
2N5378A	0791
2N5379	0037
2N5379A	0791
2N5380	0079
2N5381	0079
2N5382	0037
2N5382A	0791
2N5383	0037
2N5383A	0791
2N5385	2964
2N5390	OEM
2N5391	0325
2N5392	0325
2N5393	0325
2N5394	0325
2N5395	0325
2N5396	0325
2N5397	2888
2N5398	2888
2N5399	0493
2N5400	0342
2N5401	0342
2N5402	OEM
2N5410(TRW)	0105
2N5415	0439
2N5415J	OEM
2N5415TX	OEM
2N5416	0439
2N5416J	OEM
2N5416TX	OEM
2N5417	0016
2N5418	0155
2N5419	0155
2N5420	0155
2N5421	2041
2N5422	0560
2N5423	0560
2N5424	0042
2N5425	OEM
2N5427	0423
2N5428	0423
2N5429	0423
2N5430	0423
2N5431	1669
2N5432	0325
2N5433	0325
2N5434	0325
2N5435	0085
2N5438	0085
2N5441	1064
2N5442	1317
2N5443	1891
2N5444	2015
2N5445	2017
2N5446	2018
2N5447	0037
2N5447A	0791
2N5448	0133
2N5448A	0791
2N5449	0016
2N5450	0086
2N5451	0086
2N5452	2945
2N5453	2945
2N5454	2945
2N5455	0126
2N5456	0079
2N5457	2950
2N5458	2950
2N5459	3133
2N5460	2987
2N5461	2987

If replacement code is OEM, contact original manufacturer for replacement.

DEVICE TYPE	REPL CODE
2N5462	2987
2N5463	2987
2N5464	2987
2N5465	OEM
2N5466	0802
2N5467	0363
2N5470	0423
2N5471	OEM
2N5472	OEM
2N5473	OEM
2N5474	OEM
2N5475	OEM
2N5476	OEM
2N5477	2873
2N5478	2873
2N5479	2873
2N5480	2873
2N5481	0560
2N5482	0560
2N5483	0042
2N5484	0325
2N5485	1392
2N5485-1	1392
2N5486	0325
2N5487-1	0144
2N5487-3	0144
2N5489	0560
2N5490	0423
2N5491	0423
2N5492	0423
2N5493	0423
2N5494	0423
2N5495	0423
2N5496	0042
2N5497	0423
2N5505	OEM
2N5506	OEM
2N5507	OEM
2N5508	OEM
2N5509	OEM
2N5510	OEM
2N5511	OEM
2N5512	OEM
2N5513	OEM
2N5514	OEM
2N5515	2945
2N5516	2945
2N5517	2945
2N5518	2945
2N5519	2945
2N5520	2945
2N5521	2945
2N5522	2945
2N5523	2945
2N5524	2945
2N5525	0400
2N5526	0400
2N5539	3489
2N5541	0168
2N5542	2662
2N5543	3133
2N5544	3133
2N5545	2945
2N5546	2945
2N5547	2945
2N5548	OEM
2N5549	0325
2N5550	0076
2N5551	0076
2N5555	0325
2N5556	0325
2N5557	0325
2N5558	0325
2N5559	0074
2N5561	0325
2N5562	0325
2N5563	0325
2N5564	2945
2N5565	2945
2N5566	2945
2N5567	1064
2N5568	1317
2N5569	0154
2N5570	0147
2N5571	1064
2N5572	1317
2N5573	0154
2N5574	0147
2N5581	0079
2N5582	0079
2N5589	3576
2N5590	1233
2N5591	1977
2N5592	3133
2N5593	3133
2N5594	3133
2N5597	0925
2N5598	0178
2N5600	0178
2N5601	0178
2N5602	0178
2N5603	3362
2N5604	0178
2N5606	0042
2N5608	0560
2N5610	0560
2N5612	0560
2N5614	0620
2N5616	0620

DEVICE TYPE	REPL CODE
2N5618	0620
2N5620	0620
2N5621	1681
2N5622	0620
2N5623	1681
2N5624	0620
2N5625	1681
2N5626	0620
2N5628	0620
2N5629	0543
2N5630	0543
2N5631	0543
2N5632	1966
2N5633	1966
2N5634	1966
2N5637	0042
2N5638	3130
2N5639	3130
2N5640	0325
2N5641	3586
2N5642	3587
2N5643	2506
2N5644	3560
2N5645	2091
2N5646	1233
2N5647	OEM
2N5648	0325
2N5649	0325
2N5650	0259
2N5651	0259
2N5652	0259
2N5653	3130
2N5654	1978
2N5655	0276
2N5656	0276
2N5657	0168
2N5658	2662
2N5659	0939
2N5660	2096
2N5661	2096
2N5662	0168
2N5663	0168
2N5664	0429
2N5665	0429
2N5668	2950
2N5669	0325
2N5670	0325
2N5671	2485
2N5671J	OEM
2N5671TX	OEM
2N5672	2485
2N5672J	OEM
2N5672TX	OEM
2N5679	0892
2N5680	0892
2N5681	0086
2N5682	0644
2N5683	4344
2N5684	4344
2N5685	4345
2N5686	4345
2N5687	2041
2N5688	0560
2N5689	3568
2N5690	3569
2N5692	0085
2N5693	0085
2N5697	0689
2N5698	3560
2N5699	0560
2N5700	0042
2N5701	0042
2N5702	0493
2N5703	0560
2N5704	0042
2N5705	0042
2N5710	0698
2N5711	3586
2N5712	3587
2N5713	3587
2N5715	0622
2N5716	0325
2N5717	0325
2N5718	2950
2N5719	OEM
2N5720	OEM
2N5721	OEM
2N5722	OEM
2N5723	OEM
2N5724	OEM
2N5725	OEM
2N5726	OEM
2N5727	OEM
2N5728	OEM
2N5730	0103
2N5731	0939
2N5733	3489
2N5735	0079
2N5736	0079
2N5737	0491
2N5738	2013
2N5741	1681
2N5742	1681
2N5745	1681
2N5754	2383
2N5755	2394
2N5756	1413
2N5757	2387
2N5758	0074

DEVICE TYPE	REPL CODE
2N5759	1966
2N5760	1966
2N5763	0150
2N5764	0560
2N5765	0042
2N5766	0560
2N5767	0560
2N5768	0042
2N5769	0079
2N5770	0144
2N5771	0342
2N5772	0079
2N5777	3185
2N5778	3185
2N5779	3185
2N5780	3185
2N5781	4365
2N5782	1267
2N5783	1267
2N5784	0644
2N5785	0644
2N5786	0630
2N5787	1137
2N5788	0344
2N5789	0901
2N5790	0058
2N5791	OEM
2N5792	OEM
2N5793	0861
2N5794	2045
2N5795	0037
2N5795JTX	OEM
2N5796	2468
2N5797	OEM
2N5798	OEM
2N5799	OEM
2N5800	OEM
2N5801	OEM
2N5802	OEM
2N5803	OEM
2N5804	0363
2N5805	0363
2N5806	0469
2N5806J	OEM
2N5807	0722
2N5807J	OEM
2N5808	0778
2N5808J	OEM
2N5809	0725
2N5809J	OEM
2N5810	0079
2N5811	0037
2N5812	0079
2N5813	0532
2N5814	0086
2N5815	0532
2N5816	0086
2N5817	0532
2N5818	0086
2N5819	0532
2N5820	0076
2N5820HS	0595
2N5821	0126
2N5821HS	0791
2N5822	0086
2N5823	0791
2N5824	0076
2N5825	0076
2N5826	0079
2N5827	0079
2N5827A	0079
2N5828	0079
2N5828A	0079
2N5829	3607
2N5830	0076
2N5830A	0224
2N5831	0715
2N5832	0715
2N5833	0715
2N5834	0126
2N5835	0007
2N5836	0630
2N5837	0418
2N5838	0270
2N5839	0270
2N5840	0220
2N5841	0861
2N5842	0861
2N5843	OEM
2N5844	OEM
2N5845	0079
2N5845A	0079
2N5846	0622
2N5847	3568
2N5848	3569
2N5849	3570
2N5851	0007
2N5852	0007
2N5855	0791
2N5856	0086
2N5857	0037
2N5857TX	OEM
2N5858	0076
2N5859	0086
2N5861	0622
2N5864	0892
2N5865	0126
2N5867	0491
2N5868	2013

DEVICE TYPE	REPL CODE
2N5869	0363
2N5870	0363
2N5871	2013
2N5872	2013
2N5873	0103
2N5874	0177
2N5875	0491
2N5876	1598
2N5877	0103
2N5878	0543
2N5879	1598
2N5880	1598
2N5881	0620
2N5882	0620
2N5883	1681
2N5884	1681
2N5885	0130
2N5886	0130
2N5887	0222
2N5888	0222
2N5889	0222
2N5890	0222
2N5891	0222
2N5892	0222
2N5893	0222
2N5894	0222
2N5895	0222
2N5896	0222
2N5897	0222
2N5898	0222
2N5899	0222
2N5900	0222
2N5901	0222
2N5902	OEM
2N5903	OEM
2N5904	OEM
2N5905	OEM
2N5906	OEM
2N5907	OEM
2N5908	OEM
2N5909	OEM
2N5910	0037
2N5911	OEM
2N5912	OEM
2N5913	3421
2N5914	3560
2N5915	1420
2N5916	OEM
2N5917	1974
2N5920	2700
2N5922	2170
2N5923	2170
2N5924	1198
2N5925	1198
2N5930	3554
2N5933	3554
2N5941	2545
2N5942	2545
2N5943	2208
2N5944	3560
2N5945	2091
2N5946	1420
2N5947	2070
2N5949	0325
2N5950	0325
2N5951	0325
2N5952	0325
2N5953	0325
2N5954	1199
2N5955	1199
2N5956	1199
2N5961	0079
2N5962	0079
2N5963	0079
2N5964	0086
2N5965	0233
2N5970	0543
2N5971	0543
2N5972	0543
2N5973	0543
2N5974	1199
2N5975	1199
2N5976	1199
2N5977	0561
2N5978	0561
2N5979	0561
2N5980	1199
2N5981	1199
2N5982	1199
2N5983	0561
2N5984	0561
2N5985	0561
2N5986	1199
2N5987	1199
2N5988	1199
2N5989	0561
2N5991	0561
2N5992	1198
2N5993	1974
2N5994	2506
2N5995	1974
2N5996	1233
2N5998	0155
2N5999	0006
2N6000	0079
2N6001	0532
2N6002	0079
2N6003	0037
2N6004	0016

DEVICE TYPE	REPL CODE
2N6005	0037
2N6006	0079
2N6007	0532
2N6008	0155
2N6009	0532
2N6010	0086
2N6011	0532
2N6012	0018
2N6013	0037
2N6014	0086
2N6015	0791
2N6016	0018
2N6017	0037
2N6021	0925
2N6022	0681
2N6023	0681
2N6024	0681
2N6025	0681
2N6026	0681
2N6027	0316
2N6028	0316
2N6029	1598
2N6030	1598
2N6031	1598
2N6032	2433
2N6032J	OEM
2N6032TX	OEM
2N6033	2433
2N6033J	OEM
2N6033TX	OEM
2N6034	2897
2N6035	2897
2N6036	2897
2N6037	0558
2N6038	0558
2N6039	0558
2N6040	3530
2N6041	0602
2N6042	0602
2N6043	3529
2N6044	1212
2N6045	1212
2N6049	0905
2N6050	2432
2N6051	2432
2N6052	2432
2N6053	2277
2N6054	2277
2N6055	2427
2N6056	2427
2N6057	2439
2N6058	2439
2N6059	2439
2N6067	0037
2N6068	4412
2N6068A	0593
2N6068B	0593
2N6069	1755
2N6069A	0593
2N6069B	0593
2N6070	4414
2N6070A	0593
2N6070B	0593
2N6071	3499
2N6071A	0593
2N6071B	0593
2N6072	4121
2N6072A	OEM
2N6072B	OEM
2N6073	0485
2N6073A	0485
2N6073B	OEM
2N6074	4416
2N6074A	OEM
2N6074B	OEM
2N6075	4417
2N6075A	OEM
2N6075B	OEM
2N6076	0037
2N6077	0429
2N6078	0429
2N6079	0429
2N6080	3578
2N6081	1233
2N6082	1977
2N6083	1977
2N6084	1974
2N6085	OEM
2N6086	OEM
2N6087	OEM
2N6088	OEM
2N6089	OEM
2N6090	OEM
2N6091	OEM
2N6092	OEM
2N6093	2545
2N6094	3580
2N6095	3582
2N6096	3583
2N6097	3584
2N6098	0060
2N6099	0482
2N6100	0482
2N6101	0482
2N6102	0482
2N6103	0482
2N6105	1974
2N6106	0854
2N6107	0854

DEVICE TYPE	REPL CODE
2N6108	0854
2N6109	0925
2N6110	0854
2N6111	0854
2N6112	0111
2N6114	OEM
2N6115	OEM
2N6116	0316
2N6117	OEM
2N6118	OEM
2N6119	OEM
2N6120	OEM
2N6121	0236
2N6122	0060
2N6123	0060
2N6124	0681
2N6125	4430
2N6126	0681
2N6128	0939
2N6129	0144
2N6130	0144
2N6131	0144
2N6132	0854
2N6133	0854
2N6134	0854
2N6135	2070
2N6137	OEM
2N6138	OEM
2N6139	0422
2N6140	0422
2N6141	0422
2N6142	OEM
2N6143	OEM
2N6144	OEM
2N6145	3199
2N6146	0147
2N6147	3222
2N6148	OEM
2N6149	OEM
2N6150	OEM
2N6151	4440
2N6152	0960
2N6153	0966
2N6154	0411
2N6155	0415
2N6156	1100
2N6157	2015
2N6158	2017
2N6159	1891
2N6160	2015
2N6161	2017
2N6162	2018
2N6163	3199
2N6164	3207
2N6165	3222
2N6167	1848
2N6168	1848
2N6169	1855
2N6170	3215
2N6171	1848
2N6172	1848
2N6173	1855
2N6174	3215
2N6175	1708
2N6176	1708
2N6177	1708
2N6178	0060
2N6179	0161
2N6180	1308
2N6181	0460
2N6186	1199
2N6187	1199
2N6188	0482
2N6189	0482
2N6190	0892
2N6191	0892
2N6198	3586
2N6199	3587
2N6200	2506
2N6207	1974
2N6208	1974
2N6211	0074
2N6211JK	OEM
2N6212	0074
2N6212J	OEM
2N6213	0074
2N6213J	OEM
2N6214	2741
2N6215	3489
2N6216	0074
2N6218	0144
2N6219	0715
2N6220	0144
2N6221	0144
2N6222	0144
2N6223	0037
2N6224	0037
2N6225	0861
2N6226	1598
2N6227	1598
2N6228	1598
2N6229	1598
2N6230	1598
2N6231	1598
2N6232	0144
2N6232-4	0144
2N6233	0178
2N6234	0429
2N6235	0429

DEVICE TYPE	REPL CODE
2N6236	4458
2N6237	0174
2N6238	3854
2N6239	3620
2N6240	3324
2N6241	1502
2N6246	0491
2N6247	0491
2N6248	1681
2N6249	1991
2N6250	1991
2N6251	1991
2N6253	0103
2N6254	0103
2N6255	4464
2N6256	1212
2N6257	0130
2N6258	2433
2N6259	2433
2N6260	0178
2N6261	0429
2N6262	0543
2N6263	0178
2N6264	0178
2N6271	3554
2N6274	2433
2N6275	2433
2N6276	2433
2N6277	2433
2N6278	3489
2N6279	3489
2N6280	3489
2N6281	3489
2N6282	3525
2N6283	3525
2N6283J	OEM
2N6283TX	OEM
2N6284	3525
2N6284J	OEM
2N6284TX	OEM
2N6285	2448
2N6286	3526
2N6287	3526
2N6288	0042
2N6289	0042
2N6290	0042
2N6291	0423
2N6292	0423
2N6293	0423
2N6294	3370
2N6295	3370
2N6296	3540
2N6297	3540
2N6298	0602
2N6299	0602
2N6300	1212
2N6301	1212
2N6302	0543
2N6303	0126
2N6304	0259
2N6305	0259
2N6306	0363
2N6306J	OEM
2N6306TX	OEM
2N6307	0363
2N6307M	0074
2N6308	0363
2N6308J	OEM
2N6308M	0074
2N6308TX	OEM
2N6312	0681
2N6313	0681
2N6314	0681
2N6315	0423
2N6316	0423
2N6317	0854
2N6318	0854
2N6322	0543
2N6322MP	0543
2N6326	0130
2N6327	0130
2N6328	0130
2N6329	1681
2N6330	1681
2N6331	1681
2N6332	OEM
2N6333	OEM
2N6334	OEM
2N6335	OEM
2N6336	OEM
2N6337	OEM
2N6338	2485
2N6339	2485
2N6340	2485
2N6341	2485
2N6342	0772
2N6342A	0772
2N6343	0744
2N6343A	0744
2N6344	0617
2N6344A	0617
2N6345	0875
2N6345A	0875
2N6346	0772
2N6346A	0772
2N6347	0744
2N6347A	0744
2N6348	0617
2N6348A	0617

If replacement code is OEM, contact original manufacturer for replacement.

DEVICE TYPE	REPL CODE
2N6349	0875
2N6349A	0875
2N6354	0620
2N6355	3525
2N6356	3525
2N6357	2439
2N6358	3525
2N6359	0543
2N6360	0543
2N6365	0050
2N6365A	0050
2N6366	1591
2N6367	0144
2N6368	2946
2N6370	0144
2N6371	0103
2N6372	0144
2N6373	0429
2N6374	0429
2N6375	0698
2N6383	3373
2N6383J	OEM
2N6383TX	OEM
2N6384	3373
2N6384J	OEM
2N6384TX	OEM
2N6385	3373
2N6385J	OEM
2N6385TX	OEM
2N6386	2235
2N6387	2235
2N6388	2235
2N6389	0259
2N6394	2521
2N6395	4490
2N6396	4491
2N6397	4492
2N6398	4493
2N6399	1512
2N6400	4494
2N6401	4491
2N6402	4491
2N6403	4492
2N6404	4493
2N6405	1512
2N6413	0161
2N6420	0792
2N6421	0792
2N6422	0792
2N6423	0792
2N6424	0142
2N6425	0142
2N6426	3799
2N6427	3799
2N6428	0079
2N6428A	0079
2N6429	0079
2N6429A	0079
2N6430	0187
2N6431	0187
2N6432	0342
2N6433	0342
2N6436	1681
2N6437	1681
2N6438	2485
2N6441	OEM
2N6442	OEM
2N6443	OEM
2N6444	OEM
2N6445	OEM
2N6446	OEM
2N6447	OEM
2N6448	OEM
2N6449	OEM
2N6450	OEM
2N6451	OEM
2N6452	OEM
2N6453	OEM
2N6454	OEM
2N6455	2312
2N6456	2312
2N6457	2312
2N6458	0620
2N6459	3565
2N6465	0178
2N6466	0934
2N6467	0681
2N6468	0681
2N6469	0491
2N6470	0103
2N6471	0103
2N6472	0130
2N6473	0236
2N6474	0236
2N6475	0681
2N6476	0681
2N6477	0728
2N6478	0728
2N6483	OEM
2N6484	OEM
2N6485	OEM
2N6486	0482
2N6487	0482
2N6488	1369
2N6489	1369
2N6490	1369
2N6491	1369
2N6492	2427
2N6493	2427
2N6494	2427
2N6496	0620
2N6497	0728
2N6498	0728
2N6499	3014
2N6500	0178
2N6504	2521
2N6505	0397
2N6506	0397
2N6507	0611
2N6508	0952
2N6509	1512
2N6510	0363
2N6511	0363
2N6512	0363
2N6513	0363
2N6514	0363
2N6515	0715
2N6516	0715
2N6518	0342
2N6519	0342
2N6526	OEM
2N6530	2235
2N6531	1212
2N6532	1212
2N6534	1212
2N6535	1212
2N6536	1212
2N6538	0079
2N6539	0079
2N6540	0861
2N6541	0861
2N6542	0363
2N6543	0363
2N6544	0363
2N6545	3267
2N6546	1341
2N6546J	OEM
2N6546TX	OEM
2N6547	3038
2N6548	3535
2N6549	3535
2N6550	OEM
2N6551	0560
2N6552	0824
2N6553	0824
2N6554	3327
2N6555	3327
2N6556	3327
2N6557	0285
2N6558	0643
2N6558L	0285
2N6560	1852
2N6561	1966
2N6564	OEM
2N6565	OEM
2N6569	0103
2N6573	0270
2N6574	1852
2N6575	1852
2N6576	1394
2N6577	1394
2N6578	2428
2N6579	OEM
2N6580	OEM
2N6581	OEM
2N6582	1966
2N6583	1966
2N6584	OEM
2N6585	OEM
2N6586	OEM
2N6587	OEM
2N6588	OEM
2N6589	OEM
2N6590	OEM
2N6591	0285
2N6592	0285
2N6593	0285
2N6594	0491
2N6601	0007
2N6605	OEM
2N6606	OEM
2N6607	OEM
2N6608	OEM
2N6609	4534
2N6617	OEM
2N6618	OEM
2N6619	OEM
2N6620	OEM
2N6621	OEM
2N6623	OEM
2N6624	OEM
2N6625	OEM
2N6626	OEM
2N6627	OEM
2N6628	OEM
2N6629	OEM
2N6630	OEM
2N6631	OEM
2N6632	OEM
2N6633	OEM
2N6634	OEM
2N6635	OEM
2N6636	OEM
2N6637	OEM
2N6638	OEM
2N6639	OEM
2N6640	OEM
2N6641	OEM
2N6642	OEM
2N6643	OEM
2N6644	OEM
2N6645	OEM
2N6646	OEM
2N6648	3374
2N6648J	OEM
2N6648TX	OEM
2N6649	3374
2N6649J	OEM
2N6650	3374
2N6650J	OEM
2N6650TX	OEM
2N6653	1852
2N6654	1852
2N6655	1852
2N6656	OEM
2N6657	OEM
2N6658	OEM
2N6659	OEM
2N6660	OEM
2N6661	OEM
2N6665	OEM
2N6666	2237
2N6667	2237
2N6668	2237
2N6669	4539
2N6671	1966
2N6672	1966
2N6673	3388
2N6674	1341
2N6675	1341
2N6676	4540
2N6677	4540
2N6678	4540
2N6679	OEM
2N6680	OEM
2N6681	OEM
2N6682	OEM
2N6683	OEM
2N6684	OEM
2N6685	OEM
2N6686	OEM
2N6687	OEM
2N6688	OEM
2N6689	OEM
2N6690	OEM
2N6691	0482
2N6692	OEM
2N6693	OEM
2N6701	OEM
2N6702	1165
2N6703	1165
2N6704	OEM
2N6705	3938
2N6706	3938
2N6707	3938
2N6708	4548
2N6709	4548
2N6710	4548
2N6711	0261
2N6712	0261
2N6713	0261
2N6714	1526
2N6715	1526
2N6716	1526
2N6717	0086
2N6718	0086
2N6719	0261
2N6720	3764
2N6721	3764
2N6722	3764
2N6723	3764
2N6724	3799
2N6725	3799
2N6726	1597
2N6727	1597
2N6728	1597
2N6729	1597
2N6730	1597
2N6731	1597
2N6732	1597
2N6733	3754
2N6734	3754
2N6735	3754
2N6736	OEM
2N6737	1526
2N6738	0728
2N6740H	0728
2N6751	4567
2N6752	4567
2N6753	3060
2N6754	2993
2N6755	OEM
2N6756	OEM
2N6757	OEM
2N6758	OEM
2N6759	OEM
2N6760	OEM
2N6761	OEM
2N6762	OEM
2N6763	4572
2N6764	OEM
2N6764H	5363
2N6765	OEM
2N6766	OEM
2N6767	OEM
2N6768	OEM
2N6769	OEM
2N6770	OEM
2N6781	OEM
2N6782	OEM
2N6783	OEM
2N6784	OEM
2N6785	OEM
2N6786	OEM
2N6787	OEM
2N6788	OEM
2N6789	OEM
2N6790	OEM
2N6791	OEM
2N6792	OEM
2N6793	OEM
2N6794	OEM
2N6795	OEM
2N6796	OEM
2N6797	OEM
2N6798	OEM
2N6799	OEM
2N6800	OEM
2N6801	OEM
2N6802	OEM
2N6826	OEM
2N6832	OEM
2N6833	4590
2N6834	4591
2N6835	OEM
2N6836	3388
2N6837	OEM
2N6838	OEM
2N6839	OEM
2N7000	OEM
2N7002	OEM
2N7008	OEM
2N19616	0050
2N29260RN	0079
2N54433	0050
2N67425	OEM
2N91446	0079
2NFP4339	OEM
2NFP4340	OEM
2NJ5A	0280
2NJ5D	0211
2NJ6	0280
2NJ8A	0280
2NJ9A	0211
2NJ9D	0211
2NJ50	0136
2NJ51	0136
2NJ52	0136
2NJ53	0136
2NJ59D	0136
2NJ224	OEM
2NJ233B	0325
2NJ584	1027
2NJ770A	OEM
2NJ771A	OEM
2NL48	0595
2NS31	0211
2NS32	0211
2NS121	0211
2NSM-1	0071
2NU/9	0847
2NU/9WHT	0847
2NU/9YEL	0847
2OA72	OEM
2OA79	OEM
2OA90	0123
2OA90A-M	0123
2OA90M	0123
2OA99	0143
2OA99A	0143
2OC6	OEM
2OC26	OEM
2OC28	OEM
2OC29	OEM
2OC30	OEM
2OC35	OEM
2OC36	OEM
2OC72	0164
2OC74	OEM
2OC83	OEM
2OC84	OEM
2P05M	1259
2P1M	2270
2P2M	0940
2P4M	0095
2P5M	1243
2P6M	1243
2P389	OEM
2P389A	OEM
2P424	OEM
2P424A	OEM
2PB187	0004
2PD56	OEM
2PD57	OEM
2PD58	OEM
2PD-358	0143
2PL.75-1.5	OEM
2PL.75-1.5S	OEM
2PS428	0079
2PT6	OEM
2P.5-1	OEM
2P.5-1S	OEM
2QT2222	2905
2QT2905	0282
2R0/18	OEM
2R1/18	OEM
2R1/52	OEM
2R2/18	OEM
2R2B	OEM
2R2JA	OEM
2R2S	OEM
2R3B	OEM
2R3S	OEM
2R4C	OEM
2R4S	OEM
2R5C	OEM
2R5S	OEM
2R6C	OEM
2R6S	OEM
2R7CD	OEM
2R7S	OEM
2R8CD	OEM
2R8S	OEM
2R9.1	0684
2R9.1A	0684
2R9.1B	0684
2R10	0225
2R10A	0225
2R10B	0225
2R11	0230
2R11A	0230
2R11B	0230
2R12	0234
2R12A	0234
2R12B	0234
2R13	0237
2R13A	0237
2R13B	0237
2R15	0247
2R15A	0247
2R15B	0247
2R16	0251
2R16A	0251
2R16B	0251
2R18	0256
2R18A	0256
2R18B	0256
2R20	0262
2R20A	0262
2R20B	0262
2R22	0269
2R22A	0269
2R22B	0269
2R24	0274
2R24A	0274
2R24B	0274
2R27	0294
2R27A	0294
2R27B	0294
2R30	0309
2R30A	0309
2R30B	0309
2R33	0318
2R33A	0318
2R33B	0318
2R36	0320
2R36A	0320
2R36B	0320
2R39	0326
2R39A	0326
2R39B	0326
2R43	0337
2R43A	0337
2R43B	0337
2R47	0347
2R47A	0347
2R47B	0347
2R51	0027
2R51A	0027
2R51B	0027
2R52	OEM
2R56	0266
2R56A	0266
2R56B	0266
2R68	0405
2R68A	0405
2R68B	0405
2R75	0426
2R75A	0426
2R75B	0426
2R91	0238
2R91A	0238
2R91B	0238
2R100	1180
2R100A	1180
2R100B	1180
2R110	1190
2R110A	1190
2R110B	1190
2R120	1207
2R120A	1207
2R120B	1207
2R130	1218
2R130A	1218
2R130B	1218
2R140	1881
2R140A	1881
2R140B	1881
2R150	0647
2R150A	0647
2R150B	0647
2R160	1255
2R160A	1255
2R160B	1255
2R180	1279
2R180A	1279
2R180B	1279
2R200	0605
2R200A	0605
2R200B	0605
2R430M	OEM
2RD402B1X16	OEM
2RD402B1X32	OEM
2RD402B1X36	OEM
2RD402B1X72	OEM
2RE52	OEM
2S001	0016
2S002	0016
2S003	0016
2S004	0016
2S005	0016
2S006	0144
2S012	0042
2S013	OEM
2S014	0086
2S017	0086
2S018	0086
2S019	0086
2S020	0086
2S021	0126
2S022	0126
2S023	0126
2S033	0103
2S034	0103
2S035	0074
2S036	0074
2S043	OEM
2S044	OEM
2S045	OEM
2S046	OEM
2S095A	0079
2S0781	0563
2S2-4	OEM
2S2-4S	OEM
2S2M	OEM
2S3M	OEM
2S4M	OEM
2S12	0280
2S13	0280
2S14	0211
2S15	0211
2S15A	0211
2S15X90	0037
2S22	0211
2S24	0211
2S25	0280
2S26	0160
2S26A	0160
2S30	0280
2S31	0280
2S32	0211
2S33	0211
2S34	0211
2S35	0136
2S36	0136
2S37	0211
2S38	0211
2S39	0211
2S40	0211
2S41	0085
2S41A	0085
2S42	0085
2S43	0211
2S44	0211
2S45	0280
2S46	0211
2S47	0211
2S49	0280
2S51	0280
2S52	0280
2S53	0280
2S54	0211
2S56	0211
2S57	0050
2S58	0136
2S60	0280
2S82	0127
2S91	0280
2S92	0280
2S92A	0280
2S93	0280
2S93A	0280
2S95A	0016
2S96	0136
2S97	0136
2S98	0136
2S101	0016
2S102	0016
2S103	0016
2S104	0016
2S109	0136
2S110	0136
2S111	0280
2S112	0136
2S128Y	0208
2S131	0016
2S134	0847
2S141	0136
2S142	0136
2S143	0136
2S144	0050
2S145	0136
2S146	0136
2S148	0050
2S155	[illegible]
2S159	[illegible]
2S160	[illegible]
2S163	0
2S167	0
2S174	0
2S174R	0
2S175	00
2S176	01
2S178	02
2S179	02
2S189	00
2S201	00
2S257	12
2S273	00
2S277C	02
2S301	012
2S302	012
2S302A	012
2S303	012
2S304	012
2S305	003
2S306	003
2S307	003
2S307A	0791
2S321	0037
2S321A	0791
2S322	0037
2S322A	0037
2S322AB	0791
2S323	0037
2S323A	0791
2S324	0037
2S324A	0791
2S325	0861
2S326	0037
2S326A	0791
2S327	0037
2S327A	0791
2S351B	0050
2S363	0079
2S375B	0211
2S471-1	0050
2S494-YE	0791
2S501	0016
2S502	0016
2S503	0016
2S512	0016
2S564R	0150
2S644(S)	0086
2S645	0079
2S701	0198
2S702	0198
2S703	0198
2S711	0016
2S712	0016
2S720	OEM
2S731	0016
2S732	0016
2S733	0016
2S741	0198
2S741A	0016
2S742	0086
2S742A	0086
2S743	0233
2S743A	0233
2S744	0198
2S744A	0016
2S745	0086
2S745A	0086
2S746	0233
2S746A	0233
2S858	0342
2S1182D	0079
2S1505	OEM
2S1685QR	OEM
2S1760	0211
2S3010	0126
2S3020	0037
2S3020A	0791
2S3021	0037
2S3021A	0791
2S3030	0037
2S3030A	0791
2S3040	0037
2S3040A	0791
2S3187	0280
2S3210	0037
2S3210A	0791
2S3220	0037
2S3220A	0791
2S3221	0037
2S3221A	0791
2S3230	0037
2S3230A	0791
2S3240	0037
2S3240A	0791
2S3324	0280
2S3370	0136
2S6856	0285
2S20141	0079
2SA007H	0164
2SA076F	0050
2SA081C	0164
2SA098R	0050
2SA12	0004
2SA12(V)	0004
2SA12A	0004
2SA12B	0004

If replacement code is OEM, contact original manufacturer for replacement.

DEVICE TYPE	REPL CODE	DEVICE TYPE	REPL CODE	DEVICE TYPE	REPL CODE	DEVICE TYPE	REPL CODE	DEVICE TYPE	REPL CODE	DEVICE TYPE	REPL CODE	DEVICE TYPE	REPL CODE
	0004	2SA20A	0164	2SA32X	0050	2SA44G	0136	2SA55R	0136	2SA65X	0050	2SA73X	0050
	0004	2SA20B	0164	2SA32Y	0050	2SA44L	0136	2SA55X	0136	2SA65Y	0050	2SA73Y	0050
	0004	2SA20C	0164	2SA33	0004	2SA44M	0136	2SA55Y	0136	2SA66	0050	2SA74	0136
	0004	2SA20D	0164	2SA35	0004	2SA44OR	0136	2SA56	0136	2SA66A	0050	2SA74A	0136
	0050	2SA20E	0164	2SA35A	0004	2SA44R	0136	2SA56A	0136	2SA66B	0050	2SA74B	0136
	0050	2SA20F	0164	2SA35B	0004	2SA44X	0136	2SA57	0136	2SA66C	0050	2SA74C	0136
	0050	2SA20G	0164	2SA35C	0004	2SA44Y	0136	2SA57A	0136	2SA66D	0050	2SA74D	0136
	0050	2SA20L	0164	2SA35D	0004	2SA45	0136	2SA57B	0136	2SA66E	0050	2SA74E	0136
	0050	2SA20M	0164	2SA35E	0004	2SA45-1	0136	2SA57C	0136	2SA66F	0050	2SA74F	0136
	0050	2SA20OR	0164	2SA35F	0004	2SA45-2	0136	2SA57D	0136	2SA66G	0050	2SA74G	0136
	0050	2SA20R	0164	2SA35G	0004	2SA45-3	0136	2SA57E	0136	2SA66K	0050	2SA74H	0136
	0050	2SA20X	0164	2SA35L	0004	2SA46	0280	2SA57F	0136	2SA66L	0050	2SA74K	0136
	0050	2SA20Y	0164	2SA35M	0004	2SA47	0050	2SA57G	0136	2SA66M	0050	2SA74L	0136
	0050	2SA21	0164	2SA35OR	0004	2SA48	0050	2SA57L	0136	2SA66OR	0050	2SA74M	0136
	0050	2SA21A	0164	2SA35X	0004	2SA49	0633	2SA57M	0136	2SA66R	0050	2SA74OR	0136
	0050	2SA21B	0164	2SA35Y	0004	2SA49A	0633	2SA57OR	0136	2SA66X	0050	2SA74R	0136
	0050	2SA21C	0164	2SA36	0050	2SA49B	0633	2SA57R	0136	2SA66Y	0050	2SA74X	0136
	0050	2SA21D	0164	2SA36A	0050	2SA49C	0633	2SA57X	0136	2SA67	0004	2SA74Y	0136
	0050	2SA21E	0164	2SA36B	0050	2SA49D	0633	2SA57Y	0136	2SA67A	0004	2SA75	0136
	0050	2SA21F	0164	2SA36C	0050	2SA49E	0633	2SA58	0050	2SA67B	0004	2SA75A	0136
	0050	2SA21G	0164	2SA36D	0050	2SA49F	0633	2SA58A	0050	2SA67C	0004	2SA75B	0136
	0050	2SA21L	0164	2SA36E	0050	2SA49G	0633	2SA58B	0050	2SA67D	0004	2SA75C	0136
	0050	2SA21M	0164	2SA36F	0050	2SA49L	0633	2SA58C	0050	2SA67E	0004	2SA75D	0136
	0050	2SA21OR	0164	2SA36G	0050	2SA49M	0633	2SA58D	0050	2SA67F	0004	2SA75E	0136
	0050	2SA21R	0164	2SA36L	0050	2SA49OR	0633	2SA58E	0050	2SA67G	0004	2SA75F	0136
R	0050	2SA21X	0164	2SA36M	0050	2SA49R	0633	2SA58F	0050	2SA67H	0004	2SA75G	0136
	0050	2SA21Y	0164	2SA36OR	0050	2SA49X	0633	2SA58G	0050	2SA67K	0004	2SA75H	0136
X	0050	2SA22	0050	2SA36R	0050	2SA49Y	0633	2SA58H	0050	2SA67L	0004	2SA75K	0136
Y	0050	2SA23	0050	2SA36X	0050	2SA50	0050	2SA58J	0050	2SA67OR	0004	2SA75L	0136
5	0004	2SA24	0050	2SA36Y	0050	2SA50A	0050	2SA58K	0050	2SA67R	0004	2SA75M	0136
/5-6	0004	2SA25	0050	2SA37	0050	2SA50B	0050	2SA58L	0050	2SA67X	0004	2SA75OR	0136
15A	0004	2SA26	0050	2SA37A	0050	2SA50C	0050	2SA58M	0050	2SA67Y	0004	2SA75R	0136
15B	0004	2SA26A	0050	2SA37B	0050	2SA50D	0050	2SA58OR	0050	2SA68	0050	2SA75X	0136
A15BK	0004	2SA26B	0050	2SA37C	0050	2SA50E	0050	2SA58R	0050	2SA69	0136	2SA75Y	0136
SA15C	0004	2SA26C	0050	2SA37D	0050	2SA50F	0050	2SA58X	0050	2SA69A	0136	2SA76	0050
SA15D	0004	2SA26D	0050	2SA37E	0050	2SA50G	0050	2SA58Y	0050	2SA69B	0136	2SA76A	0050
SA15E	0004	2SA26E	0050	2SA37F	0050	2SA50L	0050	2SA59	0050	2SA69C	0136	2SA76B	0050
2SA15F	0004	2SA26F	0050	2SA37G	0050	2SA50M	0050	2SA59A	0050	2SA69D	0136	2SA76C	0050
2SA15G	0004	2SA26G	0050	2SA37L	0050	2SA50OR	0050	2SA59B	0050	2SA69E	0136	2SA76D	0050
2SA15H	0004	2SA26L	0050	2SA37M	0050	2SA50R	0050	2SA59C	0050	2SA69F	0136	2SA76E	0050
2SA15K	0004	2SA26M	0050	2SA37OR	0050	2SA50X	0050	2SA59D	0050	2SA69G	0136	2SA76F	0050
2SA15L	0004	2SA26OR	0050	2SA37R	0050	2SA50Y	0050	2SA59E	0050	2SA69H	0136	2SA76G	0050
2SA15M	0004	2SA26R	0050	2SA37X	0050	2SA51	0136	2SA59F	0050	2SA69K	0136	2SA76H	0050
2SA15OR	0004	2SA26X	0050	2SA37Y	0050	2SA51A	0136	2SA59G	0050	2SA69L	0136	2SA76K	0050
2SA15R	0004	2SA26Y	0050	2SA38	0050	2SA51B	0136	2SA59L	0050	2SA69M	0136	2SA76L	0050
2SA15RD	0004	2SA27	0050	2SA38A	0050	2SA51C	0136	2SA59M	0050	2SA69OR	0136	2SA76OR	0050
2SA15U	0004	2SA28	0050	2SA38B	0050	2SA51D	0136	2SA59OR	0050	2SA69R	0136	2SA76R	0050
2SA15V	0004	2SA28A	0050	2SA38C	0050	2SA51E	0136	2SA59R	0050	2SA69X	0136	2SA76X	0050
2SA15V/R	0004	2SA28B	0050	2SA38D	0050	2SA51F	0136	2SA59X	0050	2SA69Y	0136	2SA76Y	0050
2SA15VR	0004	2SA28C	0050	2SA38E	0050	2SA51G	0136	2SA59Y	0050	2SA70	0050	2SA77	0633
2SA15X	0004	2SA28D	0050	2SA38F	0050	2SA51L	0136	2SA60	0050	2SA70-08	0050	2SA77A	0633
2SA15Y	0004	2SA28E	0050	2SA38G	0050	2SA51M	0136	2SA60A	0050	2SA70-0B	0050	2SA77B	0633
2SA16	0004	2SA28F	0050	2SA38L	0050	2SA51OR	0136	2SA60B	0050	2SA70-0B	0050	2SA77C	0633
2SA16A	0004	2SA28G	0050	2SA38M	0050	2SA51R	0136	2SA60C	0050	2SA70A	0050	2SA77D	0633
2SA16B	0004	2SA28L	0050	2SA38OR	0050	2SA51X	0136	2SA60D	0050	2SA70B	0050	2SA77E	0633
2SA16C	0004	2SA28M	0050	2SA38R	0050	2SA51Y	0136	2SA60E	0050	2SA70C	0050	2SA77F	0633
2SA16D	0004	2SA28OR	0050	2SA38X	0050	2SA52	0136	2SA60F	0050	2SA70D	0050	2SA77G	0633
2SA16E	0004	2SA28X	0050	2SA38Y	0050	2SA52A	0136	2SA60G	0050	2SA70E	0050	2SA77H	0633
2SA16F	0004	2SA28Y	0050	2SA39	0050	2SA52B	0136	2SA60H	0050	2SA70F	0050	2SA77K	0633
2SA16L	0004	2SA29	0164	2SA39A	0050	2SA52C	0136	2SA60K	0050	2SA70G	0050	2SA77L	0633
2SA16M	0004	2SA29A	0164	2SA39B	0050	2SA52D	0136	2SA60L	0050	2SA70H	0050	2SA77M	0633
2SA16OR	0004	2SA29B	0164	2SA39C	0050	2SA52E	0136	2SA60M	0050	2SA70K	0050	2SA77OR	0633
2SA16R	0004	2SA29C	0164	2SA39D	0050	2SA52F	0136	2SA60OR	0050	2SA70L	0050	2SA77R	0633
2SA16X	0004	2SA29D	0164	2SA39E	0050	2SA52G	0136	2SA60R	0050	2SA70MA	0050	2SA77X	0633
2SA16Y	0004	2SA29E	0164	2SA39F	0050	2SA52L	0136	2SA60X	0050	2SA70OA	0050	2SA77Y	0633
2SA17	0050	2SA29F	0164	2SA39G	0050	2SA52M	0136	2SA60Y	0050	2SA70OR	0050	2SA78	0004
2SA17A	0050	2SA29G	0164	2SA39L	0050	2SA52OR	0136	2SA61	0050	2SA70R	0050	2SA79	0004
2SA17B	0050	2SA29L	0164	2SA39M	0050	2SA52R	0136	2SA61A	0050	2SA70X	0050	2SA80	0164
2SA17C	0050	2SA29M	0164	2SA39OR	0050	2SA52X	0136	2SA61C	0050	2SA70Y	0050	2SA80A	0164
2SA17D	0050	2SA29R	0164	2SA39R	0050	2SA52Y	0136	2SA61D	0050	2SA71	0050	2SA80B	0164
2SA17E	0050	2SA29X	0164	2SA39X	0050	2SA53	0633	2SA61E	0050	2SA71A	0050	2SA80D	0164
2SA17F	0050	2SA29Y	0164	2SA39Y	0050	2SA53A	0633	2SA61G	0050	2SA71AB	0050	2SA80E	0164
2SA17G	0050	2SA30	0136	2SA40	0280	2SA53B	0633	2SA61K	0050	2SA71AC	0050	2SA80F	0164
2SA17H	0050	2SA30A	0136	2SA40A	0280	2SA53C	0633	2SA61L	0050	2SA71B	0050	2SA80G	0164
2SA17L	0050	2SA30B	0136	2SA40B	0280	2SA53D	0633	2SA61OR	0050	2SA71BS	0050	2SA80H	0164
2SA17OR	0050	2SA30C	0136	2SA40C	0280	2SA53E	0633	2SA61R	0050	2SA71C	0050	2SA80K	0164
2SA17R	0050	2SA30D	0136	2SA40D	0280	2SA53F	0633	2SA61X	0050	2SA71D	0050	2SA80L	0164
2SA17X	0050	2SA30E	0136	2SA40E	0280	2SA53G	0633	2SA61Y	0050	2SA71E	0050	2SA80M	0164
2SA17Y	0050	2SA30F	0136	2SA40F	0280	2SA53L	0633	2SA64	0164	2SA71F	0050	2SA80OR	0164
2SA18	0050	2SA30G	0136	2SA40G	0280	2SA53M	0633	2SA64A	0164	2SA71G	0050	2SA80R	0164
2SA18A	0050	2SA30L	0136	2SA40L	0280	2SA53OR	0633	2SA64B	0164	2SA71H	0050	2SA80X	0164
2SA18B	0050	2SA30M	0136	2SA40M	0280	2SA53R	0633	2SA64C	0164	2SA71K	0050	2SA80Y	0164
2SA18C	0050	2SA30OR	0136	2SA40OR	0280	2SA53X	0633	2SA64D	0164	2SA71L	0050	2SA81	0050
2SA18D	0050	2SA30X	0136	2SA40R	0280	2SA53Y	0633	2SA64E	0164	2SA71M	0050	2SA82	0164
2SA18E	0050	2SA30Y	0136	2SA40X	0280	2SA54	0050	2SA64F	0164	2SA71OR	0050	2SA82A	0164
2SA18F	0050	2SA31	0280	2SA40Y	0280	2SA54A	0050	2SA64G	0164	2SA71R	0050	2SA82B	0164
2SA18G	0050	2SA31A	0280	2SA41	0050	2SA54B	0050	2SA64GN	0164	2SA71X	0050	2SA82C	0164
2SA18H	0050	2SA31B	0280	2SA41A	0050	2SA54C	0050	2SA64H	0164	2SA71Y	0050	2SA82D	0164
2SA18L	0050	2SA31C	0280	2SA41B	0050	2SA54D	0050	2SA64J	0164	2SA71YA	0050	2SA82E	0164
2SA18M	0050	2SA31D	0280	2SA41C	0050	2SA54E	0050	2SA64K	0164	2SA72	0136	2SA82F	0164
2SA18OR	0050	2SA31E	0280	2SA41D	0050	2SA54F	0050	2SA64L	0164	2SA72BLU	0136	2SA82G	0164
2SA18R	0050	2SA31F	0280	2SA41E	0050	2SA54G	0050	2SA64M	0164	2SA72BLU-BLU	0136	2SA82H	0164
2SA18X	0050	2SA31G	0280	2SA41F	0050	2SA54L	0050	2SA64OR	0164	2SA72BRN	0136	2SA82K	0164
2SA18Y	0050	2SA31L	0280	2SA41G	0050	2SA54M	0050	2SA64R	0164	2SA72ORN	0136	2SA82M	0164
2SA19	0164	2SA31M	0280	2SA41L	0050	2SA54OR	0050	2SA64X	0164	2SA72WHT	0136	2SA82OR	0164
2SA19A	0164	2SA31OR	0280	2SA41M	0050	2SA54R	0050	2SA64Y	0164	2SA73	0050	2SA82R	0164
2SA19B	0164	2SA31X	0280	2SA41OR	0050	2SA54X	0050	2SA65	0050	2SA73A	0050	2SA82X	0164
2SA19C	0164	2SA31Y	0280	2SA41R	0050	2SA54Y	0050	2SA65A	0050	2SA73B	0050	2SA82Y	0164
2SA19D	0164	2SA32	0050	2SA41X	0050	2SA55	0136	2SA65B	0050	2SA73C	0050	2SA83	0050
2SA19E	0164	2SA32A	0050	2SA41Y	0050	2SA55A	0136	2SA65C	0050	2SA73D	0050	2SA83A	0050
2SA19F	0164	2SA32B	0050	2SA42	0050	2SA55B	0136	2SA65D	0050	2SA73E	0050	2SA83B	0050
2SA19G	0164	2SA32C	0050	2SA43	0136	2SA55C	0136	2SA65E	0050	2SA73F	0050	2SA83C	0050
2SA19L	0164	2SA32D	0050	2SA44	0136	2SA55D	0136	2SA65F	0050	2SA73G	0050	2SA83D	0050
2SA19M	0164	2SA32E	0050	2SA44A	0136	2SA55E	0136	2SA65G	0050	2SA73H	0050	2SA83E	0050
2SA19OR	0164	2SA32F	0050	2SA44B	0136	2SA55F	0136	2SA65K	0050	2SA73K	0050	2SA83F	0050
2SA19R	0164	2SA32G	0050	2SA44C	0136	2SA55G	0136	2SA65L	0050	2SA73L	0050	2SA83G	0050
2SA19X	0164	2SA32L	0050	2SA44D	0136	2SA55L	0136	2SA65M	0050	2SA73M	0050	2SA83H	0050
2SA19Y	0164	2SA32M	0050	2SA44E	0136	2SA55M	0136	2SA65OR	0050	2SA73OR	0050	2SA83K	0050
2SA20	0164	2SA32OR	0050	2SA44F	0136	2SA55OR	0136	2SA65R	0050	2SA73R	0050	2SA83L	0050

If replacement code is OEM, contact original manufacturer for replacement.

DEVICE TYPE	REPL CODE
2SA83M	0050
2SA83OR	0050
2SA83R	0050
2SA83X	0050
2SA83Y	0050
2SA84	0050
2SA84A	0050
2SA84B	0050
2SA84C	0050
2SA84D	0050
2SA84E	0050
2SA84F	0050
2SA84G	0050
2SA84H	0050
2SA84K	0050
2SA84L	0050
2SA84M	0050
2SA84OR	0050
2SA84R	0050
2SA84X	0050
2SA84Y	0050
2SA85	0136
2SA85L	0136
2SA86	0050
2SA87	0050
2SA87A	0050
2SA87B	0050
2SA87C	0050
2SA87D	0050
2SA87E	0050
2SA87F	0050
2SA87G	0050
2SA87H	0050
2SA87K	0050
2SA87L	0050
2SA87M	0050
2SA87OR	0050
2SA87R	0050
2SA87X	0050
2SA87Y	0050
2SA88	0050
2SA89	0050
2SA90	0136
2SA92	0050
2SA92A	0050
2SA92B	0050
2SA92C	0050
2SA92D	0050
2SA92E	0050
2SA92F	0050
2SA92G	0050
2SA92H	0050
2SA92K	0050
2SA92L	0050
2SA92M	0050
2SA92OR	0050
2SA92R	0050
2SA92X	0050
2SA92Y	0050
2SA93	0050
2SA93A	0050
2SA93B	0050
2SA93C	0050
2SA93E	0050
2SA93F	0050
2SA93G	0050
2SA93H	0050
2SA93K	0050
2SA93L	0050
2SA93M	0050
2SA93OR	0050
2SA93R	0050
2SA93X	0050
2SA93Y	0050
2SA94	0164
2SA94A	0164
2SA94C	0164
2SA94D	0164
2SA94E	0164
2SA94F	0164
2SA94G	0164
2SA94H	0164
2SA94K	0164
2SA94L	0164
2SA94M	0164
2SA94OR	0164
2SA94R	0164
2SA94X	0164
2SA94Y	0164
2SA95	0050
2SA96	0050
2SA97	0050
2SA98	0050
2SA99	0050
2SA100	0136
2SA100A	0136
2SA100B	0136
2SA100C	0136
2SA100D	0136
2SA100E	0136
2SA100F	0136
2SA100G	0136
2SA100H	0136
2SA100J	0136
2SA100K	0136
2SA100M	0136
2SA100OR	0136
2SA100R	0136
2SA100X	0136

DEVICE TYPE	REPL CODE
2SA100Y	0136
2SA101	0136
2SA101(C)	0136
2SA101(V)	0136
2SA101-OR	0136
2SA101A	0136
2SA101AA	0136
2SA101AY	0136
2SA101B	0136
2SA101BA	0136
2SA101BB	0136
2SA101BC	0136
2SA101BX	0136
2SA101C	0136
2SA101CA	0136
2SA101CV	0136
2SA101CX	0136
2SA101D	0136
2SA101E	0136
2SA101F	0136
2SA101G	0136
2SA101H	0136
2SA101K	0136
2SA101L	0136
2SA101M	0136
2SA101OR	0136
2SA101QA	0136
2SA101R	0136
2SA101X	0136
2SA101XBX	0136
2SA101Y	0136
2SA101YA	0136
2SA101Z	0136
2SA102	0136
2SA102(BA)	0136
2SA102-OR	0136
2SA102A	0136
2SA102AA	0136
2SA102AB	0136
2SA102B	0136
2SA102BA	0136
2SA102BA-2	0136
2SA102BN	0136
2SA102C	0136
2SA102CA	0136
2SA102CA-1	0136
2SA102D	0136
2SA102E	0136
2SA102F	0136
2SA102G	0136
2SA102H	0136
2SA102K	0136
2SA102L	0136
2SA102M	0136
2SA102OR	0136
2SA102TV	0136
2SA102TV-2	0136
2SA102X	0136
2SA102Y	0136
2SA103	0136
2SA103(CA)	0136
2SA103A	0136
2SA103B	0136
2SA103C	0136
2SA103CA	0136
2SA103CAK	0136
2SA103CB	0136
2SA103CG	0136
2SA103D	0136
2SA103DA	0136
2SA103E	0136
2SA103F	0136
2SA103G	0136
2SA103GA	0136
2SA103K	0136
2SA103L	0136
2SA103M	0136
2SA103OR	0136
2SA103R	0136
2SA103X	0136
2SA103Y	0136
2SA104	0050
2SA104A	0050
2SA104B	0050
2SA104C	0050
2SA104D	0050
2SA104E	0050
2SA104F	0050
2SA104G	0050
2SA104H	0050
2SA104K	0050
2SA104L	0050
2SA104M	0050
2SA104OR	0050
2SA104P	0050
2SA104R	0050
2SA104X	0050
2SA104Y	0050
2SA105	0136
2SA105A	0136
2SA105B	0136
2SA105C	0136
2SA105D	0136
2SA105E	0136
2SA105G	0136
2SA105H	0136
2SA105K	0136
2SA105L	0136
2SA105M	0136

DEVICE TYPE	REPL CODE
2SA105OR	0136
2SA105R	0136
2SA105X	0136
2SA105Y	0136
2SA106	0136
2SA106A	0136
2SA106B	0136
2SA106C	0136
2SA106D	0136
2SA106E	0136
2SA106F	0136
2SA106G	0136
2SA106H	0136
2SA106K	0136
2SA106L	0136
2SA106M	0136
2SA106OR	0136
2SA106R	0136
2SA106X	0136
2SA106Y	0136
2SA107	0136
2SA107A	0136
2SA107B	0136
2SA107C	0136
2SA107D	0136
2SA107E	0136
2SA107F	0136
2SA107G	0136
2SA107H	0136
2SA107K	0136
2SA107L	0136
2SA107M	0136
2SA107OR	0136
2SA107R	0136
2SA107X	0136
2SA107Y	0136
2SA108	0136
2SA108A	0136
2SA108B	0136
2SA108C	0136
2SA108D	0136
2SA108E	0136
2SA108F	0136
2SA108G	0136
2SA108H	0136
2SA108K	0136
2SA108L	0136
2SA108M	0136
2SA108OR	0136
2SA108R	0136
2SA108X	0136
2SA108Y	0136
2SA109	0136
2SA109A	0136
2SA109B	0136
2SA109C	0136
2SA109D	0136
2SA109E	0136
2SA109F	0136
2SA109G	0136
2SA109K	0136
2SA109L	0136
2SA109M	0136
2SA109OR	0136
2SA109R	0136
2SA109X	0136
2SA109Y	0136
2SA110	0136
2SA110A	0136
2SA110B	0136
2SA110C	0136
2SA110D	0136
2SA110E	0136
2SA110F	0136
2SA110G	0136
2SA110K	0136
2SA110L	0136
2SA110M	0136
2SA110OR	0136
2SA110R	0136
2SA110X	0136
2SA110Y	0136
2SA111	0136
2SA111A	0136
2SA111B	0136
2SA111C	0136
2SA111D	0136
2SA111E	0136
2SA111F	0136
2SA111G	0136
2SA111K	0136
2SA111L	0136
2SA111M	0136
2SA111OR	0136
2SA111R	0136
2SA111X	0136
2SA111Y	0136
2SA112	0136
2SA112A	0136
2SA112B	0136
2SA112C	0136
2SA112D	0136
2SA112E	0136
2SA112F	0136
2SA112G	0136
2SA112GN	0136
2SA112H	0136
2SA112K	0136
2SA112L	0136

DEVICE TYPE	REPL CODE
2SA112M	0136
2SA112OR	0136
2SA112R	0136
2SA112X	0136
2SA112Y	0136
2SA113	0136
2SA113A	0136
2SA113B	0136
2SA113C	0136
2SA113D	0136
2SA113E	0136
2SA113F	0136
2SA113G	0136
2SA113GN	0136
2SA113H	0136
2SA113J	0136
2SA113L	0136
2SA113M	0136
2SA113R	0136
2SA113X	0136
2SA113Y	0136
2SA114	0136
2SA114A	0136
2SA114B	0136
2SA114C	0136
2SA114D	0136
2SA114E	0136
2SA114F	0136
2SA114G	0136
2SA114H	0136
2SA114K	0136
2SA114L	0136
2SA114M	0136
2SA114OR	0136
2SA114R	0136
2SA114X	0136
2SA114Y	0136
2SA115	0136
2SA115A	0136
2SA115B	0136
2SA115C	0136
2SA115D	0136
2SA115E	0136
2SA115F	0136
2SA115G	0136
2SA115GN	0136
2SA115H	0136
2SA115J	0136
2SA115K	0136
2SA115L	0136
2SA115M	0136
2SA115OR	0136
2SA115R	0136
2SA115X	0136
2SA115Y	0136
2SA116	0136
2SA116A	0136
2SA116B	0136
2SA116C	0136
2SA116D	0136
2SA116E	0136
2SA116F	0136
2SA116G	0136
2SA116GN	0136
2SA116H	0136
2SA116J	0136
2SA116K	0136
2SA116L	0136
2SA116M	0136
2SA116OR	0136
2SA116R	0136
2SA116X	0136
2SA116Y	0136
2SA117	0136
2SA117A	0136
2SA117B	0136
2SA117C	0136
2SA117D	0136
2SA117E	0136
2SA117F	0136
2SA117G	0136
2SA117GN	0136
2SA117H	0136
2SA117J	0136
2SA117K	0136
2SA117L	0136
2SA117M	0136
2SA117OR	0136
2SA117R	0136
2SA117X	0136
2SA117Y	0136
2SA118	0136
2SA118A	0136
2SA118B	0136
2SA118C	0136
2SA118D	0136
2SA118E	0136
2SA118F	0136
2SA118G	0136
2SA118GN	0136
2SA118H	0136
2SA118J	0136
2SA118K	0136
2SA118L	0136
2SA118M	0136
2SA118OR	0136
2SA118R	0136
2SA118X	0136
2SA118Y	0136

DEVICE TYPE	REPL CODE
2SA119	0126
2SA120	0126
2SA120A	0126
2SA121	0136
2SA121A	0136
2SA121B	0136
2SA121C	0136
2SA121D	0136
2SA121E	0136
2SA121F	0136
2SA121G	0136
2SA121GN	0136
2SA121H	0136
2SA121J	0136
2SA121K	0136
2SA121L	0136
2SA121M	0136
2SA121OR	0136
2SA121R	0136
2SA121X	0136
2SA121Y	0136
2SA122	0136
2SA122A	0136
2SA122B	0136
2SA122C	0136
2SA122D	0136
2SA122E	0136
2SA122F	0136
2SA122G	0136
2SA122GN	0136
2SA122H	0136
2SA122K	0136
2SA122L	0136
2SA122M	0136
2SA122OR	0136
2SA122R	0136
2SA122X	0136
2SA122Y	0136
2SA123	0136
2SA123A	0136
2SA123B	0136
2SA123C	0136
2SA123D	0136
2SA123E	0136
2SA123F	0136
2SA123G	0136
2SA123GN	0136
2SA123H	0136
2SA123J	0136
2SA123K	0136
2SA123L	0136
2SA123M	0136
2SA123OR	0136
2SA123R	0136
2SA123X	0136
2SA123Y	0136
2SA124	0136
2SA124A	0136
2SA124B	0136
2SA124C	0136
2SA124D	0136
2SA124E	0136
2SA124F	0136
2SA124G	0136
2SA124GN	0136
2SA124H	0136
2SA124J	0136
2SA124K	0136
2SA124L	0136
2SA124M	0136
2SA124OR	0136
2SA124R	0136
2SA124X	0136
2SA124Y	0136
2SA125	0136
2SA125A	0136
2SA125B	0136
2SA125C	0136
2SA125D	0136
2SA125E	0136
2SA125F	0136
2SA125G	0136
2SA125GN	0136
2SA125H	0136
2SA125J	0136
2SA125K	0136
2SA125L	0136
2SA125M	0136
2SA125OR	0136
2SA125R	0136
2SA125X	0136
2SA125Y	0136
2SA126	0136
2SA127	0136
2SA128	0164
2SA128A	0164
2SA128B	0164
2SA128C	0164
2SA128D	0164
2SA128E	0164
2SA128F	0164
2SA128G	0164
2SA128GN	0164
2SA128H	0164
2SA128J	0164
2SA128K	0164
2SA128L	0164
2SA128M	0164
2SA128OR	0164

DEVICE TYPE	REPL CODE
2SA128R	0164
2SA128X	0164
2SA128Y	0164
2SA129	0164
2SA129A	0164
2SA129B	0164
2SA129C	0164
2SA129D	0164
2SA129E	0164
2SA129F	0164
2SA129G	0164
2SA129GN	0164
2SA129H	0164
2SA129J	0164
2SA129K	0164
2SA129L	0164
2SA129M	0164
2SA129OR	0164
2SA129R	0164
2SA129X	0164
2SA129Y	0164
2SA130	0050
2SA130A	0050
2SA130B	0050
2SA130C	0050
2SA130D	0050
2SA130E	0050
2SA130F	0050
2SA130G	0050
2SA130GN	0050
2SA130H	0050
2SA130J	0050
2SA130K	0050
2SA130L	0050
2SA130M	0050
2SA130OR	0050
2SA130R	0050
2SA130X	0050
2SA130Y	0050
2SA131	0136
2SA131A	0136
2SA131B	0136
2SA131C	0136
2SA131D	0136
2SA131E	0136
2SA131F	0136
2SA131G	0136
2SA131GN	0136
2SA131H	0136
2SA131J	0136
2SA131K	0136
2SA131L	0136
2SA131M	0136
2SA131OR	0136
2SA131R	0136
2SA131X	0136
2SA131Y	0136
2SA132	0136
2SA132A	0136
2SA132B	0136
2SA132C	0136
2SA132D	0136
2SA132E	0136
2SA132F	0136
2SA132G	0136
2SA132H	0136
2SA132J	0136
2SA132K	0136
2SA132L	0136
2SA132M	0136
2SA132OR	0136
2SA132X	0136
2SA132Y	0136
2SA133	0136
2SA133A	0136
2SA133B	0136
2SA133C	0136
2SA133D	0136
2SA133E	0136
2SA133F	0136
2SA133G	0136
2SA133GN	0136
2SA133H	0136
2SA133J	0136
2SA133K	0136
2SA133L	0136
2SA133M	0136
2SA133OR	0136
2SA133R	0136
2SA133X	0136
2SA133Y	0136
2SA134	0136
2SA134A	0136
2SA134B	0136
2SA134C	0136
2SA134D	0136
2SA134E	0136
2SA134F	0136
2SA134G	0136
2SA134H	0136
2SA134J	0136
2SA134K	0136
2SA134L	0136
2SA134OR	0136
2SA134R	0136
2SA134X	0136
2SA134Y	0136
2SA135	0050
2SA135A	0050

DEVICE TYPE	REPL CODE
2SA135B	0050
2SA135C	0050
2SA135D	0050
2SA135E	0050
2SA135F	0050
2SA135G	0050
2SA135GN	0050
2SA135H	0050
2SA135J	0050
2SA135K	0050
2SA135L	0050
2SA135M	0050
2SA135OR	0050
2SA135R	0050
2SA135X	0050
2SA135Y	0050
2SA136	0050
2SA136A	0050
2SA136B	0050
2SA136C	0050
2SA136D	0050
2SA136E	0050
2SA136F	0050
2SA136G	0050
2SA136GN	0050
2SA136H	0050
2SA136J	0050
2SA136K	0050
2SA136L	0050
2SA136M	0050
2SA136OR	0050
2SA136R	0050
2SA136X	0050
2SA136Y	0050
2SA137	0136
2SA137A	0136
2SA137B	0136
2SA137C	0136
2SA137D	0136
2SA137E	0136
2SA137F	0136
2SA137G	0136
2SA137GN	0136
2SA137H	0136
2SA137J	0136
2SA137K	0136
2SA137L	0136
2SA137M	0136
2SA137OR	0136
2SA137R	0136
2SA137X	0136
2SA137Y	0136
2SA138	0004
2SA139	0136
2SA139A	0136
2SA139B	0136
2SA139C	0136
2SA139D	0136
2SA139E	0136
2SA139F	0136
2SA139G	0136
2SA139GN	0136
2SA139J	0136
2SA139K	0136
2SA139L	0136
2SA139M	0136
2SA139OR	0136
2SA139R	0136
2SA139X	0136
2SA139Y	0136
2SA141	0136
2SA141A	0136
2SA141B	0136
2SA141C	0136
2SA141D	0136
2SA141E	0136
2SA141F	0136
2SA141G	0136
2SA141GN	0136
2SA141H	0136
2SA141K	0136
2SA141L	0136
2SA141M	0136
2SA141OR	0136
2SA141R	0136
2SA141X	0136
2SA141Y	0136
2SA142	0136
2SA142A	0136
2SA142B	0136
2SA142C	0136
2SA142D	0136
2SA142E	0136
2SA142F	0136
2SA142G	0136
2SA142GN	0136
2SA142H	0136
2SA142J	0136
2SA142K	0136
2SA142L	0136
2SA142M	0136
2SA142OR	0136
2SA142R	0136
2SA142X	0136
2SA142Y	0136
2SA143	0136
2SA143A	0136
2SA143B	0136
2SA143C	0136

Original Device Types

DEVICE TYPE	REPL CODE
2SA143D	0136
2SA143E	0136
2SA143F	0136
2SA143G	0136
2SA143GN	0136
2SA143H	0136
2SA143J	0136
2SA143K	0136
2SA143L	0136
2SA143M	0136
2SA143OR	0136
2SA143R	0136
2SA143X	0136
2SA143Y	0136
2SA144	0136
2SA144A	0136
2SA144B	0136
2SA144C	0136
2SA144D	0136
2SA144E	0136
2SA144F	0136
2SA144G	0136
2SA144GN	0136
2SA144H	0136
2SA144J	0136
2SA144K	0136
2SA144L	0136
2SA144M	0136
2SA144OR	0136
2SA144R	0136
2SA144X	0136
2SA144Y	0136
2SA145	0136
2SA145A	0136
2SA145B	0136
2SA145C	0136
2SA145D	0136
2SA145E	0136
2SA145G	0136
2SA145GN	0136
2SA145K	0136
2SA145M	0136
2SA145OR	0136
2SA145R	0136
2SA145X	0136
2SA145Y	0136
2SA146	0136
2SA146A	0136
2SA146B	0136
2SA146C	0136
2SA146D	0136
2SA146E	0136
2SA146F	0136
2SA146G	0136
2SA146GN	0136
2SA146H	0136
2SA146J	0136
2SA146K	0136
2SA146L	0136
2SA146M	0136
2SA146OR	0136
2SA146X	0136
2SA146Y	0136
2SA147	0136
2SA147A	0136
2SA147B	0136
2SA147C	0136
2SA147D	0136
2SA147E	0136
2SA147F	0136
2SA147G	0136
2SA147H	0136
2SA147J	0136
2SA147K	0136
2SA147L	0136
2SA147M	0136
2SA147OR	0136
2SA147R	0136
2SA147X	0136
2SA147Y	0136
2SA148	0136
2SA148A	0136
2SA148B	0136
2SA148C	0136
2SA148D	0136
2SA148E	0136
2SA148F	0136
2SA148G	0136
2SA148GN	0136
2SA148H	0136
2SA148J	0136
2SA148K	0136
2SA148L	0136
2SA148M	0136
2SA148OR	0136
2SA148R	0136
2SA148X	0136
2SA148Y	0136
2SA149	0136
2SA149A	0136
2SA149B	0136
2SA149C	0136
2SA149D	0136
2SA149E	0136
2SA149F	0136
2SA149G	0136
2SA149GN	0136
2SA149H	0136
2SA149J	0136

DEVICE TYPE	REPL CODE
2SA149K	0136
2SA149L	0136
2SA149M	0136
2SA149OR	0136
2SA149R	0136
2SA149X	0136
2SA149Y	0136
2SA150	0136
2SA150A	0136
2SA150B	0136
2SA150C	0136
2SA150D	0136
2SA150E	0136
2SA150F	0136
2SA150G	0136
2SA150GN	0136
2SA150H	0136
2SA150J	0136
2SA150K	0136
2SA150L	0136
2SA150M	0136
2SA150OR	0136
2SA150R	0136
2SA150X	0136
2SA150Y	0136
2SA151	0136
2SA151A	0136
2SA151B	0136
2SA151C	0136
2SA151D	0136
2SA151E	0136
2SA151F	0136
2SA151G	0136
2SA151GN	0136
2SA151H	0136
2SA151J	0136
2SA151K	0136
2SA151L	0136
2SA151M	0136
2SA151OR	0136
2SA151R	0136
2SA151X	0136
2SA151Y	0136
2SA152	0136
2SA152A	0136
2SA152B	0136
2SA152C	0136
2SA152D	0136
2SA152E	0136
2SA152F	0136
2SA152G	0136
2SA152GN	0136
2SA152H	0136
2SA152J	0136
2SA152K	0136
2SA152L	0136
2SA152M	0136
2SA152OR	0136
2SA152R	0136
2SA152X	0136
2SA152Y	0136
2SA153	0136
2SA153A	0136
2SA153B	0136
2SA153C	0136
2SA153D	0136
2SA153E	0136
2SA153F	0136
2SA153G	0136
2SA153GN	0136
2SA153H	0136
2SA153J	0136
2SA153K	0136
2SA153L	0136
2SA153M	0136
2SA153OR	0136
2SA153R	0136
2SA153X	0136
2SA153Y	0136
2SA154	0136
2SA154A	0136
2SA154B	0136
2SA154C	0136
2SA154D	0136
2SA154E	0136
2SA154F	0136
2SA154G	0136
2SA154GN	0136
2SA154H	0136
2SA154J	0136
2SA154K	0136
2SA154L	0136
2SA154M	0136
2SA154OR	0136
2SA154R	0136
2SA154X	0136
2SA154Y	0136
2SA155	0136
2SA155A	0136
2SA155B	0136
2SA155C	0136
2SA155D	0136
2SA155E	0136
2SA155F	0136
2SA155G	0136
2SA155GN	0136
2SA155H	0136
2SA155J	0136
2SA155K	0136

DEVICE TYPE	REPL CODE
2SA155L	0136
2SA155M	0136
2SA155OR	0136
2SA155R	0136
2SA155X	0136
2SA155Y	0136
2SA156	0136
2SA156A	0136
2SA156B	0136
2SA156C	0136
2SA156D	0136
2SA156E	0136
2SA156F	0136
2SA156G	0136
2SA156GN	0136
2SA156H	0136
2SA156J	0136
2SA156K	0136
2SA156L	0136
2SA156M	0136
2SA156OR	0136
2SA156R	0136
2SA156X	0136
2SA156Y	0136
2SA157	0136
2SA157A	0136
2SA157B	0136
2SA157C	0136
2SA157D	0136
2SA157E	0136
2SA157F	0136
2SA157G	0136
2SA157GN	0136
2SA157H	0136
2SA157J	0136
2SA157K	0136
2SA157L	0136
2SA157M	0136
2SA157OR	0136
2SA157R	0136
2SA157X	0136
2SA157Y	0136
2SA159	0136
2SA159A	0136
2SA159B	0136
2SA159C	0136
2SA159D	0136
2SA159E	0136
2SA159F	0136
2SA159G	0136
2SA159GN	0136
2SA159H	0136
2SA159J	0136
2SA159K	0136
2SA159L	0136
2SA159M	0136
2SA159OR	0136
2SA159R	0136
2SA159X	0136
2SA159Y	0136
2SA160	0136
2SA160A	0136
2SA160B	0136
2SA160C	0136
2SA160D	0136
2SA160E	0136
2SA160F	0136
2SA160GN	0136
2SA160H	0136
2SA160J	0136
2SA160K	0136
2SA160L	0136
2SA160M	0136
2SA160OR	0136
2SA160R	0136
2SA160X	0136
2SA160Y	0136
2SA161	0136
2SA161A	0136
2SA161B	0136
2SA161C	0136
2SA161D	0136
2SA161E	0136
2SA161F	0136
2SA161G	0136
2SA161GN	0136
2SA161H	0136
2SA161J	0136
2SA161K	0136
2SA161L	0136
2SA161M	0136
2SA161OR	0136
2SA161R	0136
2SA161X	0136
2SA161Y	0136
2SA162	0136
2SA162A	0136
2SA162D	0136
2SA162C	0136
2SA162D	0136
2SA162E	0136
2SA162F	0136
2SA162G	0136
2SA162GN	0136
2SA162H	0136
2SA162J	0136
2SA162K	0136
2SA162L	0136
2SA162M	0136

DEVICE TYPE	REPL CODE
2SA162OR	0136
2SA162R	0136
2SA162X	0136
2SA162Y	0136
2SA163	0136
2SA163A	0136
2SA163B	0136
2SA163C	0136
2SA163D	0136
2SA163E	0136
2SA163F	0136
2SA163G	0136
2SA163GN	0136
2SA163H	0136
2SA163J	0136
2SA163K	0136
2SA163L	0136
2SA163M	0136
2SA163OR	0136
2SA163R	0136
2SA163X	0136
2SA163Y	0136
2SA164	0136
2SA164A	0136
2SA164B	0136
2SA164C	0136
2SA164D	0136
2SA164E	0136
2SA164F	0136
2SA164G	0136
2SA164GN	0136
2SA164H	0136
2SA164J	0136
2SA164K	0136
2SA164L	0136
2SA164M	0136
2SA164OR	0136
2SA164R	0136
2SA164X	0136
2SA164Y	0136
2SA165	0136
2SA165A	0136
2SA165B	0136
2SA165C	0136
2SA165D	0136
2SA165E	0136
2SA165F	0136
2SA165G	0136
2SA165GN	0136
2SA165H	0136
2SA165J	0136
2SA165K	0136
2SA165L	0136
2SA165M	0136
2SA165OR	0136
2SA165R	0136
2SA165X	0136
2SA165Y	0136
2SA166	0136
2SA166A	0136
2SA166B	0136
2SA166C	0136
2SA166D	0136
2SA166E	0136
2SA166F	0136
2SA166G	0136
2SA166GN	0136
2SA166H	0136
2SA166J	0136
2SA166K	0136
2SA166L	0136
2SA166M	0136
2SA166OR	0136
2SA166R	0136
2SA166X	0136
2SA166Y	0136
2SA167	0280
2SA167A	0280
2SA167B	0280
2SA167C	0280
2SA167D	0280
2SA167E	0280
2SA167F	0280
2SA167G	0280
2SA167GN	0280
2SA167H	0280
2SA167J	0280
2SA167K	0280
2SA167L	0280
2SA167M	0280
2SA167OR	0280
2SA167R	0280
2SA167X	0280
2SA167Y	0280
2SA168	0280
2SA168A	0280
2SA168B	0280
2SA168C	0280
2SA168D	0280
2SA168E	0280
2SA168F	0280
2SA168G	0280
2SA168GN	0280
2SA168H	0280
2SA168J	0280
2SA168K	0280
2SA168L	0280
2SA168M	0280
2SA168OR	0280

DEVICE TYPE	REPL CODE
2SA168R	0280
2SA168X	0280
2SA168Y	0280
2SA169	0280
2SA169A	0280
2SA169B	0280
2SA169C	0280
2SA169D	0280
2SA169E	0280
2SA169F	0280
2SA169G	0280
2SA169GN	0280
2SA169H	0280
2SA169J	0280
2SA169K	0280
2SA169L	0280
2SA169M	0280
2SA169OR	0280
2SA169R	0280
2SA169X	0280
2SA169Y	0280
2SA170	0280
2SA170A	0280
2SA170B	0280
2SA170C	0280
2SA170D	0280
2SA170E	0280
2SA170F	0280
2SA170G	0280
2SA170GN	0280
2SA170H	0280
2SA170J	0280
2SA170K	0280
2SA170L	0280
2SA170M	0280
2SA170OR	0280
2SA170R	0280
2SA170X	0280
2SA170Y	0280
2SA171	0280
2SA171A	0280
2SA171B	0280
2SA171C	0280
2SA171D	0280
2SA171E	0280
2SA171F	0280
2SA171G	0280
2SA171GN	0280
2SA171H	0280
2SA171J	0280
2SA171K	0280
2SA171L	0280
2SA171M	0280
2SA171OR	0280
2SA171R	0280
2SA171X	0280
2SA171Y	0280
2SA172	0280
2SA172A	0280
2SA172B	0280
2SA172C	0280
2SA172D	0280
2SA172E	0280
2SA172F	0280
2SA172G	0280
2SA172GN	0280
2SA172H	0280
2SA172J	0280
2SA172K	0280
2SA172L	0280
2SA172M	0280
2SA172OR	0280
2SA172R	0280
2SA172X	0280
2SA172Y	0280
2SA173	0004
2SA173A	0004
2SA173B	0004
2SA173C	0004
2SA173D	0004
2SA173E	0004
2SA173F	0004
2SA173G	0004
2SA173GN	0004
2SA173H	0004
2SA173J	0004
2SA173K	0004
2SA173L	0004
2SA173M	0004
2SA173OR	0004
2SA173R	0004
2SA173X	0004
2SA173Y	0004
2SA174	0211
2SA174A	0211
2SA174B	0211
2SA174C	0211
2SA174D	0211
2SA174E	0211
2SA174F	0211
2SA174G	0211
2SA174GN	0211
2SA174H	0211
2SA174J	0211
2SA174K	0211
2SA174L	0211
2SA174M	0211
2SA174OR	0211
2SA174R	0211

DEVICE TYPE	REPL CODE
2SA174X	0211
2SA174Y	0211
2SA175	0136
2SA175A	0136
2SA175B	0136
2SA175C	0136
2SA175D	0136
2SA175E	0136
2SA175F	0136
2SA175G	0136
2SA175GN	0136
2SA175H	0136
2SA175J	0136
2SA175K	0136
2SA175L	0136
2SA175M	0136
2SA175OR	0136
2SA175R	0136
2SA175X	0136
2SA175Y	0136
2SA176	0136
2SA176A	0136
2SA176B	0136
2SA176C	0136
2SA176D	0136
2SA176E	0136
2SA176F	0136
2SA176G	0136
2SA176GN	0136
2SA176H	0136
2SA176J	0136
2SA176K	0136
2SA176L	0136
2SA176M	0136
2SA176OR	0136
2SA176R	0136
2SA176X	0136
2SA176Y	0136
2SA178	0050
2SA180	0136
2SA180A	0136
2SA180B	0136
2SA180C	0136
2SA180D	0136
2SA180E	0136
2SA180F	0136
2SA180G	0136
2SA180GN	0136
2SA180H	0136
2SA180J	0136
2SA180K	0136
2SA180L	0136
2SA180M	0136
2SA180OR	0136
2SA180R	0136
2SA180X	0136
2SA180Y	0136
2SA181	0136
2SA181A	0136
2SA181B	0136
2SA181C	0136
2SA181D	0136
2SA181E	0136
2SA181F	0136
2SA181G	0136
2SA181GN	0136
2SA181H	0136
2SA181J	0136
2SA181K	0136
2SA181L	0136
2SA181M	0136
2SA181OR	0136
2SA181R	0136
2SA181X	0136
2SA181Y	0136
2SA182	0050
2SA182A	0050
2SA182B	0050
2SA182C	0050
2SA182D	0050
2SA182E	0050
2SA182F	0050
2SA182G	0050
2SA182GN	0050
2SA182H	0050
2SA182J	0050
2SA182K	0050
2SA182L	0050
2SA182M	0050
2SA182OR	0050
2SA182R	0050
2SA182X	0050
2SA182Y	0050
2SA183	0136
2SA183A	0136
2SA183B	0136
2SA183C	0136
2SA183D	0136
2SA183E	0136
2SA183F	0136
2SA183G	0136
2SA183GN	0136
2SA183H	0136
2SA183J	0136
2SA183K	0136
2SA183L	0136
2SA183M	0136
2SA183OR	0136
2SA183R	0136

DEVICE TYPE	REPL CODE
2SA183X	0136
2SA183Y	0136
2SA184	0050
2SA186	0164
2SA187TV	0004
2SA188	0136
2SA188A	0136
2SA188B	0136
2SA188C	0136
2SA188D	0136
2SA188E	0136
2SA188F	0136
2SA188G	0136
2SA188GN	0136
2SA188H	0136
2SA188J	0136
2SA188K	0136
2SA188L	0136
2SA188M	0136
2SA188OR	0136
2SA188R	0136
2SA188X	0136
2SA188Y	0136
2SA189	0136
2SA189A	0136
2SA189B	0136
2SA189C	0136
2SA189D	0136
2SA189E	0136
2SA189F	0136
2SA189G	0136
2SA189GN	0136
2SA189H	0136
2SA189J	0136
2SA189K	0136
2SA189L	0136
2SA189M	0136
2SA189OR	0136
2SA189R	0136
2SA189X	0136
2SA189Y	0136
2SA190	0050
2SA191	0050
2SA192	0050
2SA193	0050
2SA194	0050
2SA195	0050
2SA196	0050
2SA197	0136
2SA197A	0136
2SA197B	0136
2SA197C	0136
2SA197D	0136
2SA197E	0136
2SA197F	0136
2SA197G	0136
2SA197GN	0136
2SA197H	0136
2SA197J	0136
2SA197K	0136
2SA197L	0136
2SA197M	0136
2SA197OR	0136
2SA197R	0136
2SA197X	0136
2SA197Y	0136
2SA198	0050
2SA198A	0050
2SA198B	0050
2SA198C	0050
2SA198D	0050
2SA198E	0050
2SA198F	0050
2SA198G	0050
2SA198GN	0050
2SA198H	0050
2SA198J	0050
2SA198K	0050
2SA198L	0050
2SA198M	0050
2SA198OR	0050
2SA198R	0050
2SA198X	0050
2SA198Y	0050
2SA199	0050
2SA200	0136
2SA201	0136
2SA201(0)	0136
2SA201-0	0136
2SA201-N	0136
2SA201-0	0136
2SA201-OR	0136
2SA201A	0136
2SA201B	0136
2SA201CL	0136
2SA201D	0136
2SA201E	0136
2SA201F	0136
2SA201G	0136
2SA201GN	0136
2SA201H	0136
2SA201J	0136
2SA201K	0136
2SA201L	0136
2SA201M	0136
2SA201N	0136
2SA201O	0136
2SA201OR	0136
2SA201R	0136

If replacement code is OEM, contact original manufacturer for replacement.

DEVICE TYPE	REPL CODE
2SA201TV	0136
2SA201TVO	0136
2SA201X	0136
2SA201Y	0136
2SA202	0633
2SA202-OR	0633
2SA202A	0633
2SA202AP	0633
2SA202B	0633
2SA202C	0633
2SA202D	0633
2SA202D-4	0633
2SA202E	0633
2SA202F	0633
2SA202G	0633
2SA202GN	0633
2SA202H	0633
2SA202J	0633
2SA202K	0633
2SA202L	0633
2SA202M	0633
2SA202OR	0633
2SA202R	0633
2SA202X	0633
2SA202Y	0633
2SA203	0633
2SA203A	0633
2SA203AA	0633
2SA203B	0633
2SA203C	0633
2SA203D	0633
2SA203P	0633
2SA204	0280
2SA204A	0280
2SA204B	0280
2SA204C	0280
2SA204D	0280
2SA204E	0280
2SA204F	0280
2SA204G	0280
2SA204GN	0280
2SA204H	0280
2SA204J	0280
2SA204K	0280
2SA204L	0280
2SA204M	0280
2SA204OR	0280
2SA204R	0280
2SA204X	0280
2SA204Y	0280
2SA205	0280
2SA205A	0280
2SA205B	0280
2SA205C	0280
2SA205D	0280
2SA205E	0280
2SA205F	0280
2SA205G	0280
2SA205GN	0280
2SA205H	0280
2SA205J	0280
2SA205K	0280
2SA205L	0280
2SA205M	0280
2SA205OR	0280
2SA205R	0280
2SA205X	0280
2SA205Y	0280
2SA206	0280
2SA206A	0280
2SA206B	0280
2SA206C	0280
2SA206D	0280
2SA206E	0280
2SA206F	0280
2SA206G	0280
2SA206GN	0280
2SA206H	0280
2SA206J	0280
2SA206K	0280
2SA206L	0280
2SA206M	0280
2SA206OR	0280
2SA206R	0280
2SA206X	0280
2SA206Y	0280
2SA207	0280
2SA207A	0280
2SA207B	0280
2SA207C	0280
2SA207D	0280
2SA207E	0280
2SA207F	0280
2SA207G	0280
2SA207GN	0280
2SA207H	0280
2SA207J	0280
2SA207K	0280
2SA207L	0280
2SA207M	0280
2SA207OR	0280
2SA207R	0280
2SA207X	0280
2SA207Y	0280
2SA208	0847
2SA208A	0847
2SA208B	0847
2SA208C	0847
2SA208D	0847
2SA208E	0847
2SA208F	0847
2SA208G	0847
2SA208GN	0847
2SA208H	0847
2SA208J	0847
2SA208K	0847
2SA208L	0847
2SA208M	0847
2SA208OR	0847
2SA208R	0847
2SA208X	0847
2SA208Y	0847
2SA209	0847
2SA209A	0847
2SA209B	0847
2SA209C	0847
2SA209D	0847
2SA209E	0847
2SA209F	0847
2SA209G	0847
2SA209GN	0847
2SA209H	0847
2SA209J	0847
2SA209K	0847
2SA209L	0847
2SA209M	0847
2SA209OR	0847
2SA209R	0847
2SA209X	0847
2SA209Y	0847
2SA210	0847
2SA210A	0847
2SA210B	0847
2SA210C	0847
2SA210D	0847
2SA210E	0847
2SA210F	0847
2SA210G	0847
2SA210GN	0847
2SA210H	0847
2SA210J	0847
2SA210K	0847
2SA210L	0847
2SA210M	0847
2SA210OR	0847
2SA210R	0847
2SA210X	0847
2SA210Y	0847
2SA211	0280
2SA211A	0280
2SA211B	0280
2SA211C	0280
2SA211D	0280
2SA211E	0280
2SA211F	0280
2SA211G	0280
2SA211GN	0280
2SA211H	0280
2SA211J	0280
2SA211K	0280
2SA211L	0280
2SA211M	0280
2SA211OR	0280
2SA211R	0280
2SA211X	0280
2SA211Y	0280
2SA212	0050
2SA212A	0050
2SA212B	0050
2SA212C	0050
2SA212D	0050
2SA212E	0050
2SA212F	0050
2SA212G	0050
2SA212GN	0050
2SA212H	0050
2SA212J	0050
2SA212K	0050
2SA212L	0050
2SA212M	0050
2SA212OR	0050
2SA212R	0050
2SA212X	0050
2SA212Y	0050
2SA213	0136
2SA213A	0136
2SA213B	0136
2SA213C	0136
2SA213D	0136
2SA213E	0136
2SA213F	0136
2SA213G	0136
2SA213GN	0136
2SA213H	0136
2SA213J	0136
2SA213K	0136
2SA213L	0136
2SA213M	0136
2SA213OR	0136
2SA213R	0136
2SA213X	0136
2SA213Y	0136
2SA214	0136
2SA214A	0136
2SA214B	0136
2SA214C	0136
2SA214D	0136
2SA214E	0136
2SA214F	0136
2SA214G	0136
2SA214GN	0136
2SA214H	0136
2SA214J	0136
2SA214K	0136
2SA214L	0136
2SA214M	0136
2SA214OR	0136
2SA214R	0136
2SA214X	0136
2SA214Y	0136
2SA215	0136
2SA215A	0136
2SA215B	0136
2SA215C	0136
2SA215D	0136
2SA215E	0136
2SA215F	0136
2SA215G	0136
2SA215GN	0136
2SA215H	0136
2SA215J	0136
2SA215K	0136
2SA215L	0136
2SA215M	0136
2SA215OR	0136
2SA215R	0136
2SA215X	0136
2SA215Y	0136
2SA216	0136
2SA216A	0136
2SA216B	0136
2SA216C	0136
2SA216D	0136
2SA216E	0136
2SA216F	0136
2SA216G	0136
2SA216GN	0136
2SA216H	0136
2SA216J	0136
2SA216K	0136
2SA216L	0136
2SA216M	0136
2SA216OR	0136
2SA216R	0136
2SA216X	0136
2SA216Y	0136
2SA217	0280
2SA217A	0280
2SA217B	0280
2SA217C	0280
2SA217D	0280
2SA217E	0280
2SA217F	0280
2SA217G	0280
2SA217GN	0280
2SA217H	0280
2SA217J	0280
2SA217K	0280
2SA217L	0280
2SA217M	0280
2SA217OR	0280
2SA217R	0280
2SA217X	0280
2SA217Y	0280
2SA218	0050
2SA218A	0050
2SA218B	0050
2SA218C	0050
2SA218D	0050
2SA218E	0050
2SA218F	0050
2SA218G	0050
2SA218GN	0050
2SA218H	0050
2SA218J	0050
2SA218K	0050
2SA218L	0050
2SA218M	0050
2SA218OR	0050
2SA218R	0050
2SA218X	0050
2SA218Y	0050
2SA219	0136
2SA219A	0136
2SA219B	0136
2SA219C	0136
2SA219D	0136
2SA219E	0136
2SA219F	0136
2SA219G	0136
2SA219GN	0136
2SA219H	0136
2SA219J	0136
2SA219K	0136
2SA219L	0136
2SA219M	0136
2SA219OR	0136
2SA219R	0136
2SA219X	0136
2SA219Y	0136
2SA220	0136
2SA220A	0136
2SA220B	0136
2SA220C	0136
2SA220D	0136
2SA220E	0136
2SA220F	0136
2SA220G	0136
2SA220GN	0136
2SA220H	0136
2SA220J	0136
2SA220K	0136
2SA220L	0136
2SA220M	0136
2SA220OR	0136
2SA220R	0136
2SA220X	0136
2SA220Y	0136
2SA221	0136
2SA221-OR	0136
2SA221A	0136
2SA221B	0136
2SA221C	0136
2SA221D	0136
2SA221E	0136
2SA221F	0136
2SA221G	0136
2SA221GN	0136
2SA221H	0136
2SA221J	0136
2SA221K	0136
2SA221L	0136
2SA221M	0136
2SA221OR	0136
2SA221R	0136
2SA221X	0136
2SA221Y	0136
2SA222	0136
2SA222A	0136
2SA222B	0136
2SA222C	0136
2SA222D	0136
2SA222E	0136
2SA222F	0136
2SA222G	0136
2SA222GN	0136
2SA222H	0136
2SA222J	0136
2SA222K	0136
2SA222L	0136
2SA222M	0136
2SA222OR	0136
2SA222R	0136
2SA222X	0136
2SA222Y	0136
2SA223	0136
2SA223A	0136
2SA223B	0136
2SA223C	0136
2SA223D	0136
2SA223E	0136
2SA223F	0136
2SA223G	0136
2SA223GN	0136
2SA223H	0136
2SA223J	0136
2SA223K	0136
2SA223L	0136
2SA223M	0136
2SA223OR	0136
2SA223R	0136
2SA223X	0136
2SA223Y	0136
2SA224	0050
2SA224A	0050
2SA224B	0050
2SA224C	0050
2SA224D	0050
2SA224E	0050
2SA224F	0050
2SA224G	0050
2SA224GN	0050
2SA224H	0050
2SA224J	0050
2SA224K	0050
2SA224L	0050
2SA224M	0050
2SA224OR	0050
2SA224R	0050
2SA224X	0050
2SA224Y	0050
2SA225	0633
2SA225A	0633
2SA225B	0633
2SA225C	0633
2SA225D	0633
2SA225E	0633
2SA225F	0633
2SA225G	0633
2SA225GN	0633
2SA225H	0633
2SA225J	0633
2SA225K	0633
2SA225L	0633
2SA225M	0633
2SA225OR	0633
2SA225R	0633
2SA225X	0633
2SA225Y	0633
2SA226	0050
2SA227	0050
2SA227A	0050
2SA227B	0050
2SA227C	0050
2SA227D	0050
2SA227E	0050
2SA227F	0050
2SA227G	0050
2SA227GN	0050
2SA227H	0050
2SA227J	0050
2SA227K	0050
2SA227L	0050
2SA227M	0050
2SA227OR	0050
2SA227R	0050
2SA227X	0050
2SA227Y	0050
2SA228	0050
2SA229	0050
2SA229A	0050
2SA229B	0050
2SA229C	0050
2SA229D	0050
2SA229E	0050
2SA229F	0050
2SA229G	0050
2SA229GN	0050
2SA229H	0050
2SA229J	0050
2SA229K	0050
2SA229L	0050
2SA229OR	0050
2SA229R	0050
2SA229X	0050
2SA229Y	0050
2SA230	0050
2SA230A	0050
2SA230B	0050
2SA230C	0050
2SA230D	0050
2SA230E	0050
2SA230F	0050
2SA230G	0050
2SA230GN	0050
2SA230H	0050
2SA230J	0050
2SA230K	0050
2SA230L	0050
2SA230M	0050
2SA230OR	0050
2SA230R	0050
2SA230X	0050
2SA230Y	0050
2SA231	0847
2SA232	0847
2SA233	0050
2SA233A	0050
2SA233B	0050
2SA233C	0050
2SA233D	0050
2SA233E	0050
2SA233F	0050
2SA233G	0050
2SA233GN	0050
2SA233H	0050
2SA233J	0050
2SA233K	0050
2SA233L	0050
2SA233M	0050
2SA233OR	0050
2SA233R	0050
2SA233X	0050
2SA233Y	0050
2SA234	0050
2SA234A	0050
2SA234B	0050
2SA234C	0050
2SA234D	0050
2SA234E	0050
2SA234F	0050
2SA234G	0050
2SA234GN	0050
2SA234H	0050
2SA234J	0050
2SA234K	0050
2SA234L	0050
2SA234M	0050
2SA234OR	0050
2SA234R	0050
2SA234X	0050
2SA234Y	0050
2SA235	0050
2SA235A	0050
2SA235A57B2-14	0050
2SA235B	0050
2SA235C	0050
2SA235C57B2-13	0050
2SA235D	0050
2SA235E	0050
2SA235F	0050
2SA235G	0050
2SA235GN	0050
2SA235H	0050
2SA235K	0050
2SA235M	0050
2SA235OR	0050
2SA235R	0050
2SA235X	0050
2SA235Y	0050
2SA236	0136
2SA236A	0136
2SA236B	0136
2SA236C	0136
2SA236D	0136
2SA236E	0136
2SA236F	0136
2SA236G	0136
2SA236GN	0136
2SA236H	0136
2SA236J	0136
2SA236K	0136
2SA236L	0136
2SA236M	0136
2SA236OR	0136
2SA236R	0136
2SA236X	0136
2SA236Y	0136
2SA237	0136
2SA237A	0136
2SA237B	0136
2SA237C	0136
2SA237D	0136
2SA237E	0136
2SA237F	0136
2SA237G	0136
2SA237GN	0136
2SA237H	0136
2SA237J	0136
2SA237K	0136
2SA237L	0136
2SA237M	0136
2SA237OR	0136
2SA237R	0136
2SA237X	0136
2SA237Y	0136
2SA238	0136
2SA238A	0136
2SA238B	0136
2SA238C	0136
2SA238D	0136
2SA238E	0136
2SA238F	0136
2SA238G	0136
2SA238GN	0136
2SA238H	0136
2SA238J	0136
2SA238K	0136
2SA238L	0136
2SA239	0050
2SA239A	0050
2SA239B	0050
2SA239C	0050
2SA239D	0050
2SA239E	0050
2SA239F	0050
2SA239G	0050
2SA239GN	0050
2SA239GREEN	0050
2SA239H	0050
2SA239J	0050
2SA239K	0050
2SA239L	0050
2SA239M	0050
2SA239OR	0050
2SA239R	0050
2SA239RED	0050
2SA239X	0050
2SA239Y	0050
2SA240	0050
2SA240-BL	OEM
2SA240-OR	OEM
2SA240A	0050
2SA240B	0050
2SA240B2	0050
2SA240BL	0050
2SA240C	0050
2SA240D	0050
2SA240E	0050
2SA240F	0050
2SA240G	0050
2SA240GN	0050
2SA240GREEN	0050
2SA240H	0050
2SA240J	0050
2SA240K	0050
2SA240L	0050
2SA240M	0050
2SA240OR	0050
2SA240R	0050
2SA240RED	0050
2SA240X	0050
2SA240Y	0050
2SA241	0050
2SA241A	0050
2SA241B	0050
2SA241C	0050
2SA241D	0050
2SA241E	0050
2SA241F	0050
2SA241G	0050
2SA241GN	0050
2SA241H	0050
2SA241J	0050
2SA241K	0050
2SA241L	0050
2SA241M	0050
2SA241OR	0050
2SA241R	0050
2SA241X	0050
2SA241Y	0050
2SA242	0050
2SA243	0050
2SA243A	0050
2SA244	0050
2SA244A	0050
2SA245	0050
2SA246	0136
2SA246A	0136
2SA246B	0136
2SA246C	0136
2SA246D	0136
2SA246E	0136
2SA246F	0136
2SA246G	0136
2SA246GN	0136
2SA246H	0136
2SA246J	0136
2SA246K	0136
2SA246L	0136
2SA246M	0136
2SA246OR	0136
2SA246R	0136
2SA246V	0136
2SA246X	0136
2SA246Y	0136
2SA247	0136
2SA247A	0136
2SA247B	0136
2SA247C	0136
2SA247D	0136
2SA247E	0136
2SA247F	0136
2SA247G	0136
2SA247GN	0136
2SA247H	0136
2SA247J	0136
2SA247K	0136
2SA247L	0136
2SA247M	0136
2SA247OR	0136
2SA247R	0136
2SA247X	0136
2SA247Y	0136
2SA248	0004
2SA248A	0004
2SA248B	0004
2SA248C	0004
2SA248D	0004
2SA248E	0004
2SA248F	0004
2SA248G	0004
2SA248GN	0004
2SA248H	0004
2SA248J	0004
2SA248K	0004
2SA248L	0004
2SA248M	0004
2SA248OR	0004
2SA248R	0004
2SA248X	0004
2SA248Y	0004
2SA249	0164
2SA250	0136
2SA250A	0136
2SA250B	0136
2SA250C	0136
2SA250D	0136
2SA250E	0136
2SA250F	0136
2SA250G	0136
2SA250GN	0136
2SA250H	0136
2SA250J	0136
2SA250L	0136
2SA250M	0136
2SA250OR	0136
2SA250X	0136
2SA250Y	0136
2SA251	0136
2SA251A	0136
2SA251B	0136
2SA251C	0136
2SA251D	0136
2SA251E	0136
2SA251F	0136
2SA251G	0136
2SA251H	0136
2SA251J	0136
2SA251K	0136
2SA251L	0136
2SA251M	0136
2SA251OR	0136
2SA251R	0136
2SA251X	0136
2SA251Y	0136
2SA252	0136
2SA252A	0136
2SA252B	0136
2SA252C	0136
2SA252D	0136
2SA252E	0136
2SA252F	0136
2SA252G	0136
2SA252H	0136
2SA252J	0136
2SA252K	0136
2SA252L	0136
2SA252M	0136
2SA252OR	0136
2SA252R	0136
2SA252X	0136
2SA252Y	0136

Original Device Types

DEVICE TYPE	REPL CODE
2SA253	0136
2SA253A	0136
2SA253B	0136
2SA253C	0136
2SA253D	0136
2SA253E	0136
2SA253F	0136
2SA253G	0136
2SA253GN	0136
2SA253H	0136
2SA253J	0136
2SA253K	0136
2SA253L	0136
2SA253M	0136
2SA253OR	0136
2SA253R	0136
2SA253X	0136
2SA253Y	0136
2SA254	0136
2SA254A	0136
2SA254B	0136
2SA254C	0136
2SA254D	0136
2SA254E	0136
2SA254F	0136
2SA254G	0136
2SA254H	0136
2SA254J	0136
2SA254K	0136
2SA254L	0136
2SA254M	0136
2SA254OR	0136
2SA254R	0136
2SA254X	0136
2SA254Y	0136
2SA255	0136
2SA255A	0136
2SA255B	0136
2SA255C	0136
2SA255D	0136
2SA255E	0136
2SA255F	0136
2SA255G	0136
2SA255GN	0136
2SA255H	0136
2SA255J	0136
2SA255K	0136
2SA255L	0136
2SA255M	0136
2SA255OR	0136
2SA255R	0136
2SA255X	0136
2SA255Y	0136
2SA256	0136
2SA256A	0136
2SA256B	0136
2SA256C	0136
2SA256D	0136
2SA256E	0136
2SA256F	0136
2SA256G	0136
2SA256GN	0136
2SA256H	0136
2SA256J	0136
2SA256K	0136
2SA256L	0136
2SA256M	0136
2SA256OR	0136
2SA256R	0136
2SA256X	0136
2SA256Y	0136
2SA257	0136
2SA257A	0136
2SA257B	0136
2SA257C	0136
2SA257D	0136
2SA257E	0136
2SA257F	0136
2SA257G	0136
2SA257GN	0136
2SA257H	0136
2SA257J	0136
2SA257K	0136
2SA257L	0136
2SA257OR	0136
2SA257R	0136
2SA257X	0136
2SA257Y	0136
2SA258	0136
2SA258A	0136
2SA258B	0136
2SA258C	0136
2SA258D	0136
2SA258E	0136
2SA258F	0136
2SA258G	0136
2SA258GN	0136
2SA258H	0136
2SA258J	0136
2SA258K	0136
2SA258L	0136
2SA258M	0136
2SA258OR	0136
2SA258R	0136
2SA258X	0136
2SA258Y	0136
2SA259	0136
2SA259B	0136
2SA259C	0136
2SA259D	0136
2SA259E	0136
2SA259F	0136
2SA259G	0136
2SA259GN	0136
2SA259H	0136
2SA259J	0136
2SA259L	0136
2SA259M	0136
2SA259OR	0136
2SA259R	0136
2SA259X	0136
2SA259Y	0136
2SA260	0050
2SA260A	0050
2SA260B	0050
2SA260C	0050
2SA260D	0050
2SA260E	0050
2SA260F	0050
2SA260G	0050
2SA260GN	0050
2SA260H	0050
2SA260J	0050
2SA260K	0050
2SA260L	0050
2SA260M	0050
2SA260OR	0050
2SA260R	0050
2SA260X	0050
2SA260Y	0050
2SA261	0050
2SA261A	0050
2SA261B	0050
2SA261C	0050
2SA261D	0050
2SA261E	0050
2SA261F	0050
2SA261G	0050
2SA261GN	0050
2SA261H	0050
2SA261K	0050
2SA261L	0050
2SA261M	0050
2SA261OR	0050
2SA261R	0050
2SA261X	0050
2SA261Y	0050
2SA262	0050
2SA262A	0050
2SA262B	0050
2SA262C	0050
2SA262D	0050
2SA262E	0050
2SA262F	0050
2SA262G	0050
2SA262GN	0050
2SA262H	0050
2SA262J	0050
2SA262K	0050
2SA262L	0050
2SA262M	0050
2SA262OR	0050
2SA262R	0050
2SA262X	0050
2SA262Y	0050
2SA263	0050
2SA263A	0050
2SA263B	0050
2SA263C	0050
2SA263D	0050
2SA263E	0050
2SA263F	0050
2SA263G	0050
2SA263GN	0050
2SA263H	0050
2SA263J	0050
2SA263K	0050
2SA263L	0050
2SA263M	0050
2SA263OR	0050
2SA263R	0050
2SA263X	0050
2SA263Y	0050
2SA264	0050
2SA264(1)	0050
2SA264A	0050
2SA264B	0050
2SA264C	0050
2SA264D	0050
2SA264E	0050
2SA264F	0050
2SA264G	0050
2SA264GN	0050
2SA264H	0050
2SA264K	0050
2SA264L	0050
2SA264M	0050
2SA264OR	0050
2SA264R	0050
2SA264X	0050
2SA264Y	0050
2SA265	0050
2SA265A	0050
2SA265B	0050
2SA265C	0050
2SA265D	0050
2SA265E	0050
2SA265F	0050
2SA265G	0050
2SA265GN	0050
2SA265H	0050
2SA265J	0050
2SA265K	0050
2SA265L	0050
2SA265M	0050
2SA265OR	0050
2SA265R	0050
2SA265X	0050
2SA265Y	0050
2SA266	0136
2SA266A	0136
2SA266B	0136
2SA266C	0136
2SA266D	0136
2SA266E	0136
2SA266F	0136
2SA266G	0136
2SA266GN	0136
2SA266GREEN	0136
2SA266H	0136
2SA266J	0136
2SA266K	0136
2SA266L	0136
2SA266M	0136
2SA266OR	0136
2SA266R	0136
2SA266X	0136
2SA266Y	0136
2SA267	0136
2SA267A	0136
2SA267B	0136
2SA267C	0136
2SA267D	0136
2SA267E	0136
2SA267F	0136
2SA267G	0136
2SA267GN	0136
2SA267H	0136
2SA267J	0136
2SA267K	0136
2SA267L	0136
2SA267M	0136
2SA267OR	0136
2SA267R	0136
2SA267X	0136
2SA267Y	0136
2SA268	0136
2SA268A	0136
2SA268B	0136
2SA268C	0136
2SA268D	0136
2SA268E	0136
2SA268F	0136
2SA268G	0136
2SA268GN	0136
2SA268H	0136
2SA268J	0136
2SA268K	0136
2SA268L	0136
2SA268M	0136
2SA268OR	0136
2SA268R	0136
2SA268X	0136
2SA268Y	0136
2SA269	0136
2SA269A	0136
2SA269B	0136
2SA269C	0136
2SA269D	0136
2SA269E	0136
2SA269F	0136
2SA269GN	0136
2SA269H	0136
2SA269J	0136
2SA269K	0136
2SA269L	0136
2SA269M	0136
2SA269OR	0136
2SA269X	0136
2SA269Y	0136
2SA270	0136
2SA270A	0136
2SA270B	0136
2SA270C	0136
2SA270D	0136
2SA270E	0136
2SA270F	0136
2SA270G	0136
2SA270GN	0136
2SA270H	0136
2SA270J	0136
2SA270K	0136
2SA270L	0136
2SA270M	0136
2SA270OR	0136
2SA270R	0136
2SA270X	0136
2SA270Y	0136
2SA271	0136
2SA271(2)	0136
2SA271(3)	0136
2SA271A	0136
2SA271B	0136
2SA271C	0136
2SA271D	0136
2SA271E	0136
2SA271F	0136
2SA271G	0136
2SA271GN	0136
2SA271H	0136
2SA271J	0136
2SA271K	0136
2SA271L	0136
2SA271M	0136
2SA271OR	0136
2SA271R	0136
2SA271X	0136
2SA271Y	0136
2SA272	0136
2SA272A	0136
2SA272B	0136
2SA272C	0136
2SA272D	0136
2SA272E	0136
2SA272F	0136
2SA272G	0136
2SA272GN	0136
2SA272H	0136
2SA272J	0136
2SA272K	0136
2SA272L	0136
2SA272M	0136
2SA272OR	0136
2SA272R	0136
2SA272X	0136
2SA272Y	0136
2SA273	0136
2SA273A	0136
2SA273B	0136
2SA273C	0136
2SA273D	0136
2SA273E	0136
2SA273F	0136
2SA273G	0136
2SA273GN	0136
2SA273H	0136
2SA273J	0136
2SA273K	0136
2SA273L	0136
2SA273M	0136
2SA273OR	0136
2SA273X	0136
2SA273Y	0136
2SA274	0136
2SA274A	0136
2SA274B	0136
2SA274C	0136
2SA274D	0136
2SA274E	0136
2SA274F	0136
2SA274G	0136
2SA274GN	0136
2SA274H	0136
2SA274J	0136
2SA274K	0136
2SA274L	0136
2SA274M	0136
2SA274OR	0136
2SA274R	0136
2SA274X	0136
2SA274Y	0136
2SA275	0136
2SA275A	0136
2SA275B	0136
2SA275C	0136
2SA275D	0136
2SA275E	0136
2SA275F	0136
2SA275G	0136
2SA275GN	0136
2SA275H	0136
2SA275J	0136
2SA275L	0136
2SA275M	0136
2SA275OR	0136
2SA275R	0136
2SA275X	0136
2SA275Y	0136
2SA276	0050
2SA277	0280
2SA277A	0280
2SA277B	0280
2SA277C	0280
2SA277D	0280
2SA277E	0280
2SA277F	0280
2SA277G	0280
2SA277GN	0280
2SA277H	0280
2SA277J	0280
2SA277K	0280
2SA277L	0280
2SA277M	0280
2SA277OR	0280
2SA277X	0280
2SA277Y	0280
2SA278	0280
2SA278A	0280
2SA278B	0280
2SA278C	0280
2SA278D	0280
2SA278E	0280
2SA278F	0280
2SA278G	0280
2SA278GN	0280
2SA278J	0280
2SA278K	0280
2SA278L	0280
2SA278M	0280
2SA278OR	0280
2SA278R	0280
2SA278X	0280
2SA278Y	0280
2SA279	0280
2SA279A	0280
2SA279B	0280
2SA279C	0280
2SA279D	0280
2SA279E	0280
2SA279F	0280
2SA279G	0280
2SA279GN	0280
2SA279H	0280
2SA279J	0280
2SA279K	0280
2SA279L	0280
2SA279OR	0280
2SA279R	0280
2SA279X	0280
2SA279Y	0280
2SA280	0280
2SA280A	0280
2SA280B	0280
2SA280C	0280
2SA280D	0280
2SA280E	0280
2SA280F	0280
2SA280G	0280
2SA280GN	0280
2SA280H	0280
2SA280J	0280
2SA280K	0280
2SA280L	0280
2SA280M	0280
2SA280OR	0280
2SA280R	0280
2SA280X	0280
2SA280Y	0280
2SA281	0280
2SA281A	0280
2SA281B	0280
2SA281C	0280
2SA281D	0280
2SA281E	0280
2SA281F	0280
2SA281G	0280
2SA281GN	0280
2SA281H	0280
2SA281J	0280
2SA281K	0280
2SA281L	0280
2SA281M	0280
2SA281OR	0280
2SA281R	0280
2SA281X	0280
2SA281Y	0280
2SA282	0004
2SA282A	0004
2SA282B	0004
2SA282C	0004
2SA282D	0004
2SA282E	0004
2SA282F	0004
2SA282G	0004
2SA282GN	0004
2SA282H	0004
2SA282J	0004
2SA282K	0004
2SA282L	0004
2SA282M	0004
2SA282OR	0004
2SA282R	0004
2SA282X	0004
2SA282Y	0004
2SA283	0280
2SA283A	0280
2SA283B	0280
2SA283C	0280
2SA283D	0280
2SA283E	0280
2SA283F	0280
2SA283G	0280
2SA283GN	0280
2SA283H	0280
2SA283J	0280
2SA283K	0280
2SA283L	0280
2SA283M	0280
2SA283OR	0280
2SA283R	0280
2SA283X	0280
2SA283Y	0280
2SA284	0280
2SA284A	0280
2SA284B	0280
2SA284C	0280
2SA284D	0280
2SA284E	0280
2SA284F	0280
2SA284G	0280
2SA284GN	0280
2SA284J	0280
2SA284K	0280
2SA284L	0280
2SA284M	0280
2SA284OR	0280
2SA284R	0280
2SA284X	0280
2SA284Y	0280
2SA285	0050
2SA285B	0050
2SA285C	0050
2SA285E	0050
2SA285F	0050
2SA285G	0050
2SA285GN	0050
2SA285H	0050
2SA285J	0050
2SA285L	0050
2SA285M	0050
2SA285OR	0050
2SA285R	0050
2SA285X	0050
2SA285Y	0050
2SA286	0050
2SA286A	0050
2SA286B	0050
2SA286C	0050
2SA286D	0050
2SA286E	0050
2SA286F	0050
2SA286G	0050
2SA286GN	0050
2SA286H	0050
2SA286K	0050
2SA286M	0050
2SA286OR	0050
2SA286R	0050
2SA286X	0050
2SA286Y	0050
2SA287	0050
2SA287A	0050
2SA287B	0050
2SA287C	0050
2SA287D	0050
2SA287E	0050
2SA287F	0050
2SA287G	0050
2SA287GN	0050
2SA287H	0050
2SA287J	0050
2SA287K	0050
2SA287L	0050
2SA287M	0050
2SA287OR	0050
2SA287X	0050
2SA287Y	0050
2SA288	0136
2SA288A	0136
2SA288B	0136
2SA288C	0136
2SA288D	0136
2SA288E	0136
2SA288F	0136
2SA288G	0136
2SA288GN	0136
2SA288J	0136
2SA288K	0136
2SA288L	0136
2SA288M	0136
2SA288OR	0136
2SA288X	0136
2SA288Y	0136
2SA289	0136
2SA289A	0136
2SA289B	0136
2SA289C	0136
2SA289D	0136
2SA289E	0136
2SA289F	0136
2SA289G	0136
2SA289GN	0136
2SA289H	0136
2SA289J	0136
2SA289K	0136
2SA289L	0136
2SA289M	0136
2SA289OR	0136
2SA289R	0136
2SA289X	0136
2SA289Y	0136
2SA290	0136
2SA290A	0136
2SA290B	0136
2SA290C	0136
2SA290D	0136
2SA290E	0136
2SA290F	0136
2SA290G	0136
2SA290GN	0136
2SA290H	0136
2SA290J	0136
2SA290K	0136
2SA290L	0136
2SA290M	0136
2SA290OR	0136
2SA290R	0136
2SA290X	0136
2SA290Y	0136
2SA291	0136
2SA291A	0136
2SA291B	0136
2SA291C	0136
2SA291D	0136
2SA291E	0136
2SA291F	0136
2SA291G	0136
2SA291GN	0136
2SA291H	0136
2SA291J	0136
2SA291K	0136
2SA291L	0136
2SA291M	0136
2SA291OR	0136
2SA291R	0136
2SA291X	0136
2SA291Y	0136
2SA292	0136
2SA292A	0136
2SA292B	0136
2SA292C	0136
2SA292D	0136
2SA292E	0136
2SA292F	0136
2SA292G	0136
2SA292GN	0136
2SA292H	0136
2SA292J	0136
2SA292K	0136
2SA292L	0136
2SA292M	0136
2SA292OR	0136
2SA292R	0136
2SA292X	0136
2SA292Y	0136
2SA293	0136
2SA293A	0136
2SA293B	0136
2SA293C	0136
2SA293D	0136
2SA293E	0136
2SA293G	0136
2SA293GN	0136
2SA293H	0136
2SA293J	0136
2SA293K	0136
2SA293L	0136
2SA293M	0136
2SA293OR	0136
2SA293R	0136
2SA293X	0136
2SA293Y	0136
2SA294	0136
2SA294A	0136
2SA294B	0136
2SA294C	0136
2SA294D	0136
2SA294E	0136
2SA294F	0136
2SA294GN	0136
2SA294H	0136
2SA294J	0136
2SA294K	0136
2SA294L	0136
2SA294M	0136
2SA294OR	0136
2SA294R	0136
2SA294X	0136
2SA294Y	0136
2SA295	0136
2SA295A	0136
2SA295B	0136
2SA295C	0136
2SA295D	0136
2SA295E	0136
2SA295F	0136
2SA295G	0136
2SA295GN	0136
2SA295H	0136
2SA295J	0136
2SA295K	0136
2SA295L	0136
2SA295OR	0136
2SA295R	0136
2SA295Y	0136
2SA296	0136
2SA296A	0136
2SA296B	0136
2SA296C	0136
2SA296D	0136
2SA296E	0136
2SA296F	0136
2SA296G	0136
2SA296GN	0136
2SA296H	0136
2SA296J	0136
2SA296K	0136
2SA296L	0136
2SA296M	0136
2SA296OR	0136
2SA296R	0136
2SA296X	0136
2SA296Y	0136
2SA297	0136
2SA297A	0136
2SA297B	0136
2SA297C	0136
2SA297D	0136
2SA297E	0136
2SA297F	0136
2SA297G	0136
2SA297H	0136
2SA297J	0136

If replacement code is OEM, contact original manufacturer for replacement.

DEVICE TYPE	REPL CODE
2SA297K	0136
2SA297L	0136
2SA297M	0136
2SA297OR	0136
2SA297R	0136
2SA297X	0136
2SA297Y	0136
2SA298	0050
2SA298A	0050
2SA298B	0050
2SA298C	0050
2SA298D	0050
2SA298E	0050
2SA298F	0050
2SA298G	0050
2SA298GN	0050
2SA298H	0050
2SA298J	0050
2SA298K	0050
2SA298L	0050
2SA298M	0050
2SA298OR	0050
2SA298R	0050
2SA298X	0050
2SA298Y	0050
2SA299	0280
2SA300	0050
2SA301	0164
2SA301A	0164
2SA301B	0164
2SA301C	0164
2SA301D	0164
2SA301E	0164
2SA301F	0164
2SA301G	0164
2SA301GN	0164
2SA301H	0164
2SA301J	0164
2SA301K	0164
2SA301L	0164
2SA301M	0164
2SA301OR	0164
2SA301R	0164
2SA301X	0164
2SA301Y	0164
2SA302	0004
2SA303	0004
2SA304	0280
2SA304-GREEN	0280
2SA304-RED	0280
2SA304-YELLOW	0280
2SA304A	0280
2SA304B	0280
2SA304C	0280
2SA304D	0280
2SA304F	0280
2SA304G	0280
2SA304GN	0280
2SA304J	0280
2SA304K	0280
2SA304L	0280
2SA304M	0280
2SA304OR	0280
2SA304R	0280
2SA304X	0280
2SA304Y	0280
2SA305	0280
2SA305-GREEN	0280
2SA305-RED	0136
2SA305-YELLOW	0280
2SA305A	0280
2SA305B	0280
2SA305C	0280
2SA305D	0280
2SA305E	0280
2SA305F	0280
2SA305G	0280
2SA305GN	0280
2SA305H	0280
2SA305J	0280
2SA305K	0280
2SA305L	0280
2SA305M	0280
2SA305OR	0280
2SA305R	0280
2SA305X	0280
2SA305Y	0280
2SA306	0050
2SA307	0050
2SA307A	0050
2SA307B	0050
2SA307C	0050
2SA307D	0050
2SA307E	0050
2SA307F	0050
2SA307G	0050
2SA307GN	0050
2SA307J	0050
2SA307K	0050
2SA307L	0050
2SA307M	0050
2SA307OR	0050
2SA307R	0050
2SA307X	0050
2SA307Y	0050
2SA308	0136
2SA308A	0136
2SA308B	0136
2SA308C	0136
2SA308D	0136
2SA308E	0136
2SA308F	0136
2SA308G	0136
2SA308GN	0136
2SA308H	0136
2SA308J	0136
2SA308K	0136
2SA308L	0136
2SA308M	0136
2SA308OR	0136
2SA308R	0136
2SA308X	0136
2SA308Y	0136
2SA309	1785
2SA309A	1785
2SA309B	1785
2SA309C	1785
2SA309D	1785
2SA309E	1785
2SA309F	1785
2SA309G	1785
2SA309GN	1785
2SA309H	1785
2SA309J	1785
2SA309K	1785
2SA309L	1785
2SA309M	1785
2SA309OR	1785
2SA309R	1785
2SA309X	1785
2SA309Y	1785
2SA310	0136
2SA310A	0136
2SA310B	0136
2SA310C	0136
2SA310D	0136
2SA310E	0136
2SA310F	0136
2SA310G	0136
2SA310GN	0136
2SA310H	0136
2SA310J	0136
2SA310K	0136
2SA310L	0136
2SA310M	0136
2SA310OR	0136
2SA310R	0136
2SA310X	0136
2SA310Y	0136
2SA311	0004
2SA311A	0004
2SA311B	0004
2SA311C	0004
2SA311D	0004
2SA311E	0004
2SA311F	0004
2SA311G	0004
2SA311GN	0004
2SA311H	0004
2SA311J	0004
2SA311K	0004
2SA311L	0004
2SA311M	0004
2SA311OR	0004
2SA311R	0004
2SA311X	0004
2SA311Y	0004
2SA312	0004
2SA312A	0004
2SA312B	0004
2SA312C	0004
2SA312D	0004
2SA312E	0004
2SA312F	0004
2SA312G	0004
2SA312GN	0004
2SA312H	0004
2SA312J	0004
2SA312K	0004
2SA312L	0004
2SA312M	0004
2SA312OR	0004
2SA312R	0004
2SA312X	0004
2SA312Y	0004
2SA313	0136
2SA313-BLUE	0136
2SA313-GREEN	0136
2SA313-RED	0136
2SA313-YELLOW	0136
2SA313A	0136
2SA313B	0136
2SA313C	0136
2SA313D	0136
2SA313E	0136
2SA313F	0136
2SA313G	0136
2SA313GN	0136
2SA313H	0136
2SA313J	0136
2SA313K	0136
2SA313L	0136
2SA313M	0136
2SA313OR	0136
2SA313R	0136
2SA313X	0136
2SA313Y	0136
2SA314	0136
2SA314-GREEN	0136
2SA314-RED	0136
2SA314-YELLOW	0136
2SA314A	0136
2SA314B	0136
2SA314C	0136
2SA314D	0136
2SA314E	0136
2SA314F	0136
2SA314G	0136
2SA314GN	0136
2SA314H	0136
2SA314J	0136
2SA314K	0136
2SA314L	0136
2SA314M	0136
2SA314OR	0136
2SA314R	0136
2SA314X	0136
2SA314Y	0136
2SA315	0136
2SA315-GREEN	0136
2SA315-RED	0136
2SA315-YELLOW	0136
2SA315A	0136
2SA315B	0136
2SA315C	0136
2SA315D	0136
2SA315E	0136
2SA315F	0136
2SA315G	0136
2SA315GN	0136
2SA315H	0136
2SA315J	0136
2SA315K	0136
2SA315L	0136
2SA315M	0136
2SA315OR	0136
2SA315R	0136
2SA315X	0136
2SA315Y	0136
2SA316	0136
2SA316-GREEN	0136
2SA316-RED	0136
2SA316-YELLOW	0136
2SA316A	0136
2SA316B	0136
2SA316C	0136
2SA316D	0136
2SA316E	0136
2SA316F	0136
2SA316G	0136
2SA316GN	0136
2SA316H	0136
2SA316J	0136
2SA316K	0136
2SA316L	0136
2SA316M	0136
2SA316OR	0136
2SA316R	0136
2SA316X	0136
2SA316Y	0136
2SA321	0004
2SA321-1	0004
2SA321A	0004
2SA321B	0004
2SA321C	0004
2SA321D	0004
2SA321E	0004
2SA321F	0004
2SA321G	0004
2SA321GN	0004
2SA321H	0004
2SA321J	0004
2SA321K	0004
2SA321L	0004
2SA321M	0004
2SA321OR	0004
2SA321R	0004
2SA321X	0004
2SA321Y	0004
2SA322	0004
2SA322A	0004
2SA322B	0004
2SA322C	0004
2SA322D	0004
2SA322E	0004
2SA322F	0004
2SA322G	0004
2SA322GN	0004
2SA322H	0004
2SA322K	0004
2SA322L	0004
2SA322M	0004
2SA322OR	0004
2SA322R	0004
2SA322X	0004
2SA322Y	0004
2SA323	0050
2SA323A	0050
2SA323B	0050
2SA323C	0050
2SA323D	0050
2SA323E	0050
2SA323F	0050
2SA323G	0050
2SA323GN	0050
2SA323J	0050
2SA323K	0050
2SA323L	0050
2SA323M	0050
2SA323R	0050
2SA323X	0050
2SA323Y	0050
2SA324	0136
2SA324A	0136
2SA324B	0136
2SA324C	0136
2SA324D	0136
2SA324E	0136
2SA324F	0136
2SA324G	0136
2SA324GN	0136
2SA324H	0136
2SA324K	0136
2SA324L	0136
2SA324M	0136
2SA324OR	0136
2SA324R	0136
2SA324X	0136
2SA324Y	0136
2SA325	0136
2SA325A	0136
2SA325B	0136
2SA325D	0136
2SA325E	0136
2SA325G	0136
2SA325GN	0136
2SA325H	0136
2SA325J	0136
2SA325K	0136
2SA325L	0136
2SA325M	0136
2SA325OR	0136
2SA325R	0136
2SA325X	0136
2SA325Y	0136
2SA326	0136
2SA326A	0136
2SA326B	0136
2SA326C	0136
2SA326D	0136
2SA326E	0136
2SA326F	0136
2SA326G	0136
2SA326GN	0136
2SA326H	0136
2SA326J	0136
2SA326K	0136
2SA326L	0136
2SA326M	0136
2SA326OR	0136
2SA326R	0136
2SA326X	0136
2SA326Y	0136
2SA327	0050
2SA328	0280
2SA329	0136
2SA329A	0136
2SA329B	0136
2SA329C	0136
2SA329D	0136
2SA329E	0136
2SA329F	0136
2SA329G	0136
2SA329GN	0136
2SA329H	0136
2SA329J	0136
2SA329K	0136
2SA329L	0136
2SA329M	0136
2SA329OR	0136
2SA329R	0136
2SA329X	0136
2SA329Y	0136
2SA330	0136
2SA330A	0136
2SA330B	0136
2SA330C	0136
2SA330D	0136
2SA330E	0136
2SA330F	0136
2SA330G	0136
2SA330GN	0136
2SA330H	0136
2SA330J	0136
2SA330K	0136
2SA330L	0136
2SA330M	0136
2SA330OR	0136
2SA330X	0136
2SA330Y	0136
2SA331	0136
2SA331A	0136
2SA331B	0136
2SA331C	0136
2SA331D	0136
2SA331E	0136
2SA331F	0136
2SA331G	0136
2SA331GN	0136
2SA331H	0136
2SA331J	0136
2SA331K	0136
2SA331L	0136
2SA331M	0136
2SA331OR	0136
2SA331R	0136
2SA331X	0136
2SA331Y	0136
2SA332	0050
2SA332A	0050
2SA332B	0050
2SA332C	0050
2SA332D	0050
2SA332E	0050
2SA332F	0050
2SA332G	0050
2SA332GN	0050
2SA332H	0050
2SA332J	0050
2SA332K	0050
2SA332L	0050
2SA332M	0050
2SA332OR	0050
2SA332R	0050
2SA332X	0050
2SA332Y	0050
2SA333	5213
2SA334	5213
2SA335	0136
2SA335A	0136
2SA335B	0136
2SA335C	0136
2SA335D	0136
2SA335E	0136
2SA335F	0136
2SA335G	0136
2SA335GN	0136
2SA335H	0136
2SA335J	0136
2SA335K	0136
2SA335L	0136
2SA335M	0136
2SA335OR	0136
2SA335R	0136
2SA335X	0136
2SA335Y	0136
2SA336	0050
2SA337	0136
2SA337A	0136
2SA337B	0136
2SA337C	0136
2SA337D	0136
2SA337E	0136
2SA337F	0136
2SA337G	0136
2SA337GN	0136
2SA337H	0136
2SA337J	0136
2SA337K	0136
2SA337L	0136
2SA337M	0136
2SA337OR	0136
2SA337R	0136
2SA337X	0136
2SA337Y	0136
2SA338	0136
2SA338A	0136
2SA338B	0136
2SA338C	0136
2SA338D	0136
2SA338E	0136
2SA338F	0136
2SA338G	0136
2SA338GN	0136
2SA338H	0136
2SA338J	0136
2SA338K	0136
2SA338L	0136
2SA338M	0136
2SA338OR	0136
2SA338R	0136
2SA338X	0136
2SA338Y	0136
2SA339	0136
2SA339A	0136
2SA339B	0136
2SA339C	0136
2SA339D	0136
2SA339E	0136
2SA339F	0136
2SA339G	0136
2SA339GN	0136
2SA339H	0136
2SA339J	0136
2SA339K	0136
2SA339L	0136
2SA339M	0136
2SA339OR	0136
2SA339R	0136
2SA339X	0136
2SA339Y	0136
2SA340	0050
2SA341	0050
2SA341-NB	OEM
2SA341-0A	0050
2SA341-0B	0050
2SA341A	0050
2SA341B	0050
2SA341C	0050
2SA341D	0050
2SA341E	0050
2SA341F	0050
2SA341G	0050
2SA341GN	0050
2SA341H	0050
2SA341J	0050
2SA341K	0050
2SA341L	0050
2SA341M	0050
2SA341OR	0050
2SA341R	0050
2SA341X	0050
2SA341Y	0050
2SA342	0050
2SA342A	0050
2SA342B	0050
2SA342C	0050
2SA342D	0050
2SA342E	0050
2SA342F	0050
2SA342G	0050
2SA342GN	0050
2SA342H	0050
2SA342J	0050
2SA342K	0050
2SA342L	0050
2SA342M	0050
2SA342OR	0050
2SA342R	0050
2SA342X	0050
2SA342Y	0050
2SA343	0050
2SA343A	0050
2SA343B	0050
2SA343C	0050
2SA343D	0050
2SA343E	0050
2SA343F	0050
2SA343G	0050
2SA343H	0050
2SA343J	0050
2SA343K	0050
2SA343L	0050
2SA343M	0050
2SA343OR	0050
2SA343R	0050
2SA343X	0050
2SA343Y	0050
2SA344	0136
2SA344A	0136
2SA344B	0136
2SA344C	0136
2SA344D	0136
2SA344E	0136
2SA344F	0136
2SA344G	0136
2SA344GN	0136
2SA344H	0136
2SA344J	0136
2SA344K	0136
2SA344L	0136
2SA344M	0136
2SA344OR	0136
2SA344R	0136
2SA344X	0136
2SA344Y	0136
2SA345	0050
2SA346	0050
2SA347	0050
2SA348	0050
2SA348A	0050
2SA348B	0050
2SA348C	0050
2SA348D	0050
2SA348E	0050
2SA348F	0050
2SA348G	0050
2SA348GN	0050
2SA348H	0050
2SA348J	0050
2SA348K	0050
2SA348L	0050
2SA348M	0050
2SA348OR	0050
2SA348R	0050
2SA348X	0050
2SA348Y	0050
2SA349	0050
2SA350	0136
2SA350A	0136
2SA350AV	0136
2SA350B	0136
2SA350BK	0136
2SA350C	0136
2SA350D	0136
2SA350E	0136
2SA350F	0136
2SA350G	0136
2SA350GN	0136
2SA350H	0136
2SA350J	0136
2SA350K	0136
2SA350L	0136
2SA350M	0136
2SA350OR	0136
2SA350R	0136
2SA350T	0136
2SA350TY	0136
2SA350X	0136
2SA350Y	0136
2SA351	0136
2SA351(B)	0136
2SA351(GR)	0136
2SA351A	0136
2SA351A-2	0136
2SA351B	0136
2SA351C	0136
2SA351D	0136
2SA351E	0136
2SA351F	0136
2SA351G	0136
2SA351GN	0136
2SA351GR	0136
2SA351K	0136
2SA351L	0136
2SA351M	0136
2SA351OR	0136
2SA351R	0136
2SA351X	0136
2SA351Y	0136
2SA352	0136
2SA352A	0136
2SA352B	0136
2SA352C	0136
2SA352D	0136
2SA352E	0136
2SA352F	0136
2SA352G	0136
2SA352GN	0136
2SA352H	0136
2SA352J	0136
2SA352K	0136
2SA352L	0136
2SA352M	0136
2SA352OR	0136
2SA352R	0136
2SA352X	0136
2SA352Y	0136
2SA353	0136
2SA353(A)	0136
2SA353(C)	0136
2SA353-AC	0136
2SA353A	0136
2SA353AL	0136
2SA353B	0136
2SA353C	0136
2SA353CL	0136
2SA353D	0136
2SA353E	0136
2SA353F	0136
2SA353G	0136
2SA353GN	0136
2SA353H	0136
2SA353J	0136
2SA353K	0136
2SA353L	0136
2SA353M	0136
2SA353OR	0136
2SA353R	0136
2SA353X	0136
2SA353Y	0136
2SA354	0136
2SA354-B	0136
2SA354A	0136
2SA354B	0136
2SA354BK	0136
2SA354C	0136
2SA354D	0136
2SA354E	0136
2SA354F	0136
2SA354G	0136
2SA354GN	0136
2SA354H	0136
2SA354J	0136
2SA354K	0136
2SA354L	0136
2SA354M	0136
2SA354OR	0136
2SA354R	0136
2SA354X	0136
2SA354Y	0136
2SA355	0136
2SA355A	0136
2SA355B	0136
2SA355C	0136
2SA355D	0136
2SA355E	0136
2SA355F	0136
2SA355G	0136
2SA355H	0136
2SA355J	0136
2SA355K	0136
2SA355L	0136
2SA355M	0136
2SA355OR	0136
2SA355R	0136
2SA355X	0136
2SA355Y	0136
2SA356	0136
2SA356A	0136
2SA356B	0136
2SA356C	0136
2SA356D	0136
2SA356E	0136
2SA356F	0136
2SA356G	0136
2SA356GN	0136
2SA356H	0136
2SA356J	0136
2SA356K	0136
2SA356L	0136
2SA356M	0136
2SA356OR	0136

DEVICE TYPE	REPL CODE
2SA356R	0136
2SA357	0136
2SA357A	0136
2SA357B	0136
2SA357C	0136
2SA357D	0136
2SA357E	0136
2SA357F	0136
2SA357G	0136
2SA357GN	0136
2SA357H	0136
2SA357J	0136
2SA357K	0136
2SA357L	0136
2SA357M	0136
2SA357OR	0136
2SA357R	0136
2SA357X	0136
2SA357Y	0136
2SA358	0136
2SA358-3	0136
2SA358A	0136
2SA358B	0136
2SA358C	0136
2SA358D	0136
2SA358E	0136
2SA358F	0136
2SA358G	0136
2SA358GN	0136
2SA358H	0136
2SA358J	0136
2SA358K	0136
2SA358L	0136
2SA358M	0136
2SA358OR	0136
2SA358R	0136
2SA358X	0136
2SA358Y	0136
2SA359	0136
2SA359A	0136
2SA359B	0136
2SA359C	0136
2SA359D	0136
2SA359E	0136
2SA359F	0136
2SA359G	0136
2SA359GN	0136
2SA359H	0136
2SA359J	0136
2SA359K	0136
2SA359L	0136
2SA359M	0136
2SA359OR	0136
2SA359R	0136
2SA359X	0136
2SA359Y	0136
2SA360	0050
2SA360A	0050
2SA360B	0050
2SA360C	0050
2SA360E	0050
2SA360F	0050
2SA360G	0050
2SA360GN	0050
2SA360H	0050
2SA360J	0050
2SA360K	0050
2SA360L	0050
2SA360M	0050
2SA360OR	0050
2SA360R	0050
2SA360X	0050
2SA360Y	0050
2SA361	0050
2SA361A	0050
2SA361B	0050
2SA361C	0050
2SA361D	0050
2SA361E	0050
2SA361F	0050
2SA361G	0050
2SA361GN	0050
2SA361H	0050
2SA361J	0050
2SA361K	0050
2SA361L	0050
2SA361M	0050
2SA361OR	0050
2SA361R	0050
2SA361X	0050
2SA361Y	0050
2SA362	0050
2SA363	0050
2SA364	0136
2SA364A	0136
2SA364B	0136
2SA364C	0136
2SA364D	0136
2SA364E	0136
2SA364F	0136
2SA364G	0136
2SA364GN	0136
2SA364H	0136
2SA364J	0136
2SA364K	0136
2SA364L	0136
2SA364M	0136
2SA364OR	0136
2SA364R	0136

DEVICE TYPE	REPL CODE
2SA364X	0136
2SA364Y	0136
2SA365	0136
2SA365A	0136
2SA365B	0136
2SA365C	0136
2SA365D	0136
2SA365E	0136
2SA365F	0136
2SA365G	0136
2SA365GN	0136
2SA365H	0136
2SA365J	0136
2SA365K	0136
2SA365L	0136
2SA365M	0136
2SA365OR	0136
2SA365R	0136
2SA365X	0136
2SA365Y	0136
2SA366	0136
2SA366A	0136
2SA366B	0136
2SA366C	0136
2SA366D	0136
2SA366E	0136
2SA366F	0136
2SA366G	0136
2SA366GN	0136
2SA366H	0136
2SA366J	0136
2SA366K	0136
2SA366L	0136
2SA366M	0136
2SA366OR	0136
2SA366R	0136
2SA366X	0136
2SA366Y	0136
2SA367	0136
2SA367A	0136
2SA367B	0136
2SA367C	0136
2SA367D	0136
2SA367E	0136
2SA367F	0136
2SA367G	0136
2SA367GN	0136
2SA367H	0136
2SA367J	0136
2SA367K	0136
2SA367L	0136
2SA367M	0136
2SA367OR	0136
2SA367R	0136
2SA367X	0136
2SA367Y	0136
2SA368	0136
2SA368A	0136
2SA368B	0136
2SA368C	0136
2SA368D	0136
2SA368E	0136
2SA368F	0136
2SA368G	0136
2SA368GN	0136
2SA368H	0136
2SA368J	0136
2SA368K	0136
2SA368L	0136
2SA368M	0136
2SA368OR	0136
2SA368R	0136
2SA368X	0136
2SA368Y	0136
2SA369	0136
2SA369A	0136
2SA369B	0136
2SA369C	0136
2SA369D	0136
2SA369E	0136
2SA369F	0136
2SA369G	0136
2SA369GN	0136
2SA369H	0136
2SA369J	0136
2SA369K	0136
2SA369L	0136
2SA369M	0136
2SA369OR	0136
2SA369R	0136
2SA369X	0136
2SA369Y	0136
2SA370	0164
2SA371	0164
2SA372	0050
2SA373	0004
2SA373A	0004
2SA374	0004
2SA375	0004
2SA375A	0004
2SA375B	0004
2SA375C	0004
2SA375D	0004
2SA375E	0004
2SA375F	0004
2SA375G	0004
2SA375GN	0004
2SA375H	0004
2SA375J	0004

DEVICE TYPE	REPL CODE
2SA375K	0004
2SA375L	0004
2SA375M	0004
2SA375OR	0004
2SA375R	0004
2SA375X	0004
2SA375Y	0004
2SA376	0136
2SA376A	0136
2SA376B	0136
2SA376C	0136
2SA376D	0136
2SA376E	0136
2SA376F	0136
2SA376GN	0136
2SA376H	0136
2SA376J	0136
2SA376K	0136
2SA376L	0136
2SA376OR	0136
2SA376R	0136
2SA376X	0136
2SA376Y	0136
2SA377	0050
2SA377A	0050
2SA377B	0050
2SA377C	0050
2SA377D	0050
2SA377E	0050
2SA377F	0050
2SA377G	0050
2SA377GN	0050
2SA377H	0050
2SA377J	0050
2SA377K	0050
2SA377L	0050
2SA377M	0050
2SA377OR	0050
2SA377R	0050
2SA377X	0050
2SA377Y	0050
2SA378	0050
2SA379	0050
2SA380	0136
2SA380A	0136
2SA380B	0136
2SA380C	0136
2SA380D	0136
2SA380E	0136
2SA380F	0136
2SA380G	0136
2SA380GN	0136
2SA380H	0136
2SA380J	0136
2SA380K	0136
2SA380L	0136
2SA380M	0136
2SA380OR	0136
2SA380R	0136
2SA380X	0136
2SA380Y	0136
2SA381	0136
2SA381A	0136
2SA381B	0136
2SA381D	0136
2SA381E	0136
2SA381F	0136
2SA381G	0136
2SA381GN	0136
2SA381H	0136
2SA381K	0136
2SA381L	0136
2SA381M	0136
2SA381OR	0136
2SA381R	0136
2SA381X	0136
2SA381Y	0136
2SA382	0136
2SA382A	0136
2SA382B	0136
2SA382C	0136
2SA382D	0136
2SA382E	0136
2SA382F	0136
2SA382G	0136
2SA382GN	0136
2SA382H	0136
2SA382J	0136
2SA382K	0136
2SA382L	0136
2SA382M	0136
2SA382OR	0136
2SA382R	0136
2SA382X	0136
2SA382Y	0136
2SA383	0136
2SA383A	0136
2SA383B	0136
2SA383C	0136
2SA383D	0136
2SA383E	0136
2SA383F	0136
2SA383G	0136
2SA383GN	0136
2SA383H	0136
2SA383J	0136
2SA383K	0136
2SA383L	0136
2SA383M	0136

DEVICE TYPE	REPL CODE
2SA383OR	0136
2SA383R	0136
2SA383X	0136
2SA383Y	0136
2SA384	0136
2SA384A	0136
2SA384B	0136
2SA384C	0136
2SA384D	0136
2SA384E	0136
2SA384F	0136
2SA384G	0136
2SA384GN	0136
2SA384H	0136
2SA384J	0136
2SA384K	0136
2SA384L	0136
2SA384M	0136
2SA384OR	0136
2SA384R	0136
2SA384X	0136
2SA384Y	0136
2SA385	0136
2SA385A	0136
2SA385B	0136
2SA385C	0136
2SA385D	0136
2SA385E	0136
2SA385F	0136
2SA385G	0136
2SA385GN	0136
2SA385H	0136
2SA385J	0136
2SA385K	0136
2SA385L	0136
2SA385M	0136
2SA385OR	0136
2SA385R	0136
2SA385X	0136
2SA385Y	0136
2SA386	0050
2SA387	0050
2SA389	0050
2SA390	0050
2SA391	0280
2SA391A	0280
2SA391B	0280
2SA391C	0280
2SA391D	0280
2SA391E	0280
2SA391F	0280
2SA391G	0280
2SA391GN	0280
2SA391H	0280
2SA391J	0280
2SA391K	0280
2SA391L	0280
2SA391M	0280
2SA391OR	0280
2SA391R	0280
2SA391X	0280
2SA391Y	0280
2SA392	0280
2SA392A	0280
2SA392B	0280
2SA392C	0280
2SA392D	0280
2SA392E	0280
2SA392F	0280
2SA392G	0280
2SA392GN	0280
2SA392H	0280
2SA392J	0280
2SA392K	0280
2SA392L	0280
2SA392M	0280
2SA392OR	0280
2SA392R	0280
2SA392X	0280
2SA392Y	0280
2SA393	0280
2SA393A	0280
2SA394	0280
2SA394A	0280
2SA394B	0280
2SA394C	0280
2SA394D	0280
2SA394E	0280
2SA394F	0280
2SA394G	0280
2SA394GN	0280
2SA394H	0280
2SA394J	0280
2SA394K	0280
2SA394L	0280
2SA394M	0280
2SA394OR	0280
2SA394R	0280
2SA394X	0280
2SA394Y	0280
2SA395	0280
2SA395A	0280
2SA395B	0280
2SA395C	0280
2SA395D	0280
2SA395E	0280
2SA395F	0280
2SA395G	0280
2SA395GN	0280

DEVICE TYPE	REPL CODE
2SA395H	0280
2SA395J	0280
2SA395K	0280
2SA395L	0280
2SA395M	0280
2SA395OR	0280
2SA395R	0280
2SA395X	0280
2SA395Y	0280
2SA396	0211
2SA397	0211
2SA398	0211
2SA398A	0211
2SA398B	0211
2SA398C	0211
2SA398D	0211
2SA398E	0211
2SA398F	0211
2SA398G	0211
2SA398GN	0211
2SA398H	0211
2SA398J	0211
2SA398K	0211
2SA398L	0211
2SA398M	0211
2SA398OR	0211
2SA398X	0211
2SA398Y	0211
2SA399	0211
2SA399A	0211
2SA399B	0211
2SA399C	0211
2SA399D	0211
2SA399E	0211
2SA399F	0211
2SA399G	0211
2SA399GN	0211
2SA399H	0211
2SA399J	0211
2SA399K	0211
2SA399L	0211
2SA399M	0211
2SA399OR	0211
2SA399X	0211
2SA399Y	0211
2SA400	0136
2SA400A	0136
2SA400B	0136
2SA400C	0136
2SA400D	0136
2SA400E	0136
2SA400F	0136
2SA400G	0136
2SA400GN	0136
2SA400H	0136
2SA400J	0136
2SA400K	0136
2SA400L	0136
2SA400M	0136
2SA400OR	0136
2SA400R	0136
2SA400X	0136
2SA400Y	0136
2SA401	0136
2SA402	0037
2SA403	0050
2SA403A	0050
2SA403B	0050
2SA403C	0050
2SA403D	0050
2SA403E	0050
2SA403F	0050
2SA403G	0050
2SA403GN	0050
2SA403H	0050
2SA403J	0050
2SA403K	0050
2SA403L	0050
2SA403M	0050
2SA403OR	0050
2SA403R	0050
2SA403X	0050
2SA403Y	0050
2SA404	0050
2SA404A	0050
2SA404B	0050
2SA404C	0050
2SA404D	0050
2SA404E	0050
2SA404F	0050
2SA404G	0050
2SA404GN	0050
2SA404H	0050
2SA404K	0050
2SA404L	0050
2SA404M	0050
2SA404OR	0050
2SA404R	0050
2SA404X	0050
2SA404Y	0050
2SA405	0050
2SA405-O	0050
2SA406	0004
2SA406A	0004
2SA406B	0004
2SA406C	0004
2SA406D	0004
2SA406E	0004
2SA406G	0004

DEVICE TYPE	REPL CODE
2SA406GN	0004
2SA406H	0004
2SA406J	0004
2SA406K	0004
2SA406L	0004
2SA406M	0004
2SA406OR	0004
2SA406R	0004
2SA406X	0004
2SA406Y	0004
2SA407	0004
2SA407A	0004
2SA407B	0004
2SA407C	0004
2SA407D	0004
2SA407E	0004
2SA407F	0004
2SA407G	0004
2SA407GN	0004
2SA407H	0004
2SA407J	0004
2SA407K	0004
2SA407L	0004
2SA407M	0004
2SA407OR	0004
2SA407R	0004
2SA407X	0004
2SA407Y	0004
2SA408	0136
2SA408A	0136
2SA408B	0136
2SA408D	0136
2SA408E	0136
2SA408F	0136
2SA408GN	0136
2SA408H	0136
2SA408J	0136
2SA408K	0136
2SA408L	0136
2SA408M	0136
2SA408OR	0136
2SA408R	0136
2SA408X	0136
2SA408Y	0136
2SA409	0136
2SA409A	0136
2SA409B	0136
2SA409C	0136
2SA409D	0136
2SA409E	0136
2SA409F	0136
2SA409G	0136
2SA409GN	0136
2SA409H	0136
2SA409J	0136
2SA409K	0136
2SA409L	0136
2SA409M	0136
2SA409OR	0136
2SA409R	0136
2SA409X	0136
2SA409Y	0136
2SA410	0050
2SA411	0050
2SA412	0136
2SA412A	0136
2SA412B	0136
2SA412C	0136
2SA412D	0136
2SA412E	0136
2SA412F	0136
2SA412G	0136
2SA412GN	0136
2SA412H	0136
2SA412J	0136
2SA412K	0136
2SA412L	0136
2SA412M	0136
2SA412OR	0136
2SA412R	0136
2SA412X	0136
2SA412Y	0136
2SA413	0050
2SA414	0280
2SA414A	0280
2SA414B	0280
2SA414C	0280
2SA414D	0280
2SA414E	0280
2SA414F	0280
2SA414G	0280
2SA414GN	0280
2SA414H	0280
2SA414J	0280
2SA414K	0280
2SA414L	0280
2SA414M	0280
2SA414OR	0280
2SA414R	0280
2SA414X	0280
2SA415	0280
2SA415A	0280
2SA415B	0280
2SA415C	0280
2SA415D	0280
2SA415E	0280
2SA415F	0280
2SA415G	0280
2SA415H	0280

DEVICE TYPE	REPL CODE
2SA415J	0280
2SA415K	0280
2SA415L	0280
2SA415M	0280
2SA415OR	0280
2SA415R	0280
2SA415X	0280
2SA415Y	0280
2SA416	0160
2SA416A	0160
2SA416B	0160
2SA416C	0160
2SA416D	0160
2SA416E	0160
2SA416F	0160
2SA416G	0160
2SA416GN	0160
2SA416H	0160
2SA416J	0160
2SA416K	0160
2SA416L	0160
2SA416M	0160
2SA416OR	0160
2SA416R	0160
2SA416X	0160
2SA416Y	0160
2SA417	0211
2SA419	0050
2SA420	0050
2SA420A	0050
2SA420B	0050
2SA420C	0050
2SA420D	0050
2SA420E	0050
2SA420F	0050
2SA420G	0050
2SA420H	0050
2SA420J	0050
2SA420K	0050
2SA420L	0050
2SA420M	0050
2SA420OR	0050
2SA420R	0050
2SA420X	0050
2SA420Y	0050
2SA421	0050
2SA422	0050
2SA423	0280
2SA424	0280
2SA425	0050
2SA426	0050
2SA426GN	0050
2SA427	0136
2SA427A	0136
2SA427B	0136
2SA427C	0136
2SA427D	0136
2SA427E	0136
2SA427F	0136
2SA427G	0136
2SA427GN	0136
2SA427H	0136
2SA427J	0136
2SA427K	0136
2SA427L	0136
2SA427M	0136
2SA427OR	0136
2SA427R	0136
2SA427X	0136
2SA427Y	0136
2SA428	0136
2SA428A	0136
2SA428B	0136
2SA428C	0136
2SA428D	0136
2SA428E	0136
2SA428F	0136
2SA428G	0136
2SA428GN	0136
2SA428H	0136
2SA428J	0136
2SA428K	0136
2SA428L	0136
2SA428M	0136
2SA428OR	0136
2SA428R	0136
2SA428X	0136
2SA428Y	0136
2SA429	2741
2SA429-O	2741
2SA429A	2741
2SA429B	2741
2SA429C	2741
2SA429D	2741
2SA429E	2741
2SA429F	2741
2SA429G	2741
2SA429GN	2741
2SA429G-O	2741
2SA429G-R	2741
2SA429G-Y	2741
2SA429H	2741
2SA429J	2741
2SA429K	2741
2SA429L	2741
2SA429M	2741
2SA429OR	2741
2SA429R	2741
2SA429X	2741

If replacement code is OEM, contact original manufacturer for replacement.

DEVICE TYPE	REPL CODE
2SA430	0050
2SA431	0050
2SA431A	0050
2SA432	0050
2SA432A	0050
2SA432B	0050
2SA432C	0050
2SA432D	0050
2SA432E	0050
2SA432F	0050
2SA432G	0050
2SA432GN	0050
2SA432H	0050
2SA432K	0050
2SA432L	0050
2SA432M	0050
2SA432OR	0050
2SA432R	0050
2SA432X	0050
2SA432Y	0050
2SA433	0050
2SA433A	0050
2SA433B	0050
2SA433C	0050
2SA433D	0050
2SA433E	0050
2SA433F	0050
2SA433G	0050
2SA433GN	0050
2SA433H	0050
2SA433K	0050
2SA433L	0050
2SA433M	0050
2SA433OR	0050
2SA433R	0050
2SA433X	0050
2SA433Y	0050
2SA434	0050
2SA434A	0050
2SA434B	0050
2SA434C	0050
2SA434D	0050
2SA434E	0050
2SA434F	0050
2SA434G	0050
2SA434GN	0050
2SA434H	0050
2SA434J	0050
2SA434K	0050
2SA434L	0050
2SA434M	0050
2SA434OR	0050
2SA434R	0050
2SA434X	0050
2SA434Y	0050
2SA435	0050
2SA435A	0050
2SA435B	0050
2SA435C	0050
2SA435D	0050
2SA435E	0050
2SA435F	0050
2SA435G	0050
2SA435GN	0050
2SA435H	0050
2SA435J	0050
2SA435L	0050
2SA435M	0050
2SA435OR	0050
2SA435R	0050
2SA435X	0050
2SA435Y	0050
2SA436	0136
2SA436A	0136
2SA436B	0136
2SA436C	0136
2SA436D	0136
2SA436E	0136
2SA436F	0136
2SA436G	0136
2SA436GN	0136
2SA436H	0136
2SA436J	0136
2SA436K	0136
2SA437	0136
2SA437A	0136
2SA437B	0136
2SA437C	0136
2SA437D	0136
2SA437E	0136
2SA437F	0136
2SA437G	0136
2SA437GN	0136
2SA437H	0136
2SA437J	0136
2SA437K	0136
2SA437L	0136
2SA437M	0136
2SA437OR	0136
2SA437R	0136
2SA437X	0136
2SA437Y	0136
2SA438	0136
2SA438A	0136
2SA438B	0136
2SA438C	0136
2SA438D	0136
2SA438E	0136
2SA438F	0136

DEVICE TYPE	REPL CODE
2SA438G	0136
2SA438GN	0136
2SA438H	0136
2SA438J	0136
2SA438K	0136
2SA438L	0136
2SA438M	0136
2SA438OR	0136
2SA438R	0136
2SA438X	0136
2SA438Y	0136
2SA440	0050
2SA440A	0050
2SA440AL	0050
2SA440B	0050
2SA440C	0050
2SA440D	0050
2SA440E	0050
2SA440F	0050
2SA440G	0050
2SA440GN	0050
2SA440H	0050
2SA440J	0050
2SA440K	0050
2SA440L	0050
2SA440M	0050
2SA440OR	0050
2SA440R	0050
2SA440X	0050
2SA440Y	0050
2SA441	0050
2SA443	0004
2SA444	0004
2SA445	0004
2SA446	0211
2SA446A	0211
2SA446B	0211
2SA446C	0211
2SA446D	0211
2SA446E	0211
2SA446F	0211
2SA446G	0211
2SA446GN	0211
2SA446H	0211
2SA446J	0211
2SA446K	0211
2SA446L	0211
2SA446M	0211
2SA446OR	0211
2SA446R	0211
2SA446X	0211
2SA446Y	0211
2SA447	0136
2SA447A	0136
2SA447B	0136
2SA447C	0136
2SA447D	0136
2SA447E	0136
2SA447F	0136
2SA447G	0136
2SA447GN	0136
2SA447H	0136
2SA447J	0136
2SA447K	0136
2SA447L	0136
2SA447M	0136
2SA447OR	0136
2SA447R	0136
2SA447X	0136
2SA447Y	0136
2SA448	0050
2SA449	0136
2SA450	0050
2SA450H	0050
2SA451	0050
2SA451H	0050
2SA452	0050
2SA452H	0050
2SA453	0136
2SA453A	0136
2SA453B	0136
2SA453C	0136
2SA453D	0136
2SA453E	0136
2SA453F	0136
2SA453G	0136
2SA453GN	0136
2SA453H	0136
2SA453J	0136
2SA453K	0136
2SA453L	0136
2SA453M	0136
2SA453OR	0136
2SA453R	0136
2SA453X	0136
2SA453Y	0136
2SA454	0136
2SA454A	0136
2SA454B	0136
2SA454C	0136
2SA454D	0136
2SA454E	0136
2SA454F	0136
2SA454G	0136
2SA454GN	0136
2SA454H	0136
2SA454J	0136
2SA454K	0136
2SA454L	0136

DEVICE TYPE	REPL CODE
2SA454OR	0136
2SA454R	0136
2SA454X	0136
2SA454Y	0136
2SA455	0136
2SA455A	0136
2SA455B	0136
2SA455C	0136
2SA455D	0136
2SA455E	0136
2SA455F	0136
2SA455G	0136
2SA455GN	0136
2SA455H	0136
2SA455J	0136
2SA455K	0136
2SA455L	0136
2SA455M	0136
2SA455OR	0136
2SA455R	0136
2SA455X	0136
2SA455Y	0136
2SA456	0136
2SA456A	0136
2SA456B	0136
2SA456C	0136
2SA456D	0136
2SA456E	0136
2SA456F	0136
2SA456G	0136
2SA456GN	0136
2SA456H	0136
2SA456J	0136
2SA456K	0136
2SA456L	0136
2SA456M	0136
2SA456OR	0136
2SA456R	0136
2SA456X	0136
2SA456Y	0136
2SA457	0136
2SA457A	0136
2SA457B	0136
2SA457C	0136
2SA457D	0136
2SA457E	0136
2SA457F	0136
2SA457G	0136
2SA457GN	0136
2SA457H	0136
2SA457J	0136
2SA457K	0136
2SA457L	0136
2SA457M	0136
2SA457OR	0136
2SA457R	0136
2SA457X	0136
2SA457Y	0136
2SA458	1062
2SA458C	1062
2SA459	0004
2SA459Y	0004
2SA460	0050
2SA461	0050
2SA462	0050
2SA463	0050
2SA463C	0050
2SA464	0050
2SA465	0150
2SA466	0136
2SA466-2	0136
2SA466-3	0136
2SA466A	0136
2SA466B	0136
2SA466BLK	0136
2SA466BLU	0136
2SA466C	0136
2SA466D	0136
2SA466E	0136
2SA466F	0136
2SA466G	0136
2SA466GN	0136
2SA466H	0136
2SA466J	0136
2SA466K	0136
2SA466L	0136
2SA466M	0136
2SA466OR	0136
2SA466R	0136
2SA466X	0136
2SA466Y	0136
2SA466YEL	0136
2SA467	0006
2SA467A	0006
2SA467B	0006
2SA467C	0006
2SA467D	0006
2SA467E	0006
2SA467F	0006
2SA467G	0006
2SA467GN	0006
2SA467G-O	0006
2SA467G-R	0006
2SA467G-Y	0006
2SA467H	0006
2SA467J	0006
2SA467K	0006
2SA467L	0006
2SA467M	0006

DEVICE TYPE	REPL CODE
2SA467OR	0006
2SA467R	0006
2SA467X	0006
2SA467Y	0006
2SA468	0136
2SA468A	0136
2SA468B	0136
2SA468C	0136
2SA468D	0136
2SA468E	0136
2SA468F	0136
2SA468G	0136
2SA468GN	0136
2SA468H	0136
2SA468J	0136
2SA468K	0136
2SA468L	0136
2SA468M	0136
2SA468OR	0136
2SA468R	0136
2SA468X	0136
2SA468Y	0136
2SA469	0136
2SA469A	0136
2SA469B	0136
2SA469C	0136
2SA469D	0136
2SA469E	0136
2SA469F	0136
2SA469G	0136
2SA469GN	0136
2SA469H	0136
2SA469J	0136
2SA469K	0136
2SA469L	0136
2SA469M	0136
2SA469OR	0136
2SA469R	0136
2SA469X	0136
2SA469Y	0136
2SA470	0136
2SA470A	0136
2SA470B	0136
2SA470C	0136
2SA470D	0136
2SA470E	0136
2SA470F	0136
2SA470G	0136
2SA470GN	0136
2SA470H	0136
2SA470J	0136
2SA470K	0136
2SA470L	0136
2SA470M	0136
2SA470OR	0136
2SA470R	0136
2SA470X	0136
2SA470Y	0136
2SA471	0136
2SA471-1	0136
2SA471-2	0136
2SA471-3	0136
2SA471A	0136
2SA471B	0136
2SA471C	0136
2SA471D	0136
2SA471E	0136
2SA471F	0136
2SA471G	0136
2SA471GN	0136
2SA471H	0136
2SA471J	0136
2SA471K	0136
2SA471L	0136
2SA471M	0136
2SA471OR	0136
2SA471R	0136
2SA471X	0136
2SA471Y	0136
2SA472	0136
2SA472-1	0136
2SA472-2	0136
2SA472-3	0136
2SA472-4	0136
2SA472-5	0136
2SA472-6	0136
2SA472A	0136
2SA472B	0136
2SA472C	0136
2SA472D	0136
2SA472E	0136
2SA472F	0136
2SA472G	0136
2SA472GN	0136
2SA472H	0136
2SA472J	0136
2SA472K	0136
2SA472L	0136
2SA472M	0136
2SA472OR	0136
2SA472R	0136
2SA472X	0136
2SA472Y	0136
2SA473	0925
2SA473(0)	0925
2SA473(0)	0925
2SA473-GR	0925
2SA473-O	0925
2SA473-R	0925

DEVICE TYPE	REPL CODE
2SA473-Y	0925
2SA473B	0925
2SA473D	0925
2SA473GR	0925
2SA473R	0925
2SA473Y	0925
2SA474	0136
2SA474-G	0136
2SA474A	0136
2SA474B	0136
2SA474C	0136
2SA474D	0136
2SA474E	0136
2SA474F	0136
2SA474G	0136
2SA474GN	0136
2SA474H	0136
2SA474OR	0136
2SA474R	0136
2SA474X	0136
2SA474Y	0136
2SA475	0050
2SA476	0136
2SA476A	0136
2SA476B	0136
2SA476C	0136
2SA476D	0136
2SA476E	0136
2SA476F	0136
2SA476G	0136
2SA476GN	0136
2SA476H	0136
2SA476J	0136
2SA476K	0136
2SA476L	0136
2SA476M	0136
2SA476OR	0136
2SA476R	0136
2SA476X	0136
2SA476Y	0136
2SA477	0136
2SA477A	0136
2SA477B	0136
2SA477C	0136
2SA477D	0136
2SA477E	0136
2SA477F	0136
2SA477G	0136
2SA477GN	0136
2SA477H	0136
2SA477J	0136
2SA477K	0136
2SA477L	0136
2SA477M	0136
2SA477OR	0136
2SA477R	0136
2SA477X	0136
2SA477Y	0136
2SA478	0004
2SA478-G	0004
2SA478A	0004
2SA478B	0004
2SA478C	0004
2SA478D	0004
2SA478E	0004
2SA478F	0004
2SA478G	0004
2SA478GN	0004
2SA478H	0004
2SA478J	0004
2SA478K	0004
2SA478L	0004
2SA478M	0004
2SA478OR	0004
2SA478R	0004
2SA478X	0004
2SA478Y	0004
2SA479	0004
2SA479A	0004
2SA479B	0004
2SA479C	0004
2SA479D	0004
2SA479E	0004
2SA479F	0004
2SA479G	0004
2SA479GN	0004
2SA479H	0004
2SA479J	0004
2SA479K	0004
2SA479L	0004
2SA479M	0004
2SA479OR	0004
2SA479R	0004
2SA479X	0004
2SA479Y	0004
2SA480	0037
2SA480-OR	0037
2SA480A	0037
2SA480B	0037
2SA480C	0037
2SA480D	0037
2SA480E	0037
2SA480F	0037
2SA480G	0037
2SA480GN	0037
2SA480H	0037
2SA480J	0037
2SA480K	0037
2SA480L	0037

DEVICE TYPE	REPL CODE
2SA480M	0037
2SA480OR	0037
2SA480R	0037
2SA480X	0037
2SA480Y	0037
2SA482	0037
2SA482A	0037
2SA482B	0037
2SA482C	0037
2SA482D	0037
2SA482E	0037
2SA482F	0037
2SA482G	0037
2SA482GN	0037
2SA482H	0037
2SA482J	0037
2SA482K	0037
2SA482L	0037
2SA482M	0037
2SA482OR	0037
2SA482R	0037
2SA482X	0037
2SA482Y	0037
2SA483	0905
2SA483-O	0905
2SA483-R	0905
2SA483-Y	0905
2SA484	0892
2SA484-BL	0892
2SA484-R	0892
2SA484-Y	0892
2SA485	0892
2SA485-BL	0892
2SA485-BLU	0892
2SA485-R	0892
2SA485-RED	0892
2SA485-Y	0892
2SA485-YEL	0892
2SA485Y	0892
2SA486	0892
2SA486-BL	0892
2SA486-BLU	0892
2SA486-R	0892
2SA486-RED	0892
2SA486-Y	0892
2SA486-YEL	0892
2SA489	0925
2SA489-0	0925
2SA489-O	0925
2SA489-R	0925
2SA489-Y	0925
2SA490	1498
2SA490(POWER)	1498
2SA490-0	1498
2SA490-O	1498
2SA490-R	1498
2SA490-Y	1498
2SA490A	1498
2SA490B	1498
2SA490C	1498
2SA490D	1498
2SA490E	1498
2SA490F	1498
2SA490G	1498
2SA490GN	1498
2SA490H	1498
2SA490J	1498
2SA490K	1498
2SA490L	1498
2SA490LBG1	1498
2SA490M	1498
2SA490OR	1498
2SA490R	1498
2SA490X	1498
2SA490Y	1498
2SA490YA	1498
2SA490YLBG11	1498
2SA490YLBGL1	1498
2SA493	0006
2SA493-GR	0006
2SA493-O	0006
2SA493-Y	0006
2SA493A	0006
2SA493B	0006
2SA493C	0006
2SA493D	0006
2SA493E	0006
2SA493F	0006
2SA493G	0006
2SA493G-GR	0006
2SA493GN	0006
2SA493GR	3597
2SA493G-Y	0006
2SA493H	0006
2SA493J	0006
2SA493K	0006
2SA493L	0006
2SA493M	0006
2SA493OR	0006
2SA493R	0006
2SA493X	0006
2SA493Y	0006
2SA494	0693
2SA494(Y)	0693
2SA494-GR	0693
2SA494-GR-1	0693
2SA494-O	0693
2SA494-OR	0693
2SA494-Y	0693

DEVICE TYPE	REPL CODE
2SA494A	0693
2SA494B	0693
2SA494C	0693
2SA494D	0693
2SA494E	0693
2SA494F	0693
2SA494G	0693
2SA494GN	0693
2SA494GR	0693
2SA494H	0693
2SA494J	0693
2SA494K	0693
2SA494L	0693
2SA494M	0693
2SA494O	0693
2SA494OR	0693
2SA494R	0693
2SA494X	0693
2SA494Y	0693
2SA495	0006
2SA495(O)	0006
2SA495(R)	0006
2SA495(Y)	0006
2SA495-0	0006
2SA495-1	0006
2SA495-GN	0006
2SA495-O	0006
2SA495-OF	0006
2SA495-OR	0006
2SA495-ORG	0006
2SA495-ORG-G	0006
2SA495-Q	0006
2SA495-R	0006
2SA495-RD	0006
2SA495-RED	0006
2SA495-RED-G	0006
2SA495-Y	0006
2SA495-YEL	0006
2SA495-YEL-G	0006
2SA495-YL	0006
2SA495A	0006
2SA495B	0006
2SA495C	0006
2SA495D	0006
2SA495E	0006
2SA495F	0006
2SA495G	0006
2SA495G-GR	0006
2SA495GN	0006
2SA495G-O	0006
2SA495G-R	0006
2SA495G-Y	0006
2SA495H	0006
2SA495J	0006
2SA495K	0006
2SA495L	0006
2SA495M	0006
2SA495O	0006
2SA495OF	0006
2SA495OR	0006
2SA495Q	0006
2SA495R	0006
2SA495RD	0006
2SA495RED	0006
2SA495RED-G	0006
2SA495RO	0006
2SA495W	0006
2SA495W1	0006
2SA495WI	0006
2SA495X	0006
2SA495Y	0006
2SA495Y(FA-1)	OEM
2SA495YEL	0006
2SA495YEL-G	0006
2SA495YL	0006
2SA496	0525
2SA496(O)	0525
2SA496(Y)	0525
2SA496-0	0525
2SA496-O	0525
2SA496-ORG	0525
2SA496-R	0525
2SA496-RED	0525
2SA496-Y	0525
2SA496-YEL	0525
2SA496O	0525
2SA496ORG	0525
2SA496R	0525
2SA496RED	0525
2SA496Y	0525
2SA496YEL	0525
2SA497	0126
2SA497-O	0126
2SA497-ORG	0126
2SA497-R	0126
2SA497-RED	0126
2SA497-Y	0126
2SA497-YEL	0126
2SA497R	0126
2SA497RED	0126
2SA497Y	0126
2SA498	0126
2SA498(Y)	0126
2SA498-O	0126
2SA498-ORG	0126
2SA498-R	0126
2SA498-RED	0126
2SA498-Y	0126
2SA498-YEL	0126

If replacement code is OEM, contact original manufacturer for replacement.

DEVICE TYPE	REPL CODE
2SA498R	0126
2SA498RED	0126
2SA498Y	0126
2SA499	0150
2SA499-O	0150
2SA499-ORG	0150
2SA499-R	0150
2SA499-RED	0150
2SA499-Y	0150
2SA499-YEL	0150
2SA499A	0150
2SA499B	0150
2SA499C	0150
2SA499D	0150
2SA499E	0150
2SA499F	0150
2SA499G	0150
2SA499GN	0150
2SA499H	0150
2SA499J	0150
2SA499K	0150
2SA499L	0150
2SA499M	0150
2SA499O	0150
2SA499OR	0150
2SA499R	0150
2SA499X	0150
2SA499Y	0150
2SA500	0150
2SA500-O	0150
2SA500-OR	0150
2SA500-ORG	0150
2SA500-R	0150
2SA500-RED	0150
2SA500-Y	0150
2SA500-YEL	0150
2SA500A	0150
2SA500B	0150
2SA500C	0150
2SA500D	0150
2SA500E	0150
2SA500F	0150
2SA500G	0150
2SA500GN	0150
2SA500H	0150
2SA500J	0150
2SA500K	0150
2SA500L	0150
2SA500M	0150
2SA500O	0150
2SA500OR	0150
2SA500R	0150
2SA500X	0150
2SA500Y	0150
2SA501	0126
2SA502	0006
2SA502-O	0006
2SA502-OR	0006
2SA502-R	0006
2SA502-Y	0006
2SA502A	0006
2SA502B	0006
2SA502C	0006
2SA502D	0006
2SA502E	0006
2SA502F	0006
2SA502G	0006
2SA502GN	0006
2SA502H	0006
2SA502J	0006
2SA502K	0006
2SA502L	0006
2SA502M	0006
2SA502OR	0006
2SA502R	0006
2SA502X	0006
2SA502Y	0006
2SA503	0126
2SA503-GR	0126
2SA503-GRN	0126
2SA503-O	0126
2SA503-ORG	0126
2SA503-R	0126
2SA503-Y	0126
2SA503-YEL	0126
2SA503G	0126
2SA503O	0126
2SA503Y	0126
2SA504	0126
2SA504-GR	0126
2SA504-GRN	0126
2SA504-O	0126
2SA504-ORG	0126
2SA504-R	0126
2SA504-Y	0126
2SA504-YEL	0126
2SA504G	0126
2SA504O	0126
2SA504Y	0126
2SA505	0525
2SA505-O	0525
2SA505-ORG	0525
2SA505-R	0525
2SA505-RED	0525
2SA505-Y	0525
2SA505-YEL	0525
2SA505O	0525
2SA505ORG	0525
2SA505R	0525
2SA505RED	0525
2SA505Y	0525
2SA505YEL	0525
2SA506	0050
2SA506R	OEM
2SA506Y	OEM
2SA507	0050
2SA507A	0050
2SA507B	0050
2SA507C	0050
2SA507D	0050
2SA507E	0050
2SA507F	0050
2SA507G	0050
2SA507GN	0050
2SA507H	0050
2SA507J	0050
2SA507K	0050
2SA507L	0050
2SA507OR	0050
2SA507R	0050
2SA507X	0050
2SA507Y	0050
2SA508	0050
2SA508D	0050
2SA508F	0050
2SA509	0006
2SA509(A)	0006
2SA509(O)	0006
2SA509-0	0006
2SA509-O	0006
2SA509-OR	0006
2SA509-R	0006
2SA509-RD	0006
2SA509-Y	0006
2SA509-YE	0006
2SA509A	0006
2SA509B	0006
2SA509BL	0006
2SA509C	0006
2SA509D	0006
2SA509E	0006
2SA509F	0006
2SA509G	0006
2SA509GN	0006
2SA509G-O	0006
2SA509GR	0006
2SA509GR-1	0006
2SA509G-Y	0006
2SA509H	0006
2SA509J	0006
2SA509K	0006
2SA509L	0006
2SA509M	0006
2SA509OR	0006
2SA509Q	0006
2SA509R	0006
2SA509RD	0006
2SA509T	0006
2SA509V	0006
2SA509X	0006
2SA509Y	0006
2SA509YE	0006
2SA510	0892
2SA510-O	0892
2SA510-OR	0892
2SA510-ORG	0892
2SA510-R	0892
2SA510-RD	0892
2SA510-RED	0892
2SA510A	0892
2SA510B	0892
2SA510C	0892
2SA510D	0892
2SA510E	0892
2SA510F	0892
2SA510G	0892
2SA510GN	0892
2SA510H	0892
2SA510J	0892
2SA510K	0892
2SA510L	0892
2SA510M	0892
2SA510OR	0892
2SA510R	0892
2SA510X	0892
2SA510Y	0892
2SA511	0892
2SA511-G	0892
2SA511-O	0892
2SA511-OR	0892
2SA511-ORG	0892
2SA511-R	0892
2SA511-RD	0892
2SA511-RED	0892
2SA511A	0892
2SA511B	0892
2SA511C	0892
2SA511D	0892
2SA511E	0892
2SA511F	0892
2SA511G	0892
2SA511GN	0892
2SA511H	0892
2SA511J	0892
2SA511K	0892
2SA511L	0892
2SA511M	0892
2SA511OR	0892
2SA511R	0892
2SA511X	0892
2SA511Y	0892
2SA512	0892
2SA512-O	0892
2SA512-OR	0892
2SA512-OR1	0892
2SA512-ORG	0892
2SA512-R	0892
2SA512-RD	0892
2SA512-RED	0892
2SA512A	0892
2SA512B	0892
2SA512C	0892
2SA512D	0892
2SA512E	0892
2SA512F	0892
2SA512G	0892
2SA512GN	0892
2SA512H	0892
2SA512J	0892
2SA512K	0892
2SA512L	0892
2SA512M	0892
2SA512OR	0892
2SA512R	0892
2SA512X	0892
2SA512Y	0892
2SA513	0892
2SA513-O	0892
2SA513-OR	0892
2SA513-ORG	0892
2SA513-R	0892
2SA513-RD	0892
2SA513-RED	0892
2SA513A	0892
2SA513B	0892
2SA513C	0892
2SA513D	0892
2SA513E	0892
2SA513F	0892
2SA513G	0892
2SA513GN	0892
2SA513H	0892
2SA513J	0892
2SA513K	0892
2SA513L	0892
2SA513M	0892
2SA513OR	0892
2SA513R	0892
2SA513X	0892
2SA513Y	0892
2SA514A-Q	OEM
2SA514A-QR	OEM
2SA514A-R	OEM
2SA516	0892
2SA516A	0892
2SA517	0136
2SA518	0136
2SA518-G	0136
2SA518A	0136
2SA518B	0136
2SA518C	0136
2SA518D	0136
2SA518E	0136
2SA518F	0136
2SA518G	0136
2SA518GN	0136
2SA518H	0136
2SA518J	0136
2SA518K	0136
2SA518L	0136
2SA518M	0136
2SA518OR	0136
2SA518R	0136
2SA518X	0136
2SA518Y	0136
2SA522	0037
2SA522A	0037
2SA522AL	0037
2SA522B	0037
2SA522C	0037
2SA522D	0037
2SA522E	0037
2SA522F	0037
2SA522G	0037
2SA522GN	0037
2SA522H	0037
2SA522J	0037
2SA522K	0037
2SA522L	0037
2SA522M	0037
2SA522OR	0037
2SA522R	0037
2SA522X	0037
2SA522Y	0037
2SA523	0126
2SA523A	0126
2SA524	0037
2SA525	0050
2SA525A	0050
2SA525B	0050
2SA525C	0050
2SA525D	0050
2SA525E	0050
2SA525F	0050
2SA525GN	0050
2SA525H	0050
2SA525J	0050
2SA525K	0050
2SA525L	0050
2SA525M	0050
2SA525OR	0050
2SA525R	0050
2SA525X	0050
2SA525Y	0050
2SA526	0037
2SA527	0126
2SA528	0126
2SA529	0126
2SA530	0037
2SA530A	0037
2SA530B	0037
2SA530C	0037
2SA530D	0037
2SA530E	0037
2SA530F	0037
2SA530G	0037
2SA530GN	0037
2SA530GR	0037
2SA530H	0037
2SA530H1	0037
2SA530HA	0037
2SA530HB	0037
2SA530HC	0037
2SA530J	0037
2SA530K	0037
2SA530L	0037
2SA530M	0037
2SA530OR	0037
2SA530R	0037
2SA530X	0037
2SA530Y	0037
2SA532	0126
2SA532A	0126
2SA532B	0126
2SA532C	0126
2SA532D	0126
2SA532E	0126
2SA532F	0126
2SA535	0037
2SA536	0050
2SA537	0126
2SA537A	0126
2SA537AA	0126
2SA537AB	0126
2SA537AC	0126
2SA537AH	0126
2SA537AHA	0126
2SA537AHB	0126
2SA537AHC	0126
2SA537H	0126
2SA537HA	0126
2SA537HB	0126
2SA537HC	0126
2SA537K	0126
2SA537L	0126
2SA538	0004
2SA538-G	0004
2SA538G	0004
2SA539	0006
2SA539(K)	0006
2SA539(L)	0006
2SA539(M)	0006
2SA539-Y	0006
2SA539A	0006
2SA539B	0006
2SA539C	0006
2SA539D	0006
2SA539E	0006
2SA539F	0006
2SA539G	0006
2SA539GN	0006
2SA539H	0006
2SA539J	0006
2SA539K	0006
2SA539L	0006
2SA539M	0006
2SA539OR	0006
2SA539R	0006
2SA539X	0006
2SA539Y	0006
2SA542	0037
2SA542A	0037
2SA542B	0037
2SA542C	0037
2SA542D	0037
2SA542E	0037
2SA542F	0037
2SA542G	0037
2SA542GN	0037
2SA542H	0037
2SA542J	0037
2SA542K	0037
2SA542L	0037
2SA542M	0037
2SA542OR	0037
2SA542R	0037
2SA542X	0037
2SA542Y	0037
2SA543	0693
2SA544	0126
2SA544A	0126
2SA544B	0126
2SA544C	0126
2SA544D	0126
2SA544E	0126
2SA544F	0126
2SA544G	0126
2SA544GN	0126
2SA544H	0126
2SA544J	0126
2SA544K	0126
2SA544L	0126
2SA544M	0126
2SA544OR	0126
2SA544R	0126
2SA544X	0126
2SA544Y	0126
2SA545	0791
2SA545(K)	0791
2SA545(L)	0791
2SA545A	0791
2SA545B	0791
2SA545C	0791
2SA545D	0791
2SA545E	0791
2SA545F	0791
2SA545G	0791
2SA545GN	0791
2SA545H	0791
2SA545J	0791
2SA545K	0791
2SA545KLM	0791
2SA545L	0791
2SA545LM	0791
2SA545M	0791
2SA545OR	0791
2SA545R	0791
2SA545X	0791
2SA545Y	0791
2SA546	0126
2SA546A	0126
2SA546B	0126
2SA546E	0126
2SA546H	0126
2SA547	0791
2SA547A	0791
2SA548	0150
2SA548G	0150
2SA548GN	0150
2SA548H	0150
2SA548HA	0150
2SA548HB	0150
2SA548HC	0150
2SA548OR	0150
2SA548R	0150
2SA548Y	0150
2SA549	0342
2SA549A	0342
2SA549AH	0342
2SA550	0037
2SA550(Q)	0037
2SA550(R)	0037
2SA550(S)	0037
2SA550A	0037
2SA550A(Q)	0037
2SA550A(R)	0037
2SA550A(R,Q,S	0037
2SA550A(S)	0037
2SA550AB	0037
2SA550AQ	0037
2SA550AR	0037
2SA550AS	0037
2SA550B	0037
2SA550BC	0037
2SA550BL	0037
2SA550C	0037
2SA550D	0037
2SA550P	0037
2SA550Q	0037
2SA550R	0037
2SA550S	0037
2SA550Y	0037
2SA551	0126
2SA551C	0126
2SA551D	0126
2SA551E	0126
2SA552	0126
2SA553	0037
2SA554	0037
2SA554A	0037
2SA555	0006
2SA556	0006
2SA557	0150
2SA558	0037
2SA559	0037
2SA559A	0037
2SA560	0126
2SA560A	0126
2SA561	0006
2SA561(O)	0006
2SA561-0	0006
2SA561-GR	0006
2SA561-GRN	0006
2SA561-O	0006
2SA561-OR	0006
2SA561-ORG	0006
2SA561-R	0006
2SA561-RD	0006
2SA561-RED	0006
2SA561-Y	0006
2SA561-YEL	0006
2SA561-YL	0006
2SA561GRN	0006
2SA561R	0006
2SA561RD	0006
2SA561RED	0006
2SA561Y	0006
2SA561YEL	0006
2SA561YL	0006
2SA562	0006
2SA562(O)	0006
2SA562(T)	0006
2SA562(Y)	0006
2SA562-0	0006
2SA562-GR	0006
2SA562-GRN	0006
2SA562-O	0006
2SA562-OR	0006
2SA562-ORG	0006
2SA562-OTM	0006
2SA562-R	0006
2SA562-RD	0006
2SA562-RED	0006
2SA562-TMO	0006
2SA562-Y	0006
2SA562-YE	0006
2SA562-YEL	0006
2SA562-Y-TM	0006
2SA562D	0006
2SA562E	0006
2SA562G	0006
2SA562GR	0006
2SA562GRN	0006
2SA562O	0006
2SA562Q	0006
2SA562R	0006
2SA562RD	0006
2SA562RED	0006
2SA562T	0006
2SA562T(O)	0006
2SA562T(Q)	0006
2SA562TM	0006
2SA562TM(Y)	0006
2SA562TM(Y)-T	3323
2SA562TM-0	0006
2SA562TML	0006
2SA562TMM-O	0006
2SA562TM-O	0006
2SA562TMO	0006
2SA562TM-O(FA)	0006
2SA562TM-OFA	0006
2SA562TM-T	5954
2SA562TM-Y	0006
2SA562TMY	0006
2SA562TM-Y(FA)	0006
2SA562TM-Y(T)	5954
2SA562TM-YFA	0006
2SA562TO	0006
2SA562TQ	0006
2SA562V	0006
2SA562VO	0006
2SA562Y	0006
2SA562YE	0006
2SA562YEL	0006
2SA562Y-TM	0006
2SA562YTM	0006
2SA564	0203
2SA564(0)	0203
2SA564(L)	0203
2SA564(O)	0203
2SA564(P)	0203
2SA564(Q)	0203
2SA564(R)	0203
2SA564(S)	0203
2SA564(S,T)	0203
2SA564(T)	0203
2SA564(Y)	0203
2SA564-0	0203
2SA564-1	0203
2SA564-O	0203
2SA564-OGD	0203
2SA564-OR	0203
2SA564-P	0203
2SA564-PQR	0203
2SA564-Q	0203
2SA564-QR	0203
2SA564-Q.R	0203
2SA564-R	0203
2SA564-S	0203
2SA564A	0203
2SA564A(P)	0203
2SA564A(Q)	0203
2SA564A(R)	0203
2SA564A(S)	0203
2SA564A2	0203
2SA564ABQ-1	0203
2SA564AG	0203
2SA564AK	0203
2SA564AL	0203
2SA564A-P	0203
2SA564AP	0203
2SA564A-P(CU)	0203
2SA564A-PQR	0203
2SA564APQR	0203
2SA564A-Q	0203
2SA564AQ	0203
2SA564A-Q(CU)	0203
2SA564AQ-(HD)	0203
2SA564A-QCU	0203
2SA564AQR	0203
2SA564AQRS	0203
2SA564A-Q.R	0203
2SA564A-R	0203
2SA564AR	0203
2SA564A-R(CU)	0203
2SA564ARS	0203
2SA564A-S	0203
2SA564AS	0203
2SA564AT	0203
2SA564ATAQR	0203
2SA564ATAQRS	0203
2SA564B	0203
2SA564BL	0203
2SA564C	0203
2SA564D	0203
2SA564E	0203
2SA564F	0203
2SA564FQ	0203
2SA564FQ-1	0203
2SA564FR	5404
2SA564FR-1	0203
2SA564G	0203
2SA564GN	0203
2SA564H	0203
2SA564J	0203
2SA564K	0203
2SA564L	0203
2SA564M	0203
2SA564OR	0203
2SA564P	0203
2SA564PQR	0203
2SA564P.A	0203
2SA564Q	0203
2SA564QP	0203
2SA564QR	0203
2SA564QRS	0203
2SA564R	0203
2SA564RQS	0203
2SA564RS	0203
2SA564S	0203
2SA564T	0203
2SA564X	0203
2SA564XL	0203
2SA564Y	0203
2SA565	0037
2SA565(C)	0037
2SA565(D)	0037
2SA565(D,C)	0037
2SA565A	0037
2SA565AB	0037
2SA565B	0037
2SA565BA	0037
2SA565C	0037
2SA565D	0037
2SA565E	0037
2SA565F	0037
2SA565G	0037
2SA565GN	0037
2SA565H	0037
2SA565J	0037
2SA565K	0037
2SA565L	0037
2SA565M	0037
2SA565OR	0037
2SA565R	0037
2SA565X	0037
2SA565Y	0037
2SA566	0905
2SA566A	0905
2SA566B	0905
2SA566C	0905
2SA566H	0905
2SA567	0037
2SA567A	0037
2SA567B	0037
2SA567C	0037
2SA567D	0037
2SA567E	0037
2SA567F	0037
2SA567G	0037
2SA567GN	0037
2SA567GR	0037
2SA567H	0037
2SA567J	0037
2SA567K	0037
2SA567L	0037
2SA567M	0037
2SA567OR	0037
2SA567R	0037
2SA567X	0037
2SA567Y	0037
2SA568	0006
2SA568-OR	0006
2SA568A	0006
2SA568B	0006
2SA568C	0006
2SA568D	0006
2SA568E	0006
2SA568F	0006
2SA568G	0006
2SA568GN	0006
2SA568H	0006
2SA568J	0006
2SA568K	0006
2SA568L	0006
2SA568M	0006
2SA568OR	0006
2SA568R	0006
2SA568X	0006
2SA568Y	0006
2SA569	0006
2SA569A	0006
2SA569B	0006

If replacement code is OEM, contact original manufacturer for replacement.

Original Device Types

DEVICE TYPE	REPL CODE
2SA569C	0006
2SA569D	0006
2SA569E	0006
2SA569F	0006
2SA569G	0006
2SA569GN	0006
2SA569H	0006
2SA569J	0006
2SA569K	0006
2SA569L	0006
2SA569M	0006
2SA569OR	0006
2SA569R	0006
2SA569X	0006
2SA569Y	0006
2SA570	0006
2SA570A	0006
2SA570B	0006
2SA570C	0006
2SA570D	0006
2SA570E	0006
2SA570F	0006
2SA570G	0006
2SA570GN	0006
2SA570H	0006
2SA570J	0006
2SA570K	0006
2SA570L	0006
2SA570M	0006
2SA570OR	0006
2SA570R	0006
2SA570X	0006
2SA570Y	0006
2SA571	0126
2SA572	0693
2SA572Y	0693
2SA573	0006
2SA574	0436
2SA575	0006
2SA576	0150
2SA577	0436
2SA578	0693
2SA579	0693
2SA580	0126
2SA581	0126
2SA594	0126
2SA594-O	0126
2SA594-R	0126
2SA594-Y	0126
2SA594N	0126
2SA597	0126
2SA598	OEM
2SA599	0791
2SA599(Y)	0791
2SA599Y	0791
2SA603	0037
2SA603A	0037
2SA603B	0037
2SA603C	0037
2SA603D	0037
2SA603E	0037
2SA603F	0037
2SA603G	0037
2SA603GN	0037
2SA603H	0037
2SA603J	0037
2SA603K	0037
2SA603L	0037
2SA603M	0037
2SA603OR	0037
2SA603R	0037
2SA603X	0037
2SA603Y	0037
2SA604	0126
2SA605	0126
2SA606	0126
2SA607	5008
2SA607A	5008
2SA607B	5008
2SA607C	5008
2SA607D	5008
2SA607E	5008
2SA607F	5008
2SA607G	5008
2SA607GN	5008
2SA607H	5008
2SA607J	5008
2SA607K	5008
2SA607L	5008
2SA607M	5008
2SA607OR	5008
2SA607R	5008
2SA607S	5008
2SA607SA	5008
2SA607SB	5008
2SA607SC	5008
2SA607SD	5008
2SA607X	5008
2SA607Y	5008
2SA608	0006
2SA608(C)	0006
2SA608(D)	0006
2SA608(E)	0006
2SA608(F)	0006
2SA608-0	0006
2SA608-C	0006
2SA608-E	0006
2SA608-E-CTV-NP	0006
2SA608-F	0006
2SA608-F-CTV-NP	0006
2SA608-F-NP-AA	0006
2SA608-F-SPA-AC	0006
2SA608-G-NP-AA	0006
2SA608-G-SPA-AC	0006
2SA608-O	0006
2SA608-OR	0006
2SA608A	0006
2SA608A(E)	0006
2SA608A(F)	0006
2SA608AE	0006
2SA608AF	0006
2SA608B	0006
2SA608BL	0006
2SA608C	0006
2SA608D	0006
2SA608D(F)	0006
2SA608E	0006
2SA608E-SP	OEM
2SA608EZ	0006
2SA608F	0006
2SA608F-NP	OEM
2SA608F-SP	OEM
2SA608FZ	0006
2SA608G	0006
2SA608GN	0006
2SA608G-SP	OEM
2SA608H	0006
2SA608J	0006
2SA608K	0006
2SA608K(E)	0006
2SA608K(F)	0006
2SA608K-A-NP	OEM
2SA608KE	0006
2SA608KF	0006
2SA608KG	0006
2SA608KNP	3108
2SA608KNPET	0006
2SA608KNPFT	0006
2SA608KNP-T	0006
2SA608L	0006
2SA608M	0006
2SA608NP	0006
2SA608NP(F)	0006
2SA608NP(G)	0006
2SA608NP-F	OEM
2SA608NP-G	OEM
2SA608NP-T	0006
2SA608OR	0006
2SA608P	0006
2SA608R	0006
2SA608SE	0006
2SA608SEZ	0006
2SA608SF	0006
2SA608SFZ	0006
2SA608SP	0006
2SA608SP(F)	0006
2SA608SP(G)	0006
2SA608SP-E	OEM
2SA608SPE	0006
2SA608SP-F	OEM
2SA608SPF	0006
2SA608SP-G	OEM
2SA608X	0006
2SA608Y	0006
2SA609	0037
2SA609A	0037
2SA609B	0037
2SA609C	0037
2SA609D	0037
2SA609E	0037
2SA609F	0037
2SA609G	0037
2SA609GN	0037
2SA609J	0037
2SA609K	0037
2SA609L	0037
2SA609M	0037
2SA609OR	0037
2SA609R	0037
2SA609Y	0037
2SA610	0006
2SA611	0006
2SA612	0126
2SA613	0905
2SA614	0905
2SA615	0905
2SA616	0905
2SA616(1)	0905
2SA616(1,2)	0905
2SA616(2)	0905
2SA617	0342
2SA617K	0342
2SA618	0342
2SA618K	0342
2SA620	0006
2SA621	0037
2SA622	3519
2SA623	1430
2SA623-O	1430
2SA624	1430
2SA625	0126
2SA626	3500
2SA626L	3500
2SA627	3500
2SA628	0006
2SA628(EF)	0006
2SA628(F)	0006
2SA628-E	0006
2SA628-O	0006
2SA628-OR	0006
2SA628A	0006
2SA628A(E)	0006
2SA628AA	0006
2SA628AD	0006
2SA628A-E	0006
2SA628AE	0006
2SA628A-F	0006
2SA628AF	0006
2SA628B	0006
2SA628C	0006
2SA628D	0006
2SA628E	0006
2SA628E,F	0006
2SA628EF	0006
2SA628F	0006
2SA628G	0006
2SA628GN	0006
2SA628H	0006
2SA628J	0006
2SA628K	0006
2SA628L	0006
2SA628M	0006
2SA628OR	0006
2SA628R	0006
2SA628X	0006
2SA628Y	0006
2SA629	0693
2SA629A	0693
2SA629B	0693
2SA629C	0693
2SA629D	0693
2SA629E	0693
2SA629F	0693
2SA629G	0693
2SA629GN	0693
2SA629H	0693
2SA629J	0693
2SA629K	0693
2SA629L	0693
2SA629M	0693
2SA629OR	0693
2SA629R	0693
2SA629X	0693
2SA629Y	0693
2SA633	1051
2SA633A	1051
2SA634	0681
2SA634(4)K	0681
2SA634(4)L	0681
2SA634(L)	0681
2SA634(L,K)	0681
2SA634(M)	0681
2SA634A	0681
2SA634K	0681
2SA634L	0681
2SA634M	0681
2SA635	0382
2SA636	2021
2SA636(4)K	2021
2SA636(4)L	2021
2SA636(L)	2021
2SA636(M)	2021
2SA636-4K	2021
2SA636-4L	2021
2SA636A	2021
2SA636B	2021
2SA636C	2021
2SA636D	2021
2SA636K	2021
2SA636L	2021
2SA636M	2021
2SA637	0342
2SA637C	0342
2SA637D	0342
2SA637M(Q)	0342
2SA637M(R)	0342
2SA637M(S)	0342
2SA638	0342
2SA638E	0342
2SA638E,F	0342
2SA638F	0342
2SA638S	0342
2SA639	0342
2SA639L	0342
2SA639S	0342
2SA640	0693
2SA640(M)	0693
2SA640A	0693
2SA640B	0693
2SA640C	0693
2SA640D	0693
2SA640M	0693
2SA641	0693
2SA641A	0693
2SA641B	0693
2SA641BL	0693
2SA641C	0693
2SA641D	0693
2SA641G	0693
2SA641GR	0693
2SA641K	0693
2SA641L	0693
2SA641M	0693
2SA641O	0693
2SA641OR	0693
2SA641R	0693
2SA641Y	0693
2SA642	0006
2SA642-O	5954
2SA642A	0006
2SA642B	0006
2SA642C	0006
2SA642D	0006
2SA642E	0006
2SA642F	0006
2SA642G	0006
2SA642GN	0006
2SA642H	0006
2SA642J	0006
2SA642K	0006
2SA642L	0006
2SA642M	0006
2SA642OR	0006
2SA642R	0006
2SA642S	0006
2SA642V	0006
2SA642W	0006
2SA642X	0006
2SA642Y	0006
2SA643	1242
2SA643(R)	1242
2SA643(V)	1242
2SA643(V,R)	1242
2SA643(W)	1242
2SA643R	1242
2SA643S	1242
2SA643V	1242
2SA643W	1242
2SA645	1367
2SA646	1367
2SA647	0126
2SA648	2013
2SA649	2013
2SA649A	2013
2SA650	2013
2SA651	2013
2SA652	0792
2SA653	0792
2SA653A	0792
2SA653L	0792
2SA654ARS	OEM
2SA656	3500
2SA656A	3500
2SA656L	3500
2SA656M	3500
2SA657	3500
2SA657A	3500
2SA658	3500
2SA658A	3500
2SA659	0006
2SA659(D)	0006
2SA659(E)	0006
2SA659A	0006
2SA659B	0006
2SA659D	0006
2SA659E	0006
2SA659F	0006
2SA659G	0006
2SA659L	0006
2SA659P	0006
2SA659R	0006
2SA659Y	0006
2SA661	0006
2SA661-GR	0006
2SA661-O	0006
2SA661-R	0006
2SA661-Y	0006
2SA661Y	0006
2SA663	0491
2SA663-BL	0491
2SA663-R	0491
2SA663-Y	0491
2SA664	0532
2SA666	0006
2SA666A	0006
2SA666B	0006
2SA666BL	0006
2SA666C	0006
2SA666D	0006
2SA666E	0006
2SA666H	0006
2SA666HR	0006
2SA666I	0006
2SA666IQRS	0006
2SA666QRS	0006
2SA666R	0006
2SA666S	0006
2SA666Y	0006
2SA668	0126
2SA669	0342
2SA670	0925
2SA670A	0925
2SA670B	0925
2SA670C	0925
2SA671	1308
2SA671A	1308
2SA671B	1308
2SA671C	1308
2SA671K	1308
2SA671KA	1308
2SA671KB	1308
2SA671KC	1308
2SA671TD	1308
2SA672	0037
2SA672(B)	0037
2SA672A	0037
2SA672B	0037
2SA672C	0037
2SA673	0148
2SA673(B)	0148
2SA673(C)	0148
2SA673(C)-Y	0148
2SA673(D)	0148
2SA673-C	0148
2SA673-D	0148
2SA673-O	0148
2SA673-OR	0148
2SA673-Y	0148
2SA673A	0148
2SA673A(C)	0148
2SA673AB	0148
2SA673A-C	0148
2SA673AC	0148
2SA673AD	0148
2SA673AK	0148
2SA673AKA	0148
2SA673AKB	0148
2SA673AKC	0148
2SA673AKD	0148
2SA673AS	0148
2SA673AS(C)	0148
2SA673B	0148
2SA673BC	4432
2SA673C	0148
2SA673C2	0148
2SA673C26TC	4432
2SA673C26TZ	4432
2SA673CD	4432
2SA673C-TB	0148
2SA673CTZ	4432
2SA673D	0148
2SA673D26TZ	4432
2SA673DTZ	4432
2SA673E	0148
2SA673F	0148
2SA673G	0148
2SA673GN	0148
2SA673H	0148
2SA673J	0148
2SA673K	0148
2SA673L	0148
2SA673M	0148
2SA673OR	0148
2SA673R	0148
2SA673X	0148
2SA673Y	0148
2SA675	0006
2SA677	0006
2SA677-0	0006
2SA677-O	0006
2SA677-OR	0006
2SA677A	0006
2SA677B	0006
2SA677C	0006
2SA677D	0006
2SA677E	0006
2SA677F	0006
2SA677G	0006
2SA677GN	0006
2SA677H	0006
2SA677HL	0006
2SA677J	0006
2SA677K	0006
2SA677L	0006
2SA677M	0006
2SA677OR	0006
2SA677R	0006
2SA677X	0006
2SA677Y	0006
2SA678	0006
2SA678(C)	0006
2SA678(SONY)	0006
2SA678-O	0006
2SA678-OR	0006
2SA678A	0006
2SA678B	0006
2SA678C	0006
2SA678D	0006
2SA678E	0006
2SA678F	0006
2SA678G	0006
2SA678GN	0006
2SA678H	0006
2SA678J	0006
2SA678K	0006
2SA678L	0006
2SA678M	0006
2SA678OR	0006
2SA678R	0006
2SA678X	0006
2SA678Y	0006
2SA679	2013
2SA679-R	2013
2SA679-Y	2013
2SA680	2013
2SA680-R	2013
2SA680-Y	2013
2SA681	0126
2SA682	0525
2SA682-O	0525
2SA682-Y	0525
2SA682O	0525
2SA683	0532
2SA683(Q)	0532
2SA683(R)	0532
2SA683NC	0532
2SA683O	0532
2SA683P	0532
2SA683Q	0532
2SA683R	0532
2SA683S	0532
2SA683.4	0532
2SA684	0532
2SA684(Q)	0532
2SA684(R)	0532
2SA684(S)	0532
2SA684NC	0532
2SA684Q	0532
2SA684R	0532
2SA684RS	0532
2SA684S	0532
2SA685	0006
2SA693	0006
2SA693C	0006
2SA695	0436
2SA695-F	0436
2SA695C	0436
2SA695D	0436
2SA696	0006
2SA696D	0006
2SA697	0006
2SA697C	0006
2SA698	0681
2SA699	0681
2SA699(AP)	0681
2SA699(H)	0681
2SA699(Q)	0681
2SA699(R)	0681
2SA699A	0681
2SA699A(P)	0681
2SA699A(R)	0681
2SA699A(RPQ)	0681
2SA699AP	0681
2SA699AQ	0681
2SA699AR	0681
2SA699CF	0681
2SA699P	0681
2SA699Q	0681
2SA699R	0681
2SA700	0925
2SA700(B)	0925
2SA700B	0925
2SA700Y	0925
2SA701	0037
2SA701F	0037
2SA701FJ	0037
2SA701FO	0037
2SA702	0693
2SA702F	3107
2SA703	1430
2SA704	0006
2SA704A	0006
2SA704B	0006
2SA704C	0006
2SA704D	0006
2SA704E	0006
2SA704F	0006
2SA704G	0006
2SA704GN	0006
2SA704H	0006
2SA704J	0006
2SA704K	0006
2SA704L	0006
2SA704M	0006
2SA704OR	0006
2SA704R	0006
2SA704X	0006
2SA704Y	0006
2SA705	0006
2SA705A	0006
2SA705B	0006
2SA705C	0006
2SA705D	0006
2SA705E	0006
2SA705F	0006
2SA705G	0006
2SA705GN	0006
2SA705H	0006
2SA705J	0006
2SA705K	0006
2SA705L	0006
2SA705M	0006
2SA705OR	0006
2SA705R	0006
2SA705X	0006
2SA705Y	0006
2SA706	0382
2SA706(3)	0382
2SA706-2	0382
2SA706-3	0382
2SA706-4	0382
2SA706H	0382
2SA706J	0382
2SA707	0436
2SA707(V)	0436
2SA708	0126
2SA708A	0126
2SA709	0037
2SA710	0150
2SA711	0150
2SA712	0439
2SA713	3519
2SA713A	3519
2SA714	2013
2SA714-L	2013
2SA714L	2013
2SA715	0525
2SA715B	0525
2SA715C	0525
2SA715D	0525
2SA715WBP	0525
2SA715WT	0525
2SA715WT(C,B)	0525
2SA715WT-B	0525
2SA715WTB	0525
2SA715WTBC	0525
2SA715WT-C	0525
2SA715WTC	0525
2SA717	0126
2SA718	0037
2SA719	0006
2SA719(Q)	0006
2SA719(R)	0006
2SA719(RS)	0006
2SA719(S)	0006
2SA719-P	0006
2SA719-PQR	0006
2SA719-Q	0006
2SA719-R	0006
2SA719-S	0006
2SA719,R	0006
2SA719A	0006
2SA719K	0006
2SA719NC	0006
2SA719O	0006
2SA719P	0006
2SA719PQR	0006
2SA719Q	0006
2SA719QR	0006
2SA719R	0006
2SA719RS	0006
2SA719S	0006
2SA719TAQR	0006
2SA720	0006
2SA720-G	0006
2SA720-O	0006
2SA720-Q	0006
2SA720-R	0006
2SA720A	0006
2SA720A(R)	0006
2SA720A(S)	0006
2SA720B	0006
2SA720C	0006
2SA720D	0006
2SA720L	0006
2SA720NC	0006
2SA720Q	0006
2SA720QR	0006
2SA720QS	0006
2SA720R	0006
2SA720RST	0006
2SA720S	0006
2SA720Y	0006
2SA721	0693
2SA721Q	0693
2SA721R	0693
2SA721S	0693
2SA721T	0693
2SA721U	0693
2SA722	0693
2SA722-S	OEM
2SA722S	0693
2SA723	0006
2SA723R	0006
2SA724	0148
2SA725	0693
2SA725Y	0693
2SA726	0693
2SA726Y	0693
2SA728	0006
2SA728A	0006
2SA730	1242
2SA731	1242
2SA732	0126
2SA733	0006
2SA733(C)Q	0006
2SA733(C)-T	0006
2SA733(E)	0006
2SA733(K)	0006
2SA733(P)	0006
2SA733(Q)	0006
2SA733-P	0006
2SA733-Q	0006
2SA733-T	0006
2SA733A	0006
2SA733A-K	OEM
2SA733AK	0006
2SA733A-KY	OEM
2SA733AP	0006
2SA733APT	0006
2SA733A-Q	OEM
2SA733AQ	0006
2SA733AR	0006
2SA733H	0006
2SA733IO	0006
2SA733IQ	0006
2SA733K	0006
2SA733P	0006
2SA733PQ	0006
2SA733Q	0006
2SA733Q1P	0006
2SA733QP	0006

If replacement code is OEM, contact original manufacturer for replacement.

DEVICE TYPE	REPL CODE
2SA733QT	0006
2SA733R	0006
2SA733TP	0006
2SA734	0126
2SA735	0037
2SA736	0126
2SA738	0460
2SA738(B)	0460
2SA738A	0460
2SA738C	0460
2SA738D	0460
2SA739	3606
2SA740	1648
2SA740A	1648
2SA740AB	1648
2SA741	0150
2SA741H	0150
2SA742	0126
2SA742H	0126
2SA743	0525
2SA743A	0525
2SA743AB	0525
2SA743AC	0525
2SA743C	0525
2SA744	0491
2SA745	2013
2SA745A	2013
2SA746	2013
2SA747	2013
2SA747A	2013
2SA748	0681
2SA748Q	0681
2SA749	0342
2SA749A	0342
2SA750	0693
2SA751	0477
2SA751(P)	0477
2SA751(Q)	0477
2SA751(R)	0477
2SA751Q	0477
2SA751R	0477
2SA752	0477
2SA753	2013
2SA754	0925
2SA755	0925
2SA755B	0925
2SA755C	0925
2SA756	3500
2SA757	3500
2SA758	2013
2SA759	0037
2SA760	0693
2SA761	0126
2SA761-1	0126
2SA761-2	0126
2SA762	0792
2SA762-1	0792
2SA762-2	0792
2SA763	0693
2SA764	0905
2SA765	0905
2SA766	1648
2SA766S	1648
2SA767	0037
2SA768	0925
2SA768G	0925
2SA768O	0925
2SA768Q	0925
2SA768Y	0925
2SA768Y-G	OEM
2SA769	0854
2SA770	0854
2SA771	1369
2SA772	0532
2SA772-1	0532
2SA772-2	0532
2SA772-23	0532
2SA772J	0532
2SA773	2038
2SA773-1	2038
2SA773-2	2038
2SA773H	2038
2SA774	0693
2SA774A	0693
2SA775	0681
2SA775A	0681
2SA775AA	0681
2SA775AB	0681
2SA775AC	0681
2SA775B	0681
2SA775C	0681
2SA776	0150
2SA776A	0150
2SA777	0436
2SA777NC	0436
2SA777Q	0436
2SA777R	5605
2SA778	0342
2SA778(A-K)	0342
2SA778A	0342
2SA778AF-02	0342
2SA778AK	0342
2SA778AK-02	0342
2SA778K	0342
2SA778K-02	0342
2SA779	1051
2SA779K	1051
2SA779KB	1051
2SA780	1367
2SA780A	1367
2SA780AK	1367
2SA780AKB	1367
2SA781	0150
2SA781K	0150
2SA782	0037
2SA783	0148
2SA784	3519
2SA785	5061
2SA786	0037
2SA786-Q	0037
2SA786Q	0037
2SA786QL	0037
2SA786R	0037
2SA787	0693
2SA788	0037
2SA789	0148
2SA790	0635
2SA790M	0635
2SA791	3519
2SA793	0126
2SA794	0525
2SA794A	0525
2SA795	0525
2SA795A	0525
2SA796	3519
2SA797	OEM
2SA798	0693
2SA798-G-DU	0693
2SA798G	6226
2SA799	0126
2SA800	3607
2SA802	0342
2SA803	1522
2SA804	1522
2SA805	0342
2SA806	0342
2SA807	3500
2SA808	3500
2SA808A	3500
2SA809	0342
2SA810	0342
2SA811	0037
2SA811A	0037
2SA811C5	0037
2SA811C6	0037
2SA811C7	0037
2SA811C8	0037
2SA812	1742
2SA812-M51E	1742
2SA812M3	1742
2SA812M4	1742
2SA812M5	1742
2SA812M6	1742
2SA812M7	1742
2SA813	0150
2SA813S2	0150
2SA813S3	0150
2SA813S4	0150
2SA814	0681
2SA814-O	0681
2SA814-Y	0681
2SA815	0681
2SA815-O	0681
2SA815-Y	0681
2SA816	1648
2SA816-0	1648
2SA816-O	1648
2SA816-Y	1648
2SA817	0436
2SA817(Y)	0436
2SA817-Y	0436
2SA817A	0436
2SA817A-Y	OEM
2SA817AY	0436
2SA817Y	0436
2SA818	0681
2SA818-O	0681
2SA818-Y	0681
2SA819	0006
2SA820	0342
2SA821	0342
2SA822	0436
2SA823	0532
2SA824	0532
2SA825	0006
2SA825Q	0006
2SA825R	0006
2SA825S	0006
2SA826	0037
2SA826P	0037
2SA826Q	0037
2SA826R	0037
2SA826S	0037
2SA827	0148
2SA828	0532
2SA828A	0532
2SA829	0532
2SA830	3519
2SA831	0635
2SA832	0342
2SA833	OEM
2SA834	1522
2SA835	0342
2SA835H	0342
2SA836	0693
2SA836(D)	0693
2SA836(E)	0693
2SA836C	0693
2SA836D	0693
2SA836DE	0693
2SA836E	0693
2SA836F	0693
2SA837	3500
2SA838	5378
2SA838A	6239
2SA838B	5378
2SA838C	5378
2SA839	1648
2SA839-O	1648
2SA839-R	1648
2SA839-Y	1648
2SA840	1648
2SA840H	1648
2SA840JC	1648
2SA840JM	1648
2SA840K	1648
2SA841	0693
2SA841-BL	0693
2SA841-GR	0693
2SA842	0693
2SA842-BL	0693
2SA842-GR	0693
2SA843	1648
2SA844	0006
2SA844(D)	0693
2SA844(E)	0693
2SA844C	0693
2SA844CD	0006
2SA844CD-TB	0006
2SA844D	0693
2SA844DE	0693
2SA844E	0693
2SA844F	0693
2SA844O	0693
2SA845	0342
2SA845A	0342
2SA845AH	0342
2SA845H	0342
2SA847	0648
2SA847A	0648
2SA848	0439
2SA849	0439
2SA850	0006
2SA851	3617
2SA852	3617
2SA853	OEM
2SA-854	1348
2SA854	1348
2SA854(Q)	1348
2SA854-Q	1348
2SA854G	1348
2SA854PQ	1348
2SA-854Q	1348
2SA854Q	1348
2SA854QE	1348
2SA854QR	1348
2SA854QV	1348
2SA854QY	1348
2SA854QZ	1348
2SA854R	1348
2SA854RA	OEM
2SA854S	1348
2SA854SQ	1348
2SA854SQR	OEM
2SA854SR	1348
2SA855	0693
2SA856	0693
2SA856A	0693
2SA857	0342
2SA858	0342
2SA859	0037
2SA860	0342
2SA861	1051
2SA861(3)	1051
2SA861JK	1051
2SA866	0037
2SA866V	0037
2SA866VG	0037
2SA867	0150
2SA868	0037
2SA869	0037
2SA870	0037
2SA871	5367
2SA872	0648
2SA872A	0648
2SA872AD	0648
2SA872AE	0648
2SA872AF	0648
2SA872C	0648
2SA872D	0648
2SA872E	0648
2SA872F	0648
2SA873	0150
2SA874	0359
2SA874M	0359
2SA874P	0359
2SA874PQ	0359
2SA874Q	0359
2SA876	0037
2SA876HA	0037
2SA876HB	0037
2SA876HC	0037
2SA877	3500
2SA878	3500
2SA879	0342
2SA879NC	0342
2SA879P	0342
2SA879Q	0342
2SA880	0693
2SA881	0006
2SA881Q	0006
2SA881R	0006
2SA882	2013
2SA883	0532
2SA884	0532
2SA885	0525
2SA886	0525
2SA886Q	0525
2SA886V	0525
2SA886VQ	0525
2SA886VR	0525
2SA886VS	0525
2SA887	1267
2SA887P	1267
2SA887Q	1267
2SA887R	1267
2SA888	0037
2SA889	0037
2SA890	0037
2SA891	0037
2SA892	0602
2SA893	0648
2SA893A	0648
2SA893AC	0648
2SA893AD	0648
2SA893AE	0648
2SA893AF	0648
2SA893D	0648
2SA893E	0648
2SA894	0436
2SA895	0150
2SA896	0342
2SA896-1	0342
2SA896-2	0342
2SA897	0648
2SA898	0525
2SA899	5382
2SA900	0525
2SA900-R	0525
2SA900-S	0525
2SA900-T	0525
2SA901	0693
2SA902	0693
2SA903	0693
2SA904	0648
2SA904A	0648
2SA905	0648
2SA906	0693
2SA907	1681
2SA908	1598
2SA909	1598
2SA911	OEM
2SA912	0342
2SA912Q	0342
2SA912R	0342
2SA912S	0342
2SA913	1648
2SA913A	1648
2SA913A(Q)	1648
2SA913A(R)	1648
2SA913AQ	1648
2SA913ARQ	OEM
2SA914	0525
2SA915	0342
2SA916	0342
2SA916K	0342
2SA916L	0342
2SA916M	0342
2SA917	0477
2SA918	0150
2SA919	OEM
2SA920	3758
2SA921	0648
2SA921A	0648
2SA921AS	0648
2SA922	0126
2SA923	0342
2SA923-1	0342
2SA923-2	0342
2SA924	3617
2SA925	0150
2SA925-1	0150
2SA925-2	0150
2SA927	0006
2SA928	1603
2SA928A	OEM
2SA929	0693
2SA929G	0693
2SA929H	0693
2SA930	0693
2SA931	5367
2SA932	5367
2SA933	0148
2SA933(Q)	0148
2SA933(Q)-Y	0148
2SA933(Q,R)	0148
2SA933(R)	0148
2SA933(R)-Y	0148
2SA933(S)	0148
2SA933-Q	0148
2SA933-R	0148
2SA933-S	0148
2SA933-T	0148
2SA933-T-93-R	0148
2SA933-T-93-S	0148
2SA933-Y	0148
2SA933A(Q)	0148
2SA933A(Q)-Y	0148
2SA933A(R)	0148
2SA933A(R)-Y	0148
2SA933AQ	0148
2SA933AR	0148
2SA933LN	0148
2SA933P	0148
2SA933Q	0148
2SA933QL	0148
2SA933QR	0148
2SA933QV	0148
2SA933QY	0148
2SA933QZ	0148
2SA933R	0148
2SA933RA	0148
2SA933RC	0148
2SA933RK	0148
2SA933RZ	0148
2SA933S	0148
2SA933S(Q)	0148
2SA933S(Q)-Y	0148
2SA933S(QR)	0148
2SA933S(QR)-T	0148
2SA933S(QS)-T	4421
2SA933S(R)	0148
2SA933S(R)-Y	0148
2SA933S(S)	0148
2SA933SP	0148
2SA933S-Q	0148
2SA933SQ	0148
2SA933SQR	0148
2SA933SQRSTA	OEM
2SA933SQT	0148
2SA933SQZ	0148
2SA933S-R	0148
2SA933SR	0148
2SA933SRC	0148
2SA933SRD	0148
2SA933SRF	0148
2SA933SRN	0148
2SA933SRQ	0148
2SA933SRR	0148
2SA933SRU	0148
2SA933SRX	0148
2SA933SRY	0148
2SA933S-S	0148
2SA933SS	0148
2SA933SSA	0148
2SA933SSD	0148
2SA933SSJ	0148
2SA933SSK	0148
2SA933SSU	0148
2SA933XTPR	0148
2SA933XTPS	0148
2SA933Z	OEM
2SA934	0532
2SA934Q	0532
2SA934R	0532
2SA935	0436
2SA935R	0436
2SA935T103	0436
2SA936	3519
2SA-937	2484
2SA937	2484
2SA937(R)	2484
2SA937-Q	2484
2SA937-R	2484
2SA937F	OEM
2SA937LN	2484
2SA937M	2484
2SA937M(Q)	2484
2SA937M(R)	2484
2SA937M(S)	2484
2SA937MQRS	2484
2SA937M-R	OEM
2SA937MR	2484
2SA937MRS	2484
2SA937M-S	OEM
2SA937MS	2484
2SA937P	2484
2SA937Q	2484
2SA937R	2484
2SA937RQ	2484
2SA937RZ	2484
2SA937S	2484
2SA938	3183
2SA939	4479
2SA940	1911
2SA940-AR(D)	1911
2SA940-AR(E)	1911
2SA940-O	1911
2SA940-R-D(LB-SAN-5)	1911
2SA940-R-E(LB-SAN-5)	1911
2SA940-Z	1911
2SA940AD	1911
2SA940AR-D	OEM
2SA940AR-E	OEM
2SA940D	1911
2SA940E	1911
2SA940GL2	1911
2SA940LB	1911
2SA940LBGL2	1911
2SA940LBGLR	1911
2SA940R	1911
2SA940R(D)	1911
2SA940R(E)	1911
2SA940RD	1911
2SA940RE	1911
2SA940T-D	OEM
2SA940T-E	OEM
2SA940Z	3543
2SA941	0006
2SA942	0006
2SA945	0164
2SA945-0	0164
2SA945-O	0164
2SA945-Q	0164
2SA948	0126
2SA948E	0126
2SA949	0477
2SA949(Y)C1	6747
2SA949-Y(C)	0477
2SA949Y	0477
2SA949Y(FA-1)	OEM
2SA950	1348
2SA950(GR)	1348
2SA950(O)	1348
2SA950(Y)	1348
2SA950-0	1348
2SA950-O	1348
2SA950-Y	1348
2SA950FA	1348
2SA950GR	3108
2SA950Y	1348
2SA950Y(FA)	OEM
2SA951	0681
2SA951-10	0681
2SA951KA	0681
2SA951KD	0681
2SA951LC	0681
2SA951M	0681
2SA951MU	0681
2SA952	0006
2SA952(C)-T	0006
2SA952(C)-TL	0006
2SA952(L)	0006
2SA952(M)	0006
2SA952C	0006
2SA952K	0006
2SA952L	0006
2SA952LK	0006
2SA952LT	0006
2SA952M	0006
2SA952ML	0006
2SA952ML2	0006
2SA952NL2	OEM
2SA953	1348
2SA953(K)	1348
2SA953(L)	1348
2SA953K	1348
2SA953L	1348
2SA954	0006
2SA956	0150
2SA956H3	0150
2SA956H4	0150
2SA956H5	0150
2SA956H6	0150
2SA957	1648
2SA957Y	1648
2SA958	1911
2SA958-O	OEM
2SA958-O-N03	1911
2SA958-Y-N03	1911
2SA958Y	1911
2SA959	OEM
2SA962	0382
2SA963	0525
2SA964	3758
2SA964A	3758
2SA965	5329
2SA965(Y)	5329
2SA965-O	0477
2SA965Y	0477
2SA966	0532
2SA966(Y)	0532
2SA966-O	0532
2SA966O	0532
2SA966Y	0532
2SA967	OEM
2SA968	1648
2SA968(Y)	1648
2SA968A	1648
2SA968B	1648
2SA968Y	1648
2SA969	0792
2SA970	0006
2SA970BL	0006
2SA970GR	0006
2SA970NEW-BL	0006
2SA971	1598
2SA972	0532
2SA973	0693
2SA977	0525
2SA977A	0525
2SA978	0693
2SA979	5057
2SA979G	6378
2SA980	2013
2SA981	2013
2SA982	2013
2SA983	OEM
2SA984	0006
2SA984(E)	0006
2SA984(F)	0006
2SA984-E-AA	0006
2SA984-F-AA	0006
2SA984E	0006
2SA984F	3108
2SA984K	0006
2SA984K(E)	0006
2SA984K(F)	0006
2SA984K-E	OEM
2SA984KE	0006
2SA984K-F	OEM
2SA984K-T	0006
2SA985	5387
2SA985A	5387
2SA985P	3543
2SA986	3571
2SA987	0006
2SA988	0648
2SA988E	0648
2SA988F	0648
2SA988FE	0648
2SA989	0532
2SA990	0693
2SA991	0693
2SA992	0648
2SA992E	0648
2SA992EF	OEM
2SA992F	0648
2SA993	0037
2SA994	0006
2SA995	5059
2SA997	0532
2SA998	0532
2SA999	0006
2SA999-O	0006
2SA999F	OEM
2SA999L	0006
2SA1001	2013
2SA1002	2013
2SA1003	1598
2SA1004	0037
2SA1005	0037
2SA1005L	0037
2SA1006	3543
2SA1006A	3543
2SA1006B	OEM
2SA1007	1598
2SA1007A	1598
2SA1008	3544
2SA1009	OEM
2SA1009A	OEM
2SA1010	1369
2SA1010V	1369
2SA1010Y	1369
2SA1011	1648
2SA1011(D)	1648
2SA1011(E)	1648
2SA1011D	1648
2SA1011E	1648
2SA1011RA	OEM
2SA1012	3548
2SA1012-O	3548
2SA1012-Y	3548
2SA1012Y	3548
2SA-1013	1522
2SA1013	1522
2SA1013-O	1522
2SA1013-0	1522
2SA1013-O(C)	1522
2SA1013-R	1522
2SA1013R	1522
2SA1013Y	1522
2SA1014	OEM
2SA-1015	0148
2SA1015	0148
2SA1015(G)	0148
2SA1015(GR)	0148
2SA1015(GR)L	0148
2SA1015(GR)Y	0148
2SA1015(H)-Y	0148
2SA1015(J)-Y	0148
2SA1015(O)	0148
2SA1015(Y)	0148
2SA1015(Y)L	0148
2SA1015(Y)Y	0148
2SA1015(Y,GR)Y	0148
2SA1015(YG)	0148
2SA1015(YG)-LB	0148
2SA1015(YG)-T	0148
2SA1015-O	0148
2SA1015-GR	0148
2SA1015-GR-(SAN)	0148
2SA1015-GR(SAN)	0148
2SA1015-O	0148
2SA1015-O(FA)	0148
2SA1015-O(SAN)	0148
2SA1015-O(TE)	4432
2SA1015-OFA	0148
2SA1015-TPE2	0148
2SA1015-Y	0148
2SA1015-Y(FA)	0148
2SA1015-Y(SAN)	0148
2SA1015-Y(TE)	0148
2SA1015-YFA	0148
2SA1015ABL	0148
2SA1015FA	0148
2SA1015G	0148
2SA1015GR	0148
2SA1015L	0148
2SA1015O	0148
2SA1015TPE2	0148
2SA-1015Y	0148
2SA1015Y	0148

If replacement code is OEM, contact original manufacturer for replacement.

DEVICE TYPE	REPL CODE
2SA1015Y(FA)	0148
2SA1015Y-TYPE1	0148
2SA1016	0693
2SA1016(F)	0693
2SA1016(G)	0693
2SA1016E	OEM
2SA1016F	0693
2SA1016G	0693
2SA1016K	0693
2SA1016K-F	0693
2SA1016KF	0693
2SA1016KFG	0693
2SA1017	0342
2SA1017G	6771
2SA1018	0342
2SA1018-V	0342
2SA1018Q	6773
2SA1018R	6773
2SA1018V	0342
2SA1019	0648
2SA1020	0434
2SA1020(Y)	0434
2SA1020-O	0434
2SA1020-Y	0434
2SA1020O	0434
2SA1020Y	0434
2SA1020YTPE6	0434
2SA1021	0525
2SA1021-O	0525
2SA1021-R	0525
2SA1021R	2555
2SA1021Y	0525
2SA1022	1752
2SA1022(B)	1752
2SA1022(BC)-W	OEM
2SA1022(C)	1752
2SA1022C	6781
2SA1023	1603
2SA1023P	1603
2SA1023Q	1603
2SA1024	OEM
2SA1024H	OEM
2SA1025	0037
2SA1025D	0037
2SA1025E	0037
2SA1026	0006
2SA1027	0037
2SA1027(R)	0037
2SA1027R	0037
2SA1028	2013
2SA1029	0006
2SA1029B	0006
2SA1029C	0006
2SA1029D	0006
2SA1030	0006
2SA1030B	0006
2SA1030C	0006
2SA1031	0037
2SA1031B	0037
2SA1031C	0037
2SA1031D	0037
2SA1032	0037
2SA1032B	0037
2SA1032C	0037
2SA1033	0037
2SA1034	1742
2SA1035	1742
2SA1036	1752
2SA1036K	1752
2SA1036KQ	OEM
2SA1037	1742
2SA1037(R)	1742
2SA1037AK	OEM
2SA1037AK-T146-R	OEM
2SA1037K	1742
2SA1037K(FR)	1742
2SA1037K(QR)-W	OEM
2SA1037KA	1742
2SA1037KERS	OEM
2SA1037K-Q	1742
2SA1037KQ	1742
2SA1037K-QR	1742
2SA1037K-R	1742
2SA1037KR	1742
2SA1037KR-T146TP	OEM
2SA1037K-S	1742
2SA1037KS	1742
2SA1037KT97	1742
2SA1037KT147	1742
2SA1038	0648
2SA1039	0648
2SA1039(Q)	0648
2SA1039(R)	0648
2SA1039(S)	0648
2SA1039AR	0648
2SA1040	0006
2SA1041	1598
2SA1042	0491
2SA1043	1681
2SA1044	1681
2SA1045	0491
2SA1046	0491
2SA1047	0525
2SA1047F	2555
2SA1048	0013
2SA1048(GR)	0013
2SA1048-0	0013
2SA1048-GR	0013
2SA1048-O	0013
2SA1048G	0013
2SA1048GR	0013
2SA10480	0013
2SA1048Y	0013
2SA1049	0006
2SA1050	0148
2SA1050A	0148
2SA1051	1598
2SA1051A	1598
2SA1052	1742
2SA1052B	1742
2SA1052C	1742
2SA1052D	1742
2SA1053	0037
2SA1054	0037
2SA1055	0037
2SA1056	0037
2SA1057	OEM
2SA1058	OEM
2SA1059	OEM
2SA1060	0859
2SA1060P	0859
2SA1060Q	0859
2SA1061	0859
2SA1061P	0859
2SA1061Q	0859
2SA1061R	0859
2SA1062	3571
2SA1063	1598
2SA1064	1598
2SA1065	1598
2SA1066	0477
2SA1067	1598
2SA1068	1598
2SA1069	3577
2SA1069A	3577
2SA1069C	3577
2SA1069L	3577
2SA1071-O	OEM
2SA1072	1598
2SA1072A	1598
2SA1073	1598
2SA1075	3581
2SA1076	3581
2SA1077	0718
2SA1078	3585
2SA1079	0718
2SA1080	0681
2SA1081	0037
2SA1081D	0037
2SA1081E	0037
2SA1082	0037
2SA1082D	0037
2SA1082E	0037
2SA1083	0037
2SA1083D	0037
2SA1083E	0037
2SA1084	1597
2SA1084D	1597
2SA1084E	1597
2SA1085	1597
2SA1085D	1597
2SA1085E	1597
2SA1090	0037
2SA1091	0342
2SA1091-O	0342
2SA1091R	0342
2SA1092	0693
2SA1093	3571
2SA1094	3581
2SA1095	3581
2SA1096	0525
2SA1096A	0525
2SA1097-1	OEM
2SA1097-2	OEM
2SA1100	0006
2SA1100D	0006
2SA1100F	0006
2SA1100L	0006
2SA1102	0859
2SA1103	0859
2SA1104	3111
2SA1104A	OEM
2SA1105	3111
2SA1106	3111
2SA1107	3630
2SA1107A	3630
2SA1108	3630
2SA1108A	3630
2SA1109	3646
2SA1110	0525
2SA1111	1648
2SA1112	1648
2SA1113	OEM
2SA1114	OEM
2SA1115	3314
2SA1115(E)	3314
2SA1115(F)	3314
2SA1115(G)	3314
2SA1115-E	3314
2SA1115-EF	3314
2SA1115-F	3314
2SA1115E	3314
2SA1115F	3314
2SA1115G	3314
2SA1116	3606
2SA1117	3606
2SA1120	0525
2SA1121	1752
2SA1121(SC)	OEM
2SA1121C(SC)	OEM
2SA1121SB	1752
2SA1121SC	1752
2SA1121SD	1752
2SA1122	1742
2SA1122(CD)	1742
2SA1122CC	1742
2SA1122CD	1742
2SA1122CE	1742
2SA1123	1522
2SA1124	1522
2SA1125	1648
2SA1126	OEM
2SA1126H	OEM
2SA1127	0693
2SA1127NC	0693
2SA1128	0006
2SA1129	1308
2SA1133	1911
2SA1133A	1911
2SA1135	0859
2SA1136	0477
2SA1137	0648
2SA1138	0648
2SA1141	3111
2SA1142	0525
2SA1143	0693
2SA1144	0136
2SA1145	1522
2SA1146	3571
2SA1147	3606
2SA1150	0006
2SA1150Y	0006
2SA1151	0532
2SA1152	0037
2SA1153	3617
2SA1154	0532
2SA1155	OEM
2SA1156	4479
2SA1158	0436
2SA1160	0532
2SA1161	OEM
2SA1162	1742
2SA1162-G	1742
2SA1162-YG	OEM
2SA1162G	1742
2SA1162Y	1742
2SA1163	5454
2SA1164	0037
2SA1166	3630
2SA1166A	3630
2SA1169	3581
2SA1170	3646
2SA1171	5454
2SA1171PD	5454
2SA1171PE	5454
2SA1173	1522
2SA1174	0693
2SA1175	3625
2SA1175(E)	3625
2SA1175(F)	3625
2SA1175(H)	3625
2SA1175(H)-Y	3625
2SA1175(J)	3625
2SA1175(J)-Y	3625
2SA1175(J,H)-Y	3625
2SA1175-E	3625
2SA1175-F	3625
2SA1175-FEK	5910
2SA1175-HFE	3625
2SA1175-Y	3625
2SA1175E	3625
2SA1175F	3625
2SA1175F(H)	3625
2SA1175FH	3625
2SA1175H	3625
2SA1175HO	3625
2SA1175J	3625
2SA1175JHFE	3625
2SA1175TP	3625
2SA1177	OEM
2SA1177E	OEM
2SA1178	3111
2SA1179	1752
2SA1179-M5	OEM
2SA1179-M5TA	OEM
2SA1179-M6TA	OEM
2SA1179-TA	1752
2SA1180	1598
2SA1180A	1598
2SA1182	1752
2SA1183	0859
2SA1184	0525
2SA1185	0859
2SA1186	3571
2SA1187	3630
2SA1188	0477
2SA1188D	0477
2SA1188E	0477
2SA1189	0477
2SA1189D	0477
2SA1189E	0477
2SA1190	0477
2SA1190D	0477
2SA1190E	0477
2SA1191	0477
2SA1191D	0477
2SA1191E	0477
2SA1193	3890
2SA1193K	3890
2SA1194	2897
2SA1194K	2897
2SA1195	1051
2SA1195-O	1051
2SA1195O	1051
2SA1195R	1051
2SA1196	OEM
2SA1197	OEM
2SA1198	0693
2SA1198(E)	0693
2SA1198(S)	0693
2SA1198R	OEM
2SA1198RS	0693
2SA1198S	0693
2SA1199	0037
2SA1200	OEM
2SA1201	OEM
2SA1202	3645
2SA1203	OEM
2SA1204	3645
2SA1205	3111
2SA1206	5462
2SA1206K	5462
2SA1207	1522
2SA1207S	6771
2SA1207ST	OEM
2SA1207T	OEM
2SA1208	1522
2SA1208-R	OEM
2SA1208-S	OEM
2SA1208S	6771
2SA1208T	6771
2SA1209	0525
2SA1209-T	OEM
2SA1209S	2555
2SA1210	0525
2SA1211	3617
2SA1213	OEM
2SA1214	0525
2SA1215	3581
2SA1216	3646
2SA1217	0525
2SA1218	0037
2SA1219	0693
2SA1220	0525
2SA1220A	0525
2SA1220A-P	0525
2SA1220AP	0525
2SA1220A-Q	0525
2SA1220AQ	0525
2SA1221	0525
2SA1221-L	0525
2SA1221L	0525
2SA1222	5367
2SA1223	OEM
2SA1224	OEM
2SA1225	5464
2SA1226	1752
2SA1227	3111
2SA1227A	3111
2SA1228	OEM
2SA1229	OEM
2SA1230	OEM
2SA1231	OEM
2SA1232	3111
2SA1233	OEM
2SA1233L	OEM
2SA1235	1752
2SA1235-E	1752
2SA1235-T12-A-1E	1752
2SA1235-T12-A-E1	1752
2SA1236	OEM
2SA1237	OEM
2SA1238	OEM
2SA1239	OEM
2SA1240	OEM
2SA1241	5464
2SA1242	5464
2SA1243	5464
2SA1244	5464
2SA1245	4498
2SA1246	3654
2SA1246(R)	3654
2SA1246(S)	3654
2SA1246(T)	3654
2SA1246(U)	3654
2SA1246R	3654
2SA1246S	3654
2SA1246T	3654
2SA1246U	3654
2SA1247	5454
2SA1248	0525
2SA1249	0525
2SA1249-R	OEM
2SA1249-S	OEM
2SA1249-T	OEM
2SA1249R	0525
2SA1249S	0525
2SA1250	OEM
2SA1251	OEM
2SA1252	1742
2SA1252-D6-TB	OEM
2SA1252-D7-TB	OEM
2SA1252D4	1742
2SA1252D5	1742
2SA1253	0477
2SA1254	0148
2SA1255	5368
2SA1256	1742
2SA1257	5454
2SA1258	0602
2SA1259	0602
2SA1260	0602
2SA1261	1199
2SA1262	1308
2SA1262-O	OEM
2SA1262R	OEM
2SA1262Y	OEM
2SA1263	3111
2SA1264	3111
2SA1264N	OEM
2SA1264O	OEM
2SA1264T	3111
2SA1265	3111
2SA1265-O	OEM
2SA1265N	3111
2SA1265NFA-1	3111
2SA1266	0148
2SA1266GR	0148
2SA1266Y	0148
2SA1267	0148
2SA1268	1522
2SA1269	0477
2SA1270	0006
2SA1271	5472
2SA1272	5472
2SA1273	3183
2SA1274	0436
2SA1275	1522
2SA1276	0854
2SA1277	1308
2SA1278	3543
2SA1279	1308
2SA1280	3543
2SA1281	5367
2SA1282	3183
2SA1282A	3183
2SA1283	0477
2SA1283(DE)-T	0477
2SA1284	0477
2SA1285	0477
2SA1285A	0477
2SA1286	OEM
2SA1287	OEM
2SA1288	1308
2SA1289	1308
2SA1290	1308
2SA1291	1308
2SA1292	3111
2SA1293	1199
2SA1294	OEM
2SA1295	OEM
2SA1296	4011
2SA1297	3183
2SA1298	1752
2SA1299	OEM
2SA1300	4011
2SA1301	3673
2SA1301-O	OEM
2SA1302	5349
2SA1302-0	OEM
2SA1302-O	OEM
2SA13020	5349
2SA1302R	5349
2SA1303	3673
2SA1304	1648
2SA1304(OEC.1)	1648
2SA1305	0925
2SA1305Y	0925
2SA1306	OEM
2SA1306(Y)	OEM
2SA1306-Y	OEM
2SA1306A	OEM
2SA1306A-O	OEM
2SA1306A-O(LB-SAN)	OEM
2SA1306A-Y	OEM
2SA1306AY	OEM
2SA1306A-Y(LB-SAN)	OEM
2SA1306Y	OEM
2SA1307	1308
2SA1307Y	OEM
2SA1308	5475
2SA1309	0013
2SA1309(Q)	0013
2SA1309(R)	0013
2SA1309(S)	0013
2SA1309A	0013
2SA1309A(Q)	0013
2SA1309A(Q)Y	0013
2SA1309A(QR)-T	6527
2SA1309A(R)	0013
2SA1309A(R)Y	0013
2SA1309AQ	0013
2SA1309A-QRS	0013
2SA1309AQSTA	OEM
2SA1309AR	0013
2SA1309AS	0013
2SA1309Q	5420
2SA1309QRS	OEM
2SA1309R	0013
2SA1309S	0013
2SA1310	OEM
2SA1311	OEM
2SA1312	5454
2SA1313	1752
2SA1314	OEM
2SA1315	OEM
2SA1316	0648
2SA1317	0013
2SA1317(S)	0013
2SA1317S	0013
2SA1317SZ	0013
2SA1317T	0013
2SA1317T-AC	OEM
2SA1318	0037
2SA1318(T)-AA-NP	0037
2SA1318S	0037
2SA1318T	0037
2SA1319	OEM
2SA1319R	OEM
2SA1319S	OEM
2SA1319T	OEM
2SA1320	0342
2SA1321	0342
2SA1321(TPE6)	0342
2SA1322	OEM
2SA1323	OEM
2SA1324	OEM
2SA1325	OEM
2SA1326	OEM
2SA1327	1308
2SA1328	OEM
2SA1329	OEM
2SA1330	OEM
2SA1330-06	OEM
2SA1331	1752
2SA1332	OEM
2SA1333	OEM
2SA1334	OEM
2SA1335	OEM
2SA1336	OEM
2SA1337	0013
2SA1338	1752
2SA1339	0013
2SA1339ST	0013
2SA1340	OEM
2SA1341	3272
2SA1342	0703
2SA1343	OEM
2SA1344	1892
2SA1344-TA	1892
2SA1345	4130
2SA1346	3143
2SA1346-AC	3143
2SA1346Z	3143
2SA1347	OEM
2SA1348	4172
2SA1349	OEM
2SA1350	0013
2SA1351	OEM
2SA1352	OEM
2SA1353	OEM
2SA1354	OEM
2SA1355	OEM
2SA1355-O	OEM
2SA1355G	OEM
2SA1355Y	OEM
2SA1356	0525
2SA1357	OEM
2SA1358	0525
2SA1359	OEM
2SA1360	0525
2SA1361	OEM
2SA1362	1752
2SA1363	OEM
2SA1364	3645
2SA1365	1752
2SA1366	1752
2SA1367	OEM
2SA1368	OEM
2SA1369	OEM
2SA1370	OEM
2SA1371	OEM
2SA1371(D)-AE	OEM
2SA1371(E)-AE	OEM
2SA1371-AE	OEM
2SA1371D	OEM
2SA1371E	OEM
2SA1372	OEM
2SA1373	OEM
2SA1374	0013
2SA1375	OEM
2SA1376	OEM
2SA1376A	OEM
2SA1376A-K	OEM
2SA1376A-L	OEM
2SA1376A-L/K	OEM
2SA1377	OEM
2SA1378	0013
2SA1379	OEM
2SA1380	OEM
2SA1381	4479
2SA1382	OEM
2SA1383	OEM
2SA1384	OEM
2SA1385	OEM
2SA1385L	OEM
2SA1386	5480
2SA1386A	OEM
2SA1387	OEM
2SA1388	OEM
2SA1389	3630
2SA1390	0013
2SA1390B	0013
2SA1390C	0013
2SA1390D-TZ	0013
2SA1391	OEM
2SA1392	OEM
2SA1393	OEM
2SA1394	OEM
2SA1395	0681
2SA1396	1369
2SA1397	OEM
2SA1398	OEM
2SA1399	OEM
2SA1400	OEM
2SA1401	OEM
2SA1402	OEM
2SA1403	OEM
2SA1404	OEM
2SA1405	OEM
2SA1406	OEM
2SA1407	OEM
2SA1408	0525
2SA1408(O)	0525
2SA1408(R)	0525
2SA1408-O	0525
2SA1408-O(LB-SAN-2)	0525
2SA1408-O(SAN-1)	0525
2SA1408-OY	OEM
2SA1408-R(LB-SAN-2)	0525
2SA1408-R(SAN-1)	0525
2SA1408R	0525
2SA1408RY	OEM
2SA1408Y(O)	0525
2SA1408Y(R)	0525
2SA1409	OEM
2SA1410	OEM
2SA1411	OEM
2SA1412	OEM
2SA1413	OEM
2SA1414	OEM
2SA1415	5371
2SA1416	OEM
2SA1417	OEM
2SA1418	5371
2SA1420	4130
2SA1421	1032
2SA1422	OEM
2SA1423	4172
2SA1424	OEM
2SA1425	OEM
2SA1426	4104
2SA1426Y	OEM
2SA1427	4104
2SA1428	OEM
2SA1429	OEM
2SA1430	4104
2SA1431	4011
2SA1432	OEM
2SA1433	OEM
2SA1441	5485
2SA1441L	5485
2SA1442	0718
2SA1443	OEM
2SA1444	OEM
2SA1450	0477
2SA1450R	0477
2SA1450S	0477
2SA1450T	0477
2SA1451	OEM
2SA1452	OEM
2SA1453	OEM
2SA1458	0013
2SA1459	OEM
2SA1460	OEM
2SA1461	OEM
2SA1462	OEM
2SA1463	OEM
2SA1464	OEM
2SA1465	0477
2SA1466	OEM
2SA1467	OEM
2SA1468	OEM
2SA1469	1308
2SA1469R	1308
2SA1470	OEM
2SA1470S	OEM
2SA1471	OEM
2SA1474	OEM
2SA1475	OEM
2SA1476	OEM
2SA1477	OEM
2SA1478	OEM
2SA1479	OEM
2SA1480	OEM
2SA1481	0013
2SA1482	1522
2SA1483	OEM
2SA1484	OEM
2SA1485	OEM
2SA1488	0681
2SA1489	0859
2SA1490	3111
2SA1491	3111
2SA1492	5494
2SA1493	3581
2SA1494	3646
2SA1495	OEM
2SA1498	OEM
2SA1499	OEM
2SA1500	OEM
2SA1501	OEM
2SA1503	OEM

DEVICE TYPE	REPL CODE
2SA1504	1742
2SA1505	OEM
2SA1506	OEM
2SA1507-R-RA	5901
2SA1507-S-CTV-YA	0525
2SA1507-S-RA	OEM
2SA1507-T-CTV-YA	0525
2SA1507S	5901
2SA1507S-CTV-YA	2110
2SA1507T-CTV-YA	2110
2SA1515	0037
2SA1516	OEM
2SA1517	OEM
2SA1524	OEM
2SA1532	OEM
2SA1532BC	OEM
2SA1546L	OEM
2SA1546M	OEM
2SA1547QSTV2	OEM
2SA1561	2484
2SA1561Q	OEM
2SA1561R	OEM
2SA1567	OEM
2SA1606	0718
2SA1606A	0718
2SA1606D	0718
2SA1606E	0718
2SA1624	0342
2SA1624(D)-AA	0342
2SA1624(E)-AA	0342
2SA1624D	0342
2SA1624E	0342
2SA1706	OEM
2SA1706-AN	OEM
2SA1706F	OEM
2SA1738QR	OEM
2SA1767Q	OEM
2SA1837	OEM
2SA2109	OEM
2SA2408	OEM
2SA3410A	0050
2SA3410B	0050
2SA-4551	0050
2SA-4561	0050
2SA4728B	0050
2SA4930	0791
2SA4940	0791
2SA4940LBGL2	0791
2SA4950	0006
2SA4950R	0006
2SA4964	0460
2SA5610	0037
2SA5620	0791
2SA5670	0791
2SA5670R	0791
2SA5680	0791
2SA5680R	0791
2SA5690	0791
2SA5690R	0791
2SA6111	0150
2SA6361	1267
2SA6361,K	1267
2SA6361K	1267
2SA24082	0050
2SA492057B2-60	0211
2SA4712057B2-65	0050
2SA4712057B2-66	0050
2SAD7160	OEM
2SAJ15GN	0050
2SA-NJ-101	0037
2SANJ101	0037
2SAU/3H	0050
2SAU03	0050
2SAU03H	0050
2SAUUJ	0847
2SB12	0150
2SB13	0004
2SB14	0050
2SB15	0164
2SB15D	0164
2SB15E	0164
2SB16	0440
2SB16A	0440
2SB16B	0440
2SB16C	0440
2SB16D	0440
2SB16E	0440
2SB16F	0440
2SB16G	0440
2SB16GN	0440
2SB16H	0440
2SB16J	0440
2SB16K	0440
2SB16L	0440
2SB16M	0440
2SB16OR	0440
2SB16R	0440
2SB16X	0440
2SB16Y	0440
2SB17	0440
2SB17A	0440
2SB17B	0440
2SB17C	0440
2SB17D	0440
2SB17E	0440
2SB17F	0440
2SB17G	0440
2SB17GN	0440
2SB17H	0440
2SB17J	0440
2SB17K	0440
2SB17L	0440
2SB17M	0440
2SB17OR	0440
2SB17R	0440
2SB17X	0440
2SB17Y	0440
2SB18	0440
2SB18A	0440
2SB18B	0440
2SB18C	0440
2SB18D	0440
2SB18E	0440
2SB18F	0440
2SB18G	0440
2SB18GN	0440
2SB18H	0440
2SB18J	0440
2SB18K	0440
2SB18L	0440
2SB18LA	0440
2SB18M	0440
2SB18OR	0440
2SB18R	0440
2SB18X	0440
2SB18Y	0440
2SB19	0975
2SB19A	0975
2SB19B	0975
2SB19C	0975
2SB19D	0975
2SB19E	0975
2SB19F	0975
2SB19G	0975
2SB19GN	0975
2SB19H	0975
2SB19J	0975
2SB19K	0975
2SB19L	0975
2SB19M	0975
2SB19OR	0975
2SB19R	0975
2SB19X	0975
2SB19Y	0975
2SB20	0975
2SB20A	0975
2SB20B	0975
2SB20C	0975
2SB20D	0975
2SB20E	0975
2SB20F	0975
2SB20G	0975
2SB20GN	0975
2SB20H	0975
2SB20J	0975
2SB20K	0975
2SB20L	0975
2SB20M	0975
2SB20OR	0975
2SB20R	0975
2SB20X	0975
2SB20Y	0975
2SB21	0975
2SB21A	0975
2SB21B	0975
2SB21C	0975
2SB21D	0975
2SB21E	0975
2SB21F	0975
2SB21G	0975
2SB21GN	0975
2SB21H	0975
2SB21J	0975
2SB21K	0975
2SB21L	0975
2SB21M	0975
2SB21OR	0975
2SB21R	0975
2SB21X	0975
2SB21Y	0975
2SB22	0164
2SB22/09-30100	0164
2SB22A	0164
2SB22B	0164
2SB22C	0164
2SB22D	0164
2SB22E	0164
2SB22F	0164
2SB22G	0164
2SB22GN	0164
2SB22H	0164
2SB22I	0164
2SB22J	0164
2SB22K	0164
2SB22L	0164
2SB22M	0164
2SB22OR	0164
2SB22P	0164
2SB22R	0164
2SB22X	0164
2SB22Y	0164
2SB23	0004
2SB23A	0004
2SB23B	0004
2SB23C	0004
2SB23D	0004
2SB23E	0004
2SB23F	0004
2SB23G	0004
2SB23GN	0004
2SB23H	0004
2SB23J	0004
2SB23K	0004
2SB23L	0004
2SB23M	0004
2SB23OR	0004
2SB23R	0004
2SB23X	0004
2SB23Y	0004
2SB24	0164
2SB24A	0164
2SB24B	0164
2SB24C	0164
2SB24D	0164
2SB24E	0164
2SB24F	0164
2SB24G	0164
2SB24GN	0164
2SB24H	0164
2SB24J	0164
2SB24K	0164
2SB24L	0164
2SB24M	0164
2SB24OR	0164
2SB24R	0164
2SB24X	0164
2SB24Y	0164
2SB25	0160
2SB25-N	0160
2SB25A	0160
2SB25B	0160
2SB25C	0160
2SB25D	0160
2SB25E	0160
2SB25F	0160
2SB25G	0160
2SB25GN	0160
2SB25H	0160
2SB25J	0160
2SB25K	0160
2SB25L	0160
2SB25M	0160
2SB25N	0160
2SB25OR	0160
2SB25R	0160
2SB25X	0160
2SB25Y	0160
2SB26	0085
2SB26A	0085
2SB26B	0085
2SB26C	0085
2SB26D	0085
2SB26E	0085
2SB26F	0085
2SB26G	0085
2SB26GN	0085
2SB26H	0085
2SB26J	0085
2SB26K	0085
2SB26L	0085
2SB26M	0085
2SB26OR	0085
2SB26R	0085
2SB26X	0085
2SB26Y	0085
2SB27	0085
2SB27A	0085
2SB27B	0085
2SB27C	0085
2SB27D	0085
2SB27E	0085
2SB27F	0085
2SB27G	0085
2SB27GN	0085
2SB27H	0085
2SB27J	0085
2SB27K	0085
2SB27L	0085
2SB27M	0085
2SB27OR	0085
2SB27R	0085
2SB27X	0085
2SB27Y	0085
2SB28	0085
2SB28A	0085
2SB28B	0085
2SB28C	0085
2SB28D	0085
2SB28E	0085
2SB28F	0085
2SB28G	0085
2SB28GN	0085
2SB28H	0085
2SB28J	0085
2SB28K	0085
2SB28L	0085
2SB28M	0085
2SB28OR	0085
2SB28R	0085
2SB28X	0085
2SB28Y	0085
2SB29	0085
2SB29A	0085
2SB29B	0085
2SB29C	0085
2SB29D	0085
2SB29E	0085
2SB29F	0085
2SB29G	0085
2SB29GN	0085
2SB29H	0085
2SB29J	0085
2SB29K	0085
2SB29L	0085
2SB29M	0085
2SB29OR	0085
2SB29R	0085
2SB29X	0085
2SB29Y	0085
2SB30	0085
2SB30A	0085
2SB30B	0085
2SB30C	0085
2SB30D	0085
2SB30E	0085
2SB30F	0085
2SB30G	0085
2SB30GN	0085
2SB30H	0085
2SB30J	0085
2SB30K	0085
2SB30L	0085
2SB30M	0085
2SB30OR	0085
2SB30R	0085
2SB30X	0085
2SB30Y	0085
2SB31	0085
2SB31A	0085
2SB31B	0085
2SB31C	0085
2SB31D	0085
2SB31E	0085
2SB31F	0085
2SB31G	0085
2SB31GN	0085
2SB31H	0085
2SB31J	0085
2SB31K	0085
2SB31L	0085
2SB31M	0085
2SB31OR	0085
2SB31R	0085
2SB31X	0085
2SB31Y	0085
2SB32	0004
2SB32(3)	0004
2SB32(N)	0004
2SB32-0	0004
2SB32-1	0004
2SB32-2	0004
2SB32-4	0004
2SB32A	0004
2SB32B	0004
2SB32C	0004
2SB32D	0004
2SB32E	0004
2SB32F	0004
2SB32G	0004
2SB32GN	0004
2SB32H	0004
2SB32J	0004
2SB32K	0004
2SB32M	0004
2SB32OR	0004
2SB32R	0004
2SB32X	0004
2SB32Y	0004
2SB33	0004
2SB33(3)	0004
2SB33(4)	0004
2SB33(5)	0004
2SB33-4	0004
2SB33-5	0004
2SB33A	0004
2SB33B	0004
2SB33BK	0004
2SB33C	0004
2SB33D	0004
2SB33E	0004
2SB33F	0004
2SB33G	0004
2SB33GN	0004
2SB33H	0004
2SB33J	0004
2SB33K	0004
2SB33L	0004
2SB33M	0004
2SB33OR	0004
2SB33R	0004
2SB33X	0004
2SB33Y	0004
2SB34	0211
2SB34A	0211
2SB34B	0211
2SB34C	0211
2SB34D	0211
2SB34E	0211
2SB34F	0211
2SB34G	0211
2SB34GN	0211
2SB34H	0211
2SB34J	0211
2SB34K	0211
2SB34L	0211
2SB34M	0211
2SB34N	0211
2SB34OR	0211
2SB34R	0211
2SB34X	0211
2SB34Y	0211
2SB35	0164
2SB35A	0164
2SB35B	0164
2SB35C	0164
2SB35D	0164
2SB35E	0164
2SB35F	0164
2SB35G	0164
2SB35GN	0164
2SB35H	0164
2SB35J	0164
2SB35K	0164
2SB35L	0164
2SB35M	0164
2SB35OR	0164
2SB35R	0164
2SB35X	0164
2SB35Y	0164
2SB37	0004
2SB37A	0004
2SB37B	0004
2SB37C	0004
2SB37D	0004
2SB37E	0004
2SB37F	0004
2SB37G	0004
2SB37GN	0004
2SB37H	0004
2SB37J	0004
2SB37K	0004
2SB37L	0004
2SB37M	0004
2SB37OR	0004
2SB37R	0004
2SB37X	0004
2SB37Y	0004
2SB38	0211
2SB38A	0211
2SB38B	0211
2SB38C	0211
2SB38D	0211
2SB38E	0211
2SB38F	0211
2SB38G	0211
2SB38GN	0211
2SB38H	0211
2SB38J	0211
2SB38K	0211
2SB38L	0211
2SB38M	0211
2SB38OR	0211
2SB38R	0211
2SB38X	0211
2SB38Y	0211
2SB39	0004
2SB39A	0004
2SB39B	0004
2SB39C	0004
2SB39D	0004
2SB39E	0004
2SB39F	0004
2SB39G	0004
2SB39GN	0004
2SB39H	0004
2SB39J	0004
2SB39K	0004
2SB39L	0004
2SB39M	0004
2SB39OR	0004
2SB39R	0004
2SB39X	0004
2SB39Y	0004
2SB40	0004
2SB40A	0004
2SB40B	0004
2SB40C	0004
2SB40D	0004
2SB40E	0004
2SB40F	0004
2SB40G	0004
2SB40GN	0004
2SB40H	0004
2SB40J	0004
2SB40K	0004
2SB40L	0004
2SB40M	0004
2SB40OR	0004
2SB40R	0004
2SB40X	0004
2SB40Y	0004
2SB41	0085
2SB41A	0085
2SB41B	0085
2SB41C	0085
2SB41D	0085
2SB41E	0085
2SB41F	0085
2SB41G	0085
2SB41GN	0085
2SB41H	0085
2SB41J	0085
2SB41K	0085
2SB41L	0085
2SB41M	0085
2SB41OR	0085
2SB41R	0085
2SB41X	0085
2SB41Y	0085
2SB42	0085
2SB42A	0085
2SB42B	0085
2SB42C	0085
2SB42D	0085
2SB42E	0085
2SB42F	0085
2SB42G	0085
2SB42GN	0085
2SB42H	0085
2SB42J	0085
2SB42K	0085
2SB42L	0085
2SB42M	0085
2SB42OR	0085
2SB42R	0085
2SB42X	0085
2SB42Y	0085
2SB43	0004
2SB43A	0004
2SB43B	0004
2SB43C	0004
2SB43D	0004
2SB43E	0004
2SB43F	0004
2SB43G	0004
2SB43GN	0004
2SB43H	0004
2SB43J	0004
2SB43K	0004
2SB43L	0004
2SB43M	0004
2SB43OR	0004
2SB43R	0004
2SB43X	0004
2SB43Y	0004
2SB44	0004
2SB44A	0004
2SB44B	0004
2SB44C	0004
2SB44D	0004
2SB44E	0004
2SB44F	0004
2SB44G	0004
2SB44GN	0004
2SB44H	0004
2SB44J	0004
2SB44K	0004
2SB44L	0004
2SB44M	0004
2SB44OR	0004
2SB44R	0004
2SB44X	0004
2SB44Y	0004
2SB46	0004
2SB46A	0004
2SB46B	0004
2SB46C	0004
2SB46D	0004
2SB46E	0004
2SB46F	0004
2SB46G	0004
2SB46GN	0004
2SB46H	0004
2SB46J	0004
2SB46K	0004
2SB46L	0004
2SB46M	0004
2SB46OR	0004
2SB46R	0004
2SB46X	0004
2SB46Y	0004
2SB47	0004
2SB47A	0004
2SB47B	0004
2SB47C	0004
2SB47D	0004
2SB47E	0004
2SB47F	0004
2SB47G	0004
2SB47GN	0004
2SB47H	0004
2SB47J	0004
2SB47K	0004
2SB47L	0004
2SB47M	0004
2SB47OR	0004
2SB47R	0004
2SB47X	0004
2SB47Y	0004
2SB48	0211
2SB48A	0211
2SB48B	0211
2SB48C	0211
2SB48D	0211
2SB48E	0211
2SB48F	0211
2SB48G	0211
2SB48GN	0211
2SB48H	0211
2SB48J	0211
2SB48K	0211
2SB48L	0211
2SB48M	0211
2SB48OR	0211
2SB48R	0211
2SB48X	0211
2SB48Y	0211
2SB49	0211
2SB49A	0211
2SB49B	0211
2SB49C	0211
2SB49D	0211
2SB49E	0211
2SB49F	0211
2SB49G	0211
2SB49GN	0211
2SB49H	0211
2SB49J	0211
2SB49K	0211
2SB49L	0211
2SB49M	0211
2SB49OR	0211
2SB49R	0211
2SB49X	0211
2SB49Y	0211
2SB50	0211
2SB50(RECT)	0323
2SB50A	0211
2SB50B	0211
2SB50C	0211
2SB50D	0211
2SB50E	0211
2SB50F	0211
2SB50G	0211
2SB50GN	0211
2SB50H	0211
2SB50J	0211
2SB50K	0211
2SB50L	0211
2SB50M	0211
2SB50OR	0211
2SB50R	0211
2SB50X	0211
2SB50Y	0211
2SB51	0211
2SB51A	0211
2SB51B	0211
2SB51C	0211
2SB51D	0211
2SB51E	0211
2SB51F	0211
2SB51G	0211
2SB51GN	0211
2SB51H	0211
2SB51J	0211
2SB51K	0211
2SB51L	0211
2SB51M	0211
2SB51OR	0211
2SB51R	0211
2SB51X	0211
2SB51Y	0211
2SB52	0211
2SB52A	0211
2SB52B	0211
2SB52C	0211
2SB52D	0211
2SB52E	0211
2SB52F	0211
2SB52G	0211
2SB52GN	0211
2SB52H	0211
2SB52J	0211
2SB52K	0211
2SB52L	0211
2SB52M	0211
2SB52OR	0211
2SB52R	0211
2SB52X	0211
2SB52Y	0211
2SB53	0211
2SB53A	0211
2SB53B	0211
2SB53C	0211
2SB53D	0211
2SB53E	0211
2SB53F	0211
2SB53G	0211
2SB53GN	0211
2SB53H	0211
2SB53J	0211
2SB53K	0211
2SB53L	0211
2SB53M	0211
2SB53OR	0211
2SB53R	0211
2SB53X	0211
2SB53Y	0211
2SB54	0004
2SB54A	0004
2SB54B	0004
2SB54BA	0004
2SB54C	0004
2SB54D	0004
2SB54E	0004
2SB54F	0004
2SB54G	0004
2SB54GN	0004
2SB54H	0004
2SB54J	0004
2SB54K	0004
2SB54L	0004
2SB54L1	0004

If replacement code is OEM, contact original manufacturer for replacement.

DEVICE TYPE	REPL CODE
2SB54M	0004
2SB54OR	0004
2SB54R	0004
2SB54X	0004
2SB54Y	0004
2SB55	0004
2SB55A	0004
2SB55B	0004
2SB55C	0004
2SB55D	0004
2SB55E	0004
2SB55F	0004
2SB55G	0004
2SB55GN	0004
2SB55H	0004
2SB55J	0004
2SB55K	0004
2SB55L	0004
2SB55M	0004
2SB55OR	0004
2SB55R	0004
2SB55X	0004
2SB55Y	0004
2SB56	0004
2SB56(C)	0004
2SB56(F1)	OEM
2SB56/8020	0004
2SB56/117728	0004
2SB56/80205600	0004
2SB56A	0004
2SB56B	0004
2SB56C	0004
2SB56CK	0004
2SB56D	0004
2SB56E	0004
2SB56F	0004
2SB56G	0004
2SB56GN	0004
2SB56H	0004
2SB56J	0004
2SB56K	0004
2SB56L	0004
2SB56OR	0004
2SB56R	0004
2SB56X	0004
2SB56Y	0004
2SB57	0004
2SB57A	0004
2SB57B	0004
2SB57C	0004
2SB57D	0004
2SB57E	0004
2SB57F	0004
2SB57G	0004
2SB57GN	0004
2SB57H	0004
2SB57J	0004
2SB57K	0004
2SB57L	0004
2SB57M	0004
2SB57OR	0004
2SB57R	0004
2SB57X	0004
2SB57Y	0004
2SB58	0164
2SB59	0004
2SB59A	0004
2SB59B	0004
2SB59C	0004
2SB59D	0004
2SB59E	0004
2SB59F	0004
2SB59G	0004
2SB59GN	0004
2SB59J	0004
2SB59K	0004
2SB59L	0004
2SB59M	0004
2SB59OR	0004
2SB59R	0004
2SB59X	0004
2SB59Y	0004
2SB60	0004
2SB60A	0004
2SB60B	0004
2SB60C	0004
2SB60D	0004
2SB60E	0004
2SB60F	0004
2SB60G	0004
2SB60GN	0004
2SB60H	0004
2SB60J	0004
2SB60K	0004
2SB60L	0004
2SB60M	0004
2SB60OR	0004
2SB60R	0004
2SB60X	0004
2SB60Y	0004
2SB61	0004
2SB61A	0004
2SB61B	0004
2SB61C	0004
2SB61D	0004
2SB61E	0004
2SB61F	0004
2SB61G	0004
2SB61GN	0004
2SB61H	0004
2SB61J	0004
2SB61K	0004
2SB61L	0004
2SB61M	0004
2SB61OR	0004
2SB61R	0004
2SB61X	0004
2SB61Y	0004
2SB62	1010
2SB62A	1010
2SB62B	1010
2SB62C	1010
2SB62D	1010
2SB62E	1010
2SB62F	1010
2SB62G	1010
2SB62GN	1010
2SB62H	1010
2SB62J	1010
2SB62K	1010
2SB62L	1010
2SB62M	1010
2SB62OR	1010
2SB62R	1010
2SB62X	1010
2SB62Y	1010
2SB63	0222
2SB63A	0222
2SB63B	0222
2SB63C	0222
2SB63D	0222
2SB63E	0222
2SB63F	0222
2SB63G	0222
2SB63GN	0222
2SB63H	0222
2SB63K	0222
2SB63L	0222
2SB63M	0222
2SB63OR	0222
2SB63R	0222
2SB63X	0222
2SB63Y	0222
2SB64	0975
2SB64A	0975
2SB64B	0975
2SB64C	0975
2SB64D	0975
2SB64E	0975
2SB64F	0975
2SB64G	0975
2SB64H	0975
2SB64J	0975
2SB64K	0975
2SB64L	0975
2SB64M	0975
2SB64OR	0975
2SB64R	0975
2SB64X	0975
2SB64Y	0975
2SB65	0004
2SB65A	0004
2SB65B	0004
2SB65C	0004
2SB65D	0004
2SB65F	0004
2SB65G	0004
2SB65GN	0004
2SB65H	0004
2SB65J	0004
2SB65K	0004
2SB65L	0004
2SB65M	0004
2SB65OR	0004
2SB65R	0004
2SB65X	0004
2SB65Y	0004
2SB66	0004
2SB66A	0004
2SB66B	0004
2SB66C	0004
2SB66D	0004
2SB66E	0004
2SB66F	0004
2SB66G	0004
2SB66GN	0004
2SB66H	0004
2SB66J	0004
2SB66K	0004
2SB66L	0004
2SB66M	0004
2SB66OR	0004
2SB66R	0004
2SB66X	0004
2SB66Y	0004
2SB67	0004
2SB67A	0004
2SB67AH	0004
2SB67B	0004
2SB67C	0004
2SB67D	0004
2SB67E	0004
2SB67F	0004
2SB67G	0004
2SB67GN	0004
2SB67H	0004
2SB67J	0004
2SB67K	0004
2SB67L	0004
2SB67M	0004
2SB67OR	0004
2SB67R	0004
2SB67X	0004
2SB67Y	0004
2SB68	0004
2SB68A	0004
2SB68B	0004
2SB68C	0004
2SB68D	0004
2SB68E	0004
2SB68F	0004
2SB68G	0004
2SB68GN	0004
2SB68H	0004
2SB68J	0004
2SB68K	0004
2SB68L	0004
2SB68M	0004
2SB68OR	0004
2SB68R	0004
2SB68X	0004
2SB68Y	0004
2SB69	0160
2SB69A	0160
2SB69B	0160
2SB69C	0160
2SB69D	0160
2SB69E	0160
2SB69F	0160
2SB69G	0160
2SB69GN	0160
2SB69H	0160
2SB69J	0160
2SB69K	0160
2SB69L	0160
2SB69M	0160
2SB69OR	0160
2SB69R	0160
2SB69X	0160
2SB69Y	0160
2SB70	0164
2SB71	0004
2SB71A	0004
2SB71B	0004
2SB71C	0004
2SB71D	0004
2SB71E	0004
2SB71F	0004
2SB71G	0004
2SB71GN	0004
2SB71H	0004
2SB71J	0004
2SB71K	0004
2SB71L	0004
2SB71M	0004
2SB71OR	0004
2SB71R	0004
2SB71X	0004
2SB71Y	0004
2SB72	0211
2SB72A	0211
2SB72B	0211
2SB72C	0211
2SB72D	0211
2SB72E	0211
2SB72F	0211
2SB72G	0211
2SB72GN	0211
2SB72H	0211
2SB72J	0211
2SB72K	0211
2SB72L	0211
2SB72M	0211
2SB72OR	0211
2SB72R	0211
2SB72X	0211
2SB72Y	0211
2SB73	0004
2SB73/123	0004
2SB73/123806	0004
2SB73A	0004
2SB73A-1	0004
2SB73B	0004
2SB73B/0573018	0004
2SB73C	0004
2SB73D	0004
2SB73E	0004
2SB73F	0004
2SB73G	0004
2SB73GN	0004
2SB73GR	0004
2SB73H	0004
2SB73J	0004
2SB73K	0004
2SB73L	0004
2SB73M	0004
2SB73OR	0004
2SB73R	0004
2SB73S	0004
2SB73X	0004
2SB73Y	0004
2SB74	0164
2SB74A	0164
2SB74B	0164
2SB74C	0164
2SB74D	0164
2SB74E	0164
2SB74F	0164
2SB74G	0164
2SB74GN	0164
2SB74H	0164
2SB74J	0164
2SB74K	0164
2SB74L	0164
2SB74M	0164
2SB74OR	0164
2SB74R	0164
2SB74X	0164
2SB74Y	0164
2SB75	0004
2SB75(C)	0004
2SB75A	0004
2SB75AH	0004
2SB75B	0004
2SB75C	0004
2SB75C/0573153	0004
2SB75C1	0004
2SB75C-4	0004
2SB75D	0004
2SB75E	0004
2SB75F	0004
2SB75F/0573125	0004
2SB75G	0004
2SB75GN	0004
2SB75H	0004
2SB75J	0004
2SB75L	0004
2SB75LB	0004
2SB75M	0004
2SB75OR	0004
2SB75R	0004
2SB75X	0004
2SB75Y	0004
2SB76	0004
2SB76A	0004
2SB76B	0004
2SB76C	0004
2SB76D	0004
2SB76E	0004
2SB76F	0004
2SB76G	0004
2SB76GN	0004
2SB76H	0004
2SB76J	0004
2SB76K	0004
2SB76L	0004
2SB76M	0004
2SB76OR	0004
2SB76R	0004
2SB76X	0004
2SB76Y	0004
2SB77	0004
2SB77(B)	0004
2SB77(C)	0004
2SB77(C)SHARP	0004
2SB77(D)	0004
2SB77(V)	0004
2SB77/1222	0004
2SB77/122244	0004
2SB77-A	0004
2SB77-C	0004
2SB77-OR	0004
2SB77A	0004
2SB77A/P	0004
2SB77AA	0004
2SB77AB	0004
2SB77AC	0004
2SB77AD	0004
2SB77AH	0004
2SB77AP	0004
2SB77B	0004
2SB77B/0573114	0004
2SB77B2	0004
2SB77B-11	0004
2SB77B57B2-24	0004
2SB77C	0004
2SB77D	0004
2SB77E	0004
2SB77F	0004
2SB77G	0004
2SB77GN	0004
2SB77H	0004
2SB77K	0004
2SB77L	0004
2SB77M	0004
2SB77OR	0004
2SB77P	0004
2SB77PD	0004
2SB77R	0004
2SB77RED	0004
2SB77V	0004
2SB77VRED	0004
2SB77X	0004
2SB77Y	0004
2SB78	0004
2SB78A	0004
2SB78B	0004
2SB78C	0004
2SB78D	0004
2SB78E	0004
2SB78F	0004
2SB78G	0004
2SB78GN	0004
2SB78H	0004
2SB78J	0004
2SB78K	0004
2SB78L	0004
2SB78M	0004
2SB78OR	0004
2SB78R	0004
2SB78X	0004
2SB78Y	0004
2SB79	0004
2SB79A	0004
2SB79B	0004
2SB79C	0004
2SB79D	0004
2SB79E	0004
2SB79F	0004
2SB79G	0004
2SB79GN	0004
2SB79H	0004
2SB79J	0004
2SB79K	0004
2SB79L	0004
2SB79M	0004
2SB79OR	0004
2SB79R	0004
2SB79X	0004
2SB79Y	0004
2SB80	0222
2SB80A	0222
2SB80B	0222
2SB80C	0222
2SB80D	0222
2SB80E	0222
2SB80F	0222
2SB80G	0222
2SB80GN	0222
2SB80H	0222
2SB80J	0222
2SB80K	0222
2SB80L	0222
2SB80M	0222
2SB80OR	0222
2SB80R	0222
2SB80X	0222
2SB80Y	0222
2SB81	0975
2SB81A	0975
2SB81B	0975
2SB81C	0975
2SB81D	0975
2SB81E	0975
2SB81F	0975
2SB81GN	0975
2SB81H	0975
2SB81J	0975
2SB81K	0975
2SB81L	0975
2SB81M	0975
2SB81OR	0975
2SB81R	0975
2SB81X	0975
2SB81Y	0975
2SB82	0975
2SB82A	0975
2SB82B	0975
2SB82C	0975
2SB82D	0975
2SB82E	0975
2SB82F	0975
2SB82G	0975
2SB82GN	0975
2SB82H	0975
2SB82J	0975
2SB82K	0975
2SB82L	0975
2SB82M	0975
2SB82OR	0975
2SB82R	0975
2SB82X	0975
2SB82Y	0975
2SB83	0975
2SB83A	0975
2SB83B	0975
2SB83C	0975
2SB83D	0975
2SB83E	0975
2SB83F	0975
2SB83G	0975
2SB83GN	0975
2SB83H	0975
2SB83J	0975
2SB83K	0975
2SB83L	0975
2SB83M	0975
2SB83OR	0975
2SB83R	0975
2SB83X	0975
2SB83Y	0975
2SB84	0085
2SB84A	0085
2SB84B	0085
2SB84C	0085
2SB84D	0085
2SB84E	0085
2SB84F	0085
2SB84G	0085
2SB84GN	0085
2SB84H	0085
2SB84J	0085
2SB84K	0085
2SB84L	0085
2SB84M	0085
2SB84OR	0085
2SB84R	0085
2SB84X	0085
2SB84Y	0085
2SB85	0164
2SB85A	0164
2SB85B	0164
2SB85C	0164
2SB85D	0164
2SB85E	0164
2SB85F	0164
2SB85G	0164
2SB85GN	0164
2SB85H	0164
2SB85J	0164
2SB85K	0164
2SB85L	0164
2SB85M	0164
2SB85OR	0164
2SB85R	0164
2SB85X	0164
2SB85Y	0164
2SB86	0160
2SB87	0164
2SB87A	0164
2SB87B	0164
2SB87C	0164
2SB87D	0164
2SB87E	0164
2SB87F	0164
2SB87G	0164
2SB87GN	0164
2SB87H	0164
2SB87J	0164
2SB87K	0164
2SB87L	0164
2SB87M	0164
2SB87OR	0164
2SB87R	0164
2SB87X	0164
2SB87Y	0164
2SB88	0975
2SB89	0004
2SB89A	0004
2SB89AH	0004
2SB89C	0004
2SB89D	0004
2SB89E	0004
2SB89F	0004
2SB89G	0004
2SB89GN	0004
2SB89H	0004
2SB89J	0004
2SB89K	0004
2SB89L	0004
2SB89M	0004
2SB89OR	0004
2SB89R	0004
2SB89X	0004
2SB89Y	0004
2SB90	0004
2SB90A	0004
2SB90B	0004
2SB90C	0004
2SB90D	0004
2SB90E	0004
2SB90F	0004
2SB90G	0004
2SB90GN	0004
2SB90H	0004
2SB90J	0004
2SB90K	0004
2SB90L	0004
2SB90M	0004
2SB90OR	0004
2SB90R	0004
2SB90X	0004
2SB90Y	0004
2SB91	0004
2SB91A	0004
2SB91B	0004
2SB91C	0004
2SB91D	0004
2SB91E	0004
2SB91F	0004
2SB91G	0004
2SB91GN	0004
2SB91H	0004
2SB91J	0004
2SB91K	0004
2SB91L	0004
2SB91M	0004
2SB91OR	0004
2SB91R	0004
2SB91X	0004
2SB91Y	0004
2SB92	0004
2SB92A	0004
2SB92B	0004
2SB92C	0004
2SB92D	0004
2SB92E	0004
2SB92F	0004
2SB92G	0004
2SB92GN	0004
2SB92H	0004
2SB92J	0004
2SB92K	0004
2SB92L	0004
2SB92M	0004
2SB92X	0004
2SB92Y	0004
2SB93	0004
2SB94	0004
2SB94A	0004
2SB94B	0004
2SB94C	0004
2SB94D	0004
2SB94E	0004
2SB94F	0004
2SB94G	0004
2SB94GN	0004
2SB94H	0004
2SB94J	0004
2SB94K	0004
2SB94M	0004
2SB94OR	0004
2SB94R	0004
2SB94X	0004
2SB94Y	0004
2SB95	0004
2SB95A	0004
2SB95B	0004
2SB95C	0004
2SB95D	0004
2SB95E	0004
2SB95F	0004
2SB95G	0004
2SB95GN	0004
2SB95H	0004
2SB95J	0004
2SB95L	0004
2SB95M	0004
2SB95OR	0004
2SB95R	0004
2SB95X	0004
2SB95Y	0004
2SB96	0211
2SB97	0004
2SB97A	0004
2SB97B	0004
2SB97C	0004
2SB97D	0004
2SB97F	0004
2SB97G	0004
2SB97GN	0004
2SB97H	0004
2SB97K	0004
2SB97L	0004
2SB97M	0004
2SB97OR	0004
2SB97R	0004
2SB97X	0004
2SB97Y	0004
2SB98	0211
2SB98A	0211
2SB98B	0211
2SB98C	0211
2SB98D	0211
2SB98E	0211
2SB98F	0211
2SB98G	0211
2SB98GN	0211
2SB98H	0211
2SB98J	0211
2SB98K	0211
2SB98L	0211
2SB98M	0211
2SB98OR	0211
2SB98R	0211
2SB98X	0211
2SB98Y	0211
2SB99	0164
2SB99A	0164
2SB99B	0164
2SB99C	0164
2SB99D	0164
2SB99E	0164
2SB99F	0164
2SB99G	0164
2SB99GN	0164
2SB99H	0164
2SB99J	0164
2SB99K	0164
2SB99L	0164
2SB99M	0164
2SB99OR	0164
2SB99R	0164
2SB99X	0164
2SB99Y	0164
2SB100	0211
2SB100(RECT)	0323
2SB100A	0211
2SB100B	0211
2SB100C	0211
2SB100D	0211
2SB100E	0211
2SB100F	0211
2SB100G	0211
2SB100GN	0211
2SB100H	0211
2SB100J	0211
2SB100K	0211
2SB100L	0211
2SB100M	0211
2SB100OR	0211
2SB100R	0211
2SB100X	0211

DEVICE TYPE	REPL CODE	DEVICE TYPE	REPL CODE	DEVICE TYPE	REPL CODE	DEVICE TYPE	REPL CODE	DEVICE TYPE	REPL CODE	DEVICE TYPE	REPL CODE	DEVICE TYPE	REPL CODE
2SB100Y	0211	2SB108	0164	2SB115	0004	2SB123	0085	2SB129D	0975	2SB136D	0004	2SB144GN	0085
2SB101	0211	2SB108A	0164	2SB115A	0004	2SB123A	0085	2SB129E	0975	2SB136E	0004	2SB144H	0085
2SB101A	0211	2SB108B	0164	2SB115B	0004	2SB123B	0085	2SB129F	0975	2SB136F	0004	2SB144J	0085
2SB101B	0211	2SB108C	0164	2SB115C	0004	2SB123C	0085	2SB129G	0975	2SB136G	0004	2SB144K	0085
2SB101C	0211	2SB108D	0164	2SB115D	0004	2SB123D	0085	2SB129GN	0975	2SB136GN	0004	2SB144L	0085
2SB101D	0211	2SB108E	0164	2SB115E	0004	2SB123E	0085	2SB129H	0975	2SB136H	0004	2SB144M	0085
2SB101E	0211	2SB108F	0164	2SB115F	0004	2SB123F	0085	2SB129J	0975	2SB136J	0004	2SB144OR	0085
2SB101F	0211	2SB108G	0164	2SB115G	0004	2SB123G	0085	2SB129K	0975	2SB136K	0004	2SB144P	0085
2SB101G	0211	2SB108GN	0164	2SB115GN	0004	2SB123GN	0085	2SB129L	0975	2SB136L	0004	2SB144R	0085
2SB101GN	0211	2SB108H	0164	2SB115H	0004	2SB123H	0085	2SB129M	0975	2SB136M	0004	2SB144X	0085
2SB101H	0211	2SB108J	0164	2SB115J	0004	2SB123J	0085	2SB129OR	0975	2SB136OR	0004	2SB144Y	0085
2SB101J	0211	2SB108K	0164	2SB115K	0004	2SB123K	0085	2SB129R	0975	2SB136R	0004	2SB145	0085
2SB101K	0211	2SB108L	0164	2SB115L	0004	2SB123L	0085	2SB129X	0975	2SB136U	0004	2SB145A	0085
2SB101L	0211	2SB108M	0164	2SB115M	0004	2SB123M	0085	2SB129Y	0975	2SB136X	0004	2SB145B	0085
2SB101M	0211	2SB108OR	0164	2SB115OR	0004	2SB123OR	0085	2SB130	0222	2SB137	0160	2SB145C	0085
2SB101OR	0211	2SB108X	0164	2SB115R	0004	2SB123R	0085	2SB130A	0222	2SB137A	0160	2SB145D	0085
2SB101R	0211	2SB108Y	0164	2SB115X	0004	2SB123X	0085	2SB130B	0222	2SB137B	0160	2SB145E	0085
2SB101X	0211	2SB109	0164	2SB115Y	0004	2SB123Y	0085	2SB130C	0222	2SB137C	0160	2SB145F	0085
2SB101Y	0211	2SB109A	0164	2SB116	0004	2SB124	0160	2SB130D	0222	2SB137D	0160	2SB145G	0085
2SB102	0211	2SB109B	0164	2SB116A	0004	2SB124A	0160	2SB130E	0222	2SB137E	0160	2SB145GN	0085
2SB102A	0211	2SB110	0004	2SB116B	0004	2SB124B	0160	2SB130F	0222	2SB137F	0160	2SB145H	0085
2SB102B	0211	2SB110A	0004	2SB116C	0004	2SB124C	0160	2SB130G	0222	2SB137G	0160	2SB145J	0085
2SB102C	0211	2SB110B	0004	2SB116D	0004	2SB124D	0160	2SB130GN	0222	2SB137GN	0160	2SB145K	0085
2SB102D	0211	2SB110C	0004	2SB116E	0004	2SB124E	0160	2SB130H	0222	2SB137H	0160	2SB145L	0085
2SB102E	0211	2SB110D	0004	2SB116F	0004	2SB124F	0160	2SB130J	0222	2SB137K	0160	2SB145M	0085
2SB102F	0211	2SB110E	0004	2SB116G	0004	2SB124G	0160	2SB130K	0222	2SB137L	0160	2SB145OR	0085
2SB102G	0211	2SB110F	0004	2SB116GN	0004	2SB124H	0160	2SB130L	0222	2SB137OR	0160	2SB145R	0085
2SB102GN	0211	2SB110G	0004	2SB116H	0004	2SB124J	0160	2SB130M	0222	2SB137R	0160	2SB145X	0085
2SB102H	0211	2SB110GN	0004	2SB116J	0004	2SB124K	0160	2SB130OR	0222	2SB137X	0160	2SB145Y	0085
2SB102J	0211	2SB110H	0004	2SB116K	0004	2SB124L	0160	2SB130R	0222	2SB137Y	0160	2SB146	0085
2SB102K	0211	2SB110J	0004	2SB116L	0004	2SB124M	0160	2SB130X	0222	2SB138	0160	2SB146A	0085
2SB102L	0211	2SB110K	0004	2SB116M	0004	2SB124OR	0160	2SB130Y	0222	2SB138A	0160	2SB146B	0085
2SB102M	0211	2SB110L	0004	2SB116OR	0004	2SB124R	0160	2SB131	0160	2SB138B	0160	2SB146C	0085
2SB102OR	0211	2SB110M	0004	2SB116R	0004	2SB124X	0160	2SB131A	0160	2SB138C	0160	2SB146D	0085
2SB102R	0211	2SB110OR	0004	2SB116X	0004	2SB124Y	0160	2SB131B	0160	2SB138D	0160	2SB146E	0085
2SB102X	0211	2SB110R	0004	2SB116Y	0004	2SB125	0604	2SB131C	0160	2SB138E	0160	2SB146F	0085
2SB102Y	0211	2SB110X	0004	2SB117	0004	2SB126	0160	2SB131D	0160	2SB138G	0160	2SB146G	0085
2SB103	0211	2SB110Y	0004	2SB117A	0004	2SB126(F)	0160	2SB131E	0160	2SB138GN	0160	2SB146GN	0085
2SB103A	0211	2SB111	0004	2SB117B	0004	2SB126(V)	0160	2SB131F	0160	2SB138H	0160	2SB146H	0085
2SB103B	0211	2SB111A	0004	2SB117C	0004	2SB126A	0160	2SB131G	0160	2SB138J	0160	2SB146J	0085
2SB103C	0211	2SB111B	0004	2SB117D	0004	2SB126AA	0160	2SB131GN	0160	2SB138K	0160	2SB146K	0085
2SB103D	0211	2SB111C	0004	2SB117E	0004	2SB126AB	0160	2SB131H	0160	2SB138L	0160	2SB146L	0085
2SB103E	0211	2SB111D	0004	2SB117F	0004	2SB126AC	0160	2SB131J	0160	2SB138M	0160	2SB146M	0085
2SB103F	0211	2SB111E	0004	2SB117G	0004	2SB126AD	0160	2SB131K	0160	2SB138OR	0160	2SB146OR	0085
2SB103G	0211	2SB111F	0004	2SB117GN	0004	2SB126AE	0160	2SB131L	0160	2SB138R	0160	2SB146R	0085
2SB103GN	0211	2SB111G	0004	2SB117H	0004	2SB126AF	0160	2SB131M	0160	2SB138X	0160	2SB146X	0085
2SB103H	0211	2SB111GN	0004	2SB117J	0004	2SB126AG	0160	2SB131OR	0160	2SB138Y	0160	2SB146Y	0085
2SB103J	0211	2SB111H	0004	2SB117K	0004	2SB126AH	0160	2SB131R	0160	2SB140	0085	2SB147	0085
2SB103K	0211	2SB111J	0004	2SB117L	0004	2SB126B	0160	2SB131X	0160	2SB140A	0085	2SB147A	0085
2SB103L	0211	2SB111K	0004	2SB117M	0004	2SB126C	0160	2SB131Y	0160	2SB140B	0085	2SB147B	0085
2SB103M	0211	2SB111L	0004	2SB117OR	0004	2SB126D	0160	2SB132	0160	2SB140C	0085	2SB147C	0085
2SB103OR	0211	2SB111M	0004	2SB117R	0004	2SB126E	0160	2SB132A	0160	2SB140D	0085	2SB147D	0085
2SB103R	0211	2SB111OR	0004	2SB117X	0004	2SB126F	0160	2SB132B	0160	2SB140E	0085	2SB147E	0085
2SB103X	0211	2SB111R	0004	2SB117Y	0004	2SB126FV	0160	2SB132C	0160	2SB140F	0085	2SB147F	0085
2SB103Y	0211	2SB111X	0004	2SB118	0164	2SB126G	0160	2SB132D	0160	2SB140G	0085	2SB147G	0085
2SB104	0164	2SB111Y	0004	2SB119	0160	2SB126GN	0160	2SB132E	0160	2SB140GN	0085	2SB147H	0085
2SB104A	0164	2SB112	0004	2SB119A	0160	2SB126H	0160	2SB132F	0160	2SB140H	0085	2SB147J	0085
2SB104B	0164	2SB112A	0004	2SB119B	0160	2SB126J	0160	2SB132G	0160	2SB140J	0085	2SB147K	0085
2SB104C	0164	2SB112B	0004	2SB119C	0160	2SB126K	0160	2SB132GN	0160	2SB140K	0085	2SB147L	0085
2SB104D	0164	2SB112C	0004	2SB119D	0160	2SB126L	0160	2SB132H	0160	2SB140L	0085	2SB147M	0085
2SB104E	0164	2SB112D	0004	2SB119E	0160	2SB126M	0160	2SB132J	0160	2SB140M	0085	2SB147OR	0085
2SB104F	0164	2SB112E	0004	2SB119F	0160	2SB126OR	0160	2SB132L	0160	2SB140OR	0085	2SB147X	0085
2SB104G	0164	2SB112F	0004	2SB119G	0160	2SB126P	0160	2SB132M	0160	2SB140X	0085	2SB147Y	0085
2SB104GN	0164	2SB112G	0004	2SB119GN	0160	2SB126R	0160	2SB132OR	0160	2SB140Y	0085	2SB148	0604
2SB104H	0164	2SB112H	0004	2SB119H	0160	2SB126V	0160	2SB132R	0160	2SB141	0085	2SB149	0085
2SB104J	0164	2SB112J	0004	2SB119J	0160	2SB126X	0160	2SB132X	0160	2SB142	0085	2SB149-N	0085
2SB104K	0164	2SB112K	0004	2SB119K	0160	2SB126Y	0160	2SB132Y	0160	2SB142A	0085	2SB149A	0085
2SB104L	0164	2SB112L	0004	2SB119L	0160	2SB127	0160	2SB134	0004	2SB142B	0085	2SB149B	0085
2SB104M	0164	2SB112M	0004	2SB119M	0160	2SB127A	0160	2SB134A	0004	2SB142C	0085	2SB149C	0085
2SB104OR	0164	2SB112OR	0004	2SB119OR	0160	2SB127B	0160	2SB134B	0004	2SB142D	0085	2SB149D	0085
2SB104R	0164	2SB112R	0004	2SB119R	0160	2SB127C	0160	2SB134C	0004	2SB142E	0085	2SB149E	0085
2SB104X	0164	2SB112X	0004	2SB119X	0160	2SB127D	0160	2SB134D	0004	2SB142F	0085	2SB149F	0085
2SB104Y	0164	2SB112Y	0004	2SB119Y	0160	2SB127E	0160	2SB134E	0004	2SB142G	0085	2SB149G	0085
2SB105	0164	2SB113	0004	2SB120	0004	2SB127F	0160	2SB134F	0004	2SB142GN	0085	2SB149GN	0085
2SB105A	0164	2SB113A	0004	2SB120A	0004	2SB127G	0160	2SB134G	0004	2SB142H	0085	2SB149H	0085
2SB105B	0164	2SB113B	0004	2SB120B	0004	2SB127GN	0160	2SB134GN	0004	2SB142J	0085	2SB149J	0085
2SB105C	0164	2SB113C	0004	2SB120C	0004	2SB127H	0160	2SB134H	0004	2SB142K	0085	2SB149K	0085
2SB105D	0164	2SB113D	0004	2SB120D	0004	2SB127J	0160	2SB134J	0004	2SB142L	0085	2SB149L	0085
2SB105E	0164	2SB113E	0004	2SB120E	0004	2SB127K	0160	2SB134K	0004	2SB142M	0085	2SB149M	0085
2SB105F	0164	2SB113F	0004	2SB120F	0004	2SB127L	0160	2SB134L	0004	2SB142OR	0085	2SB149OR	0085
2SB105G	0164	2SB113G	0004	2SB120G	0004	2SB127M	0160	2SB134M	0004	2SB142R	0085	2SB149R	0085
2SB105GN	0164	2SB113GN	0004	2SB120GN	0004	2SB127OR	0160	2SB134OR	0004	2SB142X	0085	2SB149X	0085
2SB105H	0164	2SB113H	0004	2SB120H	0004	2SB127R	0160	2SB134R	0004	2SB142Y	0085	2SB149Y	0085
2SB105K	0164	2SB113J	0004	2SB120J	0004	2SB127X	0160	2SB134X	0004	2SB143	0085	2SB150	0164
2SB105L	0164	2SB113K	0004	2SB120K	0004	2SB127Y	0160	2SB134Y	0004	2SB143A	0085	2SB150A	0164
2SB105M	0164	2SB113L	0004	2SB120L	0004	2SB128	0975	2SB135	0004	2SB143B	0085	2SB150B	0164
2SB105OR	0164	2SB113M	0004	2SB120M	0004	2SB128(V)	0975	2SB135(C)	0004	2SB143C	0085	2SB150C	0164
2SB105R	0164	2SB113OR	0004	2SB120OR	0004	2SB128A	0975	2SB135A	0004	2SB143D	0085	2SB150D	0164
2SB105X	0164	2SB113R	0004	2SB120R	0004	2SB128B	0975	2SB135B	0004	2SB143E	0085	2SB150E	0164
2SB105Y	0164	2SB113X	0004	2SB120X	0004	2SB128C	0975	2SB135C	0004	2SB143F	0085	2SB150F	0164
2SB106	0164	2SB113Y	0004	2SB120Y	0004	2SB128D	0975	2SB135D	0004	2SB143G	0085	2SB150G	0164
2SB107	0160	2SB114	0004	2SB122	0160	2SB128E	0975	2SB135E	0004	2SB143GN	0085	2SB150GN	0164
2SB107A	0160	2SB114A	0004	2SB122A	0160	2SB128F	0975	2SB135F	0004	2SB143H	0085	2SB150H	0164
2SB107B	0160	2SB114B	0004	2SB122B	0160	2SB128G	0975	2SB135G	0004	2SB143J	0085	2SB150J	0164
2SB107C	0160	2SB114C	0004	2SB122C	0160	2SB128GN	0975	2SB135GN	0004	2SB143K	0085	2SB150K	0164
2SB107D	0160	2SB114D	0004	2SB122D	0160	2SB128H	0975	2SB135H	0004	2SB143L	0085	2SB150L	0164
2SB107E	0160	2SB114E	0004	2SB122E	0160	2SB128J	0975	2SB135J	0004	2SB143M	0085	2SB150M	0164
2SB107F	0160	2SB114F	0004	2SB122F	0160	2SB128K	0975	2SB135K	0004	2SB143OR	0085	2SB150OR	0164
2SB107G	0160	2SB114G	0004	2SB122G	0160	2SB128L	0975	2SB135L	0004	2SB143P	0085	2SB150R	0164
2SB107GN	0160	2SB114GN	0004	2SB122GN	0160	2SB128M	0975	2SB135M	0004	2SB143X	0085	2SB150X	0164
2SB107H	0160	2SB114H	0004	2SB122H	0160	2SB128OR	0975	2SB135OR	0004	2SB143Y	0085	2SB150Y	0164
2SB107J	0160	2SB114J	0004	2SB122J	0160	2SB128R	0975	2SB135R	0004	2SB144	0085	2SB151	0975
2SB107K	0160	2SB114K	0004	2SB122K	0160	2SB128V	0975	2SB135X	0004	2SB144A	0085	2SB151A	0975
2SB107L	0160	2SB114L	0004	2SB122L	0160	2SB128X	0975	2SB135Y	0004	2SB144B	0085	2SB151B	0975
2SB107M	0160	2SB114M	0004	2SB122M	0160	2SB128Y	0975	2SB136	0004	2SB144C	0085	2SB151C	0975
2SB107OR	0160	2SB114OR	0004	2SB122OR	0160	2SB129	0975	2SB136(C)	0004	2SB144D	0085	2SB151D	0975
2SB107R	0160	2SB114R	0004	2SB122R	0160	2SB129A	0975	2SB136A	0004	2SB144E	0085	2SB151E	0975
2SB107X	0160	2SB114X	0004	2SB122X	0160	2SB129B	0975	2SB136B	0004	2SB144F	0085	2SB151F	0975
2SB107Y	0160	2SB114Y	0004	2SB122Y	0160	2SB129C	0975	2SB136C	0004	2SB144G	0085	2SB151G	0975

If replacement code is OEM, contact original manufacturer for replacement.

DEVICE TYPE	REPL CODE
2SB151GN	0975
2SB151H	0975
2SB151J	0975
2SB151K	0975
2SB151L	0975
2SB151M	0975
2SB151OR	0975
2SB151R	0975
2SB151X	0975
2SB151Y	0975
2SB152	0975
2SB152A	0975
2SB152B	0975
2SB152C	0975
2SB152D	0975
2SB152E	0975
2SB152F	0975
2SB152G	0975
2SB152GN	0975
2SB152H	0975
2SB152J	0975
2SB152K	0975
2SB152L	0975
2SB152M	0975
2SB152OR	0975
2SB152R	0975
2SB152X	0975
2SB152Y	0975
2SB153	0004
2SB153A	0004
2SB153B	0004
2SB153C	0004
2SB153D	0004
2SB153E	0004
2SB153F	0004
2SB153G	0004
2SB153GN	0004
2SB153H	0004
2SB153J	0004
2SB153K	0004
2SB153L	0004
2SB153M	0004
2SB153OR	0004
2SB153R	0004
2SB153X	0004
2SB153Y	0004
2SB154	0004
2SB154A	0004
2SB154B	0004
2SB154C	0004
2SB154D	0004
2SB154E	0004
2SB154F	0004
2SB154G	0004
2SB154GN	0004
2SB154H	0004
2SB154J	0004
2SB154K	0004
2SB154L	0004
2SB154M	0004
2SB154OR	0004
2SB154R	0004
2SB154X	0004
2SB154Y	0004
2SB155	0004
2SB155A	0004
2SB155B	0004
2SB155C	0004
2SB155D	0004
2SB155E	0004
2SB155F	0004
2SB155G	0004
2SB155GN	0004
2SB155H	0004
2SB155J	0004
2SB155K	0004
2SB155L	0004
2SB155M	0004
2SB155OR	0004
2SB155R	0004
2SB155X	0004
2SB155Y	0004
2SB156	0004
2SB156-A	0004
2SB156A	0004
2SB156AA	0004
2SB156AB	0004
2SB156AC	0004
2SB156AD	0004
2SB156B	0004
2SB156B3	0004
2SB156BK	0004
2SB156C	0004
2SB156D	0004
2SB156E	0004
2SB156F	0004
2SB156G	0004
2SB156GN	0004
2SB156H	0004
2SB156J	0004
2SB156K	0004
2SB156L	0004
2SB156M	0004
2SB156OR	0004
2SB156P	0004
2SB156R	0004
2SB156X	0004
2SB156Y	0004
2SB157	0136
2SB157A	0136
2SB157B	0136
2SB157C	0136
2SB157D	0136
2SB157E	0136
2SB157F	0136
2SB157G	0136
2SB157GN	0136
2SB157H	0136
2SB157J	0136
2SB157K	0136
2SB157L	0136
2SB157M	0136
2SB157OR	0136
2SB157R	0136
2SB157X	0136
2SB157Y	0136
2SB158	0136
2SB158A	0136
2SB158B	0136
2SB158C	0136
2SB158D	0136
2SB158E	0136
2SB158F	0136
2SB158G	0136
2SB158GN	0136
2SB158H	0136
2SB158J	0136
2SB158K	0136
2SB158L	0136
2SB158M	0136
2SB158OR	0136
2SB158R	0136
2SB158X	0136
2SB158Y	0136
2SB159	0136
2SB159A	0136
2SB159B	0136
2SB159C	0136
2SB159D	0136
2SB159E	0136
2SB159F	0136
2SB159G	0136
2SB159GN	0136
2SB159H	0136
2SB159J	0136
2SB159K	0136
2SB159L	0136
2SB159M	0136
2SB159OR	0136
2SB159R	0136
2SB159X	0136
2SB159Y	0136
2SB160	0136
2SB160A	0136
2SB160B	0136
2SB160C	0136
2SB160D	0136
2SB160E	0136
2SB160F	0136
2SB160G	0136
2SB160GN	0136
2SB160H	0136
2SB160J	0136
2SB160K	0136
2SB160L	0136
2SB160M	0136
2SB160OR	0136
2SB160R	0136
2SB160X	0136
2SB160Y	0136
2SB161	0211
2SB161A	0211
2SB161B	0211
2SB161C	0211
2SB161D	0211
2SB161E	0211
2SB161F	0211
2SB161G	0211
2SB161GN	0211
2SB161H	0211
2SB161J	0211
2SB161K	0211
2SB161L	0211
2SB161M	0211
2SB161OR	0211
2SB161R	0211
2SB161X	0211
2SB161Y	0211
2SB162	0004
2SB162A	0004
2SB162B	0004
2SB162C	0004
2SB162D	0004
2SB162E	0004
2SB162F	0004
2SB162G	0004
2SB162GN	0004
2SB162H	0004
2SB162J	0004
2SB162K	0004
2SB162L	0004
2SB162M	0004
2SB162OR	0004
2SB162R	0004
2SB162X	0004
2SB162Y	0004
2SB163	0211
2SB163A	0211
2SB163B	0211
2SB163C	0211
2SB163D	0211
2SB163E	0211
2SB163F	0211
2SB163G	0211
2SB163GN	0211
2SB163H	0211
2SB163J	0211
2SB163K	0211
2SB163L	0211
2SB163M	0211
2SB163OR	0211
2SB163R	0211
2SB163X	0211
2SB163Y	0211
2SB164	0211
2SB164A	0211
2SB164B	0211
2SB164C	0211
2SB164D	0211
2SB164E	0211
2SB164F	0211
2SB164G	0211
2SB164GN	0211
2SB164H	0211
2SB164J	0211
2SB164L	0211
2SB164M	0211
2SB164OR	0211
2SB164R	0211
2SB164X	0211
2SB164Y	0211
2SB165	0211
2SB165A	0211
2SB165B	0211
2SB165C	0211
2SB165D	0211
2SB165E	0211
2SB165F	0211
2SB165G	0211
2SB165GN	0211
2SB165H	0211
2SB165J	0211
2SB165K	0211
2SB165M	0211
2SB165OR	0211
2SB165R	0211
2SB165X	0211
2SB165Y	0211
2SB166	0004
2SB166A	0004
2SB166B	0004
2SB166C	0004
2SB166D	0004
2SB166E	0004
2SB166F	0004
2SB166G	0004
2SB166GN	0004
2SB166H	0004
2SB166J	0004
2SB166K	0004
2SB166L	0004
2SB166M	0004
2SB166OR	0004
2SB166R	0004
2SB166X	0004
2SB166Y	0004
2SB167	0164
2SB167A	0164
2SB167B	0164
2SB167BK	0164
2SB167C	0164
2SB167D	0164
2SB167E	0164
2SB167F	0164
2SB167G	0164
2SB167GN	0164
2SB167H	0164
2SB167J	0164
2SB167K	0164
2SB167L	0164
2SB167M	0164
2SB167OR	0164
2SB167R	0164
2SB167X	0164
2SB167Y	0164
2SB168	0004
2SB168A	0004
2SB168B	0004
2SB168C	0004
2SB168D	0004
2SB168E	0004
2SB168F	0004
2SB168G	0004
2SB168GN	0004
2SB168H	0004
2SB168J	0004
2SB168K	0004
2SB168L	0004
2SB168M	0004
2SB168OR	0004
2SB168R	0004
2SB168X	0004
2SB168Y	0004
2SB169	0004
2SB169A	0004
2SB169B	0004
2SB169C	0004
2SB169D	0004
2SB169E	0004
2SB169F	0004
2SB169G	0004
2SB169GN	0004
2SB169H	0004
2SB169J	0004
2SB169K	0004
2SB169L	0004
2SB169M	0004
2SB169OR	0004
2SB169R	0004
2SB169X	0004
2SB169Y	0004
2SB170	0004
2SB170A	0004
2SB170B	0004
2SB170C	0004
2SB170D	0004
2SB170E	0004
2SB170F	0004
2SB170G	0004
2SB170GN	0004
2SB170H	0004
2SB170J	0004
2SB170K	0004
2SB170L	0004
2SB170M	0004
2SB170OR	0004
2SB170R	0004
2SB170X	0004
2SB170Y	0004
2SB171	0004
2SB171A	0004
2SB171B	0004
2SB171C	0004
2SB171D	0004
2SB171E	0004
2SB171F	0004
2SB171G	0004
2SB171GN	0004
2SB171H	0004
2SB171J	0004
2SB171K	0004
2SB171L	0004
2SB171M	0004
2SB171OR	0004
2SB171X	0004
2SB171Y	0004
2SB172	0004
2SB172(F)	0004
2SB172A	0004
2SB172A-1	0004
2SB172A-F	0004
2SB172AF	0004
2SB172AL	0004
2SB172B	0004
2SB172C	0004
2SB172D	0004
2SB172E	0004
2SB172F	0004
2SB172FN	0004
2SB172G	0004
2SB172GN	0004
2SB172H	0004
2SB172J	0004
2SB172K	0004
2SB172L	0004
2SB172M	0004
2SB172OR	0004
2SB172P	0004
2SB172R	0004
2SB172X	0004
2SB172Y	0004
2SB173	0004
2SB173(C)	0004
2SB173A	0004
2SB173B	0004
2SB173BL	0004
2SB173C	0004
2SB173CL	0004
2SB173D	0004
2SB173E	0004
2SB173F	0004
2SB173G	0004
2SB173GN	0004
2SB173H	0004
2SB173J	0004
2SB173K	0004
2SB173L	0004
2SB173M	0004
2SB173OR	0004
2SB173R	0004
2SB173X	0004
2SB173Y	0004
2SB174	0004
2SB174A	0004
2SB174B	0004
2SB174C	0004
2SB174D	0004
2SB174E	0004
2SB174F	0004
2SB174G	0004
2SB174GN	0004
2SB174H	0004
2SB174K	0004
2SB174L	0004
2SB174M	0004
2SB174OR	0004
2SB174R	0004
2SB174X	0004
2SB174Y	0004
2SB175	0004
2SB175(A)	0004
2SB175(B)	0004
2SB175(C)	0004
2SB175A	0004
2SB175B	0004
2SB175B-1	0004
2SB175BL	0004
2SB175C	0004
2SB175CL	0004
2SB175D	0004
2SB175E	0004
2SB175F	0004
2SB175G	0004
2SB175GN	0004
2SB175H	0004
2SB175L	0004
2SB175M	0004
2SB175OR	0004
2SB175R	0004
2SB175X	0004
2SB175Y	0004
2SB176	0004
2SB176(O)	0004
2SB176(P)	0004
2SB176-O	0004
2SB176-P	0004
2SB176-PR	0004
2SB176A	0004
2SB176B	0004
2SB176C	0004
2SB176D	0004
2SB176E	0004
2SB176F	0004
2SB176G	0004
2SB176GN	0004
2SB176H	0004
2SB176J	0004
2SB176K	0004
2SB176L	0004
2SB176M	0004
2SB176O	0004
2SB176OR	0004
2SB176P	0004
2SB176PL	0004
2SB176PR	0004
2SB176PRC	0004
2SB176R	0004
2SB176R(1)	0004
2SB176RG	0004
2SB176X	0004
2SB176Y	0004
2SB177	0004
2SB177A	0004
2SB177B	0004
2SB177C	0004
2SB177D	0004
2SB177E	0004
2SB177F	0004
2SB177G	0004
2SB177GN	0004
2SB177H	0004
2SB177K	0004
2SB177L	0004
2SB177M	0004
2SB177OR	0004
2SB177R	0004
2SB177X	0004
2SB177Y	0004
2SB178	0004
2SB178(M)	0004
2SB178-O	0004
2SB178-OR	0004
2SB178-S	0004
2SB178A	0004
2SB178B	0004
2SB178C	0004
2SB178D	0004
2SB178E	0004
2SB178F	0004
2SB178G	0004
2SB178GN	0004
2SB178H	0004
2SB178J	0004
2SB178K	0004
2SB178L	0004
2SB178M	0004
2SB178N	0004
2SB178OR	0004
2SB178R	0004
2SB178S	0004
2SB178T	0004
2SB178TC	0004
2SB178TS	0004
2SB178U	0004
2SB178V	0004
2SB178X	0004
2SB178Y	0004
2SB179	0050
2SB180	0164
2SB180A	0164
2SB181	0164
2SB181A	0164
2SB182	0050
2SB183	0004
2SB183A	0004
2SB183B	0004
2SB183C	0004
2SB183D	0004
2SB183E	0004
2SB183F	0004
2SB183G	0004
2SB183GN	0004
2SB183H	0004
2SB183J	0004
2SB183K	0004
2SB183L	0004
2SB183M	0004
2SB183OR	0004
2SB183R	0004
2SB183X	0004
2SB183Y	0004
2SB184	0004
2SB184A	0004
2SB184B	0004
2SB184C	0004
2SB184D	0004
2SB184E	0004
2SB184F	0004
2SB184G	0004
2SB184GN	0004
2SB184H	0004
2SB184J	0004
2SB184K	0004
2SB184L	0004
2SB184M	0004
2SB184OR	0004
2SB184R	0004
2SB184X	0004
2SB184Y	0004
2SB185	0004
2SB185(O)	0004
2SB185(P)	0004
2SB185-O	0004
2SB185A	0004
2SB185AA	0004
2SB185B	0004
2SB185C	0004
2SB185D	0004
2SB185E	0004
2SB185F	0004
2SB185G	0004
2SB185GN	0004
2SB185H	0004
2SB185I	0004
2SB185J	0004
2SB185L	0004
2SB185M	0004
2SB185OR	0004
2SB185P	0004
2SB185R	0004
2SB185X	0004
2SB185Y	0004
2SB186	0004
2SB186(A)	0004
2SB186(O)	0004
2SB186-1	0004
2SB186-7	0004
2SB186-K	0004
2SB186-O	0004
2SB186-OR	0004
2SB186A	0004
2SB186AG	0004
2SB186B	0004
2SB186BY	0004
2SB186C	0004
2SB186D	0004
2SB186E	0004
2SB186F	0004
2SB186G	0004
2SB186GN	0004
2SB186H	0004
2SB186J	0004
2SB186K	0004
2SB186L	0004
2SB186M	0004
2SB186OR	0004
2SB186R	0004
2SB186X	0004
2SB186Y	0004
2SB187	0004
2SB187(1)	0004
2SB187(K)	0004
2SB187-1	0004
2SB187-OR	0004
2SB187A	0004
2SB187AA	3555
2SB187B	3555
2SB187BK	0004
2SB187C	0004
2SB187D	0004
2SB187E	0004
2SB187F	0004
2SB187G	0004
2SB187GN	0004
2SB187H	0004
2SB187K	0004
2SB187L	0004
2SB187M	0004
2SB187OR	0004
2SB187R	0004
2SB187S	0004
2SB187TV	0004
2SB187X	0004
2SB187Y	0004
2SB188	0004
2SB188A	0004
2SB188B	0004
2SB188C	0004
2SB188D	0004
2SB188E	0004
2SB188F	0004
2SB188G	0004
2SB188GN	0004
2SB188H	0004
2SB188J	0004
2SB188K	0004
2SB188L	0004
2SB188M	0004
2SB188OR	0004
2SB188R	0004
2SB188X	0004
2SB188Y	0004
2SB189	0004
2SB189A	0004
2SB189B	0004
2SB189C	0004
2SB189D	0004
2SB189E	0004
2SB189F	0004
2SB189G	0004
2SB189GN	0004
2SB189H	0004
2SB189J	0004
2SB189K	0004
2SB189L	0004
2SB189M	0004
2SB189OR	0004
2SB189R	0004
2SB189X	0004
2SB189Y	0004
2SB190	0164
2SB191	0280
2SB192	0164
2SB193	0164
2SB194	0164
2SB195	0164
2SB196	0164
2SB197	0164
2SB198	0164
2SB199	0211
2SB199A	0211
2SB199B	0211
2SB199C	0211
2SB199D	0211
2SB199E	0211
2SB199F	0211
2SB199G	0211
2SB199GN	0211
2SB199H	0211
2SB199J	0211
2SB199L	0211
2SB199M	0211
2SB199OR	0211
2SB199R	0211
2SB199X	0211
2SB199Y	0211
2SB200	0004
2SB200(RECT)	1414
2SB200A	0004
2SB200B	0004
2SB200C	0004
2SB200D	0004
2SB200E	0004
2SB200F	0004
2SB200G	0004
2SB200GN	0004
2SB200H	0004
2SB200J	0004
2SB200K	0004
2SB200L	0004
2SB200M	0004
2SB200OR	0004
2SB200R	0004
2SB200X	0004
2SB200Y	0004
2SB201	0004
2SB202	0004
2SB202A	0004
2SB202B	0004
2SB202C	0004
2SB202D	0004
2SB202E	0004
2SB202F	0004
2SB202G	0004
2SB202GN	0004
2SB202H	0004
2SB202J	0004
2SB202L	0004
2SB202M	0004
2SB202OR	0004
2SB202R	0004
2SB202X	0004
2SB202Y	0004
2SB203	0604
2SB203A	0604
2SB203AA	0604
2SB203B	0604
2SB203C	0604
2SB203D	0604
2SB203E	0604
2SB203F	0604
2SB203G	0604
2SB203GN	0604

Original Device Types

DEVICE TYPE	REPL CODE
2SB203H	0604
2SB203J	0604
2SB203K	0604
2SB203L	0604
2SB203M	0604
2SB203OR	0604
2SB203R	0604
2SB203X	0604
2SB203Y	0604
2SB204	0604
2SB205	0604
2SB206	0604
2SB207	0604
2SB207A	0604
2SB208	0604
2SB208A	0604
2SB209	0604
2SB210	0604
2SB211	0604
2SB212	0604
2SB213	0604
2SB213A	0604
2SB214	0604
2SB214A	0604
2SB215	0975
2SB215A	0975
2SB215B	0975
2SB215C	0975
2SB215D	0975
2SB215E	0975
2SB215F	0975
2SB215G	0975
2SB215GN	0975
2SB215H	0975
2SB215J	0975
2SB215K	0975
2SB215M	0975
2SB215OR	0975
2SB215R	0975
2SB215X	0975
2SB215Y	0975
2SB216	0160
2SB216A	0160
2SB216C	0160
2SB216D	0160
2SB216E	0160
2SB216F	0160
2SB216G	0160
2SB216GN	0160
2SB216H	0160
2SB216J	0160
2SB216K	0160
2SB216L	0160
2SB216M	0160
2SB216OR	0160
2SB216R	0160
2SB216X	0160
2SB216Y	0160
2SB217	0160
2SB217A	0160
2SB217B	0160
2SB217C	0160
2SB217D	0160
2SB217E	0160
2SB217F	0160
2SB217G	0160
2SB217GN	0160
2SB217H	0160
2SB217K	0160
2SB217L	0160
2SB217M	0160
2SB217OR	0160
2SB217R	0160
2SB217U	0160
2SB217X	0160
2SB217Y	0160
2SB218	0211
2SB218A	0211
2SB218B	0211
2SB218C	0211
2SB218D	0211
2SB218E	0211
2SB218F	0211
2SB218G	0211
2SB218GN	0211
2SB218H	0211
2SB218J	0211
2SB218K	0211
2SB218L	0211
2SB218M	0211
2SB218OR	0211
2SB218R	0211
2SB218X	0211
2SB218Y	0211
2SB219	0211
2SB219A	0211
2SB219B	0211
2SB219E	0211
2SB219F	0211
2SB219G	0211
2SB219H	0211
2SB219J	0211
2SB219K	0211
2SB219L	0211
2SB219R	0211
2SB219X	0211
2SB219Y	0211
2SB220	0211
2SB220A	0211
2SB220B	0211
2SB220C	0211
2SB220D	0211
2SB220E	0211
2SB220F	0211
2SB220GN	0211
2SB220H	0211
2SB220J	0211
2SB220K	0211
2SB220L	0211
2SB220M	0211
2SB220OR	0211
2SB220R	0211
2SB220X	0211
2SB220Y	0211
2SB221	0211
2SB221A	0211
2SB221B	0211
2SB221C	0211
2SB221D	0211
2SB221E	0211
2SB221F	0211
2SB221G	0211
2SB221GN	0211
2SB221H	0211
2SB221J	0211
2SB221K	0211
2SB221L	0211
2SB221M	0211
2SB221OR	0211
2SB221R	0211
2SB221X	0211
2SB221Y	0211
2SB222	0211
2SB222A	0211
2SB222B	0211
2SB222C	0211
2SB222D	0211
2SB222E	0211
2SB222F	0211
2SB222G	0211
2SB222GN	0211
2SB222H	0211
2SB222J	0211
2SB222K	0211
2SB222L	0211
2SB222M	0211
2SB222OR	0211
2SB222R	0211
2SB222X	0211
2SB222Y	0211
2SB223	0211
2SB223A	0211
2SB223B	0211
2SB223C	0211
2SB223D	0211
2SB223E	0211
2SB223F	0211
2SB223G	0211
2SB223Y	0211
2SB224	0004
2SB224A	0004
2SB224B	0004
2SB224C	0004
2SB224D	0004
2SB224E	0004
2SB224F	0004
2SB224G	0004
2SB224GN	0004
2SB224H	0004
2SB224K	0004
2SB224L	0004
2SB224M	0004
2SB224OR	0004
2SB224R	0004
2SB224X	0004
2SB224Y	0004
2SB225	0211
2SB225A	0211
2SB225B	0211
2SB225C	0211
2SB225D	0211
2SB225E	0211
2SB225F	0211
2SB225G	0211
2SB225GN	0211
2SB225H	0211
2SB225K	0211
2SB225L	0211
2SB225M	0211
2SB225OR	0211
2SB225R	0211
2SB225X	0211
2SB225Y	0211
2SB226	0211
2SB226A	0211
2SB226B	0211
2SB226C	0211
2SB226D	0211
2SB226E	0211
2SB226F	0211
2SB226G	0211
2SB226GN	0211
2SB226J	0211
2SB226K	0211
2SB226L	0211
2SB226M	0211
2SB226OR	0211
2SB226R	0211
2SB226X	0211
2SB226Y	0211
2SB227	0004
2SB227A	0004
2SB227B	0004
2SB227C	0004
2SB227D	0004
2SB227E	0004
2SB227F	0004
2SB227G	0004
2SB227GN	0004
2SB227H	0004
2SB227J	0004
2SB227K	0004
2SB227L	0004
2SB227M	0004
2SB227OR	0004
2SB227R	0004
2SB227X	0004
2SB227Y	0004
2SB228	0975
2SB228A	0975
2SB228B	0975
2SB228C	0975
2SB228D	0975
2SB228E	0975
2SB228F	0975
2SB228G	0975
2SB228GN	0975
2SB228H	0975
2SB228J	0975
2SB228K	0975
2SB228L	0975
2SB228M	0975
2SB228OR	0975
2SB228R	0975
2SB228X	0975
2SB228Y	0975
2SB229	0975
2SB229A	0975
2SB229B	0975
2SB229C	0975
2SB229D	0975
2SB229E	0975
2SB229F	0975
2SB229G	0975
2SB229GN	0975
2SB229J	0975
2SB229K	0975
2SB229L	0975
2SB229M	0975
2SB229OR	0975
2SB229R	0975
2SB229X	0975
2SB229Y	0975
2SB230	0975
2SB230A	0975
2SB230B	0975
2SB230C	0975
2SB230D	0975
2SB230E	0975
2SB230F	0975
2SB230G	0975
2SB230GN	0975
2SB230H	0975
2SB230J	0975
2SB230K	0975
2SB230L	0975
2SB230M	0975
2SB230OR	0975
2SB230R	0975
2SB230X	0975
2SB230Y	0975
2SB231	0975
2SB231A	0975
2SB231B	0975
2SB231C	0975
2SB231D	0975
2SB231E	0975
2SB231F	0975
2SB231G	0975
2SB231GN	0975
2SB231H	0975
2SB231J	0975
2SB231K	0975
2SB231L	0975
2SB231M	0975
2SB231OR	0975
2SB231R	0975
2SB231X	0975
2SB231Y	0975
2SB232	0975
2SB232A	0975
2SB232B	0975
2SB232C	0975
2SB232D	0975
2SB232E	0975
2SB232F	0975
2SB232G	0975
2SB232GN	0975
2SB232H	0975
2SB232K	0975
2SB232L	0975
2SB232M	0975
2SB232OR	0975
2SB232R	0975
2SB232X	0975
2SB232Y	0975
2SB233	0975
2SB233A	0975
2SB233B	0975
2SB233C	0975
2SB233D	0975
2SB233E	0975
2SB233F	0975
2SB233G	0975
2SB233GN	0975
2SB233H	0975
2SB233J	0975
2SB233K	0975
2SB233L	0975
2SB233M	0975
2SB233OR	0975
2SB233R	0975
2SB233X	0975
2SB233Y	0975
2SB234	0975
2SB234A	0975
2SB234B	0975
2SB234C	0975
2SB234D	0975
2SB234E	0975
2SB234F	0975
2SB234G	0975
2SB234GN	0975
2SB234H	0975
2SB234J	0975
2SB234K	0975
2SB234L	0975
2SB234M	0975
2SB234N	0975
2SB234OR	0975
2SB234R	0975
2SB234X	0975
2SB234Y	0975
2SB235	0040
2SB235A	0040
2SB235B	0040
2SB235C	0040
2SB235D	0040
2SB235E	0040
2SB235F	0040
2SB235G	0040
2SB235GN	0040
2SB235H	0040
2SB235L	0040
2SB235M	0040
2SB235OR	0040
2SB235R	0040
2SB235X	0040
2SB235Y	0040
2SB236	0440
2SB236A	0440
2SB236B	0440
2SB236C	0440
2SB236D	0440
2SB236E	0440
2SB236F	0440
2SB236G	0440
2SB236GN	0440
2SB236H	0440
2SB236J	0440
2SB236K	0440
2SB236L	0440
2SB236M	0440
2SB236OR	0440
2SB236R	0440
2SB236X	0440
2SB236Y	0440
2SB237	0440
2SB237-12A	0440
2SB237-12B	0440
2SB237A	0440
2SB237B	0440
2SB237C	0440
2SB237D	0440
2SB237E	0440
2SB237F	0440
2SB237G	0440
2SB237GN	0440
2SB237H	0440
2SB237J	0440
2SB237K	0440
2SB237L	0440
2SB237M	0440
2SB237OR	0440
2SB237R	0440
2SB237X	0440
2SB237Y	0440
2SB238	0164
2SB238-12A	0164
2SB238-12B	0164
2SB238-12C	0164
2SB238A	0164
2SB238B	0164
2SB238C	0164
2SB238D	0164
2SB238E	0164
2SB238F	0164
2SB238G	0164
2SB238GN	0164
2SB238H	0164
2SB238J	0164
2SB238K	0164
2SB238L	0164
2SB238M	0164
2SB238OR	0164
2SB238R	0164
2SB238X	0164
2SB238Y	0164
2SB239	0160
2SB239A	0160
2SB239B	0160
2SB239C	0160
2SB239D	0160
2SB239E	0160
2SB239F	0160
2SB239G	0160
2SB239GN	0160
2SB239H	0160
2SB239J	0160
2SB239K	0160
2SB239L	0160
2SB239M	0160
2SB239OR	0160
2SB239R	0160
2SB239X	0160
2SB239Y	0160
2SB240	0050
2SB240A	0050
2SB240B	0050
2SB240C	0050
2SB240D	0050
2SB240E	0050
2SB240F	0050
2SB240GN	0050
2SB240H	0050
2SB240J	0050
2SB240K	0050
2SB240L	0050
2SB240M	0050
2SB240OR	0050
2SB240R	0050
2SB240X	0050
2SB240Y	0050
2SB241	0164
2SB241A	0164
2SB241B	0164
2SB241C	0164
2SB241D	0164
2SB241E	0164
2SB241F	0164
2SB241G	0164
2SB241GN	0164
2SB241H	0164
2SB241J	0164
2SB241K	0164
2SB241L	0164
2SB241M	0164
2SB241OR	0164
2SB241R	0164
2SB241V	0164
2SB241X	0164
2SB241Y	0164
2SB242	0164
2SB242A	0164
2SB242B	0164
2SB242C	0164
2SB242D	0164
2SB242E	0164
2SB242F	0164
2SB242G	0164
2SB242GN	0164
2SB242H	0164
2SB242J	0164
2SB242K	0164
2SB242L	0164
2SB242M	0164
2SB242OR	0164
2SB242R	0164
2SB242X	0164
2SB242Y	0164
2SB243	0975
2SB243A	0975
2SB243B	0975
2SB243C	0975
2SB243D	0975
2SB243E	0975
2SB243F	0975
2SB243G	0975
2SB243GN	0975
2SB243H	0975
2SB243J	0975
2SB243K	0975
2SB243L	0975
2SB243M	0975
2SB243OR	0975
2SB243R	0975
2SB243X	0975
2SB243Y	0975
2SB244	0975
2SB244A	0975
2SB244B	0975
2SB244C	0975
2SB244D	0975
2SB244E	0975
2SB244F	0975
2SB244G	0975
2SB244GN	0975
2SB244H	0975
2SB244J	0975
2SB244K	0975
2SB244L	0975
2SB244M	0975
2SB244OR	0975
2SB244R	0975
2SB244X	0975
2SB244Y	0975
2SB245	0975
2SB245A	0975
2SB245B	0975
2SB245C	0975
2SB245D	0975
2SB245E	0975
2SB245F	0975
2SB245G	0975
2SB245H	0975
2SB245J	0975
2SB245L	0975
2SB245M	0975
2SB245OR	0975
2SB245R	0975
2SB245X	0975
2SB245Y	0975
2SB246	0085
2SB246A	0085
2SB246B	0085
2SB246C	0085
2SB246D	0085
2SB246E	0085
2SB246F	0085
2SB246G	0085
2SB246GN	0085
2SB246H	0085
2SB246J	0085
2SB246K	0085
2SB246L	0085
2SB246M	0085
2SB246OR	0085
2SB246R	0085
2SB246X	0085
2SB246Y	0085
2SB247	0160
2SB247A	0160
2SB247B	0160
2SB247C	0160
2SB247D	0160
2SB247E	0160
2SB247F	0160
2SB247G	0160
2SB247GN	0160
2SB247H	0160
2SB247J	0160
2SB247K	0160
2SB247L	0160
2SB247M	0160
2SB247OR	0160
2SB247R	0160
2SB247X	0160
2SB247Y	0160
2SB248	0085
2SB248A	0085
2SB248B	0085
2SB248C	0085
2SB248D	0085
2SB248E	0085
2SB248F	0085
2SB248G	0085
2SB248GN	0085
2SB248H	0085
2SB248J	0085
2SB248K	0085
2SB248L	0085
2SB248M	0085
2SB248OR	0085
2SB248R	0085
2SB248X	0085
2SB248Y	0085
2SB249	0604
2SB249A	0604
2SB249B	0604
2SB249C	0604
2SB249D	0604
2SB249E	0604
2SB249F	0604
2SB249G	0604
2SB249GN	0604
2SB249H	0604
2SB249J	0604
2SB249K	0604
2SB249L	0604
2SB249M	0604
2SB249OR	0604
2SB249X	0604
2SB249Y	0604
2SB250	0085
2SB250A	0085
2SB250B	0085
2SB250D	0085
2SB250E	0085
2SB250F	0085
2SB250G	0085
2SB250GN	0085
2SB250H	0085
2SB250K	0085
2SB250L	0085
2SB250M	0085
2SB250OR	0085
2SB250X	0085
2SB250Y	0085
2SB251	0975
2SB251A	0975
2SB251B	0975
2SB251C	0975
2SB251D	0975
2SB251E	0975
2SB251F	0975
2SB251G	0975
2SB251GN	0975
2SB251H	0975
2SB251J	0975
2SB251K	0975
2SB251L	0975
2SB251M	0975
2SB251OR	0975
2SB251R	0975
2SB251X	0975
2SB252	0975
2SB252A	0975
2SB252B	0975
2SB252C	0975
2SB252D	0975
2SB252E	0975
2SB252F	0975
2SB252G	0975
2SB252GN	0975
2SB252H	0975
2SB252J	0975
2SB252L	0975
2SB252M	0975
2SB252OR	0975
2SB252R	0975
2SB252X	0975
2SB252Y	0975
2SB253	0975
2SB253A	0975
2SB253B	0975
2SB253C	0975
2SB253D	0975
2SB253E	0975
2SB253F	0975
2SB253G	0975
2SB253GN	0975
2SB253H	0975
2SB253K	0975
2SB253L	0975
2SB253M	0975
2SB253OR	0975
2SB253X	0975
2SB253Y	0975
2SB254	1010
2SB254(D)	1010
2SB254A	1010
2SB254B	1010
2SB254C	1010
2SB254D	1010
2SB254E	1010
2SB254F	1010
2SB254G	1010
2SB254GN	1010
2SB254H	1010
2SB254J	1010
2SB254K	1010
2SB254L	1010
2SB254M	1010
2SB254OR	1010
2SB254R	1010
2SB254X	1010
2SB254Y	1010
2SB255	1010
2SB255A	1010
2SB255B	1010
2SB255C	1010
2SB255D	1010
2SB255E	1010
2SB255F	1010
2SB255G	1010
2SB255GN	1010
2SB255H	1010
2SB255J	1010
2SB255K	1010
2SB255L	1010
2SB255M	1010
2SB255OR	1010
2SB255R	1010
2SB255X	1010
2SB255Y	1010
2SB256	1010
2SB256A	1010
2SB256B	1010
2SB256C	1010
2SB256D	1010
2SB256E	1010
2SB256F	1010
2SB256G	1010
2SB256GN	1010
2SB256H	1010
2SB256J	1010
2SB256L	1010
2SB256M	1010
2SB256OR	1010
2SB256R	1010
2SB256X	1010
2SB256Y	1010
2SB257	0004
2SB257A	0004
2SB257B	0004
2SB257C	0004
2SB257D	0004
2SB257E	0004
2SB257F	0004
2SB257G	0004
2SB257GN	0004
2SB257H	0004
2SB257K	0004

If replacement code is OEM, contact original manufacturer for replacement.

DEVICE TYPE	REPL CODE
2SB257L	0004
2SB257M	0004
2SB257OR	0004
2SB257R	0004
2SB257X	0004
2SB257Y	0004
2SB258	0437
2SB258A	0437
2SB258B	0437
2SB258C	0437
2SB258D	0437
2SB258E	0437
2SB258F	0437
2SB258G	0437
2SB258GN	0437
2SB258H	0437
2SB258J	0437
2SB258K	0437
2SB258L	0437
2SB258M	0437
2SB258OR	0437
2SB258R	0437
2SB258X	0437
2SB258Y	0437
2SB259	0437
2SB259A	0437
2SB259B	0437
2SB259C	0437
2SB259D	0437
2SB259E	0437
2SB259F	0437
2SB259GN	0437
2SB259H	0437
2SB259J	0437
2SB259K	0437
2SB259L	0437
2SB259OR	0437
2SB259R	0437
2SB259X	0437
2SB259Y	0437
2SB260	0040
2SB260A	0040
2SB260B	0040
2SB260C	0040
2SB260D	0040
2SB260E	0040
2SB260F	0040
2SB260G	0040
2SB260GN	0040
2SB260H	0040
2SB260J	0040
2SB260K	0040
2SB260L	0040
2SB260M	0040
2SB260OR	0040
2SB260R	0040
2SB260X	0040
2SB260Y	0040
2SB261	0004
2SB261A	0004
2SB261B	0004
2SB261C	0004
2SB261D	0004
2SB261E	0004
2SB261F	0004
2SB261G	0004
2SB261GN	0004
2SB261H	0004
2SB261J	0004
2SB261K	0004
2SB261L	0004
2SB261M	0004
2SB261N	0004
2SB261OR	0004
2SB261R	0004
2SB261X	0004
2SB261Y	0004
2SB262	0004
2SB262A	0004
2SB262B	0004
2SB262C	0004
2SB262D	0004
2SB262E	0004
2SB262F	0004
2SB262G	0004
2SB262GN	0004
2SB262H	0004
2SB262J	0004
2SB262K	0004
2SB262L	0004
2SB262M	0004
2SB262OR	0004
2SB262R	0004
2SB262X	0004
2SB262Y	0004
2SB263	0004
2SB263A	0004
2SB263B	0004
2SB263C	0004
2SB263D	0004
2SB263E	0004
2SB263F	0004
2SB263G	0004
2SB263GN	0004
2SB263H	0004
2SB263J	0004
2SB263K	0004
2SB263L	0004
2SB263M	0004
2SB263OR	0004
2SB263R	0004
2SB263X	0004
2SB263Y	0004
2SB264	0004
2SB264A	0004
2SB264B	0004
2SB264C	0004
2SB264D	0004
2SB264E	0004
2SB264F	0004
2SB264G	0004
2SB264GN	0004
2SB264H	0004
2SB264J	0004
2SB264K	0004
2SB264L	0004
2SB264M	0004
2SB264OR	0004
2SB264R	0004
2SB264X	0004
2SB264Y	0004
2SB265	0004
2SB265A	0004
2SB265B	0004
2SB265C	0004
2SB265D	0004
2SB265E	0004
2SB265F	0004
2SB265G	0004
2SB265GN	0004
2SB265H	0004
2SB265J	0004
2SB265K	0004
2SB265L	0004
2SB265M	0004
2SB265OR	0004
2SB265X	0004
2SB265Y	0004
2SB266	0004
2SB266A	0004
2SB266B	0004
2SB266C	0004
2SB266D	0004
2SB266E	0004
2SB266F	0004
2SB266G	0004
2SB266GN	0004
2SB266H	0004
2SB266J	0004
2SB266K	0004
2SB266L	0004
2SB266M	0004
2SB266OR	0004
2SB266P	0004
2SB266Q	0004
2SB266R	0004
2SB266X	0004
2SB266Y	0004
2SB267	0004
2SB267A	0004
2SB267B	0004
2SB267C	0004
2SB267D	0004
2SB267E	0004
2SB267F	0004
2SB267G	0004
2SB267GN	0004
2SB267H	0004
2SB267J	0004
2SB267K	0004
2SB267L	0004
2SB267M	0004
2SB267OR	0004
2SB267R	0004
2SB267X	0004
2SB267Y	0004
2SB268	0004
2SB268A	0004
2SB268B	0004
2SB268C	0004
2SB268D	0004
2SB268E	0004
2SB268F	0004
2SB268G	0004
2SB268GN	0004
2SB268H	0004
2SB268J	0004
2SB268K	0004
2SB268L	0004
2SB268M	0004
2SB268OR	0004
2SB268R	0004
2SB268X	0004
2SB268Y	0004
2SB269	0004
2SB269A	0004
2SB269B	0004
2SB269C	0004
2SB269D	0004
2SB269E	0004
2SB269F	0004
2SB269G	0004
2SB269GN	0004
2SB269J	0004
2SB269K	0004
2SB269L	0004
2SB269M	0004
2SB269OR	0004
2SB269R	0004
2SB269X	0004
2SB269Y	0004
2SB270	0004
2SB270A	0004
2SB270B	0004
2SB270C	0004
2SB270D	0004
2SB270E	0004
2SB270F	0004
2SB270G	0004
2SB270GN	0004
2SB270H	0004
2SB270J	0004
2SB270K	0004
2SB270L	0004
2SB270M	0004
2SB270OR	0004
2SB270R	0004
2SB270X	0004
2SB270Y	0004
2SB271	0004
2SB272	0004
2SB272A	0004
2SB272B	0004
2SB272C	0004
2SB272D	0004
2SB272E	0004
2SB272F	0004
2SB272G	0004
2SB272GN	0004
2SB272J	0004
2SB272K	0004
2SB272L	0004
2SB272M	0004
2SB272OR	0004
2SB272R	0004
2SB272X	0004
2SB272Y	0004
2SB273	0004
2SB273A	0004
2SB273B	0004
2SB273C	0004
2SB273D	0004
2SB273E	0004
2SB273F	0004
2SB273G	0004
2SB273GN	0004
2SB273H	0004
2SB273J	0004
2SB273K	0004
2SB273L	0004
2SB273M	0004
2SB273OR	0004
2SB273R	0004
2SB273X	0004
2SB273Y	0004
2SB274	0975
2SB274A	0975
2SB274B	0975
2SB274C	0975
2SB274D	0975
2SB274E	0975
2SB274F	0975
2SB274G	0975
2SB274H	0975
2SB274J	0975
2SB274K	0975
2SB274L	0975
2SB274M	0975
2SB274OR	0975
2SB274R	0975
2SB274V	0975
2SB274X	0975
2SB274Y	0975
2SB275	0975
2SB275A	0975
2SB275B	0975
2SB275C	0975
2SB275D	0975
2SB275E	0975
2SB275F	0975
2SB275G	0975
2SB275GN	0975
2SB275H	0975
2SB275J	0975
2SB275K	0975
2SB275L	0975
2SB275M	0975
2SB275OR	0975
2SB275R	0975
2SB275X	0975
2SB275Y	0975
2SB276	0975
2SB276A	0975
2SB276B	0975
2SB276C	0975
2SB276D	0975
2SB276E	0975
2SB276F	0975
2SB276G	0975
2SB276GN	0975
2SB276H	0975
2SB276J	0975
2SB276K	0975
2SB276L	0975
2SB276M	0975
2SB276OR	0975
2SB276R	0975
2SB276X	0975
2SB276Y	0975
2SB277	0050
2SB278	0164
2SB279	0164
2SB280	0004
2SB281	0532
2SB281C	0532
2SB282	0604
2SB282A	0604
2SB282B	0604
2SB282C	0604
2SB282D	0604
2SB282E	0604
2SB282G	0604
2SB282GN	0604
2SB282H	0604
2SB282J	0604
2SB282K	0604
2SB282L	0604
2SB282M	0604
2SB282OR	0604
2SB282R	0604
2SB282X	0604
2SB282Y	0604
2SB283	0160
2SB283A	0160
2SB283B	0160
2SB283C	0160
2SB283D	0160
2SB283E	0160
2SB283F	0160
2SB283G	0160
2SB283GN	0160
2SB283H	0160
2SB283J	0160
2SB283K	0160
2SB283L	0160
2SB283M	0160
2SB283OR	0160
2SB283R	0160
2SB283X	0160
2SB283Y	0160
2SB284	0160
2SB284A	0160
2SB284B	0160
2SB284C	0160
2SB284D	0160
2SB284E	0160
2SB284F	0160
2SB284G	0160
2SB284H	0160
2SB284J	0160
2SB284K	0160
2SB284L	0160
2SB284M	0160
2SB284OR	0160
2SB284R	0160
2SB284X	0160
2SB284Y	0160
2SB285	0975
2SB285A	0975
2SB285B	0975
2SB285C	0975
2SB285D	0975
2SB285E	0975
2SB285F	0975
2SB285G	0975
2SB285GN	0975
2SB285H	0975
2SB285J	0975
2SB285K	0975
2SB285L	0975
2SB285M	0975
2SB285OR	0975
2SB285R	0975
2SB285X	0975
2SB285Y	0975
2SB286	0975
2SB287	0975
2SB288	0050
2SB289	0050
2SB290	0280
2SB290-GREEN	0280
2SB290-YELLOW	0280
2SB290A	0280
2SB290A-BLUE	0280
2SB290B	0280
2SB290C	0280
2SB290D	0280
2SB290E	0280
2SB290F	0280
2SB290G	0280
2SB290GN	0280
2SB290H	0280
2SB290J	0280
2SB290K	0280
2SB290L	0280
2SB290M	0280
2SB290OR	0280
2SB290R	0280
2SB290X	0280
2SB290Y	0280
2SB291	0280
2SB291-GREEN	0280
2SB291-RED	0280
2SB291-YELLOW	0280
2SB291A	0280
2SB291B	0280
2SB291C	0280
2SB291D	0280
2SB291E	0280
2SB291F	0280
2SB291G	0280
2SB291GN	0280
2SB291H	0280
2SB291J	0280
2SB291K	0280
2SB291L	0280
2SB291M	0280
2SB291OR	0280
2SB291R	0280
2SB291X	0280
2SB291Y	0280
2SB292	0280
2SB292-BLUE	0280
2SB292-GREEN	0280
2SB292-ORANGE	0280
2SB292-RED	0280
2SB292-YELLOW	0280
2SB292A	0280
2SB292A-BLUE	0280
2SB292A-GREEN	0280
2SB292A-ORANG	0280
2SB292A-RED	0280
2SB292A-YELLO	0280
2SB292B	0280
2SB292C	0280
2SB292D	0280
2SB292E	0280
2SB292F	0280
2SB292G	0280
2SB292GN	0280
2SB292H	0280
2SB292J	0280
2SB292K	0280
2SB292L	0280
2SB292M	0280
2SB292OR	0280
2SB292R	0280
2SB292X	0280
2SB292Y	0280
2SB293	0004
2SB293A	0004
2SB293B	0004
2SB293C	0004
2SB293D	0004
2SB293E	0004
2SB293F	0004
2SB293G	0004
2SB293GN	0004
2SB293H	0004
2SB293J	0004
2SB293K	0004
2SB293L	0004
2SB293M	0004
2SB293OR	0004
2SB293R	0004
2SB293X	0004
2SB293Y	0004
2SB294	0004
2SB294A	0004
2SB294B	0004
2SB294C	0004
2SB294D	0004
2SB294E	0004
2SB294F	0004
2SB294G	0004
2SB294GN	0004
2SB294H	0004
2SB294J	0004
2SB294K	0004
2SB294L	0004
2SB294M	0004
2SB294OR	0004
2SB294R	0004
2SB294X	0004
2SB294Y	0004
2SB295	0975
2SB295A	0975
2SB295B	0975
2SB295C	0975
2SB295D	0975
2SB295E	0975
2SB295F	0975
2SB295G	0975
2SB295GN	0975
2SB295H	0975
2SB295J	0975
2SB295K	0975
2SB295L	0975
2SB295M	0975
2SB295OR	0975
2SB295R	0975
2SB295X	0975
2SB295Y	0975
2SB296	0975
2SB296A	0975
2SB296B	0975
2SB296C	0975
2SB296D	0975
2SB296E	0975
2SB296F	0975
2SB296G	0975
2SB296GN	0975
2SB296H	0975
2SB296J	0975
2SB296K	0975
2SB296L	0975
2SB296M	0975
2SB296OR	0975
2SB296R	0975
2SB296X	0975
2SB296Y	0975
2SB299	0004
2SB299A	0004
2SB299B	0004
2SB299C	0004
2SB299D	0004
2SB299E	0004
2SB299F	0004
2SB299G	0004
2SB299GN	0004
2SB299H	0004
2SB299J	0004
2SB299K	0004
2SB299L	0004
2SB299M	0004
2SB299OR	0004
2SB299R	0004
2SB299X	0004
2SB299Y	0004
2SB300	0975
2SB300A	0975
2SB300B	0975
2SB300C	0975
2SB300D	0975
2SB300E	0975
2SB300F	0975
2SB300G	0975
2SB300GN	0975
2SB300H	0975
2SB300J	0975
2SB300K	0975
2SB300L	0975
2SB300M	0975
2SB300OR	0975
2SB300R	0975
2SB300X	0975
2SB300Y	0975
2SB301	0975
2SB301A	0975
2SB301B	0975
2SB301C	0975
2SB301D	0975
2SB301E	0975
2SB301F	0975
2SB301G	0975
2SB301GN	0975
2SB301H	0975
2SB301J	0975
2SB301L	0975
2SB301M	0975
2SB301OR	0975
2SB301R	0975
2SB301X	0975
2SB301Y	0975
2SB302	0004
2SB302A	0004
2SB302B	0004
2SB302C	0004
2SB302D	0004
2SB302E	0004
2SB302F	0004
2SB302G	0004
2SB302J	0004
2SB302K	0004
2SB302L	0004
2SB302M	0004
2SB302OR	0004
2SB302R	0004
2SB302X	0004
2SB302Y	0004
2SB303	0004
2SB303(0)	0004
2SB303-0	0004
2SB303A	0004
2SB303B	0004
2SB303BK	0004
2SB303C	0004
2SB303D	0004
2SB303E	0004
2SB303F	0004
2SB303G	0004
2SB303GN	0004
2SB303H	0004
2SB303J	0004
2SB303K	0004
2SB303L	0004
2SB303M	0004
2SB303OR	0004
2SB303R	0004
2SB303X	0004
2SB303Y	0004
2SB304	0164
2SB304A	0164
2SB304B	0164
2SB304C	0164
2SB304D	0164
2SB304E	0164
2SB304F	0164
2SB304G	0164
2SB304GN	0164
2SB304H	0164
2SB304J	0164
2SB304K	0164
2SB304L	0164
2SB304M	0164
2SB304OR	0164
2SB304R	0164
2SB304X	0164
2SB304Y	0164
2SB309	0975
2SB309A	0975
2SB309B	0975
2SB309C	0975
2SB309D	0975
2SB309E	0975
2SB309F	0975
2SB309G	0975
2SB309GN	0975
2SB309H	0975
2SB309K	0975
2SB309L	0975
2SB309M	0975
2SB309OR	0975
2SB309R	0975
2SB309X	0975
2SB309Y	0975
2SB310	0975
2SB310A	0975
2SB310B	0975
2SB310C	0975
2SB310D	0975
2SB310E	0975
2SB310F	0975
2SB310G	0975
2SB310GN	0975
2SB310H	0975
2SB310J	0975
2SB310K	0975
2SB310L	0975
2SB310M	0975
2SB310OR	0975
2SB310R	0975
2SB310X	0975
2SB310Y	0975
2SB311	0975
2SB311A	0975
2SB311B	0975
2SB311C	0975
2SB311D	0975
2SB311E	0975
2SB311F	0975
2SB311G	0975
2SB311GN	0975
2SB311H	0975
2SB311J	0975
2SB311K	0975
2SB311L	0975
2SB311M	0975
2SB311OR	0975
2SB311R	0975
2SB311X	0975
2SB311Y	0975
2SB312	0975
2SB312A	0975
2SB312B	0975
2SB312C	0975
2SB312D	0975
2SB312E	0975
2SB312F	0975
2SB312G	0975
2SB312GN	0975
2SB312H	0975
2SB312J	0975
2SB312K	0975
2SB312L	0975
2SB312M	0975
2SB312X	0975
2SB312Y	0975
2SB313	0975
2SB313A	0975
2SB313B	0975
2SB313C	0975
2SB313D	0975
2SB313E	0975
2SB313F	0975
2SB313G	0975
2SB313GN	0975
2SB313H	0975
2SB313J	0975
2SB313K	0975
2SB313L	0975
2SB313M	0975
2SB313OR	0975
2SB313R	0975
2SB313X	0975
2SB313Y	0975
2SB314	0280
2SB314A	0280
2SB314B	0280
2SB314C	0280
2SB314D	0280
2SB314E	0280
2SB314F	0280
2SB314GN	0280
2SB314H	0280
2SB314J	0280
2SB314K	0280
2SB314L	0280
2SB314M	0280
2SB314OR	0280
2SB314R	0280
2SB314X	0280
2SB314Y	0280
2SB315	0004
2SB315A	0004

If replacement code is OEM, contact original manufacturer for replacement.

DEVICE TYPE	REPL CODE
2SB315B	0004
2SB315C	0004
2SB315D	0004
2SB315E	0004
2SB315F	0004
2SB315G	0004
2SB315GN	0004
2SB315H	0004
2SB315J	0004
2SB315L	0004
2SB315M	0004
2SB315OR	0004
2SB315R	0004
2SB315X	0004
2SB315Y	0004
2SB316	0004
2SB316A	0004
2SB316B	0004
2SB316C	0004
2SB316D	0004
2SB316E	0004
2SB316F	0004
2SB316G	0004
2SB316GN	0004
2SB316H	0004
2SB316OR	0004
2SB316R	0004
2SB316X	0004
2SB316Y	0004
2SB317	0004
2SB318	0975
2SB318A	0975
2SB318B	0975
2SB318C	0975
2SB318D	0975
2SB318E	0975
2SB318F	0975
2SB318G	0975
2SB318GN	0975
2SB318H	0975
2SB318J	0975
2SB318K	0975
2SB318L	0975
2SB318M	0975
2SB318OR	0975
2SB318R	0975
2SB318X	0975
2SB318Y	0975
2SB319	0975
2SB319A	0975
2SB319B	0975
2SB319C	0975
2SB319D	0975
2SB319E	0975
2SB319F	0975
2SB319G	0975
2SB319GN	0975
2SB319H	0975
2SB319J	0975
2SB319K	0975
2SB319L	0975
2SB319M	0975
2SB319OR	0975
2SB319R	0975
2SB319X	0975
2SB319Y	0975
2SB320	0975
2SB320A	0975
2SB320B	0975
2SB320C	0975
2SB320D	0975
2SB320E	0975
2SB320F	0975
2SB320G	0975
2SB320GN	0975
2SB320H	0975
2SB320J	0975
2SB320K	0975
2SB320L	0975
2SB320M	0975
2SB320OR	0975
2SB320R	0975
2SB320X	0975
2SB320Y	0975
2SB321	0004
2SB322	0004
2SB323	0004
2SB324	1062
2SB324(E)	1062
2SB324(F)	1062
2SB324(I)	1062
2SB324(K)	1062
2SB324(L)	1062
2SB324(N)	1062
2SB324(V)	1062
2SB324-OR	1062
2SB324,K	1062
2SB324A	1062
2SB324B	1062
2SB324C	1062
2SB324D	1062
2SB324E	1062
2SB324E-1	1062
2SB324E-L	1062
2SB324F	1062
2SB324G	1062
2SB324GN	1062
2SB324H	1062
2SB324I	1062

DEVICE TYPE	REPL CODE
2SB324J	1062
2SB324K	1062
2SB324L	1062
2SB324M	1062
2SB324N	1062
2SB324OR	1062
2SB324P	1062
2SB324R	1062
2SB324S	1062
2SB324V	1062
2SB324X	1062
2SB324Y	1062
2SB326	0004
2SB326A	0004
2SB326B	0004
2SB326C	0004
2SB326D	0004
2SB326E	0004
2SB326F	0004
2SB326G	0004
2SB326GN	0004
2SB326H	0004
2SB326J	0004
2SB326K	0004
2SB326L	0004
2SB326M	0004
2SB326OR	0004
2SB326R	0004
2SB326X	0004
2SB326Y	0004
2SB327	0004
2SB327A	0004
2SB327B	0004
2SB327C	0004
2SB327D	0004
2SB327E	0004
2SB327F	0004
2SB327G	0004
2SB327GN	0004
2SB327H	0004
2SB327J	0004
2SB327K	0004
2SB327L	0004
2SB327M	0004
2SB327OR	0004
2SB327R	0004
2SB327X	0004
2SB327Y	0004
2SB328	0211
2SB328A	0211
2SB328B	0211
2SB328C	0211
2SB328D	0211
2SB328E	0211
2SB328F	0211
2SB328G	0211
2SB328GN	0211
2SB328H	0211
2SB328J	0211
2SB328K	0211
2SB328L	0211
2SB328M	0211
2SB328OR	0211
2SB328R	0211
2SB328X	0211
2SB328Y	0211
2SB329	0004
2SB329A	0004
2SB329B	0004
2SB329C	0004
2SB329D	0004
2SB329E	0004
2SB329F	0004
2SB329G	0004
2SB329GN	0004
2SB329H	0004
2SB329J	0004
2SB329K	0004
2SB329L	0004
2SB329M	0004
2SB329OR	0004
2SB329R	0004
2SB329X	0004
2SB329Y	0004
2SB331	0040
2SB331A	0040
2SB331B	0040
2SB331C	0040
2SB331D	0040
2SB331E	0040
2SB331F	0040
2SB331G	0040
2SB331GN	0040
2SB331H	0040
2SB331J	0040
2SB331K	0040
2SB331L	0040
2SB331M	0040
2SB331OR	0040
2SB331R	0040
2SB331X	0040
2SB331Y	0040
2SB332	0040
2SB332A	0040
2SB332B	0040
2SB332C	0040
2SB332D	0040
2SB332E	0040
2SB332F	0040

DEVICE TYPE	REPL CODE
2SB332G	0040
2SB332GN	0040
2SB332H	0040
2SB332J	0040
2SB332K	0040
2SB332L	0040
2SB332M	0040
2SB332OR	0040
2SB332R	0040
2SB332X	0040
2SB332Y	0040
2SB333	0040
2SB333A	0040
2SB333B	0040
2SB333D	0040
2SB333E	0040
2SB333F	0040
2SB333G	0040
2SB333GN	0040
2SB333H	0040
2SB333HA	0040
2SB333HB	0040
2SB333J	0040
2SB333K	0040
2SB333L	0040
2SB333M	0040
2SB333OR	0040
2SB333R	0040
2SB333X	0040
2SB333Y	0040
2SB334	0440
2SB334-OR	0440
2SB334A	0440
2SB334B	0440
2SB334C	0440
2SB334D	0440
2SB334E	0440
2SB334F	0440
2SB334G	0440
2SB334GN	0440
2SB334H	0440
2SB334J	0440
2SB334K	0440
2SB334L	0440
2SB334M	0440
2SB334OR	0440
2SB334R	0440
2SB334X	0440
2SB334Y	0440
2SB335	0211
2SB336	0004
2SB337	0160
2SB337(A)	0160
2SB337(B)	0160
2SB337(C)	0160
2SB337-B	0160
2SB337-OR	0160
2SB337A	0160
2SB337B	0160
2SB337BK	0160
2SB337C	0160
2SB337D	0160
2SB337E	0160
2SB337F	0160
2SB337G	0160
2SB337GN	0160
2SB337H	0160
2SB337J	0160
2SB337K	0160
2SB337L	0160
2SB337LB	0160
2SB337M	0160
2SB337OR	0160
2SB337R	0160
2SB337X	0160
2SB337Y	0160
2SB338	0160
2SB338A	0160
2SB338B	0160
2SB338C	0160
2SB338D	0160
2SB338E	0160
2SB338F	0160
2SB338G	0160
2SB338GN	0160
2SB338H	0160
2SB338HA	0160
2SB338HB	0160
2SB338J	0160
2SB338K	0160
2SB338L	0160
2SB338M	0160
2SB338OR	0160
2SB338R	0160
2SB338X	0160
2SB338Y	0160
2SB339	0604
2SB339H	0604
2SB340	0604
2SB340H	0604
2SB341	0975
2SB341(V)	0975
2SB341A	0975
2SB341B	0975
2SB341C	0975
2SB341D	0975
2SB341E	0975
2SB341F	0975
2SB341G	0975

DEVICE TYPE	REPL CODE
2SB341GN	0975
2SB341H	0975
2SB341J	0975
2SB341K	0975
2SB341L	0975
2SB341M	0975
2SB341OR	0975
2SB341R	0975
2SB341S	0975
2SB341V	0975
2SB341V(S)	0975
2SB341X	0975
2SB341Y	0975
2SB342	0975
2SB342A	0975
2SB342B	0975
2SB342C	0975
2SB342D	0975
2SB342E	0975
2SB342F	0975
2SB342G	0975
2SB342GN	0975
2SB342H	0975
2SB342J	0975
2SB342K	0975
2SB342L	0975
2SB342M	0975
2SB342OR	0975
2SB342R	0975
2SB342X	0975
2SB342Y	0975
2SB343	0975
2SB343A	0975
2SB343B	0975
2SB343C	0975
2SB343D	0975
2SB343E	0975
2SB343F	0975
2SB343G	0975
2SB343GN	0975
2SB343H	0975
2SB343J	0975
2SB343K	0975
2SB343L	0975
2SB343M	0975
2SB343OR	0975
2SB343R	0975
2SB343X	0975
2SB343Y	0975
2SB345	0004
2SB345A	0004
2SB345B	0004
2SB345C	0004
2SB345D	0004
2SB345E	0004
2SB345F	0004
2SB345G	0004
2SB345GN	0004
2SB345H	0004
2SB345J	0004
2SB345K	0004
2SB345L	0004
2SB345M	0004
2SB345OR	0004
2SB345R	0004
2SB345X	0004
2SB345Y	0004
2SB346	0004
2SB346(Q)	0004
2SB346A	0004
2SB346B	0004
2SB346C	0004
2SB346D	0004
2SB346E	0004
2SB346F	0004
2SB346G	0004
2SB346GN	0004
2SB346H	0004
2SB346J	0004
2SB346K	0004
2SB346L	0004
2SB346M	0004
2SB346OR	0004
2SB346Q	0004
2SB346R	0004
2SB346X	0004
2SB346Y	0004
2SB347	0004
2SB347A	0004
2SB347B	0004
2SB347C	0004
2SB347D	0004
2SB347E	0004
2SB347F	0004
2SB347G	0004
2SB347GN	0004
2SB347H	0004
2SB347J	0004
2SB347K	0004
2SB347L	0004
2SB347M	0004
2SB347OR	0004
2SB347R	0004
2SB347X	0004
2SB347Y	0004
2SB348	0004
2SB348(Q)	0004
2SB348A	0004
2SB348B	0004

DEVICE TYPE	REPL CODE
2SB348C	0004
2SB348D	0004
2SB348E	0004
2SB348F	0004
2SB348G	0004
2SB348GN	0004
2SB348H	0004
2SB348J	0004
2SB348K	0004
2SB348L	0004
2SB348M	0004
2SB348OR	0004
2SB348Q	0004
2SB348R	0004
2SB348X	0004
2SB348Y	0004
2SB349	0004
2SB350	0211
2SB350A	0211
2SB350B	0211
2SB350C	0211
2SB350D	0211
2SB350E	0211
2SB350F	0211
2SB350G	0211
2SB350GN	0211
2SB350H	0211
2SB350J	0211
2SB350K	0211
2SB350L	0211
2SB350M	0211
2SB350OR	0211
2SB350R	0211
2SB350X	0211
2SB350Y	0211
2SB351	0440
2SB351A	0440
2SB351B	0440
2SB351C	0440
2SB351D	0440
2SB351E	0440
2SB351F	0440
2SB351G	0440
2SB351GN	0440
2SB351H	0440
2SB351J	0440
2SB351K	0440
2SB351L	0440
2SB351M	0440
2SB351OR	0440
2SB351R	0440
2SB351X	0440
2SB351Y	0440
2SB352	0040
2SB352A	0040
2SB352B	0040
2SB352C	0040
2SB352D	0040
2SB352E	0040
2SB352F	0040
2SB352G	0040
2SB352GN	0040
2SB352H	0040
2SB352J	0040
2SB352K	0040
2SB352L	0040
2SB352M	0040
2SB352OR	0040
2SB352R	0040
2SB352X	0040
2SB352Y	0040
2SB353	0040
2SB353A	0040
2SB353B	0040
2SB353C	0040
2SB353D	0040
2SB353E	0040
2SB353F	0040
2SB353G	0040
2SB353GN	0040
2SB353H	0040
2SB353J	0040
2SB353K	0040
2SB353L	0040
2SB353M	0040
2SB353OR	0040
2SB353R	0040
2SB353X	0040
2SB353Y	0040
2SB354	0437
2SB354A	0437
2SB354B	0437
2SB354C	0437
2SB354D	0437
2SB354E	0437
2SB354F	0437
2SB354G	0437
2SB354GN	0437
2SB354H	0437
2SB354J	0437
2SB354K	0437
2SB354L	0437
2SB354M	0437
2SB354OR	0437
2SB354R	0437
2SB354X	0437
2SB354Y	0437
2SB355	0085
2SB355A	0085

DEVICE TYPE	REPL CODE
2SB355B	0085
2SB355C	0085
2SB355D	0085
2SB355E	0085
2SB355F	0085
2SB355G	0085
2SB355GN	0085
2SB355H	0085
2SB355J	0085
2SB355K	0085
2SB355L	0085
2SB355M	0085
2SB355OR	0085
2SB355R	0085
2SB355X	0085
2SB355Y	0085
2SB356	0085
2SB356A	0085
2SB356B	0085
2SB356C	0085
2SB356D	0085
2SB356E	0085
2SB356F	0085
2SB356G	0085
2SB356GN	0085
2SB356H	0085
2SB356J	0085
2SB356K	0085
2SB356L	0085
2SB356M	0085
2SB356OR	0085
2SB356R	0085
2SB356X	0085
2SB356Y	0085
2SB357	0975
2SB357A	0975
2SB357B	0975
2SB357C	0975
2SB357D	0975
2SB357E	0975
2SB357F	0975
2SB357G	0975
2SB357GN	0975
2SB357H	0975
2SB357J	0975
2SB357K	0975
2SB357L	0975
2SB357M	0975
2SB357OR	0975
2SB357R	0975
2SB357X	0975
2SB357Y	0975
2SB358	0975
2SB358A	0975
2SB358B	0975
2SB358C	0975
2SB358D	0975
2SB358E	0975
2SB358F	0975
2SB358G	0975
2SB358GN	0975
2SB358H	0975
2SB358J	0975
2SB358K	0975
2SB358L	0975
2SB358M	0975
2SB358OR	0975
2SB358R	0975
2SB358X	0975
2SB358Y	0975
2SB359	0975
2SB359A	0975
2SB359B	0975
2SB359C	0975
2SB359D	0975
2SB359E	0975
2SB359F	0975
2SB359G	0975
2SB359GN	0975
2SB359H	0975
2SB359J	0975
2SB359K	0975
2SB359L	0975
2SB359M	0975
2SB359OR	0975
2SB359R	0975
2SB359X	0975
2SB359Y	0975
2SB360	0975
2SB360A	0975
2SB360B	0975
2SB360C	0975
2SB360D	0975
2SB360E	0975
2SB360F	0975
2SB360G	0975
2SB360GN	0975
2SB360H	0975
2SB360J	0975
2SB360K	0975
2SB360L	0975
2SB360M	0975
2SB360OR	0975
2SB360R	0975
2SB360X	0975
2SB360Y	0975
2SB361	0975
2SB361A	0975
2SB361B	0975

DEVICE TYPE	REPL CODE
2SB361C	0975
2SB361D	0975
2SB361E	0975
2SB361F	0975
2SB361G	0975
2SB361GN	0975
2SB361H	0975
2SB361J	0975
2SB361K	0975
2SB361L	0975
2SB361M	0975
2SB361OR	0975
2SB361R	0975
2SB361X	0975
2SB361Y	0975
2SB362	0975
2SB362A	0975
2SB362B	0975
2SB362C	0975
2SB362D	0975
2SB362E	0975
2SB362F	0975
2SB362G	0975
2SB362GN	0975
2SB362H	0975
2SB362J	0975
2SB362K	0975
2SB362L	0975
2SB362M	0975
2SB362OR	0975
2SB362R	0975
2SB362X	0975
2SB362Y	0975
2SB363	0854
2SB364	0164
2SB364/8023	0164
2SB364/80236400	0164
2SB364-OR	0164
2SB364A	0164
2SB364B	0164
2SB364C	0164
2SB364D	0164
2SB364E	0164
2SB364F	0164
2SB364G	0164
2SB364GN	0164
2SB364H	0164
2SB364J	0164
2SB364K	0164
2SB364L	0164
2SB364M	0164
2SB364OR	0164
2SB364R	0164
2SB364X	0164
2SB364Y	0164
2SB365	0164
2SB365A	0164
2SB365B	0164
2SB365C	0164
2SB365D	0164
2SB365E	0164
2SB365F	0164
2SB365G	0164
2SB365GN	0164
2SB365H	0164
2SB365J	0164
2SB365K	0164
2SB365L	0164
2SB365M	0164
2SB365OR	0164
2SB365R	0164
2SB365X	0164
2SB365Y	0164
2SB366	0975
2SB366A	0975
2SB366B	0975
2SB366C	0975
2SB366D	0975
2SB366E	0975
2SB366F	0975
2SB366G	0975
2SB366GN	0975
2SB366H	0975
2SB366J	0975
2SB366K	0975
2SB366L	0975
2SB366M	0975
2SB366OR	0975
2SB366R	0975
2SB366X	0975
2SB366Y	0975
2SB367	0222
2SB367(B)	0222
2SB367(B)P	0222
2SB367-4	0222
2SB367-5	0222
2SB367-OR	0222
2SB367A	0222
2SB367AL	0222
2SB367B	0222
2SB367B-2	0222
2SB367BL	0222
2SB367BP	0222
2SB367C	0222
2SB367D	0222
2SB367E	0222
2SB367F	0222
2SB367G	0222
2SB367H	0222

If replacement code is OEM, contact original manufacturer for replacement.

DEVICE TYPE	REPL CODE
2SB367J	0222
2SB367K	0222
2SB367L	0222
2SB367M	0222
2SB367OR	0222
2SB367P	0222
2SB367R	0222
2SB367X	0222
2SB367Y	0222
2SB368	0222
2SB368(B)	0222
2SB368-OR	0222
2SB368A	0222
2SB368B	0222
2SB368C	0222
2SB368D	0222
2SB368E	0222
2SB368F	0222
2SB368G	0222
2SB368GN	0222
2SB368H	0222
2SB368J	0222
2SB368K	0222
2SB368L	0222
2SB368M	0222
2SB368OR	0222
2SB368X	0222
2SB368Y	0222
2SB370	0004
2SB370(A)	0004
2SB370(B)	0004
2SB370/123	0004
2SB370/123877	0004
2SB370-O	0004
2SB370A	0004
2SB370AA	0004
2SB370AB	0004
2SB370AC	0004
2SB370AH	0004
2SB370AHA	0004
2SB370AHB	0004
2SB370B	0004
2SB370C	0004
2SB370D	0004
2SB370E	0004
2SB370F	0004
2SB370G	0004
2SB370GN	0004
2SB370H	0004
2SB370J	0004
2SB370K	0004
2SB370L	0004
2SB370M	0004
2SB370OR	0004
2SB370P	0004
2SB370R	0004
2SB370V	0004
2SB370X	0004
2SB370Y	0004
2SB371	0004
2SB371A	0004
2SB371B	0004
2SB371C	0004
2SB371D	0004
2SB371E	0004
2SB371F	0004
2SB371G	0004
2SB371GN	0004
2SB371H	0004
2SB371J	0004
2SB371K	0004
2SB371L	0004
2SB371M	0004
2SB371OR	0004
2SB371R	0004
2SB371X	0004
2SB371Y	0004
2SB372	0847
2SB373	0847
2SB373A	0847
2SB373B	0847
2SB373C	0847
2SB373D	0847
2SB373E	0847
2SB373F	0847
2SB373G	0847
2SB373GN	0847
2SB373H	0847
2SB373K	0847
2SB373L	0847
2SB373M	0847
2SB373OR	0847
2SB373R	0847
2SB373X	0847
2SB373Y	0847
2SB374	0164
2SB375	1675
2SB375-2B	1675
2SB375-5B	1675
2SB375-OR	1675
2SB375A	1675
2SB375A-2B	1675
2SB375A-5B	1675
2SB375AL	1675
2SB375A-NB	1675
2SB375ATV	1675
2SB375B	1675
2SB375C	1675
2SB375D	1675

DEVICE TYPE	REPL CODE
2SB375E	1675
2SB375F	1675
2SB375G	1675
2SB375GN	1675
2SB375H	1675
2SB375J	1675
2SB375K	1675
2SB375L	1675
2SB375M	1675
2SB375OR	1675
2SB375R	1675
2SB375X	1675
2SB375Y	1675
2SB376	0004
2SB376A	0004
2SB376B	0004
2SB376C	0004
2SB376D	0004
2SB376E	0004
2SB376F	0004
2SB376G	0004
2SB376GN	0004
2SB376J	0004
2SB376K	0004
2SB376L	0004
2SB376M	0004
2SB376OR	0004
2SB376R	0004
2SB376X	0004
2SB376Y	0004
2SB377	0211
2SB377B	0211
2SB378	0004
2SB378A	0004
2SB378B	0004
2SB378C	0004
2SB378D	0004
2SB378E	0004
2SB378F	0004
2SB378GN	0004
2SB378J	0004
2SB378K	0004
2SB378L	0004
2SB378M	0004
2SB378OR	0004
2SB378R	0004
2SB378X	0004
2SB378Y	0004
2SB379	0211
2SB379-2	0211
2SB379A	0211
2SB379B	0211
2SB379C	0211
2SB379D	0211
2SB379E	0211
2SB379F	0211
2SB379G	0211
2SB379GN	0211
2SB379H	0211
2SB379J	0211
2SB379K	0211
2SB379L	0211
2SB379M	0211
2SB379OR	0211
2SB379R	0211
2SB379Y	0211
2SB380	0211
2SB380A	0211
2SB380B	0211
2SB380C	0211
2SB380D	0211
2SB380E	0211
2SB380F	0211
2SB380GN	0211
2SB380J	0211
2SB380K	0211
2SB380L	0211
2SB380OR	0211
2SB380R	0211
2SB380X	0211
2SB380Y	0211
2SB381	0004
2SB381A	0004
2SB381B	0004
2SB381C	0004
2SB381D	0004
2SB381E	0004
2SB381F	0004
2SB381G	0004
2SB381GN	0004
2SB381H	0004
2SB381J	0004
2SB381K	0004
2SB381L	0004
2SB381M	0004
2SB381OR	0004
2SB381R	0004
2SB381X	0004
2SB381Y	0004
2SB382	0004
2SB382A	0004
2SB382B	0004
2SB382BK	0004
2SB382BN	0004
2SB382C	0004
2SB382D	0004
2SB382E	0004
2SB382F	0004
2SB382G	0004

DEVICE TYPE	REPL CODE
2SB382GN	0004
2SB382H	0004
2SB382J	0004
2SB382K	0004
2SB382L	0004
2SB382OR	0004
2SB382R	0004
2SB382X	0004
2SB382Y	0004
2SB383	0004
2SB383(2SB382)	0004
2SB383(2SB495)	0004
2SB383-1	0004
2SB383-2	0004
2SB383A	0004
2SB383B	0004
2SB383C	0004
2SB383D	0004
2SB383E	0004
2SB383F	0004
2SB383G	0004
2SB383GN	0004
2SB383H	0004
2SB383J	0004
2SB383K	0004
2SB383L	0004
2SB383M	0004
2SB383OR	0004
2SB383R	0004
2SB383X	0004
2SB383Y	0004
2SB384	0004
2SB384A	0004
2SB384B	0004
2SB384C	0004
2SB384D	0004
2SB384E	0004
2SB384F	0004
2SB384G	0004
2SB384GN	0004
2SB384H	0004
2SB384J	0004
2SB384K	0004
2SB384L	0004
2SB384M	0004
2SB384OR	0004
2SB384R	0004
2SB384X	0004
2SB384Y	0004
2SB385	0004
2SB385A	0004
2SB385B	0004
2SB385C	0004
2SB385D	0004
2SB385E	0004
2SB385F	0004
2SB385G	0004
2SB385GN	0004
2SB385H	0004
2SB385J	0004
2SB385K	0004
2SB385L	0004
2SB385M	0004
2SB385OR	0004
2SB385R	0004
2SB385X	0004
2SB385Y	0004
2SB386	0004
2SB386A	0004
2SB386B	0004
2SB386C	0004
2SB386D	0004
2SB386E	0004
2SB386F	0004
2SB386G	0004
2SB386GN	0004
2SB386H	0004
2SB386K	0004
2SB386L	0004
2SB386M	0004
2SB386OR	0004
2SB386R	0004
2SB386X	0004
2SB386Y	0004
2SB387	0004
2SB387A	0004
2SB387B	0004
2SB387D	0004
2SB387E	0004
2SB387F	0004
2SB387G	0004
2SB387GN	0004
2SB387H	0004
2SB387J	0004
2SB387K	0004
2SB387L	0004
2SB387M	0004
2SB387OR	0004
2SB387R	0004
2SB387X	0004
2SB387Y	0004
2SB389	0004
2SB389-O	0004
2SB389A	0004
2SB389B	0004
2SB389BK	0004
2SB389C	0004
2SB389D	0004
2SB389E	0004

DEVICE TYPE	REPL CODE
2SB389F	0004
2SB389G	0004
2SB389GN	0004
2SB389H	0004
2SB389J	0004
2SB389K	0004
2SB389L	0004
2SB389M	0004
2SB389OR	0004
2SB389R	0004
2SB389X	0004
2SB389Y	0004
2SB390	0975
2SB390A	0975
2SB390B	0975
2SB390C	0975
2SB390D	0975
2SB390E	0975
2SB390F	0975
2SB390G	0975
2SB390GN	0975
2SB390H	0975
2SB390J	0975
2SB390K	0975
2SB390L	0975
2SB390M	0975
2SB390OR	0975
2SB390R	0975
2SB390X	0975
2SB390Y	0975
2SB391	0160
2SB391A	0160
2SB391B	0160
2SB391C	0160
2SB391D	0160
2SB391E	0160
2SB391F	0160
2SB391G	0160
2SB391GN	0160
2SB391H	0160
2SB391J	0160
2SB391K	0160
2SB391L	0160
2SB391M	0160
2SB391OR	0160
2SB391R	0160
2SB391X	0160
2SB391Y	0160
2SB392	0280
2SB392A	0280
2SB392B	0280
2SB392C	0280
2SB392D	0280
2SB392E	0280
2SB392F	0280
2SB392G	0280
2SB392GN	0280
2SB392H	0280
2SB392J	0280
2SB392K	0280
2SB392L	0280
2SB392M	0280
2SB392OR	0280
2SB392R	0280
2SB392X	0280
2SB392Y	0280
2SB393	0280
2SB393A	0280
2SB393B	0280
2SB393C	0280
2SB393D	0280
2SB393E	0280
2SB393F	0280
2SB393G	0280
2SB393GN	0280
2SB393H	0280
2SB393J	0280
2SB393K	0280
2SB393L	0280
2SB393M	0280
2SB393OR	0280
2SB393R	0280
2SB393X	0280
2SB393Y	0280
2SB394	0280
2SB394A	0280
2SB394B	0280
2SB394C	0280
2SB394D	0280
2SB394E	0280
2SB394F	0280
2SB394G	0280
2SB394GN	0280
2SB394H	0280
2SB394K	0280
2SB394L	0280
2SB394M	0280
2SB394OR	0280
2SB394R	0280
2SB394X	0280
2SB394Y	0280
2SB395	0280
2SB395A	0280
2SB395B	0280
2SB395C	0280
2SB395D	0280
2SB395E	0280
2SB395F	0280
2SB395G	0280

DEVICE TYPE	REPL CODE
2SB395GN	0280
2SB395H	0280
2SB395J	0280
2SB395K	0280
2SB395M	0280
2SB395OR	0280
2SB395R	0280
2SB395X	0280
2SB395Y	0280
2SB396	0211
2SB396A	0211
2SB396B	0211
2SB396C	0211
2SB396D	0211
2SB396E	0211
2SB396F	0211
2SB396G	0211
2SB396GN	0211
2SB396H	0211
2SB396J	0211
2SB396K	0211
2SB396L	0211
2SB396M	0211
2SB396OR	0211
2SB396R	0211
2SB396X	0211
2SB396Y	0211
2SB397	0050
2SB398	0164
2SB400	0004
2SB400(RECT)	0473
2SB400A	0004
2SB400B	0004
2SB400BK	0004
2SB400C	0004
2SB400D	0004
2SB400E	0004
2SB400F	0004
2SB400G	0004
2SB400GN	0004
2SB400H	0004
2SB400J	0004
2SB400K	0004
2SB400L	0004
2SB400M	0004
2SB400OR	0004
2SB400R	0004
2SB400X	0004
2SB400Y	0004
2SB401	0280
2SB401A	0280
2SB401B	0280
2SB401C	0280
2SB401D	0280
2SB401E	0280
2SB401F	0280
2SB401G	0280
2SB401GN	0280
2SB401H	0280
2SB401K	0280
2SB401L	0280
2SB401M	0280
2SB401OR	0280
2SB401R	0280
2SB401X	0280
2SB401Y	0280
2SB402	0280
2SB402A	0280
2SB402B	0280
2SB402C	0280
2SB402D	0280
2SB402E	0280
2SB402F	0280
2SB402G	0280
2SB402GN	0280
2SB402H	0280
2SB402J	0280
2SB402K	0280
2SB402L	0280
2SB402M	0280
2SB402OR	0280
2SB402R	0280
2SB402X	0280
2SB402Y	0280
2SB403	0280
2SB403A	0280
2SB403B	0280
2SB403C	0280
2SB403D	0280
2SB403E	0280
2SB403F	0280
2SB403G	0280
2SB403GN	0280
2SB403H	0280
2SB403J	0280
2SB403K	0280
2SB403L	0280
2SB403M	0280
2SB403OR	0280
2SB403R	0280
2SB403X	0280
2SB403Y	0280
2SB405	0164
2SB405(K)	0164
2SB405-0	0164
2SB405-1	0164
2SB405-2C	0164
2SB405-3C	0164
2SB405-4C	0164

DEVICE TYPE	REPL CODE
2SB405-0	0164
2SB405-OR	0164
2SB405-R	0164
2SB405A	0164
2SB405AG	0164
2SB405B	0164
2SB405C	0164
2SB405D	0164
2SB405DK	0164
2SB405E	0164
2SB405EK	0164
2SB405F	0164
2SB405G	0164
2SB405GN	0164
2SB405H	0164
2SB405J	0164
2SB405K	0164
2SB405L	0164
2SB405M	0164
2SB405OR	0164
2SB405R	0164
2SB405RE	0164
2SB405X	0164
2SB405Y	0164
2SB406	0975
2SB407	0160
2SB407-0	0160
2SB407-0	0160
2SB407-OR	0160
2SB407A	0160
2SB407B	0160
2SB407BK	0160
2SB407C	0160
2SB407D	0160
2SB407E	0160
2SB407F	0160
2SB407G	0160
2SB407GN	0160
2SB407H	0160
2SB407J	0160
2SB407K	0160
2SB407M	0160
2SB407OR	0160
2SB407R	0160
2SB407TV	0160
2SB407TV-2	0160
2SB407X	0160
2SB407Y	0160
2SB408	0004
2SB408A	0004
2SB408B	0004
2SB408C	0004
2SB408D	0004
2SB408E	0004
2SB408F	0004
2SB408G	0004
2SB408GN	0004
2SB408H	0004
2SB408J	0004
2SB408K	0004
2SB408L	0004
2SB408M	0004
2SB408OR	0004
2SB408R	0004
2SB408X	0004
2SB408Y	0004
2SB409	0280
2SB410	0975
2SB411	0975
2SB411TV	0975
2SB412	0975
2SB413	0222
2SB413A	0222
2SB413B	0222
2SB413C	0222
2SB413D	0222
2SB413E	0222
2SB413F	0222
2SB413G	0222
2SB413GN	0222
2SB413H	0222
2SB413J	0222
2SB413K	0222
2SB413L	0222
2SB413M	0222
2SB413OR	0222
2SB413R	0222
2SB413X	0222
2SB413Y	0222
2SB414	1010
2SB414B	1010
2SB414C	1010
2SB414D	1010
2SB414E	1010
2SB414F	1010
2SB414G	1010
2SB414GN	1010
2SB414H	1010
2SB414J	1010
2SB414K	1010
2SB414L	1010
2SB414M	1010
2SB414OR	1010
2SB414R	1010
2SB414X	1010
2SB414Y	1010
2SB415	0164
2SB415(1)	0164
2SB415-OR	0164

DEVICE TYPE	REPL CODE
2SB415A	0164
2SB415B	0164
2SB415C	0164
2SB415D	0164
2SB415E	0164
2SB415F	0164
2SB415G	0164
2SB415GN	0164
2SB415H	0164
2SB415J	0164
2SB415K	0164
2SB415L	0164
2SB415M	0164
2SB415OR	0164
2SB415R	0164
2SB415X	0164
2SB415Y	0164
2SB416	0280
2SB416A	0280
2SB416B	0280
2SB416C	0280
2SB416D	0280
2SB416E	0280
2SB416F	0280
2SB416G	0280
2SB416GN	0280
2SB416H	0280
2SB416J	0280
2SB416K	0280
2SB416L	0280
2SB416M	0280
2SB416OR	0280
2SB416R	0280
2SB416X	0280
2SB416Y	0280
2SB417	0280
2SB417A	0280
2SB417B	0280
2SB417C	0280
2SB417D	0280
2SB417E	0280
2SB417F	0280
2SB417G	0280
2SB417GN	0280
2SB417H	0280
2SB417J	0280
2SB417K	0280
2SB417L	0280
2SB417M	0280
2SB417OR	0280
2SB417R	0280
2SB417X	0280
2SB417Y	0280
2SB419	0222
2SB421	0004
2SB422	0211
2SB422A	0211
2SB422B	0211
2SB422C	0211
2SB422D	0211
2SB422E	0211
2SB422F	0211
2SB422G	0211
2SB422GN	0211
2SB422H	0211
2SB422J	0211
2SB422K	0211
2SB422L	0211
2SB422M	0211
2SB422OR	0211
2SB422R	0211
2SB422X	0211
2SB422Y	0211
2SB423	0004
2SB423A	0004
2SB423B	0004
2SB423C	0004
2SB423D	0004
2SB423E	0004
2SB423F	0004
2SB423G	0004
2SB423GN	0004
2SB423H	0004
2SB423J	0004
2SB423K	0004
2SB423L	0004
2SB423M	0004
2SB423OR	0004
2SB423R	0004
2SB423X	0004
2SB423Y	0004
2SB424	0604
2SB424A	0604
2SB424B	0604
2SB424C	0604
2SB424D	0604
2SB424E	0604
2SB424F	0604
2SB424G	0604
2SB424GN	0604
2SB424H	0604
2SB424J	0604
2SB424K	0604
2SB424L	0604
2SB424M	0604
2SB424OR	0604
2SB424R	0604
2SB424X	0604
2SB424Y	0604

DEVICE TYPE	REPL CODE
2SB425	0160
2SB425-BL	0160
2SB425-R	0160
2SB425-Y	0160
2SB425A	0160
2SB425B	0160
2SB425C	0160
2SB425D	0160
2SB425E	0160
2SB425F	0160
2SB425G	0160
2SB425G-BL	0160
2SB425G-R	0160
2SB425G-Y	0160
2SB425H	0160
2SB425J	0160
2SB425K	0160
2SB425L	0160
2SB425M	0160
2SB425R	0160
2SB425X	0160
2SB425Y	0160
2SB426	0160
2SB426-BL	0160
2SB426-R	0160
2SB426-Y	0160
2SB426A	0160
2SB426B	0160
2SB426BL	0160
2SB426C	0160
2SB426D	0160
2SB426E	0160
2SB426F	0160
2SB426G	0160
2SB426G-BL	0160
2SB426GN	0160
2SB426G-R	0160
2SB426G-Y	0160
2SB426J	0160
2SB426K	0160
2SB426L	0160
2SB426M	0160
2SB426OR	0160
2SB426R	0160
2SB426X	0160
2SB426Y	0160
2SB427	0004
2SB427A	0004
2SB427B	0004
2SB427C	0004
2SB427D	0004
2SB427E	0004
2SB427F	0004
2SB427G	0004
2SB427GN	0004
2SB427H	0004
2SB427J	0004
2SB427K	0004
2SB427L	0004
2SB427M	0004
2SB427OR	0004
2SB427R	0004
2SB427X	0004
2SB427Y	0004
2SB428	0004
2SB428A	0004
2SB428B	0004
2SB428C	0004
2SB428D	0004
2SB428E	0004
2SB428F	0004
2SB428G	0004
2SB428GN	0004
2SB428H	0004
2SB428J	0004
2SB428K	0004
2SB428L	0004
2SB428M	0004
2SB428OR	0004
2SB428R	0004
2SB428X	0004
2SB428Y	0004
2SB429	0280
2SB430	0040
2SB431	0164
2SB432	0975
2SB432A	0975
2SB432B	0975
2SB432C	0975
2SB432D	0975
2SB432E	0975
2SB432F	0975
2SB432G	0975
2SB432GN	0975
2SB432H	0975
2SB432J	0975
2SB432K	0975
2SB432L	0975
2SB432M	0975
2SB432OR	0975
2SB432R	0975
2SB432X	0975
2SB432Y	0975
2SB433	5432
2SB434	0925
2SB434(O)	0925
2SB434-O	0925
2SB434-ORG	0925
2SB434-R	0925
2SB434-RED	0925
2SB434-Y	0925
2SB434-YEL	0925
2SB434G	0925
2SB434G-O	0925
2SB434G-R	0925
2SB434G-Y	0925
2SB434R	0925
2SB435	0925
2SB435-0	0925
2SB435-O	0925
2SB435-ORG	0925
2SB435-R	0925
2SB435-RED	0925
2SB435-Y	0925
2SB435-YEL	0925
2SB435G-O	0925
2SB435G-R	0925
2SB435G-Y	0925
2SB435O	0925
2SB435R	0925
2SB435RY	0925
2SB435Y	0925
2SB436	0050
2SB437	0050
2SB438	1062
2SB439	0004
2SB439/802	0004
2SB439/120144	0004
2SB439/80243900	0004
2SB439A	0004
2SB439B	0004
2SB439C	0004
2SB439D	0004
2SB439E	0004
2SB439F	0004
2SB439G	0004
2SB439GN	0004
2SB439H	0004
2SB439J	0004
2SB439L	0004
2SB439M	0004
2SB439OR	0004
2SB439R	0004
2SB439X	0004
2SB439Y	0004
2SB440	0004
2SB440A	0004
2SB440B	0004
2SB440C	0004
2SB440E	0004
2SB440F	0004
2SB440G	0004
2SB440GN	0004
2SB440H	0004
2SB440K	0004
2SB440L	0004
2SB440M	0004
2SB440OR	0004
2SB440R	0004
2SB440X	0004
2SB440Y	0004
2SB441	0975
2SB442	0975
2SB443	0004
2SB443A	0004
2SB443B	0004
2SB443C	0004
2SB443D	0004
2SB443E	0004
2SB443F	0004
2SB443G	0004
2SB443GN	0004
2SB443H	0004
2SB443J	0004
2SB443K	0004
2SB443L	0004
2SB443M	0004
2SB443OR	0004
2SB443R	0004
2SB443X	0004
2SB443Y	0004
2SB444	0136
2SB444A	0136
2SB444B	0136
2SB444C	0136
2SB444D	0136
2SB444E	0136
2SB444F	0136
2SB444G	0136
2SB444GN	0136
2SB444H	0136
2SB444K	0136
2SB444L	0136
2SB444M	0136
2SB444OR	0136
2SB444R	0136
2SB444X	0136
2SB444Y	0136
2SB445	1010
2SB445A	1010
2SB445B	1010
2SB445C	1010
2SB445D	1010
2SB445E	1010
2SB445F	1010
2SB445G	1010
2SB445GN	1010
2SB445H	1010
2SB445J	1010
2SB445K	1010
2SB445L	1010
2SB445M	1010
2SB445OR	1010
2SB445R	1010
2SB445X	1010
2SB445Y	1010
2SB446	1010
2SB446A	1010
2SB446B	1010
2SB446C	1010
2SB446D	1010
2SB446E	1010
2SB446F	1010
2SB446G	1010
2SB446GN	1010
2SB446H	1010
2SB446J	1010
2SB446K	1010
2SB446L	1010
2SB446M	1010
2SB446OR	1010
2SB446R	1010
2SB446X	1010
2SB446Y	1010
2SB447	0975
2SB447A	0975
2SB447B	0975
2SB447C	0975
2SB447D	0975
2SB447E	0975
2SB447F	0975
2SB447G	0975
2SB447GN	0975
2SB447H	0975
2SB447J	0975
2SB447K	0975
2SB447L	0975
2SB447OR	0975
2SB447R	0975
2SB447X	0975
2SB447Y	0975
2SB448	0222
2SB448A	0222
2SB448B	0222
2SB448C	0222
2SB448D	0222
2SB448E	0222
2SB448F	0222
2SB448G	0222
2SB448GN	0222
2SB448H	0222
2SB448K	0222
2SB448L	0222
2SB448M	0222
2SB448OR	0222
2SB448R	0222
2SB448X	0222
2SB448Y	0222
2SB449	0160
2SB449(F)	0160
2SB449A	0160
2SB449B	0160
2SB449C	0160
2SB449D	0160
2SB449E	0160
2SB449F	0160
2SB449G	0160
2SB449GN	0160
2SB449H	0160
2SB449J	0160
2SB449K	0160
2SB449L	0160
2SB449M	0160
2SB449OR	0160
2SB449P	0160
2SB449PG	0160
2SB449X	0160
2SB449Y	0160
2SB450	0164
2SB450A	0164
2SB451	0164
2SB452	0164
2SB452A	0164
2SB453	0164
2SB454	0164
2SB455	0164
2SB456	0164
2SB457	0164
2SB457A	0164
2SB457A(C)	0164
2SB457AC	0164
2SB457B	0164
2SB457C	0164
2SB457D	0164
2SB457E	0164
2SB457F	0164
2SB457G	0164
2SB457GN	0164
2SB457H	0164
2SB457J	0164
2SB457K	0164
2SB457L	0164
2SB457M	0164
2SB457OR	0164
2SB457R	0164
2SB457X	0164
2SB457Y	0164
2SB458	0222
2SB458A	0222
2SB458B	0222
2SB458BC	0222
2SB458BL	0222
2SB458C	0222
2SB458D	0222
2SB458E	0222
2SB458F	0222
2SB458G	0222
2SB458GN	0222
2SB458H	0222
2SB458J	0222
2SB458K	0222
2SB458L	0222
2SB458M	0222
2SB458OR	0222
2SB458R	0222
2SB458X	0222
2SB458Y	0222
2SB459	0004
2SB459(C)	0004
2SB459-O	0004
2SB459A	0004
2SB459B	0004
2SB459C	0004
2SB459C-2	0004
2SB459D	0004
2SB459E	0004
2SB459F	0004
2SB459G	0004
2SB459GN	0004
2SB459H	0004
2SB459J	0004
2SB459K	0004
2SB459L	0004
2SB459M	0004
2SB459OR	0004
2SB459R	0004
2SB459X	0004
2SB459Y	0004
2SB460	0004
2SB460A	0004
2SB460B	0004
2SB460C	0004
2SB460D	0004
2SB460E	0004
2SB460F	0004
2SB460G	0004
2SB460GN	0004
2SB460H	0004
2SB460J	0004
2SB460K	0004
2SB460L	0004
2SB460M	0004
2SB460OR	0004
2SB460R	0004
2SB460X	0004
2SB460Y	0004
2SB461	0847
2SB461A	0847
2SB461B	0847
2SB461BL	0847
2SB461C	0847
2SB461D	0847
2SB461E	0847
2SB461F	0847
2SB461G	0847
2SB461H	0847
2SB461J	0847
2SB461K	0847
2SB461L	0847
2SB461M	0847
2SB461OR	0847
2SB461R	0847
2SB461X	0847
2SB461Y	0847
2SB462	0222
2SB462-BL	0222
2SB462-BLU	0222
2SB462-BLU-G	0222
2SB462-R	0222
2SB462-RED	0222
2SB462-RED-G	0222
2SB462-Y	0222
2SB462-YEL	0222
2SB462-YEL-G	0222
2SB462A	0222
2SB462B	0222
2SB462C	0222
2SB462D	0222
2SB462E	0222
2SB462F	0222
2SB462G	0222
2SB462G-BL	0222
2SB462GN	0222
2SB462G-R	0222
2SB462G-Y	0222
2SB462H	0222
2SB462K	0222
2SB462L	0222
2SB462M	0222
2SB462OR	0222
2SB462R	0222
2SB462X	0222
2SB462Y	0222
2SB463	0222
2SB463(1)	0222
2SB463(Y)	0222
2SB463-BL	0222
2SB463-BLU	0222
2SB463-BLU-G	0222
2SB463-O	0222
2SB463-R	0222
2SB463-RED	0222
2SB463-RED-G	0222
2SB463-Y	0222
2SB463-YEL	0222
2SB463-YEL-G	0222
2SB463A	0222
2SB463B	0222
2SB463BL	0222
2SB463BLU	0222
2SB463BLU-G	0222
2SB463C	0222
2SB463D	0222
2SB463E	0222
2SB463F	0222
2SB463G	0222
2SB463G-BL	0222
2SB463GN	0222
2SB463G-R	0222
2SB463G-Y	0222
2SB463H	0222
2SB463J	0222
2SB463K	0222
2SB463L	0222
2SB463M	0222
2SB463OR	0222
2SB463R	5983
2SB463RED	0222
2SB463RED-G	0222
2SB463X	0222
2SB463XL	0222
2SB463Y	0222
2SB463YEL	0222
2SB463YEL-G	0222
2SB464	0975
2SB464A	0975
2SB464B	0975
2SB464C	0975
2SB464D	0975
2SB464E	0975
2SB464F	0975
2SB464G	0975
2SB464GN	0975
2SB464H	0975
2SB464J	0975
2SB464K	0975
2SB464M	0975
2SB464OR	0975
2SB464X	0975
2SB464Y	0975
2SB465	0975
2SB465A	0975
2SB465B	0975
2SB465C	0975
2SB465D	0975
2SB465E	0975
2SB465F	0975
2SB465G	0975
2SB465GN	0975
2SB465H	0975
2SB465J	0975
2SB465K	0975
2SB465L	0975
2SB465M	0975
2SB465R	0975
2SB465X	0975
2SB465Y	0975
2SB466	1010
2SB466A	1010
2SB466B	1010
2SB466C	1010
2SB466D	1010
2SB466E	1010
2SB466F	1010
2SB466G	1010
2SB466GN	1010
2SB466H	1010
2SB466J	1010
2SB466K	1010
2SB466L	1010
2SB466M	1010
2SB466OR	1010
2SB466R	1010
2SB466X	1010
2SB466Y	1010
2SB467	1010
2SB467A	1010
2SB467C	1010
2SB467D	1010
2SB467E	1010
2SB467F	1010
2SB467G	1010
2SB467GN	1010
2SB467H	1010
2SB467J	1010
2SB467K	1010
2SB467L	1010
2SB467M	1010
2SB467OR	1010
2SB467X	1010
2SB467Y	1010
2SB468	0975
2SB468-OR	0975
2SB468A	0975
2SB468B	0975
2SB468B-5	0975
2SB468C	0975
2SB468D	0975
2SB468E	0975
2SB468F	0975
2SB468G	0975
2SB468GN	0975
2SB468H	0975
2SB468J	0975
2SB468K	0975
2SB468L	0975
2SB468OR	0975
2SB468X	0975
2SB468Y	0975
2SB470	0004
2SB470A	0004
2SB470B	0004
2SB470C	0004
2SB470D	0004
2SB470E	0004
2SB470F	0004
2SB470G	0004
2SB470GN	0004
2SB470H	0004
2SB470J	0004
2SB470K	0004
2SB470L	0004
2SB470M	0004
2SB470OR	0004
2SB470R	0004
2SB470X	0004
2SB470Y	0004
2SB471	0160
2SB471-2	0160
2SB471-0	0160
2SB471A	0160
2SB471B	0160
2SB471C	0160
2SB471D	0160
2SB471E	0160
2SB471F	0160
2SB471G	0160
2SB471GN	0160
2SB471H	0160
2SB471J	0160
2SB471K	0160
2SB471L	0160
2SB471M	0160
2SB471OR	0160
2SB471R	0160
2SB471X	0160
2SB471Y	0160
2SB472	0604
2SB472(B)	0604
2SB472A	0604
2SB472B	0604
2SB472C	0604
2SB472D	0604
2SB472E	0604
2SB472F	0604
2SB472G	0604
2SB472GN	0604
2SB472H	0604
2SB472J	0604
2SB472K	0604
2SB472M	0604
2SB472OR	0604
2SB472X	0604
2SB472Y	0604
2SB473	0222
2SB473(H)	0222
2SB473A	0222
2SB473B	0222
2SB473C	0222
2SB473D	0222
2SB473E	0222
2SB473F	0222
2SB473G	0222
2SB473GN	0222
2SB473H	0222
2SB473K	0222
2SB473L	0222
2SB473M	0222
2SB473X	0222
2SB473Y	0222
2SB474	1010
2SB474(C)	1010
2SB474(MP)	1862
2SB474-2	1010
2SB474-3	1010
2SB474-4	1010
2SB474-6D	1010
2SB474-OR	1010
2SB474A	1010
2SB474B	1010
2SB474C	1010
2SB474D	1010
2SB474E	1010
2SB474F	1010
2SB474G	1010
2SB474GN	1010
2SB474H	1010
2SB474J	1010
2SB474K	1010
2SB474L	1010
2SB474M	1010
2SB474MP	1010
2SB474OR	1010
2SB474R	1010
2SB474S	1010
2SB474V4	1010
2SB474V10	1010
2SB474X	1010
2SB474Y	1010
2SB474YE1	1010
2SB474YEL	1010
2SB475	0004
2SB475A	0004
2SB475B	0004
2SB475C	0004
2SB475D	0004
2SB475E	0004
2SB475F	0004
2SB475G	0004
2SB475GN	0004
2SB475H	0004
2SB475J	0004
2SB475K	0004
2SB475L	0004
2SB475M	0004
2SB475OR	0004
2SB475P	0004
2SB475PL	0004
2SB475Q	0004
2SB475R	0004
2SB475X	0004
2SB475Y	0004
2SB476	0847
2SB476S	0847
2SB476W	0847
2SB477	0440
2SB477(C)	0440
2SB477A	0440
2SB477C	0440
2SB477D	0440
2SB477E	0440
2SB477F	0440
2SB477G	0440
2SB477GN	0440
2SB477H	0440
2SB477J	0440
2SB477K	0440
2SB477L	0440
2SB477M	0440
2SB477OR	0440
2SB477R	0440
2SB477X	0440
2SB477Y	0440
2SB478	0437
2SB478BS	0437
2SB479	0437
2SB480	0437
2SB481	0222
2SB481-OR	0222
2SB481A	0222
2SB481B	0222
2SB481C	0222
2SB481D	0222
2SB481E	0222
2SB481F	0222
2SB481G	0222
2SB481GN	0222
2SB481H	0222
2SB481J	0222
2SB481K	0222
2SB481L	0222
2SB481M	0222
2SB481OR	0222
2SB481R	0222
2SB481X	0222
2SB482	0004
2SB482A	0004
2SB482B	0004
2SB482C	0004
2SB482D	0004
2SB482E	0004
2SB482F	0004
2SB482G	0004
2SB482GN	0004
2SB482H	0004
2SB482J	0004
2SB482K	0004
2SB482L	0004
2SB482M	0004
2SB482V	0004
2SB482X	0004
2SB482Y	0004
2SB483	0604
2SB484	0604
2SB485	0604
2SB486	0004
2SB486A	0004
2SB486B	0004
2SB486C	0004
2SB486D	0004
2SB486E	0004
2SB486F	0004
2SB486G	0004
2SB486GN	0004
2SB486H	0004
2SB486J	0004
2SB486K	0004
2SB486L	0004
2SB486M	0004
2SB486OR	0004
2SB486R	0004
2SB486X	0004
2SB486Y	0004

If replacement code is OEM, contact original manufacturer for replacement.

DEVICE TYPE	REPL CODE
2SB491	OEM
2SB492	0847
2SB492A	0847
2SB492B	0847
2SB492C	0847
2SB492D	0847
2SB492E	0847
2SB492F	0847
2SB492G	0847
2SB492GN	0847
2SB492H	0847
2SB492J	0847
2SB492K	0847
2SB492L	0847
2SB492M	0847
2SB492OR	0847
2SB492R	0847
2SB492X	0847
2SB492Y	0847
2SB493	0164
2SB493W	0164
2SB494	0164
2SB495	0164
2SB495A	0164
2SB495B	0164
2SB495C	0164
2SB495D	0164
2SB495E	0164
2SB495F	0164
2SB495G	0164
2SB495GN	0164
2SB495H	0164
2SB495J	0164
2SB495K	0164
2SB495L	0164
2SB495M	0164
2SB495OR	0164
2SB495R	0164
2SB495T	0164
2SB495X	0164
2SB495Y	0164
2SB496	0004
2SB496A	0004
2SB496B	0004
2SB496C	0004
2SB496D	0004
2SB496E	0004
2SB496F	0004
2SB496G	0004
2SB496GN	0004
2SB496H	0004
2SB496J	0004
2SB496K	0004
2SB496L	0004
2SB496M	0004
2SB496OR	0004
2SB496R	0004
2SB496X	0004
2SB496Y	0004
2SB497	0004
2SB497A	0004
2SB497B	0004
2SB497C	0004
2SB497D	0004
2SB497E	0004
2SB497F	0004
2SB497G	0004
2SB497GN	0004
2SB497H	0004
2SB497J	0004
2SB497K	0004
2SB497L	0004
2SB497M	0004
2SB497OR	0004
2SB497R	0004
2SB497X	0004
2SB497Y	0004
2SB498	0004
2SB498A	0004
2SB498B	0004
2SB498C	0004
2SB498D	0004
2SB498E	0004
2SB498F	0004
2SB498G	0004
2SB498GN	0004
2SB498H	0004
2SB498J	0004
2SB498K	0004
2SB498L	0004
2SB498M	0004
2SB498OR	0004
2SB498R	0004
2SB498X	0004
2SB498Y	0004
2SB502	0905
2SB502-0	0905
2SB502-R	0905
2SB502-Y	0905
2SB502A	0905
2SB502A-0	0905
2SB502A-R	0905
2SB502A-Y	0905
2SB503	0854
2SB503-0	0854
2SB503-R	0854
2SB503-Y	0854
2SB503A	0854
2SB503A-0	0854

DEVICE TYPE	REPL CODE
2SB503A-R	0854
2SB503A-Y	0854
2SB504	0126
2SB504A	0126
2SB505	0681
2SB506	3500
2SB506A	3500
2SB507	0681
2SB507D	3752
2SB507E	0681
2SB508	0925
2SB509	0925
2SB510	0126
2SB510S	0126
2SB511	0925
2SB511-1	0925
2SB511C	0925
2SB511D	0925
2SB511E	0925
2SB512	0925
2SB512-0	0925
2SB512-0	0925
2SB512A	0925
2SB512P	0925
2SB513	0925
2SB513A	0925
2SB513P	0925
2SB513Q	0925
2SB513R	0925
2SB514	0925
2SB514D	3752
2SB514E	3752
2SB515	0925
2SB516	0004
2SB516(C)	0004
2SB516(D)	0004
2SB516C	0004
2SB516CD	0004
2SB516CD(P)	0004
2SB516D	0004
2SB516P	0004
2SB518	1681
2SB519	1681
2SB520	2013
2SB520-1	2013
2SB520-2	2013
2SB521	0854
2SB522	0854
2SB523	0925
2SB524	0925
2SB525	0532
2SB525C	0532
2SB525D	0532
2SB525E	0532
2SB526	0681
2SB526C	0681
2SB527	0681
2SB528	0681
2SB528D	2423
2SB529	0925
2SB530	2013
2SB531	2013
2SB531-0	2013
2SB531-R	2013
2SB531-Y	2013
2SB532	2013
2SB533	0854
2SB533-0	0854
2SB533O	0854
2SB533Y	0854
2SB534	0004
2SB534(A)	0004
2SB535	0164
2SB535G	0164
2SB536	0681
2SB536K	0681
2SB536L	0681
2SB536M	0681
2SB537	0681
2SB537K	0681
2SB537L	0681
2SB538	0604
2SB539	1598
2SB539A	1598
2SB539B	1598
2SB539C	1598
2SB540	0164
2SB541	3500
2SB542	0006
2SB542D	0006
2SB544	0532
2SB544(E)	0532
2SB544(F)	0532
2SB544D	0532
2SB544E	0532
2SB544E-MP	0532
2SB544F	0532
2SB544P1	0532
2SB544P2	0532
2SB546	1648
2SB546A	1648
2SB546AK	1648
2SB546AL	1648
2SB546AM	1648
2SB546C	1648
2SB546E	1648
2SB546I	1648
2SB546K	1648
2SB546L	1648

DEVICE TYPE	REPL CODE
2SB546M	1648
2SB547	1648
2SB547A	1648
2SB548	0525
2SB549	0525
2SB549P	0525
2SB549Q	0525
2SB549R	0525
2SB550	3500
2SB551	0905
2SB551H	0905
2SB552	1598
2SB553	0718
2SB553Y	0718
2SB554	1598
2SB554-0	1598
2SB554R	1598
2SB555	2013
2SB555-0	2013
2SB555-R	2013
2SB555R	2013
2SB556	2013
2SB556-0	2013
2SB556-R	2013
2SB557	2013
2SB557-0	2013
2SB557-R	2013
2SB558	2013
2SB558-0	2013
2SB558-R	2013
2SB559	0460
2SB560	0477
2SB560(E)	0477
2SB560(F)	0477
2SB560-0-MP	OEM
2SB560-F-MP-AE	0477
2SB560E	6873
2SB560F	6873
2SB561	0436
2SB561B	0436
2SB561C	0436
2SB562	0532
2SB562(C)	0532
2SB562-C	0532
2SB562A	3597
2SB562B	0532
2SB562BC	OEM
2SB562C	0532
2SB562D	0532
2SB563	0905
2SB564	0532
2SB564(1)K	0532
2SB564(1)L	0532
2SB564(L)	0532
2SB564(M)	0532
2SB564-LK-T	0532
2SB564A	3597
2SB564K	0532
2SB564L	0532
2SB564M	0532
2SB564Q	0532
2SB564QRS	0532
2SB564R	0532
2SB564S	0532
2SB565	0164
2SB565A	0164
2SB565B	0164
2SB565C	0164
2SB565D	0164
2SB565E	0164
2SB565F	0164
2SB565G	0164
2SB565GN	0164
2SB565H	0164
2SB565J	0164
2SB565K	0164
2SB565L	0164
2SB565M	0164
2SB565OR	0164
2SB565R	0164
2SB565X	0164
2SB565Y	0164
2SB566	0681
2SB566A	0681
2SB566AKB	0681
2SB566AKC	0681
2SB566B	0681
2SB566C	0681
2SB566D	0681
2SB566K	0681
2SB566KB	0681
2SB566KC	0681
2SB567	3543
2SB568	1911
2SB568-C	1911
2SB568B	1911
2SB568C	1911
2SB568D	1911
2SB569	0460
2SB570	0460
2SB571	0460
2SB572	1051
2SB573	1051
2SB574	1051
2SB575	0460
2SB576	0460
2SB577	0460
2SB578	1199
2SB579	1199

DEVICE TYPE	REPL CODE
2SB580	1199
2SB581	1199
2SB582	1199
2SB583	1199
2SB584	3530
2SB585	2277
2SB586	2277
2SB587	2432
2SB588	2432
2SB589	2432
2SB595	1369
2SB595-0	1369
2SB595-Y	1369
2SB595Y	1369
2SB596	0854
2SB596-0	0854
2SB596-0	0854
2SB596R	0854
2SB596Y	0854
2SB598	3812
2SB598(E)	3812
2SB598(F)	3812
2SB598-F-NP-AA	3812
2SB598E	3812
2SB598F	3812
2SB598F-NP	OEM
2SB598NP	OEM
2SB599	0681
2SB600	3500
2SB600(RECT)	0446
2SB601	0602
2SB601K	0602
2SB602	OEM
2SB603	OEM
2SB604	0854
2SB605	0532
2SB605K	0532
2SB606	0342
2SB607	0164
2SB608	0436
2SB609	0681
2SB611	2013
2SB611A	2013
2SB612	1598
2SB612A	1598
2SB613	3606
2SB616	3571
2SB616A	3571
2SB617	3571
2SB617A	3571
2SB618	3571
2SB618A	3571
2SB619	0164
2SB620	0150
2SB621	0532
2SB621A	0532
2SB621ARSTA	OEM
2SB621C	0532
2SB621N	0532
2SB621NC	0532
2SB621Q	0532
2SB621R	0532
2SB622	0439
2SB624	1752
2SB624-BV4	1752
2SB624-BV345	OEM
2SB624BV1	1752
2SB624BV2	1752
2SB624BV3	1752
2SB624BV4	1752
2SB624BV5	1752
2SB625	OEM
2SB626	1681
2SB627	3298
2SB628	1648
2SB628A	1648
2SB630	5467
2SB631	0525
2SB631D	2555
2SB631E	2555
2SB631F	2555
2SB631K	0525
2SB631KE	0525
2SB631KF	2555
2SB632	0460
2SB632D	2555
2SB632E	0460
2SB632F	2555
2SB632K	0460
2SB632K(E)	0460
2SB632KE	0460
2SB632KF	2555
2SB632K-SA-E	0460
2SB633	5468
2SB633E	5468
2SB633P	5468
2SB634	3500
2SB634-Y	3500
2SB634Y	3500
2SB635	0004
2SB636	0004
2SB637	0648
2SB637KC	0648
2SB637KD	0648
2SB637KE	0648
2SB638	2432
2SB638H	2432
2SB639	2432
2SB639H	2432

DEVICE TYPE	REPL CODE
2SB640	0037
2SB641	0359
2SB641(P)	0359
2SB641(Q)	0359
2SB641(R)	0359
2SB641(S)	0359
2SB641(T)	0359
2SB641-P	0359
2SB641-Q	0359
2SB641-QR	0359
2SB641-R	0359
2SB641-S	0359
2SB641B	0359
2SB641P	0359
2SB641Q	0359
2SB641QR	0359
2SB641QRS	0359
2SB641R	0359
2SB641RS	0359
2SB641S	0359
2SB641T	0359
2SB642	0825
2SB642(Q)	0825
2SB642(R)	0825
2SB642(S)	0825
2SB642(T)	0825
2SB642-0	0825
2SB642-P	0825
2SB642-PQR	0825
2SB642-Q	0825
2SB642-QR	0825
2SB642-R	0825
2SB642-S	0825
2SB642A	0825
2SB642P	0825
2SB642PQR	0825
2SB642Q	0825
2SB642QR	0825
2SB642QRS	0825
2SB642R	0825
2SB642RS	0825
2SB642RST	0825
2SB642S	0825
2SB643	0825
2SB643(Q)	0825
2SB643(R)	0825
2SB643(S)	0825
2SB643-0	0825
2SB643-P	0825
2SB643-PQR	0825
2SB643-Q	0825
2SB643-QRS	OEM
2SB643-R	0825
2SB643-S	0825
2SB643P	0825
2SB643PQR	0825
2SB643Q	0825
2SB643QR	0825
2SB643QRS	0825
2SB643R	0825
2SB643RS	0825
2SB643S	0825
2SB644	0825
2SB644(C)	0825
2SB644(Q)	0825
2SB644(R)	0825
2SB644(S)	0825
2SB644-Q	0825
2SB644-R	0825
2SB644Q	0825
2SB644QR	0825
2SB644R	0825
2SB644S	0825
2SB645	1598
2SB645E	1598
2SB646	0477
2SB646A	0477
2SB646AB	0477
2SB646B	0477
2SB646C	0477
2SB646D	0477
2SB647	0477
2SB647A	0477
2SB647AB	0477
2SB647AC	0477
2SB647B	0477
2SB647C	0477
2SB647D	0477
2SB648	0525
2SB648A	0525
2SB648B	0525
2SB648C	0525
2SB649	0525
2SB649A	0525
2SB649AB	0525
2SB649AC	0525
2SB649AD	0525
2SB649B	0525
2SB649C	0525
2SB649D	0525
2SB650	2432
2SB650H	2432
2SB653	3500
2SB653A	3500
2SB654	3500
2SB654A	3500
2SB655	1598
2SB655A	1598
2SB656	1598

DEVICE TYPE	REPL CODE
2SB656A	1598
2SB668	0602
2SB668A	0602
2SB669	0602
2SB669A	0602
2SB669Q	0602
2SB669R	0602
2SB673	0602
2SB674	0602
2SB675	0602
2SB676	0681
2SB677	0925
2SB678	OEM
2SB679	0602
2SB681	1598
2SB682	0681
2SB683	0854
2SB683C	0854
2SB683D	0854
2SB684	0532
2SB685	0073
2SB686	0859
2SB686R	OEM
2SB688	3111
2SB688-0	3111
2SB688-0(BS)	3111
2SB688R	3111
2SB689	0681
2SB690	0681
2SB691	3111
2SB692	3111
2SB693	3526
2SB693H	3526
2SB694	OEM
2SB694H	OEM
2SB695	5477
2SB696	3500
2SB696K	3500
2SB697	1598
2SB697K	1598
2SB698	0006
2SB698(F)	0006
2SB698(G)	0006
2SB698-AA	0006
2SB698-T	0006
2SB698D	3108
2SB698E	3108
2SB698F	0006
2SB698G	3108
2SB699	0925
2SB699A-B-B/C	0925
2SB699Q	0925
2SB700	3581
2SB700A	3581
2SB701	3581
2SB702	3581
2SB702A	3581
2SB703	0681
2SB703A	0681
2SB703Q	0681
2SB705	3581
2SB705A	3581
2SB705B	3581
2SB706	OEM
2SB706A	OEM
2SB707	0854
2SB708	0854
2SB709	1742
2SB709(Q)	1742
2SB709(R)	1742
2SB709(S)	1742
2SB709A	1742
2SB709AORB	1742
2SB709A-Q	1742
2SB709AQ	OEM
2SB709AQRB	1742
2SB709AQRS	1742
2SB709AQRW	1742
2SB709A-QW	1742
2SB709AQW	1742
2SB709A-RST	OEM
2SB709AW	1742
2SB709QRSTW	OEM
2SB709QR-TX	1742
2SB709R(AR)	OEM
2SB709S	1742
2SB710	1752
2SB710(Q)	1752
2SB710(R)	1752
2SB710(S)	1752
2SB710A	1752
2SB710AQRS	1752
2SB710AW	1752
2SB711	0602
2SB712	0602
2SB713	5478
2SB714	0847
2SB715	0648
2SB716	0648
2SB716A	0648
2SB716AD	0648
2SB716D	0648
2SB716E	0648
2SB717	5479
2SB717B	5479
2SB717C	5479
2SB717D	5479
2SB718	5479
2SB719	1648

DEVICE TYPE	REPL CODE
2SB720	0436
2SB721	0037
2SB722	1598
2SB723	1598
2SB724	0925
2SB725	0006
2SB726	1603
2SB726(R)	1603
2SB726(S)	1603
2SB726(T)	1603
2SB726Q	1603
2SB726R	1603
2SB726S	1603
2SB726T	1603
2SB727	1968
2SB727K	1968
2SB731	0525
2SB731K	OEM
2SB733	0532
2SB733-34	OEM
2SB733K	0532
2SB733O	0532
2SB733U	0532
2SB734	0477
2SB734-34	0532
2SB734A	0532
2SB734K	0532
2SB734U	0532
2SB736	5454
2SB736A	5454
2SB737	0648
2SB738	0477
2SB738B	0477
2SB738C	0477
2SB739	0532
2SB739B	0532
2SB739C	0532
2SB740	0477
2SB740-3	0477
2SB740B	0477
2SB740C	0477
2SB741	0477
2SB741B	0477
2SB741C	0477
2SB742	0693
2SB743	0525
2SB743Q	0525
2SB744	1613
2SB744A	1613
2SB744Q	1613
2SB744R	1613
2SB744S	1613
2SB745	4139
2SB745(S)	4139
2SB745(T)	4139
2SB745A	4139
2SB745AS	4139
2SB745S	4139
2SB745T	4139
2SB746	0148
2SB747	0854
2SB748	3581
2SB748A	3581
2SB749	3581
2SB749A	3581
2SB750	0602
2SB750A	0602
2SB750B	0602
2SB751	0602
2SB751A	0602
2SB751B	0602
2SB753	1369
2SB754	5478
2SB755	3581
2SB756K	1968
2SB757	5341
2SB758	3571
2SB758A	3571
2SB759	0648
2SB759A	0648
2SB760	0718
2SB760A	0718
2SB760B	0718
2SB761	3165
2SB761(P)	3165
2SB761(Q)	3165
2SB761A	3165
2SB761B	3165
2SB761L	3165
2SB761LB	3165
2SB761LBP	3165
2SB761LBQ	3165
2SB761P	3165
2SB761Q	3165
2SB761QLB	3165
2SB761QP	3165
2SB761R	3165
2SB762	0718
2SB762A	0718
2SB762B	0718
2SB762P	0718
2SB762Q	0718
2SB763	5478
2SB763A	5478
2SB763B	5478
2SB764	0477
2SB764(F)	0477
2SB764E	OEM
2SB764F	6873

DEVICE TYPE	REPL CODE
2SB765	1968
2SB765K	1968
2SB765SK	1968
2SB766	3645
2SB766-QR	OEM
2SB766A	3645
2SB766QR	OEM
2SB767	3645
2SB768	OEM
2SB772	0460
2SB772(P)	0460
2SB772(Q)	0460
2SB772(R)	0460
2SB772-P	0460
2SB772-Q	0460
2SB772P	0460
2SB772Q	0460
2SB772QR	0460
2SB772R	0460
2SB773	3571
2SB773A	3571
2SB774	1613
2SB774(Q)	1613
2SB774(R)	1613
2SB774(RS)-T	1613
2SB774(S)	1613
2SB774Q	1613
2SB774R	1613
2SB774S	1613
2SB774T	1613
2SB775	5478
2SB775E	5477
2SB776	5478
2SB776E	5477
2SB777	3111
2SB778	5477
2SB779	1752
2SB780	0006
2SB781	0681
2SB782	0681
2SB783	0681
2SB786	2897
2SB787	0342
2SB788	0648
2SB788(S)	0648
2SB788(T)	0648
2SB788Q	0648
2SB788S	0648
2SB788T	0648
2SB788U	0648
2SB789	5371
2SB789A	5371
2SB790	0148
2SB791	1968
2SB791K	1968
2SB792	5454
2SB792S	OEM
2SB792T	OEM
2SB793	0477
2SB793(Q)	0477
2SB793(R)	0477
2SB793(S)	0477
2SB793-QRS	0477
2SB793A	0477
2SB793Q	0477
2SB793QRS	0477
2SB793R	0477
2SB793S	0477
2SB794	2897
2SB794A	2897
2SB794K	2897
2SB794L	2897
2SB794M	2897
2SB795	2897
2SB795L	2897
2SB796	3606
2SB798	3645
2SB799	3645
2SB800	3645
2SB800(RECT)	1422
2SB801	OEM
2SB802	OEM
2SB803	OEM
2SB804	OEM
2SB804AW	OEM
2SB805	OEM
2SB806	OEM
2SB807	OEM
2SB808	0006
2SB808F	0006
2SB808F-SP	OEM
2SB808F-SPA	OEM
2SB808FT	OEM
2SB808G	6873
2SB808G1	0006
2SB810	0532
2SB810(E)	0532
2SB810(F)	0532
2SB810(H)	0532
2SB810F	0532
2SB810H	0532
2SB810HJ	0532
2SB810HJ-T	0532
2SB810HT-T	0532
2SB810J	0532
2SB811	0477
2SB812	0859
2SB812A	0859
2SB813	0681
2SB814	5454
2SB815	1752
2SB816	3111
2SB817	3111
2SB818	OEM
2SB818K	OEM
2SB819	4348
2SB819(Q)	4348
2SB819(R)	4348
2SB819AU(Q)	4348
2SB819AU(R)	4348
2SB819AUR	4348
2SB819Q	4348
2SB819QR	4348
2SB819R	4348
2SB820	OEM
2SB821	0477
2SB822	4104
2SB822(Q)	4104
2SB822(R)	4104
2SB822Q	4104
2SB823	0854
2SB824	1308
2SB824Q	1308
2SB824R	1308
2SB825	0854
2SB825R	0854
2SB825S	2580
2SB826	1308
2SB827	5478
2SB827R	7096
2SB827S	7096
2SB828	5478
2SB828Q	7096
2SB828R	7096
2SB828S	5478
2SB829	5341
2SB829R	7096
2SB829S	OEM
2SB830	0477
2SB831	1752
2SB831BB	1752
2SB831BC	1752
2SB832	OEM
2SB833	4104
2SB833(Q)	4104
2SB834	0854
2SB834(O)	0854
2SB834(Y)	0854
2SB834-O	0854
2SB834O	0854
2SB834Y	0854
2SB835	0532
2SB835(Q)	0532
2SB835(R)	0532
2SB835(S)	0532
2SB835(T)	0532
2SB835R	3597
2SB835T	0532
2SB836	3498
2SB836L	3498
2SB837	3498
2SB837B	3498
2SB837C	3498
2SB837D	3498
2SB837LB	3498
2SB837LC	3498
2SB837LD	3498
2SB838	3498
2SB838B	3498
2SB839	3498
2SB839B	3498
2SB840	3498
2SB840B	3498
2SB840C	3498
2SB840LB	3498
2SB841	3498
2SB841C	3498
2SB841LB	3498
2SB841LC	3498
2SB842	OEM
2SB843	OEM
2SB844	3498
2SB845	3498
2SB846	3498
2SB848	0859
2SB849	5478
2SB849A	5478
2SB850	1308
2SB850A	1308
2SB851	5061
2SB851P	5061
2SB851Q	5061
2SB851R	5061
2SB852	OEM
2SB853	OEM
2SB853K	OEM
2SB854	OEM
2SB854K	OEM
2SB855	5492
2SB855A	5492
2SB855B	5492
2SB856	0718
2SB856B	0718
2SB856C	0718
2SB857	2923
2SB857-F1	OEM
2SB857B	2923
2SB857C	2923
2SB857D	2923
2SB858	2923
2SB858B	2923
2SB858C	2923
2SB858D	2923
2SB859	0681
2SB859B	0681
2SB859C	0681
2SB860	1369
2SB861	1648
2SB861-02	1648
2SB861-02-C	1648
2SB861-02C	1648
2SB861B	1648
2SB861C	1648
2SB861D	1648
2SB862	OEM
2SB863	5478
2SB864	0693
2SB865	0635
2SB867	0718
2SB868	0681
2SB869	0718
2SB870	0718
2SB871	1308
2SB871A	1308
2SB872	0602
2SB872A	0602
2SB873	OEM
2SB874	0525
2SB875	0718
2SB876	2555
2SB877	0525
2SB878	OEM
2SB879	OEM
2SB880	2237
2SB881	2237
2SB882	2237
2SB883	0073
2SB884	0602
2SB885	0602
2SB886	2237
2SB887	0073
2SB888	OEM
2SB889	0460
2SB889Q	OEM
2SB890	0006
2SB891	0460
2SB892	0477
2SB892S	0477
2SB892SZ	0477
2SB892T	0477
2SB892TZ	0477
2SB893	OEM
2SB894	2484
2SB894(Q)	2484
2SB894(R)	2484
2SB894(S)	2484
2SB894S	2484
2SB895	2897
2SB895A	2897
2SB896	1308
2SB896A	1308
2SB897	3500
2SB898	0681
2SB899	0681
2SB900	1308
2SB901	0681
2SB902	1742
2SB902P	OEM
2SB902Q	OEM
2SB903	1199
2SB904	0563
2SB905	OEM
2SB906	OEM
2SB907	5353
2SB908	5353
2SB-909	4104
2SB909	4104
2SB909M	4104
2SB909MQ	OEM
2SB909MR	OEM
2SB910	0359
2SB910M	0359
2SB910MP	0359
2SB910MQ	0359
2SB910MR	0359
2SB911	4104
2SB911-M	OEM
2SB911M	4104
2SB911MP	4104
2SB911MQ	4104
2SB911MR	4104
2SB912	0073
2SB913	0073
2SB914	2448
2SB915	2448
2SB916	2448
2SB917	OEM
2SB918	0342
2SB919	0718
2SB920	0718
2SB921	0718
2SB922	5477
2SB923	4144
2SB924	4144
2SB925	1308
2SB925A	1308
2SB926	4011
2SB926-T	4011
2SB927	OEM
2SB928	OEM
2SB928A	OEM
2SB929	5464
2SB929A	5464
2SB930	5464
2SB930A	5464
2SB931	5464
2SB932	5464
2SB933	5464
2SB934	5464
2SB935	5464
2SB935A	5464
2SB936	5464
2SB936A	5464
2SB937	5353
2SB937A	5353
2SB938	5353
2SB938A	5353
2SB939	OEM
2SB939A	OEM
2SB940	1648
2SB940-P	1648
2SB940A	1648
2SB940ALBP	1648
2SB940A-P	1648
2SB940AP	1648
2SB940APLB	1648
2SB940LBP	1648
2SB940P	1648
2SB940PLB	1648
2SB941	2923
2SB941(P)	2923
2SB941(Q)	2923
2SB941A	2923
2SB941A-PLT	OEM
2SB941A-PVJ	OEM
2SB941B	2923
2SB941D	2923
2SB941P	2923
2SB941PB	2923
2SB941PD	2923
2SB941PQ	2923
2SB941Q	2923
2SB942	0718
2SB942A	0718
2SB942A-PVJ	OEM
2SB943	0718
2SB943Q	5500
2SB944	0718
2SB945	5500
2SB945P	5500
2SB946	0718
2SB947	1308
2SB947A	1308
2SB948	1308
2SB948A	1308
2SB949	0602
2SB949(P)	0602
2SB949(Q)	0602
2SB949A	0602
2SB949P	0602
2SB950	0602
2SB950A	0602
2SB951	0602
2SB951A	0602
2SB952	5464
2SB952(P)	5464
2SB952(Q)	5464
2SB952A	5464
2SB952PQ	5464
2SB952Q	5464
2SB953	1308
2SB953A	1308
2SB954	3165
2SB954A	3165
2SB955	1968
2SB955(K)	1968
2SB956	3645
2SB956S	OEM
2SB957	3498
2SB958	3498
2SB959	3498
2SB960	3498
2SB962	OEM
2SB962P	OEM
2SB962Q	OEM
2SB963	OEM
2SB964	OEM
2SB965	5477
2SB966	3111
2SB967	OEM
2SB968	3498
2SB969	OEM
2SB970	1752
2SB970-QR	OEM
2SB970QR	OEM
2SB970QRSTW	OEM
2SB971	3183
2SB972	OEM
2SB973	1752
2SB974	0602
2SB974L	0602
2SB974M	0602
2SB975	0602
2SB976	0648
2SB976(Q)	0648
2SB976(R)	0648
2SB976QR	0648
2SB976R	0648
2SB977	0635
2SB977A	0635
2SB978	3183
2SB979	3111
2SB980	3111
2SB981	3111
2SB982	3111
2SB983	1199
2SB984	1522
2SB985	0811
2SB985G	OEM
2SB985S	0811
2SB986	0460
2SB987	1522
2SB988	0925
2SB989	0681
2SB990	OEM
2SB991	1648
2SB992	1308
2SB993	1308
2SB994	1308
2SB995	0718
2SB996	3165
2SB997	0602
2SB998	2237
2SB999	2237
2SB1000	3645
2SB1000(RECT)	2443
2SB1000A	3645
2SB1001	OEM
2SB1002	3645
2SB1003	OEM
2SB1004	OEM
2SB1005	0602
2SB1007	0525
2SB1008	2897
2SB1009	0460
2SB1009P	0460
2SB1010	3183
2SB1010(Q)	3183
2SB1010Q	3183
2SB1010QZ	3183
2SB1010R	3183
2SB1010RZ	3183
2SB1011	OEM
2SB1011RL	OEM
2SB1012	OEM
2SB1012(K)	OEM
2SB1012K	OEM
2SB1013	OEM
2SB1014	OEM
2SB1015	1308
2SB1015Y	OEM
2SB1016	1369
2SB1017	0681
2SB1018	0718
2SB1019	1308
2SB1020	2237
2SB1021	0602
2SB1022	0602
2SB1023	OEM
2SB1024	OEM
2SB1025	OEM
2SB1026	OEM
2SB1027	OEM
2SB1028	OEM
2SB1029	OEM
2SB1030	0013
2SB1030AR	0013
2SB1030ARS	0013
2SB1030AS	0013
2SB1030R	0013
2SB1030RSTTA	OEM
2SB1030S	0013
2SB1031	OEM
2SB1032	OEM
2SB1033	OEM
2SB1034	2897
2SB1035	OEM
2SB1036	OEM
2SB1037	1648
2SB1038	0681
2SB1039	0681
2SB1040	OEM
2SB1040A	OEM
2SB1041	6456
2SB1042	OEM
2SB1042M	0359
2SB1043	6456
2SB1044	OEM
2SB1044M	0359
2SB1045	OEM
2SB1046	OEM
2SB1047	OEM
2SB1048	OEM
2SB1049	OEM
2SB1050	4011
2SB1050(Q)	4011
2SB1050(R)	4011
2SB1050-Q	OEM
2SB1050R	3838
2SB1051	OEM
2SB1052	OEM
2SB1052B	OEM
2SB1052D	OEM
2SB1053	OEM
2SB1054	3111
2SB1055	3111
2SB1056	OEM
2SB1057	3111
2SB1058	4011
2SB1059	0037
2SB1060	0461
2SB1061	OEM
2SB1062	OEM
2SB1062S	OEM
2SB1063	0718
2SB1064	1308
2SB1065	0460
2SB1066	OEM
2SB1067	2897
2SB1068	4011
2SB1068-K	OEM
2SB1069	OEM
2SB1070	OEM
2SB1071	OEM
2SB1072	OEM
2SB1073	OEM
2SB1074	OEM
2SB1075	OEM
2SB1076	OEM
2SB1077	0602
2SB1078	1968
2SB1079	OEM
2SB1080	3327
2SB1085	1648
2SB1086	0525
2SB1087	OEM
2SB1088	OEM
2SB1089	OEM
2SB1090	OEM
2SB1091	0602
2SB1092	OEM
2SB1093	OEM
2SB1093-T	OEM
2SB1093L	OEM
2SB1093Q	OEM
2SB1094	4011
2SB1094L	4011
2SB1095	0681
2SB1096	1648
2SB1097	OEM
2SB1098	1968
2SB1099	0602
2SB1099K	0602
2SB1099L	0602
2SB1100	OEM
2SB1101	0602
2SB1102	0602
2SB1103	0602
2SB1104	0602
2SB1105	1968
2SB1106	1968
2SB1107	OEM
2SB1108	1968
2SB1109	0525
2SB1110	OEM
2SB1111	OEM
2SB1112	OEM
2SB1113	OEM
2SB1114	OEM
2SB1115	3645
2SB1115A	3645
2SB1116	0477
2SB1116A	4090
2SB1117	OEM
2SB1118	3645
2SB1119	3645
2SB1120	OEM
2SB1121	OEM
2SB1122	3645
2SB1123	OEM
2SB1124	OEM
2SB1125	OEM
2SB1126	OEM
2SB1127	OEM
2SB1128	OEM
2SB1129	OEM
2SB1130	OEM
2SB1131	OEM
2SB1132	OEM
2SB1133	1308
2SB1133R	OEM
2SB1134	0925
2SB1134Q	OEM
2SB1134R	0925
2SB1134R-SA	0925
2SB1135	4125
2SB1135(R)	4125
2SB1135R	4124
2SB1135S	4125
2SB1136	OEM
2SB1137	OEM
2SB1140	OEM
2SB1141	0525
2SB1142	OEM
2SB1143	OEM
2SB1144	0525
2SB1145	1968
2SB1146	1968
2SB1147	1968
2SB1148	OEM
2SB1149	OEM
2SB1150	OEM
2SB1151	0460
2SB1152	OEM
2SB1153	OEM
2SB1154	OEM
2SB1155	OEM
2SB1156	OEM
2SB1157	OEM
2SB1158	OEM
2SB1159	OEM
2SB1160	OEM
2SB1161	OEM
2SB1162	5349
2SB1163	5349
2SB1164	OEM
2SB1165	OEM
2SB1165S	OEM
2SB1166	OEM
2SB1167	OEM
2SB1168	0460
2SB1168S	0460
2SB1169	OEM
2SB1170	OEM
2SB1171	OEM
2SB1172	OEM
2SB1173	OEM
2SB1174	OEM
2SB1175	OEM
2SB1176	OEM
2SB1177	OEM
2SB1178	OEM
2SB1179	OEM
2SB1180	OEM
2SB1181	OEM
2SB1182R	OEM
2SB1184	OEM
2SB1185	1308
2SB1185E	OEM
2SB1186	1648
2SB1186A	OEM
2SB1187	5475
2SB1187E	OEM
2SB1188	OEM
2SB1189	OEM
2SB1190	1648
2SB1191	OEM
2SB1192	1648
2SB1193	1968
2SB1194	OEM
2SB1195	OEM
2SB1203S	OEM
2SB1212	OEM
2SB1212-P	OEM
2SB1212P	OEM
2SB1218-QRS	OEM
2SB1218QRS	OEM
2SB1223	OEM
2SB1238	OEM
2SB1238QS	OEM
2SB1238QSTV6	OEM
2SB1240	4104
2SB1240-P	OEM
2SB1240QRTV6	OEM
2SB1240R	OEM
2SB1243	OEM
2SB1262	OEM
2SB1272	OEM
2SB1272M	OEM
2SB1274(Q)	0925
2SB1274(R)	0925
2SB1274-Q-RA	0925
2SB1274-R-RA	0925
2SB1274Q	0925
2SB1274Q-RA	0925
2SB1274R	0925
2SB1274S	0925
2SB1277	4388
2SB1277P	4388
2SB1318	OEM
2SB1318L	OEM
2SB1357	5475
2SB1357DEFTA	OEM
2SB1359	6338
2SB1364	0211
2SB1366Y	OEM
2SB1367O	OEM
2SB1367Y	OEM
2SB1370-EF	OEM
2SB1370E	OEM
2SB1370EF	OEM
2SB1375	OEM
2SB1382	OEM
2SB1420	OEM
2SB1565-E	OEM
2SB1760	0211
2SB1780	0164
2SB1780A	0164
2SB1780B	0164
2SB1780C	0164
2SB1780D	0164
2SB1780E	0164
2SB1780F	0164
2SB1780G	0164
2SB1780GN	0164
2SB1780H	0164
2SB1780J	0164
2SB1780K	0164
2SB1780L	0164
2SB1780M	0164
2SB1780OR	0164
2SB1780R	0164
2SB1780X	0164

If replacement code is OEM, contact original manufacturer for replacement.

DEVICE TYPE	REPL CODE
2SB1780Y	0164
2SB1785	0211
2SB1851	0004
2SB1860	0211
2SB2821	2484
2SB3030	0004
2SB3224E	0211
2SB3240	0164
2SB3240A	0164
2SB3240B	0164
2SB3240C	0164
2SB3240D	0164
2SB3240E	0164
2SB3240F	0164
2SB3240G	0164
2SB3240GN	0164
2SB3240H	0164
2SB3240J	0164
2SB3240K	0164
2SB3240L	0164
2SB3240M	0164
2SB3240OR	0164
2SB3240R	0164
2SB3240X	0164
2SB3240Y	0164
2SB3244	0280
2SB-3783	0164
2SB3783A	0164
2SB3783B	0164
2SB3783C	0164
2SB3783D	0164
2SB3783E	0164
2SB3783F	0164
2SB3783G	0164
2SB3783GN	0164
2SB3783H	0164
2SB3783J	0164
2SB3783K	0164
2SB3783L	0164
2SB3783M	0164
2SB3783OR	0164
2SB3783R	0164
2SB3783X	0164
2SB3783Y	0164
2SB-3812	0164
2SB3812A	0164
2SB3812B	0164
2SB3812C	0164
2SB3812D	0164
2SB3812F	0164
2SB3812G	0164
2SB3812GN	0164
2SB3812H	0164
2SB3812J	0164
2SB3812K	0164
2SB3812L	0164
2SB3812M	0164
2SB3812OR	0164
2SB3812R	0164
2SB3812X	0164
2SB3812Y	0164
2SB-3813	0164
2SB3813A	0164
2SB3813B	0164
2SB3813C	0164
2SB3813D	0164
2SB3813E	0164
2SB3813F	0164
2SB3813G	0164
2SB3813GN	0164
2SB3813H	0164
2SB3813J	0164
2SB3813K	0164
2SB3813L	0164
2SB3813M	0164
2SB3813OR	0164
2SB3813R	0164
2SB3813X	0164
2SB3813Y	0164
2SB4151	0211
2SB4340	0854
2SB4440D	0164
2SB4631	0222
2SB4718	OEM
2SB7513	0280
2SB17313	0211
2SB-C731	0015
2SBF1	0004
2SB-F1A	0004
2SBF1A	0004
2SBF2	0004
2SBF2A	0164
2SBF5	0160
2SBF20	1414
2SBF40	0473
2SBF60	0446
2SBF80	1422
2SBF100	2443
2SBM77	0164
2SB-S131	0087
2SB-S851	0277
2SBU86	0211
2SBZ	0975
2SC11	0208
2SC11A	0208
2SC11B	0208
2SC11C	0208
2SC11D	0208
2SC11E	0208

DEVICE TYPE	REPL CODE
2SC11F	0208
2SC11G	0208
2SC11GN	0208
2SC11H	0208
2SC11J	0208
2SC11K	0208
2SC11L	0208
2SC11M	0208
2SC11OR	0208
2SC11R	0208
2SC11X	0208
2SC11Y	0208
2SC12	0086
2SC12A	0086
2SC12B	0086
2SC12C	0086
2SC12D	0086
2SC12E	0086
2SC12F	0086
2SC12G	0086
2SC12GN	0086
2SC12H	0086
2SC12J	0086
2SC12K	0086
2SC12L	0086
2SC12M	0086
2SC12OR	0086
2SC12R	0086
2SC12X	0086
2SC12Y	0086
2SC13	0600
2SC13A	0600
2SC13B	0600
2SC13C	0600
2SC13D	0600
2SC13E	0600
2SC13F	0600
2SC13G	0600
2SC13GN	0600
2SC13H	0600
2SC13J	0600
2SC13K	0600
2SC13L	0600
2SC13M	0600
2SC13OR	0600
2SC13R	0600
2SC13X	0600
2SC13Y	0600
2SC14	0600
2SC14A	0600
2SC14B	0600
2SC14C	0600
2SC14D	0600
2SC14E	0600
2SC14F	0600
2SC14G	0600
2SC14GN	0600
2SC14H	0600
2SC14J	0600
2SC14K	0600
2SC14L	0600
2SC14M	0600
2SC14OR	0600
2SC14R	0600
2SC14X	0600
2SC14Y	0600
2SC15	0016
2SC15-0	0016
2SC15-1	0016
2SC15-2	0016
2SC15-3	0016
2SC15A	0016
2SC15B	0016
2SC15C	0016
2SC15D	0016
2SC15E	0016
2SC15F	0016
2SC15G	0016
2SC15GN	0016
2SC15H	0016
2SC15J	0016
2SC15K	0016
2SC15L	0016
2SC15M	0016
2SC15OR	0016
2SC15R	0016
2SC15X	0016
2SC15Y	0016
2SC16	0016
2SC16A	0016
2SC16B	0016
2SC16C	0016
2SC16D	0016
2SC16E	0016
2SC16F	0016
2SC16G	0016
2SC16GN	0016
2SC16H	0016
2SC16J	0016
2SC16K	0016
2SC16L	0016
2SC16M	0016
2SC16OR	0016
2SC16R	0016
2SC16X	0016
2SC16Y	0016
2SC17	0016
2SC17A	0016
2SC17B	0016

DEVICE TYPE	REPL CODE
2SC17C	0016
2SC17D	0016
2SC17E	0016
2SC17F	0016
2SC17G	0016
2SC17GN	0016
2SC17H	0016
2SC17J	0016
2SC17K	0016
2SC17L	0016
2SC17M	0016
2SC17OR	0016
2SC17R	0016
2SC17X	0016
2SC17Y	0016
2SC18	0079
2SC18A	0079
2SC18B	0079
2SC18C	0079
2SC18D	0079
2SC18E	0079
2SC18F	0079
2SC18G	0079
2SC18GN	0079
2SC18H	0079
2SC18J	0079
2SC18K	0079
2SC18L	0079
2SC18M	0079
2SC18OR	0079
2SC18R	0079
2SC18X	0079
2SC19	0086
2SC19A	0086
2SC19B	0086
2SC19C	0086
2SC19D	0086
2SC19E	0086
2SC19F	0086
2SC19G	0086
2SC19GN	0086
2SC19H	0086
2SC19J	0086
2SC19K	0086
2SC19L	0086
2SC19M	0086
2SC19OR	0086
2SC19R	0086
2SC19X	0086
2SC19Y	0086
2SC20	0086
2SC20A	0086
2SC20B	0086
2SC20C	0086
2SC20D	0086
2SC20E	0086
2SC20F	0086
2SC20G	0086
2SC20GN	0086
2SC20H	0086
2SC20J	0086
2SC20K	0086
2SC20L	0086
2SC20M	0086
2SC20OR	0086
2SC20R	0086
2SC20X	0086
2SC20Y	0086
2SC21	0103
2SC21A	0103
2SC21B	0103
2SC21C	0103
2SC21D	0103
2SC21E	0103
2SC21F	0103
2SC21G	0103
2SC21GN	0103
2SC21H	0103
2SC21J	0103
2SC21K	0103
2SC21L	0103
2SC21M	0103
2SC21OR	0103
2SC21R	0103
2SC21X	0103
2SC21Y	0103
2SC22	0086
2SC22A	0086
2SC22B	0086
2SC22D	0086
2SC22E	0086
2SC22F	0086
2SC22G	0086
2SC22GN	0086
2SC22H	0086
2SC22J	0086
2SC22K	0086
2SC22L	0086
2SC22M	0086
2SC22OR	0086
2SC22R	0086
2SC22X	0086
2SC22Y	0086
2SC23	0086
2SC23A	0086
2SC23B	0086
2SC23D	0086
2SC23E	0086
2SC23F	0086

DEVICE TYPE	REPL CODE
2SC23G	0086
2SC23GN	0086
2SC23H	0086
2SC23J	0086
2SC23K	0086
2SC23L	0086
2SC23M	0086
2SC23OR	0086
2SC23R	0086
2SC23X	0086
2SC23Y	0086
2SC24	0086
2SC24A	0086
2SC24B	0086
2SC24C	0086
2SC24D	0086
2SC24E	0086
2SC24F	0086
2SC24G	0086
2SC24GN	0086
2SC24H	0086
2SC24J	0086
2SC24K	0086
2SC24L	0086
2SC24M	0086
2SC24OR	0086
2SC24R	0086
2SC24X	0086
2SC24Y	0086
2SC25	1085
2SC26	0016
2SC26A	0016
2SC26B	0016
2SC26C	0016
2SC26D	0016
2SC26E	0016
2SC26F	0016
2SC26G	0016
2SC26GN	0016
2SC26H	0016
2SC26J	0016
2SC26K	0016
2SC26L	0016
2SC26M	0016
2SC26OR	0016
2SC26R	0016
2SC26X	0016
2SC26Y	0016
2SC27	0016
2SC27A	0016
2SC27B	0016
2SC27D	0016
2SC27E	0016
2SC27F	0016
2SC27G	0016
2SC27GN	0016
2SC27H	0016
2SC27J	0016
2SC27K	0016
2SC27L	0016
2SC27M	0016
2SC27OR	0016
2SC27R	0016
2SC27X	0016
2SC27Y	0016
2SC28	0198
2SC28A	0198
2SC28B	0198
2SC28C	0198
2SC28D	0198
2SC28E	0198
2SC28F	0198
2SC28G	0198
2SC28GN	0198
2SC28H	0198
2SC28J	0198
2SC28K	0198
2SC28L	0198
2SC28M	0198
2SC28OR	0198
2SC28R	0198
2SC28X	0198
2SC28Y	0198
2SC29	0198
2SC29A	0198
2SC29B	0198
2SC29C	0198
2SC29D	0198
2SC29E	0198
2SC29F	0198
2SC29G	0198
2SC29GN	0198
2SC29H	0198
2SC29J	0198
2SC29K	0198
2SC29L	0198
2SC29M	0198
2SC29OR	0198
2SC29R	0198
2SC29X	0198
2SC29Y	0198
2SC30	0086
2SC30-OR	0086
2SC30A	0086
2SC30B	0086
2SC30C	0086
2SC30D	0086
2SC30E	0086
2SC30F	0086

DEVICE TYPE	REPL CODE
2SC30G	0086
2SC30GN	0086
2SC30H	0086
2SC30J	0086
2SC30L	0086
2SC30M	0086
2SC30OR	0086
2SC30R	0086
2SC30X	0086
2SC30Y	0086
2SC31	0086
2SC31A	0086
2SC31B	0086
2SC31C	0086
2SC31D	0086
2SC31E	0086
2SC31F	0086
2SC31G	0086
2SC31GN	0086
2SC31H	0086
2SC31J	0086
2SC31K	0086
2SC31L	0086
2SC31M	0086
2SC31OR	0086
2SC31R	0086
2SC31X	0086
2SC31Y	0086
2SC32	0086
2SC32A	0086
2SC32B	0086
2SC32C	0086
2SC32D	0086
2SC32E	0086
2SC32F	0086
2SC32G	0086
2SC32GN	0086
2SC32H	0086
2SC32J	0086
2SC32K	0086
2SC32L	0086
2SC32M	0086
2SC32OR	0086
2SC32R	0086
2SC32X	0086
2SC32Y	0086
2SC33	0079
2SC33A	0079
2SC33B	0079
2SC33C	0079
2SC33D	0079
2SC33E	0079
2SC33F	0079
2SC33G	0079
2SC33GN	0079
2SC33H	0079
2SC33J	0079
2SC33K	0079
2SC33L	0079
2SC33M	0079
2SC33OR	0079
2SC33R	0079
2SC33X	0079
2SC33Y	0079
2SC34	0208
2SC34A	0208
2SC34C	0208
2SC34D	0208
2SC34E	0208
2SC34F	0208
2SC34G	0208
2SC34GN	0208
2SC34H	0208
2SC34J	0208
2SC34K	0208
2SC34L	0208
2SC34M	0208
2SC34OR	0208
2SC34R	0208
2SC34X	0208
2SC34Y	0208
2SC35	0208
2SC35A	0208
2SC35B	0208
2SC35C	0208
2SC35E	0208
2SC35F	0208
2SC35G	0208
2SC35GN	0208
2SC35H	0208
2SC35J	0208
2SC35K	0208
2SC35L	0208
2SC35M	0208
2SC35OR	0208
2SC35R	0208
2SC35X	0208
2SC35Y	0208
2SC36	0208
2SC36B	0208
2SC36C	0208
2SC36D	0208
2SC36E	0208
2SC36F	0208
2SC36G	0208
2SC36GN	0208
2SC36H	0208
2SC36J	0208
2SC36K	0208

DEVICE TYPE	REPL CODE
2SC36L	0208
2SC36M	0208
2SC36OR	0208
2SC36R	0208
2SC36X	0208
2SC36Y	0208
2SC37	0016
2SC37A	0016
2SC37B	0016
2SC37C	0016
2SC37D	0016
2SC37E	0016
2SC37F	0016
2SC37G	0016
2SC37GN	0016
2SC37H	0016
2SC37J	0016
2SC37K	0016
2SC37L	0016
2SC37M	0016
2SC37OR	0016
2SC37R	0016
2SC37X	0016
2SC37Y	0016
2SC38	0016
2SC38A	0016
2SC38B	0016
2SC38C	0016
2SC38D	0016
2SC38E	0016
2SC38F	0016
2SC38G	0016
2SC38H	0016
2SC38J	0016
2SC38K	0016
2SC38L	0016
2SC38M	0016
2SC38OR	0016
2SC38R	0016
2SC38X	0016
2SC38Y	0016
2SC39	0144
2SC39A	0144
2SC39B	0144
2SC39C	0144
2SC39D	0144
2SC39E	0144
2SC39F	0144
2SC39G	0144
2SC39GN	0144
2SC39H	0144
2SC39J	0144
2SC39K	0144
2SC39L	0144
2SC39M	0144
2SC39OR	0144
2SC39R	0144
2SC39X	0144
2SC39Y	0144
2SC40	0144
2SC40A	0144
2SC40B	0144
2SC40C	0144
2SC40D	0144
2SC40E	0144
2SC40F	0144
2SC40G	0144
2SC40GN	0144
2SC40H	0144
2SC40J	0144
2SC40K	0144
2SC40L	0144
2SC40OR	0144
2SC40R	0144
2SC40X	0144
2SC40Y	0144
2SC41	0620
2SC41B	0620
2SC41C	0620
2SC41F	0620
2SC41GN	0620
2SC41H	0620
2SC41J	0620
2SC41M	0620
2SC41OR	0620
2SC41R	0620
2SC41TV	0620
2SC41X	0620
2SC41Y	0620
2SC42	0074
2SC42A	0074
2SC42B	0074
2SC42C	0074
2SC42E	0074
2SC42F	0074
2SC42G	0074
2SC42GN	0074
2SC42H	0074
2SC42J	0074
2SC42K	0074
2SC42L	0074
2SC42M	0074
2SC42OR	0074
2SC42R	0074
2SC42X	0074
2SC42Y	0074
2SC43	0074
2SC44	0074
2SC45	0016

DEVICE TYPE	REPL CODE
2SC45B-GR	0016
2SC45B-O	0016
2SC45B-Y	0016
2SC46	0086
2SC46A	0086
2SC46B	0086
2SC46C	0086
2SC46DB	0086
2SC46E	0086
2SC46F	0086
2SC46G	0086
2SC46GN	0086
2SC46H	0086
2SC46J	0086
2SC46K	0086
2SC46L	0086
2SC46M	0086
2SC46OR	0086
2SC46X	0086
2SC46Y	0086
2SC47	0086
2SC47A	0086
2SC47B	0086
2SC47C	0086
2SC47D	0086
2SC47E	0086
2SC47G	0086
2SC47GN	0086
2SC47H	0086
2SC47J	0086
2SC47K	0086
2SC47L	0086
2SC47M	0086
2SC47OR	0086
2SC47R	0086
2SC47X	0086
2SC47Y	0086
2SC48	0086
2SC48A	0086
2SC48B	0086
2SC48C	0086
2SC48D	0086
2SC48E	0086
2SC48F	0086
2SC48G	0086
2SC48GN	0086
2SC48H	0086
2SC48J	0086
2SC48K	0086
2SC48L	0086
2SC48M	0086
2SC48OR	0086
2SC48R	0086
2SC48X	0086
2SC48Y	0086
2SC49	0086
2SC49A	0086
2SC49B	0086
2SC49C	0086
2SC49D	0086
2SC49E	0086
2SC49F	0086
2SC49G	0086
2SC49GN	0086
2SC49H	0086
2SC49J	0086
2SC49K	0086
2SC49L	0086
2SC49M	0086
2SC49OR	0086
2SC49X	0086
2SC49Y	0086
2SC50	0600
2SC50A	0600
2SC50B	0600
2SC50C	0600
2SC50E	0600
2SC50F	0600
2SC50G	0600
2SC50GN	0600
2SC50H	0600
2SC50J	0600
2SC50K	0600
2SC50L	0600
2SC50M	0600
2SC50OR	0600
2SC50R	0600
2SC50X	0600
2SC50Y	0600
2SC51	0086
2SC51A	0086
2SC51B	0086
2SC51C	0086
2SC51D	0086
2SC51E	0086
2SC51F	0086
2SC51G	0086
2SC51GN	0086
2SC51H	0086
2SC51J	0086
2SC51K	0086
2SC51L	0086
2SC51M	0086
2SC51OR	0086
2SC51R	0086
2SC51X	0086
2SC51Y	0086
2SC52	0016
2SC52A	0016

DEVICE TYPE	REPL CODE
2SC52B	0016
2SC52C	0016
2SC52D	0016
2SC52E	0016
2SC52F	0016
2SC52G	0016
2SC52GN	0016
2SC52H	0016
2SC52J	0016
2SC52K	0016
2SC52L	0016
2SC52M	0016
2SC52OR	0016
2SC52X	0016
2SC52Y	0016
2SC53	0016
2SC53A	0016
2SC53B	0016
2SC53C	0016
2SC53D	0016
2SC53E	0016
2SC53F	0016
2SC53G	0016
2SC53GN	0016
2SC53H	0016
2SC53J	0016
2SC53L	0016
2SC53OR	0016
2SC53R	0016
2SC53X	0016
2SC53Y	0016
2SC54	0016
2SC54A	0016
2SC54B	0016
2SC54C	0016
2SC54D	0016
2SC54E	0016
2SC54F	0016
2SC54G	0016
2SC54H	0016
2SC54J	0016
2SC54K	0016
2SC54L	0016
2SC54M	0016
2SC54OR	0016
2SC54X	0016
2SC54Y	0016
2SC55	0016
2SC55A	0016
2SC55B	0016
2SC55C	0016
2SC55D	0016
2SC55E	0016
2SC55F	0016
2SC55G	0016
2SC55GN	0016
2SC55H	0016
2SC55J	0016
2SC55K	0016
2SC55L	0016
2SC55M	0016
2SC55OR	0016
2SC55R	0016
2SC55X	0016
2SC55Y	0016
2SC56	0127
2SC56A	0127
2SC56B	0127
2SC56C	0127
2SC56D	0127
2SC56E	0127
2SC56F	0127
2SC56G	0127
2SC56GN	0127
2SC56H	0127
2SC56J	0127
2SC56K	0127
2SC56L	0127
2SC56M	0127
2SC56OR	0127
2SC56R	0127
2SC56X	0127
2SC56Y	0127
2SC57	0595
2SC58	0233
2SC58A	0233
2SC58AC	0233
2SC58B	0233
2SC58D	0233
2SC58E	0233
2SC58F	0233
2SC58G	0233
2SC58GN	0233
2SC58H	0233
2SC58J	0233
2SC58K	0233
2SC58L	0233
2SC58M	0233
2SC58OR	0233
2SC58R	0233
2SC58X	0233
2SC58Y	0233
2SC59	0086
2SC59A	0086
2SC59B	0086
2SC59C	0086
2SC59D	0086
2SC59E	0086
2SC59F	0086
2SC59G	0086
2SC59GN	0086
2SC59H	0086
2SC59J	0086
2SC59K	0086
2SC59L	0086
2SC59M	0086
2SC59OR	0086
2SC59R	0086
2SC59X	0086
2SC59Y	0086
2SC60	0208
2SC60A	0208
2SC60B	0208
2SC60C	0208
2SC60D	0208
2SC60E	0208
2SC60F	0208
2SC60G	0208
2SC60GN	0208
2SC60H	0208
2SC60J	0208
2SC60K	0208
2SC60L	0208
2SC60M	0208
2SC60OR	0208
2SC60R	0208
2SC60X	0208
2SC60Y	0208
2SC61	0086
2SC61A	0086
2SC61B	0086
2SC61C	0086
2SC61D	0086
2SC61E	0086
2SC61F	0086
2SC61G	0086
2SC61GN	0086
2SC61H	0086
2SC61J	0086
2SC61K	0086
2SC61L	0086
2SC61M	0086
2SC61OR	0086
2SC61R	0086
2SC61Y	0086
2SC62	0016
2SC62A	0016
2SC62B	0016
2SC62C	0016
2SC62D	0016
2SC62E	0016
2SC62F	0016
2SC62G	0016
2SC62GN	0016
2SC62H	0016
2SC62J	0016
2SC62L	0016
2SC62M	0016
2SC62OR	0016
2SC62R	0016
2SC62X	0016
2SC62Y	0016
2SC63	0144
2SC63A	0144
2SC63B	0144
2SC63D	0144
2SC63E	0144
2SC63F	0144
2SC63G	0144
2SC63GN	0144
2SC63H	0144
2SC63J	0144
2SC63K	0144
2SC63L	0144
2SC63M	0144
2SC63OR	0144
2SC63R	0144
2SC63X	0144
2SC63Y	0144
2SC64	0086
2SC64A	0086
2SC64B	0086
2SC64C	0086
2SC64D	0086
2SC64E	0086
2SC64F	0086
2SC64G	0086
2SC64GN	0086
2SC64H	0086
2SC64K	0086
2SC64L	0086
2SC64M	0086
2SC64OR	0086
2SC64R	0086
2SC64X	0086
2SC64Y	0086
2SC64Y-RST	0086
2SC65	0233
2SC65(Y)	0233
2SC65-O	0233
2SC65-OR	0233
2SC65A	0233
2SC65B	0233
2SC65C	0233
2SC65D	0233
2SC65E	0233
2SC65F	0233
2SC65G	0233
2SC65GN	0233
2SC65H	0233
2SC65K	0233
2SC65L	0233
2SC65M	0233
2SC65N	0233
2SC65OR	0233
2SC65R	0233
2SC65X	0233
2SC65Y	0233
2SC65Y(B)	0233
2SC65YA	0233
2SC65YB	0233
2SC65YTV	0233
2SC65YTV1	0233
2SC66	0233
2SC66A	0233
2SC66B	0233
2SC66C	0233
2SC66D	0233
2SC66E	0233
2SC66EV	0233
2SC66F	0233
2SC66G	0233
2SC66GN	0233
2SC66H	0233
2SC66J	0233
2SC66K	0233
2SC66L	0233
2SC66M	0233
2SC66OR	0233
2SC66R	0233
2SC66X	0233
2SC66Y	0233
2SC67	0016
2SC67A	0016
2SC67B	0016
2SC67C	0016
2SC67D	0016
2SC67E	0016
2SC67F	0016
2SC67G	0016
2SC67GN	0016
2SC67H	0016
2SC67J	0016
2SC67K	0016
2SC67L	0016
2SC67M	0016
2SC67R	0016
2SC67X	0016
2SC67Y	0016
2SC68	0016
2SC68A	0016
2SC68B	0016
2SC68C	0016
2SC68D	0016
2SC68E	0016
2SC68F	0016
2SC68G	0016
2SC68GN	0016
2SC68H	0016
2SC68J	0016
2SC68K	0016
2SC68L	0016
2SC68M	0016
2SC68OR	0016
2SC68R	0016
2SC68X	0016
2SC68Y	0016
2SC69	0086
2SC69A	0086
2SC69B	0086
2SC69C	0086
2SC69D	0086
2SC69E	0086
2SC69F	0086
2SC69G	0086
2SC69GN	0086
2SC69H	0086
2SC69J	0086
2SC69K	0086
2SC69L	0086
2SC69M	0086
2SC69OR	0086
2SC69R	0086
2SC69X	0086
2SC69Y	0086
2SC70	0233
2SC70A	0233
2SC70B	0233
2SC70C	0233
2SC70D	0233
2SC70E	0233
2SC70F	0233
2SC70G	0233
2SC70GN	0233
2SC70H	0233
2SC70J	0233
2SC70K	0233
2SC70L	0233
2SC70M	0233
2SC70OR	0233
2SC70R	0233
2SC70X	0233
2SC70Y	0233
2SC71	0600
2SC71A	0600
2SC71B	0600
2SC71C	0600
2SC71D	0600
2SC71E	0600
2SC71F	0600
2SC71G	0600
2SC71GN	0600
2SC71H	0600
2SC71J	0600
2SC71M	0600
2SC71OR	0600
2SC71R	0600
2SC71X	0600
2SC72	0600
2SC72A	0600
2SC72B	0600
2SC72C	0600
2SC72D	0600
2SC72E	0600
2SC72F	0600
2SC72G	0600
2SC72GN	0600
2SC72H	0600
2SC72J	0600
2SC72K	0600
2SC72L	0600
2SC72M	0600
2SC72OR	0600
2SC72R	0600
2SC72X	0600
2SC72Y	0600
2SC73	0600
2SC73A	0600
2SC73B	0600
2SC73C	0600
2SC73D	0600
2SC73E	0600
2SC73F	0600
2SC73G	0600
2SC73GN	0600
2SC73H	0600
2SC73J	0600
2SC73K	0600
2SC73L	0600
2SC73LBGL	0600
2SC73M	0600
2SC73OR	0600
2SC73R	0600
2SC73X	0600
2SC73Y	0600
2SC74	0198
2SC74-GR	0198
2SC74-O	0198
2SC74-R	0198
2SC74-Y	0198
2SC74A	0198
2SC74B	0198
2SC74C	0198
2SC74D	0198
2SC74E	0198
2SC74F	0198
2SC74G	0198
2SC74GN	0198
2SC74H	0198
2SC74J	0198
2SC74K	0198
2SC74L	0198
2SC74M	0198
2SC74R	0198
2SC74X	0198
2SC74Y	0198
2SC75	0600
2SC75A	0600
2SC75B	0600
2SC75B-1	0600
2SC75C	0600
2SC75E	0600
2SC75F	0600
2SC75G	0600
2SC75GN	0600
2SC75H	0600
2SC75J	0600
2SC75K	0600
2SC75L	0600
2SC75M	0600
2SC75OR	0600
2SC75R	0600
2SC75X	0600
2SC75Y	0600
2SC76	0600
2SC76A	0600
2SC76B	0600
2SC76C	0600
2SC76D	0600
2SC76E	0600
2SC76F	0600
2SC76G	0600
2SC76GN	0600
2SC76H	0600
2SC76J	0600
2SC76K	0600
2SC76L	0600
2SC76M	0600
2SC76OR	0600
2SC76R	0600
2SC76X	0600
2SC76Y	0600
2SC77	0600
2SC77A	0600
2SC77B	0600
2SC77C	0600
2SC77D	0600
2SC77F	0600
2SC77G	0600
2SC77GN	0600
2SC77H	0600
2SC77J	0600
2SC77K	0600
2SC77L	0600
2SC77M	0600
2SC77OR	0600
2SC77R	0600
2SC77X	0600
2SC77Y	0600
2SC77Z	0600
2SC78	0600
2SC78A	0600
2SC78B	0600
2SC78C	0600
2SC78D	0600
2SC78E	0600
2SC78F	0600
2SC78G	0600
2SC78GN	0600
2SC78H	0600
2SC78J	0600
2SC78K	0600
2SC78L	0600
2SC78M	0600
2SC78OR	0600
2SC78R	0600
2SC78X	0600
2SC78Y	0600
2SC79	0224
2SC79A	0224
2SC79B	0224
2SC79C	0224
2SC79D	0224
2SC79E	0224
2SC79F	0224
2SC79G	0224
2SC79GN	0224
2SC79H	0224
2SC79J	0224
2SC79K	0224
2SC79L	0224
2SC79M	0224
2SC79OR	0224
2SC79R	0224
2SC79X	0224
2SC79Y	0224
2SC80	0007
2SC80A	0007
2SC80B	0007
2SC80C	0007
2SC80D	0007
2SC80E	0007
2SC80F	0007
2SC80G	0007
2SC80GN	0007
2SC80H	0007
2SC80J	0007
2SC80K	0007
2SC80L	0007
2SC80M	0007
2SC80OR	0007
2SC80R	0007
2SC80X	0007
2SC80Y	0007
2SC81	0018
2SC81A	0018
2SC81B	0018
2SC81C	0018
2SC81D	0018
2SC81E	0018
2SC81F	0018
2SC81G	0018
2SC81GN	0018
2SC81H	0018
2SC81J	0018
2SC81K	0018
2SC81L	0018
2SC81M	0018
2SC81OR	0018
2SC81R	0018
2SC82	0127
2SC82BN	0127
2SC82R	0127
2SC83	0127
2SC84	0600
2SC85	0208
2SC85O	0208
2SC86	0208
2SC87	0016
2SC87A	0016
2SC87B	0016
2SC87C	0016
2SC87D	0016
2SC87E	0016
2SC87F	0016
2SC87G	0016
2SC87GN	0016
2SC87H	0016
2SC87J	0016
2SC87K	0016
2SC87L	0016
2SC87M	0016
2SC87OR	0016
2SC87R	0016
2SC87X	0016
2SC87Y	0016
2SC88	0086
2SC88A	0086
2SC88B	0086
2SC88C	0086
2SC88D	0086
2SC88E	0086
2SC88F	0086
2SC88G	0086
2SC88GN	0086
2SC88H	0086
2SC88J	0086
2SC88K	0086
2SC88L	0086
2SC88M	0086
2SC88OR	0086
2SC88R	0086
2SC88X	0086
2SC88Y	0086
2SC89	0600
2SC89A	0600
2SC89B	0600
2SC89C	0600
2SC89D	0600
2SC89E	0600
2SC89F	0600
2SC89G	0600
2SC89GN	0600
2SC89H	0600
2SC89J	0600
2SC89K	0600
2SC89L	0600
2SC89M	0600
2SC89OR	0600
2SC89R	0600
2SC89X	0600
2SC89Y	0600
2SC90	0600
2SC90A	0600
2SC90B	0600
2SC90C	0600
2SC90D	0600
2SC90E	0600
2SC90F	0600
2SC90G	0600
2SC90GN	0600
2SC90H	0600
2SC90J	0600
2SC90K	0600
2SC90L	0600
2SC90M	0600
2SC90OR	0600
2SC90R	0600
2SC90X	0600
2SC90Y	0600
2SC91	0600
2SC91A	0600
2SC91B	0600
2SC91C	0600
2SC91D	0600
2SC91E	0600
2SC91F	0600
2SC91G	0600
2SC91GN	0600
2SC91H	0600
2SC91J	0600
2SC91K	0600
2SC91L	0600
2SC91M	0600
2SC91OR	0600
2SC91R	0600
2SC91X	0600
2SC91Y	0600
2SC92	0042
2SC93	0042
2SC94	0042
2SC94AP	0042
2SC95	0086
2SC95A	0086
2SC95B	0086
2SC95C	0086
2SC95D	0086
2SC95E	0086
2SC95F	0086
2SC95G	0086
2SC95GN	0086
2SC95H	0086
2SC95J	0086
2SC95K	0086
2SC95L	0086
2SC95M	0086
2SC95OR	0086
2SC95R	0086
2SC95X	0086
2SC95Y	0086
2SC96	0079
2SC97	0086
2SC97A	0086
2SC97B	0086
2SC97C	0086
2SC97D	0086
2SC97E	0086
2SC97F	0086
2SC97G	0086
2SC97GN	0086
2SC97H	0086
2SC97J	0086
2SC97K	0086
2SC97L	0086
2SC97M	0086
2SC97OR	0086
2SC97R	0086
2SC97X	0086
2SC97Y	0086
2SC98	0016
2SC98A	0016
2SC98B	0016
2SC98C	0016
2SC98D	0016
2SC98E	0016
2SC98F	0016
2SC98G	0016
2SC98GN	0016
2SC98H	0016
2SC98J	0016
2SC98K	0016
2SC98L	0016
2SC98M	0016
2SC98OR	0016
2SC98R	0016
2SC98X	0016
2SC98Y	0016
2SC99	0016
2SC99A	0016
2SC99B	0016
2SC99C	0016
2SC99D	0016
2SC99E	0016
2SC99F	0016
2SC99G	0016
2SC99GN	0016
2SC99H	0016
2SC99J	0016
2SC99K	0016
2SC99L	0016
2SC99M	0016
2SC99OR	0016
2SC99R	0016
2SC99X	0016
2SC99Y	0016
2SC100	0016
2SC100-OY	0016
2SC100A	0016
2SC100B	0016
2SC100C	0016
2SC100D	0016
2SC100E	0016
2SC100F	0016
2SC100G	0016
2SC100GN	0016
2SC100J	0016
2SC100L	0016
2SC100M	0016
2SC100OR	0016
2SC100R	0016
2SC100X	0016
2SC100Y	0016
2SC101	0074
2SC101A	0074
2SC101B	0074
2SC101C	0074
2SC101D	0074
2SC101E	0074
2SC101F	0074
2SC101G	0074
2SC101GN	0074
2SC101H	0074
2SC101J	0074
2SC101K	0074
2SC101L	0074
2SC101M	0074
2SC101OR	0074
2SC101R	0074
2SC101X	0074
2SC101XL	0074
2SC101Y	0074
2SC102	OEM
2SC103	0016
2SC103A	0016
2SC103B	0016
2SC103C	0016
2SC103D	0016
2SC103E	0016
2SC103F	0016
2SC103G	0016
2SC103GN	0016
2SC103H	0016
2SC103J	0016
2SC103K	0016
2SC103L	0016
2SC103M	0016
2SC103X	0016
2SC103Y	0016
2SC104	0016
2SC104A	0016
2SC104B	0016
2SC104C	0016
2SC104D	0016
2SC104E	0016
2SC104F	0016
2SC104G	0016
2SC104GN	0016
2SC104H	0016
2SC104J	0016
2SC104K	0016
2SC104L	0016
2SC104M	0016
2SC104OR	0016

If replacement code is OEM, contact original manufacturer for replacement.

DEVICE TYPE	REPL CODE
2SC104R	0016
2SC104X	0016
2SC104Y	0016
2SC105	0016
2SC105B	0016
2SC106	0622
2SC106A	0622
2SC106B	0622
2SC106C	0622
2SC106G	0622
2SC106GN	0622
2SC106H	0622
2SC106J	0622
2SC106K	0622
2SC106L	0622
2SC106M	0622
2SC106OR	0622
2SC106R	0622
2SC106X	0622
2SC106Y	0622
2SC107	0042
2SC107G4	0042
2SC107G5	0042
2SC107G6	0042
2SC108	0086
2SC108A	0086
2SC108A-O	0086
2SC108A-R	0086
2SC108B	0086
2SC108C	0086
2SC108D	0086
2SC108E	0086
2SC108F	0086
2SC108G	0086
2SC108GN	0086
2SC108H	0086
2SC108J	0086
2SC108K	0086
2SC108L	0086
2SC108M	0086
2SC108OR	0086
2SC108R	0086
2SC108X	0086
2SC108Y	0086
2SC109	0086
2SC109A	0086
2SC109A-O	0086
2SC109A-R	0086
2SC109A-Y	0086
2SC109B	0086
2SC109C	0086
2SC109D	0086
2SC109E	0086
2SC109F	0086
2SC109G	0086
2SC109G1	0086
2SC109GN	0086
2SC109J	0086
2SC109K	0086
2SC109L	0086
2SC109OR	0086
2SC109R	0086
2SC109X	0086
2SC109Y	0086
2SC110	0016
2SC110A	0016
2SC110B	0016
2SC110C	0016
2SC110D	0016
2SC110E	0016
2SC110F	0016
2SC110G	0016
2SC110GN	0016
2SC110H	0016
2SC110J	0016
2SC110K	0016
2SC110L	0016
2SC110M	0016
2SC110OR	0016
2SC110R	0016
2SC110X	0016
2SC110Y	0016
2SC111	0016
2SC111A	0016
2SC111B	0016
2SC111C	0016
2SC111D	0016
2SC111E	0016
2SC111F	0016
2SC111G	0016
2SC111GN	0016
2SC111H	0016
2SC111J	0016
2SC111K	0016
2SC111L	0016
2SC111M	0016
2SC111OR	0016
2SC111R	0016
2SC111X	0016
2SC111Y	0016
2SC112	0086
2SC112A	0086
2SC112B	0086
2SC112C	0086
2SC112D	0086
2SC112E	0086
2SC112F	0086
2SC112G	0086
2SC112GN	0086
2SC112H	0086
2SC112J	0086
2SC112K	0086
2SC112L	0086
2SC112M	0086
2SC112OR	0086
2SC112R	0086
2SC112X	0086
2SC112Y	0086
2SC113	0086
2SC113A	0086
2SC113B	0086
2SC113C	0086
2SC113D	0086
2SC113E	0086
2SC113F	0086
2SC113G	0086
2SC113GN	0086
2SC113H	0086
2SC113J	0086
2SC113K	0086
2SC113L	0086
2SC113M	0086
2SC113OR	0086
2SC113R	0086
2SC113X	0086
2SC113Y	0086
2SC114	0086
2SC114A	0086
2SC114B	0086
2SC114C	0086
2SC114D	0086
2SC114E	0086
2SC114EF	0086
2SC114F	0086
2SC114G	0086
2SC114GN	0086
2SC114H	0086
2SC114J	0086
2SC114K	0086
2SC114L	0086
2SC114M	0086
2SC114OR	0086
2SC114R	0086
2SC114X	0086
2SC114Y	0086
2SC115	0086
2SC115-1	0086
2SC115-2	0086
2SC115-3	0086
2SC115-43	0086
2SC115A	0086
2SC115B	0086
2SC115C	0086
2SC115D	0086
2SC115E	0086
2SC115F	0086
2SC115G	0086
2SC115GN	0086
2SC115H	0086
2SC115J	0086
2SC115K	0086
2SC115L	0086
2SC115M	0086
2SC115OR	0086
2SC115R	0086
2SC115X	0086
2SC115Y	0086
2SC116	2061
2SC116-OR	2061
2SC116A	2061
2SC116B	2061
2SC116C	2061
2SC116D	2061
2SC116E	2061
2SC116F	2061
2SC116G	2061
2SC116GN	2061
2SC116H	2061
2SC116J	2061
2SC116K	2061
2SC116L	2061
2SC116M	2061
2SC116OR	2061
2SC116R	2061
2SC116T	2061
2SC116X	2061
2SC116Y	2061
2SC117	0086
2SC117A	0086
2SC117B	0086
2SC117C	0086
2SC117D	0086
2SC117F	0086
2SC117G	0086
2SC117GN	0086
2SC117H	0086
2SC117J	0086
2SC117K	0086
2SC117L	0086
2SC117M	0086
2SC117OR	0086
2SC117R	0086
2SC117X	0086
2SC117Y	0086
2SC118	0086
2SC118A	0086
2SC118B	0086
2SC118C	0086
2SC118D	0086
2SC118E	0086
2SC118F	0086
2SC118G	0086
2SC118GN	0086
2SC118H	0086
2SC118J	0086
2SC118L	0086
2SC118M	0086
2SC118OR	0086
2SC118R	0086
2SC118X	0086
2SC118Y	0086
2SC119	0086
2SC119A	0086
2SC119B	0086
2SC119C	0086
2SC119D	0086
2SC119E	0086
2SC119F	0086
2SC119G	0086
2SC119GN	0086
2SC119H	0086
2SC119J	0086
2SC119K	0086
2SC119L	0086
2SC119M	0086
2SC119OR	0086
2SC119R	0086
2SC119X	0086
2SC119Y	0086
2SC120	0016
2SC120A	0016
2SC120B	0016
2SC120C	0016
2SC120D	0016
2SC120E	0016
2SC120F	0016
2SC120G	0016
2SC120GN	0016
2SC120H	0016
2SC120J	0016
2SC120K	0016
2SC120M	0016
2SC120R	0016
2SC120Y	0016
2SC121	0086
2SC121A	0086
2SC121B	0086
2SC121C	0086
2SC121D	0086
2SC121E	0086
2SC121F	0086
2SC121G	0086
2SC121GN	0086
2SC121H	0086
2SC121J	0086
2SC121L	0086
2SC121M	0086
2SC121OR	0086
2SC121R	0086
2SC121X	0086
2SC121Y	0086
2SC122	0086
2SC122A	0086
2SC122B	0086
2SC122C	0086
2SC122D	0086
2SC122E	0086
2SC122F	0086
2SC122G	0086
2SC122GN	0086
2SC122H	0086
2SC122J	0086
2SC122K	0086
2SC122L	0086
2SC122M	0086
2SC122OR	0086
2SC122R	0086
2SC122X	0086
2SC122Y	0086
2SC123	0086
2SC123A	0086
2SC123B	0086
2SC123C	0086
2SC123D	0086
2SC123E	0086
2SC123F	0086
2SC123G	0086
2SC123GN	0086
2SC123H	0086
2SC123J	0086
2SC123K	0086
2SC123L	0086
2SC123M	0086
2SC123OR	0086
2SC123R	0086
2SC123X	0086
2SC123Y	0086
2SC124	0086
2SC124A	0086
2SC124B	0086
2SC124C	0086
2SC124D	0086
2SC124E	0086
2SC124F	0086
2SC124G	0086
2SC124GN	0086
2SC124H	0086
2SC124J	0086
2SC124K	0086
2SC124L	0086
2SC124M	0086
2SC124OR	0086
2SC124R	0086
2SC124X	0086
2SC124Y	0086
2SC125	0086
2SC126	0285
2SC127	0016
2SC127A	0016
2SC127B	0016
2SC127C	0016
2SC127D	0016
2SC127E	0016
2SC127F	0016
2SC127G	0016
2SC127GN	0016
2SC127H	0016
2SC127J	0016
2SC127K	0016
2SC127L	0016
2SC127M	0016
2SC127OR	0016
2SC127R	0016
2SC127X	0016
2SC127Y	0016
2SC128	0600
2SC128A	0600
2SC128B	0600
2SC128C	0600
2SC128D	0600
2SC128E	0600
2SC128F	0600
2SC128G	0600
2SC128GN	0600
2SC128H	0600
2SC128J	0600
2SC128K	0600
2SC128L	0600
2SC128M	0600
2SC128OR	0600
2SC128R	0600
2SC128X	0600
2SC128Y	0600
2SC129	0600
2SC129A	0600
2SC129B	0600
2SC129C	0600
2SC129D	0600
2SC129E	0600
2SC129F	0600
2SC129GN	0600
2SC129H	0600
2SC129J	0600
2SC129K	0600
2SC129L	0600
2SC129M	0600
2SC129OR	0600
2SC129R	0600
2SC129X	0600
2SC129Y	0600
2SC130	0086
2SC130A	0086
2SC130B	0086
2SC130C	0086
2SC130D	0086
2SC130E	0086
2SC130F	0086
2SC130GN	0086
2SC130H	0086
2SC130J	0086
2SC130K	0086
2SC130L	0086
2SC130M	0086
2SC130OR	0086
2SC130R	0086
2SC130X	0086
2SC130Y	0086
2SC131	0016
2SC131A	0016
2SC131B	0016
2SC131C	0016
2SC131D	0016
2SC131E	0016
2SC131F	0016
2SC131G	0016
2SC131GN	0016
2SC131H	0016
2SC131J	0016
2SC131L	0016
2SC131M	0016
2SC131OR	0016
2SC131R	0016
2SC131T	0016
2SC131Y	0016
2SC132	0016
2SC132A	0016
2SC132B	0016
2SC132C	0016
2SC132D	0016
2SC132E	0016
2SC132F	0016
2SC132G	0016
2SC132GN	0016
2SC132H	0016
2SC132J	0016
2SC132K	0016
2SC132L	0016
2SC132M	0016
2SC132OR	0016
2SC132R	0016
2SC132X	0016
2SC132Y	0016
2SC133	0016
2SC133A	0016
2SC133C	0016
2SC133D	0016
2SC133E	0016
2SC133F	0016
2SC133G	0016
2SC133GN	0016
2SC133H	0016
2SC133J	0016
2SC133K	0016
2SC133L	0016
2SC133M	0016
2SC133OR	0016
2SC133X	0016
2SC133Y	0016
2SC134	0016
2SC134A	0016
2SC134B	0016
2SC134C	0016
2SC134D	0016
2SC134E	0016
2SC134F	0016
2SC134G	0016
2SC134GN	0016
2SC134H	0016
2SC134J	0016
2SC134K	0016
2SC134L	0016
2SC134M	0016
2SC134OR	0016
2SC134R	0016
2SC134X	0016
2SC134Y	0016
2SC135	0016
2SC135A	0016
2SC135B	0016
2SC135C	0016
2SC135D	0016
2SC135E	0016
2SC135F	0016
2SC135G	0016
2SC135GN	0016
2SC135H	0016
2SC135J	0016
2SC135L	0016
2SC135M	0016
2SC135OR	0016
2SC135R	0016
2SC135X	0016
2SC135Y	0016
2SC136	0016
2SC136D	0016
2SC137	0016
2SC137A	0016
2SC137B	0016
2SC137C	0016
2SC137D	0016
2SC137E	0016
2SC137F	0016
2SC137G	0016
2SC137GN	0016
2SC137H	0016
2SC137J	0016
2SC137K	0016
2SC137L	0016
2SC137M	0016
2SC137OR	0016
2SC137R	0016
2SC137X	0016
2SC137Y	0016
2SC138	0016
2SC138A	0016
2SC138B	0016
2SC138C	0016
2SC138D	0016
2SC138E	0016
2SC138F	0016
2SC138G	0016
2SC138GN	0016
2SC138H	0016
2SC138J	0016
2SC138L	0016
2SC138M	0016
2SC138OR	0016
2SC138R	0016
2SC138S	0016
2SC138X	0016
2SC139	0016
2SC139A	0016
2SC139B	0016
2SC139C	0016
2SC139D	0016
2SC139E	0016
2SC139F	0016
2SC139G	0016
2SC139GN	0016
2SC139H	0016
2SC139J	0016
2SC139K	0016
2SC139M	0016
2SC139OR	0016
2SC139R	0016
2SC139X	0016
2SC139Y	0016
2SC140	0086
2SC140A	0086
2SC140C	0086
2SC140D	0086
2SC140E	0086
2SC140F	0086
2SC140G	0086
2SC140GN	0086
2SC140H	0086
2SC140J	0086
2SC140K	0086
2SC140L	0086
2SC140M	0086
2SC140OR	0086
2SC140R	0086
2SC140X	0086
2SC140Y	0086
2SC141	0079
2SC142	0079
2SC143	0079
2SC144	0079
2SC144A	0079
2SC145	0079
2SC146	0079
2SC146A	0079
2SC147	2354
2SC147A	2354
2SC147B	2354
2SC147C	2354
2SC147D	2354
2SC147F	2354
2SC147G	2354
2SC147GN	2354
2SC147H	2354
2SC147J	2354
2SC147K	2354
2SC147L	2354
2SC147M	2354
2SC147R	2354
2SC147X	2354
2SC147Y	2354
2SC148	0144
2SC148A	0144
2SC148B	0144
2SC148D	0144
2SC148E	0144
2SC148F	0144
2SC148G	0144
2SC148GN	0144
2SC148H	0144
2SC148J	0144
2SC148K	0144
2SC148L	0144
2SC148M	0144
2SC148OR	0144
2SC148R	0144
2SC148X	0144
2SC148Y	0144
2SC149	0086
2SC150	2377
2SC150-OR	2377
2SC150A	2377
2SC150B	2377
2SC150C	2377
2SC150D	2377
2SC150E	2377
2SC150F	2377
2SC150G	2377
2SC150GN	2377
2SC150H	2377
2SC150J	2377
2SC150K	2377
2SC150L	2377
2SC150M	2377
2SC150OR	2377
2SC150R	2377
2SC150T	2377
2SC150X	2377
2SC150Y	2377
2SC151	0198
2SC151A	0198
2SC151B	0198
2SC151C	0198
2SC151D	0198
2SC151E	0198
2SC151G	0198
2SC151GN	0198
2SC151H	0198
2SC151HA	0198
2SC151HB	0198
2SC151HC	0198
2SC151J	0198
2SC151K	0198
2SC151L	0198
2SC151M	0198
2SC151OR	0198
2SC151R	0198
2SC151X	0198
2SC151Y	0198
2SC152	0198
2SC152A	0198
2SC152B	0198
2SC152C	0198
2SC152D	0198
2SC152E	0198
2SC152F	0198
2SC152G	0198
2SC152GN	0198
2SC152H	0198
2SC152HA	0198
2SC152HB	0198
2SC152HC	0198
2SC152J	0198
2SC152K	0198
2SC152L	0198
2SC152M	0198
2SC152OR	0198
2SC152R	0198
2SC152X	0198
2SC152Y	0198
2SC153	0079
2SC154	0716
2SC154(C)	0716
2SC154-OR	0716
2SC154A	0716
2SC154B	0716
2SC154C	0716
2SC154D	0716
2SC154E	0716
2SC154F	0716
2SC154G	0716
2SC154GN	0716
2SC154H	0716
2SC154HA	0716
2SC154HB	0716
2SC154HC	0716
2SC154J	0716
2SC154K	0716
2SC154L	0716
2SC154M	0716
2SC154OR	0716
2SC154R	0716
2SC154X	0716
2SC154Y	0716
2SC155	0127
2SC155A	0127
2SC155B	0127
2SC155C	0127
2SC155D	0127
2SC155E	0127
2SC155F	0127
2SC155GN	0127
2SC155H	0127
2SC155J	0127
2SC155K	0127
2SC155L	0127
2SC155M	0127
2SC155OR	0127
2SC155R	0127
2SC155X	0127
2SC155Y	0127
2SC156	0127
2SC156A	0127
2SC156B	0127
2SC156C	0127
2SC156D	0127
2SC156E	0127
2SC156F	0127
2SC156G	0127
2SC156GN	0127
2SC156H	0127
2SC156J	0127
2SC156K	0127
2SC156L	0127
2SC156M	0127
2SC156OR	0127
2SC156R	0127
2SC156X	0127
2SC156Y	0127
2SC157	0198
2SC157A	0198
2SC157B	0198
2SC157C	0198
2SC157D	0198
2SC157E	0198
2SC157F	0198
2SC157G	0198
2SC157GN	0198
2SC157J	0198
2SC157K	0198
2SC157L	0198
2SC157M	0198
2SC157OR	0198
2SC157R	0198
2SC157Y	0198
2SC158	0198
2SC158A	0198
2SC158B	0198
2SC158C	0198
2SC158D	0198
2SC158E	0198
2SC158F	0198
2SC158G	0198
2SC158GN	0198
2SC158H	0198
2SC158J	0198
2SC158K	0198
2SC158L	0198
2SC158M	0198
2SC158OR	0198
2SC158R	0198
2SC158X	0198
2SC158Y	0198
2SC159	0016
2SC159A	0016
2SC159B	0016

DEVICE TYPE	REPL CODE
2SC159C	0016
2SC159D	0016
2SC159E	0016
2SC159F	0016
2SC159GN	0016
2SC159H	0016
2SC159J	0016
2SC159K	0016
2SC159L	0016
2SC159M	0016
2SC159OR	0016
2SC159R	0016
2SC159X	0016
2SC159Y	0016
2SC160	0016
2SC160A	0016
2SC160B	0016
2SC160C	0016
2SC160D	0016
2SC160E	0016
2SC160F	0016
2SC160G	0016
2SC160GN	0016
2SC160H	0016
2SC160J	0016
2SC160K	0016
2SC160L	0016
2SC160M	0016
2SC160OR	0016
2SC160R	0016
2SC160X	0016
2SC160Y	0016
2SC161	0543
2SC162	0079
2SC163	0493
2SC163A	0493
2SC163B	0493
2SC163C	0493
2SC163D	0493
2SC163E	0493
2SC163F	0493
2SC163G	0493
2SC163GN	0493
2SC163H	0493
2SC163J	0493
2SC163K	0493
2SC163L	0493
2SC163M	0493
2SC163OR	0493
2SC163R	0493
2SC163X	0493
2SC163Y	0493
2SC164	0018
2SC164(E)	0018
2SC164A	0018
2SC164B	0018
2SC164C	0018
2SC164D	0018
2SC164E	0018
2SC164F	0018
2SC164G	0018
2SC164GN	0018
2SC164H	0018
2SC164J	0018
2SC164K	0018
2SC164L	0018
2SC164OR	0018
2SC164R	0018
2SC164X	0018
2SC164Y	0018
2SC165	0595
2SC166	0016
2SC166A	0016
2SC166B	0016
2SC166C	0016
2SC166D	0016
2SC166E	0016
2SC166F	0016
2SC166G	0016
2SC166GN	0016
2SC166H	0016
2SC166J	0016
2SC166K	0016
2SC166L	0016
2SC166M	0016
2SC166OR	0016
2SC166R	0016
2SC166X	0016
2SC166Y	0016
2SC167	0016
2SC167A	0016
2SC167B	0016
2SC167C	0016
2SC167D	0016
2SC167E	0016
2SC167F	0016
2SC167G	0016
2SC167GN	0016
2SC167H	0016
2SC167K	0016
2SC167L	0016
2SC167M	0016
2SC167OR	0016
2SC167X	0016
2SC167Y	0016
2SC168	0079
2SC169	0016
2SC170	0016
2SC170A	0016

DEVICE TYPE	REPL CODE
2SC170B	0016
2SC170C	0016
2SC170D	0016
2SC170E	0016
2SC170F	0016
2SC170G	0016
2SC170GN	0016
2SC170H	0016
2SC170J	0016
2SC170K	0016
2SC170L	0016
2SC170M	0016
2SC170OR	0016
2SC170R	0016
2SC170X	0016
2SC170Y	0016
2SC171	0016
2SC171A	0016
2SC171B	0016
2SC171C	0016
2SC171D	0016
2SC171E	0016
2SC171F	0016
2SC171G	0016
2SC171GN	0016
2SC171H	0016
2SC171J	0016
2SC171K	0016
2SC171L	0016
2SC171M	0016
2SC171OR	0016
2SC171R	0016
2SC171X	0016
2SC171Y	0016
2SC172	0016
2SC172A	0016
2SC172B	0016
2SC172C	0016
2SC172D	0016
2SC172E	0016
2SC172F	0016
2SC172G	0016
2SC172GN	0016
2SC172H	0016
2SC172J	0016
2SC172K	0016
2SC172L	0016
2SC172M	0016
2SC172OR	0016
2SC172R	0016
2SC172Y	0016
2SC173	0600
2SC173A	0600
2SC173B	0600
2SC173C	0600
2SC173D	0600
2SC173E	0600
2SC173F	0600
2SC173G	0600
2SC173GN	0600
2SC173H	0600
2SC173J	0600
2SC173K	0600
2SC173L	0600
2SC173M	0600
2SC173OR	0600
2SC173R	0600
2SC173X	0600
2SC173Y	0600
2SC174	0016
2SC174A	0016
2SC174B	0016
2SC174C	0016
2SC174D	0016
2SC174E	0016
2SC174F	0016
2SC174G	0016
2SC174GN	0016
2SC174H	0016
2SC174J	0016
2SC174K	0016
2SC174L	0016
2SC174M	0016
2SC174OR	0016
2SC174Q	0016
2SC174R	0016
2SC174S	0016
2SC174X	0016
2SC174Y	0016
2SC175	0600
2SC175A	0600
2SC175B	0600
2SC175BL	0600
2SC175C	0600
2SC175D	0600
2SC175E	0600
2SC175F	0600
2SC175G	0600
2SC175GN	0600
2SC175H	0600
2SC175J	0600
2SC175K	0600
2SC175L	0600
2SC175M	0600
2SC175OR	0600
2SC175R	0600
2SC175X	0600
2SC175Y	0600
2SC176	0600

DEVICE TYPE	REPL CODE
2SC176A	0600
2SC176B	0600
2SC176C	0600
2SC176D	0600
2SC176E	0600
2SC176F	0600
2SC176G	0600
2SC176GN	0600
2SC176H	0600
2SC176J	0600
2SC176K	0600
2SC176L	0600
2SC176OR	0600
2SC176R	0600
2SC176X	0600
2SC176Y	0600
2SC177	0600
2SC177A	0600
2SC177B	0600
2SC177C	0600
2SC177D	0600
2SC177E	0600
2SC177F	0600
2SC177G	0600
2SC177GN	0600
2SC177H	0600
2SC177J	0600
2SC177K	0600
2SC177L	0600
2SC177M	0600
2SC177OR	0600
2SC177R	0600
2SC177X	0600
2SC177Y	0600
2SC178	0600
2SC178A	0600
2SC178B	0600
2SC178C	0600
2SC178D	0600
2SC178E	0600
2SC178F	0600
2SC178G	0600
2SC178H	0600
2SC178J	0600
2SC178K	0600
2SC178L	0600
2SC178M	0600
2SC178OR	0600
2SC178R	0600
2SC178X	0600
2SC178Y	0600
2SC179	0208
2SC179A	0208
2SC179B	0208
2SC179C	0208
2SC179D	0208
2SC179E	0208
2SC179F	0208
2SC179G	0208
2SC179GN	0208
2SC179H	0208
2SC179J	0208
2SC179K	0208
2SC179L	0208
2SC179M	0208
2SC179OR	0208
2SC179R	0208
2SC179X	0208
2SC179Y	0208
2SC180	0208
2SC180A	0208
2SC180B	0208
2SC180C	0208
2SC180D	0208
2SC180E	0208
2SC180F	0208
2SC180G	0208
2SC180GN	0208
2SC180H	0208
2SC180J	0208
2SC180K	0208
2SC180L	0208
2SC180M	0208
2SC180OR	0208
2SC180R	0208
2SC180X	0208
2SC180Y	0208
2SC181	0208
2SC181A	0208
2SC181B	0208
2SC181C	0208
2SC181D	0208
2SC181E	0208
2SC181F	0208
2SC181G	0208
2SC181GN	0208
2SC181H	0208
2SC181J	0208
2SC181L	0208
2SC181M	0208
2SC181OR	0208
2SC181R	0208
2SC181X	0208
2SC181Y	0208
2SC182	0475
2SC182(Q)	0475
2SC182(V)	0475
2SC182A	0475
2SC182B	0475

DEVICE TYPE	REPL CODE
2SC182C	0475
2SC182D	0475
2SC182E	0475
2SC182F	0475
2SC182G	0475
2SC182GN	0475
2SC182H	0475
2SC182J	0475
2SC182K	0475
2SC182L	0475
2SC182M	0475
2SC182Q	0475
2SC182R	0475
2SC182V	0475
2SC182X	0475
2SC182Y	0475
2SC183	0475
2SC183(P)	0475
2SC183(Q)(R)	0475
2SC183(R)	0475
2SC183-1	0475
2SC183-OR	0475
2SC183A	0475
2SC183AP	0475
2SC183B	0475
2SC183BK	0475
2SC183C	0475
2SC183D	0475
2SC183E	0475
2SC183F	0475
2SC183G	0475
2SC183GN	0475
2SC183H	0475
2SC183J	0475
2SC183K	0475
2SC183L	0475
2SC183M	0475
2SC183OR	0475
2SC183P	0475
2SC183Q	0475
2SC183R	0475
2SC183S	0475
2SC183W	0475
2SC183X	0475
2SC183Y	0475
2SC184	0475
2SC184(R)	0475
2SC184-OR	0475
2SC184A	0475
2SC184AP	0475
2SC184B	0475
2SC184BK	0475
2SC184C	0475
2SC184D	0475
2SC184E	0475
2SC184F	0475
2SC184G	0475
2SC184GN	0475
2SC184H	0475
2SC184J	0475
2SC184K	0475
2SC184L	0475
2SC184M	0475
2SC184OR	0475
2SC184P	0475
2SC184Q	0475
2SC184R	0475
2SC184X	0475
2SC184Y	0475
2SC185	0475
2SC185A	0475
2SC185B	0475
2SC185C	0475
2SC185E	0475
2SC185F	0475
2SC185G	0475
2SC185GN	0475
2SC185H	0475
2SC185J	0475
2SC185K	0475
2SC185L	0475
2SC185M	0475
2SC185OR	0475
2SC185Q	0475
2SC185R	0475
2SC185V	0475
2SC185X	0475
2SC185Y	0475
2SC186	0127
2SC186A	0127
2SC186B	0127
2SC186C	0127
2SC186D	0127
2SC186E	0127
2SC186F	0127
2SC186G	0127
2SC186GN	0127
2SC186H	0127
2SC186J	0127
2SC186K	0127
2SC186L	0127
2SC186M	0127
2SC186OR	0127
2SC186R	0127
2SC186X	0127
2SC186Y	0127
2SC187	0674
2SC187(I)	0674
2SC187-OR	0674

DEVICE TYPE	REPL CODE
2SC187A	0674
2SC187B	0674
2SC187C	0674
2SC187D	0674
2SC187E	0674
2SC187F	0674
2SC187G	0674
2SC187H	0674
2SC187I	0674
2SC187J	0674
2SC187K	0674
2SC187L	0674
2SC187M	0674
2SC187OR	0674
2SC187R	0674
2SC187X	0674
2SC187Y	0674
2SC188	0086
2SC188A	0086
2SC188AB	0086
2SC188B	0086
2SC188C	0086
2SC188D	0086
2SC188E	0086
2SC188F	0086
2SC188G	0086
2SC188GN	0086
2SC188J	0086
2SC188K	0086
2SC188L	0086
2SC188M	0086
2SC188OR	0086
2SC188R	0086
2SC188X	0086
2SC188Y	0086
2SC189	0086
2SC189A	0086
2SC189B	0086
2SC189C	0086
2SC189D	0086
2SC189E	0086
2SC189F	0086
2SC189G	0086
2SC189GN	0086
2SC189H	0086
2SC189J	0086
2SC189K	0086
2SC189L	0086
2SC189M	0086
2SC189OR	0086
2SC189R	0086
2SC189X	0086
2SC189Y	0086
2SC190	0086
2SC190A	0086
2SC190B	0086
2SC190C	0086
2SC190D	0086
2SC190E	0086
2SC190F	0086
2SC190GN	0086
2SC190H	0086
2SC190J	0086
2SC190K	0086
2SC190L	0086
2SC190M	0086
2SC190OR	0086
2SC190R	0086
2SC190X	0086
2SC190Y	0086
2SC191	0016
2SC191A	0016
2SC191B	0016
2SC191C	0016
2SC191D	0016
2SC191E	0016
2SC191F	0016
2SC191G	0016
2SC191GN	0016
2SC191H	0016
2SC191J	0016
2SC191K	0016
2SC191L	0016
2SC191M	0016
2SC191OR	0016
2SC191R	0016
2SC191X	0016
2SC191Y	0016
2SC192	0016
2SC192A	0016
2SC192B	0016
2SC192C	0016
2SC192D	0016
2SC192E	0016
2SC192F	0016
2SC192G	0016
2SC192GN	0016
2SC192H	0016
2SC192J	0016
2SC192K	0016
2SC192L	0016
2SC192M	0016
2SC192OR	0016
2SC192R	0016
2SC192X	0016
2SC192Y	0016
2SC193	0016
2SC193A	0016
2SC193B	0016

DEVICE TYPE	REPL CODE
2SC193C	0016
2SC193D	0016
2SC193E	0016
2SC193F	0016
2SC193G	0016
2SC193GN	0016
2SC193H	0016
2SC193J	0016
2SC193K	0016
2SC193L	0016
2SC193M	0016
2SC193OR	0016
2SC193R	0016
2SC193X	0016
2SC194	0016
2SC194A	0016
2SC194B	0016
2SC194C	0016
2SC194D	0016
2SC194E	0016
2SC194F	0016
2SC194G	0016
2SC194GN	0016
2SC194H	0016
2SC194J	0016
2SC194K	0016
2SC194L	0016
2SC194M	0016
2SC194OR	0016
2SC194X	0016
2SC194Y	0016
2SC195	0016
2SC195A	0016
2SC195B	0016
2SC195C	0016
2SC195D	0016
2SC195E	0016
2SC195F	0016
2SC195G	0016
2SC195GN	0016
2SC195H	0016
2SC195J	0016
2SC195K	0016
2SC195L	0016
2SC195M	0016
2SC195OR	0016
2SC195R	0016
2SC195X	0016
2SC195Y	0016
2SC196	0016
2SC196A	0016
2SC196B	0016
2SC196C	0016
2SC196D	0016
2SC196E	0016
2SC196F	0016
2SC196G	0016
2SC196GN	0016
2SC196H	0016
2SC196J	0016
2SC196K	0016
2SC196L	0016
2SC196M	0016
2SC196OR	0016
2SC196R	0016
2SC196X	0016
2SC196Y	0016
2SC197	0016
2SC197A	0016
2SC197B	0016
2SC197C	0016
2SC197D	0016
2SC197E	0016
2SC197F	0016
2SC197G	0016
2SC197GN	0016
2SC197H	0016
2SC197J	0016
2SC197K	0016
2SC197L	0016
2SC197M	0016
2SC197OR	0016
2SC197R	0016
2SC197X	0016
2SC197Y	0016
2SC198	0530
2SC198A	0530
2SC198H	0530
2SC198S	0530
2SC198T	0530
2SC199	0016
2SC199A	0016
2SC199B	0016
2SC199C	0016
2SC199D	0016
2SC199E	0016
2SC199F	0016
2SC199G	0016
2SC199GN	0016
2SC199H	0016
2SC199J	0016
2SC199K	0016
2SC199L	0016
2SC199M	0016
2SC199OR	0016
2SC199R	0016
2SC199X	0016
2SC199Y	0016
2SC200	0016

DEVICE TYPE	REPL CODE
2SC200A	0016
2SC200B	0016
2SC200C	0016
2SC200D	0016
2SC200E	0016
2SC200F	0016
2SC200G	0016
2SC200GN	0016
2SC200H	0016
2SC200J	0016
2SC200K	0016
2SC200L	0016
2SC200M	0016
2SC200OR	0016
2SC200R	0016
2SC200X	0016
2SC200Y	0016
2SC201	0016
2SC201A	0016
2SC201B	0016
2SC201C	0016
2SC201D	0016
2SC201E	0016
2SC201F	0016
2SC201G	0016
2SC201GN	0016
2SC201H	0016
2SC201J	0016
2SC201K	0016
2SC201L	0016
2SC201M	0016
2SC201R	0016
2SC201X	0016
2SC201Y	0016
2SC202	0016
2SC202A	0016
2SC202B	0016
2SC202C	0016
2SC202D	0016
2SC202E	0016
2SC202F	0016
2SC202G	0016
2SC202GN	0016
2SC202H	0016
2SC202J	0016
2SC202K	0016
2SC202L	0016
2SC202M	0016
2SC202OR	0016
2SC202R	0016
2SC202X	0016
2SC202Y	0016
2SC203	0595
2SC203A	0595
2SC203AA	0595
2SC203B	0595
2SC203C	0595
2SC203D	0595
2SC203E	0595
2SC203F	0595
2SC203G	0595
2SC203GN	0595
2SC203H	0595
2SC203J	0595
2SC203K	0595
2SC203L	0595
2SC203M	0595
2SC203OR	0595
2SC203R	0595
2SC203X	0595
2SC203Y	0595
2SC204	0016
2SC204A	0016
2SC204B	0016
2SC204C	0016
2SC204D	0016
2SC204E	0016
2SC204F	0016
2SC204G	0016
2SC204GN	0016
2SC204H	0016
2SC204J	0016
2SC204K	0016
2SC204L	0016
2SC204M	0016
2SC204OR	0016
2SC204R	0016
2SC204X	0016
2SC204Y	0016
2SC205	0016
2SC205A	0016
2SC205B	0016
2SC205C	0016
2SC205D	0016
2SC205F	0016
2SC205G	0016
2SC205GN	0016
2SC205H	0016
2SC205J	0016
2SC205K	0016
2SC205L	0016
2SC205M	0016
2SC205OR	0016
2SC205R	0016
2SC205X	0016
2SC205Y	0016
2SC206	0216
2SC206-OR	0216
2SC206A	0216

If replacement code is OEM, contact original manufacturer for replacement.

DEVICE TYPE	REPL CODE
2SC206B	0216
2SC206C	0216
2SC206D	0216
2SC206E	0216
2SC206G	0216
2SC206GN	0216
2SC206H	0216
2SC206J	0216
2SC206K	0216
2SC206L	0216
2SC206M	0216
2SC206OR	0216
2SC206R	0216
2SC206RED	0216
2SC206WHITE	0216
2SC206X	0216
2SC206Y	0216
2SC207	0259
2SC208	0007
2SC208A	0007
2SC208B	0007
2SC208C	0007
2SC208D	0007
2SC208E	0007
2SC208F	0007
2SC208G	0007
2SC208GN	0007
2SC208H	0007
2SC208J	0007
2SC208K	0007
2SC208L	0007
2SC208M	0007
2SC208OR	0007
2SC208R	0007
2SC208X	0007
2SC208Y	0007
2SC209	0259
2SC210	0086
2SC210A	0086
2SC210B	0086
2SC210C	0086
2SC210D	0086
2SC210E	0086
2SC210F	0086
2SC210G	0086
2SC210GN	0086
2SC210H	0086
2SC210J	0086
2SC210K	0086
2SC210L	0086
2SC210M	0086
2SC210OR	0086
2SC210R	0086
2SC210X	0086
2SC210Y	0086
2SC211	0086
2SC211A	0086
2SC211B	0086
2SC211C	0086
2SC211D	0086
2SC211F	0086
2SC211G	0086
2SC211GN	0086
2SC211H	0086
2SC211J	0086
2SC211K	0086
2SC211L	0086
2SC211M	0086
2SC211OR	0086
2SC211R	0086
2SC211X	0086
2SC211Y	0086
2SC212	0086
2SC212A	0086
2SC212B	0086
2SC212C	0086
2SC212D	0086
2SC212E	0086
2SC212F	0086
2SC212G	0086
2SC212GN	0086
2SC212H	0086
2SC212J	0086
2SC212K	0086
2SC212L	0086
2SC212M	0086
2SC212OR	0086
2SC212R	0086
2SC212X	0086
2SC212Y	0086
2SC213	0622
2SC213A	0622
2SC213B	0622
2SC213C	0622
2SC213D	0622
2SC213E	0622
2SC213F	0622
2SC213G	0622
2SC213GN	0622
2SC213H	0622
2SC213J	0622
2SC213K	0622
2SC213L	0622
2SC213M	0622
2SC213OR	0622
2SC213R	0622
2SC213X	0622
2SC213Y	0622
2SC214	0622
2SC214A	0622
2SC214B	0622
2SC214C	0622
2SC214D	0622
2SC214E	0622
2SC214F	0622
2SC214G	0622
2SC214GN	0622
2SC214H	0622
2SC214J	0622
2SC214K	0622
2SC214L	0622
2SC214M	0622
2SC214OR	0622
2SC214X	0622
2SC214Y	0622
2SC215	0622
2SC215A	0622
2SC215B	0622
2SC215C	0622
2SC215E	0622
2SC215F	0622
2SC215G	0622
2SC215GN	0622
2SC215H	0622
2SC215J	0622
2SC215K	0622
2SC215L	0622
2SC215M	0622
2SC215OR	0622
2SC215R	0622
2SC215X	0622
2SC215Y	0622
2SC216	0086
2SC216A	0086
2SC216B	0086
2SC216C	0086
2SC216D	0086
2SC216E	0086
2SC216F	0086
2SC216G	0086
2SC216GN	0086
2SC216H	0086
2SC216J	0086
2SC216K	0086
2SC216L	0086
2SC216M	0086
2SC216OR	0086
2SC216R	0086
2SC216X	0086
2SC216Y	0086
2SC217	0086
2SC217A	0086
2SC217B	0086
2SC217C	0086
2SC217D	0086
2SC217E	0086
2SC217F	0086
2SC217G	0086
2SC217GN	0086
2SC217H	0086
2SC217J	0086
2SC217K	0086
2SC217L	0086
2SC217M	0086
2SC217OR	0086
2SC217R	0086
2SC217X	0086
2SC217Y	0086
2SC218	0086
2SC218A	0086
2SC218B	0086
2SC218C	0086
2SC218D	0086
2SC218E	0086
2SC218F	0086
2SC218G	0086
2SC218GN	0086
2SC218H	0086
2SC218J	0086
2SC218L	0086
2SC218M	0086
2SC218OR	0086
2SC218R	0086
2SC218X	0086
2SC218Y	0086
2SC219	0007
2SC220	0086
2SC220A	0086
2SC220B	0086
2SC220C	0086
2SC220D	0086
2SC220E	0086
2SC220F	0086
2SC220G	0086
2SC220GN	0086
2SC220H	0086
2SC220K	0086
2SC220L	0086
2SC220M	0086
2SC220OR	0086
2SC220R	0086
2SC220X	0086
2SC220Y	0086
2SC221	0086
2SC221A	0086
2SC221B	0086
2SC221C	0086
2SC221D	0086
2SC221E	0086
2SC221F	0086
2SC221G	0086
2SC221GN	0086
2SC221H	0086
2SC221J	0086
2SC221K	0086
2SC221L	0086
2SC221M	0086
2SC221OR	0086
2SC221R	0086
2SC221X	0086
2SC221Y	0086
2SC222	0086
2SC222A	0086
2SC222B	0086
2SC222C	0086
2SC222D	0086
2SC222E	0086
2SC222F	0086
2SC222G	0086
2SC222GN	0086
2SC222H	0086
2SC222J	0086
2SC222L	0086
2SC222M	0086
2SC222OR	0086
2SC222R	0086
2SC222X	0086
2SC222Y	0086
2SC223	0622
2SC223A	0622
2SC223B	0622
2SC223C	0622
2SC223D	0622
2SC223E	0622
2SC223F	0622
2SC223G	0622
2SC223GN	0622
2SC223H	0622
2SC223J	0622
2SC223L	0622
2SC223M	0622
2SC223OR	0622
2SC223R	0622
2SC223X	0622
2SC223Y	0622
2SC224	0622
2SC224A	0622
2SC224B	0622
2SC224C	0622
2SC224D	0622
2SC224E	0622
2SC224F	0622
2SC224G	0622
2SC224GN	0622
2SC224GR-GL	0622
2SC224H	0622
2SC224J	0622
2SC224K	0622
2SC224L	0622
2SC224M	0622
2SC224OR	0622
2SC224R	0622
2SC224X	0622
2SC224Y	0622
2SC225	0086
2SC225A	0086
2SC225B	0086
2SC225C	0086
2SC225D	0086
2SC225E	0086
2SC225F	0086
2SC225G	0086
2SC225GN	0086
2SC225H	0086
2SC225J	0086
2SC225L	0086
2SC225M	0086
2SC225OR	0086
2SC225R	0086
2SC225X	0086
2SC225Y	0086
2SC226	0086
2SC226A	0086
2SC226B	0086
2SC226C	0086
2SC226D	0086
2SC226E	0086
2SC226F	0086
2SC226G	0086
2SC226GN	0086
2SC226H	0086
2SC226J	0086
2SC226K	0086
2SC226L	0086
2SC226M	0086
2SC226OR	0086
2SC226R	0086
2SC226X	0086
2SC226Y	0086
2SC227	0086
2SC227A	0086
2SC227B	0086
2SC227C	0086
2SC227D	0086
2SC227E	0086
2SC227F	0086
2SC227G	0086
2SC227GN	0086
2SC227H	0086
2SC227J	0086
2SC227K	0086
2SC227L	0086
2SC227M	0086
2SC227OR	0086
2SC227R	0086
2SC227X	0086
2SC227Y	0086
2SC228	0086
2SC228A	0086
2SC228B	0086
2SC228C	0086
2SC228D	0086
2SC228E	0086
2SC228F	0086
2SC228G	0086
2SC228GN	0086
2SC228H	0086
2SC228J	0086
2SC228K	0086
2SC228L	0086
2SC228M	0086
2SC228OR	0086
2SC228R	0086
2SC228X	0086
2SC228Y	0086
2SC229	0622
2SC229A	0622
2SC229B	0622
2SC229C	0622
2SC229D	0622
2SC229E	0622
2SC229F	0622
2SC229G	0622
2SC229GN	0622
2SC229H	0622
2SC229J	0622
2SC229K	0622
2SC229L	0622
2SC229M	0622
2SC229OR	0622
2SC229R	0622
2SC229X	0622
2SC229Y	0622
2SC230	2988
2SC230A	2988
2SC230B	2988
2SC230C	2988
2SC230D	2988
2SC230E	2988
2SC230F	2988
2SC230G	2988
2SC230GN	2988
2SC230H	2988
2SC230J	2988
2SC230K	2988
2SC230L	2988
2SC230M	2988
2SC230OR	2988
2SC230R	2988
2SC230X	2988
2SC230Y	2988
2SC231	2996
2SC231A	2996
2SC231B	2996
2SC231C	2996
2SC231D	2996
2SC231E	2996
2SC231F	2996
2SC231G	2996
2SC231GN	2996
2SC231H	2996
2SC231J	2996
2SC231K	2996
2SC231L	2996
2SC231M	2996
2SC231OR	2996
2SC231R	2996
2SC231X	2996
2SC231Y	2996
2SC232	3009
2SC232A	3009
2SC232B	3009
2SC232C	3009
2SC232D	3009
2SC232E	3009
2SC232F	3009
2SC232G	3009
2SC232GN	3009
2SC232H	3009
2SC232J	3009
2SC232K	3009
2SC232L	3009
2SC232M	3009
2SC232OR	3009
2SC232R	3009
2SC232X	3009
2SC232Y	3009
2SC233	0622
2SC233A	0622
2SC233B	0622
2SC233C	0622
2SC233D	0622
2SC233F	0622
2SC233G	0622
2SC233GN	0622
2SC233H	0622
2SC233J	0622
2SC233L	0622
2SC233M	0622
2SC233OR	0622
2SC233R	0622
2SC233X	0622
2SC233Y	0622
2SC234	0086
2SC234A	0086
2SC234B	0086
2SC234C	0086
2SC234D	0086
2SC234E	0086
2SC234F	0086
2SC234G	0086
2SC234GN	0086
2SC234H	0086
2SC234J	0086
2SC234K	0086
2SC234L	0086
2SC234M	0086
2SC234OR	0086
2SC234R	0086
2SC234X	0086
2SC234Y	0086
2SC235	0018
2SC235(0)	0018
2SC235-0	0018
2SC235-O	0018
2SC235A	0018
2SC235B	0018
2SC235C	0018
2SC235D	0018
2SC235E	0018
2SC235F	0018
2SC235G	0018
2SC235GN	0018
2SC235H	0018
2SC235J	0018
2SC235K	0018
2SC235L	0018
2SC235M	0018
2SC235OR	0018
2SC235R	0018
2SC235X	0018
2SC235Y	0018
2SC236	0086
2SC236A	0086
2SC236B	0086
2SC236C	0086
2SC236D	0086
2SC236E	0086
2SC236F	0086
2SC236G	0086
2SC236GN	0086
2SC236H	0086
2SC236J	0086
2SC236K	0086
2SC236L	0086
2SC236M	0086
2SC236OR	0086
2SC236R	0086
2SC236X	0086
2SC236Y	0086
2SC237	0016
2SC237A	0016
2SC237B	0016
2SC237C	0016
2SC237D	0016
2SC237E	0016
2SC237F	0016
2SC237G	0016
2SC237GN	0016
2SC237H	0016
2SC237J	0016
2SC237K	0016
2SC237L	0016
2SC237M	0016
2SC237OR	0016
2SC237R	0016
2SC237X	0016
2SC237Y	0016
2SC238	0016
2SC238A	0016
2SC238B	0016
2SC238C	0016
2SC238D	0016
2SC238E	0016
2SC238F	0016
2SC238G	0016
2SC238GN	0016
2SC238H	0016
2SC238J	0016
2SC238K	0016
2SC238L	0016
2SC238M	0016
2SC238OR	0016
2SC238R	0016
2SC238X	0016
2SC238Y	0016
2SC239	0016
2SC239A	0016
2SC239B	0016
2SC239C	0016
2SC239D	0016
2SC239E	0016
2SC239F	0016
2SC239G	0016
2SC239GN	0016
2SC239H	0016
2SC239J	0016
2SC239K	0016
2SC239L	0016
2SC239M	0016
2SC239OR	0016
2SC239R	0016
2SC239X	0016
2SC239Y	0016
2SC240	0103
2SC240A	0103
2SC240B	0103
2SC240C	0103
2SC240D	0103
2SC240E	0103
2SC240F	0103
2SC240G	0103
2SC240GN	0103
2SC240H	0103
2SC240J	0103
2SC240K	0103
2SC240L	0103
2SC240M	0103
2SC240OR	0103
2SC240R	0103
2SC240X	0103
2SC240Y	0103
2SC241	0103
2SC241A	0103
2SC241B	0103
2SC241C	0103
2SC241D	0103
2SC241E	0103
2SC241F	0103
2SC241G	0103
2SC241GN	0103
2SC241H	0103
2SC241K	0103
2SC241L	0103
2SC241M	0103
2SC241OR	0103
2SC241R	0103
2SC241X	0103
2SC241Y	0103
2SC242	0103
2SC242A	0103
2SC242B	0103
2SC242C	0103
2SC242D	0103
2SC242E	0103
2SC242F	0103
2SC242G	0103
2SC242GN	0103
2SC242H	0103
2SC242J	0103
2SC242K	0103
2SC242L	0103
2SC242M	0103
2SC242OR	0103
2SC242R	0103
2SC242X	0103
2SC242Y	0103
2SC243	0177
2SC244	1966
2SC244A	1966
2SC244B	1966
2SC244C	1966
2SC244D	1966
2SC244E	1966
2SC244F	1966
2SC244G	1966
2SC244GN	1966
2SC244J	1966
2SC244K	1966
2SC244L	1966
2SC244M	1966
2SC244OR	1966
2SC244R	1966
2SC244X	1966
2SC244Y	1966
2SC245	1966
2SC246	1966
2SC247	0086
2SC247A	0086
2SC247B	0086
2SC247C	0086
2SC247D	0086
2SC247E	0086
2SC247F	0086
2SC247G	0086
2SC247GN	0086
2SC247H	0086
2SC247J	0086
2SC247L	0086
2SC247M	0086
2SC247OR	0086
2SC247R	0086
2SC247X	0086
2SC247Y	0086
2SC248	0016
2SC248(0)	0016
2SC248A	0016
2SC248B	0016
2SC248C	0016
2SC248D	0016
2SC248E	0016
2SC248F	0016
2SC248G	0016
2SC248J	0016
2SC248K	0016
2SC248L	0016
2SC248M	0016
2SC248OR	0016
2SC248R	0016
2SC248X	0016
2SC248Y	0016
2SC249	0086
2SC249A	0086
2SC249B	0086
2SC249C	0086
2SC249D	0086
2SC249E	0086
2SC249F	0086
2SC249G	0086
2SC249GN	0086
2SC249H	0086
2SC249J	0086
2SC249K	0086
2SC249L	0086
2SC249M	0086
2SC249OR	0086
2SC249R	0086
2SC249X	0086
2SC249Y	0086
2SC250	0016
2SC250A	0016
2SC250B	0016
2SC250C	0016
2SC250D	0016
2SC250E	0016
2SC250F	0016
2SC250G	0016
2SC250GN	0016
2SC250H	0016
2SC250J	0016
2SC250K	0016
2SC250L	0016
2SC250M	0016
2SC250OR	0016
2SC250R	0016
2SC250X	0016
2SC250Y	0016
2SC251	0259
2SC251A	0259
2SC251B	0259
2SC251C	0259
2SC251D	0259
2SC251E	0259
2SC251F	0259
2SC251G	0259
2SC251GN	0259
2SC251H	0259
2SC251J	0259
2SC251K	0259
2SC251L	0259
2SC251M	0259
2SC251OR	0259
2SC251R	0259
2SC251X	0259
2SC251Y	0259
2SC252	0259
2SC252A	0259
2SC252B	0259
2SC252C	0259
2SC252D	0259
2SC252E	0259
2SC252F	0259
2SC252G	0259
2SC252GN	0259
2SC252H	0259
2SC252J	0259
2SC252K	0259
2SC252L	0259
2SC252M	0259
2SC252OR	0259
2SC252R	0259
2SC252X	0259
2SC252Y	0259
2SC253	0259
2SC253A	0259
2SC253B	0259
2SC253C	0259
2SC253D	0259
2SC253E	0259
2SC253F	0259
2SC253G	0259
2SC253GN	0259
2SC253H	0259
2SC253J	0259
2SC253K	0259
2SC253L	0259
2SC253M	0259
2SC253OR	0259
2SC253R	0259
2SC253RS	0259
2SC253X	0259
2SC253Y	0259
2SC254	0086
2SC255	0086
2SC256	0086
2SC257	1984
2SC258	0264
2SC259	0264
2SC260	0595
2SC260D	0595
2SC260E	0595
2SC261	0698
2SC262	0595

DEVICE TYPE	REPL CODE
2SC263	0144
2SC263A	0144
2SC263B	0144
2SC263C	0144
2SC263D	0144
2SC263E	0144
2SC263F	0144
2SC263G	0144
2SC263GN	0144
2SC263H	0144
2SC263J	0144
2SC263K	0144
2SC263L	0144
2SC263M	0144
2SC263OR	0144
2SC263R	0144
2SC263X	0144
2SC263Y	0144
2SC264	0079
2SC265	0079
2SC266	0475
2SC266A	0475
2SC266B	0475
2SC266C	0475
2SC266D	0475
2SC266E	0475
2SC266F	0475
2SC266G	0475
2SC266GN	0475
2SC266H	0475
2SC266J	0475
2SC266K	0475
2SC266L	0475
2SC266M	0475
2SC266OR	0475
2SC266R	0475
2SC266X	0475
2SC266Y	0475
2SC267	0016
2SC267A	0016
2SC267B	0016
2SC267C	0016
2SC267D	0016
2SC267E	0016
2SC267F	0016
2SC267G	0016
2SC267GN	0016
2SC267H	0016
2SC267J	0016
2SC267K	0016
2SC267L	0016
2SC267M	0016
2SC267OR	0016
2SC267R	0016
2SC267X	0016
2SC267Y	0016
2SC268	0086
2SC268-OR	0086
2SC268A	0086
2SC268B	0086
2SC268C	0086
2SC268D	0086
2SC268E	0086
2SC268F	0086
2SC268G	0086
2SC268GN	0086
2SC268H	0086
2SC268J	0086
2SC268K	0086
2SC268L	0086
2SC268M	0086
2SC268OR	0086
2SC268R	0086
2SC268X	0086
2SC268Y	0086
2SC269	0144
2SC269A	0144
2SC269B	0144
2SC269C	0144
2SC269D	0144
2SC269E	0144
2SC269F	0144
2SC269G	0144
2SC269GN	0144
2SC269H	0144
2SC269J	0144
2SC269K	0144
2SC269L	0144
2SC269M	0144
2SC269OR	0144
2SC269R	0144
2SC269X	0144
2SC269Y	0144
2SC270	0074
2SC271	0144
2SC271A	0144
2SC271B	0144
2SC271C	0144
2SC271D	0144
2SC271E	0144
2SC271F	0144
2SC271G	0144
2SC271GN	0144
2SC271H	0144
2SC271J	0144
2SC271K	0144
2SC271L	0144
2SC271M	0144
2SC271OR	0144
2SC271R	0144
2SC271X	0144
2SC271Y	0144
2SC272	0144
2SC272A	0144
2SC272B	0144
2SC272C	0144
2SC272D	0144
2SC272E	0144
2SC272F	0144
2SC272G	0144
2SC272GN	0144
2SC272H	0144
2SC272J	0144
2SC272K	0144
2SC272L	0144
2SC272M	0144
2SC272OR	0144
2SC272R	0144
2SC272X	0144
2SC272Y	0144
2SC273	0233
2SC277C	0208
2SC278-OR	0475
2SC280	0111
2SC280AO	0111
2SC281	0536
2SC281(B)	0536
2SC281-OR	0536
2SC281A	0536
2SC281B	0536
2SC281BL	0536
2SC281C	0536
2SC281C-EP	0536
2SC281D	0536
2SC281E	0536
2SC281EP	0536
2SC281F	0536
2SC281G	0536
2SC281GN	0536
2SC281H	0536
2SC281J	0536
2SC281K	0536
2SC281L	0536
2SC281M	0536
2SC281OR	0536
2SC281R	0536
2SC281X	0536
2SC281Y	0536
2SC282	0016
2SC282A	0016
2SC282B	0016
2SC282C	0016
2SC282D	0016
2SC282E	0016
2SC282F	0016
2SC282G	0016
2SC282GN	0016
2SC282H	0016
2SC282J	0016
2SC282K	0016
2SC282L	0016
2SC282M	0016
2SC282OR	0016
2SC282R	0016
2SC282X	0016
2SC282Y	0016
2SC283	1562
2SC283A	1562
2SC283B	1562
2SC283C	1562
2SC283D	1562
2SC283E	1562
2SC283F	1562
2SC283G	1562
2SC283GN	1562
2SC283H	1562
2SC283J	1562
2SC283K	1562
2SC283L	1562
2SC283M	1562
2SC283OR	1562
2SC283R	1562
2SC283X	1562
2SC283Y	1562
2SC284	0016
2SC284A	0016
2SC284B	0016
2SC284C	0016
2SC284D	0016
2SC284E	0016
2SC284F	0016
2SC284G	0016
2SC284GN	0016
2SC284H	0016
2SC284J	0016
2SC284K	0016
2SC284L	0016
2SC284M	0016
2SC284OR	0016
2SC284R	0016
2SC284X	0016
2SC284Y	0016
2SC285	0493
2SC285A	0493
2SC285B	0493
2SC285C	0493
2SC285D	0493
2SC285E	0493
2SC285F	0493
2SC285G	0493
2SC285GN	0493
2SC285H	0493
2SC285J	0493
2SC285K	0493
2SC285L	0493
2SC285M	0493
2SC285OR	0493
2SC285R	0493
2SC285X	0493
2SC285Y	0493
2SC286	0144
2SC286A	0144
2SC286B	0144
2SC286C	0144
2SC286D	0144
2SC286E	0144
2SC286F	0144
2SC286G	0144
2SC286GN	0144
2SC286H	0144
2SC286J	0144
2SC286K	0144
2SC286L	0144
2SC286M	0144
2SC286OR	0144
2SC286R	0144
2SC286X	0144
2SC286Y	0144
2SC287	0475
2SC287-OR	0475
2SC287A	0475
2SC287A-B	OEM
2SC287B	0475
2SC287C	0475
2SC287D	0475
2SC287E	0475
2SC287F	0475
2SC287G	0475
2SC287GN	0475
2SC287H	0475
2SC287J	0475
2SC287K	0475
2SC287L	0475
2SC287M	0475
2SC287OR	0475
2SC287R	0475
2SC287X	0475
2SC287Y	0475
2SC288	1342
2SC288A	1342
2SC288A(5B)	1342
2SC288A1	1342
2SC288A1B	1342
2SC288A5-B	1342
2SC288AB	1342
2SC288B	1342
2SC288C	1342
2SC288D	1342
2SC288E	1342
2SC288F	1342
2SC288G	1342
2SC288GN	1342
2SC288H	1342
2SC288J	1342
2SC288K	1342
2SC288L	1342
2SC288M	1342
2SC288OR	1342
2SC288R	1342
2SC288X	1342
2SC288Y	1342
2SC289	0144
2SC289A	0144
2SC289B	0144
2SC289C	0144
2SC289D	0144
2SC289E	0144
2SC289F	0144
2SC289G	0144
2SC289GN	0144
2SC289H	0144
2SC289J	0144
2SC289K	0144
2SC289L	0144
2SC289M	0144
2SC289OR	0144
2SC289R	0144
2SC289X	0144
2SC289Y	0144
2SC290	0698
2SC291	0086
2SC291A	0086
2SC291B	0086
2SC291C	0086
2SC291D	0086
2SC291E	0086
2SC291F	0086
2SC291G	0086
2SC291GN	0086
2SC291H	0086
2SC291J	0086
2SC291K	0086
2SC291M	0086
2SC291OR	0086
2SC291R	0086
2SC291X	0086
2SC291Y	0086
2SC292	0622
2SC292A	0622
2SC292B	0622
2SC292C	0622
2SC292D	0622
2SC292E	0622
2SC292F	0622
2SC292G	0622
2SC292GN	0622
2SC292H	0622
2SC292J	0622
2SC292K	0622
2SC292L	0622
2SC292M	0622
2SC292OR	0622
2SC292R	0622
2SC292X	0622
2SC292Y	0622
2SC293	0622
2SC293A	0622
2SC293B	0622
2SC293C	0622
2SC293D	0622
2SC293E	0622
2SC293F	0622
2SC293G	0622
2SC293GN	0622
2SC293H	0622
2SC293J	0622
2SC293K	0622
2SC293L	0622
2SC293M	0622
2SC293OR	0622
2SC293R	0622
2SC293X	0622
2SC293Y	0622
2SC294	0285
2SC294X	0285
2SC295	0079
2SC296	0423
2SC297	0560
2SC298	0622
2SC298-4	0622
2SC299	3336
2SC299A	3336
2SC299B	3336
2SC299C	3336
2SC299D	3336
2SC299E	3336
2SC299F	3336
2SC299G	3336
2SC299GN	3336
2SC299H	3336
2SC299J	3336
2SC299K	3336
2SC299L	3336
2SC299M	3336
2SC299OR	3336
2SC299R	3336
2SC299X	3336
2SC299Y	3336
2SC300	0016
2SC300A	0016
2SC300B	0016
2SC300C	0016
2SC300D	0016
2SC300E	0016
2SC300F	0016
2SC300G	0016
2SC300GN	0016
2SC300H	0016
2SC300J	0016
2SC300K	0016
2SC300L	0016
2SC300M	0016
2SC300OR	0016
2SC300R	0016
2SC300X	0016
2SC300Y	0016
2SC301	0016
2SC301A	0016
2SC301B	0016
2SC301C	0016
2SC301D	0016
2SC301E	0016
2SC301F	0016
2SC301G	0016
2SC301GN	0016
2SC301H	0016
2SC301J	0016
2SC301K	0016
2SC301L	0016
2SC301M	0016
2SC301OR	0016
2SC301R	0016
2SC301X	0016
2SC301Y	0016
2SC302	0016
2SC302A	0016
2SC302B	0016
2SC302C	0016
2SC302D	0016
2SC302E	0016
2SC302F	0016
2SC302G	0016
2SC302GN	0016
2SC302H	0016
2SC302J	0016
2SC302K	0016
2SC302L	0016
2SC302M	0016
2SC302OR	0016
2SC302R	0016
2SC302X	0016
2SC302Y	0016
2SC303	0595
2SC304	0595
2SC305	0595
2SC305A	0595
2SC305B	0595
2SC305C	0595
2SC305D	0595
2SC305E	0595
2SC305F	0595
2SC305G	0595
2SC305GN	0595
2SC305H	0595
2SC305J	0595
2SC305K	0595
2SC305L	0595
2SC305M	0595
2SC305OR	0595
2SC305R	0595
2SC305X	0595
2SC305Y	0595
2SC306	0086
2SC306A	0086
2SC306B	0086
2SC306C	0086
2SC306D	0086
2SC306E	0086
2SC306F	0086
2SC306G	0086
2SC306GN	0086
2SC306H	0086
2SC306J	0086
2SC306K	0086
2SC306L	0086
2SC306M	0086
2SC306OR	0086
2SC306R	0086
2SC306X	0086
2SC306Y	0086
2SC307	0086
2SC307-OR	0086
2SC307A	0086
2SC307B	0086
2SC307C	0086
2SC307D	0086
2SC307E	0086
2SC307F	0086
2SC307G	0086
2SC307GN	0086
2SC307H	0086
2SC307J	0086
2SC307K	0086
2SC307L	0086
2SC307M	0086
2SC307OR	0086
2SC307R	0086
2SC307X	0086
2SC307Y	0086
2SC308	0086
2SC308A	0086
2SC308B	0086
2SC308C	0086
2SC308D	0086
2SC308E	0086
2SC308F	0086
2SC308G	0086
2SC308GN	0086
2SC308H	0086
2SC308J	0086
2SC308K	0086
2SC308L	0086
2SC308M	0086
2SC308OR	0086
2SC308R	0086
2SC308X	0086
2SC308Y	0086
2SC309	0086
2SC309A	0086
2SC309B	0086
2SC309C	0086
2SC309D	0086
2SC309E	0086
2SC309F	0086
2SC309G	0086
2SC309GN	0086
2SC309H	0086
2SC309J	0086
2SC309K	0086
2SC309L	0086
2SC309M	0086
2SC309OR	0086
2SC309R	0086
2SC309X	0086
2SC309Y	0086
2SC310	0086
2SC310A	0086
2SC310B	0086
2SC310C	0086
2SC310D	0086
2SC310E	0086
2SC310F	0086
2SC310G	0086
2SC310GN	0086
2SC310H	0086
2SC310J	0086
2SC310K	0086
2SC310L	0086
2SC310M	0086
2SC310OR	0086
2SC310R	0086
2SC310X	0086
2SC310Y	0086
2SC311	0698
2SC311A	OEM
2SC311B	0698
2SC311C	0698
2SC311D	0698
2SC311E	0698
2SC311F	0698
2SC311G	0698
2SC311GN	0698
2SC311H	0698
2SC311J	0698
2SC311K	0698
2SC311L	0698
2SC311M	0698
2SC311OR	0698
2SC311R	0698
2SC311X	0698
2SC311Y	0698
2SC312	0630
2SC312A	0630
2SC312B	0630
2SC312C	0630
2SC312D	0630
2SC312E	0630
2SC312F	0630
2SC312G	0630
2SC312GN	0630
2SC312H	0630
2SC312J	0630
2SC312K	0630
2SC312L	0630
2SC312M	0630
2SC312OR	0630
2SC312R	0630
2SC312X	0630
2SC312Y	0630
2SC-313	3390
2SC313	3390
2SC313-OR	3390
2SC313A	3390
2SC313B	3390
2SC313C	3390
2SC313D	3390
2SC313E	3390
2SC313F	3390
2SC313G	3390
2SC313GN	3390
2SC313H	3390
2SC313J	3390
2SC313K	3390
2SC313L	3390
2SC313M	3390
2SC313OR	3390
2SC313R	3390
2SC313X	3390
2SC313Y	3390
2SC314	0630
2SC314TA	OEM
2SC315	0016
2SC315A	0016
2SC315B	0016
2SC315C	0016
2SC315D	0016
2SC315E	0016
2SC315F	0016
2SC315G	0016
2SC315GN	0016
2SC315H	0016
2SC315J	0016
2SC315K	0016
2SC315L	0016
2SC315M	0016
2SC315OR	0016
2SC315R	0016
2SC315X	0016
2SC315Y	0016
2SC316	0016
2SC316A	0016
2SC316B	0016
2SC316C	0016
2SC316D	0016
2SC316E	0016
2SC316F	0016
2SC316G	0016
2SC316GN	0016
2SC316H	0016
2SC316J	0016
2SC316K	0016
2SC316L	0016
2SC316M	0016
2SC316OR	0016
2SC316R	0016
2SC316X	0016
2SC316Y	0016
2SC317	0016
2SC317A	0016
2SC317B	0016
2SC317C	0016
2SC317D	0016
2SC317E	0016
2SC317F	0016
2SC317G	0016
2SC317GN	0016
2SC317H	0016
2SC317J	0016
2SC317K	0016
2SC317L	0016
2SC317M	0016
2SC317OR	0016
2SC317R	0016
2SC317X	0016
2SC317Y	0016
2SC318	0016
2SC318A	0016
2SC318AB	0016
2SC318B	0016
2SC318C	0016
2SC318D	0016
2SC318E	0016
2SC318F	0016
2SC318G	0016
2SC318GN	0016
2SC318H	0016
2SC318J	0016
2SC318K	0016
2SC318L	0016
2SC318M	0016
2SC318OR	0016
2SC318R	0016
2SC318X	0016
2SC318Y	0016
2SC319	0493
2SC319A	0493
2SC319B	0493
2SC319C	0493
2SC319D	0493
2SC319E	0493
2SC319F	0493
2SC319G	0493
2SC319GN	0493
2SC319H	0493
2SC319J	0493
2SC319K	0493
2SC319L	0493
2SC319M	0493
2SC319OR	0493
2SC319R	0493
2SC319X	0493
2SC319Y	0493
2SC320	3421
2SC320A	3421
2SC320B	3421
2SC320C	3421
2SC320D	3421
2SC320E	3421
2SC320F	3421
2SC320G	3421
2SC320GN	3421
2SC320H	3421
2SC320J	3421
2SC320K	3421
2SC320L	3421
2SC320M	3421
2SC320OR	3421
2SC320R	3421
2SC320X	3421
2SC320Y	3421
2SC321	0016
2SC321A	0016
2SC321B	0016
2SC321C	0016
2SC321D	0016
2SC321E	0016
2SC321F	0016
2SC321G	0016
2SC321GN	0016
2SC321H	0016
2SC321HA	0016
2SC321HB	0016
2SC321HC	0016
2SC321J	0016
2SC321K	0016
2SC321L	0016
2SC321M	0016
2SC321OR	0016
2SC321R	0016
2SC321X	0016
2SC321Y	0016
2SC322	0079
2SC323	0016
2SC323A	0016
2SC323B	0016
2SC323C	0016
2SC323D	0016
2SC323E	0016
2SC323F	0016
2SC323G	0016
2SC323GN	0016
2SC323H	0016
2SC323J	0016
2SC323K	0016
2SC323L	0016
2SC323M	0016
2SC323OR	0016
2SC323R	0016
2SC323X	0016
2SC323Y	0016
2SC324	0016
2SC324(H)	0016
2SC324A	0016

If replacement code is OEM, contact original manufacturer for replacement.

DEVICE TYPE	REPL CODE
2SC324B	0016
2SC324C	0016
2SC324D	0016
2SC324E	0016
2SC324F	0016
2SC324G	0016
2SC324GN	0016
2SC324H	0016
2SC324HA	0016
2SC324J	0016
2SC324K	0016
2SC324L	0016
2SC324M	0016
2SC324OR	0016
2SC324R	0016
2SC324X	0016
2SC324Y	0016
2SC325	0259
2SC325C	0259
2SC326	0418
2SC327	0127
2SC328	0259
2SC329	0259
2SC330	2844
2SC331	2844
2SC332	0079
2SC333	0016
2SC334	0016
2SC335	0536
2SC336	0016
2SC337	0016
2SC338	0076
2SC339	0079
2SC340	0475
2SC340H	OEM
2SC341	0324
2SC342	0079
2SC343	0016
2SC344	0127
2SC344(Y)	0127
2SC344Y	0127
2SC345	0324
2SC346	0016
2SC347	0016
2SC348	0016
2SC348A	0016
2SC348B	0016
2SC348C	0016
2SC348D	0016
2SC348E	0016
2SC348F	0016
2SC348G	0016
2SC348GN	0016
2SC348H	0016
2SC348J	0016
2SC348K	0016
2SC348L	0016
2SC348M	0016
2SC348OR	0016
2SC348X	0016
2SC348Y	0016
2SC349	0127
2SC349R	0127
2SC350	0016
2SC350A	0016
2SC350B	0016
2SC350C	0016
2SC350D	0016
2SC350E	0016
2SC350F	0016
2SC350G	0016
2SC350GN	0016
2SC350H	0016
2SC350J	0016
2SC350K	0016
2SC350L	0016
2SC350M	0016
2SC350OR	0016
2SC350R	0016
2SC350X	0016
2SC350Y	0016
2SC351	0127
2SC351(FA)	0127
2SC351(FA-1)	OEM
2SC351A	0127
2SC351B	0127
2SC351C	0127
2SC351D	0127
2SC351E	0127
2SC351F	0127
2SC351FA1	0127
2SC351G	0127
2SC351GN	0127
2SC351H	0127
2SC351J	0127
2SC351K	0127
2SC351L	0127
2SC351M	0127
2SC351OR	0127
2SC351R	0127
2SC351X	0127
2SC351Y	0127
2SC352	0454
2SC352-OR	0454
2SC352A	0454
2SC352AC	0454
2SC352B	0454
2SC352C	0454
2SC352D	0454
2SC352E	0454
2SC352F	0454
2SC352G	0454
2SC352GN	0454
2SC352H	0454
2SC352J	0454
2SC352K	0454
2SC352L	0454
2SC352M	0454
2SC352OR	0454
2SC352R	0454
2SC352X	0454
2SC352Y	0454
2SC353	0086
2SC353A	0086
2SC353AC	0086
2SC353B	0086
2SC353C	0086
2SC353D	0086
2SC353E	0086
2SC353F	0086
2SC353G	0086
2SC353GN	0086
2SC353H	0086
2SC353J	0086
2SC353K	0086
2SC353L	0086
2SC353M	0086
2SC353OR	0086
2SC353R	0086
2SC353X	0086
2SC353Y	0086
2SC354	0936
2SC355	0936
2SC356	0079
2SC356A	0079
2SC356B	0079
2SC356C	0079
2SC356D	0079
2SC356E	0079
2SC356F	0079
2SC356G	0079
2SC356GN	0079
2SC356H	0079
2SC356J	0079
2SC356K	0079
2SC356L	0079
2SC356M	0079
2SC356OR	0079
2SC356R	0079
2SC356X	0079
2SC356Y	0079
2SC360	0016
2SC360-OR	0016
2SC360A	0016
2SC360B	0016
2SC360C	0016
2SC360D	0016
2SC360E	0016
2SC360F	0016
2SC360G	0016
2SC360GN	0016
2SC360H	0016
2SC360J	0016
2SC360K	0016
2SC360L	0016
2SC360M	0016
2SC360OR	0016
2SC360R	0016
2SC360X	0016
2SC360Y	0016
2SC361	0155
2SC361A	0155
2SC361B	0155
2SC361C	0155
2SC361D	0155
2SC361E	0155
2SC361F	0155
2SC361G	0155
2SC361GN	0155
2SC361H	0155
2SC361J	0155
2SC361K	0155
2SC361L	0155
2SC361M	0155
2SC361OR	0155
2SC361R	0155
2SC361X	0155
2SC361Y	0155
2SC362	0155
2SC362A	0155
2SC362B	0155
2SC362C	0155
2SC362D	0155
2SC362E	0155
2SC362F	0155
2SC362G	0155
2SC362GN	0155
2SC362H	0155
2SC362J	0155
2SC362K	0155
2SC362L	0155
2SC362M	0155
2SC362OR	0155
2SC362R	0155
2SC362X	0155
2SC362Y	0155
2SC363	0016
2SC363-OR	0016
2SC363A	0016
2SC363B	0016
2SC363C	0016
2SC363D	0016
2SC363E	0016
2SC363F	0016
2SC363G	0016
2SC363GN	0016
2SC363H	0016
2SC363J	0016
2SC363K	0016
2SC363L	0016
2SC363M	0016
2SC363OR	0016
2SC363R	0016
2SC363X	0016
2SC363Y	0016
2SC364	0530
2SC366	0286
2SC366-0	0286
2SC366-ORG-G	0286
2SC366-RED-G	0286
2SC366A	0286
2SC366B	0286
2SC366C	0286
2SC366D	0286
2SC366E	0286
2SC366F	0286
2SC366G	0286
2SC366GN	0286
2SC366G-O	0286
2SC366G-R	0286
2SC366G-Y	0286
2SC366H	0286
2SC366J	0286
2SC366K	0286
2SC366L	0286
2SC366M	0286
2SC366OR	0286
2SC366X	0286
2SC367	0286
2SC367-O	0286
2SC367-ORG-G	0286
2SC367-R	0286
2SC367-RED-G	0286
2SC367-Y	0286
2SC367-YEL-G	0286
2SC367A	0286
2SC367B	0286
2SC367C	0286
2SC367D	0286
2SC367E	0286
2SC367F	0286
2SC367G	0286
2SC367GN	0286
2SC367G-O	0286
2SC367G-R	0286
2SC367G-Y	0286
2SC367H	0286
2SC367J	0286
2SC367K	0286
2SC367L	0286
2SC367M	0286
2SC367OR	0286
2SC367R	0286
2SC367X	0286
2SC367Y	0286
2SC368	0111
2SC368-BL	0111
2SC368-GR	0111
2SC368A	0111
2SC368B	0111
2SC368C	0111
2SC368D	0111
2SC368E	0111
2SC368F	0111
2SC368G	0111
2SC368GN	0111
2SC368H	0111
2SC368J	0111
2SC368K	0111
2SC368L	0111
2SC368M	0111
2SC368OR	0111
2SC368R	0111
2SC368V	0111
2SC368X	0111
2SC368Y	0111
2SC369	0111
2SC369-BL	0111
2SC369-BLU-G	0111
2SC369-GR	0111
2SC369-GRN-G	0111
2SC369-V	0111
2SC369A	0111
2SC369B	0111
2SC369BL	0111
2SC369C	0111
2SC369D	0111
2SC369E	0111
2SC369F	0111
2SC369G	0111
2SC369G/BL	0111
2SC369G/GR	0111
2SC369G-BL	0111
2SC369GBL	0111
2SC369G-GR	0111
2SC369GGR	0111
2SC369GN	0111
2SC369GR	0111
2SC369G-V	0111
2SC369H	0111
2SC369J	0111
2SC369K	0111
2SC369L	0111
2SC369M	0111
2SC369OR	0111
2SC369R	0111
2SC369V	0111
2SC369X	0111
2SC369Y	0111
2SC370	0155
2SC370-0	0155
2SC370-G	0155
2SC370-O	0155
2SC370-T	0155
2SC370A	0155
2SC370B	0155
2SC370C	0155
2SC370D	0155
2SC370E	0155
2SC370F	0155
2SC370G	0155
2SC370GN	0155
2SC370H	0155
2SC370I	0155
2SC370J	0155
2SC370K	0155
2SC370L	0155
2SC370M	0155
2SC370OR	0155
2SC370R	0155
2SC370X	0155
2SC370Y	0155
2SC371	0191
2SC371(0)	0191
2SC371(O)	0191
2SC371-0	0191
2SC371-O	0191
2SC371-OR	0191
2SC371-ORG-G	0191
2SC371-R	0191
2SC371-RED-G	0191
2SC371-T	0191
2SC371A	0191
2SC371B	0191
2SC371C	0191
2SC371D	0191
2SC371E	0191
2SC371F	0191
2SC371G	0191
2SC371GN	0191
2SC371G-O	0191
2SC371G-R	0191
2SC371H	0191
2SC371J	0191
2SC371K	0191
2SC371L	0191
2SC371M	0191
2SC371O	0191
2SC371OR	0191
2SC371R	0191
2SC371R-1	0191
2SC371RED-G	0191
2SC371T	0191
2SC371X	0191
2SC371Y	0191
2SC372	0076
2SC372(H)	0076
2SC372(O)	0076
2SC372(Y)	0076
2SC372-0	0076
2SC372-1	0076
2SC372-2	0076
2SC372-O	0076
2SC372-OR	0076
2SC372-ORG	0076
2SC372-ORG-G	0076
2SC372-Y	0076
2SC372-YEL-G	0076
2SC372A	0076
2SC372AR	0076
2SC372B	0076
2SC372C	0076
2SC372D	0076
2SC372E	0076
2SC372F	0076
2SC372G	0076
2SC372GN	0076
2SC372G-O	0076
2SC372G-Y	0076
2SC372J	0076
2SC372K	0076
2SC372L	0076
2SC372M	0076
2SC372O	0076
2SC372OR	0076
2SC372R	0076
2SC372X	0076
2SC372Y	0076
2SC372Y1	0076
2SC372YEL	0076
2SC372YEL-G	0076
2SC372Z	0076
2SC373	0076
2SC373(GR)	0076
2SC373-0	0076
2SC373-14	0076
2SC373-G	0076
2SC373-O	0076
2SC373-OR	0076
2SC373A	0076
2SC373AL	0076
2SC373B	0076
2SC373C	0076
2SC373D	0076
2SC373E	0076
2SC373F	0076
2SC373G	0076
2SC373GN	0076
2SC373GR	0076
2SC373H	0076
2SC373J	0076
2SC373K	0076
2SC373L	0076
2SC373M	0076
2SC373OR	0076
2SC373R	0076
2SC373W	0076
2SC373X	0076
2SC373Y	0076
2SC374	0552
2SC374(BL)	0552
2SC374(V)	0552
2SC374-BL	0552
2SC374-OR	0552
2SC374-V	0552
2SC374A	0552
2SC374B	0552
2SC374BL	0552
2SC374BLK	0552
2SC374C	0552
2SC374D	0552
2SC374E	0552
2SC374F	0552
2SC374G	0552
2SC374GN	0552
2SC374H	0552
2SC374J	0552
2SC374K	0552
2SC374L	0552
2SC374M	0552
2SC374OR	0552
2SC374R	0552
2SC374V	0552
2SC374X	0552
2SC374Y	0552
2SC375	0127
2SC375-0	0127
2SC375-O	0127
2SC375-Y	0127
2SC375A	0127
2SC375B	0127
2SC375C	0127
2SC375D	0127
2SC375E	0127
2SC375F	0127
2SC375G	0127
2SC375GN	0127
2SC375H	0127
2SC375J	0127
2SC375K	0127
2SC375L	0127
2SC375M	0127
2SC375OR	0127
2SC375R	0127
2SC375X	0127
2SC376	0286
2SC376A	0286
2SC376B	0286
2SC376C	0286
2SC376D	0286
2SC376E	0286
2SC376F	0286
2SC376G	0286
2SC376GN	0286
2SC376H	0286
2SC376J	0286
2SC376K	0286
2SC376L	0286
2SC376M	0286
2SC376OR	0286
2SC376R	0286
2SC376X	0286
2SC376Y	0286
2SC377	0076
2SC377-0	0076
2SC377-BN	0076
2SC377-BRN	0076
2SC377-O	0076
2SC377-OR	0076
2SC377-ORG	0076
2SC377-R	0076
2SC377-RED	0076
2SC377A	0076
2SC377B	0076
2SC377BN	0076
2SC377BRN	0076
2SC377C	0076
2SC377D	0076
2SC377E	0076
2SC377F	0076
2SC377G	0076
2SC377GN	0076
2SC377H	0076
2SC377J	0076
2SC377K	0076
2SC377L	0076
2SC377M	0076
2SC377O	0076
2SC377OR	0076
2SC377R	0076
2SC377RED	0076
2SC377X	0076
2SC377Y	0076
2SC378	0155
2SC378-O	0155
2SC378-ORG	0155
2SC378-R	0155
2SC378-RED	0155
2SC378-Y	0155
2SC378-YEL	0155
2SC378A	0155
2SC378B	0155
2SC378C	0155
2SC378D	0155
2SC378E	0155
2SC378F	0155
2SC378G	0155
2SC378GN	0155
2SC378H	0155
2SC378J	0155
2SC378K	0155
2SC378L	0155
2SC378M	0155
2SC378O	0155
2SC378OR	0155
2SC378R	0155
2SC378X	0155
2SC378Y	0155
2SC379	0286
2SC379A	0286
2SC379B	0286
2SC379C	0286
2SC379D	0286
2SC379E	0286
2SC379F	0286
2SC379G	0286
2SC379GN	0286
2SC379H	0286
2SC379J	0286
2SC379K	0286
2SC379L	0286
2SC379M	0286
2SC379OR	0286
2SC379R	0286
2SC379X	0286
2SC379Y	0286
2SC380	0286
2SC380(O)	0286
2SC380(R)	0286
2SC380(Y)	0286
2SC380-0	0286
2SC380-0/4454C	0286
2SC380-BRN	0286
2SC380-O	0286
2SC380-O/4454C	0286
2SC380-OR	0286
2SC380-ORG	0286
2SC380-R	0286
2SC380-RED	0286
2SC380-Y	0286
2SC380-YEL	0286
2SC380A	0286
2SC380A(D)	0286
2SC380A(O)	0286
2SC380A(R)	0286
2SC380A(Y)	0286
2SC380A-0	0286
2SC380A0	0286
2SC380A-JA-8	OEM
2SC380A-JA-Y	OEM
2SC380A-O	0286
2SC380AO	0286
2SC380A-O(TV)	0286
2SC380A-R	0286
2SC380AR	0286
2SC380ATV	0286
2SC380A-Y	0286
2SC380AY	0286
2SC380B	0286
2SC380B-Y	0286
2SC380BY	0286
2SC380C	0286
2SC380C-Y	0286
2SC380CY	0286
2SC380D	0286
2SC380D-Y	0286
2SC380DY	0286
2SC380E	0286
2SC380E-Y	0286
2SC380EY	0286
2SC380F	0286
2SC380F-Y	0286
2SC380FY	0286
2SC380G	0286
2SC380GN	0286
2SC380H	0286
2SC380J	0286
2SC380K	0286
2SC380L	0286
2SC380M	0286
2SC380O	0286
2SC380OR	0286
2SC380R	0286
2SC380RED	0286
2SC380T	0286
2SC380TM-O	0286
2SC380TM-Y	0286
2SC380V	0286
2SC380X	0286
2SC380Y	0286
2SC380YEL	0286
2SC381	0113
2SC381(BN)	0113
2SC381-0	0113
2SC381-BN	0113
2SC381-BRN	0113
2SC381-O	0113
2SC381-OR	0113
2SC381-ORG	0113
2SC381-R	0113
2SC381-RED	0113
2SC381A	0113
2SC381B	0113
2SC381BN	0113
2SC381BN-1	0113
2SC381C	0113
2SC381D	0113
2SC381E	0113
2SC381F	0113
2SC381G	0113
2SC381GN	0113
2SC381H	0113
2SC381J	0113
2SC381K	0113
2SC381L	0113
2SC381M	0113
2SC381O	0113
2SC381OR	0113
2SC381R	0113
2SC381RED	0113
2SC381RL	0113
2SC381X	0113
2SC381Y	0113
2SC382	0184
2SC382(BL)	0184
2SC382(BN)	0184
2SC382(R)	0184
2SC382-BK	OEM
2SC382-BK(1)	0184
2SC382-BK(2)	0184
2SC382-BL	OEM
2SC382-BR	OEM
2SC382-G	0184
2SC382-GR	0184
2SC382-GY	0184
2SC382-OR	0184
2SC382-R	0184
2SC382-V	0184
2SC382A	0184
2SC382B	0184
2SC382BK	0184
2SC382BK1	0184
2SC382BK2	0184
2SC382BL	0184
2SC382BN	0184
2SC382BR	0184
2SC382C	0184
2SC382D	0184
2SC382E	0184
2SC382F	0184
2SC382G	0184
2SC382GN	0184
2SC382GR	0184
2SC382GY	0184
2SC382H	0184
2SC382J	0184
2SC382K	0184
2SC382L	0184
2SC382M	0184
2SC382OR	0184
2SC382R	0184
2SC382TM	0184
2SC382V	0184
2SC382W	0184
2SC382W,R	0184
2SC382X	0184
2SC382Y	0184
2SC383	0076
2SC383(T)	0076
2SC383-OR	0076
2SC383A	0076
2SC383B	0076
2SC383C	0076
2SC383D	0076
2SC383E	0076
2SC383F	0076
2SC383G	0076
2SC383GN	0076
2SC383H	0076
2SC383J	0076
2SC383K	0076
2SC383L	0076
2SC383M	0076
2SC383OR	0076
2SC383R	0076
2SC383T	0076
2SC383TM	0076
2SC383W	0076
2SC383WT	0076
2SC383X	0076
2SC383Y	0076
2SC384	0127
2SC384(O)	0127

Original Device Types

DEVICE TYPE	REPL CODE
2SC384(Y)	0127
2SC384-0	0127
2SC384A	0127
2SC384B	0127
2SC384C	0127
2SC384D	0127
2SC384E	0127
2SC384F	0127
2SC384G	0127
2SC384GN	0127
2SC384H	0127
2SC384J	0127
2SC384K	0127
2SC384L	0127
2SC384M	0127
2SC384OR	0127
2SC384R	0127
2SC384X	0127
2SC384Y	0127
2SC385	0549
2SC385A	0549
2SC385B	0549
2SC385C	0549
2SC385D	0549
2SC385E	0549
2SC385F	0549
2SC385G	0549
2SC385GN	0549
2SC385H	0549
2SC385J	0549
2SC385K	0549
2SC385L	0549
2SC385M	0549
2SC385OR	0549
2SC385R	0549
2SC385X	0549
2SC385Y	0549
2SC386	0127
2SC386-0	0127
2SC386A	0127
2SC386A-0(TV)	0127
2SC386AO	0127
2SC386B	0127
2SC386C	0127
2SC386D	0127
2SC386E	0127
2SC386F	0127
2SC386G	0127
2SC386GN	0127
2SC386H	0127
2SC386J	0127
2SC386K	0127
2SC386L	0127
2SC386M	0127
2SC386OR	0127
2SC386R	0127
2SC386X	0127
2SC386Y	0127
2SC387	0549
2SC387(FA-3)	0549
2SC387-G	0549
2SC387-OR	0549
2SC387A	0549
2SC387A(FA-3)	0549
2SC387AG	0549
2SC387B	0549
2SC387C	0549
2SC387D	0549
2SC387E	0549
2SC387F	0549
2SC387FA3	0549
2SC387G	0549
2SC387GN	0549
2SC387H	0549
2SC387J	0549
2SC387K	0549
2SC387L	0549
2SC387M	0549
2SC387OR	0549
2SC387R	0549
2SC387X	0549
2SC387Y	0549
2SC388	0842
2SC388(FA-1)	OEM
2SC388-A	0842
2SC388-0	0842
2SC388-OR	0842
2SC388A	0127
2SC388ABC	0842
2SC388ACC	0842
2SC388A-TM	0842
2SC388ATM	0842
2SC388ATM(FA)	0842
2SC388ATM(FA-1)	OEM
2SC388ATMFA	0842
2SC388B	0842
2SC388C	0842
2SC388D	0842
2SC388E	0842
2SC388F	0842
2SC388FA	0842
2SC388G	0842
2SC388GN	0842
2SC388H	0842
2SC388J	0842
2SC388K	0842
2SC388L	0842
2SC388M	0842
2SC388OR	0842
2SC388R	0842
2SC388TM	0842
2SC388X	0842
2SC388Y	0842
2SC389	0007
2SC389-0	0007
2SC389-OR	0007
2SC389A	0007
2SC389AFP	0007
2SC389B	0007
2SC389BLB-0	0007
2SC389C	0007
2SC389D	0007
2SC389E	0007
2SC389F	0007
2SC389G	0007
2SC389GN	0007
2SC389H	0007
2SC389J	0007
2SC389K	0007
2SC389L	0007
2SC389LP	0007
2SC389M	0007
2SC389O	0007
2SC389OR	0007
2SC389P	0007
2SC389R	0007
2SC389X	0007
2SC389Y	0007
2SC390	0259
2SC390A	0259
2SC390B	0259
2SC390C	0259
2SC390D	0259
2SC390E	0259
2SC390F	0259
2SC390G	0259
2SC390GN	0259
2SC390H	0259
2SC390J	0259
2SC390K	0259
2SC390L	0259
2SC390M	0259
2SC390OR	0259
2SC390R	0259
2SC390X	0259
2SC390Y	0259
2SC391	0008
2SC391A	0008
2SC391B	0008
2SC391C	0008
2SC391D	0008
2SC391E	0008
2SC391F	0008
2SC391G	0008
2SC391GN	0008
2SC391H	0008
2SC391J	0008
2SC391K	0008
2SC391L	0008
2SC391M	0008
2SC391OR	0008
2SC391R	0008
2SC391X	0008
2SC391Y	0008
2SC392	0259
2SC392A	0259
2SC392B	0259
2SC392C	0259
2SC392D	0259
2SC392E	0259
2SC392F	0259
2SC392G	0259
2SC392GN	0259
2SC392H	0259
2SC392J	0259
2SC392K	0259
2SC392L	0259
2SC392M	0259
2SC392OR	0259
2SC392R	0259
2SC392X	0259
2SC392Y	0259
2SC393	0079
2SC394	0155
2SC394(0)	0155
2SC394(O)	0155
2SC394-0	0155
2SC394-GR	0155
2SC394-GRN	0155
2SC394-O	0155
2SC394-OR	0155
2SC394-ORG	0155
2SC394-R	0155
2SC394-RED	0155
2SC394-Y	0155
2SC394-YEL	0155
2SC394A	0155
2SC394AP	0155
2SC394B	0155
2SC394C	0155
2SC394D	0155
2SC394E	0155
2SC394F	0155
2SC394G	0155
2SC394GN	0155
2SC394GR	0155
2SC394GRN	0155
2SC394H	0155
2SC394J	0155
2SC394K	0155
2SC394L	0155
2SC394M	0155
2SC394O	0155
2SC394OR	0155
2SC394R	0155
2SC394RED	0155
2SC394W	0155
2SC394X	0155
2SC394Y	0155
2SC394YEL	0155
2SC395	0016
2SC395A	0016
2SC395A-O	0016
2SC395A-ORG	0016
2SC395A-R	0016
2SC395A-RED	0016
2SC395A-Y	0016
2SC395A-YEL	0016
2SC395B	0016
2SC395C	0016
2SC395D	0016
2SC395E	0016
2SC395F	0016
2SC395G	0016
2SC395GN	0016
2SC395H	0016
2SC395J	0016
2SC395K	0016
2SC395L	0016
2SC395M	0016
2SC395OR	0016
2SC395R	0016
2SC395X	0016
2SC395Y	0016
2SC396	0493
2SC396A	0493
2SC396B	0493
2SC396C	0493
2SC396D	0493
2SC396E	0493
2SC396F	0493
2SC396G	0493
2SC396GN	0493
2SC396GR	0493
2SC396H	0493
2SC396J	0493
2SC396K	0493
2SC396L	0493
2SC396M	0493
2SC396OR	0493
2SC396R	0493
2SC396X	0493
2SC396Y	0493
2SC397	0007
2SC397A	0007
2SC397B	0007
2SC397C	0007
2SC397D	0007
2SC397E	0007
2SC397F	0007
2SC397G	0007
2SC397GN	0007
2SC397H	0007
2SC397J	0007
2SC397K	0007
2SC397L	0007
2SC397M	0007
2SC397OR	0007
2SC397R	0007
2SC397X	0007
2SC397Y	0007
2SC398	0008
2SC398(FA-1)	0008
2SC398A	0008
2SC398B	0008
2SC398C	0008
2SC398D	0008
2SC398E	0008
2SC398F	0008
2SC398FA1	0008
2SC398G	0008
2SC398GN	0008
2SC398H	0008
2SC398J	0008
2SC398K	0008
2SC398L	0008
2SC398M	0008
2SC398OR	0008
2SC398R	0008
2SC398X	0008
2SC398Y	0008
2SC399	0008
2SC399A	0008
2SC399B	0008
2SC399C	0008
2SC399D	0008
2SC399E	0008
2SC399F	0008
2SC399FA1	0008
2SC399G	0008
2SC399GN	0008
2SC399H	0008
2SC399J	0008
2SC399K	0008
2SC399L	0008
2SC399M	0008
2SC399OR	0008
2SC399R	0008
2SC399X	0008
2SC399Y	0008
2SC400	0016
2SC400-0	0016
2SC400-GR	0016
2SC400-O	0016
2SC400-R	0016
2SC400-Y	0016
2SC400A	0016
2SC400B	0016
2SC400C	0016
2SC400D	0016
2SC400E	0016
2SC400F	0016
2SC400G	0016
2SC400GN	0016
2SC400H	0016
2SC400J	0016
2SC400K	0016
2SC400L	0016
2SC400M	0016
2SC400OR	0016
2SC400R	0016
2SC400X	0016
2SC400Y	0016
2SC401	0155
2SC401A	0155
2SC401B	0155
2SC401C	0155
2SC401D	0155
2SC401E	0155
2SC401F	0155
2SC401G	0155
2SC401GN	0155
2SC401H	0155
2SC401J	0155
2SC401K	0155
2SC401L	0155
2SC401M	0155
2SC401OR	0155
2SC401R	0155
2SC401X	0155
2SC401Y	0155
2SC402	0155
2SC402A	0155
2SC402B	0155
2SC402C	0155
2SC402D	0155
2SC402E	0155
2SC402F	0155
2SC402G	0155
2SC402GN	0155
2SC402H	0155
2SC402J	0155
2SC402K	0155
2SC402L	0155
2SC402M	0155
2SC402OR	0155
2SC402R	0155
2SC402X	0155
2SC402Y	0155
2SC403	0155
2SC403(C)	0155
2SC403(SONY)	0155
2SC403-OR	0155
2SC403A	0155
2SC403AL	0155
2SC403B	0155
2SC403C	0155
2SC403C4	OEM
2SC403CG	0155
2SC403D	0155
2SC403E	0155
2SC403F	0155
2SC403G	0155
2SC403GN	0155
2SC403H	0155
2SC403J	0155
2SC403K	0155
2SC403L	0155
2SC403M	0155
2SC403OR	0155
2SC403R	0155
2SC403SP3	OEM
2SC403X	0155
2SC403Y	0155
2SC404	0155
2SC404A	0155
2SC404B	0155
2SC404C	0155
2SC404D	0155
2SC404E	0155
2SC404F	0155
2SC404G	0155
2SC404GN	0155
2SC404H	0155
2SC404J	0155
2SC404K	0155
2SC404L	0155
2SC404M	0155
2SC404OR	0155
2SC404R	0155
2SC404X	0155
2SC404Y	0155
2SC405	0016
2SC405A	0016
2SC405B	0016
2SC405C	0016
2SC405D	0016
2SC405E	0016
2SC405F	0016
2SC405G	0016
2SC405GN	0016
2SC405H	0016
2SC405J	0016
2SC405K	0016
2SC405L	0016
2SC405M	0016
2SC405OR	0016
2SC405R	0016
2SC405X	0016
2SC405Y	0016
2SC406	0016
2SC406A	0016
2SC406B	0016
2SC406C	0016
2SC406D	0016
2SC406E	0016
2SC406F	0016
2SC406G	0016
2SC406GN	0016
2SC406H	0016
2SC406J	0016
2SC406K	0016
2SC406L	0016
2SC406M	0016
2SC406OR	0016
2SC406R	0016
2SC406X	0016
2SC406Y	0016
2SC407	1966
2SC408	1966
2SC409	1966
2SC410	1966
2SC410A	1966
2SC411	1966
2SC412	0642
2SC413	0630
2SC414	0630
2SC415	0630
2SC416	0630
2SC420	0630
2SC423	0016
2SC423-0	0016
2SC423A	0016
2SC423B	0016
2SC423C	0016
2SC423D	0016
2SC423E	0016
2SC423F	0016
2SC423G	0016
2SC423GN	0016
2SC423H	0016
2SC423J	0016
2SC423K	0016
2SC423L	0016
2SC423M	0016
2SC423OR	0016
2SC423R	0016
2SC423X	0016
2SC423Y	0016
2SC424	0016
2SC424D	0016
2SC425	0016
2SC425A	0016
2SC425B	0016
2SC425C	0016
2SC425D	0016
2SC425E	0016
2SC425F	0016
2SC425G	0016
2SC425GN	0016
2SC425H	0016
2SC425K	0016
2SC425L	0016
2SC425M	0016
2SC425OR	0016
2SC425R	0016
2SC425X	0016
2SC425Y	0016
2SC426	0079
2SC427	0079
2SC428	0079
2SC429	0475
2SC429A	0475
2SC429B	0475
2SC429C	0475
2SC429D	0475
2SC429E	0475
2SC429F	0475
2SC429G	0475
2SC429GN	0475
2SC429H	0475
2SC429J	0475
2SC429K	0475
2SC429L	0475
2SC429M	0475
2SC429OR	0475
2SC429R	0475
2SC429X	0475
2SC429Y	0475
2SC430	0475
2SC430A	0475
2SC430B	0475
2SC430C	0475
2SC430D	0475
2SC430E	0475
2SC430F	0475
2SC430G	0475
2SC430GN	0475
2SC430H	0475
2SC430J	0475
2SC430K	0475
2SC430L	0475
2SC430M	0475
2SC430OR	0475
2SC430R	0475
2SC430W	0475
2SC430X	0475
2SC430Y	0475
2SC433	0079
2SC437	0630
2SC438	0698
2SC439	0018
2SC440	0493
2SC441	0493
2SC441A	0493
2SC441B	0493
2SC441C	0493
2SC441D	0493
2SC441E	0493
2SC441F	0493
2SC441G	0493
2SC441GN	0493
2SC441H	0493
2SC441J	0493
2SC441K	0493
2SC441L	0493
2SC441M	0493
2SC441OR	0493
2SC441R	0493
2SC441X	0493
2SC441Y	0493
2SC442	0493
2SC442A	0493
2SC442B	0493
2SC442C	0493
2SC442D	0493
2SC442E	0493
2SC442F	0493
2SC442G	0493
2SC442GN	0493
2SC442H	0493
2SC442J	0493
2SC442K	0493
2SC442L	0493
2SC442M	0493
2SC442OR	0493
2SC442R	0493
2SC442X	0493
2SC442Y	0493
2SC443	0086
2SC443A	0086
2SC443B	0086
2SC443C	0086
2SC443D	0086
2SC443E	0086
2SC443F	0086
2SC443G	0086
2SC443GN	0086
2SC443H	0086
2SC443J	0086
2SC443K	0086
2SC443L	0086
2SC443M	0086
2SC443OR	0086
2SC443R	0086
2SC443X	0086
2SC443Y	0086
2SC444	0595
2SC445	0168
2SC445A	0168
2SC445B	0168
2SC445C	0168
2SC445D	0168
2SC445E	0168
2SC445F	0168
2SC445G	0168
2SC445GN	0168
2SC445H	0168
2SC445J	0168
2SC445K	0168
2SC445L	0168
2SC445M	0168
2SC445OR	0168
2SC445R	0168
2SC445X	0168
2SC445Y	0168
2SC446	5568
2SC454	0076
2SC454(B)	0076
2SC454(C)	0076
2SC454-3	0076
2SC454-5	0076
2SC454-C	0076
2SC454-OR	0076
2SC454A	0076
2SC454B	0076
2SC454B-6	0076
2SC454BL	0076
2SC454C	0076
2SC454D	0076
2SC454E	0076
2SC454F	0076
2SC454G	0076
2SC454GN	0076
2SC454H	0076
2SC454J	0076
2SC454K	0076
2SC454L	0076
2SC454LA	0076
2SC454M	0076
2SC454OR	0076
2SC454R	0076
2SC454X	0076
2SC454Y	0076
2SC455	0155
2SC455-OR	0155
2SC455A	0155
2SC455B	0155
2SC455C	0155
2SC455D	0155
2SC455E	0155
2SC455F	0155
2SC455G	0155
2SC455GN	0155
2SC455H	0155
2SC455J	0155
2SC455K	0155
2SC455L	0155
2SC455M	0155
2SC455OR	0155
2SC455R	0155
2SC455X	0155
2SC455Y	0155
2SC456	0698
2SC456A	0698
2SC456B	0698
2SC456C	0698
2SC456D	0698
2SC456E	0698
2SC456F	0698
2SC456G	0698
2SC456GN	0698
2SC456H	0698
2SC456J	0698
2SC456K	0698
2SC456L	0698
2SC456M	0698
2SC456OR	0698
2SC456R	0698
2SC456X	0698
2SC456Y	0698
2SC457	0079
2SC458	0076
2SC458(A)	0076
2SC458(B)	0076
2SC458(C)	0076
2SC458(C,D)	0076
2SC458(D)	0076
2SC458(LG)	0076
2SC458-4	0076
2SC458-5	0076
2SC458-B	0076
2SC458-C	0076
2SC458-D	0076
2SC458-GR	0076
2SC458-O	0076
2SC458-OR	0076
2SC458-Y	0076
2SC458A	0076
2SC458AD	0076
2SC458AK	0076
2SC458B	0076
2SC458BC	0076
2SC458BCD	1419
2SC458BD	0076
2SC458B-GR	0076
2SC458BL	0076
2SC458BM	0076
2SC458B-O	0076
2SC458B-Y	0076
2SC458C	0076
2SC458C/L6	0076
2SC458CD	0076
2SC458CD-TB	0076
2SC458CL	0076
2SC458CLG	0076
2SC458CM	0076
2SC458CTZ	1419
2SC458D	0076
2SC458DTZ	1419
2SC458E	0076
2SC458F	0076
2SC458G	0076
2SC458GN	0076
2SC458H	0076
2SC458J	0076
2SC458K	0076
2SC458KB	0076
2SC458L	0076
2SC458L(C)	0076
2SC458L6	0076
2SC458LB	0076
2SC458LC	0076
2SC458LD	0076
2SC458LG	0076
2SC458LG(B)	0076
2SC458LG(C)	0076
2SC458LG(D)	0076
2SC458LGA	0076
2SC458LGB	0076
2SC458LGBM	0076
2SC458LGC	0076
2SC458LGC-6	0076

If replacement code is OEM, contact original manufacturer for replacement.

DEVICE TYPE	REPL CODE
2SC458LGD	0076
2SC458LGO	0076
2SC458LGR	0076
2SC458LGS	0076
2SC458M	0076
2SC458OR	0076
2SC458R	0076
2SC458RGS	0076
2SC458X	0076
2SC458Y	0076
2SC459	0286
2SC459A	0286
2SC459B	0286
2SC459C	0286
2SC459D	0286
2SC459E	0286
2SC459F	0286
2SC459G	0286
2SC459GN	0286
2SC459H	0286
2SC459J	0286
2SC459K	0286
2SC459L	0286
2SC459M	0286
2SC459OR	0286
2SC459R	0286
2SC459X	0286
2SC459Y	0286
2SC460	0151
2SC460(A)	0151
2SC460(B)	0151
2SC460(C)	0151
2SC460-5	0151
2SC460-B	0151
2SC460-C	0151
2SC460-OR	0151
2SC460A	0151
2SC460A2302-17	0151
2SC460B	0151
2SC460BL	0151
2SC460C	0151
2SC460D	0151
2SC460E	0151
2SC460F	0151
2SC460G	0151
2SC460GB	0151
2SC460GN	0151
2SC460H	0151
2SC460J	0151
2SC460K	0151
2SC460L	0151
2SC460M	0151
2SC460OR	0151
2SC460R	0151
2SC460X	0151
2SC460Y	0151
2SC461	0151
2SC461(8F)	0151
2SC461-8F	0151
2SC461-A	0151
2SC461-B	0151
2SC461A	0151
2SC461AL	0151
2SC461B	0151
2SC461BF	0151
2SC461BK	0151
2SC461BL	0151
2SC461C	0151
2SC461E	0151
2SC461EP	0151
2SC461F	0151
2SC461L	0151
2SC462	0079
2SC463	0007
2SC463A	0007
2SC463B	0007
2SC463C	0007
2SC463D	0007
2SC463E	0007
2SC463F	0007
2SC463G	0007
2SC463GN	0007
2SC463H	0007
2SC463J	0007
2SC463K	0007
2SC463L	0007
2SC463M	0007
2SC463OR	0007
2SC463R	0007
2SC463X	0007
2SC463Y	0007
2SC464	0259
2SC464A	0259
2SC464B	0259
2SC464C	0259
2SC464D	0259
2SC464E	0259
2SC464F	0259
2SC464G	0259
2SC464GN	0259
2SC464H	0259
2SC464J	0259
2SC464K	0259
2SC464L	0259
2SC464M	0259
2SC464OR	0259
2SC464R	0259
2SC464X	0259
2SC464Y	0259
2SC465	0259
2SC465A	0259
2SC465B	0259
2SC465C	0259
2SC465D	0259
2SC465E	0259
2SC465F	0259
2SC465G	0259
2SC465GN	0259
2SC465H	0259
2SC465J	0259
2SC465K	0259
2SC465L	0259
2SC465M	0259
2SC465OR	0259
2SC465R	0259
2SC465X	0259
2SC465Y	0259
2SC466	0259
2SC466A	0259
2SC466B	0259
2SC466C	0259
2SC466D	0259
2SC466E	0259
2SC466F	0259
2SC466G	0259
2SC466GN	0259
2SC466H	0259
2SC466J	0259
2SC466K	0259
2SC466L	0259
2SC466M	0259
2SC466OR	0259
2SC466R	0259
2SC466X	0259
2SC466Y	0259
2SC467	0144
2SC468	0016
2SC468A	0016
2SC468B	0016
2SC468H	0016
2SC468LGR	0016
2SC469	0475
2SC469-OR	0475
2SC469A	0475
2SC469B	0475
2SC469C	0475
2SC469D	0475
2SC469E	0475
2SC469F	0475
2SC469G	0475
2SC469GN	0475
2SC469H	0475
2SC469J	0475
2SC469K	0475
2SC469L	0475
2SC469M	0475
2SC469OR	0475
2SC469Q	0475
2SC469R	0475
2SC469X	0475
2SC469Y	0475
2SC470	0233
2SC470-3	0233
2SC470-4	0233
2SC470-5	0233
2SC470-6	0233
2SC470A	0233
2SC470B	0233
2SC470C	0233
2SC470D	0233
2SC470E	0233
2SC470F	0233
2SC470G	0233
2SC470GN	0233
2SC470H	0233
2SC470J	0233
2SC470K	0233
2SC470L	0233
2SC470M	0233
2SC470OR	0233
2SC470R	0233
2SC470X	0233
2SC470Y	0233
2SC471	0018
2SC472	0224
2SC472A	0224
2SC472B	0224
2SC472C	0224
2SC472D	0224
2SC472E	0224
2SC472F	0224
2SC472G	0224
2SC472GN	0224
2SC472H	0224
2SC472J	0224
2SC472K	0224
2SC472L	0224
2SC472M	0224
2SC472OR	0224
2SC472R	0224
2SC472X	0224
2SC472Y	0224
2SC473	0016
2SC473AC	0016
2SC474	0079
2SC475	0475
2SC475A	0475
2SC475B	0475
2SC475C	0475
2SC475D	0475
2SC475E	0475
2SC475F	0475
2SC475G	0475
2SC475GN	0475
2SC475H.TEMP	0475
2SC475M	0475
2SC475OR	0475
2SC475R	0475
2SC475X	0475
2SC475Y	0475
2SC476	0475
2SC476C	0475
2SC476D	0475
2SC476E	0475
2SC476F	0475
2SC476G	0475
2SC476GN	0475
2SC476H	0475
2SC476J	0475
2SC476K	0475
2SC476L	0475
2SC476M	0475
2SC476OR	0475
2SC476R	0475
2SC476X	0475
2SC476Y	0475
2SC477	0007
2SC477A	0007
2SC477B	0007
2SC477C	0007
2SC477D	0007
2SC477E	0007
2SC477F	0007
2SC477G	0007
2SC477GN	0007
2SC477H	0007
2SC477J	0007
2SC477K	0007
2SC477L	0007
2SC477M	0007
2SC477OR	0007
2SC477R	0007
2SC477X	0007
2SC477Y	0007
2SC478	0016
2SC478(D)	0016
2SC478-4	0016
2SC478-O	0016
2SC478A	0016
2SC478B	0016
2SC478C	0016
2SC478D	0016
2SC478E	0016
2SC478F	0016
2SC478G	0016
2SC478GN	0016
2SC478H	0016
2SC478J	0016
2SC478K	0016
2SC478L	0016
2SC478M	0016
2SC478OR	0016
2SC478R	0016
2SC478X	0016
2SC478Y	0016
2SC479	0086
2SC479A	0086
2SC479B	0086
2SC479C	0086
2SC479D	0086
2SC479E	0086
2SC479F	0086
2SC479G	0086
2SC479GN	0086
2SC479H	0086
2SC479J	0086
2SC479K	0086
2SC479L	0086
2SC479M	0086
2SC479OR	0086
2SC479R	0086
2SC479X	0086
2SC479Y	0086
2SC480	0324
2SC481	0698
2SC481-OR	0698
2SC481A	0698
2SC481B	0698
2SC481C	0698
2SC481D	0698
2SC481E	0698
2SC481F	0698
2SC481G	0698
2SC481GN	0698
2SC481H	0698
2SC481J	0698
2SC481K	0698
2SC481L	0698
2SC481M	0698
2SC481OR	0698
2SC481R	0698
2SC481X	0698
2SC481Y	0698
2SC482	1593
2SC482(Y)	1593
2SC482-0	1593
2SC482-GR	1593
2SC482-GRN	1593
2SC482-O	1593
2SC482-OR	1593
2SC482-ORG	1593
2SC482-Y	1593
2SC482-YEL	1593
2SC482A	1593
2SC482B	1593
2SC482C	1593
2SC482D	1593
2SC482E	1593
2SC482F	1593
2SC482G	1593
2SC482GN	1593
2SC482GR	1593
2SC482GRN	1593
2SC482GRY	1593
2SC482H	1593
2SC482J	1593
2SC482K	1593
2SC482L	1593
2SC482M	1593
2SC482OR	1593
2SC482R	1593
2SC482X	1593
2SC482Y	1593
2SC482YEL	1593
2SC483	1027
2SC484	1480
2SC484-BL	1480
2SC484-BLU	1480
2SC484-R	1480
2SC484-RED	1480
2SC484-Y	1480
2SC484-YEL	1480
2SC484A	1480
2SC484B	1480
2SC484BL	1480
2SC484C	1480
2SC484D	1480
2SC484E	1480
2SC484F	1480
2SC484G	1480
2SC484GN	1480
2SC484H	1480
2SC484J	1480
2SC484K	1480
2SC484L	1480
2SC484M	1480
2SC484OR	1480
2SC484R	1480
2SC484X	1480
2SC484Y	1480
2SC485	1480
2SC485(Y)	1480
2SC485-BL	1480
2SC485-BLU	1480
2SC485-OR	1480
2SC485-R	1480
2SC485-RED	1480
2SC485-Y	1480
2SC485-YEL	1480
2SC485A	1480
2SC485B	1480
2SC485BL	1480
2SC485C	1480
2SC485D	1480
2SC485E	1480
2SC485F	1480
2SC485G	1480
2SC485GN	1480
2SC485H	1480
2SC485J	1480
2SC485K	1480
2SC485L	1480
2SC485M	1480
2SC485OR	1480
2SC485R	1480
2SC485X	1480
2SC485Y	1480
2SC486	1480
2SC486(Y)	1480
2SC486-BL	1480
2SC486-BLU	1480
2SC486-R	1480
2SC486-RED	1480
2SC486-Y	1480
2SC486-YEL	1480
2SC486A	1480
2SC486B	1480
2SC486BL	1480
2SC486C	1480
2SC486D	1480
2SC486E	1480
2SC486F	1480
2SC486G	1480
2SC486GN	1480
2SC486H	1480
2SC486J	1480
2SC486K	1480
2SC486L	1480
2SC486M	1480
2SC486OR	1480
2SC486R	1480
2SC486X	1480
2SC486Y	1480
2SC487	0178
2SC487A	0178
2SC488	0178
2SC488H	0178
2SC489	0178
2SC489-BL	0178
2SC489-BLU	0178
2SC489-R	0178
2SC489-RED	0178
2SC489-Y	0178
2SC489-YEL	0178
2SC489A	0178
2SC489B	0178
2SC489C	0178
2SC489D	0178
2SC489E	0178
2SC489F	0178
2SC489G	0178
2SC489GN	0178
2SC489H	0178
2SC489J	0178
2SC489K	0178
2SC489L	0178
2SC489M	0178
2SC489OR	0178
2SC489R	0178
2SC489X	0178
2SC489Y	0178
2SC490	0178
2SC490-BL	0178
2SC490-BLU	0178
2SC490-R	0178
2SC490-RED	0178
2SC490-Y	0178
2SC490-YEL	0178
2SC491	0178
2SC491-BL	0178
2SC491-BLU	0178
2SC491-R	0178
2SC491-RED	0178
2SC491-Y	0178
2SC491-YEL	0178
2SC491A	0178
2SC491B	0178
2SC491BL	0178
2SC491C	0178
2SC491D	0178
2SC491E	0178
2SC491F	0178
2SC491G	0178
2SC491GN	0178
2SC491H	0178
2SC491J	0178
2SC491K	0178
2SC491L	0178
2SC491M	0178
2SC491OR	0178
2SC491R	0178
2SC491X	0178
2SC491Y	0178
2SC492	0177
2SC493	0867
2SC493-BL	0867
2SC493-R	0867
2SC493-Y	0867
2SC493A	0867
2SC493B	0867
2SC493C	0867
2SC493D	0867
2SC493E	0867
2SC493F	0867
2SC493G	0867
2SC493GN	0867
2SC493H	0867
2SC493J	0867
2SC493K	0867
2SC493L	0867
2SC493M	0867
2SC493OR	0867
2SC493R	0867
2SC493X	0867
2SC493Y	0867
2SC494	0620
2SC494-BL	0620
2SC494-R	0620
2SC494-Y	0620
2SC494A	0620
2SC494B	0620
2SC494BL	0620
2SC494C	0620
2SC494D	0620
2SC494E	0620
2SC494F	0620
2SC494G	0620
2SC494GN	0620
2SC494H	0620
2SC494J	0620
2SC494K	0620
2SC494L	0620
2SC494M	0620
2SC494OR	0620
2SC494R	0620
2SC494X	0620
2SC494Y	0620
2SC495	0786
2SC495-0	0786
2SC495-O	0786
2SC495-OR	0786
2SC495-ORG	0786
2SC495-R	0786
2SC495-RED	0786
2SC495-Y	0786
2SC495-YEL	0786
2SC495A	0786
2SC495B	0786
2SC495C	0786
2SC495D	0786
2SC495E	0786
2SC495F	0786
2SC495G	0786
2SC495GN	0786
2SC495H	0786
2SC495J	0786
2SC495K	0786
2SC495L	0786
2SC495M	0786
2SC495O	0786
2SC495OR	0786
2SC495P	0786
2SC495R	0786
2SC495RED	0786
2SC495T	0786
2SC495X	0786
2SC495Y	0786
2SC495YEL	0786
2SC495Y-X	OEM
2SC496	0786
2SC496(O)	0786
2SC496(Y)	0786
2SC496-0	0786
2SC496-O	0786
2SC496-OR	0786
2SC496-ORG	0786
2SC496-R	0786
2SC496-RED	0786
2SC496-Y	0786
2SC496-YEL	0786
2SC496A	0786
2SC496B	0786
2SC496C	0786
2SC496D	0786
2SC496E	0786
2SC496F	0786
2SC496G	0786
2SC496GN	0786
2SC496H	0786
2SC496J	0786
2SC496K	0786
2SC496L	0786
2SC496M	0786
2SC496OR	0786
2SC496R	0786
2SC496RED	0786
2SC496X	0786
2SC496Y	0786
2SC496YEL	0786
2SC497	0086
2SC497-0	0086
2SC497-O	0086
2SC497-OR	0086
2SC497-ORG	0086
2SC497-R	0086
2SC497-RED	0086
2SC497-Y	0086
2SC497-YEL	0086
2SC497A	0086
2SC497B	0086
2SC497C	0086
2SC497D	0086
2SC497E	0086
2SC497F	0086
2SC497G	0086
2SC497GN	0086
2SC497H	0086
2SC497J	0086
2SC497K	0086
2SC497L	0086
2SC497M	0086
2SC497OR	0086
2SC497R	0086
2SC497RED	0086
2SC497X	0086
2SC497Y	0086
2SC498	0086
2SC498-0	0086
2SC498-O	0086
2SC498-OR	0086
2SC498-ORG	0086
2SC498-R	0086
2SC498-RED	0086
2SC498-Y	0086
2SC498-YEL	0086
2SC498A	0086
2SC498B	0086
2SC498C	0086
2SC498D	0086
2SC498E	0086
2SC498F	0086
2SC498G	0086
2SC498GN	0086
2SC498H	0086
2SC498J	0086
2SC498K	0086
2SC498L	0086
2SC498M	0086
2SC498OR	0086
2SC498R	0086
2SC498RED	0086
2SC498X	0086
2SC498Y	0086
2SC498YEL	0086
2SC499	2375
2SC499(A)	2375
2SC499(R)	2375
2SC499(Y)	2375
2SC499-OR	2375
2SC499-R	2375
2SC499-R(FA-1)	2375
2SC499-RED	2375
2SC499-RY	2375
2SC499-Y	2375
2SC499-Y(FA-1)	2375
2SC499-YEL	2375
2SC499A	2375
2SC499B	2375
2SC499C	2375
2SC499D	2375
2SC499E	2375
2SC499F	2375
2SC499FA1	2375
2SC499G	2375
2SC499GN	2375
2SC499H	2375
2SC499J	2375
2SC499K	2375
2SC499L	2375
2SC499M	2375
2SC499OR	2375
2SC499R	2375
2SC499RED	2375
2SC499RY	2375
2SC499X	2375
2SC499Y	2375
2SC499YEL	2375
2SC500	0233
2SC500(N)	0233
2SC500(Y)	0233
2SC500A	0233
2SC500B	0233
2SC500C	0233
2SC500D	0233
2SC500E	0233
2SC500F	0233
2SC500G	0233
2SC500GN	0233
2SC500H	0233
2SC500J	0233
2SC500K	0233
2SC500L	0233
2SC500M	0233
2SC500OR	0233
2SC500R	0233
2SC500X	0233
2SC500Y	0233
2SC501	0086
2SC501-0	0086
2SC501-ORG	0086
2SC501-R	0086
2SC501-RED	0086
2SC501-Y	0086
2SC501-YEL	0086
2SC501A	0086
2SC501B	0086
2SC501C	0086
2SC501D	0086
2SC501E	0086
2SC501F	0086
2SC501G	0086
2SC501GN	0086
2SC501H	0086
2SC501J	0086
2SC501K	0086
2SC501L	0086
2SC501M	0086
2SC501O	0086
2SC501OR	0086
2SC501R	0086
2SC501X	0086
2SC501Y	0086
2SC502	0698
2SC502A	0698
2SC502B	0698
2SC502C	0698
2SC502D	0698
2SC502E	0698
2SC502F	0698
2SC502G	0698
2SC502GN	0698
2SC502H	0698
2SC502J	0698
2SC502K	0698
2SC502L	0698
2SC502M	0698
2SC502OR	0698
2SC502R	0698
2SC502X	0698
2SC502Y	0698
2SC503	0086
2SC503-GR	0086
2SC503-O	0086
2SC503-Y	0086
2SC503A	0086
2SC503B	0086
2SC503C	0086
2SC503D	0086
2SC503E	0086
2SC503F	0086
2SC503G	0086
2SC503GN	0086
2SC503GR	0086

DEVICE TYPE	REPL CODE
2SC503H	0086
2SC503J	0086
2SC503K	0086
2SC503L	0086
2SC503M	0086
2SC503OR	0086
2SC503R	0086
2SC503X	0086
2SC504	0086
2SC504-GR	0086
2SC504-O	0086
2SC504-Y	0086
2SC504A	0086
2SC504B	0086
2SC504C	0086
2SC504D	0086
2SC504E	0086
2SC504F	0086
2SC504G	0086
2SC504GN	0086
2SC504GR	0086
2SC504H	0086
2SC504J	0086
2SC504K	0086
2SC504L	0086
2SC504M	0086
2SC504OR	0086
2SC504R	0086
2SC504X	0086
2SC505	0233
2SC505-O	0233
2SC505-ORG	0233
2SC505-R	0233
2SC505-RED	0233
2SC505A	0233
2SC505B	0233
2SC505C	0233
2SC505D	0233
2SC505E	0233
2SC505F	0233
2SC505G	0233
2SC505GN	0233
2SC505H	0233
2SC505J	0233
2SC505K	0233
2SC505L	0233
2SC505M	0233
2SC505O	0233
2SC505OR	0233
2SC505R	0233
2SC505X	0233
2SC505Y	0233
2SC506	0233
2SC506-O	0233
2SC506-ORG	0233
2SC506-R	0233
2SC506-RED	0233
2SC506A	0233
2SC506B	0233
2SC506C	0233
2SC506D	0233
2SC506E	0233
2SC506F	0233
2SC506G	0233
2SC506GN	0233
2SC506H	0233
2SC506J	0233
2SC506K	0233
2SC506L	0233
2SC506M	0233
2SC506O	0233
2SC506OR	0233
2SC506R	0233
2SC506X	0233
2SC506Y	0233
2SC507	0233
2SC507-0	0233
2SC507-O	0233
2SC507-R	0233
2SC507-Y	0233
2SC507A	0233
2SC507B	0233
2SC507C	0233
2SC507D	0233
2SC507E	0233
2SC507F	0233
2SC507G	0233
2SC507GN	0233
2SC507H	0233
2SC507J	0233
2SC507K	0233
2SC507L	0233
2SC507M	0233
2SC507OR	0233
2SC507R	0233
2SC507X	0233
2SC507Y	0233
2SC508	0178
2SC508A	0178
2SC508B	0178
2SC508C	0178
2SC508D	0178
2SC508E	0178
2SC508F	0178
2SC508G	0178
2SC508GN	0178
2SC508OR	0178
2SC508X	0178
2SC508Y	0178
2SC509	0155
2SC509(0)	0155
2SC509(O)	0155
2SC509(Y)	0155
2SC509-0	0155
2SC509-O	0155
2SC509-Y	0155
2SC509G	0155
2SC509G-O	0155
2SC509G-Y	0155
2SC5090	0155
2SC509Y	0155
2SC510	1480
2SC510-O	1480
2SC510-ORG	1480
2SC510-R	1480
2SC510-RED	1480
2SC510A	1480
2SC510B	1480
2SC510C	1480
2SC510D	1480
2SC510E	1480
2SC510F	1480
2SC510G	1480
2SC510GN	1480
2SC510H	1480
2SC510J	1480
2SC510K	1480
2SC510L	1480
2SC510M	1480
2SC5100	1480
2SC5100R	1480
2SC510R	1480
2SC510X	1480
2SC510Y	1480
2SC511	1480
2SC511-O	1480
2SC511-ORG	1480
2SC511-R	1480
2SC511-RED	1480
2SC511A	1480
2SC511B	1480
2SC511C	1480
2SC511D	1480
2SC511E	1480
2SC511F	1480
2SC511G	1480
2SC511GN	1480
2SC511H	1480
2SC511J	1480
2SC511K	1480
2SC511L	1480
2SC511M	1480
2SC5110	1480
2SC5110R	1480
2SC511R	1480
2SC511X	1480
2SC511Y	1480
2SC512	1480
2SC512-0	1480
2SC512-O	1480
2SC512-ORG	1480
2SC512-R	1480
2SC512-RED	1480
2SC512A	1480
2SC512B	1480
2SC512C	1480
2SC512D	1480
2SC512E	1480
2SC512F	1480
2SC512G	1480
2SC512GN	1480
2SC512H	1480
2SC512J	1480
2SC512K	1480
2SC512L	1480
2SC512M	1480
2SC5120	1480
2SC5120R	1480
2SC512R	1480
2SC512X	1480
2SC512Y	1480
2SC513	1480
2SC513-0	1480
2SC513-O	1480
2SC513-ORG	1480
2SC513-R	1480
2SC513-RED	1480
2SC513A	1480
2SC513B	1480
2SC513C	1480
2SC513D	1480
2SC513E	1480
2SC513F	1480
2SC513G	1480
2SC513GN	1480
2SC513H	1480
2SC513J	1480
2SC513K	1480
2SC513L	1480
2SC513M	1480
2SC5130	1480
2SC5130R	1480
2SC513R	1480
2SC513X	1480
2SC513Y	1480
2SC514	0142
2SC514A	0142
2SC514B	0142
2SC514C	0142
2SC514D	0142
2SC514E	0142
2SC514F	0142
2SC514G	0142
2SC514GN	0142
2SC514H	0142
2SC514J	0142
2SC514K	0142
2SC514L	0142
2SC514M	0142
2SC514OR	0142
2SC514R	0142
2SC514X	0142
2SC514Y	0142
2SC515	0142
2SC515-OR	0142
2SC515A	0142
2SC515A(BK)	0142
2SC515AM	0142
2SC515AX	0142
2SC515AY	0142
2SC515B	0142
2SC515BK	0142
2SC515C	0142
2SC515D	0142
2SC515E	0142
2SC515F	0142
2SC515G	0142
2SC515GN	0142
2SC515H	0142
2SC515J	0142
2SC515K	0142
2SC515L	0142
2SC515M	0142
2SC515OR	0142
2SC515R	0142
2SC515X	0142
2SC515Y	0142
2SC516	0086
2SC516A	0086
2SC516B	0086
2SC516C	0086
2SC516D	0086
2SC516E	0086
2SC516F	0086
2SC516G	0086
2SC516GN	0086
2SC516H	0086
2SC516J	0086
2SC516K	0086
2SC516L	0086
2SC516M	0086
2SC516OR	0086
2SC516R	0086
2SC516X	0086
2SC516Y	0086
2SC517	1411
2SC517-OR	1411
2SC517A	1411
2SC517B	1411
2SC517C	1411
2SC517D	1411
2SC517E	1411
2SC517F	1411
2SC517G	1411
2SC517GN	1411
2SC517H	1411
2SC517J	1411
2SC517K	1411
2SC517L	1411
2SC517M	1411
2SC517OR	1411
2SC517R	1411
2SC517S	1411
2SC517X	1411
2SC517Y	1411
2SC518	0168
2SC518A	0168
2SC519	0867
2SC519A	0867
2SC520	0867
2SC520A	0867
2SC521	0867
2SC521A	0867
2SC522	2314
2SC522-0	2314
2SC522-O	2314
2SC522-ORG	2314
2SC522-R	2314
2SC522-RED	2314
2SC5220	2314
2SC522R	2314
2SC523	2314
2SC523-0	2314
2SC523-O	2314
2SC523-ORG	2314
2SC523-R	2314
2SC523-RED	2314
2SC5230	2314
2SC523R	2314
2SC524	3336
2SC524-0	3336
2SC524-O	3336
2SC524-ORG	3336
2SC524-R	3336
2SC524-RED	3336
2SC5240	3336
2SC524R	3336
2SC525	1411
2SC525-0	1411
2SC525-O	1411
2SC525-ORG	1411
2SC525-R	1411
2SC525-RED	1411
2SC5250	1411
2SC525R	1411
2SC526	0233
2SC526-0	0233
2SC526A	0233
2SC526B	0233
2SC526C	0233
2SC526D	0233
2SC526E	0233
2SC526F	0233
2SC526G	0233
2SC526GN	0233
2SC526H	0233
2SC526J	0233
2SC526K	0233
2SC526L	0233
2SC526M	0233
2SC526OR	0233
2SC526R	0233
2SC526X	0233
2SC526Y	0233
2SC527	0007
2SC528	0111
2SC529	0286
2SC529A	0286
2SC529B	0286
2SC529C	0286
2SC529D	0286
2SC529E	0286
2SC529F	0286
2SC529G	0286
2SC529GN	0286
2SC529H	0286
2SC529J	0286
2SC529K	0286
2SC529L	0286
2SC529M	0286
2SC529OR	0286
2SC529R	0286
2SC529X	0286
2SC529Y	0286
2SC530	0079
2SC531	0155
2SC532	0079
2SC5320	0079
2SC533	0079
2SC534	0079
2SC535	0127
2SC535(B)	0127
2SC535(C)	0127
2SC535-OR	0127
2SC535A	0127
2SC535ABC	0127
2SC535AL	0127
2SC535B	0127
2SC535C	0127
2SC535D	0127
2SC535E	0127
2SC535F	0127
2SC535G	0127
2SC535GN	0127
2SC535H	0127
2SC535J	0127
2SC535K	0127
2SC535L	0127
2SC535M	0127
2SC535NB	OEM
2SC535OR	0127
2SC535R	0127
2SC535X	0127
2SC535Y	0127
2SC536	0537
2SC536(C)	0537
2SC536(D)	0537
2SC536(E)	0537
2SC536(F)	0537
2SC536(G)	0537
2SC536(RED)	OEM
2SC536-AUD-SPA	OEM
2SC536-D	0537
2SC536-E	0537
2SC536-E-NP	0537
2SC536-F	0537
2SC536-F-NP	0537
2SC536-F-NP-AA	0537
2SC536-F-SPA-AC	0537
2SC536-G	0537
2SC536-G-NP	0537
2SC536-G-NP-AA	0537
2SC536-G-SP	OEM
2SC536-G-SPT-AC	0537
2SC536-OR	0537
2SC536A	0507
2SC536AE-NP	OEM
2SC536AE-SP	OEM
2SC536AF	0537
2SC536AF-SP	OEM
2SC536AG	0537
2SC536AG-SP	OEM
2SC536AUD	OEM
2SC536AUD-G	OEM
2SC536AUD-H	OEM
2SC536AUD-KG	OEM
2SC536B	0537
2SC536C	0537
2SC536D	0537
2SC536DK	0537
2SC536E	0537
2SC536ED	0537
2SC536EH	0537
2SC536EJ	0537
2SC536EN	0537
2SC536EP	0537
2SC536ER	0537
2SC536ESP	OEM
2SC536ET	0537
2SC536EW	OEM
2SC536EZ	0537
2SC536F	0537
2SC536F1	0537
2SC536F2	0537
2SC536FC	0537
2SC536F-NP	0537
2SC536FP	0537
2SC536FS	0537
2SC536FS6	0537
2SC536FSP	OEM
2SC536FZ	0537
2SC536G	0537
2SC536G-1	0537
2SC536G1	0537
2SC536G2	0537
2SC536GF	0537
2SC536GJ	0537
2SC536GK	0537
2SC536GL	0537
2SC536GM	0537
2SC536GN	0537
2SC536G-NP	0537
2SC536GP	0537
2SC536GT	0537
2SC536GV	0537
2SC536GY	0537
2SC536GZ	0537
2SC536H	0537
2SC536J	0537
2SC536K	0537
2SC536K(F)	0537
2SC536K(G)	0537
2SC536KE	0537
2SC536KF	0537
2SC536K-F-NP-AA	0537
2SC536KF-NP-AA	4366
2SC536KG	0537
2SC536K-G-NP-AA	0537
2SC536KN(PG)	0537
2SC536KNP	0537
2SC536KNP-AA	0537
2SC536K-NP-E	OEM
2SC536K-NP-F	OEM
2SC536KNPFT	0537
2SC536KNP-T	0537
2SC536L	0537
2SC536M	0537
2SC536NP	0537
2SC536NP(E)	0537
2SC536NP(F)	0537
2SC536NP(G)	0537
2SC536NP-E	0537
2SC536NPE	0537
2SC536NP-F	OEM
2SC536NPF	0537
2SC536NP-G	OEM
2SC536NPG	0537
2SC536NP-T	0537
2SC536OR	0537
2SC536R	0537
2SC536SE	0537
2SC536SEZ	0537
2SC536SF	0537
2SC536SFZ	0537
2SC536SP	0537
2SC536SP(F)	0537
2SC536SP(G)	0537
2SC536SP-AC	0537
2SC536SP-E	OEM
2SC536SPE	0537
2SC536SP-F	OEM
2SC536SPF	0537
2SC536SP-G	OEM
2SC536W	0537
2SC536X	0537
2SC536XL	0537
2SC536Y	0507
2SC536YF	0537
2SC537	0191
2SC537(F)	0191
2SC537(G)	0191
2SC537-C7	0191
2SC537-EV	0191
2SC537ALC	0191
2SC537B	0191
2SC537BK	0191
2SC537C	0191
2SC537C7	0191
2SC537D	0191
2SC537D1	0191
2SC537D2	0191
2SC537E	0191
2SC537EF	0191
2SC537EH	0191
2SC537EJ	0191
2SC537EV	0191
2SC537F	0191
2SC537F1	0191
2SC537F2	0191
2SC537FC	0191
2SC537F-C7	0191
2SC537FC7	0191
2SC537FJ	0191
2SC537FK	0191
2SC537FV	0191
2SC537G	0191
2SC537G1	0191
2SC537G2	0191
2SC537GFL	0191
2SC537GI	0191
2SC537H	0191
2SC537HT	0191
2SC537WF	0191
2SC538	0079
2SC538(A)	0079
2SC538(P)	0079
2SC538(R)	0079
2SC538-Q	0079
2SC538A	0079
2SC538A(Q)	0079
2SC538A-P	0079
2SC538AQ	0079
2SC538A-R	0079
2SC538AR	0079
2SC538AS	0079
2SC538K	0079
2SC538P	0079
2SC538Q	0079
2SC538R	0079
2SC538S	0079
2SC538T	0079
2SC539	0016
2SC539(L)(K)	0016
2SC539(R)	0016
2SC539R	0016
2SC539S	0016
2SC539T	0016
2SC540	0475
2SC540A	0475
2SC540B	0475
2SC540C	0475
2SC540D	0475
2SC540E	0475
2SC540F	0475
2SC540G	0475
2SC540GN	0475
2SC540H	0475
2SC540J	0475
2SC540K	0475
2SC540L	0475
2SC540M	0475
2SC540OR	0475
2SC540R	0475
2SC540X	0475
2SC540Y	0475
2SC541	0560
2SC542	0560
2SC543	0042
2SC543A	0042
2SC543B	0042
2SC543C	0042
2SC543D	0042
2SC543E	0042
2SC543F	0042
2SC543G	0042
2SC543GN	0042
2SC543H	0042
2SC543J	0042
2SC543K	0042
2SC543L	0042
2SC543M	0042
2SC543OR	0042
2SC543R	0042
2SC543X	0042
2SC543Y	0042
2SC544	2209
2SC544A	2209
2SC544AG	2209
2SC544B	2209
2SC544C	2209
2SC544D	0127
2SC544D(VHF)	0127
2SC544E	2209
2SC544F	2209
2SC544G	2209
2SC544GN	2209
2SC544H	2209
2SC544J	2209
2SC544K	2209
2SC544L	2209
2SC544M	2209
2SC544OR	2209
2SC544R	2209
2SC544X	2209
2SC544Y	2209
2SC545	0127
2SC545A	0127
2SC545B	0127
2SC545C	0127
2SC545D	0127
2SC545E	0127
2SC546	0007
2SC546K	0007
2SC547	0111
2SC547B	OEM
2SC547C	OEM
2SC548	0560
2SC548C	OEM
2SC549	OEM
2SC550	0560
2SC550A	0560
2SC551	0042
2SC552	0042
2SC553	0042
2SC554	0042
2SC555	0493
2SC556	0224
2SC556B	OEM
2SC557	OEM
2SC557C	OEM
2SC558	4928
2SC558A	OEM
2SC558B	OEM
2SC559	0018
2SC559B	OEM
2SC560	0086
2SC560A	0086
2SC560B	0086
2SC560C	0086
2SC560D	0086
2SC560E	0086
2SC560F	0086
2SC560G	0086
2SC560GN	0086
2SC560H	0086
2SC560J	0086
2SC560K	0086
2SC560L	0086
2SC560M	0086
2SC560OR	0086
2SC560R	0086
2SC560X	0086
2SC560Y	0086
2SC561	1431
2SC561-OR	1431
2SC561A	1431
2SC561B	1431
2SC561C	1431
2SC561D	1431
2SC561E	1431
2SC561F	1431
2SC561G	1431
2SC561GN	1431
2SC561H	1431
2SC561J	1431
2SC561K	1431
2SC561L	1431
2SC561M	1431
2SC561OR	1431
2SC561R	1431
2SC561X	1431
2SC561Y	1431
2SC562	0216
2SC562(O)	0216
2SC562-OR	0216
2SC562A	0216
2SC562B	0216
2SC562C	0216
2SC562D	0216
2SC562E	0216
2SC562F	0216
2SC562G	0216
2SC562GN	0216
2SC562H	0216
2SC562J	0216
2SC562K	0216
2SC562L	0216
2SC562M	0216
2SC562OR	0216
2SC562R	0216
2SC562X	0216
2SC562Y	0216
2SC563	0835
2SC563(A)	0835
2SC563-F	0835
2SC563-G	0835
2SC563-OR	0835
2SC563A	0835
2SC563B	0835
2SC563C	0835
2SC563D	0835
2SC563E	0835
2SC563F	0835
2SC563G	0835
2SC563GN	0835
2SC563H	0835
2SC563J	0835
2SC563K	0835
2SC563L	0835
2SC563M	0835
2SC563OR	0835
2SC563R	0835
2SC563X	0835
2SC563Y	0835
2SC564	0016
2SC564(Q)	0016
2SC564(Q)(R)	0016
2SC564(S)	0016
2SC564A	0016
2SC564AP	0016
2SC564B	0016
2SC564C	0016
2SC564D	0016

If replacement code is OEM, contact original manufacturer for replacement.

DEVICE TYPE	REPL CODE
2SC564E	0016
2SC564F	0016
2SC564G	0016
2SC564GN	0016
2SC564H	0016
2SC564J	0016
2SC564K	0016
2SC564L	0016
2SC564M	0016
2SC564OR	0016
2SC564P	0016
2SC564PL	0016
2SC564Q	0016
2SC564QC	0016
2SC564R	0016
2SC564S	0016
2SC564T	0016
2SC564X	0016
2SC564Y	0016
2SC565	0016
2SC566	0259
2SC566A	0259
2SC566B	0259
2SC566C	0259
2SC566D	0259
2SC566E	0259
2SC566F	0259
2SC566G	0259
2SC566GN	0259
2SC566H	0259
2SC566J	0259
2SC566K	0259
2SC566L	0259
2SC566M	0259
2SC566OR	0259
2SC566R	0259
2SC566X	0259
2SC566Y	0259
2SC567	0259
2SC567A	0259
2SC567B	0259
2SC567C	0259
2SC567D	0259
2SC567E	0259
2SC567F	0259
2SC567G	0259
2SC567GN	0259
2SC567H	0259
2SC567J	0259
2SC567K	0259
2SC567L	0259
2SC567M	0259
2SC567OR	0259
2SC567R	0259
2SC567X	0259
2SC567Y	0259
2SC568	0259
2SC568A	0259
2SC568B	0259
2SC568C	0259
2SC568D	0259
2SC568E	0259
2SC568F	0259
2SC568G	0259
2SC568GN	0259
2SC568H	0259
2SC568J	0259
2SC568K	0259
2SC568L	0259
2SC568M	0259
2SC568OR	0259
2SC568R	0259
2SC568X	0259
2SC568Y	0259
2SC569	0016
2SC570	0016
2SC571	2041
2SC572	3321
2SC573	3322
2SC575	0086
2SC576	0016
2SC576E	0016
2SC576F	0016
2SC576G	0016
2SC577	0016
2SC578	0861
2SC579	0400
2SC580	0086
2SC580A	0086
2SC580B	0086
2SC580C	0086
2SC580D	0086
2SC580E	0086
2SC580F	0086
2SC580G	0086
2SC580GN	0086
2SC580H	0086
2SC580J	0086
2SC580K	0086
2SC580L	0086
2SC580M	0086
2SC580OR	0086
2SC580R	0086
2SC580T	0086
2SC580X	0086
2SC580Y	0086
2SC581	0007
2SC582	2658
2SC582(BX)	2658

DEVICE TYPE	REPL CODE
2SC582(BX)(BY)	2658
2SC582(BY)	2658
2SC582-OR	2658
2SC582,B	2658
2SC582A	2658
2SC582B	2658
2SC582BC	2658
2SC582BX	2658
2SC582BY	2658
2SC582C	2658
2SC582D	2658
2SC582E	2658
2SC582EA	2658
2SC582EH	2658
2SC582F	2658
2SC582G	2658
2SC582GN	2658
2SC582H	2658
2SC582J	2658
2SC582K	2658
2SC582L	2658
2SC582M	2658
2SC582OR	2658
2SC582R	2658
2SC582X	2658
2SC582Y	2658
2SC583	0259
2SC584	0219
2SC585	0042
2SC586	0177
2SC587	0016
2SC587A	0016
2SC587B	0016
2SC587C	0016
2SC587D	0016
2SC587E	0016
2SC587F	0016
2SC587G	0016
2SC587GN	0016
2SC587H	0016
2SC587J	0016
2SC587K	0016
2SC587L	0016
2SC587M	0016
2SC587OR	0016
2SC587R	0016
2SC587X	0016
2SC587Y	0016
2SC588	0016
2SC588A	0016
2SC588B	0016
2SC588C	0016
2SC588D	0016
2SC588E	0016
2SC588F	0016
2SC588G	0016
2SC588GN	0016
2SC588H	0016
2SC588J	0016
2SC588K	0016
2SC588L	0016
2SC588M	0016
2SC588OR	0016
2SC588R	0016
2SC588X	0016
2SC588Y	0016
2SC589	0233
2SC589A	0233
2SC589B	0233
2SC589C	0233
2SC589D	0233
2SC589E	0233
2SC589F	0233
2SC589G	0233
2SC589GN	0233
2SC589H	0233
2SC589J	0233
2SC589K	0233
2SC589L	0233
2SC589M	0233
2SC589OR	0233
2SC589R	0233
2SC589X	0233
2SC589Y	0233
2SC590	0233
2SC590A	0233
2SC590B	0233
2SC590C	0233
2SC590D	0233
2SC590E	0233
2SC590F	0233
2SC590G	0233
2SC590GN	0233
2SC590H	0233
2SC590J	0233
2SC590K	0233
2SC590L	0233
2SC590M	0233
2SC590OR	0233
2SC590X	0233
2SC590Y	0233
2SC591	0168
2SC591A	0168
2SC591B	0168
2SC591C	0168
2SC591D	0168
2SC591E	0168
2SC591F	0168
2SC591G	0168

DEVICE TYPE	REPL CODE
2SC591GN	0168
2SC591H	0168
2SC591J	0168
2SC591K	0168
2SC591L	0168
2SC591M	0168
2SC591OR	0168
2SC591R	0168
2SC591X	0168
2SC591Y	0168
2SC592	0560
2SC593	0016
2SC594	0016
2SC594-O	0016
2SC594-R	0016
2SC594-Y	0016
2SC594A	0016
2SC594B	0016
2SC594C	0016
2SC594D	0016
2SC594E	0016
2SC594F	0016
2SC594G	0016
2SC594GN	0016
2SC594H	0016
2SC594J	0016
2SC594K	0016
2SC594L	0016
2SC594M	0016
2SC594OR	0016
2SC594R	0016
2SC594Y	0016
2SC595	0016
2SC595A	0016
2SC595B	0016
2SC595C	0016
2SC595D	0016
2SC595E	0016
2SC595F	0016
2SC595G	0016
2SC595GN	0016
2SC595H	0016
2SC595J	0016
2SC595K	0016
2SC595L	0016
2SC595M	0016
2SC595OR	0016
2SC595X	0016
2SC595Y	0016
2SC596	0016
2SC596A	0016
2SC596B	0016
2SC596C	0016
2SC596E	0016
2SC596F	0016
2SC596G	0016
2SC596GN	0016
2SC596H	0016
2SC596J	0016
2SC596K	0016
2SC596L	0016
2SC596M	0016
2SC596OR	0016
2SC596R	0016
2SC596X	0016
2SC596Y	0016
2SC597	0936
2SC598	0560
2SC599	0042
2SC600	0042
2SC601	0493
2SC601N	0493
2SC602	0007
2SC603	0400
2SC604	0127
2SC605	0475
2SC605(B)	0475
2SC605(L)	0475
2SC605-B	0475
2SC605-OR	0475
2SC605A	0475
2SC605B	0475
2SC605C	0475
2SC605D	0475
2SC605E	0475
2SC605F	0475
2SC605G	0475
2SC605GN	0475
2SC605H	0475
2SC605J	0475
2SC605K	0475
2SC605L	0475
2SC605M	0475
2SC605OR	0475
2SC605Q	0475
2SC605R	0475
2SC605TW	0475
2SC605X	0475
2SC605Y	0475
2SC606	0475
2SC606(B)	0475
2SC606(VHF)	0475
2SC606-B	0475
2SC606A	0475
2SC606B	0475
2SC606C	0475
2SC606D	0475
2SC606E	0475
2SC606F	0475

DEVICE TYPE	REPL CODE
2SC606G	0475
2SC606GN	0475
2SC606H	0475
2SC606J	0475
2SC606K	0475
2SC606L	0475
2SC606M	0475
2SC606N	0475
2SC606OR	0475
2SC606R	0475
2SC606X	0475
2SC606Y	0475
2SC607	0086
2SC608	1411
2SC608A	1411
2SC608AA	1411
2SC608B	1411
2SC608C	1411
2SC608D	1411
2SC608E	1411
2SC608F	1411
2SC608G	1411
2SC608GN	1411
2SC608H	1411
2SC608J	1411
2SC608K	1411
2SC608L	1411
2SC608M	1411
2SC608OR	1411
2SC608R	1411
2SC608T	1411
2SC608X	1411
2SC608Y	1411
2SC609	1411
2SC609-OR	1411
2SC609A	1411
2SC609B	1411
2SC609C	1411
2SC609D	1411
2SC609E	1411
2SC609F	1411
2SC609G	1411
2SC609GN	1411
2SC609H	1411
2SC609J	1411
2SC609K	1411
2SC609L	1411
2SC609M	1411
2SC609OR	1411
2SC609R	1411
2SC609T	1411
2SC609X	1411
2SC609Y	1411
2SC610	0086
2SC610A	0086
2SC610B	0086
2SC610C	0086
2SC610D	0086
2SC610E	0086
2SC610F	0086
2SC610G	0086
2SC610GN	0086
2SC610H	0086
2SC610J	0086
2SC610K	0086
2SC610L	0086
2SC610M	0086
2SC610OR	0086
2SC610R	0086
2SC610X	0086
2SC610Y	0086
2SC611	0259
2SC611A	0259
2SC611B	0259
2SC611C	0259
2SC611D	0259
2SC611E	0259
2SC611F	0259
2SC611G	0259
2SC611GN	0259
2SC611H	0259
2SC611J	0259
2SC611K	0259
2SC611L	0259
2SC611M	0259
2SC611OR	0259
2SC611R	0259
2SC611X	0259
2SC611Y	0259
2SC612	0259
2SC612A	0259
2SC612B	0259
2SC612C	0259
2SC612D	0259
2SC612E	0259
2SC612F	0259
2SC612G	0259
2SC612GN	0259
2SC612H	0259
2SC612J	0259
2SC612K	0259
2SC612L	0259
2SC612M	0259
2SC612OR	0259
2SC612R	0259
2SC612X	0259
2SC612Y	0259
2SC613	0144
2SC613A	0144

DEVICE TYPE	REPL CODE
2SC613B	0144
2SC613C	0144
2SC613D	0144
2SC613E	0144
2SC613F	0144
2SC613G	0144
2SC613GN	0144
2SC613H	0144
2SC613J	0144
2SC613K	0144
2SC613L	0144
2SC613M	0144
2SC613OR	0144
2SC613R	0144
2SC613X	0144
2SC613Y	0144
2SC614	0698
2SC614A	0698
2SC614B	0698
2SC614C	0698
2SC614D	0698
2SC614E	0698
2SC614F	0698
2SC614G	0698
2SC614GN	0698
2SC614H	0698
2SC614J	0698
2SC614K	0698
2SC614L	0698
2SC614M	0698
2SC614OR	0698
2SC614R	0698
2SC614X	0698
2SC614Y	0698
2SC615	0698
2SC615-OR	0698
2SC615A	0698
2SC615B	0698
2SC615C	0698
2SC615D	0698
2SC615E	0698
2SC615F	0698
2SC615G	0698
2SC615GN	0698
2SC615H	0698
2SC615J	0698
2SC615K	0698
2SC615L	0698
2SC615M	0698
2SC615OR	0698
2SC615R	0698
2SC615X	0698
2SC615Y	0698
2SC616	0630
2SC617	0630
2SC618	0007
2SC618A	0007
2SC619	0286
2SC619(B)	0286
2SC619(C)	0286
2SC619A	0286
2SC619B	0286
2SC619C	0286
2SC619D	0286
2SC619E	0286
2SC619F	0286
2SC619G	0286
2SC619GN	0286
2SC619H	0286
2SC619J	0286
2SC619K	0286
2SC619L	0286
2SC619M	0286
2SC619OR	0286
2SC619R	0286
2SC619X	0286
2SC619Y	0286
2SC620	0076
2SC620(C)	0076
2SC620(D)	0076
2SC620(E)	0076
2SC620-D	0076
2SC620-E	0076
2SC620-OR	0076
2SC620A	0076
2SC620B	0076
2SC620C	0076
2SC620CD	0076
2SC620D	0076
2SC620DE	0076
2SC620E	0076
2SC620F	0076
2SC620G	0076
2SC620GN	0076
2SC620H	0076
2SC620J	0076
2SC620K	0076
2SC620L	0076
2SC620M	0076
2SC620OR	0076
2SC620R	0076
2SC620X	0076
2SC620Y	0076
2SC621	0286
2SC621A	0286
2SC621B	0286
2SC621C	0286
2SC621D	0286
2SC621E	0286

DEVICE TYPE	REPL CODE
2SC621F	0286
2SC621G	0286
2SC621GN	0286
2SC621H	0286
2SC621J	0286
2SC621K	0286
2SC621L	0286
2SC621M	0286
2SC621OR	0286
2SC621R	0286
2SC621X	0286
2SC621Y	0286
2SC622	0016
2SC622A	0016
2SC622B	0016
2SC622C	0016
2SC622D	0016
2SC622E	0016
2SC622F	0016
2SC622G	0016
2SC622GN	0016
2SC622H	0016
2SC622J	0016
2SC622K	0016
2SC622L	0016
2SC622M	0016
2SC622OR	0016
2SC622R	0016
2SC622X	0016
2SC622Y	0016
2SC623	0016
2SC624	0016
2SC626	0016
2SC627	0233
2SC628	0493
2SC628E	0493
2SC628F	0493
2SC629	0224
2SC629-31	0224
2SC629-41	0224
2SC629A	0224
2SC629B	0224
2SC629C	0224
2SC629D	0224
2SC629E	0224
2SC629F	0224
2SC629G	0224
2SC629GN	0224
2SC629H	0224
2SC629J	0224
2SC629K	0224
2SC629L	0224
2SC629M	0224
2SC629OR	0224
2SC629R	0224
2SC629X	0224
2SC629Y	0224
2SC631	0286
2SC631A	0286
2SC631B	0286
2SC631C	0286
2SC631D	0286
2SC631E	0286
2SC631F	0286
2SC631G	0286
2SC631GN	0286
2SC631H	0286
2SC631J	0286
2SC631K	0286
2SC631L	0286
2SC631M	0286
2SC631OR	0286
2SC631R	0286
2SC631X	0286
2SC631Y	0286
2SC632	2075
2SC632(1)	2075
2SC632-OR	2075
2SC632A	2075
2SC632B	2075
2SC632C	2075
2SC632D	2075
2SC632E	2075
2SC632F	2075
2SC632G	2075
2SC632GN	2075
2SC632H	2075
2SC632J	2075
2SC632K	2075
2SC632L	2075
2SC632M	2075
2SC632OR	2075
2SC632R	2075
2SC632X	2075
2SC632Y	2075
2SC633	2075
2SC633-7	2075
2SC633-OR	2075
2SC633A	2075
2SC633B	2075
2SC633C	2075
2SC633D	2075
2SC633E	2075
2SC633F	2075
2SC633G	2075
2SC633GN	2075
2SC633H	2075
2SC633J	2075
2SC633K	2075

DEVICE TYPE	REPL CODE
2SC633L	2075
2SC633M	2075
2SC633OR	2075
2SC633R	2075
2SC633X	2075
2SC633Y	2075
2SC634	2075
2SC634(2)	2075
2SC634-O	2075
2SC634-OR	2075
2SC634-SP	2075
2SC634-SP8	2075
2SC634A	2075
2SC634AK	2075
2SC634AL	2075
2SC634AXL	2075
2SC634B	2075
2SC634C	2075
2SC634D	2075
2SC634E	2075
2SC634F	2075
2SC634G	2075
2SC634GN	2075
2SC634H	2075
2SC634J	2075
2SC634K	2075
2SC634L	2075
2SC634M	2075
2SC634OR	2075
2SC634R	2075
2SC634SP	2075
2SC634SP-8	2075
2SC634SP8	2075
2SC634X	2075
2SC634Y	2075
2SC635	0079
2SC635A	0079
2SC636	0042
2SC637	3321
2SC638	3322
2SC638C	3322
2SC639	0016
2SC640	0475
2SC640A	0475
2SC640B	0475
2SC640C	0475
2SC640D	0475
2SC640E	0475
2SC640F	0475
2SC640G	0475
2SC640GN	0475
2SC640J	0475
2SC640K	0475
2SC640L	0475
2SC640M	0475
2SC640R	0475
2SC640X	0475
2SC641	0286
2SC641A	0286
2SC641B	0286
2SC641C	0286
2SC641D	0286
2SC641E	0286
2SC641F	0286
2SC641G	0286
2SC641GN	0286
2SC641H	0286
2SC641J	0286
2SC641K	0286
2SC641K-B	OEM
2SC641L	0286
2SC641M	0286
2SC641OR	0286
2SC641R	0286
2SC641X	0286
2SC641Y	0286
2SC642	0003
2SC642A	0003
2SC643	1150
2SC643-OR	1150
2SC643A	1150
2SC643B	1150
2SC643C	1150
2SC643D	1150
2SC643E	1150
2SC643F	1150
2SC643G	1150
2SC643GN	1150
2SC643H	1150
2SC643J	1150
2SC643K	1150
2SC643L	1150
2SC643M	1150
2SC643OR	1150
2SC643R	1150
2SC643X	1150
2SC643Y	1150
2SC644	0111
2SC644(F)	0111
2SC644(H)	0111
2SC644(R)	0111
2SC644(R,S)	0111
2SC644(S)	0111
2SC644(T)	0111
2SC644-OR	0111
2SC644A	0111
2SC644B	0111
2SC644C	0111
2SC644D	0111

If replacement code is OEM, contact original manufacturer for replacement.

DEVICE TYPE	REPL CODE
2SC644E	0111
2SC644F	0111
2SC644F(H)(S)	0111
2SC644FH	0111
2SC644FHS	0111
2SC644FR	0111
2SC644FS	0111
2SC644G	0111
2SC644GN	0111
2SC644H	0111
2SC644H(S)	0111
2SC644J	0111
2SC644K	0111
2SC644L	0111
2SC644M	0111
2SC644OR	0111
2SC644P	0111
2SC644PJ	0111
2SC644Q	0111
2SC644R	0111
2SC644S	0111
2SC644T	0111
2SC644X	0111
2SC644Y	0111
2SC645	0674
2SC645(B)	0674
2SC645-OR	0674
2SC645A	0674
2SC645B	0674
2SC645B-1	0674
2SC645C	0674
2SC645D	0674
2SC645E	0674
2SC645F	0674
2SC645G	0674
2SC645GN	0674
2SC645GR	0674
2SC645H	0674
2SC645J	0674
2SC645K	0674
2SC645L	0674
2SC645M	0674
2SC645N	0674
2SC645OR	0674
2SC645R	0674
2SC645V	0674
2SC645X	0674
2SC645Y	0674
2SC646	0620
2SC647	4387
2SC647Q	4387
2SC647R	4387
2SC648	0016
2SC648A	0016
2SC648B	0016
2SC648C	0016
2SC648D	0016
2SC648E	0016
2SC648F	0016
2SC648G	0016
2SC648H	0016
2SC648J	0016
2SC648K	0016
2SC648L	0016
2SC648M	0016
2SC648OR	0016
2SC648R	0016
2SC648X	0016
2SC648Y	0016
2SC649	0016
2SC649A	0016
2SC649B	0016
2SC649C	0016
2SC649D	0016
2SC649E	0016
2SC649F	0016
2SC649G	0016
2SC649GN	0016
2SC649H	0016
2SC649J	0016
2SC649K	0016
2SC649L	0016
2SC649M	0016
2SC649OR	0016
2SC649R	0016
2SC649X	0016
2SC649Y	0016
2SC650	0536
2SC650-OR	0536
2SC650-Y	0536
2SC650A	0536
2SC650B	0536
2SC650C	0536
2SC650D	0536
2SC650E	0536
2SC650F	0536
2SC650G	0536
2SC650GN	0536
2SC650H	0536
2SC650J	0536
2SC650K	0536
2SC650L	0536
2SC650M	0536
2SC650OR	0536
2SC650R	0536
2SC650X	0536
2SC650Y	0536
2SC651	0493
2SC652	0493
2SC653	0259
2SC654	0493
2SC654A	0493
2SC654B	0493
2SC654C	0493
2SC654D	0493
2SC654E	0493
2SC654F	0493
2SC654G	0493
2SC654GN	0493
2SC654H	0493
2SC654J	0493
2SC654K	0493
2SC654L	0493
2SC654M	0493
2SC654OR	0493
2SC654R	0493
2SC654X	0493
2SC654Y	0493
2SC655	0016
2SC655A	0016
2SC655B	0016
2SC655C	0016
2SC655D	0016
2SC655E	0016
2SC655F	0016
2SC655G	0016
2SC655GN	0016
2SC655H	0016
2SC655J	0016
2SC655K	0016
2SC655L	0016
2SC655M	0016
2SC655OR	0016
2SC655R	0016
2SC655X	0016
2SC655Y	0016
2SC656	0259
2SC656A	0259
2SC656B	0259
2SC656C	0259
2SC656D	0259
2SC656E	0259
2SC656F	0259
2SC656G	0259
2SC656GN	0259
2SC656H	0259
2SC656J	0259
2SC656K	0259
2SC656L	0259
2SC656M	0259
2SC656OR	0259
2SC656R	0259
2SC656X	0259
2SC656Y	0259
2SC657	0127
2SC657A	0127
2SC657B	0127
2SC657C	0127
2SC657D	0127
2SC657E	0127
2SC657F	0127
2SC657G	0127
2SC657GN	0127
2SC657H	0127
2SC657J	0127
2SC657K	0127
2SC657L	0127
2SC657M	0127
2SC657OR	0127
2SC657R	0127
2SC657X	0127
2SC657Y	0127
2SC658	0127
2SC658A	0127
2SC658B	0127
2SC658C	0127
2SC658D	0127
2SC658E	0127
2SC658F	0127
2SC658G	0127
2SC658GN	0127
2SC658H	0127
2SC658J	0127
2SC658K	0127
2SC658L	0127
2SC658M	0127
2SC658OR	0127
2SC658R	0127
2SC658X	0127
2SC658Y	0127
2SC659	0127
2SC659A	0127
2SC659B	0127
2SC659C	0127
2SC659D	0127
2SC659E	0127
2SC659F	0127
2SC659G	0127
2SC659GN	0127
2SC659H	0127
2SC659J	0127
2SC659K	0127
2SC659L	0127
2SC659M	0127
2SC659OR	0127
2SC659R	0127
2SC659X	0127
2SC659Y	0127
2SC660	0007
2SC661	0007
2SC662	0144
2SC662A	0144
2SC662B	0144
2SC662C	0144
2SC662D	0144
2SC662E	0144
2SC662F	0144
2SC662G	0144
2SC662GN	0144
2SC662H	0144
2SC662J	0144
2SC662K	0144
2SC662L	0144
2SC662M	0144
2SC662OR	0144
2SC662R	0144
2SC662X	0144
2SC662Y	0144
2SC663	0007
2SC663A	0007
2SC663B	0007
2SC663C	0007
2SC663D	0007
2SC663E	0007
2SC663F	0007
2SC663G	0007
2SC663GN	0007
2SC663H	0007
2SC663J	0007
2SC663K	0007
2SC663L	0007
2SC663M	0007
2SC663OR	0007
2SC663R	0007
2SC663X	0007
2SC663Y	0007
2SC664	0103
2SC664(B)	0103
2SC664(C)	0103
2SC664-OR	0103
2SC664A	0103
2SC664B	0103
2SC664C	0103
2SC664D	0103
2SC664E	0103
2SC664F	0103
2SC664L	0103
2SC664MNP-F	OEM
2SC664OR	0103
2SC664R	0103
2SC664Y	0103
2SC665	0867
2SC665H	0867
2SC665HA	0867
2SC665HB	0867
2SC666	0177
2SC666C	0177
2SC667	0007
2SC668	0127
2SC668(AUDIO)	0127
2SC668(C)	0127
2SC668(D)	0127
2SC668-OR	0127
2SC668-SP	0127
2SC668A	0127
2SC668B	0127
2SC668B1	0127
2SC668BC2	0127
2SC668C	0127
2SC668C1	0127
2SC668C2	0127
2SC668CD	0127
2SC668D	0127
2SC668D0	0127
2SC668D1	0127
2SC668DE	0127
2SC668D0	0127
2SC668DV	0127
2SC668DX	0127
2SC668DZ	0127
2SC668E	0127
2SC668E1	0127
2SC668E2	0127
2SC668EP	0127
2SC668EV	0127
2SC668EX	0127
2SC668F	0127
2SC668G	0127
2SC668GN	0127
2SC668H	0127
2SC668K	0127
2SC668L	0127
2SC668M	0127
2SC668OR	0127
2SC668R	0127
2SC668SP	0127
2SC668X	0127
2SC668Y	0127
2SC669	0622
2SC669A	0622
2SC669AB	0622
2SC669C	0622
2SC669D	0622
2SC673	0007
2SC673(B)	0007
2SC673B	0007
2SC673C	0007
2SC673C2	0007
2SC673D	0007
2SC674	0127
2SC674(C)	0127
2SC674(D)	0127
2SC674(F)	0127
2SC674(G)	0127
2SC674-B	0127
2SC674-F	0127
2SC674B	0127
2SC674C	0127
2SC674CK	0127
2SC674CL	0127
2SC674D	0127
2SC674E	0127
2SC674F	0127
2SC674G	0127
2SC675	0363
2SC676	0168
2SC677	0867
2SC678	0103
2SC679	1027
2SC679H	1027
2SC680	1504
2SC680(A)	1504
2SC680A	1504
2SC680B	1504
2SC680C	1504
2SC680G	1504
2SC680GN	1504
2SC680H	1504
2SC680J	1504
2SC680K	1504
2SC680L	1504
2SC680M	1504
2SC680R	1504
2SC680X	1504
2SC681	2437
2SC681(A)	2437
2SC681(B)	2437
2SC681A	2437
2SC681AYL	2437
2SC681AYL-12	OEM
2SC681L	2437
2SC681Y	2437
2SC681YL	2437
2SC682	0047
2SC682(B)	0047
2SC682-OR	0047
2SC682A	0047
2SC682B	0047
2SC682C	0047
2SC682D	0047
2SC682E	0047
2SC682F	0047
2SC682G	0047
2SC682GN	0047
2SC682H	0047
2SC682J	0047
2SC682K	0047
2SC682L	0047
2SC682M	0047
2SC682OR	0047
2SC682R	0047
2SC682X	0047
2SC682Y	0047
2SC683	0047
2SC683(B)	0047
2SC683-0	0047
2SC683-OR	0047
2SC683A	0047
2SC683B	0047
2SC683C	0047
2SC683D	0047
2SC683E	0047
2SC683F	0047
2SC683G	0047
2SC683GN	0047
2SC683H	0047
2SC683J	0047
2SC683K	0047
2SC683L	0047
2SC683M	0047
2SC683OR	0047
2SC683P	0047
2SC683Q	0047
2SC683R	0047
2SC683S	0047
2SC683V	0047
2SC683X	0047
2SC683Y	0047
2SC684	0127
2SC684-OR	0127
2SC684A	0127
2SC684B	0127
2SC684BK	0127
2SC684C	0127
2SC684D	0127
2SC684E	0127
2SC684F	0127
2SC684G	0127
2SC684GN	0127
2SC684H	0127
2SC684J	0127
2SC684K	0127
2SC684L	0127
2SC684M	0127
2SC684OR	0127
2SC684R	0127
2SC684TM	0127
2SC684X	0127
2SC684Y	0127
2SC685	0142
2SC685(0)	0142
2SC685-0	0142
2SC685-O	0142
2SC685A	0142
2SC685ABK	0142
2SC685AL	0142
2SC685B	0142
2SC685BK	0142
2SC685C	0142
2SC685D	0142
2SC685E	0142
2SC685F	0142
2SC685G	0142
2SC685GN	0142
2SC685H	0142
2SC685J	0142
2SC685K	0142
2SC685L	0142
2SC685M	0142
2SC685OR	0142
2SC685P	0142
2SC685R	0142
2SC685SY	0142
2SC685X	0142
2SC685Y	0142
2SC686	0233
2SC686A	0233
2SC686B	0233
2SC686C	0233
2SC686D	0233
2SC686E	0233
2SC686F	0233
2SC686G	0233
2SC686GN	0233
2SC686H	0233
2SC686J	0233
2SC686K	0233
2SC686L	0233
2SC686M	0233
2SC686OR	0233
2SC686R	0233
2SC686X	0233
2SC686Y	0233
2SC687	0177
2SC687A	0177
2SC687B	0177
2SC687C	0177
2SC687D	0177
2SC687E	0177
2SC687F	0177
2SC687G	0177
2SC687GN	0177
2SC687H	0177
2SC687J	0177
2SC687K	0177
2SC687L	0177
2SC687M	0177
2SC687OR	0177
2SC687R	0177
2SC687X	0177
2SC687Y	0177
2SC688	0224
2SC688-SP	0224
2SC688A	0224
2SC688B	0224
2SC688C	0224
2SC688D	0224
2SC688E	0224
2SC688F	0224
2SC688G	0224
2SC688GN	0224
2SC688H	0224
2SC688J	0224
2SC688K	0224
2SC688L	0224
2SC688M	0224
2SC688OR	0224
2SC688R	0224
2SC688X	0224
2SC688Y	0224
2SC689	0016
2SC689A	0016
2SC689B	0016
2SC689C	0016
2SC689D	0016
2SC689E	0016
2SC689G	0016
2SC689GN	0016
2SC689H	0016
2SC689J	0016
2SC689K	0016
2SC689L	0016
2SC689M	0016
2SC689OR	0016
2SC689R	0016
2SC689X	0016
2SC689Y	0016
2SC690	3587
2SC690A	3587
2SC691	0560
2SC692	0560
2SC693	0552
2SC693-OR	0552
2SC693A	0552
2SC693B	0552
2SC693C	0552
2SC693D	0552
2SC693E	0552
2SC693EB	0552
2SC693ET	0552
2SC693F	0552
2SC693FC	0552
2SC693FL	0552
2SC693FP	0552
2SC693FU	0552
2SC693G	0552
2SC693GL	0552
2SC693GN	0552
2SC693GS	0552
2SC693GU	0552
2SC693GZ	0552
2SC693H	0552
2SC693J	0552
2SC693K	0552
2SC693L	0552
2SC693M	0552
2SC693NP	0552
2SC693OR	0552
2SC693R	0552
2SC693U	0552
2SC693X	0552
2SC693Y	0552
2SC694	0111
2SC694A	0111
2SC694B	0111
2SC694C	0111
2SC694D	0111
2SC694E	0111
2SC694F	0111
2SC694G	0111
2SC694GN	0111
2SC694H	0111
2SC694J	0111
2SC694K	0111
2SC694L	0111
2SC694M	0111
2SC694OR	0111
2SC694R	0111
2SC694X	0111
2SC694Y	0111
2SC694Z	0111
2SC695	0475
2SC696	0644
2SC696(B)	0644
2SC696(D)	0644
2SC696(E)	0644
2SC696(H)	0644
2SC696-4	0644
2SC696-OR	0644
2SC696A	0644
2SC696AA	0644
2SC696AB	0644
2SC696AD	0644
2SC696AE	0644
2SC696AF	0644
2SC696AG	0644
2SC696AH	0644
2SC696AI	0644
2SC696B	0644
2SC696BL	0644
2SC696C	0644
2SC696D	0644
2SC696E	0644
2SC696F	0644
2SC696G	0644
2SC696GN	0644
2SC696GU	0644
2SC696H	0644
2SC696I	0644
2SC696J	0644
2SC696K	0644
2SC696L	0644
2SC696M	0644
2SC696OR	0644
2SC696R	0644
2SC696U	0644
2SC696X	0644
2SC696Y	0644
2SC697	1908
2SC697-OR	1908
2SC697A	1908
2SC697B	1908
2SC697C	1908
2SC697D	1908
2SC697E	1908
2SC697F	1908
2SC697G	1908
2SC697GN	1908
2SC697H	1908
2SC697J	1908
2SC697K	1908
2SC697L	1908
2SC697M	1908
2SC697OR	1908
2SC697R	1908
2SC697X	1908
2SC697Y	1908
2SC698	0630
2SC699	0595
2SC700	0595
2SC701	0018
2SC701A	0018
2SC701B	0018
2SC701C	0018
2SC701D	0018
2SC701E	0018
2SC701F	0018
2SC701G	0018
2SC701GN	0018
2SC701H	0018
2SC701J	0018
2SC701K	0018
2SC701L	0018
2SC701M	0018
2SC701OR	0018
2SC701R	0018
2SC701X	0018
2SC701Y	0018
2SC702	0018
2SC702A	0018
2SC702B	0018
2SC702C	0018
2SC702D	0018
2SC702E	0018
2SC702F	0018
2SC702G	0018
2SC702GN	0018
2SC702H	0018
2SC702J	0018
2SC702K	0018
2SC702L	0018
2SC702M	0018
2SC702OR	0018
2SC702R	0018
2SC702X	0018
2SC702Y	0018
2SC703	0042
2SC704	0042
2SC705	0127
2SC705A	0127
2SC705B	0127
2SC705C	0127
2SC705D	0127
2SC705E	0127
2SC705F	0127
2SC705G	0127
2SC705GN	0127
2SC705J	0127
2SC705K	0127
2SC705L	0127
2SC705M	0127
2SC705OR	0127
2SC705R	0127
2SC705TV	0127
2SC705TVV	0127
2SC705TW	0127
2SC705X	0127
2SC705Y	0127
2SC706	0007
2SC707	0259
2SC707A	0259
2SC707B	0259
2SC707C	0259
2SC707D	0259
2SC707E	0259
2SC707F	0259
2SC707G	0259
2SC707GN	0259
2SC707H	0259
2SC707K	0259
2SC707L	0259
2SC707M	0259
2SC707OR	0259
2SC707R	0259
2SC707X	0259
2SC707Y	0259
2SC708	0086
2SC708(A)	0086
2SC708(B)	0086
2SC708(C)	0086
2SC708-OR	0086
2SC708A	0086
2SC708AA	0086
2SC708AB	0086
2SC708AC	0086
2SC708AH	0086
2SC708B	0086
2SC708C	0086
2SC708D	0086
2SC708E	0086
2SC708F	0086
2SC708G	0086
2SC708GN	0086
2SC708H	0086
2SC708HA	0086
2SC708HB	0086
2SC708L	0086
2SC708M	0086
2SC708OR	0086
2SC708R	0086
2SC708X	0086
2SC708Y	0086
2SC709	0286
2SC709(B)(C)	0286
2SC709(C)	0286
2SC709A	0286
2SC709B	0286
2SC709C	0286
2SC709CD	0286
2SC709D	0286
2SC709E	0286
2SC709F	0286
2SC709G	0286

If replacement code is OEM, contact original manufacturer for replacement.

DEVICE TYPE	REPL CODE
2SC709GN	0286
2SC709H	0286
2SC709J	0286
2SC709L	0286
2SC709M	0286
2SC709OR	0286
2SC709X	0286
2SC709Y	0286
2SC710	0368
2SC710(B)	0368
2SC710(C)	0368
2SC710(D)	0368
2SC710(E)	0368
2SC710-0	0368
2SC710-3	0368
2SC710-5	0368
2SC710-13	0368
2SC710-14	0368
2SC710-D	0368
2SC710-E	0368
2SC710-OR	0368
2SC710AL	0368
2SC710B	0368
2SC710B2	0368
2SC710BC	0368
2SC710C	0368
2SC710D	0368
2SC710DB	0368
2SC710DE	0368
2SC710E	0368
2SC710F	0368
2SC710G	0368
2SC710GN	0368
2SC710H	0368
2SC710K	0368
2SC710L	0368
2SC710M	0368
2SC710OR	0368
2SC710R	0368
2SC710S	0368
2SC710X	0368
2SC710XL	0368
2SC710Y	0368
2SC711	0380
2SC711(D)	0380
2SC711(E)	0380
2SC711(F)	0380
2SC711-OR	0380
2SC711A	0380
2SC711A(E)	0380
2SC711A(F)	0380
2SC711A-E	0380
2SC711AE	0380
2SC711A-F	0380
2SC711AF	0380
2SC711AG	0380
2SC711AN	0380
2SC711B	0380
2SC711C	0380
2SC711D	0380
2SC711E	0380
2SC711EF	0380
2SC711F	0380
2SC711G	2465
2SC711GN	0380
2SC711H	0380
2SC711J	0380
2SC711L	0380
2SC711M	0380
2SC711OR	0380
2SC711R	0380
2SC711X	0380
2SC711Y	0380
2SC712	0076
2SC712(D)	0076
2SC712A	0076
2SC712B	0076
2SC712BC	0076
2SC712C	0076
2SC712CD	0076
2SC712D	0076
2SC712DC	0076
2SC712E	0076
2SC712F	0076
2SC712G	0076
2SC712GN	0076
2SC712H	0076
2SC712J	0076
2SC712K	0076
2SC712L	0076
2SC712M	0076
2SC712OR	0076
2SC712R	0076
2SC712W	0076
2SC712X	0076
2SC712Y	0076
2SC713	0286
2SC713A	0286
2SC713B	0286
2SC713C	0286
2SC713D	0286
2SC713E	0286
2SC713F	0286
2SC713G	0286
2SC713GN	0286
2SC713H	0286
2SC713J	0286
2SC713K	0286
2SC713L	0286
2SC713M	0286
2SC713MC	0286
2SC713MD	0286
2SC713ME	0286
2SC713OR	0286
2SC713R	0286
2SC713W	0286
2SC713X	0286
2SC713Y	0286
2SC714	0155
2SC714A	0155
2SC714B	0155
2SC714C	0155
2SC714D	0155
2SC714E	0155
2SC714F	0155
2SC714G	0155
2SC714GN	0155
2SC714H	0155
2SC714J	0155
2SC714K	0155
2SC714L	0155
2SC714M	0155
2SC714OR	0155
2SC714R	0155
2SC714X	0155
2SC714Y	0155
2SC715	0286
2SC715(D)	0286
2SC715(E)	0286
2SC715-OR	0286
2SC715A	0286
2SC715B	0286
2SC715C	0286
2SC715D	0286
2SC715E	0286
2SC715EJ	0286
2SC715EV	0286
2SC715F	0286
2SC715G	0286
2SC715GN	0286
2SC715H	0286
2SC715J	0286
2SC715K	0286
2SC715L	0286
2SC715M	0286
2SC715OR	0286
2SC715R	0286
2SC715X	0286
2SC715XL	0286
2SC715Y	0286
2SC716	0155
2SC716A	0155
2SC716B	0155
2SC716C	0155
2SC716D	0155
2SC716E	0155
2SC716F	0155
2SC716G	0155
2SC716GN	0155
2SC716H	0155
2SC716J	0155
2SC716K	0155
2SC716L	0155
2SC716M	0155
2SC716OR	0155
2SC716R	0155
2SC716X	0155
2SC716Y	0155
2SC717	0127
2SC717-TM	0127
2SC717B	0127
2SC717BK	0127
2SC717BLK	0127
2SC717F	0127
2SC717G	0127
2SC717GN	0127
2SC717H	0127
2SC717K	0127
2SC717L	0127
2SC717M	0127
2SC717TM	0127
2SC717X	0127
2SC718	0079
2SC719	0155
2SC719Q	0155
2SC720	0007
2SC721	0079
2SC722	0127
2SC723	0127
2SC723BL	0127
2SC724	0155
2SC725	0155
2SC725-06	0155
2SC726	0079
2SC727	0715
2SC727A	0715
2SC727B	0715
2SC727C	0715
2SC727D	0715
2SC727E	0715
2SC727F	0715
2SC727G	0715
2SC727GN	0715
2SC727H	0715
2SC727J	0715
2SC727K	0715
2SC727L	0715
2SC727M	0715
2SC727R	0715
2SC727X	0715
2SC727Y	0715
2SC728	0715
2SC728A	0715
2SC728B	0715
2SC728D	0715
2SC728E	0715
2SC728F	0715
2SC728G	0715
2SC728GN	0715
2SC728H	0715
2SC728J	0715
2SC728K	0715
2SC728L	0715
2SC728M	0715
2SC728OR	0715
2SC728R	0715
2SC728X	0715
2SC728Y	0715
2SC729	0418
2SC730	2041
2SC731	0493
2SC731R	0493
2SC732	0111
2SC732(BL)	0111
2SC732-B	0111
2SC732-BL	0111
2SC732-BLU	0111
2SC732-G	0111
2SC732-GR	0111
2SC732-GRN	0111
2SC732-OR	0111
2SC732-V	0111
2SC732-V10	0111
2SC732-VIO	0111
2SC732A	0111
2SC732B	0111
2SC732BL	0111
2SC732BL-1	0111
2SC732BLU	0111
2SC732C	0111
2SC732D	0111
2SC732E	0111
2SC732F	0111
2SC732G	0111
2SC732GN	0111
2SC732GR	0111
2SC732GR/4454C	0111
2SC732GRB	0111
2SC732GRN	0111
2SC732H	0111
2SC732J	0111
2SC732L	0111
2SC732M	0111
2SC732NEW-BL	OEM
2SC732OR	0111
2SC732R	0111
2SC732S	0111
2SC732TM	OEM
2SC732V	0111
2SC732V10	0111
2SC732VIO	0111
2SC732X	0111
2SC732Y	0111
2SC733	0111
2SC733(BL)	0111
2SC733(GR)	0111
2SC733(Y)	0111
2SC733-0	0111
2SC733-B	0111
2SC733-BL	0111
2SC733-BLU	0111
2SC733-G	0111
2SC733-GR	0111
2SC733-GRN	0111
2SC733-O	0111
2SC733-OR	0111
2SC733-ORG	0111
2SC733-Y	0111
2SC733-YEL	0111
2SC733A	OEM
2SC733B	0111
2SC733BL	0111
2SC733BLK	0111
2SC733BLU	0111
2SC733C	0111
2SC733D	0111
2SC733E	0111
2SC733ER	0111
2SC733F	0111
2SC733G	0111
2SC733GN	0111
2SC733GR	0111
2SC733GRN	0111
2SC733H	0111
2SC733J	0111
2SC733K	0111
2SC733L	0111
2SC733M	0111
2SC733O	0111
2SC733OR	0111
2SC733Q	0111
2SC733R	0111
2SC733S	0111
2SC733S-BL	0111
2SC733V	0111
2SC733X	0111
2SC733Y	0111
2SC733YEL	0111
2SC734	0076
2SC734(O)	0076
2SC734(R)	0076
2SC734(Y)	0076
2SC734-0	0076
2SC734-G	0076
2SC734-GR	0076
2SC734-GRN	0076
2SC734-O	0076
2SC734-OR	0076
2SC734-ORG	0076
2SC734-OY	0076
2SC734-R	0076
2SC734-RED	0076
2SC734-Y	0076
2SC734-YEL	0076
2SC734A	0076
2SC734B	0076
2SC734C	0076
2SC734D	0076
2SC734E	0076
2SC734F	0076
2SC734G	0076
2SC734GN	0076
2SC734GR	0076
2SC734GRN	0076
2SC734H	0076
2SC734J	0076
2SC734K	0076
2SC734K/GR	0076
2SC734L	0076
2SC734M	0076
2SC734OR	0076
2SC734R	0076
2SC734RED	0076
2SC734X	0076
2SC734Y	0076
2SC734YEL	0076
2SC735	0191
2SC735(0)	0191
2SC735(FA-4)	OEM
2SC735(O)	0191
2SC735(Y)	0191
2SC735-0	0191
2SC735-GR	0191
2SC735-GRN	0191
2SC735-O	0191
2SC735-OR	0191
2SC735-ORG	0191
2SC735-ORN	0191
2SC735-OY	0191
2SC735-R	0191
2SC735-RED	0191
2SC735-Y	0191
2SC735-YEL	0191
2SC735A	0191
2SC735B	0191
2SC735C	0191
2SC735D	0191
2SC735E	0191
2SC735F	0191
2SC735FA3	0191
2SC735G	0191
2SC735GN	0191
2SC735GR	0191
2SC735GRN	0191
2SC735H	0191
2SC735J	0191
2SC735K	0191
2SC735L	0191
2SC735M	0191
2SC735O	0191
2SC735OR	0191
2SC735RED	0191
2SC735X	0191
2SC735Y	0191
2SC735YEL	0191
2SC735YFA-5	0191
2SC736	0867
2SC736A	0867
2SC736B	0867
2SC736C	0867
2SC736D	0867
2SC736E	0867
2SC736F	0867
2SC736G	0867
2SC736GN	0867
2SC736H	0867
2SC736J	0867
2SC736K	0867
2SC736L	0867
2SC736M	0867
2SC736OR	0867
2SC736R	0867
2SC736X	0867
2SC736Y	0867
2SC737	0079
2SC737Y	0079
2SC738	0895
2SC738A	0895
2SC738B	0895
2SC738C	0895
2SC738D	0895
2SC738E	0895
2SC738F	0895
2SC738G	0895
2SC738GN	0895
2SC738H	0895
2SC738J	0895
2SC738K	0895
2SC738L	0895
2SC738M	0895
2SC738OR	0895
2SC738R	0895
2SC738X	0895
2SC738Y	0895
2SC739	0127
2SC739A	0127
2SC739B	0127
2SC739C	0127
2SC739D	0127
2SC739E	0127
2SC739F	0127
2SC739G	0127
2SC739GN	0127
2SC739H	0127
2SC739K	0127
2SC739L	0127
2SC739M	0127
2SC739OR	0127
2SC739R	0127
2SC739Y	0127
2SC740	0144
2SC740A	0144
2SC740B	0144
2SC740C	0144
2SC740D	0144
2SC740E	0144
2SC740F	0144
2SC740G	0144
2SC740GN	0144
2SC740H	0144
2SC740J	0144
2SC740K	0144
2SC740L	0144
2SC740M	0144
2SC740OR	0144
2SC740R	0144
2SC740X	0144
2SC740Y	0144
2SC741	0689
2SC741B	0689
2SC741C	0689
2SC741D	0689
2SC741E	0689
2SC741F	0689
2SC741GN	0689
2SC741H	0689
2SC741J	0689
2SC741K	0689
2SC741L	0689
2SC741M	0689
2SC741OR	0689
2SC741R	0689
2SC741X	0689
2SC741Y	0689
2SC742	0936
2SC743	0936
2SC743A	0936
2SC744	0086
2SC744A	0086
2SC745	0936
2SC746	0168
2SC746A	0168
2SC746B	0168
2SC746C	0168
2SC746D	0168
2SC746E	0168
2SC746F	0168
2SC746GN	0168
2SC746H	0168
2SC746J	0168
2SC746K	0168
2SC746L	0168
2SC746M	0168
2SC746OR	0168
2SC746R	0168
2SC746X	0168
2SC746Y	0168
2SC748	0144
2SC748A	0144
2SC748C	0144
2SC748D	0144
2SC748E	0144
2SC748F	0144
2SC748G	0144
2SC748GN	0144
2SC748H	0144
2SC748J	0144
2SC748K	0144
2SC748L	0144
2SC748M	0144
2SC748OR	0144
2SC748R	0144
2SC748X	0144
2SC748Y	0144
2SC749	0936
2SC751	0111
2SC752	0551
2SC752(G)TM-Y	0551
2SC752-ORG-G	0551
2SC752-RED-G	0551
2SC752-YEL-G	0551
2SC752A	0551
2SC752B	0551
2SC752C	0551
2SC752D	0551
2SC752E	0551
2SC752F	0551
2SC752G	0551
2SC752GA	0551
2SC752GN	0551
2SC752G-O	0551
2SC752G-R	0551
2SC752GTM-O	0551
2SC752G-TM-Y	0551
2SC752GTM-Y	0551
2SC752GTMY	0551
2SC752G-Y	0551
2SC752H	0551
2SC752J	0551
2SC752K	0551
2SC752L	0551
2SC752M	0551
2SC752OR	0551
2SC752R	0551
2SC752TM-YFA	0551
2SC752X	0551
2SC752Y	0551
2SC753	0475
2SC753GTM-Y	0475
2SC754	0079
2SC755	0016
2SC756	0622
2SC756(4-4)	0622
2SC756(44)	0622
2SC756-1	0622
2SC756-2	0622
2SC756-2-4	0622
2SC756-2-5	0622
2SC756-3	0622
2SC756-4	0622
2SC756-OR	0622
2SC756A	0622
2SC756B	0622
2SC756C	0622
2SC756D	0622
2SC756E	0622
2SC756F	0622
2SC756G	0622
2SC756GN	0622
2SC756H	0622
2SC756J	0622
2SC756K	0622
2SC756L	0622
2SC756M	0622
2SC756OR	0622
2SC756R	0622
2SC756X	0622
2SC756Y	0622
2SC757	0475
2SC758	0074
2SC758OR	0074
2SC759	0543
2SC760	0103
2SC761	1154
2SC761(Y)	1154
2SC761A	1154
2SC761B	1154
2SC761C	1154
2SC761D	1154
2SC761E	1154
2SC761F	1154
2SC761G	1154
2SC761GN	1154
2SC761H	1154
2SC761J	1154
2SC761K	1154
2SC761L	1154
2SC761M	1154
2SC761OR	1154
2SC761P	1154
2SC761R	1154
2SC761X	1154
2SC761Y	1154
2SC761Z	1154
2SC762	1154
2SC762B	1154
2SC762C	1154
2SC762D	1154
2SC762E	1154
2SC762F	1154
2SC762G	1154
2SC762GN	1154
2SC762H	1154
2SC762J	1154
2SC762K	1154
2SC762L	1154
2SC762M	1154
2SC762R	1154
2SC762X	1154
2SC762Y	1154
2SC763	0127
2SC763(C)	0127
2SC763-C	0127
2SC763-OR	0127
2SC763A	0127
2SC763B	0127
2SC763C	0127
2SC763D	0127
2SC763E	0127
2SC763F	0127
2SC763G	0127
2SC763GN	0127
2SC763H	0127
2SC763J	0127
2SC763K	0127
2SC763L	0127
2SC763M	0127
2SC763OR	0127
2SC763R	0127
2SC763S	0127
2SC763T	0127
2SC763X	0127
2SC763Y	0127
2SC764	0079
2SC765	0130
2SC765A	0130
2SC765B	0130
2SC765C	0130
2SC765D	0130
2SC765E	0130
2SC765F	0130
2SC765G	0130
2SC765GN	0130
2SC765H	0130
2SC765J	0130
2SC765K	0130
2SC765L	0130
2SC765M	0130
2SC765OR	0130
2SC765R	0130
2SC765X	0130
2SC765Y	0130
2SC766	0103
2SC767	0103
2SC768	0620
2SC768A	0620
2SC768B	0620
2SC768C	0620
2SC768D	0620
2SC768E	0620
2SC768F	0620
2SC768G	0620
2SC768GN	0620
2SC768H	0620
2SC768J	0620
2SC768K	0620
2SC768L	0620
2SC768M	0620
2SC768OR	0620
2SC768R	0620
2SC768X	0620
2SC768Y	0620
2SC769	0620
2SC770	0074
2SC771	0224
2SC771A	0224
2SC771B	0224
2SC771BX	0224
2SC771C	0224
2SC771D	0224
2SC771E	0224
2SC771F	0224
2SC771G	0224
2SC771GN	0224
2SC771H	0224
2SC771J	0224
2SC771K	0224
2SC771L	0224
2SC771M	0224
2SC771OR	0224
2SC771R	0224
2SC771X	0224
2SC771Y	0224
2SC772	1144
2SC772-OR	1144
2SC772A	1144
2SC772B	1144
2SC772BG	1144
2SC772BH	1144
2SC772BV	1144
2SC772BX	1144
2SC772BY	1144
2SC772C	1144
2SC772C1	1144
2SC772C2	1144
2SC772CA	1144
2SC772CK	1144
2SC772CL	1144
2SC772CS	1144
2SC772CU	1144
2SC772CV	1144
2SC772CX	1144
2SC772D	1144
2SC772DJ	1144
2SC772DU	1144
2SC772DV	1144
2SC772DX	1144
2SC772DY	1144
2SC772E	1144
2SC772F	1144
2SC772G	1144
2SC772GN	1144
2SC772H	1144
2SC772J	1144
2SC772K	1144
2SC772KB	1144
2SC772KC	1144
2SC772KD	1144
2SC772KD1	1144
2SC772KD2	1144
2SC772L	1144
2SC772M	1144
2SC772OR	1144

DEVICE TYPE	REPL CODE
2SC772R	1144
2SC772RB-D	1144
2SC772RD	1144
2SC772RS-D	1144
2SC772X	1144
2SC772Y	1144
2SC773	0788
2SC773(E)	0788
2SC773A	0788
2SC773B	0788
2SC773C	0788
2SC773D	0788
2SC773E	0788
2SC773F	0788
2SC773G	0788
2SC773GN	0788
2SC773H	0788
2SC773J	0788
2SC773K	0788
2SC773L	0788
2SC773M	0788
2SC773OR	0788
2SC773R	0788
2SC773X	0788
2SC773Y	0788
2SC774	0698
2SC774B	0698
2SC774C	0698
2SC774D	0698
2SC774E	0698
2SC774F	0698
2SC774G	0698
2SC774GN	0698
2SC774H	0698
2SC774J	0698
2SC774K	0698
2SC774L	0698
2SC774M	0698
2SC774R	0698
2SC774X	0698
2SC774Y	0698
2SC775	0698
2SC775A	0698
2SC775B	0698
2SC775C	0698
2SC775D	0698
2SC775E	0698
2SC775F	0698
2SC775G	0698
2SC775GN	0698
2SC775H	0698
2SC775J	0698
2SC775K	0698
2SC775L	0698
2SC775M	0698
2SC775OR	0698
2SC775R	0698
2SC775X	0698
2SC775Y	0698
2SC776	1400
2SC776(Y)	1400
2SC776A	1400
2SC776B	1400
2SC776C	1400
2SC776D	1400
2SC776E	1400
2SC776F	1400
2SC776G	1400
2SC776GN	1400
2SC776H	1400
2SC776J	1400
2SC776K	1400
2SC776L	1400
2SC776M	1400
2SC776OR	1400
2SC776R	1400
2SC776X	1400
2SC776Y	1400
2SC777	1411
2SC777-OR	1411
2SC777A	1411
2SC777AP	1411
2SC777B	1411
2SC777C	1411
2SC777D	1411
2SC777E	1411
2SC777F	1411
2SC777G	1411
2SC777GN	1411
2SC777H	1411
2SC777J	1411
2SC777K	1411
2SC777L	1411
2SC777M	1411
2SC777OR	1411
2SC777R	1411
2SC777X	1411
2SC777Y	1411
2SC778	1411
2SC778A	1411
2SC778B	1411
2SC778C	1411
2SC778D	1411
2SC778E	1411
2SC778F	1411
2SC778G	1411
2SC778GN	1411
2SC778H	1411
2SC778J	1411

DEVICE TYPE	REPL CODE
2SC778K	1411
2SC778L	1411
2SC778M	1411
2SC778OR	1411
2SC778R	1411
2SC778X	1411
2SC778Y	1411
2SC779	1027
2SC779-O	1027
2SC779-R	1027
2SC779-Y	1027
2SC779O	1027
2SC780	0861
2SC780-O	0861
2SC780-ORG-G	0861
2SC780-R	0861
2SC780-RED-G	0861
2SC780-Y	0861
2SC780-YEL-G	0861
2SC780A	0861
2SC780A/G	0861
2SC780AG	0861
2SC780AG-O	0861
2SC780AG-R	0861
2SC780AG-Y	0861
2SC780B	0861
2SC780C	0861
2SC780D	0861
2SC780E	0861
2SC780F	0861
2SC780G	0861
2SC780GA	0861
2SC780GN	0861
2SC780G-O	0861
2SC780G-R	0861
2SC780G-Y	0861
2SC780H	0861
2SC780J	0861
2SC780K	0861
2SC780L	0861
2SC780M	0861
2SC780O	0861
2SC780OR	0861
2SC780R	0861
2SC780X	0861
2SC780Y	0861
2SC781	0698
2SC781A	0698
2SC781AK	0698
2SC781B	0698
2SC781C	0698
2SC781D	0698
2SC781E	0698
2SC781F	0698
2SC781G	0698
2SC781H	0698
2SC781J	0698
2SC781K	0698
2SC781M	0698
2SC781OR	0698
2SC781R	0698
2SC781X	0698
2SC781Y	0698
2SC782	1469
2SC782(FA6)	OEM
2SC782(FAZ-1)	OEM
2SC782-OR	1469
2SC782A	1469
2SC782B	1469
2SC782C	1469
2SC782D	1469
2SC782E	1469
2SC782F	1469
2SC782G	1469
2SC782GN	1469
2SC782H	1469
2SC782J	1469
2SC782K	1469
2SC782L	1469
2SC782OR	1469
2SC782R	1469
2SC782X	1469
2SC782Y	1469
2SC783	1027
2SC783-O	1027
2SC783-R	1027
2SC783-Y	1027
2SC784	1144
2SC784(BN)	1144
2SC784(R)	1144
2SC784-0	1144
2SC784-6	1144
2SC784-B	1144
2SC784-BN	1144
2SC784-BRN	1144
2SC784-O	1144
2SC784-OR	1144
2SC784-ORG	1144
2SC784-R	1144
2SC784-RED	1144
2SC784-Y	1144
2SC784A	1144
2SC784B	1144
2SC784BN	1144
2SC784BN-1	1144
2SC784BRN	1144
2SC784C	1144
2SC784D	1144
2SC784E	1144

DEVICE TYPE	REPL CODE
2SC784F	1144
2SC784G	1144
2SC784GN	1144
2SC784H	1144
2SC784J	1144
2SC784K	1144
2SC784L	1144
2SC784M	1144
2SC784OR	1144
2SC784P	1144
2SC784Q	1144
2SC784R	1144
2SC784RA	1144
2SC784RED	1144
2SC784X	1144
2SC785	1144
2SC785(E)(D)	1144
2SC785(R)	1144
2SC785-0	1144
2SC785-B	1144
2SC785-BN	1144
2SC785-BRN	1144
2SC785-O	1144
2SC785-ORG	1144
2SC785-O.JA	1144
2SC785-R	1144
2SC785-RED	1144
2SC785-Y	1144
2SC785-YEL	1144
2SC785A	1144
2SC785B	1144
2SC785BL	1144
2SC785BN	1144
2SC785BR	1144
2SC785BRN	1144
2SC785C	1144
2SC785D	1144
2SC785E	1144
2SC785F	1144
2SC785G	1144
2SC785GN	1144
2SC785GR	1144
2SC785H	1144
2SC785JA	1144
2SC785K	1144
2SC785L	1144
2SC785M	1144
2SC785O	1144
2SC785R	1144
2SC785RA	1144
2SC785RED	1144
2SC785V	1144
2SC785X	1144
2SC785Y	1144
2SC785YEL	1144
2SC786	0216
2SC786A	0216
2SC786B	0216
2SC786C	0216
2SC786D	0216
2SC786E	0216
2SC786F	0216
2SC786G	0216
2SC786GN	0216
2SC786H	0216
2SC786J	0216
2SC786K	0216
2SC786L	0216
2SC786M	0216
2SC786OR	0216
2SC786R	0216
2SC786X	0216
2SC786Y	0216
2SC787	1154
2SC787A	1154
2SC787B	1154
2SC787C	1154
2SC787D	1154
2SC787E	1154
2SC787F	1154
2SC787G	1154
2SC787GN	1154
2SC787H	1154
2SC787K	1154
2SC787L	1154
2SC787M	1154
2SC787OR	1154
2SC787R	1154
2SC787X	1154
2SC787Y	1154
2SC788	0233
2SC788B	0233
2SC788C	0233
2SC788D	0233
2SC788E	0233
2SC788F	0233
2SC788G	0233
2SC788GN	0233
2SC788J	0233
2SC788K	0233
2SC788L	0233
2SC788M	0233
2SC788OR	0233
2SC788R	0233
2SC788X	0233
2SC788Y	0233
2SC789	0228
2SC789-0	0228
2SC789-O	0228

DEVICE TYPE	REPL CODE
2SC789-R	0228
2SC789-Y	0228
2SC789A	0228
2SC789B	0228
2SC789C	0228
2SC789D	0228
2SC789E	0228
2SC789F	0228
2SC789G	0228
2SC789GN	0228
2SC789H	0228
2SC789J	0228
2SC789K	0228
2SC789L	0228
2SC789M	0228
2SC789OR	0228
2SC789R	0228
2SC789X	0228
2SC789Y	0228
2SC790	1496
2SC790(O)	1496
2SC790-O	1496
2SC790-R	1496
2SC790-Y	1496
2SC790Y	1496
2SC791	1147
2SC791-OR	1147
2SC791A	1147
2SC791B	1147
2SC791C	1147
2SC791D	1147
2SC791E	1147
2SC791F	1147
2SC791FA1	1147
2SC791G	1147
2SC791GN	1147
2SC791H	1147
2SC791J	1147
2SC791K	1147
2SC791M	1147
2SC791OR	1147
2SC791R	1147
2SC791X	1147
2SC791Y	1147
2SC792	0220
2SC792(FA-3)	OEM
2SC792FA-3	0220
2SC793	0103
2SC793-BL	0103
2SC793-BLU	0103
2SC793-R	0103
2SC793-RED	0103
2SC793-Y	0103
2SC793-YEL	0103
2SC793A	0103
2SC793B	0103
2SC793BL	0103
2SC793C	0103
2SC793E	0103
2SC793F	0103
2SC793G	0103
2SC793GN	0103
2SC793H	0103
2SC793J	0103
2SC793K	0103
2SC793L	0103
2SC793M	0103
2SC793OR	0103
2SC793R	0103
2SC793X	0103
2SC793Y	0103
2SC794	0103
2SC794A	0103
2SC794B	0103
2SC794C	0103
2SC794D	0103
2SC794E	0103
2SC794F	0103
2SC794G	0103
2SC794GN	0103
2SC794H	0103
2SC794J	0103
2SC794K	0103
2SC794L	0103
2SC794OR	0103
2SC794R	0103
2SC794RA	0103
2SC794X	0103
2SC794Y	0103
2SC795	0142
2SC795(BN)	0142
2SC795A	0142
2SC795B	0142
2SC795BN	0142
2SC795C	0142
2SC795D	0142
2SC795F	0142
2SC795G	0142
2SC795GN	0142
2SC795H	0142
2SC795J	0142
2SC795L	0142
2SC795M	0142
2SC795OR	0142
2SC795R	0142
2SC795X	0142
2SC795Y	0142
2SC796	0016
2SC796A	0016

DEVICE TYPE	REPL CODE
2SC796B	0016
2SC796C	0016
2SC796D	0016
2SC796E	0016
2SC796F	0016
2SC796G	0016
2SC796GN	0016
2SC796H	0016
2SC796J	0016
2SC796K	0016
2SC796L	0016
2SC796M	0016
2SC796OR	0016
2SC796R	0016
2SC796X	0016
2SC796Y	0016
2SC797	0079
2SC797A	0079
2SC797B	0079
2SC797C	0079
2SC797D	0079
2SC797E	0079
2SC797F	0079
2SC797G	0079
2SC797GN	0079
2SC797H	0079
2SC797J	0079
2SC797K	0079
2SC797L	0079
2SC797M	0079
2SC797OR	0079
2SC797R	0079
2SC797X	0079
2SC797Y	0079
2SC798	0698
2SC798A	0698
2SC798B	0698
2SC798C	0698
2SC798D	0698
2SC798E	0698
2SC798G	0698
2SC798GN	0698
2SC798H	0698
2SC798J	0698
2SC798K	0698
2SC798L	0698
2SC798M	0698
2SC798OR	0698
2SC798R	0698
2SC798X	0698
2SC798Y	0698
2SC799	1411
2SC799-4	1411
2SC799A	1411
2SC799AP	1411
2SC799B	1411
2SC799C	1411
2SC799D	1411
2SC799E	1411
2SC799F	1411
2SC799G	1411
2SC799GN	1411
2SC799H	1411
2SC799J	1411
2SC799K	1411
2SC799L	1411
2SC799M	1411
2SC799OR	1411
2SC799R	1411
2SC799X	1411
2SC799Y	1411
2SC800	0630
2SC800A	0630
2SC800B	0630
2SC800C	0630
2SC800D	0630
2SC800E	0630
2SC800F	0630
2SC800G	0630
2SC800GN	0630
2SC800H	0630
2SC800J	0630
2SC800K	0630
2SC800L	0630
2SC800M	0630
2SC800OR	0630
2SC800R	0630
2SC800X	0630
2SC800Y	0630
2SC801	0086
2SC802	0086
2SC803	0698
2SC803A	0698
2SC803B	0698
2SC803C	0698
2SC803D	0698
2SC803E	0698
2SC803F	0698
2SC803G	0698
2SC803H	0698
2SC803J	0698
2SC803K	0698
2SC803L	0698
2SC803M	0698
2SC803OR	0698
2SC803R	0698
2SC803X	0698
2SC803Y	0698
2SC804	0475

DEVICE TYPE	REPL CODE
2SC804H	0475
2SC805	0716
2SC805A	0716
2SC805A1	0716
2SC805A2	0716
2SC805B	0716
2SC805C	0716
2SC805D	0716
2SC805E	0716
2SC805F	0716
2SC805G	0716
2SC805GN	0716
2SC805H	0716
2SC805J	0716
2SC805L	0716
2SC805OR	0716
2SC805R	0716
2SC805X	0716
2SC805Y	0716
2SC806	0363
2SC806(A)	0363
2SC806A	0363
2SC806B	0363
2SC806C	0363
2SC806D	0363
2SC806E	0363
2SC806F	0363
2SC806G	0363
2SC806J	0363
2SC806L	0363
2SC806M	0363
2SC806OR	0363
2SC806R	0363
2SC806X	0363
2SC806Y	0363
2SC807	1966
2SC807(A)	1966
2SC807A	1966
2SC807AK	1966
2SC808	2414
2SC809	0259
2SC810	0493
2SC811	0007
2SC812	0630
2SC812A	0630
2SC812B	0630
2SC812C	0630
2SC812D	0630
2SC812E	0630
2SC812F	0630
2SC812G	0630
2SC812GN	0630
2SC812H	0630
2SC812J	0630
2SC812K	0630
2SC812L	0630
2SC812M	0630
2SC812OR	0630
2SC812R	0630
2SC812Y	0630
2SC814	0079
2SC814A	0079
2SC814B	0079
2SC814C	0079
2SC814D	0079
2SC814E	0079
2SC814F	0079
2SC814G	0079
2SC814GN	0079
2SC814H	0079
2SC814J	0079
2SC814K	0079
2SC814M	0079
2SC814OR	0079
2SC814R	0079
2SC814X	0079
2SC814Y	0079
2SC815	0155
2SC815(K)	0155
2SC815(L)	0155
2SC815-1	0155
2SC815-Y	6844
2SC815A	0155
2SC815B	0155
2SC815BK	0155
2SC815C	0155
2SC815D	0155
2SC815E	0155
2SC815F	0155
2SC815G	0155
2SC815GN	0155
2SC815H	0155
2SC815J	0155
2SC815K	0155
2SC815K,L	0155
2SC815L	0155
2SC815LJ	0155
2SC815M	0155
2SC815OR	0155
2SC815R	0155
2SC815S	0155
2SC815SA	0155
2SC815SC	0155
2SC815X	0155
2SC815Y	0155
2SC816	0086
2SC816A	0086
2SC816B	0086
2SC816C	0086

DEVICE TYPE	REPL CODE
2SC816D	0086
2SC816E	0086
2SC816F	0086
2SC816G	0086
2SC816GN	0086
2SC816H	0086
2SC816HL	0086
2SC816K	0086
2SC816L	0086
2SC816M	0086
2SC816OR	0086
2SC816R	0086
2SC816X	0086
2SC816Y	0086
2SC817	0259
2SC818	0168
2SC818A	0168
2SC818B	0168
2SC818C	0168
2SC818D	0168
2SC818E	0168
2SC818F	0168
2SC818G	0168
2SC818GN	0168
2SC818H	0168
2SC818J	0168
2SC818K	0168
2SC818L	0168
2SC818M	0168
2SC818OR	0168
2SC818R	0168
2SC818X	0168
2SC818Y	0168
2SC819	2700
2SC821	2041
2SC822	2041
2SC823	0418
2SC824	0418
2SC825	1027
2SC825A	1027
2SC825B	1027
2SC825C	1027
2SC825D	1027
2SC825E	1027
2SC825F	1027
2SC825G	1027
2SC825GN	1027
2SC825H	1027
2SC825J	1027
2SC825K	1027
2SC825L	1027
2SC825M	1027
2SC825OR	1027
2SC825R	1027
2SC825X	1027
2SC825Y	1027
2SC826	0086
2SC826A	0086
2SC826B	0086
2SC826C	0086
2SC826D	0086
2SC826E	0086
2SC826F	0086
2SC826G	0086
2SC826H	0086
2SC826J	0086
2SC826K	0086
2SC826L	0086
2SC826M	0086
2SC826OR	0086
2SC826R	0086
2SC826X	0086
2SC826Y	0086
2SC827	0086
2SC827A	0086
2SC827B	0086
2SC827C	0086
2SC827D	0086
2SC827E	0086
2SC827F	0086
2SC827G	0086
2SC827GN	0086
2SC827H	0086
2SC827J	0086
2SC827K	0086
2SC827M	0086
2SC827OR	0086
2SC827R	0086
2SC827X	0086
2SC827Y	0086
2SC828	1220
2SC828(A)	1220
2SC828(H)	1220
2SC828(K)	1220
2SC828(N)	1220
2SC828(O)	1220
2SC828(P)	1220
2SC828(P)(Q)	1220
2SC828(Q)	1220
2SC828(R)	1220
2SC828(R)(Q)	1220
2SC828(R)(S)	1220
2SC828(R,Q,P)	1220
2SC828(R,S,T)	1220
2SC828(S)	1220
2SC828(S)(T)	1220
2SC828(T)	1220
2SC828-0	1220
2SC828-O	1220

If replacement code is OEM, contact original manufacturer for replacement.

DEVICE TYPE	REPL CODE
2SC828-OR	1220
2SC828A	1220
2SC828A(P)	1220
2SC828A(Q)	1220
2SC828A(R)	1220
2SC828A(R.S.)	1220
2SC828A(S)	1220
2SC828A(T)	1220
2SC828AO	1220
2SC828AP	1220
2SC828AQ	1220
2SC828AR	1220
2SC828AS	1220
2SC828AT	1220
2SC828B	1220
2SC828C	1220
2SC828D	1220
2SC828E	1220
2SC828F	1220
2SC828G	1220
2SC828GN	1220
2SC828H	1220
2SC828HR	1220
2SC828K	1220
2SC828L	1220
2SC828LS	1220
2SC828M	1220
2SC828OR	1220
2SC828P	1220
2SC828PQ	1220
2SC828Q	1220
2SC828Q-6	1220
2SC828QR	1220
2SC828QRS	1220
2SC828R	1220
2SC828R-1	1220
2SC828RA	1220
2SC828RH	1220
2SC828RS	1220
2SC828RST	1220
2SC828RT	1220
2SC828S	1220
2SC828T	1220
2SC828X	1220
2SC828Y	1220
2SC828YL	1220
2SC829	0151
2SC829(B)	0151
2SC829(C)	0151
2SC829(Y)	0151
2SC829/4454C	0151
2SC829-OR	0151
2SC829A	0151
2SC829AK	0151
2SC829B	0151
2SC829BC	0151
2SC829BJ	0151
2SC829BK	0151
2SC829C	0151
2SC829CL	0151
2SC829D	0151
2SC829E	0151
2SC829G	0151
2SC829GN	0151
2SC829H	0151
2SC829K	0151
2SC829L	0151
2SC829M	0151
2SC829OR	0151
2SC829R	0151
2SC829X	0151
2SC829Y	0151
2SC830	1147
2SC830A	1147
2SC830B	1147
2SC830C	1147
2SC830D	1147
2SC830E	1147
2SC830F	1147
2SC830G	1147
2SC830GN	1147
2SC830H	1147
2SC830J	1147
2SC830K	1147
2SC830L	1147
2SC830M	1147
2SC830OR	1147
2SC830R	1147
2SC830X	1147
2SC830Y	1147
2SC831	0042
2SC833	0429
2SC833BL	0429
2SC834L	0429
2SC835	0127
2SC836	0127
2SC836M	0127
2SC837	0224
2SC837(K)	0224
2SC837(KL)	0224
2SC837(L)	0224
2SC837A	0224
2SC837B	0224
2SC837C	0224
2SC837D	0224
2SC837E	0224
2SC837F	0224
2SC837G	0224
2SC837GN	0224
2SC837H	0224
2SC837J	0224
2SC837K	0224
2SC837KL	0224
2SC837L	0224
2SC837M	0224
2SC837OR	0224
2SC837R	0224
2SC837WF	0224
2SC837X	0224
2SC837Y	0224
2SC838	0191
2SC838(A)	0191
2SC838(E)	0191
2SC838(F)	0191
2SC838(H)	0191
2SC838(J)	0191
2SC838(K)	0191
2SC838(L)	0191
2SC838(M)	0191
2SC838(O)	0191
2SC838(S)	0191
2SC838-2	0191
2SC838-O	0191
2SC838A	0191
2SC838BL	0191
2SC838C	0191
2SC838E	0191
2SC838F	0191
2SC838H	0191
2SC838HF	0191
2SC838I	0191
2SC838J	0191
2SC838K	0191
2SC838L	0191
2SC838M	0191
2SC839	0076
2SC839(F)	0076
2SC839(H)	0076
2SC839(J)	0076
2SC839(JI)	0076
2SC839(L)	0076
2SC839(M)	0076
2SC839-E	0076
2SC839-F	0076
2SC839A	0076
2SC839B	0076
2SC839C	0076
2SC839D	0076
2SC839E	0076
2SC839F	0076
2SC839G	0076
2SC839GN	0076
2SC839H	0076
2SC839J	0076
2SC839J1	0076
2SC839JH	0076
2SC839K	0076
2SC839L	0076
2SC839M	0076
2SC839OR	0076
2SC839R	0076
2SC839X	0076
2SC839Y	0076
2SC840	2028
2SC840(P)	2028
2SC840/A	2028
2SC840A	2028
2SC840AC	2028
2SC840B	2028
2SC840C	2028
2SC840D	2028
2SC840E	2028
2SC840F	2028
2SC840G	2028
2SC840GN	2028
2SC840H	2028
2SC840HP	2028
2SC840J	2028
2SC840K	2028
2SC840L	2028
2SC840M	2028
2SC840OR	2028
2SC840P	2028
2SC840Q	2028
2SC840R	2028
2SC840X	2028
2SC840Y	2028
2SC841	2041
2SC841H	2041
2SC844	0493
2SC844D	0493
2SC845	0493
2SC847	0016
2SC847A	0016
2SC847B	0016
2SC847C	0016
2SC847D	0016
2SC847E	0016
2SC847F	0016
2SC847G	0016
2SC847GN	0016
2SC847H	0016
2SC847J	0016
2SC847K	0016
2SC847L	0016
2SC847M	0016
2SC847OR	0016
2SC847R	0016
2SC847X	0016
2SC848	0016
2SC848A	0016
2SC848B	0016
2SC848C	0016
2SC848D	0016
2SC848E	0016
2SC848F	0016
2SC848G	0016
2SC848GN	0016
2SC848H	0016
2SC848J	0016
2SC848K	0016
2SC848L	0016
2SC848M	0016
2SC848OR	0016
2SC848R	0016
2SC848X	0016
2SC848Y	0016
2SC849	0016
2SC849A	0016
2SC849B	0016
2SC849C	0016
2SC849D	0016
2SC849E	0016
2SC849F	0016
2SC849G	0016
2SC849GN	0016
2SC849H	0016
2SC849J	0016
2SC849K	0016
2SC849L	0016
2SC849M	0016
2SC849OR	0016
2SC849R	0016
2SC849X	0016
2SC849Y	0016
2SC850	0016
2SC850A	0016
2SC850B	0016
2SC850C	0016
2SC850D	0016
2SC850E	0016
2SC850F	0016
2SC850G	0016
2SC850GN	0016
2SC850H	0016
2SC850J	0016
2SC850K	0016
2SC850L	0016
2SC850M	0016
2SC850OR	0016
2SC850R	0016
2SC850X	0016
2SC850Y	0016
2SC851	0130
2SC851A	0130
2SC851B	0130
2SC851C	0130
2SC851D	0130
2SC851E	0130
2SC851F	0130
2SC851G	0130
2SC851GN	0130
2SC851H	0130
2SC851K	0130
2SC851L	0130
2SC851M	0130
2SC851OR	0130
2SC851R	0130
2SC851X	0130
2SC851Y	0130
2SC852	0493
2SC852A	0493
2SC853	0218
2SC853-OR	0218
2SC853A	0218
2SC853B	0218
2SC853C	0218
2SC853D	0218
2SC853E	0218
2SC853F	0218
2SC853G	0218
2SC853GN	0218
2SC853H	0218
2SC853J	0218
2SC853K	0218
2SC853KLM	0218
2SC853L	0218
2SC853M	0218
2SC853OR	0218
2SC853R	0218
2SC853X	0218
2SC853Y	0218
2SC854	0418
2SC855	0418
2SC856	0730
2SC856-02	0730
2SC856-OR	0730
2SC856A	0730
2SC856B	0730
2SC856C	0730
2SC856D	0730
2SC856E	0730
2SC856F	0730
2SC856G	0730
2SC856GN	0730
2SC856J	0730
2SC856K	0730
2SC856L	0730
2SC856M	0730
2SC856OR	0730
2SC856R	0730
2SC856X	0730
2SC856Y	0730
2SC857	0233
2SC857H	0233
2SC857K	0233
2SC858	0111
2SC858A	0111
2SC858B	0111
2SC858C	0111
2SC858D	0111
2SC858E	0111
2SC858F	0111
2SC858FG	0111
2SC858G	0111
2SC858GA	0111
2SC858GN	0111
2SC858H	0111
2SC858J	0111
2SC858K	0111
2SC858L	0111
2SC858M	0111
2SC858OR	0111
2SC858R	0111
2SC858X	0111
2SC858Y	0111
2SC859	0111
2SC859A	0111
2SC859B	0111
2SC859D	0111
2SC859E	0111
2SC859F	0111
2SC859FG	0111
2SC859G	0111
2SC859GK	0111
2SC859GL	0111
2SC859GM	0111
2SC859GN	0111
2SC859H	0111
2SC859J	0111
2SC859K	0111
2SC859L	0111
2SC859M	0111
2SC859OR	0111
2SC859R	0111
2SC859X	0111
2SC859Y	0111
2SC860	0007
2SC860A	0007
2SC860B	0007
2SC860C	0007
2SC860D	0007
2SC860E	0007
2SC860F	0007
2SC860G	0007
2SC860GN	0007
2SC860H	0007
2SC860J	0007
2SC860K	0007
2SC860L	0007
2SC860M	0007
2SC860OR	0007
2SC860R	0007
2SC860X	0007
2SC860Y	0007
2SC861	0642
2SC862	0363
2SC863	0007
2SC863A	0007
2SC863B	0007
2SC863C	0007
2SC863D	0007
2SC863E	0007
2SC863F	0007
2SC863G	0007
2SC863GN	0007
2SC863H	0007
2SC863J	0007
2SC863K	0007
2SC863L	0007
2SC863M	0007
2SC863OR	0007
2SC863R	0007
2SC863X	0007
2SC863Y	0007
2SC864	0007
2SC865	0698
2SC866	0630
2SC866A	0630
2SC866B	0630
2SC866C	0630
2SC866D	0630
2SC866E	0630
2SC866F	0630
2SC866G	0630
2SC866GN	0630
2SC866H	0630
2SC866J	0630
2SC866K	0630
2SC866L	0630
2SC866M	0630
2SC866OR	0630
2SC866R	0630
2SC866X	0630
2SC866Y	0630
2SC867	1360
2SC867-0R	1360
2SC867-2	1360
2SC867-3	1360
2SC867-4	1360
2SC867-5	1360
2SC867-OR	1360
2SC867A	1360
2SC867B	1360
2SC867C	1360
2SC867D	1360
2SC867E	1360
2SC867F	1360
2SC867G	1360
2SC867GN	1360
2SC867H	1360
2SC867J	1360
2SC867K	1360
2SC867L	1360
2SC867M	1360
2SC867OR	1360
2SC867R	1360
2SC867X	1360
2SC867Y	1360
2SC868	0861
2SC868A	0861
2SC868B	0861
2SC868C	0861
2SC868D	0861
2SC868E	0861
2SC868F	0861
2SC868G	0861
2SC868GN	0861
2SC868H	0861
2SC868J	0861
2SC868K	0861
2SC868L	0861
2SC868M	0861
2SC868OR	0861
2SC868R	0861
2SC868X	0861
2SC868Y	0861
2SC869	0861
2SC869A	0861
2SC869B	0861
2SC869C	0861
2SC869D	0861
2SC869E	0861
2SC869F	0861
2SC869G	0861
2SC869GN	0861
2SC869J	0861
2SC869K	0861
2SC869L	0861
2SC869M	0861
2SC869OR	0861
2SC869R	0861
2SC869X	0861
2SC869Y	0861
2SC870	0076
2SC870A	0076
2SC870B	0076
2SC870C	0076
2SC870D	0076
2SC870E	0076
2SC870F	0076
2SC870FL	0076
2SC870G	0076
2SC870GN	0076
2SC870H	0076
2SC870J	0076
2SC870K	0076
2SC870L	0076
2SC870M	0076
2SC870OR	0076
2SC870R	0076
2SC870X	0076
2SC870Y	0076
2SC871	0151
2SC871(F)	0151
2SC871-G	0151
2SC871A	0151
2SC871AM	0151
2SC871B	0151
2SC871C	0151
2SC871D	0151
2SC871E	0151
2SC871F	0151
2SC871G	0151
2SC871GN	0151
2SC871H	0151
2SC871J	0151
2SC871K	0151
2SC871L	0151
2SC871M	0151
2SC871OR	0151
2SC871R	0151
2SC871X	0151
2SC871Y	0151
2SC872	0418
2SC873	4463
2SC874	0018
2SC875	0086
2SC875(D)	0086
2SC875(E)	0086
2SC875(F)	0086
2SC875-1	0086
2SC875-1C	0086
2SC875-1D	0086
2SC875-1E	0086
2SC875-1F	0086
2SC875-2	0086
2SC875-2C	0086
2SC875-2D	0086
2SC875-2E	0086
2SC875-2F	0086
2SC875-3	0086
2SC875-3C	0086
2SC875-3D	0086
2SC875-3E	0086
2SC875-3F	0086
2SC875B	0086
2SC875C	0086
2SC875D	3426
2SC875DL	0086
2SC875E	3426
2SC875EL	0086
2SC875F	0086
2SC875G	0086
2SC875GN	0086
2SC875J	0086
2SC875K	0086
2SC875L	0086
2SC875M	0086
2SC875OR	0086
2SC875X	0086
2SC875Y	0086
2SC876	0086
2SC876(F)	0086
2SC876-E	0086
2SC876-F	0086
2SC876A	0086
2SC876B	0086
2SC876C	0086
2SC876D	0086
2SC876E	0086
2SC876F	0086
2SC876G	0086
2SC876GN	0086
2SC876H	0086
2SC876J	0086
2SC876K	0086
2SC876L	0086
2SC876M	0086
2SC876OR	0086
2SC876R	0086
2SC876TV	0086
2SC876TV(D)	0086
2SC876TV(E)	0086
2SC876TVD	0086
2SC876TV-E	0086
2SC876TVE	0086
2SC876TVEF	0086
2SC876TV-F	0086
2SC876TVF	0086
2SC876X	0086
2SC876Y	0086
2SC877	0016
2SC878	0016
2SC879	0127
2SC880	0324
2SC880R	0324
2SC881	0218
2SC881A	0218
2SC881B	0218
2SC881C	0218
2SC881D	0218
2SC881E	0218
2SC881F	0218
2SC881G	0218
2SC881GN	0218
2SC881H	0218
2SC881K	0218
2SC881L	0218
2SC881M	0218
2SC881OR	0218
2SC881R	0218
2SC881X	0218
2SC881Y	0218
2SC882	0620
2SC883	0423
2SC884	0178
2SC885	0074
2SC886	0074
2SC887	0074
2SC888	0074
2SC889	0103
2SC890	3846
2SC891	2091
2SC892	0042
2SC893	2314
2SC894	0155
2SC894A	0155
2SC894B	0155
2SC894C	0155
2SC894D	0155
2SC894E	0155
2SC894F	0155
2SC894G	0155
2SC894GN	0155
2SC894H	0155
2SC894J	0155
2SC894K	0155
2SC894L	0155
2SC894M	0155
2SC894OR	0155
2SC894R	0155
2SC894X	0155
2SC894Y	0155
2SC895	0074
2SC895-2	0074
2SC895-3	0074
2SC895-4	0074
2SC895-5	0074
2SC896	0016
2SC896A	0016
2SC896B	0016
2SC896C	0016
2SC896D	0016
2SC896E	0016
2SC896F	0016
2SC896G	0016
2SC896GN	0016
2SC896J	0016
2SC896K	0016
2SC896L	0016
2SC896M	0016
2SC896OR	0016
2SC896R	0016
2SC896X	0016
2SC896Y	0016
2SC897	0177
2SC897A	0177
2SC897B	0177
2SC898	0177
2SC898A	0177
2SC898B	0177
2SC898D	0177
2SC898E	0177
2SC898F	0177
2SC898G	0177
2SC898GN	0177
2SC898H	0177
2SC898J	0177
2SC898K	0177
2SC898L	0177
2SC898M	0177
2SC898OR	0177
2SC898X	0177
2SC898Y	0177
2SC899	2402
2SC899A	2402
2SC899B	2402
2SC899C	2402
2SC899D	2402
2SC899E	2402
2SC899F	2402
2SC899G	2402
2SC899GN	2402
2SC899H	2402
2SC899J	2402
2SC899K	2402
2SC899L	2402
2SC899M	2402
2SC899OR	2402
2SC899R	2402
2SC899X	2402
2SC899Y	2402
2SC900	1221
2SC900(E)	1221
2SC900(E)(L)	1221
2SC900(F)	1221
2SC900(U)	1221
2SC900-OR	1221
2SC900A	1221
2SC900AF	1221
2SC900B	1221
2SC900C	1221
2SC900D	1221
2SC900E	1221
2SC900F	1221
2SC900G	1221
2SC900J	1221
2SC900K	1221
2SC900L	1221
2SC900M	1221
2SC900OR	1221
2SC900R	1221
2SC900S	1221
2SC900SA	1221
2SC900SB	1221
2SC900SC	1221
2SC900SD	1221
2SC900U	1221
2SC900UE	1221
2SC900V	1221
2SC900VE	1221
2SC900X	1221
2SC900Y	1221
2SC901	0642
2SC901A	0642
2SC901B	0642
2SC901C	0642
2SC901D	0642
2SC901E	0642
2SC901F	0642
2SC901G	0642
2SC901GN	0642
2SC901H	0642
2SC901J	0642
2SC901K	0642
2SC901L	0642
2SC901M	0642
2SC901OR	0642
2SC901R	0642
2SC901X	0642
2SC901Y	0642
2SC902	0620

If replacement code is OEM, contact original manufacturer for replacement.

DEVICE TYPE	REPL CODE
2SC903	0155
2SC904	0155
2SC905	0861
2SC906	0016
2SC906(F)	0016
2SC906F	0016
2SC907	0016
2SC907A	0016
2SC907AC	0016
2SC907AD	0016
2SC907AH	0016
2SC907B	0016
2SC907C	0016
2SC907D	0016
2SC907E	0016
2SC907F	0016
2SC907G	0016
2SC907GN	0016
2SC907H	0016
2SC907HA	0016
2SC907J	0016
2SC907K	0016
2SC907L	0016
2SC907M	0016
2SC907OR	0016
2SC907R	0016
2SC907X	0016
2SC907Y	0016
2SC908	1354
2SC909	0560
2SC910	OEM
2SC911	0560
2SC911A	0560
2SC912	0144
2SC912M	0144
2SC913	0861
2SC913A	0861
2SC913B	0861
2SC913C	0861
2SC913D	0861
2SC913E	0861
2SC913F	0861
2SC913G	0861
2SC913GN	0861
2SC913H	0861
2SC913J	0861
2SC913K	0861
2SC913L	0861
2SC913M	0861
2SC913OR	0861
2SC913R	0861
2SC913X	0861
2SC913Y	0861
2SC914	0016
2SC914A	0016
2SC915	0016
2SC915A	0016
2SC916	0042
2SC917	0007
2SC917(K)	0007
2SC917A	0007
2SC917B	0007
2SC917C	0007
2SC917D	0007
2SC917E	0007
2SC917F	0007
2SC917G	0007
2SC917GN	0007
2SC917H	0007
2SC917J	0007
2SC917K	0007
2SC917L	0007
2SC917M	0007
2SC917OR	0007
2SC917R	0007
2SC917X	0007
2SC917Y	0007
2SC918	0224
2SC918A	0224
2SC918AL	0224
2SC918B	0224
2SC918C	0224
2SC918D	0224
2SC918E	0224
2SC918F	0224
2SC918G	0224
2SC918GN	0224
2SC918H	0224
2SC918J	0224
2SC918K	0224
2SC918L	0224
2SC918LF	0224
2SC918M	0224
2SC918OR	0224
2SC918R	0224
2SC918X	0224
2SC918XL	0224
2SC918Y	0224
2SC920	1213
2SC920-0Q	1213
2SC920-OR	1213
2SC920A	1213
2SC920B	1213
2SC920C	1213
2SC920CL	1213
2SC920D	1213
2SC920E	1213
2SC920F	1213
2SC920G	1213

DEVICE TYPE	REPL CODE
2SC920GN	1213
2SC920H	1213
2SC920L	1213
2SC920M	1213
2SC920OR	1213
2SC920Q	1213
2SC920R	1213
2SC920X	1213
2SC920Y	1213
2SC921	2381
2SC921(L)	2381
2SC921(VHF)	2381
2SC921A	2381
2SC921B	2381
2SC921C	2381
2SC921C1	2381
2SC921CL	2381
2SC921D	2381
2SC921E	2381
2SC921F	2381
2SC921G	2381
2SC921GN	2381
2SC921H	2381
2SC921J	2381
2SC921K	2381
2SC921L	2381
2SC921M	2381
2SC921OR	2381
2SC921R	2381
2SC921W	2381
2SC921X	2381
2SC921Y	2381
2SC922	0937
2SC922A	0937
2SC922B	0937
2SC922C	0937
2SC922K	0937
2SC922L	0937
2SC922M	0937
2SC-923	0552
2SC923	0552
2SC923(E)	0552
2SC923(E)(F)	0552
2SC923(F)	0552
2SC923A	0552
2SC923B	0552
2SC923C	0552
2SC923D	0552
2SC923E	0552
2SC923F	0552
2SC923G	0552
2SC923GN	0552
2SC923H	0552
2SC923K	0552
2SC923L	0552
2SC923M	0552
2SC923OR	0552
2SC923R	0552
2SC923X	0552
2SC923Y	0552
2SC924	0549
2SC924A	0549
2SC924B	0549
2SC924C	0549
2SC924D	0549
2SC924E	0549
2SC924F	0549
2SC924G	0549
2SC924GN	0549
2SC924H	0549
2SC924J	0549
2SC924K	0549
2SC924L	0549
2SC924M	0549
2SC924OR	0549
2SC924R	0549
2SC924X	0549
2SC924Y	0549
2SC925	0286
2SC925(M)	0286
2SC925A	0286
2SC925B	0286
2SC925C	0286
2SC925D	0286
2SC925E	0286
2SC925F	0286
2SC925G	0286
2SC925GN	0286
2SC925H	0286
2SC925J	0286
2SC925K	0286
2SC925L	0286
2SC925M	0286
2SC925OR	0286
2SC925R	0286
2SC925X	0286
2SC925Y	0286
2SC926	0261
2SC926(A)	0261
2SC926-OR	0261
2SC926A	0261
2SC926B	0261
2SC926C	0261
2SC926D	0261
2SC926E	0261
2SC926F	0261
2SC926G	0261
2SC926GN	0261
2SC926H	0261

DEVICE TYPE	REPL CODE
2SC926J	0261
2SC926K	0261
2SC926L	0261
2SC926M	0261
2SC926OR	0261
2SC926R	0261
2SC926X	0261
2SC926Y	0261
2SC927	0007
2SC927(C)	0007
2SC927(D)	0007
2SC927(E)	0007
2SC927A	0007
2SC927B	0007
2SC927C	0007
2SC927C(E)	0007
2SC927C(K)	0007
2SC927CJ	0007
2SC927CK	0007
2SC927CT	0007
2SC927CU	0007
2SC927CW	0007
2SC927D	0007
2SC927E	0007
2SC927E,Z	0007
2SC927F	0007
2SC927G	0007
2SC927GN	0007
2SC927H	0007
2SC927J	0007
2SC927K	0007
2SC927L	0007
2SC927M	0007
2SC927OR	0007
2SC927R	0007
2SC927X	0007
2SC927XL	0007
2SC927Y	0007
2SC927Z	0007
2SC928	0007
2SC928A	0007
2SC928B	0007
2SC928C	0007
2SC928D	0007
2SC928E	0007
2SC928F	0007
2SC928G	0007
2SC928GN	0007
2SC928H	0007
2SC928J	0007
2SC928K	0007
2SC928L	0007
2SC928M	0007
2SC928OR	0007
2SC928R	0007
2SC928X	0007
2SC928Y	0007
2SC929	1066
2SC929(0)	1066
2SC929(E)	1066
2SC929-0	1066
2SC929-O	1066
2SC929A	1066
2SC929B	1066
2SC929C	1066
2SC929C1	1066
2SC929D	1066
2SC929D1	1066
2SC929DE	1066
2SC929DP	1066
2SC929DU	1066
2SC929DV	1066
2SC929E	1497
2SC929ED	1066
2SC929EZ	1066
2SC929F	1066
2SC929FK	1066
2SC929G	1066
2SC929GN	1066
2SC929H	1066
2SC929J	1066
2SC929K	1066
2SC929L	1066
2SC929M	1066
2SC929NP	1066
2SC929OR	1066
2SC929R	1066
2SC929SP	1066
2SC929X	1066
2SC929Y	1066
2SC930	2209
2SC930(C)	2209
2SC930(D)	2209
2SC930(E)	2209
2SC930(F)	2209
2SC930-E-SP	OEM
2SC930-E-SPA-AC	OEM
2SC930-F-SPA-AC	OEM
2SC930-OR	2209
2SC930A	2209
2SC930B	2209
2SC930BK	2209
2SC930C	2209
2SC930C-IF	2209
2SC930CK	2209
2SC930CL	2209
2SC930CS	2209
2SC930D	2209
2SC930DB	2209

DEVICE TYPE	REPL CODE
2SC930DC	2209
2SC930DE	2209
2SC930DK	2209
2SC930DS	2209
2SC930DT	2209
2SC930DT-2	2209
2SC930DX	2209
2SC930DZ	2209
2SC930E	2209
2SC930E1	4517
2SC930E2	OEM
2SC930EP	2209
2SC930E-SP	OEM
2SC930ET	2209
2SC930EV	2209
2SC930EX	2209
2SC930F	2209
2SC930F-SP	OEM
2SC930G	2209
2SC930GN	2209
2SC930H	2209
2SC930J	2209
2SC930K	2209
2SC930L	2209
2SC930M	2209
2SC930NP	2209
2SC930OR	2209
2SC930R	2209
2SC930SP	2209
2SC930SPE	4517
2SC930X	2209
2SC930Y	2209
2SC931	0219
2SC931C	0219
2SC931D	0219
2SC931E	0219
2SC932	0560
2SC932(E)	0560
2SC932A	0560
2SC932B	0560
2SC932BK	0560
2SC932C	0560
2SC932D	0560
2SC932E	0560
2SC932F	0560
2SC932G	0560
2SC932GN	0560
2SC932H	0560
2SC932J	0560
2SC932K	0560
2SC932L	0560
2SC932M	0560
2SC932OR	0560
2SC932R	0560
2SC932X	0560
2SC932Y	0560
2SC933	0191
2SC933(D)	0191
2SC933(F)	0191
2SC933(G)	0191
2SC933A	0191
2SC933B	0191
2SC933BB	0191
2SC933C	0191
2SC933D	0191
2SC933D(F)	0191
2SC933E	0191
2SC933E(F)	0191
2SC933F	0191
2SC933FB	0191
2SC933FP	0191
2SC933G	0191
2SC933GN	0191
2SC933H	0191
2SC933J	0191
2SC933K	0191
2SC933L	0191
2SC933M	0191
2SC933OR	0191
2SC933R	0191
2SC933X	0191
2SC933Y	0191
2SC934	0016
2SC934-0	0016
2SC934A	0016
2SC934B	0016
2SC934C	0016
2SC934D	0016
2SC934E	0016
2SC934F	0016
2SC934G	0016
2SC934GN	0016
2SC934H	0016
2SC934J	0016
2SC934K	0016
2SC934L	0016
2SC934M	0016
2SC934OR	0016
2SC934Q	0016
2SC934R	0016
2SC934X	0016
2SC934Y	0016
2SC935	0074
2SC935-OR	0074
2SC935A	0074
2SC935B	0074
2SC935C	0074
2SC935D	0074
2SC935E	0074

DEVICE TYPE	REPL CODE
2SC935F	0074
2SC935G	0074
2SC935GN	0074
2SC935H	0074
2SC935J	0074
2SC935K	0074
2SC935M	0074
2SC935OR	0074
2SC935R	0074
2SC935X	0074
2SC935Y	0074
2SC936	0003
2SC936(BK)	0003
2SC936A	0003
2SC936B	0003
2SC936BK	0003
2SC936C	0003
2SC936D	0003
2SC936E	0003
2SC936F	0003
2SC936G	0003
2SC936GN	0003
2SC936H	0003
2SC936J	0003
2SC936K	0003
2SC936L	0003
2SC936M	0003
2SC936OR	0003
2SC936R	0003
2SC936X	0003
2SC936Y	0003
2SC937	0103
2SC937(BK)	0103
2SC937(XL)	0103
2SC937(YL,BK)	0103
2SC937-01	0103
2SC937A	0103
2SC937B	0103
2SC937BK	0103
2SC937C	0103
2SC937D	0103
2SC937E	0103
2SC937F	0103
2SC937G	0103
2SC937GN	0103
2SC937H	0103
2SC937J	0103
2SC937K	0103
2SC937L	0103
2SC937M	0103
2SC937OR	0103
2SC937R	OEM
2SC937X	0103
2SC937Y	0103
2SC938	0155
2SC938-O	0155
2SC938A	0155
2SC938B	0155
2SC938C	0155
2SC938D	0155
2SC938E	0155
2SC938F	0155
2SC938G	0155
2SC938GN	0155
2SC938H	0155
2SC938J	0155
2SC938K	0155
2SC938L	0155
2SC938M	0155
2SC938OR	0155
2SC938R	0155
2SC938X	0155
2SC938Y	0155
2SC939	0642
2SC939(L)	0642
2SC939I	0642
2SC939L	0642
2SC940	0826
2SC940(L)(M)	0826
2SC940(M)	0826
2SC940(P)	0826
2SC940(Q)	0826
2SC940(R)	0826
2SC940L	0826
2SC940M	0826
2SC940Q	0826
2SC941	0155
2SC941(O)	0155
2SC941(O),(R)	0155
2SC941(R)	0155
2SC941-04	0155
2SC941-0Y	0155
2SC941-O	0155
2SC941-R	0155
2SC941-TM.O.JA	0155
2SC941-Y	0155
2SC941K	0155
2SC941O	0155
2SC941R	0155
2SC941TM	0155
2SC941TM-0	OEM
2SC941TM-O	0155
2SC941Y	0155
2SC942	0016
2SC943	0016
2SC943A	0016
2SC943B	0016
2SC943C	0016
2SC943D	0016

DEVICE TYPE	REPL CODE
2SC943E	0016
2SC943F	0016
2SC943G	0016
2SC943GN	0016
2SC943H	0016
2SC943J	0016
2SC943K	0016
2SC943L	0016
2SC943M	0016
2SC943OR	0016
2SC943R	0016
2SC943X	0016
2SC943Y	0016
2SC944	0076
2SC944S	0076
2SC-945	0076
2SC945	0076
2SC945(C)-T	0076
2SC945(C)T	0076
2SC945(K)	0076
2SC945(L)	0076
2SC945(L)P	1419
2SC945(L)Q	0076
2SC945(P)	0076
2SC945(Q)	0076
2SC945(R)	0076
2SC945(TK)	0076
2SC945(TK,P)	0076
2SC945(TP)	0076
2SC945(TQ)	0076
2SC945(TQ,Q)	0076
2SC945-0	0076
2SC945-K	0076
2SC945-O	0076
2SC945-OR	0076
2SC945-P	0076
2SC945-Q	0076
2SC945-T	0076
2SC945-TQ	0076
2SC945-YX	0076
2SC-945A	0076
2SC945A	0076
2SC945A(C)	0076
2SC945A(C)P	0076
2SC945A(C)Q	0076
2SC945A(C)-T	0076
2SC945A(C)T	0076
2SC945A(K)	0076
2SC945A(P)	0076
2SC945A(Q)	0076
2SC945A(Q,R)	0076
2SC945A(R)	0076
2SC945A/D	0076
2SC945A-AQ	0076
2SC945A-D	OEM
2SC945A-K	OEM
2SC945AK	0076
2SC945AKT	0076
2SC945A-P	OEM
2SC945AP	0076
2SC945A-PA	0076
2SC945APT	0076
2SC945A-Q	0076
2SC945AQ	0076
2SC945A-QA	0076
2SC945A-R	OEM
2SC945AR	0076
2SC945A-RA	0076
2SC945B	0076
2SC945C	0076
2SC945D	0076
2SC945E	0076
2SC945F	0076
2SC945G	0076
2SC945GN	0076
2SC945H	0076
2SC945J	0076
2SC945K	0076
2SC945L	0076
2SC945LP	0076
2SC945LQ	0076
2SC945M	0076
2SC945OR	0076
2SC945P	0076
2SC945P1	0076
2SC945PA	0076
2SC945PJ	0076
2SC945PQ	0076
2SC945Q	0076
2SC945QL	0076
2SC945QP	0076
2SC945QZ	0076
2SC945R	0076
2SC945RA	0076
2SC945S	0076
2SC945T	0076
2SC945TK	0076
2SC945TP	0076
2SC945TQ	0076
2SC945X	0076
2SC945Y	0076
2SC945Z	OEM
2SC945ZA	0076
2SC947	0007
2SC947A	0007
2SC947B	0007
2SC947C	0007
2SC947D	0007
2SC947E	0007

DEVICE TYPE	REPL CODE
2SC947F	0007
2SC947G	0007
2SC947GN	0007
2SC947H	0007
2SC947J	0007
2SC947K	0007
2SC947L	0007
2SC947M	0007
2SC947OR	0007
2SC947R	0007
2SC947X	0007
2SC947Y	0007
2SC948	0216
2SC948A	0216
2SC948B	0216
2SC948D	0216
2SC948E	0216
2SC948F	0216
2SC948G	0216
2SC948GN	0216
2SC948H	0216
2SC948J	0216
2SC948K	0216
2SC948L	0216
2SC948M	0216
2SC948OR	0216
2SC948R	0216
2SC948X	0216
2SC948Y	0216
2SC949	0216
2SC950	0016
2SC951	0079
2SC952	0079
2SC953	0016
2SC954	0016
2SC955	0016
2SC956	0016
2SC957	0144
2SC957A	0144
2SC957AL	0144
2SC957B	0144
2SC957C	0144
2SC957D	0144
2SC957E	0144
2SC957F	0144
2SC957G	0144
2SC957GN	0144
2SC957H	0144
2SC957J	0144
2SC957K	0144
2SC957L	0144
2SC957M	0144
2SC957OR	0144
2SC957R	0144
2SC957X	0144
2SC957XL	0144
2SC957Y	0144
2SC959	0086
2SC959A	0086
2SC959B	0086
2SC959C	0086
2SC959D	0086
2SC959E	0086
2SC959F	0086
2SC959G	0086
2SC959GN	0086
2SC959H	0086
2SC959J	0086
2SC959K	0086
2SC959L	0086
2SC959M	0086
2SC959OR	0086
2SC959R	0086
2SC959S	0086
2SC959SA	0086
2SC959SB	0086
2SC959SC	0086
2SC959SD	0086
2SC959X	0086
2SC959Y	0086
2SC960	5623
2SC961	0620
2SC962	0620
2SC963	0016
2SC964	0016
2SC965	0016
2SC966	0016
2SC966A	0016
2SC966B	0016
2SC966C	0016
2SC966D	0016
2SC966E	0016
2SC966F	0016
2SC966G	0016
2SC966GN	0016
2SC966H	0016
2SC966J	0016
2SC966K	0016
2SC966L	0016
2SC966M	0016
2SC966OR	0016
2SC966R	0016
2SC966X	0016
2SC966Y	0016
2SC967	0016
2SC967A	0016
2SC967B	0016
2SC967C	0016
2SC967D	0016

If replacement code is OEM, contact original manufacturer for replacement.

DEVICE TYPE	REPL CODE
2SC967E	0016
2SC967F	0016
2SC967G	0016
2SC967GN	0016
2SC967H	0016
2SC967J	0016
2SC967K	0016
2SC967L	0016
2SC967M	0016
2SC967OR	0016
2SC967R	0016
2SC967X	0016
2SC967Y	0016
2SC968	2929
2SC968A	2929
2SC968B	2929
2SC968C	2929
2SC968D	2929
2SC968E	2929
2SC968F	2929
2SC968G	2929
2SC968GN	2929
2SC968H	2929
2SC968J	2929
2SC968K	2929
2SC968L	2929
2SC968M	2929
2SC968OR	2929
2SC968P	2929
2SC968R	2929
2SC968X	2929
2SC968Y	2929
2SC969	0016
2SC970	0016
2SC971	2952
2SC971A	2952
2SC971B	2952
2SC971BK	2952
2SC971C	2952
2SC971D	2952
2SC971E	2952
2SC971F	2952
2SC971G	2952
2SC971GN	2952
2SC971H	2952
2SC971J	2952
2SC971K	2952
2SC971L	2952
2SC971M	2952
2SC971OR	2952
2SC971R	2952
2SC971X	2952
2SC971Y	2952
2SC972	0086
2SC972A	0086
2SC972B	0086
2SC972C	0086
2SC972D	0086
2SC972E	0086
2SC972F	0086
2SC972G	0086
2SC972GN	0086
2SC972H	0086
2SC972J	0086
2SC972K	0086
2SC972L	0086
2SC972M	0086
2SC972OR	0086
2SC972R	0086
2SC972X	0086
2SC972Y	0086
2SC973	2091
2SC973A	2091
2SC975	1420
2SC975A	1420
2SC976	0086
2SC976TV	0086
2SC979	0016
2SC979-O	0016
2SC979-R	0016
2SC979-Y	0016
2SC979A	0016
2SC979A-O	0016
2SC979A-R	0016
2SC980	0286
2SC980A	0286
2SC980A/G	0286
2SC980AG	0286
2SC980AG-O	0286
2SC980AG-R	0286
2SC980G	0286
2SC980G-O	0286
2SC980G-R	0286
2SC980G-Y	0286
2SC981	0178
2SC982	0400
2SC983	0066
2SC983(FA-1)	OEM
2SC983(O)	0066
2SC983-0	0066
2SC983-O	0066
2SC983-R	0066
2SC983-Y	0066
2SC983D	0066
2SC983FA-1	0066
2SC983FH-1	0066
2SC983O	0066
2SC983R	0066
2SC983S	0066
2SC983Y	0066
2SC983YFA-1	0066
2SC984	0536
2SC984A	0536
2SC984B	0536
2SC984C	0536
2SC984D	0536
2SC984E	0536
2SC984F	0536
2SC984G	0536
2SC984GN	0536
2SC984H	0536
2SC984J	0536
2SC984L	0536
2SC984M	0536
2SC984OR	0536
2SC984R	0536
2SC984X	0536
2SC984Y	0536
2SC985	0111
2SC985A	0111
2SC987	0007
2SC987A	0007
2SC988	0007
2SC988A	0007
2SC988B	0007
2SC990	0042
2SC991	0493
2SC991A	0493
2SC991B	0493
2SC991C	0493
2SC991D	0493
2SC991E	0493
2SC991F	0493
2SC991G	0493
2SC991GN	0493
2SC991H	0493
2SC991J	0493
2SC991K	0493
2SC991L	0493
2SC991M	0493
2SC991OR	0493
2SC991R	0493
2SC991X	0493
2SC991Y	0493
2SC992	0493
2SC992A	0493
2SC992B	0493
2SC992C	0493
2SC992D	0493
2SC992E	0493
2SC992F	0493
2SC992G	0493
2SC992GN	0493
2SC992H	0493
2SC992J	0493
2SC992K	0493
2SC992L	0493
2SC992M	0493
2SC992OR	0493
2SC992R	0493
2SC992X	0493
2SC992Y	0493
2SC993	0016
2SC993D	0016
2SC993E	0016
2SC994	0079
2SC995	0168
2SC995A	0168
2SC995B	0168
2SC995C	0168
2SC995E	0168
2SC995F	0168
2SC995G	0168
2SC995GN	0168
2SC995H	0168
2SC995J	0168
2SC995K	0168
2SC995L	0168
2SC995M	0168
2SC995OR	0168
2SC995R	0168
2SC995X	0168
2SC995Y	0168
2SC996	3280
2SC997	0224
2SC997A	0224
2SC997B	0224
2SC997C	0224
2SC997D	0224
2SC997E	0224
2SC997F	0224
2SC997G	0224
2SC997GN	0224
2SC997H	0224
2SC997J	0224
2SC997K	0224
2SC997L	0224
2SC997M	0224
2SC997OR	0224
2SC997R	0224
2SC997X	0224
2SC997Y	0224
2SC998	3846
2SC999	0065
2SC999A	0065
2SC999B	0065
2SC999C	0065
2SC999D	0065
2SC999E	0065
2SC999F	0065
2SC999G	0065
2SC999GN	0065
2SC999H	0065
2SC999J	0065
2SC999K	0065
2SC999L	0065
2SC999M	0065
2SC999OR	0065
2SC999R	0065
2SC999X	0065
2SC999Y	0065
2SC1000	0111
2SC1000(BL)	0111
2SC1000(GR)	0111
2SC1000-BL	0111
2SC1000-GR	0111
2SC1000-Y	0111
2SC1000A	0111
2SC1000B	0111
2SC1000BL	0111
2SC1000C	0111
2SC1000D	0111
2SC1000E	0111
2SC1000F	0111
2SC1000G	0111
2SC1000G-BL	0111
2SC1000GBL	0111
2SC1000G-GR	0111
2SC1000GN	0111
2SC1000GR	0111
2SC1000GR(FA-1)	OEM
2SC1000H	0111
2SC1000J	0111
2SC1000K	0111
2SC1000L	0111
2SC1000M	0111
2SC1000OR	0111
2SC1000R	0111
2SC1000X	0111
2SC1000Y	0111
2SC1001	2039
2SC1004	0003
2SC1004A	0003
2SC1004B	0003
2SC1004C	0003
2SC1004D	0003
2SC1004E	0003
2SC1004F	0003
2SC1004G	0003
2SC1004GN	0003
2SC1004H	0003
2SC1004J	0003
2SC1004K	0003
2SC1004L	0003
2SC1004M	0003
2SC1004OR	0003
2SC1004R	0003
2SC1004X	0003
2SC1004Y	0003
2SC1005	0065
2SC1005A	0065
2SC1005B	0065
2SC1005C	0065
2SC1005D	0065
2SC1005E	0065
2SC1005F	0065
2SC1005G	0065
2SC1005GN	0065
2SC1005H	0065
2SC1005J	0065
2SC1005K	0065
2SC1005L	0065
2SC1005M	0065
2SC1005OR	0065
2SC1005R	0065
2SC1005X	0065
2SC1005Y	0065
2SC1006	0111
2SC1006A	0111
2SC1006B	0111
2SC1006C	0111
2SC1007	0016
2SC1008	0086
2SC1008A	0086
2SC1009	0224
2SC1009A	0224
2SC1009F1	0224
2SC1009F2	0224
2SC1009F3	0224
2SC1009F4	0224
2SC1009F5	0224
2SC1010	0111
2SC1010A	0111
2SC1010B	0111
2SC1010C	0111
2SC1010D	0111
2SC1010E	0111
2SC1010F	0111
2SC1010G	0111
2SC1010GN	0111
2SC1010H	0111
2SC1010J	0111
2SC1010K	0111
2SC1010L	0111
2SC1010M	0111
2SC1010OR	0111
2SC1010R	0111
2SC1010X	0111
2SC1010Y	0111
2SC1011	0698
2SC1012	0233
2SC1012A	0233
2SC1012B	0233
2SC1012C	0233
2SC1012D	0233
2SC1012E	0233
2SC1012F	0233
2SC1012G	0233
2SC1012GN	0233
2SC1012H	0233
2SC1012J	0233
2SC1012K	0233
2SC1012L	0233
2SC1012M	0233
2SC1012OR	0233
2SC1012R	0233
2SC1012X	0233
2SC1012Y	0233
2SC1013	2046
2SC1013-OR	2046
2SC1013A	2046
2SC1013B	2046
2SC1013C	2046
2SC1013D	2046
2SC1013E	2046
2SC1013F	2046
2SC1013G	2046
2SC1013GN	2046
2SC1013H	2046
2SC1013J	2046
2SC1013K	2046
2SC1013L	2046
2SC1013LJ	2046
2SC1013M	2046
2SC1013OR	2046
2SC1013PJ	2046
2SC1013R	2046
2SC1013X	2046
2SC1013Y	2046
2SC1014	2046
2SC1014-1	2046
2SC1014-2	2046
2SC1014A	2046
2SC1014B	2046
2SC1014BY	2046
2SC1014C	2046
2SC1014CD	2046
2SC1014D	2046
2SC1014D1	2046
2SC1014E	2046
2SC1014F	2046
2SC1014G	2046
2SC1014GA	2046
2SC1014GN	2046
2SC1014H	2046
2SC1014J	2046
2SC1014K	2046
2SC1014L	2046
2SC1014LG	2046
2SC1014LR	2046
2SC1014M	2046
2SC1014OR	2046
2SC1014R	2046
2SC1014W	2046
2SC1014X	2046
2SC1014Y	2046
2SC1015	OEM
2SC1015-O	OEM
2SC1015Y	OEM
2SC1016	OEM
2SC1017	2050
2SC1018	2050
2SC1018A	2050
2SC1018B	2050
2SC1018C	2050
2SC1018D	2050
2SC1018E	2050
2SC1018F	2050
2SC1018G	2050
2SC1018GN	2050
2SC1018H	2050
2SC1018J	2050
2SC1018K	2050
2SC1018L	2050
2SC1018M	2050
2SC1018R	2050
2SC1018X	2050
2SC1018Y	2050
2SC1019	OEM
2SC1019C	0042
2SC1020	OEM
2SC1021Q	OEM
2SC1023	0127
2SC1023(O)	0127
2SC1023(O)	0127
2SC1023-O	0127
2SC1023-Y	0127
2SC1023A	0127
2SC1023B	0127
2SC1023C	0127
2SC1023D	0127
2SC1023E	0127
2SC1023F	0127
2SC1023G	0127
2SC1023GN	0127
2SC1023H	0127
2SC1023J	0127
2SC1023K	0127
2SC1023L	0127
2SC1023M	0127
2SC1023OR	0127
2SC1023R	0127
2SC1023X	0127
2SC1023Y	0127
2SC1024	0178
2SC1024(D)	0178
2SC1024(F)	0178
2SC1024-E	0178
2SC1024A	0178
2SC1024B	0178
2SC1024C	0178
2SC1024D	0178
2SC1024E	0178
2SC1024F	0178
2SC1024G	0178
2SC1024L	0178
2SC1024Y	0178
2SC1025	0934
2SC1025(D)	0934
2SC1025C	OEM
2SC1025CTV	0934
2SC1025D	0934
2SC1025E	0934
2SC1025J	0934
2SC1025MT	0934
2SC1026	0127
2SC1026(G)	0127
2SC1026-O	0127
2SC1026-R	0127
2SC1026A	0127
2SC1026B	0127
2SC1026BL	0127
2SC1026C	0127
2SC1026D	0127
2SC1026E	0127
2SC1026F	0127
2SC1026G	0127
2SC1026GN	0127
2SC1026GR	0127
2SC1026H	0127
2SC1026J	0127
2SC1026K	0127
2SC1026L	0127
2SC1026M	0127
2SC1026OR	0127
2SC1026R	0127
2SC1026X	0127
2SC1026Y	0127
2SC1027	0642
2SC1030	2059
2SC1030-OR	2059
2SC1030A	2059
2SC1030B	2059
2SC1030B2C	2059
2SC1030C	2059
2SC1030D	2059
2SC1030E	2059
2SC1030F	2059
2SC1030G	2059
2SC1030H	2059
2SC1030J	2059
2SC1030K	2059
2SC1030L	2059
2SC1030M	2059
2SC1030OR	2059
2SC1030R	2059
2SC1030X	2059
2SC1030Y	2059
2SC1031	1027
2SC1032	0127
2SC1032(Y)	0127
2SC1032A	0127
2SC1032B	0127
2SC1032BL	0127
2SC1032C	0127
2SC1032D	0127
2SC1032E	0127
2SC1032F	0127
2SC1032G	0127
2SC1032GN	0127
2SC1032H	0127
2SC1032J	0127
2SC1032K	0127
2SC1032L	0127
2SC1032M	0127
2SC1032OR	0127
2SC1032R	0127
2SC1032X	0127
2SC1032Y	0127
2SC1033	0715
2SC1033A	0715
2SC1033B	0715
2SC1033C	0715
2SC1033D	0715
2SC1033E	0715
2SC1033F	0715
2SC1033G	0715
2SC1033GN	0715
2SC1033H	0715
2SC1033J	0715
2SC1033K	0715
2SC1033L	0715
2SC1033OR	0715
2SC1033R	0715
2SC1033X	0715
2SC1033Y	0715
2SC1034	2064
2SC1034-3	2064
2SC1034-4	2064
2SC1034-5	2064
2SC1035	0007
2SC1035A	0007
2SC1035B	0007
2SC1035C	0007
2SC1035D	0007
2SC1035E	0007
2SC1035F	0007
2SC1035GN	0007
2SC1035H	0007
2SC1035J	0007
2SC1035L	0007
2SC1035M	0007
2SC1035OR	0007
2SC1035R	0007
2SC1035X	0007
2SC1035Y	0007
2SC1036	0007
2SC1036A	0007
2SC1036B	0007
2SC1036C	0007
2SC1036D	0007
2SC1036E	0007
2SC1036F	0007
2SC1036G	0007
2SC1036GN	0007
2SC1036H	0007
2SC1036J	0007
2SC1036K	0007
2SC1036L	0007
2SC1036M	0007
2SC1036OR	0007
2SC1036R	0007
2SC1036X	0007
2SC1036Y	0007
2SC1037	OEM
2SC1042	1591
2SC1043	2070
2SC1044	0007
2SC1045	0003
2SC1045(B)	0003
2SC1045B	0003
2SC1045C	0003
2SC1045D	0003
2SC1045E	0003
2SC1045F	0003
2SC1045G	0003
2SC1045GN	0003
2SC1045H	0003
2SC1045J	0003
2SC1045K	0003
2SC1045L	0003
2SC1045M	0003
2SC1045OR	0003
2SC1045R	0003
2SC1045X	0003
2SC1045Y	0003
2SC1046	0065
2SC1046-OR	0065
2SC1046A	0065
2SC1046B	0065
2SC1046C	0065
2SC1046D	0065
2SC1046E	0065
2SC1046F	0065
2SC1046G	0065
2SC1046GN	0065
2SC1046H	0065
2SC1046J	0065
2SC1046K	0065
2SC1046L	0065
2SC1046M	0065
2SC1046N	0065
2SC1046OR	0065
2SC1046R	0065
2SC1046X	0065
2SC1046Y	0065
2SC1047	0113
2SC1047(B)	0113
2SC1047(C)	0113
2SC1047(CD)	0113
2SC1047-C	0113
2SC1047-D	0113
2SC1047A	0113
2SC1047B	0113
2SC1047BC	0113
2SC1047BCD	0113
2SC1047C	0113
2SC1047D	0113
2SC1047E	0113
2SC1047F	0113
2SC1047G	0113
2SC1047GN	0113
2SC1047GR	0113
2SC1047H	0113
2SC1047J	0113
2SC1047K	0113
2SC1047L	0113
2SC1047M	0113
2SC1047R	0113
2SC1047X	0113
2SC1047Y	0113
2SC1048	0233
2SC1048(C)	0233
2SC1048(D)	0233
2SC1048B	0233
2SC1048C	0233
2SC1048D	0233
2SC1048D,C	0233
2SC1048DC	0233
2SC1048E	0233
2SC1048F	0233
2SC1050	0220
2SC1050C	0220
2SC1050D	0220
2SC1050E	0220
2SC1050F	0220
2SC1051	0177
2SC1051L	0177
2SC1052	0086
2SC1053	0086
2SC1054	0007
2SC1055	0429
2SC1055H	0429
2SC1055U	0429
2SC1056	0233
2SC1056A	0233
2SC1056B	0233
2SC1056C	0233
2SC1056D	0233
2SC1056E	0233
2SC1056F	0233
2SC1056G	0233
2SC1056GN	0233
2SC1056H	0233
2SC1056J	0233
2SC1056K	0233
2SC1056L	0233
2SC1056M	0233
2SC1056OR	0233
2SC1056R	0233
2SC1056X	0233
2SC1056Y	0233
2SC1057	OEM
2SC1058	OEM
2SC1059	0142
2SC1059A	0142
2SC1059B	0142
2SC1059C	0142
2SC1059D	0142
2SC1059E	0142
2SC1059F	0142
2SC1059G	0142
2SC1059GN	0142
2SC1059H	0142
2SC1059J	0142
2SC1059K	0142
2SC1059L	0142
2SC1059M	0142
2SC1059OR	0142
2SC1059R	0142
2SC1059X	0142
2SC1059Y	0142
2SC1060	0042
2SC1060(C,D)	0042
2SC1060A	0042
2SC1060B	0042
2SC1060BL	0042
2SC1060BM	0042
2SC1060BY	0042
2SC1060C	0042
2SC1060D	0042
2SC1060E	0042
2SC1060F	0042
2SC1060G	0042
2SC1060GN	0042
2SC1060H	0042
2SC1060J	0042
2SC1060K	0042
2SC1060L	0042
2SC1060M	0042
2SC1060OR	0042
2SC1060R	0042
2SC1060X	0042
2SC1060Y	0042
2SC1061	0042
2SC1061(B)	0042
2SC1061(C)	0042
2SC1061-C	0042
2SC1061A	0042
2SC1061B	0042
2SC1061BM	0042
2SC1061C	0042
2SC1061D	0042
2SC1061K	0042
2SC1061KA	0042
2SC1061KB	0042
2SC1061KC	0042
2SC1061SC	0042
2SC1061T	0042
2SC1061TB	0042
2SC1062	0233
2SC1063	0018
2SC1064	0324
2SC1066	0259
2SC1067	0007
2SC1068	0418
2SC1069	0630
2SC1070	0007
2SC1070-1	0007
2SC1070-2	0007
2SC1070-B	0007
2SC1071	0016
2SC1072	0086

DEVICE TYPE	REPL CODE
2SC1072A	0086
2SC1072B	0086
2SC1072C	0086
2SC1072D	0086
2SC1072E	0086
2SC1072F	0086
2SC1072G	0086
2SC1072GN	0086
2SC1072H	0086
2SC1072J	0086
2SC1072K	0086
2SC1072L	0086
2SC1072M	0086
2SC1072OR	0086
2SC1072R	0086
2SC1072X	0086
2SC1072Y	0086
2SC1073BGS2	OEM
2SC1074	2091
2SC1075	1420
2SC1076	2093
2SC1078	2096
2SC1079	0177
2SC1079-R	0177
2SC1079-Y	0177
2SC1080	0177
2SC1080-R	0177
2SC1080-Y	0177
2SC1080B	0177
2SC1080LB	0177
2SC1082	OEM
2SC1083	2041
2SC1084	0224
2SC1085	0178
2SC1086	0223
2SC1086A	0223
2SC1086B	0223
2SC1086C	0223
2SC1086D	0223
2SC1086E	0223
2SC1086F	0223
2SC1086G	0223
2SC1086GN	0223
2SC1086H	0223
2SC1086J	0223
2SC1086K	0223
2SC1086L	0223
2SC1086M	0223
2SC1086OR	0223
2SC1086R	0223
2SC1086X	0223
2SC1086Y	0223
2SC1087	OEM
2SC1088	0285
2SC1089	0285
2SC1089(B)	0285
2SC1089(B,D)	0285
2SC1089(D)	0285
2SC1089B	0285
2SC1089C	0285
2SC1089D	0285
2SC1090	0233
2SC1091A	0560
2SC1092	OEM
2SC1092FA	OEM
2SC1093	OEM
2SC1095	0219
2SC1095-K	0219
2SC1096	0390
2SC1096(M)	0390
2SC1096-4Z-L	0390
2SC1096-4ZL	0390
2SC1096-OR	0390
2SC1096A	0390
2SC1096B	0390
2SC1096C	0390
2SC1096D	0390
2SC1096E	0390
2SC1096F	0390
2SC1096G	0390
2SC1096GN	0390
2SC1096H	0390
2SC1096J	0390
2SC1096K	0390
2SC1096L	0390
2SC1096LM	0390
2SC1096M	0390
2SC1096N	0390
2SC1096OR	0390
2SC1096Q	0390
2SC1096R	0390
2SC1096X	0390
2SC1096Y	0390
2SC1097	OEM
2SC1098	0560
2SC1098(4)K	0560
2SC1098(4)L	0560
2SC1098(L)	0560
2SC1098(M)	0560
2SC1098-4K	0560
2SC1098-4L	0560
2SC1098-42M	0560
2SC1098A	0560
2SC1098B	0560
2SC1098C	0560
2SC1098D	0560
2SC1098L	0560
2SC1098M	0560
2SC1099	1146
2SC1099(K)	1146
2SC1099K	1146
2SC1100	1150
2SC1100A	1150
2SC1100B	1150
2SC1100C	1150
2SC1100D	1150
2SC1100E	1150
2SC1100F	1150
2SC1100G	1150
2SC1100GN	1150
2SC1100H	1150
2SC1100J	1150
2SC1100K	1150
2SC1100L	1150
2SC1100M	1150
2SC1100OR	1150
2SC1100R	1150
2SC1100X	1150
2SC1100Y	1150
2SC1101	0003
2SC1101A	0003
2SC1101B	0003
2SC1101C	0003
2SC1101D	0003
2SC1101E	0003
2SC1101F	0003
2SC1101G	0003
2SC1101GN	0003
2SC1101H	0003
2SC1101J	0003
2SC1101K	0003
2SC1101L	0003
2SC1101M	0003
2SC1101OR	0003
2SC1101R	0003
2SC1101X	0003
2SC1101Y	0003
2SC1102	0142
2SC1102(L)	0142
2SC1102(M)	0142
2SC1102A	0142
2SC1102B	0142
2SC1102C	0142
2SC1102K	0142
2SC1102L	0142
2SC1102M	0142
2SC1103	0233
2SC1103(A)	0233
2SC1103(L)	0233
2SC1103A	0233
2SC1103B	0233
2SC1103C	0233
2SC1103L	0233
2SC1104	1209
2SC1104A	1209
2SC1104B	1209
2SC1104C	1209
2SC1104L	1209
2SC1105	0142
2SC1105(K)	0142
2SC1105(L)	0142
2SC1105(M)	0142
2SC1105A	0142
2SC1105B	0142
2SC1105C	0142
2SC1105D	0142
2SC1105E	0142
2SC1105F	0142
2SC1105G	0142
2SC1105GN	0142
2SC1105H	0142
2SC1105J	0142
2SC1105K	0142
2SC1105L	0142
2SC1105M	0142
2SC1105OR	0142
2SC1105R	0142
2SC1105X	0142
2SC1105Y	0142
2SC1106	0220
2SC1106(K)	0220
2SC1106(L)	0220
2SC1106(M)	0220
2SC1106A	0220
2SC1106B	0220
2SC1106C	0220
2SC1106K	0220
2SC1106L	0220
2SC1106M	0220
2SC1106P	0220
2SC1106PQ	0220
2SC1106Q	0220
2SC1107	0236
2SC1107Q	0236
2SC1107YG	0236
2SC1108	0236
2SC1109	0236
2SC1110	0236
2SC1111	0867
2SC1112	0867
2SC1113	0429
2SC1114	2125
2SC1114-O	OEM
2SC1115	0177
2SC1116	0543
2SC1116-0	0543
2SC1116-O	0543
2SC1116A	0543
2SC1117	0007
2SC1117A	0007
2SC1117B	0007
2SC1117C	0007
2SC1117D	0007
2SC1117E	0007
2SC1117F	0007
2SC1117G	0007
2SC1117GN	0007
2SC1117H	0007
2SC1117J	0007
2SC1117K	0007
2SC1117L	0007
2SC1117M	0007
2SC1117OR	0007
2SC1117R	0007
2SC1117X	0007
2SC1117Y	0007
2SC1122	1420
2SC1122A	1420
2SC1123	0079
2SC1123A	0079
2SC1123B	0079
2SC1123C	0079
2SC1123D	0079
2SC1123E	0079
2SC1123F	0079
2SC1123GN	0079
2SC1123H	0079
2SC1123J	0079
2SC1123K	0079
2SC1123L	0079
2SC1123M	0079
2SC1123OR	0079
2SC1123R	0079
2SC1123X	0079
2SC1123Y	0079
2SC1124	0264
2SC1124(13)	0264
2SC1124-2	0575
2SC1124-OR	0264
2SC1124A	0264
2SC1124B	0264
2SC1124C	0264
2SC1124D	0264
2SC1124E	0264
2SC1124F	0264
2SC1124G	0264
2SC1124GN	0264
2SC1124H	0264
2SC1124J	0264
2SC1124K	0264
2SC1124L	0264
2SC1124M	0264
2SC1124OR	0264
2SC1124R	0264
2SC1124X	0264
2SC1124Y	0264
2SC1126	0224
2SC1126A	0224
2SC1126B	0224
2SC1126E	0224
2SC1126F	0224
2SC1126GN	0224
2SC1126J	0224
2SC1126K	0224
2SC1126L	0224
2SC1126M	0224
2SC1126OR	0224
2SC1126R	0224
2SC1126X	0224
2SC1126Y	0224
2SC1127	1068
2SC1127-1	1068
2SC1127-2	1068
2SC1127-O	1068
2SC1127-OR	1068
2SC1127A	1068
2SC1127B	1068
2SC1127C	1068
2SC1127D	1068
2SC1127E	1068
2SC1127F	1068
2SC1127G	1068
2SC1127GA	1068
2SC1127GN	1068
2SC1127H	1068
2SC1127J	1068
2SC1127JR	1068
2SC1127K	1068
2SC1127L	1068
2SC1127M	1068
2SC1127OR	1068
2SC1127R	1068
2SC1127X	1068
2SC1127Y	1068
2SC1128	0224
2SC1128(M)	0224
2SC1128(S)	0224
2SC1128-O	0224
2SC1128A	0224
2SC1128B	0224
2SC1128BL	0224
2SC1128C	0224
2SC1128D	0224
2SC1128G	0224
2SC1128H	0224
2SC1128M	0224
2SC1128R	0224
2SC1128S	0224
2SC1128Y	0224
2SC1129	0224
2SC1129(M)	0224
2SC1129(R)	0224
2SC1129-0	0224
2SC1129A	0224
2SC1129B	0224
2SC1129BL	0224
2SC1129C	0224
2SC1129G	0224
2SC1129M	0224
2SC1129R	0224
2SC1129Y	0224
2SC1130	0003
2SC1131	0642
2SC1132	0065
2SC1133	0065
2SC1133D	0065
2SC1138	2146
2SC1140	1852
2SC1141	1852
2SC1142	0363
2SC1143	0363
2SC1150	0086
2SC1151	0003
2SC1151A	0003
2SC1152	0074
2SC1152(F)	0074
2SC1152F	0074
2SC1152G	0074
2SC1153	1150
2SC1153A	1150
2SC1154	0223
2SC1155	0264
2SC1156	0264
2SC1156A	0264
2SC1156B	0264
2SC1156C	0264
2SC1156D	0264
2SC1156E	0264
2SC1156F	0264
2SC1156G	0264
2SC1156GN	0264
2SC1156H	0264
2SC1156K	0264
2SC1156L	0264
2SC1156M	0264
2SC1156OR	0264
2SC1156R	0264
2SC1156X	0264
2SC1156Y	0264
2SC1157	0264
2SC1157A	0264
2SC1157B	0264
2SC1157C	0264
2SC1157E	0264
2SC1157F	0264
2SC1157G	0264
2SC1157GN	0264
2SC1157H	0264
2SC1157J	0264
2SC1157K	0264
2SC1157L	0264
2SC1157M	0264
2SC1157OR	0264
2SC1157R	0264
2SC1157X	0264
2SC1157Y	0264
2SC1158	0475
2SC1159	0007
2SC1160	0178
2SC1160(L)	0178
2SC1160-O	0178
2SC1160K	0178
2SC1160L	0178
2SC1161	0178
2SC1162	0563
2SC1162(C)	0563
2SC1162A	0563
2SC1162B	0563
2SC1162C	0563
2SC1162CP	0563
2SC1162D	0563
2SC1162W	0563
2SC1162W(C)	0563
2SC1162W(D)	0563
2SC1162WBP	0563
2SC1162WT	0563
2SC1162WTB	0563
2SC1162WTC	0563
2SC1163	0276
2SC1164	0418
2SC1164-O	0418
2SC1164-R	0418
2SC1165	0689
2SC1166	0733
2SC1166(O)	0733
2SC1166-GR	0733
2SC1166-O	0733
2SC1166-R	0733
2SC1166-Y	0733
2SC1166A	0733
2SC1166B	0733
2SC1166C	0733
2SC1166E	0733
2SC1166F	0733
2SC1166G	0733
2SC1166GN	0733
2SC1166H	0733
2SC1166K	0733
2SC1166L	0733
2SC1166M	0733
2SC1166O	0733
2SC1166OR	0733
2SC1166U	0733
2SC1166Y	0733
2SC1167	0313
2SC1168	0142
2SC1168X	0142
2SC1169	0630
2SC1170	1150
2SC1170(B)	1150
2SC1170-OR	1150
2SC1170A	1150
2SC1170B	1150
2SC1170B(FA-2)	OEM
2SC1170BFA-2	1150
2SC1170C	1150
2SC1170D	1150
2SC1170E	1150
2SC1170F	1150
2SC1170FA-2	1150
2SC1170G	1150
2SC1170GN	1150
2SC1170H	1150
2SC1170J	1150
2SC1170K	1150
2SC1170L	1150
2SC1170M	1150
2SC1170OR	1150
2SC1170R	1150
2SC1170X	1150
2SC1170Y	1150
2SC1171	0065
2SC1172	0313
2SC1172A	0313
2SC1172B	0313
2SC1172C	0313
2SC1172D	0313
2SC1172E	0313
2SC1172F	0313
2SC1172G	0313
2SC1172GN	0313
2SC1172H	0313
2SC1172J	0313
2SC1172K	0313
2SC1172L	0313
2SC1172M	0313
2SC1172R	0313
2SC1172X	0313
2SC1172Y	0313
2SC1173	0042
2SC1173(O)	0042
2SC1173(X)	0042
2SC1173-0	0042
2SC1173-B	0042
2SC1173-C	0042
2SC1173-GR	0042
2SC1173-O	0042
2SC1173-R	0042
2SC1173-Y	0042
2SC1173A	0042
2SC1173B	0042
2SC1173GR	0042
2SC1173R	0042
2SC1173X	0042
2SC1173X(Y)	0042
2SC1173XO	0042
2SC1173XY	0042
2SC1173Y	0042
2SC1173YX	OEM
2SC1174	0313
2SC1174-OR	0313
2SC1174A	0313
2SC1174B	0313
2SC1174C	0313
2SC1174D	0313
2SC1174E	0313
2SC1174F	0313
2SC1174G	0313
2SC1174GN	0313
2SC1174H	0313
2SC1174J	0313
2SC1174K	0313
2SC1174L	0313
2SC1174M	0313
2SC1174OR	0313
2SC1174R	0313
2SC1174X	0313
2SC1174Y	0313
2SC1175	0191
2SC1175(C)	0191
2SC1175(D,E,F)	0191
2SC1175C	0191
2SC1175CTV	0191
2SC1175CTV-D	OEM
2SC1175CTV-E	OEM
2SC1175CTV-F	OEM
2SC1175D	0191
2SC1175E	0191
2SC1175E,D,F	0191
2SC1175F	0191
2SC1175F(H)	0191
2SC1176	2170
2SC1177	1198
2SC1178	OEM
2SC1178A	1977
2SC1180	0007
2SC1181	0047
2SC1182	0007
2SC1182B	0007
2SC1182D	0007
2SC1184	0003
2SC1184A	0003
2SC1184B	0003
2SC1184C	0003
2SC1184D	0003
2SC1184E	0003
2SC1185	0074
2SC1185A	0074
2SC1185B	0074
2SC1185C	0074
2SC1185K	0074
2SC1185L	0074
2SC1185M	0074
2SC1187	0224
2SC1188	0224
2SC1188-O	0224
2SC1189	0224
2SC1189L	0224
2SC1190	1233
2SC1195	0103
2SC1195(FA-1)	0103
2SC1195A	0103
2SC1195B	0103
2SC1195C	0103
2SC1195D	0103
2SC1195E	0103
2SC1195F	0103
2SC1195FA-1	0103
2SC1195FA-5	0103
2SC1195G	0103
2SC1195GN	0103
2SC1195GR	0103
2SC1195H	0103
2SC1195J	0103
2SC1195K	0103
2SC1195L	0103
2SC1195M	0103
2SC1195O	0103
2SC1195OR	0103
2SC1195R	0103
2SC1195X	0103
2SC1195Y	0103
2SC1196A	OEM
2SC1197A	OEM
2SC1199	0418
2SC1204	0155
2SC1204B	6035
2SC1204C	0155
2SC1204D	0155
2SC1205	2183
2SC1205A	2183
2SC1205B	2183
2SC1205C	2183
2SC1206	0086
2SC1206A	0086
2SC1206B	0086
2SC1207	OEM
2SC1207A	OEM
2SC1207B	OEM
2SC1208	OEM
2SC1209	0076
2SC1209(C)	0076
2SC1209(D)	0076
2SC1209C	0076
2SC1209D	0076
2SC1209E	0076
2SC1210	0951
2SC1210Y	0951
2SC1211	0155
2SC1212	0563
2SC1212A	0563
2SC1212AB	0563
2SC1212ABWT	0563
2SC1212AC	0563
2SC1212ACWT	0563
2SC1212B	0563
2SC1212C	0563
2SC1213	0191
2SC1213(A)	0191
2SC1213(B)	0191
2SC1213(C)	0191
2SC1213(D)	0191
2SC1213-OR	0191
2SC1213A	0191
2SC1213A(C)	0191
2SC1213AA	0191
2SC1213AB	0191
2SC1213A-C	0191
2SC1213AC	0191
2SC1213AD	0191
2SC1213AK	0191
2SC1213AKB	0191
2SC1213AKC	0191
2SC1213AKD	0191
2SC1213B	0191
2SC1213BC	0191
2SC1213C	0191
2SC1213C21TZ	3131
2SC1213CD	0191
2SC1213D	0191
2SC1213D21TZ	3131
2SC1213D-24	0191
2SC1213E	0191
2SC1213F	0191
2SC1213G	0191
2SC1213GN	0191
2SC1213H	0191
2SC1213J	0191
2SC1213K	0191
2SC1213L	0191
2SC1213M	0191
2SC1213OR	0191
2SC1213Q	0191
2SC1213X	0191
2SC1213Y	0191
2SC1214	0155
2SC1214(B)	0155
2SC1214A	0155
2SC1214B	0155
2SC1214C	0155
2SC1214D	0155
2SC1215	0127
2SC1215(R)	0127
2SC1215C	0127
2SC1215D	0127
2SC1215E	0127
2SC1215F	0127
2SC1215G	0127
2SC1215GN	0127
2SC1215H	0127
2SC1215J	0127
2SC1215K	0127
2SC1215L	0127
2SC1215M	0127
2SC1215OR	0127
2SC1215R	0127
2SC1215X	0127
2SC1215Y	0127
2SC1216	0224
2SC1217	0285
2SC1218	0086
2SC1219	0155
2SC1220	0155
2SC1220(E)	0155
2SC1220-003	0155
2SC1220A	0155
2SC1220A(QPR)	0155
2SC1220AP	0155
2SC1220AQ	0155
2SC1220AR	0155
2SC1220E	0155
2SC1220P	0155
2SC1220Q	0155
2SC1220R	0155
2SC1222	2190
2SC1222(E)	2190
2SC1222A	2190
2SC1222B	2190
2SC1222C	2190
2SC1222D	2190
2SC1222E	2190
2SC1222H	2190
2SC1222U	2190
2SC1223	0418
2SC1224	0338
2SC1226	0219
2SC1226(A)	0219
2SC1226(AP)	0219
2SC1226(P)	0219
2SC1226(Q)	0219
2SC1226(R)	0219
2SC1226-O	0219
2SC1226A	0219
2SC1226A(P)	0219
2SC1226A(Q)	0219
2SC1226A(QPR)	0219
2SC1226A(R)	0219
2SC1226AC	0219
2SC1226ACF	0219
2SC1226AF	0219
2SC1226AP	0219
2SC1226AQ	0219
2SC1226AR	0219
2SC1226ARL	0219
2SC1226ARL(P)	0219
2SC1226ARL(Q)	0219
2SC1226ARL(R)	0219
2SC1226B	0219
2SC1226BL	0219
2SC1226C	0219
2SC1226CF	0219
2SC1226D	0219
2SC1226E	0219
2SC1226F	0219
2SC1226G	0219
2SC1226H	0219
2SC1226L	0219
2SC1226OR	0219
2SC1226P	0219
2SC1226Q	0219
2SC1226R	0219
2SC1226RL	0219
2SC1226RL(P)	0219
2SC1226RL(Q)	0219
2SC1226RL(QR)	0219
2SC1226RL(R)	0219
2SC1226RL/P	0219
2SC1226RLP	0219
2SC1226RLQ	0219
2SC1226RLR	0219
2SC1226Y	0219
2SC1227	0074

If replacement code is OEM, contact original manufacturer for replacement.

DEVICE TYPE	REPL CODE
2SC1228	1966
2SC1229	0168
2SC1229Q	0168
2SC1229R	0168
2SC1230	1966
2SC1231	0861
2SC1232	OEM
2SC1233	OEM
2SC1235	0142
2SC1235A	0142
2SC1235AL	0142
2SC1235AM	0142
2SC1235G	0142
2SC1236	0224
2SC1237	0936
2SC1237E	0936
2SC1239	1908
2SC1240	0224
2SC1241	1198
2SC1241A	1198
2SC1243	2016
2SC1243-24	2016
2SC1243C1	2016
2SC1243C2	2016
2SC1243D1	2016
2SC1243D2	2016
2SC1244	0079
2SC1246	0155
2SC1246A	0155
2SC1246R	0155
2SC1246S	0155
2SC1246T	0155
2SC1247	0155
2SC1247A	0155
2SC1247AF	0155
2SC1252	2208
2SC1253	0418
2SC1254	0007
2SC1256	2212
2SC1257	2170
2SC1258	1233
2SC1259	1974
2SC1260	0259
2SC1266	0086
2SC1267	2170
2SC1268	1165
2SC1271A	OEM
2SC1273	OEM
2SC1274	0016
2SC1275	0259
2SC1276	0111
2SC1277	0324
2SC1278	0861
2SC1278S	0861
2SC1279	0861
2SC1279S	0861
2SC1280	0400
2SC1280A	0400
2SC1280AS	0400
2SC1280S	0400
2SC1281	0079
2SC1282	0324
2SC1283	0079
2SC1284	0127
2SC1285	0155
2SC1290	2229
2SC1292	0642
2SC1293	1066
2SC1293(A)	1066
2SC1293A	1066
2SC1293B	1066
2SC1293C	1066
2SC1293D	1066
2SC1295	1150
2SC1295(O)	1150
2SC1295-O	1150
2SC1295-0	1150
2SC1296	0065
2SC1298	1974
2SC1300	0065
2SC1302	OEM
2SC1303	1865
2SC1304	1504
2SC1306	0839
2SC1306A	0839
2SC1306I	0839
2SC1306K	0839
2SC-1307	0836
2SC1307	0836
2SC1307-1	0836
2SC1308	0313
2SC1308(K)	0313
2SC1308K	0313
2SC1308L	0313
2SC1308N	0313
2SC1309	0065
2SC1310	0286
2SC1311	0286
2SC1312	1221
2SC1312A	1221
2SC1312BC	1221
2SC1312C	1221
2SC1312D	1221
2SC1312E	1221
2SC1312F	1221
2SC1312G	1221
2SC1312GN	1221
2SC1312H	1221
2SC1312J	1221
2SC1312K	1221
2SC1312L	1221
2SC1312M	1221
2SC1312OR	1221
2SC1312R	1221
2SC1312X	1221
2SC1312Y	1221
2SC1313	0111
2SC1313(G)	0111
2SC1313B	0111
2SC1313Y	0111
2SC1314	1974
2SC1315	0233
2SC1316	2245
2SC1316-3	2245
2SC1316T	OEM
2SC1317	0155
2SC1317(P)	0155
2SC1317(Q)	0155
2SC1317(R)	0155
2SC1317(S)	0155
2SC1317-OR	0155
2SC1317-Q	0155
2SC1317-R	0155
2SC1317-S	0155
2SC1317-T	0155
2SC1317A	0155
2SC1317BC	0155
2SC1317C	0155
2SC1317E	0155
2SC1317G	0155
2SC1317GR	0155
2SC1317L	0155
2SC1317NC	0155
2SC1317NC(Q)	0155
2SC1317OR	0155
2SC1317Q	0155
2SC1317R	0155
2SC1317S	0155
2SC1317T	0155
2SC1317V	0155
2SC1317Y	0155
2SC1318	0155
2SC1318(P)	0155
2SC1318(P,R)	0155
2SC1318(Q)	0155
2SC1318(R)	0155
2SC1318(S)	0155
2SC1318-R	0155
2SC1318-S	0155
2SC1318A	0155
2SC1318B	0155
2SC1318C	0155
2SC1318E	0155
2SC1318F	0155
2SC1318G	0155
2SC1318GN	0155
2SC1318H	0155
2SC1318J	0155
2SC1318K	0155
2SC1318L	0155
2SC1318M	0155
2SC1318P	0155
2SC1318PR	0155
2SC1318Q	0155
2SC1318QP	0155
2SC1318QR	0155
2SC1318R	0155
2SC1318S	0155
2SC1318S,R	0155
2SC1318Y	0155
2SC1319	0007
2SC1319C	0007
2SC1320	0224
2SC1320(K)	0224
2SC1320A	0224
2SC1320B	0224
2SC1320C	0224
2SC1320D	0224
2SC1320E	0224
2SC1320F	0224
2SC1320G	0224
2SC1320GN	0224
2SC1320H	0224
2SC1320J	0224
2SC1320K	0224
2SC1320L	0224
2SC1320M	0224
2SC1320OR	0224
2SC1320R	0224
2SC1320X	0224
2SC1320Y	0224
2SC1321	0007
2SC1321Q2	0007
2SC1321Q3	0007
2SC1321Q4	0007
2SC1321Q5	0007
2SC1322	0177
2SC1324	0418
2SC1324(C)	0418
2SC1324C	0418
2SC1325	0891
2SC1325A	0891
2SC1325AK	0891
2SC1325AL	0891
2SC1326	0493
2SC1327	1221
2SC1327FS	1221
2SC1327R	1221
2SC1327S	1221
2SC1327T	1221
2SC1327TU	1221
2SC1327TV	1221
2SC1327U	1221
2SC1328	0111
2SC1328(U)	0111
2SC1328(U)(T)	0111
2SC1328T	0111
2SC1328U	0111
2SC1330	0218
2SC1331	0007
2SC1335	1221
2SC1335(B)	1221
2SC1335(C)	1221
2SC1335(D)	1221
2SC1335(E)	1221
2SC1335-OR	1221
2SC1335A	1221
2SC1335B	1221
2SC1335C	1221
2SC1335D	1221
2SC1335E	1221
2SC1335F	1221
2SC1335G	1221
2SC1335GN	1221
2SC1335H	1221
2SC1335J	1221
2SC1335K	1221
2SC1335L	1221
2SC1335M	1221
2SC1335OR	1221
2SC1335R	1221
2SC1335X	1221
2SC1335Y	1221
2SC1336	0144
2SC1336JK	0144
2SC1337	1420
2SC1337A	1420
2SC1338A	OEM
2SC1339	OEM
2SC1340	2170
2SC1342	0127
2SC1342(A)	0127
2SC1342(B)	0127
2SC1342(C)	0127
2SC1342-OR	0127
2SC1342A	0127
2SC1342B	0127
2SC1342C	0127
2SC1342D	0127
2SC1342E	0127
2SC1342F	0127
2SC1342G	0127
2SC1342GN	0127
2SC1342J	0127
2SC1342K	0127
2SC1342L	0127
2SC1342M	0127
2SC1342OR	0127
2SC1342R	0127
2SC1342X	0127
2SC1342Y	0127
2SC1343	0171
2SC1343-05	0171
2SC1343B	0171
2SC1343BL	0171
2SC1343D	0171
2SC1343E	0171
2SC1343F	0171
2SC1343G	0171
2SC1343GN	0171
2SC1343G-R	0171
2SC1343GR	0171
2SC1343H	0171
2SC1343J	0171
2SC1343K	0171
2SC1343L	0171
2SC1343M	0171
2SC1343O	0171
2SC1343OR	0171
2SC1343R	0171
2SC1343X	0171
2SC1343Y	0171
2SC1344	0552
2SC1344(E)	0552
2SC1344C	0552
2SC1344D	0552
2SC1344E	0552
2SC1344F	0552
2SC1345	0552
2SC1345(E)	0552
2SC1345C	0552
2SC1345D	0552
2SC1345E	0552
2SC1345F	0552
2SC1345KD	0552
2SC1345KE	0552
2SC1345KF	0552
2SC1346	0218
2SC1346(Q)	0218
2SC1346(R)	0218
2SC1346R	0218
2SC1346S	0218
2SC1347	0218
2SC1347(Q)	0218
2SC1347(R)	0218
2SC1347(S)	0218
2SC1347A	0218
2SC1347B	0218
2SC1347C	0218
2SC1347D	0218
2SC1347F	0218
2SC1347G	0218
2SC1347L	0218
2SC1347Q	0218
2SC1347R	0218
2SC1347RQ	0218
2SC1347X	0218
2SC1347Y	0218
2SC1348	0223
2SC1348-1	0223
2SC1348-2	0223
2SC1348-3	0223
2SC1349	0861
2SC1350	OEM
2SC1351	0630
2SC1352	0079
2SC1353	0630
2SC1355	OEM
2SC1356	OEM
2SC1358	2051
2SC1358A	2051
2SC1358K	2051
2SC1358K1	2051
2SC1358K2	2051
2SC1358K3	2051
2SC1358L	2051
2SC1358M	2051
2SC1359	0151
2SC1359(A)	0151
2SC1359(B)	0151
2SC1359(C)	0151
2SC1359(C,B)	0151
2SC1359-C	0151
2SC1359A	0151
2SC1359B	0151
2SC1359BC	0151
2SC1359C	0151
2SC1359Q	0151
2SC1360	1663
2SC1360-C	6723
2SC1360A	1663
2SC1360ANC	1663
2SC1361	0155
2SC1362	0155
2SC1362(47)	0155
2SC1363	0076
2SC1364	0076
2SC1364-6	0076
2SC1364-7	0076
2SC1364-8	0076
2SC1364-OR	0076
2SC1364A	0076
2SC1364B	0076
2SC1364C	0076
2SC1364D	0076
2SC1364E	0076
2SC1364H	0076
2SC1364K	0076
2SC1364L	0076
2SC1364M	0076
2SC1364OR	0076
2SC1364R	0076
2SC1364X	0076
2SC1364Y	0076
2SC1365	0418
2SC1366	0418
2SC1367	0003
2SC1367A	0003
2SC1368	1790
2SC1368(B)	1790
2SC1368B	1790
2SC1368C	1790
2SC1368D	1790
2SC1372	0155
2SC1372Y	0155
2SC1373	0111
2SC1374	2272
2SC1374H	2272
2SC1375	0111
2SC1375H	0111
2SC1376	0079
2SC1376H	0079
2SC1377	0836
2SC1380	0016
2SC1380-BL	0016
2SC1380-GR	0016
2SC1380A	0016
2SC1380A-BL	0016
2SC1380A-GR	0016
2SC1381	1591
2SC1382	0563
2SC1382-0	0563
2SC1382-O	0563
2SC1382-Y	0563
2SC1383	0018
2SC1383(P,Q,R)	0018
2SC1383(R)	0018
2SC1383(S)	0018
2SC1383-P	0018
2SC1383-Q	0018
2SC1383-QR	0018
2SC1383-R	0018
2SC1383,RS	0018
2SC1383NC	0018
2SC1383NC(R)	0018
2SC1383NC-P	0018
2SC1383NCP	0018
2SC1383NCPQR	0018
2SC1383NC-Q	0018
2SC1383NCQ	0018
2SC1383NC-R	0018
2SC1383NCR	0018
2SC1383NOPQR	0018
2SC1383P	0018
2SC1383PQR	0018
2SC1383Q	0018
2SC1383QR	0018
2SC1383R	0018
2SC1383RS	0018
2SC1383S	0018
2SC1383X	0018
2SC1384	0018
2SC1384(Q)	0018
2SC1384-OR	0018
2SC1384A	0018
2SC1384B	0018
2SC1384C	0018
2SC1384D	0018
2SC1384E	0018
2SC1384F	0018
2SC1384G	0018
2SC1384GN	0018
2SC1384H	0018
2SC1384J	0018
2SC1384K	0018
2SC1384L	0018
2SC1384M	0018
2SC1384NC	0018
2SC1384NC-R	OEM
2SC1384OR	0018
2SC1384Q	0018
2SC1384Q,R	0018
2SC1384R	0018
2SC1384RS	0018
2SC1384S	0018
2SC1384X	0018
2SC1384Y	0018
2SC1385	0016
2SC1385H	0016
2SC1386	0086
2SC1386H	0086
2SC1387	0418
2SC1388	0086
2SC1390	0597
2SC1390(L)	0597
2SC1390(L,Y)	0597
2SC1390(V)	0597
2SC1390(W)	0597
2SC1390(X)	0597
2SC1390(Y)	0597
2SC1390I(W)	0597
2SC1390IW	0597
2SC1390J(X)	0597
2SC1390JX	0597
2SC1390W	0597
2SC1390WH	0597
2SC1390WX	0597
2SC1390X	0597
2SC1390XK	0597
2SC1390YM	0597
2SC1391	0142
2SC1391VL	0142
2SC1393	0224
2SC1393K	0224
2SC1393L	0224
2SC1393M	0224
2SC1394	0224
2SC1395	0007
2SC1396	0007
2SC1398	0042
2SC1398(Q)	0042
2SC1398A	0042
2SC1398P	0042
2SC1398Q	0042
2SC1399	0563
2SC1399E	0563
2SC1400	0530
2SC1402	0177
2SC1403	0543
2SC1403A	0543
2SC1404	OEM
2SC1406	2287
2SC1406(P)	2287
2SC1406Q	2287
2SC1407	2287
2SC1407(Q)	2287
2SC1407B	2287
2SC1407Q	2287
2SC1407X	2287
2SC1409	0392
2SC1409(B)	0392
2SC1409(C)	0392
2SC1409A	0392
2SC1409A(B)	0392
2SC1409AA	0392
2SC1409AB	0392
2SC1409AC	0392
2SC1409B	0392
2SC1409C	0392
2SC1410	0392
2SC1410A	0392
2SC1410AA	0392
2SC1410AB	0392
2SC1410AC	0392
2SC1410B	0392
2SC1410C	0392
2SC1411	0016
2SC1412	0861
2SC1413	0065
2SC1413A	0065
2SC1413AH	0065
2SC1413B	0065
2SC1413C	0065
2SC1413D	0065
2SC1413E	0065
2SC1413F	0065
2SC1413G	0065
2SC1413GN	0065
2SC1413K	0065
2SC1413L	0065
2SC1413R	0065
2SC1413X	0065
2SC1413Y	0065
2SC1414	OEM
2SC1415	1085
2SC1415-05	1085
2SC1416	0016
2SC1416A	0016
2SC1417	0127
2SC1417(G)	0127
2SC1417(V)	0127
2SC1417(V,G)	0127
2SC1417(W)	0127
2SC1417D(V)	0127
2SC1417F	0127
2SC1417V	0127
2SC1417VF	0127
2SC1417VW	0127
2SC1417W	0127
2SC1418	0042
2SC1419	2292
2SC1419-C	2292
2SC1419B	2292
2SC1419C	2292
2SC1424	0259
2SC1426	0630
2SC1427	0224
2SC1428	0111
2SC1429	0551
2SC1429-1	0551
2SC1429-2	0551
2SC1430	0622
2SC1431	1027
2SC1431-1	1027
2SC1431-2	1027
2SC1432	OEM
2SC1433	0642
2SC1434	1852
2SC1436	1852
2SC1437-NC	0261
2SC1438	0079
2SC1439	0079
2SC1440	1852
2SC1441	1852
2SC1444	0423
2SC1445	0423
2SC1446	0643
2SC1446(LA)(P)	0643
2SC1446(LA)(Q)	0643
2SC1446(LB)(P)	0643
2SC1446(LB)(Q)	0643
2SC1446(P)	0643
2SC1446(Q)	0643
2SC1446-O	0643
2SC1446LB	0643
2SC1446LB(Q)	0643
2SC1446LB(R)	0643
2SC1446LBQ	0643
2SC1446LBR	0643
2SC1446P	0643
2SC1446Q	0643
2SC1447	0643
2SC1447(R)	0643
2SC1447-0	0643
2SC1447-O	0643
2SC1447FA-2	0643
2SC1447GL	0643
2SC1447GL3	0643
2SC1448	1284
2SC1448A	1284
2SC1448A(FA-1)	OEM
2SC1448C	1284
2SC1448LB	1284
2SC1449	0161
2SC1449(K)	0161
2SC1449(L)	0161
2SC1449CB	0161
2SC1449M	0161
2SC1450	1027
2SC1450S	1027
2SC1451	0715
2SC1452	0715
2SC1453	0111
2SC1454	0220
2SC1454-0	0220
2SC1454-O	0220
2SC1456	0142
2SC1456(L)	0142
2SC1456(M)	0142
2SC1457	0861
2SC1459-E	OEM
2SC1463	0642
2SC1464	0630
2SC1465	OEM
2SC1466	0065
2SC1467	0065
2SC1468	1966
2SC1469	0363
2SC1469A	0363
2SC1471	0286
2SC1472	0400
2SC1472K	0400
2SC1472KA	0400
2SC1472KB	0400
2SC1473	0261
2SC1473(NC)	0261
2SC1473(Q)	0261
2SC1473(R)	0261
2SC1473-Q	0261
2SC1473-QNC	0261
2SC1473-QRNC	0261
2SC1473-R	0261
2SC1473-RNC	0261
2SC1473A	0261
2SC1473A(Q)	0261
2SC1473A(R)	0261
2SC1473AE	0261
2SC1473AH	0261
2SC1473AQ	0261
2SC1473AQH	0261
2SC1473AR	0261
2SC1473C	0261
2SC1473CN	0261
2SC1473N	0261
2SC1473NC	0261
2SC1473NC(P)	0261
2SC1473NC(Q)	0261
2SC1473NC(R)	0261
2SC1473NC-P	0261
2SC1473NC-Q	0261
2SC1473NCQ	0261
2SC1473NCQR	0261
2SC1473NC-R	0261
2SC1473NCR	0261
2SC1473NE	0261
2SC1473P	0261
2SC1473Q	0261
2SC1473QNC	0261
2SC1473QR	0261
2SC1473QRNC	0261
2SC1473R	0261
2SC1473RNC	0261
2SC1474	0018
2SC1474-3	0018
2SC1474-4	0018
2SC1474J	0018
2SC1474S	0018
2SC1475	1978
2SC1475(13)	1978
2SC1475-13	1978
2SC1475A	1978
2SC1475D	1978
2SC1475H	1978
2SC1475K	1978
2SC1477	0363
2SC1478	0224
2SC1478A	0224
2SC1479	0418
2SC1480	OEM
2SC1481	OEM
2SC1482	2335
2SC1485	0715
2SC1486	OEM
2SC1487	OEM
2SC1488	OEM
2SC1489	OEM
2SC1490	2170
2SC1491	1198
2SC1492	1977
2SC1493	0715
2SC1493NC	0715
2SC1493P4	0715
2SC1494	1974
2SC1495	OEM
2SC1498	OEM
2SC1499	OEM
2SC1500	OEM
2SC1501	0276
2SC1501(Q)	0276
2SC1501(R)	0276
2SC1501P	0276
2SC1501Q	0276
2SC1501R	0276
2SC1502L	0168
2SC1504	2096
2SC1504E	OEM
2SC1505	0955
2SC1505(1)	0955
2SC1505(1)K	0955
2SC1505(1)L	0955
2SC1505(I)	0955
2SC1505(K)	0955
2SC1505-1	0955
2SC1505A	0955
2SC1505H(LB)	0955
2SC1505I	0955
2SC1505K	0955
2SC1505L	0955
2SC1505LA	0955
2SC1505M	6256
2SC1506	0168
2SC1506K	0168
2SC1506L	0168

DEVICE TYPE	REPL CODE
2SC1506M	0168
2SC1507	0955
2SC1507(1)K	OEM
2SC1507(1)L	OEM
2SC1507(1)M	OEM
2SC1507(2)	OEM
2SC1507(K)	0955
2SC1507(L)	0955
2SC1507-1	OEM
2SC1507-K	0955
2SC1507-L	0955
2SC1507-O	0955
2SC1507-Y	0955
2SC1507H	0955
2SC1507J	0955
2SC1507K	0955
2SC1507K,L,M	0955
2SC1507L	0955
2SC1507LK	0955
2SC1507M	0955
2SC1507Y	0955
2SC1509	0324
2SC1509-O	0324
2SC1509NC	0324
2SC1509O	0324
2SC1509Q	0324
2SC1509R	0324
2SC1509Y	0324
2SC1511	OEM
2SC1512	OEM
2SC1513	0418
2SC1514	1085
2SC1514-05	1085
2SC1514-14	1085
2SC1514-15	1085
2SC1514BK	1085
2SC1514BVC	1085
2SC1514CS	1085
2SC1514CS-CVC	1085
2SC1514CVC	1085
2SC1514VC	1085
2SC1515	0715
2SC1515AX	0715
2SC1515C	0715
2SC1515K	0715
2SC1516	1946
2SC1516K	1946
2SC1516K-B	1946
2SC1516KC	1946
2SC1517	1708
2SC1517A	1708
2SC1517AK	1708
2SC1517AKB	1708
2SC1517AKC	1708
2SC1518	0018
2SC1518NC	0018
2SC1519	0168
2SC1520	0168
2SC1520-1	0168
2SC1520C	0168
2SC1520I	0168
2SC1520K	0168
2SC1520K-1	0168
2SC1520L	0168
2SC1520L-1	0168
2SC1520M	0168
2SC1520M-1	0168
2SC1521	2359
2SC1521K	2359
2SC1521L	2359
2SC1522	1085
2SC1523	OEM
2SC1524	OEM
2SC1525	OEM
2SC1526	OEM
2SC1529	0259
2SC1533	OEM
2SC1534	OEM
2SC1535	OEM
2SC1536	OEM
2SC1537	0111
2SC1537(S)	0111
2SC1537-0	0111
2SC1537-O	0111
2SC1537B	0111
2SC1538	0111
2SC1538S	0111
2SC1538S(A)	0111
2SC1538SA	0111
2SC1539	0079
2SC1540	0079
2SC1541	0155
2SC1542	0111
2SC1543	0224
2SC1544	0079
2SC1545	0400
2SC1545M	0400
2SC1546	0400
2SC1547	0007
2SC1549	0861
2SC1550	0276
2SC1550(K)	0276
2SC1550(L)	0276
2SC1550(M)	0276
2SC1550-1	0276
2SC1552	2382
2SC1553	0007
2SC1553A	0007
2SC1556	0698

DEVICE TYPE	REPL CODE
2SC1561	OEM
2SC1562	OEM
2SC1563	OEM
2SC1564	OEM
2SC1565	0563
2SC1565A	OEM
2SC1566	0276
2SC1566F	0276
2SC1567	0563
2SC1567A	0563
2SC1567A-Q	OEM
2SC1567R	0563
2SC1568	1790
2SC1568(R)	1790
2SC1568-R	1790
2SC1568-S	1790
2SC1568R	1790
2SC1568S	1790
2SC1569	0643
2SC1569(FA-2)	OEM
2SC1569(FA-5)	0643
2SC1569(K)	0643
2SC1569-0	0643
2SC1569-O	0643
2SC1569BK	0643
2SC1569FA-1	0643
2SC1569FA-3	0643
2SC1569FA-5	0643
2SC1569FA5	0643
2SC1569K	0643
2SC1569L	0643
2SC1569LB	0643
2SC1569LB-O	OEM
2SC1569LBO	0643
2SC1569LB-R	OEM
2SC1569LBR	0643
2SC1569LB-Y	OEM
2SC1569LBY	0643
2SC1569O	0643
2SC1569R	0643
2SC1569Y	0643
2SC1570	0111
2SC1570F	0507
2SC1570FL	0111
2SC1570G	0507
2SC1570GL	0111
2SC1570HL	0111
2SC1570LH	0111
2SC1571	0111
2SC1571F	0507
2SC1571G	0111
2SC1571H	0111
2SC1571L	0111
2SC1571L-F	OEM
2SC1571L-G	OEM
2SC1571L-H	OEM
2SC1572	OEM
2SC1573	0261
2SC1573(P)	0261
2SC1573(Q)	0261
2SC1573(TO126)	0261
2SC1573-O	0261
2SC1573A	0261
2SC1573A(Q)	0261
2SC1573A(R)	0261
2SC1573AH	0261
2SC1573AQ	0261
2SC1573AQH	0261
2SC1573AQR	0261
2SC1573AR	0261
2SC1573B	0261
2SC1573B(R)	0261
2SC1573BR	0261
2SC1573NC	0261
2SC1573NC(Q)	0261
2SC1573NC-Q	0261
2SC1573NC-R	0261
2SC1573P	0261
2SC1573P(TO126)	0261
2SC1573PQ	0261
2SC1573PQ(TO126)	0261
2SC1573Q	0261
2SC1573Q(TO126)	0261
2SC1573QNC	0261
2SC1573QR	0261
2SC1573R	0261
2SC1574	0103
2SC1576	0363
2SC1577	0363
2SC1577A	0363
2SC1578	0363
2SC1579	0363
2SC1580	0363
2SC1581	2312
2SC1582	2312
2SC1583	2411
2SC1583F	2411
2SC1583G	2411
2SC1583H	2411
2SC1584	0543
2SC1585	2414
2SC1585F	2414
2SC1585F,H	2414
2SC1585H	2414
2SC1586	2414
2SC1587	OEM
2SC1588	0418
2SC1589	2050
2SC1592	OEM

DEVICE TYPE	REPL CODE
2SC1593	OEM
2SC1594	OEM
2SC1595	OEM
2SC1596	0079
2SC1597	0861
2SC1600	OEM
2SC1601	0224
2SC1602	0079
2SC1603	OEM
2SC1604	OEM
2SC1605	1977
2SC1605A	1977
2SC1606	0560
2SC1607	0330
2SC1608	0630
2SC1609	2433
2SC1610	0620
2SC1613	0715
2SC1613A	0715
2SC1614	0715
2SC1615	0715
2SC1617	2437
2SC1618	2438
2SC1618B	2438
2SC1619	0867
2SC1619A	0867
2SC1620-0	OEM
2SC1620-Y	OEM
2SC1621	0144
2SC1621-0	0144
2SC1621B2	0144
2SC1621B3	0144
2SC1621B4	0144
2SC1622	0007
2SC1622A	0007
2SC1622D6	0007
2SC1622D7	0007
2SC1622D8	0007
2SC1622O	0007
2SC1622Y	0007
2SC1623	0724
2SC1623(L6)	0724
2SC1623-L5L6	OEM
2SC1623-L6	0724
2SC1623L	OEM
2SC1623L3	0724
2SC1623L4	0724
2SC1623L5	0724
2SC1623L5L6	0724
2SC1623L6	0724
2SC1623L7	0724
2SC1623L51E	0724
2SC1624	0236
2SC1624-0	0236
2SC1624-Y	0236
2SC1625	0236
2SC1625-0	0236
2SC1625-O	0236
2SC1625-Y	0236
2SC1625Y	0236
2SC1625YLBGH	0236
2SC1625YLBGL	0236
2SC1625YLBGL1	0236
2SC1626	0236
2SC1626-0	0236
2SC1626-Y	0236
2SC1626O	0236
2SC1626Y	0236
2SC1627	0733
2SC1627(Y)	0733
2SC1627-0	0733
2SC1627-Y	0733
2SC1627-Y(FA)	0733
2SC1627-YFA	0733
2SC1627A	0733
2SC1627AO	0733
2SC1627A-Y	0733
2SC1627AY	0733
2SC1627Y	0733
2SC1627Y(FA)	0733
2SC1628	0264
2SC1628-0	0264
2SC1628-Y	0264
2SC1628-Y.JA	0264
2SC1629	0130
2SC1629A	0130
2SC1629A(O)	0130
2SC1629A(R)	0130
2SC1629AO	0130
2SC1629AR	0130
2SC1629M	0130
2SC1629R	0130
2SC1630	0233
2SC1631	0018
2SC1632	0286
2SC1633	2456
2SC1634	2459
2SC1634-7	0076
2SC1635	0086
2SC1636	2460
2SC1636(22)	2460
2SC1637	0111
2SC1638	0418
2SC1639	0016
2SC1640	2465
2SC1641	0016
2SC1641Q	0016
2SC1641R	0016
2SC1642	0111

DEVICE TYPE	REPL CODE
2SC1643	0155
2SC1644	0155
2SC1645	0400
2SC1645B	0400
2SC1646	0400
2SC1646A	0400
2SC1646B	0400
2SC1647	0111
2SC1647Q	0111
2SC1647RY	0111
2SC1648	0111
2SC1648E	0111
2SC1648EF	0111
2SC1648SH	0111
2SC1649	0715
2SC1650	0715
2SC1651	0715
2SC-1652	2471
2SC1652	2471
2SC1652(Q)	2471
2SC1652(R)	2471
2SC1652-O	2471
2SC1652-Q	2471
2SC1652M	2471
2SC1652Q	5920
2SC1652QY	2471
2SC1652R	2471
2SC1652RE	2471
2SC1652RQ	2471
2SC1652S	2471
2SC1653	0861
2SC1653N2	0861
2SC1653N3	0861
2SC1653N4	0861
2SC1654	0861
2SC1654N5	0861
2SC1654N6	0861
2SC1654N7	0861
2SC1655	OEM
2SC1655A	OEM
2SC1657	OEM
2SC1658	OEM
2SC1658(C)	0582
2SC1658(R)	0582
2SC1659	OEM
2SC1661	OEM
2SC1662	OEM
2SC1663	0264
2SC1663H	0264
2SC1664	0629
2SC1664A	0629
2SC1664O	0629
2SC1664OR	0629
2SC1664Q	0629
2SC1664R	0629
2SC1665	0259
2SC1666	0363
2SC1667	0103
2SC1668	1977
2SC1669	0392
2SC1669-O	0392
2SC1669-R	0392
2SC1669-Y	0392
2SC1670	2483
2SC1670H	2483
2SC1670J	2483
2SC1670JA	2483
2SC1670JD	2483
2SC1670JW	2483
2SC1672	2485
2SC1673	OEM
2SC1674	1066
2SC1674(L)	1066
2SC1674(M)	1066
2SC1674-M	1066
2SC1674AL	1066
2SC1674K	1066
2SC1674L	1066
2SC1674M	1066
2SC1675	0076
2SC1675B	0076
2SC1675K	0076
2SC1675L	0076
2SC1675L1	OEM
2SC1675L1TA	OEM
2SC1675L-FM	OEM
2SC1675M	0076
2SC1676	1974
2SC1678	2495
2SC1678E	2495
2SC1679	0836
2SC1680	OEM
2SC1681	1757
2SC1681-BL	1757
2SC1681-GR	1757
2SC1681A	1757
2SC1681B	1757
2SC1681B6	1757
2SC1681BL	1757
2SC1681C	1757
2SC1681F	1757
2SC1681G	1757
2SC1681GN	1757
2SC1681GR	1757
2SC1681H	1757
2SC1681J	1757
2SC1681K	1757
2SC1681L	1757
2SC1681M	1757

DEVICE TYPE	REPL CODE
2SC1681O	1757
2SC1681OR	1757
2SC1681R	1757
2SC1681V	1757
2SC1682	0111
2SC1682-BL	0111
2SC1682-GR	0111
2SC1682V	0111
2SC1683	0392
2SC1683(LB)	0392
2SC1683(LB)P	0392
2SC1683(LB)Q	0392
2SC1683-0	0392
2SC1683-O	0392
2SC1683-Q	0392
2SC1683A	0392
2SC1683LA	0392
2SC1683LB	0392
2SC1683LD	0392
2SC1683LDP	0392
2SC1683LDQ	0392
2SC1683P	0392
2SC1683Q	0392
2SC1683Q(LB)	0392
2SC1683R	0392
2SC1684	0155
2SC1684(Q)	0155
2SC1684(QR)	0155
2SC1684(Q.R)	0155
2SC1684(R)	0155
2SC1684(S)	0155
2SC1684(T)	0155
2SC1684-P	0155
2SC1684-Q	0155
2SC1684-Q.R	0155
2SC1684-R	0155
2SC1684-S	0155
2SC1684A(Q)	0155
2SC1684A(R)	0155
2SC1684A(S)	0155
2SC1684BL	0155
2SC1684H(R)	0155
2SC1684HE	0155
2SC1684K	0155
2SC1684P	0155
2SC1684Q	0155
2SC1684QR	0155
2SC1684QRS	0155
2SC1684R	0155
2SC1684RSTTA	OEM
2SC1684S	0155
2SC1684T	0155
2SC1685	0286
2SC1685(C)	0286
2SC1685(Q)	0286
2SC1685(R)	0286
2SC1685(S)	0286
2SC1685-O	0286
2SC1685-P	0286
2SC1685-PQR	0286
2SC1685-Q	0286
2SC1685-QR	0286
2SC1685-QRS	0286
2SC1685-R	0286
2SC1685-RS	0286
2SC1685-S	0286
2SC1685A(Q)	0286
2SC1685A(R)	0286
2SC1685A(S)	0286
2SC1685AQR	OEM
2SC1685AQRS	0383
2SC1685C	0286
2SC1685CR	0286
2SC1685CR-Q	0286
2SC1685NCQRS	OEM
2SC1685P	0286
2SC1685PQ	0286
2SC1685PQR	0286
2SC1685PQR(M)	0286
2SC1685PR	0286
2SC1685Q	0286
2SC1685QL	0286
2SC1685QR	0286
2SC1685QRS	0286
2SC1685R	0286
2SC1685RS	0286
2SC1685RST	0286
2SC1685S	0286
2SC1685ST	0286
2SC1685T	0286
2SC1685TAQR	0286
2SC1685TAQRS	0286
2SC1685Y	0383
2SC1686	0224
2SC1686B	0224
2SC1686V	0224
2SC1687	0224
2SC1688	0016
2SC1688D	OEM
2SC1689	2506
2SC1698	OEM
2SC1703	OEM
2SC1706	0715
2SC1706H	0715
2SC1707	0016
2SC1707A	0016
2SC1707AH	0016
2SC1707AHB	0016
2SC1707H	0016

DEVICE TYPE	REPL CODE
2SC1707HB	0016
2SC1707HC	0016
2SC1707HD	0016
2SC1708	0530
2SC1708A	0530
2SC1709	0079
2SC1711	OEM
2SC1711A	OEM
2SC1717	0630
2SC1718	OEM
2SC1719	0861
2SC1720	0861
2SC1721	0018
2SC1722	1085
2SC1722BK	1085
2SC1722BKS	1085
2SC1722S	1085
2SC1723	1085
2SC1723-02	1085
2SC1723A	1085
2SC1724	0630
2SC1725	OEM
2SC1726	OEM
2SC1727	2525
2SC1728	1173
2SC1728-3	1173
2SC1728D	1173
2SC1728F	1173
2SC1728H	1173
2SC1729	2526
2SC1730	0127
2SC1730K	0127
2SC1730L	0127
2SC1731	OEM
2SC1732	OEM
2SC1733	OEM
2SC1734	0016
2SC1734H	0016
2SC1735	0286
2SC1736	0111
2SC1737	0111
2SC1738	0111
2SC1739	0016
2SC1740	0151
2SC1740(Q)	0151
2SC1740(Q)-L	0151
2SC1740(Q)-Y	0151
2SC1740(Q)Y	0151
2SC1740(QR)-T	0151
2SC1740(R)	0151
2SC1740(R)-L	0151
2SC1740(R)-Y	0151
2SC1740(R)Y	0151
2SC1740(S)	0151
2SC1740-E	OEM
2SC1740-Q	0151
2SC1740-R	0151
2SC1740-S	0151
2SC1740-T-93-R	0151
2SC1740-T-93-S	0151
2SC1740-Y	0151
2SC1740C	0151
2SC1740D	0151
2SC1740EF	0151
2SC1740FF	0151
2SC1740HF	0151
2SC1740JF	0151
2SC1740L	0151
2SC1740LN	0151
2SC1740LNCS	0151
2SC1740LN-E	OEM
2SC1740LNE	0151
2SC1740LN-R	OEM
2SC1740LN-S	OEM
2SC1740LNS	0151
2SC1740LNT	OEM
2SC1740LN-Z	OEM
2SC1740P	0151
2SC1740Q	0151
2SC1740QH	0151
2SC1740QJ	0151
2SC1740QN	0151
2SC1740QP	0151
2SC1740QQ	0151
2SC1740QR	0151
2SC1740QU	0151
2SC1740QV	0151
2SC1740QW	0151
2SC1740QZ	0151
2SC1740R	0151
2SC1740RE	0151
2SC1740RH	0151
2SC1740RL	0151
2SC1740RM	0151
2SC1740RQ	0151
2SC1740RR	0151
2SC1740RS	0151
2SC1740RS(RS)	0151
2SC1740RS(S)	0151
2SC1740RS-TB	0151
2SC1740RU	0151
2SC1740RV	0151
2SC1740RZ	0151
2SC1740S	0151
2SC1740S(Q)	0151
2SC1740S(Q)Y	0151
2SC1740S(QR)	0151
2SC1740S(QR)-T	0151
2SC1740S(QR)-Y	0151

DEVICE TYPE	REPL CODE
2SC1740S(QRS)	0151
2SC1740S(R)	0151
2SC1740S(R)-T	0151
2SC1740S(R)Y	0151
2SC1740S(RS)	0151
2SC1740S(S)	0151
2SC1740S(S35)	0151
2SC1740S(SR)-T	2814
2SC1740S,B40	0151
2SC1740S-E	0151
2SC1740SE	0151
2SC1740SED	0151
2SC1740SEE	0151
2SC1740SEH	0151
2SC1740SER	0151
2SC1740SES	0151
2SC1740SEY	0151
2SC1740SLNST	OEM
2SC1740S-Q	0151
2SC1740SQ	0151
2SC1740SQA	0151
2SC1740S-QF	0151
2SC1740SQR	0151
2SC1740SQT	0151
2SC1740SQY	0151
2SC1740S-R	0151
2SC1740SR	0151
2SC1740SRC	0151
2SC1740SRF	0151
2SC1740SRJ	0151
2SC1740SRK	0151
2SC1740SRM	0151
2SC1740SRP	0151
2SC1740SRQ	0151
2SC1740SRR	0151
2SC1740S-RS	0151
2SC1740SRS	0151
2SC1740S-RSF	OEM
2SC1740SRT	0151
2SC1740SRY	0151
2SC1740S-S	0151
2SC1740S-SF	OEM
2SC1740SSJ	0151
2SC1740SSK	0151
2SC1740SSN	0151
2SC1740SSS	0151
2SC1740SST	0151
2SC1740SSU	0151
2SC1740SSX	0151
2SC1740SSZ	0151
2SC1740ST-93	0151
2SC1740TPQ	0151
2SC1740TPR	0151
2SC1740TPS	0151
2SC1740XTPQ	0151
2SC1740XTPQR	0151
2SC1740XTPQS	0151
2SC1740XTPR	0151
2SC1740XTPS	0151
2SC1740Z	OEM
2SC1741	1513
2SC1741A	1513
2SC1741AP	1513
2SC1741AQ	1513
2SC1741P	1513
2SC1741Q	1513
2SC1741QR	1513
2SC1741R	1513
2SC1741S	1513
2SC1741SQ	1513
2SC1741SR	1513
2SC1742	OEM
2SC1743	0261
2SC1743NCR	0261
2SC1744	OEM
2SC1745	OEM
2SC1746	OEM
2SC1746A	0079
2SC1747	2525
2SC1748	0187
2SC1748Q	0187
2SC1748R	0187
2SC1749	1085
2SC1749XD	1085
2SC1752	0715
2SC1753	0715
2SC1755	1085
2SC1755C	1085
2SC1755D	1085
2SC1755E	1085
2SC1756	0643
2SC1756(C)	0643
2SC1756(D)	0643
2SC1756(E)	0643
2SC1756-0	0643
2SC1756-F	0643
2SC1756C	0643
2SC1756C,D,E	0643
2SC1756D	0643
2SC1756E	0643
2SC1756HP	0643
2SC1756K	0643
2SC1756M	0643
2SC1756M-C	OEM
2SC1756M-D	OEM
2SC1756M-E	OEM
2SC1757	1085
2SC1758	0285
2SC1760	1946

DEVICE TYPE	REPL CODE
2SC1760-2	1946
2SC1760-3	1946
2SC1760H	1946
2SC1760K	1946
2SC1761	1946
2SC1761KJ	1946
2SC1761LE	1946
2SC1761MC	1946
2SC1762-1	OEM
2SC1762-2	OEM
2SC1763	2544
2SC1764	2545
2SC1765	0630
2SC1766	0380
2SC1766B	0380
2SC1768	2548
2SC1769	2550
2SC1770	0079
2SC1771	0016
2SC1772	0111
2SC1773	2556
2SC1774	0079
2SC1775	0530
2SC1775A	0530
2SC1775AE	0530
2SC1775E	0530
2SC1775F	0530
2SC1776	0079
2SC1777	2556
2SC1778	0224
2SC1779	0224
2SC1780	0086
2SC1781	0016
2SC1781H	0016
2SC1781HA	0016
2SC1781HB	0016
2SC1781HC	0016
2SC1782	0867
2SC1783	0867
2SC1784	0543
2SC1785	2414
2SC1786	2414
2SC1787	1270
2SC1788	0155
2SC1788A	0155
2SC1788R	0155
2SC1789	0007
2SC1790	0259
2SC1790JD	0259
2SC1791	0142
2SC1792	OEM
2SC1797	OEM
2SC1798	OEM
2SC1799	OEM
2SC1800	OEM
2SC1801	0259
2SC1802	0224
2SC1803	OEM
2SC1804	OEM
2SC1805	OEM
2SC1806	OEM
2SC1807	0493
2SC1808	OEM
2SC1809	0127
2SC1810	2598
2SC1810HU	2598
2SC1810J	2598
2SC1810JL	2598
2SC1811	0261
2SC1811-1	0261
2SC1811-2	0261
2SC1811KC	0261
2SC1812	OEM
2SC1812-P	OEM
2SC1813	0016
2SC-1815	0076
2SC1815	0076
2SC1815(FA)	OEM
2SC1815(G)	0076
2SC1815(G)-T	0076
2SC1815(GR)	0076
2SC1815(GR)L	0076
2SC1815(GR)-Y	0076
2SC1815(GR)Y	0076
2SC1815(O)	0076
2SC1815(Y)	0076
2SC1815(Y)L	0076
2SC1815(Y)-T	4366
2SC1815(Y)Y	0076
2SC1815(Y,GR)Y	0076
2SC1815(YG)	0076
2SC1815(YG)-T	0076
2SC1815-0	0076
2SC1815-BL	0076
2SC1815-BL(T)	0076
2SC1815-GR	0076
2SC1815-GR(FA)	0076
2SC1815-GRFA	0076
2SC1815-O	0076
2SC1815-OFA	0076
2SC1815-Q	2814
2SC1815-Y	0076
2SC1815-Y(FA)	0076
2SC1815-YFA	0076
2SC1815BL	0076
2SC1815BL(T)	1419
2SC1815FA	0076
2SC1815G	0076
2SC1815GR	0076
2SC1815GR(FA-2)	OEM
2SC1815N(GR)	0076
2SC1815N-BL	0076
2SC1815NEW-BL	0076
2SC1815NEW-GR	0076
2SC1815NEW-O	0076
2SC1815NEW-Y	0076
2SC1815N-GR	0076
2SC1815N-Y	0076
2SC1815O	0076
2SC1815Q	1419
2SC1815TPE2	0076
2SC-1815Y	0076
2SC1815Y	0076
2SC1815Y(FA)	0076
2SC1815Y(GR)	0076
2SC1816	2495
2SC1816H	2495
2SC1816HL	2495
2SC1817	0836
2SC1818	0177
2SC1819	0643
2SC1819M	0643
2SC1819MIS	0643
2SC1819ML	0643
2SC1819MR	0643
2SC1819RL	0643
2SC1820	0233
2SC1821	OEM
2SC1822	OEM
2SC1823	OEM
2SC1824	OEM
2SC1825	OEM
2SC1826	0423
2SC1826(P)	0423
2SC1826(Q)	0423
2SC1826(R)	0423
2SC1826(Y)	0423
2SC1826-0	0423
2SC1826-G	0423
2SC1826-O	0423
2SC1826-Y	0423
2SC1826D	0423
2SC1826G	0423
2SC1826O	0423
2SC1826P	0423
2SC1826Q	0423
2SC1826R	0423
2SC1826Y	0423
2SC1827	0236
2SC1827(Y)	0236
2SC1828	2096
2SC1828B	2096
2SC1828R	2096
2SC1829	2620
2SC1829(FA-2)	OEM
2SC1829C	2620
2SC1829FA-1	2620
2SC1830	0543
2SC1831	0561
2SC1832	2626
2SC1833	0286
2SC1834	0286
2SC1835	OEM
2SC1836	0936
2SC1837	OEM
2SC1838	OEM
2SC1840	0111
2SC1840E	0111
2SC1840F	0111
2SC1840P	0111
2SC1841	0530
2SC1841(E)	0530
2SC1841E	0530
2SC1841F	0530
2SC1841FE	0530
2SC1841FP	0530
2SC1841P	0530
2SC1841Q	0530
2SC1841R	0530
2SC1841U	0530
2SC1842	0111
2SC1842E	0111
2SC1843	0111
2SC1844	0111
2SC1844E	0111
2SC1844F	0111
2SC1845	0530
2SC1845(E)	0530
2SC1845(U)	0530
2SC1845E	0530
2SC1845EF	0530
2SC1845F	0530
2SC1845Q	0530
2SC1845R	0530
2SC1845S	0530
2SC1845U	0530
2SC1846	0786
2SC1846(Q)	0786
2SC1846(R)	0786
2SC1846-Q	OEM
2SC1846-R	0786
2SC1846-S	0786
2SC1846B	0786
2SC1846P	0786
2SC1846Q	0786
2SC1846Q/R/S	0786
2SC1846QRS	0786
2SC1846R	0786
2SC1846RS	0786
2SC1846RSLB	0786
2SC1846RSRL	0786
2SC1846S	0786
2SC1847	0563
2SC1847(Q)	0563
2SC1847(R)	0563
2SC1847-0	0563
2SC1847A	0563
2SC1847B	0563
2SC1847C	0563
2SC1847D	0563
2SC1847E	0563
2SC1847F	0563
2SC1847G	0563
2SC1847GN	0563
2SC1847LG	0563
2SC1847NV	0563
2SC1847P	0563
2SC1847Q	0563
2SC1847R	0563
2SC1847V	0563
2SC1847V(Q)	0563
2SC1847V(R)	0563
2SC1847VG	0563
2SC1847VQ	0563
2SC1847VR	0563
2SC1847X	0563
2SC1847Y	0563
2SC1848	0219
2SC1848(Q)	0219
2SC1848(R)	0219
2SC1848P	0219
2SC1848Q	0219
2SC1848R	0219
2SC1848V	0219
2SC1849	0079
2SC1850	0079
2SC1851	0079
2SC1852	0079
2SC1853	1220
2SC1853B	1220
2SC1853C	1220
2SC1854	0286
2SC1854C	0286
2SC1854R	0286
2SC1854S	0286
2SC1855	0224
2SC1856	0224
2SC1856-02	0224
2SC1856M	0224
2SC1857	0079
2SC1859	0630
2SC1860	2650
2SC1861	2650
2SC1862	0187
2SC1863	0429
2SC1864	0429
2SC1864A	0429
2SC1865	0728
2SC1866	0620
2SC1867	1966
2SC1867A	1966
2SC1868	1966
2SC1869	0620
2SC1870	2414
2SC1871	1966
2SC1871A	1966
2SC1872	OEM
2SC1873	OEM
2SC1874	OEM
2SC1875	2660
2SC1875(K)	2660
2SC1875K	2660
2SC1875L	2660
2SC1875R	2660
2SC1876	OEM
2SC1879	OEM
2SC1880	OEM
2SC1880K	OEM
2SC1881	1212
2SC1881K	1212
2SC1882	OEM
2SC1883	OEM
2SC1883K	OEM
2SC1884	OEM
2SC1885	0261
2SC1886	0259
2SC1887	0259
2SC1888	0630
2SC-1890	0530
2SC1890	0530
2SC1890(E)	0530
2SC1890A	0530
2SC1890A(D)	0530
2SC1890A(E)	0530
2SC1890A(E)L.P.F.	0530
2SC1890A-D	0530
2SC1890AD	0530
2SC1890A-E	0530
2SC1890AE	0530
2SC1890A-EV	0530
2SC1890A-F	OEM
2SC1890AF	0530
2SC1890AL	0530
2SC1890D	0530
2SC-1890E	0530
2SC1890E	0530
2SC1890F	0530
2SC1890F(E)	0530
2SC1890P	0530
2SC1891	0223
2SC1892	2660
2SC1893	2668
2SC1893-O	2668
2SC1893FA-1	2668
2SC1894	0313
2SC1894(ARV)	0313
2SC1894K	0313
2SC1894N	0313
2SC1895	0313
2SC1896	0313
2SC1897	0086
2SC1898	0008
2SC1899	0008
2SC1900	0530
2SC1901	0418
2SC1902	2041
2SC1903	0563
2SC1904	0563
2SC1905	0955
2SC1905(1)	0955
2SC1905(H)B	0955
2SC1905(H)L	0955
2SC1905(LB)	0955
2SC1905(M)	0955
2SC1905H	0955
2SC1905H(LB)	0955
2SC1905HLB	0955
2SC1905LB	0955
2SC1905M	0955
2SC-1906	0127
2SC1906	0127
2SC1906-T	0127
2SC1906-Y	0127
2SC1906F	0127
2SC1907	0127
2SC1908	2679
2SC1908E	2679
2SC1908H	2679
2SC1909	0936
2SC1909K	0936
2SC1909R	0936
2SC1913	0392
2SC1913A	0392
2SC1913A(Q)	0392
2SC1913A(R)	0392
2SC1913AQ	3113
2SC1913AR	0392
2SC1914	0530
2SC1914A	0530
2SC1915	0530
2SC1915-T	OEM
2SC1916	1978
2SC1917	1978
2SC1918	1978
2SC1919	0127
2SC1919C	0127
2SC1921	0261
2SC1921-03	0261
2SC1922	2668
2SC1923	2691
2SC1923(O)	2691
2SC1923(Y)	2691
2SC1923-0	2691
2SC1923-02	OEM
2SC1923-O	2691
2SC1923A	2691
2SC1923BN	2691
2SC1923R	1497
2SC1923Y	2691
2SC1924	OEM
2SC1925	OEM
2SC1926	OEM
2SC1927	OEM
2SC1928	1978
2SC1929	0168
2SC1929(Q)	0168
2SC1929(R)	0168
2SC1929LB	0168
2SC1929Q	0168
2SC1929R	0168
2SC1940	0079
2SC1941	1327
2SC1941-OY	1327
2SC1941K	1327
2SC1941L	1327
2SC1941M	1327
2SC1942	2660
2SC1942-03	2660
2SC1942-10	2660
2SC1942-10/03	0048
2SC1943	OEM
2SC1944	0836
2SC1945	0836
2SC1946	2526
2SC1946A	2526
2SC1947	2702
2SC1948	OEM
2SC1948-1	OEM
2SC1948-2	OEM
2SC1948-5	OEM
2SC1949	OEM
2SC1950	OEM
2SC1951	1386
2SC1952	OEM
2SC1953	0563
2SC1954	0127
2SC1955	2702
2SC1956	0630
2SC1957	2710
2SC1957K	2710
2SC1958Y	OEM
2SC1959	0286
2SC1959(OYG)	0286
2SC1959(Y)	0286
2SC1959(Y)-T	0286
2SC1959(Y)-Y	0286
2SC1959-O	0286
2SC1959-Y	0286
2SC1959-Y(FA)	0286
2SC1959-YFA	0286
2SC1959GR	0286
2SC1959NEW-Y	0286
2SC1959Y	0286
2SC1959Y(FA)	0286
2SC1960	0286
2SC1961	0224
2SC1962	0338
2SC1962-O	0338
2SC1963	1946
2SC1963-1	1946
2SC1963Y	1946
2SC1964	2495
2SC1964D	2495
2SC1965	2702
2SC1965A	2702
2SC1966	OEM
2SC1967	OEM
2SC1968	OEM
2SC1968A	OEM
2SC1969	0836
2SC1969B	0836
2SC1969BH	0836
2SC1969H	0836
2SC1970	2719
2SC1971	2719
2SC1972	2720
2SC1973	0018
2SC1974	0836
2SC1974C	0836
2SC1975	0936
2SC1976	2725
2SC1977	OEM
2SC1978	2719
2SC1980	1270
2SC1980(S)	1270
2SC1980(T)	1270
2SC1980R	1270
2SC1980S	1270
2SC1980T	1270
2SC1981	OEM
2SC1982	0264
2SC-1983	2058
2SC1983	2058
2SC1983(O)	2058
2SC1983(R)	2058
2SC1983(Y)	2058
2SC1983-O	2058
2SC1983D	2058
2SC1983K	2058
2SC1983O	2058
2SC1983P	2319
2SC1983Q	2058
2SC1983R	2058
2SC1983Y	2058
2SC1984	1212
2SC1985	0423
2SC1986	0482
2SC1987	0429
2SC1989	0007
2SC1990	0127
2SC1990B	0127
2SC1991	0079
2SC1992	0016
2SC1993	0016
2SC1994	0016
2SC1995	0016
2SC1996	0016
2SC1997	0016
2SC1998	0079
2SC1999	0111
2SC2000	0286
2SC2000L	0286
2SC2001	1513
2SC2001(L)	1513
2SC2001-L	1513
2SC2001F	1513
2SC2001G	1513
2SC2001I	1513
2SC2001K	1513
2SC2001KTA	OEM
2SC2001L	1513
2SC2001L1	1513
2SC2001LK	1513
2SC2001M	1513
2SC2001R	1513
2SC-2002	0951
2SC2002	0951
2SC2002(K)	0951
2SC2002(L)	0951
2SC2002(M)	0951
2SC2002-L	0951
2SC2002-M	0951
2SC2002B	0951
2SC2002K	0951
2SC2002L	0951
2SC2002M	0951
2SC2002Y	OEM
2SC2003	0286
2SC2003K	0286
2SC2008	OEM
2SC2009	0224
2SC2010	0224
2SC2011	0224
2SC2011L	0224
2SC2012	2757
2SC2013	0224
2SC2014	1386
2SC2017	1852
2SC2018	1852
2SC2019	1852
2SC2020	0936
2SC-2021	1140
2SC2021	1140
2SC2021-Q	1140
2SC2021-R	1140
2SC2021E	1140
2SC2021F	OEM
2SC2021L	1140
2SC2021LN	1140
2SC2021LR	OEM
2SC2021LS	OEM
2SC2021M	1140
2SC2021M(Q)	1140
2SC2021M(R)	1140
2SC2021M(S)	1140
2SC2021MLN	1140
2SC2021MQR	1140
2SC2021M-QRS	1140
2SC2021MQRS	1140
2SC2021M-R	OEM
2SC2021MR	1140
2SC2021MRS	1140
2SC2021M-S	OEM
2SC2021MS	1140
2SC2021P	1140
2SC2021Q	1140
2SC2021R	1140
2SC2021RJ	1140
2SC2021RS	1140
2SC2021S	1140
2SC2021SE	OEM
2SC2021Y	1140
2SC2022	0955
2SC2023	2613
2SC2024	1591
2SC2025	0259
2SC2026	1806
2SC2026K	1806
2SC2027	0313
2SC2028	0890
2SC2028-B/20	0890
2SC2028B	0890
2SC2028B/20	0890
2SC2028B20	0890
2SC2028D	0890
2SC2028L	0890
2SC2029	0936
2SC2029/1	0936
2SC2029/3	0936
2SC2029-1	0936
2SC2029-3	0936
2SC2029-B/10	0936
2SC2029B	0936
2SC2029B/10	0936
2SC2029B10	0936
2SC2029C	0936
2SC2031	OEM
2SC2032	OEM
2SC2033	OEM
2SC2034	2769
2SC2035	0079
2SC2036	0563
2SC2037	2770
2SC2038	OEM
2SC2039	OEM
2SC2040	2774
2SC2043	0836
2SC2044	OEM
2SC2050	0936
2SC2051	0286
2SC2053	0079
2SC2055	0079
2SC2056	2702
2SC2057	0224
2SC2057-C	0224
2SC2057C	0224
2SC2057D	0224
2SC2057E	0224
2SC2057E1	0224
2SC2057F	0224
2SC2058	0286
2SC2058(P)	0286
2SC2058P-SP	OEM
2SC2058PZ	0286
2SC2058Q	6197
2SC2058Q-SP	OEM
2SC2058QZ	0286
2SC2058S	0286
2SC2058S-P	0286
2SC2058Z	OEM
2SC2059	0724
2SC2059K	0724
2SC2059KN	0724
2SC2060	0018
2SC2060(Q)	0018
2SC2060(R)	0018
2SC2060-O	OEM
2SC2060K	0018
2SC2060Q	0018
2SC2060QC	0018
2SC2060QZ	0018
2SC2060R	0018
2SC2060RD	0018
2SC2060RQ	0018
2SC2060RZ	0018
2SC2060Y	OEM
2SC2061	0324
2SC2061QU	OEM
2SC2062	2797
2SC2063	0582
2SC2063M	0582
2SC2063P	0582
2SC2065	OEM
2SC2066	OEM
2SC2067	1978
2SC-2068	0643
2SC2068	0643
2SC2068(FA-1)	0643
2SC2068(LB)	0643
2SC2068-GS-1	0643
2SC2068-LB	0643
2SC2068-O	OEM
2SC2068BK	0643
2SC2068BKLB	0643
2SC2068BXLB	0643
2SC2068FA-1	0643
2SC2068GS-1	0643
2SC-2068LB	0643
2SC2068LB	0643
2SC2068LBBK	0643
2SC2068R	6256
2SC2068S	6256
2SC2068Y	OEM
2SC2069	0016
2SC2070	0016
2SC2071	0276
2SC2072	OEM
2SC2072-1	OEM
2SC2072-2	OEM
2SC2073	1284
2SC2073(B)	1284
2SC2073(C)	1284
2SC2073(FA-1)	1284
2SC2073(FA-3)	1284
2SC2073-AR(D)	1284
2SC2073-AR(E)	1284
2SC2073-B	1284
2SC2073-C	1284
2SC2073-LBGL2	1284
2SC2073-Z	1284
2SC2073A	1284
2SC2073A-D	OEM
2SC2073AD	1284
2SC2073A-E	OEM
2SC2073AE	1284
2SC2073AR	OEM
2SC2073AR-D	OEM
2SC2073AR-E	OEM
2SC2073B	1284
2SC2073BGS2	1284
2SC2073BGSC	1284
2SC2073BLGL2	1284
2SC2073C	1284
2SC2073D	1284
2SC2073E	1284
2SC2073FA-1	1284
2SC2073FA-4	1284
2SC2073GL2	1284
2SC2073J	1284
2SC2073L	1284
2SC2073LB	1284
2SC2073LBGL	1284
2SC2073LBGL2	1284
2SC2073LBGS2	1284
2SC2073LBL2	1284
2SC2073R	1284
2SC2073R(D)	1284
2SC2073R(E)	1284
2SC2073R-D	OEM
2SC2073RD	1284
2SC2073R-E	OEM
2SC2073RE	1284
2SC2073SO	1284
2SC2073T	1284
2SC2073Y	1284
2SC2073Z	3113
2SC2074	1984
2SC2074C	1984
2SC2074Y	1984
2SC2075	2495
2SC2076	2814
2SC2076B	2814
2SC2076C	2814
2SC2076CB	2814
2SC2076CD	2814
2SC2076D	2814
2SC2078	2495
2SC2079	1591
2SC2079E	1591
2SC2080	0161
2SC2081	2821
2SC2082	OEM
2SC2083	OEM

If replacement code is OEM, contact original manufacturer for replacement.

DEVICE TYPE	REPL CODE
2SC2085	0643
2SC2085P	0643
2SC2085Q	0643
2SC2086	2825
2SC2087	0079
2SC2088	1562
2SC2089	0530
2SC2091	1591
2SC2092	0936
2SC2093	0836
2SC2094	2526
2SC2097	OEM
2SC2098	0836
2SC2099	2835
2SC2100	2312
2SC2101	1198
2SC2102	1233
2SC2103	1974
2SC2103A	1974
2SC2104	OEM
2SC2105	1420
2SC2106	OEM
2SC2107	0079
2SC2107G3	0079
2SC2108	OEM
2SC2109	0079
2SC2110	0086
2SC2111	0016
2SC2113	0563
2SC2114	OEM
2SC2115	2844
2SC2116	OEM
2SC2117	2702
2SC2118	2702
2SC2119	OEM
2SC2120	0866
2SC2120(O)	4993
2SC2120(Y)	0866
2SC2120-0	0866
2SC2120-O	0866
2SC2120-TE	4993
2SC2120-Y	0866
2SC2120FA	0866
2SC-2120Y	0866
2SC2120Y	0866
2SC2120Y-FA	OEM
2SC2120YFA	0866
2SC2121	0363
2SC2122	0363
2SC2122A	0363
2SC2123	0065
2SC2124	2847
2SC2125	2847
2SC2126	0178
2SC2126A	0178
2SC2127	0543
2SC2127A	0543
2SC2128	OEM
2SC2128A	OEM
2SC2129	1978
2SC2130	1978
2SC2131	0187
2SC2132	2852
2SC2133	OEM
2SC2134	OEM
2SC2135	2860
2SC2137	1966
2SC2138	0168
2SC2139	1852
2SC2139A	1852
2SC2140	0074
2SC2141	0264
2SC2141-10	0264
2SC2141KC	0264
2SC2141KD	0264
2SC2141LC	0264
2SC2141M	0264
2SC2141NE	0264
2SC2142	OEM
2SC2143	2844
2SC2144	OEM
2SC2145	0698
2SC2146	0936
2SC2147	OEM
2SC2148	OEM
2SC2149	OEM
2SC2150	OEM
2SC2151	1852
2SC2152	OEM
2SC2153	0079
2SC2159	OEM
2SC2160	OEM
2SC2161	OEM
2SC2162	OEM
2SC2163	OEM
2SC2164	OEM
2SC2165	OEM
2SC2166	2495
2SC2167	1284
2SC2167-Y	1284
2SC2167Y	1284
2SC2168	1284
2SC2168(O)	1284
2SC2168(Y)	1284
2SC2168-O	1284
2SC2168-O-N03	1284
2SC2168-OY	1284
2SC2168-Y	1284
2SC2168-Y-N03	1284
2SC2168Y	1284
2SC2169	OEM
2SC2172	OEM
2SC2173	OEM
2SC2174	OEM
2SC2175	1966
2SC2176	OEM
2SC2177	OEM
2SC2178	0392
2SC2178Y	0392
2SC2179	OEM
2SC2180	1974
2SC2181	2884
2SC2182	2506
2SC2183	OEM
2SC2184	0836
2SC2185	OEM
2SC2186	OEM
2SC2188	2052
2SC2189	0620
2SC2190	0065
2SC2191	0363
2SC2192	OEM
2SC2193	0715
2SC2194	0551
2SC2195	0836
2SC2196	0630
2SC2197	1974
2SC2198	0629
2SC2199	0620
2SC2199(FA-1)	OEM
2SC2200	0728
2SC2204	OEM
2SC2205	OEM
2SC2206	2223
2SC2206(B)	2223
2SC2206(C)	2223
2SC2206(D)	2223
2SC2206(R)	2223
2SC2206(S)	2223
2SC2206-C	2223
2SC2206B	2223
2SC2206BC	2223
2SC2206C	2223
2SC2206CD	2223
2SC2206Q	2223
2SC2206R	2223
2SC2206S	2223
2SC2207	0836
2SC2208	OEM
2SC2209	0161
2SC2209R	0161
2SC2210	0224
2SC2210E	2536
2SC2210F	2536
2SC2212	0144
2SC2212A	0144
2SC2213	0144
2SC2214	OEM
2SC2215	0224
2SC2216	2902
2SC2217	0224
2SC2218	OEM
2SC2219	OEM
2SC2220	OEM
2SC2221	1984
2SC2222	1984
2SC2223	OEM
2SC2223F14	OEM
2SC2224	0338
2SC2224A	0338
2SC2227	OEM
2SC2228	1327
2SC2228(D)	1327
2SC2228(E)	1327
2SC2228A	1327
2SC2228A-C	OEM
2SC2228A-D	OEM
2SC2228A-E	OEM
2SC2228C	1327
2SC2228D	1327
2SC2228E	1327
2SC2228F	1327
2SC2228K	1327
2SC2228K(D)	1327
2SC2228K(E)	1327
2SC2228K-D	OEM
2SC2228K-E	OEM
2SC2228M	1327
2SC2228Y	1327
2SC2228Y(E)	1327
2SC-2229	0066
2SC2229	0066
2SC2229(FA-2)	OEM
2SC2229(O)	0066
2SC2229(Y)	0066
2SC2229-0	0066
2SC2229-C	0066
2SC2229-M	0066
2SC2229-M(SAN-1)	0066
2SC2229-O	0066
2SC2229-O(FA)	0066
2SC2229-O(FA-1)	OEM
2SC2229-OY	0066
2SC2229-Y	0066
2SC2229D	0066
2SC2229E	0066
2SC2229M	0066
2SC22290	0066
2SC2229Y	0066
2SC2229Y-TYPE6	0066
2SC2230	0261
2SC2230(G)	0261
2SC2230(GR)	0261
2SC2230(GR)Y	0261
2SC2230(Y)	0261
2SC2230-A	0261
2SC2230-GR	0261
2SC2230-Y	0261
2SC2230A	0261
2SC2230A(G)	0261
2SC2230A(GR)	0261
2SC2230A(Y)	0261
2SC2230AG	0261
2SC2230A-GR	0261
2SC2230AGR	0261
2SC2230A-GR(C)	0261
2SC2230AP	0261
2SC2230A-Y	0261
2SC2230AY	0261
2SC2230AYG	0261
2SC2230G	0261
2SC2230GR	0261
2SC2230Y	0261
2SC2230YGR	0261
2SC2231	0168
2SC2231A	0168
2SC2231Y	0168
2SC2233	1538
2SC2233LBGL2	1538
2SC2234	1974
2SC2235	1386
2SC2235-O	1386
2SC-2236	2910
2SC2236	2910
2SC2236(D)	2910
2SC2236(O)	2910
2SC2236(Y)	2910
2SC2236-CY	2910
2SC2236-O	2910
2SC2236-OY	2910
2SC2236-Y	2910
2SC2236A	OEM
2SC2236GR	2910
2SC2236Q	2910
2SC2236Y	2910
2SC2237	2526
2SC2238	2911
2SC2238(Y)	3113
2SC2238A	2911
2SC2238B	3113
2SC2238Y	3113
2SC2239	1027
2SC2240	2131
2SC2240BL	2131
2SC2240GL	2131
2SC2240GR	2131
2SC2240NEW-BL	2131
2SC2241	1085
2SC2242	1085
2SC2242BK	1085
2SC2242GL2	1085
2SC2242LB	1085
2SC2242LBGL2	1085
2SC2243	2912
2SC2244	0074
2SC2245	0074
2SC2246	0363
2SC2247	0728
2SC2248	0728
2SC2249	OEM
2SC2250	OEM
2SC2251	OEM
2SC2251-O	0261
2SC2252	OEM
2SC2253	OEM
2SC2254	OEM
2SC2255	OEM
2SC2256	2414
2SC2257	0276
2SC2257A	0276
2SC2258	0276
2SC2258A	0276
2SC2258B	0276
2SC2259	2924
2SC2259C	2924
2SC2259G	2924
2SC2260	0177
2SC2261	0177
2SC2262	0177
2SC2263	0111
2SC2264	1978
2SC2265	0007
2SC2266	2928
2SC2267	0187
2SC2267H	0187
2SC2270	OEM
2SC2270B	OEM
2SC2271	0261
2SC2271(D)	0261
2SC2271(D)-AE	0261
2SC2271(E)	0261
2SC2271(E)-AE	0261
2SC2271(M)	0261
2SC2271(N)	0261
2SC2271-AE	0261
2SC2271-B-CTV	0261
2SC2271-D-CTV	0261
2SC2271-E-CTV	0261
2SC2271-WB	0261
2SC2271C	0261
2SC2271CTV	OEM
2SC2271D	0261
2SC2271E	0261
2SC2271M	0261
2SC2271N	0261
2SC2271Q	0261
2SC2271Z	0261
2SC2272	OEM
2SC2273	OEM
2SC2274	0155
2SC2274(D)	0155
2SC2274(E)	0155
2SC2274(F)	0155
2SC2274-D	0155
2SC2274-DE	0155
2SC2274-E	0155
2SC2274-E-AA	0155
2SC2274-F	0155
2SC2274-F-AA	0155
2SC2274D	2536
2SC2274E	0155
2SC2274F	0155
2SC2274K	0155
2SC2274K-AA-E	0826
2SC2274K-D	OEM
2SC2274KE	0155
2SC2274K-T	0155
2SC2275	1284
2SC2275A	1284
2SC2276	2939
2SC2277	0079
2SC2278	0285
2SC2279	0338
2SC2281	1198
2SC2282	OEM
2SC2283	OEM
2SC2284	OEM
2SC2284A	OEM
2SC2285	OEM
2SC2285A	OEM
2SC2286	OEM
2SC2287	OEM
2SC2287K	OEM
2SC2287M	OEM
2SC2288	OEM
2SC2288K	OEM
2SC2288M	OEM
2SC2289	OEM
2SC2289K	OEM
2SC2289M	OEM
2SC2290	2946
2SC2291	2947
2SC2292	2928
2SC2293	2928
2SC2294	0127
2SC2295	1733
2SC2296	0530
2SC2297	0423
2SC2298	0161
2SC2298A	0161
2SC2298B	0161
2SC2299	1978
2SC2300	0936
2SC2301	OEM
2SC2302	OEM
2SC2303	1852
2SC2304	1852
2SC2305	2185
2SC2306	1852
2SC2307	1506
2SC2308	0155
2SC2308B	0155
2SC2308C	0155
2SC2309	0111
2SC2309(E)	0111
2SC2309(F)	0111
2SC2309D	0111
2SC2309E	0111
2SC2309F	0111
2SC2309L	0111
2SC2310	0286
2SC2311	0161
2SC2312	0836
2SC2313	OEM
2SC2314	1591
2SC2314D	1591
2SC2315	0423
2SC2316	OEM
2SC2317	0392
2SC2318	OEM
2SC2319	OEM
2SC2320	0151
2SC2320GR	OEM
2SC2320L	0151
2SC2321	0620
2SC2322	0620
2SC2323	0620
2SC2324	2970
2SC2324K	2970
2SC2325	OEM
2SC2326	OEM
2SC2327	OEM
2SC2328	OEM
2SC2329	2700
2SC2330	0261
2SC2330-O	0261
2SC2330-Y	OEM
2SC2330A	0261
2SC2330A-GR	0261
2SC2330A-Y	0261
2SC2330GR	OEM
2SC2330Y	0261
2SC2331	0392
2SC2331-Y	1624
2SC2331Y	0392
2SC2333	0261
2SC2334	2268
2SC2334K	2268
2SC2334L	2268
2SC2335	0728
2SC2336	0392
2SC2336A	0392
2SC2336B	0392
2SC2337	0867
2SC2337A	0867
2SC2338	OEM
2SC2339	OEM
2SC2340	0261
2SC2341	OEM
2SC2342	OEM
2SC2344	1284
2SC2344(D)	1284
2SC2344(E)	1284
2SC2344D	1284
2SC2344E	1284
2SC2344RA	OEM
2SC2345	OEM
2SC2347	1591
2SC2348	0224
2SC2349	0127
2SC2350	0042
2SC2351	OEM
2SC2352	0224
2SC2353	2991
2SC2354	0429
2SC2356	1852
2SC2357	2993
2SC2358	2993
2SC2359	0728
2SC2360	OEM
2SC2361	2997
2SC2361A	2997
2SC2362	1227
2SC2362-T	1227
2SC2362G	1227
2SC2362H	1227
2SC2362K	1227
2SC2362K(F)-AA	1227
2SC2362K(G)-AA	1227
2SC2362K-AA	1227
2SC2362KD	1227
2SC2362KE	1227
2SC2362KF	1227
2SC2362K-T	1227
2SC2363	0079
2SC2363G	7000
2SC2364	3001
2SC2365	0642
2SC2366	OEM
2SC2367	OEM
2SC2368	3007
2SC2369	3007
2SC2370	OEM
2SC2371	0276
2SC2371(1)	0276
2SC2371(1)K	0276
2SC2371(1)L	0276
2SC2371(1)LD	0276
2SC2371(1)M	0276
2SC2371(1)MK	0276
2SC2371(8)K	0276
2SC2371(8)L	0276
2SC2371(8)M	0276
2SC2371(K)	0276
2SC2371(L)	0276
2SC2371(N)-L	0276
2SC2371(P)-L	0276
2SC2371K	0276
2SC2371L	0276
2SC2371M	0276
2SC2372(L)	0728
2SC2373	0728
2SC2373(1)K	0728
2SC2373(1)L	0728
2SC2373(1)M	0728
2SC2373(L)	0728
2SC2373K	0728
2SC2373L	0728
2SC2373M	0728
2SC2375	0261
2SC2376	0276
2SC2377	2223
2SC2377(B)	2223
2SC2377(C)	2223
2SC2377(CD)	2223
2SC2377(D)	2223
2SC2377-C	2223
2SC2377C	2223
2SC2377D	2223
2SC2378	1386
2SC2379	OEM
2SC2380	OEM
2SC2381	OEM
2SC2382	1974
2SC-2383	1562
2SC2383	1562
2SC2383(O)	1562
2SC2383(Q)	1562
2SC2383(R)	1562
2SC2383-0	1562
2SC2383-O	1562
2SC2383-O(SAN)	1562
2SC2383-R(SAN)	1562
2SC2383-Y	1562
2SC2383-Y(C)	1562
2SC2383O	1562
2SC2383R	1562
2SC2383Y	1562
2SC2384	0338
2SC2385	0111
2SC2386	1978
2SC2387	2928
2SC2388	1852
2SC2388A	1852
2SC2389	0530
2SC2389Z	OEM
2SC2390	0530
2SC2391	OEM
2SC2392	OEM
2SC2393	0261
2SC2393D	0261
2SC2394	0836
2SC2394D	0836
2SC2395	OEM
2SC2396	0111
2SC2397	OEM
2SC2398	0177
2SC2399	0144
2SC2400	OEM
2SC2401	3025
2SC2402	0543
2SC2403	OEM
2SC2404	OEM
2SC2404(C)	OEM
2SC2404(D)	OEM
2SC2404C	OEM
2SC2405	1733
2SC2406	OEM
2SC2407	OEM
2SC2407-1	OEM
2SC2408	OEM
2SC2409	0111
2SC2410	0008
2SC2411	1733
2SC2412	0724
2SC2412K	0724
2SC2412K(BR)	0724
2SC2412K(QR)-W	OEM
2SC2412KBRST	OEM
2SC2412K-Q	0724
2SC2412KQ	0724
2SC2412K-QR	0724
2SC2412K-R	0724
2SC2412KR	0724
2SC2412KR(BS)	0724
2SC2412KS	0724
2SC2412K-T-146-R	0724
2SC2412KT147	0724
2SC2412Q	0724
2SC2412R	0724
2SC2413	0724
2SC2414	3037
2SC2415	3038
2SC2416	3038
2SC2417	OEM
2SC2419	OEM
2SC2420	3041
2SC2421	2548
2SC2422	2548
2SC2422A	2548
2SC2423	OEM
2SC2424	OEM
2SC2425	0955
2SC2427	0728
2SC2428	0543
2SC2429	1852
2SC2429A	1852
2SC2430	0620
2SC2431	0620
2SC2432	0620
2SC2433	2433
2SC2434	2433
2SC2435	1991
2SC2436	2626
2SC2437	3038
2SC2438	OEM
2SC2439	OEM
2SC2440	0728
2SC2441	0563
2SC2441E	2551
2SC2442	OEM
2SC2443	OEM
2SC2443-24	0551
2SC2444	OEM
2SC2445	OEM
2SC2447	0259
2SC2448	3058
2SC2449	3058
2SC2450	1852
2SC2451	1852
2SC2452	3060
2SC2453	3060
2SC2454	OEM
2SC2455	OEM
2SC2456	0276
2SC2456(LB-SAN-1)	0276
2SC2456LB	0276
2SC2456LB(Z)	0276
2SC2456Z	0276
2SC2458	0076
2SC2458(G)	0076
2SC2458(GR)	0076
2SC2458(LB-SAN-1)	0076
2SC2458(Y)	0076
2SC2458-GR	0076
2SC2458-O	0076
2SC2458BL	0076
2SC2458GR	0076
2SC2458GR-T	OEM
2SC2458O	0076
2SC2458Y	0076
2SC2459	1270
2SC2459BL	1270
2SC2459GR	1270
2SC2460	0177
2SC2460A	0177
2SC2461	0543
2SC2461B	0543
2SC2462	0724
2SC2462(LD)	0724
2SC2462-C	0724
2SC2462B	0724
2SC2462C	0724
2SC2462CD	OEM
2SC2462D	0724
2SC2462L	OEM
2SC2462LD	OEM
2SC2463	0724
2SC2463D	0724
2SC2463E	0724
2SC2463E(DE)	OEM
2SC2463F	0724
2SC2464	OEM
2SC2465	3007
2SC2466	OEM
2SC2467	3007
2SC2468	OEM
2SC2469	OEM
2SC2470	3007
2SC2471	0224
2SC2472	1806
2SC2473	2851
2SC2474	2272
2SC2475	0861
2SC2476	2272
2SC2477	2272
2SC2480	OEM
2SC2481	0563
2SC2481-0	0563
2SC2481-O	0563
2SC2481-OLBGL1	0563
2SC2481-Y	0563
2SC2481R	0563
2SC2481RLBGL1	0563
2SC2481Y	0563
2SC-2482	0261
2SC2482	0261
2SC2482(C)	0261
2SC2482(C1)	0261
2SC2482(C1)-T	0261
2SC2482(FA-1)	0261
2SC2482(Y)-T	0261
2SC2482-1	0261
2SC2482A	0261
2SC2482BK	0261
2SC2482C	OEM
2SC2482FA-1	0261
2SC2482FA1	0261
2SC2482K	0261
2SC2482V	0261
2SC2482Z	0261
2SC2483	0219
2SC2483(O)	0219
2SC2483(R)	0219
2SC2483(Y)	0219
2SC2483-O	0219
2SC2483-R	0219
2SC2483-Y	0219
2SC2483O	0219
2SC2483R	0219
2SC2483Y	0219
2SC2484	3081
2SC2484P	3081
2SC2485	3081
2SC2486	3082
2SC2487	0543
2SC2488	0543
2SC2489	0543
2SC2490	OEM
2SC2491	1165
2SC2492	0543
2SC2493	0543
2SC2494	OEM
2SC2494K	OEM
2SC2494M	OEM
2SC2495	OEM
2SC2495K	OEM
2SC2495M	OEM
2SC2496	OEM
2SC2496A	OEM
2SC2497	0563
2SC2497A	0563
2SC2497Q	0563

If replacement code is OEM, contact original manufacturer for replacement.

DEVICE TYPE	REPL CODE
2SC2497R	0563
2SC2498	1806
2SC2499	2851
2SC2500	3093
2SC2501	0728
2SC2502	0728
2SC2503	OEM
2SC2504	OEM
2SC2505	0728
2SC2506	0728
2SC2507	1852
2SC2508	2884
2SC2509	3101
2SC2510	OEM
2SC2511	0155
2SC2512	0224
2SC2512-03	0224
2SC2512-04	0224
2SC2516	1165
2SC2516A	1165
2SC2516AK	1165
2SC2516AL	1165
2SC2516AM	1165
2SC2516K	1165
2SC2516L	1165
2SC2516M	1165
2SC2517	1165
2SC2518	0728
2SC2519	0127
2SC2522	0543
2SC2522A	0543
2SC2523	0543
2SC2525	2276
2SC2526	2276
2SC2527	1165
2SC2528	3112
2SC2529	3113
2SC2530	3115
2SC2532	OEM
2SC2534	0728
2SC2535	0177
2SC2536	1506
2SC2537	2185
2SC2538	OEM
2SC2539	OEM
2SC2540	3119
2SC2541	2185
2SC2542	0728
2SC2543	3120
2SC2544	3120
2SC2544E	OEM
2SC2545	0111
2SC2545D	0111
2SC2545E	0111
2SC2545F	0111
2SC2545T44	OEM
2SC2546	3120
2SC2546D	OEM
2SC2546E	OEM
2SC2546F	OEM
2SC2547	3120
2SC2547D	OEM
2SC2547E	OEM
2SC2548	1506
2SC2548(FA)	OEM
2SC2549	0224
2SC2550	0016
2SC2551	0261
2SC2551(O)	0261
2SC2551-0	0261
2SC2551-O	0261
2SC25510	0261
2SC2551R	0261
2SC2551RO	0261
2SC2552	0728
2SC2553	0728
2SC-2555	2185
2SC2555	2185
2SC2555-2	2185
2SC2555Z	2185
2SC2556	0563
2SC2556A	0563
2SC2557	0111
2SC2558	OEM
2SC2559	OEM
2SC2561	0127
2SC2562	3125
2SC2563	3082
2SC2564	2276
2SC2565	2276
2SC2566	OEM
2SC2567	OEM
2SC2567A	OEM
2SC2568	0276
2SC2568*1*	0276
2SC2568*1*(K)	0276
2SC2568*1*(L)	0276
2SC2568*1*(M)	0276
2SC2568(1)	0276
2SC2568(1)(K)	0276
2SC2568(1)(L)	0276
2SC2568(1)(M)	0276
2SC2568(1)-K	0276
2SC2568(1)K	0276
2SC2568(1)KD	0276
2SC2568(1)-L	0276
2SC2568(1)L	0276
2SC2568(1)LD	0276
2SC2568(1)-M	0276
2SC2568(1)M	0276
2SC2568(K)	0276
2SC2568(L)	0276
2SC2568(M)	0276
2SC2568-1	0276
2SC2568-1-(K)	0276
2SC2568-1-(L)	0276
2SC2568-1-(M)	0276
2SC2568K	0276
2SC2568KD	0276
2SC2568L	0276
2SC2568LD	0276
2SC2568LM	0276
2SC2568M	0276
2SC2569	0728
2SC2570	2851
2SC2570A	2851
2SC2570AE	2851
2SC2571	OEM
2SC2571-1	OEM
2SC2571-2	OEM
2SC2572	0016
2SC2575	0286
2SC2575L	0286
2SC2577	0194
2SC2578	0194
2SC2579	3082
2SC2580	3082
2SC2581	3082
2SC2582	0563
2SC2584	OEM
2SC2585	OEM
2SC2586	OEM
2SC2587	0046
2SC2587A	0046
2SC2588	0046
2SC2588A	0046
2SC2589	0867
2SC2590	0563
2SC2591	0392
2SC2592	0392
2SC2593	OEM
2SC2594	0161
2SC2594V	0161
2SC2595	OEM
2SC2596	OEM
2SC2597	OEM
2SC2598	OEM
2SC2599	OEM
2SC2600	OEM
2SC2601	OEM
2SC2602	0111
2SC2602G	0507
2SC2603	0518
2SC2603(D)	0518
2SC2603(E)	0518
2SC2603(F)	0518
2SC2603(G)	0518
2SC2603-D	0518
2SC2603-E	0518
2SC2603-F	0518
2SC2603-G	0518
2SC2603E	0518
2SC2603F	0518
2SC2603G	0518
2SC2603R	0518
2SC2603S	0518
2SC2604	0191
2SC2605	0191
2SC2606	0285
2SC2607	2414
2SC2608	2414
2SC2609	OEM
2SC-2610	0261
2SC2610	0261
2SC2610(BK)	0261
2SC2610(BR)	0261
2SC2610(Y)	0261
2SC2610-05	0261
2SC2610-5	0261
2SC2610B	0261
2SC2610BK	0261
2SC2610K	0261
2SC2610KK	0261
2SC2611	0276
2SC2611-06	0276
2SC2611-5	0276
2SC2611BK	OEM
2SC2612	0168
2SC2613	0728
2SC2613K	0728
2SC2614	OEM
2SC2615	OEM
2SC2615K	OEM
2SC2616	2928
2SC2617	2928
2SC2618	1733
2SC2619	0724
2SC2619A	0724
2SC2619B	0724
2SC2619C	0724
2SC2620	OEM
2SC2620A	OEM
2SC2620B	OEM
2SC2620C	OEM
2SC2621	0276
2SC2621(C)	0276
2SC2621(D)	0276
2SC2621(D)-RAC	0276
2SC2621(E)	0276
2SC2621(E)-RAC	0276
2SC2621-C	0276
2SC2621-C-RA	0276
2SC2621-D	0276
2SC2621-D-RA	0276
2SC2621-E	0276
2SC2621-E-RA	0276
2SC2621-RA	0276
2SC2621-RA(C)	0276
2SC2621-RA(D)	0276
2SC2621-RA(E)	0276
2SC2621-RAC	0276
2SC2621-RAD	0276
2SC2621-RAE	0276
2SC2621C	0276
2SC2621D	0276
2SC2621E	0276
2SC2621RA	0276
2SC2621RA(C)	0276
2SC2621RA(D)	0276
2SC2621RA(E)	0276
2SC2621RA-C	OEM
2SC2621RA-D	OEM
2SC2621RA-E	OEM
2SC2621S	OEM
2SC2623	1852
2SC2624	2185
2SC2625	2185
2SC2626	1506
2SC2627	3160
2SC2628	OEM
2SC2629	3164
2SC2630	3119
2SC2631	0261
2SC2631QRS	OEM
2SC2632	0261
2SC2633	1085
2SC2634	0111
2SC2634-R	0111
2SC2634-RS	0111
2SC2634-S	0111
2SC2634NC	0111
2SC2634NC-T	OEM
2SC2634R	0111
2SC2634RS	0111
2SC2634S	0111
2SC2635	OEM
2SC2635H	OEM
2SC2636	0127
2SC2636(R)	0127
2SC2636S	0127
2SC2636T	0127
2SC2637	1085
2SC2637K	1085
2SC2637L	1085
2SC2637Q	1085
2SC2637R	1085
2SC2637RL	1085
2SC2638	OEM
2SC2639	OEM
2SC2640	3170
2SC2641	OEM
2SC2642	OEM
2SC2643	OEM
2SC2644	OEM
2SC2644S	OEM
2SC2645	OEM
2SC2646	0127
2SC2647	0224
2SC2647C	0224
2SC2648	OEM
2SC2649	OEM
2SC2650	OEM
2SC2651	OEM
2SC2652	OEM
2SC2653	0168
2SC2653(H)	0168
2SC2653(H)CL	0168
2SC2653(H)CL(LB)	0168
2SC2653(H)LB	0168
2SC2653(LB)	0168
2SC2653B	0168
2SC2653C	0168
2SC2653CL	0168
2SC2653H	0168
2SC2653H(C)	0168
2SC2653H(L)	0168
2SC2653H(LB)	0168
2SC2653H-CL	0168
2SC2653HCL	0168
2SC2653HCL(LB)	0168
2SC2653HLB	0168
2SC2653H.C	0168
2SC2653H.L	0168
2SC2653L	0168
2SC2653LB	0168
2SC2653LBA	0168
2SC2653Q	0168
2SC2654	3178
2SC2655	0018
2SC2655(Y)	0018
2SC2655(Y)-T	0018
2SC2655(Y)-Y	0018
2SC2655-O	0018
2SC2655-Y	0018
2SC2655-Y(C)	0018
2SC26550	0018
2SC26550YTPE6	0018
2SC2655Y	0018
2SC2655YTPE6	0018
2SC2656	2185
2SC2657	1325
2SC2657A	1325
2SC2658	0276
2SC2658A	0276
2SC2659	0313
2SC2659A	0313
2SC2660	0392
2SC2660A	0392
2SC2662	0622
2SC2664	0622
2SC2665	3081
2SC2666	OEM
2SC2667	OEM
2SC2668	1144
2SC2668-O	1497
2SC2668Y	1497
2SC2669	0155
2SC2669-O	0155
2SC2669-Y	0155
2SC2669Y	1518
2SC2670	0155
2SC2671	2851
2SC2672	OEM
2SC2673	0009
2SC2673-Q	0009
2SC2673-R	0009
2SC2673Q	1566
2SC2673R	1569
2SC2674	1386
2SC2675	0155
2SC2676	0530
2SC2677	0330
2SC2678	0330
2SC2679	0144
2SC2680	OEM
2SC2681	0194
2SC2682	0563
2SC2684	OEM
2SC2685	OEM
2SC2686	OEM
2SC2687	0261
2SC2688	0276
2SC2688(1)-K	OEM
2SC2688(1)-L	OEM
2SC2688(1)-M	0276
2SC2688(2)	0276
2SC2688(K)	0276
2SC2688(L)	0276
2SC2688(M)	0276
2SC2688(N)	0276
2SC2688-K	0276
2SC2688-L	0276
2SC2688-LA	0276
2SC2688-LK	0276
2SC2688-M	0276
2SC2688-N	0276
2SC2688A	0276
2SC2688K	0276
2SC2688KA	0276
2SC2688L	0276
2SC2688LA	0276
2SC2688LK	0276
2SC2688LM	0276
2SC2688M	0276
2SC2688MA	0276
2SC2688N	0276
2SC2689	OEM
2SC2690	0563
2SC2690A	0563
2SC2690A-Q	0563
2SC2691	0155
2SC2692	OEM
2SC2693	OEM
2SC2694	3194
2SC2695	OEM
2SC2700	OEM
2SC2701	OEM
2SC2702	OEM
2SC-2703	1562
2SC2703	1562
2SC2703Y	1562
2SC2704	0563
2SC2705	1562
2SC2705P	OEM
2SC2706	3082
2SC2707	3200
2SC2710	0155
2SC2710(Y)	0155
2SC2710-O	0155
2SC2710-Y	0155
2SC2710Y	0155
2SC2711	OEM
2SC2712	0724
2SC2712(Y)	0724
2SC2712-YG	0724
2SC2712G	0724
2SC2712GR	0724
2SC2712Y	0724
2SC2713	2332
2SC2713-L	5010
2SC2714	OEM
2SC2714-Y	OEM
2SC2715	0724
2SC2716	0724
2SC2717	0313
2SC2717(FA)	OEM
2SC2717FA	0313
2SC2718	0018
2SC2719	0155
2SC2719K	0155
2SC2719L	0155
2SC2720	3203
2SC2721	0018
2SC2722	0046
2SC2723	1506
2SC2724	0191
2SC2724(D)	0191
2SC2724(E)	0191
2SC2724-C	0191
2SC2724-CD	0191
2SC2724-D	0191
2SC2724-E	0191
2SC2724C	0191
2SC2724D	0191
2SC2724E	0191
2SC2724T	0191
2SC2725	OEM
2SC2726	OEM
2SC2727	2844
2SC2728	OEM
2SC2729	2844
2SC2730	OEM
2SC2731	OEM
2SC2732	OEM
2SC2733	OEM
2SC2734	2919
2SC2735	2919
2SC2735J-L	2919
2SC2735J-TL	2919
2SC2736	2919
2SC2737	2851
2SC2738	3209
2SC2739	0728
2SC2740	3211
2SC2741	OEM
2SC2743	OEM
2SC2744	OEM
2SC2745	OEM
2SC2746	OEM
2SC2748	OEM
2SC2749	2185
2SC2749A	2185
2SC2750	0194
2SC2751	1506
2SC2751A	1506
2SC2752	3217
2SC2753	1506
2SC2754	0261
2SC2755	OEM
2SC2756	OEM
2SC2757	OEM
2SC2757R	OEM
2SC2758	OEM
2SC2759	OEM
2SC2760	OEM
2SC2761	OEM
2SC2762	0493
2SC2763	OEM
2SC2766	2276
2SC2766A	2276
2SC2767	0728
2SC2768	0728
2SC2768A	0728
2SC2769	2185
2SC2770	OEM
2SC2771	OEM
2SC2773	3226
2SC2774	3226
2SC2775	2332
2SC2776	1733
2SC2776VA	1733
2SC2776VB	1733
2SC2776VC	1733
2SC2778	OEM
2SC2778(BC)-W	OEM
2SC2780	1562
2SC2781	OEM
2SC2782	OEM
2SC2783	OEM
2SC2784	1270
2SC2784E	1270
2SC2784EU	OEM
2SC2784F	1270
2SC2785	0249
2SC2785(E)	0249
2SC2785(F)	0249
2SC2785(H)	0249
2SC2785(H)-Y	0249
2SC2785(J)	0249
2SC2785(J)-Y	0249
2SC2785(J,H)-Y	0249
2SC2785(JH)-T	0249
2SC2785-E	0249
2SC2785-F	0249
2SC2785-FEK	OEM
2SC2785-HFE	0249
2SC2785-K	0249
2SC2785-Y	0249
2SC2785E	0249
2SC2785EE	0249
2SC2785EF	0249
2SC2785F	0249
2SC2785F(H)	0249
2SC2785FF	0249
2SC2785H	0249
2SC2785J	0249
2SC2785JHFE	0249
2SC2785K	0249
2SC2785TP	0249
2SC2785Y	0249
2SC2786	0286
2SC2786L	0286
2SC2786MTA	OEM
2SC2786SP	OEM
2SC2787	0127
2SC2787L	0127
2SC2788	2185
2SC2789	2185
2SC2789C	0276
2SC2789CD	0276
2SC2790	2441
2SC2790A	2441
2SC2791	3229
2SC2792	3231
2SC2793	OEM
2SC2794	0563
2SC2796	OEM
2SC2797	OEM
2SC2798	OEM
2SC2799	OEM
2SC2800	OEM
2SC2802	OEM
2SC2803	2784
2SC2804	OEM
2SC2805	OEM
2SC2806	OEM
2SC2808	0530
2SC2809	2613
2SC2809-O	2613
2SC28090	2613
2SC2809Q	2613
2SC2809R	OEM
2SC2809V	2613
2SC2810	0728
2SC2811	OEM
2SC2812	0724
2SC2812-L5	0724
2SC2812-L5TA	0724
2SC2812-L5TA(E)	0724
2SC2812-L6	0724
2SC2812-L6TA	0724
2SC2812-L6TA(F)	0724
2SC2812-L7	0724
2SC2812-L7-TA	0724
2SC2812-L7TA	0724
2SC2812-L7TA(G)	0724
2SC2812-TA	0724
2SC2812L6	0724
2SC2812TB	OEM
2SC2813	OEM
2SC2814	0724
2SC2814-F5TA	OEM
2SC2814-TA	0724
2SC2814F5-TA	OEM
2SC2815	0728
2SC2816	0728
2SC2817	OEM
2SC2818	0079
2SC2818H	0079
2SC2819	1852
2SC2819H	1852
2SC2820	1852
2SC2821	2332
2SC2821UD	2332
2SC2821UE	2332
2SC2821UF	2332
2SC2822	0016
2SC2823	OEM
2SC2824	0563
2SC2825	3242
2SC2826	3014
2SC2827	0728
2SC2828	2185
2SC2829	1966
2SC2830	1852
2SC2831	2766
2SC2831A	2766
2SC2832	0168
2SC2832A	0168
2SC2832K	0168
2SC2832L	0168
2SC2833	2784
2SC2833A	2784
2SC2834	3248
2SC2834-AM	2795
2SC2834A	2795
2SC2834AM	2795
2SC2837	0194
2SC2838	0046
2SC2839	0151
2SC2839-D	0151
2SC2839-F-SPA-AC	0151
2SC2839D	0151
2SC2839E	0151
2SC2839EZ	0151
2SC2839F	0151
2SC2839FZ	0151
2SC2840	0127
2SC2840D	0127
2SC2840E	2848
2SC2841	2185
2SC2844	OEM
2SC2845	OEM
2SC2847	OEM
2SC2850	OEM
2SC2851	OEM
2SC2852	0418
2SC2853	1386
2SC2853D	1386
2SC2853E	1386
2SC2854	1386
2SC2854D	1386
2SC2854E	1386
2SC2855	1386
2SC2855D	1386
2SC2855E	1386
2SC2856	1386
2SC2856E	1386
2SC2857	0261
2SC2857D	0261
2SC2857E	0261
2SC2858	OEM
2SC2858K	OEM
2SC2859	1733
2SC2860	1497
2SC2865	1852
2SC2865A	1852
2SC2867	0728
2SC2868	0155
2SC2869	OEM
2SC2870	OEM
2SC2870L	OEM
2SC2871	OEM
2SC2872	0324
2SC2873	OEM
2SC2875	OEM
2SC2876	2844
2SC2877	0161
2SC2878	3094
2SC2878(A)	3094
2SC2878(B)	3094
2SC2878(B)-T	3094
2SC2878(B)-Y	3094
2SC2878-A	3094
2SC2878A	3094
2SC-2878B	3094
2SC2878B	3094
2SC2879	OEM
2SC2880	OEM
2SC2881	OEM
2SC2882	0667
2SC2883	OEM
2SC2884	0127
2SC2885	OEM
2SC2885K	OEM
2SC2885L	OEM
2SC2885M	OEM
2SC2886	OEM
2SC2887	OEM
2SC2888	OEM
2SC2889	OEM
2SC2890	OEM
2SC2891	OEM
2SC2892	OEM
2SC2893	OEM
2SC2894	OEM
2SC2895	OEM
2SC2896	OEM
2SC2897	OEM
2SC2898	0728
2SC2899	3217
2SC2900	2613
2SC2901	3274
2SC2902	1852
2SC2903	1852
2SC2904	OEM
2SC2905	OEM
2SC2906	OEM
2SC2906A	OEM
2SC2907	OEM
2SC2908	3279
2SC2909	0261
2SC2909R	0261
2SC2909S	0261
2SC2909T	OEM
2SC2910	1562
2SC2910S	OEM
2SC2910T	OEM
2SC2911	0563
2SC2911-T	OEM
2SC2911S	0563
2SC2912	0563
2SC2913	0728
2SC2914	3267
2SC2914(Z)	3267
2SC2914Z	3267
2SC2915	OEM
2SC2917	OEM
2SC2918	OEM
2SC2920	1341
2SC2921	3286
2SC2922	3226
2SC2923	0276
2SC2923-RL	0276
2SC2923RL	0276
2SC2924	OEM
2SC2925	0400
2SC2926	0127
2SC2927	2598
2SC2928	3292
2SC2928-01	3292
2SC2929	3014
2SC2930	1852

DEVICE TYPE	REPL CODE
2SC2931	OEM
2SC2932	OEM
2SC2933	OEM
2SC2934	0276
2SC2935	0276
2SC2936	OEM
2SC2937	2185
2SC2938	2185
2SC2939	1506
2SC2940	OEM
2SC2941	OEM
2SC2942	OEM
2SC2943	1506
2SC2944	1506
2SC2946	OEM
2SC2946-1K	OEM
2SC2946-1L	OEM
2SC2946-1M	OEM
2SC2946-1N	OEM
2SC2946K	OEM
2SC2946L	OEM
2SC2946M	OEM
2SC2946N	OEM
2SC2947	0155
2SC2948	OEM
2SC2949	OEM
2SC2950	OEM
2SC2951	OEM
2SC2952	OEM
2SC2953	OEM
2SC2954	OEM
2SC2955	OEM
2SC2956	OEM
2SC2957	OEM
2SC2958	1386
2SC2958-L	OEM
2SC2958K	1386
2SC2959	1562
2SC2960	3302
2SC2960E	3302
2SC2960F	3302
2SC2960G	3302
2SC2961	2185
2SC2962	2784
2SC2963	0060
2SC2964	2928
2SC2965	2928
2SC2966	OEM
2SC2967	0127
2SC2968	0530
2SC2969	OEM
2SC2970	OEM
2SC2971	1341
2SC2972	1341
2SC2973	OEM
2SC2974	OEM
2SC2975	OEM
2SC2976	3060
2SC2977	OEM
2SC2978	OEM
2SC2979	2096
2SC2980	OEM
2SC2981	3060
2SC2982	OEM
2SC2983	OEM
2SC2984	OEM
2SC2985	OEM
2SC2986	0127
2SC2987	0194
2SC2987A	0194
2SC2988	OEM
2SC2989	OEM
2SC2991	OEM
2SC2992	OEM
2SC2995	0530
2SC2996	1733
2SC2997	OEM
2SC2998	0286
2SC2999	0127
2SC2999(E)	0127
2SC2999D	0127
2SC2999E	0127
2SC3000	0286
2SC3000(D)-AA	0286
2SC3000(E)-AA	0286
2SC3000-AA	0286
2SC3000-D	0286
2SC3000-E	0286
2SC3000-F-AA	0286
2SC3000D	0286
2SC3000E	0286
2SC3000EZ	0286
2SC3000F	0286
2SC3001	OEM
2SC3004	OEM
2SC3005	OEM
2SC3006	OEM
2SC3007	3329
2SC3008	OEM
2SC3009	OEM
2SC3010	OEM
2SC3011	OEM
2SC3011MA	OEM
2SC3012	0194
2SC3013	OEM
2SC3014	OEM
2SC3015	2919
2SC3016	2919
2SC3017	0698
2SC3018	OEM
2SC3019	OEM
2SC3020	OEM
2SC3021	OEM
2SC3022	OEM
2SC3023	0223
2SC3024	2847
2SC3025	0065
2SC3026	2847
2SC3027	0065
2SC3028	OEM
2SC3029	2844
2SC3030	1506
2SC3031	3060
2SC3032	1506
2SC3033	3060
2SC3034	3058
2SC3035	0429
2SC3036	2096
2SC3037	2851
2SC3038	3014
2SC3038N	3014
2SC3039	0728
2SC3040	3211
2SC3041	3038
2SC3042	2185
2SC3043	3038
2SC3044	1341
2SC3044A	1341
2SC3045	1341
2SC3046	3338
2SC3047	3339
2SC3048	3060
2SC3049	3060
2SC3050	0223
2SC3051	3217
2SC3052	1733
2SC3052E	1733
2SC3053	1733
2SC3054	3345
2SC3055	0728
2SC3056	0728
2SC3056A	0728
2SC3057	0728
2SC3058	OEM
2SC3058A	OEM
2SC3059	0223
2SC3060	0223
2SC3061	OEM
2SC3062	OEM
2SC3063	0276
2SC3063-RL	0276
2SC3063RL	0276
2SC3064	OEM
2SC3064-E	OEM
2SC3064E	OEM
2SC3065	OEM
2SC3066	OEM
2SC3066-F	OEM
2SC3067	OEM
2SC3068	OEM
2SC3068-AA	OEM
2SC3069	OEM
2SC3070	OEM
2SC3071	OEM
2SC3072	OEM
2SC3073	3350
2SC3073Y	3350
2SC3074	OEM
2SC3075	OEM
2SC3076	3350
2SC3077	OEM
2SC3078	0582
2SC3078M	0582
2SC3079	1497
2SC3079M	1497
2SC3080	2223
2SC3080M	2223
2SC3080MM	2223
2SC3080MN	2223
2SC3080MP	2223
2SC3080MQ	2223
2SC3081	0127
2SC3082	OEM
2SC3082I	OEM
2SC3083	2784
2SC3084	3038
2SC3085	0259
2SC3086	3357
2SC3087	OEM
2SC3088	2784
2SC3089	3360
2SC3090	1506
2SC3091	1852
2SC3092	1852
2SC3093	1852
2SC3094	0259
2SC3095	OEM
2SC3096	OEM
2SC3098	OEM
2SC3099	2919
2SC3101	0698
2SC3102	OEM
2SC3103	OEM
2SC3104	OEM
2SC3105	OEM
2SC3106	OEM
2SC3107	OEM
2SC3108	OEM
2SC3109	OEM
2SC3110	2919
2SC3111	OEM
2SC3112	0111
2SC3112(A)	0111
2SC3112(B)	0111
2SC3112A	0111
2SC3112AB	0111
2SC3113	OEM
2SC3114	0155
2SC3114(S)	0155
2SC3114(T)	0155
2SC3114-R	0155
2SC3114-S	0155
2SC3114-T	0155
2SC3114R	0155
2SC3114S	OEM
2SC3114T	0155
2SC3114U	OEM
2SC3115	2332
2SC3116	0563
2SC3116(T)	0563
2SC3116S	0563
2SC3116ST	0563
2SC3116T	0563
2SC3117	0563
2SC3117-R	OEM
2SC3117-S	OEM
2SC3117R	0563
2SC3117S	0563
2SC3118	OEM
2SC3119	2919
2SC3120	3372
2SC3121	OEM
2SC3122	OEM
2SC3123	OEM
2SC3124	3372
2SC3125	3372
2SC3125HH	OEM
2SC3126	2844
2SC3127	2919
2SC3128	2851
2SC3129	OEM
2SC3130	2919
2SC3131	3375
2SC3132	OEM
2SC3133	OEM
2SC3134	0724
2SC3135	1386
2SC3136	1806
2SC3137	OEM
2SC3138	3379
2SC3139	OEM
2SC3140	OEM
2SC3141	OEM
2SC3142	OEM
2SC3143	2332
2SC3144	1212
2SC3145	1212
2SC3146	2235
2SC3147	OEM
2SC3148	2766
2SC3149	3357
2SC3149L	3357
2SC3150	3357
2SC3151	2784
2SC3152	2784
2SC3153	2784
2SC3153K	2784
2SC3153L	2784
2SC3153M	2784
2SC3154	0223
2SC3155	0223
2SC3156	3385
2SC3157	3060
2SC3157K	3060
2SC3157L	3060
2SC3158	0728
2SC3159	0728
2SC3159L	0728
2SC3160	0728
2SC3161	OEM
2SC3162	3014
2SC3163	0728
2SC3164	3387
2SC3165	3014
2SC3166	0728
2SC3167	3388
2SC3168	0259
2SC3169	3014
2SC3169KT	3014
2SC3170	0728
2SC3170-P	0728
2SC3171	2185
2SC3172	OEM
2SC3173	1165
2SC3174	0728
2SC3174R	0728
2SC3174S	0728
2SC3175	0728
2SC3176	0728
2SC3177	0127
2SC3178	OEM
2SC3179	0060
2SC3179(G)	0060
2SC3179(Y)	0060
2SC3179-O	OEM
2SC3179G	0060
2SC3179R	OEM
2SC3179Y	0060
2SC3180	3082
2SC3180N	OEM
2SC3181	0194
2SC3181-0	OEM
2SC3181N	OEM
2SC3182	0194
2SC3182N	0194
2SC3182NFA-1	0194
2SC3183	3357
2SC3184	3357
2SC3185	2964
2SC3186	OEM
2SC3187	3392
2SC3187V	3392
2SC3189	0728
2SC3190	0191
2SC3191	0191
2SC3192	0127
2SC3193	0127
2SC3194	3396
2SC3195	1497
2SC3196	1497
2SC3197	1066
2SC3198	0286
2SC3198GR	0286
2SC3198Y	0286
2SC3199	0286
2SC3199GR	0286
2SC3200	OEM
2SC3201	OEM
2SC3202	0155
2SC3203	3398
2SC3204	0286
2SC3205	3093
2SC3206	0086
2SC3207	0086
2SC3208	1085
2SC3209	0261
2SC3209K	0261
2SC3209L	0261
2SC3209LK	0261
2SC3209M	0261
2SC3210	2185
2SC3211	2784
2SC3211A	2784
2SC3212	3360
2SC3212A	3360
2SC3213	OEM
2SC3214	0065
2SC3215	OEM
2SC3216	OEM
2SC3217	OEM
2SC3218	OEM
2SC3219	0728
2SC3220	3387
2SC3221	OEM
2SC3222	3388
2SC3223	OEM
2SC3224	OEM
2SC3225	OEM
2SC3226	3406
2SC3227	0324
2SC3228	1562
2SC3229	2598
2SC3230	3115
2SC3231	3014
2SC3232	1506
2SC3233	OEM
2SC3234	3113
2SC3235	3014
2SC3236	0728
2SC3237	3411
2SC3238	1085
2SC3239	0060
2SC3240	OEM
2SC3241	OEM
2SC3242	3093
2SC3242A	3093
2SC3242G	3093
2SC3243	1386
2SC3243(DE)-T	1386
2SC3243D	1386
2SC3243DE	1386
2SC3243E	1386
2SC3244	1386
2SC3245	1386
2SC3245A	1386
2SC3246	OEM
2SC3246-T11-A-H/J	OEM
2SC3246H	OEM
2SC3247	0400
2SC3248	0261
2SC3249	0261
2SC3250	OEM
2SC3251	OEM
2SC3252	1506
2SC3253	0060
2SC3254	0060
2SC3255	0060
2SC3256	1361
2SC3257	0728
2SC3258	0060
2SC3259	3357
2SC3260	2784
2SC3261	2784
2SC3262	OEM
2SC3263	1506
2SC3264	OEM
2SC3265	1733
2SC3265Y	OEM
2SC3266	1500
2SC3267	3093
2SC3268	OEM
2SC3269	0261
2SC3270	0261
2SC3271	0276
2SC3271(N)	0276
2SC3271(N)-L	0276
2SC3271(NP)	0276
2SC3271(NP)-C1	0276
2SC3271(NP)-L	0276
2SC3271(P)	0276
2SC3271(P)-L	0276
2SC3271-N	0276
2SC3271F	0276
2SC3271F-N	0276
2SC3271N	0276
2SC3271P	0276
2SC3272	0276
2SC3273	2851
2SC3274	OEM
2SC3275	OEM
2SC3276	OEM
2SC3277	2185
2SC3278	1966
2SC3279	1500
2SC3280	3435
2SC3280-0	3702
2SC3280R	OEM
2SC3281	3436
2SC3281-0	3702
2SC3281-0	3436
2SC3281-R	3436
2SC3281G	OEM
2SC3282	OEM
2SC3282A	OEM
2SC3283	OEM
2SC3283A	OEM
2SC3283R	1562
2SC3284	3435
2SC3284-0	OEM
2SC3285	3438
2SC3286	OEM
2SC3287	OEM
2SC3288	OEM
2SC3289	OEM
2SC3290	OEM
2SC3291	OEM
2SC3292	0278
2SC3293	0278
2SC3294	2366
2SC3295	OEM
2SC3296	0392
2SC3296D	OEM
2SC3296E	OEM
2SC3297	0042
2SC3298	OEM
2SC3298(Y)	OEM
2SC3298A	OEM
2SC3298A-O	OEM
2SC3298A-O(LB-SAN)	OEM
2SC3298A-T(LB-SAN)	OEM
2SC3298A-Y	OEM
2SC3298AY	OEM
2SC3298A-Y(LB-SAN)	OEM
2SC3298B	OEM
2SC3298B-Y	OEM
2SC3298Y	OEM
2SC3299	0060
2SC3299(Y)	0060
2SC3299Y	0060
2SC3300	2367
2SC3300A	2367
2SC3300ALF	2367
2SC3301	OEM
2SC3301G	OEM
2SC3302	2844
2SC3303	OEM
2SC3303Y	OEM
2SC3304	OEM
2SC3305	OEM
2SC3306	2185
2SC3306S	OEM
2SC3307	OEM
2SC3308	3446
2SC3309	3014
2SC3310	0728
2SC3311	2954
2SC3311(Q)	2954
2SC3311(R)	2954
2SC3311(S)	2954
2SC3311-Q	OEM
2SC3311-QRS	OEM
2SC3311A	2954
2SC3311A(Q)	2954
2SC3311A(Q)Y	2954
2SC3311A(Q,R)-Y	2954
2SC3311A(QR)-T	2954
2SC3311A(R)	2954
2SC3311A(R)Y	2954
2SC3311A-Q	OEM
2SC3311AQ	2954
2SC3311A-QRS	2954
2SC3311AQRS	2954
2SC3311AQSTA	OEM
2SC3311A-R	2954
2SC3311AR	2954
2SC3311ARS	2954
2SC3311A-RTA	2954
2SC3311AS	2954
2SC3311Q	2954
2SC3311QR	2954
2SC3311QRS	OEM
2SC3311QRSTA	OEM
2SC3311QS	2954
2SC3311R	2954
2SC3311S	2954
2SC3312	3447
2SC3313	OEM
2SC3314	OEM
2SC3315	OEM
2SC3315C	4452
2SC3316	3449
2SC3317	3014
2SC3317-S	3014
2SC3317S	3014
2SC3318	3387
2SC3319	3388
2SC3320	OEM
2SC3321	3388
2SC3322	2784
2SC3323	OEM
2SC3324	2332
2SC3325	1733
2SC3326	OEM
2SC3327	OEM
2SC3328	OEM
2SC3329	0530
2SC3330	2954
2SC3330-S-AC	2954
2SC3330-T	2954
2SC3330T	2954
2SC3331	0018
2SC3331-V	OEM
2SC3331S	0018
2SC3331ST	0018
2SC3331T	0018
2SC3331U	0018
2SC3332	OEM
2SC3333	0261
2SC3334	0261
2SC3334-T	0261
2SC3335	3217
2SC3336	1506
2SC3337	3454
2SC3338	OEM
2SC3339	OEM
2SC3340	OEM
2SC3341	OEM
2SC3342	OEM
2SC3343	OEM
2SC3344	0728
2SC3345	OEM
2SC3346	OEM
2SC3347	OEM
2SC3348	OEM
2SC3349	OEM
2SC3350	OEM
2SC3351	OEM
2SC3352	3460
2SC3353	3462
2SC3354	OEM
2SC3354S	OEM
2SC3354T	OEM
2SC3355	OEM
2SC3356	OEM
2SC3357	OEM
2SC3358	OEM
2SC3359	0155
2SC3360	OEM
2SC3361	1733
2SC3362	OEM
2SC3363	OEM
2SC3364	OEM
2SC3365	2185
2SC3366	OEM
2SC3367	OEM
2SC3368	OEM
2SC3369	OEM
2SC3370	OEM
2SC3371	OEM
2SC3372	OEM
2SC3372-5	OEM
2SC3373	OEM
2SC3374	3468
2SC3374-O	OEM
2SC3375	OEM
2SC3376	2784
2SC3377	OEM
2SC3377P	OEM
2SC3377Q	OEM
2SC3377R	OEM
2SC3378	OEM
2SC3379	OEM
2SC3380	3471
2SC3381	OEM
2SC3382	OEM
2SC3383	OEM
2SC3384	3388
2SC3385	3388
2SC3386	OEM
2SC3387	3476
2SC3387-01AB	3476
2SC3387-SY	3476
2SC3388	OEM
2SC3389	OEM
2SC3390	2954
2SC3391	OEM
2SC3391C	OEM
2SC3392	1733
2SC3393	2954
2SC3393ST	2954
2SC3394	OEM
2SC3395	3478
2SC3396	0981
2SC3396(CY)-TA	0981
2SC3396TA	OEM
2SC3397	OEM
2SC3398	3482
2SC3398-TA	3482
2SC3399	0898
2SC3399-AC	0898
2SC3400	2323
2SC3400-AC	2323
2SC3400-T	2323
2SC3400Z	2323
2SC3401	OEM
2SC3402	0832
2SC3402-T	0832
2SC3403	OEM
2SC3404	OEM
2SC3405	OEM
2SC3406	OEM
2SC3407	OEM
2SC3408	OEM
2SC3409	2784
2SC3410	OEM
2SC3411	3388
2SC3412	3486
2SC3413	2954
2SC3413(C)	2954
2SC3413(D)	2954
2SC3413B	2954
2SC3413BC	2954
2SC3413C	2954
2SC3413CD	OEM
2SC3413D-TZ	2954
2SC3414	OEM
2SC3415	3487
2SC3416	OEM
2SC3417	0276
2SC3418	OEM
2SC3419	0563
2SC3420	OEM
2SC3421	0563
2SC3422	OEM
2SC3422Y	OEM
2SC3423	0563
2SC3424	OEM
2SC3425	OEM
2SC3426	1733
2SC3427	OEM
2SC3428	OEM
2SC3429	OEM
2SC3430	OEM
2SC3431	OEM
2SC3432	OEM
2SC3433	OEM
2SC3434	1506
2SC3435	1506
2SC3436	OEM
2SC3437	OEM
2SC3438	OEM
2SC3439	OEM
2SC3440	1733
2SC3441	1733
2SC3442	OEM
2SC3443	OEM
2SC3444	0667
2SC3445	OEM
2SC3446	3460
2SC3447	OEM
2SC3448	3493
2SC3449	3494
2SC3450	OEM
2SC3450M	OEM
2SC3451	OEM
2SC3452	OEM
2SC3453	OEM
2SC3454	OEM
2SC3455	OEM
2SC3456	3497
2SC3457	OEM
2SC3458	3438
2SC3459	3438
2SC3460	3360
2SC3461	3360
2SC3462	OEM
2SC3463	OEM
2SC3464	OEM
2SC3465	OEM
2SC3466	3360
2SC3467	0261
2SC3467D	0261
2SC3467E	0261
2SC3468	0261
2SC3468(D)-AE	0261
2SC3468(E)-AE	0261
2SC3468-AE	0261
2SC3468-D	0261
2SC3468D	0261
2SC3468E	0261
2SC3468F	0261
2SC3469	OEM
2SC3470	2954

If replacement code is OEM, contact original manufacturer for replacement.

Original Device Types

DEVICE TYPE	REPL CODE
2SC3471	OEM
2SC3472	OEM
2SC3473	OEM
2SC3474	OEM
2SC3475	OEM
2SC3476	OEM
2SC3477	OEM
2SC3478	OEM
2SC3478A	OEM
2SC3479	3507
2SC3480	3507
2SC3481	3507
2SC3482	3507
2SC3483	3511
2SC3484	3511
2SC3485	3513
2SC3486	3514
2SC3487	OEM
2SC3488	2954
2SC3489	OEM
2SC3490	3357
2SC3491	OEM
2SC3492	OEM
2SC3493	2919
2SC3494	2954
2SC3495	0530
2SC3496	OEM
2SC3497	OEM
2SC3498	OEM
2SC3499	OEM
2SC3500	OEM
2SC3501	OEM
2SC3502	0276
2SC3503	0276
2SC3503-RA	0276
2SC3503CDEF	OEM
2SC3503D	0276
2SC3503E	0276
2SC3503F	0276
2SC3504	OEM
2SC3505	2784
2SC3506	3438
2SC3507	3438
2SC3508	OEM
2SC3509	OEM
2SC3510	2851
2SC3511	OEM
2SC3512	3454
2SC3513	OEM
2SC3514	OEM
2SC3515	3471
2SC3516	OEM
2SC3517	OEM
2SC3518	OEM
2SC3518L	OEM
2SC3519	3524
2SC3519A	OEM
2SC3520	OEM
2SC3521	1733
2SC3522	OEM
2SC3523	OEM
2SC3524	OEM
2SC3525	OEM
2SC3526	OEM
2SC3527	1506
2SC3528	OEM
2SC3529	OEM
2SC3531	OEM
2SC3532	OEM
2SC3533	3494
2SC3534	3494
2SC3535	3494
2SC3536	3494
2SC3537	OEM
2SC3538	OEM
2SC3539	OEM
2SC3540	OEM
2SC3541	OEM
2SC3542	OEM
2SC3544	OEM
2SC3545	OEM
2SC3545T44	OEM
2SC3546	0046
2SC3547	2919
2SC3548	OEM
2SC3549	3357
2SC3550	2784
2SC3551	2784
2SC3552	3542
2SC3552L	3542
2SC3553	2129
2SC3553BC	2129
2SC3553CD	2954
2SC3554	OEM
2SC3555	OEM
2SC3556	OEM
2SC3557	OEM
2SC3558	OEM
2SC3559	3460
2SC3560	3547
2SC3561	3547
2SC3562	OEM
2SC3563	OEM
2SC3564	OEM
2SC3565	OEM
2SC3566	1165
2SC3567	1165
2SC3568	0728
2SC3568K	0728
2SC3568L	0728
2SC3569	3014
2SC3570	3462
2SC3571	3462
2SC3572	0728
2SC3573	OEM
2SC3574	OEM
2SC3575	OEM
2SC3576	OEM
2SC3577	3360
2SC3578	OEM
2SC3579	OEM
2SC3580	OEM
2SC3581	OEM
2SC3582	OEM
2SC3583	OEM
2SC3584	OEM
2SC3585	OEM
2SC3586	OEM
2SC3587	OEM
2SC3588	OEM
2SC3589	OEM
2SC3590	OEM
2SC3591	OEM
2SC3592	OEM
2SC3593	OEM
2SC3594	OEM
2SC3595	OEM
2SC3596	OEM
2SC3597	OEM
2SC3598	OEM
2SC3599	OEM
2SC3600	OEM
2SC3601	OEM
2SC3602	OEM
2SC3603	3562
2SC3604	OEM
2SC3605	3454
2SC3606	OEM
2SC3607	OEM
2SC3608	OEM
2SC3609	OEM
2SC3610	OEM
2SC3611	OEM
2SC3612	OEM
2SC3613	OEM
2SC3614	OEM
2SC3615	OEM
2SC3616	OEM
2SC3617	OEM
2SC3618	OEM
2SC3619	0276
2SC3619LB	5285
2SC3620	0276
2SC3620(LB)	0276
2SC3620(LB-SAN-1)	0276
2SC3620EQ	0276
2SC3620LB	0276
2SC3621	0563
2SC3621(O)	0563
2SC3621(R)	0563
2SC3621-0(LB-SAN-2)	0563
2SC3621-O	0563
2SC3621-O(LB-SAN-2)	0563
2SC3621-O(SAN-1)	0563
2SC3621-OY	OEM
2SC3621-R	0563
2SC3621-R(LB-SAN-2)	0563
2SC3621-R(SAN-1)	0563
2SC3621R	0563
2SC3621RY	OEM
2SC3621Y	0563
2SC3621Y(O)	0563
2SC3621Y(R)	0563
2SC3622	OEM
2SC3622A	OEM
2SC3622AK	OEM
2SC3622AL	OEM
2SC3623	OEM
2SC3623-K	OEM
2SC3623A	OEM
2SC3623A-L	OEM
2SC3623K	OEM
2SC3623L	OEM
2SC3624	OEM
2SC3624A	OEM
2SC3625	0728
2SC3626	2908
2SC3627	0728
2SC3628	OEM
2SC3629	OEM
2SC3630	OEM
2SC3631	OEM
2SC3632	OEM
2SC3633	OEM
2SC3634	OEM
2SC3635	OEM
2SC3636	3494
2SC3637	OEM
2SC3638	OEM
2SC3639	OEM
2SC3640	OEM
2SC3641	OEM
2SC3642	3494
2SC3643	3494
2SC3644	OEM
2SC3645	3471
2SC3646	OEM
2SC3647	OEM
2SC3648	3471
2SC3649	OEM
2SC3650	OEM
2SC3651	OEM
2SC3652	OEM
2SC3653	0898
2SC3657	2784
2SC3658	0055
2SC3659	0055
2SC3660	OEM
2SC3661	OEM
2SC3662	OEM
2SC3663	OEM
2SC3664	OEM
2SC3665	OEM
2SC3666	3608
2SC3667	0009
2SC3668	OEM
2SC3669	OEM
2SC3670	0009
2SC3671	3471
2SC3672	OEM
2SC3673	OEM
2SC3674	OEM
2SC3675	OEM
2SC3676	OEM
2SC3677	OEM
2SC3678	2784
2SC3679	2784
2SC3680	3486
2SC3681	OEM
2SC3682	OEM
2SC3683	OEM
2SC3684	OEM
2SC3685	3494
2SC3686	3494
2SC3687	3494
2SC3688	3494
2SC3689	OEM
2SC3690	OEM
2SC3691	0060
2SC3691K	0060
2SC3691L	0060
2SC3692	OEM
2SC3693	OEM
2SC3694	OEM
2SC3695	OEM
2SC3696	OEM
2SC3697	OEM
2SC3698	OEM
2SC3699	OEM
2SC3700	OEM
2SC3701	OEM
2SC3702	OEM
2SC3703	OEM
2SC3704	OEM
2SC3705	1061
2SC3706	OEM
2SC3707	2919
2SC3708	1386
2SC3708-R	1386
2SC3708-S	1386
2SC3708R	1386
2SC3708S	1386
2SC3708T	OEM
2SC3709	OEM
2SC3710	OEM
2SC3711	OEM
2SC3712	OEM
2SC3713	OEM
2SC3714	OEM
2SC3715	OEM
2SC3716	OEM
2SC3717	0127
2SC3718	OEM
2SC3719	OEM
2SC3720	0224
2SC3720A	0224
2SC3720B	0224
2SC3720C	0224
2SC3720D	0224
2SC3720F	0224
2SC3720G	0224
2SC3720GN	0224
2SC3720H	0224
2SC3720J	0224
2SC3720K	0224
2SC3720L	0224
2SC3720OR	0224
2SC3720R	0224
2SC3720X	0224
2SC3720Y	0224
2SC3721	0079
2SC3722	OEM
2SC3723	OEM
2SC3724	1506
2SC3725	1506
2SC3726	OEM
2SC3727	OEM
2SC3728	OEM
2SC3729	3486
2SC3730	OEM
2SC3731	2954
2SC3732	OEM
2SC3733	OEM
2SC3734	OEM
2SC3735	OEM
2SC3736	OEM
2SC3737	3494
2SC3738	OEM
2SC3739	OEM
2SC3740	1500
2SC3741	OEM
2SC3742	OEM
2SC3743	3357
2SC3744	OEM
2SC3745	OEM
2SC3746	0060
2SC3746R	0060
2SC3747	0060
2SC3748	0060
2SC3748R	0060
2SC3749	3460
2SC3750	3462
2SC3751	3639
2SC3752	3497
2SC3753	OEM
2SC3754	OEM
2SC3755	OEM
2SC3756	OEM
2SC3757	OEM
2SC3757QR	OEM
2SC3758	OEM
2SC3759	OEM
2SC3760	OEM
2SC3761	OEM
2SC3762	OEM
2SC3763	OEM
2SC3764	OEM
2SC3765	OEM
2SC3766	OEM
2SC3769	OEM
2SC3770	OEM
2SC3771	OEM
2SC3772	OEM
2SC3773	OEM
2SC3773(3-4)-W	OEM
2SC3774	OEM
2SC3775	OEM
2SC3776	2851
2SC3777	2851
2SC3778	2851
2SC3779	3454
2SC3780	OEM
2SC3781	OEM
2SC3782	OEM
2SC3783	2784
2SC3784	1061
2SC3785	1061
2SC3786	OEM
2SC3787	OEM
2SC3788	OEM
2SC3789	0276
2SC3789-C	0276
2SC3789-CD	0276
2SC3789-D	0276
2SC3789-DE	0276
2SC3789-E	5090
2SC3789-YA	5090
2SC3789D	0276
2SC3789E	5090
2SC3790	0276
2SC3790D	0276
2SC3791	OEM
2SC3792	OEM
2SC3793	OEM
2SC3794	3357
2SC3795	3653
2SC3795B	3653
2SC3796	2784
2SC3797	3494
2SC3798	2784
2SC3799	3494
2SC3800	0224
2SC3800A	0224
2SC3800B	0224
2SC3800C	0224
2SC3800D	0224
2SC3800E	0224
2SC3800F	0224
2SC3800G	0224
2SC3800GN	0224
2SC3800H	0224
2SC3800J	0224
2SC3800K	0224
2SC3800L	0224
2SC3800M	0224
2SC3800OR	0224
2SC3800R	0224
2SC3800X	0224
2SC3800Y	0224
2SC3801	OEM
2SC3802	OEM
2SC3803	OEM
2SC3804	OEM
2SC3805	OEM
2SC3806	OEM
2SC3807	OEM
2SC3807-R-CTV-YA	OEM
2SC3807K	OEM
2SC3807R	OEM
2SC3808	OEM
2SC3809	OEM
2SC3810	OEM
2SC3811	0739
2SC3811-QRS	OEM
2SC3812	OEM
2SC3813	2185
2SC3814	OEM
2SC3815	OEM
2SC3816	OEM
2SC3817	OEM
2SC3818	OEM
2SC3819	OEM
2SC3820	OEM
2SC3821	2908
2SC3822	OEM
2SC3823	OEM
2SC3824	OEM
2SC3825	OEM
2SC3826	OEM
2SC3827	OEM
2SC3828	OEM
2SC3829	OEM
2SC3830	3668
2SC3831	3387
2SC3832	2908
2SC3833	1506
2SC3833LF-128	1506
2SC3834	OEM
2SC3835	OEM
2SC3836	OEM
2SC3840K	OEM
2SC3841	OEM
2SC3842	OEM
2SC3843	OEM
2SC3844	OEM
2SC3845	OEM
2SC3846	OEM
2SC3847	OEM
2SC3850	OEM
2SC3851	0236
2SC3851(A)	7120
2SC3852	3674
2SC3853	3675
2SC3854	0127
2SC3855	0194
2SC3856	3678
2SC3857	2276
2SC3858	3226
2SC3868	3460
2SC3869	3462
2SC3870	3462
2SC3871	3682
2SC3871LP	3682
2SC3872	2185
2SC3873	2185
2SC3874	OEM
2SC3875	0724
2SC3875S	0724
2SC3875S(LO)	0724
2SC3876	1435
2SC3877	OEM
2SC3878	OEM
2SC3879	0724
2SC3880	OEM
2SC3881	OEM
2SC3886A	5621
2SC3890	0728
2SC3892A	OEM
2SC3898	OEM
2SC3902	0563
2SC3902-R-RA	0563
2SC3902-S-CTV-YA	0563
2SC3902-S-RA	OEM
2SC3902-T-CTV-YA	0563
2SC3902S	0563
2SC3902S-CTV-YA	4243
2SC3902T	0563
2SC3902T-CTV-YA	0563
2SC3907	3702
2SC3931	OEM
2SC3938	OEM
2SC3939R	OEM
2SC3940	0224
2SC3940A	0224
2SC3940AQSTA	OEM
2SC3940B	0224
2SC3940C	0224
2SC3940D	0224
2SC3940E	0224
2SC3940F	0224
2SC3940G	0224
2SC3940GN	0224
2SC3940H	0224
2SC3940J	0224
2SC3940K	0224
2SC3940L	0224
2SC3940M	0224
2SC3940OR	0224
2SC3940R	0224
2SC3940X	0224
2SC3940Y	0224
2SC3942	3709
2SC3942M	5579
2SC3942RL	5579
2SC3987	3723
2SC4006	3723
2SC-4012	0018
2SC4012A	0018
2SC4012B	0018
2SC4012C	0018
2SC4012D	0018
2SC4012E	0018
2SC4012F	0018
2SC4012G	0018
2SC4012GN	0018
2SC4012H	0018
2SC4012J	0018
2SC4012L	0018
2SC4012M	0018
2SC4012X	0018
2SC4012Y	0018
2SC4026	OEM
2SC-4033	0224
2SC4033A	0224
2SC4033B	0224
2SC4033C	0224
2SC4033D	0224
2SC4033E	0224
2SC4033F	0224
2SC4033G	0224
2SC4033GN	0224
2SC4033H	0224
2SC4033J	0224
2SC4033K	0224
2SC4033L	0224
2SC4033M	0224
2SC4033OR	0224
2SC4033R	0224
2SC4033X	0224
2SC4033Y	0224
2SC4038LN	OEM
2SC4038Q-TL2	OEM
2SC4038R-TL2	OEM
2SC4048	OEM
2SC4052	OEM
2SC4054	OEM
2SC4059	1506
2SC4073	OEM
2SC4075	1085
2SC4075(D)-YAC	1085
2SC4075(E)-YAC	1085
2SC4075-YAC	1085
2SC4075D	1085
2SC4075E	1085
2SC4108	1506
2SC4108-CTV-YB	1506
2SC4108N	1506
2SC4108Z	1506
2SC4109	OEM
2SC4115S	1500
2SC4115SQR	1500
2SC4116	0042
2SC4116A	0042
2SC4116B	0042
2SC4116C	0042
2SC4116D	0042
2SC4116E	0042
2SC4116F	0042
2SC4116G	0042
2SC4116GN	0042
2SC4116H	0042
2SC4116J	0042
2SC4116K	0042
2SC4116L	0042
2SC4116M	0042
2SC4116OR	0042
2SC4116R	0042
2SC4116X	0042
2SC4116Y	0042
2SC4130	3794
2SC4133	3796
2SC4159	1165
2SC4159-E	1165
2SC4159D	1165
2SC4159E	1165
2SC4161L	OEM
2SC4161M	OEM
2SC4161MN-RA11	OEM
2SC4169	1061
2SC4170	OEM
2SC4212(H)	OEM
2SC4212-C1	OEM
2SC4212H	OEM
2SC4212HL	OEM
2SC4212HLB	OEM
2SC4212HLBS	OEM
2SC4217	0276
2SC4217(D)-RAC	0276
2SC4217(E)-RAC	0276
2SC4217-RAC	0276
2SC4217D	0276
2SC4217E	0276
2SC4226	OEM
2SC4237	OEM
2SC4256	OEM
2SC4274	0728
2SC4274-02	0728
2SC4274-02F9	0728
2SC4304	3357
2SC4417	OEM
2SC4418	2908
2SC4418LF608	2908
2SC4418Y	2908
2SC4418Y-F	2908
2SC4421LP	OEM
2SC4423	2185
2SC4423-CTV	2185
2SC4423-CTV-YB	2185
2SC4424	3098
2SC4424-L-YB	3098
2SC4424-M-YB	3098
2SC4424L	3098
2SC4424M	3098
2SC4440	OEM
2SC4475	OEM
2SC4502	OEM
2SC4502-T	OEM
2SC4502-Y	OEM
2SC4533	OEM
2SC4533LP	OEM
2SC4544	3926
2SC4547	OEM
2SC4589	3941
2SC4601B	OEM
2SC4634	OEM
2SC4636	OEM
2SC4663	OEM
2SC4663NPR-F	OEM
2SC4664	OEM
2SC4664MNP-F	OEM
2SC4664NPR-F	OEM
2SC4721P	OEM
2SC4721Q	OEM
2SC4722(NP)	OEM
2SC4722N	OEM
2SC4746	4030
2SC4763	OEM
2SC4793	OEM
2SC4797	OEM
2SC4833MNP	OEM
2SC4833N	OEM
2SC4834MNP	OEM
2SC4927-01	OEM
2SC5083	OEM
2SC5083(L-P)-T	OEM
2SC5100	0168
2SC5100A	0168
2SC5100B	0168
2SC5100C	0168
2SC5100D	0168
2SC5100E	0168
2SC5100F	0168
2SC5100G	0168
2SC5100GN	0168
2SC5100H	0168
2SC5100J	0168
2SC5100K	0168
2SC5100L	0168
2SC5100M	0168
2SC5100OR	0168
2SC5100R	0168
2SC5100Y	0168
2SC5110	0168
2SC5110A	0168
2SC5110B	0168
2SC5110C	0168
2SC5110D	0168
2SC5110E	0168
2SC5110F	0168
2SC5110G	0168
2SC5110GN	0168
2SC5110H	0168
2SC5110J	0168
2SC5110K	0168
2SC5110L	0168
2SC5110M	0168
2SC5110OR	0168
2SC5110R	0168
2SC5110X	0168
2SC5110Y	0168
2SC5120	0560
2SC5130	0560
2SC5370	0018
2SC5370A	0018
2SC5370B	0018
2SC5370C	0018
2SC5370D	0018
2SC5370E	0018
2SC5370F	0018
2SC5370G	0018
2SC5370GN	0018
2SC5370H	0018
2SC5370J	0018
2SC5370K	0018
2SC5370L	0018
2SC5370M	0018
2SC5370OR	0018
2SC5370R	0018
2SC5370X	0018
2SC5370Y	0018
2SC5430OR	0224
2SC5488	OEM
2SC5598	OEM
2SC5884	OEM
2SC5886	OEM
2SC5944	OEM
2SC5945	OEM
2SC5946	OEM
2SC5947	OEM
2SC5988	OEM
2SC5991	OEM
2SC6031	OEM
2SC6036	OEM
2SC6039	OEM
2SC6042	OEM
2SC6045	OEM
2SC6052	OEM
2SC6056	OEM
2SC6059	OEM
2SC6080	OEM
2SC6081	OEM
2SC6082	OEM
2SC6094	OEM

If replacement code is OEM, contact original manufacturer for replacement.

DEVICE TYPE	REPL CODE
2SC6095	OEM
2SC6096	OEM
2SC6193	OEM
2SC6228	OEM
2SC6235	OEM
2SC6251	OEM
2SC6277	OEM
2SC6284	OEM
2SC6287	OEM
2SC6292	OEM
2SC6295	OEM
2SC6297	OEM
2SC6299	OEM
2SC6301	OEM
2SC6304	OEM
2SC6316	OEM
2SC6318	OEM
2SC6341	OEM
2SC6379	OEM
2SC6423	OEM
2SC6438	OEM
2SC6488	OEM
2SC6491	OEM
2SC6543	OEM
2SC6545	OEM
2SC6547	OEM
2SC6556	OEM
2SC6603	OEM
2SC6604	OEM
2SC8146	0018
2SC8290	0079
2SC9011E	0079
2SC9011F	0079
2SC9011H	0079
2SC9012A	OEM
2SC9013	OEM
2SC9633	OEM
2SC10148	0560
2SC11070	0560
2SC14505	0042
2SC15211	0168
2SC17401	0198
2SC17405	0198
2SC27354	OEM
2SC32816	OEM
2SC32840	OEM
2SC82915	OEM
2SCD78	0042
2SCF1	0144
2SCF1A	0224
2SCF1B	0224
2SCF1C	0224
2SCF1D	0224
2SCF1E	0224
2SCF1F	0224
2SCF1G	0224
2SCF1GN	0224
2SCF1H	0224
2SCF1J	0224
2SCF1K	0224
2SCF1L	0224
2SCF1M	0224
2SCF1OR	0224
2SCF1R	0224
2SCF1X	0224
2SCF1Y	0224
2SCF-2	0016
2SCF2A	0224
2SCF2B	0224
2SCF2C	0224
2SCF2D	0224
2SCF2E	0224
2SCF2G	0224
2SCF2GN	0224
2SCF2H	0224
2SCF2J	0224
2SCF2K	0224
2SCF2L	0224
2SCF2M	0224
2SCF2R	0224
2SCF2X	0224
2SCF2Y	0224
2SCF3	0630
2SCF3A	0630
2SCF3B	0630
2SCF3C	0630
2SCF3D	0630
2SCF3E	0630
2SCF3F	0630
2SCF3G	0630
2SCF3GN	0630
2SCF3H	0630
2SCF3J	0630
2SCF3K	0630
2SCF3L	0630
2SCF3M	0630
2SCF3OR	0630
2SCF3R	0630
2SCF3X	0630
2SCF3Y	0630
2SCF5	0016
2SCF5A	0079
2SCF5B	0079
2SCF5C	0079
2SCF5D	0079
2SCF5E	0079
2SCF5F	0079
2SCF5G	0079
2SCF5GN	0079
2SCF5H	0079
2SCF5J	0079
2SCF5K	0079
2SCF5L	0079
2SCF5M	0079
2SCF5OR	0079
2SCF5R	0079
2SCF5X	0079
2SCF5Y	0079
2SC-F6	0079
2SCF6	1400
2SCF6A	0086
2SCF6B	0086
2SCF6C	0086
2SCF6D	0086
2SCF6E	0086
2SCF6F	0086
2SCF6G	0086
2SCF6GN	0086
2SCF6H	0086
2SCF6J	0086
2SCF6K	0086
2SCF6L	0086
2SCF6M	0086
2SCF6OR	0086
2SCF6R	0086
2SCF6X	0086
2SCF6Y	0086
2SCF-8	4208
2SCF8	4208
2SCF8A	4208
2SCF8B	4208
2SCF8C	4208
2SCF8D	4208
2SCF8E	4208
2SCF8F	4208
2SCF8G	4208
2SCF8GN	4208
2SCF8H	4208
2SCF8J	4208
2SCF8K	4208
2SCF8L	4208
2SCF8OR	4208
2SCF8R	4208
2SCF8X	4208
2SCF8Y	4208
2SC-F11	0224
2SCF11	0475
2SC-F11A	0224
2SC-F11B	0224
2SC-F11C	0224
2SC-F11D	0224
2SC-F11E	0224
2SC-F11F	0224
2SC-F11G	0224
2SC-F11GN	0224
2SC-F11H	0224
2SC-F11J	0224
2SC-F11K	0224
2SC-F11L	0224
2SC-F11M	0224
2SC-F11OR	0224
2SC-F11R	0224
2SC-F11X	0224
2SC-F11Y	0224
2SCF12A	0630
2SCF12B	0630
2SCF12C	0630
2SCF12D	0630
2SCF12E	0630
2SCF12F	0630
2SCF12G	0630
2SCF12GN	0630
2SCF12H	0630
2SCF12J	0630
2SCF12K	0630
2SCF12L	0630
2SCF12M	0630
2SCF12OR	0630
2SCF12R	0630
2SCF12X	0630
2SCF12Y	0630
2SC-F14	0224
2SC-F14A	0224
2SC-F14B	0224
2SC-F14C	0224
2SC-F14D	0224
2SC-F14E	0224
2SC-F14F	0224
2SC-F14G	0224
2SC-F14GN	0224
2SC-F14H	0224
2SC-F14J	0224
2SC-F14K	0224
2SC-F14L	0224
2SC-F14M	0224
2SC-F14OR	0224
2SC-F14R	0224
2SC-F14X	0224
2SC-F14Y	0224
2SCF812	0630
2SCI090B	0233
2SCI520K	0168
2SCM39J	0224
2SCM39X	0224
2SCM93D	0103
2SCM95K	0142
2SCM98F	0086
2SC-NJ-100	0111
2SC-NJ100	0111
2SCNJ100	0111
2SC-NJ-107	0086
2SCNJ107	0086
2SCS183A	0018
2SCS183B	0018
2SCS183C	0018
2SCS183D	0018
2SCS183E	0198
2SCS183F	0018
2SCS183G	0018
2SCS183GN	0018
2SCS183H	0018
2SCS183J	0018
2SCS183K	0018
2SCS183L	0018
2SCS183M	0018
2SCS183OR	0018
2SCS183R	0018
2SCS183Y	0018
2SCS184	0018
2SCS184A	0018
2SCS184B	0018
2SCS184C	0018
2SCS184D	0018
2SCS184E	0198
2SCS184F	0018
2SCS184G	0018
2SCS184GN	0018
2SCS184H	0018
2SCS184J	0144
2SCS184K	0018
2SCS184L	0018
2SCS184M	0018
2SCS184OR	0018
2SCS184R	0018
2SCS184X	0018
2SCS184Y	0018
2SCS429	0224
2SCS429A	0224
2SCS429B	0224
2SCS429C	0224
2SCS429D	0224
2SCS429E	0224
2SCS429F	0224
2SCS429G	0224
2SCS429GN	0224
2SCS429H	0224
2SCS429J	0144
2SCS429K	0224
2SCS429L	0224
2SCS429M	0224
2SCS429OR	0224
2SCS429R	0224
2SCS429X	0224
2SCS429Y	0224
2SCS430	0224
2SCS430A	0224
2SCS430B	0224
2SCS430C	0224
2SCS430D	0224
2SCS430E	0224
2SCS430F	0224
2SCS430GN	0224
2SCS430H	0144
2SCS430J	0224
2SCS430K	0224
2SCS430L	0224
2SCS430M	0224
2SCS430R	0224
2SCS430X	0224
2SCS430Y	0224
2SCS461	0224
2SCS461A	0224
2SCS461B	0224
2SCS461C	0224
2SCS461D	0224
2SCS461E	0224
2SCS461F	0144
2SCS461G	0224
2SCS461GN	0224
2SCS461H	0224
2SCS461J	0224
2SCS461K	0224
2SCS461L	0224
2SCS461M	0224
2SCS461OR	0224
2SCS461R	0224
2SCS461X	0224
2SCS461Y	0224
2SCS469	0224
2SCS469A	0224
2SCS469B	0224
2SCS469C	0224
2SCS469D	0224
2SCS469E	0224
2SCS469F	0144
2SCS469G	0224
2SCS469GN	0224
2SCS469H	0224
2SCS469J	0224
2SCS469K	0224
2SCS469L	0224
2SCS469M	0224
2SCS469R	0224
2SCS469X	0224
2SCS469Y	0224
2SCU64M	1400
2SD2A	OEM
2SD11	0208
2SD11A	0208
2SD11B	0208
2SD11C	0208
2SD11D	0208
2SD11E	0208
2SD11F	0208
2SD11G	0208
2SD11GN	0208
2SD11H	0208
2SD11J	0208
2SD11K	0208
2SD11L	0208
2SD11M	0208
2SD11OR	0208
2SD11R	0208
2SD11X	0208
2SD11Y	0208
2SD12	0074
2SD12A	0074
2SD12B	0074
2SD12C	0074
2SD12E	0074
2SD12F	0074
2SD12G	0074
2SD12GN	0074
2SD12H	0074
2SD12J	0074
2SD12K	0074
2SD12L	0074
2SD12M	0074
2SD12OR	0074
2SD12R	0074
2SD12X	0074
2SD12Y	0074
2SD13	OEM
2SD14	OEM
2SD15	0130
2SD15A	0130
2SD15B	0130
2SD15C	0130
2SD15D	0130
2SD15E	0130
2SD15F	0130
2SD15G	0130
2SD15GN	0130
2SD15H	0130
2SD15J	0130
2SD15K	0130
2SD15L	0130
2SD15M	0130
2SD15OR	0130
2SD15R	0130
2SD15X	0130
2SD15Y	0130
2SD16	0130
2SD16A	0130
2SD16B	0130
2SD16C	0130
2SD16D	0130
2SD16E	0130
2SD16F	0130
2SD16G	0130
2SD16GN	0130
2SD16H	0130
2SD16J	0130
2SD16K	0130
2SD16L	0130
2SD16M	0130
2SD16OR	0130
2SD16R	0130
2SD16X	0130
2SD16Y	0130
2SD17	0130
2SD17A	0130
2SD17C	0130
2SD17D	0130
2SD17E	0130
2SD17F	0130
2SD17GN	0130
2SD17H	0130
2SD17J	0130
2SD17K	0130
2SD17L	0130
2SD17M	0130
2SD17R	0130
2SD17X	0130
2SD17Y	0130
2SD18	0867
2SD19	0208
2SD19A	0208
2SD19B	0208
2SD19C	0208
2SD19E	0208
2SD19F	0208
2SD19G	0208
2SD19GN	0208
2SD19H	0208
2SD19J	0208
2SD19K	0208
2SD19M	0208
2SD19Y	0208
2SD20	0208
2SD20A	0208
2SD20B	0208
2SD20C	0208
2SD20D	0208
2SD20E	0208
2SD20F	0208
2SD20G	0208
2SD20GN	0208
2SD20H	0208
2SD20J	0208
2SD20K	0208
2SD20L	0208
2SD20M	0208
2SD20OR	0208
2SD20R	0208
2SD20X	0208
2SD20Y	0208
2SD21	0208
2SD21A	0208
2SD21B	0208
2SD21C	0208
2SD21D	0208
2SD21E	0208
2SD21F	0208
2SD21G	0208
2SD21GN	0208
2SD21H	0208
2SD21J	0208
2SD21K	0208
2SD21L	0208
2SD21M	0208
2SD21OR	0208
2SD21R	0208
2SD21X	0208
2SD21Y	0208
2SD22	0208
2SD22A	0208
2SD22B	0208
2SD22C	0208
2SD22D	0208
2SD22E	0208
2SD22F	0208
2SD22G	0208
2SD22GN	0208
2SD22H	0208
2SD22J	0208
2SD22K	0208
2SD22L	0208
2SD22M	0208
2SD22OR	0208
2SD22R	0208
2SD22X	0208
2SD22Y	0208
2SD23	0208
2SD23A	0208
2SD23B	0208
2SD23C	0208
2SD23D	0208
2SD23E	0208
2SD23F	0208
2SD23G	0208
2SD23GN	0208
2SD23H	0208
2SD23J	0208
2SD23K	0208
2SD23L	0208
2SD23M	0208
2SD23OR	0208
2SD23R	0208
2SD23X	0208
2SD23Y	0208
2SD24	0142
2SD24(D)	0142
2SD24(F)	0142
2SD24A	0142
2SD24B	0142
2SD24C	0142
2SD24CK	0142
2SD24D	0142
2SD24DR	0142
2SD24E	0142
2SD24F	0142
2SD24G	0142
2SD24GN	0142
2SD24H	0142
2SD24J	0142
2SD24K	OEM
2SD24K(C)	0142
2SD24K(E)	0142
2SD24KC	0142
2SD24KD	0142
2SD24KE	0142
2SD24L	0142
2SD24M	0142
2SD24OR	0142
2SD24R	0142
2SD24X	0142
2SD24Y	0142
2SD24Y(C)	0142
2SD24Y(D,E)	0142
2SD24Y(E,F)	0142
2SD24Y(KC)	0142
2SD24YC	0142
2SD24YD	0142
2SD24YE	0142
2SD24YF	0142
2SD24YK	0142
2SD24YKC	0142
2SD24YL	0142
2SD24YLC	0142
2SD24YLD	0142
2SD24YLE	0142
2SD24YLF	0142
2SD24YM	0142
2SD24YM(C-E)	0142
2SD24YMC	0142
2SD24YME	0142
2SD25	0208
2SD25A	0208
2SD25B	0208
2SD25C	0208
2SD25D	0208
2SD25E	0208
2SD25G	0208
2SD25GN	0208
2SD25H	0208
2SD25J	0208
2SD25K	0208
2SD25L	0208
2SD25M	0208
2SD25OR	0208
2SD25R	0208
2SD25X	0208
2SD25Y	0208
2SD26	0867
2SD26A	0867
2SD26B	0867
2SD26C	0867
2SD26D	0867
2SD26E	0867
2SD26F	0867
2SD26G	0867
2SD26GN	0867
2SD26H	0867
2SD26J	0867
2SD26K	0867
2SD26L	0867
2SD26M	0867
2SD26OR	0867
2SD26R	0867
2SD26X	0867
2SD26Y	0867
2SD27	0208
2SD28	0042
2SD28A	0042
2SD28B	0042
2SD28C	0042
2SD28D	0042
2SD28E	0042
2SD28F	0042
2SD28G	0042
2SD28GN	0042
2SD28H	0042
2SD28J	0042
2SD28K	0042
2SD28L	0042
2SD28M	0042
2SD28OR	0042
2SD28R	0042
2SD28X	0042
2SD28Y	0042
2SD29	0042
2SD29A	0042
2SD29B	0042
2SD29E	0042
2SD29F	0042
2SD29GN	0042
2SD29J	0042
2SD29K	0042
2SD29L	0042
2SD29M	0042
2SD29OR	0042
2SD29R	0042
2SD29X	0042
2SD29Y	0042
2SD30	0208
2SD30-0	0208
2SD30-0R	0208
2SD30-N	0208
2SD30-O	0208
2SD30-OR	0208
2SD30A	0208
2SD30B	0208
2SD30C	0208
2SD30D	0208
2SD30E	0208
2SD30F	0208
2SD30G	0208
2SD30GN	0208
2SD30H	0208
2SD30J	0208
2SD30K	0208
2SD30L	0208
2SD30M	0208
2SD30N	0208
2SD30OR	0208
2SD30P	0208
2SD30R	0208
2SD30X	0208
2SD30Y	0208
2SD31	0208
2SD31B	0208
2SD31C	0208
2SD31D	0208
2SD31E	0208
2SD31F	0208
2SD31G	0208
2SD31GN	0208
2SD31H	0208
2SD31J	0208
2SD31K	0208
2SD31L	0208
2SD31M	0208
2SD31OR	0208
2SD31R	0208
2SD31X	0208
2SD31Y	0208
2SD32	0208
2SD32A	0208
2SD32B	0208
2SD32C	0208
2SD32D	0208
2SD32E	0208
2SD32F	0208
2SD32G	0208
2SD32GN	0208
2SD32H	0208
2SD32J	0208
2SD32K	0208
2SD32L	0208
2SD32M	0208
2SD32OR	0208
2SD32R	0208
2SD32X	0208
2SD32Y	0208
2SD33	0007
2SD33A	0007
2SD33B	0007
2SD33C	0007
2SD33D	0007
2SD33E	0007
2SD33F	0007
2SD33G	0007
2SD33GN	0007
2SD33H	0007
2SD33J	0007
2SD33K	0007
2SD33L	0007
2SD33OR	0007
2SD33X	0007
2SD33Y	0007
2SD34	0208
2SD34A	0208
2SD34B	0208
2SD34C	0208
2SD34D	0208
2SD34E	0208
2SD34F	0208
2SD34G	0208
2SD34GN	0208
2SD34H	0208
2SD34J	0208
2SD34K	0208
2SD34L	0208
2SD34M	0208
2SD34OR	0208
2SD34R	0208
2SD34X	0208
2SD34Y	0208
2SD35	0208
2SD36	0208
2SD36A	0208
2SD36B	0208
2SD36C	0208
2SD36D	0208
2SD36E	0208
2SD36F	0208
2SD36G	0208
2SD36GN	0208
2SD36H	0208
2SD36J	0208
2SD36K	0208
2SD36L	0208
2SD36M	0208
2SD36OR	0208
2SD36R	0208
2SD36X	0208
2SD36Y	0208
2SD37	0208
2SD37A	0208
2SD37B	0208
2SD37C	0208
2SD37D	0208
2SD37E	0208
2SD37F	0208
2SD37G	0208
2SD37GN	0208
2SD37H	0208
2SD37K	0208
2SD37L	0208
2SD37M	0208
2SD37OR	0208
2SD37R	0208
2SD37X	0208
2SD37Y	0208
2SD38	0208
2SD38A	0208
2SD38B	0208
2SD38C	0208
2SD38D	0208
2SD38E	0208
2SD38F	0208
2SD38G	0208
2SD38GN	0208
2SD38H	0208
2SD38J	0208
2SD38K	0208
2SD38L	0208
2SD38M	0208
2SD38OR	0208
2SD38R	0208
2SD38X	0208
2SD38Y	0208

If replacement code is OEM, contact original manufacturer for replacement.

DEVICE TYPE	REPL CODE
2SD40V5	OEM
2SD41	0130
2SD41A	0130
2SD41B	0130
2SD41C	0130
2SD41D	0130
2SD41E	0130
2SD41F	0130
2SD41G	0130
2SD41GN	0130
2SD41H	0130
2SD41J	0130
2SD41K	0130
2SD41L	0130
2SD41M	0130
2SD41OR	0130
2SD41R	0130
2SD41X	0130
2SD41Y	0130
2SD43	0208
2SD43A	0208
2SD43B	0208
2SD43C	0208
2SD43D	0208
2SD43E	0208
2SD43F	0208
2SD43G	0208
2SD43GN	0208
2SD43H	0208
2SD43J	0208
2SD43K	0208
2SD43L	0208
2SD43M	0208
2SD43OR	0208
2SD43R	0208
2SD43X	0208
2SD43Y	0208
2SD44	0208
2SD44A	0208
2SD44B	0208
2SD44C	0208
2SD44D	0208
2SD44E	0208
2SD44F	0208
2SD44G	0208
2SD44GN	0208
2SD44H	0208
2SD44J	0208
2SD44K	0208
2SD44L	0208
2SD44M	0208
2SD44R	0208
2SD44X	0208
2SD44Y	0208
2SD45	0074
2SD46	0074
2SD47	0620
2SD48	0042
2SD49	0042
2SD49A	0042
2SD49B	0042
2SD49C	0042
2SD49D	0042
2SD49E	0042
2SD49F	0042
2SD49G	0042
2SD49GN	0042
2SD49J	0042
2SD49K	0042
2SD49L	0042
2SD49M	0042
2SD49OR	0042
2SD49R	0042
2SD49X	0042
2SD49Y	0042
2SD50	0867
2SD50A	0867
2SD50B	0867
2SD50C	0867
2SD50D	0867
2SD50E	0867
2SD50F	0867
2SD50G	0867
2SD50GN	0867
2SD50H	0867
2SD50J	0867
2SD50K	0867
2SD50L	0867
2SD50M	0867
2SD50OR	0867
2SD50R	0867
2SD50X	0867
2SD50Y	0867
2SD51	0867
2SD51A	0867
2SD51B	0867
2SD51C	0867
2SD51D	0867
2SD51E	0867
2SD51F	0867
2SD51G	0867
2SD51GN	0867
2SD51H	0867
2SD51J	0867
2SD51K	0867
2SD51L	0867
2SD51M	0867
2SD51OR	0867
2SD51R	0867
2SD51X	0867
2SD51Y	0867
2SD52	0103
2SD53	0130
2SD53A	0130
2SD53B	0130
2SD53C	0130
2SD53E	0130
2SD53F	0130
2SD53G	0130
2SD53GN	0130
2SD53H	0130
2SD53J	0130
2SD53K	0130
2SD53L	0130
2SD53M	0130
2SD53OR	0130
2SD53R	0130
2SD53X	0130
2SD53Y	0130
2SD54	0130
2SD55	0130
2SD55A	0130
2SD56	4474
2SD56A	4474
2SD56B	4474
2SD56C	4474
2SD56D	4474
2SD56E	4474
2SD56F	4474
2SD56G	4474
2SD56GN	4474
2SD56H	4474
2SD56J	4474
2SD56K	4474
2SD56L	4474
2SD56M	4474
2SD56OR	4474
2SD56R	4474
2SD56X	4474
2SD56Y	4474
2SD57	0178
2SD58	0178
2SD59	0074
2SD60	0074
2SD61	0208
2SD61A	0208
2SD61B	0208
2SD61C	0208
2SD61D	0208
2SD61E	0208
2SD61F	0208
2SD61G	0208
2SD61GN	0208
2SD61H	0208
2SD61J	0208
2SD61K	0208
2SD61L	0208
2SD61M	0208
2SD61OR	0208
2SD61R	0208
2SD61X	0208
2SD61Y	0208
2SD62	0208
2SD62A	0208
2SD62B	0208
2SD62C	0208
2SD62D	0208
2SD62E	0208
2SD62G	0208
2SD62GN	0208
2SD62H	0208
2SD62J	0208
2SD62K	0208
2SD62L	0208
2SD62M	0208
2SD62R	0208
2SD62X	0208
2SD62Y	0208
2SD63	0208
2SD63A	0208
2SD63B	0208
2SD63C	0208
2SD63E	0208
2SD63F	0208
2SD63G	0208
2SD63GN	0208
2SD63H	0208
2SD63J	0208
2SD63K	0208
2SD63L	0208
2SD63M	0208
2SD63OR	0208
2SD63R	0208
2SD63X	0208
2SD63Y	0208
2SD64	0208
2SD64A	0208
2SD64B	0208
2SD64C	0208
2SD64D	0208
2SD64E	0208
2SD64F	0208
2SD64G	0208
2SD64GN	0208
2SD64H	0208
2SD64J	0208
2SD64K	0208
2SD64M	0208
2SD64OR	0208
2SD64R	0208
2SD64X	0208
2SD64Y	0208
2SD65	0208
2SD65-1	0208
2SD65A	0208
2SD65B	0208
2SD65C	0208
2SD65D	0208
2SD65E	0208
2SD65F	0208
2SD65G	0208
2SD65GN	0208
2SD65H	0208
2SD65J	0208
2SD65K	0208
2SD65L	0208
2SD65M	0208
2SD65OR	0208
2SD65R	0208
2SD65X	0208
2SD65Y	0208
2SD66	0208
2SD66A	0208
2SD66B	0208
2SD66C	0208
2SD66D	0208
2SD66E	0208
2SD66F	0208
2SD66G	0208
2SD66GN	0208
2SD66H	0208
2SD66J	0208
2SD66K	0208
2SD66L	0208
2SD66M	0208
2SD66OR	0208
2SD66R	0208
2SD66X	0208
2SD66Y	0208
2SD67	0177
2SD67B	0177
2SD67C	0177
2SD67D	0177
2SD67E	0177
2SD68	0130
2SD68A	0130
2SD68B	0130
2SD68C	0130
2SD68D	0130
2SD68E	0130
2SD68F	0130
2SD68G	0130
2SD68GN	0130
2SD68H	0130
2SD68J	0130
2SD68L	0130
2SD68M	0130
2SD68OR	0130
2SD68R	0130
2SD68X	0130
2SD68Y	0130
2SD69	6140
2SD70	0130
2SD70A	0130
2SD70B	0130
2SD70C	0130
2SD70D	0130
2SD70E	0130
2SD70F	0130
2SD70G	0130
2SD70GN	0130
2SD70H	0130
2SD70J	0130
2SD70K	0130
2SD70L	0130
2SD70M	0130
2SD70OR	0130
2SD70R	0130
2SD70X	0130
2SD70Y	0130
2SD71	0178
2SD72	0208
2SD72(K)	0208
2SD72-2C	0208
2SD72-3C	0208
2SD72-4C	0208
2SD72-6	0208
2SD72-OR	0208
2SD72A	0208
2SD72B	0208
2SD72C	0208
2SD72D	0208
2SD72E	0208
2SD72EJ	0208
2SD72F	0208
2SD72G	0208
2SD72GA	0208
2SD72H	0208
2SD72J	0208
2SD72K	0208
2SD72L	0208
2SD72M	0208
2SD72OR	0208
2SD72P	0208
2SD72R	0208
2SD72RE	0208
2SD72X	0208
2SD72Y	0208
2SD73	0867
2SD74	0286
2SD75	0208
2SD75(A)	0208
2SD75(B)	0208
2SD75A	0208
2SD75AH	0208
2SD75B	0208
2SD75C	0208
2SD75D	0208
2SD75E	0208
2SD75F	0208
2SD75G	0208
2SD75GN	0208
2SD75H	0208
2SD75J	0208
2SD75K	0208
2SD75L	0208
2SD75M	0208
2SD75OR	0208
2SD77	0208
2SD77(C)	0208
2SD77-A	0208
2SD77A	0208
2SD77AH	0208
2SD77B	0208
2SD77C	0208
2SD77D	0208
2SD77E	0208
2SD77F	0208
2SD77G	0208
2SD77GN	0208
2SD77H	0208
2SD77J	0208
2SD77L	0208
2SD77M	0208
2SD77OR	0208
2SD77P	0208
2SD77R	0208
2SD77X	0208
2SD77Y	0208
2SD78	3009
2SD78A	3009
2SD79	6829
2SD80	0130
2SD80A	0130
2SD80B	0130
2SD80C	0130
2SD80D	0130
2SD80E	0130
2SD80F	0130
2SD80G	0130
2SD80GN	0130
2SD80H	0130
2SD80J	0130
2SD80K	0130
2SD80L	0130
2SD80OR	0130
2SD80R	0130
2SD80X	0130
2SD80Y	0130
2SD81	0867
2SD81A	0867
2SD81B	0867
2SD81C	0867
2SD81E	0867
2SD81F	0867
2SD81G	0867
2SD81GN	0867
2SD81H	0867
2SD81J	0867
2SD81K	0867
2SD81L	0867
2SD81M	0867
2SD81OR	0867
2SD81R	0867
2SD81X	0867
2SD81Y	0867
2SD82	0867
2SD82-OR	0867
2SD82A	0867
2SD82B	0867
2SD82C	0867
2SD82D	0867
2SD82E	0867
2SD82F	0867
2SD82G	0867
2SD82GN	0867
2SD82H	0867
2SD82J	0867
2SD82K	0867
2SD82L	0867
2SD82M	0867
2SD82OR	0867
2SD82R	0867
2SD82X	0867
2SD82Y	0867
2SD83	0867
2SD84	0867
2SD85GR	0728
2SD88	0867
2SD88A	0867
2SD88OY	0867
2SD90	0042
2SD90A	0042
2SD90B	0042
2SD90C	0042
2SD90D	0042
2SD90E	0042
2SD90F	0042
2SD90G	0042
2SD90GN	0042
2SD90H	0042
2SD90J	0042
2SD90K	0042
2SD90L	0042
2SD90M	0042
2SD90OR	0042
2SD90X	0042
2SD90Y	0042
2SD91	0042
2SD91A	0042
2SD91B	0042
2SD91E	0042
2SD91F	0042
2SD91G	0042
2SD91GN	0042
2SD91H	0042
2SD91J	0042
2SD91L	0042
2SD91M	0042
2SD91OR	0042
2SD91R	0042
2SD91X	0042
2SD91Y	0042
2SD92	0178
2SD92D	0178
2SD93	0074
2SD94	0392
2SD96	0208
2SD96A	0208
2SD96B	0208
2SD96C	0208
2SD96D	0208
2SD96E	0208
2SD96F	0208
2SD96G	0208
2SD96GN	0208
2SD96H	0208
2SD96J	0208
2SD96K	0208
2SD96L	0208
2SD96OR	0208
2SD96R	0208
2SD96X	0208
2SD96Y	0208
2SD100	0208
2SD100A	0208
2SD100B	0208
2SD100C	0208
2SD100D	0208
2SD100E	0208
2SD100F	0208
2SD100G	0208
2SD100GN	0208
2SD100H	0208
2SD100J	0208
2SD100K	0208
2SD100L	0208
2SD100M	0208
2SD100OR	0208
2SD100R	0208
2SD100X	0208
2SD100Y	0208
2SD101	0600
2SD101AK	0600
2SD102	0178
2SD102-O	0178
2SD102-R	0178
2SD102-Y	0178
2SD102A	0178
2SD102B	0178
2SD102C	0178
2SD102D	0178
2SD102E	0178
2SD102F	0178
2SD102G	0178
2SD102GN	0178
2SD102H	0178
2SD102J	0178
2SD102K	0178
2SD102L	0178
2SD102M	0178
2SD102OR	0178
2SD102R	0178
2SD102X	0178
2SD102Y	0178
2SD103	0178
2SD103-O	0178
2SD103-R	0178
2SD103-Y	0178
2SD104	0208
2SD104A	0208
2SD104B	0208
2SD104C	0208
2SD104D	0208
2SD104E	0208
2SD104F	0208
2SD104G	0208
2SD104GN	0208
2SD104H	0208
2SD104K	0208
2SD104L	0208
2SD104M	0208
2SD104OR	0208
2SD104R	0208
2SD104X	0208
2SD104Y	0208
2SD105	0208
2SD105A	0208
2SD105B	0208
2SD105C	0208
2SD105D	0208
2SD105E	0208
2SD105F	0208
2SD105G	0208
2SD105GN	0208
2SD105H	0208
2SD105J	0208
2SD105K	0208
2SD105L	0208
2SD105M	0208
2SD105OR	0208
2SD105R	0208
2SD105X	0208
2SD105Y	0208
2SD107	0103
2SD108	0103
2SD110	0177
2SD110-0	0177
2SD110-O	0177
2SD110-ORG	0177
2SD110-R	0177
2SD110-RED	0177
2SD110-Y	0177
2SD110-YEL	0177
2SD111	0177
2SD111-O	0177
2SD111-ORG	0177
2SD111-R	0177
2SD111-RED	0177
2SD111-Y	0177
2SD111-YEL	0177
2SD111R	0177
2SD113	0130
2SD113-O	0130
2SD113-ORG	0130
2SD113-R	0130
2SD113-RED	0130
2SD113-Y	0130
2SD113-YEL	0130
2SD113R	0130
2SD114	0130
2SD114-O	0130
2SD114-ORG	0130
2SD114-R	0130
2SD114-RED	0130
2SD114-Y	0130
2SD114-YEL	0130
2SD114R	0130
2SD116	0103
2SD117	0620
2SD117C	0620
2SD118	0177
2SD118-B	0177
2SD118-BL	0177
2SD118-BLU	0177
2SD118-R	0177
2SD118-RED	0177
2SD118-Y	0177
2SD118-YEL	0177
2SD118A	0177
2SD118BL	0177
2SD118C	0177
2SD118D	0177
2SD118R	0177
2SD118Y	0177
2SD119	0177
2SD119-BL	0177
2SD119-BLU	0177
2SD119-R	0177
2SD119-RED	0177
2SD119-Y	0177
2SD119-YEL	0177
2SD119A	0177
2SD119B	0177
2SD119BL	0177
2SD119C	0177
2SD119D	0177
2SD119R	0177
2SD119Y	0177
2SD120	1480
2SD120D	1480
2SD120E	1480
2SD120F	1480
2SD120G	1480
2SD120GN	1480
2SD120H	1480
2SD120J	1480
2SD120K	1480
2SD120L	1480
2SD120M	1480
2SD120OR	1480
2SD120R	1480
2SD120X	1480
2SD120Y	1480
2SD121	1480
2SD121C	1480
2SD121D	1480
2SD121E	1480
2SD121F	1480
2SD121G	1480
2SD121GN	1480
2SD121H	1480
2SD121J	1480
2SD121K	1480
2SD121L	1480
2SD121M	1480
2SD121OR	1480
2SD121R	1480
2SD121X	1480
2SD121Y	1480
2SD122	0042
2SD123	0236
2SD123(P)	OEM
2SD124	0867
2SD124A	0867
2SD124AH	0867
2SD124B	0867
2SD124C	0867
2SD124E	0867
2SD124F	0867
2SD124G	0867
2SD124GN	0867
2SD124H	0867
2SD124J	0867
2SD124K	0867
2SD124L	0867
2SD124M	0867
2SD124OR	0867
2SD124R	0867
2SD124X	0867
2SD124Y	0867
2SD125	0867
2SD125A	0867
2SD125AH	0867
2SD125B	0867
2SD125C	0867
2SD125E	0867
2SD125F	0867
2SD125G	0867
2SD125GN	0867
2SD125H	0867
2SD125J	0867
2SD125K	0867
2SD125L	0867
2SD125M	0867
2SD125OR	0867
2SD125R	0867
2SD125X	0867
2SD125Y	0867
2SD126	0867
2SD126H	0867
2SD126HA	0867
2SD126HB	0867
2SD126S	0867
2SD127	0208
2SD127A	0208
2SD127B	0208
2SD127C	0208
2SD127D	0208
2SD127E	0208
2SD127F	0208
2SD127G	0208
2SD127GN	0208
2SD127H	0208
2SD127J	0208
2SD127K	0208
2SD127L	0208
2SD127M	0208
2SD127OR	0208
2SD127R	0208
2SD127X	0208
2SD127Y	0208
2SD128	0208
2SD128A	0208
2SD128B	0208
2SD128C	0208
2SD128D	0208
2SD128E	0208
2SD128F	0208
2SD128G	0208
2SD128GN	0208
2SD128H	0208
2SD128J	0208
2SD128K	0208
2SD128L	0208
2SD128M	0208
2SD128OR	0208
2SD128R	0208
2SD128X	0208
2SD128Y	0208
2SD129	0178
2SD129-BL	0178
2SD129-BLU	0178
2SD129-R	0178
2SD129-RED	0178
2SD129-Y	0178
2SD129-YEL	0178
2SD130	1147
2SD130(BL)	1147
2SD130(FA-1)	OEM
2SD130(Y)	1147
2SD130-BL	1147
2SD130-BLU	1147
2SD130-R	1147
2SD130-RED	1147
2SD130-Y	1147
2SD130-YEL	1147
2SD130A	1147
2SD130B	1147
2SD130BL	1147
2SD130C	1147
2SD130D	1147
2SD130E	1147

DEVICE TYPE	REPL CODE
2SD130F	1147
2SD130G	1147
2SD130GN	1147
2SD130H	1147
2SD130J	1147
2SD130K	1147
2SD130OR	1147
2SD130R	1147
2SD130X	1147
2SD130Y	1147
2SD131	0620
2SD132	0130
2SD134	0018
2SD134A	0018
2SD134C	0018
2SD134D	0018
2SD134F	0018
2SD134G	0018
2SD134GN	0018
2SD134H	0018
2SD134J	0018
2SD134K	0018
2SD134L	0018
2SD134M	0018
2SD134OR	0018
2SD134R	0018
2SD134X	0018
2SD134Y	0018
2SD136	0142
2SD136D	0142
2SD136E	0142
2SD136F	0142
2SD136G	0142
2SD136GN	0142
2SD136J	0142
2SD136K	0142
2SD136L	0142
2SD136M	0142
2SD136OR	0142
2SD136R	0142
2SD136X	0142
2SD136Y	0142
2SD137	0142
2SD137A	0142
2SD137B	0142
2SD137C	0142
2SD137D	0142
2SD137F	0142
2SD137G	0142
2SD137H	0142
2SD137J	0142
2SD137K	0142
2SD137L	0142
2SD137M	0142
2SD137OR	0142
2SD137R	0142
2SD137X	0142
2SD137Y	0142
2SD138	0168
2SD139	0168
2SD141	0042
2SD141A	0042
2SD141B	0042
2SD141C	0042
2SD141E	0042
2SD141F	0042
2SD141G	0042
2SD141GN	0042
2SD141H	0042
2SD141H01	0042
2SD141J	0042
2SD141K	0042
2SD141L	0042
2SD141M	0042
2SD141OR	0042
2SD141R	0042
2SD141X	0042
2SD141Y	0042
2SD142	0178
2SD142A	0178
2SD142B	0178
2SD142C	0178
2SD142D	0178
2SD142E	0178
2SD142F	0178
2SD142G	0178
2SD142GN	0178
2SD142H	0178
2SD142J	0178
2SD142K	0178
2SD142L	0178
2SD142M	0178
2SD142OR	0178
2SD142R	0178
2SD142X	0178
2SD142Y	0178
2SD143	0178
2SD144	0178
2SD145	0178
2SD146	0042
2SD146A	0042
2SD146B	0042
2SD146C	0042
2SD146D	0042
2SD146E	0042
2SD146F	0042
2SD146G	0042
2SD146GN	0042
2SD146H	0042

DEVICE TYPE	REPL CODE
2SD146J	0042
2SD146K	0042
2SD146L	0042
2SD146OR	0042
2SD146R	0042
2SD146UK	0042
2SD146VK	0042
2SD146X	0042
2SD146Y	0042
2SD147	0042
2SD147A	0042
2SD147B	0042
2SD147C	0042
2SD147D	0042
2SD147E	0042
2SD147F	0042
2SD147GN	0042
2SD147H	0042
2SD147J	0042
2SD147K	0042
2SD147L	0042
2SD147M	0042
2SD147OR	0042
2SD147R	0042
2SD147X	0042
2SD147Y	0042
2SD148	0042
2SD149	0086
2SD150	0178
2SD150A	0178
2SD150B	0178
2SD150C	0178
2SD150E	0178
2SD150F	0178
2SD150G	0178
2SD150GN	0178
2SD150H	0178
2SD150J	0178
2SD150K	0178
2SD150L	0178
2SD150M	0178
2SD150OR	0178
2SD150R	0178
2SD150X	0178
2SD150Y	0178
2SD151	0620
2SD151A	0620
2SD151B	0620
2SD151C	0620
2SD151E	0620
2SD151F	0620
2SD151G	0620
2SD151GN	0620
2SD151H	0620
2SD151J	0620
2SD151K	0620
2SD151L	0620
2SD151M	0620
2SD151OR	0620
2SD151R	0620
2SD151Y	0620
2SD152	0392
2SD152(L)	0392
2SD152(M)	0392
2SD152A	0392
2SD152B	0392
2SD152C	0392
2SD152D	0392
2SD152E	0392
2SD152F	0392
2SD152G	0392
2SD152GN	0392
2SD152H	0392
2SD152J	0392
2SD152K	0392
2SD152L	0392
2SD152M	0392
2SD152OR	0392
2SD152R	0392
2SD152X	0392
2SD152Y	0392
2SD154	0236
2SD154(L)	0236
2SD154A	0236
2SD154B	0236
2SD154C	0236
2SD154E	0236
2SD154F	0236
2SD154G	0236
2SD154GN	0236
2SD154H	0236
2SD154J	0236
2SD154K	0236
2SD154L	0236
2SD154M	0236
2SD154R	0236
2SD154X	0236
2SD154Y	0236
2SD155	0178
2SD155(K)	0178
2SD155(L)	0178
2SD155-H	0178
2SD155-K	0178
2SD155H	0178
2SD155K	0178
2SD155L	0178
2SD156	0142
2SD156B	0142
2SD156C	0142

DEVICE TYPE	REPL CODE
2SD156D	0142
2SD156E	0142
2SD156F	0142
2SD156G	0142
2SD156GN	0142
2SD156H	0142
2SD156K	0142
2SD156L	0142
2SD156M	0142
2SD156OR	0142
2SD156R	0142
2SD156X	0142
2SD156Y	0142
2SD157	0142
2SD157A	0142
2SD157B	0142
2SD157C	0142
2SD157D	0142
2SD157G	0142
2SD157GN	0142
2SD157H	0142
2SD157J	0142
2SD157K	0142
2SD157L	0142
2SD157OR	0142
2SD157R	0142
2SD157X	0142
2SD157Y	0142
2SD158	1027
2SD158A	1027
2SD158B	1027
2SD158C	1027
2SD158D	1027
2SD158E	1027
2SD158F	1027
2SD158GN	1027
2SD158H	1027
2SD158J	1027
2SD158K	1027
2SD158L	1027
2SD158M	1027
2SD158OR	1027
2SD158R	1027
2SD158X	1027
2SD158Y	1027
2SD159	1027
2SD159A	1027
2SD159B	1027
2SD159C	1027
2SD159D	1027
2SD159E	1027
2SD159F	1027
2SD159G	1027
2SD159GN	1027
2SD159H	1027
2SD159J	1027
2SD159K	1027
2SD159L	1027
2SD159M	1027
2SD159OR	1027
2SD159R	1027
2SD159X	1027
2SD159Y	1027
2SD160	0630
2SD161	0620
2SD161A	0620
2SD161B	0620
2SD161C	0620
2SD161D	0620
2SD161E	0620
2SD161F	0620
2SD161G	0620
2SD161GN	0620
2SD161H	0620
2SD161J	0620
2SD161K	0620
2SD161L	0620
2SD161M	0620
2SD161OR	0620
2SD161R	0620
2SD161X	0620
2SD161Y	0620
2SD162	0208
2SD162A	0208
2SD162B	0208
2SD162C	0208
2SD162D	0208
2SD162E	0208
2SD162F	0208
2SD162G	0208
2SD162GN	0208
2SD162H	0208
2SD162J	0208
2SD162K	0208
2SD162L	0208
2SD162M	0208
2SD162OR	0208
2SD162R	0208
2SD162X	0208
2SD162Y	0208
2SD163	0103
2SD163A	0103
2SD163B	0103
2SD163C	0103
2SD163D	0103
2SD163E	0103
2SD163F	0103
2SD163G	0103
2SD163GN	0103

DEVICE TYPE	REPL CODE
2SD163H	0103
2SD163J	0103
2SD163K	0103
2SD163L	0103
2SD163M	0103
2SD163OR	0103
2SD163R	0103
2SD163X	0103
2SD163Y	0103
2SD164	0103
2SD164A	0103
2SD164B	0103
2SD164C	0103
2SD164D	0103
2SD164E	0103
2SD164F	0103
2SD164G	0103
2SD164GN	0103
2SD164H	0103
2SD164J	0103
2SD164K	0103
2SD164L	0103
2SD164M	0103
2SD164OR	0103
2SD164R	0103
2SD164X	0103
2SD164Y	0103
2SD165	1966
2SD166	1966
2SD167	0208
2SD167A	0208
2SD167B	0208
2SD167C	0208
2SD167E	0208
2SD167F	0208
2SD167G	0208
2SD167GN	0208
2SD167H	0208
2SD167J	0208
2SD167K	0208
2SD167L	0208
2SD167M	0208
2SD167OR	0208
2SD167R	0208
2SD167X	0208
2SD167Y	0208
2SD168	3373
2SD170	0208
2SD170(A)	0208
2SD170(B)	0208
2SD170A	0208
2SD170A/PB	0208
2SD170AA	0208
2SD170AB	0208
2SD170AC	0208
2SD170B	0208
2SD170BC	0208
2SD170C	0208
2SD171	0642
2SD172	0103
2SD172A	0103
2SD172B	0103
2SD172C	0103
2SD172D	0103
2SD172E	0103
2SD172F	0103
2SD172G	0103
2SD172GN	0103
2SD172H	0103
2SD172J	0103
2SD172K	0103
2SD172L	0103
2SD172M	0103
2SD172OR	0103
2SD172R	0103
2SD172X	0103
2SD172Y	0103
2SD173	0130
2SD173A	0130
2SD173B	0130
2SD173C	0130
2SD173D	0130
2SD173E	0130
2SD173F	0130
2SD173G	0130
2SD173GN	0130
2SD173H	0130
2SD173J	0130
2SD173K	0130
2SD173M	0130
2SD173OR	0130
2SD173X	0130
2SD173Y	0130
2SD174	0103
2SD174A	0103
2SD174B	0103
2SD174C	0103
2SD174D	0103
2SD174E	0103
2SD174F	0103
2SD174G	0103
2SD174GN	0103
2SD174H	0103
2SD174J	0103
2SD174K	0103
2SD174L	0103
2SD174M	0103
2SD174R	0103
2SD174X	0103

DEVICE TYPE	REPL CODE
2SD174Y	0103
2SD175	0103
2SD175A	0103
2SD175B	0103
2SD175C	0103
2SD175D	0103
2SD175E	0103
2SD175F	0103
2SD175G	0103
2SD175GN	0103
2SD175H	0103
2SD175J	0103
2SD175K	0103
2SD175L	0103
2SD175M	0103
2SD175OR	0103
2SD175R	0103
2SD175X	0103
2SD175Y	0103
2SD176	0130
2SD176A	0130
2SD176B	0130
2SD176C	0130
2SD176D	0130
2SD176E	0130
2SD176F	0130
2SD176G	0130
2SD176GN	0130
2SD176H	0130
2SD176J	0130
2SD176K	0130
2SD176L	0130
2SD176M	0130
2SD176OR	0130
2SD176X	0130
2SD176Y	0130
2SD177	0620
2SD178	0208
2SD178A	0208
2SD178B	0208
2SD178C	0208
2SD178D	0208
2SD178E	0208
2SD178F	0208
2SD178G	0208
2SD178GN	0208
2SD178H	0208
2SD178J	0208
2SD178K	0208
2SD178L	0208
2SD178M	0208
2SD178Q	0208
2SD178T	0208
2SD178X	0208
2SD178Y	0208
2SD180	2793
2SD180A	2793
2SD180B	2793
2SD180C	2793
2SD180D	2793
2SD180F	2793
2SD180G	2793
2SD180GN	2793
2SD180H	2793
2SD180J	2793
2SD180K	2793
2SD180L	2793
2SD180M	2793
2SD180OR	2793
2SD180R	2793
2SD180X	2793
2SD180Y	2793
2SD181	1966
2SD181A	1966
2SD182	0086
2SD182A	0086
2SD182B	0086
2SD182C	0086
2SD182D	0086
2SD182E	0086
2SD182F	0086
2SD182G	0086
2SD182GN	0086
2SD182H	0086
2SD182J	0086
2SD182K	0086
2SD182L	0086
2SD182OR	0086
2SD182R	0086
2SD182X	0086
2SD182Y	0086
2SD183	0086
2SD183B	0086
2SD183E	0086
2SD183GN	0086
2SD183K	0086
2SD183L	0086
2SD183OR	0086
2SD183R	0086
2SD184	0042
2SD184A	0042
2SD184B	0042
2SD184C	0042
2SD184D	0042
2SD184E	0042
2SD184F	0042
2SD184H	0042
2SD184J	0042
2SD184K	0042

DEVICE TYPE	REPL CODE
2SD184L	0042
2SD184M	0042
2SD184OR	0042
2SD184Y	0042
2SD185	0042
2SD186	0208
2SD186A	0208
2SD186B	0208
2SD186C	0208
2SD186D	0208
2SD186E	0208
2SD186F	0208
2SD186G	0208
2SD186GN	0208
2SD186H	0208
2SD186J	0208
2SD186K	0208
2SD186L	0208
2SD186M	0208
2SD186OR	0208
2SD186R	0208
2SD186X	0208
2SD186Y	0208
2SD187	0208
2SD187-OR	0208
2SD187A	0208
2SD187B	0208
2SD187C	0208
2SD187D	0208
2SD187E	0208
2SD187F	0208
2SD187G	0208
2SD187GN	0208
2SD187H	0208
2SD187J	0208
2SD187K	0208
2SD187L	0208
2SD187M	0208
2SD187OR	0208
2SD187R	0208
2SD187X	0208
2SD187Y	0208
2SD188	2556
2SD188A	2556
2SD188B	2556
2SD188C	2556
2SD188D	2556
2SD188E	2556
2SD188F	2556
2SD188G	2556
2SD188GN	2556
2SD188H	2556
2SD188J	2556
2SD188L	2556
2SD188M	2556
2SD188OR	2556
2SD188R	2556
2SD188X	2556
2SD188Y	2556
2SD189	0177
2SD189A	0177
2SD189B	0177
2SD189C	0177
2SD189D	0177
2SD189E	0177
2SD189F	0177
2SD189G	0177
2SD189GN	0177
2SD189H	0177
2SD189J	0177
2SD189K	0177
2SD189L	0177
2SD189M	0177
2SD189OR	0177
2SD189R	0177
2SD189X	0177
2SD189Y	0177
2SD190	0142
2SD190B	0142
2SD190C	0142
2SD190D	0142
2SD190E	0142
2SD190F	0142
2SD190GN	0142
2SD190H	0142
2SD190J	0142
2SD190K	0142
2SD190L	0142
2SD190M	0142
2SD190OR	0142
2SD190R	0142
2SD190X	0142
2SD190Y	0142
2SD191	0208
2SD192	0208
2SD193	0208
2SD194	0208
2SD195	0208
2SD195A	0208
2SD195B	0208
2SD195C	0208
2SD195D	0208
2SD195E	0208
2SD195F	0208
2SD195G	0208
2SD195GN	0208
2SD195H	0208
2SD195J	0208
2SD195K	0208

DEVICE TYPE	REPL CODE
2SD195L	0208
2SD195M	0208
2SD195OR	0208
2SD195R	0208
2SD195X	0208
2SD195Y	0208
2SD198	0042
2SD198(H)	0042
2SD198(R)	0042
2SD198A	0042
2SD198AP	0042
2SD198A-Q	OEM
2SD198AQ	0042
2SD198A-R	OEM
2SD198AR	0042
2SD198B	0042
2SD198C	0042
2SD198D	0042
2SD198F	0042
2SD198G	0042
2SD198H	0042
2SD198HQ	0042
2SD198HR	0042
2SD198J	0042
2SD198K	0042
2SD198OR	0042
2SD198P	0042
2SD198Q	0042
2SD198R	0042
2SD198S	0042
2SD198V	0042
2SD198X	0042
2SD198Y	0042
2SD199	0003
2SD200	0223
2SD200-OR	0223
2SD200A	0223
2SD200B	0223
2SD200C	0223
2SD200D	0223
2SD200E	0223
2SD200F	0223
2SD200G	0223
2SD200GN	0223
2SD200H	0223
2SD200J	0223
2SD200K	0223
2SD200L	0223
2SD200M	0223
2SD200OR	0223
2SD200X	0223
2SD200Y	0223
2SD201	0867
2SD201-O	0867
2SD201A	0867
2SD201B	0867
2SD201C	0867
2SD201F	0867
2SD201G	0867
2SD201GN	0867
2SD201H	0867
2SD201J	0867
2SD201K	0867
2SD201L	0867
2SD201M	0867
2SD201M(O)	0867
2SD201M(O,Y)	0867
2SD201M(Y)	0867
2SD201MO	0867
2SD201MY	0867
2SD201Q	0867
2SD201X	0867
2SD201Y	0867
2SD202	0867
2SD203	0867
2SD204	0086
2SD204(L)	0086
2SD204A	0086
2SD204B	0086
2SD204BL	0086
2SD204C	0086
2SD204D	0086
2SD204E	0086
2SD204F	0086
2SD204G	0086
2SD204GA	0086
2SD204GN	0086
2SD204H	0086
2SD204K	0086
2SD204L	0086
2SD204M	0086
2SD204R	0086
2SD204Y	0086
2SD205	0086
2SD205(L)	0086
2SD205(M)	0086
2SD205A	0086
2SD205B	0086
2SD205C	0086
2SD205D	0086
2SD205E	0086
2SD205F	0086
2SD205G	0086
2SD205GN	0086
2SD205H	0086
2SD205J	0086
2SD205L	0086
2SD205M	0086
2SD205OR	0086

If replacement code is OEM, contact original manufacturer for replacement.

DEVICE TYPE	REPL CODE
2SD205R	0086
2SD205X	0086
2SD205Y	0086
2SD206	0543
2SD206A	0543
2SD207	0103
2SD207A	0103
2SD208	0620
2SD208A	0620
2SD211	0103
2SD211-O	0103
2SD211-OR	0103
2SD211A	0103
2SD211B	0103
2SD211C	0103
2SD211D	0103
2SD211E	0103
2SD211F	0103
2SD211G	0103
2SD211GN	0103
2SD211H	0103
2SD211J	0103
2SD211K	0103
2SD211L	0103
2SD211M	0103
2SD211M(O)	0103
2SD211M(Y)	0103
2SD211OR	0103
2SD211R	0103
2SD211X	0103
2SD211Y	0103
2SD212	0103
2SD212A	0103
2SD212B	0103
2SD212C	0103
2SD212D	0103
2SD212E	0103
2SD212F	0103
2SD212G	0103
2SD212GN	0103
2SD212H	0103
2SD212J	0103
2SD212K	0103
2SD212L	0103
2SD212M	0103
2SD212OR	0103
2SD212R	0103
2SD212X	0103
2SD212Y	0103
2SD213	0177
2SD214	0177
2SD215	0086
2SD216	0086
2SD217	0867
2SD218	0867
2SD219	0086
2SD219A	0086
2SD219B	0086
2SD219C	0086
2SD219D	0086
2SD219E	0086
2SD219F	0086
2SD219G	0086
2SD219GN	0086
2SD219H	0086
2SD219J	0086
2SD219K	0086
2SD219L	0086
2SD219M	0086
2SD219OR	0086
2SD219R	0086
2SD219X	0086
2SD219Y	0086
2SD220	0086
2SD220A	0086
2SD220B	0086
2SD220C	0086
2SD220D	0086
2SD220E	0086
2SD220F	0086
2SD220G	0086
2SD220GN	0086
2SD220H	0086
2SD220J	0086
2SD220K	0086
2SD220L	0086
2SD220M	0086
2SD220OR	0086
2SD220R	0086
2SD220X	0086
2SD220Y	0086
2SD221	0086
2SD221F	0086
2SD222	1480
2SD222A	1480
2SD222B	1480
2SD222C	1480
2SD222D	1480
2SD222E	1480
2SD222F	1480
2SD222G	1480
2SD222GN	1480
2SD222H	1480
2SD222J	1480
2SD222K	1480
2SD222L	1480
2SD222M	1480
2SD222OR	1480
2SD222R	1480
2SD222X	1480
2SD222Y	1480
2SD223	1480
2SD223A	1480
2SD223B	1480
2SD223C	1480
2SD223D	1480
2SD223E	1480
2SD223F	1480
2SD223G	1480
2SD223GN	1480
2SD223H	1480
2SD223J	1480
2SD223K	1480
2SD223L	1480
2SD223M	1480
2SD223OR	1480
2SD223R	1480
2SD223X	1480
2SD223Y	1480
2SD224	1480
2SD226	0178
2SD226-0	0178
2SD226-O	0178
2SD226A	0178
2SD226A(O)	0178
2SD226AO	0178
2SD226AP	0178
2SD226B	0178
2SD226BP	0178
2SD226C	0178
2SD226E	0178
2SD226F	0178
2SD226G	0178
2SD226GN	0178
2SD226H	0178
2SD226J	0178
2SD226K	0178
2SD226L	0178
2SD226M	0178
2SD226OR	0178
2SD226P	0178
2SD226Q	0178
2SD226R	0178
2SD226X	0178
2SD226Y	0178
2SD227	1419
2SD227(PANASONIC)	1419
2SD227(R)	1419
2SD227-175	1419
2SD227-OR	1419
2SD227A	1419
2SD227B	1419
2SD227C	1419
2SD227D	1419
2SD227E	1419
2SD227G	1419
2SD227GN	1419
2SD227H	1419
2SD227J	1419
2SD227K	1419
2SD227L	1419
2SD227LF	1419
2SD227M	1419
2SD227OR	1419
2SD227R	1419
2SD227V	1419
2SD227X	1419
2SD227Y	1419
2SD228	0155
2SD228F	0155
2SD228G	0155
2SD228GN	0155
2SD228H	0155
2SD228J	0155
2SD228K	0155
2SD228L	0155
2SD228M	0155
2SD228OR	0155
2SD228R	0155
2SD228X	0155
2SD228Y	0155
2SD231	0130
2SD232	0130
2SD232A	0130
2SD234	0228
2SD234(O)	0228
2SD234-0	0228
2SD234-O	0228
2SD234-ORG	0228
2SD234-R	0228
2SD234-RED	0228
2SD234-Y	0228
2SD234-YEL	0228
2SD234A	0228
2SD234B	0228
2SD234C	0228
2SD234D	0228
2SD234E	0228
2SD234F	0228
2SD234G	0228
2SD234GA	0228
2SD234GN	0228
2SD234G-O	0228
2SD234G-R	0228
2SD234GR	0228
2SD234G-Y	0228
2SD234H	0228
2SD234J	0228
2SD234K	0228
2SD234L	0228
2SD234M	0228
2SD234N	0228
2SD234O	0228
2SD234OR	0228
2SD234R	0228
2SD234X	0228
2SD234Y	0228
2SD235	0228
2SD235(O)	0228
2SD235(R)	0228
2SD235(Y)	0228
2SD235-0	0228
2SD235-O	0228
2SD235-OR	0228
2SD235-ORG	0228
2SD235-R	0228
2SD235-RED	0228
2SD235-Y	0228
2SD235-YEL	0228
2SD235-Y.JA	0228
2SD235A	0228
2SD235B	0228
2SD235C	0228
2SD235D	0228
2SD235E	0228
2SD235F	0228
2SD235G	0228
2SD235GN	0228
2SD235G-O	0228
2SD235G-R	0228
2SD235G-Y	0228
2SD235H	0228
2SD235J	0228
2SD235L	0228
2SD235LBY	0228
2SD235M	0228
2SD235O	0228
2SD235OR	0228
2SD235R	0228
2SD235RED	0228
2SD235RY	0228
2SD235X	0228
2SD235Y	0228
2SD236	4098
2SD236(02Y)	4098
2SD236A	4098
2SD236B	4098
2SD236C	4098
2SD236D	4098
2SD236E	4098
2SD236F	4098
2SD236G	4098
2SD236GN	4098
2SD236H	4098
2SD236J	4098
2SD236K	4098
2SD236L	4098
2SD236M	4098
2SD236R	4098
2SD236X	4098
2SD236Y	4098
2SD237	0178
2SD238	0178
2SD238(F)	0178
2SD238F	0178
2SD241	0042
2SD242	0423
2SD242YLC	0423
2SD243	0236
2SD244	0236
2SD245K(C)	OEM
2SD246	0223
2SD246A	0223
2SD246B	0223
2SD246C	0223
2SD246D	0223
2SD246E	0223
2SD246F	0223
2SD246G	0223
2SD246GN	0223
2SD246H	0223
2SD246J	0223
2SD246K	0223
2SD246L	0223
2SD246M	0223
2SD246OR	0223
2SD246R	0223
2SD246X	0223
2SD246Y	0223
2SD247	0620
2SD247K	0620
2SD249	0130
2SD250	0130
2SD251	1027
2SD254	0178
2SD255	0178
2SD256	0042
2SD257	0423
2SD258	0236
2SD259	0236
2SD260	0620
2SD261	0218
2SD261(L)	0218
2SD261(O)	0218
2SD261(O,Q)	0218
2SD261(Q)	0218
2SD261(R)	0218
2SD261(V)	0218
2SD261-0	0218
2SD261-O	0218
2SD261A	0218
2SD261B	0218
2SD261C	0218
2SD261D	0218
2SD261E	0218
2SD261F	0218
2SD261G	0218
2SD261GN	0218
2SD261H	0218
2SD261J	0218
2SD261K	0218
2SD261L	0218
2SD261M	0218
2SD261OR	0218
2SD261P	0218
2SD261Q	0218
2SD261R	0218
2SD261S	0218
2SD261U	0218
2SD261V	0218
2SD261W	0218
2SD261X	0218
2SD261Y	0218
2SD262	0363
2SD265	0363
2SD266	0363
2SD271R	OEM
2SD272	0363
2SD273	0363
2SD274	0363
2SD280	OEM
2SD283	0074
2SD284	0429
2SD285	0074
2SD286	0543
2SD287	0543
2SD287A	0543
2SD287B	0543
2SD287C	0543
2SD288	0423
2SD288(O)	0423
2SD288A	0423
2SD288B	0423
2SD288C	0423
2SD288K	0423
2SD288L	0423
2SD289	0423
2SD289A	0423
2SD289B	0423
2SD289C	0423
2SD290	0178
2SD291	4098
2SD291(R)	4098
2SD291-0	4098
2SD291B	4098
2SD291BL	4098
2SD291C	4098
2SD291D	4098
2SD291E	4098
2SD291F	4098
2SD291G	4098
2SD291GA	4098
2SD291GN	4098
2SD291H	4098
2SD291J	4098
2SD291K	4098
2SD291L	4098
2SD291M	4098
2SD291OR	4098
2SD291R	4098
2SD291X	4098
2SD291Y	4098
2SD292	0042
2SD292-0	0042
2SD292A	0042
2SD292B	0042
2SD292BL	0042
2SD292C	0042
2SD292D	0042
2SD292E	0042
2SD292F	0042
2SD292G	0042
2SD292GA	0042
2SD292GN	0042
2SD292H	0042
2SD292J	0042
2SD292K	0042
2SD292L	0042
2SD292M	0042
2SD292OR	0042
2SD292R	0042
2SD292X	0042
2SD292Y	0042
2SD293	0363
2SD294	0363
2SD297	0178
2SD299	0223
2SD299(V)	0223
2SD299A	0223
2SD299B	0223
2SD299C	0223
2SD299E	0223
2SD299F	0223
2SD299G	0223
2SD299GN	0223
2SD299H	0223
2SD299J	0223
2SD299K	0223
2SD299M	0223
2SD299OR	0223
2SD299R	0223
2SD299SL	0223
2SD299X	0223
2SD299Y	0223
2SD300	0223
2SD300A	0223
2SD300B	0223
2SD300C	0223
2SD300D	0223
2SD300G	0223
2SD300H	0223
2SD300J	0223
2SD301	4013
2SD310	1852
2SD311	1852
2SD312	0003
2SD312A	0003
2SD312B	0003
2SD312C	0003
2SD312D	0003
2SD312E	0003
2SD312F	0003
2SD312G	0003
2SD312GN	0003
2SD312H	0003
2SD312J	0003
2SD312K	0003
2SD312L	0003
2SD312M	0003
2SD312X	0003
2SD312Y	0003
2SD313	0042
2SD313(D)	0042
2SD313(D,E)	0042
2SD313(DE)	0042
2SD313(E)	0042
2SD313(EF)	0042
2SD313(F)	0042
2SD313A	0042
2SD313B	0042
2SD313C	0042
2SD313D	0042
2SD313DE	0042
2SD313D-RA	OEM
2SD313E	0042
2SD313E-RA	OEM
2SD313F	0042
2SD313G	0042
2SD313GN	0042
2SD313H	0042
2SD313HP	0042
2SD313HP-D	0042
2SD313HP-E	0042
2SD313L	0042
2SD313M	0042
2SD313N	0042
2SD313R	0042
2SD313RA	0042
2SD313RA(D)	0042
2SD313RA(E)	0042
2SD313RA(F)	0042
2SD313Y	0042
2SD314	0042
2SD314A	0042
2SD314B	0042
2SD314C	0042
2SD314D	0042
2SD314E	0042
2SD314F	0042
2SD314GN	0042
2SD314H	0042
2SD314L	0042
2SD314M	0042
2SD314N	0042
2SD314R	0042
2SD314Y	0042
2SD315	4251
2SD315C	4251
2SD315D	4251
2SD315E	4251
2SD316	0177
2SD316-1	0177
2SD316-2	0177
2SD317	0423
2SD317A	0423
2SD317A(F)	0423
2SD317A(F)(P)	0423
2SD317A(P)	0423
2SD317AF	0423
2SD317AP	0423
2SD317F	0423
2SD317P	0423
2SD318	0423
2SD318-O	0423
2SD318A	0423
2SD318B	0423
2SD318P	0423
2SD318Q	0423
2SD319	0543
2SD320	0074
2SD321	0363
2SD322	0177
2SD322A	0177
2SD322B	0177
2SD322C	0177
2SD323	0177
2SD323A	0177
2SD323B	0177
2SD323C	0177
2SD324	0142
2SD324E	0142
2SD325	0560
2SD325-1	0560
2SD325-OR	0560
2SD325A	0560
2SD325B	0560
2SD325C	0560
2SD325D	0560
2SD325DS	OEM
2SD325E	0560
2SD325ES	OEM
2SD325F	0560
2SD325G	0560
2SD325GN	0560
2SD325H	0560
2SD325J	0560
2SD325K	0560
2SD325L	0560
2SD325M	0560
2SD325OR	0560
2SD325R	OEM
2SD325X	0560
2SD325Y	0560
2SD326	0142
2SD326A	0142
2SD326B	0142
2SD326C	0142
2SD326D	0142
2SD326E	0142
2SD326L	0142
2SD326M	0142
2SD326OR	0142
2SD326X	0142
2SD326Y	0142
2SD327	0286
2SD327-V	0286
2SD327A	0286
2SD327B	0286
2SD327C	0286
2SD327D	0286
2SD327E	0286
2SD327F	0286
2SD327L	0286
2SD327OR	0286
2SD327R	0286
2SD328	0086
2SD328S	0086
2SD329	0187
2SD330	1607
2SD330(E)	1607
2SD330D	1607
2SD330E	1607
2SD330F	OEM
2SD331	0042
2SD332	0867
2SD334	0177
2SD334A	0177
2SD334R	0177
2SD335	0130
2SD336	0324
2SD336(R)	0324
2SD336(Y)	0324
2SD338	0130
2SD339	0177
2SD339-1	0177
2SD339-2	0177
2SD340	0177
2SD340-1	0177
2SD340-2	0177
2SD341	0103
2SD341H	0103
2SD342	0423
2SD343	0228
2SD343A	0228
2SD343B	0228
2SD343C	0228
2SD343D	0228
2SD343H	0228
2SD343J	0228
2SD343K	0228
2SD344	0423
2SD345	0423
2SD346	0423
2SD347	0423
2SD348	0313
2SD349	0400
2SD350	2051
2SD350(Q)	2051
2SD350A	2051
2SD350Q	2051
2SD350T	2051
2SD351	0363
2SD352	0208
2SD352D	0208
2SD352H	0208
2SD353	0220
2SD355	0018
2SD355C	0018
2SD355D	0018
2SD356	0236
2SD356E	0236
2SD357	0236
2SD357E	0236
2SD358	0236
2SD358E	0236
2SD359	1607
2SD359C	1607
2SD359C2	1607
2SD359D	1607
2SD359D1	1607
2SD359D2	1607
2SD360	0042
2SD360D	0042
2SD360E	0042
2SD361	0042
2SD361C	0042
2SD361D	0042
2SD362	1165
2SD365	0042
2SD365A	0042
2SD365B	0042
2SD365H	0042
2SD365HP	0042
2SD365P	0042
2SD366	0042
2SD366A	0042
2SD367	0208
2SD367A	0208
2SD367B	0208
2SD367C	0208
2SD367E	0208
2SD367F	0208
2SD367H	0208
2SD367J	0208
2SD367K	0208
2SD367L	0208
2SD367M	0208
2SD367OR	0208
2SD367R	0208
2SD367X	0208
2SD367Y	0208
2SD368	0313
2SD368A-D	0313
2SD368A-E	0313
2SD369	0543
2SD369-0	0543
2SD369-O	0543
2SD369C	0543
2SD369Y	0543
2SD369Y(FA)	OEM
2SD370	0177
2SD371	0867
2SD371-O	0867
2SD371-R	0867
2SD371-Y	0867
2SD375	0620
2SD376	1966
2SD376A	1966
2SD377	1966
2SD378	0630
2SD379	0177
2SD380	0313
2SD380A	0313
2SD381	0236
2SD381CDTA	OEM
2SD381K	0236
2SD381L	0236
2SD382	0236
2SD383	0363
2SD384	3370
2SD385	1212
2SD386	0392
2SD386A	0392
2SD386A(D)	0392
2SD386A(E)	0392
2SD386A-D	0392
2SD386AD	0392
2SD386A-E	0392
2SD386AE	0392
2SD386A-Y	OEM
2SD386AY	2146
2SD386A-YD	OEM
2SD386A-YE	OEM
2SD386E	0392
2SD386Y	0392
2SD387	0392
2SD387A	0392
2SD387AC	0392
2SD387AD	0392
2SD387A-E	OEM
2SD387AE	0392
2SD387A-F	OEM
2SD387AF	0392
2SD387AS	0392
2SD387D	0392
2SD387E	0392
2SD387S	0392
2SD388	0867
2SD389	0042
2SD389(LP)	0042
2SD389(O)	0042
2SD389(O,LP,P)	0042
2SD389(P)	0042
2SD389-0	0042
2SD389-O	0042
2SD389A	0042
2SD389A(F)	0042
2SD389A(O)	0042
2SD389A(P)	0042
2SD389AF	0042
2SD389AFO	0042
2SD389AFO	0042
2SD389AFP	0042

DEVICE TYPE	REPL CODE
2SD389AFPQ	0042
2SD389AO	0042
2SD389AP	0042
2SD389AQ	0042
2SD389B	0042
2SD389B(0)	0042
2SD389B(P)	0042
2SD389BLB	0042
2SD389BLB-0	0042
2SD389BLB-P	0042
2SD389BP	0042
2SD389BQ	0042
2SD389L	0042
2SD389LB	0042
2SD389LBP	0042
2SD389LP	0042
2SD389P	0042
2SD389Q	0042
2SD389R	0042
2SD390	0042
2SD390A	0042
2SD390P	0042
2SD390Q	0042
2SD392	0286
2SD393	0363
2SD394	0313
2SD395	0363
2SD396	0363
2SD400	2910
2SD400(E)	2910
2SD400(F)	2910
2SD400-D-MP	2910
2SD400-E-MP	2910
2SD400-F-MP	2910
2SD400D	1569
2SD400E	2910
2SD400E-MP	OEM
2SD400F	2910
2SD400F-MP	OEM
2SD400FZ	2910
2SD400G	OEM
2SD400P1	2910
2SD400P2	2910
2SD401	1284
2SD401(A)	1284
2SD401(K)	1284
2SD401(L)	1284
2SD401-O	1284
2SD401-Y	1284
2SD401A	1284
2SD401A(K)	1284
2SD401A(L)	1284
2SD401A-K	1284
2SD401AK	1284
2SD401A-L	1284
2SD401AL	1284
2SD401A-M	OEM
2SD401AM	1284
2SD401E	1284
2SD401EK	1284
2SD401H	1284
2SD401I	1284
2SD401IJ	1284
2SD401K	1284
2SD401KL	1284
2SD401L	1284
2SD401LA	1284
2SD401LM	1284
2SD401M	1284
2SD401R	1284
2SD401Y	1284
2SD402	0392
2SD402A	0392
2SD402A-K	OEM
2SD402AK	0392
2SD402A-L	OEM
2SD402AL	0392
2SD402A-M	OEM
2SD402AM	0392
2SD402K	0392
2SD402L	0392
2SD402M	0392
2SD404	0423
2SD404G	0423
2SD405	1212
2SD407	OEM
2SD408	OEM
2SD409	1212
2SD410	OEM
2SD411	2439
2SD412	OEM
2SD413	0233
2SD414	0563
2SD414Q	0563
2SD415	0563
2SD415P	0563
2SD415Q	0563
2SD415R	0563
2SD415Y	0563
2SD416	0313
2SD417	0867
2SD418	0065
2SD419	1212
2SD422	0429
2SD423	0429
2SD424	0543
2SD424-O	0543
2SD424R	0543
2SD425	0177

DEVICE TYPE	REPL CODE
2SD425-O	0177
2SD425-R	0177
2SD426	0177
2SD426-O	0177
2SD426-R	0177
2SD427	0177
2SD427-O	0177
2SD427-R	0177
2SD428	0867
2SD428-O	0867
2SD428-R	0867
2SD429	0363
2SD430	0620
2SD431	0620
2SD432	0620
2SD433	2414
2SD434	2414
2SD435	2414
2SD436	2414
2SD437	0363
2SD437W	0363
2SD438	0191
2SD438(E)	0191
2SD438(F)	0191
2SD438D	5920
2SD438E	0191
2SD438F	0191
2SD438TV	OEM
2SD439	0563
2SD439E	0563
2SD439F	2551
2SD445	OEM
2SD457	OEM
2SD458	0363
2SD458C	0363
2SD459	2235
2SD460	1212
2SD461	0642
2SD463	2427
2SD464	2439
2SD467	0951
2SD467(B)	0951
2SD467(C)	0951
2SD467-C	0951
2SD467B	0951
2SD467BC	0951
2SD467C	0951
2SD468	0018
2SD468(C)	0018
2SD468A	0018
2SD468AC	0018
2SD468B	0018
2SD468BC	0018
2SD468C	0018
2SD468D	0018
2SD468E	0018
2SD468F	0018
2SD468G	0018
2SD468GN	0018
2SD468H	0018
2SD468L	0018
2SD468LN	0018
2SD468Y	0018
2SD469	0177
2SD470	1751
2SD470-A	1751
2SD470-B	1751
2SD470A	1751
2SD470B	1751
2SD471	0018
2SD471(1)L	0018
2SD471(A)	0018
2SD471(K)	0018
2SD471(L)	0018
2SD471-L	0018
2SD471A	0018
2SD471B	0018
2SD471C	0018
2SD471D	0018
2SD471E	0018
2SD471G	0018
2SD471GN	0018
2SD471K	0018
2SD471KA	0018
2SD471L	0018
2SD471LA	0018
2SD471M	0018
2SD471N	0018
2SD471OR	0018
2SD471P	0018
2SD471Q	0018
2SD471R	0018
2SD471X	0018
2SD471Y	0018
2SD472	OEM
2SD472H	OEM
2SD473	0018
2SD473H	0018
2SD474K	0007
2SD475	0042
2SD475A	0042
2SD476	0423
2SD476A	0423
2SD476AKB	0423
2SD476AKC	0423
2SD476B	0423
2SD476C	0423
2SD476D	0423
2SD476G	0423

DEVICE TYPE	REPL CODE
2SD476KB	0423
2SD476KC	0423
2SD4760	0423
2SD476YL	0423
2SD477	0042
2SD478	0392
2SD478-08	0392
2SD478-C	0392
2SD478A-01	0392
2SD478B	0392
2SD478BS	0392
2SD478C	0392
2SD478CS	0392
2SD478D	0392
2SD478Y	0392
2SD478YL	0392
2SD479	0161
2SD480	0161
2SD481	0161
2SD482	0276
2SD483	0276
2SD484	0168
2SD485	0161
2SD486	0161
2SD487	0161
2SD488	0161
2SD489	0161
2SD490	0161
2SD491	0561
2SD492	0103
2SD493	0561
2SD494	0561
2SD495	0561
2SD496	3529
2SD497	3529
2SD498	3529
2SD499	0561
2SD500	0561
2SD501	0561
2SD502	2427
2SD503	2427
2SD504	2439
2SD505	2439
2SD506	2439
2SD511A	OEM
2SD517	0313
2SD518	0178
2SD519	0429
2SD520	0130
2SD522	0103
2SD523	2427
2SD524	1394
2SD525	4247
2SD525-O	4247
2SD525-Y	4247
2SD525R	4247
2SD525Y	4247
2SD526	2997
2SD526(O)	2997
2SD526-O	2997
2SD528	OEM
2SD529	1852
2SD530	OEM
2SD531	0423
2SD532	5709
2SD533	0363
2SD534	0543
2SD535	2414
2SD536	1966
2SD537	1966
2SD538	1852
2SD538A	1852
2SD539	0363
2SD539A	0363
2SD540	OEM
2SD541	OEM
2SD542	OEM
2SD543	OEM
2SD544	0423
2SD545	0018
2SD545(E)	0018
2SD545(F)	0018
2SD545(G)	0018
2SD545E	0018
2SD545F	0018
2SD545G	0018
2SD545NP	OEM
2SD546	2096
2SD546K	2096
2SD547	OEM
2SD548	OEM
2SD549	0558
2SD550	3362
2SD551	0543
2SD552	2414
2SD553	4887
2SD553-Y	4887
2SD553Y	4887
2SD554	1027
2SD555	0867
2SD556	0543
2SD557	0543
2SD558	OEM
2SD560	1212
2SD560L	OEM
2SD560M	OEM
2SD562	0836
2SD562A	0836
2SD562B	0836

DEVICE TYPE	REPL CODE
2SD562C	0836
2SD562D	0836
2SD562E	0836
2SD562K	0836
2SD562L	0836
2SD562OR	0836
2SD562R	0836
2SD562X	0836
2SD562Y	0836
2SD564	4246
2SD565	1991
2SD568	0423
2SD568(K)	0423
2SD569	0423
2SD570	0423
2SD571	1978
2SD571(L)	1978
2SD571-L	OEM
2SD571K	1978
2SD571L	1978
2SD572	0363
2SD573	0363
2SD575	4928
2SD575L	4928
2SD576	0233
2SD577	0065
2SD578	OEM
2SD579	0178
2SD580	0747
2SD581	0177
2SD581A	0177
2SD582	4944
2SD582A	4944
2SD583	2414
2SD586	2939
2SD586A	2939
2SD586R	2939
2SD587	2939
2SD587A	2939
2SD588	2939
2SD588A	2939
2SD589	0065
2SD590	0086
2SD591	0111
2SD591R	0111
2SD592	0286
2SD592A	0286
2SD592ANC	0286
2SD592AQ	1569
2SD592AR	1569
2SD592NC	0286
2SD592R	0286
2SD593	0168
2SD594	OEM
2SD596	1733
2SD596DV1	1733
2SD596DV2	1733
2SD596DV3	1733
2SD596DV4	1733
2SD596DV5	1733
2SD597	0867
2SD598	0867
2SD599	0111
2SD600	0563
2SD600D	2551
2SD600E	2551
2SD600F	2551
2SD600K	0563
2SD600KE	2551
2SD600KF	2551
2SD601	0724
2SD601(Q)	0724
2SD601(R)	0724
2SD601(S)	0724
2SD601-QRS	OEM
2SD601A	0724
2SD601A-Q	0724
2SD601AQRS	0724
2SD601AQRW	0724
2SD601A-QW	0724
2SD601AQW	0724
2SD601AR	0724
2SD601ATW	0724
2SD601AW	0724
2SD601Q	0724
2SD601QR	OEM
2SD601QRSTW	OEM
2SD601R	0724
2SD601R(TR)	OEM
2SD601S	0724
2SD602	1733
2SD602(R)	1733
2SD602(S)	1733
2SD602A	1733
2SD602R	OEM
2SD602R(W-R)	OEM
2SD602S	1733
2SD603	0155
2SD604	0074
2SD605	0363
2SD606	0363
2SD608	0392
2SD608A	0392
2SD610	1027
2SD611	0429
2SD612	1790
2SD612(E)	1790
2SD612(F)	1790
2SD612E	1790

DEVICE TYPE	REPL CODE
2SD612F	2551
2SD612K	1790
2SD612K(E)	1790
2SD612KE	1790
2SD612KF	2551
2SD612K-SA-E	1790
2SD613	0060
2SD613(D)	0060
2SD613(E)	0060
2SD613E	0060
2SD613P	0060
2SD614	OEM
2SD615	OEM
2SD616	0867
2SD617	2439
2SD619	1946
2SD620	0161
2SD621	2847
2SD622	0429
2SD624	0233
2SD625	5727
2SD626	OEM
2SD627	2660
2SD628	2439
2SD628H	2439
2SD629	2439
2SD629H	2439
2SD630	0130
2SD631	2433
2SD632	0270
2SD632P	0270
2SD632Q	0270
2SD632R	0270
2SD633	1212
2SD634	1212
2SD635	1212
2SD636	0582
2SD636(O)	0582
2SD636(P)	0582
2SD636(Q)	0582
2SD636(R)	0582
2SD636(R.S.T)	0582
2SD636(S)	0582
2SD636(T)	0582
2SD636-0	0582
2SD636-O	0582
2SD636-P	0582
2SD636-Q	0582
2SD636-QRS	0582
2SD636-Q.R	0582
2SD636-R	0582
2SD636-R.S.T	0582
2SD636-S	0582
2SD636P	0582
2SD636Q	0582
2SD636QR	0582
2SD636QRS	0582
2SD636R	0582
2SD636RS	0582
2SD636S	0582
2SD636T	0582
2SD637	0582
2SD637(C)	0582
2SD637(CR)	0582
2SD637(Q)	0582
2SD637(R)	0582
2SD637(S)	0582
2SD637(T)	0582
2SD637-O	0582
2SD637-P	0582
2SD637-PQR	0582
2SD637-Q	0582
2SD637-QR	0582
2SD637-QRS	0582
2SD637-R	0582
2SD637-RS	0582
2SD637-S	0582
2SD637C	0582
2SD637CR	0582
2SD637O	0582
2SD637P	0582
2SD637PQ	0582
2SD637PQR	0582
2SD637PR	0582
2SD637Q	0582
2SD637QR	0582
2SD637QRS	0582
2SD637R	0582
2SD637RS	0582
2SD637RST	0582
2SD637S	0582
2SD637ST	0582
2SD637T	0582
2SD638	0672
2SD638(Q)	0672
2SD638(R)	0672
2SD638(S)	0672
2SD638-Q	0672
2SD638-R	0672
2SD638-RS	0672
2SD638-S	0672
2SD638P	0672
2SD638PQR	0672
2SD638Q	0672
2SD638QR	0672
2SD638QRS	0672
2SD638R	0672
2SD638RS	0672
2SD638S	0672

DEVICE TYPE	REPL CODE
2SD638T	0672
2SD639	0672
2SD639(Q)	0672
2SD639(R)	0672
2SD639(S)	0672
2SD639-R	0672
2SD639-S	0672
2SD639P	0672
2SD639Q	0672
2SD639QR	0672
2SD639QRST	0672
2SD639R	0672
2SD639S	0672
2SD639T	0672
2SD640	0363
2SD641	1852
2SD642	OEM
2SD643	OEM
2SD644	OEM
2SD645	OEM
2SD646	OEM
2SD646A	OEM
2SD647	OEM
2SD647A	1386
2SD648	OEM
2SD648A	OEM
2SD649	0065
2SD649(Q)	0065
2SD649Q	0065
2SD650	OEM
2SD650H	OEM
2SD651	OEM
2SD652	OEM
2SD654	0086
2SD655	1978
2SD655(D)	1978
2SD655(E)	1978
2SD655(F)	1978
2SD655D	1978
2SD655E	1978
2SD655EF	1978
2SD655F	1978
2SD656	0178
2SD657	0074
2SD658	0065
2SD660	0630
2SD661	0582
2SD661(R)	0582
2SD661(S)	0582
2SD661(T)	0582
2SD661-R	0582
2SD661-S	0582
2SD661RS	0582
2SD661S	0582
2SD661ST	0582
2SD661T	0582
2SD662	0715
2SD662(Q)	0715
2SD662(R)	0715
2SD663	0363
2SD664	1212
2SD665	0543
2SD666	1386
2SD666A	1386
2SD666AB	1386
2SD666AC	1386
2SD666B	1386
2SD666C	1386
2SD666D	1386
2SD667	1386
2SD667(A)	1386
2SD667(B)	1386
2SD667A	1386
2SD667AB	1386
2SD667AC	1386
2SD667B	1386
2SD667C	1386
2SD667D	1386
2SD668	0563
2SD668A	0563
2SD668AB	0563
2SD668AC	0563
2SD668B	0563
2SD668D	0563
2SD669	0563
2SD669A	0563
2SD669AB	0563
2SD669AC	0563
2SD669B	0563
2SD669C	0563
2SD669D	0563
2SD670	1394
2SD670H	1394
2SD671	0155
2SD671(L)	0155
2SD672	0074
2SD673	0867
2SD673A	0867
2SD674	0867
2SD674A	0867
2SD675	0543
2SD675A	0543
2SD676	0543
2SD676A	0543
2SD677	0074
2SD678	1212
2SD678A	1212
2SD679	1212
2SD679A	1212

DEVICE TYPE	REPL CODE
2SD683	1852
2SD683A	1852
2SD683S	1852
2SD684	OEM
2SD684A	OEM
2SD685	1852
2SD686	1212
2SD687	1212
2SD688	1480
2SD689	0055
2SD690	0429
2SD691	1212
2SD692	2427
2SD692(Q)	2427
2SD692M	2427
2SD692P	2427
2SD692Q	2427
2SD693	0074
2SD694	OEM
2SD695	OEM
2SD696	0014
2SD696A	0014
2SD696AB	OEM
2SD696AC	OEM
2SD697	OEM
2SD698	OEM
2SD698B	OEM
2SD699	OEM
2SD700	OEM
2SD701QR-TW	OEM
2SD702	OEM
2SD703	OEM
2SD704	0060
2SD704D	0060
2SD704E	0060
2SD705	0363
2SD706	0074
2SD707	0363
2SD708	OEM
2SD709	OEM
2SD710	1733
2SD711	2626
2SD711A	2626
2SD712	0482
2SD712A	0482
2SD713	0482
2SD715	0134
2SD716	3081
2SD716R	OEM
2SD717	3081
2SD718	0194
2SD720	OEM
2SD720D	0208
2SD720E	0208
2SD720PJ	0208
2SD721	1212
2SD722	5748
2SD723	0236
2SD724	0392
2SD725	0065
2SD725-06	0065
2SD726	0236
2SD726B	6170
2SD726C	6170
2SD727	3602
2SD728	3081
2SD729	3525
2SD729H	3525
2SD730	1394
2SD730H	1394
2SD731	3279
2SD732	0867
2SD732K	0867
2SD733	0543
2SD733K	0543
2SD733U	0543
2SD734	1513
2SD734(E)-AA	1513
2SD734(F)	1513
2SD734(F)-AA	1513
2SD734(G)	1513
2SD734(G)-AA	1513
2SD734-AA	1513
2SD734-F-AA	1513
2SD734-G-AA	1513
2SD734-NP	1513
2SD734-T	1513
2SD734E	1513
2SD734F	1513
2SD734G	1513
2SD735	2276
2SD735-O	2276
2SD735A-B-B/C	2276
2SD736	2276
2SD736A	2276
2SD737	2276
2SD738	2276
2SD738A	2276
2SD741	OEM
2SD743	1284
2SD743A	1284
2SD743R	1284
2SD745	2276
2SD745A	2276
2SD745B	2276
2SD746	OEM
2SD746A	OEM
2SD746A-01	OEM
2SD747	0079

If replacement code is OEM, contact original manufacturer for replacement.

DEVICE TYPE	REPL CODE
2SD748	3227
2SD748A	3227
2SD748A-01	3227
2SD748A-01A	3227
2SD748B	3227
2SD749	0363
2SD749P	0363
2SD750	0543
2SD751	0194
2SD752	0543
2SD753	0543
2SD754	0324
2SD755	0530
2SD755D	0530
2SD755E	0530
2SD755F	0530
2SD756	0530
2SD756A	0530
2SD756AF	0530
2SD756D	0530
2SD756E	0530
2SD757	0168
2SD757C	0168
2SD757D	0168
2SD758	1085
2SD758C	1085
2SD758D	1085
2SD759	0392
2SD760	0392
2SD761	0392
2SD761D	0392
2SD761E	0392
2SD761V	0392
2SD761VL	0392
2SD761VL-D	OEM
2SD761VL-E	OEM
2SD761VR	0392
2SD761VR-D	OEM
2SD761VRD	0392
2SD761VR-E	OEM
2SD761VRE	0392
2SD762	0695
2SD762-O	0695
2SD762-OP	0695
2SD762-P	0695
2SD762A	0695
2SD762LB	0695
2SD762OP	0695
2SD762OPLB	0695
2SD762P	0695
2SD762PLB	0695
2SD762Q	0695
2SD763	1386
2SD764	0223
2SD765	0223
2SD766	2096
2SD766D	2096
2SD766P	2096
2SD766Q	2096
2SD766R	2096
2SD767	0155
2SD767(R)	0155
2SD767(S)	0155
2SD767R	0155
2SD767S	0155
2SD768	1959
2SD768K	1959
2SD769	0286
2SD770	0155
2SD771	0111
2SD772	0392
2SD772A	0392
2SD772B	0392
2SD772BLB	0392
2SD773	0018
2SD773-4	0018
2SD773-34	0018
2SD773K	0018
2SD773L	0018
2SD773U	0018
2SD774	2030
2SD774(U2)	2030
2SD774-3	2030
2SD774-4	2030
2SD774-34	2030
2SD774-K	2030
2SD774K	2030
2SD774R	2030
2SD774U	2030
2SD774U2	2030
2SD776	2548
2SD777	0270
2SD777(FA-4)	OEM
2SD777(FA-5)	0270
2SD777-1	0270
2SD777FA-1	0270
2SD777FA-5	0270
2SD777FA5	0270
2SD778	0286
2SD778(Q)	0286
2SD778(R)	0286
2SD778Q	0286
2SD778QR	0286
2SD778R	0286
2SD779	0286
2SD780	2332
2SD780A	2332
2SD780DW1	2332
2SD780DW2	2332

DEVICE TYPE	REPL CODE
2SD780DW3	2332
2SD780DW4	2332
2SD780DW5	2332
2SD781	0563
2SD782	0728
2SD783	0313
2SD784	2847
2SD785	2847
2SD786	0530
2SD787	0715
2SD787B	0715
2SD787C	0715
2SD787D	0715
2SD787E	0715
2SD788	1500
2SD788-3	0018
2SD788-4	0018
2SD788-5	0018
2SD788-34TZ	1500
2SD788-C	0018
2SD788B	0018
2SD788C	1500
2SD788D	0018
2SD788E	0018
2SD789	2561
2SD789(B)	2561
2SD789(C)	2561
2SD789(D)	2561
2SD789(E)	2561
2SD789-03C	2561
2SD789-3	2561
2SD789-4	2561
2SD789B	2561
2SD789BC	2561
2SD789C	2561
2SD789D	2561
2SD789DE	2561
2SD789E	2561
2SD789M	2561
2SD789P	2561
2SD790	1386
2SD790B	1386
2SD790C	1386
2SD792	0313
2SD792S	0313
2SD792T	0313
2SD793	0563
2SD793Q	0563
2SD794	0161
2SD794(2)	0161
2SD794A	0161
2SD794AD	0161
2SD794AQ	0161
2SD794D	0161
2SD794P	0161
2SD794Q	0161
2SD794R	0161
2SD795	0836
2SD795A	0836
2SD795AQ	0836
2SD795P	0836
2SD795Q	0836
2SD796	OEM
2SD797	2433
2SD798	OEM
2SD799	OEM
2SD800	1852
2SD801	1852
2SD802	3060
2SD803	2439
2SD803Q	2439
2SD804	0042
2SD804B	0042
2SD804H	0042
2SD804HLB	0042
2SD804HP	0042
2SD804L	0042
2SD804P	0042
2SD805	OEM
2SD806	OEM
2SD807	1493
2SD807M	1493
2SD808	0191
2SD809	0563
2SD809D	OEM
2SD809E	OEM
2SD810	4450
2SD811	5758
2SD812	1165
2SD812-Q	1165
2SD812Q	1165
2SD813	1733
2SD814	2332
2SD814A	2332
2SD814S	OEM
2SD815	OEM
2SD816	5307
2SD817	OEM
2SD818	2660
2SD819	2668
2SD820	2051
2SD820(FA-1)	2051
2SD820FA-1	2051
2SD821	0313
2SD822	5407
2SD822P	5407
2SD822Q	5407
2SD823	1165
2SD824	2276

DEVICE TYPE	REPL CODE
2SD824A	2276
2SD825	2276
2SD825A	2276
2SD826	OEM
2SD827	OEM
2SD827K	OEM
2SD828	OEM
2SD829	OEM
2SD829K	OEM
2SD830	OEM
2SD832	OEM
2SD833	1212
2SD834	5307
2SD835	5307
2SD836	1212
2SD836A	1212
2SD836B	1212
2SD836Q	1212
2SD836R	1212
2SD837	1212
2SD837A	1212
2SD837AB	1212
2SD837AD	1212
2SD837AQ	1212
2SD837AR	1212
2SD837B	1212
2SD837LB	1212
2SD837LBPQ	1212
2SD837O	1212
2SD837P	1212
2SD837PQ	1212
2SD837Q	1212
2SD837R	1212
2SD838	2847
2SD838B	2847
2SD839	1212
2SD840	1212
2SD841	3014
2SD842	2428
2SD843	0060
2SD843(Y)	0060
2SD844	0194
2SD844E	0194
2SD845	2276
2SD845O	2276
2SD847	2367
2SD848	2939
2SD848A	2939
2SD849	0065
2SD850	2668
2SD851	OEM
2SD852	OEM
2SD854	0622
2SD855	1165
2SD855A	1165
2SD855B	1165
2SD856	0060
2SD856(P)	0060
2SD856(Q)	0060
2SD856(R)	0060
2SD856A	0060
2SD856B	0060
2SD856L	0060
2SD856LB	0060
2SD856M	0060
2SD856M(LB)	0060
2SD856MLB	0060
2SD856MLD	0060
2SD856O	0060
2SD856P	0060
2SD856PLB	0060
2SD856PQ	0060
2SD856PQ(LB)	0060
2SD856PQR	0060
2SD856Q	0060
2SD856Q(LB)	0060
2SD856R	0060
2SD856S	0060
2SD857	0060
2SD857A	0060
2SD857B	0060
2SD858	5451
2SD858A	5451
2SD858B	5451
2SD859	0168
2SD859A	0168
2SD859B	0168
2SD860	0168
2SD860A	0168
2SD860B	0168
2SD861	0168
2SD861A	0168
2SD861B	0168
2SD862	OEM
2SD863	1386
2SD863E	1386
2SD863E-AE	1386
2SD863F	1386
2SD864	1959
2SD864K	1959
2SD865	OEM
2SD866	3125
2SD866(P)	3125
2SD866(Q)	3125
2SD866A	3125
2SD866B	3125
2SD866D	3125
2SD866P	3125
2SD866Q	3125

DEVICE TYPE	REPL CODE
2SD866R	3125
2SD866S	3125
2SD867	0177
2SD868	0055
2SD868(FA)	OEM
2SD869	0055
2SD869(L)	0055
2SD869-14	0055
2SD869-L	0055
2SD869B	0055
2SD869FA	0055
2SD869L	0055
2SD870	0055
2SD870(FA)	0055
2SD870A	0055
2SD870FA	0055
2SD870Z	0055
2SD871	0055
2SD871(Z)	0055
2SD871Z	0055
2SD872	0728
2SD873	4786
2SD874	0667
2SD874(R)	0667
2SD874-QR	OEM
2SD874A	0667
2SD874QR	OEM
2SD874QRS	OEM
2SD874R	0667
2SD875	0667
2SD876	0392
2SD877	0178
2SD877Y	2361
2SD878	0103
2SD878(FA-1)	OEM
2SD879	1500
2SD880	0461
2SD880-0	0461
2SD880-O	0461
2SD880-OLBGL2	0461
2SD880-Y	0461
2SD880GR	0461
2SD880OLBGL2	0461
2SD880Y	0461
2SD880YLBGL2	0461
2SD882	0161
2SD882(P)	0161
2SD882-P	0161
2SD882P	0161
2SD882PM	0161
2SD882Q	0161
2SD882R	0161
2SD883	2939
2SD883A	2939
2SD884	0728
2SD885	0728
2SD886	0060
2SD886A	0060
2SD886F	0060
2SD886P	0060
2SD886Q	0060
2SD887	1212
2SD888	1212
2SD889	0111
2SD889Q	0111
2SD889R	0111
2SD889S	0507
2SD889T	0111
2SD890	0530
2SD891	0400
2SD891A	0400
2SD892	2797
2SD892A	2797
2SD892A-Q	2797
2SD892AQ	2797
2SD892AR	2797
2SD893	0400
2SD893A	0400
2SD893Q	0400
2SD893R	0400
2SD894	0558
2SD894E	OEM
2SD894K	OEM
2SD895	3082
2SD896	3082
2SD897	0055
2SD897A	0055
2SD898	0055
2SD898-B13	0055
2SD898A	0055
2SD898B	0055
2SD899	0055
2SD899A	0055
2SD900	0055
2SD900A	0055
2SD900B	0055
2SD900B-06	0055
2SD900D	0055
2SD901	0392
2SD901D	OEM
2SD901E	OEM
2SD902	0074
2SD903	0055
2SD904	0055
2SD904K	0055
2SD904P	0055
2SD905	0313
2SD906	0055
2SD907	3081

DEVICE TYPE	REPL CODE
2SD908	0194
2SD909	3435
2SD910	0194
2SD911	0543
2SD912	0543
2SD913	2433
2SD914	2433
2SD915	OEM
2SD916	1212
2SD917	0728
2SD918	0482
2SD919	0155
2SD920	0074
2SD920B	0074
2SD921	3001
2SD921-03	OEM
2SD922	OEM
2SD923	OEM
2SD926	0236
2SD927	0236
2SD928	0236
2SD929	2548
2SD929A	OEM
2SD929B	6341
2SD930	OEM
2SD931	3001
2SD932	OEM
2SD933	5419
2SD933A	5419
2SD933H	5419
2SD934	OEM
2SD934K	OEM
2SD935	2613
2SD936	OEM
2SD937	OEM
2SD938	OEM
2SD939	OEM
2SD940	OEM
2SD941	OEM
2SD942	OEM
2SD943	OEM
2SD944	1061
2SD946	0558
2SD946-Q	0558
2SD946-Q.R	0558
2SD946-R	0558
2SD946A	0558
2SD946B	0558
2SD946S	OEM
2SD947	0558
2SD948	0016
2SD949	0313
2SD950	2668
2SD950Q	2668
2SD951	0055
2SD952	0223
2SD953	0223
2SD954	0223
2SD955	0715
2SD956	0055
2SD957	0055
2SD957A	0055
2SD957A-03	0055
2SD958	0530
2SD958(R)	0530
2SD958(S)	0530
2SD958(T)	0530
2SD958R	0530
2SD958S	0530
2SD958T	0530
2SD958U	0530
2SD959	1165
2SD959LB	1165
2SD959Q	1165
2SD960	0236
2SD961	1165
2SD962	OEM
2SD963	OEM
2SD965	1500
2SD966	OEM
2SD966(R)	OEM
2SD967	0400
2SD968	3471
2SD968A	3471
2SD968A-RS	OEM
2SD969	0672
2SD970	1959
2SD970K	1959
2SD971	OEM
2SD972	2366
2SD973	0009
2SD973(Q)	0009
2SD973(R)	0009
2SD973(S)	0009
2SD973A	0009
2SD973A-QRS	OEM
2SD973AR	1386
2SD973ARS	OEM
2SD973Q	0009
2SD973QRS	0009
2SD973R	0009
2SD973RS	0009
2SD973S	0009
2SD974	1386
2SD975	0563
2SD975C	0563
2SD976	0728
2SD976A	0728
2SD977	5307

DEVICE TYPE	REPL CODE
2SD978	5307
2SD979	5342
2SD980	OEM
2SD981	OEM
2SD981-03	OEM
2SD982	OEM
2SD983	OEM
2SD985	3703
2SD985-L	3703
2SD985K	3703
2SD985L	3703
2SD985M	3703
2SD985T	3703
2SD986	2970
2SD986K	2970
2SD987	0728
2SD990	OEM
2SD990K	OEM
2SD991	5307
2SD991K	5307
2SD992	3350
2SD992M	3350
2SD993	0055
2SD993A	0055
2SD994	0055
2SD995	2847
2SD997	2414
2SD998	3703
2SD999	0667
2SD999-CLCK	0667
2SD1000	0667
2SD1001	0667
2SD1002	OEM
2SD1003	OEM
2SD1004	OEM
2SD1005	OEM
2SD1006	OEM
2SD1007	OEM
2SD1009	OEM
2SD1010	0111
2SD1010R	6758
2SD1010S	0111
2SD1011	0530
2SD1011R	0530
2SD1011S	0530
2SD1011T	0530
2SD1012	0286
2SD1012F	0286
2SD1012G	5920
2SD1012G1	0286
2SD1012G-SPA	OEM
2SD1012H	5920
2SD1014	OEM
2SD1015	OEM
2SD1015C	OEM
2SD1016	0055
2SD1017	2613
2SD1018	2613
2SD1020	0018
2SD1020H	0018
2SD1021	0018
2SD1022	1212
2SD1023	5417
2SD1024	1212
2SD1025	2096
2SD1026	OEM
2SD1027	OEM
2SD1029	0042
2SD1030	OEM
2SD1031	1959
2SD1032	3081
2SD1032A	3081
2SD1032B	3081
2SD1033	3350
2SD1034	OEM
2SD1034A	OEM
2SD1035	1165
2SD1036	2276
2SD1037	OEM
2SD1038	OEM
2SD1039	3553
2SD1040	0543
2SD1040A	0543
2SD1041	3554
2SD1042	OEM
2SD1043	5419
2SD1044	0134
2SD1044A	0134
2SD1045	OEM
2SD1046	0194
2SD1046D-L	OEM
2SD1046E-L	OEM
2SD1047	0194
2SD1048	1733
2SD1048-X6	1733
2SD1048-X6-TA	1733
2SD1048-X6TA	1733
2SD1048-X6TA(F)	1733
2SD1048-X7	1733
2SD1048-X7-TA	1733
2SD1048-X7-TA(G)	1733
2SD1048-X7TA	1733
2SD1048-X7TA(G)	1733
2SD1048-X8	1733
2SD1048-X8-TA	1733
2SD1048-X8-TA(H)	1733
2SD1048-X8TA	1733
2SD1048-X8TA(H)	1733
2SD1049	3602

DEVICE TYPE	REPL CODE
2SD1050	OEM
2SD1051	0018
2SD1051(Q)	0018
2SD1051(R)	0018
2SD1051Q	0018
2SD1051QR	0018
2SD1051R	0018
2SD1052	2069
2SD1052-A	2069
2SD1052A	2069
2SD1052A(FA)	2069
2SD1052A-FA	2069
2SD1052AFA	2069
2SD1052FA	2069
2SD1053	3350
2SD1054	5422
2SD1055	0009
2SD1055Q	0009
2SD1055R	0009
2SD1056	OEM
2SD1059	2201
2SD1060	0060
2SD1060K	0060
2SD1061	0060
2SD1062	1966
2SD1063	3081
2SD1064	3081
2SD1065	2367
2SD1065Q	2367
2SD1065R	2367
2SD1065S	2367
2SD1066	OEM
2SD1067	OEM
2SD1068	2993
2SD1069	0728
2SD1070	3081
2SD1071	OEM
2SD1072	OEM
2SD1073	OEM
2SD1074	3350
2SD1075	3350
2SD1076	3350
2SD1076L	3350
2SD1077	3350
2SD1078	3350
2SD1078B	3350
2SD1079	3350
2SD1079C	3350
2SD1079LB	3350
2SD1079LC	3350
2SD1080	3350
2SD1080C	3350
2SD1080LC	3350
2SD1081	3350
2SD1081B	3350
2SD1081C	3350
2SD1081LB	3350
2SD1081LC	3350
2SD1082	OEM
2SD1083	3350
2SD1083L	3350
2SD1084	OEM
2SD1085	5307
2SD1085K	5307
2SD1087	OEM
2SD1088	OEM
2SD1089	OEM
2SD1090	3001
2SD1091	OEM
2SD1092	3516
2SD1092(FA)	OEM
2SD1092(FA-1)	OEM
2SD1092FA	3516
2SD1093	2993
2SD1094	3588
2SD1095	0065
2SD1096	3350
2SD1097	0065
2SD1098	0065
2SD1098-42M	0065
2SD1099	0065
2SD1100	OEM
2SD1101	1733
2SD1101AB	1733
2SD1101AC	1733
2SD1102	0065
2SD1103	0065
2SD1104	0065
2SD1105	2414
2SD1106	OEM
2SD1107	OEM
2SD1109	3081
2SD1109A	3081
2SD1110	0130
2SD1110A	0130
2SD1111	OEM
2SD1111-T	OEM
2SD1112	OEM
2SD1113	OEM
2SD1113K	OEM
2SD1114	OEM
2SD1114K	OEM
2SD1115	5307
2SD1115K	5307
2SD1116	OEM
2SD1116K	OEM
2SD1117	0060
2SD1117A	0060
2SD1118	OEM

If replacement code is OEM, contact original manufacturer for replacement.

DEVICE TYPE	REPL CODE
2SD1119	OEM
2SD1120	5307
2SD1121	5307
2SD1122	OEM
2SD1123	OEM
2SD1124	5307
2SD1125	OEM
2SD1126	1959
2SD1126K	1959
2SD1127	1959
2SD1127K	1959
2SD1128	OEM
2SD1129	OEM
2SD1129K	OEM
2SD1130	0130
2SD1130K	0130
2SD1131	0236
2SD1131B	0236
2SD1131C	0236
2SD1131D	0236
2SD1132	0236
2SD1132B	0236
2SD1132C	0236
2SD1132D	0236
2SD1133	0236
2SD1133(D)	0236
2SD1133B	0236
2SD1133C	0236
2SD1133D	0236
2SD1134	0236
2SD1134-C	0236
2SD1134C	0236
2SD1135	0236
2SD1136	3014
2SD1137	0236
2SD1138	0392
2SD1138(C)	0392
2SD1138(D)	0392
2SD1138(G)	0392
2SD1138-02-C	0392
2SD1138-02C	0392
2SD1138-C	0392
2SD1138-D	0392
2SD1138-O	0392
2SD1138-Y	0392
2SD1138B	0392
2SD1138C	0392
2SD1138D	0392
2SD1138Y	0392
2SD1139	0392
2SD1139B	0392
2SD1140	0130
2SD1141	OEM
2SD1141K	OEM
2SD1142	1541
2SD1143	0065
2SD1145	OEM
2SD1146	3093
2SD1147	OEM
2SD1148	3082
2SD1149	OEM
2SD1150	0728
2SD1151	0223
2SD1152	OEM
2SD1153	0400
2SD1154	0363
2SD1155	OEM
2SD1156	OEM
2SD1157	2069
2SD1158	OEM
2SD1159	3014
2SD1160	5450
2SD1160-O	OEM
2SD1160Y	5450
2SD1161	OEM
2SD1162	5307
2SD1162K	5307
2SD1162L	5307
2SD1162M	5307
2SD1163	2819
2SD1163A	2819
2SD1164	3225
2SD1164K	3225
2SD1164L	3225
2SD1164M	3225
2SD1165	OEM
2SD1165A	OEM
2SD1166	OEM
2SD1168	0065
2SD1169	5452
2SD1170	5453
2SD1171	0055
2SD1172	0055
2SD1173	0055
2SD1174	0055
2SD1175	0055
2SD1176	1212
2SD1176A	1212
2SD1177	0563
2SD1177C	0563
2SD1178	0563
2SD1178C	0563
2SD1179	0563
2SD1180	0563
2SD1181	OEM
2SD1182	OEM
2SD1183	0313
2SD1184	0313
2SD1185	0313
2SD1186	0313
2SD1187	5456
2SD1189	0161
2SD1190	5457
2SD1191	2235
2SD1192	2235
2SD1193	3627
2SD1194	1212
2SD1195	1212
2SD1196	2235
2SD1197	1573
2SD1198	0400
2SD1198A	0400
2SD1199	0582
2SD1199-R	0582
2SD1200	0161
2SD1201	1852
2SD1202	3038
2SD1203	3038
2SD1204	3038
2SD1205	0400
2SD1205(P)	0400
2SD1205(Q)	0400
2SD1205(R)	0400
2SD1205A	0400
2SD1205PQR	0400
2SD1205R	0400
2SD1206	0672
2SD1206(Q)	0672
2SD1206(R)	0672
2SD1206(S)	0672
2SD1206Q	0672
2SD1206R	0672
2SD1206S	0672
2SD1207	3938
2SD1207-F	3938
2SD1207-T	3938
2SD1207S	3938
2SD1207SZ	3938
2SD1207T	3938
2SD1207TZ	3938
2SD1208	2427
2SD1208(FA)	2427
2SD1208FA	2427
2SD1209	3892
2SD1210	3627
2SD1211	1562
2SD1212	3115
2SD1213	3602
2SD1213C	OEM
2SD1214	1212
2SD1215	1212
2SD1216	1212
2SD1217	0278
2SD1218	2366
2SD1219	1212
2SD1220	OEM
2SD1220-OY	OEM
2SD1221	5450
2SD1222	5352
2SD1223	5352
2SD1224	5463
2SD1225	0009
2SD1225M	0009
2SD1225M(Q)	0009
2SD1225M(R)	0009
2SD1225MQ	1566
2SD1225MR	1566
2SD1226	2471
2SD1226M	2471
2SD1226MP	2471
2SD1226MQ	2471
2SD1226MR	2471
2SD1226Q	2471
2SD1227	0009
2SD1227M	0009
2SD1227MF	OEM
2SD1227MP	0009
2SD1227MQ	0009
2SD1227MR	0009
2SD1228	2471
2SD1228M	2471
2SD1229	3746
2SD1230	3748
2SD1231	1394
2SD1232	1394
2SD1233	1394
2SD1234	1394
2SD1235	1165
2SD1236	1165
2SD1237	1165
2SD1237R	1165
2SD1238	3279
2SD1239	2485
2SD1240	2485
2SD1241	3081
2SD1241A	3081
2SD1242	3081
2SD1242A	3081
2SD1243	0194
2SD1243A	0194
2SD1244	1500
2SD1244R	OEM
2SD1245	5307
2SD1246	1500
2SD1246-T	1500
2SD1246-U	1500
2SD1246T	1500
2SD1246U	6912
2SD1247	OEM
2SD1247S	OEM
2SD1247T	OEM
2SD1248	1959
2SD1249	OEM
2SD1249A	OEM
2SD1250	OEM
2SD1250A	OEM
2SD1251	5450
2SD1251A	5450
2SD1252	5450
2SD1252A	5450
2SD1253	5450
2SD1253A	5450
2SD1254	5450
2SD1255	5450
2SD1256	5450
2SD1257	5450
2SD1257A	5450
2SD1258	OEM
2SD1259	0558
2SD1259A	0558
2SD1260	5352
2SD1260A	5352
2SD1261	5352
2SD1261A	5352
2SD1262	OEM
2SD1262A	OEM
2SD1263	0168
2SD1263A	0168
2SD1264	0392
2SD1264-D	0392
2SD1264-P	0392
2SD1264-PLB	0392
2SD1264A	0392
2SD1264ALBP	0392
2SD1264A-P	0392
2SD1264AP	0392
2SD1264APLB	0392
2SD1264LBP	0392
2SD1264P	0392
2SD1264PLB	0392
2SD1265	1162
2SD1265(LB)	1162
2SD1265(LB)O	1162
2SD1265(P)	1162
2SD1265(Q)	1162
2SD1265(QP)	1162
2SD1265-0	1162
2SD1265-0P	1162
2SD1265-LB	1162
2SD1265-O	1162
2SD1265-OP	1162
2SD1265-OP(LB)	1162
2SD1265-P	1162
2SD1265-Q	1162
2SD1265-R	1162
2SD1265A	1162
2SD1265B	1162
2SD1265L	1162
2SD1265LB	1162
2SD1265LB-O	1162
2SD1265LB-P	1162
2SD1265O	1162
2SD1265OLB	1162
2SD1265OP	1162
2SD1265OPQ	1162
2SD1265P	1162
2SD1266	0060
2SD1266(P)	0060
2SD1266(Q)	0060
2SD1266(R)	0060
2SD1266-P	0060
2SD1266-PQ	0060
2SD1266-Q	0060
2SD1266A	0060
2SD1266A(P)	0060
2SD1266A(P,Q)	0060
2SD1266A(Q)	0060
2SD1266A-P	OEM
2SD1266AP	0060
2SD1266A-PLT	OEM
2SD1266A-PVJ	OEM
2SD1266A-Q	0060
2SD1266AQ	0060
2SD1266LB	0060
2SD1266LB(P)	0060
2SD1266LB(Q)	0060
2SD1266LBPQ	0060
2SD1266LVPQ	0060
2SD1266P	0060
2SD1266PQ	0060
2SD1266PQR	0060
2SD1266Q	0060
2SD1267	0060
2SD1267(P)	0060
2SD1267A	0060
2SD1267ALBPQ	0060
2SD1267AP	OEM
2SD1267APQ	4866
2SD1267A-PVJ	OEM
2SD1267P	0060
2SD1267Q	6170
2SD1268	1165
2SD1269	1165
2SD1270	1165
2SD1271	1165
2SD1271A	1165
2SD1271B	OEM
2SD1271BP	OEM
2SD1272	OEM
2SD1273	0928
2SD1273(P)	0928
2SD1273(Q)	0928
2SD1273A	0928
2SD1273LB-O	0928
2SD1273LB-P	0928
2SD1273P	0928
2SD1273Q	0928
2SD1273QP	0928
2SD1274	4007
2SD1274A	4007
2SD1274B	4007
2SD1275	1212
2SD1275A	1212
2SD1276	1212
2SD1276(LB)	1212
2SD1276-AP	1212
2SD1276-LB	1212
2SD1276A	1212
2SD1276A(LB)P	1212
2SD1276A(LP)	1212
2SD1276AP	1212
2SD1276B	1212
2SD1276LB	1212
2SD1276P	1212
2SD1276PLB	1212
2SD1276Q	6940
2SD1276QLB	1212
2SD1277	1212
2SD1277A	1212
2SD1277B	1212
2SD1277BP	1212
2SD1277R	1212
2SD1278	5471
2SD1279	0313
2SD1279FA	0313
2SD1280	0667
2SD1280R	0667
2SD1281	3350
2SD1282	3350
2SD-1283	0208
2SD1283	0208
2SD1284	3350
2SD1286	3225
2SD1287	4013
2SD1288	OEM
2SD1289	0194
2SD1290	1541
2SD1291	1541
2SD1292	1562
2SD1292-R-T103	OEM
2SD1292R	OEM
2SD1292T103Q	1562
2SD1292T103R	1562
2SD1293	3608
2SD1293M	3608
2SD1293MP	3608
2SD1293MQ	3608
2SD1293MR	3608
2SD1294	2989
2SD1294(FA)	2989
2SD1294FA	2989
2SD1295	3350
2SD1296	OEM
2SD1297	OEM
2SD1298	OEM
2SD1299	OEM
2SD1300	0313
2SD1301	0055
2SD1302	1500
2SD1302STTA	OEM
2SD1302T	OEM
2SD1303	1386
2SD1304	OEM
2SD1305	OEM
2SD1306	1733
2SD1306E	1733
2SD1307	OEM
2SD1308	3136
2SD1308L	3136
2SD1309	3136
2SD1310	0236
2SD1311	0236
2SD1312	3938
2SD1313	OEM
2SD1314	1506
2SD1315	5452
2SD1316	0558
2SD1317	0558
2SD1318	0086
2SD1319	5350
2SD1320	5350
2SD1321	5350
2SD1322	2235
2SD1323	2235
2SD1324	2235
2SD1325	3723
2SD1326	3723
2SD1327	3723
2SD1328	1733
2SD1328(DS)	1733
2SD1328STTW	OEM
2SD1329	5350
2SD1330	0014
2SD1330(R)	0014
2SD1330(S)	0014
2SD1330(T)	0014
2SD1330-RS	0014
2SD1330R	0014
2SD1330RST	0014
2SD1330S	0014
2SD1330ST	0014
2SD1330T	0014
2SD1331	OEM
2SD1332	0194
2SD1333	0194
2SD1334	0194
2SD1335	0194
2SD1336	3136
2SD1336A	3136
2SD1337	OEM
2SD1338	OEM
2SD1339	OEM
2SD1340	OEM
2SD1341	0065
2SD1341P	OEM
2SD1342	0055
2SD1343	OEM
2SD1344	0055
2SD1344E	0055
2SD1345	0060
2SD1346	OEM
2SD1347	0018
2SD1348	0161
2SD1348S	OEM
2SD1348T	OEM
2SD1349	OEM
2SD1350	OEM
2SD1351	0042
2SD1352	0236
2SD1353	3691
2SD1354	3692
2SD1355	3693
2SD1356	3694
2SD1357	1212
2SD1358	2235
2SD1359	2235
2SD1360	OEM
2SD1361	OEM
2SD1362	3178
2SD1363	0060
2SD1363-Y	0060
2SD1364	0728
2SD1365	OEM
2SD1366	1733
2SD1366A	1733
2SD1367	OEM
2SD1368	0667
2SD1369	2235
2SD1370	1212
2SD1371	OEM
2SD1372	3014
2SD1373	2613
2SD1374	0270
2SD1375	0270
2SD1376	3703
2SD1376K	3703
2SD1377	1959
2SD1378	0563
2SD1379	0161
2SD1380	0563
2SD1381	0563
2SD1382	2970
2SD1383	5364
2SD1384	3093
2SD1384(R)	3093
2SD1384Q	3093
2SD1384QZ	3093
2SD1384R	3093
2SD1384RZ	3093
2SD1385	OEM
2SD1386	OEM
2SD1387	OEM
2SD1387-2	OEM
2SD1387-3	OEM
2SD1387-4	OEM
2SD1388	OEM
2SD1388-4	OEM
2SD1389	5471
2SD1390	2766
2SD1391	1541
2SD1392	2366
2SD1392L	2366
2SD1393	0278
2SD1394	2366
2SD1395	2366
2SD1396	1541
2SD1397	1541
2SD1397-OR1-YB	1541
2SD1397B	1541
2SD1398	1541
2SD1398-CTV-YB	1541
2SD1398-OR1-YB	1541
2SD1398-ORI-YB	1541
2SD1398S	1541
2SD1398S-CA	1541
2SD1399	1541
2SD1400	3438
2SD1401	3438
2SD1402	5481
2SD1403	3360
2SD1404	0728
2SD1405	2069
2SD1405BL	OEM
2SD1406	3125
2SD1406-O	3125
2SD1406-Y	3125
2SD1406-YGR	3125
2SD1406GR	3125
2SD1406OR	OEM
2SD1406Y	3125
2SD1406YR	OEM
2SD1407	0482
2SD1408	0236
2SD1408(OY)-LB	6535
2SD1408-Y	0236
2SD1409	OEM
2SD1410	OEM
2SD1411	1165
2SD1412	3125
2SD1413	OEM
2SD1414	OEM
2SD1415	3136
2SD1416	3136
2SD1417	3136
2SD1418	OEM
2SD1419	OEM
2SD1420	OEM
2SD1421	OEM
2SD1422	OEM
2SD1423	2954
2SD1423AS	2954
2SD1423Q	OEM
2SD1424	OEM
2SD1425	1541
2SD-1426	1541
2SD1426	1541
2SD1426(E)	1541
2SD1426(F)	1541
2SD1426FA	1541
2SD1426Y	OEM
2SD1427	1541
2SD1427(FA-1)	1541
2SD1427-LB	1541
2SD1427FA-1	1541
2SD1428	5482
2SD1428E	5482
2SD1428FA	1541
2SD1428FA-3	1541
2SD1428Y	OEM
2SD1429	3438
2SD1430	3438
2SD1431	3476
2SD1432	3360
2SD1433	3360
2SD1434	OEM
2SD1435	OEM
2SD1436	OEM
2SD1437	OEM
2SD1438	3703
2SD1439	1541
2SD1439-Q	1541
2SD1439-Q(LB)	1541
2SD1439-QLB	1541
2SD1439LB	1541
2SD1439P	1541
2SD1439PLB	1541
2SD1439Q	1541
2SD1439R	1541
2SD1439RL	1541
2SD1440	1541
2SD1441	1541
2SD1441(B)	1541
2SD1441(L)	1541
2SD1441-LB	1541
2SD1441LB	1541
2SD1441R	1541
2SD1441RL	1541
2SD1442	0060
2SD1443	0060
2SD1444	0060
2SD1445	0060
2SD1446	OEM
2SD1447	OEM
2SD1448	OEM
2SD1449	OEM
2SD1449S	OEM
2SD1449T	OEM
2SD1450	3447
2SD1450QRSTA	OEM
2SD1450R	OEM
2SD1450RS	OEM
2SD1450S	OEM
2SD1450T	OEM
2SD1451	3438
2SD1452	3438
2SD1453	1541
2SD1453-04	1541
2SD1453-06	1541
2SD1454	3438
2SD1455	1541
2SD1455-04	1541
2SD1455-05	1541
2SD1456	3360
2SD1457	OEM
2SD1457A	OEM
2SD1457AKU	OEM
2SD1457A-P	OEM
2SD1457AP	OEM
2SD1458	0645
2SD1458(Q)	0645
2SD1458(R)	0645
2SD1458(S)	OEM
2SD1459	3728
2SD1459D	3728
2SD1459Q	3728
2SD1459R	3728
2SD1460	OEM
2SD1461	OEM
2SD1462	OEM
2SD1463	OEM
2SD1464	OEM
2SD1465	OEM
2SD1466	OEM
2SD1468	1500
2SD1468S	1500
2SD1468SQ	OEM
2SD1468SR	OEM
2SD1468SS	OEM
2SD1469	0009
2SD1469F	OEM
2SD1469Q	OEM
2SD1469R	OEM
2SD1469S	OEM
2SD1470	OEM
2SD1471	5369
2SD1472	OEM
2SD1473	OEM
2SD1474	OEM
2SD1475	1165
2SD1475(Q)	1165
2SD1475(R)	1165
2SD1475Q	1165
2SD1476	1165
2SD1476(Q)	1165
2SD1476(R)	1165
2SD1477	OEM
2SD1478	5364
2SD1479	3438
2SD1480	0236
2SD1481	0278
2SD1482	OEM
2SD1483	OEM
2SD1484	OEM
2SD1485	0194
2SD1486	0194
2SD1487	0194
2SD1488	0194
2SD1489	1500
2SD1490	1386
2SD1491	1061
2SD1492	3438
2SD1493	3438
2SD1494	3438
2SD1495	3438
2SD1496	3438
2SD1497	3360
2SD1497-06	3360
2SD1498	3360
2SD1499	1165
2SD1500	OEM
2SD1501	OEM
2SD1501Q	OEM
2SD1502	OEM
2SD1503	2784
2SD1504	OEM
2SD1504E	OEM
2SD1505	0060
2SD1506	0161
2SD1507	OEM
2SD1508	OEM
2SD1509	5352
2SD1510	1212
2SD1511	OEM
2SD1512	OEM
2SD1513	1500
2SD1513(Q)	1500
2SD1513(R)	1500
2SD1513R	1500
2SD1514	OEM
2SD1515	OEM
2SD1516	OEM
2SD1517	OEM
2SD1518	4450
2SD1519	OEM
2SD1520	OEM
2SD1521	1061
2SD1522	5342
2SD1523	4334
2SD1524	5342
2SD1525	OEM
2SD1526	OEM
2SD1527	OEM
2SD1528	OEM
2SD1529	OEM
2SD1530	OEM
2SD1531	5497
2SD1531(Q)	OEM
2SD1531(R)	OEM
2SD1531R	5497
2SD1532	OEM
2SD1533	OEM
2SD1534	OEM
2SD1535	OEM
2SD1536	OEM
2SD1537	OEM
2SD1538	OEM
2SD1539	0060
2SD1540	OEM
2SD1541	1541
2SD1542	3438
2SD1543	3438
2SD1544	3438
2SD1545	3438

If replacement code is OEM, contact original manufacturer for replacement.

DEVICE TYPE	REPL CODE
2SD1546	3360
2SD1547	3360
2SD1548	OEM
2SD1549	4450
2SD1550	1506
2SD1551	3438
2SD1552	3438
2SD1553	1541
2SD1554	1541
2SD1554-C1	1541
2SD1554-LB	1541
2SD1554FA	1541
2SD1555	0556
2SD1555(E)	0556
2SD1555(LBOEC)	0556
2SD1555(Y)-T	0556
2SD1555-1	0556
2SD1555-C1	0556
2SD1555-LB	0556
2SD1555-LB-S1	0556
2SD1555LBS1	0556
2SD1556	2129
2SD1556(E)	1541
2SD1556-LB	2129
2SD1556LBMA	2129
2SD1556MA	2129
2SD1557	OEM
2SD1558	1212
2SD1559	OEM
2SD1562	0392
2SD1563	0563
2SD1564	OEM
2SD1565	5417
2SD1566	OEM
2SD1567	OEM
2SD1568	OEM
2SD1569	OEM
2SD1569FA-5	OEM
2SD1571	3357
2SD1572	1212
2SD1573	OEM
2SD1574	OEM
2SD1575	2766
2SD1576	3438
2SD1577	3360
2SD1578	OEM
2SD1579	OEM
2SD1579K	OEM
2SD1580	0060
2SD1581	OEM
2SD1581L	OEM
2SD1582	OEM
2SD1583	OEM
2SD1584	OEM
2SD1585	0236
2SD1585-K	0236
2SD1585K	0236
2SD1585L	OEM
2SD1585LK	0236
2SD1586	0236
2SD1587	0392
2SD1588	OEM
2SD1589	3136
2SD1590	3136
2SD1591	OEM
2SD1592	OEM
2SD1593	OEM
2SD1593KL	OEM
2SD1593L	OEM
2SD1593M	OEM
2SD1594	OEM
2SD1595	3723
2SD1597	OEM
2SD1598	OEM
2SD1599	5504
2SD1600	5504
2SD1601	2235
2SD1602	2235
2SD1603	2235
2SD1604	2235
2SD1605	1959
2SD1606	1959
2SD1607	OEM
2SD1608	1959
2SD1609	0563
2SD1610	OEM
2SD1611	OEM
2SD1612	OEM
2SD1613	OEM
2SD1614	OEM
2SD1615	0667
2SD1615A	0667
2SD1615A-GP	7145
2SD1615Y-GP	OEM
2SD1616	0563
2SD1616A	0563
2SD1617	5508
2SD1618	0667
2SD1619	0667
2SD1620	OEM
2SD1621	OEM
2SD1622	5510
2SD1623	OEM
2SD1624	OEM
2SD1625	OEM
2SD1626	OEM
2SD1627	OEM
2SD1628	OEM
2SD1629	1061

DEVICE TYPE	REPL CODE
2SD1630	1061
2SD1631	OEM
2SD1632	1541
2SD1633	3136
2SD1634	3136
2SD1635	OEM
2SD1636	OEM
2SD1637	1061
2SD1637Y	1061
2SD1638	OEM
2SD1639	OEM
2SD1640	OEM
2SD1641	3516
2SD1642	OEM
2SD1643	OEM
2SD1644	OEM
2SD1645	1061
2SD1646	5417
2SD1647	0278
2SD1647C3	0278
2SD1648	OEM
2SD1649	1541
2SD1649-CA	1541
2SD1649-CTV-YB	1541
2SD-1650	1541
2SD1650	2129
2SD1650-CTV-YB	2129
2SD1650YD	1541
2SD1651	0556
2SD1651-CTV-TB	0556
2SD1651-CTV-YB	0556
2SD1651-KR	0556
2SD1651YB	0556
2SD1651YD	0556
2SD1652	1541
2SD1652-CTV-YB	1541
2SD1653	3494
2SD1654	3494
2SD1655	3494
2SD1656	3494
2SD1657	OEM
2SD1658	1061
2SD1659	OEM
2SD1660	OEM
2SD1661	OEM
2SD1662	OEM
2SD1663	3494
2SD1664	OEM
2SD1664R	OEM
2SD1665	OEM
2SD1666	0060
2SD1666Q	0060
2SD1666R	0060
2SD1666S	0060
2SD1667	0060
2SD1668	0060
2SD1669	OEM
2SD1670	OEM
2SD1671	OEM
2SD1672	OEM
2SD1673	OEM
2SD1676	OEM
2SD1677	OEM
2SD1678	OEM
2SD1679	OEM
2SD1680	4450
2SD1681	0563
2SD1682	OEM
2SD1682S	OEM
2SD1683	OEM
2SD1684	0563
2SD1685	0286
2SD1685RS	0286
2SD1686	OEM
2SD1687	OEM
2SD1688	OEM
2SD1689	OEM
2SD1690	OEM
2SD1691	OEM
2SD1692	OEM
2SD1693	OEM
2SD1694	OEM
2SD1695	OEM
2SD1696	OEM
2SD1697	OEM
2SD1698	OEM
2SD1699	OEM
2SD1700	OEM
2SD1701	OEM
2SD1702	OEM
2SD1703	OEM
2SD1704	OEM
2SD1705	0194
2SD1706	OEM
2SD1706Q	OEM
2SD1707	OEM
2SD1708	OEM
2SD1709	1541
2SD1710	OEM
2SD1711	1541
2SD1712	OEM
2SD1713	OEM
2SD1714	OEM
2SD1715	OEM
2SD1716	OEM
2SD1717	3436
2SD1718	3436
2SD1719	OEM
2SD1720	0042

DEVICE TYPE	REPL CODE
2SD1721	OEM
2SD1722	OEM
2SD1723	OEM
2SD1724	OEM
2SD1725	0161
2SD1725S	0161
2SD1726	OEM
2SD1727	1541
2SD1728	1541
2SD1729	1541
2SD1730	1541
2SD1731	1541
2SD1732	5345
2SD1733	OEM
2SD1734	2766
2SD1735	3494
2SD1736	3494
2SD1737	3494
2SD1738	3494
2SD1739	3494
2SD1740	OEM
2SD1741	OEM
2SD1742	OEM
2SD1743	OEM
2SD1744	OEM
2SD1745	OEM
2SD1746	OEM
2SD1747	OEM
2SD1748	5352
2SD1749	5352
2SD1750	OEM
2SD1751	OEM
2SD1752	OEM
2SD1753	OEM
2SD1754	OEM
2SD1755	OEM
2SD1756	OEM
2SD1757	OEM
2SD1757KQSTW	OEM
2SD1757KR	OEM
2SD1757KS	OEM
2SD1758F5	OEM
2SD1758R	OEM
2SD1760	OEM
2SD1761	OEM
2SD1761-E	OEM
2SD1761-F	OEM
2SD1761DEF	OEM
2SD1761F	OEM
2SD1762	0060
2SD1763	0392
2SD1764	0278
2SD1765	1212
2SD1765S	OEM
2SD1766	OEM
2SD1767	OEM
2SD1768	OEM
2SD1768S	OEM
2SD1769	1959
2SD1770	0392
2SD1771	OEM
2SD1772	0392
2SD1773	1959
2SD1774	2069
2SD1775	OEM
2SD1776	2069
2SD1776(P)	2069
2SD1776(Q)	2069
2SD1776-PQ0	OEM
2SD1776P	2319
2SD1776Q	2319
2SD1777	OEM
2SD1779	OEM
2SD1781K	2349
2SD1781KAFR	OEM
2SD1783L	2366
2SD1785	OEM
2SD1788	3136
2SD1796	OEM
2SD1812P	OEM
2SD1812Q	OEM
2SD1819	OEM
2SD1819A	OEM
2SD1823	OEM
2SD1825	3136
2SD1825-LU	3136
2SD1832	0042
2SD1832C4	0042
2SD1833	0042
2SD1833C4	0042
2SD1833F	0042
2SD1843	2849
2SD1843-T	OEM
2SD1843L	2849
2SD1843T	2849
2SD1846	2129
2SD1847	3507
2SD1855	0060
2SD1855E	0060
2SD1856	3723
2SD1858Q	OEM
2SD1858R2	OEM
2SD1859	2471
2SD1862	0009
2SD1862-P	OEM
2SD1876	2129
2SD1876-CA	2129
2SD1877	2129
2SD1877S-SONY-CA	2129

DEVICE TYPE	REPL CODE
2SD1878	2129
2SD1878-CA	2129
2SD1878CA	2129
2SD1878S	2129
2SD1879	2129
2SD1879-CTV-YB	2129
2SD1880	5330
2SD1880YD	5330
2SD1881	OEM
2SD1885	5431
2SD1886	3494
2SD1886CA	3494
2SD1887	2490
2SD1888	OEM
2SD1894	0065
2SD1900	OEM
2SD1910	1541
2SD1911	1541
2SD1913	0042
2SD1913-O	0042
2SD1913-Q-RA	0042
2SD1913-R-RA	0042
2SD1913Q	0042
2SD1913QRA	0042
2SD1913R	0042
2SD1913RRA	6852
2SD1913S	OEM
2SD1915T	OEM
2SD1919	0009
2SD1920	0009
2SD1929	1061
2SD1929Q	1061
2SD1929TP	1061
2SD1941	3360
2SD1941-06	3360
2SD1941-07	3360
2SD1944	2069
2SD1959	3494
2SD1959-02	3494
2SD1978	OEM
2SD1981	OEM
2SD1991	OEM
2SD2005Q	OEM
2SD2005R	OEM
2SD2006	1386
2SD2010	5552
2SD2010P	OEM
2SD2012	1165
2SD2021	OEM
2SD2021M	OEM
2SD2023E	0461
2SD2023F18E	0461
2SD2037	3446
2SD2037EFTA	OEM
2SD2045	5557
2SD2046	OEM
2SD2057	1541
2SD2057LB	1541
2SD2059Y	OEM
2SD2061	5563
2SD2061DE	0042
2SD2061EF	OEM
2SD2089	2129
2SD2089-LBSONY	2129
2SD2095	2129
2SD2096	OEM
2SD2096-EF	OEM
2SD2096EL	OEM
2SD2096F	OEM
2SD2125	2129
2SD2137-OP	OEM
2SD2144S	OEM
2SD2148	OEM
2SD2148-C1	OEM
2SD2157P	OEM
2SD2271D	OEM
2SD2340	0042
2SD2348	OEM
2SD2348-LB	OEM
2SD2350	0042
2SD2394-EF	OEM
2SD2394-F	OEM
2SD2394E	OEM
2SD2394EF	OEM
2SD2724D	OEM
2SD3243DE	OEM
2SD3311(R)	OEM
2SD5254	OEM
2SD7180	OEM
2SD7620P	0768
2SD9008	OEM
2SD-F1	0600
2SDF1	0600
2SD-F1A	0600
2SDF1A	0600
2SD-F1B	0208
2SD-F1C	0208
2SD-F1D	0208
2SD-F1E	0208
2SD-F1F	0208
2SD-F1G	0208
2SD-F1GN	0208
2SD-F1H	0208
2SD-F1J	0208
2SD-F1K	0208
2SD-F1L	0208
2SD-F1M	0208
2SD-F10R	0208
2SD-F1R	0208

DEVICE TYPE	REPL CODE
2SD-F1X	0208
2SD-F1Y	0208
2SDF2A	0130
2SDF2B	0130
2SDF2C	0130
2SDF2D	0130
2SDF2E	0130
2SDF2F	0130
2SDF2G	0130
2SDF2GN	0130
2SDF2H	0130
2SDF2J	0130
2SDF2K	0130
2SDF2L	0130
2SDF2M	0130
2SDF20R	0130
2SDF2R	0130
2SDF2X	0130
2SDF2Y	0130
2SDF946A	OEM
2SDL2F	0208
2SD026	0142
2SDQ22Q	0161
2SDU36C	0142
2SDU37E	0142
2SDU47X	0130
2SDU57M	0142
2SDU780R	0018
2SDU84GN	0086
2SDU90R	0208
2SE629	0144
2SE4002	0111
2SE4002-1	0208
2SF1	OEM
2SF2	OEM
2SF4	OEM
2SF11	1103
2SF11Q	2336
2SF12	2491
2SF13	0435
2SF14	0240
2SF15	1487
2SF16	2659
2SF18	0676
2SF21	2519
2SF22	0741
2SF23	3105
2SF24	0745
2SF25	3109
2SF26	2917
2SF28	0747
2SF31A	0731
2SF32	0712
2SF32A	0712
2SF33	0469
2SF33A	0469
2SF34	0469
2SF34A	0469
2SF35	0721
2SF35A	0721
2SF36	0721
2SF36A	0721
2SF38	0722
2SF38A	0722
2SF41	OEM
2SF41F	0325
2SF42	OEM
2SF43	OEM
2SF44	OEM
2SF45	OEM
2SF46	OEM
2SF47	OEM
2SF48	OEM
2SF49	OEM
2SF50	OEM
2SF51	OEM
2SF52	OEM
2SF53	OEM
2SF54	OEM
2SF55	2188
2SF56	2188
2SF57	0567
2SF58	0762
2SF59	0762
2SF60	0740
2SF61	0740
2SF62	0740
2SF63	0741
2SF64	0741
2SF65	0741
2SF66	0745
2SF67	0745
2SF68	0747
2SF69	0747
2SF70	0747
2SF71	3307
2SF72	2188
2SF73	0567
2SF74	3258
2SF75	0762
2SF76	3271
2SF77	0740
2SF101	0344
2SF101B	0344
2SF102	0901
2SF102A	0901
2SF102B	0901
2SF103	2342
2SF104	0058

DEVICE TYPE	REPL CODE
2SF104A	0058
2SF104B	0058
2SF105	0407
2SF106	0407
2SF106A	0407
2SF106B	0407
2SF108	0407
2SF108A	0407
2SF108B	0407
2SF110B	1683
2SF111	OEM
2SF112	OEM
2SF113	OEM
2SF114	OEM
2SF115	OEM
2SF116	0710
2SF116A	0710
2SF118	OEM
2SF120	OEM
2SF121	OEM
2SF123	OEM
2SF125	OEM
2SF131	3277
2SF132	0731
2SF133	0712
2SF134	0469
2SF135	0469
2SF136	0721
2SF137	0722
2SF138	0778
2SF139	3291
2SF200	2809
2SF201	0336
2SF205	3243
2SF206	0764
2SF207	2875
2SF210	0778
2SF210A	0778
2SF211	0725
2SF211A	0764
2SF212	2875
2SF212A	2875
2SF215	0764
2SF216	0764
2SF217	0725
2SF218	0725
2SF219	OEM
2SF220	OEM
2SF221	3419
2SF222	1103
2SF223	2491
2SF224	0240
2SF225	0240
2SF226	2659
2SF227	0676
2SF228	2809
2SF229	2342
2SF230	2188
2SF231	2188
2SF232	2188
2SF233	0762
2SF234	0762
2SF235	0740
2SF236	0740
2SF237	0740
2SF239	0731
2SF240	0712
2SF241	0469
2SF242	0469
2SF243	0721
2SF244	0721
2SF245	0722
2SF248	1825
2SF248A	1825
2SF248T	0042
2SF261	1103
2SF261A	1103
2SF262	2491
2SF262A	2491
2SF263	0240
2SF263A	0240
2SF264	0240
2SF264A	0240
2SF265	2659
2SF265A	2659
2SF266	2659
2SF266A	2659
2SF267	0676
2SF267A	0676
2SF268	0676
2SF268A	0676
2SF269	2809
2SF269A	2809
2SF271	2449
2SF271A	2449
2SF272	2491
2SF272A	2491
2SF273	0240
2SF273A	0240
2SF274	0435
2SF274A	0435
2SF275	0435
2SF275A	0435
2SF276	2659
2SF276A	2659
2SF277	1487
2SF277A	1487
2SF278	0747
2SF278A	0747

DEVICE TYPE	REPL CODE
2SF279	3243
2SF279A	3243
2SF281	2188
2SF281A	2188
2SF282	0567
2SF282A	0567
2SF283	3258
2SF283A	3258
2SF284	0762
2SF284A	0762
2SF285	3268
2SF285A	3268
2SF286	3271
2SF286A	3271
2SF287	0740
2SF287A	0740
2SF288	0740
2SF288A	0740
2SF289	3291
2SF289A	3291
2SF291	0731
2SF291A	0731
2SF292	0712
2SF292A	0712
2SF293	0469
2SF293A	0469
2SF294	0469
2SF294A	0469
2SF295	0721
2SF295A	0721
2SF296	0721
2SF296A	0721
2SF297	0722
2SF297A	0722
2SF298	0722
2SF298A	0722
2SF299	0778
2SF299A	0778
2SF301	OEM
2SF302	OEM
2SF303	OEM
2SF304	OEM
2SF305	OEM
2SF306	OEM
2SF307	OEM
2SF308	OEM
2SF309	OEM
2SF316	OEM
2SF321	3419
2SF321A	3419
2SF322	1103
2SF322A	1103
2SF323	2491
2SF323A	2491
2SF324	0240
2SF324A	0240
2SF325	0240
2SF325A	0240
2SF326	2659
2SF326A	2659
2SF327	2659
2SF327A	2659
2SF328	0676
2SF328A	0676
2SF329	2809
2SF329A	2809
2SF330	0336
2SF330A	0336
2SF331	OEM
2SF332	OEM
2SF333	OEM
2SF334	OEM
2SF335	OEM
2SF336	OEM
2SF337	OEM
2SF338	OEM
2SF339	OEM
2SF340	OEM
2SF341	OEM
2SF342	OEM
2SF343	OEM
2SF344	OEM
2SF345	OEM
2SF346	OEM
2SF347	OEM
2SF348	OEM
2SF352	OEM
2SF353	OEM
2SF354	OEM
2SF355	OEM
2SF356	OEM
2SF357	0447
2SF358	OEM
2SF359	OEM
2SF360	OEM
2SF361	OEM
2SF362	OEM
2SF363	OEM
2SF364	OEM
2SF365	OEM
2SF366	OEM
2SF367	OEM
2SF371	OEM
2SF373	OEM
2SF374	OEM
2SF375	OEM
2SF376	OEM
2SF377	OEM
2SF378	OEM

DEVICE TYPE	REPL CODE
2SF379	OEM
2SF401	OEM
2SF402	OEM
2SF403	OEM
2SF404	OEM
2SF405	OEM
2SF406	OEM
2SF407	OEM
2SF408	OEM
2SF416	OEM
2SF421	OEM
2SF422	OEM
2SF426	OEM
2SF427	OEM
2SF431	0764
2SF432	0764
2SF433	2875
2SF434	0766
2SF435	OEM
2SF436	OEM
2SF437	OEM
2SF438	2875
2SF439	0766
2SF440	OEM
2SF441	0505
2SF442	0505
2SF443	OEM
2SF448	0336
2SF448A	0336
2SF449	2875
2SF450	OEM
2SF451	0687
2SF452	0766
2SF453	OEM
2SF454	0764
2SF454A	0764
2SF455	0766
2SF456	OEM
2SF457	0764
2SF458	0766
2SF459	0750
2SF460	0712
2SF461	0469
2SF462	0721
2SF463	0740
2SF464	3291
2SF478	OEM
2SF479	OEM
2SF480	OEM
2SF481	OEM
2SF482	OEM
2SF483	OEM
2SF484	OEM
2SF485	OEM
2SF486	OEM
2SF487	OEM
2SF490	OEM
2SF491	OEM
2SF492	OEM
2SF493	OEM
2SF494	OEM
2SF495	OEM
2SF496	OEM
2SF497	OEM
2SF511	OEM
2SF512	OEM
2SF513	OEM
2SF514	OEM
2SF515	OEM
2SF516	OEM
2SF517	OEM
2SF518	OEM
2SF521	0344
2SF522	0447
2SF523	0058
2SF524	0407
2SF525	0407
2SF526	1683
2SF527	1683
2SF528	2095
2SF529	OEM
2SF530	0505
2SF531	0863
2SF532	0710
2SF533	OEM
2SF534	OEM
2SF535	OEM
2SF536	OEM
2SF537	OEM
2SF538	OEM
2SF539	OEM
2SF540	OEM
2SF541	OEM
2SF542	OEM
2SF543	OEM
2SF544	OEM
2SF562	OEM
2SF563	OEM
2SF564	OEM
2SF565	OEM
2SF566	OEM
2SF605	OEM
2SF644S	OEM
2SF656	0901
2SF656ZS	0901
2SF657	0901
2SF658	0058
2SF660	1110
2SF661	2932
2SF662	0963
2SF664	0901
2SF664S	0901
2SF665	1103
2SF666	2491
2SF667	0240
2SF668	0676
2SF674S	0901
2SF676S	0901
2SF694	OEM
2SF695	OEM
2SF696	OEM
2SF711	0567
2SF713	3271
2SF714	0747
2SF716	OEM
2SF717	0766
2SF721	0712
2SF722	0469
2SF724	0722
2SF725	0778
2SF726	0725
2SF727	0766
2SF732	0469
2SF733	0721
2SF734	0722
2SF743	OEM
2SF744	OEM
2SF745	OEM
2SF746	OEM
2SF747	OEM
2SF748	OEM
2SF752	0762
2SF752A	0745
2SF755	3291
2SF762	0762
2SF762A	0762
2SF775	0698
2SF812	OEM
2SF813	OEM
2SF814	OEM
2SF815	OEM
2SF816	OEM
2SF817	OEM
2SF832	OEM
2SF833	OEM
2SF834	OEM
2SF835	OEM
2SF836	OEM
2SF837	OEM
2SF861	OEM
2SF862	OEM
2SF863	OEM
2SF875	OEM
2SF876	OEM
2SF877	OEM
2SF894	OEM
2SF932	OEM
2SF933	OEM
2SF934	OEM
2SF935	OEM
2SF936	OEM
2SF937	OEM
2SF938	OEM
2SF939	OEM
2SF940	0505
2SF941	0447
2SF942	0940
2SF1060	0505
2SF1110	0344
2SF1111	0344
2SF1112	0058
2SF1113	0407
2SF1114	0407
2SF1115	2095
2SF1116	2089
2SF1117	0505
2SF1118	0710
2SF1119	OEM
2SF1125	OEM
2SF1126	OEM
2SF1127	OEM
2SF1128	OEM
2SF1129	OEM
2SF1130	OEM
2SF1141	OEM
2SF1145	1900
2SF1146	1900
2SF1147	0738
2SF1148	0738
2SF1149	2147
2SF1150	OEM
2SF1151	OEM
2SF1152	OEM
2SF1168	0239
2SF1168A	0239
2SF1188	0239
2SF1188A	0239
2SF1188B	0239
2SF1188C	0239
2SF1188D	0239
2SF1188E	0239
2SF1188F	0239
2SF1188G	0239
2SF1188H	0239
2SF1188K	0239
2SF1188L	0239
2SF1188M	0239
2SF1188N	0239
2SF1188R	0239
2SF1188Y	0239
2SF1189	0061
2SF1189A	0061
2SF1189C	0061
2SF1189D	0061
2SF1189E	0061
2SF1189F	0061
2SF1189G	0061
2SF1189H	0061
2SF1189K	0061
2SF1189L	0061
2SF1189M	0061
2SF1189N	0061
2SF1189R	0061
2SF1189Y	0061
2SF1365	OEM
2SF1366	OEM
2SF1367	OEM
2SF1368	OEM
2SF1369	OEM
2SF1370	OEM
2SF1418C	0505
2SF1418D	0863
2SF1418E	0710
2SF1419	0710
2SF1420	0710
2SF1420D	0239
2SF1420E	0710
2SF1420F	0710
2SF1421	0710
2SF1422	0239
2SF1422A	0239
2SF1422B	0239
2SF1422C	0239
2SF1422D	0239
2SFD111	0143
2SFF05	OEM
2SFF10	OEM
2SFF15	OEM
2SFT212	0160
2SF.T212	0085
2SG536GN	0016
2SH11	2136
2SH12	1669
2SH13	OEM
2SH14	OEM
2SH18	2136
2SH18K	2136
2SH18L	2136
2SH18M	2136
2SH18N	2136
2SH19	2136
2SH19K	2136
2SH19L	2136
2SH19M	2136
2SH19N	2136
2SH20	2136
2SH21	1669
2SH22	1175
2SH203	0280
2SH460	0440
2SH643	0791
2SH678	0150
2SJ2A	0015
2SJ4A	0015
2SJ8A	0071
2SJ11	0325
2SJ12	0325
2SJ13	OEM
2SJ15	OEM
2SJ16	OEM
2SJ17	OEM
2SJ18	OEM
2SJ19	OEM
2SJ20	OEM
2SJ22	OEM
2SJ29	OEM
2SJ32	OEM
2SJ33	OEM
2SJ39	OEM
2SJ40	OEM
2SJ40CD	OEM
2SJ40CTA	OEM
2SJ43	OEM
2SJ44	OEM
2SJ45	OEM
2SJ47	OEM
2SJ48	OEM
2SJ49	OEM
2SJ49(H)	OEM
2SJ49H	OEM
2SJ50	OEM
2SJ50H	OEM
2SJ51	OEM
2SJ55	OEM
2SJ56	OEM
2SJ56(H)	OEM
2SJ56H	OEM
2SJ59(H)	OEM
2SJ60A	0015
2SJ68	OEM
2SJ69	OEM
2SJ70	OEM
2SJ72	OEM
2SJ73	OEM
2SJ74	OEM
2SJ75	OEM
2SJ76	OEM
2SJ77	OEM
2SJ77(K)	OEM
2SJ77K	OEM
2SJ78	OEM
2SJ79	OEM
2SJ79(K)	OEM
2SJ79K	OEM
2SJ81	OEM
2SJ82	OEM
2SJ83	OEM
2SJ84	OEM
2SJ90	OEM
2SJ91	OEM
2SJ92	OEM
2SJ96	OEM
2SJ99	OEM
2SJ100	OEM
2SJ101	OEM
2SJ102	2271
2SJ103	2987
2SJ103GR	OEM
2SJ104	OEM
2SJ104V	OEM
2SJ105	2987
2SJ105-Y	2987
2SJ105Y	2987
2SJ106	OEM
2SJ106G-TE85L	OEM
2SJ107	OEM
2SJ108	OEM
2SJ109	OEM
2SJ110	OEM
2SJ111	OEM
2SJ112	OEM
2SJ113	OEM
2SJ114	OEM
2SJ115	OEM
2SJ116	OEM
2SJ117	5360
2SJ118	OEM
2SJ119	OEM
2SJ120(L)	OEM
2SJ120(S)	OEM
2SJ120L	OEM
2SJ120S	OEM
2SJ122	OEM
2SJ123	OEM
2SJ125	OEM
2SJ126	OEM
2SJ164PQRTA	OEM
2SK2SET	0325
2SK11	0325
2SK12	0325
2SK13	0325
2SK15	3384
2SK16(H)	0325
2SK16HA	0325
2SK17	0325
2SK17(O)	0325
2SK17-0	0325
2SK17-OR	0325
2SK17A	0325
2SK17B	0325
2SK17BL	0325
2SK17GR	0325
2SK170	0325
2SK17R	0325
2SK17Y	0325
2SK18	OEM
2SK18A	OEM
2SK18L	OEM
2SK19	0688
2SK19(BL)	0688
2SK19(GN)	0688
2SK19(GR)	0688
2SK19-14	0688
2SK19-BL	0688
2SK19-GR	0688
2SK19-Y	0688
2SK19A	0688
2SK19B	0688
2SK19BB	0688
2SK19BL	0688
2SK19FET	0688
2SK19GB	0688
2SK19GE	0688
2SK19GR	0688
2SK19H	0688
2SK19K	0688
2SK19TM	0688
2SK19V	0688
2SK19Y	0688
2SK22Y	0325
2SK23	3717
2SK23A	3717
2SK23A(812)	3717
2SK23A-8	3717
2SK23A8	3717
2SK23A-9	3717
2SK23A9	3717
2SK23A-540	3717
2SK23A540	3717
2SK23AL	3717
2SK23G	3717
2SK24	0325
2SK24C	0325
2SK24D	0325
2SK24DR	0325
2SK24E	0325
2SK24F	0325
2SK24G	0325
2SK25	0325
2SK25C	0325
2SK25D	0325
2SK25E	0325
2SK25ET	0325
2SK25F	0325
2SK25G	0325
2SK30	3046
2SK30(O)	3046
2SK30-O	3046
2SK30-Y	3046
2SK30A	3046
2SK30A(D)	3046
2SK30A(FA-1)	3046
2SK30A(G)	3046
2SK30A(G)F	3046
2SK30A(G)L	3046
2SK30A(G)P	3046
2SK30A(GR)	3046
2SK30A-1	3046
2SK30A-GR	3046
2SK30AGR	3046
2SK30A-O	3046
2SK30AO	3046
2SK30A-R	3046
2SK30ATM	3046
2SK30A-TM(FA-1)	3046
2SK30ATM(FA-1)	OEM
2SK30ATM(O)	3046
2SK30ATM(Y)	3046
2SK30ATMFA-1	3046
2SK30ATMFA1	3046
2SK30ATMFA-I	3046
2SK30ATM-G	OEM
2SK30ATM-GR	3046
2SK30ATMGR	3046
2SK30ATM-O	3046
2SK30ATM-R	3046
2SK30ATM-Y	3046
2SK30A-Y	3046
2SK30AY	3046
2SK30D	3046
2SK30GR	3046
2SK30O	3046
2SK30R	3046
2SK30Y	3046
2SK31(C)	0325
2SK31C	0325
2SK32	0325
2SK32B	0325
2SK33	1280
2SK33D	1280
2SK33E	1280
2SK33F	1280
2SK33H	1280
2SK34	3155
2SK34(C)	3155
2SK34(D)	3155
2SK34(D)(E)	3155
2SK34(E)	3155
2SK34-D	3155
2SK34A	3155
2SK34B	3155
2SK34C	3155
2SK34D	3155
2SK34E	3155
2SK35	0325
2SK35-0	0325
2SK35-1	0325
2SK35-2	0325
2SK35A	0325
2SK35BL	0325
2SK35C	0325
2SK35GN	0325
2SK35R	0325
2SK35Y	0325
2SK37	0325
2SK37(K)	0325
2SK37H	0325
2SK38	OEM
2SK38A	OEM
2SK39	0212
2SK39(P)	0212
2SK39A	0212
2SK39B	0212
2SK39P	0212
2SK39Q	0212
2SK40	3354
2SK40(C)	3354
2SK40(D)	3354
2SK40-3	3354
2SK40A	3354
2SK40B	3354
2SK40C	3354
2SK40D	3354
2SK41	1211
2SK41C	1211
2SK41D	1211
2SK41E	1211
2SK41E1	1211
2SK41F	1211
2SK42	0325
2SK42-CM1	0325
2SK42-CMI	0325
2SK43	1758
2SK43(03)	1758
2SK43(2)	1758
2SK43(3A)	1758
2SK43(S)	1758
2SK43(S)-D	1758
2SK43-3	1758
2SK43-4	1758
2SK43-OR	1758
2SK43A	1758
2SK43B	1758
2SK43C	1758
2SK43D	1758
2SK43E	1758
2SK43F	1758
2SK43GN	1758
2SK43H	1758
2SK43J	1758
2SK43OR	1758
2SK43R	1758
2SK43S	1758
2SK43X	1758
2SK43Y	1758
2SK44	0325
2SK44(D)	0325
2SK45	0849
2SK45B	0849
2SK46	OEM
2SK47	2950
2SK48	3249
2SK49	1280
2SK49E2	1280
2SK49F	1280
2SK49H	1280
2SK49H1	1280
2SK49H2	1280
2SK49I	1280
2SK49M	1280
2SK50	OEM
2SK54	1280
2SK54B	1280
2SK54C	1280
2SK55	0325
2SK55(CD)	0325
2SK55C	0325
2SK55CD	0325
2SK55D	0325
2SK55DE	0325
2SK55R	0325
2SK56	OEM
2SK57	OEM
2SK58	OEM
2SK59	1758
2SK60	OEM
2SK61	1211
2SK61(Y)	1211
2SK61-LV	OEM
2SK61-O	1211
2SK61GR	1211
2SK61Y	1211
2SK63	OEM
2SK65	OEM
2SK66	3133
2SK67	OEM
2SK67A	OEM
2SK67J2	OEM
2SK67J3	OEM
2SK67J4	OEM
2SK67J5	OEM
2SK67J6	OEM
2SK67J7	OEM
2SK67J8	OEM
2SK68	3342
2SK68-L	3342
2SK68-M	3342
2SK68A	3342
2SK68A(L)	3342
2SK68A(M)	3342
2SK68AK	3342
2SK68AL	3342
2SK68A-M	3342
2SK68AM	3342
2SK68L	3342
2SK68M	3342
2SK68Q	3342
2SK69	OEM
2SK70	OEM
2SK72	2945
2SK73	OEM
2SK74	OEM
2SK79	OEM
2SK80	OEM
2SK81	OEM
2SK83	0325
2SK83R	0325
2SK84	3133
2SK84R	OEM
2SK85	OEM
2SK87(H)	3133
2SK87HB	3133
2SK87HD	3133
2SK88	OEM
2SK92	OEM
2SK93	OEM
2SK94	OEM
2SK94X1	OEM
2SK95	OEM
2SK96	OEM
2SK97	OEM
2SK101	OEM
2SK102	OEM
2SK103	OEM
2SK104	0325
2SK104E	0325
2SK104F	0325
2SK104H	0325
2SK105	1758
2SK105(E)	1758
2SK105(F)	1758
2SK105A	1758
2SK105A-10	1758
2SK105A-30	1758
2SK105E	1758
2SK105F	1758
2SK106	0325
2SK106(D)	0325
2SK106D	0325
2SK107	0325
2SK107(3)	0325
2SK107-1	0325
2SK107-2	0325
2SK107-4	0325
2SK108	OEM
2SK108-C	OEM
2SK108C	OEM
2SK109	OEM
2SK109A	OEM
2SK110	OEM
2SK111	OEM
2SK112	OEM
2SK113	1155
2SK117	3133
2SK117(BL)	3133
2SK117-BL	3133
2SK117-GR	3133
2SK117-GR(TPE2)	3133
2SK117-GR-FA-2	3133
2SK117-GRFA-2	3133
2SK117-Y	3133
2SK117A	3133
2SK117B(C)	3133
2SK117BL	3133
2SK117GR	3133
2SK117Y	3133
2SK118	1758
2SK118(O)	1758
2SK118-O	1758
2SK118-R	1758
2SK118NEW-R	OEM
2SK118Q	1758
2SK118R	1758
2SK119	0325
2SK120	1392
2SK121	1758
2SK121-2	1758
2SK121-3	OEM
2SK121-4	OEM
2SK121-5	OEM
2SK121-6	OEM
2SK123	OEM
2SK124	OEM
2SK125	OEM
2SK127	1758
2SK127A	OEM
2SK128	OEM
2SK130	OEM
2SK130A	OEM
2SK131	OEM
2SK132	OEM
2SK133	OEM
2SK134	OEM
2SK134(H)	OEM
2SK134H	OEM
2SK135	OEM
2SK135(H)	OEM
2SK135H	OEM
2SK136	3717
2SK137	OEM
2SK137A	OEM
2SK138	OEM
2SK140	OEM
2SK141	1155
2SK141A	1155
2SK146	OEM
2SK147	OEM
2SK148	0325
2SK149	OEM
2SK150	OEM
2SK151	OEM
2SK152	0325
2SK152(2)	0325
2SK152(4)	0325
2SK152-1	0325
2SK152-2	0325
2SK152-3	0325
2SK152-4	OEM
2SK152L	0325
2SK152LA	0325
2SK154	OEM
2SK155	OEM
2SK156	OEM
2SK157	OEM
2SK158	OEM
2SK159	OEM
2SK160	OEM
2SK160A	OEM
2SK160K4	OEM
2SK160K5	OEM
2SK160K6	OEM
2SK160K7	OEM
2SK161	3622

If replacement code is OEM, contact original manufacturer for replacement.

DEVICE TYPE	REPL CODE
2SK161-O	OEM
2SK161Y	OEM
2SK162	OEM
2SK163	OEM
2SK163L	OEM
2SK163M	OEM
2SK165	OEM
2SK168	0325
2SK168D	0325
2SK168F	0325
2SK169	OEM
2SK170	0325
2SK170V	0325
2SK171	OEM
2SK173	OEM
2SK175	OEM
2SK176	OEM
2SK176(H)	OEM
2SK176H	OEM
2SK184	OEM
2SK185	OEM
2SK186	3133
2SK186B	3133
2SK186C	3133
2SK186D	3133
2SK186E	3133
2SK187	OEM
2SK187C	OEM
2SK187D	OEM
2SK187E	OEM
2SK187F	OEM
2SK190	OEM
2SK190E	OEM
2SK190F	OEM
2SK190G	OEM
2SK190H	OEM
2SK191	3249
2SK192	0325
2SK192A	0325
2SK193	OEM
2SK194	0325
2SK195	1392
2SK195H	1392
2SK196	3249
2SK196(H)	OEM
2SK196R	3249
2SK197	0325
2SK197YC	0325
2SK197YD	0325
2SK197YE	0325
2SK198	OEM
2SK198L	OEM
2SK199	OEM
2SK201	OEM
2SK203	OEM
2SK208	OEM
2SK209	OEM
2SK209-G	OEM
2SK209-GR	OEM
2SK209-Y	OEM
2SK209G	OEM
2SK210	OEM
2SK211	OEM
2SK212	1392
2SK212A	1392
2SK212A-E	OEM
2SK212AE	1392
2SK212E	1392
2SK212F	OEM
2SK213	OEM
2SK214	OEM
2SK214(K)	OEM
2SK214K	OEM
2SK215	OEM
2SK216	OEM
2SK216(K)	OEM
2SK216K	OEM
2SK217	OEM
2SK217ZD	OEM
2SK217ZE	OEM
2SK218	OEM
2SK218(Q)	OEM
2SK218Q	OEM
2SK218R	OEM
2SK220(H)	OEM
2SK220H	OEM
2SK221(H)	OEM
2SK221H	OEM
2SK222	OEM
2SK222E	OEM
2SK223	OEM
2SK225	OEM
2SK226	OEM
2SK227	OEM
2SK238	OEM
2SK240	OEM
2SK241	OEM
2SK242	OEM
2SK246	OEM
2SK246-Y	OEM
2SK246BL	OEM
2SK246GR	OEM
2SK246GR3	OEM
2SK246Y	OEM
2SK247	OEM
2SK258(H)	OEM
2SK258H	OEM
2SK259(H)	OEM
2SK259H	OEM
2SK260(H)	OEM
2SK260H	OEM
2SK266	OEM
2SK270	OEM
2SK271	OEM
2SK272	OEM
2SK273	OEM
2SK274	OEM
2SK275	OEM
2SK276	OEM
2SK277	OEM
2SK278	OEM
2SK279	OEM
2SK280	OEM
2SK281	OEM
2SK283	OEM
2SK286	OEM
2SK287(K)	OEM
2SK287K	OEM
2SK288(K)	OEM
2SK288K	OEM
2SK289(H)	OEM
2SK290(H)	OEM
2SK291	OEM
2SK291E	OEM
2SK291F	OEM
2SK291G	OEM
2SK291H	OEM
2SK292	OEM
2SK293	OEM
2SK293A	OEM
2SK294	6385
2SK295	5060
2SK296	1466
2SK298	OEM
2SK299	OEM
2SK300	OEM
2SK301	3133
2SK301(O)	3133
2SK301(P)	3133
2SK301(Q)	3133
2SK301(R)	3133
2SK301Q	3133
2SK301QR	3133
2SK301QTA	OEM
2SK301R	3133
2SK302	OEM
2SK303	OEM
2SK304	0325
2SK304D	OEM
2SK304E	OEM
2SK308	OEM
2SK310	1465
2SK311	OEM
2SK312	OEM
2SK313	OEM
2SK314	OEM
2SK315	OEM
2SK315E	OEM
2SK316	OEM
2SK317	OEM
2SK318	OEM
2SK319	1465
2SK320	OEM
2SK321	OEM
2SK322	OEM
2SK322WE	OEM
2SK322WF	OEM
2SK322WG	OEM
2SK322WH	OEM
2SK323	OEM
2SK323KB	OEM
2SK323KC	OEM
2SK323KD	OEM
2SK323KE	OEM
2SK324	OEM
2SK325	OEM
2SK330	OEM
2SK331	OEM
2SK332	OEM
2SK333	OEM
2SK334	OEM
2SK336	OEM
2SK337	OEM
2SK343	OEM
2SK344	OEM
2SK345	5361
2SK346	5361
2SK347	OEM
2SK349	OEM
2SK350	OEM
2SK351	5899
2SK352	OEM
2SK353	OEM
2SK354	OEM
2SK354A	OEM
2SK355	OEM
2SK356	OEM
2SK357	OEM
2SK358	OEM
2SK359	OEM
2SK360	OEM
2SK362	1758
2SK363	OEM
2SK364	OEM
2SK365	OEM
2SK366	OEM
2SK367	OEM
2SK368	OEM
2SK369	OEM
2SK370	OEM
2SK371	OEM
2SK372	OEM
2SK373	OEM
2SK374	OEM
2SK375(L)	OEM
2SK375(S)	OEM
2SK376	OEM
2SK377	OEM
2SK378	OEM
2SK379	OEM
2SK380	OEM
2SK381	1758
2SK381-B	1758
2SK381-E	1758
2SK381B	1758
2SK381C	4282
2SK381D	4282
2SK381TP-C	1758
2SK382	1466
2SK383	5060
2SK384(L)	OEM
2SK384(S)	OEM
2SK385	OEM
2SK386	OEM
2SK387	OEM
2SK388	OEM
2SK389	OEM
2SK398	5363
2SK399	OEM
2SK400	OEM
2SK401	OEM
2SK402	OEM
2SK403	OEM
2SK404	OEM
2SK405	OEM
2SK406	OEM
2SK407	OEM
2SK408	OEM
2SK409	OEM
2SK410	OEM
2SK411	1668
2SK412	OEM
2SK413	OEM
2SK414	OEM
2SK415	OEM
2SK416	OEM
2SK416(L)	OEM
2SK416(S)	OEM
2SK417	OEM
2SK418	OEM
2SK419	OEM
2SK420	OEM
2SK421	OEM
2SK422	OEM
2SK423	OEM
2SK424	1668
2SK425	OEM
2SK426	OEM
2SK427	OEM
2SK428	5060
2SK429	OEM
2SK429(L)	OEM
2SK429(S)	OEM
2SK430	OEM
2SK430(L)	OEM
2SK430(S)	OEM
2SK431	OEM
2SK433	OEM
2SK435	OEM
2SK436	OEM
2SK437	OEM
2SK437(H)	OEM
2SK438	OEM
2SK439	OEM
2SK440	2176
2SK441	OEM
2SK442	OEM
2SK443	OEM
2SK444	OEM
2SK445	OEM
2SK446	OEM
2SK447	OEM
2SK448	OEM
2SK449	OEM
2SK453	OEM
2SK454	OEM
2SK455	OEM
2SK456	OEM
2SK457	OEM
2SK458	OEM
2SK459	OEM
2SK462	OEM
2SK463	OEM
2SK464	OEM
2SK468	OEM
2SK470	OEM
2SK477	OEM
2SK479	OEM
2SK482	OEM
2SK484	OEM
2SK493G	OEM
2SK494	OEM
2SK503A	OEM
2SK505	OEM
2SK507	OEM
2SK508	OEM
2SK511	OEM
2SK512	OEM
2SK513	3718
2SK518	OEM
2SK519	OEM
2SK520	OEM
2SK521	OEM
2SK522	OEM
2SK525	OEM
2SK526	OEM
2SK527	OEM
2SK528	OEM
2SK529	OEM
2SK530	OEM
2SK531	OEM
2SK532	OEM
2SK534	OEM
2SK535	OEM
2SK535(L)	OEM
2SK535(S)	OEM
2SK537	OEM
2SK538	OEM
2SK539	OEM
2SK541	OEM
2SK543	OEM
2SK544	OEM
2SK544F	OEM
2SK545	OEM
2SK546	OEM
2SK549	OEM
2SK551	OEM
2SK552	OEM
2SK553	OEM
2SK554	5769
2SK555	5769
2SK556	OEM
2SK557	OEM
2SK559	OEM
2SK560	OEM
2SK562	OEM
2SK564	OEM
2SK565	OEM
2SK566	OEM
2SK568	OEM
2SK570	OEM
2SK570S	OEM
2SK572	OEM
2SK573	OEM
2SK575	OEM
2SK578	OEM
2SK579(L)	OEM
2SK579(S)	OEM
2SK580(L)	OEM
2SK580(S)	OEM
2SK581	OEM
2SK582	OEM
2SK583	OEM
2SK585	OEM
2SK586	OEM
2SK587	OEM
2SK590	OEM
2SK593	OEM
2SK600	OEM
2SK601	OEM
2SK602	OEM
2SK603	OEM
2SK604	OEM
2SK606	0325
2SK607	OEM
2SK608	OEM
2SK609	OEM
2SK610	OEM
2SK614	OEM
2SK615	OEM
2SK616	OEM
2SK617	OEM
2SK618	OEM
2SK619	OEM
2SK621	OEM
2SK624	OEM
2SK625	OEM
2SK626	OEM
2SK627	OEM
2SK628	OEM
2SK629	OEM
2SK630	OEM
2SK631	OEM
2SK632	OEM
2SK633	OEM
2SK634	OEM
2SK635	OEM
2SK636	OEM
2SK637	OEM
2SK638	OEM
2SK646	OEM
2SK652Q	OEM
2SK652R	OEM
2SK656	OEM
2SK705	OEM
2SK739	OEM
2SK774	OEM
2SK786	OEM
2SK796	OEM
2SK817	OEM
2SK824	OEM
2SK824(1)	OEM
2SK1033B	0325
2SK1059	OEM
2SK1105	OEM
2SK1199	OEM
2SK1206-01	OEM
2SK1429	OEM
2SK1521A	0325
2SK1916-53-F87	OEM
2SK1917	OEM
2SK2050F	OEM
2SK2050F121	OEM
2SK2255	OEM
2SK2255-01M	OEM
2SL17	OEM
2SL24F	OEM
2SM1	0015
2SM3	0087
2SM6	OEM
2SM11	OEM
2SM12	OEM
2SM13	OEM
2SM15	0422
2SM20	OEM
2SM21	OEM
2SM22	OEM
2SM25	0422
2SM40	OEM
2SM41	OEM
2SM42	OEM
2SM43	OEM
2SM55	OEM
2SM58	OEM
2SM63	OEM
2SM64	OEM
2SM65	OEM
2SM66	OEM
2SM67	OEM
2SM68	OEM
2SM69	OEM
2SM70	OEM
2SM71	0408
2SM72	0422
2SM73	0147
2SM74	0147
2SM75	OEM
2SM79	OEM
2SM100	3152
2SM125	OEM
2SM125A	OEM
2SM136	OEM
2SM137	OEM
2SM138	OEM
2SM139	OEM
2SM140	OEM
2SM141	OEM
2SM142	OEM
2SM143	OEM
2SM144	0154
2SM145	0147
2SM146	0147
2SM147	0279
2SM150	OEM
2SM151	OEM
2SM200	OEM
2SM610B	0037
2SM1020A4	OEM
2SM1020A4PL	OEM
2SM1020A4T	OEM
2SM1020A5	OEM
2SM1020A5PL	OEM
2SM1020A5T	OEM
2SM1020A6	OEM
2SM1020A6PL	OEM
2SM1020A6T	OEM
2SM1020A7	OEM
2SM1020A7PL	OEM
2SM1020A7T	OEM
2SM1020A8PL	OEM
2SM1020A8T	OEM
2SM1020AB	OEM
2SM1020E4	OEM
2SM1020E4PL	OEM
2SM1020E4T	OEM
2SM1020E5	OEM
2SM1020E5PL	OEM
2SM1020E5T	OEM
2SM1020E6	OEM
2SM1020E6PL	OEM
2SM1020E6T	OEM
2SM1020E7	OEM
2SM1020E7T	OEM
2SM1020GE8	OEM
2SM1020GE8PL	OEM
2SM1020GE8T	OEM
2SM1020GE9	OEM
2SM1020GE9PL	OEM
2SM1020GE9T	OEM
2SM1020GE10	OEM
2SM1020GE10PL	OEM
2SM1020GE10T	OEM
2SN733	0079
2SN784	0493
2SN3841	0037
2SO12	0042
2SO14	0007
2SO17	0630
2SO18	0630
2SO20	0630
2SO33	0103
2SO34	0103
2SO35	0103
2SO36	0103
2SO95A	0079
2S002	0861
2SP405	0086
2SQ371	0144
2SR05K	0015
2SR1K	0015
2SR1K-2	OEM
2SR1K-4	OEM
2SR24	0325
2SR54	0086
2SR68AM	0325
2SR173	0086
2SR677	0791
2SV341V	0975
2SVB	0323
2SVB10	1414
2T1	OEM
2T2P	0211
2T3	0004
2T11	0164
2T12	0164
2T13	0164
2T14	0164
2T14A	0136
2T15	0164
2T15X3	0127
2T16	0164
2T17	0164
2T20	0050
2T21	0164
2T22	0164
2T23	0164
2T24	0164
2T25	0164
2T26	0164
2T40	0079
2T41	0079
2T42	0079
2T43	0079
2T44	0079
2T51	0038
2T52	0600
2T53	0600
2T54	0600
2T55	0600
2T56	0600
2T57	0600
2T58	0600
2T61	0038
2T62	0038
2T63	0038
2T64	0038
2T64R	0038
2T65	0038
2T65R	0038
2T66	0038
2T66R	0038
2T67	0600
2T69	0038
2T71	0600
2T72	0600
2T73	0600
2T73R	0600
2T74	0600
2T75	0600
2T75R	0600
2T76	0600
2T76R	0600
2T77	0600
2T77R	0600
2T78	0600
2T78R	0600
2T82	0600
2T83	0600
2T84	0038
2T85	0038
2T85A	0038
2T86	0038
2T89	0038
2T172	0079
2T201	0050
2T202	0079
2T203	0050
2T204	0050
2T204A	0050
2T205	0050
2T205A	0050
2T230	0211
2T231	0211
2T311	0164
2T312	0164
2T313	0164
2T314	0164
2T315	0164
2T321	0164
2T322	0164
2T323	0164
2T324	0164
2T383	0164
2T402	0198
2T403	0079
2T404	0079
2T501	0015
2T502	0015
2T503	0015
2T504	0015
2T505	0015
2T506	0015
2T507	0071
2T508	0071
2T509	0071
2T510	0071
2T513	0600
2T520	0600
2T521	0600
2T522	0038
2T523	0038
2T524	0600
2T551	0600
2T552	0038
2T650	0600
2T681	0038
2T682	0038
2T918	0079
2T919	0127
2T2001	0211
2T2102	0595
2T2708	0079
2T2785	0079
2T2857	0079
2T3011	0160
2T3021	0160
2T3022	0160
2T3030	0160
2T3031	0160
2T3032	0160
2T3033	0160
2T3041	0160
2T3042	0160
2T3043	0160
2TDA1054M	OEM
2TN15	0280
2TN32	0280
2TN45A	0211
2TN48	0280
2TN49	0280
2TN52	0280
2TN53	0280
2TN56	0211
2TN95	0211
2TN95A	0211
2U4	0023
2U8	0023
2U12	0023
2U16	0017
2U20	0017
2V58.5B	OEM
2V110A	OEM
2V110B	OEM
2V205	OEM
2V362	0211
2V363	0211
2V435	OEM
2V464	0280
2V465	0280
2V466	0280
2V467	0280
2V482	0280
2V483	0280
2V484	0280
2V485	0050
2V486	0280
2V559	0050
2V560	0050
2V561	0050
2V562	0050
2V563	0050
2V631	0280
2V632	0280
2V633	0280
2VR5.6	0876
2VR5.6A	0876
2VR5.6B	0876
2VR6	3127
2VR6A	3127
2VR6B	3127
2VR6.2	3127
2VR6.2A	3127
2VR6.2B	3127
2VR6.8	0205
2VR6.8A	0205
2VR6.8B	0205
2VR7	0205
2VR7A	0205
2VR7B	0205
2VR7.5	0480
2VR7.5A	0480
2VR7.5B	0480
2VR8.2	0504
2VR8.2A	0504
2VR8.2B	0504
2VR8.5	3317
2VR8.5A	3317
2VR8.5B	3317
2VR9.1	0684
2VR9.1A	0684
2VR9.1B	0684
2VR10	0225
2VR10A	0225
2VR10B	0225
2VR11	0230
2VR11A	0230
2VR11B	0230
2VR12	0234
2VR12A	0234
2VR12B	0234
2VR13	0237
2VR13A	0237
2VR13B	0237

DEVICE TYPE	REPL CODE
2VR13R	0237
2VR14	1397
2VR14A	1397
2VR14B	1397
2VR15	0247
2VR15A	0247
2VR15B	0247
2VR16	0251
2VR16A	0251
2VR16B	0251
2VR18	0256
2VR18A	0256
2VR18B	0256
2VR20	0262
2VR20A	0262
2VR20B	0262
2VR22	0269
2VR22A	0269
2VR22B	0269
2VR24	0274
2VR24A	0274
2VR24B	0274
2VR27	0294
2VR27A	0294
2VR27B	0294
2VR28	1177
2VR28A	1177
2VR28B	1177
2VR30	0309
2VR30A	0309
2VR30B	0309
2VR33	0318
2VR33A	0318
2VR33B	0318
2VR36	0320
2VR36A	0320
2VR36B	0320
2VR39	0326
2VR39A	0326
2VR39B	0326
2VR43	0337
2VR43A	0337
2VR43B	0337
2VR47	0347
2VR47A	0347
2VR47B	0347
2VR51	0027
2VR51A	0027
2VR51B	0027
2VR56	0266
2VR56A	0266
2VR56B	0266
2VR62	0386
2VR62A	0386
2VR62B	0386
2VR67	0405
2VR67A	0405
2VR67B	0405
2VR68	0405
2VR68A	0405
2VR68B	0405
2VR75	0426
2VR75A	0426
2VR75B	0426
2VR80	0444
2VR80A	0444
2VR80B	0444
2VR82	0444
2VR82A	0444
2VR82B	0444
2VR90	0238
2VR90A	0238
2VR90B	0238
2VR91	0238
2VR91A	0238
2VR91B	0238
2VR100	1180
2VR100A	1180
2VR100B	1180
2VR105	1180
2VR105A	1180
2VR105B	1180
2VR110	1190
2VR110A	1190
2VR110B	1190
2VR120	1207
2VR120A	1207
2VR120B	1207
2VR130	1218
2VR130A	1218
2VR130B	1218
2VR150	0647
2VR150A	0647
2VR150B	0647
2VR160	1255
2VR160A	1255
2VR160B	1255
2VR180	1279
2VR180A	1279
2VR180B	1279
2VR200	0605
2VR200A	0605
2VR200B	0605
2W005	0106
2W005G	OEM
2W005M	1858
2W01	0106
2W01G	OEM
2W01M	2785

DEVICE TYPE	REPL CODE
2W02	0106
2W02-5008-L	OEM
2W02-5008L	OEM
2W02G	OEM
2W02M	2785
2W04	0106
2W04G	OEM
2W04M	2785
2W06	0677
2W06G	OEM
2W06M	2189
2W08	0118
2W08G	OEM
2W08M	3736
2W010	0787
2W3A	0015
2W4A	0015
2W5A	0015
2W6A	0071
2W7A	0071
2W9A	0071
2W10	0787
2W10G	5725
2W10M	5725
2W12A	OEM
2W15A	OEM
2W20A	OEM
2WCDR152-1203B	OEM
2WDCR7K603B35	OEM
2WDCR7K703B35	OEM
2WDCR7K803B35	OEM
2WDCR7K903B35	OEM
2WDCR7K1003B35	OEM
2WDCR7K1103B4	OEM
2WDCR7K1103B35	OEM
2WDCR7K1203B35	OEM
2WDCR7K1203BR	OEM
2WDCR7K1303B4	OEM
2WDCR7K1303B35	OEM
2WDCR7K1403B4	OEM
2WDCR7K1503B4	OEM
2WDCR152-603B	OEM
2WDCR152-703B	OEM
2WDCR152-803B	OEM
2WDCR152-903B	OEM
2WDCR152-1003B	OEM
2WDCR152-1103B	OEM
2WDCR152-1303B	OEM
2WMT1	1250
2WMT2	1577
2WMT4	1581
2WMT6	1586
2WMT8	1132
2WMT10	0044
2WMT12	OEM
2X2N3055	OEM
2X9A116	0015
2XAA111	0143
2XAA112	0143
2XAA113	0143
2XAA119	0143
2XBD181	OEM
2XBD182	OEM
2XBD183	OEM
2XBD184	OEM
2XBDY20	OEM
2XR440M	OEM
2XT95423C	OEM
2YQ	OEM
2Z	OEM
2Z16A	0251
2Z18A	0256
2Z27A	0294
2.2B	OEM
2.2Z22D10	OEM
2.3L15	OEM
2.3L30	OEM
2.3L60	OEM
2.5Z6.8D	0205
2.5Z6.8D5	0205
2.5Z6.8D10	0205
2.5Z7.5D	0480
2.5Z7.5D5	0480
2.5Z7.5D10	0480
2.5Z8.2D	0504
2.5Z8.2D5	0504
2.5Z8.2D10	0504
2.5Z9.1D	0684
2.5Z9.1D5	0684
2.5Z9.1D10	0684
2.5Z10	0225
2.5Z10D5	0225
2.5Z10D10	0225
2.5Z11D	0230
2.5Z11D5	0230
2.5Z11D10	0230
2.5Z12D	0234
2.5Z12D5	0234
2.5Z12D10	0234
2.5Z13D	0237
2.5Z13D5	0237
2.5Z13D10	0237
2.5Z14D	1397
2.5Z14D5	1397
2.5Z14D10	1397
2.5Z15D	0247
2.5Z15D5	0247
2.5Z15D10	0247
2.5Z16D	0251

DEVICE TYPE	REPL CODE
2.5Z16D5	0251
2.5Z16D10	0251
2.5Z17D	1178
2.5Z17D5	1178
2.5Z17D10	1178
2.5Z18D	0256
2.5Z18D5	0256
2.5Z18D10	0256
2.5Z19D	2395
2.5Z19D5	2395
2.5Z19D10	2395
2.5Z20D	0262
2.5Z20D5	0262
2.5Z20D10	0262
2.5Z22D	0269
2.5Z22D5	0269
2.5Z24D	0269
2.5Z24D5	0274
2.5Z24D10	0274
2.5Z25D	2399
2.5Z25D5	2399
2.5Z25D10	2399
2.5Z27D	0294
2.5Z27D5	0294
2.5Z27D10	0294
2.5Z30D	0309
2.5Z30D5	0309
2.5Z30D10	0309
2.5Z33D	0318
2.5Z33D5	0318
2.5Z33D10	0318
2.5Z36D	0320
2.5Z36D5	0320
2.5Z36D10	0320
2.5Z39D	0326
2.5Z39D5	0326
2.5Z39D10	0326
2.5Z43D	0337
2.5Z43D5	0337
2.5Z43D10	0337
2.5Z45	0347
2.5Z45D5	0347
2.5Z45D10	0347
2.5Z47D	0347
2.5Z47D5	0347
2.5Z47D10	0347
2.5Z50D	0027
2.5Z50D5	0027
2.5Z50D10	0027
2.5Z51D	0027
2.5Z51D5	0027
2.5Z51D10	0027
2.5Z52D	0027
2.5Z52D5	0027
2.5Z52D10	0027
2.5Z56D	0266
2.5Z56D5	0266
2.5Z56D10	0266
2.5Z62D	0386
2.5Z62D5	0386
2.5Z62D10	0386
2.5Z68D	0405
2.5Z68D5	0405
2.5Z68D10	0405
2.5Z75D	0426
2.5Z75D5	0426
2.5Z75D10	0426
2.5Z82D	0444
2.5Z82D5	0444
2.5Z82D10	0444
2.5Z91D	0238
2.5Z91D5	0238
2.5Z91D10	0238
2.5Z100D	1180
2.5Z100D5	1180
2.5Z100D10	1180
2.7B	0760
2.7B2	0760
2.7B3	0760
2.7L1	0760
2.4341.0018	0160
2.V008M01	0280
2.V0038H03	0211
2x0C318	OEM
2x0C308	OEM
3-1477	0540
3-DL2	0023
3A03	OEM
3A05	OEM
3A1	0087
3A2	OEM
3A4	OEM
3A5R5	OEM
3A5R12-12	OEM
3A5R15-15	OEM
3A6	OEM
3A8	OEM
3A12R5	OEM
3A12R12-12	OEM
3A12R15-15	OEM
3A15	0110
3A24R5	OEM
3A24R12-12	OEM
3A24R15-15	OEM
3A30	0110
3A50	0087
3A50FS	0087
3A69-34000-001	OEM
3A100	0087

DEVICE TYPE	REPL CODE
3A100FS	0087
3A120	OEM
3A133	OEM
3A152	0015
3A154	0015
3A156	0015
3A158	0071
3A-200	0071
3A200	0015
3A200FS	0087
3A252	0015
3A254	0015
3A256	0015
3A258	0071
3A300	1747
3A300FS	OEM
3A400	0087
3A400FS	0087
3A500	1771
3A500FS	0087
3A600	0087
3A600FS	0087
3A700	0087
3A700FS	0087
3A800	0087
3A800FS	0087
3A900	0087
3A900FS	0087
3A1000	0087
3A1000FS	0087
3A1100	OEM
3A1100FS	OEM
3A1200	OEM
3A1200FS	OEM
3A1300	OEM
3A1300FS	OEM
3A1510	0071
3A2510	0071
3A3030	OEM
3A3030A	OEM
3A3060	OEM
3A3060A	OEM
3A3100	OEM
3A3100A	OEM
3A3150	OEM
3A3150A	OEM
3A3200	OEM
3A3200A	OEM
3AF2	0087
3AF4	1747
3AF6	0965
3AF8	0817
3ALB	OEM
3AS1	0015
3AS2	0015
3B4B41	0323
3B4.1	0323
3B15	0211
3B15-1	0211
3B30S	OEM
3B60S	OEM
3B100S	OEM
3B150S	OEM
3B200S	OEM
3B1034	OEM
3B3015	OEM
3B3030	OEM
3B3060	OEM
3B3100	OEM
3B3150	OEM
3B3200	OEM
3BB-20B01	0277
3BB-20801	0015
3BH61	0087
3BS1	0015
3BS2	0015
3BZ5.6	0876
3BZ5.6A	0876
3BZ5.6B	0876
3BZ6.2	0185
3BZ6.2A	0185
3BZ6.2B	0185
3BZ6.8	0205
3BZ6.8A	0205
3BZ6.8B	0205
3BZ7.5	0480
3BZ7.5A	0480
3BZ7.5B	0480
3BZ8.2	0504
3BZ8.2A	0504
3BZ8.2B	0504
3BZ9.1	0684
3BZ9.1A	0684
3BZ9.1B	0684
3BZ10	0225
3BZ10A	0225
3BZ10B	0225
3BZ11	0230
3BZ11A	0230
3BZ11B	0230
3BZ12	0234
3BZ12A	0234
3BZ12B	0234
3BZ13	0237
3BZ13A	0237
3BZ13B	0237
3BZ15	0247
3BZ15A	0247
3BZ15B	0247

DEVICE TYPE	REPL CODE
3BZ16	0251
3BZ16A	0251
3BZ16B	0251
3BZ18	0256
3BZ18A	0256
3BZ18B	0256
3BZ20	0262
3BZ20A	0262
3BZ20B	0262
3BZ22	0269
3BZ22A	0269
3BZ22B	0269
3BZ24	0274
3BZ24A	0274
3BZ24B	0274
3BZ27	0294
3BZ27A	0294
3BZ27B	0294
3BZ30	0309
3BZ30A	0309
3BZ30B	0309
3BZ33	0318
3BZ33A	0318
3BZ33B	0318
3BZ36	0320
3BZ36A	0320
3BZ36B	0320
3BZ39	0326
3BZ39A	0326
3BZ39B	0326
3BZ43	0337
3BZ43A	0337
3BZ43B	0337
3BZ47	0347
3BZ47A	0347
3BZ47B	0347
3BZ51	0027
3BZ51A	0027
3BZ51B	0027
3BZ56	0266
3BZ56A	0266
3BZ56B	0266
3BZ-61	0087
3BZ61	0564
3BZ62	0386
3BZ62A	0386
3BZ62B	0386
3BZ68	0405
3BZ68A	0405
3BZ68B	0405
3BZ75	0426
3BZ75A	0426
3BZ75B	0426
3BZ82	0444
3BZ82A	0444
3BZ82B	0444
3BZ91	0238
3BZ91A	0238
3BZ91B	0238
3BZ100	1180
3BZ100A	1180
3BZ100B	1180
3BZ110	1190
3BZ110A	1190
3BZ110B	1190
3BZ120	1207
3BZ120A	1207
3BZ120B	1207
3BZ130	1218
3BZ130A	1218
3BZ130B	1218
3BZ150	0647
3BZ150A	0647
3BZ150B	0647
3BZ160	1255
3BZ160A	1255
3BZ160B	1255
3BZ180	1279
3BZ180A	1279
3BZ180B	1279
3BZ200	0605
3BZ200A	0605
3BZ200B	0605
3BZD5.6	OEM
3BZD5.6A	OEM
3BZD5.6B	OEM
3BZD6.2	OEM
3BZD6.2A	OEM
3BZD6.2B	OEM
3BZD6.8	OEM
3BZD6.8A	OEM
3BZD6.8B	OEM
3BZD7.5	OEM
3BZD7.5A	OEM
3BZD7.5B	OEM
3BZD8.2	OEM
3BZD8.2A	OEM
3BZD8.2B	OEM
3BZD9.1	OEM
3BZD9.1A	OEM
3BZD9.1B	OEM
3BZD10	OEM
3BZD10A	OEM
3BZD10B	OEM
3BZD11	OEM
3BZD11A	OEM
3BZD11B	OEM
3BZD12	OEM
3BZD12A	OEM

DEVICE TYPE	REPL CODE
3BZD12B	OEM
3BZD13	OEM
3BZD13A	OEM
3BZD13B	OEM
3BZD15	OEM
3BZD15A	OEM
3BZD15B	OEM
3BZD16	OEM
3BZD16A	OEM
3BZD16B	OEM
3BZD18	OEM
3BZD18A	OEM
3BZD18B	OEM
3BZD20	OEM
3BZD20A	OEM
3BZD20B	OEM
3BZD22	OEM
3BZD22A	OEM
3BZD22B	OEM
3BZD24	OEM
3BZD24A	OEM
3BZD24B	OEM
3BZD27	OEM
3BZD27A	OEM
3BZD27B	OEM
3BZD30	OEM
3BZD30A	OEM
3BZD30B	OEM
3BZD33	OEM
3BZD33A	OEM
3BZD33B	OEM
3BZD36	OEM
3BZD36A	OEM
3BZD36B	OEM
3BZD39	OEM
3BZD39A	OEM
3BZD39B	OEM
3BZD43	OEM
3BZD43A	OEM
3BZD43B	OEM
3BZD47	OEM
3BZD47A	OEM
3BZD47B	OEM
3BZD51	OEM
3BZD51A	OEM
3BZD51B	OEM
3BZD56	OEM
3BZD56A	OEM
3BZD56B	OEM
3BZD62	OEM
3BZD62A	OEM
3BZD62B	OEM
3BZD68	OEM
3BZD68A	OEM
3BZD68B	OEM
3BZD75	OEM
3BZD75A	OEM
3BZD75B	OEM
3BZD82	OEM
3BZD82A	OEM
3BZD82B	OEM
3BZD91	OEM
3BZD91A	OEM
3BZD91B	OEM
3BZD100	OEM
3BZD100A	OEM
3BZD100B	OEM
3BZD110	OEM
3BZD110A	OEM
3BZD110B	OEM
3BZD120	OEM
3BZD120A	OEM
3BZD120B	OEM
3BZD130	OEM
3BZD130A	OEM
3BZD130B	OEM
3BZD150	OEM
3BZD150A	OEM
3BZD150B	OEM
3BZD160	OEM
3BZD160A	OEM
3BZD160B	OEM
3BZD180	OEM
3BZD180A	OEM
3BZD180B	OEM
3BZD200	OEM
3BZD200A	OEM
3BZD200B	OEM
3C05	0708
3C05R	0933
3C1	2901
3C2	1124
3C4	0805
3C6	0319
3C6RH	0997
3C8	1132
3C10	0044
3C10R	0947
3C12	OEM
3C14	OEM
3C15	0580
3C15A	OEM
3C15R	0947
3C16	OEM
3C20	0580
3C20R	0947
3C30	2060
3C30A	OEM
3C30R	4523

DEVICE TYPE	REPL CODE
3C40	1000
3C40R	1012
3C50	2076
3C50R	5567
3C51	OEM
3C60	2081
3C60A	OEM
3C60R	1073
3C63B	OEM
3C63C	OEM
3C70	2088
3C70R	1138
3C80	2088
3C80R	1138
3C90	0612
3C90R	1188
3C91B	OEM
3C91C	OEM
3C92B	OEM
3C92C	OEM
3C100	0612
3C100A	OEM
3C100R	1188
3C128	OEM
3C169	OEM
3C200	OEM
3C200A	OEM
3C201	OEM
3C211	OEM
3C300	OEM
3C400	OEM
3C500	OEM
3C600	OEM
3C800	OEM
3C1000	OEM
3C1030	OEM
3C1030A	OEM
3C1060	OEM
3C1060A	OEM
3C1100	OEM
3C1100A	OEM
3C1150	OEM
3C1150A	OEM
3C1200	OEM
3C1200A	OEM
3C2030	OEM
3C2030A	OEM
3C2060	OEM
3C2060A	OEM
3C2100	OEM
3C2100A	OEM
3C2200	OEM
3C2200A	OEM
3C6030	1137
3C6060	OEM
3C6100	0447
3C6150	OEM
3C6200	OEM
3CC11	OEM
3CC12	0871
3CC13	0871
3CD12	0871
3CD13	0871
3CGOR	1073
3CGOR	1073
3CS1	0015
3CS2	0015
3CZ6.8	OEM
3CZ6.8A	OEM
3CZ6.8B	OEM
3CZ7.5	OEM
3CZ7.5A	OEM
3CZ7.5B	OEM
3CZ8.2	OEM
3CZ8.2A	OEM
3CZ8.2B	OEM
3CZ9.1	OEM
3CZ9.1A	OEM
3CZ9.1B	OEM
3CZ10	OEM
3CZ10A	OEM
3CZ10B	OEM
3CZ11	OEM
3CZ11A	OEM
3CZ11B	OEM
3CZ12	OEM
3CZ12A	OEM
3CZ12B	OEM
3CZ13	OEM
3CZ13A	OEM
3CZ13B	OEM
3CZ15	OEM
3CZ15A	OEM
3CZ15B	OEM
3CZ16	OEM
3CZ16A	OEM
3CZ16B	OEM
3CZ18	OEM
3CZ18A	OEM
3CZ18B	OEM
3CZ20	OEM
3CZ20A	OEM
3CZ20B	OEM
3CZ22	OEM
3CZ22A	OEM
3CZ22B	OEM
3CZ24	OEM
3CZ24A	OEM
3CZ24B	OEM

If replacement code is OEM, contact original manufacturer for replacement.

DEVICE TYPE	REPL CODE
3CZ27	OEM
3CZ27A	OEM
3CZ27B	OEM
3CZ30	OEM
3CZ30A	OEM
3CZ30B	OEM
3CZ33	OEM
3CZ33A	OEM
3CZ33B	OEM
3CZ36	OEM
3CZ36A	OEM
3CZ36B	OEM
3CZ39	OEM
3CZ39A	OEM
3CZ39B	OEM
3CZ43	OEM
3CZ43A	OEM
3CZ43B	OEM
3CZ47	OEM
3CZ47A	OEM
3CZ47B	OEM
3CZ51	OEM
3CZ51A	OEM
3CZ51B	OEM
3CZ56	OEM
3CZ56A	OEM
3CZ56B	OEM
3CZ62	OEM
3CZ62A	OEM
3CZ62B	OEM
3CZ66	OEM
3CZ66A	OEM
3CZ66B	OEM
3CZ75	OEM
3CZ75A	OEM
3CZ75B	OEM
3CZ82	OEM
3CZ82A	OEM
3CZ82B	OEM
3CZ91	OEM
3CZ91A	OEM
3CZ91B	OEM
3CZ100	OEM
3CZ100A	OEM
3CZ100B	OEM
3CZ110	OEM
3CZ110A	OEM
3CZ110B	OEM
3CZ120	OEM
3CZ120A	OEM
3CZ120B	OEM
3CZ130	OEM
3CZ130A	OEM
3CZ130B	OEM
3CZ150	OEM
3CZ150A	OEM
3CZ150B	OEM
3CZ160	OEM
3CZ160A	OEM
3CZ160B	OEM
3CZ180	OEM
3CZ180A	OEM
3CZ180B	OEM
3CZ200	OEM
3CZ200A	OEM
3CZ200B	OEM
3D1.E	0323
3D4B41	1414
3D5R5	OEM
3D5R12-12	OEM
3D5R15-15	OEM
3D24R5	OEM
3D24R12-12	OEM
3D24R15-15	OEM
3D-702	0402
3DC11	OEM
3DC12	0871
3DC13	0871
3DD10	0373
3DD12	0871
3DD15	OEM
3DD20	OEM
3DD25	OEM
3DD30	OEM
3DE15	OEM
3DF10	OEM
3DF20	OEM
3DF25	OEM
3DF30	OEM
3DH61	0087
3DL2	0023
3DL4C	0023
3DS1	0015
3DS2	0015
3DS3	0087
3DZ61	0087
3E05	0110
3E-1	0016
3E1	0953
3E-2	0016
3E2	0242
3E-3	0016
3E-4	0143
3E4	0540
3E6	0965
3E8	0817
3E10	0922
3E25	OEM
3E-27	0004
3E-28	0164
3E-29	0164
3E50A	OEM
3E-64	0015
3E-65	0015
3ES1	0015
3ES2	0015
3EZ3.9D	0806
3EZ3.9D1	OEM
3EZ3.9D2	OEM
3EZ3.9D3	OEM
3EZ3.9D4	OEM
3EZ3.9D5	0806
3EZ3.9D10	0806
3EZ4.3D	0821
3EZ4.3D1	OEM
3EZ4.3D2	OEM
3EZ4.3D3	OEM
3EZ4.3D4	OEM
3EZ4.3D5	0821
3EZ4.3D10	0821
3EZ4.7D	0833
3EZ4.7D1	OEM
3EZ4.7D2	OEM
3EZ4.7D3	OEM
3EZ4.7D4	OEM
3EZ4.7D5	0833
3EZ4.7D10	0833
3EZ5.1D	0442
3EZ5.1D1	OEM
3EZ5.1D2	OEM
3EZ5.1D3	OEM
3EZ5.1D4	OEM
3EZ5.1D5	0442
3EZ5.1D10	0442
3EZ5.6D	0876
3EZ5.6D1	OEM
3EZ5.6D2	OEM
3EZ5.6D3	OEM
3EZ5.6D4	OEM
3EZ5.6D5	0876
3EZ5.6D10	0876
3EZ6.2D	0185
3EZ6.2D1	OEM
3EZ6.2D2	OEM
3EZ6.2D3	OEM
3EZ6.2D4	OEM
3EZ6.2D5	0185
3EZ6.2D10	0185
3EZ6.8D	0205
3EZ6.8D1	OEM
3EZ6.8D2	OEM
3EZ6.8D3	OEM
3EZ6.8D4	OEM
3EZ6.8D5	0205
3EZ6.8D10	0205
3EZ7.5D	0480
3EZ7.5D1	OEM
3EZ7.5D2	OEM
3EZ7.5D3	OEM
3EZ7.5D4	OEM
3EZ7.5D5	0480
3EZ7.5D10	0480
3EZ8.2D	0504
3EZ8.2D1	OEM
3EZ8.2D2	OEM
3EZ8.2D3	OEM
3EZ8.2D4	OEM
3EZ8.2D5	0504
3EZ8.2D10	0504
3EZ9.1D	0684
3EZ9.1D1	OEM
3EZ9.1D2	OEM
3EZ9.1D3	OEM
3EZ9.1D4	OEM
3EZ9.1D5	0684
3EZ9.1D10	0684
3EZ10D	0225
3EZ10D1	OEM
3EZ10D2	OEM
3EZ10D3	OEM
3EZ10D4	OEM
3EZ10D5	0225
3EZ10D10	0225
3EZ11D	0230
3EZ11D1	OEM
3EZ11D2	OEM
3EZ11D3	OEM
3EZ11D4	OEM
3EZ11D5	0230
3EZ11D10	0230
3EZ12D	0234
3EZ12D1	OEM
3EZ12D2	OEM
3EZ12D3	OEM
3EZ12D4	OEM
3EZ12D5	0234
3EZ12D10	0234
3EZ13D	0237
3EZ13D1	OEM
3EZ13D2	OEM
3EZ13D3	OEM
3EZ13D4	OEM
3EZ13D5	0237
3EZ13D10	0237
3EZ14D	1397
3EZ14D1	OEM
3EZ14D2	OEM
3EZ14D3	OEM
3EZ14D4	OEM
3EZ14D5	1397
3EZ14D10	1397
3EZ15D	0247
3EZ15D1	OEM
3EZ15D2	OEM
3EZ15D3	OEM
3EZ15D4	OEM
3EZ15D5	0247
3EZ15D10	0247
3EZ16D	0251
3EZ16D1	OEM
3EZ16D2	OEM
3EZ16D3	OEM
3EZ16D4	OEM
3EZ16D5	0251
3EZ16D10	0251
3EZ17	1178
3EZ17D	1178
3EZ17D1	OEM
3EZ17D2	OEM
3EZ17D3	OEM
3EZ17D4	OEM
3EZ17D5	1178
3EZ17D10	1178
3EZ18D	0256
3EZ18D1	OEM
3EZ18D2	OEM
3EZ18D3	OEM
3EZ18D4	OEM
3EZ18D5	0256
3EZ18D10	0256
3EZ19	2395
3EZ19D	OEM
3EZ19D1	OEM
3EZ19D2	OEM
3EZ19D3	OEM
3EZ19D4	OEM
3EZ19D5	2395
3EZ19D10	2395
3EZ20D	0262
3EZ20D1	OEM
3EZ20D2	OEM
3EZ20D3	OEM
3EZ20D4	OEM
3EZ20D5	0262
3EZ20D10	0262
3EZ22D	0269
3EZ22D1	OEM
3EZ22D2	OEM
3EZ22D3	OEM
3EZ22D4	OEM
3EZ22D5	0269
3EZ22D10	0269
3EZ24D	0274
3EZ24D1	OEM
3EZ24D2	OEM
3EZ24D3	OEM
3EZ24D4	OEM
3EZ24D5	0274
3EZ24D10	0274
3EZ27D	0294
3EZ27D1	OEM
3EZ27D2	OEM
3EZ27D3	OEM
3EZ27D4	OEM
3EZ27D5	0294
3EZ27D10	0294
3EZ28D	1177
3EZ28D1	OEM
3EZ28D2	OEM
3EZ28D3	OEM
3EZ28D4	OEM
3EZ28D5	1177
3EZ28D10	1177
3EZ30D	0309
3EZ30D1	OEM
3EZ30D2	OEM
3EZ30D3	OEM
3EZ30D4	OEM
3EZ30D5	0309
3EZ30D10	0309
3EZ33D	0318
3EZ33D1	OEM
3EZ33D2	OEM
3EZ33D3	OEM
3EZ33D4	OEM
3EZ33D5	0318
3EZ33D10	0318
3EZ36D	0320
3EZ36D1	OEM
3EZ36D2	OEM
3EZ36D3	OEM
3EZ36D4	OEM
3EZ36D5	0320
3EZ36D10	0320
3EZ39D	0326
3EZ39D1	OEM
3EZ39D2	OEM
3EZ39D3	OEM
3EZ39D4	OEM
3EZ39D5	0326
3EZ39D10	0326
3EZ43D	0337
3EZ43D1	OEM
3EZ43D2	OEM
3EZ43D3	OEM
3EZ43D4	OEM
3EZ43D5	0337
3EZ43D10	0337
3EZ47D	0347
3EZ47D1	OEM
3EZ47D2	OEM
3EZ47D3	OEM
3EZ47D4	OEM
3EZ47D5	0347
3EZ47D10	0347
3EZ51D	0027
3EZ51D1	OEM
3EZ51D2	OEM
3EZ51D3	OEM
3EZ51D4	OEM
3EZ51D5	0027
3EZ51D10	0027
3EZ56D	0266
3EZ56D1	OEM
3EZ56D2	OEM
3EZ56D3	OEM
3EZ56D4	OEM
3EZ56D5	0266
3EZ56D10	0266
3EZ62D	0386
3EZ62D1	OEM
3EZ62D2	OEM
3EZ62D3	OEM
3EZ62D4	OEM
3EZ62D5	0386
3EZ62D10	0386
3EZ68D	0405
3EZ68D1	OEM
3EZ68D2	OEM
3EZ68D3	OEM
3EZ68D4	OEM
3EZ68D5	0405
3EZ68D10	0405
3EZ75D	0426
3EZ75D1	OEM
3EZ75D2	OEM
3EZ75D3	OEM
3EZ75D4	OEM
3EZ75D5	0426
3EZ75D10	0426
3EZ82D	0444
3EZ82D1	OEM
3EZ82D2	OEM
3EZ82D3	OEM
3EZ82D4	OEM
3EZ82D5	0444
3EZ82D10	0444
3EZ91D	0238
3EZ91D1	OEM
3EZ91D2	OEM
3EZ91D3	OEM
3EZ91D4	OEM
3EZ91D5	0238
3EZ91D10	0238
3EZ100D	1180
3EZ100D1	OEM
3EZ100D2	OEM
3EZ100D3	OEM
3EZ100D4	OEM
3EZ100D5	1180
3EZ100D10	1180
3EZ110D	1190
3EZ110D1	OEM
3EZ110D2	OEM
3EZ110D3	OEM
3EZ110D4	OEM
3EZ110D5	1190
3EZ110D10	1190
3EZ120D	1207
3EZ120D1	OEM
3EZ120D2	OEM
3EZ120D3	OEM
3EZ120D4	OEM
3EZ120D5	1207
3EZ120D10	1207
3EZ130D	1218
3EZ130D1	OEM
3EZ130D2	OEM
3EZ130D3	OEM
3EZ130D4	OEM
3EZ130D5	1218
3EZ130D10	1218
3EZ140D	1881
3EZ140D1	OEM
3EZ140D2	OEM
3EZ140D3	OEM
3EZ140D4	OEM
3EZ140D5	1881
3EZ140D10	1881
3EZ150D	0647
3EZ150D1	OEM
3EZ150D2	OEM
3EZ150D3	OEM
3EZ150D4	OEM
3EZ150D5	0647
3EZ150D10	0647
3EZ160D	1255
3EZ160D1	OEM
3EZ160D2	OEM
3EZ160D3	OEM
3EZ160D4	OEM
3EZ160D5	1255
3EZ160D10	1255
3EZ170D	2102
3EZ170D1	OEM
3EZ170D2	OEM
3EZ170D3	OEM
3EZ170D4	OEM
3EZ170D5	2102
3EZ170D10	2102
3EZ180D	1279
3EZ180D1	OEM
3EZ180D2	OEM
3EZ180D3	OEM
3EZ180D4	OEM
3EZ180D5	1279
3EZ180D10	1279
3EZ190D	2225
3EZ190D1	OEM
3EZ190D2	OEM
3EZ190D3	OEM
3EZ190D4	OEM
3EZ190D5	2225
3EZ190D10	2225
3EZ200D	0605
3EZ200D1	OEM
3EZ200D2	OEM
3EZ200D3	OEM
3EZ200D4	OEM
3EZ200D5	0605
3EZ200D10	0605
3E.5	OEM
3F5	0708
3F5-D	0708
3F5D	0097
3F5R	2291
3F10	0580
3F10D	0580
3F10D-C	0580
3F10R	2291
3F15	0580
3F20	0580
3F20D	0580
3F20D-C	0580
3F20R	2291
3F30	2060
3F30(SCR)	1137
3F30D	2060
3F30R	0476
3F40	0580
3F40D	1000
3F40D-C	1000
3F40R	0476
3F50	2076
3F50D	2076
3F50R	0476
3F60	2081
3F60(SCR)	0344
3F60D	2081
3F60R	0476
3F80	2088
3F80D	2088
3F80R	0449
3F100	0612
3F100(SCR)	0901
3F100D	0612
3F100R	0449
3F120	OEM
3F150	0058
3F150(SCR)	2342
3F200	0058
3F200(SCR)	0058
3FB050R	0729
3FB100R	0729
3FB200R	0729
3FB400R	0737
3FB600R	0737
3FB800R	0742
3FB1000R	0742
3FC11	OEM
3FC12	0853
3FC13	1000
3FD	OEM
3FD12	0853
3FD13	1012
3FF05	OEM
3FF10	OEM
3FF15	OEM
3FF30	OEM
3FF40	OEM
3FF50	OEM
3FOR3B42(C1)-T	OEM
3FR5	0933
3FR10	0947
3FR15	0947
3FR20	0947
3FR30	4523
3FR40	1012
3FR50	5567
3FR60	1073
3FR80	1138
3FR100	1188
3FS1	0015
3FS2	0015
3G1	0087
3G2	2950
3G2(RECT)	0087
3G4	0087
3G4B41	0473
3G6	0087
3G8	0087
3G10	0087
3G15	OEM
3G30	OEM
3G42IG5	OEM
3G60	OEM
3G100	OEM
3G150	OEM
3G152	0015
3G154	0015
3G156	0015
3G158	0071
3G200	OEM
3G252	0015
3G254	0015
3G256	0015
3G258	0071
3G421G5	0407
3G1510	0071
3G2510	0071
3GA	0015
3GC11	OEM
3GC12	1000
3GCRCX	OEM
3GCRCY	OEM
3GCRCZ	OEM
3GCRDX	OEM
3GCRDY	OEM
3GCRDZ	OEM
3GCREX	OEM
3GCREY	OEM
3GCREZ	OEM
3GCRFX	OEM
3GCRFY	OEM
3GCRFZ	OEM
3GCRGX	OEM
3GCRGY	OEM
3GCRGZ	OEM
3GCRHX	OEM
3GCRHY	OEM
3GCRHZ	OEM
3GCRIX	OEM
3GCRIY	OEM
3GCRIZ	OEM
3GCRJX	OEM
3GCRJY	OEM
3GCRJZ	OEM
3GD12	0853
3GH61	0087
3GS1	0015
3GS2	0015
3GZ61	0540
3H77-00001-070	OEM
3H81-00050-001	OEM
3H82-00020-000	0334
3H82-00030-003	1165
3H83-00050-000	2178
3H83-00050-002	OEM
3H83-00160-000	2289
3H84-00040-004	0643
3H750M	OEM
3H7500H	OEM
3HC11	OEM
3HS1	0015
3HS2	0015
3J	OEM
3J4B41	0446
3J15	1137
3J30	1137
3J60	0344
3J100	0901
3J301	OEM
3JC11	OEM
3JC12	2081
3JH41	OEM
3JH-61	0023
3JH61	0023
3JH61(FA-2)	0023
3JH61(FA2)	0023
3JH61-FA-2	0023
3JH61FA-2	0023
3JH441	OEM
3JM61	OEM
3JU21.95B	OEM
3JUS1.62A	OEM
3JUS1.62B	OEM
3JUS1.85A	OEM
3JUS1.85B	OEM
3JUS1.95A	OEM
3JUS2.19A	OEM
3JUS2.19B	OEM
3JUS2.29A	OEM
3JUS2.29B	OEM
3JZ61	0965
3K	OEM
3K15	1137
3K101	OEM
3K121	OEM
3K301	OEM
3K3000LO	OEM
3K5797	0037
3K210000LO	OEM
3KC11	OEM
3KCB10	0729
3KCB20	0729
3KCB40	0737
3KCB60	0737
3KD8	OEM
3KM3LB	OEM
3KM300LA	OEM
3KM3000LA	OEM
3KM50000PA2	OEM
3KS110	OEM
3KZ6.8	0205
3KZ6.8A	OEM
3KZ6.8B	OEM
3KZ7.5	0480
3KZ7.5A	OEM
3KZ7.5B	OEM
3KZ8.2	0504
3KZ8.2A	OEM
3KZ8.2B	OEM
3KZ9.1	0684
3KZ9.1A	OEM
3KZ9.1B	OEM
3KZ10	0225
3KZ10A	OEM
3KZ10B	OEM
3KZ11	0230
3KZ11A	OEM
3KZ11B	OEM
3KZ12	0234
3KZ12A	OEM
3KZ12B	OEM
3KZ13	0237
3KZ13A	OEM
3KZ13B	OEM
3KZ14	OEM
3KZ14A	OEM
3KZ14B	OEM
3KZ15	0247
3KZ15A	OEM
3KZ15B	OEM
3KZ16	OEM
3KZ16A	OEM
3KZ16B	OEM
3KZ17	OEM
3KZ17A	OEM
3KZ17B	OEM
3KZ18	OEM
3KZ18A	OEM
3KZ18B	OEM
3KZ19	OEM
3KZ19A	OEM
3KZ19B	OEM
3KZ20	0262
3KZ20A	OEM
3KZ20B	OEM
3KZ22	0269
3KZ22A	OEM
3KZ22B	OEM
3KZ24	0274
3KZ24A	OEM
3KZ24B	OEM
3KZ27	0294
3KZ27A	OEM
3KZ27B	OEM
3KZ28	OEM
3KZ28A	OEM
3KZ28B	OEM
3KZ30	0309
3KZ30A	OEM
3KZ30B	OEM
3KZ33	0318
3KZ33A	OEM
3KZ33B	OEM
3KZ36	0320
3KZ36A	OEM
3KZ36B	OEM
3KZ39	0326
3KZ39A	OEM
3KZ39B	OEM
3KZ43	0318
3KZ43A	OEM
3KZ43B	OEM
3KZ47	0347
3KZ47A	OEM
3KZ47B	OEM
3KZ51	0027
3KZ51A	OEM
3KZ51B	OEM
3KZ56	0266
3KZ56A	OEM
3KZ56B	OEM
3KZ62	0386
3KZ62A	OEM
3KZ62B	OEM
3KZ66	0405
3KZ68	OEM
3KZ68A	OEM
3KZ68B	OEM
3KZ75	0426
3KZ75A	OEM
3KZ75B	OEM
3KZ82	0444
3KZ82A	OEM
3KZ82B	OEM
3KZ91	0238
3KZ91A	OEM
3KZ91B	OEM
3KZ100	1180
3KZ100A	OEM
3KZ100B	OEM
3KZ110	OEM
3KZ110A	OEM
3KZ110B	OEM
3KZ120	1207
3KZ120A	OEM
3KZ120B	OEM
3KZ130	1218
3KZ130A	OEM
3KZ130B	OEM
3KZ140	OEM

If replacement code is OEM, contact original manufacturer for replacement.

DEVICE TYPE	REPL CODE
3KZ140A	OEM
3KZ140B	OEM
3KZ150	0647
3KZ150A	OEM
3KZ150B	OEM
3KZ160	1255
3KZ160A	OEM
3KZ160B	OEM
3KZ170	OEM
3KZ170A	OEM
3KZ170B	OEM
3KZ180	1279
3KZ180A	OEM
3KZ180B	OEM
3KZ190	OEM
3KZ190A	OEM
3KZ190B	OEM
3KZ200	0605
3KZ200A	OEM
3KZ200B	OEM
3L1.4-1.6	OEM
3L1.8-2.1	OEM
3L4-2001	0143
3L4-2001-1	0019
3L4-2001-1A	0143
3L4-2001-3	0133
3L4-2001-4	0133
3L4-2003-1	0019
3L4-2003-3	0143
3L4-2003-4	0143
3L4-2003-7	0019
3L4-3001-1	0133
3L4-3001-5	0023
3L4-3001-7	0133
3L4-3001-8	0023
3L4-3002-7	0133
3L4-3002-10	0133
3L4-3002-13	0795
3L4-3002-20	0133
3L4-3002-25	0133
3L4-3002-30	0015
3L4-3002-31	0133
3L4-3002-32	0133
3L4-3002-33	0133
3L4-3501-1	OEM
3L4-3503-1	1029
3L4-3503-5	0030
3L4-3503-6	0030
3L4-3504-1	3687
3L4-3504-1(G)	OEM
3L4-3504-1G	OEM
3L4-3504-1GREEN	OEM
3L4-3504-2	3687
3L4-3504-2(B)	3687
3L4-3504-2B	3687
3L4-3504-2BLUE	3687
3L4-3504-3	3687
3L4-3504-3(W)	3687
3L4-3504-3W	3687
3L4-3504-3WHITE	3687
3L4-3504-4	3687
3L4-3504-4(R)	3687
3L4-3504-4R	3687
3L4-3504-9	OEM
3L4-3504-11	OEM
3L4-3504-12	OEM
3L4-3504-RED	OEM
3L4-3505-1	0244
3L4-3505-2	0025
3L4-3505-3	0649
3L4-3505-4	0317
3L4-3506-1	OEM
3L4-3506-2	0025
3L4-3506-3	0649
3L4-3506-4	OEM
3L4-3506-7	0244
3L4-3506-12	0025
3L4-3506-21	0157
3L4-3506-29	0002
3L4-3506-31	0162
3L4-3506-40	0420
3L4-3506-43	0025
3L4-3506-45	0495
3L4-3506-49	0441
3L4-3508-1	3687
3L4-3508-2	3687
3L4-5007-3	0144
3L4-6001-01	0103
3L4-6004	0178
3L4-6005-1	0178
3L4-6005-3	0161
3L4-6005-5	0178
3L4-6006-1	0142
3L4-6007-1	0178
3L4-6007-2	0016
3L4-6007-3	0144
3L4-6007-4	0595
3L4-6007-9	OEM
3L4-6007-10	0007
3L4-6007-11	0007
3L4-6007-12	0007
3L4-6007-13	0144
3L4-6007-14	0007
3L4-6007-15	0111
3L4-6007-19	0595
3L4-6007-20	0007
3L4-6007-21	0007
3L4-6007-22	0007
3L4-6007-23	0007
3L4-6007-34	0037
3L4-6007-35	0007
3L4-6007-41	0224
3L4-6007-51	0144
3L4-6010-1	0423
3L4-6010-3	0079
3L4-6010-4	0791
3L4-6010-6	0086
3L4-6010-8	0126
3L4-6011-1	0423
3L4-6011-2	1367
3L4-6011-3	1367
3L4-6011-9	0925
3L4-6011-11	0925
3L4-6011-12	0925
3L4-6011-14	0925
3L4-6011-52	1367
3L4-6011-53	1367
3L4-6012-2	0423
3L4-6012-3	0423
3L4-6012-4	0161
3L4-6012-5	0423
3L4-6012-6	0423
3L4-6012-8	0423
3L4-6012-55	0560
3L4-6012-56	0423
3L4-6012-58	0423
3L4-6013-2	0854
3L4-6013-4	0460
3L4-6013-5	0854
3L4-6013-6	0925
3L4-6013-8	0854
3L4-6013-15	0925
3L4-6013-55	0854
3L4-6013-56	0854
3L4-6013-58	0854
3L4-6021-01	2439
3L4-6024-1	3536
3L4-6503-1	0212
3L4-6503-2	0212
3L4-6504-2	OEM
3L4-7004-1	0316
3L4-9002-1	2714
3L4-9004-1	1395
3L4-9004-3	1395
3L4-9004-4	1395
3L4-9004-6	1395
3L4-9004-51	1395
3L4-9006-1	2617
3L4-9006-51	2617
3L4-9007-0	1345
3L4-9007-1	1443
3L4-9007-51	1443
3L4-9008-01	1843
3L4-9008-51	1843
3L4-9009-1	OEM
3L4-9013-01	0141
3L4-9014-1	OEM
3L4-9015-1	3079
3L4-9020-1	1048
3L4-9021-1	OEM
3L4-9022-1	OEM
3L4-9024-1	OEM
3L4-9025-1	OEM
3L5-0011-4	OEM
3L5-0016-01	OEM
3L5-0016-02	OEM
3L5-0016-1	OEM
3L5-0016-2	OEM
3L5-1002-02	OEM
3L5-1002-04	OEM
3L5-1002-05	OEM
3L5-1002-2	OEM
3L5-1002-4	OEM
3L5-1002-5	OEM
3L1015	OEM
3L1030	1137
3L1060	OEM
3L2015	1137
3L2100	OEM
3L46007-1	0079
3L46007-2	0079
3LA-6007-4	0595
3LC11	OEM
3LC12	2088
3LF11	0816
3LH61	0087
3LK-2001-1	OEM
3LR	OEM
3LZ61	0817
3M5U12-5	OEM
3M90	OEM
3M95	OEM
3MA	0774
3MC	0050
3MC11	OEM
3MCCV102	OEM
3MS5	0015
3MS10	0015
3MS20	0015
3MS30	0015
3MS40	0015
3MS50	0015
3N	OEM
3N0128	OEM
3N0138	OEM
3N0139	OEM
3N0140	OEM
3N0141	OEM
3N0142	OEM
3N0143	OEM
3N0152	OEM
3N0153	OEM
3N0154	OEM
3N0159	OEM
3N0187	OEM
3N0200	OEM
3N0204	OEM
3N0205	OEM
3N0206	OEM
3N0211	OEM
3N0212	OEM
3N0213	OEM
3N21	0136
3N22	0600
3N23	0600
3N23A	0600
3N23B	0600
3N23C	0600
3N25	OEM
3N25/501	0050
3N26	OEM
3N27	OEM
3N29	0600
3N30	0600
3N31	0600
3N32	OEM
3N33	OEM
3N34	0050
3N35	0016
3N35A	0050
3N36	0600
3N37	0600
3N39	OEM
3N40	OEM
3N41	OEM
3N42	OEM
3N43	OEM
3N44	OEM
3N44A	OEM
3N49	0440
3N50	0440
3N51	0440
3N52	0440
3N56	OEM
3N57	OEM
3N58	OEM
3N59	OEM
3N60	OEM
3N62	OEM
3N63	OEM
3N64	OEM
3N65	OEM
3N66	OEM
3N67	OEM
3N68	OEM
3N68A	OEM
3N69	OEM
3N70	OEM
3N71	0127
3N72	0127
3N73	0127
3N74	0111
3N75	0111
3N76	0111
3N77	0111
3N78	0111
3N79	0111
3N80	OEM
3N81	OEM
3N82	OEM
3N83	2336
3N84	OEM
3N85	OEM
3N86	OEM
3N87	0127
3N88	0127
3N90	0037
3N91	0037
3N92	0037
3N93	0037
3N94	0037
3N95	0037
3N96	OEM
3N97	OEM
3N98	OEM
3N99	OEM
3N100	OEM
3N101	OEM
3N102	OEM
3N103	OEM
3N104	OEM
3N105	OEM
3N106	OEM
3N107	OEM
3N112	0150
3N113	0150
3N114	0037
3N115	0037
3N116	0037
3N117	0037
3N118	0037
3N119	0037
3N120	0127
3N121	0127
3N123	0693
3N124	OEM
3N125	OEM
3N126	OEM
3N127	0007
3N128	0849
3N129	0037
3N130	0037
3N131	0037
3N132	0037
3N133	0037
3N134	0037
3N135	0037
3N136	0037
3N138	OEM
3N139	OEM
3N140	0353
3N141	0212
3N142	0849
3N143	0849
3N152	0849
3N153	0849
3N154	0849
3N155	OEM
3N155A	OEM
3N156	OEM
3N156A	OEM
3N157	OEM
3N157A	OEM
3N158	OEM
3N158A	OEM
3N159	0212
3N161	2458
3N163	OEM
3N164	OEM
3N165	OEM
3N166	OEM
3N169	OEM
3N170	OEM
3N171	OEM
3N172	OEM
3N173	OEM
3N175	0983
3N176	0983
3N182	OEM
3N187	0212
3N188	OEM
3N189	OEM
3N190	OEM
3N191	OEM
3N200	2458
3N201	0212
3N201A	0212
3N202	2458
3N203	2458
3N203A	OEM
3N204	2458
3N205	2458
3N206	2458
3N208	3839
3N209	2458
3N211	2458
3N212	4511
3N213	0212
3N219	OEM
3N220	OEM
3N221	OEM
3N222	OEM
3N225	OEM
3N225A	OEM
3N243	OEM
3N244	OEM
3N245	OEM
3N246	0277
3N247	0277
3N248	0289
3N249	0296
3N250	0302
3N251	0250
3N252	0250
3N253	0277
3N254	0289
3N255	0289
3N256	0296
3N257	0302
3N258	0250
3N259	0250
3N261	OEM
3N262	OEM
3N263	OEM
3N2603	OEM
3N3536	OEM
3N6285	3525
3NC11	OEM
3NC12	0612
3NF11	OEM
3NV10	0017
3NV10F	0017
3NV15	0017
3NV15F	0017
3NV20	OEM
3NV20F	OEM
3NV30	OEM
3NV30F	OEM
3NVF10	0017
3NVF12	0017
3NVF15	0017
3NZ61	0087
3P1M	OEM
3P1MH	0447
3P2M	OEM
3P2MH	0346
3P2MM	OEM
3P4M	OEM
3P4MH	0095
3P5M	OEM
3P6M	OEM
3PT40	OEM
3PT60	OEM
3PT80	OEM
3PT100	OEM
3QF11	OEM
3R0/18	OEM
3R1/18	OEM
3R2/18	OEM
3R2B	OEM
3R2S	OEM
3R3B	OEM
3R3S	OEM
3R4C	OEM
3R4S	OEM
3R5C	OEM
3R5S	OEM
3R6C	OEM
3R6S	OEM
3R7CD	OEM
3R7.5	0480
3R7.5A	0480
3R7.5B	0480
3R8CD	OEM
3R8S	OEM
3R8.2	0504
3R8.2A	0504
3R8.2B	0504
3R9.1	0684
3R9.1A	0684
3R9.1B	0684
3R10	0225
3R10A	0225
3R10B	0225
3R11	0230
3R11A	0230
3R11B	0230
3R12	0234
3R12A	0234
3R12B	0234
3R13	0237
3R13B	0237
3R15	0247
3R15A	0247
3R15B	0247
3R16	0251
3R16B	0251
3R18	0256
3R18A	0256
3R18B	0256
3R20	0262
3R20A	0262
3R20B	0262
3R22	0269
3R22A	0269
3R22B	0269
3R24	0274
3R24A	0274
3R24B	0274
3R27	0294
3R27A	0294
3R27B	0294
3R30	0309
3R30A	0309
3R30B	0309
3R33	0318
3R33A	0318
3R33B	0318
3R35	0320
3R35A	0320
3R35B	0320
3R36	0320
3R36A	0320
3R36B	0320
3R39	0326
3R39A	0326
3R39B	0326
3R43	0337
3R43A	0337
3R43B	0337
3R47	0347
3R47A	0347
3R47B	0347
3R51	0027
3R51A	0027
3R51B	0027
3R52	OEM
3R56	0266
3R56A	0266
3R56B	0266
3R62	0386
3R62A	0386
3R62B	0386
3R68	0405
3R68A	0405
3R68B	0405
3R75	0426
3R75A	0426
3R75B	0426
3R82	0444
3R82A	0444
3R82B	0444
3R91	0238
3R91A	0238
3R91B	0238
3R100	1180
3R100A	1180
3R100B	1180
3R110	1190
3R110A	1190
3R110B	1190
3R120	1207
3R120A	1207
3R120B	1207
3R130	1218
3R130A	1218
3R130B	1218
3R140	1881
3R140A	1881
3R140B	1881
3R150	0647
3R150A	0647
3R150B	0647
3R160	1255
3R160A	1255
3R180	1279
3R180A	1279
3R180B	1279
3R200	0605
3R200A	0605
3R200B	0605
3RB20/6	OEM
3RB20/10	OEM
3RB80/6	OEM
3RB80/10	OEM
3RC2	1110
3RC5	1110
3RC5A	1110
3RC10	2932
3RC10A	2932
3RC15	0963
3RC15A	0963
3RC20	0963
3RC20A	0963
3RC25	2933
3RC30	2933
3RC30A	2933
3RC40	2936
3RC40A	2936
3RC50	3671
3RC50A	3671
3RC60	3439
3RC60A	3439
3RCF	1110
3RL5A	1110
3RL10A	2932
3RL20A	0963
3RL30A	2933
3RL40A	2936
3RL50A	3671
3RL60A	3439
3RO/18	OEM
3RUS2120	OEM
3RUS2150	OEM
3RUS2180	OEM
3S001	OEM
3S002	OEM
3S003	OEM
3S004	0016
3S05E	0110
3S1E	0953
3S2E	0242
3S2.8-3.1	OEM
3S3E	1747
3S4E	0540
3S4M	0095
3S4M//LB	0095
3S4M//LB1E	0095
3S4M/LB1E	0095
3S5E	1771
3S5R15-15	OEM
3S6E	0965
3S11	OEM
3S12	OEM
3S12R15-15	OEM
3S14	OEM
3S16	OEM
3S24R15-15	OEM
3S28R15-15	OEM
3S30B	0325
3S105	0087
3S271	OEM
3S331	OEM
3SA324	0050
3SB05	OEM
3SB10	OEM
3SB20	OEM
3SB40	OEM
3SB60	OEM
3SB80	OEM
3SB100	6438
3SB629	0015
3SBMA1F	OEM
3SBMA2	0729
3SBMA2F	0729
3SBMA4	0729
3SBMA4F	0729
3SBMA6	0737
3SBMA8	0742
3SBMA05F	OEM
3SBMB1F	OEM
3SBMB2	0729
3SBMB2F	0729
3SBMB4	0729
3SBMB4F	0729
3SBMB6	0737
3SBMB8	0737
3SBMB05F	OEM
3SBMC05F	OEM
3SBMC1F	OEM
3SBMC2	0729
3SBMC2F	0729
3SBMC4	0729
3SBMC4F	0729
3SBMC6	0737
3SBMC8	0737
3SC1358K1	0363
3SC1415	1085
3SF1	OEM
3SF2	3382
3SF4	OEM
3SF11	2336
3SF11(Q)	2336
3SF11P	2336
3SF11Q	2336
3SFR0	OEM
3SJ11	0844
3SJ11A	0844
3SK072	OEM
3SK3E	0212
3SK14	0849
3SK15	OEM
3SK15A	OEM
3SK16	OEM
3SK17	OEM
3SK18	OEM
3SK19	OEM
3SK20	0353
3SK20(H)	OEM
3SK20H	0353
3SK20HW	0353
3SK20HY	0353
3SK21	0212
3SK21(H)	OEM
3SK21H	0212
3SK22	0212
3SK22-Y	0212
3SK22GR	0849
3SK22Y	0212
3SK23	0325
3SK28	0212
3SK29	0849
3SK30	0325
3SK30(B)	0212
3SK30A	0325
3SK-30B	0849
3SK30B	0325
3SK30C	0325
3SK32	0353
3SK32(B)	0212
3SK32A	0353
3SK32B	0353
3SK32B-6	0353
3SK32C	0353
3SK32D	0353
3SK32E	0353
3SK32E-4	0353
3SK33	0849
3SK34	0212
3SK34C	0325
3SK-35	0212
3SK35	0212
3SK35-BL	0212
3SK35-GR	0212
3SK35-Y	0212
3SK35BL	0212
3SK35G	0212
3SK37	0212
3SK38	0983
3SK38A	0983
3SK-39	0212
3SK39	0353
3SK39(E)	0212
3SK39(P)	0212
3SK39(Z)	OEM
3SK39E	0353
3SK39P	0353
3SK39Q	0353
3SK39R	0353
3SK40	0353
3SK40I	0212
3SK40M	0212
3SK-41	OEM
3SK41	1031
3SK41(L)	1031
3SK41C	1031
3SK41L	1031
3SK41M	1031
3SK44	0212
3SK45	0414
3SK45-B04	0414
3SK45-B09	0414
3SK45B	0414
3SK45B-09	0414
3SK45B09	0414
3SK47	2458
3SK48	0353
3SK49	0353
3SK49(Z)	OEM
3SK49E2	0212
3SK49NC	0212
3SK49Q	0212
3SK51	0353

DEVICE TYPE	REPL CODE
3SK53	0212
3SK55	OEM
3SK59	0212
3SK59(GR)	0212
3SK59BL	0212
3SK59GR	0212
3SK60	2458
3SK61	2458
3SK63	2458
3SK66	OEM
3SK70	2458
3SK71	OEM
3SK72	0371
3SK73	0212
3SK73(Y)	0212
3SK73-GR	0212
3SK73Y	1450
3SK74	0371
3SK74L	OEM
3SK76	OEM
3SK77	OEM
3SK78	2458
3SK79	OEM
3SK80	0371
3SK81	4237
3SK82	OEM
3SK83	0371
3SK85	0371
3SK87	0371
3SK87(K)	0371
3SK88	0371
3SK88(K)	0371
3SK90	OEM
3SK95	0371
3SK96	OEM
3SK97	OEM
3SK100	OEM
3SK101	0369
3SK102	0371
3SK103	OEM
3SK104	OEM
3SK104V	OEM
3SK107	OEM
3SK108	0384
3SK112	OEM
3SK113	0371
3SK114	0371
3SK115	0371
3SK116	OEM
3SK118	OEM
3SK119	OEM
3SK120	OEM
3SK121	OEM
3SK122	OEM
3SK122L	OEM
3SK122M	OEM
3SK123	OEM
3SK125	OEM
3SK126	OEM
3SK127	OEM
3SK128	OEM
3SK129	OEM
3SK131	OEM
3SK132	OEM
3SK133	OEM
3SK134	OEM
3SK135	OEM
3SK136	OEM
3SK137	OEM
3SK138	OEM
3SK139	OEM
3SK140	OEM
3SK140-0	OEM
3SK140-O	OEM
3SK141	OEM
3SK142	OEM
3SK143	OEM
3SK144	OEM
3SK145	OEM
3SK146	OEM
3SK150	OEM
3SK151	OEM
3SK152	OEM
3SK153	OEM
3SK156	OEM
3SK159	OEM
3SK160	OEM
3SK162	OEM
3SK164	OEM
3SK165	OEM
3SK166	OEM
3SK169	OEM
3SK179	OEM
3SK183	OEM
3SK184	OEM
3SK186	OEM
3SK411	0212
3SK458	OEM
3SM0	OEM
3SM2	3382
3SM4	0095
3SM6	0031
3SM8	OEM
3SM1020A4	OEM
3SM1020A4PL	OEM
3SM1020A4T	OEM
3SM1020A5	OEM
3SM1020A5PL	OEM
3SM1020A5T	OEM
3SM1020A6	OEM
3SM1020A6PL	OEM
3SM1020A6T	OEM
3SM1020A7	OEM
3SM1020A7PL	OEM
3SM1020A7T	OEM
3SM1020A8	OEM
3SM1020A8PL	OEM
3SM1020A8T	OEM
3SM1020E4	OEM
3SM1020E4PL	OEM
3SM1020E4T	OEM
3SM1020E5	OEM
3SM1020E5PL	OEM
3SM1020E5T	OEM
3SM1020E6	OEM
3SM1020E6PL	OEM
3SM1020E6T	OEM
3SM1020E7	OEM
3SM1020E7PL	OEM
3SM1020E7T	OEM
3SM1020GE8	OEM
3SM1020GE8PL	OEM
3SM1020GE8T	OEM
3SM1020GE9	OEM
3SM1020GE9PL	OEM
3SM1020GE9T	OEM
3SM1020GE10	OEM
3SM1020GE10PL	OEM
3SM1020GE10T	OEM
3SS4M	0095
3T6.2	2221
3T6.2A	2221
3T6.2B	2221
3T7.5	0221
3T7.5A	0221
3T7.5B	0221
3T9.1	1618
3T9.1A	1618
3T9.1B	1618
3T11	0691
3T11A	0691
3T11B	0691
3T12	0870
3T12A	0870
3T12B	0870
3T13	1020
3T13A	1020
3T13B	1020
3T16	1402
3T16A	1402
3T16B	1402
3T20	1717
3T20A	1717
3T20B	1717
3T24	1736
3T24A	1736
3T24B	1736
3T30	1772
3T30A	1772
3T30B	1772
3T33	1788
3T33A	1788
3T33B	1788
3T36	1796
3T36A	1796
3T36B	1796
3T39	1804
3T39A	1804
3T39B	1804
3T43	1194
3T43A	1194
3T43B	1194
3T47	0022
3T47A	0022
3T47B	0022
3T51	0132
3T51A	0132
3T51B	0132
3T56	0207
3T56A	0207
3T56B	0207
3T62	0263
3T62A	0263
3T62B	0263
3T68	0310
3T68A	0310
3T68B	0310
3T75	0329
3T75A	0329
3T75B	0329
3T82	0356
3T82A	0356
3T82B	0356
3T91	0412
3T91A	0412
3T91B	0412
3T100	0438
3T100A	0438
3T100B	0438
3T110	0488
3T110A	0488
3T110B	0488
3T120	0509
3T120A	0509
3T120B	0509
3T130	0524
3T130A	0524
3T130B	0524
3T150	0063
3T150A	0063
3T150B	0063
3T160	0401
3T160A	0401
3T160B	0401
3T180	0634
3T180A	0634
3T180B	0634
3T200	0668
3T200A	0668
3T200B	0668
3T201	0600
3T202	0600
3T203	0600
3T501	0015
3T502	0015
3T503	0015
3T504	0015
3T505	0015
3T506	0015
3T507	0071
3T508	0071
3T509	0071
3T510	0071
3TCRA	1110
3TCRB	1110
3TCRC	2932
3TCRD	0963
3TCRE	0963
3TCRF	2933
3TCRG	2933
3TCRH	2936
3TCRI	3671
3TCRJ	3439
3TCRK	OEM
3TCRL	OEM
3TE120	0103
3TE130	OEM
3TE140	0103
3TE150	0595
3TE160	0595
3TE220	0065
3TE225	OEM
3TE230	0103
3TE240	0103
3TE250	0595
3TE260	0595
3TE467	OEM
3TE477	OEM
3TH62	0087
3TRC10A	0641
3TX002	0065
3TX003	0103
3TX004	0103
3TX621	3560
3TX622	2091
3TX820	3560
3TX822	2091
3TZ3.6	0796
3TZ3.6A	0796
3TZ3.6B	0796
3TZ3.6C	0796
3TZ3.6D	0796
3TZ3.9	0806
3TZ3.9A	0806
3TZ3.9B	0806
3TZ3.9C	0806
3TZ3.9D	0806
3TZ4.3	0821
3TZ4.3A	0821
3TZ4.3B	0821
3TZ4.3C	0821
3TZ4.3D	0821
3TZ4.7	0833
3TZ4.7A	0833
3TZ4.7B	0833
3TZ4.7C	0833
3TZ4.7D	0833
3TZ5.1	0442
3TZ5.1A	0442
3TZ5.1B	0442
3TZ5.1C	0442
3TZ5.1D	0442
3TZ5.6	0876
3TZ5.6A	0876
3TZ5.6B	0876
3TZ5.6C	0876
3TZ5.6D	0876
3TZ6.2	0185
3TZ6.2A	0185
3TZ6.2B	0185
3TZ6.2C	0185
3TZ6.2D	0185
3TZ6.8	0205
3TZ6.8A	0205
3TZ6.8B	0205
3TZ6.8C	0205
3TZ6.8D	0205
3TZ7.5	0480
3TZ7.5A	0480
3TZ7.5B	0480
3TZ7.5C	0480
3TZ7.5D	0480
3TZ8.2	0504
3TZ8.2A	0504
3TZ8.2B	0504
3TZ8.2C	0504
3TZ8.2D	0504
3TZ9.1	0684
3TZ9.1A	0684
3TZ9.1B	0684
3TZ9.1C	0684
3TZ9.1D	0684
3TZ10	0225
3TZ10A	0225
3TZ10B	0225
3TZ10C	0225
3TZ10D	0225
3TZ11	0230
3TZ11A	0230
3TZ11B	0230
3TZ11C	0230
3TZ11D	0230
3TZ12	0234
3TZ12A	0234
3TZ12B	0234
3TZ12C	0234
3TZ12D	0234
3TZ13	0237
3TZ13A	0237
3TZ13B	0237
3TZ13C	0237
3TZ13D	0237
3TZ14	1397
3TZ14A	1397
3TZ14B	1397
3TZ14C	1397
3TZ14D	1397
3TZ15	0247
3TZ15A	0247
3TZ15B	0247
3TZ15C	0247
3TZ15D	0247
3TZ16	0251
3TZ16A	0251
3TZ16B	0251
3TZ16C	0251
3TZ16D	0251
3TZ17	1178
3TZ17A	1178
3TZ17B	1178
3TZ17C	1178
3TZ17D	1178
3TZ18	0256
3TZ18A	0256
3TZ18B	0256
3TZ18C	0256
3TZ18D	0256
3TZ19	2395
3TZ19A	2395
3TZ19B	2395
3TZ19C	2395
3TZ19D	2395
3TZ20	0262
3TZ20A	0262
3TZ20B	0262
3TZ20C	0262
3TZ20D	0262
3TZ22	0269
3TZ22A	0269
3TZ22B	0269
3TZ22C	0269
3TZ22D	0269
3TZ24	0274
3TZ24A	0274
3TZ24B	0274
3TZ24C	0274
3TZ24D	0274
3TZ27	0294
3TZ27A	0294
3TZ27B	0294
3TZ27C	0294
3TZ27D	0294
3TZ30	0309
3TZ30A	0309
3TZ30B	0309
3TZ30C	0309
3TZ30D	0309
3TZ33	0318
3TZ33A	0318
3TZ33B	0318
3TZ33C	0318
3TZ33D	0318
3TZ34C	0318
3TZ34D	OEM
3TZ36	0320
3TZ36A	0320
3TZ36B	0320
3TZ36C	0320
3TZ36D	0320
3TZ39	0326
3TZ39A	0326
3TZ39B	0326
3TZ39C	0326
3TZ39D	0326
3TZ43	0337
3TZ43A	0337
3TZ43B	0337
3TZ43C	0337
3TZ43D	0337
3TZ45A	0347
3TZ45B	0347
3TZ45C	0347
3TZ45D	OEM
3TZ47	0347
3TZ47A	0347
3TZ47B	0347
3TZ47C	0347
3TZ47D	0347
3TZ50	0027
3TZ50A	0027
3TZ50B	0027
3TZ50C	0027
3TZ50D	0027
3TZ51	0027
3TZ51A	0027
3TZ51B	0027
3TZ51C	0027
3TZ51D	0027
3TZ52	0027
3TZ52A	0027
3TZ52B	0027
3TZ52C	0027
3TZ52D	OEM
3TZ56	0266
3TZ56A	0266
3TZ56B	0266
3TZ56C	0266
3TZ56D	0266
3TZ62	0386
3TZ62A	0386
3TZ62B	0386
3TZ62C	0386
3TZ62D	0386
3TZ68	0405
3TZ68A	0405
3TZ68B	0405
3TZ68C	0405
3TZ68D	0405
3TZ75	0426
3TZ75A	0426
3TZ75B	0426
3TZ75C	0426
3TZ75D	0426
3TZ82	0444
3TZ82A	0444
3TZ82B	0444
3TZ82C	0444
3TZ82D	0444
3TZ91	0238
3TZ91A	0238
3TZ91B	0238
3TZ91C	0238
3TZ91D	0238
3TZ100	1180
3TZ100A	1180
3TZ100B	1180
3TZ100C	1180
3TZ100D	1180
3TZ105	1190
3TZ105A	1190
3TZ105B	1190
3TZ105C	1190
3TZ105D	OEM
3TZ110	1190
3TZ110A	1190
3TZ110B	1190
3TZ110C	1190
3TZ110D	1190
3TZ120	1207
3TZ120A	1207
3TZ120B	1207
3TZ120C	1207
3TZ120D	1207
3TZ130	1218
3TZ130A	1218
3TZ130B	1218
3TZ130C	1218
3TZ130D	1218
3TZ140	1881
3TZ140A	1881
3TZ140B	1881
3TZ140C	1881
3TZ140D	1881
3TZ150	0647
3TZ150A	0647
3TZ150B	0647
3TZ150C	0647
3TZ150D	0647
3TZ160	1255
3TZ160A	1255
3TZ160A0A	OEM
3TZ160B	1255
3TZ160C	1255
3TZ160D	1255
3TZ175	1279
3TZ175A	1279
3TZ175B	1279
3TZ175C	1279
3TZ175D	OEM
3TZ180	1279
3TZ180A	1279
3TZ180B	1279
3TZ180C	1279
3TZ180D	1279
3TZ200	0605
3TZ200A	0605
3TZ200B	0605
3TZ200C	0605
3TZ200D	0605
3TZ220	OEM
3TZ220A	OEM
3TZ220B	OEM
3TZ220C	OEM
3TZ220D	OEM
3TZ455	OEM
3UT40	OEM
3VR5.6	0876
3VR5.6A	0876
3VR5.6B	0876
3VR6	3127
3VR6A	3127
3VR6B	3127
3VR6.2	0185
3VR6.2A	0185
3VR6.2B	0185
3VR6.8	0205
3VR6.8A	0205
3VR6.8B	0205
3VR7.5	0480
3VR7.5A	0480
3VR7.5B	0480
3VR8.2	0504
3VR8.2A	0504
3VR8.2B	0504
3VR8.5	3317
3VR8.5A	3317
3VR8.5B	3317
3VR9.1	0684
3VR9.1A	0684
3VR9.1B	0684
3VR10	0225
3VR10A	0225
3VR10B	0225
3VR11	0230
3VR11A	0230
3VR11B	0230
3VR12	0234
3VR12A	0234
3VR12B	0234
3VR13	0237
3VR13A	0237
3VR13B	0237
3VR15	0247
3VR15A	0247
3VR15B	0247
3VR16	0251
3VR16A	0251
3VR16B	0251
3VR18	0256
3VR18A	0256
3VR18B	0256
3VR20	0262
3VR20A	0262
3VR20B	0262
3VR22	0269
3VR22A	0269
3VR22B	0269
3VR24	0274
3VR24A	0274
3VR24B	0274
3VR27	0294
3VR27A	0294
3VR27B	0294
3VR30	0309
3VR30A	0309
3VR30B	0309
3VR33	0318
3VR33A	0318
3VR33B	0318
3VR36	0320
3VR36A	0320
3VR36B	0320
3VR39	0326
3VR39A	0326
3VR39B	0326
3VR43	0337
3VR43A	0337
3VR43B	0337
3VR47	0347
3VR47A	0347
3VR47B	0347
3VR51	0027
3VR51A	0027
3VR51B	0027
3VR56	0266
3VR56A	0266
3VR56B	0266
3VR62	0386
3VR62A	0386
3VR62B	0386
3VR67	0405
3VR67A	0405
3VR67B	0405
3VR68	0405
3VR68A	0405
3VR68B	0405
3VR75	0426
3VR75A	0426
3VR75B	0426
3VR82	0444
3VR82A	0444
3VR82B	0444
3VR90	0238
3VR90A	0238
3VR90B	0238
3VR91	0238
3VR91A	0238
3VR91B	0238
3VR100	1180
3VR100A	1180
3VR100B	1180
3VR110	1190
3VR110A	1190
3VR110B	1190
3VR120	1207
3VR120A	1207
3VR120B	1207
3VR130	1218
3VR130A	1218
3VR130B	1218
3VR150	0647
3VR150A	0647
3VR150B	0647
3VR160	1255
3VR160A	1255
3VR160B	1255
3VR180	1279
3VR180A	1279
3VR180B	1279
3VR200	0605
3VR200A	0605
3VR200B	0605
3W5R5	OEM
3W5R12-12	OEM
3W5R15-15	OEM
3W12R5	OEM
3W12R15-15	OEM
3W24R5	OEM
3W24R12-12	OEM
3W24R15-15	OEM
3W28R5	OEM
3W28R12-12	OEM
3W28R15-15	OEM
3W5412-5	OEM
3W12412-12	OEM
3WB-5A	1438
3WB-6A	1445
3WB-7A	1458
3WB-9D	1490
3WB-11D	0510
3WB-13	0237
3WB-13D	0870
3WB-15D	0247
3WB-17D	1078
3WB-25D	2399
3WB-30D	1788
3WB-55D	0207
3WB-85D	0356
3WB-100D	0438
3WB-120D	0509
3WC-13D	0237
3WC-15D	0247
3WC-25D	2399
3WM1	OEM
3WM2	OEM
3WM4	OEM
3WM6	OEM
3WM8	OEM
3WM10	OEM
3WM12	OEM
3X11/1	0015
3Z3.9M	2035
3Z3.9T5	6327
3Z3.9T10	0547
3Z3.9T20	0547
3Z4.3T5	2403
3Z4.3T10	2403
3Z4.3T20	2403
3Z4.7M	1438
3Z4.7T5	7132
3Z4.7T10	2113
3Z4.7T20	2113
3Z5.1T5	2410
3Z5.1T10	2410
3Z5.1T20	2410
3Z5.6M	1901
3Z5.6T5	1901
3Z5.6T10	1901
3Z5.6T20	1901
3Z6.2T5	0696
3Z6.2T10	0696
3Z6.2T20	0696
3Z6.8	0205
3Z6.8A	0205
3Z6.8B	0205
3Z6.8M	1601
3Z6.8T5	1601
3Z6.8T10	1601
3Z6.8T20	1601
3Z7.5T5	1616
3Z7.5T10	1616
3Z7.5T20	1616
3Z8.2M	1622
3Z8.2T5	1622
3Z8.2T10	1622
3Z8.2T20	1622
3Z9.0	0684
3Z9.0A	0684
3Z9.0B	0684
3Z9.1T5	0627
3Z9.1T10	0627
3Z9.1T20	0627
3Z10M	0510
3Z10T5	0992
3Z10T10	0992
3Z10T20	0992
3Z11T5	0995
3Z11T10	0995
3Z11T20	0995
3Z12	0234
3Z12A	0234
3Z12B	0234
3Z12M	0870

DEVICE TYPE	REPL CODE
3Z12T5	1263
3Z12T10	1263
3Z12T20	1263
3Z13T5	1249
3Z13T10	1249
3Z13T20	1249
3Z15	0247
3Z15A	0247
3Z15B	0247
3Z15M	1274
3Z15T5	1639
3Z15T10	1639
3Z15T20	1639
3Z16T5	1703
3Z16T10	1703
3Z16T20	1703
3Z18	0256
3Z18A	0256
3Z18B	0256
3Z18M	1078
3Z18T5	1716
3Z18T10	1716
3Z18T20	1716
3Z20T5	5729
3Z20T10	1731
3Z20T20	1731
3Z21	0262
3Z21A	0262
3Z21B	0262
3Z22M	1723
3Z22T5	0727
3Z22T10	0727
3Z22T20	0727
3Z24	0274
3Z24A	0274
3Z24B	0274
3Z24T5	1756
3Z24T10	1756
3Z24T20	1756
3Z27	0294
3Z27A	0294
3Z27B	0294
3Z27M	1761
3Z27T5	1782
3Z27T10	1782
3Z27T20	1782
3Z30	0309
3Z30A	0309
3Z30B	0309
3Z30T5	1794
3Z30T10	1794
3Z30T20	1794
3Z33	0318
3Z33A	0318
3Z33B	0318
3Z33T5	1788
3Z33T10	1788
3Z36	0320
3Z36A	0320
3Z36B	0320
3Z524A	OEM
3.0B2	0118
3.3B1	1080
3.3B1(ZENER)	0299
3.3B2	OEM
3.3Y	0299
3.6B	0376
3.6B1	0188
3.6C	0188
3.6CP	0376
3.6E25	OEM
3.6E50A	OEM
3.6E100	OEM
3.6E250	OEM
3.6EB1	0376
3.6L1	0376
3.6N1	0376
3.9B2	0036
3.9N2	0036
3.9Y	0036
3.58MC	4465
4-009	0443
4-08018-667	1257
4-4A-1A7-1	0160
4-6B-3A7-1	0265
4-8P-2A7-1	0004
4-9L-4A7-1	0004
4-12-1A7-1	0016
4-14A17-1	0160
4-46	0111
4-47	0191
4-47(SEARS)	0016
4-48	1862
4-48(SEARS)	1862
4-50	0015
4-65-1A7-1	0050
4-65-2A7-1	0208
4-65-4A7-1	0600
4-65A17-1	0050
4-65B17-1	0050
4-65C17-1	0160
4-66-1A7-1	0004
4-66-2A7-1	0004
4-66-3A7-1	0004
4-74-3A7-1	0280
4-77A17-1	0440
4-77C17-1	0160
4-88A17-1	0160
4-88B17-1	0050
4-88C17-1	0280
4-92-1A7-1	0440
4-142	0222
4-142-1(SEARS)	0847
4-202A16	0143
4-202R101	0015
4-203-5100-69372	0552
4-221R806	OEM
4-221R807	OEM
4-227R801	OEM
4-279	0136
4-280	0136
4-282	0143
4-324	0325
4-397	0086
4-398	0086
4-399	0144
4-400	0144
4-432	0050
4-433	0144
4-434	0144
4-435	0085
4-436	0157
4-436(SEARS)	0157
4-437	0030
4-443	0127
4-850	0127
4-851	0016
4-852	0143
4-853	0143
4-854	0143
4-855	0143
4-856	0244
4-857	0143
4-882	0143
4-1544	0111
4-1545	0016
4-1546	0847
4-1723	0456
4-1724	0133
4-1726	0133
4-1790	0016
4-1791	0111
4-1792	0595
4-1807	0015
4-1848	0822
4-2020	0143
4-2020-03173	0015
4-2020-03200	0015
4-2020-03500	0143
4-2020-03571	0143
4-2020-03700	0102
4-2020-03800	0102
4-2020-03900	0071
4-2020-05000	0196
4-2020-05200	0015
4-2020-05400	0196
4-2020-05600	0123
4-2020-05800	0133
4-2020-06100	0133
4-2020-06200	0133
4-2020-06300	1303
4-2020-06400	0133
4-2020-06500	0102
4-2020-06600	0053
4-2020-06700	0102
4-2020-06800	0015
4-2020-07300	0015
4-2020-07500	0636
4-2020-07600	0015
4-2020-07601	0015
4-2020-07700	0015
4-2020-07800	0102
4-2020-07801	0102
4-2020-07802	0102
4-2020-07900	0196
4-2020-08000	0102
4-2020-08001	0133
4-2020-08200	0133
4-2020-08500	0015
4-2020-08600	0123
4-2020-08700	0196
4-2020-08900	0015
4-2020-09200	0495
4-2020-09400	1303
4-2020-09600	OEM
4-2020-10100	0133
4-2020-10500	0015
4-2020-11300	0025
4-2020-11600	OEM
4-2020-12000	0495
4-2020-12300	0317
4-2020-12400	0137
4-2020-13300	0820
4-2020-14400	0102
4-2020-14500	0015
4-2020-14600	0102
4-2020-15100	0102
4-2020-15600	0133
4-2021	0924
4-2021-04170	0277
4-2021-04470	0071
4-2021-04570	0102
4-2021-04770	0071
4-2021-04870	0102
4-2021-04970	0015
4-2021-05000	1149
4-2021-05070	0975
4-2021-05170	0917
4-2021-05470	0649
4-2021-05870	0143
4-2021-06970	0102
4-2021-07470	0133
4-2021-07570	0757
4-2021-07670	0124
4-2021-08070	0757
4-2021-08270	0015
4-2021-08570	0565
4-2021-09070	0317
4-2021-09370	0023
4-2021-10270	0023
4-2021-10470	0023
4-2021-10870	0317
4-2021-14970	0017
4-2029-70791	0030
4-2029-70860	0124
4-2029-71590	0133
4-2039-70380	1066
4-2039-70431	0076
4-2040-08000	0133
4-2060-02300	0878
4-2060-02400	0878
4-2060-02600	0855
4-2060-02900	0855
4-2060-03900	0855
4-2060-04000	0167
4-2060-04200	0043
4-2060-04300	0389
4-2060-04600	0167
4-2060-04800	0607
4-2060-04900	0607
4-2060-05200	0884
4-2060-07200	0043
4-2060-07300	0389
4-2060-07500	0881
4-2060-09100	3705
4-2060-09200	0856
4-2061-05170	0395
4-2061-05370	2639
4-2069-70232	2926
4-2069-70430	OEM
4-2069-71660	4410
4-2069-71710	4834
4-2069-71730	3725
4-2069-71842	OEM
4-2073	0050
4-3033	0015
4-3034	0133
4-3036	2262
4-5145	0016
4-18341	0030
4-68681-2	0004
4-68681-3	0004
4-68682-3	0016
4-68687-3	0229
4-68689-3	0071
4-68695-3	0144
4-68697-3	0071
4-684120-3	0144
4-685285-3	0144
4-686105-3	0071
4-686106-3	0071
4-686107-3	0144
4-686108-3	0144
4-686112-3	0144
4-686114-3	0144
4-686116-3	0229
4-686118-3	0144
4-686119-3	0144
4-686120-3	0144
4-686126-3	0144
4-686127-3	0144
4-686130-3	0178
4-686131-3	0144
4-686132-3	0016
4-686139-3	0071
4-686140-3	0144
4-686143-3	0016
4-686144-3	0016
4-686145-3	0233
4-686147-3	0229
4-686148-3	0229
4-686149-3	0015
4-686150-3	0229
4-686151-3	0229
4-686163-3	0004
4-686169-3	0144
4-686170-3	0126
4-686171-3	0144
4-686172-3	0144
4-686173-3	0016
4-686177-3	0229
4-686179-3	0071
4-686182-3	0086
4-686183-3	0016
4-686184-3	0071
4-686186-3	0071
4-686189-3	0071
4-686195-3	0004
4-686196-3	0004
4-686199-3	0071
4-686201-3	0071
4-686207-3	0144
4-686208-3	0144
4-686209-3	0144
4-686212-3	0071
4-686213-3	0160
4-686224-3	0144
4-686226-3	0178
4-686227-3	0201
4-686228-3	0144
4-686229-3	0126
4-686230-3	0126
4-686231-3	0016
4-686232-3	0233
4-686234-3	0423
4-686235-3	0126
4-686238-3	0126
4-686244-3	0144
4-686251-3	0144
4-686252-3	0338
4-686256-1	0136
4-686256-2	0136
4-686256-3	0136
4-686257-3	0016
4-3022861	0144
4-3023190	0130
4-3023212	0016
4-3023221	0016
4-3023222	0126
4-3023223	0086
4-3023843	0103
4-3023844	0086
4-3025763	0144
4-3025764	0144
4-3025765	0144
4-3025766	0016
4-3025767	0144
4-3540012	0015
4-8134842	0338
4-30203845	0126
4-202003571	0143
4-202104170	0015
4-202104570	0102
4-202104770	0164
4-202104970	0133
4-202105070	0924
4-202105470	0649
4-202115770	0102
4-2020035000	0143
4-AAZ10	0143
4A-1	0160
4A-1-70	0160
4A-1-70-12	0160
4A-1-70-12-7	0160
4A-1-A-7B	0160
4A-1A	0160
4A-1A0	0160
4A-1A0R	0160
4A-1A1	0160
4A-1A2	0160
4A-1A3	0160
4A-1A3P	0160
4A-1A4	0160
4A-1A4-7	0160
4A-1A5	0160
4A-1A5L	0160
4A-1A6	0160
4A-1A6-4	0160
4A-1A7	0160
4A-1A7-1	0160
4A-1A8	0160
4A-1A9	0160
4A-1A9G	0160
4A-1A19	0160
4A-1A21	0160
4A-1A82	0160
4A2	0376
4A-10	0160
4A-11	0160
4A-12	0160
4A-13	0160
4A-14	0160
4A-15	0160
4A-16	0160
4A-17	0160
4A-18	0160
4A-19	0160
4A132	1250
4A162	0319
4A232	0319
4A262	0319
4A600	0319
4A1122	0319
4A2122	0319
4AF05NLH	OEM
4AF05NLV	OEM
4AF05NPP	OEM
4AF1	OEM
4AF1LH	OEM
4AF1LV	OEM
4AF1NLH	OEM
4AF1NLV	OEM
4AF1NPP	OEM
4AF2	OEM
4AF2LH	OEM
4AF2LV	OEM
4AF2NLH	OEM
4AF2NLV	OEM
4AF2NPP	OEM
4AF4	OEM
4AF4LH	OEM
4AF4LV	OEM
4AF4NLH	OEM
4AF4NLV	OEM
4AF4NPP	OEM
4AF6NLH	OEM
4AF6NLV	OEM
4AF6NPP	OEM
4AJ4DX52D	0015
4AJ4DX520	0015
4B1	0036
4B4B41	0323
4B10Y	OEM
4B20Y	OEM
4B30Y	OEM
4B40Y	OEM
4BA4B41	OEM
4BV60	OEM
4BV80	OEM
4BV100	OEM
4BV120	OEM
4BV130	OEM
4C1	2901
4C2	1124
4C4	0805
4C6	0319
4C8	1132
4C10	0044
4C12	OEM
4C14	OEM
4C16	OEM
4C28	0086
4C29	0086
4C30	0086
4C31	0086
4C43	0079
4D4	0015
4D4B41	1414
4D4B42	OEM
4D4B44	OEM
4D6	0015
4D8	0015
4D20	0016
4D21	0016
4D22	0016
4D24	0016
4D25	0016
4D26	0016
4D48B41	OEM
4E20	OEM
4E20-8	OEM
4E20-28	OEM
4E20A	OEM
4E20M-8	OEM
4E20M8	OEM
4E20M-28	OEM
4E20M28	OEM
4E30-8	OEM
4E30-28	OEM
4E30A	OEM
4E30M-8	OEM
4E30M8	OEM
4E30M-28	OEM
4E40-8	OEM
4E40-28	OEM
4E40A	OEM
4E40M-8	OEM
4E40M8	OEM
4E40M-28	OEM
4E40M28	OEM
4E50-8	OEM
4E50-28	OEM
4E50A	OEM
4E50M-8	OEM
4E50M8	OEM
4E50M-28	OEM
4E50M28	OEM
4E80-8	OEM
4E80-28	OEM
4E100-8	OEM
4E100-28	OEM
4E100A	OEM
4E100M-8	OEM
4E100M8	OEM
4E100M-28	OEM
4E100M28	OEM
4E200-8	OEM
4E200-28	OEM
4E200A	OEM
4E200M-28	OEM
4EP30-3	OEM
4EP30-8	OEM
4EP30-28	OEM
4EP30A	OEM
4EP30M3	OEM
4EP30M8	OEM
4EP30M28	OEM
4EX580	OEM
4EX581	OEM
4EX582	OEM
4EX583	OEM
4E.5	OEM
4F05	OEM
4F1	OEM
4F2	OEM
4F3	OEM
4F4	OEM
4F5	OEM
4F6	OEM
4F8	OEM
4F10	OEM
4F12	OEM
4FB5	0015
4FB10	0015
4FB20	0015
4FB30	0015
4FB40	0015
4FC5	OEM
4FC10	OEM
4FC20	OEM
4FC30	OEM
4FC40	OEM
4FDC	OEM
4G2	3249
4G4	OEM
4G4B41	0473
4G5	OEM
4G8	0071
4G50	OEM
4G100	OEM
4G132	0319
4G162	0319
4G200	OEM
4G232	0319
4G262	0319
4G1122	0319
4G2122	0319
4GA	0015
4GZ10A	0510
4GZ10B	0510
4GZ12A	0870
4GZ12B	0509
4GZ15A	1274
4GZ15B	0063
4GZ18A	1078
4GZ18B	0634
4GZ22A	1723
4GZ27A	1761
4GZ33A	1788
4GZ39A	1804
4GZ47A	0022
4GZ56A	0207
4GZ68A	0310
4GZ82A	0356
4H1HA	OEM
4H-130-30302	0015
4H-130-30312	0123
4H-130-30702	0015
4H-130-30845	0554
4H-130-30862	0057
4H-130-30869	0123
4H-130-30993	1414
4H-130-31012	0019
4H-130-31438	0023
4H-130-31907	0554
4H-130-34173	0253
4H-130-40936	0079
4H-130-40937	0079
4H-130-41356	3807
4H-130-41395	3617
4H-130-41448	0808
4H-130-41461	3807
4H-130-44121	0079
4H-130-44196	0079
4H-209-80751	0496
4H-209-80786	5616
4H-209-80821	1260
4H-209-80967	2926
4H-209-81252	5172
4H-209-81473	4573
4H1034-4196	0079
4H1034-4283	0477
4H1303-0219	0143
4H1303-0613	0124
4H1303-1201	0023
4H1303-1356	0015
4H1303-1393	0023
4H1304	OEM
4H1304-0938	0079
4H1304-0941	0037
4H1304-1025	0224
4H1304-1084	0261
4H1304-1558	2766
4H1304-4104	0006
4H1304-4121	0079
4H1304-4154	0224
4H1304-4196	0079
4H1304-4283	0477
4H1304-4349	1386
4H1304-4358	0037
4H2098-0357	3451
4H13030621	0133
4H13030847	0124
4H13030862	0057
4H13030983	0124
4H13030984	0124
4H13031168	0071
4H13031173	0087
4H13031174	0023
4H13031201	0102
4H13032214	OEM
4H13034167	0471
4H13034197	1005
4H13034278	0062
4H13034328	OEM
4H13040938	0079
4H13040941	0037
4H13040995	0925
4H13041025	OEM
4H13041041	0324
4H13041085	0728
4H13041275	0071
4H13041436	2860
4H13041487	0031
4H13041773	0276
4H13041782	0261
4H13041803	0261
4H13041817	OEM
4H13044196	0079
4H13044197	0037
4H13044283	0477
4H13044349	1386
4H13044358	0037
4H13044503	0079
4H20980587	1074
4H20981194	1057
4H20981472	OEM
4H20981787	1780
4HEADIC	OEM
4I29200602	0123
4IC-402	4485
4IC-407	0579
4IC-409	2305
4J4B41	0446
4J24X539	0137
4J31	OEM
4J32	OEM
4J33	OEM
4J34	OEM
4J35	OEM
4J36	OEM
4J37	OEM
4J43	OEM
4J44	OEM
4J50	OEM
4J50-5	OEM
4J50A	OEM
4J50M5	OEM
4J50M25	OEM
4J50TO	OEM
4J50TR	OEM
4J52A	OEM
4J53	OEM
4J57	OEM
4J58	OEM
4J59	OEM
4J100-5	OEM
4J100M5	OEM
4J100M25	OEM
4J200-5	OEM
4J200-25	OEM
4J200M5	OEM
4J200M25	OEM
4JA2FX355	0015
4JA2X355	0015
4JA4DR700	0015
4JA4DX520	0015
4JA6MR700	0015
4JA10DX3	0015
4JA10DX32	0015
4JA10EX3	0015
4JA11BX4	0097
4JA12C101	OEM
4JA16MR700M	0071
4JA27DR700	0109
4JA38DR700	1132
4JA70MR700	1986
4JA211A	0015
4JB2C6	0917
4JBC12	0917
4JC12X070	OEM
4JD1A17	0280
4JD1A73	0211
4JD1B2	OEM
4JD1B3	OEM
4JD1B4	OEM
4JD3B1	0600
4JD4A2	OEM
4JD4A3	OEM
4JD4A4	OEM
4JD4A5	OEM
4JD5E29	1669
4JD7A35	OEM
4JD12C102	OEM
4JD12X009	OEM
4JD12X010	OEM
4JD12X011	OEM
4JD12X012	OEM
4JD12X013	OEM
4JD12X014	OEM
4JD12X043	OEM
4JD12X047	OEM
4JD12X132	OEM
4JD20AB	OEM
4JFBD1	OEM
4JFBD2	OEM
4JFBD3	OEM
4JFBD4	OEM
4JFBD5	OEM
4JFBD6	OEM
4JFBD7	OEM
4JUS2.16A	OEM
4JUS2.16B	OEM
4JUS2.46A	OEM
4JUS2.46B	OEM
4JUS2.60A	OEM
4JUS2.60B	OEM
4JUS2.92A	OEM
4JUS2.92B	OEM
4JUS3.05A	OEM
4JUS3.05B	OEM
4JX1A520	0211

DEVICE TYPE	REPL CODE
4JX1A520B	0211
4JX1A520C	0211
4JX1A520D	0280
4JX1A520E	0280
4JX1A813	0050
4JX1C707	0050
4JX1C850	0211
4JX1C850A	0280
4JX1C1224	0211
4JX1D925	0004
4JX1E596	0016
4JX1E821	0280
4JX1E850	0038
4JX2A60	0280
4JX2A601	0038
4JX2A616	0208
4JX2A801	0600
4JX2A816	0038
4JX2A822	0038
4JX5E670	1669
4JX7A972	0016
4JX8D404	0160
4JX8P404	0160
4JX8P409	0160
4JX11C2848	0086
4JX16A567	0016
4JX16A569	0038
4JX16A667	0016
4JX16A667/G	0079
4JX16A667/O	0079
4JX16A667/R	0079
4JX16A667/Y	0079
4JX16A667G	0016
4JX16A667O	0016
4JX16A667R	0016
4JX16A667Y	0016
4JX16A668	0016
4JX16A668/G	0079
4JX16A668/O	0079
4JX16A668/Y	0079
4JX16A668G	0016
4JX16A668O	0079
4JX16A668Y	0016
4JX16A669	0016
4JX16A669/G	1518
4JX16A669/Y	1518
4JX16A669G	0016
4JX16A669Y	0016
4JX16A670	0016
4JX16A670G	0016
4JX16A6670	0079
4JX16A6680	0016
4JX16B670/B	0079
4JX16B670/G	1518
4JX16B670/R	0079
4JX16B670/Y	0079
4JX16B670B	0016
4JX16B670G	0016
4JX16B670R	0016
4JX16B670Y	0016
4JX16E3860	0016
4JX16E3890	0016
4JX16E3960	0016
4JX17A567	OEM
4JX24X539	0137
4JX29A826	0037
4JX29A829	0037
4JX2816	0038
4JX2825	0038
4JZ4X539	0137
4JZ4XL12	0137
4K3CB	OEM
4K3CC	OEM
4K3SJ	OEM
4K3SK	OEM
4K3SL	OEM
4K3SN	OEM
4K5SL1	OEM
4K5SL3	OEM
4K30	1137
4K60	0344
4K100	0901
4K200LW	OEM
4K200LX	OEM
4KD20A7	OEM
4KM50LB	OEM
4KM50LC	OEM
4KM50SI	OEM
4KM50SJ	OEM
4KM50SK	OEM
4KM70LH	OEM
4KM70LH2	OEM
4KM70SJ	OEM
4KM70SK	OEM
4KM100LA	OEM
4KM100LF	OEM
4KM100LH	OEM
4KM150LA	OEM
4KM150LB	OEM
4KM150LF	OEM
4KM150LH	OEM
4KM150LH1	OEM
4KM3000LR	OEM
4KM50000LA3	OEM
4KM50000LA5	OEM
4KM50000LQ	OEM
4KM50000LR	OEM
4KMP300LU	OEM
4KMP10000LF	OEM
4KMV150LH1	OEM
4KP3SN	OEM
4KZ	OEM
4L30	1137
4L60	OEM
4L100	0901
4N22	OEM
4N22A	OEM
4N23	OEM
4N23A	OEM
4N24	OEM
4N24A	OEM
4N25	0541
4N25A	0541
4N26	0541
4N27	0541
4N28	0541
4N29	1053
4N29A	1053
4N30	1053
4N31	1053
4N32	1109
4N32A	1109
4N33	1109
4N35	0315
4N36	0315
4N37	0315
4N38	OEM
4N38A	0541
4N39	OEM
4N40	OEM
4N41	OEM
4N45	OEM
4N46	OEM
4N47	OEM
4N48	OEM
4N49	OEM
4N51	OEM
4N52	OEM
4N53	OEM
4N54	OEM
4N55	OEM
4N55TXV	OEM
4N55TXVB	OEM
4NT100	OEM
4NT175	OEM
409C	0030
4R0/18	OEM
4R1/18	OEM
4R2/18	OEM
4R2B	OEM
4R2S	OEM
4R3B	OEM
4R3S	OEM
4R4C	OEM
4R4S	OEM
4R5C	OEM
4R5S	OEM
4R6C	OEM
4R6S	OEM
4R7CD	OEM
4R8CD	OEM
4R8S	OEM
4R52	OEM
4RCM5	1103
4RCM10	OEM
4RCM20	OEM
4RCM30	2659
4RCM40	OEM
4RCM50	OEM
4RCM60	OEM
4RO/18	OEM
4SD46-2	0143
4SM1	0015
4SM3	0087
4SM6	OEM
4SM15	OEM
4SM25	OEM
4SM40	OEM
4SM55	OEM
4SM70	OEM
4SM100	OEM
4SM150	OEM
4SM200	OEM
4ST41	OEM
4T6.2	0636
4T6.2A	0636
4T10T	OEM
4T20T	OEM
4T40T	OEM
4T60T	OEM
4T80T	OEM
4T100T	OEM
4T501	0015
4T502	0015
4T503	0015
4T504	0015
4T505	0015
4T506	0015
4T507	0071
4T508	0071
4T509	0071
4T510	0071
4TP001	OEM
4V27	OEM
4W01-13	0137
4W70	OEM
4W80	OEM
4W81	OEM
4W82	OEM
4X11	OEM
4Z9-4Z12	OEM
4.3Y	0275
4.7B	OEM
4.7B1	0140
4.7B2	0140
4.7B3	0140
4.7EB1	0140
4.7EB3	0140
4.7X	0140
4.7Z	0041
5	OEM
5-8	0016
5-30082.3	0071
5-30082.4	0071
5-30086.1	0015
5-30088.1	0015
5-30088.2	0229
5-30088.3	0071
5-30094.1	0015
5-30095.1	0015
5-30098.1	0015
5-30099.1	0015
5-30099.3	0015
5-30099.4	0015
5-30106.1	0015
5-30109.1	0015
5-30111.1	0229
5-30113.1	0015
5-30119.1	0288
5-30120.1	0015
5-30122.1	0229
5-30132.1	0774
5-113641	0141
5-70004503	0016
5-70005452	0016
5-70005503	0016
5-7000901504	0016
5A1	0015
5A1/18	OEM
5A1/52	OEM
5A2	0015
5A2/18	OEM
5A2/52	OEM
5A3	0015
5A4	0015
5A4D	0015
5A4D-C	0015
5A5	0015
5A5/18	OEM
5A5/52	OEM
5A5D	0015
5A5R5	OEM
5A5R12-12	OEM
5A5R15-15	OEM
5A6	0015
5A6D	0015
5A6D-C	0015
5A8	0071
5A8D	2637
5A8DC	0071
5A10	0071
5A10C	0071
5A10D	2637
5A10D-C	0071
5A12	OEM
5A12/18	OEM
5A12R5	OEM
5A12R12-12	OEM
5A12R15-15	OEM
5A13	OEM
5A13/18	OEM
5A14	OEM
5A14/18	OEM
5A15	1137
5A15/18	OEM
5A16	OEM
5A16/18	OEM
5A22	OEM
5A22/18	OEM
5A23	OEM
5A23/18	OEM
5A24	OEM
5A24/18	OEM
5A24R5	OEM
5A24R12-12	OEM
5A24R15-15	OEM
5A25	OEM
5A25/18	OEM
5A26	OEM
5A26/18	OEM
5A30	OEM
5A50	OEM
5A52	OEM
5A52/18	OEM
5A53	OEM
5A53/18	OEM
5A54	OEM
5A54/18	OEM
5A55	OEM
5A55/18	OEM
5A56	OEM
5A56/18	OEM
5A60	OEM
5A100	OEM
5A200	OEM
5A400	OEM
5A600	OEM
5A800	OEM
5A1000	OEM
5AA5-1	0440
5A-D	0015
5AD8	0087
5AD8DC	0087
5AD10	0087
5AD10C	0087
5AD10D	0087
5AGD	OEM
5AV20	OEM
5AV30	OEM
5AV50	OEM
5AV60	OEM
5AV65	OEM
5AV80	OEM
5AV100	OEM
5AV120	OEM
5AV130	OEM
5AX90518A72	OEM
5B05	OEM
5B-1	0015
5B1	0015
5B1/18	OEM
5B1/52	OEM
5B-2	0015
5B2	0087
5B2/18	OEM
5B2/52	OEM
5B-2-H5W	0015
5B2C11	OEM
5B2Z11	OEM
5B3	0015
5B4	0302
5B5/18	OEM
5B5/52	OEM
5B7	0253
5B12	OEM
5B12/18	OEM
5B13	OEM
5B13/18	OEM
5B14	OEM
5B14/18	OEM
5B15	OEM
5B15/18	OEM
5B-15H	0015
5B16	OEM
5B16/18	OEM
5B22	OEM
5B22/18	OEM
5B23	OEM
5B23/18	OEM
5B24	OEM
5B24/18	OEM
5B25	OEM
5B25/18	OEM
5B26	OEM
5B26/18	OEM
5B52	OEM
5B52/18	OEM
5B53	OEM
5B53/18	OEM
5B54	OEM
5B54/18	OEM
5B55	OEM
5B55/18	OEM
5B56	OEM
5B56/18	OEM
5B2012	OEM
5BG2C11	OEM
5BG2C41	1236
5BG2Z11	OEM
5BG2Z41	1727
5BL2C41	1236
5BL41	1127
5C	OEM
5C(C)1	0589
5C05/18	OEM
5C05/52	OEM
5C054	OEM
5C054/18	OEM
5C055	OEM
5C055/18	OEM
5C056	OEM
5C056/18	OEM
5C057	OEM
5C057/18	OEM
5C1/52	OEM
5C1R10	OEM
5C1R20	OEM
5C1R30	OEM
5C1S10	OEM
5C1S20	OEM
5C1S30	OEM
5C2/18	OEM
5C2/52	OEM
5C3	0589
5C4-5S	OEM
5C4.5-5.5S	OEM
5C5.4-5.9S	OEM
5C5.5-6.5S	OEM
5C5.6S	OEM
5C6.7S	OEM
5C14	OEM
5C14/18	OEM
5C15	OEM
5C15/18	OEM
5C16	OEM
5C16/18	OEM
5C17	OEM
5C17/18	OEM
5C24	OEM
5C24/18	OEM
5C25	OEM
5C25/18	OEM
5C26	OEM
5C26/18	OEM
5C27	OEM
5C27/18	OEM
5C28	OEM
5C29	OEM
5C30	OEM
5C50	OEM
5C100	OEM
5C200	OEM
5C400	OEM
5C600	OEM
5C800	OEM
5C1000	OEM
5CA/18	OEM
5CH1M	1236
5CH1MU	1236
5CH2M	1236
5CH2MU	1236
5CH2SM	1236
5CH3M	OEM
5CH4M	OEM
5CL2C41	1236
5CL41	1127
5CS04SM	1236
5D1	0015
5D2	0015
5D2C11	OEM
5D2Z11	OEM
5D4	OEM
5D5R5	OEM
5D5R12-12	OEM
5D5R15-15	OEM
5D6	OEM
5D8	0071
5D10	0071
5D12R5	OEM
5D12R12-12	OEM
5D12R15-15	OEM
5D15	OEM
5D20	OEM
5D20-1	0802
5D24R5	OEM
5D24R12-12	OEM
5D24R15-15	OEM
5D25	OEM
5D30	OEM
5D40	OEM
5D50	OEM
5D50FS	OEM
5D100	OEM
5D100FS	OEM
5D200	OEM
5D200FS	OEM
5D400	OEM
5D400FS	OEM
5D600	OEM
5D600FS	OEM
5D800	OEM
5D800FS	OEM
5D1000	OEM
5DG2C11	OEM
5DG2C41	1236
5DG2Z11	OEM
5DG2Z41	1727
5DH1M	OEM
5DH2M	OEM
5DH3M	OEM
5DH4M	OEM
5DL2C41	1236
5DL2CZ41A-LB	OEM
5DL41	1127
5E1	0015
5E1R20	OEM
5E1R40	OEM
5E1R60	OEM
5E1S20	OEM
5E1S40	OEM
5E1S60	OEM
5E2	0015
5E3	0015
5E4	0015
5E5	0015
5E6	0015
5E8	0071
5E25	OEM
5E29	OEM
5E50A	OEM
5E100	OEM
5E150	OEM
5E200	OEM
5E250	OEM
5EH1M	1727
5EH2M	1727
5EH3M	1727
5EH4M	1727
5EZ3.3D	0782
5EZ3.6D	0796
5EZ3.6D5	0796
5EZ3.6D10	0796
5EZ3.9D	0806
5EZ3.9D5	0806
5EZ3.9D10	0806
5EZ4.3D5	0821
5EZ4.3D10	0821
5EZ4.7D	0833
5EZ4.7D5	0833
5EZ4.7D10	0833
5EZ5.1D	0442
5EZ5.1D5	0442
5EZ5.1D10	0442
5EZ5.6D	0876
5EZ5.6D5	0876
5EZ5.6D10	0876
5EZ6.2D	0185
5EZ6.2D5	0185
5EZ6.2D10	0185
5EZ6.8D	0205
5EZ6.8D5	0205
5EZ6.8D10	0205
5EZ7.5D	0480
5EZ7.5D5	0480
5EZ7.5D10	0480
5EZ8.2D	0504
5EZ8.2D5	0504
5EZ8.2D10	0504
5EZ8.7D	0504
5EZ8.7D5	3317
5EZ8.7D10	3317
5EZ9.1D	0684
5EZ9.1D5	0684
5EZ9.1D10	0684
5EZ10D	0225
5EZ10D5	0225
5EZ10D10	0225
5EZ11D	0230
5EZ11D5	0230
5EZ11D10	0230
5EZ12D	0234
5EZ12D5	0234
5EZ12D10	0234
5EZ13D	0237
5EZ13D5	0237
5EZ13D10	0237
5EZ14D	1397
5EZ14D5	1397
5EZ14D10	1397
5EZ15D	0247
5EZ15D5	OEM
5EZ15D10	0247
5EZ16D	0251
5EZ16D5	0251
5EZ16D10	0251
5EZ17D	1178
5EZ17D5	1178
5EZ17D10	1178
5EZ18D	0256
5EZ18D5	0256
5EZ18D10	0256
5EZ19D	2395
5EZ19D5	2395
5EZ19D10	2395
5EZ20D5	0262
5EZ20D10	0262
5EZ22D	0269
5EZ22D5	0269
5EZ22D10	0269
5EZ24D	0274
5EZ24D5	0274
5EZ24D10	0274
5EZ25D	2399
5EZ25D5	2399
5EZ25D10	2399
5EZ27D	0294
5EZ27D5	0294
5EZ27D10	0294
5EZ28D	1177
5EZ28D5	1177
5EZ28D10	1177
5EZ30D	0309
5EZ30D5	0309
5EZ30D10	0309
5EZ33D	0318
5EZ33D5	0318
5EZ33D10	0318
5EZ36D	0320
5EZ36D5	0320
5EZ36D10	0318
5EZ39D	0326
5EZ39D5	0326
5EZ39D10	0326
5EZ43D	0347
5EZ43D5	0337
5EZ43D10	0337
5EZ47D	0347
5EZ47D10	0347
5EZ51D	0027
5EZ51D5	0027
5EZ51D10	0027
5EZ56D	0266
5EZ56D5	0266
5EZ56D10	0266
5EZ60D	2856
5EZ60D5	2856
5EZ60D10	2856
5EZ62D5	0386
5EZ75D	0426
5EZ75D10	0426
5EZ82D	0444
5EZ82D5	0444
5EZ82D10	0444
5EZ87D	3028
5EZ87D5	3028
5EZ87D10	3028
5EZ91D	0238
5EZ91D5	0238
5EZ91D10	0238
5EZ100D	1180
5EZ100D5	1180
5EZ100D10	1180
5EZ110D5	1190
5EZ110D10	1190
5EZ120D	1207
5EZ120D5	1207
5EZ130D	1218
5EZ130D10	OEM
5EZ140D	1881
5EZ140D5	1881
5EZ140D10	1881
5EZ150D	0647
5EZ150D5	0647
5EZ150D10	0647
5EZ160D	1255
5EZ160D5	1255
5EZ160D10	1255
5EZ170D	2102
5EZ170D5	2102
5EZ170D10	2102
5EZ180D	1279
5EZ180D5	1279
5EZ180D10	1279
5EZ190D	2225
5EZ190D5	2225
5EZ190D10	2225
5EZ200D	0605
5EZ200D5	0605
5EZ200D10	0605
5EZ1200D10	OEM
5E.5	OEM
5F1	0708
5F1R07	OEM
5F1R15	OEM
5F1S07	OEM
5F1S15	OEM
5F1S07	OEM
5F5	0970
5F10	0871
5F15	0319
5F15E2	OEM
5F15E3	OEM
5F20	1250
5F25	OEM
5F30	0853
5F40	0764
5F50	OEM
5F60	OEM
5F80	OEM
5FF05	OEM
5FF10	OEM
5FF15	OEM
5FF30	OEM
5FF40	OEM
5FF50	OEM
5FWJ2C41	1236
5G1	OEM
5G2	4288
5G2C11	OEM
5G2Z11	OEM
5G3	OEM
5G4	OEM
5G5	OEM
5G8	0071
5G514	OEM
5G515	OEM
5G516	OEM
5GA	0015
5G-D	0015
5GD	0015
5GF	0071
5GFH	0015
5GG2C11	OEM
5GG2C41	OEM
5GG2Z11	OEM
5GG2Z41	OEM
5GJ	0071
5GJ/FR1N	0015
5GWJ2C41	1236
5H	0015
5H3P	0970
5H3PN	0985
5H4D1	0015
5H300RT	OEM
5H500RT	OEM
5H750M	0015
5H20984841	OEM
5J2C11	OEM
5J2Z11	OEM
5J3P	0084
5J26	OEM
5J-F1	0015
5JG2C11	OEM
5JG2Z11	OEM
5JG2Z41	OEM
5JH3	OEM
5JUS2.70A	OEM
5JUS2.70B	OEM
5JUS3.08A	OEM
5JUS3.08B	OEM
5JUS3.25A	OEM
5JUS3.25B	OEM
5JUS3.65A	OEM

DEVICE TYPE	REPL CODE
5JUS3.65B	OEM
5JUS3.81A	OEM
5JUS3.81B	OEM
5K090B	OEM
5K10	0071
5K20	OEM
5K40	OEM
5K50	OEM
5K50SE	OEM
5K50SG	OEM
5K60	OEM
5K70SG	OEM
5K70SG-WB	OEM
5K70SH1	OEM
5K100	OEM
5K300SK	OEM
5K4164ANP-15	2357
5KF10(B)	OEM
5KF20(B)	OEM
5KF20B	1236
5KF30	1664
5KF30B	5796
5KF40	1664
5KF40B	5796
5KFU10	OEM
5KFU10B	OEM
5KFU20	OEM
5KFU20B	OEM
5KM50SJ	OEM
5KM70SI	OEM
5KM70SJ	OEM
5KM300SI	OEM
5KM1000SG	OEM
5KP5.0	2396
5KP5.0A	2396
5KP6.0	3114
5KP6.0A	3114
5KP6.5	OEM
5KP6.5A	OEM
5KP7.0	OEM
5KP7.0A	OEM
5KP7.5	OEM
5KP7.5A	OEM
5KP8.0	OEM
5KP8.0A	OEM
5KP8.5	OEM
5KP8.5A	OEM
5KP9.0	OEM
5KP9.0A	OEM
5KP10	OEM
5KP10A	OEM
5KP11	3172
5KP11A	3172
5KP12	OEM
5KP12A	OEM
5KP13	OEM
5KP13A	OEM
5KP14	OEM
5KP14A	OEM
5KP15	3191
5KP15A	3191
5KP16	OEM
5KP16A	OEM
5KP17	OEM
5KP17A	3201
5KP18	OEM
5KP18A	OEM
5KP20	OEM
5KP20A	OEM
5KP22	1915
5KP22A	1915
5KP24	OEM
5KP24A	OEM
5KP26	OEM
5KP26A	OEM
5KP28	OEM
5KP28A	OEM
5KP30	OEM
5KP30A	OEM
5KP33	OEM
5KP33A	OEM
5KP36	OEM
5KP36A	OEM
5KP40	OEM
5KP40A	OEM
5KP43	0568
5KP43A	0568
5KP45	OEM
5KP45A	OEM
5KP48	OEM
5KP48A	OEM
5KP51	OEM
5KP51A	OEM
5KP54	OEM
5KP54A	OEM
5KP58	0831
5KP58A	0831
5KP60	OEM
5KP60A	OEM
5KP64	OEM
5KP64A	OEM
5KP64P	OEM
5KP70	OEM
5KP70A	OEM
5KP75	OEM
5KP75A	OEM
5KP78	OEM
5KP78A	OEM
5KP85	OEM

DEVICE TYPE	REPL CODE
5KP85A	OEM
5KP90	OEM
5KP90A	OEM
5KP100	OEM
5KP100A	OEM
5KP110	OEM
5KP110A	OEM
5KQ30	OEM
5KQ30(B)	OEM
5KQ30B	OEM
5KQ40	OEM
5KQ40(B)	OEM
5KQ40B	OEM
5KQ50	OEM
5KQ50(B)	OEM
5KQ50B	OEM
5KQ60	OEM
5KQ60(B)	OEM
5KQ60B	OEM
5KQ90	OEM
5KQ90B	OEM
5KQ100	OEM
5KQ100B	OEM
5L1-2	OEM
5L1-2S	OEM
5L15	0348
5L20	OEM
5L20(1)	OEM
5L25	OEM
5L30	OEM
5L30(1)	OEM
5L35	OEM
5L40	OEM
5L40(1)	OEM
5L45	OEM
5L50	OEM
5L55	OEM
5L60	OEM
5L60(1)	OEM
5L65	OEM
5L70	OEM
5L75	OEM
5L75(1)	OEM
5L80	OEM
5L85	OEM
5L90	OEM
5L95	OEM
5L100	OEM
5L100(1)	OEM
5L110	OEM
5L120	OEM
5L130	OEM
5L140	OEM
5L150	OEM
5L150(1)	OEM
5L160	OEM
5L170	OEM
5L180	OEM
5L190	OEM
5L200	OEM
5L200(1)	OEM
5L209	OEM
5L210	OEM
5L220	OEM
5L230	OEM
5L240	OEM
5L250	OEM
5L260	OEM
5L270	OEM
5L280	OEM
5L290	OEM
5L300	OEM
5LT32	OEM
5LT36ZY1	OEM
5LT61ZK	OEM
5M10	OEM
5M15	OEM
5M20	OEM
5M30	OEM
5M40	3751
5M47ZS5	6801
5M47ZS10	OEM
5M47ZS10.5	OEM
5M50	3757
5M60	1112
5M80	3768
5M80/E3517	OEM
5M81	OEM
5M100	3011
5MA1	OEM
5MA2	0015
5MA3	OEM
5MA4	0015
5MA4D	0015
5MA5	0015
5MA6	0015
5MA8	0071
5MA10	0071
5MF1	0015
5MS5	0015
5MS10	0015
5MS20	0015
5MS30	0015
5MS40	0015
5MS50	0015
5N02F1-A	OEM
5N05M1	OEM
5N1	0015
5N2	0097

DEVICE TYPE	REPL CODE
5N3	0097
5N10K1	OEM
5N15A	OEM
5N15A9	OEM
5N20A	OEM
5N20A9	OEM
5N25A	OEM
5N25A9	OEM
5N30A	OEM
5N30A9	OEM
5N35A	OEM
5N35A9	OEM
5N40A	OEM
5N40A9	OEM
5N45A	OEM
5N45A9	OEM
5N50A	OEM
5N50A9	OEM
5N55A	OEM
5N55A9	OEM
5N60A	OEM
5N60A9	OEM
5N65A	OEM
5N65A9	OEM
5N70A	OEM
5N70A9	OEM
5N75A	OEM
5N75A9	OEM
5N80A	OEM
5N80A9	OEM
5N85A	OEM
5N85A9	OEM
5N90A	OEM
5N90A9	OEM
5N95A	OEM
5N95A9	OEM
5N100A	OEM
5N100A9	OEM
5N110A	OEM
5N120A	OEM
5N130A	OEM
5N140A	OEM
5N150A	OEM
5N160A	OEM
5N170A	OEM
5N190A	OEM
5N200A	OEM
5N210A	OEM
5N220A	OEM
5N230A	OEM
5N240A	OEM
5N250A	OEM
5N260A	OEM
50	0143
5P-01	0015
5P05M	2095
5P1M	3869
5P2	0097
5P2M	3876
5P3	0097
5P4M	0710
5P5M	0863
5P6M	0863
5P15	OEM
5P30	OEM
5P60	OEM
5P100	OEM
5PL.6-1.2	OEM
5PL.6-1.2S	OEM
5P.4-.8	OEM
5P.4-.8S	OEM
5P.35-.5	OEM
5P.35-.5S	OEM
5Q3	1630
5R	OEM
5R0/18	OEM
5R1/18	OEM
5R1/52	OEM
5R2/4	OEM
5R2/18	OEM
5R2/66	OEM
5R2B	OEM
5R2S	OEM
5R3/4	OEM
5R3/66	OEM
5R3B	OEM
5R3P	OEM
5R3S	OEM
5R4/3	OEM
5R4/3C	OEM
5R4/4	OEM
5R4/59	OEM
5R4C	OEM
5R4S	OEM
5R5/3	OEM
5R5/3D	OEM
5R5/4	OEM
5R5/61	OEM
5R5C	OEM
5R5S	OEM
5R6/3	OEM
5R6/3D	OEM
5R6/5	OEM
5R6/61	OEM
5R6C	OEM
5R6S	OEM
5R7/3	OEM
5R7/3D	OEM
5R7/5	OEM

DEVICE TYPE	REPL CODE
5R7/61	OEM
5R7CD	OEM
5R7S	OEM
5R8/3	OEM
5R8/3D	OEM
5R8/5	OEM
5R8/63	OEM
5R8CD	OEM
5R8S	OEM
5R52	OEM
5RC2	3419
5RC5	1103
5RC5A	1103
5RC10	2491
5RC10A	2491
5RC10G	0240
5RC15	0240
5RC15A	0240
5RC20	0240
5RC20A	0240
5RC25	2659
5RC30	2659
5RC30A	2659
5RC40	0676
5RC40A	0676
5RC50	2809
5RC50A	2809
5RC50B	2809
5RC60	0336
5RC60A	0336
5RCF5A	OEM
5RCF10A	OEM
5RCF20A	OEM
5RCF25A	OEM
5RCF30A	OEM
5RCF40A	OEM
5RCF50A	OEM
5RCF60A	OEM
5RCL2	3419
5RCL5	1103
5RCL10	2491
5RCL15	0240
5RCL20	0240
5RE52	OEM
5RKB1	OEM
5RKB2	OEM
5RKB3	OEM
5RKB4	OEM
5RKB5	OEM
5RKB6	OEM
5RKB7	OEM
5RKB8	OEM
5R0/18	OEM
5S1S03	OEM
5S1S04	OEM
5S2.8-2.85	OEM
5S3P	OEM
5S3.4-3.7	OEM
5S4M	OEM
5SB05	OEM
5SB10	OEM
5SB20	OEM
5SB40	OEM
5SB60	OEM
5SB80	OEM
5SB100	OEM
5SKB1	OEM
5SKB2	OEM
5SKB3	OEM
5SKB4	OEM
5SKB5	OEM
5SKB6	OEM
5SKB7	OEM
5SKB8	OEM
5SM1020A4	OEM
5SM1020A4PL	OEM
5SM1020A4T	OEM
5SM1020A5	OEM
5SM1020A5PL	OEM
5SM1020A5T	OEM
5SM1020A6	OEM
5SM1020A6PL	OEM
5SM1020A6T	OEM
5SM1020A7	OEM
5SM1020A7PL	OEM
5SM1020A7T	OEM
5SM1020A8	OEM
5SM1020A8PL	OEM
5SM1020A8T	OEM
5SM1020E4	OEM
5SM1020E4PL	OEM
5SM1020E4T	OEM
5SM1020E5	OEM
5SM1020E5PL	OEM
5SM1020E5T	OEM
5SM1020E6	OEM
5SM1020E6PL	OEM
5SM1020E6T	OEM
5SM1020E7	OEM
5SM1020E7PL	OEM
5SM1020E7T	OEM
5SM1020GE8PL	OEM
5SM1020GE8T	OEM
5SM1020GE9	OEM
5SM1020GE9PL	OEM
5SM1020GE9T	OEM
5SM1020GE10	OEM
5SM1020GE10PL	OEM
5SM1020GE10T	OEM

DEVICE TYPE	REPL CODE
5SP510-6	OEM
5SPT10-4	OEM
5SPT10-5	OEM
5SPT12-4	OEM
5SPT12-5	OEM
5T3P	OEM
5TB1	2491
5TB1-CE00	OEM
5TB1-CH00	OEM
5TB1-EE00	OEM
5TB1-EH00	OEM
5TB1-GE00	OEM
5TB1-GH00	OEM
5TB1-JE00	OEM
5TB2	0240
5TB2-CE00	OEM
5TB2-CH00	OEM
5TB2-EE00	OEM
5TB2-EH00	OEM
5TB2-GE00	OEM
5TB2-GH00	OEM
5TB2-JE00	OEM
5TB3	2659
5TB3-CE00	OEM
5TB3-CH00	OEM
5TB3-EE00	OEM
5TB3-EH00	OEM
5TB3-GE00	OEM
5TB3-GH00	OEM
5TB3-JE00	OEM
5TB4	0676
5TB4-CE00	OEM
5TB4-CH00	OEM
5TB4-EE00	OEM
5TB4-EH00	OEM
5TB4-GE00	OEM
5TB4-GH00	OEM
5TB4-JE00	OEM
5TB5	2809
5TB5-CE00	OEM
5TB5-CH00	OEM
5TB5-EE00	OEM
5TB5-EH00	OEM
5TB5-GE00	OEM
5TB5-GH00	OEM
5TB5-JE00	OEM
5TB6	0336
5TB6-CE00	OEM
5TB6-CH00	OEM
5TB6-EE00	OEM
5TB6-EH00	OEM
5TB6-GE00	OEM
5TB6-GH00	OEM
5TB6-JE00	OEM
5TB7	OEM
5TB7-CE00	OEM
5TB7-CH00	OEM
5TB7-EE00	OEM
5TB7-EH00	OEM
5TB7-GE00	OEM
5TB7-GH00	OEM
5TB7-JE00	OEM
5TB8	OEM
5TC05	OEM
5TC025	OEM
5TC10	OEM
5TC15	OEM
5TC20	OEM
5TC25	OEM
5TC30	OEM
5TC35	OEM
5TC40	OEM
5TC50	OEM
5TC60	OEM
5TC70	OEM
5TC80	OEM
5TC90	OEM
5TC100	OEM
5TC110	OEM
5TC120	OEM
5TCA10	OEM
5TCA20	OEM
5TCA30	OEM
5TCA40	OEM
5TCA50	OEM
5TCA60	OEM
5TCA70	OEM
5TCA80	OEM
5TCA90	OEM
5TCA100	OEM
5TCA110	OEM
5TCA120	OEM
5TCRA	3419
5TCRB	1103
5TCRC	2491
5TCRD	OEM
5TCRE	0240
5TCRF	OEM
5TCRG	6942
5TCRH	0676
5TCRI	0336
5TCRJ	0336
5TCRK	OEM
5TCRL	OEM
5V	OEM
5V3P	OEM
5W3P	OEM
5W5R5	OEM
5W5R12-12	OEM

DEVICE TYPE	REPL CODE
5W5R15-15	OEM
5W12R5	OEM
5W12R12-12	OEM
5W12R15-15	OEM
5W24R5	OEM
5W24R12-12	OEM
5W24R15-15	OEM
5W28R5	OEM
5W28R12-12	OEM
5W28R15-15	OEM
5W70	OEM
5W71	OEM
5WM1020GE8	OEM
5X3P	OEM
5X10	OEM
5Y3P	OEM
5Z27	0294
5Z30	0309
5Z82B	0444
5ZB	OEM
5ZS3.3	0782
5ZS3.3A	0782
5ZS3.3B	0782
5ZS3.6	0796
5ZS3.6A	0796
5ZS3.6B	0796
5ZS3.9	0806
5ZS3.9A	0806
5ZS3.9B	0806
5ZS4.3	0821
5ZS4.3A	0821
5ZS4.3B	0821
5ZS5.1	0442
5ZS5.1A	0442
5ZS5.1B	0442
5ZS5.6	0876
5ZS5.6A	0876
5ZS5.6B	0876
5ZS6.0A	3127
5ZS6.0B	3127
5ZS6.2	0185
5ZS6.2A	0185
5ZS6.2B	0185
5ZS7.5	0480
5ZS7.5A	0480
5ZS7.5B	0480
5ZS8.2	0504
5ZS8.2A	0504
5ZS8.2B	0504
5ZS8.7	3317
5ZS8.7B	1490
5ZS9.1	0684
5ZS9.1A	0684
5ZS9.1B	0684
5ZS10	0225
5ZS10A	0225
5ZS10B	0225
5ZS11	0230
5ZS11A	0230
5ZS12	0234
5ZS12A	0234
5ZS12B	0234
5ZS13	0237
5ZS13B	0237
5ZS14	1397
5ZS14B	1397
5ZS15	0247
5ZS15A	0247
5ZS15B	0247
5ZS18	0256
5ZS18A	0256
5ZS18B	0256
5ZS20	0262
5ZS20A	0262
5ZS20B	0262
5ZS22	0269
5ZS22A	0269
5ZS22B	0269
5ZS24	0274
5ZS24A	0274
5ZS24B	0274
5ZS27	0294
5ZS27A	0294
5ZS27B	0294
5ZS30	0309
5ZS30A	0309
5ZS30B	0309
5ZS33	0318
5ZS33A	0318
5ZS33B	0318
5ZS36	0320
5ZS36A	0320
5ZS36B	0320
5ZS39	0326
5ZS39A	0326
5ZS39B	0326
5ZS43	0337
5ZS43A	0337
5ZS43B	0337
5ZS47	0347
5ZS47A	0347
5ZS47B	0347
5ZS51	0027
5ZS51A	0027
5ZS51B	0027
5ZS68	0405
5ZS68A	0405
5ZS68B	0405
5ZS75	0426

DEVICE TYPE	REPL CODE
5ZS75A	0426
5ZS75B	0426
5ZS82	0444
5ZS82A	0444
5ZS87	0444
5ZS87A	0444
5ZS87B	0444
5ZS91	0238
5ZS91A	0238
5ZS91B	0238
5ZS100	1180
5ZS100A	1180
5ZS100B	1180
5.1	0041
5.1B	0041
5.1B1	0041
5.1B2	0041
5.1B2,,	0041
5.1EB(3)	0041
5.1EB2	0041
5.1EB3	0041
5.1N1	0041
5.1N2	0041
5.1N3	0041
5.1X	0041
5.6B	0253
5.6B1	0253
5.6B2	0253
5.6B3	0253
5.6EB	0253
5.6EB1	0883
5.6L2	0253
5.6M	OEM
5.6N1	0253
5.6N2	0253
5.6V	0157
5.6X	0041
5.6Y	0253
5.6Z	0253
6	0079
6-0004799	0576
6-02	0079
6-04	0016
6-04GRN	0016
6-04ORN	0016
6-04S1	0016
6-04S2	0016
6-05	0016
6-05F	0016
6-0451	0016
6-0452	0016
6-11	0016
6-13	0004
6-19	0016
6-24	0178
6-30	0016
6-31	0037
6-31A	0037
6-38	0037
6-38A	0037
6-49	0693
6-50	0136
6-53	0164
6-53/63	0211
6-53A	0004
6-53F	0004
6-60	0050
6-60-P	0211
6-60A	0050
6-60B	0136
6-60C	0050
6-60D	0136
6-60E	0050
6-60F	0050
6-60P	0004
6-60T	0050
6-60X	0050
6-61	0136
6-61-P	0211
6-61A	0050
6-61B	0004
6-61D	0136
6-61E	0050
6-61F	0050
6-61P	0004
6-61T	0050
6-61X	0050
6-62	0136
6-62-P	0211
6-62A	0004
6-62B	0004
6-62C	0050
6-62D	0004
6-62E	0050
6-62F	0136
6-62P	0136
6-62X	0136
6-63	0004
6-63A	0004
6-63T	0004
6-65	0004
6-65T	0004
6-66	0004
6-66T	0004
6-67	0004
6-67T	0004
6-69	0050
6-69X	0050
6-70	0136

If replacement code is OEM, contact original manufacturer for replacement.

DEVICE TYPE	REPL CODE
6-71	0136
6-72	0136
6-84F	0050
6-85F	0050
6-87	0004
6-88	0004
6-88(AUTOMATIC)	0222
6-89	0136
6-89X	0136
6-90	0007
6-93	1221
6-223-636	1965
6-2708	0144
6-4799	0576
6-5363	0004
6-13759A02	OEM
6-13780A01	OEM
6-59010	0015
6-1260039	0050
6-1260039A	0050
6-6490004	0103
6A05	3657
6A05G	3657
6A1	0589
6A1(RECT)	1282
6A1G	1282
6A2	1282
6A2G	1287
6A4	1287
6A4G	1287
6A6	1293
6A6F	0711
6A6F1	OEM
6A6G	1293
6A8	1296
6A8-1A5L	0050
6A8F	OEM
6A8F1	OEM
6A8G	1296
6A10	1296
6A10F	OEM
6A10F1	OEM
6A10G	1296
6A15	0970
6A30	0970
6A50	0970
6A100	3733
6A150	OEM
6A200	0989
6A300	3743
6A400	0197
6A500	0200
6A600	0204
6A700	0206
6A753	OEM
6A800	0206
6A900	0586
6A1000	0586
6A1200	OEM
6A10227	0016
6A10228	0086
6A10229	0222
6A10422	0016
6A10423	0016
6A10520	0016
6A10622	0164
6A10624	0004
6A10851	0016
6A10855	0016
6A11180	0016
6A11223	0007
6A11301	0633
6A11665	0004
6A11668	0004
6A12515	0164
6A12516	0004
6A12517	0164
6A12677	0144
6A12678	0280
6A12679	0144
6A12680	0136
6A12681	0016
6A12682	0079
6A12683	0016
6A12684	0004
6A12685	0004
6A12725	0016
6A12788	0016
6A12789	0016
6A12889	0050
6A12988	0233
6A12989	0004
6A12990	0004
6A12992	0208
6A12993	0038
6A16399	0076
6AKB7	OEM
6ALZ1	OEM
6ALZ6.8	OEM
6ALZ6.8A	OEM
6ALZ6.8B	OEM
6ALZ7.5	OEM
6ALZ7.5A	OEM
6ALZ7.5B	OEM
6ALZ8.2	OEM
6ALZ8.2A	OEM
6ALZ8.2B	OEM
6ALZ8.7	OEM
6ALZ8.7A	OEM
6ALZ8.7B	OEM
6ALZ9.1	OEM
6ALZ9.1A	OEM
6ALZ9.1B	OEM
6ALZ10	OEM
6ALZ10A	OEM
6ALZ10B	OEM
6ALZ11	OEM
6ALZ11A	OEM
6ALZ11B	OEM
6ALZ12	OEM
6ALZ12A	OEM
6ALZ12B	OEM
6ALZ13	OEM
6ALZ13A	OEM
6ALZ13B	OEM
6ALZ14	OEM
6ALZ14A	OEM
6ALZ14B	OEM
6ALZ15	OEM
6ALZ15A	OEM
6ALZ15B	OEM
6ALZ16	OEM
6ALZ16A	OEM
6ALZ16B	OEM
6ALZ17	OEM
6ALZ17A	OEM
6ALZ17B	OEM
6ALZ18	OEM
6ALZ18A	OEM
6ALZ18B	OEM
6ALZ19	OEM
6ALZ19A	OEM
6ALZ19B	OEM
6ALZ20	OEM
6ALZ20A	OEM
6ALZ20B	OEM
6ALZ22	OEM
6ALZ22A	OEM
6ALZ22B	OEM
6ALZ23	OEM
6ALZ23A	OEM
6ALZ23B	OEM
6ALZ24	OEM
6ALZ24A	OEM
6ALZ24B	OEM
6ALZ25	OEM
6ALZ25A	OEM
6ALZ25B	OEM
6ALZ27	OEM
6ALZ27A	OEM
6ALZ27B	OEM
6ALZ28	OEM
6ALZ28A	OEM
6ALZ28B	OEM
6ALZ30	OEM
6ALZ30A	OEM
6ALZ30B	OEM
6ALZ33	OEM
6ALZ33A	OEM
6ALZ33B	OEM
6ALZ36	OEM
6ALZ36A	OEM
6ALZ36B	OEM
6ALZ39	OEM
6ALZ39A	OEM
6ALZ39B	OEM
6ALZ43	OEM
6ALZ43A	OEM
6ALZ43B	OEM
6ALZ47	OEM
6ALZ47A	OEM
6ALZ47B	OEM
6ALZ51	OEM
6ALZ51A	OEM
6ALZ51B	OEM
6ALZ56	OEM
6ALZ56A	OEM
6ALZ56B	OEM
6ALZ60	OEM
6ALZ60A	OEM
6ALZ60B	OEM
6ALZ62	OEM
6ALZ62A	OEM
6ALZ62B	OEM
6ALZ68	OEM
6ALZ68A	OEM
6ALZ68B	OEM
6ALZ75	OEM
6ALZ75A	OEM
6ALZ75B	OEM
6ALZ82	OEM
6ALZ82A	OEM
6ALZ82B	OEM
6ALZ87	OEM
6ALZ87A	OEM
6ALZ87B	OEM
6ALZ91	OEM
6ALZ91A	OEM
6ALZ91B	OEM
6ALZ100	OEM
6ALZ100A	OEM
6ALZ100B	OEM
6ALZ110	OEM
6ALZ110A	OEM
6ALZ110B	OEM
6ALZ120	OEM
6ALZ120A	OEM
6ALZ120B	OEM
6ALZ130	OEM
6ALZ130A	OEM
6ALZ130B	OEM
6ALZ140	OEM
6ALZ140A	OEM
6ALZ140B	OEM
6ALZ150	OEM
6ALZ150A	OEM
6ALZ150B	OEM
6ALZ160	OEM
6ALZ160A	OEM
6ALZ160B	OEM
6ALZ170	OEM
6ALZ170A	OEM
6ALZ170B	OEM
6ALZ180	OEM
6ALZ180A	OEM
6ALZ180B	OEM
6ALZ190	OEM
6ALZ190A	OEM
6ALZ190B	OEM
6ALZ200	OEM
6ALZ200A	OEM
6ALZ200B	OEM
6ALZ561	OEM
6B	OEM
6B1	0253
6B2	0636
6B2(3%)	0157
6B2CTA	OEM
6B3	0253
6B4B41	4073
6B4CTA	OEM
6B4DM	OEM
6B4HA	OEM
6B8CTA	OEM
6B8DM	OEM
6B8HA	OEM
6B10	OEM
6B12DM	OEM
6B12MA	OEM
6B16DM	OEM
6B16MA	OEM
6B20	OEM
6B40	OEM
6B60	0446
6B82	OEM
6BCTA	OEM
6BG11	1567
6BL6	OEM
6BM6	OEM
6BM61	OEM
6C	OEM
6C05	0970
6C05R	0985
6C1	0091
6C2	0471
6C3	0295
6C7	OEM
6C8	OEM
6C9	OEM
6C10	3733
6C10R	0910
6C15	0989
6C15R	0990
6C20	3740
6C20R	0990
6C30	3743
6C30R	0993
6C40	0197
6C40R	0997
6C50	0200
6C50R	1001
6C60	0204
6C60R	0515
6C70	0206
6C70R	1008
6C80	0206
6C80R	1008
6C90	0586
6C90R	0948
6C100	0586
6C100R	0948
6C153S	OEM
6CC11	1250
6CC12	0871
6CC13	0989
6CD12	0871
6CD13	0989
6CG12	0197
6CM7	OEM
6CZ5.0	OEM
6CZ5.0A	OEM
6CZ5.0B	OEM
6CZ6.0	OEM
6CZ6.0A	OEM
6CZ6.0B	OEM
6CZ6.5	OEM
6CZ6.5A	OEM
6CZ6.5B	OEM
6CZ7.0	OEM
6CZ7.0A	OEM
6CZ7.0B	OEM
6CZ7.5	OEM
6CZ7.5A	OEM
6CZ7.5B	OEM
6CZ8.0	OEM
6CZ8.0A	OEM
6CZ8.0B	OEM
6CZ8.5	OEM
6CZ8.5A	OEM
6CZ8.5B	OEM
6CZ9.0	OEM
6CZ9.0A	OEM
6CZ9.0B	OEM
6CZ10	OEM
6CZ10A	OEM
6CZ10B	OEM
6CZ11	OEM
6CZ11A	OEM
6CZ11B	OEM
6CZ12	OEM
6CZ12A	OEM
6CZ12B	OEM
6CZ13	OEM
6CZ13A	OEM
6CZ13B	OEM
6CZ14	OEM
6CZ14A	OEM
6CZ14B	OEM
6CZ15	OEM
6CZ15A	OEM
6CZ15B	OEM
6CZ16	OEM
6CZ16A	OEM
6CZ16B	OEM
6CZ17	OEM
6CZ17A	OEM
6CZ17B	OEM
6CZ18	OEM
6CZ18A	OEM
6CZ18B	OEM
6CZ20	OEM
6CZ20A	OEM
6CZ20B	OEM
6CZ22	OEM
6CZ22A	OEM
6CZ22B	OEM
6CZ24	OEM
6CZ24A	OEM
6CZ24B	OEM
6CZ26	OEM
6CZ26A	OEM
6CZ26B	OEM
6CZ28	OEM
6CZ28A	OEM
6CZ28B	OEM
6CZ30	OEM
6CZ30A	OEM
6CZ30B	OEM
6CZ33	OEM
6CZ33A	OEM
6CZ33B	OEM
6CZ36	OEM
6CZ36A	OEM
6CZ36B	OEM
6CZ40	OEM
6CZ40A	OEM
6CZ40B	OEM
6CZ43	OEM
6CZ43A	OEM
6CZ43B	OEM
6CZ45	OEM
6CZ45A	OEM
6CZ45B	OEM
6CZ48	OEM
6CZ48A	OEM
6CZ48B	OEM
6CZ51	OEM
6CZ51A	OEM
6CZ51B	OEM
6CZ54	OEM
6CZ54A	OEM
6CZ54B	OEM
6CZ58	OEM
6CZ58A	OEM
6CZ58B	OEM
6CZ60	OEM
6CZ60A	OEM
6CZ60B	OEM
6CZ64	OEM
6CZ64A	OEM
6CZ64B	OEM
6CZ70	OEM
6CZ70A	OEM
6CZ70B	OEM
6CZ75	OEM
6CZ75A	OEM
6CZ75B	OEM
6CZ78	OEM
6CZ78A	OEM
6CZ78B	OEM
6CZ85	OEM
6CZ85A	OEM
6CZ85B	OEM
6CZ90	OEM
6CZ90A	OEM
6CZ90B	OEM
6CZ100	OEM
6CZ100A	OEM
6CZ100B	OEM
6CZ110	OEM
6CZ110A	OEM
6CZ110B	OEM
6D02M1-A	OEM
6D05M1	OEM
6D4B41	1414
6D10K1	OEM
6D20	OEM
6D50	OEM
6D50FS	OEM
6D72	OEM
6D73	OEM
6D100	OEM
6D100FS	OEM
6D122	0004
6D122R	0004
6D122T	0004
6D122TC	0004
6D122TH	0004
6D122U	0004
6D122V	0004
6D122W	0004
6D122Y	0004
6D200	OEM
6D200FS	OEM
6D300A	0023
6D400	OEM
6D400FS	OEM
6D450A	OEM
6D600	OEM
6D600FS	OEM
6D800	OEM
6D800FS	OEM
6D1000	OEM
6D67800A03	0774
6DC11	0319
6DC12	0871
6DD12	0871
6DG11	1567
6DM	0244
6E	OEM
6E20	OEM
6E40A	OEM
6E55	OEM
6E100	OEM
6E175	OEM
6F	OEM
6F4	OEM
6F5	0970
6F5-D	0970
6F5A	0970
6F5B	0970
6F5D	0097
6F5R	2291
6F5RA	2291
6F5RB	2291
6F8	OEM
6F10	3733
6F10-D	3733
6F10A	3733
6F10B	3733
6F10D	0097
6F10R	2291
6F10RA	2291
6F10RB	2291
6F12	OEM
6F15	0989
6F15-D	0989
6F15A	0989
6F15B	0989
6F15D	0989
6F15R	2291
6F15RA	2291
6F15RB	2291
6F16	OEM
6F20	0989
6F20-D	0989
6F20A	0989
6F20B	0989
6F20D	0097
6F20R	2291
6F20RA	2291
6F20RB	2291
6F30	3743
6F30-D	3743
6F30A	3743
6F30B	3743
6F30R	0476
6F30RA	0476
6F30RB	0476
6F40	0197
6F40-D	0197
6F40A	0197
6F40B	0197
6F40D	0805
6F40R	0476
6F40RA	0476
6F40RB	0476
6F50	0200
6F50-D	0200
6F50A	0200
6F50B	0200
6F50R	0476
6F50RA	0476
6F50RB	0476
6F60	0204
6F60-D	0204
6F60A	0204
6F60B	0204
6F60D	0204
6F60R	0476
6F60RA	0476
6F60RB	0476
6F70	0206
6F70A	0206
6F70B	0206
6F70R	0449
6F70RA	0449
6F70RB	0449
6F80	0206
6F80-D	0206
6F80A	0206
6F80B	0206
6F80D	0206
6F80R	0449
6F80RA	0449
6F80RB	0449
6F90	0586
6F90A	0586
6F90B	0586
6F90R	0449
6F90RA	0449
6F90RB	0449
6F100	0586
6F100-D	0586
6F100A	0586
6F100B	0586
6F100D	0586
6F100R	0449
6F100RA	0449
6F100RB	0449
6F120	OEM
6FA10-D	0910
6FB050	1042
6FB1L	0323
6FB2L	1414
6FB4L	0473
6FB6L	0446
6FB100	1042
6FB200	1042
6FB400	1045
6FB600	1045
6FB800	OEM
6FB1000	OEM
6FC11	OEM
6FC12	0853
6FC13	0197
6FD12	0853
6FD13	0997
6FG11	2153
6FH05S	1567
6FH1S	1567
6FH2S	1567
6FH3S	2153
6FH4S	2153
6FL10S02	OEM
6FL20S02	OEM
6FL40S02	5327
6FL50	OEM
6FL60	OEM
6FL60S02	OEM
6FR5	0985
6FR5-D	0985
6FR5A	0985
6FR10	0910
6FR10-D	0910
6FR10A	0910
6FR10B	0910
6FR15	0990
6FR15A	0990
6FR15B	0990
6FR20	0990
6FR20-D	0990
6FR20A	0990
6FR20B	0990
6FR30	0993
6FR30A	0993
6FR30B	0993
6FR40	0997
6FR40-D	0997
6FR40A	0997
6FR40B	0997
6FR50	1001
6FR50A	1001
6FR50B	1001
6FR58	0985
6FR60	0515
6FR60-D	0515
6FR60A	0515
6FR60B	0515
6FR70	1008
6FR70A	1008
6FR70B	1008
6FR80	1008
6FR80-D	1008
6FR80A	1008
6FR80B	1008
6FR90	0948
6FR90A	0948
6FR90B	0948
6FR100	0948
6FR100A	0948
6FR100B	0948
6FXF11	OEM
6FXF12	OEM
6G2	OEM
6G4	0196
6G4B41	0473
6G8	0015
6GA175D	0015
6GA1750	0015
6GC1	0196
6GC1BY1	3598
6GC11	OEM
6GC12	0197
6GD1	0484
6GD12	0853
6GG11	2153
6GX1	1080
6H65F	OEM
6H65FW	OEM
6H65H	OEM
6H65HW	OEM
6H75AE	OEM
6H100AE	OEM
6H150AE	OEM
6H200A	OEM
6H250AR	OEM
6H300AE	OEM
6H350A	OEM
6H400AE	OEM
6H450A	OEM
6H500A	OEM
6H600A	OEM
6H740AC	1919
6H750A	OEM
6HC11	OEM
6J4B41	0446
6J7L	OEM
6JB41	OEM
6JC11	OEM
6JC12	0204
6JG11	0711
6JUS3.24	OEM
6JUS3.24B	OEM
6JUS3.70A	OEM
6JUS3.70B	OEM
6JUS3.90A	OEM
6JUS3.90B	OEM
6JUS4.38A	OEM
6JUS4.38B	OEM
6JUS4.57A	OEM
6JUS4.57B	OEM
6KB7	OEM
6KC11	OEM
6KZ5.0	OEM
6KZ5.0A	OEM
6KZ5.0B	OEM
6KZ6.0	OEM
6KZ6.0A	OEM
6KZ6.0B	OEM
6KZ6.5	OEM
6KZ6.5A	OEM
6KZ6.5B	OEM
6KZ7.0	OEM
6KZ7.0A	OEM
6KZ7.0B	OEM
6KZ7.5	OEM
6KZ7.5A	OEM
6KZ7.5B	OEM
6KZ8.0	OEM
6KZ8.0A	OEM
6KZ8.0B	OEM
6KZ8.5	OEM
6KZ8.5A	OEM
6KZ8.5B	OEM
6KZ9.0	OEM
6KZ9.0A	OEM
6KZ9.0B	OEM
6KZ10	OEM
6KZ10A	OEM
6KZ10B	OEM
6KZ11	OEM
6KZ11A	OEM
6KZ11B	OEM
6KZ12	OEM
6KZ12A	OEM
6KZ12B	OEM
6KZ13	OEM
6KZ13A	OEM
6KZ13B	OEM
6KZ14	OEM
6KZ14A	OEM
6KZ14B	OEM
6KZ15	OEM
6KZ15A	OEM
6KZ15B	OEM
6KZ16	OEM
6KZ16A	OEM
6KZ16B	OEM
6KZ17	OEM
6KZ17A	OEM
6KZ17B	OEM
6KZ18	OEM
6KZ18A	OEM
6KZ18B	OEM
6KZ20	OEM
6KZ20A	OEM
6KZ20B	OEM
6KZ22	OEM
6KZ22A	OEM
6KZ22B	OEM
6KZ24	OEM
6KZ24A	OEM
6KZ24B	OEM
6KZ26	OEM
6KZ26A	OEM
6KZ26B	OEM
6KZ28	OEM
6KZ28A	OEM
6KZ28B	OEM
6KZ30	OEM
6KZ30A	OEM
6KZ30B	OEM

DEVICE TYPE	REPL CODE
6KZ33	OEM
6KZ33A	OEM
6KZ33B	OEM
6KZ36	OEM
6KZ36A	OEM
6KZ36B	OEM
6KZ40	OEM
6KZ40A	OEM
6KZ40B	OEM
6KZ43	OEM
6KZ43A	OEM
6KZ43B	OEM
6KZ45	OEM
6KZ45A	OEM
6KZ45B	OEM
6KZ48	OEM
6KZ48A	OEM
6KZ48B	OEM
6KZ51	OEM
6KZ51A	OEM
6KZ51B	OEM
6KZ54	OEM
6KZ54A	OEM
6KZ54B	OEM
6KZ58	OEM
6KZ58A	OEM
6KZ58B	OEM
6KZ60	OEM
6KZ60A	OEM
6KZ60B	OEM
6KZ64	OEM
6KZ64A	OEM
6KZ65B	OEM
6KZ70	OEM
6KZ70A	OEM
6KZ70B	OEM
6KZ75	OEM
6KZ75A	OEM
6KZ75B	OEM
6KZ78	OEM
6KZ78A	OEM
6KZ78B	OEM
6KZ85	OEM
6KZ85A	OEM
6KZ85B	OEM
6KZ90	OEM
6KZ90A	OEM
6KZ90B	OEM
6KZ100	OEM
6KZ100A	OEM
6KZ100B	OEM
6KZ110	OEM
6KZ110A	OEM
6KZ110B	OEM
6L122	0004
6LC11	OEM
6LC12	0206
6LF11	1008
6M4	0015
6M10	0871
6M20	0871
6M30	1126
6M40	0853
6M60	0319
6M70	OEM
6M72	OEM
6M80	1132
6M100	0044
6M404-1	0015
6M404-2	0015
6M404-3	0015
6M404-4	0015
6M404-5	0015
6M404-6	0015
6M404-7	0015
6MA10	0871
6MA20	0871
6MA40	0853
6MA60	0267
6MA80	1119
6MA100	0281
6MC	0050
6MC11	OEM
6N02F1	OEM
6N05M1	OEM
6N10K1	OEM
6N126	OEM
6N127	OEM
6N128	OEM
6N129	OEM
6N130	OEM
6N131	OEM
6N132	OEM
6N133	OEM
6N134	OEM
6N135	5426
6N136	1291
6N136-020	1291
6N136-HP	1291
6N137	5423
6N138	5427
6N139	5427
6N140	OEM
6N-1039E-32	OEM
6NC11	OEM
6NC12	0586
6NF11	0948
6P05M1A	OEM
6P10K1	OEM
6P02F1-A	OEM
6QF11	OEM
6RH05S	1547
6RH1S	1567
6RH2S	1567
6RH3S	2153
6RH4S	3139
6RKB1	OEM
6RKB2	OEM
6RKB3	OEM
6RKB4	OEM
6RKB5	OEM
6RKB6	OEM
6RKB7	OEM
6RKB8	OEM
6RM36	OEM
6RM40	OEM
6RM42	OEM
6RM48	OEM
6RM56	OEM
6RM60	OEM
6RM64	OEM
6RM72	OEM
6RM80	OEM
6RM100	OEM
6RM120	OEM
6RM150	OEM
6RM200	OEM
6RM250	OEM
6RS1DH4	OEM
6RS1DH6	OEM
6RS5AJ1B1	OEM
6RS5SJ5B5	OEM
6RS5SP1B1	OEM
6RS5SP2B2	OEM
6RS5SP3B3	OEM
6RS5SP4B4	OEM
6RS5SP5B5	4408
6RS5SP6B6	OEM
6RS5SP7B7	OEM
6RS5SP8B8	OEM
6RS5SP9B9	OEM
6RS5SP10B10	4471
6RS5SP11B11	3257
6RS5SP12B12	3257
6RS5SP13B13	3257
6RS5SP14B14	3257
6RS5SP15B15	OEM
6RS5SP16B16	1101
6RS5SP17B17	OEM
6RS5SP18B18	OEM
6RS5SP19B19	OEM
6RS5SP20B20	OEM
6RS5VP1B	OEM
6RS5VP2B	OEM
6RS5VP3B	OEM
6RS5VP4B	OEM
6RS5VP5B	OEM
6RS5VP6B	OEM
6RS5VP7B	OEM
6RS5VP8B	OEM
6RS5VP9B	OEM
6RS5VP10B	OEM
6RS5VP11B	OEM
6RS5VP12B	OEM
6RS5VP13B	OEM
6RS5VP14B	OEM
6RS5VP15B	OEM
6RS5VP16B	OEM
6RS5VP17B	OEM
6RS5VP18B	OEM
6RS5VP19B	OEM
6RS5VP20B	OEM
6RS5VP21B	OEM
6RS5VP22B	OEM
6RS5VP23B	OEM
6RS5VP24B	OEM
6RS6PH13BCJ1	1303
6RS6PH13BJJ1	1303
6RS6PH13BKJ1	1303
6RS6PH13BMJ1	1303
6RS7PH13BKJ1	1303
6RS7PH30BCB1	0774
6RS7PH130BCB1	0774
6RS17DH4	OEM
6RS17DH6	OEM
6RS18PH110BEB1	0774
6RS18PH110BHB1	0774
6RS18PH110BMB1	0774
6RS18PH110BNB1	0774
6RS20AP62B2	0246
6RS20AP181	OEM
6RS20PH6RGD1	OEM
6RS20SJ3BB	1374
6RS20SJ5B5	OEM
6RS20SP1B1	OEM
6RS20SP2B2	OEM
6RS20SP3B3	OEM
6RS20SP4B4	OEM
6RS20SP5B5	3415
6RS20SP6B6	OEM
6RS20SP7B7	OEM
6RS20SP8B8	OEM
6RS20SP10B10	OEM
6RS20SP11B11	OEM
6RS20SP12B12	OEM
6RS20SP13B13	OEM
6RS20SP14B14	OEM
6RS20SP15B15	OEM
6RS20SP16B16	OEM
6RS20SP17B17	OEM
6RS20SP18B18	OEM
6RS20SP19B19	OEM
6RS20SP20B20	OEM
6RS20VJ10B	0246
6RS20VP1B	OEM
6RS20VP2B	OEM
6RS20VP3B	OEM
6RS20VP4B	OEM
6RS20VP5B	OEM
6RS20VP6B	OEM
6RS20VP7B	OEM
6RS20VP8B	OEM
6RS20VP9B	OEM
6RS20VP10B	OEM
6RS20VP11B	OEM
6RS20VP12B	OEM
6RS20VP13B	OEM
6RS20VP14B	OEM
6RS20VP15B	OEM
6RS20VP16B	OEM
6RS20VP17B	OEM
6RS20VP18B	OEM
6RS20VP19B	OEM
6RS20VP20B	OEM
6RS20VP21B	OEM
6RS20VP22B	OEM
6RS20VP23B	OEM
6RS20VP24B	OEM
6RS36PH13BJJ1	1303
6RS36PH13BJK1	1303
6RS36PH13BKJ1	1303
6RS36PH13BLJ1	1303
6RS51GX12	0474
6RS56PHL3BKJ1	1303
6RW32DEB3	1347
6RW32FEB3	1347
6RW32HEB3	1905
6RW32KEB3	1905
6RW32MEB3	1905
6RW32PEB3	0193
6RW32REB3	0193
6RW56AC	OEM
6RW56BC	OEM
6RW56CC	OEM
6RW56DC	OEM
6RW56EC	OEM
6RW56FC	OEM
6RW56GC	OEM
6RW56HC	OEM
6RW56KC	OEM
6RW59PY	OEM
6RW59RY	OEM
6RW59SY	OEM
6RW59UY	OEM
6RW59VY	OEM
6RW59WY	OEM
6RW62HY	0015
6RW71AY	OEM
6RW71B	OEM
6RW71BY	OEM
6RW71C	OEM
6RW71CY	OEM
6RW71D	OEM
6RW71DY	OEM
6RW71E	OEM
6RW71EY	OEM
6RW71F	OEM
6RW71FY	OEM
6RW71G	OEM
6RW71GY	OEM
6RW71H	OEM
6RW71HY	OEM
6RW71KY	OEM
6RW71MY	OEM
6RW71PY	OEM
6RW71RY	OEM
6Rs20SP9B9	OEM
6S	OEM
6S08	OEM
6S010	OEM
6S015	OEM
6S020	OEM
6S2	1669
6S4	1287
6S6	1293
6S8	1296
6S10	1296
6S18	OEM
6S28	OEM
6S38	OEM
6S48	OEM
6S58	OEM
6S68	OEM
6S110	OEM
6S115	OEM
6S120	OEM
6S210	OEM
6S215	OEM
6S220	OEM
6S310	OEM
6S315	OEM
6S320	OEM
6S410	OEM
6S415	OEM
6S420	OEM
6S510	OEM
6S515	OEM
6S520	OEM
6S610	OEM
6S615	OEM
6S620	OEM
6S10729A01	OEM
6S13759A01	OEM
6S13759A02	OEM
6S13761-A01	OEM
6S13761A01	OEM
6SB050	0323
6SB50	0323
6SB100	0323
6SB200	1414
6SB400	0473
6SB600	0446
6SB800	1422
6SB1000	2443
6SBF20	OEM
6SBF40	OEM
6SBF60	OEM
6SBF80	OEM
6SBF100	OEM
6SKB1	OEM
6SKB2	OEM
6SKB3	OEM
6SKB4	OEM
6SKB5	OEM
6SKB6	OEM
6SKB7	OEM
6SKB8	OEM
6SM1	OEM
6SM3	OEM
6SM6	OEM
6SM15	OEM
6SM25	OEM
6SM40	OEM
6SM55	OEM
6SM70	OEM
6SM100	OEM
6SM150	OEM
6SM200	OEM
6T06	4244
6T06A	OEM
6T06B	OEM
6T06C	OEM
6T06HA	OEM
6T06HX	OEM
6T010	OEM
6T010A	OEM
6T010B	OEM
6T010C	OEM
6T010HA	OEM
6T010HX	OEM
6T015	OEM
6T015A	OEM
6T015B	OEM
6T015C	OEM
6T015HA	OEM
6T015HX	OEM
6T10T	OEM
6T16	4244
6T16A	OEM
6T16B	OEM
6T16C	OEM
6T16HA	OEM
6T16HX	OEM
6T20T	OEM
6T26	4244
6T26A	OEM
6T26B	OEM
6T26C	OEM
6T26HA	OEM
6T26HX	OEM
6T36	OEM
6T36A	OEM
6T36B	OEM
6T36C	OEM
6T36HA	OEM
6T36HX	OEM
6T40T	OEM
6T46	OEM
6T46A	OEM
6T46B	OEM
6T46C	OEM
6T46HA	OEM
6T46HX	OEM
6T56	OEM
6T56A	OEM
6T56B	OEM
6T56C	OEM
6T56HA	OEM
6T56HX	OEM
6T60T	OEM
6T66	OEM
6T66A	OEM
6T66B	OEM
6T66C	OEM
6T66HA	OEM
6T66HX	OEM
6T80T	OEM
6T100T	OEM
6T110	OEM
6T110A	OEM
6T110B	OEM
6T110C	OEM
6T110HA	OEM
6T110HX	OEM
6T115	OEM
6T115A	OEM
6T115B	OEM
6T115C	OEM
6T115HA	OEM
6T115HX	OEM
6T210	OEM
6T210A	OEM
6T210B	OEM
6T210C	OEM
6T210HA	OEM
6T210HX	OEM
6T215	OEM
6T215A	OEM
6T215B	OEM
6T215C	OEM
6T215HA	OEM
6T215HX	OEM
6T310	OEM
6T310A	OEM
6T310B	OEM
6T310C	OEM
6T310HA	OEM
6T310HX	OEM
6T315	OEM
6T315A	OEM
6T315B	OEM
6T315C	OEM
6T315HA	OEM
6T315HX	OEM
6T410	OEM
6T410A	OEM
6T410B	OEM
6T410C	OEM
6T410HA	OEM
6T410HX	OEM
6T415	OEM
6T415A	OEM
6T415B	OEM
6T415C	OEM
6T415HA	OEM
6T415HX	OEM
6T510	OEM
6T510A	OEM
6T510B	OEM
6T510C	OEM
6T510HA	OEM
6T510HX	OEM
6T515	OEM
6T515A	OEM
6T515B	OEM
6T515C	OEM
6T515HA	OEM
6T515HX	OEM
6T610	OEM
6T610A	OEM
6T610B	OEM
6T610C	OEM
6T610HA	OEM
6T610HX	OEM
6T615	OEM
6T615A	OEM
6T615B	OEM
6T615C	OEM
6T615HA	OEM
6T615HX	OEM
6TC10	OEM
6TC20	OEM
6TC30	OEM
6TC40	OEM
6TC50	OEM
6TC60	OEM
6TC70	OEM
6TC80	OEM
6TC90	OEM
6TC100	OEM
6TC110	OEM
6TC120	OEM
6TCA10	OEM
6TCA20	OEM
6TCA30	OEM
6TCA40	OEM
6TCA50	OEM
6TCA60	OEM
6TCA70	OEM
6TCA80	OEM
6TCA90	OEM
6TCA100	OEM
6TCA110	OEM
6TCA120	OEM
6TW1	OEM
6TW2	OEM
6TW3	OEM
6TW4	OEM
6TW5	OEM
6TW6	OEM
6U4	OEM
6U8	OEM
6U12	OEM
6V23	OEM
6V-200	0157
6V200	0157
6V203	OEM
6V211	OEM
6V221	OEM
6W52	OEM
6W70	OEM
6W73	OEM
6W411	OEM
6W411V	OEM
6WM1	OEM
6WM2	OEM
6WM4	OEM
6WM6	OEM
6WM10	OEM
6WM12	OEM
6X2C	OEM
6X97047A01	0016
6X97047A02	0004
6X97174A01	0133
6X97174XA08	0133
6Z0345	OEM
6.2B	0471
6.2B1	0091
6.2B1(1WATT)	0636
6.2B2	0471
6.2B3	0471
6.2EB3	0471
6.2N1	0471
6.2N2	0471
6.2Y	0471
6.2Z	0471
6.8	0062
6.8B	0062
6.8B1	0062
6.8B2	0062
6.8B2(ZENER)	3369
6.8B2(ZENER),,	0062
6.8B3	0062
6.8EB1	0062
6.8EB3	0062
6.8EN2	0062
6.8L	OEM
6.8M	OEM
6.8N2	0062
6.8SC20	OEM
6.8V	0062
6.8X	0062
6.8Y	0062
6.8Z	0062
7	OEM
7-0002	0015
7-0003	0277
7-0004	0015
7-0005	0143
7-0006	0143
7-0007	0063
7-0008	0015
7-0009	0438
7-0013	0133
7-0014	0037
7-0015	0016
7-1(SARKES)	0016
7-1(STANDEL)	0160
7-2	0160
7-2(SARKES)	0016
7-2(STANDEL)	3033
7-3(SARKES)	0016
7-3(STANDEL)	0126
7-4	0144
7-4(SARKES)	0016
7-4(STANDEL)	0126
7-6	0144
7-6(SARKES)	0079
7-7	0144
7-7(SARKES)	0016
7-8	0144
7-8(SARKES)	0016
7-8(STANDEL)	0296
7-9	0007
7-9(SARKES)	0007
7-10	0007
7-10(SARKES)	0007
7-11(SARKES)	0007
7-12	0103
7-12(STANDEL)	0103
7-13	0103
7-13(STANDEL)	0103
7-14A	0126
7-15(SARKES)	0016
7-15(STANDEL)	0071
7-16	0127
7-16(SARKES)	0016
7-17	0016
7-17(SARKES)	0016
7-18(SARKES)	0016
7-19	0142
7-19(SARKES)	0016
7-19(STANDEL)	0142
7-20	0127
7-20(SARKES)	0016
7-21(SARKES)	0007
7-22	0127
7-22(SARKES)	0007
7-23	0127
7-23(SARKES)	0007
7-24(SARKES)	0007
7-25(SARKES)	0007
7-26	0127
7-27	OEM
7-28	0007
7-28(STANDEL)	0905
7-29	0030
7-29(STANDEL)	0178
7-36	0127
7-40	0212
7-43	0127
7-44	0144
7-45	0127
7-59-001/3477	0143
7-59-005/3477	0277
7-59-0013477	0143
7-59-0053477	0277
7-59-010/3477	0211
7-59-019/3477	0079
7-59-020/3477	0079
7-59-021/3477	0079
7-59-022/3477	0079
7-59-023/3477	0079
7-59-024/3477	0079
7-59-029/3477	0280
7-59-060/3477	0211
7-59-068	0016
7-59-0103477	0164
7-59-0193477	0144
7-59-0203477	0144
7-59-0213477	0144
7-59-0223477	0144
7-59-0233477	0144
7-59-0243477	0016
7-59-0293477	0004
7-59-0603477	0004
7-117-02	0353
7-759-651-35	0878
7-5806	0725
7-5851	0147
7-5851-02	0147
7-5851V	0147
7-5853	2353
7-5853A	1413
7-5853B	1413
7-6006-00	0953
7-25806-01	0725
7-73004-02	0847
7-73004-03	0847
7-73004-04	0847
7-73004-1	0847
7-466201	0130
7-1585122	0147
7-2580504	1487
7-2585301-01	2353
7-2585301-02	1413
7-2585303	1413
7A1/18	OEM
7A1/52	OEM
7A2	0062
7A2/18	OEM
7A2/52	OEM
7A5/18	OEM
7A5/52	OEM
7A12	OEM
7A12/18	OEM
7A13	OEM
7A13/18	OEM
7A14	OEM
7A14/18	OEM
7A15	OEM
7A15/18	OEM
7A16	OEM
7A16/18	OEM
7A22	OEM
7A22/18	OEM
7A23	OEM
7A23/18	OEM
7A24	OEM
7A24/18	OEM
7A25	OEM
7A25/18	OEM
7A26	OEM
7A26/18	OEM
7A30	0127
7A30(GE)	0086
7A30(SHERWOOD)	0016
7A31	0127
7A31(GE)	0086
7A31(SHERWOOD)	0016
7A32	0127
7A32(GE)	0086
7A32(SHERWOOD)	0086
7A35	0595
7A35(GE)	0086
7A52	OEM
7A52/18	OEM
7A53	OEM
7A53/18	OEM
7A54	OEM
7A54/18	OEM
7A55	OEM
7A55/18	OEM
7A56	OEM
7A56/18	OEM
7A995	0086
7A995(GE)	0086
7A995(SHERWOOD)	0086
7A1011	0086
7A1011(GE)	0086
7A1011(SHERWOOD)	0086
7AAC299M	OEM
7B1	0142
7B1(2%)	OEM
7B1(ZENER)	OEM
7B1/18	OEM
7B1/52	OEM
7B2	0142
7B2(ZENER)	0062
7B2/18	OEM
7B2/52	OEM
7B3	OEM
7B-3B1	0302

If replacement code is OEM, contact original manufacturer for replacement.

DEVICE TYPE	REPL CODE
7B4	OEM
7B5/18	OEM
7B5/52	OEM
7B12	OEM
7B12/18	OEM
7B13	0142
7B13/18	OEM
7B14	OEM
7B14/18	OEM
7B15	OEM
7B15/18	OEM
7B16	OEM
7B16/18	OEM
7B20T	OEM
7B22	OEM
7B22/18	OEM
7B23	OEM
7B23/18	OEM
7B24	OEM
7B24/18	OEM
7B25	OEM
7B25/18	OEM
7B26	OEM
7B26/18	OEM
7B33	0142
7B34	0142
7B40T	OEM
7B52	OEM
7B52/18	OEM
7B53	OEM
7B53/18	OEM
7B54	OEM
7B54/18	OEM
7B55	OEM
7B55/18	OEM
7B56	OEM
7B56/18	OEM
7B60T	OEM
7B80T	OEM
7BM10	OEM
7BM10A	OEM
7BT13ZRK	OEM
7BT25ZYK	OEM
7C05/18	OEM
7C05/52	OEM
7C054	OEM
7C054/18	OEM
7C055	OEM
7C055/18	OEM
7C056	OEM
7C056/18	OEM
7C057	OEM
7C057/18	OEM
7C1	0142
7C1(ZENER)	0077
7C1/18	OEM
7C1/52	OEM
7C2	0142
7C2/18	OEM
7C2/52	OEM
7C3	0142
7C4	OEM
7C13	0595
7C14	OEM
7C14/18	OEM
7C15	OEM
7C15/18	OEM
7C16	OEM
7C16/18	OEM
7C17	OEM
7C17/18	OEM
7C24	OEM
7C24/18	OEM
7C25	OEM
7C25/18	OEM
7C26	OEM
7C26/18	OEM
7C27	OEM
7C27/18	OEM
7CM8	OEM
7D	0015
7D1	0142
7D2	0142
7D3	0142
7D4	OEM
7D8	OEM
7D10	OEM
7D12	OEM
7D13	0595
7D15	OEM
7D33	0142
7D34	0142
7D75A	OEM
7D150A	OEM
7D210	0071
7D210A	0071
7D300A	OEM
7E1	0142
7E2	0142
7E3	0142
7E4	OEM
7E13	0595
7E17	OEM
7E34A	OEM
7E45	OEM
7E90	OEM
7E115	OEM
7EM11	OEM
7E.5	OEM
7F13	0595
7G1	0142
7G2	0142
7G3	0142
7G4	0142
7G13	0595
7G33	OEM
7G34	OEM
7GM7	OEM
7GM10	OEM
7H75AE	OEM
7H150AE	OEM
7H300AE	OEM
7H450A	OEM
7JUS3.78A	OEM
7JUS3.78B	OEM
7JUS4.31A	OEM
7JUS4.31B	OEM
7JUS4.55A	OEM
7JUS4.55B	OEM
7JUS5.11A	OEM
7JUS5.11B	OEM
7JUS5.33A	OEM
7JUS5.33B	OEM
7K705M	0196
7L	OEM
7L6-0105	0178
7L6-0444-1	4293
7L6-0495-14	0015
7L6-0531-1	4293
7L6-0531-2	4293
7L6-0531-19	4095
7L6-0592-1	4293
7L6-0592-1(NPN)	OEM
7L6-0592-1(PNP)	OEM
7MA60	0015
7OR2S	0145
7R0/18	OEM
7R1/18	OEM
7R1/52	OEM
7R2/4	OEM
7R2/18	OEM
7R2/66	OEM
7R2B	OEM
7R2S	OEM
7R3/4	OEM
7R3/66	OEM
7R3B	OEM
7R3S	OEM
7R4/3	OEM
7R4/3C	OEM
7R4/59	OEM
7R4C	OEM
7R4S	OEM
7R5/3	OEM
7R5/3D	OEM
7R5/5	OEM
7R5/61	OEM
7R5C	OEM
7R5S	OEM
7R6/3	OEM
7R6/3D	OEM
7R6/5	OEM
7R6/61	OEM
7R6C	OEM
7R6S	OEM
7R7/3	OEM
7R7/3D	OEM
7R7/5	OEM
7R7/61	OEM
7R7CD	OEM
7R7S	OEM
7R8/3	OEM
7R8/3D	OEM
7R8/5	OEM
7R8/63	OEM
7R8CD	OEM
7R8S	OEM
7R52	OEM
7RE52	OEM
7RO/18	OEM
7S133DC	OEM
7SA15	0050
7SB33E	0211
7TB1	2449
7TB1-CE00	OEM
7TB1-CH00	OEM
7TB1-EE00	OEM
7TB1-EH00	OEM
7TB1-GE00	OEM
7TB1-JE00	OEM
7TB2	0240
7TB2-CE00	OEM
7TB2-CH00	OEM
7TB2-EE00	OEM
7TB2-EH00	OEM
7TB2-GE00	OEM
7TB2-GH00	OEM
7TB2-JE00	OEM
7TB3	2659
7TB3-CE00	OEM
7TB3-CH00	OEM
7TB3-EE00	OEM
7TB3-EH00	OEM
7TB3-GE00	OEM
7TB3-JE00	OEM
7TB4	0747
7TB4-CE00	OEM
7TB4-CH00	OEM
7TB4-EE00	OEM
7TB4-EH00	OEM
7TB4-GE00	OEM
7TB4-GH00	OEM
7TB4-JE00	OEM
7TB5	3291
7TB5-CE00	OEM
7TB5-CH00	OEM
7TB5-EE00	OEM
7TB5-EH00	OEM
7TB5-GE00	OEM
7TB5-GH00	OEM
7TB5-JE00	OEM
7TB6	0687
7TB6-CE00	OEM
7TB6-CH00	OEM
7TB6-EE00	OEM
7TB6-EH00	OEM
7TB6-GE00	OEM
7TB6-GH00	OEM
7TB6-JE00	OEM
7TB7	OEM
7TB7-CE00	OEM
7TB7-CH00	OEM
7TB7-EE00	OEM
7TB7-EH00	OEM
7TB7-GE00	OEM
7TB7-GH00	OEM
7TB8	0766
7TB8-CE00	OEM
7TB8-CH00	OEM
7TB8-EE00	OEM
7TB8-EH00	OEM
7TB8-GE00	OEM
7TB8-GH00	OEM
7TB8-JE00	OEM
7TCRA	OEM
7TCRB	OEM
7TCRC	OEM
7TCRD	OEM
7TCRE	OEM
7TCRF	OEM
7TCRG	OEM
7TCRH	OEM
7TCRI	OEM
7TCRJ	OEM
7TCRK	OEM
7TCRL	OEM
7UV10	OEM
7V10T	OEM
7V204	OEM
7V205	OEM
7V224	OEM
7V242	OEM
7V254	OEM
7V272	OEM
7V273	OEM
7V312	OEM
7VM705M	0196
7W412	OEM
7W412V	OEM
7XF	OEM
7Z3	OEM
7.5B	0077
7.5B1	0062
7.5B2	0077
7.5JB2	0077
8-0001300	0071
8-0001400	0071
8-0024-1	0144
8-0024-2	0144
8-0024-3	0198
8-0060	0280
8-0062	0211
8-00243	0016
8-005202	0016
8-0050100	0016
8-0050300	0208
8-0050400	0050
8-0050500	0050
8-0050600	0208
8-0050700	0160
8-0051500	0016
8-0051600	0126
8-0052102	0016
8-0052302	0079
8-0052402	0103
8-0052600	0016
8-0052700	0126
8-0052800	0208
8-0053001	0016
8-0053300	0086
8-0053400	0016
8-0053600	0016
8-0053702	1514
8-0104900	0050
8-0105200	0050
8-0105300	0050
8-0205400	0004
8-0205600	0004
8-0222631U	0004
8-0236400	0004
8-0236430	0004
8-0243900	0004
8-0318250	0016
8-0337390	0016
8-0338030	0144
8-0338040	0144
8-0339430	0144
8-0339440	0144
8-0383840	0144
8-0383930	0144
8-0383940	0016
8-0389910	0016
8-0389930	0016
8-0414120	0103
8-0414130	0103
8-0421980	0016
8-1	0178
8-1(BENDIX)	0178
8-4(BENDIX)	0016
8-9V	0244
8-22	0015
8-25	0015
8-38	0015
8-179-918-77	0017
8-543-640-00	OEM
8-619-030-007	0004
8-619-030-008	0004
8-619-030-009	0004
8-619-030-011	0143
8-619-030-012	0015
8-619-030-014	0004
8-619-030-015	0222
8-619-030-016	0004
8-619-030-017	0004
8-639-001-095	0015
8-697-020-567	0004
8-697-020-568	0004
8-697-020-569	0004
8-697-020-570	0016
8-697-020-571	0143
8-709-801-71	0102
8-710-222-21	0015
8-710-815-55	0133
8-712-540-00	OEM
8-712-600-00	0143
8-713-300-57	0039
8-717-105-28	OEM
8-719-000-04	2203
8-719-000-05	0323
8-719-000-08	OEM
8-719-000-24	0058
8-719-000-28	0407
8-719-018-47	OEM
8-719-022-97	OEM
8-719-026-11	0143
8-719-027-43	OEM
8-719-028-72	OEM
8-719-031-68	OEM
8-719-031-80	OEM
8-719-032-13	OEM
8-719-100-06	1981
8-719-100-23	0275
8-719-100-30	0041
8-719-100-35	0253
8-719-100-37	0471
8-719-100-38	0471
8-719-100-43	0077
8-719-100-44	0077
8-719-100-55	0057
8-719-100-64	0052
8-719-100-65	0052
8-719-100-68	0053
8-719-100-72	0686
8-719-100-76	0375
8-719-100-80	0671
8-719-101-03	0166
8-719-101-04	0166
8-719-101-08	OEM
8-719-101-14	0010
8-719-101-38	0376
8-719-101-39	0376
8-719-101-40	0036
8-719-101-58	0253
8-719-101-67	0077
8-719-102-47	0760
8-719-102-57	0376
8-719-102-58	0376
8-719-102-60	0036
8-719-102-61	0036
8-719-102-64	0140
8-719-102-67	0041
8-719-102-68	0041
8-719-102-69	0041
8-719-102-70	0253
8-719-102-71	0253
8-719-102-72	0253
8-719-102-73	0471
8-719-102-74	0471
8-719-102-76	0062
8-719-102-88	0057
8-719-102-89	0064
8-719-102-90	0064
8-719-102-91	0064
8-719-102-99	0053
8-719-103-06	0053
8-719-103-09	0686
8-719-103-11	0686
8-719-103-16	0420
8-719-105-51	OEM
8-719-105-82	OEM
8-719-105-91	OEM
8-719-106-07	0010
8-719-106-08	OEM
8-719-106-16	OEM
8-719-106-34	OEM
8-719-106-43	OEM
8-719-106-63	OEM
8-719-106-70	OEM
8-719-107-16	OEM
8-719-107-82	OEM
8-719-108-32	0253
8-719-109-54	OEM
8-719-109-66	OEM
8-719-109-68	0376
8-719-109-69	0376
8-719-109-71	OEM
8-719-109-72	0036
8-719-109-74	0275
8-719-109-81	0140
8-719-109-84	0041
8-719-109-85	0041
8-719-109-86	0589
8-719-109-88	0253
8-719-109-89	0253
8-719-109-90	0253
8-719-109-92	0471
8-719-109-93	0471
8-719-109-96	0295
8-719-109-97	0062
8-719-109-98	0062
8-719-110-06	OEM
8-719-110-09	0165
8-719-110-12	OEM
8-719-110-13	0057
8-719-110-14	OEM
8-719-110-16	0064
8-719-110-17	0064
8-719-110-18	0064
8-719-110-26	0064
8-719-110-31	0052
8-719-110-32	OEM
8-719-110-33	0052
8-719-110-34	0053
8-719-110-35	OEM
8-719-110-36	0053
8-719-110-48	0210
8-719-110-49	0375
8-719-110-52	0671
8-719-110-53	0671
8-719-110-61	0494
8-719-110-66	0714
8-719-110-72	0195
8-719-110-76	OEM
8-719-110-78	0166
8-719-110-90	OEM
8-719-111-07	0181
8-719-112-01	0052
8-719-112-07	0052
8-719-112-24	0052
8-719-112-25	0052
8-719-113-07	0053
8-719-113-21	0053
8-719-113-25	0053
8-719-113-63	0683
8-719-114-34	OEM
8-719-119-97	OEM
8-719-120-53	OEM
8-719-121-98	0917
8-719-122-00	0120
8-719-122-03	0120
8-719-122-07	0705
8-719-123-25	OEM
8-719-124-07	0494
8-719-124-08	0402
8-719-127-07	0760
8-719-127-25	0455
8-719-130-07	0292
8-719-130-25	0195
8-719-133-07	0299
8-719-136-07	0188
8-719-139-07	0036
8-719-143-07	0275
8-719-151-07	0041
8-719-151-77	0041
8-719-156-07	0253
8-719-156-23	0253
8-719-156-25	0253
8-719-156-27	0253
8-719-156-73	OEM
8-719-158-39	OEM
8-719-160-24	0162
8-719-160-44	0057
8-719-162-07	0471
8-719-168-07	0062
8-719-168-08	OEM
8-719-168-77	0062
8-719-175-07	0077
8-719-175-26	0077
8-719-182-07	0165
8-719-182-25	0165
8-719-190-00	0494
8-719-191-81	2939
8-719-193-03	OEM
8-719-200-01	0015
8-719-200-02	0015
8-719-200-2	0015
8-719-200-10	0015
8-719-205-10	0015
8-719-210-10	0071
8-719-300-07	0446
8-719-300-09	1236
8-719-300-33	0283
8-719-300-38	0023
8-719-300-45	0023
8-719-300-59	1236
8-719-300-65	0017
8-719-300-70	0102
8-719-300-76	0023
8-719-300-91	0609
8-719-301-14	0348
8-719-301-18	0071
8-719-301-45	2542
8-719-301-64	0102
8-719-302-00	0023
8-719-302-06	0023
8-719-302-43	0017
8-719-302-44	0023
8-719-303-21	0023
8-719-304-63	0071
8-719-305-07	1045
8-719-305-15	0158
8-719-311-23	OEM
8-719-311-31	0023
8-719-311-72	1045
8-719-311-87	OEM
8-719-311-89	OEM
8-719-312-10	0031
8-719-312-61	0023
8-719-312-62	0023
8-719-312-71	OEM
8-719-312-72	OEM
8-719-320-11	0102
8-719-320-31	0102
8-719-331-10	0023
8-719-400-18	0907
8-719-404-46	OEM
8-719-422-21	0143
8-719-500-01	0241
8-719-500-04	4581
8-719-500-26	OEM
8-719-500-34	0199
8-719-500-41	1236
8-719-500-67	1236
8-719-500-69	OEM
8-719-501-34	1015
8-719-502-20	1414
8-719-503-06	0737
8-719-503-40	0087
8-719-504-10	0323
8-719-510-02	OEM
8-719-510-09	OEM
8-719-510-26	1090
8-719-510-48	OEM
8-719-510-53	OEM
8-719-510-63	OEM
8-719-510-64	OEM
8-719-511-10	0277
8-719-511-20	0289
8-719-511-40	0296
8-719-512-20	2234
8-719-521-10	0277
8-719-549-41	0017
8-719-713-93	0720
8-719-768-71	0720
8-719-800-34	0199
8-719-800-43	1291
8-719-800-69	OEM
8-719-800-83	0315
8-719-801-02	OEM
8-719-801-35	OEM
8-719-801-70	0102
8-719-801-71	0102
8-719-802-30	0133
8-719-803-06	OEM
8-719-810-20	OEM
8-719-812-41	3019
8-719-812-43	0841
8-719-815-55	0133
8-719-815-80	0124
8-719-815-85	0124
8-719-820-13	OEM
8-719-900-02	0015
8-719-900-26	1217
8-719-900-63	0023
8-719-900-93	0023
8-719-900-95	0023
8-719-901-02	0080
8-719-901-03	0023
8-719-901-09	0023
8-719-901-13	0015
8-719-901-19	0023
8-719-901-21	0120
8-719-901-24	0095
8-719-901-31	OEM
8-719-901-39	0102
8-719-901-58	0031
8-719-901-83	0133
8-719-901-92	0023
8-719-901-93	0023
8-719-901-94	0023
8-719-901-95	0023
8-719-901-96	OEM
8-719-901-98	0315
8-719-902-85	0031
8-719-903-06	3604
8-719-903-09	0023
8-719-903-29	0023
8-719-906-15	0102
8-719-906-24	0402
8-719-907-30	0031
8-719-907-40	0023
8-719-907-50	OEM
8-719-908-03	0087
8-719-908-06	OEM
8-719-911-06	3173
8-719-911-19	0124
8-719-911-53	0965
8-719-911-54	0540
8-719-911-55	0540
8-719-912-54	0087
8-719-918-77	0017
8-719-920-04	0023
8-719-920-62	OEM
8-719-921-53	0031
8-719-923-64	3688
8-719-923-76	0124
8-719-924-06	0023
8-719-924-73	0133
8-719-925-06	0023
8-719-926-06	0023
8-719-927-74	1166
8-719-928-04	0031
8-719-928-08	0102
8-719-930-11	0317
8-719-930-12	0137
8-719-931-05	0162
8-719-931-06	0301
8-719-931-07	0649
8-719-931-08	0244
8-719-931-10	0170
8-719-931-13	0365
8-719-933-13	0031
8-719-933-73	OEM
8-719-934-00	OEM
8-719-935-08	0077
8-719-935-86	OEM
8-719-936-12	1005
8-719-936-82	OEM
8-719-936-83	0087
8-719-936-84	OEM
8-719-936-85	0023
8-719-936-96	1166
8-719-937-08	0165
8-719-937-10	0064
8-719-938-36	0023
8-719-938-40	0609
8-719-938-68	OEM
8-719-939-02	OEM
8-719-941-13	0087
8-719-941-17	OEM
8-719-941-74	1090
8-719-945-80	0348
8-719-945-84	OEM
8-719-945-91	0039
8-719-946-33	OEM
8-719-948-59	0546
8-719-950-10	OEM
8-719-951-28	OEM
8-719-961-03	0023
8-719-971-20	3887
8-719-974-81	0628
8-719-975-20	OEM
8-719-975-21	OEM
8-719-976-64	OEM
8-719-979-85	0031
8-719-980-78	OEM
8-719-981-00	0031
8-719-981-01	0023
8-719-982-04	2542
8-719-987-06	OEM
8-719-988-55	OEM
8-719-988-57	2088
8-719-989-22	OEM
8-719-991-18	OEM
8-720-177-54	2030
8-721-323-00	0004
8-722-381-20	3717
8-722-383	3717
8-722-923-00	0622
8-723-302-00	1758
8-723-303	3342
8-723-303-03	1758
8-723-303-20	1758
8-723-304-00	1758
8-723-650	0038
8-724-034-00	0155
8-724-375-01	0155
8-724-733-30	0233
8-725-412-00	0264
8-725-413-00	0313
8-725-800-00	0224
8-725-923-00	0224
8-726-357-10	0286
8-726-368-10	0155
8-726-388-00	0076
8-727-225-00	OEM
8-727-632-00	0382
8-727-786-00	0006
8-727-786-01	0037
8-729-016-15	OEM
8-729-016-32	OEM
8-729-019-01	OEM
8-729-019-51	OEM
8-729-100-09	2916
8-729-100-66	0724
8-729-101-13	0316
8-729-101-31	0316
8-729-103-41	0532
8-729-103-42	0532
8-729-103-43	0532
8-729-104-33	1284
8-729-107-26	0236

DEVICE TYPE	REPL CODE
8-729-107-78	OEM
8-729-107-84	OEM
8-729-108-53	0249
8-729-109-53	0836
8-729-113-32	0532
8-729-115-10	1758
8-729-115-30	1758
8-729-116-42	3225
8-729-117-43	2030
8-729-117-54	3625
8-729-117-63	0191
8-729-118-76	2660
8-729-119-76	3625
8-729-119-78	0249
8-729-119-79	4264
8-729-119-80	0276
8-729-120-28	0724
8-729-122-01	0525
8-729-122-03	0525
8-729-122-12	0525
8-729-133-40	2268
8-729-133-53	0728
8-729-137-32	0728
8-729-138-72	0371
8-729-138-82	0371
8-729-140-50	0261
8-729-140-75	0667
8-729-140-96	2030
8-729-140-97	0532
8-729-140-98	0018
8-729-141-48	OEM
8-729-141-89	0236
8-729-157-11	1978
8-729-167-42	1066
8-729-168-82	0276
8-729-169-01	0563
8-729-169-02	OEM
8-729-173-37	0006
8-729-177-22	0460
8-729-177-32	0018
8-729-177-33	0018
8-729-177-41	2030
8-729-177-42	2030
8-729-177-43	2030
8-729-178-54	0249
8-729-178-56	0249
8-729-188-23	0161
8-729-191-81	0482
8-729-194-58	0076
8-729-198-52	3703
8-729-200-02	0015
8-729-200-17	0342
8-729-200-33	0042
8-729-201-32	1522
8-729-201-62	2185
8-729-201-78	3125
8-729-201-96	0286
8-729-201-98	1541
8-729-202-03	0236
8-729-202-53	1541
8-729-202-58	0556
8-729-203-80	0556
8-729-204-82	0013
8-729-204-83	0013
8-729-207-35	2987
8-729-208-39	OEM
8-729-208-72	OEM
8-729-209-03	0261
8-729-212-02	0866
8-729-212-12	0261
8-729-213-01	0261
8-729-213-02	0261
8-729-213-11	0261
8-729-213-12	0261
8-729-216-22	1742
8-729-220-93	OEM
8-729-230-49	0724
8-729-231-60	3125
8-729-231-95	2129
8-729-238-32	1562
8-729-245-83	0076
8-729-245-84	0076
8-729-255-12	0261
8-729-265-52	0018
8-729-266-93	0155
8-729-271-02	0155
8-729-271-32	2332
8-729-286-90	0055
8-729-287-00	0055
8-729-287-10	0055
8-729-288-02	0461
8-729-300-90	3360
8-729-301-11	3476
8-729-301-46	0261
8-729-301-56	3476
8-729-304-00	1758
8-729-304-50	3360
8-729-304-92	0525
8-729-305-01	3360
8-729-305-42	1280
8-729-306-92	0563
8-729-307-62	0423
8-729-307-82	0392
8-729-308-92	2561
8-729-309-06	0530
8-729-309-08	0530
8-729-309-36	0648
8-729-311-42	0270
8-729-313-42	0236
8-729-313-82	0392
8-729-316-12	0042
8-729-316-16	0042
8-729-317-12	1308
8-729-320-61	1500
8-729-322-78	0285
8-729-323-82	0392
8-729-326-11	0276
8-729-326-82	1911
8-729-331-53	0423
8-729-336-11	1648
8-729-341-34	0065
8-729-341-93	2292
8-729-345-42	0220
8-729-348-47	0693
8-729-364-12	0286
8-729-372-31	1085
8-729-372-52	0155
8-729-374-02	0477
8-729-375-01	0155
8-729-377-12	1369
8-729-378-82	0018
8-729-378-84	0018
8-729-378-91	2561
8-729-378-92	2561
8-729-378-93	2561
8-729-382-64	0423
8-729-383-03	0194
8-729-386-12	1648
8-729-390-06	0055
8-729-395-70	0055
8-729-395-73	0055
8-729-398-09	0055
8-729-400-81	0060
8-729-400-82	0060
8-729-403-27	OEM
8-729-420-00	1652
8-729-422-27	0724
8-729-422-36	OEM
8-729-423-35	2954
8-729-423-37	2954
8-729-423-44	0013
8-729-424-00	1062
8-729-432-44	OEM
8-729-447-53	0004
8-729-468-43	0532
8-729-600-08	1758
8-729-600-12	OEM
8-729-600-21	1752
8-729-600-27	2075
8-729-600-28	2075
8-729-609-06	0076
8-729-612-77	0037
8-729-617-77	0037
8-729-660-37	0003
8-729-663-47	0076
8-729-665-47	0155
8-729-671-13	0368
8-729-671-14	0368
8-729-671-15	0368
8-729-800-35	1541
8-729-800-87	1541
8-729-801-22	0286
8-729-801-45	OEM
8-729-801-71	1541
8-729-801-76	1541
8-729-802-50	1541
8-729-803-04	2209
8-729-803-81	0261
8-729-806-84	0127
8-729-809-29	1165
8-729-809-41	2129
8-729-810-49	2129
8-729-820-24	OEM
8-729-820-50	0693
8-729-821-87	2129
8-729-822-65	3494
8-729-831-33	0042
8-729-880-82	0006
8-729-883-91	0076
8-729-883-92	0151
8-729-900-36	0887
8-729-900-52	OEM
8-729-900-61	4172
8-729-900-65	4130
8-729-900-89	0898
8-729-901-01	3478
8-729-901-02	OEM
8-729-901-04	1892
8-729-901-07	OEM
8-729-901-81	0724
8-729-902-07	OEM
8-729-902-10	3014
8-729-902-11	1140
8-729-903-10	OEM
8-729-903-30	OEM
8-729-903-80	5060
8-729-905-52	0009
8-729-906-24	5307
8-729-906-38	0276
8-729-906-39	0276
8-729-906-45	OEM
8-729-907-46	OEM
8-729-920-71	1742
8-729-920-74	0724
8-729-920-90	0728
8-729-920-92	OEM
8-729-922-68	0013
8-729-922-69	2954
8-729-924-82	OEM
8-729-924-83	1165
8-729-924-86	OEM
8-729-924-90	OEM
8-729-925-79	OEM
8-729-926-73	OEM
8-729-927-12	1500
8-729-927-22	OEM
8-729-931-43	0728
8-729-965-22	2471
8-729-967-32	0009
8-729-978-62	0037
8-729-982-22	4104
8-729-993-72	2484
8-729-37501	0155
8-739-100-00	0371
8-739-178-54	OEM
8-741-013-00	OEM
8-741-013-10	OEM
8-741-100-62	OEM
8-741-101-60	OEM
8-741-103-00	OEM
8-741-103-80	OEM
8-741-103-90	OEM
8-741-104-10	OEM
8-741-104-30	OEM
8-741-104-40	OEM
8-741-104-41	OEM
8-741-104-50	OEM
8-741-104-60	OEM
8-741-104-70	OEM
8-741-108-30	OEM
8-741-108-40	OEM
8-741-108-60	OEM
8-741-108-70	OEM
8-741-108-80	OEM
8-741-109-00	OEM
8-741-109-10	OEM
8-741-109-20	OEM
8-741-109-30	OEM
8-741-109-40	OEM
8-741-109-50	OEM
8-741-115-30	OEM
8-741-131-70	OEM
8-741-132-30	OEM
8-741-135-70	OEM
8-741-138-70	OEM
8-741-139-80	OEM
8-741-148-33	OEM
8-741-156-80	OEM
8-741-618-11	OEM
8-741-637-11	OEM
8-742-230	OEM
8-743-350-00	OEM
8-743-420-00	OEM
8-743-580-00	OEM
8-743-600-00	OEM
8-743-610-00	OEM
8-743-640-00	OEM
8-743-690-00	OEM
8-743-760-00	OEM
8-743-780-00	OEM
8-743-790-00	OEM
8-743-800-00	OEM
8-743-830-00	OEM
8-743-840-00	OEM
8-743-901-00	OEM
8-743-903-00	OEM
8-743-904-00	OEM
8-743-912-00	OEM
8-743-921-00	OEM
8-743-945-10	OEM
8-743-946-10	OEM
8-743-947-20	OEM
8-743-948-10	OEM
8-743-949-00	OEM
8-743-950-10	OEM
8-743-951-20	OEM
8-743-952-00	OEM
8-743-953-00	OEM
8-743-954-00	OEM
8-743-955-00	OEM
8-743-956-10	OEM
8-743-958-10	OEM
8-743-959-10	OEM
8-743-960-00	OEM
8-743-969-10	OEM
8-743-970-10	OEM
8-743-971-00	OEM
8-743-972-00	OEM
8-743-973-00	OEM
8-743-976-00	OEM
8-743-977-00	OEM
8-743-978-00	OEM
8-743-980-00	OEM
8-743-981-00	OEM
8-743-982-10	OEM
8-743-982-20	OEM
8-743-983-00	OEM
8-743-984-00	OEM
8-743-988-00	OEM
8-743-999-00	OEM
8-743-999-10	OEM
8-749-20-81	OEM
8-749-900-15	5276
8-749-900-36	OEM
8-749-900-80	OEM
8-749-901-35	3913
8-749-901-43	OEM
8-749-920-57	OEM
8-749-920-61	OEM
8-749-920-62	OEM
8-749-920-65	OEM
8-749-920-81	OEM
8-749-921-10	OEM
8-749-921-99	OEM
8-749-922-13	OEM
8-749-930-35	3445
8-749-933-35	OEM
8-749-938-30	OEM
8-749-939-69	OEM
8-749-939-70	OEM
8-749-939-71	OEM
8-749-953-14	OEM
8-749-956-10	1533
8-749-956-11	OEM
8-749-956-30	1572
8-749-958-30	1572
8-749-960-11	OEM
8-749-960-21	OEM
8-750-105-11	1634
8-750-253-20	OEM
8-750-930-00	OEM
8-750-990-00	OEM
8-751-001-00	4434
8-751-300-00	3961
8-751-310-00	4261
8-751-340-00	3707
8-751-350-00	3964
8-751-360-00	3966
8-751-370-00	4294
8-751-380-00	3968
8-751-390-00	4242
8-751-410-00	3970
8-751-430-00	3973
8-751-450-00	3975
8-751-500-00	OEM
8-751-700-10	3992
8-751-771-00	4000
8-751-771-10	4000
8-751-830-02	OEM
8-751-860-00	OEM
8-751-870-00	OEM
8-751-880-00	OEM
8-751-940-02	OEM
8-751-941-01	OEM
8-751-960-00	OEM
8-751-3300-00	3697
8-752-000-80	OEM
8-752-001-30	OEM
8-752-001-31	OEM
8-752-001-40	OEM
8-752-001-41	OEM
8-752-006-10	OEM
8-752-006-12	OEM
8-752-007-30	OEM
8-752-008-20	OEM
8-752-010-00	OEM
8-752-010-60	OEM
8-752-011-20	2870
8-752-012-52	OEM
8-752-012-91	OEM
8-752-013-90	OEM
8-752-019-20	OEM
8-752-019-30	OEM
8-752-030-26	3083
8-752-030-69	OEM
8-752-031-72	OEM
8-752-032-27	OEM
8-752-033-32	OEM
8-752-034-95	OEM
8-752-035-29	OEM
8-752-035-32	OEM
8-752-035-39	OEM
8-752-035-52	OEM
8-752-035-53	OEM
8-752-035-54	OEM
8-752-036-29	OEM
8-752-036-60	OEM
8-752-037-15	OEM
8-752-037-24	OEM
8-752-052-88	OEM
8-752-053-17	OEM
8-752-053-30	OEM
8-752-053-38	OEM
8-752-059-67	OEM
8-752-062-86	OEM
8-752-321-18	OEM
8-752-332-83	OEM
8-752-602-54	OEM
8-752-813-89	OEM
8-752-817-58	OEM
8-752-830-22	OEM
8-752-841-16	OEM
8-757-280-00	OEM
8-757-604-00	OEM
8-757-605-00	OEM
8-757-611-00	OEM
8-757-660-00	OEM
8-757-959-00	OEM
8-758-040-00	OEM
8-758-100-00	OEM
8-758-140-00	OEM
8-758-150-00	OEM
8-758-160-00	OEM
8-758-220-00	OEM
8-758-320-00	OEM
8-758-480-00	4166
8-758-660-00	OEM
8-758-690-00	OEM
8-758-740-00	OEM
8-758-850-00	OEM
8-758-851-00	OEM
8-758-852-00	OEM
8-759-000-49	0101
8-759-001-20	6205
8-759-006-10	OEM
8-759-011-65	OEM
8-759-013-06	0624
8-759-013-09	0334
8-759-014-40	OEM
8-759-030-99	1346
8-759-038-39	OEM
8-759-040-00	OEM
8-759-040-46	3428
8-759-040-53	0035
8-759-043-86	OEM
8-759-044-56	OEM
8-759-045-38	1063
8-759-054-08	OEM
8-759-069-14	OEM
8-759-084-09	OEM
8-759-088-00	OEM
8-759-089-13	OEM
8-759-090-21	OEM
8-759-093-26	OEM
8-759-093-28	OEM
8-759-093-29	OEM
8-759-10-31	0904
8-759-32-89	OEM
8-759-100-06	OEM
8-759-100-07	2916
8-759-100-09	2916
8-759-100-38	OEM
8-759-100-54	OEM
8-759-100-56	OEM
8-759-100-60	5209
8-759-100-75	5048
8-759-101-60	1750
8-759-101-72	OEM
8-759-101-77	OEM
8-759-101-80	4665
8-759-102-12	OEM
8-759-102-28	OEM
8-759-103-00	0553
8-759-103-68	0675
8-759-103-93	0631
8-759-104-05	1880
8-759-105-56	2623
8-759-105-57	2625
8-759-105-59	OEM
8-759-105-82	2042
8-759-106-61	3250
8-759-108-05	1292
8-759-110-17	2926
8-759-110-31	0904
8-759-111-44	OEM
8-759-111-85	4665
8-759-111-88	4796
8-759-112-06	OEM
8-759-112-22	5023
8-759-112-38	1048
8-759-112-77	4840
8-759-112-93	OEM
8-759-113-53	4635
8-759-113-58	0553
8-759-113-63	0683
8-759-113-65	OEM
8-759-113-68	0675
8-759-113-73	2026
8-759-113-78	2042
8-759-115-17	OEM
8-759-115-18	OEM
8-759-115-56	2623
8-759-116-18	OEM
8-759-125-56	0490
8-759-131-11	2104
8-759-131-49	OEM
8-759-132-40	0625
8-759-133-90	0176
8-759-135-80	0770
8-759-140-01	0478
8-759-140-11	0215
8-759-140-29	2233
8-759-140-53	0035
8-759-140-66	0101
8-759-142-04	OEM
8-759-145-27	3254
8-759-145-58	0360
8-759-146-17	OEM
8-759-146-37	OEM
8-759-147-49	OEM
8-759-147-60	OEM
8-759-148-68	OEM
8-759-148-69	OEM
8-759-149-90	OEM
8-759-150-61	1292
8-759-153-20	OEM
8-759-153-42	OEM
8-759-154-62	OEM
8-759-155-21	OEM
8-759-157-40	1329
8-759-157-41	1329
8-759-157-52	2872
8-759-157-60	1479
8-759-164-18	OEM
8-759-170-08	1196
8-759-170-12	OEM
8-759-171-05	0624
8-759-171-12	OEM
8-759-178-05	1292
8-759-178-12	1828
8-759-192-70	OEM
8-759-200-93	OEM
8-759-204-96	OEM
8-759-206-12	OEM
8-759-206-28	3905
8-759-208-67	OEM
8-759-208-92	OEM
8-759-208-94	OEM
8-759-220-00	OEM
8-759-220-04	OEM
8-759-231-56	OEM
8-759-234-63	OEM
8-759-240-01	0478
8-759-240-11	0215
8-759-240-12	0498
8-759-240-13	0413
8-759-240-16	0101
8-759-240-27	1949
8-759-240-29	2233
8-759-240-30	0500
8-759-240-49	0001
8-759-240-51	0366
8-759-240-52	0024
8-759-240-53	0035
8-759-240-66	0101
8-759-240-69	0119
8-759-240-71	0129
8-759-240-81	0332
8-759-243-19	OEM
8-759-245-20	2675
8-759-245-28	3198
8-759-250-26	OEM
8-759-271-20	1018
8-759-271-33	OEM
8-759-272-15	0087
8-759-273-20	3804
8-759-275-18	OEM
8-759-276-07	0912
8-759-276-14	5205
8-759-276-30	3772
8-759-276-58	5172
8-759-312-21	4808
8-759-312-44	4854
8-759-400-01	4743
8-759-400-88	3437
8-759-402-35	4129
8-759-403-42	OEM
8-759-403-44	OEM
8-759-420-37	OEM
8-759-420-48	OEM
8-759-424	4038
8-759-424-00	0167
8-759-425	4038
8-759-425-00	0167
8-759-500-31	OEM
8-759-510-89	OEM
8-759-510-90	OEM
8-759-512-50	OEM
8-759-512-85	OEM
8-759-512-86	OEM
8-759-517-74	OEM
8-759-518-39	OEM
8-759-600-05	0767
8-759-600-08	OEM
8-759-600-43	4165
8-759-600-50	OEM
8-759-600-95	3948
8-759-601-12	OEM
8-759-601-77	OEM
8-759-601-86	OEM
8-759-601-87	OEM
8-759-601-95	3948
8-759-602-54	OEM
8-759-603-94	OEM
8-759-604-29	0624
8-759-604-37	OEM
8-759-604-83	OEM
8-759-605-14	OEM
8-759-605-15	OEM
8-759-605-39	OEM
8-759-605-55	0767
8-759-608-07	OEM
8-759-608-11	OEM
8-759-608-43	4165
8-759-608-64	OEM
8-759-608-65	OEM
8-759-608-82	OEM
8-759-608-83	OEM
8-759-610-95	3948
8-759-618-30	OEM
8-759-618-48	3956
8-759-619-03	0810
8-759-630-78	OEM
8-759-631-22	OEM
8-759-632-89	OEM
8-759-633-67	OEM
8-759-634-45	OEM
8-759-634-46	OEM
8-759-634-50	OEM
8-759-634-54	OEM
8-759-634-62	OEM
8-759-634-69	OEM
8-759-635-34	OEM
8-759-636-30	OEM
8-759-636-31	OEM
8-759-636-45	OEM
8-759-645-19	0969
8-759-651-34	1634
8-759-651-35	0878
8-759-651-42	1634
8-759-660-00	OEM
8-759-679-32	OEM
8-759-684-78	4112
8-759-701-59	3332
8-759-701-79	6352
8-759-701-92	OEM
8-759-710-04	OEM
8-759-710-40	OEM
8-759-710-68	OEM
8-759-710-69	OEM
8-759-711-23	OEM
8-759-720-74	OEM
8-759-748-69	OEM
8-759-800-12	4384
8-759-800-15	3333
8-759-800-18	OEM
8-759-800-65	2666
8-759-800-81	1028
8-759-801-25	1951
8-759-801-98	0732
8-759-802-10	2063
8-759-803-24	OEM
8-759-803-25	OEM
8-759-803-29	3097
8-759-805-37	OEM
8-759-812-01	0579
8-759-814-05	4830
8-759-820-63	OEM
8-759-820-92	2346
8-759-820-93	4094
8-759-822-02	2346
8-759-822-03	OEM
8-759-826-01	3103
8-759-832-10	2063
8-759-833-01	1215
8-759-833-61	1260
8-759-841-25	1951
8-759-841-40	0362
8-759-878-02	1590
8-759-878-03	1590
8-759-879-03	1590
8-759-900-04	1595
8-759-900-09	1642
8-759-900-10	1662
8-759-900-45	OEM
8-759-900-46	OEM
8-759-900-70	OEM
8-759-900-74	0243
8-759-900-86	0291
8-759-900-93	1888
8-759-901-03	OEM
8-759-901-23	0979
8-759-901-38	0427
8-759-902-22	OEM
8-759-903-86	3063
8-759-904-69	0119
8-759-904-94	4422
8-759-905-15	OEM
8-759-905-16	OEM
8-759-905-17	OEM
8-759-905-19	OEM
8-759-905-35	OEM
8-759-905-45	OEM
8-759-905-86	OEM
8-759-905-94	OEM
8-759-906-09	OEM
8-759-907-16	OEM
8-759-907-22	OEM
8-759-907-76	OEM
8-759-907-79	OEM
8-759-907-86	OEM
8-759-908-13	OEM
8-759-908-24	OEM
8-759-908-25	OEM
8-759-908-41	OEM
8-759-908-68	OEM
8-759-908-75	OEM
8-759-909-15	OEM
8-759-909-49	OEM
8-759-909-50	OEM
8-759-910-21	OEM
8-759-910-39	OEM
8-759-911-66	OEM
8-759-913-11	OEM
8-759-914-09	OEM
8-759-914-66	OEM
8-759-915-31	OEM
8-759-918-29	OEM
8-759-918-75	OEM
8-759-919-05	OEM
8-759-920-81	OEM
8-759-922-85	OEM
8-759-923-41	OEM
8-759-924-12	1812
8-759-925-74	OEM
8-759-927-51	OEM
8-759-928-10	OEM
8-759-929-62	6021
8-759-932-33	0101
8-759-932-80	2339
8-759-933-00	OEM
8-759-934-19	OEM
8-759-937-59	1824

If replacement code is OEM, contact original manufacturer for replacement.

DEVICE TYPE	REPL CODE
8-759-940-88	OEM
8-759-945-58	0360
8-759-947-18	OEM
8-759-947-49	OEM
8-759-951-02	2105
8-759-953-87	2295
8-759-955-50	OEM
8-759-965-60	OEM
8-759-966-81	OEM
8-759-971-56	OEM
8-759-972-43	OEM
8-759-978-66	OEM
8-759-980-10	OEM
8-759-980-43	4375
8-759-980-58	4385
8-759-980-59	OEM
8-759-981-40	OEM
8-759-981-41	OEM
8-759-981-69	2881
8-759-981-80	OEM
8-759-982-10	OEM
8-759-982-13	0334
8-759-982-21	1292
8-759-982-25	OEM
8-759-982-26	1828
8-759-982-31	0624
8-759-982-34	OEM
8-759-982-37	OEM
8-759-983-38	OEM
8-759-983-44	OEM
8-759-983-69	OEM
8-759-984-06	OEM
8-759-985-20	2675
8-759-985-27	OEM
8-759-985-28	OEM
8-759-986-39	OEM
8-759-987-16	0631
8-759-987-47	OEM
8-759-987-58	OEM
8-759-987-89	OEM
8-759-989-67	OEM
8-759-994-51	0976
8-759-996-43	OEM
8-759-997-64	OEM
8-760-335-10	0018
8-760-343-10	0018
8-760-413	1978
8-760-413-10	1978
8-760-514-10	0532
8-760-523-10	0532
8-761-622-00	2460
8-762-020-00	0342
8-763-113-00	1946
8-763-213-00	1051
8-764-803-00	2495
8-765-170-01	0338
8-765-212-20	0150
8-765-222-20	1946
8-765-401-00	0144
8-765-422-00	0325
8-765-423-00	0325
8-765-424-00	0325
8-765-440-00	OEM
8-765-500-00	0264
8-765-510-00	0681
8-765-620-00	OEM
8-769-132-00	1758
8-769-194-00	1758
8-769-200-00	0325
8-769-200-10	0325
8-769-200-30	0325
8-769-200-40	0325
8-795-145-58	0360
8-795-270-70	0878
8-795-600-95	3948
8-851-340-00	3707
8-902-0706-071	0016
8-905-013-752	0015
8-905-013-759	0015
8-905-013-760	0015
8-905-014-017	0016
8-905-198-001	0015
8-905-198-004	0015
8-905-198-005	0015
8-905-198-007	0015
8-905-198-008	0015
8-905-198-010	0015
8-905-198-034	0015
8-905-305-004	0123
8-905-305-007	0143
8-905-305-020	0143
8-905-305-023	0123
8-905-305-055	0143
8-905-305-318	0143
8-905-305-327	0143
8-905-305-330	0143
8-905-305-336	0143
8-905-305-338	0143
8-905-305-339	0143
8-905-305-342	0143
8-905-305-348	0143
8-905-305-400	0015
8-905-305-405	0143
8-905-305-555	0143
8-905-305-561	0143
8-905-305-580	0143
8-905-305-635	0143
8-905-313-007	0123
8-905-313-008	0123
8-905-313-010	0143
8-905-313-011	0143
8-905-313-018	0123
8-905-313-100	0143
8-905-313-101	0143
8-905-313-120	0143
8-905-405-002	0015
8-905-405-026	0015
8-905-405-069	0015
8-905-405-077	0143
8-905-405-098	0133
8-905-405-105	0071
8-905-405-134	0015
8-905-405-146	0015
8-905-405-160	0071
8-905-405-170	0071
8-905-405-206	0015
8-905-405-838	0143
8-905-406-020	0133
8-905-413-092	0015
8-905-421-109	0157
8-905-421-118	0012
8-905-421-128	0002
8-905-421-215	0451
8-905-421-228	0012
8-905-421-234	0137
8-905-421-239	0002
8-905-421-300	0133
8-905-421-315	0331
8-905-421-319	0137
8-905-421-715	0002
8-905-605-016	0004
8-905-605-030	0004
8-905-605-032	0004
8-905-605-050	0004
8-905-605-051	0004
8-905-605-075	0004
8-905-605-090	0004
8-905-605-091	0004
8-905-605-105	0208
8-905-605-108	0208
8-905-605-109	0208
8-905-605-110	0208
8-905-605-111	0208
8-905-605-112	0208
8-905-605-113	0208
8-905-605-120	0164
8-905-605-123	0164
8-905-605-124	0164
8-905-605-125	0164
8-905-605-126	0164
8-905-605-127	0164
8-905-605-128	0164
8-905-605-129	0164
8-905-605-230	0004
8-905-605-232	0004
8-905-605-234	0004
8-905-605-250	0164
8-905-605-255	0164
8-905-605-260	0164
8-905-605-264	0164
8-905-605-266	0164
8-905-605-268	0164
8-905-605-269	0164
8-905-605-292	0004
8-905-605-305	0004
8-905-605-320	0050
8-905-605-365	0208
8-905-605-384	0208
8-905-605-390	0208
8-905-605-607	0222
8-905-605-624	0160
8-905-605-635	0160
8-905-605-636	0160
8-905-605-637	0160
8-905-605-644	0007
8-905-605-650	0222
8-905-605-775	0975
8-905-605-908	0975
8-905-606-001	0050
8-905-606-003	0050
8-905-606-007	0050
8-905-606-008	0050
8-905-606-010	0050
8-905-606-016	0050
8-905-606-051	0050
8-905-606-075	0050
8-905-606-077	0050
8-905-606-090	0050
8-905-606-105	0050
8-905-606-106	0050
8-905-606-120	0050
8-905-606-142	0050
8-905-606-152	0050
8-905-606-153	0050
8-905-606-154	0050
8-905-606-155	0050
8-905-606-158	0050
8-905-606-165	0050
8-905-606-168	0050
8-905-606-180	0050
8-905-606-211	0050
8-905-606-225	0050
8-905-606-241	0050
8-905-606-255	0050
8-905-606-256	0050
8-905-606-349	0050
8-905-606-350	0050
8-905-606-351	0050
8-905-606-352	0050
8-905-606-360	0050
8-905-606-375	0050
8-905-606-390	0050
8-905-606-391	0050
8-905-606-392	0050
8-905-606-405	0050
8-905-606-419	0050
8-905-606-420	0050
8-905-606-423	0050
8-905-606-720	0160
8-905-606-750	0004
8-905-606-800	0004
8-905-606-815	0004
8-905-606-817	0004
8-905-606-885	0004
8-905-613-010	0004
8-905-613-015	0208
8-905-613-062	0208
8-905-613-070	0164
8-905-613-071	0164
8-905-613-131	0164
8-905-613-132	0164
8-905-613-133	0164
8-905-613-150	2866
8-905-613-160	0164
8-905-613-210	0160
8-905-613-215	0160
8-905-613-232	0975
8-905-613-240	0222
8-905-613-241	0222
8-905-613-242	0222
8-905-613-245	0222
8-905-613-250	0160
8-905-613-265	0222
8-905-613-266	0222
8-905-613-277	0222
8-905-613-282	0222
8-905-613-283	0222
8-905-613-284	0222
8-905-613-295	0975
8-905-613-555	0222
8-905-613-640	0004
8-905-613-710	0004
8-905-613-955	0004
8-905-615-156	0004
8-905-705-112	0016
8-905-705-403	0016
8-905-705-405	0016
8-905-705-410	0086
8-905-706-010	0285
8-905-706-044	0007
8-905-706-055	0007
8-905-706-060	0007
8-905-706-067	0285
8-905-706-068	0285
8-905-706-070	0007
8-905-706-071	0007
8-905-706-075	0007
8-905-706-080	0007
8-905-706-101	0007
8-905-706-104	0016
8-905-706-110	0007
8-905-706-112	0007
8-905-706-201	0016
8-905-706-202	0016
8-905-706-203	0016
8-905-706-206	0016
8-905-706-208	0016
8-905-706-211	0016
8-905-706-215	0016
8-905-706-235	0016
8-905-706-236	0016
8-905-706-238	0016
8-905-706-239	0016
8-905-706-240	0016
8-905-706-242	0016
8-905-706-244	0016
8-905-706-245	0016
8-905-706-246	0016
8-905-706-247	0037
8-905-706-250	0016
8-905-706-251	0037
8-905-706-253	0037
8-905-706-254	0037
8-905-706-255	0037
8-905-706-256	0037
8-905-706-257	0016
8-905-706-260	0016
8-905-706-263	0016
8-905-706-280	0037
8-905-706-286	0037
8-905-706-287	0037
8-905-706-288	0037
8-905-706-289	0037
8-905-706-290	0037
8-905-706-336	0016
8-905-706-545	0126
8-905-706-555	0103
8-905-706-556	0103
8-905-706-557	0103
8-905-706-606	0016
8-905-706-730	0007
8-905-706-790	0050
8-905-706-801	0161
8-905-706-901	0325
8-905-707-254	0016
8-905-707-265	0016
8-905-707-313	0016
8-905-713-058	0037
8-905-713-101	0103
8-905-713-110	0161
8-905-713-556	0103
8-905-713-810	0126
8-917-300-70	0023
8-2409501	0016
8-2410300	0126
8-81250108	0111
8-81250109	0016
8-B	OEM
8-C	OEM
8A	OEM
8A01	0143
8A1ABG0	OEM
8A1ABG1	OEM
8A1HA1	OEM
8A1HS0	OEM
8A1HS1	OEM
8A1PA1	OEM
8A1PG0	OEM
8A1PG1	OEM
8A1PS0	OEM
8A1PS1	OEM
8A1QBG0	OEM
8A1QBG1	OEM
8A2ABG0	OEM
8A2ABG2	OEM
8A2HA2	OEM
8A2HS0	OEM
8A2HS1	OEM
8A2PA2	OEM
8A2PG0	OEM
8A2PG2	OEM
8A2PS0	OEM
8A2PS1	OEM
8A2QBG0	OEM
8A2QBG2	OEM
8A3ABG0	OEM
8A3ABG3	OEM
8A3HA3	OEM
8A3HS0	OEM
8A3HS1	OEM
8A3PA3	OEM
8A3PG0	OEM
8A3PG3	OEM
8A3PS0	OEM
8A3PS1	OEM
8A3QBG0	OEM
8A3QBG3	OEM
8A4ABG0	OEM
8A4ABG4	OEM
8A4HA4	OEM
8A4HS0	OEM
8A4HS1	OEM
8A4PA4	OEM
8A4PG0	OEM
8A4PG4	OEM
8A4PS0	OEM
8A4PS1	OEM
8A4QBG0	OEM
8A4QBG4	OEM
8A5ABG0	OEM
8A5ABG5	OEM
8A5HA5	OEM
8A5HS0	OEM
8A5HS1	OEM
8A5PA5	OEM
8A5PG0	OEM
8A5PG5	OEM
8A5PS0	OEM
8A5PS1	OEM
8A5QBG0	OEM
8A5QBG5	OEM
8A6ABG0	OEM
8A6ABG6	OEM
8A6HS0	OEM
8A6HS1	OEM
8A6PG0	OEM
8A6PG6	OEM
8A6PS0	OEM
8A6PS1	OEM
8A6QBG0	OEM
8A6QBG6	OEM
8A7ABG0	OEM
8A7ABG7	OEM
8A7HS0	OEM
8A7HS1	OEM
8A7PG0	OEM
8A7PG7	OEM
8A7PS0	OEM
8A7PS1	OEM
8A7QBG0	OEM
8A7QBG7	OEM
8A8ABG0	OEM
8A8ABG8	OEM
8A8HS0	OEM
8A8HS1	OEM
8A8PG0	OEM
8A8PG8	OEM
8A8PS0	OEM
8A8PS1	OEM
8A8QBG0	OEM
8A8QBG8	OEM
8A9ABG0	OEM
8A9ABG9	OEM
8A9HS0	OEM
8A9HS1	OEM
8A9PG0	OEM
8A9PG9	OEM
8A9PS0	OEM
8A9PS1	OEM
8A9QBG0	OEM
8A9QBG9	OEM
8A10ABG0	OEM
8A10ABG10	OEM
8A10HS0	OEM
8A10HS1	OEM
8A10PG0	OEM
8A10PG10	OEM
8A10PS0	OEM
8A10PS1	OEM
8A10QBG0	OEM
8A10QBG10	OEM
8A11ABG0	OEM
8A11ABG11	OEM
8A11HS0	OEM
8A11HS1	OEM
8A11PG0	OEM
8A11PG11	OEM
8A11PS0	OEM
8A11PS1	OEM
8A11QBG0	OEM
8A11QBG11	OEM
8A12ABG0	OEM
8A12ABG12	OEM
8A12HS0	OEM
8A12HS1	OEM
8A12PG0	OEM
8A12PG12	OEM
8A12PS0	OEM
8A12PS1	OEM
8A12QBG0	OEM
8A12QBG12	OEM
8A13ABG0	OEM
8A13ABG13	OEM
8A13PG0	OEM
8A13PG13	OEM
8A13QBG0	OEM
8A13QBG13	OEM
8A14ABG0	OEM
8A14ABG14	OEM
8A14PG0	OEM
8A14PG14	OEM
8A14QBG0	OEM
8A14QBG14	OEM
8A15ABG0	OEM
8A15ABG15	OEM
8A15PG0	OEM
8A15PG15	OEM
8A15QBG0	OEM
8A15QBG15	OEM
8A16ABG0	OEM
8A16ABG16	OEM
8A16PG0	OEM
8A16PG16	OEM
8A16QBG0	OEM
8A16QBG16	OEM
8A17ABG0	OEM
8A17ABG17	OEM
8A17PG0	OEM
8A17PG17	OEM
8A17QBG0	OEM
8A17QBG17	OEM
8A18ABG0	OEM
8A18ABG18	OEM
8A18PG0	OEM
8A18PG18	OEM
8A18QBG0	OEM
8A18QBG18	OEM
8A19ABG0	OEM
8A19ABG19	OEM
8A19PG0	OEM
8A19PG19	OEM
8A19QBG0	OEM
8A19QBG19	OEM
8A20ABG0	OEM
8A20ABG20	OEM
8A20PG0	OEM
8A20PG20	OEM
8A20QBG0	OEM
8A20QBG20	OEM
8A1002	0400
8A1003	0400
8A10521	0222
8A10625	0222
8A11083	0222
8A11667	0015
8A11721	0222
8A12359	0222
8A12789	0016
8A12991	0085
8A13164	0222
8A13718	0164
8AF05NLH	OEM
8AF05NLV	OEM
8AF05NPP	OEM
8AF1NLH	OEM
8AF1NLV	OEM
8AF1NPP	OEM
8AF2NLH	OEM
8AF2NLV	OEM
8AF2NPP	OEM
8AF4NLH	OEM
8AF4NLV	OEM
8AF4NPP	OEM
8AN10	1577
8AN20	1577
8AN30	0109
8AN40	3751
8AN50	OEM
8AN60	OEM
8AN80	OEM
8AN100	3011
8AN120	OEM
8AN140	OEM
8AN160	OEM
8B	OEM
8BT30ZK	OEM
8C1	OEM
8C2	OEM
8C4	OEM
8C6	OEM
8C8	OEM
8C12	OEM
8C14	OEM
8C16	OEM
8C91	0012
8C200	0037
8C201	0037
8C202	0037
8C203	0037
8C204	0037
8C205	0037
8C206	0037
8C207	0037
8C327-25	OEM
8C430	0037
8C430K	0037
8C440	0037
8C440K	0037
8C443	0037
8C443K	0037
8C445	0037
8C445K	0037
8C449	0037
8C449K	0037
8C450	0037
8C460	0037
8C460K	0037
8C463	0037
8C463K	0037
8C465	0037
8C465K	0037
8C466	0037
8C466K	0037
8C467	0037
8C467K	0037
8C468	0037
8C468K	0037
8C469	0037
8C469K	0037
8C470	0037
8C470K	0037
8C700	0037
8C700A	0037
8C700B	0037
8C702	0037
8C702A	0037
8C702B	0037
8C704	0037
8C740	0037
8C740G	0037
8C740M	0037
8C742	0037
8C742G	0037
8C742M	0037
8C7400	0037
8C7420	0037
8CG15	0015
8CG15RE	0015
8D	1283
8D4	0015
8D6	0015
8D8	0071
8D10	0071
8E	0050
8E(AUTOMATIC)	0136
8E0	OEM
8E1	OEM
8E7H	OEM
8E15	OEM
8E30A	OEM
8E70	OEM
8E110	OEM
8F	0050
8F(AUTOMATIC)	0136
8F5H	OEM
8F5H1	OEM
8F6H	OEM
8F6H1	OEM
8FR	OEM
8G6H	OEM
8G6H1	OEM
8G7	0071
8GA	0071
8H303	0160
8J5091	OEM
8JUS4.32A	OEM
8JUS4.32B	OEM
8JUS5.20A	OEM
8JUS5.20B	OEM
8JUS5.84A	OEM
8JUS5.84B	OEM
8JUS6.10A	OEM
8JUS6.10B	OEM
8JV30	OEM
8JV60	OEM
8JV100	OEM
8JV120	OEM
8JV150	OEM
8JV160	OEM
8JV180	OEM
8KBS	OEM
8L	0050
8L201	0160
8L201B	0160
8L201C	0160
8L201R	0160
8L201V	0160
8L301V	0085
8L404	0160
8M-26102	1030
8MT05ZYK	OEM
8N	OEM
8NA	0196
8P	0050
8P1M	2089
8P-2	0004
8P-2-70	0004
8P-2-70-12	0004
8P-2-70-12-7	0004
8P-2A	0004
8P-2A0	0004
8P-2A0R	0004
8P-2A1	0004
8P-2A2	0004
8P-2A3	0004
8P-2A3P	0004
8P-2A4	0004
8P-2A4-7	0004
8P-2A4-7B	0004
8P-2A5	0004
8P-2A5L	0004
8P-2A6	0004
8P-2A6-2	0004
8P-2A7	0004
8P-2A7-1	0004
8P-2A8	0004
8P-2A9	0004
8P-2A9G	0004
8P-2A19	0211
8P-2A21	0004
8P-2A82	0004
8P-2AO	0164
8P-2AOR	0164
8P2M	0505
8P70BLU	0042
8P10	OEM
8P-20	0004
8P-21	0004
8P-22	0004
8P-23	0004
8P-24	0004
8P-25	0004
8P-26	0004
8P-27	0004
8P-28	0004
8P-29	0004
8P40	0085
8P50S	0222
8P73BLU	0042
8P111	0325
8P202	0222
8P345	0042
8P-404	0160
8P404	0160
8P404B	0160
8P404F	0085
8P404M	0160
8P404M-1	0160
8P404N	0160
8P404ORN	0085
8P-404R	0160
8P404R	0160
8P404T	0222
8P404V	0160
8P415C	0160
8P416C	0160
8P-505	0222
8P505	0222
8P880	0440
8P880B	0440
8P9253	0149
8PC60	0160
8PS60	0160
8Q-3-01	0111
8Q-3-02	0150
8Q-3-04	0208
8Q-3-10	0111
8Q-3-11	0037
8Q-3-12	0079
8Q-3-13	0086
8Q-3-14	0037
8Q-3-23	2763
8Q-7-01	0015
8Q-7-02	0015
8Q-7-03	0015
8RCM5	2519
8RCM10	OEM
8RCM20	0745
8RCM30	OEM
8RCM40	OEM
8RCM50	OEM
8RCM60	OEM
8RCU5	OEM

If replacement code is OEM, contact original manufacturer for replacement.

Original Device Types

DEVICE TYPE	REPL CODE
8RCU10	OEM
8RCU20	OEM
8RCU30	OEM
8RCU40	OEM
8RCU50	OEM
8RCU60	OEM
8RCV5	2519
8RCV20	0745
8S	OEM
8S191A04	0015
8SB050	OEM
8SB100	0323
8SB200	1414
8SB400	0473
8SB600	0446
8SM1	OEM
8SM3	OEM
8SM6	OEM
8SM15	OEM
8SM25	OEM
8SM40	OEM
8SM55	OEM
8SM70	OEM
8SM100	OEM
8SM150	OEM
8SM200	OEM
8T04A	2300
8T04HA	2300
8T14A	2300
8T14HA	2300
8T24A	2300
8T24HA	2300
8T28	0581
8T32F	OEM
8T33	OEM
8T34A	2300
8T34HA	2300
8T35F	OEM
8T36F	OEM
8T44A	2300
8T44HA	2300
8T54A	OEM
8T54HA	OEM
8T64A	OEM
8T64HA	OEM
8T95	4647
8T3404F	OEM
8T3404N	OEM
8TW1	OEM
8TW2	OEM
8TW3	OEM
8TW4	OEM
8TW5	OEM
8TW6	OEM
8U4	0795
8U8	0015
8U12	0072
8V228	OEM
8V257	OEM
8V274	OEM
8V301	OEM
8VT01	OEM
8VT02	OEM
8W413	OEM
8W413V	OEM
8.2B1	0077
8.2M	OEM
8.2Z	0165
8.75VZENER	0012
9-719-000-04	2203
9-900-285-01	OEM
9-900-615-01	1386
9-900-616-01	0477
9-900-617-01	1414
9-900-618-01	0023
9-900-619-01	0686
9-900-843-01	OEM
9-900-910-01	OEM
9-900-929-01	OEM
9-900-930-01	OEM
9-900-931-01	0023
9-901-134-01	2919
9-901-371-01	1329
9-901-372-01	OEM
9-901-373-01	OEM
9-901-374-01	OEM
9-901-375-01	OEM
9-901-376-01	OEM
9-901-385-01	OEM
9-901-386-01	2919
9-901-387-01	OEM
9-901-403-01	OEM
9-901-496-01	OEM
9-901-499-01	0546
9-901-500-01	OEM
9-901-504-01	0462
9-901-505-01	1028
9-901-506-01	OEM
9-901-507-01	2346
9-901-518-01	OEM
9-901-519-01	OEM
9-901-548-01	0546
9-901-549-01	OEM
9-901-550-01	3604
9-901-551-01	2400
9-901-557-01	OEM
9-904-341-01	OEM
9-904-343-01	OEM
9-904-344-01	OEM
9-904-366-01	0826
9-904-457-01	OEM
9-904-517-01	OEM
9-904-519-01	OEM
9-904-528-01	OEM
9-904-554-01	5090
9-904-656-01	OEM
9-905-606-001	0050
9-906-318-01	OEM
9-906-466-01	OEM
9-906-605-01	OEM
9-907-498-01	OEM
9-908-789-01	OEM
9-908-790-01	OEM
9-908-794-01	OEM
9-908-796-01	OEM
9-908-797-01	OEM
9-982-051-00	OEM
9-982-071-00	OEM
9-982-071-02	OEM
9-5108	0050
9-5110	0050
9-5111	0050
9-5112	0600
9-5113	0600
9-5114	0600
9-5116	0050
9-5117	0050
9-5118	0050
9-5119	0050
9-5120	0050
9-5120A	0280
9-5121	0050
9-5122	0050
9-5123	0050
9-5124	0050
9-5125	0144
9-5126	0144
9-5127	0144
9-5128	0144
9-5129	0144
9-5130	0144
9-5131	0144
9-5201	0004
9-5202	0208
9-5203	0004
9-5204	0004
9-5208	0004
9-5209	0004
9-5212	0004
9-5213	0004
9-5214	0004
9-5216	0016
9-5217	0004
9-5218	0004
9-5220	0086
9-5221	0079
9-5222-1	0004
9-5222-2	0208
9-5223	0144
9-5224-1	0004
9-5224-2	0208
9-5225	0016
9-5226-003	0126
9-5226-004	0086
9-5226-1	0126
9-5226-2	0016
9-5226-3	0126
9-5226-4	0086
9-5227	0016
9-5250	0160
9-5251	0160
9-5252	0142
9-5252-1	0142
9-5252-2	0142
9-5252-3	0142
9-5252-4	0142
9-5257	0265
9-5296	0016
9-9101	0136
9-9102	0136
9-9103	0136
9-9104	0004
9-9105	0050
9-9106	0050
9-9107	0050
9-9108	0050
9-9109-1	0016
9-9109-2	0016
9-9120	0050
9-9121	0050
9-9201	0004
9-9202	0004
9-9203	0004
9-51141400	0160
9-511410100	0004
9-511410200	0004
9-511410900	0004
9-511413500	0004
9-511511500	0071
9A1	1030
9A1(3%)	OEM
9A2	0165
9A8-1A64	0050
9A1000	OEM
9ACW	OEM
9AM10	OEM
9B7	OEM
9BM10	OEM
9C3	0057
9CR12U02G	OEM
9CR12U04G	OEM
9CR12U06G	OEM
9CR12U08G	OEM
9CR12U10G	OEM
9CR12U12G	OEM
9CR20U08G	OEM
9CR20U10G	OEM
9CR20U11G	OEM
9CR30U02G	OEM
9CR30U04G	OEM
9CR30U06G	OEM
9CR30U08G	OEM
9D02M1	OEM
9D05M1	OEM
9D10K1	OEM
9D11	0133
9D12	0143
9D13	0015
9D14	0139
9D15	0012
9D16	0143
9D141003-12	0012
9DI	0133
9DI1	0133
9DI2	0143
9DI3	0015
9DI4	0139
9DI5	0012
9DT1100310	0143
9E13	OEM
9E26A	OEM
9E45	OEM
9E85	OEM
9E150	OEM
9GR2	0016
9H00DC	0682
9H00PC	0682
9H01DC	5338
9H01PC	5338
9H04DC	1907
9H04PC	1907
9H05DC	3251
9H05PC	3251
9H08DC	5355
9H08PC	5355
9H10DC	0685
9H10PC	0685
9H11DC	2398
9H11PC	2398
9H20DC	3715
9H20PC	3715
9H21DC	4856
9H21PC	4856
9H22	4596
9H22DC	4596
9H22PC	4596
9H30DC	5383
9H30PC	5383
9H40DC	0559
9H40PC	0559
9H50DC	1792
9H50PC	1792
9H51DC	1944
9H51PC	1944
9H52DC	2020
9H52PC	2020
9H53DC	2101
9H53PC	2101
9H54DC	2172
9H54PC	2172
9H55DC	3158
9H55PC	3158
9H60DC	5410
9H60PC	5410
9H61DC	2663
9H61PC	2663
9H62DC	2731
9H62PC	2731
9H71DC	3264
9H71PC	3264
9H72DC	3313
9H72PC	3313
9H73DC	2463
9H73PC	2463
9H74DC	2492
9H74PC	2492
9H76DC	5302
9H76PC	5302
9H78DC	5418
9H78PC	5418
9H101DC	5521
9H101PC	5521
9H102DC	5523
9H102PC	5523
9H103DC	2969
9H103PC	2969
9H106DC	5253
9H106PC	5253
9H108DC	0180
9H108PC	0180
9H450A	OEM
9H600A	OEM
9HM7	OEM
9JUS4.86A	OEM
9JUS4.86B	OEM
9JUS5.54A	OEM
9JUS5.54B	OEM
9JUS5.85A	OEM
9JUS5.85B	OEM
9JUS6.57A	OEM
9JUS6.57B	OEM
9JUS6.86A	OEM
9JUS6.86B	OEM
9K	OEM
9L	OEM
9L-4	0004
9L-4-70	0004
9L-4-70-12	0004
9L-4-70-12-7	0004
9L-4A	0004
9L-4A0	0004
9L-4A0R	0004
9L-4A1	0004
9L-4A2	0004
9L-4A3	0004
9L-4A3P	0004
9L-4A4	0004
9L-4A4-7	0004
9L-4A4-7B	0004
9L-4A5	0004
9L-4A5L	0004
9L-4A6	0004
9L-4A6-1	0004
9L-4A7	0004
9L-4A7-1	0004
9L-4A8	0004
9L-4A9	0004
9L-4A9G	0004
9L-4A19	0004
9L-4A21	0004
9L-4A82	0004
9L-40	0004
9L-41	0004
9L-42	0004
9L-43	0004
9L-44	0004
9L-45	0004
9L-46	0004
9L-47	0004
9L-48	0004
9L-49	0004
9LR2	0196
9LR2-1	0196
9LR2-1S	OEM
9LR2-2	0484
9LR2-3	0196
9LR2-4	1080
9LR2-24	0196
9LR2-385	OEM
9LR2-S	0196
9LR2-S1	0196
9LR21	0196
9LS17TDM	OEM
9LS95DC	OEM
9LS95DM	OEM
9LS95FC	OEM
9LS95FM	OEM
9LS95PC	OEM
9LS164DC	OEM
9LS164DM	OEM
9LS164FC	OEM
9LS164FM	OEM
9LS164PC	4497
9LS170DC	OEM
9LS170DM	OEM
9LS170FC	OEM
9LS170FM	OEM
9LS170PC	OEM
9LS174DC	0260
9LS174DM	0260
9LS174FM	OEM
9LS174PC	0260
9LS175DC	1672
9LS175FC	OEM
9LS175FM	OEM
9LS175PC	1672
9LS194DC	1304
9LS194DM	1304
9LS194FC	OEM
9LS194FM	OEM
9LS194PC	1304
9LS195DC	OEM
9LS195DM	OEM
9LS195FC	OEM
9LS195FM	OEM
9LS195PC	1315
9LS295DC	OEM
9LS295DM	OEM
9LS295FC	OEM
9LS295FM	OEM
9LS295PC	OEM
9LS670DC	OEM
9LS670DM	OEM
9LS670FC	OEM
9LS670FM	OEM
9LS670PC	OEM
9M	OEM
9M12	OEM
9M13	OEM
9M31	OEM
9M40	OEM
9M61	OEM
9M62A	OEM
9M64A	OEM
9M65	OEM
9M65A	OEM
9M66	OEM
9M72	OEM
9M73	OEM
9M80	OEM
9M80A	OEM
9M90	OEM
9M301	OEM
9M302	OEM
9M303	OEM
9M501	OEM
9M502	OEM
9M503	OEM
9M601	OEM
9M602	OEM
9M603	OEM
9M611	OEM
9M612	OEM
9M613	OEM
9M751	OEM
9M752	OEM
9M753	OEM
9MF2A100	OEM
9MF2A200	OEM
9MF2A400	OEM
9MF2B100	OEM
9MF2B200	OEM
9MF2B400	OEM
9MF2B600	OEM
9MF2B800	OEM
9MF10B100	OEM
9MF10B100R	OEM
9MF10B200	OEM
9MF10B200R	OEM
9MF10B400	OEM
9MF10B400R	OEM
9MF10B600	OEM
9MF10B600R	OEM
9MF10B800	OEM
9MF10B800R	OEM
9MT19ZK	OEM
9N00	0232
9N00DC	0232
9N00PC	0232
9N01DC	0268
9N01PC	0268
9N02	0314
9N02DC	0314
9N02F1	OEM
9N02PC	0314
9N03DC	0335
9N03PC	0335
9N04	0361
9N04DC	0361
9N04PC	0361
9N05DC	0385
9N05M1	OEM
9N05PC	0385
9N06	1206
9N06DC	1206
9N06PC	1206
9N07DC	1339
9N07PC	1339
9N08	0467
9N08DC	0467
9N08PC	0467
9N09DC	0492
9N09PC	0492
9N10	0512
9N10DC	0512
9N10K1	OEM
9N10PC	0512
9N11DC	0527
9N11PC	0527
9N12DC	2242
9N12PC	2242
9N13	1441
9N13DC	1441
9N13PC	1441
9N14DC	2243
9N14PC	2243
9N16DC	1349
9N16PC	1349
9N17DC	1352
9N17PC	1352
9N20	0697
9N20DC	0697
9N20PC	0697
9N21DC	1357
9N21PC	1357
9N23DC	3466
9N23PC	3466
9N25DC	3477
9N25PC	3477
9N26DC	0803
9N26PC	0803
9N27	0818
9N27DC	0818
9N27PC	0818
9N30DC	0873
9N30PC	0873
9N32	0899
9N32DC	0899
9N32PC	0899
9N37DC	3520
9N37PC	3520
9N38DC	0996
9N39DC	5836
9N39PC	5836
9N40DC	1024
9N40PC	1024
9N50DC	0743
9N50PC	0743
9N51	1168
9N51DC	1168
9N51PC	1168
9N53DC	1185
9N53PC	1185
9N54DC	1202
9N54PC	1202
9N60DC	1275
9N60PC	1275
9N70DC	1404
9N70PC	1404
9N72DC	1426
9N72PC	1426
9N73	1172
9N73DC	1172
9N73PC	1172
9N74	1313
9N74DC	1313
9N74PC	1313
9N76	1158
9N76DC	1158
9N76PC	1158
9N86	1368
9N86DC	1368
9N86PC	1368
9N107DC	0942
9N107PC	0942
9N122DC	1139
9N122PC	1139
9N123	1157
9N123DC	1157
9N123PC	1157
9N132DC	1271
9N132PC	1271
90C69-1	0015
9P10K1	OEM
9PO2F1-A	OEM
9RE1	0015
9S00DC	0704
9S00PC	0704
9S03DC	2217
9S03PC	2217
9S04DC	2263
9S04PC	2263
9S05DC	2321
9S05PC	2321
9S011	0004
9S037	0016
9S10DC	2445
9S10PC	2445
9S11DC	2447
9S11PC	2447
9S15DC	2451
9S15PC	2451
9S20DC	1017
9S20PC	1017
9S22DC	2461
9S22PC	2461
9S40DC	2475
9S40PC	2475
9S64DC	2496
9S64PC	2496
9S65DC	2497
9S65PC	2497
9S74DC	2504
9S74PC	2504
9S112DC	1617
9S112PC	1617
9S113DC	1623
9S113PC	1623
9S114DC	1629
9S114PC	1629
9S133DC	1819
9S133PC	1819
9S134DC	1827
9S134PC	1827
9S140DC	1886
9S140PC	1886
9T10T	OEM
9T20T	OEM
9T40T	OEM
9T60T	OEM
9T80T	OEM
9T100T	OEM
9TR1	0127
9TR2	0368
9TR3	0368
9TR4	1400
9TR5	1908
9TR6	1400
9TR7	0155
9TR8	0236
9TR9	2276
9TR10	0016
9TR11	1400
9TR11001-01	0127
9TR21001-02	0016
9TR31001-03	0016
9TR61001-07	0086
9TR91001-09	0103
9TRZ1001-02	0016
9V54	OEM
9V58	OEM
9VT01	OEM
9VT02	OEM
9X64	OEM
9XT2107-501	OEM
9.0B2	OEM
9.1B	0057
9.1B2	6397
9.1B3	0057
9.1BL	OEM
9.1M	OEM
9.1N3	0057
9.1Z	0057
9.S037	0079
10	0016
10-0007	0936
10-0009	0390
10-001	3815
10-002	0076
10-003	0076
10-004	1211
10-004#	1540
10-004(IC)	1540
10-004IC	1540
10-005	4042
10-005(IC)	4042
10-006	0890
10-007	0936
10-008	0155
10-009	0936
10-010	0133
10-010#	1196
10-010(IC)	1196
10-010IC	1196
10-012	0023
10-013	0012
10-015	0030
10-016	0253
10-024	2939
10-080009	0144
10-080010	0016
10-085001	0143
10-085004	0123
10-085005	0133
10-085006	0015
10-085009	0015
10-085010	0015
10-085013	0030
10-085014	0143
10-085018	0143
10-085025	0133
10-085026	0015
10-085027	0162
10-085030	0015
10-0800010	0016
10-1	0007
10-2	0007
10-2SA49	0136
10-2SB54	0004
10-2SB56	0004
10-2SC80	0144
10-2SC94	0144
10-2SC380	0127
10-7	0015
10-12	0015
10-42	0015
10-102005	0071
10-I3002-004	0103
10A	0050
10A1/18	OEM
10A1/52	OEM
10A2/18	OEM
10A2/52	OEM
10A5/18	OEM
10A5/52	OEM
10A12	OEM
10A12/18	OEM
10A13	OEM
10A13/18	OEM
10A14	OEM
10A14/18	OEM
10A15	OEM
10A15/18	OEM
10A16	OEM
10A16/18	OEM
10A22	OEM
10A22/18	OEM
10A23	OEM
10A23/18	OEM
10A24	OEM
10A24/18	OEM
10A25	OEM
10A25/18	OEM
10A26	OEM
10A26/18	OEM
10A50	OEM
10A52	OEM
10A52/18	OEM
10A53	OEM
10A53/18	OEM
10A54	OEM
10A54/18	OEM
10A55	OEM
10A55/18	OEM
10A56	OEM
10A56/18	OEM
10A100	OEM
10A200	OEM
10A400	OEM
10A590B	0015
10A600	OEM
10A800	OEM
10AG2	0015

If replacement code is OEM, contact original manufacturer for replacement.

DEVICE TYPE	REPL CODE
10AG4	0015
10AG6	0015
10AG8	0071
10AG10	0071
10AL2	0015
10AL6	0015
10AM10	OEM
10AS	0015
10AT2	0015
10AT4	0015
10AT6	0015
10AT8	0071
10AT10	0071
10B	0050
10B05T	OEM
10B1	0015
10B1(ZENER)	0057
10B1(ZENER),,	0064
10B1/18	OEM
10B1/52	OEM
10B-2	0015
10B2	0064
10B2(ZENER)	OEM
10B2/18	OEM
10B2/52	OEM
10B-2-B1W	0015
10B2-B1W	0277
10B-2-N1W	0015
10B2FR	OEM
10B3	0015
10B3(1W-ZENER)	0170
10B3(ZENER)	0064
10B-4	0474
10B4	0015
10B-4-C4	0474
10B4B41	OEM
10B4FR	OEM
10B5	0015
10B5/18	OEM
10B5/52	OEM
10B6	0015
10B6FR	OEM
10B8	0071
10B10	0087
10B10T	OEM
10B12	OEM
10B12/18	OEM
10B13	OEM
10B13/18	OEM
10B14	OEM
10B14/18	OEM
10B15	OEM
10B15/18	OEM
10B16	OEM
10B16/18	OEM
10B20T	OEM
10B22	OEM
10B22/18	OEM
10B23	OEM
10B23/18	OEM
10B24	OEM
10B24/18	OEM
10B25	OEM
10B25/18	OEM
10B26	OEM
10B26/18	OEM
10B40T	OEM
10B52	OEM
10B52/18	OEM
10B53	OEM
10B53/18	OEM
10B54	OEM
10B54/18	OEM
10B55	OEM
10B55/18	OEM
10B56	OEM
10B56/18	OEM
10B62	0170
10B80T	OEM
10B551	0144
10B551-2	0007
10B551-3	0007
10B553	0144
10B553-2	0007
10B553-2,3	OEM
10B553-3	0007
10B555	0144
10B555-2	0144
10B555-2,3	OEM
10B555-3	0144
10B556	0144
10B556-2	0144
10B556-2,3	OEM
10B556-3	0144
10B701	OEM
10B705	OEM
10B1051	0144
10B1055	0144
10BG2C11	OEM
10BG2Z11	OEM
10BL2C41	0909
10BT16ZK	OEM
10B-Y	0015
10BYZ1400	OEM
10BYZ1600	OEM
10BZX79C	0064
10C	0015
10C05	0015
10C05/18	OEM

DEVICE TYPE	REPL CODE
10C05/52	OEM
10C054	OEM
10C054/18	OEM
10C055	OEM
10C055/18	OEM
10C056	OEM
10C056/18	OEM
10C057	OEM
10C057/18	OEM
10C1	0015
10C1R10	OEM
10C1R20	OEM
10C1R30	OEM
10C1S10	OEM
10C1S20	OEM
10C1S30	OEM
10C2	0015
10C2/18	OEM
10C2/52	OEM
10C2R10	OEM
10C2R11	OEM
10C2R20	OEM
10C2R21	OEM
10C2R30	OEM
10C2R31	OEM
10C2S10	OEM
10C2S11	OEM
10C2S20	OEM
10C2S21	OEM
10C2S30	OEM
10C2S31	OEM
10C2SR11	OEM
10C2SR20	OEM
10C2SR21	OEM
10C2SR30	OEM
10C2SR31	OEM
10C2SS11	OEM
10C2SS21	OEM
10C2SS30	OEM
10C2SS31	OEM
10C3	0015
10C4	0015
10C4D	0015
10C4.5	OEM
10C4.5-5.5	OEM
10C5	0015
10C5.5-6.5	OEM
10C5.6	OEM
10C6	0015
10C6.5-7.5	OEM
10C6.7	OEM
10C8	0071
10C10	0071
10C11/18	OEM
10C11/52	OEM
10C12	OEM
10C14/18	OEM
10C15	OEM
10C15/18	OEM
10C16/18	OEM
10C17	OEM
10C17/18	OEM
10C24	OEM
10C24/18	OEM
10C25	OEM
10C25/18	OEM
10C26	OEM
10C26/18	OEM
10C27	OEM
10C27/18	OEM
10C573	0144
10C573-2	0007
10C573-2,3	OEM
10C573-3	0007
10C574	0144
10C574-2	0007
10C574-2,3	OEM
10C574-3	0007
10CC11	OEM
10CH2SM	1236
10CL2C41	0909
10CS04SM	OEM
10CZ6.8	0248
10CZ6.8A	OEM
10CZ6.8B	OEM
10CZ7.5	OEM
10CZ7.5A	OEM
10CZ7.5B	OEM
10CZ8.2	OEM
10CZ8.2A	OEM
10CZ8.2B	OEM
10CZ9.1	OEM
10CZ9.1A	OEM
10CZ9.1B	OEM
10CZ10	OEM
10CZ10A	OEM
10CZ10B	OEM
10CZ11	OEM
10CZ11A	OEM
10CZ11B	OEM
10CZ12	OEM
10CZ12A	OEM
10CZ12B	OEM
10CZ13	OEM
10CZ13A	OEM
10CZ13B	OEM
10CZ15	OEM
10CZ15A	OEM
10CZ15B	OEM

DEVICE TYPE	REPL CODE
10CZ16	OEM
10CZ16A	OEM
10CZ16B	OEM
10CZ18	OEM
10CZ18A	OEM
10CZ18B	OEM
10CZ20	OEM
10CZ20A	OEM
10CZ20B	OEM
10CZ22	OEM
10CZ22A	OEM
10CZ22B	OEM
10CZ24	OEM
10CZ24A	OEM
10CZ24B	OEM
10CZ27	OEM
10CZ27A	OEM
10CZ27B	OEM
10CZ30	OEM
10CZ30A	OEM
10CZ30B	OEM
10CZ33	OEM
10CZ33A	OEM
10CZ33B	OEM
10CZ36	OEM
10CZ36A	OEM
10CZ36B	OEM
10CZ39	OEM
10CZ39A	OEM
10CZ39B	OEM
10CZ43	OEM
10CZ43A	OEM
10CZ43B	OEM
10CZ47	OEM
10CZ47A	OEM
10CZ47B	OEM
10CZ51	OEM
10CZ51A	OEM
10CZ51B	OEM
10CZ56	OEM
10CZ56A	OEM
10CZ56B	OEM
10CZ62	OEM
10CZ62A	OEM
10CZ62B	OEM
10CZ66	OEM
10CZ66A	OEM
10CZ66B	OEM
10CZ75	OEM
10CZ75A	OEM
10CZ75B	OEM
10CZ82	OEM
10CZ82A	OEM
10CZ82B	OEM
10CZ91	OEM
10CZ91A	OEM
10CZ91B	OEM
10CZ100	OEM
10CZ100A	OEM
10CZ100B	OEM
10CZ103	OEM
10CZ110	OEM
10CZ110A	OEM
10CZ110B	OEM
10CZ120	OEM
10CZ120A	OEM
10CZ120B	OEM
10CZ130A	OEM
10CZ130B	OEM
10CZ150	OEM
10CZ150A	OEM
10CZ150B	OEM
10CZ160	OEM
10CZ160A	OEM
10CZ160B	OEM
10CZ180	OEM
10CZ180A	OEM
10CZ180B	OEM
10CZ200	OEM
10CZ200A	OEM
10CZ200B	OEM
10D	0015
10D(ZENER)	0248
10D-02	0015
10D-05	0015
10D05	0015
10D-06	0015
10D06	0540
10D0.5	0015
10D-1	0015
10D1	0023
10D-2	0023
10D2	0023
10D-2B	0015
10D-2B(-4)	0015
10D-2B(4)	0015
10D-2B-4	0102
10D3	0015
10D3G	0015
10D-4	0071
10D4	0023
10D4A	0023
10D4B	0023
10D4B41	OEM
10D4E	0015
10D4L	0102
10D-5	0102
10D5	0015
10D5A	0102

DEVICE TYPE	REPL CODE
10D5B	0015
10D5D	0102
10D5E	0015
10D-6	0015
10D6	0023
10D6D	0015
10D6E	0015
10D6FD	OEM
10D7	0071
10D7E	0071
10D7F	0071
10D-7K	0071
10D8	0023
10D-10	0071
10D10	0071
10D556-2,3	OEM
10D701	OEM
10D702	OEM
10DB	0015
10DB1	0277
10DB2	0289
10DB2A	0289
10DB2P	0296
10DB4	0296
10DB4A	0296
10DB4P	0302
10DB6	0302
10DB6A	0296
10DB6A-C	0296
10DB6P	2010
10DB8	0250
10DB10	0250
10DC	0015
10DC05	0424
10DC0B	0015
10DC0H	0015
10DC0.5	0015
10DC-1	OEM
10DC1	0015
10DC1BLACK	0015
10DC-1N	0196
10DC-1R	0015
10DC1R	0071
10DC-2	0023
10DC2	0015
10DC-2B	0015
10DC-2C	0015
10DC-2F	0015
10DC2F	0015
10DC-2J	0015
10DC-4	0015
10DC4	0015
10DC-4R	0015
10DC4R	0015
10DC-4R-202	0015
10DC5	0015
10DC8	0071
10DC8R	0071
10DC11	OEM
10DCIN	0015
10DCIR	0015
10DCOB	0015
10DCOH	0015
10DCO.5	0015
10DCR	4977
10DF1	0609
10DF2	0795
10DF4	0015
10DF6	0072
10DF8	0071
10DG2C11	OEM
10DG2Z11	OEM
10DI	0071
10DK241U	OEM
10DL1	0015
10DL2	0015
10DL2C41	0909
10D-V	0015
10DX2	0015
10DZ	0015
10E-1	0015
10E1	0015
10E1LF	0015
10E1R20	OEM
10E1R40	OEM
10E1R60	OEM
10E1S20	OEM
10E1S40	OEM
10E1S60	OEM
10E-2	0015
10E2	0015
10E2-TA2B	0015
10E2N	OEM
10E2R20	OEM
10E2R21	OEM
10E2R40	OEM
10E2R41	OEM
10E2R60	OEM
10E2R61	OEM
10E2S20	OEM
10E2S21	OEM
10E2S40	OEM
10E2S41	OEM
10E2S60	OEM
10E2S61	OEM
10E4	0015
10E4A	0015
10E4B	0015
10E6	0015

DEVICE TYPE	REPL CODE
10E-7L	0015
10E8	0071
10E10	0071
10E10(FA-8)	0071
10E10-TA2B5	0071
10E-10P	0015
10E10P	0015
10E11F	0015
10E12	OEM
10E24A	OEM
10E40	OEM
10E75	OEM
10E120	OEM
10E1051	OEM
10EB1	0170
10ELS4	0023
10ELS4-TA-1	0023
10ELS6	0023
10ELS6TA1	0023
10EN2	OEM
10EZ3.3D	OEM
10EZ3.3D5	OEM
10EZ3.3D10	OEM
10EZ3.6D	OEM
10EZ3.6D5	OEM
10EZ3.6D10	OEM
10EZ3.9D	OEM
10EZ3.9D5	OEM
10EZ3.9D10	OEM
10EZ4.3D	OEM
10EZ4.3D5	OEM
10EZ4.3D10	OEM
10EZ4.7D	OEM
10EZ4.7D5	OEM
10EZ4.7D10	OEM
10EZ5.1D	OEM
10EZ5.1D5	OEM
10EZ5.1D10	OEM
10EZ5.6D	OEM
10EZ5.6D5	OEM
10EZ5.6D10	OEM
10EZ6.2D	OEM
10EZ6.2D5	OEM
10EZ6.2D10	OEM
10EZ6.8D	OEM
10EZ6.8D5	OEM
10EZ6.8D10	OEM
10EZ7.5D	OEM
10EZ7.5D5	OEM
10EZ7.5D10	OEM
10EZ8.2D	OEM
10EZ8.2D5	OEM
10EZ8.2D10	OEM
10EZ9.1D	OEM
10EZ9.1D5	OEM
10EZ9.1D10	OEM
10EZ10D	OEM
10EZ10D5	OEM
10EZ10D10	OEM
10EZ11D	OEM
10EZ11D5	OEM
10EZ11D10	OEM
10EZ12D	OEM
10EZ12D5	OEM
10EZ12D10	OEM
10EZ13D	OEM
10EZ13D5	OEM
10EZ13D10	OEM
10EZ14D	OEM
10EZ14D5	OEM
10EZ14D10	OEM
10EZ15D	OEM
10EZ15D5	OEM
10EZ15D10	OEM
10EZ16D	OEM
10EZ16D5	OEM
10EZ16D10	OEM
10EZ17D	OEM
10EZ17D5	OEM
10EZ17D10	OEM
10EZ18D	OEM
10EZ18D5	OEM
10EZ18D10	OEM
10EZ19D	OEM
10EZ19D5	OEM
10EZ19D10	OEM
10EZ20D	OEM
10EZ20D5	OEM
10EZ20D10	OEM
10EZ22D	OEM
10EZ22D5	OEM
10EZ22D10	OEM
10EZ24D	OEM
10EZ24D5	OEM
10EZ24D10	OEM
10EZ25D	OEM
10EZ25D5	OEM
10EZ25D10	OEM
10EZ27D	OEM
10EZ27D5	OEM
10EZ27D10	OEM
10EZ30D	OEM
10EZ30D5	OEM
10EZ30D10	OEM
10EZ33D	OEM
10EZ33D5	OEM
10EZ33D10	OEM
10EZ36D	OEM
10EZ36D5	OEM

DEVICE TYPE	REPL CODE
10EZ36D10	OEM
10EZ39D	OEM
10EZ39D5	OEM
10EZ39D10	OEM
10EZ43D	OEM
10EZ43D5	OEM
10EZ43D10	OEM
10EZ45D	OEM
10EZ45D5	OEM
10EZ45D10	OEM
10EZ47D	OEM
10EZ47D5	OEM
10EZ47D10	OEM
10EZ50D	OEM
10EZ50D5	OEM
10EZ50D10	OEM
10EZ51D	OEM
10EZ51D5	OEM
10EZ51D10	OEM
10EZ52D	OEM
10EZ52D5	OEM
10EZ52D10	OEM
10EZ56D	OEM
10EZ56D5	OEM
10EZ56D10	OEM
10EZ62D	OEM
10EZ62D5	OEM
10EZ62D10	OEM
10EZ68D	OEM
10EZ68D5	OEM
10EZ68D10	OEM
10EZ82D	OEM
10EZ82D5	OEM
10EZ82D10	OEM
10EZ91D	OEM
10EZ91D5	OEM
10EZ91D10	OEM
10EZ100D	OEM
10EZ100D5	OEM
10EZ100D10	OEM
10EZ105D	OEM
10EZ105D5	OEM
10EZ105D10	OEM
10EZ110D	OEM
10EZ110D5	OEM
10EZ110D10	OEM
10EZ120D	OEM
10EZ120D5	OEM
10EZ120D10	OEM
10EZ130D	OEM
10EZ130D5	OEM
10EZ130D10	OEM
10EZ140D	OEM
10EZ140D5	OEM
10EZ140D10	OEM
10EZ150D	OEM
10EZ150D5	OEM
10EZ150D10	OEM
10EZ160D	OEM
10EZ160D5	OEM
10EZ160D10	OEM
10EZ175D	OEM
10EZ175D5	OEM
10EZ175D10	OEM
10EZ180D	OEM
10EZ180D5	OEM
10EZ180D10	OEM
10EZ200D	OEM
10EZ200D5	OEM
10EZ200D10	OEM
10E.5	OEM
10F1	0015
10F1R07	OEM
10F1R15	OEM
10F1S07	OEM
10F1S15	OEM
10F2R07	OEM
10F2R15	OEM
10F2S07	OEM
10F2S15	OEM
10F2S07	OEM
10F5	3733
10F10	0871
10F15	0871
10F20	0871
10F30	0853
10F40	0853
10F50	1609
10FC11	OEM
10FCRA	1825
10FCRB	1825
10FCRC	1825
10FCRD	1825
10FCRE	1825
10FCRF	1825
10FCRG	1825
10FCRH	1825
10FCRI	OEM
10FCRJ	OEM
10FCRK	OEM
10FCRL	OEM
10FIR15	OEM
10FWJ2C11	3266
10FWJ2C41	3266
10G4	0071
10G4B41	OEM
10G1051	0144
10G1052	0144
10GA	0071

DEVICE TYPE	REPL CODE
10GC11	0853
10GD11	0853
10GG2C11	OEM
10GG2Z11	OEM
10GW2C11	OEM
10GW2C41	OEM
10GWJ2C11	3266
10GWJ2C41	3266
10H	0015
10H3	0097
10H3N	2291
10H3P	3733
10H3PN	0910
10H380RT	OEM
10H551	0144
10H551-2,3	OEM
10H553	0144
10H553-2,3	OEM
10H1051	0144
10H1053	0144
10HB05	2363
10HB10	2363
10HB20	2363
10HB40	2369
10HB60	2370
10HB80	2372
10HC11	OEM
10I10	0071
10IP005	OEM
10IP005D	OEM
10IP005N	OEM
10IP01	OEM
10IP01D	OEM
10IP01N	OEM
10IP02	OEM
10IP02D	OEM
10IP02N	OEM
10IP04	OEM
10IP04D	OEM
10IP04N	OEM
10IP06	OEM
10IP06D	OEM
10IP06N	OEM
10IP08	OEM
10IP08D	OEM
10IP08N	OEM
10IP10	OEM
10IP10D	OEM
10IP10N	OEM
10J1	1228
10J2	0015
10J2F	0015
10J3P	0090
10J4	OEM
10J4B41	OEM
10JC11	OEM
10JF1	OEM
10JF2	OEM
10JG2C11	OEM
10JG2Z11	OEM
10JH3	1560
10JUS5.04A	OEM
10JUS5.04B	OEM
10JUS6.16A	OEM
10JUS6.16B	OEM
10JUS6.50A	OEM
10JUS6.50AB	OEM
10JUS7.30A	OEM
10JUS7.30B	OEM
10JUS7.62A	OEM
10JUS7.62B	OEM
10K	0015
10K241U	1993
10KB7	OEM
10KC11	OEM
10KM7	OEM
10KZ6.8	OEM
10KZ6.8A	OEM
10KZ6.8B	OEM
10KZ7.5	OEM
10KZ7.5A	OEM
10KZ7.5B	OEM
10KZ8.2	OEM
10KZ8.2A	OEM
10KZ8.2B	OEM
10KZ9.1	OEM
10KZ9.1A	OEM
10KZ9.1B	OEM
10KZ10	OEM
10KZ10A	OEM
10KZ10B	OEM
10KZ11	OEM
10KZ11A	OEM
10KZ11B	OEM
10KZ12	OEM
10KZ12A	OEM
10KZ12B	OEM
10KZ13	OEM
10KZ13A	OEM
10KZ13B	OEM
10KZ15	OEM
10KZ15A	OEM
10KZ15B	OEM
10KZ16	OEM
10KZ16A	OEM
10KZ16B	OEM
10KZ18	OEM
10KZ18A	OEM
10KZ18B	OEM

DEVICE TYPE	REPL CODE
10KZ20	OEM
10KZ20A	OEM
10KZ20B	OEM
10KZ22	OEM
10KZ22A	OEM
10KZ22B	OEM
10KZ24	OEM
10KZ24A	OEM
10KZ24B	OEM
10KZ27	OEM
10KZ27A	OEM
10KZ27B	OEM
10KZ30	OEM
10KZ30A	OEM
10KZ30B	OEM
10KZ33	OEM
10KZ33A	OEM
10KZ33B	OEM
10KZ36	OEM
10KZ36A	OEM
10KZ36B	OEM
10KZ39	OEM
10KZ39A	OEM
10KZ39B	OEM
10KZ43	OEM
10KZ43A	OEM
10KZ43B	OEM
10KZ47	OEM
10KZ47A	OEM
10KZ47B	OEM
10KZ51	OEM
10KZ51A	OEM
10KZ51B	OEM
10KZ56	OEM
10KZ56A	OEM
10KZ56B	OEM
10KZ62	OEM
10KZ62A	OEM
10KZ62B	OEM
10KZ66	OEM
10KZ66A	OEM
10KZ66B	OEM
10KZ75	OEM
10KZ75A	OEM
10KZ75B	OEM
10KZ82	OEM
10KZ82A	OEM
10KZ82B	OEM
10KZ91	OEM
10KZ91A	OEM
10KZ91B	OEM
10KZ100	OEM
10KZ100A	OEM
10KZ100B	OEM
10KZ110	OEM
10KZ110A	OEM
10KZ110B	OEM
10KZ120	OEM
10KZ120A	OEM
10KZ120B	OEM
10KZ130	OEM
10KZ130A	OEM
10KZ130B	OEM
10KZ150	OEM
10KZ150A	OEM
10KZ150B	OEM
10KZ160	OEM
10KZ160A	OEM
10KZ160B	OEM
10KZ180	OEM
10KZ180A	OEM
10KZ180B	OEM
10KZ200	OEM
10KZ200A	OEM
10KZ200B	OEM
10L1-1.5	OEM
10L1-1.5S	OEM
10L1.5-2	OEM
10L1.5-2S	OEM
10L15	0348
10L20	OEM
10L25	OEM
10L30	OEM
10L35	OEM
10L40	OEM
10L45	OEM
10L50	OEM
10L55	OEM
10L60	OEM
10L65	OEM
10L70	OEM
10L75	OEM
10L80	OEM
10L85	OEM
10L90	OEM
10L95	OEM
10L100	OEM
10L110	OEM
10L120	OEM
10L130	OEM
10L140	OEM
10L150	OEM
10L160	OEM
10L170	OEM
10L180	OEM
10L190	OEM
10L200	OEM
10L210	OEM
10L220	OEM
10L230	OEM
10L240	OEM
10L250	OEM
10L260	OEM
10L270	OEM
10L280	OEM
10L290	OEM
10L300	OEM
10LC11	OEM
10LZ3.3D5	OEM
10LZ3.3D10	OEM
10LZ3.6D5	OEM
10LZ3.6D10	OEM
10M	0015
10M10	2901
10M15	1124
10M17Z	1532
10M17Z5	1532
10M17Z10	1532
10M20	1124
10M25Z	1748
10M25Z5	1748
10M25Z10	1748
10M30	1126
10M40	0805
10M50	1195
10M60	0319
10M80	1132
10M100	0044
10MA10	0254
10MA20	1107
10MA30	1111
10MA40	0258
10MA60	0267
10MA80	1119
10MA100	0281
10MA120	OEM
10MA140	OEM
10MA160	OEM
10MC11	OEM
10N1	0015
10N2	0097
10N2(ZENER)	0064
10N3	0064
10N15A	OEM
10N15A9	OEM
10N20A	OEM
10N20A9	OEM
10N25A	OEM
10N25A9	OEM
10N30A	OEM
10N30A9	OEM
10N35A	OEM
10N35A9	OEM
10N40A	OEM
10N40A9	OEM
10N45A	OEM
10N45A9	OEM
10N50A	OEM
10N50A9	OEM
10N55A	OEM
10N55A9	OEM
10N60A	OEM
10N60A9	OEM
10N65A	OEM
10N65A9	OEM
10N70A	OEM
10N70A9	OEM
10N75A	OEM
10N75A9	OEM
10N80A	OEM
10N80A9	OEM
10N85A	OEM
10N85A9	OEM
10N90A	OEM
10N90A9	OEM
10N95A	OEM
10N95A9	OEM
10N100A	OEM
10N100A9	OEM
10N110A	OEM
10N130A	OEM
10N140A	OEM
10N150A	OEM
10N160A	OEM
10N170A	OEM
10N180A	OEM
10N190A	OEM
10N200A	OEM
10N210A	OEM
10N220A	OEM
10N230A	OEM
10N240A	OEM
10N250A	OEM
10N260A	OEM
10NC11	OEM
100-202	0143
10P1	0037
10P1A	0037
10P1S	0567
10P1SG	0567
10P2	0097
10P2S	0762
10P2SG	0762
10P3	0097
10P3S	3271
10P3SG	0740
10P4	OEM
10P4S	0740
10P4SG	0740
10P5S	3291
10P5SG	0764
10P6S	0764
10P6SG	0764
10P8S	0766
10P8SG	0766
10P10SG	OEM
10P12SG	OEM
10PCRA	1651
10PCRB	1651
10PCRC	1651
10PCRD	1651
10PCRE	1651
10PCRF	1584
10PCRG	1584
10PCRH	1584
10PCRI	1665
10PCRJ	1665
10PCRK	OEM
10PCRL	OEM
10PCRNA	OEM
10PCRNB	OEM
10PCRNC	OEM
10PCRND	OEM
10PCRNE	OEM
10PCRNF	OEM
10PCRNG	OEM
10PCRNH	OEM
10PCRNI	OEM
10PCRNJ	OEM
10PCRNK	OEM
10PCRNL	OEM
10PCRSA	2519
10PCRSB	2519
10PCRSC	0741
10PCRSD	0745
10PCRSE	0745
10PCRSF	0747
10PCRSG	0747
10PCRSH	0747
10PCRSI	0752
10PCRSJ	0752
10PCRSK	OEM
10PCRSL	OEM
10PF	0030
10PM8	OEM
10PZ6.8	1458
10PZ6.8A	1458
10PZ6.8B	1458
10PZ6.8C	1458
10PZ6.8D	1458
10PZ7.5	0221
10PZ7.5A	0221
10PZ7.5B	0221
10PZ7.5C	0221
10PZ7.5D	0221
10PZ8.2	1490
10PZ8.2A	1490
10PZ8.2B	1490
10PZ8.2C	1490
10PZ8.2D	1490
10PZ9.1	1618
10PZ9.1A	1618
10PZ9.1B	1618
10PZ9.1C	1618
10PZ9.1D	1618
10PZ10	0510
10PZ10A	0510
10PZ10B	0510
10PZ10C	0510
10PZ10D	0510
10PZ11	0691
10PZ11A	0691
10PZ11B	0691
10PZ11C	0691
10PZ11D	0691
10PZ12	0870
10PZ12A	0870
10PZ12B	0870
10PZ12C	0870
10PZ12D	0870
10PZ13	1020
10PZ13A	1020
10PZ13B	1020
10PZ13C	1020
10PZ13D	1020
10PZ14	1153
10PZ14A	1153
10PZ14B	1153
10PZ14C	1153
10PZ14D	1153
10PZ15	1274
10PZ15A	1274
10PZ15B	1274
10PZ15C	1274
10PZ15D	1274
10PZ16	1402
10PZ16A	1402
10PZ16B	1402
10PZ16C	1402
10PZ16D	1402
10PZ18	1078
10PZ18A	1078
10PZ18B	1078
10PZ18C	1078
10PZ18D	1078
10PZ20	1717
10PZ20A	1717
10PZ20B	1717
10PZ20C	1717
10PZ20D	1717
10PZ22	1723
10PZ22A	1723
10PZ22B	1723
10PZ22C	1723
10PZ22D	1723
10PZ24	1736
10PZ24A	1736
10PZ24B	1736
10PZ24C	1736
10PZ24D	1736
10PZ27	1761
10PZ27A	1761
10PZ27B	1761
10PZ27C	1761
10PZ27D	1761
10PZ30	1772
10PZ30A	1772
10PZ30B	1772
10PZ30C	1772
10PZ30D	1772
10PZ33	1788
10PZ33A	1788
10PZ33B	1788
10PZ33C	1788
10PZ33D	1788
10PZ36	1796
10PZ36A	1796
10PZ36B	1796
10PZ36C	1796
10PZ36D	1796
10PZ39	1804
10PZ39A	1804
10PZ39B	1804
10PZ39C	1804
10PZ39D	1804
10PZ43	1194
10PZ43A	1194
10PZ43B	1194
10PZ43C	1194
10PZ43D	1194
10PZ47	0022
10PZ47A	0022
10PZ47B	0022
10PZ47C	0022
10PZ47D	0022
10PZ51	0132
10PZ51A	0132
10PZ51B	0132
10PZ51C	0132
10PZ51D	0132
10PZ56	0207
10PZ56A	0207
10PZ56B	0207
10PZ56C	0207
10PZ56D	0207
10PZ62	0263
10PZ62A	0263
10PZ62B	0263
10PZ62C	0263
10PZ62D	0263
10PZ68	0310
10PZ68A	0310
10PZ68B	0310
10PZ68C	0310
10PZ68D	0310
10PZ75	0329
10PZ75A	0329
10PZ75B	0329
10PZ75C	0329
10PZ75D	0356
10PZ82	0356
10PZ82A	0356
10PZ82B	0356
10PZ82C	0356
10PZ82D	0356
10PZ91	0412
10PZ91A	0412
10PZ91B	0412
10PZ91C	0412
10PZ91D	0412
10PZ100	0438
10PZ100A	0438
10PZ100B	0438
10PZ100C	0438
10PZ100D	0438
10PZ110	0488
10PZ110A	0488
10PZ110B	0488
10PZ110C	0488
10PZ110D	0488
10PZ120	0509
10PZ120A	0509
10PZ120B	0509
10PZ120C	0509
10PZ120D	0509
10PZ130	0524
10PZ130A	0524
10PZ130B	0524
10PZ130C	0524
10PZ130D	0524
10PZ150	0063
10PZ150A	0063
10PZ150B	0063
10PZ150C	0063
10PZ150D	0063
10PZ160	0401
10PZ160A	0401
10PZ160B	0401
10PZ160C	0401
10PZ160D	0401
10PZ180	0634
10PZ180A	0634
10PZ180B	0634
10PZ180C	0634
10PZ180D	0634
10PZ200	0668
10PZ200A	0668
10PZ200B	0668
10PZ200C	0668
10PZ200D	0668
10PZ220	OEM
10PZ220A	OEM
10PZ220B	OEM
10PZ220C	OEM
10PZ220D	OEM
10P.5-.75	OEM
10P.5-.75S	OEM
10P.75-1	OEM
10P.75-1S	OEM
10Q3	1250
10R0/18	OEM
10R1/18	OEM
10R1/52	OEM
10R1A10	OEM
10R1A10M	OEM
10R1A20	OEM
10R1A20M	OEM
10R1A40	OEM
10R1A40M	OEM
10R1A60	OEM
10R1A60M	OEM
10R1A80	OEM
10R1A80M	OEM
10R1A100	OEM
10R1A100M	OEM
10R1A110	OEM
10R1A110M	OEM
10R1A120	OEM
10R1A120M	OEM
10R1B	0015
10R2	5005
10R2/4	OEM
10R2/18	OEM
10R2/66	OEM
10R2B	0015
10R2S	OEM
10R3/4	OEM
10R3/66	OEM
10R3B	0015
10R3P	OEM
10R3S	OEM
10R4/3	OEM
10R4/3C	OEM
10R4/4	OEM
10R4/59	OEM
10R4B	0015
10R4C	OEM
10R4S	OEM
10R5/3	OEM
10R5/3D	OEM
10R5/5	OEM
10R5/61	OEM
10R5B	0015
10R5C	OEM
10R5S	OEM
10R6/3	OEM
10R6/3D	OEM
10R6/5	OEM
10R6/61	OEM
10R6B	0015
10R6C	OEM
10R6S	OEM
10R7/3	OEM
10R7/3D	OEM
10R7/5	OEM
10R7/61	OEM
10R7B	0071
10R7CD	OEM
10R7S	OEM
10R8/3	OEM
10R8/3D	OEM
10R8/5	OEM
10R8/63	OEM
10R8B	0071
10R8CD	OEM
10R8S	OEM
10R9B	0071
10R10B	0071
10R11B	0102
10R12B	0102
10R13B	0102
10R14B	0102
10R15B	0348
10R16B	OEM
10R18B	OEM
10R20B	OEM
10R22B	OEM
10R24B	OEM
10R26B	OEM
10R27.5	OEM
10R27.5A	OEM
10R27.5B	OEM
10R27.5C	OEM
10R27.5D	OEM
10R28B	OEM
10R30B	OEM
10R32B	OEM
10R34B	OEM
10R36B	OEM
10R38B	OEM
10R40B	OEM
10R42B	OEM
10R44B	OEM
10R46B	OEM
10R48B	OEM
10R50B	OEM
10RC2	3307
10RC2A	3307
10RC5	2188
10RC5A	2188
10RC10	0567
10RC10A	0567
10RC15	3258
10RC15A	3258
10RC20	0762
10RC20A	0762
10RC25	3268
10RC25A	0721
10RC30	3271
10RC30A	3271
10RC40	0740
10RC40A	0740
10RC50	3291
10RC50A	3291
10RC60	0764
10RC60A	0764
10RC70	2875
10RC70A	2875
10RC80	0766
10RC80A	0766
10RC90	OEM
10RC100	0679
10RC110	0679
10RC110A	0679
10RC120	0679
10RCF2A	OEM
10RCF5A	OEM
10RCF10A	OEM
10RCF15A	OEM
10RCF20A	OEM
10RCF25A	OEM
10RCF30A	OEM
10RCF40A	OEM
10RCF50A	OEM
10RCF60A	OEM
10RCF70A	OEM
10RCF80A	OEM
10RCU5	OEM
10RCU10	OEM
10RCU20	OEM
10RCU30	OEM
10RCU40	OEM
10RCU50	OEM
10RCU60	OEM
10RE52	OEM
10RIA10	0469
10RIA20	0469
10RIA40	0722
10RIA60	0725
10RIA80	0679
10RIA100	0679
10RIA110	0679
10RIA120	0679
10RM60	OEM
10RM80	OEM
10RM100	OEM
10RM120	OEM
10RM150	OEM
10RM200	OEM
10RM220	OEM
10RM250	OEM
10RO/18	OEM
10RZ6.8	1601
10RZ6.8A	1601
10RZ6.8B	1601
10RZ6.8C	1458
10RZ6.8D	1458
10RZ7.1C	OEM
10RZ7.1D	OEM
10RZ7.5	1616
10RZ7.5A	1616
10RZ7.5B	1616
10RZ7.5C	0221
10RZ7.5D	0221
10RZ8.2	1622
10RZ8.2A	1490
10RZ8.2B	1622
10RZ8.2C	1490
10RZ8.2D	1490
10RZ9.1	0627
10RZ9.1A	0627
10RZ9.1B	0627
10RZ9.1C	1618
10RZ9.1D	1618
10RZ10	0510
10RZ10A	0992
10RZ10B	0992
10RZ10C	0510
10RZ10D	0510
10RZ11	0995
10RZ11A	0995
10RZ11B	0691
10RZ11C	0691
10RZ11D	0691
10RZ12	1263
10RZ12A	1263
10RZ12B	1263
10RZ12C	0870
10RZ12D	0870
10RZ13	1249
10RZ13A	1249
10RZ13B	1249
10RZ13C	1020
10RZ13D	1020
10RZ15	1639
10RZ15A	1639
10RZ15B	1639
10RZ15C	1274
10RZ15D	1274
10RZ16	1703
10RZ16A	1703
10RZ16B	1703
10RZ16C	1402
10RZ16D	1402
10RZ18	1716
10RZ18A	1716
10RZ18B	1716
10RZ18C	1078
10RZ18D	1078
10RZ20	1731
10RZ20A	1731
10RZ20B	1731
10RZ20C	1717
10RZ20D	1717
10RZ22	0727
10RZ22A	0727
10RZ22B	0727
10RZ22C	1723
10RZ22D	1723
10RZ24	1756
10RZ24A	1756
10RZ24B	1756
10RZ24C	1736
10RZ24D	1736
10RZ27	1782
10RZ27A	1782
10RZ27B	1782
10RZ27C	1761
10RZ27D	1761
10RZ30	1794
10RZ30A	1794
10RZ30B	1794
10RZ30C	1772
10RZ30D	1772
10RZ33	1799
10RZ33A	1799
10RZ33B	1799
10RZ33C	1788
10RZ33D	1788
10RZ36	1796
10RZ36A	1809
10RZ36B	1809
10RZ36C	1796
10RZ36D	1796
10RZ39	1817
10RZ39A	1817
10RZ39B	1817
10RZ39C	1804
10RZ39D	1804
10RZ43	1826
10RZ43A	1826
10RZ43B	1826
10RZ43C	6989
10RZ43D	6989
10RZ47	1853
10RZ47A	1853
10RZ47B	1853
10RZ47C	0022
10RZ47D	0022
10RZ51	1866
10RZ51A	1866
10RZ51B	1866
10RZ51C	0132
10RZ51D	0132
10RZ56	1884
10RZ56A	1884
10RZ56B	1884
10RZ56C	0207
10RZ56D	0207
10RZ62	0263
10RZ62A	0263
10RZ62B	0263
10RZ62C	0263
10RZ62D	0263
10RZ68	1902
10RZ68A	1902
10RZ68B	1902
10RZ68C	0310
10RZ68D	0310
10RZ75	0736
10RZ75A	0736
10RZ75B	0736
10RZ75C	0329
10RZ75D	0329
10RZ82	1909
10RZ82A	1909
10RZ82B	1909
10RZ82C	0356
10RZ82D	0356
10RZ91	1914
10RZ91A	1914
10RZ91B	1914
10RZ91C	0412

If replacement code is OEM, contact original manufacturer for replacement.

DEVICE TYPE	REPL CODE
10RZ91D	0412
10RZ100	1163
10RZ100A	1163
10RZ100B	1163
10RZ100C	0438
10RZ100D	0438
10RZ110	1933
10RZ110A	1933
10RZ110B	1933
10RZ110C	0488
10RZ110D	0488
10RZ120	1941
10RZ120A	1941
10RZ120B	0509
10RZ120C	0509
10RZ120D	0509
10RZ130	1947
10RZ130A	1947
10RZ130B	1947
10RZ130C	0524
10RZ130D	0524
10RZ150	1961
10RZ150A	1961
10RZ150B	1961
10RZ150C	0063
10RZ150D	0063
10RZ160	0357
10RZ160A	0357
10RZ160B	0357
10RZ160C	0401
10RZ160D	0401
10RZ180	0776
10RZ180A	0776
10RZ180B	0776
10RZ180C	0634
10RZ180D	0634
10RZ200	1071
10RZ200A	1071
10RZ200B	1071
10RZ200C	0668
10RZ200D	0668
10RZ220	OEM
10RZ220A	OEM
10RZ220B	OEM
10RZ220C	OEM
10RZ220D	OEM
10S1S03	OEM
10S1S04	OEM
10S2-3	OEM
10S2-3S	OEM
10S2S03	OEM
10S2S04	OEM
10S2SC	OEM
10S2.5-3R	OEM
10S2.5-3.5	OEM
10S2.5-3.5S	OEM
10S3-4	OEM
10S3-4S	OEM
10S3P	OEM
10S3SC	OEM
10S3.5R	OEM
10S4SC	OEM
10S5SC	OEM
10S6SC	OEM
10S20	OEM
10S30	OEM
10S40	OEM
10S50	OEM
10SB050	OEM
10SB100	OEM
10SB200	OEM
10SB400	OEM
10SB600	OEM
10SC05	OEM
10SC1	OEM
10SC2	OEM
10SC3	OEM
10SC3.5-3.5S	OEM
10SC3.5-4.5	OEM
10SC4	OEM
10SC6	OEM
10SL101	OEM
10SM1	OEM
10SM3	OEM
10SM6	OEM
10SM15	OEM
10SM25	OEM
10SM40	OEM
10SM55	OEM
10SM70	OEM
10SM100	OEM
10SM150	OEM
10SM200	OEM
10SP005	OEM
10SP005D	OEM
10SP005N	OEM
10SP01	OEM
10SP01D	OEM
10SP01N	OEM
10SP02	OEM
10SP02D	OEM
10SP02N	OEM
10SP04	OEM
10SP04D	OEM
10SP04N	OEM
10SP06	OEM
10SP06D	OEM
10SP06N	OEM
10SP08	OEM
10SP08D	OEM
10SP08N	OEM
10SP10	OEM
10SP10D	OEM
10SP10N	OEM
10SS2SC	OEM
10SS3SC	OEM
10SS4SC	OEM
10SS5SC	OEM
10SS6SC	OEM
10T2	OEM
10T2FR	OEM
10T3P	OEM
10T4	OEM
10T4FR	OEM
10T4S	OEM
10T5.6	1445
10T6FR	OEM
10T6.2	2221
10T6.8	1458
10T7.5	0221
10T8.2	1490
10T9.1	1618
10T10	0510
10T11	0691
10T12	0870
10T13	1020
10T15	1274
10T16	1402
10T18	1078
10T20	1717
10T22	1723
10T24	1736
10T27	1761
10T30	1772
10T33	1788
10T36	1796
10T39	1804
10T43	1194
10T47	0022
10T51	0132
10T56	0207
10T62	0263
10T68	0310
10T75	0329
10T82	0356
10T91	0412
10T100	0438
10T110	0488
10T120	0509
10T130	0524
10T150	0063
10T160	0401
10T180	0634
10T200	0668
10TB050	OEM
10TB1	0741
10TB1-CE00	OEM
10TB1-CH00	OEM
10TB1-EE00	OEM
10TB1-EH00	OEM
10TB1-GE00	OEM
10TB1-GH00	OEM
10TB1-JE00	OEM
10TB2	0745
10TB2-CE00	OEM
10TB2-CH00	OEM
10TB2-EE00	OEM
10TB2-EH00	OEM
10TB2-GE00	OEM
10TB2-GH00	OEM
10TB2-JE00	OEM
10TB3	2917
10TB3-CE00	OEM
10TB3-CH00	OEM
10TB3-EE00	OEM
10TB3-EH00	OEM
10TB3-GE00	OEM
10TB3-GH00	OEM
10TB3-JE00	OEM
10TB4	0747
10TB4-CE00	OEM
10TB4-CH00	OEM
10TB4-EE00	OEM
10TB4-EH00	OEM
10TB4-GE00	OEM
10TB4-GH00	OEM
10TB4-J300	OEM
10TB5	3291
10TB5-CE00	OEM
10TB5-CH00	OEM
10TB5-EE00	OEM
10TB5-EH00	OEM
10TB5-GE00	OEM
10TB5-GH00	OEM
10TB5-JE00	OEM
10TB6	0764
10TB6-CE00	OEM
10TB6-CH00	OEM
10TB6-EE00	OEM
10TB6-EH00	OEM
10TB6-GE00	OEM
10TB6-GH00	OEM
10TB6-JE00	OEM
10TB7	2875
10TB7-CE00	OEM
10TB7-CH00	OEM
10TB7-EE00	OEM
10TB7-EH00	OEM
10TB7-GE00	OEM
10TB7-GH00	OEM
10TB7-JE00	OEM
10TB8	0766
10TB8-CE00	OEM
10TB8-CH00	OEM
10TB8-EE00	OEM
10TB8-EH00	OEM
10TB8-GE00	OEM
10TB8-GH00	OEM
10TB8-JE00	OEM
10TB100	OEM
10TB200	OEM
10TB400	OEM
10TB600	OEM
10TC05	OEM
10TC05B	OEM
10TC10	OEM
10TC10B	OEM
10TC15	OEM
10TC15B	OEM
10TC20	OEM
10TC20B	OEM
10TC25	OEM
10TC30	OEM
10TC30B	OEM
10TC35	OEM
10TC40	OEM
10TC40B	OEM
10TC50	OEM
10TC50B	OEM
10TC60	OEM
10TC60B	OEM
10TC70	OEM
10TC80	OEM
10TC90	OEM
10TC100	OEM
10TC110	OEM
10TC120	OEM
10TCA10	OEM
10TCA20	OEM
10TCA30	OEM
10TCA40	OEM
10TCA50	OEM
10TCA60	OEM
10TCA70	OEM
10TCA80	OEM
10TCA90	OEM
10TCA100	OEM
10TCA110	OEM
10TCA120	OEM
10TQ030	OEM
10TQ035	OEM
10TQ040	OEM
10TQ045	OEM
10V3P	OEM
10V13	OEM
10V-50	0080
10V-100	0609
10V-200	0795
10VJ	0170
10VT01	OEM
10VZENER	0170
10W3P	OEM
10W5R12-12	OEM
10W5R15-15	OEM
10W12R5	OEM
10W12R12-12	OEM
10W12R15-15	OEM
10W24R12-12	OEM
10W24R15-15	OEM
10WM1	OEM
10WM2	OEM
10WM4	OEM
10WM5	OEM
10WM8	OEM
10WM10	OEM
10X3P	OEM
10X7-8	OEM
10X7.5-8.5	OEM
10X8-9	OEM
10X8.5-9.6	OEM
10X9-10	OEM
10XT26	0211
10Y	0064
10Y3P	OEM
10YHB10	OEM
10Z3.3T	4231
10Z3.3T5	1006
10Z3.3T10	1006
10Z3.6T	2397
10Z3.6T5	1380
10Z3.6T10	1380
10Z3.9M	2035
10Z3.9T5	0547
10Z3.9T10	0547
10Z3.9T20	0547
10Z4.3T5	2403
10Z4.3T10	2403
10Z4.3T20	2403
10Z4.7M	1438
10Z4.7T5	2113
10Z4.7T10	2113
10Z4.7T20	2113
10Z5.1T5	2410
10Z5.1T10	2410
10Z5.1T20	2410
10Z5.6M	1445
10Z5.6T5	1901
10Z5.6T10	1901
10Z5.6T20	1901
10Z6	0015
10Z6.2T5	0696
10Z6.2T10	0696
10Z6.2T20	0696
10Z6.8M	1458
10Z6.8T5	1601
10Z6.8T10	1601
10Z6.8T20	1601
10Z7.5T5	1616
10Z7.5T10	1616
10Z7.5T20	1616
10Z8.2M	1490
10Z8.2T5	1622
10Z8.2T10	1622
10Z8.2T20	1622
10Z9.1T5	0627
10Z9.1T10	0627
10Z9.1T20	0627
10Z10	0510
10Z10M	0510
10Z10T5	0992
10Z10T10	0992
10Z10T20	0992
10Z11T5	0995
10Z11T10	0995
10Z11T20	0995
10Z12	0870
10Z12M	0870
10Z12T5	1263
10Z12T10	1263
10Z12T20	1263
10Z13T5	1249
10Z13T10	1249
10Z13T20	1249
10Z15M	1274
10Z15T5	1639
10Z15T10	1639
10Z15T20	1639
10Z16T5	1703
10Z16T10	1703
10Z16T20	1703
10Z18M	1078
10Z18T10	1716
10Z18T20	1716
10Z20T5	1731
10Z20T10	1731
10Z20T20	1731
10Z22M	1723
10Z22T5	0727
10Z22T10	0727
10Z22T20	0727
10Z24T5	1756
10Z24T10	1756
10Z24T20	1756
10Z27M	1761
10Z27T5	1782
10Z27T10	1782
10Z27T20	1782
10Z30T5	1772
10Z30T20	1794
10Z33T5	1799
10Z33T10	1799
10ZB	OEM
11/1	0133
11/1+12/1	0015
11/1&12/1	0015
11/10	0102
11/15	0015
11/20	1303
11/1592	0015
11-0399	0160
11-0400	0160
11-0422	0155
11-0423	0400
11-0429	0015
11-0430	0133
11-0769	0015
11-0770	0460
11-0771	0015
11-0772	0161
11-0773	0460
11-0774	0155
11-0775	0400
11-0778	0155
11-0781	0133
11-085001	0143
11-085003	0015
11-085004	0143
11-085005	0143
11-085007	0143
11-085008	0143
11-085010	0016
11-085012	0133
11-085013	0015
11-085014	0143
11-085015	0143
11-085022	0143
11-085024	0015
11-60	2301
11-102	4640
11-102-001	0277
11-103	0626
11-104	OEM
11-113	1540
11-115	4291
11-116	OEM
11-117	1786
11-1592	0015
11-11911-1	0130
11-102001	0015
11-102003	0277
11-108002	0277
11-120007	0015
11A1	0181
11B	0181
11B1	0248
11B2	0181
11B3	0181
11B551	0144
11B551-2	0111
11B551-2,3	OEM
11B551-3	0111
11B552	0144
11B552-2	0111
11B552-2,3	OEM
11B552-3	0111
11B554	0144
11B554-2	0111
11B554-2,3	OEM
11B554-3	0111
11B555	0144
11B555-2	0111
11B555-2,3	OEM
11B555-3	0111
11B556	0595
11B556-2	0861
11B556-2,3	OEM
11B556-3	0861
11B560	0595
11B560-2	0861
11B560-2,3	OEM
11B560-3	0861
11B1052	0144
11B1055	0144
11B1257	0861
11B1258	0861
11B1259	0861
11B1260	0861
11BT23GK	OEM
11BT33ZK	OEM
11C1B1	0142
11C2	0181
11C3B1	0142
11C3B3	0142
11C5B1	0142
11C7B1	0142
11C10B1	0142
11C11B1	0142
11C11B20	0630
11C44DC	0011
11C211B20	0595
11C551	0144
11C551-2	0111
11C551-2,3	OEM
11C551-3	0111
11C553	0144
11C553-2	0111
11C553-2,3	OEM
11C553-3	0111
11C557	0144
11C557-2	0007
11C557-2,3	OEM
11C557-3	0007
11C702	0018
11C1051	0144
11C1053	0144
11C1057	0144
11C1536	0086
11CB1	OEM
11CB2	OEM
11CB3	OEM
11CB4	OEM
11CB5	OEM
11CB6	OEM
11CB7	OEM
11CB8	OEM
11CF1	OEM
11CF2	OEM
11CF3	OEM
11CF4	OEM
11CF5	OEM
11CF6	OEM
11CF7	OEM
11CF8	OEM
11CM6	OEM
11DF1	0080
11DF2	0080
11DF3	0795
11DF4	0795
11DM6	OEM
11DQ03	0735
11DQ04	1335
11DQ05	1335
11DQ06	3888
11DQ09	3888
11DQ10	OEM
11E1	0015
11E1TA1	0015
11E1TA1-T	0015
11E2	0015
11E2TA-1	0015
11E2TA1	0015
11E4	0015
11E11	OEM
11E22A	OEM
11E35	OEM
11E60	OEM
11E100	OEM
11EB2	OEM
11EQS04	6808
11ES1	0015
11ES1TA1	0015
11ES1TB3	OEM
11ES2	0015
11F1SC	0871
11F2SC	0871
11F3SC	0853
11FA30S	OEM
11G702	OEM
11G703	OEM
11G1052	0007
11G1053	0007
11HM6	OEM
11J2	0015
11J2F	0015
11JUS5.94A	OEM
11JUS5.94B	OEM
11JUS6.78A	OEM
11JUS6.78B	OEM
11JUS7.15A	OEM
11JUS7.15B	OEM
11JUS8.03A	OEM
11JUS8.03B	OEM
11JUS8.38A	OEM
11JUS8.38B	OEM
11K60	0201
11N1	0181
11N267	0143
11PD18M	OEM
11PD18P	OEM
11PH18M	OEM
11PS18M	OEM
11R05S	0084
11R055	0097
11R1S	0090
11R1SC	0097
11R2	OEM
11R2S	0097
11R2SC	0097
11R3S	0105
11R3SC	0109
11R4S	0109
11R5S	0116
11R6S	0122
11R8S	0131
11R10S	0145
11R12S	OEM
11R45	0109
11R105	0145
11R106	0142
11RA10S	OEM
11RC10	0567
11RC20	0762
11RC30	0740
11RC40	0740
11RC50	0764
11RC60	0764
11RC80	0766
11RC100	OEM
11RT1	0150
11T1	OEM
11T2	OEM
11T4	0179
11T4S	1259
11V53A	OEM
11V54	OEM
11V64	OEM
11V651	OEM
11V652	OEM
11W71	OEM
11W72	OEM
11Y	0181
11YM6	OEM
11Z4	OEM
11Z6	OEM
11Z6A	OEM
11Z6AF	0292
11Z6F	0292
11Z6FAF	OEM
12/1	0196
12/1N	1080
12/1N10	1149
12/3	0196
12/100	0196
12-085005	0143
12-085006	0143
12-085009	0143
12-085029	0143
12-085031	0015
12-085034	0143
12-085035	0143
12-085038	0143
12-085040	0015
12-085041	0123
12-087003	0143
12-087004	0133
12-1	0233
12-1-70	0016
12-1-70-12	0016
12-1-70-12-7	0016
12-1-73	0280
12-1-74	0280
12-1-75	0280
12-1-76	0280
12-1-78	0280
12-1-83	0280
12-1-91	0280
12-1-92	0280
12-1-93	0280
12-1-95	0004
12-1-96	0004
12-1-100	0280
12-1-102	0280
12-1-103	0280
12-1-104	0280
12-1-105	0280
12-1-106	0004
12-1-107	0004
12-1-120	0004
12-1-128	0280
12-1-135	0050
12-1-137	0050
12-1-138	0050
12-1-139	0050
12-1-148	0004
12-1-150	0050
12-1-157	0050
12-1-161	0280
12-1-162	0280
12-1-164	0004
12-1-179	0280
12-1-180	0280
12-1-184	0004
12-1-186	0280
12-1-189	0136
12-1-190	0050
12-1-191	0004
12-1-226	0004
12-1-227	0004
12-1-228	0050
12-1-229	0050
12-1-230	0050
12-1-231	0050
12-1-232	0004
12-1-233	0050
12-1-234	0280
12-1-235	0280
12-1-236	0280
12-1-240	0280
12-1-241	0280
12-1-242	0136
12-1-243	0136
12-1-244	0136
12-1-246	0004
12-1-254	0280
12-1-256	0050
12-1-257	0136
12-1-258	0050
12-1-259	0050
12-1-260	0050
12-1-266	0004
12-1-267	0004
12-1-270	0160
12-1-271	0160
12-1-272	0004
12-1-273	0280
12-1-274	0004
12-1-275	0280
12-1-276	0016
12-1-277	0016
12-1-278	0016
12-1-279	0016
12-1-289	0280
12-1A	0016
12-1A0	0016
12-1A0R	0016
12-1A1	0016
12-1A2	0016
12-1A3	0016
12-1A3P	0016
12-1A4	0016
12-1A4-7	0016
12-1A4-7B	0016
12-1A5	0016
12-1A5L	0016
12-1A6	0016
12-1A6A	0016
12-1A7	0016
12-1A7-1	0016
12-1A8	0016
12-1A9	0016
12-1A9G	0016
12-1A19	0016
12-1A21	0016
12-1A82	0016
12-4	0016
12-10	0016
12-11	0016
12-12	0016
12-13	0016
12-14	0016
12-15	0016
12-16	0016
12-17	0016
12-18	0016
12-19	0016
12-1032-41	2525
12-23163-3	0113
12-100001	0015
12-100003	0015
12-102001	0015
12-679115-01	0725
12-680701-05	0725
12A	0053

If replacement code is OEM, contact original manufacturer for replacement.

DEVICE TYPE	REPL CODE
12A1	0052
12A2	1005
12A3	1005
12A6F	0711
12A6F1	OEM
12A8	OEM
12A8F	OEM
12A8F1	OEM
12A10F	OEM
12A10F1	OEM
12A50	OEM
12A100	OEM
12A104	OEM
12A105	OEM
12A107	OEM
12A108	OEM
12A200	OEM
12A300	OEM
12A304	OEM
12A308	OEM
12A400	OEM
12A500	OEM
12A600	OEM
12A700	0131
12A800	0131
12A900	0145
12A904	OEM
12A1000	0145
12A9275	0211
12A9275-1	0211
12AL240	0280
12B	0052
12B(1WATT)	0137
12B,,	0052
12B1	1005
12B1(1-WATT)	0137
12B1(1WATT)	0137
12B-2	0015
12B2	0137
12B2(.5W)	1005
12B-2-BIP-M	0071
12B-2B1P-M	0015
12B3	0052
12B20	OEM
12B40	OEM
12B490100	OEM
12BG11	1530
12BH11	1520
12BL2C41	2274
12BM	1216
12C05	0084
12C05R	0534
12C1	1005
12C2	0015
12C2(ZENER)	0052
12C2P-114	0277
12C3	0052
12C5Y	0137
12C70R	0550
12C10	0090
12C10R	0748
12C15	0097
12C15R	0765
12C20	0097
12C20R	0765
12C30	0105
12C30R	0777
12C40	0105
12C40R	0538
12C50	0116
12C50R	0801
12C60	0122
12C60R	0816
12C70	0131
12C70R	0545
12C80	0131
12C80R	0545
12C90	0145
12C90R	0550
12C100	0145
12C100R	0550
12C101	OEM
12C102	OEM
12CC12	1250
12CD12	1577
12CF11	0545
12CG11	0853
12CH1FM	OEM
12CH1M	2234
12CH2FM	OEM
12CH2M	2234
12CH3FM	OEM
12CH3M	OEM
12CH4FM	OEM
12CH4M	OEM
12CLN	0016
12CTQ030	3266
12CTQ035	3266
12CTQ040	3266
12CTQ045	6890
12D02M1	OEM
12D05M1	OEM
12D2C41	OEM
12D10K1-A	OEM
12DF15E1	OEM
12DF15E2	OEM
12DF15E3	OEM
12DG11	1530
12DH11	1520
12E10	OEM
12E20A	OEM
12E40	OEM
12E70	OEM
12E109	OEM
12E120	OEM
12F5	0084
12F5A	0084
12F5B	0084
12F5R	2291
12F5RA	2291
12F5RB	2291
12F10	0090
12F10A	0090
12F10B	0090
12F10R	2291
12F10RA	2291
12F10RB	2291
12F15	0097
12F15A	0097
12F15B	0097
12F15E1	OEM
12F15E3	OEM
12F15R	2291
12F15RA	2291
12F15RB	2291
12F20	0097
12F20A	0097
12F20B	0097
12F20R	2291
12F20RA	2291
12F20RB	2291
12F30	0105
12F30A	0105
12F30B	0105
12F30R	0476
12F30RA	0476
12F30RB	0476
12F40	0109
12F40A	0109
12F40B	0109
12F40R	0476
12F40RA	0476
12F40RB	0476
12F50	0116
12F50A	0116
12F50B	0116
12F50R	0476
12F50RA	0476
12F50RB	0476
12F60	0122
12F60A	0122
12F60B	0122
12F60R	0476
12F60RA	0476
12F60RB	0476
12F70	3011
12F70A	3011
12F70B	0131
12F70R	3011
12F70RA	0449
12F70RB	0449
12F80	0131
12F80A	0131
12F80B	0131
12F80R	0449
12F80RA	0449
12F80RB	0449
12F90	0145
12F90A	0145
12F90B	0145
12F90R	0449
12F90RA	0449
12F90RB	0449
12F100	0145
12F100A	0145
12F100B	0145
12F100R	0449
12F100RA	0449
12F100RB	0449
12F120	OEM
12FC12	1581
12FD12	3281
12FG11	0029
12FH05S	1567
12FH1S	1567
12FH2S	1567
12FH3S	2153
12FH4S	2153
12FH11	1847
12FL5S02	1567
12FL5S05	1567
12FL5S10	1567
12FL10S02	1567
12FL10S05	1567
12FL10S10	1567
12FL20S02	1567
12FL20S05	1567
12FL20S10	1567
12FL40S02	2153
12FL40S05	2153
12FL40S10	2153
12FL60S02	0711
12FL60S05	0711
12FL60S10	OEM
12FL80S05	OEM
12FL80S10	OEM
12FL100S05	OEM
12FL100S10	OEM
12FR5	0534
12FR5A	0534
12FR5B	0534
12FR10	0748
12FR10A	0748
12FR10B	5038
12FR15	0765
12FR15A	0765
12FR15B	0765
12FR20	0765
12FR20A	0765
12FR20B	0765
12FR30	0777
12FR30A	0777
12FR30B	0777
12FR40	0538
12FR40A	0538
12FR40B	0538
12FR50	0801
12FR50A	0801
12FR50B	0801
12FR60	0816
12FR60A	0816
12FR60B	0816
12FR70B	0545
12FR80	0545
12FR80A	0545
12FR80B	0545
12FR90	0550
12FR90A	0550
12FR90B	0550
12FR100	0550
12FR100A	0550
12FR100B	0550
12FX11	OEM
12FX12	OEM
12FXF11	OEM
12G1	OEM
12G2	OEM
12G4	0102
12G101	OEM
12G301	OEM
12G302	OEM
12GC11	1581
12GG11	0029
12GH11	1847
12H301	OEM
12H302	OEM
12H303	OEM
12H901	OEM
12H902	OEM
12J2	0015
12J2F	0015
12J301	OEM
12J302	OEM
12J303	OEM
12J901	OEM
12J902	OEM
12J905	OEM
12JC11	1586
12JG11	0601
12JH11	1851
12JUS6.48A	OEM
12JUS6.48B	OEM
12JUS7.39A	OEM
12JUS7.39B	OEM
12JUS7.80A	OEM
12JUS7.80B	OEM
12JUS8.76A	OEM
12JUS8.76B	OEM
12JUS9.14A	OEM
12JUS9.14B	OEM
12K60	0288
12KM7	OEM
12L2C41	OEM
12LC11	1132
12LF11	0545
12M2	0160
12MC	0050
12MF5	1567
12MF10	1567
12MF15	1567
12MF20	1567
12MF30	OEM
12MF40	OEM
12MZ	0050
12N02F1	OEM
12N05M1	OEM
12N10K1	OEM
12NC11	0044
12NF11	0550
12P02F1-A	OEM
12P05M1-A	OEM
12P2	0133
12P10K1	OEM
12PR40B	0538
12PR60B	0816
12QF11	OEM
12R0/18	OEM
12R1/18	OEM
12R1/52	OEM
12R2	OEM
12R2/4	OEM
12R2/66	OEM
12R2B	OEM
12R2S	OEM
12R3/4	OEM
12R3/66	OEM
12R3B	OEM
12R3S	OEM
12R4/3	OEM
12R4/3C	OEM
12R4/4	OEM
12R4/59	OEM
12R4C	OEM
12R4S	OEM
12R5/3	OEM
12R5/3D	OEM
12R5/5	OEM
12R5/61	OEM
12R5C	OEM
12R5S	OEM
12R6/3	OEM
12R6/3D	OEM
12R6/5	OEM
12R6/61	OEM
12R6C	OEM
12R6S	OEM
12R7/3	OEM
12R7/3D	OEM
12R7/5	OEM
12R7/61	OEM
12R7CD	OEM
12R7S	OEM
12R8/3	OEM
12R8/3D	OEM
12R8/5	OEM
12R8/63	OEM
12R8CD	OEM
12R8S	OEM
12R52	OEM
12RCM5	2188
12RCM10	0567
12RCM20	OEM
12RCM30	OEM
12RCM40	OEM
12RCM50	OEM
12RCM60	OEM
12RH05S	1567
12RH1S	1567
12RH2S	1547
12RH3S	3139
12RH4S	2153
12R0/18	OEM
12S1573	OEM
12S1831	OEM
12SM1	OEM
12SM3	OEM
12SM6	OEM
12SM15	OEM
12SM25	OEM
12SM40	OEM
12SM55	OEM
12SM70	OEM
12SM100	OEM
12SM150	OEM
12SM200	OEM
12SZ-EQB01-12A	0137
12T05T	OEM
12T1	OEM
12T2	OEM
12T4	OEM
12T4S	OEM
12T10T	OEM
12T20T	OEM
12T40T	OEM
12T60T	OEM
12T80T	OEM
12TB050	OEM
12TB100	OEM
12TB200	OEM
12TB400	OEM
12TB600	OEM
12TC10	OEM
12TC20	OEM
12TC30	OEM
12TC40	OEM
12TC50	OEM
12TC60	OEM
12TC70	OEM
12TC80	OEM
12TC90	OEM
12TC100	OEM
12TC110	OEM
12TC120	OEM
12TCA10	OEM
12TCA20	OEM
12TCA30	OEM
12TCA40	OEM
12TCA50	OEM
12TCA60	OEM
12TCA70	OEM
12TCA80	OEM
12TCA90	OEM
12TCA100	OEM
12TCA110	OEM
12TCA120	OEM
12V	0137
12V1W	0137
12V66	OEM
12X	0052
12X006	OEM
12X008	OEM
12X010	OEM
12X011	OEM
12X012	OEM
12X013	OEM
12X014	OEM
12X015	OEM
12X040	OEM
12X043	OEM
12X047	0016
12X058	OEM
12X059	OEM
12X070	OEM
12X084A	OEM
12X084B	OEM
12X165	OEM
12Y	0137
12Z	0365
12Z4	OEM
12Z6	OEM
12Z6A	OEM
12Z6F	0188
12Z6FAF	3947
13-0002	0170
13-0003	0133
13-0003(PACE)	0124
13-0004	0143
13-0006	0006
13-0006A	0037
13-0009	0155
13-0010	0155
13-0014	0429
13-0015	0015
13-0017	0320
13-0020	0144
13-0021	0016
13-0022	0076
13-0024	0016
13-0028	0698
13-0029	0156
13-0032	0103
13-0035	0079
13-0040	0016
13-0041	0037
13-0043	0037
13-0043A	0037
13-0044	0006
13-0044A	0037
13-0048	0037
13-0049	0161
13-0050	0023
13-0058	0155
13-0061	0037
13-0061A	0037
13-0062	0144
13-0062-1	0224
13-0063	0076
13-0064	0079
13-0065	0144
13-0065(1)	0007
13-0078	0247
13-0079	5568
13-0079A	5568
13-0097	0561
13-001200-00X	0123
13-0022345-00X	OEM
13-0023511-04X	0532
13-0023534-03X	OEM
13-0023535-03X	OEM
13-0023713-08X	OEM
13-0106	1268
13-0117	0030
13-0118	0212
13-0129	5062
13-0130	3797
13-0158	OEM
13-0161	0083
13-0164	0451
13-0165	0212
13-0167	OEM
13-0178	0018
13-0321-5	0016
13-0321-6	0016
13-0321-7	0016
13-0321-8	0016
13-0321-9	0016
13-0321-10	0016
13-0321-11	0016
13-0321-12	0016
13-0321-14	0144
13-0321-15	0144
13-0321-16	0144
13-0321-17	0144
13-0321-21	0144
13-0321-81	0016
13-076003	OEM
13-085002	0133
13-085012	0123
13-085015	0133
13-085022	0133
13-085023	0133
13-085024	0071
13-085026	0133
13-085027	0071
13-085028	0012
13-085029	0123
13-085039	0015
13-085042	0012
13-086028	1925
13-087005	0133
13-087005(DIODE)	0133
13-087005(TRANS)	0004
13-087027	0015
13-1-6	0632
13-3-6	7091
13-4P63-I	0042
13-9-6	0632
13-10-6	0632
13-11-6	0632
13-12-03-0	0143
13-17-6	0016
13-17-6(SEARS)*	0079
13-26-6	0654
13-27-6	2715
13-28-6	2571
13-29-5	0167
13-29-6	0167
13-30-6	0364
13-36-6	0654
13-40-6	0354
13-41-6	0349
13-42-6	0352
13-43-6	OEM
13-46-6	OEM
13-50-6	0623
13-52-6	0701
13-53-6	4239
13-54-6	OEM
13-56-6	0856
13-57-6	0751
13-59-6	2557
13-60-6	0701
13-61-6	2557
13-62-6	OEM
13-64-6	0167
13-66-9	OEM
13-67-6	0751
13-72-6	OEM
13-73-6	0802
13-74-6	6412
13-100-6	6653
13-101-6	OEM
13-103-6	OEM
13-105-6	5300
13-106-6	OEM
13-107-6	OEM
13-108-6	OEM
13-109-6	OEM
13-110-6	OEM
13-111-6	OEM
13-112-6	OEM
13-113-6	OEM
13-114-6	OEM
13-115-6	OEM
13-116-6	OEM
13-117-6	1672
13-118-6	OEM
13-119-6	OEM
13-120-6	OEM
13-121-6	OEM
13-122-6	OEM
13-123-6	OEM
13-124-6	OEM
13-125-6	OEM
13-126-6	OEM
13-127-6	OEM
13-128-6	OEM
13-129-6	OEM
13-131-6	OEM
13-132-6	OEM
13-133-6	OEM
13-135-6	OEM
13-136-6	OEM
13-137-6	OEM
13-138-6	OEM
13-139-6	OEM
13-140-6	OEM
13-141-6	OEM
13-142-6	OEM
13-143-6	OEM
13-144-6	OEM
13-145-6	OEM
13-146-6	OEM
13-147-6	OEM
13-148-6	OEM
13-149-6	OEM
13-150-6	OEM
13-151-6	OEM
13-152-6	OEM
13-153-6	OEM
13-155-6	OEM
13-156-6	OEM
13-157-6	OEM
13-158-6	OEM
13-159-6	OEM
13-160-6	OEM
13-161-6	OEM
13-162-6	OEM
13-163-6	OEM
13-164-6	OEM
13-165-6	OEM
13-166-6	OEM
13-167-6	OEM
13-168-6	OEM
13-169-6	OEM
13-170-6	OEM
13-171-6	OEM
13-172-6	OEM
13-173-6	OEM
13-900-6	OEM
13-901-6	OEM
13-902-6	OEM
13-903-6	OEM
13-905-6	OEM
13-906-6	OEM
13-907-6	OEM
13-908-6	OEM
13-912-6	OEM
13-915-6	OEM
13-917-6	OEM
13-920-6	OEM
13-922-6	OEM
13-923-6	OEM
13-925-6	OEM
13-926-6	OEM
13-928-6	OEM
13-931-6	OEM
13-932-6	OEM
13-933-6	OEM
13-935-6	OEM
13-936-6	OEM
13-937-6	OEM
13-938-6	OEM
13-939-6	OEM
13-941-6	OEM
13-945-6	OEM
13-946-6	OEM
13-949-6	OEM
13-950-6	OEM
13-951-6	OEM
13-952-6	OEM
13-954-6	OEM
13-956-6	OEM
13-957-6	OEM
13-959-6	OEM
13-960-6	OEM
13-961-6	OEM
13-963-6	OEM
13-964-6	OEM
13-965-6	OEM
13-967-6	OEM
13-968-6	OEM
13-970-6	OEM
13-971-6	OEM
13-972-6	OEM
13-973-6	OEM
13-974-6	OEM
13-975-6	OEM
13-977-6	OEM
13-979-6	OEM
13-981-6	OEM
13-982-6	OEM
13-984-6	OEM
13-985-6	OEM
13-986-6	OEM
13-987-6	OEM
13-1032-5	0144
13-1032-51	0224
13-1836-1	0016
13-1847-1	0423
13-2384-2	0007
13-6247-3	0015
13-10102-1	0015
13-10320-14	0007
13-10321-1	0144
13-10321-2	0144
13-10321-3	0015
13-10321-5	0144
13-10321-6	0007
13-10321-7	0144
13-10321-8	0144
13-10321-9	0144
13-10321-10	0007
13-10321-11	0007
13-10321-12	0144
13-10321-13	OEM
13-10321-14	0144
13-10321-15	0144
13-10321-16	0144
13-10321-17	0144
13-10321-20	0144
13-10321-21	0144
13-10321-24	0917
13-10321-26	0144
13-10321-29	0127
13-10321-30	0144
13-10321-31	0127
13-10321-32	0127
13-10321-34	2518
13-10321-35	0224
13-10321-36	0224
13-10321-37	0212
13-10321-41	0224
13-10321-42	0917
13-10321-43	0144
13-10321-46	0224
13-10321-47	0007
13-10321-48	0628
13-10321-50	0259
13-10321-51	0144
13-10321-52	OEM
13-10321-53	0212
13-10321-54	0133
13-10321-55	0133
13-10321-56	0111
13-10321-59	0079
13-10321-62	0224
13-10321-65	0127
13-10321-66	0224
13-10321-67	0144
13-10321-70	2458
13-10321-71	0224
13-10321-72	0127

If replacement code is OEM, contact original manufacturer for replacement.

DEVICE TYPE	REPL CODE
13-10321-75	0008
13-10321-76	0007
13-10321-79	0144
13-10321-151	OEM
13-10321-152	OEM
13-10321-153	2518
13-10321-154	1029
13-10321-155	OEM
13-10321-156	0224
13-10321-157	0224
13-10321-158	3607
13-10321-159	0371
13-10321-160	0127
13-10321-161	0371
13-12001-0	0123
13-12002-0	0143
13-12003-0	0019
13-13021-15	0127
13-13543-1	0142
13-13543-2	0142
13-14065-77	0079
13-14085-1	0144
13-14085-2	0144
13-14085-3	0127
13-14085-4	0127
13-14085-6	0016
13-14085-7	0016
13-14085-9	0004
13-14085-10	0004
13-14085-11	0164
13-14085-12	0164
13-14085-13	0037
13-14085-14	0400
13-14085-15	0016
13-14085-15A	0086
13-14085-16	0127
13-14085-17	0127
13-14085-18	0004
13-14085-23	0004
13-14085-24	0144
13-14085-25	0164
13-14085-26	0224
13-14085-27	0144
13-14085-28	0050
13-14085-29	0142
13-14085-30	0050
13-14085-31	0136
13-14085-32	0136
13-14085-33	0136
13-14085-34	0016
13-14085-35	0004
13-14085-41	0111
13-14085-49	0111
13-14085-50	0016
13-14085-54	0076
13-14085-60	0004
13-14085-71	0004
13-14085-72	0016
13-14085-74	0144
13-14085-75	0144
13-14085-76	0144
13-14085-77	0144
13-14085-83	0076
13-14085-84	0016
13-14085-85	0111
13-14085-86	1267
13-14085-87	0037
13-14085-88	0042
13-14085-89	0016
13-14085-91	0191
13-14085-92	0006
13-14085-93	0042
13-14085-94	0086
13-14085-95	0111
13-14085-96	0111
13-14085-97	0111
13-14085-121	0042
13-14085-122	0079
13-14085-125	0286
13-14085-126	1790
13-14094-1	0143
13-14094-2	0143
13-14094-3	0143
13-14094-5	0143
13-14094-8	0123
13-14094-9	0143
13-14094-11	0143
13-14094-12	0015
13-14094-13	0030
13-14094-14	0143
13-14094-15	0143
13-14094-16	0015
13-14094-17	0015
13-14094-24	0080
13-14094-31	0030
13-14094-33	0133
13-14094-35	OEM
13-14094-36	0139
13-14094-38	0015
13-14094-39	0015
13-14094-42	0015
13-14094-54	0015
13-14097-7	0133
13-14261-1	0015
13-14261-3	0015
13-14278-1	0030
13-14278-2	0030
13-14279-1	0208
13-14604-1	0265
13-14605-1	0086
13-14606-1	0016
13-14627-1	0015
13-14627-4	0015
13-14735	0160
13-14735-1	0160
13-14735A	0160
13-14778-1	0265
13-14858-1	0015
13-14879-1	0002
13-14879-2	0137
13-14879-3	0002
13-14879-4	0649
13-14879-5	0420
13-14879-6	0531
13-14879-7	0531
13-14886-1	0136
13-14887-1	0136
13-14888-3	0211
13-14889-1	0136
13-14890-1	0143
13-15465-1	6563
13-15804-1	0016
13-15805-1	0211
13-15806-1	0160
13-15808-1	0076
13-15808-2	0016
13-15809-1	0233
13-15810-1	0007
13-15833-1	0086
13-15835-1	0224
13-15836-1	0211
13-15840-1	0016
13-15840-2	0016
13-15841-1	0076
13-15842-1	0324
13-15865-1	0016
13-16104-8	0015
13-16104-9	0015
13-16105-3	1303
13-16106-1	0774
13-16219-7	0196
13-16235-8	0143
13-16247-3	0015
13-16570-1	0037
13-16570-1A	0037
13-16570-2	0037
13-16570-2A	0037
13-16592-1	0975
13-16607-1	0975
13-16608-1	0975
13-16744-1	0144
13-16769-1	0016
13-17174-1	0071
13-17174-2	0015
13-17174-3	0015
13-17174-4	0087
13-17174-5	0071
13-17204	0162
13-17204-1	0133
13-17546-2	0133
13-17557-1	0015
13-17569	0474
13-17569-1	0604
13-17569-2	0133
13-17595-2	0133
13-17596-1	0474
13-17596-2	0124
13-17596-2(BIAS)	0917
13-17596-2(DET/AGC)	0133
13-17596-2(RECT)	0015
13-17596-3	0124
13-17596-4	0124
13-17596-5	0133
13-17596-6	0124
13-17596-7	0124
13-17596-8	0124
13-17596-9	0124
13-17596-10	0124
13-17607-1	0975
13-17607B	0975
13-17608-1	0975
13-17608-2	0975
13-17608B	0975
13-17608C	0975
13-17609-1	0975
13-17825-1	0015
13-17918-1	0103
13-18032-1	0004
13-18033-1	0004
13-18034	0160
13-18034-1	0160
13-18034A	0160
13-18087-1	0016
13-18087-2	0016
13-18158-1	0016
13-18198-1	0604
13-18282	0142
13-18282-1	0142
13-18304-1	0004
13-18359	0142
13-18359-1	0142
13-18359-3	0142
13-18359A	0142
13-18363	0198
13-18363-1	0155
13-18363-1A	0198
13-18364-1	0016
13-18365-1	0111
13-18458-1	0015
13-18481-1	0242
13-18481-2	0110
13-18481-3	0242
13-18642-1	0604
13-18642-2	0604
13-18642-2D	0604
13-18642-2E	0604
13-18642-2F	0604
13-18642-3	0604
13-18642-3A	0604
13-18642-3B	0604
13-18642-20	0604
13-18654-1	0600
13-18671-1	0004
13-18671-1A	0004
13-18671-1B	0004
13-18671-1C	0004
13-18924-1	0940
13-18924-1(LARGE-CAS	1396
13-18924-1(SMALL-CAS	1137
13-18924-2	0344
13-18924-3	1137
13-18924-4	1137
13-18924-5	1137
13-18924-7	0058
13-18924-8	1137
13-18927-1	0286
13-18927-1A	0086
13-18944-1	0280
13-18944-2	0280
13-18946-1	0050
13-18946-2	0050
13-18947-1	0050
13-18948-1	0136
13-18948-2	0050
13-18949-1	0144
13-18950-1	0007
13-18950-2	0050
13-18951-1	0136
13-18951-2	0136
13-18956-2	0280
13-19776-1	0037
13-21606-1	0975
13-22017-0	0133
13-22154-0	0030
13-22154-500	0554
13-22156-0	0005
13-22319-0	0317
13-22452-0	0015
13-22463-0	0015
13-22463-1	0023
13-22581	0016
13-22581-1	0016
13-22582-1	0037
13-22582-1A	0037
13-22606-0	0124
13-22609-0	0015
13-22690-1	0325
13-22692-1	0325
13-22692-2	0325
13-22739-1	0160
13-22741	0160
13-22741-1	0265
13-23001-2	0224
13-23002-2	0224
13-23013-2	1144
13-23160-2	0330
13-23160-3	0127
13-23160-4	0079
13-23160-5	0127
13-23163-2	0113
13-23309-5	0076
13-23323-4	0079
13-23323-6	0144
13-23324-6	0079
13-23325-5	0079
13-23326-6	2050
13-23327-4	0111
13-23338-3	0079
13-23338-4	0111
13-23339-2	0111
13-23339-3	0286
13-23505-2	0086
13-23506-2	0436
13-23508-0	0460
13-23510-4	0018
13-23543-1	0142
13-23543-2	0142
13-23594-1	0103
13-23785-1	0004
13-23822	0008
13-23822-1	0008
13-23824	0079
13-23824-1	0008
13-23824-2	0007
13-23824-3	0330
13-23825-1	0233
13-23826-1	0037
13-23826-1A	0037
13-23826-2	0037
13-23826-2A	0037
13-23826-3	0037
13-23826-3A	0037
13-23826-Q	0006
13-23840-1	0086
13-23892-5	0595
13-23916-1	0016
13-23917-1	0133
13-25226-1	0079
13-25343-1	0142
13-26009-1	0007
13-26377-1	0222
13-26377-2	2763
13-26386-1	0037
13-26386-1A	0037
13-26386-2	0037
13-26386-2A	0037
13-26386-3	0037
13-26386-4	0037
13-26576-1	0007
13-26576-2	0007
13-26577-1	0007
13-26577-2	0007
13-26577-3	0007
13-26614-1	0087
13-26666	0086
13-26666-1	0086
13-27050-1	0038
13-27404-1	0079
13-27404-2	0016
13-27432-1	0086
13-27432-2	0079
13-27433-1	0016
13-27443-1	0086
13-27596-5	0133
13-27974-1	0276
13-27974-2	0276
13-27974-3	0276
13-27974-4	0276
13-28222-1	1199
13-28222-2	0561
13-28222-3	0561
13-28222-4	0561
13-28336-1	0460
13-28336-2	0161
13-28336-20	1657
13-28386-1	0126
13-28391-1	0037
13-28391-1A	0037
13-28391-2	0037
13-28391-2A	0037
13-28392-1	0086
13-28392-2	0086
13-28392-3	0086
13-28392-5	0595
13-28393-1	0037
13-28393-1A	0037
13-28393-2	0037
13-28393-2A	0037
13-28393-3	0086
13-28394-1	0086
13-28394-2	0086
13-28394-3	0086
13-28394-4	0126
13-28394-4A	0126
13-28394-5	0126
13-28394-5A	0126
13-28394-6	0126
13-28432-1	0861
13-28432-2	0233
13-28469-2	0042
13-28471-1	0086
13-28471-1(METAL)	0086
13-28532-1	0561
13-28584	0144
13-28584-1	0144
13-28654-1	0325
13-28654-2	0325
13-28654-3	2888
13-28654-4	2888
13-28654-5	0325
13-29033-1	0079
13-29033-2	0079
13-29033-3	0079
13-29033-4	0111
13-29033-5	0079
13-29033-6	0079
13-29033-7	OEM
13-29165-1	0087
13-29165-2	0087
13-29392-2	0144
13-29432-1	0016
13-29437-1	0338
13-29656-1	0774
13-29663-1	0102
13-29687-2	0133
13-29775-1	0400
13-29775-2	0400
13-29775-3	0400
13-29776-1	0037
13-29776-1A	0037
13-29776-2	0037
13-29776-3	0037
13-29777-1	0030
13-29867-1	0133
13-29867-2	0133
13-29867-2(SUPP)	0015
13-29867-2(SW)	0133
13-29947-1	0079
13-29974-1	0007
13-29974-4	0276
13-30281	0143
13-31013-1	0715
13-31013-1/2	0037
13-31013-2	0037
13-31013-3	0196
13-31013-4	0144
13-31013-5	0047
13-31013-6	0015
13-31013-7	0224
13-31013-8	0144
13-31013-9	0006
13-31013-10	0079
13-31013-11	0715
13-31013-13	0595
13-31013-17	1051
13-31013-18	1165
13-31013-19	0532
13-31013-20	0224
13-31013-24	0436
13-31013-25	0324
13-31013-26	0595
13-31013-27	0563
13-31013-28	0155
13-31013-30	0006
13-31013-43	2660
13-31013-44	0955
13-31013-45	0076
13-31013-49	0086
13-31013-50	0126
13-31013-52	0006
13-31013-53	1066
13-31013-54	0224
13-31013-59	1284
13-31014-1	0015
13-31014-2	0196
13-31014-3	0015
13-31014-4	0917
13-31014-6	0023
13-31014-7	0023
13-31014-32	0348
13-31014-41	0378
13-32362-1	0111
13-32363-1	OEM
13-32364-1	0037
13-32366-1	0144
13-32366-2	0144
13-32630-1	0079
13-32630-2	0595
13-32630-3	0086
13-32630-4	0595
13-32631	0791
13-32631-1	0037
13-32631-2	0791
13-32631-3	0126
13-32632-1	0551
13-32634-1	0560
13-32635-1	1267
13-32636-1	0042
13-32638-1	0161
13-32640-1	0042
13-32642-1	0161
13-33172-1	0182
13-33172-2	0102
13-33173-1	0849
13-33174-1	0285
13-33174-2	0285
13-33174-8	OEM
13-33175-1	0400
13-33175-2	0400
13-33176-1	0285
13-33177-1	0030
13-33178-1	0791
13-33179	OEM
13-33179-1	0245
13-33179-2	0002
13-33179-3	0213
13-33179-4	0170
13-33179-5	0531
13-33179-6	0137
13-33179-7	0186
13-33179-8	1606
13-33179-9	OEM
13-33179-12	3026
13-33179-14	0137
13-33180-1	0551
13-33181-1	0223
13-33181-2	0223
13-33181-3	0223
13-33182-1	0074
13-33182-2	0074
13-33183-0	1396
13-33183-1	6169
13-33184-1	0316
13-33185-1	0566
13-33186-1	0039
13-33187-1	0049
13-33187-2	0189
13-33187-3	0700
13-33187-4	0494
13-33187-5	0104
13-33187-6	0064
13-33187-7	0064
13-33187-8	0287
13-33187-9	0705
13-33187-10	0340
13-33187-11	0052
13-33187-12	0157
13-33187-13	0089
13-33187-14	0376
13-33187-15	0714
13-33187-18	0686
13-33187-19	0275
13-33187-20	0213
13-33187-22	OEM
13-33187-23	0455
13-33187-26	0077
13-33187-27	1817
13-33187-28	OEM
13-33187-30	OEM
13-33187-31	0700
13-33187-34	0186
13-33187-38	0165
13-33187-39	0295
13-33188-2	0103
13-33189-1	0920
13-33190-1	0296
13-33350-1	0016
13-33376-1	0015
13-33595-1	0016
13-33595-2	0016
13-33595-3	0016
13-33742-1	0560
13-33742-1METAL	0086
13-33742-1PLASTIC	0551
13-33925-1	0161
13-33959-1	1137
13-34001-1	0400
13-34002-1	0423
13-34002-2	0423
13-34002-3	2997
13-34002-4	0423
13-34002-5	2997
13-34003-1	0218
13-34003-1(METAL)	0086
13-34004-1	0382
13-34004-1(METAL)	0126
13-34045-1	0144
13-34045-2	0144
13-34046-1	0560
13-34046-2	0560
13-34046-3	0560
13-34046-4	0560
13-34046-5	0236
13-34047-1	1267
13-34047-2	1267
13-34047-3	1267
13-34047-4	0681
13-34048-1	OEM
13-34056-1	0133
13-34057-1	0015
13-34089-2	0276
13-34089-4	0276
13-34338-1	OEM
13-34367-1	0037
13-34368-1	0242
13-34369-1	0150
13-34371-1	0086
13-34372-1	0264
13-34372-2	0560
13-34373-1	0382
13-34373-2	1267
13-34374-1	0130
13-34375-1	0325
13-34375-2	0325
13-34378-1	0325
13-34378-2	0325
13-34378-3	0325
13-34381	0079
13-34381-1	0111
13-34381-2	0111
13-34616-1	0423
13-34617-1	0854
13-34684-1	0130
13-34838-1	0423
13-34839-1	0854
13-34901-1	1287
13-34940-1	0037
13-35059-1	0167
13-35089-1	1708
13-35089-2	1708
13-35089-3	1708
13-35089-4	1708
13-35224-1	OEM
13-35226-1	0079
13-35257-1	0276
13-35257-2	0276
13-35257-3	0276
13-35257-4	0276
13-35324	3534
13-35324-1	2258
13-35550	0144
13-35550-1	0861
13-35621-1	0143
13-35792-1	0004
13-35807-1	0861
13-35807-2	0079
13-35894-4	0190
13-36386-1	0037
13-36444-3	0161
13-36445-3	0460
13-37526-1	0551
13-37527-1	0382
13-37708-1	0543
13-37833-2	OEM
13-37868-1	0023
13-37869-1	0715
13-37870-1	0065
13-37870-2	0065
13-37900-1	0937
13-37905-1	0039
13-37933-1	OEM
13-37933-2	0378
13-39004-1	0042
13-39004-2	0042
13-39046-3	0560
13-39047-3	1267
13-39072-1	0182
13-39072-2	0182
13-39073-1	0102
13-39074-1	0236
13-39098-1	0560
13-39099-1	0042
13-39099-2	0423
13-39100-1	0925
13-39106-1	OEM
13-39114-1	0079
13-39114-2	0079
13-39114-3	0595
13-39115-1	0342
13-39115-2	0791
13-39115-3	0037
13-39146-1	0338
13-39174-1	0338
13-39607-4	4121
13-39678-1	4121
13-39678-3	4121
13-39678-4	4121
13-39819-1	0925
13-39819-2	0925
13-39851-1	0861
13-39860-1	0015
13-39863-1	0039
13-39867-2	0102
13-39884-1	0042
13-39884-2	0042
13-39884-3	0042
13-39970-1	0037
13-40083-1	0037
13-40083-2	0037
13-40312-1	0111
13-40340-1	2997
13-40341-1	0681
13-40342-1	2485
13-40343-1	1598
13-40344-1	0563
13-40345-1	0525
13-40346-1	0130
13-40347-1	1681
13-41122-1	0071
13-41122-2	0023
13-41122-4	0071
13-41123-2	0071
13-41123-4	0071
13-41628-2	0236
13-41628-3	0042
13-41628-5	0219
13-41629-4	0681
13-41738-1	0543
13-43005-1	0236
13-43112-1	0325
13-43250-1	0133
13-43382-1	0288
13-43463-1	0313
13-43463-2	0065
13-43633-1	0363
13-43634-1	0037
13-43635-1	0423
13-43766-1	0015
13-43773-1	0079
13-43777-1	0023
13-43777-2	0023
13-43790-1	0560
13-43791-1	1267
13-43956-1	1217
13-43956-2	1217
13-44290	0325
13-44291	0325
13-45016-1	0223
13-45018-1	1708
13-45147-3	0348
13-45321-1	0285
13-50385-1	0525
13-50482-1	0460
13-50484-1	0280
13-50486-1	0280
13-50528-1	0111
13-50631-1	0280
13-50944-1	0280
13-55009-1	0975
13-55009-2	0975
13-55010-1	0288
13-55018-4	0086
13-55020-1	0144
13-55029-1	0015
13-55030-1	0102
13-55031-1	0102
13-55031-2	0102
13-55031-3	0102
13-55046-1	0143
13-55061-1	0016
13-55061-1(AGC)	0086
13-55061-1(HOR-MUL)	0079
13-55061-1(SOUND)	0079
13-55061-2	0016
13-55061-2(SYL)	0079
13-55061-2(WARDS)	0086
13-55062-1	0233
13-55063-1	0144
13-55064-1	0086
13-55064-1(SYL)	0178
13-55064-1(WARDS)	0595
13-55065-1	0144

If replacement code is OEM, contact original manufacturer for replacement.

DEVICE TYPE	REPL CODE
13-55066-1	0016
13-55066-2	0016
13-55066-2(SYL)	0086
13-55066-2(WARDS)	0086
13-55067-1	0016
13-55068-1	0016
13-55069-1	0037
13-55069-1A	0037
13-55078-1	0102
13-55166-1	0143
13-55166-1/2439-	0143
13-55166-1/2439-2	0143
13-55323-1	OEM
13-55332-1	0137
13-55333-1	0913
13-67539-1	0277
13-67539-1/3464	0277
13-67544-1	0157
13-67583-5	0144
13-67583-6	0016
13-67584-4/3464	0007
13-67585-4	0016
13-67585-4/2439-	0861
13-67585-4/2439-2	0079
13-67585-5	0016
13-67585-5/2439-	0861
13-67585-5/2439-2	0144
13-67585-6/2439-	0079
13-67585-7	0079
13-67585-7/2439-	0861
13-67585-7/2439-2	0016
13-67586-3	0016
13-67590-1	0133
13-67590-1/2439-	0143
13-67590-1/2439-2	0143
13-67599-3	0004
13-67599-3/2439-	1303
13-67599-3/2439-2	0004
13-67599-9	0136
13-67600-3	0030
13-67600-4	OEM
13-67600-4/2439-2	0030
13-67600-5	0030
13-68617-1	0016
13-85943-1	0196
13-85943-2	0196
13-85943-3	0196
13-85962-1	0143
13-86416-1	0038
13-86420-1	0600
13-87433-1	0600
13-87539-4	0015
13-87539-4(REALTONE)	0277
13-94096-2	0004
13-100000	2557
13-100001	OEM
13-102001	0023
13-105698-1	0086
13-200003	OEM
13-200004	OEM
13-200005	OEM
13-200007	OEM
13-200008	OEM
13-200009	OEM
13-200010	OEM
13-200011	OEM
13-200013	OEM
13-200014	OEM
13-200015	OEM
13-200016	OEM
13-200017	OEM
13-200018	OEM
13-200019	OEM
13-200020	OEM
13-200021	OEM
13-200022	OEM
13-200023	OEM
13-200024	OEM
13-200025	OEM
13-200026	OEM
13-200028	OEM
13-200029	OEM
13-200030	OEM
13-200031	OEM
13-200032	OEM
13-200033	OEM
13-200034	OEM
13-200035	OEM
13-200036	OEM
13-200037	OEM
13-200038	OEM
13-200039	OEM
13-200040	OEM
13-200041	OEM
13-200042	OEM
13-200045	OEM
13-200050	OEM
13-200052	OEM
13-200056	OEM
13-200057	OEM
13-200058	OEM
13-200059	OEM
13-200060	OEM
13-200061	OEM
13-200062	OEM
13-200068	OEM
13-200069	OEM
13-200070	OEM
13-200072	OEM
13-200073	OEM
13-200074	OEM
13-200076	OEM
13-200077	OEM
13-200078	OEM
13-200079	OEM
13-200080	OEM
13-200081	OEM
13-200083	OEM
13-200084	OEM
13-200085	OEM
13-200089	OEM
13-200090	OEM
13-200091	OEM
13-200092	OEM
13-200093	OEM
13-200094	OEM
13-200095	OEM
13-200096	OEM
13-200100	OEM
13-200101	OEM
13-200102	OEM
13-200103	OEM
13-200104	OEM
13-200114	OEM
13-200117	OEM
13-200118	OEM
13-200119	OEM
13-200122	OEM
13-200123	OEM
13-200124	OEM
13-200126	OEM
13-200133	OEM
13-200134	OEM
13-200136	OEM
13-200138	OEM
13-200139	OEM
13-200140	OEM
13-200143	OEM
13-200148	OEM
13-200154	OEM
13-200155	OEM
13-200162	OEM
13-200165	OEM
13-200172	OEM
13-200173	OEM
13-200176	OEM
13-200177	OEM
13-200178	OEM
13-200179	OEM
13-200180	OEM
13-200181	OEM
13-200182	OEM
13-200183	OEM
13-200184	OEM
13-200185	OEM
13-200186	OEM
13-200187	OEM
13-200188	OEM
13-200191	OEM
13-200192	OEM
13-200193	OEM
13-200194	OEM
13-200195	OEM
13-200197	OEM
13-200198	OEM
13-200199	OEM
13-200200	OEM
13-200201	OEM
13-200202	OEM
13-200203	OEM
13-200206	OEM
13-200207	OEM
13-200208	OEM
13-200210	OEM
13-200211	OEM
13-200216	OEM
13-200218	OEM
13-200220	OEM
13-200222	OEM
13-200223	OEM
13-200226	OEM
13-200227	OEM
13-200228	OEM
13-200229	OEM
13-200230	OEM
13-200231	OEM
13-200232	OEM
13-200238	OEM
13-200240	OEM
13-200244	OEM
13-200254	OEM
13-200263	OEM
13-200264	OEM
13-200265	OEM
13-350591	0167
13-1000000	2557
13-1000001	2557
13-3015121-1	0313
13-3015122-1	0079
13-3015123-1	0037
13-3015124-1	0439
13-3015125-1	1852
13-3015127-1	0473
13-3015128-1	0715
13-3015132-1	0861
13-3015733-1	0037
13-3015735-1	2395
13-3015739-1	OEM
13-3017119-2	2022
13-3017148-1	4337
13-3017149-1	0079
13-3017150-1	0037
13-3017629-1	0264
13-3017630-1	0426
13A	0053
13A8	OEM
13A565	OEM
13B	0053
13B,,	0053
13B1	0053
13B2	0053
13B3	0053
13BKJ1	1303
13D1	OEM
13D2	OEM
13D4	0015
13DD02F	0170
13E1	OEM
13E2	OEM
13E10	OEM
13E20A	OEM
13E35	OEM
13E60	OEM
13E100	OEM
13F64	OEM
13J2	0015
13J2F	0015
13K3	OEM
13M85	OEM
13N1	0053
13N2	0053
13P1	0015
13P2	0133
13P2(RECTIFIER)	0071
13PT030	0908
13R2	OEM
13RC2	3277
13RC5	0731
13RC10	0712
13RC15	0469
13RC20	0469
13RC25	0721
13RC30	0721
13RC40	0722
13RC50	0725
13RC60	0725
13RC70	0679
13RC80	0679
13RECT-SI-1014	0023
13S	0201
13SD-1N60	0019
13SD-1N60S	0019
13SI-TMS3453N2L	OEM
13ST-2SC2073SO	1284
13T1	OEM
13T4	0142
13V64	OEM
13V66/LD932	OEM
13V92	OEM
13Y	0053
13Z4	OEM
13Z6	OEM
13Z6A	OEM
13Z6F	0456
13Z6FAF	OEM
13Z-1005	0143
14-0072-1	0015
14-0072-1(PHILCO)	0015
14-0072-2	0015
14-0072-2(PHILCO)	0015
14-0072-3	0015
14-0072-3(PHILCO)	0015
14-0104-1	0042
14-0104-2	0925
14-0104-3	0126
14-0104-4	0086
14-0104-5	0400
14-0104-7	0016
14-0110-1	0007
14-0110-2	0007
14-1	0016
14-2	0016
14-3	0016
14-201I-02	0372
14-501-01	0196
14-501-02	0196
14-502-01	0143
14-503-01	0484
14-503-02	0484
14-503-03	0484
14-503-04	0484
14-503-08	0484
14-504-01	0143
14-504-04	0484
14-507-01	0133
14-508-01	0030
14-509-01	0137
14-509-02	0162
14-510-01	0143
14-511-01	0143
14-512-01	0143
14-513-01	0143
14-514-01	0143
14-514-02	0133
14-514-03	0102
14-514-04	0123
14-514-04/64	0123
14-514-05	0143
14-514-06	0143
14-514-08	0143
14-514-09	0143
14-514-10	0143
14-514-11	0123
14-514-11/61	0123
14-514-12	0124
14-514-13	0123
14-514-13/63	0123
14-514-14	0102
14-514-15	0133
14-514-16	0123
14-514-16/66	0123
14-514-17	0133
14-514-18	0102
14-514-19	0133
14-514-20	1149
14-514-21	0143
14-514-22	0143
14-514-46	OEM
14-514-55	0143
14-514-61	0143
14-514-62	0124
14-514-64	0124
14-514-65	1149
14-514-66	0102
14-514-70	0133
14-514-72	0143
14-515-01	0137
14-515-02	0162
14-515-03	0137
14-515-04	0649
14-515-05	0495
14-515-06	0510
14-515-07	0002
14-515-08	0870
14-515-09	0002
14-515-11	0002
14-515-12	0012
14-515-13	0649
14-515-14	0495
14-515-15	0162
14-515-16	0012
14-515-17	0649
14-515-18	0794
14-515-19	0012
14-515-23	0686
14-515-24	0012
14-515-25	0137
14-515-26	0649
14-515-73	0002
14-557-10	0004
14-564-08	0164
14-566-08	0136
14-569-09	0050
14-572-10	0079
14-573-10	0160
14-574-10	0160
14-575-10	0016
14-576-10	0164
14-577-10	0004
14-578-10	0160
14-579-10	0160
14-580-01	0050
14-581-01	0050
14-582-01	0136
14-583-01	0016
14-584-01	0004
14-585-01	0050
14-586-01	0085
14-587-01	0050
14-588-01	0050
14-589-01	0160
14-590-01	0160
14-591-01	0050
14-593-01	1669
14-593-03	1669
14-600-01	0136
14-600-02	0136
14-600-04	0136
14-600-07	0143
14-600-10	0050
14-600-11	0050
14-600-13	0050
14-600-16	0136
14-600-19	0050
14-600-20	0050
14-600-22	0050
14-601-01	0160
14-601-02	0604
14-601-03	0160
14-601-04	0160
14-601-05	0160
14-601-06	0160
14-601-07	0160
14-601-08	0160
14-601-09	0160
14-601-10	0103
14-601-11	0160
14-601-12	0103
14-601-13	0103
14-601-14	0074
14-601-15	0103
14-601-15A	0103
14-601-16	0103
14-601-16A	0103
14-601-17	0161
14-601-18	0103
14-601-20	0103
14-602-01	0286
14-602-02	0016
14-602-03	0127
14-602-04	0004
14-602-05	0004
14-602-05A	0004
14-602-06	0164
14-602-07	0164
14-602-08	0164
14-602-09	0164
14-602-3	2866
14-602-10	0004
14-602-11	0037
14-602-11A	0037
14-602-12	0016
14-602-13	0037
14-602-14	0016
14-602-15	0004
14-602-16	0016
14-602-17	0016
14-602-18	0086
14-602-19	0233
14-602-20	0037
14-602-20A	0037
14-602-21	0208
14-602-22	0016
14-602-23	0016
14-602-24	0086
14-602-25	0111
14-602-26	0016
14-602-27	2866
14-602-28	0126
14-602-28A	0126
14-602-29	0086
14-602-30	0086
14-602-31	0007
14-602-32	0037
14-602-32A	0037
14-602-33	2866
14-602-34	0007
14-602-35	0016
14-602-36	0233
14-602-37	0233
14-602-38	2866
14-602-39	2866
14-602-40	2866
14-602-41	0144
14-602-42	0037
14-602-42A	0037
14-602-43	0086
14-602-44	0037
14-602-44A	0037
14-602-45	0007
14-602-46	0111
14-602-46A	0111
14-602-47	0037
14-602-47A	0037
14-602-48	0016
14-602-49	0079
14-602-50	0016
14-602-51	0164
14-602-52	0208
14-602-54	0791
14-602-54A	0037
14-602-55	0595
14-602-55A	0016
14-602-56	0037
14-602-56A	0037
14-602-57	4163
14-602-58	0037
14-602-58A	0037
14-602-59	0086
14-602-60	0126
14-602-61	0016
14-602-62	0016
14-602-63	0111
14-602-64	0693
14-602-65	0086
14-602-66	0126
14-602-67	0178
14-602-68	0037
14-602-69	0016
14-602-70	0595
14-602-71	0791
14-602-72	0086
14-602-73	0126
14-602-74	0595
14-602-75	0791
14-602-76	0168
14-602-77	0127
14-602-77B	0127
14-602-78	0016
14-602-79	0126
14-602-79A	0126
14-602-80	0016
14-602-81	0016
14-602-85	0126
14-602-580	0037
14-602-600	0037
14-603-01	0086
14-603-02	0007
14-603-02A	0086
14-603-03	0007
14-603-04	0016
14-603-05	0127
14-603-05-2	0127
14-603-06	0127
14-603-07	0086
14-603-08	0007
14-603-09	0007
14-603-3	OEM
14-603-10	0016
14-603-11	0016
14-603-12	0144
14-603-13	0007
14-604-01	2763
14-604-02	0222
14-604-03	3545
14-604-07	0160
14-604-08	0160
14-607-29	0126
14-607-29A	0126
14-608-01	0161
14-608-02	0161
14-608-02A	0423
14-609-01	0338
14-609-01A	0338
14-609-02	0285
14-609-02A	0338
14-609-03	0042
14-609-03A	0423
14-609-04	0042
14-609-05	0338
14-609-49A	0144
14-700-01	0325
14-700-02	0325
14-700-03	0325
14-700-04	0325
14-700-05	0325
14-800-32	0016
14-801-23	0111
14-802-12	0016
14-803-12	0037
14-804-12	0037
14-805-12	0016
14-807-12	0150
14-808-12	0037
14-850-12	0007
14-851-12	0007
14-853-23	0016
14-854-12	0016
14-856-23	0037
14-857-12	0037
14-858-12	0016
14-862-32	0016
14-863-23	0037
14-865-12	0016
14-866-32	0016
14-867-32	0037
14-901-12	0285
14-2000-01	0400
14-2000-02	0400
14-2000-03	0400
14-2000-04	0400
14-2000-05	0400
14-2002-01	0325
14-2007	2161
14-2007-00	2161
14-2007-00B	0753
14-2007-01	0753
14-2007-02	2161
14-2007-03	0753
14-2008-01	0664
14-2010-01	0666
14-2010-03	0349
14-2011-01	0372
14-40325A	0103
14-40363A	0103
14-40369A	0103
14-40421A	0103
14-40464A	0103
14-40465A	0103
14-40466A	0103
14-40471	0103
14-40934-1	0103
14-60237	0285
14-SAM	1892
14A	0160
14A0	0160
14A1	0160
14A1-A82	0160
14A2	0160
14A3	0160
14A4	0160
14A5	0160
14A6	0160
14A7	0160
14A8	0160
14A8-1	0050
14A8-1-12	0050
14A9	0160
14A10	0160
14A10R	0160
14A11	0160
14A12	0160
14A13	0160
14A13P	0160
14A14	0160
14A14-7	0160
14A14-7B	0160
14A15	0160
14A15L	0160
14A16	0160
14A16-5	0160
14A17	0160
14A17-1	0160
14A18	0160
14A19	0160
14A19G	0160
14A502	OEM
14B338-001	OEM
14B348-003	1067
14B348-1	1067
14B348-2	1067
14B348-3	1067
14B348-4	1067
14B348-5	1067
14DM8	OEM
14DN160	3521
14E20A	OEM
14E30	OEM
14E50	OEM
14E100	OEM
14J2	0015
14J2F	0015
14L0035	3729
14LN033	2581
14LN034	2593
14LQ007	0579
14LV233	3254
14MW69	0004
14P1	OEM
14P2	0015
14R2	OEM
14SI-HEF4011UBP	OEM
14SP-LT-3211B	OEM
14T-009-003	OEM
14T1	OEM
14T4	0179
14T4S	OEM
14Z4	OEM
14Z6	OEM
14Z6A	OEM
14Z6F	0533
14Z6FAF	OEM
15	OEM
15-0075603-04X	OEM
15-033-0	0015
15-082019	0079
15-085002	0143
15-085003	0143
15-085004	0196
15-085005	0133
15-085006	0015
15-085007	0015
15-085008	0133
15-085009	0143
15-085015	0015
15-085016	0015
15-085018	0917
15-085027	0143
15-085032	0143
15-085033	0071
15-085037	0143
15-085038	0196
15-085039	0196
15-085040	0015
15-085041	0071
15-085042	0924
15-085043	0015
15-085047	0196
15-085061	0143
15-088002	0037
15-088002A	0037
15-088003	0127
15-088004	0144
15-1	0016
15-2	0016
15-2R51-1	0701
15-3	0037
15-4	0037
15-5	0037
15-15	OEM
15-19CC19	0028
15-20A70	0123
15-20A90	0123
15-20A-90M	0143
15-30	0037
15-40	0037
15-50	0037
15-166N	0127
15-5016T/IC7313AP	0362
15-5021/ICLM8363	1247
15-7142C-1	0701
15-14504-1	3380
15-26587-1	0632
15-31015-1	6671
15-31015-7	0167
15-31015-13	OEM
15-33201-1	0167
15-33201-2	0167
15-34005-1	2529
15-34048	1940
15-34048-1	0664
15-34049-1	0701
15-34049-3	0701
15-34202-1	2529
15-34379-1	0443
15-34401-1	2282
15-34408-1	0664
15-34452	0678
15-34452-1	0678
15-34502-1	0360
15-34502-2	0360
15-34502-3	0360
15-34503	0701

If replacement code is OEM, contact original manufacturer for replacement.

DEVICE TYPE	REPL CODE
15-34503-1	0701
15-34503-2	0701
15-34906-1	1611
15-35059	0167
15-35059-1	0167
15-35059-2	0167
15-36446-1	1395
15-36647-1	3954
15-36994-1	2755
15-36995-1	1421
15-37534-1	3380
15-37534-2	3380
15-37700-1	2167
15-37701-1	2043
15-37701-2	2043
15-37702-1	0352
15-37703-1	0354
15-37704-1	0349
15-37833-2	1705
15-39060-1	1808
15-39061-1	3195
15-39075-1	0856
15-39098-1	0560
15-39207-1	2608
15-39208-1	2688
15-39209-1	1337
15-39600-1	3380
15-39600-2	3380
15-40140-2	0334
15-40183-1	0520
15-41545-1	3380
15-41627-1	2831
15-41627-2	2831
15-41627-3	2831
15-41764-1	2812
15-41764-2	2812
15-41856-1	3195
15-43098-1	1878
15-43251-1	1808
15-43251-2	1808
15-43312-1	2817
15-43312-2	2817
15-43636-1	1170
15-43637-1	4661
15-43638-1	2743
15-43703-1	0856
15-43704-1	2233
15-43705-1	0215
15-43706-1	2384
15-45141-1	0215
15-45184-1	1957
15-45185-1	1143
15-45186-1	0625
15-45300-1	0167
15-45693-1	3981
15-53201-1	1940
15-53201-2	1940
15-71420-1	0701
15-100001	0015
15-100002	0015
15-100003	2130
15-100004	0015
15-103022	0015
15-108001	2903
15-108002	0196
15-108003	0015
15-108004	0015
15-108005	0015
15-108006	0015
15-108007	0474
15-108008	0133
15-108009	0196
15-108010	0015
15-108011	0015
15-108012	0774
15-108013	0102
15-108014	0474
15-108015	0015
15-108016	0015
15-108017	0774
15-108020	5083
15-108021	0015
15-108022	0015
15-108023	0196
15-108024	0015
15-108025	0133
15-108026	0102
15-108027	0102
15-108028	1303
15-108029	0102
15-108030	0201
15-108031	0015
15-108032	0015
15-108033	0649
15-108034	0015
15-108035	0102
15-108036	0015
15-108037	0015
15-108038	0474
15-108039	OEM
15-108040	0102
15-108041	0102
15-108042	0133
15-108043	0102
15-108044	0102
15-108045	0201
15-108046	0071
15-108047	0102
15-108048	0102
15-108049	0015
15-108050	0015
15-108107	0774
15-123060	1054
15-123065	0042
15-123100	0826
15-123101	0133
15-123102	0023
15-123103	0102
15-123104	0288
15-123105	0015
15-123106	0649
15-123202	0236
15-123230	1150
15-123231	0288
15-123242	0023
15-123243	0023
15-123300	0015
15-123303	1051
15-123353	OEM
15-123412	0392
15-123413	OEM
15-123425	0080
15-301513-1	1048
15-392007-1	2608
15-2210921	0050
15-2221011	0004
15-3012472-01X	OEM
15-3013086-01X	OEM
15-3013885-01X	OEM
15-3013886-01X	OEM
15-3013887-01X	OEM
15-3013888-01X	OEM
15-3015129-1	0521
15-3015130-1	0379
15-3015131-1	3711
15-3015131-2	3710
15-3015131-3	3711
15-3015132-1	0622
15-3015569-1	0478
15-3015570-1	3611
15-3015734-1	0069
15-3015740-1	5702
15-3015807-1	OEM
15-3015808-1	OEM
15-3017045-1	1057
15-3017119-1	2022
15-3017119-2	2022
15-3017148-1	OEM
15-3017630-1	0426
15-22210111	0004
15-22210131	0050
15-22210300	0050
15-22210921	0050
15-22211021	0050
15-22211200	0004
15-22211328	0004
15-22211921	0050
15-22214400	0050
15-22214411	0050
15-22214435	0050
15-22214821	0050
15-22214831	0050
15-22216500	0004
15-22216600	0050
15-22217400	0050
15-22223720	0016
15-30115734-1	OEM
15A1HA1	OEM
15A1HS0	OEM
15A1HS1	OEM
15A1PA1	OEM
15A1PS0	OEM
15A1PS1	OEM
15A2HA2	OEM
15A2HS0	OEM
15A2HS1	OEM
15A2PA2	OEM
15A2PS0	OEM
15A2PS1	OEM
15A3HA3	OEM
15A3HS0	OEM
15A3HS1	OEM
15A3PA3	OEM
15A3PS0	OEM
15A3PS1	OEM
15A4HA4	OEM
15A4HS0	OEM
15A4HS1	OEM
15A4PA4	OEM
15A4PS0	OEM
15A4PS1	OEM
15A5HA5	OEM
15A5HS0	OEM
15A5HS1	OEM
15A5PA5	OEM
15A5PS0	OEM
15A5PS1	OEM
15A6HS0	OEM
15A6HS1	OEM
15A6PS0	OEM
15A6PS1	OEM
15A7HS0	OEM
15A7HS1	OEM
15A7PS0	OEM
15A7PS1	OEM
15A8HS0	OEM
15A8HS1	OEM
15A8PS0	OEM
15A8PS1	OEM
15A9HS0	OEM
15A9HS1	OEM
15A9PS0	OEM
15A9PS1	OEM
15A10HS0	OEM
15A10HS1	OEM
15A10PS0	OEM
15A10PS1	OEM
15A11HS0	OEM
15A11HS1	OEM
15A11PS0	OEM
15A11PS1	OEM
15A12HS0	OEM
15A12HS1	OEM
15A12PS0	OEM
15A12PS1	OEM
15A50	OEM
15A100	OEM
15A200	OEM
15A400	OEM
15A600	OEM
15A800	OEM
15A14663	6818
15A14664	0261
15A14665	OEM
15A14724	OEM
15A14741	5500
15A14742	5500
15A17125	0532
15A17126	1541
15AGD	OEM
15AK60	OEM
15AK80	OEM
15AK100	OEM
15AN20	OEM
15AN30	OEM
15AN40	OEM
15AN50	OEM
15AN60	OEM
15AN80	OEM
15AN90	OEM
15AN100	OEM
15AN120	OEM
15AN140	OEM
15AN160	OEM
15B011	OEM
15B1	0953
15B1(ZENER)	0879
15B2	OEM
15B2(1WATT)	0002
15B2(ZENER)	0879
15B2,,	0686
15B3	0686
15B3(ZENER)	1105
15B3(ZENER),,	0686
15B4	0540
15B4B41	0323
15B6	0965
15B8	0817
15B10	0922
15BC11	OEM
15BD11	0015
15BD11(FA-1)	OEM
15BG15	OEM
15BL11	OEM
15C05	0110
15C1	0953
15C2	0242
15C2D	0242
15C4	0540
15C6	0965
15C8	0817
15C10	0922
15C12	OEM
15C14	OEM
15C16	OEM
15CC11	OEM
15CD11	OEM
15CG11	OEM
15CG15	OEM
15CL11	OEM
15D02M1	OEM
15D05M1	OEM
15D4B41	5400
15D4D41	OEM
15D10K1	OEM
15D48B41	OEM
15DD02F	0170
15DF4	0023
15DF6	0023
15DF8	OEM
15DF15E1	OEM
15DF15E2	OEM
15DF15E3	OEM
15DG15	OEM
15DL11	OEM
15E	2312
15E6	0965
15E7	0817
15E8	0817
15E10	0922
15E20A	OEM
15E40	OEM
15E60	OEM
15E100	OEM
15F6	0965
15F8	0817
15F10	0922
15F15E1	OEM
15F15E2	OEM
15F15E3	OEM
15FC11	OEM
15FD11	OEM
15FT1N	OEM
15FT2N	OEM
15FT3N	OEM
15FT4N	OEM
15FT5N	OEM
15FT6N	OEM
15FWJ11	OEM
15G4B41	5402
15GQJ11	OEM
15GS1	OEM
15GS2	OEM
15GS4	OEM
15GS6	OEM
15GS8	OEM
15GS10	OEM
15GS12	OEM
15GWJ11	OEM
15H3P	0989
15H3PN	0990
15IP005	OEM
15IP005D	OEM
15IP005N	OEM
15IP01	OEM
15IP01D	OEM
15IP01N	OEM
15IP02	OEM
15IP02D	OEM
15IP02N	OEM
15IP04	OEM
15IP04D	OEM
15IP04N	OEM
15IP06	OEM
15IP06D	OEM
15IP06N	OEM
15IP08	OEM
15IP08D	OEM
15IP08N	OEM
15IP10	OEM
15IP10D	OEM
15IP10N	OEM
15J2	0015
15J2F	0015
15J4B41	5403
15KP17	OEM
15KP17A	OEM
15KP18	OEM
15KP18A	OEM
15KP20	OEM
15KP20A	OEM
15KP22	OEM
15KP22A	OEM
15KP24	OEM
15KP24A	OEM
15KP26	OEM
15KP26A	OEM
15KP28	OEM
15KP28A	OEM
15KP30	OEM
15KP30A	OEM
15KP33	OEM
15KP33A	OEM
15KP36	OEM
15KP36A	OEM
15KP40	OEM
15KP40A	OEM
15KP43	OEM
15KP43A	OEM
15KP45	OEM
15KP45A	OEM
15KP48	OEM
15KP48A	OEM
15KP51	OEM
15KP51A	OEM
15KP54	OEM
15KP54A	OEM
15KP58	OEM
15KP58A	OEM
15KP60	OEM
15KP60A	OEM
15KP64	OEM
15KP64A	OEM
15KP70	OEM
15KP70A	OEM
15KP75	OEM
15KP75A	OEM
15KP78	OEM
15KP78A	OEM
15KP85	OEM
15KP85A	OEM
15KP90	OEM
15KP90A	OEM
15KP100	OEM
15KP100A	OEM
15KP110	OEM
15KP110A	OEM
15KP120	OEM
15KP120A	OEM
15KP130	OEM
15KP130A	OEM
15KP150	OEM
15KP150A	OEM
15KP160	OEM
15KP160A	OEM
15KP170	OEM
15KP170A	OEM
15KP180	OEM
15KP180A	OEM
15KP200	OEM
15KP200A	OEM
15KP220	OEM
15KP220A	OEM
15KP240	OEM
15KP240A	OEM
15KP260	OEM
15KP260A	OEM
15KP280	OEM
15KP280A	OEM
15KS17	OEM
15KS17A	OEM
15KS18	OEM
15KS18A	OEM
15KS20	OEM
15KS20A	OEM
15KS22	OEM
15KS22A	OEM
15KS24	OEM
15KS24A	OEM
15KS26	OEM
15KS26A	OEM
15KS28	OEM
15KS28A	OEM
15KS30	OEM
15KS30A	OEM
15KS33	OEM
15KS33A	OEM
15KS36	OEM
15KS36A	OEM
15KS40	OEM
15KS40A	OEM
15KS43	OEM
15KS43A	OEM
15KS45	OEM
15KS45A	OEM
15KS48	OEM
15KS48A	OEM
15KS51	OEM
15KS51A	OEM
15KS54	OEM
15KS54A	OEM
15KS58	OEM
15KS58A	OEM
15KS60	OEM
15KS60A	OEM
15KS64	OEM
15KS64A	OEM
15KS70	OEM
15KS70A	OEM
15KS75	OEM
15KS75A	OEM
15KS78	OEM
15KS78A	OEM
15KS85	OEM
15KS85A	OEM
15KS90	OEM
15KS90A	OEM
15KS100	OEM
15KS100A	OEM
15KS110	OEM
15KS110A	OEM
15KS120	OEM
15KS120A	OEM
15KS130	OEM
15KS130A	OEM
15KS150	OEM
15KS150A	OEM
15KS160	OEM
15KS160A	OEM
15KS170	OEM
15KS170A	OEM
15KS180	OEM
15KS180A	OEM
15KS200	OEM
15KS200A	OEM
15KS220	OEM
15KS220A	OEM
15KS240	OEM
15KS240A	OEM
15KS260	OEM
15KS260A	OEM
15KS280	OEM
15KS280A	OEM
15L15	OEM
15L20	OEM
15L25	OEM
15L30	OEM
15L35	OEM
15L40	OEM
15L45	OEM
15L50	OEM
15L55	OEM
15L60	OEM
15L65	OEM
15L70	OEM
15L75	OEM
15L80	OEM
15L85	OEM
15L90	OEM
15L95	OEM
15L100	OEM
15L110	OEM
15L120	OEM
15L130	OEM
15L140	OEM
15L150	OEM
15L160	OEM
15L170	OEM
15L180	OEM
15L190	OEM
15L200	OEM
15L210	OEM
15L220	OEM
15L230	OEM
15L240	OEM
15L250	OEM
15L260	OEM
15L270	OEM
15L280	OEM
15L290	OEM
15L300	OEM
15LC11	OEM
15M3PN	0990
15M4N	OEM
15M45	0686
15MA300	OEM
15MA400	OEM
15MLA160	OEM
15MLA180	OEM
15MLA200	OEM
15MQ20	1600
15MQ30	1600
15MQ40	1600
15MT15ZK	OEM
15MT20ZA	OEM
15N02F1	OEM
15N05M1	OEM
15N1	0015
15N2	0097
15N2(POWER)	0097
15N3	0097
15N3(POWER)	0097
15N10K1	OEM
15N15A	OEM
15N15A9	OEM
15N20A	OEM
15N20A9	OEM
15N25A	OEM
15N25A9	OEM
15N30A	OEM
15N30A9	OEM
15N35A	OEM
15N35A9	OEM
15N40A	OEM
15N40A9	OEM
15N45A	OEM
15N45A9	OEM
15N50A	OEM
15N50A9	OEM
15N55A	OEM
15N55A9	OEM
15N60A	OEM
15N60A9	OEM
15N65A	OEM
15N65A9	OEM
15N70A	OEM
15N70A9	OEM
15N75A	OEM
15N75A9	OEM
15N80A	OEM
15N80A9	OEM
15N85A	OEM
15N85A9	OEM
15N90A	OEM
15N90A9	OEM
15N95A	OEM
15N95A9	OEM
15N100A	OEM
15N100A9	OEM
15N110A	OEM
15N120A	OEM
15N130A	OEM
15N140A	OEM
15N160A	OEM
15N170A	OEM
15N180A	OEM
15N190A	OEM
15N200A	OEM
15N210A	OEM
15N220A	OEM
15N230A	OEM
15N240A	OEM
15N250A	OEM
15N260A	OEM
15P02F1-A	OEM
15P05M1-A	OEM
15P1	OEM
15P2	0133
15P2(POWER)	0097
15P3	0097
15P3(POWER)	0097
15P10K1	OEM
15PK17A	3201
15PK18A	OEM
15PK20A	OEM
15PK22A	1915
15PK24A	OEM
15PK26A	OEM
15PK28A	OEM
15PK30A	OEM
15PK33A	OEM
15PK36A	OEM
15PK40A	OEM
15PK43A	0568
15PK45A	OEM
15PK48A	OEM
15PK51A	OEM
15PK54A	OEM
15PK58A	0831
15PK60A	OEM
15PK64A	OEM
15PK70A	OEM
15PK75A	OEM
15PK78A	OEM
15PK85A	OEM
15PK90A	OEM
15PK100A	OEM
15PK110A	OEM
15PK120A	OEM
15PK130A	OEM
15PK150A	OEM
15PK160A	OEM
15PK170A	1405
15Q3	1250
15R0/18	OEM
15R1/18	OEM
15R1/52	OEM
15R2	OEM
15R2/4	OEM
15R2/66	OEM
15R2B	OEM
15R2S	OEM
15R3/4	OEM
15R3B	OEM
15R3S	OEM
15R4/3	OEM
15R4/3C	OEM
15R4/4	OEM
15R4/59	OEM
15R4C	OEM
15R4S	OEM
15R5/3	OEM
15R5/3D	OEM
15R5/5	OEM
15R5/61	OEM
15R5C	OEM
15R5S	OEM
15R6/3	OEM
15R6/3D	OEM
15R6/5	OEM
15R6/61	OEM
15R6C	OEM
15R6S	OEM
15R7/3	OEM
15R7/3D	OEM
15R7/5	OEM
15R7/61	OEM
15R7CD	OEM
15R7S	OEM
15R8/3	OEM
15R8/3D	OEM
15R8/5	OEM
15R8/63	OEM
15R8CD	OEM
15R8S	OEM
15R52	OEM
15R0/18	OEM
15S05	0110
15S1	0953
15S2	0242
15S4	0540
15S6	0965
15SB03S	1600
15SB04M	OEM
15SB04S	1600
15SI-CA3064E	0802
15SM6	OEM
15SM15	OEM
15SM25	OEM
15SM40	OEM
15SM55	OEM
15SM70	OEM
15SM100	OEM
15SM150	OEM
15SM200	OEM
15SP-005	OEM
15SP-01	OEM
15SP-02	OEM
15SP-04	OEM
15SP-06	OEM
15SP-08	OEM
15SP-10	OEM
15ST-2SB561C(AU)E	0436
15SZ-RD24EB	0494
15T1	OEM
15T4	OEM
15TB050	OEM
15TB100	OEM
15TB200	OEM
15TB400	OEM
15TB600	OEM
15V66	OEM
15W71	OEM
15Y	0686
15Z3	0037
15Z4	OEM
15Z6	OEM
15Z6A	OEM
15Z6AF	0451
15Z6F	0451
15Z6FAF	OEM
16	OEM
16-2	0015

DEVICE TYPE	REPL CODE
16-17	0325
16-19(SYMPHONIC)	0126
16-20(SYMPHONIC)	0086
16-500	0774
16-736	0144
16-3006	0077
16-21426	0007
16-147191229	0016
16-171191368	0016
16-207190405	0004
16-501190016	0071
16A1	0038
16A1(FLEETWOOD)	0016
16A2	0038
16A2(FLEETWOOD)	0016
16A132	OEM
16A545-7	0016
16A667-GRN	0111
16A667-ORG	0111
16A667-RED	0111
16A667-YEL	0111
16A667G	0079
16A667R	0086
16A667Y	0079
16A668-GRN	0111
16A668-ORG	0111
16A668-YEL	0111
16A668G	0079
16A668Y	0079
16A669-GRN	0111
16A669-YEL	0111
16A669G	0079
16A787	0211
16A1938	0016
16A6670	0086
16A6680	0086
16A14692	0133
16A14693	0124
16A14694	0133
16A14695	OEM
16A14696	OEM
16A14697	OEM
16A14698	OEM
16A14699	OEM
16A14721	OEM
16A14739	OEM
16A17050	OEM
16A17051	OEM
16A17052	0023
16A17130	OEM
16A17131	OEM
16A17132	OEM
16A17133	OEM
16A17134	OEM
16A17135	OEM
16A17136	OEM
16A17137	OEM
16ATCRA	0469
16ATCRB	0469
16ATCRC	0469
16ATCRD	0469
16ATCRE	0469
16ATCRF	0722
16ATCRG	0722
16ATCRH	0722
16ATCRI	0725
16ATCRJ	0725
16ATCRK	0679
16ATCRL	0679
16B1	0861
16B1(ZENER)	0445
16B2	0079
16B20	OEM
16B40	OEM
16B670-GRN	0400
16B670-RED	0861
16B670-YEL	0861
16B670R	0715
16B670Y	0715
16C025	0469
16C025B	0469
16C025C	3307
16C050	0469
16C050B	0469
16C050C	2188
16C-4	0015
16C4	0015
16C4B1P	0277
16C-4P	0277
16C10	0469
16C10B	0469
16C10C	0567
16C15	0469
16C15B	0469
16C15C	3258
16C20	0469
16C20B	0469
16C20C	0762
16C20R	OEM
16C25	0722
16C25B	0722
16C25C	3268
16C30	0722
16C30B	0722
16C30C	3271
16C35	0722
16C35B	0722
16C35C	0722
16C40	0722
16C40B	0722
16C40C	0740
16C40R	OEM
16C45	0725
16C45B	0725
16C45C	0725
16C50	0725
16C50B	0725
16C50C	3291
16C60	0725
16C60B	0725
16C60C	0764
16C70	0679
16C70B	0679
16C70C	2875
16C80	0679
16C80B	0679
16C80C	0766
16C90	0679
16C90B	0679
16C90C	0679
16C100	0679
16C100B	0679
16C100C	0679
16C110	0679
16C110B	0679
16C110C	0679
16C120	0679
16C120B	0679
16C120C	0679
16E08	OEM
16E4	OEM
16E5	OEM
16E6	OEM
16E16A	OEM
16E35	OEM
16E50	OEM
16E90	OEM
16E1330	0079
16E1330(GE)	0016
16F5	5514
16F5R	2291
16F10	5582
16F10B	2900
16F10R	2291
16F15	2900
16F20	2900
16F20R	2291
16F30	5583
16F30R	0476
16F40	3751
16F40R	0476
16F50	3757
16F50R	0476
16F60	1112
16F60R	0476
16F80	1610
16F80R	0449
16F100	3011
16F100R	0449
16F120	OEM
16F-B1	OEM
16FCRA	OEM
16FCRB	OEM
16FCRC	OEM
16FCRD	OEM
16FCRE	OEM
16FCRF	OEM
16FCRG	OEM
16FCRH	OEM
16FCRI	OEM
16FCRJ	OEM
16FCRK	OEM
16FCRL	OEM
16FDC	OEM
16FL5S02	OEM
16FL5S05	OEM
16FL5S10	OEM
16FL10S02	OEM
16FL10S05	OEM
16FL10S10	OEM
16FL20S02	OEM
16FL20S05	OEM
16FL20S10	OEM
16FL40S02	OEM
16FL40S10	OEM
16FL60S02	OEM
16FL60S05	OEM
16FL60S10	OEM
16FL80S05	OEM
16FL80S10	OEM
16FL100S05	OEM
16FL100S10	OEM
16FR10	2343
16FR20	1241
16FR30	OEM
16FR40	2365
16FR60	2376
16FR80	1264
16FR100	0449
16G010F	OEM
16G011F	OEM
16G011L	OEM
16G2	0016
16G5	0015
16G27	0325
16G28	0133
16G28A	0133
16GN	0127
16J1	0144
16J2	0144
16J2(DIODE)	0015
16J2F	0015
16J3	0079
16K1	0144
16K2	0144
16K3	0144
16KPR	OEM
16KQ30	OEM
16KQ30B	OEM
16KQ40	OEM
16KQ40B	OEM
16KQ50	OEM
16KQ50B	OEM
16KQ60	OEM
16KQ60B	OEM
16KQ90	OEM
16KQ90B	OEM
16KQ100	OEM
16KQ100B	OEM
16KZ	OEM
16L2	0144
16L3	0144
16L4	0111
16L5	0144
16L22	0144
16L23	0144
16L24	0111
16L25	0111
16L42	0016
16L43	0016
16L44	0016
16L45	0111
16L62	0016
16L63	0016
16L64	0127
16L65	0111
16M10	OEM
16M11	OEM
16M12	OEM
16MQ30	OEM
16MQ30(B)	OEM
16MQ40	OEM
16MQ40(B)	OEM
16MQ50	OEM
16MQ50(B)	OEM
16MQ60	OEM
16MQ60(B)	OEM
16MS1	OEM
16MS1A	OEM
16MS1B	OEM
16MS1C	OEM
16MS1D	OEM
16N2	0420
16NT100	OEM
16P1	OEM
16P2	2130
16P2881	0400
16P3367	0400
16PCRA	1651
16PCRB	1651
16PCRC	1651
16PCRD	1651
16PCRE	1651
16PCRF	1584
16PCRG	1584
16PCRH	1584
16PCRI	1665
16PCRJ	1665
16PCRK	OEM
16PCRL	OEM
16PCRNA	OEM
16PCRNB	OEM
16PCRNC	OEM
16PCRND	OEM
16PCRNE	OEM
16PCRNF	OEM
16PCRNG	OEM
16PCRNH	OEM
16PCRNI	OEM
16PCRNJ	OEM
16PCRNK	OEM
16PCRNL	OEM
16PCRSA	2519
16PCRSB	2519
16PCRSC	0741
16PCRSD	0745
16PCRSE	0745
16PCRSF	0747
16PCRSG	0747
16PCRSH	0747
16PCRSI	0752
16PCRSJ	0752
16PCRSK	OEM
16PCRSL	OEM
16RC2	0722
16RC2A	3277
16RC5	0722
16RC5A	0731
16RC10	0722
16RC10A	0712
16RC10AS24	0804
16RC15	0722
16RC15A	0469
16RC20	0722
16RC20A	0469
16RC20AS24	0804
16RC25	0722
16RC25A	0721
16RC30	0722
16RC30A	0722
16RC30AS24	0804
16RC40	0722
16RC40A	0722
16RC40AS24	0804
16RC50A	0778
16RC50AS24	OEM
16RC60	0725
16RC60A	0725
16RC60AS24	0804
16RC70	OEM
16RC70A	OEM
16RC80	OEM
16RC80A	0750
16RC90	OEM
16RC90A	OEM
16RC100	OEM
16RC100A	OEM
16RC110A	0679
16RC120A	OEM
16RCF5A	OEM
16RCF10A	OEM
16RCF15A	OEM
16RCF20A	OEM
16RCF25A	OEM
16RCF30A	OEM
16RCF40A	OEM
16RCF50A	OEM
16RCF60A	OEM
16RCF70A	OEM
16RCF80A	OEM
16RIA10	0469
16RIA10M	OEM
16RIA20	0469
16RIA20M	OEM
16RIA30	OEM
16RIA40	0725
16RIA40M	OEM
16RIA60	0725
16RIA60M	OEM
16RIA80	0679
16RIA80M	OEM
16RIA100	0679
16RIA100M	OEM
16RIA110	0679
16RIA110M	OEM
16RIA120	0679
16RIA120M	OEM
16S2-AW01-11	0317
16SL107	OEM
16SL108	OEM
16ST-2SA733P	0006
16ST-2SC945Q	0076
16ST-2SC945Q/R	0076
16ST-2SC945Q/R(AU)E	0076
16ST-2SC945R	0076
16ST-2SD994	0055
16SZ-AW01-11	0317
16T1	OEM
16T4	0346
16T4S	OEM
16T17	OEM
16T20	OEM
16T37-RE00	OEM
16T40	OEM
16TCRA	0469
16TCRB	0469
16TCRC	0469
16TCRD	0469
16TCRE	0469
16TCRF	0722
16TCRG	0722
16TCRH	0725
16TCRI	0725
16TCRJ	0725
16TCRK	0679
16TCRL	0679
16U1	0079
16X	0445
16X1	0016
16X2	0016
16X39	0133
16Z4	0137
16Z6	0162
16Z6A	0162
16Z6F	0162
16Z6FAF	OEM
17-10	0015
17-410	0015
17-443	0086
17-451	0016
17-457	0079
17-458	0126
17-459	0037
17-459A	0037
17-12054-1	1823
17-12056-1	1041
17-12057-1	2097
17-12058-1	0141
17-12064-1	0562
17-12065-1	0081
17-12089-1	2078
17A4422-1	0160
17A17068	OEM
17E07	OEM
17E14A	OEM
17E33	OEM
17E45	OEM
17E75	OEM
17P1	OEM
17P2	0133
17RCG5	OEM
17REB60	OEM
17S04	OEM
17S06	OEM
17S08	1046
17S08R	1789
17S10	1046
17S10R	1789
17S12	OEM
17S14	OEM
17S16	OEM
17S50	0143
17SP-LT-3211B	OEM
17SP-P722-7R	OEM
17T1	OEM
17T4	0346
17T4S	OEM
17TB1	0567
17TB1-CE00	OEM
17TB1-CH00	OEM
17TB1-EE00	OEM
17TB1-EH00	OEM
17TB1-GE00	OEM
17TB1-GH00	OEM
17TB1-JE00	OEM
17TB1-JH00	OEM
17TB1-LE00	OEM
17TB1-LH00	OEM
17TB1-NE00	OEM
17TB1-NH00	OEM
17TB1-RE00	OEM
17TB2	0762
17TB2-CE00	OEM
17TB2-CH00	OEM
17TB2-EE00	OEM
17TB2-EH00	OEM
17TB2-GE00	OEM
17TB2-GH00	OEM
17TB2-JE00	OEM
17TB2-JH00	OEM
17TB2-LE00	OEM
17TB2-LH00	OEM
17TB2-NE00	OEM
17TB2-NH00	OEM
17TB2-RE00	OEM
17TB3	3271
17TB3-CE00	OEM
17TB3-CH00	OEM
17TB3-EE00	OEM
17TB3-EH00	OEM
17TB3-GE00	OEM
17TB3-GH00	OEM
17TB3-JE00	OEM
17TB3-JH00	OEM
17TB3-LE00	OEM
17TB3-LH00	OEM
17TB3-NE00	OEM
17TB3-NH00	OEM
17TB3-RE00	OEM
17TB4	0740
17TB4-CE00	OEM
17TB4-CH00	OEM
17TB4-EE00	OEM
17TB4-EH00	OEM
17TB4-GE00	OEM
17TB4-GH00	OEM
17TB4-JE00	OEM
17TB4-JH00	OEM
17TB4-LE00	OEM
17TB4-LH00	OEM
17TB4-NE00	OEM
17TB4-NH00	OEM
17TB4-RE00	OEM
17TB5	3291
17TB5-CH00	OEM
17TB5-EE00	OEM
17TB5-EH00	OEM
17TB5-GE00	OEM
17TB5-GH00	OEM
17TB5-JE00	OEM
17TB5-JH00	OEM
17TB5-LE00	OEM
17TB5-LH00	OEM
17TB5-NE00	OEM
17TB5-NH00	OEM
17TB5-RE00	OEM
17TB6	0764
17TB6-CE00	OEM
17TB6-CH00	OEM
17TB6-EE00	OEM
17TB6-EH00	OEM
17TB6-GE00	OEM
17TB6-GH00	OEM
17TB6-JE00	OEM
17TB6-JH00	OEM
17TB6-LE00	OEM
17TB6-LH00	OEM
17TB6-NE00	OEM
17TB6-NH00	OEM
17TB6-RE00	OEM
17TB7	2875
17TB7-CE00	OEM
17TB7-CH00	OEM
17TB7-EE00	OEM
17TB7-EH00	OEM
17TB7-GE00	OEM
17TB7-GH00	OEM
17TB7-JE00	OEM
17TB7-JH00	OEM
17TB7-LE00	OEM
17TB7-LH00	OEM
17TB7-NE00	OEM
17TB7-NH00	OEM
17TB7-RE00	OEM
17TB8	0766
17TB8-CE00	OEM
17TB8-CH00	OEM
17TB8-EE00	OEM
17TB8-EH00	OEM
17TB8-GE00	OEM
17TB8-GH00	OEM
17TB8-JE00	OEM
17TB8-JH00	OEM
17TB8-LE00	OEM
17TB8-LH00	OEM
17TB8-NE00	OEM
17TB8-NH00	OEM
17TB8-RE00	OEM
17TB9	OEM
17TB10	OEM
17TB11	OEM
17TB12	OEM
17TB16-CH00	OEM
17TB16-GH00	OEM
17V10	OEM
17Z4	OEM
17Z6	0157
17Z6A	0157
17Z6AF	0157
17Z6F	0157
17Z6FAF	OEM
18-085001	0015
18-085002	0137
18-3	0375
18-607-2	0076
18-3539-1	0086
18A	0375
18A8-1	0050
18A8-1-12	0050
18A8-1-127	0050
18A8-1L	0050
18A8-1L8	0050
18AA-1-82	0160
18B1	0375
18BT03Z	OEM
18BT12Z	OEM
18DB2A	0289
18DB2A-C	0289
18DB4A	0296
18DB4A-C	0296
18DB6A	0302
18DB6A-C	0250
18DB8A	0250
18DB8A-C	0250
18DB10A	0250
18DB10A-C	0250
18E06	OEM
18E12A	OEM
18E27	OEM
18E40	OEM
18E55	OEM
18EB3	0375
18J2	0071
18J2F	0071
18L62	OEM
18NT100	OEM
18P2	1335
18P-2A82	0004
18RC2	3277
18RC5	0731
18RC10	0712
18RC15	0722
18RC20	0469
18RC25	0721
18RC30	0721
18RC40	0722
18RC50	0778
18RC60	0725
18RM90	OEM
18SI-SN76650N	0395
18ST-2SC1815Y(AP)	0076
18ST-2SC1815Y(AP)E	0076
18ST-2SC2371M	0276
18ST-2SD478C	0392
18ST-BU205-01	0065
18ST-MPSH81	0037
18T1	OEM
18Z6	0636
18Z6A	0636
18Z6AF	0636
18Z6F	0636
18Z6FAF	OEM
19-003935	0103
19-020-001	0136
19-020-002	0136
19-020-003	0004
19-020-005	0136
19-020-007	0004
19-020-015	0211
19-020-019	0040
19-020-031	0050
19-020-032	0050
19-020-033	0280
19-020-034	0004
19-020-035	0004
19-020-036	0004
19-020-037	0144
19-020-038	0086
19-020-043	5677
19-020-043A	0016
19-020-044	0007
19-020-045	0178
19-020-048	0007
19-020-052	0144
19-020-056	0042
19-020-058	0016
19-020-066	0042
19-020-070	0224
19-020-071	0127
19-020-072	0493
19-020-073	0016
19-020-074	0016
19-020-075	0086
19-020-076	0698
19-020-077	0698
19-020-079	0632
19-020-081	1669
19-020-44	0144
19-020-100	0126
19-020-101	0042
19-020-102	1267
19-020-111	2170
19-020-112	1977
19-020-114	0006
19-020-115	0325
19-040-002	0015
19-040-003	0015
19-040-004	0015
19-080-001	0133
19-080-002	0015
19-080-008	0133
19-080-009	0143
19-080-014	0041
19-090-007	0760
19-090-008	0317
19-090-008A	0170
19-090-014	0041
19-090-015	0188
19-05589	1404
19-09973-0	0268
19-076001	0579
19-085005	0143
19-085010	0015
19-085016	0030
19-085017	0649
19-085018	0143
19-085022	0015
19-1	0037
19-1-7	OEM
19-2	0037
19-3	0037
19-6-7	OEM
19-10	0037
19-20	0037
19-30	0037
19-130-004	0619
19-130-005	1208
19-3415	0164
19-3416	0164
19-3692	0086
19-3934-643	0086
19-3935-641	0264
19-8049	OEM
19-9072	OEM
19-10476-0	5338
19-100001	0071
19-C	1303
19A11552B-P1	0636
19A11644P1	0664
19A115024-P4	0015
19A115024-P6	0071
19A115056-P1	0975
19A115061-P1	0016
19A115061-P2	0016
19A115077-P1	0211
19A115077-P2	0211
19A115086-P1	0143
19A115087-P1	0136
19A115089	0097
19A115094-P1	0440
19A115098	0136
19A115098-P1	0136
19A115099-P1	0136
19A115100-P1	0015
19A115101	0160
19A115101-P1	0160
19A115102-P1	0016
19A115103-P1	0600
19A115108-P1	0016
19A115108-P2	0016
19A115123-P1	0016
19A115123-P2	0016
19A115129-2	0038
19A115129-P1	0038
19A115140-P1	0050
19A115140-P2	0050
19A115142-P1	0016
19A115142-P2	0016
19A115145-P3	0015
19A115145-P4	0015
19A115178-P1	0037
19A115178-P2	0037
19A115180-2	0126

If replacement code is OEM, contact original manufacturer for replacement.

DEVICE TYPE	REPL CODE
19A115184-P1	0160
19A115192-P1	0050
19A115192-P2	0050
19A115200-P1	0042
19A115201-P1	0600
19A115201-P2	0600
19A115208	0280
19A115208-P1	0280
19A115208-P2	0280
19A115249-1	0144
19A115253-P1	0016
19A115253-P2	0016
19A115267P1	0160
19A115281-P1	0211
19A115300-1	0086
19A115300-2	0086
19A115300-P1	0560
19A115300-P2	0560
19A115300-P3	0560
19A115301-P1	0280
19A115301-P2	0280
19A115304-2	0086
19A115315-P1	0016
19A115315-P2	0016
19A115322	0917
19A115322-P1	0917
19A115328-P1	0016
19A115328-P2	0016
19A115341P1	0160
19A115342-1	0144
19A115342-P1	0016
19A115342-P2	0016
19A115359-P1	0016
19A115359-P2	0016
19A115361-P1	0160
19A115362-P1	0016
19A115362-P2	0016
19A115385-P1	0160
19A115410-P1	0016
19A115410-P2	0016
19A115440-1	0144
19A115441-1	0144
19A115458-P1	0037
19A115458-P2	0037
19A115460-P1	0722
19A115487-P1	0440
19A115527	0178
19A115527-P1	0178
19A115528-P3	0012
19A115531-P1	0975
19A115540-P1	0440
19A115546-P1	0600
19A115546-P2	0600
19A115548-P1	0280
19A115552-P1	0079
19A115552-P2	0079
19A115553-P1	0050
19A115554-P1	0050
19A115554-P2	0050
19A115556-P1	0280
19A115561	0160
19A115567-P1	0050
19A115567-P2	0050
19A115569-P1	0015
19A115569-P2	0015
19A115623-P1	0142
19A115623-P2	0142
19A115628-P1	0050
19A115628-P2	0050
19A115635-1	0050
19A115635-P1	0050
19A115636-P1	0050
19A115653-P1	0037
19A115653-P2	0037
19A115654-P1	0037
19A115654-P2	0037
19A115665-P1	0050
19A115665-P2	0050
19A115666-1	0144
19A115673-P1	0600
19A115673-P2	0600
19A115674-P1	0004
19A115674-P2	0004
19A115683-P1	0722
19A115683-P2	0722
19A115688-P1	0037
19A115688-P2	0037
19A115706-P1	0037
19A115706-P2	0037
19A115720-1	0016
19A115728-1	0016
19A115747-P1	0722
19A115768-P1	0037
19A115768-P2	0037
19A115786	0016
19A115786A	0016
19A115913-1	1823
19A115913-2	2078
19A115913-3	1041
19A115913-4	2097
19A115913-10	0358
19A115913-19	0141
19A115913-20	0562
19A115913P2	2078
19A115913P14	1831
19A116180-7	1024
19A116180-8	0743
19A116180-11	1202
19A116180-18	1368
19A116180-22	0385
19A116180-24	1208
19A116180-25	0980
19A116180-27	0834
19A116180-29	1486
19A116180-P13	1404
19A116180P1	0232
19A116180P2	0268
19A116180P3	0314
19A116180P4	0512
19A116180P5	0697
19A116180P7	1024
19A116180P8	0743
19A116180P11	1202
19A116180P15	1172
19A116180P16	1313
19A116180P17	1368
19A116180P20	0361
19A123160-1	0144
19A126813	0103
19A126813A	0103
19A126826-P2	0130
19AI15527-1	0178
19AR2	0015
19AR3	0143
19AR4	0484
19AR4-1	0484
19AR5	0015
19AR6-1	0208
19AR6-2	0208
19AR6-3	0208
19AR7-1	0004
19AR7-2	0004
19AR8	0123
19AR11	0015
19AR12	0015
19AR13-1	0050
19AR13-2	0050
19AR13-3	0050
19AR13-4	0050
19AR14-1	0004
19AR14-2	0004
19AR16-1	0004
19AR16-2	0004
19AR17	0015
19AR18	0050
19AR19-1	0004
19AR19-2	0004
19AR20	0016
19AR22	0123
19AR24	0050
19AR25	0004
19AR26	0004
19AR27	0004
19AR29	0196
19AR29-1	0196
19AR30	0774
19AR31	0160
19AR32	0004
19AR34	0015
19AR35	0142
19AR36	0016
19B200011-P1	0097
19B200011-P2	0097
19B200011-P3	0097
19B200011-P4	0097
19B200011-P5	0015
19B200054-P1	0004
19B200061-P1	0004
19B200061-P2	0004
19B200061-P3	0004
19B200061-P4	0004
19B200063-P1	0004
19B200065-P1	0600
19B200065-P2	0600
19B200129-P1	0280
19B200130-P1	0050
19B200130-P2	0050
19B200132-P1	0211
19B200132-P2	0211
19B200132-P3	0211
19B200132-P4	0211
19B200210-P1	0004
19B200210-P2	0004
19B200210-P3	0004
19B200248-P1	0940
19B200248-P2	0940
19B200248-P3	0940
19B200249-P1	0133
19B200249-P2	0133
19B200249-P3	0133
19B200379-P1	0012
19B2000129-P1	0280
19B2000130-P1	0050
19B2000130-P2	0050
19B2000132-P1	0004
19B2000132-P2	0004
19B2000132-P3	0004
19B2000132-P4	0004
19BA009	OEM
19C300LL5-1	0086
19C300073-P1	0211
19C300073-P2	0211
19C300073-P3	0211
19C300073-P4	0211
19C300073-P5	0211
19C300073-P6	0211
19C300076-P1	0015
19C300076-P2	0015
19C300076-P3	0015
19C300076-P4	0015
19C300076-P5	0015
19C300076-P6	0015
19C300113-P1	0160
19C300114-P1	0016
19C300114-P2	0016
19C300115-P1	0086
19C300128-P1	0211
19C300128-P2	0211
19C300128-P3	0211
19C300128-P4	0211
19C300128-P5	0211
19C300128-P6	0211
19C300128-P7	0211
19C300128-P8	0211
19C300138-P4	0211
19C300138-P8	0211
19C300216-P1	0050
19C300216-P2	0050
19C307022	0030
19C307022-P1	0030
19C3001414P2	0079
19CC19	0649
19E06	OEM
19E12A	OEM
19E15-1	3714
19E18-1	3833
19E19-2	3861
19E25	OEM
19E40	OEM
19E70	OEM
19L-4A82	0004
19P1	0133
19P2	1335
19SZ-WZ-120	0052
19Z6	OEM
19Z6A	OEM
19Z6AF	0025
19Z6F	0025
19Z6FAF	OEM
20	0262
20-0352	1854
20-08048-433	OEM
20-1	0144
20-1(ZENER)	0700
20-3	0700
20-16-3	0143
20-22-08	0015
20-1680-143	0015
20-1680-174	0086
20-1680-175	0143
20-1680-189	0164
20-9062	0110
20A0007	0004
20A0009	0004
20A0015	0004
20A0017	0160
20A0041	0160
20A0042	0160
20A0053	0016
20A0054	0015
20A0059	0382
20A0060	0382
20A0073	0016
20A0074	0160
20A0075	0126
20A0076	0086
20A05	0071
20A1	0953
20A2	0242
20A3	1747
20A4	0540
20A5	1771
20A6	0965
20A6F	0601
20A8	0817
20A8F	OEM
20A10	0922
20A10F	OEM
20A11	0133
20A12	OEM
20A13	0133
20A50	OEM
20A-70	0143
20A70	0143
20A79	0143
20A-90	0123
20A90	0123
20A-90H	0143
20A90H	0019
20A90HMP	0123
20A90LF	0143
20A-90M	0143
20A90M	0143
20A90MLF	0123
20A90Z	0123
20A100	OEM
20A200	OEM
20A400	OEM
20A600	OEM
20A800	OEM
20A10849	0016
20AS	0015
20B1	OEM
20B1M	0632
20B2FR	OEM
20B4FR	OEM
20B6FR	OEM
20B20	OEM
20B40	OEM
20B409	0162
20B410	0162
20BG2C11	6551
20BG2Z11	OEM
20BL2C41	2234
20BS	0015
20C	0015
20C05	0071
20C1	0953
20C1R10	OEM
20C1R20	OEM
20C1R30	OEM
20C1S10	OEM
20C1S20	OEM
20C1S30	OEM
20C2	0242
20C2R10	OEM
20C2R20	OEM
20C2R30	OEM
20C2S10	OEM
20C2S20	OEM
20C2S30	OEM
20C3	1747
20C4	0540
20C5	1771
20C6	0965
20C8	0965
20C9	0817
20C10	0817
20C12	OEM
20C14	OEM
20C16	OEM
20C18	OEM
20C71	0004
20C72	0004
20C2073BGSZ	OEM
20CL2C41	2234
20CS04FM	OEM
20CS04H	2234
20CS04M	1942
20CT030	OEM
20CT035	OEM
20CT040	OEM
20CT045	OEM
20D02M1	OEM
20D05	0071
20D05MA	OEM
20D1	OEM
20D2	0023
20D2FM	OEM
20D2L	0071
20D4	0071
20D4FM	OEM
20D6	0071
20D6FM	OEM
20D8	OEM
20D10	0071
20D10K1	OEM
20DCS10	0015
20DCS10R	0015
20E06	OEM
20E1	0071
20E1R20	OEM
20E1R40	OEM
20E1R60	OEM
20E1S20	OEM
20E1S40	OEM
20E1S60	OEM
20E2	0071
20E2R20	OEM
20E2R40	OEM
20E2R60	OEM
20E2S20	OEM
20E2S40	OEM
20E2S60	OEM
20E4	0071
20E4FD	OEM
20E6	0071
20E8	0071
20E10	0071
20E10FA13	0071
20E12A	OEM
20E20	OEM
20E40	OEM
20E-54-5411	2017
20E70	OEM
20E1114	1030
20EB	0700
20F1	0580
20F1R07	OEM
20F1R15	OEM
20F1S07	OEM
20F1S15	OEM
20F2R07	OEM
20F2R15	OEM
20F2S07	OEM
20F2S15	OEM
20F5	1600
20F5R	2291
20F10	5282
20F10R	2291
20F15	0871
20F20	0871
20F20R	2291
20F30	5283
20F30R	0476
20F40	0853
20F40R	0476
20F50	OEM
20F60	1609
20F80	1610
20F100	1614
20FQ020	0615
20FQ025	OEM
20FQ030	0615
20FQ035	0615
20FQ040	0615
20FQ045	0615
20FQS07	OEM
20FR5	5005
20FR10	5491
20FR20	1635
20FR30	5495
20FR40	1251
20FR50	5496
20FR60	1205
20FR80	2137
20FR100	2251
20FT1K	OEM
20FT1L	OEM
20FT2K	OEM
20FT2L	OEM
20FT3K	OEM
20FT3L	OEM
20FT4K	OEM
20FT4L	OEM
20FT5K	OEM
20FT5L	OEM
20FT6K	OEM
20FT6L	OEM
20FT7K	OEM
20FT8K	OEM
20G2C11	OEM
20G2Z11	OEM
20G3Z11	OEM
20GG2C11	5985
20GG2Z11	OEM
20H	0015
20H2	0097
20H3	0097
20H3N	2291
20H3P	0989
20H3PN	0990
20H260RT	OEM
20HA3	0319
20HB5	1600
20HB20	0871
20HB40	0853
20J2	0097
20J3P	0097
20JG2C11	2709
20JG2Z11	OEM
20JH3	2657
20K	0015
20L2C41	OEM
20L15	OEM
20L20	OEM
20L25	OEM
20L30	OEM
20L35	OEM
20L40	OEM
20L45	OEM
20L50	OEM
20L55	OEM
20L60	OEM
20L65	OEM
20L70	OEM
20L75	OEM
20L80	OEM
20L85	OEM
20L90	OEM
20L95	OEM
20L100	OEM
20L110	OEM
20L120	OEM
20L130	OEM
20L140	OEM
20L150	OEM
20L160	OEM
20L170	OEM
20L180	OEM
20L190	OEM
20L200	OEM
20L210	OEM
20L220	OEM
20L230	OEM
20L240	OEM
20L250	OEM
20L260	OEM
20L270	OEM
20L280	OEM
20L290	OEM
20L300	OEM
20LF	OEM
20M	0015
20M10	2653
20M15	OEM
20M20	2657
20M30	2664
20M40	2006
20M50	2677
20M60	2682
20M80	3900
20M100	2655
20MA10	2695
20MA20	2657
20MA30	2006
20MA40	2006
20MA60	2682
20MA80	3900
20MA100	3208
20MA120	OEM
20MA140	OEM
20MA160	OEM
20MC	0050
20MLA60	OEM
20MLA80	OEM
20MLA100	OEM
20MLA120	OEM
20N02F1	OEM
20N05M1	OEM
20N1	0015
20N2	0097
20N3	0097
20N10K1	OEM
20N15A	OEM
20N15A9	OEM
20N20A	OEM
20N20A9	OEM
20N25A	OEM
20N25A9	OEM
20N30A	OEM
20N30A9	OEM
20N35A	OEM
20N35A9	OEM
20N40A	OEM
20N40A9	OEM
20N45A	OEM
20N45A9	OEM
20N50A	OEM
20N50A9	OEM
20N55A	OEM
20N55A9	OEM
20N60A	OEM
20N60A9	OEM
20N65A	OEM
20N65A9	OEM
20N70A	OEM
20N70A9	OEM
20N75A	OEM
20N75A9	OEM
20N80A	OEM
20N80A9	OEM
20N85A	OEM
20N85A9	OEM
20N90A	OEM
20N90A9	OEM
20N95A	OEM
20N95A9	OEM
20N100A	OEM
20N100A9	OEM
20N110A	OEM
20N120A	OEM
20N130A	OEM
20N140A	OEM
20N150A	OEM
20N160A	OEM
20N170A	OEM
20N180A	OEM
20N190A	OEM
20N200A	OEM
20N210A	OEM
20N220A	OEM
20N230A	OEM
20N240A	OEM
20N250A	OEM
20N260A	OEM
20P02F1-A	OEM
20P05M1-A	OEM
20P1S	0469
20P1SG	0712
20P2	0097
20P2S	0469
20P2SG	0469
20P3	0097
20P3S	0721
20P3SG	0722
20P4S	0722
20P4SG	0722
20P5S	0778
20P5SG	0725
20P6S	0725
20P6SG	0725
20P8SG	0750
20P10K1	OEM
20P10SG	OEM
20P12SG	OEM
20PM10	OEM
20Q3	1241
20R2	3189
20R3	0325
20R3P	OEM
20RM60	OEM
20RM80	OEM
20RM100	OEM
20RM120	OEM
20RM150	OEM
20RM200	OEM
20RM220	OEM
20RM250	OEM
20S05	0071
20S1	OEM
20S1S03	OEM
20S1S04	OEM
20S2	OEM
20S2S03	OEM
20S2S04	OEM
20S2SC	OEM
20S3	0805
20S3P	OEM
20S3SC	OEM
20S4	0805
20S4SC	OEM
20S5	0580
20S5SC	OEM
20S6	OEM
20S6SC	OEM
20SI-TA7607AP	0912
20SS2SC	OEM
20SS3SC	OEM
20SS4SC	OEM
20SS5SC	OEM
20SS6SC	OEM
20T2FR	OEM
20T3P	OEM
20T4FR	OEM
20T6FR	OEM
20TB1	0712
20TB1-CE00	OEM
20TB1-CH00	OEM
20TB1-EE00	OEM
20TB1-EH00	OEM
20TB1-GE00	OEM
20TB1-GH00	OEM
20TB1-JH00	OEM
20TB1-LE00	OEM
20TB1-LH00	OEM
20TB1-NE00	OEM
20TB1-NH00	OEM
20TB1-RE00	OEM
20TB2	0469
20TB2-CE00	OEM
20TB2-CH00	OEM
20TB2-EE00	OEM
20TB2-EH00	OEM
20TB2-GH00	OEM
20TB2-JH00	OEM
20TB2-LE00	OEM
20TB2-LH00	OEM
20TB2-NE00	OEM
20TB2-NH00	OEM
20TB2-RE00	OEM
20TB3	0721
20TB3-CE00	OEM
20TB3-CH00	OEM
20TB3-EE00	OEM
20TB3-EH00	OEM
20TB3-GE00	OEM
20TB3-GH00	OEM
20TB3-JH00	OEM
20TB3-LE00	OEM
20TB3-LH00	OEM
20TB3-NE00	OEM
20TB3-NH00	OEM
20TB3-RE00	OEM
20TB4	0722
20TB4-CE00	OEM
20TB4-CH00	OEM
20TB4-EE00	OEM
20TB4-EH00	OEM
20TB4-GE00	OEM
20TB4-GH00	OEM
20TB4-JE00	OEM
20TB4-JH00	OEM
20TB4-LE00	OEM
20TB4-LH00	OEM
20TB4-NE00	OEM
20TB4-NH00	OEM
20TB4-RE00	OEM
20TB5	0778
20TB5-CE00	OEM
20TB5-CH00	OEM
20TB5-EH00	OEM
20TB5-GE00	OEM
20TB5-GH00	OEM
20TB5-JE00	OEM
20TB5-JH00	OEM
20TB5-LE00	OEM
20TB5-LH00	OEM
20TB5-NE00	OEM
20TB5-NH00	OEM
20TB5-RE00	OEM
20TB6	0725
20TB6-CE00	OEM
20TB6-CH00	OEM
20TB6-EE00	OEM
20TB6-EH00	OEM
20TB6-GE00	OEM
20TB6-GH00	OEM
20TB6-JE00	OEM
20TB6-JH00	OEM
20TB6-LE00	OEM
20TB6-LH00	OEM
20TB6-NE00	OEM
20TB6-NH00	OEM
20TB6-RE00	OEM
20TB7	OEM
20TB7-CE00	OEM
20TB7-CH00	OEM
20TB7-EE00	OEM
20TB7-EH00	OEM
20TB7-GE00	OEM
20TB7-GH00	OEM

DEVICE TYPE	REPL CODE
20TB7-JE00	OEM
20TB7-JH00	OEM
20TB7-LE00	OEM
20TB7-LH00	OEM
20TB7-NE00	OEM
20TB7-NH00	OEM
20TB7-RE00	OEM
20TB8	OEM
20TB8-CE00	OEM
20TB8-CH00	OEM
20TB8-EE00	OEM
20TB8-EH00	OEM
20TB8-GE00	OEM
20TB8-GH00	OEM
20TB8-JE00	OEM
20TB8-JH00	OEM
20TB8-LE00	OEM
20TB8-LH00	OEM
20TB8-NE00	OEM
20TB8-NH00	OEM
20TB8-RE00	OEM
20TB9	OEM
20TB10	OEM
20TB11	OEM
20TB12	OEM
20TC05	OEM
20TC025	OEM
20TC10	OEM
20TC15	OEM
20TC20	OEM
20TC25	OEM
20TC30	OEM
20TC40	OEM
20TC50	OEM
20TC60	OEM
20TC70	OEM
20TC80	OEM
20TC90	OEM
20TC100	OEM
20TC110	OEM
20TC120	OEM
20TCA10	OEM
20TCA20	OEM
20TCA30	OEM
20TCA40	OEM
20TCA50	OEM
20TCA60	OEM
20TCA70	OEM
20TCA80	OEM
20TCA90	OEM
20TCA100	OEM
20TCA110	OEM
20TCA120	OEM
20V3P	OEM
20V10	OEM
20V-HG	0004
20V-MG	0280
20W3P	OEM
20X	0671
20X3P	OEM
20Y	0700
20Y3P	OEM
20Z6	OEM
20Z6A	OEM
20Z6AF	0649
20Z6F	0649
20Z6FAF	OEM
20Z8F	OEM
20ZB	OEM
21/3	0484
21/3.92	0484
21-0101	0595
21-1	0016
21-1AA	0316
21-1AA(MAGNAVOX)	0316
21-1L	0016
21-2	0127
21-4	0127
21-6	0127
21-7	0127
21-28	0160
21-32	0050
21-33	0050
21-34	0004
21-35	0142
21-36	0004
21-37	0004
21-810	0015
21-810-2	0015
21A001-000	0015
21A001-00	0015
21A002	0196
21A002-000	0196
21A003-016	0143
21A004-000	0102
21A005	0474
21A005-000	0004
21A006-000	0015
21A007-000	0015
21A008-000	0015
21A008-001	0015
21A008-002	0102
21A008-003	0015
21A008-008	0133
21A008-016	0474
21A009	0133
21A009-000	0143
21A009-002	0123
21A009-008	0133
21A009-009	0133
21A015-001	0004
21A015-002	2763
21A015-003	0222
21A015-004	0144
21A015-005	0208
21A015-006	0164
21A015-008	0150
21A015-008A	0037
21A015-009	0037
21A015-009A	0037
21A015-011	0037
21A015-011A	0037
21A015-012	0037
21A015-012A	0037
21A015-013	0050
21A015-014	0144
21A015-016	0144
21A015-018	0086
21A015-019	0086
21A015-020	0016
21A015-021	2763
21A015-022	0222
21A015-025	0037
21A015-026	0086
21A015-027	0016
21A020-001	0015
21A020-005	0143
21A020-006	0015
21A037-003	0137
21A037-006	0028
21A037-008	0137
21A037-009	0531
21A037-012	0165
21A037-016	0098
21A037-017	1649
21A037-018	0365
21A037-020	0317
21A037-021	0213
21A038-000	0004
21A039-000	0004
21A040-000	0004
21A040-001	0143
21A040-003	0144
21A040-004	0144
21A040-005	0004
21A040-007	0144
21A040-010	0144
21A040-014	0004
21A040-015	0740
21A040-016	0144
21A040-017	0144
21A040-019	0144
21A040-020	0016
21A040-021	0004
21A040-022	0004
21A040-023	0127
21A040-024	0127
21A040-025	0127
21A040-031	0136
21A040-032	0155
21A040-033	0016
21A040-033A	0016
21A040-034	0016
21A040-035	0847
21A040-036	0164
21A040-037	0016
21A040-045	0224
21A040-046	0330
21A040-047	0330
21A040-049	0126
21A040-050	0126
21A040-051	0086
21A040-052	0042
21A040-053	0224
21A040-054	0127
21A040-055	0127
21A040-056	0016
21A040-057	0004
21A040-058	0004
21A040-059	0037
21A040-060	0164
21A040-061	0164
21A040-063	0127
21A040-064	0111
21A040-065	0111
21A040-066	0111
21A040-067	0111
21A040-068	0470
21A040-077	0016
21A040-078	0016
21A040-079	0004
21A040-081	0164
21A040-082	0111
21A040-083	0111
21A040-091	0224
21A040-092	0016
21A040-36	0004
21A040-37	0016
21A040-44	0015
21A040-54	0224
21A041-002	0030
21A041-003	0030
21A045-000	0050
21A048-000	0050
21A048-00	OEM
21A049-000	0050
21A050-000	0050
21A050-001	0050
21A050-004	0144
21A051-000	0280
21A053-000	0004
21A054-000	0004
21A055-000	0004
21A062-000	0150
21A063-000	0211
21A064-000	0160
21A074-000	0004
21A079-000	0015
21A097-000	0160
21A1	0474
21A2	0196
21A3	0774
21A4	1303
21A6	0015
21A7	0015
21A11-002	0139
21A13-041	0030
21A101-001	2571
21A101-001(IC)	2571
21A101-001(RECT)	0015
21A101-002	2715
21A101-004	2882
21A101-005	0528
21A101-006	0638
21A101-007	0673
21A101-008	0930
21A101-009	0946
21A101-010	0964
21A101-011	2886
21A101-012	2887
21A101-013	2889
21A101-015	1808
21A101-016	0354
21A101-017	0352
21A101-018	2891
21A101-14	OEM
21A102-001	0015
21A102-002	0250
21A103-005	0030
21A103-006	0123
21A103-007	0015
21A103-010	0143
21A103-011	0143
21A103-012	0071
21A103-013	0071
21A103-015	0071
21A103-016	0123
21A103-017	0143
21A103-018	0015
21A103-019	0143
21A103-021	0015
21A103-022	0143
21A103-043	1925
21A103-044	0015
21A103-045	0317
21A103-047	0030
21A103-048	0123
21A103-049	0649
21A103-050	0015
21A103-052	0143
21A103-055	0143
21A103-057	OEM
21A103-058	0015
21A103-060	0030
21A103-064	0012
21A103-065	0133
21A103-069	0030
21A103-070	0015
21A103-104	0071
21A103-064	0012
21A105-001	0144
21A105-004	0086
21A105-006	0086
21A108-001	0133
21A108-002	0012
21A108-003	0133
21A108-004	0133
21A109-001	0143
21A109-002	0143
21A109-003	0123
21A109-022	0143
21A110-001	0015
21A110-002	0015
21A110-003	0015
21A110-004	0071
21A110-005	0102
21A110-006	0023
21A110-007	0015
21A110-008	0102
21A110-009	0102
21A110-012	0015
21A110-013	0102
21A110-014	0102
21A110-071	0023
21A110-072	0071
21A111-001	0133
21A111-002	0139
21A112-001	0037
21A112-002	0126
21A112-003	0006
21A112-004	0791
21A112-006	0102
21A112-007	0127
21A112-010	0007
21A112-013	0155
21A112-015	0016
21A112-017	0016
21A112-018	0016
21A112-019	1220
21A112-020	0076
21A112-023	0003
21A112-025	0142
21A112-029	0142
21A112-031	0074
21A112-033	0142
21A112-036	2051
21A112-045	0111
21A112-046	0076
21A112-047	0037
21A112-048	1267
21A112-049	0560
21A112-050	0016
21A112-058	0076
21A112-062	1220
21A112-063	1220
21A112-065	0037
21A112-070	0168
21A112-071	0168
21A112-074	0861
21A112-084	0224
21A112-085	0016
21A112-086	0144
21A112-087	0144
21A112-088	0016
21A112-089	0016
21A112-090	0016
21A112-091	0016
21A112-092	0016
21A112-093	0037
21A112-094	0925
21A112-095	0042
21A112-096	1209
21A112-098	0168
21A112-100	0037
21A112-101	0151
21A112-102	0006
21A112-103	0065
21A112-104	0016
21A112-120	1327
21A112-124	2125
21A112-125	0681
21A112-126	1648
21A113-002	0325
21A118-008	0168
21A118-029	0042
21A118-031	0018
21A118-032	0532
21A118-036	2051
21A118-049	0423
21A118-063	1284
21A118-096	0261
21A118-12	OEM
21A118-100	0955
21A118-114	0220
21A118-122	1648
21A118-124	2125
21A118-131	1806
21A118-137	0076
21A118-139	0006
21A118-140	0076
21A118-153	0276
21A118-201	0951
21A118-233	0560
21A118-234	0728
21A118-235	2268
21A118-236	1978
21A118-237	0161
21A118-238	5407
21A118-863	0085
21A119-005	0123
21A119-008	0053
21A119-030	0015
21A119-034	1005
21A119-038	0057
21A119-040	0120
21A119-041	0157
21A119-045	1329
21A119-049	0700
21A119-063	1284
21A119-064	0471
21A119-068	0023
21A119-073	0165
21A119-075	0015
21A119-077	2637
21A119-079	0015
21A119-080	0133
21A119-108	0041
21A119-123	0123
21A119-187	0023
21A119-189	0015
21A119-190	0023
21A120-001	0395
21A120-002	1808
21A120-008	0167
21A120-009	4642
21A120-016	0232
21A120-017	0467
21A120-044	1055
21A120-045	4641
21A120-049	0490
21A120-050	4060
21A120-051	0912
21A120-052	2625
21A120-071	2231
21A120-074	4549
21A120-099	0912
21A120-178	0767
21A120-185	2623
21A120-204	1429
21A120-208	4635
21A500	0474
21A500-000	0474
21A112095	0042
21B1AH	2457
21B1M	0654
21B1Z	0654
21B-14	0015
21B-17	0015
21D1FM	OEM
21D2FM	OEM
21D3FM	OEM
21D4FM	OEM
21D6FM	OEM
21DQ03	0031
21DQ04	0031
21DQ05	0031
21DQ06	0031
21DQ09	0031
21DQ10	0031
21E06	OEM
21E12A	OEM
21E18	OEM
21E38A	OEM
21E60	OEM
21ECG5123	3317
21F02-2F	OEM
21F02-2N	OEM
21F02-4F	OEM
21F02-4I	OEM
21F02-4N	OEM
21F02-21	OEM
21F02B	OEM
21F02F	OEM
21F02I	OEM
21F02N	OEM
21F1SC	0871
21F2SC	0871
21F3SC	0805
21FQ030	0615
21FQ035	0615
21FQ040	0615
21FQ045	0615
21J-23840	0015
21K60	0143
21L02-1B	OEM
21L02-1F	OEM
21L02-1I	OEM
21L02-1N	OEM
21L02-2F	OEM
21L02-2I	OEM
21L02-2N	OEM
21L02-3B	OEM
21L02-3F	OEM
21L02-3I	OEM
21L02-3N	OEM
21L02B	4973
21L02F	OEM
21L02FDC	OEM
21L02FFC	OEM
21L02FPC	OEM
21L02HDC	OEM
21L02HFC	OEM
21L02HPC	OEM
21L02I	OEM
21L02N	OEM
21L021DC	OEM
21L021FC	OEM
21L021PC	OEM
21L022DC	OEM
21L022FC	OEM
21L022PC	OEM
21M006	0136
21M007	0280
21M018	0030
21M020	0006
21M022	0037
21M025	2021
21M026	2021
21M027	1242
21M028	1267
21M084	0016
21M086	0016
21M091	0111
21M093	0224
21M094	0127
21M095	0111
21M099	0224
21M122	0016
21M123	0016
21M124	0111
21M125	0016
21M137	0111
21M138	0111
21M139	0016
21M140	0127
21M146	0016
21M149	0016
21M150	0016
21M151	0127
21M152	0127
21M153	0127
21M154	0127
21M160	0111
21M161	0111
21M174	0111
21M178	0127
21M179	0127
21M180	0560
21M181	0560
21M182	0127
21M183	0560
21M184	0560
21M185	0086
21M186	0016
21M188	0127
21M192	0218
21M193	0595
21M196	0688
21M200	0016
21M214	0012
21M224	0325
21M228	0086
21M248	0015
21M252	OEM
21M283	0015
21M286	0560
21M288	0143
21M289	0123
21M302	0015
21M307	0012
21M312	0015
21M315	0015
21M316	0030
21M317	0015
21M323	0143
21M325	0123
21M327	OEM
21M330	0133
21M345	0681
21M355	0037
21M369	0390
21M386	0015
21M387	0086
21M408	0111
21M417	0015
21M419	0015
21M430	OEM
21M431	OEM
21M432	0143
21M433	0015
21M434	0015
21M435	0015
21M436	0015
21M437	0015
21M443	1267
21M446	0111
21M448	0086
21M455	0086
21M459	1242
21M465	0460
21M466	0161
21M469	0015
21M485	1478
21M487	0015
21M488	0016
21M492	0030
21M493	0170
21M502	0030
21M506	1611
21M519	0015
21M520	0016
21M526	0030
21M532	1215
21M534	0325
21M541	0324
21M545	0015
21M556	0560
21M562	0133
21M563	0016
21M568	0123
21M577	0224
21M578	0016
21M579	0016
21M581	0037
21M582	4543
21M583	0030
21M584	0012
21M585	0012
21M586	1149
21M588	0701
21M594	0143
21M028	1267
21MW132	0211
21P560	OEM
21PT5	OEM
21PT10	OEM
21PT20	OEM
21PT40	OEM
21PT60	OEM
21R05S	3189
21R1S	2901
21R1SC	2901
21R2S	1124
21R2SC	1124
21R3S	1126
21R3SC	1126
21R4S	0805
21R5S	1195
21R6S	0319
21R8S	1132
21R10S	0044
21R12S	OEM
21R35	0805
21RC10	0567
21RC20	0762
21RC30	0740
21RC40	0740
21RC50	0764
21RC60	0764
21RC80	0766
21RC100	OEM
21RECT-SI-1039(AU)	0795
21SZ-RD13EB1	0053
21Z6	OEM
21Z6A	OEM
22-00920-000	OEM
22-001001	0016
22-001002	0144
22-001003	0144
22-001004	0144
22-001005	0144
22-001006	0016
22-001007	0016
22-001008	0178
22-001009	0178
22-001010	0037
22-002001	0222
22-002006	0211
22-002007	0004
22-002008	0222
22-002009	0222
22-004003	0143
22-004004	0071
22-1	0127
22-1(ZENER)	0705
22-1-005	0019
22-1-044	0133
22-1-075	0023
22-1-5	0019
22-1-70	0133
22-1-127	0030
22-1-129	0143
22-1-130	0030
22-1-131	0012
22-1-132	0091
22-1-138	0015
22-908-03A	OEM
22-908-03B	OEM
22-950-3B	OEM
22A50	OEM
22A100	OEM
22A200	OEM
22A400	OEM
22A600	OEM
22A800	OEM
22BC11	OEM
22BD11	OEM
22BH18M	OEM
22BH18P	OEM
22CC11	OEM
22CD11	OEM
22E05	OEM
22E8	0280
22E10A	OEM
22E15	OEM
22E30	OEM
22E50	OEM
22E89	0847
22FC11	OEM
22FD11	OEM
22MPG5	OEM
22MPS5	OEM
22NT100	OEM
22P1	OEM
22PB18M	OEM
22PD18M	OEM
22PD18P	OEM
22PH18M	OEM
22PS18M	OEM
22R2	0871
22R4	OEM
22R22	1124
22RC2	3277
22RC5	0731
22RC10	0712
22RC15	0469
22RC20	0469
22RC25	0721
22RC30	0721
22RC40	0722
22RC50	0778
22RC60	0725
22RC70	OEM
22RIA10	5895
22RIA10M	OEM
22RIA20	0469
22RIA20M	OEM
22RIA40	0722
22RIA40M	OEM
22RIA60	0725
22RIA60M	OEM
22RIA80	0750
22RIA80M	OEM
22RIA100	4855
22RIA100M	OEM
22RIA110	OEM
22RIA110M	OEM
22RIA120	0679
22RIA120M	OEM
22ST-2SA1015-O	0148
22ST-2SA10150	0148
22ST-2SD1177C	0563
22V10	OEM

If replacement code is OEM, contact original manufacturer for replacement.

DEVICE TYPE	REPL CODE
22Z6	0012
22Z6A	0012
22Z6AF	0012
22Z6F	0012
22Z6FAF	OEM
23	0016
23-0003	0242
23-0004	0015
23-0010	0015
23-0017	0015
23-0018	0015
23-1	0037
23-2	0037
23-3	0037
23-10	0037
23-20	0037
23-30	0037
23-5009	0604
23-5014	0004
23-5017	0004
23-5020	0155
23-5021	0155
23-5022	0155
23-5023	0155
23-5024	0155
23-5025	0155
23-5026	0155
23-5027	0155
23-5029	0155
23-5031	0178
23-5032	3545
23-5033	0016
23-5034	0847
23-5035	0103
23-5037	0086
23-5038	0103
23-5039	0086
23-5042	0160
23-5044	0400
23-5045	0126
23-5048	2866
23-5051	3280
23-5052	0079
23-6001-16	0164
23-6001-17	0164
23-6001-20	0164
23-6001-21	0164
23-6001-23	0164
23-PT274-120	0007
23-PT274-121	0016
23-PT274-122	0037
23-PT274-123	0007
23-PT274-125	0133
23-PT275-121	0007
23-PT275-122	0144
23-PT275-124	0917
23-PT283-122	0007
23-PT283-124	0007
23-PT283-125	0133
23-PT284-122	0050
23-PT284-123	0050
23-ST-2SA733Q(AU)E	0006
23B-210-025	0160
23B-210-230-2	0975
23B114044	0016
23B114053	0086
23B114054	0086
23B-210067-001	4465
23B-210067-002	4465
23B210679-1	0774
23BSC101	0015
23C05C	OEM
23C025	0469
23C025C	0469
23C050	0469
23C050C	0469
23C10	0469
23C10C	0469
23C15	0469
23C15C	0469
23C20	0469
23C20C	0469
23C25	0722
23C25C	0722
23C30	0722
23C30C	0722
23C35	0722
23C35C	0722
23C40	0722
23C40C	0722
23C45	0725
23C45C	0725
23C50	0725
23C50B	0725
23C50C	0725
23C60	0725
23C60B	0725
23C60C	0725
23C70	0679
23C70B	0750
23C70C	0679
23C80	0679
23C80B	0750
23C80C	0679
23C90B	OEM
23C100B	OEM
23C110B	OEM
23C120B	OEM
23C1000-1-539	OEM
23D	0400
23E001-1	0016
23E05	OEM
23E10A	OEM
23E20	OEM
23E30	OEM
23E60	OEM
23J2	0133
23J2C	OEM
23K3	OEM
23R2	OEM
23RA5	0469
23RA10	0469
23RA20	0469
23RA30	0725
23RA40	0725
23RA50	0725
23RA60	0725
23RA80	OEM
23SD-1SS151	0917
23SD-BAW62	0124
23ST-2SA733Q(AU)E	0006
23ST-2SA733R(AU)E	0006
23ST-MPS9682J	0037
23T1	OEM
23T2	OEM
23T3	OEM
23T4	OEM
23T5	OEM
23T6	OEM
23T7	OEM
23T8	OEM
23T9	OEM
23T10	OEM
23T11	OEM
23T12	OEM
23TB1	0712
23TB2	0469
23TB3	0721
23TB4	0722
23TB5	0778
23TB6	0725
23TB7	OEM
23TB8	OEM
23TB9	OEM
23TB10	OEM
23TB11	OEM
23TB12	OEM
23Z6	OEM
23Z6A	OEM
23Z6AF	0170
23Z6F	0170
23Z6FAF	OEM
24	OEM
24-002	0016
24-016	0127
24-016-001	0150
24-016-005	0127
24-2	0494
24-198	0015
24-602-25	0016
24-3564	0144
24-28201	0133
24-DP1	0015
24A	0016
24A-015-003	OEM
24A1	0144
24B	0016
24B(ZENER)	0494
24B1	0016
24B2	0494
24B4	0494
24C01A	OEM
24C01AI/P	OEM
24C01CB1	OEM
24C02AB1	OEM
24C02AIP	OEM
24C04AI/P	OEM
24C04B1	OEM
24C04PC	OEM
24CH1FM	OEM
24CH1FN	OEM
24CH2FM	OEM
24CH3FM	OEM
24CH4FM	OEM
24C01BI/P	OEM
24D400P1	OEM
24DP1	0015
24E-000	0030
24E-001	0133
24E-001C	0030
24E-002	0917
24E-002-0	0917
24E-002-C	0917
24E-005	OEM
24E005	OEM
24E-006	0133
24E05	OEM
24E-022	0917
24E10A	OEM
24E20	OEM
24E35	OEM
24E60	OEM
24F15E1	OEM
24J2	0133
24J2C	OEM
24K570S	OEM
24LC01B/P	OEM
24LC01BI/P	OEM
24LC02BI/P	OEM
24LC01BI/P	OEM
24M	OEM
24MW11	0280
24MW15	0211
24MW16	0211
24MW27	0004
24MW28	0004
24MW29	0211
24MW34	0004
24MW43	0211
24MW44	0050
24MW55	0050
24MW59	0050
24MW60	0211
24MW61	0136
24MW69	0211
24MW70	0211
24MW74	0136
24MW76	0030
24MW77	0280
24MW78	0211
24MW83	0211
24MW84	0211
24MW87	0143
24MW107	0211
24MW111	0004
24MW115	0211
24MW116	0211
24MW119	0076
24MW122	0123
24MW130	0038
24MW132	0211
24MW152	0136
24MW156	0211
24MW157	0050
24MW175	0229
24MW178	0004
24MW179	0004
24MW185	0004
24MW186	0211
24MW187	0004
24MW192	0277
24MW196	0229
24MW197	0015
24MW199	0143
24MW205	0136
24MW207	0649
24MW208	0015
24MW227	0015
24MW241	0015
24MW243	0143
24MW244	0123
24MW246	0071
24MW256	0164
24MW263	0211
24MW267	0015
24MW268	0015
24MW269	0015
24MW271	0136
24MW287	0127
24MW303	0136
24MW331	0012
24MW333	0016
24MW339	0050
24MW340	0050
24MW350	0050
24MW351	0050
24MW352	0136
24MW353	0136
24MW361	0127
24MW368	0136
24MW370	0004
24MW372	0016
24MW384	0004
24MW405	0050
24MW406	0050
24MW407	0050
24MW408	0211
24MW441	0004
24MW454	0016
24MW458	0016
24MW460	0016
24MW461	0016
24MW525	0050
24MW535	0127
24MW593	0127
24MW594	0127
24MW595	0127
24MW596	0127
24MW597	0127
24MW598	0004
24MW599	0004
24MW600	0004
24MW601	0004
24MW602	0071
24MW603	0143
24MW605	0071
24MW607	1080
24MW608	0004
24MW609	0016
24MW613	0004
24MW614	0004
24MW615	0004
24MW618	0822
24MW619	0015
24MW652	0688
24MW653	0127
24MW654	0151
24MW655	0536
24MW-656	0127
24MW656	0127
24MW657	0076
24MW658	0111
24MW659	0111
24MW660	0076
24MW661	0006
24MW662	0042
24MW663	0086
24MW664	0030
24MW665	0123
24MW667	0133
24MW669	0015
24MW670	1083
24MW671	0015
24MW673	0151
24MW674	0536
24MW675	0144
24MW676	0111
24MW677	0076
24MW700	0127
24MW714	1480
24MW721	0015
24MW723	0325
24MW724	0144
24MW725	0144
24MW726	0030
24MW727	0892
24MW734	0030
24MW736	0688
24MW737	0155
24MW738	0286
24MW739	0286
24MW740	0191
24MW741	0164
24MW742	0636
24MW743	0012
24MW744	1925
24MW746	0211
24MW747	0280
24MW748	0211
24MW760	0016
24MW763	0211
24MW764	0211
24MW765	0211
24MW768	0015
24MW771	0143
24MW772	0015
24MW773	0016
24MW774	0016
24MW775	0016
24MW776	0016
24MW777	0004
24MW778	0042
24MW779	0071
24MW780	0004
24MW781	0004
24MW782	0004
24MW783	0164
24MW785	0133
24MW789	0004
24MW790	0016
24MW793	0286
24MW795	0016
24MW796	0016
24MW797	0016
24MW799	0004
24MW801	0016
24MW805	0127
24MW807	0016
24MW808	0016
24MW809	0016
24MW812	0127
24MW813	0127
24MW814	0151
24MW815	0127
24MW816	0136
24MW817	0016
24MW818	0016
24MW819	0164
24MW820	0123
24MW823	0016
24MW824	0004
24MW825	0133
24MW826	0111
24MW827	0144
24MW828	0847
24MW829	0071
24MW851	0015
24MW852	0144
24MW853	0004
24MW854	0016
24MW855	0016
24MW856	0004
24MW857	0004
24MW858	0133
24MW860	0143
24MW861	0133
24MW862	0015
24MW863	0127
24MW864	0071
24MW865	0144
24MW867	0015
24MW871	0015
24MW874	0076
24MW892	0004
24MW893	0004
24MW894	0133
24MW895	0211
24MW896	0211
24MW899	0016
24MW924	0071
24MW950	0030
24MW953	0127
24MW954	0016
24MW955	0030
24MW956	0133
24MW957	0007
24MW958	0007
24MW961	0016
24MW964	0111
24MW965	0111
24MW967	0143
24MW973	0164
24MW974	0229
24MW975	0229
24MW976	0037
24MW977	0042
24MW978	0161
24MW988	0016
24MW989	0325
24MW990	0111
24MW991	0136
24MW992	0016
24MW994	0222
24MW995	0015
24MW996	2155
24MW997	3550
24MW998	0012
24MW1022	0111
24MW1023	0016
24MW1024	0016
24MW1025	0111
24MW1028	0579
24MW1029	0143
24MW1030	0143
24MW1031	0037
24MW1038	0144
24MW1040	0164
24MW1043	0143
24MW1049	0037
24MW1051	0143
24MW1052	0030
24MW1057	0127
24MW1058	0127
24MW1059	0016
24MW1060	0111
24MW1061	0006
24MW1062	5883
24MW1062(NPN)	OEM
24MW1062(PNP)	OEM
24MW1063	0030
24MW1065	0170
24MW1066	0015
24MW1067	0143
24MW1068	0016
24MW1069	0016
24MW1071	0015
24MW1081	0127
24MW1082	0144
24MW1083	0004
24MW1084	0004
24MW1089	0016
24MW1092	0143
24MW1096	0016
24MW1105	0720
24MW1106	0127
24MW1107	0451
24MW1108	0023
24MW1109	0133
24MW1110	4563
24MW1112	0030
24MW1113	0139
24MW1115	0004
24MW1116	2866
24MW1118	0087
24MW1120	0155
24MW1122	0212
24MW1123	0015
24MW1124	0023
24MW1125	0170
24MW1126	OEM
24MW1141	0016
24MW1143	0563
24MW1144	0015
24MW1146	0015
24MW1147	0016
24MW1152	0076
24MW1161	0086
24MW1162	0015
24MW7796	0079
24MY716	OEM
24P02F1-A	OEM
24P05M1-A	OEM
24P10K1	OEM
24R	0079
24R2	0805
24S04	OEM
24S06	OEM
24S08	OEM
24S10	2224
24S12	OEM
24S14	OEM
24S16	OEM
24SD-1SV70	1029
24ST-MPS9410AJ	0086
24ST-MPS9700G	0079
24SZ-HZ4B3	0036
24T-002	0144
24T002	0127
24T-011-008	0144
24T011-008	0144
24T-011-011	0037
24T011-012	0007
24T-011-013	0224
24T-011-015	0127
24T-013-003	0007
24T-013-005	0144
24T-015-013	0079
24T-016	0127
24T016	0127
24T-016-(016)	0111
24T-016-001	0007
24T-016-005	0144
24T-016-010	0127
24T-016-011	0144
24T-016-013	0144
24T-016-015	0144
24T-016-016	0144
24T-016-024	0127
24T-016-0B	0127
24T021	0007
24T-026-001	0325
24T-029	OEM
24T-029-001	OEM
24T013003	0127
24T013005	0127
24T016001	0127
24T016005	0127
24T1	OEM
24T2	OEM
24T3	OEM
24T4	OEM
24T5	OEM
24T6	OEM
24T7	OEM
24T8	OEM
24T9	OEM
24T10	OEM
24T11	OEM
24T12	OEM
24TB1	0567
24TB2	0762
24TB3	3271
24TB4	0740
24TB5	3291
24TB6	0764
24TB7	2875
24TB8	0766
24TB9	OEM
24TB10	OEM
24TB11	OEM
24TB12	OEM
24V10A	OEM
24V11	OEM
24Z6	OEM
24Z6A	OEM
24Z6F	0317
24Z6FAF	OEM
25	OEM
25-06D	0913
25-3	0143
25-5	0015
25-100015	0144
25-108550-03	OEM
25A	0144
25A1	0144
25A1ABG0	OEM
25A1ABG1	OEM
25A1PG0	OEM
25A1PG1	OEM
25A1QBG0	OEM
25A1QBG1	OEM
25A2	0016
25A2ABG0	OEM
25A2ABG2	OEM
25A2PG0	OEM
25A2PG2	OEM
25A2QBG0	OEM
25A2QBG2	OEM
25A3ABG0	OEM
25A3ABG3	OEM
25A3PG0	OEM
25A3PG3	OEM
25A3QBG0	OEM
25A3QBG3	OEM
25A4ABG0	OEM
25A4ABG4	OEM
25A4PG0	OEM
25A4PG4	OEM
25A4QBG0	OEM
25A4QBG4	OEM
25A5ABG0	OEM
25A5ABG5	OEM
25A5PG0	OEM
25A5PG5	OEM
25A5QBG0	OEM
25A5QBG5	OEM
25A6ABG0	OEM
25A6ABG6	OEM
25A6PG0	OEM
25A6PG6	OEM
25A6QBG0	OEM
25A7ABG0	OEM
25A7ABG7	OEM
25A7PG0	OEM
25A7PG7	OEM
25A7QBG0	OEM
25A7QBG7	OEM
25A8ABG0	OEM
25A8ABG8	OEM
25A8PG0	OEM
25A8PG8	OEM
25A8QBG0	OEM
25A8QBG8	OEM
25A9ABG0	OEM
25A9ABG9	OEM
25A9PG0	OEM
25A9PG9	OEM
25A9QBG0	OEM
25A9QBG9	OEM
25A10ABG0	OEM
25A10ABG10	OEM
25A10PG0	OEM
25A10PG10	OEM
25A10QBG0	OEM
25A10QBG10	OEM
25A11ABG0	OEM
25A11ABG11	OEM
25A11PG0	OEM
25A11PG11	OEM
25A11QBG0	OEM
25A11QBG11	OEM
25A12ABG0	OEM
25A12ABG12	OEM
25A12PG0	OEM
25A12PG12	OEM
25A12QBG0	OEM
25A12QBG12	OEM
25A13ABG0	OEM
25A13ABG13	OEM
25A13PG0	OEM
25A13PG13	OEM
25A13QBG0	OEM
25A13QBG13	OEM
25A14ABG0	OEM
25A14ABG14	OEM
25A14PG0	OEM
25A14PG14	OEM
25A14QBG0	OEM
25A14QBG14	OEM
25A15ABG0	OEM
25A15ABG15	OEM
25A15PG0	OEM
25A15PG15	OEM
25A15QBG0	OEM
25A15QBG15	OEM
25A16ABG0	OEM
25A16ABG16	OEM
25A16PG0	OEM
25A16PG16	OEM
25A16QBG0	OEM
25A16QBG16	OEM
25A17ABG0	OEM
25A17ABG17	OEM
25A17PG0	OEM
25A17PG17	OEM
25A17QBG0	OEM
25A17QBG17	OEM
25A18ABG0	OEM
25A18ABG18	OEM
25A18PG0	OEM
25A18PG18	OEM
25A18QBG0	OEM
25A18QBG18	OEM
25A19ABG0	OEM
25A19ABG19	OEM
25A19PG0	OEM
25A19PG19	OEM
25A19QBG0	OEM
25A19QBG19	OEM
25A20ABG0	OEM
25A20ABG20	OEM
25A20PG0	OEM
25A20PG20	OEM
25A20QBG0	OEM
25A20QBG20	OEM
25A321	0050
25A324	0050
25A440A	0050
25A473Y	0925
25A561Y	0037
25A1262-005	0144
25A1273-001	0016
25A1281-001	0144
25AC3SL	3152
25AC3SR	3152
25AC5SL	4100
25AC5SR	4100
25AC6SL	4100
25AC6SR	4100
25AK20	OEM
25AK30	OEM
25AK50	OEM
25AM624	0144
25AN20	OEM
25AN30	OEM
25AN40	OEM
25AN50	OEM
25AN60	OEM
25AN80	OEM
25AN100	OEM
25AN120	OEM
25AN140	OEM

DEVICE TYPE	REPL CODE	DEVICE TYPE	REPL CODE	DEVICE TYPE	REPL CODE	DEVICE TYPE	REPL CODE	DEVICE TYPE	REPL CODE	DEVICE TYPE	REPL CODE	DEVICE TYPE	REPL CODE
25AN160	OEM	25H5A	1560	25L60	OEM	25PCRC	1650	25T9	OEM	25TB9-EH00	OEM	26T3	OEM
25ANS	OEM	25H10	1560	25L65	OEM	25PCRD	1650	25T10	OEM	25TB9-GE00	OEM	26T4	OEM
25B	0144	25H10A	1560	25L70	OEM	25PCRE	1650	25T11	OEM	25TB9-GH00	OEM	26T5	OEM
25B05T	OEM	25H15	2840	25L75	OEM	25PCRF	2647	25T12	OEM	25TB9-JE00	OEM	26T6	OEM
25B-1	0144	25H15A	2840	25L80	OEM	25PCRG	2647	25T20	OEM	25TB9-JH00	OEM	26T7	OEM
25B1	0144	25H20A	2840	25L85	OEM	25PCRH	2647	25T40	OEM	25TB9-LE00	OEM	26T8	OEM
25B1C	2743	25H25	4308	25L90	OEM	25PCRI	2649	25TB1-CE00	OEM	25TB9-NE00	OEM	26T9	OEM
25B1T	2743	25H25A	4308	25L95	OEM	25PCRJ	2649	25TB1-CH00	OEM	25TB9-RE00	OEM	26T10	OEM
25B2	0224	25H30	4308	25L100	OEM	25PCRK	OEM	25TB1-EE00	OEM	25TB10-CE00	OEM	26T11	OEM
25B2C	2743	25H30A	4308	25L110	OEM	25PCRL	OEM	25TB1-EH00	OEM	25TB10-CH00	OEM	26T12	OEM
25B2T	2743	25H40	2850	25L120	OEM	25PCRNA	OEM	25TB1-GE00	OEM	25TB10-EE00	OEM	26TB1-CE00	OEM
25B4B41	OEM	25H40A	2850	25L130	OEM	25PCRNB	OEM	25TB1-GH00	OEM	25TB10-EH00	OEM	26TB1-CH00	OEM
25B10T	OEM	25H50	2871	25L140	OEM	25PCRNC	OEM	25TB1-JE00	OEM	25TB10-GE00	OEM	26TB1-EE00	OEM
25B20T	OEM	25H50A	2871	25L150	OEM	25PCRND	OEM	25TB1-LE00	OEM	25TB10-GH00	OEM	26TB1-EH00	OEM
25B21	0079	25H60	2871	25L160	OEM	25PCRNE	OEM	25TB1-LH00	OEM	25TB10-JE00	OEM	26TB1-GE00	OEM
25B40T	OEM	25H60A	2871	25L170	OEM	25PCRNF	OEM	25TB1-NE00	OEM	25TB10-JH00	OEM	26TB1-GH00	OEM
25B60T	OEM	25H70	2833	25L180	OEM	25PCRNG	OEM	25TB1-NH00	OEM	25TB10-LE00	OEM	26TB1-JE00	OEM
25B186	0211	25H70A	2833	25L190	OEM	25PCRNH	OEM	25TB1-RE00	OEM	25TB10-LH00	OEM	26TB1-JH00	OEM
25B375A-NB	0975	25H80	2833	25L200	OEM	25PCRNI	OEM	25TB2-CE00	OEM	25TB10-NE00	OEM	26TB1-LE00	OEM
25B378	0004	25H80A	2833	25L210	OEM	25PCRNJ	OEM	25TB2-CH00	OEM	25TB10-NH00	OEM	26TB1-LH00	OEM
25B496	0280	25H90	2473	25L220	OEM	25PCRNK	OEM	25TB2-EE00	OEM	25TB10-RE00	OEM	26TB1-NE00	OEM
25C05	3189	25H90A	2473	25L230	OEM	25PCRNL	OEM	25TB2-EH00	OEM	25TB11-CE00	OEM	26TB1-NH00	OEM
25C05R	1630	25H100	2473	25L240	OEM	25PCRSA	2188	25TB2-GE00	OEM	25TB11-CH00	OEM	26TB1-RE00	OEM
25C1	OEM	25H100A	2473	25L250	OEM	25PCRSB	2188	25TB2-GH00	OEM	25TB11-EE00	OEM	26TB2-CE00	OEM
25C2	OEM	25HB5	3762	25L260	OEM	25PCRSC	0567	25TB2-JE00	OEM	25TB11-EH00	OEM	26TB2-CH00	OEM
25C4	OEM	25HB10	2653	25L270	OEM	25PCRSD	0762	25TB2-JH00	OEM	25TB11-GE00	OEM	26TB2-EE00	OEM
25C6	OEM	25HB15	2657	25L280	OEM	25PCRSE	0762	25TB2-LE00	OEM	25TB11-GH00	OEM	26TB2-EH00	OEM
25C8	OEM	25HB20	2657	25L290	OEM	25PCRSF	0740	25TB2-LH00	OEM	25TB11-JE00	OEM	26TB2-GE00	OEM
25C10	2901	25HB25	2664	25L300	OEM	25PCRSG	0740	25TB2-NE00	OEM	25TB11-JH00	OEM	26TB2-GH00	OEM
25C10R	0254	25HB30	2664	25LC11	OEM	25PCRSH	0740	25TB2-NH00	OEM	25TB11-LE00	OEM	26TB2-JE00	OEM
25C12	OEM	25HB35	2006	25LC12	3900	25PCRSI	0764	25TB2-RE00	OEM	25TB11-LH00	OEM	26TB2-JH00	OEM
25C14	OEM	25HB40	2006	25LF11	1119	25PCRSJ	0764	25TB3-CE00	OEM	25TB11-NE00	OEM	26TB2-LE00	OEM
25C15	1124	25HB50	2677	25LS2521	OEM	25PCRSK	0766	25TB3-CH00	OEM	25TB11-NH00	OEM	26TB2-LH00	OEM
25C15R	1107	25HB60	2682	25M10	OEM	25PCRSL	OEM	25TB3-EE00	OEM	25TB11-RE00	OEM	26TB2-NE00	OEM
25C16	OEM	25HBR5	1783	25MC11	OEM	25PDU5M	OEM	25TB3-EH00	OEM	25TB12-CE00	OEM	26TB2-NH00	OEM
25C20	1124	25HBR10	1783	25N15	0002	25PW5	OEM	25TB3-GE00	OEM	25TB12-CH00	OEM	26TB2-RE00	OEM
25C30	1126	25HBR20	1783	25N15A	OEM	25PW10	OEM	25TB3-GH00	OEM	25TB12-EE00	OEM	26TB3-CE00	OEM
25C30R	0258	25HBR30	1783	25N15A9	OEM	25PW20	OEM	25TB3-JE00	OEM	25TB12-EH00	OEM	26TB3-CH00	OEM
25C40	0805	25HBR40	1783	25N20A	OEM	25PW30	OEM	25TB3-JH00	OEM	25TB12-GE00	OEM	26TB3-EE00	OEM
25C40R	0258	25HBR50	1783	25N20A9	OEM	25PW40	OEM	25TB3-LE00	OEM	25TB12-GH00	OEM	26TB3-EH00	OEM
25C50	1195	25HBR60	1783	25N25A	OEM	25PW50	OEM	25TB3-LH00	OEM	25TB12-JE00	OEM	26TB3-GE00	OEM
25C50R	1644	25HC11	OEM	25N25A9	OEM	25PW60	OEM	25TB3-NE00	OEM	25TB12-JH00	OEM	26TB3-GH00	OEM
25C60	0319	25HR5	1783	25N27	0441	25QF11	OEM	25TB3-NH00	OEM	25TB12-LE00	OEM	26TB3-JE00	OEM
25C60R	0267	25HR5A	1783	25N30A	OEM	25R	0144	25TB3-RE00	OEM	25TB12-LH00	OEM	26TB3-JH00	OEM
25C70R	1119	25HR10	1783	25N30A9	OEM	25R2	OEM	25TB4-CE00	OEM	25TB12-NE00	OEM	26TB3-LE00	OEM
25C80	1132	25HR10A	1783	25N35A	OEM	25RIA10	OEM	25TB4-CH00	OEM	25TB12-RE00	OEM	26TB3-LH00	OEM
25C80R	1119	25HR15	1783	25N35A9	OEM	25RIA10M	OEM	25TB4-EH00	OEM	25TCRA	0469	26TB3-NE00	OEM
25C90	0044	25HR15A	1783	25N40A	OEM	25RIA20	OEM	25TB4-GE00	OEM	25TCRB	0469	26TB3-NH00	OEM
25C90R	0281	25HR20	1783	25N40A9	OEM	25RIA20M	OEM	25TB4-GH00	OEM	25TCRC	0469	26TB3-RE00	OEM
25C100	0044	25HR20A	1783	25N45A	OEM	25RIA40	OEM	25TB4-JE00	OEM	25TCRD	0469	26TB4-CE00	OEM
25C100R	0281	25HR30	1783	25N45A9	OEM	25RIA40M	OEM	25TB4-JH00	OEM	25TCRE	0469	26TB4-CH00	OEM
25C206	0144	25HR30A	1783	25N50A	OEM	25RIA60	OEM	25TB4-LE00	OEM	25TCRF	0722	26TB4-EE00	OEM
25C536E	0004	25HR40	1783	25N50A9	OEM	25RIA60M	OEM	25TB4-LH00	OEM	25TCRG	0722	26TB4-EH00	OEM
25CC11	OEM	25HR40A	1783	25N55A	OEM	25RIA80	OEM	25TB4-NE00	OEM	25TCRH	0722	26TB4-GE00	OEM
25CC13	1124	25HR50	1783	25N55A9	OEM	25RIA80M	OEM	25TB4-NH00	OEM	25TCRI	0725	26TB4-GH00	OEM
25CD13	1107	25HR50A	1783	25N60A	OEM	25RIA100	OEM	25TB4-RE00	OEM	25TCRJ	0725	26TB4-JE00	OEM
25CS58LGBM	0016	25HR60	1783	25N60A9	OEM	25RIA100M	OEM	25TB5-CE00	OEM	25TCRK	0679	26TB4-JH00	OEM
25D4B41	OEM	25HR60A	1783	25N65A	OEM	25RIA110	OEM	25TB5-CH00	OEM	25TCRL	0679	26TB4-LE00	OEM
25D10	0071	25HR80	1818	25N65A9	OEM	25RIA110M	OEM	25TB5-EE00	OEM	25VC02	OEM	26TB4-LH00	OEM
25D15	OEM	25HR80A	1818	25N70A	OEM	25RIA120	OEM	25TB5-EH00	OEM	25VC05	OEM	26TB4-NE00	OEM
25D20	OEM	25HR100	1818	25N70A9	OEM	25RIA120M	OEM	25TB5-GE00	OEM	25VC1	OEM	26TB4-NH00	OEM
25D25	OEM	25HR100A	1818	25N75A	OEM	25SD-1N60-TMS	0019	25TB5-GH00	OEM	25VC2	OEM	26TB4-RE00	OEM
25D30	OEM	25IP005	OEM	25N75A9	OEM	25SDG1	OEM	25TB5-JE00	OEM	25VC3	OEM	26TB5-CE00	OEM
25D40	OEM	25IP005D	OEM	25N80A	OEM	25SDG2	OEM	25TB5-JH00	OEM	25VC4	OEM	26TB5-CH00	OEM
25D50	OEM	25IP005N	OEM	25N80A9	OEM	25SDG2JHKO	OEM	25TB5-LE00	OEM	25VC5	OEM	26TB5-EE00	OEM
25DC11	OEM	25IP01	OEM	25N85A	OEM	25SDG2JHNO	OEM	25TB5-LH00	OEM	25VC6	OEM	26TB5-EH00	OEM
25DF10	OEM	25IP01D	OEM	25N85A9	OEM	25SDG3	OEM	25TB5-NE00	OEM	25VC7	OEM	26TB5-GE00	OEM
25DF15	OEM	25IP01N	OEM	25N90A	OEM	25SDG4	OEM	25TB5-NH00	OEM	25VC8	OEM	26TB5-GH00	OEM
25DF20	OEM	25IP02	OEM	25N90A9	OEM	25SDG4JHKO	OEM	25TB5-RE00	OEM	25Z6	0137	26TB5-JE00	OEM
25DF25	OEM	25IP02D	OEM	25N95A	OEM	25SDG4JHNO	OEM	25TB6-CE00	OEM	25Z6A	0137	26TB5-JH00	OEM
25DF30	OEM	25IP02N	OEM	25N95A9	OEM	25SDG5	OEM	25TB6-CH00	OEM	25Z6AF	0137	26TB5-LE00	OEM
25E05	OEM	25IP04	OEM	25N100A	OEM	25SDG5JHKO	OEM	25TB6-EE00	OEM	25Z6F	0137	26TB5-LH00	OEM
25E10A	OEM	25IP04D	OEM	25N100A9	OEM	25SDG5JHNO	OEM	25TB6-EH00	OEM	25Z6FAF	OEM	26TB5-NE00	OEM
25E19	OEM	25IP04N	OEM	25N110A	OEM	25SDG6	OEM	25TB6-GE00	OEM	26	OEM	26TB5-NH00	OEM
25E33	OEM	25IP06	OEM	25N120A	OEM	25SDG6JHKO	OEM	25TB6-GH00	OEM	26/P19503	0086	26TB5-RE00	OEM
25E55	OEM	25IP06D	OEM	25N130A	OEM	25SDG6JHNO	OEM	25TB6-JE00	OEM	26-16162-1	1030	26TB6-CE00	OEM
25F-B1F	0289	25IP06N	OEM	25N140A	OEM	25SP005	OEM	25TB6-JH00	OEM	26-4701508	0071	26TB6-CH00	OEM
25FC11	OEM	25IP08	OEM	25N150A	OEM	25SP005D	OEM	25TB6-LE00	OEM	26E04	OEM	26TB6-EE00	OEM
25FC13	0805	25IP08D	OEM	25N160A	OEM	25SP005N	OEM	25TB6-LH00	OEM	26E08A	OEM	26TB6-EH00	OEM
25FCRA	OEM	25IP08N	OEM	25N170A	OEM	25SP01	OEM	25TB6-NE00	OEM	26E17	OEM	26TB6-GE00	OEM
25FCRB	OEM	25IP10	OEM	25N180A	OEM	25SP01D	OEM	25TB6-NH00	OEM	26E30	OEM	26TB6-GH00	OEM
25FCRC	OEM	25IP10D	OEM	25N190A	OEM	25SP01N	OEM	25TB6-RE00	OEM	26E45	OEM	26TB6-JE00	OEM
25FCRD	OEM	25IP10N	OEM	25N200A	OEM	25SP02	OEM	25TB7-CE00	OEM	26G10	OEM	26TB6-JH00	OEM
25FCRE	OEM	25J2	0133	25N210A	OEM	25SP02D	OEM	25TB7-EE00	OEM	26G20	OEM	26TB6-LE00	OEM
25FCRF	OEM	25J2C	0133	25N220A	OEM	25SP02N	OEM	25TB7-EH00	OEM	26G40	OEM	26TB6-LH00	OEM
25FCRG	OEM	25J4B41	OEM	25N230A	OEM	25SP04	OEM	25TB7-GE00	OEM	26G60	OEM	26TB6-NE00	OEM
25FCRH	OEM	25JB1L	2363	25N240A	OEM	25SP04D	OEM	25TB7-GH00	OEM	26G80	OEM	26TB6-NH00	OEM
25FCRI	OEM	25JB2L	2363	25N250A	OEM	25SP04N	OEM	25TB7-JE00	OEM	26G100	OEM	26TB6-RE00	OEM
25FCRJ	OEM	25JB4L	2369	25N260A	OEM	25SP-06	OEM	25TB7-JH00	OEM	26G120	OEM	26TB7-CE00	OEM
25FCRK	OEM	25JC11	OEM	25NC12	2655	25SP06	OEM	25TB7-LE00	OEM	26J2	0133	26TB7-CH00	OEM
25FCRL	OEM	25JC12	2682	25NF11	0281	25SP06D	OEM	25TB7-LH00	OEM	26J2C	0133	26TB7-EH00	OEM
25FXF11	OEM	25K10	0071	25P1	0143	25SP06N	OEM	25TB7-NE00	OEM	26MB5A	OEM	26TB7-GE00	OEM
25FXF12	OEM	25K20	OEM	25P1SG	OEM	25SP08	OEM	25TB7-NH00	OEM	26MB10A	5400	26TB7-GH00	OEM
25G4B41	OEM	25K30	OEM	25P2SG	OEM	25SP08D	OEM	25TB7-RE00	OEM	26MB20A	5400	26TB7-JH00	OEM
25G5	3762	25K40	OEM	25P3SG	OEM	25SP08N	OEM	25TB8-CE00	OEM	26MB40A	3486	26TB7-LE00	OEM
25G5R	2665	25K50	OEM	25P4SG	OEM	25SP10	OEM	25TB8-CH00	OEM	26MB60A	5403	26TB7-LH00	OEM
25G10	2653	25KB7	OEM	25P5	OEM	25SP10D	OEM	25TB8-EE00	OEM	26MB80A	6680	26TB7-NE00	OEM
25G10R	2695	25KC11	OEM	25P6SG	OEM	25SP10N	OEM	25TB8-EH00	OEM	26MB100A	6086	26TB7-NH00	OEM
25G20	2657	25L01I	OEM	25P8SG	OEM	25ST-2SC18150	0076	25TB8-GE00	OEM	26MB120A	OEM	26TB7-RE00	OEM
25G20R	2768	25L01N	OEM	25P10	OEM	25ST-2SD781	0563	25TB8-GH00	OEM	26MW613	0004	26TB8-CE00	OEM
25G40	2006	25L15	OEM	25P12SG	OEM	25T-002	0127	25TB8-JE00	OEM	260D15901	0111	26TB8-CH00	OEM
25G40R	2907	25L20	OEM	25P20	OEM	25T1	0211	25TB8-JH00	OEM	26P1	OEM	26TB8-EE00	OEM
25G60	2682	25L25	OEM	25P20SG	OEM	25T2	OEM	25TB8-LE00	OEM	26R2	0319	26TB8-EH00	OEM
25G80	3900	25L30	OEM	25P30	OEM	25T3	OEM	25TB8-LH00	OEM	26R2S	0319	26TB8-GE00	OEM
25G100	2655	25L35	OEM	25P40	OEM	25T4	OEM	25TB8-NE00	OEM	26ST-2SC383TM	0076	26TB8-GH00	OEM
25G120	OEM	25L40	OEM	25P50	OEM	25T5	OEM	25TB8-NH00	OEM	26ST-JE9014C	4898	26TB8-JH00	OEM
25GC11	OEM	25L45	OEM	25P60	OEM	25T6	OEM	25TB8-RE00	OEM	26T1	0211	26TB8-LE00	OEM
25GC12	2006	25L50	OEM	25PCRA	1650	25T7	OEM	25TB9-CH00	OEM	26T2	OEM	26TB8-LH00	OEM
25H5	1560	25L55	OEM	25PCRB	1650	25T8	OEM	25TB9-EE00	OEM	26T2C	OEM	26TB8-NE00	OEM

If replacement code is OEM, contact original manufacturer for replacement.

DEVICE TYPE	REPL CODE	DEVICE TYPE	REPL CODE	DEVICE TYPE	REPL CODE	DEVICE TYPE	REPL CODE	DEVICE TYPE	REPL CODE	DEVICE TYPE	REPL CODE	DEVICE TYPE	REPL CODE
26TB8-NH00	OEM	27TB1-CH00	OEM	27TB9-CE00	OEM	28-35-00	1501	28TB6-EH00	OEM	29-180B	0738	29TB6-EH00	OEM
26TB8-RE00	OEM	27TB1-EE00	OEM	27TB9-CH00	OEM	28-35-01	2804	28TB6-GE00	OEM	29-180C	0738	29TB6-GE00	OEM
26TB9-CE00	OEM	27TB1-EH00	OEM	27TB9-EE00	OEM	28-37-00	1067	28TB6-GH00	OEM	29-180D	0738	29TB6-GH00	OEM
26TB9-CH00	OEM	27TB1-GE00	OEM	27TB9-EH00	OEM	28-39-01	2546	28TB6-JE00	OEM	29-180E	0738	29TB6-JE00	OEM
26TB9-EH00	OEM	27TB1-GH00	OEM	27TB9-GE00	OEM	28-65-01	0015	28TB6-JH00	OEM	29-180M	0738	29TB6-JH00	OEM
26TB9-GE00	OEM	27TB1-JE00	OEM	27TB9-GH00	OEM	28-254566-1	0015	28TB6-LE00	OEM	29-180N	0738	29TB6-LE00	OEM
26TB9-GH00	OEM	27TB1-JH00	OEM	27TB9-JE00	OEM	28A477	0136	28TB6-LH00	OEM	29-180P	0738	29TB6-LH00	OEM
26TB9-JH00	OEM	27TB1-LE00	OEM	27TB9-JH00	OEM	28A608E	0150	28TB6-NE00	OEM	29-180PA	0738	29TB6-RE00	OEM
26TB9-LE00	OEM	27TB1-LH00	OEM	27TB9-LE00	OEM	28BR2	OEM	28TB6-NH00	OEM	29-180PB	0738	29TB7	OEM
26TB9-LH00	OEM	27TB1-NE00	OEM	27TB9-LH00	OEM	28BR2S	OEM	28TB6-RE00	OEM	29-180S	0738	29TB7-CE00	OEM
26TB9-NE00	OEM	27TB1-NH00	OEM	27TB9-NE00	OEM	28C16-15	OEM	28TB7-CE00	OEM	29-180T	0738	29TB7-CH00	OEM
26TB9-NH00	OEM	27TB1-RE00	OEM	27TB9-NH00	OEM	28C620	OEM	28TB7-CH00	OEM	29-218A	0738	29TB7-EE00	OEM
26TB9-RE00	OEM	27TB2	OEM	27TB9-RE00	OEM	28C774	OEM	28TB7-EE00	OEM	29-218B	0738	29TB7-EH00	OEM
26TB10-CE00	OEM	27TB2-CE00	OEM	27TB10	OEM	28C777	OEM	28TB7-EH00	OEM	29-218C	0738	29TB7-GE00	OEM
26TB10-CH00	OEM	27TB2-CH00	OEM	27TB10-CE00	OEM	28C1014	OEM	28TB7-GE00	OEM	29-218D	0738	29TB7-GH00	OEM
26TB10-EE00	OEM	27TB2-EE00	OEM	27TB10-CH00	OEM	28CPQ030	1942	28TB7-GH00	OEM	29-218F	0738	29TB7-JE00	OEM
26TB10-EH00	OEM	27TB2-EH00	OEM	27TB10-EE00	OEM	28CPQ040	1942	28TB7-JE00	OEM	29-218H	0738	29TB7-JH00	OEM
26TB10-GE00	OEM	27TB2-GE00	OEM	27TB10-EH00	OEM	28CPQ050	OEM	28TB7-JH00	OEM	29-218K	0738	29TB7-LE00	OEM
26TB10-GH00	OEM	27TB2-GH00	OEM	27TB10-GE00	OEM	28CPQ060	OEM	28TB7-LE00	OEM	29-218M	0738	29TB7-LH00	OEM
26TB10-JH00	OEM	27TB2-JE00	OEM	27TB10-GH00	OEM	28E04	OEM	28TB7-LH00	OEM	29-218P	0738	29TB7-RE00	OEM
26TB10-LE00	OEM	27TB2-JH00	OEM	27TB10-JE00	OEM	28E08A	OEM	28TB7-NE00	OEM	29-218S	0738	29TB8	OEM
26TB10-LH00	OEM	27TB2-LE00	OEM	27TB10-JH00	OEM	28E15	OEM	28TB7-NH00	OEM	29-218V	0738	29TB8-CE00	OEM
26TB10-NE00	OEM	27TB2-LH00	OEM	27TB10-LE00	OEM	28E30	OEM	28TB7-RE00	OEM	29-218Z	0738	29TB8-CH00	OEM
26TB10-NH00	OEM	27TB2-NE00	OEM	27TB10-LH00	OEM	28E50	OEM	28TB8-CE00	OEM	29-218ZB	0738	29TB8-EE00	OEM
26TB10-RE00	OEM	27TB2-NH00	OEM	27TB10-NE00	OEM	28J2	1335	28TB8-CH00	OEM	29-218ZD	0738	29TB8-EH00	OEM
26TB11-CE00	OEM	27TB2-RE00	OEM	27TB10-NH00	OEM	28J2C	0133	28TB8-EE00	OEM	29B1B	0349	29TB8-GE00	OEM
26TB11-CH00	OEM	27TB3	OEM	27TB10-RE00	OEM	28L22	OEM	28TB8-EH00	OEM	29B1Z	0349	29TB8-GH00	OEM
26TB11-EE00	OEM	27TB3-CE00	OEM	27TB11	OEM	28M018	0030	28TB8-GE00	OEM	29D5R19	OEM	29TB8-JE00	OEM
26TB11-EH00	OEM	27TB3-CH00	OEM	27TB11-CE00	OEM	28P1	0143	28TB8-GH00	OEM	29D9R19GH	OEM	29TB8-JH00	OEM
26TB11-GE00	OEM	27TB3-EE00	OEM	27TB11-CH00	OEM	28R2	1132	28TB8-JE00	OEM	29M01201	OEM	29TB8-LE00	OEM
26TB11-GH00	OEM	27TB3-EH00	OEM	27TB11-EE00	OEM	28R25	1132	28TB8-JH00	OEM	29P1	0188	29TB8-LH00	OEM
26TB11-JH00	OEM	27TB3-GE00	OEM	27TB11-EH00	OEM	28ST-2SC1213C	0191	28TB8-LE00	OEM	29ST-3SK80	0371	29TB8-RE00	OEM
26TB11-LE00	OEM	27TB3-GH00	OEM	27TB11-GE00	OEM	28ST-2SC2230AY	0261	28TB8-LH00	OEM	29SZ-RD11EB2	0181	29TB9	OEM
26TB11-LH00	OEM	27TB3-JE00	OEM	27TB11-GH00	OEM	28ST-2SD467(C)	0951	28TB8-NE00	OEM	29T1	OEM	29TB9-CE00	OEM
26TB11-NE00	OEM	27TB3-JH00	OEM	27TB11-JE00	OEM	28ST-2SD467C	0951	28TB8-NH00	OEM	29T2	OEM	29TB9-CH00	OEM
26TB11-NH00	OEM	27TB3-LE00	OEM	27TB11-JH00	OEM	28ST-MPS9633C	0079	28TB8-RE00	OEM	29T2C	OEM	29TB9-EE00	OEM
26TB11-RE00	OEM	27TB3-LH00	OEM	27TB11-LE00	OEM	28T1	OEM	28TB9-CH00	OEM	29T3	OEM	29TB9-EH00	OEM
26TB12-CE00	OEM	27TB3-NE00	OEM	27TB11-LH00	OEM	28T2	OEM	28TB9-EE00	OEM	29T4	OEM	29TB9-GE00	OEM
26TB12-CH00	OEM	27TB3-NH00	OEM	27TB11-NE00	OEM	28T3	OEM	28TB9-EH00	OEM	29T5	OEM	29TB9-GH00	OEM
26TB12-EE00	OEM	27TB3-RE00	OEM	27TB11-NH00	OEM	28T4	OEM	28TB9-GE00	OEM	29T6	OEM	29TB9-JE00	OEM
26TB12-EH00	OEM	27TB4	OEM	27TB11-RE00	OEM	28T5	OEM	28TB9-GH00	OEM	29T7	OEM	29TB9-JH00	OEM
26TB12-GE00	OEM	27TB4-CE00	OEM	27TB12	OEM	28T6	OEM	28TB9-JE00	OEM	29T8	OEM	29TB9-LE00	OEM
26TB12-GH00	OEM	27TB4-CH00	OEM	27TB12-CE00	OEM	28T7	OEM	28TB9-JH00	OEM	29T9	OEM	29TB9-LH00	OEM
26TB12-JH00	OEM	27TB4-EE00	OEM	27TB12-CH00	OEM	28T8	OEM	28TB9-LE00	OEM	29T10	OEM	29TB9-RE00	OEM
26TB12-LE00	OEM	27TB4-EH00	OEM	27TB12-EE00	OEM	28T9	OEM	28TB9-LH00	OEM	29T11	OEM	29TB10	OEM
26TB12-NE00	OEM	27TB4-GE00	OEM	27TB12-EH00	OEM	28T10	OEM	28TB9-NE00	OEM	29T12	OEM	29TB10-CE00	OEM
26TB12-NH00	OEM	27TB4-GH00	OEM	27TB12-GE00	OEM	28T11	OEM	28TB9-NH00	OEM	29TB1	OEM	29TB10-CH00	OEM
26TB12-RE00	OEM	27TB4-JE00	OEM	27TB12-GH00	OEM	28T12	OEM	28TB9-RE00	OEM	29TB1-CE00	OEM	29TB10-EE00	OEM
26Z6	OEM	27TB4-JH00	OEM	27TB12-JH00	OEM	28TB1-CE00	OEM	28TB10-CE00	OEM	29TB1-CH00	OEM	29TB10-EH00	OEM
26Z6A	OEM	27TB4-LE00	OEM	27TB12-LE00	OEM	28TB1-CH00	OEM	28TB10-CH00	OEM	29TB1-EE00	OEM	29TB10-GE00	OEM
26Z6F	0365	27TB4-LH00	OEM	27TB12-LH00	OEM	28TB1-EE00	OEM	28TB10-EE00	OEM	29TB1-EH00	OEM	29TB10-GH00	OEM
26Z6FAF	OEM	27TB4-NE00	OEM	27TB12-NE00	OEM	28TB1-EH00	OEM	28TB10-EH00	OEM	29TB1-GE00	OEM	29TB10-JE00	OEM
27-0432-00-00	OEM	27TB4-NH00	OEM	27TB12-NH00	OEM	28TB1-GE00	OEM	28TB10-GE00	OEM	29TB1-GH00	OEM	29TB10-JH00	OEM
27-0435-00	OEM	27TB4-RE00	OEM	27TB12-RE00	OEM	28TB1-GH00	OEM	28TB10-GH00	OEM	29TB1-JE00	OEM	29TB10-LE00	OEM
27-0460-00	OEM	27TB5	OEM	27TB13	OEM	28TB1-JE00	OEM	28TB10-JE00	OEM	29TB1-JH00	OEM	29TB10-LH00	OEM
27-0473-01	OEM	27TB5-CE00	OEM	27TB14	OEM	28TB1-JH00	OEM	28TB10-JH00	OEM	29TB1-LH00	OEM	29TB10-RE00	OEM
27-0473-03	OEM	27TB5-CH00	OEM	27TB15	OEM	28TB1-LE00	OEM	28TB10-LE00	OEM	29TB1-RE00	OEM	29TB11	OEM
27-0501-00	OEM	27TB5-EE00	OEM	27X	0455	28TB1-LH00	OEM	28TB10-LH00	OEM	29TB2	OEM	29TB11-CE00	OEM
27-0532-03	OEM	27TB5-EH00	OEM	27Y	0455	28TB1-NE00	OEM	28TB10-NE00	OEM	29TB2-CE00	OEM	29TB11-CH00	OEM
27-0554-01	OEM	27TB5-GE00	OEM	27Z6	0100	28TB1-RE00	OEM	28TB10-NH00	OEM	29TB2-CH00	OEM	29TB11-EE00	OEM
27-0602-01	OEM	27TB5-GH00	OEM	27Z6A	0100	28TB2-CE00	OEM	28TB10-RE00	OEM	29TB2-EE00	OEM	29TB11-EH00	OEM
27-0605-00	OEM	27TB5-JE00	OEM	27Z6F	0100	28TB2-CH00	OEM	28TB11-CE00	OEM	29TB2-EH00	OEM	29TB11-GE00	OEM
27-0690-00	OEM	27TB5-JH00	OEM	27Z6FAF	OEM	28TB2-EE00	OEM	28TB11-CH00	OEM	29TB2-GE00	OEM	29TB11-GH00	OEM
27-0692-00	OEM	27TB5-LE00	OEM	28-1-01	0015	28TB2-EH00	OEM	28TB11-EE00	OEM	29TB2-GH00	OEM	29TB11-JE00	OEM
27-06900000	OEM	27TB5-LH00	OEM	28-1-02	0015	28TB2-GE00	OEM	28TB11-EH00	OEM	29TB2-JE00	OEM	29TB11-JH00	OEM
27-226	0196	27TB5-NE00	OEM	28-6-01	0015	28TB2-GH00	OEM	28TB11-GE00	OEM	29TB2-JH00	OEM	29TB11-LE00	OEM
27-C226	0196	27TB5-NH00	OEM	28-7-01	0015	28TB2-JE00	OEM	28TB11-GH00	OEM	29TB2-LE00	OEM	29TB11-LH00	OEM
27A10446-101-11	0333	27TB5-RE00	OEM	28-13-01	0015	28TB2-JH00	OEM	28TB11-JE00	OEM	29TB2-LH00	OEM	29TB11-NE00	OEM
27B1(1WATT)	1606	27TB6	OEM	28-14-01	0015	28TB2-LE00	OEM	28TB11-JH00	OEM	29TB2-RE00	OEM	29TB11-RE00	OEM
27C226	0196	27TB6-CE00	OEM	28-15-01	0015	28TB2-LH00	OEM	28TB11-LE00	OEM	29TB3	OEM	29TB12	OEM
27C256-20	OEM	27TB6-EE00	OEM	28-15-02	0015	28TB2-NE00	OEM	28TB11-LH00	OEM	29TB3-CE00	OEM	29TB12-CE00	OEM
27C256-20FA	OEM	27TB6-EH00	OEM	28-18-01	0015	28TB2-RE00	OEM	28TB11-NE00	OEM	29TB3-CH00	OEM	29TB12-CH00	OEM
27C256-XX	OEM	27TB6-GE00	OEM	28-19-01	0015	28TB3-CE00	OEM	28TB11-NH00	OEM	29TB3-EE00	OEM	29TB12-EE00	OEM
27E04	OEM	27TB6-GH00	OEM	28-20-01	0015	28TB3-CH00	OEM	28TB11-RE00	OEM	29TB3-EH00	OEM	29TB12-EH00	OEM
27E08A	OEM	27TB6-JE00	OEM	28-20-02	0015	28TB3-EE00	OEM	28TB12-CE00	OEM	29TB3-GE00	OEM	29TB12-GE00	OEM
27E16	OEM	27TB6-JH00	OEM	28-21-01	0015	28TB3-EH00	OEM	28TB12-CH00	OEM	29TB3-GH00	OEM	29TB12-GH00	OEM
27E30	OEM	27TB6-LE00	OEM	28-22-01	0015	28TB3-GE00	OEM	28TB12-EE00	OEM	29TB3-JE00	OEM	29TB12-JE00	OEM
27E50	OEM	27TB6-LH00	OEM	28-22-02	0015	28TB3-GH00	OEM	28TB12-EH00	OEM	29TB3-JH00	OEM	29TB12-JH00	OEM
27J2	0133	27TB6-NE00	OEM	28-22-03	0015	28TB3-JE00	OEM	28TB12-GE00	OEM	29TB3-LE00	OEM	29TB12-LE00	OEM
27P1	0143	27TB6-NH00	OEM	28-22-04	0015	28TB3-JH00	OEM	28TB12-GH00	OEM	29TB3-LH00	OEM	29TB12-LH00	OEM
27ST-2SA673C	0148	27TB6-RE00	OEM	28-22-05	0015	28TB3-LE00	OEM	28TB12-JE00	OEM	29TB3-RE00	OEM	29TB12-NE00	OEM
27ST-2SD871	0055	27TB7	OEM	28-22-06	0015	28TB3-LH00	OEM	28TB12-JH00	OEM	29TB4	OEM	29TB12-RE00	OEM
27ST-LM2682	0079	27TB7-CE00	OEM	28-22-07	0015	28TB3-NE00	OEM	28TB12-LE00	OEM	29TB4-CE00	OEM	29TB13	OEM
27T1	OEM	27TB7-CH00	OEM	28-22-10	0015	28TB3-RE00	OEM	28TB12-LH00	OEM	29TB4-CH00	OEM	29TB14	OEM
27T2	OEM	27TB7-EE00	OEM	28-22-11	0087	28TB4-CE00	OEM	28TB12-NE00	OEM	29TB4-EE00	OEM	29TB15	OEM
27T3	OEM	27TB7-EH00	OEM	28-22-12	0015	28TB4-CH00	OEM	28TB12-NH00	OEM	29TB4-EH00	OEM	29V008M01	0280
27T4	OEM	27TB7-GE00	OEM	28-22-13	0102	28TB4-EE00	OEM	28TB12-RE00	OEM	29TB4-GE00	OEM	29V0038H03	0211
27T5	OEM	27TB7-GH00	OEM	28-22-14	0015	28TB4-EH00	OEM	28TB13	OEM	29TB4-GH00	OEM	29V011H01	0280
27T6	OEM	27TB7-JE00	OEM	28-22-15	0015	28TB4-GE00	OEM	28TB14	OEM	29TB4-JE00	OEM	29V012H01	0280
27T7	OEM	27TB7-JH00	OEM	28-22-16	0087	28TB4-GH00	OEM	28TB15	OEM	29TB4-JH00	OEM	30	OEM
27T8	OEM	27TB7-LE00	OEM	28-22-17	0015	28TB4-JE00	OEM	28V12	OEM	29TB4-LE00	OEM	30-004-001	0604
27T9	OEM	27TB7-LH00	OEM	28-22-18	0087	28TB4-JH00	OEM	28Z6	0002	29TB4-LH00	OEM	30-007-0	0693
27T10	OEM	27TB7-NE00	OEM	28-22-19	1303	28TB4-LE00	OEM	28Z6A	0002	29TB4-RE00	OEM	30-005072	0103
27T11	OEM	27TB7-NH00	OEM	28-22-19(ELECTROHOME		28TB4-LH00	OEM	28Z6AF	0002	29TB5	OEM	30-090	0161
27T12	OEM	27TB7-RE00	OEM		0102	28TB4-NE00	OEM	28Z6F	0002	29TB5-CE00	OEM	30-30	OEM
27T401	0280	27TB8	OEM	28-22-20	0087	28TB4-RE00	OEM	28Z6FAF	OEM	29TB5-CH00	OEM	30-219A	0468
27T402	0280	27TB8-CE00	OEM	28-22-21	0015	28TB5-CE00	OEM	29-178A	0738	29TB5-EE00	OEM	30-219B	0468
27T403	0211	27TB8-CH00	OEM	28-22-22	0102	28TB5-CH00	OEM	29-178B	0738	29TB5-EH00	OEM	30-219C	0468
27T404	0211	27TB8-EE00	OEM	28-23-01	0774	28TB5-EE00	OEM	29-178C	0738	29TB5-GE00	OEM	30-219D	0468
27T405	0211	27TB8-EH00	OEM	28-24-01	1303	28TB5-EH00	OEM	29-178D	0738	29TB5-GH00	OEM	30-219F	0468
27T406	0160	27TB8-GE00	OEM	28-25-01	0015	28TB5-GE00	OEM	29-178E	0738	29TB5-JE00	OEM	30-219H	0468
27T407	0440	27TB8-GH00	OEM	28-26-01	0474	28TB5-GH00	OEM	29-178M	0738	29TB5-JH00	OEM	30-219K	0468
27T408	0600	27TB8-JH00	OEM	28-29-01	0015	28TB5-JE00	OEM	29-178N	0738	29TB5-LE00	OEM	30-219M	0468
27T409	0198	27TB8-LE00	OEM	28-31-00	1323	28TB5-LE00	OEM	29-178P	0738	29TB5-LH00	OEM	30-219P	0468
27T410	0038	27TB8-LH00	OEM	28-32-00	0190	28TB5-LH00	OEM	29-178PA	0738	29TB5-RE00	OEM	30-219S	0468
27T411	0198	27TB8-NE00	OEM	28-32-0X	0190	28TB5-NE00	OEM	29-178PB	0738	29TB6	OEM	30-219V	0468
27T412	0136	27TB8-NH00	OEM	28-32-OX	0190	28TB6-CE00	OEM	29-178S	0738	29TB6-CE00	OEM	30-219Z	0468
27TB1	OEM	27TB8-RE00	OEM	28-33-01	0409	28TB6-CH00	OEM	29-178T	0738	29TB6-CH00	OEM	30-219ZB	0468
27TB1-CE00	OEM	27TB9	OEM	28-33-02	0409	28TB6-EE00	OEM	29-180A	0738	29TB6-EE00	OEM	30-219ZD	0468

DEVICE TYPE	REPL CODE
30-254A	0468
30-254B	0468
30-254C	0468
30-254D	0468
30-254E	0468
30-254F	0468
30-254K	0468
30-254P	0468
30-254V	0468
30-254ZB	0468
30-576-4	OEM
30-8054-7	0015
30-8057-13	0913
30A6F	0601
30A8F	OEM
30A10F	OEM
30AD50	OEM
30AD100	OEM
30AD200	OEM
30AD400	OEM
30AD600	OEM
30AD800	OEM
30AS	0015
30AV60	OEM
30AV80	OEM
30AV100	OEM
30B	0195
30B2	0195
30B2C11	OEM
30B2FR	OEM
30B2Z11	OEM
30B3	0195
30B4FR	OEM
30B5	0015
30B6FR	OEM
30BG2C11	OEM
30BG2C15	OEM
30BG2Z11	OEM
30BG11	1530
30BG15	1530
30BL2C11	OEM
30BL11	OEM
30BS	0015
30C	0015
30C1	OEM
30C2	OEM
30C4	OEM
30C6	OEM
30C8	OEM
30C10	OEM
30C12	OEM
30C14	OEM
30C16	OEM
30CB	0015
30CG2C15	OEM
30CG15	OEM
30CL2C11	OEM
30CL11	OEM
30CS03F	OEM
30CS04F	OEM
30CTQ30	OEM
30CTQ35	OEM
30CTQ40	OEM
30CTQ45	OEM
30D1	0087
30D1FC	OEM
30D2	0071
30D2C11	OEM
30D2Z11	OEM
30D4	0087
30DF	0031
30DF1	0564
30DF2	0087
30DF2-FC	0087
30DF2-FIN	0031
30DF6	0031
30DF6-FC	0031
30DG2C11	OEM
30DG2C15	OEM
30DG2Z11	OEM
30DG11	1530
30DG15	OEM
30DL1	OEM
30DL2	OEM
30DL2C11	OEM
30DL4	OEM
30DL11	OEM
30E08A	OEM
30F1	0947
30FG11	0029
30FH05S	1520
30FH1S	1520
30FH2S	1520
30FH3S	1847
30FH4S	1847
30FQ030	0615
30FQ030A	0615
30FQ035	0615
30FQ035A	0615
30FQ040	0615
30FQ040A	0615
30FQ045	0615
30FQ045A	0615
30FWJ2C11	OEM
30FWJ2C12	0584
30FWJ11	OEM
30G2C11	OEM
30G6P41	OEM
30G6P42	OEM
30GC15	OEM
30GG2C11	OEM
30GG2Z11	OEM
30GG11	0029
30GWJ2C11	OEM
30GWJ2C12	0584
30GWJ11	OEM
30H	0015
30H3	0097
30H3N	0476
30H3P	3743
30H3PN	0993
30HB050	5569
30HB100	5569
30HB200	1643
30HB400	5604
30HB600	1673
30HB800	1589
30J2	0097
30J2C11	OEM
30J2Z11	OEM
30J3P	0105
30J6P41	OEM
30J6P42	OEM
30JG2C11	OEM
30JG2Z11	OEM
30JG11	0601
30JH3	4308
30K	0015
30L6P41	OEM
30L15	OEM
30L20	OEM
30L25	OEM
30L29	OEM
30L30	OEM
30L35	OEM
30L40	OEM
30L45	OEM
30L50	OEM
30L55	OEM
30L60	OEM
30L65	OEM
30L70	OEM
30L75	OEM
30L80	OEM
30L85	OEM
30L90	OEM
30L95	OEM
30L100	OEM
30L110	OEM
30L120	OEM
30L130	OEM
30L140	OEM
30L150	OEM
30L160	OEM
30L170	OEM
30L180	OEM
30L190	OEM
30L200	OEM
30L210	OEM
30L220	OEM
30L230	OEM
30L240	OEM
30L250	OEM
30L260	OEM
30L270	OEM
30L280	OEM
30L290	OEM
30L300	OEM
30LF	OEM
30LSP42	OEM
30M	0015
30MF5	0029
30MF10	1530
30MF15	1530
30MF20	1530
30MF30	0029
30MF40	0029
30N2	0097
30N3	0097
30N15A	OEM
30N15A9	OEM
30N20A	OEM
30N20A9	OEM
30N25A	OEM
30N25A9	OEM
30N30A	OEM
30N30A9	OEM
30N35A	OEM
30N35A9	OEM
30N40A	OEM
30N40A9	OEM
30N45A	OEM
30N45A9	OEM
30N50A	OEM
30N50A9	OEM
30N55A	OEM
30N55A9	OEM
30N60A	OEM
30N60A9	OEM
30N65A	OEM
30N65A9	OEM
30N70A	OEM
30N70A9	OEM
30N75A	OEM
30N75A9	OEM
30N80A	OEM
30N80A9	OEM
30N85A	OEM
30N85A9	OEM
30N90A	OEM
30N90A9	OEM
30N95A	OEM
30N95A9	OEM
30N100A9	OEM
30N110A	OEM
30N120A	OEM
30N130A	OEM
30N150A	OEM
30N160A	OEM
30N170A	OEM
30N180A	OEM
30N190A	OEM
30N200A	OEM
30N210A	OEM
30N220A	OEM
30N230A	OEM
30N240A	OEM
30N260A	OEM
30NS	OEM
30P1	0143
30P2	0097
30P3	0097
30P4	OEM
30Q3	1581
30Q6P42	OEM
30QHC030	OEM
30QHC045	OEM
30R1	0953
30R2	OEM
30R2SW	OEM
30R3	0540
30R3P	OEM
30R4	OEM
30R6	0965
30R8	0087
30R10	0087
30R12	OEM
30R25	0044
30RECT-SI-1020	0023
30RECT-SI-1044(AU)	0015
30RH05S	1530
30RH1S	1530
30RH2S	1530
30RH3S	0029
30RH4S	0029
30S05	0110
30S1	0953
30S1S03	0086
30S1S04	OEM
30S2	1771
30S3	1747
30S3P	OEM
30S4	0540
30S5	1771
30S6	0965
30S8	0817
30S10	0922
30SB02S	1545
30SB02SA	1545
30SB03S	1545
30SB03SA	1545
30SB04S	1545
30SB04SA	1545
30SB025S	1545
30SP-LT-6540G	OEM
30ST-2SC2611	0276
30SZ-HZ5C1	0248
30T2	OEM
30T2FR	OEM
30T3P	OEM
30T4FR	OEM
30T6FR	OEM
30TB050	OEM
30TB100	OEM
30TB200	OEM
30TB400	OEM
30TB600	OEM
30TC10	OEM
30TC20	OEM
30TC30	OEM
30TC40	OEM
30TC50	OEM
30TC60	OEM
30TC70	OEM
30TC80	OEM
30TC90	OEM
30TC100	OEM
30TC110	OEM
30TC120	OEM
30TCA10	OEM
30TCA20	OEM
30TCA30	OEM
30TCA40	OEM
30TCA50	OEM
30TCA60	OEM
30TCA70	OEM
30TCA80	OEM
30TCA90	OEM
30TCA100	OEM
30TCA110	OEM
30TCA120	OEM
30U6P42	OEM
30V3P	OEM
30V10	OEM
30V11	OEM
30V12	OEM
30V-H6	0004
30V-HG	0004
30W3P	OEM
30X3P	OEM
30X5PV-CM	OEM
30X10PV-CM	OEM
30Y	0195
30Y3P	OEM
30n140A	OEM
30n250A	OEM
31-0001	0050
31-0002	0050
31-0003	0050
31-0004	0050
31-0005	0136
31-0006	0004
31-0007	0016
31-0008	0004
31-0009	0016
31-002-0	0076
31-0010	0178
31-0012	0111
31-0013	0111
31-0015	0050
31-0016	0050
31-0017	0004
31-0018	0004
31-0025	0004
31-0026	0004
31-0033	0004
31-0035	0164
31-0041	0136
31-0042	0136
31-0048	0127
31-0049	0127
31-0050	0127
31-0051	0144
31-0052	0111
31-0053	0004
31-0054	0127
31-0055	0791
31-0065	0136
31-0066	0560
31-0068	0016
31-0069	0016
31-0070	0004
31-0075	0164
31-0080	0016
31-0081	0016
31-0082	0016
31-0083	0086
31-0084	0016
31-0085	0016
31-0097	0144
31-0098	0127
31-0099	0111
31-011	0133
31-015	0190
31-025	0211
31-025-0	0368
31-027-0	0380
31-058-0	2046
31-069-0	0839
31-091-3	0836
31-093	0133
31-0100	0111
31-0101	0218
31-0102	1242
31-0103	0144
31-0104	0016
31-0105	0164
31-0106	0016
31-0107	0004
31-0108	0136
31-0115	0016
31-0116	0016
31-0123	0050
31-0124	0050
31-0132	0136
31-0134	0136
31-0135	0136
31-0139	0136
31-0141	0136
31-0148	0004
31-0150	0050
31-0153	0004
31-0161	0136
31-0163	0136
31-0165	0050
31-0166	0050
31-0168	0050
31-0170	0050
31-0171	0136
31-0172	0004
31-0175	1010
31-0177	0007
31-0178	0136
31-0180	0050
31-0181	0050
31-0182	0164
31-0183	0164
31-0184	0136
31-0187	0016
31-0188	0004
31-0189	0164
31-0190	0136
31-0191	0136
31-0192	0222
31-0196	0975
31-0205	0211
31-0206	0050
31-0217	0136
31-0228	0136
31-0229	0164
31-0230	0016
31-0239	0016
31-0240	0160
31-0241	0050
31-0241-1	0050
31-0242	0144
31-0243	0144
31-0246	0016
31-0247	0211
31-0248	0822
31-0253	0280
31-1	0016
31-1N914	0124
31-16	0016
31-101	OEM
31-194	0071
31-195	0015
31-2104733	0050
31-21004900	0050
31-21007744	0050
31-21024033	0050
31-21024044	0050
31-21047111	0050
31-21050611	0050
31-21050622	0050
31-22005400	0004
31DF1	0087
31DF2	0087
31DF3	1362
31DF4	1362
31DQ03	2542
31DQ04	0564
31DQ05	3604
31DQ06	3604
31DQ09	3604
31DQ10	OEM
31K111-002	OEM
31K111-02	OEM
31MQ30	OEM
31MQ40	OEM
31N2	OEM
31P1	OEM
31P4	OEM
31R2	1124
31SI-UPC1360C	1627
31ST-2SA733Q	0006
31ST-2SA733Q/R	0006
31ST-2SA733Q/R(AU)E	0006
31ST-2SA733R	0006
31ST-2SC1923-0	2691
31ST-2SC1923-8	OEM
31ST-2SC1923-O	2691
31Z6	0299
31Z6A	0299
32-0000	0143
32-0001	0143
32-0002	0143
32-0003	0143
32-0004	0143
32-0005	0012
32-0007	0123
32-0008	0143
32-006-0	0558
32-0013	0143
32-0022	0133
32-0023	0143
32-0025	0244
32-0026	0015
32-0029	0143
32-0033	1925
32-0035	0143
32-0036	0143
32-0037	0015
32-0038	0871
32-0039	0123
32-0042	0015
32-0043	0030
32-0044	0030
32-0045	0015
32-0046	0015
32-0047	0015
32-0048	0229
32-0049	0170
32-0050	0015
32-0057	0133
32-0059	0015
32-0060	0015
32-0061	0015
32-0062	0196
32-0063	0133
32-209A	0217
32-209B	0217
32-209C	0217
32-209D	0217
32-209F	0217
32-209H	0217
32-209K	0217
32-209M	0217
32-209P	0658
32-209S	0658
32-251A	0217
32-251B	0217
32-251C	0217
32-251D	0217
32-251F	0217
32-251H	0217
32-251K	0217
32-251M	0217
32-251P	0658
32-251S	0658
32-3305-1	0190
32-3305-4	0190
32-13843-2	0016
32-16591	3033
32-16599	0265
32-18537	0143
32-18539	0123
32-20738	0016
32-20739	0037
32-23555-1	2624
32-23555-2	2624
32-23555-3	2624
32-23555-4	2624
32-29778	0190
32-29778-1	0190
32-29778-2	0190
32-29778-3	0190
32-29778-4	0190
32-33051-1	0190
32-33051-5	0190
32-33057-1	0190
32-33057-2	1501
32-33057-3	0190
32-33057-4	0190
32-33057-5	0190
32-33057-6	0190
32-33091-6	1067
32-33091-9	1067
32-33094-1	1501
32-33094-2	1501
32-33094-3	1501
32-33094-4	1501
32-33094-5	1501
32-33094-6	1501
32-35894-1	0190
32-35894-2	1501
32-35894-3	1501
32-35894-4	1501
32-35894-5	0190
32-35894-6	1501
32-35894-6X7	1501
32-35894-7	1501
32-39091-1	1067
32-39091-2	1067
32-39091-3	1706
32-39091-4	1067
32-39091-5	1067
32-39091-6	1067
32-39091-7	1706
32-39091-8	1706
32-39091-9	1067
32-39091-15	1067
32-39091-132-39091-5	OEM
32-39704-1	1501
32-39704-2	1501
32-43737-1	3666
32-3309094-5	OEM
32E07A	OEM
32E64	0126
32KBS	OEM
32N2	OEM
32P4	OEM
32R2	1124
32ST-HZ20-04(AU)	0700
32ST-JE9018H	0144
32SZ-HZ20-04	0700
32SZ-HZ20-04(AU)	0700
32Z6	0188
32Z6A	0188
33-0002	0015
33-0005	OEM
33-0006	0015
33-0023	0015
33-0024	0015
33-0025	0649
33-0026	0015
33-0029	0015
33-0030	0015
33-0031	0015
33-0033	0774
33-0036	0102
33-0039	0440
33-070	0016
33-071	0016
33-090	0161
33-096	0460
33-2N3906	0006
33-3H22P	OEM
33-4-3	1030
33-4-3A	1030
33-7-3	1030
33-8-3	1030
33-1000-00	0211
33-1001-00	0211
33-1002-00	0822
33-1004-00	0160
33-1009-01	0004
33-1019-00	0004
33-1020-00	0004
33-1021-00	0004
33A14666	OEM
33A14667	OEM
33A14668	3515
33A14670	OEM
33A14725	OEM
33A17056	4422
33A17057	4927
33A17058	1346
33A17059	6433
33A17060	OEM
33A17061	OEM
33A17062	3955
33A17063	OEM
33B	0166
33B1	0195
33B1/1W	OEM
33B1/2.5W	OEM
33B2	0166
33BH05M	OEM
33BH05P	OEM
33C	0166
33D5R02	OEM
33D9R02GH	OEM
33G59019	0196
33G59024	0015
33G59113	1303
33G59121	0015
33G59122	0015
33H50	0144
33K3	OEM
33K59	0325
33M10	OEM
33P1	OEM
33PB05M	OEM
33PD05M	OEM
33PH05M	OEM
33RECT-SI-1066	0271
33SI-TA7335P	4406
33ST-2SC710D	0368
33ST-2SC2610	0261
33ST-2SD478A-01	0392
33ST-2SD748-01	3227
33ST-2SD748A-01	3227
33T1	OEM
33T2	OEM
33T3	OEM
33T4	OEM
33T5	OEM
33T6	OEM
33T7	OEM
33T8	OEM
33T9	OEM
33T10	OEM
33T11	OEM
33T12	OEM
33V10	OEM
33Y	0166
33Z6	0188
33Z6A	0188
34	0023
34-0012	1030
34-028-0	0019
34-029-0	0030
34-032-0	0124
34-0661-85	OEM
34-2N3563	0144
34-2N3904	0076
34-6	0050
34-6-16-47	0037
34-34-6015-43	0016
34-119	0050
34-220	0050
34-221	0050
34-298	0050
34-1000	0103
34-1000A	0103
34-1002	0042
34-1003	0925
34-1006	0276
34-1026	0178
34-1027	0905
34-1028	0103
34-1028A	0103
34-2001-1	0143
34-3022-7	0143
34-3057-48	OEM
34-5082-0012	0023
34-5170-0020	2848
34-5171-0016	0693
34-5172-0076	1221
34-5261-0141	0006
34-5262-0128	0286
34-6000-3	0050
34-6000-4	0004
34-6000-5	0004
34-6000-6	0004
34-6000-7	0004
34-6000-8	0004
34-6000-9	0050
34-6000-10	0050
34-6000-11	0050
34-6000-12	0050
34-6000-13	0050
34-6000-14	0050
34-6000-15	0004
34-6000-16	0050
34-6000-17	0050
34-6000-18	0050
34-6000-19	0050
34-6000-20	0050
34-6000-25	0050
34-6000-26	0050
34-6000-27	0004

If replacement code is OEM, contact original manufacturer for replacement.

DEVICE TYPE	REPL CODE
34-6000-28	0004
34-6000-29	0004
34-6000-30	0004
34-6000-31	0004
34-6000-32	0004
34-6000-33	0004
34-6000-34	0004
34-6000-50	OEM
34-6000-52	OEM
34-6000-54	OEM
34-6000-55	OEM
34-6000-56	OEM
34-6000-58	0050
34-6000-59	0050
34-6000-60	0050
34-6000-61	0050
34-6000-62	0050
34-6000-63	0050
34-6000-64	0016
34-6000-65	0050
34-6000-66	0050
34-6000-67	0050
34-6000-68	0050
34-6000-69	0144
34-6000-70	0144
34-6000-71	0854
34-6000-72	0330
34-6000-76	0050
34-6000-77	0050
34-6000-78	0050
34-6000-79	0050
34-6000-80	0050
34-6000-81	0050
34-6000-82	0050
34-6000-83	0050
34-6000-84	0050
34-6000-85	0050
34-6000-85HR52	OEM
34-6001-1	0085
34-6001-3	0144
34-6001-5	0016
34-6001-6	0144
34-6001-7	0004
34-6001-8	0004
34-6001-9	0004
34-6001-10	0004
34-6001-11	0004
34-6001-12	0004
34-6001-13	0004
34-6001-14	0004
34-6001-15	0037
34-6001-16	0004
34-6001-18	0004
34-6001-19	0004
34-6001-20	0004
34-6001-21	0004
34-6001-22	0004
34-6001-23	0004
34-6001-26	0004
34-6001-28	0164
34-6001-29	0004
34-6001-30	0004
34-6001-31	0004
34-6001-33	0004
34-6001-34	0016
34-6001-41	0004
34-6001-42	0004
34-6001-43	0004
34-6001-44	0004
34-6001-47	0004
34-6001-48	0016
34-6001-49	0016
34-6001-50	0016
34-6001-51	0086
34-6001-52	0016
34-6001-53	0079
34-6001-54	0136
34-6001-55	0079
34-6001-56	0079
34-6001-57	0079
34-6001-58	0079
34-6001-60	0079
34-6001-61	0050
34-6001-62	0016
34-6001-63	0016
34-6001-64	0144
34-6001-65	0233
34-6001-66	0164
34-6001-69	0016
34-6001-70	0016
34-6001-71	0198
34-6001-72	0004
34-6001-73	0111
34-6001-74	0016
34-6001-76	0004
34-6001-77	0016
34-6001-78	0086
34-6001-79	0160
34-6001-80	0086
34-6001-82	0086
34-6001-83	0086
34-6001-84	0208
34-6001-85	0086
34-6001-86	0126
34-6001-7434-6001-76	OEM
34-6002-1	0178
34-6002-2	0160
34-6002-3	0160
34-6002-4	0160

DEVICE TYPE	REPL CODE
34-6002-5	0160
34-6002-6	0160
34-6002-7	0160
34-6002-8	0160
34-6002-9	0160
34-6002-10	0160
34-6002-11	0160
34-6002-13	0160
34-6002-14	0160
34-6002-17	0085
34-6002-18	0160
34-6002-18A	0085
34-6002-19	0160
34-6002-20	0160
34-6002-21	0142
34-6002-22	0085
34-6002-22A	0085
34-6002-23	1681
34-6002-24	0975
34-6002-25	0924
34-6002-26	0142
34-6002-27	0103
34-6002-28	0178
34-6002-29	0178
34-6002-30	2013
34-6002-31	0975
34-6002-32	0103
34-6002-32A	0103
34-6002-33	0604
34-6002-34	0160
34-6002-35	0130
34-6002-37	0130
34-6002-41	0566
34-6002-42	1367
34-6002-43	0423
34-6002-45	0285
34-6002-46	0142
34-6002-49	0975
34-6002-54	0079
34-6002-55	0079
34-6002-56	0042
34-6002-57	0925
34-6002-58	0642
34-6002-61	0074
34-6002-62	0142
34-6002-63	0065
34-6002-64	0065
34-6005-1	0050
34-6005-2	0178
34-6005-3	0178
34-6006-1	0042
34-6007-1	0016
34-6007-2	0016
34-6007-3	0016
34-6007-4	0595
34-6007-5	0079
34-6007-6	0079
34-6007-7	0079
34-6007-8	0595
34-6007-9	0079
34-6007-10	0007
34-6007-11	0007
34-6007-12	0007
34-6007-13	0007
34-6007-14	0111
34-6008	0004
34-6009	0004
34-6010	4163
34-6015-1	0016
34-6015-2	0016
34-6015-3	0016
34-6015-4	0016
34-6015-5	0016
34-6015-6	0016
34-6015-7	0016
34-6015-8	0127
34-6015-9	0016
34-6015-10	0016
34-6015-11	0016
34-6015-12	0079
34-6015-13	0016
34-6015-14	0016
34-6015-15	0008
34-6015-15(PHILCO)	0007
34-6015-16	0008
34-6015-17	0008
34-6015-18	0007
34-6015-19	0008
34-6015-20	0016
34-6015-21	0016
34-6015-22	0007
34-6015-23	0086
34-6015-24	0086
34-6015-25	0079
34-6015-26	0037
34-6015-27	0224
34-6015-28	0338
34-6015-29	0008
34-6015-30	0086
34-6015-31	0007
34-6015-32	0050
34-6015-33	0050
34-6015-34	0050
34-6015-35	0050
34-6015-36	0050
34-6015-37	0330
34-6015-38	0330
34-6015-39	0050
34-6015-40	0050

DEVICE TYPE	REPL CODE
34-6015-41	0016
34-6015-42	0037
34-6015-43	0079
34-6015-43A	0016
34-6015-44	0016
34-6015-46	0224
34-6015-47	0224
34-6015-48	0224
34-6015-49	0144
34-6015-50	0127
34-6015-51	0086
34-6015-52	0224
34-6015-54	0016
34-6015-62	0330
34-6015-64	0285
34-6015-80	0016
34-6016-2	0016
34-6016-3	0016
34-6016-4	0016
34-6016-6	0079
34-6016-7	0016
34-6016-8	0016
34-6016-11	0004
34-6016-12	0126
34-6016-13	1267
34-6016-14	0016
34-6016-15	0037
34-6016-15A	0037
34-6016-16	0086
34-6016-17	0144
34-6016-18	0016
34-6016-19	0016
34-6016-22	0086
34-6016-23	0126
34-6016-23A	0126
34-6016-24	0016
34-6016-25	0016
34-6016-26	0086
34-6016-27	0086
34-6016-28	0050
34-6016-29	0050
34-6016-30	0086
34-6016-31	0566
34-6016-32	0037
34-6016-32A	0126
34-6016-33	0086
34-6016-41	0861
34-6016-44	0595
34-6016-45	0042
34-6016-46	0925
34-6016-47	0037
34-6016-49	0016
34-6016-49A	0016
34-6016-50	0004
34-6016-51	0595
34-6016-53	0566
34-6016-54	1367
34-6016-56	0861
34-6016-59	0791
34-6016-60	0037
34-6016-63	0016
34-6016-64	0693
34-6016-65	1267
34-6016-1734-6016-18	OEM
34-6017-3	0400
34-6018-2	0325
34-6075-46	0079
34-8001-43	0136
34-8002-1	0143
34-8002-2	0143
34-8002-3	0143
34-8002-4	0143
34-8002-5	0143
34-8002-6	0143
34-8002-7	0143
34-8002-22	OEM
34-8002-23	OEM
34-8002-27	OEM
34-8003	0015
34-8015-12	OEM
34-8022	0143
34-8022-1	0143
34-8022-2	0143
34-8022-3	0143
34-8022-4	0143
34-8022-5	0143
34-8022-6	0143
34-8022-6(PHILCO)	0123
34-8022-7	0143
34-8022-9	OEM
34-8022-77	0143
34-8024-24	0102
34-8024-25	0102
34-8026-1	0015
34-8026-2	0015
34-8026-3	0015
34-8026-4	0015
34-8027	0917
34-8028	1030
34-8034	0196
34-8034-1	0015
34-8034-2	0015
34-8034-3	0015
34-8034-4	0015
34-8034-7	0196
34-8036-1	0015
34-8036-2	0015
34-8036-3	0015
34-8036-4	0015

DEVICE TYPE	REPL CODE
34-8037	0196
34-8037-1	0196
34-8037-2	0196
34-8037-3	0196
34-8037-4	0196
34-8040-2	0015
34-8042-1	0015
34-8042-2	0015
34-8042-3	0015
34-8043-4	1030
34-8047-1	0913
34-8047-2	0015
34-8047-13	0143
34-8048-1	0015
34-8048-2	0229
34-8048-3	0015
34-8048-4	0015
34-8048-5	0015
34-8050-1	0015
34-8050-2	0015
34-8050-5	0015
34-8050-6	0015
34-8050-7	0015
34-8050-8	0015
34-8050-9	0015
34-8050-10	0071
34-8050-14	0015
34-8051	0015
34-8052-1	OEM
34-8053-2	0774
34-8053-3	0774
34-8053-4	0774
34-8053-6	OEM
34-8053-7	0774
34-8054-1	0015
34-8054-2	0015
34-8054-3	0015
34-8054-4	0015
34-8054-5	0015
34-8054-6	0015
34-8054-7	0015
34-8054-8	0102
34-8054-9	0015
34-8054-10	0015
34-8054-11	0015
34-8054-12	0015
34-8054-13	0015
34-8054-14	0015
34-8054-15	0015
34-8054-16	0920
34-8054-17	1217
34-8054-18	0015
34-8054-19	OEM
34-8054-23	0015
34-8054-25	0102
34-8054-27	0015
34-8055-2	0015
34-8055-3	2637
34-8056-1	1303
34-8057-1	0030
34-8057-2	OEM
34-8057-3	0123
34-8057-4	1649
34-8057-5	0133
34-8057-6	0133
34-8057-7	0162
34-8057-8	2510
34-8057-9	0649
34-8057-10	0025
34-8057-11	0015
34-8057-12	0002
34-8057-13	0133
34-8057-14	0137
34-8057-15	1303
34-8057-16	0649
34-8057-18	0015
34-8057-19	0030
34-8057-21	OEM
34-8057-22	0102
34-8057-23	0143
34-8057-24	0102
34-8057-25	0143
34-8057-26	0143
34-8057-27	0649
34-8057-28	0015
34-8057-29	0143
34-8057-30	0143
34-8057-31	0565
34-8057-32	0157
34-8057-33	0137
34-8057-34	0913
34-8057-36	OEM
34-8057-37	0162
34-8057-38	0039
34-8057-39	0102
34-8057-46	0326
34-8057-47	0256
34-8057-53	0137
34-8057-56	0317
34-8058	0474
34-8058-1	0474
34-8058-1#(2)	0474
34-8058-2	0474
34-8058-7	0474
34-8059-1	0015
34-8059-2	0015
34-8059-6	0095
34-8059-7	0095
34-8061-1	0201

DEVICE TYPE	REPL CODE
34-8062-1	0757
34-8062-4	0378
34-9037-1	0196
34-9037-2	0913
34-11016	0201
34-60001-63	0079
34D5R02	OEM
34D9R02GH	OEM
34E06A	OEM
34E3L	0144
34E31	0144
34G3098-2	0534
34GA	0985
34H31	0144
34ID	0989
34MT3	OEM
34MW117	OEM
34N2	OEM
34P1AA	0037
34P4	0133
34R2	0805
34SI-UPC1373HA	2026
34ST-2N6558	0643
34ST-2SD468C	0018
34ST-MPS9418AT	0324
34SZ-UZ-12B	0052
34SZ-UZ12B	0052
34TB4	OEM
34TB5	OEM
34TB6	OEM
34TB7	OEM
34TB8	OEM
34TB9	OEM
34TB10	OEM
34V12	OEM
34Z6	OEM
34Z6A	OEM
35	0023
35-003-001	0015
35-013-0	0005
35-43	OEM
35-44	OEM
35-1000	0071
35-1003	0071
35-1004	0015
35-1005	0015
35-1006	5514
35-1008	0242
35-1014	0133
35-1029	0071
35-210631-01	0576
35-210631-02	0576
35-39306001	0016
35-39306002	0016
35-39306003	0016
35AD50	OEM
35AD100	OEM
35AD200	OEM
35AD400	OEM
35AD600	OEM
35AD800	OEM
35B611	0289
35BL611	0289
35C	0010
35C05	3189
35C05B	OEM
35C05BF	OEM
35C05F	OEM
35C05R	1630
35C025	0641
35C025B	0641
35C025BF	0608
35C025F	0608
35C050	0641
35C050B	0641
35C050BF	0608
35C050F	0608
35C10	2901
35C10B	0641
35C10BF	0608
35C10F	0608
35C10R	0254
35C15	1124
35C15B	0641
35C15BF	0608
35C15F	0608
35C15R	1107
35C20	1124
35C20B	0641
35C20BF	0608
35C20F	0608
35C20R	1107
35C25	0217
35C25B	0217
35C25BF	0610
35C25F	0610
35C30	1126
35C30B	0217
35C30BF	0610
35C30F	0610
35C30R	0258
35C35	0468
35C35B	0468
35C35BF	OEM
35C35F	OEM
35C40	0805
35C40B	0217
35C40BF	0610
35C40F	0610

DEVICE TYPE	REPL CODE
35C40R	0258
35C45	0468
35C45B	0468
35C45BF	OEM
35C45F	OEM
35C50	1195
35C50B	0217
35C50BF	0610
35C50F	0610
35C50R	1644
35C60	0319
35C60B	0217
35C60BF	0610
35C60F	0610
35C60R	0267
35C70	1132
35C70(RECT.)	1132
35C70B	0658
35C70BF	0468
35C70F	0468
35C70R	1119
35C80	1132
35C80B	0658
35C80BF	0468
35C80F	0468
35C80R	1119
35C90	0044
35C90B	0658
35C90BF	0468
35C90F	0468
35C90R	0281
35C100	0044
35C100B	0658
35C100BF	0468
35C100F	0468
35C100R	0281
35C110	0658
35C110B	0658
35C110BF	0468
35C110F	0468
35C120	0658
35C120B	0658
35C120BF	0468
35C120F	0468
35D5R02	OEM
35D9R02GH	OEM
35E05A	OEM
35F70	OEM
35F80	OEM
35F90	OEM
35F100	OEM
35F110	OEM
35F120	OEM
35FCRA	OEM
35FCRB	OEM
35FCRC	OEM
35FCRD	OEM
35FCRE	OEM
35FCRF	OEM
35FCRG	OEM
35FCRH	OEM
35FCRI	OEM
35FCRJ	OEM
35FCRK	OEM
35FCRL	OEM
35H5	3189
35H10	1124
35H10R	1107
35H20	1124
35H30	0805
35H40	0805
35HR20	1107
35HR30	0258
35HR40	0258
35M10	OEM
35N2	OEM
35P1	0211
35P2	0211
35P2C	0211
35P4	0133
35PCRA	1650
35PCRB	1650
35PCRC	1650
35PCRD	1650
35PCRE	1650
35PCRF	2647
35PCRG	2647
35PCRH	2647
35PCRI	2649
35PCRJ	2649
35PCRK	OEM
35PCRL	OEM
35PCRNA	OEM
35PCRNB	OEM
35PCRNC	OEM
35PCRND	OEM
35PCRNE	OEM
35PCRNF	OEM
35PCRNG	OEM
35PCRNH	OEM
35PCRNI	OEM
35PCRNJ	OEM
35PCRNK	OEM
35PCRNL	OEM
35PCRSA	0731
35PCRSB	0731
35PCRSC	0712
35PCRSD	0469
35PCRSE	0469

DEVICE TYPE	REPL CODE
35PCRSF	0722
35PCRSG	0722
35PCRSH	0722
35PCRSI	0725
35PCRSJ	0725
35PCRSK	0750
35PCRSL	0750
35RCS10A	OEM
35RCS20A	OEM
35RCS30A	OEM
35RCS40A	OEM
35RCS50A	OEM
35RCS60A	OEM
35RCS80A	OEM
35RE60	OEM
35RE90	OEM
35RE100	OEM
35RE110	OEM
35RE120	OEM
35RE130	OEM
35SI-UPC574J	1329
35SI-UPC580C	1055
35SP005	OEM
35SP005D	OEM
35SP005N	OEM
35SP01	OEM
35SP01D	OEM
35SP01N	OEM
35SP02	OEM
35SP02D	OEM
35SP02N	OEM
35SP04	OEM
35SP04D	OEM
35SP04N	OEM
35SP06	OEM
35SP06D	OEM
35SP06N	OEM
35SP08	OEM
35SP08D	OEM
35SP08N	OEM
35SP10	OEM
35SP10D	OEM
35SP10N	OEM
35SZ-UPC574J	1329
35T1	0211
35T2	OEM
35T3	OEM
35T4	OEM
35T5	OEM
35T6	OEM
35T7	OEM
35T8	OEM
35T9	OEM
35T10	OEM
35T11	OEM
35T12	OEM
35TC4I	OEM
35TCRA	0469
35TCRB	0469
35TCRC	0469
35TCRD	0469
35TCRE	0469
35TCRF	0722
35TCRG	0722
35TCRH	0722
35TCRI	0725
35TCRJ	0725
35TCRK	0679
35TCRL	0679
35TE18	OEM
35TE20	OEM
35TE22	OEM
35TE24	OEM
35TE26	OEM
35TE28	OEM
35TE30	OEM
35TE35	OEM
35TE40	OEM
35V10	OEM
35V11	OEM
35V12	OEM
35Z6	0140
35Z6A	0140
36-0041	1478
36-0083	2872
36-6343	0137
36B2(2%)	OEM
36B3	0010
36D5R37	OEM
36D9R37GH	OEM
36D-32	0133
36D32	0133
36E004-1	0143
36GRA	0985
36J003-1	0004
36MB5A	OEM
36MB10A	2716
36MB20A	2800
36MB40A	2925
36MB60A	2925
36MB80A	3184
36MB100A	3184
36MB120A	OEM
36P1	0211
36P1C	0211
36P1F	0211
36P2F	0211
36P3	0211
36P3A	0211

DEVICE TYPE	REPL CODE	DEVICE TYPE	REPL CODE	DEVICE TYPE	REPL CODE	DEVICE TYPE	REPL CODE	DEVICE TYPE	REPL CODE	DEVICE TYPE	REPL CODE	DEVICE TYPE	REPL CODE
36P3C	0211	36TB3-JH00	OEM	37-8-01	1030	37TB4-GE00	OEM	37TB13	OEM	38TB6-JH00	OEM	39P2C	0280
36P4	0211	36TB3-LE00	OEM	37-8-1	1030	37TB4-GH00	OEM	37TB14	OEM	38TB6-LE00	OEM	39PC1	0085
36P4C	0211	36TB3-LH00	OEM	37-210	2018	37TB4-JE00	OEM	37TB15	OEM	38TB6-LH00	OEM	39R-26	0441
36P5	0211	36TB3-NE00	OEM	37H01	1402	37TB4-JH00	OEM	37TB16-JH00	OEM	38TB6-NE00	OEM	39RECT-SI-1006	0023
36P5C	0211	36TB3-NH00	OEM	37H02	1761	37TB4-LE00	OEM	37TB30EH00	OEM	38TB6-NH00	OEM	39SD-1N4140	OEM
36P7	0004	36TB3-RE00	OEM	37H03	0668	37TB4-LH00	OEM	37Z6	0157	38TB6-RE00	OEM	39SD-1N4148	0124
36P7C	0004	36TB4-CE00	OEM	37H04	0547	37TB4-NE00	OEM	37Z6A	0157	38TB7-CE00	OEM	39ST-2SD820	2051
36P7T	0004	36TB4-CH00	OEM	37H05	2403	37TB4-NH00	OEM	38-11016	0201	38TB7-EE00	OEM	39ST-2SD898B	0055
36P8	0004	36TB4-EE00	OEM	37H06	0870	37TB4-RE00	OEM	38A-64C	0649	38TB7-EH00	OEM	39SZ-RD12FB	0137
36P8C	0004	36TB4-EH00	OEM	37H07	0524	37TB5-CE00	OEM	38A64C	0649	38TB7-GE00	OEM	39T1	0211
36R2	0319	36TB4-GE00	OEM	37H08	0510	37TB5-CH00	OEM	38A4148-000	0128	38TB7-GH00	OEM	39TB7	OEM
36R2S	0319	36TB4-GH00	OEM	37H09	0616	37TB5-EE00	OEM	38A4148-001	0128	38TB7-JE00	OEM	39TB9	OEM
36RA50	0468	36TB4-JH00	OEM	37P4	OEM	37TB5-EH00	OEM	38A4157-000	0128	38TB7-JH00	OEM	39TB11	OEM
36RA60	0468	36TB4-LE00	OEM	37RA60	0217	37TB5-GE00	OEM	38A4871-000	OEM	38TB7-LE00	OEM	39TB13	OEM
36RA70	OEM	36TB4-LH00	OEM	37RA80	0658	37TB5-GH00	OEM	38A5576-000	0473	38TB7-LH00	OEM	39Z6	0062
36RA80	0468	36TB4-NE00	OEM	37RA100	0658	37TB5-JE00	OEM	38A5577-000	0551	38TB7-NE00	OEM	39Z6A	0062
36RA90	OEM	36TB4-NH00	OEM	37RA120	0658	37TB5-JH00	OEM	38A5578-000	0382	38TB7-NH00	OEM	40	OEM
36RA100	0468	36TB4-RE00	OEM	37RA140	OEM	37TB5-LE00	OEM	38E04A	OEM	38TB7-RE00	OEM	40-013-0	OEM
36RA110	0468	36TB5-CE00	OEM	37RA150	OEM	37TB5-LH00	OEM	38N2	OEM	38TB8-CE00	OEM	40-035-0	0215
36RA120	0468	36TB5-CH00	OEM	37RA160	OEM	37TB5-NE00	OEM	38P1	0160	38TB8-CH00	OEM	40-063-0	2828
36RA140	OEM	36TB5-EE00	OEM	37RC10	0217	37TB5-NH00	OEM	38P1C	0160	38TB8-EE00	OEM	40-065-19-001	1859
36RA150	OEM	36TB5-EH00	OEM	37RC10A	0217	37TB5-RE00	OEM	38R2	1132	38TB8-EH00	OEM	40-065-19-002	0341
36RA160	OEM	36TB5-GE00	OEM	37RC20	0217	37TB6-CE00	OEM	38R2S	1132	38TB8-GE00	OEM	40-065-19-003	1844
36RC10	0217	36TB5-GH00	OEM	37RC20A	0217	37TB6-CH00	OEM	38SI-SN76730N	OEM	38TB8-GH00	OEM	40-065-19-004	1835
36RC10A	0217	36TB5-JH00	OEM	37RC30	0217	37TB6-EE00	OEM	38SI-UPC1373H	2026	38TB8-JH00	OEM	40-065-19-005	0033
36RC20	0217	36TB5-LE00	OEM	37RC30A	0217	37TB6-EH00	OEM	38ST-2SA10150(AP)	0148	38TB8-LE00	OEM	40-065-19-006	1041
36RC20A	0217	36TB5-LH00	OEM	37RC40	0217	37TB6-GE00	OEM	38ST-2SA10150(AP)B	0148	38TB8-LH00	OEM	40-065-19-007	3399
36RC30	0217	36TB5-NE00	OEM	37RC40A	0217	37TB6-GH00	OEM	38ST-2SC1855	0224	38TB8-NE00	OEM	40-065-19-008	2097
36RC30A	0217	36TB5-NH00	OEM	37RC60A	0217	37TB6-JE00	OEM	38T1	0211	38TB8-NH00	OEM	40-065-19-012	0559
36RC40	0217	36TB5-RE00	OEM	37RC70A	OEM	37TB6-JH00	OEM	38T2	OEM	38TB8-RE00	OEM	40-065-19-013	0141
36RC40A	0217	36TB6-CE00	OEM	37RC80	0658	37TB6-LE00	OEM	38T3	OEM	38TB9-CE00	OEM	40-065-19-014	1481
36RC50A	OEM	36TB6-CH00	OEM	37RC80A	OEM	37TB6-LH00	OEM	38T4	OEM	38TB9-CH00	OEM	40-065-19-016	1024
36RC60	0217	36TB6-EE00	OEM	37RC100	0658	37TB6-NE00	OEM	38T5	OEM	38TB9-EE00	OEM	40-065-19-027	1432
36RC60A	0217	36TB6-EH00	OEM	37RC100A	OEM	37TB6-NH00	OEM	38T6	OEM	38TB9-EH00	OEM	40-065-19-029	1368
36RC80	0468	36TB6-GE00	OEM	37RC110	0658	37TB6-RE00	OEM	38T7	OEM	38TB9-GE00	OEM	40-065-19-030	1486
36RC80A	0468	36TB6-GH00	OEM	37RC110A	OEM	37TB7-CE00	OEM	38T8	OEM	38TB9-GH00	OEM	40-0502	0229
36RC100	0468	36TB6-JE00	OEM	37RC120	0658	37TB7-CH00	OEM	38T9	OEM	38TB9-JE00	OEM	40-601	0211
36RC100A	OEM	36TB6-JH00	OEM	37RC120A	OEM	37TB7-EE00	OEM	38T10	OEM	38TB9-JH00	OEM	40-13184-3	0852
36RC110	0468	36TB6-LE00	OEM	37RE60	0217	37TB7-EH00	OEM	38T11	OEM	38TB9-LE00	OEM	40A50	3189
36RC110A	OEM	36TB6-LH00	OEM	37RE70	0658	37TB7-GE00	OEM	38T12	OEM	38TB9-LH00	OEM	40A100	2901
36RC120	0468	36TB6-NE00	OEM	37RE80	0658	37TB7-GH00	OEM	38TB1-CE00	OEM	38TB9-NE00	OEM	40A150	1124
36RC120A	OEM	36TB6-NH00	OEM	37RE90	0658	37TB7-JE00	OEM	38TB1-EE00	OEM	38TB9-NH00	OEM	40A200	1124
36RCS10A	OEM	36TB6-RE00	OEM	37RE100	OEM	37TB7-JH00	OEM	38TB1-EH00	OEM	38TB9-RE00	OEM	40A300	1126
36RCS20A	OEM	36TB7-CE00	OEM	37RE100M	OEM	37TB7-LE00	OEM	38TB1-GE00	OEM	38TB10-CE00	OEM	40A400	0805
36RCS30A	OEM	36TB7-CH00	OEM	37RE110	OEM	37TB7-LH00	OEM	38TB1-GH00	OEM	38TB10-CH00	OEM	40A500	1195
36RCS40A	OEM	36TB7-EE00	OEM	37RE120	OEM	37TB7-NE00	OEM	38TB1-JE00	OEM	38TB10-EE00	OEM	40A600	0319
36RCS50A	OEM	36TB7-EH00	OEM	37RE130	OEM	37TB7-NH00	OEM	38TB1-JH00	OEM	38TB10-EH00	OEM	40AD50	OEM
36RCS60A	OEM	36TB7-GE00	OEM	37RE130M	OEM	37TB7-RE00	OEM	38TB1-LE00	OEM	38TB10-GE00	OEM	40AD100	OEM
36RCS80A	OEM	36TB7-GH00	OEM	37REH60	0217	37TB8-CE00	OEM	38TB1-LH00	OEM	38TB10-GH00	OEM	40AD200	OEM
36RE50	0468	36TB7-JE00	OEM	37REH80	0658	37TB8-CH00	OEM	38TB1-NE00	OEM	38TB10-JE00	OEM	40AD400	OEM
36RE60	0468	36TB7-JH00	OEM	37REH100	0658	37TB8-EE00	OEM	38TB1-NH00	OEM	38TB10-JH00	OEM	40AD600	OEM
36RE70	OEM	36TB7-LE00	OEM	37REH120	0658	37TB8-EH00	OEM	38TB1-RE00	OEM	38TB10-LE00	OEM	40AD800	OEM
36RE80	0468	36TB7-LH00	OEM	37REH130	OEM	37TB8-GE00	OEM	38TB2-CE00	OEM	38TB10-LH00	OEM	40AS	0015
36RE90	OEM	36TB7-NH00	OEM	37SE-RD10EB	0064	37TB8-GH00	OEM	38TB2-CH00	OEM	38TB10-NE00	OEM	40AV20	OEM
36RE100	0468	36TB8-CE00	OEM	37SZ-BZ-120	0137	37TB8-JE00	OEM	38TB2-EE00	OEM	38TB10-NH00	OEM	40AV30	OEM
36RE110	0468	36TB8-CH00	OEM	37SZ-RD10EB	0064	37TB8-JH00	OEM	38TB2-EH00	OEM	38TB10-RE00	OEM	40AV50	OEM
36RE120	0738	36TB8-EE00	OEM	37T1	0136	37TB8-LE00	OEM	38TB2-GE00	OEM	38TB11-CE00	OEM	40AV60	OEM
36RE130	OEM	36TB8-EH00	OEM	37T2	OEM	37TB8-LH00	OEM	38TB2-GH00	OEM	38TB11-CH00	OEM	40B05T	OEM
36RE130M	OEM	36TB8-GE00	OEM	37T3	OEM	37TB8-NE00	OEM	38TB2-JE00	OEM	38TB11-EE00	OEM	40B5	0015
36REH60	0610	36TB8-LE00	OEM	37T4	OEM	37TB8-NH00	OEM	38TB2-JH00	OEM	38TB11-EH00	OEM	40B10T	OEM
36REH80	0468	36TB8-LH00	OEM	37T5	OEM	37TB8-RE00	OEM	38TB2-LE00	OEM	38TB11-GE00	OEM	40B20T	OEM
36REH100	0468	36TB8-RE00	OEM	37T6	OEM	37TB9-CE00	OEM	38TB2-LH00	OEM	38TB11-GH00	OEM	40B40T	OEM
36REH120	0468	36TB9-CE00	OEM	37T7	OEM	37TB9-CH00	OEM	38TB2-NE00	OEM	38TB11-JE00	OEM	40B605	OEM
36REH130	OEM	36TB9-CH00	OEM	37T8	OEM	37TB9-EE00	OEM	38TB2-NH00	OEM	38TB11-JH00	OEM	40BS	0015
36SI-CA3065E	0167	36TB9-EE00	OEM	37T9	OEM	37TB9-EH00	OEM	38TB2-RE00	OEM	38TB11-LE00	OEM	40C	0015
36SI-UPD1705C-012	OEM	36TB9-EH00	OEM	37T10	OEM	37TB9-GE00	OEM	38TB3-CE00	OEM	38TB11-LH00	OEM	40C05	4031
36ST-2SB857C	2923	36TB9-GE00	OEM	37T11	OEM	37TB9-GH00	OEM	38TB3-CH00	OEM	38TB11-NE00	OEM	40C05B	4031
36T1	0211	36TB9-GH00	OEM	37T12	OEM	37TB9-JE00	OEM	38TB3-EE00	OEM	38TB11-NH00	OEM	40C025	1704
36T2	OEM	36TB9-JE00	OEM	37TB1-CE00	OEM	37TB9-JH00	OEM	38TB3-EH00	OEM	38TB11-RE00	OEM	40C025B	1704
36T3	OEM	36TB9-JH00	OEM	37TB1-CH00	OEM	37TB9-LE00	OEM	38TB3-GE00	OEM	38TB12-CE00	OEM	40C050	1704
36T4	OEM	36TB9-LE00	OEM	37TB1-EE00	OEM	37TB9-LH00	OEM	38TB3-GH00	OEM	38TB12-CH00	OEM	40C050B	1704
36T5	OEM	36TB9-LH00	OEM	37TB1-EH00	OEM	37TB9-NE00	OEM	38TB3-JE00	OEM	38TB12-EE00	OEM	40C1	OEM
36T6	OEM	36TB9-RE00	OEM	37TB1-GE00	OEM	37TB9-NH00	OEM	38TB3-JH00	OEM	38TB12-EH00	OEM	40C1R10	OEM
36T7	OEM	36TB10-CE00	OEM	37TB1-GH00	OEM	37TB9-RE00	OEM	38TB3-LE00	OEM	38TB12-GE00	OEM	40C1R20	OEM
36T8	OEM	36TB10-CH00	OEM	37TB1-JE00	OEM	37TB10-CE00	OEM	38TB3-LH00	OEM	38TB12-GH00	OEM	40C1R30	OEM
36T9	OEM	36TB10-EE00	OEM	37TB1-JH00	OEM	37TB10-CH00	OEM	38TB3-NE00	OEM	38TB12-JE00	OEM	40C1S10	OEM
36T10	OEM	36TB10-EH00	OEM	37TB1-LE00	OEM	37TB10-EE00	OEM	38TB3-NH00	OEM	38TB12-JH00	OEM	40C1S20	OEM
36T11	OEM	36TB10-GE00	OEM	37TB1-LH00	OEM	37TB10-EH00	OEM	38TB3-RE00	OEM	38TB12-LE00	OEM	40C1S30	OEM
36T12	OEM	36TB10-GH00	OEM	37TB1-NE00	OEM	37TB10-GE00	OEM	38TB4-CE00	OEM	38TB12-LH00	OEM	40C2	OEM
36TB1-CE00	OEM	36TB10-JE00	OEM	37TB1-NH00	OEM	37TB10-GH00	OEM	38TB4-CH00	OEM	38TB12-NE00	OEM	40C2FM	OEM
36TB1-CH00	OEM	36TB10-JH00	OEM	37TB1-RE00	OEM	37TB10-JH00	OEM	38TB4-EE00	OEM	38TB12-NH00	OEM	40C2PW8V1SP	0150
36TB1-EE00	OEM	36TB10-LE00	OEM	37TB2-CE00	OEM	37TB10-LE00	OEM	38TB4-EH00	OEM	38TB12-RE00	OEM	40C2R10	OEM
36TB1-EH00	OEM	36TB10-LH00	OEM	37TB2-CH00	OEM	37TB10-LH00	OEM	38TB4-GE00	OEM	38TB13	OEM	40C2R20	OEM
36TB1-GE00	OEM	36TB10-RE00	OEM	37TB2-EE00	OEM	37TB10-NE00	OEM	38TB4-GH00	OEM	38Z6	0471	40C2R30	OEM
36TB1-GH00	OEM	36TB11-CH00	OEM	37TB2-EH00	OEM	37TB10-NH00	OEM	38TB4-JE00	OEM	38Z6A	0471	40C2R110	OEM
36TB1-JE00	OEM	36TB11-EE00	OEM	37TB2-GE00	OEM	37TB10-RE00	OEM	38TB4-JH00	OEM	39-033-0	2301	40C2S10	OEM
36TB1-JH00	OEM	36TB11-EH00	OEM	37TB2-GH00	OEM	37TB11-CE00	OEM	38TB4-LE00	OEM	39-033-2	2301	40C2S20	OEM
36TB1-LE00	OEM	36TB11-GE00	OEM	37TB2-JE00	OEM	37TB11-CH00	OEM	38TB4-LH00	OEM	39-047-1	1899	40C2S30	OEM
36TB1-LH00	OEM	36TB11-GH00	OEM	37TB2-JH00	OEM	37TB11-EE00	OEM	38TB4-NE00	OEM	39-054-1	1786	40C4	OEM
36TB1-NE00	OEM	36TB11-JE00	OEM	37TB2-LE00	OEM	37TB11-EH00	OEM	38TB4-NH00	OEM	39-059-0	0651	40C4FM	OEM
36TB1-NH00	OEM	36TB11-JH00	OEM	37TB2-LH00	OEM	37TB11-GE00	OEM	38TB4-RE00	OEM	39-060-0	4644	40C6	OEM
36TB2-CE00	OEM	36TB11-LE00	OEM	37TB2-NE00	OEM	37TB11-GH00	OEM	38TB5-CE00	OEM	39-060-1	1128	40C6FM	OEM
36TB2-CH00	OEM	36TB11-LH00	OEM	37TB2-RE00	OEM	37TB11-JH00	OEM	38TB5-EE00	OEM	39-061-0	1540	40C8	OEM
36TB2-EE00	OEM	36TB11-RE00	OEM	37TB3-CE00	OEM	37TB11-LE00	OEM	38TB5-EH00	OEM	39-062-0	OEM	40C10	1704
36TB2-EH00	OEM	36TB12-CH00	OEM	37TB3-CH00	OEM	37TB11-LH00	OEM	38TB5-GE00	OEM	39-076-0	1910	40C10B	1704
36TB2-GE00	OEM	36TB12-EE00	OEM	37TB3-EE00	OEM	37TB11-NE00	OEM	38TB5-GH00	OEM	39-077-0	0176	40C12	OEM
36TB2-GH00	OEM	36TB12-EH00	OEM	37TB3-GE00	OEM	37TB11-NH00	OEM	38TB5-JE00	OEM	39-11	OEM	40C14	OEM
36TB2-JH00	OEM	36TB12-GE00	OEM	37TB3-GH00	OEM	37TB11-RE00	OEM	38TB5-JH00	OEM	39-13	0137	40C15	1704
36TB2-LE00	OEM	36TB12-GH00	OEM	37TB3-JE00	OEM	37TB12-CE00	OEM	38TB5-LE00	OEM	39A9	0211	40C15B	1704
36TB2-LH00	OEM	36TB12-JE00	OEM	37TB3-JH00	OEM	37TB12-CH00	OEM	38TB5-LH00	OEM	39A69-2	0133	40C16	OEM
36TB2-NE00	OEM	36TB12-JH00	OEM	37TB3-LE00	OEM	37TB12-EE00	OEM	38TB5-NE00	OEM	39B2	0533	40C20	1704
36TB2-NH00	OEM	36TB12-LH00	OEM	37TB3-LH00	OEM	37TB12-EH00	OEM	38TB5-NH00	OEM	39B3	OEM	40C20B	1704
36TB2-RE00	OEM	36TB13	OEM	37TB3-NE00	OEM	37TB12-GH00	OEM	38TB5-RE00	OEM	39B6	OEM	40C25	4031
36TB3-CE00	OEM	36TBA-RE00	OEM	37TB3-NH00	OEM	37TB12-JH00	OEM	38TB6-CE00	OEM	39B-9E	OEM	40C25B	4031
36TB3-CH00	OEM	36V12	OEM	37TB3-RE00	OEM	37TB12-LE00	OEM	38TB6-EE00	OEM	39J2F	OEM	40C30	4031
36TB3-EE00	OEM	36W4A	2997	37TB4-CE00	OEM	37TB12-LH00	OEM	38TB6-EH00	OEM	39N2	OEM	40C30B	4031
36TB3-EH00	OEM	36Z6	0162	37TB4-CH00	OEM	37TB12-NE00	OEM	38TB6-GE00	OEM	39P1	0160	40C35	4031
36TB3-GE00	OEM	36Z6A	0162	37TB4-EE00	OEM	37TB12-NH00	OEM	38TB6-GH00	OEM	39P1C	0160	40C35B	4031
36TB3-GH00	OEM	36ZW38	OEM	37TB4-EH00	OEM	37TB12-RE00	OEM	38TB6-JE00	OEM	39P1K	OEM	40C40	4031

DEVICE TYPE	REPL CODE
40C40B	4031
40C45	4031
40C45B	4031
40C50	4031
40C50B	4031
40C60	4031
40C60B	4031
40C70	0679
40C70B	0679
40C80	0679
40C80B	0679
40C90	0679
40C90B	0679
40C100	0679
40C100B	0679
40C110	0679
40C110B	0679
40C120	0679
40C120B	0679
40CDQ020	OEM
40CDQ030	OEM
40CDQ035	OEM
40CDQ040	OEM
40CDQ045	OEM
40CPQ040	OEM
40CPQ045	OEM
40D1	3657
40D2	7006
40D4	7006
40D6	7007
40D8	7007
40D1547	0050
40D6665A03	0015
40E03A	OEM
40E1R20	OEM
40E1R40	OEM
40E1R60	OEM
40E1S20	OEM
40E1S40	OEM
40E1S60	OEM
40E2FM	OEM
40E2R20	OEM
40E2R40	OEM
40E2R60	OEM
40E2S20	OEM
40E2S40	OEM
40E2S60	OEM
40E4FM	OEM
40E6FM	OEM
40F05	OEM
40F1R07	OEM
40F1R15	OEM
40F1S07	OEM
40F1S15	OEM
40F2R07	OEM
40F2R15	OEM
40F2S07	OEM
40F2S15	OEM
40F10	OEM
40F15	OEM
40F20	OEM
40F25	OEM
40F30	OEM
40F35	OEM
40F40	OEM
40F45	OEM
40F50	OEM
40F60	OEM
40H	0015
40H000F	OEM
40H000P	OEM
40H002F	OEM
40H3	0097
40H3N	0476
40H3P	0197
40H3PN	0997
40HA10	OEM
40HA20	OEM
40HA40	OEM
40HA60	OEM
40HA80	OEM
40HA100	OEM
40HA120	OEM
40HA140	OEM
40HA160	OEM
40HC032F	OEM
40HF5	3189
40HF5R	0267
40HF10	2901
40HF10R	0267
40HF15	1124
40HF20	1124
40HF20R	0267
40HF30	1126
40HF30R	0267
40HF40	0805
40HF40R	0258
40HF50	1195
40HF50R	0267
40HF60	0319
40HF60R	0267
40HF70	1132
40HF80	1132
40HF80R	0281
40HF90	0044
40HF100	0044
40HF100R	0281
40HF120	OEM
40HF140	OEM
40HF160	OEM
40HFL5S02	1530
40HFL5S05	1530
40HFL10S02	1530
40HFL10S05	1530
40HFL10S10	4895
40HFL20S02	1530
40HFL20S05	1530
40HFL20S10	4895
40HFL40S02	0029
40HFL40S05	0029
40HFL40S10	7089
40HFL60S02	0601
40HFL60S05	0601
40HFL60S10	6231
40HFL80S05	OEM
40HFL80S10	OEM
40HFL100S05	OEM
40HFL100S10	OEM
40HFR5	1630
40HFR10	0254
40HFR15	1107
40HFR20	1107
40HFR30	1111
40HFR40	0258
40HFR50	1644
40HFR60	0267
40HFR70	1119
40HFR80	1119
40HFR90	0281
40HFR100	0281
40HFR120	OEM
40HFR140	OEM
40HFR160	OEM
40J	OEM
40J2	0015
40J2F	OEM
40J3P	0109
40JH3	2850
40JH3R	2191
40JP3	OEM
40JZ	0015
40K	0015
40KR	0015
40L15	OEM
40L20	OEM
40L25	OEM
40L30	OEM
40L35	OEM
40L40	OEM
40L45	OEM
40L50	OEM
40L55	OEM
40L60	OEM
40L65	OEM
40L70	OEM
40L75	OEM
40L80	OEM
40L85	OEM
40L90	OEM
40L95	OEM
40L100	OEM
40L110	OEM
40L120	OEM
40L130	OEM
40L140	OEM
40L150	OEM
40L160	OEM
40L170	OEM
40L180	OEM
40L190	OEM
40L200	OEM
40L210	OEM
40L220	OEM
40L230	OEM
40L240	OEM
40L250	OEM
40L260	OEM
40L270	OEM
40L280	OEM
40L290	OEM
40L300	OEM
40LF	OEM
40M	0015
40N1	0015
40N2	0097
40N3	0097
40N15A	OEM
40N15A9	OEM
40N20A	OEM
40N20A9	OEM
40N25A	OEM
40N25A9	OEM
40N30A	OEM
40N30A9	OEM
40N35A	OEM
40N35A9	OEM
40N40A	OEM
40N40A9	OEM
40N45A	OEM
40N45A9	OEM
40N50A	OEM
40N50A9	OEM
40N55A	OEM
40N55A9	OEM
40N60A	OEM
40N60A9	OEM
40N65A	OEM
40N65A9	OEM
40N70A	OEM
40N70A9	OEM
40N75A	OEM
40N75A9	OEM
40N80A	OEM
40N80A9A9	OEM
40N85A	OEM
40N85A9	OEM
40N90A	OEM
40N90A9	OEM
40N95A	OEM
40N95A9	OEM
40N100A	OEM
40N100A9	OEM
40N110A	OEM
40N120A	OEM
40N130A	OEM
40N140A	OEM
40N150A	OEM
40N160A	OEM
40N170A	OEM
40N180A	OEM
40N190A	OEM
40N200A	OEM
40N210A	OEM
40N220A	OEM
40N230A	OEM
40N240A	OEM
40N250A	OEM
40NJ	0015
40P1	0211
40P2	0211
40P2(POWER)	0097
40P2RECT	0097
40P3	0097
40Q3	0853
40Q4	3751
40R2S	0044
40R3	0805
40R3P	OEM
40RCS5	1704
40RCS10	1704
40RCS15	OEM
40RCS20	1704
40RCS25	OEM
40RCS30	4031
40RCS40	4031
40RCS50	4031
40RCS60	4031
40RCS70	0679
40RCS80	0679
40RCS90	0679
40RCS100	0679
40RCS110	0679
40RCS120	0679
40S2S03	OEM
40S2S04	OEM
40S3	0805
40S3P	OEM
40S5	1000
40SB02S	1545
40SB02SA	1545
40SB03S	1545
40SB03SA	1545
40SB04SA	1545
40SB025S	1545
40SG1	5058
40SG2	0609
40SG3	0133
40SG4	0795
40SG5	0795
40SG6	0015
40SG7	0373
40SG8	0373
40SI-MC1364P	0802
40SI-UPB562C	OEM
40SL05	OEM
40SL1	OEM
40SL2	OEM
40SL3	OEM
40SL4	OEM
40SL5	OEM
40SL6	OEM
40SP-BP-104	OEM
40ST-2SC2371L	0276
40T1	OEM
40T2	OEM
40T3	OEM
40T3P	OEM
40T4	OEM
40T5	OEM
40T6	OEM
40T7	OEM
40T8	OEM
40T9	OEM
40T10	OEM
40T11	OEM
40T12	OEM
40T13	OEM
40T14	OEM
40T15	OEM
40TC10	OEM
40TC20	OEM
40TC30	OEM
40TC40	OEM
40TC50	OEM
40TC60	OEM
40TC70	OEM
40TC80	OEM
40TC90	OEM
40TC100	OEM
40TC120	OEM
40TCA10	OEM
40TCA20	OEM
40TCA30	OEM
40TCA40	OEM
40TCA50	OEM
40TCA60	OEM
40TCA70	OEM
40TCA80	OEM
40TCA90	OEM
40TCA100	OEM
40TCA110	OEM
40TCA120	OEM
40V3P	OEM
40V10	OEM
40V12	OEM
40W3P	OEM
40X3P	OEM
40Y3P	0015
40Z6	OEM
40Z6A	OEM
41-001	0474
41-0318	0103
41-0318A	0103
41-0499	0016
41-0500	0126
41-0500A	0126
41-0606	0168
41-0905	0137
41-0909	0178
41-J2	0015
41B581014	0086
41C1FM	OEM
41C2FM	OEM
41C3FM	OEM
41C4FM	OEM
41C6FM	OEM
41C-402	4485
41C-405	OEM
41C-406	3263
41C-407	0579
41C-409	2305
41D5R03	OEM
41D9R03GH	OEM
41E1FM	OEM
41E2FM	OEM
41E3-1	0143
41E3FM	OEM
41E4FM	OEM
41E5R03	OEM
41E6FM	OEM
41F1FM	OEM
41F2FM	OEM
41F3FM	OEM
41F4FM	OEM
41F6FM	OEM
41GJ298	OEM
41HA40	OEM
41HA60	OEM
41HA80	OEM
41HA100	OEM
41HF10	OEM
41HF20	OEM
41HF40	OEM
41HF60	OEM
41HF80	OEM
41HF100	OEM
41HF120	OEM
41HF140	OEM
41HF160	OEM
41HFR10	OEM
41HFR20	OEM
41HFR40	OEM
41HFR60	OEM
41HFR80	OEM
41HFR90	0281
41HFR100	OEM
41HFR120	OEM
41HFR140	OEM
41HFR160	OEM
41J2	0015
41J2F	OEM
41MQ30	OEM
41MQ40	OEM
41MQ50	OEM
41MQ60	OEM
41N	0007
41N1	0144
41N2	0144
41N2A	0144
41N2AA	0144
41N2B	0144
41N3	0144
41R1FM	OEM
41R2FM	OEM
41R3FM	OEM
41R4FM	OEM
41R6FM	OEM
41SI-HA1128	0167
41SI-LM3065N	0167
41Z6	0165
41Z6A	0165
42-051	0071
42-1	2482
42-2	2472
42-7	0015
42-14027	0015
42-16599	0160
42-17143	0211
42-17443	0015
42-17443A	0015
42-17444	0198
42-18109	0847
42-18111	0198
42-18310	0142
42-19642	0086
42-19643	0126
42-19644	0198
42-19645	0071
42-19670	0198
42-19671	0004
42-19681	0123
42-19682	0050
42-19683	0144
42-19792	0050
42-19840	0016
42-19862	0208
42-19862A	0164
42-19863	0004
42-19863A	0004
42-19864	0004
42-19864A	0004
42-19865	0015
42-20222	0004
42-20738	0198
42-20739	0126
42-20960	0178
42-20961	0103
42-20961A	0103
42-21232	0126
42-21233	0086
42-21234	0016
42-21362	0123
42-21400	0015
42-21401	0007
42-21402	0007
42-21403	0136
42-21404	0208
42-21405	0164
42-21406	0004
42-21407	0198
42-21408	0015
42-21443	0265
42-21866	0015
42-22008	0037
42-22008A	0037
42-22009	0400
42-22154	0595
42-22158	0016
42-22532	0127
42-22533	0016
42-22534	0004
42-22535	0164
42-22535Q	0211
42-22536	0030
42-22537	0143
42-22538	0133
42-22539	0143
42-22540	1303
42-22755	0143
42-22778	0050
42-22779	0050
42-22780	0050
42-22781	0050
42-22784	0050
42-22785	0144
42-22786	0016
42-22787	0016
42-22809	0016
42-22810	0037
42-22810A	0037
42-22811	0016
42-22812	0016
42-22834	0265
42-22835	0015
42-22847	0016
42-23348	0016
42-23349	0016
42-23350	0015
42-23350A	0015
42-23459	0042
42-23541	0037
42-23541A	0037
42-23542	0016
42-23622	0004
42-23960	0127
42-23960P	0127
42-23961	0127
42-23961P	0127
42-23962	0127
42-23962P	0127
42-23963	0127
42-23963P	0127
42-23964	0016
42-23964P	0016
42-23965	0136
42-23965P	0050
42-23966	0016
42-23966P	0016
42-23967	0164
42-23967P	0164
42-23968	0160
42-23968P	0160
42-23969	0143
42-23970	0030
42-23972	0123
42-23975	0015
42-23977	0922
42-24263	0400
42-24387	0133
42-27202	0015
42-27277	0111
42-27278	0015
42-27372	0007
42-27373	0007
42-27374	0079
42-27375	0079
42-27376	0211
42-27377	0579
42-27378	0143
42-27379	0030
42-27380	0143
42-27381	0143
42-27463	0015
42-27529	1144
42-27530	0155
42-27533	0155
42-27534	0076
42-27535	0111
42-27536	0006
42-27537	0191
42-27538	0925
42-27539	0042
42-27540	0005
42-27541	0137
42-27542	0133
42-27543	0019
42-27544	0123
42-28056	0016
42-28057	0164
42-28058	0015
42-28199	0143
42-28200	0030
42-28201	0133
42-28202	0015
42-28203	0144
42-28204	0144
42-28205	0016
42-28206	0144
42-28207	0016
42-28208	0037
42-28210	0016
42-28211	0037
42-30092	0111
42A11	0015
42A14	0143
42A23	0015
42B2	0015
42B16	0015
42D5R03	OEM
42D9R03GH	OEM
42E03A	OEM
42J	OEM
42J2	0015
42K	OEM
42L	OEM
42N2	OEM
42NT100	OEM
42OT1	0164
42P1	0280
42P15	0050
42P83	0016
42R2	0097
42ST-2SC930E	2209
42ST-2SD666C	1386
42ST-LBC547C	OEM
42T1	OEM
42T2	OEM
42T3	OEM
42T4	OEM
42T5	OEM
42T6	OEM
42T7	OEM
42T8	OEM
42T9	OEM
42T10	OEM
42T11	OEM
42T12	OEM
42T13	OEM
42T14	OEM
42T15	OEM
42TB1-CE00	OEM
42TB1-EE00	OEM
42TB1-EH00	OEM
42TB1-GE00	OEM
42TB1-GH00	OEM
42TB1-JE00	OEM
42TB1-JH00	OEM
42TB1-LE00	OEM
42TB1-LH00	OEM
42TB1-NE00	OEM
42TB1-NH00	OEM
42TB1-RE00	OEM
42TB2-CE00	OEM
42TB2-EE00	OEM
42TB2-EH00	OEM
42TB2-GE00	OEM
42TB2-GH00	OEM
42TB2-JE00	OEM
42TB2-JH00	OEM
42TB2-LE00	OEM
42TB2-LH00	OEM
42TB2-NE00	OEM
42TB2-NH00	OEM
42TB2-RE00	OEM
42TB3-CE00	OEM
42TB3-EE00	OEM
42TB3-EH00	OEM
42TB3-GE00	OEM
42TB3-GH00	OEM
42TB3-JE00	OEM
42TB3-JH00	OEM
42TB3-LE00	OEM
42TB3-LH00	OEM
42TB3-NE00	OEM
42TB3-NH00	OEM
42TB3-RE00	OEM
42TB4-CE00	OEM
42TB4-EE00	OEM
42TB4-EH00	OEM
42TB4-GE00	OEM
42TB4-GH00	OEM
42TB4-JE00	OEM
42TB4-JH00	OEM
42TB4-LE00	OEM
42TB4-LH00	OEM
42TB4-NE00	OEM
42TB4-NH00	OEM
42TB4-RE00	OEM
42TB5-CE00	OEM
42TB5-EE00	OEM
42TB5-EH00	OEM
42TB5-GE00	OEM
42TB5-GH00	OEM
42TB5-JE00	OEM
42TB5-JH00	OEM
42TB5-LE00	OEM
42TB5-LH00	OEM
42TB5-NE00	OEM
42TB5-NH00	OEM
42TB5-RE00	OEM
42TB6-CE00	OEM
42TB6-EE00	OEM
42TB6-EH00	OEM
42TB6-GE00	OEM
42TB6-GH00	OEM
42TB6-JE00	OEM
42TB6-JH00	OEM
42TB6-LE00	OEM
42TB6-LH00	OEM
42TB6-NE00	OEM
42TB6-NH00	OEM
42TB6-RE00	OEM
42TB7-CE00	OEM
42TB7-EE00	OEM
42TB7-EH00	OEM
42TB7-GE00	OEM
42TB7-GH00	OEM
42TB7-JE00	OEM
42TB7-JH00	OEM
42TB7-LE00	OEM
42TB7-LH00	OEM
42TB7-NE00	OEM
42TB7-NH00	OEM
42TB7-RE00	OEM
42TB8-CE00	OEM
42TB8-EE00	OEM
42TB8-EH00	OEM
42TB8-GE00	OEM
42TB8-GH00	OEM
42TB8-JE00	OEM
42TB8-JH00	OEM
42TB8-LE00	OEM
42TB8-LH00	OEM
42TB8-NE00	OEM
42TB8-NH00	OEM
42TB8-RE00	OEM
42TB9-CE00	OEM
42TB9-EE00	OEM
42TB9-EH00	OEM
42TB9-GE00	OEM
42TB9-GH00	OEM
42TB9-JE00	OEM
42TB9-JH00	OEM
42TB9-LE00	OEM
42TB9-LH00	OEM
42TB9-NE00	OEM
42TB9-NH00	OEM
42TB9-RE00	OEM
42TB10-CE00	OEM
42TB10-EE00	OEM
42TB10-EH00	OEM
42TB10-GE00	OEM
42TB10-GH00	OEM
42TB10-JE00	OEM
42TB10-JH00	OEM
42TB10-LE00	OEM
42TB10-LH00	OEM
42TB10-NE00	OEM
42TB10-NH00	OEM
42TB10-RE00	OEM
42TB11-CE00	OEM
42TB11-EE00	OEM
42TB11-EH00	OEM
42TB11-GE00	OEM
42TB11-GH00	OEM
42TB11-JE00	OEM
42TB11-JH00	OEM
42TB11-LE00	OEM
42TB11-LH00	OEM
42TB11-NE00	OEM
42TB11-NH00	OEM
42TB11-RE00	OEM
42TB12-CE00	OEM

DEVICE TYPE	REPL CODE
42TB12-EE00	OEM
42TB12-EH00	OEM
42TB12-GE00	OEM
42TB12-GH00	OEM
42TB12-JE00	OEM
42TB12-JH00	OEM
42TB12-LE00	OEM
42TB12-LH00	OEM
42TB12-NE00	OEM
42TB12-NH00	OEM
42TB12-RE00	OEM
42TB13-CE00	OEM
42TB13-EE00	OEM
42TB13-EH00	OEM
42TB13-GE00	OEM
42TB13-GH00	OEM
42TB13-JE00	OEM
42TB13-JH00	OEM
42TB13-LE00	OEM
42TB13-LH00	OEM
42TB13-NE00	OEM
42TB13-NH00	OEM
42TB13-RE00	OEM
42TB14-CE00	OEM
42TB14-EE00	OEM
42TB14-EH00	OEM
42TB14-GE00	OEM
42TB14-GH00	OEM
42TB14-JE00	OEM
42TB14-JH00	OEM
42TB14-LE00	OEM
42TB14-LH00	OEM
42TB14-NE00	OEM
42TB14-NH00	OEM
42TB14-RE00	OEM
42TB15-CE00	OEM
42TB15-EE00	OEM
42TB15-EH00	OEM
42TB15-GE00	OEM
42TB15-GH00	OEM
42TB15-JE00	OEM
42TB15-JH00	OEM
42TB15-LE00	OEM
42TB15-LH00	OEM
42TB15-NH00	OEM
42TB15-RE00	OEM
42X25	0015
42X32	0015
42X210	0038
42X230	0004
42X233	0211
42X244	0015
42X244B	0015
42X245	0015
42X245B	0015
42X308	0211
42X309	0211
42X310	0038
42X311	0211
42Z6	0057
42Z6A	0057
43	OEM
43-022861	0144
43-023190	0130
43-023212	0079
43-023221	0016
43-023222	0126
43-023223	0086
43-023843	0103
43-023844	0086
43-025763	0144
43-025764	0144
43-025765	0144
43-025766	0016
43-025767	0144
43-025834	0222
43-0203845	0126
43-540012	0015
43A144188-5	0147
43A168135-1	1041
43A168135-2	1823
43A168135-3	2097
43A168135-4	0141
43A168135-6	1831
43A168135-7	0033
43A168135-8	0081
43A168135-9	0358
43A168135-10	0562
43A223006P1	0682
43A223007	0268
43A223008	5338
43A223009	0314
43A223012	0685
43A223015	0873
43A223017	0559
43A223018	0559
43A223025	1172
43A223026P1	1313
43A223028	1158
43A223029P1	1052
43A223030	1486
43A223031	1715
43A223033P1	0117
43A223034P1	0569
43B	0054
43C216408P1	0682
43C216409P1	5338
43C216410P1	1907
43C216411P1	3715

DEVICE TYPE	REPL CODE
43C216414P1	1944
43C216447	1926
43C216447P1	1926
43D5R03	OEM
43D9R03GH	OEM
43J	OEM
43K	OEM
43N3	0198
43N6	0198
43P1	0280
43P2	0211
43P3	0280
43P4	0211
43P4C	0211
43P6	0211
43P6A	0211
43P6C	0280
43P7	0211
43P7A	0211
43P7C	0211
43P13	0050
43R2	OEM
43ST-2SC535B	0127
43ST-2SC1514-05	1085
43ST-2SC1514-15	1085
43SZ-RD6.2E	0471
43SZ-YZ-063	0471
43SZ-YZ063	0471
43T1	OEM
43T2	OEM
43T3	OEM
43T4	OEM
43T5	OEM
43T6	OEM
43T7	OEM
43T8	OEM
43T9	OEM
43T10	OEM
43TB1-JH00	OEM
43TB1-LE00	OEM
43TB1-LH00	OEM
43TB1-NE00	OEM
43TB1-RE00	OEM
43TB2-JH00	OEM
43TB2-LE00	OEM
43TB2-LH00	OEM
43TB2-NE00	OEM
43TB2-NH00	OEM
43TB2-RE00	OEM
43TB3-JH00	OEM
43TB3-LE00	OEM
43TB3-LH00	OEM
43TB3-NE00	OEM
43TB3-NH00	OEM
43TB4-JH00	OEM
43TB4-LE00	OEM
43TB4-LH00	OEM
43TB4-NE00	OEM
43TB4-NH00	OEM
43TB4-RE00	OEM
43TB5-JH00	OEM
43TB5-LE00	OEM
43TB5-LH00	OEM
43TB5-NE00	OEM
43TB5-NH00	OEM
43TB5-RE00	OEM
43TB6-JH00	OEM
43TB6-LE00	OEM
43TB6-LH00	OEM
43TB6-NE00	OEM
43TB6-NH00	OEM
43TB6-RE00	OEM
43TB7-JH00	OEM
43TB7-LE00	OEM
43TB7-LH00	OEM
43TB7-NE00	OEM
43TB7-NH00	OEM
43TB7-RE00	OEM
43TB8-JH00	OEM
43TB8-LE00	OEM
43TB8-LH00	OEM
43TB8-NE00	OEM
43TB8-NH00	OEM
43TB8-RE00	OEM
43TB9-JH00	OEM
43TB9-LE00	OEM
43TB9-LH00	OEM
43TB9-NE00	OEM
43TB9-NH00	OEM
43TB9-RE00	OEM
43TB10-JH00	OEM
43TB10-LE00	OEM
43TB10-LH00	OEM
43TB10-NE00	OEM
43TB10-NH00	OEM
43TB10-RE00	OEM
43TB11-JH00	OEM
43TB11-LE00	OEM
43TB11-LH00	OEM
43TB11-NE00	OEM
43TB11-NH00	OEM
43TB11-RE00	OEM
43TB12-JH00	OEM
43TB12-LE00	OEM
43TB12-LH00	OEM
43TB12-NE00	OEM
43TB12-NH00	OEM
43TB12-RE00	OEM
43TB13	OEM

DEVICE TYPE	REPL CODE
43TB13-JH00	OEM
43TB13-LE00	OEM
43TB13-LH00	OEM
43TB13-NE00	OEM
43TB13-NH00	OEM
43TB13-RE00	OEM
43TB14-JH00	OEM
43TB14-LE00	OEM
43TB14-LH00	OEM
43TB14-NE00	OEM
43TB14-NH00	OEM
43TB14-RE00	OEM
43TB15-CE00	OEM
43TB15-EE00	OEM
43TB15-EH00	OEM
43TB15-GE00	OEM
43TB15-GH00	OEM
43TB15-JE00	OEM
43TB15-JH00	OEM
43TB15-LE00	OEM
43TB15-LH00	OEM
43TB15-NE00	OEM
43TB15-NH00	OEM
43TB15-RE00	OEM
43X16A567	0198
43Z6	0170
43Z6A	0170
44	OEM
44-13	0136
44-530	0015
44-44886G01	0155
44A-1A5	0160
44A354637-001	0178
44A417779-001	0312
44B1	0856
44B1Z	0856
44BH05M	OEM
44BH05P	OEM
44D5R03	OEM
44D9R03GH	OEM
44E03A	OEM
44J	OEM
44K	OEM
44P1	0050
44PB05M	OEM
44PBF7C	OEM
44PD05M	OEM
44PD05P	OEM
44PH05M	OEM
44PH05P	OEM
44PS05M	OEM
44R2	0197
44R2R	0997
44SD-CDG00	0133
44SP-LD271	0516
44SP-LP271	OEM
44ST-2SB568C	1911
44ST-2SB649AC	0525
44SZ-WZ040	0036
44SZ-WZ-063	0471
44T1	OEM
44T-100-118	OEM
44T-100-120	1050
44T-300-91	0030
44T-300-92	0041
44T-300-93	0064
44T-300-94	0077
44T-300-95	0015
44T-300-96	0133
44T-300-97	0019
44T-300-98	OEM
44T-300-99	2057
44T-300-100	0083
44T-300-101	1899
44T-300-102	1050
44T-300-103	0151
44T-300-104	0076
44T-300-105	0325
44T-300-106	0076
44T-300-107	0161
44T-300-108	2710
44T-300-109	0936
44T-300-110	1066
44T-300-111	1221
44T-300-112	0076
44T-300-113	0325
44T-300-358	OEM
44T-300-359	4830
44TB7	OEM
44TB9	OEM
44TB11	OEM
45A2FX355	0015
45AN30	OEM
45AN40	OEM
45AN50	OEM
45AN60	OEM
45AN80	OEM
45AN100	OEM
45AN120	OEM
45AN140	OEM
45AN160	OEM
45ANS	OEM
45AR20	OEM
45AR30	OEM
45AR50	OEM
45AR60	OEM
45AR80	OEM
45AR100	OEM
45AR120	OEM

DEVICE TYPE	REPL CODE
45AR140	OEM
45AR160	OEM
45AR4040	OEM
45ARS	OEM
45D5R03	OEM
45D9R03GH	OEM
45E03A	OEM
45F10	OEM
45F15	OEM
45F20	OEM
45F25	OEM
45F30	OEM
45F40	OEM
45F50	OEM
45F60	OEM
45F80	OEM
45J	OEM
45K	OEM
45L5	0599
45L5R	1347
45L10	0599
45L10R	1347
45L15	0599
45L15R	1347
45L20	0599
45L20R	1347
45L25	0599
45L25R	1347
45L30	0599
45L30R	1347
45L35	1986
45L35R	1905
45L40	1986
45L40R	1905
45L45	1986
45L45R	1905
45L50	1986
45L50R	1905
45L60	1986
45L60R	1905
45L70	0657
45L70R	0193
45L80	0657
45L80R	0193
45L90	0657
45L90R	0193
45L100	0657
45L100R	0193
45L120	OEM
45LK10A	0599
45LK10R	1347
45LK20A	0599
45LK20R	1347
45LK30A	0599
45LK30R	1347
45LK40A	1986
45LK40R	1905
45LK50A	1986
45LK50R	1905
45LK60A	1986
45LK60R	1905
45LK80A	0657
45LK80R	0193
45LK100A	0657
45LK100R	0193
45M10	0599
45M15	0599
45M20	0599
45M30	0599
45M40	0599
45M50	1986
45M60	1986
45M80	0657
45M100	0657
45MA10	1347
45MA20	1347
45MA30	1347
45MA40	1347
45MA60	1905
45MA80	0193
45MA100	0193
45MA120	0202
45MA140	0202
45MA160	OEM
45MLA60	OEM
45MLA80	OEM
45MLA100	OEM
45MLA120	OEM
45MN-1	OEM
45MR	OEM
45MR-1	OEM
45MS2	OEM
45N1	0038
45N2	0038
45N2A	0208
45N2M	0016
45N3	0016
45N4	0016
45N4M	0016
45NP	0144
45P1U5M	OEM
45R2	OEM
45R2S	OEM
45SD-W005M	0106
45ST-2SC454C	0076
45ST-2SC1106	0220
45SZ-RD5.1FB	0041
45TB7	OEM
45V10	OEM

DEVICE TYPE	REPL CODE
45V12	OEM
45X1A502C	0211
45X1A520C	0004
45X2	0280
45X11PV-CM	OEM
45ZE1	OEM
46	OEM
46-06311-3	0127
46-4	0917
46-29-6	0167
46-34-3	0015
46-36-3	0015
46-86-3	0015
46-119-3	0007
46-163-3	0004
46-530-3	0010
46-840-3	0004
46-861-3	0917
46-862	0917
46-862-3	0050
46-864-3	0144
46-865-3	0050
46-866-3	0050
46-867-3	0233
46-868-3	1283
46-869-3	0004
46-1325-3	4196
46-1340-3	2715
46-1343-3	0579
46-1344-3	OEM
46-1346-3	2715
46-1347-3	0829
46-1348-3	0829
46-1352-3	0753
46-1354-3	OEM
46-1356-3	0829
46-1357-3	0878
46-1361-3	0167
46-1362-3	2634
46-1363-3	4582
46-1364-3	0403
46-1365-3	0395
46-1366-3	0789
46-1369-3	0878
46-1370-3	2639
46-1370-9	OEM
46-1382-3	OEM
46-1383-3	2476
46-1391-3	2669
46-1392-3	0864
46-1393-3	1326
46-1394-3	4594
46-1395-3	2949
46-1396-3	3915
46-1397-3	4595
46-1398-3	2231
46-1399-3	OEM
46-5002-1	0632
46-5002-3	3724
46-5002-4	0632
46-5002-5	0377
46-5002-6	2715
46-5002-7	0632
46-5002-8	0789
46-5002-11	0349
46-5002-12	0352
46-5002-13	0354
46-5002-15	0167
46-5002-16	0364
46-5002-19	0372
46-5002-21	0377
46-5002-23	0379
46-5002-26	0167
46-5002-27	0372
46-5002-28	0395
46-5002-31	0403
46-5290-0042	0362
46-5290-0178	2063
46-5291-0164	2295
46-6661-2	0229
46-6829	0127
46-8257-3	0016
46-8477-3	0111
46-8601-3	0015
46-8607-3	3046
46-8610-3	0004
46-8611	0004
46-8611-3	0004
46-8611-4	0196
46-8612-3	0050
46-8613	0917
46-8613-3	0136
46-8613-3(HOR-OUT)	0975
46-8613-3(SYNC)	0050
46-8614	0917
46-8614-3	0600
46-8615-3	0160
46-8616-3	0123
46-8617-3	0160
46-8619	0015
46-8619-3	0143
46-8620-3	0102
46-8621	0030
46-8621-3	0277
46-8622	0917
46-8622-3	0975
46-8623	0143
46-8623-3	0143
46-8624-3	0191

DEVICE TYPE	REPL CODE
46-8625	0917
46-8625-1	0136
46-8625-2	0136
46-8625-3	0136
46-8625-4	0136
46-8625-5	0136
46-8625-6	0136
46-8627	0917
46-8629	0127
46-8630	0143
46-8631-3	0004
46-8634-3	0160
46-8634-9	OEM
46-8636-3	0136
46-8638	0325
46-8638-3	0160
46-8638P-3	0086
46-8642	0030
46-8643	0917
46-8643-3	0917
46-8644	0030
46-8644-3	0143
46-8645	OEM
46-8646	0212
46-8646-3	0143
46-8647	0127
46-8647-3	0015
46-8648	0325
46-8649	0325
46-8650	0127
46-8651	0079
46-8652	0127
46-8653	0127
46-8654	0127
46-8655	OEM
46-8656	OEM
46-8660-3	0004
46-8661	0015
46-8661-3	0015
46-8663-3	0917
46-8664-3	3555
46-8665-3	0004
46-8666-3	0004
46-8668-3	0004
46-8671-3	0208
46-8672-3	0007
46-8676-3	0133
46-8677-2	0144
46-8677-3	0259
46-8679-1	0004
46-8679-2	0004
46-8679-3	0004
46-8680-1	0004
46-8680-2	0004
46-8680-3	0004
46-8681-1	0004
46-8681-2	0004
46-8681-3	0164
46-8682-2	0016
46-8682-3	0111
46-8687-3	0015
46-8688-3	0917
46-8689-3	0015
46-8694-3	1080
46-8695-3	0016
46-8697-3	0015
46-13101-3	0855
46-13103-3	0043
46-13104-3	0389
46-13104-5	OEM
46-13105-3	0607
46-13110-3	OEM
46-13118-3	OEM
46-13119-3	0884
46-13120-3	OEM
46-13121-3	2284
46-13122-3	OEM
46-13123-3	OEM
46-13124-3	0856
46-13125-3	0886
46-13126-3	2671
46-13127-3	OEM
46-13128-3	OEM
46-13129-3	OEM
46-13130-3	OEM
46-13131-3	0843
46-13133-3	3144
46-13134-3	OEM
46-13142-3	OEM
46-13143-3	1055
46-13145-3	0167
46-13148-3	OEM
46-13149-3	OEM
46-13150-3	OEM
46-13152-3	3228
46-13153-3	3945
46-13154-3	2671
46-13155-3	OEM
46-13156-3	0462
46-13162-3	3707
46-13163-3	4226
46-13164-3	4242
46-13165-3	4261
46-13166-3	3697
46-13167-3	3966
46-13168-3	4294
46-13169-3	4319
46-13170-3	OEM
46-13171-3	3964

DEVICE TYPE	REPL CODE
46-13172-3	2960
46-13173-3	1018
46-13174-3	3970
46-13174-9	OEM
46-13175-3	4434
46-13176-3	3246
46-13177-3	2295
46-13178-3	0312
46-13180-3	0916
46-13184-3	0852
46-13185-3	0857
46-13187-3	OEM
46-13188-3	0912
46-13190-3	4134
46-13193-3	OEM
46-13196-3	OEM
46-13197-3	OEM
46-13200-3	OEM
46-13203-3	OEM
46-13204-3	OEM
46-13208-3	0918
46-13212-3	OEM
46-13215-3	OEM
46-13216-3	2054
46-13217-3	OEM
46-13219-3	OEM
46-13221-3	OEM
46-13222-3	OEM
46-13223-3	0904
46-13225-3	5159
46-13227-3	OEM
46-13228-3	OEM
46-13230-3	OEM
46-13231-3	OEM
46-13232-3	OEM
46-13233-3	3968
46-13234-3	5635
46-13237-3	1446
46-13238-3	2054
46-13239-3	5717
46-13243-3	OEM
46-13244-3	OEM
46-13245-3	OEM
46-13246-3	OEM
46-13247-3	OEM
46-13248-3	OEM
46-13249-3	OEM
46-13250-3	OEM
46-13251-3	OEM
46-13252-3	OEM
46-13253-3	OEM
46-13254-3	OEM
46-13255-3	OEM
46-13256-3	OEM
46-13257-3	0332
46-13258-3	0215
46-13259-3	OEM
46-13260-3	OEM
46-13261-3	OEM
46-13262-3	OEM
46-13263-3	OEM
46-13264-3	6030
46-13267-3	0918
46-13268-3	OEM
46-13270-3	OEM
46-13271-3	OEM
46-13272-3	0478
46-13273-3	OEM
46-13274-3	OEM
46-13275-3	OEM
46-13279-3	0770
46-13290-3	OEM
46-13291-3	OEM
46-13292-3	OEM
46-13294-3	OEM
46-13295-3	OEM
46-13296-3	0101
46-13297-3	0770
46-13298-3	2865
46-13299-3	3973
46-13301-3	3980
46-13302-3	0865
46-13303-3	OEM
46-13304-3	OEM
46-13305-3	OEM
46-13306-3	OEM
46-13307-3	4963
46-13309-3	4138
46-13310-3	0360
46-13311-3	1321
46-13312-3	4750
46-13313-3	0413
46-13314-3	0056
46-13315-3	0413
46-13316-3	0300
46-13317-3	0478
46-13318-3	OEM
46-13319-3	0119
46-13320-3	OEM
46-13321-3	OEM
46-13322-3	2063
46-13323-3	0865
46-13324-3	4338
46-13326-3	OEM
46-13327-3	1026
46-13329-3	OEM
46-13334-3	OEM
46-13337-3	OEM
46-13338-3	5775

DEVICE TYPE	REPL CODE
46-13339-3	5159
46-13340-3	0167
46-13341-3	3246
46-13342-3	OEM
46-13343-3	OEM
46-13344-3	OEM
46-13345-3	0360
46-13346-3	OEM
46-13349-3	OEM
46-13350-3	OEM
46-13351-3	OEM
46-13353-3	OEM
46-13354-3	OEM
46-13356-3	0663
46-13358-3	OEM
46-13359-3	OEM
46-13360-3	OEM
46-13361-3	OEM
46-13368-3	OEM
46-13371-3	OEM
46-13379-3	0865
46-13380-3	OEM
46-13381-3	OEM
46-13383-3	OEM
46-13384-3	OEM
46-13385-3	OEM
46-13386-3	OEM
46-13389-3	OEM
46-13392-3	2926
46-13393-3	OEM
46-13395-3	OEM
46-13398-3	OEM
46-13399-3	OEM
46-13400-3	OEM
46-13401-3	OEM
46-13402-3	OEM
46-13404-3	0360
46-13405-3	OEM
46-13406-3	OEM
46-13411	1446
46-13411-3	1446
46-13413-3	OEM
46-13414-3	OEM
46-13420-3	OEM
46-13421-3	OEM
46-13422-3	OEM
46-13423-3	OEM
46-13424-3	OEM
46-13427-3	OEM
46-13428-3	1446
46-13434-3	OEM
46-13435-3	OEM
46-13436-3	OEM
46-13437-3	OEM
46-13439-3	OEM
46-13440-3	OEM
46-13441-3	OEM
46-13442-3	OEM
46-13443-3	OEM
46-13444-3	OEM
46-13445-3	OEM
46-13446-3	OEM
46-13447-3	OEM
46-13448-3	3776
46-13449-3	OEM
46-13450-3	OEM
46-13452-3	OEM
46-13453-3	OEM
46-13454-3	OEM
46-13455-3	0979
46-13459-3	OEM
46-13461-3	OEM
46-13462-3	OEM
46-13463-3	OEM
46-13464-3	OEM
46-13465-3	OEM
46-13466-3	4276
46-13469-3	OEM
46-13473-3	OEM
46-13474-3	OEM
46-13478-3	OEM
46-13479-3	OEM
46-13480-3	OEM
46-13481-3	OEM
46-13482-3	OEM
46-13483-3	OEM
46-13484-3	2105
46-13485-3	OEM
46-13486-3	OEM
46-13487-3	2228
46-13488-3	OEM
46-13489-3	OEM
46-13491-3	2675
46-13493-3	4422
46-13494-3	OEM
46-13495-3	0631
46-13502-3	4962
46-13503-3	OEM
46-13504-3	OEM
46-13505-3	OEM
46-13507-3	OEM
46-13508-3	OEM
46-13509-3	OEM
46-13513-3	OEM
46-13514-3	5399
46-13515-3	OEM
46-13516-3	OEM
46-13517-3	OEM
46-13518-3	OEM
46-13519-3	OEM
46-13520-3	OEM
46-13521-3	OEM
46-13522-3	OEM
46-13523-3	OEM
46-13524-3	OEM
46-13527-3	0332
46-13528-3	4857
46-13531-3	OEM
46-13534-3	OEM
46-13537-3	OEM
46-13538-3	1201
46-13539-3	OEM
46-13540-3	OEM
46-13541-3	OEM
46-13542-3	OEM
46-13543-3	OEM
46-13545-3	OEM
46-13546-3	3518
46-13547-3	5874
46-13547-9	OEM
46-13548-3	1988
46-13550-3	3738
46-13551-3	2623
46-13552-3	5095
46-13554-3	OEM
46-13557-3	OEM
46-13558-3	OEM
46-13559-3	OEM
46-13561-3	2284
46-13567-3	OEM
46-13571-3	OEM
46-13572-3	OEM
46-13573-3	6092
46-13574-3	OEM
46-13575-3	OEM
46-13577-3	OEM
46-13580-3	OEM
46-13581-3	2666
46-13582-3	OEM
46-13583-3	OEM
46-13585-3	OEM
46-13586-3	OEM
46-13587-3	OEM
46-13588-3	3502
46-13588-9	3502
46-13590-3	3772
46-13591-3	OEM
46-13593-3	0243
46-13594-3	OEM
46-13595-3	OEM
46-13596-3	OEM
46-13597-3	OEM
46-13598-3	OEM
46-13599-3	OEM
46-13600-3	OEM
46-13601-3	OEM
46-13604	1988
46-13604--3	OEM
46-13604-3	1988
46-13607-3	OEM
46-13608-3	OEM
46-13609-3	OEM
46-13611-3	OEM
46-13612-3	OEM
46-13613-3	OEM
46-13616-3	OEM
46-13617-3	OEM
46-13618-3	OEM
46-13620-3	OEM
46-13621-3	OEM
46-13622-3	OEM
46-13623-3	OEM
46-13624-3	OEM
46-13625-3	OEM
46-13626-3	OEM
46-13627-3	OEM
46-13628-3	OEM
46-13629-3	OEM
46-13630-3	OEM
46-13631-3	OEM
46-13632-3	OEM
46-13633-3	OEM
46-13634-3	OEM
46-13639-3	OEM
46-13642-3	OEM
46-13643-3	OEM
46-13644-3	OEM
46-13645-3	4410
46-13646-3	OEM
46-13647-3	OEM
46-13648-3	OEM
46-13649-3	5216
46-13650-3	OEM
46-13651-3	OEM
46-13652-3	OEM
46-13653-3	OEM
46-13654-3	OEM
46-13659-3	OEM
46-13661-3	OEM
46-13662-3	0035
46-13663-3	OEM
46-13664-3	OEM
46-13665-3	OEM
46-13666-3	OEM
46-13667-3	OEM
46-13668-3	OEM
46-13669-3	OEM
46-13674-4	OEM
46-13675-3	OEM
46-13676-3	OEM
46-13679-3	OEM
46-13686-3	1028
46-13687-3	OEM
46-13688-3	OEM
46-13689-3	3828
46-13690-3	OEM
46-13691-3	OEM
46-13692-3	OEM
46-13699-3	OEM
46-13700-3	OEM
46-13704-3	0332
46-13706-3	OEM
46-13708-3	OEM
46-13709-3	OEM
46-13714-3	OEM
46-13715-3	OEM
46-13719-3	OEM
46-13720-3	OEM
46-13727-3	OEM
46-13728-3	OEM
46-13729-3	OEM
46-13730-3	0024
46-13731-3	OEM
46-13732-3	OEM
46-13738-3	OEM
46-13739-3	OEM
46-13740-3	OEM
46-13741-3	OEM
46-13742-3	OEM
46-13743-3	OEM
46-13744-3	OEM
46-13745-3	OEM
46-13746-3	OEM
46-13747-3	OEM
46-13748-3	OEM
46-13749-3	OEM
46-13750-3	OEM
46-13751-3	OEM
46-13752-3	OEM
46-13753-3	OEM
46-13754-3	OEM
46-13755-3	OEM
46-13758-3	OEM
46-13759-3	OEM
46-13762-3	OEM
46-13763-3	OEM
46-13764-3	OEM
46-13766-3	OEM
46-13767-3	OEM
46-13768-3	OEM
46-13769-3	OEM
46-13771-3	4349
46-13773-3	OEM
46-13774-3	OEM
46-13775-3	OEM
46-13776-3	OEM
46-13780-3	OEM
46-13781-3	OEM
46-13782-3	OEM
46-13784-3	OEM
46-13785-3	OEM
46-13786-3	OEM
46-13787-3	OEM
46-13788-3	OEM
46-13789-3	OEM
46-13790-3	OEM
46-13791-3	OEM
46-13792-3	OEM
46-13793-3	OEM
46-13794-3	OEM
46-13795-3	OEM
46-13796-3	OEM
46-13797-3	OEM
46-13798-3	OEM
46-13799-3	OEM
46-13800-3	OEM
46-13801-3	OEM
46-13802-3	OEM
46-13803-3	OEM
46-13804-3	OEM
46-13805-3	OEM
46-13806-3	OEM
46-13807-3	OEM
46-13808-3	OEM
46-13809-3	OEM
46-13810-3	OEM
46-13815-3	2026
46-13816-3	0129
46-13817-3	OEM
46-13818-3	OEM
46-13819-3	OEM
46-13820-3	OEM
46-13827-3	OEM
46-13828-3	OEM
46-13829-3	4137
46-13830-3	1676
46-13832-3	OEM
46-13833-3	OEM
46-13834-3	OEM
46-13835-3	OEM
46-13839-3	4307
46-13840-3	OEM
46-13841-3	OEM
46-13842-3	OEM
46-13844-3	OEM
46-13845-3	OEM
46-13846-3	OEM
46-13847-3	OEM
46-13848-3	OEM
46-13849-3	OEM
46-13850-3	OEM
46-13851-3	OEM
46-13852-3	OEM
46-13853-3	OEM
46-13855-3	OEM
46-13856-3	OEM
46-13857-3	OEM
46-13858-3	OEM
46-13859-3	OEM
46-13860-3	OEM
46-13861-3	OEM
46-13862-3	OEM
46-13863-3	OEM
46-13869-3	OEM
46-13870-3	OEM
46-13871-3	OEM
46-13872-3	OEM
46-13873-3	OEM
46-13874-3	OEM
46-13875-3	OEM
46-13876-3	OEM
46-13877-3	4377
46-13879-3	OEM
46-13887-3	OEM
46-13888-3	OEM
46-13889-3	OEM
46-13890-3	OEM
46-13891-3	OEM
46-13892-3	OEM
46-13893-3	OEM
46-13894-3	OEM
46-13895-3	OEM
46-13896-3	OEM
46-13899-3	OEM
46-13900-3	OEM
46-13901-3	OEM
46-13902-3	OEM
46-13903-3	OEM
46-13904-3	OEM
46-13905-3	OEM
46-13906-3	OEM
46-13907-3	OEM
46-13908-3	OEM
46-13909-3	OEM
46-13910-3	OEM
46-13911-3	4650
46-13912-3	1442
46-13916-3	OEM
46-13917-3	OEM
46-13921-3	1529
46-13922-3	3916
46-13923-3	OEM
46-13924-3	5011
46-13925-3	6560
46-13926-3	OEM
46-13927-3	OEM
46-13928-3	OEM
46-13930-3	OEM
46-13931-3	OEM
46-13932-3	OEM
46-13933-3	OEM
46-13934-3	OEM
46-13935-3	OEM
46-13936-3	OEM
46-13937-3	OEM
46-13938-3	1988
46-13939-3	OEM
46-13941-3	OEM
46-13942-3	OEM
46-13943-3	OEM
46-13944-3	OEM
46-13946-3	OEM
46-13947-3	OEM
46-13949-3	OEM
46-13951-3	OEM
46-13953-3	OEM
46-13954-3	OEM
46-13956-3	OEM
46-13957-3	OEM
46-13958-3	OEM
46-13959-3	OEM
46-13960-3	OEM
46-13962-3	4102
46-13963-3	1988
46-13967-3	OEM
46-13969-3	3253
46-13970-9	OEM
46-13972-3	OEM
46-13973-3	OEM
46-13975-3	OEM
46-13977-3	OEM
46-13978-3	OEM
46-13979-3	OEM
46-13980-3	6991
46-13983-3	OEM
46-13984-3	OEM
46-13985-3	OEM
46-13986-3	OEM
46-13987-3	OEM
46-13988-3	OEM
46-13989-3	OEM
46-13990-3	OEM
46-13991-3	OEM
46-13992-3	OEM
46-13993-3	OEM
46-13994-3	OEM
46-13995-3	OEM
46-13996-3	OEM
46-13997-3	OEM
46-13998-3	OEM
46-13999-3	OEM
46-14367-3	0332
46-16261	0015
46-16261-3	0015
46-16428-3	OEM
46-16513-4	OEM
46-18583-3	OEM
46-25156-3	OEM
46-35986-3	OEM
46-37328-3	0556
46-61249-3	0015
46-61267-3	0133
46-61294-3	OEM
46-61307-3	0133
46-61951-3	0120
46-61961-3	0120
46-61992-3	OEM
46-67120A13	0015
46-80309-3	0133
46-81187-3	0285
46-83183-3	OEM
46-83395-3	OEM
46-83619-3	OEM
46-83648-3	OEM
46-83649-3	OEM
46-84120-3	0144
46-85285-3	0144
46-86101-3	0127
46-86102-3	0136
46-86104-3	0924
46-86104-9	OEM
46-86105-3	1303
46-86106-3	0102
46-86107-3	0007
46-86108-3	0007
46-86109-3	0127
46-86110-3	3613
46-86112-2	0007
46-86112-3	0184
46-86113-3	0007
46-86114-3	0007
46-86115-3	0208
46-86116-3	0077
46-86117-3	0007
46-86118-3	0007
46-86119-3	0007
46-86120-3	0127
46-86121-3	0016
46-86122-3	0016
46-86123-3	0050
46-86125-3	0222
46-86126-3	0127
46-86127-3	0127
46-86130-3	0178
46-86130-9	OEM
46-86131-3	0330
46-86132-3	0191
46-86133-3	0127
46-86134-3	0917
46-86135-3	0222
46-86135-9	OEM
46-86136-3	0160
46-86138-3	1303
46-86139-3	0015
46-86140-3	0144
46-86141-3	0071
46-86143-3	0016
46-86144-3	0016
46-86145-3	0016
46-86146-3	0071
46-86147-3	0071
46-86148	0071
46-86148-3	0015
46-86149-3	0015
46-86150-3	0229
46-86151-3	0229
46-86152-3	0016
46-86163-3	0004
46-86165-3	0004
46-86166-3	0211
46-86168-3	0133
46-86169-3	0016
46-86170-3	0037
46-86171-3	0016
46-86172-3	0127
46-86173-3	0233
46-86177-3	0015
46-86178-3	1149
46-86179-3	0102
46-86180-3	0774
46-86182-3	0233
46-86183-3	0233
46-86184	0133
46-86184-3	0133
46-86185-3	1080
46-86186-3	0133
46-86187-3	0133
46-86189-3	0178
46-86192-3	0191
46-86194-3	1083
46-86195-3	0004
46-86196-3	0164
46-86198-3	0847
46-86198-9	OEM
46-86199-3	0015
46-86200-3	0123
46-86201-3	0015
46-86201-3(CENTERING	0920
46-86207-3	0127
46-86208-3	0144
46-86209-3	0127
46-86210-3	0233
46-86211-3	0208
46-86212-3	0015
46-86213-3	0160
46-86214-3	0143
46-86220-3	0196
46-86224-3	0007
46-86225-3	0649
46-86226-3	0178
46-86227-3	0201
46-86228-3	0016
46-86229-3	0037
46-86229-3A	0037
46-86230-3	0037
46-86231-3	0016
46-86232-3	0233
46-86233-3	0086
46-86234-3	0161
46-86235-3	0460
46-86238-3	0037
46-86238-3A	0037
46-86239-3	0127
46-86240-3	0144
46-86244-3	0144
46-86247-2	0016
46-86247-3	0191
46-86249	0030
46-86249-3	0760
46-86250-3	0133
46-86251-3	0127
46-86252-3	0155
46-86253-3	0143
46-86254-3	0913
46-86256-1	0136
46-86256-2	0136
46-86256-3	0004
46-86257-3	0016
46-86261-3	0015
46-86262-3	0144
46-86264-3	0015
46-86265-3	0127
46-86266-3	0143
46-86267-3	0015
46-86268-3	0016
46-86269-3	0144
46-86270-3	0015
46-86271-3	0015
46-86272-3	0276
46-86274-3	0016
46-86279-3	0201
46-86280-3	0102
46-86281-3	0102
46-86282-3	0102
46-86283-3	0037
46-86284-3	OEM
46-86285-3	0007
46-86286-3	0917
46-86287-3	OEM
46-86288-3	0917
46-86289-3	0030
46-86290-3	0015
46-86291-3	0201
46-86292-3	0378
46-86293-3	0037
46-86294-3	0757
46-86295-3	0127
46-86296-3	0365
46-86297-3	0720
46-86298-3	0554
46-86299-3	0127
46-86300-3	0050
46-86301-3	0127
46-86302-3	0127
46-86303-3	0015
46-86304-3	0484
46-86305-3	0030
46-86307-3	0015
46-86308-3	0102
46-86309-3	0133
46-86309-3(ZENER)	0137
46-86310-3	0016
46-86311-3	0127
46-86312-3	0196
46-86313-3	0288
46-86314-3	0144
46-86314-3A	0144
46-86315-3	0086
46-86316-3	0325
46-86317-3	0042
46-86318-3	0233
46-86319-3	0276
46-86320-3	0102
46-86321-3	0015
46-86322-3	0015
46-86323-3	0170
46-86324-3	0102
46-86326-3	0015
46-86327-3	0285
46-86328-3	0102
46-86329-3	0375
46-86330-3	0128
46-86331-3	0102
46-86332-3	0196
46-86334-3	0378
46-86335-3	0042
46-86336-3	0196
46-86337-3	0015
46-86338-3	0133
46-86339-3	0015
46-86340-3	1303
46-86341-3	0100
46-86342-3	0150
46-86343-3	0133
46-86344-3	0285
46-86345-3	0066
46-86346-3	0285
46-86347-3	0423
46-86348-3	0042
46-86349-3	0178
46-86349-9	2361
46-86350-3	0178
46-86351-3	0015
46-86352-3	0127
46-86353-3	0111
46-86354-3	0127
46-86355-2	0015
46-86355-3	0015
46-86357-3	0144
46-86358-3	0015
46-86360-3	0560
46-86364-3	0015
46-86365-3	0023
46-86371-3	0086
46-86372-3	OEM
46-86373-3	0004
46-86374-3	0042
46-86375-3	0016
46-86376-3	0151
46-86377-3	0037
46-86378-3	0016
46-86379-3	0137
46-86380-3	0913
46-86381	0086
46-86381-3	0086
46-86382-3	4056
46-86383-3	0015
46-86384-3	0142
46-86386-3	0142
46-86387-3	0142
46-86388-3	0103
46-86388-9	OEM
46-86389-3	0065
46-86390-3	0003
46-86391-3	0102
46-86392-3	0495
46-86393-2	0102
46-86393-3	0102
46-86394-3	0636
46-86394-3(10)	0636
46-86395-3	0102
46-86396-3	0353
46-86397-3	0144
46-86398-3	0595
46-86399-3	0693
46-86400-3	0042
46-86401-3	0100
46-86402-3	0015
46-86403-3	0126
46-86404-3	0016
46-86405-3	0086
46-86406-3	0126
46-86407-3	0016
46-86408-3	0016
46-86409-3	0016
46-86411-3	0460
46-86412-3	0786
46-86414-3	OEM
46-86415-3	0065
46-86416-3	0015
46-86419-3	0016
46-86420-3	0015
46-86421-3	0012
46-86422-3	0133
46-86424-3	0037
46-86425-3	0126
46-86426-3	0086
46-86427-3	0086
46-86428-3	0133
46-86429-3	0203
46-86430-3	0004
46-86431-3	0133
46-86432-3	0098
46-86433-3	0015
46-86434-3	0111
46-86435-3	0127
46-86436-3	0133
46-86437-3	0015
46-86438-3	0015
46-86439-3	0142
46-86439-9	OEM
46-86440-3	0102
46-86442-3	0244
46-86444-3	1323
46-86445-3	0720
46-86446-3	0142
46-86446-9	0142
46-86447-3	1674
46-86448-3	3383
46-86451-3	0378
46-86452-3	OEM
46-86454-3	0378
46-86455-3	0220

DEVICE TYPE	REPL CODE
46-86456-3	0649
46-86457-3	0471
46-86458-3	0317
46-86459-3	0018
46-86460-3	0091
46-86461-3	0065
46-86462-3	0317
46-86463-3	0015
46-86464-3	0002
46-86465-3	1329
46-86466-3	0955
46-86467-3	0098
46-86467-5	0098
46-86468-3	OEM
46-86469-3	0168
46-86472-9	OEM
46-86473-3	OEM
46-86474-3	1329
46-86475-3	0220
46-86476-3	0934
46-86476-9	0934
46-86477-3	2051
46-86478-3	OEM
46-86479-3	OEM
46-86481-3	0133
46-86482-3	0643
46-86483-3	0244
46-86484-3	0143
46-86485-3	0191
46-86486-3	0065
46-86487-3	0317
46-86488-3	1228
46-86489-3	0023
46-86492-3	1150
46-86492-9	OEM
46-86493-3	0074
46-86494	0071
46-86494-3	0071
46-86496-3	0178
46-86496-9	OEM
46-86497-3	0015
46-86498-3	OEM
46-86499-3	0071
46-86500-3	1284
46-86501-3	0471
46-86502-3	OEM
46-86503-3	0015
46-86504-3	0181
46-86505-3	0643
46-86505-9	0643
46-86506-3	0168
46-86507-3	0023
46-86508-3	0955
46-86509-3	0643
46-86511-3	0017
46-86512-3	0016
46-86513-3	0076
46-86513-4	0537
46-86514-3	0006
46-86515-3	0925
46-86516-3	0560
46-86517-3	0006
46-86518-3	1675
46-86518-9	OEM
46-86519-3	0023
46-86520-3	0288
46-86521-3	0077
46-86522-3	0074
46-86522-9	0220
46-86523-3	0137
46-86524-3	OEM
46-86525-1	0102
46-86525-3	0102
46-86526-3	0392
46-86529-3	0402
46-86530-3	0010
46-86530-3(RECT)	0348
46-86531-3	0842
46-86533-3	1284
46-86534-3	1911
46-86535-3	0842
46-86536-3	0537
46-86537-3	5606
46-86538-3	0224
46-86540-3	0155
46-86543-3	0552
46-86544-3	0127
46-86545-3	0595
46-86546-3	0006
46-86547-3	0378
46-86550-3	0191
46-86551-3	0006
46-86551-9	OEM
46-86552	0392
46-86552-3	0392
46-86553-3	0071
46-86554-3	0010
46-86555-3	OEM
46-86556-3	0955
46-86557-3	0313
46-86558-3	0087
46-86559-3	0091
46-86559-3(DIODE)	0133
46-86560-3	0348
46-86561-3	0270
46-86561-9	OEM
46-86562-3	0071
46-86563-3	2051
46-86565-3	1348

DEVICE TYPE	REPL CODE
46-86566-3	0062
46-86567-3	0643
46-86568-3	0178
46-86569-3	0710
46-86569-9	OEM
46-86570-3	0052
46-86571-3	0316
46-86572-3	0066
46-86573-3	0342
46-86574-3	0532
46-86574-9	2613
46-86577-3	6208
46-86578-3	0006
46-86579-3	0537
46-86581-3	0681
46-86582-3	6258
46-86583-3	0551
46-86584-3	0382
46-86585-9	OEM
46-86587-3	1005
46-86589-3	6853
46-86590-3	0042
46-86591-3	0066
46-86592-3	OEM
46-86595-3	OEM
46-86596-3	0076
46-86598-3	6936
46-86599-3	1607
46-86600-3	3108
46-86602-3	0400
46-86605-3	0471
46-86606-3	OEM
46-86609-3	OEM
46-86610-3	0564
46-86612-3	2536
46-86614-3	0191
46-86615-3	OEM
46-86616-3	0423
46-86617-3	3316
46-86618-3	6027
46-86619-3	0041
46-86620-9	2660
46-86621-3	7116
46-86622-3	5196
46-86623-3	OEM
46-86625-3	0693
46-86626-3	0053
46-86627-3	0445
46-86628-3	0261
46-86629-3	0629
46-86630-3	0066
46-86631-3	OEM
46-86633-3	OEM
46-86634-3	OEM
46-86635-3	1092
46-86637-3	0203
46-86639-3	OEM
46-86641-3	OEM
46-86642-3	0140
46-86643-3	OEM
46-86644-3	OEM
46-86645-3	OEM
46-86646-3	1284
46-86646-9	1284
46-86647-3	0261
46-86648-3	0261
46-86649	0064
46-86649-3	0064
46-86652-3	OEM
46-86653-3	OEM
46-86654-3	0494
46-86660-3	0507
46-86663-9	OEM
46-86664-3	3328
46-86665-3	3369
46-86666-3	0257
46-86667-3	OEM
46-86668-3	OEM
46-86669-3	OEM
46-86670-3	0516
46-86671-3	0161
46-86672-3	3342
46-86673-3	1513
46-86674-3	0275
46-86677-3	0052
46-86678-3	0166
46-86679-3	0148
46-86680-3	2902
46-86681-3	0055
46-86681-9	0055
46-86682-3	0076
46-86683-3	0537
46-86684-3	2548
46-86687-3	2484
46-86689-3	OEM
46-86691-3	OEM
46-86693-3	OEM
46-86694-3	OEM
46-86695-3	OEM
46-86696-3	OEM
46-86700-3	0018
46-86701-3	OEM
46-86702-3	0879
46-86703-3	OEM
46-86704-3	OEM
46-86708-3	OEM
46-86709-3	0344
46-86710-3	OEM
46-86711-3	OEM

DEVICE TYPE	REPL CODE
46-86712-3	0057
46-86713-3	OEM
46-86715-3	0925
46-86716-3	0715
46-86717-3	3717
46-86718-3	1211
46-86719-3	1348
46-86720-3	4999
46-86721-3	0400
46-86722-3	1348
46-86724-3	0018
46-86725-3	0693
46-86726-3	1140
46-86727-3	0037
46-86728-3	3812
46-86729-3	0582
46-86730-3	1140
46-86732-3	OEM
46-86733-3	OEM
46-86734-3	0071
46-86735-3	0253
46-86736-3	0076
46-86737-3	0023
46-86738-3	0253
46-86739-3	0023
46-86740-3	0015
46-86741-3	0019
46-86742-3	0299
46-86743-3	0055
46-86744-3	0018
46-86746-3	5458
46-86747-3	OEM
46-86748-3	1911
46-86749-3	OEM
46-86750-3	OEM
46-86751-3	0261
46-86752-3	0283
46-86753-3	0055
46-86753-9	0055
46-86754-3	2613
46-86755-3	OEM
46-86756-3	0530
46-86760-3	OEM
46-86762-3	OEM
46-86763-3	OEM
46-86767-3	OEM
46-86768-3	OEM
46-86770-3	0461
46-86771-3	OEM
46-86772-3	OEM
46-86773-3	OEM
46-86774-3	0019
46-86774-4	0019
46-86778-3	OEM
46-86781-3	0686
46-86782-3	OEM
46-86783-3	OEM
46-86784-3	OEM
46-86786-3	OEM
46-86787-3	OEM
46-86788-3	OEM
46-86789-3	OEM
46-86791-3	OEM
46-86792-3	1165
46-86793-3	0551
46-86794-3	0471
46-86797-3	0062
46-86800-3	OEM
46-86801-3	OEM
46-86802-3	OEM
46-86803-3	OEM
46-86804-3	2910
46-86805-3	0006
46-86807-3	3046
46-86808-3	OEM
46-86809-3	OEM
46-86810-3	OEM
46-86811-3	6197
46-86813-3	OEM
46-86814-3	OEM
46-86815-3	OEM
46-86816-3	OEM
46-86817-3	OEM
46-86818-3	OEM
46-86820-3	1076
46-86823-3	0283
46-86825-3	OEM
46-86826-3	OEM
46-86827-3	OEM
46-86828-3	OEM
46-86830-3	0883
46-86831-3	2209
46-86832-3	OEM
46-86834-3	OEM
46-86835-3	OEM
46-86836-3	0064
46-86837-3	OEM
46-86838-3	OEM
46-86840-3	OEM
46-86841-3	OEM
46-86843-3	0525
46-86844-3	OEM
46-86845-3	OEM
46-86846-3	OEM
46-86847-3	OEM
46-86848-3	OEM
46-86849-9	OEM
46-86850-3	OEM
46-86852-3	OEM

DEVICE TYPE	REPL CODE
46-86856-3	OEM
46-86858-3	OEM
46-86859-3	OEM
46-86860-3	0558
46-86861-3	OEM
46-86862-3	0111
46-86863-3	3046
46-86864-3	OEM
46-86865-3	OEM
46-86866-3	OEM
46-86867-3	0015
46-86870-3	0076
46-86872-3	OEM
46-86873-3	OEM
46-86874-3	OEM
46-86875-3	OEM
46-86877-3	OEM
46-86878-3	OEM
46-86879-3	0275
46-86880-3	OEM
46-86881-3	OEM
46-86883-3	0376
46-86884-3	OEM
46-86885-3	OEM
46-86886-3	OEM
46-86887-3	2484
46-86888-3	2058
46-86889-3	0018
46-86891-3	OEM
46-86893-3	OEM
46-86894-3	OEM
46-86895-3	OEM
46-86896-3	5174
46-86897-3	0118
46-86898-3	OEM
46-86899-3	OEM
46-86900-3	OEM
46-86901-3	OEM
46-86902-3	OEM
46-86903-3	OEM
46-86904-3	OEM
46-86906-3	0325
46-86906-9	OEM
46-86907-3	0530
46-86908-3	0693
46-86909-3	2551
46-86910-3	2555
46-86912-3	0253
46-86913-3	OEM
46-86914-3	5653
46-86915-3	0041
46-86916-3	7090
46-86917-3	4977
46-86918-3	0023
46-86919-3	OEM
46-86920-3	0825
46-86921-3	0062
46-86922-3	0181
46-86923-3	5478
46-86924-3	0006
46-86925-3	0062
46-86926-3	OEM
46-86927-3	OEM
46-86929-3	OEM
46-86930-3	0006
46-86931-3	0375
46-86932-3	OEM
46-86933-3	0471
46-86934-3	OEM
46-86935-3	OEM
46-86936-3	OEM
46-86937-3	0079
46-86938-3	0037
46-86940-3	OEM
46-86941-3	OEM
46-86942-3	0062
46-86944-3	OEM
46-86946-3	OEM
46-86948-3	OEM
46-86949-3	OEM
46-86950-3	2423
46-86951-3	0728
46-86952-3	2580
46-86953-3	0057
46-86954-3	0041
46-86955-3	OEM
46-86956-3	0532
46-86957-3	OEM
46-86958-3	OEM
46-86959-3	OEM
46-86961-3	OEM
46-86962-3	0219
46-86963-3	1051
46-86964-3	OEM
46-86965-3	OEM
46-86967-3	OEM
46-86968-3	0148
46-86969-3	3516
46-86970-3	OEM
46-86971-3	OEM
46-86972-3	OEM
46-86973-3	OEM
46-86974-3	OEM
46-86975-3	OEM
46-86976-3	OEM
46-86977-3	OEM
46-86978-3	OEM
46-86979-3	OEM
46-86980-3	OEM

DEVICE TYPE	REPL CODE
46-86982-3	0242
46-86983-3	OEM
46-86984-3	OEM
46-86985-3	OEM
46-86986-3	OEM
46-86987-3	OEM
46-86988-3	OEM
46-86989-3	OEM
46-86990-3	OEM
46-86991-3	OEM
46-86992-3	OEM
46-86993-3	0436
46-86994-3	OEM
46-86995-3	OEM
46-86996-3	0062
46-86997-3	0471
46-86998-3	0041
46-86999-3	0375
46-88676-3	0178
46-96679-3	0148
46-101701-3	OEM
46-101910-3	OEM
46-131000-3	OEM
46-131008-3	OEM
46-131009-3	OEM
46-131012-3	OEM
46-131013-3	OEM
46-131014-3	0107
46-131015-3	OEM
46-131018-3	OEM
46-131019-3	OEM
46-131020-3	OEM
46-131021-3	OEM
46-131022-3	OEM
46-131023-3	OEM
46-131024-3	OEM
46-131025-3	OEM
46-131026-3	OEM
46-131027-3	OEM
46-131028-3	OEM
46-131029-3	OEM
46-131030-3	2186
46-131031-3	OEM
46-131032-3	OEM
46-131033-3	OEM
46-131034-3	OEM
46-131046-3	OEM
46-131047-3	OEM
46-131048-3	OEM
46-131050-3	OEM
46-131051-3	OEM
46-131052-3	OEM
46-131053-3	3522
46-131055-3	OEM
46-131058-3	OEM
46-131059-3	OEM
46-131060-3	2913
46-131065-3	OEM
46-131066-3	OEM
46-131067-3	OEM
46-131068-3	OEM
46-131070-3	OEM
46-131071-3	OEM
46-131072-3	OEM
46-131073-3	0360
46-131074-3	OEM
46-131075-3	OEM
46-131076-3	OEM
46-131077-3	OEM
46-131078-3	OEM
46-131079-3	OEM
46-131080-3	OEM
46-131081-3	OEM
46-131082-3	OEM
46-131083-3	OEM
46-131084-3	OEM
46-131085-3	OEM
46-131086-3	OEM
46-131087-3	OEM
46-131088-3	OEM
46-131089-3	OEM
46-131092-3	OEM
46-131093-3	OEM
46-131095-3	OEM
46-131096-3	OEM
46-131097-3	OEM
46-131098-3	4869
46-131099-3	4353
46-131100-3	4129
46-131101-3	OEM
46-131102-3	OEM
46-131103-3	OEM
46-131104-3	OEM
46-131105-3	OEM
46-131108-3	OEM
46-131110-3	OEM
46-131111-3	OEM
46-131112-3	OEM
46-131113-3	OEM
46-131114-3	OEM
46-131115-3	OEM
46-131116-3	OEM
46-131117-3	OEM
46-131118-3	OEM
46-131119-3	OEM
46-131120-3	OEM
46-131121-3	OEM
46-131122-3	OEM
46-131123-3	OEM

DEVICE TYPE	REPL CODE
46-131124-3	OEM
46-131125-3	OEM
46-131126-3	0291
46-131127-3	OEM
46-131128-3	OEM
46-131129-3	OEM
46-131130-3	OEM
46-131131-3	4310
46-131132-9	OEM
46-131133-3	OEM
46-131134-3	OEM
46-131135-3	OEM
46-131136-3	5529
46-131137-3	OEM
46-131138-3	OEM
46-131139-3	OEM
46-131141-3	OEM
46-131142-3	OEM
46-131145-3	OEM
46-131146-3	OEM
46-131147-3	OEM
46-131148-3	OEM
46-131149-3	OEM
46-131152-3	OEM
46-131154-3	OEM
46-131155-3	OEM
46-131156-3	OEM
46-131158-3	5881
46-131171-3	OEM
46-131172-3	0024
46-131173-3	OEM
46-131174-3	OEM
46-131175-3	OEM
46-131176-3	OEM
46-131177-3	OEM
46-131178-3	OEM
46-131179-3	OEM
46-131180-3	OEM
46-131181-3	OEM
46-131182-3	OEM
46-131183-3	OEM
46-131185-3	OEM
46-131186-3	OEM
46-131187-3	OEM
46-131188-3	6144
46-131189-3	OEM
46-131190-3	OEM
46-131191-3	OEM
46-131192-3	OEM
46-131195-3	OEM
46-131196-3	OEM
46-131197-3	OEM
46-131198-3	OEM
46-131199-3	OEM
46-131200-3	OEM
46-131202-3	OEM
46-131203-3	5011
46-131204-3	4266
46-131205-3	OEM
46-131206-3	0539
46-131207-3	0557
46-131208-3	0732
46-131209-3	OEM
46-131210-3	OEM
46-131211-3	OEM
46-131212-3	OEM
46-131213-3	OEM
46-131214-3	OEM
46-131215-3	OEM
46-131216-3	OEM
46-131217-9	OEM
46-131219-3	OEM
46-131220-3	OEM
46-131221-3	OEM
46-131222-3	OEM
46-131224-3	OEM
46-131225-3	OEM
46-131226-3	OEM
46-131227-3	OEM
46-131228-3	OEM
46-131229-3	OEM
46-131230-3	0621
46-131231-3	OEM
46-131232-3	OEM
46-131233-3	OEM
46-131234-3	OEM
46-131235-3	OEM
46-131236-3	OEM
46-131237-3	3437
46-131238-3	0416
46-131239-3	OEM
46-131240-3	0057
46-131241-3	OEM
46-131242-3	OEM
46-131243-3	OEM
46-131245-3	4927
46-131246-3	OEM
46-131248-3	OEM
46-131249-3	OEM
46-131251-3	OEM
46-131252-3	OEM
46-131253-3	OEM
46-131254-3	OEM
46-131255-3	OEM
46-131256-3	OEM
46-131257-3	OEM
46-131258-3	OEM
46-131259-3	OEM
46-131260-3	OEM

DEVICE TYPE	REPL CODE
46-131261-3	OEM
46-131262-3	0334
46-131263-3	OEM
46-131264-3	OEM
46-131265-3	4194
46-131266-3	5256
46-131267-3	OEM
46-131268-3	OEM
46-131269-3	OEM
46-131270-3	OEM
46-131271-3	OEM
46-131272-3	OEM
46-131273-3	OEM
46-131274-3	3159
46-131275-3	OEM
46-131276-3	OEM
46-131277-3	6887
46-131278-3	OEM
46-131279-3	OEM
46-131280-3	OEM
46-131281-3	OEM
46-131283-3	OEM
46-131284-3	OEM
46-131285-3	OEM
46-131286-3	OEM
46-131287-3	OEM
46-131288-3	OEM
46-131289-3	OEM
46-131290-3	OEM
46-131291-3	OEM
46-131292-3	OEM
46-131295-3	OEM
46-131296-3	OEM
46-131298-3	OEM
46-131299-3	OEM
46-131302-3	OEM
46-131304-3	OEM
46-131305-3	OEM
46-131306-3	OEM
46-131307-3	OEM
46-131308-3	OEM
46-131313-3	OEM
46-131314-3	OEM
46-131315-3	0391
46-131316-3	OEM
46-131321-3	OEM
46-131322-3	OEM
46-131323-3	OEM
46-131324-3	OEM
46-131326-3	OEM
46-131327-3	OEM
46-131328-3	OEM
46-131329-3	OEM
46-131330-3	OEM
46-131331-3	OEM
46-131332-3	OEM
46-131333-3	OEM
46-131334-3	OEM
46-131335-3	OEM
46-131336-3	OEM
46-131339-3	OEM
46-131340-3	OEM
46-131341-3	2957
46-131342-3	3027
46-131343-3	3097
46-131344-3	OEM
46-131345-3	OEM
46-131347-3	OEM
46-131348-3	OEM
46-131349-3	3367
46-131350-3	OEM
46-131351-3	3445
46-131352-3	OEM
46-131353-3	OEM
46-131354-3	OEM
46-131357-3	1571
46-131358-3	OEM
46-131360-3	OEM
46-131361-3	OEM
46-131362-3	OEM
46-131363-3	OEM
46-131364-3	OEM
46-131365-3	OEM
46-131366-3	OEM
46-131367-3	OEM
46-131368-3	OEM
46-131369-3	OEM
46-131370-3	OEM
46-131371-3	OEM
46-131373-3	OEM
46-131378-3	OEM
46-131382-3	OEM
46-131383-3	OEM
46-131384-3	OEM
46-131385-3	OEM
46-131386-3	OEM
46-131387-3	OEM
46-131388-3	OEM
46-131389-3	OEM
46-131390-3	OEM
46-131391-3	OEM
46-131393-3	OEM
46-131395-3	OEM
46-131396-3	1291
46-131398-3	OEM
46-131399-3	OEM
46-131400-3	OEM
46-131401-3	4381
46-131404-3	OEM

DEVICE TYPE	REPL CODE	DEVICE TYPE	REPL CODE	DEVICE TYPE	REPL CODE	DEVICE TYPE	REPL CODE	DEVICE TYPE	REPL CODE	DEVICE TYPE	REPL CODE	DEVICE TYPE	REPL CODE
46-131405-3	OEM	46-131553-3	OEM	46-131685-3	OEM	46-131853-3	OEM	46-132005-3	OEM	46-132190-3	OEM	46-132339-3	OEM
46-131406-3	OEM	46-131554-3	OEM	46-131687-3	OEM	46-131854-3	OEM	46-132006-3	OEM	46-132191-3	OEM	46-132340-3	OEM
46-131407-3	OEM	46-131555-3	OEM	46-131688-3	OEM	46-131857-3	OEM	46-132007-3	OEM	46-132192-3	OEM	46-132341-3	OEM
46-131408-3	OEM	46-131561-3	OEM	46-131689-3	OEM	46-131858-3	OEM	46-132009-3	OEM	46-132193-3	OEM	46-132342-3	OEM
46-131409-3	OEM	46-131562-3	OEM	46-131693-3	OEM	46-131859-3	OEM	46-132022-3	OEM	46-132197-3	OEM	46-132343-3	OEM
46-131410-3	OEM	46-131563-3	OEM	46-131694-3	OEM	46-131860-3	OEM	46-132023-3	1807	46-132201-3	OEM	46-132344-3	OEM
46-131411-3	OEM	46-131564-3	OEM	46-131695-3	OEM	46-131861-3	OEM	46-132024-3	OEM	46-132203-3	OEM	46-132347-3	OEM
46-131412-3	OEM	46-131565-3	OEM	46-131696-3	OEM	46-131862-3	OEM	46-132039-3	OEM	46-132204-3	OEM	46-132348-3	OEM
46-131413-3	OEM	46-131566-3	OEM	46-131697-3	OEM	46-131863-3	OEM	46-132040-3	OEM	46-132205-3	OEM	46-132349-3	OEM
46-131414-3	OEM	46-131567-3	OEM	46-131699-3	OEM	46-131864-3	OEM	46-132041-3	OEM	46-132206-3	OEM	46-132350-3	OEM
46-131415-3	OEM	46-131568-3	OEM	46-131700-3	OEM	46-131865-3	OEM	46-132042-3	OEM	46-132207-3	OEM	46-132351-3	OEM
46-131416-3	OEM	46-131569-3	OEM	46-131701-3	OEM	46-131866-3	OEM	46-132043-3	OEM	46-132208-3	OEM	46-132352-3	OEM
46-131418-3	OEM	46-131570-3	OEM	46-131702-3	OEM	46-131868-3	OEM	46-132044-3	OEM	46-132209-3	OEM	46-132354-3	OEM
46-131419-3	OEM	46-131571-3	OEM	46-131703-3	OEM	46-131869-3	OEM	46-132045-3	OEM	46-132210-3	OEM	46-132355-3	OEM
46-131420-3	OEM	46-131572-3	OEM	46-131704-3	OEM	46-131870-3	OEM	46-132046-3	OEM	46-132211-3	OEM	46-132356-3	OEM
46-131421-3	OEM	46-131573-3	OEM	46-131705-3	OEM	46-131871-3	OEM	46-132048-3	OEM	46-132213-3	OEM	46-132358-3	OEM
46-131423-3	OEM	46-131575-3	OEM	46-131712-3	OEM	46-131872-3	OEM	46-132050-3	OEM	46-132214-3	OEM	46-132359-3	OEM
46-131424-3	OEM	46-131576-3	OEM	46-131713-3	OEM	46-131873-3	OEM	46-132051-3	OEM	46-132215-3	OEM	46-132360-3	OEM
46-131428-3	OEM	46-131577-3	3378	46-131715-3	OEM	46-131874-3	OEM	46-132052-3	OEM	46-132216-3	OEM	46-132361-3	OEM
46-131430-3	OEM	46-131578-3	OEM	46-131716-3	OEM	46-131875-3	OEM	46-132053-3	OEM	46-132220-3	OEM	46-132362-3	OEM
46-131431-3	OEM	46-131579-3	OEM	46-131717-3	OEM	46-131876-3	OEM	46-132054-3	OEM	46-132221-3	OEM	46-132364-3	OEM
46-131432-3	3236	46-131580-3	OEM	46-131718-3	OEM	46-131877-3	OEM	46-132056-3	OEM	46-132222-3	OEM	46-132365-3	OEM
46-131433-3	3959	46-131581-3	OEM	46-131719-3	OEM	46-131878-3	OEM	46-132059-3	OEM	46-132228-3	OEM	46-132367-3	OEM
46-131434-3	OEM	46-131582-3	OEM	46-131721-3	OEM	46-131882-3	OEM	46-132061-3	OEM	46-132231-3	OEM	46-132368-3	OEM
46-131435-3	OEM	46-131583-3	OEM	46-131723-3	OEM	46-131885-3	OEM	46-132062-3	OEM	46-132232-3	OEM	46-132369-3	OEM
46-131436-3	OEM	46-131584-3	OEM	46-131724-3	OEM	46-131895-3	OEM	46-132063-3	OEM	46-132233-3	OEM	46-132371-3	OEM
46-131438-9	3913	46-131585-3	1641	46-131725-3	OEM	46-131896-3	OEM	46-132064-3	OEM	46-132234-3	OEM	46-132372-3	OEM
46-131439-3	OEM	46-131586-3	OEM	46-131726-3	OEM	46-131897-3	OEM	46-132065-3	OEM	46-132235-3	OEM	46-132373-3	OEM
46-131452-3	OEM	46-131587-3	OEM	46-131727-3	OEM	46-131898-3	OEM	46-132066-3	OEM	46-132237-3	OEM	46-132374-3	OEM
46-131453-3	OEM	46-131588-3	OEM	46-131729-3	OEM	46-131899-3	OEM	46-132067-3	OEM	46-132238-3	OEM	46-132375-3	OEM
46-131454-3	OEM	46-131589-3	OEM	46-131730-3	OEM	46-131900-3	OEM	46-132068-3	OEM	46-132242-3	OEM	46-132376-3	OEM
46-131455-3	OEM	46-131590-3	OEM	46-131734-3	OEM	46-131901-3	OEM	46-132069-3	OEM	46-132243-3	OEM	46-132377-3	OEM
46-131456-3	OEM	46-131591-3	OEM	46-131735-3	OEM	46-131902-3	OEM	46-132070-3	2334	46-132244-3	OEM	46-132378-3	OEM
46-131457-3	OEM	46-131592-3	OEM	46-131736-3	OEM	46-131903-3	OEM	46-132071-3	OEM	46-132245-3	OEM	46-132379-3	OEM
46-131458-3	OEM	46-131593-3	OEM	46-131737-3	OEM	46-131905-3	0683	46-132072-3	OEM	46-132246-3	OEM	46-132381-3	OEM
46-131459-3	OEM	46-131594-3	OEM	46-131739-3	OEM	46-131906-3	OEM	46-132075-3	OEM	46-132247-3	OEM	46-132382-3	OEM
46-131460-3	OEM	46-131595-3	OEM	46-131740-3	OEM	46-131914-3	OEM	46-132076-3	OEM	46-132248-3	OEM	46-132383-3	OEM
46-131461-3	OEM	46-131596-3	OEM	46-131741-3	OEM	46-131915-3	OEM	46-132077-3	OEM	46-132249-3	OEM	46-132386-3	OEM
46-131462-3	OEM	46-131597-3	OEM	46-131744-3	OEM	46-131916-3	OEM	46-132079-3	OEM	46-132250-3	OEM	46-132389-3	OEM
46-131463-3	OEM	46-131598-3	OEM	46-131745-3	0315	46-131917-3	OEM	46-132080-3	OEM	46-132252-3	OEM	46-132390-3	OEM
46-131464-3	OEM	46-131599-3	OEM	46-131746-3	OEM	46-131918-3	OEM	46-132083-3	OEM	46-132253-3	OEM	46-132391-3	OEM
46-131465-3	OEM	46-131600-3	OEM	46-131747-3	OEM	46-131919-3	OEM	46-132084-3	OEM	46-132256-3	OEM	46-132392-3	OEM
46-131466-3	OEM	46-131601-3	OEM	46-131751-3	OEM	46-131920-3	OEM	46-132085-3	OEM	46-132257-3	OEM	46-132393-3	OEM
46-131467-3	OEM	46-131602-3	OEM	46-131755-3	OEM	46-131921-3	OEM	46-132086-3	OEM	46-132259-3	OEM	46-132394-3	0101
46-131469-3	OEM	46-131604-3	OEM	46-131756-3	OEM	46-131922-3	OEM	46-132087-3	OEM	46-132260-3	OEM	46-132395-3	OEM
46-131470-3	OEM	46-131605-3	OEM	46-131757-3	OEM	46-131923-3	OEM	46-132088-3	OEM	46-132261-3	OEM	46-132396-3	OEM
46-131471-3	OEM	46-131606-3	OEM	46-131758-3	OEM	46-131924-3	OEM	46-132090-3	OEM	46-132263-3	OEM	46-132397-3	OEM
46-131472-3	OEM	46-131609-3	OEM	46-131759-3	OEM	46-131926-3	OEM	46-132092-3	OEM	46-132264-3	OEM	46-132398-3	2870
46-131473-3	OEM	46-131610-3	OEM	46-131760-3	OEM	46-131927-3	OEM	46-132093-3	OEM	46-132265-3	OEM	46-132399-3	OEM
46-131474-3	OEM	46-131611-3	OEM	46-131761-3	OEM	46-131928-3	OEM	46-132094-3	OEM	46-132266-3	OEM	46-132404-3	OEM
46-131475-3	OEM	46-131613-3	OEM	46-131762-3	OEM	46-131929-3	OEM	46-132099-3	OEM	46-132268-3	OEM	46-132405-3	OEM
46-131476-3	OEM	46-131614-3	OEM	46-131766-3	OEM	46-131931-3	OEM	46-132101-3	OEM	46-132269-3	OEM	46-132407-3	OEM
46-131477-3	OEM	46-131615-3	OEM	46-131767-3	OEM	46-131932-3	OEM	46-132104-3	OEM	46-132270-3	OEM	46-132408-3	OEM
46-131478-3	OEM	46-131616-3	OEM	46-131768-3	OEM	46-131933-3	OEM	46-132105-3	OEM	46-132271-3	OEM	46-132409-3	OEM
46-131479-3	OEM	46-131617-3	OEM	46-131770-3	OEM	46-131934-3	OEM	46-132106-3	OEM	46-132272-3	OEM	46-132411-3	OEM
46-131480-3	OEM	46-131618-3	OEM	46-131771-3	OEM	46-131935-3	OEM	46-132108-3	OEM	46-132273-3	OEM	46-132412-3	OEM
46-131481-3	OEM	46-131619-3	OEM	46-131772-3	4382	46-131937-3	OEM	46-132109-3	OEM	46-132274-3	OEM	46-132413-3	OEM
46-131482-3	OEM	46-131620-3	OEM	46-131773-3	OEM	46-131938-3	OEM	46-132111-3	OEM	46-132275-3	OEM	46-132415-3	OEM
46-131483-3	OEM	46-131621-3	OEM	46-131774-3	1166	46-131940-3	OEM	46-132112-3	OEM	46-132277-3	OEM	46-132416-3	OEM
46-131484-3	OEM	46-131622-3	OEM	46-131776-3	OEM	46-131941-3	OEM	46-132113-3	OEM	46-132279-3	OEM	46-132417-3	OEM
46-131485-3	OEM	46-131623-3	OEM	46-131777-3	OEM	46-131942-3	OEM	46-132115-3	OEM	46-132280-3	OEM	46-132418-3	OEM
46-131486-3	OEM	46-131624-3	OEM	46-131778-3	OEM	46-131943-3	OEM	46-132116-3	OEM	46-132281-3	OEM	46-132419-3	OEM
46-131488-3	OEM	46-131625-3	OEM	46-131779-3	OEM	46-131944-3	OEM	46-132117-3	OEM	46-132283-3	OEM	46-132420-3	OEM
46-131491-3	OEM	46-131626-3	OEM	46-131780-3	OEM	46-131946-3	OEM	46-132118-3	OEM	46-132284-3	OEM	46-132422-3	0035
46-131492-3	OEM	46-131627-3	OEM	46-131781-3	OEM	46-131947-3	OEM	46-132119-3	OEM	46-132285-3	OEM	46-132423-3	OEM
46-131493-3	OEM	46-131629-3	2111	46-131782-3	OEM	46-131948-3	OEM	46-132120-3	OEM	46-132286-3	2090	46-132424-3	OEM
46-131494-9	OEM	46-131631-3	OEM	46-131783-3	OEM	46-131950-3	OEM	46-132122-3	OEM	46-132287-3	3953	46-132425-3	OEM
46-131495-3	OEM	46-131632-3	OEM	46-131785-3	OEM	46-131951-3	OEM	46-132124-3	OEM	46-132288-3	OEM	46-132426-3	OEM
46-131500-3	OEM	46-131633-3	OEM	46-131786-3	OEM	46-131952-3	OEM	46-132127-3	OEM	46-132289-3	OEM	46-132428-3	OEM
46-131503-3	OEM	46-131634-3	OEM	46-131787-3	3236	46-131954-3	OEM	46-132129-3	OEM	46-132290-3	OEM	46-132430-3	OEM
46-131504-3	OEM	46-131635-3	OEM	46-131788-3	OEM	46-131955-3	OEM	46-132130-3	OEM	46-132291-3	OEM	46-132433-3	OEM
46-131505-3	OEM	46-131636-3	OEM	46-131789-3	OEM	46-131956-3	OEM	46-132131-3	OEM	46-132293-3	OEM	46-132434-3	OEM
46-131507-3	0391	46-131637-3	OEM	46-131790-3	OEM	46-131958-3	OEM	46-132132-3	OEM	46-132295-3	OEM	46-132435-3	OEM
46-131508-3	OEM	46-131638-3	OEM	46-131793-3	0624	46-131960-3	OEM	46-132133-3	OEM	46-132296-3	OEM	46-132437-3	OEM
46-131509-3	OEM	46-131639-3	OEM	46-131794-3	OEM	46-131962-3	OEM	46-132134-3	OEM	46-132298-3	OEM	46-132438-3	OEM
46-131510-3	OEM	46-131640-3	OEM	46-131798-3	OEM	46-131963-3	OEM	46-132135-3	OEM	46-132299-3	OEM	46-132440-3	OEM
46-131511-3	OEM	46-131641-3	OEM	46-131799-3	OEM	46-131964-3	OEM	46-132137-3	OEM	46-132300-3	OEM	46-132441-3	OEM
46-131512-3	OEM	46-131642-3	OEM	46-131800-3	OEM	46-131968-3	OEM	46-132139-3	OEM	46-132301-3	OEM	46-132442-3	OEM
46-131513-3	OEM	46-131644-3	OEM	46-131801-3	0624	46-131969-3	4117	46-132142-3	OEM	46-132302-3	OEM	46-132443-3	OEM
46-131514-3	2870	46-131645-3	OEM	46-131802-3	OEM	46-131970-3	OEM	46-132143-3	OEM	46-132303-3	3690	46-132444-3	OEM
46-131515-3	3083	46-131646-3	OEM	46-131804-3	OEM	46-131971-3	OEM	46-132149-3	OEM	46-132304-3	OEM	46-132445-3	OEM
46-131516-3	1786	46-131647-3	OEM	46-131806-3	OEM	46-131972-3	OEM	46-132150-3	OEM	46-132305-3	3254	46-132446-3	OEM
46-131519-3	6397	46-131648-3	OEM	46-131807-3	OEM	46-131974-3	OEM	46-132152-3	OEM	46-132307-3	OEM	46-132447-3	OEM
46-131520-3	OEM	46-131649-3	OEM	46-131808-3	OEM	46-131975-3	OEM	46-132153-3	OEM	46-132308-3	OEM	46-132448-3	OEM
46-131521-3	OEM	46-131650-3	OEM	46-131809-3	OEM	46-131976-3	OEM	46-132154-3	OEM	46-132309-3	OEM	46-132449-3	OEM
46-131522-3	OEM	46-131651-3	OEM	46-131812-3	OEM	46-131977-3	OEM	46-132155-3	OEM	46-132310-3	OEM	46-132450-3	OEM
46-131523-3	OEM	46-131652-3	OEM	46-131813-3	OEM	46-131978-3	OEM	46-132156-3	OEM	46-132311-3	OEM	46-132451-3	OEM
46-131524-3	OEM	46-131655-3	OEM	46-131814-3	OEM	46-131979-3	OEM	46-132157-3	OEM	46-132312-3	OEM	46-132452-3	OEM
46-131525-3	OEM	46-131656-3	0462	46-131816-3	OEM	46-131980-3	OEM	46-132158-3	OEM	46-132313-3	OEM	46-132453-3	OEM
46-131526-3	OEM	46-131658-3	OEM	46-131817-3	OEM	46-131981-3	OEM	46-132159-3	OEM	46-132314-3	OEM	46-132454-3	OEM
46-131531-3	OEM	46-131659-3	OEM	46-131818-3	OEM	46-131982-3	OEM	46-132161-3	OEM	46-132315-3	OEM	46-132455-3	OEM
46-131532-3	OEM	46-131660-3	OEM	46-131822-3	OEM	46-131986-3	OEM	46-132162-3	OEM	46-132316-3	OEM	46-132456-3	OEM
46-131533-3	OEM	46-131662-3	OEM	46-131825-3	OEM	46-131987-3	OEM	46-132163-3	OEM	46-132317-3	OEM	46-132459-3	OEM
46-131535-3	OEM	46-131664-3	OEM	46-131826-3	OEM	46-131988-3	OEM	46-132164-3	OEM	46-132318-3	OEM	46-132460-3	OEM
46-131536-3	OEM	46-131665-3	OEM	46-131827-3	OEM	46-131989-3	OEM	46-132165-3	OEM	46-132320-3	OEM	46-132461-3	OEM
46-131538-3	OEM	46-131666-3	OEM	46-131828-3	OEM	46-131990-3	OEM	46-132166-3	OEM	46-132321-3	OEM	46-132462-3	OEM
46-131539-3	OEM	46-131667-3	OEM	46-131831-3	OEM	46-131992-3	OEM	46-132167-3	OEM	46-132323-3	OEM	46-132463-3	OEM
46-131540-3	OEM	46-131669-3	OEM	46-131832-3	OEM	46-131993-3	OEM	46-132169-3	OEM	46-132324-3	OEM	46-132464-3	OEM
46-131541-3	OEM	46-131670-3	OEM	46-131837-3	OEM	46-131994-3	OEM	46-132171-3	OEM	46-132325-3	OEM	46-132465-3	OEM
46-131542-3	OEM	46-131671-3	OEM	46-131838-3	OEM	46-131995-3	OEM	46-132172-3	OEM	46-132326-3	OEM	46-132473-3	OEM
46-131543-3	OEM	46-131672-3	OEM	46-131839-3	OEM	46-131996-3	OEM	46-132175-3	OEM	46-132327-3	OEM	46-132474-3	OEM
46-131544-3	OEM	46-131675-3	OEM	46-131840-3	OEM	46-131997-3	OEM	46-132177-3	OEM	46-132328-3	OEM	46-132475-3	OEM
46-131545-3	OEM	46-131676-3	OEM	46-131841-3	OEM	46-131998-3	OEM	46-132178-3	OEM	46-132329-3	OEM	46-132476-3	OEM
46-131546-3	OEM	46-131677-3	OEM	46-131844-3	OEM	46-131999-3	OEM	46-132179-3	OEM	46-132330-3	OEM	46-132480-3	OEM
46-131547-3	OEM	46-131678-3	0035	46-131847-3	OEM	46-132000-3	OEM	46-132180-3	OEM	46-132333-3	OEM	46-132483-3	OEM
46-131548-3	OEM	46-131680-3	OEM	46-131848-3	OEM	46-132001-3	OEM	46-132182-3	OEM	46-132334-3	OEM	46-132485-3	OEM
46-131549-3	OEM	46-131681-3	OEM	46-131849-3	OEM	46-132002-3	OEM	46-132183-3	OEM	46-132335-3	OEM	46-132489-3	OEM
46-131551-3	OEM	46-131682-3	OEM	46-131850-3	OEM	46-132003-3	OEM	46-132185-3	OEM	46-132336-3	OEM	46-132490-3	OEM
46-131552-3	OEM	46-131683-3	2413	46-131851-3	OEM	46-132004-3	OEM	46-132188-3	OEM	46-132338-3	OEM	46-132498-3	OEM

DEVICE TYPE	REPL CODE
46-132499-3	OEM
46-132500-3	OEM
46-132501-3	OEM
46-132502-3	OEM
46-132506-3	OEM
46-132507-3	OEM
46-132510-3	OEM
46-132511-3	OEM
46-132512-3	OEM
46-132513-3	OEM
46-132514-3	OEM
46-132515-3	OEM
46-132517-3	OEM
46-132518-3	OEM
46-132519-3	OEM
46-132520-3	OEM
46-132521-3	OEM
46-132525-3	OEM
46-132526-3	OEM
46-132527-3	OEM
46-132528-3	OEM
46-132529-3	OEM
46-132532-3	OEM
46-132534-3	OEM
46-132535-3	OEM
46-132536-3	OEM
46-132539-3	OEM
46-132540-3	OEM
46-132542-3	OEM
46-132543-3	OEM
46-132544-3	OEM
46-132545-3	OEM
46-132546-3	OEM
46-132547-3	OEM
46-132548-3	OEM
46-132549-3	OEM
46-132550-3	OEM
46-132551-3	OEM
46-132552-3	OEM
46-132553-3	OEM
46-132554-3	OEM
46-132555-3	OEM
46-132556-3	OEM
46-132557-3	OEM
46-132558-3	OEM
46-132560-3	OEM
46-132561-3	OEM
46-132562-3	OEM
46-132563-3	OEM
46-132564-3	OEM
46-132565-3	OEM
46-132566-3	OEM
46-132567-3	OEM
46-132568-3	OEM
46-132569-3	OEM
46-132572-3	OEM
46-132574-3	OEM
46-132575-3	OEM
46-132578-3	OEM
46-132579-3	OEM
46-132580-3	OEM
46-132582-3	OEM
46-132583-3	OEM
46-132588-3	OEM
46-132589-3	OEM
46-132590-3	OEM
46-132593-3	OEM
46-132595-3	OEM
46-132598-3	OEM
46-132600-3	OEM
46-132604-3	OEM
46-132605-3	OEM
46-132607-3	OEM
46-132608-3	OEM
46-132609-3	OEM
46-132610-3	OEM
46-132611-3	OEM
46-132612-3	OEM
46-132617-3	OEM
46-132618-3	OEM
46-132619-3	OEM
46-132620-3	OEM
46-132621-3	OEM
46-132622-3	OEM
46-132623-3	OEM
46-132624-3	OEM
46-132625-3	OEM
46-132627-3	OEM
46-132628-3	OEM
46-132629-3	OEM
46-132630-3	OEM
46-132631-3	OEM
46-132634-3	OEM
46-132635-3	OEM
46-132636-3	OEM
46-132637-3	2116
46-132638-3	2122
46-132639-3	OEM
46-132640-3	OEM
46-132645-3	OEM
46-132647-3	4094
46-132648-3	2346
46-132649-3	4930
46-132650-3	OEM
46-132651-3	OEM
46-132652-3	OEM
46-132653-3	OEM
46-132654-3	OEM
46-132655-3	OEM
46-132657-3	OEM
46-132660-3	OEM
46-132661-3	OEM
46-132662-3	OEM
46-132663-3	OEM
46-132664-3	OEM
46-132665-3	OEM
46-132666-3	OEM
46-132667-3	OEM
46-132668-3	OEM
46-132669-3	OEM
46-132670-3	OEM
46-132671-3	OEM
46-132672-3	OEM
46-132673-3	OEM
46-132674-3	OEM
46-132675-3	OEM
46-132677-3	OEM
46-132678-3	OEM
46-132679-3	OEM
46-132680-3	OEM
46-132681-3	OEM
46-132682-3	OEM
46-132683-3	OEM
46-132684-3	OEM
46-132689-3	OEM
46-132690-3	OEM
46-132691-3	OEM
46-132692-3	OEM
46-132694-3	OEM
46-132695-3	OEM
46-132697-3	OEM
46-132698-3	OEM
46-132699-3	OEM
46-132702-3	OEM
46-132703-3	OEM
46-132704-3	OEM
46-132705-3	OEM
46-132706-3	OEM
46-132708-3	OEM
46-132709-3	OEM
46-132710-3	OEM
46-132711-3	OEM
46-132712-3	OEM
46-132713-3	OEM
46-132714-3	OEM
46-132715-3	OEM
46-132716-3	OEM
46-132717-3	OEM
46-132718-3	OEM
46-132719-3	OEM
46-132720-3	OEM
46-132721-3	OEM
46-132722-3	OEM
46-132726-3	OEM
46-132727-3	OEM
46-132728-3	OEM
46-132729-3	OEM
46-132730-3	OEM
46-132731-3	OEM
46-132732-3	OEM
46-132733-3	OEM
46-132734-3	OEM
46-132735-3	OEM
46-132736-3	OEM
46-132737-3	OEM
46-132738-3	OEM
46-132739-3	OEM
46-132740-3	OEM
46-132741-3	OEM
46-132744-3	OEM
46-132745-3	OEM
46-132747-3	OEM
46-132748-3	OEM
46-132750-3	OEM
46-132751-3	OEM
46-132754-3	OEM
46-132758-3	OEM
46-132760-3	OEM
46-132763-3	OEM
46-132765-3	OEM
46-132766-3	OEM
46-132767-3	OEM
46-132768-3	OEM
46-132769-3	OEM
46-132770-3	OEM
46-132771-3	OEM
46-132772-3	OEM
46-132773-3	OEM
46-132774-3	OEM
46-132775-3	OEM
46-132778-3	OEM
46-132779-3	OEM
46-132781-3	OEM
46-132783-3	OEM
46-132784-3	OEM
46-132786-3	OEM
46-132787-3	OEM
46-132792-3	OEM
46-132793-3	OEM
46-132794-3	OEM
46-132796-3	OEM
46-132797-3	OEM
46-132798-3	OEM
46-132799-3	OEM
46-132802-3	OEM
46-132803-3	OEM
46-132804-3	OEM
46-132805-3	OEM
46-132806-3	OEM
46-132810-3	OEM
46-132811-3	OEM
46-132812-3	OEM
46-132813-3	OEM
46-132814-3	OEM
46-132815-3	OEM
46-132816-3	OEM
46-132817-3	OEM
46-132820-3	OEM
46-132821-3	OEM
46-132826-3	OEM
46-132827-3	OEM
46-132828-3	OEM
46-132829-3	OEM
46-132830-3	OEM
46-132831-3	OEM
46-132832-3	OEM
46-132833-3	OEM
46-132834-3	OEM
46-132835-3	OEM
46-132836-3	OEM
46-132837-3	OEM
46-132838-3	OEM
46-132839-3	OEM
46-132840-3	OEM
46-132843-3	OEM
46-132844-3	OEM
46-132845-3	OEM
46-132846-3	OEM
46-132848-3	OEM
46-132849-3	OEM
46-132850-3	0624
46-132852-3	OEM
46-132853-3	OEM
46-132854-3	OEM
46-132855-3	OEM
46-132856-3	OEM
46-132857-3	OEM
46-132858-3	OEM
46-132859-3	OEM
46-132860-3	OEM
46-132865-3	OEM
46-132866-3	OEM
46-132867-3	OEM
46-132868-3	OEM
46-132869-3	OEM
46-132870-3	OEM
46-132871-3	OEM
46-132872-3	OEM
46-132873-3	OEM
46-132874-3	OEM
46-132875-3	OEM
46-132876-3	OEM
46-132877-3	OEM
46-132878-3	OEM
46-132879-3	OEM
46-132882-3	OEM
46-132883-3	OEM
46-132884-3	1654
46-132885-3	OEM
46-132886-3	4422
46-132888-3	OEM
46-132891-3	OEM
46-132893-3	OEM
46-132894-3	OEM
46-132895-3	OEM
46-132896-3	OEM
46-132897-3	OEM
46-132898-3	OEM
46-132899-3	OEM
46-132900-3	OEM
46-132902-3	OEM
46-132904-3	OEM
46-132906-3	OEM
46-132907-3	OEM
46-132908-3	OEM
46-132909-3	OEM
46-132912-3	OEM
46-132914-3	OEM
46-132915-3	OEM
46-132917-3	OEM
46-132918-3	OEM
46-132919-3	OEM
46-132920-3	OEM
46-132921-3	OEM
46-132922-3	OEM
46-132923-3	OEM
46-132924-3	OEM
46-132925-3	OEM
46-132929-3	OEM
46-132930-3	OEM
46-132931-3	3515
46-132932-3	OEM
46-132934-3	OEM
46-132935-3	OEM
46-132939-3	OEM
46-132940-3	OEM
46-132941-3	OEM
46-132943-3	OEM
46-132947-3	OEM
46-132949-3	OEM
46-132950-3	OEM
46-132951-3	OEM
46-132952-3	OEM
46-132953-3	OEM
46-132954-3	OEM
46-132955-3	OEM
46-132956-3	OEM
46-132957-3	OEM
46-132958-3	OEM
46-132959-3	OEM
46-132960-3	OEM
46-132961-3	OEM
46-132962-3	OEM
46-132963-3	OEM
46-132964-3	OEM
46-132965-3	OEM
46-132966-3	OEM
46-132967-3	OEM
46-132968-3	OEM
46-132969-3	OEM
46-132970-3	OEM
46-132972-3	OEM
46-132973-3	OEM
46-132974-3	OEM
46-132975-3	OEM
46-132976-3	OEM
46-132977-3	OEM
46-132978-3	OEM
46-132979-3	OEM
46-132986-3	OEM
46-132988-3	OEM
46-132992-3	OEM
46-132997-3	OEM
46-133000-3	OEM
46-133001-3	OEM
46-133004-3	OEM
46-133005-3	OEM
46-133009-3	OEM
46-133010-3	OEM
46-133011-3	OEM
46-133012-3	OEM
46-133013-3	OEM
46-133017-3	OEM
46-133018-3	OEM
46-133019-3	OEM
46-133021-3	OEM
46-133022-3	OEM
46-133024-3	OEM
46-133025-3	OEM
46-133027-3	OEM
46-133028-3	OEM
46-133029-3	OEM
46-133034-3	OEM
46-133035-3	OEM
46-133036-3	OEM
46-133037-3	OEM
46-133038-3	OEM
46-133039-3	OEM
46-133040-3	OEM
46-133042-3	OEM
46-133043-3	OEM
46-133044-3	OEM
46-133045-3	OEM
46-133046-3	OEM
46-133047-3	OEM
46-133048-3	OEM
46-133050-3	OEM
46-133051-3	OEM
46-133052-3	OEM
46-133057-3	OEM
46-133058-3	OEM
46-133059-3	OEM
46-133061-3	OEM
46-133062-3	OEM
46-133063-3	OEM
46-133064-3	OEM
46-133066-3	OEM
46-133067-3	OEM
46-133068-3	OEM
46-133069-3	OEM
46-133071-3	OEM
46-133072-3	OEM
46-133073-3	OEM
46-133074-3	OEM
46-133075-3	OEM
46-133076-3	OEM
46-133077-3	OEM
46-133078-3	OEM
46-133079-3	OEM
46-133080-3	OEM
46-133081-3	OEM
46-133082-3	OEM
46-133083-3	OEM
46-133084-3	OEM
46-133085-3	OEM
46-133086-3	OEM
46-133087-3	OEM
46-133088-3	OEM
46-133089-3	OEM
46-133090-3	OEM
46-133091-3	OEM
46-133092-3	OEM
46-133093-3	OEM
46-133094-3	OEM
46-133095-3	OEM
46-133103-3	OEM
46-133105-3	OEM
46-133106-3	OEM
46-133107-3	OEM
46-133108-3	OEM
46-133109-3	OEM
46-133110-3	OEM
46-133111-3	OEM
46-133112-3	OEM
46-133113-3	OEM
46-133114-3	OEM
46-133116-3	OEM
46-133119-3	OEM
46-133120-3	OEM
46-133121-3	OEM
46-133122-3	OEM
46-133123-3	OEM
46-133124-3	OEM
46-133125-3	OEM
46-133126-3	OEM
46-133127-3	OEM
46-133136-3	OEM
46-133137-3	OEM
46-133142-3	OEM
46-133143-3	OEM
46-133144-3	OEM
46-133145-3	OEM
46-133147-3	OEM
46-133148-3	OEM
46-133149-3	OEM
46-133154-3	OEM
46-133155-3	OEM
46-133156-3	OEM
46-133158-3	OEM
46-133159-3	OEM
46-133160-3	OEM
46-133161-3	OEM
46-133166-3	OEM
46-133169-3	OEM
46-133182-3	OEM
46-133183-3	OEM
46-133184-3	OEM
46-133185-3	OEM
46-133186-3	OEM
46-133187-3	OEM
46-133188-3	OEM
46-133190-3	OEM
46-133192-3	OEM
46-133193-3	OEM
46-133194-3	OEM
46-133195-3	OEM
46-133200-3	OEM
46-133201-3	OEM
46-133206-3	OEM
46-133207-3	OEM
46-133208-3	OEM
46-133209-3	OEM
46-133210-3	OEM
46-133211-3	OEM
46-133212-3	OEM
46-133213-3	OEM
46-133214-3	OEM
46-133216-3	OEM
46-133221-3	OEM
46-133240-3	OEM
46-133244-3	OEM
46-133245-3	OEM
46-133246-3	OEM
46-133253-3	OEM
46-133261-3	OEM
46-133262-3	OEM
46-133263-3	OEM
46-133270-3	OEM
46-133274-3	OEM
46-133275-3	2194
46-133276-3	OEM
46-133278-3	OEM
46-133283-3	OEM
46-133284-3	OEM
46-133285-3	OEM
46-133293-3	6183
46-133300-3	OEM
46-133301-3	OEM
46-133302-3	OEM
46-133304-3	OEM
46-133305-3	OEM
46-133306-3	OEM
46-133307-3	OEM
46-133315-3	OEM
46-133316-3	OEM
46-133321-3	OEM
46-133322-3	OEM
46-133323-3	OEM
46-133325-3	3440
46-133326-3	OEM
46-133327-3	OEM
46-133328-3	OEM
46-133329-3	OEM
46-133330-3	OEM
46-133331-3	OEM
46-133332-3	OEM
46-133333-3	OEM
46-133334-3	OEM
46-133336-3	OEM
46-133337-3	OEM
46-133339-3	OEM
46-133340-3	OEM
46-133341-3	OEM
46-133342-3	OEM
46-133343-3	OEM
46-133344-3	3515
46-133346-3	OEM
46-133347-3	OEM
46-133351-3	OEM
46-133352-3	OEM
46-133353-3	OEM
46-133354-3	OEM
46-133358-3	OEM
46-133359-3	OEM
46-133362-3	OEM
46-133363-3	OEM
46-133365-3	OEM
46-133367-3	OEM
46-133368-3	OEM
46-133371-3	OEM
46-133378-3	OEM
46-133379-3	OEM
46-133380-3	OEM
46-133381-3	OEM
46-133382-3	OEM
46-133383-3	OEM
46-133384-3	OEM
46-133386-3	OEM
46-133387-3	OEM
46-133388-3	1166
46-133389-3	1291
46-133402-3	OEM
46-133403-3	OEM
46-133404-3	OEM
46-133405-3	OEM
46-133406-3	OEM
46-133407-3	OEM
46-133408-3	OEM
46-133409-3	OEM
46-133410-3	OEM
46-133411-3	OEM
46-133412-3	OEM
46-133413-3	OEM
46-133414-3	OEM
46-133415-3	OEM
46-133416-3	OEM
46-133417-3	OEM
46-133418-3	OEM
46-133419-3	OEM
46-133420-3	OEM
46-133421-3	OEM
46-133422-3	OEM
46-133423-3	OEM
46-133424-3	OEM
46-133425-3	OEM
46-133426-3	OEM
46-133427-3	OEM
46-133428-3	OEM
46-133430-3	OEM
46-133432-3	OEM
46-133437-3	OEM
46-133438-3	OEM
46-133441-3	OEM
46-133444-3	OEM
46-133445-3	OEM
46-133446-3	OEM
46-133447-3	OEM
46-133448-3	OEM
46-133449-3	OEM
46-133450-3	OEM
46-133451-3	OEM
46-133452-3	OEM
46-133453-3	OEM
46-133454-3	OEM
46-133455-3	OEM
46-133456-3	OEM
46-133458-3	OEM
46-133459-3	OEM
46-133460-3	OEM
46-133462-3	OEM
46-133463-3	OEM
46-133464-3	OEM
46-133465-3	OEM
46-133471-3	OEM
46-133472-3	OEM
46-133473-3	OEM
46-133475-3	OEM
46-133476-3	OEM
46-133477-3	OEM
46-133478-3	OEM
46-133479-3	OEM
46-133480-3	OEM
46-133481-3	OEM
46-133483-3	OEM
46-133489-3	OEM
46-133491-3	OEM
46-133492-3	OEM
46-133493-3	OEM
46-133495-3	OEM
46-133501-3	OEM
46-133503-3	OEM
46-133504-3	OEM
46-133506-3	OEM
46-133507-3	OEM
46-133508-3	OEM
46-133509-3	OEM
46-133510-3	OEM
46-133515-3	OEM
46-133517-3	OEM
46-133522-3	OEM
46-133523-3	OEM
46-133524-3	OEM
46-133528-3	OEM
46-133535-3	OEM
46-133537-3	OEM
46-133543-3	OEM
46-133544-3	OEM
46-133545-3	OEM
46-133546-3	OEM
46-133547-3	OEM
46-133548-3	OEM
46-133549-3	OEM
46-133550-3	OEM
46-133551-3	OEM
46-133552-3	OEM
46-133553-3	OEM
46-133554-3	OEM
46-133555-3	OEM
46-133558-3	OEM
46-133565-3	OEM
46-133566-3	OEM
46-133570-3	OEM
46-133571-3	OEM
46-133578-3	OEM
46-133593-3	OEM
46-133691-3	OEM
46-133692-3	5642
46-133693-3	OEM
46-133694-3	OEM
46-133695-3	OEM
46-133696-3	OEM
46-133697-3	0315
46-133801-3	OEM
46-133831-3	1291
46-133832-3	OEM
46-133912-3	OEM
46-133967-3	OEM
46-134039-3	OEM
46-134041-3	OEM
46-351284-3	OEM
46-351285-3	OEM
46-353671-3	OEM
46-611117-3	1993
46-720336-3	OEM
46-721199-3	OEM
46-721209-3	OEM
46-721236-3	OEM
46-721243-3	OEM
46-721246-3	OEM
46-721267-3	OEM
46-721342-3	OEM
46-721344-3	OEM
46-721348-3	OEM
46-721352-3	OEM
46-721360-3	OEM
46-721806-3	OEM
46-721807-3	OEM
46-721809-3	OEM
46-721811-3	OEM
46-721813-3	OEM
46-721815-3	OEM
46-721821-3	OEM
46-721822-3	OEM
46-721823-3	OEM
46-721824-3	OEM
46-721826-3	OEM
46-721828-3	OEM
46-721829-3	OEM
46-721894-3	OEM
46-721898-3	OEM
46-721903-3	OEM
46-721905-3	OEM
46-721906-3	OEM
46-721908-3	OEM
46-721909-3	OEM
46-721910-3	OEM
46-721911-3	OEM
46-721913-3	OEM
46-721914-3	OEM
46-721916-3	OEM
46-721917-3	OEM
46-721924-3	OEM
46-721925-3	OEM
46-721926-3	OEM
46-721927-3	OEM
46-721928-3	OEM
46-721929-3	OEM
46-721930-3	OEM
46-721931-3	OEM
46-721933-3	OEM
46-721934-3	OEM
46-721936-3	OEM
46-721938-3	OEM
46-721940-3	OEM
46-721942-3	OEM
46-721945-3	OEM
46-721946-3	OEM
46-721947-3	OEM
46-721951-3	OEM
46-721952-3	OEM
46-721953-3	OEM
46-721954-3	OEM
46-721955-3	OEM
46-721964-3	OEM
46-721968-3	OEM
46-721969-3	OEM
46-721971-3	OEM
46-721972-3	OEM
46-721975-3	OEM
46-721982-3	OEM
46-722012-3	OEM
46-722013-3	OEM
46-722016-3	OEM
46-722017-3	OEM
46-722020-3	OEM
46-722021-3	OEM
46-722022-3	OEM
46-722023-3	OEM
46-722024-3	OEM
46-722025-3	OEM
46-722026-3	OEM
46-722029-3	OEM
46-722038-3	OEM

If replacement code is OEM, contact original manufacturer for replacement.

DEVICE TYPE	REPL CODE
46-722039-3	OEM
46-722040-3	OEM
46-722042-3	OEM
46-722043-3	OEM
46-722044-3	OEM
46-722047-3	OEM
46-722050-3	OEM
46-722051-3	OEM
46-722053-3	OEM
46-722054-3	OEM
46-722057-3	OEM
46-722059-3	OEM
46-722060-3	OEM
46-722062-3	OEM
46-722065-3	OEM
46-722066-3	1122
46-722068-3	OEM
46-722070-3	OEM
46-722071-3	OEM
46-722072-3	OEM
46-722073-3	OEM
46-722074-3	OEM
46-722076-3	OEM
46-722079-3	OEM
46-722080-3	OEM
46-722083-3	OEM
46-722086-3	OEM
46-722088-3	OEM
46-722089-3	OEM
46-722090-3	OEM
46-722091-3	OEM
46-722092-3	OEM
46-722093-3	OEM
46-722095-3	OEM
46-722096-3	OEM
46-722097-3	OEM
46-722098-3	OEM
46-722099-3	OEM
46-722100-3	OEM
46-722101-3	OEM
46-722104-3	OEM
46-722105-3	OEM
46-722106-3	OEM
46-722107-3	OEM
46-722108-3	OEM
46-722111-3	OEM
46-722112-3	OEM
46-722113-3	OEM
46-722114-3	OEM
46-722115-3	OEM
46-722116-3	OEM
46-722118-3	OEM
46-722121-3	OEM
46-722122-3	OEM
46-722123-3	OEM
46-722124-3	OEM
46-722126-3	OEM
46-722128-3	OEM
46-722129-3	OEM
46-722130-3	OEM
46-722131-3	OEM
46-722133-3	OEM
46-722134-3	OEM
46-722135-3	OEM
46-722420-3	OEM
46-722421-3	OEM
46-722540-3	OEM
46-722543-3	OEM
46-722547-3	OEM
46-722608-3	OEM
46-722633-3	OEM
46-750033-3	OEM
46-750034-3	OEM
46-750035-3	OEM
46-750036-3	OEM
46-750037-3	OEM
46-750038-3	OEM
46-750052-3	OEM
46-750055-3	OEM
46-750058-3	OEM
46-751270-3	OEM
46-751271-3	OEM
46-751633-3	OEM
46-751634-3	OEM
46-751652-3	OEM
46-751656-3	OEM
46-751657-3	OEM
46-751658-3	OEM
46-751659-3	OEM
46-751660-3	OEM
46-752091-3	OEM
46-752111-3	OEM
46-752113-3	OEM
46-752116-3	OEM
46-752117-3	OEM
46-752182-3	OEM
46-752183-3	OEM
46-752185-3	OEM
46-752189-3	OEM
46-753126-3	OEM
46-753127-3	OEM
46-753128-3	OEM
46-753129-3	OEM
46-753134-3	OEM
46-753135-3	OEM
46-753138-3	OEM
46-754471-3	OEM
46-754472-3	OEM
46-754547-3	OEM
46-754548-3	OEM
46-754549-3	OEM
46-754550-3	OEM
46-754552-3	OEM
46-754557-3	OEM
46-754559-3	OEM
46-754560-3	OEM
46-754561-3	OEM
46-754562-3	OEM
46-754564-3	OEM
46-754565-3	OEM
46-754566-3	OEM
46-754567-3	OEM
46-754568-3	OEM
46-754569-3	OEM
46-754570-3	OEM
46-754571-3	OEM
46-754572-3	OEM
46-754573-3	OEM
46-754574-3	OEM
46-754575-3	OEM
46-754577-3	OEM
46-754579-3	OEM
46-754580-3	OEM
46-754581-3	OEM
46-754583-3	OEM
46-754584-3	OEM
46-754586-3	OEM
46-754587-3	OEM
46-754588-3	OEM
46-754589-3	OEM
46-754590-3	OEM
46-754591-3	OEM
46-754592-3	OEM
46-754593-3	OEM
46-754594-3	OEM
46-754595-3	OEM
46-754596-3	OEM
46-754599-3	OEM
46-754601-3	OEM
46-754602-3	OEM
46-754603-3	OEM
46-754605-3	OEM
46-754606-3	OEM
46-754607-3	OEM
46-754608-3	OEM
46-754610-3	OEM
46-754611-3	OEM
46-754612-3	OEM
46-754650-3	OEM
46-754661-3	OEM
46-754709-3	OEM
46-754710-3	OEM
46-754711-3	OEM
46-754712-3	OEM
46-754713-3	OEM
46-754714-3	OEM
46-754715-3	OEM
46-756047-3	OEM
46-756054-3	OEM
46-756055-3	OEM
46-756065-3	OEM
46-756085-3	OEM
46-756118-3	OEM
46-756127-3	OEM
46-756142-3	OEM
46-756148-3	OEM
46-756149-3	OEM
46-756157-3	OEM
46-756384-3	OEM
46-756385-3	OEM
46-756386-3	OEM
46-756388-3	OEM
46-756389-3	OEM
46-756390-3	OEM
46-756391-3	OEM
46-756392-3	OEM
46-756410-3	OEM
46-756411-3	OEM
46-756413-3	OEM
46-756414-3	OEM
46-756421-3	OEM
46-756484-3	OEM
46-756525-3	OEM
46-756527-3	OEM
46-756528-3	OEM
46-756529-3	OEM
46-756530-3	OEM
46-756533-3	OEM
46-756534-3	OEM
46-756688-3	OEM
46-756690-3	OEM
46-801344-3	OEM
46-801386-3	OEM
46-852644-3	OEM
46-861000-3	OEM
46-861002-3	OEM
46-861005-3	OEM
46-861007-3	OEM
46-861008-3	OEM
46-861009-3	1369
46-861010-3	0728
46-861011-3	0563
46-861012-3	0525
46-861014-3	5378
46-861015-3	0502
46-861016-3	0087
46-861017-3	0955
46-861018-3	0955
46-861019-3	2668
46-861020-3	0261
46-861021-3	0130
46-861022-3	0023
46-861023-3	0392
46-861024-3	OEM
46-861025-3	3880
46-861027-3	3094
46-861028-3	OEM
46-861029-3	OEM
46-861030-3	OEM
46-861031-3	1911
46-861032-3	OEM
46-861034-3	OEM
46-861036-3	OEM
46-861037-2	0286
46-861037-3	3302
46-861038-3	0301
46-861039-3	1284
46-861040-3	1911
46-861041-3	0023
46-861042-3	2912
46-861043-3	0261
46-861044-3	OEM
46-861045-3	0471
46-861046-3	0057
46-861048-3	OEM
46-861049-3	OEM
46-861051-3	OEM
46-861053-3	OEM
46-861054-3	OEM
46-861055-3	OEM
46-861056-3	OEM
46-861057-3	OEM
46-861058-3	OEM
46-861059-3	OEM
46-861060-3	OEM
46-861061-3	OEM
46-861062-3	OEM
46-861063-3	OEM
46-861064-3	OEM
46-861065-3	OEM
46-861066-3	OEM
46-861067-3	0365
46-861068-3	OEM
46-861070-3	OEM
46-861071-3	OEM
46-861072-3	OEM
46-861074-3	OEM
46-861075-3	OEM
46-861076-3	OEM
46-861077-3	1165
46-861078-3	OEM
46-861080-3	OEM
46-861084-3	OEM
46-861086-3	OEM
46-861087-3	OEM
46-861088-3	OEM
46-861089-3	0140
46-861090-3	OEM
46-861091-3	OEM
46-861092-3	OEM
46-861094-3	0166
46-861095-3	OEM
46-861096-3	3046
46-861097-3	2069
46-861098-3	OEM
46-861102-3	OEM
46-861103-3	OEM
46-861107-3	OEM
46-861111-3	0012
46-861113-3	0253
46-861114-3	2910
46-861114-3(DIODE)	0283
46-861115-3	0724
46-861116-3	0077
46-861117-3	OEM
46-861118-3	0170
46-861119-3	0726
46-861120-3	OEM
46-861122-3	OEM
46-861123-3	OEM
46-861124-3	OEM
46-861125-3	OEM
46-861126-3	OEM
46-861128-3	OEM
46-861129-3	OEM
46-861130-3	OEM
46-861131-3	OEM
46-861132-3	OEM
46-861133-3	OEM
46-861134-3	OEM
46-861135-3	0686
46-861136-3	OEM
46-861137-3	OEM
46-861139-3	OEM
46-861140-3	OEM
46-861141-3	0195
46-861142-3	0283
46-861144-3	OEM
46-861145-3	0257
46-861146-3	OEM
46-861148-3	3654
46-861149-3	0102
46-861149-9	OEM
46-861151-3	0276
46-861153-3	OEM
46-861154-3	OEM
46-861155-3	OEM
46-861156-3	OEM
46-861157-3	OEM
46-861158-3	0064
46-861159-3	0057
46-861160-3	OEM
46-861161-3	2703
46-861162-3	OEM
46-861163-3	OEM
46-861166-3	OEM
46-861168-3	OEM
46-861169-3	OEM
46-861170-3	0057
46-861171-3	OEM
46-861172-3	0064
46-861173-3	OEM
46-861174-3	2542
46-861175-3	3688
46-861176-3	0181
46-861178-3	2848
46-861179-3	0041
46-861180-3	OEM
46-861181-3	OEM
46-861182-3	OEM
46-861183-9	OEM
46-861184-3	OEM
46-861185-3	OEM
46-861186-3	0471
46-861187-3	0396
46-861188-3	OEM
46-861189-3	OEM
46-861190-3	OEM
46-861191-3	OEM
46-861192-3	OEM
46-861193-3	OEM
46-861194-3	OEM
46-861195-3	OEM
46-861196-3	OEM
46-861197-3	OEM
46-861199-3	OEM
46-861200-3	1131
46-861201-3	OEM
46-861202-3	OEM
46-861203-3	2536
46-861204-3	OEM
46-861205-3	OEM
46-861206-9	OEM
46-861207-3	OEM
46-861209-3	2989
46-861210-3	OEM
46-861211-3	0700
46-861212-3	0434
46-861214-3	OEM
46-861215-3	OEM
46-861216-3	1541
46-861217-3	OEM
46-861218-3	OEM
46-861219-3	OEM
46-861220-3	OEM
46-861222-3	2427
46-861222-9	2427
46-861223-3	OEM
46-861224-3	OEM
46-861225-3	OEM
46-861226-3	OEM
46-861227-3	OEM
46-861228-3	OEM
46-861229-3	OEM
46-861230-3	OEM
46-861231-3	0018
46-861232-3	OEM
46-861233-3	OEM
46-861235-3	OEM
46-861236-3	OEM
46-861237-3	OEM
46-861238-3	OEM
46-861239-3	OEM
46-861240-3	OEM
46-861241-3	OEM
46-861243-3	OEM
46-861244-3	OEM
46-861245-3	0898
46-861246-3	OEM
46-861247-3	OEM
46-861248-3	OEM
46-861249-3	OEM
46-861250-3	OEM
46-861251-3	OEM
46-861252-3	4404
46-861253-3	1541
46-861254-3	OEM
46-861255-3	OEM
46-861256-3	0516
46-861257-3	OEM
46-861258-3	OEM
46-861259-3	OEM
46-861261-3	OEM
46-861262-3	OEM
46-861263-3	OEM
46-861265-3	0023
46-861266-3	4877
46-861267-3	OEM
46-861268-3	OEM
46-861269-3	3491
46-861270-3	OEM
46-861271-3	1541
46-861272-3	0276
46-861274-3	0124
46-861275-3	OEM
46-861276-3	OEM
46-861277-3	OEM
46-861278-3	OEM
46-861279-3	OEM
46-861280-3	OEM
46-861281-3	0162
46-861282-3	OEM
46-861283-3	OEM
46-861284-3	OEM
46-861285-3	0253
46-861286-3	OEM
46-861287-3	OEM
46-861288-3	OEM
46-861289-3	2185
46-861290-3	5132
46-861291-3	OEM
46-861292-3	OEM
46-861293-3	OEM
46-861294-3	0062
46-861295-3	2628
46-861296-3	5420
46-861297-3	5449
46-861298-3	5449
46-861299-3	0148
46-861300-3	OEM
46-861301-3	0887
46-861302-3	OEM
46-861303-3	OEM
46-861304-3	0071
46-861305-3	OEM
46-861306-3	1978
46-861307-3	OEM
46-861308-3	OEM
46-861309-3	OEM
46-861310-3	OEM
46-861311-3	OEM
46-861312-3	OEM
46-861313-3	OEM
46-861314-3	OEM
46-861315-3	OEM
46-861316-3	OEM
46-861317-3	0705
46-861318-3	OEM
46-861319-3	OEM
46-861320-3	OEM
46-861321-3	OEM
46-861323-3	OEM
46-861324-3	OEM
46-861325-3	OEM
46-861326-3	OEM
46-861327-3	1541
46-861328-3	OEM
46-861329-3	OEM
46-861330-3	OEM
46-861331-3	OEM
46-861332-3	0253
46-861333-3	1005
46-861336-3	5989
46-861337-3	OEM
46-861338-3	OEM
46-861339-3	OEM
46-861340-3	0053
46-861341-3	OEM
46-861342-3	OEM
46-861343-3	OEM
46-861344-3	5586
46-861345-3	1032
46-861346-3	6054
46-861347-3	OEM
46-861348-3	OEM
46-861349-3	OEM
46-861350-3	OEM
46-861352-3	OEM
46-861353-3	OEM
46-861354-3	OEM
46-861355-3	OEM
46-861356-3	OEM
46-861357-3	OEM
46-861359-3	OEM
46-861360-3	OEM
46-861361-3	OEM
46-861362-3	4468
46-861363-3	OEM
46-861365-3	0832
46-861366-3	OEM
46-861367-3	0485
46-861368-3	OEM
46-861369-3	OEM
46-861370-3	OEM
46-861371-3	OEM
46-861372-3	OEM
46-861373-3	OEM
46-861374-3	OEM
46-861375-3	OEM
46-861376-3	OEM
46-861378-3	OEM
46-861379-3	1032
46-861380-3	OEM
46-861381-1	0052
46-861381-3	0053
46-861382-3	0166
46-861383-3	OEM
46-861384-3	OEM
46-861385-3	OEM
46-861386-3	OEM
46-861387-3	OEM
46-861388-3	OEM
46-861389-9	OEM
46-861390-3	OEM
46-861391-3	OEM
46-861392-3	OEM
46-861393-3	OEM
46-861394-3	OEM
46-861395-3	OEM
46-861396-3	0317
46-861397-3	0677
46-861398-3	OEM
46-861401-3	OEM
46-861402-3	OEM
46-861403-3	OEM
46-861404-3	OEM
46-861405-3	OEM
46-861406-3	OEM
46-861407-3	0155
46-861408-3	OEM
46-861409-3	OEM
46-861410-3	0071
46-861411-3	OEM
46-861412-3	0286
46-861413-3	1541
46-861414-3	OEM
46-861415-3	OEM
46-861416-3	OEM
46-861417-3	OEM
46-861418-3	OEM
46-861419-3	OEM
46-861420-3	OEM
46-861421-3	OEM
46-861422-3	1541
46-861423-3	OEM
46-861424-3	OEM
46-861425-3	OEM
46-861426-3	OEM
46-861427-3	OEM
46-861428-3	OEM
46-861429-3	OEM
46-861430-3	OEM
46-861431-3	OEM
46-861432-3	OEM
46-861433-3	OEM
46-861434-3	OEM
46-861435-3	OEM
46-861436-3	OEM
46-861437-3	OEM
46-861438-3	OEM
46-861439-3	0151
46-861440-3	OEM
46-861441-3	OEM
46-861442-3	OEM
46-861443-3	OEM
46-861444-3	OEM
46-861445-3	0342
46-861446-3	0261
46-861447-3	OEM
46-861448-3	OEM
46-861449-3	OEM
46-861450-3	OEM
46-861451-3	OEM
46-861452-3	OEM
46-861453-3	0556
46-861454-3	OEM
46-861455-3	OEM
46-861456-3	OEM
46-861457-3	OEM
46-861458-3	OEM
46-861459-3	1083
46-861460-3	OEM
46-861461-3	5695
46-861462-3	3185
46-861463-3	1562
46-861464-3	3360
46-861465-3	0854
46-861466-3	0053
46-861467-3	0023
46-861468-3	OEM
46-861469-3	OEM
46-861470-3	OEM
46-861471-3	OEM
46-861472-3	OEM
46-861473-3	OEM
46-861474-3	OEM
46-861475-3	0556
46-861476-3	OEM
46-861477-3	OEM
46-861478-3	OEM
46-861479-3	OEM
46-861480-3	OEM
46-861481-3	OEM
46-861482-3	OEM
46-861483-3	OEM
46-861484-3	OEM
46-861485-3	OEM
46-861486-3	OEM
46-861487-3	OEM
46-861490-3	1522
46-861491-3	OEM
46-861492-3	OEM
46-861493-3	OEM
46-861494-3	OEM
46-861495-3	OEM
46-861496-3	OEM
46-861497-3	OEM
46-861498-3	OEM
46-861499-3	OEM
46-861500-3	0140
46-861501-3	OEM
46-861502-3	OEM
46-861503-3	OEM
46-861504-3	0671
46-861505-3	OEM
46-861506-3	OEM
46-861507-3	OEM
46-861508-3	0036
46-861510-3	OEM
46-861511-3	OEM
46-861513-3	OEM
46-861514-3	OEM
46-861515-3	OEM
46-861516-3	OEM
46-861517-3	OEM
46-861518-3	OEM
46-861519-3	OEM
46-861520-3	OEM
46-861521-3	OEM
46-861522-3	OEM
46-861523-3	OEM
46-861524-3	OEM
46-861525-3	OEM
46-861527-3	OEM
46-861529-3	OEM
46-861530-3	OEM
46-861531-3	OEM
46-861532-3	OEM
46-861533-3	OEM
46-861534-3	OEM
46-861535-3	OEM
46-861536-3	OEM
46-861537-3	OEM
46-861538-3	0485
46-861540-3	OEM
46-861543-3	OEM
46-861544-3	OEM
46-861545-3	2851
46-861546-3	OEM
46-861547-3	0181
46-861548-3	OEM
46-861549-3	OEM
46-861550-3	OEM
46-861551-3	0376
46-861552-3	OEM
46-861553-3	OEM
46-861554-3	OEM
46-861555-3	0077
46-861556-3	OEM
46-861557-3	OEM
46-861558-3	OEM
46-861559-3	OEM
46-861560-3	0907
46-861561-3	OEM
46-861562-3	OEM
46-861565-3	OEM
46-861566-3	0907
46-861567-3	OEM
46-861568-3	OEM
46-861569-3	OEM
46-861570-3	OEM
46-861571-3	OEM
46-861572-3	OEM
46-861573-3	OEM
46-861574-3	0724
46-861575-3	OEM
46-861576-3	OEM
46-861577-3	OEM
46-861578-3	OEM
46-861579-3	OEM
46-861580-3	OEM
46-861581-3	OEM
46-861582-3	OEM
46-861583-3	OEM
46-861584-3	OEM
46-861585-3	OEM
46-861586-3	OEM
46-861587-3	OEM
46-861588-3	OEM
46-861589-3	OEM
46-861590-3	OEM
46-861592-3	OEM
46-861593-3	OEM
46-861594-3	OEM
46-861596-3	OEM
46-861597-3	OEM
46-861598-3	OEM
46-861600-3	1733
46-861603-3	OEM
46-861604-3	OEM
46-861605-3	0525
46-861606-1	5556
46-861606-3	0563
46-861607-3	OEM
46-861608-3	OEM
46-861609-3	OEM
46-861610-3	OEM
46-861611-3	OEM
46-861612-3	OEM
46-861613-3	OEM
46-861614-3	OEM
46-861615-3	OEM
46-861617-3	OEM
46-861618-3	OEM
46-861619-3	OEM
46-861621-3	OEM
46-861622-3	OEM
46-861627-3	1541
46-861628-3	OEM
46-861629-3	OEM
46-861630-3	OEM
46-861631-3	OEM
46-861632-3	OEM

If replacement code is OEM, contact original manufacturer for replacement.

DEVICE TYPE	REPL CODE
46-861633-3	1875
46-861634-3	0023
46-861636-3	OEM
46-861637-3	OEM
46-861639-3	OEM
46-861642-3	OEM
46-861643-3	OEM
46-861644-3	OEM
46-861645-3	3192
46-861646-3	0729
46-861647-3	5586
46-861648-3	1032
46-861649-3	2129
46-861650-3	OEM
46-861651-3	OEM
46-861652-3	6104
46-861653-3	OEM
46-861654-3	0124
46-861657-3	OEM
46-861659-3	OEM
46-861660-3	OEM
46-861661-3	OEM
46-861662-3	OEM
46-861664-3	OEM
46-861665-3	OEM
46-861666-3	OEM
46-861667-3	OEM
46-861668-3	OEM
46-861669-3	OEM
46-861672-3	OEM
46-861673-3	OEM
46-861677-3	OEM
46-861678-3	OEM
46-861680-3	0276
46-861684-3	1386
46-861685-3	OEM
46-861686-3	OEM
46-861687-3	OEM
46-861691-3	0009
46-861692-3	OEM
46-861693-3	OEM
46-861694-3	OEM
46-861695-3	OEM
46-861696-3	OEM
46-861699-3	0446
46-861700-3	OEM
46-861701-3	OEM
46-861702-3	OEM
46-861703-3	OEM
46-861704-3	OEM
46-861705-3	OEM
46-861707-3	OEM
46-861708-3	OEM
46-861709-3	OEM
46-861711-3	OEM
46-861712-3	0261
46-861713-3	OEM
46-861714-3	OEM
46-861716-3	OEM
46-861717-3	0023
46-861718-3	0031
46-861720-3	0376
46-861721-3	OEM
46-861722-3	OEM
46-861723-3	OEM
46-861724-3	OEM
46-861725-3	OEM
46-861728-3	OEM
46-861729-3	OEM
46-861730-3	OEM
46-861734-3	OEM
46-861735-3	OEM
46-861737-3	0795
46-861738-3	0407
46-861739-3	3305
46-861741-3	OEM
46-861744-3	OEM
46-861745-3	OEM
46-861747-3	OEM
46-861748-3	OEM
46-861752-3	OEM
46-861753-3	OEM
46-861754-3	OEM
46-861756-3	OEM
46-861757-3	OEM
46-861758-3	OEM
46-861759-3	OEM
46-861761-3	OEM
46-861762-3	OEM
46-861763-3	OEM
46-861767-3	OEM
46-861768-3	OEM
46-861773-3	OEM
46-861776-3	OEM
46-861777-3	OEM
46-861780-3	OEM
46-861781-3	OEM
46-861783-3	OEM
46-861786-3	OEM
46-861787-3	OEM
46-861788-3	OEM
46-861789-3	OEM
46-861790-3	OEM
46-861791-3	OEM
46-861801-3	OEM
46-861803-3	OEM
46-861807-3	OEM
46-861810-3	OEM
46-861811-3	OEM
46-861812-3	OEM
46-861816-3	OEM
46-861817-3	OEM
46-861825-3	OEM
46-861826-3	OEM
46-861829-3	OEM
46-861832-3	OEM
46-861833-3	OEM
46-861834-3	OEM
46-861836-3	OEM
46-861837-3	OEM
46-861838-3	OEM
46-861839-3	OEM
46-861840-3	OEM
46-861842-3	OEM
46-861843-3	OEM
46-861845-3	OEM
46-861851-3	0042
46-861854-3	OEM
46-861857-3	0042
46-861861-3	1189
46-861862-3	OEM
46-861863-3	OEM
46-861864-3	4299
46-861865-3	OEM
46-861866-3	OEM
46-861867-3	2323
46-861868-3	2478
46-861869-3	1276
46-861870-3	0879
46-861871-3	OEM
46-861872-3	OEM
46-861874-3	OEM
46-861875-3	OEM
46-861878-3	OEM
46-861883-3	OEM
46-861884-3	OEM
46-861888-3	OEM
46-861889-3	OEM
46-861890-3	OEM
46-861891-3	OEM
46-861893-3	OEM
46-861895-3	OEM
46-861896-3	OEM
46-861897-3	OEM
46-861898-3	OEM
46-861899-3	OEM
46-861902-3	OEM
46-861904-3	OEM
46-861906-3	OEM
46-861907-3	OEM
46-861908-3	OEM
46-861909-3	OEM
46-861911-3	OEM
46-861912-3	0339
46-861913-3	0589
46-861914-3	OEM
46-861915-3	OEM
46-861916-3	OEM
46-861917-3	OEM
46-861918-3	OEM
46-861919-3	OEM
46-861920-3	OEM
46-861923-3	OEM
46-861924-3	0062
46-861925-3	OEM
46-861926-3	OEM
46-861927-3	OEM
46-861930-3	OEM
46-861931-3	OEM
46-861932-3	OEM
46-861933-3	OEM
46-861934-3	OEM
46-861935-3	OEM
46-861936-3	OEM
46-861937-3	OEM
46-861938-3	OEM
46-861940-3	OEM
46-861942-3	OEM
46-861943-3	OEM
46-861944-3	1386
46-861945-3	OEM
46-861946-3	OEM
46-861948-3	OEM
46-861949-3	OEM
46-861950-3	OEM
46-861951-3	OEM
46-861952-3	OEM
46-861954-3	OEM
46-861955-3	OEM
46-861957-3	OEM
46-861958-3	OEM
46-861959-3	OEM
46-861961-3	OEM
46-861962-3	OEM
46-861963-3	OEM
46-861964-3	OEM
46-861966-3	OEM
46-861967-3	OEM
46-861968-3	OEM
46-861969-3	OEM
46-861970-3	OEM
46-861971-3	OEM
46-861981-3	OEM
46-861982-3	OEM
46-861983-3	OEM
46-861984-3	OEM
46-861986-3	OEM
46-861987-3	OEM
46-861988-3	OEM
46-861989-3	OEM
46-861990-3	OEM
46-861991-3	OEM
46-861992-3	OEM
46-861993-3	OEM
46-861994-3	OEM
46-861995-3	OEM
46-861997-3	OEM
46-861998-3	OEM
46-861999-3	OEM
46-862000-3	OEM
46-862001-3	OEM
46-862002-3	4767
46-862003-3	OEM
46-862004-3	OEM
46-862007-3	OEM
46-862009-3	OEM
46-862010-3	OEM
46-862012-3	OEM
46-862013-3	OEM
46-862014-3	OEM
46-862015-3	OEM
46-862016-3	OEM
46-862017-3	OEM
46-862019-3	OEM
46-862020-3	OEM
46-862021-3	OEM
46-862022-3	OEM
46-862023-3	OEM
46-862024-3	OEM
46-862025-3	OEM
46-862026-3	OEM
46-862027-3	OEM
46-862028-3	OEM
46-862029-3	OEM
46-862030-3	OEM
46-862031-3	OEM
46-862033-3	OEM
46-862034-3	OEM
46-862035-3	OEM
46-862036-3	OEM
46-862037-3	OEM
46-862038-3	OEM
46-862039-3	OEM
46-862040-3	OEM
46-862041-3	OEM
46-862042-3	OEM
46-862043-3	OEM
46-862045-3	0077
46-862046-3	OEM
46-862047-3	OEM
46-862048-3	OEM
46-862049-3	OEM
46-862055-3	OEM
46-862057-3	OEM
46-862058-3	OEM
46-862066-3	OEM
46-862067-3	OEM
46-862068-3	0155
46-862069-3	OEM
46-862070-3	OEM
46-862072-3	OEM
46-862073-3	OEM
46-862074-3	OEM
46-862075-3	OEM
46-862077-3	OEM
46-862078-3	OEM
46-862079-3	OEM
46-862080-3	OEM
46-862081-3	5420
46-862083-3	OEM
46-862087-3	OEM
46-862088-3	OEM
46-862089-3	0166
46-862090-3	OEM
46-862091-3	OEM
46-862092-3	1506
46-862094-3	OEM
46-862096-3	OEM
46-862097-3	1541
46-862098-3	OEM
46-862099-3	OEM
46-862100-3	1541
46-862103-3	OEM
46-862107-3	OEM
46-862108-3	OEM
46-862109-3	OEM
46-862111-3	OEM
46-862112-3	0031
46-862113-3	OEM
46-862116-3	OEM
46-862118-3	OEM
46-862119-3	OEM
46-862120-3	OEM
46-862121-3	0925
46-862122-3	OEM
46-862124-3	OEM
46-862125-3	OEM
46-862126-3	OEM
46-862129-3	OEM
46-862131-3	OEM
46-862132-3	OEM
46-862133-3	OEM
46-862134-3	OEM
46-862135-3	OEM
46-862136-3	OEM
46-862138-3	OEM
46-862139-3	OEM
46-862140-3	OEM
46-862142-3	OEM
46-862145-3	OEM
46-862146-3	OEM
46-862151-3	OEM
46-862152-3	0546
46-862153-3	0017
46-862155-3	OEM
46-862156-3	OEM
46-862157	OEM
46-862157-3	4052
46-862158-3	OEM
46-862159-3	OEM
46-862161-3	OEM
46-862162-3	OEM
46-862163-3	OEM
46-862164-3	OEM
46-862165-3	OEM
46-862166-3	OEM
46-862167-3	0102
46-862168-3	0023
46-862169-3	OEM
46-862170-3	2129
46-862173-3	OEM
46-862176-3	OEM
46-862177-3	OEM
46-862178-3	OEM
46-862179-3	OEM
46-862180-3	OEM
46-862181-3	OEM
46-862182-3	OEM
46-862183-3	OEM
46-862184-3	OEM
46-862186-3	OEM
46-862187-3	OEM
46-862188-3	OEM
46-862189-3	OEM
46-862191-3	OEM
46-862192-3	OEM
46-862193-3	OEM
46-862194-3	OEM
46-862195-3	OEM
46-862196-3	OEM
46-862209-3	OEM
46-862211-3	OEM
46-862212-3	OEM
46-862217-3	OEM
46-862219-3	OEM
46-862220-3	OEM
46-862221-3	OEM
46-862223-3	OEM
46-862224-3	OEM
46-862227-3	OEM
46-862228-3	OEM
46-862229-3	OEM
46-862230-3	OEM
46-862232-3	OEM
46-862233-3	OEM
46-862234-3	OEM
46-862236-3	OEM
46-862237-3	OEM
46-862238-3	OEM
46-862239-3	OEM
46-862240-3	OEM
46-862241-3	OEM
46-862244-3	OEM
46-862245-3	OEM
46-862248-3	OEM
46-862249-3	OEM
46-862253-3	OEM
46-862254-3	OEM
46-862255-3	OEM
46-862257-3	OEM
46-862259-3	OEM
46-862260-3	OEM
46-862262-3	OEM
46-862269-3	OEM
46-862270-3	OEM
46-862271-3	OEM
46-862276-3	OEM
46-862277-3	OEM
46-862282-3	OEM
46-862283-3	OEM
46-862286-3	OEM
46-862287-3	OEM
46-862288-3	OEM
46-862289-3	0079
46-862290-3	0168
46-862291-3	0643
46-862292-3	OEM
46-862293-3	OEM
46-862294-3	OEM
46-862295-3	1541
46-862297-3	OEM
46-862298-3	OEM
46-862299-3	OEM
46-862300-3	OEM
46-862301-3	0023
46-862302-3	OEM
46-862303-3	OEM
46-862304-3	OEM
46-862305-3	OEM
46-862306-3	OEM
46-862307-3	OEM
46-862308-3	OEM
46-862309-3	OEM
46-862318-3	0275
46-862321-3	1045
46-862322-3	0031
46-862323-3	0031
46-862325-3	OEM
46-862326-3	OEM
46-862327-3	OEM
46-862328-3	0276
46-862329-3	5462
46-862330-3	OEM
46-862331-3	OEM
46-862332-3	OEM
46-862333-3	OEM
46-862336-3	0261
46-862337-3	OEM
46-862339-3	OEM
46-862340-3	OEM
46-862341-3	OEM
46-862342-3	OEM
46-862343-3	OEM
46-862345-3	OEM
46-862346-3	OEM
46-862347-3	OEM
46-862348-3	OEM
46-862349-3	OEM
46-862350-3	0140
46-862351-3	OEM
46-862352-3	OEM
46-862353-3	OEM
46-862354-3	OEM
46-862355-3	OEM
46-862356-3	OEM
46-862359-3	OEM
46-862360-3	OEM
46-862361-3	OEM
46-862362-3	OEM
46-862363-3	OEM
46-862366-3	OEM
46-862367-3	OEM
46-862368-3	OEM
46-862369-3	OEM
46-862370-3	OEM
46-862371-3	OEM
46-862372-3	OEM
46-862373-3	OEM
46-862374-3	OEM
46-862375-3	OEM
46-862377-3	OEM
46-862378-3	OEM
46-862379-3	OEM
46-862380-3	OEM
46-862381-3	OEM
46-862391-3	OEM
46-862392-3	OEM
46-862393-3	OEM
46-862394-3	OEM
46-862395-3	OEM
46-862396-3	OEM
46-862398-3	OEM
46-862399-3	OEM
46-862401-3	OEM
46-862402-3	OEM
46-862406-3	OEM
46-862407-3	OEM
46-862408-3	OEM
46-862409-3	0031
46-862410-3	0471
46-862411-3	3098
46-862412-3	OEM
46-862413-3	2185
46-862414-3	OEM
46-862415-3	OEM
46-862425-3	OEM
46-862427-3	OEM
46-862429-3	OEM
46-862435-3	OEM
46-862436-3	OEM
46-862437-3	OEM
46-862438-3	OEM
46-862439-3	OEM
46-862441-3	OEM
46-862442-3	OEM
46-862446-3	1500
46-862447-3	OEM
46-862448-3	OEM
46-862449-3	OEM
46-862450-3	OEM
46-862452-3	OEM
46-862453-3	OEM
46-862456-3	1005
46-862459-3	OEM
46-862460-3	OEM
46-862461-3	OEM
46-862462-3	OEM
46-862463-3	OEM
46-862464-3	OEM
46-862465-3	OEM
46-862466-3	OEM
46-862467-3	OEM
46-862469-3	OEM
46-862470-3	OEM
46-862471-3	OEM
46-862472-3	OEM
46-862473-3	OEM
46-862474-3	OEM
46-862476-3	OEM
46-862477-3	OEM
46-862478-3	OEM
46-862479-3	OEM
46-862480-3	OEM
46-862481-3	OEM
46-862482-3	OEM
46-862483-3	OEM
46-862484-3	OEM
46-862491-3	OEM
46-862493-3	OEM
46-862494-3	OEM
46-862495-3	OEM
46-862497-3	OEM
46-862498-3	OEM
46-862500-3	OEM
46-862503-3	OEM
46-862504-3	OEM
46-862505-3	OEM
46-862506-3	OEM
46-862507-3	OEM
46-862508-3	OEM
46-862509-3	OEM
46-862510-3	OEM
46-862512-3	OEM
46-862515-3	OEM
46-862516-3	5685
46-862517-3	OEM
46-862522-3	2185
46-862528-3	4945
46-862529-3	5901
46-862530-3	0563
46-862532-3	5952
46-862541-3	OEM
46-862544-3	OEM
46-862550-3	OEM
46-862552-3	OEM
46-862553-3	OEM
46-862557-3	0049
46-862559-3	2739
46-862564-3	OEM
46-862565-3	OEM
46-862566-3	OEM
46-862567-3	OEM
46-862568-3	OEM
46-862569-3	OEM
46-862571-3	OEM
46-862573-3	OEM
46-862574-3	OEM
46-862575-3	OEM
46-862576-3	OEM
46-862577-3	OEM
46-862578-3	OEM
46-862579-3	OEM
46-862580-3	OEM
46-862581-3	OEM
46-862582-3	1541
46-862585-3	OEM
46-862586-3	OEM
46-862587-3	OEM
46-862588-3	OEM
46-862589-3	OEM
46-862590-3	OEM
46-862591-3	OEM
46-862593-3	OEM
46-862594-3	OEM
46-862595-3	2185
46-862598-3	OEM
46-862599-3	OEM
46-862600-3	OEM
46-862601-3	OEM
46-862602-3	OEM
46-862603-3	OEM
46-862604-3	OEM
46-862605-3	OEM
46-862606-3	OEM
46-862607-3	OEM
46-862608-3	OEM
46-862609-3	OEM
46-862611-3	0002
46-862612-3	0052
46-862614-3	0253
46-862615-3	0062
46-862616-3	0041
46-862620-3	OEM
46-862621-3	OEM
46-862622-3	OEM
46-862623-3	OEM
46-862624-3	OEM
46-862625-3	OEM
46-862630-3	OEM
46-862631-3	OEM
46-862633-3	1875
46-862634-3	0064
46-862639-3	0165
46-862642-3	0052
46-862650-3	OEM
46-862651-3	0062
46-862652-3	0077
46-862653-3	0036
46-862654-3	OEM
46-862655-3	OEM
46-862656-3	OEM
46-862657-3	OEM
46-862658-3	OEM
46-862659-3	OEM
46-862660-3	OEM
46-862661-3	OEM
46-862662-3	OEM
46-862663-3	OEM
46-862664-3	OEM
46-862665-3	OEM
46-862666-3	OEM
46-862667-3	OEM
46-862668-3	OEM
46-862669-3	OEM
46-862670-3	OEM
46-862671-3	OEM
46-862672-3	OEM
46-862673-3	OEM
46-862674-3	OEM
46-862676-3	OEM
46-862677-3	OEM
46-862678-3	OEM
46-862679-3	OEM
46-862680-3	OEM
46-862681-3	OEM
46-862682-3	OEM
46-862683-3	OEM
46-862684-3	OEM
46-862685-3	OEM
46-862686-3	OEM
46-862687-3	OEM
46-862688-3	OEM
46-862689-3	OEM
46-862690-3	OEM
46-862691-3	OEM
46-862693-3	OEM
46-862694-3	OEM
46-862695-3	OEM
46-862696-3	OEM
46-862697-3	OEM
46-862764-3	OEM
46-862770-3	OEM
46-862771-3	OEM
46-862772-3	OEM
46-862775-3	OEM
46-862776-3	OEM
46-862777-3	OEM
46-862778-3	OEM
46-862782-3	OEM
46-862784-3	OEM
46-862786-3	OEM
46-862787-3	OEM
46-862788-3	OEM
46-862791-3	OEM
46-862794-3	OEM
46-862795-3	OEM
46-862796-3	0195
46-862857-3	6153
46-862858-3	3494
46-862994-3	0031
46-863230-3	1145
46-863232-3	OEM
46-863233-3	OEM
46-863234-3	0037
46-863235-3	0079
46-863236-3	OEM
46-863237-3	OEM
46-863238-3	0261
46-863239-3	0079
46-863240-3	0037
46-863241-3	5034
46-863242-3	1597
46-866209	0313
46-961037-3	OEM
46-8691025-3	OEM
46AR1	0015
46AR2	0015
46AR3	0015
46AR4	0015
46AR5	0015
46AR6	0015
46AR7	0015
46AR8	0015
46AR9	0015
46AR10	0015
46AR11	0015
46AR12	0015
46AR13	0071
46AR15	0071
46AR16	0071
46AR18	0071
46AR21	0015
46AR27	0015
46AR28	0015
46AR29	0015
46AR35	0015
46AR50	0071
46AR52	0071
46AR59	0071
46AX1	0015
46AX2	0015
46AX3	0015
46AX4	0015
46AX5	0015
46AX7	0015
46AX8	0015
46AX9	0015
46AX10	0015
46AX11	0015
46AX12	0071
46AX13	0071
46AX14	0071
46AX16	0015
46AX17	0015
46AX19	0015
46AX21	0015
46AX30	0015
46AX34	0015
46AX52	0015
46AX54	0015
46AX55	0071
46AX56	0071
46AX59	0071

If replacement code is OEM, contact original manufacturer for replacement.

DEVICE TYPE	REPL CODE
46AX70	0071
46AX82	0071
46AX84	0071
46AX85	0071
46B-3A5	0265
46BD1	0015
46BD2	0015
46BD5	0015
46BD8	0015
46BD9	0015
46BD11	0015
46BD12	0015
46BD14	0015
46BD19	0015
46BD25	0015
46BD27	0071
46BD30	0071
46BD32	0071
46BD33	0071
46BD34	0071
46BD38	0071
46BD39	0071
46BD52	0071
46BD101	0071
46BR5	0015
46BR7	0015
46BR9	0015
46BR10	0015
46BR11	0015
46BR15	0015
46BR17	0015
46BR18	0071
46BR21	0071
46BR27	0071
46BR62	0071
46BR63	0071
46BR64	0071
46BR68	0071
46BX2	0015
46BX3	0015
46C-42	0144
46D5R03	OEM
46D9R03GH	OEM
46J	OEM
46K	OEM
46R2	0122
46R2R	0515
46R2S	0204
46RECT-SI-1002	0023
46SI-SAF1039P	0950
46ST-MPS9700E	0079
46ST-MPS9700F	0079
46TB1-CE00	OEM
46TB1-CH00	OEM
46TB1-EE00	OEM
46TB1-EH00	OEM
46TB1-JH00	OEM
46TB1-LE00	OEM
46TB1-LH00	OEM
46TB1-NE00	OEM
46TB1-NH00	OEM
46TB2-CE00	OEM
46TB2-CH00	OEM
46TB2-EE00	OEM
46TB2-EH00	OEM
46TB2-JH00	OEM
46TB2-LE00	OEM
46TB2-LH00	OEM
46TB2-NE00	OEM
46TB2-NH00	OEM
46TB3-CE00	OEM
46TB3-CH00	OEM
46TB3-EE00	OEM
46TB3-EH00	OEM
46TB3-JH00	OEM
46TB3-LE00	OEM
46TB3-LH00	OEM
46TB3-NE00	OEM
46TB3-NH00	OEM
46TB4-CE00	OEM
46TB4-CH00	OEM
46TB4-EE00	OEM
46TB4-EH00	OEM
46TB4-JH00	OEM
46TB4-LE00	OEM
46TB4-LH00	OEM
46TB4-NE00	OEM
46TB4-NH00	OEM
46TB5-CE00	OEM
46TB5-CH00	OEM
46TB5-EE00	OEM
46TB5-EH00	OEM
46TB5-GH00	OEM
46TB5-JH00	OEM
46TB5-LE00	OEM
46TB5-LH00	OEM
46TB5-NE00	OEM
46TB5-NH00	OEM
46TB6-CE00	OEM
46TB6-CH00	OEM
46TB6-EE00	OEM
46TB6-EH00	OEM
46TB6-GH00	OEM
46TB6-JH00	OEM
46TB6-LE00	OEM
46TB6-LH00	OEM
46TB6-NE00	OEM
46TB6-NH00	OEM
46TB7-JE00	OEM
46TB7-JH00	OEM
46TB7-LE00	OEM
46TB7-LH00	OEM
46TB7-NE00	OEM
46TB7-NH00	OEM
46TB7-RE00	OEM
46TB8-CHG0	OEM
46TB8-JE00	OEM
46TB8-JH00	OEM
46TB8-LE00	OEM
46TB8-LH00	OEM
46TB8-NE00	OEM
46TB8-NH00	OEM
46TB8-RE00	OEM
46TB9-JE00	OEM
46TB9-JH00	OEM
46TB9-LE00	OEM
46TB9-LH00	OEM
46TB9-NE00	OEM
46TB9-NH00	OEM
46TB9-RE00	OEM
46TB10-JE00	OEM
46TB10-JH00	OEM
46TB10-LE00	OEM
46TB10-LH00	OEM
46TB10-NE00	OEM
46TB10-NH00	OEM
46TB10-RE00	OEM
46TB11-JE00	OEM
46TB11-JH00	OEM
46TB11-LE00	OEM
46TB11-LH00	OEM
46TB11-NE00	OEM
46TB11-NH00	OEM
46TB11-RE00	OEM
46TB12-JE00	OEM
46TB12-JH00	OEM
46TB12-LE00	OEM
46TB12-LH00	OEM
46TB12-NH00	OEM
46TB12-RE00	OEM
46TB13	OEM
46TB13-JH00	OEM
46TB13-LE00	OEM
46TB13-LH00	OEM
46TB13-NE00	OEM
46TB13-NH00	OEM
46TB13-RE00	OEM
46TB14-JH00	OEM
46TB14-LE00	OEM
46TB14-LH00	OEM
46TB14-NE00	OEM
46TB14-NH00	OEM
46TB14-RE00	OEM
46TB15-JH00	OEM
46TB15-LE00	OEM
46TB15-LH00	OEM
46TB15-NE00	OEM
46TB15-NH00	OEM
46TB15-RE00	OEM
47-1	0917
47-2	0917
47-2(DIO)	0917
47-2(XSTR)	0127
47-4	0917
47-866882-3	OEM
47B	0999
47C23-3	0150
47C221AF4241	OEM
47C231AN4943	OEM
47C232AN4989	OEM
47C337ANR801	OEM
47C420AF8582	OEM
47C432AN-8088	OEM
47C432AN-8088Z	OEM
47C432AN-8098	OEM
47C432AN-8497	OEM
47C432AN8693	OEM
47C432AN-8734	OEM
47C432AN8939	OEM
47C432AN8989	OEM
47C434N-3509Z	OEM
47C634N2427	OEM
47C634N2438	OEM
47C634N2453	OEM
47C634N-2454	OEM
47C634N2454	OEM
47C670N1264	OEM
47C870N4627	OEM
47C1638NU313	OEM
47D5R03	OEM
47D9R03GH	OEM
47P1	0160
47R2	OEM
47S155275	0316
47TB1-CE00	OEM
47TB1-CH00	OEM
47TB1-EE00	OEM
47TB1-JE00	OEM
47TB1-JH00	OEM
47TB1-LE00	OEM
47TB1-LH00	OEM
47TB1-NE00	OEM
47TB1-NH00	OEM
47TB2-CE00	OEM
47TB2-CH00	OEM
47TB2-EE00	OEM
47TB2-EH00	OEM
47TB2-JE00	OEM
47TB2-JH00	OEM
47TB2-LE00	OEM
47TB2-LH00	OEM
47TB2-NE00	OEM
47TB2-NH00	OEM
47TB3-CE00	OEM
47TB3-CH00	OEM
47TB3-EE00	OEM
47TB3-EH00	OEM
47TB3-JH00	OEM
47TB3-LE00	OEM
47TB3-LH00	OEM
47TB3-NE00	OEM
47TB3-NH00	OEM
47TB4-CE00	OEM
47TB4-CH00	OEM
47TB4-EE00	OEM
47TB4-EH00	OEM
47TB4-JH00	OEM
47TB4-LE00	OEM
47TB4-LH00	OEM
47TB4-NE00	OEM
47TB4-NH00	OEM
47TB5-CE00	OEM
47TB5-CH00	OEM
47TB5-EE00	OEM
47TB5-EH00	OEM
47TB5-JH00	OEM
47TB5-LE00	OEM
47TB5-NE00	OEM
47TB5-NH00	OEM
47TB6-CE00	OEM
47TB6-CH00	OEM
47TB6-EE00	OEM
47TB6-EH00	OEM
47TB6-JH00	OEM
47TB6-LE00	OEM
47TB6-NE00	OEM
47TB6-NH00	OEM
47TB7	OEM
47TB7-CE00	OEM
47TB7-CH00	OEM
47TB7-EE00	OEM
47TB7-EH00	OEM
47TB7-JH00	OEM
47TB7-LE00	OEM
47TB7-NE00	OEM
47TB7-NH00	OEM
47TB8-CE00	OEM
47TB8-CH00	OEM
47TB8-EE00	OEM
47TB8-EH00	OEM
47TB8-JH00	OEM
47TB8-LE00	OEM
47TB8-NE00	OEM
47TB8-NH00	OEM
47TB9	OEM
47TB9-CE00	OEM
47TB9-CH00	OEM
47TB9-EE00	OEM
47TB9-EH00	OEM
47TB9-JH00	OEM
47TB9-LE00	OEM
47TB9-NE00	OEM
47TB9-NH00	OEM
47TB10-CE00	OEM
47TB10-CH00	OEM
47TB10-EE00	OEM
47TB10-EH00	OEM
47TB10-JH00	OEM
47TB10-LE00	OEM
47TB10-NE00	OEM
47TB10-NH00	OEM
47TB11	OEM
47TB11-CE00	OEM
47TB11-CH00	OEM
47TB11-EE00	OEM
47TB11-EH00	OEM
47TB11-JH00	OEM
47TB11-LE00	OEM
47TB11-NE00	OEM
47TB11-NH00	OEM
47TB12-CE00	OEM
47TB12-EE00	OEM
47TB12-EH00	OEM
47TB12-JH00	OEM
47TB12-LE00	OEM
47TB12-NE00	OEM
47TB12-NH00	OEM
47V10	OEM
47V11	OEM
47V12	OEM
47Z102536-P1	0133
48-00155077	0133
48-00155119	0155
48-00155128	0023
48-00155192	1005
48-00155222	0261
48-01	0015
48-01-041	1233
48-01-043	2170
48-01-048	1354
48-01-055	3572
48-01-094	1233
48-01+	0015
48-0305A01	OEM
48-03005A01	0133
48-03005A03	0143
48-03005A05	0133
48-03005A06	0124
48-03005A07	0015
48-03005A08	0030
48-03073A06	0244
48-03073A08	0931
48-03073A09	0012
48-03073A10	1276
48-21-0644-012	1198
48-35P1	0211
48-36P1	0211
48-36P3	0211
48-39P1	0160
48-39P3	0211
48-43P3	0211
48-43P4	0211
48-45N2	0111
48-56P1	0280
48-57B2	0160
48-57B42	0160
48-77	1030
48-110	0196
48-171-A06	0004
48-191A01	0015
48-191A01-9	0015
48-191A02	0015
48-191A03	0015
48-191A04	0015
48-191A05	0102
48-191A05A	0102
48-191A06	0015
48-191A07	0015
48-191A07A	0015
48-191A08	0015
48-191A09	0015
48-971-A95	0016
48-971A04	0144
48-971A05	0016
48-971A13	0142
48-971A203	0050
48-1005	0015
48-1050-SL12318	1823
48-1050-SL12319	1041
48-1050-SL12320	2097
48-1050-SL12321	1831
48-1050-SL12322	0033
48-1050-SL12323	0081
48-1050-SL12324	0141
48-1050-SL12325	0562
48-1050-SL12326	0358
48-1050-SL12327	0333
48-3003A02	0018
48-3003A03	0016
48-3003A04	0006
48-3003A05	0016
48-3003A06	0532
48-3003A09	0325
48-3003A10	0212
48-3003A11	0079
48-3003A12	0016
48-6712	0196
48-6712A02	0913
48-8213H07	0524
48-8213H08	0510
48-8375D02	0940
48-8613	0917
48-8613-3	0136
48-8619-3	0143
48-10001-A01	0015
48-10001-A03	0015
48-10001-A030-1	0015
48-10062A01	0015
48-10062A01A	0015
48-10062A02	0015
48-10062A04	0015
48-10062A05	0015
48-10062A05A	0015
48-10073A01	0004
48-10073A02	0211
48-10074A01	0004
48-10074A02	0211
48-10074A03	0004
48-10075A01	0265
48-10075A02	0265
48-10075A03	0265
48-10075A04	0265
48-10075A05	0265
48-10075A06	0265
48-10075A07	0265
48-10075A08	0265
48-10079A01	0050
48-10079A02	0050
48-10103A01	0265
48-10103A02	0265
48-10103A03	0265
48-10103A04	0265
48-10103A05	0265
48-10103A06	0085
48-10103A07	0265
48-10103A08	0265
48-10103A09	0265
48-10103A10	0265
48-10103A11	0085
48-10346A01	0162
48-10346A02	0920
48-10516C02	0030
48-10577A01	0133
48-10577A13	0133
48-12443	0004
48-13466	0038
48-13470	0144
48-13481	0016
48-13494A1G	0111
48-13707	0849
48-17162A06	0004
48-17162A10	0164
48-17162A13	0142
48-17162A17	0004
48-17162A22	0004
48-17271A03	0004
48-21598B01	0004
48-32000	0137
48-34816	0133
48-40004S05	2050
48-40004S06	0936
48-40118B01	0037
48-40118B01A	0037
48-40170-G01	0111
48-40170G01	0111
48-40171G01	0079
48-40172G01	0222
48-40235C02	OEM
48-40235G01	0015
48-40235G02	0023
48-40246-G01	0111
48-40246G01	0111
48-40246G02	0016
48-40247G01	0127
48-40247G02	0111
48-40458A04	0157
48-40458A06	0244
48-40458A064	0244
48-40516C01	0247
48-40516C02	0030
48-40734P01	0786
48-40738P01	0133
48-40739P01	0015
48-40764P01	2495
48-41508A01	0015
48-41508A02	0015
48-41763C01	0244
48-41763C02	0244
48-41763C03	0244
48-41768G01	0019
48-41784J03	1790
48-41784J04	1790
48-41785J03	0460
48-41785J04	0460
48-41815J02	0127
48-41816J01	0127
48-41816J02	0151
48-41873J02	0057
48-41873J03	0170
48-41884J03	0161
48-42098B01	0037
48-42098B01A	0037
48-42383A0	0030
48-42383A01	0030
48-42485B01	0030
48-42503P01	0005
48-42884P01	1591
48-42885P01	0936
48-42899J01	0139
48-43238G01	0037
48-43265G01	0015
48-43283A01	0030
48-43351A01	0004
48-43351A02	0127
48-43351A03	0144
48-43351A04	0144
48-43351A05	0144
48-43354A81	0016
48-43354A82	0016
48-43354A83	0164
48-43467J01	3046
48-43992J01	0224
48-44080J05	0012
48-44885G01	1220
48-44885G02	1220
48-44886G01	0155
48-45323G01	0720
48-47389C01	OEM
48-57120A01	0015
48-60022A13	0016
48-60022A14	0178
48-60022A97	0143
48-60022A98	0015
48-60077A06	0143
48-60154A01	0123
48-61074B01	0143
48-61767B01	0143
48-62334A01	0123
48-62334A02	0143
48-63005A66	0016
48-63005A72	0016
48-63006A56	0143
48-63026A45	0142
48-63026A46	0144
48-63026A47	0016
48-63026A48	0016
48-63029A16	0050
48-63029A17	0164
48-63029A18	0211
48-63029A19	0211
48-63029A20	0143
48-63029A60	0050
48-63029A90	0136
48-63029A91	0211
48-63029A92	0280
48-63029A93	0211
48-63029A94	0211
48-63044A05	0004
48-63075A72	0050
48-63075A73	0050
48-63075A74	0136
48-63075A75	0050
48-63075A76	0280
48-63075A78	0143
48-63076A52	0016
48-63076A81	0127
48-63076A82	0127
48-63076A83	0016
48-63077A03	0280
48-63077A10	0016
48-63077A11	0143
48-63077A29	0144
48-63077A30	0016
48-63077A31	0016
48-63077A32	0143
48-63078A52	0127
48-63078A54	0127
48-63078A59	0280
48-63078A60	0280
48-63078A61	0280
48-63078A62	0164
48-63078A63	0050
48-63078A64	0280
48-63078A65	0136
48-63078A66	0164
48-63078A68	0164
48-63078A69	0164
48-63078A70	0016
48-63078A71	0016
48-63078A86	2866
48-63079A97	0016
48-63081A82	0050
48-63082A15	0280
48-63082A16	0164
48-63082A24	0136
48-63082A25	0016
48-63082A26	0016
48-63082A27	0016
48-63082A45	0016
48-63082A71	0016
48-63084A03	0004
48-63084A04	0004
48-63084A05	0004
48-63084A06	0143
48-63086A16	0015
48-63086A19	0164
48-63090A01	0143
48-63590A01	0143
48-64169	0015
48-64978A10	0265
48-64978A11	0265
48-64978A24	0265
48-64978A27	0050
48-64978A28	0050
48-64978A29	0050
48-64978A39	0212
48-64978A40	0037
48-64978A40A	0037
48-64978A41	0037
48-64978A41A	0037
48-65108A23	0004
48-65108A62	0004
48-65112A65	0144
48-65112A67	0144
48-65112A68	0007
48-65112A73	0917
48-65113A84	0917
48-65113A88	0144
48-65118A64	0144
48-65123A67	0144
48-65123A94	0016
48-65123A95	0144
48-65132A79	0050
48-65144A72	0144
48-65145A74	0015
48-65146A61	0007
48-65146A62	0007
48-65146A63	0007
48-65147A72	0016
48-65173A78	0007
48-65174A24	0007
48-65177A77	0126
48-65177A77A	0126
48-65831A02	1080
48-65837A02	0143
48-65937A02	0143
48-66037A03	0015
48-66037A04	0015
48-66037A05	0015
48-66037A08	0015
48-66037A10	0015
48-66037A12	0015
48-66544A88	1030
48-66629A02	0015
48-66629A03	0015
48-66629A05	0015
48-66629A06	0015
48-66653A001	0015
48-66653A002	0015
48-66653A003	0774
48-66653A005	0774
48-66653A0053	0774
48-66653A02	0474
48-66653APT001	0474
48-66653APT015	0774
48-66653APT0U1	0474
48-66653APT1003	0774
48-66653APTOU1	0474
48-66654A02	0015
48-66865A01	1030
48-66865A02	5036
48-66865A03	5036
48-66865A04	5036
48-67020A11	0143
48-67120A01	0015
48-67120A02	0133
48-67120A03	0133
48-67120A04	0071
48-67120A05	3812
48-67120A06	0015
48-67120A07	0015
48-67120A08	1303
48-67120A09	0143
48-67120A0607	0015
48-67120A10	0071
48-67120A11	0133
48-67926A01	0015
48-68688A79	0015
48-68688A79B	0015
48-69394A01	0016
48-82137H01	1402
48-82137H02	1761
48-82137H03	0668
48-82137H04	0547
48-82137H05	2403
48-82137H06	0870
48-82137H09	0616
48-82256C33	0760
48-82292A03	0143
48-82732C04	1250
48-82965F	0576
48-83461E17	2874
48-83461E23	2084
48-83461E28	0309
48-83461E34	0876
48-83461E39	0262
48-83461E44	0833
48-83741C01	0710
48-83875D01	0447
48-83875D03	0469
48-83875D04	3283
48-83875D05	3284
48-83875D06	0940
48-84973CB1	0725
48-86148	0015
48-86168-3	0143
48-86193	0143
48-86200-3	0143
48-86289-3	0133
48-86343-3	0143
48-86376-3	0016
48-86444-2	1323
48-86797-3	0062
48-86904GF	0050
48-90066A01	0196
48-90068A01	0474
48-90158A01	0102
48-90165A01	0037
48-90165A01A	0037
48-90172A01	0016
48-90210A01	0143
48-90222A08	0143
48-90229A01	0015
48-90232A01	0007
48-90232A03	0144
48-90232A04	0144
48-90232A05	0016
48-90232A06	0126
48-90232A06A	0126
48-90232A07	0142
48-90232A08	0086
48-90232A09	0126
48-90232A09A	0126
48-90232A10	0144
48-90232A11	0016
48-90232A12	0126
48-90232A12A	0126
48-90232A13	0016
48-90232A14	0325
48-90232A15	0126
48-90232A15A	0126
48-90232A16	0142
48-90232A17	0007
48-90232A18	0007
48-90232A19	0144
48-90233A01	0143
48-90233A04	0015
48-90233A05	0917
48-90233A06	0143
48-90233A08	0143
48-90234A01	0015
48-90234A02	0229
48-90234A03	0474
48-90234A11	0934
48-90234A12	0102
48-90234A13	0155
48-90234A14	0006
48-90234A36	0563
48-90234A38	0693
48-90234A39	0325
48-90234A58	3442
48-90234A59	0151
48-90234A60	0720

DEVICE TYPE	REPL CODE
48-90234A61	2874
48-90234A62	0091
48-90234A64	0111
48-90234A66	0917
48-90234A86	0168
48-90234A97	0140
48-90234A98	0375
48-90234A99	0693
48-90235A01	0196
48-90334A67	0133
48-90343A02	0127
48-90343A05	1220
48-90343A06	1220
48-90343A07	0203
48-90343A08	0203
48-90343A17	0155
48-90343A30	0181
48-90343A50	0076
48-90343A52	0065
48-90343A53	0102
48-90343A54	0182
48-90343A55	0162
48-90343A56	2028
48-90343A57	0261
48-90343A58	0342
48-90343A59	0203
48-90343A61	0392
48-90343A62	0071
48-90343A64	0023
48-90343A66	0143
48-90343A67	0155
48-90343A68	0006
48-90343A70	0525
48-90343A72	1220
48-90343A73	1220
48-90343A75	0113
48-90343A76	0219
48-90343A77	0155
48-90343A79	1221
48-90343A80	0151
48-90343A83	0155
48-90343A85	0563
48-90343A88	0139
48-90343A89	0120
48-90343A91	0015
48-90343A92	0015
48-90343A93	0015
48-90343A94	0471
48-90420A01	0582
48-90420A02	0681
48-90420A03	0124
48-90420A04	0133
48-90420A05	0532
48-90420A06	0925
48-90420A11	1220
48-90420A17	0036
48-90420A18	0052
48-90420A19	0582
48-90420A29	0359
48-90420A30	1220
48-90420A31	2223
48-90420A34	4139
48-90420A36	0151
48-90420A42	0825
48-90420A60	0530
48-90420A61	0042
48-90420A71	0133
48-90420A72	0720
48-90420A73	0261
48-90420A76	0023
48-90420A77	0023
48-90420A79	1802
48-90420A80	1097
48-90420A82	0276
48-90420A83	0648
48-90420A84	0392
48-90420A85	0015
48-90420A86	0015
48-90420A87	0015
48-90420A88	0455
48-90420A92	0276
48-90420A97	0023
48-90420A98	0023
48-90420A99	0155
48-90423A93	0286
48-90432A01	0155
48-90432A03	1015
48-90432A04	0199
48-90432A06	0006
48-90432A08	0525
48-90432A15	0023
48-90432A19	0124
48-90432A28	0786
48-90432A32	2223
48-90432A33	0563
48-90432A34	0582
48-90432A35	0672
48-90432A37	0582
48-90432A40	0057
48-90432A42	0253
48-90432A43	0471
48-90432A55	0825
48-90432A63	0036
48-90432A69	1085
48-90432A70	0392
48-90432A71	0955
48-90432A72	0348
48-90432A73	1613
48-90432A75	0317
48-90432A76	0219
48-90432A77	0825
48-90432A78	0062
48-90432A79	0023
48-90432A80	0071
48-90432A85	2427
48-90432A90	0532
48-90432A91	0111
48-90432A92	0286
48-90432A93	0286
48-90432A94	1603
48-90432A95	1220
48-90432A96	0124
48-90432A97	0124
48-90432A99	0100
48-90445A01	0039
48-90445A05	1220
48-90445A28	0325
48-90445A32	0455
48-90445A44	1270
48-90445A48	0276
48-90445A50	0091
48-90445A53	0261
48-90445A54	0563
48-90445A56	1284
48-90445A57	0786
48-90445A60	0017
48-90445A61	0071
48-90445A62	0681
48-90445A63	0475
48-90445A64	0127
48-90445A65	0276
48-90445A66	0686
48-90445A67	0441
48-90445A68	0203
48-90445A69	0643
48-90445A70	0276
48-90445A71	0276
48-90445A72	1220
48-90445A73	1220
48-90445A74	1220
48-90445A75	1220
48-90445A90	0859
48-90445A92	0602
48-90445A95	0071
48-90445A96	0023
48-90445A97	1329
48-97046A02	0004
48-97046A03	0004
48-97046A04	0144
48-97046A05	0127
48-97046A06	0127
48-97046A07	0127
48-97046A08	0136
48-97046A09	0136
48-97046A1S	0222
48-97046A10	0004
48-97046A14	0451
48-97046A15	0222
48-97046A16	0136
48-97046A17	0127
48-97046A18	0144
48-97046A20	0127
48-97046A21	0127
48-97046A22	0016
48-97046A23	0016
48-97046A24	0016
48-97046A25	0007
48-97046A26	0037
48-97046A27	0037
48-97046A28	0016
48-97046A29	0086
48-97046A30	0042
48-97046A31	0222
48-97046A32	0004
48-97046A33	0004
48-97046A34	0004
48-97046A36	0126
48-97046A37	0164
48-97046A38	0042
48-97046A39	0126
48-97046A40	0086
48-97046A42	0016
48-97046A43	0016
48-97046A45	0007
48-97046A46	0016
48-97046A47	0325
48-97046A48	0325
48-97046A50	0016
48-97046A51	0144
48-97046A52	0079
48-97046A53	0211
48-97046A54	0004
48-97046A55	0004
48-97046A56	0004
48-97046A57	0016
48-97048A01	0133
48-97048A02	0143
48-97048A04	0015
48-97048A05	0123
48-97048A06	0143
48-97048A07	0015
48-97048A08	0137
48-97048A10	0015
48-97048A16	0137
48-97048A17	0133
48-97048A18	0015
48-97048A19	0143
48-97048A20	0071
48-97127A01	0015
48-97127A02	0144
48-97127A03	0144
48-97127A04	0086
48-97127A05	0164
48-97127A06	0127
48-97127A09	0164
48-97127A012	0016
48-97127A013	0016
48-97127A015	0150
48-97127A018	0016
48-97127A019	0086
48-97127A12	0016
48-97127A13	0016
48-97127A15	0150
48-97127A18	0016
48-97127A19	0016
48-97127A20	0004
48-97127A22	0004
48-97127A23	0164
48-97127A24	0016
48-97127A29	0016
48-97127A30	0004
48-97127A31	0004
48-97127A32	0211
48-97127A33	0016
48-97162A01	0144
48-97162A02	0144
48-97162A03	0136
48-97162A04	0016
48-97162A05	0016
48-97162A06	0164
48-97162A07	0164
48-97162A08	0164
48-97162A09	0016
48-97162A11	0004
48-97162A12	0079
48-97162A15	0016
48-97162A16	0004
48-97162A18	0004
48-97162A19	0004
48-97162A20	0004
48-97162A21	0016
48-97162A23	0016
48-97162A24	0004
48-97162A25	0004
48-97162A26	0127
48-97162A28	0007
48-97162A30	0007
48-97162A31	0007
48-97162A32	0007
48-97162A33	0016
48-97162A34	0164
48-97168A01	0143
48-97168A02	0015
48-97168A03	0123
48-97168A04	0143
48-97168A06	0015
48-97168A07	0143
48-97168A09	0143
48-97168A10	0015
48-97168A11	0015
48-97168A13	0143
48-97168A14	0015
48-97172A01	0015
48-97177A01	0325
48-97177A02	0007
48-97177A03	0007
48-97177A04	0144
48-97177A06	0325
48-97177A07	0144
48-97177A08	0144
48-97177A09	0016
48-97177A010	0460
48-97177A011	0460
48-97177A015	0143
48-97177A10	0551
48-97177A11	0382
48-97177A12	0016
48-97177A13	0016
48-97177A14	0037
48-97177A14A	0037
48-97177A15	0143
48-97177H01	0325
48-97221A01	0004
48-97221A02	0004
48-97221A03	0004
48-97221A04	0004
48-97221A05	0004
48-97222A01	0143
48-97222A02	0015
48-97238A01	0136
48-97238A02	0136
48-97238A03	0136
48-97238A04	0016
48-97238A05	0004
48-97238A06	0222
48-97238A07	0004
48-97239A01	0143
48-97270A01	0015
48-97270A02	0143
48-97271A01	0004
48-97271A02	0004
48-97271A03	0004
48-97271A04	0004
48-97271A05	0050
48-97271A06	0050
48-97271A3	0004
48-97271A4	0004
48-97271A5	0004
48-97271A6	0004
48-97305A02	0157
48-97305A03	0277
48-97305A05	0157
48-97762A02	0144
48-97768A06	0071
48-103083	0440
48-115107	0023
48-123173	0079
48-123522	0136
48-123536	0136
48-123802	0016
48-123803	0016
48-124158	0004
48-124159	0004
48-124175	0004
48-124204	0160
48-124216	0600
48-124217	0600
48-124218	0600
48-124219	0211
48-124220	0600
48-124221	0600
48-124246	0160
48-124247	0160
48-124255	0136
48-124256	0136
48-124258	0164
48-124259	0164
48-124275	0164
48-124276	0164
48-124279	0164
48-124285	0160
48-124286	0211
48-124296	0050
48-124297	0211
48-124300	0164
48-124302	0160
48-124303	0211
48-124304	0211
48-124305	0136
48-124306	0211
48-124307	0280
48-124308	0280
48-124309	0211
48-124310	0136
48-124311	0136
48-124312	0136
48-124314	0280
48-124315	0280
48-124316	0136
48-124318	0211
48-124319	0211
48-124322	0164
48-124327	0280
48-124328	0280
48-124329	0440
48-124332	0160
48-124343	0280
48-124344	0280
48-124345	0280
48-124346	0136
48-124347	0136
48-124348	0136
48-124349	0136
48-124350	0136
48-124351	0136
48-124352	0136
48-124353	0211
48-124354	0211
48-124355	0211
48-124356	0160
48-124357	0280
48-124358	0280
48-124359	0280
48-124360	0136
48-124363	0050
48-124364	0136
48-124365	0136
48-124366	0136
48-124367	0136
48-124368	0050
48-124370	0280
48-124371	0280
48-124373	0004
48-124377	0136
48-124378	0280
48-124379	0280
48-124380	0280
48-124388	0136
48-124389	0280
48-124398	0280
48-124443	0280
48-124444	0280
48-124445	0280
48-124446	0280
48-124804	0144
48-124805	0144
48-124808	0144
48-125204	0160
48-125208	0160
48-125228	0136
48-125229	0280
48-125230	0280
48-125231	0280
48-125232	0280
48-125233	0600
48-125234	0600
48-125235	0600
48-125236	0600
48-125237	0280
48-125238	0280
48-125239	0280
48-125240	0280
48-125242	0280
48-125252	0440
48-125267	0160
48-125271	0164
48-125276	0164
48-125278	0136
48-125282	0004
48-125285	0004
48-125286	0440
48-125288	0160
48-125294	0004
48-125296	0280
48-125299	0440
48-125332	0160
48-127021	0025
48-128093	0136
48-128094	0164
48-128095	0136
48-128096	0136
48-128219	0050
48-128239	0600
48-128303	0280
48-129934	0160
48-129935	0160
48-129936	0160
48-129937	0160
48-134101	0136
48-134145	0560
48-134173	0016
48-134190A1G	0007
48-134302	0160
48-134372	0136
48-134387	0143
48-134404	0136
48-134405	0136
48-134406	0136
48-134407	0164
48-134408	0004
48-134411	0050
48-134412	0050
48-134413	0050
48-134414	0136
48-134415	0280
48-134416	0280
48-134417	0280
48-134418	0280
48-134419	0280
48-134420	0280
48-134421	0280
48-134422	0280
48-134423	0280
48-134424	0280
48-134425	0280
48-134426	0280
48-134427	0280
48-134428	0280
48-134429	0280
48-134430	0440
48-134431	0847
48-134432	0280
48-134433	0280
48-134434	0136
48-134439	0050
48-134443	0280
48-134444	0280
48-134445	0280
48-134446	0280
48-134447	0160
48-134448	0160
48-134449	0160
48-134450	0280
48-134454	0136
48-134456	0136
48-134457	0136
48-134458	0280
48-134459	0975
48-134462	0280
48-134463	0160
48-134464	0016
48-134465	0016
48-134466	0280
48-134467	2900
48-134468	0280
48-134469	0280
48-134470	0280
48-134471	0280
48-134472	0280
48-134473	0280
48-134474	0280
48-134475	0280
48-134476	0280
48-134477	0280
48-134478	0126
48-134478A	0126
48-134479	0136
48-134480	0136
48-134481	0136
48-134482	0164
48-134483	0164
48-134484	0050
48-134485	0050
48-134486	0050
48-134487	0160
48-134488	0160
48-134493	0160
48-134494	0280
48-134495	0280
48-134496	0280
48-134499	0280
48-134500	0280
48-134501	0280
48-134504	0050
48-134506	0050
48-134507	0050
48-134508	0136
48-134509	0280
48-134510	0280
48-134512	0280
48-134514	0136
48-134519	0160
48-134520	0600
48-134521	0136
48-134522	0136
48-134524	0050
48-134525	0037
48-134525A	0037
48-134526	0050
48-134535	0280
48-134536	0136
48-134537	0143
48-134538	0280
48-134539	0280
48-134540	0280
48-134541	0280
48-134542	0280
48-134543	0280
48-134544	0280
48-134545	0136
48-134547	0136
48-134552	0100
48-134553	0280
48-134554	0280
48-134555	0280
48-134556	0280
48-134557	0280
48-134558	0280
48-134559	0280
48-134560	0160
48-134561	0136
48-134562	0280
48-134563	0280
48-134564	0280
48-134565	0280
48-134567	0280
48-134570	0160
48-134572	0004
48-134573	0211
48-134574	0160
48-134575	0160
48-134576	0136
48-134577	0136
48-134578	0136
48-134579	0050
48-134584	0847
48-134585	0847
48-134587	0143
48-134588	0143
48-134591	0136
48-134592	0160
48-134600	0136
48-134601	0136
48-134602	0136
48-134603	0280
48-134604	0280
48-134605	0136
48-134606	0160
48-134610	0280
48-134611	0160
48-134612	0160
48-134613	0160
48-134621	0004
48-134622	0440
48-134623	0975
48-134625	0280
48-134626	0280
48-134631	0280
48-134632	0004
48-134633	0164
48-134634	0160
48-134635	0136
48-134636	0280
48-134637	0280
48-134638	0160
48-134639	0160
48-134640	0604
48-134641	0280
48-134644	0160
48-134645	0160
48-134646	0160
48-134647	0160
48-134648	0233
48-134649	0160
48-134651	0160
48-134652	0975
48-134653	0012
48-134654	0016
48-134655	0280
48-134656	0280
48-134657	0280
48-134659	0100
48-134663	0039
48-134664	0016
48-134665	0016
48-134666	0016
48-134667	0016
48-134668	0016
48-134669	0016
48-134670	0160
48-134672	0160
48-134673	0016
48-134674	0016
48-134675	0016
48-134676	0050
48-134677	0050
48-134678	0050
48-134679	0050
48-134680	0136
48-134681	0136
48-134682	0136
48-134683	0136
48-134684	0136
48-134689	0086
48-134690	0016
48-134691	0016
48-134692	0604
48-134693	0050
48-134694	0050
48-134695	0604
48-134696	0160
48-134697	0136
48-134698	0162
48-134699	0331
48-134700	0600
48-134701	0103
48-134701A	0103
48-134702	0037
48-134702A	0037
48-134703	0016
48-134704	0331
48-134705	0016
48-134706	0144
48-134709	0144
48-134711	0136
48-134713	0144
48-134714	0016
48-134715	0103
48-134715A	0103
48-134717	0144
48-134718	0016
48-134719	0144
48-134720	0016
48-134721	0016
48-134722	0160
48-134723	0160
48-134724	0144
48-134725	0144
48-134726	0016
48-134727	0085
48-134728	0149
48-134729	0604
48-134730	0160
48-134731	0085
48-134732	0016
48-134733	0016
48-134733A	0016
48-134734	0016
48-134734A	0016
48-134737	0016
48-134738	0160
48-134739	0016
48-134739A	0142
48-134740	0604
48-134741	0604
48-134742	0604
48-134743	0604
48-134744	0160
48-134745	0037
48-134745A	0037
48-134746	0160
48-134747	0160
48-134748	0604
48-134749	0160
48-134750	0160
48-134751	0160
48-134752	0604
48-134753	0604
48-134756	0007
48-134757	0160
48-134758	0160
48-134759	0160
48-134760	0604
48-134761	0604
48-134763	0160
48-134764	0160
48-134765	0016
48-134766	0160
48-134767	0160
48-134768	0016
48-134769	0015
48-134772	0144
48-134773	0144
48-134774	0144
48-134775	0016
48-134776	0016
48-134777	0144
48-134779	0144
48-134780	0144
48-134781	0133
48-134782	0016
48-134783	0144
48-134784	0144

If replacement code is OEM, contact original manufacturer for replacement.

DEVICE TYPE	REPL CODE
48-134785	0016
48-134786	0144
48-134787	0144
48-134788	0604
48-134789	0127
48-134790	0015
48-134791	0016
48-134792	1669
48-134795	0136
48-134796	0136
48-134797	0136
48-134798	0136
48-134800	0144
48-134801	0016
48-134802	0016
48-134803	0079
48-134804	0016
48-134805	0224
48-134806	0144
48-134807	0016
48-134808	0079
48-134809	0016
48-134810	0715
48-134811	0016
48-134814	0144
48-134815	0037
48-134815A	0037
48-134816	0133
48-134817	0016
48-134818	0144
48-134819	0233
48-134820	0144
48-134821	0144
48-134822	0016
48-134823	0079
48-134824	0016
48-134825	0224
48-134826	0144
48-134827	0144
48-134828	0144
48-134829	0037
48-134829A	0037
48-134830	0037
48-134830A	0693
48-134831	0037
48-134831A	0693
48-134832	0037
48-134832A	0693
48-134833	0037
48-134833A	0037
48-134837	0144
48-134838	0086
48-134839	0016
48-134840	0016
48-134841	0016
48-134842	0016
48-134843	0233
48-134844	0016
48-134845	0127
48-134846	0715
48-134847	0016
48-134848	0016
48-134850	0794
48-134851	0794
48-134852	0016
48-134853	0233
48-134854	0016
48-134855	0144
48-134856	0847
48-134857	0224
48-134858	0039
48-134859	0136
48-134860	0136
48-134861	0136
48-134862	0136
48-134865	0037
48-134865A	0037
48-134866	0037
48-134866A	0037
48-134867	0037
48-134867A	0037
48-134868	0037
48-134868A	0037
48-134869	0037
48-134869A	0037
48-134870	0037
48-134870A	0037
48-134871	0037
48-134871A	0037
48-134872	0142
48-134879	0144
48-134880	0136
48-134882	0103
48-134884	0103
48-134885	0142
48-134888	0160
48-134889	0016
48-134891	0144
48-134892	0144
48-134893	0144
48-134894	0016
48-134895	0016
48-134896	0016
48-134897	0016
48-134898	0233
48-134899	0016
48-134900	0074
48-134901	0642
48-134902	0144

DEVICE TYPE	REPL CODE
48-134903	0016
48-134904	0127
48-134904A1G	0127
48-134904F	0127
48-134905	0016
48-134906	0016
48-134907	0085
48-134908	0144
48-134909	0037
48-134909A	0037
48-134910	0037
48-134910A	0037
48-134910F	0037
48-134911	0037
48-134912	0244
48-134913	0037
48-134913A	0037
48-134914	0037
48-134914A	0037
48-134915	0037
48-134915A	0037
48-134916	0913
48-134917	0913
48-134918	0016
48-134919	0233
48-134920	0142
48-134921	0102
48-134922	0127
48-134923	0127
48-134924	0127
48-134925	0127
48-134926	0127
48-134927	0233
48-134928	0016
48-134929	0016
48-134930	0160
48-134931	0600
48-134932	0330
48-134933	0016
48-134933E	0016
48-134934	0975
48-134935	0016
48-134936	0178
48-134937	0144
48-134938	0160
48-134939	0102
48-134940	0037
48-134940A	0037
48-134941	0086
48-134942	0016
48-134943	0037
48-134943A	0037
48-134944	0325
48-134945	0127
48-134946	0144
48-134947	0160
48-134948	0144
48-134949	0144
48-134950	0144
48-134951	0126
48-134951A	0126
48-134952	0016
48-134953	0086
48-134954	0123
48-134956	0280
48-134957	0012
48-134958	0015
48-134959	0015
48-134960	0144
48-134961	0127
48-134962	0127
48-134963	0127
48-134964	0127
48-134965	0127
48-134966	0127
48-134967	0037
48-134967A	0037
48-134969	0103
48-134970	0016
48-134972	0142
48-134973	0037
48-134973A	0037
48-134974	0160
48-134975	0037
48-134975A	0037
48-134977	0160
48-134978	0102
48-134979	0144
48-134980	0016
48-134981	0127
48-134982	0086
48-134983	0144
48-134985	0144
48-134987	0460
48-134988	0016
48-134989	0037
48-134989A	0037
48-134990	0015
48-134991	0162
48-134992	0016
48-134993	0157
48-134994	0016
48-134995	0642
48-134996	0016
48-134997	0111
48-134998	0276
48-136665	0016
48-137000	0365
48-137001	0975

DEVICE TYPE	REPL CODE
48-137002	0233
48-137003	0016
48-137004	0144
48-137005	0086
48-137006	0144
48-137007	0016
48-137008	0103
48-137008A	0103
48-137010	0016
48-137011	0264
48-137013	0016
48-137014	0016
48-137015	0111
48-137017	0100
48-137019	0016
48-137020	0037
48-137020A	0037
48-137021	0037
48-137022	0016
48-137023	0325
48-137024	1137
48-137025	0085
48-137026	0085
48-137027	0103
48-137027A	0103
48-137029	0015
48-137030	2997
48-137031	0160
48-137032	0037
48-137032A	0037
48-137033	0144
48-137034	0441
48-137035	0233
48-137036	0103
48-137036A	0103
48-137037	0161
48-137039	0847
48-137040	0127
48-137041	0086
48-137043	0016
48-137044	0016
48-137045	0037
48-137045A	0037
48-137046	0037
48-137047	0016
48-137048	0100
48-137053	0103
48-137053A	0103
48-137055	0144
48-137056	0016
48-137057	0016
48-137058	1669
48-137059	0127
48-137061	0037
48-137062	0455
48-137065	3331
48-137067	0037
48-137067A	0037
48-137068	0037
48-137068A	0037
48-137069	0037
48-137069A	0037
48-137070	0849
48-137071	0127
48-137072	0016
48-137073	0016
48-137074	0015
48-137075	0127
48-137076	0127
48-137077	0127
48-137078	0085
48-137079	0103
48-137079A	0103
48-137080	0423
48-137083	0016
48-137088	0086
48-137089	0016
48-137090	0037
48-137090A	0037
48-137091	0551
48-137092	0551
48-137093	0264
48-137096	0016
48-137098	0110
48-137101	0016
48-137102	0085
48-137104	0144
48-137105	0144
48-137106	0016
48-137107	0016
48-137108	0016
48-137109	0016
48-137110	0016
48-137111	0016
48-137112	0102
48-137115	0016
48-137116	0130
48-137118	0160
48-137119	0160
48-137120	0160
48-137121	0604
48-137122	0160
48-137123	0160
48-137124	0160
48-137125	0604
48-137126	0144
48-137127	0037
48-137127A	0037
48-137128	0161

DEVICE TYPE	REPL CODE
48-137130	0012
48-137132	0441
48-137133	0920
48-137134	0642
48-137136	0144
48-137137	0016
48-137138	0016
48-137139	0016
48-137140	0144
48-137142	2270
48-137143	0015
48-137144	0144
48-137145	0161
48-137146	0161
48-137147	0161
48-137148	0161
48-137149	0551
48-137153	0460
48-137154	0460
48-137155	0460
48-137156	0460
48-137157	0460
48-137158	0144
48-137160	0382
48-137164	0012
48-137165	1669
48-137166	0144
48-137167	0913
48-137168	0382
48-137169	0551
48-137171	0079
48-137171D	0016
48-137172	0016
48-137173	0037
48-137173A	0037
48-137174	0016
48-137175	0103
48-137175A	0103
48-137176	0037
48-137176A	0037
48-137177	0137
48-137178	0975
48-137179	0642
48-137180	0103
48-137180A	0103
48-137183	0122
48-137184	2659
48-137190	0144
48-137191	0144
48-137192	0016
48-137193	0164
48-137194	0144
48-137195	0037
48-137196	0144
48-137197	0007
48-137198	0015
48-137199	0211
48-137200	0161
48-137202	0551
48-137203	0642
48-137205	0071
48-137206	0016
48-137207	0142
48-137208	0015
48-137212	0071
48-137213	0085
48-137214	0222
48-137215	0085
48-137216	0085
48-137217	0085
48-137218	0085
48-137219	0085
48-137220	0085
48-137234	0975
48-137235	0975
48-137238	0861
48-137239	0338
48-137240	0382
48-137251	0103
48-137256	0460
48-137257	0016
48-137258	0460
48-137259	0460
48-137260	0016
48-137265	0016
48-137266	0039
48-137267	0222
48-137268	0222
48-137269	0222
48-137270	0222
48-137271	0222
48-137272	0137
48-137279	0062
48-137280	1137
48-137281	0058
48-137282	1669
48-137290	0071
48-137291	0015
48-137295	0058
48-137299	0143
48-137300	0111
48-137301	0071
48-137302	0071
48-137303	0460
48-137304	0460
48-137306	0705
48-137307	0086
48-137308	0222
48-137309	0423

DEVICE TYPE	REPL CODE
48-137310	0854
48-137311	2997
48-137312	3165
48-137314	0382
48-137315	0086
48-137316	0071
48-137318	0037
48-137319	0551
48-137320	0382
48-137321	0791
48-137322	0253
48-137324	0037
48-137325	0111
48-137326	0074
48-137327	0540
48-137329	0085
48-137330	0686
48-137331	0460
48-137333	0103
48-137336	0016
48-137338	2809
48-137339	0144
48-137340	0540
48-137341	0065
48-137342	0975
48-137343	0325
48-137344	0103
48-137347	0023
48-137348	0023
48-137349	4588
48-137350	0016
48-137351	0224
48-137352	0144
48-137353	0016
48-137354	0144
48-137355	0144
48-137364	0233
48-137366	0037
48-137367	0975
48-137368	0103
48-137369	2997
48-137370	3165
48-137371	0144
48-137372	0144
48-137373	0016
48-137374	0016
48-137375	0144
48-137376	0144
48-137377	0016
48-137378	0016
48-137379	0037
48-137380	0037
48-137381	0037
48-137382	0037
48-137383	0037
48-137384	0016
48-137385	0133
48-137388	0144
48-137390	0111
48-137391	0037
48-137392	0400
48-137393	0064
48-137394	0686
48-137395	0058
48-137396	0042
48-137398	0016
48-137399	0016
48-137400	0127
48-137421	0244
48-137437	0042
48-137472	0460
48-137473	0161
48-137483	0144
48-137487	0030
48-137488	0212
48-137491	0007
48-137495	0143
48-137501	0460
48-137502	0037
48-137504	0037
48-137505	0551
48-137506	0042
48-137507	0925
48-137509	0016
48-137514	0133
48-137526	2997
48-137527	3165
48-137531	2270
48-137540	0925
48-137549	0161
48-137550	0460
48-137552	0560
48-137553	1267
48-137554	OEM
48-137562	3165
48-137563	0365
48-137566	0925
48-137567	0212
48-137573	0133
48-137577	1083
48-137601	OEM
48-137605	0143
48-137606	OEM
48-137607	0649
48-137610	0382
48-137612	0127
48-137621	1083
48-137855	0016
48-137978	0160

DEVICE TYPE	REPL CODE
48-137988	0086
48-155035	0006
48-155039	0019
48-155041	0080
48-155042	0219
48-155044	2096
48-155045	0203
48-155046	0155
48-155047	0124
48-155048	0317
48-155051	0261
48-155058	2051
48-155059	0955
48-155060	0133
48-155061	0143
48-155063	0296
48-155065	0203
48-155068	0593
48-155070	0276
48-155071	0286
48-155073	0286
48-155074	0219
48-155077	0133
48-155081	0181
48-155083	0071
48-155087	0224
48-155088	0016
48-155089	1220
48-155093	0286
48-155095	0006
48-155097	0219
48-155099	0023
48-155107	0023
48-155108	0102
48-155110	0955
48-155113	0058
48-155114	0019
48-155116	0681
48-155119	0155
48-155124	0223
48-155125	0102
48-155126	0023
48-155128	0023
48-155130	0643
48-155131	0120
48-155136	0023
48-155140	0130
48-155145	0041
48-155146	0636
48-155148	0139
48-155150	0155
48-155152	0348
48-155153	0168
48-155154	0155
48-155156	0006
48-155157	2439
48-155158	1268
48-155159	0124
48-155173	0139
48-155177	0015
48-155178	0023
48-155179	0023
48-155182	0495
48-155184	0471
48-155185	0041
48-155189	2051
48-155192	1005
48-155193	0015
48-155195	0023
48-155196	0137
48-155197	0065
48-155198	0071
48-155207	0041
48-155209	0643
48-155210	0052
48-155212	0023
48-155213	0643
48-155221	0130
48-155222	0261
48-155223	0102
48-155224	0065
48-155225	0301
48-155235	0015
48-155236	0015
48-155253	0502
48-155258	0124
48-155259	1085
48-155263	0002
48-155279	0120
48-155284	0203
48-155289	0182
48-155293	0582
48-155294	0825
48-155295	0025
48-155296	0023
48-155298	0064
48-155299	0883
48-155334	0124
48-155343	0825
48-355008	0123
48-355029	0058
48-355046	0139
48-355048	0023
48-355052	0224
48-355053	0151
48-355054	0276
48-355056	1751
48-355059	0042
48-443006A06	0317

DEVICE TYPE	REPL CODE
48-644587	0143
48-644676	0050
48-644677	0050
48-644678	0004
48-644679	0164
48-644681	0143
48-645867	0050
48-646954	0015
48-647311	0143
48-647313	0143
48-647713	0143
48-647829	0015
48-660370A05	0015
48-674297U	0917
48-674970	0917
48-711052	0143
48-732230	1030
48-733746	0015
48-739300	0143
48-741255	0913
48-741280	0143
48-741656	0913
48-741724	0913
48-741752	0484
48-742698	0484
48-742970	0917
48-746831	0015
48-751656	0196
48-751724	1080
48-752497	0015
48-754153	0484
48-822321E07	1401
48-822321E08	1537
48-859248	0086
48-859428	0086
48-865539	0123
48-869001	0004
48-869087B	0085
48-869090	0975
48-869099B	0085
48-869138	0086
48-869148	0211
48-869170	0086
48-869182	0160
48-869184	0086
48-869198	0004
48-869205	0604
48-869206	1669
48-869225	0178
48-869228	0086
48-869237	0160
48-869244	0103
48-869248	0016
48-869249	0211
48-869253	0211
48-869254	0038
48-869256	2136
48-869259	0103
48-869263	0086
48-869264	1669
48-869266	0007
48-869273C	0642
48-869274	0178
48-869278	0103
48-869279C	0074
48-869282	0211
48-869283	0038
48-869301	0178
48-869302	0103
48-869306	1825
48-869308	0126
48-869321	0103
48-869337	0074
48-869380	0086
48-869393	0178
48-869400	0126
48-869408	0074
48-869426	0126
48-869427	0604
48-869450	0007
48-869464	0086
48-869475	0211
48-869475A	0211
48-869476	0038
48-869476A	0038
48-869481	0007
48-869561	6348
48-1370893	0460
48-8690468	0050
48-44885601	1220
48-46785104	0460
48-V34816	0133
48A40458A06	0165
48A40516C02	0030
48A41508A01	0015
48A41508A02	0015
48A42383A01	0030
48A62692A01	0143
48A90456A80	OEM
48A124315	0211
48AA0516C02	0030
48B41266G01	0102
48B41768G01	0019
48B43265G01	0015
48B62334A01	0143
48B63494A01	0030
48B63494H01	0030
48B66629A01	0143
48B66629A02	0143

DEVICE TYPE	REPL CODE
48B66629A03	0015
48B66629A05	0015
48B67020A11	0143
48B90158A01	0540
48B732230	1030
48B7322230	1030
48C40235C01	0023
48C40235G01	0015
48C40235G02	0102
48C40235G1	0015
48C40524A02	0133
48C42428A01	0019
48C61074B01	0143
48C61767B01	0143
48C65831A02	1080
48C65831A03	0484
48C65832A02	0143
48C65837A01	0143
48C65837A02	0143
48C66037A03	0015
48C66037A04	0015
48C66037A05	0015
48C66037A10	0015
48C66037A12	0015
48C66629A02	0015
48C-66653A-001	0474
48C66653A02	0015
48C66865A01	1030
48C67120A01	0143
48C67120A02	0133
48C67120A05	0015
48C67926A01	0015
48C125233	0038
48C125235	0038
48C125236	0038
48C125237	0136
48C134587	0143
48C134816	0133
48C134840	0127
48C666534A02	0474
48C674970	0917
48C742970	0917
48C751656	0196
48D5R03	OEM
48D9R03GH	OEM
48D67	0133
48D6653APT.005	0774
48D8375D02	0940
48D60154A01	0143
48D63590A01	0143
48D66037A03	0015
48D66037A04	0015
48D66037A05	0015
48D66037A08	0015
48D66653-APT.00	0774
48D66653A02	0474
48D66653A05	0774
48D66653APT001	0474
48D66653APT003	0774
48D66653APT005	0774
48D66653APT015	0774
48D66653APT.005	0774
48D66654A02	0015
48D67120A01	0015
48D67120A02	0015
48D67120A05	0015
48D67120A06	0015
48D67120A07	0015
48D67120A08	1303
48D67120A09	0143
48D67120A011	0079
48D67120A11	0133
48D67120A11(DIO)	0143
48D67120A11(XSTR)	0079
48D67120A13	0133
48D67120A13(C)	0133
48D67120AB	0023
48D67120H02	0015
48D69723A01	1323
48D69723A02	1323
48D69723A02-185	1323
48D69723B01	1323
48D83875D01	0447
48D83875D03	0469
48D83875D04	3283
48D83875D05	3284
48D83875D06	0940
48D90068A01	0474
48D90210A01	0143
48DL7120A02	0913
48E03A	OEM
48E05A	OEM
48G10346A01	0143
48G10346A02	0133
48G13001A01	1027
48J	OEM
48K	OEM
48K35P1	0136
48K36P1	0136
48K36P3	0136
48K39P1	0085
48K43P3	0136
48K43P4	0136
48K45N2	0038
48K56P1	0211
48K57B2	0085
48K57B42	0085
48K64169	0015
48K125230	0085
48K134450	0211
48K134458	0211
48K134494	0050
48K134495	0050
48K134496	0050
48K134587	0143
48K134601	0050
48K134796	0050
48K134798	0050
48K544539	0019
48K640675	0019
48K644681	0019
48K646594	0015
48K646954	0015
48K647311	0143
48K647713	0143
48K647769	0143
48K647829	0015
48K741255	0196
48K741752	0484
48K742698	0484
48K746831	0015
48K751724	1080
48K752297	0015
48K752497	0015
48K754153	0484
48K865539	0123
48K869001	0280
48M355002	0224
48M355004	1220
48M355005	0643
48M355006	0203
48M355007	0006
48M355008	0143
48M355009	0019
48M355012	0261
48M355013	0102
48M355014	0133
48M355016	0023
48M355021	0182
48M355023	0023
48M355025	0080
48M355029	0058
48M355035	0124
48M355036	0120
48M355037	0155
48M355038	0955
48M355039	0219
48M355040	0042
48M355041	0867
48M355042	2096
48M355043	2051
48M355044	0955
48M355045	0015
48M355046	0139
48M355047	0757
48M355048	0102
48M355049	0015
48M355050	1606
48M355051	0074
48M355052	0224
48M355053	0151
48M355054	0276
48M355055	0525
48M355056	1751
48M355057	0276
48M355058	0342
48M355059	0042
48M355062	0317
48M355065	1834
48NSP1035	0160
48P1	0004
48P-2A5	0004
48P60022A97	0143
48P60022A98	0030
48P60077A06	0143
48P63005A72	0079
48P63006A56	0143
48P63076A81	0079
48P63076A82	0079
48P63077A03	0211
48P63077A11	0143
48P63077A31	0079
48P63077A32	0143
48P63077A52	0143
48P63078A52	0079
48P63078A62	0211
48P63078A69	0127
48P63078A70	0079
48P63078A71	0016
48P63078A86	7077
48P63079A97	0079
48P63082A24	0079
48P63082A25	0079
48P63082A26	0079
48P63082A27	0086
48P63082A45	0144
48P63082A71	0144
48P63082A74	OEM
48P63086A18	0079
48P64978A27	0050
48P64978A28	0050
48P64978A29	0050
48P65112A65	0079
48P65112A73	0917
48P65113A88	0079
48P65118A64	0079
48P65123A67	0007
48P65123A95	0127
48P65137A77	OEM
48P65144A72	0127
48P65145A74	OEM
48P65146A61	0127
48P65146A62	0127
48P65146A63	0144
48P65148A02	0005
48P65148A04	OEM
48P65173A77	0015
48P65173A78	0144
48P65174A24	0007
48P65175A12	0285
48P65175A24	OEM
48P65193A55	0917
48P65193A56	0127
48P65194A92	0127
48Q134722	0160
48R2	0206
48R2S	OEM
48R25	0206
48R10001-A01	0015
48R10001-A03	0015
48R10001-A030-1	0015
48R10062A01	0015
48R10062A02	0015
48R10062A04	0015
48R10062A05	0015
48R10073A02	0136
48R10074A02	0136
48R100620A02	0015
48R100620A04	0015
48R100620A05	0015
48R134407	0211
48R134573	0211
48R134587	0143
48R134621	0004
48R134632	0004
48R134665	0038
48R134666	0086
48R134671	0015
48R134722	0160
48R660370A05	0015
48R859428	0086
48R869138	0086
48R869148	0211
48R869170	0086
48R869206	1669
48R869248	0079
48R869249	0211
48R869253	0211
48R869254	0038
48R869256	2136
48R869264	1669
48R869282	0211
48R869283	0038
48R869306	1825
48R869426	0150
48R869464	0086
48R869475	0211
48R869475A	0211
48R869476	0038
48R869476A	0038
48S00155035	0006
48S00155039	0019
48S00155060	0124
48S00155077	0133
48S00155087	0224
48S00155093	0286
48S00155114	0019
48S00155131	0120
48S00155137	0023
48S00155138	0015
48S00155144	1329
48S00155146	0471
48S00155160	0133
48S00155177	0015
48S00155222	0261
48S00155258	0124
48S00155284	0203
48S00155334	0124
48S00155343	0825
48S00155365	0286
48S00155400	0582
48S00155403	0582
48S00355052	0224
48S50	0905
48S191A02	0015
48S191A04	0015
48S191A04(A)	0015
48S191A05	0102
48S191A06	0015
48S191A07	0015
48S191A08	0102
48S191A11	0015
48S10062A01	0015
48S10062A02	0015
48S10062A05	0015
48S10062A05A	0015
48S10346A02	0143
48S10577A01	0133
48S10577A02	0133
48S10577A04	0102
48S10577A11	0133
48S10577A13	0133
48S10641D62	0636
48S32000	0137
48S40149P01	1221
48S40170G01	0076
48S40171G01	0086
48S40172G01	0222
48S40235G02	0015
48S40241G01	0111
48S40246G01	0127
48S40247G01	0111
48S40247G02	0079
48S40382J01	0560
48S40383J01	1267
48S40606G01	0079
48S40606G02	0111
48S40606J02	0111
48S40607G01	0111
48S40607J01	0111
48S40662G02	0228
48S40662G04	0042
48S40929P01	2190
48S41508A01	0015
48S43239G0Z	0233
48S43240G01	0042
48S43241G01	0042
48S43467J01	3046
48S43991J01	0144
48S43992J01	0144
48S44883G01	0042
48S44884G01	0925
48S44885G01	0016
48S44885G02	0016
48S44886G01	0155
48S44887G01	0133
48S65123A67	0127
48S67120A11	0102
48S67120A13	0133
48S90158A01	0540
48S90233A01	0143
48S90233A02	0143
48S90233A04	0015
48S90233A06	0143
48S90233A07	0137
48S90233A08	0143
48S90234A01	0540
48S90234A02	0015
48S90234A03	0474
48S90235A01	0196
48S97048A04	0540
48S97127A01	0296
48S97168A01	0143
48S97168A04	0143
48S97168A07	0143
48S97172A01	0540
48S97305A02	0157
48S97305A03	0540
48S134404	0211
48S134405	0050
48S134406	0050
48S134407	0050
48S134408	0211
48S134458	0136
48S134587	0143
48S134666	0079
48S134695	0160
48S134718	0079
48S134719	0079
48S134720	0079
48S134721	0079
48S134732	0079
48S134733	0079
48S134733A	0079
48S134734	0016
48S134734A	0079
48S134736	0015
48S134737	0160
48S134739	0142
48S134739A	0142
48S134746	0160
48S134747	0160
48S134751	0160
48S134756	0127
48S134758	0160
48S134759	0160
48S134760	0160
48S134761	0160
48S134765	0079
48S134766	0160
48S134767	0160
48S134768	0079
48S134773	0079
48S134774	0079
48S134775	0079
48S134776	0079
48S134783	0079
48S134784	0079
48S134785	0079
48S134789	0079
48S134790	0015
48S134797	0136
48S134804	0144
48S134805	0079
48S134807	0016
48S134809	0079
48S134810	0079
48S134811	0079
48S134814	0037
48S134815	0037
48S134816	0133
48S134819	0233
48S134820	0007
48S134821	0150
48S134823	0079
48S134825	0127
48S134826	0079
48S134827	0079
48S134830	0037
48S134831	0037
48S134832	0037
48S134837	0079
48S134838	0233
48S134840	0079
48S134841	0079
48S134842	0079
48S134843	0285
48S134844	0079
48S134845	0144
48S134846	0079
48S134850	0137
48S134851	0012
48S134853	0086
48S134854	0016
48S134855	0144
48S134857	0079
48S134858	0039
48S134860	0280
48S134861	0280
48S134862	0280
48S134872	0142
48S134879	0079
48S134888	0160
48S134889	0079
48S134894	0079
48S134898	0079
48S134899	0079
48S134900	0074
48S134901	0642
48S134902	0144
48S134903	0016
48S134904	0079
48S134905	0016
48S134906	0016
48S134908	0079
48S134909	0150
48S134910	0150
48S134912	0244
48S134913	0037
48S134915	0037
48S134916	0196
48S134917	0196
48S134918	0079
48S134919	0233
48S134920	0142
48S134921	0015
48S134922	0079
48S134923	0079
48S134924	0079
48S134925	0079
48S134926	0079
48S134927	0079
48S134932	0007
48S134933	0016
48S134934	0975
48S134935	0086
48S134936	0423
48S134937	0330
48S134938	0975
48S134939	0015
48S134941	0079
48S134942	0086
48S134943	0037
48S134944	0325
48S134945(A2C)	0079
48S134946	0144
48S134947	0085
48S134948	0144
48S134949	0007
48S134950	0144
48S134952	0079
48S134953	0086
48S134954	0143
48S134956	0050
48S134957	0012
48S134958	0015
48S134959	0015
48S134960	0144
48S134961	0079
48S134962	0079
48S134963	0079
48S134964	0079
48S134970	0144
48S134972	0142
48S134974	0085
48S134978	1303
48S134979	0144
48S134981	0127
48S134988	0016
48S134989	0037
48S134990	0015
48S134992	0079
48S134995	0642
48S134997	0016
48S134998	0142
48S137000	0137
48S137001	0975
48S137002	0285
48S137003	0079
48S137006	0016
48S137006(A3S)	0079
48S137007	0595
48S137014	0086
48S137015	0111
48S137017	0002
48S137021	0244
48S137021(10)	0244
48S137022	0086
48S137029	0015
48S137031	0160
48S137032	0037
48S137033	0079
48S137034	0441
48S137040	0015
48S137041	0016
48S137044	0079
48S137045	0037
48S137047	0086
48S137055	0144
48S137056	0016
48S137057	0016
48S137063	0039
48S137065	OEM
48S137070	0325
48S137074	0015
48S137076	OEM
48S137081	0201
48S137082	0288
48S137093	0264
48S137094	OEM
48S137101	0143
48S137106	0086
48S137107	0016
48S137108	0016
48S137109	0079
48S137110	0016
48S137111	0079
48S137113	0338
48S137114	0757
48S137115	0016
48S137127	0037
48S137133	0133
48S137134	0642
48S137145	0560
48S137158	0127
48S137160	0382
48S137164	0012
48S137167	1149
48S137168	0382
48S137169	0551
48S137170	0636
48S137171	0079
48S137171(D)	0079
48S137172	0016
48S137173	0037
48S137174	0016
48S137190	0079
48S137191	0079
48S137192	0079
48S137203	0642
48S137206	0086
48S137207	0142
48S137208	0015
48S137260	0079
48S137266	0039
48S137270	0222
48S137272	0039
48S137281	0058
48S137295	0058
48S137299	0143
48S137300	0016
48S137308	0222
48S137309	0423
48S137310	0854
48S137311	0042
48S137312	3165
48S137314	0037
48S137315	0016
48S137321	0037
48S137323	0423
48S137330	0002
48S137331	1267
48S137337	OEM
48S137341	0642
48S137342	0975
48S137343	0325
48S137344	0103
48S137347	0023
48S137348	0023
48S137350	0079
48S137351	0079
48S137364	0338
48S137364(H)	0233
48S137369	2997
48S137370	3165
48S137386	0861
48S137387	0636
48S137389	0402
48S137397	1323
48S137415	0285
48S137415(I)	0233
48S137442	0565
48S137472	1267
48S137473	0560
48S137476	0338
48S137495	0143
48S137498	0016
48S137512	0079
48S137524	0363
48S137528	1027
48S137530	0016
48S137531	2270
48S137533	0102
48S137535	0178
48S137539	0065
48S137543	0016
48S137546	0102
48S137548	0074
48S137551	0102
48S137560	OEM
48S137572	0551
48S137855	0079
48S144258	0124
48S155001	0004
48S155002	0015
48S155005	0178
48S155006	0233
48S155013	0423
48S155014	0854
48S155030	0338
48S155034	0955
48S155035	0006
48S155037	0023
48S155039	0019
48S155040	0023
48S155041	0080
48S155042	0219
48S155043	0643
48S155044	2096
48S155044(15)	0142
48S155045	0203
48S155046	0155
48S155047	0124
48S155048	0317
48S155050	0058
48S155051	0261
48S155052	0120
48S155053	0867
48S155054	0502
48S155054(ZENER)	0170
48S155056	0182
48S155057	0023
48S155058	2051
48S155059	0955
48S155060	0133
48S155061	0143
48S155062	0042
48S155063	0296
48S155066	0925
48S155067	0103
48S155068	0593
48S155069	0955
48S155070	0276
48S155071	0286
48S155072	0643
48S155073	0286
48S155074	0219
48S155075	2336
48S155076	0065
48S155077	0133
48S155078	0143
48S155079	0023
48S155080	0091
48S155081	0181
48S155082	5186
48S155083	0071
48S155084	0317
48S155085	0071
48S155086	0629
48S155087	0224
48S155087(SCR)	0058
48S155088	0016
48S155089	1220
48S155090	0223
48S155091	1284
48S155092	0058
48S155093	0286
48S155094	0286
48S155095	0006
48S155096	0643
48S155097	0219
48S155098	1267
48S155099	0023
48S155100	0023
48S155103	0137
48S155104	0002
48S155105	0042
48S155106	0015
48S155107	0023
48S155108	0102
48S155110	0643
48S155111	0219
48S155114	0019
48S155116	0681
48S155117	0155
48S155118	0826
48S155121	0018
48S155122	0532
48S155123	0218
48S155126	0023
48S155127	0451
48S155128	0023
48S155129	0124
48S155131	0120
48S155136	0023
48S155137	0023
48S155138	0023
48S155139	0348
48S155140	0130
48S155141	0079
48S155144	1329
48S155150	0155
48S155160	0133

If replacement code is OEM, contact original manufacturer for replacement.

DEVICE TYPE	REPL CODE
48S155169	0140
48S155177	0015
48S155178	0023
48S155179	0023
48S155180	5224
48S155183	0124
48S155189	2051
48S155193	0023
48S155195	0023
48S155199	0392
48S155199(ZENER)	0471
48S155201	0023
48S155207	0041
48S155213	0643
48S155222	0261
48S155223	0102
48S155224	0065
48S155225	0301
48S155238	0015
48S155239	1097
48S155241	2439
48S155242	0155
48S155252	5051
48S155258	0124
48S155259	1085
48S155267	5255
48S155275	0316
48S155279	0120
48S155280	0883
48S155281	0091
48S155282	0062
48S155284	0203
48S155285	0825
48S155286	0060
48S155287	0006
48S155289	0182
48S155292	0023
48S155293	0582
48S155294	0825
48S155295	0025
48S155296	0023
48S155298	0064
48S155299	0041
48S155302	0168
48S155305	0955
48S155307	0286
48S155310	0071
48S155311	0025
48S155312	0317
48S155313	0102
48S155314	0502
48S155317	0170
48S155320	0582
48S155323	0064
48S155324	0071
48S155325	0165
48S155333	0392
48S155336	0023
48S155337	0060
48S155343	0825
48S155349	0002
48S155366	0120
48S155376	0609
48S155377	0315
48S155378	0705
48S155380	0203
48S155382	1212
48S155383	2336
48S155387	0445
48S164844	OEM
48S317113	0233
48S317476	0233
48S623334A01	0143
48S1552132	0643
48SP134804	0079
48SP134826	0127
48SP134837	0127
48SP134855	0079
48SP134857	0127
48SP134894	0079
48SP134897	0079
48SP134903	0079
48SP134904	0127
48SP134905	0079
48SP134906	0079
48SP134933	0079
48SP134937	0127
48ST-2SD870	0055
48ST-2SD871	0055
48T43394P01	0530
48T43688P01	OEM
48T66544A88	1030
48V68629A40	OEM
48X00155093	OEM
48X90	OEM
48X9271A03	0211
48X90223A02	0155
48X90223A03	0091
48X90223A04	0471
48X90229A02	0672
48X90229A03	2471
48X90229A04	0261
48X90229A05	0055
48X90229A06	0018
48X90229A07	0620
48X90229A08	OEM
48X90229A09	0041
48X90229A15	0530
48X90229A16	2008

DEVICE TYPE	REPL CODE
48X90229A17	0286
48X90229A18	0203
48X90229A19	1603
48X90229A20	0155
48X90229A36	0064
48X90229A37	0261
48X90229A38	0400
48X90229A45	0359
48X90229A46	7147
48X90229A47	0134
48X90229A48	0602
48X90229A52	0203
48X90229A56	0672
48X90229A61	0460
48X90229A66	2923
48X90229A67	0053
48X90229A68	0053
48X90229A69	0006
48X90229A70	0275
48X90229A94	0530
48X90229A97	2223
48X90229A98	0118
48X90229A99	OEM
48X90232A01	0079
48X90232A02	0079
48X90232A03	0079
48X90232A04	0079
48X90232A05	0086
48X90232A06	0150
48X90232A07	0086
48X90232A08	0079
48X90232A09	0150
48X90232A10	0079
48X90232A11	0079
48X90232A12	0150
48X90232A13	0079
48X90232A14	0325
48X90232A15	0150
48X90232A16	0142
48X90232A17	0127
48X90232A18	0127
48X90232A19	0127
48X90232A20	0127
48X90232A40	1220
48X90232A47	0928
48X90232A67	0124
48X90232A68	0091
48X90232A69	0091
48X90232A72	0686
48X90232A74	0077
48X90232A75	2923
48X90232A76	OEM
48X90232A80	2058
48X90232A83	0124
48X90232A84	0636
48X90232A89	0879
48X90232A90	0343
48X90232A91	0494
48X90232A92	0165
48X90232A93	0137
48X90232A94	0071
48X90232A95	0017
48X90232A96	0041
48X90232A97	0760
48X90232A98	0705
48X90232A99	0261
48X90233A01	0143
48X90233A02	0143
48X90233A03	OEM
48X90233A04	0015
48X90233A05	0917
48X90233A06	0143
48X90233A07	0137
48X90233A08	0030
48X90233A09	0554
48X90233A17	0133
48X90233A25	0060
48X90233A26	0392
48X90233A27	0168
48X90233A28	0060
48X90233A39	0053
48X90233A40	0825
48X90233A41	0253
48X90233A50	0041
48X90233A55	2008
48X90233A67	0917
48X90234A01	0015
48X90234A02	0015
48X90234A03	0015
48X90234A05	0288
48X90234A59	0151
48X90234A66	0917
48X90234A67	0133
48X90235A01	0196
48X90235A05	0057
48X90235A09	0343
48X90235A70	0009
48X90235A72	0133
48X90237A18	0286
48X90237A19	0155
48X90237A20	0155
48X90237A62	0071
48X90237A63	0077
48X90237A64	0062
48X90237A65	0052
48X90237A66	0133
48X90343A02	0127
48X90343A15	0203
48X90343A39	0471

DEVICE TYPE	REPL CODE
48X90343A54	0182
48X90343A83	0155
48X90380A03	0392
48X90380A04	0532
48X90380A05	0023
48X90380A06	0140
48X90380A07	0494
48X90380A08	0017
48X90380A09	0041
48X90380A10	0620
48X90380A29	0343
48X90380A30	0686
48X90380A46	0120
48X90380A74	0582
48X90380A76	0203
48X90380A77	0532
48X90380A78	0582
48X90380A80	0672
48X90380A85	0057
48X90380A86	0071
48X90380A87	0163
48X90420A01	0582
48X90420A24	0582
48X90420A28	0825
48X90420A29	0359
48X90420A31	2223
48X90420A42	0825
48X90420A43	0018
48X90420A53	0120
48X90420A56	0060
48X90420A74	0955
48X90420A83	0648
48X90420A85	0015
48X90420A86	0015
48X90420A87	0015
48X90420A88	0455
48X90420A90	0313
48X90420A92	0276
48X90420A98	0023
48X90432A01	0155
48X90432A11	1778
48X90432A14	1603
48X90432A15	0023
48X90432A32	2223
48X90432A35	0672
48X90432A36	0672
48X90432A37	0582
48X90432A41	0879
48X90432A42	0253
48X90432A54	0359
48X90432A55	0825
48X90432A57	0582
48X90445A11	2223
48X90445A17	0582
48X90445A33	0127
48X90445A34	0151
48X90445A40	0299
48X90445A52	0181
48X90445A60	2528
48X90445A63	0042
48X90445A71	0276
48X90445A74	1220
48X90445A91	0672
48X90445A93	0558
48X90445A94	0445
48X90445A96	0023
48X90445A97	1329
48X90456A04	0359
48X90456A07	4348
48X90456A11	0672
48X90456A12	0111
48X90456A16	0253
48X90456A24	0124
48X90456A26	0124
48X90456A32	0695
48X90456A34	1329
48X90456A35	0672
48X90456A47	0286
48X90456A70	0023
48X90456A71	0052
48X90456A72	0133
48X90456A73	0999
48X90456A75	0955
48X90456A76	0161
48X90456A77	2660
48X90456A80	0700
48X90456A94	0023
48X90456A95	0023
48X90456A96	0062
48X90478A04	0582
48X90478A05	1270
48X90478A06	3165
48X90478A10	0023
48X90478A13	0062
48X90478A15	0859
48X90478A16	0859
48X90478A17	3165
48X90478A18	0648
48X90478A20	1212
48X90478A21	0155
48X90478A22	1212
48X90478A23	3125
48X90478A24	4883
48X90478A26	1414
48X90478A27	0071
48X90478A27(MI-152A)	OEM
48X90478A28	1015
48X90478A29	0053
48X90478A30	0041

DEVICE TYPE	REPL CODE
48X90478A31	OEM
48X90478A45	0203
48X90478A46	0532
48X90478A47	0359
48X90478A48	0825
48X90478A50	0423
48X90478A51	0563
48X90478A52	0582
48X90478A53	0582
48X90478A54	0672
48X90478A55	0042
48X90478A57	0023
48X90478A60	0023
48X90478A61	0041
48X90478A68	0582
48X90478A69	0672
48X90478A70	0120
48X90478A71	0041
48X90478A72	0062
48X90478A73	0077
48X90478A75	0023
48X90478A77	0700
48X90478A78	0181
48X90478A91	0041
48X90478A94	4348
48X90478A98	0018
48X90488A01	0124
48X90488A16	0359
48X90488A17	0582
48X90488A42	0261
48X90488A44	0445
48X90488A45	0023
48X90488A46	0077
48X90488A47	0015
48X90488A48	0015
48X90488A49	0261
48X90488A50	0363
48X90488A51	2668
48X90488A61	1751
48X90488A62	0695
48X90488A63	0686
48X90488A66	2668
48X90488A68	0181
48X90488A69	0071
48X90488A71	OEM
48X90488A74	0359
48X90532A36	0672
48X97046A15	0222
48X97046A16	0136
48X97046A17	0079
48X97046A18	0079
48X97046A19	0151
48X97046A20	0151
48X97046A21	0151
48X97046A22	1220
48X97046A23	1220
48X97046A24	0111
48X97046A25	0079
48X97046A31	0222
48X97046A32	0440
48X97046A34	0211
48X97046A36	0203
48X97046A48	1062
48X97046A50	0079
48X97046A51	0144
48X97046A52	0086
48X97046A53	0004
48X97046A54	0004
48X97046A55	0004
48X97046A60	0016
48X97046A61	0016
48X97046A62	0016
48X97048A02	0143
48X97048A04	0015
48X97048A05	0143
48X97048A06	0143
48X97048A07	0015
48X97048A08	0137
48X97048A09	0030
48X97048A10	0015
48X97048A18	0079
48X97048A19	0143
48X97127A01	0015
48X97127A05	0211
48X97127A22	0211
48X97162A01	0144
48X97162A02	0144
48X97162A03	0050
48X97162A04	0144
48X97162A05	0016
48X97162A06	0004
48X97162A07	0004
48X97162A09	0144
48X97162A10	0144
48X97162A17	0211
48X97162A19	0211
48X97162A21	0016
48X97162A36	0164
48X97162A37	0164
48X97168A01	0143
48X97168A02	0030
48X97168A03	0143
48X97168A04	0143
48X97168A06	0015
48X97168A07	0143
48X97168A10	0015
48X97168A-11	OEM
48X97168A11	0030
48X97168A16	0015

DEVICE TYPE	REPL CODE
48X97172A01	0015
48X97177A03	0079
48X97177A10	0551
48X97177A11	0382
48X97177A12	0086
48X97177A13	0079
48X97177A14	0150
48X97177A15	0143
48X97177H01	0325
48X97221A01	0211
48X97221A02	0050
48X97221A03	0211
48X97221A04	0211
48X97221A05	0211
48X97222A01	0143
48X97222A02	0015
48X97238A01	0136
48X97238A02	0136
48X97238A03	0136
48X97238A04	0016
48X97238A05	0211
48X97238A06	0822
48X97239A01	0143
48X97271A01	0015
48X97271A02	0143
48X97271A03	0211
48X97305A02	0157
48X97305A03	0787
48X134902	0127
48X134970	0079
48X155172	0825
48X155222	0261
48X155238	1802
48X155241	2439
48X355052	0224
48X644681	0143
48X9048844	OEM
49-1	0144
49-1042	0015
49-3112	0015
49-61963	2270
49-B-1-AH	OEM
49A0000	1432
49A0002-000	1172
49A0005-000	0512
49A0006-000	0697
49A0010	0033
49A0012-000	1313
49A0510	0033
49P1C	0160
49RECT-SI-154	0023
49ST-2SC1213D-24	0191
49X90232A05	0079
50	OEM
50-30307-01	OEM
50-30307-07	0532
50-30702-06	0006
50-30702-07	0006
50-30715-06	0148
50-30715-07	0148
50-30715-08	0148
50-30718-07	0532
50-30727-05	1911
50-30737-07	1522
50-30740-06	1648
50-40101-04	0127
50-40101-05	0144
50-40102-04	0127
50-40102-05	0079
50-40105-08	0286
50-40106-09	0037
50-40201-08	0016
50-40201-09	0076
50-40201-10	0079
50-40204-10	0150
50-40205-09	0037
50-40301-02	2668
50-40306-07	0018
50-40708-06	0076
50-40708-07	0076
50-40708-09	0076
50-40711-06	0066
50-40711-07	0066
50-40712-08	0866
50-40724-03	0055
50-40725-06	0313
50-40726-05	1284
50-40727-08	0261
50-40733-04	0261
50-40734-08	0270
50-40735-07	1562
50-40741-12	2069
50-40743-01	0556
50-40744-06	0392
50-41001-01	2660
50-41302-01	1493
50-50704-11	3046
50A	0015
50A52	0136
50A102	0136
50A103	0126
50A103K	0136
50AC40A	OEM
50AC60A	OEM
50AC80A	OEM
50AC100A	OEM
50AC120A	OEM
50AD50	OEM
50AD100	OEM

DEVICE TYPE	REPL CODE
50AD200	OEM
50AD400	OEM
50AD600	OEM
50AD800	OEM
50AS	0015
50B5	0015
50B54	0004
50B173-C	0004
50B173-S	0004
50B173C	0211
50B175-B	0280
50B175A	0004
50B175B	0004
50B175C	0004
50B324	0164
50B364	0164
50B415	0164
50B423	0004
50BU75-C	0004
50C	0015
50C05	3762
50C05R	2665
50C5R	2182
50C6-7	OEM
50C6.5-7.5	OEM
50C10	2653
50C10R	2695
50C20	2657
50C20R	2768
50C30	2664
50C30R	2855
50C40	2006
50C40R	2907
50C50	2677
50C50R	2974
50C60	2682
50C60R	3051
50C70	3900
50C70R	3117
50C80	3900
50C80R	3117
50C90	2655
50C90R	3208
50C100	2655
50C100R	3208
50C371	0016
50C372	0016
50C373	0016
50C374	0016
50C380-0	0127
50C380-0R	0127
50C380-OR	0127
50C394-0	0127
50C394-R	0127
50C538	0016
50C644	0016
50C784	0144
50C784-R	0127
50C828	0016
50C829	0144
50C829B	0144
50C829C	0144
50C838	0016
50C1047	0144
50CJ139	0079
50CZ5.0	OEM
50CZ5.0A	OEM
50CZ5.0B	OEM
50CZ6.0	OEM
50CZ6.0A	OEM
50CZ6.0B	OEM
50CZ6.5	OEM
50CZ6.5A	OEM
50CZ6.5B	OEM
50CZ7.0	OEM
50CZ7.0A	OEM
50CZ7.0B	OEM
50CZ7.5	OEM
50CZ7.5A	OEM
50CZ7.5B	OEM
50CZ8.0	OEM
50CZ8.0A	OEM
50CZ8.0B	OEM
50CZ8.5	OEM
50CZ8.5A	OEM
50CZ8.5B	OEM
50CZ9.0	OEM
50CZ9.0A	OEM
50CZ9.0B	OEM
50CZ10	OEM
50CZ10A	OEM
50CZ10B	OEM
50CZ11	OEM
50CZ11A	OEM
50CZ11B	OEM
50CZ12	OEM
50CZ12A	OEM
50CZ12B	OEM
50CZ13	OEM
50CZ13A	OEM
50CZ13B	OEM
50CZ14	OEM
50CZ14A	OEM
50CZ14B	OEM
50CZ15	OEM
50CZ15A	OEM
50CZ15B	OEM
50CZ16	OEM
50CZ16A	OEM

DEVICE TYPE	REPL CODE
50CZ16B	OEM
50CZ17	OEM
50CZ17A	OEM
50CZ17B	OEM
50CZ18	OEM
50CZ18A	OEM
50CZ18B	OEM
50CZ20	OEM
50CZ20A	OEM
50CZ20B	OEM
50CZ22	OEM
50CZ22A	OEM
50CZ22B	OEM
50CZ24	OEM
50CZ24A	OEM
50CZ24B	OEM
50CZ26	OEM
50CZ26A	OEM
50CZ26B	OEM
50CZ28	OEM
50CZ28A	OEM
50CZ28B	OEM
50CZ30	OEM
50CZ30A	OEM
50CZ30B	OEM
50CZ33	OEM
50CZ33A	OEM
50CZ33B	OEM
50CZ36	OEM
50CZ36A	OEM
50CZ36B	OEM
50CZ40	OEM
50CZ40A	OEM
50CZ40B	OEM
50CZ43	OEM
50CZ43A	OEM
50CZ43B	OEM
50CZ45	OEM
50CZ45A	OEM
50CZ45B	OEM
50CZ48	OEM
50CZ48A	OEM
50CZ48B	OEM
50CZ51	OEM
50CZ51A	OEM
50CZ51B	OEM
50CZ54	OEM
50CZ54A	OEM
50CZ54B	OEM
50CZ58	OEM
50CZ58A	OEM
50CZ58B	OEM
50CZ60	OEM
50CZ60A	OEM
50CZ60B	OEM
50CZ64	OEM
50CZ64A	OEM
50CZ64B	OEM
50CZ70	OEM
50CZ70A	OEM
50CZ70B	OEM
50CZ75	OEM
50CZ75A	OEM
50CZ75B	OEM
50CZ78	OEM
50CZ78A	OEM
50CZ78B	OEM
50CZ85	OEM
50CZ85A	OEM
50CZ85B	OEM
50CZ90	OEM
50CZ90A	OEM
50CZ90B	OEM
50CZ100	OEM
50CZ100A	OEM
50CZ100B	OEM
50CZ110	OEM
50CZ110A	OEM
50CZ110B	OEM
50D1	OEM
50D2	0015
50D4	0015
50D6	OEM
50D8	0071
50D10	0071
50D12	OEM
50D14	OEM
50D15	OEM
50D16	OEM
50D18	OEM
50D20	OEM
50D25	OEM
50D30	OEM
50D40	OEM
50D50	OEM
50E03A	OEM
50E05	0015
50E05A	OEM
50E1	0015
50E2	0015
50E3	0015
50E4	0015
50E5	0015
50E6	0015
50E7	0071
50E8	0071
50E10	0071
50E12	OEM
50E14	OEM

If replacement code is OEM, contact original manufacturer for replacement.

DEVICE TYPE	REPL CODE
50E16	5351
50E18	OEM
50E20	OEM
50F1	2076
50F5	0200
50F20	OEM
50F30	OEM
50F40	OEM
50F50	OEM
50G6P41	OEM
50H3	OEM
50H3N	0476
50H3P	0200
50H3PN	1001
50HQ020	1545
50HQ030	1545
50HQ035	1545
50HQ040	1545
50HQ045	1545
50HV100AV10L	OEM
50J	OEM
50J1	1771
50J2P	0116
50J3P	0116
50J6P41	OEM
50JH3	2871
50K	OEM
50K10	OEM
50K20	OEM
50K30	OEM
50K40	OEM
50K50	OEM
50KZ5.0	OEM
50KZ5.0A	OEM
50KZ5.0B	OEM
50KZ6.0	OEM
50KZ6.0A	OEM
50KZ6.0B	OEM
50KZ6.5	OEM
50KZ6.5A	OEM
50KZ6.5B	OEM
50KZ7.0	OEM
50KZ7.0A	OEM
50KZ7.0B	OEM
50KZ7.5	OEM
50KZ7.5A	OEM
50KZ7.5B	OEM
50KZ8.0	OEM
50KZ8.0A	OEM
50KZ8.0B	OEM
50KZ8.5	OEM
50KZ8.5A	OEM
50KZ8.5B	OEM
50KZ9.0	OEM
50KZ9.0A	OEM
50KZ9.0B	OEM
50KZ10	OEM
50KZ10A	OEM
50KZ10B	OEM
50KZ11	OEM
50KZ11A	OEM
50KZ11B	OEM
50KZ12	OEM
50KZ12A	OEM
50KZ12B	OEM
50KZ13	OEM
50KZ13A	OEM
50KZ13B	OEM
50KZ14	OEM
50KZ14A	OEM
50KZ14B	OEM
50KZ15	OEM
50KZ15A	OEM
50KZ15B	OEM
50KZ16	OEM
50KZ16A	OEM
50KZ16B	OEM
50KZ17	OEM
50KZ17A	OEM
50KZ17B	OEM
50KZ18	OEM
50KZ18A	OEM
50KZ18B	OEM
50KZ20	OEM
50KZ20A	OEM
50KZ20B	OEM
50KZ22	OEM
50KZ22A	OEM
50KZ22B	OEM
50KZ24	OEM
50KZ24A	OEM
50KZ24B	OEM
50KZ26	OEM
50KZ26A	OEM
50KZ26B	OEM
50KZ28	OEM
50KZ28A	OEM
50KZ28B	OEM
50KZ30	OEM
50KZ30A	OEM
50KZ30B	OEM
50KZ33	OEM
50KZ33A	OEM
50KZ33B	OEM
50KZ36	OEM
50KZ36A	OEM
50KZ36B	OEM
50KZ40	OEM
50KZ40A	OEM
50KZ40B	OEM
50KZ43	OEM
50KZ43A	OEM
50KZ43B	OEM
50KZ45	OEM
50KZ45A	OEM
50KZ45B	OEM
50KZ48	OEM
50KZ48A	OEM
50KZ48B	OEM
50KZ51	OEM
50KZ51A	OEM
50KZ51B	OEM
50KZ54	OEM
50KZ54A	OEM
50KZ54B	OEM
50KZ58	OEM
50KZ58A	OEM
50KZ58B	OEM
50KZ60	OEM
50KZ60A	OEM
50KZ60B	OEM
50KZ64	OEM
50KZ64A	OEM
50KZ64B	OEM
50KZ70	OEM
50KZ70A	OEM
50KZ70B	OEM
50KZ75	OEM
50KZ75A	OEM
50KZ75B	OEM
50KZ78	OEM
50KZ78A	OEM
50KZ78B	OEM
50KZ85	OEM
50KZ85A	OEM
50KZ85B	OEM
50KZ90	OEM
50KZ90A	OEM
50KZ90B	OEM
50KZ100	OEM
50KZ100A	OEM
50KZ100B	OEM
50KZ110	OEM
50KZ110A	OEM
50KZ110B	OEM
50L6P41	OEM
50L15	OEM
50L20	OEM
50L25	OEM
50L30	OEM
50L35	OEM
50L40	OEM
50L45	OEM
50L50	OEM
50L55	OEM
50L60	OEM
50L65	OEM
50L70	OEM
50L75	OEM
50L80	OEM
50L85	OEM
50L90	OEM
50L95	OEM
50L100	OEM
50L110	OEM
50L120	OEM
50L130	OEM
50L140	OEM
50L150	OEM
50L160	OEM
50L170	OEM
50L180	OEM
50L190	OEM
50L200	OEM
50L210	OEM
50L220	OEM
50L230	OEM
50L240	OEM
50L250	OEM
50L260	OEM
50L270	OEM
50L280	OEM
50L290	OEM
50L300	OEM
50LA	OEM
50LF	OEM
50LF11	OEM
50M	0015
50M9.1SZ	OEM
50M9.1SZ5	0994
50M24Z5	0274
50M33Z5	0318
50M62Z5	0386
50MV80	OEM
50MV100	OEM
50MV120	OEM
50N15A	OEM
50N15A9	OEM
50N20A	OEM
50N20A9	OEM
50N25A	OEM
50N25A9	OEM
50N30A	OEM
50N30A9	OEM
50N35A	OEM
50N35A9	OEM
50N40A	OEM
50N40A9	OEM
50N45A	OEM
50N45A9	OEM
50N50A	OEM
50N50A9	OEM
50N55A	OEM
50N55A9	OEM
50N60A	OEM
50N60A9	OEM
50N65A	OEM
50N65A9	OEM
50N70A	OEM
50N70A9	OEM
50N75A	OEM
50N75A9	OEM
50N80A	OEM
50N80A9	OEM
50N85A	OEM
50N85A9	OEM
50N90A	OEM
50N90A9	OEM
50N95A	OEM
50N95A9	OEM
50N100A	OEM
50N100A9	OEM
50N110A	OEM
50N120A	OEM
50N130A	OEM
50N140A	OEM
50N150A	OEM
50N160A	OEM
50N170A	OEM
50N180A	OEM
50N190A	OEM
50N200A	OEM
50N210A	OEM
50N220A	OEM
50N230A	OEM
50N240A	OEM
50NF11	OEM
50P1SP	OEM
50P2	0136
50P2SG	OEM
50P2SP	OEM
50P3	0136
50P3SP	OEM
50P4SG	OEM
50P4SP	OEM
50P5SP	OEM
50P6SG	OEM
50P6SP	OEM
50P8SG	OEM
50P8SP	OEM
50P10SG	OEM
50P10SP	OEM
50P12SP	OEM
50P16SP	OEM
50Q6P42	OEM
50QF11	OEM
50R2	OEM
50R2S	0586
50R3P	OEM
50RCS5	1704
50RCS5S60	1704
50RCS10	1704
50RCS10S60	1704
50RCS20	1704
50RCS20S60	1704
50RCS30	4031
50RCS30S60	4031
50RCS40	4031
50RCS40S60	4031
50RCS50	OEM
50RCS60	4031
50RCS60S60	4031
50RCS70	OEM
50RCS80	0679
50RCS80S60	0679
50RCS90	OEM
50RCS100	0679
50RCS100S60	0679
50RCS110	0679
50RCS110S60	0679
50RCS120	0679
50RCS120S60	0679
50RIA5	OEM
50RIA10	1704
50RIA10M	OEM
50RIA20	1704
50RIA20M	OEM
50RIA30	OEM
50RIA40	4031
50RIA40M	OEM
50RIA60	4031
50RIA60M	OEM
50RIA80	0679
50RIA80M	OEM
50RIA100	0679
50RIA100M	OEM
50RIA110	0679
50RIA110M	OEM
50RIA120	0679
50RIA120M	OEM
50S2SP	OEM
50S2.5-3.5	OEM
50S3P	OEM
50S3SP	OEM
50S3.4	OEM
50S4SP	OEM
50S5	2076
50S5SP	OEM
50S6SP	OEM
50S8SP	OEM
50S20	OEM
50S30	OEM
50S40	OEM
50S50	OEM
50SI-HA11446	2122
50SQ030	OEM
50SQ035	OEM
50SQ040	OEM
50SQ045	OEM
50SQ060	OEM
50SQ080	OEM
50SQ090	OEM
50SQ0100	OEM
50SS2SP	OEM
50SS3SP	OEM
50SS4SP	OEM
50SS5SP	OEM
50SS6SP	OEM
50SS8SP	OEM
50ST-2SC945R	0076
50ST-2SD467(B)	0951
50ST-2SD467B	0951
50SZ-HZ12A2	1005
50T3P	OEM
50T6.8	0926
50T6.8A	0926
50T6.8B	0926
50T7.5	0944
50T7.5A	0944
50T7.5B	0944
50T8.2	0958
50T8.2A	0958
50T8.2B	0958
50T9.1	0994
50T9.1A	0994
50T9.1B	0994
50T10	1009
50T10A	1009
50T10B	1009
50T11	1019
50T11A	1019
50T11B	1019
50T12	0889
50T12A	0889
50T12B	0889
50T13	1049
50T13A	1049
50T13B	1049
50T15	0932
50T15A	0932
50T15B	0932
50T16	1079
50T16A	1079
50T16B	1079
50T18	1106
50T18A	1106
50T18B	1106
50T20	1135
50T20A	1135
50T20B	1135
50T22	1152
50T22A	1152
50T22B	1152
50T24	1164
50T24A	1164
50T24B	1164
50T27	1184
50T27A	1184
50T27B	1184
50T30	1210
50T30A	1210
50T30B	1210
50T33	1223
50T33A	1223
50T33B	1223
50T36	1232
50T36A	1232
50T36B	1232
50T39	1246
50T39A	1246
50T39B	1246
50T43	1265
50T43A	1265
50T43B	1265
50T45	1273
50T47	1290
50T47A	1290
50T47B	1290
50T51	1307
50T51A	1307
50T51B	1307
50T62	1353
50T62A	1353
50T62B	1353
50T68	1365
50T68A	1365
50T68B	1365
50T75	1384
50T75A	1384
50T75B	1384
50T82	1401
50T82A	1401
50T82B	1401
50T91	1412
50T91A	1412
50T91B	1412
50T100	1423
50T100A	1423
50T100B	1423
50T110	1444
50T110A	1444
50T110B	1444
50T120	1457
50T120A	1457
50T120B	1457
50T130	1470
50T130A	1470
50T130B	1470
50T150	1505
50T150A	1505
50T150B	1505
50T160	1521
50T160A	1521
50T160B	1521
50T180	1548
50T180A	1548
50T180B	1548
50T200	1587
50T200A	1587
50T200B	1587
50T801	OEM
50U6P42	OEM
50V3P	OEM
50V10	OEM
50V11	OEM
50W3P	OEM
50X3P	OEM
50X7-8	OEM
50X7.5-8.5	OEM
50X8-9	OEM
50X8.5-9.5	OEM
50X9-10A	OEM
50XV	OEM
50Y3P	OEM
50Z6.2	0914
50Z6.2A	0914
50Z6.2B	0914
50Z6.8	0926
50Z6.8A	0926
50Z6.8B	0926
50Z7.5	0944
50Z7.5A	0944
50Z7.5B	0944
50Z8.2	0958
50Z8.2A	0958
50Z8.2B	0958
50Z9.1	0012
50Z9.1A	0012
50Z9.1B	0012
50Z9.1C	0012
50Z10	1009
50Z10A	1009
50Z10B	1009
50Z11	0794
50Z11A	0794
50Z11B	0794
50Z12	0137
50Z12A	0137
50Z12B	0137
50Z12C	0137
50Z13	1049
50Z13A	1049
50Z13B	1049
50Z14	0100
50Z14A	0100
50Z14B	0100
50Z15	0002
50Z15A	0002
50Z15B	0002
50Z15C	0002
50Z16	1079
50Z16A	1079
50Z16B	1079
50Z17	1096
50Z17A	1096
50Z17B	1096
50Z18	1106
50Z18A	1106
50Z18B	1106
50Z19	1123
50Z19A	1123
50Z19B	1123
50Z20	1135
50Z20A	1135
50Z20B	1135
50Z22	1152
50Z22A	1152
50Z22B	1152
50Z24	1164
50Z24A	1164
50Z24B	1164
50Z25	1174
50Z25A	1174
50Z25B	1174
50Z27	0441
50Z27A	0441
50Z27B	0441
50Z27C	0441
50Z30	1210
50Z30A	1210
50Z30B	1210
50Z33	0039
50Z33A	0039
50Z33B	0039
50Z33C	0039
50Z36	1232
50Z36A	1232
50Z36B	1232
50Z39	1246
50Z39A	1246
50Z39B	1246
50Z43	1265
50Z43A	1265
50Z43B	1265
50Z45	1273
50Z45A	1273
50Z45B	1273
50Z47	1290
50Z47A	1290
50Z47B	1290
50Z50	1299
50Z50A	1299
50Z50B	1299
50Z51	1307
50Z51A	1307
50Z51B	1307
50Z52	1322
50Z52A	1322
50Z52B	1322
50Z52D	1322
50Z52D5	1322
50Z52D10	1322
50Z56	0783
50Z56A	0783
50Z56B	0783
50Z62	0783
50Z62A	0783
50Z62B	0783
50Z68	1365
50Z68A	1365
50Z68B	1365
50Z75	1384
50Z75A	1384
50Z75B	1384
50Z82	0331
50Z82A	0331
50Z82B	0331
50Z82C	0331
50Z91	1412
50Z91A	1412
50Z91B	1412
50Z100	1423
50Z100A	1423
50Z100B	1423
50Z105	OEM
50Z105A	OEM
50Z105B	OEM
50Z110	0149
50Z110A	0149
50Z110B	0149
50Z120	1457
50Z120A	1457
50Z120B	1457
50Z130	1470
50Z130A	1470
50Z130B	1470
50Z140	1484
50Z140A	1484
50Z140B	1484
50Z140D	1484
50Z140D10	1484
50Z150	1505
50Z150A	1505
50Z150B	1505
50Z160	1505
50Z160A	1505
50Z160B	1505
50Z175	1531
50Z175A	1531
50Z175B	1531
50Z175D	1531
50Z175D5	1531
50Z175D10	1531
50Z180	1548
50Z180A	1548
50Z180B	1548
50Z200	1587
50Z200A	1587
50Z200B	1587
50Z1040D5	OEM
51	0144
51-01003-04	0133
51-02004-12	0041
51-02006-12	0165
51-02007-12	0057
51-02008-12	0062
51-02020-05	0064
51-02025-04	0091
51-02029-04	0137
51-02032-04	0077
51-03007-06	0023
51-03009A02	0785
51-03015-02	0080
51-03017-01	0795
51-03020-02	0023
51-03021-02	0023
51-03024-02	0023
51-03025-04	0133
51-03026-04	0023
51-03029-04	0023
51-03030-04	0023
51-03035-04	0087
51-04001-01	0123
51-06002-00	0554
51-08001-05	0124
51-08001-11	0124
51-08004-06	0124
51-08006-02	0124
51-4	0276
51-903	1143
51-1059A01	1345
51-1375A05	OEM
51-3009A02	OEM
51-3009A07	OEM
51-10276A01	2624
51-10302A01	0632
51-10302A1	0632
51-10305A43	OEM
51-10422A01	0654
51-10422A02	0654
51-10437A01	0654
51-10534A01	3008
51-10534A03	2027
51-10534A04	3008
51-10542A01	0623
51-10566A01	0701
51-10594A01	1345
51-10600A	6845
51-10600A01	4926
51-10611A09	2100
51-10611A10	OEM
51-10611A11	0232
51-10611A12	0361
51-10611A15	0743
51-10611A16	1432
51-10619A01	1345
51-10631A01	0664
51-10634A01	OEM
51-10636A	1980
51-10636A01	1980
51-10637A01	2747
51-10638A01	1843
51-10650A01	0360
51-10655A03	0856
51-10655A17	0478
51-10655A19	0413
51-10655A21	1143
51-10655B02	0789
51-10658A01	2755
51-10658A02	2755
51-10658A03	2755
51-10672A01	1759
51-10678A01	1843
51-10679A13	0789
51-10711A01	1395
51-10711A02	1395
51-10726A01	OEM
51-10726A02	OEM
51-10730A01	OEM
51-10740A02	OEM
51-13753-A19	0521
51-13753A01	1055
51-13753A04	4038
51-13753A10	3269
51-13753A11	0167
51-13753A18	3273
51-13753A19	0521
51-13753A20	2789
51-13753A21	3787
51-13753A22	2623
51-13753A23	4495
51-13753A24	5095
51-13753A25	3609
51-13753A26	3738
51-13753A27	4549
51-13753A28	2623
51-13753A29	4060
51-13753A35	3738
51-13753A39	3738
51-13753A40	1170
51-13753A46	2049
51-13753A47	3270
51-13753A48	3351
51-13753A49	3359
51-13753A50	3089
51-13753A52	3502
51-13753A53	3282
51-13753A54	2623
51-13753A55	1429
51-13763A22	3566
51-13763A27	2623
51-15066A01	0701
51-40000S08	1048
51-40000S09	OEM
51-40464P01	0520
51-40464P01	OEM
51-41850J01	6202
51-42211P01	2534
51-42211P01	OEM
51-42908J01	0520
51-43639B02	2027
51-43684B01	3008
51-43684B02	2027
51-43684B0P	2027
51-44789J01	2019
51-44789J02	2019
51-44837J04	3262
51-70177A07	0623
51-84320A09	1980
51-90305A04	0852
51-90305A05	0857
51-90305A6	0167
51-90305A17	OEM

If replacement code is OEM, contact original manufacturer for replacement.

DEVICE TYPE	REPL CODE
51-90305A20	4034
51-90305A21	2960
51-90305A24	4067
51-90305A27	4079
51-90305A30	4154
51-90305A31	4158
51-90305A37	4759
51-90305A39	4744
51-90305A40	1580
51-90305A41	4758
51-90305A43	4747
51-90305A44	4751
51-90305A45	5071
51-90305A46	4162
51-90305A50	0395
51-90305A59	1329
51-90305A60	2581
51-90305A61	0167
51-90305A62	4746
51-90305A72	2295
51-90305A73	4752
51-90305A77	4729
51-90305A78	1260
51-90305A87	0167
51-90305A88	4857
51-90305A89	1429
51-90305A90	0395
51-90305A91	0167
51-90305A92	4060
51-90433A07	3862
51-90433A08	2259
51-90433A09	1029
51-90433A10	0119
51-90433A11	0398
51-90433A12	1599
51-90433A13	1143
51-90433A27	2259
51-90433A30	2259
51-90433A32	1580
51-90433A33	1582
51-90433A34	1585
51-90433A35	4751
51-90433A36	OEM
51-90433A52	6865
51-90433A53	3609
51-90433A54	6868
51-90433A55	6150
51-90433A56	OEM
51-90433A60	4038
51-90433A63	2295
51-90433A64	OEM
51-90433A66	0857
51-90433A86	OEM
51-90433A87	0679
51-355051	OEM
51-10422801	0654
51-40000508	OEM
51-40000509	OEM
51-43639802	2027
51-44789302	OEM
51-44837504	OEM
51A	OEM
51B	OEM
51C43684B02	2027
51C436884B01	2027
51D170	0211
51D176	0004
51D188	0280
51D189	0280
51D70177A01	1753
51D70177A02	0789
51D70177A02(C)	2393
51D70177A07	0167
51D70177B02	0789
51G10679A03	0856
51G10679A13	0789
51G10679A14	OEM
51GX3	0474
51GX14	0474
51HQ045	1545
51HV100AV14L	OEM
51HV100AV16L	OEM
51IN60P	0143
51IS34S	0143
51LN60P	0143
51LS34S	0143
51M33801A01	4038
51M33801A04	1055
51M33801A05	3273
51M33801A06	2942
51M33801A07	3269
51M70177A01	1753
51M70177A03	0856
51M70177A05	0395
51M70177A07	0623
51M70177B02	0789
51N3M	1681
51P	OEM
51P2	0136
51P4	0136
51PB	OEM
51PBH	OEM
51PBHL	OEM
51PBL	OEM
51PL	OEM
51R8432A09	1980
51R84320A09	1980
51RC10	0608
51RC20	0608
51RC30	0610
51RC40	0610
51RC50	0610
51RC60	0610
51RC70	0468
51RC80	0468
51RC100	0468
51RC120	0468
51RCG5	OEM
51RCG10	OEM
51RCG15	OEM
51RCG20	OEM
51RCG25	OEM
51RCG30	OEM
51RCG40	OEM
51RCG50	OEM
51RCG80	OEM
51S02	OEM
51S1S34	0143
51S105FA01	0701
51S137A48	3351
51S137A50	3089
51S1048A01	0167
51S1056GA01	0701
51S1059A01	0701
51S1373A40	1170
51S3753A48	3351
51S10276A01	2624
51S10302A01	0632
51S10382A	0654
51S10408A01	0829
51S10422A	0654
51S10422A01	0654
51S10432A01	3724
51S10437A01	0654
51S10534A01	2027
51S10534A03	2027
51S10566A01	0701
51S10585A01	OEM
51S10592A01	0701
51S10594A01	1345
51S10600A	2457
51S10600A01	3779
51S10611A09	2100
51S10611A10	OEM
51S10611A11	0232
51S10611A12	0361
51S10611A15	0743
51S10611A16	1432
51S10636A	1980
51S10636A01	1980
51S10638A01	1843
51S10655A01	1753
51S10655A15	OEM
51S10655A17	0478
51S10655A18	0215
51S10655A19	0413
51S10655B01	1753
51S10655B03	0856
51S10655B05	OEM
51S10655B13	0167
51S10655C05	0395
51S13747A03	2546
51S13749A46	2049
51S13752A47	3270
51S13753A01	1055
51S13753A02	3269
51S13753A03	2942
51S13753A06	3273
51S13753A07	0167
51S13753A08	2602
51S13753A09	0167
51S13753A10	3269
51S13753A11	0167
51S13753A23	4495
51S13753A28	2623
51S13753A37	5215
51S13753A39	3738
51S13753A40	1170
51S13753A46	2049
51S13753A47	3270
51S13753A48	3351
51S13753A49	3359
51S13753A50	3089
51S13753A52	3502
51S13753A53	3282
51S13753A55	1429
51S13753A56	2623
51S13753A60	7104
51S13753A64	3089
51S13753A65	2122
51S13753A66	OEM
51S13753A67	3916
51S13753A68	5208
51S23753A47	3270
51S33753A47	3270
51S44789J01	2019
51S44789J02	2019
51S90518A12	6784
51S1059401	1337
51SI13753A40	OEM
51SZ-UZ-4.0C	0275
51T40113T01	4933
51X2	OEM
51X9050A78	OEM
51X13753A24	5095
51X90226A72	OEM
51X90266A15	0631
51X90266A26	OEM
51X90266A45	0770
51X90305A21	2960
51X90305A41	4758
51X90305A45	5071
51X90305A46	4162
51X90305A59	1329
51X90305A61	0167
51X90305A88	4857
51X90433A12	1599
51X90433A28	4750
51X90433A33	1582
51X90433A35	4751
51X90433A47	OEM
51X90433A52	5193
51X90433A53	3609
51X90433A54	2019
51X90433A55	6150
51X90433A56	OEM
51X90433A99	5194
51X90458A01	1585
51X90458A02	5546
51X90458A03	5227
51X90458A04	OEM
51X90458A05	5547
51X90458A06	5192
51X90458A07	0176
51X90458A15	2053
51X90458A30	3738
51X90458A32	OEM
51X90458A33	OEM
51X90458A34	OEM
51X90458A50	3738
51X90458A51	0167
51X90458A52	5094
51X90458A55	3503
51X90458A56	4759
51X90458A71	3074
51X90458A72	3084
51X90458A73	2019
51X90458A74	OEM
51X90458A75	3759
51X90458A76	OEM
51X90458A77	OEM
51X90458A78	5562
51X90458A79	OEM
51X90458A80	3753
51X90458A81	5225
51X90458A82	5169
51X90458A83	0360
51X90458A84	0360
51X90458A88	0101
51X90458A89	0921
51X90458A90	3198
51X90458A91	OEM
51X90458A92	OEM
51X90458A93	0176
51X90458A94	0176
51X90458A95	1349
51X90458A96	1206
51X90458A97	OEM
51X90458A98	0923
51X90480A02	OEM
51X90480A03	OEM
51X90480A07	0624
51X90480A08	0624
51X90480A09	0334
51X90480A16	0101
51X90480A17	0921
51X90480A18	2053
51X90480A20	0683
51X90480A32	3730
51X90480A41	3731
51X90480A46	0923
51X90480A49	OEM
51X90480A57	2623
51X90480A64	3744
51X90480A65	3747
51X90480A66	0334
51X90480A92	0489
51X90480A93	3759
51X90480A94	1170
51X90480A95	3503
51X90480A96	0976
51X90495A14	3503
51X90507A20	2019
51X90507A21	OEM
51X90507A24	0923
51X90507A28	2025
51X90507A29	2026
51X90507A30	OEM
51X90507A39	0770
51X90507A40	0770
51X90507A46	0699
51X90507A77	2066
51X90507A78	2067
51X90507A79	OEM
51X90507A80	2068
51X90507A81	OEM
51X90507A82	OEM
51X90507A83	2025
51X90507A84	0478
51X90507A85	0129
51X90507A86	OEM
51X90507A87	2073
51X90507A88	OEM
51X90507A91	OEM
51X90507A92	1014
51X90507A93	1957
51X90507A94	OEM
51X90507A95	OEM
51X90507A97	2055
51X90507A98	0498
51X90507A99	0119
51X90518A01	OEM
51X90518A02	OEM
51X90518A03	OEM
51X90518A07	0624
51X90518A08	0056
51X90518A09	5229
51X90518A13	5544
51X90518A19	2553
51X90518A28	3198
51X90518A34	2053
51X90518A35	1599
51X90518A36	0215
51X90518A55	OEM
51X90518A68	OEM
51X90518A69	0973
51X90518A71	0969
51X90518A72	OEM
51X90518A73	OEM
51X90518A74	OEM
51X90518A89	OEM
51X90518A90	OEM
51X90518A95	OEM
51X90518A97	5191
51X90580A41	OEM
51X90597A88	OEM
51X904080A96	0976
51X9048095	OEM
51Z6	0782
51Z6A	0782
52-010-106-0	0532
52-010-109-0	0342
52-010-151-0	0622
52-020-108-0	0016
52-020-166-0	0224
52-020-167-0	0224
52-020-169-0	0224
52-020-173-0	2050
52-025-004-0	0103
52-040-009-0	0579
52-040-010-0	1215
52-045-001-0	1235
52-050-021-0	0143
52-051-008-0	0143
52-051-009-0	0124
52-051-017-0	0143
52-052-004-0	0123
52-053-005-0	0039
52-053-013-0	0012
52-062-007-0	0139
52-1	0015
52-4	0276
52A011	0015
52A011-1	0071
52A15	0143
52A26	0636
52BBIA	0015
52BBLA	0015
52C04	0004
52D189	0164
52DS-18	0015
52GT	OEM
52HQ030	1545
52HQ035	1545
52HQ040	1545
52HQ045	1545
52HV100AV18L	OEM
52J	OEM
52L	OEM
52P	OEM
52PB	OEM
52PBH	OEM
52PBHL	OEM
52PBL	OEM
52PL	OEM
52RCG5	OEM
52RCG10	OEM
52RCG15	OEM
52RCG20	OEM
52RCG25	OEM
52RCG30	OEM
52RCG40	OEM
52RCG50	OEM
52RCG60	OEM
52SD-1	0015
52SD1	0015
52SI-SAA1061	OEM
52ST-2SD467C	0951
52ST-JE9018G	0144
52TA7611AP-S	0069
52X2	OEM
52Z4	OEM
52Z6	2397
52Z6A	2397
53-0051-2	0015
53-0082-3	0071
53-0082-1003	0229
53-0086-1	0015
53-0088-1	0015
53-0088-2	0229
53-0088-3	0071
53-0088-1002	0071
53-0088-1003	0071
53-0088-1004	0071
53-0094-1	0015
53-0095-1	0015
53-0098-1	0015
53-0099-1	0015
53-0099-3	0015
53-0099-4	0015
53-0106-1	0015
53-0106-1001	0071
53-0109-1	0015
53-0111-1	0229
53-0111-1001	0071
53-0113-1	0015
53-0120-1	0015
53-0122-1	0229
53-0136T	0071
53-1110	0016
53-1173	0400
53-1362	0560
53-1487	2158
53-1516	0037
53-1517	0940
53-1967	0560
53A001-1	0143
53A001-2	0143
53A001-3	0015
53A001-4	0071
53A001-5	0025
53A001-6	0030
53A001-9	0015
53A001-10	0137
53A001-12	0015
53A001-13	OEM
53A001-31	OEM
53A001-33	0100
53A001-34	0071
53A001-35	0015
53A006-1	0143
53A008-1	0143
53A009-1	0196
53A0016	0554
53A0081	0143
53A010-1	0087
53A011-1	0015
53A014-1	0102
53A015-1	0102
53A016-1	0098
53A017-1	0102
53A018-1	1149
53A019-1	0133
53A020-1	0133
53A022-1	0015
53A022-2	0143
53A022-3	0133
53A022-4	0015
53A022-5	0102
53A022-6	0012
53A022-7	0012
53A022-8	OEM
53A022-10	0201
53A030-1	0133
53B001-1	0143
53B001-2	0143
53B001-3	0143
53B001-5	0015
53B001-6	0015
53B001-7	0133
53B001-8	0030
53B001-9	0133
53B003-1	0015
53B003-2	0015
53B004-1	0143
53B004-2	0143
53B005-2	0143
53B006-1	0015
53B007-1	0143
53B008-1	0917
53B010-1	0196
53B010-2	0196
53B010-3	0015
53B010-4	0015
53B010-5	0087
53B010-6	0133
53B010-7	0123
53B010-8	0920
53B010-9	0920
53B010-10	0102
53B011-1	0015
53B011-2	0015
53B011-3	0774
53B012-1	0065
53B013-1	0133
53B014-1	0133
53B015-1	0062
53B017-1	0924
53B018-1	0277
53B019-1	0201
53B019-2	0015
53B019-3	0196
53B020-2	0015
53B020-3	0030
53C001	0143
53C001-1	0143
53C002-1	0302
53C003-1	0015
53C005-3	0015
53C006-1	0143
53C006-2	0143
53C006-52	0143
53C007-1	0015
53C008-1	0196
53C009-1	0015
53C009-2	0143
53C011-4	0201
53C012-1	0015
53C013-1	0188
53C014-1	0015
53C015-1	0182
53C016-1	0015
53C017-1	0102
53C020-1	0143
53C021-1	0302
53C022-1	0015
53CD01-1	0143
53D001-1	0030
53D002-1	0143
53D002-2	0143
53D003-1	0015
53D003-2	0015
53D5R08	OEM
53D9R08GH	OEM
53E001-1	0143
53E003-1	0143
53E003-2	0133
53E006-1	0143
53E009-1	0030
53E010-1	0015
53E011-1	0277
53E011-2	0636
53F001-1	0015
53F002-1	0015
53H001-1	0015
53H001-2	0143
53J001-1	0087
53J002-1	0087
53J002-2	0229
53J003-1	0015
53J003-2	0133
53J004-1	0636
53J004-2	0365
53J004-3	0649
53K001-2	0143
53K001-3	0030
53K001-4	0301
53K001-5	0143
53K001-6	0015
53K001-6(5,7)	0015
53K001-7	0015
53K001-7(6,8)	0015
53K001-9	0229
53K001-10	OEM
53K001-11	0015
53K001-13	OEM
53K001-14	0012
53K001-15	0030
53K001-16	OEM
53K001-18	0012
53L001-1	0123
53L001-5	0023
53L001-10	0019
53L001-14	0023
53L001-15	0133
53LS140J16	OEM
53LS140N16	OEM
53LS141J16	OEM
53LS141N16	OEM
53LS240J16	OEM
53LS240N16	OEM
53LS241J16	OEM
53LS241N16	OEM
53N001-1	0143
53N001-2	0015
53N001-3	0015
53N001-4	0565
53N001-5	0133
53N001-6	0030
53N001-7	0143
53N001-9	0402
53N002-2	0015
53N003-1	0143
53N003-2	0143
53N003-3	0123
53N004-1	0030
53N004-5	0143
53N004-6	0143
53N004-7	0015
53N004-8	0015
53N004-9	0015
53N004-10	0123
53N004-11	0143
53N004-12	0133
53N004-13	0030
53N004-14	0143
53N49	0440
53P151	0198
53P153	0160
53P157	0211
53P158	0198
53P159	0198
53P161	1221
53P162	1221
53P163	0198
53P165	0076
53P166	0006
53P169	0086
53P170	0126
53RA441J	OEM
53RA441N	OEM
53S080J	OEM
53S080N	OEM
53S081J	OEM
53S081N	OEM
53S140J16	OEM
53S140N16	OEM
53S141J16	OEM
53S141N16	OEM
53S240J16	OEM
53S240N16	OEM
53S241J16	OEM
53S241N16	OEM
53SC-15	0030
53ST-2SC2344	1284
53SZ-BZ-110	0317
53SZ-HZ9B3	0322
53SZ-RD11FB	0317
53T001-1	0143
53T001-2	0030
53T001-3	0015
53T001-4	0143
53T001-5	0143
53T001-6	0143
53T001-7	0133
53U001-1	0143
53U001-2	0133
53W001-6	0015
53W001-7	0124
53Y001-1	0015
53YT	OEM
53Z4	OEM
53Z6	2035
53Z6A	2035
53Z20D	OEM
53Z47D5	OEM
53Z62D	OEM
53Z62D10	OEM
53Z68D	OEM
53Z68D5	OEM
53Z68D10	OEM
53Z75D5	OEM
53Z110D	OEM
53Z130D5	OEM
54	0144
54-1	0016
54A	0079
54B	0016
54B2C	OEM
54BLK	0144
54BLU	0016
54BRN	0144
54C	0016
54CAY	OEM
54D	0016
54D5R08	OEM
54D9408GH	OEM
54E	0715
54F	0016
54F00DM	OEM
54F00FM	OEM
54F02DM	OEM
54F02FM	OEM
54F04DM	OEM
54F04FM	OEM
54F08DM	OEM
54F08FM	OEM
54F10DM	OEM
54F10FM	OEM
54F11DM	OEM
54F11FM	OEM
54F20DM	OEM
54F20FM	OEM
54F32DM	OEM
54F32FM	OEM
54F64DM	OEM
54F64FM	OEM
54F74DM	OEM
54F74FM	OEM
54F109DM	OEM
54F109FM	OEM
54F138DM	OEM
54F151DM	OEM
54F151FM	OEM
54F153DM	OEM
54F153FM	OEM
54F157DM	OEM
54F157FM	OEM
54F158DM	OEM
54F158FM	OEM
54F160DM	OEM
54F160FM	OEM
54F161DM	OEM
54F161FM	OEM
54F162DM	OEM
54F162FM	OEM
54F163DM	OEM
54F163FM	OEM
54F175DM	OEM
54F175FM	OEM
54F181DM	OEM
54F181FM	OEM
54F182DM	OEM
54F182FM	OEM
54F189DM	OEM
54F189FM	OEM
54F240DM	OEM
54F240FM	OEM
54F241DM	OEM
54F241FM	OEM
54F242DM	OEM
54F242FM	OEM
54F243DM	OEM

If replacement code is OEM, contact original manufacturer for replacement.

DEVICE TYPE	REPL CODE
54F243FM	OEM
54F244DM	OEM
54F244FM	OEM
54F245DM	OEM
54F251DM	OEM
54F251FM	OEM
54F253DM	OEM
54F253FM	OEM
54F257DM	OEM
54F257FM	OEM
54F258DM	OEM
54F258FM	OEM
54F280DM	OEM
54F280FM	OEM
54F283DM	OEM
54F283FM	OEM
54F289DM	OEM
54F289FM	OEM
54F352DM	OEM
54F352FM	OEM
54F353DM	OEM
54F353FM	OEM
54F373DM	OEM
54F373FM	OEM
54F374DM	OEM
54F374FM	OEM
54F500DM	OEM
54F533DM	OEM
54F533FM	OEM
54F534DM	OEM
54F534FM	OEM
54F537DM	OEM
54F537FM	OEM
54F538DM	OEM
54F538FM	OEM
54F539DM	OEM
54F545DM	OEM
54F588DM	OEM
54GRN	0016
54H00DM	OEM
54H00FM	OEM
54H01DM	OEM
54H01FM	OEM
54H04DM	OEM
54H04FM	OEM
54H05DM	OEM
54H05FM	OEM
54H08DM	OEM
54H08FM	OEM
54H10DM	OEM
54H10FM	OEM
54H11DM	OEM
54H11FM	OEM
54H20DM	OEM
54H20FM	OEM
54H21DM	OEM
54H21FM	OEM
54H22DM	OEM
54H22FM	OEM
54H30DM	OEM
54H30FM	OEM
54H40DM	OEM
54H40FM	OEM
54H50DM	OEM
54H50FM	OEM
54H51DM	OEM
54H51FM	OEM
54H52DM	OEM
54H52FM	OEM
54H53DM	OEM
54H53FM	OEM
54H54DM	OEM
54H54FM	OEM
54H55FM	OEM
54H60FM	OEM
54H61DM	OEM
54H61FM	OEM
54H62FM	OEM
54H71FM	OEM
54H72DM	OEM
54H72FM	OEM
54H73DM	OEM
54H73FM	OEM
54H74DM	OEM
54H74FM	OEM
54H76FM	OEM
54H78FM	OEM
54H87DM	OEM
54H87FM	OEM
54H101FM	OEM
54H102FM	OEM
54H103DM	OEM
54H103FM	OEM
54H106DM	OEM
54H106FM	OEM
54H108DM	OEM
54H108FM	OEM
54H183FM	OEM
54HC00F	OEM
54HC02F	OEM
54HC08F	OEM
54HC20F	OEM
54HC30F	OEM
54HCT00F	OEM
54HCT02F	OEM
54HCT04F	OEM
54HCT08F	REPL
54HCT10F	REPL
54HCT11F	OEM

DEVICE TYPE	REPL CODE
54HCT27F	OEM
54HCT30F	OEM
54HCT32F	OEM
54HCT74F	OEM
54HCT258F	OEM
54LS00DM	OEM
54LS00FM	OEM
54LS02	1559
54LS02DM	1559
54LS02FM	OEM
54LS03DM	OEM
54LS03FM	OEM
54LS04DM	OEM
54LS04FM	OEM
54LS05DM	OEM
54LS05FM	OEM
54LS08DM	OEM
54LS08FM	OEM
54LS09FM	OEM
54LS10DM	OEM
54LS10FM	OEM
54LS11DM	OEM
54LS11FM	OEM
54LS13DM	OEM
54LS13FM	OEM
54LS14DM	OEM
54LS14FM	OEM
54LS15FM	OEM
54LS20DM	OEM
54LS20FM	OEM
54LS21DM	OEM
54LS21FM	OEM
54LS22FM	OEM
54LS26DM	OEM
54LS26FM	OEM
54LS27DM	OEM
54LS27FM	OEM
54LS28DM	OEM
54LS28FM	OEM
54LS30DM	OEM
54LS30FM	OEM
54LS32DM	OEM
54LS32FM	OEM
54LS33FM	OEM
54LS37DM	OEM
54LS37FM	OEM
54LS38	OEM
54LS38DM	OEM
54LS38FM	OEM
54LS40DM	OEM
54LS40FM	OEM
54LS42DM	OEM
54LS42FM	OEM
54LS47DM	OEM
54LS47FM	OEM
54LS48DM	OEM
54LS48FM	OEM
54LS49DM	OEM
54LS49FM	OEM
54LS51DM	OEM
54LS51FM	OEM
54LS54FM	OEM
54LS55FM	OEM
54LS74DM	OEM
54LS74FM	OEM
54LS76DM	OEM
54LS76FM	OEM
54LS78DM	OEM
54LS78FM	OEM
54LS83ADM	OEM
54LS83AFM	OEM
54LS85DM	OEM
54LS85FM	OEM
54LS89DM	OEM
54LS89FM	OEM
54LS90FM	OEM
54LS92FM	OEM
54LS93DM	OEM
54LS93FM	OEM
54LS95BDM	OEM
54LS95BFM	OEM
54LS107DM	OEM
54LS107FM	OEM
54LS109DM	OEM
54LS109FM	OEM
54LS112DM	OEM
54LS112FM	OEM
54LS113FM	OEM
54LS114FM	OEM
54LS125A/BCBJA	OEM
54LS125A/BCBJC	OEM
54LS125ADM	OEM
54LS125AFM	OEM
54LS126DM	OEM
54LS126FM	OEM
54LS133DM	OEM
54LS133FM	OEM
54LS136DM	OEM
54LS136FM	OEM
54LS138DM	OEM
54LS138FM	OEM
54LS139DM	OEM
54LS139FM	OEM
54LS151DM	OEM
54LS151FM	OEM
54LS152FM	OEM
54LS153DM	OEM
54LS153FM	OEM
54LS155DM	OEM

DEVICE TYPE	REPL CODE
54LS155FM	OEM
54LS156DM	OEM
54LS156FM	OEM
54LS157DM	OEM
54LS157FM	OEM
54LS158DM	OEM
54LS158FM	OEM
54LS160FM	OEM
54LS161DM	OEM
54LS161FM	OEM
54LS162FM	OEM
54LS163DM	OEM
54LS163FM	OEM
54LS164DM	OEM
54LS164FM	OEM
54LS165DM	OEM
54LS165FM	OEM
54LS168DM	OEM
54LS168FM	OEM
54LS169DM	OEM
54LS169FM	OEM
54LS170DM	OEM
54LS170FM	OEM
54LS175DM	OEM
54LS175FM	OEM
54LS181FM	OEM
54LS189DM	OEM
54LS189FM	OEM
54LS190FM	OEM
54LS191DM	OEM
54LS191FM	OEM
54LS192DM	OEM
54LS192FM	OEM
54LS193FM	OEM
54LS194ADM	OEM
54LS194AFM	OEM
54LS195ADM	OEM
54LS195AFM	OEM
54LS196FM	OEM
54LS240DM	OEM
54LS240FM	OEM
54LS241DM	OEM
54LS241FM	OEM
54LS244DM	OEM
54LS244FM	OEM
54LS245DM	OEM
54LS245FM	OEM
54LS247DM	OEM
54LS247FM	OEM
54LS251DM	OEM
54LS251FM	OEM
54LS253DM	OEM
54LS253FM	OEM
54LS256DM	OEM
54LS256FM	OEM
54LS257DM	OEM
54LS257FM	OEM
54LS258ADM	OEM
54LS258AFM	OEM
54LS258DM	OEM
54LS258FM	OEM
54LS259DM	OEM
54LS259FM	OEM
54LS260DM	OEM
54LS260FM	OEM
54LS266DM	OEM
54LS266FM	OEM
54LS273DM	OEM
54LS273FM	OEM
54LS279DM	OEM
54LS279FM	OEM
54LS283DM	OEM
54LS283FM	OEM
54LS283J	OEM
54LS283W	OEM
54LS289DM	OEM
54LS289FM	OEM
54LS290DM	OEM
54LS290FM	OEM
54LS293DM	OEM
54LS293FM	OEM
54LS295ADM	OEM
54LS295AFM	OEM
54LS298DM	OEM
54LS298FM	OEM
54LS299DM	OEM
54LS299FM	OEM
54LS322DM	OEM
54LS322FM	OEM
54LS323DM	OEM
54LS323FM	OEM
54LS347DM	OEM
54LS347FM	OEM
54LS352DM	OEM
54LS352FM	OEM
54LS353DM	OEM
54LS353FM	OEM
54LS365ADM	OEM
54LS365AFM	OEM
54LS366ADM	OEM
54LS366AFM	OEM
54LS367ADM	OEM
54LS367AFM	OEM
54LS368ADM	OEM
54LS368AFM	OEM
54LS373DM	OEM
54LS373FM	OEM
54LS374DM	OEM
54LS374FM	OEM

DEVICE TYPE	REPL CODE
54LS375DM	OEM
54LS375FM	OEM
54LS377DM	OEM
54LS377FM	OEM
54LS378DM	OEM
54LS378FM	OEM
54LS379DM	OEM
54LS379FM	OEM
54LS384DM	OEM
54LS384FM	OEM
54LS386DM	OEM
54LS386FM	OEM
54LS390DM	OEM
54LS390FM	OEM
54LS393DM	OEM
54LS393FM	OEM
54LS395DM	OEM
54LS395FM	OEM
54LS447DM	OEM
54LS447FM	OEM
54LS490DM	OEM
54LS490FM	OEM
54LS533DM	OEM
54LS533FM	OEM
54LS534DM	OEM
54LS534FM	OEM
54LS563DC	OEM
54LS563DM	OEM
54LS563FM	OEM
54LS563PC	OEM
54LS564DM	OEM
54LS564FM	OEM
54LS573DM	OEM
54LS573FM	OEM
54LS574DM	OEM
54LS574FM	OEM
54LS670DM	OEM
54LS670FM	OEM
54LS670W	OEM
54ORN	0144
54R	OEM
54RED	0144
54S00DM	OEM
54S00FM	OEM
54S03DM	OEM
54S04ADM	OEM
54S04AFM	OEM
54S04DM	OEM
54S04FM	OEM
54S05ADM	OEM
54S05AFM	OEM
54S05DM	OEM
54S05FM	OEM
54S09FM	OEM
54S10DM	OEM
54S10FM	OEM
54S11DM	OEM
54S11FM	OEM
54S15DM	OEM
54S15FM	OEM
54S20DM	OEM
54S20FM	OEM
54S22DM	OEM
54S22FM	OEM
54S30DM	OEM
54S30FM	OEM
54S32DM	OEM
54S32FM	OEM
54S40DM	OEM
54S40FM	OEM
54S51DM	OEM
54S51FM	OEM
54S64DM	OEM
54S64FM	OEM
54S65FM	OEM
54S74DM	OEM
54S74FM	OEM
54S86DM	OEM
54S86FM	OEM
54S109DM	OEM
54S109FM	OEM
54S112DM	OEM
54S112FM	OEM
54S113FM	OEM
54S114FM	OEM
54S132DM	OEM
54S132FM	OEM
54S133DM	OEM
54S133FM	OEM
54S134FM	OEM
54S135FM	OEM
54S137DM	OEM
54S137FM	OEM
54S138DM	OEM
54S138FM	OEM
54S139DM	OEM
54S139FM	OEM
54S140DM	OEM
54S140FM	OEM
54S151DM	OEM
54S151FM	OEM
54S153DM	OEM
54S153FM	OEM
54S157DM	OEM
54S157FM	OEM
54S158DM	OEM
54S158FM	OEM
54S175DM	OEM
54S175FM	OEM

DEVICE TYPE	REPL CODE
54S181DM	OEM
54S181FM	OEM
54S182DM	OEM
54S182FM	OEM
54S189FM	OEM
54S194DM	OEM
54S194FM	OEM
54S253DM	OEM
54S253FM	OEM
54S258DM	OEM
54S258FM	OEM
54S260DM	OEM
54S260FM	OEM
54S280DM	OEM
54S280FM	OEM
54S289DM	OEM
54S289FM	OEM
54SC137C(A)	OEM
54SC138C(A)	OEM
54SC139C(A)	OEM
54SC237C(A)	OEM
54SC238C(A)	OEM
54SC239C(A)	OEM
54SC240C(A)	OEM
54SC241C(MA)	OEM
54SC242C(MA)	OEM
54SC243C(MA)	OEM
54SC244C(MA)	OEM
54SC245C(MA)	OEM
54SC540C(MA)	OEM
54SC541C(MA)	OEM
54SD-BA282	0163
54SI-TA7075P	0855
54SP-LT-527R	OEM
54ST-2SC1815Y	0076
54ST-2SC1815Y(AU)E	0076
54WHT	0016
54YEL	0016
54Z4	OEM
54Z6	2401
54Z6A	2401
55-0014	OEM
55-1	0143
55-641	0400
55-642	0086
55-643	0126
55-1016	0004
55-1026	0016
55-1027	0208
55-1029	0164
55-1031	0164
55-1032	0038
55-1034	0050
55-1082	0016
55-1083	0037
55-1083A	0037
55-1084	0086
55-1085	0037
55-1085A	0037
55C05	0217
55C05B	0217
55C05BF	0217
55C05F	0217
55C025	0217
55C025B	0217
55C025BF	0217
55C025F	0217
55C050	0608
55C050B	0608
55C050BF	0641
55C050F	0641
55C10	0217
55C10B	0217
55C10BF	0217
55C10F	0217
55C15	0217
55C15B	0217
55C15BF	0217
55C15F	0217
55C20	0217
55C20B	0217
55C20BF	0217
55C20F	0217
55C25	0217
55C25B	0217
55C25BF	0217
55C25F	0217
55C30	0217
55C30B	0217
55C30BF	0217
55C30F	0217
55C35	0217
55C35B	0468
55C35BF	0217
55C35F	0217
55C40	0217
55C40B	0468
55C40BF	0217
55C40F	0217
55C45	0217
55C45B	0468
55C45BF	0217
55C45F	0217
55C50	0217
55C50B	0468
55C50BF	0217
55C50F	0217
55C60	0217
55C60B	0468

DEVICE TYPE	REPL CODE
55C60BF	0217
55C60F	0217
55C70	0217
55C70B	0468
55C70BF	0658
55C70F	0658
55C80	0468
55C80B	0468
55C80BF	0658
55C80F	0658
55C90	0468
55C90B	0468
55C90BF	0658
55C90F	0658
55C100	0468
55C100B	0468
55C100BF	0658
55C100F	0658
55C110	0468
55C110B	0468
55C110BF	0658
55C110F	0658
55C120	0468
55C120B	0468
55C120BF	0658
55C120F	0658
55C130	OEM
55C130B	OEM
55C130BF	OEM
55C130F	OEM
55C150	OEM
55C150B	OEM
55C150BF	OEM
55C150F	OEM
55C160	OEM
55C160B	OEM
55C160BF	OEM
55C160F	OEM
55C170	OEM
55C170B	OEM
55C170BF	OEM
55C170F	OEM
55C180	OEM
55C180B	OEM
55C180BF	OEM
55C180F	OEM
55C05BF	OEM
55D5R02	OEM
55D9R08GH	OEM
55E04A	OEM
55HQ015	OEM
55HQ020	OEM
55HQ025	OEM
55HQ030	OEM
55P	OEM
55P2	0136
55P3	0136
55PB	OEM
55PBH	OEM
55PBHL	OEM
55PBL	OEM
55PL	OEM
55SD-1N60S	0019
55SI-MC14011BCP	0215
55ST-2SD401AK	1284
55SZ-RD3.3EB	0299
55SZ-RD12EB	1005
55V10	OEM
55V11	OEM
55Z4	0170
55Z6	1438
55Z6A	1438
56-1	0143
56-2	0143
56-3	0143
56-4	0143
56-5	0133
56-6	0062
56-7	3683
56-8	0143
56-10	0143
56-11	0143
56-15	0102
56-16	0041
56-19	0012
56-20	0143
56-21	0917
56-23	0030
56-24	0133
56-25	0002
56-26	0143
56-27	0133
56-28	0133
56-31	0649
56-32	0365
56-33	0015
56-35	0016
56-36	0420
56-44	0162
56-45	0531
56-46	0012
56-47	0441
56-49	0030
56-51	0137
56-52	0015
56-53	0794
56-54	0244
56-55	0350
56-56	0124

DEVICE TYPE	REPL CODE
56-57	0137
56-58	0471
56-59	0140
56-62	0012
56-63	0025
56-64	0726
56-66	0350
56-67	0170
56-68	1268
56-70	0188
56-71	0025
56-72	0931
56-73	0133
56-74	1276
56-78	0229
56-89	0143
56-93	0133
56-4780	OEM
56-4826	0151
56-4827	0111
56-4829	1220
56-4830	0018
56-4831	0532
56-4832	0133
56-4833	2596
56-4834	0520
56-4835	0030
56-4836	0139
56-4837	0052
56-4839	0015
56-4885	0244
56-4886	0123
56-8086	0127
56-8086A	0127
56-8086B	0127
56-8086C	0127
56-8087	0127
56-8087B	0127
56-8087C	0127
56-8088	0127
56-8088A	0127
56-8088C	0127
56-8089	0016
56-8089A	0016
56-8089C	0016
56-8090	0016
56-8090A	0016
56-8090C	0016
56-8091	0004
56-8091A	0004
56-8091B	0004
56-8091C	0004
56-8091D	0004
56-8092	0004
56-8092A	0004
56-8092B	0004
56-8093	0143
56-8094	2866
56-8095	0123
56-8096	0649
56-8097	0015
56-8098	0037
56-8098A	0037
56-8098B	0037
56-8098C	0037
56-8099	0164
56-8100	0208
56-8100A	0208
56-8100B	0208
56-8100C	0208
56-8100D	0208
56-8101	0164
56-8196	0111
56-8197	0111
56-8198	0015
56-8199	0015
56-861041-3	0023
56A1	0632
56A1-1	0632
56A3-1	0167
56A4-1	0352
56A5-1	0354
56A6-1	0349
56A7-1	0039
56A7-26	OEM
56A7-45	OEM
56A9-1	0789
56A11-1	0697
56A13-1	1082
56A14-1	1426
56A15-1	1208
56A16-1	1793
56A17-1	1337
56A19-1	0857
56A20-1	0802
56A21-1	2239
56A22-1	0016
56A23-1	0848
56A24-1	2043
56A25-1	0848
56A29-2	OEM
56A34-1	3921
56A42-1	4147
56A46-1	2457
56A49-1	3986
56A49-2	3986
56A53-1	OEM
56A55-1	1328
56A60-1	4194

If replacement code is OEM, contact original manufacturer for replacement.

DEVICE TYPE	REPL CODE
56A65-1	1170
56A74-1	3738
56A74LS-00	OEM
56A75-1	0490
56A75-1A	0490
56A101-1	3451
56A102-1	4134
56A107-1	2826
56A111-05	OEM
56A111-1	OEM
56A112-1	OEM
56A113-1	1329
56A120-1	2533
56A121-1	0904
56A122-1	1057
56A123-1	OEM
56A124-1	OEM
56A125-1	OEM
56A126-1	OEM
56A127-1	OEM
56A128-1	OEM
56A137-1	0663
56A138-1	4109
56A141-1	3445
56A141-4	3953
56A161-1-2	OEM
56A161-2	OEM
56A221-1	0079
56A223-2	4550
56A232-1	OEM
56A234-1	OEM
56A236-1	OEM
56A249-1	OEM
56A250-1	OEM
56A265-1	0101
56A355-1	2666
56A357-1	OEM
56A359-1	4094
56A360-1	2346
56A402-L	1066
56A404-1	OEM
56A551	1328
56B74-1	3738
56B75-1	2623
56C1	0632
56C1-1	0632
56C7-1	0039
56C7-26	OEM
56C7-45	OEM
56C17-1	1337
56C49-1	3986
56D3-1	0167
56D4-1	0352
56D5-1	0354
56D5R08	OEM
56D6-1	0349
56D9-1	0789
56D9R08GH	OEM
56D17-1	1337
56D20-1	0802
56D49-1	3986
56D49-2	3986
56D53-1	OEM
56D65-1	1170
56D75-1	2623
56D75-1A	0490
56P1	0050
56P2	0050
56P3	0050
56P4	0050
56P4P	0050
56SI-TA7644BP	0663
56ST-LM2152	0079
56ST-MJE9742	0276
56SZ-WZ120	0052
56SZ-WZ120(AU)	0052
56TB13	OEM
56TB14	OEM
56TB15	OEM
56TB16	OEM
56TB16-CE00	OEM
56TB16-EE00	OEM
56TB16-JE00	OEM
56TB16-JH00	OEM
56TB16-LE00	OEM
56TB16-LH00	OEM
56TB16-NE00	OEM
56TB17	OEM
56TB17-CE00	OEM
56TB17-EE00	OEM
56TB17-EH00	OEM
56TB17-JE00	OEM
56TB17-JH00	OEM
56TB17-LE00	OEM
56TB17-LH00	OEM
56TB17-NE00	OEM
56TB18	OEM
56TB18-CE00	OEM
56TB18-CH00	OEM
56TB18-EE00	OEM
56TB18-EH00	OEM
56TB18-JE00	OEM
56TB18-JH00	OEM
56TB18-LE00	OEM
56TB18-NE00	OEM
56TB18-NH00	OEM
56TB19	OEM
56TB19-CE00	OEM
56TB19-CH00	OEM
56TB19-EE00	OEM
56TB19-JE00	OEM
56TB19-JH00	OEM
56TB19-LE00	OEM
56TB19-LH00	OEM
56TB19-NE00	OEM
56TB19-NH00	OEM
56TB20	OEM
56TB20-CE00	OEM
56TB20-CH00	OEM
56TB20-EE00	OEM
56TB20-EH00	OEM
56TB20-JE00	OEM
56TB20-JH00	OEM
56TB20-LE00	OEM
56TB20-NE00	OEM
56Z4	0137
56Z6	0442
56Z6A	0442
57-0006	0015
57-0004503	0016
57-0005452	0016
57-000901504	0016
57-00901504	0079
57-1,,	0064
57-2	0015
57-6	0015
57-12	0015
57-13	0015
57-15	0015
57-17	0015
57-18	3189
57-20	0015
57-21	0015
57-22	0015
57-23	0015
57-24	0015
57-25	0229
57-26	0229
57-27	0015
57-28	0015
57-29	0229
57-31	0015
57-32	0196
57-33	0015
57-36	0774
57-38	0015
57-39	0762
57-42	0087
57-42A	0087
57-43	0229
57-46	0015
57-49	0229
57-52	0293
57-55	0901
57-56	0293
57-58	0015
57-59	0015
57-60	0015
57-61	0061
57-62	2327
57-64	0071
57-65	0080
57-83	0409
57-90	3655
57-98	3655
57-100	0064
57-110	0181
57-120	0052
57-150	OEM
57-160	0445
57-200	0700
57-220	0705
57-249	1276
57-330	0166
57-439	0275
57-519	OEM
57-569	0253
57-629	0471
57-689	0062
57-829	0165
57-919	0057
57A1-55	1381
57A2-1	0136
57A2-4	0164
57A2-15	0004
57A2-19	0136
57A2-24	0004
57A2-49	2866
57A2.4	0004
57A3-1	1381
57A3-2	1381
57A3-3	2866
57A3-13	1381
57A5-1	0136
57A5-2	0136
57A5-3	0004
57A5-4	0136
57A5-5	0004
57A5-6	0144
57A5-7	0144
57A5-8	0144
57A6-1	0211
57A6-2	0160
57A6-3	0160
57A6-4	0016
57A6-5	0208
57A6-6	0038
57A6-7	0198
57A6-8	0160
57A6-9	0016
57A6-11	0016
57A6-12	0160
57A6-15	0126
57A6-18	1381
57A6-28	1381
57A6-34	4163
57A7-1	0144
57A10-1	0144
57A10A-8-6	0144
57A12-1	0233
57A12-2	0233
57A14-2	0086
57A15-2	0016
57A15-3	0016
57A19-1	0037
57A20-1	0144
57A20G14	0905
57A21-2	0144
57A21-4	0144
57A21-5	0144
57A21-6	0144
57A21-7	0144
57A21-8	0016
57A21-9	0144
57A21-10	0144
57A21-11	0127
57A21-12	0127
57A21-13	0127
57A21-14	0127
57A21-15	0144
57A21-16	0007
57A21-17	0007
57A21-18	0007
57A21-20	OEM
57A21-21	0127
57A21-26	0127
57A21-45	0127
57A23-1	0126
57A23-2	0126
57A23-3	0126
57A24-1	0016
57A24-2	0016
57A24-3	0016
57A29-2	2571
57A30	OEM
57A31-1	0325
57A31-2	0688
57A31-3	0688
57A31-4	0325
57A32-1	0823
57A32-2	1478
57A32-3	1621
57A32-5	OEM
57A32-6	4570
57A32-7	OEM
57A32-8	OEM
57A32-9	OEM
57A32-10	0701
57A32-11	0470
57A32-12	0701
57A32-13	OEM
57A32-14	OEM
57A32-15	OEM
57A32-16	0701
57A32-17	4561
57A32-19	1994
57A32-21	1478
57A32-22	0443
57A32-25	4543
57A32-27	4543
57A32-28	1478
57A32-29	1478
57A32-32	0520
57A33-1	1381
57A63-11	0313
57A100-3	2866
57A100-5	2866
57A100-385	2866
57A101-4	0144
57A102-4	0007
57A103-4	0007
57A104-1	0233
57A104-2	0233
57A104-3	0233
57A104-4	0233
57A104-5	0233
57A104-6	0233
57A104-7	0233
57A104-8	0233
57A104-8-6	0086
57A105-12	0016
57A106-12	0037
57A107-1	0144
57A107-2	0144
57A107-3	0144
57A107-4	0144
57A107-5	0144
57A107-6	0144
57A107-8	0144
57A108-1	0150
57A108-2	0150
57A108-3	0150
57A108-4	0150
57A108-5	0150
57A108-6	0150
57A108-6-8	0126
57A108-7	0150
57A108-8	0150
57A111-9	4163
57A126-1	0111
57A126-12	0127
57A128-9	0086
57A129-9	0086
57A130-9	0126
57A131-10	0086
57A132-10	0126
57A133-12	0037
57A134-12	0144
57A135-12	0016
57A136-12	0016
57A137-12	0037
57A137-12A	0037
57A138-4	0008
57A138-4-6	0008
57A139-1	0007
57A139-2	0007
57A139-3	0007
57A139-4	0008
57A139-4-6	0144
57A140-12	0016
57A141-4	0007
57A142-4	0330
57A142-4(3RDLF)	0330
57A143-12	0111
57A144-12	0111
57A145-12	0037
57A145-12A	0037
57A146-12	0144
57A147-12	0037
57A148-12	0037
57A149-12	0325
57A150-12	0325
57A151-6	0144
57A152-12	0144
57A153-6	0016
57A153-9	0079
57A155-10	0178
57A156	0086
57A156-9	0016
57A157	0037
57A157-9	0037
57A158-10	0142
57A159	0037
57A159-2	OEM
57A159-12	0037
57A160-8	0233
57A164-4	0007
57A166-12	0016
57A167-9	0595
57A168-9	0791
57A169-12	0325
57A172-8	0264
57A174-8	0037
57A175-12	0037
57A177-12	0007
57A178-12	0037
57A179-4	0007
57A180-4	0330
57A181-12	0016
57A182-10	0079
57A182-12	0079
57A183-11	0233
57A184-11	0007
57A184-12	0016
57A185-11	0065
57A185-12	0037
57A186-11	0065
57A187-12	0236
57A188-12	0681
57A189-8	0037
57A190-12	OEM
57A191-12	0016
57A192-10	0142
57A193-11	0285
57A193-12	0079
57A194-1L	0233
57A194-11	0233
57A194-12	0058
57A195-10	0142
57A196-10	0270
57A196-11	0065
57A197-12	0037
57A198-11	0065
57A199-4	0016
57A200-12	0016
57A201-14	0037
57A202-13	0016
57A203-14	0016
57A204-14	0016
57A205-14	0728
57A206-14	0925
57A207-8	0233
57A208-8	0285
57A211-8	0264
57A213-11	0065
57A214-2	0161
57A214-12	0551
57A215-12	0037
57A216-12	0037
57A219-14	0155
57A220-14	0006
57A235-12	0037
57A236-11	0285
57A240-14	0006
57A241-14	0155
57A243-10	0058
57A244-14	0423
57A245-14	0854
57A249-4	0007
57A250-14	0042
57A251-14	0854
57A252-1	0016
57A253-14	0016
57A256-10	0103
57A258-8	0037
57A259-10	0058
57A261-10	0233
57A263-11	0313
57A264-12	OEM
57A265-4	0016
57A267-4	0212
57A268-9	0016
57A273-14	2235
57A274-14	0602
57A277-14	1648
57A278-14	1284
57A279-14	0042
57A280-14	0155
57A281-11	0233
57A281-14	0006
57A282-12	0016
57A283-11	0233
57A286-10	0042
57A290-12	0325
57A293-11	0233
57A294-12	0079
57A295-8	0285
57A296-10	0392
57A297-12	OEM
57A301-12	OEM
57A302-12	OEM
57A305-12	0037
57A309-12	2945
57A310-14	0681
57A311-14	0236
57A312	0313
57A312-11	0285
57A313-11	0065
57A401-11	0728
57A402-4	0007
57A402-L	1066
57A403-12	0037
57A404-12	0324
57A405-12	0436
57A411-1	OEM
57A411-3	3046
57A413-1	0261
57A419-GR	0076
57A419-GY	0076
57A419-Y	0076
57A420-GR	0148
57A420-Q	0148
57A421-11	0728
57A422-8	1327
57A423-8	0261
57A423-8T	0261
57A424-11	0055
57A428-12	1284
57A429-16	OEM
57A431-2	OEM
57A431-J	OEM
57A446-2	0866
57A458-0	0006
57A458-O	0006
57A459-Y	0261
57A460-11	3507
57A460-12	1541
57A460-15	1541
57A485-1	1648
57A493-10	0715
57B2-1	0050
57B2-2	0050
57B2-3	0164
57B2-4	0164
57B2-5	0038
57B2-6	0004
57B2-7	0164
57B2-8	0208
57B2-9	0050
57B2-10	0164
57B2-11	0050
57B2-12	0004
57B2-13	0050
57B2-14	0050
57B2-15	0164
57B2-16	0164
57B2-17	0050
57B2-18	0050
57B2-19	0050
57B2-20	0050
57B2-21	0164
57B2-22	0050
57B2-23	0004
57B2-24	0004
57B2-25	0164
57B2-26	0050
57B2-27	0016
57B2-28	0016
57B2-29	0004
57B2-30	0050
57B2-31	0050
57B2-32	0004
57B2-33	0004
57B2-34	0004
57B2-35	0050
57B2-36	0004
57B2-37	0050
57B2-38	0086
57B2-39	0004
57B2-40	0050
57B2-41	0050
57B2-42	0050
57B2-43	0004
57B2-44	0004
57B2-45	0004
57B2-46	0004
57B2-47	0178
57B2-48	0050
57B2-49	2866
57B2-50	0050
57B2-51	0050
57B2-52	0004
57B2-57	0086
57B2-58	0178
57B2-59	0016
57B2-60	0004
57B2-61	0164
57B2-62	0016
57B2-63	0016
57B2-64	0016
57B2-65	0050
57B2-66	0050
57B2-67	0050
57B2-68	0050
57B2-70	0037
57B2-70A	0037
57B2-71	0037
57B2-71A	0037
57B2-72	0004
57B2-73	0016
57B2-75	0050
57B2-77	0050
57B2-78	0004
57B2-79	0079
57B2-80	0050
57B2-83	0004
57B2-84	0178
57B2-85	0079
57B2-87	0016
57B2-88	0004
57B2-89	0050
57B2-90	0050
57B2-93	0050
57B2-97	0016
57B2-101	0016
57B2-102	0016
57B2-103	0016
57B2-104	0050
57B2-105	0050
57B2-113	0016
57B2-116	0016
57B2-126	0016
57B2-149	0050
57B2-153	0016
57B2-157	0050
57B2-158	0050
57B2-159	0050
57B2-192	0016
57B3-1	1381
57B3-2	1381
57B3-3	2866
57B3-4	0004
57B3-5	0004
57B3-6	0004
57B3-7	3033
57B3-8	3033
57B3-9	3033
57B3-10	3033
57B3-11	3033
57B3-12	0265
57B3-13	1381
57B3-105	0050
57B4-1	0160
57B4-2	0160
57B4-4	0160
57B5-6	0111
57B5-7	0111
57B5-8	0111
57B6	0211
57B6-4	0079
57B6-9	0079
57B6-11	0079
57B6-12	0085
57B6-19	0079
57B12-4	0086
57B21	0144
57B21-2	0144
57B21-4	0144
57B21-5	0144
57B21-6	0007
57B21-7	0007
57B21-9	0127
57B21-12	0144
57B21-13	0144
57B21-14	0144
57B21-15	0144
57B21-16	0144
57B21-17	0007
57B21-18	0144
57B27-2	0079
57B30-12	0127
57B31-14	0681
57B100-3	2866
57B100-5	2866
57B100-7	0004
57B100-11	0222
57B100-385	2866
57B101-4	0144
57B-102-4	0079
57B102-4	0007
57B103-4	0007
57B104-8	0233
57B105-12	0016
57B106-12	0037
57B107-8	0016
57B108-6	0037
57B108-6A	0037
57B109-9	0086
57B110-9	0126
57B111-9	4163
57B112-9	0126
57B112-9A	0126
57B113-9	0086
57B114-9	0126
57B114-9A	0126
57B115-9	0126
57B115-9A	0126
57B116-9	0126
57B116-9A	0126
57B117-9	0016
57B118-12	0016
57B119-2	0007
57B119-12	0016
57B120-12	0016
57B121-9	0016
57B122-9	0037
57B122-9A	0037
57B123-10	0178
57B124-10	0085
57B125-9	0016
57B126-12	0016
57B128-9	0086
57B129-9	0016
57B130-9	0037
57B130-9A	0037
57B131-10	0042
57B132-10	0925
57B133-12	0037
57B134-12	0144
57B135-12	0016
57B136-1	0233
57B136-2	0233
57B136-3	0233
57B136-4	0233
57B136-5	0233
57B136-6	0233
57B136-7	0233
57B136-8	0233
57B136-9	0233
57B136-10	0233
57B136-11	0233
57B136-12	0016
57B137-12	0037
57B137-12A	0037
57B138-4	0008
57B139-4	0008
57B140-12	0016
57B141-1	0144
57B141-2	0144
57B141-3	0144
57B141-4	0007
57B142-1	0144
57B142-2	0144
57B142-3	0144
57B142-4	0330
57B143-1	0144
57B143-2	0144
57B143-3	0144
57B143-4	0144
57B143-5	0144
57B143-6	0144
57B143-7	0144
57B143-8	0144
57B143-9	0144
57B143-10	0144
57B143-11	0144
57B143-12	0016
57B144-12	0016
57B145-12	0037
57B145-12A	0037
57B146-12	0016
57B147-12	0037
57B147-12A	0037
57B148-1	0126
57B148-2	0126
57B148-3	0126
57B148-4	0126
57B148-5	0126
57B148-6	0126
57B148-7	0126
57B148-8	0126
57B148-9	0126
57B148-10	0126
57B148-11	0126
57B148-12	0126
57B148-12A	0126
57B149-2	0325
57B149-12	0325
57B150-12	0325
57B151-6	0144
57B152-1	0016
57B152-2	0016
57B152-3	0016

Original Device Types

DEVICE TYPE	REPL CODE
57B152-4	0016
57B152-5	0016
57B152-6	0016
57B152-7	0016
57B152-8	0016
57B152-9	0016
57B152-10	0016
57B152-11	0016
57B152-12	0144
57B153-1	0016
57B153-2	0079
57B153-3	0016
57B153-4	0016
57B153-5	0016
57B153-6	0016
57B153-7	0016
57B153-8	0016
57B153-9	0016
57B155-10	0178
57B156-9	0016
57B157-9	0037
57B157-9A	0037
57B158-1	0142
57B158-2	0142
57B158-3	0142
57B158-4	0142
57B158-5	0142
57B158-6	0142
57B158-7	0142
57B158-8	0142
57B158-9	0142
57B158-10	0178
57B159-12	0037
57B159-12A	0037
57B160-1	0144
57B160-2	0144
57B160-3	0144
57B160-4	0144
57B160-5	0144
57B160-6	0144
57B160-7	0144
57B160-8	0144
57B162-12	0595
57B163-12	0126
57B163-12A	0126
57B166-12	0144
57B167-9	0086
57B168	0004
57B168-9	0126
57B169	0004
57B169-2	0325
57B169-12	0325
57B170	0004
57B170-9	0086
57B171-9	0126
57B175-9	0103
57B175-9A	0103
57B175-12	0037
57B179-4	0127
57B179-12	OEM
57B180	0050
57B181-12	0079
57B182-12	0076
57B184	0050
57B184-1	0079
57B184-12	0007
57B185-12	0037
57B186	0050
57B186-11	0065
57B187	0050
57B187-12	0461
57B188	0050
57B188-12	0681
57B191-12	0016
57B192-10	0142
57B192-11	0142
57B194-11	0016
57B196-10	0270
57B198-11	0065
57B208-8	0285
57B211-8	0264
57B213-11	0065
57B214-12	0551
57B243-10	0058
57B245-14	0854
57B258-8	0037
57B263-11	0313
57B277-14	1648
57B278-14	1284
57B279-4	1085
57B279-14	1085
57B280-14	0155
57B281-14	0006
57B282-12	0016
57B283-11	0233
57B286-10	0423
57B296-10	0392
57B310-14	0681
57B311-14	0236
57B312-11	0285
57B313-11	0055
57B316-10	2620
57B401-11	0728
57B402-4	1066
57B402-L	1066
57B404-12	0622
57B413-1	0715
57B418-10	2548
57B419-GR	0076
57B419-P	0076
57B419-Q	0076
57B420-GR	0148
57B420-P	0006
57B420-Q	0006
57B421-11	1165
57B423-8	0261
57B424-11	0055
57B425-F	OEM
57B426-F	OEM
57B427-1	0168
57B428-8	0392
57B428-12	1284
57B428-12-P	0392
57B428-12P	0392
57B429-8	1648
57B429-12	1911
57B429-12-P	1648
57B429-12P	1648
57B432-1	OEM
57B433-1	0280
57B434-1	0016
57B446-2	0866
57B459-Y	0261
57B460-12	1541
57B483-11	0233
57B1127	0004
57B1130	0050
57B1131	0050
57B1143	0004
57B1186	0050
57C5	0038
57C5-1	0136
57C5-2	0050
57C5-3	0004
57C5-4	0136
57C5-5	0004
57C5-6	0144
57C5-7	0144
57C5-8	0144
57C5-9	0136
57C5-10	0136
57C6-1	0211
57C6-2	0160
57C6-3	0160
57C6-4	0016
57C6-5	0208
57C6-6	0038
57C6-6A	0280
57C6-6B	0280
57C6-6C	0280
57C6-7	0198
57C6-8	0160
57C6-9	0016
57C6-10	0142
57C6-11	0016
57C6-12	0160
57C6-14	0142
57C6-15	0126
57C6-16	0136
57C6-17	0198
57C6-18	1381
57C6-19	0198
57C6-20	0208
57C6-21	0208
57C6-22	0211
57C6-23	0160
57C6-24	0142
57C6-25	0004
57C6-26	0126
57C6-26A	0126
57C6-27	0198
57C6-28	1381
57C6-29	0198
57C6-30	0198
57C6-31	0126
57C6-31A	0126
57C6-32	0198
57C6-33	0126
57C6-33A	0126
57C6-34	4163
57C7-1	0144
57C7-2	0144
57C7-3	0144
57C7-4	0144
57C7-5	0144
57C7-6	0144
57C7-7	0144
57C7-8	0086
57C7-9	0016
57C7-10	0016
57C7-15	0016
57C7-17	0016
57C7-18	0016
57C7-20	0016
57C9-2	0160
57C10-1	0144
57C10-2	0144
57C11-1	0016
57C12-1	0233
57C12-2	0233
57C12-3	0178
57C12-4	0086
57C12-5	0178
57C14-1	0086
57C14-2	0086
57C14-3	0086
57C15-1	0079
57C15-2	0016
57C15-3	0016
57C15-4	0016
57C15-5	0037
57C15-50	0037
57C16	0136
57C16-1	0016
57C19-1	0037
57C19-1A	0037
57C20-1	0144
57C22-1	0604
57C22-2	0604
57C23-1	0126
57C23-2	0126
57C23-3	0126
57C24-1	0016
57C24-2	0016
57C24-3	0016
57C24-4	0016
57C27-1	0016
57C27-2	0144
57C28	2624
57C29-1	2571
57C29-2	2571
57C68	0004
57C104-8	0007
57C105-12	0079
57C109-9	0086
57C110-9	0126
57C121-9	0016
57C122-9	0126
57C142-4	0007
57C148-12	0126
57C148-12A	0126
57C156-9	0016
57C157-9	0037
57C157-90	0037
57C164-4	0007
57C179-4	1066
57C182-12	0079
57C191-12	0016
57C208-8	0261
57C279-14	0042
57C423-8	0261
57D1-1	0143
57D1-2	0143
57D1-3	0208
57D1-4	0208
57D1-5	0208
57D1-6	0208
57D1-7	0164
57D1-8	0004
57D1-9	0050
57D1-10	0050
57D1-11	0211
57D1-12	0050
57D1-13	0050
57D1-14	0004
57D1-15	0050
57D1-16	0050
57D1-17	0164
57D1-18	0164
57D1-19	0164
57D1-20	0164
57D1-21	0164
57D1-22	0050
57D1-23	0004
57D1-24	0050
57D1-25	0050
57D1-26	0004
57D1-27	0004
57D1-28	0004
57D1-30	0050
57D1-31	0050
57D1-32	0050
57D1-33	0050
57D1-34	0004
57D1-35	0050
57D1-36	0050
57D1-37	0050
57D1-38	0164
57D1-39	0050
57D1-40	0004
57D1-41	0050
57D1-42	0164
57D1-43	0004
57D1-44	0004
57D1-45	0050
57D1-46	0050
57D1-47	0050
57D1-48	0050
57D1-49	0050
57D1-50	0050
57D1-51	0016
57D1-52	0150
57D1-53	0004
57D1-54	0143
57D1-55	1381
57D1-56	0164
57D1-57	0050
57D1-58	0004
57D1-59	0004
57D1-60	0164
57D1-61	0050
57D1-62	0143
57D1-63	0229
57D1-64	0143
57D1-65	0917
57D1-66	0004
57D1-67	0050
57D1-68	0004
57D1-69	0050
57D1-70	0004
57D1-71	0004
57D1-72	0050
57D1-73	0050
57D1-74	0050
57D1-75	0016
57D1-76	0037
57D1-77	0164
57D1-78	0208
57D1-79	0004
57D1-80	0050
57D1-81	0050
57D1-82	0004
57D1-83	0004
57D1-84	0050
57D1-85	0050
57D1-86	0050
57D1-87	0050
57D1-88	0050
57D1-89	0050
57D1-90	0004
57D1-91	0004
57D1-92	0004
57D1-93	0004
57D1-94	0050
57D1-95	0004
57D1-96	0050
57D1-97	0050
57D1-98	0050
57D1-99	0050
57D1-100	0050
57D1-101	0050
57D1-102	0050
57D1-103	0050
57D1-104	0004
57D1-105	0004
57D1-106	0004
57D1-107	0136
57D1-108	0050
57D1-109	0050
57D1-110	0050
57D1-111	0004
57D1-112	0050
57D1-113	0050
57D1-114	0136
57D1-115	0050
57D1-116	0004
57D1-117	0004
57D1-118	0004
57D1-119	0160
57D1-120	0050
57D1-121	0004
57D1-122	0233
57D1-123	0016
57D1-124	0016
57D3-6	0211
57D4-1	0160
57D4-2	0160
57D5-1	0136
57D5-2	0136
57D5-4	0136
57D5R21	OEM
57D6-4	0016
57D6-10	0142
57D6-12	0160
57D6-19	0198
57D9-1	0050
57D9-2	0160
57D9R21GH	OEM
57D14-1	0016
57D14-2	0016
57D14-3	0016
57D17-1	1337
57D19-1	0037
57D19-2	0037
57D19-3	0037
57D19-10	0037
57D19-20	0037
57D19-30	0037
57D24-1	0144
57D24-2	0144
57D24-3	0144
57D68	0211
57D107-8	0144
57D108-8	OEM
57D126	0211
57D127	0276
57D130	0136
57D131	0136
57D132	0136
57D132-9	0050
57D136-12	0016
57D143	0211
57D156	0211
57D159-12	0126
57D166-12	0144
57D169	0211
57D170	0211
57D180	0280
57D184	0280
57D186	0136
57D187	0136
57D188	0136
57D189	0211
57D1127	0211
57D1130	0050
57D1131	0050
57D1132	0050
57D1143	0211
57D1186	0050
57DG-23	0160
57DG-32	0160
57L1-1	0050
57L1-2	0050
57L1-3	0050
57L1-4	0050
57L1-5	0164
57L1-6	0004
57L1-7	0164
57L1-8	0004
57L1-9	0050
57L1-10	0050
57L1-11	0050
57L1-12	0050
57L2-1	0086
57L2-2	0016
57L3-1	0016
57L3-4	0016
57L5-1	0085
57L106-9	0325
57LS300J	OEM
57LS301J	OEM
57LS304J	OEM
57LS306J	OEM
57LS307J	OEM
57LS376J	OEM
57LS380J	OEM
57M1-1	0050
57M1-2	0050
57M1-3	0050
57M1-4	0050
57M1-5	0004
57M1-6	0004
57M1-7	0164
57M1-8	0164
57M1-9	0050
57M1-10	0050
57M1-11	0050
57M1-12	0050
57M1-13	0126
57M1-13A	0126
57M1-14	0016
57M1-15	0016
57M1-16	0208
57M1-17	0050
57M1-18	0208
57M1-19	0016
57M1-20	0016
57M1-21	0126
57M1-21A	0126
57M1-22	0126
57M1-23	0016
57M1-24	0016
57M1-25	0126
57M1-25A	0126
57M1-26	0016
57M1-27	0016
57M1-28	0016
57M1-29	0016
57M1-30	0016
57M1-31	0016
57M1-32	0016
57M1-33	0086
57M1-34	0086
57M1-35	0164
57M2-1	0208
57M2-2	0208
57M2-3	0164
57M2-4	0164
57M2-6	0208
57M2-7	0086
57M2-8	0164
57M2-9	0208
57M2-10	0126
57M2-10A	0126
57M2-11	0086
57M2-14	0086
57M2-15	0126
57M2-15A	0126
57M2-16	0233
57M2-17	0233
57M2-18	0086
57M2-506	2866
57M3-1	0103
57M3-1A	0103
57M3-2	0103
57M3-3	0086
57M3-4	0103
57M3-4A	0103
57M3-5	0178
57M3-6	0103
57M3-6A	0103
57M3-7	0085
57M3-8	0085
57M3-9P	0222
57M3-10P	0222
57M3-12	0222
57P1	0211
57RECT-SI-154(AU)	0023
57RECT-SI-1021(AO)	OEM
57RECT-SI-1021(AU)	0023
57S-06	0301
57S300J	OEM
57S301J	OEM
57S304J	OEM
57S306J	OEM
57S307J	OEM
57S373J	OEM
57S374J	OEM
57S376J	OEM
57S378J	OEM
57S380J	OEM
57S382J	OEM
57SI-HA11423	1201
57ST-3SK60	2458
57TB11	OEM
57TB12	OEM
57TB13	OEM
57TB14	OEM
57TB15	OEM
57TB16	OEM
57TB16-CE00	OEM
57TB16-GE00	OEM
57TB16-RE00	OEM
57TB17	OEM
57TB17-CE00	OEM
57TB17-GE00	OEM
57TB17-RE00	OEM
57TB18	OEM
57TB18-CE00	OEM
57TB18-GE00	OEM
57TB18-RE00	OEM
57TB19	OEM
57TB19-CE00	OEM
57TB19-GE00	OEM
57TB19-RE00	OEM
57TB20	OEM
57TB20-CE00	OEM
57TB20-GE00	OEM
57TB20-RE00	OEM
57X14	0015
57Z4	OEM
57Z6	1445
57Z6A	1445
58-01	1250
58-1	0224
58A290-12	0325
58B2-14	0050
58P	OEM
58PB	OEM
58PBH	OEM
58PBHL	OEM
58PBL	OEM
58PL	OEM
58SI-LM393N	0631
58SI-LM393P	0631
58ST-2SA1029C	0006
58ST-2SC2344E	1284
58ST-MPS9418AS	0324
58SZ-RD6.2EB3	0471
58TB9	OEM
58TB10	OEM
58TB11	OEM
58TB12	OEM
58TB13	OEM
58TB13-CE00	OEM
58TB13-GE00	OEM
58TB13-RE00	OEM
58TB14	OEM
58TB14-CE00	OEM
58TB14-GE00	OEM
58TB14-RE00	OEM
58TB15	OEM
58TB15-CE00	OEM
58TB15-GE00	OEM
58TB15-RE00	OEM
58TB16	OEM
58TB16-CE00	OEM
58TB16-GE00	OEM
58TB16-RE00	OEM
58TB17	OEM
58TB17-CE00	OEM
58TB17-GE00	OEM
58TB17-RE00	OEM
58TB18	OEM
58TB18-CE00	OEM
58TB18-GE00	OEM
58TB18-RE00	OEM
58TB19	OEM
58TB19-CE00	OEM
58TB19-GE00	OEM
58TB19-RE00	OEM
58TB20	OEM
58TB20-CE00	OEM
58TB20-GE00	OEM
58TB20-RE00	OEM
58Z6	2221
58Z6A	2221
59-1	0917
59-5035	OEM
59-5055	OEM
59-5635	OEM
59-5637	OEM
59-5655	OEM
59-5657	OEM
59-5735	OEM
59-5737	OEM
59-5755	OEM
59-5757	OEM
59-5800	OEM
59-5802	OEM
59-5804	OEM
59-5840	OEM
59-5842	OEM
59-5844	OEM
59B402781	OEM
59B402787	1069
59B402788	1810
59E001-1	0133
59SD-1N34A(AU)	0123
59SD-134A(AU)	OEM
59SI-LA1210	0603
59SI-LM393N	0631
59SI-UPA53C	4138
59SI-UPD1913C	OEM
59ST-BC548B	0079
59Z6	0205
59Z6A	0205
60	OEM
60-3	0015
60-20	OEM
60-1045	OEM
60-1046	OEM
60-1069	OEM
60-3012	OEM
60-3013	OEM
60-3014	OEM
60-3015	OEM
60-3016	OEM
60-3039-4	OEM
60-3039-6	OEM
60-3039-12	OEM
60-3040-12	OEM
60-3040-24	OEM
60-3053-12	OEM
60-3053-24	OEM
60AC80	OEM
60BC15	2657
60C	0015
60CDQ020	OEM
60CDQ030	OEM
60CDQ035	OEM
60CDQ040	OEM
60CDQ045	OEM
60D	0015
60D5R05	OEM
60D9R05GH	OEM
60DC15	2657
60DE10	0071
60DE12	OEM
60DE16	OEM
60DE18	OEM
60DE20	OEM
60DE22	OEM
60DE24	OEM
60DE25	OEM
60DE1414	OEM
60EQ5A	OEM
60F1	2081
60F5	0204
60F80	OEM
60FC15	2006
60FWJ2C11	OEM
60FWJ11	OEM
60GC15	2006
60GWJ2C11	OEM
60GWJ11	OEM
60H	0015
60H3	OEM
60H3N	0476
60H3P	0204
60H3PN	0515
60HF5	3762
60HF10	2653
60HF10R	1783
60HF15	2657
60HF20	2657
60HF25	2664
60HF30	2664
60HF30R	1783
60HF35	2006
60HF40	2006
60HF40R	1783
60HF45	2677
60HF50	2677
60HF50R	1783
60HF60	2682
60HF60R	1783
60HF80	3900
60HF100	2655
60HF100R	1818
60HFR10	2768
60HFR20	2768
60HFR60	6548
60HQ030	OEM
60HQ035	OEM
60HQ045	OEM
60HQ060	OEM
60HQ080	OEM
60HQ090	OEM
60HQ100	OEM
60J1	0965
60J2	0015
60J2P	0122
60J3P	0122
60JC15	2682
60JH3	2871
60KS200C	OEM
60L15	OEM
60L20	6250
60L25	OEM
60L30	OEM
60L35	OEM
60L40	OEM

DEVICE TYPE	REPL CODE
60L45	OEM
60L50	OEM
60L55	OEM
60L60	OEM
60L65	OEM
60L70	OEM
60L75	OEM
60L80	OEM
60L85	OEM
60L90	OEM
60L95	OEM
60L100	OEM
60L110	OEM
60L120	OEM
60L130	OEM
60L140	OEM
60L150	OEM
60L160	OEM
60L170	OEM
60L180	OEM
60L190	OEM
60L200	OEM
60L210	OEM
60L220	OEM
60L230	OEM
60L240	OEM
60L250	OEM
60L260	OEM
60L270	OEM
60L280	OEM
60L290	OEM
60L300	OEM
60LA	0087
60LC15	3900
60LF	OEM
60M	0015
60N15A	OEM
60N15A9	OEM
60N20A	OEM
60N20A9	OEM
60N25A	OEM
60N25A9	OEM
60N30A	OEM
60N30A9	OEM
60N35A9	OEM
60N40A	OEM
60N40A9	OEM
60N45A	OEM
60N45A9	OEM
60N50A	OEM
60N50A9	OEM
60N55A	OEM
60N55A9	OEM
60N60A	OEM
60N60A9	OEM
60N65A	OEM
60N65A9	OEM
60N70A	OEM
60N70A9	OEM
60N75A	OEM
60N75A9	OEM
60N80A	OEM
60N80A9	OEM
60N85A	OEM
60N85A9	OEM
60N90A	OEM
60N90A9	OEM
60N95A	OEM
60N95A9	OEM
60N100A	OEM
60N100A9	OEM
60N110A	OEM
60N120A	OEM
60N130A	OEM
60N140A	OEM
60N150A	OEM
60N160A	OEM
60N170A	OEM
60N180A	OEM
60N190A	OEM
60N200A	OEM
60N220A	OEM
60N230A	OEM
60N240A	OEM
60N250A	OEM
60NC15	3900
60QC15	OEM
60R3P	OEM
60S05	1282
60S1	1282
60S2	1287
60S2D	1669
60S3	1287
60S3P	OEM
60S4	1287
60S4D	1287
60S5	2081
60S6	1293
60S6D	1293
60S8	1296
60S10	1296
60SB03S	1545
60SB04S	1545
60SP-PD-41P1	OEM
60SP-PD-41PI	OEM
60SP-RD-41PI	OEM
60SP-SD-41PI	OEM
60ST-2SC945R	0076
60SZ-RD9.1EB2	0057
60T3P	OEM
60T4	1137
60V3P	OEM
60V10	OEM
60V12	OEM
60W3P	OEM
60X3P	OEM
60Y3P	OEM
60Z6	0221
60Z6A	0221
61	OEM
61-259	0143
61-607	0004
61-608	0004
61-654	0004
61-655	0004
61-656	0004
61-746	0016
61-747	0086
61-751	0016
61-754	0016
61-755	0016
61-756	0229
61-782	0160
61-813	0086
61-814	0016
61-815	0016
61-820	0229
61-926	0229
61-928	0004
61-929	0004
61-1053-1	0423
61-1130	0004
61-1131	0004
61-1215	0004
61-1320	0015
61-1400	0016
61-1401	0016
61-1402	0016
61-1403	0016
61-1404	0016
61-1763	0016
61-1764	0086
61-1765	0015
61-1906	0222
61-1907	0004
61-1934	0004
61-1935	0004
61-3096-90	0854
61-7728	0015
61-8968	1303
61-8968M	1303
61-8969	0774
61-8969M	0774
61-59395	0143
61-260039	0136
61-260039A	0136
61A0001-10	2261
61A0001-11	1478
61A0001-12	0433
61A001-10	2261
61A001-11	1478
61A001-12	0433
61A023-2	2393
61A030-6	0167
61A030-9	2854
61A0306	0167
61B002-1	0050
61B003-1	0136
61B004-1	0004
61B005-1	0004
61B006-1	0004
61B007-1	0144
61B007-2	0144
61B009-1	0004
61B0015-1	0004
61B015-1	0004
61B016-1	0004
61B017-1	0004
61B018-1	0004
61B019-1	0004
61B020-1	0004
61B021-1	0004
61B022-2	0004
61B022-3	0004
61B023-1	0004
61B026-1	0004
61B027-1	0164
61B042-9	0579
61B1C	0802
61C001-1	0198
61C001-2	4034
61C001-4	2960
61C001-7	4067
61C001-11	4079
61C001-12	1524
61C001-30	4744
61C001-32	1580
61C001-33	4758
61C001-34	4752
61C001-35	4747
61C001-36	4751
61C001-38	4746
61C002-1	0211
61C002-2	4759
61C002-3	4162
61C003-1	0211
61C004-1	0086
61C005-1	0160
61C66-1	2977
61C66-2	2977
61C66-3	2977
61C66-4	2977
61C66-5	2977
61C66-6	2977
61C66-7	2977
61C66-8	2977
61C66-9	2977
61C66-10	2977
61E17	2874
61E34	0876
61E39	0262
61E44	0833
61F1FM	OEM
61F2FM	OEM
61F3FM	OEM
61F4FM	OEM
61F6FM	OEM
61H20B27F-8803	OEM
61H20B48F-8M1	OEM
61H20B56F	OEM
61J2	0015
61K001-8	OEM
61K001-9	0882
61K001-10	0520
61K001-12	0882
61K001-13	1479
61L001-3	4627
61L001-4	OEM
61L001-5	4641
61L001-6	4642
61L001-7	OEM
61L001-8	4637
61MQ30	OEM
61MQ40	OEM
61MQ50	OEM
61MQ60	OEM
61N004-18	OEM
61N004-19	OEM
61P1	0136
61P1D	0050
61P10	0136
61R1FM	OEM
61R2FM	OEM
61R3FM	OEM
61R4FM	OEM
61R6FM	OEM
61SI-SAB1018P	OEM
61SI-SN74145N	0619
61SP-LT-4234G	OEM
61ST-BC557B	0037
61SV	OEM
61T1	OEM
61T1C	OEM
61T2	OEM
61T2C	OEM
61T3	OEM
61T3C	OEM
61T4	0344
61T5	OEM
61T6	OEM
61T7	OEM
61T8	OEM
61T9	OEM
61T10	OEM
61T11	OEM
61T12	OEM
61T13	OEM
61T14	OEM
61T15	OEM
61T16	OEM
61T18	OEM
61T19	OEM
61T20	OEM
61T20-EE00	OEM
61T20-RE00	OEM
61T21	OEM
61T21-EE00	OEM
61T21-RE00	OEM
61T22	OEM
61T22-EE00	OEM
61T22-RE00	OEM
61T23	OEM
61T23-EE00	OEM
61T23-RE00	OEM
61T24	OEM
61T24-EE00	OEM
61T24-RE00	OEM
61T25-EE00	OEM
61T25-RE00	OEM
61T26	OEM
61T26-EE00	OEM
61T26-RE00	OEM
61T27-EE00	OEM
61T27-RE00	OEM
61T28	OEM
61T28-EE00	OEM
61T28-RE00	OEM
61T29-EE00	OEM
61T29-RE00	OEM
61T30	OEM
61T30-EE00	OEM
61T30-RE00	OEM
61T31-CE00	OEM
61T31-GE00	OEM
61T31-RE00	OEM
61T32	OEM
61T32-CE00	OEM
61T32-GE00	OEM
61T32-RE00	OEM
61T33-CE00	OEM
61T33-GE00	OEM
61T33-RE00	OEM
61T34	OEM
61T34-CE00	OEM
61T34-GE00	OEM
61T34-RE00	OEM
61T35-CE00	OEM
61T35-GE00	OEM
61T35-RE00	OEM
61T36	OEM
61T36-CE00	OEM
61T36-GE00	OEM
61T36-RE00	OEM
61T37	OEM
61T37-CE00	OEM
61T37-GE00	OEM
61T38	OEM
61T38-CE00	OEM
61T38-GE00	OEM
61T38-RE00	OEM
61T39	OEM
61T39-CE00	OEM
61T39-GE00	OEM
61T39-RE00	OEM
61T40-CE00	OEM
61T40-GE00	OEM
61T40-RE00	OEM
61TB1-JH00	OEM
61TB2-JH00	OEM
61TB3-JH00	OEM
61TB4-JH00	OEM
61TB5-JH00	OEM
61TB6-JH00	OEM
61TB7	OEM
61TB7-JH00	OEM
61TB8-JH00	OEM
61TB9	OEM
61TB9-EE00	OEM
61TB9-RE00	OEM
61TB10-EE00	OEM
61TB10-RE00	OEM
61TB11	OEM
61TB11-EE00	OEM
61TB11-RE00	OEM
61TB12-EE00	OEM
61TB12-RE00	OEM
61TB13	OEM
61TB13-EE00	OEM
61TB13-RE00	OEM
61TB14-EE00	OEM
61TB14-RE00	OEM
61TB15	OEM
61TB15-EE00	OEM
61TB15-RE00	OEM
61TB16-EE00	OEM
61TB16-RE00	OEM
61TB17	OEM
61TB17-EE00	OEM
61TB17-RE00	OEM
61TB18-EE00	OEM
61TB18-RE00	OEM
61TB19	OEM
61TB19-EE00	OEM
61TB19-RE00	OEM
61TB20-EE00	OEM
61TB20-RE00	OEM
61TC	OEM
61Z6	0504
62-3597-1	0150
62-3597-2	0150
62-7567	0016
62-8555	0079
62-8781	0050
62-10234	0143
62-10655	0143
62-12034	0143
62-13258	0004
62-13259	0208
62-13261	0015
62-13477	0015
62-13494	0004
62-15318	0143
62-15483	0015
62-16013	0196
62-16711	0015
62-16712	0196
62-16769	0143
62-16841	0143
62-16905	0016
62-16918	0004
62-16919	0178
62-17232	0917
62-17390	0050
62-17391	0050
62-17550	0016
62-18135	0015
62-18337	0196
62-18415	0280
62-18416	0280
62-18417	0280
62-18418	0050
62-18419	0050
62-18420	0004
62-18421	0004
62-18422	0050
62-18423	0280
62-18424	0280
62-18425	0016
62-18426	0233
62-18427	0085
62-18428	0160
62-18429	0604
62-18430	0004
62-18431	0015
62-18434	0015
62-18435	0015
62-18436	0015
62-18438	0015
62-18641	0016
62-18642	0016
62-18643	0016
62-18782	0133
62-18828	0016
62-19115	0015
62-19260	0917
62-19280	0016
62-19452	0037
62-19516	0016
62-19548	0016
62-19581	0127
62-19620	0917
62-19734	0196
62-19749	0015
62-19814	0015
62-19837	0016
62-19838	0016
62-19846	0143
62-20154	0037
62-20154A	0037
62-20155	0016
62-20223	0133
62-20240	0016
62-20241	0016
62-20242	0016
62-20243	0016
62-20244	0037
62-20244A	0037
62-20319	0133
62-20360	0016
62-20437	0133
62-20565	1303
62-20597	0133
62-20643	0002
62-21369	0015
62-21496	0016
62-21552	0133
62-21573	0367
62-21574	0495
62-21683	0150
62-21699	1030
62-22038	0016
62-22039	0016
62-22250	0016
62-22251	0016
62-22524	0150
62-22529	0150
62-26597	0133
62-26851	0050
62-109474	0196
62-112524	0143
62-113998	0015
62-114267	0127
62-118825	0071
62-125528	0133
62-126321	0133
62-126856	0015
62-127532	0917
62-128343	0280
62-129556	0917
62-129604	0127
62-130045	0133
62-130046	0133
62-130047	0012
62-130139	0150
62-130761	0002
62-130762	0002
62-132497	0150
62-134074	0917
62A01	0015
62A02	0015
62A04	0015
62A05	0015
62B40B20F-8752	OEM
62B40B32F-8810	OEM
62B40B32F-8827	OEM
62J2	0015
62LS141J16	OEM
62R2	OEM
62R2R	1250
62ST-2SC458	0076
62ST-2SC458C	0076
62ST-LBC547B	OEM
62SV	OEM
62T1	OEM
62T2	OEM
62T3	OEM
62T4	0573
62T5	OEM
62T6	OEM
62T7	OEM
62T8	OEM
62T9	OEM
62T10	OEM
62T11	OEM
62T12	OEM
62T13	OEM
62T14	OEM
62T15	OEM
62T16	OEM
62T17	OEM
62T18	OEM
62T19	OEM
62T20	OEM
62T20-LH00	OEM
62T21	OEM
62T21-LH00	OEM
62T22	OEM
62T22-LH00	OEM
62T23	OEM
62T23-LH00	OEM
62T24	OEM
62T24-LH00	OEM
62T25-LH00	OEM
62T26	OEM
62T26-LH00	OEM
62T27-LH00	OEM
62T28	OEM
62T28-LH00	OEM
62T29-LH00	OEM
62T29-NH00	OEM
62T30	OEM
62T30-LH00	OEM
62T30-NH00	OEM
62T31-LH00	OEM
62T31-NH00	OEM
62T32	OEM
62T32-LH00	OEM
62T32-NH00	OEM
62T33-LH00	OEM
62T33-NH00	OEM
62T34	OEM
62T34-LH00	OEM
62T34-NH00	OEM
62T35-LH00	OEM
62T35-NH00	OEM
62T36	OEM
62T36-LH00	OEM
62T36-NH00	OEM
62T37	OEM
62T37-LH00	OEM
62T37-NH00	OEM
62T38	OEM
62T38-LH00	OEM
62T38-NH00	OEM
62T39-LH00	OEM
62T39-NH00	OEM
62T40	OEM
62T40-LH00	OEM
62T40-NH00	OEM
62TB7	OEM
62TB9	OEM
62TB11	OEM
62TB13	OEM
62TB15	OEM
62TB17	OEM
62TB19	OEM
62Z6	0684
62Z6A	0684
63-00-84381-2	0211
63-3954	0050
63-7246	0164
63-7247	0004
63-7248	0208
63-7396	0004
63-7397	0004
63-7398	0004
63-7399	0004
63-7420	0004
63-7421	0016
63-7433	0015
63-7538	0050
63-7541	0050
63-7547	0280
63-7548	0050
63-7549	0208
63-7564	0004
63-7565	0208
63-7567	0016
63-7579	0050
63-7580	0050
63-7581	0050
63-7582	0050
63-7596	0004
63-7660	0050
63-7670	0016
63-7871	0004
63-7872	0004
63-7873	0004
63-8119	0050
63-8120	0164
63-8376	0050
63-8377	0050
63-8378	0050
63-8379	0050
63-8380	0004
63-8381	0143
63-8473	0208
63-8512	0178
63-8555	0016
63-8590	0160
63-8685	0015
63-8699	0050
63-8700	0050
63-8701	0016
63-8702	0016
63-8703	0004
63-8704	0004
63-8705	0208
63-8706	0160
63-8707	0103
63-8707A	0103
63-8819	0229
63-8824	0015
63-8825	0244
63-8945	0178
63-8954	0050
63-8955	0143
63-9072	0050
63-9337	0016
63-9338	0016
63-9339	0016
63-9340	0208
63-9341	0016
63-9516	0016
63-9517	0050
63-9518	0016
63-9519	0004
63-9520	0004
63-9521	0004
63-9522	0030
63-9523	0143
63-9659	0004
63-9664	0050
63-9665	0050
63-9782	0030
63-9783	0244
63-9787	0229
63-9829	0016
63-9830	0016
63-9831	0016
63-9832	0016
63-9833	0016
63-9847	0016
63-9876	0050
63-9877	0050
63-9941	0050
63-9942	0137
63-10035	0050
63-10036	0050
63-10037	0004
63-10038	0004
63-10062	0178
63-10064	0015
63-10145	0050
63-10146	0050
63-10147	0004
63-10148	0050
63-10149	0050
63-10150	0050
63-10151	0004
63-10152	0004
63-10153	0004
63-10154	0004
63-10156	0004
63-10158	0004
63-10159	0004
63-10188	0016
63-10195	0050
63-10196	0050
63-10200	0280
63-10235-64	OEM
63-10375	0050
63-10376	0050
63-10377	0016
63-10378	0160
63-10383	0208
63-10384	0164
63-10408	0004
63-10708	0016
63-10709	0015
63-10725	0016
63-10732	0016
63-10733	0016
63-10734	0016
63-10735	0016
63-10736	0016
63-10737	0016
63-10739	0143
63-10860	0016
63-11025	0016
63-11055	0050
63-11073	0164
63-11074	0143
63-11143	0016
63-11144	0004
63-11147	0025
63-11148	0289
63-11215	0229
63-11289	0016
63-11290	0178
63-11291	0015
63-11468	0016
63-11469	0016
63-11470	0016
63-11471	0016
63-11472	0016
63-11474	0004
63-11496	0050
63-11497	0164
63-11582	0050
63-11584	0050
63-11585	0280
63-11586	0004

DEVICE TYPE	REPL CODE
63-11659	0025
63-11660	0016
63-11661	0004
63-11757	0016
63-11758	0016
63-11759	0016
63-11762	0071
63-11825	0079
63-11831	0016
63-11832	0016
63-11833	0016
63-11834	0030
63-11878	0178
63-11879	0143
63-11881	0015
63-11934	0016
63-11935	0016
63-11936	0086
63-11937	0016
63-11938	0178
63-11957	0015
63-11989	0086
63-11991	0103
63-11991A	0103
63-12003	0016
63-12004	0016
63-12062	0016
63-12077	0229
63-12110	0025
63-12154	0037
63-12154A	0037
63-12156	0037
63-12156A	0037
63-12157	0037
63-12157A	0037
63-12158	0143
63-12272	0016
63-12273	0074
63-12287	0289
63-12316	0004
63-12317	0004
63-12366	0071
63-12605	0016
63-12607	0143
63-12608	0016
63-12609	0016
63-12610	0050
63-12641	0016
63-12642	0016
63-12645	0143
63-12669	0164
63-12670	0164
63-12696	0016
63-12697	0016
63-12698	0164
63-12706	0016
63-12707	0016
63-12750	0016
63-12751	0016
63-12752	0016
63-12753	0016
63-12754	0143
63-12755	0143
63-12756	0143
63-12757	0143
63-12874	0016
63-12875	0016
63-12876	0004
63-12877	0016
63-12878	0016
63-12879	0016
63-12880	0004
63-12881	0004
63-12933	0016
63-12940	0016
63-12941	0016
63-12942	0016
63-12943	0016
63-12944	0079
63-12945	0164
63-12946	0016
63-12947	0164
63-12948	0016
63-12949	0016
63-12950	0016
63-12951	0016
63-12952	0016
63-12953	0016
63-12954	0178
63-12989	0086
63-12990	0086
63-13025	0050
63-13080	0143
63-13214	0086
63-13215	0086
63-13216	0074
63-13322	0037
63-13322A	0037
63-13323	0164
63-13419	0016
63-13438	0016
63-13440	0016
63-13441	0016
63-13839	0050
63-13840	0004
63-13842	0143
63-13864	0016
63-13899	0050
63-13903	0015

DEVICE TYPE	REPL CODE
63-13919	0015
63-13926	0325
63-13927	0016
63-13954	0074
63-14032	0016
63-14051	0016
63-14052	0016
63-14057	0016
63-14195	0015
63-15483	0015
63-16918	0004
63-17390	0050
63-18135	0229
63-18416	0280
63-18418	0050
63-18419	0050
63-18420	0004
63-18421	0004
63-18423	0050
63-18424	0050
63-18426	0285
63-18427	0085
63-18430	0004
63-18643	0016
63-19173	0293
63-19280	0016
63-19282	0016
63-22724	0143
63-23041	0004
63-25179	0004
63-25180	0004
63-25181	0004
63-25182	0004
63-25261	0164
63-25281	0004
63-25282	0004
63-25342	0123
63-25720	0004
63-25726	0050
63-25727	0004
63-25728	0004
63-25729	0004
63-25933	0143
63-25942	0004
63-25944	0004
63-25946	0280
63-26382	0143
63-26597	0015
63-26849	0280
63-26850	0050
63-26851	0004
63-27278	0280
63-27279	0280
63-27280	0280
63-27281	0004
63-27366	0050
63-27367	0280
63-27483	0229
63-27500	0050
63-27622	0015
63-28250	0143
63-28348	0050
63-28358	0050
63-28390	0004
63-28399	0004
63-28426	0233
63-28888	0143
63-29383	0229
63-29451	0160
63-29459	0160
63-29461	0016
63-29661	0050
63-29662	0280
63-29663	0280
63-29664	0280
63-29665	0004
63-29666	0004
63-29819	0050
63-29820	0050
63-29821	0050
63-29862	0050
63-29863	0280
63-250128-2	0917
63-250128-8	0917
63-250128-12	0917
63-250128-14	0917
63-250128-16	0917
63-259128-3	OEM
63A03Y	OEM
63J2	0015
63LS10J16	OEM
63LS140N16	OEM
63LS141N16	OEM
63LS240J16	OEM
63LS240N16	OEM
63LS241J16	OEM
63LS241N16	OEM
63N1	0642
63N50	0440
63P3	0211
63R2	OEM
63RA441J	OEM
63RA441N	OEM
63RECT-SI-1021	0023
63S080J	OEM
63S080N	OEM
63S081J	OEM
63S081N	OEM
63S140J16	OEM

DEVICE TYPE	REPL CODE
63S140N16	OEM
63S141J16	OEM
63S141N16	OEM
63S240J16	OEM
63S240N16	OEM
63S241J16	OEM
63S241N16	OEM
63ST-2SC1162C	0563
63Z6	0510
63Z6A	0510
64-1	0143
64-8054-6	0015
64-J2	0015
64EPA	OEM
64EPB	OEM
64J2	0015
64KZ	OEM
64N1	0074
64R2	0109
64RECT-SI-1001(AU)	0795
64RECT-U-1012	2010
64SP-LT-6740R	OEM
64T1	0280
64T4	0058
64Z6	0691
64Z6A	0691
65	0015
65-08001	0822
65-080001	1862
65-085002	0143
65-085003	0143
65-085004	0137
65-085010	0143
65-085012	0143
65-085013	0015
65-1	0016
65-1-70	0050
65-1-70-12	0050
65-1-70-12-7	0050
65-1A	0050
65-1A0	0050
65-1A0R	0050
65-1A1	0050
65-1A2	0050
65-1A3	0050
65-1A3P	0050
65-1A4	0050
65-1A4-7	0050
65-1A4-7B	0050
65-1A5	0050
65-1A5L	0050
65-1A6	0050
65-1A6-5	0050
65-1A7	0050
65-1A7-1	0050
65-1A8	0050
65-1A9	0050
65-1A9G	0050
65-1A19	0050
65-1A21	0050
65-1A82	0050
65-2	0208
65-2-70	0208
65-2-70-12	0208
65-2-70-12-7	0208
65-2A	0208
65-2A0	0208
65-2A0R	0208
65-2A1	0208
65-2A2	0208
65-2A3	0208
65-2A3P	0208
65-2A4	0208
65-2A4-7	0208
65-2A4-7B	0208
65-2A5	0208
65-2A5L	0208
65-2A6	0208
65-2A6-1	0208
65-2A7	0208
65-2A7-1	0208
65-2A8	0208
65-2A9	0208
65-2A9G	0208
65-2A19	0208
65-2A21	0208
65-2A82	0208
65-4	0600
65-4-70	0600
65-4-70-12	0600
65-4-70-12-7	0600
65-4A	0600
65-4A0	0600
65-4A0R	0600
65-4A1	0600
65-4A2	0600
65-4A3	0600
65-4A3P	0600
65-4A4	0600
65-4A4-7	0600
65-4A4-7B	0600
65-4A5	0600
65-4A5L	0600
65-4A6	0600
65-4A6-2	0600
65-4A7	0600
65-4A7-1	0600
65-4A8	0600
65-4A9	0600

DEVICE TYPE	REPL CODE
65-4A9G	0600
65-4A19	0600
65-4A21	0600
65-4A82	0600
65-10	0050
65-11	0050
65-12	0050
65-13	0050
65-14	0050
65-15	0050
65-16	0050
65-17	0050
65-18	0050
65-19	0050
65-20	0208
65-21	0208
65-22	0208
65-23	0208
65-24	0208
65-25	0208
65-26	0208
65-27	0208
65-28	0208
65-29	0208
65-40	0600
65-41	0600
65-42	0600
65-43	0600
65-44	0600
65-45	0600
65-46	0600
65-47	0600
65-48	0600
65-49	0600
65-11137-23	2755
65-11502-03	0079
65-20820-00	4003
65-22009-00	OEM
65-27328-01	4818
65-33361-00	1260
65-37323-164	0799
65-45402-00	2295
65-45430-00	OEM
65-48629-00	OEM
65-80001	1010
65-250128-12	0917
65-744238	0484
65A	0037
65A0	0050
65A1	0050
65A2	0050
65A3	0050
65A4	0050
65A5	0050
65A6	0050
65A7	0050
65A8	0050
65A9	0050
65A10	0050
65A10R	0050
65A12	0050
65A13	0050
65A13P	0050
65A14	0050
65A14-7	0050
65A14-7B	0050
65A15	0050
65A15L	0050
65A16	0050
65A16-3	0050
65A17	0050
65A17-1	0050
65A18	0050
65A19	0050
65A19G	0050
65A-70	0050
65A-70-12	0050
65A-70-12-7	0050
65A119	0050
65A121	0050
65A182	0050
65B	0037
65B0	0136
65B1	0136
65B3	0136
65B4	0136
65B5	0136
65B6	0136
65B7	0136
65B8	0136
65B9	0136
65B10	0136
65B10R	0136
65B11	0136
65B12	0136
65B13	0136
65B13P	0136
65B14	0136
65B14-7	0136
65B14-7B	0136
65B15	0136
65B15L	0136
65B16	0136
65B16-2	0136
65B17	0136
65B17-1	0136
65B18	0136
65B19	0136
65B19G	0136

DEVICE TYPE	REPL CODE
65B-70	0136
65B-70-12	0136
65B-70-12-7	0136
65B119	0136
65B121	0136
65B182	0136
65BO	0136
65C	0037
65C0	0160
65C01	0050
65C02	0050
65C02A	OEM
65C03	0050
65C1	0037
65C2	0160
65C3	0160
65C4	0160
65C5	0160
65C6	0160
65C7	0160
65C8	0160
65C9	0160
65C10	0160
65C10R	0160
65C11	0160
65C12	0160
65C13	0160
65C13P	0160
65C14	0160
65C14-7	0160
65C14-7B	0160
65C15	0160
65C15L	0160
65C16	0160
65C16-4	0160
65C17	0160
65C17-1	0160
65C18	0160
65C19	0160
65C19G	0160
65C22	OEM
65C-70	0160
65C-70-12	0160
65C-70-12-7	0160
65C119	0160
65C121	0160
65C182	0160
65D	0037
65D1	0037
65D5R09	OEM
65D9R09GH	OEM
65E	0037
65E05A	OEM
65E1	0037
65F	0037
65F1	0037
65J2	0015
65P117	0015
65P124	0015
65P124-1	0015
65P124-2	0015
65P153	0015
65P155	0015
65P206	0015
65P284	0015
65P297	0015
65R2	OEM
65R2S	OEM
65SI-HA11580	1055
65ST-2SC454B	0076
65ST-2SC535B	0127
65T1	0004
65Z6	0870
65Z6A	0870
66-1	0004
66-1-70	0004
66-1-70-12	0004
66-1-70-12-7	0004
66-1A	0004
66-1A0	0004
66-1A0R	0004
66-1A1	0004
66-1A2	0004
66-1A3	0004
66-1A3P	0004
66-1A4	0004
66-1A4-7	0004
66-1A4-7B	0004
66-1A5	0004
66-1A5L	0004
66-1A6	0004
66-1A6-3	0004
66-1A7	0004
66-1A7-1	0004
66-1A8	0004
66-1A9	0004
66-1A9G	0004
66-1A19	0004
66-1A21	0004
66-1A82	0004
66-2	0004
66-2-70	0004
66-2-70-12	0004
66-2-70-12-7	0004
66-2A	0004
66-2A0	0004
66-2A0R	0004
66-2A1	0004
66-2A2	0004

DEVICE TYPE	REPL CODE
66-2A3	0004
66-2A3P	0004
66-2A4	0004
66-2A4-7	0004
66-2A4-7B	0004
66-2A5	0004
66-2A5L	0004
66-2A6	0004
66-2A6-4	0004
66-2A7	0004
66-2A7-1	0004
66-2A8	0004
66-2A9	0004
66-2A9G	0004
66-2A19	0004
66-2A21	0004
66-2A82	0004
66-3	0004
66-3-70	0004
66-3-70-12	0004
66-3-70-12-7	0004
66-3A	0004
66-3A0	0004
66-3A0R	0004
66-3A1	0004
66-3A2	0004
66-3A3	0004
66-3A3P	0004
66-3A4	0004
66-3A4-7	0004
66-3A4-7B	0004
66-3A5	0004
66-3A5L	0004
66-3A6	0004
66-3A6C	0004
66-3A7	0004
66-3A7-1	0004
66-3A8	0004
66-3A9	0004
66-3A9G	0004
66-3A19	0004
66-3A21	0004
66-3A82	0004
66-10	0004
66-11	0004
66-12	0004
66-13	0004
66-14	0004
66-15	0004
66-16	0004
66-17	0004
66-18	0004
66-19	0004
66-20	0004
66-21	0004
66-22	0004
66-23	0004
66-24	0004
66-25	0004
66-26	0004
66-27	0004
66-28	0004
66-29	0004
66-30	0004
66-31	0004
66-32	0004
66-33	0004
66-34	0004
66-35	0004
66-36	0004
66-37	0004
66-38	0004
66-2246	0015
66-6023	0004
66-6023-00	0004
66-6024-00	0004
66-6025-00	0004
66-6026-00	0004
66-6027-00	0004
66-6028-00	0004
66-6030-00	0015
66-6031-00	0015
66-6033	0004
66-8504	0071
66-11502-02	0079
66-11502-03	0079
66-11702-11	0079
66-18050-03	OEM
66-19013-07J	0079
66-19014-03J	4898
66-19014-035	4898
66-19400-00	0161
66-19400-01	0161
66-20324-00	0808
66-20438-00	0925
66-29015-03J	0037
66-29015-035	0037
66-F29-1	0016
66-P11I20	0037
66-P1120	0037
66A52	OEM
66A96-1	0190
66B-020A	2494
66C0039-001	OEM
66F-001	0015
66F001	0015
66F-001-1	0015
66F001-1	0015
66F-010	0042

DEVICE TYPE	REPL CODE
66F016-1	0133
66F016-2	0133
66F017	0913
66F018	0133
66F020-1	0276
66F020-2	0276
66F021-1	0007
66F022-1	0007
66F023-1	0037
66F024-1	0037
66F025-1	0086
66F026-1	0111
66F027-1	0016
66F028-1	0016
66F029-1	0016
66F039-1	0400
66F041-1	0037
66F042-1	0007
66F-054-1	1501
66F-054-2	1501
66F-054-3	1501
66F054-3	0190
66F054-4	0190
66F057-1	0016
66F057-2	0016
66F058-2	0551
66F074-1	0285
66F074-2	0285
66F074-3	0285
66F074-4	0285
66F-084-1	0325
66F-112-1	1501
66F112-1	1501
66F-112-2	1501
66F112-2	1501
66F125-1	2167
66F136-1	2043
66F159-1	1067
66F159-2	1067
66F159-3	1067
66F175-1	2608
66F176-1	2688
66F-181-1	1501
66F181-1	1501
66F1271	0349
66F1551	0856
66F1751	2608
66F1761	2688
66J2	0015
66R2	0122
66R2S	0816
66ST-2SC1514BK	1085
66ST-2SD819	2668
66SZ-RD10EB3	0064
66WM8	OEM
66X0003	1030
66X0003-001	1030
66X0003-1	1030
66X002-001	0913
66X0020-000	0143
66X0020-001	0143
66X0023-001	0015
66X0023-002	2637
66X0023-003	0229
66X0023-004	0015
66X0023-005	0015
66X0023-006	0015
66X0023-007	0015
66X0023-008	0015
66X0023-009	0071
66X0023-1	0015
66X0024-000	0196
66X0025-000	0196
66X0025-000-001	0196
66X0025-000,001	0196
66X0025-001	0196
66X0028-001	0015
66X0028-008	0071
66X0033-000	0015
66X0033-001	0015
66X0035-000	0774
66X0035-001	0774
66X0036-001	1303
66X0036-002	0102
66X0037-001	0015
66X0038-001	0102
66X0039-001	0143
66X0040-001	OEM
66X0040-003	0636
66X0040-004	0726
66X0040-005	0064
66X0040-007	0402
66X0040-008	0028
66X0040-009	0002
66X0040-011	0533
66X0040-012	0783
66X0040-019	0036
66X0040-020	0057
66X0040-021	0152
66X0041-001	0474
66X0043-001	0133
66X0044-001	0133
66X0044-100	0133
66X0045-001	0128
66X0046-001	0124
66X0047-001	0143
66X0047-901	0143
66X0048-001	0015
66X0049-001	0861

If replacement code is OEM, contact original manufacturer for replacement.

DEVICE TYPE	REPL CODE
66X0049-001(DIODE)	0123
66X0049-002	0143
66X0049-004	0028
66X0049-100	0123
66X0050-001	0030
66X0051-001	0143
66X0053-001	1217
66X0054-001	1217
66X0054-002	1217
66X0055-001	1217
66X0056-001	0133
66X0060-001	1501
66X0060-002	1501
66X0061-001	0128
66X0062-001	0133
66X0067-002	OEM
66X0068-002	0102
66X0069-001	0133
66X0070-001	0124
66X0071-001	0023
66X0073-001	0087
66X002000	0143
66X041-001	0015
66X053-001	0015
66X060-001	OEM
66X14	0015
66X14-C	0015
66X14C	0015
66X16	0015
66X17	0015
66X19	0015
66X20	0143
66X21	0196
66X23	0015
66X24	0015
66X25	0196
66X25-0	0196
66X26	0015
66X29	0133
66X41-1	0474
66X45-1	0128
66X218	0196
66XZ18	0196
66Z6	0237
66Z6A	0237
67-01603-01	2879
67-1000-00	0015
67-1003-00	0015
67-10430-01	1329
67-20315-01	OEM
67-30201-05	0030
67-30321-05	3688
67-31043-05	0554
67-31212-04	0052
67-31603-01	OEM
67-31703-01	OEM
67-32009-01	0663
67-32109-02	4109
67-32706-03	OEM
67-32706-04	OEM
67-32720-01	2024
67-32720-02	1949
67-32720-03	OEM
67-32720-04	OEM
67-32808-02	5011
67-33105-01	OEM
67-33215-01	3378
67-34002-06	0080
67-34003-06	0023
67-34148-07	0124
67-35402-06	0242
67-39221-04P	0057
67-40060-01	0019
67-40060-02	0019
67-51703-02	3862
67-53315-01	1529
67-90430-01	1329
67-93401-01	3365
67-93501-01	3884
67-RECT-U-1002	OEM
67C015H20LSS	OEM
67C024H20LSS	OEM
67C030H20LSS	OEM
67C050H20LSS	OEM
67C075H20LSS	OEM
67C100H20LSS	OEM
67C120H20LSS	OEM
67C150H20LSS	OEM
67C200H20LSS	OEM
67C250H20LSS	OEM
67C300H20LSS	OEM
67C400H20LSS	OEM
67C500H20LSS	OEM
67C750H20LSS	OEM
67D006H53PNN	OEM
67D010H53PNN	OEM
67D015H53PNN	OEM
67D015H55FNN	OEM
67D018H55FNN	OEM
67D020H53PNN	OEM
67D024H55FNN	OEM
67D030H53PNN	OEM
67D036H55FNN	OEM
67D048H55FNN	OEM
67D050H04TNN	OEM
67D050H20TTS	OEM
67D050H53PNN	OEM
67D050H55FNN	OEM
67D075H04TNN	OEM
67D075H20TTS	OEM
67D075H55FNN	OEM
67D080H53PNN	OEM
67D100H04TNN	OEM
67D100H20TTS	OEM
67D100H53PNN	OEM
67D100H55FNN	OEM
67D120H04TNN	OEM
67D120H20TTS	OEM
67D120H53PNN	OEM
67D120H55FNN	OEM
67D150H04TNN	OEM
67D150H20TTS	OEM
67D150H53PNN	OEM
67D150H55FNN	OEM
67D200H04TNN	OEM
67D200H20TTS	OEM
67D200H53PNN	OEM
67D200H55FNN	OEM
67D250H20TTS	OEM
67D250H53PNN	OEM
67D250H55FNN	OEM
67D300H04TNN	OEM
67D300H20TTS	OEM
67D300H53PNN	OEM
67D300H55FNN	OEM
67D400H04TNN	OEM
67D400H20TTS	OEM
67D400H55FNN	OEM
67D500H20TTS	OEM
67D500H55FNN	OEM
67D600H20TTS	OEM
67D750H20TTS	OEM
67J2	0071
67J2A	0071
67LS300J	OEM
67LS301J	OEM
67LS304J	OEM
67LS306J	OEM
67LS307J	OEM
67LS376J	OEM
67LS380J	OEM
67P1	0160
67P2	0160
67P3	0160
67R2	OEM
67RECT-U-1002	OEM
67S300J	OEM
67S301J	OEM
67S304J	OEM
67S306J	OEM
67S307J	OEM
67S373J	OEM
67S374J	OEM
67S376J	OEM
67S378J	OEM
67S380J	OEM
67S382J	OEM
67SD-1N34A	0123
67Z6	1397
67Z6A	1397
68-A-8318-P1	0037
68A831-P1	0037
68A7349-D32	1041
68A7349-D46	0141
68A7349-D62	0562
68A7349-PD32	1041
68A7349PD30	1823
68A7349PD36	1831
68A7349PD45	0081
68A7349PD46	0141
68A7349PD62	0562
68A7652P1	1350
68A7652P2	2899
68A8225P001	0415
68A9025	0232
68A9026	0682
68A9027	0314
68A9028	0361
68A9030	0512
68A9031	0685
68A9032	1206
68A9033	0697
68A9034	3466
68A9035	0873
68A9036	3520
68A9037	0996
68A9038	0743
68A9040	1275
68A9041	1432
68A9042	1158
68A9047	5253
68A9048	1495
68A9049	1539
68HDS	OEM
68P1	0160
68P1B	0160
68PB08M	OEM
68PBF9C	OEM
68PD08M	OEM
68PH08M	OEM
68PS08M	OEM
68R2	0131
68R2S	0131
68SP-CQY89A	OEM
68ST-2SD819	2668
68ST-MPS9700E	0079
68SZ-RD12FB3	0137
68X0003	1030
68X0003-001	1030
68X003-001	1030
68X0040-004	0726
68X0040-005	0170
68Z6	0247
68Z6A	0247
69-001	0239
69-01014-01	OEM
69-02005-01	OEM
69-02006-02	0112
69-02007-01	OEM
69-1810	0079
69-1811	0079
69-1812	0079
69-1813	0079
69-1814	0086
69-1815	0150
69-1816	0111
69-1817	0150
69-1818	0161
69-1819	0460
69-1820	0143
69-1821	0030
69-1822	0133
69-1823	0015
69-2246	0015
69-2922	0019
69-3116	2557
69AJ110	0905
69B1M	0354
69N1	0144
69SP112	0126
69ST-2SD313F	0042
69SZ-HZ12A3(AU)	1005
69V097H62	0079
70	OEM
70ATB7	OEM
70ATB9	OEM
70ATB11	OEM
70B1C	0352
70B1Z	0352
70BN20	OEM
70BN30	OEM
70BN40	OEM
70BN50	OEM
70BN60	OEM
70BN80	OEM
70BN100	OEM
70BN120	OEM
70BN140	OEM
70BN160	OEM
70BNS	OEM
70C05	OEM
70C05B	OEM
70C05BF	OEM
70C05C	OEM
70C05CF	OEM
70C05F	OEM
70C025	0217
70C025B	0217
70C025BF	0217
70C025F	0217
70C050	0217
70C050B	0217
70C050BF	0217
70C050F	0217
70C10	0217
70C10B	0217
70C10BF	0217
70C10C	OEM
70C10CF	OEM
70C10F	0217
70C15	0217
70C15B	0217
70C15BF	0217
70C15C	OEM
70C15CF	OEM
70C15F	0217
70C20	0217
70C20B	0217
70C20BF	0217
70C20C	OEM
70C20CF	OEM
70C20F	0217
70C25	0217
70C25B	0217
70C25BF	0217
70C25C	OEM
70C25CF	OEM
70C25F	0217
70C30	0217
70C30B	0217
70C30BF	0217
70C30C	OEM
70C30CF	OEM
70C30F	0217
70C35	0217
70C35B	0468
70C35BF	0217
70C35C	0217
70C35CF	OEM
70C35F	0217
70C40	0217
70C40B	0468
70C40BF	0217
70C40C	0217
70C40CF	OEM
70C40F	0217
70C45	0217
70C45B	0468
70C45BF	0217
70C45C	OEM
70C45CF	OEM
70C45F	0217
70C50	0217
70C50B	0468
70C50BF	0217
70C50C	OEM
70C50CF	OEM
70C50F	0217
70C60	0217
70C60B	0468
70C60BF	0217
70C60C	OEM
70C60CF	OEM
70C60F	0217
70C70	0468
70C70B	0468
70C70BF	0658
70C70C	OEM
70C70CF	OEM
70C70F	0658
70C80	0468
70C80B	0468
70C80BF	0658
70C80C	OEM
70C80CF	OEM
70C80F	0658
70C90	0468
70C90B	0468
70C90BF	0658
70C90C	OEM
70C90CF	OEM
70C90F	0658
70C100	0468
70C100B	0468
70C100BF	0658
70C100C	OEM
70C100CF	OEM
70C100F	1725
70C110	0468
70C110B	0468
70C110BF	0658
70C110C	OEM
70C110CF	OEM
70C110F	0658
70C120	0468
70C120B	0468
70C120BF	0658
70C120C	OEM
70C120CF	OEM
70C120F	0658
70C130	OEM
70C130B	OEM
70C130BF	OEM
70C130F	OEM
70C140	OEM
70C140B	OEM
70C140BF	OEM
70C140F	OEM
70C150	OEM
70C150B	OEM
70C150BF	OEM
70C150F	OEM
70C160	OEM
70C160B	OEM
70C160BF	OEM
70C160F	OEM
70C170	OEM
70C170B	OEM
70C170BF	OEM
70C170F	OEM
70C180	OEM
70C180B	OEM
70C180BF	OEM
70C180F	OEM
70D5R10	OEM
70D9R10GH	OEM
70E04A	OEM
70F10	OEM
70F15	OEM
70F20	OEM
70F25	OEM
70F30	OEM
70F40	0015
70F50	OEM
70F60	OEM
70F100	OEM
70H5	1560
70H5A	1560
70H5AR	2179
70H5R	2179
70H10	1560
70H10A	1560
70H10AR	2179
70H10R	2179
70H15	2840
70H15A	2840
70H15AR	2182
70H15R	2182
70H20	2840
70H20A	2840
70H20AR	2182
70H20R	2182
70H25	4308
70H25A	4308
70H25AR	3601
70H25R	3601
70H30	4308
70H30A	4308
70H30AR	3601
70H30R	3601
70H40	2850
70H40A	2850
70H40AR	2191
70H40R	2191
70H50	2871
70H50A	2871
70H50AR	2197
70H50R	2197
70H60	2871
70H60A	2871
70H60AR	2197
70H60R	2197
70H70	2833
70H70A	2833
70H70AR	2216
70H70R	2216
70H80	2833
70H80A	2833
70H80AR	2216
70H80R	2216
70H90	2473
70H90A	2473
70H90AR	2340
70H90R	2340
70H100	2473
70H100A	2473
70H100AR	2340
70H100R	2340
70H120A	2643
70HA40	OEM
70HA60	OEM
70HA80	OEM
70HA100	OEM
70HA120	OEM
70HA140	OEM
70HA160	OEM
70HF10	1560
70HF20	2840
70HF40	2850
70HF60	2871
70HF60A	OEM
70HF80	2833
70HF80T	2845
70HF100	2473
70HF100A	OEM
70HF100T	2845
70HF120	2643
70HF140	OEM
70HF160	OEM
70HFL5S02	4212
70HFL5S05	4212
70HFL5S10	4212
70HFL10S02	4212
70HFL10S05	4212
70HFL10S10	4212
70HFL20S02	4212
70HFL20S05	4212
70HFL20S10	4212
70HFL40S02	OEM
70HFL40S05	OEM
70HFL40S10	OEM
70HFL60S02	5622
70HFL60S05	5622
70HFL60S10	5622
70HFL80S05	OEM
70HFL80S10	OEM
70HFL100S05	OEM
70HFL100S10	OEM
70HFR10	2179
70HFR20	2182
70HFR30	OEM
70HFR40	2191
70HFR50	OEM
70HFR60	2197
70HFR80	2216
70HFR80T	1818
70HFR100	2340
70HFR100T	1818
70HFR120	5325
70HFR140	OEM
70HFR160	OEM
70HG10	OEM
70HG20	OEM
70HG40	OEM
70HG60	OEM
70HG80	OEM
70HG100	OEM
70HG120	OEM
70HR5A	1745
70HR10A	1745
70HR15	2182
70HR15A	2182
70HR20	2182
70HR20A	2182
70HR30A	1783
70HR40	2191
70HR40A	2191
70HR50	2197
70HR50A	2197
70HR60	2197
70HR60A	2197
70HR90	2473
70HR90A	2473
70HR100	2473
70HR100A	2473
70HR120A	5325
70KS200C	OEM
70M10	1023
70M15	1023
70M20	1023
70M30	1023
70M40	1023
70M50	1046
70M60	1036
70M80	1046
70M100	1046
70MA10	0501
70MA20	0501
70MA30	0501
70MA40	0501
70MA60	1777
70MA80	2832
70MA100	1789
70MA120	3167
70MA140	3167
70MA160	OEM
70MLA60	OEM
70MLA80	OEM
70MLA100	OEM
70MLA120	OEM
70MLAB160	OEM
70MLAB180	OEM
70MLAB200	OEM
70MLAB250	OEM
70N1	0016
70N1M	0111
70N2	0198
70N3	0127
70N4	0016
70R2	OEM
70R2S	0145
70RC2	OEM
70RC5	OEM
70RC10	OEM
70RC15	OEM
70RC20	OEM
70RC30	OEM
70RC40	OEM
70RC50	OEM
70RC50A	OEM
70RC60	OEM
70RC60A	OEM
70RC70A	OEM
70RC80A	OEM
70RCF5A	OEM
70RCF10A	OEM
70RCF15A	OEM
70RCF20A	OEM
70RCF25A	OEM
70RCF30A	OEM
70RCF40A	OEM
70RCF60A	OEM
70RCF70A	OEM
70RCF80A	OEM
70RCS10A	OEM
70RCS20A	OEM
70RCS30A	OEM
70RCS40A	OEM
70RCS50A	OEM
70RCS60A	OEM
70RE60	0468
70RE60A	0468
70RE70	0468
70RE70A	0468
70RE80	0468
70RE80A	0468
70RE90	0468
70RE90A	0468
70RE100	0468
70RE100A	0468
70RE110	0468
70RE110A	0468
70RE120	0468
70RE120A	0468
70RE130	OEM
70REB60	0468
70REB70	0468
70REB80	0468
70REB90	0468
70REB100	0468
70REB110	0468
70REB120	0468
70REB130	OEM
70S5	OEM
70S10	OEM
70S15	OEM
70S25	OEM
70S30	OEM
70S35	OEM
70S40	0015
70S45	OEM
70S50	OEM
70S60	OEM
70S70	OEM
70S2020	OEM
70SD-1SV124	0628
70ST-2SC945Q	0076
70STB7	OEM
70STB9	OEM
70STB11	OEM
70STB13	OEM
70T1	OEM
70T2	OEM
70T3	OEM
70T4	OEM
70T5	OEM
70T6	OEM
70T7	OEM
70T8	OEM
70T9	OEM
70T10	OEM
70T11	OEM
70T12	OEM
70T13	OEM
70T14	OEM
70T15	OEM
70T16	OEM
70T17	OEM
70T18	OEM
70T19	OEM
70T20	OEM
70T25	OEM
70T30	OEM
70T35	OEM
70T40	0015
70T45	OEM
70T50	OEM
70T60	OEM
70T70	OEM
70TB1	OEM
70TB1-CE00	OEM
70TB1-CH00	OEM
70TB1-GH00	OEM
70TB1-JE00	OEM
70TB1-JH00	OEM
70TB1-LE00	OEM
70TB1-LH00	OEM
70TB1-NE00	OEM
70TB1-NH00	OEM
70TB1-RE00	OEM
70TB2	OEM
70TB2-CE00	OEM
70TB2-CH00	OEM
70TB2-GH00	OEM
70TB2-JE00	OEM
70TB2-JH00	OEM
70TB2-LE00	OEM
70TB2-LH00	OEM
70TB2-NE00	OEM
70TB2-NH00	OEM
70TB2-RE00	OEM
70TB3	OEM
70TB3-CE00	OEM
70TB3-CH00	OEM
70TB3-GH00	OEM
70TB3-JE00	OEM
70TB3-JH00	OEM
70TB3-LE00	OEM
70TB3-LH00	OEM
70TB3-NE00	OEM
70TB3-NH00	OEM
70TB3-RE00	OEM
70TB4	OEM
70TB4-CE00	OEM
70TB4-CH00	OEM
70TB4-GE00	OEM
70TB4-GH00	OEM
70TB4-JE00	OEM
70TB4-JH00	OEM
70TB4-LE00	OEM
70TB4-LH00	OEM
70TB4-NE00	OEM
70TB4-NH00	OEM
70TB4-RE00	OEM
70TB5	OEM
70TB5-CE00	OEM
70TB5-CH00	OEM
70TB5-GE00	OEM
70TB5-GH00	OEM
70TB5-JE00	OEM
70TB5-JH00	OEM
70TB5-LE00	OEM
70TB5-LH00	OEM
70TB5-NE00	OEM
70TB5-NH00	OEM
70TB6	OEM
70TB6-CE00	OEM
70TB6-CH00	OEM
70TB6-GE00	OEM
70TB6-GH00	OEM
70TB6-JE00	OEM
70TB6-JH00	OEM
70TB6-LE00	OEM
70TB6-LH00	OEM
70TB6-NE00	OEM
70TB6-NH00	OEM
70TB6-RE00	OEM
70TB7	OEM
70TB7-CH00	OEM
70TB7-GE00	OEM
70TB7-GH00	OEM
70TB7-JE00	OEM
70TB7-JH00	OEM
70TB7-LE00	OEM
70TB7-LH00	OEM
70TB7-NE00	OEM
70TB7-NH00	OEM
70TB7-RE00	OEM
70TB8	OEM
70TB8-CE00	OEM
70TB8-CH00	OEM
70TB8-GH00	OEM
70TB8-JE00	OEM

DEVICE TYPE	REPL CODE
70TB8-JH00	OEM
70TB8-LE00	OEM
70TB8-LH00	OEM
70TB8-NE00	OEM
70TB8-NH00	OEM
70TB8-RE00	OEM
70TB9	OEM
70TB9-CE00	OEM
70TB9-CH00	OEM
70TB9-GH00	OEM
70TB9-JE00	OEM
70TB9-JH00	OEM
70TB9-LE00	OEM
70TB9-LH00	OEM
70TB9-NE00	OEM
70TB9-NH00	OEM
70TB9-RE00	OEM
70TB10	OEM
70TB10-CE00	OEM
70TB10-CH00	OEM
70TB10-GH00	OEM
70TB10-JE00	OEM
70TB10-JH00	OEM
70TB10-LE00	OEM
70TB10-LH00	OEM
70TB10-NE00	OEM
70TB10-NH00	OEM
70TB10-RE00	OEM
70TB11	OEM
70TB11-CE00	OEM
70TB11-CH00	OEM
70TB11-GH00	OEM
70TB11-JE00	OEM
70TB11-JH00	OEM
70TB11-LE00	OEM
70TB11-LH00	OEM
70TB11-NE00	OEM
70TB11-NH00	OEM
70TB12	OEM
70TB12-CE00	OEM
70TB12-CH00	OEM
70TB12-GH00	OEM
70TB12-JE00	OEM
70TB12-JH00	OEM
70TB12-LE00	OEM
70TB12-LH00	OEM
70TB12-NE00	OEM
70TB12-NH00	OEM
70TB12-RE00	OEM
70TB13	OEM
70TB13-CE00	OEM
70TB13-CH00	OEM
70TB13-GH00	OEM
70TB13-JE00	OEM
70TB13-JH00	OEM
70TB13-LH00	OEM
70TB13-NE00	OEM
70TB13-NH00	OEM
70TB13-RE00	OEM
70TB14	OEM
70TB14-CE00	OEM
70TB14-CH00	OEM
70TB14-GH00	OEM
70TB14-JE00	OEM
70TB14-JH00	OEM
70TB14-LE00	OEM
70TB14-LH00	OEM
70TB14-NE00	OEM
70TB14-NH00	OEM
70TB14-RE00	OEM
70TB15	OEM
70TB15-CE00	OEM
70TB15-CH00	OEM
70TB15-GH00	OEM
70TB15-JE00	OEM
70TB15-JH00	OEM
70TB15-LE00	OEM
70TB15-LH00	OEM
70TB15-NE00	OEM
70TB15-NH00	OEM
70TB15-RE00	OEM
70TB16	OEM
70TB16-CE00	OEM
70TB16-CH00	OEM
70TB16-GH00	OEM
70TB16-JE00	OEM
70TB16-JH00	OEM
70TB16-LE00	OEM
70TB16-LH00	OEM
70TB16-NE00	OEM
70TB16-NH00	OEM
70TB16-RE00	OEM
70TB17	OEM
70TB17-CE00	OEM
70TB17-CH00	OEM
70TB17-GH00	OEM
70TB17-JE00	OEM
70TB17-JH00	OEM
70TB17-LE00	OEM
70TB17-LH00	OEM
70TB17-NE00	OEM
70TB17-NH00	OEM
70TB17-RE00	OEM
70TB18	OEM
70TB19	OEM
70TB20	OEM
70U5	1023
70U5R	0501
70U10	1023
70U10R	0501
70U15	1023
70U15A	1023
70U15AR	0501
70U15R	0501
70U20	1023
70U20R	0501
70U25	1023
70U25A	1023
70U25AR	0501
70U25R	0501
70U30	1023
70U30R	0501
70U35	1036
70U35R	1777
70U40	1023
70U40R	0501
70U45	1036
70U45R	1777
70U50	1036
70U50R	1777
70U60	1036
70U60R	1777
70U70	1046
70U70A	1046
70U70AR	1789
70U70R	1789
70U80	1046
70U80R	1789
70U90	1046
70U90A	1046
70U90AR	1789
70U90R	1789
70U100	1046
70U100R	1789
70U120	OEM
70UW15	1023
70UW15R	0501
70UW25	1023
70UW25R	0501
70UW35	1023
70UW35R	0501
70UW45	1036
70UW45R	1777
70UW65	1046
70UW65R	1789
70V10	OEM
70V10A	OEM
70V11A	OEM
70.00.730	0136
70.01.704	0016
71-126268	0016
71HA60	OEM
71HF5	OEM
71HF10	OEM
71HF20	OEM
71HF40	OEM
71HF60	OEM
71HF80	OEM
71HF100	OEM
71HF120	OEM
71HF140	OEM
71HF160	OEM
71HFR10	OEM
71HFR20	OEM
71HFR40	OEM
71HFR60	OEM
71HFR80	OEM
71HFR100	OEM
71HFR120	OEM
71HFR140	OEM
71HFR160	OEM
71N1	0142
71N1B	0016
71N1T	0142
71N2	0142
71N2T	0142
71RA30	0468
71RA50	0468
71RA60	0468
71RA80	0468
71RA100	0468
71RA110	0468
71RA120	0468
71RA140	OEM
71RA150	OEM
71RA160	OEM
71RB50	0468
71RB60	0468
71RB80	0468
71RB100	0468
71RB110	0468
71RB120	0468
71RB140	OEM
71RB150	OEM
71RB160	OEM
71RC2B	OEM
71RC5A	0217
71RC5B	0608
71RC10	OEM
71RC10A	0217
71RC10B	0608
71RC15A	0217
71RC20	OEM
71RC20A	0217
71RC20B	0608
71RC25A	0217
71RC30	OEM
71RC30A	0217
71RC30AS60	0217
71RC30B	0610
71RC40	OEM
71RC40A	0217
71RC40B	0610
71RC50	OEM
71RC50A	0217
71RC50B	OEM
71RC60	OEM
71RC60A	0217
71RC60AS50	0217
71RC60B	OEM
71RC80	OEM
71RC80A	0468
71RC80AS50	0468
71RC100	OEM
71RC100A	OEM
71RC110A	OEM
71RC120	OEM
71RC120A	OEM
71RCF70A	OEM
71RCF80A	OEM
71RCG10	OEM
71RCG15	OEM
71RCG20	OEM
71RCG25	OEM
71RCG30	OEM
71RCG40	OEM
71RCG50	OEM
71RCG60	OEM
71RCS10A	OEM
71RCS20A	OEM
71RCS30A	OEM
71RCS40A	OEM
71RCS50A	OEM
71RCS60A	OEM
71RCS80A	OEM
71RE60	0468
71RE70	0468
71RE80	0468
71RE90	0468
71RE90A	OEM
71RE100	0468
71RE100A	OEM
71RE110	0468
71RE120	0468
71RE130	OEM
71RE130A	OEM
71REA50	0468
71REA60	0468
71REA70	OEM
71REA80	0468
71REA90	OEM
71REA100	0468
71REA110	0468
71REA120	0468
71REA130	OEM
71REB50	0468
71REB60	0468
71REB70	OEM
71REB80	0468
71REB90	OEM
71REB100	0468
71REB110	0468
71REB120	0468
71REB130	OEM
71RECT-SI-1019	0348
71REH60	0610
71REH80	0468
71REH100	0468
71REH110	0468
71REH120	0468
71REH130	OEM
71RIA10	OEM
71RIA20	OEM
71RIA40	OEM
71RIA60	OEM
71RIA80	OEM
71RIA100	OEM
71RIA110	OEM
71RIA120	OEM
71ST-2SC1505L	0955
71SZ-RD12B(AU)	0052
71SZ-UZ-12AB(AU)	OEM
71SZ-UZ-12B(AU)	0052
71T1	OEM
71T1C	OEM
71T2	0285
71T2C	OEM
71T3	OEM
71T3C	OEM
71T4	OEM
71T5	OEM
71T6	OEM
71T7	OEM
71T8	OEM
71T9	OEM
71T10	OEM
71T11	OEM
71T12	OEM
71T13	OEM
71T14	OEM
71T15	OEM
71T16	OEM
71T17	OEM
71T18	OEM
71T19	OEM
71T20	OEM
71T20-CE00	OEM
71T20-CH00	OEM
71T20-EE00	OEM
71T20-EH00	OEM
71T20-GE00	OEM
71T20-GH00	OEM
71T20-JE00	OEM
71T20-JH00	OEM
71T20-LE00	OEM
71T20-LH00	OEM
71T20-NE00	OEM
71T20-NH00	OEM
71T20-RE00	OEM
71T21	OEM
71T21-CE00	OEM
71T21-CH00	OEM
71T21-EE00	OEM
71T21-EH00	OEM
71T21-GE00	OEM
71T21-GH00	OEM
71T21-JE00	OEM
71T21-JH00	OEM
71T21-LE00	OEM
71T21-LH00	OEM
71T21-NE00	OEM
71T21-NH00	OEM
71T21-RE00	OEM
71T22	OEM
71T22-CE00	OEM
71T22-CH00	OEM
71T22-EE00	OEM
71T22-EH00	OEM
71T22-GE00	OEM
71T22-GH00	OEM
71T22-JE00	OEM
71T22-JH00	OEM
71T22-LE00	OEM
71T22-LH00	OEM
71T22-NE00	OEM
71T22-NH00	OEM
71T22-RE00	OEM
71T23	OEM
71T23-CE00	OEM
71T23-CH00	OEM
71T23-EE00	OEM
71T23-EH00	OEM
71T23-GE00	OEM
71T23-GH00	OEM
71T23-JE00	OEM
71T23-JH00	OEM
71T23-LE00	OEM
71T23-LH00	OEM
71T23-NE00	OEM
71T23-NH00	OEM
71T23-RE00	OEM
71T24	OEM
71T24-CE00	OEM
71T24-CH00	OEM
71T24-EE00	OEM
71T24-EH00	OEM
71T24-GE00	OEM
71T24-GH00	OEM
71T24-JE00	OEM
71T24-JH00	OEM
71T24-LE00	OEM
71T24-LH00	OEM
71T24-NE00	OEM
71T24-NH00	OEM
71T24-RE00	OEM
71T25	OEM
71T25-CE00	OEM
71T25-CH00	OEM
71T25-EE00	OEM
71T25-EH00	OEM
71T25-GE00	OEM
71T25-GH00	OEM
71T25-JE00	OEM
71T25-JH00	OEM
71T25-LE00	OEM
71T25-LH00	OEM
71T25-NE00	OEM
71T25-NH00	OEM
71T25-RE00	OEM
71T26	OEM
71T26-CE00	OEM
71T26-CH00	OEM
71T26-EE00	OEM
71T26-EH00	OEM
71T26-GE00	OEM
71T26-GH00	OEM
71T26-JE00	OEM
71T26-JH00	OEM
71T26-LE00	OEM
71T26-LH00	OEM
71T26-NE00	OEM
71T26-NH00	OEM
71T26-RE00	OEM
71T27-CE00	OEM
71T27-CH00	OEM
71T27-EE00	OEM
71T27-EH00	OEM
71T27-GE00	OEM
71T27-GH00	OEM
71T27-JE00	OEM
71T27-JH00	OEM
71T27-LE00	OEM
71T27-LH00	OEM
71T27-NE00	OEM
71T27-NH00	OEM
71T27-RE00	OEM
71T28	OEM
71T28-CE00	OEM
71T28-CH00	OEM
71T28-EE00	OEM
71T28-EH00	OEM
71T28-GE00	OEM
71T28-GH00	OEM
71T28-JE00	OEM
71T28-JH00	OEM
71T28-LE00	OEM
71T28-LH00	OEM
71T28-NE00	OEM
71T28-NH00	OEM
71T28-RE00	OEM
71T29-CE00	OEM
71T29-CH00	OEM
71T29-EE00	OEM
71T29-EH00	OEM
71T29-GE00	OEM
71T29-GH00	OEM
71T29-JE00	OEM
71T29-JH00	OEM
71T29-LE00	OEM
71T29-LH00	OEM
71T29-NE00	OEM
71T29-NH00	OEM
71T29-RE00	OEM
71T30-CE00	OEM
71T30-CH00	OEM
71T30-EE00	OEM
71T30-EH00	OEM
71T30-GE00	OEM
71T30-GH00	OEM
71T30-JE00	OEM
71T30-JH00	OEM
71T30-LE00	OEM
71T30-LH00	OEM
71T30-NE00	OEM
71T30-NH00	OEM
71T30-RE00	OEM
71T31-CE00	OEM
71T31-CH00	OEM
71T31-EE00	OEM
71T31-EH00	OEM
71T31-GE00	OEM
71T31-GH00	OEM
71T31-JE00	OEM
71T31-JH00	OEM
71T31-LE00	OEM
71T31-LH00	OEM
71T31-NE00	OEM
71T31-NH00	OEM
71T31-RE00	OEM
71T32	OEM
71T32-CE00	OEM
71T32-CH00	OEM
71T32-EE00	OEM
71T32-EH00	OEM
71T32-GE00	OEM
71T32-GH00	OEM
71T32-JE00	OEM
71T32-JH00	OEM
71T32-LE00	OEM
71T32-LH00	OEM
71T32-NE00	OEM
71T32-NH00	OEM
71T32-RE00	OEM
71T33-CE00	OEM
71T33-CH00	OEM
71T33-EE00	OEM
71T33-EH00	OEM
71T33-GE00	OEM
71T33-GH00	OEM
71T33-JE00	OEM
71T33-JH00	OEM
71T33-LE00	OEM
71T33-LH00	OEM
71T33-NE00	OEM
71T33-NH00	OEM
71T33-RE00	OEM
71T34	OEM
71T34-CE00	OEM
71T34-CH00	OEM
71T34-EE00	OEM
71T34-EH00	OEM
71T34-GE00	OEM
71T34-GH00	OEM
71T34-JE00	OEM
71T34-JH00	OEM
71T34-LE00	OEM
71T34-LH00	OEM
71T34-NE00	OEM
71T34-NH00	OEM
71T34-RE00	OEM
71T35-GE00	OEM
71T35-GH00	OEM
71T35-JE00	OEM
71T35-JH00	OEM
71T35-LE00	OEM
71T35-LH00	OEM
71T35-NE00	OEM
71T35-RE00	OEM
71T36	OEM
71T36-CE00	OEM
71T36-GE00	OEM
71T36-GH00	OEM
71T36-JE00	OEM
71T36-JH00	OEM
71T36-LE00	OEM
71T36-LH00	OEM
71T36-NE00	OEM
71T36-NH00	OEM
71T36-RE00	OEM
71T37	OEM
71T37-CE00	OEM
71T37-GE00	OEM
71T37-GH00	OEM
71T37-JE00	OEM
71T37-JH00	OEM
71T37-LE00	OEM
71T37-LH00	OEM
71T37-NE00	OEM
71T37-NH00	OEM
71T37-RE00	OEM
71T38	OEM
71T38-CE00	OEM
71T38-GE00	OEM
71T38-GH00	OEM
71T38-JE00	OEM
71T38-JH00	OEM
71T38-LE00	OEM
71T38-LH00	OEM
71T38-NE00	OEM
71T38-NH00	OEM
71T38-RE00	OEM
71T39	OEM
71T39-CE00	OEM
71T39-CH00	OEM
71T39-GE00	OEM
71T39-GH00	OEM
71T39-JE00	OEM
71T39-JH00	OEM
71T39-LE00	OEM
71T39-LH00	OEM
71T39-NE00	OEM
71T39-NH00	OEM
71T39-RE00	OEM
71T40-CE00	OEM
71T40-CH00	OEM
71T40-GE00	OEM
71T40-GH00	OEM
71T40-JE00	OEM
71T40-JH00	OEM
71T40-LE00	OEM
71T40-LH00	OEM
71T40-NE00	OEM
71T40-NH00	OEM
71T40-RE00	OEM
71TB1-CE00	OEM
71TB1-CH00	OEM
71TB1-EE00	OEM
71TB1-EH00	OEM
71TB1-GE00	OEM
71TB1-GH00	OEM
71TB1-JH00	OEM
71TB1-LE00	OEM
71TB1-LH00	OEM
71TB1-RE00	OEM
71TB2-CE00	OEM
71TB2-CH00	OEM
71TB2-EE00	OEM
71TB2-EH00	OEM
71TB2-GE00	OEM
71TB2-GH00	OEM
71TB2-JE00	OEM
71TB2-JH00	OEM
71TB2-LH00	OEM
71TB2-RE00	OEM
71TB3-CE00	OEM
71TB3-CH00	OEM
71TB3-EE00	OEM
71TB3-EH00	OEM
71TB3-GE00	OEM
71TB3-GH00	OEM
71TB3-JE00	OEM
71TB3-JH00	OEM
71TB3-LE00	OEM
71TB3-LH00	OEM
71TB3-RE00	OEM
71TB4-CE00	OEM
71TB4-CH00	OEM
71TB4-EE00	OEM
71TB4-EH00	OEM
71TB4-GE00	OEM
71TB4-GH00	OEM
71TB4-JE00	OEM
71TB4-JH00	OEM
71TB4-LE00	OEM
71TB4-LH00	OEM
71TB4-RE00	OEM
71TB5-CE00	OEM
71TB5-CH00	OEM
71TB5-EH00	OEM
71TB5-GE00	OEM
71TB5-GH00	OEM
71TB5-JE00	OEM
71TB5-JH00	OEM
71TB5-LE00	OEM
71TB5-LH00	OEM
71TB5-RE00	OEM
71TB6-CE00	OEM
71TB6-CH00	OEM
71TB6-EE00	OEM
71TB6-EH00	OEM
71TB6-GE00	OEM
71TB6-GH00	OEM
71TB6-JE00	OEM
71TB6-JH00	OEM
71TB6-LE00	OEM
71TB6-LH00	OEM
71TB7	OEM
71TB7-CE00	OEM
71TB7-CH00	OEM
71TB7-EE00	OEM
71TB7-EH00	OEM
71TB7-GE00	OEM
71TB7-GH00	OEM
71TB7-JE00	OEM
71TB7-JH00	OEM
71TB7-LE00	OEM
71TB7-LH00	OEM
71TB8-CE00	OEM
71TB8-CH00	OEM
71TB8-EE00	OEM
71TB8-EH00	OEM
71TB8-GE00	OEM
71TB8-GH00	OEM
71TB8-JE00	OEM
71TB8-JH00	OEM
71TB8-LE00	OEM
71TB8-LH00	OEM
71TB9	OEM
71TB9-CE00	OEM
71TB9-CH00	OEM
71TB9-EE00	OEM
71TB9-EH00	OEM
71TB9-GE00	OEM
71TB9-GH00	OEM
71TB9-JE00	OEM
71TB9-JH00	OEM
71TB9-LH00	OEM
71TB10-CE00	OEM
71TB10-CH00	OEM
71TB10-EE00	OEM
71TB10-EH00	OEM
71TB10-GE00	OEM
71TB10-GH00	OEM
71TB10-JE00	OEM
71TB10-JH00	OEM
71TB10-LE00	OEM
71TB10-LH00	OEM
71TB11	OEM
71TB11-CE00	OEM
71TB11-CH00	OEM
71TB11-EE00	OEM
71TB11-EH00	OEM
71TB11-GE00	OEM
71TB11-GH00	OEM
71TB11-JE00	OEM
71TB11-JH00	OEM
71TB11-LE00	OEM
71TB11-LH00	OEM
71TB12-CE00	OEM
71TB12-CH00	OEM
71TB12-EE00	OEM
71TB12-EH00	OEM
71TB12-GE00	OEM
71TB12-GH00	OEM
71TB12-JE00	OEM
71TB12-JH00	OEM
71TB12-LE00	OEM
71TB12-LH00	OEM
71TB13	OEM
71TB13-CE00	OEM
71TB13-CH00	OEM
71TB13-EE00	OEM
71TB13-EH00	OEM
71TB13-GE00	OEM
71TB13-GH00	OEM
71TB13-JE00	OEM
71TB13-JH00	OEM
71TB13-LE00	OEM
71TB13-LH00	OEM
71TB13-NE00	OEM
71TB13-NH00	OEM
71TB13-RE00	OEM
71TB14-CE00	OEM
71TB14-CH00	OEM
71TB14-EE00	OEM
71TB14-EH00	OEM
71TB14-GE00	OEM
71TB14-GH00	OEM
71TB14-JE00	OEM
71TB14-JH00	OEM
71TB14-LE00	OEM
71TB14-LH00	OEM
71TB14-NE00	OEM
71TB14-NH00	OEM
71TB14-RE00	OEM
71TB15	OEM
71TB15-CE00	OEM
71TB15-CH00	OEM
71TB15-EE00	OEM
71TB15-EH00	OEM
71TB15-GE00	OEM
71TB15-GH00	OEM
71TB15-JE00	OEM
71TB15-JH00	OEM
71TB15-LE00	OEM
71TB15-LH00	OEM
71TB15-NE00	OEM
71TB15-NH00	OEM
71TB15-RE00	OEM
71TB16-CE00	OEM
71TB16-EE00	OEM
71TB16-EH00	OEM
71TB16-GE00	OEM
71TB16-JE00	OEM

If replacement code is OEM, contact original manufacturer for replacement.

DEVICE TYPE	REPL CODE
71TB16-JH00	OEM
71TB16-LE00	OEM
71TB16-LH00	OEM
71TB16-NE00	OEM
71TB16-NH00	OEM
71TB16-RE00	OEM
71TB17	OEM
71TB17-CE00	OEM
71TB17-CH00	OEM
71TB17-EE00	OEM
71TB17-EH00	OEM
71TB17-GE00	OEM
71TB17-GH00	OEM
71TB17-JE00	OEM
71TB17-JH00	OEM
71TB17-LE00	OEM
71TB17-LH00	OEM
71TB17-NE00	OEM
71TB17-NH00	OEM
71TB17-RE00	OEM
71TB18-CE00	OEM
71TB18-CH00	OEM
71TB18-EE00	OEM
71TB18-EH00	OEM
71TB18-GE00	OEM
71TB18-GH00	OEM
71TB18-JE00	OEM
71TB18-JH00	OEM
71TB18-LE00	OEM
71TB18-LH00	OEM
71TB18-NE00	OEM
71TB18-NH00	OEM
71TB18-RE00	OEM
71TB19	OEM
71TB19-CE00	OEM
71TB19-CH00	OEM
71TB19-EH00	OEM
71TB19-GE00	OEM
71TB19-GH00	OEM
71TB19-JE00	OEM
71TB19-JH00	OEM
71TB19-LE00	OEM
71TB19-LH00	OEM
71TB19-NE00	OEM
71TB19-RE00	OEM
71TB20-CE00	OEM
71TB20-CH00	OEM
71TB20-EE00	OEM
71TB20-EH00	OEM
71TB20-GE00	OEM
71TB20-GH00	OEM
71TB20-JE00	OEM
71TB20-JH00	OEM
71TB20-LE00	OEM
71TB20-LH00	OEM
71TB20-NE00	OEM
71TB20-RE00	OEM
71TC	OEM
71Z6	OEM
71Z6A	OEM
72-6	OEM
72-9	0102
72-11	0015
72-14	OEM
72-15	0015
72-18	0133
72B1Z	0751
72HG10	OEM
72HG20	OEM
72HG40	OEM
72HG60	OEM
72HG80	OEM
72HG100	OEM
72HG120	OEM
72N1	0144
72N1B	0086
72N2	0144
72N2B	0016
72R2S	OEM
72RA30	OEM
72RA50	0217
72RA60	0217
72RA80	0658
72RA100	0658
72RA120	0658
72RA140	OEM
72RA150	OEM
72RA160	OEM
72RB50	0217
72RB60	0217
72RB80	0658
72RB100	0658
72RB120	0658
72RB140	OEM
72RB150	OEM
72RB160	OEM
72RC2B	0658
72RC5A	0217
72RC5B	0641
72RC10A	0217
72RC10B	0641
72RC15A	0217
72RC20A	0217
72RC20B	0641
72RC25A	0217
72RC30A	0217
72RC30AS60	0217
72RC30B	0217
72RC40A	0217

DEVICE TYPE	REPL CODE
72RC40B	0217
72RC50A	0217
72RC50B	OEM
72RC60A	0217
72RC60AS50	0217
72RC60B	OEM
72RC80A	0658
72RC80AS50	OEM
72RC100A	OEM
72RC110A	OEM
72RC120A	OEM
72RCF70A	OEM
72RCF80A	OEM
72RCG5	OEM
72RCG10	OEM
72RCG15	OEM
72RCG20	OEM
72RCG25	OEM
72RCG30	OEM
72RCG40	OEM
72RCG50	OEM
72RCG60	OEM
72REA60	OEM
72REA70	OEM
72REA80	OEM
72REA90	OEM
72REA100	OEM
72REA110	OEM
72REA120	OEM
72REA130	OEM
72REB60	OEM
72REB70	OEM
72REB80	OEM
72REB90	OEM
72REB100	OEM
72REB110	OEM
72REB120	OEM
72REB130	OEM
72REH60	0217
72REH80	0658
72REH100	0658
72REH110	0658
72REH120	0658
72REH130	OEM
72RIA10	OEM
72RIA20	OEM
72RIA40	OEM
72RIA60	OEM
72RIA80	OEM
72RIA100	OEM
72RIA110	OEM
72RIA120	OEM
72SI-TA7074P	0395
72ST-2SD900B	0055
72ST-MPS9750F	0037
72ST-MPS9750G	0037
72T1	OEM
72T2	0236
72T3	OEM
72T4	OEM
72T6	OEM
72T7	OEM
72T8	OEM
72X7377ESD	OEM
72X7385	OEM
72X7437	OEM
72X8287ESD	OEM
72X8290	OEM
72X8299	OEM
72X8319	OEM
72Z	0015
72Z4	OEM
72Z6	OEM
72Z6A	OEM
73-15	0002
73-17	0420
73-22	0495
73-31	0157
73-44	0253
73-125-B-003	OEM
73A01	0847
73A02	0847
73A03	0136
73A60-11	0102
73C180475	0167
73C180476-5	0167
73C180837-1	0666
73C180837-2	0666
73C180837-3	0666
73C180843	0802
73C180843-1	0802
73C180843-2	0802
73C180843-3	0802
73C180843-4	0802
73C181254	1501
73C182051	1501
73C182051-1	1501
73C182051-2	1501
73C182051-3	1501
73C182186	0167
73C182763-1	0751
73C182763-2	0751
73C182764-1	0856
73C182764-2	0856
73C182764-4	0856
73N1	0007
73N1B	0016
73N51	0440
73SI-MC1353P	0855

DEVICE TYPE	REPL CODE
73SI-MC1358P	0167
73ST-2SC945R(AU)E	0076
73ST-2SC1921-03	0261
73T1	OEM
73T2	0285
73T3	OEM
73T4	OEM
73T5	OEM
73T6	OEM
73T7	OEM
73T8	OEM
73T9	OEM
73T10	OEM
73T11	OEM
73T12	OEM
73T13	OEM
73T14	OEM
73T15	OEM
73T16	OEM
73T17	OEM
73T18	OEM
73T19	OEM
73T20	OEM
73T20-CE00	OEM
73T20-CH00	OEM
73T20-EE00	OEM
73T20-EH00	OEM
73T20-GE00	OEM
73T20-GH00	OEM
73T20-JE00	OEM
73T20-JH00	OEM
73T20-LE00	OEM
73T20-RE00	OEM
73T21	OEM
73T21-CE00	OEM
73T21-CH00	OEM
73T21-EE00	OEM
73T21-EH00	OEM
73T21-GE00	OEM
73T21-GH00	OEM
73T21-JE00	OEM
73T21-JH00	OEM
73T21-LE00	OEM
73T21-RE00	OEM
73T22	OEM
73T22-CE00	OEM
73T22-CH00	OEM
73T22-EE00	OEM
73T22-EH00	OEM
73T22-GE00	OEM
73T22-GH00	OEM
73T22-JE00	OEM
73T22-JH00	OEM
73T22-LE00	OEM
73T22-RE00	OEM
73T23	OEM
73T23-CE00	OEM
73T23-CH00	OEM
73T23-EE00	OEM
73T23-EH00	OEM
73T23-GE00	OEM
73T23-GH00	OEM
73T23-JE00	OEM
73T23-JH00	OEM
73T23-LE00	OEM
73T23-RE00	OEM
73T24	OEM
73T24-CE00	OEM
73T24-CH00	OEM
73T24-EE00	OEM
73T24-EH00	OEM
73T24-GE00	OEM
73T24-GH00	OEM
73T24-JE00	OEM
73T24-JH00	OEM
73T24-LE00	OEM
73T24-RE00	OEM
73T25-CE00	OEM
73T25-CH00	OEM
73T25-EE00	OEM
73T25-EH00	OEM
73T25-GE00	OEM
73T25-GH00	OEM
73T25-JE00	OEM
73T25-JH00	OEM
73T25-LE00	OEM
73T25-LH00	OEM
73T25-RE00	OEM
73T26	OEM
73T26-CE00	OEM
73T26-CH00	OEM
73T26-EE00	OEM
73T26-EH00	OEM
73T26-GE00	OEM
73T26-GH00	OEM
73T26-JE00	OEM
73T26-JH00	OEM
73T26-LE00	OEM
73T26-LH00	OEM
73T26-RE00	OEM
73T27-CE00	OEM
73T27-CH00	OEM
73T27-EE00	OEM
73T27-EH00	OEM
73T27-GE00	OEM
73T27-GH00	OEM
73T27-JE00	OEM
73T27-JH00	OEM
73T27-LE00	OEM

DEVICE TYPE	REPL CODE
73T27-LH00	OEM
73T27-RE00	OEM
73T28	OEM
73T28-CE00	OEM
73T28-CH00	OEM
73T28-EE00	OEM
73T28-EH00	OEM
73T28-GE00	OEM
73T28-GH00	OEM
73T28-JE00	OEM
73T28-JH00	OEM
73T28-LE00	OEM
73T28-LH00	OEM
73T28-RE00	OEM
73T29-CE00	OEM
73T29-CH00	OEM
73T29-EE00	OEM
73T29-EH00	OEM
73T29-GE00	OEM
73T29-GH00	OEM
73T29-JE00	OEM
73T29-JH00	OEM
73T29-LE00	OEM
73T29-LH00	OEM
73T29-RE00	OEM
73T30-CE00	OEM
73T30-CH00	OEM
73T30-EE00	OEM
73T30-EH00	OEM
73T30-GE00	OEM
73T30-GH00	OEM
73T30-JE00	OEM
73T30-JH00	OEM
73T30-L300	OEM
73T30-LH00	OEM
73T30-RE00	OEM
73TB1-CE00	OEM
73TB1-CH00	OEM
73TB1-EE00	OEM
73TB1-EH00	OEM
73TB1-GE00	OEM
73TB1-GH00	OEM
73TB1-JE00	OEM
73TB1-JH00	OEM
73TB1-LE00	OEM
73TB1-LH00	OEM
73TB1-NE00	OEM
73TB1-NH00	OEM
73TB1-RE00	OEM
73TB2-CE00	OEM
73TB2-CH00	OEM
73TB2-EE00	OEM
73TB2-EH00	OEM
73TB2-GE00	OEM
73TB2-GH00	OEM
73TB2-JE00	OEM
73TB2-JH00	OEM
73TB2-LE00	OEM
73TB2-LH00	OEM
73TB2-NE00	OEM
73TB2-NH00	OEM
73TB2-RE00	OEM
73TB3-CE00	OEM
73TB3-CH00	OEM
73TB3-EE00	OEM
73TB3-EH00	OEM
73TB3-GE00	OEM
73TB3-GH00	OEM
73TB3-JE00	OEM
73TB3-JH00	OEM
73TB3-LE00	OEM
73TB3-LH00	OEM
73TB3-NE00	OEM
73TB3-NH00	OEM
73TB3-RE00	OEM
73TB4-CE00	OEM
73TB4-CH00	OEM
73TB4-EE00	OEM
73TB4-EH00	OEM
73TB4-GE00	OEM
73TB4-GH00	OEM
73TB4-JE00	OEM
73TB4-JH00	OEM
73TB4-LE00	OEM
73TB4-LH00	OEM
73TB4-NE00	OEM
73TB4-NH00	OEM
73TB4-RE00	OEM
73TB5-CE00	OEM
73TB5-CH00	OEM
73TB5-EE00	OEM
73TB5-EH00	OEM
73TB5-GE00	OEM
73TB5-GH00	OEM
73TB5-JE00	OEM
73TB5-JH00	OEM
73TB5-LE00	OEM
73TB5-LH00	OEM
73TB5-NE00	OEM
73TB5-NH00	OEM
73TB5-RE00	OEM
73TB6-CE00	OEM
73TB6-CH00	OEM
73TB6-EE00	OEM
73TB6-EH00	OEM
73TB6-GE00	OEM
73TB6-GH00	OEM
73TB6-JE00	OEM
73TB6-JH00	OEM

DEVICE TYPE	REPL CODE
73TB6-LE00	OEM
73TB6-LH00	OEM
73TB6-NE00	OEM
73TB6-NH00	OEM
73TB6-RE00	OEM
73TB7	OEM
73TB7-CE00	OEM
73TB7-CH00	OEM
73TB7-EE00	OEM
73TB7-EH00	OEM
73TB7-GH00	OEM
73TB7-JE00	OEM
73TB7-JH00	OEM
73TB7-LE00	OEM
73TB7-LH00	OEM
73TB7-NE00	OEM
73TB7-NH00	OEM
73TB7-RE00	OEM
73TB8-CH00	OEM
73TB8-EE00	OEM
73TB8-EH00	OEM
73TB8-GE00	OEM
73TB8-GH00	OEM
73TB8-JE00	OEM
73TB8-JH00	OEM
73TB8-LE00	OEM
73TB8-LH00	OEM
73TB8-NE00	OEM
73TB8-NH00	OEM
73TB8-RE00	OEM
73TB9	OEM
73TB9-CE00	OEM
73TB9-CH00	OEM
73TB9-EE00	OEM
73TB9-EH00	OEM
73TB9-GE00	OEM
73TB9-GH00	OEM
73TB9-JE00	OEM
73TB9-JH00	OEM
73TB9-LE00	OEM
73TB9-LH00	OEM
73TB9-NH00	OEM
73TB9-RE00	OEM
73TB10-CE00	OEM
73TB10-CH00	OEM
73TB10-EE00	OEM
73TB10-EH00	OEM
73TB10-GE00	OEM
73TB10-GH00	OEM
73TB10-JE00	OEM
73TB10-JH00	OEM
73TB10-LE00	OEM
73TB10-LH00	OEM
73TB10-NE00	OEM
73TB10-NH00	OEM
73TB10-RE00	OEM
73TB11	OEM
73TB11-CE00	OEM
73TB11-CH00	OEM
73TB11-EE00	OEM
73TB11-EH00	OEM
73TB11-GE00	OEM
73TB11-GH00	OEM
73TB11-JE00	OEM
73TB11-JH00	OEM
73TB11-LE00	OEM
73TB11-LH00	OEM
73TB11-NE00	OEM
73TB11-NH00	OEM
73TB11-RE00	OEM
73TB12-CE00	OEM
73TB12-CH00	OEM
73TB12-EE00	OEM
73TB12-EH00	OEM
73TB12-GE00	OEM
73TB12-GH00	OEM
73TB12-JE00	OEM
73TB12-JH00	OEM
73TB12-LE00	OEM
73TB12-LH00	OEM
73TB12-NE00	OEM
73TB12-NH00	OEM
73TB12-RE00	OEM
73TB13	OEM
73TB13-CE00	OEM
73TB13-CH00	OEM
73TB13-EE00	OEM
73TB13-EH00	OEM
73TB13-GE00	OEM
73TB13-GH00	OEM
73TB13-JE00	OEM
73TB13-JH00	OEM
73TB13-LE00	OEM
73TB13-LH00	OEM
73TB13-NE00	OEM
73TB13-NH00	OEM
73TB13-RE00	OEM
73TB14-CE00	OEM
73TB14-CH00	OEM
73TB14-EE00	OEM
73TB14-EH00	OEM
73TB14-GE00	OEM
73TB14-GH00	OEM
73TB14-JE00	OEM
73TB14-JH00	OEM
73TB14-LE00	OEM
73TB14-LH00	OEM
73TB14-NE00	OEM
73TB14-NH00	OEM

DEVICE TYPE	REPL CODE
73TB14-RE00	OEM
73TB15	OEM
73TB15-CE00	OEM
73TB15-CH00	OEM
73TB15-EE00	OEM
73TB15-EH00	OEM
73TB15-GE00	OEM
73TB15-GH00	OEM
73TB15-JE00	OEM
73TB15-JH00	OEM
73TB15-LE00	OEM
73TB15-LH00	OEM
73TB15-NE00	OEM
73TB15-NH00	OEM
73TB15-RE00	OEM
73TB16-CE00	OEM
73TB16-CH00	OEM
73TB16-EE00	OEM
73TB16-EH00	OEM
73TB16-GE00	OEM
73TB16-GH00	OEM
73TB16-JE00	OEM
73TB16-LE00	OEM
73TB16-LH00	OEM
73TB16-NE00	OEM
73TB16-NH00	OEM
73TB16-RE00	OEM
73TB17	OEM
73TB17-CE00	OEM
73TB17-CH00	OEM
73TB17-EE00	OEM
73TB17-EH00	OEM
73TB17-GE00	OEM
73TB17-GH00	OEM
73TB17-JE00	OEM
73TB17-JH00	OEM
73TB17-LE00	OEM
73TB17-LH00	OEM
73TB17-NE00	OEM
73TB17-NH00	OEM
73TB17-RE00	OEM
73TB18-CE00	OEM
73TB18-CH00	OEM
73TB18-EE00	OEM
73TB18-EH00	OEM
73TB18-GE00	OEM
73TB18-GH00	OEM
73TB18-JE00	OEM
73TB18-JH00	OEM
73TB18-LE00	OEM
73TB18-LH00	OEM
73TB18-NE00	OEM
73TB18-NH00	OEM
73TB18-RE00	OEM
73TB19	OEM
73TB19-CE00	OEM
73TB19-CH00	OEM
73TB19-EE00	OEM
73TB19-EH00	OEM
73TB19-GE00	OEM
73TB19-GH00	OEM
73TB19-JE00	OEM
73TB19-JH00	OEM
73TB19-LE00	OEM
73TB19-LH00	OEM
73TB19-NE00	OEM
73TB19-NH00	OEM
73TB19-RE00	OEM
73TB20-CE00	OEM
73TB20-CH00	OEM
73TB20-EE00	OEM
73TB20-EH00	OEM
73TB20-GE00	OEM
73TB20-GH00	OEM
73TB20-JH00	OEM
73TB20-LE00	OEM
73TB20-LH00	OEM
73TB20-NE00	OEM
73TB20-NH00	OEM
73TB20-RE00	OEM
73W00124	0467
73Z4	OEM
73Z6	0815
73Z6A	0815
74	0144
74-3	0280
74-3-70	0280
74-3-70-12	0280
74-3-70-12-7	0280
74-3A	0280
74-3A0	0280
74-3A0R	0280
74-3A1	0280
74-3A2	0280
74-3A3	0280
74-3A3P	0280
74-3A4	0280
74-3A4-7	0280
74-3A4-7B	0280
74-3A5	0280
74-3A5L	0280
74-3A6	0280
74-3A6-3	0280
74-3A7	0280
74-3A7-1	0280
74-3A8	0280
74-3A9	0280
74-3A9G	0280
74-3A19	0280

DEVICE TYPE	REPL CODE
74-3A21	0280
74-3A82	0280
74-30	0280
74-31	0280
74-32	0280
74-33	0280
74-34	0280
74-35	0280
74-36	0280
74-37	0280
74-38	0280
74-39	0280
74A01	0847
74A02	0847
74A03	0847
74ACT00E	OEM
74ACT04E	OEM
74ACT04M	OEM
74ACT05E	OEM
74ACT08E	OEM
74ACT32E	OEM
74ACT32M	OEM
74ACT193M	OEM
74ALS00	OEM
74ALS00A	OEM
74ALS00AD(SM)	OEM
74ALS00AN	OEM
74ALS02	OEM
74ALS02D(SM)	OEM
74ALS02N	OEM
74ALS04A	OEM
74ALS04AN	OEM
74ALS04BD(SM)	OEM
74ALS08	OEM
74ALS08D(SM)	OEM
74ALS08N	OEM
74ALS09	OEM
74ALS10	OEM
74ALS10AD(SM)	OEM
74ALS10N	OEM
74ALS11AD(SM)	OEM
74ALS27	OEM
74ALS27N	OEM
74ALS32	OEM
74ALS32D(SM)	OEM
74ALS32N	OEM
74ALS38	OEM
74ALS74	OEM
74ALS74A	OEM
74ALS74AD(SM)	OEM
74ALS74AN	OEM
74ALS109	OEM
74ALS109AN	OEM
74ALS138	OEM
74ALS138N	OEM
74ALS174D(SM)	OEM
74ALS175	OEM
74ALS175N	OEM
74ALS240ADW(SM)	OEM
74ALS244A	OEM
74ALS244ADW(SM)	OEM
74ALS244AN	OEM
74ALS245A	OEM
74ALS245ADW(SM)	OEM
74ALS257	OEM
74ALS373	OEM
74ALS573	OEM
74ALS573BDW(SM)	OEM
74ALS573N	OEM
74ALS652DW(SM)	OEM
74AS573DW(SM)	OEM
74AS646DW(SM)	OEM
74AS760DW(SM)	OEM
74C04	2958
74CH646EN	OEM
74CH648EN	OEM
74CH7038EN	OEM
74F00	OEM
74F00(SM)	OEM
74F00DC	OEM
74F00N	OEM
74F00PC	OEM
74F02DC	OEM
74F02PC	OEM
74F04	OEM
74F04(SM)	OEM
74F04DC	OEM
74F04PC	OEM
74F08	OEM
74F08(SM)	OEM
74F08DC	OEM
74F08N	OEM
74F08PC	OEM
74F10	OEM
74F10(SM)	OEM
74F10DC	OEM
74F10N	OEM
74F10PC	OEM
74F11	OEM
74F11DC	OEM
74F11N	OEM
74F11PC	OEM
74F20	OEM
74F20DC	OEM
74F20N	OEM
74F20PC	OEM
74F27N	OEM
74F30D(SM)	OEM
74F32	OEM

DEVICE TYPE	REPL CODE
74F32(SM)	OEM
74F32DC	OEM
74F32PC	OEM
74F64DC	OEM
74F64PC	OEM
74F74	OEM
74F74(SM)	OEM
74F74DC	OEM
74F74N	OEM
74F74PC	OEM
74F86PC	OEM
74F109	OEM
74F109DC	OEM
74F109PC	OEM
74F125(SM)	OEM
74F138(SM)	OEM
74F138DC	OEM
74F138PC	OEM
74F139	OEM
74F139DC	OEM
74F139N	OEM
74F139PC	OEM
74F151	OEM
74F151DC	OEM
74F151FC	OEM
74F151PC	OEM
74F153	OEM
74F153DC	OEM
74F153FC	OEM
74F153PC	OEM
74F157DC	OEM
74F157FC	OEM
74F157PC	OEM
74F158	OEM
74F158A	OEM
74F158A(SM)	OEM
74F158APC	OEM
74F158DC	OEM
74F158FC	OEM
74F158N	OEM
74F158PC	OEM
74F160DC	OEM
74F160PC	OEM
74F161DC	OEM
74F161PC	OEM
74F162DC	OEM
74F162PC	OEM
74F163DC	OEM
74F163PC	OEM
74F174	OEM
74F174(SM)	OEM
74F174N	OEM
74F174PC	OEM
74F175	OEM
74F175DC	OEM
74F175N	OEM
74F175PC	OEM
74F181DC	OEM
74F181PC	OEM
74F182DC	OEM
74F182PC	OEM
74F189DC	OEM
74F189FC	OEM
74F189PC	OEM
74F191PC	OEM
74F240DC	OEM
74F240FC	OEM
74F240PC	OEM
74F241	OEM
74F241DC	OEM
74F241FC	OEM
74F241N	OEM
74F241PC	OEM
74F242DC	OEM
74F242FC	OEM
74F242PC	OEM
74F243DC	OEM
74F243FC	OEM
74F243PC	OEM
74F244(SM)	OEM
74F244DC	OEM
74F244FC	OEM
74F244PC	OEM
74F245	OEM
74F245(SM)	OEM
74F245DC	OEM
74F245PC	OEM
74F251DC	OEM
74F251FC	OEM
74F251PC	OEM
74F253DC	OEM
74F253FC	OEM
74F253PC	OEM
74F257	OEM
74F257DC	OEM
74F257FC	OEM
74F257N	OEM
74F257PC	OEM
74F258DC	OEM
74F258FC	OEM
74F258PC	OEM
74F280(SM)	OEM
74F280DC	OEM
74F280PC	OEM
74F283(SM)	OEM
74F283DC	OEM
74F283PC	OEM
74F289DC	OEM
74F289FC	OEM
74F289PC	OEM
74F352DC	OEM
74F352FC	OEM
74F352PC	OEM
74F353DC	OEM
74F353FC	OEM
74F353PC	OEM
74F373DC	OEM
74F373PC	OEM
74F374	OEM
74F374DC	OEM
74F374PC	OEM
74F379	OEM
74F500DC	OEM
74F500PC	OEM
74F533DC	OEM
74F533PC	OEM
74F534DC	OEM
74F534PC	OEM
74F537DC	OEM
74F537FC	OEM
74F537PC	OEM
74F538DC	OEM
74F538FC	OEM
74F538PC	OEM
74F539DC	OEM
74F539FC	OEM
74F539PC	OEM
74F545PC	OEM
74F588DC	OEM
74F588PC	OEM
74H00DC	0682
74H00PC	0682
74H01DC	5338
74H01PC	5338
74H04DC	1907
74H04PC	1907
74H05DC	3251
74H05PC	3251
74H08PC	5355
74H10DC	0685
74H10PC	0685
74H11DC	2398
74H11PC	2398
74H20DC	3715
74H20PC	3715
74H21DC	4856
74H21PC	4856
74H22	4596
74H22/9H22	4596
74H22/94H22	4596
74H22DC	4596
74H22PC	4596
74H30DC	5383
74H30PC	5383
74H40DC	0559
74H40PC	0559
74H50DC	1792
74H50FC	1792
74H50PC	1792
74H51DC	1944
74H51PC	1944
74H52DC	2020
74H52PC	2020
74H53DC	2101
74H53PC	2101
74H54DC	2172
74H54PC	2172
74H55DC	3158
74H55PC	3158
74H60DC	5410
74H60PC	5410
74H61DC	2663
74H61PC	2663
74H62DC	2731
74H62PC	2731
74H71DC	3264
74H71PC	3264
74H72DC	3313
74H72PC	3313
74H73DC	2463
74H73PC	2463
74H74DC	2492
74H74PC	2492
74H76DC	5302
74H76PC	5302
74H78DC	5418
74H78PC	5418
74H87DC	2579
74H87PC	2579
74H101DC	5521
74H101PC	5521
74H102DC	5523
74H102PC	5523
74H103DC	2969
74H103PC	2969
74H106DC	5253
74H106PC	5253
74H108DC	0180
74H108PC	0180
74H183DC	4399
74H183PC	4399
74HC00	OEM
74HC00/FP	OEM
74HC00D	OEM
74HC00N	2344
74HC02	OEM
74HC02D	OEM
74HC02N	1452
74HC03D	OEM
74HC03N	1453
74HC04	OEM
74HC04AP	1455
74HC04D	OEM
74HC04N	1561
74HC05	OEM
74HC08	OEM
74HC08D	OEM
74HC08M	OEM
74HC08N	2408
74HC10	OEM
74HC10D	OEM
74HC10N	2429
74HC11D	OEM
74HC11N	2435
74HC14	OEM
74HC14AP	OEM
74HC14D	OEM
74HC14N	2466
74HC20D	OEM
74HC20N	1475
74HC21D	OEM
74HC21N	1477
74HC27	OEM
74HC27D	OEM
74HC27N	1485
74HC30	OEM
74HC30D	OEM
74HC30N	1489
74HC32	OEM
74HC32D	OEM
74HC32N	2610
74HC42D	OEM
74HC42N	1515
74HC58D	OEM
74HC58N	OEM
74HC73D	OEM
74HC73N	1558
74HC74	OEM
74HC74D	OEM
74HC74N	1561
74HC75D	OEM
74HC75N	1565
74HC85D	OEM
74HC85E	OEM
74HC85N	1583
74HC86	OEM
74HC86D	OEM
74HC86N	2999
74HC93D	OEM
74HC93N	1594
74HC107D	OEM
74HC107N	4692
74HC109D	OEM
74HC109N	6333
74HC112D	OEM
74HC112N	0214
74HC123D	OEM
74HC123N	3907
74HC125D	OEM
74HC125N	6344
74HC126D	OEM
74HC126N	6346
74HC132D	OEM
74HC132N	6349
74HC137D	OEM
74HC137N	0784
74HC138	OEM
74HC138D	OEM
74HC138N	6353
74HC139	OEM
74HC139D	OEM
74HC139N	3950
74HC147D	OEM
74HC147N	0938
74HC151D	OEM
74HC151N	6358
74HC153D	OEM
74HC153N	6360
74HC154	OEM
74HC154D	OEM
74HC154EN	OEM
74HC154N	6361
74HC157	OEM
74HC157D	OEM
74HC157N	OEM
74HC158D	OEM
74HC158N	1088
74HC160D	OEM
74HC160N	1117
74HC161D	OEM
74HC161N	1129
74HC162D	OEM
74HC162N	1142
74HC163D	OEM
74HC163N	4057
74HC164D	OEM
74HC164N	6368
74HC165D	OEM
74HC165N	6370
74HC166	OEM
74HC166D	OEM
74HC166N	4741
74HC173D	OEM
74HC173N	6380
74HC174D	OEM
74HC174N	6383
74HC175D	OEM
74HC175N	6384
74HC181D	OEM
74HC181N	4748
74HC182D	OEM
74HC182N	4749
74HC190D	OEM
74HC190N	4753
74HC191D	OEM
74HC191N	4754
74HC192D	OEM
74HC192N	4755
74HC193D	OEM
74HC193N	4756
74HC194D	OEM
74HC194N	1557
74HC195D	OEM
74HC195N	1578
74HC221D	OEM
74HC221N	4774
74HC237D	OEM
74HC237N	1958
74HC238D	OEM
74HC238N	4784
74HC240D	OEM
74HC240N	1979
74HC241D	OEM
74HC241N	1990
74HC242D	OEM
74HC242N	1996
74HC243D	OEM
74HC243N	2001
74HC244	OEM
74HC244D	OEM
74HC244N	4792
74HC245D	OEM
74HC245N	2014
74HC251D	OEM
74HC251N	2065
74HC253D	OEM
74HC253N	4800
74HC257D	OEM
74HC257N	4802
74HC258D	OEM
74HC258N	4804
74HC259	OEM
74HC259D	OEM
74HC259N	4806
74HC273	OEM
74HC273D	OEM
74HC273N	4822
74HC280D	OEM
74HC280N	4822
74HC283D	OEM
74HC283N	4828
74HC297D	OEM
74HC297N	4843
74HC299D	OEM
74HC299N	6464
74HC354D	OEM
74HC354N	4862
74HC356D	OEM
74HC356N	4863
74HC365D	OEM
74HC365N	2880
74HC366D	OEM
74HC366N	2885
74HC367	OEM
74HC367D	OEM
74HC367N	2894
74HC368	OEM
74HC368(SM)	OEM
74HC368D	OEM
74HC368N	2898
74HC373	OEM
74HC373D	OEM
74HC373N	4462
74HC374	OEM
74HC374D	OEM
74HC374N	6579
74HC377D	OEM
74HC377N	6581
74HC390D	OEM
74HC390N	6595
74HC393D	OEM
74HC393N	6598
74HC423D	OEM
74HC423N	4876
74HC533D	OEM
74HC533N	3677
74HC534D	OEM
74HC534N	3681
74HC540D	OEM
74HC540N	4888
74HC541D	OEM
74HC541N	4889
74HC563D	OEM
74HC563N	4892
74HC564D	OEM
74HC564N	4893
74HC573D	OEM
74HC573N	6765
74HC574D	OEM
74HC574N	6767
74HC583D	OEM
74HC583N	4899
74HC597D	OEM
74HC597N	4901
74HC640D	OEM
74HC640N	4912
74HC643D	OEM
74HC643N	4914
74HC646D	OEM
74HC646N	4915
74HC648D	OEM
74HC648N	4916
74HC670D	OEM
74HC670N	4919
74HC688D	OEM
74HC688N	4923
74HC4002D	OEM
74HC4002N	5488
74HC4015D	OEM
74HC4015N	5506
74HC4016D	OEM
74HC4016M	OEM
74HC4016N	5507
74HC4017D	OEM
74HC4017N	5511
74HC4020D	OEM
74HC4020N	5512
74HC4024D	OEM
74HC4024N	5515
74HC4040D	OEM
74HC4040N	5532
74HC4046AD	OEM
74HC4046AN	5152
74HC4046AP	OEM
74HC4049D	OEM
74HC4049N	5538
74HC4050D	OEM
74HC4050N	5540
74HC4051D	OEM
74HC4051N	5543
74HC4052D	OEM
74HC4052N	5545
74HC4053	OEM
74HC4053AP	OEM
74HC4053D	OEM
74HC4053N	5548
74HC4059D	OEM
74HC4059N	5558
74HC4060D	OEM
74HC4060N	5561
74HC4066	OEM
74HC4066D	OEM
74HC4066N	5571
74HC4066S	OEM
74HC4067D	OEM
74HC4067N	5572
74HC4075D	OEM
74HC4075N	5576
74HC4094AP	OEM
74HC4094D	OEM
74HC4094N	5596
74HC4316D	OEM
74HC4316N	5955
74HC4351D	OEM
74HC4351N	5981
74HC4352D	OEM
74HC4352N	5982
74HC4353D	OEM
74HC4353N	5984
74HC4510D	OEM
74HC4510N	6087
74HC4511D	OEM
74HC4511N	6088
74HC4514D	OEM
74HC4514N	6093
74HC4515D	OEM
74HC4515N	6094
74HC4516D	OEM
74HC4516N	6097
74HC4518D	OEM
74HC4518N	6102
74HC4520D	OEM
74HC4520N	6106
74HC4538D	OEM
74HC4538N	6126
74HC4543D	OEM
74HC4543N	6137
74HC7030D	OEM
74HC7030N	5519
74HC7038EN	OEM
74HC7046AD	OEM
74HC7046AN	5148
74HC7046D	OEM
74HC7046N	5148
74HC7266D	OEM
74HC7266N	5889
74HC7597D	OEM
74HC7597N	OEM
74HC40102D	OEM
74HC40102N	5072
74HC40103D	OEM
74HC40103N	5073
74HC40104D	OEM
74HC40104N	5074
74HC40105D	OEM
74HC40105N	6084
74HCT00D	OEM
74HCT00N	6904
74HCT02	4910
74HCT02D	OEM
74HCT02N	4910
74HCT03D	OEM
74HCT03N	5917
74HCT04	3296
74HCT04D	OEM
74HCT04N	3296
74HCT08	3299
74HCT08D	OEM
74HCT08N	3299
74HCT10D	OEM
74HCT10N	5918
74HCT11D	OEM
74HCT11N	5919
74HCT14	3300
74HCT14D	OEM
74HCT14N	3300
74HCT20D	OEM
74HCT20N	5921
74HCT21D	OEM
74HCT21N	5922
74HCT27D	OEM
74HCT27N	5923
74HCT30D	OEM
74HCT30N	5925
74HCT32	3306
74HCT32D	OEM
74HCT32N	3306
74HCT42D	OEM
74HCT42N	5928
74HCT73D	OEM
74HCT73N	5933
74HCT74D	OEM
74HCT74N	5934
74HCT75D	OEM
74HCT75N	5935
74HCT85D	OEM
74HCT85N	5936
74HCT86D	OEM
74HCT86N	5938
74HCT93D	OEM
74HCT93N	5940
74HCT107D	OEM
74HCT107N	6610
74HCT109D	OEM
74HCT109N	6611
74HCT112D	OEM
74HCT112N	6613
74HCT123D	OEM
74HCT123N	6617
74HCT125D	OEM
74HCT125N	6618
74HCT126D	OEM
74HCT126N	6619
74HCT132D	OEM
74HCT132N	6623
74HCT137D	OEM
74HCT137N	4398
74HCT138	6090
74HCT138D	OEM
74HCT138N	6090
74HCT139D	OEM
74HCT139N	6628
74HCT147D	OEM
74HCT147N	6645
74HCT151D	OEM
74HCT151N	6649
74HCT153D	OEM
74HCT153N	6651
74HCT154D	OEM
74HCT154EN	OEM
74HCT154N	5045
74HCT157D	OEM
74HCT157N	6656
74HCT158D	OEM
74HCT158N	6095
74HCT160D	OEM
74HCT161N	6660
74HCT162D	OEM
74HCT162N	6661
74HCT163D	OEM
74HCT163N	6664
74HCT164D	OEM
74HCT164N	6666
74HCT165D	OEM
74HCT165N	6667
74HCT166D	OEM
74HCT166N	6669
74HCT173D	OEM
74HCT173N	6673
74HCT174D	OEM
74HCT174N	6674
74HCT175D	OEM
74HCT175N	6675
74HCT181D	OEM
74HCT181N	6679
74HCT182D	OEM
74HCT182N	6120
74HCT190D	OEM
74HCT190N	6686
74HCT191D	OEM
74HCT191N	6687
74HCT192D	OEM
74HCT192N	6690
74HCT193D	OEM
74HCT193N	6691
74HCT194D	OEM
74HCT194N	6693
74HCT195	6100
74HCT195D	OEM
74HCT195N	6100
74HCT221D	OEM
74HCT221N	6722
74HCT237D	OEM
74HCT237N	6733
74HCT238D	OEM
74HCT238N	6734
74HCT240-2N	OEM
74HCT240D	OEM
74HCT240N	4050
74HCT241D	OEM
74HCT241N	6737
74HCT242D	OEM
74HCT242N	6738
74HCT243D	OEM
74HCT243N	6741
74HCT244	6108
74HCT244(SM)	OEM
74HCT244D	OEM
74HCT244N	6108
74HCT245	4075
74HCT245-2N	OEM
74HCT245D	OEM
74HCT245N	4075
74HCT251D	OEM
74HCT251N	6749
74HCT253D	OEM
74HCT253N	6752
74HCT257D	OEM
74HCT257N	6754
74HCT258D	OEM
74HCT258N	6755
74HCT259D	OEM
74HCT259N	6757
74HCT273-2N	OEM
74HCT273D	OEM
74HCT273N	4126
74HCT280D	OEM
74HCT280N	6779
74HCT283D	OEM
74HCT283N	6782
74HCT297D	OEM
74HCT297N	6793
74HCT299D	OEM
74HCT299N	6794
74HCT354D	OEM
74HCT354N	0790
74HCT356D	OEM
74HCT356N	0813
74HCT365D	OEM
74HCT365N	6821
74HCT366D	OEM
74HCT366N	6824
74HCT367D	OEM
74HCT367N	6825
74HCT368D	OEM
74HCT368N	6826
74HCT373D	OEM
74HCT373N	6123
74HCT374	OEM
74HCT374-2N	OEM
74HCT374D	OEM
74HCT374N	4311
74HCT374P	OEM
74HCT377D	OEM
74HCT377N	6831
74HCT390D	OEM
74HCT390N	6837
74HCT393D	OEM
74HCT393N	6840
74HCT423D	OEM
74HCT423N	4460
74HCT533D	OEM
74HCT533N	5719
74HCT534D	OEM
74HCT534N	5737
74HCT540D	OEM
74HCT540N	6954
74HCT541D	OEM
74HCT541N	6955
74HCT563D	OEM
74HCT563N	6965
74HCT564D	OEM
74HCT564N	6966
74HCT573D	OEM
74HCT573N	6971
74HCT574D	OEM
74HCT574N	6178
74HCT583D	OEM
74HCT583N	6981
74HCT597D	OEM
74HCT597N	6987
74HCT640D	OEM
74HCT640N	7029
74HCT643D	OEM
74HCT643N	7036
74HCT646D	OEM
74HCT646N	7039
74HCT648D	OEM
74HCT648N	4507
74HCT670D	OEM
74HCT670N	7069
74HCT688D	OEM
74HCT688N	3412
74HCT4002D	OEM
74HCT4002N	3364
74HCT4015D	OEM
74HCT4015N	3405
74HCT4016D	OEM
74HCT4016N	3409
74HCT4017D	OEM
74HCT4017N	3417
74HCT4020D	OEM
74HCT4020N	3430

DEVICE TYPE	REPL CODE
74HCT4024D	OEM
74HCT4024N	3448
74HCT4040D	OEM
74HCT4040N	3506
74HCT4046AD	OEM
74HCT4046AN	5146
74HCT4051D	OEM
74HCT4051N	3556
74HCT4052D	OEM
74HCT4052N	3558
74HCT4053D	OEM
74HCT4053N	3564
74HCT4059D	OEM
74HCT4059N	3605
74HCT4060D	OEM
74HCT4060N	3614
74HCT4066D	OEM
74HCT4066N	3629
74HCT4067D	OEM
74HCT4067N	3633
74HCT4075D	OEM
74HCT4075N	3665
74HCT4094D	OEM
74HCT4094N	3737
74HCT4316D	OEM
74HCT4316N	4324
74HCT4351D	OEM
74HCT4351N	4392
74HCT4352D	OEM
74HCT4352N	4394
74HCT4353D	OEM
74HCT4353N	4396
74HCT4510D	OEM
74HCT4510N	4891
74HCT4511D	OEM
74HCT4511N	4894
74HCT4514D	OEM
74HCT4514N	4903
74HCT4515D	OEM
74HCT4515N	4908
74HCT4516D	OEM
74HCT4516N	4909
74HCT4518D	OEM
74HCT4518N	4913
74HCT4520D	OEM
74HCT4520N	4918
74HCT4538D	OEM
74HCT4538N	4958
74HCT4543D	OEM
74HCT4543N	4979
74HCT7030D	OEM
74HCT7030N	3464
74HCT7038N	OEM
74HCT7046AD	OEM
74HCT7046AN	3527
74HCT7046D	OEM
74HCT7046N	3527
74HCT7266D	OEM
74HCT7266N	4248
74HCT7597D	OEM
74HCT7597N	OEM
74HCT40102D	OEM
74HCT40102N	5067
74HCT40103D	OEM
74HCT40103N	5068
74HCT40104D	OEM
74HCT40104N	5069
74HCT40105D	OEM
74HCT40105N	5070
74HCU04	OEM
74HCU04A	OEM
74HCU04AP	OEM
74HCU04D	OEM
74HCU04N	4602
74L03	1579
74LS00	1527
74LS00CH	1527
74LS00DC	1527
74LS00J	1527
74LS00N	1527
74LS00NA	1527
74LS00PC	1527
74LS00W	1527
74LS02	1559
74LS02CH	1559
74LS02DC	1559
74LS02J	1559
74LS02N	1559
74LS02PC	1559
74LS02W	1559
74LS03	1579
74LS03CH	1579
74LS03DC	1579
74LS03J	1579
74LS03P	1579
74LS03PC	1579
74LS03W	1579
74LS04	1595
74LS04(SM)	OEM
74LS04DC	1595
74LS04N	6179
74LS04NA	1595
74LS04P	1595
74LS04PC	1595
74LS05	1608
74LS05(SM)	OEM
74LS05DC	1608
74LS05N	6180
74LS05PC	1608
74LS06	OEM
74LS06(MIN)	OEM
74LS06(SM)	OEM
74LS07	OEM
74LS08	1633
74LS08CH	1633
74LS08DC	1633
74LS08J	1633
74LS08N	1633
74LS08NA	1633
74LS08PC	1633
74LS08W	1633
74LS09	1642
74LS09N	OEM
74LS09PC	1642
74LS10	1662
74LS10(SM)	OEM
74LS10CH	1662
74LS10DC	1662
74LS10J	1662
74LS10N	1662
74LS10PC	1662
74LS10W	1662
74LS11	1667
74LS11CH	1667
74LS11DC	1667
74LS11J	1667
74LS11N	1667
74LS11PC	1667
74LS11W	1667
74LS12	1679
74LS13DC	1688
74LS13N	1688
74LS13PC	1688
74LS14	1698
74LS14(SM)	OEM
74LS14D(SM)	OEM
74LS14DC	1698
74LS14N	1698
74LS14PC	1698
74LS14PF	1698
74LS15	1707
74LS15PC	1707
74LS16-1	OEM
74LS17	OEM
74LS20	0034
74LS20CH	0034
74LS20DC	0034
74LS20J	0034
74LS20N	0034
74LS20PC	0034
74LS20W	0034
74LS21	1763
74LS21DC	1763
74LS21PC	1763
74LS22	1775
74LS22PC	1775
74LS26	1382
74LS26DC	1382
74LS26PC	1382
74LS27	0183
74LS27CH	0183
74LS27DC	0183
74LS27J	0183
74LS27N	7050
74LS27PC	0183
74LS27W	0183
74LS28	OEM
74LS28DC	0472
74LS28PC	0472
74LS30	0828
74LS30CH	0828
74LS30DC	0828
74LS30J	0828
74LS30N	7057
74LS30PC	0828
74LS30W	0828
74LS32	0088
74LS32(SM)	OEM
74LS32CH	0088
74LS32DC	0088
74LS32J	0088
74LS32L	0088
74LS32N	0088
74LS32P	0088
74LS32PC	0088
74LS32W	0088
74LS33	1832
74LS33PC	1832
74LS37	1730
74LS37DC	1730
74LS37FC	1730
74LS37PC	1730
74LS38	1839
74LS38-1	1839
74LS38CH	1839
74LS38DC	1839
74LS38FC	1839
74LS38J	1839
74LS38N	1839
74LS38PC	1839
74LS38W	1839
74LS40DC	0135
74LS40PC	0135
74LS42	1841
74LS42CH	1841
74LS42DC	1841
74LS42J	1841
74LS42PC	1841
74LS42W	1841
74LS47DC	1845
74LS47PC	1845
74LS48DC	1849
74LS48PC	1849
74LS49DC	1850
74LS49PC	1850
74LS51	1033
74LS51CH	1033
74LS51DC	1033
74LS51J	1033
74LS51PC	1033
74LS51W	1033
74LS54DC	1857
74LS54PC	1857
74LS55PC	0457
74LS73	1867
74LS73CH	1867
74LS73J	1867
74LS73W	1867
74LS74	0243
74LS74A	0243
74LS74AN	0243
74LS74ANB	OEM
74LS74APC	0243
74LS74CH	0243
74LS74J	0243
74LS74PC	0243
74LS74W	0243
74LS75	1870
74LS75(SM)	OEM
74LS75CH	1870
74LS75J	1870
74LS75N	7093
74LS75W	1870
74LS76	2180
74LS76DC	2180
74LS76PC	2180
74LS77	1872
74LS83ACH	2218
74LS83ADC	2218
74LS83AJ	2218
74LS83APC	2218
74LS83AW	2218
74LS85	0431
74LS85CH	0431
74LS85DC	0431
74LS85J	0431
74LS85PC	0431
74LS85W	0431
74LS86	0291
74LS86CH	0291
74LS86DC	0291
74LS86J	0291
74LS86N	0291
74LS86PC	0291
74LS86W	0291
74LS89DC	OEM
74LS89FC	OEM
74LS89PC	OEM
74LS90	1882
74LS90DC	1882
74LS90PC	1882
74LS91	1885
74LS92	1887
74LS92PC	1887
74LS93	1888
74LS93CH	1888
74LS93DC	1888
74LS93J	1888
74LS93N	1888
74LS93PC	1888
74LS93W	1888
74LS95	OEM
74LS95BDC	0771
74LS95BFC	0771
74LS95BPC	0771
74LS95N	OEM
74LS107	7013
74LS107CH	1602
74LS107DC	1602
74LS107J	1602
74LS107PC	1602
74LS107W	1602
74LS109	1906
74LS109A	4972
74LS109CH	1906
74LS109DC	1906
74LS109J	1906
74LS109PC	1906
74LS109W	1906
74LS112	2128
74LS112A	OEM
74LS112DC	2128
74LS112PC	2128
74LS113PC	2256
74LS114PC	2302
74LS122	5084
74LS122N	5084
74LS123	0979
74LS123CH	0979
74LS123J	0979
74LS123N	0979
74LS123W	0979
74LS125	0075
74LS125A	0075
74LS125AD(SM)	OEM
74LS125ADC	0075
74LS125AM	OEM
74LS125AN	0075
74LS125APC	0075
74LS126	2877
74LS126A	OEM
74LS126AN	OEM
74LS126DC	2877
74LS126PC	2877
74LS132	1625
74LS132DC	1625
74LS132N	1625
74LS132PC	1625
74LS133DC	3400
74LS133PC	3400
74LS136CH	1628
74LS136DC	1628
74LS136J	1628
74LS136PC	1628
74LS136W	1628
74LS138	0427
74LS138DC	0427
74LS138N	0427
74LS138NA	OEM
74LS138P	0427
74LS138PC	0427
74LS139	0153
74LS139A	OEM
74LS139AN	OEM
74LS139DC	0153
74LS139N	0153
74LS139PC	0153
74LS145	1564
74LS145PC	OEM
74LS148	3912
74LS151	1646
74LS151DC	1646
74LS151N	5136
74LS151PC	1646
74LS152	OEM
74LS153	0959
74LS153DC	0959
74LS153N	0959
74LS153NB	OEM
74LS153PC	0959
74LS154	5045
74LS155	0209
74LS155DC	0209
74LS155N	4206
74LS155NA	4206
74LS155PC	0209
74LS156	1654
74LS156DC	1654
74LS156N	1654
74LS156PC	1654
74LS157	1161
74LS157DC	1161
74LS157N	1161
74LS157PC	1161
74LS158	1656
74LS158DC	1656
74LS158PC	1656
74LS160PC	0837
74LS161	0858
74LS161A	5166
74LS161AN	0858
74LS161APC	5166
74LS161CH	0858
74LS161DC	0858
74LS161J	0858
74LS161PC	0858
74LS161W	0858
74LS162	OEM
74LS162N	OEM
74LS162PC	0880
74LS163	0893
74LS163CH	0893
74LS163DC	0893
74LS163J	0893
74LS163N	5184
74LS163PC	0893
74LS163W	0893
74LS164	4339
74LS164DC	4339
74LS164FC	4339
74LS164N	4339
74LS164PC	4339
74LS165DC	4356
74LS165FC	4356
74LS165PC	4356
74LS166	4371
74LS166A	OEM
74LS168DC	0967
74LS168PC	0967
74LS169DC	0986
74LS169PC	0986
74LS170DC	2629
74LS170FC	2629
74LS170PC	2629
74LS173A	5217
74LS174	0260
74LS174B	OEM
74LS174N	0260
74LS174NA	OEM
74LS174PC	0260
74LS175	1672
74LS175DC	1672
74LS175N	1672
74LS175PC	1672
74LS181PC	1678
74LS189DC	OEM
74LS189FC	OEM
74LS189PC	OEM
74LS190PC	1686
74LS191	7026
74LS191CH	1687
74LS191DC	1687
74LS191J	1687
74LS191PC	1687
74LS191W	1687
74LS192	1689
74LS192DC	1689
74LS192PC	1689
74LS193	1692
74LS193CH	1692
74LS193DC	1692
74LS193J	1692
74LS193N	1692
74LS193PC	1692
74LS193W	1692
74LS194	1304
74LS194ADC	1304
74LS194AFC	1304
74LS194APC	1304
74LS195	1519
74LS195ADC	1315
74LS195AFC	1315
74LS195APC	1315
74LS195N	1519
74LS196	2834
74LS196PC	2834
74LS197	2469
74LS197DC	2469
74LS197N	OEM
74LS197PC	2469
74LS221	1239
74LS221CH	1239
74LS221J	1239
74LS221PC	OEM
74LS221W	1239
74LS240	0452
74LS240DC	0452
74LS240N	0452
74LS240PC	0452
74LS241	1726
74LS241DC	1726
74LS241N	1726
74LS241PC	1726
74LS242	1728
74LS242DC	1728
74LS242PC	1728
74LS243	7038
74LS243APC	OEM
74LS243DC	0906
74LS243N	0906
74LS243PC	0906
74LS244	0458
74LS244(SM)	OEM
74LS244DC	0458
74LS244N	0458
74LS244PC	0458
74LS245	0463
74LS245DC	0463
74LS245N	0463
74LS245NA	0463
74LS245PC	0463
74LS247DC	1732
74LS247PC	1732
74LS248DC	1732
74LS248PC	1734
74LS249DC	1735
74LS249PC	1735
74LS251(SM)	OEM
74LS251DC	1737
74LS251PC	1737
74LS253DC	1739
74LS253PC	1739
74LS256DC	OEM
74LS256PC	OEM
74LS257	1744
74LS257A	1744
74LS257AN	1744
74LS257APC	1744
74LS257DC	1744
74LS257N	1744
74LS257PC	1744
74LS258	1746
74LS258APC	1746
74LS258DC	1746
74LS258N	1746
74LS258PC	1746
74LS259	3205
74LS259DC	3205
74LS259PC	3205
74LS260DC	5995
74LS260PC	5995
74LS266	0592
74LS266CH	0592
74LS266DC	0592
74LS266J	0592
74LS266PC	0592
74LS266W	0592
74LS273	0894
74LS273DC	0894
74LS273N	0894
74LS273PC	0894
74LS279	5401
74LS279DC	3290
74LS279N	5401
74LS279PC	3290
74LS280	1773
74LS283	1779
74LS283DC	1779
74LS283N	OEM
74LS283PC	1779
74LS289DC	OEM
74LS289PC	OEM
74LS290	4425
74LS290DC	4425
74LS290PC	4425
74LS293	0082
74LS293DC	0082
74LS293PC	0082
74LS295ADC	2227
74LS295AFC	2227
74LS295APC	2227
74LS298DC	3371
74LS298PC	3371
74LS299DC	4426
74LS299FC	4426
74LS322	1805
74LS322A	OEM
74LS322DC	1805
74LS322FC	1805
74LS322PC	1805
74LS323DC	OEM
74LS323FC	OEM
74LS323PC	OEM
74LS324	6000
74LS327	6002
74LS347DC	OEM
74LS347PC	OEM
74LS352	0761
74LS352DC	0761
74LS352PC	0761
74LS353DC	0773
74LS353PC	0773
74LS365	0943
74LS365A	0943
74LS365ADC	0943
74LS365AN	OEM
74LS365APC	0943
74LS366ADC	0956
74LS366APC	0956
74LS367	0977
74LS367A	0977
74LS367ADC	0977
74LS367APC	0977
74LS368	0991
74LS368A	0991
74LS368ADC	0991
74LS368AN	0991
74LS368APC	0991
74LS373	0709
74LS373DC	0709
74LS373N	0709
74LS373PC	0709
74LS374	0713
74LS374DC	0713
74LS374N	0713
74LS374P	0713
74LS374PC	0713
74LS375	OEM
74LS375DC	OEM
74LS375N	OEM
74LS375PC	OEM
74LS377DC	1120
74LS377PC	1120
74LS378	1133
74LS378DC	1133
74LS378FC	1133
74LS378PC	1133
74LS379DC	1151
74LS379FC	1151
74LS379PC	1151
74LS390	1288
74LS390DC	1288
74LS390PC	1288
74LS393	0819
74LS393DC	0819
74LS393N	0819
74LS393NA	0819
74LS393PC	0819
74LS395A	1330
74LS395DC	1330
74LS395FC	1330
74LS395PC	1330
74LS447DC	OEM
74LS447PC	OEM
74LS471	OEM
74LS490	2213
74LS490PC	2213
74LS502DC	OEM
74LS502PC	OEM
74LS503DC	OEM
74LS503PC	OEM
74LS504DC	OEM
74LS504PC	OEM
74LS533DC	5719
74LS533PC	5719
74LS534DC	5737
74LS534PC	5737
74LS540DC	2541
74LS540PC	2541
74LS541	2547
74LS541DC	2547
74LS541N	2547
74LS541PC	2547
74LS574PC	6178
74LS590	OEM
74LS590N	OEM
74LS612	OEM
74LS612N	OEM
74LS624	3141
74LS629	OEM
74LS629N	3175
74LS645	0775
74LS646	OEM
74LS646NT	OEM
74LS670	1130
74LS670DC	1130
74LS670FC	1130
74LS670N	OEM
74LS670PC	1130
74LS688	3412
74LS783	OEM
74LS785	OEM
74LS2578C	OEM
74LSP14APC	1304
74N1	0016
74P1	0037
74P1M	0126
74Q1262	0050
74Q22881	1862
74S00	0704
74S00DC	0704
74S00N	0704
74S00PC	0704
74S02	2238
74S02PC	2238
74S03DC	2217
74S03PC	2217
74S04	2263
74S04ADC	2263
74S04APC	2263
74S04DC	2263
74S04N	6201
74S04PC	2263
74S05	2321
74S05ADC	2321
74S05APC	2321
74S05D(SM)	OEM
74S05DC	2321
74S05N	OEM
74S05PC	2321
74S08	2569
74S08DC	2569
74S08N	OEM
74S08PC	2680
74S09DC	2667
74S09PC	2667
74S10	2445
74S10DC	2445
74S10N	OEM
74S10PC	2445
74S11	2447
74S11DC	2447
74S11N	6206
74S11PC	2447
74S15DC	2451
74S15PC	2451
74S20	5754
74S20DC	1017
74S20N	5754
74S20PC	1017
74S21	OEM
74S22DC	2461
74S22PC	2461
74S30DC	3726
74S30PC	3726
74S32	3847
74S32DC	OEM
74S32N	3847
74S32PC	3847
74S37	5756
74S38	0780
74S40	2475
74S40DC	OEM
74S40PC	2475
74S51	4305
74S51DC	4305
74S51N	OEM
74S51NA	4305
74S51PC	4305
74S64	2496
74S64DC	2496
74S64PC	2496
74S65	2497
74S65DC	2497
74S65PC	2497
74S74	2504
74S74DC	2504
74S74PC	2504
74S85	5773
74S86	2511
74S86DC	2511
74S86PC	2511
74S109DC	OEM
74S109PC	OEM
74S112	1617
74S112A	1617
74S112DC	1617
74S112N	1617
74S112PC	1617
74S113DC	1623
74S113PC	1623
74S114DC	1629
74S114PC	1629

DEVICE TYPE	REPL CODE
74S132	2134
74S132DC	2134
74S132PC	2134
74S133	OEM
74S133DC	1819
74S133N	OEM
74S133PC	1819
74S134DC	1827
74S134PC	1827
74S135PC	OEM
74S137DC	OEM
74S137PC	OEM
74S138	2138
74S138DC	2138
74S138N	OEM
74S138PC	2138
74S139	OEM
74S139DC	OEM
74S139PC	OEM
74S140DC	1886
74S140PC	1886
74S151	1955
74S151DC	1955
74S151N	OEM
74S151PC	1955
74S153	2151
74S153DC	2151
74S153PC	2151
74S157	1695
74S157DC	1695
74S157PC	1695
74S158	OEM
74S158DC	2154
74S158N	OEM
74S158PC	2154
74S163	0893
74S163N	OEM
74S174	2132
74S174N	2132
74S174NA	2132
74S174PC	2132
74S175	2141
74S175DC	2141
74S175N	2141
74S175PC	2141
74S181DC	2165
74S181PC	2165
74S182	2166
74S182DC	2166
74S182PC	2166
74S189DC	OEM
74S189FC	OEM
74S189PC	OEM
74S194DC	1931
74S194FC	1931
74S194PC	1931
74S195	OEM
74S195A	OEM
74S240	OEM
74S240DC	2475
74S240PC	OEM
74S241	OEM
74S241DC	OEM
74S241N	OEM
74S241PC	OEM
74S251DC	2198
74S253DC	OEM
74S253PC	OEM
74S257DC	OEM
74S257PC	OEM
74S258	OEM
74S258DC	2205
74S258PC	2205
74S280	2220
74S280PC	OEM
74S288	2175
74S289DC	OEM
74S289FC	OEM
74S289PC	OEM
74S299	OEM
74S373	2264
74S374	2266
74SC137C(A)	OEM
74SC137D(A)	OEM
74SC137P(A)	OEM
74SC138C(A)	OEM
74SC138D(A)	OEM
74SC138P(A)	OEM
74SC139C(A)	OEM
74SC139D(A)	OEM
74SC139P(A)	OEM
74SC237C(A)	OEM
74SC237P(A)	OEM
74SC238C(A)	OEM
74SC238D(A)	OEM
74SC239C(A)	OEM
74SC239D(A)	OEM
74SC239P(A)	OEM
74SC240C(MA)	OEM
74SC240D(MA)	OEM
74SC240P(MA)	OEM
74SC241C(MA)	OEM
74SC241D(MA)	OEM
74SC241P(MA)	OEM
74SC242C(MA)	OEM
74SC242D(MA)	OEM
74SC242P(MA)	OEM
74SC243C(MA)	OEM
74SC243D(MA)	OEM
74SC243P(MA)	OEM
74SC244C(MA)	OEM
74SC244D(MA)	OEM
74SC244P(MA)	OEM
74SC540C(MA)	OEM
74SC540D(MA)	OEM
74SC540P(MA)	OEM
74SC541C(MA)	OEM
74SC541D(MA)	OEM
74SC541P(MA)	OEM
74SL258ADC	OEM
74SL266DC	OEM
74SL299PC	OEM
74ST-2SA10150(AU)E	0148
74ST-2SC763D	0127
74ST-2SC785-0	1144
74ST-2SC1906(AP)	0127
74ST-2SD1138C	0392
74T1	OEM
74T2	0236
74T3	OEM
74T4	OEM
74T5	OEM
74T6	OEM
74T7	OEM
74T8	OEM
74T9	OEM
74T10	OEM
74T11	OEM
74T12	OEM
74T13	OEM
74T14	OEM
74T15	OEM
74T20	OEM
74T20-CE00	OEM
74T20-CH00	OEM
74T20-EE00	OEM
74T20-GE00	OEM
74T20-GH00	OEM
74T20-JE00	OEM
74T20-JH00	OEM
74T20-LE00	OEM
74T20-LH00	OEM
74T20-NE00	OEM
74T20-NH00	OEM
74T20-RE00	OEM
74T21	OEM
74T21-CE00	OEM
74T21-CH00	OEM
74T21-EE00	OEM
74T21-GH00	OEM
74T21-JE00	OEM
74T21-JH00	OEM
74T21-LE00	OEM
74T21-LH00	OEM
74T21-NE00	OEM
74T21-NH00	OEM
74T21-RE00	OEM
74T22	OEM
74T22-CE00	OEM
74T22-CH00	OEM
74T22-EE00	OEM
74T22-GE00	OEM
74T22-GH00	OEM
74T22-JE00	OEM
74T22-JH00	OEM
74T22-LE00	OEM
74T22-LH00	OEM
74T22-NE00	OEM
74T22-NH00	OEM
74T22-RE00	OEM
74T23	OEM
74T23-CE00	OEM
74T23-CH00	OEM
74T23-EE00	OEM
74T23-EH00	OEM
74T23-GE00	OEM
74T23-GH00	OEM
74T23-JE00	OEM
74T23-JH00	OEM
74T23-LE00	OEM
74T23-LH00	OEM
74T23-NE00	OEM
74T23-NH00	OEM
74T23-RE00	OEM
74T24	OEM
74T24-CE00	OEM
74T24-CH00	OEM
74T24-EE00	OEM
74T24-EH00	OEM
74T24-GE00	OEM
74T24-GH00	OEM
74T24-JE00	OEM
74T24-JH00	OEM
74T24-LE00	OEM
74T24-LH00	OEM
74T24-NE00	OEM
74T24-NH00	OEM
74T24-RE00	OEM
74T25-CE00	OEM
74T25-CH00	OEM
74T25-EE00	OEM
74T25-EH00	OEM
74T25-GE00	OEM
74T25-GH00	OEM
74T25-JE00	OEM
74T25-LE00	OEM
74T25-LH00	OEM
74T25-NE00	OEM
74T25-NH00	OEM
74TB1-CE00	OEM
74TB1-CH00	OEM
74TB1-EE00	OEM
74TB1-GE00	OEM
74TB1-GH00	OEM
74TB1-JE00	OEM
74TB1-LE00	OEM
74TB1-LH00	OEM
74TB1-NE00	OEM
74TB1-NH00	OEM
74TB1-RE00	OEM
74TB2-CE00	OEM
74TB2-CH00	OEM
74TB2-GE00	OEM
74TB2-JE00	OEM
74TB2-JH00	OEM
74TB2-LE00	OEM
74TB2-LH00	OEM
74TB2-NE00	OEM
74TB2-NH00	OEM
74TB2-RE00	OEM
74TB3-CE00	OEM
74TB3-CH00	OEM
74TB3-EE00	OEM
74TB3-GE00	OEM
74TB3-GH00	OEM
74TB3-JE00	OEM
74TB3-JH00	OEM
74TB3-LE00	OEM
74TB3-LH00	OEM
74TB3-NE00	OEM
74TB3-NH00	OEM
74TB3-RE00	OEM
74TB4-CE00	OEM
74TB4-CH00	OEM
74TB4-EE00	OEM
74TB4-GE00	OEM
74TB4-GH00	OEM
74TB4-JE00	OEM
74TB4-JH00	OEM
74TB4-LE00	OEM
74TB4-LH00	OEM
74TB4-NE00	OEM
74TB4-NH00	OEM
74TB4-RE00	OEM
74TB5-CE00	OEM
74TB5-CH00	OEM
74TB5-EE00	OEM
74TB5-GE00	OEM
74TB5-GH00	OEM
74TB5-JE00	OEM
74TB5-JH00	OEM
74TB5-LE00	OEM
74TB5-LH00	OEM
74TB5-NE00	OEM
74TB5-NH00	OEM
74TB5-RE00	OEM
74TB6-CE00	OEM
74TB6-CH00	OEM
74TB6-EE00	OEM
74TB6-GE00	OEM
74TB6-GH00	OEM
74TB6-JE00	OEM
74TB6-JH00	OEM
74TB6-LE00	OEM
74TB6-NE00	OEM
74TB6-NH00	OEM
74TB6-RE00	OEM
74TB7-CE00	OEM
74TB7-CH00	OEM
74TB7-EE00	OEM
74TB7-GE00	OEM
74TB7-GH00	OEM
74TB7-JE00	OEM
74TB7-JH00	OEM
74TB7-LE00	OEM
74TB7-LH00	OEM
74TB7-NE00	OEM
74TB7-NH00	OEM
74TB7-RE00	OEM
74TB8-CE00	OEM
74TB8-CH00	OEM
74TB8-EE00	OEM
74TB8-GE00	OEM
74TB8-GH00	OEM
74TB8-JE00	OEM
74TB8-JH00	OEM
74TB8-LE00	OEM
74TB8-LH00	OEM
74TB8-NE00	OEM
74TB8-NH00	OEM
74TB8-RE00	OEM
74TB9-CE00	OEM
74TB9-CH00	OEM
74TB9-EE00	OEM
74TB9-GE00	OEM
74TB9-GH00	OEM
74TB9-JE00	OEM
74TB9-JH00	OEM
74TB9-LE00	OEM
74TB9-LH00	OEM
74TB9-NE00	OEM
74TB9-NH00	OEM
74TB9-RE00	OEM
74TB10-CE00	OEM
74TB10-CH00	OEM
74TB10-EE00	OEM
74TB10-GE00	OEM
74TB10-GH00	OEM
74TB10-JE00	OEM
74TB10-JH00	OEM
74TB10-LE00	OEM
74TB10-LH00	OEM
74TB10-NE00	OEM
74TB10-NH00	OEM
74TB10-RE00	OEM
74TB11-CE00	OEM
74TB11-CH00	OEM
74TB11-EE00	OEM
74TB11-GE00	OEM
74TB11-GH00	OEM
74TB11-JE00	OEM
74TB11-JH00	OEM
74TB11-LE00	OEM
74TB11-LH00	OEM
74TB11-NE00	OEM
74TB11-NH00	OEM
74TB11-RE00	OEM
74TB12-CE00	OEM
74TB12-CH00	OEM
74TB12-EE00	OEM
74TB12-GE00	OEM
74TB12-GH00	OEM
74TB12-JE00	OEM
74TB12-JH00	OEM
74TB12-LE00	OEM
74TB12-LH00	OEM
74TB12-NE00	OEM
74TB12-NH00	OEM
74TB12-RE00	OEM
74TB13	OEM
74TB13-CE00	OEM
74TB13-CH00	OEM
74TB13-EE00	OEM
74TB13-GE00	OEM
74TB13-GH00	OEM
74TB13-JE00	OEM
74TB13-JH00	OEM
74TB13-LE00	OEM
74TB13-LH00	OEM
74TB13-NE00	OEM
74TB13-NH00	OEM
74TB13-RE00	OEM
74TB14-CE00	OEM
74TB14-CH00	OEM
74TB14-EE00	OEM
74TB14-GE00	OEM
74TB14-GH00	OEM
74TB14-JE00	OEM
74TB14-JH00	OEM
74TB14-LE00	OEM
74TB14-LH00	OEM
74TB14-NE00	OEM
74TB14-NH00	OEM
74TB14-RE00	OEM
74TB15	OEM
74TB15-CE00	OEM
74TB15-CH00	OEM
74TB15-EE00	OEM
74TB15-GE00	OEM
74TB15-GH00	OEM
74TB15-JE00	OEM
74TB15-JH00	OEM
74TB15-LE00	OEM
74TB15-LH00	OEM
74TB15-NE00	OEM
74TB15-NH00	OEM
74TB15-RE00	OEM
74TB16-CE00	OEM
74TB16-CH00	OEM
74TB16-EE00	OEM
74TB16-GE00	OEM
74TB16-GH00	OEM
74TB16-JE00	OEM
74TB16-JH00	OEM
74TB16-LE00	OEM
74TB16-LH00	OEM
74TB16-NE00	OEM
74TB16-NH00	OEM
74TB16-RE00	OEM
74TB17	OEM
74TB17-CH00	OEM
74TB17-EE00	OEM
74TB17-GE00	OEM
74TB17-GH00	OEM
74TB17-JE00	OEM
74TB17-JH00	OEM
74TB17-LE00	OEM
74TB17-LH00	OEM
74TB17-NE00	OEM
74TB17-NH00	OEM
74TB17-RE00	OEM
74TB18-CE00	OEM
74TB18-CH00	OEM
74TB18-GE00	OEM
74TB18-GH00	OEM
74TB18-JE00	OEM
74TB18-LE00	OEM
74TB18-LH00	OEM
74TB18-NH00	OEM
74TB18-RE00	OEM
74TB19	OEM
74TB19-CE00	OEM
74TB19-CH00	OEM
74TB19-EE00	OEM
74TB19-GE00	OEM
74TB19-GH00	OEM
74TB19-JE00	OEM
74TB19-JH00	OEM
74TB19-LE00	OEM
74TB19-LH00	OEM
74TB19-NE00	OEM
74TB19-NH00	OEM
74TB19-RE00	OEM
74TB20-CE00	OEM
74TB20-CH00	OEM
74TB20-EE00	OEM
74TB20-GE00	OEM
74TB20-GH00	OEM
74TB20-JE00	OEM
74TB20-JH00	OEM
74TB20-LE00	OEM
74TB20-LH00	OEM
74TB20-NE00	OEM
74TB20-NH00	OEM
74TB20-RE00	OEM
74Z4	OEM
74Z6	0827
74Z6A	0827
75	0015
75(HEP)	0493
75-0-010	OEM
75-30	OEM
75-30-01	OEM
75-30-02	OEM
75-31	OEM
75-31-01	OEM
75-31-02	OEM
75-43	OEM
75-100-001	0124
75-461	0004
75A198-11	0065
75B3	OEM
75BV50H	OEM
75BV100H	OEM
75BV150H	OEM
75D01	0447
75D02	0940
75D03	0469
75D04	3283
75D05	3284
75D06	0940
75D1	0015
75D2	0015
75D4	OEM
75D6	OEM
75D8	0071
75D10	0071
75D12	OEM
75D14	OEM
75D16	OEM
75D18	OEM
75D20	OEM
75E03A	OEM
75E1	0015
75E2	0015
75E3	0015
75E4	0015
75E5	0015
75E6	0015
75E7	0071
75E8	0071
75E10	0071
75E12	OEM
75E14	OEM
75E16	OEM
75E18	OEM
75E20	1023
75E211	0491
75F05	0015
75F545DC	OEM
75G6P41	OEM
75HQ030	OEM
75HQ035	OEM
75HQ040	OEM
75HQ045	OEM
75J6P41	OEM
75L6P41	OEM
75N1	0233
75N5	0002
75N5AA	0016
75N5.1	0162
75N5.6	0157
75N6.2	0636
75N12	0137
75N27	0441
75R1B	0015
75R2B	0015
75R3B	0015
75R4B	0015
75R5B	0015
75R6B	0015
75R7B	0071
75R8B	0071
75R9B	0071
75R10B	0071
75R11B	0102
75R12B	0102
75R13B	0102
75R14B	0102
75R15B	OEM
75R16B	OEM
75R18B	OEM
75R20B	OEM
75R22B	OEM
75R24B	OEM
75R26B	OEM
75R28B	OEM
75R30B	OEM
75R32B	OEM
75R34B	OEM
75R36B	OEM
75R38B	OEM
75R40B	OEM
75R42B	OEM
75R44B	OEM
75R46B	OEM
75R48B	OEM
75R50B	OEM
75S107F	OEM
75S107N	OEM
75S108F	OEM
75S108N	OEM
75S207F	OEM
75S207N	OEM
75S208F	OEM
75S208N	OEM
75ST-2SC2229Y	0066
75ST-2SD467C(AU)E	0951
75ST-BC547B	0079
75SZ-HZ6C3	0471
75TE3.9	OEM
75TE3.9A	OEM
75TE4.7	OEM
75TE4.7A	OEM
75TE5.6	OEM
75TE5.6A	OEM
75TE6.8	OEM
75TE6.8A	OEM
75TE8.2	OEM
75TE8.2A	OEM
75TE10	OEM
75TE10A	OEM
75TE12	OEM
75TE12A	OEM
75TE15	OEM
75TE15A	OEM
75TE18	OEM
75TE18A	OEM
75TE22	OEM
75TE22A	OEM
75TE27	OEM
75TE27A	OEM
75TE33	OEM
75TE33A	OEM
75TE39	OEM
75TE39A	OEM
75TE47	OEM
75TE47A	OEM
75TE56	OEM
75TE56A	OEM
75W-005	0106
75W005	0106
75Z05	OEM
75Z4	OEM
75Z6	0846
75Z6A	0846
75Z6.8	0025
75Z6.8A	0025
75Z6.8B	0025
75Z6.8C	OEM
75Z7.5	0649
75Z7.5A	0649
75Z7.5B	0649
75Z7.5C	OEM
75Z8.2	0244
75Z8.2A	0244
75Z8.2B	0244
75Z8.2C	OEM
75Z9.1	0012
75Z9.1A	0012
75Z9.1B	0012
75Z9.1C	0012
75Z10	0170
75Z10A	0170
75Z10B	0170
75Z10C	OEM
75Z11	0794
75Z11A	0794
75Z11B	0794
75Z11C	0794
75Z12	0137
75Z12A	0137
75Z12B	0137
75Z12C	0137
75Z13	0365
75Z13A	0365
75Z13B	0365
75Z13C	0365
75Z14	0100
75Z14A	0100
75Z14B	0100
75Z14C	0100
75Z15	0002
75Z15A	0002
75Z15B	0002
75Z15C	0002
75Z16	0420
75Z16A	0420
75Z16B	0420
75Z16C	OEM
75Z17	1649
75Z18	0495
75Z18A	0495
75Z18B	0495
75Z18C	OEM
75Z19	0949
75Z19A	0949
75Z19B	0949
75Z19C	OEM
75Z20	0531
75Z20A	0531
75Z20B	0531
75Z20C	OEM
75Z22	0565
75Z22A	0565
75Z22B	0565
75Z22C	OEM
75Z24	0402
75Z24A	0402
75Z24B	0402
75Z24C	OEM
75Z25	1606
75Z25A	1606
75Z25B	1606
75Z25C	OEM
75Z27	0441
75Z27A	0441
75Z27B	0441
75Z27C	0441
75Z30	0726
75Z30A	0726
75Z30B	0726
75Z30C	OEM
75Z33	0039
75Z33A	0039
75Z33B	0039
75Z33C	0039
75Z36	0820
75Z36A	0820
75Z36B	0820
75Z36C	OEM
75Z39	0350
75Z39A	0350
75Z39B	0350
75Z39C	OEM
75Z43	0931
75Z43A	0931
75Z43B	0931
75Z43C	OEM
75Z45	OEM
75Z45A	OEM
75Z45B	OEM
75Z45C	OEM
75Z47	0999
75Z47A	0999
75Z47B	0999
75Z47C	OEM
75Z50	OEM
75Z50A	OEM
75Z50B	OEM
75Z50C	OEM
75Z51	0502
75Z51A	0502
75Z51B	0502
75Z51C	OEM
75Z52	OEM
75Z52A2A	OEM
75Z52B	OEM
75Z52C	OEM
75Z56	1834
75Z56A	1834
75Z56B	1834
75Z56C	1834
75Z62	0783
75Z62A	0783
75Z62B	0783
75Z62C	0783
75Z68	2157
75Z68A	2157
75Z68B	2157
75Z68C	OEM
75Z71A	OEM
75Z71B	OEM
75Z71C	OEM
75Z75	1189
75Z75A	1189
75Z75B	1189
75Z75C	OEM
75Z82	0331
75Z82A	0331
75Z82B	0331
75Z82C	0331
75Z91	1311
75Z91A	1311
75Z91B	1311
75Z91C	OEM
75Z100	0098
75Z100A	0098
75Z100B	0098
75Z100C	OEM
75Z105	OEM
75Z105A	OEM
75Z105B	OEM
75Z105C	OEM
75Z110	0149
75Z110A	0149
75Z110B	0149
75Z110C	0149
75Z120	0186
75Z120A	0186
75Z120B	0186
75Z120C	OEM

If replacement code is OEM, contact original manufacturer for replacement.

DEVICE TYPE	REPL CODE
75Z130	0213
75Z130A	0213
75Z130B	0213
75Z130C	OEM
75Z140	0245
75Z140A	0245
75Z140B	0245
75Z140C	OEM
75Z150	0028
75Z150A	0028
75Z150B	0028
75Z150C	OEM
75Z160	0255
75Z160A	0255
75Z160B	0255
75Z160C	OEM
75Z175	OEM
75Z175A	OEM
75Z175B	OEM
75Z175C	OEM
75Z180	0367
75Z180A	0367
75Z180B	0367
75Z180C	OEM
75Z200	0421
75Z200A	0421
75Z200B	0421
75Z200C	OEM
76	0136
76-005-003	1608
76-005-005	1382
76-005-007	0243
76-005-013	0977
76-0105	0178
76-1	0086
76-100-001	1118
76-120-001	0973
76-140-001	2318
76-190-001	2759
76-500-001	1134
76-1400-001	OEM
76-11770	0085
76-13570-37	OEM
76-13570-39	0144
76-13570-59	0144
76-13570-63	OEM
76-13570-65	0917
76-13570-85	0917
76-13848-23	0917
76-13866-17	0144
76-13866-18	0076
76-13866-19	0144
76-13866-19(VHF)	0127
76-13866-20	0144
76-13866-59	0144
76-13866-62	0007
76-14090-1	0086
76-14196-1	0123
76-14327-1	0190
76-14327-2	0190
76-14327-3	0190
76-14327-4	0190
76-14327-5	0190
76-14327-6	0190
76-14327-7	0190
76-14327-8	0190
76-I(SYLVANIA)	0037
76-L(SYLVANIA)	0037
76G1450J	0015
76N1	0016
76N1(REMOTE)	0079
76N1(VID)	0111
76N1B	0086
76N1M	0016
76N2	0016
76N2B	0086
76N3B	0086
76RECT-SI-176	0091
76SD-1N4148(AU)	0124
76SI-HD14011BP	0215
76SK-1N4148(AU)	0124
76SO-1N4148(AU)	0124
76ST-2SC1514-15	1085
76ST-2SC1815GR	0076
76X0076-001	OEM
76X0077-001	OEM
76Z4	0052
76Z6	0868
76Z6A	0868
77	0280
77(DIO)	0795
77-001	0239
77-27166-2	0050
77-270877-2	0160
77-270878-2	0160
77-270993-1	0229
77-271025-1	0004
77-271026-1	0004
77-271027-1	0004
77-271029-1	0050
77-271029-2	0050
77-271031-1	0143
77-271032-1	0143
77-271036-1	0004
77-271037-1	0004
77-271038-1	0050
77-271039-1	0004
77-271166-2	0050
77-271166-3	0050

DEVICE TYPE	REPL CODE
77-271334-1	0015
77-271374-1	0015
77-271453-1	0016
77-271490	0086
77-271490-1	0086
77-271491-1	0160
77-271798-1	0161
77-271798-3	0161
77-271818-1	0126
77-271967-1	0016
77-272913-1	0042
77-272914-1	0925
77-272999-1	0086
77-273001-2	0016
77-273001-3	0050
77-273004-1	0164
77-273715-1	0042
77-273716-1	0925
77-273738-1	0042
77-273739-1	0925
77A	0440
77A0	0440
77A01	0133
77A02	0133
77A1	0440
77A2	0440
77A3	0440
77A4	0440
77A5	0440
77A6	0440
77A7	0440
77A8	0440
77A9	0440
77A10	0440
77A10R	0440
77A11	0133
77A12	0440
77A13	0133
77A13P	0440
77A14	0440
77A14-7	0440
77A14-7B	0440
77A15	0440
77A15L	0440
77A16	0440
77A16-1	0440
77A17	0440
77A17-1	0440
77A18	0440
77A19	0440
77A19G	0440
77A-70	0440
77A-70-12	0440
77A-70-12-7	0440
77A119	0440
77A121	0440
77A182	0440
77B0	0265
77B2C	OEM
77C	0160
77C0	0160
77C1	0160
77C2	0160
77C3	0160
77C4	0160
77C5	0160
77C6	0160
77C7	0160
77C8	0160
77C9	0160
77C10	0160
77C10R	0160
77C11	0160
77C12	0160
77C13	0160
77C13P	0160
77C14	0160
77C14-7	0160
77C14-7B	0160
77C15	0160
77C15L	0160
77C16	0160
77C16-3	0160
77C17	0160
77C17-1	0160
77C18	0160
77C19	0160
77C19G	0160
77C-70	0160
77C-70-12	0160
77C-70-12-7	0160
77C119	0160
77C121	0160
77C182	0160
77C800-005	1823
77C800-007	1041
77C800-008	0141
77C813-002	0081
77N1	0016
77N2	0016
77N2B	0086
77N3	0198
77N4	0016
77N5	0016
77N6	0016
77SS-2N5062	0901
77ST-2SD669AC	0563
77ST-2SD869	0055
77Z4	OEM

DEVICE TYPE	REPL CODE
77Z6	0885
77Z6A	0885
78-001	0061
78-900-001	OEM
78-900-002	0006
78-900-003	0382
78-900-004	4330
78-5009	0160
78-254566-1	0015
78-254566-4	0015
78-271030-1	0229
78-271031-1	0123
78-271032-2	0123
78-271143-1	0015
78-271199-1	0143
78-271228-1	0143
78-271383-1	0137
78-271383-2	0441
78-271383-3	0002
78-271383-4	0002
78-272160-1	0015
78-272212-1	0160
78-273002-1	0143
78-273008	0015
78-273008-1	0015
78-273085	0015
78A200010P4	0232
78BLK	0280
78C01	0144
78C02	0144
78D5R10	OEM
78D9R10GH	OEM
78D56	OEM
78EP	OEM
78GRN	0280
78H05-KC	2863
78H05KC	2863
78H12	OEM
78H05KC	2863
78HV05CDA	1916
78HV05CU	0624
78HV05DA	OEM
78HV06CDA	OEM
78HV06CU	OEM
78HV06DA	OEM
78HV08CDA	OEM
78HV08CU	OEM
78HV08DA	OEM
78HV12CDA	6745
78HV12CU	3485
78HV12DA	OEM
78HV14CDA	OEM
78HV14CU	OEM
78HV14DA	OEM
78HV15CDA	OEM
78HV15CU	OEM
78HV15DA	OEM
78HV18CDA	OEM
78HV18CU	OEM
78HV18DA	OEM
78HV24CDA	OEM
78HV24CU	OEM
78HV24DA	OEM
78L02AA	OEM
78L05	1292
78L05-AN	1292
78L05-AV	1292
78L05A	1292
78L05ACP	1292
78L05AV	1292
78L05AVP	1292
78L05CLP	OEM
78L05H	OEM
78L05J	1292
78L05V	1292
78L06	2301
78L06A	2301
78L06AC	2301
78L06C	2301
78L06CLP	OEM
78L08	0083
78L08A	0083
78L08AC	0083
78L08AWC	0083
78L09	1786
78L09A	1786
78L09CLP	1786
78L09K	OEM
78L09T	OEM
78L010AP	OEM
78L12	1828
78L12A	1828
78L12ACDB	OEM
78L12ACS	OEM
78L12ACZ	1828
78L12CDB	OEM
78L12CLP	OEM
78L12CS	OEM
78L15A	OEM
78L15ACDB	OEM
78L15ACS	OEM
78L15CDB	OEM
78L15CLP	OEM
78L15CS	OEM
78L18CLP	OEM
78L18J	OEM
78L24	OEM
78L24CLP	OEM
78L56	OEM

DEVICE TYPE	REPL CODE
78L62	2301
78L62AC	2301
78L62AWC	2301
78L62WV	2301
78L82AC	0083
78L82AWC	OEM
78L02ACDB	OEM
78L02ACS	OEM
78L02CDB	OEM
78L02CS	OEM
78L05ACDB	OEM
78L05ACS	OEM
78L05CDB	OEM
78L05CS	OEM
78L06ACS	OEM
78L06CS	OEM
78LR05	OEM
78M05	0624
78M05(SM)	OEM
78M05A	0624
78M05C	0624
78M05CDB	OEM
78M05CU	OEM
78M05DB	OEM
78M05E	OEM
78M05HC	OEM
78M05HF	0624
78M06CDB	OEM
78M06CU	OEM
78M06DB	OEM
78M08	1196
78M08CDB	OEM
78M08CU	OEM
78M08DB	OEM
78M09	1786
78M09A	OEM
78M09E	OEM
78M09P	OEM
78M12	0334
78M12A	OEM
78M12CDB	OEM
78M12CU	OEM
78M12DB	OEM
78M12KC	0334
78M13A	OEM
78M15CDB	OEM
78M15CU	OEM
78M15DB	OEM
78M15H	OEM
78M18A	2259
78M18M	OEM
78M20CDB	OEM
78M20CU	OEM
78M20DB	OEM
78M24CDB	OEM
78M24CU	OEM
78M24DB	OEM
78M93D	OEM
78MHV05CDB	OEM
78MHV05CU	OEM
78MHV05DB	OEM
78MHV06CDB	OEM
78MHV06CU	OEM
78MHV06DB	OEM
78MHV08CDB	OEM
78MHV08CU	OEM
78MHV08DB	OEM
78MHV12CDB	OEM
78MHV12DB	OEM
78MHV13CU	OEM
78MHV15CDB	OEM
78MHV15CU	OEM
78MHV15DB	OEM
78MHV18CU	OEM
78MHV20CDB	OEM
78MHV20DB	OEM
78MHV24CDB	OEM
78MHV24CU	OEM
78MHV24DB	OEM
78N1	0016
78N2B	0016
78RED	0280
78SZ-UZP-12B	0137
78YEL	0280
78Z4	OEM
78Z6	0914
78Z6A	0914
79F015	0143
79F114-1	0086
79F114-2A	0126
79F114-3	0086
79F114-4	0126
79F114-4A	0126
79F150	1030
79F150-2	1030
79F153	0474
79L05	4508
79L05AC	4508
79L05CLP	OEM
79L12	1836
79L12CLP	OEM
79L15	OEM
79L15CLP	OEM
79L18CZ	OEM
79LL05	OEM
79LS05	OEM
79M05	1285
79M05A	OEM
79M05C	1285

DEVICE TYPE	REPL CODE
79M05CDB	OEM
79M05CT	1285
79M05CU	OEM
79M05DB	OEM
79M05.2CDB	OEM
79M05.2CU	OEM
79M05.2DB	OEM
79M06CDB	OEM
79M06CU	OEM
79M06DB	OEM
79M08CDB	OEM
79M08CU	OEM
79M08DB	OEM
79M12	1838
79M12A	OEM
79M12C	1838
79M12CDB	OEM
79M12CKC	1838
79M12CU	OEM
79M12DB	OEM
79M15CDB	OEM
79M15CU	OEM
79M15DB	OEM
79M18CDB	OEM
79M18CU	OEM
79M18DB	OEM
79M24CDB	OEM
79M24CU	OEM
79M24DB	OEM
79P1	0004
79SI-UPC393C	0631
79Z4	OEM
79Z6	0926
79Z6A	0926
80	1217
80-001300	0071
80-001400	0071
80-0256-18	OEM
80-050100	0016
80-050300	0208
80-050400	0050
80-050500	0050
80-050600	0208
80-050700	0160
80-051500	0016
80-051600	0126
80-052102	0016
80-052202	0016
80-052302	0016
80-052402	0103
80-052600	0016
80-052700	0126
80-052800	0208
80-053001	0016
80-053300	0086
80-053400	0016
80-053600	0144
80-053702	1514
80-60-1	0143
80-308-2	0016
80-1059	0139
80-4464-04	OEM
80-104900	0136
80-105200	0136
80-105300	0136
80-205400	0004
80-205600	0004
80-236400	0004
80-236430	0004
80-243900	0004
80-318250	0016
80-337390	0016
80-338030	0144
80-338040	0144
80-339440	0144
80-383840	0144
80-383930	0144
80-383940	0016
80-389910	0016
80-389930	0016
80-414120	0103
80-414130	0103
80-421980	0016
80-2226314	0004
80A5	0071
80AS	0071
80B1AM	OEM
80B-US	OEM
80C48C098	OEM
80C51FV568	OEM
80C88	OEM
80C97	3180
80C888-2	OEM
80E-US	OEM
80H	0071
80H3	0206
80H3N	0449
80H3P	3011
80L15	OEM
80L20	6250
80L25	OEM
80L30	OEM
80L35	OEM
80L40	OEM
80L45	OEM
80L50	OEM
80L55	OEM
80L60	OEM
80L65	OEM

DEVICE TYPE	REPL CODE
80L70	OEM
80L75	OEM
80L80	OEM
80L85	OEM
80L90	OEM
80L95	OEM
80L100	OEM
80L110	OEM
80L120	OEM
80L130	OEM
80L140	OEM
80L150	OEM
80L160	OEM
80L170	OEM
80L180	OEM
80L190	OEM
80L200	OEM
80L210	OEM
80L220	OEM
80L230	OEM
80L240	OEM
80L250	OEM
80L260	OEM
80L270	OEM
80L280	OEM
80L290	OEM
80L300	OEM
80N9A	OEM
80N15A	OEM
80N15A9	OEM
80N20A	OEM
80N20A9	OEM
80N25A	OEM
80N25A9	OEM
80N30A	OEM
80N30A9	OEM
80N35A	OEM
80N35A9	OEM
80N40A	OEM
80N40A9	OEM
80N45A	OEM
80N45A9	OEM
80N50A	OEM
80N50A9	OEM
80N55A	OEM
80N55A9	OEM
80N60A	OEM
80N60A9	OEM
80N65A	OEM
80N65A9	OEM
80N70A	OEM
80N70A9	OEM
80N75A	OEM
80N75A9	OEM
80N80A	OEM
80N80A9	OEM
80N85A	OEM
80N85A9	OEM
80N90A9	OEM
80N95A	OEM
80N95A9	OEM
80N100A	OEM
80N100A9	OEM
80N110A	OEM
80N120A	OEM
80N130A	OEM
80N140A	OEM
80N150A	OEM
80N160A	OEM
80N170A	OEM
80N180A	OEM
80N190A	OEM
80N200A	OEM
80N210A	OEM
80N220A	OEM
80N230A	OEM
80N240A	OEM
80N250A	OEM
80P1	0004
80P1,610080-1	0004
80P1SP	OEM
80P2	0126
80P2B	0126
80P2SP	OEM
80P3	0126
80P3B	0126
80P3SP	OEM
80P4SP	OEM
80P5SP	OEM
80P6SP	OEM
80P8SP	OEM
80P10SP	OEM
80P12SP	OEM
80P16SP	OEM
80S0030	OEM
80S2SP	OEM
80S3SP	OEM
80S4SP	OEM
80S5SP	OEM
80S6SP	OEM
80S8SP	OEM
80SI-M51358B	0912
80SI-M51358P	0912
80SI-MAB8021	OEM
80SQ030	OEM
80SQ035	OEM
80SQ040	OEM
80SQ045	OEM
80SS2SP	OEM

DEVICE TYPE	REPL CODE
80SS3SP	OEM
80SS4SP	OEM
80SS5SP	OEM
80SS6SP	OEM
80SS8SP	OEM
80ST-2SC2271E	0261
80ST-2SC2753	1506
80T1	OEM
80T2	0018
80T3	OEM
80T4	OEM
80T5	OEM
80T6	OEM
80T7	OEM
80T8	OEM
80T9	OEM
80T10	OEM
80T11	OEM
80T12	OEM
80T20	OEM
80T20-CE00	OEM
80T20-CH00	OEM
80T20-EE00	OEM
80T20-EH00	OEM
80T21	OEM
80T21-CE00	OEM
80T21-CH00	OEM
80T21-EE00	OEM
80T21-EH00	OEM
80T22	OEM
80T22-CE00	OEM
80T22-CH00	OEM
80T22-EE00	OEM
80T22-EH00	OEM
80T23	OEM
80T23-CE00	OEM
80T23-CH00	OEM
80T23-EE00	OEM
80T23-EH00	OEM
80T24	OEM
80T24-CH00	OEM
80T24-EE00	OEM
80T24-EH00	OEM
80T25-CH00	OEM
80T25-EE00	OEM
80T25-EH00	OEM
80TB1-NE00	OEM
80TB1-NH00	OEM
80TB1-RE00	OEM
80TB2-NE00	OEM
80TB2-NH00	OEM
80TB2-RE00	OEM
80TB3-NE00	OEM
80TB3-NH00	OEM
80TB3-RE00	OEM
80TB4-NE00	OEM
80TB4-NH00	OEM
80TB4-RE00	OEM
80TB5-NE00	OEM
80TB5-NH00	OEM
80TB5-RE00	OEM
80TB6-NE00	OEM
80TB6-NH00	OEM
80TB6-RE00	OEM
80TB7-CE00	OEM
80TB7-CH00	OEM
80TB7-EE00	OEM
80TB7-EH00	OEM
80TB8-CE00	OEM
80TB8-CH00	OEM
80TB8-EE00	OEM
80TB8-EH00	OEM
80TB9-CE00	OEM
80TB9-CH00	OEM
80TB9-EE00	OEM
80TB9-EH00	OEM
80TB10-CE00	OEM
80TB10-CH00	OEM
80TB10-EE00	OEM
80TB10-EH00	OEM
80TB11-CE00	OEM
80TB11-CH00	OEM
80TB11-EE00	OEM
80TB11-EH00	OEM
80TB12-CE00	OEM
80TB12-CH00	OEM
80TB12-EH00	OEM
80TB13	OEM
80TB13-CE00	OEM
80TB13-CH00	OEM
80TB13-EH00	OEM
80TB14-CH00	OEM
80TB14-EH00	OEM
80TB15	OEM
80TB15-CE00	OEM
80TB15-CH00	OEM
80TB15-EE00	OEM
80TB15-EH00	OEM
80TB16-CE00	OEM
80TB16-CH00	OEM
80TB16-EE00	OEM
80TB16-EH00	OEM
80TB17	OEM
80TB17-CE00	OEM
80TB17-CH00	OEM
80TB17-EE00	OEM
80TB17-EH00	OEM
80TB18-CE00	OEM
80TB18-CH00	OEM

If replacement code is OEM, contact original manufacturer for replacement.

DEVICE TYPE	REPL CODE
80TB18-EE00	OEM
80TB18-EH00	OEM
80TB19	OEM
80TB19-CE00	OEM
80TB19-CH00	OEM
80TB19-EE00	OEM
80TB19-EH00	OEM
80TB20-CE00	OEM
80TB20-CH00	OEM
80TB20-EE00	OEM
80TB20-EH00	OEM
80TF3.9	OEM
80TF3.9A	OEM
80TF4.7	OEM
80TF4.7A	OEM
80TF5.6	OEM
80TF5.6A	OEM
80TF6.8	OEM
80TF6.8A	OEM
80TF8.2	OEM
80TF8.2A	OEM
80TF10	OEM
80TF10A	OEM
80TF12	OEM
80TF12A	OEM
80TF15	OEM
80TF15A	OEM
80TF18	OEM
80TF18A	OEM
80TF22	OEM
80TF22A	OEM
80TF27	OEM
80TF27A	OEM
80TF33	OEM
80TF33A	OEM
80TF39	OEM
80TF39A	OEM
80TF47	OEM
80TF47A	OEM
80TF56	OEM
80TF56A	OEM
80V10A	OEM
80V11	OEM
80X8-8.5	OEM
80X8.5-9	OEM
80X9-9.5	OEM
80X9.5-10	OEM
80Z4	OEM
80Z6	0944
80Z6A	0944
81-1007	OEM
81-23860400-3	0551
81-23860400A	0551
81-23860400B	0551
81-27123100-3	0015
81-27123150-8	0143
81-27123300-3	0015
81-27123307-3	0015
81-27125140-7	0016
81-27125140-7A	0016
81-27125140-7B	0016
81-27125160-5	0016
81-27125160-5A	0016
81-27125160-5B	0016
81-27125270-2	0016
81-27125270-2A	0016
81-27125270-2B	0016
81-27125300-7	0016
81-27125530-9	0551
81-27125530-9A	0551
81-27125530-9B	0551
81-27126100-0	0551
81-27126130-7	0160
81-27126130-7A	0160
81-27126130-7B	0160
81-46123001-3	0143
81-46123002-1	0030
81-46123004-5	0015
81-46123004-7	0015
81-46123005	0015
81-46123006-2	0143
81-46123010-4	2270
81-46123013-8	0143
81-46123014-6	0015
81-46123015-3	0143
81-46123018-7	0015
81-46123029-4	0143
81-46125001-1	0136
81-46125002-9	0004
81-46125003-7	0004
81-46125004-5	0004
81-46125005-2	0164
81-46125006-0	0144
81-46125007-8	0127
81-46125009-4	0004
81-46125010-2	0004
81-46125011-0	0004
81-46125012-8	0127
81-46125013-6	0127
81-46125016-9	0016
81-46125018-5	0164
81-46125019-3	0016
81-46125026-8	0016
81-46125027-6	0016
81-46125028-4	0160
81-46125029-2	0004
81-46125030-0	0127
81-46125032-6	0127
81-46125033-4	0127
81-46125034-2	0004
81-46125036-7	0004
81-46125065-6	0325
81-46128001-8	3263
81-46128002-6	2305
81-46143002-7	0120
81A4-1	OEM
81A5-2	OEM
81BUS	OEM
81F1FM	OEM
81F1FMA	OEM
81F2FM	OEM
81F2FMA	OEM
81F3FM	OEM
81F3FMA	OEM
81P3	0126
81R1FM	OEM
81R1FMA	OEM
81R2FM	OEM
81R2FMA	OEM
81R3FM	OEM
81R3FMA	OEM
81RK10	OEM
81RK10M	OEM
81RK20	OEM
81RK20M	OEM
81RK40	OEM
81RK40M	OEM
81RK60	OEM
81RK60M	OEM
81RK80	OEM
81RK80M	OEM
81RK100	OEM
81RK120	OEM
81RK120M	OEM
81RK140	OEM
81RK140M	OEM
81RK160	OEM
81RK160M	OEM
81RL10	0526
81RL20	0526
81RL30	0526
81RL40	0526
81RL50	0526
81RL60	0526
81RL80	0108
81RL100	0108
81RL110	0108
81RL120	0108
81RLA50	0610
81RLA60	0610
81RLA80	0468
81RLA100	0468
81RLA120	0468
81RLA140	OEM
81RLA160	OEM
81RLB50	0526
81RLB60	0526
81RLB80	0108
81RLB100	0108
81RLB110	0108
81RLB120	0108
81RM10	0526
81RM30	0526
81RM40	0526
81RM50	0526
81RM60	0526
81RM70	0108
81RM80	0108
81RM90	0108
81RM100	0108
81SD-FV1043	0554
81ST-2SA10150(AU)B	0148
81ST-2SA1050(AU)B	0148
81ST-2SB561C	0436
81SV-KB262	0120
81T1	OEM
81T2	0086
81T3	OEM
81T4	OEM
81T5	OEM
81T6	OEM
81T7	OEM
81T8	OEM
81T9	OEM
81T10	OEM
81T11	OEM
81T12	OEM
81T20	OEM
81T20-NE00	OEM
81T20-NH00	OEM
81T20-RE00	OEM
81T21	OEM
81T21-NE00	OEM
81T21-NH00	OEM
81T22	OEM
81T22-NE00	OEM
81T22-NH00	OEM
81T22-RE00	OEM
81T23	OEM
81T23-NE00	OEM
81T23-NH00	OEM
81T23-RE00	OEM
81T24	OEM
81T24-NE00	OEM
81T24-NH00	OEM
81T24-RE00	OEM
81T25-NE00	OEM
81T25-NH00	OEM
81T25-RE00	OEM
81TB1-CE00	OEM
81TB1-CH00	OEM
81TB1-EE00	OEM
81TB1-LE00	OEM
81TB1-LH00	OEM
81TB1-NE00	OEM
81TB2-CE00	OEM
81TB2-CH00	OEM
81TB2-EE00	OEM
81TB2-LE00	OEM
81TB2-LH00	OEM
81TB2-NE00	OEM
81TB3-CE00	OEM
81TB3-CH00	OEM
81TB3-EE00	OEM
81TB3-LE00	OEM
81TB3-LH00	OEM
81TB3-NE00	OEM
81TB4-CE00	OEM
81TB4-CH00	OEM
81TB4-EE00	OEM
81TB4-LE00	OEM
81TB4-LH00	OEM
81TB4-NE00	OEM
81TB5-CE00	OEM
81TB5-CH00	OEM
81TB5-EE00	OEM
81TB5-LE00	OEM
81TB5-LH00	OEM
81TB5-NE00	OEM
81TB6-CE00	OEM
81TB6-CH00	OEM
81TB6-EE00	OEM
81TB6-LE00	OEM
81TB6-LH00	OEM
81TB6-NE00	OEM
81TB7	OEM
81TB7-CE00	OEM
81TB7-CH00	OEM
81TB7-EE00	OEM
81TB7-LE00	OEM
81TB7-LH00	OEM
81TB7-NE00	OEM
81TB8-CE00	OEM
81TB8-CH00	OEM
81TB8-EE00	OEM
81TB8-LE00	OEM
81TB8-LH00	OEM
81TB8-NE00	OEM
81TB9	OEM
81TB9-CE00	OEM
81TB9-CH00	OEM
81TB9-LE00	OEM
81TB9-LH00	OEM
81TB9-NE00	OEM
81TB10-NH00	OEM
81TB10-RE00	OEM
81TB11	OEM
81TB11-NH00	OEM
81TB11-RE00	OEM
81TB12-NH00	OEM
81TB12-RE00	OEM
81TB13	OEM
81TB13-NH00	OEM
81TB13-RE00	OEM
81TB14-NH00	OEM
81TB14-RE00	OEM
81TB15	OEM
81TB15-NH00	OEM
81TB15-RE00	OEM
81TB16-NH00	OEM
81TB16-RE00	OEM
81TB17	OEM
81TB17-NH00	OEM
81TB17-RE00	OEM
81TB18-NH00	OEM
81TB18-RE00	OEM
81TB19	OEM
81TB19-NH00	OEM
81TB19-RE00	OEM
81TB20-NE00	OEM
81TB20-NH00	OEM
81TB20-RE00	OEM
81X0042-100	1657
81Z6	0958
81Z6A	0958
82	0049
82-2	OEM
82-3	OEM
82-4	0015
82-409501	0016
82-410300	0126
82A205	OEM
82HM181CJG	OEM
82HM181CPG	OEM
82HM181MJG	OEM
82HM191CJG	OEM
82HM191CPG	OEM
82HM191MJG	OEM
82M432B2	1554
82N1	0127
82P3	OEM
82RECT-SI-154(AU)	0023
82RL10	0655
82RL20	0655
82RL30	0655
82RL40	0655
82RL50	0655
82RL60	0655
82RL80	0020
82RL100	0020
82RL110	0020
82RL120	0020
82RLA50	0217
82RLA60	0217
82RLA80	0658
82RLA100	0658
82RLA110	0658
82RLA120	0658
82RLA130	OEM
82RLA140	OEM
82RLA150	OEM
82RLA160	OEM
82RLB50	0655
82RLB60	0655
82RLB80	0655
82RLB100	0020
82RLB110	0020
82RLB120	0020
82RM10	0655
82RM20	0655
82RM30	0655
82RM40	0655
82RM50	0655
82RM60	0655
82RM70	OEM
82RM80	0020
82RM90	OEM
82RM100	0020
82S100	OEM
82S100N	OEM
82S100N/7700	OEM
82S123	2175
82S210F	OEM
82S210N	OEM
82SP-LT-211	OEM
82ST-2SC288A	1342
82T2	0042
82Z6	OEM
82Z6A	OEM
83	0126
83-1	0037
83-2	0102
83-11	0379
83-28	2041
83-43-3	0774
83-829	0015
83-880	0015
83-1056	0160
83A30-1	0015
83A40-1	1303
83B38-1	0143
83C6V8	0025
83C30-1	0015
83C40-1	1303
83D5R11	OEM
83D9R11GH	OEM
83N52	0440
83P1	0037
83P1A	0037
83P1B	0037
83P1BC	0037
83P1M	0037
83P1MC	0037
83P2	0037
83P2A	0037
83P2AA	0037
83P2AA1	0037
83P2B	0126
83P2M	0037
83P2M1	0037
83P2N	0037
83P3	0037
83P3A	0037
83P3AA	0037
83P3AA1	0037
83P3B	0037
83P3B1	0037
83P3M	0037
83P3M1	0037
83P4	0037
83X3202	OEM
83Z6	1009
83Z6A	1009
84	0160
84-44-3	0774
84A	0160
84A0	0160
84A1	0160
84A2	0160
84A3	0160
84A4	0160
84A5	0160
84A6	0160
84A7	0160
84A8	0160
84A9	0160
84A10	0160
84A10R	0160
84A12	0160
84A13	0160
84A13P	0160
84A14	0160
84A14-7	0160
84A14-7B	0160
84A15	0160
84A15L	0160
84A16	0160
84A16B	0160
84A17	0160
84A17-1	0160
84A18	0160
84A19	0160
84A19G	0160
84A20	0133
84A-70	0160
84A-70-12	0160
84A-70-12-7	0160
84A121	0160
84A182	0160
84AA1	0160
84AA19	0160
84B	0160
84C01	0276
84D5R11	OEM
84D9R11GH	OEM
84G01	0276
84S157PC	1695
84SZ-BZV46-1VR	OEM
84SZ-RD12EB3	0052
84TB1-JH00	OEM
84TB1-LE00	OEM
84TB1-LH00	OEM
84TB1-NE00	OEM
84TB1-NH00	OEM
84TB2-CE00	OEM
84TB2-CH00	OEM
84TB2-EE00	OEM
84TB2-EH00	OEM
84TB2-LH00	OEM
84TB2-NE00	OEM
84TB2-NH00	OEM
84TB3-CE00	OEM
84TB3-CH00	OEM
84TB3-EH00	OEM
84TB3-LH00	OEM
84TB3-NE00	OEM
84TB3-NH00	OEM
84TB4-CE00	OEM
84TB4-CH00	OEM
84TB4-EE00	OEM
84TB4-EH00	OEM
84TB4-LH00	OEM
84TB4-NE00	OEM
84TB4-NH00	OEM
84TB5-CE00	OEM
84TB5-CH00	OEM
84TB5-EE00	OEM
84TB5-EH00	OEM
84TB5-LH00	OEM
84TB5-NE00	OEM
84TB5-NH00	OEM
84TB6-CE00	OEM
84TB6-CH00	OEM
84TB6-EE00	OEM
84TB6-EH00	OEM
84TB6-LH00	OEM
84TB6-NE00	OEM
84TB6-NH00	OEM
84TB7	OEM
84TB7-CE00	OEM
84TB7-CH00	OEM
84TB7-EE00	OEM
84TB7-EH00	OEM
84TB7-LH00	OEM
84TB7-NE00	OEM
84TB8-CE00	OEM
84TB8-CH00	OEM
84TB8-EE00	OEM
84TB8-EH00	OEM
84TB8-LH00	OEM
84TB8-NE00	OEM
84TB9	OEM
84TB9-CE00	OEM
84TB9-CH00	OEM
84TB9-EE00	OEM
84TB9-EH00	OEM
84TB9-LH00	OEM
84TB9-NE00	OEM
84TB10-CE00	OEM
84TB10-CH00	OEM
84TB10-EE00	OEM
84TB10-EH00	OEM
84TB10-LH00	OEM
84TB10-NE00	OEM
84TB11	OEM
84TB11-CE00	OEM
84TB11-CH00	OEM
84TB11-EE00	OEM
84TB11-LH00	OEM
84TB11-NE00	OEM
84TB12-CE00	OEM
84TB12-CH00	OEM
84TB12-EE00	OEM
84TB12-LH00	OEM
84TB12-NE00	OEM
84TB13	OEM
84TB13-CE00	OEM
84TB13-CH00	OEM
84TB13-EE00	OEM
84TB13-LH00	OEM
84TB13-NE00	OEM
84TB14-CE00	OEM
84TB14-CH00	OEM
84TB14-EE00	OEM
84TB14-LH00	OEM
84TB14-NE00	OEM
84TB15-CE00	OEM
84TB15-CH00	OEM
84TB15-EE00	OEM
84TB15-LH00	OEM
84TB15-NE00	OEM
84Z6	OEM
84Z6A	OEM
85	0609
85-5	0015
85-370-2-BLU	0160
85-557-2	OEM
85A500161	0415
85D5R11	OEM
85D9R11GH	OEM
85HA40	OEM
85HA60	OEM
85HA80	OEM
85HA100	OEM
85HA120	OEM
85HA140	OEM
85HA160	OEM
85HF10	2778
85HF20	2778
85HF30	OEM
85HF30A	OEM
85HF40	2813
85HF60	2813
85HF60A	OEM
85HF80	2845
85HF100	2845
85HF100A	OEM
85HF120	2643
85HF140	OEM
85HF160	OEM
85HFL10S02	OEM
85HFL10S05	OEM
85HFL10S10	OEM
85HFL20S02	OEM
85HFL20S05	OEM
85HFL20S10	OEM
85HFL40S02	OEM
85HFL40S05	OEM
85HFL40S10	OEM
85HFL60S02	OEM
85HFL60S05	OEM
85HFL60S10	OEM
85HFL80S02	OEM
85HFL80S05	OEM
85HFL80S10	OEM
85HFL100S05	OEM
85HFL100S10	OEM
85HFR10	1745
85HFR20	1745
85HFR30	OEM
85HFR30A	OEM
85HFR40	1783
85HFR50	OEM
85HFR60	1783
85HFR60A	OEM
85HFR80	1818
85HFR100	1818
85HFR100A	OEM
85HFR120	5325
85HFR140	OEM
85HFR160	OEM
85HQ030	OEM
85HQ035	OEM
85HQ040	OEM
85HQ045	OEM
85P1	OEM
85ST-2SA733Q	0006
85ST-2SC458C(AU)E	0076
85Z6	0889
85Z6A	0889
86	0023
86-0001	0030
86-0002	0143
86-0005	0012
86-0006	0015
86-0007	1080
86-0007-004	0016
86-0008	0019
86-0016-01	0015
86-0022-001	0016
86-0029-001	0016
86-0031-001	0016
86-0033-007	0822
86-0036-001	0037
86-0046-001	OEM
86-0509	0030
86-0510	0062
86-0511	0015
86-0513	0123
86-0514	0030
86-0515	0139
86-0516	0080
86-0518	0005
86-0519	0133
86-0520	0139
86-0521	0052
86-0522	0133
86-1-3	0015
86-3-1	0484
86-3-3	0015
86-4-1	0015
86-4-2	0600
86-5-2	0208
86-6-2	0208
86-7-1	0015
86-8-2	0160
86-9-1	0196
86-9-3	0015
86-10-01	0143
86-10-1	0143
86-10-2	0600
86-11-2	0600
86-12-1	0917
86-12-2	0600
86-12-3	0917
86-13-2	0208
86-14-1	0123
86-14-2	0208
86-15-1	0123
86-16-2	0004
86-18-1	0196
86-18-1A	0196
86-18-2	0050
86-19-2	0160
86-20-1	0143
86-20-2	0050
86-21-1	0015
86-21-2	0004
86-22-1	0143
86-22-2	0004
86-22-3	0015
86-23-1	0920
86-23-2	0004
86-23-3	0015
86-24-2	0208
86-25-1	0137
86-25-2	0208
86-26-1	0015
86-26-2	0600
86-27-1	0133
86-27-2	0004
86-28-2	0004
86-28-3	0015
86-29-2	0004
86-30-1	0015
86-30-2	0004
86-30-3	0015
86-31-1	0137
86-31-2	0600
86-32-1	0015
86-32-2	0004
86-33-2	0004
86-34-3	0015
86-35-1	0012
86-35-2	0208
86-35-3	0015
86-36-1	0015
86-36-2	0050
86-37-1	6564
86-37-2	0050
86-37-3	0474
86-37-18	OEM
86-38-1	0015
86-38-2	0050
86-39-1	OEM
86-39-2	0004
86-40-3	0015
86-41-1	1149
86-42-3	0015
86-43-1	0484
86-43-3	0015
86-44-2	0600
86-44-3	6811
86-45-1	0143
86-45-2	0004
86-45-3	1303
86-46-2	0004
86-46-3	0015
86-47-2	0004
86-47-3	0474
86-48-1	0123
86-48-2	0004
86-49-1	0143
86-49-2	0004
86-49-3	0015
86-50-2	0004
86-50-3	0015
86-51-2	0015
86-51-3	0015
86-52-1	0015
86-52-9	OEM
86-53-3	0133
86-54-2	0004
86-54-3	0015
86-55-3	0474
86-56-3	0474
86-56-3B	0474
86-56-3C	0474
86-56-3F	0474
86-57-3	0015
86-58-1	0012
86-58-2	0016
86-58-3	0015
86-59-1	0015
86-59-2	0004
86-59-3	0015
86-60-1	0143
86-60-2	0004
86-60-3	0015
86-61-1	0137
86-61-2	0004
86-62-1	0133
86-62-2	0160

DEVICE TYPE	REPL CODE
86-62-3	0015
86-63-2	0160
86-63-3	0015
86-64-1	0143
86-65-1	0137
86-65-3	0015
86-65-4	0133
86-65-9	0015
86-66-1	0531
86-67-0	0015
86-67-1	0365
86-67-2	0071
86-67-3	0023
86-67-8	0015
86-67-9	0015
86-68-3	6976
86-68-3A	0277
86-70-1	0133
86-72-2	0004
86-72-3	0071
86-73-1	0531
86-73-2	0004
86-74-1	0133
86-74-2	0004
86-74-9	0133
86-75-1	0402
86-75-2	0004
86-75-3	0015
86-75-3(SEARS)	0015
86-76-1	0133
86-76-2	0208
86-77-1	0133
86-77-2	0004
86-78-1	0133
86-78-2	0004
86-78-3	0015
86-79-2	0004
86-80-1	0015
86-80-2	0004
86-80-3	0015
86-81-2	0208
86-82-2	0004
86-83-2	0004
86-84-1	0015
86-84-2	0004
86-85-1	0137
86-86-2	0050
86-87-2	0050
86-87-3	0030
86-88-1	0531
86-88-2	0050
86-88-3	0143
86-89-1	0015
86-89-2	0050
86-89-3	0015
86-90-2	0050
86-91-1	0402
86-91-2	0050
86-92-1	0137
86-92-2	0050
86-93-1	0420
86-93-2	0004
86-93-3	OEM
86-94-1	0039
86-95-1	0274
86-95-2	0004
86-96-1	0605
86-96-3	0474
86-97-1	0196
86-98-2	0004
86-99-2	0050
86-100-2	0050
86-101-2	0050
86-102-2	0050
86-102-3	0344
86-103-2	0004
86-103-3	0863
86-103-9	3404
86-104-3	0182
86-105-3	0102
86-106-1	1501
86-106-2	1501
86-106-3	1501
86-106-3-1	1501
86-106-3-2	1501
86-106-3-3	1501
86-106-3-4	1501
86-106-3-5	1501
86-106-4	1501
86-106-5	1501
86-107-2	0050
86-108-2	0136
86-109-1	0451
86-109-2	0136
86-109-3	0572
86-109-5	OEM
86-110-01	OEM
86-110-1	3465
86-110-2	0016
86-110-3	0901
86-111-1	0565
86-111-2	0136
86-111-3	0015
86-112-2	0050
86-113-1	0402
86-113-2	OEM
86-113-3	3402
86-114-1	0244
86-114-2	0004

DEVICE TYPE	REPL CODE
86-114-3	1323
86-115-2	0004
86-115-3	OEM
86-116-1	0030
86-116-2	0136
86-116-3	0023
86-117-1	1606
86-117-2	0050
86-117-3	0102
86-118-1	0170
86-118-3	0102
86-119-1	0820
86-119-2	0016
86-120-2	0160
86-121-2	0086
86-123-2	0016
86-125-1	0143
86-126-2	0004
86-126-3	0564
86-127-2	0160
86-127-3	2546
86-128-2	0004
86-128-3	0564
86-129-2	0004
86-130-2	0004
86-131-2	0004
86-132-2	0004
86-133-2	0004
86-135-1	0244
86-135-2	0050
86-136-2	0050
86-138-2	0144
86-138-3	0127
86-139-1	0531
86-139-2	0016
86-139-3	0015
86-141-1	OEM
86-141-2	0160
86-142-2	0160
86-143-2	0016
86-144-2	0016
86-145-1	0649
86-146-1	0143
86-146-2	0160
86-146-3	0015
86-147-1	0015
86-147-2	0160
86-147-3	0015
86-148-1	0256
86-149-1	0999
86-149-2	0050
86-150-1	OEM
86-150-2	0050
86-150-3	0098
86-151-2	0050
86-152-1	OEM
86-152-2	0004
86-153-1	OEM
86-155-2	0016
86-156-2	0004
86-156-2A	0211
86-157-2	0079
86-158-2	0016
86-159-2	0004
86-160-2	0126
86-161-2	0086
86-162-2	0050
86-163-2	0050
86-164-2	0050
86-164-9	OEM
86-165-2	0276
86-166-2	0016
86-169-2	0004
86-170-2	0086
86-171-2	0016
86-172-2	0004
86-173-2	0160
86-173-9	0160
86-175-2	0016
86-176-2	0004
86-177-2	0276
86-178-2	0037
86-178-20	0037
86-179-2	0050
86-180-2	0050
86-181-2	0050
86-182-2	0016
86-183-2	0037
86-183-20	0037
86-185-2	0007
86-186-2	0007
86-188-2	0016
86-189-2	0016
86-191-2	0016
86-192-2	0016
86-194-2	0016
86-195-2	0016
86-196-2	0016
86-198-2	0016
86-199-2	0016
86-201-2	0016
86-202-2	0016
86-204-2	0007
86-205-2	0007
86-207-2	0086
86-208-2	0086
86-210-2	0086
86-211-2	0086
86-213-2	0233

DEVICE TYPE	REPL CODE
86-214-2	0233
86-215-2	0233
86-217-2	0037
86-217-20	0037
86-218-2	0037
86-218-20	0037
86-219-2	0037
86-221-2	0642
86-222-2	0642
86-224-2	0642
86-224-9	OEM
86-225-2	0642
86-227-2	0142
86-228-2	0276
86-230-2	0160
86-231-2	0160
86-232-2	0160
86-233-2	0037
86-234-2	0086
86-235-2	0160
86-236-2	0142
86-237-2	0016
86-238-2	0016
86-243-2	0144
86-244-2	0144
86-245-2	0144
86-246-2	0037
86-246-20	0037
86-247-2	0016
86-248-2	0160
86-249-2	0004
86-249-9	0142
86-250-2	0016
86-251-2	0037
86-251-20	0037
86-253-2	0050
86-254-2	0050
86-255-2	0016
86-256-2	0016
86-257-2	0276
86-259-2	0142
86-260-2	0142
86-261-2	0142
86-262-2	0007
86-262-9	0065
86-263-2	0007
86-264-2	0016
86-265-2	0016
86-266-2	0086
86-267-2	0086
86-271-2	0178
86-272-2	0178
86-273-2	0086
86-275-2	0142
86-276-2	0037
86-276-20	0037
86-277-2	0016
86-278-2	0050
86-279-2	0050
86-280-2	0136
86-281-2	0136
86-282-2	0136
86-283-2	0004
86-284-2	0551
86-286-2	0037
86-286-20	0037
86-287-2	0276
86-287-9	OEM
86-289-2	0007
86-290-2	0007
86-291-2	0086
86-291-7	OEM
86-291-9	0016
86-292-2	0975
86-293-2	0016
86-294-2	0037
86-294-20	0037
86-295-2	0004
86-296-2	0050
86-297-2	0004
86-298-2	0037
86-298-20	0037
86-300-2	0004
86-301-2	0208
86-303-2	0004
86-304-2	0004
86-305-2	0004
86-306-2	0086
86-308-2	0016
86-309-2	0016
86-310-2	0016
86-311-2	0136
86-312-2	0050
86-313-2	0160
86-316-2	0233
86-317-2	0160
86-319-2	0160
86-320-2	0050
86-321-2	0050
86-322-2	0050
86-323-2	0016
86-324-2	0016
86-327-2	0016
86-328-2	0016
86-329-2	0126
86-330-2	0086
86-334-2	0126
86-336-2	0086
86-339-2	0016

DEVICE TYPE	REPL CODE
86-339-9	0016
86-340-2	0037
86-340-20	0037
86-342-2	0016
86-344-2	0161
86-344-9	OEM
86-347-2	0050
86-347-7	0050
86-348-2	0050
86-353-2	0160
86-354-2	0160
86-359-2	0016
86-362-2	0016
86-363-2	0050
86-365-2	0016
86-366-2	0050
86-367-2	0050
86-368-2	0050
86-370-2	0604
86-370-2YEL	0160
86-373-2	0050
86-374-2	0050
86-376-2	0050
86-379-2	0016
86-381-2	0224
86-386-2	0144
86-389-2	0198
86-390-2	0016
86-391-2	0016
86-392-2	0004
86-393-2	0086
86-396-2	0460
86-396-9	OEM
86-399-2	0016
86-399-9	0016
86-400-2	1221
86-403-2	0016
86-406-2	0037
86-407-2	0037
86-412-2	0178
86-416-2	0144
86-417-2	0144
86-419-2	0004
86-420-2	0016
86-421-2	0004
86-422-2	0018
86-422-3	0551
86-423-2	0532
86-423-3	0382
86-428-2	0086
86-428-9	0086
86-431-2	0126
86-431-9	0126
86-440-2	0086
86-441-2	0086
86-442-2	0007
86-444-2	0111
86-445-2	0016
86-449-2	0136
86-449-9	0050
86-452-2	0086
86-455-3	0363
86-457-2	0016
86-458-2	0016
86-459-2	0037
86-460-2	0016
86-461-2	0016
86-462-2	0016
86-463-2	3426
86-463-3	0086
86-464-2	0212
86-465-2	0079
86-467-2	0144
86-472-2	0016
86-475-2	0037
86-476-2	0004
86-477-2	0325
86-479-2	OEM
86-480-9	0222
86-481-1	0016
86-481-2	0016
86-482-2	0037
86-483-2	0016
86-483-3	0016
86-484-2	0016
86-485-2	0016
86-486-2	0016
86-486-9	OEM
86-487-2	0142
86-487-3	0142
86-488-2	0144
86-490-2	0144
86-491-2	0144
86-493-2	0016
86-494-2	0016
86-495-2	0016
86-496-2	0016
86-497-2	0004
86-500-2	0325
86-501-2	0037
86-502-2	0016
86-506-2	0161
86-507-2	0460
86-509-2	0004
86-510-2	0086
86-511-9	0047
86-512-2	0440
86-513-2	0330
86-515-2	0861

DEVICE TYPE	REPL CODE
86-515-2(SEARS)	0016
86-520-2	0016
86-521-2	0861
86-525-2	0127
86-526-2	0111
86-527-2	0037
86-528-2	0037
86-529-2	0042
86-530-2	0925
86-533-2	0006
86-534-2	0016
86-536-2	0079
86-537-2	0079
86-538-2	0079
86-540-2	0353
86-541-2	0400
86-547-2	0037
86-549-2	0400
86-550-2	0861
86-551-2	0016
86-552-2	0037
86-554-2	0016
86-555-2	0006
86-556-2	0285
86-557-2	0861
86-558-2	0400
86-561-2	0016
86-563-2	0065
86-563-9	0065
86-564-2	0363
86-564-3	0065
86-564-9	0065
86-565-1	OEM
86-565-2	0016
86-566-2	0861
86-567-2	0086
86-568-2	0423
86-568-9	0042
86-569-2	OEM
86-570-2	0006
86-570-3	0006
86-572-2	0316
86-572-3	0316
86-573-2	0016
86-574-2	0133
86-590-2	6830
86-593-2	0224
86-593-9	1144
86-594-2	0224
86-595-2	0016
86-596-2	0127
86-597-2	0127
86-598-2	0016
86-599-2	0016
86-600-2	0037
86-601-2	0595
86-601-9	OEM
86-602-2	0791
86-602-9	OEM
86-604-2	0042
86-605-2	0212
86-606-2	0212
86-607-02	0076
86-607-2	0191
86-608-2	0693
86-609-2	0218
86-609-9	0324
86-610-2	0436
86-610-9	0006
86-611-2	0861
86-612-2	0233
86-613-2	0236
86-614-2	0681
86-615-2	0861
86-619-2	0224
86-620-2	0224
86-621-2	0224
86-622-2	0037
86-624-2	0142
86-624-9	0142
86-625-2	0212
86-626-2	0313
86-626-9	0065
86-628-2	0285
86-628-9	6052
86-629-2	0233
86-629-9	OEM
86-630-2	0861
86-631-2	0285
86-632-2	0212
86-633-2	0065
86-633-9	0313
86-646-2	0016
86-648-2	0168
86-649-2	0861
86-650-2	0086
86-651-2	0715
86-653-2	0285
86-655-2	0198
86-659-2	0532
86-660-2	0551
86-661-2	0016
86-663-2	0042
86-664-2	0285
86-664-9	3016
86-665-2	0103
86-665-9	0103
86-668-2	2136
86-668-9	OEM

DEVICE TYPE	REPL CODE
86-669-2	0037
86-670-2	OEM
86-671-2	2235
86-672-2	0285
86-673-2	0566
86-674-2	0086
86-675-2	0086
86-702-2	0016
86-801-2	OEM
86-802-2	OEM
86-803-2	OEM
86-804-9	1284
86-805-2	OEM
86-807-2	OEM
86-809-2	OEM
86-810-9	1284
86-900-2	OEM
86-901-2	OEM
86-902-2	OEM
86-903-2	OEM
86-904-2	OEM
86-905-2	OEM
86-906-2	OEM
86-907-2	OEM
86-909-2	OEM
86-910-2	OEM
86-911-2	OEM
86-912-2	OEM
86-914-2	OEM
86-916-2	OEM
86-918-2	OEM
86-920-2	OEM
86-922-2	OEM
86-923-2	OEM
86-925-2	OEM
86-926-2	OEM
86-927-2	OEM
86-928-2	OEM
86-929-2	OEM
86-931-2	OEM
86-932-2	OEM
86-933-2	OEM
86-934-2	OEM
86-936-2	OEM
86-937-2	OEM
86-1392	0016
86-5000-2	0004
86-5000-3	0015
86-5001-2	0004
86-5001-3	0015
86-5002-3	0015
86-5003-2	0208
86-5003-3	0015
86-5004-2	0208
86-5005-2	0208
86-5006-2	0004
86-5006-3	0015
86-5007-2	0208
86-5007-3	0143
86-5008-2	0208
86-5009-2	0623
86-5009-3	0015
86-5010-3	0133
86-5011-2	0208
86-5011-3	0133
86-5012-2	0208
86-5012-3	0015
86-5013-2	0208
86-5015-2	0208
86-5015-3	0133
86-5016-2	0208
86-5017-2	0208
86-5018-2	0016
86-5024-3	0015
86-5026-2	0208
86-5027-2	0004
86-5027-3	0133
86-5028-3	0242
86-5029-2	0208
86-5029-3	0015
86-5032-3	0242
86-5034-2	0208
86-5037-3	0133
86-5039-2	0160
86-5040-2	0016
86-5041-2	0016
86-5042-2	0004
86-5043-2	0160
86-5044-2	0016
86-5045-2	0016
86-5046-2	0016
86-5047-2	0208
86-5048-2	0208
86-5049-2	0016
86-5050-2	0016
86-5051-2	0016
86-5052-2	0164
86-5055-2	0016
86-5056-2	0016
86-5057-2	0160
86-5058-2	0160
86-5060-2	0208
86-5061-2	0208
86-5062-2	0208
86-5063-2	0004
86-5064-2	0126
86-5064-2A	0126
86-5065-2	0198
86-5067-2	0004

DEVICE TYPE	REPL CODE
86-5070-2	0086
86-5073-2	0086
86-5074-2	0086
86-5075-2	0595
86-5079-2	0037
86-5080-2	0208
86-5081-2	0016
86-5082-2	0126
86-5082-2A	0126
86-5083-2	0160
86-5084-2	0103
86-5084-2A	0103
86-5085-2	0178
86-5086-2	0208
86-5087-2	0208
86-5088-2	0160
86-5089-2	0160
86-5090-2	0160
86-5091-2	0004
86-5093-2	0086
86-5095-2	0325
86-5096-2	0325
86-5097-2	0016
86-5099-2	0016
86-5100-2	0161
86-5101-2	0103
86-5101-2A	0103
86-5102-2	0042
86-5103-2	0016
86-5104-2	0791
86-5105-2	0400
86-5106-2	0178
86-5107-2	0042
86-5108-2	0042
86-5109-2	0086
86-5110-2	0016
86-5111-2	0016
86-5112-2	0103
86-5113-2	0160
86-5114-2	0016
86-5122-2	0325
86-5125-2	0160
86-5911-3	0015
86-5943-2	0160
86-10003	0224
86-86525-3	0348
86-86560-3	0348
86-100002	0224
86-100003	0079
86-100004	0224
86-100005	0016
86-100006	0224
86-100007	0224
86-100008	0016
86-100009	0037
86-100010	0018
86-100011	0532
86-100012	0030
86-100013	0133
86-100014	0133
86-200003	OEM
86-200055	OEM
86A318	0004
86A327	0111
86A332	0103
86A338	0178
86A339	0905
86C04	0460
86HF10	OEM
86HF20	OEM
86HF40	OEM
86HF60	OEM
86HF80	OEM
86HF100	OEM
86HF120	OEM
86HF140	OEM
86HF160	OEM
86HFR10	OEM
86HFR20	OEM
86HFR40	OEM
86HFR60	OEM
86HFR80	OEM
86HFR100	OEM
86HFR120	OEM
86HFR140	OEM
86HFR160	OEM
86SI-74145N	0619
86ST-MPS9750D	0037
86X0006-001	0076
86X0007-001	1221
86X0007-004	0144
86X0007-104	0016
86X0007-204	0144
86X0008-001	0016
86X0009-001	0160
86X00011-001	0211
86X006-001	0016
86X007-004	0016
86X007-034	0016
86X0011-001	0136
86X0012-001	0111
86X0013-001	0136
86X0014-001	0050
86X0015-001	0160
86X0016-001	0037
86X0016-001A	0037
86X0017-001	0004
86X0018-001	0164
86X0019-001	0016

DEVICE TYPE	REPL CODE
86X0022-001	0016
86X0024-001	2624
86X0025-001	0198
86X0025-001(AFO)	0142
86X0025-001(TIK)	0079
86X0027-001	2624
86X0028-001	0142
86X0029-001	0016
86X0030-001	0265
86X0031-001	0016
86X0031-002	0016
86X0031-003	0016
86X0032-001	0016
86X0033-001	0222
86X0033-01	0222
86X0034-001	0330
86X0035-001	0086
86X0036-001	0037
86X0036-001(DIODE)	0102
86X0036-001(TRANSIST	0037
86X0036-001A	0037
86X0037-001	2866
86X0037-002	0004
86X0037-100	2866
86X0038-001	0007
86X0040-00	0016
86X0040-001	0016
86X0041-001	0037
86X0041-001A	0037
86X0042-001	3165
86X0042-002	2997
86X0042-100	1657
86X0043-001	0144
86X0044-001	0037
86X0044-001A	0037
86X0045-001	0016
86X0046-001	0037
86X0047-001	0037
86X0048-001	0016
86X0049-001	0076
86X0050-001	0016
86X0051-001	0079
86X0052-001	0224
86X0053-001	0167
86X0054-00	0079
86X0054-001	0016
86X0055-001	2549
86X0056-001	0328
86X0056-002	0328
86X0058-001	0198
86X0058-002	0198
86X0058-003	0198
86X0059-001	0161
86X0059-002	0561
86X0060-001	0127
86X0061-001	0127
86X0062-001	0127
86X0063-001	0016
86X0064-001	1611
86X0065-001	0285
86X0066-001	0006
86X0066-003	0006
86X0067-001	2329
86X0067-002	4606
86X0068-001	1712
86X0068-002	4615
86X0069-001	0239
86X0070-001	0316
86X0071-001	0079
86X0072-001	0037
86X0073-001	0551
86X0073-002	0563
86X0074-001	0161
86X0075-001	0460
86X0075-002	0925
86X0076-001	1708
86X0077-001	0239
86X0077-002	0061
86X0078-001	0061
86X0080-001	3165
86X0080-002	2997
86X0080-100	3732
86X0081-001	0095
86X0081-002	0095
86X0082-001	0521
86X0083-001	0161
86X0084-001	0848
86X0085-001	OEM
86X0086-001	2786
86X0087-001	OEM
86X0088-001	OEM
86X0089-001	0400
86X0090-001	1221
86X037-001	0038
86X0113-001	0076
86X0114-001	0006
86X0115-001	0285
86X0116-001	0285
86X0123-001	3873
86X0124-001	3909
86X0127-001	0861
86X0128-001	0728
86X0129-001	4247
86X2	0604
86X3	0280
86X6	0016
86X6-1	0016
86X6-4-518	0016
86X7-2	0016

DEVICE TYPE	REPL CODE
86X7-3	0016
86X7-4	0016
86X7-6	0127
86X7-6013	0144
86X8-1	0016
86X8-2	0016
86X8-3	0016
86X8-4	0016
86X34-1	0016
86X46	0037
86X47	0037
86X53	OEM
86X53-1	0167
86X56-2	0328
86X65-1	0285
86X74-1	0161
86X75-2	0925
86X82-1	0521
86X86-1	2786
86X87-1	OEM
86X88-1	7117
86X115	0285
86X483-2	0086
86X6029-001	0144
86Z6	1049
86Z6A	1049
87	0023
87-0001	1211
87-0002	1213
87-0002-1	2209
87-0003	1066
87-0004	1215
87-0005	0076
87-0006	0218
87-0009	1221
87-0013	0552
87-0014	0076
87-0015	0633
87-0016	0136
87-0017	0633
87-0018	0004
87-0019	0004
87-0020	0004
87-0021	0004
87-0022	1235
87-0023-7	0224
87-0023-T	1220
87-0027	0127
87-0028	1242
87-0029	0218
87-0203-1	0155
87-0212-1	0111
87-0216	1478
87-0217	1479
87-0218-U	1221
87-0227	0429
87-0228	1496
87-0229	1498
87-0230	0111
87-0230-1	0111
87-0233	1509
87-0234	0520
87-0235	0151
87-0235-C	0224
87-0235A	0151
87-0235B	0151
87-0235C	0151
87-0236-Q	0018
87-0236-R	0018
87-0236-S	0018
87-0236Q	0018
87-0236R	0018
87-0236S	0018
87-0237-R	0532
87-0237-S	0532
87-0238	0111
87-0239-A	0018
87-0246	1524
87-5-3	0474
87-10-0	0143
87-10-1	0143
87-56-3	0474
87-67-3	0023
87-71	OEM
87-104-3	0182
87-218-U	0111
87-423-2	0006
87-593-2	1144
87-724-375-01	0155
87A	0023
87B	OEM
87B02	0551
87C48	OEM
87C201-25-T	OEM
87RECT-SI-154	0023
87RECT-SI-154(AU)	0023
87SI-TA7680AP	4109
87Z6	1058
87Z6A	1058
88	OEM
88-2	OEM
88-3	0015
88-125	0015
88-831	0015
88-832	0015
88-833	0015
88-9132	4586
88-9132F	4586
88-9302	1421

DEVICE TYPE	REPL CODE
88-9302RS	1421
88-9302S	1421
88-9574	1443
88-9774	OEM
88-9774F	OEM
88-9779	1759
88-9779F	1759
88-9824R	OEM
88-9841	OEM
88-9841R	1843
88-9842	OEM
88-9842F	1345
88-9842R	1345
88-9842RS	1345
88-9842S	1345
88-18920	2124
88-20372	2755
88-20404	0520
88-98302	OEM
88-98302R	OEM
88-1250108	0111
88-1250109	0016
88-B-7-258	0678
88-B-7258	0678
88A752271	2552
88B	0050
88B0	0050
88B1	0050
88B2	0050
88B3	0050
88B4	0050
88B5	0050
88B6	0050
88B7	0050
88B8	0050
88B9	0050
88B10	0050
88B10R	0050
88B11	0050
88B12	0050
88B13	0050
88B13P	0050
88B14	0050
88B14-7	0050
88B14-7B	0050
88B15	0050
88B15L	0050
88B16	0050
88B16E	0050
88B17	0050
88B17-1	0050
88B18	0050
88B19	0050
88B19G	0050
88B-70	0050
88B-70-12	0050
88B-70-12-7	0050
88B119	0050
88B121	0050
88B182	0050
88B7258	0678
88C	0280
88C0	0280
88C1	0280
88C2	0280
88C3	0280
88C4	0280
88C5	0280
88C6	0280
88C7	0280
88C8	0280
88C9	0280
88C10	0280
88C10R	0280
88C11	0280
88C12	0280
88C13	0280
88C13P	0280
88C14	0280
88C14-7	0280
88C14-7B	0280
88C15	0280
88C15L	0280
88C16	0280
88C16D	0280
88C17	0280
88C17-1	0280
88C18	0280
88C19	0280
88C19G	0280
88C-70	0280
88C-70-12	0280
88C-70-12-7	0280
88C119	0280
88C121	0280
88C182	0280
88CC-70-12	0280
88ST-2SA1015Y	0148
88SZ-RD12FB2	0137
88X0053-001	2393
88X8841	OEM
88Z6	0932
88Z6A	0932
89-REST-U-1081	OEM
89RECT-U-1001	OEM
89ST-MPS9410AK	0086
89WJ75/46N	0143
90-30	0076
90-31	0155

DEVICE TYPE	REPL CODE
90-32	0076
90-33	1221
90-35	2632
90-36	0911
90-37	2057
90-38	0219
90-39	1313
90-45	0007
90-46	1342
90-47	1342
90-48	0076
90-49	0549
90-50	3717
90-54	0050
90-55	3717
90-56	0191
90-57	0076
90-58	0353
90-59	0050
90-60	1144
90-61	0191
90-62	3717
90-65	0076
90-66	0155
90-67	1313
90-68	0268
90-69	0076
90-70	1221
90-71	0191
90-72	2632
90-73	0911
90-74	2057
90-75	0390
90-140	0076
90-175	4208
90-176	0390
90-177	1978
90-178	0212
90-179	3717
90-180	0151
90-181	1220
90-450	0839
90-451	0390
90-452	0076
90-453	0076
90-454	1978
90-455	0330
90-456	0076
90-457	0076
90-458	0076
90-459	0076
90-600	0228
90-601	0076
90-602	0368
90-603	0111
90-604	1144
90-605	0076
90-606	3717
90-607	1280
90-608	1280
90-609	0839
90-610	0836
90-612	0076
90-613	3717
90-614	1221
90-1648	OEM
90T2	0127
90T112-1	OEM
90V10A	OEM
90V11	OEM
90X9.5-10.5	OEM
91-3	0143
91-4	0050
91-46	0143
91-50	0030
91-US	OEM
91A	0144
91A01	0015
91A02	0015
91A03	0015
91A04	0071
91A05	0015
91A06	0015
91A08	0102
91A018	0015
91A11	0015
91AJ150	0178
91B	0144
91B(ZENER)	1311
91BGRN	0144
91C	0016
91C-US	OEM
91D	0016
91D5R11	OEM
91D9R11GH	OEM
91E	0016
91F	0144
91H	OEM
91L11PC	OEM
91N1	0123
91N1B	0007
91RC5	0608
91RC10	0608
91RC20	0608
91RC30	0610
91RC40	0610
91RC60	0610
91RC80	0468
91RC100	0468

DEVICE TYPE	REPL CODE
91RC110	0468
91RC120	0468
91RL10	OEM
91RL20	OEM
91RL30	OEM
91RL40	OEM
91RL50	OEM
91RL60	OEM
91RM10	OEM
91RM20	OEM
91RM30	OEM
91RM40	OEM
91RM50	OEM
91RM60	OEM
91T1	OEM
91T2	OEM
91T3	OEM
91T4	OEM
91T5	OEM
91T6	0111
91TB1-JE00	OEM
91TB1-JH00	OEM
91TB1-LE00	OEM
91TB1-LH00	OEM
91TB1-NE00	OEM
91TB1-NH00	OEM
91TB2-JE00	OEM
91TB2-JH00	OEM
91TB2-LE00	OEM
91TB2-LH00	OEM
91TB2-NE00	OEM
91TB2-NH00	OEM
91TB3-JE00	OEM
91TB3-JH00	OEM
91TB3-LE00	OEM
91TB3-LH00	OEM
91TB3-NE00	OEM
91TB3-NH00	OEM
91TB4-JE00	OEM
91TB4-JH00	OEM
91TB4-LE00	OEM
91TB4-LH00	OEM
91TB4-NE00	OEM
91TB4-NH00	OEM
91TB5-JE00	OEM
91TB5-JH00	OEM
91TB5-LE00	OEM
91TB5-LH00	OEM
91TB5-NE00	OEM
91TB5-NH00	OEM
91TB6-JE00	OEM
91TB6-JH00	OEM
91TB6-LE00	OEM
91TB6-LH00	OEM
91TB6-NE00	OEM
91TB6-NH00	OEM
91TB7	OEM
91TB7-JE00	OEM
91TB7-JH00	OEM
91TB7-LE00	OEM
91TB7-LH00	OEM
91TB7-NE00	OEM
91TB7-NH00	OEM
91TB8-JE00	OEM
91TB8-JH00	OEM
91TB8-LE00	OEM
91TB8-LH00	OEM
91TB8-NE00	OEM
91TB8-NH00	OEM
91TB9	OEM
91TB9-JE00	OEM
91TB9-JH00	OEM
91TB9-LE00	OEM
91TB9-LH00	OEM
91TB9-NE00	OEM
91TB9-NH00	OEM
91TB10-JE00	OEM
91TB10-JH00	OEM
91TB10-LE00	OEM
91TB10-LH00	OEM
91TB10-NE00	OEM
91TB10-NH00	OEM
91TB11	OEM
91TB11-JE00	OEM
91TB11-JH00	OEM
91TB11-LE00	OEM
91TB11-LH00	OEM
91TB11-NE00	OEM
91TB11-NH00	OEM
91TB12-JE00	OEM
91TB12-JH00	OEM
91TB12-LE00	OEM
91TB12-LH00	OEM
91TB12-NE00	OEM
91TB12-NH00	OEM
92	OEM
92-1	0143
92-1-70	0440
92-1-70-12	0440
92-1-70-12-7	0440
92-1A	0440
92-1A0	0440
92-1A0R	0440
92-1A1	0440
92-1A2	0440
92-1A3	0440
92-1A3P	0440
92-1A4	0440
92-1A4-7	0440

DEVICE TYPE	REPL CODE
92-1A4-7B	0440
92-1A5	0440
92-1A5L	0440
92-1A6	0440
92-1A6-1	0440
92-1A7	0440
92-1A7-1	0440
92-1A8	0440
92-1A9	0440
92-1A9G	0440
92-1A19	0440
92-1A21	0440
92-1A82	0440
92-2	OEM
92-10	0440
92-11	0440
92-11-1	0015
92-12	0440
92-13	0440
92-14	0440
92-15	0440
92-16	0440
92-17	0440
92-18	0440
92-19	0440
92-64-1	0913
92-1001	0143
92A11-1	0015
92B12-2	0015
92BIC	1345
92BLC	1345
92D5R11	OEM
92D7R11	OEM
92D9R11GH	OEM
92GU01	5034
92GU01A	5034
92GU05	5034
92GU06	5034
92GU45	3799
92GU45A	3799
92GU51	1597
92GU51A	1597
92GU55	1597
92GU56	1597
92L102-2	0015
92N1	0144
92N1B	0144
92PE37A	3938
92PE37B	3938
92PE37C	3938
92PE77A	4548
92PE77B	4548
92PE77C	4548
92PE487	0261
92PE488	0261
92PE489	0261
92PU01	5027
92PU01A	5027
92PU05	5034
92PU06	6134
92PU07	5034
92PU10	0261
92PU36	3764
92PU36A	3764
92PU36B	3764
92PU36C	3764
92PU45	5093
92PU45A	5093
92PU51	1597
92PU51A	1597
92PU55	1597
92PU56	1597
92PU57	1597
92PU100	5034
92PU200	1597
92PU391	3754
92PU392	3754
92PU393	3754
92RC5	OEM
92RC10	OEM
92RC20	OEM
92RC30	0217
92RC40	OEM
92RC60	0217
92RC80	OEM
92RC100	OEM
92RC110	OEM
92RC120	OEM
92RL10	OEM
92RL20	OEM
92RL30	OEM
92RL40	OEM
92RL50	OEM
92RL60	OEM
92RM10	OEM
92RM20	OEM
92RM30	OEM
92RM40	OEM
92RM50	OEM
92RM60	OEM
92T6	0111
93-13	0133
93-188	2015
93-302	0015
93A1-20	0015
93A1-21	0015
93A1D-1	0015
93A2	0229
93A3D-2	0015

DEVICE TYPE	REPL CODE
93A4-2	0015
93A4-4	0015
93A5-2	0196
93A5-8	0196
93A5-9	0196
93A5-10	0196
93A6-1	0015
93A6-2	0015
93A8-1	0143
93A8-4	OEM
93A9	0211
93A9-1	0004
93A9-2	0004
93A9-3	0280
93A9-4	0280
93A10-1	0015
93A11	0015
93A12	0015
93A12-1	0015
93A12-3	0015
93A15-1	0229
93A15-10	0196
93A25-1	0143
93A25-2	0123
93A25-3	0143
93A25-B	0143
93A27-1	0143
93A27-2	0015
93A27-3	0133
93A27-8	0133
93A30-1	0015
93A30-3	0015
93A31-1	0133
93A33-1	0143
93A36-1	0030
93A38-1	0143
93A39-1	0402
93A39-2	0157
93A39-3	0012
93A39-5	0531
93A39-6	0170
93A39-7	0531
93A39-8	0402
93A39-11	0025
93A39-12	0137
93A39-13	0317
93A39-14	1606
93A39-15	0016
93A39-17	0127
93A39-19	0157
93A39-21	OEM
93A39-24	0002
93A39-25	0783
93A39-26	0157
93A39-28	0213
93A39-30	0170
93A39-31	0365
93A39-34	0456
93A39-35	0350
93A39-37	0213
93A39-40	0039
93A39-43	0137
93A39-44	0317
93A39-45	0402
93A39-48	1649
93A39-49	0299
93A39-50	0420
93A40-1	0102
93A40-2	OEM
93A41-2	0019
93A41-5	0143
93A42-1	0015
93A42-2	0015
93A42-7	0015
93A42-8	0015
93A43-1	0917
93A43-2	0917
93A43-4	0917
93A44-2	0917
93A45-1	0015
93A45-2	0015
93A47	0015
93A48-1	0133
93A48-2	0015
93A51-3	0015
93A52-1	0015
93A52-2	0133
93A52-7	0102
93A53-1	0474
93A53-2	0474
93A53-3	0015
93A55-1	0317
93A56-1	0015
93A57-1	0774
93A58-1	0015
93A59-1	0917
93A59-2	0917
93A59-10	OEM
93A60-1	0102
93A60-2	0015
93A60-3	0102
93A60-5	0133
93A60-6	0133
93A60-7	0023
93A60-8	0015
93A60-9	0015
93A60-10	0102
93A60-11	0102
93A60-14	0015

If replacement code is OEM, contact original manufacturer for replacement.

DEVICE TYPE	REPL CODE
93A60-80	0015
93A61-1	OEM
93A61-2	OEM
93A63-1	0030
93A63-2	0917
93A63-3	OEM
93A63-4	OEM
93A64-1	0133
93A64-2	0133
93A64-3	0133
93A64-4	OEM
93A64-5	0133
93A64-7	0133
93A64-8	0133
93A65-1	OEM
93A67-1	0015
93A68	OEM
93A69-1	0102
93A69-2	0133
93A70-1	OEM
93A71-1	0102
93A75-1	0474
93A76-1	0030
93A76-6	OEM
93A77-1	0143
93A78-1	0015
93A79-6	0071
93A80-1	0137
93A83-1	0123
93A91-1	1501
93A91-2	1501
93A91-3	1501
93A91-4	1501
93A93-1	0378
93A93-2	0378
93A96-1	0190
93A96-2	0190
93A96-3	0190
93A97-1	0015
93A97-3	0015
93A98-1	2984
93A99-2	2984
93A99-3	2984
93A99-7	1054
93A99-8	1054
93A102-2	0365
93A104-1	0015
93A104-2	4423
93A105-1	0143
93A106-1	OEM
93A106-2	OEM
93A106-3	OEM
93A107-2	0149
93A110-1	0143
93A111-1	6588
93A112-1	0071
93A120	0015
93A664-2	0133
93B1-1	0015
93B1-2	0015
93B1-3	0015
93B1-4	0015
93B1-5	0015
93B1-6	0015
93B1-7	0015
93B1-8	0015
93B1-9	0015
93B1-10	0015
93B1-11	0015
93B1-12	0015
93B1-13	0015
93B1-14	0015
93B1-15	0015
93B1-16	0015
93B1-17	0015
93B1-18	0015
93B1-20	0015
93B1-21	0015
93B2-1	0015
93B3-3	1030
93B3-3,4	1030
93B3-4	1030
93B5	0196
93B5-1	0196
93B5-2	0196
93B5-3	0196
93B5-3-6	0196
93B5-3-8	0196
93B5-3-9	0196
93B5-4	0196
93B5-5	0196
93B5-6	0196
93B5-7	0196
93B5-8	0196
93B5-9	0196
93B5-10	0196
93B6	0196
93B8	0196
93B8-1	0143
93B9	0196
93B12	0015
93B12-1	0015
93B12-2	0015
93B12-3	0015
93B15-1	0229
93B19-1	0474
93B20-1	0229
93B20-2	0229
93B20-3	0229

DEVICE TYPE	REPL CODE
93B21-5	OEM
93B22-3	1030
93B24-1	0015
93B24-2	0015
93B24-3	0015
93B25-1	0143
93B25-2	0143
93B25-3	0143
93B27-1	0143
93B27-2	0015
93B27-3	0123
93B27-4	0133
93B27-5	0133
93B30-1	0015
93B30-3	0015
93B34-2	0015
93B36-1	0030
93B38-1	0143
93B38-5	0143
93B39-1	0012
93B39-2	0162
93B39-3	0012
93B39-4	0441
93B39-5	0531
93B39-6	0170
93B39-7	0531
93B39-12	0137
93B39-13	0137
93B39-31	0365
93B41	0143
93B41-1	0143
93B41-2	0143
93B41-3	0143
93B41-4	0015
93B41-5	0030
93B41-6	0015
93B41-8	0015
93B41-12	0015
93B41-14	0015
93B41-15	0030
93B41-20	0229
93B41-29	0143
93B42-1	0015
93B42-2	0015
93B42-3	0015
93B42-4	0015
93B42-5	0015
93B42-6	0015
93B42-7	0015
93B42-8	0015
93B42-9	0015
93B42-10	0015
93B42-11	0015
93B42-12	0015
93B44-1	0917
93B45-1	0023
93B45-2	0015
93B45-3	0071
93B46-1	0143
93B47-1	0015
93B48-1	0133
93B48-2	0133
93B48-3	0133
93B48-4	0133
93B51-3	0015
93B52-1	0102
93B52-1-55	OEM
93B52-2	0102
93B52-27	0072
93B53-1	0015
93B53-2	0474
93B57-1	0774
93B58-1	0102
93B59-1	0917
93B60-3	0102
93B60-8	0133
93B60-10	0102
93B60-11	0102
93B63-1	OEM
93B63-2	OEM
93B64-1	0133
93B64-2	0133
93B65-1	0071
93B65-2	0071
93B67-1	0102
93B68-14	0917
93B71-1	0071
93B77-1	0143
93B91-1	1501
93B96-1	0190
93B96-2	0190
93B97-1	0015
93B122	0015
93B123	0015
93B130-1	OEM
93BLC	2257
93C	3061
93C1-20	0015
93C1-21	0015
93C1-2021	0474
93C2-6	0143
93C5-1	0484
93C5-2	0484
93C5-3	0484
93C5-4	0484
93C5-5	0196
93C5-6	0196
93C5-7	0196
93C5-8	0196

DEVICE TYPE	REPL CODE
93C5-9	0196
93C5-10	0196
93C6-7	0196
93C6-9	0196
93C7-1	0143
93C7-2	0133
93C7-3	0133
93C8-1	0143
93C12-1	0015
93C12-2	0015
93C12-3	0015
93C16-2	0015
93C18-1	0071
93C18-2	0071
93C19-1	0015
93C21-1	0143
93C21-2	0143
93C21-3	0143
93C21-4	0143
93C21-5	0143
93C21-6	0143
93C21-7	0143
93C21-8	0143
93C22-1	1030
93C22-3	1030
93C24-1	0015
93C24-2	0015
93C24-3	0015
93C24-4	0015
93C25-2	0123
93C25-3	0143
93C25-4	0015
93C26-1	0015
93C26-2	0015
93C26-3	0015
93C26-4	0015
93C26-5	0015
93C26-6	1080
93C26-7	0196
93C26-7,9	0196
93C26-8	0015
93C26-9	0196
93C26-10	0196
93C26-11	0196
93C27-1	0143
93C27-2	0133
93C27-3	0133
93C27-4	0015
93C27-5	0143
93C27-6	0143
93C27-7	0133
93C28-4	0015
93C30-1	0015
93C30-3	0015
93C39-1	0188
93C39-2	0052
93C39-3	0012
93C39-4	0441
93C39-5	0531
93C39-6	0137
93C39-7	0531
93C39-8	0402
93C39-9	0649
93C39-10	0649
93C39-11	0198
93C39-12	0137
93C39-13	0137
93C39-14	0649
93C39-19	0157
93C39-24	0002
93C39-25	0783
93C39-28	0213
93C39-43	0137
93C39-45	0402
93C39-49	0299
93C39-51	0471
93C39-52	OEM
93C39-58	0402
93C39-77	0041
93C39-79	0471
93C39-80	0057
93C39-106	OEM
93C40	1303
93C40-1	1303
93C42-2	0015
93C42-7	0015
93C46	OEM
93C46AN	OEM
93C46N	OEM
93C51-3	0484
93C52-1	0102
93C52-1-55	OEM
93C52-5	0015
93C52-9	0015
93C52-9-52T	0015
93C52-9-55	0071
93C52-10	0015
93C52-14	0023
93C52-15	0015
93C52-27	0071
93C52-952T	OEM
93C53-2	0474
93C55-1	0143
93C59	0196
93C59-2	0196
93C60-3	0102
93C60-5	0133
93C60-6	0133
93C60-7	0023

DEVICE TYPE	REPL CODE
93C60-9	0031
93C60-9-55	0023
93C60-10	0102
93C60-21	0023
93C60-21-52T	0023
93C60-21-55	0023
93C60-21-64	0023
93C60-26	0023
93C60-27	0023
93C60-27-52T	0023
93C64-1	0124
93C64-2	0124
93C64-3	0124
93C64-5	0133
93C64-10	0023
93C64-11	0124
93C64-11H	0124
93C64-11H-52T	0124
93C65-11H	OEM
93C69-1	0102
93C77-1	0143
93C91-2	1501
93C93-1	0378
93C97-1	0015
93C112-1	0502
93C118-1	0071
93C118-2	0071
93C218	0143
93C267	0484
93CG9-1	0102
93D8-1	0143
93D39-7	0531
93D39-43	0137
93D39-43-52T	0137
93D39-51	0471
93D39-52-52T	OEM
93D39-54	0053
93D39-54-52T	0053
93D39-67	OEM
93D39-73	0157
93D39-77-52T	0041
93D39-79-52T	0471
93D39-80	0057
93D39-80-52T	0057
93D39-106-52T	OEM
93D52-1	0015
93D60-7	0023
93D60-21-52T	0023
93D60-21-55	0023
93D60-21-64	0023
93D60-27-52T	0023
93D64-2	0124
93D91-1	1501
93D91-2	1501
93D91-3	1501
93D91-4	1501
93D91-5	1501
93D91-6	1501
93D96-1	0190
93D96-2	0190
93D96-3	0190
93D98-1	2984
93D99-1	2984
93D99-2	2984
93D99-3	2984
93D99-4	2984
93D99-5	1054
93D99-6	1054
93D99-7	1054
93D99-8	1054
93D99-9	2984
93D112-57	0917
93ERH-1X1128	0167
93ERH-DX0155//	0023
93ERH-DX0156//	0015
93EVHD1N4148//	0124
93H00DC	OEM
93H00DM	OEM
93H00FC	OEM
93H00FM	OEM
93H00PC	OEM
93H72DC	OEM
93H72DM	OEM
93H72FC	OEM
93H72FM	OEM
93H72PC	OEM
93H87DC	2579
93H87PC	2579
93H183DC	4399
93H193PC	4399
93K2-1	0196
93L00DC	1315
93L00DM	1315
93L00FC	OEM
93L00FM	OEM
93L00PC	1519
93L01DC	OEM
93L01FM	OEM
93L01PC	OEM
93L02DM	OEM
93L08DC	2127
93L08DM	2127
93L08FM	OEM
93L08PC	2127
93L09DC	OEM
93L09DM	OEM
93L09FM	OEM
93L09PC	OEM
93L3-1	0229

DEVICE TYPE	REPL CODE
93L3-2	0229
93L5-1	0015
93L5-2	0229
93L5-3	0087
93L5-4	0015
93L5-5	0087
93L5-6	0015
93L5-7	0015
93L5-8	0087
93L10DC	OEM
93L10DM	OEM
93L10FM	OEM
93L10PC	OEM
93L11DC	OEM
93L11DM	OEM
93L11FM	OEM
93L12DC	OEM
93L12DM	OEM
93L12FM	OEM
93L12PC	OEM
93L14DC	OEM
93L14DM	OEM
93L14FM	OEM
93L14PC	OEM
93L16DC	2685
93L16DM	OEM
93L16FM	OEM
93L16PC	2685
93L18DC	OEM
93L18DM	OEM
93L18FM	OEM
93L18PC	OEM
93L21DC	OEM
93L21DM	OEM
93L21FM	OEM
93L21PC	OEM
93L22DC	OEM
93L22DM	OEM
93L22FM	OEM
93L22PC	OEM
93L24DC	OEM
93L24DM	OEM
93L24FM	OEM
93L24PC	OEM
93L28DC	OEM
93L28DM	OEM
93L28FC	OEM
93L28FM	OEM
93L28PC	OEM
93L34DC	OEM
93L34DM	OEM
93L34FM	OEM
93L34PC	OEM
93L38DC	OEM
93L38DM	OEM
93L38FC	OEM
93L38FM	OEM
93L38PC	OEM
93L40DC	OEM
93L40DM	OEM
93L40FM	OEM
93L40PC	OEM
93L41DC	OEM
93L41DM	OEM
93L41FM	OEM
93L41PC	OEM
93L60DC	OEM
93L60DM	OEM
93L60FM	OEM
93L60PC	OEM
93L66DC	OEM
93L66DM	OEM
93L66FM	OEM
93L66PC	OEM
93L101-2	0015
93L102-2	0015
93L103-4	0015
93L104-5	0087
93L107-2	0015
93L415DC	OEM
93L415DM	OEM
93L415FC	OEM
93L415FM	OEM
93L415PC	OEM
93L422A	OEM
93L422DC	OEM
93L422DM	OEM
93L422FC	OEM
93L422FM	OEM
93L422N	OEM
93L425DC	OEM
93L425DM	OEM
93L425FM	OEM
93L425PC	OEM
93M8-1	0229
93N9-1	1501
93P1AA	0037
93S00DC	OEM
93S00DM	OEM
93S00FC	OEM
93S00FM	OEM
93S00PC	OEM
93S10DC	OEM
93S10DM	OEM
93S10FM	OEM
93S10PC	OEM
93S12DC	OEM
93S12DM	OEM
93S12FM	OEM

DEVICE TYPE	REPL CODE
93S12PC	OEM
93S16DC	OEM
93S16DM	OEM
93S16FM	OEM
93S16PC	OEM
93S21DC	OEM
93S21DM	OEM
93S21FM	OEM
93S21PC	OEM
93S41DC	OEM
93S41DM	OEM
93S41FM	OEM
93S41PC	OEM
93S42DC	OEM
93S42DM	OEM
93S42FM	OEM
93S42PC	OEM
93S43DC	OEM
93S43DM	OEM
93S43FM	OEM
93S43PC	OEM
93S46DC	OEM
93S46DM	OEM
93S46FM	OEM
93S46PC	OEM
93S47DC	OEM
93S47DM	OEM
93S47FM	OEM
93S47PC	OEM
93S48DC	OEM
93S48DM	OEM
93S48FM	OEM
93S48PC	OEM
93S62DC	OEM
93S62DM	OEM
93S62FM	OEM
93S62PC	OEM
93S153DC	2151
93S157DC	1695
93S157PC	1695
93S158DC	2154
93S158PC	2154
93S174DC	2132
93S174PC	2132
93S175DC	2141
93S175PC	2141
93S194DC	1931
93S194PC	1931
93SC165	0103
93SC165133	0103
93SC165133A	0103
93SE165	0130
93T6	0111
93T4100203100	0151
93T4105002132	0461
93T4120265210	0057
93T4120280156	0143
93T4120290136	0143
93T4120482210	0030
93T4125005150	1005
93T4130060052	0015
93T4150133000	4147
93T4150182072	2063
93X60-9	OEM
93Z510DC	OEM
93Z510DM	OEM
93Z510FC	OEM
93Z510FM	OEM
93Z510LC	OEM
93Z510LM	OEM
93Z510PC	OEM
93Z511DC	OEM
93Z511DM	OEM
93Z511FC	OEM
93Z511FM	OEM
93Z511LM	OEM
93Z511PC	OEM
93.20.709	0289
93.20.714	0289
93.24.401	0143
93.24.601	0143
93.24.604	0143
94-42-9	0143
94-1066-1	0015
94A-1A6-4	0160
94A60-3	OEM
94A64-1	OEM
94A80-1	0137
94BLC	0823
94N1	0016
94N1B	0016
94N1R	0016
94N1V	0086
94N2	0016
94T1	0280
94T2	OEM
94T3	OEM
94T4	OEM
94T5	OEM
94TB1-JH00	OEM
94TB1-LE00	OEM
94TB1-NE00	OEM
94TB1-NH00	OEM
94TB2-JE00	OEM
94TB2-JH00	OEM
94TB2-LE00	OEM
94TB2-LH00	OEM
94TB2-NE00	OEM
94TB2-NH00	OEM

DEVICE TYPE	REPL CODE
94TB3-JE00	OEM
94TB3-JH00	OEM
94TB3-LE00	OEM
94TB3-LH00	OEM
94TB3-NE00	OEM
94TB3-NH00	OEM
94TB4-JE00	OEM
94TB4-JH00	OEM
94TB4-LE00	OEM
94TB4-LH00	OEM
94TB4-NH00	OEM
94TB5-JE00	OEM
94TB5-JH00	OEM
94TB5-LE00	OEM
94TB5-LH00	OEM
94TB5-NE00	OEM
94TB5-NH00	OEM
94TB6-JE00	OEM
94TB6-JH00	OEM
94TB6-LE00	OEM
94TB6-LH00	OEM
94TB6-NE00	OEM
94TB6-NH00	OEM
94TB7	OEM
94TB7-JE00	OEM
94TB7-JH00	OEM
94TB7-LE00	OEM
94TB7-LH00	OEM
94TB7-NE00	OEM
94TB7-NH00	OEM
94TB8-JE00	OEM
94TB8-JH00	OEM
94TB8-LE00	OEM
94TB8-LH00	OEM
94TB8-NE00	OEM
94TB8-NH00	OEM
94TB9	OEM
94TB9-JE00	OEM
94TB9-JH00	OEM
94TB9-LE00	OEM
94TB9-LH00	OEM
94TB9-NE00	OEM
94TB9-NH00	OEM
94TB10-JE00	OEM
94TB10-JH00	OEM
94TB10-LE00	OEM
94TB10-LH00	OEM
94TB10-NE00	OEM
94TB10-NH00	OEM
95	0124
95/1016	OEM
95/1032	OEM
95/4000/2	OEM
95/4000/4	OEM
95/4005/2	OEM
95/4005/3	OEM
95/5032	OEM
95/6011	OEM
95/6110	OEM
95/6440	OEM
95-108	0050
95-110	0050
95-111	0136
95-112	0600
95-113	0600
95-114	0600
95-116	0050
95-117	0050
95-118	0050
95-119	0050
95-120	0050
95-120A	0280
95-121	0050
95-122	0050
95-123	0050
95-124	0050
95-125	0144
95-126	0144
95-127	0144
95-128	0144
95-129	0144
95-130	0144
95-131	0144
95-201	0004
95-202	0208
95-203	0004
95-204	0004
95-208	0004
95-209	0004
95-212	0004
95-213	0004
95-214	0004
95-216	0016
95-217	0004
95-218	0004
95-220	0086
95-221	0016
95-222-1	0004
95-222-2	0208
95-223	0144
95-224-1	0004
95-224-2	0208
95-225	0016
95-226-003	0126
95-226-004	0086
95-226-1	0126
95-226-2	0016
95-226-3	0126
95-226-4	0086

DEVICE TYPE	REPL CODE
95-227	0016
95-250	0160
95-251	0160
95-252	0142
95-252-1	0142
95-252-2	0142
95-252-3	0142
95-252-4	0142
95-296	0016
95-11410100	0004
95-11410200	0004
95-11410900	0004
95-11413500	0004
95-11414000	0211
95-11511500	0071
95A5-10	OEM
95B5-8	0196
95ERRMCG044115	OEM
95ERRMCG044117	OEM
95ERRMCG044118	OEM
95ERRMCG044119	OEM
95ERRMCG526A15	OEM
95ERRMCG526A17	0133
95ERRMCG526A18	OEM
95ERRMCG526A21	0866
95ERRMCG575A15	OEM
95ERRMCG575A17	0133
95ERRMCG575A18	OEM
95ERRMCG575A21	0866
95G	0023
95H00DC	OEM
95KUAB0015AM	0681
95KUAC0004AZ	0076
95KUAD0031SZ	1212
95KUAD0036MZ	3136
95KUB0429BZ	OEM
95KUBC0112AZ	0015
95KUBC0150BZ	0015
95KUBC0215BK	0080
95KUBC0215BZ	0080
95KUBC0216CZ	OEM
95KUBC112AZ	0015
95KUBC216CZ	OEM
95KUBD0115AZ	0253
95KUBD0128AZ	0375
95KUBD0429BZ	3910
95KUBDAC5R6B	2121
95KUBDAS5R6B	3910
95KUCB0027AZ	0624
95KUCC0013DZ	OEM
95KUCZ0051ZZ	OEM
95KUCZ0097ZZ	OEM
96-0008	0143
96-138-2	0144
96-5005-01	0285
96-5007-01	0143
96-5007-02	0123
96-5022-01	0015
96-5022-1	0015
96-5023-01	0015
96-5026-01	0160
96-5032-01	0004
96-5033-01	0004
96-5033-02	0004
96-5033-03	0004
96-5033-04	0004
96-5045-01	0160
96-5046-01	0015
96-5059-01	0143
96-5062-01	0050
96-5064-01	0160
96-5076-01	0164
96-5080-02	0016
96-5081-01	0160
96-5082-01	0015
96-5085-01	0004
96-5085-02	0004
96-5086-02	0160
96-5088-01	0015
96-5091-01	0137
96-5093-01	0123
96-5093-02	0123
96-5093-03	0123
96-5094-01	0123
96-5094-01(TRANSISTO	0136
96-5095-01	0050
96-5096-01	0015
96-5098-01	0004
96-5099-01	0050
96-5100-01	0160
96-5101-01	0004
96-5102-01	0004
96-5103-01	0015
96-5106-01	0242
96-5107-01	0086
96-5107-02	0086
96-5109-01	0015
96-5109-02	0015
96-5110-02	0002
96-5110-03	0002
96-5112-01	0087
96-5113-01	0015
96-5115-01	0016
96-5115-02	0016
96-5115-03	0016
96-5115-04	0016
96-5115-05	0016
96-5116-01	0170
96-5117-01	0103
96-5117-91	0103
96-5118-01	0133
96-5120-01	0302
96-5124-02	0649
96-5125-01	0160
96-5131-01	0007
96-5132-01	0142
96-5133-01-02	0137
96-5133-02	0137
96-5135-01	0142
96-5138-01	0050
96-5139-01	0050
96-5140-01	0050
96-5141-01	0050
96-5143-01	0160
96-5143-02	0160
96-5148-01	0160
96-5149-01	0242
96-5152-01	0016
96-5152-03	0016
96-5153-01	0016
96-5153-03	0016
96-5155-01	0160
96-5161-01	0595
96-5162-03	0103
96-5162-04	0103
96-5163-01	0007
96-5164-02	0103
96-5164-03	0103
96-5165-01	0126
96-5166-01	0015
96-5170-01	0086
96-5174-01	0007
96-5175-01	0007
96-5176-01	0126
96-5177-01	0016
96-5178-01	0071
96-5180-01	0086
96-5180-02	0086
96-5184-01	0242
96-5187-01	0016
96-5190-01	0178
96-5191-01	0178
96-5192-01	0160
96-5193-01	0087
96-5194-01	0087
96-5195-01	0922
96-5196-01	0015
96-5198-01	0007
96-5199-01	0007
96-5201-01	0103
96-5203-01	0086
96-5204-01	0086
96-5205-01	0208
96-5207-01	0103
96-5208-01	0086
96-5209-01	0126
96-5213-01	0016
96-5214-01	1137
96-5215-01	0037
96-5219-01	0285
96-5220-01	4163
96-5228-01	0016
96-5229-01	0016
96-5231-01	0086
96-5235-01	0007
96-5236-01	0007
96-5237-01	0016
96-5238-01	0632
96-5238-02	0632
96-5241-01	0922
96-5244-01	0086
96-5246-01	1124
96-5248-04	0137
96-5248-09	0456
96-5248-11	0456
96-5249-01	0234
96-5249-02	0205
96-5249-03	0256
96-5250-01	0234
96-5250-03	0309
96-5250-06	0247
96-5250-07	0225
96-5252-01	0086
96-5254-01	0133
96-5255-01	0016
96-5256-01	0086
96-5257-01	0016
96-5258-01	0693
96-5259-01	0007
96-5260-01	0007
96-5269-01	2136
96-5284-01	0074
96-5334-04	0256
96-5334-05	0262
96-5344-02	0870
96-5344-03	0247
96-51430-02	0160
96E1S953	0133
96E1S2076A	0124
96E2SA733	0006
96E2SC945	0076
96E2SC1213	0191
96E2SC2827	0728
96EP	OEM
96ERD5.6EB2	0253
96ESM-1A-04FR	0023
96ESM-3-02FR	0031
96L02	1468
96L02DC	1468
96L02DM	1468
96L02FM	OEM
96L02PC	1468
96LS02DC	2934
96LS02DM	2934
96LS02FM	OEM
96LS02PC	2934
96LS32DC	OEM
96LS32DM	OEM
96LS32FM	OEM
96LS32PC	OEM
96LS42DC	OEM
96LS42DM	OEM
96LS42PC	OEM
96N(AIRLINE)	0144
96N1	0142
96N927	0144
96N932	0144
96NPT	0144
96S02DC	4292
96S02PC	4292
96XZ0778-44N	0143
96XZ0868/15X	0143
96XZ077844N	0143
96XZ778/21N	0143
96XZ778/27N	0196
96XZ778/44N	0143
96XZ801/06N	0160
96XZ801/10N	0160
96XZ801/14N	0016
96XZ801/34X	0160
96XZ801/37N	0050
96XZ801/50N	0004
96XZ6050/25N	0007
96XZ6051-28N	0136
96XZ6051-35N	0136
96XZ6051-36N	0136
96XZ6052/52N	0016
96XZ6052-52N	0079
96XZ6053/11N	0016
96XZ6053/24N	0164
96XZ6053/27N	0004
96XZ6053/35N	0016
96XZ6053/36N	0016
96XZ6053/38N	0086
96XZ6053/51N	0004
96XZ6053-09N	0211
96XZ6053-10N	0136
96XZ6053-11N	0079
96XZ6053-24N	0136
96XZ6053-27N	0211
96XZ6053-51N	0164
96XZ6054/45X	0822
97	OEM
97-3	1303
97A83	0085
97A283-11	OEM
97EPA	OEM
97EPB	0600
97N2	0600
97N2U	0208
97P1	0211
97P1U	0164
98-1	0004
98-2	0004
98-3	0004
98-4	0004
98-5	0004
98-6	0004
98-7	0004
98-301	0015
98-302	0229
98A40-2	OEM
98A40-7	OEM
98A40-17	OEM
98A12518	0015
98B5-4	0484
98DM	OEM
98M2SA1162Y/G	1742
98M2SC1318-R/	0155
98M2SC2712Y/G	0724
98M2SD601A//	0724
98M2SD601A///	0724
98MDT144EK//	3478
98M-HSM2838C/	OEM
98M-IMT1/////	OEM
98MMA157/////	OEM
98MMA223/////	OEM
98N2P	2763
98P1P	0222
98T2	0079
99-101	0136
99-102	0136
99-103	0136
99-104	0004
99-105	0050
99-106	0050
99-107	0050
99-108	0050
99-109-1	0016
99-109-2	0016
99-120	0050
99-121	0050
99-201	0004
99-202	0004
99-203	0004
99-PWR	0085
99AT6	0211
99B5	0211
99BA6	0280
99BE6	0280
99E16-1	3418
99K7	0600
99L-4A6-1	0004
99L6	0038
99L6(SHARP)	0086
99P1	0037
99P1M	0037
99P2	0126
99P2B	0126
99P3AA	0004
99P6	0693
99P10	0037
99P117	0126
99PLM	0126
99S001	0160
99S002	0004
99S003	0004
99S004	0164
99S004A	0004
99S005	0004
99S006	0050
99S007	0050
99S07	0600
99S010	0004
99S010A	0004
99S011	0004
99S011A	0004
99S012	0016
99S012A	0016
99S012E	0016
99S014	0160
99S014A	0160
99S015	0160
99S016	0144
99S016-1	0144
99S017	0144
99S018	0144
99S018A	0144
99S019	0144
99S019A	0144
99S019B	0144
99S020	0198
99S022	0632
99S025	0155
99S025A	0016
99S031	0007
99S032	0007
99S032(3RD-LF)	0330
99S033	0111
99S033-40	1381
99S034	0861
99S035	0016
99S036	0016
99S036(TELEDYNE)	0086
99S037	0144
99S038	0016
99S039	0037
99S040	0861
99S041	0212
99S042	0829
99S044	0007
99S045	0353
99S045A	0212
99S046	0212
99S047	0178
99S049	2617
99S053	0701
99S055	0224
99S056	0224
99S057	0604
99S060-1	0861
99S061-1	0861
99S063-1	1708
99S067-1	0212
99S070-1	0861
99S073	0126
99S074	0086
99S075	0423
99S077-1	0861
99S079	0065
99S079-1	0065
99S082-1	0167
99S083-1	0423
99S084-1	0037
99S085	0016
99S087-1	0233
99S090-1	0144
99S091-1	0551
99S092-1	0382
99S094-1	0352
99S095-1	0354
99S096-1	0666
99S097-1	0802
99S099-1	0925
99S0102-1	0285
99S100-1	0042
99S101-1	0233
99S102-1	0264
99S103-1	0130
99S103-2	0130
99S103-3	0130
99S105-1	0161
99SA7	0208
99SK5	0600
99SK7	0600
99S022	0632
99SQ7	0600
100-001-01/2228-3	0229
100-003-40/228-3	0143
100-003-40/2228-3	0143
100-007-10/2228/3	0133
100-007-13/228-3	0143
100-007-13/2228-3	0143
100-008-53/2228-3	0030
100-0051	0143
100-00310-09	0133
100-00340-00	0143
100-00910-07	0019
100-00914-10	0123
100-011-20	0133
100-011-50/2228-3	0133
100-0110-01	OEM
100-0124	0143
100-0125	0143
100-0495-15	0006
100-0673-04	0148
100-01110-01	0133
100-01120-09	0133
100-1	0551
100-1A	OEM
100-5	OEM
100-10	0091
100-11	0133
100-12	0019
100-13	0015
100-14	0012
100-15	0015
100-120	0133
100-125	0133
100-130	0133
100-132-00	0015
100-135	0091
100-136	0019
100-137	0133
100-138	0015
100-139	0012
100-140	OEM
100-180	0019
100-181	0123
100-184	0133
100-215	0019
100-216	0133
100-217	0023
100-218	0025
100-219	0244
100-435	0133
100-436	0019
100-437	0062
100-438	0023
100-439	0143
100-520	0023
100-521	0165
100-522	0012
100-523	0012
100-524	0139
100-525	0402
100-526	0030
100-527	0023
100-914-10	OEM
100-929-60/2228-3	OEM
100-10121-05	0015
100-10132-00	0015
100-22363-08	0030
100A	0015
100B63	0211
100C	0071
100C05	1084
100C05B	1084
100C025	0738
100C025B	0738
100C050	0738
100C050B	0738
100C3.8-4.4	OEM
100C-4R	0015
100C4.4-5	OEM
100C4.8-5.4	OEM
100C5.4-5.9	OEM
100C5.9-6.4	OEM
100C6-7	OEM
100C6B	OEM
100C6.2-6.8	OEM
100C6.8-7.4	OEM
100C10	0738
100C10B	0738
100C15	0738
100C15B	0738
100C20	0738
100C20B	0738
100C25	0738
100C25B	0738
100C30	0738
100C30B	0738
100C35	0738
100C35B	0738
100C40	0738
100C40B	0738
100C45	0738
100C45B	0738
100C50	0738
100C50B	0738
100C60	0738
100C60B	0738
100C70	0738
100C70B	0738
100C80	0738
100C80B	0738
100C90	0738
100C90B	0738
100C100	0738
100C100B	0738
100C110	0738
100C110B	0738
100C120	0738
100C120B	0738
100C130	1889
100C130B	1889
100C140	1889
100C140B	1889
100C150	1889
100C150B	1889
100C160	1889
100C160B	1889
100C170	OEM
100C170B	OEM
100C180	OEM
100C180B	OEM
100D10	0071
100D20	OEM
100D30	OEM
100D40	OEM
100EXD21	OEM
100F20	OEM
100F30	OEM
100F40	OEM
100F50	OEM
100H3	OEM
100HF20	OEM
100HF20F	OEM
100HF20M	OEM
100HF40	OEM
100HF40F	OEM
100HF40M	OEM
100HF60	OEM
100HF60F	OEM
100HF60M	OEM
100HF80	OEM
100HF80F	OEM
100HF80M	OEM
100HF100	OEM
100HF100F	OEM
100HF100M	OEM
100HF120	OEM
100HF120F	OEM
100HF120M	OEM
100HF140	OEM
100HF140F	OEM
100HF140M	OEM
100HF160	OEM
100HF160F	OEM
100HF160M	OEM
100HHFR20	OEM
100HHFR20HFR	OEM
100HHFR20M	OEM
100HHFR40	OEM
100HHFR40HFR	OEM
100HHFR40M	OEM
100HHFR60	OEM
100HHFR60HFR	OEM
100HHFR60M	OEM
100HHFR80	OEM
100HHFR80HFR	OEM
100HHFR80M	OEM
100HHFR100	OEM
100HHFR100HFR	OEM
100HHFR100M	OEM
100HHFR120	OEM
100HHFR120HFR	OEM
100HHFR120M	OEM
100HHFR140	OEM
100HHFR140HFR	OEM
100HHFR140M	OEM
100HHFR160	OEM
100HHFR160HFR	OEM
100HHFR160M	OEM
100J6P41	OEM
100JB05L	5807
100JB1L	5807
100JB2L	6056
100JB4L	6057
100JB6L	6060
100JB8L	6061
100JB10L	4900
100JB12L	OEM
100JD12	OEM
100JH21	OEM
100K10	0071
100K10A	0599
100K10R	1347
100K20	OEM
100K20A	0599
100K20R	1347
100K30	OEM
100K30A	0599
100K30R	1347
100K40A	1986
100K40R	1905
100K50A	1986
100K50R	1905
100K60A	1986
100K60R	1905
100K80A	0657
100K80R	0193
100K100A	0657
100K100R	0193
100KF10	OEM
100KF20	OEM
100KF30	OEM
100L1-1.2	OEM
100L1.05-1.25	OEM
100L1.05-2.5	OEM
100L1.5-1.8	OEM
100L1.6-2	OEM
100L1.25-1.5	OEM
100L6P41	OEM
100L15	OEM
100L20	0348
100L25	OEM
100L30	OEM
100L35	OEM
100L40	OEM
100L45	OEM
100L50	OEM
100L55	OEM
100L60	OEM
100L65	OEM
100L70	OEM
100L75	OEM
100L80	OEM
100L85	OEM
100L90	OEM
100L95	OEM
100L100	OEM
100L110	OEM
100L120	OEM
100L130	OEM
100L140	OEM
100L150	OEM
100L160	OEM
100L170	OEM
100L180	OEM
100L190	OEM
100L200	OEM
100L210	OEM
100L220	OEM
100L230	OEM
100L240	OEM
100L250	OEM
100L260	OEM
100L270	OEM
100L280	OEM
100L290	OEM
100L300	OEM
100MAB180	OEM
100MAB200	OEM
100MAB250	OEM
100MLAB160	OEM
100MLAB180	OEM
100MLAB200	OEM
100N1	0007
100N1AS	0007
100N1P	0007
100N1P(SOUND)	0079
100N1P(VID)	0007
100N3	0008
100N3P	0007
100N15A	OEM
100N15A(	OEM
100N20A	OEM
100N20A9	OEM
100N25A	OEM
100N25A9	OEM
100N30A	OEM
100N30A9	OEM
100N35A	OEM
100N35A9	OEM
100N39	OEM
100N40A	OEM
100N40A9	OEM
100N45A9	OEM
100N50A	OEM
100N50A9	OEM
100N55A	OEM
100N55A9	OEM
100N60A	OEM
100N60A9	OEM
100N65A	OEM
100N65A9	OEM
100N70A	OEM
100N70A9	OEM
100N75A	OEM
100N75A9	OEM
100N80A	OEM
100N80A9	OEM
100N85A	OEM
100N85A9	OEM
100N90A	OEM
100N90A9	OEM
100N95A	OEM
100N95A9	OEM
100N100A	OEM
100N100A9	OEM
100N110A	OEM
100N120A	OEM
100N130A	OEM
100N140A	OEM
100N150A	OEM
100N160A	OEM
100N170A	OEM
100N180A	OEM
100N190A	OEM
100N200A	OEM
100N210A	OEM
100N220A	OEM

If replacement code is OEM, contact original manufacturer for replacement.

DEVICE TYPE	REPL CODE
100N230A	OEM
100N240A	OEM
100N250A	OEM
100P4H	OEM
100P6H	OEM
100P8H	OEM
100P10H	OEM
100P12H	OEM
100P16H	OEM
100PBH	OEM
100PIN-BNC	OEM
100PIN-PP	OEM
100PIN-RM	OEM
100PV-BNC	OEM
100PV-PP	OEM
100PV-RM	OEM
100Q6P41	OEM
100QD21	OEM
100QH21	OEM
100Q-PIN-RM	OEM
100R1B	0015
100R2B	0015
100R3B	0015
100R4B	0015
100R5B	0015
100R6B	0015
100R7B	0071
100R8B	0071
100R9B	0071
100R10B	0071
100R11B	0102
100R12B	0102
100R13B	0102
100R14B	0102
100R15B	0017
100R16B	0348
100R18B	0348
100R20B	0348
100R22B	OEM
100R24B	OEM
100R26B	OEM
100R28B	OEM
100R30B	OEM
100R32B	OEM
100R34B	OEM
100R36B	OEM
100R38B	OEM
100R40B	OEM
100R42B	OEM
100R44B	OEM
100R46B	OEM
100R48B	OEM
100R50B	OEM
100S2-2.6	OEM
100S2.5-3	OEM
100S3-3.5	OEM
100S3.2-3.8	OEM
100S5SP	OEM
100S6SP	OEM
100S8SP	OEM
100S10SP	OEM
100S12SP	OEM
100S20	OEM
100S30	OEM
100S40	OEM
100S50	OEM
100SS3-07DC	OEM
100SS3-07SC	OEM
100SS4-07SC	OEM
100SS5-07SC	OEM
100SS5SP	OEM
100SS6SP	OEM
100SS8SP	OEM
100SS10SP	OEM
100SS12SP	OEM
100Series	OEM
100T2	0103
100T2A	0103
100U6P41	OEM
100UV-PV-RM	OEM
100V10A	OEM
100V11	OEM
100W1	0144
100X6	0103
100X6A	0103
100X7-8	OEM
100X7.4-8.0	OEM
100X8-9	OEM
100X9-10	OEM
100YD21	OEM
100Z4	OEM
100d15	OEM
100d25	OEM
101-0373-00	0076
101-2(ADMIRAL)	0144
101-3(ADMIRAL)	0144
101-4(ADMIRAL)	0144
101-12	0280
101-15	0004
101-6742	0600
101-6742(RCA)	0293
101-6744(RCA)	0293
101A	0050
101A1AX1	OEM
101A1AY1	OEM
101A1AYF1	OEM
101A1EX1	OEM
101A1EY1	OEM
101A1EYF1	OEM
101A2AX2	OEM
101A2AY2	OEM
101A2AYF2	OEM
101A2EX2	OEM
101A2EY2	OEM
101A2EYF2	OEM
101A3AX3	OEM
101A3AY3	OEM
101A3AYF3	OEM
101A3EX3	OEM
101A3EY3	OEM
101A3EYF3	OEM
101A4AX4	OEM
101A4AY4	OEM
101A4AYF4	OEM
101A4EX4	OEM
101A4EY4	OEM
101A4EYF4	OEM
101A5AX5	OEM
101A5AY5	OEM
101A5AYF5	OEM
101A5EX5	OEM
101A5EY5	OEM
101A5EYF5	OEM
101B	0050
101B1-11	OEM
101B1-12	OEM
101B1-13	OEM
101B1-14	OEM
101B1-26	OEM
101B1-27	OEM
101B1-31	OEM
101B1-32	OEM
101B1-41	OEM
101B1-42	OEM
101B1-51	OEM
101B1-52	OEM
101B3-11	OEM
101B3-12	OEM
101B3-13	OEM
101B3-14	OEM
101B3-26	OEM
101B3-27	OEM
101B3-31	OEM
101B3-32	OEM
101B3-41	OEM
101B3-42	OEM
101B3-51	OEM
101B6	0015
101B7-11	OEM
101B7-12	OEM
101B7-13	OEM
101B7-14	OEM
101B7-26	OEM
101B7-27	OEM
101B7-31	OEM
101B7-32	OEM
101B7-41	OEM
101B7-42	OEM
101B7-51	OEM
101B7-52	OEM
101BE-52	OEM
101F1-11	OEM
101F1-12	OEM
101F1-13	OEM
101F1-14	OEM
101F1-26	OEM
101F1-27	OEM
101F1-31	OEM
101F1-32	OEM
101F1-41	OEM
101F1-42	OEM
101F1-51	OEM
101F1-52	OEM
101F1FM	OEM
101F1S	1347
101F2FM	OEM
101F2S	1347
101F3-11	OEM
101F3-12	OEM
101F3-13	OEM
101F3-14	OEM
101F3-26	OEM
101F3-27	OEM
101F3-31	OEM
101F3-32	OEM
101F3-41	OEM
101F3-42	OEM
101F3-51	OEM
101F3-52	OEM
101F3FM	OEM
101F3S	1347
101F4S	1905
101F5S	1905
101F6S	1905
101F7-11	OEM
101F7-12	OEM
101F7-13	OEM
101F7-14	OEM
101F7-26	OEM
101F7-27	OEM
101F7-31	OEM
101F7-32	OEM
101F7-41	OEM
101F7-42	OEM
101F7-51	OEM
101F7-52	OEM
101F8S	0193
101F10S	0193
101F12S	0202
101KL20S15	OEM
101KL20S20	OEM
101KL40S10	OEM
101KL40S15	OEM
101KL40S20	OEM
101KL60S10	OEM
101KL60S15	OEM
101KL60S20	OEM
101KL80S10	OEM
101KL80S15	OEM
101KL80S20	OEM
101KL80S25	OEM
101KL100S10	OEM
101KL100S15	OEM
101KL100S20	OEM
101KL100S25	OEM
101KL100S30	OEM
101KL120S10	OEM
101KL120S15	OEM
101KL120S20	OEM
101KL120S25	OEM
101KL120S30	OEM
101KL130S10	OEM
101KL130S15	OEM
101KL130S20	OEM
101KL130S30	OEM
101KL140S10	OEM
101KL140S15	OEM
101KL140S20	OEM
101KL140S25	OEM
101KL140S30	OEM
101KL150S10	OEM
101KL150S15	OEM
101KL150S20	OEM
101KL150S25	OEM
101KL150S30	OEM
101KL160S10	OEM
101KL160S15	OEM
101KL160S20	OEM
101KL160S25	OEM
101KL160S30	OEM
101KLR40S10	OEM
101KLR40S15	OEM
101KLR40S20	OEM
101KLR40S25	OEM
101KLR40S30	OEM
101KLR60S10	OEM
101KLR60S15	OEM
101KLR60S20	OEM
101KLR60S25	OEM
101KLR60S30	OEM
101KLR80S10	OEM
101KLR80S15	OEM
101KLR80S20	OEM
101KLR80S25	OEM
101KLR80S30	OEM
101KLR100S10	OEM
101KLR100S15	OEM
101KLR100S20	OEM
101KLR100S25	OEM
101KLR100S30	OEM
101KLR120S10	OEM
101KLR120S15	OEM
101KLR120S20	OEM
101KLR120S25	OEM
101KLR120S30	OEM
101KLR130S10	OEM
101KLR130S15	OEM
101KLR130S20	OEM
101KLR130S25	OEM
101KLR130S30	OEM
101KLR140S10	OEM
101KLR140S15	OEM
101KLR140S20	OEM
101KLR140S25	OEM
101KLR140S30	OEM
101KLR150S10	OEM
101KLR150S15	OEM
101KLR150S20	OEM
101KLR150S25	OEM
101KLR150S30	OEM
101KLR160S10	OEM
101KLR160S15	OEM
101KLR160S20	OEM
101KLR160S25	OEM
101KLR160S30	OEM
101LLS	OEM
101M	0050
101P1	0037
101P10	0037
101R1FM	OEM
101R1S	OEM
101R2FM	OEM
101R2S	OEM
101R3FM	OEM
101R3S	OEM
101R4S	OEM
101RA10	0738
101RA20	0738
101RA30	0738
101RA50	0738
101RA60	0738
101RA80	0738
101RA100	0738
101RA110	0738
101RA120	0738
101RA130	1889
101RC10	0738
101RC20	0738
101RC30	0738
101RC40	0738
101RC50	0738
101RC60	0738
101RC80	0738
101RC120	OEM
101RE50	0738
101RE60	0738
101RE60M	OEM
101RE70	OEM
101RE70M	OEM
101RE80	0738
101RE80M	OEM
101RE90	OEM
101RE90M	OEM
101RE100	0738
101RE100M	OEM
101RE110	0738
101RE110M	OEM
101RE120	0738
101RE120M	OEM
101RE130	OEM
101RE130M	OEM
101S1AX0	OEM
101S1AX1	OEM
101S1AY0	OEM
101S1AY1	OEM
101S1AYF0	OEM
101S1AYF1	OEM
101S1EX0	OEM
101S1EX1	OEM
101S1EY0	OEM
101S1EY1	OEM
101S1EYF0	OEM
101S1EYF1	OEM
101S2AX0	OEM
101S2AX1	OEM
101S2AY0	OEM
101S2AY1	OEM
101S2AYF0	OEM
101S2AYF1	OEM
101S2EX0	OEM
101S2EX1	OEM
101S2EY0	OEM
101S2EY1	OEM
101S2EYF0	OEM
101S2EYF1	OEM
101S3AX0	OEM
101S3AX1	OEM
101S3AY0	OEM
101S3AY1	OEM
101S3AYF0	OEM
101S3AYF1	OEM
101S3EX0	OEM
101S3EX1	OEM
101S3EY0	OEM
101S3EY1	OEM
101S3EYF0	OEM
101S3EYF1	OEM
101S4AX0	OEM
101S4AY0	OEM
101S4AY1	OEM
101S4AYF0	OEM
101S4AYF1	OEM
101S4EX0	OEM
101S4EX1	OEM
101S4EY0	OEM
101S4EY1	OEM
101S4EYF0	OEM
101S4EYF1	OEM
101S5AX0	OEM
101S5AX1	OEM
101S5AY0	OEM
101S5AY1	OEM
101S5AYF0	OEM
101S5AYF1	OEM
101S5EX0	OEM
101S5EX1	OEM
101S5EY0	OEM
101S5EY1	OEM
101S5EYF0	OEM
101S5EYF1	OEM
101S6AX0	OEM
101S6AX1	OEM
101S6AY0	OEM
101S6AY1	OEM
101S6AYF0	OEM
101S6AYF1	OEM
101S6EX0	OEM
101S6EX1	OEM
101S6EY0	OEM
101S6EY1	OEM
101S6EYF0	OEM
101S6EYF1	OEM
101S7AX0	OEM
101S7AX1	OEM
101S7AY0	OEM
101S7AY1	OEM
101S7AYF0	OEM
101S7AYF1	OEM
101S7EX0	OEM
101S7EX1	OEM
101S7EY0	OEM
101S7EY1	OEM
101S7EYF0	OEM
101S7EYF1	OEM
101S8AX0	OEM
101S8AX1	OEM
101S8AY0	OEM
101S8AY1	OEM
101S8AYF0	OEM
101S8AYF1	OEM
101S8EX0	OEM
101S8EX1	OEM
101S8EY0	OEM
101S8EY1	OEM
101S8EYF0	OEM
101S8EYF1	OEM
101S9AX0	OEM
101S9AX1	OEM
101S9AY0	OEM
101S9AY1	OEM
101S9AYF0	OEM
101S9AYF1	OEM
101S9EX0	OEM
101S9EX1	OEM
101S9EY0	OEM
101S9EY1	OEM
101S9EYF0	OEM
101S9EYF1	OEM
101S10AX0	OEM
101S10AX1	OEM
101S10AY0	OEM
101S10AY1	OEM
101S10AYF0	OEM
101S10AYF1	OEM
101S10EX0	OEM
101S10EY0	OEM
101S10EYF0	OEM
101S11AX0	OEM
101S11AX1	OEM
101S11AY0	OEM
101S11AY1	OEM
101S11AYF0	OEM
101S11AYF1	OEM
101S12AX0	OEM
101S12AX1	OEM
101S12AY0	OEM
101S12AY1	OEM
101S12AYF0	OEM
101S12AYF1	OEM
101SS2-07DC	OEM
101SS2-07SC	OEM
101SS5-07SC	OEM
101XLS	OEM
102-02	0143
102-0145-16	OEM
102-0373-00	0076
102-0394--25	OEM
102-0394-25	0155
102-0454-02	0076
102-0460-02	0151
102-0461-02	0151
102-0495-00	0786
102-0495-20	0786
102-0535-02	0127
102-0732-28	0111
102-0735-02	0191
102-0735-25	0191
102-0828-17	1220
102-0945-16	0076
102-0945-17	0076
102-0945-38	0076
102-0945-39	0076
102-4	0007
102-207	0143
102-339	0133
102-412	0124
102-1047-03	0113
102-1061-01	0042
102-1166-25	0733
102-1317-18	0155
102-1335-04	1221
102-1342-02	0127
102-1343-02	0171
102-1384-17	0018
102-1675-02	0224
102-1675-11	0076
102-1675-12	0076
102-1678-00	2495
102A1AX1	OEM
102A1AY1	OEM
102A1AYF1	OEM
102A1EX1	OEM
102A1EY1	OEM
102A1EYF1	OEM
102A2AX2	OEM
102A2AY2	OEM
102A2AYF2	OEM
102A2EX2	OEM
102A2EY2	OEM
102A2EYF2	OEM
102A3AX3	OEM
102A3AY3	OEM
102A3AYF3	OEM
102A3EX3	OEM
102A3EY3	OEM
102A3EYF3	OEM
102A4AX4	OEM
102A4AY4	OEM
102A4AYF4	OEM
102A4EX4	OEM
102A4EY4	OEM
102A4EYF4	OEM
102A5AX5	OEM
102A5AY5	OEM
102A5AYF5	OEM
102A5EX5	OEM
102A5EY5	OEM
102A5EYF5	OEM
102B1-11	OEM
102B1-12	OEM
102B1-13	OEM
102B1-14	OEM
102B1-26	OEM
102B1-27	OEM
102B1-31	OEM
102B1-32	OEM
102B1-41	OEM
102B1-42	OEM
102B1-51	OEM
102B1-52	OEM
102B3-11	OEM
102B3-12	OEM
102B3-13	OEM
102B3-14	OEM
102B3-26	OEM
102B3-27	OEM
102B3-31	OEM
102B3-32	OEM
102B3-41	OEM
102B3-42	OEM
102B3-51	OEM
102B3-52	OEM
102B6	0015
102B7-11	OEM
102B7-12	OEM
102B7-13	OEM
102B7-14	OEM
102B7-26	OEM
102B7-27	OEM
102B7-31	OEM
102B7-32	OEM
102B7-41	OEM
102B7-42	OEM
102B7-51	OEM
102B7-52	OEM
102D	0015
102DE14020	0583
102F1-11	OEM
102F1-12	OEM
102F1-13	OEM
102F1-14	OEM
102F1-26	OEM
102F1-27	OEM
102F1-31	OEM
102F1-32	OEM
102F1-41	OEM
102F1-42	OEM
102F1-51	OEM
102F1-52	OEM
102F3-11	OEM
102F3-12	OEM
102F3-13	OEM
102F3-14	OEM
102F3-26	OEM
102F3-27	OEM
102F3-31	OEM
102F3-32	OEM
102F3-41	OEM
102F3-42	OEM
102F3-51	OEM
102F3-52	OEM
102F7-11	OEM
102F7-12	OEM
102F7-13	OEM
102F7-14	OEM
102F7-26	OEM
102F7-27	OEM
102F7-31	OEM
102F7-32	OEM
102F7-41	OEM
102F7-42	OEM
102F7-51	OEM
102F7-52	OEM
102G1BX0	OEM
102G1BX1	OEM
102G1BY0	OEM
102G1BY1	OEM
102G1BYF0	OEM
102G1BYF1	OEM
102G1EX0	OEM
102G1EX1	OEM
102G1EY0	OEM
102G1EY1	OEM
102G1EYF0	OEM
102G1EYF1	OEM
102G2BX0	OEM
102G2BX2	OEM
102G2BY0	OEM
102G2BY2	OEM
102G2BYF0	OEM
102G2BYF2	OEM
102G2EX0	OEM
102G2EX2	OEM
102G2EY0	OEM
102G2EY2	OEM
102G2EYF0	OEM
102G2EYF2	OEM
102G3BX0	OEM
102G3BX3	OEM
102G3BY0	OEM
102G3BY3	OEM
102G3BYF0	OEM
102G3BYF3	OEM
102G3EX0	OEM
102G3EX3	OEM
102G3EY0	OEM
102G3EY3	OEM
102G3EYF0	OEM
102G3EYF3	OEM
102G4BX0	OEM
102G4BX4	OEM
102G4BY0	OEM
102G4BY4	OEM
102G4BYF0	OEM
102G4BYF4	OEM
102G4EX0	OEM
102G4EX4	OEM
102G4EY0	OEM
102G4EY4	OEM
102G4EYF0	OEM
102G4EYF4	OEM
102G5BX0	OEM
102G5BX5	OEM
102G5BY0	OEM
102G5BY5	OEM
102G5BYF0	OEM
102G5BYF5	OEM
102G5EX0	OEM
102G5EX5	OEM
102G5EY0	OEM
102G5EY5	OEM
102G5EYF0	OEM
102G5EYF5	OEM
102G6BX0	OEM
102G6BX6	OEM
102G6BY0	OEM
102G6BY6	OEM
102G6BYF0	OEM
102G6BYF6	OEM
102G6EX0	OEM
102G6EX6	OEM
102G6EY0	OEM
102G6EY6	OEM
102G6EYF0	OEM
102G6EYF6	OEM
102G7BX0	OEM
102G7BX7	OEM
102G7BY0	OEM
102G7BY7	OEM
102G7BYF0	OEM
102G7BYF7	OEM
102G7EX0	OEM
102G7EX7	OEM
102G7EY0	OEM
102G7EY7	OEM
102G7EYF0	OEM
102G7EYF7	OEM
102G8BX0	OEM
102G8BX8	OEM
102G8BY0	OEM
102G8BY8	OEM
102G8BYF0	OEM
102G8BYF8	OEM
102G8EX0	OEM
102G8EX8	OEM
102G8EY0	OEM
102G8EY8	OEM
102G8EYF0	OEM
102G8EYF8	OEM
102G9BX0	OEM
102G9BX9	OEM
102G9BY0	OEM
102G9BY9	OEM
102G9BYF0	OEM
102G9BYF9	OEM
102G9EX0	OEM
102G9EY0	OEM
102G9EYF0	OEM
102G10BX0	OEM
102G10BX10	OEM
102G10BY0	OEM
102G10BY10	OEM
102G10BYF0	OEM
102G10BYF10	OEM
102G10EX0	OEM
102G10EY0	OEM
102G10EYF0	OEM
102G11BX0	OEM
102G11BX11	OEM
102G11BY0	OEM
102G11BY11	OEM
102G11BYF0	OEM
102G11BYF11	OEM
102G11EX0	OEM
102G11EY0	OEM
102G11EYF0	OEM
102G12BX0	OEM
102G12BX12	OEM
102G12BY0	OEM
102G12BY12	OEM
102G12BYF0	OEM
102G12BYF12	OEM
102G12EX0	OEM
102G12EY0	OEM
102G12EYF0	OEM
102G13BX0	OEM
102G13BX13	OEM
102G13BY0	OEM
102G13BY13	OEM
102G13BYF0	OEM
102G13BYF13	OEM
102G13EX0	OEM

If replacement code is OEM, contact original manufacturer for replacement.

DEVICE TYPE	REPL CODE
102G13EY0	OEM
102G13EYF0	OEM
102G14BX0	OEM
102G14BX14	OEM
102G14BY0	OEM
102G14BY14	OEM
102G14BYF0	OEM
102G14BYF14	OEM
102G14EX0	OEM
102G14EY0	OEM
102G14EYF0	OEM
102G15BX0	OEM
102G15BX15	OEM
102G15BY0	OEM
102G15BY15	OEM
102G15BYF0	OEM
102G15BYF15	OEM
102G15EX0	OEM
102G15EY0	OEM
102G15EYF0	OEM
102G16BX0	OEM
102G16BX16	OEM
102G16BY0	OEM
102G16BY16	OEM
102G16BYF0	OEM
102G16BYF16	OEM
102G16EX0	OEM
102G16EY0	OEM
102G16EYF0	OEM
102G17BX17	OEM
102G17BY17	OEM
102G17BYF17	OEM
102G18BX18	OEM
102G18BY18	OEM
102G18BYF18	OEM
102G19BX19	OEM
102G19BY19	OEM
102G19BYF19	OEM
102G20BX20	OEM
102G20BY20	OEM
102G20BYF20	OEM
102P1	0037
102P10	0037
102S1AX0	OEM
102S1AX1	OEM
102S1AY0	OEM
102S1AY1	OEM
102S1AYF0	OEM
102S1AYF1	OEM
102S1EX0	OEM
102S1EX1	OEM
102S1EY0	OEM
102S1EY1	OEM
102S1EYF0	OEM
102S1EYF1	OEM
102S2AX0	OEM
102S2AX1	OEM
102S2AY0	OEM
102S2AY1	OEM
102S2AYF0	OEM
102S2AYF1	OEM
102S2EX0	OEM
102S2EX1	OEM
102S2EY0	OEM
102S2EY1	OEM
102S2EYF0	OEM
102S2EYF1	OEM
102S3AX0	OEM
102S3AX1	OEM
102S3AY0	OEM
102S3AY1	OEM
102S3AYF0	OEM
102S3AYF1	OEM
102S3EX0	OEM
102S3EX1	OEM
102S3EY0	OEM
102S3EY1	OEM
102S3EYF0	OEM
102S3EYF1	OEM
102S4AX0	OEM
102S4AX1	OEM
102S4AY0	OEM
102S4AY1	OEM
102S4AYF0	OEM
102S4AYF1	OEM
102S4EX0	OEM
102S4EX1	OEM
102S4EY0	OEM
102S4EY1	OEM
102S4EYF0	OEM
102S4EYF1	OEM
102S5AX0	OEM
102S5AX1	OEM
102S5AY0	OEM
102S5AY1	OEM
102S5AYF0	OEM
102S5AYF1	OEM
102S5EX0	OEM
102S5EX1	OEM
102S5EY0	OEM
102S5EY1	OEM
102S5EYF0	OEM
102S5EYF1	OEM
102S6AX0	OEM
102S6AX1	OEM
102S6AY0	OEM
102S6AY1	OEM
102S6AYF0	OEM
102S6AYF1	OEM
102S6EX0	OEM
102S6EX1	OEM
102S6EY0	OEM
102S6EY1	OEM
102S6EYF0	OEM
102S6EYF1	OEM
102S7AX0	OEM
102S7AX1	OEM
102S7AY0	OEM
102S7AY1	OEM
102S7AYF0	OEM
102S7AYF1	OEM
102S7EX0	OEM
102S7EX1	OEM
102S7EY0	OEM
102S7EY1	OEM
102S7EYF0	OEM
102S7EYF1	OEM
102S8AX0	OEM
102S8AX1	OEM
102S8AY0	OEM
102S8AY1	OEM
102S8AYF0	OEM
102S8AYF1	OEM
102S8EX0	OEM
102S8EX1	OEM
102S8EY0	OEM
102S8EY1	OEM
102S8EYF0	OEM
102S8EYF1	OEM
102S9AX0	OEM
102S9AX1	OEM
102S9AY0	OEM
102S9AY1	OEM
102S9AYF0	OEM
102S9AYF1	OEM
102S9EX0	OEM
102S9EX1	OEM
102S9EY0	OEM
102S9EY1	OEM
102S9EYF0	OEM
102S9EYF1	OEM
102S10AX0	OEM
102S10AX1	OEM
102S10AY0	OEM
102S10AY1	OEM
102S10AYF0	OEM
102S10AYF1	OEM
102S10EX0	OEM
102S10EY0	OEM
102S10EYF0	OEM
102S11AX0	OEM
102S11AX1	OEM
102S11AY0	OEM
102S11AY1	OEM
102S11AYF0	OEM
102S11AYF1	OEM
102S12AX0	OEM
102S12AX1	OEM
102S12AY0	OEM
102S12AY1	OEM
102S12AYF0	OEM
102S12AYF1	OEM
102SD13780	2042
103	OEM
103-00315-06	OEM
103-0227-18	1419
103-0235-85	0228
103-0461-02	OEM
103-0571-11	1978
103-0571-12	1946
103-1	OEM
103-3	0143
103-4	0144
103-9	OEM
103-15	OEM
103-18	OEM
103-19	0143
103-20	0196
103-21	OEM
103-22	0025
103-23	0143
103-23-01	0143
103-23-1	0143
103-24	OEM
103-28	1030
103-29	0015
103-30	OEM
103-31	0143
103-32	0196
103-34	0143
103-39	0030
103-42	0133
103-43	0196
103-44	0143
103-46	OEM
103-47	0030
103-47-01	0030
103-49	0917
103-50	OEM
103-51	0133
103-59	0015
103-60	0917
103-61	0917
103-62	0917
103-65	0917
103-70	1030
103-71	1030
103-72	OEM
103-73	0143
103-74	0143
103-76	0015
103-79	0143
103-82	0015
103-83	0365
103-84	0100
103-85	0365
103-87	0143
103-89	1030
103-90	0123
103-93	0924
103-96	0023
103-96-01	0365
103-101	2903
103-102	0123
103-104	0030
103-105	0402
103-105-01	0402
103-111	0072
103-112	0102
103-114	0143
103-123	OEM
103-126	0565
103-128	OEM
103-131	0133
103-135	0820
103-136	0002
103-137	0565
103-140	0299
103-140A	0062
103-141	0124
103-141-01	6867
103-142	0133
103-142-01	0133
103-144	0274
103-144-01	0274
103-144-03	0309
103-145	0133
103-145-01	0015
103-146	0030
103-150	0441
103-151	0309
103-152	1030
103-152-01	1030
103-157	0421
103-158	0137
103-159	0133
103-160	0102
103-166	OEM
103-176	0030
103-176-01	0720
103-176-01A	1029
103-178	0133
103-178-02	0133
103-178-02A	0133
103-179	OEM
103-185	0015
103-189	0030
103-191	0015
103-192	5332
103-193	0102
103-193-01	0023
103-194	0077
103-196	0102
103-196-01	OEM
103-197	OEM
103-202	0143
103-202-01	0917
103-202-GE	0917
103-203	0015
103-206	1255
103-207	0213
103-208	0213
103-211	5334
103-212	0565
103-212-76	0023
103-212-B	OEM
103-214	0102
103-215	1323
103-216	0015
103-219	0030
103-222	0133
103-222-01	0133
103-227	OEM
103-228	0015
103-231	0445
103-235	0030
103-235-01	0030
103-235-01A	OEM
103-236	0039
103-237	1329
103-238	1881
103-239	0288
103-239-01	0288
103-239-02	0378
103-239-02B	0378
103-239-02103-240	OEM
103-240	0133
103-244	0102
103-245	0015
103-246	0317
103-247	0102
103-248	0402
103-249	OEM
103-251	1606
103-252	0495
103-253	OEM
103-253-01	OEM
103-253-02	OEM
103-254	0015
103-254-01	0015
103-255	OEM
103-256	0237
103-257	OEM
103-258	0757
103-259	OEM
103-260	2874
103-261	0015
103-261-02	0348
103-261-02A	0015
103-261-03	0182
103-261-04	0023
103-261-04A	0023
103-263	0102
103-263A	0102
103-266	OEM
103-266-01	OEM
103-266-02#	OEM
103-269	OEM
103-269-01	OEM
103-269-02	OEM
103-270	0350
103-271	0019
103-272	0012
103-274	OEM
103-275	0288
103-275-01	0288
103-276	0402
103-278	0565
103-278-01	OEM
103-279	1276
103-279-01	0137
103-279-02	0760
103-279-03	0677
103-279-04	0118
103-279-05	0292
103-279-06	0188
103-279-07	0456
103-279-08	0533
103-279-08A	0275
103-279-09	0140
103-279-09A	0140
103-279-10	0162
103-279-10A	0041
103-279-11	0157
103-279-11A	0253
103-279-12	0091
103-279-12A	0091
103-279-13	0471
103-279-13A	0471
103-279-14	0025
103-279-14A	0062
103-279-14R	0062
103-279-15	0649
103-279-15A	0649
103-279-16	0165
103-279-16A	0165
103-279-17	1083
103-279-18	0012
103-279-18A	OEM
103-279-19	0064
103-279-19A	0064
103-279-20	0317
103-279-20A	0181
103-279-21	0137
103-279-21A	0052
103-279-22	0053
103-279-22A	0053
103-279-23	0100
103-279-23A	0879
103-279-24	0686
103-279-24A	0686
103-279-25	0420
103-279-26	1649
103-279-26A	0210
103-279-27	0495
103-279-28	0949
103-279-28A	0949
103-279-29	0700
103-279-29A	0700
103-279-30	0565
103-279-31	0402
103-279-31A	0471
103-279-32	1606
103-279-33	0441
103-279-34	0257
103-279-35	0726
103-279-36	0820
103-279-36A	OEM
103-279-37	0820
103-279-37A	0820
103-279-38	0350
103-279-39	0931
103-279-40	0999
103-279-41	0502
103-279-42	0869
103-279-43	1156
103-279-43A	1156
103-279-44	0783
103-279-45	2157
103-279-46	1189
103-279-46A	1189
103-279-47	0331
103-279-48	3026
103-279-49	1311
103-279-50	0098
103-279-51	0149
103-279-52	0186
103-279-53	0213
103-279-54	0245
103-279-55	0028
103-279-56	0255
103-279-57	0877
103-279-58	0367
103-279-59	2858
103-279-60	0421
103-279A	1276
103-280	OEM
103-280-01	OEM
103-280-02	0023
103-280-03	OEM
103-281	1029
103-284	0023
103-284A	0023
103-285	0378
103-287	0102
103-288	0039
103-288-01	0039
103-289	0402
103-290	0162
103-292	0402
103-293	1029
103-293-02	0378
103-295	0023
103-295-01	0133
103-295-01A	0133
103-295-02	0133
103-295-02A	0133
103-295-03	0023
103-295A	0133
103-298-04	OEM
103-298-05	0023
103-298-05A	0023
103-301	OEM
103-301-06A	OEM
103-301-09	0025
103-301-13	0012
103-301-13A	0012
103-301-17A	0365
103-301-19A	0002
103-301-22	0531
103-301-22A	0531
103-301-24	0402
103-301-24A	0494
103-301-27	OEM
103-304-08	OEM
103-305	0348
103-305A	0348
103-306	5351
103-306A	3328
103-308	0052
103-308A	0052
103-309-01	0064
103-309-01A	0064
103-312	0031
103-315-01	0015
103-315-01A	0015
103-315-03	0015
103-315-03A	0015
103-315-04	0015
103-315-06	0015
103-315-06A	0015
103-316-04	0017
103-317	OEM
103-326	0023
103-326A	0023
103-327	OEM
103-328	OEM
103-328-01	OEM
103-328-03	OEM
103-329	OEM
103-330	0795
103-330A	0795
103-334	0195
103-334-01	OEM
103-336-11	0253
103-336-15	0077
103-336-15A	0077
103-336-18	0057
103-336-18A	0057
103-336-19	0064
103-336-19A	0064
103-336-20A	OEM
103-339	0102
103-339-01	0031
103-339-01A	0031
103-339-02A	0031
103-339-03	0102
103-339-04	0102
103-339-04A	0031
103-339-4	0102
103-339A	0031
103-344	0015
103-344-01	0015
103-344-01A	0015
103-344-02	0015
103-344-02A	0015
103-344-03A	OEM
103-344-04A	OEM
103-344-05	OEM
103-344-06	0015
103-344-06A	0015
103-344A	0015
103-345-06	0015
103-347-03A	OEM
103-350	OEM
103-352-01	OEM
103-355-06	0071
103-355-06A	0071
103-360-01	OEM
103-360-02	OEM
103-361	OEM
103-368-03	0023
103-372	OEM
103-376-02	OEM
103-376-02A	OEM
103-380	OEM
103-381	OEM
103-387	OEM
103-387B	OEM
103-389	OEM
103-389A	OEM
103-392	OEM
103-393	OEM
103-394	0017
103-398	OEM
103-398A	OEM
103-399-12	OEM
103-399-29	OEM
103-402-03	OEM
103-404	OEM
103-408	0471
103-408A	0471
103-409	0137
103-409A	0052
103-410	OEM
103-412	5374
103-412A	OEM
103-417-01	OEM
103-417-02	OEM
103-417-03	0031
103-417-03A	0031
103-417-04	0023
103-417-04A	0031
103-417-05	0031
103-417-05A	0031
103-417A	0031
103-429	OEM
103-433	0124
103-433A	OEM
103-443	OEM
103-447	OEM
103-461	OEM
103-461A	OEM
103-462	OEM
103-462-01	OEM
103-462-03	OEM
103-467A	OEM
103-469A	OEM
103-471-01	OEM
103-471A	OEM
103-472A	OEM
103-475A	OEM
103-476-15A	OEM
103-476-16A	OEM
103-476-18A	OEM
103-476-19A	OEM
103-476-20A	OEM
103-598	OEM
103-765-01	0037
103-902	OEM
103-1344	OEM
103-29000	OEM
103-Z9000	0402
103-Z9001	0143
103-Z9002	0649
103-Z9003	0137
103-Z9004	0039
103-Z9005	0188
103-Z9006	0162
103-Z9007	0157
103-Z9008	0636
103-Z9009	0062
103-Z9010	0170
103-Z9011	0794
103-Z9012	0365
103-Z9013	0002
103-Z9014	0441
103-Z9015	0869
103-Z9016	0783
103-Z9017	0331
103-Z9018	0149
103-Z9019	0165
103-Z9020	0025
103-Z9021	0244
103-Z9022	0495
103-Z9023	0531
103-Z9024	0726
103-Z9025	0502
103-Z9026	0173
103-Z9027	0186
103-Z9028	0225
103-Z9029	0234
103-Z9030	0247
103-Z9031	0256
103-Z9032	0269
103-Z9033	0294
103-Z9034	0318
103-Z9035	0320
103-Z9036	0347
103-Z9037	0263
103-Z9038	2874
103-Z9039	0139
103-Z9040	3683
103-Z9041	0533
103-Z9042	0317
103-Z9043	1180
103-Z9044	0120
103-Z9045	1925
103A1AX1	OEM
103A1AY1	OEM
103A1AYF1	OEM
103A1EX1	OEM
103A1EY1	OEM
103A1EYF1	OEM
103A2AX2	OEM
103A2AY2	OEM
103A2AYF2	OEM
103A2EX2	OEM
103A2EY2	OEM
103A2EYF2	OEM
103A3AX3	OEM
103A3AY3	OEM
103A3AYF3	OEM
103A3EX3	OEM
103A3EY3	OEM
103A3EYF3	OEM
103A4AX4	OEM
103A4AY4	OEM
103A4AYF4	OEM
103A4EX4	OEM
103A4EY4	OEM
103A4EYF4	OEM
103A5AX5	OEM
103A5AY5	OEM
103A5AYF5	OEM
103A5EX5	OEM
103A5EY5	OEM
103A5EYF5	OEM
103B1-11	OEM
103B1-12	OEM
103B1-13	OEM
103B1-14	OEM
103B1-26	OEM
103B1-27	OEM
103B1-31	OEM
103B1-32	OEM
103B1-41	OEM
103B1-42	OEM
103B1-51	OEM
103B1-52	OEM
103B3-11	OEM
103B3-12	OEM
103B3-13	OEM
103B3-14	OEM
103B3-26	OEM
103B3-27	OEM
103B3-31	OEM
103B3-32	OEM
103B3-41	OEM
103B3-42	OEM
103B3-51	OEM
103B3-52	OEM
103B6	1335
103B7-11	OEM
103B7-12	OEM
103B7-13	OEM
103B7-14	OEM
103B7-26	OEM
103B7-27	OEM
103B7-31	OEM
103B7-32	OEM
103B7-41	OEM
103B7-42	OEM
103B7-51	OEM
103B7-52	OEM
103B98M090	1786
103B98N060	0923
103D079400	0683
103DE75200	0539
103DE76710	OEM
103EP	OEM
103F1-11	OEM
103F1-12	OEM
103F1-13	OEM
103F1-14	OEM
103F1-26	OEM
103F1-27	OEM
103F1-31	OEM
103F1-32	OEM
103F1-41	OEM
103F1-42	OEM
103F1-51	OEM
103F1-52	OEM
103F3-11	OEM
103F3-12	OEM
103F3-13	OEM
103F3-14	OEM
103F3-26	OEM
103F3-27	OEM
103F3-31	OEM
103F3-32	OEM
103F3-41	OEM
103F3-42	OEM
103F3-51	OEM
103F3-52	OEM
103F7-11	OEM
103F7-12	OEM
103F7-13	OEM
103F7-14	OEM
103F7-26	OEM
103F7-27	OEM
103F7-31	OEM
103F7-32	OEM

DEVICE TYPE	REPL CODE
103F7-41	OEM
103F7-42	OEM
103F7-51	OEM
103F7-52	OEM
103G3125	0015
103G3125A	0015
103LLS	OEM
103MLA60	OEM
103MLA80	OEM
103MLA100	OEM
103MLA120	OEM
103P(AIRLINE)	0037
103PJLA160	OEM
103PJLA180	OEM
103PJLA200	OEM
103PJLA250	OEM
103S1AX0	OEM
103S1AX1	OEM
103S1AY0	OEM
103S1AY1	OEM
103S1AYF0	OEM
103S1AYF1	OEM
103S1EX0	OEM
103S1EX1	OEM
103S1EY0	OEM
103S1EY1	OEM
103S1EYF0	OEM
103S1EYF1	OEM
103S2AX0	OEM
103S2AX1	OEM
103S2AY0	OEM
103S2AY1	OEM
103S2AYF0	OEM
103S2AYF1	OEM
103S2EX0	OEM
103S2EX1	OEM
103S2EY0	OEM
103S2EY1	OEM
103S2EYF0	OEM
103S2EYF1	OEM
103S3AX0	OEM
103S3AX1	OEM
103S3AY0	OEM
103S3AY1	OEM
103S3AYF0	OEM
103S3AYF1	OEM
103S3EX0	OEM
103S3EX1	OEM
103S3EY0	OEM
103S3EY1	OEM
103S3EYF0	OEM
103S3EYF1	OEM
103S4AX0	OEM
103S4AX1	OEM
103S4AY0	OEM
103S4AY1	OEM
103S4AYF0	OEM
103S4AYF1	OEM
103S4EX0	OEM
103S4EX1	OEM
103S4EY0	OEM
103S4EY1	OEM
103S4EYF0	OEM
103S4EYF1	OEM
103S5AX0	OEM
103S5AX1	OEM
103S5AY0	OEM
103S5AY1	OEM
103S5AYF0	OEM
103S5AYF1	OEM
103S5EX0	OEM
103S5EX1	OEM
103S5EY0	OEM
103S5EY1	OEM
103S5EYF0	OEM
103S5EYF1	OEM
103S6AX0	OEM
103S6AX1	OEM
103S6AY0	OEM
103S6AY1	OEM
103S6AYF0	OEM
103S6AYF1	OEM
103S6EX0	OEM
103S6EX1	OEM
103S6EY0	OEM
103S6EY1	OEM
103S6EYF0	OEM
103S6EYF1	OEM
103S7AX0	OEM
103S7AX1	OEM
103S7AY0	OEM
103S7AY1	OEM
103S7AYF0	OEM
103S7AYF1	OEM
103S7EX0	OEM
103S7EX1	OEM
103S7EY0	OEM
103S7EY1	OEM
103S7EYF0	OEM
103S7EYF1	OEM
103S8AX0	OEM
103S8AX1	OEM
103S8AY0	OEM
103S8AY1	OEM
103S8AYF0	OEM
103S8AYF1	OEM
103S8EX0	OEM
103S8EX1	OEM
103S8EY0	OEM
103S8EY1	OEM
103S8EYF0	OEM
103S8EYF1	OEM
103S9AX0	OEM
103S9AX1	OEM
103S9AY0	OEM
103S9AY1	OEM
103S9AYF0	OEM
103S9AYF1	OEM
103S9EX0	OEM
103S9EX1	OEM
103S9EY0	OEM
103S9EY1	OEM
103S9EYF0	OEM
103S9EYF1	OEM
103S10AX0	OEM
103S10AX1	OEM
103S10AY0	OEM
103S10AY1	OEM
103S10AYF0	OEM
103S10AYF1	OEM
103S10EX0	OEM
103S10EY0	OEM
103S10EYF0	OEM
103S11AX0	OEM
103S11AX1	OEM
103S11AY0	OEM
103S11AY1	OEM
103S11AYF0	OEM
103S11AYF1	OEM
103S12AX0	OEM
103S12AX1	OEM
103S12AY0	OEM
103S12AY1	OEM
103S12AYF0	OEM
103S12AYF1	OEM
103SD78370	2346
103TB7	OEM
103TB9	OEM
103TB11	OEM
104-1	0030
104-2	OEM
104-17	0136
104-17(RCA)	0037
104-19	0136
104-21	0136
104A1AX1	OEM
104A1AY1	OEM
104A1AYF1	OEM
104A1EX1	OEM
104A1EY1	OEM
104A1EYF1	OEM
104A2AX2	OEM
104A2AY2	OEM
104A2AYF2	OEM
104A2EX2	OEM
104A2EY2	OEM
104A2EYF2	OEM
104A3AX3	OEM
104A3AY3	OEM
104A3AYF3	OEM
104A3EX3	OEM
104A3EY3	OEM
104A3EYF3	OEM
104A4AX4	OEM
104A4AY4	OEM
104A4AYF4	OEM
104A4EX4	OEM
104A4EY4	OEM
104A4EYF4	OEM
104A5AX5	OEM
104A5AY5	OEM
104A5AYF5	OEM
104A5EX5	OEM
104A5EY5	OEM
104A5EYF5	OEM
104AHU7K	OEM
104B1-11	OEM
104B1-12	OEM
104B1-13	OEM
104B1-14	OEM
104B1-26	OEM
104B1-27	OEM
104B1-31	OEM
104B1-32	OEM
104B1-41	OEM
104B1-42	OEM
104B1-51	OEM
104B1-52	OEM
104B3-11	OEM
104B3-12	OEM
104B3-13	OEM
104B3-14	OEM
104B3-26	OEM
104B3-27	OEM
104B3-31	OEM
104B3-32	OEM
104B3-41	OEM
104B3-42	OEM
104B3-51	OEM
104B3-52	OEM
104B6	0015
104B7-11	OEM
104B7-12	OEM
104B7-13	OEM
104B7-14	OEM
104B7-26	OEM
104B7-27	OEM
104B7-31	OEM
104B7-32	OEM
104B7-41	OEM
104B7-42	OEM
104B7-51	OEM
104B7-52	OEM
104F1-11	OEM
104F1-12	OEM
104F1-13	OEM
104F1-14	OEM
104F1-26	OEM
104F1-27	OEM
104F1-31	OEM
104F1-32	OEM
104F1-41	OEM
104F1-42	OEM
104F1-51	OEM
104F1-52	OEM
104F3-11	OEM
104F3-12	OEM
104F3-13	OEM
104F3-14	OEM
104F3-26	OEM
104F3-27	OEM
104F3-31	OEM
104F3-32	OEM
104F3-41	OEM
104F3-42	OEM
104F3-51	OEM
104F3-52	OEM
104F7-11	OEM
104F7-12	OEM
104F7-13	OEM
104F7-14	OEM
104F7-26	OEM
104F7-27	OEM
104F7-31	OEM
104F7-32	OEM
104F7-41	OEM
104F7-42	OEM
104F7-51	OEM
104F7-52	OEM
104H01	0086
104K	OEM
104S1AX0	OEM
104S1AX1	OEM
104S1AY0	OEM
104S1AY1	OEM
104S1AYF0	OEM
104S1AYF1	OEM
104S1EX0	OEM
104S1EX1	OEM
104S1EY0	OEM
104S1EY1	OEM
104S1EYF0	OEM
104S1EYF1	OEM
104S2AX0	OEM
104S2AX1	OEM
104S2AY0	OEM
104S2AY1	OEM
104S2AYF0	OEM
104S2AYF1	OEM
104S2EX0	OEM
104S2EX1	OEM
104S2EY0	OEM
104S2EY1	OEM
104S2EYF0	OEM
104S2EYF1	OEM
104S3AX0	OEM
104S3AX1	OEM
104S3AY0	OEM
104S3AY1	OEM
104S3AYF0	OEM
104S3AYF1	OEM
104S3EX0	OEM
104S3EX1	OEM
104S3EY0	OEM
104S3EY1	OEM
104S3EYF0	OEM
104S3EYF1	OEM
104S4AX0	OEM
104S4AX1	OEM
104S4AY0	OEM
104S4AY1	OEM
104S4AYF0	OEM
104S4AYF1	OEM
104S4EX0	OEM
104S4EX1	OEM
104S4EY0	OEM
104S4EY1	OEM
104S4EYF0	OEM
104S4EYF1	OEM
104S5AX0	OEM
104S5AX1	OEM
104S5AY0	OEM
104S5AY1	OEM
104S5AYF0	OEM
104S5AYF1	OEM
104S5EX0	OEM
104S5EX1	OEM
104S5EY0	OEM
104S5EY1	OEM
104S5EYF0	OEM
104S5EYF1	OEM
104S6AX0	OEM
104S6AX1	OEM
104S6AY0	OEM
104S6AY1	OEM
104S6AYF0	OEM
104S6AYF1	OEM
104S6EX0	OEM
104S6EX1	OEM
104S6EY0	OEM
104S6EY1	OEM
104S6EYF0	OEM
104S6EYF1	OEM
104S7AX0	OEM
104S7AX1	OEM
104S7AY0	OEM
104S7AY1	OEM
104S7AYF0	OEM
104S7AYF1	OEM
104S7EX0	OEM
104S7EX1	OEM
104S7EY0	OEM
104S7EY1	OEM
104S7EYF0	OEM
104S7EYF1	OEM
104S8AX0	OEM
104S8AX1	OEM
104S8AY0	OEM
104S8AY1	OEM
104S8AYF0	OEM
104S8AYF1	OEM
104S8EX0	OEM
104S8EX1	OEM
104S8EY0	OEM
104S8EY1	OEM
104S8EYF0	OEM
104S8EYF1	OEM
104S9AX0	OEM
104S9AX1	OEM
104S9AY0	OEM
104S9AY1	OEM
104S9AYF0	OEM
104S9AYF1	OEM
104S9EX0	OEM
104S9EX1	OEM
104S9EY0	OEM
104S9EY1	OEM
104S9EYF0	OEM
104S9EYF1	OEM
104S10AX0	OEM
104S10AX1	OEM
104S10AY0	OEM
104S10AY1	OEM
104S10AYF0	OEM
104S10AYF1	OEM
104S10EX0	OEM
104S10EY0	OEM
104S10EYF0	OEM
104S11AX0	OEM
104S11AX1	OEM
104S11AY0	OEM
104S11AY1	OEM
104S11AYF0	OEM
104S11AYF1	OEM
104S12AX0	OEM
104S12AX1	OEM
104S12AY0	OEM
104S12AY1	OEM
104S12AYF0	OEM
104S12AYF1	OEM
104T2	0103
104TB7	OEM
104TB9	OEM
104TB11	OEM
104Z4	OEM
105	0143
105(ADMIRAL)	0016
105-001-04	0079
105-001-05	0079
105-001-07	0330
105-001-08	0144
105-005-04/2228-3	0144
105-005-12	0144
105-006-08	0016
105-008-04/2228-3	0144
105-009-21/2228-3	0144
105-00106-00	0144
105-00107-09	0127
105-00108-07	0144
105-060-09	0016
105-085-33	0016
105-02004-09	0144
105-02005-07	0127
105-02006-05	0127
105-02008-01	0127
105-06004-00	0144
105-06007-05	0016
105-08243-05	0079
105-110	OEM
105-114	OEM
105-904-85	0079
105-904-86	0079
105-904-87	0079
105-931-91/2228-3	0016
105-941-97/2228-3	0144
105-24191-04	0144
105-28196-07	0016
105B6	0015
105B7-11	OEM
105B7-12	OEM
105B7-13	OEM
105B7-14	OEM
105B7-26	OEM
105B7-27	OEM
105B7-31	OEM
105B7-32	OEM
105B7-41	OEM
105B7-42	OEM
105B7-51	OEM
105B7-52	OEM
105DE87820	OEM
105S925080	2194
105T4	OEM
105V10	OEM
105Z4	OEM
106	0015
106-001	0113
106-002	0076
106-003	0076
106-004	0390
106-005	0786
106-006	2495
106-007	0133
106-008	0019
106-009	0139
106-010	0057
106-011	0023
106-012	5117
106-013	0087
106-1	0015
106-120	0037
106-351	1066
106A	0447
106B	0940
106B6	0015
106B7-11	OEM
106B7-12	OEM
106B7-13	OEM
106B7-14	OEM
106B7-26	OEM
106B7-27	OEM
106B7-31	OEM
106B7-32	OEM
106B7-41	OEM
106B7-42	OEM
106B7-51	OEM
106B7-52	OEM
106C	1222
106D	OEM
106E	OEM
106F	2095
106F4	1259
106KB0	0126
106KBA	0126
106LL2	OEM
106M	0325
106P1	0160
106P1AG	0160
106P1T	0160
106Q	2095
106RED	0037
106Y	2095
106Z4	OEM
107-0021-00	1669
107-1	OEM
107-2	OEM
107-3	OEM
107-4	OEM
107-8	0016
107-3088	0079
107A	0050
107A(SCR)	0447
107B	0050
107B(SCR)	0940
107B6	0015
107B7-11	OEM
107B7-12	OEM
107B7-13	OEM
107B7-14	OEM
107B7-26	OEM
107B7-27	OEM
107B7-31	OEM
107B7-32	OEM
107B7-41	OEM
107B7-42	OEM
107B7-51	OEM
107B7-52	OEM
107BRN	0016
107C	1222
107D	0095
107E	OEM
107F	2095
107M	0050
107N1	0086
107N2	0016
107P341	0576
107Q	2095
107Y	2095
107Z4	OEM
108	0015
108(FARFISA)	0143
108-0049-08	0325
108-0068-12	0325
108-1	0050
108-2	0050
108-3	0050
108-4	0050
108-6	0037
108-60	0037
108-74	0143
108-116	OEM
108A	0447
108A4	0015
108B	0940
108B1-11	OEM
108B1-12	OEM
108B1-13	OEM
108B1-14	OEM
108B1-26	OEM
108B1-27	OEM
108B1-31	OEM
108B1-32	OEM
108B1-41	OEM
108B1-42	OEM
108B1-51	OEM
108B1-52	OEM
108B3-11	OEM
108B3-12	OEM
108B3-13	OEM
108B3-14	OEM
108B3-26	OEM
108B3-27	OEM
108B3-31	OEM
108B3-32	OEM
108B3-41	OEM
108B3-42	OEM
108B3-51	OEM
108B3-52	OEM
108B4	0015
108B6	0015
108B7-11	OEM
108B7-12	OEM
108B7-13	OEM
108B7-14	OEM
108B7-26	OEM
108B7-27	OEM
108B7-31	OEM
108B7-32	OEM
108B7-41	OEM
108B7-42	OEM
108B7-51	OEM
108B7-52	OEM
108C	1222
108D	0095
108E-E2	0015
108F	1259
108F1-11	OEM
108F1-12	OEM
108F1-13	OEM
108F1-14	OEM
108F1-26	OEM
108F1-27	OEM
108F1-31	OEM
108F1-32	OEM
108F1-41	OEM
108F1-42	OEM
108F1-51	OEM
108F1-52	OEM
108F3-11	OEM
108F3-12	OEM
108F3-13	OEM
108F3-14	OEM
108F3-26	OEM
108F3-31	OEM
108F3-32	OEM
108F3-41	OEM
108F3-42	OEM
108F3-51	OEM
108F3-52	OEM
108F7-11	OEM
108F7-12	OEM
108F7-13	OEM
108F7-14	OEM
108F7-26	OEM
108F7-27	OEM
108F7-31	OEM
108F7-32	OEM
108F7-41	OEM
108F7-42	OEM
108F7-51	OEM
108F7-52	OEM
108GRN	0037
108Q	1396
108T2	2485
108Y	1396
108Z4	OEM
109	0016
109-036500	0143
109-1	0079
109-1(RCA)	0016
109-192	0143
109A949CE	2015
109A949CM-P1	0411
109B6	0015
109B7-11	OEM
109B7-12	OEM
109B7-13	OEM
109B7-14	OEM
109B7-26	OEM
109B7-27	OEM
109B7-31	OEM
109B7-32	OEM
109B7-41	OEM
109B7-42	OEM
109B7-51	OEM
109B7-52	OEM
109LL2	OEM
109SLS	OEM
109T2	2485
109UA	OEM
109UB	OEM
109UC	OEM
109UD	OEM
109WA	OEM
109WB	OEM
109WC	OEM
109WD	OEM
109XA	OEM
109XB	OEM
109XC	OEM
109XD	OEM
109Z4	OEM
110	0037
110-01563-00	0004
110-629	0015
110-635	0015
110-636	0071
110-672	0015
110-684	0015
110-763	0143
110-US	OEM
110B6	0071
110BH1	0774
110LL2	OEM
110P	OEM
110P1	0037
110P1AA	0037
110P1M	0037
110P2	0150
110PB	OEM
110PB9M	OEM
110PBF9	OEM
110PBHL	OEM
110PBL	OEM
110PBM	OEM
110PD9M	OEM
110PH9M	OEM
110PL	OEM
110PS9M	OEM
110T4	OEM
110Z4	OEM
111-0-9170-00700	OEM
111-4-2020-0860	0143
111-4-2020-04600	0167
111-4-2020-05600	0123
111-4-2020-06100	0133
111-4-2020-06200	0102
111-4-2020-08000	0015
111-4-2020-08600	0143
111-4-2020-11300	0025
111-4-2020-11500	0317
111-4-2020-13300	0820
111-4-2020-14400	0102
111-4-2020-14500	0015
111-4-2020-14600	0015
111-4-2020-15100	0102
111-4-2060-0400	0167
111-4-2060-02300	0878
111-4-2060-02400	0878
111-4-2060-03900	0855
111-4-2060-04000	0167
111-4-2060-05100	0881
111-4-2060-05200	0884
111-731	0325
111-6191	0725
111-6510	0208
111-6910	0050
111-6935	0136
111N1	0491
111N2	0103
111N4	0103
111N4A	0103
111N4B	0103
111N5	OEM
111P1	0491
111P3B	0905
111P7	0491
111RK10	OEM
111RK10M	OEM
111RK20	OEM
111RK20M	OEM
111RK40	OEM
111RK40M	OEM
111RK60	OEM
111RK60M	OEM
111RK80	OEM
111RK80M	OEM
111RK100	OEM
111RK100M	OEM
111RK120	OEM
111RK120M	OEM
111RLA60	OEM
111RLA80	OEM
111RLA100	OEM
111RLA120	OEM
111T2	0233
111Z4	3930
112-000172	0037
112-000185	0037
112-000187	0037
112-000267	0050
112-001	0280
112-002	0136
112-003	0211
112-004	0211
112-0011-A	0079
112-034923	0004
112-1A82	0016
112-2	0126
112-2A	0126
112-3	3698
112-7	0037
112-8	0037

DEVICE TYPE	REPL CODE
112-10	0037
112-361	0086
112-362	0630
112-363	0103
112-500-0-50	0133
112-500-0-501	0133
112-520	0144
112-521	0144
112-522	0144
112-523	0016
112-524	0160
112-525	0086
112-601-0-102	0015
112-826	0015
112-200525	0004
112-202147	0160
112-203053	0086
112-203055	0103
112-203391	0111
112-7292955	0604
112P1	OEM
112RLA60	OEM
112RLA80	OEM
112RLA100	OEM
112RLA120	OEM
113-039	0015
113-118	0198
113-321	0015
113-392	0015
113-398	0144
113-938	0144
113-998	0229
113A7739	0015
113N1	OEM
113N1AG	0400
113N2	0400
113Z4	OEM
114-013	0015
114-1	0079
114-4-2020-14500	0015
114-118	0144
114-267	0144
114N4P	0208
114P3P	0164
115	3314
115-039	0015
115-063	0160
115-1	0198
115-4	0198
115-13	0079
115-225	0016
115-227	0050
115-228	0050
115-229	0050
115-268	0160
115-269	0160
115-275	0136
115-559	0015
115-599	0015
115-867	0229
115-875	0016
115B7T	OEM
115E	3314
115E(2SA1115E)	3314
115F	3314
115T4	OEM
115UA	OEM
115UB	OEM
115UC	OEM
115UD	OEM
115UE	OEM
115UF	OEM
115WA	OEM
115WB	OEM
115WC	OEM
115WD	OEM
115XA	OEM
115XB	OEM
115XC	OEM
115XD	OEM
115YA	OEM
115YB	OEM
115YC	OEM
115YD	OEM
115Z4	0002
116-052	0015
116-068	0975
116-072	0136
116-073	0144
116-074	0016
116-075	0142
116-078	0016
116-079	0144
116-080	0144
116-082	0144
116-083	0144
116-085	0016
116-086	0975
116-087	0975
116-088	0975
116-089	0975
116-091	0004
116-092	0016
116-1	3844
116-198	0144
116-199	0144
116-200	0144
116-201	0004
116-202	0050
116-203	0004
116-206	0004
116-207	0050
116-208	0050
116-209	0050
116-687	0208
116-756	0136
116-757	0004
116-875	0016
116-997	0004
116BT300R	OEM
116BT500R	OEM
117-1	0142
117-2	0164
117-6	2866
117-9	1657
117-134-11	0015
117-A40	0378
118-1	0144
118-2	0144
118-3	0144
118-4	0144
118-7470	OEM
118J2	OEM
118UA	OEM
118UB	OEM
118UC	OEM
118UD	OEM
118XA	OEM
118XB	OEM
118XC	OEM
118XD	OEM
119-0016	0440
119-0054	0016
119-0055	0037
119-0056	0016
119-0068	0178
119-0075	0130
119-0077	0086
119-6511	0015
120-000190	0050
120-00-19	0211
120-001-300	0015
120-00190	0136
120-00195	0004
120-00213	0136
120-001190	0050
120-001192	0280
120-001195	0211
120-001300	0015
120-001301	0123
120-001795	0211
120-001798	0698
120-002012	0004
120-002013	0004
120-002014	0004
120-002213	0050
120-002214	0050
120-002216	0050
120-002513	0136
120-002515	0136
120-002518	0136
120-002520	0164
120-002521	0004
120-002656	0136
120-002748	0004
120-003147	0133
120-003148	0015
120-003149	0071
120-003150	0222
120-003151	0698
120-004048	0136
120-004061	0071
120-004480	0016
120-004482	0016
120-004483	0016
120-004492	0136
120-004493	0164
120-004494	0164
120-004495	0164
120-004496	0144
120-004497	0144
120-004498	0143
120-004499	0143
120-004503	0015
120-004629	0913
120-004722	0050
120-004723	0144
120-004724	0144
120-004725	0144
120-004727	0004
120-004728	0164
120-004729	0004
120-004730	0143
120-004877	0133
120-004878	0015
120-004879	0012
120-004880	0016
120-004881	0144
120-004882	0016
120-004883	0016
120-004884	0338
120-004887	0222
120-004888	0600
120-005291	0144
120-005292	0144
120-005293	0144
120-005294	0144
120-005295	0144
120-005296	0144
120-005297	0144
120-005298	0144
120-005299	0143
120-005300	0030
120-006604	0037
120-01193	0004
120-02213	0050
120-1	0016
120-2	0016
120-3	0016
120-7	0016
120-8	0016
120-8A	0016
120-190	0004
120-7396	0031
120A11	0133
120BLU	0016
120P1	0037
120P1M	0037
120T4	OEM
120V10	OEM
121	0212
121(DIO)	0023
121(SEARS)	0016
121-1	0037
121-1RED	0037
121-6	0208
121-7	0208
121-8	0208
121-9	0004
121-10	0004
121-11	0004
121-12	0004
121-14	0280
121-14(COLUMBIA)	0086
121-14(ZENITH)	0015
121-15	0600
121-16	0600
121-17	0600
121-18	0142
121-18(ZENITH)	0164
121-19	0004
121-21	0600
121-22	0600
121-24	0600
121-25	0600
121-26	0600
121-27	0211
121-31	0143
121-33	0004
121-34	0004
121-43	0004
121-44	0136
121-45	0136
121-46	0004
121-47	0004
121-48	0050
121-49	0050
121-50	0600
121-51	0600
121-52	0004
121-53	0280
121-54	0136
121-59	0208
121-60	0208
121-61	0004
121-62	0136
121-63	0050
121-64	0004
121-65	0136
121-66	0136
121-67	0136
121-68	0004
121-69	0004
121-70	0600
121-71	0600
121-72	0004
121-73	0050
121-74	0050
121-75	0050
121-76	0050
121-78	0050
121-79	0050
121-80	0280
121-81	0280
121-82	0280
121-83	0280
121-84	0280
121-85	0280
121-86	0280
121-87	0280
121-88	0280
121-89	0280
121-90	0280
121-91	0280
121-92	0280
121-93	0280
121-94	0280
121-95	0211
121-96	0211
121-99-1A	OEM
121-100	0600
121-101	0050
121-102	0050
121-103	0050
121-104	0050
121-105	0050
121-106	0211
121-107	0211
121-113	0144
121-119	0050
121-120	0164
121-128	0280
121-132	0050
121-134	0050
121-135	0050
121-136	0050
121-137	0050
121-138	0050
121-139	0050
121-145	0280
121-146	0280
121-147	0280
121-148	0004
121-150	0136
121-151	0004
121-152	0004
121-153	0136
121-154	0136
121-157	0050
121-160	0280
121-161	0136
121-162	0136
121-163	0004
121-164	0004
121-165	0600
121-166	0600
121-167	0164
121-170	0037
121-171	0160
121-179	0050
121-180	0050
121-181	0050
121-184	0004
121-185	0050
121-186	0136
121-187	0136
121-189	0136
121-190	0004
121-191	0004
121-192	0004
121-193	0004
121-195B	0016
121-200	0211
121-205	0280
121-206	0280
121-207	0280
121-208	0280
121-209	0280
121-210	0280
121-211	0280
121-212	0280
121-213	0280
121-219	0280
121-220	0280
121-221	0280
121-222	0280
121-225	0280
121-226	0004
121-227	0004
121-228	0136
121-229	0136
121-230	0136
121-231	0136
121-232	0136
121-233	0136
121-234	0280
121-235	0280
121-236	0280
121-237	0208
121-238	0208
121-239	0004
121-240	0004
121-240X	0004
121-241	0136
121-241(ZENITH)	0136
121-242	0050
121-243	0050
121-244	0050
121-245	0004
121-246	0004
121-247	0208
121-247BLK	0208
121-247VIL	0600
121-247VIO	0600
121-247WHT	0600
121-248	0208
121-248BLU	0208
121-248VIL	0600
121-248VIO	0600
121-254	0280
121-256	0136
121-257	0050
121-258	0050
121-259	0050
121-260	0050
121-261	0136
121-262	0136
121-263	0136
121-266	0211
121-267	0211
121-268	0050
121-269	0050
121-270	0160
121-271	0160
121-272	0004
121-273	0004
121-274	0004
121-275	0004
121-276	0198
121-277	0198
121-278	0198
121-279	0086
121-283	0007
121-284	0050
121-286	0079
121-287	0004
121-288	0079
121-289	0136
121-290	0136
121-291	0004
121-292	0136
121-293	0136
121-294	0050
121-295	0050
121-296	0136
121-297	0136
121-298	0050
121-299	0136
121-300	0004
121-301	0004
121-302	0600
121-303	0007
121-304	0136
121-305	0004
121-306	0004
121-307	0004
121-308	0160
121-309	0004
121-310	0004
121-311	0164
121-311B	0164
121-311C	0164
121-311D	0164
121-311E	0164
121-311F	0164
121-312	0050
121-313	0050
121-314	0004
121-315	0142
121-316	0144
121-317	0144
121-318	0144
121-318L	0144
121-319	0004
121-320	0004
121-321	0144
121-327	0004
121-328	0004
121-329	0050
121-330	0050
121-331	0050
121-332	0050
121-333	0050
121-334	0050
121-335	0050
121-336	0050
121-345	0144
121-347	0136
121-348	0004
121-349	0050
121-350	0136
121-351	5347
121-352	0136
121-353	0136
121-354	0280
121-356	0050
121-357	0050
121-358	0050
121-359	0050
121-360	0847
121-361	0233
121-362	0280
121-363	0160
121-364	0016
121-365	0038
121-366	0016
121-367	0016
121-368	0050
121-369	0076
121-370	0975
121-370-X	OEM
121-371	0160
121-372	0211
121-373	0164
121-373B	0164
121-373C	0164
121-373C(ORANGE)	0164
121-373D	0164
121-373D(RED)	0164
121-373E	0164
121-373E(ORANGE)	0164
121-373F	0164
121-373F(YELLOW)	0164
121-373G	0164
121-373H	0164
121-373H(YELLOW)	0164
121-373I	0164
121-373J	0164
121-373J(YELLOW)	0164
121-373K	0164
121-373K(GREEN)	0164
121-373L	0164
121-373M	0164
121-373M(GREEN)	0164
121-373N	0164
121-373N(BLUE)	0164
121-373O	0164
121-373P	0164
121-373P(BLUE)	0164
121-373Q	0164
121-373Q(VIOLET)	0164
121-374	0004
121-375	0004
121-377	0007
121-378	0007
121-379	0007
121-380	0007
121-381	0136
121-382	0160
121-383	0050
121-384	0050
121-385	0136
121-388	0847
121-389	0160
121-389-X	OEM
121-395	0004
121-395-X	OEM
121-396	0004
121-397	0280
121-398	0160
121-399	0004
121-400	0164
121-401	0847
121-401BROWN	0847
121-401ORANGE	0847
121-401RED	0847
121-401YELLOW	0847
121-403	0164
121-404	0016
121-406	0604
121-408	0004
121-409	0004
121-410	0004
121-411	0050
121-412	0050
121-413	0050
121-414	0050
121-415	0050
121-415B	0050
121-416	0211
121-417	0037
121-418	0604
121-418BLUE	0604
121-418GREEN	0604
121-418ORANGE	0604
121-418VIOLET	0604
121-418YELLOW	0604
121-419	0006
121-420	0004
121-421	0164
121-422	0198
121-423	0016
121-425	0164
121-425A	0164
121-425A(BROWN)	0164
121-425B	0164
121-425B(RED)	0164
121-425C	0164
121-425D	0164
121-425D(RECT)	0164
121-425E	0164
121-425E(ORANGE)	0164
121-425F	0164
121-425F(YELLOW)	0164
121-425G	0164
121-425G(ORANGE)	0164
121-425H	0164
121-425H(YELLOW)	0164
121-425I	0164
121-425I(GREEN)	0164
121-425J	0164
121-425J(YELLOW)	0164
121-425K	0164
121-425K(GREEN)	0164
121-425L	0164
121-425L(BLUE)	0164
121-425M	0164
121-425M(GREEN)	0164
121-425N	0164
121-425N(BLUE)	0164
121-425O	0164
121-425P	0164
121-425Q	0164
121-425Q(VIOLET)	0164
121-426	0050
121-427	0050
121-428	0050
121-429	0050
121-430	0086
121-430B	0079
121-430CL	0079
121-431	0086
121-432	0136
121-433	0079
121-433A	0079
121-433CL	0079
121-434	0079
121-434H	0079
121-435	0079
121-436	0142
121-437	0004
121-441	0037
121-442	0079
121-443	OEM
121-443A	OEM
121-444	0006
121-445	0233
121-446	0006
121-447	0079
121-447-04	OEM
121-448	0016
121-449	0313
121-450	0016
121-451	0142
121-452	0074
121-453	0016
121-460	0136
121-461	0007
121-462	0007
121-470	0006
121-471	0007
121-472	0144
121-473	0233
121-477	OEM
121-480	0144
121-481	0007
121-482	0127
121-483	0144
121-490	0004
121-491	0136
121-492	0136
121-493	0136
121-494	0136
121-495	0037
121-496	0006
121-497	0037
121-497WHT	0006
121-498	0079
121-499	0079
121-499-01	0079
121-499-01A	0079
121-499-01C	0079
121-499-1A	0079
121-499A	0079
121-500	0086
121-501	0224
121-502	0224
121-503	0224
121-504	0079
121-505	0144
121-506	0144
121-507	0224
121-508	0007
121-509	0016
121-510	0144
121-520	0076
121-521	0224
121-521(MOTOROLA)	0079
121-521(ZENITH)	0224
121-522	0076
121-523	0016
121-524	0016
121-524A	0224
121-526	0007
121-538	0050
121-538B	0050
121-539	0050
121-540	0050
121-540B	0050
121-541	0050
121-541B	0050
121-542	0050
121-542B	0050
121-543	0004
121-544	0004
121-546	0224
121-546B	0144
121-547	0224
121-551	0224
121-552	0050
121-553	0050
121-554	0136
121-555	0136
121-557	0208
121-558	0208
121-560	0144
121-580	0007
121-581	0016
121-582	0142
121-582GREEN	0142
121-583	0144
121-585	0259
121-585B	0259
121-587	0144
121-588	0178
121-600	0079
121-600(ZENITH)	0079
121-601	0050
121-602	0037
121-603	0037
121-610	0016
121-612	0144
121-612-16	0144
121-613	0144
121-613-16	0144
121-614	0144
121-614-9	0144
121-615	0150
121-616	0144
121-617	OEM
121-619	OEM
121-620	OEM
121-622	OEM

If replacement code is OEM, contact original manufacturer for replacement.

DEVICE TYPE	REPL CODE
121-626	OEM
121-627	OEM
121-630	0905
121-632	0004
121-633	0004
121-634	0004
121-635	0004
121-636	0004
121-636B	0004
121-636C	0004
121-636D	0004
121-637	0144
121-638	0144
121-638B	0144
121-639	0016
121-639CL	0086
121-640	0004
121-641	0208
121-641BLU	0208
121-641GRN	0208
121-641ORN	0208
121-641RED	0208
121-641YEL	0208
121-642	0006
121-643	0006
121-644	0144
121-645	0144
121-646	0016
121-647	0016
121-648	0016
121-649	0016
121-650	0038
121-651	0144
121-652	0016
121-653	0144
121-654	0037
121-655	0076
121-656	0076
121-657	0016
121-658	0144
121-661	0037
121-662	0076
121-663	0211
121-664	0037
121-665	0103
121-667	0006
121-668	0076
121-671	0076
121-672	0016
121-673	0208
121-674	0211
121-675	1221
121-676	0076
121-677	1221
121-678	0155
121-678GREEN	0155
121-678YELLOW	0155
121-679	0037
121-679GREEN	0037
121-679YELLOW	0037
121-680	0037
121-681	0037
121-682	0037
121-683	0037
121-684	0076
121-687	0079
121-692	0007
121-695	0079
121-697	0050
121-698	0050
121-699	0037
121-699-01	0037
121-699-02	0037
121-699A	0037
121-701	0079
121-702	0076
121-703	0076
121-704	0006
121-705	0211
121-706	0079
121-707	0460
121-708	0161
121-709	0460
121-710	0161
121-711	0076
121-712	0276
121-713	0142
121-714	0037
121-716	0037
121-719	0042
121-722	0086
121-723	0007
121-725	0006
121-726	0103
121-726A	0103
121-730	0144
121-731	0325
121-732	0144
121-733	0076
121-734	0130
121-735	0127
121-737	0086
121-737CL	0086
121-739	0127
121-741	0259
121-742	0127
121-743	0233
121-743-01	0338
121-743-01(T05)	0233
121-743-01(X51)	0338
121-744	0233
121-745	0016
121-746	0037
121-748	0007
121-751	0016
121-752	3799
121-753	0144
121-754	0224
121-755	0264
121-756	0325
121-758	0003
121-758-X	0003
121-758X	0003
121-759	1150
121-759X	0065
121-760	0007
121-761	0007
121-762	0600
121-762CL	0208
121-764	0016
121-765	0126
121-765-01	0037
121-766	0086
121-766-01	0595
121-767	0111
121-767CL	0016
121-768	0111
121-768CL	0016
121-770	0423
121-770-X	0042
121-770CL	0423
121-770X	0042
121-772	0042
121-772CL	0423
121-773	0079
121-773CL	0086
121-774	0037
121-774CL	0126
121-775	0007
121-776	0187
121-777	0187
121-777-01	0233
121-779	0127
121-782	0212
121-783	0212
121-784	0212
121-785	0212
121-786	0212
121-787	0212
121-787-01	0079
121-792	0187
121-793	0160
121-801	0037
121-802	0007
121-803	0681
121-804	0042
121-805	0161
121-806	0042
121-807	0144
121-808	0161
121-809	0236
121-812	0016
121-819	0144
121-819-01	0861
121-821	0003
121-822	0285
121-823	0007
121-824	0007
121-825	0079
121-826	0212
121-827	0144
121-828	OEM
121-829	0103
121-830	0280
121-831	0065
121-831-X	0065
121-834	0127
121-835	0076
121-836	0079
121-840	0224
121-841	0144
121-843	0264
121-843A	0595
121-844	0086
121-845	0126
121-846	0076
121-847	0007
121-848	0144
121-849	0127
121-850	0079
121-851	0144
121-853	0423
121-853-X	OEM
121-853X	0560
121-854	0042
121-855	0224
121-855-01	0224
121-856	0016
121-857	0144
121-858	0325
121-859	OEM
121-860	0325
121-861	0037
121-862	0079
121-863	0079
121-865	0037
121-866	OEM
121-868	0285
121-868-01	0285
121-868-02	0285
121-869	0144
121-871	0861
121-872	0224
121-873	0925
121-875	0037
121-876	0111
121-877	0079
121-878	0086
121-879	0037
121-880	0423
121-880-X	OEM
121-881	0016
121-883	0016
121-884	0144
121-885	0224
121-886	0925
121-887	0228
121-888	0079
121-889	0016
121-895	0079
121-895A	0079
121-896	0037
121-897	OEM
121-899	0144
121-900	0076
121-906	0224
121-907	0224
121-908	0224
121-909	0144
121-910	0144
121-911	0187
121-911X	0187
121-912	5409
121-913	0016
121-914	0016
121-915	0016
121-916	0016
121-921	OEM
121-921-01	0042
121-921-02	OEM
121-924	0224
121-925	0144
121-926	0925
121-926-01	0925
121-926-02	OEM
121-927	0042
121-927-01	0042
121-927-02	OEM
121-928	0111
121-929	0224
121-930	0224
121-931	0079
121-932	0144
121-933	1027
121-943	0224
121-944	0079
121-945	0224
121-946	0127
121-950	0224
121-950-01	0224
121-951	0224
121-952	0037
121-953	0212
121-954	0224
121-956	0042
121-957	0015
121-958	0144
121-964	OEM
121-965	OEM
121-966	0042
121-966-01	0042
121-966-02	0042
121-966-03	0042
121-967	0423
121-968	0144
121-969	0925
121-969-01	OEM
121-969-02	0925
121-970	0042
121-970-02	0236
121-972	0079
121-972-01	0079
121-973	0037
121-973A	0037
121-974	0127
121-975	0079
121-975A	0079
121-976	0042
121-976-01	0042
121-977	0925
121-977-01	0854
121-978	0037
121-978-01	0037
121-978-01A	0807
121-978-03	0037
121-978-03A	0037
121-978-03C	0037
121-978A	0037
121-980	0382
121-980-01	0382
121-982	0079
121-983	0127
121-984	0127
121-984-01	OEM
121-985	0065
121-985-01	OEM
121-985-X	OEM
121-986	0037
121-986A	0037
121-987	0236
121-987-01	2997
121-987-02	0423
121-987-03	0768
121-988	0681
121-988-01	OEM
121-988-02	0854
121-988-03	0925
121-989	0285
121-990	0285
121-992	0236
121-992-01	0236
121-993	1027
121-994	0681
121-995	0127
121-996	0642
121-997	0925
121-998	0917
121-999	0475
121-1001	0224
121-1002	0127
121-1003	0313
121-1004	0079
121-1005	0037
121-1006	0042
121-1007	0382
121-1008	0236
121-1008-01	0429
121-1009	0681
121-1010	0144
121-1012	0602
121-1013	1212
121-1014	0338
121-1015	0065
121-1016	0006
121-1017	0155
121-1018	0127
121-1019	0037
121-1019-01	0037
121-1019-01A	0037
121-1019-01C	0037
121-1019A	0037
121-1020	0861
121-1020A	0861
121-1021	0436
121-1023	0224
121-1024	0212
121-1026	OEM
121-1027	1514
121-1028	1027
121-1028-01	0392
121-1029	0065
121-1029-01	0065
121-1029-02	0065
121-1030	0212
121-1031	0178
121-1032	0211
121-1033	0313
121-1034	0285
121-1034-01	0285
121-1034-1	0285
121-1034A	0285
121-1035	0324
121-1035A	0324
121-1036	0436
121-1037	0285
121-1040	0079
121-1040-01	0079
121-1040-01A	0079
121-1040A	0079
121-1042	3892
121-1042-01	3892
121-1042A	3892
121-1043	0037
121-1043A	0037
121-1044	0127
121-1045	0286
121-1048	0325
121-1049	0236
121-1051	0537
121-1059	0037
121-1059A	0037
121-1060	OEM
121-1061	0236
121-1061-01	0392
121-1062	0037
121-1063	0285
121-1064	0037
121-1064A	0037
121-1065	0079
121-1065A	0079
121-1066	0371
121-1070	0728
121-1072	0236
121-1072-01	0236
121-1082-02	0042
121-1084	0079
121-1084A	0079
121-1086-01	2908
121-1089	OEM
121-1090	OEM
121-1091	OEM
121-1092	0065
121-1094	OEM
121-1096	0079
121-1101	0342
121-1101-02	0037
121-1101A	0037
121-1102	0006
121-1102A	0006
121-1103	OEM
121-1109	2908
121-1109-01	2908
121-1110	OEM
121-1112	0681
121-1112-01	0681
121-1112-1	OEM
121-1114	0055
121-1124	0160
121-1125	OEM
121-1125-01	OEM
121-1125-01A	OEM
121-1125B	OEM
121-1126	OEM
121-1126A	OEM
121-1127-01	OEM
121-1127-01A	OEM
121-1128	OEM
121-1130	OEM
121-1130A	OEM
121-1130B	OEM
121-1133	1742
121-1133A	1752
121-1133B	OEM
121-1134	0160
121-1136-01	OEM
121-1136-1	4450
121-1139A	OEM
121-1140	0261
121-1140A	0261
121-1140C	0261
121-1141	1541
121-1141-01	1541
121-1141-02	1541
121-1145	0006
121-1145A	0006
121-1146	0715
121-1146A	0079
121-1148	1541
121-1148-01	OEM
121-1149	OEM
121-1153	1500
121-1158	1806
121-1158A	1806
121-1160	OEM
121-1162-01A	OEM
121-1167	3514
121-1167-01	OEM
121-1169	OEM
121-1187	OEM
121-1189-01	OEM
121-1190	OEM
121-1201	OEM
121-1210	OEM
121-1216-01	OEM
121-1225-01	OEM
121-1226B	OEM
121-1227A	OEM
121-1232	0079
121-1232A	0079
121-1237	OEM
121-1256	OEM
121-1256A	OEM
121-1256Z	OEM
121-1261A	OEM
121-1262A	OEM
121-1263	OEM
121-1263A	OEM
121-1264	OEM
121-1264-01	OEM
121-1264-01A	OEM
121-1264A	OEM
121-1290	OEM
121-1291-01	OEM
121-1292	0079
121-1295	OEM
121-1299	OEM
121-1300-01A	OEM
121-1300A	OEM
121-1301-01A	OEM
121-1301A	OEM
121-1325	OEM
121-1330	0280
121-1350	0280
121-1360	0280
121-1390	0280
121-1400	0280
121-1410	0600
121-5065	0016
121-9003	0436
121-505501	0016
121-Z9000	0079
121-Z9000-A	0016
121-Z9000A	0016
121-Z9001	0313
121-Z9002	0460
121-Z9003	0037
121-Z9004	0004
121-Z9005	0126
121-Z9006	0160
121-Z9007	0164
121-Z9008	0560
121-Z9009	0975
121-Z9010	0178
121-Z9011	0103
121-Z9012	0142
121-Z9013	0136
121-Z9014	0218
121-Z9015	0208
121-Z9016	0276
121-Z9017	0222
121-Z9018	0074
121-Z9019	1242
121-Z9020	0150
121-Z9021	0224
121-Z9022	0642
121-Z9023	2763
121-Z9024	0400
121-Z9025	0604
121-Z9026	0561
121-Z9027	0698
121-Z9028	0168
121-Z9029	0004
121-Z9030	0208
121-Z9031	0038
121-Z9032	0160
121-Z9033	0440
121-Z9034	0130
121-Z9035	0161
121-Z9036	0551
121-Z9037	0630
121-Z9038	3519
121-Z9039	0936
121-Z9040	0836
121-Z9041	1411
121-Z9042	0177
121-Z9043	0543
121-Z9044	0003
121-Z9045	0715
121-Z9046	0342
121-Z9047	0236
121-Z9048	0681
121-Z9049	1708
121-Z9050	0037
121-Z9051	0847
121-Z9052	1681
121-Z9053	0382
121-Z9054	0264
121-Z9055	0566
121-Z9056	1367
121-Z9057	0905
121-Z9058	0491
121-Z9059	0602
121-Z9060	2258
121-Z9061	3370
121-Z9062	2013
121-Z9063	1598
121-Z9064	1411
121-Z9065	0076
121-Z9066	0018
121-Z9067	0532
121-Z9068	0324
121-Z9069	1946
121-Z9070	0259
121-Z9071	0316
121-Z9072	0040
121-Z9073	2427
121-Z9074	2277
121-Z9075	3373
121-Z9076	3374
121-Z9077	2439
121-Z9078	2432
121-Z9079	1394
121-Z9080	2448
121-Z9081	3525
121-Z9082	3526
121-Z9083	0558
121-Z9084	2897
121-Z9085	1212
121-Z9086	2235
121-Z9087	2237
121-Z9088	0418
121-Z9089	0177
121-Z9090	0363
121-Z9091	0543
121-Z9092	2016
121-Z9093	0259
121-Z9094	0892
121-Z9095	1480
121-Z9096	2485
121-Z9097	2700
121-Z9098	0050
121-Z9099	1199
121-Z9100	1031
121-Z9101	0353
121-Z9102	3280
121-Z9103	0436
121-Z9104	1430
121-Z9105	0525
121-Z9106	0392
121-Z9107	1085
121-Z9108	2214
121-Z9109	1751
121-Z9110	2096
121-Z9111	0728
121-Z9112	0055
121-Z9113	3081
121-Z9114	0859
121-Z9115	1648
121-Z9119	0859
121A	0023
121A-US	OEM
121G1241	0102
121G3019	0016
121G3020	0016
121GI241	0102
121GL241	0102
121N1	OEM
121S2	0316
121T2	0144
121U1	OEM
121U2	OEM
121U2AT	0316
121U3AC	0316
122	0023
122-1	0016
122-2	0079
122-2(DIO)	0071
122-2N1B	0363
122-3	OEM
122-6	0016
122-80	0015
122-229	0280
122-522	1497
122-1028	0440
122-1028A	0440
122-1625	0160
122-1648	0136
122-1962	0038
122-A484	0144
122A	0023
122A-US	OEM
122GRN	0037
122N2	0363
122T2	0144
122YEL	0037
123-001	0823
123-002	0212
123-003	0325
123-004	0016
123-005	0111
123-007	0111
123-008	0836
123-009	0839
123-010	0016
123-011	0219
123-011A	0219
123-012	0155
123-015	0143
123-016	0133
123-017	0400
123-017(DIODE)	0124
123-018	0102
123-019	0244
123-020	0012
123-021	0015
123-022	0133
123-024	0124
123-025	0133
123A-US	OEM
123B-001	0103
123B-002	0219
123B-003	0133
123B-004	0244
123B-005	0636
123B-006	0420
123N1	0003
123S425	0086
123S-437	0086
123T2	0144
124	0023
124-0028	0071
124-0165	0319
124-0178	0015
124-1	0160
124-N16	0600
124A	0023
124J490	0015
124N1	0600
124N16	0016
125-1	0150
125-4	0975
125-4(RCA)	0975
125-121	0086
125-402	0160
125-403	0160
125-410	0178
125-425C	0164
125-655	0224
125-A000-F01	0816
125-B415	0042
125A	OEM
125A000F01	0816
125A134	0004
125A137	0086
125A137A	0086
125AS251	0178
125B	OEM
125B132	0016
125B133	0037
125B139	0086
125B410	0178
125BN20	OEM
125BN30	OEM
125BN40	OEM
125BN50	OEM
125BN60	OEM
125BN80	OEM
125BN100	OEM
125BN120	OEM
125BN140	OEM
125BN160	OEM
125BNS	OEM
125C3	2699

If replacement code is OEM, contact original manufacturer for replacement.

DEVICE TYPE	REPL CODE
125C3RB	2699
125C211	0016
125CN30	OEM
125CN40	OEM
125CN50	OEM
125CN60	OEM
125CN80	OEM
125CN100	OEM
125CN120	OEM
125CN140	OEM
125CN160	OEM
125CNS	OEM
125P1	0037
125P1M	0037
125P116	0037
125PALB**	0766
125PAM**	0766
125PI	0037
125PL	0037
125PL50	OEM
125PL60	OEM
125PL80	OEM
125PL100	OEM
125PLB50	OEM
125PLB60	OEM
125PLB80	OEM
125PLB100	OEM
125PLB110	OEM
125PLB120	OEM
125PM10	OEM
125PM20	OEM
125PM30	OEM
125PM40	OEM
125PM50	OEM
125PM60	OEM
125PM80	OEM
125PM100	OEM
125T1	OEM
125T4	OEM
125UA	OEM
125UB	OEM
125UC	OEM
125UD	OEM
125UE	OEM
125UF	OEM
125V10	OEM
125WA	OEM
125WB	OEM
125WC	OEM
125WD	OEM
125XA	OEM
125XB	OEM
125XC	OEM
125XD	OEM
125YA	OEM
125YB	OEM
125YC	OEM
125YD	OEM
126	0023
126-1	2826
126-4	0015
126-7	0015
126-7(ARVIN)	0015
126-12	0016
126-40	1301
126A	0023
126N1	0208
126N2	0208
126P1	0164
126T1	OEM
127	0016
127-001	0861
127-1	0473
127-7	0600
127-115	0111
127A905F01	0267
127T1	OEM
128	0144
128-853	0560
128-9050	0103
128A157F01	0267
128A157F02	0267
128C212H01	2078
128C213H01	1797
128C830H05	0569
128N2	0086
128N3	OEM
128N4	0144
128WHT	0016
129-4	0164
129-5	0160
129-6	0085
129-7	0085
129-8	0004
129-8-1	0004
129-8-1A	0004
129-8-2	0004
129-9	0160
129-10	0160
129-11	0136
129-13	0160
129-14	0198
129-15	0198
129-16	0079
129-17	0208
129-18	0164
129-20	0037
129-21	0016
129-23	0178
129-27	0007
129-28	0086
129-29	0126
129-30	0208
129-31	0004
129-32	0004
129-33	0042
129-33(PILOT)	0016
129-34	0037
129-34(PILOT)	0037
129BRN	0016
129N1	0007
129WHT	0016
130	0127
130-04	OEM
130-06	OEM
130-08	OEM
130-013	2017
130-10	OEM
130-104	0160
130-105	0402
130-138	0144
130-144	2208
130-146	0103
130-149	0037
130-150	2208
130-152	0259
130-152-00	0259
130-172	2208
130-174	2208
130-185	0007
130-185-98	0259
130-240	0007
130-245	1400
130-338-00	0015
130-398-99	0015
130-30189	0133
130-30192	0015
130-30256	0071
130-30261	0277
130-30265	0133
130-30266	0133
130-30274	0133
130-30281	0143
130-30301	0143
130-30313	0015
130-30702	0133
130-40089	0600
130-40095	0004
130-40096	0208
130-40214	0016
130-40215	0016
130-40216	0111
130-40229	0143
130-40236	0004
130-40294	0016
130-40304	0144
130-40311	0016
130-40312	0016
130-40313	0016
130-40314	0208
130-40315	0037
130-40317	0016
130-40318	0016
130-40347	0208
130-40349	3545
130-40352	0004
130-40357	0016
130-40362	0144
130-40421	0144
130-40429	0037
130-40439	3545
130-40456	0164
130-40459	0144
130-40883	0111
130-40896	0016
130-40901	0111
130-40922	0016
130-410012-00	OEM
130-410033-00	OEM
130-410043-11	OEM
130-410051-01	OEM
130-410062-11	OEM
130-410068-01	OEM
130-410075-00	OEM
130-411091-00	OEM
130-V10	0127
130-VLO	0127
130A17268	0015
130HF20	OEM
130HF20F	OEM
130HF20M	OEM
130HF40	OEM
130HF40F	OEM
130HF40M	OEM
130HF60	OEM
130HF60F	OEM
130HF60M	OEM
130HF80	OEM
130HF80F	OEM
130HF80M	OEM
130HF120	OEM
130HF120F	OEM
130HF120M	OEM
130HF130	OEM
130HF130F	OEM
130HF130M	OEM
130HF140	OEM
130HF140F	OEM
130HF140M	OEM
130HF160	OEM
130HF160F	OEM
130HF160M	OEM
130HHFR20	OEM
130HHFR20HFR	OEM
130HHFR20M	OEM
130HHFR40	OEM
130HHFR40HFR	OEM
130HHFR40M	OEM
130HHFR60	OEM
130HHFR60HFR	OEM
130HHFR60M	OEM
130HHFR80	OEM
130HHFR80HFR	OEM
130HHFR80M	OEM
130HHFR120	OEM
130HHFR120HFR	OEM
130HHFR120M	OEM
130HHFR130	OEM
130HHFR130HFR	OEM
130HHFR130M	OEM
130HHFR140	OEM
130HHFR140HFR	OEM
130HHFR140M	OEM
130HHFR160	OEM
130HHFR160HFR	OEM
130HHFR160M	OEM
130ORN	0127
130P4	OEM
130T4	OEM
130YE	0015
131-2	0560
131A	0196
131A246F01	0267
131A246F02	0267
131N2	0338
131N2G	0042
131P4	OEM
132-001	0004
132-002	0016
132-003	0285
132-004	0016
132-005	0016
132-007	0126
132-008	0007
132-009	0007
132-010	0004
132-011	0016
132-014	0285
132-015	0144
132-017	0016
132-018	0016
132-019	0050
132-020	0050
132-021	0016
132-021B	0086
132-022	0086
132-023	0016
132-024	0460
132-025	0074
132-026	0016
132-027	0050
132-028	0142
132-029	0693
132-030	0016
132-031	0693
132-032	0126
132-033	0142
132-038	0086
132-039	0126
132-041	0016
132-042	0016
132-045	0353
132-047	0353
132-048	0849
132-049	0325
132-050	0016
132-051	0016
132-052A	0400
132-054	0016
132-055	0016
132-056	0037
132-057	0016
132-059	0142
132-062	0016
132-063	0016
132-065	0178
132-066	0086
132-069	0016
132-070	0103
132-072	0161
132-074	0037
132-075	0016
132-076	0007
132-077	0016
132-078	0276
132-082	0007
132-085	0103
132-087	0007
132-090	0004
132-3	0042
132-5	0423
132-90	0004
132-185	0007
132-501	0016
132-502	0016
132-503	0016
132-504	0016
132-515	0142
132-516	0142
132-521	0142
132-522	0142
132-523	0142
132-524	0142
132-525	0142
132-526	0142
132-539	0016
132-540	0016
132-541	0103
132-542	1514
132-B3511	0196
132B3511	0196
132N1	0086
132P4	OEM
133-002	0829
133-0021-0	1854
133-1	0551
133-3	0042
133P100	1350
134-1	0382
134-1P	0150
134-4P	0037
134B1038-4	0111
134B1038-8	0111
134B1038-13	0111
134B1038-21	0079
134B1038-22	0111
134B1040-7	0595
134P1	0693
134P1(REMOTE)	0791
134P1(VID)	0150
134P1A	0037
134P1AA	0126
134P1M	0037
134P2	0126
134P4	0037
134P4AA	0037
134P4M	0037
134P6	0150
135-3	0015
135B1N	4220
135C	0609
135C4L	0609
135H1M	0285
135N1	0233
135N1M	0233
135P4	0229
136-000100	1030
136-000200	1030
136-000300	1030
136-12	0086
136P1	0150
136P4	OEM
136RED	0016
137	0150
137-003	2018
137-12	0150
137-684	0071
137-718	0015
137-737	0242
137-759	0002
137-824	0143
137-828	0071
137P4	OEM
137Q31	0190
138-4	0016
139-4	0144
139N1	0224
139N1D	0224
140-013	0015
140BV50H	OEM
140BV100H	OEM
140BV150H	OEM
140N1	0103
140N2	0103
140PL10	OEM
140PL20	OEM
140PL30	OEM
140PL40	OEM
140PL50	OEM
140PL60	OEM
140PM10	OEM
140PM20	OEM
140PM30	OEM
140PM40	OEM
140PM50	OEM
140PM60	OEM
140T4	OEM
141-003	0143
141-4	0079
141-402	0007
141-430	0079
142	0133
142-001	0016
142-002	0111
142-003	0111
142-004	0042
142-005	0155
142-006	0219
142-007	0622
142-008	0042
142-009	0133
142-010	0133
142-011	0143
142-012	0012
142-014	0015
142-015	0420
142-016	0025
142-3	0016
142-4	0016
142N1	0155
142N1P	0007
142N3	0079
142N3P	0079
142N3T	0016
142N4	0016
142N5	0016
142N6	0144
142N7	OEM
142N8	OEM
142N10	OEM
143	0222
143-12	0079
144-3	0133
144-4	0233
144-6	OEM
144-12	0079
144A-1	0160
144A-1-12	0160
144A-1-12-3	0160
144A-1-12-8	0160
144N1	0285
144N1G	0285
144N1G(AMP)	0285
144N1G(VID)	0285
144N1P	0285
144N2	0285
144N3	0285
144N4	0233
145-2	0285
145-6	OEM
145-12	0150
145-100	0170
145-470	0999
145-2023DN	OEM
145-2023DNT	OEM
145-2023SN	OEM
145-2023SNT	OEM
145-2025	OEM
145-2051	OEM
145-2068	OEM
145-2071	OEM
145-2071R	OEM
145-2072-1	OEM
145-2072-2	OEM
145-2072-4	OEM
145-2073	OEM
145-2076	OEM
145-T1B	0136
145A9254	0015
145A9786	0015
145N1	0007
145N1(LAST-LF)	0330
145N1P	0007
145NIP	0079
145T1B	0050
146-1	0285
146-12	0079
146-T1	0160
146D-1	0133
146D1	2130
146D1B112	0913
146N3	0016
146N5	0016
146T1	0085
147-7031-01	0079
147-7040-01	1248
147-7043-01	1048
147-T1	0160
147N1	0018
147P2	0532
147T1	0085
148-1	0086
148-3	0102
148-3(XSTR)	0079
148-12	0150
148-134622	0440
148N1	0086
148N2	0016
148N3	0086
148N212	0016
148P-2	0004
148P-2-12	0004
148P-2-12-8	0004
149-3	OEM
149-4	0042
149-10	0042
149-12	0325
149-142-G1	0133
149L-4	0004
149L-4-12	0004
149L-4-12-8	0004
149N1	0161
149N2	0161
149N2B	2997
149N2D	0042
149N4B	2997
149N2002D	0042
149N2004	0042
149N20020	3132
149P1	0460
149P1B	3165
149P1D	0042
149P4B	3165
149P2001D	0925
149P2003	0925
149P20010	OEM
150	0015
150-001-9-005	0019
150-001-9-007	0143
150-002-9-001	0143
150-004-9-001	0143
150-005-9-001	0143
150-006-9-001	0143
150-012-9-001	0019
150-013-7-001	OEM
150-013-9-001	0019
150-014-9-001	0019
150-014-9001	OEM
150-015-9-001	0123
150-016-9-001	0143
150-018-9-001	0015
150-022-9-001	0015
150-066-9-001	0124
150-1	0079
150-2	OEM
150-12	0325
150BC15	0599
150C05	OEM
150C05B	OEM
150C025	0738
150C025B	0738
150C050	0738
150C050B	0738
150C4.7-5.3	OEM
150C6.2-6.8	OEM
150C10	0738
150C10B	0738
150C15	0738
150C15B	0738
150C20	0738
150C20B	0738
150C25	0738
150C25B	0738
150C30	0738
150C30B	0738
150C35	0738
150C35B	0738
150C40	0738
150C40B	0738
150C45	0738
150C45B	0738
150C50	0738
150C50B	0738
150C60	0738
150C60B	0738
150C70	0738
150C70B	0738
150C80	0738
150C80B	0738
150C90	0738
150C90B	0738
150C100	0738
150C100B	0738
150C110	0738
150C110B	0738
150C120	0738
150C120B	0738
150C130	1889
150C130B	1889
150C140	1889
150C140B	1889
150C150	1889
150C150B	1889
150C160	1889
150C160B	1889
150C170	OEM
150C170B	OEM
150C180	OEM
150C180B	OEM
150D	0002
150DC15	0599
150FC15	0599
150GC15	0599
150JC15	1986
150K**	0599
150K40A	OEM
150K5A	0599
150K5R	1347
150K10A	0599
150K10AM	OEM
150K10R	1347
150K15A	0599
150K15R	1347
150K20A	0599
150K20AM	OEM
150K20R	1347
150K25A	0599
150K25R	1347
150K30A	0599
150K30AM	OEM
150K30R	1347
150K40A	0599
150K40AM	OEM
150K40R	1905
150K50A	1986
150K50R	1905
150K60A	1986
150K60AM	OEM
150K60R	1905
150K70A	0657
150K70R	0193
150K80A	0657
150K80AM	OEM
150K80R	0193
150K90A	3218
150K100A	0657
150K100AM	OEM
150K100R	0193
150K120A	3309
150K120AM	OEM
150K140A	OEM
150KR5A	4978
150KR10A	1347
150KR10AM	OEM
150KR20A	1347
150KR20AM	OEM
150KR30A	1347
150KR30AM	OEM
150KR40A	1347
150KR40AM	OEM
150KR50A	7094
150KR60A	1905
150KR60AM	OEM
150KR80A	0193
150KR80AM	OEM
150KR100A	0193
150KR100AM	OEM
150KR120A	OEM
150KR120AM	OEM
150KS5	OEM
150KS10	OEM
150KS15	OEM
150KS20	OEM
150KS25	OEM
150KS30	OEM
150KS40	OEM
150KS50	OEM
150KS60	OEM
150KS70	OEM
150KS80	OEM
150KS90	OEM
150KS100	OEM
150KS120	OEM
150L5A	0599
150L5R	1347
150L10A	0599
150L10R	1347
150L15A	0599
150L15R	1347
150L20A	0599
150L20R	1347
150L25A	0599
150L25R	1347
150L30A	0599
150L30R	1347
150L40A	1986
150L40R	1905
150L50A	1986
150L50R	1905
150L60A	1986
150L60R	1905
150L70A	0657
150L70R	0193
150L80A	0657
150L80R	0193
150L90A	0657
150L90R	0193
150L100A	0657
150L100R	0193
150L120A	3309
150L140A	OEM
150LC15	0657
150LD11	1789
150LD13	OEM
150MA60	OEM
150MA80	OEM
150MA100	OEM
150MA120	OEM
150MA140	OEM
150MA160	OEM
150MA180	OEM
150MLAB160	OEM
150MLAB180	OEM
150MLAB200	OEM
150MLAB250	OEM
150N1	0007
150N2	0016
150N3	0007
150N15A	OEM
150N15A9	OEM
150N20A	OEM
150N20A9	OEM
150N25A	OEM
150N25A9	OEM
150N30A	OEM
150N30A9	OEM
150N35A	OEM
150N35A9	OEM
150N40A	OEM
150N40A9	OEM
150N45A	OEM
150N45A9	OEM
150N50A	OEM
150N50A9	OEM
150N55A	OEM
150N55A9	OEM
150N60A	OEM
150N60A9	OEM
150N65A	OEM
150N65A9	OEM
150N70A	OEM
150N70A9	OEM
150N75A	OEM

If replacement code is OEM, contact original manufacturer for replacement.

DEVICE TYPE	REPL CODE
150N75A9	OEM
150N80A	OEM
150N80A9	OEM
150N85A	OEM
150N85A9	OEM
150N90A	OEM
150N90A9	OEM
150N95A	OEM
150N95A9	OEM
150N100A	OEM
150N100A9	OEM
150N110A	OEM
150N120A	OEM
150N130A	OEM
150N140A	OEM
150N150A	OEM
150N160A	OEM
150N170A	OEM
150N180A	OEM
150N190A	OEM
150N200A	OEM
150N210A	OEM
150N220A	OEM
150N230A	OEM
150N240A	OEM
150N250A	OEM
150NC15	0657
150ND13	OEM
150P1SP	OEM
150P2SP	OEM
150P3SP	OEM
150P4SP	OEM
150P5SP	OEM
150P6SP	OEM
150P8SP	OEM
150P10SP	OEM
150P12SP	OEM
150P16SP	OEM
150QC15	3309
150QD13	OEM
150R1B	0087
150R2B	0242
150R3B	1747
150R4B	0540
150R5B	0965
150R6B	0965
150R7B	0817
150R8B	0087
150R9B	0087
150R10B	0087
150R11B	OEM
150R12B	OEM
150R13B	OEM
150R14B	OEM
150R15B	OEM
150R16B	OEM
150R18B	OEM
150R20B	OEM
150R22B	OEM
150R24B	OEM
150R26B	OEM
150R28B	OEM
150R30B	OEM
150R32B	OEM
150R34B	OEM
150R36B	OEM
150R38B	OEM
150R40B	OEM
150R42B	OEM
150R44B	OEM
150R46B	OEM
150R48B	OEM
150R50B	OEM
150RC5	OEM
150RC10	0738
150RC15	OEM
150RC20	0738
150RC25	OEM
150RC30	0738
150RC35	OEM
150RC40	0738
150RC50	0738
150RC60	0738
150RC60A	OEM
150RC70	0738
150RC70A	OEM
150RC80	0738
150RC80A	OEM
150RE50	0738
150RE60	0738
150RE80	0738
150RE90	OEM
150RE100	0738
150RE110	0738
150RE120	0738
150RE130	OEM
150S2SP	OEM
150S3SP	OEM
150S4SP	OEM
150S5SP	OEM
150S6SP	OEM
150S8SP	OEM
150S10H	OEM
150S12H	OEM
150SS2SP	OEM
150SS3SP	OEM
150SS4SP	OEM
150SS5SP	OEM
150SS6SP	OEM
150SS8SP	OEM
150SS10H	OEM
150SS12H	OEM
150T4	OEM
150TD13	OEM
150X9-9.75	OEM
150YD13	OEM
151	0023
151-001-9-001	0133
151-002-9-001	0133
151-006-9-001	0039
151-0021-9-001	OEM
151-04SPC	OEM
151-06SPC	OEM
151-08SPC	OEM
151-011-9-001	0133
151-011-9-011	0015
151-013-9-001	0913
151-014-9-001	0920
151-015-9-001	0124
151-018-9-001	0015
151-021-001	0133
151-021-9-001	0133
151-022-9-001	0133
151-023-9-001	0015
151-024-9-001	0133
151-025-9-001	0229
151-029-9-003	0015
151-030-9-001	0133
151-030-9-002	1149
151-030-9-003	0636
151-030-9-004	0030
151-030-9-005	0023
151-030-9-006	0133
151-031-9-001	0636
151-031-9-006	0012
151-032-004	0133
151-032-9-001	0133
151-032-9-002	0133
151-032-9-004	0133
151-034-9-001	0133
151-035-9-001	0133
151-035-9-004	0133
151-039-9-003	0030
151-040-7-003	0023
151-040-9-001	0133
151-040-9-002	0133
151-040-9-003	0023
151-045-9-001	0015
151-045-9-003	0102
151-049-9-001	0133
151-049-9-002	0133
151-051-9-001	0133
151-059-9-001	0133
151-060-9-001	0124
151-061-9-001	0133
151-062-9-001	1029
151-064-9-001	0133
151-064-9-002	0133
151-066-9-001	0124
151-067-9-001	0133
151-069-9-001	0124
151-072-9-001	0163
151-0149	0178
151-1	0023
151-3	0023
151-4	0023
151-6	0079
151-9	0348
151-10SPC	OEM
151-12SPC	OEM
151-14SPC	OEM
151-16SPC	OEM
151-18SPC	OEM
151-20SPC	OEM
151-22SPC	OEM
151-26SPC	OEM
151-28SPC	OEM
151-30SPC	OEM
151-32-9-004	0133
151-267-9-001	0133
151A	0023
151F1FM	OEM
151F2FM	OEM
151F3FM	OEM
151M11	0144
151MQ30	OEM
151MQ40	OEM
151N1	0144
151N2	0016
151N4	0016
151N5	0079
151N11	0144
151N116	0144
151R1FM	OEM
151R2FM	OEM
151R3FM	OEM
151RA10	0738
151RA20	0738
151RA30	0738
151RA40	0738
151RA50	0738
151RA60	0738
151RA80	0738
151RA100	0738
151RA110	0738
151RA120	0738
151RA140	1889
151RA150	1889
151RA160	1889
151RB10	0738
151RB20	0738
151RB30	0738
151RB40	0738
151RB50	0738
151RB60	0738
151RB80	0738
151RB100	0738
151RB110	0738
151RB120	0738
151RB140	1889
151RB150	1889
151RB160	1889
151RC10	1084
151RC10A	0738
151RC15	OEM
151RC15A	OEM
151RC20	1084
151RC20A	0738
151RC30	1086
151RC30A	0738
151RC40	1086
151RC40A	0738
151RC50	1086
151RC50A	0738
151RC60	1086
151RC60A	0738
151RC80	1102
151RC80A	0738
151RC100	OEM
151RC120	OEM
151RCF15A	OEM
151RCF25A	OEM
151RCF60A	OEM
151RCF70A	OEM
151RCF80A	OEM
151RE50	0738
151RE60	0738
151RE80	0738
151RE100	0738
151RE110	0738
151RE120	0738
151REA50	0738
151REA60	0738
151REA70	OEM
151REA80	0738
151REA90	OEM
151REA100	0738
151REA110	0738
151REA120	0738
151REA130	OEM
151REB50	0738
151REB60	0738
151REB70	OEM
151REB80	0738
151REB90	OEM
151REB100	0738
151REB110	0738
151REB120	0738
151REB130	OEM
151RL50	0428
151RL60	0428
151RL80	1871
151RL100	1871
151RL110	1871
151RL120	1871
151RM10	0428
151RM20	0428
151RM30	0428
151RM40	0428
151RM50	0428
151RM60	0428
151RM80	1871
151RM100	1871
152	0023
152-006-9-001	0039
152-008-7-001	0030
152-008-9-001	0012
152-0047	0540
152-0047-00	0540
152-04SPC	OEM
152-06SPC	OEM
152-08SPC	OEM
152-012-9-001	0012
152-019-9-001	0636
152-021-9-001	0794
152-029-9-002	0039
152-042-9-001	0414
152-042-9-001(FET)	0212
152-042-9-002	0025
152-047	0015
152-047-9-001	0471
152-047-9-004	0471
152-051-9-001	0012
152-051-9-002	0636
152-052-9-002	0091
152-054-9-001	0244
152-057-9-001	0420
152-079-001	0162
152-079-9-001	0041
152-079-9-002	0636
152-079-9-003	0091
152-079-9-004	0649
152-082-9-001	0012
152-0154	1413
152-10SPC	OEM
152-12	OEM
152-12SPC	OEM
152-14SPC	OEM
152-16SPC	OEM
152-18SPC	OEM
152-20SPC	OEM
152-22SPC	OEM
152-26SPC	OEM
152-28SPC	OEM
152-30SPC	OEM
152A	OEM
152CB	0415
152ET	OEM
152K5A	OEM
152K10A	OEM
152K10AM	OEM
152K15A	OEM
152K20A	OEM
152K20AM	OEM
152K25A	OEM
152K30A	OEM
152K30AM	OEM
152K40A	OEM
152K40AM	OEM
152K50A	OEM
152K60A	OEM
152K60AM	OEM
152K70A	OEM
152K80A	OEM
152K80AM	OEM
152K90A	OEM
152K100A	OEM
152K100AM	OEM
152K120A	OEM
152K120AM	OEM
152L5A	OEM
152L10A	OEM
152L15A	OEM
152L20A	OEM
152L25A	OEM
152L30A	OEM
152L40A	OEM
152L50A	OEM
152L60A	OEM
152L70A	OEM
152L80A	OEM
152L90A	OEM
152L100A	OEM
152L120A	OEM
152N-2	0130
152N2C	0130
152P1B	0905
152PJA180	OEM
152PJA200	OEM
152PJA250	OEM
152PJAL60	OEM
152PJAL80	OEM
152PJAL100	OEM
152PJAL120	OEM
152SC711(D)	0025
153-004-9-001	0030
153-008-9-001	0720
153-008-9-002	0139
153-04	3512
153-04SPC	OEM
153-05	3512
153-06	3512
153-06SPC	OEM
153-07	3512
153-08	3512
153-08SPC	OEM
153-09	3512
153-011-9-001	0120
153-9	0086
153-10	3512
153-10SPC	OEM
153-12	3512
153-12SPC	OEM
153-14	3512
153-14SPC	OEM
153-16	3489
153-16SPC	OEM
153-18	3489
153-18SPC	OEM
153-20SPC	OEM
153-22SPC	OEM
153-24SPC	OEM
153-26SPC	OEM
153-28SPC	OEM
153-30SPC	OEM
153E	OEM
153N1	0423
153N5	OEM
153N5C	0042
154	0023
154-001-9-001	0030
154-004-9-001	0030
154-04	3512
154-04SPC	OEM
154-05	3512
154-06	3512
154-06SPC	OEM
154-07	3512
154-08	3512
154-08SPC	OEM
154-09	3512
154-10	3512
154-10SPC	OEM
154-12	3512
154-12SPC	OEM
154-14	3512
154-16	3489
154-16SPC	OEM
154-18	3489
154-18SPC	OEM
154-20SPC	OEM
154-22SPC	OEM
154-24SPC	OEM
154-26SPC	OEM
154-28SPC	OEM
154-30SPC	OEM
154A	0023
154A3675-105	0211
154A3676	0136
154A3676-205	0211
154A3677	0136
154A3679	0211
154A3679-5110	0136
154A3680	0160
154A3992	0015
154A5943	0142
154A5943-1	0142
154A5944-410	0079
154A5944-413	0111
154A5945-519	0111
154A5946	0016
154A5946-622	0086
154A5946-624	0079
154A5946-667	0126
154A5947-7732	0150
154A8681	0211
154E	OEM
154K5A	OEM
154K10A	OEM
154K10AM	OEM
154K15A	OEM
154K20A	OEM
154K20AM	OEM
154K25A	OEM
154K30A	OEM
154K30AM	OEM
154K40A	OEM
154K40AM	OEM
154K50A	OEM
154K60A	OEM
154K60AM	OEM
154K70A	OEM
154K80A	OEM
154K80AM	OEM
154K90A	OEM
154K100A	OEM
154K100AM	OEM
154K120A	OEM
154K120AM	OEM
154L5A	OEM
154L10A	OEM
154L15A	OEM
154L20A	OEM
154L25A	OEM
154L30A	OEM
154L40A	OEM
154L50A	OEM
154L51	OEM
154L60A	OEM
154L70A	OEM
154L80A	OEM
154L90A	OEM
154L100A	OEM
154L120A	OEM
154T1	0050
154T1A	0050
154T1B	0050
154WA	2203
154WK	0694
155D06010B	OEM
155N1	0423
155T1	0050
156	0016
156(DIO)	0023
156-0011-00	1069
156-0012-00	OEM
156-0017	2537
156-0017-00	2537
156-04	0130
156-06	0130
156-08	0130
156-032	0103
156-043	0103
156-043A	0103
156-044	0103
156-053	0103
156-063	0103
156-064	0103
156-083	0130
156-084	0130
156-0148-00	0361
156-0151-00	1554
156-0176-00	1922
156-10	1966
156-104	0130
156-123	0130
156-124	0130
156-143	OEM
156A	0023
156B	OEM
156C	OEM
156D	OEM
156E	OEM
156T1	0050
156WHT	0016
157	0037
157-009-9-001	0165
157-100	0064
157-120	0052
157-180	0375
157-439	0275
157-829	OEM
157-870	0331
157-1120	0052
157A	OEM
157B	OEM
157C	OEM
157D	OEM
157E	OEM
157N3	0551
157P4	0382
157T1	0050
157T1A	0050
157YEL	0037
158-3	0037
158-10	0142
158P1	0126
158P1M	0126
158P2	0037
158P2M	0037
158P3	0037
159-12	0150
159T1	0050
160-8	0079
160E15	OEM
160F15	OEM
160G2G41	OEM
160J2G41	OEM
160L2G41	OEM
160Q2G41	OEM
160T1	0050
160U2G41	OEM
161-001I	0079
161-006-0001	3714
161-006-0002	3714
161-009I	0079
161-009J	0079
161-011-0001	3861
161-011-0002	3861
161-011H	0286
161-011J	0079
161-012-0002	4236
161-012H	0037
161-012J	0037
161-014H	0076
161-015G	0037
161-015H	0342
161-015I	0342
161-016(H)	0079
161-016H	0079
161-016K	0016
161-017I	0286
161-017J	0286
161-017K	0286
161-037H	0079
161-037J	0286
161-039G	0079
161-039H	0079
161-039I	0079
161-039J	0079
161-040G	0168
161-040H	0342
161-040I	0342
161-040J	0342
161-041H	0715
161-041I	0079
161-041J	0079
161-042G	0042
161-042H	0042
161-043G	0925
161-043H	0925
161-044D	0392
161-044E	0392
161-046G	0168
161-046H	0168
161-046I	0168
161-047(H)	0042
161-047G	0461
161-047H	0461
161-050G	0037
161-050H	0037
161-051G	0261
161-051H	0261
161-053G	1085
161-053H	0261
161-054A	0392
161-061H	0079
161-0161	0079
161-0161H	0079
161-1	OEM
161-5	0423
161-118-0001	1257
161-152-0101	1069
161-152-0102	1069
161-LNC	0103
161N2	1514
161N4C	0103
161RL10	OEM
161RL20	OEM
161RL30	OEM
161RL40	OEM
161RL50	OEM
161RL60	OEM
161RM10	OEM
161RM20	OEM
161RM30	OEM
161RM40	OEM
161RM50	OEM
161RM60	OEM
161T1	0050
161T2	0144
162-005A	0023
162-028A	0516
162-038A	OEM
162-041A	OEM
162-045A	2400
162-047A	OEM
162-3	3327
162-4	OEM
162-10	OEM
162-12	OEM
162-16	OEM
162-18	0315
162-18-0	0315
162-26-01	OEM
162-28	OEM
162J2	0015
162N2	0161
162P1	0460
162T1	0050
162T2	0144
163-04	3512
163-04A	OEM
163-04SPC	OEM
163-05	3512
163-06	3512
163-06A	OEM
163-06SPC	OEM
163-07	3512
163-08	3512
163-08A	OEM
163-08SPC	OEM
163-09	3512
163-10	3512
163-10SPC	OEM
163-12	3512
163-12A	OEM
163-12SPC	OEM
163-14	3512
163-14A	OEM
163-14SPC	OEM
163-16	3489
163-16A	OEM
163-18	3489
163-18A	OEM
163-18SPC	OEM
163-20A	OEM
163-20SPC	OEM
163-22A	OEM
163-22SPC	OEM
163-24SPC	OEM
163-28SPC	OEM
163-30SPC	OEM
163J2	0015
164-003A	0120
164-006A	OEM
164-04	3512
164-04A	OEM
164-04SPC	OEM
164-05	3512
164-06	3512
164-06A	OEM
164-06SPC	OEM
164-07	3512
164-08	3512
164-08A	OEM
164-08SPC	OEM
164-09	3512
164-10	3512
164-10A	OEM
164-10SPC	OEM
164-12	3512
164-12A	OEM
164-12SPC	OEM
164-14	3512
164-14A	OEM
164-14SPC	OEM
164-16	3489
164-16A	OEM
164-18	3489
164-18A	OEM
164-20A	OEM
164-20SPC	OEM
164-22A	OEM
164-22SPC	OEM
164-24SPC	OEM
164-26SPC	OEM
164-28SPC	OEM
164-30SPC	OEM
164J2	0015
165	0196
165-1A82	0050
165-2A82	0208
165-4A82	0600
165-432-2-48-1	0124
165A-182	0050
165B-182	0050
165C-182	0160
165J2	0015
165N1	0079
166	1080
166-1A82	0004
166-2A82	0004

DEVICE TYPE	REPL CODE
166-3A82	0004
166J2	0015
167-001B	0167
167-006A	1329
167-006B	1329
167-013A	0395
167-014A	1808
167-018A	0167
167-019A	0167
167-024B	OEM
167-028A	OEM
167-030A	OEM
167-035A	OEM
167-9	0595
167J2	0071
167N1	0016
167N2	0016
168-9	0791
168J2	0087
168N1	0016
169-257	0142
169-284	0142
169J2	OEM
170(RCA)	0007
170-1	0133
170-9	0086
170J2	OEM
171-001-9-001	0160
171-003-9-001	0142
171-015-9-001	0160
171-016-9-001	0222
171-1	0155
171-2	0609
171-3	OEM
171-4	OEM
171-9	0126
171J	OEM
171K	OEM
171RC10A	OEM
171RC10AM	OEM
171RC20A	OEM
171RC20AM	OEM
171RC40A	OEM
171RC40AM	OEM
171RC60A	OEM
171RC60AM	OEM
171RC80A	OEM
171RC80AM	OEM
171RK80A	OEM
171RK80AM	OEM
171RK100A	OEM
171RK100AM	OEM
171RK120A	OEM
171RK120AM	OEM
171RK140A	OEM
171RK140AM	OEM
171RK160A	OEM
171RK160AM	OEM
172-001-9-001	0630
172-003-9-001	0103
172-003-9-001A	0103
172-006-9-001	0560
172-007-9-001	1411
172-008-9-001	1411
172-009-9-001	1411
172-010-9-001	0228
172-011-9-001	0219
172-013-9-001	0836
172-014-9-001	0042
172-014-9-002	2296
172-014-9-003	0042
172-024-9-004	0390
172-028-9-001	2495
172-028-9-002	1946
172-031-9-003	0228
172-038-9-001	0890
172-038-9-002	0936
172-038-9-003	0390
172-044-9-001	0042
173-1	0325
173-1(SYLVANIA)	0325
173-15	0420
173A419-2	0160
173A3936	0160
173A3963	0160
173A3970	0211
173A3981	0015
173A3981-1	0015
173A4348	0004
173A4349	0004
173A4389-1	0004
173A4390	0004
173A4391	0086
173A4393	0015
173A4394	0102
173A4394-1	0133
173A4399	0016
173A4416	0016
173A4419	0160
173A4419-1	0160
173A4419-2	0160
173A4419-3	0160
173A4419-4	0160
173A4419-5	0160
173A4419-6	0160
173A4419-7	0160
173A4419-8	0160
173A4419-9	0160
173A4419-10	0160
173A4420	0160
173A4420-1	0160
173A4420-5	0160
173A4421-1	0160
173A4422-1	0160
173A4424	0474
173A4436	0160
173A4469	0160
173A4470-11	0016
173A4470-13	0079
173A4470-32	0016
173A4473-5	0111
173A4489-2	0086
173A4490-2	0560
173A4490-5	0086
173A4490-7	0042
173A4491-2	0103
173A4491-2A	0103
173H	OEM
174-002-9-001	0208
174-1	0015
174-2	0015
174-3	0015
174-3A82	0280
174-25566-01	0126
174A4424	0474
175-006-9-001	1062
175-007-9-001	0004
175-008-9-001	0004
175-010-9-001	0925
175-2	0400
175-8349	0343
175E15	OEM
175F15	OEM
175PA50	0159
175PA60	0159
175PA80	0096
175PA100	0096
175PA110	0096
175PA120	0096
175PA130	OEM
175PA140	OEM
175PA150	OEM
175PA160	OEM
175PAK50	OEM
175PAK60	OEM
175PAK80	OEM
175PAK100	OEM
175PAK110	OEM
175PAK120	OEM
175PAK130	OEM
175PAK140	OEM
175PAK150	OEM
175PAK160	OEM
175RA60	1900
175RA80	0738
175RA100	0738
175RA110	0738
175RA120	0738
175RA140	OEM
175RA150	OEM
175RA160	OEM
175RC10	1900
175RC20	1900
175RC30	1900
175RC40	1900
175RC60	1900
175RC80	0738
176-003	0144
176-003-9-001	0144
176-004	0086
176-004-9-001	0144
176-004-9-002	0839
176-005	0144
176-005-9-001	0144
176-006	0144
176-006-9-001	0144
176-007	0144
176-007-9-001	0144
176-008-9-001	0016
176-014-9-001	0016
176-016-9-001	0155
176-017-9-001	0016
176-018-9-001	1400
176-024-9-001	0016
176-024-9-002	0839
176-025-9-001	0076
176-025-9-002	0127
176-026-9-001	0224
176-029-9-001	0155
176-029-9-002	1400
176-029-9-003	1400
176-029-9-004	1591
176-031-9-001	0155
176-031-9-002	0111
176-037-9-001	0224
176-037-9-002	0733
176-037-9-003	0111
176-040-9-001	2556
176-042-9-001	0155
176-042-9-002	0076
176-042-9-003	0076
176-042-9-004	0111
176-042-9-005	0219
176-042-9-006	0191
176-042-9-007	0228
176-043-9-002	0076
176-044-9-001	0836
176-044-9-002	0549
176-047-9-001	0191
176-048-9-002	0076
176-049-9-002	0076
176-054-9-001	0076
176-055-9-004	2292
176-056-9-001	1144
176-056-9-003	1144
176-056-9-005	0733
176-060-9-001	2209
176-060-9-002	0076
176-060-9-003	0076
176-060-9-004	2814
176-062-9-001	0076
176-065-9-001	0076
176-070-9-001	1984
176-072-9-005	0151
176-073-9-001	0127
176-073-9-002	0836
176-073-9-003	0018
176-073-9-012	0836
176-074-9-001	0171
176-074-9-002	1591
176-074-9-003	0936
176-074-9-004	0951
176-075-9-001	0113
176-075-9-003	0151
176-075-9-004	1220
176-075-9-006	0786
176-075-9-008	0786
176-087-9-001	0839
176-087-9-002	0836
176T1	OEM
176T2	OEM
176T3	OEM
176T4	OEM
176T5	OEM
176T6	OEM
176T7	OEM
176T8	OEM
176T9	OEM
176T10	OEM
176T11	OEM
176T12	OEM
177-001	0126
177-001-9-001	0126
177-006-9-001	0006
177-006-9-002	0006
177-007-9-001	0203
177-012-9-001	0006
177-018-9-001	0006
177-019-9-003	0006
177-020-9-001	0006
177-023-9-001	0925
177-025-9-001	0148
177-025-9-002	0693
177A07	0623
177E	OEM
179-1	0133
179-2	0133
179-4	0170
179-16	OEM
179-46444-01	1257
179-46444-02	1069
179-46444-04	OEM
179-46444-05	3901
179-46445-01	1041
179-46445-02	2097
179-46445-03	1835
179-46445-05	3399
179-46445-06	1844
179-46445-08	1859
179-46445-09	0341
179-46445-10	1632
179-46447-21	1554
179N2	5465
179P1	4466
180-1	0015
180-253	OEM
180N1	0330
180N1P	0224
180P1SP	OEM
180P2SP	OEM
180P3SP	OEM
180P4SP	OEM
180P5SP	OEM
180P6SP	OEM
180P8SP	OEM
180P10SP	OEM
180P12SP	OEM
180P16SP	OEM
180T2	0103
180T2A	0103
180T2B	0103
180T2C	0103
181(WORKMAN)	0190
181-000100	2100
181-000200	0654
181-003-9-001	2730
181-1	0133
181-1F	0133
181-2	0124
181-2F	OEM
181-3	0124
181-1002	0124
181-1003	0124
181N1	0007
181N1D	0007
181N2	0007
181N2D	0007
181T2	0130
181T2A	0103
181T2B	0177
181T2C	0103
182-009-9-001	0688
182-014-9-002	3046
182-014-9-003	0325
182-015-9-001	3155
182-029-9-001	0688
182-037-9-001	0688
182-038-9-001	0414
182-039-9-001	0688
182-044-1-001	OEM
182-044-9-001	3342
182-044-9-002	0325
182-045-9-001	0325
182-046-9-001	3342
182-056-9-001	0688
182-138-9-001	0414
182-144-9-002	OEM
182T2A	0543
182T2B	0543
182T2C	0543
183-1(SYLVANIA)	6169
183P1	0126
183T2A	0363
183T2B	0363
184-1	0102
184-1G033	1080
184-2	0102
184A-1	0160
184A-1-12	0160
184A-1-12-7	0160
184A-1L	0160
184A-1L8	0160
184T2A	0363
184T2B	0363
184T2C	0074
185-008	2296
185-009-01	OEM
185-4	OEM
185-5	OEM
185-6	0015
185-6(RECT)	0015
185-6-01	OEM
185-8	0505
185-9	0016
185-9-01	0576
185-9-02	0574
185-9-1	OEM
185-16	OEM
185-84	OEM
185-736	0178
185-Z9000	2336
185-Z9001	3960
185-Z9002	0061
185-Z9003	0764
185-Z9004	1137
185-Z9005	0344
185-Z9006	0901
185-Z9007	0058
185-Z9008	0940
185-Z9009	0505
185-Z9010	0710
185-Z9011	3696
185-Z9012	0725
185-Z9013	1652
185-Z9014	1651
185-Z9015	1650
185-Z9016	1712
185-Z9017	2329
185-Z9018	2329
185-Z9019	1791
185-Z9020	1683
185-Z9021	1502
185-Z9022	0863
185T2A	0363
185T2B	0363
186-001	0155
186-002	0076
186-003	0076
186-004	0076
186-005	0111
186-006	0191
186-007	0076
186-008	0390
186-009	0219
186-011	0133
186-012	0133
186-013	0143
186-015	0133
186-016	0012
186-1	1143
186N1	0007
187-1	0331
187-6	0170
187-7	0402
188-5	0147
188-68-35	0474
188-70-48	0474
188-826	0160
188B-1-82	0050
188C-1-82	0280
188P-2	0004
188P-2-12	0004
188P-2-127	0004
188P-2L	0004
188P-2L8	0004
189	0144
189-1	0065
189L-4	0004
189L-4-12	0004
189L-4-127	0004
189L-4L	0004
189L-4L8	0004
189N1	0313
189N1G	0313
190N1	0264
190N1C	0161
190N1D	0264
190N3	0161
190N3C	0161
190V039H18	0774
192-1A82	0440
192-120	0137
192-156	OEM
192-160	OEM
192-200	0137
192-330	0039
192-519	OEM
192-569	0157
192-1120	0137
192N2	0065
194N1	0065
194N1D	1966
195N1	0042
195N1C	0042
195N1D	0042
195N3	0042
195P2	0925
195P2C	0854
195P4	0925
196-654	0015
199-POWER	0160
200	0015
200-007	0144
200-010	0144
200-011	0086
200-015	0144
200-016	0198
200-018	0178
200-052	0037
200-053	0325
200-055	0144
200-056	0144
200-057	0016
200-058	0016
200-064	0325
200-076	0042
200-846	0016
200-856	2866
200-862	0016
200-863	0016
200-3172-208	OEM
200-3174-006	OEM
200-4302	OEM
200-6582	0087
200-6582-22	0229
200A	0038
200A#	0738
200B	0738
200B40	OEM
200B45	OEM
200B50	OEM
200BN20	OEM
200BN30	OEM
200BN40	OEM
200BN50	OEM
200BN60	OEM
200BN80	OEM
200BN100	OEM
200BN120	OEM
200BN140	OEM
200BN160	OEM
200BNS	OEM
200C	0738
200C4.7-5.3	OEM
200CNQ020	OEM
200CNQ030	OEM
200CNQ035	OEM
200CNQ040	OEM
200CNQ045	OEM
200D	0738
200E	0738
200EXD21	OEM
200F	0738
200H	0738
200JH21	OEM
200K	0738
200M	0738
200MAB180	OEM
200MAB200	OEM
200MAB250	OEM
200N15A	OEM
200N15A9	OEM
200N20A	OEM
200N20A9	OEM
200N25A	OEM
200N25A9	OEM
200N30A	OEM
200N30A9	OEM
200N35A	OEM
200N35A9	OEM
200N40A	OEM
200N40A9	OEM
200N45A	OEM
200N45A9	OEM
200N50A	OEM
200N50A9	OEM
200N55A	OEM
200N55A9	OEM
200N60A	OEM
200N60A9	OEM
200N65A	OEM
200N65A9	OEM
200N70A	OEM
200N70A9	OEM
200N75A	OEM
200N75A9	OEM
200N80A	OEM
200N80A9	OEM
200N85A	OEM
200N85A9	OEM
200N90A	OEM
200N90A9	OEM
200N95A	OEM
200N95A9	OEM
200N100A	OEM
200N100A9	OEM
200N110A	OEM
200N120A	OEM
200N130A	OEM
200N140A	OEM
200N150A	OEM
200N160A	OEM
200N170A	OEM
200N180A	OEM
200N190A	OEM
200N200A	OEM
200N210A	OEM
200N220A	OEM
200N230A	OEM
200N240A	OEM
200N250A	OEM
200PIN-BNC	OEM
200PIN-PP	OEM
200PIN-RM	OEM
200PV-BNC	OEM
200PV-PP	OEM
200PV-RM	OEM
200QD1	OEM
200QD21	OEM
200QH21	OEM
200R1B	0087
200R2B	2645
200R3B	1747
200R4B	5570
200R5B	1771
200R6B	0965
200R7B	0087
200R8B	0087
200R9B	0087
200R10B	0087
200R11B	OEM
200R12B	OEM
200R13B	OEM
200R14B	OEM
200R15B	OEM
200R16B	OEM
200R18B	OEM
200R20B	OEM
200R22B	OEM
200R24B	OEM
200R26B	OEM
200R28B	OEM
200R30B	OEM
200R32B	OEM
200R34B	OEM
200R36B	OEM
200R38B	OEM
200R40B	OEM
200R42B	OEM
200R44B	OEM
200R46B	OEM
200R48B	OEM
200R50B	OEM
200RB4	0540
200RS45	OEM
200RS50	OEM
200S	0738
200S2-2.65	OEM
200S40	OEM
200S45	OEM
200S50	OEM
200UB5	1023
200UB5R	0501
200UB10	1023
200UB10R	0501
200UB20	1023
200UB20R	0501
200UB30	1023
200UB30R	0501
200UB40	1036
200UB40R	1777
200UB50	1036
200UB50R	1777
200X2100-022	0878
200X2110-269	0167
200X2120-012	0395
200X2120-033	0855
200X2174-006	0151
200X2300-033	1201
200X2501-708	1055
200X2600-103	OEM
200X2600-183	3503
200X3110-607	0220
200X3145-404	0220
200X3151-432	1085
200X3162-538	0236
200X3172-208	1085
200X3174-006	0151
200X3174-014	0151
200X3174-021	0151
200X3189-304	2668
200X3190-604	0127
200X3192-101	0261
200X3206-800	0643
200X3207-306	1284
200X3222-907	0066
200X3223-025	0261
200X3224-007	2131
200X3224-204	1085
200X4049-081	1498
200X4082-614	0037
200X4085-415	1348
200X4094-001	1911
200X4101-500	0148
200X4547-806	0392
200X4547-814	0392
200X4589-802	0055
200X8000-026	0019
200X8010-094	0133
200X8010-102	0120
200X8010-165	0133
200X8100-130	0023
200X8130-171	0348
200X8220-531	0686
200X8220-851	0057
200X8220-878	0057
200X9120-224	0023
200YD21	OEM
201	0037
201-15	0004
201-25-4343-12	0144
201-211	OEM
201-254323-12	0144
201-254323-13	0144
201-254343-12	0144
201-254343-13	0127
201-254343-34	0007
201-254343-49	0144
201-283818-1	0212
201-283818-2	0212
201-283818-3	0212
201A	0050
201A#	0469
201A0723	0244
201AP	OEM
201B	0050
201B#	0469
201C	0762
201CNQ020	OEM
201CNQ030	OEM
201CNQ035	OEM
201CNQ040	OEM
201CNQ045	OEM
201D	0762
201E	3268
201F	3271
201H	3291
201K	3291
201M	0764
201P	2875
201S	0766
201U	2188
201UL140S20	OEM
201V	4855
201X2000-118	0019
201X2010-131	0124
201X2010-144	0124
201X2010-159	0124
201X2010-165	0133
201X2100-119	0348
201X2100-126	0023
201X2100-164	0348
201X2120-009	0023
201X2130-234	0023
201X2220-118	0057
201X2220-878	0057
201X2230-042	0301
201X3120-255	0023
201X3130-109	0071
201Z	4855
201ZB	OEM
201ZD	OEM
201ZF	OEM
201ZH	OEM
201ZK	OEM
202	0164
202(RECT)	2900
202-1	3247
202-2	0149
202-3	OEM
202-5-1260-32123	3688
202-5-2300-01710	0196
202-5-2300-01810	0023
202-5-2300-01910	0609
202-5-2450-13540	0023
202-5-3200-03010	0118
202-5-3210-05110	0041
202-5-3210-06220	0471
202-5-3210-13020	0053
202-5-3220-30010	0195
202-5-3220-36010	0010
202-5-9110-18820	0143
202-5-9531-01010	0133

DEVICE TYPE	REPL CODE
202-207	OEM
202-237	OEM
202A	0038
202A#	0469
202B	0741
202C	0745
202D	0745
202E	0747
202F	0747
202H	0747
202K	0752
202M	0764
202N1	0382
2020(RECT)	2900
202P	2875
202P1	0382
202S	0766
202S03	OEM
202U	2519
202V	4855
202Z	4855
202ZB	OEM
202ZD	OEM
202ZF	OEM
202ZH	OEM
202ZK	OEM
203	0015
203(ZENER)	0700
203-1	0212
203-2	0015
203-3	0609
203-4	0212
203-5-4570-73452	1513
203-5-4570-73462	1513
203-5-4921-01270	5372
203-5-5000-53660	0537
203-5-5000-53664	0537
203-5-5083-33160	0018
203-5-5100-53660	0537
203-5-5251-57069	0111
203-5-5251-57079	0111
203-5-6580-40050	2910
203-5-6850-40050	2910
203-5-7000-92970	0693
203-5-8620-32550	0560
203-6	0212
203A	3102
203A6137	0636
203B	0741
203C	0745
203D	0745
203E	0747
203F	0747
203H	0747
203J	OEM
203JPLA160	OEM
203JPLA180	OEM
203JPLA200	OEM
203JPLA250	OEM
203K	0752
203M	0764
203P	2875
203S	0766
203S4D	0212
203U	3102
203V	4855
203X3189-408	0313
203Z	4855
203ZB	OEM
203ZD	OEM
203ZF	OEM
203ZH	OEM
203ZK	OEM
204A	OEM
204A(SCR)	2188
204B	OEM
204C	OEM
204D	OEM
204F	OEM
204H	OEM
204K	OEM
204M	OEM
204Z4	OEM
205	OEM
205-5-9040-44210	0133
205A2-333	0155
205Z4	OEM
206-5-0131-22210	0621
206-5-0723-20110	2572
206-5-0783-21011	2063
206-5-0824-10110	2614
206-5-0834-10214	3190
206-5-1384-192210	OEM
206-5-2341-41610	4280
206A	OEM
206B	OEM
206C	OEM
206D	OEM
206E	OEM
206F	OEM
206H	OEM
206K	OEM
206M	OEM
206Z4	OEM
207A	0050
207A(SCR)	1110
207A1	0050
207A3	0004
207A7	0004
207A9	0144
207A10	0016
207A16	0178
207A16A	0178
207A20	0085
207A20A	0085
207A26	0160
207A31	0079
207A32-B	1413
207B	0050
207B(SCR)	2932
207C	OEM
207C(SCR)	0963
207D	OEM
207D(SCR)	0963
207E	OEM
207E(SCR)	2933
207H	OEM
207H(SCR)	0676
207K	OEM
207K(SCR)	2809
207M	0050
207M(SCR)	0336
207V073C04	0037
207Z4	OEM
208M	OEM
208Z4	OEM
209-1	0037
209-30	0042
209-31	0143
209-32	0133
209-846	0016
209-856	2866
209-862	0016
209-863	0016
209A	OEM
209B	OEM
209B-FLAG	0217
209B-FLEX	0217
209C	OEM
209D-FLAG	0217
209D-FLEX	0217
209F-FLAG	0217
209F-FLEX	0217
209H	OEM
209H-FLAG	0217
209H-FLEX	0217
209K	OEM
209K-FLAG	0217
209K-FLEX	0217
209M	OEM
209M-FLAG	0217
209M-FLEX	0217
209P	OEM
209S	OEM
209S-FLAG	0658
209S-FLEX	0468
209V	OEM
209Z	OEM
209Z4	OEM
210X2100-126	OEM
210Z4	OEM
211-58	0540
211-141	OEM
211A	0280
211A6381-I	0178
211B	0280
211B-FLAG	0217
211B-FLEX	0217
211C	0280
211D	0280
211D-FLAG	0217
211D-FLEX	0217
211E	0280
211F	0280
211F-FLAG	0217
211F-FLEX	0217
211H	OEM
211H-FLAG	0217
211H-FLEX	0217
211K	OEM
211K-FLAG	0217
211K-FLEX	0217
211M	OEM
211M-FLAG	0217
211M-FLEX	0217
211P	OEM
211S	OEM
211S-FLAG	0658
211S-FLEX	0468
211V	OEM
211Z	OEM
211Z4	0794
212-00102	0190
212-00104	0190
212-012	0190
212-013	0190
212-2	OEM
212-4	OEM
212-5	OEM
212-7	0015
212-18	0015
212-21	0015
212-22	0015
212-23	0015
212-25	0015
212-26	OEM
212-27	0015
212-33	0015
212-35	0015
212-36	0015
212-37	0015
212-38	0015
212-39	0015
212-40	0580
212-41	0015
212-42	0015
212-46	0474
212-47	0015
212-48	0015
212-49	0015
212-50	0015
212-51	0015
212-57	0015
212-58	0015
212-59	0015
212-60	0023
212-61	0015
212-62	0708
212-63	0474
212-64	0015
212-65	0015
212-66	0201
212-68	0102
212-70	0015
212-71	0015
212-72	0015
212-75	0015
212-76	0015
212-76-02	0015
212-77	0015
212-79	0015
212-80	0102
212-85	0774
212-85B	0774
212-86	0102
212-90	0378
212-92	0015
212-93	1791
212-94	0015
212-94B	0015
212-95	0378
212-96	0795
212-102	3858
212-103	3858
212-104	3858
212-105	3858
212-106	3858
212-108	0128
212-109	0128
212-110	0128
212-128	3858
212-129	3858
212-130	3858
212-130-X	0128
212-130X	3858
212-131	3858
212-132	3858
212-133	3858
212-134	3858
212-135	3858
212-136	3858
212-137	3858
212-138	3858
212-139	3858
212-139-01	3858
212-139-02	3858
212-139-03	1501
212-140	3858
212-140-01	3858
212-141	1067
212-141-01	1067
212-141-02	1706
212-141-03	1706
212-141-04	1706
212-141-1	OEM
212-142	1501
212-142-01	1501
212-142-02	1067
212-143	3655
212-144	3655
212-145	2109
212-145-01	1067
212-145-02	1067
212-146	2109
212-146-01	2109
212-146-02	2109
212-147	1067
212-148	1067
212-148-01	3358
212-149	1067
212-150	1067
212-151	1501
212-168-04	OEM
212-192	0015
212-241	1067
212-245	1067
212-505	0079
212-507	0079
212-695	0016
212-699	0037
212-741-005	2659
212-Z9000	0087
212-Z9001	0302
212-Z9002	0250
212-Z9003	1303
212-Z9004	0293
212-Z9005	0258
212-Z9006	0805
212-Z9007	1217
212-Z9008	0288
212-Z9009	OEM
212-Z9010	0348
212-Z9011	0538
212-Z9012	0122
212-Z9013	0131
212-Z9014	0145
212-Z9015	0319
212-Z9016	1023
212-Z9017	2363
212-Z9018	4417
212-Z9019	0966
212-Z9020	0577
212-Z9021	2387
212-Z9500A	0190
212-Z9521	2983
212-Z9522	1501
212-Z9523	1067
212-Z9526A	1706
212-Z9529	2546
212-Z9530	3655
212-Z9531	2804
212-Z9532	0409
212-Z9533	1710
212-Z9534	2982
212-Z9535	2985
212-Z9536-A	1997
212-Z9537	1197
212-Z9538	2984
212-Z9539	1054
212A	OEM
212B	OEM
212C	OEM
212D	OEM
212E	OEM
212F	OEM
212M	OEM
212P	OEM
212R2S	OEM
212S	OEM
212Z4	OEM
213-1	0086
213A	OEM
213B	OEM
213C	OEM
213D	OEM
213E	OEM
213F	OEM
213H	OEM
213K	OEM
213M	OEM
213Z4	OEM
214H	OEM
215-51	0015
215-58	0015
215-76(GE)	0015
215-76GE	0015
215T4	OEM
215Z4	0002
217-1	0144
217-76-02	0015
217N1	0642
217N1D	0177
217N2	0642
217N2C	0177
217N4	0642
218-22	0016
218-23	0016
218-24	0016
218-25	0016
218-26	0086
218-A-464-PM	2017
218A	OEM
218A4164P1M	2017
218B	0738
218D	0738
218F	0738
218H	0738
218K	0738
218M	0738
218P	0738
218S	0738
218V	0738
218Z	0738
218ZB	0738
218ZD	0738
218ZF	OEM
218ZH	OEM
219-103-2006	0167
219-303-4305	OEM
219B-FLAG	0217
219B-FLEX	0217
219BLY-A	OEM
219D-FLAG	0217
219D-FLEX	0217
219F-FLAG	0217
219F-FLEX	0217
219H	OEM
219H-FLAG	0217
219H-FLEX	0217
219K-FLAG	0217
219K-FLEX	0217
219M-FLAG	0217
219M-FLEX	0217
219S-FLAG	0658
219S-FLEX	0468
219ZB-FLEX	0468
219ZD-FLEX	0468
219Z-FLEX	0468
220-001001	0016
220-001002	0016
220-001011	0144
220-001012	0144
220-002001	0222
220-003001	0015
220-008001	0325
220C	OEM
220T4	OEM
221	0565
221-0048	0167
221-00267	OEM
221-00563	OEM
221-00586	OEM
221-00590	OEM
221-00707	OEM
221-31	0632
221-32	0632
221-34	0664
221-35	OEM
221-36	2161
221-37	0753
221-39	0753
221-40	0410
221-41	0753
221-42	0352
221-43	0354
221-45	2457
221-45-01	2457
221-46	0666
221-48	0167
221-48-01	0167
221-62	0349
221-64	2472
221-64(IC)	2482
221-64(RECT)	0102
221-65	0701
221-69	1337
221-69-01	1337
221-76	0623
221-76-01	0623
221-77	2480
221-78	1337
221-79	0701
221-79-01	0701
221-80	2482
221-81	OEM
221-82	0666
221-83	OEM
221-83-01	OEM
221-84	OEM
221-84-01	OEM
221-86	2493
221-87	0352
221-87-01	2830
221-89	1759
221-90	1345
221-91	0520
221-91-01	0520
221-92	OEM
221-93	0312
221-94	2499
221-95	2499
221-96	2501
221-96C	2501
221-97	2502
221-97-01	2502
221-98	1170
221-98-05	1170
221-98A05	1170
221-100	OEM
221-101-01	0334
221-102	OEM
221-102-01	5890
221-102A	OEM
221-103	3788
221-103-01A	OEM
221-103A	3788
221-104	3719
221-104D	6428
221-105	3720
221-105-01	3720
221-106	2801
221-106-01	2801
221-107	1421
221-108	2755
221-111-02	OEM
221-116	3922
221-121	0176
221-129	0625
221-129A-01	0625
221-130	OEM
221-131	2975
221-132	3990
221-134	OEM
221-134-01	OEM
221-140	3978
221-141	0848
221-143	OEM
221-144	OEM
221-146	6363
221-147	6364
221-148	OEM
221-149	OEM
221-149-01	OEM
221-153	OEM
221-154	OEM
221-155	OEM
221-155-01	OEM
221-157-02	OEM
221-158	1057
221-158-01	1057
221-158-02	1057
221-158-03	1057
221-158-02001	1057
221-159	OEM
221-166	0624
221-166-00	0624
221-166-0	1812
221-166-02	1196
221-166-04	0334
221-166-09	OEM
221-166-4	0334
221-166-9	OEM
221-167-01	1292
221-167-04	OEM
221-167-05	1828
221-171	3381
221-175	4002
221-175-01	4002
221-177	3711
221-178	4274
221-179-01	2107
221-179-01-E-01	2107
221-179-02	2107
221-179-03	2107
221-179-05	OEM
221-179-1	2107
221-182	OEM
221-182-01	1957
221-187	5215
221-187A	5215
221-188	0625
221-190	1425
221-190-01	1104
221-190-01A	1104
221-190-E-01	1104
221-193	3788
221-193A	3788
221-201	OEM
221-201-03	OEM
221-202	4006
221-213	0624
221-213-00	0624
221-213-02	1196
221-213-04	0334
221-213-4	0334
221-217	6369
221-230	3772
221-240	0360
221-241	OEM
221-242-01	OEM
221-250-02	5564
221-250-03	OEM
221-261	4008
221-261-01	4008
221-261B03	4008
221-264	OEM
221-267	OEM
221-285-01	OEM
221-285-02	OEM
221-285-03	OEM
221-290	OEM
221-290-01	OEM
221-291	4189
221-296	OEM
221-296-00	OEM
221-296-04	0334
221-297	OEM
221-300	OEM
221-302	OEM
221-303	3437
221-313-01	OEM
221-314	OEM
221-316	4727
221-343	OEM
221-346	6163
221-347	4385
221-347-01	4385
221-347A01	OEM
221-348	OEM
221-364	0522
221-364-01	OEM
221-365	OEM
221-365-01	OEM
221-365-02	OEM
221-366	OEM
221-369	OEM
221-370	OEM
221-371-01	OEM
221-371-02	OEM
221-373-02	OEM
221-376	OEM
221-377-01	OEM
221-383	3254
221-384A01	OEM
221-394	OEM
221-418	OEM
221-418-01	OEM
221-418-02	OEM
221-419	OEM
221-422-01	3710
221-446	OEM
221-446-00	OEM
221-465	OEM
221-466	OEM
221-467	3236
221-468	6394
221-473	OEM
221-475	3083
221-475C	3083
221-476	2870
221-476D	2870
221-479-06	OEM
221-482	OEM
221-493	OEM
221-496-01	OEM
221-516	3635
221-517-01	OEM
221-517-1	OEM
221-518	OEM
221-522	OEM
221-524	OEM
221-524-01	OEM
221-528	OEM
221-537	OEM
221-538	OEM
221-539	OEM
221-540	OEM
221-544	OEM
221-544-01	OEM
221-550	OEM
221-550-01	OEM
221-554-01	OEM
221-556	OEM
221-560	OEM
221-561	4094
221-563	OEM
221-565	OEM
221-566	OEM
221-568	OEM
221-569	OEM
221-570	OEM
221-571	OEM
221-571-01	OEM
221-572	OEM
221-573	OEM
221-578	OEM
221-579	OEM
221-580-04	OEM
221-586	OEM
221-587	OEM
221-589	OEM
221-590	OEM
221-598	3253
221-600	OEM
221-604	OEM
221-604-02	OEM
221-605	OEM
221-628	OEM
221-636	OEM
221-637	OEM
221-637-01	4360
221-648	OEM
221-650	OEM
221-650-03	OEM
221-652-01	OEM
221-656	OEM
221-679	4326
221-679-01	OEM
221-682-01	OEM
221-683	OEM
221-683A	OEM
221-684-01	OEM
221-685	OEM
221-685-02	OEM
221-701	OEM
221-703	OEM
221-704	OEM
221-705	OEM
221-707	OEM
221-708	OEM
221-708-01	OEM
221-749	OEM
221-776	OEM
221-776-01	OEM
221-782	OEM
221-791	OEM
221-796	6419
221-806	OEM
221-812	OEM
221-826	OEM
221-899	OEM
221-948-01	OEM
221-949-01	OEM
221-4757D04	3083
221-4767D02	2870
221-Z9000	2624
221-Z9001	2571
221-Z9002	0678
221-Z9003	0829
221-Z9004	2715
221-Z9005	0654
221-Z9006	0443
221-Z9007	2549
221-Z9008	0328
221-Z9009	2743
221-Z9010	0856
221-Z9011	1395
221-Z9012	0789
221-Z9013	0395
221-Z9014	0364
221-Z9015	1843
221-Z9016	2557
221-Z9017	1301

DEVICE TYPE	REPL CODE
221-Z9018	1696
221-Z9019	1192
221-Z9020	0026
221-Z9021	0579
221-Z9022	0878
221-Z9023	0911
221-Z9024	1478
221-Z9025	2617
221-Z9026	1215
221-Z9027	0520
221-Z9028	0785
221-Z9029	1808
221-Z9030	0723
221-Z9031	3380
221-Z9032	0751
221-Z9033	3735
221-Z9034	0360
221-Z9035	5670
221-Z9036	0802
221-Z9037	2755
221-Z9038	2167
221-Z9039	2043
221-Z9040	3195
221-Z9041	0521
221-Z9042	0973
221-Z9043	0624
221-Z9044	1292
221-Z9045	0083
221-Z9046	0864
221-Z9047	0528
221-Z9048	2639
221-Z9049	4003
221-Z9050	2949
221-Z9051	1050
221-Z9052	2942
221-Z9053	3269
221-Z9054	3273
221-Z9055	2049
221-Z9056	2786
221-Z9057	2789
221-Z9058	0886
221-Z9059	0852
221-Z9060	0857
221-Z9061	1540
221-Z9062	3318
221-Z9063	1816
221-Z9064	1055
221-Z9065	4224
221-Z9066	2019
221-Z9067	4069
221-Z9068	4042
221-Z9069	4072
221-Z9070	3815
221-Z9071	1128
221-Z9072	2072
221-Z9073	2233
221-Z9074	0001
221-Z9075	0232
221-Z9076	0361
221-Z9077	0697
221-Z9078	0569
221-Z9079	1823
221-Z9080	1041
221-Z9081	1831
221-Z9082	2622
221-Z9083	0081
221-Z9084	0141
221-Z9085	0562
221-Z9086	1157
221-Z9087	1345
221-Z9088	2930
221-Z9089	3219
221-Z9090	2027
221-Z9091	5679
221-Z9092	OEM
221-Z9093	1611
221-Z9094	3761
221-Z9095	3763
221-Z9096	0848
221-Z9097	1077
221-Z9098	0417
221-Z9099	2262
221-Z9100	4561
221-Z9101	4565
221-Z9102	4566
221-Z9103	4093
221-Z9104	1621
221-Z9105	4570
221-Z9106	4571
221-Z9107	2124
221-Z9108	2582
221-Z9109	2599
221-Z9110	2603
221-Z9111	2591
221-Z9112	0470
221-Z9113	4582
221-Z9114	2305
221-Z9115	1248
221-Z9116	4420
221-Z9117	4594
221-Z9118	4595
221-Z9119	3366
221-Z9120	4600
221-Z9121	0881
221-Z9122	0884
221-Z9123	4147
221-Z9124	2602
221-Z9125	4401
221-Z9126	2798

DEVICE TYPE	REPL CODE
221-Z9127	4395
221-Z9128	4616
221-Z9129	4087
221-Z9130	4086
221-Z9131	4018
221-Z9132	1260
221-Z9133	0215
221-Z9134	5684
221-Z9135	2872
221-Z9136	2279
221-Z9137	3915
221-Z9138	4543
221-Z9139	2699
221-Z9140	3766
221-Z9141	1854
221-Z9142	2705
221-Z9143	1797
221-Z9144	0410
221-Z9145	1975
221-Z9146	4147
221-Z9147	2368
221-Z9148	2358
221-Z9149	3352
221-Z9150	0923
221-Z9151	2581
221-Z9152	3285
221-Z9153	1878
221-Z9154	2831
221-Z9155	2812
221-Z9156	0673
221-Z9157	0638
221-Z9158	4034
221-Z9159	2602
221-Z9160	1479
221-Z9161	1524
221-Z9162	2960
221-Z9163	0398
221-Z9164	0101
221-Z9165	OEM
221-Z9166	0916
221-Z9167	2817
221-Z9168	5689
221-Z9169	3397
221-Z9170	3976
221-Z9171	3977
221-Z9172	2022
221-Z9173	2823
221-Z9174	3270
221-Z9175	3832
221-Z9176	3089
221-Z9177	2231
221-Z9178	5691
221-Z9179	4635
221-Z9180	4067
221-Z9181	4079
221-Z9182	5562
221A	OEM
221C	OEM
221E	OEM
221G3114	2537
221G3114-1	2537
222-1	0267
222-2	0267
222A	OEM
222B	OEM
222C	OEM
222D	OEM
222F	OEM
222H	OEM
222K	OEM
222M	OEM
222N1	0551
222P	OEM
222S	OEM
222V	OEM
222Z	OEM
223	0144
223-1	0037
223-3	OEM
223-4	OEM
223-5	OEM
223-6	OEM
223-10-01	OEM
223-18	3502
223-18-01	3955
223-20-01	OEM
223-28	3574
223-40	2364
223-Z9000	2854
223-Z9001	2857
223-Z9002	2859
223-Z9003	2861
223-Z9004	2634
223-Z9005	2632
223-Z9006	1994
223-Z9007	2057
223-Z9008	2864
223A	OEM
223B	OEM
223D	OEM
223F	OEM
223H	OEM
223K	OEM
223M	OEM
223P	OEM
223P1	0037
223P1AA	0037
223S	OEM
223U	OEM

DEVICE TYPE	REPL CODE
223Z	OEM
223ZB	OEM
223ZD	OEM
224-1	0111
224B007-1	OEM
224N1	0111
225	OEM
225A6946-P000	0232
225A6946-P003	0335
225A6946-P004	0361
225A6946-P010	0512
225A6946-P020	0697
225A6946-P050	0743
225A6946-P093	0569
225A6976-P095	1486
225ITTN	OEM
225T4	OEM
226-1	0079
226-1(SYLVANIA)	0016
226-2	0079
226-2(DIO)	OEM
226-2(DIODE)	0102
226-3	3698
226-4	3426
226-8	OEM
226N1	0079
226P10601	0918
226P30104	OEM
227-3	3327
227-2000D1	0196
227-200001	0196
228-1	0824
228A	OEM
228B	OEM
228D	OEM
228F	OEM
228H	OEM
228ITTJ	OEM
228K	OEM
228M	OEM
228N1	0551
228N5	0824
228P	OEM
228S	OEM
228V	OEM
228Z	OEM
228ZB	OEM
228ZD	OEM
228ZF	OEM
228ZH	OEM
228ZK	OEM
229-0011	0196
229-0014	0143
229-0026	0050
229-0027	0164
229-0028	0164
229-0029	0164
229-0030	0164
229-0030-8	OEM
229-0035	0143
229-0037	0143
229-0038	0050
229-0041	0164
229-0049	0143
229-0050-8	OEM
229-0050-13	0016
229-0050-14	0016
229-0050-15	0016
229-0055	0164
229-0056	0164
229-0057	0143
229-0062	0164
229-0077	0050
229-0079	0050
229-0080	0164
229-0082	0050
229-0083	0050
229-0085	0164
229-0086	0050
229-0087	0050
229-0088	0164
229-0089	0050
229-0090	0050
229-0091	0164
229-0092	0164
229-0093	0143
229-0095	0050
229-0096	0030
229-0097	0164
229-0098	0050
229-0099	0050
229-0-180-33	0127
229-0-180-34	0127
229-0100	0164
229-0102	0015
229-0105	0143
229-0106	0050
229-0107	0143
229-0108	0143
229-0110	0050
229-0111	0050
229-0112	0050
229-0114	0917
229-0116	0085
229-0117	0015
229-0119	0015
229-0120	0015
229-0121	0050
229-0122	2866

DEVICE TYPE	REPL CODE
229-0122-20	0325
229-0123	0050
229-0124	0164
229-0125	0164
229-0129	0050
229-0130	0164
229-0131	0050
229-0132	0050
229-0133	0164
229-0134	2866
229-0135	0015
229-0136	0050
229-0137	0164
229-0138	0164
229-0139	0164
229-0140	0164
229-0141	0143
229-0142	0164
229-0143	0164
229-0144	0079
229-0145	0050
229-0146	0164
229-0147	0015
229-0149	0079
229-0150	0079
229-0151	0079
229-0151-3	0144
229-0152	0079
229-0154	0079
229-0156-12	OEM
229-0162	0015
229-0180-26	OEM
229-0180-31	OEM
229-0180-32	0007
229-0180-33	0007
229-0180-34	0144
229-0180-119	0007
229-0180-123	0016
229-0180-124	0144
229-0180-130	OEM
229-0180-149	0144
229-0182-65	0143
229-0185-2	0144
229-0185-3	0144
229-0190-29	0144
229-0190-30	0007
229-0190-31	0127
229-0190-90	0016
229-0191-29	0127
229-0191-30	0127
229-0191-31	OEM
229-0192-18	0212
229-0192-19	0144
229-0192-20	0325
229-0204-4	0144
229-0204-6	0007
229-0204-6(VHF)	0007
229-0204-23	0144
229-0210-14	0144
229-0210-19	0127
229-0214-17	OEM
229-0214-40	0144
229-0214-41	OEM
229-0220-9	0144
229-0220-14	0917
229-0220-19	0144
229-0240-19	OEM
229-0240-20	0917
229-0240-25	0007
229-0248-45	0144
229-0248-46	OEM
229-0250-10	0144
229-0250-11	0917
229-0260-18	0127
229-180-32	0007
229-485-2	0127
229-1054-5	0015
229-1054-9	0015
229-1054-47	OEM
229-1054-52	OEM
229-1054-82	0015
229-1054-85	0015
229-1200-29	0917
229-1200-36	0016
229-1301-21	1137
229-1301-32	0330
229-1301-39	3380
229-1301-40	3735
229-1301-41	0349
229-1301-42	1808
229-1301-43	3195
229-1301-44	0856
229-1513-46	0086
229-5100-15U	0144
229-5100-15V	0144
229-5100-31V	0007
229-5100-32	0007
229-5100-32V	0007
229-5100-33V	0144
229-5100-164	0917
229-5100-224	0144
229-5100-225	0144
229-5100-226	0144
229-5100-227	0136
229-5100-228	0144
229-5100-231	0143
229-5100-232	0015
229-5100-233	0015
229-5100-234	0015

DEVICE TYPE	REPL CODE
229-5100-235	0102
229-6011	0086
229-9191-31	OEM
229A	OEM
229B	OEM
229D	OEM
229F	OEM
229H	OEM
229K	OEM
229M	OEM
229P	OEM
229S	OEM
229V	OEM
229Z	OEM
229ZB	OEM
229ZD	OEM
229ZF	OEM
229ZH	OEM
229ZK	OEM
230-0006	0019
230-006	0143
230-0014	0133
230-0023	0057
230H	OEM
230T4	OEM
231-0000-01	0164
231-0004	0103
231-0004-01	0111
231-0004-03	0111
231-0006-03	0604
231-0006B	0160
231-0008	0178
231-0009	0004
231-006B	0160
231-0011	0160
231-0013	0086
231-0015	0160
231RC10A	OEM
231RC10AM	OEM
231RC20A	OEM
231RC20AM	OEM
231RC40A	OEM
231RC40AM	OEM
231RC60A	OEM
231RC60AM	OEM
231RC80A	OEM
231RC80AM	OEM
231RK80A	OEM
231RK100A	OEM
231RK100AM	OEM
231RK120A	OEM
231RK120AM	OEM
231TB7	OEM
232	OEM
232-0001	0015
232-0007	0748
232-0008	0254
232-0012	0691
232-0014	0097
232-1	0079
232-2	0076
232-1006	0015
232-1009	0015
232-1011	0015
232B2	0079
232N1	0079
232N2	0079
232TB7	OEM
233(SEARS)	0086
233J	OEM
233K	OEM
233L	OEM
233TB7	OEM
234-1	2455
234J	OEM
234K	OEM
234L	OEM
234TB7	OEM
235	0280
235-1	OEM
235H	OEM
235J	OEM
235K	OEM
235L	OEM
236-0002	1041
236-0003	0141
236-0005	0232
236-0006	0335
236-0007	0361
236-0008	0385
236-0009	1172
236-0012	0962
236-0017	0562
239	OEM
240-US	OEM
240A	0469
240B	0469
240C	0469
240D	0469
240E	0722
240E15	OEM
240F	0722
240F15	OEM
240H	0722
240K	0725
240M	0725
240P	0679
240PAC10	OEM
240PAC20	OEM

DEVICE TYPE	REPL CODE
240PAC30	OEM
240PAC40	OEM
240PAC50	OEM
240PAC60	OEM
240PAC70	OEM
240PAC80	OEM
240PAC90	OEM
240PAC100	OEM
240PAL50	0459
240PAL60	0459
240PAL80	0587
240PAL100	0587
240PAL110	0587
240PAL120	0587
240PAM10	0459
240PAM20	0459
240PAM30	0459
240PAM40	0459
240PAM50	0459
240PAM60	0459
240PAM80	0587
240PAM100	0587
240S	0679
240T4	OEM
240U	0469
240V	0679
240Z	0679
240ZB	0679
240ZD	0679
241-15A	0050
241C025	OEM
241C025B	OEM
241C25	OEM
241C25B	OEM
241C30	OEM
241C30B	OEM
241C35	OEM
241C35B	OEM
241C40	OEM
241C40B	OEM
241C45	OEM
241C45B	OEM
241C50	OEM
241C50B	OEM
241C60	OEM
241C60B	OEM
241C70	OEM
241C70B	OEM
241C80	OEM
241C80B	OEM
241C90	OEM
241C90B	OEM
241C100	OEM
241C100B	OEM
241C110	OEM
241C110B	OEM
241C120	OEM
241C120B	OEM
241TB7	OEM
241UA	OEM
241UB	OEM
241UC	OEM
241UD	OEM
241UF	OEM
241UH	OEM
241UK	OEM
241UM	OEM
241WA	OEM
241WB	OEM
241WC	OEM
241WD	OEM
241WF	OEM
241WH	OEM
241WK	OEM
241WM	OEM
242	0494
242-1	OEM
242-997	2136
242B7	OEM
242H	OEM
242XB	OEM
242XD	OEM
242XF	OEM
242XH	OEM
242XK	OEM
242XM	OEM
242XP	OEM
242XS	OEM
242YB	OEM
242YD	OEM
242YF	OEM
242YH	OEM
242YK	OEM
242YM	OEM
242YP	OEM
242YS	OEM
242ZB	OEM
242ZD	OEM
242ZF	OEM
242ZH	OEM
242ZK	OEM
242ZM	OEM
242ZP	OEM
242ZS	OEM
243A-US	OEM
244	OEM
244H	OEM
244TB7	OEM
245N1B	0642

DEVICE TYPE	REPL CODE
245N2B	0642
246-3	OEM
246H	OEM
246P1	0037
246P463060	0057
246P463070	0057
246P464030	0064
246P464060	0064
247-1	0917
247-257	0016
247-621	0015
247-623	0004
247-624	0160
247-625	0086
247-626	0086
247-629	0016
248-38104-1	0086
249-1L	0224
249-140	OEM
249-160	OEM
249-399	OEM
249-519	0041
249-569	0883
249-629	0295
249-759	0077
249-7867-3331-504	OEM
249-7868-3331-504	OEM
249-7869-3331-504	OEM
249-7870-3331-504	OEM
249-7871-3331-504	OEM
249-7872-3331-504	OEM
249-7967-3332-504	OEM
249-7968-3332-504	OEM
249-7969-3332-504	OEM
249-7970-3332-504	OEM
249-7971-3332-504	OEM
249-7972-3332-504	OEM
249-8067-3333-504	OEM
249-8068-3333-504	OEM
249-8069-3333-504	OEM
249-8070-3333-504	OEM
249-8071-3333-504	OEM
249-8072-3333-504	OEM
249-N1	OEM
249H	OEM
249HA	OEM
249N1	0224
250-0359	1607
250-0373	0076
250-0380	0286
250-0700	0111
250-0711	0380
250-0712	0076
250-1	0285
250-3	0285
250-4	OEM
250-1213	0191
250-1312	1221
250A-FLAG	0217
250A-FLEX	0217
250B35	OEM
250BC15	1023
250B-FLAG	0217
250B-FLEX	0217
250DC15	1023
250D-FLAG	0217
250D-FLEX	0217
250FC15	1023
250F-FLAG	0217
250F-FLEX	0217
250GC15	1023
250GD15	OEM
250H	OEM
250H-FLEX	0468
250JB05L	2363
250JB1L	2363
250JB2L	2363
250JB3L	2369
250JB4L	2369
250JB5L	2370
250JB6L	2370
250JB8L	2372
250JB10L	4900
250JB12L	OEM
250JC15	1036
250K-FLEX	0468
250LC15	1046
250MA60	OEM
250MA80	OEM
250MA100	OEM
250MA120	OEM
250MA140	OEM
250MA160	OEM
250MA180	OEM
250M-FLEX	0468
250MLAB160	OEM
250MLAB180	OEM
250MLAB200	OEM
250MLAB250	OEM
250N3	0285
250N15A	OEM
250N15A9	OEM
250N20A	OEM
250N20A9	OEM
250N25A	OEM
250N25A9	OEM
250N30A	OEM
250N30A9	OEM
250N35A	OEM

If replacement code is OEM, contact original manufacturer for replacement.

DEVICE TYPE	REPL CODE	DEVICE TYPE	REPL CODE	DEVICE TYPE	REPL CODE	DEVICE TYPE	REPL CODE	DEVICE TYPE	REPL CODE	DEVICE TYPE	REPL CODE	DEVICE TYPE	REPL CODE
250N35A9	OEM	250R38B	OEM	251RC10	1900	254F-FLAG	0217	260P04503	0079	260P17105	0368	260P356010	0127
250N40A	OEM	250R40	OEM	251RC20	1900	254F-FLEX	0217	260P04505	0079	260P17106	0368	260P360030	0006
250N40A9	OEM	250R40B	OEM	251RC30	1900	254G	OEM	260P05402	0127	260P17201	0144	260P385010	0066
250N45A	OEM	250R42	OEM	251RC40	1900	254GI	0015	260P05402A	0127	260P17501	0380	260P387010	2910
250N45A9	OEM	250R42B	OEM	251RC50	1900	254H	OEM	260P05801	0144	260P17502	0380	260P387030	2910
250N50A	OEM	250R44	OEM	251RC60	1900	254H-FLAG	0217	260P05901	0007	260P17503	0380	260P416020	0155
250N50A9	OEM	250R44B	OEM	251RC80	0738	254H-FLEX	0217	260P05901A	0007	260P17601	0127	260P416030	0155
250N55A	OEM	250R46	OEM	251RC100	OEM	254K-FLAG	0217	260P06901	0144	260P17602	0144	260P416040	0155
250N55A9	OEM	250R46B	OEM	251RC120	OEM	254K-FLEX	0217	260P06902	0144	260P17603	0127	260P419030	0191
250N60A	OEM	250R48	OEM	251RL60	1700	254M-FLAG	0217	260P06903	0144	260P17701	0380	260P419040	0191
250N60A9	OEM	250R48B	OEM	251RL80	0345	254M-FLEX	0217	260P06904	0079	260P17702	0007	260P422010	0261
250N65A	OEM	250R50	OEM	251RL100	0345	254P-FLAG	0658	260P07001	0079	260P17704	0380	260P428020	1284
250N65A9	OEM	250R50B	OEM	251RL110	0345	254P-FLEX	0658	260P07002	0079	260P19009	0074	260P428070	1284
250N70A	OEM	250RA10	1900	251RL120	0345	254S-FLAG	0217	260P07004	0144	260P19101	0016	260P469010	0342
250N70A9	OEM	250RA20	1900	251RM60	1700	254S-FLEX	0217	260P07301	0079	260P19108	0003	260P469020	0342
250N75A	OEM	250RA30	1900	251RM80	0345	254V-FLEX	0468	260P07502	0222	260P19208	0103	260P559010	0151
250N75A9	OEM	250RA40	0738	251RM100	0345	254ZB-FLEX	0468	260P07601	0164	260P19501	0016	260P559030	0151
250N80A	OEM	250RA50	0738	251RM110	0345	254ZD-FLEX	0468	260P07701	0079	260P19503	0111	260P559050	0151
250N80A9	OEM	250RA60	0738	251RM120	0345	254Z-FLEX	0468	260P07702	0079	260P19909	0074	260P560040	0148
250N85A	OEM	250RA70	OEM	251RM130	OEM	255B7-CH00	OEM	260P07703	0076	260P20101	0178	260P560050	0148
250N85A9	OEM	250RA80	0738	251RM140	OEM	255P13701	OEM	260P07704	0079	260P21001	0280	260P561010	OEM
250N90A	OEM	250RA100	0738	251RM150	OEM	256M	OEM	260P07705	0076	260P21002	0280	260P571010	0276
250N90A9	OEM	250RA110	0738	251RM160	OEM	256M11	OEM	260P07707	0079	260P21102	0560	260P573020	1648
250N95A	OEM	250RA120	0738	251RM170	OEM	256TB7	OEM	260P07901	0144	260P21106	0566	260P574020	0392
250N95A9	OEM	250RA140	OEM	251S-FLEX	0468	257-1	0494	260P08001	0144	260P21208	0142	260P582010	OEM
250N100A	OEM	250RA150	OEM	251UL20S10	OEM	257-IF	OEM	260P08201	0037	260P21308	0560	260P603010	1032
250N100A9	OEM	250RA160	OEM	251UL20S15	OEM	257A-US	OEM	260P08401	0144	260P21608	0065	260P604010	0887
250N110A	OEM	250RA170	OEM	251UL40S10	OEM	258-1	OEM	260P08601	0178	260P21901	0074	260P606010	2129
250N120A	OEM	250RA180	OEM	251UL40S15	OEM	258-2	0382	260P08801	0079	260P21908	0220	260P607010	2129
250N130A	OEM	250RA190	OEM	251UL40S20	OEM	258TB7	OEM	260P08801A	0079	260P22001	0325	260P608010	2129
250N140A	OEM	250RA200	OEM	251UL60S10	OEM	259	OEM	260P08901	0065	260P22002	0325	260P632010	0887
250N150A	OEM	250RC5	OEM	251UL60S15	OEM	260	OEM	260P08908	0826	260P22003	3155	260P641030	OEM
250N160A	OEM	250RC10	OEM	251UL60S20	OEM	260-00301	OEM	260P09201	0007	260P22004	3155	260P654020	0286
250N170A	OEM	250RC15	OEM	251UL80S10	OEM	260-2	0023	260P09402	0142	260P22101	0233	260P664010	OEM
250N180A	OEM	250RC20	OEM	251UL80S15	OEM	260-10-006	0688	260P09501	0003	260P22203	0285	260P664030	OEM
250N190A	OEM	250RC30	OEM	251UL80S20	OEM	260-10-016	0006	260P09508	0003	260P22204	0285	260Z00109	0144
250N200A	OEM	250RC40	OEM	251UL100S10	OEM	260-10-020	0076	260P09701	0065	260P22801	0390	260Z00209	0144
250N210A	OEM	250RC50	OEM	251UL100S15	OEM	260-10-021	0191	260P09902	0079	260P24008	0103	260Z00309	0144
250N220A	OEM	250RC60	OEM	251UL100S20	OEM	260-10-031	0030	260P045040	0124	260P24308	0003	260Z00401	0191
250N230A	OEM	250RL60	0345	251UL100S25	OEM	260-10-033	0023	260P106A1	0155	260P24408	0642	260Z00402	0016
250N240A	OEM	250RL80	1700	251UL100S30	OEM	260-10-036	0911	260P1003	0086	260P24803	0178	260Z00407	OEM
250N250A	OEM	250RL100	0345	251UL120S20	OEM	260-10-039	0216	260P1030	0233	260P24901	0007	260Z00703	0164
250NC15	1046	250RL110	0345	251UL120S25	OEM	260-10-042	0111	260P1060	0144	260P25504	1348	260Z01201	0164
250P4H	OEM	250RL120	0345	251UL120S30	OEM	260-10-044	0165	260P1601	0007	260P25601	3314	261-02	OEM
250P6H	OEM	250RM60	1700	251UL130S20	OEM	260-10-046	0636	260P1610	0007	260P25604	3314	261-03	OEM
250P8H	OEM	250RM80	0345	251UL130S30	OEM	260-10-048	0123	260P1760	0127	260P26201	0342	261A	0738
250P10H	OEM	250RM100	0345	251UL140S20	OEM	260-10-049	0139	260P4002	0079	260P26301	0681	261B	0738
250P12H	OEM	250RM110	0345	251UL140S25	OEM	260-10-050	0229	260P10003	0086	260P28107	0191	261C	0738
250P16H	OEM	250RM120	0345	251UL140S30	OEM	260-10-052	0733	260P10003A	0086	260P28401	0042	261D	0738
250PA60	0159	250RM130	OEM	251UL150S30	OEM	260-10-053	0936	260P10005	0076	260P28701	0560	261F	0738
250PA80	0096	250RM140	OEM	251UL160S25	OEM	260-10-054	0042	260P10301	0233	260P30701	0261	261H	0738
250PA100	0096	250RM150	OEM	251UL160S30	OEM	260-10-055	0836	260P10403	0144	260P31303	0155	261J	OEM
250PA120	0096	250RM160	OEM	251UL180S25	OEM	260-10-056	0549	260P10501	0144	260P31402	0076	261K	0738
250PA140	OEM	250RM170	OEM	251UL180S30	OEM	260-61-011	0133	260P10502	0144	260P33108	0065	261M	0738
250PA150	OEM	250RS35	OEM	251UL260S30	OEM	260-61-047	0133	260P10601	0007	260P33408	0003	261S	0738
250PA160	OEM	250RS40	OEM	251ZB-FLEX	0468	260-P10305	OEM	260P10602	0144	260P33802	0518	261Z	0738
250PAC10	0192	250S2H	OEM	251ZD-FLEX	0468	260C33801	0518	260P10707	0595	260P33803	0518	261ZB	0738
250PAC20	0192	250S4H	OEM	251Z-FLEX	0468	260D00401	0004	260P10801	2375	260P33804	0518	261ZD	0738
250PAC30	0159	250S6H	OEM	252-2	0133	260D00402	0004	260P11101	0144	260P33805	0518	262-1	2528
250PAC40	0159	250S30	OEM	252GT	OEM	260D00403	0004	260P11101A	0144	260P33806	0518	263H	OEM
250PAC50	0159	250S35	OEM	253-5	0531	260D00404	0004	260P11102	0330	260P34008	2660	263P05209	0024
250PAC60	0159	250S40	OEM	253B	OEM	260D00507	0143	260P11209	0065	260P34202	0392	263P06602	0101
250PAL10	0459	250S-FLEX	0468	253PJA60	OEM	260D0404	0136	260P11302	0079	260P34602	1085	263P17402	1551
250PAL20	0459	250T4	OEM	253PJA80	OEM	260D02501	0004	260P11303	0079	260P35101	1085	263P82401	OEM
250PAL30	0459	250V-FLEX	0468	253PJA100	OEM	260D02601	0004	260P11304	0016	260P35301	0715	263P82501	OEM
250PAL40	0459	250ZB-FLEX	0468	253PJA120	OEM	260D02701	0164	260P11305	0016	260P35401	0955	263P89601	OEM
250PAL50	0459	250ZD-FLEX	0468	253PJA140	OEM	260D04501	0136	260P11403	0037	260P35601	0127	263P142030	OEM
250PAL60	0459	250Z-FLEX	0468	253PJA160	OEM	260D04701	0164	260P11502	0079	260P35908	2660	263P142040	OEM
250PAM10	0459	251-2	0124	253PJLA60	OEM	260D05701	0144	260P11503	0079	260P36001	0006	263P170030	OEM
250PAM20	0459	251-250-03	OEM	253PJLA80	OEM	260D05704	0127	260P11504	0079	260P36003	0006	263P228040	OEM
250PAM30	0459	251A-FLAG	0217	253PJLA100	OEM	260D05707	0144	260P11505	0079	260P36501	0224	263P228050	OEM
250PAM40	0459	251A-FLEX	0217	253PJLA120	OEM	260D05709	0079	260P12001	0079	260P37108	0236	263P229010	OEM
250PAM50	0459	251B-FLAG	0217	253RA10	OEM	260D07201	0233	260P12002	0079	260P38501	0066	263P229020	OEM
250PAM60	0459	251B-FLEX	0217	253RA20	OEM	260D07412	0016	260P12401	0086	260P38701	2910	263P299010	6930
250P-FLEX	0468	251D-FLAG	0217	253RA30	OEM	260D07901	0086	260P12701	0042	260P41604	0155	263P545020	OEM
250R1B	0953	251D-FLEX	0217	253RA40	OEM	260D08013	0127	260P13001	0136	260P41605	0155	263P546010	OEM
250R2B	0242	251F2S	OEM	253RA60	OEM	260D08201	0079	260P13604	0951	260P41802	0563	263P548010	OEM
250R3B	1747	251F3S	OEM	253RA80	OEM	260D08214	0566	260P13701	0079	260P41904	0191	263P684010	OEM
250R4B	0540	251F4S	1777	253RA100	OEM	260D08514	0004	260P13702	0079	260P41905	0191	263P701010	3806
250R5B	1771	251F5S	1777	253RA120	OEM	260D08601	0086	260P13704	0436	260P42002	1284	263P755020	OEM
250R6B	0965	251F6S	1777	253RA140	OEM	260D08701	0086	260P14101	0076	260P42201	0261	263P757010	OEM
250R7B	0817	251F8S	1789	253RA150	OEM	260D08801	0016	260P14102	0016	260P42502	0276	263P792020	OEM
250R8B	0817	251F10S	1789	253RA160	OEM	260D08912	0164	260P14103	0016	260P42504	0276	263P798010	OEM
250R9B	0922	251F12S	OEM	253RA170	OEM	260D09001	0016	260P14104	0086	260P42706	0423	263P872040	3521
250R10B	0922	251F-FLAG	0217	253RA180	OEM	260D09301	0103	260P14202	0423	260P42807	1284	264-701508	0071
250R11B	OEM	251F-FLEX	0217	253RA190	OEM	260D09314	0016	260P14407	1284	260P43201	0055	264D00101	0133
250R12B	OEM	251H	OEM	253RA200	OEM	260D09413	0004	260P15100	0142	260P43501	0055	264D00209	0133
250R13B	OEM	251H-FLEX	0468	253RL60	OEM	260D09612	0086	260P15108	0142	260P44501	OEM	264D00503	0030
250R14B	OEM	251K-FLEX	0468	253RL80	OEM	260D106A1	0016	260P15201	0037	260P44701	1541	264D00505	0015
250R15B	OEM	251M	OEM	253RL100	OEM	260D13701	0111	260P15202	0037	260P45902	1758	264D00507	0301
250R16B	OEM	251M1	0038	253RL110	OEM	260D13702	0016	260P15203	0037	260P46901	0342	264D00612	0143
250R18B	OEM	251M-FLEX	0468	253RL120	OEM	260D13704	0004	260P15701	0380	260P52301	0261	264D00701	0143
250R20B	OEM	251R2S	OEM	253RM60	OEM	260D15901	0076	260P16009	0178	260P57101	0276	264D00801	0123
250R22B	OEM	251R3S	OEM	253RM80	OEM	260D15902	0191	260P16101	0047	260P58201	OEM	264D00901	0143
250R24	OEM	251R4S	OEM	253RM100	OEM	260J	OEM	260P16202	0142	260P60401	0887	264D0701	0123
250R24B	OEM	251RA10	OEM	253RM110	OEM	260K	OEM	260P16208	0142	260P60601	2129	264D01001	0143
250R26	OEM	251RA20	OEM	253RM120	OEM	260P0770	0079	260P16301	0144	260P70403	0144	264D01112	0015
250R26B	OEM	251RA30	OEM	254-0102-139	0144	260P01209	0160	260P16302	0144	260P70501	0144	264D02602	OEM
250R28	OEM	251RA40	OEM	254-10	0340	260P02903	0079	260P16304	0951	260P70502	0144	264H	OEM
250R28B	OEM	251RA60	1900	254-11	0052	260P02903A	0079	260P16502	0006	260P141103	0016	264P0080	0143
250R30	OEM	251RA80	0738	254-27	3946	260P02908	0079	260P16503	0006	260P166040	0148	264P00401	0143
250R30B	OEM	251RA100	0738	254-34	OEM	260P03001	0164	260P16504	0006	260P254010	3314	264P00501	0196
250R32	OEM	251RA120	0738	254-36	0375	260P03201	0007	260P16603	0148	260P255040	1348	264P00502	0196
250R32B	OEM	251RA140	OEM	254A-FLAG	0217	260P03201A	0007	260P16802	0436	260P256010	3314	264P00506	0196
250R34	OEM	251RA150	OEM	254A-FLEX	0217	260P04001	0079	260P17002	0086	260P256030	3314	264P00601	0015
250R34B	OEM	251RA160	OEM	254B-FLAG	0217	260P04002	0079	260P17101	0368	260P256040	3314	264P00602	0071
250R36	OEM	251RA180	OEM	254B-FLEX	0217	260P04003	0076	260P17102	0079	260P325030	0018	264P00801	0123
250R36B	OEM	251RA190	OEM	254D-FLAG	0217	260P04004	0079	260P17103	0079	260P338040	0518	264P01011	0015
250R38	OEM	251RA200	OEM	254D-FLEX	0217	260P04502	0079	260P17104	0016	260P338050	0518	264P01012	0015

DEVICE TYPE	REPL CODE
264P01301	0143
264P01305	0143
264P01306	0123
264P01350	0143
264P01508	1303
264P01701	0102
264P02001	0015
264P02104	0774
264P02301	0102
264P02402	0102
264P02501	0162
264P02502	0165
264P02503	0057
264P03001	0015
264P03301	0301
264P03302	0317
264P03303	0137
264P03401	0917
264P03501	OEM
264P03601	0102
264P03603	0102
264P03604	0017
264P03605	0102
264P03606	0102
264P03607	0015
264P03701	0015
264P03703	0087
264P03802	0143
264P04003	0002
264P04005	0157
264P04108	OEM
264P04109	OEM
264P04206	0378
264P04301	0015
264P04303	0071
264P04402	0015
264P04501	0133
264P04502	0124
264P04504	0133
264P04507	0133
264P04701	0102
264P04702	0015
264P04703	0015
264P04705	0102
264P04801	0015
264P04901	0015
264P05001	0015
264P05002	0015
264P05404	0124
264P05501	OEM
264P05902	0030
264P06301	0004
264P06601	0015
264P06603	0102
264P06606	0015
264P08002	0015
264P08801	0023
264P08901	0201
264P08902	0288
264P09001	0071
264P09101	0102
264P09301	0137
264P09501	0102
264P09707	0456
264P045040	0124
264P045080	0124
264P10F02	0317
264P10102	0023
264P10103	0102
264P10105	0015
264P10201	0023
264P10202	0640
264P10308	0686
264P10403	0248
264P10502	0317
264P10906	0420
264P11003	0062
264P11007	0010
264P11009	0195
264P12303	0917
264P13002	0102
264P14002	0288
264P14701	0071
264P17401	0062
264P17402	0052
264P19303	0041
264P19304	0118
264P19305	0118
264P19306	0052
264P19307	0375
264P19308	0057
264P20006	0323
264P20301	0841
264P22001	0062
264P22006	0064
264P22008	0036
264P22103	0052
264P22104	0494
264P22105	0091
264P22202	0879
264P22203	0295
264P22205	0064
264P22206	0726
264P22208	0064
264P22404	OEM
264P22501	OEM
264P22601	OEM
264P23101	0023
264P24401	3452
264P27202	OEM
264P28501	0015
264P29004	0052
264P29009	OEM
264P29101	0494
264P29102	0091
264P29104	0062
264P29205	0091
264P29206	0376
264P29209	0376
264P33004	0100
264P33009	0100
264P34109	0140
264P34202	OEM
264P34206	1276
264P34702	4327
264P39401	OEM
264P39402	OEM
264P39802	0106
264P46004	0140
264P46006	0041
264P46106	0091
264P46302	0165
264P46402	0064
264P46502	0052
264P47003	0195
264P50203	0589
264P51201	0742
264P102020	0640
264P157010	0348
264P157040	0017
264P203020	OEM
264P225020	OEM
264P231010	0023
264P244010	3452
264P244020	1329
264P285010	0015
264P295020	0023
264P295030	0017
264P341020	0064
264P341090	0140
264P358040	0102
264P358070	0031
264P358090	0546
264P389010	0917
264P457080	0299
264P458050	OEM
264P460040	0140
264P460050	0140
264P460060	0041
264P460090	0041
264P461040	0253
264P462020	0062
264P462080	0165
264P463010	0077
264P463020	0165
264P463050	0057
264P463060	0057
264P463080	0057
264P463090	0057
264P464060	0064
264P465010	0052
264P465060	0053
264P467090	0705
264P469050	0455
264P469080	OEM
264P470040	0166
264P471010	0010
264P471020	0166
264P483080	0041
264P484020	0157
264P485050	OEM
264P485060	0649
264P486020	0244
264P486080	OEM
264P487070	0137
264P487090	0137
264P488020	0053
264P488080	0002
264P490040	OEM
264P491090	0726
264P501040	0677
264P501050	0118
264P501090	OEM
264P502010	OEM
264P502030	0589
264P508080	0241
264P512010	0742
264P512020	0742
264P512030	0729
264P521030	0023
264P521040	0023
264P522010	0023
264P528010	OEM
264P528020	OEM
264P535010	OEM
264P543010	0795
264P544010	0106
264P544090	0106
264P550010	0841
264P566010	OEM
264P613010	OEM
264P825010	0023
264PA71010	0062
264Z00103	0030
264Z00201	0015
264Z00701	0143
265-1	0016
265D00702	0133
265H	OEM
265P03301	0139
265Z00101	0133
266-1	0133
266-1F	0133
266A30N	OEM
266P00101	2624
266P00101(XSTR)	0079
266P00102	2624
266P00103	OEM
266P00301	1634
266P00602	4260
266P00801	1634
266P0501	3359
266P01002	1329
266P01102	OEM
266P01802	OEM
266P03901	OEM
266P05201	OEM
266P05202	OEM
266P05301	OEM
266P05302	OEM
266P016010	1028
266P037010	OEM
266P064010	OEM
266P305A1	OEM
266P10101	0395
266P10103	0395
266P10201	1808
266P10202	1808
266P10601	0918
266P11301	0912
266P11501	OEM
266P11601	0069
266P12101	0539
266P13003	OEM
266P13401	OEM
266P13501	OEM
266P13701	OEM
266P15401	0631
266P16201	4120
266P19201	2666
266P19701	OEM
266P19801	4377
266P20402	4672
266P21401	OEM
266P21409	OEM
266P21501	4102
266P22101	OEM
266P24901	OEM
266P28101	3437
266P30102	0167
266P30103	0167
266P30109	0167
266P30201	0470
266P30501	OEM
266P30601	2262
266P30706	1994
266P30801	OEM
266P32301	0167
266P32302	0167
266P32402	1128
266P32501	1540
266P32801	2755
266P35001	0416
266P35801	1612
266P36402	5501
266P38802	1529
266P41801	OEM
266P50101	3359
266P57001	2879
266P59201	OEM
266P60301	OEM
266P60304	4196
266P60502	0856
266P61401	4389
266P62103	0312
266P63001	0663
266P67301	OEM
266P68801	OEM
266P71301	0692
266P71401	OEM
266P75201	2722
266P82803	OEM
266P87803	OEM
266P92304	1292
266P93402	0334
266P130030	OEM
266P154010	0631
266P197010	OEM
266P922010	0624
266P922020	0624
266P922030	4480
266P923020	1786
266P931010	1786
266P934020	0334
266P934040	0334
266P934060	0624
266P963030	OEM
267-3	OEM
267P90201	OEM
267P90202	3502
267P90205	3503
267P90206	OEM
267P91002	2364
267P91003	3955
267P91006	2364
267P916020	OEM
267P925010	OEM
268H	OEM
268P02902	OEM
268P03301	0112
268P033010	0112
268P058010	OEM
269M01201	0133
269P33804	0518
269P559030	OEM
269V004-H01	0196
270-Y30B60	0738
270-Y30B70	0738
270-Y30B80	0738
270-Y30B90	0738
270-Y30C100	0738
270-Y30C110	0738
270-Y30C120	0738
270PAF10V20	OEM
270PAF10V25	OEM
270PAF10V30	OEM
270PAF20V20	OEM
270PAF20V25	OEM
270PAF20V30	OEM
270PAF40V20	OEM
270PAF40V25	OEM
270PAF40V30	OEM
270PAF60V20	OEM
270PAF60V25	OEM
270PAF60V30	OEM
270PAF80V20	OEM
270PAF80V25	OEM
270PAF80V30	OEM
270PAF100V20	OEM
270PAF100V25	OEM
270PAF100V30	OEM
270PAF120V20	OEM
270PAF120V25	OEM
270PAF120V30	OEM
270PE2-35	OEM
270XY5-35SC00	OEM
271H	OEM
271XY5-35SC00	OEM
272H	OEM
272K	OEM
272P01302	OEM
272P01401	3884
272P026010	2111
272P13801	OEM
272P13901	OEM
272P14001	3097
272P15901	4927
272P132010	OEM
272P138010	OEM
272P139010	OEM
272P140010	3097
272P165030	OEM
272P181010	OEM
272P184010	OEM
272P187010	OEM
272P188010	OEM
272P189010	OEM
272P237010	OEM
272P238010	4094
272P238020	4094
272P239010	2346
272P239020	OEM
272P239030	OEM
272P239040	OEM
272P240010	OEM
272P255010	OEM
272P255020	OEM
272P351010	OEM
272P351020	OEM
272P394010	OEM
272P440010	OEM
272P460010	OEM
272P461010	3515
272P469010	OEM
272P490010	OEM
272P491010	OEM
272P492010	OEM
272P493010	OEM
272P628010	OEM
272P629010	OEM
272P630010	OEM
272P631010	OEM
272P648010	OEM
272P657010	OEM
272P658010	OEM
272P837020	OEM
272P860010	OEM
272P860030	OEM
274-54-B-IC	OEM
274-69BIC	0354
274P008030	OEM
274P008040	OEM
274P157010	OEM
274P177010	OEM
275-0-3110-04600	OEM
275A-US	OEM
275H	OEM
275J	OEM
275K	OEM
275L	OEM
275RF5	OEM
275RF10	OEM
275RF20	OEM
275RF30	OEM
275RF40	OEM
275RF50	OEM
275RF60	OEM
275U5A	1023
275U5AR	0501
275U10A	1023
275U10AR	0501
275U15A	1023
275U15AR	0501
275U20A	1023
275U20AR	0501
275U25A	1036
275U25AR	1777
275U30A	1036
275U30AR	1777
275U40A	1036
275U40AR	1777
275U50A	1036
275U50AR	1777
275U60A	1036
275U60AR	1777
275U70A	2832
275U70AR	1789
275U80A	2832
275U80AR	1789
275U90A	2832
275U90AR	1789
275U100A	2832
275U100AR	1789
276-007	0312
276-009	1192
276-010	0410
276-018	5666
276-019	1778
276-021	0862
276-022	4679
276-025	2195
276-026	3019
276-030	5016
276-033	5116
276-035	5386
276-036	4023
276-036B	4023
276-037	0841
276-038	0360
276-041	1962
276-053	4989
276-065	4300
276-066	4304
276-066A	4304
276-068	7155
276-069	7156
276-070	4052
276-073	OEM
276-075	4873
276-081	5153
276-0225	OEM
276-0707	OEM
276-113	7158
276-116	OEM
276-124	7158
276-130	5123
276-134	5051
276-142	7159
276-143	6872
276-145	5006
276-561	0636
276-562	0012
276-563	0137
276-564	0002
276-565	0162
276-568	0830
276-569	OEM
276-570	0838
276-571	0840
276-572	0649
276-703	1048
276-705	1050
276-706	0379
276-1000	7095
276-1001	0593
276-1002	0593
276-1003	0593
276-1020	7099
276-1050	2730
276-1067	7102
276-1079	1396
276-1089	2095
276-1090	2089
276-1101	0023
276-1102	0023
276-1103	0795
276-1104	0015
276-1114	0087
276-1122	0124
276-1123	0123
276-1124	0917
276-1141	0110
276-1142	0953
276-1143	0242
276-1144	0540
276-1146	7106
276-1151	0277
276-1152	0277
276-1161	4192
276-1165	0735
276-1171	7107
276-1172	1414
276-1173	7108
276-1180	7109
276-1181	5694
276-1185	5483
276-1251	0812
276-1252	OEM
276-1301	OEM
276-1302	OEM
276-1303	OEM
276-1307	OEM
276-1308	OEM
276-1337	OEM
276-1603	0076
276-1604	0006
276-1617	0016
276-1656	OEM
276-1704	OEM
276-1705	3062
276-1707	3068
276-1708	3069
276-1709	3071
276-1710	OEM
276-1711	0625
276-1712	0176
276-1713	2247
276-1714	3391
276-1715	3741
276-1717	2718
276-1718	0585
276-1720	3381
276-1721	3781
276-1723	0973
276-1725	0379
276-1728	3285
276-1729	2279
276-1731	3063
276-1735	2082
276-1737	3079
276-1740	0026
276-1743	1543
276-1748	OEM
276-1749	OEM
276-1750	OEM
276-1757	0443
276-1758	0785
276-1759	0167
276-1762	5280
276-1763	OEM
276-1764	OEM
276-1767	OEM
276-1770	0624
276-1771	0334
276-1772	1321
276-1773	1285
276-1774	1838
276-1777	2838
276-1778	2564
276-1782	OEM
276-1783	OEM
276-1784	OEM
276-1786	OEM
276-1787	OEM
276-1792	3792
276-1795	OEM
276-1796	OEM
276-1797	OEM
276-1801	0232
276-1802	0361
276-1803	1172
276-1805	1108
276-1808	1208
276-1811	0314
276-1813	1158
276-1818	1313
276-1822	0467
276-1828	0619
276-1835	0977
276-1915	0088
276-2001	0600
276-2001/RS2001	0600
276-2002	0600
276-2002/RS2002	0600
276-2003	0050
276-2003/RS2003	0050
276-2004	0004
276-2004/RS2004	0004
276-2005	0208
276-2005/RS2005	0208
276-2006	0085
276-2006/RS2006	0085
276-2007	0211
276-2007/RS2007	0164
276-2008	0715
276-2008/RS2008	0715
276-2009	0079
276-2009/RS2009	0079
276-2010	0127
276-2010/RS2010	0079
276-2011	0144
276-2011/RS2011	0144
276-2012	0715
276-2012/RS2012	0715
276-2013	0016
276-2013/RS2013	0016
276-2014	0079
276-2014/RS2014	0079
276-2015	0144
276-2015/RS2015	0144
276-2016	4376
276-2016/RS2016	0079
276-2017	0042
276-2017/RS2017	0042
276-2018	0566
276-2018/RS2018	0566
276-2019	1361
276-2019/RS2019	1361
276-2020	0482
276-2021	0150
276-2021/RS2021	0150
276-2022	0037
276-2022/RS2022	0037
276-2023	0037
276-2023/RS2023	0037
276-2024	0037
276-2024/RS2024	0037
276-2025	0925
276-2025/RS2025	0925
276-2026	1367
276-2026/RS2026	1367
276-2027	0925
276-2027/RS2027	1369
276-2028	0325
276-2029	1893
276-2030	0076
276-2030/RS2030	0079
276-2031	0330
276-2031/RS2031	0330
276-2032	0006
276-2032/RS2032	0037
276-2033	0079
276-2033/RS2033	0079
276-2034	0037
276-2034/RS2034	0037
276-2035	0325
276-2036	0325
276-2038	0418
276-2038/RS2038	0418
276-2039	0177
276-2039/RS2039	0177
276-2040	2013
276-2040/RS2040	2013
276-2041	0103
276-2041/RS2041	0103
276-2042	1394
276-2042/RS2042	3373
276-2043	0491
276-2043/RS2043	0491
276-2044	3007
276-2048	0042
276-2051	0076
276-2055	0313
276-2057	0076
276-2058	0076
276-2059	0079
276-2060	0400
276-2061	0715
276-2062	1392
276-2068	2235
276-2072	2341
276-2073	OEM
276-2074	OEM
276-2329	OEM
276-2334	0304
276-2335	OEM
276-2336	OEM
276-2337	OEM
276-2401	0478
276-2411	0215
276-2413	0413
276-2417	0513
276-2423	0519
276-2449	0001
276-2466	0101
276-2481	0626
276-2503	0522
276-2505	0522
276-2520	0508
276-2521	0511
276H	OEM
277-14	1911
277-221	OEM
277-1012	OEM
277A	OEM
277G	OEM
277H	OEM
277J	OEM
277K	OEM
278-2	0628
278-14	1284
278H	OEM
279-02	OEM
279-14ITT	0064
279-22ITT744	OEM
279H	OEM
280-0001	0911
280-0002	0520
280-07	0725
280H	OEM
281	0016
283H	OEM
284H	OEM
284J	OEM
285H	OEM
285J	OEM
285K	OEM
285L	OEM
286-1BFY	OEM
286H	OEM
286HM	OEM
286HP	OEM
286J	OEM

DEVICE TYPE	REPL CODE
287H	OEM
288(SEARS)	0015
288-0550R	4127
288H	OEM
288HC	OEM
289H	OEM
289J	OEM
289K	OEM
289L	OEM
290-1003	0133
290A	OEM
290V02H69	0144
290V034C01	0071
291-04	0015
291-20	0015
291H	OEM
292-10	0015
292A	OEM
292H	OEM
293H	OEM
294	0037
294-42-9	0019
294-92-2	0079
294H	OEM
295	OEM
295-01	0133
295-03	0023
295-1	0133
295-95-1	0176
295H	OEM
295L001H01	0015
295L001M01	0015
295L001M02	0015
295L002H01	0015
295L002M03	0229
295L003M01	0015
295V002H01	0015
295V003H01	0774
295V003M01	0774
295V005H02	0015
295V005H03	0143
295V006H01	0015
295V006H02	0015
295V006H03	0015
295V006H05	0015
295V006H06	0229
295V006H07	0015
295V006H08	0229
295V006H09	0015
295V007H02	0015
295V008H01	0015
295V012H01	0015
295V012H02	0015
295V012H03	0015
295V012H06	0015
295V014H01	0015
295V014H04	0015
295V014H7	0015
295V015H02	0015
295V016H01	0015
295V017H01	0015
295V020B01	0015
295V020H01	0015
295V023H01	0015
295V027C01	0015
295V027C01-1	0015
295V028C01	0015
295V028C02	0015
295V028C03	0015
295V028C04	0071
295V029C01	0015
295V029C02	0015
295V031B	1303
295V031B02	0474
295V031C02	0474
295V033B01	0774
295V034C01	1303
295V035C01	0015
295V041H04	0160
296	2520
296-18-9	0004
296-19-9	0004
296-42-9	0019
296-46-9	0050
296-50-9	0016
296-51-9	0016
296-55-9	0537
296-56-9	0016
296-58-9	0086
296-59-9	0016
296-60-9	0004
296-61-9	0222
296-62-9	0004
296-64-9-1	0164
296-77-9	0155
296-81-9	1400
296-86	0224
296-98-9	0076
296H	OEM
296L002B01	0124
296L003B01	0137
296V001H01	1030
296V001M01	1030
296V002H01	0143
296V002H02	0143
296V002H05	0143
296V002H06	0143
296V002H07	0143
296V002H08	0143

DEVICE TYPE	REPL CODE
296V002M01	0143
296V004H01	0196
296V006H02	0143
296V006H03	0015
296V006H07	0015
296V007H02	0143
296V011H03	0365
296V012H01	0143
296V015B01	0143
296V015H01	0143
296V017H01	0137
296V018B01	0137
296V018B02	0649
296V019B01	0137
296V019B04	0794
296V020B01	0133
296V020B02	0015
296V024B01	0143
296V024B02	0133
296V034C01	1303
297C011H01	0050
297H	OEM
297L001H01	0208
297L001H02	0208
297L001H03	0004
297L001M01	0208
297L001M02	0004
297L001M03	0004
297L002H01	0004
297L005H01	0004
297L006H01	0016
297L006H02	0016
297L007C02	0016
297L007H01	0016
297L007H02	0016
297L007H03	0016
297L007H03/C03	0079
297L008C02	0086
297L008H01	0086
297L010C01	0693
297L012C01	0037
297L013B01	0016
297L013B02	0037
297V002H03	0208
297V002H04	0208
297V002H05	0208
297V002M04	0208
297V002M05	0208
297V003	0211
297V003H02	0004
297V003H03	0211
297V003H06	0004
297V003H08	0004
297V003H09	0004
297V003M01	0211
297V003M07	0211
297V004H01	0004
297V004H03	0004
297V004H04	0004
297V004H06	0004
297V004H08	0004
297V004H09	0004
297V004H010	0211
297V004H10	0211
297V004H11	0004
297V004H14	0211
297V004H15	0211
297V004H16	0211
297V004M01	0211
297V005H01	0004
297V008H01	0280
297V008M01	0050
297V07K01	OEM
297V010H01	0004
297V010M01	0004
297V011	0280
297V011H01	0136
297V011H02	0280
297V012	0280
297V012H01	0136
297V012H02	0280
297V012H03	0280
297V012H04	0050
297V012H05	0280
297V012H06	0280
297V012H07	0050
297V012H08	0280
297V012H09	0280
297V012H10	0136
297V012H11	0050
297V012H12	0050
297V012H13	0050
297V012H14	0211
297V012H15	0136
297V017H01	0280
297V017H02	0280
297V018H01	0004
297V019B01	0280
297V019B04	0794
297V019H01	0280
297V020H01	0050
297V020H02	0280
297V020M01	0280
297V021H01	0280
297V021H02	0280
297V021H03	0280
297V022H01	0280
297V024H01	0050
297V024H03	0050

DEVICE TYPE	REPL CODE
297V025H02	0211
297V025H03	0004
297V025H04	0211
297V025H05	0211
297V025H15	0211
297V026H01	0136
297V026H03	0280
297V027C01	0015
297V027H01	0211
297V032H01	0211
297V033H01	0211
297V034H01	0136
297V035H01	0136
297V036H01	0136
297V036H02	0136
297V037B02	0004
297V037H01	0211
297V037H02	0211
297V038H01	0280
297V038H02	0050
297V038H03	0050
297V038H04	0050
297V038H05	0280
297V038H06	0136
297V038H07	0211
297V038H09	0280
297V038H10	0136
297V038H11	0136
297V038H12	0136
297V040H01	0211
297V040H08	0211
297V040H09	0164
297V040H10	0211
297V040H11	0211
297V040H12	0211
297V040H13	0211
297V040H15	0160
297V040H16	0211
297V041H01	0265
297V041H02	0160
297V041H03	0160
297V041H04	0160
297V041H05	0085
297V041H06	0160
297V041H07	0085
297V041H15	0085
297V042C01	0211
297V042C02	0211
297V042C03	0211
297V042C04	0211
297V042H01	0211
297V042H02	0211
297V042H03	0136
297V042H04	0136
297V043H01	0211
297V044H01	0280
297V045H01	0136
297V045H02	0136
297V049H01	0198
297V049H03	0198
297V049H04	0198
297V049H05	0016
297V049H06	0086
297V050C02	0004
297V050H01	0474
297V050H02	0474
297V050H03	0211
297V051C03	0211
297V051C04	0211
297V051H01	0211
297V051H02	0211
297V051H03	0211
297V051H04	0004
297V052C01	0211
297V052H01	0004
297V052H02	0211
297V052H04	0004
297V053C01	0211
297V053H01	0004
297V053H02	0004
297V054C01	0280
297V054C02	0280
297V054H01	0136
297V054H02	0136
297V055C01	0280
297V055H01	0136
297V057H01	0164
297V057H02	0164
297V059H01	0198
297V059H02	0198
297V059H03	0198
297V060H01	0142
297V060H02	0142
297V060H03	0142
297V061C01	0103
297V061C01A	0103
297V061C02	0103
297V061C02A	0103
297V061C03	0198
297V061C04	0198
297V061C05	0086
297V061C06	0198
297V061C07	0016
297V061H01	0016
297V061H02	0016
297V061H03	0016
297V062C01	0160
297V062C05	0160
297V062C06	0086

DEVICE TYPE	REPL CODE
297V063C01	0136
297V063H01	0050
297V064B01	0050
297V064H01	0050
297V065C01	0136
297V065C02	0136
297V065C03	0136
297V065H01	0280
297V065H02	0280
297V065H03	0280
297V068C01	OEM
297V069C01	0642
297V070C01	0136
297V070H49	0144
297V071C02	OEM
297V071C03	0142
297V071H03	0142
297V072C01	0144
297V072C03	0144
297V072C04	0144
297V072C05	0086
297V072C06	0016
297V073C01	0126
297V073C02	0126
297V073C03	0037
297V073C04	0037
297V074C01	0007
297V074C02	0198
297V074C03	0198
297V074C04	0198
297V074C05	OEM
297V074C06	0016
297V074C07	0016
297V074C08	0076
297V074C09	0144
297V074C10	0086
297V074C11	0126
297V074C12	0086
297V075C01	OEM
297V076B01	0211
297V076C01	0211
297V077C01	0136
297V078C01	0144
297V078C02	0144
297V080C01	0126
297V081C01	0004
297V082B01	0086
297V082B02	0126
297V082B03	0126
297V083C01	0037
297V083C02	0016
297V083C03	0126
297V083C04	0086
297V084C01	0276
297V085C01	0016
297V085C02	0016
297V085C03	0016
297V085C04	0016
297V086C01	0037
297V086C02	0016
297V086C03	0016
297V086C04	0460
297V087B02	0551
297V86C02	OEM
298H	OEM
299-POWER	0440
299B09001	0124
299D09001	0124
299D09002	0041
299D09005	0124
299D13101	0879
299H	OEM
299POWER	0440
300	0211
300A	0599
300AR	1347
300B	0599
300BR	1347
300C	0599
300CR	1347
300D	0599
300DR	1347
300E	0599
300ER	1347
300EXD13	OEM
300EXH21	OEM
300F	0599
300FR	1347
300FXD13	OEM
300G	1986
300GR	1905
300H	1986
300HR	1905
300JB05L	4152
300JB1L	4152
300JB2L	4152
300JB3L	4159
300JB4L	4159
300JB5L	4159
300JB6L	4159
300JH21	OEM
300K	1986
300KR	1905
300LD11A	OEM
300LD13	OEM
300M	1986
300MR	1905
300N15A	OEM
300N15A9	OEM

DEVICE TYPE	REPL CODE
300N20A	OEM
300N20A9	OEM
300N25A	OEM
300N25A9	OEM
300N30A	OEM
300N30A9	OEM
300N35A	OEM
300N35A9	OEM
300N40A	OEM
300N40A9	OEM
300N45A	OEM
300N45A9	OEM
300N50A	OEM
300N50A9	OEM
300N55A	OEM
300N55A9	OEM
300N60A	OEM
300N60A9	OEM
300N65A	OEM
300N65A9	OEM
300N70A	OEM
300N70A9	OEM
300N75A	OEM
300N75A9	OEM
300N80A	OEM
300N80A9	OEM
300N85A	OEM
300N85A9	OEM
300N90A	OEM
300N90A9	OEM
300N95A	OEM
300N95A9	OEM
300N100A	OEM
300N100A9	OEM
300N110A	OEM
300N120A	OEM
300N130A	OEM
300N140A	OEM
300N150A	OEM
300N160A	OEM
300N170A	OEM
300N180A	OEM
300N190A	OEM
300N200A	OEM
300N210A	OEM
300N220A	OEM
300N230A	OEM
300N240A	OEM
300N250A	OEM
300ND11A	OEM
300ND13	OEM
300P2S	OEM
300PA50	OEM
300PA60	0159
300PA80	0096
300PA100	0096
300PA120	0096
300PA140	OEM
300PA150	OEM
300PA160	OEM
300PAC10	0192
300PAC20	0192
300PAC30	0159
300PAC40	0159
300PAC50	0159
300PAC60	0159
300QD13	OEM
300QH21	OEM
300R1B	0953
300R2B	0242
300R3B	1747
300R4B	OEM
300R5B	1771
300R6B	2645
300R7B	0817
300R8B	0817
300R9B	2645
300R10B	OEM
300R11B	OEM
300R12B	OEM
300R13B	OEM
300R14B	OEM
300R15B	OEM
300R16B	OEM
300R18B	OEM
300R20B	OEM
300R22B	OEM
300R24B	OEM
300R26B	OEM
300R28B	OEM
300R30B	OEM
300R32B	OEM
300R34B	OEM
300R36B	OEM
300R38B	OEM
300R40B	OEM
300R42B	OEM
300R44B	OEM
300R46B	OEM
300R48B	OEM
300R50B	OEM
300RA10	1900
300RA20	1900
300RA30	1900
300RA40	1900
300RA50	0738
300RA60	0738
300RA80	0738
300RA100	0738

DEVICE TYPE	REPL CODE
300RA110	0738
300RA120	0738
300RA130	OEM
300RA140	OEM
300RA150	OEM
300RA160	OEM
300RA170	OEM
300RA180	OEM
300RA190	OEM
300RA200	OEM
300RB10	1900
300RB20	1900
300RB30	1900
300RB40	0738
300RB50	0738
300RB60	0738
300RB70	OEM
300RB80	0738
300RB90	OEM
300RB100	0738
300RB110	0738
300RB120	0738
300RB130	OEM
300RB140	OEM
300RB150	OEM
300RB160	OEM
300RB170	OEM
300RB180	OEM
300RB190	OEM
300RB200	OEM
300RC5	OEM
300RC10	OEM
300RC15	OEM
300RC20	OEM
300RC30	OEM
300RC40	OEM
300RC50	OEM
300RC60	OEM
300S5H	OEM
300S6H	OEM
300S8H	OEM
300S10H	OEM
300S12H	OEM
300SS5H	OEM
300SS6H	OEM
300SS8H	OEM
300SS10H	OEM
300SS12H	OEM
300TD11A	OEM
300TD13	OEM
300U5	1023
300U5A	1023
300U5AR	0501
300U5R	0501
300U10	1023
300U10A	1023
300U10AR	0501
300U10R	0501
300U15	1023
300U15A	1023
300U15AR	0501
300U15R	0501
300U20	1023
300U20A	1023
300U20AR	0501
300U20R	0501
300U25	1023
300U25A	1023
300U25AR	0501
300U25R	0501
300U30	1023
300U30A	1023
300U30AR	0501
300U30R	0501
300U40	1036
300U40A	1023
300U40AR	0501
300U40R	1777
300U50	1036
300U50A	1036
300U50AR	1777
300U50R	1777
300U60	1036
300U60A	1036
300U60AR	1777
300U60R	1777
300U70	1046
300U70A	1046
300U70AR	1789
300U70R	1789
300U80	1046
300U80A	1046
300U80AR	1789
300U80R	1789
300U90	1046
300U90A	1046
300U90AR	1789
300U90R	1789
300U100	1046
300U100A	1046
300U100AR	1789
300U100R	1789
300U120A	3928
300UR20A	0501
300UR30A	0501
300UR40A	0501
300UR60A	1777
300UR80A	1789
300UR100A	1789

DEVICE TYPE	REPL CODE
300UR120A	6729
300WD11A	OEM
300WD13	OEM
300YD13	OEM
301	0211
301-1	0474
301-576-2	0466
301-576-3	0664
301-576-4	0232
301-576-6	0086
301-576-14	0379
301-679-1	1913
301-680-1	3779
301-695-1	1233
301-696-1	1354
301AL	6467
301C025	OEM
301C025B	OEM
301C25	OEM
301C25B	OEM
301C30	OEM
301C30B	OEM
301C35	OEM
301C35B	OEM
301C40	OEM
301C40B	OEM
301C45	OEM
301C45B	OEM
301C50	OEM
301C50B	OEM
301C60	OEM
301C60B	OEM
301C70	OEM
301C70B	OEM
301C80	OEM
301C80B	OEM
301C90	OEM
301C90B	OEM
301C100	OEM
301C100B	OEM
301C110	OEM
301C110B	OEM
301C120	OEM
301C120B	OEM
301CL	6718
301F1FM	OEM
301F2FM	OEM
301F3FM	OEM
301R1FM	OEM
301R2FM	OEM
301R3FM	OEM
301RA10	OEM
301RA20	OEM
301RA30	OEM
301RA40	OEM
301RA60	1900
301RA80	0738
301RA100	0738
301RA120	0738
301RA130	OEM
301RA140	OEM
301RA150	OEM
301RA160	OEM
301RA170	OEM
301RA180	OEM
301RA190	OEM
301RA200	OEM
301RB10	OEM
301RB20	OEM
301RB30	OEM
301RB40	OEM
301RB50	OEM
301RB60	OEM
301RB80	OEM
301RB100	OEM
301RB110	OEM
301RB120	OEM
301RB130	OEM
301RB140	OEM
301RB150	OEM
301RB160	OEM
301RB170	OEM
301RB180	OEM
301RB190	OEM
301RB200	OEM
301RC10	1900
301RC20	1900
301RC30	6678
301RC40	1900
301RC50	1900
301RC60	1900
301RC80	1900
301RC100	OEM
301RC120	OEM
301U80	1046
301U100	1046
301U120	3928
301U120A	4558
301U140	2832
301U160	OEM
301U160A	OEM
301U180	OEM
301U200	OEM
301U210	OEM
301U220	OEM
301U230	OEM
301U240	OEM
301U250	OEM
301UR120	3531

If replacement code is OEM, contact original manufacturer for replacement.

DEVICE TYPE	REPL CODE
301UR140	3167
301UR160	OEM
301UR160A	OEM
301UR180	OEM
301UR200	OEM
302	0211
302(OPTO)	OEM
302-2	0253
302-679-1	2050
302-680	2050
302-681	OEM
302-K	0319
302A	1124
302AL	6468
302AR	0267
302B	2901
302BR	0267
302C	1124
302CL	6468
302CR	0267
302D	1124
302DR	0267
302E	1126
302E-RT	OEM
302F	1126
302FR	0267
302G	0805
302GR	0267
302G-RT	OEM
302H	0805
302HR	0267
302K	1195
302KR	0267
302M	0319
302MR	0267
302P	1132
302S	1132
302U5A	OEM
302U10A	OEM
302U15A	OEM
302U20A	OEM
302U25A	OEM
302U30A	OEM
302U40A	OEM
302U50A	OEM
302U60A	OEM
302U70A	OEM
302U80A	OEM
302U90A	OEM
302U100A	OEM
302U120A	OEM
302V	0044
302Z	0044
303-1	0037
303-1(IC)	0490
303-2	0126
303-4	OEM
303-5	OEM
303-7	OEM
303-K	0319
303A	1124
303AL	6469
303AR	0267
303B	2901
303BR	0267
303C	1124
303CL	6469
303CR	0267
303D	1124
303DR	0267
303E	1126
303E-RT	OEM
303F	1126
303FR	0267
303G	0805
303GR	0267
303G-RT	OEM
303H	0805
303HR	0267
303K	1195
303KR	0267
303M	0319
303MR	0267
303P	1132
303PJA180	OEM
303PJA200	OEM
303PJA250	OEM
303RA10	OEM
303RA20	OEM
303RA30	OEM
303RA40	OEM
303RA60	OEM
303RA80	OEM
303RA100	OEM
303RA120	OEM
303RA130	OEM
303RA140	OEM
303RA150	OEM
303RA170	OEM
303RA180	OEM
303RA190	OEM
303RA200	OEM
303RB10	OEM
303RB20	OEM
303RB30	OEM
303RB40	OEM
303RB60	OEM
303RB80	OEM
303RB100	OEM
303RB120	OEM
303RB130	OEM
303RB140	OEM
303RB150	OEM
303RB160	OEM
303RB170	OEM
303RB180	OEM
303RB190	OEM
303RB200	OEM
303S	1132
303U80	OEM
303U100	OEM
303U120	OEM
303U140	OEM
303U160	OEM
303U180	OEM
303U200	OEM
303U210	OEM
303U220	OEM
303U230	OEM
303U240	OEM
303U250	OEM
303V	0044
303Z	0044
303Z4	OEM
304	OEM
304A	0097
304AL	6470
304AR	2291
304B	0090
304BR	2291
304C	0109
304CL	6470
304CR	2291
304D	0109
304DR	2291
304F	0105
304FR	0476
304H	0109
304HR	0476
304K	0116
304KR	0476
304M	0122
304MR	0476
304P	0131
304PR	0449
304RA60	OEM
304RA80	OEM
304RA100	OEM
304RA120	OEM
304RA130	OEM
304RA140	OEM
304RA150	OEM
304RA160	OEM
304RA170	OEM
304S	0131
304SR	0449
304V	0145
304Z	0145
304Z4	OEM
305-1	3753
305-2	OEM
305-3	OEM
305-4	OEM
305A	0708
305AR	2291
305B	4140
305BR	2291
305C	0580
305CR	2291
305D	0580
305DR	2291
305F	2060
305FR	0476
305GI418C	0348
305H	1000
305HR	0476
305K	2076
305KR	0476
305M	2081
305MR	0476
305P	2088
305PR	0449
305S	2088
305SR	0449
305T4	OEM
305U80	OEM
305U100	OEM
305U120	OEM
305U140	OEM
305U160	OEM
305U180	OEM
305U200	OEM
305U210	OEM
305U220	OEM
305U230	OEM
305U240	OEM
305U250	OEM
305V	0612
305Z	0612
305Z4	OEM
305ZR	0449
306-1	0016
306-8	0079
306-12	OEM
306AL	6471
306CL	6471
306Z4	OEM
307-005-9-001	0579
307-007-9-001	0911
307-007-9-002	2632
307-008-9-001	0823
307-029-1-001	0911
307-047-9-001	0417
307-059-9-001	0229
307-095-9-001	OEM
307-095-9-002	0215
307-095-9-003	1292
307-095-9-004	2253
307-107-9-001	4633
307-107-9-002	OEM
307-107-9-003	1050
307-107-9-004	4624
307-107-9-005	4620
307-107-9-006	4700
307-108-9-001	2380
307-112-9-001	1816
307-112-9-002	1292
307-112-9-003	1292
307-112-9-005	0417
307-112-9-006	4209
307-112-9-007	2261
307-112-9-009	2253
307-113-9-001	0215
307-113-9-002	2306
307-113-9-003	1292
307-113-9-004	4209
307-115-9-001	2261
307-120-9-001	4069
307-128-9-002	OEM
307-131-9-001	2306
307-131-9-002	1128
307-131-9-003	4644
307-131-9-005	OEM
307-131-9-006	0706
307-131-9001	2306
307-133-9-004	1540
307-133-9-005	4648
307-133-9001	4648
307-143-9-001	4645
307-143-9-002	4042
307-143-9-003	4652
307-143-9-004	0325
307-143-9-007	OEM
307-145-9-001	2328
307-152-9-001	OEM
307-152-9-004	0626
307-152-9-005	2834
307-152-9-007	1527
307-152-9-008	OEM
307-152-9-009	1888
307-152-9-010	6494
307-152-9-011	OEM
307-152-9-012	0215
307A	0015
307AL	6472
307B	0015
307C	0015
307CL	6472
307D	0015
307F	0015
307H	0015
307K	0015
307M	0015
307Z4	OEM
308-1	OEM
308A	2900
308AR	2291
308B	2900
308BR	2291
308C	2900
308CR	2291
308D	2900
308DR	2291
308F	1112
308FR	0476
308H	1112
308HR	0476
308K	1112
308KR	0476
308Z4	OEM
309-01ITT	0064
309-033-0	2301
309-3	OEM
309-4	OEM
309-6	OEM
309-324-616	0143
309-327-601	0649
309-327-608	0649
309-327-802	0649
309-327-803	0071
309-327-910	0015
309-327-916	0143
309-327-926	0016
309-327-927	0015
309-327-931	0004
309-327-932	0015
309U10	OEM
309U80	OEM
309U100	OEM
309U120	OEM
309U140	OEM
309U160	OEM
309U180	OEM
309U200	OEM
309U210	OEM
309U220	OEM
309U230	OEM
309U240	OEM
309U250	OEM
309Z4	OEM
310	0015
310-068	0136
310-4	0015
310-68	0136
310-123	0136
310-124	0136
310-139	0136
310-187	0016
310-188	0004
310-189	0164
310-190	0136
310-191	0050
310-192	0222
310-8028-001	0740
310J	OEM
310L	OEM
310T4	OEM
310VF	OEM
310Z4	OEM
311AL	6474
311CL	6474
311Z4	0794
312A12	OEM
312AJ	6475
312AL	6475
312CJ	6475
312CL	6475
312J	OEM
312K	OEM
312L	OEM
312R2S	OEM
312Z4	OEM
313AJ	OEM
313AL	OEM
313CJ	OEM
313CL	OEM
313Z4	OEM
314-6006	0178
314-6007-1	0016
314-6007-2	0016
314-6007-3	0016
314-6010-3	0079
315	0015
315-01A	OEM
315-6CGI	0015
315S02	OEM
315T4	OEM
315Z4	OEM
316Z4	OEM
317-2627-1	0141
317K	2838
317T	2564
317Z4	OEM
318-1	6035
318-2	3076
318-3	OEM
318-7	OEM
319A	0599
319B	0599
319D	0599
319E	0599
319F	0599
319G	1986
319G06	OEM
319H	1986
319K	1986
319Z4	OEM
320	OEM
320A	0015
320B	0015
320C	0015
320D	0015
320F	0015
320H	0015
320K	0015
320M	0015
320P	0071
320S	0071
320T4	OEM
320VF	OEM
320Z	0071
320Z4	OEM
321	1724
321-919	OEM
321A	OEM
321AJ	6478
321AL	6478
321CJ	6478
321CL	6478
321Z4	OEM
322	OEM
322-0147	0015
322-1	0178
322A	OEM
322AJ	6479
322AL	6479
322CJ	6479
322CL	6479
322E	OEM
322G	OEM
322T1	0004
323AJ	6480
323AL	6480
323CJ	6480
323CL	6480
323T1	0004
324	0004
324(DIO)	OEM
324(IC)	0625
324-0011	0196
324-0012	1030
324-0014	0143
324-0015	0474
324-0016	0050
324-0026	0050
324-0027	0136
324-0028	0136
324-0029	0004
324-0030	0164
324-0035	0143
324-0037	0143
324-0038	0136
324-0041	0004
324-0049	0143
324-0055	0211
324-0056	0164
324-0057	0143
324-0062	0164
324-0074	0004
324-0077	0050
324-0079	0136
324-0080	0164
324-0082	0050
324-0083	0050
324-0085	0164
324-0086	0050
324-0087	0050
324-0088	0211
324-0089	0004
324-0090	0280
324-0091	0004
324-0092	0004
324-0093	0004
324-0093(PHILCO)	0143
324-0095	0050
324-0097	0164
324-0098	0136
324-0099	0136
324-012	0050
324-019	0086
324-0100	0004
324-0102	0229
324-0105	0143
324-0106	0136
324-0107	0123
324-0107-01	0143
324-0108	0143
324-0110	0050
324-0111	0050
324-0112	0050
324-0113	0917
324-0114	0917
324-0115	0604
324-0116	0975
324-0117	0015
324-0118	2900
324-0119	0015
324-0120	0015
324-0121	0136
324-0122	0208
324-0123	0050
324-0124	0164
324-0125	0164
324-0126	0222
324-0128	0975
324-0129	0050
324-0130	0136
324-0131	0050
324-0132	0050
324-0133	0211
324-0134	0038
324-0135	0015
324-0136	0050
324-0137	0633
324-0138	0633
324-0139	0004
324-0140	0164
324-0141	0143
324-0142	0004
324-0143	0211
324-0144	0211
324-0145	0136
324-0146	0211
324-0147	0015
324-0149	0144
324-0150	0144
324-0151	0198
324-0152	0198
324-0154	0198
324-0155	0050
324-0160	0143
324-0162	0015
324-0166	OEM
324-0187	0050
324-0611	4163
324-1	0144
324-132	0050
324-144	0211
324-6005-5	0016
324-6011	0086
324-6013	0086
324A	OEM
324AL	6481
324CL	6481
324T1	0004
324T2	0004
325-0025-327	0143
325-0025-329	0004
325-0025-330	0004
325-0025-331	0004
325-0028-79	0004
325-0028-80	0004
325-0028-81	0004
325-0028-82	0136
325-0028-83	0004
325-0028-84	0144
325-0028-85	0050
325-0028-86	0143
325-0028-87	0143
325-0028-89	0015
325-0030-315	0004
325-0030-317	0004
325-0030-318	0004
325-0030-319	0004
325-0031-303	0016
325-0031-304	0198
325-0031-305	0198
325-0031-306	0164
325-0031-310	0016
325-0031-335	0143
325-0031-338	0015
325-0036-536	0004
325-0036-562	0143
325-0036-564	0136
325-0036-565	0136
325-0042-311	0277
325-0042-351	0016
325-0047-516	0004
325-0047-517	0015
325-0054-310	0004
325-0054-311	0004
325-0054-312	0015
325-0076-306	0016
325-0076-307	0016
325-0076-308	0016
325-0076-315	0015
325-0081-100	0016
325-0081-101	0016
325-0081-102	0164
325-0081-109	0289
325-0081-110	0015
325-0135-B	0015
325-0137-6	0004
325-0141-23	0015
325-0500-12	0111
325-0500-13	0111
325-0574-30	0016
325-0574-31	0016
325-0574-32	0030
325-0670	0004
325-0670-1	0004
325-0670-7	0004
325-0670-16	0015
325-0670A	0004
325-1370-18	0004
325-1370-19	0016
325-1370-20	0016
325-1375-10	0136
325-1375-11	0136
325-1375-12	0136
325-1376-53	0004
325-1376-54	0136
325-1376-55	0136
325-1376-56	0004
325-1376-57	0004
325-1376-58	0004
325-1376-60	0143
325-1378-18	0127
325-1378-19	0127
325-1378-20	0004
325-1378-21	0164
325-1378-22	0004
325-1441-10	0015
325-1441-11	0015
325-1442-8	0211
325-1442-9	0142
325-1446-26	0086
325-1446-27	0086
325-1446-28	0233
325-1446-29	0015
325-1513-29	0198
325-1513-30	0198
325-1513-46	0086
325-1771-15	0016
325-1771-16	0127
325-4610-100	0071
325AJ	6482
325AL	6482
325CJ	6482
325CL	6482
325EQ	OEM
325L	OEM
325PAH50	OEM
325PAH60	OEM
325PAH80	OEM
325PAH100	OEM
325PAH110	OEM
325PAH120	OEM
325RA5	OEM
325RA10	OEM
325RA20	OEM
325RA30	OEM
325RA40	OEM
325RA50	OEM
325RA60	OEM
325T1	0004
325T4	OEM
326	OEM
326A	OEM
326AJ	6483
326AL	6483
326CJ	6483
326CL	6483
326T1	0004
327-1	4601
327-3	OEM
328-3	OEM
329-1	3713
329-100	OEM
329-130	OEM
329-200	OEM
329-629	OEM
329A	1023
329AR	0501
329B	1023
329BR	0501
329C	1023
329CR	0501
329D	1023
329DR	0501
329E	1023
329ER	0501
329F	1023
329FR	0501
329G	1036
329GR	1777
329H	1036
329HR	1777
329K	1036
329KR	1777
329M	1036
329MR	1777
329P	1046
329PR	1789
329S	1046
329SR	1789
329V	1046
329VR	1789
329Z	1046
329ZR	1789
330-2	OEM
330-5	OEM
330-10	OEM
330-11	OEM
330D05P	OEM
330GI	0133
330T4	OEM
330VF	OEM
331-1	0423
331AL	OEM
331CL	6484
331H-15	OEM
331TB7	OEM
331TB9	OEM
331TB11	OEM
331TB13	OEM
332AJ	6485
332AL	6485
332TB7	OEM
332TB9	OEM
332TB11	OEM
332TB13	OEM
333-1	0150
333-2	OEM
333AJ	6486
333AL	6486
333CJ	6486
333CL	6486
333TB7	OEM
333TB9	OEM
333TB11	OEM
333TB13	OEM
334-1	OEM
334-377	0074
334AL	6487
334CL	6487
334TB7	OEM
334TB9	OEM
334TB11	OEM
335-1	0065
335A	1630
335AJ	6488
335AL	6488
335B	0254
335C	1107
335CJ	6488
335CL	6488
335D	1107
335F	1111
335G	0267
335H	0258
335K	1644
335M	0267
335P	1119
335S	1119
335V	0281
335Z	0281
336A	1630
336B	0254
336C	1107
336D	1107
336F	1111
336G	0267

If replacement code is OEM, contact original manufacturer for replacement.

DEVICE TYPE	REPL CODE
336H	0258
336K	1644
336M	0267
336S	1119
336V	0281
336Z	0281
337-120	OEM
337-220	OEM
337-330	OEM
337-829	OEM
337-919	OEM
337A	0534
337B	0748
337C	0538
337D	0538
337F	0777
337H	0538
337K	0801
337M	0816
337P	0545
337S	0545
337V	0550
337Z	0550
338-6551-A	OEM
339	0176
339-04	0102
339-5	OEM
339-6	2033
339-7	OEM
339-529-001	0143
339-529-002	0170
339A	1023
339AR	0501
339B	1023
339BR	0501
339C	1023
339CR	0501
339D	1023
339DR	0501
339E	1023
339ER	0501
339F	1023
339FR	0501
339G	1036
339GI	OEM
339GR	1777
339H	1036
339HR	1777
339K	1036
339KR	1777
339M	1036
339MR	1777
339P	1046
339PR	1789
339S	1046
339SR	1789
339V	1046
339VR	1789
339Z	1046
339ZR	1789
340-3	OEM
340RA10	6678
340RA20	6678
340RA30	6678
340RA40	6678
340RA60	6678
340T4	OEM
341-0127-A	OEM
341-0128-A	OEM
341A	0970
341AR	2291
341B	3733
341BR	2291
341C	0989
341CR	2291
341D	0989
341DR	2291
341F	3743
341FR	0476
341H	3743
341HR	0476
341K	0200
341KR	0476
341M	0204
341MR	0476
341P	0206
341PR	0449
341S	0206
341SR	0449
341V	0586
341Y	0586
341Z	0586
342-0132-B	OEM
342-0133-A	OEM
342-0134-B	OEM
342-0135-B	OEM
342-0265-A	OEM
342-0272-A	OEM
342-1	0086
342AJ	6489
342AL	6489
342CJ	6489
342CL	6489
343-1001	OEM
343-1002	OEM
343AJ	6490
343AL	6490
343CJ	6490
343CL	6490
343RA10	OEM
343RA20	OEM
343RA30	OEM
343RA40	OEM
343RA60	OEM
344-0010-B	OEM
344-0011	OEM
344-0020-A	OEM
344-0021	OEM
344-0041-A	OEM
344-1	0086
344-6000-2	0016
344-6000-3	0144
344-6000-3A	0144
344-6000-4	0016
344-6000-5	0016
344-6000-5A	0016
344-6001-1	0086
344-6001-2	0086
344-6002-3	0016
344-6005-1	0016
344-6005-2	0079
344-6005-5	0016
344-6005-6	0133
344-6005-7	OEM
344-6005-8	OEM
344-6006-6	0143
344-6011-1	0086
344-6011-2	0086
344-6012-1	0126
344-6013-1B	0086
344-6013-4	0086
344-6014-1B	0126
344-6015-7	0144
344-6015-7A	0144
344-6015-8	0127
344-6015-9	0127
344-6015-10	0144
344-6015-11	0144
344-6017-1	0037
344-6017-2	0016
344-6017-3	0126
344-6017-4	0086
344-6017-5	0016
344-6017-6	0144
344AL	OEM
344CL	OEM
344TB7	OEM
344TB9	OEM
345-2	0086
346-3	0079
346-4	0079
346A	0985
346B	0910
346C	0990
346D	0990
346F	0993
346H	0993
346K	1001
346M	0515
346P	1008
346S	1008
346V	0948
346Z	0948
347-2	OEM
347-4	OEM
347AL	6492
347CL	OEM
349-1	0086
349-2	0079
349-7	OEM
349-113-011	1823
349-113-012	1041
349-113-013	0081
349-113-014	0141
349-113-015	2078
349-113-022	0081
349-113-024	2619
349-113-025	2622
349AL	OEM
349CL	OEM
350	0280
350-1	OEM
350-2	OEM
350-3	OEM
350-4	OEM
350-5	OEM
350B20	OEM
350B25	OEM
350B30	OEM
350D7216	0076
350D7217	0076
350J	OEM
350K	OEM
350K3516	0006
350K3517	0006
350K5717	0148
350L	OEM
350P	OEM
350P2SP	OEM
350P4SP	OEM
350P6SP	OEM
350P8SP	OEM
350P10SP	OEM
350P12SP	OEM
350P16SP	OEM
350PAF10V08	OEM
350PAF10V10	OEM
350PAF10V12	OEM
350PAF20V08	OEM
350PAF20V10	OEM
350PAF20V12	OEM
350PAF40V08	OEM
350PAF40V10	OEM
350PAF40V12	OEM
350PAF60V08	OEM
350PAF60V10	OEM
350PAF60V12	OEM
350R1B	OEM
350R2B	OEM
350R3B	OEM
350R4B	OEM
350R5B	OEM
350R6B	OEM
350R7B	OEM
350R8B	OEM
350R9B	OEM
350R10B	OEM
350R11B	OEM
350R12B	OEM
350R13B	OEM
350R14B	OEM
350R15B	OEM
350R16B	OEM
350R18B	OEM
350R20B	OEM
350R22B	OEM
350R24B	OEM
350R26B	OEM
350R28B	OEM
350R30B	OEM
350R32B	OEM
350R34B	OEM
350R36B	OEM
350R38B	OEM
350R40B	OEM
350R42B	OEM
350R44B	OEM
350R46B	OEM
350R48B	OEM
350R50B	OEM
350RA10	1900
350RA20	1900
350RA30	1900
350RA40	1900
350RA60	1900
350RA80	0738
350RA100	0738
350RA120	0738
350RA130	OEM
350RA140	OEM
350RA150	OEM
350RA160	OEM
350RA170	OEM
350RA180	OEM
350RB50	OEM
350RB60	OEM
350RB70	OEM
350RB80	OEM
350RB90	OEM
350RB100	OEM
350RB110	OEM
350RB120	OEM
350RC10	OEM
350RC20	OEM
350RC30	OEM
350RC40	OEM
350RC60	OEM
350RS25	OEM
350RS30	OEM
350S20	OEM
350S25	OEM
350S30	OEM
350T4	OEM
351	0031
351-07	OEM
351-029-020	0410
351-2	OEM
351-3006	1069
351-3031	0133
351-7008-010	1257
351-7008-020	3901
351-7011-020	3418
351-7011-030	3861
351-7011-040	3861
351-7011-050	3714
351-7011-060	3714
351-7011-070	1257
351-7011-080	1257
351-7011-090	1364
351-7011-100	1069
351-7011-110	1069
351-7011-120	1069
351-7011-150	3337
351-7011-160	3337
351-7011-170	OEM
351-7011-180	OEM
351-7015-010	3845
351-7015-020	3850
351-7015-030	1257
351-7015-040	1364
351-7015-050	3897
351-7015-060	3901
351-7015-070	3337
351-7025-010	4236
351-7025-020	1069
351-7026-010	3845
351-7026-030	1257
351-7026-060	3901
351-7026-070	3337
351-7121-010	OEM
351-7121-020	1069
351-7206-080	3845
351-7206-090	3850
351-7206-100	1257
351-7206-110	1364
351-7206-140	3337
351-7206-150	3845
351-7206-160	3850
351-7206-170	1257
351-7206-180	1364
351-7206-190	3897
351-7206-200	3901
351-7206-210	3337
351-7577-010	1350
351A	0031
351RA10	OEM
351RA20	OEM
351RA30	OEM
351RA40	OEM
351RA60	1900
351RA80	0738
351RA100	0738
351RA120	0738
351RA130	OEM
351RA140	OEM
351RA150	OEM
351RA160	OEM
351RA170	OEM
351RA180	OEM
352	0211
352(DIO)	0031
352-0023-001	1041
352-0197-000	0016
352-0677-040	0103
352-0677-051	0103
352-5	0551
352-6	0566
352-7	OEM
352A	0031
352AR	2291
352B	2291
352BR	2291
352C	2291
352CR	2291
352D	2291
352DR	2291
352F	0476
352FR	0476
352H	0476
352HR	0476
352K	0476
352KR	0476
352M	0476
352MR	0476
352YT	OEM
353	0211
353(IC)	3741
353-6	OEM
353-9001-001	0136
353-9001-002	0136
353-9001-003	0136
353-9002-002	0136
353-9008-001	0126
353-9012-001	0004
353-9201-001	0265
353-9203-001	0042
353-9301-001	0086
353-9301-002	0050
353-9301-004	0126
353-9304-001	0037
353-9304-004	0126
353-9306-001	0016
353-9306-002	0016
353-9306-003	0016
353-9310-001	0016
353-9312-001	0004
353-9314-001	0016
353-9315-001	0016
353-9317-001	0126
353-9318-001	0111
353-9318-002	0111
353-9319-001	0016
353-9319-002	0016
353-9502-001	0042
353PJLA160	OEM
353PJLA180	OEM
353PJLA200	OEM
353PJLA250	OEM
353RA10	OEM
353RA20	OEM
353RA30	OEM
353RA40	OEM
353RA60	OEM
353RA80	OEM
353RA100	OEM
353RA120	OEM
353RA130	OEM
353RA140	OEM
353RA150	OEM
353RA160	OEM
353RA170	OEM
353RA180	OEM
353Y	OEM
354	0031
354(CHRYSLER)	1345
354-3052	0211
354-9001-001	0019
354-9101-002	0123
354-9101-006	0015
354-9102-001	0015
354-9110-001	0015
354A	0031
354Y	OEM
355	OEM
355AJ	OEM
355AL	OEM
355CJ	OEM
355CL	OEM
355D6	0007
355D7	0212
355D8	0007
355D9	0016
356	0031
356A	0031
357A-10	2170
357A-22	1233
358-03	OEM
358MHZ	1030
358P007A1	OEM
359A	0015
359B	0015
359C	0015
359D	0015
359F	0015
359H	0015
359K	0015
359M	0015
359P	0071
359S	0071
359V	0071
359Z	0071
360-1	0224
360-2	0342
360-32	0133
360BM	OEM
360J	OEM
360K	OEM
360K1010	0133
360K1015	0124
360K1027	0124
360K3039	0052
360K3068	0253
360K3098	0052
360K3100	0041
360K3106	0471
360K3121	0062
360K3139	0253
360K7160	0015
360M	OEM
361-1	0560
361K7086	0795
361K7090	0015
361K7091	0071
361K7093	0015
361K7094	0015
361K7160	0015
361K7287	0023
361Q8003	0106
362-1	3016
362-1R	OEM
362-3	0285
362-6	0285
362-7	3016
363A	OEM
363AL	6497
363B	OEM
363CL	6497
363D	OEM
363F	OEM
363H	OEM
363M	OEM
363S	OEM
363Z	OEM
364-1	0037
364-1(SYLVANIA)	0037
364-3	OEM
364-2001	0019
364-10048	0325
364-B12	0725
364B14	0687
365	OEM
365-1	0016
365L	OEM
365T	0280
365T1	0280
366-1	0079
366-1(SYLVANIA)	0144
366-2	0079
366-2(SYLVANIA)	0144
366A	0970
366A-RT	OEM
366B	3733
366B-RT	OEM
366C	0989
366C-RT	OEM
366D	0989
366D-RT	OEM
366F	3743
366F-RT	OEM
366H	0197
366H-RT	OEM
366K	0200
366K-RT	OEM
366M	0204
366M-RT	OEM
366P	0258
366RA	0985
366RB	0910
366RC	0990
366RD	0990
366RF	0993
366RH	0997
366RK	1001
366RM	0515
366RP	1008
366RS	1008
366RV	0948
366RZ	0948
366S	0206
366V	0586
366Z	0586
367A	0097
367A-RT	OEM
367B	0090
367B-RT	OEM
367C	0109
367C-RT	OEM
367D	0109
367D-RT	OEM
367F	0105
367F-RT	OEM
367H	0109
367H-RT	OEM
367K	0116
367K-RT	OEM
367M	0122
367M-RT	OEM
367P	0131
367RA	0765
367RB	0748
367RC	0538
367RD	0538
367RF	0777
367RH	0538
367RK	0801
367RM	0816
367RP	0545
367RS	0545
367RV	0550
367RZ	0550
367S	0131
367V	0145
367Z	0145
368A	5514
368B	5582
368BPY	OEM
368C	2900
368D	2900
368F	5583
368H	3751
368K	3757
368M	1112
368P	3768
368RA	1238
368RB	1240
368RC	1241
368RD	1241
368RF	1245
368RH	1251
368RK	1253
368RM	1205
368RP	1264
368RS	1264
368RV	0449
368RZ	0449
368S	3768
368V	3011
368Z	3011
369-2	0015
369-3	0015
369HHD	4300
370	OEM
370-1	0065
370-2	0351
370AJ	1295
370AL	1295
370CJ	1295
370CL	1295
370J	OEM
370K	OEM
370L	OEM
371A	3189
371AJ	1416
371AL	1416
371B	2901
371C	1124
371CJ	1416
371CL	1416
371D	1124
371F	1126
371H	0805
371K	1195
371M	0319
371P	1132
371RA	1630
371RB	0254
371RC	1107
371RD	1107
371RF	1111
371RH	0258
371RK	1644
371RM	0267
371RP	1119
371RS	1119
371RV	0281
371RY	0281
371RZ	0281
371S	1132
371V	0044
371Z	0044
372AL	1556
372CL	1556
372Z	0550
373AL	OEM
373CL	OEM
374-1	0325
374A	OEM
374AL	OEM
374B	OEM
374C	OEM
374CL	OEM
374D	OEM
374F	OEM
374H	OEM
374K	OEM
374M	OEM
375CQY	OEM
375RA5	OEM
375RA10	OEM
375RA20	OEM
375RA30	OEM
375RA40	OEM
375RA50	OEM
375RA60	OEM
376-0062	0850
376A	OEM
376B	OEM
376C	OEM
376D	OEM
376E	OEM
376F	OEM
376G	OEM
376H	OEM
376K	OEM
376M	OEM
377-1	OEM
377A	OEM
377B	OEM
377C	OEM
377D	OEM
377E	OEM
377F	OEM
377G	OEM
377H	OEM
377K	OEM
377M	OEM
377Y	OEM
378-1	0325
378-44	0103
378-44A	0103
380	OEM
380(IC)	0379
380-2	0313
380-3	OEM
380A/C	0379
380B/M	0379
380H61	0015
380J	OEM
380K	OEM
383	0050
383(IC)	1048
384A	0015
384B	0015
384C	0015
384D	0015
384F	0015
384H	0015
384K	0071
384M	0071
384P	0071
384S	0071
384V	0071
384Z	0071
385A	0015
385B	0015
385C	0015
385D	0015
385F	0015
385H	0015
385K	0015
385KW	0015
385M	0015
385P	0071
385S	0071
385Z	0071
386	3063
386-1	0037
386-1(SYLVANIA)	0037
386-1AY	0015
386-1CY	0015
386-1FY	0015
386-7118P1	0144
386-7178P1	0016
386-7183P1	0103
386-7183P1A	0103
386-7184P1	0126
386-7185P1	0016
386-7188P1	0144
386-7243-P001	0007
386AK	0015
386AW	0015
386AX	0015
386AY	0015

If replacement code is OEM, contact original manufacturer for replacement.

DEVICE TYPE	REPL CODE
386BW	0015
386BY	0015
386CW	0015
386CX	0015
386CY	0015
386D	OEM
386DW	0015
386DY	0015
386FW	0015
386FX	0015
386FY	0015
386K	0015
386KX	0015
386KY	0015
386MW	0015
386MY	0015
387-1	OEM
387E	OEM
388A	0920
388B	0102
388C	1090
388D	1090
388F	OEM
388H	OEM
388K	OEM
388M	OEM
388S	OEM
388Z	OEM
389-1	OEM
389C	OEM
389E	OEM
390-1	0808
390-1,,	0037
390J	OEM
390K	OEM
392-1	0144
392-1(SYLVANIA)	0086
392-2	0313
393-1	0791
393-1(SYLVANIA)	0126
394-1	0069
394-2	2826
394-3074-2	0211
394-3074-5	0211
394-3097-1	0211
394-3097-2	0211
394-3102-1	0600
394-3127-1	0086
394-3127-2	0086
394-3127-3	0086
394-3135	0162
394-3135A	0103
396-7178P1	0144
398-2	OEM
398-8418-1	3833
398-8972-1	1275
398-13223-1	0232
398-13224-1	0361
398-13225-1	0385
398-13226-1	0268
398-13632-1	0361
398B	0953
398P	6972
398V	0927
398Z	0087
399-3	3132
399-4	3132
399-5	0236
399-8	0236
399A	0110
399B	0242
399C	0242
399D	0242
400-2	OEM
400-1735	3899
400-1736	3924
400A	0143
400B	0599
400C	0599
400C-11958	0015
400D	0143
400E	OEM
400E15	OEM
400E120	0749
400EXD21	OEM
400F	0599
400F15	OEM
400G	OEM
400H	1986
400K	1986
400M	1986
400MAB180	OEM
400MAB200	OEM
400MAB250	OEM
400P	0657
400P4SP	OEM
400P6SP	OEM
400P8SP	OEM
400P12S	OEM
400PIN-RM	OEM
400PV-RM	OEM
400QD21	OEM
400R1B	OEM
400R2B	OEM
400R3B	OEM
400R4B	OEM
400R5B	OEM
400R6B	OEM
400R7B	OEM
400R8B	OEM
400R9B	OEM
400R10B	OEM
400R11B	OEM
400R12B	OEM
400R13B	OEM
400R14B	OEM
400R15B	OEM
400R16B	OEM
400R18B	OEM
400R20B	OEM
400R22B	OEM
400R24B	OEM
400R26B	OEM
400R28B	OEM
400R30B	OEM
400R32B	OEM
400R34B	OEM
400R36B	OEM
400R38B	OEM
400R40B	OEM
400R42B	OEM
400R44B	OEM
400R46B	OEM
400R48B	OEM
400R50B	OEM
400S	0657
400S2-2.6	OEM
400S2SP	OEM
400S3SP	OEM
400S4SP	OEM
400S5SP	OEM
400S6SP	OEM
400S8SP	OEM
400SS2SP	OEM
400SS3SP	OEM
400SS4SP	OEM
400SS5SP	OEM
400SS6SP	OEM
400SS8SP	OEM
400V	0657
400YD21	OEM
400Z	0657
401-041-6507	0276
401A	0599
401B	0599
401C	0599
401D	0599
401F	0599
401H	1986
401K	1986
401M	1986
401P	0657
401PDA20L15	OEM
401PDA40L15	OEM
401PDA40L20	OEM
401PDA60L15	OEM
401PDA60L20	OEM
401PDA80L15	OEM
401PDA80L20	OEM
401PDA100L20	OEM
401PDA100L25	OEM
401PDA100L30	OEM
401PDA120L20	OEM
401PDA120L25	OEM
401PDA120L30	OEM
401PDA130L20	OEM
401PDA130L30	OEM
401PDA140L20	OEM
401PDA140L25	OEM
401PDA140L30	OEM
401PDA160L25	OEM
401PDA160L30	OEM
401PDA180L25	OEM
401PDA180L30	OEM
401PDL40S15	OEM
401PDL40S20	OEM
401PDL60S15	OEM
401PDL60S20	OEM
401PDL80S15	OEM
401PDL80S20	OEM
401PDL100S20	OEM
401PDL100S30	OEM
401PDL120S20	OEM
401PDL120S30	OEM
401PDL130S20	OEM
401PDL130S30	OEM
401PDL140S30	OEM
401PDL150S30	OEM
401PDL160S30	OEM
401S	0657
401V	0657
401Z	0657
402	OEM
402A	3189
402B	2901
402C	1124
402D	1124
402F	1126
402H	0805
402K	1195
402M	0319
402P	1132
402S	1132
402V	0044
402Z	0044
403-009/07	0103
403-1	3316
403-2	OEM
403-4	0917
403-6	OEM
403A	3189
403B	1124
403C	1124
403D	1124
403F	1126
403H	0805
403K	1195
403M	0319
403P	1132
403PJA60	OEM
403PJA80	OEM
403PJA100	OEM
403PJA120	OEM
403PJA140	OEM
403PJA160	OEM
403S	1132
403V	0044
403Z	0044
404	OEM
404-1(SYL)	0079
404-2	0127
404-2(SYLVANIA)	0016
404A	0097
404B	0090
404B(NCR)	0126
404C	0109
404CM	OEM
404D	0109
404F	0105
404H	0109
404K	0116
404M	0122
404P	0131
404S	0131
404V	0145
404VM	OEM
404Z	0145
405-000-4407	0887
405-000-6104	0898
405-000-6302	0898
405-000-9907	2094
405-001-7407	0148
405-001-7605	0148
405-004-3109	0203
405-004-3208	0203
405-004-4205	0006
405-004-4502	0006
405-004-4601	0006
405-004-4809	0006
405-004-5004	0006
405-004-5103	0006
405-006-1103	0148
405-006-1202	0148
405-006-1301	0148
405-006-1707	0148
405-006-1806	0148
405-006-4906	1911
405-006-5101	1911
405-006-6702	0006
405-007-5308	0477
405-007-6701	3812
405-010-6507	0261
405-010-6606	0261
405-010-6705	0261
405-011-7305	0151
405-011-7404	0151
405-011-7503	0151
405-011-7513	0151
405-011-8401	0151
405-011-8500	0151
405-011-8609	0151
405-012-2002	0076
405-012-2101	0076
405-012-2309	0076
405-012-7908	0951
405-012-8004	0951
405-013-2001	1284
405-013-2209	1284
405-013-4500	0261
405-013-4609	0261
405-013-6207	0261
405-013-6306	0261
405-013-6801	0155
405-013-7006	0155
405-014-2406	1562
405-014-2505	1562
405-014-8408	0276
405-014-8409	0276
405-014-8507	0276
405-014-8606	0276
405-014-9207	2851
405-015-2900	0009
405-015-3006	0009
405-016-1100	0151
405-016-6204	0286
405-016-6402	0286
405-016-6600	0286
405-016-8109	0276
405-016-9403	OEM
405-017-1901	0155
405-017-9501	2954
405-018-2501	0898
405-018-2600	2323
405-018-7100	1386
405-018-7209	1386
405-019-1909	0537
405-019-2708	0537
405-019-2807	0537
405-019-2817	0537
405-019-3804	0537
405-019-4504	0537
405-019-4505	0537
405-019-6706	0537
405-019-6805	0537
405-020-2001	OEM
405-020-5200	0076
405-020-5220	0076
405-020-5705	0076
405-020-6900	0076
405-020-7501	0076
405-020-7709	0076
405-020-7907	0076
405-021-5605	0400
405-022-0005	1500
405-022-4409	1541
405-022-6601	1541
405-022-6700	3596
405-022-6809	0556
405-023-4903	2910
405-023-5009	2910
405-023-5306	2910
405-024-9907	1513
405-025-0200	1513
405-027-0604	OEM
405-027-0802	3046
405-027-1007	3046
405-031-1000	OEM
405-033-1701	OEM
405-039-3303	0155
405-039-3402	0155
405-039-8001	1500
405-040-6102	1327
405-041-6507	0276
405-041-6705	0276
405-053-0504	0275
405-056-8701	1541
405-056-8800	1541
405-058-0208	OEM
405-058-0802	0203
405-058-4800	1506
405-059-9804	0042
405-059-9903	0042
405-064-7307	0925
405-064-7406	0925
405-066-4304	0276
405-082-2407	2129
405-083-1300	2185
405-095-9004	2185
405-2	0224
405-3	OEM
405-4	OEM
405-114-6700	2110
405-114-6908	6429
405-114-7103	0563
405-114-7301	0563
405A	2002
405AR	0267
405B	1250
405BR	0267
405C	1250
405CR	0267
405D	1250
405DR	0267
405E	0319
405ER	0267
405F	0319
405FR	0267
405H	0319
405HR	0476
406	OEM
406-000-2806	0563
406-000-2905	0563
406-000-3001	0563
406-000-3100	0563
406-000-3209	0525
406-000-3308	0525
406-000-3407	0525
406-000-3506	0525
406-000-3605	0276
406-000-5302	0066
406-000-5708	1284
406-000-6309	1284
406-000-6804	0148
406-000-7108	1911
406-000-7603	1911
406-001-0900	0276
406-001-7404	5624
406-001-7503	5624
406-001-7602	6608
406-001-7701	6608
406-004-9702	1506
406-006-2701	2185
406-006-6504	0006
406-3	OEM
407-002-6809	1166
407-002-9602	1166
407-004-8306	0071
407-004-9105	0023
407-005-2303	0023
407-005-4208	0133
407-005-4505	0124
407-005-7605	0071
407-005-8602	0023
407-006-0407	0023
407-006-0902	0023
407-006-3903	0023
407-006-4108	0023
407-006-5709	0564
407-006-6706	0283
407-007-6605	0023
407-007-6606	0023
407-007-6903	0023
407-007-7405	0023
407-007-7603	0023
407-007-9904	0124
407-008-0207	0133
407-008-6001	0907
407-008-6704	0124
407-009-8905	0283
407-009-9100	0031
407-011-3004	0023
407-011-3103	0023
407-011-4308	0023
407-011-4407	0023
407-012-4208	0124
407-012-4406	0133
407-012-5809	0133
407-012-6202	0907
407-012-6212	0907
407-013-1008	0133
407-013-1206	0133
407-013-1909	0023
407-013-2609	0143
407-013-3200	0023
407-013-4207	0133
407-013-4306	0124
407-013-6508	0124
407-013-7109	0133
407-016-0107	2703
407-016-7205	2703
407-018-4202	OEM
407-020-6607	OEM
407-026-8407	OEM
407-027-4606	5174
407-027-5207	5132
407-028-2205	OEM
407-028-8900	OEM
407-028-9105	4052
407-029-1207	OEM
407-029-7308	OEM
407-030-2507	1076
407-031-0205	OEM
407-031-7204	OEM
407-033-1507	OEM
407-040-2009	OEM
407-040-4003	OEM
407-042-7309	3163
407-043-0200	4878
407-048-1103	0041
407-048-1130	0041
407-048-1202	4111
407-048-1608	0041
407-048-2001	0253
407-048-3206	0062
407-048-3404	0062
407-048-3503	0062
407-048-3701	0077
407-048-4005	0077
407-048-4203	0077
407-048-5101	0057
407-048-5200	0057
407-048-5606	0064
407-048-5903	0064
407-048-6603	0181
407-048-6900	0053
407-048-7105	0053
407-048-7303	0052
407-048-7600	0053
407-048-7808	0053
407-048-7907	0376
407-048-8003	0879
407-049-2901	1083
407-049-3403	0137
407-049-8507	OEM
407-050-2501	0166
407-051-5303	0494
407-051-6409	0275
407-051-6706	OEM
407-052-3506	0010
407-053-0702	0589
407-053-4700	0494
407-053-6001	0589
407-054-0008	0064
407-054-3306	0052
407-054-4709	0053
407-054-4808	0053
407-054-5805	0879
407-055-7907	0376
407-056-7906	0041
407-056-8002	0041
407-057-4102	0062
407-057-4201	0062
407-057-6304	0077
407-057-9602	0057
407-057-9701	0057
407-057-9800	0057
407-059-3400	0137
407-065-0905	OEM
407-070-8408	0057
407-075-3903	OEM
407-075-9509	0062
407-076-2608	0102
407-076-3803	OEM
407-079-2605	0031
407-088-6011	0907
407-088-6502	0023
407-095-8001	0102
407-096-9809	0471
407-102-6006	0143
407-102-6501	OEM
407-102-6600	OEM
407-103-1604	0546
407-103-2502	OEM
407-103-2601	OEM
407-103-6500	0031
407-104-2204	1166
407-104-2402	1166
407-124-5209	1191
407-124-5506	0023
407-124-5605	1090
407-124-5803	1191
407-124-6404	0023
407-124-6406	0023
407-124-6503	OEM
407-143-2708	0166
407-147-5705	OEM
407A	0970
407AR	2900
407B	3733
407BR	2291
407C	0989
407CR	2900
407D	0989
407DR	2291
407F	3743
407FR	0476
407H	0197
407HR	0476
407K	0200
407KR	1112
407M	0204
407MR	0476
407P	0206
407RA	0985
407RB	0910
407RC	0990
407RD	0990
407RF	0993
407RH	0997
407RK	1001
407RM	0515
407RP	1008
407RS	1008
407RV	0948
407RZ	0948
407S	0206
407V	0586
407Z	0586
408	OEM
408-000-0301	1291
408-008-2406	0124
408-008-8606	0071
408-009-3501	OEM
408-009-3600	OEM
408-009-3709	OEM
408-009-3808	OEM
408-009-9008	0023
408-009-9107	0140
408-009-9206	0031
408-009-9305	0023
408-009-9404	0080
408-010-0902	0062
408-014-8408	0276
408-1	OEM
408-5	OEM
408A	0097
408AR	2900
408B	0090
408BR	2291
408C	0109
408CR	2291
408D	0109
408DR	2291
408F	0105
408FR	0476
408H	0109
408HR	0476
408K	0116
408KR	0476
408M	0122
408MR	0476
408P	0131
408RA	0765
408RB	0748
408RC	0765
408RD	0538
408RF	0777
408RH	0777
408RK	0801
408RM	0816
408RP	0545
408RS	0545
408RV	0550
408RZ	0550
408S	0131
408V	0145
408Z	0145
409-001-0604	1529
409-004-5705	3083
409-004-8102	OEM
409-004-8409	2870
409-006-0203	1654
409-009-2709	0035
409-011-1608	OEM
409-015-1208	OEM
409-017-7208	3253
409-017-7505	3097
409-018-9706	OEM
409-019-0801	4266
409-019-1709	OEM
409-019-3109	0539
409-019-4205	0462
409-019-4700	0332
409-019-5608	0732
409-019-5707	2957
409-019-6209	2666
409-019-6407	4377
409-019-6605	4381
409-022-2205	OEM
409-032-3407	OEM
409-036-0105	2111
409-037-5406	OEM
409-040-8401	OEM
409-042-9000	1654
409-046-7705	OEM
409-047-5601	5817
409-047-8602	3913
409-047-9104	1641
409-047-9203	3445
409-051-2801	0024
409-051-3006	0035
409-052-4408	OEM
409-053-7309	5011
409-058-4105	OEM
409-063-4206	OEM
409-067-3304	OEM
409-074-9900	OEM
409-083-0103	6028
409-093-3804	OEM
409-107-6708	OEM
409-108-0200	OEM
409-108-2402	OEM
409-108-2600	OEM
409-109-4108	OEM
409-109-4405	OEM
409-110-4807	OEM
409-111-3106	OEM
409-117-9607	3690
409-118-2102	OEM
409-121-4209	OEM
409-121-4407	OEM
409-132-7800	OEM
409-132-7909	0035
409-135-5407	OEM
409-135-5506	4930
409-142-4905	OEM
409-145-6500	OEM
409-146-7001	4094
409-147-9905	OEM
409-151-7904	OEM
409-156-6605	4094
409-169-5305	OEM
409-169-5800	OEM
409-172-8003	OEM
409-172-8102	OEM
409-173-2703	OEM
409-173-2802	OEM
409-173-5209	OEM
409-174-1408	OEM
409-184-9807	OEM
409-185-0803	OEM
409-187-7701	OEM
409-193-2905	OEM
409-194-8401	OEM
409-199-5306	OEM
409-219-9406	OEM
409-220-8009	OEM
409-225-9704	OEM
409-228-5604	OEM
409-242-6007	OEM
409-243-0806	2364
409-246-8809	OEM
409-272-3205	OEM
409-274-3302	OEM
409-289-3403	OEM
409-301-2803	OEM
409-301-9604	OEM
409A	5514
409AR	2291
409B	5582
409BR	2291
409C	2900
409CR	2291
409D	2900
409DR	2291
409F	5583
409FR	0476
409H	3751
409HR	0476
409K	3757
409KR	0476
409M	1112
409MR	0476
409P	3768
409RA	1238
409RB	1240
409RC	1241
409RD	1241
409RF	1245
409RH	2365
409RK	1253
409RM	1205
409RP	1264

If replacement code is OEM, contact original manufacturer for replacement.

DEVICE TYPE	REPL CODE
409RS	1264
409RV	0449
409RY	2251
409RZ	0449
409S	3768
409V	3011
409Z	3011
410	0363
410-012-0150	0004
410-013-0240	0004
410-015-2409	OEM
410-019-4706	OEM
410-022-6506	OEM
410-024-9307	OEM
410-051-5808	1654
410-067-3300	1654
410-070-2000	OEM
410-070-9801	OEM
410-2	OEM
410-165-3202	OEM
410-214-7502	OEM
410A	0599
410AR	1347
410B	0599
410BR	1347
410C	0599
410CR	1347
410D	0599
410DR	1347
410E	0599
410ER	1347
410F	0599
410FR	1347
410H	1986
410HR	1905
410K	1986
410KF	OEM
410KR	1905
410M	1986
410MR	1905
410PE2-35	OEM
410XY5-35SC00	OEM
411	0363
411-1	OEM
411-6	OEM
411-237	0086
411A	OEM
411B	OEM
411C	OEM
411D	OEM
411E	OEM
411F	OEM
411H	1986
411HR	1905
411K	1986
411KR	1905
411M	1986
411MR	1905
411P	0657
411PR	0193
411S	0657
411SR	0193
411V	0657
411VR	0193
411Z	0657
411ZD	OEM
411ZH	OEM
411ZR	0193
412	0280
412-1A5	0016
412-2	3161
412H	OEM
413	0363
414	0079
414-2	OEM
414A-15	0160
414H	OEM
415	OEM
415C01111P	OEM
415T4	OEM
416	0522
416-1D400-26	OEM
416-1D400-115	OEM
416A	1560
416AR	1783
416B	1560
416BR	1783
416C	2840
416CR	1783
416D	2840
416F	4308
416FR	1783
416H	2850
416K	2871
416KR	1783
416M	2871
416P	2833
416PR	1818
416RA	1560
416RB	1560
416RC	2182
416RD	2182
416RF	2191
416RH	2191
416RK	2197
416RM	2197
416RP	2216
416RS	2216
416RV	2340
416RZ	2340
416S	2833
416SR	1818
416V	2473
416VR	1818
416Z	2473
416ZR	1818
417-1	3880
417-2	0050
417-5	0004
417-6	0164
417-7	0016
417-11	0050
417-12	0050
417-13	0050
417-14	0050
417-16	0050
417-17	0004
417-18	0164
417-19	0127
417-20	0160
417-21	0164
417-22	0050
417-23	0050
417-25	0050
417-27	0050
417-28	0164
417-29	0604
417-29BLK	0604
417-29GRN	0604
417-29WHT	0604
417-30	0160
417-31	0050
417-32	0160
417-33	0050
417-35	0050
417-36	0050
417-37	0050
417-38	0050
417-39	0050
417-40	0004
417-41	0004
417-42	0604
417-43	0126
417-44	0160
417-45	0160
417-46	0160
417-47	0004
417-48	0004
417-49	0086
417-50	0050
417-51	0164
417-52	0004
417-53	0050
417-54	0050
417-56	0050
417-57	0050
417-58	0050
417-59	0086
417-60	0050
417-62	0085
417-66	0050
417-67	0076
417-68	0050
417-69	0016
417-70	0050
417-71	0050
417-72	0050
417-73	0050
417-74	0004
417-75	0004
417-76	0050
417-77	0076
417-78	0004
417-79	0136
417-81	1669
417-84	0144
417-85	0144
417-87	0086
417-88	0086
417-89	0086
417-90	0160
417-91	0111
417-92	0016
417-93	0016
417-94	0155
417-99	0160
417-101	0103
417-102	0150
417-103	0004
417-105	0007
417-106	0007
417-107	0144
417-108	0016
417-109	0076
417-110	0016
417-111	0126
417-112	0975
417-113	0975
417-114	0076
417-115	0233
417-116	0037
417-118	0155
417-119	2624
417-120	0604
417-121	0208
417-122	0164
417-124	0144
417-125	0144
417-126	0016
417-127	0198
417-128	0086
417-129	0016
417-132	0037
417-133	0007
417-134	0079
417-135	0155
417-136	0126
417-137	0086
417-138	0126
417-139	0130
417-139A	0103
417-140	0325
417-141	0160
417-142	0604
417-143	0050
417-144	0161
417-145	0460
417-146	0004
417-147	0164
417-148	0164
417-149	0164
417-150	0164
417-151	0164
417-152	0004
417-153	0037
417-153-13431	0693
417-154	0016
417-155	0086
417-158	0074
417-159	0276
417-160	0222
417-161	0400
417-167	0325
417-168	0037
417-169	0325
417-170	0126
417-171	0079
417-172	0016
417-175	0042
417-176	0037
417-177	0440
417-182	0037
417-183	1669
417-184	0037
417-185	0016
417-187	1669
417-190	0007
417-192	0016
417-194	0325
417-195	0276
417-196	0037
417-197	0016
417-199	0178
417-200	0037
417-201	0037
417-203	0561
417-204	0074
417-205	0127
417-206	0849
417-207	0849
417-211	0325
417-212	0103
417-212A	0103
417-213	0111
417-215	0103
417-216	0160
417-217	0016
417-218	0016
417-219	0079
417-220	0264
417-221	1367
417-222	0400
417-223	0212
417-224	0018
417-226-1	0016
417-228	0016
417-229	0016
417-231	0325
417-233	0016
417-235	0037
417-237	0086
417-237-13163	0086
417-239	0642
417-240	0212
417-241	0325
417-243	0007
417-244	0016
417-245	0285
417-246	0325
417-247	0086
417-248	0642
417-250	0086
417-252	0325
417-253	0325
417-254	0130
417-255	0126
417-257	0086
417-258	0007
417-260	0126
417-262	0007
417-272	2258
417-282	0543
417-283	0079
417-284	0150
417-286	0177
417-294	0715
417-298	1284
417-801	0198
417-811	0076
417-821	0079
417-822	0037
417-823	0316
417-834	0338
417-836	0342
417-852	0042
417-863	0212
417-864	0086
417-865	0037
417-874	0006
417-875	0076
417-923	0313
417-926	0551
417-932	0561
417-937	0079
417A	1124
417B	2901
417C	1124
417D	1124
417F	1126
417H	0805
417HD	OEM
417K	1195
417M	0319
417P	1132
417RA	1107
417RB	0254
417RC	1107
417RD	1107
417RF	1111
417RH	0258
417RK	1644
417RM	0267
417RP	1119
417RS	1119
417RV	0281
417RZ	0281
417S	1132
417V	0044
417Z	0044
418A	1124
418B	2901
418C	1124
418D	1124
418E	0805
418F	1126
418H	0805
418K	1195
418M	0319
418P	1132
418RA	1107
418RB	0254
418RC	1107
418RD	1107
418RF	1111
418RH	0258
418RK	1644
418RM	0267
418RP	1119
418RS	1119
418RV	0281
418RZ	0281
418S	1132
418V	0044
418Z	0044
419-1	2309
419-2	3937
419-3	OEM
419A	1124
419B	2901
419C	1124
419D	1124
419F	1126
419H	0805
419K	1195
419M	0319
419P	1132
419RA	1107
419RB	0254
419RC	1107
419RD	1107
419RF	1111
419RH	0258
419RK	1644
419RM	0267
419RP	1119
419RS	1119
419RV	0281
419S	1132
419V	0044
420	OEM
420-1	0037
420-301	OEM
420-2003-173	0071
420-2005-000	0196
420-2104-570	0071
420KF	OEM
420PA40	OEM
420PB30	1765
420PB50	1765
420PB60	1765
420PB80	1814
420PB100	5318
420PB110	6580
420PB120	1814
420PB130	5319
420PB140	5319
420PB150	5319
420PB160	5319
420PB170	OEM
420PB180	OEM
420PB190	OEM
420PB200	OEM
420PBL50	0769
420PBL60	0769
420PBL80	0483
420PBL100	0483
420PBL110	0483
420PBL120	0483
420PBM50	0769
420PBM60	0769
420PBM80	0483
420PBM100	0483
420PBM110	0483
420PBM120	0483
420PL60	OEM
420PL80	OEM
420PL100	OEM
420PL110	OEM
420PL120	OEM
420PM60	OEM
420PM80	OEM
420PM100	OEM
420PM110	OEM
420PM120	OEM
420PM130	OEM
420PM140	OEM
420PM150	OEM
420PM160	OEM
420PM170	OEM
420T1	0004
420T4	OEM
421	OEM
421-1	0042
421-2	0925
421-3	0042
421-4	0925
421-5	0482
421-6	0050
421-6B	0050
421-7	0050
421-7B	0050
421-8	0050
421-8(MAG)	1369
421-8B	0050
421-9	0004
421-10	0004
421-11	0004
421-11B	0004
421-12	0004
421-12B	0004
421-13	0004
421-13B	0004
421-13C	0211
421-14	0004
421-14B	0004
421-15	0004
421-15B	0004
421-16	0050
421-17	0050
421-18	0142
421-19	0004
421-20	0136
421-20B	0136
421-21	0136
421-21B	0136
421-22	0136
421-22B	0136
421-24	0160
421-25	0160
421-26	0136
421-4027	0015
421-6599	0160
421-7143	0004
421-7443	0015
421-7443A	0015
421-7444	0016
421-8109	0004
421-8111	0016
421-8310	0142
421-9644	0016
421-9670	0016
421-9671	0004
421-9682	0280
421-9683	0144
421-9792	0050
421-9840	0016
421-9862	0208
421-9862A	0004
421-9863	0004
421-9863A	0004
421-9864	0004
421-9864A	0004
421-9865	0015
421T1	0004
422	OEM
422-0222	0004
422-0738	0016
422-0739	0126
422-0960	0178
422-0961	0103
422-1232	0126
422-1233	0086
422-1234	0016
422-1362	0015
422-1401	0144
422-1402	0144
422-1403	0136
422-1404	0086
422-1405	0126
422-1406	0004
422-1407	0016
422-1408	0071
422-1443	0265
422-1866	0015
422-2008	0126
422-2158	0086
422-2532	0144
422-2533	0016
422-2534	0016
422-2535	0004
422-2540	0015
422-2778	0050
422-2779	0050
422-2780	0050
422B	OEM
423	4160
423-1	OEM
423-800-235	2097
423-800175	1041
423-800176	0562
423-800177	1041
423-800178	0141
423-800194	1844
423-800202	1831
423-800223	1481
423-800234	1823
423-800236	0358
423AB1A11	OEM
423AB1AB1	0277
423AB1AF1	OEM
423AB1AI1	OEM
423AC1A11	OEM
423AC1AI1	OEM
423AF1AH1	OEM
423BB1A11	OEM
423BB1AB1	0289
423BB1AF1	OEM
423BB1AI1	OEM
423BC1A11	OEM
423BC1AI1	OEM
423BF1AH1	OEM
423DB1A11	OEM
423DB1AB1	0296
423DB1AF1	OEM
423DB1AI1	OEM
423DC1A11	OEM
423DC1AI1	OEM
423DF1AH1	OEM
423EB2AF1	OEM
423EF2AH1	OEM
423EF3AH1	OEM
423FB1AB1	0277
423FB1AF1	OEM
423FB1AI1	OEM
423FC1A11	OEM
423FC1AI1	OEM
423FF1AH1	OEM
423MB1A11	OEM
423MB1AB1	0302
423MB1AF1	OEM
423MB1AI1	OEM
423MB2A11	OEM
423MB2AI1	OEM
423MB6AH1	OEM
423MC1A11	OEM
423MC1AI1	OEM
423MC2A11	OEM
423MC2AI1	OEM
423MC5AE1	OEM
423MC6AH1	OEM
423MC11AH1	OEM
423MD5AE1	OEM
423MD6AH1	OEM
423MD11AH1	OEM
423MF1AH1	OEM
423MH5AE1	OEM
423MH6AH1	OEM
423MH10AE1	OEM
423MH11AH1	OEM
423NB1A11	OEM
423NB1AB1	0250
423NB1AF1	OEM
423NB1AI1	OEM
423NB2AF1	OEM
423NB2AI1	OEM
423NC1A11	OEM
423NC1AI1	OEM
423NC2A11	OEM
423NC2AI1	OEM
423NC7AE1	OEM
423NC17AH1	OEM
423ND17AH1	OEM
423NF1AH1	OEM
423NF2AH1	OEM
423NH10AE1	OEM
423NH16AE1	OEM
423NH26AH1	OEM
423PB1AB1	0250
423PB10AH1	OEM
423PC8AE1	OEM
423PC10AE1	OEM
423PC13AE1	OEM
423PC15AE1	OEM
423PD8AE1	OEM
423PD10AE1	OEM
423PD13AE1	OEM
423PD15AE1	OEM
423PH5AS1	OEM
423PH8AS1	OEM
423PH10AE1	OEM
423PH10AS1	OEM
423PH13AS1	OEM
423PH15AE1	OEM
423PH15AS1	OEM
423PH20AS1	OEM
423PH25AS1	OEM
423SC14AH1	OEM
423SD14AH1	OEM
423SH14AH1	OEM
423SH19AH1	OEM
423SH24AH1	OEM
423TB8AH1	OEM
424	OEM
424AB1AB1	OEM
424AB1AF1	OEM
424AF1AH1	OEM
424BB1AB1	OEM
424BB1AF1	OEM
424BF1AH1	OEM
424DB1AB1	OEM
424DB1AF1	OEM
424DF1AH1	OEM
424DF2AH1	OEM
424EF2AH1	OEM
424EF3AH1	OEM
424FB1AB1	OEM
424FB1AF1	OEM
424FF1AH1	OEM
424K	OEM
424M	OEM
424MB1AB1	OEM
424MB1AF1	OEM
424MC5AE1	OEM
424MC10AE1	OEM
424MC13AE1	OEM
424MD5AE1	OEM
424MD10AE1	OEM
424MD13AE1	OEM
424MF1AH1	OEM
424MH5AE1	OEM
424MH5AS1	OEM
424MH9AS1	OEM
424MH10AE1	OEM
424MH13AE1	OEM
424MH13AS1	OEM
424MH17AE1	OEM
424MH17AS1	OEM
424MH21AE1	OEM
424MH21AS1	OEM
424MH25AE1	OEM
424PFE1	OEM
424PFS1	OEM
425	OEM
425K	OEM
425M	OEM
425NH9AB1	OEM
425NH10AB1	OEM
425NH14AB1	OEM
425NH17AB1	OEM
425Q	OEM
425SH2AB1	0182
425SH3AB1	OEM
425SH4AB1	0348
425SH6AB1	OEM
425SH8AB1	OEM
425SH23AB1	OEM
425SH29AB1	OEM
425SH38AB1	OEM
425T	OEM
425T4	OEM
426	OEM
426-1	1966
426EH2AB1	0017
426EH3AB1	0017
426EH4AB1	OEM
426EH6AB1	OEM
426EH8AB1	OEM
426EH10AB1	OEM
426EH12AB1	OEM
426EH16AB1	OEM
426EH20AB1	OEM
427	OEM
427-1	0264
427-2	0566
427-3	0566
427N2	0264
428	OEM
428-1	0824
428-2	0824
428N2	0824
429	OEM
429-0092-1	0050
429-0092-2	0004
429-0092-3	0004
429-0092-56	0143
429-0093-69	0136
429-0093-71	0071
429-0094-39	0136
429-0910-50	0050
429-0910-51	0004
429-0910-52	0004
429-0910-53	0123
429-0910-54	0143

DEVICE TYPE	REPL CODE
429-0958-41	0277
429-0958-42	0016
429-0958-43	0636
429-0981-12	0127
429-0985-12	0127
429-0986-12	0144
429-0989-68	0015
429-10001-0A	0084
429-10002-0A	0097
429A	1023
429AR	0501
429B	1023
429BR	0501
429C	1023
429CR	0501
429D	1023
429DR	0501
429E	1023
429ER	0501
429F	1023
429FR	0501
429G	1036
429GR	1777
429H	1036
429HR	1777
429K	1036
429M	1036
429MR	1777
429RR	1777
430	0079
430-31	0015
430-85	0164
430-86	0016
430-87	0016
430-22861	0144
430-23190	0130
430-23212	0016
430-23221	0016
430-23222	0126
430-23223	0086
430-23843	0103
430-23844	0086
430-25762	0325
430-25763	0144
430-25764	0144
430-25765	0144
430-25766	0016
430-25767	0144
430-25834	0222
430-203845	0126
430CL	0016
430KF	OEM
430T4	OEM
431	4160
431-1B400-26	OEM
431-1B400-26R	OEM
431-1B400-115	OEM
431-1B400-115R	OEM
431-2A60-26	OEM
431-2A60-26R	OEM
431-2A60-26SR	OEM
431-2A60-115	OEM
431-2A60-115R	OEM
431-2A60-115SR	OEM
431-2A400-26	OEM
431-2A400-26R	OEM
431-2A400-26SR	OEM
431-2A400-115R	OEM
431-2A400-115SR	OEM
431-2B60-26	OEM
431-2B60-26R	OEM
431-2B60-26SR	OEM
431-2B60-115	OEM
431-2B60-115R	OEM
431-2B60-115SR	OEM
431-3A400-26R	OEM
431-3A400-115R	OEM
431-3B60-115	OEM
431-3B400-26	OEM
431-3B400-115	OEM
431-26551A	0701
431L	OEM
432	OEM
432-1	0086
433	0079
433(ZENITH)	0016
433-1	1270
433-1(SYLVANIA)	0079
433-2	1541
433-2,,	1541
433-4	OEM
433-5	OEM
433B	OEM
433CL	0016
433J	OEM
433M852	0126
434	0111
434-1	0037
434-1,,	0037
434-2	OEM
434-3	OEM
434-4	0037
434A	OEM
434B	OEM
434TS1	OEM
435-1	0079
435-1,,	0079
435-2	0079
435-2,,	0079
435-4	OEM
435-21026-0A	0232
435-21027-0A	0314
435-21028-0A	0361
435-21029-0A	0467
435-21030-0A	0512
435-21033-0A	0697
435-21034-0A	1168
435-21035-0A	1368
435-23006-0A	1172
435-23007-0A	1313
435-40012	0015
435A	1630
435B	0254
435C	1107
435D	1107
435F	1111
435H	0258
435J	OEM
435K	1644
435L	OEM
435M	0267
435P	2216
435S	2216
435V	2340
435Z	2340
436-10010-0A	0834
436-10011-0A	1715
436A	1630
436B	1107
436C	1107
436D	1107
436F	0258
436H	0258
436K	0267
436M	0267
436P	1119
436S	1119
436V	0281
436Z	0281
437-1	0261
437A	0534
437B	0748
437C	0538
437D	0538
437F	0777
437H	0538
437K	0801
437M	0816
437P	0131
437S	0131
437V	0145
437Z	0449
438-1	OEM
439A	1023
439AR	0501
439B	1023
439BR	0501
439C	1023
439CR	0501
439D	1023
439DR	0501
439E	1023
439ER	0501
439F	1023
439FR	0501
439G	1036
439GR	1777
439H	1036
439HR	1777
439K	1036
439KR	1777
439M	1036
439MR	1777
440	OEM
440C-C	OEM
440C-E	OEM
440KF	OEM
440T4	OEM
441	OEM
441-1	OEM
441A	0970
441B	3733
441C	0989
441D	0989
441F	3743
441H	0197
441K	0200
441M	0204
441P	0206
441S	0206
441V	0586
441Z	0586
442-1	0008
442-2	1705
442-5	0364
442-9	0443
442-10	2161
442-16	0701
442-18	0785
442-22	0312
442-24	2169
442-25	2222
442-28	0664
442-30	1922
442-30-2897	1922
442-33	0666
442-39	1300
442-48	OEM
442-50	OEM
442-54	0624
442-57	0352
442-58	0354
442-60	2239
442-63	1321
442-66	2246
442-71	2247
442-73	OEM
442-74	0372
442-75	2104
442-82	0666
442-96	0417
442-99	1143
442-602	0625
442-604	3729
442-612	3063
442-613	3829
442-615	0692
442-622	OEM
442-623	2782
442-624	OEM
442-630	1285
442-635	3972
442-636	4799
442-640	2748
442-644	1828
442-647	3428
442-648	2517
442-650	OEM
442-651	2863
442-654	3381
442-655	OEM
442-659	3920
442-663	0334
442-664	1838
442-665	4508
442-672	4106
442-673	4798
442-674	0334
442-675	1838
442-676	OEM
442-677	4727
442-682	3893
442-686	0379
442-687	OEM
442-688	3781
442-691	1196
442-693	OEM
442-702	1916
442-703	OEM
442-704	2564
442-705	2736
442-713	1048
442-715	2782
442-728	0770
442C-A	OEM
442C-D	OEM
442J	OEM
442K	OEM
442L	OEM
442R	0997
442TB1	OEM
442TB2	OEM
442TB3	OEM
442TB4	OEM
442TB5	OEM
442TB6	OEM
442TB7	OEM
442TB8	OEM
442TB9	OEM
442TB10	OEM
442TB11	OEM
442TB12	OEM
442TB13	OEM
442TB14	OEM
442TB15	OEM
442TB16	OEM
442TB17	OEM
443-1	0232
443-2	0697
443-3	0873
443-4	1426
443-5	1172
443-6	1313
443-7	1208
443-7-16088	1208
443-9	OEM
443-10	OEM
443-11	OEM
443-12	0512
443-13	1432
443-15	0743
443-16	1158
443-17	1377
443-18	0361
443-21	OEM
443-22	0175
443-23	1139
443-25	1495
443-26	0704
443-27	0358
443-32	0175
443-34	0834
443-35	1038
443-36	1108
443-42	1617
443-43	5523
443-44	1441
443-44-2854	1441
443-45	0467
443-46	0314
443-53	1052
443-54	0335
443-58	2398
443-59	5791
443-60	0498
443-61	1143
443-62	0011
443-65	0818
443-66	1921
443-67	2267
443-68	0685
443-70	2969
443-71	0682
443-72	1352
443-73	1349
443-74	1262
443-77	0996
443-87	0619
443-89	0492
443-90	1157
443-162	1926
443-603	0215
443-604	2846
443-606	1949
443-607	0413
443-612	1926
443-622	1451
443-623	0527
443-623-51644	0527
443-625	1271
443-628	1950
443-629	1208
443-640	0569
443-642	0385
443-680	1486
443-683	OEM
443-698	1368
443-703	0478
443-704	2055
443-706	0129
443-707	3600
443-708	2347
443-709	2177
443-711	1732
443-712	2072
443-713	2228
443-717	1193
443-718	1315
443-719	0592
443-720	3180
443-721	OEM
443-722	2267
443-727	1468
443-728	1527
443-729	1688
443-730	0243
443-731	4425
443-732	0828
443-733	0082
443-737	1043
443-738	3431
443-745	1579
443-751	0626
443-752	1672
443-753	OEM
443-754	0452
443-755	1595
443-757	0858
443-760	0056
443-762	4547
443-764	2048
443-777	3611
443-778	2384
443-779	1559
443-780	1633
443-781	1870
443-782	0209
443-783	1256
443-784	0500
443-785	3047
443-792	1625
443-794	0508
443-795	0511
443-797	1662
443-798	0034
443-799	1161
443-800	0183
443-801	2834
443-802	1744
443-804	3205
443-805	0894
443-807	1841
443-811	0075
443-813	1882
443-815	1692
443-816	1642
443-817	1689
443-818	1608
443-819	OEM
443-822	0153
443-824	1726
443-827	0388
443-828	1867
443-829	2180
443-836	1551
443-837	0709
443-839	0906
443-840	4181
443-841	1116
443-842	OEM
443-843	0448
443-854	3290
443-855	1779
443-856	OEM
443-857	0977
443-858	2243
443-860	OEM
443-863	0713
443-864	1667
443-871	1830
443-872	1698
443-874	OEM
443-875	0088
443-877	0427
443-879	0260
443-881	3481
443-884	1728
443-885	0463
443-886	0498
443-887	0519
443-888	3767
443-889	1564
443-891	0291
443-892	4371
443-893	1288
443-896	2238
443-897	2263
443-898	0780
443-899	2497
443-900	2504
443-904	0522
443-906	OEM
443-909	0521
443-912	3912
443-913	OEM
443-915	2511
443-916	3611
443-919	2877
443-920	0431
443-921	1288
443-927	3308
443-928	0988
443-929	0513
443-930	2331
443-934	0893
443-935	2151
443-942	0979
443-948	2128
443-951	1033
443-955	0959
443-958	0146
443-961	1763
443-964	1695
443-965	OEM
443-971	3412
443-972	2264
443-973	0819
443-974	5756
443-976	2569
443-979	2266
443-981	2138
443-982	2475
443-983	2141
443-999	3141
444-1	0055
444-1(IC)	2178
444-229-8	1638
444T20	OEM
444T21	OEM
444T22	OEM
444T23	OEM
444T24	OEM
444T26	OEM
444T28	OEM
444TB7	OEM
444TB9	OEM
444TB11	OEM
444TB13	OEM
444TB15	OEM
444TB17	OEM
444TB19	OEM
445-0028-P1	0725
445-0028-P2	0725
445-0300P3	0415
445-0303-001	1413
445-0303-002	1413
445-1	1541
445N1	OEM
446A	0985
446B	0989
446C	0990
446D	0990
446F	0997
446H	0997
446K	1001
446M	0515
446P	0206
446S	0206
446V	0586
446Z	0586
447	0079
447(ZENITH)	0016
447-1	0008
448A662	0086
449	0161
449-1	3523
450	OEM
450-1	0079
450-2	0111
450-3	OEM
450-4	OEM
450-5	OEM
450E15	OEM
450E120	OEM
450F15	OEM
450J	OEM
450K	OEM
450KF	OEM
450PF5	OEM
450PF10	OEM
450PF20	OEM
450PF30	OEM
450PF50	OEM
450PF60	OEM
450Pf40	OEM
450R1B	OEM
450R2B	OEM
450R3B	OEM
450R4B	OEM
450R5B	OEM
450R6B	OEM
450R7B	OEM
450R8B	OEM
450R9B	OEM
450R10B	OEM
450R11B	OEM
450R12B	OEM
450R13B	OEM
450R14B	OEM
450R15B	OEM
450R16B	OEM
450R18B	OEM
450R20B	OEM
450R22B	OEM
450R24B	OEM
450R26B	OEM
450R28B	OEM
450R30B	OEM
450R32B	OEM
450R34B	OEM
450R36B	OEM
450R38B	OEM
450R40B	OEM
450R42B	OEM
450R44B	OEM
450R46B	OEM
450R48B	OEM
450R50B	OEM
450T4	OEM
450VB5A	OEM
450VB10A	OEM
450VB15A	OEM
450VB20A	OEM
450VB25A	OEM
450VB30A	OEM
450VB40A	OEM
450VB50A	OEM
451	OEM
451-3	0037
451D	1250
451J	OEM
451K	OEM
451L	OEM
451PF5	OEM
451PF10	OEM
451PF20	OEM
451PF30	OEM
451PF40	OEM
451PF50	OEM
451PF60	OEM
453J	OEM
453K	OEM
453L	OEM
454-A2534-1	0015
454-A2534-10	0015
454A104	0086
454J	OEM
454K	OEM
454L	OEM
455-1	1283
456-1	OEM
456-2	0079
456J	OEM
456K	OEM
456T20	OEM
456T21	OEM
456T22	OEM
456T23	OEM
456T24	OEM
456TB7	OEM
456TB9	OEM
456TB11	OEM
456TB13	OEM
456TB15	OEM
456TB17	OEM
456TB19	OEM
458-1	OEM
458-3	0595
458J	OEM
458K	OEM
458L	OEM
458T20	OEM
458T21	OEM
458T22	OEM
458T23	OEM
458T24	OEM
458TB7	OEM
458TB9	OEM
458TB11	OEM
458TB13	OEM
458TB15	OEM
458TB17	OEM
458TB19	OEM
460	OEM
460-1	0037
460J	OEM
460K	OEM
460KF	OEM
460L	OEM
462-1	OEM
462-3	0423
462-11	OEM
462TB7	OEM
462TB9	OEM
462TB11	OEM
462TB13	OEM
464-1	3360
464-100-19	0143
464-103-19	0143
464-106-19	0143
464-110-19	0123
464-111-19	0143
464-113-19	0143
464-119-19	0123
464-280-19	0229
464-285-15	0015
464-285-19	0229
464-311-19	0170
464TB7	OEM
464TB9	OEM
464TB11	OEM
464.062.15	0636
465-005-19	0004
465-032-19	0050
465-036-19	0004
465-042-19	0050
465-045-19	0050
465-049-19	0050
465-061-19	0050
465-067-19	0164
465-072-19	0004
465-073-19	0004
465-075-19	0004
465-080-19	0004
465-082-19	0004
465-086-19	0050
465-1	OEM
465-1A5	0050
465-2A5	0208
465-4A5	0600
465-106-19	0016
465-108-19	0164
465-115-19	0004
465-132-19	0004
465-137-19	0222
465-146-19	0050
465-163-19	0004
465-165-19	0004
465-166-19	0822
465-181-19	0127
465-191-15	0004
465-199-19	0012
465-206-19	0222
465-223-19	0050
465-260-15	3545
465A-15	0050
465B-15	0050
465C-15	0160
466-1A5	0004
466-2A5	0004
466-3A5	0004
466TB7	OEM
466TB9	OEM
467C	0436
468-1	1708
469-1	OEM
470	OEM
470KF	OEM
470PB30	1765
470PB40	OEM
470PB50	1765
470PB60	1765
470PB80	1814
470PB100	1814
470PB110	1814
470PB120	5318
470PB130	5319
470PB140	5319
470PB150	5319
470PB160	5319
470PB170	OEM
470PB180	OEM
470PB190	OEM
470PB200	OEM
470PDA10	2159
470PDA20	2159
470PDA40	2159
470PDA60	2159
470PDA80	0479
470PDA100	0479
470PDAR10	OEM
470PDAR20	OEM
470PDAR40	OEM

If replacement code is OEM, contact original manufacturer for replacement.

DEVICE TYPE	REPL CODE
470PDAR60	OEM
471-010	0015
471-1	0285
471-1(IC)	OEM
471-1(METAL)	0086
471-1(PLASTIC)	0551
471PD140	OEM
471PD160	OEM
471PD180	OEM
471PD200	OEM
471PDA120	1204
471PDA140	OEM
471PDA150	OEM
471PDA160	OEM
471PDA180	OEM
471PDA200	OEM
471PDA210	OEM
471PDA220	OEM
471PDA230	OEM
471PDA240	OEM
471PDA250	OEM
472-1	OEM
472-H12	OEM
472US1	OEM
473-1	OEM
473A31	0160
473B5	0004
473B6-2	0211
473B6-2A	0211
473B6-4	0211
473B6-5	0211
473B6-7	0211
474	OEM
474-004	0015
474-025	0015
474-1	0079
474-3A5	0280
475	OEM
475-018	0242
475-1	0037
476-1	0439
476T1	OEM
476T2	OEM
476T3	OEM
476T4	OEM
476T5	OEM
476T6	OEM
476T7	OEM
476T8	OEM
476T9	OEM
476T10	OEM
476T11	OEM
476T12	OEM
477-054-5805	4526
477-0375-001	1041
477-0377-001	1823
477-0379-001	0562
477-0380-001	0081
477-0381-001	0141
477-0412-004	1172
477-0415-002	0743
477-0417-002	1185
477-1	1852
477-2	OEM
477A15	0440
478-1	OEM
478C	0097
478E	OEM
479-1	0715
479-1006-002	0764
479-1066	0764
479C	OEM
479E	OEM
479K	OEM
479M	OEM
480	OEM
480-1	0861
480-9(SEARS)	0222
480KF	OEM
482-1	0037
482-US1	OEM
482US2	OEM
484A15	0160
488B15	0050
488C15	0280
490	OEM
490-2(SEARS)	0144
491-2(SEARS)	0144
491A948	0086
492-1A5	0440
499	0079
499-1	0016
500-1	0079
500-4	0079
500B10	0071
500B15	OEM
500B20	OEM
500EXH21	OEM
500F	OEM
500HXD25	OEM
500P8H	OEM
500P12H	OEM
500P16H	OEM
500P20H	OEM
500P25H	OEM
500PA5	OEM
500PA10	OEM
500PA20	OEM
500PA30	OEM
500PA40	OEM
500PA50	OEM
500PA60	OEM
500R1B	OEM
500R2B	OEM
500R3B	OEM
500R4B	OEM
500R5B	OEM
500R6B	OEM
500R7B	OEM
500R8B	OEM
500R9B	OEM
500R10B	OEM
500R11B	OEM
500R12B	OEM
500R13B	OEM
500R14B	OEM
500R15B	OEM
500R16B	OEM
500R18B	OEM
500R20B	OEM
500R22B	OEM
500R24B	OEM
500R26B	OEM
500R28B	OEM
500R30B	OEM
500R32B	OEM
500R34B	OEM
500R36B	OEM
500R38B	OEM
500R40B	OEM
500R42B	OEM
500R44B	OEM
500R46B	OEM
500R48B	OEM
500R50B	OEM
500RS20	0348
500S2.8-3.2	OEM
500S10	0071
500S10H	OEM
500S12H	OEM
500S15	0182
500S20	0348
500SS10H	OEM
500SS12H	OEM
500V5A	OEM
500V10A	OEM
500V20A	OEM
500V30A	OEM
500V40A	OEM
500V50A	OEM
500V60A	OEM
501	OEM
501-363-2	0242
501ES001M	0144
501P	OEM
501PA5	OEM
501PA10	OEM
501PA20	OEM
501PA30	OEM
501PA40	OEM
501PA50	OEM
501PA60	OEM
501PBA60	OEM
501PBQ50	OEM
501PBQ70	OEM
501PBQ80	OEM
501PBQ90	OEM
501PBQ100	OEM
501PBQ110	OEM
501PBQ120	OEM
501T1	0050
501V60	5046
501V60B	5046
501V60S	OEM
501V80	0749
501V80B	0749
501V100	0749
501V100B	0749
501V120	0749
501V120B	0749
501V140	OEM
501V140B	OEM
501V160	OEM
501V180	OEM
501V200	OEM
501V210	OEM
501V220	OEM
501V230	OEM
501V240	OEM
501V250	OEM
501VF60B	OEM
501VF80B	OEM
501VF100B	OEM
501VF120	OEM
501VF120B	OEM
501VF140	OEM
501VF140B	OEM
501VF160	OEM
501VF180	OEM
501VF200	OEM
501VF210	OEM
501VF220	OEM
501VF230	OEM
501VF240	OEM
501VR60B	2315
501VR80B	2792
501VR100B	2792
501VR120	2792
501VR120B	2792
501VR140	OEM
501VR140B	OEM
501VR160	OEM
501VR180	OEM
501VR200	OEM
501VR210	OEM
501VR220	OEM
501VR230	OEM
501VR240	OEM
502	0079
502V5A	OEM
502V10A	OEM
502V20A	OEM
502V30A	OEM
502V40A	OEM
502V50A	OEM
502V60A	OEM
503	OEM
503-T10192	0087
503-T21271	0143
503-T21472	0143
503-T21651	0030
503PJA180	OEM
503PJA200	OEM
503PJA250	OEM
503PJLA60	OEM
503PJLA80	OEM
503PJLA100	OEM
503PJLA120	OEM
503T1	0050
503V60	OEM
503V60B	OEM
503V80	OEM
503V80B	OEM
503V100	OEM
503V100B	OEM
503V120	OEM
503V120B	OEM
503V140	OEM
503V140B	OEM
503V160	OEM
503V180	OEM
503V200	OEM
503V210	OEM
503V220	OEM
503V230	OEM
503V240	OEM
503V250	OEM
504T1	0050
505ES105	0847
505T1	0050
506T1	0050
507-4756-3331-500	OEM
507-4757-3331-500	5990
507-4758-3331-500	5990
507-4759-3331-500	OEM
507-4760-3331-500	OEM
507-4761-3331-500	OEM
507-4856-3332-500	OEM
507-4857-3332-500	2708
507-4858-3332-500	2708
507-4859-3332-500	OEM
507-4860-3332-500	OEM
507-4861-3332-500	OEM
507-4956-3333-500	OEM
507-4957-3333-500	2753
507-4958-3333-500	2753
507-4959-3333-500	OEM
507-4960-3333-500	OEM
507-4961-3333-500	OEM
507T1	0050
508	OEM
508-1(IC)	OEM
508ES020P	0150
508ES021P	0150
508T1	0050
509	0079
509-01	OEM
509-1	0313
509-10	OEM
509-50	OEM
509ES025P	0211
510A-90	0143
510A90	0143
510ED46	0143
510ES030M	0595
510ES031M	0595
510IN60	0143
510IS34	0143
510LN60	0143
510LS34	0143
511-1	2860
511-3	0766
511-515	0198
511-519	0198
511-898	0015
511ES035P	0595
511ES036P	0595
511S345	0143
512ES040P	0847
512M	OEM
512M11	OEM
512RED	0016
513-891	0484
513-1500-067	OEM
513ES045P	0847
514-048	1876
514-053	0723
515	0198
515-074	0698
515-299	0015
515-521	0144
515ES045M	0150
515ES046M	0150
516	0198
516ES047M	0150
516ES048M	0150
517-0021	0015
517-0025	0015
517-0031	0015
517-0033	0015
517-1	4450
517-3	OEM
517-518	0198
518	OEM
518-499	0015
519-1	0079
519-528A43	OEM
519ES067M	0079
519ES068M	0079
520ES070M	0150
521-081	0123
521-094	0015
521-145	0143
521-9165	OEM
521-9166	OEM
521-9173	OEM
521-9174	OEM
521-9175	OEM
521-9176	OEM
521-9177	OEM
521-9178	OEM
521-9180	OEM
521-9181	OEM
521-9183	OEM
521-9184	OEM
521-9185	1981
521-9186	1989
521-9189	3019
521-9190	OEM
521-9195	1989
521-9200	OEM
521-9206	0841
521-9207	3124
521-9208	OEM
521-9210	OEM
521-9211	OEM
521-9212	OEM
521-9214	OEM
521-9215	OEM
521-9216	OEM
521-9217	1358
521-9222	OEM
521-9223	OEM
521-9224	1778
521-9225	3157
521-9240	OEM
521-9246	1962
521-9247	OEM
521-9248	3157
521-9249	1962
521-9250	1778
521-9251	OEM
521-9253	OEM
521-9254	OEM
521-9256	1962
521-9258	3157
521-9260	1778
521-9264	OEM
521-9265	OEM
521-9266	OEM
521ES071M	0150
521T1	0280
522	0079
522(ZENITH)	0127
522-726	0015
522-893	0196
522-958	0250
522ES075M	0595
522ES076M	0595
523	OEM
523-0001-002	0015
523-0001-003	0015
523-0006-002	0124
523-0007-001	0124
523-0009-041	0554
523-0009-049	0030
523-006-002	0124
523-0013-002	0015
523-0013-201	0071
523-0017-001	0087
523-0501-002	0023
523-1000-001	0015
523-1000-067	0143
523-1000-295	0143
523-1000-326	2487
523-1000-881	0015
523-1000-882	0015
523-1000-883	0015
523-1002-326	6555
523-1500-067	0143
523-1500-803	0124
523-1500-881	0015
523-1500-883	0124
523-2001-100	0170
523-2003-001	0143
523-2003-100	0170
523-2003-150	0002
523-2003-519	0162
523-2004-100	0225
523-2005-100	0170
523-2005-279	1312
523-2503-100	0170
523-2503-150	0002
523-2503-519	0162
523-2503-689	0025
523-2509-100	0170
523-4001-001	0087
523AB1A11	OEM
523AB1AF1	OEM
523AB1AI1	OEM
523AC1AD1	OEM
523AC1AI1	OEM
523AD1AD1	OEM
523BB1A11	OEM
523BB1AF1	OEM
523BB1AI1	OEM
523BC1A11	OEM
523BC1AD1	OEM
523BC1AI1	OEM
523BD1AD1	OEM
523DB1A11	OEM
523DB1AF1	OEM
523DB1AI1	OEM
523DC1A11	OEM
523DC1AD1	OEM
523DC1AI1	OEM
523DD1AD1	OEM
523ES077M	0595
523ES078M	0595
523FB1A11	OEM
523FB1AF1	OEM
523FB1AI1	OEM
523FC1A11	OEM
523FC1AD1	OEM
523FC1AI1	OEM
523FD1AD1	OEM
523MB1A11	OEM
523MB1AF1	OEM
523MB1AI1	OEM
523MB2A11	OEM
523MB2AI1	OEM
523MC1A11	OEM
523MC1AD1	OEM
523MC1AI1	OEM
523MC2A11	OEM
523MC2AI1	OEM
523MD1AD1	OEM
523MH5AE1	OEM
523MH5AS1	OEM
523MH9AS1	OEM
523MH10AE1	OEM
523MH13AS1	OEM
523MH17AS1	OEM
523MH21AS1	OEM
523MH25AS1	OEM
523NB1A11	OEM
523NB1AF1	OEM
523NB1AI1	OEM
523NB2A11	OEM
523NB2AI1	OEM
523NC1A11	OEM
523NC1AD1	OEM
523NC1AI1	OEM
523NC2A11	OEM
523NC2AI2	OEM
523NC4AE1	OEM
523NC7AE1	OEM
523NC10AE1	OEM
523ND1AD1	OEM
523ND4AE1	OEM
523ND7AE1	OEM
523ND10AE1	OEM
523NH10AE1	OEM
523NH13AE1	OEM
523NH16AE1	OEM
523NH20AE1	OEM
523R	OEM
524-457	0143
524AB1AF1	OEM
524AC1AD1	OEM
524AD1AD1	OEM
524BB1AF1	OEM
524BC1AD1	OEM
524BD1AD1	OEM
524DB1AF1	OEM
524DC1AD1	OEM
524DD1AD1	OEM
524FB1AF1	OEM
524FC1AD1	OEM
524FD1AD1	OEM
524MB1AF1	OEM
524MC1AD1	OEM
524MC5AE1	OEM
524MC10AE1	OEM
524MD1AD1	OEM
524MD5AE1	OEM
524MD10AE1	OEM
524MH5AE1	OEM
524MH5AS1	OEM
524MH9AS1	OEM
524MH10AE1	OEM
524MH13AE1	OEM
524MH13AS1	OEM
524MH17AE1	OEM
524MH17AS1	OEM
524MH21AE1	OEM
524MH21AS1	OEM
524MH25AS1	OEM
524WHT	0016
525	OEM
525-24	0015
525-212	0071
525-498	0015
525-877	0143
525BF	0240
526	OEM
526-1	0532
526-376	0015
527-062	0196
527-1	OEM
527-798	0229
528-325	0229
528BSY	OEM
529-01-1	OEM
529-01-5	OEM
529-1	0224
529-2-1	OEM
529-2-5	OEM
530	0015
530-073-31	0157
530-082-3	0071
530-082-4	0071
530-082-1003	0229
530-086-1	0015
530-088-1	0015
530-088-2	0229
530-088-3	0071
530-088-1002	0071
530-088-1003	0071
530-088-1004	0071
530-094-1	0015
530-095-1	0015
530-098-1	0015
530-099-1	0015
530-099-3	0015
530-099-4	0015
530-106-1	0015
530-106-1001	0071
530-109-1	0015
530-111-1	0015
530-111-1001	0071
530-113-1	0015
530-120-1	0015
530-122-1	0229
530-136-T	0071
530-165-8	2982
531	OEM
531US	OEM
532-1	1511
532-3	1511
532-5	OEM
532-341	0015
532-341A	0015
532BXY	OEM
535-1	6066
536	OEM
536-1	0007
536-2	0007
536DP	0224
536FS	0016
536FU	0016
536GT	0111
536J2F	0015
537FS	0144
537J2F	0015
538J2F	0015
539-003-1	OEM
539-003-5	OEM
539-005-1	OEM
539-005-5	OEM
539-01-1	OEM
539-01-5	OEM
539J2F	0015
540-008	0540
540-010	0015
540-013	0100
540-014	0071
540-015	0242
540J2F	0015
540PB30	OEM
540PB50	OEM
540PB60	OEM
543H	OEM
544-0003-043	3971
544-1	OEM
544-2002-004	2617
544-2002-008	1696
544-2003-002	0789
544-2003-005	1321
544-2003-006	OEM
544-2004-001	4586
544-2006-001	1248
544-2006-011	1048
544-2006-11	3769
544-2007-001	0030
544-2008-001	OEM
544-2009-555	0973
544-2020-002	0625
544-3001-001	2867
544-3001-002	OEM
544-3001-003	OEM
544-3001-102	OEM
544-3001-103	0478
544-3001-117	0513
544-3001-140	0001
544-3001-143	1544
544-3001-201	3594
544-3001-213	OEM
544H	OEM
545-1	5034
545-2	OEM
545-3	OEM
545-5	OEM
545-6	OEM
546	0144
546-1	1597
547-2001	OEM
547-2002	OEM
547-2003	OEM
547-2004	OEM
547-2005	OEM
547-2006	OEM
547-2007	OEM
547-2008	OEM
547J2F	0071
548D	OEM
549-0101	OEM
549-0104	OEM
549-0201	OEM
549-0301	OEM
549-1	0150
549-2	0037
549-4001-002	OEM
550	OEM
550-0204	OEM
550-0205	OEM
550-0206	OEM
550-0206-004	OEM
550-0304	OEM
550-0305	OEM
550-0306	OEM
550-0306-004	OEM
550-0404	OEM
550-0405	OEM
550-0406	OEM
550-0406-004	OEM
550-0504	OEM
550-0505	OEM
550-0506	OEM
550-0604	OEM
550-0605	OEM
550-0606	OEM
550-2204	OEM
550-2205	OEM
550-2206	OEM
550-2304	OEM
550-2305	OEM
550-2306	OEM
550-2404	OEM
550-2405	OEM
550-2406	OEM
550-3004	OEM
550E	OEM
550H	OEM
550P4H	OEM
550P6H	OEM
550P8H	OEM
550P10H	OEM
550P12H	OEM
550P16H	OEM
550PA5	2843
550PA10	2843
550PA20	2843
550PA30	1765
550PA40	1765
550PA50	1765
550PA60	1765
550PB30	1765
550PB50	1765
550PB60	1765
550PB80	1814
550PB100	1814
550PB120	1814
550PB140	6834
550PB150	5319
550PB160	5319
550PB180	OEM
550R1B	OEM
550R2B	OEM
550R3B	OEM
550R4B	OEM
550R5B	OEM
550R6B	OEM
550R7B	OEM
550R8B	OEM
550R9B	OEM
550R10B	OEM
550R11B	OEM
550R12B	OEM
550R13B	OEM
550R14B	OEM
550R15B	OEM
550R16B	OEM
550R18B	OEM
550R20B	OEM
550R22B	OEM
550R24B	OEM
550R26B	OEM
550R28B	OEM
550R30B	OEM
550R32B	OEM
550R34B	OEM

If replacement code is OEM, contact original manufacturer for replacement.

DEVICE TYPE	REPL CODE
550R36B	OEM
550R38B	OEM
550R40B	OEM
550R42B	OEM
550R44B	OEM
550R46B	OEM
550R48B	OEM
550R50B	OEM
551	0144
551-0002	0273
551-006	0723
551-010-00	0141
551-013-00	0081
551-0205	OEM
551-0305	OEM
551-0405	OEM
551-0505	OEM
551A	1876
551B	OEM
551E	OEM
551F	OEM
551H	OEM
551M	0224
551PA5	OEM
551PA10	OEM
551PA20	OEM
551PA30	OEM
551PA40	OEM
551PA50	OEM
551PA60	OEM
552	0845
552E	OEM
552F	OEM
552H	OEM
553	OEM
553E	OEM
553F	OEM
553H	OEM
554-1	0261
554E	OEM
554F	OEM
554H	OEM
555	0973
555-3	0086
555-2001	OEM
555-2002	OEM
555-2003	OEM
555-2004	OEM
555-2005	OEM
555-2006	OEM
555-2007	OEM
555-2008	OEM
555-2301	OEM
555-2303	OEM
555-2401	OEM
555-3001	OEM
555-3002	OEM
555-3003	OEM
555-3004	OEM
555-3005	OEM
555-3006	OEM
555-3007	OEM
555-3008	OEM
555-4001	OEM
555-4003	OEM
555-4007	OEM
555-4301	OEM
555-4303	OEM
555-4401	OEM
555-4403	OEM
555D	0973
555H	OEM
555PB	OEM
556	3285
556-1	OEM
556D	3285
556E	OEM
556F	OEM
556H	OEM
558-0101-001	OEM
558-0101-003	OEM
558-0101-004	OEM
558-0101-005	OEM
558-0101-006	OEM
558-0101-007	OEM
558-0102-001	OEM
558-0102-003	OEM
558-0102-004	OEM
558-0102-005	OEM
558-0102-006	OEM
558-0102-007	OEM
558-0201-001	OEM
558-0201-003	OEM
558-0201-004	OEM
558-0201-005	OEM
558-0201-006	OEM
558-0201-007	OEM
558-0202-001	OEM
558-0202-003	OEM
558-0202-004	OEM
558-0202-005	OEM
558-0202-006	OEM
558-0202-007	OEM
558-0301-001	OEM
558-0301-003	OEM
558-0301-004	OEM
558-0301-005	OEM
558-0301-006	OEM
558-0301-007	OEM
558-0302-001	OEM
558-0302-003	OEM
558-0302-004	OEM
558-0302-005	OEM
558-0302-006	OEM
558-0302-007	OEM
558CP	3798
558E	OEM
558F	OEM
558H	OEM
559-0101-001	OEM
559-0101-003	OEM
559-0101-004	OEM
559-0101-005	OEM
559-0101-006	OEM
559-0101-007	OEM
559-0102-001	OEM
559-0102-003	OEM
559-0102-004	OEM
559-0102-005	OEM
559-0102-006	OEM
559-0102-007	OEM
559-0103-001	OEM
559-0103-003	OEM
559-0103-004	OEM
559-0103-005	OEM
559-0103-006	OEM
559-0103-007	OEM
559-0201-001	OEM
559-0201-003	OEM
559-0201-004	OEM
559-0201-005	OEM
559-0201-006	OEM
559-0201-007	OEM
559-0202-001	OEM
559-0202-003	OEM
559-0202-004	OEM
559-0202-006	OEM
559-0202-007	OEM
559-0301-001	OEM
559-0301-003	OEM
559-0301-004	OEM
559-0301-005	OEM
559-0301-006	OEM
559-0301-007	OEM
559-0302-001	OEM
559-0302-003	OEM
559-0302-004	OEM
559-0302-005	OEM
559-0302-006	OEM
559-0302-007	OEM
559-1	0037
559-1492-001	1364
559-1493-001	3337
559-1494-001	3901
559-1495-001	3850
559-1496-001	1257
559-2101-001	OEM
559-2101-003	OEM
559-2101-004	OEM
559-2101-005	OEM
559-2101-006	OEM
559-2101-007	OEM
559-2201-001	OEM
559-2201-003	OEM
559-2201-004	OEM
559-2201-005	OEM
559-2201-006	OEM
559-2201-007	OEM
559-2301-001	OEM
559-2301-003	OEM
559-2301-004	OEM
559-2301-005	OEM
559-2301-006	OEM
559-2301-007	OEM
559-3001-001	OEM
559-3001-003	OEM
559H	OEM
560	OEM
560-0101	OEM
560-0102	OEM
560-0103	OEM
560-0104	OEM
560-0105	OEM
560-0106	OEM
560-0107	OEM
560-0108	OEM
560-0201	OEM
560-0202	OEM
560-0203	OEM
560-0204	OEM
560-0205	OEM
560-0206	OEM
560-0207	OEM
560-0208	OEM
560-0301	OEM
560-0302	OEM
560-0304	OEM
560-0305	OEM
560-0306	OEM
560-0307	OEM
560-0308	OEM
560-0401	OEM
560-0402	OEM
560-0403	OEM
560-0404	OEM
560-0405	OEM
560-0406	OEM
560-0407	OEM
560-0408	OEM
560-0501	OEM
560-0502	OEM
560-0503	OEM
560-0504	OEM
560-0505	OEM
560-0506	OEM
560-0507	OEM
560-0508	OEM
560-0601	OEM
560-0602	OEM
560-0603	OEM
560-0604	OEM
560-0605	OEM
560-0606	OEM
560-0607	OEM
560-0608	OEM
560-2	0016
560E	OEM
560F	OEM
560H	OEM
561	OEM
561-0884-006	1800
561-0884-041	1975
561-0884-043	0312
561-0884-047	2358
561-1	1511
561-2	OEM
561-4	OEM
561-2511-106	0143
561-2511-120	0143
561-2511-190	0143
561-2511-210	0143
561-2511-240	0143
561-2511-290	0143
561-2511-400	0143
561-2511-503	0143
561-2511-510	0143
561-2511-530	0143
561-2511-535	0015
561-2511-538	0143
561-2511-540	0143
561-2511-544	0143
561-2511-555	0143
561-2511-558	0143
561-2511-564	0104
561-2511-570	0143
561-2511-575	0143
561-2511-580	0133
561-2511-700	0036
561-2511-900	0062
561-2554-560	0133
561-2554-570	0133
561-2554-571	0133
561-2554-577	0133
561-2554-578	0133
561-2554-610	0133
561-2554-620	0133
561-2554-820	0133
561-2554-821	0133
561-2554-824	0133
561-2554-830	0133
561-2554-841	0133
561-2554-842	0133
561-2554-850	0133
561-2554-880	0102
561-2556-250	0133
561-2556-270	0133
561-2556-290	0133
561-2556-360	0143
561-2556-450	0015
561-2556-590	0133
561-2556-620	0133
561-2556-950	0143
561-2557-032	0376
561-2557-042	0036
561-2557-050	0140
561-2557-060	0091
561-2557-062	0205
561-2557-070	0062
561-2557-092	0636
561-2557-142	0170
561-2557-152	0247
561-2557-161	0137
561-2557-191	0420
561-2560-460	0299
561-2560-461	0292
561-2560-480	0456
561-2560-491	0533
561-2560-501	0451
561-2560-520	0157
561-2560-522	0157
561-2560-530	0471
561-2560-531	0636
561-2560-680	0671
561-2560-754	0025
561-2560-755	0077
561-2560-756	0244
561-2560-757	0012
561-2560-758	0170
561-2560-816	0301
561-2560-820	0636
561-2560-823	0636
561-2560-825	0636
561-2560-914	0124
561-2560-915	0133
561-2560-935	0012
561-2560-957	0025
561-2560-958	0649
561-2560-961	0170
561-2560-962	0170
561-2560-963	0137
561-2560-964	0441
561-2560-966	0375
561-2560-967	0375
561-2560-969	0565
561-2560-970	0402
561-2560-971	0441
561-2561-313	0322
561-2561-321	0032
561-2561-322	0068
561-2561-358	1717
561-2561-359	1723
561-2561-370	0263
561-2561-590	1901
561-2561-604	0992
561-2561-766	0636
561-2561-776	0420
561-2561-780	0531
561-2561-788	0502
561-2568-191	1078
561-2568-280	1194
561-2568-300	0132
561-2568-312	0207
561-2573-260	3683
561-2579-707	1601
561-2580-003	0736
561-2580-040	0438
561-2580-152	0668
561-2580-392	0783
561-2583-110	0889
561-2583-141	0932
561-2583-606	0133
561-2583-608	0133
561-2583-791	0234
561-2583-793	0234
561-2588-092	0405
561-2588-185	0533
561-2588-930	0442
561-2589-930	0547
561-2589-971	1901
561-2590-090	0133
561-2591-480	0124
561-2591-540	0124
561-2593-701	1276
561-2594-480	0124
561-2597-180	0256
561-2597-280	0292
561-2597-291	0188
561-2597-310	0533
561-2597-321	0451
561-2597-341	0157
561-2597-350	0636
561-2597-351	0636
561-2597-361	0025
561-2597-381	0244
561-2597-383	0012
561-2597-420	0137
561-2597-421	0137
561-2597-440	0002
561-2597-441	0002
561-2597-464	0495
561-2597-471	0531
561-2597-570	0502
561-2598-000	0299
561-2598-250	0041
561-2599-340	0102
561-2599-360	0102
561-2600-230	0012
561-2602-210	1276
561-2602-300	0041
561-2602-310	0012
561-2602-350	0025
561-2602-430	0365
561-2602-440	0100
561-2603-383	0442
561-2603-530	0251
561-2654-756	0091
561-2706-000	0102
561-2707-770	0124
561-2709-560	0869
561-4800-350	0154
561-4800-402	0154
561-4800-403	0147
561-4800-425	0576
561-4800-440	2353
561-4800-442	2353
561-4800-443	2356
561-4800-445	2017
561-4800-446	0576
561-4800-448	2353
561-4800-452	0154
561-4800-502	0154
561-4800-520	4233
561-5004-412	0015
561-5005-360	0015
561-5005-380	0015
561-5005-390	0015
561-5005-400	0015
561-5005-470	0015
561-5005-890	0133
561-5006-460	0015
561-5006-470	0015
561-5008-810	0015
561-5010-820	0015
561-5010-840	0015
561-5010-950	0015
561-5010-960	0015
561-5011-190	2081
561-5011-200	2081
561-5011-842	1124
561-5011-850	1124
561-5011-860	1124
561-5011-870	0805
561-5011-876	0258
561-5011-880	0805
561-5011-900	0319
561-5011-902	0319
561-5011-970	0319
561-5011-976	1107
561-5012-021	0097
561-5012-027	0765
561-5012-042	0109
561-5012-180	0015
561-5012-210	0015
561-5012-220	0015
561-5013-422	0989
561-5013-432	0989
561-5015-660	0540
561-5015-662	0015
561-5015-860	2081
561-5016-140	0989
561-5016-160	0515
561-5016-920	0015
561-5016-930	0015
561-5016-940	0015
561-5016-950	0015
561-5016-960	0015
561-5017-640	0015
561-5020-690	0015
561-5020-700	0015
561-5020-701	0015
561-5020-710	0015
561-5021-292	2657
561-5021-576	0258
561-5021-600	0319
561-5024-830	0015
561-5026-100	0015
561-5026-130	0015
561-5031-930	0015
561-5031-940	0015
561-5031-950	0015
561-5032-080	2002
561-5032-086	2003
561-5032-090	1250
561-5032-140	1586
561-5032-530	0015
561-5032-770	0015
561-5032-790	0015
561-5032-890	1347
561-5032-891	0599
561-5032-930	1986
561-5034-951	0133
561-5037-400	1036
561-5037-680	0044
561-5040-010	0015
561-5040-020	0015
561-5040-030	0015
561-5040-040	0015
561-5040-050	0015
561-5040-060	0071
561-5040-070	0071
561-5040-440	1023
561-5040-480	1023
561-5040-600	0015
561-5040-821	0540
561-5040-830	0015
561-5040-850	0015
561-5045-270	0319
561-5047-210	0242
561-5053-970	0965
561-5106-830	0567
561-5106-880	0740
561-5106-890	0764
561-5106-900	0764
561-5106-920	0766
561-5117-750	0676
561-5138-980	0740
561-5138-990	0725
561-5141-720	0676
561-5144-420	2270
561-5144-440	3404
561-5150-600	1137
561-5150-610	0344
561-5150-620	0901
561-6160-368	1600
561-6160-550	0277
561-6160-650	0323
561-6160-680	0344
561-6161-090	0953
561-6162-030	0240
561-6162-246	0289
561-6162-247	0087
561-6162-248	0940
561-6162-250	0240
561-6162-252	0015
561-6162-295	0469
561-6163-025	0015
561-6163-047	0805
561-6163-050	0740
561-6163-070	0805
561-6163-072	0258
561-6163-080	2850
561-6163-094	0740
561-6164-002	0473
561-6164-003	0015
561-6164-025	0015
561-6164-160	0687
561-6164-180	0109
561-6164-185	0853
561-6164-250	0805
561-6164-252	0722
561-6164-253	0267
561-6164-255	2850
561-6164-275	0015
561-6164-280	0015
561-6164-295	1644
561-6164-297	2871
561-6164-410	1609
561-6164-420	1205
561-6165-080	0336
561-6165-085	0740
561-6166-010	0015
561-6166-020	0965
561-6168-800	3658
561-6561-860	0015
561-6800-035	0208
561-6800-109	0004
561-6800-123	0280
561-6800-167	0600
561-6800-169	0600
561-6800-170	0600
561-6800-174	0440
561-6800-176	0085
561-6800-188	0211
561-6800-212	0208
561-6800-214	0208
561-6800-215	0004
561-6800-217	0004
561-6800-228	0208
561-6800-241	0211
561-6800-242	0085
561-6800-250	0085
561-6800-270	0211
561-6800-277	0440
561-6800-301	0160
561-6800-327	0280
561-6800-335	0198
561-6800-336	0198
561-6800-337	0127
561-6800-362	0211
561-6800-376	0085
561-6800-380	0085
561-6800-396	0280
561-6800-404	0004
561-6800-405	0211
561-6800-406	0004
561-6800-414	0280
561-6800-441	0440
561-6800-444	0038
561-6800-456	0085
561-6800-466	0211
561-6800-472	0086
561-6800-482	0016
561-6800-484	0016
561-6800-495	0016
561-6800-497	0086
561-6800-503	0280
561-6800-508	0211
561-6800-519	0280
561-6800-520	0280
561-6800-524	0211
561-6800-526	0211
561-6800-541	0198
561-6800-555	0160
561-6800-586	0280
561-6800-591	0004
561-6800-597	0280
561-6800-600	0280
561-6800-652	0211
561-6800-656	0086
561-6800-657	0086
561-6800-662	0847
561-6800-669	0160
561-6800-679	0600
561-6800-685	0600
561-6800-687	0740
561-6800-696	0198
561-6800-697	0086
561-6800-700	0050
561-6800-706	0144
561-6800-718	0144
561-6800-741	0050
561-6800-914	0144
561-6800-916	0144
561-6800-929	0016
561-6800-930	0016
561-6801-007	0211
561-6801-008	0211
561-6801-034	0126
561-6801-038	0847
561-6801-040	0085
561-6801-100	0280
561-6801-102	0208
561-6801-136	0160
561-6801-175	0211
561-6801-178	0050
561-6801-182	0847
561-6801-183	0847
561-6801-184	0847
561-6801-185	0280
561-6801-301	0050
561-6801-303	0164
561-6801-304	0600
561-6801-305	0164
561-6801-308	0600
561-6801-309	0280
561-6801-311	0600
561-6801-363	0160
561-6801-365	0160
561-6801-370	0211
561-6801-372	0211
561-6801-373	0211
561-6801-378	0211
561-6801-379	0211
561-6801-381	0211
561-6801-404	0211
561-6801-415	0211
561-6801-420	0086
561-6801-502	0085
561-6801-537	0160
561-6801-542	0604
561-6801-544	0604
561-6801-545	0604
561-6801-556	0604
561-6801-557	0604
561-6801-592	0127
561-6801-605	0600
561-6801-606	0600
561-6801-613	0086
561-6801-623	0126
561-6801-666	0160
561-6801-670	1669
561-6801-711	0086
561-6801-755	0085
561-6801-922	0126
561-6801-926	0004
561-6801-990	0086
561-6801-997	0280
561-6801-998	0280
561-6802-076	0440
561-6802-078	0440
561-6802-137	0085
561-6802-147	0085
561-6802-153	0437
561-6802-156	0437
561-6802-188	0050
561-6802-218	0198
561-6802-221	0016
561-6802-223	0016
561-6802-270	0086
561-6802-369	0016
561-6802-375	0211
561-6802-382	0050
561-6802-484	0016
561-6802-501	0016
561-6802-564	0211
561-6802-646	2136
561-6802-714	0127
561-6802-869	0160
561-6802-904	0126
561-6802-905	0126
561-6802-907	0037
561-6802-908	0037
561-6802-923	0111
561-6802-924	0111
561-6802-925	0111
561-6802-996	0050
561-6803-011	0016
561-6803-012	0037
561-6803-013	0016
561-6803-053	0086
561-6803-054	0178
561-6803-055	0103
561-6803-115	0016
561-6803-133	0126
561-6803-135	0037
561-6803-214	0160
561-6803-215	0160
561-6803-250	0037
561-6803-300	0493
561-6803-371	0050
561-6803-391	0111
561-6803-393	0111
561-6803-397	0111
561-6803-403	0218
561-6803-416	0155
561-6803-417	0155
561-6803-440	0187
561-6803-441	1027
561-6803-442	0543
561-6803-448	0103
561-6803-460	0325
561-6803-502	0126
561-6803-504	0037
561-6803-565	0079
561-6803-566	0079
561-6803-567	0324
561-6803-568	0086
561-6803-569	0086
561-6803-585	0142
561-6803-638	0037
561-6803-639	0037
561-6803-644	0037
561-6803-645	0126
561-6803-646	0079
561-6803-702	0037
561-6803-703	0037
561-6803-705	0324
561-6803-711	0111
561-6803-715	0103
561-6803-721	0127
561-6803-732	0975
561-6803-739	0142
561-6803-766	0178

If replacement code is OEM, contact original manufacturer for replacement.

DEVICE TYPE	REPL CODE
561-6803-771	0130
561-6803-772	0130
561-6803-773	0543
561-6803-789	0491
561-6803-819	0325
561-6803-822	3133
561-6803-858	0111
561-6803-902	0363
561-6803-903	0079
561-6803-904	0079
561-6803-905	0037
561-6803-906	0037
561-6804-030	0126
561-6804-037	0126
561-6804-092	1155
561-6804-123	0079
561-6804-124	0079
561-6804-125	0037
561-6804-126	0037
561-6804-234	0126
561-6804-235	0126
561-6804-237	0086
561-6804-240	0142
561-6804-249	0037
561-6804-275	0144
561-6804-348	0177
561-6804-356	0037
561-6804-360	2987
561-6804-387	0925
561-6804-389	0925
561-6804-400	0079
561-6804-401	0079
561-6804-402	0037
561-6804-403	0037
561-6804-424	0155
561-6804-851	2136
561-6804-855	1155
561-6804-905	2013
561-6804-921	0161
561-6804-946	0086
561-6805-190	0161
561-6805-195	0460
561-6805-240	0363
561-6805-247	0325
561-6805-294	0042
561-6805-296	0042
561-6805-302	0130
561-6805-308	0400
561-6805-320	0086
561-6805-322	0126
561-6805-354	0037
561-6805-415	0439
561-6805-485	1392
561-6805-486	0325
561-6805-670	0177
561-6805-830	0715
561-6806-027	0316
561-6806-028	0316
561-6806-307	0363
561-6901-000	0280
561-6901-006	0600
561-6901-008	0038
561-6901-010	0975
561-6901-110	0975
561-6901-410	0363
561-6901-411	1966
561-6901-440	0065
561-6901-473	0791
561-6901-527	0273
561-6901-826	2427
561-6901-850	0178
561-6902-105	1199
561-6902-521	0161
561-6936-220	0103
561-6940-251	0178
561-6940-347	0086
561-6952-923	0016
561-6953/020	0016
561-6953-003	0004
561-6953-004	0004
561-6953-006	0050
561-6953-007	0050
561-6953-009	0160
561-6953-010	0208
561-6953-011	0600
561-6953-020	0016
561-6953-021	0142
561-6953-024	0086
561-6953-025	0126
561-6953-041	0042
561-6954-030	0681
561-6954-099	0127
561-6955-935	0325
561-6956-373	0975
561-6956-515	0155
561-6956-531	0127
561-6956-534	0037
561-6956-550	0324
561-6956-553	0400
561-6956-555	0037
561-6956-602	0551
561-6956-651	0382
561-6956-652	0382
561-6956-740	2258
561-6956-942	0560
561-6958-405	0382
561C	0324
561E	OEM
561F	OEM
561H	OEM
562	OEM
562-1	0143
562E	OEM
562F	OEM
562H	OEM
564HD	OEM
564HDS	OEM
565	3381
565-072	0076
565-073	2046
565-074	0016
565-075	1411
565-076	1411
565-1	0086
567	3781
568H	OEM
569-0320-809	0410
569-0320-813	0692
569-0320-814	0358
569-0320-815	1929
569-0320-818	1831
569-0320-820	0141
569-0320-821	0562
569-0320-822	0141
569-0320-826	1831
569-0320-831	0026
569-0540-355	0973
569-0540-358	0360
569-0540-623	0026
569-0541-000	0618
569-0541-006	2992
569-0541-014	3066
569-0541-018	2434
569-0541-024	0078
569-0541-030	4164
569-0541-031	4170
569-0541-036	4175
569-0541-037	4178
569-0541-038	3045
569-0541-041	4187
569-0541-042	4191
569-0541-043	4195
569-0542-744	0011
569-0544-300	1823
569-0544-304	1041
569-0544-305	2097
569-0544-308	1831
569-0544-312	1176
569-0544-318	0033
569-0544-322	0141
569-0544-328	0333
569-0544-348	0562
569-0690-300	2687
569-0690-301	1696
569-0690-500	0478
569-0690-501	0215
569-0690-502	0413
569-0690-550	0398
569-0690-681	2749
569-0773-600	0973
569-0773-610	0360
569-0773-800	0232
569-0773-811	0527
569-0880-430	1823
569-0880-444	0033
569-0880-445	0081
569-0880-446	0141
569-0880-450	2619
569-0880-458	0466
569-0880-462	0562
569-0881-801	5338
569-0881-804	1907
569-0881-810	0685
569-0881-811	2398
569-0881-820	3715
569-0881-821	4856
569-0881-822	4596
569-0881-830	5383
569-0881-840	0559
569-0881-850	1792
569-0881-851	1944
569-0881-852	2020
569-0881-853	2101
569-0881-855	3158
569-0881-860	5410
569-0881-874	2492
569-0881-908	0180
569-0882-600	0232
569-0882-601	0268
569-0882-602	0314
569-0882-603	0335
569-0882-604	0361
569-0882-605	0385
569-0882-606	1206
569-0882-607	1339
569-0882-608	0467
569-0882-609	0492
569-0882-610	0512
569-0882-612	2242
569-0882-613	1441
569-0882-614	2243
569-0882-616	1349
569-0882-617	1352
569-0882-620	0697
569-0882-625	3477
569-0882-627	0818
569-0882-630	0873
569-0882-632	0899
569-0882-633	4193
569-0882-637	3520
569-0882-638	0996
569-0882-640	1024
569-0882-641	1038
569-0882-642	1052
569-0882-647	1108
569-0882-650	0743
569-0882-651	1168
569-0882-653	1185
569-0882-654	1202
569-0882-660	1275
569-0882-672	1426
569-0882-673	1172
569-0882-674	1313
569-0882-675	1432
569-0882-676	1158
569-0882-680	1535
569-0882-682	1574
569-0882-685	0374
569-0882-686	1368
569-0882-689	5455
569-0882-690	1208
569-0882-692	0834
569-0882-693	0569
569-0882-696	1715
569-0882-707	0942
569-0882-709	0968
569-0882-721	0175
569-0882-722	1139
569-0882-723	1157
569-0882-726	1193
569-0882-732	1271
569-0882-741	1377
569-0882-745	0619
569-0882-750	1492
569-0882-751	1495
569-0882-753	1539
569-0882-754	1555
569-0882-755	1576
569-0882-760	1631
569-0882-761	1645
569-0882-764	0734
569-0882-765	1685
569-0882-774	1770
569-0882-775	1787
569-0882-776	1795
569-0882-777	1803
569-0882-779	1820
569-0882-780	1829
569-0882-781	1842
569-0882-782	1856
569-0882-790	1912
569-0882-791	1917
569-0882-792	1921
569-0882-793	1926
569-0882-796	1950
569-0882-797	1956
569-0883-200	0704
569-0883-203	2217
569-0883-204	2263
569-0883-205	2321
569-0883-210	2445
569-0883-211	2447
569-0883-215	2451
569-0883-220	1017
569-0883-222	2461
569-0883-240	2475
569-0883-264	2496
569-0883-265	2497
569-0883-274	2504
569-0883-312	1617
569-0883-338	2138
569-0883-340	1886
569-0883-353	2151
569-0883-357	1695
569-0883-358	2154
569-0883-374	2132
569-0885-850	1231
569-0885-853	1272
569-0885-854	1262
569-0886-031	1940
569-0886-800	0243
569-0964-001	1705
569-0964-002	1300
569-0965-006	1039
569-1000-020	0876
569-2142-100	0015
569-8800-090	0042
569-9440-102	4987
570-004503	0016
570-005503	0016
570-1	0037
570H	OEM
571-844	0103
571-844A	0103
572-0040-051	0160
572-1	0079
572-683	0016
573-469	0016
573-472	0144
573-474	0144
573-474A	0144
573-475	0144
573-479	0016
573-480	0016
573-481	0016
573-491	0144
573-494	0144
573-495	0144
573-507	0144
573-509	0144
573-515	0142
573-518	0050
573-529	0004
573-532	0086
574	0037
574-1	0150
574-844	0074
575-028	0015
575-042	0015
575-048	0015
575-050	0015
576-0001-002	0037
576-0001-002(SEARS)	0004
576-0001-003	0004
576-0001-004	0198
576-0001-005	0198
576-0001-006	0144
576-0001-008	0016
576-0001-009	0004
576-0001-012	0016
576-0001-013	0037
576-0001-014	0079
576-0002-001	2997
576-0002-002	0160
576-0002-003	0103
576-0002-004	0004
576-0002-005	0160
576-0002-006	0016
576-0002-008	0037
576-0002-009	0440
576-0002-010	0440
576-0002-011	0161
576-0002-012	0004
576-0002-013	0208
576-0002-026	2997
576-0002-029	2997
576-0003-001	0144
576-0003-002	0144
576-0003-003	0144
576-0003-004	0144
576-0003-005	0144
576-0003-006	0144
576-0003-007	0144
576-0003-008	0136
576-0003-009	0136
576-0003-010	0050
576-0003-011	0079
576-0003-012	0150
576-0003-013	0136
576-0003-014	0136
576-0003-015	0136
576-0003-017	0037
576-0003-018	0144
576-0003-019	0037
576-0003-020	0144
576-0003-021	0144
576-0003-022	0111
576-0003-023	0007
576-0003-024	0136
576-0003-026	0007
576-0003-027	0144
576-0003-028	0144
576-0003-029	0127
576-0003-12	1381
576-0003-224	0212
576-0004-004	0325
576-0004-005	0630
576-0004-006	2354
576-0004-007	2700
576-0004-008	0493
576-0004-009	2700
576-0004-010	0016
576-0004-011	2700
576-0004-012	2700
576-0004-013	0698
576-0004-021	2312
576-0004-035	0264
576-0004-035(IC)	4003
576-0004-051	1198
576-0004-053	2312
576-0004-104	2700
576-0004-105	0936
576-0005-001	1669
576-0005-004	0086
576-0006-003	0325
576-0006-011	0144
576-0006-221	0353
576-0006-222	0212
576-0006-227	0212
576-0007-001	0400
576-001	0004
576-001-013	0150
576-002-001	0560
576-002-004	0211
576-003-009	0050
576-004-051	1198
576-005	0004
576-0036-212	0698
576-0036-213	0698
576-0036-847	0198
576-0036-913	0698
576-0036-916	0016
576-0036-917	0016
576-0036-918	0144
576-0036-919	0144
576-0036-920	0016
576-0036-921	0016
576-0040-051	0160
576-0040-251	0103
576-0040-253	0004
576-0040-254	0160
576-1	0330
576-14	0379
576-2000-278	0440
576-2000-990	0136
576-2000-993	0136
576-2001-970	0440
577B196	0576
577R819H01	0086
579-0001-009	0280
580-029	0015
580H	OEM
581B83-7	2018
586-2	0016
586-151	1823
586-152	1831
586-153	0141
586-154	0333
586-155	0562
586-187	0081
586-303	1041
586-308	1835
586-331	0945
586-412	1039
586-415	1433
586-425	0358
586-442	0466
586-517	0919
586-528	0888
586-546	0850
586-547	0982
586-780	1011
587-033	1722
588-40-202	1478
588U	0325
589H	OEM
590	1077
591	OEM
592	OEM
592-027	1257
592-028	1364
592-029-0	1069
592-032-0	OEM
592-032-1	OEM
592-079-0	OEM
592-081-0	1069
594-1	OEM
595-1	0016
595-1(SYLVANIA)	0016
595-2	0016
595-2(SYLVANIA)	0016
596-2	0133
596-5	0133
597-1	0150
597-2	OEM
598H	OEM
600	0015
600/8EDA-XX01-4	OEM
600/11DA-1A01-4	OEM
600/11DA-1C01-4	OEM
600/11DA-1D01-4	OEM
600/11DA-2A01-4	OEM
600/11DA-2B01-4	OEM
600/11DA-2C01-4	OEM
600/11DA-2D01-4	OEM
600/11DA-3A01-4	OEM
600/11DA-3B01-4	OEM
600/11DA-3C01-4	OEM
600/11DA-3D01-4	OEM
600/11DA-4A01-4	OEM
600/11DA-4B01-4	OEM
600/11DA-4C01-4	OEM
600/11DA-4D01-4	OEM
600-104-308	1062
600-207-801	0076
600-224-605	1221
600-229-201	0191
600-301-801	1419
600C	0133
600HT1U	3745
600HT2U	3745
600HT3U	3745
600HT4U	3745
600HT6U	3745
600HT8U	3782
600HT10U	5749
600PBQ120	OEM
600PBQ130	OEM
600PBQ140	OEM
600PBQ150	OEM
600PBQ160	OEM
600PBQ170	OEM
600PBQ180	OEM
600PIN-RM	OEM
600PV-RM	OEM
600R1B	OEM
600R2B	OEM
600R3B	OEM
600R4B	OEM
600R5B	OEM
600R6B	OEM
600R7B	OEM
600R8B	OEM
600R9B	OEM
600R10B	OEM
600R11B	OEM
600R12B	OEM
600R13B	OEM
600R14B	OEM
600R15B	OEM
600R16B	OEM
600R18B	OEM
600R20B	OEM
600R22B	OEM
600R24B	OEM
600R26B	OEM
600R28B	OEM
600R30B	OEM
600R32B	OEM
600R34B	OEM
600R36B	OEM
600R38B	OEM
600R40B	OEM
600R42B	OEM
600R44B	OEM
600R46B	OEM
600R48B	OEM
600R50B	OEM
600X0091-086	0016
600X0092-086	0144
600X0093-086	0007
600X0094-086	0007
600X0095-086	0037
600X0096-066	0143
600X0097-066	0143
600X0098-066	0196
600X0099-066	0015
600X0100-066	0133
600X0101-066	0137
600X0141-000	0127
600X0143-000	0127
600X0175-000	0127
600X0186-000	0917
600X0195-000	0127
601	0015
601-0226-0666	0365
601-0100865	1441
601-1	0016
601-2	0016
601C	0133
601X0048-066	0087
601X0049-066	0015
601X0149-086	0016
601X0150-066	0143
601X0151-066	0143
601X0152-066	0015
601X0224-066	0133
601X0225-066	0133
601X0226-066	0365
601X0226-0666	0137
601X0227-066	0015
601X0375-006	0102
601X0402-038	0365
601X0417-086	0037
601X0432-066	OEM
602	OEM
602-032	0160
602-040	0004
602-075	0050
602-56	0037
602-60	0037
602-61	0144
602-113	0144
602X0006-000	OEM
602X0008-002	0111
602X-0018-000	OEM
602X0018-000	0016
602X0018-002	0111
602X0019-000	0015
603	0518
603-020	0050
603-030	0050
603-040	0050
603-5	OEM
603-113	0144
603E	0518
603F	0518
603G	0518
603G(2SC2603G)	0518
604	0016
604-030	0050
604-080	0050
604-3	0400
604-113	0144
604B	0015
604C	0133
605	0198
605-030	0004
605-113	0144
606-020	0004
606-113	0015
606AJ24	OEM
606C	OEM
606J	OEM
606K	OEM
606L	OEM
606M	OEM
607-030	0016
607-113	0015
607AJ24	OEM
608	0015
608-030	0015
608-2	0086
608-3	0086
608-101	0015
608-112	0975
608-113	0015
608C	OEM
609-020	0050
609-030	0015
609-0371923	OEM
609-112	0016
609-113	0015
609-2497000	OEM
609-2497001	OEM
610	0015
610-001-103	0015
610-009-2402	0057
610-017-706	0133
610-035	0004
610-035-1	0004
610-036	0004
610-036-1	0004
610-036-2	0004
610-036-3	0004
610-036-4	0004
610-036-5	0004
610-036-7	0004
610-036-8	0004
610-039	0160
610-039-1	0160
610-039-103	OEM
610-040	0004
610-040-1	0004
610-040-2	0004
610-041	0144
610-041-1	0144
610-041-2	0144
610-041-3	0144
610-042	0144
610-042-1	0144
610-043	0004
610-043-1	0004
610-043-2	0004
610-043-3	0004
610-043-4	0004
610-043-6	0004
610-043-7	0004
610-045	0144
610-045-1	0144
610-045-2	0144
610-045-3	0016
610-045-4	0016
610-046-7	0004
610-050	0050
610-050-1	0050
610-050-2	0050
610-050-3	0050
610-051	0050
610-051-1	0050
610-051-2	0050
610-051-4	0050
610-052	0136
610-052-1	0136
610-053	0050
610-053-1	0050
610-053-2	0050
610-055	0050
610-055-1	0050
610-055-2	0050
610-055-3	0050
610-056	0136
610-056-1	0136
610-056-2	0136
610-056-3	0136
610-056-4	0136
610-067	0160
610-067-1	0160
610-067-2	0160
610-067-3	0160
610-068	0160
610-068-1	0160
610-069	0144
610-069-1	0144
610-070	0016
610-070-1	0016
610-070-2	0016
610-070-3	0016
610-071	0142
610-071-1	0142
610-072	0144
610-072-1	0144
610-072-2	0144
610-073	0144
610-073-1	0144
610-074	0136
610-074-1	0136
610-076	0016
610-076-1	0016
610-076-2	0016
610-077	0016
610-077-1	0016
610-077-2	0016
610-077-3	0016
610-077-4	0016
610-077-5	0016
610-078	0016
610-078-1	0016
610-079	0004
610-079-1	0004
610-080	0004
610-080-1	0004
610-083	0126

DEVICE TYPE	REPL CODE
610-083-1	0126
610-083-2	0126
610-083-3	0126
610-11	OEM
610BB	OEM
610C	0015
610N	OEM
612	0015
612-1	0560
612-16	0079
612-16(ZENITH)	0144
612-16A	0144
612-60039	0136
612-60039A	0136
612C	0133
613	0079
613(ZENITH)	0144
613-4	0233
613-72	0144
614	0015
614(ZENITH)	0144
614-1	0086
614-2	0079
614-3	0016
614-12	0144
614-118	0133
614A-1-5L	0160
614A1-5L	0160
614C	0133
614H	OEM
614X	OEM
614X1	0004
614X2	0004
614X3	0004
614X4	0004
614X5	0004
614X6	0004
614X7	0004
614X8	0208
614X9	0004
614X10	0004
615-1	0127
616	0015
616-1	0150
616-501B	OEM
616C	0133
617-7	OEM
617-10	0016
617-15	0143
617-17	0143
617-46	0015
617-50	0004
617-52	0004
617-53	0015
617-54	0050
617-55	0050
617-56	0136
617-57	0050
617-58	0050
617-62	0015
617-63	0016
617-64	0016
617-65	0007
617-66	0015
617-67	0016
617-68	0016
617-69	0164
617-70	0004
617-71	0016
617-87	0198
617-117	0042
617-156	0143
617-161	0016
617-162	0015
617-163	0015
618	0015
618C	0133
619	1413
619-2	1413
619-3	2394
619-3(RCA)	2394
620	0015
620-1	0037
620-2	0037
620-56	0150
620C	0133
622	0015
622-1	0086
622-1(RCA)	0086
622C	0133
622H	OEM
623(RCA)	0042
623-1	0560
624	0015
624(RCA)	0925
624-0005	0196
624-0006	0196
624-0007	0196
624-0009	0015
624-0010	0015
624-0011	0143
624-1	0460
624C	0133
625-1	0079
625-1(RCA)	0086
626	0016
626-1	0016
626N002A	OEM
627-1	0037
628-3	0551
629-3	0382
629A02	0015
630-002	0143
630-002-801	0102
630-052	0015
630-076	0016
630-077	0170
630-079	0143
631-1	0126
631-3(SYLVANIA)	0126
632	OEM
632-1(SYLVANIA)	0551
632-3	0423
634	2075
634-1	0127
634H	OEM
635	0037
635-1	0212
635-1(RCA)	0353
635H	OEM
636H	OEM
637	0003
637-1	0074
637-1(RCA)	0074
638	0016
638-1	0086
638H	0178
638HJ	0178
639	0079
639(ZENITH)	0016
639CL	0086
639H	OEM
640-1(SYLVANIA)	0042
640H	OEM
641H	OEM
642-028	0143
642-068	0196
642-098	0123
642-102	0143
642-116	0050
642-117	0004
642-119	0143
642-126	0133
642-132	0143
642-147	0050
642-150	0004
642-151	0164
642-152	0222
642-173	0050
642-174	0016
642-176	0160
642-186	0649
642-199	0143
642-202	0050
642-206	0085
642-207	0050
642-216	0277
642-217	0222
642-219	0015
642-221	0143
642-229	0127
642-230	0127
642-236	0025
642-242	0016
642-246	0016
642-254	0127
642-255	0002
642-260	0127
642-261	0161
642-264	0085
642-266	0460
642-268	0127
642-269	0127
642-270	0127
642-271	0265
642-272	0160
642-274	0127
642-275	0133
642-277	0208
642-279	0123
642-280	0030
642-281	0133
642-304	0229
642-306	0086
642-316	0085
642-319	0127
642AS4076-101	0127
643RLX	0791
644-1	0460
644-405E	0286
646H	OEM
647-1	0065
648	OEM
648HD	OEM
649-1	0065
649-3	5785
650	0143
650-1	OEM
650-105	0211
650-106	0211
650-107	0211
650-108	0211
650-109	0038
650-110	0015
650C	OEM
650C0	0376
650C1	0036
650C2	0036
650C3	OEM
650C4	OEM
650C5	0275
650C6	0275
650C7	OEM
650PBQ50	OEM
650PBQ60	OEM
650PBQ70	OEM
650PBQ80	OEM
650PBQ90	OEM
650PBQ100	OEM
650PBQ110	OEM
650PBQ120	OEM
651	0140
651-001C	0019
651-022A	0143
651-023A	0143
651C	OEM
651C0	0140
651C1	0140
651C2	0140
651C3	0140
651C4	0041
651C5	0041
651C6	0041
651C7	0041
651C8	0041
651C9	0253
651PDB40L20	OEM
651PDB40L25	OEM
651PDB50L20	OEM
651PDB50L25	OEM
651PDB60L20	OEM
651PDB60L25	OEM
651PDB80L20	OEM
651PDB80L25	OEM
651PDB80L30	OEM
651PDB100L25	OEM
651PDB100L30	OEM
651PDB110L25	OEM
651PDB120L25	OEM
651PDB120L30	OEM
651PDB130L30	OEM
651PDB140L30	OEM
651PDB160L30	OEM
652	0091
652-003A	0023
652-005A	0023
652-005B	0080
652-013A	0106
652-015B	0133
652C	OEM
652C0	0157
652C1	0253
652C2	0253
652C3	0091
652C4	0091
652C5	0091
652C6	0091
652C7	0471
652C8	0471
652C9	OEM
653	OEM
653-010A	OEM
653-022A	1615
653-023A	1778
653-2	OEM
653-2F	OEM
653-3	OEM
653-3F	OEM
653-4	OEM
653-4F	OEM
653-5	OEM
653-503A	OEM
653C	OEM
653C0	OEM
653C1	0062
653C2	0062
653C3	0062
653C4	0062
653C5	OEM
653C6	0077
653C7	0077
653C8	OEM
653C9	0165
654-1	0325
654-1(SYLVANIA)	0325
654-202B	0636
654-215A	0636
654-222A	0402
654-401A	0030
654-404A	OEM
654-411B	0365
654C9	0057
655-60CP20	OEM
655-60MP20	OEM
655-CP4	OEM
655-MP4	OEM
655-MP5	OEM
655-NV-90C	OEM
655-NV-90C-P883	OEM
655-NV-90M	OEM
655C9	0064
655NV-60C	OEM
655NV-60C-P883	OEM
655NV-60M	OEM
655NV-60M-P883	OEM
655NV-90M-883	OEM
655NV-400C	OEM
655NV-400C-P883	OEM
655NV-400M	OEM
655NV-400M-883	OEM
656-1	0285
656-2	0561
656-136	0004
656-137	0004
656-138	0164
656-139	0004
656-141	0015
656-142	0143
657-010A	OEM
657H	OEM
658H	OEM
660-126	0016
660-127	0144
660-128	0086
660-131	0016
660-134	0198
660-145	0086
660-220	0016
660-221	0016
660-222	0016
660-224	0164
660-225	0016
660-227	0004
660-228	0208
660-230	0015
660B	0211
662-009R	0079
662-010E	0037
662-039C	OEM
662-048B	0079
662-048C	OEM
662HA	OEM
662N004S	0037
662N007B	0079
662N009S	0079
662N010I	0037
662N031B	6705
662N033C	0715
662N046A	OEM
662N047A	OEM
662N607C	OEM
662T26	OEM
662T28	OEM
662T30	OEM
662T32	OEM
662T34	OEM
663-1	OEM
663-3	OEM
664-015B	2997
664-016B	0733
664-409B	0111
664-410C	0111
664-410D	0111
664-412C	0191
664-413B	1144
664-420C	0224
664-423B	0006
664-614B	1144
664-615B	0111
664H	OEM
664N427B	OEM
665-1A5L	0050
665-2A5L	0208
665-4A5L	0600
665-833B	0436
665A-1-5L	0050
665B-1-5L	0136
665C-1-5L	0160
666/BL60	OEM
666-1	0037
666-1-A-5L	0004
666-2-A-5L	0004
666-3A-5L	0004
666T20	OEM
666T21	OEM
666T22	OEM
666T23	OEM
666T24	OEM
666TB19	OEM
667-004B	0688
667-008B	0212
667-009A	1211
667-012A	OEM
668-009A	2755
668-028B	1571
668-032A	OEM
668-033A	0520
668-037A	3475
668-045A	0176
668-056A	OEM
668-160B	OEM
668-606A	OEM
668-614A	0651
668-616A	1612
668TB7	OEM
668TB9	OEM
668TB11	OEM
668TB13	OEM
668TB15	OEM
669	0037
669-1	0642
669A464H01	1041
669A471H01	1426
669A492H01	0562
670HA	OEM
671A290H01	1823
671A291H01	0033
671A292H01	0081
671A293H01	0141
673-1	OEM
673-1S	OEM
673-2	OEM
673-2S	OEM
673-3	OEM
673-3S	OEM
673-4	OEM
673-4S	OEM
673-5	OEM
673-5S	OEM
673-6	OEM
673-6S	OEM
673-7	OEM
673-8	OEM
673-9	OEM
673-10	OEM
673-11	OEM
673-12	OEM
673-75	OEM
673-85	OEM
674-3A5L	0280
675-153	0164
675-154	0004
675-155	0004
675-156	0004
675-158	0143
675-206	0822
676-1	0330
676-1(DIO)	OEM
676-1S	OEM
676-2	OEM
676-2S	OEM
676-3	OEM
676-3S	OEM
676-4	OEM
676-4S	OEM
676-5	OEM
676-5S	OEM
676-6	OEM
676-6S	OEM
676-12	OEM
676-18	OEM
676-24	OEM
676-30	OEM
676-36	OEM
676-42	OEM
676-48	OEM
676-50	OEM
676-RS	OEM
676H	OEM
677A-1-5L	0440
677H	OEM
678-1	OEM
678-2	OEM
678-3	OEM
678-4	OEM
678-5	OEM
678-6	OEM
679-1	1684
679-1(DIO)	1643
679-2	1643
679-3	1673
679-4	1673
679-5	1673
679-6	1673
680-1	0007
680-1(DIO)	1643
680-1(IC)	3779
680-2	1643
680-3	OEM
680-4	OEM
680-5	OEM
680-6	OEM
681-1D	OEM
681-1N	OEM
681-1P	OEM
681-2D	OEM
681-2N	OEM
681-2P	OEM
681-3D	OEM
681-3N	OEM
681-3P	OEM
681-4D	OEM
681-4N	OEM
681-4P	OEM
681-5D	OEM
681-5N	OEM
681-5P	OEM
681-6D	OEM
681-6N	OEM
681-6P	OEM
682-1	OEM
682-2	OEM
682-3	OEM
682-4	OEM
682-5	OEM
682-6	OEM
683	OEM
683-1	OEM
683-2	OEM
683-3	OEM
683-4	OEM
683-5	OEM
683-6	OEM
684-1	2092
684-2	OEM
684-3	OEM
684-4	OEM
684-5	OEM
684-6	OEM
684-652423-1	0015
684A-1-5L	0160
686-0012	0086
686-0112	0086
686-0130	0086
686-0210	0178
686-0243	0103
686-143	0103
686-143A	0103
688-1	0058
688-10	OEM
688-12	OEM
688-15	OEM
688-18	OEM
688-20	OEM
688-25	OEM
688-255A	OEM
688B-1-5L	0050
688C	OEM
688C-1-5L	0280
689-1	OEM
689-1D	OEM
689-1N	OEM
689-1P	OEM
689-2D	OEM
689-2N	OEM
689-2P	OEM
689-3D	OEM
689-3N	OEM
689-3P	OEM
689-4D	OEM
689-4N	OEM
689-4P	OEM
689-5D	OEM
689-5N	OEM
689-5P	OEM
689-6D	OEM
689-6N	OEM
689-6P	OEM
690C092H88	0015
690L270H02	0160
690L297H01	0050
690L297H02	0160
690V02H69	0144
690V08H36	0079
690V010H40	0007
690V010H41	0144
690V010H42	0136
690V010H43	0030
690V028H28	0144
690V028H48	0144
690V028H49	0917
690V028H69	0007
690V028H89	0144
690V031H33	0015
690V034H29	0136
690V034H30	0004
690V034H31	0004
690V034H32	0143
690V034H39	0211
690V034H74	0143
690V037H91	0143
690V037H92	0474
690V038H22	0087
690V038H23	0196
690V039H17	1303
690V039H18	0774
690V039H22	0087
690V039H52	0015
690V039H54	0474
690V040H56	0050
690V040H57	0136
690V040H58	0136
690V040H59	0136
690V040H60	0136
690V040H61	0211
690V040H62	0211
690V040H63	0143
690V041H08	0015
690V043H62	0164
690V043H63	0164
690V047H54	0136
690V047H55	0136
690V047H56	0280
690V047H57	0280
690V047H58	0211
690V047H59	0004
690V047H60	0211
690V047H61	0143
690V047H97	0016
690V049H81	0144
690V052H23	0211
690V052H24	0211
690V052H50	0143
690V052H63	0050
690V052H68	0143
690V053H57	0015
690V054H20	0004
690V054H21	0211
690V056H27	0050
690V056H29	0050
690V056H30	0050
690V056H31	0050
690V056H32	0050
690V056H33	0004
690V056H34	0004
690V056H89	0136
690V056H90	0211
690V057H25	0050
690V057H27	0211
690V057H28	0211
690V057H59	0136
690V057H62	0136
690V059H20	0004
690V059H21	0164
690V059H52	0004
690V059H55	0004
690V059H63	0143
690V060H58	0144
690V060H59	0144
690V061H98	0211
690V061H99	0211
690V062H47	0004
690V063H14	0136
690V063H15	0136
690V063H16	0211
690V063H17	0211
690V063H50	0004
690V063H51	0211
690V066H44	0050
690V066H45	0136
690V066H46	0004
690V066H47	0004
690V066H48	0143
690V066H49	0136
690V066H89	0050
690V067H09	0143
690V067H35	0208
690V067H69	OEM
690V068H29	0136
690V068H30	0211
690V068H31	0211
690V068H32	0143
690V069H39	0015
690V070H49	0144
690V070H98	0144
690V073H59	0136
690V073H60	0143
690V073H61	0030
690V073H85	0136
690V075H62	0086
690V075H68	0144
690V077H34	0136
690V077H35	0136
690V077H36	0136
690V077H37	0211
690V077H73	0004
690V080H36	0007
690V080H37	0136
690V080H38	0233
690V080H39	0211
690V080H40	0136
690V080H41	0198
690V080H42	0975
690V080H43	0975
690V080H44	0211
690V080H45	0642
690V080H47	0015
690V080H49	0102
690V080H50	0102
690V080H51	0102
690V080H52	0015
690V080H53	0289
690V080H54	0087
690V080H91	0015
690V081H07	0144
690V081H08	0050
690V081H09	0160
690V081H40	0277
690V081H91	0102
690V081H92	0002
690V081H96	0208
690V081H97	0085
690V082H40	0015
690V082H47	0164
690V083H89	0143
690V084H60	0136
690V084H61	0004
690V084H62	0016
690V084H63	0004
690V084H94	0144
690V084H95	0144
690V084H96	0144
690V085	0164
690V085H42	0136
690V085H44	0164
690V086H39	0211
690V086H51	0016
690V086H52	0144
690V086H86	0037
690V086H87	0144
690V086H88	0016
690V086H89	0126
690V086H90	0086
690V086H91	0015
690V086H92	0015
690V086H94	0007
690V086H95	0007
690V086H96	0144
690V088H20	0133
690V088H44	0144
690V088H45	0144
690V088H46	0127
690V088H47	0127

If replacement code is OEM, contact original manufacturer for replacement.

DEVICE TYPE	REPL CODE
690V088H48	0144
690V088H49	0127
690V088H50	0016
690V088H51	0016
690V088H52	0004
690V089H46	0127
690V089H86	0127
690V089H89	0016
690V089H90	0004
690V089H91	0015
690V092H52	0016
690V092H54	0016
690V092H81	0016
690V092H84	0016
690V092H85	0143
690V092H88	0229
690V092H96	0016
690V092H97	0016
690V094H17	0103
690V094H18	0004
690V094H20	0004
690V094H21	0016
690V094H35	0015
690V097H59	0004
690V097H62	0016
690V098H48	0016
690V098H49	0016
690V098H50	0016
690V098H51	0222
690V098H52	0143
690V098H53	0015
690V099H59	0164
690V099H79	0016
690V0103H27	0016
690V06199	0280
690V38H25	1030
690V68H32	0143
690V73H60	0143
690V102H39	0208
690V102H40	0133
690V102H71	0016
690V102H96	0136
690V103H23	0144
690V103H24	0144
690V103H25	0144
690V103H26	0144
690V103H27	0144
690V103H28	0127
690V103H29	0079
690V103H30	0086
690V103H31	0016
690V103H32	0079
690V103H33	0016
690V103H53	0137
690V103H54	0143
690V104H53	0004
690V104H54	0004
690V105H19	0151
690V105H-21	0050
690V105H21	0136
690V105H-22	0079
690V105H22	1480
690V105H-24	0136
690V105H24	0004
690V105H-25	0079
690V105H25	0076
690V105H26	0136
690V105H27	0286
690V105H28	1144
690V105H29	1144
690V105H30	0030
690V105H31	0015
690V105H32	0012
690V109H44	0015
690V109H46	0127
690V109H72	0030
690V110H30	0144
690V110H31	0144
690V110H32	0144
690V110H33	0144
690V110H34	0086
690V110H36	0086
690V110H55	0126
690V110H88	0143
690V110H89	0560
690V111H63	0030
690V114H29	0144
690V114H30	0016
690V114H31	0144
690V114H33	0016
690V114H36	0015
690V116H19	0144
690V116H20	0127
690V116H21	0016
690V116H22	0325
690V116H23	0037
690V116H24	0086
690V116H25	0460
690V116H26	0161
690V116H40	0137
690V116H41	0015
690V118H57	0133
690V118H58	0015
690V118H59	0144
690V118H60	0037
690V118H61	0037
690V118H62	0004
690V119H13	0030

DEVICE TYPE	REPL CODE
690V119H14	0015
690V119H15	0133
690V119H54	0164
690V119H94	0136
690V119H95	0050
690V119H96	0050
690V119H97	0050
690V120H89	0079
690V037H91	0143
692-0014-00	OEM
692-0016-00	2867
692-0020-00	3429
693EP	0016
693FS	0016
693G	0016
693GT	0016
694D	0016
694E	0016
695-1	OEM
695-2	OEM
695-3	OEM
695-4	OEM
695-5	2883
695-6	OEM
695C	OEM
695M	OEM
696-1	OEM
696-2	OEM
696-3	OEM
696-4	OEM
696-5	OEM
696-6	OEM
697-1	OEM
697-2	OEM
697-3	OEM
697-4	OEM
697-5	OEM
697-6	OEM
697-15CP2	OEM
697-CP4	OEM
697NV-400C	OEM
697NV-400C-P883	OEM
697NV-400M	OEM
697NV-400M-P883	OEM
698-1	OEM
698-2	OEM
698-3	OEM
698-4	OEM
698-5	OEM
698-6	OEM
698V102H39	0038
699	0016
700	OEM
700-04	0325
700-1	OEM
700-2	OEM
700-3	OEM
700-4	OEM
700-5	OEM
700-6	OEM
700-110	0086
700-113	0103
700-133	0037
700-134	0079
700-135	0079
700-136	0126
700-137	0015
700-154	0286
700-155	0111
700-156	0079
700-159	0012
700-325	0286
700A858-249	OEM
700A858-285	0015
700A858-286	0015
700A858-318	0076
700A858-319	0076
700A858-322	0133
700A858-328	0111
700M	OEM
700PA50	4148
700PA60	4148
700PA80	1725
700PA100	1725
700PA110	1725
700PA120	1725
700PA130	OEM
700PA140	OEM
700PA150	OEM
700PA160	OEM
700PA170	OEM
700PA180	OEM
700PA190	OEM
700PA200	OEM
700PA210	OEM
700PBQ10	OEM
700PBQ20	OEM
700PBQ30	OEM
700PBQ40	OEM
700PBQ50	OEM
700PBQ60	OEM
700PK50	4148
700PK60	4148
700PK80	1725
700PK100	1725
700PK120	1725
700PK140	OEM
700PK150	OEM
700PK160	OEM

DEVICE TYPE	REPL CODE
700PK180	OEM
700PK200	OEM
700PK210	OEM
700PK220	OEM
700PK230	OEM
700PK240	OEM
700PK250	OEM
700PK260	OEM
700PK270	OEM
700PK280	OEM
700PK290	OEM
700PK300	OEM
700PKL160S686	OEM
700PKL160T60	OEM
700PKL160T606	OEM
700PKL160T626	OEM
700PKL160T663	OEM
700PKL160T673	OEM
700PKL160T683	OEM
700PKL180S686	OEM
700PKL180T60	OEM
700PKL180T606	OEM
700PKL180T626	OEM
700PKL180T663	OEM
700PKL180T673	OEM
700PKL180T683	OEM
700PKL190S686	OEM
700PKL190T60	OEM
700PKL190T606	OEM
700PKL190T626	OEM
700PKL190T663	OEM
700PKL190T673	OEM
700PKL190T683	OEM
700PKL200S686	OEM
700PKL200T60	OEM
700PKL200T606	OEM
700PKL200T626	OEM
700PKL200T663	OEM
700PKL200T673	OEM
700PKL200T683	OEM
700PKL210S686	OEM
700PKL210T60	OEM
700PKL210T606	OEM
700PKL210T626	OEM
700PKL210T663	OEM
700PKL210T673	OEM
700PKL210T683	OEM
700PKL220S686	OEM
700PKL220T60	OEM
700PKL220T606	OEM
700PKL220T626	OEM
700PKL220T663	OEM
700PKL220T673	OEM
700PKL220T683	OEM
700U	OEM
700UM	OEM
701	OEM
701-1	OEM
701-2	OEM
701-3	OEM
701-4	OEM
701-5	OEM
701-6	OEM
701F6S	OEM
701F8S	OEM
701F10S	OEM
701F12S	OEM
701F14S	OEM
701F16S	OEM
701F18S	OEM
702-810	0196
703-056-(4)	0079
703-1	0086
703-2	0086
703A	OEM
703B	0086
703C	OEM
703D	OEM
703F	OEM
703H	0016
703PJA60	OEM
703PJA80	OEM
703PJA100	OEM
703PJA120	OEM
703PJA140	OEM
703PJA160	OEM
704-15K36	OEM
704-15K36T	OEM
704-34	OEM
704-34A	OEM
704-34B	OEM
704-45A	OEM
704-56A	OEM
704-190A	OEM
704-1549	OEM
705-600007	0593
705A	OEM
706-1	OEM
707	2997
708-1	OEM
708-2	OEM
708-6	OEM
708-13	OEM
709CE	1301
710	0015
710-0	OEM
710-0300-015	OEM
710-0300-025	OEM
710-0300-075	OEM

DEVICE TYPE	REPL CODE
710-0301-015	OEM
710-0301-025	OEM
710-0301-075	OEM
710-0302-015	OEM
710-0302-025	OEM
710-0302-075	OEM
710-0304-015	OEM
710-0304-025	OEM
710-0304-075	OEM
710-0305-015	OEM
710-0305-025	OEM
710-0305-075	OEM
710-1	3613
710-2	OEM
710-3	OEM
710A	1023
710AR	0501
710B	1023
710BR	0501
710C	1023
710CE	1797
710CR	0501
710D	1023
710DR	0501
710E	1023
710ER	0501
710F	1023
710FR	0501
710H	1036
710HR	1777
710K	1036
710KR	1777
710M	1036
710MR	1777
710P	1046
710PR	1789
710S	1046
710SR	1789
710V	1046
710VR	1789
710Z	1046
710ZR	1789
711	OEM
711-1	OEM
711-6	OEM
713-2	OEM
713A	0319
713AR	0267
713B	0319
713BR	0267
713C	0319
713CR	0267
713D	0319
713DR	0267
713F	0319
713FR	0267
713H	0319
713HR	0267
713K	0319
713KR	0267
713M	0319
713MR	0267
713PC	OEM
714A	0319
714AR	0267
714B	0319
714BR	0267
714C	0319
714CR	0267
714D	0319
714DR	0267
714F	0319
714FR	0267
714H	0319
714HR	0267
714K	0319
714KR	0267
714M	0319
714MR	0267
715EN	0016
715FB	0127
715HC	1554
717A	1023
717B	1023
717C	1023
717D	1023
717E	1023
717F	1023
717H	1036
717K	1036
717M	1036
717P	1046
717S	1046
717V	1046
717Z	1046
718-1	OEM
719-1	OEM
719H	OEM
720-35019	0211
720DC	1421
722	OEM
723	0026
723A/B	OEM
723BE	1192
723C	0026
723CE	1192
723CJ	0026
724	OEM
724C	0191

DEVICE TYPE	REPL CODE
724D	0191
725A	OEM
725H	OEM
726B	OEM
726C	OEM
729-3	0086
730	OEM
730-6001	OEM
730-6002	OEM
730-6004	OEM
730-6005	OEM
730-6006	OEM
730-6007	OEM
730-6009	OEM
730-6010	OEM
730-6011	OEM
730-6012	OEM
730-6014	OEM
730-6016	OEM
730-6020	OEM
730-6023	OEM
730-6025	OEM
730-6029	OEM
730-6030	OEM
730-6031	OEM
733-1	OEM
733W00024	1823
733W00025	1041
733W00026	2097
733W00027	1831
733W00028	0141
733W00029	0562
733W00030	0358
733W00039	1208
733W00042	0232
733W00043	0697
733W00048	1835
733W00049	1844
733W00050	0341
733W00064	0361
733W00065	0512
733W00067	1313
733W00068	1368
733W00069	1158
733W00076	1926
733W00126	0899
733W00133	0569
733W00183	1492
733W00211	0215
734EU	0325
735	OEM
735DAC	OEM
736	OEM
737	0079
737(ZENITH)	0016
737-1-1	OEM
737-1-2	OEM
737-1-3	OEM
737-1-4	OEM
737-1-5	OEM
739	0692
739-0261-601	OEM
739-0261-602	OEM
739-0361-601	OEM
739-0361-602	OEM
739-0461-601	OEM
739-0461-602	OEM
739-0561-601	OEM
739-0561-602	OEM
739-0661-601	OEM
739-0661-602	OEM
739-0761-601	OEM
739-0761-602	OEM
739-0861-601	OEM
739-0861-602	OEM
739-0961-601	OEM
739-0961-602	OEM
739-2-1	OEM
739-2-2	OEM
739-2-3	OEM
739-2-4	OEM
739-2-5	OEM
739-3-1	OEM
739-3-2	OEM
739-3-3	OEM
739-3-4	OEM
739-3-5	OEM
739-4-1	OEM
739-4-2	OEM
739-4-3	OEM
739-4-4	OEM
739-4-5	OEM
739-52-1	OEM
739-52-2	OEM
739-52-3	OEM
739-52-4	OEM
739-52-5	OEM
739-53-1	OEM
739-53-3	OEM
739-53-4	OEM
739-53-5	OEM
739-54-1	OEM
739-54-2	OEM
739-54-3	OEM
739-54-4	OEM
739-54-5	OEM
739-1061-601	OEM
739-1061-602	OEM
739DC	0692

DEVICE TYPE	REPL CODE
739H01	0016
740	OEM
740-56-2-1	OEM
740-56-5-1	OEM
740-937-120	1018
740-1003-150	2155
740-2001-110	0355
740-2001-306	2262
740-2001-307	0701
740-2002-111	0664
740-2003-150	2155
740-2007-120	1018
740-3006	OEM
740-3012	OEM
740-3014	OEM
740-5853-350	OEM
740-5903-301	1215
740-8120-160	2568
740-8160-190	0579
740-9000-554	1478
740-9000-566	0433
740-9000-754	1478
740-9003-301	1215
740-9003-350	0416
740-9007-046	1393
740-9007-205	1050
740-9011-322	4093
740-9016-105	0443
740-9017-092	4557
740-9037-120	1018
740-9037-204	3366
740-9607-205	1050
741	0312
741-3-1	OEM
741-3-2	OEM
741-3-3	OEM
741-3-4	OEM
741-3-5	OEM
741-4-1	OEM
741-4-2	OEM
741-4-3	OEM
741-4-4	OEM
741-4-5	OEM
741C	0312
742	0143
742-1	0551
742-58-2-1	OEM
742-58-5-1	OEM
743	0233
744	0086
745-0005	OEM
745-0007	OEM
745-0008	OEM
745-0009	OEM
745-0017	OEM
745-0018	OEM
745H	OEM
746H	OEM
746HC	0753
748	0079
748(ZENITH)	0016
749-1501	OEM
749-1601	OEM
749-1701	OEM
749-1801	OEM
749-1901	OEM
749-2001	OEM
749-8160-190	0579
749DHC	2552
750	OEM
750-045	0178
750-137	0211
750-138	0211
750-139	0211
750-140	0211
750-141	0015
750-35019	0004
750A858-285	0023
750A858-319	0079
750A858-328	0111
750A858-448	0042
750C838-123	0016
750C838-124	0016
750C838-125	0016
750D858-211	0071
750D858-212	0016
750D858-213	0144
750H	OEM
750M63-104	0050
750M63-105	0164
750M63-115	0004
750M63-116	0050
750M63-117	0050
750M63-119	0016
750M63-120	0016
750M63-146	0016
750M63-147	0164
750M63-148	0164
750M63-149	0015
751-2001-212	0015
751-4000-88	0143
751-6300-001	0015
751H	OEM
752-1	OEM
752RT	OEM
753-0100-069	1267
753-0101-047	0144
753-0101-226	0560
753-0160-699	1267

DEVICE TYPE	REPL CODE
753-1303-801	0286
753-1372-100	0076
753-1644-100	0111
753-1828-001	0016
753-2000-003	0016
753-2000-004	0016
753-2000-006	0178
753-2000-007	0151
753-2000-008	0016
753-2000-009	0016
753-2000-011	0016
753-2000-100	0111
753-2000-101	0037
753-2000-107	0086
753-2000-460	0127
753-2000-463	0822
753-2000-535	0224
753-2000-710	0144
753-2000-711	0380
753-2000-735	0016
753-2000-870	0016
753-2000-871	0016
753-2001-173	0042
753-2100-001	0016
753-2100-002	0222
753-2100-008	0016
753-3000-535	0224
753-4000-010	0111
753-4000-011	0016
753-4000-024	0325
753-4000-101	0016
753-4000-537	0016
753-4000-668	0127
753-4000-929	0127
753-4001-474	1010
753-4001-932	0042
753-4004-248	0037
753-5751-359	0151
753-5851-359	0151
753-6000--002	OEM
753-6000-002	1220
753-6000-019	0688
753-8400-230	0111
753-8500-380	0076
753-8510-470	0086
753-9000-019	0688
753-9000-839	0076
753-9000-922	0937
753-9001-674	1066
753-9001-675	0076
753-9010-021	1669
753-9010-235	0042
753-9020-784	1144
753-9050-785	1144
753R	OEM
754-0102-139	0005
754-1003-030	0143
754-2000-001	0133
754-2000-002	0015
754-2000-005	0015
754-2000-009	0143
754-2000-011	0133
754-2009-150	0133
754-2509-150	0133
754-2720-007	0062
754-2720-021	0133
754-2900-105	OEM
754-2900-106	OEM
754-4000-088	0143
754-4000-188	0143
754-4000-410	0143
754-4000-553	0760
754-5000-021	0015
754-5700-282	0244
754-5710-219	0244
754-5750-282	0244
754-5750-283	0133
754-5750-284	0015
754-5900-090	0143
754-8130-140	0005
754-9000-082	0165
754-9000-351	0030
754-9000-473	0133
754-9001-124	0139
754-9002-687	0720
754-9030-009	1083
754-9052-473	0133
754-9053-090	0012
754R	OEM
755N	OEM
755P	OEM
756	0015
756-1	OEM
756H	OEM
757N	OEM
757P	OEM
759H	OEM
759P	OEM
760H	OEM
761	OEM
762-2	0061
765-1	OEM
767	0079
767(ZENITH)	0016
767CL	0016
767H	OEM
768	2188
768-1	OEM
770-045	0042

If replacement code is OEM, contact original manufacturer for replacement.

DEVICE TYPE	REPL CODE
770A	1023
770B	1023
770C	1023
770D	1023
770E	1023
770F	1023
770H	1036
770K	1036
770M	1036
770P	1046
770S	1046
770V	1046
770Z	1046
771-12	OEM
771-14	OEM
771-16	OEM
771A	0501
771B	0501
771C	0501
771D	0501
771E	0501
771F	0501
771H	1777
771HD	OEM
771HDS	OEM
771K	1777
771M	1777
771P	1789
771S	1789
771V	1789
772-724	0211
772A	0127
772B	0127
772B1	0127
772BJ	0127
772BL	0127
772BM	0127
772BN	0127
772BY	0127
772C	0127
772CC	0127
772D	0127
772D1	0127
772DC	0127
772DG	0127
772E	0127
772EH	0127
772F	0127
772FE	0127
772G	0127
773	0079
773(ZENITH)	0016
773-3000-535	OEM
773CL	0086
773RED	0144
774	0150
774(ZENITH)	0037
774CL	0126
774H	OEM
775-1	0400
775-1(SYLVANIA)	0400
775BRN	0144
776-1	0126
776-1(SYLVANIA)	0037
776-2	0037
776-3	0037
776GRN	0016
777-2	0023
777R	OEM
778-1	5379
779-1	OEM
779BLU	0144
780-4	OEM
780-6	OEM
780-7	5396
780DA	OEM
780WHT	0016
781-1	1389
781D	OEM
781H	OEM
783B	OEM
783D	OEM
783F	OEM
783H	OEM
783K	OEM
783M	OEM
783RED	0144
783S	OEM
783Z	OEM
783ZD	OEM
783ZH	OEM
783ZM	OEM
783ZS	OEM
783ZZ	OEM
784A	OEM
784B	OEM
784C	OEM
784D	OEM
784E	OEM
784F	OEM
784H	OEM
784K	OEM
784M	OEM
784ORN	0144
784S	OEM
784Z	OEM
784ZB	OEM
784ZD	OEM
785H	OEM
785YEL	0016
786	0144
786H	OEM
787BLU	0144
788A	1023
788AR	0501
788B	1023
788BR	0501
788C	1023
788CR	0501
788D	1023
788DR	0501
788F	1023
788FR	0501
788H	1036
788HR	1777
788K	1036
788KR	1777
788M	1036
788MR	1777
788S	1046
788SR	1789
788Z	1046
788ZR	1789
790	OEM
790H	OEM
791	0016
792-238	0015
792-286	0211
792-287	0211
792-288	0211
792-289	0211
792-290	0211
792-292	0143
792-414	0023
792H	OEM
796H	OEM
800	OEM
800-000-025	0190
800-002-00	0123
800-003-00	0143
800-004-00	0137
800-005-00	0143
800-006-00	0015
800-007-00	0196
800-008-00	0474
800-0016	1349
800-010-00	0087
800-011-00	1303
800-012-00	0774
800-013-00	0015
800-014-00	0015
800-014-01	0015
800-016-00	0133
800-017-00	0015
800-018-00	0087
800-018-02	0087
800-020-00	0143
800-021-00	0133
800-022-00	0143
800-023-00	0137
800-024-00	0015
800-028-00	0420
800-029-30	0102
800-030-00	0025
800-032-00	0015
800-034-00	0205
800-035-00	0137
800-036-00	0133
800-037-00	0087
800-038-00	0015
800-039-00	0143
800-040-00	0133
800-041-00	0015
800-042-00	0133
800-043-00	0087
800-045-00	3620
800-046-00	0015
800-050-00	0087
800-051-00	0087
800-01300-504	0015
800-01400	0015
800-1	OEM
800-2	OEM
800-3	OEM
800-4	OEM
800-5	OEM
800-6	OEM
800-122	0142
800-158	0142
800-172	0142
800-180	0142
800-181	0142
800-196	3033
800-198	OEM
800-203	0142
800-204	0142
800-205	0164
800-208	2866
800-219	2866
800-245	4163
800-250-102	0198
800-253	0265
800-256	0142
800-256(METAL)	0142
800-256(PLASTIC)	0276
800-270	1657
800-282	0102
800-284	0126
800-289	1657
800-292	0126
800-294	4163
800-305	4163
800-310	2866
800-310(CP)	OEM
800-310CP	0208
800-315	4163
800-321	0142
800-328	4163
800-329	0085
800-356	OEM
800-362	3732
800-364	OEM
800-366	3732
800-398	OEM
800-401	0142
800-501-00	0076
800-501-01	0016
800-501-02	0016
800-501-03	0016
800-501-04	0016
800-501-11	0016
800-501-22	0016
800-502-00	0004
800-503-00	2866
800-504-00	0136
800-505-00	0164
800-506-00	0600
800-507-00	0160
800-508-00	0016
800-509-00	0076
800-510-00	0103
800-510-01	0103
800-511-00	0126
800-512-00	0086
800-513-00	0086
800-514-00	0016
800-515-00	0086
800-516-00	0126
800-517-00	0143
800-518-00	0160
800-521-01	0016
800-521-02	0086
800-521-03	0086
800-522-01	0016
800-522-02	0016
800-522-03	0086
800-522-04	0016
800-523-01	0037
800-523-02	0037
800-524-02	0103
800-524-02A	0103
800-524-03	0103
800-524-03A	0103
800-524-04A	0103
800-525-03	0126
800-525-04	0037
800-525-04A	0103
800-526-00	0016
800-527-00	0037
800-528	0208
800-528-00	0208
800-529-00	0016
800-530-00	0016
800-530-01	0016
800-533-00	0042
800-534-00	0016
800-534-00(DIO)	0030
800-534-00(XSTR)	0079
800-534-01	0016
800-535-00	0325
800-535-01	0325
800-536-00	0144
800-537-01	0208
800-537-02	1514
800-537-03	0164
800-538-00	0016
800-544-00	0016
800-544-00(U.J.T.)	1669
800-544-10	0016
800-544-20	0016
800-544-30	0016
800-546-00	0042
800-547-00	0037
800-548-00	0016
800-550-00	0074
800-550-10	0642
800-552-00	0338
800-557-00	0007
800-601	OEM
800-608	OEM
800-619	2480
800-669B	3919
800-670	OEM
800-670B	4082
800-671B	OEM
800-672	1381
800-672-B	OEM
800-763	0065
800-764	0065
800-765	OEM
800-766	OEM
800-767	1591
800-771	0925
800-791	3858
800-817	OEM
800-818	6236
800-50100	0016
800-50300	0208
800-50400	0050
800-50500	0136
800-50600	0208
800-50700	0160
800-51500	0016
800-51600	0126
800-52102	0016
800-52202	0016
800-52302	0016
800-52402	0103
800-52600	0016
800-52700	0126
800-52800	0208
800-53001	0198
800-53300	0086
800-53400	0016
800-53600	0144
800-53702	1514
800EXD25	OEM
800EXD26	OEM
800EXH21	OEM
800FXD25	OEM
800FXD26	OEM
800HB1U	5738
800HB2U	5738
800HB3U	5738
800HB4U	5738
800HB6U	5738
800P12H	OEM
800PDB10	OEM
800PDB20	OEM
800PDB30	OEM
800PDB40	OEM
800PDB50	OEM
800PDB60	OEM
800PIN-RM	OEM
800PV-RM	OEM
800UD25	OEM
800UD26	OEM
800YD25	OEM
800YD26	OEM
800YKD25	OEM
800YKD26	OEM
801	OEM
801-04900	0136
801-05200	0136
801-05300	0136
801-1	OEM
801-2	OEM
801-3	OEM
801-4	OEM
801-5	OEM
801-6	OEM
801B	0016
801F8H	OEM
801F12H	OEM
801F16H	OEM
801F20H	OEM
801F25H	OEM
801F30H	OEM
801PDB60B	2159
801PDB80B	0479
801PDB100B	0479
801PDB120	0479
801PDB120B	0479
801PDB140	OEM
801PDB140B	OEM
801PDB160	OEM
801PDB180	OEM
801PDB200	OEM
801PDB210	OEM
801PDB220	OEM
801PDB230	OEM
801PDB240	OEM
801PDB250	OEM
801PJA250	OEM
801PJA300	OEM
801PJA400	OEM
802	OEM
802-0027	0183
802-0038	1839
802-0245	0463
802-0260	5995
802-0266	0592
802-05400	0004
802-05600	0004
802-1	OEM
802-2	OEM
802-3	OEM
802-4	OEM
802-22631U	0004
802-36400	0004
802-36430	0004
802-43900	0004
802CPY	OEM
803-0073	1536
803-1	OEM
803-2	OEM
803-3	OEM
803-4	OEM
803-18250	0016
803-37390	0016
803-38030	0144
803-38040	0144
803-39430	0144
803-39440	0144
803-83840	0144
803-83930	0144
803-83940	0016
803-89910	0016
803-89930	0016
804	0016
804-1D	OEM
804-1N	OEM
804-1P	OEM
804-2D	OEM
804-2N	OEM
804-2P	OEM
804-3D	OEM
804-3N	OEM
804-3P	OEM
804-4D	OEM
804-4N	OEM
804-4P	OEM
804-14120	0103
804-14130	0103
804-21980	0016
804A765-2	0922
805-1060	0143
805-10600	0143
807-1	0086
808-3	OEM
808-206	0015
808-304	0004
808-305	0004
808-306	0004
808-307	0004
808-308	0004
808-309	0208
808-310	0004
808-311	0004
808-312	0143
809-1	OEM
809A	OEM
809B	OEM
809B-FLAG	0217
809B-FLEX	0217
809C	OEM
809D	OEM
809D-FLAG	0217
809D-FLEX	0217
809E	OEM
809F	OEM
809F-FLAG	0217
809F-FLEX	0217
809H-FLAG	0217
809H-FLEX	0217
809K	OEM
809K-FLAG	0217
809K-FLEX	0217
809M-FLAG	0217
809M-FLEX	0217
809P	OEM
809S	OEM
809S-FLAG	0658
809S-FLEX	0468
810	2755
810-4	OEM
810-6	OEM
810-8	OEM
812H	OEM
813H	OEM
814	OEM
814H	OEM
815-003	0740
815-181D	0211
815-1810	0004
818WHT	0016
819H	OEM
820H	OEM
821	OEM
821CXYA	OEM
821CXYB	OEM
821CXYC	OEM
821CXYD	OEM
822	0016
822-1	0015
822-1(SYLVANIA)	0007
822-2	1228
822-3	0015
822-4	0015
822-5	1228
822A	0016
822ABLU	0016
822B	0016
823-1	OEM
823B	0016
823WHT	0016
824-09501	0016
824-1	0086
824-1(SYLVANIA)	0008
824-10300	0126
825CPY	OEM
826	0042
826-1	2033
826BAY	OEM
827BRN	0016
828GRN	0016
829	0037
829A	0037
829B	0037
829C	0037
829D	0037
829DE	0037
829E	0037
829F	0037
830	0004
830H	OEM
832-1	OEM
832H	OEM
833	0037
833-3	OEM
835-300-0002	0725
835H	OEM
837-1	OEM
837H	OEM
838-1	OEM
838H	OEM
839H	OEM
840PE2-98	OEM
840XY2-35SC	OEM
845H	OEM
846H	OEM
848	4166
848HA	OEM
848HD	OEM
848HDS	OEM
850PA50	6330
850PA60	6330
850PA80	6167
850PA100	1725
850PA110	1725
850PA120	1725
850PA130	OEM
850PA140	OEM
850PA150	OEM
850PA160	OEM
850PA170	OEM
850PA180	OEM
850PA190	OEM
850PA200	OEM
850PK50	4148
850PK60	4148
850PK80	1725
850PK100	1725
850PK120	1725
850PK140	OEM
850PK150	OEM
850PK160	OEM
850PK180	OEM
850PK200	OEM
850PK210	OEM
850PK220	OEM
850PK230	OEM
850PK240	OEM
851-0372-130	0015
851H	OEM
851PDE40L20	OEM
851PDE40L25	OEM
851PDE40L30	OEM
851PDE50L20	OEM
851PDE50L25	OEM
851PDE50L30	OEM
851PDE60L20	OEM
851PDE60L25	OEM
851PDE60L30	OEM
851PDE80L20	OEM
851PDE80L25	OEM
851PDE80L30	OEM
851PDE100L20	OEM
851PDE100L25	OEM
851PDE100L30	OEM
851PDE120L20	OEM
851PDE120L25	OEM
851PDE120L30	OEM
851PDE130L20	OEM
851PDE130L25	OEM
851PDE130L30	OEM
851PDE140L20	OEM
851PDE140L25	OEM
851PDE140L30	OEM
851PDE160L20	OEM
851PDE160L25	OEM
851PDE160L30	OEM
852-92-43	0460
853-0300-632	0016
853-0300-634	1267
853-0300-643	0791
853-0300-644	0111
853-0300-900	0016
853-0300-923	0016
853-0301-096	0390
853-0301-317	0155
853-0373-110	0016
854-0372-020	0015
854H	OEM
856H	OEM
857-1	OEM
858GS	0016
859GK	0111
860-022-01	0130
860-1GI	OEM
861H	OEM
862-1	OEM
863-1	OEM
863-254B	0133
863-567B	0133
863-776B	0133
863H	OEM
866-6	0103
866-6(BENDIX)	0103
866H	OEM
868-1	OEM
868A-01	OEM
868A-02	OEM
868A-09	OEM
868A-10	OEM
868A-11	OEM
868A-12	OEM
868A-14	OEM
868A-15	OEM
868A-16	OEM
868A-18	OEM
868A-20	OEM
868A-21	OEM
868A-22	OEM
868A-23A	OEM
868A-23B	OEM
868A-23C	OEM
868A-24A	OEM
868A-25	OEM
868A-27	OEM
868A-32	OEM
868A-33	OEM
868A-40	OEM
868A-41	OEM
868A-58	OEM
868A-59	OEM
868A-60	OEM
868A-61	OEM
868A-62	OEM
868A-63	OEM
868A-64	OEM
868A-65	OEM
868A-67	OEM
868A-68	OEM
868A-69	OEM
868A-70	OEM
868A-71	OEM
868A-80	OEM
868G-50	OEM
868G-51	OEM
868G-55	OEM
868G-56	OEM
868G-57	OEM
869-1	OEM
869H	OEM
870H	OEM
873-1	OEM
874H	OEM
876H	OEM
880	OEM
880-1	OEM
880-2	OEM
880-101-00	0632
880-102-00	2617
880-103-00	1069
880-207-000	0015
880-250-001	0160
880-250-010	0160
880-250-011	0161
880-250-102	0016
880-250-107	0126
880-250-108	0016
880-250-109	0016
880-250108	0111
881-1	OEM
881-250-102	0016
881-250-107	0126
881-250-108	0016
881-250-109	0016
881-250108	0111
881H	OEM
882-1	OEM
884-250-001	0160
884-250-010	0160
884-250-011	0161
885-1	OEM
885B	OEM
886	OEM
886-1	OEM
887	OEM
895	0079
895-1	OEM
898	1833
900C	OEM
900HC	3418
901-000-6-51	0004
901-5553-7	0760
901F	0079
901SS1-07SC	OEM
902	OEM
902-000-2-04	1221
902-000-8-04	0076
902-001-7-18	1144
902-002-3-06	0380
902-003-0-012	0076
902-003-0-12	0076
902-003-3-17	0076
902-003-6-006	1221
902-003-6-06	1221
902-2	OEM
903	0007
903-00390	0143
903-00391	0554
903-00393	0133
903-00394	0015
903-1B	0030
903-3	0016
903-3G	0016
903-8B	0123
903-9B	0143
903-10B	0143
903-11	0030
903-12B	0143

If replacement code is OEM, contact original manufacturer for replacement.

DEVICE TYPE	REPL CODE	DEVICE TYPE	REPL CODE	DEVICE TYPE	REPL CODE	DEVICE TYPE	REPL CODE	DEVICE TYPE	REPL CODE	DEVICE TYPE	REPL CODE	DEVICE TYPE	REPL CODE
903-13	OEM	903-101	OEM	903-357	0052	903-568	OEM	905-5	3550	905-201	1018	905-388	OEM
903-13B	0229	903-101B	0030	903-358	OEM	903-574	OEM	905-5B	OEM	905-202	2104	905-391	OEM
903-14B	0015	903-102	0015	903-359	OEM	903-575	OEM	905-7	OEM	905-203	OEM	905-407	OEM
903-15B	0015	903-103	OEM	903-360	OEM	903-579	OEM	905-9-B	2876	905-204	2728	905-408	OEM
903-16	OEM	903-103B	0143	903-361	OEM	903-583	OEM	905-9B	OEM	905-205	OEM	905-409	OEM
903-16B	0143	903-104B	0015	903-362	OEM	903-600	OEM	905-13	2576	905-206	OEM	905-414	OEM
903-17	0087	903-105	OEM	903-363	OEM	903-606	OEM	905-13B	2576	905-208	OEM	905-416	OEM
903-17A	0087	903-105B	0015	903-365	OEM	903-630	OEM	905-17B	OEM	905-209	3791	905-417	OEM
903-17B	0087	903-108	0143	903-366	OEM	903-631	OEM	905-18	OEM	905-210	2295	905-418	OEM
903-17C	0087	903-108B	0143	903-375	1615	903-632	OEM	905-19	OEM	905-211	OEM	905-424	OEM
903-18	0143	903-109	0015	903-378	1778	903-639	OEM	905-19B	OEM	905-212	OEM	905-442	OEM
903-18B	0143	903-110B	0005	903-380	OEM	903-652	OEM	905-20	OEM	905-213	OEM	905-454	3804
903-19B	0030	903-112B	0133	903-385	1778	903-655	OEM	905-23	OEM	905-214	2614	905-463	0023
903-20	0133	903-113B	0143	903-386	OEM	903-664	OEM	905-24B	0701	905-215	OEM	905-465	4280
903-22	OEM	903-114	OEM	903-389	OEM	903-665	OEM	905-27	2557	905-216	OEM	905-488	OEM
903-23	0143	903-114B	0143	903-391	OEM	903-667	OEM	905-28	2557	905-217	0683	905-491	OEM
903-23A	0143	903-115	OEM	903-394	OEM	903-669	OEM	905-30	OEM	905-218	2233	905-492	OEM
903-23B	0143	903-116B	0133	903-397	0133	903-670	OEM	905-38B	0701	905-219	1292	905-529	OEM
903-23C	0143	903-117	0015	903-407	0143	903-671	OEM	905-39B	0470	905-220	0332	905-537	OEM
903-23D	0143	903-117B	0015	903-408	OEM	903-680	OEM	905-40	OEM	905-221	0969	905-538	OEM
903-23E	0143	903-118	0133	903-412	0087	903-684	OEM	905-41	OEM	905-222	OEM	905-539	OEM
903-25B	0133	903-118B	0133	903-413	OEM	903-734	0124	905-44B	OEM	905-223	3948	905-540	OEM
903-26B	0188	903-119	0012	903-414	OEM	903-753	OEM	905-46B	0701	905-226	4003	905-585	OEM
903-27	0019	903-119B	0012	903-415	7011	903-754	OEM	905-47	OEM	905-227	OEM	905-645	OEM
903-27B	0019	903-120B	0157	903-417	OEM	903-756	OEM	905-48	OEM	905-229	4933	905-658	0770
903-28	OEM	903-121	0133	903-418	OEM	903-760	OEM	905-55	1940	905-231	OEM	905-659	0631
903-28B	0015	903-121B	0133	903-419	OEM	903-762	OEM	905-58	OEM	905-232	OEM	905-662	0413
903-29B	0143	903-123B	0030	903-424	OEM	903-764	OEM	905-59	1248	905-233	2958	905-717	0413
903-30B	0143	903-125	0015	903-425	OEM	903-800	0343	905-66	2295	905-234	3948	905-738	OEM
903-32	0015	903-126	OEM	903-426	0015	903-827	OEM	905-69	2295	905-235	2063	905-830	OEM
903-33	OEM	903-128	OEM	903-427	0365	903-830	0015	905-77	OEM	905-236	OEM	905-920	OEM
903-34	0143	903-129	0015	903-428	OEM	903-852	0589	905-78	OEM	905-237	OEM	905-922	0631
903-34A	0143	903-130	OEM	903-429	OEM	903-864	0133	905-80	OEM	905-238	2751	905-923	OEM
903-34B	0143	903-133	0133	903-430	OEM	903-869	0124	905-100	OEM	905-239	6039	905-940	3635
903-34C	0143	903-136B	0133	903-435	0123	903-878	OEM	905-101	3973	905-240	OEM	905-1085	0631
903-34D	0143	903-137	0554	903-438	0015	903-885	1329	905-102	1208	905-241	3198	905-1087	OEM
903-34E	0143	903-137B	0030	903-440	1615	903-970	0323	905-103	3970	905-242	0334	905-1106	0478
903-35	OEM	903-138	OEM	903-440B	0030	903-976	0133	905-104	4434	905-243	3791	905-1123	4266
903-35B	0071	903-139	OEM	903-443	0649	903-984	0253	905-105	0878	905-244	1248	905-1126	0101
903-36	0071	903-140	OEM	903-444	0071	903-1008	OEM	905-106	3948	905-246	OEM	905-1141	OEM
903-36A	0071	903-141B	0030	903-445	0023	903-1010	OEM	905-107	4261	905-247	OEM	905-1224	2418
903-36B	0071	903-142	OEM	903-447	OEM	903-1018	OEM	905-108	3697	905-248	OEM	905-1225	OEM
903-36C	0071	903-142B	1989	903-448	0471	903-1022	0589	905-109	3964	905-249	OEM	905-1226	OEM
903-36D	0071	903-146	0015	903-449	0181	903-1071	0064	905-110	3966	905-250	OEM	905-1227	0625
903-36F	0071	903-149	0015	903-450	0015	903-1081	0883	905-111	4294	905-251	OEM	905-1228	OEM
903-37B	0143	903-150	OEM	903-451	0087	903-1084	OEM	905-112	3975	905-252	OEM	905-1229	OEM
903-39	0015	903-152	OEM	903-453	0062	903-1088	OEM	905-113	OEM	905-253	1260	905-1230	OEM
903-40	0143	903-154	0015	903-454	0760	903-1089	OEM	905-114	3707	905-254	3190	905-1231	OEM
903-41B	0133	903-155	OEM	903-455	0086	903-1113	OEM	905-115	3968	905-255	0360	905-1233	OEM
903-42	OEM	903-156	0015	903-456	0053	903-1124	0343	905-116	4242	905-256	3956	905-1233-01	OEM
903-42B	0133	903-156B	0015	903-457	0015	903-1128	0041	905-118	2295	905-257	OEM	905-1234	OEM
903-43#	OEM	903-157	0015	903-463	0023	903-1200	2436	905-118-01	2295	905-259	OEM	905-1235	OEM
903-43B	0143	903-163	0030	903-465	OEM	903-1202	OEM	905-120	2755	905-260	6173	905-1236	OEM
903-45B	0143	903-163B	0628	903-466	OEM	903-1203	OEM	905-121	0416	905-261	OEM	905-1238	0770
903-46	0015	903-164	0015	903-467	0124	903-1204	OEM	905-122	1571	905-262	2675	905-1239	0413
903-47	0760	903-164B	0015	903-468	OEM	903-1205	OEM	905-123	4390	905-263	OEM	905-1240	0413
903-47B	0720	903-166	OEM	903-470	OEM	903-1206	0041	905-124	OEM	905-264	OEM	905-1247	OEM
903-48	OEM	903-166B	0025	903-472	OEM	903-1207	0041	905-125	0478	905-265	2675	905-1247-01	OEM
903-48B	0133	903-167	OEM	903-473	OEM	903-1208	OEM	905-126	0215	905-266	OEM	905-1248	OEM
903-49	0015	903-167B	0019	903-474	OEM	903-1209	0248	905-127	OEM	905-267	1292	905-1289	3202
903-49B	0015	903-168	OEM	903-475	0917	903-1210	OEM	905-131	6166	905-268	0626	905-1298	OEM
903-50	0015	903-168B	0019	903-476	0143	903-1211	OEM	905-136	OEM	905-269	OEM	905-1339	OEM
903-51	0143	903-169B	0023	903-477	OEM	903-1213	OEM	905-137	4209	905-270	OEM	905-1366	OEM
903-51B	0143	903-171B	0030	903-479	0053	903-1214	0041	905-138	1571	905-271	OEM	905-1372	2742
903-52	0015	903-172	OEM	903-480	0036	903-1216	OEM	905-139	1612	905-275	OEM	905-1373	0413
903-52B	0015	903-173	OEM	903-481	0041	903-1217	OEM	905-140	OEM	905-276	OEM	905-1376	OEM
903-54	OEM	903-174	OEM	903-482	0041	903-1218	OEM	905-141	0520	905-277	OEM	905-1377	OEM
903-54B	0143	903-176	OEM	903-483	1329	903-1220	0907	905-142	OEM	905-278	OEM	905-1378	OEM
903-56	OEM	903-177B	0133	903-484	OEM	903-1229	OEM	905-144	1612	905-279	OEM	905-1482	OEM
903-58	0133	903-178	0133	903-486	0133	903-1233	OEM	905-145	OEM	905-280	OEM	905-10083	OEM
903-58B	0133	903-179	0012	903-487	0077	903-1241	OEM	905-147	1014	905-281	OEM	905-10084	OEM
903-59	0015	903-180	0025	903-492	0064	903-1242	OEM	905-151	OEM	905-282	OEM	905-10085	2364
903-63	OEM	903-185	OEM	903-494	OEM	903-1243	OEM	905-152	0843	905-283	OEM	905-10090	OEM
903-65	0143	903-193	OEM	903-495	0165	903-1270	0140	905-154	OEM	905-284	0362	905-10091	OEM
903-65B	0143	903-197	0023	903-496	0253	903-1271	0471	905-155	0651	905-286	OEM	905HC	3714
903-67	0554	903-198	0015	903-502	OEM	903-1273	OEM	905-156	3110	905-287	4854	905HM	OEM
903-67B	0030	903-200	OEM	903-507	0123	903-1275	0064	905-157	OEM	905-288	0176	906	OEM
903-68	0015	903-206	0447	903-508	0124	903-10053	0124	905-158	OEM	905-289	2063	906HC	4234
903-68B	0102	903-208	0106	903-509	0057	903-10054	0041	905-159	OEM	905-291	3190	907	OEM
903-69	0015	903-212	0143	903-510	0015	903-10055	0052	905-161	4378	905-293	OEM	907HC	4236
903-69B	0015	903-303	0015	903-511	0028	903-10056	1329	905-162	OEM	905-294	OEM	907HM	OEM
903-70	OEM	903-304	1282	903-512	0015	903-10057	0057	905-164	OEM	905-295	OEM	908	OEM
903-71B	0030	903-305	6646	903-513	0041	903-10058	0023	905-165	0904	905-298	4003	908D1	0143
903-72	OEM	903-306	1282	903-519	0124	903-10059	0071	905-166	OEM	905-299	OEM	908HC	3845
903-72B	0133	903-311	0133	903-523	0124	903-10060	0705	905-170	4209	905-301	2728	909	OEM
903-73	0015	903-318	OEM	903-528	0124	903-10061	2189	905-171	1612	905-303	OEM	909-27125-140	0016
903-73B	0053	903-321	OEM	903-529	OEM	903-10067	0036	905-172	4219	905-308	OEM	909-27125-160	0127
903-77	OEM	903-325	1778	903-531	0292	903HC	3861	905-173	2534	905-310	OEM	909HC	3850
903-79B	0015	903-326	1615	903-532	0023	903HM	OEM	905-174	OEM	905-311	OEM	909HM	OEM
903-80	OEM	903-327	0019	903-533	0052	903J	0079	905-175	OEM	905-323	OEM	910	0007
903-80B	0030	903-330	0023	903-534	0124	903Y00212	0229	905-177	1018	905-324	OEM	910H	OEM
903-82	0133	903-331	0636	903-535	2542	903Y00228	0012	905-178	0651	905-325	0706	910HC	1257
903-82B	0133	903-332	0124	903-537	0188	903Y002149	0016	905-180	OEM	905-326	4831	910HM	OEM
903-83	OEM	903-333	0253	903-538	OEM	903Y002150	0111	905-181	OEM	905-327	OEM	910X1	0164
903-83B	0143	903-334	0540	903-539	OEM	903Y-002151	0382	905-182	OEM	905-328	OEM	910X2	0164
903-84B	0133	903-335	0062	903-540	OEM	903Y002151	1267	905-184	OEM	905-329	4830	910X3	0004
903-85B	0030	903-337	0064	903-541	0199	903Y002152	0560	905-185	OEM	905-338	OEM	910X4	0004
903-86	OEM	903-341	0019	903-542	OEM	904	0007	905-186	0413	905-340	OEM	910X5	0004
903-91	0015	903-342	OEM	903-543	OEM	904-95	0016	905-189	4000	905-344	0243	910X6	0004
903-92	OEM	903-343	OEM	903-544	0052	904-95A	0016	905-190	1329	905-347	2105	910X7	0164
903-92B	0143	903-345	OEM	903-545	0137	904-95B	0111	905-191	0878	905-349	OEM	910X8	0164
903-93	OEM	903-347	0023	903-546	0455	904-96B	0079	905-193	OEM	905-354	0035	910X10	0164
903-95B	0133	903-348	0087	903-556	0143	904-97B	0143	905-194	OEM	905-360	OEM	911H	OEM
903-96B	0030	903-349	0019	903-557	0133	904A	0007	905-195	OEM	905-368	1201	911HC	1364
903-97	OEM	903-351	OEM	903-558	0628	904HC	3833	905-196	OEM	905-369	0101	911HM	OEM
903-97B	0137	903-352	OEM	903-559	0301	905	0007	905-197	OEM	905-375	OEM	912-1A6A	0016
903-99	0554	903-353	OEM	903-560	0301	905-00066	2295	905-198	4112	905-380	0366	912H	OEM
903-99B	0030	903-355	0019	903-561	0110	905-00160	1395	905-199	OEM	905-384	OEM	912HC	3897
903-100B	0133	903-356	0087	903-562	0023	905-00161	3557	905-200	1949	905-387	OEM	913H	OEM

DEVICE TYPE	REPL CODE
913HC	3901
913HM	OEM
914	0133
914-000-2-00	0133
914-000-4-00	0019
914-000-6-00	0133
914-000-7-00	0720
914-001-1-00	0139
914-001-7-00	0123
914A-1-6-5	0160
914F298-1	0086
914HC	1069
914HM	OEM
915	OEM
915H	OEM
915HC	3899
915HM	OEM
916-000-4-00	OEM
916-31001-1	0004
916-31001-1B	0004
916-31001-7B	0004
916-31003-5B	0004
916-31007-5	0004
916-31007-5B	0004
916-31012-6	0004
916-31012-6B	0004
916-31019-3	0136
916-31019-3B	0136
916-31024-3	0016
916-31024-3B	0127
916-31024-5	0127
916-31024-5B	0144
916-31025-4	0007
916-31025-4B	0007
916-31025-5	0016
916-31025-5B	0127
916-31026-9B	0004
916-32000-7	0143
916-32003-2	0143
916-32006-2	0143
916-33003-2	0015
917-1201-0	0579
917-12010	0579
918H-1	OEM
918H-2	OEM
919-00-1445	0157
919-00-1445-001	0649
919-00-1445-002	0137
919-00-1445-1	0649
919-00-1445-2	0137
919-00-1449-1	0087
919-00-2440-1	0015
919-00-2440-2	0015
919-00-3309	0531
919-00-3309-2	0234
919-00-3309-3	0251
919-00-4326	0133
919-00-4799	0133
919-00-5045	0229
919-00-7109	0229
919-00-7349	0162
919-00-7394	0133
919-00-7766	0015
919-00-7776	0015
919-00-9929	0133
919-001172	0133
919-001172-1	0133
919-001445-2	0137
919-001449	0242
919-001449-1	0242
919-001449-3	0242
919-002240-1	0242
919-002440	0087
919-002440-1	0087
919-002440-2	0540
919-003309	0531
919-003309-1	0495
919-004326	0133
919-004799	0015
919-005045	0242
919-007109	0015
919-007394	0015
919-007776	0953
919-008862	0242
919-01-0459	0015
919-01-0623	0015
919-01-0829	0229
919-01-0829-1	0229
919-01-0867	0143
919-01-0873	0133
919-01-1172-1	0133
919-01-1211	0015
919-01-1212	0015
919-01-1213	0012
919-01-1214	0244
919-01-1215	0133
919-01-1307	0133
919-01-1339	0229
919-01-1340	0234
919-01-3035	0012
919-01-3058	0162
919-01-3072	0133
919-03036	0087
919-010459	0242
919-010623	0242
919-010829	0953
919-010829-1	0242
919-010867	0143
919-010873	0133

DEVICE TYPE	REPL CODE
919-011211	0953
919-011212	0015
919-011215	0133
919-011307	0133
919-011339	0087
919-011340	0870
919-013036	0953
919-013044	0015
919-013058	0137
919-013059	0133
919-013060	0133
919-013061	0087
919-013067	0133
919-013072	0015
919-013079	0087
919-013081	0133
919-013082	0133
919-015618-1	0531
919-017406-011	0269
919-017406-013	0247
919-017406-056	0256
919-017406-057	0225
919-72-2	1413
920	OEM
920B03003	OEM
920B04201	OEM
920B05301	OEM
920B3003	OEM
920B19803	OEM
920B19804	OEM
920B19808	OEM
921	OEM
921-01122	0016
921-01123	0079
921-01124	0079
921-01125	0532
921-01127	0079
921-01128	3519
921-01129	0400
921-01130	0532
921-01131	0161
921-1A	0050
921-1B	0050
921-2A	0050
921-2B	0050
921-3A	0050
921-3B	0050
921-4	0004
921-4A	0050
921-4B	0136
921-5A	0208
921-5B	0208
921-6A	0208
921-6B	0208
921-7	0016
921-7B	0208
921-8	0016
921-9	0004
921-9B	0050
921-10	0050
921-10B	0050
921-11	OEM
921-11B	0050
921-12	0004
921-12B	0050
921-13	0004
921-13B	0050
921-14	0004
921-14B	0004
921-15	0050
921-15B	0050
921-16	0050
921-16B	0050
921-17	0050
921-17B	0050
921-19	0050
921-20	0144
921-20A	0144
921-20B	0144
921-20BK	0016
921-21	0144
921-21A	0144
921-21B	0144
921-21BK	0144
921-22	0144
921-22A	0144
921-22B	0144
921-22BG	0016
921-23	0144
921-23A	0144
921-23B	0144
921-23BK	0016
921-24#	OEM
921-24B	0004
921-25B	0050
921-26	0016
921-26A	0016
921-26B	0050
921-27	0004
921-27A	0004
921-27B	0016
921-28	0016
921-28-BX	OEM
921-28A	0016
921-28B	0016
921-28BLU	0016
921-29B	0037
921-30	0144
921-30A	0144

DEVICE TYPE	REPL CODE
921-30B	0144
921-31	0144
921-31A	0144
921-31B	0144
921-32	0144
921-32A	0144
921-32B	0144
921-33	0144
921-33B	0144
921-34	0144
921-34A	0144
921-34B	0144
921-35	0004
921-35A	0004
921-35B	0004
921-36	OEM
921-36B	0004
921-37B	0004
921-38	0004
921-38A	0004
921-38B	0004
921-39	0004
921-39A	0004
921-39B	0004
921-40	0004
921-40A	0004
921-40B	0004
921-41	OEM
921-41B	0004
921-43B	0016
921-43E	0079
921-44B	0164
921-45	0004
921-45A	0004
921-45B	0004
921-46	0016
921-46A	0016
921-46B	0208
921-46BK	0016
921-47	0016
921-47A	0016
921-47B	0126
921-47BL	0016
921-48	OEM
921-49B	0016
921-50	OEM
921-50B	0016
921-51	0004
921-51B	0004
921-52	0004
921-52B	0004
921-53	OEM
921-53B	0164
921-54	OEM
921-54B	0208
921-55B	0007
921-56B	0007
921-57B	0007
921-58B	0007
921-59B	0127
921-60B	0086
921-61	0142
921-61B	0142
921-62	0127
921-62A	0127
921-62B	0127
921-63	0127
921-63A	0127
921-63B	0127
921-64	0127
921-64A	0127
921-64B	0127
921-64C	0127
921-65	0136
921-65A	0136
921-65B	0136
921-66	0136
921-66A	0136
921-66B	0136
921-67	0004
921-67A	0004
921-67B	0004
921-68	0004
921-68A	0004
921-68B	0004
921-69	0164
921-69A	0164
921-69B	0164
921-70	0037
921-70A	0037
921-70B	0037
921-71	0208
921-71A	0208
921-71B	0208
921-72B	0144
921-73B	0016
921-77B	0016
921-80	OEM
921-81	0280
921-84	0127
921-84A	0127
921-84B	0127
921-85	0127
921-85A	0127
921-85B	0127
921-86	0127
921-86A	0127
921-86B	0127
921-88	0142

DEVICE TYPE	REPL CODE
921-88A	0142
921-88B	0142
921-89	OEM
921-91	OEM
921-92B	0111
921-93B	0016
921-97B	0144
921-98	OEM
921-98B	0144
921-99B	0016
921-100	3555
921-100B	0004
921-101	OEM
921-102	0127
921-102A	0127
921-102B	0127
921-103B	0150
921-104	0150
921-105B	0164
921-106B	0111
921-107	OEM
921-108	OEM
921-109B	0016
921-110B	0126
921-111B	0111
921-112B	0150
921-113	OEM
921-114	0224
921-114B	0111
921-115	0224
921-115B	2209
921-116	0224
921-116B	0111
921-117	OEM
921-117B	0016
921-118B	0164
921-119	0111
921-119B	0127
921-120B	0016
921-121	OEM
921-123B	0016
921-124B	0016
921-125B	0016
921-126B	0325
921-127	OEM
921-127B	1066
921-128B	0016
921-129B	0127
921-130	0127
921-133	0076
921-133B	0076
921-140B	0164
921-141B	0127
921-142B	0127
921-143B	0127
921-144	OEM
921-145	0007
921-145B	0127
921-147B	0004
921-148B	0004
921-150B	0004
921-151	OEM
921-152	OEM
921-152B	0111
921-153	OEM
921-153B	0004
921-154	OEM
921-154B	0111
921-155	OEM
921-155B	0111
921-156B	0142
921-157B	0849
921-158B	0144
921-159	OEM
921-159B	0016
921-160B	0037
921-161B	0016
921-163B	0560
921-164	0560
921-164B	OEM
921-165	0560
921-167	0560
921-170	OEM
921-170B	0144
921-171B	0144
921-172B	0144
921-173	OEM
921-173B	0144
921-174B	0144
921-176	0224
921-176B	0127
921-177	0144
921-177B	0127
921-179	OEM
921-181B	0127
921-182B	0126
921-188B	0086
921-189B	0016
921-191B	0016
921-193	OEM
921-194	OEM
921-195B	0016
921-196	0111
921-196B	0111
921-197B	0037
921-198	3132
921-198BX	1657
921-200B	0016
921-201	OEM

DEVICE TYPE	REPL CODE
921-202	0224
921-202B	2209
921-203	OEM
921-203B	0325
921-204	OEM
921-204B	0127
921-205B	0111
921-206	0111
921-206B	0111
921-207	1221
921-207B	0111
921-208B	0111
921-209	0286
921-209B	0076
921-210B	0127
921-211B	0127
921-212B	0144
921-213B	0127
921-214B	0016
921-215B	0016
921-216B	0004
921-217B	0004
921-222B	0004
921-223B	0004
921-224	OEM
921-224B	0164
921-225B	0016
921-226	0127
921-226B	0127
921-227B	0164
921-228B	0016
921-229B	0016
921-230B	0595
921-231B	0325
921-232	OEM
921-232B	0127
921-233	OEM
921-233B	0127
921-234	OEM
921-234B	0016
921-235B	0127
921-236B	0086
921-237	OEM
921-237B	0191
921-238B	0004
921-239	OEM
921-239B	0111
921-240B	0111
921-241B	0086
921-242B	0164
921-243	OEM
921-243B	1062
921-244	OEM
921-244B	0164
921-250	OEM
921-252B	0016
921-253	OEM
921-254B	0037
921-255B	0016
921-256	OEM
921-256X	0164
921-257B	0111
921-258B	0111
921-262	OEM
921-263	OEM
921-264B	0079
921-265	0144
921-265B	0224
921-266	0144
921-266B	0144
921-267	0144
921-267B	0144
921-268B	0016
921-269	OEM
921-269B	0016
921-270	OEM
921-270B	0126
921-271	OEM
921-272B	1220
921-273B	0004
921-274B	0004
921-275B	0155
921-276B	0155
921-281B	0111
921-282B	0164
921-283	0016
921-288	0016
921-289	OEM
921-290	OEM
921-291B	0079
921-292	0037
921-292B	0037
921-293	OEM
921-294	OEM
921-296	OEM
921-296B	0037
921-301	0224
921-301B	0224
921-304	OEM
921-304B	0079
921-305	OEM
921-305B	0016
921-306B	0079
921-307	0079
921-307B	0079
921-308	0037
921-308B	0037
921-309	0079
921-310	OEM

DEVICE TYPE	REPL CODE
921-311	OEM
921-312	OEM
921-312B	0079
921-313B	0144
921-314	0079
921-314B	0016
921-315B	0126
921-316	OEM
921-317	OEM
921-318B	0004
921-319	OEM
921-319B	0004
921-324	OEM
921-325B	0144
921-326B	0127
921-327B	0164
921-330	OEM
921-331	OEM
921-332B	0037
921-333B	2481
921-333P	2481
921-334	OEM
921-334B	0144
921-335	OEM
921-335B	0144
921-336	OEM
921-336B	0144
921-337B	0111
921-338B	0144
921-340	0086
921-341	OEM
921-345	0079
921-345B	0016
921-346	0342
921-347	0715
921-348B	0037
921-349	0224
921-350B	0224
921-351	0079
921-351B	0079
921-352	OEM
921-352B	0079
921-353	0079
921-353B	0079
921-354B	0079
921-355	0079
921-355B	0079
921-356	OEM
921-357	OEM
921-357B	2050
921-358	OEM
921-359	OEM
921-360B	0079
921-362	OEM
921-366	OEM
921-367	OEM
921-369	0016
921-370	OEM
921-371	OEM
921-372	OEM
921-376	OEM
921-377	0715
921-378	0324
921-379	0224
921-382	3535
921-383	3536
921-387	OEM
921-388	OEM
921-389	6236
921-389(PNP)	OEM
921-393	OEM
921-394	OEM
921-395	OEM
921-396	OEM
921-398	OEM
921-399	OEM
921-400	OEM
921-401	OEM
921-402	OEM
921-403	0400
921-404	3519
921-405	0037
921-407	0079
921-408	0079
921-409	OEM
921-410	0018
921-411	0313
921-412	6236
921-414	OEM
921-416	OEM
921-421	OEM
921-422	OEM
921-424	0313
921-425	1591
921-426	OEM
921-428	0144
921-429	0236
921-430	0681
921-431	0042
921-432	0925
921-433	0925
921-434	0042
921-436	OEM
921-437	OEM
921-438	OEM
921-439	OEM
921-446	OEM
921-449	0079
921-450	0079

DEVICE TYPE	REPL CODE
921-451	OEM
921-452	3519
921-453	1954
921-462	0079
921-463	0079
921-464	0079
921-465	OEM
921-467	0236
921-468	OEM
921-469	OEM
921-470	0079
921-472	1954
921-473	1954
921-474	OEM
921-476	2140
921-477	OEM
921-478	OEM
921-481	OEM
921-497	OEM
921-538	OEM
921-548	OEM
921-732	OEM
921-931	0042
921-1000	OEM
921-1001	OEM
921-1002	OEM
921-1005	2460
921-1006	0264
921-1007	0037
921-1008	1978
921-1009	0042
921-1010	1978
921-1011	0867
921-1013	0155
921-1014	0224
921-1015	0076
921-1016	0006
921-1017	0076
921-1018	1946
921-1019	1758
921-1020	0264
921-1021	0382
921-1022	0155
921-1026	OEM
921-1028	0076
921-1031	3342
921-1034	OEM
921-1037	OEM
921-1043	OEM
921-1045	0079
921-1047	0042
921-1049	0079
921-1051	0079
921-1052	0264
921-1053	0111
921-1055	0111
921-1059	0144
921-1060	0079
921-1062	0079
921-1063	0079
921-1065	0079
921-1067	2888
921-1068	0127
921-1069	0286
921-1070	0006
921-1071	0286
921-1072	0079
921-1073	0286
921-1074	0111
921-1075	0286
921-1076	0276
921-1077	0286
921-1078	0286
921-1080	OEM
921-1083	OEM
921-1084	OEM
921-1086	OEM
921-1090	0622
921-1093	OEM
921-1095	OEM
921-1102	OEM
921-1103	0079
921-1105	OEM
921-1108	OEM
921-1111	OEM
921-1114	1386
921-1115	0477
921-1116	OEM
921-1118	OEM
921-1119	OEM
921-1120	OEM
921-1121	OEM
921-1122	0079
921-1123	OEM
921-1125	OEM
921-1126	OEM
921-1127	OEM
921-1128	OEM
921-1129	OEM
921-1132	OEM
921-1133	OEM
921-1134	0037
921-1135	OEM
921-1142	0079
921-1143	0925
921-1147	OEM
921-1148	OEM
921-1149	OEM
921-1150	3072

If replacement code is OEM, contact original manufacturer for replacement.

DEVICE TYPE	REPL CODE
921-1152	0286
921-1153	OEM
921-1154	OEM
921-1156	OEM
921-1157	OEM
921-1163	OEM
921-1164	OEM
921-1165	OEM
921-1166	OEM
921-1168	0286
921-1169	0042
921-1171	0006
921-1176	OEM
921-1177	OEM
921-1180	0006
921-1182	0006
921-1185	0286
921-1186	OEM
921-1187	OEM
921-1188	0360
921-1191	0693
921-1192	OEM
921-1195	0079
921-1196	0079
921-1197	0037
921-1199	0423
921-1200	0693
921-1201	0325
921-1202	0018
921-1203	1946
921-1204	0076
921-1205	3717
921-1206	1758
921-1207	0037
921-1208	0532
921-1209	0155
921-1210	0042
921-1211	OEM
921-1214	0037
921-1216	OEM
921-1219	OEM
921-1220	OEM
921-1227	OEM
921-1228	OEM
921-1229	0325
921-1231	OEM
921-1234	OEM
921-1236	OEM
921-1237	OEM
921-1239	OEM
921-1240	OEM
921-1241	OEM
921-1244	0224
921-1245	OEM
921-1247	OEM
921-1249	OEM
921-1250	2030
921-1251	2561
921-1252	1140
921-1253	0037
921-1255	0325
921-1257	OEM
921-1258	OEM
921-1259	OEM
921-1261	OEM
921-1262	0037
921-1263	0079
921-1267	OEM
921-1268	OEM
921-1269	OEM
921-1270	OEM
921-1271	OEM
921-1273	OEM
921-1279	0286
921-1280	OEM
921-1281	0037
921-1282	1221
921-1283	OEM
921-1284	0076
921-1285	0325
921-1286	0937
921-1287	0076
921-1288	0161
921-1289	OEM
921-1290	0551
921-1291	0037
921-1292	0286
921-1293	0144
921-1294	0224
921-1295	0079
921-1298	OEM
921-1299	OEM
921-1301	OEM
921-1303	0111
921-1304	OEM
921-1305	0324
921-1307	OEM
921-1308	OEM
921-1309	3703
921-1310	0009
921-1311	3625
921-1312	0018
921-1313	0013
921-1314	0076
921-1315	0076
921-1316	4104
921-1317	0249
921-1318	OEM
921-1319	1308
921-1320	0150
921-1321	0325
921-1322	1051
921-1323	OEM
921-1324	OEM
921-1325	0086
921-1326	3625
921-1327	2030
921-1334	OEM
921-1335	0144
921-1336	0144
921-1337	0144
921-1338	0079
921-1339	0079
921-1340	0037
921-1341	0006
921-1342	OEM
921-1355	OEM
921-1356	OEM
921-1358	OEM
921-1362	0286
921-1370	OEM
921-1384	OEM
921-1400	OEM
921-1420	OEM
921-1427	OEM
921-1468	OEM
921-1469	OEM
921-1470	OEM
921-1471	OEM
921-1478	OEM
921-1483	0224
921-1509	OEM
921-1515	0825
921-1516	OEM
921-1544	0018
921-1550	0832
921-1560	0359
921-1564	0127
921-1612	OEM
921-1632	0724
921-1689	4130
921-1690	OEM
921-1693	4172
921-1701	1386
921-1704	OEM
921-1799	0887
921-1801	1032
921-1803	OEM
921-1824	OEM
921-1838	0898
921-1866	OEM
921-1867	0006
921-1868	OEM
921-1869	OEM
921-1869(2SD1449T)	OEM
921-1870	2954
921-1871	2954
921-1872	0013
921-1873	0013
921-1874	0151
921-1875	0898
921-1876	1032
921-1877	0013
921-1878	1242
921-1879	0148
921-1880	0832
921-1882	OEM
921-1883	2954
921-1884	OEM
921-1885	0018
921-1886	0532
921-1887	0013
921-1888	2954
921-1889	1758
921-1890	0434
921-1891	0013
921-1892	2954
921-1893	0155
921-1894	OEM
921-1895	OEM
921-1896	2923
921-1897	0151
921-1898	OEM
921-1899	OEM
921-1900	0832
921-1924	1562
921-1946	0151
921-1947	OEM
921-1988X	1657
921-3558	0079
921-10026	0079
921-10027	1092
921-10028	6193
921-10029	0006
921-10030	0066
921-10031	1541
921-10032	OEM
921-10033	0006
921-10034	0261
921-10035	1911
921-10036	0392
921-10037	OEM
921-10038	OEM
921HC	3337
921HM	OEM
922	OEM
923	OEM
923HC	OEM
923HM	OEM
924-2209	0400
924-16598	0004
924-17945	0160
925	OEM
925-1	OEM
925-2	3537
926	OEM
926FM	OEM
926HC	3924
926HM	OEM
927HC	4218
930(0V)	0111
930B	0111
930DC	1823
930DM	OEM
930DZ	0127
930FM	OEM
930HM	OEM
930PC	1823
930X1	0016
930X2	0016
930X3	0016
930X4	0144
930X5	0144
930X6	0136
930X7	0004
930X8	0004
930X9	0004
930X10	0004
931	0133
931DC	3061
931FM	OEM
931PC	3061
932	OEM
932DC	1041
932DM	OEM
932FM	OEM
932HM	OEM
932PC	1041
933DC	2097
933DM	OEM
933FM	OEM
933HM	OEM
933PC	2097
934	OEM
934Q5FC	OEM
935-1	2250
935DC	1176
935PC	1176
936-10	0015
936-20	0015
936-NPN	0161
936DC	1831
936PC	1831
937DC	1835
937PC	1835
940	OEM
941	OEM
941DC	2622
941DM	OEM
941FM	OEM
941HM	OEM
941PC	2622
941T1	0847
942	OEM
942-002	0723
942-337	OEM
942-341	OEM
943	OEM
944	OEM
944DC	0033
944DM	OEM
944FM	OEM
944H	OEM
944HM	OEM
944PC	0033
945	1669
945DC	0081
945DM	OEM
945FM	OEM
945HM	OEM
945PC	0081
946	OEM
946-8620-3	0015
946DC	0141
946DM	OEM
946FM	OEM
946PC	0141
947	OEM
947-1	0079
948-10300	0725
948DC	3399
948FM	OEM
948HM	OEM
948PC	3399
949	OEM
949DC	1844
949DM	OEM
949FM	OEM
949PC	1844
950	OEM
950-D10	OEM
950DC	2619
950DM	OEM
950FM	OEM
950H	OEM
950H1-S	OEM
950HA-1	OEM
950HM	OEM
950PC	2619
951	0018
951-1	0211
951-1(SYLVANIA)	0136
951DC	2078
951DM	OEM
951FM	OEM
951HM	OEM
951PC	2078
952	0086
952DC	OEM
953	0086
953DC	OEM
953H	0037
954	OEM
954DC	OEM
955	OEM
955-1	0050
955-2	0050
955-3	0050
955DC	OEM
956	OEM
956DC	OEM
957	OEM
957-7656-0	1137
957DC	OEM
958	OEM
958DC	OEM
958HM	OEM
960DC	OEM
961DC	1859
961PC	1859
962DC	0562
962DM	OEM
962FM	OEM
962PC	0562
963DC	0341
963FM	OEM
963PC	0341
964-2073B	0016
964-2209	0400
964-16598	0136
964-16599	0085
964-17142	0211
964-17443	0015
964-17444	0016
964-17887	0085
964-17945	0160
964-19862	2866
964-19862A	0004
964-19863	0004
964-19864	0004
964-19865	0015
964-20738	0016
964-20739	0126
964-21866	0015
964-22008	0126
964-22009	0400
964-22158	0086
964-24387	0086
964-24584	0016
964-25046	0086
964-27986	0566
964-174443	0015
964DC	OEM
965-1A6-5	0050
965-2A6-1	0208
965-4A6-2	0600
965A16-3	0050
965B16-2	0136
965C-16-4	0160
965T1	0211
966-1A6-3	0004
966-2A6-4	0004
966-3A6C	0004
966DC	OEM
967-MP4	OEM
967D	OEM
970	OEM
971-11	OEM
971-12	OEM
971DC	OEM
972-659E-0	0006
972-659E-O	0342
972D7	0015
972X1	0050
972X2	0050
972X3	0050
972X4	0050
972X5	0050
972X6	0136
972X7	0136
972X8	0136
972X9	0004
972X10	0004
972X11	0004
972X12	0004
973	OEM
974-1(SYLVANIA)	0276
974-2	0285
974-2(SYLVANIA)	0276
974-3	0276
974-3A6-3	0276
974-4	0285
974-4(SYLVANIA)	0276
974HC	1810
975	0079
976-0036-921	0133
977-1	0133
977-2	0015
977-2B	0015
977-3B	0071
977-4	0015
977-4B	0277
977-5	0015
977-6B	0277
977-8	0015
977-8B	0015
977-9	0015
977-10	0015
977-10B	0015
977-11	0015
977-11B	0015
977-12	0015
977-13	0015
977-13B	0015
977-14	OEM
977-14B	0196
977-15	OEM
977-16	0015
977-17	0015
977-18	0015
977-18B	0015
977-19	0023
977-19B	0023
977-20	0015
977-20B	0015
977-22	0015
977-22B	0015
977-23	0015
977-23B	0015
977-24B	0277
977-25B	0015
977-26	0015
977-27B	0015
977-28	0015
977-28B	0015
977-30	0774
977-31	3858
977-32	3858
977-33	3858
977-34	1067
977-34-01	1067
977-35	1067
977-35-02	1706
977-35-03	1067
977-36	1706
977-37	1706
977-38	1501
977-39	1501
977-40	3655
977-41	2109
977-42	2109
977-43	2109
977-44	OEM
977-45	3872
977-46	OEM
977-52	2983
977-218	0277
977-Z9500A	0190
977-Z9521	3858
977-Z9522	1501
977-Z9523	1067
977-Z9526A	1706
977-Z9529	2546
977-Z9530	3655
977-Z9531	2804
977-Z9532	0409
977-Z9533	1710
977-Z9534	2982
977-Z9535	2985
977-Z9536	1997
977-Z9536A	1997
977-Z9537	1197
977-Z9538	2984
977-Z9539	1054
977A1-6-1	0440
978-1923	0126
980H	OEM
981H	OEM
982H	OEM
983H	OEM
984A-1-6B	0160
985H	OEM
986M	0037
987T1	0211
988T1	0211
989DC	OEM
989HM	OEM
989T1	0211
990-110	OEM
990-130	OEM
990-150	OEM
990-160	OEM
990-170	OEM
990-180	OEM
990-190	OEM
990T1	0211
991-00-1172	0133
991-00-1172-1	0133
991-00-1219	0016
991-00-1221	0004
991-00-1222	0004
991-00-1449	0229
991-00-1449-1	0229
991-00-2232	0016
991-00-2248	0016
991-00-2298	0016
991-00-2356	0016
991-00-2356/K	0016
991-00-2440	0229
991-00-2440-1	0229
991-00-2440-2	0229
991-00-2873	0016
991-00-2888	0086
991-00-3144	0016
991-00-3304	0016
991-00-7394	0015
991-00-7776	0015
991-00-8393	0016
991-00-8393A	0016
991-00-8393M	0016
991-00-8394	0016
991-00-8394A	0016
991-00-8394AH	0016
991-00-8395	0016
991-002232	0155
991-002298	0111
991-002356	0111
991-002873	0111
991-003304	0155
991-008393	0155
991-01-0098	0037
991-01-0099	0160
991-01-0461	0400
991-01-0462	0037
991-01-1216	0160
991-01-1217	0164
991-01-1219	0016
991-01-1220	0016
991-01-1221	0004
991-01-1222	0004
991-01-1223	0004
991-01-1224	0004
991-01-1225	0037
991-01-1305	0086
991-01-1306	0016
991-01-1312	0016
991-01-1314	0086
991-01-1315	0126
991-01-1316	0127
991-01-1317	0103
991-01-1318	0016
991-01-1319	0037
991-01-1705	0016
991-01-1706	0325
991-01-2328	0037
991-01-2686	0126
991-01-3044	0016
991-01-3055	0325
991-01-3056	0016
991-01-3057	0016
991-01-3058	0037
991-01-3063	0103
991-01-3068	0016
991-01-3170	0423
991-01-3543	0400
991-01-3544	0016
991-01-3599	0037
991-01-3683	0016
991-01-3740	0016
991-01-5001	0423
991-01-5063	0423
991-010098	0037
991-010462	0155
991-011217	0164
991-011219	0016
991-011220	0016
991-011221	0004
991-011222	0004
991-011223	0004
991-011225	0037
991-011305	0086
991-011306	0016
991-011312	0016
991-011313	0016
991-011314	0086
991-011318	0016
991-011319	0037
991-011576	0325
991-011706	0325
991-012686	0079
991-013044	0155
991-013056	0155
991-013057	0155
991-013058	0037
991-013063	0042
991-013068	0155
991-013543	0400
991-013544	0111
991-015587	0016
991-015614	0006
991-015615	0155
991-015663	0400
991-015942	0723
991-016274	0155
991-019542-1	1876
991-025515	0215
991-025515-001	0215
991-025515-002	0215
991-025517	0626
991-026005	0500
991-026022	1949
991-026022-00	1949
991-026025	0478
991-026605	0519
991-026609	1014
991-026615	0513
991-027358	0413
991-027358-001	0413
991-028650-001	0398
991-028679-001	1143
991-028680-001	1525
991-028681-001	1957
991-028682-001	2072
991-029866	1749
991T1	0211
992-00-1192	0085
992-00-2271	0103
992-00-2298	0016
992-00-3139	0103
992-00-3139A	0103
992-00-3144	0016
992-00-3172	0142
992-00-4091	0103
992-00-4092	0103
992-00-8870	0160
992-00-8890	0604
992-00-8890L	0604
992-008-890	0265
992-00271	0103
992-00271A	0103
992-001192	0160
992-002271	0103
992-003139	0103
992-003172	0429
992-004091	0103
992-008870	0604
992-008890	0160
992-01-1216	0160
992-01-1218	0160
992-01-1317	0103
992-01-3684	0233
992-01-3705	0178
992-01-3738	0016
992-02271	0103
992-011216	0160
992-011218	0085
992-011317	0042
992-017169	0103
992-020054	2427
992-1A6-1	0440
992-10	0015
992-531-01	0277
992T1	0004
995-01-6130	0925
995-01-6131	0042
1000	OEM
1000-0948	OEM
1000-0963	OEM
1000-06971	OEM
1000-12	0012
1000-17	0143
1000-23	1395
1000-25	0632
1000-129	0015
1000-130	0133
1000-131	0143
1000-132	0012
1000-135	0151
1000-136	0076
1000-137	0076
1000-138	2046
1000-139	1400
1000-139*	0698
1000-140	2050
1000-141	1411
1000-142	0042
1000-2640	0137
1000-6914	0062
1000-6948	1260
1000-6963	5172
1000-6971	4372
1000-6989	1392
1000-7052	1513
1000-7227	3816
1000-7334	1513
1000-7359	0106
1000-9959	1066
1000DS10	OEM
1000EXD22	OEM
1000FXD22	OEM
1000HB1U	5739
1000HB2U	5739
1000HB3U	5739
1000HB4U	5739
1000HB6U	5739
1000HT1U	5736
1000HT2U	5736
1000HT3U	5736
1000HT4U	5736
1000HT6U	5736
1000HT8U	5736
1000PA50	6330
1000PA60	6330
1000PA80	6167
1000PA100	6167
1000PA110	6167
1000PA120	6167
1000PA130	OEM
1000PA140	OEM
1000PA150	OEM
1000PK50	OEM
1000PK60	OEM
1000PK80	OEM

DEVICE TYPE	REPL CODE
1000PK100	OEM
1000PK120	6167
1000PK140	OEM
1000PK150	OEM
1000PK160	OEM
1000PK170	OEM
1000PK180	OEM
1000PKL60S65	OEM
1000PKL60S646	OEM
1000PKL60S656	OEM
1000PKL60S666	OEM
1000PKL60S676	OEM
1000PKL60S683	OEM
1000PKL60T603	OEM
1000PKL60T623	OEM
1000PKL60T663	OEM
1000PKL80S65	OEM
1000PKL80S646	OEM
1000PKL80S656	OEM
1000PKL80S666	OEM
1000PKL80S676	OEM
1000PKL80S683	OEM
1000PKL80T603	OEM
1000PKL80T623	OEM
1000PKL80T663	OEM
1000PKL100S65	OEM
1000PKL100S646	OEM
1000PKL100S656	OEM
1000PKL100S666	OEM
1000PKL100S676	OEM
1000PKL100S683	OEM
1000PKL100T603	OEM
1000PKL100T623	OEM
1000PKL100T663	OEM
1000PKL110S65	OEM
1000PKL110S646	OEM
1000PKL110S656	OEM
1000PKL110S666	OEM
1000PKL110S676	OEM
1000PKL110S683	OEM
1000PKL110T603	OEM
1000PKL110T623	OEM
1000PKL110T663	OEM
1000PKL120S65	OEM
1000PKL120S646	OEM
1000PKL120S656	OEM
1000PKL120S666	OEM
1000PKL120S676	OEM
1000PKL120S683	OEM
1000PKL120T603	OEM
1000PKL120T623	OEM
1000PKL120T663	OEM
1000PKL130S65	OEM
1000PKL130S646	OEM
1000PKL130S656	OEM
1000PKL130S666	OEM
1000PKL130S676	OEM
1000PKL130S683	OEM
1000PKL130T603	OEM
1000PKL130T623	OEM
1000PKL140S65	OEM
1000PKL140S646	OEM
1000PKL140S656	OEM
1000PKL140S666	OEM
1000PKL140S676	OEM
1000PKL140S683	OEM
1000PKL140T603	OEM
1000PKL140T623	OEM
1000PKL140T663	OEM
1000PKL1305663	OEM
1000YKD22	OEM
1001	6768
1001(EF-JOHNSON)	2867
1001(TRANS)	0086
1001-0014	0143
1001-0036	0443
1001-01	0086
1001-02	0191
1001-03	0368
1001-04	0380
1001-05	0155
1001-06	0788
1001-07	0698
1001-08	1411
1001-09	0236
1001-012	0012
1001-0158	OEM
1001-0478	0120
1001-5	1413
1001-6	1413
1001-10	0143
1001-11	0015
1001-12	0012
1001-13	0102
1001-1179	0079
1001-4975	OEM
1001-5600	0120
1001-5659	0143
1001-5667	0015
1001-5683	0023
1001-5709	OEM
1001-5717	0261
1001-5766	0037
1001-5816	0925
1001-5824	0925
1001-5832	0055
1001-5840	0168
1001-5865	0079
1001-5915	0286
1001-5949	0076
1001-5956	2533
1001-5972	0167
1001-6657	0461
1001-6699	0071
1001-6731	0023
1001-6806	0133
1001-6848	0288
1001-6863	0062
1001-6939	0261
1001-6947	0715
1001-6962	0342
1001-7005	OEM
1001-7044	0079
1001-7069	0079
1001-7077	0079
1001-7085	0342
1001-7119	0392
1001-7135	0079
1001-7150	0852
1001-7168	0857
1001-7877	OEM
1001-9552	0133
1001-9602	0143
1001-9610	0023
1001-9628	0261
1001-9636	0155
1001-9644	0286
1001-9651	0392
1001-9677	0852
1001-9685	0857
1001-9693	0167
1001-9727	1429
1001-9735	1329
1001-9958	0224
1001-9966	0079
1001-9974	0079
1001-9990	0079
1001PDE120	OEM
1001PDE140	OEM
1001PDE160	OEM
1001PDE180	OEM
1001PDE200	OEM
1001PDE210	OEM
1001PDE220	OEM
1001PDE230	OEM
1001PDE240	OEM
1001PDE250	OEM
1002	0693
1002-0006	0579
1002-0014	0143
1002-01	0325
1002-02	2209
1002-02A	0144
1002-03	0016
1002-04	0016
1002-04-1	0144
1002-05	0004
1002-07	0021
1002-08	0143
1002-09	0143
1002-0219	0296
1002-0345	0362
1002-0402	0079
1002-0436	0925
1002-0519	0589
1002-17	0143
1002-17(DIODE)	0143
1002-68	2209
1002-345	OEM
1002-4404	0137
1002-4891	0037
1002-5567	0224
1002-5591	0681
1002-5625	0079
1002-5690	0077
1002-6219	0296
1002-9296	0120
1002-9668	0100
1002A	OEM
1002A(JULIETTE)	0127
1002A-1	0126
1002A-2	0086
1003	0004
1003(E.F.-JOHNSON)	0004
1003(JULIETTE)	0136
1003(TRANS)	0887
1003-01	0127
1003-02	2393
1003-0284	OEM
1003-0518	1648
1003-0567	1055
1003-11	0133
1003-13	0015
1003-2084	0301
1003-3116	0362
1003-6754	0111
1003A	OEM
1003PJA250	OEM
1003PJA300	OEM
1004	0198
1004(2SC537)	0016
1004(DIODE)	0143
1004(G.E.)	0007
1004(JULIETTE)	0144
1004(JULIETTE)*	0144
1004(TRANS)	0898
1004-01	0164
1004-02	1400
1004-03	0016
1004-0392	0037
1004-0434	0224
1004-0558	0317
1004-0780	1221
1004-0798	0239
1004-3	OEM
1004-17	0144
1004-25	OEM
1004-1986	1247
1004-1994	1260
1004-2042	1318
1004A	OEM
1005	0016
1005(2SC537)	0016
1005(JULIETTE)	0004
1005(JULIETTE)*	0004
1005(JULIETTERECTIFI	0015
1005-17	0004
1005-20	0015
1005A	OEM
1006	0224
1006(DIODE)	0143
1006(G.E.)	0007
1006(JULIETTE)	0079
1006(JULIETTE)*	0004
1006(JULIETTE-RECTIF	0015
1006/0945	0064
1006-0887	1260
1006-0895	1513
1006-0903	2063
1006-0937	0423
1006-0945	0064
1006-17	0071
1006-21	0151
1006-24	0015
1006-48	0151
1006-78	0050
1006-93	0004
1006-1737	0002
1006-2594	0006
1006-5977	0030
1006-5985	0170
1006-7445	3711
1006-9292	0019
1006A	OEM
1007	0143
1007(JULIETTE)*	0136
1007(JULIETTE-RECTIF	0015
1007-0951	0795
1007-17	0004
1007-1124	0143
1007-3054	0224
1007-3062	0224
1007-3088	0224
1007-3153	0076
1007-3229	0715
1007-3344	0400
1007-5869	0286
1007-5901	0140
1007-5919	0554
1007-6537	3982
1008	0086
1008(E.F.-JOHNSON)	0004
1008(E.F.JOHNSON)	0164
1008(JULIETTE)	0222
1008(POWER)	0222
1008-02	0016
1008-17	0085
1009	0004
1009(E.F.-JOHNSON)	0004
1009(G.E.)	0007
1009(JULIETTE-RECTIF	0015
1009-01	1280
1009-01*	0212
1009-02	0155
1009-03	1400
1009-03*	0698
1009-03-17	0086
1009-04	1411
1009-04-17	0698
1009-05	2046
1009-05A	0423
1009-06	0025
1009-06(10)	0649
1009-07	0015
1009-08	0133
1009-09	0015
1009-2	0079
1009-11	0512
1009-17	0016
1009-127	0688
1009-8403	1284
1010	0170
1010(20A90)	0123
1010/8033	0037
1010-0866	0079
1010-14	1400
1010-17	1400
1010-17(R.F.)	0698
1010-78	0050
1010-87	0050
1010-89	0136
1010-143	0133
1010-145	0143
1010-3417	0023
1010-7738	0037
1010-7928	0016
1010-7936	0423
1010-7951	0181
1010-7985	0065
1010-7993	0086
1010-8009	0006
1010-8025	0065
1010-8033	0203
1010-8041	0086
1010-8066	0330
1010-8082	0016
1010-8090	0016
1010-8116	0102
1010-8132	0565
1010-8165	0124
1010-8173	0143
1010-9486	0023
1010-9494	0017
1010-9932	0856
1010-9940	0167
1010-9965	0802
1010-9973	0666
1010A	0167
1011	0143
1011-01	0698
1011-02	0025
1011-0302	0124
1011-11	0338
1011-11(R.F.)	0144
1012	0015
1012(GE)	0037
1012(G.E.)	0037
1012-17	0143
1012-17(POWER)	0097
1012A1PGBP	OEM
1012A3PGAO	OEM
1012A10	OEM
1012B30	OEM
1012C1PGAO	OEM
1012D3PGBP	OEM
1012GE	0150
1013	0143
1013-15	2209
1013-16	0004
1014	0079
1014(E.F.-JOHNSON)	0004
1014-25	0632
1014AO	OEM
1014AP	OEM
1014BO	OEM
1014BP	OEM
1014CO	OEM
1014CP	OEM
1015	0002
1015-0518	OEM
1016	0693
1016(GE)	0037
1016-17	0015
1016-77	0143
1016-78	0015
1016-79	0015
1016-80	0632
1016-81	0042
1016-83	0144
1016-84	0144
1016-85	0475
1018-25	1837
1018-1048	0825
1018-1063	0076
1018-2400	0019
1018-3259	0133
1018-4943	0008
1018-6278	0170
1018-6963	0015
1018-7961	0071
1018-9884	0133
1019-0569	0365
1019-25	2155
1019-74	0004
1019-1203	0330
1019-1385	0102
1019-3852	0016
1019-6699	0143
1019M	0037
1020	0015
1020-01	OEM
1020-17	0016
1020-25	0470
1020G	0127
1021	0325
1021-17	0211
1021-25	0911
1021A	OEM
1022	OEM
1022-0911	0036
1022-1976	0165
1022-5548	0015
1023-17	0211
1023A	OEM
1023AD8DIA3	OEM
1023AD8DIB3	OEM
1023AD8DIC3	OEM
1023AD8DID3	OEM
1023AD16PDA3	OEM
1023AD16PDB3	OEM
1023AD16PDC3	OEM
1023AD16PDD3	OEM
1023AD16SEA3	OEM
1023AD16SEB3	OEM
1023AD16SEC3	OEM
1023AD16SED3	OEM
1023EX16DI005	OEM
1023EX16DI16AOP	OEM
1023EX16DI16BOP	OEM
1023EX32DI00P	OEM
1023EX32DI16AOP	OEM
1023EX32DI16BOP	OEM
1023EX32PD00P	OEM
1023EX32PD16AOP	OEM
1023EX32PD16BOP	OEM
1023EX32SE00P	OEM
1023EX32SE16AOP	OEM
1023EX32SE16BOP	OEM
1023G	0016
1023G(GE)	0016
1023M	OEM
1024	0015
1024-17	0085
1024G	0079
1024M11	OEM
1025	OEM
1025-2712	0167
1025-4172	0167
1025-4712	0167
1025G	0079
1026	0127
1026G	0016
1027	0441
1027(G.E.)	0144
1027-4231	0770
1027-4264	5218
1027-4389	3902
1027-4454	OEM
1027-4462	5260
1027-4850	5261
1027-4959	5262
1027-5006	4925
1027-5113	5271
1027-5394	OEM
1027-5535	5263
1027-5816	OEM
1027-5881	OEM
1027-6244	3159
1027-6319	OEM
1027-6459	OEM
1027-6533	0825
1027-6624	0648
1027-7259	1165
1027-7291	1165
1027-7366	0018
1027-7986	0434
1027-8042	0018
1027-8059	0672
1027-9974	0023
1027-9982	0609
1027-9990	0041
1027G	0086
1028	0079
1028-0006	0253
1028-0014	0253
1028-0030	0064
1028-0048	0064
1028-0055	0023
1028-0063	1778
1028-0071	0124
1028-0089	0124
1028-0097	0041
1028-0105	0062
1028-1079	OEM
1028-3984	0133
1028-4297	0111
1028-4305	2481
1028-4313	0144
1028-6680	0005
1028-6755	0023
1028A	OEM
1028G	0016
1029	0079
1029(G.E.)	0016
1029-02	0065
1029-1788	OEM
1029G	0016
1030	0037
1030(G.E.)	0065
1030-8DIA300P	OEM
1030-8DIB300P	OEM
1030-8DIC300P	OEM
1030-8DID300P	OEM
1030-16DIA300P	OEM
1030-16DIC300P	OEM
1030-16DID300P	OEM
1030-16PDA300P	OEM
1030-16PDB300P	OEM
1030-16PDC300P	OEM
1030-16PDD300P	OEM
1030-16SEA300P	OEM
1030-16SEB300P	OEM
1030-16SEC300P	OEM
1030-16SED300P	OEM
1030-17	0143
1030-25	1172
1030-32DIA300P	OEM
1030-32DIB300P	OEM
1030-32DIC300P	OEM
1030-32DID300P	OEM
1030-32PDA300P	OEM
1030-32PDB300P	OEM
1030-32PDC300P	OEM
1030-32PDD300P	OEM
1030-32SEA300P	OEM
1030-32SEB300P	OEM
1030-32SEC300P	OEM
1030-32SED300P	OEM
1030-64PDA300P	OEM
1030-64PDB300P	OEM
1030-64PDC300P	OEM
1030-64PDD300P	OEM
1030-64SEA300P	OEM
1030-64SEB300P	OEM
1030-64SEC300P	OEM
1030-64SED300P	OEM
1030-5530	0060
1031-17	0087
1031-25	0268
1032	0211
1032-25	2872
1032-3467	0139
1032-3475	0139
1032-3483	0139
1032-8243	0309
1032R40	OEM
1033	0137
1033-0911	0133
1033-0983	0133
1033-0991	0133
1033-1	0136
1033-2	0136
1033-3	0136
1033-4	0136
1033-5	0211
1033-6	0038
1033-7	0211
1033-8	0015
1033-17	0015
1033-1007	0124
1033-1247	2067
1033-1262	2068
1033-1270	0058
1033-1338	0489
1033-1874	2073
1033-1916	0019
1033-4340	0101
1033-4985	0203
1033-5313	0359
1033-5537	0359
1033-6204	0825
1033-6451	0155
1033-6493	0155
1033-6949	0286
1033-6956	0286
1033-7079	2223
1033-7400	0582
1033-7426	0582
1033-7699	0672
1033-7905	0582
1034	0164
1034-17	0016
1034-43	0600
1035	0164
1035-02	OEM
1035-80	1220
1035-7564	0624
1035Z4	OEM
1036	0164
1036A	OEM
1037	OEM
1037-0575	2411
1037-65	1220
1037-7174	3816
1038	0142
1038(G.E.)	0086
1038/9922	0203
1038-1	0079
1038-1-10	0016
1038-6	0016
1038-6CL	0016
1038-8	0016
1038-10	0016
1038-15	0016
1038-15CL	0016
1038-18	0016
1038-18CL	0016
1038-21	0016
1038-23	0016
1038-23CL	0016
1038-24	0016
1038-1697	0080
1038-1721	0015
1038-1788	0182
1038-1804	0528
1038-1853	0168
1038-1861	0629
1038-9922	0016
1039-0060	0006
1039-01	0076
1039-0433	0219
1039-0441	0155
1039-0458	0155
1039-0461	1220
1039-0482	1663
1039-0515	0168
1039-0516	0643
1039-0888	0127
1039-0961	1220
1039-1290	2051
1039-9772	0842
1040	0015
1040-01	1221
1040-02	0178
1040-03	0016
1040-04	1400
1040-05	1411
1040-06	OEM
1040-08	0143
1040-09	0181
1040-0968	OEM
1040-2	0016
1040-7	0086
1040-10	0015
1040-11	0086
1040-59	0050
1040-80	0208
1040-81	0015
1040-155	0155
1040-1373	OEM
1040-9068	0006
1040-9332	0023
1040-9373	0494
1041-63	0023
1041-66	0133
1041-68	0138
1041-69	2406
1041-71	0155
1041-72	0076
1041-76	2050
1041-77	0698
1041-4950	0582
1041-4951	0582
1041-6949	0148
1042-01	0212
1042-02	3046
1042-03	0076
1042-04	0076
1042-05	0191
1042-06	0006
1042-07	0368
1042-08	0839
1042-09	0836
1042-10	1010
1042-11	0823
1042-12	0019
1042-13	0123
1042-14	0143
1042-15	0133
1042-16	0124
1042-17	0015
1042-18	0012
1042-19	0165
1042-20	0143
1042-23	0120
1042-7938	0395
1043-0049	0133
1043-07	0076
1043-1195	0212
1043-1229	0155
1043-1260	0127
1043-1278	0037
1043-1286	0042
1043-1294	0925
1043-1310	0460
1043-1328	0551
1043-1344	0701
1043-1534	0137
1043-1583	0120
1043-7069	OEM
1043-7267	OEM
1043-7275	3772
1043-7309	0142
1043-7358	0168
1043-7374	0037
1043-7382	0015
1044	0015
1044-0295	0037
1044-6888	0436
1044-7035	3270
1044-7043	2049
1044-7049	2049
1044-7878	0423
1044-8983	0133
1044-9544	0086
1045-0518	0015
1045-0534	0015
1045-2761	0167
1045-2951	0016
1045-3082	0086
1045-4676	0037
1045-7802	0133
1045-7828	0042
1045-7836	0925
1045-7844	0103
1045-7851	0103
1045Z4	OEM
1047-25	0433
1047-715	OEM
1047-1571	0080
1047-7115	0006
1048-1307	0080
1048-6408	OEM
1048-6421	0133
1048-7361	0261
1048-9839	0244
1048-9870	0019
1048-9888	0123
1048-9904	0224
1048-9912	0079
1048-9920	0076
1048-9938	0030
1048-9946	0015

DEVICE TYPE	REPL CODE
1048-9953	0139
1048-9987	0133
1048-9995	0015
1049-0035	0532
1049-0060	0151
1049-0092	0111
1049-0100	1220
1049-0167	0018
1049-1744	0076
1049-3435	0023
1050	0015
1050-5816	0649
1050B	0150
1051-5559	0123
1052-0229	0148
1052-17	0004
1052-6390	2942
1052-6408	3269
1052-6416	3273
1054-8295	1323
1055-7726	0224
1055Z4	OEM
1056-4417	0023
1056-8269	0423
1056-8376	OEM
1056-8384	4591
1056-8392	OEM
1057-17	0004
1057-2071	0286
1057-9140	0317
1058	0285
1058-3227	3327
1058-3664	0824
1059-2848	0102
1059-7805	0139
1059-7961	0071
1059-9140	0317
1059M	0037
1060	0015
1060-17	0004
1060-17(DRIVER)	0164
1060-17(PRE-AMP)	0211
1060-3553	0143
1060-6564	0006
1060-8149	0113
1060-9428	0261
1061-4667	0143
1061-5972	0167
1061-6274	1150
1061-6282	0220
1061-6290	0017
1061-7090	OEM
1061-8312	0006
1061-8320	0076
1061-8321	OEM
1061-8338	1284
1061-8346	0643
1061-8353	0643
1061-8361	0071
1061-8379	0023
1061-8387	0023
1061-8668	0643
1061-8807	0006
1061-8908	0643
1061-8916	0023
1061-8924	0023
1061-8932	1005
1061-9068	0006
1061-9153	3915
1061-9161	2949
1061-9526	4238
1061-9666	1326
1061-9856	2231
1062-0615	0144
1062-0706	OEM
1062-1414	OEM
1062-6018	0037
1062-6414	0006
1062-7511	2668
1062GE	0133
1063-79	0015
1063-3553	0143
1063-4145	0025
1063-4806	2329
1063-4814	1712
1063-4939	2786
1063-5019	0521
1063-5027	OEM
1063-5142	0285
1063-5209	0039
1063-5381	0016
1063-5399	0079
1063-5423	0037
1063-5431	0037
1063-5449	0037
1063-5646	1648
1063-5688	1284
1063-5704	0955
1063-6454	0025
1063-6926	0037
1063-7247	OEM
1063-7411	OEM
1063-7916	0285
1063-8369	0220
1063-8435	0203
1063-8591	0019
1063-8799	OEM
1063-8963	0191
1063-8971	0071
1063-8989	0471
1064-6461	0537
1064-6545	0879
1064M	0037
1065-2055	3273
1065-2212	2026
1065-2543	2628
1065-2642	2628
1065-3225	0042
1065-3525	0042
1065-4861	0232
1065-6775	0502
1065-9928	0023
1065-9936	0955
1065-9944	2096
1065Z4	OEM
1066-9666	1326
1067	0233
1068-17	4208
1069	0042
1069-7032	0867
1070	0071
1070(XSTR)	3014
1070-0623	0023
1070-0631	0955
1070-4781	1710
1070-5481	0039
1070-8576	2209
1071	0925
1071-2081	0118
1071-4186	1308
1071-4251	1790
1071-4913	0155
1072-1	0236
1072-6677	OEM
1072-8491	0324
1072G	0791
1072K	0791
1073-6361	2811
1073-6486	0006
1074	0313
1074-117	0103
1074-120	0012
1074-1582	0037
1075	0330
1075-5122	0471
1075-6146	0392
1075Z4	OEM
1076-0999	0532
1076-1377	0007
1076-1484	0133
1076-1559	0285
1076-1674	0017
1076-4801	0642
1076-5196	0037
1077	0666
1077-07	0016
1077-0261	0080
1077-240Z	OEM
1077-2309	0018
1077-2325	0143
1077-2333	0005
1077-2341	0133
1077-2358	0120
1077-2366	0015
1077-2374	0901
1077-2382	0579
1077-2390	1215
1077-2408	0433
1077-2760	0143
1077-3828	1284
1077-3836	0015
1077-3844	1571
1077-9270	0005
1077-9296	0123
1077H11	OEM
1077H12	OEM
1078-7372	0196
1079	0802
1079-5110	0144
1079-9928	0530
1079-9955	0023
1079-9963	0023
1080	0071
1080-03	0042
1080-05	0622
1080-06	0219
1080-07	0191
1080-08	0023
1080-20	0286
1080-21	0076
1080-130	0042
1080-9978	OEM
1080G	0126
1081-1511	0324
1081-3087	0324
1081-3186	0039
1081-3285	3696
1081-3293	0344
1081-3301	0016
1081-3319	0016
1081-3343	0313
1081-3350	0264
1081-3368	0042
1081-3475	0079
1081-3541	2786
1081-3558	3089
1081-3624	0402
1081-4000	0037
1081-4010	0037
1081-4739	2537
1081-8581	0037
1081-9464	0016
1082-1056	0079
1082-1403	0079
1082-2211	0023
1082-4803	1197
1082-6717	0015
1082-6915	2953
1082-9208	0917
1082-9240	0974
1083-17	0164
1084	0071
1084(GE)	0144
1085(GE)	1501
1085-8967	0144
1085Z4	OEM
1086	1501
1086(GE)	1501
1086(G.E.)	0190
1086-1227	1029
1086-3967	0715
1086-4080	0124
1086-9931	2295
1087	0086
1087-01	2050
1087-02	0076
1087-1606	0224
1087-1630	0148
1087-1648	0693
1087-1663	0148
1087-1713	0133
1087-2901	0076
1088-1	OEM
1088-9517	0133
1089-1	OEM
1089-7619	0433
1089-9938	1911
1090	0071
1090-0033	0023
1090-0071	0133
1090-0827	0157
1090-1	OEM
1090-1981	0133
1090-19801	0133
1090A	OEM
1091	0565
1091-0826	0057
1091-1	OEM
1091-3549	0181
1092	0130
1093	0079
1093-2697	0143
1093-3034	0127
1093-3059	0286
1093A41-165	OEM
1094-7877	0042
1094Z4	OEM
1095	0037
1095-01	0325
1095-2638	0006
1095-2679	0062
1095-6324	0253
1095J2F	0015
1095Z4	OEM
1096-11	0698
1096-12	0537
1096J2F	0015
1097	0802
1097-85	0006
1098-14	1607
1098-15	0698
1098-16	OEM
1098-1975	0041
1098-8665	2528
1099-0950	0086
1099-1214	1284
1099-2949	0706
1099-2972	0023
1099-3616	1491
1099-3624	3079
1099-6346	2395
1099-6395	3790
1099-6544	2827
1099-6981	0348
1099-9050	0079
1099-9100	0219
1099-9175	1284
1099-9183	0168
1099-9217	0079
1100	1267
1100-01A	OEM
1100-75	0698
1100-1179	0079
1100-3951	0286
1100-3985	0065
1100-6962	0140
1100-9446	3807
1100-9453	0127
1100-9461	0076
1100-9479	0286
1100-9487	0124
1100-9750	0362
1100I	OEM
1101	1312
1101/01A	OEM
1101-01A	OEM
1101-1640	1369
1101-9072	3271
1101A	OEM
1101ADM	OEM
1101PDK120	OEM
1101PDK140	OEM
1101PDK160	OEM
1101PDK180	OEM
1101PDK200	OEM
1101PDK220	OEM
1101PDK240	OEM
1101PDK260	OEM
1101PDK280	OEM
1101PDK300	OEM
1102	0188
1102-17	0164
1102-17A	0004
1102-63	0600
1102-6853	0023
1102-9568	0628
1103	0188
1103-88	0015
1103-3669	1329
1103-3727	0133
1103-5557	0236
1103-7751	1055
1103I	OEM
1103N	OEM
1104	0140
1104/01A	OEM
1104-01A	OEM
1104-94	0004
1104-95	0208
1104-96	0015
1104-5960	0261
1105	0253
1105-15	0160
1105-2750	0023
1105-3063	0066
1105-4558	1220
1105Z4	OEM
1106	0062
1106-29	0015
1106-36	0015
1106-97	0144
1106-99	0086
1107	0077
1107B-938N78	OEM
1108	0165
1109	0057
1110	0064
1110-86	0015
1111	0181
1111-17	0050
1111-18	0050
1111-4469	OEM
1111P	0387
1112	0052
1112-8	0037
1112-78	0086
1112-79	0698
1113	0879
1113-03	0016
1113-13	0050
1113AD-A-CJ-O	OEM
1113ADACJO	OEM
1113AD-A-CJ-P	OEM
1113ADACJP	OEM
1113AD-A-OO-O	OEM
1113ADAOOO	OEM
1113AD-A-OO-P	OEM
1113ADAOOP	OEM
1113AD-B-CJ-O	OEM
1113ADBCJO	OEM
1113AD-B-CJ-P	OEM
1113ADBCJP	OEM
1113AD-B-OO-O	OEM
1113ADBOOO	OEM
1113AD-B-OO-P	OEM
1113ADBOOP	OEM
1113EX8C	OEM
1113EX16C	OEM
1113EX16T	OEM
1113EX80	OEM
1113EX160	OEM
1113SX8C	OEM
1113SX16C	OEM
1113SX16T	OEM
1113SX80	OEM
1113SX85	OEM
1113SX160	OEM
1114	OEM
1114-4458	2581
1114E1	OEM
1114Z4	0052
1115	0065
1115-16	0015
1116	0210
1116-42	0015
1116-3243	0079
1116-4027	0016
1116-4035	0079
1116-4084	0037
1116-4092	0037
1116-4100	0079
1116-4118	0079
1116-4126	0037
1116-4365	0133
1116-5016	0037
1116-6691	0857
1116-7004	0436
1117(MC)	0079
1117-76	0015
1117-1857	0079
1117-4422	0079
1117-5353	0066
1118	0375
1118-17	0143
1118-20	0015
1119	0285
1119-54	0050
1119-55	0050
1119-56	0050
1119-57	0004
1119-58	0208
1119-59	0004
1119-9254	2691
1120	0700
1120-17	0133
1120-18	0015
1120-6034	OEM
1120-6364	0276
1121	OEM
1121-1	1426
1121-17	0015
1122	0705
1122-96	0136
1122-2403	0079
1122-2411	0037
1122-2430	0532
1122-2437	0532
1122-8863	0362
1122Z4	OEM
1123-55	0144
1123-56	0144
1123-57	0144
1123-58	0144
1123-59	0144
1123-60	0016
1123-1388	0313
1123-1719	0162
1123-2624	0023
1124	0160
1124-0017	0162
1124-0239	0912
1124-7234	2094
1124A	0160
1124B	0160
1124C	0085
1125	0560
1125/2582	0037
1125-1188	1348
1125-2582	0037
1125-2590	0554
1125-2608	0062
1125-7722	0143
1125Z4	OEM
1126-0023	0582
1126-0031	0053
1126-0049	0023
1126-0239	0912
1126-0304	0825
1126-0312	0825
1126-0320	0825
1126-0352	0582
1126-0353	0582
1126-0395	0124
1126-5170	1348
1126-7127	0203
1126-7150	0151
1126-7226	0006
1126-7234	2094
1126-8851	0079
1127	0441
1127-8454	0582
1127-8470	0015
1127-8488	0471
1128	0037
1128-17	0004
1128-4239	4857
1128-4304	1270
1128-4437	0015
1128-7540	OEM
1128-7547	0005
1129	1750
1129-8015	0148
1133	0039
1133-0289	2542
1133-0305	0253
1133-0321	0064
1135-3646	0037
1135-7076	0324
1135-8066	0023
1135-8074	0039
1135-9098	0348
1135-9254	0313
1135-9262	0285
1135Z4	OEM
1136-4064	0133
1137	0666
1137-7058	0057
1137-7215	0124
1138	0015
1138-0789	2022
1138-0979	0069
1138-4757	0270
1138-9442	0395
1138-9459	OEM
1138-9467	0203
1138-9640	0042
1138-9657	0015
1139-17	0015
1140	OEM
1140-17	0004
1140-2302	0852
1140-2310	0212
1140-2328	0079
1140-2864	0079
1140-2898	0210
1140-3193	0079
1140-3195	OEM
1140-3201	0079
1140-3219	0392
1140-3391	0058
1140-3409	0686
1140-4118	4549
1140-4126	0181
1140-4233	3937
1140-4258	0566
1140-4597	0726
1140-7673	0023
1141-3119	1170
1143-0584	1949
1143-0741	0359
1143-0956	0735
1143-1046	0133
1143-1049	0124
1143-2903	0582
1143-4784	1962
1144-1482	0825
1144-3876	0006
1144-5749	0023
1144-9261	0042
1145	0280
1145-17	0004
1145-2695	0052
1145-2703	0079
1145-2711	0224
1145-2729	0566
1145-2737	0037
1145-2745	0079
1145-2752	0079
1145-2778	0042
1145-2786	0912
1145-2794	3977
1145-3156	0079
1145-3164	0037
1145-3370	0564
1145-4071	0212
1145-4188	1029
1145-4253	0037
1145-4741	3537
1145-4949	0079
1145-5011	1392
1145-5037	OEM
1146	0280
1146-0888	0124
1146-0896	0005
1146-0912	0019
1146-5200	0133
1146S	0181
1147	0150
1147-07	OEM
1147-08	0624
1147-09	0215
1147-0249	OEM
1147-0523	0041
1147-0531	1778
1147-10	2247
1147-11	2233
1147-12	1544
1147-161	3815
1147-161(VARACTOR)	0005
1147-3204	0062
1147-3220	0700
1147-3238	0181
1147-4738	OEM
1147-6249	0148
1147-7673	0023
1147-8815	5195
1148-0956	0735
1148-17	0164
1148-7944	0023
1148-9499	3074
1148-9523	3084
1148-9556	2019
1148-9580	OEM
1148-9648	1270
1148-9671	3165
1149-0182	0062
1149-0224	0023
1149-0729	0167
1149-0740	1597
1149-0760	0037
1149-0794	0079
1149-2253	OEM
1149-2469	0037
1149-2477	0144
1149-2493	0955
1149-2501	2660
1149-2527	0019
1149-2535	0124
1149-2543	0795
1149-2576	0133
1149-4332	0076
1149-4374	0342
1149-4390	0076
1149-4408	1327
1149-4424	0161
1149-6729	0167
1150	0095
1150-7746	0076
1150-7753	1513
1150-7761	0064
1150-8504	1221
1150-9098	2223
1151	0395
1151-3736	0076
1151-5442	OEM
1153-0128	1386
1154	6783
1154-5563	1211
1154-5613	0006
1154-5696	4052
1154-6199	0076
1154-6561	0210
1154-6579	0392
1154-6587	0006
1154-7072	0629
1154-9680	0533
1154-9880	0533
1154-9938	0176
1155	2295
1155-1785	0037
1155-1934	3960
1155-5562	0603
1155-5570	0706
1155-5588	4890
1155-5596	0076
1155-5604	0563
1155-5612	0143
1155-5620	0124
1155-5636	0064
1155-6560	0006
1155-6578	2295
1155-8228	1066
1155-8236	0076
1155-8764	0212
1155-9259	2926
1155GE	0102
1155Z4	OEM
1156	1323
1156(TC)	1816
1156-1644	0133
1157	0086
1157-0165	0442
1157-2096	0552
1157-2104	2209
1157-4712	2209
1157-4720	OEM
1157-4738	2295
1157-4746	0143
1157-4753	0143
1157-4761	0133
1157-4779	0005
1157-4787	0023
1157-4795	0124
1157-4829	0532
1157-4837	0079
1157-4845	0016
1157-4852	0079
1157-4860	0715
1157-4878	4898
1157-4886	0144
1157-8564	0215
1157-8572	2184
1157-8580	0392
1157-8598	0681
1157-8606	0127
1157-8622	0076
1157-8630	0191
1157-8648	0148
1157-8655	0133
1157-8663	2232
1157-8671	0023
1157-8689	1329
1157-8697	0275
1157-8705	0181
1157-8929	0139
1157-8937	1925
1157-8945	0120
1158	OEM
1159	OEM
1159-2318	5346
1159-2342	3983
1160	OEM
1160-7009	0471
1160-7611	0253
1160-7629	0057
1160-8015	0023
1160H	OEM
1161-2926	0253
1161-7719	0224
1163-4110	0057
1163-4243	0076
1163-5423	OEM
1163-7907	OEM
1163-9507	1367
1163-9572	0006
1163-9960	OEM
1164-4762	4511
1164-4804	0037
1164-4812	OEM
1164B	2657
1164D	2006
1164F	2682
1164H	3900

If replacement code is OEM, contact original manufacturer for replacement.

DEVICE TYPE	REPL CODE
1164K	2655
1165(GE)	0167
1166	0142
1166-3829	3977
1166-3879	3977
1166-3887	0239
1166-5197	5785
1166-5205	4601
1168-2754	0143
1168-2788	0137
1168-7944	0023
1169	0233
1169-2076	0167
1169-4478	0062
1169-7653	0681
1170-0036	2896
1171-8309	OEM
1172-0026	0865
1172-0034	1201
1172-0036	2896
1172-0042	0334
1172-1396	0055
1172-1494	0016
1172-1936	OEM
1172-5157	2019
1172-5272	0825
1172-5306	0928
1172-5330	0582
1172-5413	0053
1172-5504	0124
1172-9811	0127
1172-9829	0037
1172-9837	0286
1172-9845	0079
1172-9852	0286
1172-9860	0018
1172-9878	0362
1172-9886	1318
1172-9894	2728
1172-9902	0554
1172-9910	0062
1173-1171	0127
1173-1494	0079
1173-2575	3807
1173-4613	4406
1173-4639	4410
1173-6824	0079
1173-6832	OEM
1173-8192	1992
1173-8200	0264
1173-8220	0264
1173-8291	0133
1173-8309	0023
1173-8317	0820
1173-8325	0023
1173-8556	0064
1174	0395
1175-0056	0799
1175-0064	0286
1175-5659	OEM
1175-7879	0699
1175-7903	0759
1175-7911	0770
1175-7937	0631
1175-7945	0582
1175-8000	0770
1175-8018	0631
1175-8026	0672
1175-8190	0825
1175-8281	0582
1175-8299	0582
1175-8349	0343
1175-8364	0299
1176-5005	3759
1176-5013	1220
1176-5039	0041
1176-6367	0224
1176-7126	0079
1176-7134	0037
1176-7142	0076
1176-7159	1284
1176-7167	0261
1176-7175	0055
1176-7183	0261
1176-7191	3516
1176-7209	0066
1176-7217	0023
1176-7225	0023
1176-7233	0471
1176-7241	0023
1176-7258	OEM
1176-7266	0663
1176-7274	2284
1176-7645	0023
1176-8819	0842
1176-8827	0006
1176-72167	0023
1177	0244
1177-1144	3738
1177-1151	2623
1177-1169	5095
1177-1177	0261
1177-1185	2668
1177-1193	0276
1177-1201	0392
1177-1219	0023
1177-1227	0023
1177-1235	0023
1177-1243	0133

DEVICE TYPE	REPL CODE
1177-1250	0091
1177-1268	0441
1177-1276	1005
1177-1284	0445
1177-1292	0710
1177H01	OEM
1177H02	OEM
1177H03	OEM
1177H04	OEM
1177H09	OEM
1178	0167
1178-3842	0168
1178-3859	0079
1178-4291	0261
1178-7066	OEM
1178-7209	0066
1178-7454	0912
1178-7462	0663
1178-7470	0037
1178-7488	0006
1178-7491	0643
1178-7496	0643
1178-7512	0079
1178-7520	0842
1178-7538	1284
1178-7546	1648
1178-7553	0715
1178-7561	0270
1178-7579	0055
1178-7587	0055
1178-7603	0023
1178-7611	0133
1178-7637	0499
1178-7645	0023
1178-7652	0071
1178-7660	0023
1178-7678	0023
1178-7686	0471
1178-7694	0471
1178-7696	OEM
1178-7812	OEM
1178-7868	0471
1179	6785
1179-0334	0764
1179-1308	1051
1179-1324	1165
1179-1332	0728
1179-1340	0080
1179-1357	0023
1179-1365	0015
1179-2256	0076
1179-2264	0595
1179-2306	0316
1179-2330	0516
1179-9186	0124
1179A	OEM
1180	OEM
1180-1	OEM
1180-1859	0124
1180-1867	0006
1180-1875	0015
1180-1883	0901
1180-1891	OEM
1180-1909	0023
1180-1917	1911
1180-1925	0055
1180-1933	0041
1180-1941	2051
1180-1958	0286
1180-1966	2427
1180-1974	3046
1180-1982	0436
1180-1990	2069
1180-2006	0076
1180-2014	3133
1180-2022	0733
1180-2030	0166
1180-2048	0062
1180-2055	1676
1180-2063	OEM
1180-2071	5216
1180-2089	3772
1180-2881	0127
1180-4457	0052
1180-5520	0041
1180-5959	OEM
1180-5991	2122
1180-6296	0441
1180-6429	OEM
1180-6445	5509
1180-6734	0471
1180-6858	OEM
1180-6866	OEM
1180-9446	OEM
1181-3201	0706
1181-3268	0170
1181-3396	1392
1181-3409	0127
1181-3417	0111
1181-3425	0111
1181-3433	2910
1181-3458	0537
1181-3466	0563
1181-3474	2910
1181-3565	5162
1181-3739	0261
1181-3763	0663
1181-3771	0076
1181-3789	0261

DEVICE TYPE	REPL CODE
1181-3797	0066
1181-3805	0168
1181-3813	1284
1181-3821	0023
1181-3912	0643
1181-3920	0955
1181-4472	0925
1181-4910	6838
1181-4928	1143
1181-4944	OEM
1181-5263	0052
1181-7103	0133
1181-7111	0023
1181-7129	3738
1181-7137	2623
1181-7145	0167
1181-7152	0392
1181-7160	0643
1181-7194	1329
1181-7228	0643
1183-0494	0118
1183-17	0698
1183-1633	2786
1183-2052	0133
1183-2060	0681
1183-3258	0055
1183-3266	1005
1183-3274	0795
1183-3282	0102
1183-3290	0023
1183-3308	0473
1183-3314	0015
1183-3316	0015
1183-3340	OEM
1183-3458	0133
1183-3548	0133
1183-8216	0375
1184-4	OEM
1184-5109	0014
1184-5641	1743
1184-5658	0064
1184-5666	0064
1184-5682	0023
1184-5708	0124
1184-5716	0124
1184-5732	0181
1184-8108	0490
1184-8728	1429
1184-9056	1806
1184-9064	1005
1184-9098	0795
1184G	0126
1185-0161	2553
1185-0260	0203
1185-5	OEM
1186	0037
1186-1	0030
1186-3	0030
1186-4	0030
1186-5	0917
1186-6	OEM
1186-5573	0023
1186-5581	0494
1186-5672	2896
1186-5839	3764
1186-5888	0865
1186-5912	5209
1186-5953	0901
1186-5987	0276
1186-5995	0276
1186-6035	0285
1187	0086
1188-0101	0181
1188-0333	2628
1188-3097	0582
1188-6306	0080
1190	0065
1190-9983	0023
1190H	OEM
1191-3290	OEM
1191-3308	OEM
1191-3316	OEM
1191-3340	1945
1191-3357	0413
1191-3365	0121
1191-3373	1348
1191-3381	2223
1191-3399	0060
1191-3415	0071
1191-3423	0071
1191-3431	0140
1191-3449	0140
1191-4769	0127
1191-4793	0728
1191-4819	0133
1191-4991	4350
1191-7860	0168
1191-8547	5304
1191-8869	2703
1192	0211
1192(IC)	2393
1192(XSTR)	0211
1192-7837	0120
1192-7845	0123
1192-7852	2533
1192-7860	0168
1192-7878	0076
1192-7886	0042
1192-9288	0168

DEVICE TYPE	REPL CODE
1193-0054	2872
1193-0062	0155
1193-0070	0168
1193-0083	0076
1193-0096	0532
1193-0104	1165
1193-0112	0041
1193-0120	OEM
1193-4148	OEM
1193-4163	1165
1193-4171	0064
1193-4189	0735
1193-4197	0471
1193-4205	0118
1193-8503	OEM
1193-8511	0645
1193-8529	OEM
1193-8537	0694
1195-2679	0062
1195-4455	0564
1195-6372	0091
1195-6380	0133
1195-6398	0133
1195-6406	0023
1195-6414	0023
1195-6422	0023
1195-6430	0181
1195-6448	0120
1195-6455	0564
1195-6463	1993
1195-6497	0261
1195-6505	0286
1195-6513	1284
1195-6521	0955
1195-6539	0261
1195-6547	3282
1195-6562	1429
1195-6570	0490
1195-6893	0057
1195-6901	0276
1195-6943	0065
1195-6950	0130
1197-0027	0064
1197-1645	OEM
1197-1686	0825
1197-1777	0582
1197-1793	0053
1197-1819	0165
1197-5055	OEM
1197-5505	OEM
1197-6024	0041
1197-6032	3725
1197-6040	4520
1198-4143	0006
1198-4176	0261
1198-4200	0261
1198-4267	5133
1198-4291	0866
1198-4325	0261
1198-4358	0148
1198-4382	0124
1198-4388	0124
1198-4853	0724
1198-4887	OEM
1198-4911	OEM
1198-5181	0724
1199-7822	0023
1199-7830	0317
1199-7848	0167
1199-7855	1201
1199-7855(ZE)	0317
1199-7863	0037
1199-7871	0261
1200	2472
1200/1	OEM
1200/2	OEM
1200/11	OEM
1200/12	OEM
1200/22	OEM
1200-1913	0006
1200-2036	1911
1200-2093	0155
1200-2148	0023
1200-2184	0023
1200-2200	2090
1200-2242	0365
1200-2267	OEM
1200-2358	0976
1200-2366	OEM
1200-2382	1329
1200-2416	0176
1200-2440	3518
1200-2531	0286
1200-3412	0166
1200-3422	OEM
1200-3430	3916
1200-3448	2122
1200-3463	2666
1200-3489	OEM
1200-3497	0024
1200-3505	1359
1200-3513	OEM
1200-3530	3916
1200-3539	0006
1200-3547	0951
1200-3554	0276
1200-3562	0276
1200-3588	0055
1200-3596	3969

DEVICE TYPE	REPL CODE
1200-3604	1758
1200-3612	0166
1200-3620	0124
1200-3638	0023
1200-3646	0031
1200-3653	0106
1200-3661	OEM
1200-3679	OEM
1200-4099	OEM
1200-4123	OEM
1200-4156	OEM
1200-4883	OEM
1200-4925	4290
1200-5112	OEM
1200-8336	0313
1200-9437	1733
1200-9460	1607
1200-9494	1386
1200-9528	0124
1200-9551	5174
1200-9585	0140
1200H	OEM
1200PDE10	OEM
1200PDE20	OEM
1200PDE30	OEM
1200PDE40	OEM
1200PDE50	OEM
1200PDE60	OEM
1200PN50	OEM
1200PN60	OEM
1200PN80	OEM
1200PN100	OEM
1200PN120	OEM
1200PN140	OEM
1200PN150	OEM
1200PN160	OEM
1200PN180	OEM
1200PN200	OEM
1200PN210	OEM
1200PN220	OEM
1200PN230	OEM
1200PN240	OEM
1200PN250	OEM
1200XY2-35SC	OEM
1201	OEM
1201(TRANS)	OEM
1201-4114	0015
1201-4148	0317
1201-4171	0023
1201-4205	0795
1201-7331	0127
1201-7349	0079
1201-7372	0532
1201-7406	4052
1201-7430	0092
1201-7737	OEM
1201-7760	OEM
1201-8156	0342
1201-8180	0076
1201-8214	0286
1201-8248	0099
1201-8271	OEM
1201-9618	5819
1201-W1K	OEM
1201PDE20	OEM
1201PDE30	OEM
1201PDE40	OEM
1201PDE60	OEM
1201PDE80	OEM
1201PDE100	OEM
1201PDE120	OEM
1201PDE140	OEM
1201SS1-07SC	OEM
1202	0832
1202(SWITCH)	0832
1202(TRANS)	0832
1202-0229	0148
1202-7678	0014
1202-7702	2203
1202-7710	1091
1202-7751	0181
1202-7769	0053
1202-W4K	OEM
1202-W8K	OEM
1202H	OEM
1202M	0079
1202XY2-35SC	OEM
1203	0815
1203(TRANS)	0887
1203-6323	0359
1203-6380	0879
1203-6406	0343
1203-6414	0053
1203-7362	0143
1203-7370	0795
1203-7388	0071
1203-7396	0031
1203-7412	0224
1203-7420	0842
1203-7438	0079
1203-7446	0148
1203-7453	0286
1203-7461	1496
1203-7479	1498
1203-7487	0006
1203-7495	0286
1203-WAP	OEM
1204	0898
1204(TRANS)	0898

DEVICE TYPE	REPL CODE
1204-0010	0006
1204-0028	2437
1204-W1KAP	OEM
1205	0015
1205-7790	0053
1205-7816	0261
1205-7832	0912
1205-7840	0066
1205-7845	0006
1205-7857	0155
1205-7865	0006
1205-7873	1624
1205-7899	0261
1205-7907	3518
1205-7923	0023
1205A	0015
1206	6786
1206(TRANS)	OEM
1206-17	0123
1206-6667	OEM
1206-W1KAP	OEM
1207	0944
1207-17	0143
1207-4543	0023
1207-4553	OEM
1207-4707	1949
1207-4758	0018
1207-6519	OEM
1207-6543	0023
1207-6576	OEM
1207-6600	0133
1207-6634	0015
1207-6667	0015
1207-6691	0041
1207-6725	1284
1207-6758	0018
1207-6782	0563
1207-6816	0168
1207-6873	0450
1207-6907	1949
1207-6931	0035
1207-6964	0525
1207-7673	0018
1207-WAP	OEM
1208	0958
1208-3853	0064
1208-3861	0140
1209	0994
1209-0445	0261
1209-3027	0023
1209-3035	0181
1209-3042(DIO)	OEM
1209-3043	0181
1209-3043(DIO)	0023
1209-5873	0066
1209-7143	1624
1210	1009
1210-17	0475
1210-17(MIXER)	0079
1210-17(OSC)	0079
1210-17B	0016
1210-W23K	OEM
1210-W46K	OEM
1211	1019
1211-9327	OEM
1211-W46K	OEM
1211-W92K	OEM
1212	0889
1212(ROSS)	0133
1212-4	0936
1212-5009	0023
1213	1049
1213-6446	0133
1213-6453	OEM
1214	0037
1214-1560	OEM
1214-4915	0148
1214-4945	0912
1214-4952	0148
1214-4957	0148
1214-4960	0148
1214-4978	0866
1214-4986	0261
1214-4994	0023
1214-5009	0023
1214-5017	0091
1214-5024	0071
1214-5025	0071
1214-7952	0148
1214-W1K	OEM
1215	0015
1215-3839	0015
1215-4274	0017
1215-5693	0031
1215-5701	0041
1215-5719	OEM
1215-5727	OEM
1215-8663	4310
1215-8671	4189
1215-9620	OEM
1215RA	2291
1216	OEM
1216-2541	0246
1216-2575	3542
1217-0536	0037
1217-0544	0261
1217-2284	4650
1217-2383	0261
1217-5164	0014

DEVICE TYPE	REPL CODE
1217-8554	OEM
1217-8562	1140
1217-8570	1140
1217-8588	0155
1217-8596	0155
1217-8604	0286
1217-8612	0286
1217-8620	2484
1217-8935	4379
1217-8943	6336
1217-8950	5258
1217-8968	5259
1217-8976	5259
1217-8992	3925
1217-9008	5261
1217-9016	5262
1217-9024	0770
1217-9032	OEM
1217-9040	5263
1217-9057	OEM
1217-9065	3159
1217-9073	OEM
1217-9081	0112
1217-9099	2484
1217-9107	0648
1217-9115	1140
1217-9123	2052
1217-9131	0728
1217-9149	0728
1217-9156	0060
1217-9164	0014
1217-9172	0060
1217-9180	0023
1217-9198	0182
1217-9206	0041
1217-9214	0041
1217-9222	0253
1217-9230	0253
1217-9248	0064
1217-9255	0064
1217-9271	0031
1217-9289	0451
1217-9297	0023
1217-9305	0343
1217-9313	0062
1217-9321	0064
1217-9339	0053
1217-9347	0053
1217-9362	0283
1217-9370	0296
1217-9388	0640
1217-9396	0296
1218	OEM
1218H	OEM
1219-8685	OEM
1220	OEM
1220-3055	OEM
1220H	OEM
1221	OEM
1221-1021	OEM
1222	OEM
1222-6684	0006
1222-7104	0079
1222H	OEM
1223	0757
1223-1924	0445
1223-2229	0018
1223-2237	2367
1223-2245	0076
1223-2252	0031
1223-2260	0041
1223-2278	0023
1223-2286	0375
1223-2294	0705
1224	OEM
1224-3804	OEM
1224H	OEM
1225-4975	0101
1226H	OEM
1227	OEM
1227-0187	0076
1227-17	0016
1227-4296	0155
1227-4304	0155
1227-4312	0286
1227-4320	0286
1227-4338	1165
1227-6457	OEM
1227-8495	OEM
1227-8610	0976
1227-9142	1624
1227-9741	0556
1228-17	0016
1228-4017	0076
1228-4022	1790
1228-4410	0663
1228-4428	0842
1228-4451	1541
1228-4469	1541
1228H	OEM
1229-0862	5815
1229-1134	4852
1229-1159	4172
1229-1167	1032
1229-1175	0832
1229-1183	0887
1229-1191	0898
1229-1308	0057
1229-4617	OEM

DEVICE TYPE	REPL CODE
1229-4625	2913
1229-4633	0129
1229-4641	OEM
1229-4658	0129
1229-4690	5301
1229-4724	0825
1229-4799	0726
1229-4807	0077
1229-4815	0077
1229-6133	OEM
1229-6141	OEM
1229-8170	2897
1229-8477	0076
1229-8485	0076
1229-8709	0052
1229-8725	1329
1229-8733	0041
1229-8766	0133
1229H	0144
1230	6789
1230-9266	0165
1231-0082	OEM
1231-0108	OEM
1231-0116	2066
1231-0132	0360
1231-0140	0360
1231-0157	3828
1231-0165	2913
1231-0181	OEM
1231-0207	4496
1231-0215	4383
1231-0256	0553
1231-0280	OEM
1231-0314	OEM
1231-0348	2484
1231-0355	0825
1231-0363	2223
1231-0371	0582
1231-0389	1978
1231-0397	0060
1231-0405	0057
1231-0421	OEM
1231-0447	0057
1231-3185	0148
1231-4720	0866
1231-4787	0148
1231-8325	3828
1231-8333	0631
1231-8341	0683
1231-8358	6165
1231-8366	0631
1231-8374	OEM
1231-8390	OEM
1231-8408	1140
1231-8416	0359
1231-8424	0700
1231-8432	0700
1231-8440	0023
1231-8788	1778
1231-8903	OEM
1232-2889	0686
1232-7284	OEM
1232-7292	1529
1232-7300	3916
1232-7409	0955
1232-7573	5048
1232-7730	2797
1232-7755	3133
1232-7805	0473
1232-7862	OEM
1232-7946	1493
1233	OEM
1233H	OEM
1234-1095	1389
1234-1674	5011
1234-5849	2989
1234-5856	OEM
1234-5864	0076
1234-5880	0786
1234-5898	0023
1234-5906	0023
1234-5914	0700
1234-5922	4137
1234-5930	OEM
1234-5948	5011
1234-5955	1551
1234-5963	5256
1234H	OEM
1235-2126	2797
1235-7943	0017
1235H	OEM
1236	OEM
1236H	OEM
1237-5473	0071
1237-5739	0079
1237-6042	0037
1237-6208	OEM
1237-7912	2026
1237-9467	0041
1238-9730	0532
1238-9748	0023
1238-9755	0041
1238-9763	0006
1238-9771	1284
1238H	OEM
1239	OEM
1239-4755	0124
1239-6537	1529
1239-6545	2122
1239-6552	0015
1239-6560	0133
1239-7766	0101
1239-7774	1648
1239-7782	0148
1239-7808	0071
1239-7824	0023
1240	OEM
1240-3432	3626
1240-3440	3198
1240-3457	0513
1240H	OEM
1241	0280
1241A	0211
1241H	OEM
1242-1814	OEM
1242-7522	4486
1243	OEM
1243H	OEM
1244-3834	1541
1244H	OEM
1245H	OEM
1247	OEM
1247-7097	OEM
1247-7121	0015
1247-7139	OEM
1247-7147	OEM
1247-9523	1541
1247H	OEM
1248	0004
1248-4010	0140
1248H	OEM
1250	OEM
1250-1326	0317
1250-4320	OEM
1250-4346	0042
1250-4478	OEM
1250-4486	OEM
1250-4643	0883
1250-4650	0700
1250-5210	0024
1250-5228	OEM
1250-5855	0079
1251-1-1	0212
1252-4906	0023
1253-5F	OEM
1253-6	OEM
1253-6F	OEM
1253-7	OEM
1253-7F	OEM
1253-8	OEM
1253-8F	OEM
1253-9	OEM
1253-10	OEM
1253H	OEM
1254	0150
1254-1025	1090
1254-1074	0031
1254-1082	0102
1254-1090	0023
1254-1108	0118
1254-1413	0018
1254-1694	0334
1254-1942	1165
1254-1959	1165
1254-2106	2178
1254-2171	OEM
1254-2593	2490
1254-3583	2289
1254-3591	OEM
1254-5661	OEM
1254-5869	0523
1254-5877	OEM
1254-5885	4102
1254-5893	OEM
1254-7170	OEM
1254-7295	OEM
1254H	OEM
1255-1063	4379
1255-1081	5324
1255-1099	5328
1255-1107	OEM
1255-1123	3925
1255-1131	4117
1255-1164	OEM
1255-1198	3159
1255-1206	0112
1255-1214	1065
1255-1222	0832
1255-1289	0124
1255-1297	5358
1255-2626	0052
1255H	OEM
1256-1213	4141
1256-1221	OEM
1256-1478	OEM
1256-1486	OEM
1256-1494	OEM
1256H	OEM
1257-15	OEM
1257-3812	0643
1257-4497	OEM
1257-4505	OEM
1257-4521	OEM
1257-4537	3521
1257-4539	3521
1257-4547	OEM
1257-4554	OEM
1257-4562	OEM
1257-4570	0686
1257-4596	0023
1257-4612	0705
1257-4620	0057
1257-4638	0636
1258-2352	0091
1258-2532	0471
1258-2540	0071
1258-9198	0023
1259-12	OEM
1259-2713	OEM
1259-2721	OEM
1259-2739	1032
1259-2747	0887
1259-2754	3501
1259-2804	2234
1259-4917	3772
1259-4941	OEM
1259-4958	2026
1259-4966	4496
1259-4974	OEM
1259-5005	0500
1259-5021	OEM
1259-5039	OEM
1259-5047	1541
1259-5054	0253
1259-5062	0471
1259-5070	0124
1259-5096	0041
1259-5104	0064
1259-5112	0091
1259-5120	OEM
1259-5138	0015
1259-5146	0283
1259-5153	OEM
1259-5161	OEM
1259-5179	3898
1259-5187	3898
1259-6516	2872
1259-6524	0539
1259-6532	0107
1259-6565	0076
1259-6581	0276
1259-6615	OEM
1259-6623	2628
1259-6631	OEM
1259-7702	2203
1263A	0015
1263B	OEM
1264	0349
1264H	OEM
1266-0268	0079
1266-2053	1329
1266-2219	0006
1266-2227	0148
1266-2243	0286
1266-2250	0219
1266-2268	0079
1266-2276	0124
1266-2284	1978
1266-2292	0392
1266-2300	0055
1266-2318	0124
1266-2326	0137
1266-2334	0181
1266-2342	0023
1266-2359	0023
1266-2367	0023
1266-2375	0023
1266-2383	OEM
1266-2383(DS2(LED))	OEM
1266-2391	0080
1266-2409	0023
1266-2417	0795
1266-2425	0683
1266-2433	2090
1266-2441	2103
1266-2458	2116
1266-2466	2122
1266-2471	0795
1266-2953	1329
1266-5201	0461
1266H	OEM
1268H	OEM
1270-1751	4097
1270-1769	3365
1270-1777	3884
1270-1785	4383
1270-1793	0062
1270-1801	0023
1270-4797	5733
1270-4805	0039
1270-4813	0053
1270-4821	0471
1270-4839	OEM
1270-4847	3431
1270-5208	0006
1270-5214	0261
1270-5216	0261
1270-5224	0286
1270-5232	0261
1270-5240	0286
1270-5257	1541
1270-5265	0023
1270-5273	0023
1270-5281	0124
1270-5299	4146
1270-5588	0140
1270-6701	0795
1270-7014	0124
1270-7030	0133
1270-7048	0133
1271-2972	0643
1271-3111	3953
1272	0016
1272H	OEM
1273-8480	0023
1275-3034	0261
1276-1292	0376
1276H	OEM
1277-17	0004
1277-7074	OEM
1277-7264	OEM
1277-7272	OEM
1277-7280	OEM
1277-9302	0451
1277A	4214
1278	0395
1278(GE)	0395
1279-3147	0253
1282	0560
1285	1267
1287-5670	3537
1287-7049	0079
1287-7239	0017
1287-7247	0031
1287-7254	OEM
1287-7262	OEM
1287-7304	OEM
1287-7668	0700
1287-7734	OEM
1287-7908	3997
1287-7916	3365
1287-7924	3884
1287-7932	0001
1287-7940	3772
1287-8989	3997
1287-9987	0636
1287-9995	0157
1288-0134	0181
1288-0175	OEM
1288-0183	2564
1288-0357	OEM
1288-0779	0064
1288-0787	OEM
1288-0795	OEM
1288-0803	0002
1288-1389	0077
1288H	OEM
1289	0015
1289A	OEM
1290-3613	1327
1290-6798	OEM
1290-6806	OEM
1290-6814	0041
1290-6848	1624
1290-6855	1624
1290-6863	0037
1290-6871	0037
1290-6885	1624
1290-6954	OEM
1290-7010	OEM
1292H	OEM
1294	0150
1294-9624	0643
1294H	OEM
1296-1892	0539
1296-1900	0462
1296-1918	0101
1296-1926	3957
1296-1934	0261
1296-1942	0556
1296-1959	0556
1296-2718	1291
1296-2742	0261
1297-1404	1541
1297-1412	1541
1298-0439	OEM
1298-0447	OEM
1298-0777	0539
1298-0785	0976
1298-0793	0557
1298-0801	0079
1298-0819	0037
1298-0827	0023
1298-0835	0261
1298-0961	OEM
1298-0967	OEM
1298-3391	0144
1298-3409	0334
1298-3425	OEM
1298-3888	0334
1298-3896	OEM
1298-3904	OEM
1298-3920	5272
1298-3995	OEM
1298-6923	2232
1298-6931	0079
1298-6949	OEM
1299-7862	0334
1300PKL20S63	OEM
1300PKL20S625	OEM
1300PKL20S635	OEM
1300PKL20S645	OEM
1300PKL20S655	OEM
1300PKL20S663	OEM
1300PKL20S673	OEM
1300PKL20S683	OEM
1300PKL20T603	OEM
1300PKL30S63	OEM
1300PKL30S625	OEM
1300PKL30S635	OEM
1300PKL30S645	OEM
1300PKL30S655	OEM
1300PKL30S663	OEM
1300PKL30S673	OEM
1300PKL30S683	OEM
1300PKL30T603	OEM
1300PKL40S63	OEM
1300PKL40S625	OEM
1300PKL40S635	OEM
1300PKL40S645	OEM
1300PKL40S655	OEM
1300PKL40S666	OEM
1300PKL40S673	OEM
1300PKL40S683	OEM
1300PKL40T603	OEM
1300PKL50S63	OEM
1300PKL50S625	OEM
1300PKL50S645	OEM
1300PKL50S655	OEM
1300PKL50S666	OEM
1300PKL50S676	OEM
1300PKL50S683	OEM
1300PKL50T603	OEM
1300PKL60S63	OEM
1300PKL60S625	OEM
1300PKL60S635	OEM
1300PKL60S645	OEM
1300PKL60S655	OEM
1300PKL60S666	OEM
1300PKL60S673	OEM
1300PKL60S683	OEM
1300PKL60T603	OEM
1300PKL502635	OEM
1300RN	0127
1301	OEM
1302	0782
1303	0806
1303-1802	0392
1303-2479	0053
1304	OEM
1305	0876
1306	0205
1307	0480
1307H	0701
1308	0504
1308-1802	0392
1309	0684
1309-0915	1005
1309-1004	1551
1309-1020	OEM
1310	0225
1310-4997	0261
1310-8253	0006
1310-8261	0261
1310-8287	0842
1310-8295	0286
1310-8303	0589
1310-8311	1541
1310-8352	0057
1310-8378	OEM
1310-8386	OEM
1310-8394	0539
1310-8395	0286
1310-8402	0557
1310-8410	0732
1310-8522	0057
1310-8586	OEM
1311	0230
1311-8641	1945
1312	0234
1313	1397
1313-5496	5330
1313-5538	1911
1313-5546	2364
1313-5553	0462
1313-5561	1527
1313-5652	0168
1313-5785	OEM
1313-5793	0035
1314	0037
1314-1718	1284
1315	0016
1315-9967	2129
1315-9975	OEM
1315-9983	0643
1316	0016
1316-0908	0261
1316-0916	0042
1316-0924	0342
1316-0932	0261
1316-0951	3772
1316-0963	4189
1316-0973	3083
1316-0981	2870
1316-17	0004
1316-1003	3635
1316-1005	3635
1316-1013	2957
1316-1021	OEM
1316-1039	0523
1316-1047	OEM
1316-1054	OEM
1316-1062	OEM
1316-3167	0006
1316-3175	0023
1316-3183	0253
1316-3191	OEM
1316-3217	0035
1316-3225	0315
1316-3233	OEM
1316-3241	OEM
1316-3258	0462
1316-6137	OEM
1316-7887	3817
1316-7895	1988
1316-7903	OEM
1316-7911	1329
1316-7929	1329
1316-7937	2268
1316-7945	0728
1316-7952	0728
1316-7960	0071
1316-7978	0023
1316-7986	0071
1316-7994	0023
1317-17	0004
1318	OEM
1318-5996	0556
1318-6275	0261
1319	OEM
1319BB	OEM
1319M	OEM
1319MF	OEM
1320	0211
1321	0065
1321-01	OEM
1321-1461	0137
1321-1917	1541
1321-1925	OEM
1321-4051	0267
1322	OEM
1322-01	OEM
1323	OEM
1324	0236
1325-1186	4422
1325-1418	OEM
1327	OEM
1327-1291	2122
1327-1309	1201
1327-4922	0006
1327-4980	0261
1327-5029	0261
1327-5045	OEM
1327-5052	OEM
1327-5060	1541
1327-5078	0041
1327-5110	0019
1327-5136	0023
1327-5144	0031
1328-0003	0071
1328-0011	0053
1328-0037	OEM
1329	0211
1330	0211
1330-5313	0486
1330-5339	0053
1330-8937	0124
1330-8945	0471
1330-8952	0041
1330-8960	0023
1330-8978	0062
1330-8994	0053
1330-9018	0471
1330-9026	0023
1330-9034	0091
1330-9059	0023
1330-9067	0023
1330-9075	0041
1330-9083	0023
1330-9091	0348
1330-9109	0257
1330-9117	0195
1330-9125	0071
1330-9141	0140
1330-9158	0023
1330-9174	0124
1330-9182	OEM
1330-9190	3927
1330-9208	4241
1330-9216	4129
1330-9224	0334
1330-9232	1641
1330-9240	1699
1330-9256	0118
1330-9257	1529
1330-9265	2870
1330-9273	3083
1330-9281	3916
1330-9299	OEM
1331-7532	0825
1331-7540	0582
1331-7565	0582
1331-7573	2223
1331-7581	0825
1331-7607	0276
1331-7615	0168
1331-7623	3619
1331-7672	0786
1331-9272	0077
1332	OEM
1333	OEM
1333-0535	OEM
1336	OEM
1339	0103
1340	0280
1343	OEM
1344-6448	0253
1344-8253	0023
1344-8287	0023
1344-8477	0261
1345	1591
1346-6347	4550
1346-8897	OEM
1346-8905	2666
1346-8913	OEM
1346-8921	4094
1346-8939	2346
1346-8947	3955
1346-8954	3046
1346-8962	0248
1346-8970	0137
1346-8988	0057
1346-8996	0253
1346-9002	0057
1346-9010	0471
1346-9028	0072
1347	0751
1347-17	0004
1347-7112	2400
1348	0751
1348-1429	0037
1348-1437	0079
1348-1445	0133
1348-1452	0039
1348-1593	2178
1348-4969	0532
1348-4977	0155
1348-4985	0041
1348-4993	0057
1348-5008	0181
1348-5016	0195
1348-5024	2364
1348-5032	3515
1348-5040	OEM
1348-5057	OEM
1348A12H01	1352
1348A14H01	1158
1348A30H01	1823
1348A32H01	1041
1348A36H01	1831
1348A44H01	0033
1348A45H01	0081
1348A46H01	0141
1348A62H01	0562
1349-17	0208
1350	0280
1351	0856
1352	0490
1352-9870	1541
1355-3086	0023
1356A	OEM
1358	1591
1359-0161	0334
1359-3827	0582
1359-3835	0582
1360	0211
1362-17	0004
1362-17A	0004
1364-17	0004
1365-4991	OEM
1365-5113	0035
1365-5121	OEM
1365-6939	0041
1365-6947	OEM
1365-7192	2232
1368-1747	0031
1368-1754	0102
1368-1762	0376
1368-1796	OEM
1368-1804	OEM
1368-1812	OEM
1368-1838	OEM
1368-1846	3253
1368-1853	1028
1368-1861	OEM
1368-1879	OEM
1368-2638	1329
1368C/D	0016
1370-7229	0023
1370-7237	OEM
1370-7245	0376
1370-7252	OEM
1370-7260	OEM
1370-7351	1386
1370-7369	3494
1370-7377	0718
1371-17	0208
1371-5404	0052
1371-5412	1216
1371-5420	OEM
1373-17	0151
1373-17A	0127
1373-17AL	0144
1374-17	0155
1374-17A	0155
1374-17AC	0016
1374-6714	0343
1374-6722	0133
1374-6730	1131
1374-6748	0165
1374-6755	0091
1374-6763	0023
1374-6771	OEM

DEVICE TYPE	REPL CODE
1374-6789	0162
1374-6797	OEM
1374-6813	0071
1374-6821	2346
1374-6839	OEM
1374-6847	2364
1374-6854	OEM
1374-6896	0286
1374-6920	0261
1374-7019	0261
1374-7605	OEM
1376	OEM
1376-001	2018
1376-002	2018
1376-2257	0041
1376-6456	0165
1377-2637	0023
1377-2645	0031
1377-2652	4111
1377-2660	OEM
1377-2678	0077
1377-2686	0057
1377-2694	OEM
1377-2710	OEM
1377-2728	OEM
1377-2736	OEM
1377-2744	3344
1377-2751	OEM
1377-2769	OEM
1377-2777	6996
1377-3775	OEM
1377-3783	OEM
1377-3791	4519
1377-3809	0006
1377-3817	1386
1377-3825	OEM
1377-3833	7103
1377-3841	1092
1377-3858	0286
1377-3866	0532
1377-4591	OEM
1377-4609	OEM
1377-6208	0023
1377-6224	OEM
1377-6232	OEM
1377-6349	0477
1377-6356	OEM
1378-8633	OEM
1378-8658	OEM
1380-4679	OEM
1380-4687	4310
1380-4695	OEM
1380-4729	1786
1380-5601	0023
1380-5650	OEM
1380-5668	OEM
1380-5676	OEM
1380-5692	OEM
1380-5700	OEM
1380-5726	OEM
1380-5734	0271
1380-5742	OEM
1380-6054	0071
1380-6062	0023
1380-6070	1641
1380-6146	OEM
1380-6179	1742
1380-6187	1752
1380-6195	OEM
1380-6203	1541
1380-6211	0724
1384-3263	4415
1384-3271	OEM
1384-3289	0023
1384-3909	OEM
1384-4972	OEM
1384-5029	OEM
1384-5185	OEM
1384S1	OEM
1385-3890	0042
1390	0280
1390-6417	OEM
1392	OEM
1392-2620	OEM
1392-2687	OEM
1392-2901	OEM
1392-2921	1386
1392-3081	OEM
1393-7123	0261
1393-9210	OEM
1393-9228	OEM
1393-9236	1127
1393-9285	3772
1393-9319	OEM
1393-9327	4129
1393-9345	OEM
1393-9350	OEM
1393-9376	4329
1393-9384	OEM
1393-9574	OEM
1393-9590	5792
1393-9608	4250
1396	0600
1400	0280
1401-1225	OEM
1401-1405	OEM
1401-1407	OEM
1401-1410	OEM
1401-1415	OEM
1401-1420	OEM
1401-1425	OEM
1402	0016
1402A	OEM
1402B	0016
1402C	0079
1402CD	0079
1402D	0079
1402E	0079
1402E/D	OEM
1403A	OEM
1404A	OEM
1405-17	0015
1405A	OEM
1406T	OEM
1407T	OEM
1410	0280
1410-102	0133
1410-102(ROGERS)	0229
1410-167	0015
1410-169	0124
1410-169(ROGERS)	0133
1410-171	0143
1411	0208
1411-130	0025
1411-134	1263
1411-135	0137
1411-136	0170
1411-137	0636
1412-1	0016
1412-1-12	0016
1412-1-12-8	0016
1412-170	0015
1412-177	1396
1412-182	0242
1412-190	0242
1412DA1AP	OEM
1412DA1BP	OEM
1412DA1CP	OEM
1412DA1DP	OEM
1412DA1EP	OEM
1412DA2AP	OEM
1412DA2BP	OEM
1412DA2CP	OEM
1412DA2DP	OEM
1412DA2EP	OEM
1412DA3AP	OEM
1412DA3BP	OEM
1412DA3CP	OEM
1412DA3DP	OEM
1412DA3EP	OEM
1412DA4AP	OEM
1412DA4BP	OEM
1412DA4CP	OEM
1412DA4DP	OEM
1412DA4EP	OEM
1413-159	0040
1413-160	0211
1413-168	0604
1413-172	0604
1413-175	0211
1413-178	0160
1414-157	0086
1414-158	0037
1414-158(ROGERS)	0037
1414-173	0079
1414-174	0155
1414-176	0037
1414-179	3525
1414-180	2427
1414-183	0155
1414-184	0086
1414-185	0126
1414-186	0086
1414-187	0126
1414-188	0130
1414-189	0086
1414A	0160
1414A-12	0160
1414A-12-8	0160
1415	0016
1416	0086
1417	0224
1417-177	0440
1420-1-1	0127
1420-2-2	0007
1423	0127
1431	OEM
1434	0236
1436-17	0004
1438B	OEM
1438B-83	OEM
1439	OEM
1441-0415	OEM
1441-0420	OEM
1441-0425	OEM
1441-0615	OEM
1441-0620	OEM
1441-0625	OEM
1441-0815	OEM
1441-0820	OEM
1441-0825	OEM
1441-1015	OEM
1441-1020	OEM
1441-1025	OEM
1441-1215	OEM
1441-1220	OEM
1441-1225	OEM
1441-1415	OEM
1441-1420	OEM
1441-1425	OEM
1443D	OEM
1443D-83	OEM
1444	OEM
1445	OEM
1446-17	OEM
1452N1	OEM
1453G	OEM
1455-7-4	0127
1458	0360
1459	0211
1461C	OEM
1461C-83	OEM
1462	1157
1464-607-7	0409
1464-607-8	0409
1465	0016
1465-1	0050
1465-1-12	0050
1465-1-12-8	0050
1465-2	0208
1465-2-12	0208
1465-2-12-8	0208
1465-4	0600
1465-4-12	0600
1465-4-12-8	0600
1465A	0050
1465A-12	0050
1465A-12-8	0050
1465B	0136
1465B-12	0136
1465B-12-8	0136
1465C	0160
1465C-12	0160
1465C-12-8	0160
1466-1	0004
1466-1-12	0004
1466-1-12-8	0004
1466-2	0004
1466-2-12	0004
1466-2-12-8	0004
1466-3	0004
1466-3-12	0004
1466-3-12-8	0004
1471-4356	1831
1471-4364	0141
1471-4372	0562
1471-4380	1823
1471-4398	0033
1471-4729	0267
1471-9860	1831
1473	0006
1474-3	0280
1474-3-12	0280
1474-3-12-8	0280
1476-17	0004
1476-17-6	0004
1477-3436	1854
1477A	0440
1477A-12	0440
1477A-12-8	0440
1478-6982	1929
1479-0224	1835
1479-0240	0682
1479-0257	4856
1479-0265	3715
1479-7971	1907
1480B	OEM
1480B-83	OEM
1482	0016
1482-17	1400
1483	0102
1483(XSTR)	0086
1483E209	OEM
1483E419	0102
1484A	0160
1484A-12	0160
1484A-12-8	0160
1485-9896	1619
1486-4	OEM
1486-8780	0685
1487-17	4208
1488(SEARS)	0015
1488B	0050
1488B-12	0050
1488B-12-8	0050
1488C	0280
1488C-12	0280
1488C-12-8	0280
1489-17	0143
1490	OEM
1492-1	0440
1492-1-12	0440
1492-1-12-8	0440
1493-17	0016
1496	3801
1499-009-0202	OEM
1499-009-0309	OEM
1499-009-0901	OEM
1500PJA10	OEM
1500PJA20	OEM
1500PJA30	OEM
1500PJA40	OEM
1501	0224
1502-0229	0148
1502B	0079
1502C	0079
1502CH	OEM
1502D	0224
1502M	OEM
1502MH833	5785
1504	OEM
1505	0015
1505A	0015
1506T	OEM
1507T	OEM
1512	0143
1512,-36/1-01-883	OEM
1512C-16/1-01	OEM
1512C-36/1-01	OEM
1512C-1601	OEM
1512C-160212	OEM
1512C-160312H	OEM
1512C-160612HP	OEM
1512M-16/1-01-883	OEM
1512M-16/1-06	OEM
1512M-36/1-01	OEM
1512M-160712HP833	OEM
1512M-161112HP833	OEM
1514	0127
1515	0079
1517	OEM
1518	OEM
1521B	0905
1524	0198
1524(PNP)	OEM
1526	0050
1526(RECT)	OEM
1526/802	OEM
1527	OEM
1527-5282	1041
1527-5308	0081
1540	0016
1542	0619
1542(GE)	0619
1548-17	0178
1550	0143
1550-17	0365
1553-17	0037
1559	0574
1559(GE)	0593
1559-17	0160
1559-17A	0160
1561-0403	0103
1561-0404	0103
1561-0408	0130
1561-0410	0130
1561-0604	0103
1561-0608	0130
1561-0610	0130
1561-0615	0130
1561-0803	0130
1561-0804	0130
1561-0805	0130
1561-0808	0130
1561-0810	0130
1561-0815	0130
1561-17	0160
1561-1004	0130
1561-1005	0130
1561-1008	0130
1561-1010	0130
1561-1015	0130
1561-1208	1852
1561-1210	1852
1561-1215	1852
1561-1404	0177
1561-1410	1852
1561-1415	1852
1561-1608	1852
1561-1610	1852
1561-1615	1852
1561A603	0103
1561A608	0130
1561A615	0130
1562	0015
1567	0016
1567-0	0016
1567-2	0016
1571-0401	0178
1571-0402	0178
1571-0420	0178
1571-0425	0178
1571-0601	0178
1571-0602	0178
1571-0620	0178
1571-0801	0178
1571-0802	0178
1571-0820	0178
1571-0825	0178
1571-32	0178
1571-33	0178
1571-34	OEM
1571-35	OEM
1571-36	OEM
1571-41	0178
1571-42	0178
1571-43	0178
1571-44	OEM
1571-46	OEM
1571-1001	0178
1571-1002	0178
1571-1020	0178
1571-1025	0178
1571-1220	0178
1571-1225	0178
1571-1425	0178
1582	0037
1582-0403	0103
1582-0404	0103
1582-0405	0103
1582-0408	0130
1582-0410	0130
1582-0415	0130
1582-0508	0130
1582-0510	0130
1582-0603	0103
1582-0604	0103
1582-0605	0103
1582-0608	0130
1582-0610	0130
1582-0615	0130
1582-0803	0130
1582-0804	0130
1582-0805	0130
1582-0808	0130
1582-0810	0130
1582-0815	0130
1582-1003	0130
1582-1004	0130
1582-1005	0130
1582-1008	0130
1582-1010	0130
1582-1015	0130
1582-1208	0130
1582-1210	0130
1582-1215	0130
1582-1810	0130
1600EXD24	OEM
1600FD26	OEM
1600FXD24	OEM
1600GXD21	OEM
1600PN50	OEM
1600PN60	OEM
1600PN80	OEM
1600PN100	OEM
1600PN120	OEM
1600PN140	OEM
1600PN150	OEM
1600PN160	OEM
1601GPT	OEM
1601PDK120	2139
1601PDK140	OEM
1601PDK160	OEM
1601PDK180	OEM
1601PDK200	OEM
1601PDK210	OEM
1601PDK220	OEM
1601PDK230	OEM
1601PDK240	OEM
1601PDK250	OEM
1602	0006
1602E	OEM
1603	OEM
1604/OPI	OEM
1604/POC	OEM
1605	OEM
1606	OEM
1607	OEM
1607-6812	OEM
1607A80	1206
1608	OEM
1609	0065
1611-17	0015
1612SK24E	0325
1614	OEM
1615C	0196
1616	OEM
1616/MIC	OEM
1616/OII	OEM
1616/OIO	OEM
1616C	0484
1616CCI	OEM
1616HCO	OEM
1617	OEM
1617C	1080
1619	OEM
1620	OEM
1620BB	OEM
1620DMA	OEM
1620M	OEM
1620MF	OEM
1620TTL	OEM
1627-1413	2899
1627-2064	0358
1634-17	0475
1634-17-14A	0144
1642	OEM
1643	0624
1652H	OEM
1653H	OEM
1654H	OEM
1655H	OEM
1663C-01D	OEM
1663C-02D	OEM
1663M-01D	OEM
1663M-02D	OEM
1664TTL	OEM
1666C-01A12	OEM
1666C-02B12H	OEM
1666C-03C12HP	OEM
1666C-06A12HP833	OEM
1666M-07B	OEM
1666M-11C12	OEM
1666M-15A12H	OEM
1667M-6D	OEM
1672	OEM
1673M-5B	OEM
1674-17	0015
1682	OEM
1687-17	0475
1699-17	0133
1700	OEM
1700-5182	2537
1701	OEM
1701-01	OEM
1702	0198
1702-0	0079
1702-01	OEM
1702-0	0079
1702K	0079
1702M	0018
1702MNO	0079
1703	0103
1703-01	OEM
1706-3B	OEM
1710	0155
1710-1	1143
1710-2	4150
1710-3	OEM
1711-0402	OEM
1711-0405	OEM
1711-0602	OEM
1711-0605	OEM
1711-0802	OEM
1711-0805	OEM
1711-17	0016
1711-1002	OEM
1711-1005	OEM
1711-1202	OEM
1711-1205	OEM
1711-1402	OEM
1711-1405	OEM
1711-1602	OEM
1711-1605	OEM
1711-1802	OEM
1712-17	0016
1713-0402	OEM
1713-0405	OEM
1713-0602	OEM
1713-0605	OEM
1713-0802	OEM
1713-0805	OEM
1713-17	0015
1713-1002	OEM
1713-1005	OEM
1713-1202	OEM
1713-1205	OEM
1713-1402	OEM
1713-1405	OEM
1713-1602	OEM
1713-1605	OEM
1713-1802	OEM
1714-0402	0429
1714-0405	0429
1714-0602	0429
1714-0605	0429
1714-0802	0429
1714-0805	0429
1714-2	OEM
1714-1002	0429
1714-1005	0429
1714-1202	0429
1714-1205	0429
1714-1402	0429
1714-1405	0429
1714-1602	0429
1714-1605	0429
1714-1802	0429
1714-1805	0429
1715-3	OEM
1716-0402	OEM
1716-0405	OEM
1716-2	5609
1716-1805	OEM
1717	OEM
1717-0402	OEM
1717-0405	OEM
1717-0602	OEM
1717-0605	OEM
1717-0802	OEM
1717-0805	OEM
1717-1002	OEM
1717-1005	OEM
1717-1202	OEM
1717-1205	OEM
1717-1402	OEM
1717-1405	OEM
1717-1602	OEM
1717-1605	OEM
1717-1802	OEM
1719-1	5641
1720	OEM
1720-5	OEM
1720-6	OEM
1720R	OEM
1723-0405	2485
1723-0410	2485
1723-0605	2485
1723-0610	2485
1723-0805	2485
1723-0810	2485
1723-17	0016
1723-1005	2485
1723-1010	2485
1723-1205	2485
1723-1210	2485
1723-1405	2485
1723-1410	2485
1723-1605	2485
1723-1610	2485
1723-1805	0130
1723-1810	0130
1723R	OEM
1725	OEM
1726	OEM
1726-0405	OEM
1726-0410	OEM
1726-0605	OEM
1726-0610	OEM
1726-0805	OEM
1726-0810	OEM
1726-1005	OEM
1726-1010	OEM
1726-1205	OEM
1726-1210	OEM
1726-1405	OEM
1726-1410	OEM
1726-1605	OEM
1726-1610	OEM
1726-1805	OEM
1726-1810	OEM
1727	OEM
1731	OEM
1732	OEM
1736R	OEM
1737R	4989
1738R	4873
1739R	OEM
1740R	OEM
1741-0051	0232
1741-0069	0682
1741-0085	0268
1741-0119	0314
1741-0143	0361
1741-0150	1907
1741-0176	0385
1741-0184	3251
1741-0200	0467
1741-0234	0512
1741-0242	0685
1741-0275	2398
1741-0291	1352
1741-0325	0697
1741-0333	3715
1741-0366	4596
1741-0416	0873
1741-0424	5383
1741-0440	3520
1741-0473	1024
1741-0481	0559
1741-0564	1168
1741-0572	1944
1741-0598	1185
1741-0606	2101
1741-0622	1202
1741-0630	2172
1741-0663	3158
1741-0689	1275
1741-0697	5410
1741-0721	2492
1741-0747	1432
1741-0770	0117
1741-0804	1368
1741-0895	0569
1741-0952	1486
1741-1018	0942
1741-1042	1492
1741-1075	1495
1741-1133	0527
1741-1190	1038
1741-1224	1052
1741-1257	0175
1741-1299	4856
1741-1323	4399
1741-1349	1829
1741-1380	2020
1743-0610	2485
1743-0620	OEM
1743-0630	2485
1743-0810	OEM
1743-0820	2485
1743-0830	OEM
1743-1010	2485
1743-1020	OEM
1743-1030	2485
1743-1210	OEM
1743-1220	2485
1743-1230	OEM
1743-1410	2485
1743-1420	2485
1743-1430	2485
1743-1610	OEM
1743-1620	2485
1743-1630	OEM
1743-1810	2433
1743R	OEM
1748-0620	OEM
1748-1020	OEM
1748-1420	OEM
1750	OEM
1750-00-0-0	OEM
1750-00-WXX-WXX	OEM
1750-01-0-0	OEM

If replacement code is OEM, contact original manufacturer for replacement.

DEVICE TYPE	REPL CODE
1750-01-WXX-WXX	OEM
1750-02-0-0	OEM
1750-02-WXX-WXX	OEM
1751	6386
1751-17	1220
1751G036	0016
1753-1	OEM
1753-2	OEM
1754	OEM
1754-1	0661
1754-3	0625
1755	OEM
1756-1	0912
1756-17	0133
1757	OEM
1757-7	OEM
1760-1	6135
1760-2	6135
1760-3	OEM
1761	OEM
1761-1	6143
1761-17	0144
1762	OEM
1762-2	OEM
1763	OEM
1763-0415	0130
1763-0420	0130
1763-0425	0130
1763-0615	0130
1763-0620	0130
1763-0625	0130
1763-0815	0130
1763-0820	2201
1763-0825	0130
1763-1015	0130
1763-1020	0130
1763-1025	0130
1763-1215	0130
1763-1220	0130
1763-1225	0130
1763-1415	0130
1763-1420	0130
1763-1425	0130
1763-1615	OEM
1763-1625	OEM
1763-1815	OEM
1764	OEM
1765	OEM
1766	0137
1768	OEM
1768-0415	OEM
1768-0420	OEM
1768-0425	OEM
1768-0615	OEM
1768-0625	OEM
1768-0815	OEM
1768-0825	OEM
1768-1015	OEM
1768-1025	OEM
1768-1215	OEM
1768-1225	OEM
1768-1415	OEM
1768-1425	OEM
1768-1625	OEM
1768-1815	OEM
1771-0440	OEM
1771-0450	OEM
1771-0460	OEM
1771-0640	OEM
1771-0650	OEM
1771-0660	OEM
1771-0840	OEM
1771-0850	OEM
1771-0860	OEM
1771-1040	OEM
1771-1050	OEM
1771-1060	OEM
1771-1240	OEM
1771-1250	OEM
1771-1260	OEM
1771-1440	OEM
1771-1450	OEM
1771-1460	OEM
1771-1640	OEM
1772	OEM
1773-3	OEM
1774-2	OEM
1775-2	OEM
1776-0450	OEM
1776-0650	OEM
1776-1050	OEM
1776-1250	OEM
1776-1450	OEM
1777-1	OEM
1777-17	0004
1778-17	0143
1780-1	2040
1780-2	OEM
1785	OEM
1787	OEM
1788	OEM
1789-17	0086
1791	OEM
1792-17	0144
1793	OEM
1794	OEM
1794-17	1083
1794-17(10)	0012
1795	OEM
1798	OEM
1799-17	0016
1800	2867
1800-17	0016
1800DC	0850
1800PC	0850
1800XY2-35SC	OEM
1801DC	0874
1801PC	0874
1801SS1-07SC	OEM
1802-01	OEM
1802DC	0888
1802L	0532
1802M	0532
1802PC	0888
1802XY2-35SC	OEM
1803-1	3579
1803DC	0900
1803PC	0900
1804	0723
1804-17	0233
1804DC	0919
1804PC	0919
1805	0268
1805DC	0929
1805PC	0929
1806	0361
1806-17	0103
1806-17A	0103
1806DC	0945
1806PC	0945
1807	1052
1807DC	0962
1807PC	0962
1808	1208
1808-8	OEM
1808DC	0982
1808PC	0982
1809	0133
1809(I.C.)	0723
1809DC	0998
1809PC	0998
1810	OEM
1810DC	1011
1811	OEM
1811/4K	OEM
1811/16K	OEM
1811-110-10111	0917
1811DC	1025
1811PC	1025
1812-1	0016
1812-1-12	0016
1812-1-127	0016
1812-1L	0016
1812-1L8	0016
1812DC	1039
1812PC	1039
1813	OEM
1813DC	4538
1813PC	4538
1814	OEM
1814/4K	OEM
1814/16K	OEM
1814A	0160
1814A-12	0160
1814A-127	0160
1814AL	0160
1814AL-8	0160
1814DC	4541
1814PC	4541
1815	OEM
1816CMOS-8L	OEM
1816CMOS-8S	OEM
1816CMOS-16L	OEM
1816CMOS-16S	OEM
1816R	OEM
1817R	OEM
1818	0133
1818(I.C.)	0723
1818-1	OEM
1820	OEM
1820-0054	0232
1820-0055	1208
1820-0063	1168
1820-0068	0512
1820-0069	0697
1820-0070	0873
1820-0075	1172
1820-0077	1313
1820-0087	2619
1820-0095	3399
1820-0099	0569
1820-0122	0358
1820-0174	0361
1820-0214	1052
1820-0261	0175
1820-0301	1432
1820-0304	1426
1820-0328	0314
1820-0341	1632
1820-0352	1854
1820-0430	1922
1820-0476	1554
1820-0495	0527
1820-0512	2492
1820-0861	0141
1820-0862	3399
1820-0863	0466
1820-0864	0335
1820-0865	0033
1820-0869	1176
1820-0870	0467
1820-0894	0361
1820-1064	0734
1820-1068	5383
1820-1111	1024
1820-1172	1349
1820R	OEM
1821	OEM
1821-1	OEM
1821R	OEM
1826-1	0024
1827-17	0004
1833	OEM
1834	OEM
1835	OEM
1835-17	0191
1839-2	OEM
1840-17	0222
1841	OEM
1841-17	0016
1842	OEM
1843	OEM
1843-17	0007
1844-17	0037
1845-17	0007
1846	OEM
1846-17	0133
1847	OEM
1848-17	0015
1849	0015
1849R	0015
1850	0015
1850-2	OEM
1850-17	0037
1850R	0015
1851-17	0229
1852-17	0144
1853-0045	0126
1854-0003	0016
1854-17	0015
1855-17	0015
1858	0600
1858-1	OEM
1858-1/KIT	OEM
1858A	0012
1859	OEM
1859-14	0160
1859-16	0160
1859-17	0325
1859R	0015
1861	3108
1862	OEM
1862-001	OEM
1862-002	OEM
1865-1	0050
1865-1-12	0050
1865-1-127	0050
1865-1L	0050
1865-2	0208
1865-2-12	0208
1865-2L	0208
1865-2L8	0208
1865-4	0600
1865-4-12	0600
1865-4-127	0600
1865-4L	0600
1865-4L8	0600
1865A	0050
1865A-12	0050
1865A-127	0050
1865AL8	0050
1865B	0136
1865B-12	0136
1865B-127	0136
1865BL	0136
1865BL8	0136
1865C	0160
1865C-12	0160
1865C-127	0160
1865CL	0160
1865CL8	0160
1866-1	0004
1866-1-12	0004
1866-1-127	0004
1866-1L	0004
1866-1L8	0004
1866-2	0004
1866-2-12	0004
1866-2-127	0004
1866-2L	0004
1866-2L8	0004
1866-3	0004
1866-3-12	0004
1866-3-127	0004
1866-3L	0004
1866-3L8	0004
1866-17	0016
1867-17	0037
1868	OEM
1868-001	OEM
1868-002	OEM
1871	OEM
1872	OEM
1874-2	OEM
1874-3	0280
1874-3-12	0280
1874-3-127	0280
1874-3L	0280
1874-3L8	0280
1877A	0440
1877A-12	0440
1877A-127	0440
1877AL	0440
1877AL8	0440
1879-17	0155
1879-17A	0086
1880-17	0144
1881-17	0144
1882-17	0016
1883-17	0016
1884-17	0016
1884A	0160
1884A-12	0160
1884A-127	0160
1884AL	0160
1884AL8	0160
1885-17	0233
1886-17	OEM
1888B	0050
1888B-12	0050
1888B-127	0050
1888BL	0050
1888BL8	0050
1888C	0280
1888C-12	0280
1888C-127	0280
1888CL	0280
1888CL8	0280
1889-17	1480
1890-17	0144
1892-1	0440
1892-1-12	0440
1892-1-127	0440
1892-1L	0440
1892-1L8	0440
1893	OEM
1893-17	0016
1894	OEM
1895	OEM
1900	OEM
1901-0028	0540
1901-0036	0540
1901-0045	0242
1901-0388	0540
1901-0389	0540
1901-0557	2698
1906-17	0208
1915-17	0155
1915A	OEM
1916-17	1400
1917-17	0004
1919-17	0004
1919-17A	0004
1922-00	OEM
1923-17	0144
1923-17-1	0144
1925-17	0151
1929-17	0016
1931-17	0127
1931-17A	0144
1932-17	0016
1933-17	0086
1934-17	0325
1935-17	0203
1936-17	0178
1937-17	0133
1940-17	0037
1941-17	0023
1943-430	0079
1945-17	0222
1946-17	0004
1947-17	0123
1948-17	0920
1949-17	0023
1950	OEM
1950-17	0015
1951-17	0136
1952-17	0007
1954-17	0004
1955-17	0222
1956-17	0143
1957	OEM
1957-2	OEM
1958-17	0208
1960-17	0004
1961-17	0076
1966-17	0016
1968-17	2046
1969-17	0042
1970-16	0015
1970-17	0015
1971-17	0071
1972-17	0086
1973-17	0004
1974-17	0004
1977-17	0143
1979-17	0133
1980-17	0143
1981/1904	OEM
1981/1908	OEM
1981/1912	OEM
1981/1916	OEM
1981/1920	OEM
1981-17	0133
1982/1904	OEM
1982-17	0015
1983/1908	OEM
1983/1912	OEM
1983/1916	OEM
1983-17	0144
1985/1	OEM
1985/2	OEM
1998-17	0127
1999	0016
1999-17	0127
2000-001	2253
2000-002	0215
2000-003	2306
2000-004	1292
2000-005	1816
2000-006	4209
2000-007	2261
2000-008	1292
2000-009	3801
2000-01	OEM
2000-010	4484
2000-011	4484
2000-017	1540
2000-029	4645
2000-030	OEM
2000-031	4042
2000-032	4652
2000-033	4648
2000-034	0325
2000-035	2828
2000-036	2328
2000-037	4291
2000-038	OEM
2000-052	4639
2000-053	OEM
2000-101	0325
2000-102	0414
2000-103	0414
2000-104	3342
2000-105	0325
2000-107	0325
2000-108	0688
2000-111	0688
2000-125	0133
2000-201	0148
2000-202	0693
2000-203	1062
2000-204	0151
2000-205	0127
2000-206	2814
2000-207	1591
2000-208	0936
2000-209	0951
2000-210	0076
2000-211	0839
2000-212	0836
2000-213	0076
2000-214	0127
2000-215	0018
2000-216	2292
2000-218	0006
2000-221	2438
2000-235	0076
2000-240	1066
2000-241	0786
2000-242	0836
2000-245	0076
2000-258	0286
2000-265	0006
2000-266	0380
2000-267	0127
2000-268	0368
2000-269	0018
2000-270	0890
2000-271	0936
2000-275	0286
2000-276	0018
2000-301	0019
2000-302	0124
2000-303	0133
2000-304	0023
2000-305	0012
2000-308	0012
2000-309	0030
2000-312	0420
2000-317	0133
2000-318	0123
2000-320	0015
2000-321	0041
2000-322	0091
2000-323	0720
2000-324	0091
2000-325	0077
2000-326	0420
2000-327	0012
2000-328	0091
2000-329	0636
2000-330	0323
2000-331	0139
2000-332	0133
2000-333	0041
2000-339	0139
2000-341	0163
2000-342	0120
2000-344	0030
2000-345	0133
2000-348	0087
2000-367	0165
2001	0042
2001(DIO)	OEM
2001(E.F.JOHNSON)	0042
2001-17	0016
2001-303	0133
2001A	OEM
2001PDK60	3900
2001PDK80	2139
2001PDK100	2139
2001PDK120	2139
2001PDK140	OEM
2001PDK160	OEM
2002	0160
2002(DIO)	OEM
2002A	OEM
2003	0103
2003(DIO)	OEM
2003-01	OEM
2003-17	0016
2004	0164
2004-01	0937
2004-02	2209
2004-03	0937
2004-04	0076
2004-05	1419
2004-06	0006
2004-14	0076
2004-67	0124
2004-68	0120
2004-69	5116
2004-70	0215
2005	OEM
2005(TRANS)	OEM
2006	0016
2006-17	0053
2007-01	0160
2008	0178
2008-17	0016
2009	0440
2010	0440
2010(RECT.)	5282
2010-17	0325
2011	0161
2012	0211
2014	OEM
2014I	0168
2015	0050
2015(RECT)	5282
2015-00	0136
2015-1	0103
2015-1A	0103
2015-2	0103
2015-2A	0103
2015-3	0103
2015-3A	0103
2016	1898
2016P	1898
2017-108	0390
2017-109	0228
2017-111	0139
2017-113	0301
2017-114	0015
2017-115	0076
2018	OEM
2018-003-106	0006
2018-01	0537
2018-12	OEM
2018-401-10	0076
2018-411-100	0076
2018-413-10	0076
2018-4140102	1144
2019-40	1031
2019-45	0124
2020	0050
2020(RECT)	2900
2020(RECT.)	2900
2020-00	0136
2020-01	1392
2020-02	1154
2020-03	0549
2020-04	0325
2020-05	0537
2020-06	1419
2020-07	0006
2020-08	0005
2020-6	OEM
2020-7	OEM
2020-8	OEM
2020-9	OEM
2020R	1241
2021	0050
2021-00	0136
2021-05	0023
2022	OEM
2022-01	0150
2022-03	0076
2022-04	0936
2022-05	0016
2022-06	0212
2022-07	0143
2022-08	0015
2022-244	0076
2023	OEM
2023-34	OEM
2023-41	0012
2025	5283
2025-1	2743
2025-2	2743
2025-3	2743
2025-9110-18810	0143
2026	0016
2026-00	0016
2026-4	2997
2026-5	2997
2027	0079
2027-00	0698
2027R.F.	0079
2028	0144
2028-00	0144
2030	5283
2030R	5495
2031-17	0023
2032-33	1066
2032-34	1066
2032-35	0006
2032-36	1280
2032-37	0076
2032-38	0215
2032-40	1419
2035	0853
2035-83	OEM
2035-5100-53660	0537
2035-5100-69372	0552
2039-2	0600
2040	0853
2040-17	0170
2041-01	0151
2041-02	0155
2041-03	0219
2041-04	0936
2041-05	0133
2041-06	0012
2042-1	2482
2042-2	2472
2042-17	0004
2043	OEM
2043-17	0037
2044-17	0198
2045	5286
2045-17	0015
2045Z4	OEM
2047-52	4669
2049-01	OEM
2049-03	1292
2049-32	0030
2050	5286
2052	0015
2055A100-50	2281
2056-04	2261
2056-05	0417
2056-06	3695
2056-10	0006
2056-75	0325
2056-105	2253
2057-305	OEM
2057A2-22	OEM
2057A2-27	0079
2057A2-28	0211
2057A2-37	0136
2057A2-48	0142
2057A2-49	0050
2057A2-60	0136
2057A2-62	0016
2057A2-64	0016
2057A2-65	0136
2057A2-66	0136
2057A2-80	0136
2057A2-81	0142
2057A2-84	0142
2057A2-87	0224
2057A2-103	0016
2057A2-109	0127
2057A2-110	0127
2057A2-113	0016
2057A2-116	0127
2057A2-117	0127
2057A2-117(OSC)	0079
2057A2-117(RF-AMP)	0079
2057A2-119	0127
2057A2-120	0127
2057A2-121	0016
2057A2-122	0076
2057A2-127	0144
2057A2-128	0127
2057A2-131	0076
2057A2-133	0160
2057A2-143	0079
2057A2-145	0016
2057A2-146	0016
2057A2-147	0004
2057A2-148	0004
2057A2-149	0050
2057A2-150	0791
2057A2-151	0086
2057A2-152	0016
2057A2-153	0016
2057A2-154	0016
2057A2-155	0016
2057A2-156	0086
2057A2-157	0007
2057A2-158	0007
2057A2-159	0050
2057A2-163	0151
2057A2-165	0004
2057A2-166	0050
2057A2-167	0208
2057A2-175	OEM
2057A2-179	0144
2057A2-180	0007
2057A2-181	0007

Original Device Types

DEVICE TYPE	REPL CODE
2057A2-182	0037
2057A2-183	0037
2057A2-184	0016
2057A2-185	0007
2057A2-187	0007
2057A2-192	0007
2057A2-193	0007
2057A2-195	0007
2057A2-196	0007
2057A2-197	0007
2057A2-198	0037
2057A2-199	0086
2057A2-200	0150
2057A2-201	0144
2057A2-202	0007
2057A2-203	0150
2057A2-204	0007
2057A2-205	0050
2057A2-206	0211
2057A2-207	0007
2057A2-208	0016
2057A2-209	0016
2057A2-210	0004
2057A2-211	0222
2057A2-212	2050
2057A2-215	0016
2057A2-216	1144
2057A2-217	0127
2057A2-218	1144
2057A2-219	0113
2057A2-220	0286
2057A2-221	0127
2057A2-222	0016
2057A2-223	0142
2057A2-224	0144
2057A2-225	0151
2057A2-226	0016
2057A2-227	0076
2057A2-228	0191
2057A2-229	0006
2057A2-230	0086
2057A2-231	0050
2057A2-232	0050
2057A2-234	0086
2057A2-237	0127
2057A2-241	0004
2057A2-249	0111
2057A2-251	0007
2057A2-252	0050
2057A2-253	OEM
2057A2-254	OEM
2057A2-255	OEM
2057A2-256	0144
2057A2-257	0111
2057A2-258	0007
2057A2-259	0127
2057A2-260	0111
2057A2-261	0233
2057A2-262	0111
2057A2-263	0136
2057A2-264	0016
2057A2-265	0975
2057A2-272	0111
2057A2-273	0111
2057A2-274	0111
2057A2-275	0111
2057A2-276	0016
2057A2-277	0037
2057A2-278	0016
2057A2-279	0016
2057A2-280	0016
2057A2-281	0016
2057A2-284	0086
2057A2-285	0016
2057A2-288	0004
2057A2-289	0076
2057A2-289(4TH-LF)	0330
2057A2-290	0111
2057A2-294	0016
2057A2-295	0086
2057A2-296	0016
2057A2-297	0016
2057A2-298	0148
2057A2-300	0016
2057A2-301	0042
2057A2-302	0085
2057A2-303	0016
2057A2-304	0127
2057A2-305	0127
2057A2-306	0076
2057A2-307	0037
2057A2-309	0144
2057A2-310	0144
2057A2-311	0144
2057A2-312	OEM
2057A2-313	0144
2057A2-314	0144
2057A2-315	OEM
2057A2-316	0016
2057A2-317	0164
2057A2-318	OEM
2057A2-319	0079
2057A2-320	OEM
2057A2-322	0127
2057A2-323	0127
2057A2-324	0016
2057A2-325	0127
2057A2-326	0127
2057A2-329	0004

DEVICE TYPE	REPL CODE
2057A2-331	0127
2057A2-332	0016
2057A2-333	0111
2057A2-334	0111
2057A2-341	0016
2057A2-342	0144
2057A2-343	0037
2057A2-352	0111
2057A2-353	0037
2057A2-356	0324
2057A2-359	0037
2057A2-370	1221
2057A2-373	0111
2057A2-374	0016
2057A2-385	0111
2057A2-386	0144
2057A2-387	0016
2057A2-390	0016
2057A2-391	0111
2057A2-392	0144
2057A2-393	0144
2057A2-394	0144
2057A2-395	0127
2057A2-396	0016
2057A2-397	0037
2057A2-398	0016
2057A2-399	0016
2057A2-400	0037
2057A2-401	0016
2057A2-402	0144
2057A2-403	0037
2057A2-404	0111
2057A2-405	0111
2057A2-406	0037
2057A2-412	0016
2057A2-427	0155
2057A2-428	0111
2057A2-429	OEM
2057A2-430	0037
2057A2-432	0224
2057A2-433	0016
2057A2-434	0016
2057A2-436	1220
2057A2-445	0688
2057A2-446	0006
2057A2-448	0144
2057A2-449	0016
2057A2-452	0191
2057A2-454	0111
2057A2-457	0037
2057A2-463	0016
2057A2-464	0016
2057A2-465	0151
2057A2-466	0597
2057A2-468	0155
2057A2-475	0111
2057A2-477	0127
2057A2-478	0151
2057A2-479	0597
2057A2-480	0148
2057A2-483	0224
2057A2-484	0018
2057A2-486	0155
2057A2-487	0016
2057A2-489	0037
2057A2-501	0127
2057A2-502	0111
2057A2-503	0127
2057A2-504	0144
2057A2-505	0144
2057A2-507	0224
2057A2-508	0224
2057A2-509	0144
2057A2-510	0016
2057A2-511	0016
2057A2-518	0111
2057A2-524	0155
2057A2-526	0224
2057A2-527	0224
2057A2-529	0693
2057A2-530	0086
2057A2-539	0127
2057A2-540	0127
2057A2-541	0127
2057A2-542	0552
2057A2-543	0552
2057A2-558	0286
2057A2-559	0079
2057A2-560	0079
2057A2-561	0037
2057A10-64	0016
2057A32-25	4543
2057A32-26	1478
2057A32-33	0520
2057A100	0103
2057A100-4	2866
2057A100-8	0004
2057A100-9	0004
2057A100-10	0136
2057A100-14	2866
2057A100-15	0211
2057A100-16	1514
2057A100-17	0086
2057A100-18	2866
2057A100-21	0004
2057A100-23	0164
2057A100-24	0004
2057A100-26	1514
2057A100-30	2866

DEVICE TYPE	REPL CODE
2057A100-31	OEM
2057A100-32	OEM
2057A100-33	OEM
2057A100-34	0164
2057A100-35	2866
2057A100-40	5883
2057A100-41	0164
2057A100-44	0164
2057A100-45	4095
2057A100-47	1657
2057A100-48	0164
2057A100-49	0161
2057A100-50	4095
2057A100-51	0037
2057A100-53	0127
2057A100-55	0004
2057A100-58	1657
2057A100-60	OEM
2057A100-62	3919
2057A100-66	1657
2057A100-67	0324
2057A100-69	3689
2057A100-72	1381
2057A100-72(NPN)	OEM
2057A100-72(PNP)	OEM
2057A-288	0211
2057A289	0086
2057A-429	0144
2057A-430	0150
2057B0-3	2866
2057B-2	0280
2057B2-4	0211
2057B2-14	0144
2057B2-23	0004
2057B2-27	0475
2057B2-28	0475
2057B2-29	0004
2057B2-32	0211
2057B2-34	0211
2057B2-35	0136
2057B2-37	0136
2057B2-38	0016
2057B2-41	0136
2057B2-42	0136
2057B2-43	0211
2057B2-44	0211
2057B2-45	0211
2057B2-46	0038
2057B2-47	0142
2057B2-48	0633
2057B2-49	0004
2057B2-50	0136
2057B2-51	0136
2057B2-52	0004
2057B2-57	0211
2057B2-58	0142
2057B2-59	0286
2057B2-60	0633
2057B2-61	0164
2057B2-62	0076
2057B2-63	0111
2057B2-64	0127
2057B2-65	0136
2057B2-66	0136
2057B2-67	0136
2057B2-68	0136
2057B2-69	0016
2057B2-70	0136
2057B2-71	0136
2057B2-72	0211
2057B2-73	0198
2057B2-77	0136
2057B2-78	0211
2057B2-79	0136
2057B2-80	0136
2057B2-81	0136
2057B2-83	0211
2057B2-84	0142
2057B2-85	0144
2057B2-86	0211
2057B2-87	0144
2057B2-88	0136
2057B2-89	0050
2057B2-90	0050
2057B2-93	0050
2057B2-94	0004
2057B2-97	0016
2057B2-99	0004
2057B2-101	0127
2057B2-102	0127
2057B2-103	0144
2057B2-104	0136
2057B2-105	0136
2057B2-107	0004
2057B2-108	0144
2057B2-109	0144
2057B2-110	0144
2057B2-111	0144
2057B2-112	0144
2057B2-113	0016
2057B2-114	0127
2057B2-115	0127
2057B2-116	0127
2057B2-117	0127
2057B2-118	0136
2057B2-119	0144
2057B2-120	0144
2057B2-121	0016
2057B2-122	0016

DEVICE TYPE	REPL CODE
2057B2-123	0016
2057B2-124	0004
2057B2-125	0127
2057B2-127	0144
2057B2-128	0144
2057B2-129	0004
2057B2-130	0016
2057B2-133	0160
2057B2-134	0975
2057B2-135	0004
2057B2-136	0975
2057B2-137	0004
2057B2-138	0007
2057B2-139	0007
2057B2-140	0233
2057B2-141	0136
2057B2-142	0004
2057B2-143	0127
2057B2-149	0050
2057B2-150	0925
2057B2-151	0042
2057B2-152	0016
2057B2-153	0016
2057B2-154	0016
2057B2-155	0016
2057B2-157	0050
2057B2-158	0050
2057B2-159	0050
2057B2-160	0144
2057B2-161	0144
2057B2-162	0144
2057B2-192	0007
2057B2-206	0004
2057B2A2-118	0050
2057B45-14	0004
2057B-59	0111
2057B-84	0142
2057B-85	0016
2057B100-1	0004
2057B100-3	2866
2057B100-4	0208
2057B100-5	2866
2057B100-6	0004
2057B100-7	0211
2057B100-8	0004
2057B100-9	0004
2057B100-11	0222
2057B100-12	0144
2057B100-13	0004
2057B100-16	1514
2057B100-17	0086
2057B101-4	0016
2057B102-4	0016
2057B103-4	0016
2057B104-8	0233
2057B106-12	0037
2057B107-8	0086
2057B108-6	0037
2057B109-9	0086
2057B110-9	0126
2057B111-9	4163
2057B112-9	0126
2057B-113	0144
2057B113-9	0086
2057B114-9	0126
2057B115-9	0126
2057B116-9	0126
2057B117-9	0016
2057B118-12	0016
2057B119-2	0016
2057B120-12	0016
2057B121-9	0126
2057B122-9	0126
2057B123-10	0142
2057B124-10	0160
2057B125-9	0016
2057B126-12	0086
2057B129-9	0086
2057B141-4	0016
2057B142-4	0016
2057B143-12	0016
2057B144-12	0086
2057B145-12	0126
2057B146-12	0016
2057B147-12	0037
2057B149-12	0325
2057B151-6	0016
2057B152-12	0016
2057B153-9	0086
2057B155-10	0178
2057B156-9	0086
2057B157-9	0074
2057B158-10	0178
2057B159-12	0037
2057B163-12	0126
2057B168	0004
2057B169	0004
2057B175-9	0074
2057B186	0050
2057B206	0004
2058	0006
2058-02	3342
2058D	OEM
2060	1609
2060-024	0015
2061-42	1292
2061-44	0133
2061-45	0030
2061A45-38	0133

DEVICE TYPE	REPL CODE
2061A45-47	0211
2061A45-72	0015
2061B45-14	0164
2061B45-35	0015
2062-17	1411
2063-17	0076
2063-17-12	0016
2064	0079
2064(CROWN)	0016
2064(IC)	OEM
2064-17	0244
2064C	OEM
2065-03	0076
2065-04	0076
2065-06	0936
2065-07	2814
2065-08	0124
2065-17	0133
2065-50	4622
2065-51	0215
2065-52	1292
2065-53	2253
2065-54	0076
2065-55	0414
2066-17	0133
2068-018-104	0520
2068-022-108	2755
2068DD	OEM
2070	1610
2074	OEM
2074-17	0688
2074AD	OEM
2075	OEM
2075-3	OEM
2076	0133
2077-07	0143
2077-2	2755
2079-40	1031
2079-41	OEM
2079-42	0165
2079-93	0012
2079-129	OEM
2080	1610
2081	OEM
2081-6	0925
2081-17	0390
2082	OEM
2082-6	0042
2084-17	0015
2085-17	0161
2085Z4	OEM
2087-46	0012
2090	0285
2090(CROWN)	0233
2090A43-1	0917
2090A53-2	0474
2091-02	6351
2091-03	2710
2091-49	1215
2091-50	2568
2091-51	0076
2091A41-81	OEM
2093	OEM
2093/1097	3263
2093A2-289	0144
2093A3D-2	0229
2093A3D-20	0222
2093A4-2	0229
2093A5-2	0484
2093A5-10	0196
2093A6-1	0229
2093A6-2	0229
2093A8-1	0143
2093A9-3	0280
2093A9-4	0280
2093A12-1	0015
2093A25-3	0143
2093A33-1	0143
2093A38-1	0143
2093A38-4	0015
2093A38-5	0143
2093A38-8	0030
2093A38-10	0143
2093A38-11	0133
2093A38-13	0015
2093A38-14	0143
2093A38-15	0143
2093A38-16	0012
2093A38-21	0143
2093A38-22	0123
2093A38-23	0211
2093A38-24	0086
2093A38-27	0143
2093A38-28	0015
2093A38-29	0030
2093A38-30	0143
2093A38-31	0143
2093A38-32	0143
2093A38-33	0143
2093A38-34	0143
2093A38-35	0123
2093A38-37	0133
2093A38-40	0143
2093A39-12	0137
2093A41	0143
2093A41-2	0143
2093A41-5	0030
2093A41-6	0071
2093A41-8	0015

DEVICE TYPE	REPL CODE
2093A41-14	0143
2093A41-16	0012
2093A41-23	0030
2093A41-24	0012
2093A41-24A	0012
2093A41-25	0030
2093A41-27	0030
2093A41-28	0071
2093A41-29	0123
2093A41-37	OEM
2093A41-38	0143
2093A41-40	0211
2093A41-41	0211
2093A41-42	0071
2093A41-43	0071
2093A41-45	0015
2093A41-49	0015
2093A41-50	0143
2093A41-51	0015
2093A41-53	0649
2093A41-54	0015
2093A41-55	0015
2093A41-56	0137
2093A41-57	0015
2093A41-58	1080
2093A41-59	0143
2093A41-60	0137
2093A41-61	0015
2093A41-62	0015
2093A41-63	0162
2093A41-65	0023
2093A41-66	0015
2093A41-67	0102
2093A41-68	0143
2093A41-69	0102
2093A41-70	0025
2093A41-71	OEM
2093A41-72	0102
2093A41-73	0102
2093A41-74	0288
2093A41-75	0015
2093A41-76	0133
2093A41-77	0023
2093A41-78	0015
2093A41-79	0030
2093A41-80	OEM
2093A41-81	0015
2093A41-82	0053
2093A41-84	0015
2093A41-85	0531
2093A41-86	0015
2093A41-87	0164
2093A41-88	0015
2093A41-89	0015
2093A41-90	0162
2093A41-91	0030
2093A41-92	0143
2093A41-94	OEM
2093A41-95	0030
2093A41-96	0015
2093A41-101	0030
2093A41-102	0139
2093A41-103	0015
2093A41-104	0015
2093A41-105	0133
2093A41-108	0015
2093A41-110	0015
2093A41-112	0289
2093A41-113	0053
2093A41-115	0023
2093A41-116	0015
2093A41-117	0030
2093A41-125	0012
2093A41-126	6665
2093A41-129	0133
2093A41-130	0030
2093A41-131	0015
2093A41-139	0015
2093A41-141	0143
2093A41-155	0133
2093A41-156	0133
2093A41-157	0005
2093A41-158	0133
2093A41-159	0012
2093A41-161	0005
2093A41-162	0030
2093A41-163	0012
2093A41-164	0133
2093A41-165	0133
2093A41-166	0030
2093A41-167	0143
2093A41-169	0143
2093A41-171	0133
2093A41-172	0133
2093A41-173	0015
2093A41-180	0015
2093A41-181	0143
2093A41-182	0015
2093A41-185	0015
2093A41-186	0012
2093A41-187	0143
2093A41-188	0005
2093A41-189	0015
2093A41-193	0030
2093A41-194	0162
2093A41-196	0019
2093A41-197	0015
2093A41-198	OEM
2093A41-200	0005

DEVICE TYPE	REPL CODE
2093A42-7	0015
2093A43-2	0917
2093A52-1	0071
2093A59-1	0917
2093A69-1	0071
2093A71-1	0071
2093A75-1	0474
2093A77-1	0143
2093A78-1	0071
2093A79-6	0071
2093A80-1	0137
2093B4-6	0015
2093B11-21	0123
2093B38-4	0133
2093B38-9	0917
2093B38-14	0015
2093B41-6	0015
2093B41-8	0015
2093B41-9	0015
2093B41-10	0015
2093B41-11	0143
2093B41-12	0015
2093B41-14	0071
2093B41-15	OEM
2093B41-16	0012
2093B41-18	0015
2093B41-20	0015
2093B41-21	0015
2093B41-22	0015
2093B41-23	0030
2093B41-24	0012
2093B41-25	0015
2093B41-27	0030
2093B41-28	0015
2093B41-29	0123
2093B41-32	0015
2093B41-33	0924
2093B41-34	0015
2093B41-35	0015
2093B41-37	0015
2093B41-62	0015
2094	OEM
2095	OEM
2095Z4	OEM
2097	OEM
2097/1114	4586
2098	OEM
2101	0111
2101(DIO)	OEM
2101A	OEM
2101A-2F	OEM
2101A-2N	OEM
2101A-4F	OEM
2101A-4N	OEM
2101AF	OEM
2101AN	OEM
2101N	OEM
2102	0015
2102-010	0143
2102-014	0015
2102-017	0015
2102-025	0143
2102-028	0143
2102-029	0124
2102-031	0030
2102-032	0057
2102-074	0015
2102-1DC	OEM
2102-1DL	OEM
2102-1DM	OEM
2102-1FC	OEM
2102-1FL	OEM
2102-1FM	OEM
2102-1PC	OEM
2102-2DC	OEM
2102-2DL	OEM
2102-2DM	OEM
2102-2FC	OEM
2102-2FL	OEM
2102-2FM	OEM
2102-2PC	OEM
2102-120	0050
2102-124	0050
2102A	4973
2102A(DIO)	OEM
2102AN-4L	4973
2102DC	OEM
2102FDC	4973
2102FDL	OEM
2102FDM	OEM
2102FFC	OEM
2102FFL	OEM
2102FFM	OEM
2102FPC	4973
2102HDC	OEM
2102HDL	OEM
2102HDM	OEM
2102HFC	OEM
2102HFL	OEM
2102HFM	OEM
2102HPC	OEM
2102L1DC	OEM
2102L1DL	OEM
2102L1DM	OEM
2102L1FC	OEM
2102L1FL	OEM
2102L1FM	OEM
2102L2DC	OEM
2102L2DL	OEM

If replacement code is OEM, contact original manufacturer for replacement.

DEVICE TYPE	REPL CODE
2102L2DM	OEM
2102L2FC	OEM
2102L2FL	OEM
2102L2FM	OEM
2102L2PC	OEM
2102LFDC	OEM
2102LFDL	OEM
2102LFDM	OEM
2102LFFC	OEM
2102LFFL	OEM
2102LFFM	OEM
2102LFPC	OEM
2102LHDC	OEM
2102LHDL	OEM
2102LHDM	OEM
2102LHFC	OEM
2102LHFL	OEM
2102LHFM	OEM
2102LHPC	OEM
2103	OEM
2103A	OEM
2104	4974
2105	0030
2105Z4	OEM
2106	OEM
2106-17	0301
2106-119	0004
2106-120	0050
2106-121	0004
2106-122	0004
2106-123	0004
2106-124	0143
2106AF001	OEM
2107	4902
2108	OEM
2109	OEM
2109-101-0064	OEM
2109-101-0066	OEM
2109-101-0204	OEM
2109-101-0709	OEM
2109-101-1005	OEM
2109-101-1403	OEM
2109-101-5304	2623
2109-102-5304	2623
2109-102-5401	5095
2109-103-2006	0167
2109-103-2200	0912
2109-103-3001	0663
2109-104-9800	OEM
2109-110-1407	OEM
2109-301-0707	1527
2109-301-6008	2026
2109-301-6600	2090
2109-301-6707	4788
2109-401-1400	OEM
2110	0871
2110N-41	0133
2110N-42	0015
2110N-132	0007
2110N-133	0007
2110N-134	0178
2111-11	OEM
2111A	0664
2111A-2F	OEM
2111A-2N	OEM
2111A-4F	OEM
2111A-4N	OEM
2111AF	OEM
2111AN	OEM
2112	OEM
2112-17	0004
2112A-2F	OEM
2112A-2N	OEM
2112A-4F	OEM
2112A-4N	OEM
2112AF	OEM
2112AN	OEM
2113	0196
2114	2048
2114-0	0233
2114-17	0164
2114-30L	2048
2114L	2048
2114L-3	2048
2114L-30	2048
2115	0871
2115R	1635
2115Z4	OEM
2116-17	0015
2117	0522
2118	OEM
2118-004-10	0023
2118-3	OEM
2118-4	OEM
2118-7	OEM
2118-201-10	OEM
2119	OEM
2119-101-0500	0167
2119-101-1204	0878
2119-101-1301	3915
2119-101-1408	2949
2119-101-1505	2284
2119-101-1602	1055
2119-101-2607	0912
2119-101-3204	0176
2119-101-4306	0167
2119-101-5107	3738
2119-101-5204	0490
2119-101-9408	0912
2119-101-9602	0663
2119-101-9709	0663
2119-102-0309	0167
2119-102-0503	0539
2119-102-0600	0557
2119-102-0907	0539
2119-102-1003	0557
2119-102-2600	0912
2119-102-3003	0976
2119-102-4008	OEM
2119-102-4309	5048
2119-102-5100	3738
2119-102-5401	5095
2119-102-5702	0976
2119-102-7304	1529
2119-102-7401	3916
2119-102-7605	0334
2119-103-0603	1329
2119-103-1006	OEM
2119-103-2302	3431
2119-201-1300	0176
2119-201-3009	3431
2119-202-0803	2872
2119-401-1201	OEM
2119-601-2709	3518
2119-601-5005	4650
2119-601-5209	3518
2119-901-0201	3957
2119-901-1109	0334
2120-17	0420
2121	OEM
2121-17	0076
2122-3	0400
2122-17	0143
2124	OEM
2125	5283
2125Z4	OEM
2127-17	1144
2130	5283
2132	0560
2132-17	1221
2133	OEM
2134-17	0475
2135	0853
2135E	3008
2135Z4	OEM
2137-103-2803	0037
2137-103-4308	0037
2137-301-0106	0261
2137-301-0504	0144
2137-301-1800	0079
2137-301-5305	0261
2137-301-5402	0261
2137-301-5509	1624
2137-301-5606	1624
2137-301-7208	0079
2137-401-5304	0037
2138-17	0007
2139-101-4005	0006
2139-101-4102	0006
2139-101-4209	0148
2139-101-5301	0006
2139-103-2807	0037
2139-103-3307	0006
2139-103-3802	0006
2139-103-4001	0006
2139-103-4302	0037
2139-103-5705	0681
2139-103-7006	1597
2139-103-7103	0037
2139-103-8807	0286
2139-103-8904	0286
2139-103-9006	0286
2139-103-9307	0006
2139-103-9404	0261
2139-103-9501	0261
2139-103-9608	0148
2139-103-9705	0006
2139-301-0401	0144
2139-301-0508	0144
2139-301-0809	0842
2139-301-0906	0842
2139-301-1406	0261
2139-301-2304	0168
2139-301-2314	0168
2139-301-2401	0168
2139-301-2410	0643
2139-301-5105	0261
2139-301-5202	0261
2139-301-5309	0261
2139-301-5406	0261
2139-301-5503	1624
2139-301-5600	1624
2139-301-6508	5133
2139-301-7202	0079
2139-301-7707	0076
2139-301-8508	0155
2139-301-8702	0155
2139-301-9100	1624
2139-301-9202	0168
2139-301-9804	0079
2139-301-9901	0079
2139-302-3603	2367
2139-302-6006	0018
2139-302-6501	0076
2139-302-7406	0076
2139-302-7409	0079
2139-302-7506	0076
2139-302-8103	1284
2139-302-8501	0191
2139-302-8608	0191
2139-303-320	2209
2139-304-5609	0866
2139-304-5706	0866
2139-305-3408	0261
2139-305-9204	0066
2139-305-9301	0066
2139-307-2603	0224
2139-308-0305	0224
2139-401-0701	1493
2139-401-1007	1541
2139-401-1104	1541
2139-401-1308	1541
2139-401-5308	0037
2139-401-8507	2668
2139-401-9609	0065
2139-401-9808	0261
2141	OEM
2141-UCB	OEM
2141-UCE	OEM
2145	5286
2145-401-3502	0055
2148-027-10	0019
2148-033-10	0087
2148-034-10	0052
2148-17	0015
2148H	OEM
2148H-3	OEM
2148HL	OEM
2148HL-3	OEM
2149-040-0200	0055
2149-101-0505	0681
2149-101-8505	1911
2149-101-9005	1348
2149-101-9102	0532
2149-101-9500	0037
2149-201-4504	1648
2149-201-4601	1911
2149-202-0400	0532
2149-203-0607	3518
2149-301-0202	1284
2149-301-0309	0066
2149-301-0406	1541
2149-301-0503	0556
2149-301-409	OEM
2149-301-510	0066
2149-301-1401	2797
2149-301-1406	0261
2149-301-1605	0103
2149-301-4202	5133
2149-301-4309	0079
2149-301-4804	0313
2149-301-4901	0103
2149-301-5100	0066
2149-301-8105	1284
2149-301-8202	0066
2149-301-8901	0066
2149-301-9100	0155
2149-301-9508	0155
2149-302-7307	0955
2149-302-7404	0643
2149-302-7501	0955
2149-302-9103	0643
2149-302-9109	0643
2149-302-9608	0643
2149-302-9802	0955
2149-303-4305	2660
2149-303-5805	0276
2149-303-7300	0392
2149-303-8004	0276
2149-303-8402	0955
2149-304-3109	0629
2149-401-0405	1541
2149-401-0502	0556
2149-401-6007	0392
2149-401-6104	1165
2149-401-6308	1165
2149-401-8201	0079
2149-401-8308	0595
2149-401-9303	0392
2149-401-9808	0261
2149-401-9905	0261
2149-404-0103	0270
2149-404-0200	0055
2149-404-0307	0055
2149H	OEM
2149H-2	OEM
2149H-3	OEM
2149HL-3	OEM
2150	5286
2150-17	0012
2151-17	0123
2152	0015
2153	OEM
2153NS	OEM
2154	OEM
2155Z4	OEM
2156	OEM
2157-301-1703	0079
2159-101-0607	0148
2159-101-0207	0076
2159-301-0304	0261
2159-301-1707	0079
2159-301-1804	0079
2159-301-2096	0261
2159-301-2906	0261
2159-302-1707	0079
2159-401-2604	3133
2159-701-0207	0076
2160	1669
2160-17	0219
2161-17	0244
2163-17	0178
2164A-15	OEM
2164A-20	OEM
2165	0167
2167-17	1908
2167-201-0704	0023
2167-206-1102	0023
2167-208-5108	0023
2167-301-6305	0023
2167-401-6304	0053
2167-401-8003	0253
2167-401-8100	0471
2167-406-4804	0124
2167-408-0509	0140
2167-408-0606	0053
2167-408-0800	0471
2168-17	0390
2169	OEM
2169-101-0602	0019
2169-101-0806	0123
2169-101-1306	2232
2169-102-2605	0015
2169-103-0608	0133
2169-103-1807	0124
2169-103-2006	0167
2169-103-2608	0473
2169-201-0106	OEM
2169-201-0601	0080
2169-201-0708	0023
2169-201-0805	0795
2169-201-0902	0015
2169-201-2601	0795
2169-201-2708	0015
2169-202-1900	0143
2169-202-2002	0133
2169-202-2109	0133
2169-202-2206	0023
2169-202-2303	0102
2169-202-2400	0023
2169-202-2604	0133
2169-205-0700	0133
2169-205-0807	0124
2169-205-1006	0057
2169-205-1103	1005
2169-205-3307	0023
2169-205-3404	0499
2169-206-1009	0023
2169-206-1106	0023
2169-206-1203	0023
2169-207-0201	0023
2169-207-1109	0071
2169-207-1206	0017
2169-208-1708	0023
2169-208-5102	0023
2169-208-5306	0071
2169-208-5704	0023
2169-208-5706	0023
2169-209-1105	0080
2169-210-5808	0071
2169-213-1102	0071
2169-213-1108	0071
2169-218-1307	0023
2169-218-1404	0023
2169-218-1501	0023
2169-219-3106	0023
2169-219-4207	0023
2169-219-4208	0023
2169-219-5106	0023
2169-220-1508	0023
2169-301-0207	0076
2169-301-1100	0133
2169-301-2901	0133
2169-301-2907	0133
2169-301-6105	0031
2169-301-6309	0023
2169-304-0104	0023
2169-304-1109	0023
2169-304-1400	0023
2169-308-0504	0023
2169-401-0900	0057
2169-401-1206	1005
2169-401-4104	0301
2169-401-5808	1329
2169-401-7604	0445
2169-403-2304	0705
2169-403-2401	0441
2169-403-5600	0471
2169-403-6508	0375
2169-403-7503	1005
2169-403-7707	0445
2169-403-8508	0039
2169-404-0103	0077
2169-404-0307	0317
2169-404-1100	0317
2169-404-1807	0471
2169-404-1904	0077
2169-404-2006	0041
2169-404-4006	0041
2169-404-4404	0365
2169-406-4804	0124
2169-406-4808	0124
2169-408-0600	0053
2169-408-0707	OEM
2169-408-0804	0471
2169-408-5704	0023
2169-501-1603	OEM
2169-503-6401	OEM
2169-901-0303	OEM
2170	1610
2176	OEM
2177	OEM
2179-001-0200	0710
2179-001-1302	0316
2179-301-9207	0023
2180-17	0143
2180-151	0144
2180-152	0144
2180-153	0016
2180-154	0016
2180-155	0178
2181-17	0155
2189-106-0500	0343
2189-301-0906	0842
2189-301-9804	0079
2195	OEM
2195-17	0111
2196-17	0123
2197-17	0127
2199-17	0015
2199-101-9408	0912
2200-17	0015
2201	OEM
2201(DIO)	OEM
2201A	OEM
2202	4172
2202(DIO)	OEM
2202-17	0390
2202A	OEM
2203	0244
2203-17	0244
2203A	OEM
2204	4130
2204-17	0198
2205	0030
2205-17	0030
2206	0015
2206(TRANS)	OEM
2206-17	0015
2207	OEM
2207-17	0111
2208	OEM
2208-17	0086
2208-21	OEM
2209	OEM
2209-17	0133
2209-21	OEM
2210-17	0925
2211-17	0560
2211B	0015
2212	OEM
2212-17	0111
2213	OEM
2213-17	0127
2213-21	OEM
2214	OEM
2214-17	2209
2215	0429
2215-17	0050
2215-21	OEM
2217	OEM
2217-21	OEM
2218	OEM
2218-17	0030
2218-21	OEM
2220-17	0037
2220-29	OEM
2221	5034
2221A	5034
2222	0016
2222A	0016
2223	OEM
2223-21	OEM
2224-17	0224
2225-17	0224
2226-17	0224
2231	0037
2231-17	1149
2231A	OEM
2231B	OEM
2231D	OEM
2231F	OEM
2231H	OEM
2231K	OEM
2231M	OEM
2231P	OEM
2231S	OEM
2232-17	0143
2232A	OEM
2232B	OEM
2232D	OEM
2232F	OEM
2232H	OEM
2232K	OEM
2232M	OEM
2232P	OEM
2232S	OEM
2232V	OEM
2232Z	OEM
2233-17	0015
2243	0160
2243-21	OEM
2244-1	0681
2245-17	0236
2246-17	0218
2252	0015
2261	0079
2263	0016
2269	0037
2270	0016
2271	0111
2272	0037
2275-1	1899
2275-17	0016
2280-13	0143
2281-13	0143
2282-13	0143
2282-17(RAT.DET)	0123
2283-13	0143
2284-13	0143
2284-17	0127
2289-13	0913
2290-13	0143
2290-17	0016
2291-17	0127
2295	0178
2301	OEM
2301A	OEM
2302	0015
2302(VARACTOR)	OEM
2302-17	0151
2302A	OEM
2303	OEM
2303A	OEM
2304	0133
2304-1	1022
2305	OEM
2306	OEM
2307	OEM
2308	OEM
2309	0012
2309-110-0013	OEM
2309-110-4208	OEM
2309-110-4305	OEM
2309-110-7601	OEM
2309-115-0319	0516
2309-115-3101	0516
2309-115-6106	OEM
2309-201-0032	OEM
2309D	OEM
2312	OEM
2312-2	OEM
2312-3	OEM
2313	OEM
2314	OEM
2314K80AM	OEM
2315	OEM
2315-046	0015
2316B-3CJ	OEM
2316B-3CK	OEM
2316B-4CK	OEM
2320-17	0076
2321	OEM
2321-17	0155
2322	0079
2322-17	1978
2323-17	0157
2324-010-10	OEM
2328-17	0139
2330-021	0649
2330-191	0071
2330-201	0071
2330-252	0071
2331	OEM
2332-3CJ	OEM
2332-3CK	OEM
2332-4CJ	OEM
2332-4CK	OEM
2332-20F	OEM
2332-20I	OEM
2332-20N	OEM
2332-25F	OEM
2332-25I	OEM
2332-25N	OEM
2332-30F	OEM
2332-30N	OEM
2332-45I	OEM
2332-45N	OEM
2332A	OEM
2332B	OEM
2332E20F	OEM
2332E20I	OEM
2332E20N	OEM
2332E25F	OEM
2332E25I	OEM
2332E25N	OEM
2332E30F	OEM
2332E30I	OEM
2332E30N	OEM
2332E45F	OEM
2332E45I	OEM
2332E45N	OEM
2332M25F	OEM
2332M25I	OEM
2332M30F	OEM
2332M30I	OEM
2332M45F	OEM
2332M45I	OEM
2334-17	0390
2336-17	3717
2337-17	0076
2338-17	0076
2339-17	0244
2340-17	1978
2341-17	0025
2344	OEM
2355D	OEM
2359-17	0212
2360-17	0015
2361-17	0155
2362-1	0079
2362-1(SYLVANIA)	0144
2362-17	0157
2364	OEM
2364-3CJ	OEM
2364-4CK	OEM
2364-5CJ	OEM
2364-5CK	OEM
2364-20F	OEM
2364-20I	OEM
2364-20N	OEM
2364-25F	OEM
2364-25I	OEM
2364-25N	OEM
2364-30F	OEM
2364-30N	OEM
2364-35F	OEM
2364-35N	OEM
2364-45F	OEM
2364-130	OEM
2364A	OEM
2364E20F	OEM
2364E20N	OEM
2364E25F	OEM
2364E25N	OEM
2364E30F	OEM
2364E30N	OEM
2364E45F	OEM
2364E45N	OEM
2364M25F	OEM
2364M30F	OEM
2364M45F	OEM
2365	OEM
2365S	OEM
2381-17	0037
2382-17	0042
2392	OEM
2396-17	0002
2399-17	0839
2400	OEM
2400-17	0133
2400-23	0015
2400-27	0015
2401	OEM
2401A	OEM
2402	0015
2402(VARACTOR)	OEM
2402-17	0030
2402-453	0004
2402-454	0004
2402-455	0004
2402-456	0222
2402-457	0004
2402-459	0143
2402-461	0015
2402-462	0015
2402-463	0015
2402A	OEM
2403	0133
2403A	OEM
2405	0133
2405-453	0004
2405-454	0004
2405-455	0004
2405-456	0004
2405-457	0004
2405-458	0143
2405-459	0015
2405-462	0015
2405-463	0015
2408-17	0560
2408-326	0004
2408-328	0004
2408-329	0164
2408-330	0143
2408-331	0015
2411-17	0023
2413	OEM
2413-21	OEM
2414-1	OEM
2415	OEM
2415-21	OEM
2416	OEM
2417	0142
2418-17	1411
2419	OEM
2419-21	OEM
2420M54532P	OEM
2420NPEP159	OEM
2420SLM339N	0176
2420SN07406	1206
2420SN07407	1339
2420SN75188	0508
2420SN75189	0511
2420SNLS151	1646
2420SNNE555	0973
2420SNS412N	1860
2420T78005P	0624
2420TMM16P2	1898
2421T05Z24Y	0494
2421T05Z39X	0036
2421T1Z1300	0365

DEVICE TYPE	REPL CODE
2421T1Z1500	0002
2421T1Z2000	0531
2421T2B4B41	0323
2421T5B4B41	0323
2421TS5277B	0023
2422TA1015Y	0148
2422TC1815Y	0076
2422TD14380	3703
2422TD71700	3081
2427	0127
2427(RCA)	0144
2432	0829
2434	0829
2437-17	0836
2443	0127
2443(RCA)	0016
2444	0142
2444(RCA)	0142
2445	0144
2446	0079
2446(RCA)	0016
2447	0079
2447(RCA)	0016
2448	0127
2448(RCA)	0037
2448-17	0368
2449-17	0111
2450	0079
2450(RCA)	0144
2450-17	1280
2450S	OEM
2451-17	0165
2452-17	0012
2453-17	0296
2454-17	0031
2460	OEM
2460-13	0015
2470-1724	5338
2470-1732	3251
2473	0079
2473(RCA)	0144
2473-2109	0467
2474	0233
2474(RCA)	0233
2475	0127
2475(RCA)	0016
2476	0079
2476(RCA)	0144
2477	0079
2477(RCA)	0144
2478	0136
2478A	0136
2478B	0136
2482	0136
2482(RCA)	0126
2484	0016
2487B	0050
2488	0136
2488A	0136
2489	0136
2489A	0050
2490	0004
2490A	0004
2491	0142
2491A	0142
2491B	0142
2494	0975
2494(RCA)	0975
2495	0086
2495(RCA)	0016
2495-012	0136
2495-013	0136
2495-014	0004
2495-078	0050
2495-079	0050
2495-080	0004
2495-082	0050
2495-166-1	0144
2495-166-2	0016
2495-166-4	0144
2495-166-8	0144
2495-166-9	0144
2495-200	0136
2495-376	0050
2495-377	0050
2495-378	0050
2495-388	0004
2495-488-1	0050
2495-488-2	0050
2495-489	0071
2495-520	0144
2495-521	0144
2495-522-1	0144
2495-522-4	0016
2495-523-1	0144
2495-529	0016
2495-567-2	0004
2495-567-3	0004
2495-586-2	0004
2496	0975
2497-473	0004
2497-496	0004
2498-163	0086
2498-507-2	0144
2498-507-3	0144
2498-508-2	0144
2498-508-3	0144
2498-513	0071
2498-903-2	0144
2498-903-3	0144
2500	0975
2500(RCA)	0975
2501	OEM
2501(IC)	OEM
2501A	OEM
2501PDK20	OEM
2501PDK30	OEM
2501PDK40	OEM
2501PDK60	OEM
2501PDN120	OEM
2501PDN140	OEM
2501PDN160	OEM
2501PDN180	OEM
2501PDN200	OEM
2501PDN210	OEM
2501PDN220	OEM
2501PDN230	OEM
2501PDN240	OEM
2501PDN250	OEM
2502	0126
2502(RCA)	0004
2502A	OEM
2503	OEM
2503A	OEM
2503V	OEM
2505K	OEM
2510-31	0143
2510-32	0015
2510-101	0224
2510-102	0224
2510-103	0111
2510-104	0016
2510-105	0560
2513IXCM2140	OEM
2513IXCMXXXX	OEM
2514	1974
2514NXCMXXXX	OEM
2516	0167
2516(EPROM)	OEM
2516-1	0167
2516JL-45	5585
2516N	OEM
2517	OEM
2518	OEM
2522	0015
2523	0015
2524N	OEM
2529N	OEM
2531A	OEM
2540	OEM
2546	0127
2546(RCA)	0016
2554-1(RCA)	2743
2554-2(RCA)	2743
2554-3	2743
2554-3(RCA)	2743
2554-4	2743
2554-4(RCA)	2743
2559	2549
2559-1	2549
2560	0328
2560(J.C.PENNEY)	0328
2560-1	0328
2584	0198
2601	OEM
2601A	OEM
2601PDN120	OEM
2601PDN140	OEM
2601PDN160	OEM
2601PDN180	OEM
2601PDN200	OEM
2601PDN210	OEM
2601PDN220	OEM
2601PDN230	OEM
2601PDN240	OEM
2601PDN250	OEM
2602	OEM
2602A	OEM
2603	0518
2603-180	0004
2603-181	0004
2603-182	0004
2603-183	0136
2603-184	0016
2603-186	0143
2603A	OEM
2603E	0518
2603F	0518
2605A	OEM
2605B	OEM
2605D	OEM
2605F	OEM
2605H	OEM
2605K	OEM
2605M	OEM
2606-286	0004
2606-287	0004
2606-288	0847
2606-291	0004
2606-292	0136
2606-294	0144
2606-295	0050
2606-296	0143
2606-297	0030
2606-299	0015
2606-303	0451
2607-45F	OEM
2607-45I	OEM
2607-45N	OEM
2608I	OEM
2609-0011	OEM
2612	0004
2613-15F	OEM
2613-15I	OEM
2613-15N	OEM
2613-20F	OEM
2613-20I	OEM
2613-20N	OEM
2613-25F	OEM
2613-25I	OEM
2613-25N	OEM
2613-45F	OEM
2613-45I	OEM
2613-45N	OEM
2614-15F	OEM
2614-15I	OEM
2614-15N	OEM
2614-20F	OEM
2614-20I	OEM
2614-20N	OEM
2614-25F	OEM
2614-25I	OEM
2614-25N	OEM
2614-45F	OEM
2614-45I	OEM
2614-45N	OEM
2615A	OEM
2615B	OEM
2615D	OEM
2615F	OEM
2615H	OEM
2615K	OEM
2615M	OEM
2616-30F	OEM
2616-30I	OEM
2616-30N	OEM
2616-35F	OEM
2616-35I	OEM
2616-35N	OEM
2616-45F	OEM
2616-45I	OEM
2616-45N	OEM
2616A20F	OEM
2616A20I	OEM
2616A20N	OEM
2616A25F	OEM
2616A25I	OEM
2616A25N	OEM
2616A30F	OEM
2616A30I	OEM
2616A30N	OEM
2616A45F	OEM
2616A45I	OEM
2616A45N	OEM
2616AE20F	OEM
2616AE20I	OEM
2616AE20N	OEM
2616AE25F	OEM
2616AE25I	OEM
2616AE25N	OEM
2616AE30F	OEM
2616AE30I	OEM
2616AE30N	OEM
2616AE45F	OEM
2616AE45I	OEM
2616AE45N	OEM
2616AM20F	OEM
2616AM20I	OEM
2616AM25F	OEM
2616AM25I	OEM
2616AM30F	OEM
2616AM30I	OEM
2616AM45F	OEM
2616AM45I	OEM
2616E-35F	OEM
2616E-35I	OEM
2616E-45F	OEM
2616E-45I	OEM
2617-45F	OEM
2617-45I	OEM
2617-45N	OEM
2620	0233
2621	OEM
2622	OEM
2626-1	1696
2632-35F	OEM
2632-35N	OEM
2632-45F	OEM
2632-45I	OEM
2632-45N	OEM
2632A20F	OEM
2632A20I	OEM
2632A20N	OEM
2632A25F	OEM
2632A25I	OEM
2632A25N	OEM
2632A-30F	OEM
2632A30F	OEM
2632A-30I	OEM
2632A30I	OEM
2632A-30N	OEM
2632A30N	OEM
2632A45F	OEM
2632A45I	OEM
2632A45N	OEM
2632AE20F	OEM
2632AE20I	OEM
2632AE20N	OEM
2632AE25F	OEM
2632AE25I	OEM
2632AE25N	OEM
2632AE30F	OEM
2632AE30I	OEM
2632AE30N	OEM
2632AE45F	OEM
2632AE45I	OEM
2632AE45N	OEM
2632AM25F	OEM
2632AM25I	OEM
2632AM30F	OEM
2632AM30I	OEM
2632AM45F	OEM
2632AM45I	OEM
2632E-45F	OEM
2632E-45I	OEM
2633(RCA)	0144
2634	0127
2634(RCA)	0144
2634-1	0144
2635	OEM
2636	0144
2637	OEM
2637I	OEM
2637N	OEM
2641N	OEM
2650A1I	OEM
2650A1N	OEM
2650AI	OEM
2650AN	OEM
2650B1I	OEM
2650B1N	OEM
2650BI	OEM
2650BN	OEM
2651I	OEM
2651N	OEM
2652	OEM
2652-1I	OEM
2652-1N	OEM
2652I	OEM
2652N	OEM
2653I	OEM
2653N	OEM
2655	OEM
2656	OEM
2657	OEM
2661-1I	OEM
2661-1N	OEM
2661-2I	OEM
2661-2N	OEM
2661-3I	OEM
2661-3N	OEM
2661E	OEM
2663E	OEM
2664-45F	OEM
2664-45I	OEM
2664-45N	OEM
2664A20F	OEM
2664A20I	OEM
2664A20N	OEM
2664A25F	OEM
2664A25I	OEM
2664A25N	OEM
2664A-30F	OEM
2664A-30I	OEM
2664A-30N	OEM
2664A-45F	OEM
2664A-45I	OEM
2664A-45N	OEM
2664AE20F	OEM
2664AE20I	OEM
2664AE20N	OEM
2664AE25F	OEM
2664AE25I	OEM
2664AE25N	OEM
2664AE30F	OEM
2664AE30I	OEM
2664AE30N	OEM
2664AE45F	OEM
2664AE45I	OEM
2664AE45N	OEM
2664AM25F	OEM
2664AM25I	OEM
2664AM30F	OEM
2664AM30I	OEM
2664AM45F	OEM
2664AM45I	OEM
2664E	OEM
2664E-50F	OEM
2664E-50I	OEM
2665	OEM
2665E	OEM
2666E	OEM
2667	0006
2668E	OEM
2670I	OEM
2670N	OEM
2671	0160
2671I	OEM
2671N	OEM
2671R	4327
2672I	OEM
2672N	OEM
2673AI	OEM
2673AN	OEM
2673I	OEM
2673N	OEM
2673R	4331
2674R	4332
2675R	4333
2676R	2206
2678R	2199
2684-0002	OEM
2700	0050
2701	OEM
2701A	OEM
2702	0124
2702A	OEM
2703	OEM
2703-384	0004
2703-385	0004
2703-386	0004
2703-387	0847
2703-388	0164
2703-389	0143
2703-390	0015
2703A	OEM
2704	0143
2704-384	0004
2704-385	0004
2704-386	0004
2704-387	0164
2704-388	0143
2704-389	0015
2706	OEM
2716	2278
2716-SA2-A0	2278
2716-SC2-A1	2278
2732	2697
2732/2332	2697
2732A	2697
2732A-2	2697
2732A-2-QGE-1	OEM
2732A-3	OEM
2732A-4	OEM
2732A-20	OEM
2732A-25	OEM
2732A-30	2697
2732D	2697
2735	OEM
2737	OEM
2745B	0085
2755	OEM
2757	OEM
2762	0111
2763	0015
2764	0812
2764-2	0812
2764-25	0812
2764-30	0812
2764-250	0812
2764-450	0812
2764-FA4-A2	0812
2764-FA5-A3	0812
2764-FC5-A3	0812
2764-RX8-R1	OEM
2764-RX8-R2	OEM
2764A-2	OEM
2765	OEM
2766	0015
2767	OEM
2774	0162
2777	0015
2779	0015
2780	0160
2780-3	0265
2780-4	0160
2780-5	0160
2781	0211
2784	0015
2786	0015
2787	0016
2789	0143
2791	0050
2792	0050
2794	0015
2795A	0012
2796	0155
2797	0050
2798	0037
2799	0006
2799-1	1843
2799-2	1843
2800	OEM
2801	OEM
2801A	OEM
2802	0071
2802(VARACTOR)	OEM
2802A	OEM
2802A(VARACTOR)	OEM
2803	OEM
2803A	OEM
2804	OEM
2805	OEM
2816	OEM
2816-3	OEM
2816-4	OEM
2840	OEM
2842	OEM
2844	OEM
2845	OEM
2853	0560
2853-1	0560
2853-2	0042
2853-3	0560
2854	0016
2854-1	0560
2854-2	0042
2854-3	0560
2855-1	0560
2855-2	0042
2855-3	0560
2856-1	0560
2856-2	0042
2856-3	0560
2900-007	0144
2901	2742
2901(MIN)	OEM
2901(SM)	2742
2901-010	0160
2903D	0631
2904-003	0079
2904-008	0160
2904-014	0160
2904-016	0136
2904-029	0136
2904-030	0630
2904-032	0086
2904-033	0079
2904-034	0198
2904-035	0198
2904-037	0698
2904-038	0037
2904-038H05	0280
2904-045	0079
2904-053	0079
2904-054	0079
2904-057	0161
2904-058	0551
2904-059	0698
2904-061	1591
2904D	0770
2906	OEM
2906-004	2616
2906-005	0785
2906A	OEM
2907	0037
2907A	OEM
2910A	OEM
2910A-1	OEM
2910A-2	OEM
2910A-4	OEM
2911A	OEM
2911A-1	OEM
2911A-2	OEM
2911A-4	OEM
2912A(A)	OEM
2913(A)	OEM
2914(A)	OEM
2920	OEM
2920-6	OEM
2920-8	OEM
2920-16	OEM
2920-18	OEM
2921(A)	OEM
2925	0016
2925(E.F.-JOHNSON)	0079
2936	OEM
2957A100-25	0178
2979DRAEB	OEM
3000	OEM
3000BD21	OEM
3000D11A	OEM
3000PJA10	OEM
3001	0144
3001-201	3594
3001A	0015
3001F	0764
3001G	2875
3001H	0766
3001PDN60	OEM
3001PDN80	OEM
3001PDN100	OEM
3001PDN120	OEM
3001PDN140	OEM
3002	0144
3003	0127
3003(SEARS)	0127
3004(SEARS)	0127
3004-856	0004
3005	2536
3005(SEARS)	0016
3005-861	0016
3006	2536
3006(SEARS)	0079
3007	2536
3007(SEARS)	0016
3008	0050
3008(E.F.-JOHNSON)	0050
3009	0050
3009(E.F.-JOHNSON)	0050
3009(SEARS)	0004
3010	0136
3010(SEARS)	0004
3011	0079
3011(E.F.-JOHNSON)	0079
3011(E.F.JOHNSON)	0016
3012	0150
3012(EFJOHNSON)	0150
3012(SEARS)	0847
3013	0211
3013(SEARS)	0847
3014	0211
3014(SEARS)	0004
3015FBM	OEM
3015FBM10	OEM
3015FBM15	OEM
3015FBM20	OEM
3015FBN	OEM
3015FBN10	OEM
3015FBN15	OEM
3015FBN20	OEM
3015G	OEM
3015G10	OEM
3016B	0139
3017	0037
3018	0144
3018(E.F.-JOHNSON)	0144
3018(E.F.JOHNSON)	0144
3019	0037
3019(EFJOHNSON)	0136
3020	0144
3021	0144
3022	0086
3023	0007
3026	0079
3027	0007
3028	0144
3029	0127
3030	OEM
3034	0086
3034(RCA)	0086
3035Z4	OEM
3037	OEM
3040	OEM
3041	OEM
3045Z4	OEM
3050	OEM
3052P	5293
3052V	2539
3054	0178
3055Z4	OEM
3064TC	0364
3065Z4	OEM
3066DC	2549
3066F	OEM
3067DC	0328
3067F	OEM
3068	0079
3069	0015
3069(ARVIN)	0015
3069-1	OEM
3074A21	0015
3075DC	1345
3075Z4	OEM
3085Z4	OEM
3095Z4	OEM
3100-48	OEM
3101E	OEM
3102	OEM
3103	OEM
3104	OEM
3105	3189
3105Z4	OEM
3107-103-40501	OEM
3107-105-40501	2109
3107-108-4040X	3655
3107-108-4050X	1067
3107-108-40401	3655
3107-108-40404	3655
3107-108-40501	1067
3107-204-9000	0016
3107-204-90010	0016
3107-204-90020	0016
3107-204-90070	0160
3107-204-90080	0144
3107-204-90100	0127
3107-204-90140	0085
3107-204-90150	0111
3107-204-90190	0085
3110	2901
3110R	0254
3111	0016
3112	0004
3113	0016
3115	1124
3115R	1107
3115Z4	OEM
3119-901-110	0334
3120	1124
3120R	1107
3122P	4329
3122V	OEM
3124	OEM
3124-45	OEM
3125	1126
3125R	1111
3125Z4	OEM
3126	OEM
3130	1126
3130-0000-014	0083
3130-3167-909	0770
3130-3169-909	OEM
3130-3193-512	2257
3130-3248-801	1248
3130-3268-100	OEM
3130R	1111
3135	0805
3135/709	0664
3135Z4	OEM
3136/5072A	0244
3137-103-4308	0037
3140	0805
3140R	0258

If replacement code is OEM, contact original manufacturer for replacement.

DEVICE TYPE	REPL CODE
3145	1195
3146-977	0103
3150	1195
3150R	1644
3151	OEM
3152-159	0178
3152-170	0103
3153GM1	3270
3155Z4	OEM
3160	0319
3161	0112
3161R	0112
3161R99	OEM
3165Z4	OEM
3170	1132
3175Z4	OEM
3180	1132
3182	3076
3189	OEM
3195Z4	OEM
3198-0009	OEM
3198-0045	OEM
3202	3426
3202-5H01	0086
3202-51-01	0086
3203	OEM
3204	OEM
3205	6903
3205A	OEM
3205Z4	OEM
3206	OEM
3212	OEM
3213	3764
3214	OEM
3215	6905
3222	0126
3225	2664
3225R	2855
3227-E	0144
3233-9	OEM
3235	0805
3245	1195
3255-9-7K	OEM
3256-9-7K	OEM
3257-9-7C	OEM
3258-9-7K	OEM
3260-91-7R	OEM
3260-92-7R	OEM
3270	1132
3271/25	OEM
3280	1132
3283-00160-000	2289
3291	OEM
3291/14	OEM
3292/14	OEM
3293/14	OEM
3301(SEARS)	0127
3302	0037
3306	OEM
3307-4-5F	OEM
3322-6	0143
3324-30	OEM
3325-0500-13	OEM
3326-4-5E	OEM
3329/03	OEM
3329-9-5F	OEM
3330-9-5F	OEM
3331	OEM
3331-9-5F	OEM
3333-9-7K	OEM
3334	OEM
3341-9-7K	OEM
3342	OEM
3347	OEM
3348	OEM
3349	OEM
3353	OEM
3354/25	OEM
3355/25	OEM
3356/25	OEM
3367	0400
3370	0144
3373	0130
3374-3	1030
3377	0015
3383-9-5F	OEM
3391	0111
3391(SEARS)	0111
3391A	0111
3391A(SEARS)	0111
3400-2412-3	2537
3401DM	OEM
3402	4858
3402-110	OEM
3402-210	OEM
3402-312	OEM
3402-313	OEM
3402B	OEM
3403	6927
3403(RCA)	0079
3404-110	OEM
3404-210	OEM
3404-312	OEM
3404-313	OEM
3411	OEM
3414	OEM
3417	OEM
3420	OEM
3421	OEM
3425	0280
3430J	OEM
3430K	OEM
3431J	OEM
3431K	OEM
3434	0280
3435	0280
3450	3555
3451	3555
3452	OEM
3455	OEM
3456A	OEM
3456B	OEM
3458	0050
3461	0079
3462	OEM
3475	4256
3476	0233
3476(RCA)	0007
3491	OEM
3492	0079
3500	0280
3500A	OEM
3500B	OEM
3500C	OEM
3500E	OEM
3500MP	OEM
3500R	OEM
3500S	OEM
3500T	OEM
3500U(M)	OEM
3501	0086
3501-9-6G	OEM
3501A	OEM
3501B	OEM
3501C	OEM
3501R	OEM
3501S	OEM
3501T	OEM
3503	0142
3503(RCA)	0142
3504	0211
3504(RCA)	0050
3505	0143
3505(AIRLINE)	0913
3505(RCA)	0016
3505A	OEM
3506	0365
3506(RCA)	0016
3507	0086
3507(RCA)	0004
3507(SEARS)	0127
3507(WARDS)	0079
3507-9-6G	OEM
3507J	OEM
3508	0086
3508(RCA)	0079
3508(SEARS)	0007
3508(WARDS)	0144
3508J	OEM
3509	0144
3509(SEARS)	0079
3509(WARDS)	0079
3510	0079
3510(RCA)	0079
3510(SEARS)	0144
3510(WARDS)	0144
3510AM	OEM
3510BM	OEM
3510CM	OEM
3510SM	OEM
3510VM(M)	OEM
3511	0144
3511(SEARS)	0325
3511(WARDS)	0325
3512	0160
3512(RCA)	0325
3512(SEARS)	0325
3512(WARDS)	0325
3512-9-6G	OEM
3512-9-7C	OEM
3513	0037
3513(RCA)	0079
3513(SEARS)	0037
3513(WARDS)	0037
3513-9-7C	OEM
3514	0160
3514(RCA)	0975
3514(SEARS)	0127
3514(WARDS)	0079
3514-91-7R	OEM
3514-92-7R	OEM
3515	0085
3515(RCA)	0160
3516	0160
3516(RCA)	0007
3516(WARDS)	0144
3517	0004
3518	0127
3518(RCA)	0007
3519	0016
3519(RCA)	0016
3519-1	0016
3519-2	0016
3520	0137
3520(DIO)	0137
3520(RCA)	0142
3520(WARDS)	0137
3520(XSTR)	0142
3520-1	0142
3520-6	OEM
3520-7	OEM
3520-8	OEM
3520-9	OEM
3521	0086
3521(SEARS)	0079
3521-1	0127
3521H	OEM
3521J	OEM
3521K	OEM
3521L	OEM
3521R	OEM
3522	0150
3522(SEARS)	0037
3522J	OEM
3522K	OEM
3522L	OEM
3522S	OEM
3523	0595
3523(RCA)	0126
3523(SEARS)	0079
3523J	OEM
3523K	OEM
3523L	OEM
3524	0791
3524(RCA)	0144
3524(SEARS)	0037
3524-1	0144
3524-2	0127
3525	1267
3525(RCA)	0016
3526	0560
3526(RCA)	0016
3527	0079
3527(RCA)	0144
3527-1	0127
3527AM	OEM
3527BM	OEM
3527CM	OEM
3528AM	OEM
3528BM	OEM
3528CM	OEM
3529	0015
3529A	0087
3529B	0087
3529C	0087
3530	0127
3530(RCA)	0144
3530-1	0127
3530-2	0127
3531	0259
3531-021-000	1441
3531-022-000	1912
3531-023-000	1377
3531-033-000	2472
3532	0233
3532(RCA)	0079
3532-1	0086
3532-9A-7K	OEM
3532-9B-7K	OEM
3533	0701
3533(RCA)	0126
3533-1	0126
3534	0050
3534(RCA)	0050
3534-9-7T	OEM
3535	0008
3535(RCA)	0007
3536	0079
3536-1	0079
3536-2	0079
3537	0079
3537(RCA)	0144
3538	0079
3538(RCA)	0016
3538-7	0239
3538-8	0061
3539	0086
3539(RCA)	0144
3539-1CD	OEM
3539-1CP	OEM
3539-2CP	OEM
3539-307-001	0144
3539-307-002	0144
3539DC	2642
3539UCP	OEM
3540	0150
3540#	0468
3540(RCA)	0037
3540-1	0037
3541	0086
3541(RCA)	0016
3542J	OEM
3542S	OEM
3543	0079
3543(RCA)	0016
3544	0004
3544(RCA)	0086
3544-1	0079
3545	0233
3545(RCA)	0233
3546	0079
3546(RCA)	0016
3546-1(RCA)	0016
3546-2(RCA)	0016
3547	0285
3548	0079
3548(RCA)	0016
3549	0150
3549(RCA)	0037
3549-1	0150
3549-1(RCA)	0037
3549-2	0150
3550	0164
3550J	OEM
3550K	OEM
3550S	OEM
3551	0136
3551(RCA)	0016
3551A	0136
3551A-BLU	0136
3551A-GRN	0136
3551J	OEM
3551S	OEM
3552	3280
3552(RCA)	0233
3552-1	0233
3552-1(RCA)	0338
3553	3280
3553AM	OEM
3554	0086
3554(RCA)	0016
3554AM	OEM
3554BM	OEM
3554CM	OEM
3554SM	OEM
3555	0086
3555-01	0725
3555-02	0725
3555-03	0725
3555-3	0086
3556	0086
3557	OEM
3558	0079
3558(RCA)	0016
3559	0150
3559(RCA)	0037
3559-1	0037
3560	0079
3560-1	0086
3560-1(RCA)	0127
3560-2	0086
3561	0079
3561-1	0086
3562	0150
3562(RCA)	0037
3563	0150
3563(RCA)	0037
3564	0136
3564(RCA)	0074
3565	0142
3565(RCA)	0086
3565-1	0079
3566	0142
3566(RCA)	0086
3567	0142
3567(RCA)	0142
3567-2	0042
3567-2(RCA)	0276
3568	0007
3568(RCA)	0144
3568(WARDS)	0144
3569	0007
3569(RCA)	0016
3569-1	0079
3570	0037
3570(RCA)	0037
3570-1	0037
3570P	0037
3571	0086
3571(RCA)	0079
3571-1	0007
3571AM	OEM
3571R	0079
3572	0127
3572(RCA)	0016
3572-3	0144
3572AM	OEM
3573AM	OEM
3574	0037
3574(RCA)	0037
3574-1	0126
3574-1(RCA)	0037
3576	0330
3576(RCA)	0144
3577	0079
3577-1	0079
3578-1	0004
3579	0007
3579(RCA)	0007
3580-9-6G	OEM
3580J	OEM
3580JM	OEM
3581	0150
3581(RCA)	0037
3581J	OEM
3581JM	OEM
3582	3280
3582(RCA)	0233
3582J	OEM
3582JM	OEM
3583	0540
3583(RCA)	0975
3583AM	OEM
3583JM	OEM
3584	0142
3584(RCA)	0142
3584-9-6G	OEM
3584JM	OEM
3585	OEM
3585-5	0239
3585-6	0061
3585-7	0239
3585-8	0061
3586	0136
3586(RCA)	0016
3586-2	0007
3588	0016
3588(RCA)	0212
3588-2	0212
3589	0016
3590	0126
3591	0086
3592(RCA)	0126
3597	0037
3597-1	0037
3597-2	0037
3598(RCA)	0144
3598-2	0126
3600	0211
3600-01-020-25722-01	OEM
3600-01-020-25722-02	OEM
3600-01-020-25722-04	OEM
3600-01-020-25722-05	OEM
3600-01-020-25722-07	OEM
3600-01-020-25722-10	OEM
3600-02-040-25383-01	OEM
3600-02-040-25383-02	OEM
3600-02-040-25383-04	OEM
3600-02-040-25383-05	OEM
3600-02-040-25383-07	OEM
3600-02-040-25383-10	OEM
3600-03-080-25711-01	OEM
3600-03-080-25711-02	OEM
3600-03-080-25711-04	OEM
3600-03-080-25711-05	OEM
3600-03-080-25711-07	OEM
3600-03-080-25711-10	OEM
3600-04-080-25712-01	OEM
3600-04-080-25712-02	OEM
3600-04-080-25712-04	OEM
3600-04-080-25712-05	OEM
3600-04-080-25712-07	OEM
3600-04-080-25712-10	OEM
3600-06-240-25713-01	OEM
3600-06-240-25713-02	OEM
3600-06-240-25713-04	OEM
3600-06-240-25713-05	OEM
3600-06-240-25713-07	OEM
3600-06-240-25713-10	OEM
3600-14-020-25722-01	OEM
3600-14-020-25722-02	OEM
3600-14-020-25722-04	OEM
3600-14-020-25722-05	OEM
3600-14-020-25722-07	OEM
3600-14-020-25722-10	OEM
3600-20-040-25383-01	OEM
3600-20-040-25383-02	OEM
3600-20-040-25383-04	OEM
3600-20-040-25383-05	OEM
3600-20-040-25383-07	OEM
3600-20-040-25383-10	OEM
3600-21-020-26542-01	OEM
3600-21-020-26542-02	OEM
3600-21-020-26542-04	OEM
3600-21-020-26542-05	OEM
3600-21-020-26542-07	OEM
3600-21-020-26542-10	OEM
3601	0086
3601(RCA)	0079
3601-1	0079
3602	OEM
3603	0050
3603-1	0079
3604	0400
3604(RCA)	0144
3604-1	0400
3604-3	0079
3606-210	OEM
3606-312	OEM
3606-313	OEM
3606AG	OEM
3606AM	OEM
3606BG	OEM
3606GM	OEM
3607	0600
3607-210	OEM
3607-312	OEM
3607-313	OEM
3607BG	OEM
3608(RCA)	0086
3608-1	0086
3608-1(RCA)	0264
3608-2	0086
3608-2(J.C.-PENNEY)	0086
3608-2(J.C.PENNEY)	0086
3608-2(RCA)	0086
3608-210	OEM
3608-312	OEM
3608-313	OEM
3609	0600
3610	0127
3610(RCA)	0144
3611	0560
3612	0560
3612(RCA)	0042
3612-210	OEM
3612-312	OEM
3612-313	OEM
3613	0338
3613(RCA)	0285
3613-2	0285
3613-3	0338
3613-3(RCA)	0338
3613-210	OEM
3613-312	OEM
3613-313	OEM
3614	OEM
3614-1	0079
3614-3	0079
3614-210	OEM
3614-312	OEM
3614-313	OEM
3615	0086
3615(RCA)	0086
3615-1	0086
3616	0150
3616-1	0037
3616-1(RCA)	0126
3617	0050
3618	0353
3618(RCA)	0144
3618-0010	OEM
3618-0020	OEM
3618-1	0160
3619-1	2383
3619-2	1413
3619-3	1413
3620	0150
3620(RCA)	0037
3620-1	0037
3621(RCA)	0042
3622-1	0086
3622-2	0086
3624-06	OEM
3624-1	0460
3624-2	0460
3624-16	OEM
3624-22	OEM
3624-23	OEM
3624-24	OEM
3624-25	OEM
3624-26	OEM
3624-27	OEM
3624-28	OEM
3625	0079
3625(RCA)	0016
3626	0016
3626AP	OEM
3626BP	OEM
3626CP	OEM
3627	0037
3627(RCA)	0037
3627AM	OEM
3627BM	OEM
3628	0208
3628-2(RCA)	0285
3628-3	0551
3628A-3	OEM
3628A-4	OEM
3629	OEM
3629-3	0382
3629AM	OEM
3629AP	OEM
3629BM	OEM
3629BP	OEM
3629CM	OEM
3629CP	OEM
3629SM	OEM
3630AM	OEM
3630BM	OEM
3630CM	OEM
3630SM	OEM
3631	0037
3631-1	0016
3631-1(RCA)	0042
3632	OEM
3632-1	0423
3632-2(RCA)	0560
3633-1	0829
3633-1(IC)	0829
3634	0074
3634-1	0642
3635	0212
3635(RCA)	0212
3635-1	0212
3635-2	0212
3636	0142
3636B-2	OEM
3637	0136
3637(RCA)	0074
3637-1	0103
3638	0178
3638-1	0905
3640(RCA)	0423
3646-2(RCA)	0144
3647	0074
3648(RCA)	0975
3649	0642
3650	OEM
3650-6	OEM
3650-7	OEM
3650-8	OEM
3650-9	OEM
3650HG	OEM
3650JG	OEM
3650KG	OEM
3650MG	OEM
3651	0150
3652	0050
3652HG	OEM
3652JG	OEM
3652MG	OEM
3653	OEM
3656AG	OEM
3656BG	OEM
3657-1	0224
3657-2	0224
3665	0103
3666	0150
3666(RCA)	0086
3666-1	0037
3669	0065
3669-2	0642
3676	0330
3676-1	0007
3679	0338
3679-1	0338
3680-1	0007
3681	0042
3681-1	1854
3682	0925
3682(RCA)	0460
3683	0103
3686	0136
3686(RCA)	0004
3686-1	0167
3687	0233
3687-1	0285
3688	0058
3688-1	0058
3693(ARVINE)	0144
3697-1	0037
3700-01-010-26295-01	OEM
3700-01-010-26295-02	OEM
3700-01-010-26295-03	OEM
3700-01-010-26295-04	OEM
3700-01-010-26295-07	OEM
3700-01-010-26295-10	OEM
3700-02-012-26296-01	OEM
3700-02-012-26296-02	OEM
3700-02-012-26296-03	OEM
3700-02-012-26296-04	OEM
3700-02-012-26296-07	OEM
3700-02-012-26296-10	OEM
3700-03-016-26297-01	OEM
3700-03-016-26297-02	OEM
3700-03-016-26297-03	OEM
3700-03-016-26297-04	OEM
3700-03-016-26297-07	OEM
3700-03-016-26297-10	OEM
3700-04-016-26297-01	OEM
3700-04-016-26297-02	OEM
3700-04-016-26297-03	OEM
3700-04-016-26297-04	OEM
3700-04-016-26297-07	OEM
3700-04-016-26297-10	OEM
3700-05-020-25722-01	OEM
3700-05-020-25722-02	OEM
3700-05-020-25722-03	OEM
3700-05-020-25722-04	OEM
3700-05-020-25722-07	OEM
3700-05-020-25722-10	OEM
3700-06-020-25722-01	OEM
3700-06-020-25722-02	OEM
3700-06-020-25722-03	OEM
3700-06-020-25722-04	OEM
3700-06-020-25722-07	OEM
3700-06-020-25722-10	OEM
3700-153	0015
3702-210	OEM
3702-312	OEM
3702-313	OEM
3703	0636
3714H1	0103
3714H1A	0103
3736-8	OEM
3736-18	OEM
3746	0280
3755-1	0062
3755-2	0062
3793	0725
3798NS	OEM
3800-4-6H	OEM
3800-9-6H	OEM
3801	1062
3801-4-6H	OEM
3801-9-6H	OEM
3801-210	OEM
3801-312	OEM
3801-313	OEM
3802	OEM
3802-210	OEM
3802-312	OEM
3802-313	OEM
3803	OEM
3803-210	OEM
3803-312	OEM
3803-313	OEM
3804	3555
3804-210	OEM
3804-312	OEM
3804-313	OEM

DEVICE TYPE	REPL CODE
3805	OEM
3805-210	OEM
3805-212	OEM
3805-313	OEM
3806	5213
3806-210	OEM
3806-312	OEM
3806-313	OEM
3807	OEM
3807-210	OEM
3807-312	OEM
3807-313	OEM
3808-210	OEM
3808-312	OEM
3808-313	OEM
3809	OEM
3810	OEM
3811	OEM
3812	OEM
3819	0325
3819(RCA)	0325
3851	0136
3852	0280
3867	0016
3868	0698
3878	0233
3881	0007
3888A-D13	0062
3888A-D17	0062
3900	2247
3900-01	OEM
3900-05	OEM
3900-06	OEM
3901-01	OEM
3901-05	OEM
3901-06	OEM
3902-210	OEM
3902-312	OEM
3902-313	OEM
3903-01	OEM
3903-05	OEM
3903-06	OEM
3906-110	OEM
3906-210	OEM
3906-312	OEM
3906-313	OEM
3907	0211
3907(2N404A)	0136
3907/2N404A	0050
3907-01	OEM
3907-05	OEM
3907-06	OEM
3908-01	OEM
3908-05	OEM
3908-06	OEM
3909	3062
3909-210	OEM
3909-312	OEM
3909-313	OEM
3912-01	OEM
3912-05	OEM
3912-06	OEM
3913-01	OEM
3913-05	OEM
3915-01	OEM
3915-05	OEM
3915-06	OEM
3918-110	OEM
3918-210	OEM
3918-312	OEM
3918-313	OEM
3919-01	OEM
3919-05	OEM
3919-06	OEM
3920-01	OEM
3920-05	OEM
3920-06	OEM
3921	OEM
3922-01	OEM
3922-05	OEM
3922-06	OEM
3926-01	OEM
3926-05	OEM
3926-06	OEM
3927-110	OEM
3927-210	OEM
3927-312	OEM
3927-313	OEM
3929-210	OEM
3929-312	OEM
3929-313	OEM
3930-210	OEM
3930-312	OEM
3930-313	OEM
3931-01	OEM
3931-05	OEM
3931-06	OEM
3935-01	OEM
3935-05	OEM
3935-06	OEM
3937-01	OEM
3937-05	OEM
3937-06	OEM
3938-01	OEM
3938-05	OEM
3938-06	OEM
3939-01	OEM
3939-05	OEM
3939-06	OEM
3940-01	OEM
3940-05	OEM
3940-06	OEM
3947-210	OEM
3947-312	OEM
3947-313	OEM
3961	0136
3961(G.E.)	0050
3970	0280
3970(G.E.)	0004
3970-01	OEM
3970-05	OEM
3970-06	OEM
3970CL	0004
3999	0016
4000	2024
4000(IC)	2024
4001	0233
4001(IC)	0478
4001(MODULE)	4437
4001-151	0071
4001-222	0050
4001-223	0050
4001-224	0004
4001-225	0004
4001-226	0004
4001-228	0264
4001-230	0143
4001B	0233
4001BDC	0478
4001BDM	0478
4001BFC	0478
4001BFM	0478
4001BPC	0478
4001PDR60	OEM
4001PDR80	OEM
4001PDR100	OEM
4001PDR120	OEM
4001PDR140	OEM
4001PDR160	OEM
4001PDR180	OEM
4001PDR200	OEM
4002	6967
4002(IC)	2055
4002A	2055
4002BDC	2055
4002BDM	2055
4002BFC	2055
4002BFM	2055
4002BPC	2055
4003	OEM
4003(SEARS)	0143
4003E	0079
4004	0698
4004(E.F.JOHNSON)	0698
4004(PENNCREST)	0004
4004(SEARS)	0143
4005	0630
4005(E.F.JOHNSON)	0698
4005(E-F-JOHNSON)	0698
4005(E.F.-JOHNSON)	0630
4005(SEARS)	0143
4006	2700
4006(E.F.JOHNSON)	0086
4006(SEARS)	0143
4006BDC	0646
4006BDM	OEM
4006BFC	OEM
4006BFM	OEM
4006BPC	6735
4007	0698
4007(IC)	2846
4007(SEARS)	0143
4007UBDC	2846
4007UBDM	OEM
4007UBFC	OEM
4007UBPC	2846
4008	0493
4008(E-F-JOHNSON)	0493
4008(E.F.-JOHNSON)	0493
4008(SEARS)	0143
4008B	0493
4008BDC	0988
4008BDM	0988
4008BFC	0988
4008BFM	0988
4008BPC	0988
4009	6968
4009(IC)	1999
4009(PENNCREST)	0164
4009(SEARS)	0143
4010	0016
4010(PENNCREST)	0143
4010(SEARS)	0143
4011	0037
4011(IC)	0215
4011(SEARS)	0143
4011-PC	0215
4011A	0215
4011BDC	0215
4011BDM	0215
4011BFC	0215
4011BFM	0215
4011BPC	0215
4011PC	0215
4012	0086
4012(IC)	0498
4012A	0498
4012BDC	0498
4012BDM	0498
4012BFC	0498
4012BFM	0498
4012BPC	0498
4013	0630
4013(IC)	0413
4013(RECTIFIER)	0015
4013B	0413
4013BDC	0413
4013BDM	0413
4013BFC	0413
4013BFM	0413
4013BPC	0413
4014	0015
4014-000-10160	0004
4014-200-30110	0133
4014BDC	0860
4014BDM	OEM
4014BFC	OEM
4014BFM	OEM
4014BPC	0860
4015	1014
4015(IC)	1014
4015BDC	1014
4015BDM	OEM
4015BFC	OEM
4015BFM	OEM
4015BPC	1014
4016	1143
4016(IC)	1143
4016(RAM)	1898
4016BDC	1143
4016BDM	OEM
4016BFC	OEM
4016BFM	OEM
4016BPC	1143
4016CX	1898
4017	0513
4017(IC)	0513
4017BDC	0513
4017BDM	0513
4017BFC	0513
4017BFM	0513
4017BP	OEM
4017BPC	0513
4018BDC	1391
4018BDM	1391
4018BFC	1391
4018BFM	1391
4018BPC	1391
4019	1525
4019(IC)	1525
4019BDC	4370
4019BDM	OEM
4019BFC	OEM
4019BFM	OEM
4019BPC	1525
4020	2707
4020(CMOS)	1661
4020(IC)	1661
4020-8	OEM
4020-9	OEM
4020BDC	1661
4020BDM	1661
4020BFC	1661
4020BFM	1661
4020BPC	1661
4021	0016
4021(IC)	1749
4021BDC	1749
4021BDM	OEM
4021BFC	OEM
4021BFM	OEM
4021BPC	1749
4022	0079
4022BDC	1256
4022BDM	1256
4022BFC	1256
4022BFM	1256
4022BPC	1256
4023	0519
4023(IC)	0519
4023-25	OEM
4023A	0519
4023BDC	0519
4023BDM	0519
4023BFC	0519
4023BFM	0519
4023BPC	0519
4024	1957
4024(IC)	1957
4024BDC	1957
4024BDM	1957
4024BFC	1957
4024BFM	1957
4024BPC	1957
4024D	OEM
4024D(SM)	OEM
4025	2072
4025(IC)	2072
4025A	2072
4025BDC	2072
4025BDM	2072
4025BFC	2072
4025BFM	2072
4025BPC	2072
4027	1949
4027(IC)	1949
4027B	1949
4027BDC	1949
4027BDM	1949
4027BFC	1949
4027BFM	1949
4027BPC	1949
4028BDC	2228
4028BDM	2228
4028BFC	2228
4028BFM	2228
4028BPC	2228
4029	2233
4029BDC	2233
4029BDM	2233
4029BFC	2233
4029BFM	2233
4029BPC	2233
4030	0500
4030(IC)	0500
4030BDC	0500
4030BDM	0500
4030BFC	0500
4030BFM	0500
4030BPC	0500
4031BDC	2971
4031BDM	OEM
4031BFC	OEM
4031BFM	OEM
4031BPC	2971
4034BDC	3615
4034BDM	OEM
4034BFC	OEM
4034BFM	OEM
4034BPC	3615
4034L	0560
4034S	0560
4035	0698
4035BDC	2777
4035BDM	OEM
4035BFC	OEM
4035BFM	OEM
4035BPC	2777
4036	0126
4040	0056
4040(IC)	0056
4040BDC	0056
4040BDM	0056
4040BFC	0056
4040BFM	0056
4040BPC	0056
4041(IC)	OEM
4041-000-1018	0143
4041-000-10150	0015
4041-000-10160	0004
4041-000-20120	0004
4041-000-30180	0004
4041-000-40270	0127
4041-000-40300	0127
4041-000-60170	0127
4041-000-60200	0127
4041-000-80100	0136
4041-200-10180	0143
4041-200-30110	0162
4041-200-40100	0143
4041BDC	3174
4041BDM	OEM
4041BFC	OEM
4041BFM	OEM
4041BPC	3174
4042	0121
4042(IC)	0121
4042BDC	0121
4042BDM	0121
4042BFC	0121
4042BFM	0121
4042BPC	0121
4043BDC	1553
4043BDM	1553
4043BFC	1553
4043BFM	1553
4043BPC	1553
4044BDC	2308
4044BDM	2308
4044BFC	2308
4044BFM	2308
4044BPC	2308
4044D	OEM
4044D(SM)	OEM
4046(SEARS)	0016
4047BDC	2331
4047BDM	2331
4047BFC	2331
4047BFM	2331
4047BPC	2331
4049	0001
4049(IC)	0001
4049B	0001
4049BDC	0001
4049BDM	0001
4049BFC	OEM
4049BFM	OEM
4049BPC	0001
4050	0398
4050(IC)	0398
4050BDC	0398
4050BDM	0398
4050BFC	OEM
4050BFM	OEM
4050BPC	0398
4051	0366
4051(IC)	0366
4051-300-10150	0015
4051BDC	0366
4051BDM	0366
4051BFC	OEM
4051BFM	OEM
4051BPC	0366
4052	0024
4052(IC)	0024
4052(RCA)	0224
4052BDC	0024
4052BDM	0024
4052BFC	OEM
4052BFM	OEM
4052BPC	0024
4053	0035
4053BDC	0035
4053BDM	0035
4053BFC	OEM
4053BFM	OEM
4053BPC	0035
4053BS	OEM
4055	3304
4055(IC)	3304
4057	0016
4060DJRC8686	OEM
4066	0079
4066(IC)	0101
4066BDC	0101
4066BDM	0101
4066BFC	OEM
4066BFM	OEM
4066BPC	0101
4067-12	OEM
4067-15	OEM
4067BDC	3742
4067BDM	3742
4067BFC	OEM
4067BFM	OEM
4067BPC	3742
4068	6973
4068-83	OEM
4068BDC	2503
4068BDM	2503
4068BFC	2503
4068BFM	2503
4068BPC	2503
4069UBDC	0119
4069UBDM	0119
4069UBFC	OEM
4069UBFM	OEM
4069UBPC	0119
4070BDC	2516
4070BDM	2516
4070BFC	2516
4070BFM	2516
4070BPC	2516
4071BDC	0129
4071BDM	0129
4071BFC	0129
4071BFM	0129
4071BPC	0129
4072BDC	2524
4072BDM	2524
4072BFC	2524
4072BFM	2524
4072BPC	2524
4073BDC	1536
4073BDM	1536
4073BFC	1536
4073BFM	1536
4073BPC	1536
4075BDC	2540
4075BDM	2540
4075BFC	2540
4075BFM	2540
4075BPC	2540
4076BDC	3496
4076BDM	3496
4076BFC	OEM
4076BFM	OEM
4076BPC	3496
4077	OEM
4077BDC	2558
4077BDM	2558
4077BFC	2558
4077BFM	2558
4077BPC	2558
4078BDC	0921
4078BDM	0921
4078BFC	0921
4078BFM	0921
4078BPC	0921
4080	0325
4080-187-0507	0086
4080-838-0001	0178
4080-838-2	0178
4080-838-3	0178
4080-866-0006	0423
4080-866-1	0178
4080-866-2	0178
4080-879-0001	0423
4081	0626
4081(IC)	0626
4081BDC	0626
4081BDM	0626
4081BFC	0626
4081BFM	0626
4081BPC	0626
4082-03	OEM
4082-501-0001	0160
4082-748-0002	0157
4082BDC	0300
4082BDM	0300
4082BFC	0300
4082BFM	0300
4082BPC	0300
4085	0016
4085BDC	0303
4085BDM	0303
4085BFC	0303
4085BFM	0303
4085BM	OEM
4085BPC	0303
4085KG	OEM
4085SM	OEM
4086	0037
4086BDC	0535
4086BDM	0535
4086BFC	0535
4086BFM	0535
4086BPC	0535
4087	0037
4092BFM	OEM
4093BDC	2384
4093BDM	2384
4093BFC	OEM
4093BPC	OEM
4099A	0382
4100	OEM
4101	OEM
4101-685	0015
4102	4973
4103	OEM
4104	2700
4104BDC	OEM
4104BDM	OEM
4104BFC	OEM
4104BFM	OEM
4104BPC	OEM
4104UMC	OEM
4104USC	OEM
4115	OEM
4115-04	OEM
4116	0522
4117	OEM
4120-9	OEM
4124	0932
4125	1096
4126	1106
4127	1135
4127JG	OEM
4127KG	OEM
4128	1152
4129	1174
4130	1184
4131	1210
4132	1223
4133	1232
4134	1246
4135	1265
4136	1290
4137	1307
4138	1331
4140	1365
4141	1384
4142	1401
4143	1412
4144	1423
4145	1986
4145(RECTIFIER)	1986
4145(ZENER)	1444
4146	1457
4148	1505
4148(DIO)	0124
4149	1521
4150	1548
4151	1587
4161	0236
4162	OEM
4163	OEM
4164	2357
4164-2	2357
4164C	2357
4164P1M	2017
4165	OEM
4167	0079
4167(AIRLINE)	0007
4167(PENNCREST)	0007
4167(SEARS)	0144
4168	0079
4168(PENNCREST)	0007
4168(SEARS)	0144
4168(WARDS)	0007
4169	0079
4169(PENNCREST)	0007
4169(SEARS)	0144
4169(WARDS)	0007
4170	OEM
4189	OEM
4189-83(M)	OEM
4201	OEM
4203J	OEM
4203K	OEM
4203S	OEM
4203SQ	OEM
4204J	OEM
4204K	OEM
4204S	OEM
4205J	OEM
4205K	OEM
4205S	OEM
4206J	OEM
4206K	OEM
4207	OEM
4207-S	OEM
4208-3268-000	2050
4213AM	OEM
4213BM	OEM
4213SM	OEM
4214AP	OEM
4214BP	OEM
4214RM	OEM
4214SM	OEM
4220	1009
4220-A	OEM
4220-AS	OEM
4220-B	OEM
4220-B5	OEM
4221	1019
4221-S	OEM
4222	0889
4223	1049
4224	0932
4225	1096
4226	1106
4227	1135
4228	1152
4229	1174
4230	1184
4231	1210
4232	1223
4233	1232
4234	1246
4235	1265
4236	1290
4237	1307
4238	1331
4240	1365
4241	1384
4242	1401
4243	1412
4244	1423
4245	1444
4246	1457
4247	0085
4248	1505
4249	1521
4250	1548
4251	1587
4253	OEM
4264-15	OEM
4264-20	OEM
4295-1	1186
4295-1(RCA)	1186
4295-2	1186
4295-2(RCA)	2282
4300ACD	OEM
4300ACP	OEM
4301	OEM
4302	OEM
4309(AIRLINE)	0016
4310(AIRLINE)	0037
4310-(AIRLINE)	0037
4312	0142
4312(RCA)	0142
4313	0211
4315	0211
4320	1009
4321	1019
4322	0889
4323	1049
4324	0932
4325	1096
4326	1106
4327	1135
4328	1152
4329	1174
4330	1184
4331	0160
4331(ZENER)	1210
4332	1223
4333	1232
4334	1246
4335	1265
4336	1290
4337	1307
4338	1331
4340	1365
4341	1384
4342	1401
4343	1412
4344	1423
4345	1444
4346	1457
4347	0085
4348	0211
4348(ZENER)	1505
4349	0211
4349(ZENER)	1521
4350	1548
4351	1587
4351.0012	0143
4351.0013	0143
4351.0031	0143
4354	0143
4354.0012	0015

If replacement code is OEM, contact original manufacturer for replacement.

DEVICE TYPE	REPL CODE
4354.0013	0030
4357	OEM
4358	OEM
4360D	0042
4362	OEM
4363	0136
4363BLU	0136
4363GRN	0136
4363ORN	0136
4363WHT	0136
4364	0136
4364(IC)	OEM
4365	0136
4366	0136
4367	0136
4368	0050
4381P1	1069
4398	0211
4400	OEM
4400Y	2193
4403	0015
4404	OEM
4416	OEM
4419-4	1681
4420	1009
4421	1019
4422	0889
4423	1049
4424	0932
4425	1096
4426	1106
4427	1135
4428	1152
4429	1174
4430	1184
4431	1210
4432	1223
4433	1232
4434	1246
4435	1265
4436	1290
4437	1307
4437-3	2257
4438	1331
4438-3	2755
4439-1	1843
4439-1(RCA)	1843
4439-2	1843
4439-2(RCA)	1843
4440	1365
4441	1384
4442	0037
4442(ZENER)	1401
4442-3	0551
4442-366	0423
4443	1412
4444	1423
4445	1444
4446	1457
4448	1505
4449	1521
4450	0211
4450(ZENER)	1548
4451	0280
4451(ZENER)	1587
4453	OEM
4454	0136
4456	0136
4457	0136
4459	0975
4460-2	1843
4460-3	1843
4462	0211
4463	0160
4464	0198
4464G-20L	OEM
4465	0198
4466	0280
4466ORN	0211
4467	0847
4467(RECT)	1250
4467(RECTIFIER)	1250
4468BRN	0211
4469RED	0211
4470	0198
4470-31	0198
4470-32	0198
4470-33	0198
4470-ORN	0847
4470M-32	0198
4471	0280
4471ORN	0211
4471YEL	0211
4472	0280
4472GRN	0211
4473	0004
4473-1	0144
4473-2	0144
4473-3	0144
4473-4	0016
4473-5	0016
4473-5X	0016
4473-6	0144
4473-7	0144
4473-8	0144
4473-9	0016
4473-11	0144
4473-12	0016
4473-M-3	0079
4473-M3	0086
4473-M-12	0086
4473-N	0086
4474	0280
4474YEL	0211
4475	0280
4475GRN	0211
4476	0280
4476BLU	0211
4477	0280
4477PUR	0211
4477V10	0004
4478	0126
4483	0086
4484	0280
4484-1	0150
4484-2	0150
4485	0280
4485-1	0150
4486	0280
4491-6	0161
4491-9	0161
4501	0136
4506	3767
4509	0136
4510	0211
4510BDC	1963
4510BDM	1963
4510BFC	1963
4510BFM	1963
4510BPC	1963
4511	1544
4511BDC	1544
4511BDM	1544
4511BFC	OEM
4511BFM	OEM
4511BPC	1544
4512BDC	2120
4512BDM	2120
4512BFC	OEM
4512BFM	OEM
4512BPC	2120
4514BDC	1830
4514BDM	1830
4514BFC	OEM
4514BFM	OEM
4514BPC	1830
4515BDC	3600
4515BDM	3600
4515BFC	OEM
4515BFM	OEM
4515BPC	3600
4516BDC	2347
4516BDM	2347
4516BFC	2347
4516BFM	2347
4516BPC	2347
4518	1043
4518(IC)	1043
4518BDC	1043
4518BDM	1043
4518BFC	1043
4518BFM	1043
4518BPC	1043
4519BDC	OEM
4519BDM	OEM
4519BFC	OEM
4519BFM	OEM
4519BPC	OEM
4520BDC	2675
4520BDM	2675
4520BFC	2675
4520BFM	2675
4520BPC	2675
4528BDC	1043
4528BDM	3198
4528BFC	OEM
4528BFM	OEM
4528BPC	3198
4531BDC	3325
4531BDM	3325
4531BFC	OEM
4531BFM	OEM
4531BPC	3325
4532BDC	1016
4532BDM	1016
4532BFC	OEM
4532BFM	OEM
4532BPC	1016
4539BDC	3656
4539BDM	3656
4539BFC	OEM
4539BFM	OEM
4539BPC	3656
4543BDC	5020
4543BDM	5020
4543BFC	OEM
4543BFM	OEM
4543BPC	5020
4545	0136
4545BLU	0136
4545WHT	0136
4552	0100
4552-01	OEM
4553	0280
4553-01	OEM
4553BLU	0211
4553BRN	0211
4553GRN	0211
4553ORN	0211
4553RED	0211
4553V10	0004
4553YEL	0211
4554	0211
4554-01	OEM
4555BDC	2938
4555BDM	2938
4555BFC	OEM
4555BFM	OEM
4555BPC	2938
4556BDC	3431
4556BDM	3431
4556BFC	OEM
4556BFM	OEM
4556BPC	3431
4557BDC	OEM
4557BDM	OEM
4557BFC	OEM
4557BFM	OEM
4557BPC	OEM
4558	4035
4558C	0360
4558D	0360
4558DD	0360
4558DV	0360
4560D	OEM
4562	0211
4563	0211
4564	0211
4565	0211
4567	0211
4570	7010
4572-01	OEM
4572-05	OEM
4572-06	OEM
4573	0440
4582BDC	4660
4582BDM	4660
4582BFC	OEM
4582BFM	OEM
4582BPC	4660
4582BRN	0160
4583BDC	1297
4583BDM	1297
4583BFC	OEM
4583BFM	OEM
4583BPC	1297
4583RED	0160
4584GRN	0160
4586	0136
4587	0144
4589	0136
4590	0150
4594	0198
4595	0136
4596	0211
4597	0440
4597GRN	0440
4597RED	0440
4603	0136
4604	0136
4605	0136
4605RED	0136
4607	0211
4608	0160
4620GRN	0160
4621	0136
4622	0440
4623	0975
4624	0198
4627	0211
4630	0198
4632	0136
4633BN9503	OEM
4640	0604
4640P	0604
4648	0233
4649	0160
4652	0975
4653	0012
4663	0039
4677	0136
4684-120-3	0144
4685-285-3	0144
4685S	OEM
4686-81-2	0004
4686-81-3	0004
4686-82-3	0016
4686-87-3	0229
4686-89-3	0071
4686-95-3	0144
4686-97-3	0071
4686-105-3	0071
4686-106-3	0071
4686-107-3	0144
4686-108-3	0144
4686-112-3	0144
4686-114-3	0144
4686-116-3	0229
4686-118-3	0144
4686-119-3	0144
4686-120-3	0144
4686-126-3	0144
4686-127-3	0144
4686-130-3	0178
4686-131-3	0144
4686-132-3	0016
4686-139-3	0071
4686-140-3	0144
4686-143-3	0016
4686-144-3	0016
4686-147-3	0229
4686-148-3	0229
4686-149-3	0015
4686-150-3	0229
4686-151-3	0229
4686-163-3	0004
4686-169-3	0144
4686-170-3	0126
4686-171-3	0144
4686-172-3	0144
4686-173-3	0016
4686-177-3	0229
4686-179-3	0071
4686-182-3	0086
4686-183-3	0016
4686-184-3	0071
4686-186-3	0071
4686-189-3	0071
4686-195-3	0004
4686-196-3	0004
4686-199-3	0071
4686-201-3	0071
4686-207-3	0144
4686-208-3	0144
4686-209-3	0144
4686-212-3	0071
4686-213-3	0160
4686-224-3	0144
4686-226-3	0178
4686-227-3	0201
4686-228-3	0144
4686-229-3	0126
4686-230-3	0126
4686-231-3	0016
4686-232-3	0233
4686-234-3	0423
4686-235-3	0126
4686-238-3	0126
4686-244-3	0144
4686-251-3	0144
4686-252-3	0338
4686-256-1	0136
4686-256-2	0136
4686-256-3	0136
4686-257-3	0016
4689	0086
4699	0331
4700	0600
4701	0144
4702	0604
4703	OEM
4704	0331
4705	0198
4705-01	OEM
4706	0144
4707	OEM
4708	OEM
4709	0144
4709-02	OEM
4713	OEM
4714	0198
4715	0103
4715-01	OEM
4715A	0103
4719	OEM
4719-01	OEM
4720BDC	OEM
4720BDM	OEM
4720BFC	OEM
4720BFM	OEM
4720BPC	OEM
4721	OEM
4721BDC	OEM
4721BDM	OEM
4721BFC	OEM
4721BFM	OEM
4721BPC	OEM
4722	0160
4722BLU	0160
4722GRN	0160
4722ORN	0160
4722PUR	0160
4722RED	0160
4722YEL	0160
4723	OEM
4723BDC	OEM
4723BDM	OEM
4723BFC	OEM
4723BFM	OEM
4723BPC	OEM
4724BDC	OEM
4724BDM	OEM
4724BFC	OEM
4724BFM	OEM
4724BPC	OEM
4725	OEM
4725BDM	OEM
4725BFC	OEM
4725BFM	OEM
4725BPC	OEM
4727	0160
4727-01	OEM
4727BDC	OEM
4727BDM	OEM
4727BFC	OEM
4727BFM	OEM
4727BPC	OEM
4729	0604
4730	0160
4731	OEM
4731-83	OEM
4731BDC	OEM
4731BDM	OEM
4731BFC	OEM
4731BFM	OEM
4731BPC	OEM
4732	0016
4732-83	OEM
4733	0016
4733-83	OEM
4734	0016
4734-83	OEM
4734BDC	OEM
4734BDM	OEM
4734BPC	OEM
4735	OEM
4735-83	OEM
4735BDC	OEM
4735BDM	OEM
4735BFC	OEM
4735BFM	OEM
4735BPC	OEM
4736B	OEM
4736B-83	OEM
4736BDC	OEM
4736BDM	OEM
4736BFC	OEM
4736BFM	OEM
4736BPC	OEM
4737	0016
4737BDC	OEM
4737BDM	OEM
4737BFC	OEM
4737BFM	OEM
4737BPC	OEM
4740	0225
4741BDC	OEM
4741BDM	OEM
4741BFC	OEM
4741BFM	OEM
4741BPC	OEM
4745	0423
4746	0127
4756	0007
4765	0016
4768	0016
4778-8(RCA)	0087
4780D	OEM
4789	0127
4792	1669
4800-200	0004
4800-220	0004
4800-221	0004
4800-222	0004
4800-223	0822
4800-224	0920
4801-0000-001	0037
4801-0000-003	0111
4801-0000-010	0079
4801-0000-016	0008
4801-0000-035	0144
4801-0000-060	0037
4801-00154	0133
4801-00628	0143
4801-00629	0143
4801-00801	0012
4801-1100-0011	0079
4801-1100-011	0160
4802-0000-002	0161
4802-00002	0144
4802-00003	0016
4802-00004	0037
4802-00005	0103
4802-00005A	0103
4802-00006	0016
4802-00007	0698
4802-00008	1411
4802-00009	0016
4802-00010	0325
4802-00012	0016
4802-00014	0155
4802-00015	0016
4802-00016	2046
4802-00017	0698
4802-00019	1411
4802-3268-000	2050
4802-3274-200	0561
4804-3267-901	2050
4805-1241-200	0124
4806-0000-004	0080
4806-0000-005	0087
4808-0000-037	0869
4809-0000-001	0554
4811-0000-025	0325
4811-0000-039	3421
4815	0037
4816B	OEM
4819	0233
4820	0144
4820-0201	0124
4821	0144
4821-0000-027	1233
4821-0644-011	1233
4821-0644-012	1974
4822-130-30132	0025
4822-130-30192	0015
4822-130-30193	0157
4822-130-30256	0015
4822-130-30259	0015
4822-130-30261	0289
4822-130-30264	0162
4822-130-30281	0143
4822-130-30284	0451
4822-130-30287	0649
4822-130-30301	0143
4822-130-30311	0143
4822-130-30312	0143
4822-130-30414	0277
4822-130-34382	0165
4822-130-40095	0164
4822-130-40096	0208
4822-130-40132	0103
4822-130-40182	0133
4822-130-40184	0016
4822-130-40212	2763
4822-130-40213	0222
4822-130-40214	0127
4822-130-40215	0127
4822-130-40216	0127
4822-130-40229	0143
4822-130-40233	0085
4822-130-40235	0004
4822-130-40236	0004
4822-130-40252	0050
4822-130-40255	0050
4822-130-40304	0007
4822-130-40311	0127
4822-130-40312	0127
4822-130-40313	0127
4822-130-40314	0208
4822-130-40315	0037
4822-130-40317	0127
4822-130-40318	0127
4822-130-40319	2866
4822-130-40333	0079
4822-130-40348	0150
4822-130-40354	0016
4822-130-40356	0086
4822-130-40361	0016
4822-130-40369	0150
4822-130-40385	3545
4822-130-40441	0050
4822-130-40454	0016
4822-130-40456	0164
4822-130-40477	0150
4822-130-40508	0150
4822-130-40537	0161
4822-130-40614	0150
4822-130-41275	0071
4822-130-41782	OEM
4822-130-50221	0015
4822-130-50228	0296
4822-130-60588	0832
4822-130-81053	OEM
4825	0224
4826	0144
4827	OEM
4828-4	0143
4835-13-47796	OEM
4835-109-87542	OEM
4835-110-57006	OEM
4835-111-17023	OEM
4835-130-370-37	OEM
4835-130-7058	OEM
4835-130-27065	OEM
4835-130-27066	OEM
4835-130-34382	OEM
4835-130-35235	OEM
4835-130-37001	OEM
4835-130-37003	OEM
4835-130-37005	5699
4835-130-37006	OEM
4835-130-37007	5700
4835-130-37008	OEM
4835-130-37009	0031
4835-130-37011	0124
4835-130-37015	0533
4835-130-37016	OEM
4835-130-37045	0671
4835-130-37046	0166
4835-130-37047	0023
4835-130-37048	0124
4835-130-37049	0053
4835-130-37051	5700
4835-130-37052	0031
4835-130-37053	0023
4835-130-37054	5710
4835-130-37056	OEM
4835-130-37057	OEM
4835-130-37058	1090
4835-130-37059	0031
4835-130-37061	0031
4835-130-37062	0124
4835-130-37063	0705
4835-130-37064	0494
4835-130-37065	OEM
4835-130-37066	OEM
4835-130-37067	0012
4835-130-37068	0883
4835-130-37069	OEM
4835-130-37074	5712
4835-130-37081	0700
4835-130-37082	0195
4835-130-37084	0253
4835-130-37085	0165
4835-130-37086	0057
4835-130-37088	OEM
4835-130-37089	2628
4835-130-37091	0455
4835-130-37092	0010
4835-130-37094	0031
4835-130-37095	OEM
4835-130-37096	0057
4835-130-37098	OEM
4835-130-37099	0031
4835-130-37101	OEM
4835-130-37112	OEM
4835-130-37113	OEM
4835-130-37116	0015
4835-130-37119	OEM
4835-130-37121	0295
4835-130-37126	0248
4835-130-37145	0124
4835-130-37151	3301
4835-130-37161	5730
4835-130-37182	4415
4835-130-37203	5742
4835-130-37235	0124
4835-130-37236	5747
4835-130-37299	5753
4835-130-37304	0140
4835-130-37328	0124
4835-130-37329	0253
4835-130-37332	0133
4835-130-37334	0609
4835-130-37335	0195
4835-130-37336	0471
4835-130-37337	0023
4835-130-37355	5772
4835-130-37384	OEM
4835-130-37389	OEM
4835-130-37397	OEM
4835 130 37399	0471
4835-130-37402	0057
4835-130-37404	0445
4835-130-37405	0077
4835-130-37423	5787
4835-130-37433	4022
4835-130-37471	1414
4835-130-37472	1414
4835-130-37491	1664
4835-130-37501	5795
4835-130-37502	2106
4835-130-37503	5797
4835-130-37516	OEM
4835-130-37529	5803
4835-130-37535	5149
4835-130-37536	OEM
4835-130-37541	OEM
4835-130-37544	4407
4835-130-37547	OEM
4835-130-37548	5808
4835-130-37549	4024
4835-130-37551	4581
4835-130-37552	0471
4835-130-37553	5811
4835-130-37555	OEM
4835-130-37556	OEM
4835-130-37557	OEM
4835-130-37558	OEM
4835-130-37559	OEM
4835-130-37561	OEM
4835-130-37562	5223
4835-130-37563	1664
4835-130-37629	OEM
4835-130-37663	1145
4835-130-37664	5832
4835-130-37672	OEM
4835-130-37679	OEM
4835-130-37681	OEM
4835-130-37715	OEM
4835-130-37738	OEM
4835-130-37739	OEM
4835-130-37741	OEM
4835-130-37773	OEM
4835-130-37774	OEM
4835-130-37777	OEM
4835-130-37778	OEM
4835-130-37815	OEM
4835-130-37816	OEM
4835-130-37829	OEM
4835-130-41275	OEM
4835-130-47042	6692
4835-130-47048	2260
4835-130-47049	0037
4835-130-47051	0079
4835-130-47052	0079
4835-130-47053	0037
4835-130-47054	0037
4835-130-47055	0079
4835-130-47056	0111
4835-130-47058	3799
4835-130-47059	0261
4835-130-47061	1541
4835-130-47062	OEM
4835-130-47063	OEM
4835-130-47064	5932
4835-130-47065	1733
4835-130-47066	1733
4835-130-47067	0556
4835-130-47068	OEM
4835-130-47072	6695
4835-130-47073	0261

DEVICE TYPE	REPL CODE
4835-130-47083	OEM
4835-130-47084	1369
4835-130-47085	1733
4835-130-47086	1733
4835-130-47087	1752
4835-130-47088	4489
4835-130-47089	0236
4835-130-47091	2129
4835-130-47092	OEM
4835-130-47094	OEM
4835-130-47095	1733
4835-130-47096	4489
4835-130-47097	2129
4835-130-47098	OEM
4835-130-47099	0261
4835-130-47101	1165
4835-130-47107	6697
4835-130-47112	1752
4835-130-47116	OEM
4835-130-47117	1526
4835-130-47126	0037
4835-130-47137	0151
4835-130-47173	6700
4835-130-47191	6703
4835-130-47235	0151
4835-130-47243	1032
4835-130-47244	0887
4835-130-47275	0286
4835-130-47353	0887
4835-130-47358	0076
4835-130-47397	1032
4835-130-47413	4124
4835-130-47422	0013
4835-130-47424	0006
4835-130-47425	3183
4835-130-47426	0477
4835-130-47427	0018
4835-130-47428	2323
4835-130-47429	0537
4835-130-47431	3938
4835-130-47432	3093
4835-130-47433	2910
4835-130-47434	OEM
4835-130-47435	0151
4835-130-47436	0151
4835-130-47439	3143
4835-130-47441	0151
4835-130-47442	0148
4835-130-47458	OEM
4835-130-47479	0898
4835-130-47503	2691
4835-130-47525	4007
4835-130-47536	OEM
4835-130-47537	OEM
4835-130-47538	OEM
4835-130-47543	1348
4835-130-47544	OEM
4835-130-47545	2923
4835-130-47575	0148
4835-130-47576	1348
4835-130-47639	3906
4835-130-47705	OEM
4835-130-47744	0261
4835-130-47749	5330
4835-130-47751	2349
4835-130-47755	OEM
4835-130-47756	0866
4835-130-47757	0866
4835-130-47758	OEM
4835-130-47759	0261
4835-130-47761	3516
4835-130-47762	1541
4835-130-47763	OEM
4835-130-47764	0161
4835-130-47765	0148
4835-130-47768	5033
4835-130-47769	0276
4835-130-47771	OEM
4835-130-47796	OEM
4835-130-47892	OEM
4835-130-47896	6776
4835-130-47897	5330
4835-130-47931	OEM
4835-130-48011	6788
4835-130-57003	1243
4835-130-57004	OEM
4835-130-67001	OEM
4835-130-87001	OEM
4835-130-87055	OEM
4835-130-87056	OEM
4835-130-87263	1348
4835-130-97005	0315
4835-130-97006	1291
4835-130-97042	5642
4835-130-97043	OEM
4835-130-97058	OEM
4835-205-88384	OEM
4835-209-07822	OEM
4835-209-1702	OEM
4835-209-8708	OEM
4835-209-8752	OEM
4835-209-8755	OEM
4835-209-17001	OEM
4835-209-17002	OEM
4835-209-17003	OEM
4835-209-17004	1588
4835-209-17005	3806
4835-209-17021	OEM
4835-209-17024	OEM
4835-209-17026	OEM
4835-209-17027	OEM
4835-209-17032	3811
4835-209-17033	0035
4835-209-17148	3161
4835-209-17314	OEM
4835-209-17315	OEM
4835-209-17339	OEM
4835-209-17345	OEM
4835-209-17352	OEM
4835-209-17355	OEM
4835-209-17368	OEM
4835-209-17369	OEM
4835-209-17432	OEM
4835-209-17433	OEM
4835-209-17475	OEM
4835-209-47005	3802
4835-209-47006	OEM
4835-209-47007	OEM
4835-209-47008	OEM
4835-209-47009	OEM
4835-209-47013	OEM
4835-209-47015	OEM
4835-209-47017	OEM
4835-209-47026	OEM
4835-209-47053	OEM
4835-209-47054	OEM
4835-209-47056	2364
4835-209-47058	OEM
4835-209-47059	OEM
4835-209-47061	OEM
4835-209-47062	OEM
4835-209-47063	OEM
4835-209-47064	OEM
4835-209-47071	OEM
4835-209-47072	OEM
4835-209-47073	OEM
4835-209-47074	OEM
4835-209-47076	OEM
4835-209-47091	OEM
4835-209-47101	OEM
4835-209-47106	OEM
4835-209-47107	OEM
4835-209-47108	OEM
4835-209-47109	OEM
4835-209-47111	OEM
4835-209-47112	OEM
4835-209-47113	OEM
4835-209-47808	OEM
4835-209-67953	OEM
4835-209-87057	OEM
4835-209-87065	0351
4835-209-87066	3495
4835-209-87067	0624
4835-209-87068	OEM
4835-209-87069	0732
4835-209-87071	5805
4835-209-87072	OEM
4835-209-87073	2957
4835-209-87074	1028
4835-209-87075	OEM
4835-209-87079	0770
4835-209-87081	0631
4835-209-87082	1612
4835-209-87083	OEM
4835-209-87084	OEM
4835-209-87085	3097
4835-209-87086	OEM
4835-209-87087	OEM
4835-209-87088	OEM
4835-209-87089	4265
4835-209-87091	5824
4835-209-87092	OEM
4835-209-87094	0557
4835-209-87105	0101
4835-209-87114	4141
4835-209-87252	1292
4835-209-87259	0624
4835-209-87261	2780
4835-209-87262	2922
4835-209-87263	0176
4835-209-87264	1329
4835-209-87266	OEM
4835-209-87268	5806
4835-209-87269	OEM
4835-209-87271	0124
4835-209-87272	OEM
4835-209-87273	OEM
4835-209-87274	0624
4835-209-87275	2259
4835-209-87277	0024
4835-209-87278	OEM
4835-209-87283	0334
4835-209-87323	OEM
4835-209-87342	3521
4835-209-87398	OEM
4835-209-87399	OEM
4835-209-87406	4422
4835-209-87531	OEM
4835-209-87541	OEM
4835-209-87552	4506
4835-209-87553	OEM
4835-209-87554	OEM
4835-209-87556	OEM
4835-209-87557	OEM
4835-209-87559	OEM
4835-209-87561	OEM
4835-209-87562	OEM
4835-209-87563	OEM
4835-209-87564	OEM
4835-209-87565	OEM
4835-209-87605	OEM
4835-209-87612	0976
4835-209-87712	OEM
4835-209-87741	OEM
4835-209-87801	OEM
4835-209-87803	1824
4835-209-87806	OEM
4835-209-87807	OEM
4835-209-87808	OEM
4835-209-87821	1292
4835-209-87826	OEM
4835-209-87827	0334
4835-209-87834	5805
4835-209-87835	3495
4835-209-87836	OEM
4835-209-87837	OEM
4835-209-87838	OEM
4835-209-87904	OEM
4835-209-87947	OEM
4835-209-87948	OEM
4835-209-87949	OEM
4835-209-87982	OEM
4835-209-88002	OEM
4835-209-88003	4360
4835-209-88004	OEM
4835-209-88005	OEM
4835-209-88009	OEM
4835-209-88023	OEM
4835-209-88025	OEM
4835-209-88098	OEM
4835-209-88103	OEM
4835-209-88106	OEM
4835-209-88107	OEM
4835-209-88108	OEM
4835-209-88109	OEM
4835-209-88114	OEM
4835-209-88187	OEM
4835-209-88189	OEM
4835-209-88191	OEM
4835-209-88216	OEM
4835-209-88217	OEM
4835-209-88243	OEM
4835-209-88337	OEM
4835-209-88379	OEM
4835-209-88381	OEM
4835-209-88382	OEM
4835-209-88384	OEM
4835-209-88385	OEM
4835-209-88386	OEM
4835-209-88427	OEM
4835-209-88428	OEM
4835-209-88439	OEM
4835-210-27002	OEM
4835-210-27003	OEM
4835-218-27004	OEM
4836-130-37333	0062
4836-209-87531	OEM
4837	0144
4838	0233
4838-230-37058	OEM
4839	0016
4840	0016
4841	0016
4842	0016
4843	0233
4844	0037
4845	0144
4851	0127
4852	0016
4853	0233
4854	0016
4855	0144
4856	OEM
4857	0224
4858	0039
4872	0142
4882	0103
4888A	0160
4888B	0160
4904	0111
4907-976	0004
4927	0079
4935	0595
4935-130-37048	OEM
4937	0111
4943	0006
4989	0006
5000	1876
5001	1669
5001-002	0368
5001-010	2046
5001-014	0076
5001-020	0076
5001-021	0191
5001-032	0155
5001-037	0151
5001-038	0076
5001-043	0380
5001-044	2050
5001-046	0212
5001-047	3046
5001-050	0839
5001-064	0390
5001-068	0622
5001-070	0155
5001-074	0076
5001-080	0143
5001-083	0133
5001-089	0023
5001-107	0133
5001-112	0139
5001-117	0023
5001-125	0012
5001-128	0133
5001-135	0015
5001-141	0019
5001-142	0062
5001-144	0133
5001-152	0012
5001-156	0133
5001-160	0157
5001-161	0019
5001-162	0133
5001-163	0080
5001-164	0133
5001-165	0120
5001-196	0030
5001-197	0041
5001-508	0103
5001-509	0037
5001-510	0144
5001-511	1221
5001-512	0208
5001-513	0786
5001-514	2495
5001-517	1946
5001-539	0155
5001-540	0155
5001-541	0151
5001-542	0224
5001-543	0113
5001-544	0151
5001-545	1220
5002	OEM
5002(THOMAS)	0723
5002-030	1714
5002-031	1050
5003	OEM
5003P	OEM
5004	OEM
5005	1876
5005(THOMAS)	1876
5007	1876
5007(THOMAS)	1876
5009	0623
5009(THOMAS)	0623
5025S	OEM
5035	OEM
5051-300-10150	0015
5052	0211
5055	OEM
5055J18	OEM
5056H-116	OEM
5065	0079
5070	OEM
5070A	OEM
5082-2835	0917
5082-4350	1291
5082-4420	OEM
5082-4480	OEM
5082-4483	OEM
5082-4486	OEM
5082-4520	OEM
5082-4550	3157
5082-4550A	OEM
5082-4555	OEM
5082-4587	OEM
5082-4620	OEM
5082-4650	1962
5082-4650A	OEM
5082-4655	OEM
5082-4687	OEM
5082-4787	OEM
5082-4920	OEM
5082-4950	1778
5082-4950A	OEM
5082-4955	OEM
5082-4987	OEM
5082-7650	OEM
5082-7651	OEM
5082-7653	OEM
5082-7730	4989
5082-7731	4870
5082-7736	OEM
5082-7740	4190
5082-7750	OEM
5082-7751	OEM
5082-7760	OEM
5085	0050
5086J24	OEM
5087J24	OEM
5090A	OEM
5093	0144
5096	0325
5101N60	0019
5101S34	0143
5104	OEM
5107	OEM
5116-1JC	OEM
5116JC	OEM
5120A90	0143
5120L	OEM
5122	0137
5124	0002
5130	0441
5132	0039
5133	OEM
5136	OEM
5142	0331
5151DC	OEM
5151JC	OEM
5152	OEM
5153	OEM
5155J18	OEM
5156JC	OEM
5157	OEM
5158	7083
5158-1	0521
5159	OEM
5160	OEM
5161	OEM
5162	OEM
5163	OEM
5164	OEM
5165	OEM
5166	OEM
5167	OEM
5168	OEM
5169	OEM
5170	OEM
5171	OEM
5172	OEM
5173	OEM
5174	OEM
5175	OEM
5176	OEM
5201-5	OEM
5201-6	OEM
5201A-5	OEM
5201A-6	OEM
5203N1	2510
5203RNI	1080
5203RNL	1080
5205NI	0015
5205NL	0015
5210DC-1	0953
5213	OEM
5213-4000	0122
5217	0133
5218	OEM
5220	OEM
5222	0137
5224	0002
5224(SCR)	0740
5226-1	0037
5226-2	0016
5228B(ZENER)	0036
5230	0441
5230A	0140
5232	0039
5233-7100	0527
5242	0331
5253	0085
5255-1J24	OEM
5256-1J24	OEM
5260-1J24	OEM
5261-1J24	OEM
5275-1J24	OEM
5276-1J24	OEM
5280-1J24	OEM
5280-2J24	OEM
5281-1J24	OEM
5281-2J24	OEM
5282-1J24	OEM
5283-1J24	OEM
5291	OEM
5300-1J16	OEM
5300-2	0415
5300-3	0415
5300-82-3	0071
5300-82-4	0071
5300-82-1003	0229
5300-86-1	0015
5300-88-1	0015
5300-88-2	0229
5300-88-3	0071
5300-88-1002	0071
5300-88-1003	0071
5300-88-1004	0071
5300-94-1	0015
5300-95-1	0015
5300-98-1	0015
5300-99-1	0015
5300-99-3	0015
5300-99-4	0015
5301-06-1	0015
5301-06-1001	0071
5301-09-1	0015
5301-1J16	OEM
5301-5	OEM
5301-6	OEM
5301-11-1	0229
5301-11-1001	0071
5301-13-1	0015
5301-20-1	0015
5301-22-1	0229
5301-36-T	0071
5301A-5	OEM
5301A-6	OEM
5303-1	1413
5303-2	1413
5305-1J16	OEM
5306-1J16	OEM
5308-1J20	OEM
5309-1J20	OEM
5310	2623
5320	OEM
5320-003	0178
5321	OEM
5322	OEM
5323	OEM
5324	0247
5326	0256
5327	0262
5328	0269
5330	0294
5330-1J16	OEM
5331	0309
5331-1J16	OEM
5332	0318
5333	0320
5334	0326
5335	0337
5336	0347
5337	0027
5340	0405
5340-1J24	OEM
5340-2J	OEM
5340-2JS	OEM
5341	0426
5341-1J24	OEM
5341-2J	OEM
5341-2JS	OEM
5342	0444
5343	0238
5344	1180
5345(ZENER)	1190
5345-2	6001
5346	1207
5348	0647
5348-1J20	OEM
5349-1J20	OEM
5350	1279
5350-1J18	OEM
5351	0605
5351-1J18	OEM
5351-31	OEM
5352-1J18	OEM
5353-1J18	OEM
5361-1N60P	0123
5371	OEM
5380-1J24	OEM
5380-2J	OEM
5380-2JS	OEM
5380-21	0015
5380-71	0016
5380-72	0016
5380-73	0086
5380-73(POWER)	0178
5381-1J24	OEM
5381-2J	OEM
5381-2JS	OEM
5388-1J	OEM
5389-1J	OEM
5389-2J	OEM
5400DM	OEM
5400FM	OEM
5401	OEM
5401DM	OEM
5401FM	OEM
5402DM	OEM
5402FM	OEM
5403	OEM
5403-MS	0012
5403DM	OEM
5403FM	OEM
5404DM	OEM
5404FM	OEM
5405DM	OEM
5405FM	OEM
5406DM	OEM
5406FM	OEM
5407DM	OEM
5407FM	OEM
5408DM	OEM
5408FM	OEM
5409DM	OEM
5409FM	OEM
5410DM	OEM
5410FM	OEM
5411DM	OEM
5411FM	OEM
5412DM	OEM
5412FM	OEM
5413DM	OEM
5413FM	OEM
5414DM	OEM
5414FM	OEM
5416	0015
5416DM	OEM
5416FM	OEM
5417DM	OEM
5417F	0155
5417FM	OEM
5417G	0155
5417H	0155
5419G	OEM
5420DM	OEM
5420FM	OEM
5421DM	OEM
5421FM	OEM
5422DM	OEM
5422FM	OEM
5423FM	OEM
5425DM	OEM
5425FM	OEM
5427DM	OEM
5427FM	OEM
5430DM	OEM
5430FM	OEM
5432DM	OEM
5432FM	OEM
5437DM	OEM
5437FM	OEM
5438DM	OEM
5438FM	OEM
5439DM	OEM
5439FM	OEM
5440DM	OEM
5440FM	OEM
5442ADM	OEM
5442AFM	OEM
5443AFM	OEM
5444AFM	OEM
5445DM	OEM
5445FM	OEM
5446-4	OEM
5446ADM	OEM
5446AFM	OEM
5447ADM	OEM
5447AFM	OEM
5448DM	OEM
5448FM	OEM
5450DM	OEM
5450FM	OEM
5451DM	OEM
5451FM	OEM
5453DM	OEM
5453FM	OEM
5454DM	OEM
5454FM	OEM
5459	0325
5460DM	OEM
5460FM	OEM
5464	0004
5470FM	OEM
5472DM	OEM
5472FM	OEM
5473DM	OEM
5473FM	OEM
5474DM	OEM
5474FM	OEM
5475DM	OEM
5475FM	OEM
5476DM	OEM
5476FM	OEM
5477FM	OEM
5480FM	OEM
5482FM	OEM
5483ADM	OEM
5483AFM	OEM
5485DM	OEM
5485FM	OEM
5486DM	OEM
5486FM	OEM
5487-1	OEM
5487-3	OEM
5488-1	OEM
5488-3	OEM
5489DM	OEM
5489FM	OEM
5490ADM	OEM
5490AFM	OEM
5490DM	OEM
5491ADM	OEM
5491AFM	OEM
5492ADM	OEM
5492AFM	OEM
5493ADM	OEM
5493AFM	OEM
5494DM	OEM
5494FM	OEM
5495ADM	OEM
5495AFM	OEM
5496DM	OEM
5496FM	OEM
5506	0760
5507	0118
5508	0299
5510	0036
5512	0140
5513	0041
5514	0253
5516	0062
5520	0064
5522	0052
5522-8	0242
5524	1618
5525	0445
5526	0375
5528	0705
5528DM	OEM
5530	0455
5532	0166
5532P	OEM
5540	7105
5540	0161
5555E	OEM
5556E	OEM
5557E	OEM
5558E	OEM
5586	OEM
5601-MS	0015

If replacement code is OEM, contact original manufacturer for replacement.

DEVICE TYPE	REPL CODE	DEVICE TYPE	REPL CODE	DEVICE TYPE	REPL CODE	DEVICE TYPE	REPL CODE	DEVICE TYPE	REPL CODE	DEVICE TYPE	REPL CODE	DEVICE TYPE	REPL CODE
5601-NI	0015	5631-MA150	0133	5655	OEM	6102	1149	6349-1J20	OEM	6613PM	OEM	7016-L	OEM
5601-NL	0015	5632-1S2226	0015	5657	OEM	6115	0244	6349-2J	OEM	6615	OEM	7021	0701
5607	4417	5632-1S2227	0015	5661-C6B3	1137	6115A	OEM	6350-1J18	OEM	6615/BL312	OEM	7025	OEM
5609	OEM	5632-AN217(BB)	2581	5662-HA1325	4018	6116	OEM	6351	0212	6616	OEM	7031	0160
5609A	OEM	5632-DFA1A4	0023	5665	OEM	6116ASP-20	OEM	6351-1J18	OEM	6624	OEM	7058	OEM
5610	OEM	5632-DFM1A4	0023	5667	OEM	6117	OEM	6352-1J18	OEM	6624/MCT18	OEM	7088	OEM
5611-628(F)	0150	5632-ERB24-04	0023	5680	0126	6117/MST11	OEM	6353-1J18	OEM	6629	0071	7090	OEM
5611-673	0037	5632-ERB24-06	0023	5685-D2SBA40	0296	6133	OEM	6367-1	0079	6629-A05	0071	7108	0016
5611-673D	0037	5632-ERB28-04	0023	5685-S1VBA40	OEM	6136	0016	6378	OEM	6630	OEM	7109	0224
5611-695	0791	5632-ERC04-04	0071	5701	0037	6143H827	0079	6378/BL62	OEM	6631	OEM	7110	OEM
5611-695C	0791	5632-HZ11A	0170	5702	OEM	6151	0086	6378/MXT20	OEM	6632	OEM	7110-1(A)	OEM
5611-933(R)	0148	5632-S1B01-02	0015	5710	OEM	6151(RCA)	0086	6380-1	0103	6632/MLT24	OEM	7110-2(A)	OEM
5611-950(Y)	1348	5632-S5566B	0015	5719	OEM	6154	0136	6380-1J24	OEM	6633	OEM	7111	OEM
5611-1015(Y)	0148	5632-S5566G	0795	5720	OEM	6155	0136	6380-2J	OEM	6633/MLT17	OEM	7112	0133
5611-6370	0280	5632-SIB01-02	0080	5721	0016	6155J18	OEM	6380-2JS	OEM	6634/MLT25	OEM	7113	0079
5611-6730	0150	5632-U06G	0965	5722-3000	0050	6156J24	OEM	6381-1	0178	6635	OEM	7115	0144
5611-RN2206	OEM	5632-W03B	0023	5735	OEM	6158	0144	6381-1J24	OEM	6635/MST18	OEM	7116	0224
5612-75(C)	0280	5632-W06A	0015	5737	OEM	6158-3	0144	6381-2J	OEM	6636	OEM	7117	0079
5612-77(C)	2866	5632-W06B	0023	5755	OEM	6161J24	OEM	6381-2JS	OEM	6636/BL87	OEM	7118	0144
5612-77C	2866	5633-1S85	0720	5757	2747	6162	0136	6388-1J	OEM	6636/MST24	OEM	7122	0144
5612-370	0004	5633-1S85Y	0030	5765	OEM	6162/ATR388	OEM	6389-1J	OEM	6637	OEM	7122-5	0079
5612-370C	0004	5633-1S2790W	0628	5766-25	0211	6163	OEM	6389-2J	OEM	6637/MST29	OEM	7123	0144
5613-458	0155	5633-152790W	0554	5767	OEM	6167	OEM	6390	OEM	6638/MST30	OEM	7124	0144
5613-458B	0155	5633-SD115	0760	5780	OEM	6171-17	0102	6393	OEM	6639	OEM	7125	0144
5613-458C	0155	5635	OEM	5789	OEM	6171-18	0102	6406A	OEM	6639/BL46	OEM	7126	0144
5613-458D	0076	5635-HZ11	0181	5792	OEM	6171J24	OEM	6410	OEM	6639/MCT14	OEM	7127	0144
5613-458LGC	0155	5635-HZ11A	0170	5792/MSA16	OEM	6172J34	OEM	6410A	OEM	6640	OEM	7128	0144
5613-460	0127	5635-HZ15	0686	5793	OEM	6177A	OEM	6417	OEM	6640/BL60	OEM	7129	0079
5613-460A	0127	5635-HZT33	3452	5793/MSA17	OEM	6181-1	0233	6421S	OEM	6640/MCT16	OEM	7130A	1750
5613-460B	0127	5635-RD4R7EB	0140	5795	OEM	6185-3	0144	6432-3	0012	6641	OEM	7131	0144
5613-460C	0127	5635-RD4R7EB2	0140	5800	OEM	6214	OEM	6440	0211	6641/BL86	OEM	7132	0144
5613-461	0224	5635-RD5R1EB	0041	5802	OEM	6221	0353	6445	0280	6642	OEM	7133	0144
5613-461(B)	0079	5635-RD6R8EB	0062	5804	OEM	6227	0212	6450CX-506	OEM	6644	OEM	7134	0144
5613-461C	0224	5635-RD6R8EB2	0062	5805	OEM	6229	OEM	6452	0211	6644/BL95	OEM	7149A	0167
5613-535	0127	5635-RD7R5EB2	0077	5832-32	OEM	6230	OEM	6455	OEM	6645	OEM	7152	OEM
5613-535(B)	0079	5635-RD8R2EB	0165	5836	OEM	6233	OEM	6455/MCA12	OEM	6645/BL95H	OEM	7152/MOT16B	OEM
5613-535A	0127	5635-RD8R2EB2	0165	5837	OEM	6236	OEM	6458D	0360	6645/MXT61	OEM	7156	OEM
5613-535B	0127	5635-RD9R1EB	0057	5840	0937	6255-1J24	OEM	6464	OEM	6646	OEM	7161	0517
5613-535C	0127	5635-RD9R1EB1	0057	5842	OEM	6256-1J24	OEM	6500-1	OEM	6646/MXT70	OEM	7161(WURLITZER)	0517
5613-710B	0368	5635-RD9R1EB2	0057	5844	OEM	6260	OEM	6500-4	OEM	6647	OEM	7163	OEM
5613-711(E)	0079	5635-RD10EB	0064	5845	OEM	6260-1J24	OEM	6502	3065	6648	OEM	7166	OEM
5613-711E	0380	5635-RD11EB2	0181	5847	0042	6261-1J24	OEM	6502A	3065	6649	OEM	7166/BL933	OEM
5613-828(S)	1220	5635-RD13EB	0053	5853	OEM	6264	OEM	6502AD	3065	6650	OEM	7166/MLT10	OEM
5613-828S	1220	5635-RD15EB	0879	5853/MST23	OEM	6264-150	OEM	6502B	3065	6651-486	0086	7171	0079
5613-870(F)	0079	5635-RD20EB3	0210	5862	0007	6264LP-15	OEM	6502B/2MHZ	OEM	6664	4315	7172	0079
5613-871(F)	0111	5635-RD30EB	0195	5864-ATR321	OEM	6264P-150	OEM	6502C	3065	6685	OEM	7172-54	0440
5613-1047(B)	0113	5635-RD36EB	0010	5865	OEM	6271	OEM	6504	OEM	6762	0037	7173	0144
5613-1162	0566	5635-ZB1-10	0170	5865/MCT32	OEM	6275-1J24	OEM	6504B	OEM	6780	OEM	7174	0144
5613-1209	0595	5636-1SS133	0133	5865/TR361	OEM	6276	OEM	6507	3070	6780/MXK47	OEM	7175	0144
5613-1209C	0595	5636-RD5R1EB	0041	5866/MST15	OEM	6276-1J24	OEM	6507(AIRLINE)	0144	6781	OEM	7176	0079
5613-1213	0715	5636-RD6R8EB	0062	5872-0010	OEM	6280-1J24	OEM	6509R4	OEM	6781/MXK11	OEM	7177	0144
5613-1213D	0086	5637	OEM	5872-0020	OEM	6280-2J24	OEM	6510	OEM	6795	OEM	7178	0144
5613-1317NC(Q)	0155	5637-1N25D	OEM	5877	OEM	6281-1J24	OEM	6510A	OEM	6795A/MXT12	OEM	7182	OEM
5613-1327(T)	1221	5637-4	5924	5883	OEM	6281-2J24	OEM	6510CBM	OEM	6800	0388	7204A	2557
5613-1327T	1221	5637-GL31AR	3019	5900	OEM	6283-1J24	OEM	6514	0076	6802	0393	7206-0	0474
5613-1335	0079	5637-LN250RP	OEM	5921	OEM	6290J24	OEM	6517	0079	6804	OEM	7208	OEM
5613-1335(E)	0111	5637-LT9223A	OEM	5921/MSA18	OEM	6291J24	OEM	6522	OEM	6805	OEM	7208B	OEM
5613-1335D	0079	5638-10	7139	5922	OEM	6292J24	OEM	6523-34	0015	6810	OEM	7209A	0474
5613-1342	0224	5638-10KIT	OEM	5922/MSA29	OEM	6293J24	OEM	6523A	OEM	6818	0440	7211-9	0015
5613-1342C	0127	5641-MV11	0015	5923	0953	6300-1J16	OEM	6525B	OEM	6821	0448	7212-3	0015
5613-1359(B)	0151	5642	OEM	5927	OEM	6300-1N	1918	6526	OEM	6838	0489	7212-3A	0015
5613-1359(C)	0151	5644	OEM	5927/MST21	OEM	6301-1J16	OEM	6526A	OEM	6840	OEM	7213	0042
5613-1359B	0151	5648-1	6020	5932	1843	6304	OEM	6526A-1	OEM	6842	OEM	7213-0	0229
5613-1383NC(R)	0018	5651-2	OEM	5939A	OEM	6304/BL43	OEM	6529B	OEM	6844	OEM	7213-7	0015
5613-1473A(R)	0261	5651-C6B3	OEM	5976	OEM	6305	OEM	6531	0127	6845	OEM	7214	0144
5613-1573A(R)	0261	5652-AN217	2581	5981	OEM	6305-1J16	OEM	6532	1973	6850	0514	7214(LOWREY)	0103
5613-1573B(R)	0261	5652-AN217(BB)	2581	6000-015	OEM	6306-1J16	OEM	6533	OEM	6851	OEM	7214-6	0071
5613-1684(Q)	0155	5652-AN217BB	2581	6001	1248	6308-1J20	OEM	6543	OEM	6861	OEM	7214A	0103
5613-1684(R)	0155	5652-AN5316	5047	6001-061	OEM	6309-1J20	OEM	6543A	OEM	6889	0581	7214S	0015
5613-1684(T)	0155	5652-AN5436	3747	6002	OEM	6310	OEM	6550	OEM	6890	OEM	7215	0144
5613-1684H(R)	0155	5652-AN5512	0976	6003	0325	6310/MXK14	OEM	6551	0701	6890/BL650	OEM	7215-0	0004
5613-1684T	0155	5652-AN7218	2926	6008	0050	6312	OEM	6551A	OEM	6900K91-002	0033	7215-1	0071
5613-1740(R)	0151	5652-AN7410	1318	6011	0325	6312/MXK15	OEM	6552	0360	6900K91-003	0141	7215-2	0071
5613-1788(R)	0155	5652-AN7410NS	1318	6013	0325	6313	0050	6558S	OEM	6900K91-004	0562	7216	0144
5613-1788R	0155	5652-BA527	2788	6019	0133	6314	OEM	6560-101	OEM	6900K91-006	1831	7217	0144
5613-1815(GR)	0076	5652-BA5402	2799	6022	OEM	6314/MXK16	OEM	6562	0770	6900K91-007	1041	7218	0144
5613-1846(Q)	0786	5652-BA6124	5162	6022/MCA13	OEM	6315	OEM	6564	OEM	6900K91-009	1433	7219	0144
5613-1846Q	0786	5652-BA7001	OEM	6024	OEM	6316	OEM	6564/BL71	OEM	6900K93-001	2078	7219-3	0222
5613-1923(Y)	2691	5652-EA6124	OEM	6024/ATR387	OEM	6322	OEM	6565	OEM	6900K93-002	2622	7220	0144
5613-3153	2784	5652-HA1156	0520	6024/MSA20	OEM	6322/BL25	OEM	6567	OEM	6900K95-001	0333	7220-1(A)	OEM
5613-3620	0276	5652-HA1325	4018	6027	OEM	6322/MLT11	OEM	6567R9	OEM	6900K95-002	0358	7221	0144
5613-4581	0007	5652-LA4265	OEM	6027H	OEM	6330-1J16	OEM	6568/MCT17	OEM	6904	OEM	7221A	0167
5613-4581C	0079	5652-LA7520	0539	6033	OEM	6331-1J16	OEM	6569	OEM	6905	OEM	7225	OEM
5613-C114YS	OEM	5652-LA7911	OEM	6033YF	2536	6334	OEM	6570	0312	6905/BL613	OEM	7228	OEM
5613-RN1206	OEM	5652-M51513L	4640	6034	OEM	6334-BL27	OEM	6570-036	OEM	6905/MCT10	OEM	7230	OEM
5614-330(E)	1607	5652-TA78L005	1292	6041	OEM	6335-1J24	OEM	6572	0360	6906	OEM	7232	0144
5614-359C	1607	5652-TA7230P	4372	6043	OEM	6336-1J24	OEM	6581	OEM	6906/BL643	OEM	7233	0224
5614-360D	0042	5652-TA7658P	5172	6050	OEM	6340-001-505A	OEM	6581R4AR	OEM	6906/MCT26	OEM	7233S	0224
5614-882(P)	0161	5652-UPC1018C	2926	6055JA8	OEM	6340-002-505R	OEM	6584/OMK7000/TK69	OEM	6914	0631	7234	0144
5614-1197	1573	5652-UPC1373H	2026	6056	1293	6340-005-505G	OEM	6589	OEM	6915	OEM	7235	0144
5614-1397	1541	5653-AN5156K	OEM	6056J24	OEM	6340-1J24	OEM	6591/MCA10	OEM	6941/CONCEPT48	OEM	7236	0224
5614-1398	1541	5653-AN5265	4422	6072J24	OEM	6340-2J	OEM	6592	OEM	6942/CONCEPT48	OEM	7236S	0224
5614-1846	2129	5653-C1870CA1	OEM	6080-002-304	OEM	6340-2JS	OEM	6592/MCC10	OEM	6950	OEM	7237	0144
5614-3600	OEM	5653-M5278L56	OEM	6080-004-304	OEM	6340-301-505A	OEM	6593	OEM	6950JR	OEM	7238	0144
5617-7	4362	5653-PST520G	OEM	6080-005-304	OEM	6340-302-505R	OEM	6593/MXC11	OEM	6959	OEM	7239	0211
5622-TPS703A	OEM	5653-STR30130	2364	6080-900-304	OEM	6340-305-505G	OEM	6594	OEM	6962/BL665	OEM	7242	OEM
5623	OEM	5653-TA7777N	3236	6081	OEM	6341-1J24	OEM	6594/MCC1	OEM	6967A	OEM	7247-A137	OEM
5626-31	OEM	5653-TA7812S	6029	6081/ATR408	OEM	6342-001-505A	OEM	6595	OEM	6975	OEM	7250	OEM
5629-6	5802	5653-TA8207K	OEM	6081/MCA11	OEM	6342-002-505R	OEM	6596	OEM	6975/MXK21	OEM	7252	0042
5631-0A90	0019	5653-TA78012A	0334	6081-900-304	OEM	6342-005-505G	OEM	6596/BL317	OEM	7000	OEM	7253	0285
5631-0A90M	0143	5654-D1709C	OEM	6082-002-304	OEM	6342-301-505A	OEM	6597	OEM	7001	0079	7253(LOWREY)	0233
5631-1N34A	0123	5654-HD38991A	2728	6082-004-304	OEM	6342-302-505R	OEM	6599	OEM	7001(E.F.-JOHNSON)	0079	7254	OEM
5631-1N60	0019	5654-LC8640-6	OEM	6082-005-304	OEM	6342-305-505G	OEM	6601	OEM	7002	0079	7256	OEM
5631-1N60P	0123	5654-M50433-1	OEM	6086J24	OEM	6343-1	0016	6602	OEM	7005G(LOWREY)	0016	7258	OEM
5631-1S2076	0133	5654-M50442-1	OEM	6087J24	OEM	6344	OEM	6602/MSC10	OEM	7006	0079	7261	0224
5631-1S2473	0133	5654-M50442-2	OEM	6099-2	0423	6344-QKH235	OEM	6604	4974	7008	OEM	7262	0144
5631-20A90	0123	5654-M58473P	OEM	6100	0164	6344A	OEM	6605	4902	7014	0079	7263	0224
5631-20A90H	0019	5654-M58659P	OEM	6100-35	0164	6348-1J20	OEM	6605/MLT23	OEM	7015	0079	7264	0224
5631-20A90H(1)	0123	5654-UPD6104C	OEM	6101	OEM	6348-2J	OEM	6613/BL324	OEM	7016	OEM	7270P	OEM

DEVICE TYPE	REPL CODE
7300/MPT26	OEM
7301	OEM
7302	0079
7303	0150
7303-1	0037
7303-3	0037
7306-1	0079
7306-7	0079
7306-8	0079
7306-13	0079
7308	OEM
7309/BL693	OEM
7310	0142
7310(G.E.)	0142
7311	0142
7312	0178
7313	0666
7316	0551
7316-1	0560
7317	0142
7317-1	1267
7318	0079
7318-1	0079
7318-2	3076
7318-6	0079
7320	OEM
7321	0861
7321-1	0079
7322	0178
7322-1	0142
7324/MPT11	OEM
7335-1	0065
7346-3	0079
7349	OEM
7349-2	0079
7349-5	0079
7351	0460
7358-00	OEM
7360R7	OEM
7362-3	0285
7362-6	0285
7363-1	0037
7364-1	0423
7364-6053P1	0698
7366-1	0065
7372-1	0212
7381/BL979	OEM
7387	OEM
7393	OEM
7394	OEM
7398-6117P1	0016
7398-6118P1	0016
7398-6119P1	0016
7398-6120P	2046
7398-6120P1	2046
7400	0232
7400/9N00	0232
7400-6A	0232
7400-9A	0232
7400A	0232
7400DC	0232
7400N	2674
7400PC	0232
7401-6A	0268
7401-9A	0268
7401DC	0268
7401PC	0268
7402	0314
7402-6A	0314
7402-9A	0314
7402DC	0314
7402N	0314
7402PC	0314
7403	OEM
7403-6A	0335
7403-9A	0335
7403DC	0335
7403N	0335
7403PC	0335
7404	0361
7404-6A	0361
7404-9A	0361
7404A	0361
7404DC	0361
7404N	0361
7404PC	0361
7405	0385
7405-6A	0385
7405-9A	0385
7405DC	0385
7405PC	0385
7406	1206
7406DC	0385
7406N	1206
7406PC	1206
7407	1339
7407A	1339
7407DC	1339
7407N	1339
7407PC	1339
7408	0467
7408-6A	0467
7408-9A	0467
7408A	0467
7408DC	0467
7408N	0467
7408PC	0467
7409DC	0492
7409PC	0492
7410	0512
7410-6A	0512
7410-9A	0512
7410DC	0512
7410N	0512
7410PC	0512
7411-6A	0527
7411-9A	0527
7411DC	0527
7411PC	0527
7412DC	2242
7412PC	2242
7413	0161
7413(IC)	1441
7413DC	1441
7413PC	1441
7414	0042
7414(IC)	2243
7414DC	2243
7414N	4459
7414PC	2243
7415	OEM
7416	1349
7416DC	1349
7416N	1349
7416NA	0979
7416PC	1349
7417	1352
7417DC	1352
7417N	1352
7417NA	1352
7417PC	1352
7419	0664
7420	0925
7420(IC)	0697
7420-6A	0697
7420-9A	0697
7420DC	0697
7420PC	0697
7421	1357
7421DC	1357
7421PC	1357
7422DC	4603
7422PC	4603
7423	0042
7423DC	3466
7423PC	3466
7425	1363
7425DC	3477
7425PC	3477
7426	OEM
7426DC	0803
7426PC	0803
7427	0144
7427(IC)	0818
7427DC	0818
7427PC	0818
7428	0144
7429	0016
7430	1370
7430-6A	0873
7430-9A	0873
7430DC	0873
7430N	1370
7430PC	0873
7431	0016
7432	1376
7432(IC)	0899
7432DC	0899
7432PC	0899
7433	0079
7437	0551
7437(IC)	3520
7437DC	3520
7437PC	3520
7438	0996
7438DC	0996
7438N	0996
7438PC	0996
7439DC	5836
7439PC	5836
7440-6A	1024
7440-9A	1024
7440DC	6004
7440PC	1024
7441	1038
7441-6A	1038
7441-9A	1038
7441DC	1038
7441PC	1038
7442ADC	1052
7442APC	1052
7442DC	1052
7442PC	1052
7443	1060
7443ADC	1060
7443APC	1060
7443DC	1060
7443PC	1060
7444	1072
7444ADC	1072
7444APC	1072
7444DC	1072
7444PC	1072
7445	1082
7445DC	1082
7445PC	1082
7446	1098
7446ADC	1098
7446APC	1098
7446PC	1098
7447	1108
7447/BL325A	OEM
7447/MCT30	OEM
7447ADC	1108
7447APC	1108
7447BDC	1108
7447BPC	1108
7447DC	1108
7447PC	1108
7448	1125
7448DC	1125
7448PC	1125
7449A	OEM
7450-6A	0743
7450-9A	0743
7450DC	0743
7450PC	0743
7451	1168
7451-6A	1168
7451-9A	1168
7451DC	1168
7451N	4533
7451PC	1168
7452A	OEM
7453-6A	1185
7453-9A	1185
7453DC	1185
7453PC	1185
7454-6A	1202
7454-9A	1202
7454DC	1202
7454PC	1202
7460	1275
7460-6A	1275
7460-9A	1275
7460DC	1275
7460PC	1275
7461	OEM
7470	1427
7470-6A	1404
7470-9A	1404
7470DC	1404
7470PC	1404
7472	1426
7472-6A	1426
7472-9A	1426
7472DC	1426
7472PC	1426
7473	1172
7473-6A	1172
7473-9A	1172
7473DC	1172
7473F	1172
7473PC	1172
7474	1313
7474/9N74	1313
7474-6A	1313
7474-9A	1313
7474DC	1313
7474F	1313
7474N	1313
7474PC	1313
7475	1432
7475-6A	1432
7475-9A	1432
7475DC	1432
7475PC	1432
7476	1158
7476-6A	1158
7476-9A	1158
7476DC	1158
7476N	1158
7476PC	1158
7480DC	1535
7480PC	1535
7482-6A	1574
7482-9A	1574
7482DC	1574
7482PC	1574
7483ADC	3121
7483APC	3121
7483DC	0117
7483PC	0117
7484A	OEM
7485	0374
7485DC	0374
7485PC	0374
7486	1368
7486DC	1368
7486PC	1368
7489DC	5455
7489FC	OEM
7489PC	5455
7490	1208
7490-6A	1208
7490-9A	1208
7490ADC	1208
7490APC	1208
7490DC	1208
7490PC	1208
7491ADC	0980
7491AFC	OEM
7491APC	0980
7491DC	0980
7491PC	0980
7492	0834
7492-6A	0834
7492-9A	0834
7492ADC	0834
7492APC	0834
7492DC	0834
7492PC	0834
7493	0569
7493/9393	0569
7493-6A	0569
7493-9A	0569
7493ADC	0569
7493APC	0569
7493DC	0569
7493PC	0569
7494DC	1702
7494FC	OEM
7494PC	1702
7495	1454
7495ADC	1486
7495AFC	OEM
7495APC	1486
7495DC	1486
7495PC	1486
7496DC	1715
7496FC	OEM
7496PC	1715
7501	0155
7501R1	OEM
7502	0111
7503	0037
7504	OEM
7504-1	OEM
7505	0155
7506	0016
7507	0086
7508	OEM
7508A02	OEM
7508A37	OEM
7509	0111
7510	0429
7510(TRANSISTOR)	0178
7511	0155
7513	0086
7514	0103
7515	0016
7516	0016
7517	0016
7517S	4858
7518	0016
7519	0155
7520F	OEM
7520N	OEM
7521B	OEM
7522F	OEM
7522N	OEM
7523B	OEM
7524F	OEM
7524N	OEM
7525B	OEM
7527-5A	0649
7528F	OEM
7528N	OEM
7529	OEM
7547	OEM
7554	0012
7555	1543
7568	0015
7574	0649
7577	OEM
7585	0079
7586	0079
7587	0079
7588	0079
7589	0016
7590	0016
7601	OEM
7602	OEM
7603	0164
7604	OEM
7605	OEM
7619	OEM
7625	1634
7626	0037
7628LA	OEM
7635	0006
7636	1647
7637	0016
7638	0532
7639	0532
7640	1655
7641	0016
7642	0144
7646IT	OEM
7662	OEM
7663	OEM
7666	0167
7675	0016
7692	OEM
7700-01D	OEM
7700-010	OEM
7701	0071
7701-1.5A	OEM
7701-2A	OEM
7701-2X	OEM
7701-3A	OEM
7701-3X	OEM
7701-4A	OEM
7701-4X	OEM
7701-5B	OEM
7701-5X	OEM
7701-6A	OEM
7701-6X	OEM
7701-7A	OEM
7701-7X	OEM
7701-8X	OEM
7701-9A	OEM
7701-9X	OEM
7701-10B	OEM
7701-10X	OEM
7701-12A	OEM
7701-12X	OEM
7701-15A	OEM
7701-15X	OEM
7701-18A	OEM
7701-18X	OEM
7701-20B	OEM
7701-20X	OEM
7702	OEM
7702-1A	0071
7704	OEM
7704-1	0071
7705	OEM
7706-1	0071
7707	1713
7708	0782
7708-1	0071
7709	0796
7710	0796
7710-1	0071
7711	0821
7711-1	0071
7712	0833
7712-1	0071
7713-1	0071
7714	0876
7715	0185
7715-3	OEM
7715-3X	OEM
7715-4	OEM
7715-4X	OEM
7715-5	OEM
7715-5X	OEM
7715-6	OEM
7715-7	OEM
7715-7X	OEM
7715-8	OEM
7715-9	OEM
7715-10	OEM
7715-10X	OEM
7715-12	OEM
7715-15	OEM
7715-15X	OEM
7715-18	OEM
7715-20	OEM
7715-20X	OEM
7715-30	OEM
7715-30X	OEM
7715-40	OEM
7715-40X	OEM
7715-50	OEM
7715-50X	OEM
7717	0480
7719	0684
7720	0225
7721	0230
7722	0030
7724	0247
7726	0256
7727	0262
7728	0269
7730	0294
7732	0318
7733	0320
7733(XSTR)	2050
7734	0326
7735	0337
7736	0347
7737	0027
7738	0266
7739	0386
7740	0405
7741	0426
7753	1759
7753DMP	0103
7801	OEM
7801OBC	OEM
7802	OEM
7803	OEM
7804	OEM
7805	0624
7805A	0624
7805C	0624
7805CDA	1922
7805CK	OEM
7805CT	0624
7805H	0624
7806CDA	OEM
7806CK	OEM
7806CU	0923
7806DA	OEM
7808C	1196
7808CDA	OEM
7808CK	OEM
7808CU	1196
7808DA	OEM
7809A	OEM
7810	0144
7810(IC)	OEM
7810G	OEM
7810H	OEM
7811	0144
7812	0144
7812(IC)	0334
7812A	0334
7812C	1351
7812CDA	3393
7812CT	0334
7812CU	0334
7812DA	5502
7812P	0334
7813	0144
7814	OEM
7815	1321
7815(IC)	1321
7815CDA	2000
7815CU	1321
7815DA	1309
7816	0016
7817	0016
7818	0016
7818CDA	OEM
7818CU	2259
7818DA	OEM
7821	OEM
7821/MPT12	OEM
7823/MLT13	OEM
7824CDA	OEM
7824CU	2239
7824DA	OEM
7829A	OEM
7849/BTR140	OEM
7849/MCT23	OEM
7850CU	OEM
7874	1876
7874(WURLITZER)	0517
7880	OEM
7885	OEM
7889	OEM
7893	OEM
7901/MPT13	OEM
7902/MPT17	OEM
7905	1285
7905CDA	2004
7905CU	1285
7905DA	2004
7905UC	1285
7905.2CDA	OEM
7905.2CU	OEM
7906C	2648
7906CDA	OEM
7906CK	OEM
7906CU	2648
7906DA	OEM
7908	OEM
7908CDA	OEM
7908CK	OEM
7908CT	OEM
7908CU	2791
7908DA	OEM
7909	OEM
7911/ACIA	OEM
7911/BP2	OEM
7911/CP1-701	OEM
7911/CP3-01	OEM
7911/CP3-03	OEM
7911/CP9/B/	OEM
7911/CRC-4K	OEM
7911/CRC-8K	OEM
7911/DSC	OEM
7911/PIA	OEM
7911/RPC	OEM
7911/RPC-16	OEM
7911/RPC-16A	OEM
7911/RPC-16C	OEM
7911/RPC-32	OEM
7911/RPC-32A	OEM
7911/RPC-32C	OEM
7911/SMC	OEM
7911/SP9	OEM
7911/SP9/B/	OEM
7911/SP80	OEM
7911/SP80/A/	OEM
7911-CP80/A/	OEM
7912	1838
7912C	1838
7912CDA	3059
7912CU	3067
7912DA	3059
7915C	3829
7915CU	3220
7915DA	3214
7918CDA	OEM
7918CK	OEM
7918CU	3346
7918DA	OEM
7920	OEM
7920A	OEM
7921	0242
7924CDA	OEM
7924CK	OEM
7924CU	3599
7924DA	OEM
7950	OEM
7950.2DA	OEM
7962	0015
7977	OEM
7991	0111
7992	0079
7993	0164
8000-00001-068	0004
8000-00003-004	0006
8000-00003-033	0368
8000-00003-034	0151
8000-00003-035	0076
8000-00003-036	0076
8000-00003-037	0006
8000-00003-038	0004
8000-00003-039	0004
8000-00003-040	0222
8000-00003-041	2050
8000-00003-042	2050
8000-00003-043	2050
8000-00003-044	0015
8000-00003-045	0143
8000-00003-046	0139
8000-00004-003	0008
8000-00004-004	0037
8000-00004-060	0123
8000-00004-061	0139
8000-00004-062	0133
8000-00004-063	0123
8000-00004-064	0133
8000-00004-065	0057
8000-00004-066	0120
8000-00004-067	0139
8000-00004-068	0023
8000-00004-079	0079
8000-00004-080	3046
8000-00004-081	0688
8000-00004-082	0788
8000-00004-083	1411
8000-00004-084	1411
8000-00004-085	0363
8000-00004-086	0208
8000-00004-087	2046
8000-00004-088	0004
8000-00004-089	0037
8000-00004-85	0016
8000-00004-184	0002
8000-00004-185	0219
8000-00004-236	OEM
8000-00004-239	0091
8000-00004-241	0620
8000-00004-242	0191
8000-00004-243	0191
8000-00004-248	0030
8000-00004-298	1144
8000-00004-299	0127
8000-00004-300	0076
8000-00004-301	0076
8000-00004-303	0018
8000-00004-305	0632
8000-00004-306	2393
8000-00004-307	0470
8000-00004-308	2155
8000-00004-P079	0016
8000-00004-P081	0325
8000-00004-P082	0086
8000-00004-P083	1411
8000-00004-P084	1411
8000-00004-P085	0155
8000-00004-P086	0155
8000-00004-P088	0004
8000-00004-P089	0037
8000-00005-001	0688
8000-00005-002	0076
8000-00005-004	0111
8000-00005-005	0076
8000-00005-007	1419
8000-00005-009	1400
8000-00005-010	1400
8000-00005-011	1411
8000-00005-013	1591
8000-00005-015	0143
8000-00005-016	0143
8000-00005-017	0143
8000-00005-018	0143
8000-00005-019	0649
8000-00005-020	0636
8000-00005-021	0012
8000-00005-022	0015
8000-00005-023	0143
8000-00005-055	0076
8000-00005-152	0015
8000-00006-001	1400
8000-00006-002	1411
8000-00006-003	0368
8000-00006-004	0037
8000-00006-005	0895
8000-00006-006	2793
8000-00006-007	0019
8000-00006-008	0133
8000-00006-009	0012
8000-00006-010	0039
8000-00006-011	1235
8000-00006-146	0091
8000-00006-147	0023
8000-00006-190	0867
8000-00006-201	0080
8000-00006-230	4133
8000-00006-231	0015
8000-00006-232	0012
8000-00006-278	0786
8000-00006-279	2548
8000-00006-280	0076
8000-00006-281	0133
8000-00009-089	0076

DEVICE TYPE	REPL CODE
8000-00009-174	0191
8000-00009-177	0127
8000-00009-178	0688
8000-00009-280	2459
8000-0004-004	0037
8000-0004-086	0208
8000-0004-P060	0143
8000-0004-P-061	0015
8000-0004-P061	0133
8000-0004-P062	0133
8000-0004-P063	0143
8000-0004-P064	0133
8000-0004-P065	0012
8000-0004-P066	0133
8000-0004-P067	0015
8000-0004-P068	0015
8000-0004-P079	0368
8000-0004-P080	0325
8000-0004-P081	0688
8000-0004-P082	0788
8000-0004-P083	1411
8000-0004-P084	1411
8000-0004-P085	0076
8000-0004-P086	0076
8000-0004-P088	0600
8000-0004-P089	0037
8000-0004-P090	0823
8000-0004-P185	0219
8000-0004-P060	0143
8000-0004-P061	0133
8000-0004-P062	0133
8000-0004-P063	0143
8000-0004-P064	0133
8000-0004-P065	0012
8000-0004-P066	0133
8000-0004-P067	0015
8000-0004-P068	0015
8000-0004-P079	0016
8000-0004-P080	0325
8000-0004-P081	0325
8000-0004-P082	0016
8000-0004-P083	0086
8000-0004-P084	0086
8000-0004-P085	0016
8000-0004-P086	0208
8000-0004-P088	0004
8000-0004-P089	0037
8000-0005-001	0325
8000-0005-002	0127
8000-0005-003	0127
8000-0005-004	0076
8000-0005-007	0016
8000-0005-008	0103
8000-0005-009	0016
8000-0005-010	0086
8000-0005-011	1411
8000-0005-015	0143
8000-0005-016	0143
8000-0005-017	0143
8000-0005-018	0143
8000-0005-019	0649
8000-0005-020	0636
8000-0005-021	0012
8000-0005-022	0015
8000-0005-023	0143
8000-0009-089	0076
8000-00010-017	0325
8000-00010-109	0133
8000-00011-004	0368
8000-00011-0048	0007
8000-00011-041	0023
8000-00011-042	0133
8000-00011-043	0012
8000-00011-044	0015
8000-00011-045	0139
8000-00011-046	0143
8000-00011-047	0368
8000-00011-048	0368
8000-00011-049	0076
8000-00011-050	0042
8000-00011-051	1400
8000-00011-052	1908
8000-00011-053	0414
8000-00011-054	3155
8000-00011-055	3155
8000-00011-064	0012
8000-00011-103	0649
8000-00011-104	0015
8000-00011-116	OEM
8000-00011-166	0030
8000-00012-038	0133
8000-00012-039	0076
8000-00012-040	0218
8000-00012-041	4101
8000-00016-127	0911
8000-00028-037	0155
8000-00028-038	0191
8000-00028-039	0042
8000-00028-040	0925
8000-00028-041	0228
8000-00028-042	0361
8000-00028-043	0569
8000-00028-044	0619
8000-00028-045	0139
8000-00028-046	0143
8000-00028-047	0064
8000-00028-048	0023
8000-00028-206	0076
8000-00029-006	0076
8000-00029-007	0951
8000-00030-007	0155
8000-00030-008	1978
8000-00030-009	0208
8000-00030-010	0080
8000-00032-025	0076
8000-00032-026	1154
8000-00032-027	0208
8000-00032-028	0836
8000-00032-029	OEM
8000-00032-030	2124
8000-00032-031	0823
8000-00034-010	0015
8000-00035-001	0133
8000-00035-002	0023
8000-00035-003	0127
8000-00038-001	0936
8000-00038-002	0011
8000-00038-003	1645
8000-00038-004	0232
8000-00038-005	4596
8000-00038-006	0569
8000-00038-007	1313
8000-00038-008	0133
8000-00038-009	0143
8000-00038-010	0005
8000-00041-015	0143
8000-00041-016	0143
8000-00041-017	0139
8000-00041-018	0012
8000-00041-019	0133
8000-00041-040	2209
8000-00041-041	0076
8000-00041-042	0076
8000-00041-043	0390
8000-00041-044	0839
8000-00041-046	0076
8000-00041-047	2710
8000-00042-007	0124
8000-00042-008	0232
8000-00042-009	0512
8000-00042-010	OEM
8000-00042-011	1926
8000-00042-012	OEM
8000-00042-013	0414
8000-00042-014	0316
8000-00042-015	0344
8000-00043-019	1144
8000-00043-020	1062
8000-00043-021	0012
8000-00043-064	1173
8000-00043-065	2495
8000-00043-067	0041
8000-00043-068	0322
8000-00045-004	1645
8000-00047-002	3618
8000-00047-003	3610
8000-00047-006	2301
8000-00049-010	0080
8000-00049-020	0030
8000-00049-021	0041
8000-00049-025	3815
8000-00049-026	2411
8000-00049-053	0076
8000-00049-054	0076
8000-00049-055	0151
8000-00049-056	0037
8000-00049-057	0151
8000-00049-058	1978
8000-00049-059	2825
8000-00049-060	2495
8000-00049-061	0018
8000-00049-062	1280
8000-00053-004	0785
8000-00053-005	OEM
8000-00057-007	OEM
8000-00057-008	OEM
8000-00057-009	1544
8000-00057-011	0012
8000-00058-003	0091
8000-00058-004	0006
8000-00058-006	0785
8000-00058-008	4644
8000-00058-009	1540
8000-004-P061	0071
8000-004-P063	0143
8000-004-P064	0071
8000-004-P067	0071
8000-004-P068	0071
8000-004-P089	0126
8000-0016-127	0911
8000-0016-129	0911
8000-0028-042	0361
8000-0030-009	0208
8000-0058-006	0785
8000-0095-010	0698
8000-028-206	OEM
8001	OEM
8002	OEM
8003	1965
8003-114	0016
8003-115	0015
8004	1965
8005	OEM
8005(PENNCREST)	0130
8006	OEM
8007	OEM
8008	OEM
8010	OEM
8010-51	0030
8010-52	0133
8010-53	0143
8010-171	2393
8010-172	2031
8010-173	0325
8010-174	0007
8010-175	0007
8010-176	0016
8011	OEM
8012	OEM
8016	OEM
8016-L	OEM
8020-202	0133
8020-203	0015
8020-204	0111
8020-205	0016
8020-206	0042
8021B	OEM
8021C	OEM
8021D	OEM
8021E	OEM
8021F	OEM
8023B	OEM
8023C	OEM
8023D	OEM
8023E	OEM
8023F	OEM
8025MJP	OEM
8031	OEM
8032	OEM
8035	OEM
8035MJP	OEM
8039-6I	OEM
8039-6N	OEM
8039I	OEM
8039N	OEM
8040	OEM
8041	OEM
8041A	OEM
8042	OEM
8042-105	OEM
8042-139	OEM
8042AH	OEM
8045	OEM
8048	OEM
8049	OEM
8049HC150	OEM
8050	0155
8050C	0155
8051	OEM
8051AHL5220123	OEM
8051C	OEM
8051CPU-001	OEM
8052ALR	OEM
8052ALY	OEM
8053ALB	OEM
8054	OEM
8054ALB	OEM
8055	OEM
8058	OEM
8070-4	0004
8071-4	0004
8072-4	0004
8073-4	0004
8074-4	0016
8075-4	0016
8080A	4547
8085	OEM
8087	OEM
8087-2	OEM
8088	OEM
8098	0395
8099/MST38	OEM
8100	OEM
8100-00-00	OEM
8100-00-01	OEM
8100-00-10	OEM
8100-00-11	OEM
8100-01-00	OEM
8100-01-01	OEM
8100-01-10	OEM
8100-01-11	OEM
8100-10-00	OEM
8100-10-01	OEM
8100-10-10	OEM
8100-10-11	OEM
8100-11-00	OEM
8100-11-01	OEM
8100-11-10	OEM
8100-11-11	OEM
8100FMP	OEM
8101-01A	OEM
8102-206	0015
8102-207	0079
8102-208	0191
8102-209	0016
8102-210	0822
8103	OEM
8104-0-0	OEM
8104-16-1	OEM
8104-32-0	OEM
8104-32-1	OEM
8104-64-1	OEM
8106-207	0007
8108-85/88	OEM
8110	OEM
8111	OEM
8112S	OEM
8113	OEM
8114	2048
8115	OEM
8116	0522
8120-00-0	OEM
8120-00-1	OEM
8120-01-0	OEM
8120-01-1	OEM
8120-10-0	OEM
8120-10-1	OEM
8120-11-1	OEM
8121B	OEM
8121C	OEM
8121D	OEM
8121E	OEM
8121F	OEM
8123B	OEM
8123C	OEM
8123D	OEM
8123E	OEM
8123F	OEM
8128	1898
8131	OEM
8131-01	OEM
8131-10	OEM
8131-11	OEM
8132	OEM
8135-0	OEM
8142	OEM
8143	OEM
8149	OEM
8150-00	OEM
8150-01	OEM
8150-10	OEM
8150-11	OEM
8154	OEM
8155	OEM
8155-P	OEM
8175-7937	0631
8198	OEM
8199	OEM
8200	OEM
8200-202	0111
8200-203	0111
8200-204	1010
8201	OEM
8202	OEM
8210	OEM
8211	OEM
8212	1860
8220	OEM
8221	OEM
8222	OEM
8223	OEM
8225P1	0415
8230	OEM
8231	OEM
8232	OEM
8235	OEM
8235-041-00330	OEM
8235-041-00390	OEM
8235-041-01100	OEM
8235-041-04130	OEM
8235-041-04131	OEM
8235-041-04180	OEM
8237A	OEM
8237A-5	OEM
8237A5	OEM
8239	OEM
8242	2149
8245	OEM
8247	OEM
8249	OEM
8250	OEM
8251A	OEM
8252	OEM
8253	OEM
8253-5	OEM
8253C-5	OEM
8254-2	OEM
8255A	0051
8255A-5	0051
8255A5	OEM
8255AC-5	0051
8258	OEM
8259A	OEM
8266	OEM
8272/MLT41	OEM
8280	2173
8281-1	0079
8281H	OEM
8284A	OEM
8288	OEM
8290	OEM
8300-8	0015
8301	0037
8302	0079
8303	0126
8304	0086
8310	OEM
8313/MPT16	OEM
8318	OEM
8318A	OEM
8323	OEM
8326	OEM
8332	OEM
8332A	OEM
8333	OEM
8340X7	OEM
8341	OEM
8345	OEM
8346	OEM
8347	OEM
8348	OEM
8348WMA	OEM
8356	OEM
8360R1	OEM
8360R2	OEM
8362R8	OEM
8364R7	OEM
8366	OEM
8370	OEM
8394	0111
8400-1	0126
8400-1A	0126
8400-1B	0126
8405	0037
8406X	OEM
8409-07004	0148
8409-07004-1	0148
8409-07004-2	0148
8409-07004-3	0148
8409-07100	0532
8409-07100-1	1522
8409-07211	0076
8409-07211-1	0076
8409-07211-2	0151
8409-07211-3	0286
8409-07213	0733
8409-07219	0018
8409-07220	0728
8409-07221	3357
8409-07223	2908
8409-07318	3136
8409-07319	2849
8409-07320-1	1386
8409-07321	1061
8409-07321-1	1061
8409-07502	1166
8409-07503	4529
8409-08004	1595
8409-08006	OEM
8409-08007	OEM
8409-7320	5508
8409-10061	OEM
8409-10062	OEM
8409-10064	OEM
8409-10130	OEM
8409-10136	OEM
8409-10468	OEM
8409-10710	OEM
8409-10718	OEM
8409-11102	0133
8409-11103	0124
8409-11138	0124
8409-11151	0023
8409-11152	0133
8409-11206	1875
8409-11303	1090
8409-11303-1	1090
8409-11308	OEM
8409-11309	1335
8409-11332	0546
8409-11346	0546
8409-11389	0546
8409-11403	0015
8409-12023	0275
8409-12023-1	0140
8409-12122	2904
8409-12317	1005
8409-12417	0726
8409-12480	0253
8409-12483	0052
8409-12493	0455
8410	0512
8411	0527
8412	2242
8413	1441
8414	2243
8415	OEM
8416	1349
8417	1352
8423	OEM
8440-121	1280
8440-122	0937
8440-123	0076
8440-124	0155
8440-126	0076
8467	OEM
8468	OEM
8471	0560
8474	1313
8475	1432
8475A	OEM
8476	1158
8477	OEM
8477A	OEM
8478	OEM
8481/BLT032	OEM
8482	0917
8493/F2913	OEM
8496PT	OEM
8500-201	0004
8500-202	0004
8500-203	0004
8500-204	0222
8500-206	0015
8500R3	OEM
8501R1	OEM
8502	OEM
8502H	OEM
8502R0	OEM
8502RO	OEM
8503	0127
8504	0016
8507H	OEM
8508H	OEM
8510	2288
8510H	OEM
8511	OEM
8511H	OEM
8512	OEM
8513A	0102
8513H	OEM
8514	OEM
8514A	OEM
8515H	OEM
8516	OEM
8516C	OEM
8516H	OEM
8517H	OEM
8517HA	OEM
8518H	OEM
8519H	OEM
8520	1048
8520A-1	OEM
8520A1	OEM
8520H	OEM
8520R4	OEM
8521	0079
8522H	OEM
8523H	0079
8524E	OEM
8530	1208
8540	0037
8550	0324
8550C	0006
8554-9	0086
8563-R7	OEM
8563R9	OEM
8563R9B	OEM
8564-1	2329
8564-2	1712
8564-3	2329
8564-4	1712
8564-R4	OEM
8564R4	OEM
8564R5C	OEM
8564R6V6	OEM
8568	OEM
8574/MXT132	OEM
8578/MST60	OEM
8580R5	OEM
8582	OEM
8582A	OEM
8599/MKT39	OEM
8601	0037
8602	0155
8606	0144
8607	0144
8609	0144
8610	OEM
8611	0144
8612	OEM
8614-007-0	0079
8616MDJ	OEM
8618	OEM
8622/MST52	OEM
8622MBV	OEM
8644	OEM
8682	OEM
8691	OEM
8701	OEM
8701-4586-21	OEM
8701C	OEM
8710	OEM
8710-51	0030
8710-52	0133
8710-53	0143
8710-54	0012
8710-55	0015
8710-161	0325
8710-162	0937
8710-163	0016
8710-164	0076
8710-165	0076
8710-166	0076
8710-167	0016
8710-168	1419
8710-169	0037
8710-170	0560
8712	OEM
8713H	OEM
8715H	OEM
8721-R3	OEM
8721R3	OEM
8722-R1	OEM
8722H	OEM
8722R2	OEM
8730H	OEM
8736H	OEM
8740	OEM
8741H	OEM
8744A	OEM
8748	OEM
8753H	OEM
8754H	OEM
8757H	OEM
8763	OEM
8764	OEM
8770/BTR546	OEM
8798	OEM
8800-201	0015
8800-202	0016
8800-203	0016
8800-204	0016
8800-205	0042
8800-206	0015
8805	OEM
8806	OEM
8813	OEM
8813BA	OEM
8821	OEM
8824	OEM
8825	OEM
8826	OEM
8835	OEM
8836	OEM
8837	OEM
8840	OEM
8840-51	0015
8840-52	0030
8840-53	0133
8840-54	0019
8840-55	0057
8840-56	0139
8840-121	1280
8840-122	0937
8840-123	0076
8840-124	0155
8840-126	0076
8840-161	0325
8840-162	0127
8840-163	0076
8840-164	0076
8840-165	0076
8840-166	0076
8840-167	0079
8840-168	0079
8840-169	0560
8840-171	1478
8848	OEM
8849	OEM
8852	OEM
8855	OEM
8864-1	0015
8867	0037
8868-6	0086
8868-7	0111
8868-8	0079
8880	OEM
8880-3	0086
8883-2	0160
8883-4	0086
8886-2	0079
8895	OEM
8896	OEM
8898	OEM
8899	OEM
8900	OEM
8902	OEM
8908	OEM
8910-51	0030
8910-52	0133
8910-53	0019
8910-54	0244
8910-55	0139
8910-141	0325
8910-142	0937
8910-143	0076
8910-144	0076
8910-145	1419
8910-146	0390
8910-147	0911
8910-148	1478
8934	OEM
8945	OEM
8947	OEM
8948	OEM
8949	OEM
8955	OEM
8956	OEM
8979	OEM
8999-115	0085
8999-201	0015
8999-202	0004
8999-203	0164
8999-205	0015
8999A-1333	OEM
9000-0061	OEM
9000-0080	OEM
9000-0150	OEM
9000-0151	OEM
9000-0152	OEM
9000-0153	OEM
9000-0870	OEM
9000-0871	OEM
9000-1151	OEM
9000-1152	OEM
9000-1153	OEM
9000-1154	OEM
9000-1155	OEM
9000-1156	OEM
9000-1157	OEM
9000-1158	OEM
9000DC	OEM

DEVICE TYPE	REPL CODE
9000DM	OEM
9000FM	OEM
9001	OEM
9001-0601	OEM
9001DC	OEM
9001DM	OEM
9001FM	OEM
9002	2674
9002DC	OEM
9002DM	OEM
9002FM	OEM
9003	2676
9003-0631	OEM
9003-0636	OEM
9003DC	OEM
9003DM	OEM
9003FM	OEM
9004	2678
9004DC	OEM
9004DM	OEM
9004FM	OEM
9005	0632
9005-0	0160
9005-0151	OEM
9005-0162	OEM
9005DC	OEM
9005DM	OEM
9005FM	OEM
9006	0632
9006DC	OEM
9006DM	OEM
9006FM	OEM
9007	OEM
9007DC	OEM
9007DM	OEM
9007FM	OEM
9008	2681
9008DC	OEM
9008DM	OEM
9008FM	OEM
9009DC	OEM
9009DM	OEM
9009FM	OEM
9010	0785
9010-1101	OEM
9010-1111	OEM
9010-1140	OEM
9010-1200	OEM
9010-1210	OEM
9010-1230	OEM
9010-1324	OEM
9010-1332	OEM
9010-1334	OEM
9010-2101	OEM
9010-2410	OEM
9010-2500	OEM
9010-2550	OEM
9010-3000	OEM
9010-3100	OEM
9010-5301	OEM
9010-5302	OEM
9010-6030	OEM
9010-6040	OEM
9010-6050	OEM
9010-6070	OEM
9010-6080	OEM
9010-6085	OEM
9010-6090	OEM
9010-6210	OEM
9010-6220	OEM
9010-6320	OEM
9010-7010	OEM
9011	0224
9011(G)	0079
9011E	0224
9011F	0079
9011G	0079
9011H	0224
9012	1369
9012-001	OEM
9012DC	OEM
9012DM	OEM
9012FM	OEM
9012G	0037
9012H	0037
9012HE	0004
9012HF	0004
9012HG	0150
9012O	OEM
9013	0079
9013G	0079
9013H	0079
9013HF	0079
9013HG	0079
9013HH	0079
9014	0079
9014(D)	0079
9014B	0079
9014C	0079
9014DC	OEM
9014DM	OEM
9014FM	OEM
9014N	OEM
9015	0037
9015C	0006
9015D	0006
9015DC	OEM
9015DM	OEM
9015FM	OEM

DEVICE TYPE	REPL CODE
9016	0224
9016(F)	0079
9016(G)	0079
9016D	0224
9016DC	OEM
9016DM	OEM
9016E	0224
9016F	0224
9016FM	OEM
9017FM	OEM
9018	0224
9018D	0224
9018E	0224
9018F	0144
9018G	0144
9018H	0144
9019-6215	OEM
9019-6421	OEM
9019-6422	OEM
9019-7224	OEM
9019-7225	OEM
9020DC	OEM
9020DM	OEM
9020FM	OEM
9020H-04	OEM
9021	OEM
9022	OEM
9022DC	OEM
9022DM	OEM
9022FM	OEM
9024DC	OEM
9024DM	OEM
9024FM	OEM
9030DC	OEM
9033	0079
9033-1	0079
9033-2	0111
9033-3	0111
9033-4	0111
9033-5	0111
9033G	0086
9033G(SYLVANIA)	0079
9033GREEN	0111
9033ORANGE	0111
9033Q	0111
9033RED	0111
9033WHITE	0111
9040H-02	OEM
9040H-03	OEM
9042B	0715
9050	OEM
9081	OEM
9093-1-6A	0358
9093-9-6A	0358
9093-9-7A	0358
9093BC	OEM
9093DC	0358
9093DM	OEM
9093FC	OEM
9093FM	OEM
9093PC	0358
9094-1-6A	1632
9094-9-6A	1632
9094-9-7A	1632
9094BC	OEM
9094DC	1632
9094DM	OEM
9094FC	OEM
9094FM	OEM
9094PC	1632
9097-1-6A	1481
9097-9-6A	1481
9097-9-7A	1481
9097BC	OEM
9097DC	1481
9097DM	OEM
9097FC	OEM
9097FM	OEM
9097PC	1481
9099-1-6A	0333
9099-9-6A	0333
9099-9-7A	0333
9099BC	OEM
9099DC	0333
9099DM	OEM
9099FC	OEM
9099FM	OEM
9099PC	0333
9100-1	0015
9109DC	1350
9109DM	OEM
9110	OEM
9110DC	2899
9111	OEM
9111DC	2959
9111DM	OEM
9111FM	OEM
9112	2033
9112DC	1929
9112DM	OEM
9123	OEM
9128DS-1411	OEM
9135PC	1619
9137-C-1004	0081
9137-C-1020	0562
9157PC	1433
9158PC	0466
9170	OEM
9200	OEM

DEVICE TYPE	REPL CODE
9201	OEM
9202	OEM
9203-8	0071
9210H-02	OEM
9216	OEM
9225H-04	OEM
9240H02	OEM
9242G	OEM
9300	0144
9300-5	0015
9300-6	0015
9300-7	0015
9300A	0144
9300B	0144
9300DC	OEM
9300DM	OEM
9300FC	OEM
9300FM	OEM
9300PC	4267
9300Z	0144
9301-1	0015
9301DC	2196
9301DM	OEM
9301FM	OEM
9301PC	2196
9302-2	0229
9302-2A	0229
9302-3	0015
9302FM	OEM
9304DC	OEM
9304DM	OEM
9304FM	OEM
9304PC	OEM
9305FM	OEM
9305PC	OEM
9306DC	OEM
9306DM	OEM
9306FM	OEM
9306PC	OEM
9307DC	OEM
9307DM	OEM
9307FM	OEM
9307PC	1125
9308DC	1070
9308DM	OEM
9308FM	OEM
9308PC	1070
9309DC	5528
9309DM	OEM
9309FM	OEM
9309PC	5528
9310DC	OEM
9310DM	OEM
9310FM	OEM
9310PC	1631
9311DC	OEM
9311DM	OEM
9311FM	OEM
9311PC	1555
9312DC	OEM
9312DM	OEM
9312FM	OEM
9312PC	OEM
9313DC	OEM
9313DM	OEM
9313FM	OEM
9313PC	OEM
9314	0144
9314DC	2488
9314DM	OEM
9314FM	OEM
9314PC	2488
9315DC	1038
9315DM	OEM
9315FM	OEM
9315PC	1038
9316/74161	1645
9316B	OEM
9316DC	5957
9316DM	6659
9316FM	OEM
9316PC	6670
9317BDC	1098
9317BDM	1098
9317BFM	OEM
9317BPC	1098
9317CDC	1098
9317CDM	1098
9317CFM	OEM
9318DC	4145
9318DM	OEM
9318FM	OEM
9318PC	4145
9319FM	OEM
9319PC	OEM
9320FM	OEM
9321DC	5958
9321DM	OEM
9321FM	OEM
9321PC	5958
9322DC	OEM
9322DM	1605
9322FM	OEM
9322PC	1605
9324DC	OEM
9324DM	OEM
9324PC	OEM
9328DC	5959
9328DM	OEM

DEVICE TYPE	REPL CODE
9328FC	OEM
9328FM	OEM
9328PC	5959
9330-006-11112	0015
9330-011-70112	0004
9330-092-90112	0636
9330-228-60112	0133
9330-229-20112	0015
9330-229-60112	0127
9330-229-70112	0127
9330-688-30112	0111
9330-767-60112	0037
9330-908-10112	0037
9334DC	1744
9334DM	1744
9334FM	OEM
9334PC	3205
9338DC	OEM
9338DM	OEM
9338FC	OEM
9338FM	OEM
9338PC	OEM
9339-560-70112	OEM
9339-841-90112	OEM
9340DC	OEM
9340DM	OEM
9340FM	OEM
9340PC	OEM
9341DC	1842
9341DM	OEM
9341FM	OEM
9341PC	1842
9342DC	1856
9342DM	OEM
9342FM	OEM
9342PC	OEM
9344FM	OEM
9345DC	1082
9345PC	1082
9348-3	0015
9348DC	OEM
9348DM	OEM
9348FM	OEM
9348PC	OEM
9352DC	1052
9352FM	OEM
9352PC	1052
9353DC	1060
9353PC	1060
9354DC	1072
9354PC	1072
9357A	1098
9357ADC	1098
9357APC	1098
9357B	1108
9357BDC	1108
9357BPC	1108
9358	1125
9358DC	1125
9358PC	1125
9360DC	1921
9360DM	OEM
9360FM	OEM
9360PC	1921
9366DC	1926
9366DM	OEM
9366FM	OEM
9366PC	1926
9367-1	0086
9368DC	5963
9368PC	5963
9370DC	5964
9370PC	5964
9374DC	5965
9374PC	5965
9375DC	1432
9375PC	1432
9377	2940
9380DC	1535
9380PC	1535
9382DC	1574
9382PC	1574
9383DC	0117
9383PC	0117
9385DC	0374
9385PC	0374
9386DC	0592
9386DM	0592
9386FM	OEM
9386PC	0592
9390	0211
9390DC	1208
9390PC	1208
9391	0211
9391DC	0980
9391PC	0980
9392DC	0834
9392PC	0834
9393	0569
9393DC	0569
9393PC	0569
9394DC	1702
9394PC	1702
9395DC	1486
9395PC	1486
9396DC	1715
9396PC	1715
9400-8	0004
9400-9	0004

DEVICE TYPE	REPL CODE
9400CJ	OEM
9400CN	OEM
9401-7	0004
9401CJ	OEM
9401CN	OEM
9402CJ	OEM
9402CN	OEM
9403-2	0160
9403-3	0050
9403-6	0050
9403-7	0004
9403-8	0136
9403-9	0004
9403DC	OEM
9403DM	OEM
9404-0	0160
9404-2	0086
9404-3	0698
9404-9	0178
9405-0	0086
9405-1	0086
9405-2	0126
9406	0502
9409-4	0178
9409DC	OEM
9409DM	OEM
9409PC	OEM
9410A	0079
9412	OEM
9423DC	OEM
9423FC	OEM
9423PC	OEM
9426B	0144
9426B,C	0144
9426C	0144
9440	0133
9445	OEM
9464C-0839	OEM
9491A	OEM
9491AJ	OEM
9491AM	OEM
9491BJ	OEM
9491BM	OEM
9495AE	OEM
9495BE	OEM
9495CE	OEM
9495CJ	OEM
9495DE	OEM
9495DJ	OEM
9495EE	OEM
9495HE	OEM
9495HJ	OEM
9496AE	OEM
9496BE	OEM
9496CE	OEM
9496CJ	OEM
9496DE	OEM
9496DJ	OEM
9496EE	OEM
9496HE	OEM
9496HJ	OEM
9501	0111
9502	0111
9502DC	OEM
9503DC	OEM
9504DC	OEM
9510	OEM
9510-1	0050
9510-2	0050
9510-3	0136
9510-7	0050
9511	OEM
9513	0144
9517A-5PC	OEM
9520	OEM
9521	0701
9534DC	OEM
9550	OEM
9550-1	1899
9564	0004
9586-102/120	OEM
9586-121/124	OEM
9591-102/120	OEM
9591-121/124	OEM
9600	0079
9600-5	0079
9600A	OEM
9600C	0144
9600DC	6476
9600DM	6476
9600F	0144
9600FM	OEM
9600G	0144
9600H	0144
9600PC	6476
9601	0144
9601-12	0144
9601DC	2267
9601DM	OEM
9601FM	OEM
9601PC	2267
9602	2286
9602DC	2286
9602DM	6528
9602FM	OEM
9602PC	2286
9604E	7119
9604F	0144
9605	OEM

DEVICE TYPE	REPL CODE
9606	0062
9607	0077
9609	OEM
9611	OEM
9614C	OEM
9614DC	OEM
9614DM	OEM
9614FM	OEM
9614PC	OEM
9615C	OEM
9615DC	5316
9615DM	5316
9615FM	OEM
9615PC	5316
9616	OEM
9616DC	OEM
9616DM	OEM
9616EDC	OEM
9616EPC	OEM
9616PC	OEM
9617K	0086
9618	0224
9619(A)	OEM
9620	OEM
9620-2	OEM
9620A	OEM
9622	OEM
9622-2	OEM
9623	0079
9623F	0144
9623G	0144
9623H	0144
9625F	0144
9625H	0144
9627DC	OEM
9627DM	OEM
9627PC	OEM
9629	OEM
9629-8	OEM
9629-16	OEM
9629-24	OEM
9629A	OEM
9629A-16	OEM
9629A-24	OEM
9630C	0144
9631U	1954
9632	0079
9636ARC	OEM
9636ARM	OEM
9636ATC	OEM
9637	OEM
9637-8	OEM
9637ARC	OEM
9637ARM	OEM
9637ATC	OEM
9638	OEM
9638-32	OEM
9638-48	OEM
9638-56	OEM
9638RC	OEM
9638RM	OEM
9638TC	OEM
9640-2	OEM
9640A	OEM
9640DC/26S10	OEM
9640DM/26S10	OEM
9640PC/26S10	OEM
9643TC	OEM
9645DC/3245	OEM
9645PC/3245	OEM
9650-001	0015
9650-1DC	OEM
9650-1DM	OEM
9650-2DC	OEM
9650-2DM	OEM
9650-3DC	OEM
9650-3DM	OEM
9650A	OEM
9650A-2	OEM
9654-0330.31	3732
9654-0331.31	3732
9664PC	7044
9665PC	0845
9666DC	1007
9666PC	1007
9667DC	1134
9667PC	1134
9668DC	1261
9668PC	1261
9682	0037
9690A	OEM
9696H	0086
9697	OEM
9699	OEM
9725	OEM
9740	OEM
9740H02	OEM
9803	0133
9804	OEM
9808	OEM
9817	OEM
9818	OEM
9826	OEM
9861B-43	0143
9864DS-1176	OEM
9903HC	OEM
9905HC	OEM
9906	OEM
9907HC	OEM

DEVICE TYPE	REPL CODE
9908	OEM
9909	OEM
9910	OEM
9914HC	OEM
9915	OEM
9915HC	OEM
9916	OEM
9917	OEM
9920-3-6	0071
9920-4	0004
9920-5	0004
9920-6-1	0016
9920-6-2	0016
9920-7-2	0016
9921-7	0004
9921-8	0004
9923(REGENCY)	OEM
9923-U8A99238X	OEM
9923-U8A-992328X	OEM
9925-0	0160
9925-2	0142
9925-2-1	0142
9925-2-2	0142
9926HM	OEM
9930BC	OEM
9930DC	OEM
9930DM	OEM
9930FC	OEM
9930FM	OEM
9932BC	OEM
9932DC	OEM
9932DM	OEM
9932FC	OEM
9932FM	OEM
9933BC	OEM
9933DC	OEM
9933DM	OEM
9933FC	OEM
9933FM	OEM
9934BC	OEM
9934DC	OEM
9934FC	OEM
9942PC	1856
9944BC	OEM
9944DC	OEM
9944DM	OEM
9944FC	OEM
9944FM	OEM
9945BC	OEM
9945DC	OEM
9945DM	OEM
9945FC	OEM
9945FM	OEM
9946BC	OEM
9946DC	OEM
9946DM	OEM
9946FC	OEM
9948BC	OEM
9948DC	OEM
9948DM	OEM
9948FC	OEM
9948FM	OEM
9951BC	OEM
9951DC	OEM
9951DM	OEM
9951FC	OEM
9951FM	OEM
9962BC	OEM
9962DC	OEM
9962DM	OEM
9962FC	OEM
9962FM	OEM
9970	0118
9973	0041
9981	0796
9982	0821
9983	0442
9984	0185
9992	0827
9993	0868
9994	0914
10003	0130
10010	0084
10010IOF	OEM
10014DC	OEM
10031	0196
10032	0004
10032A	OEM
10036	0004
10036-001	0126
10037	0004
10038	0004
10039	0004
10060	OEM
10060-0	2260
10100	0071
10100B	OEM
10100BD	OEM
10100F	OEM
10101B	OEM
10101F	OEM
10101GOA	OEM
10101OOA	OEM
10102B	OEM
10102F	OEM
10102GOA	OEM
10102OOA	OEM
10103B	OEM
10103F	OEM

DEVICE TYPE	REPL CODE
10103GOA	OEM
10103OOA	OEM
10104B	OEM
10104F	OEM
10104GOA	OEM
10104OOA	OEM
10105B	OEM
10105F	OEM
10105GOA	OEM
10105OOA	OEM
10106B	OEM
10106F	OEM
10106GOA	OEM
10106OOA	OEM
10107B	OEM
10107F	OEM
10108B	OEM
10108F	OEM
10109B	OEM
10109F	OEM
10110B	OEM
10110F	OEM
10111B	OEM
10111F	OEM
10112B	OEM
10112F	OEM
10113B	OEM
10113F	OEM
10114	OEM
10114DC	OEM
10115	OEM
10115DC	OEM
10116	OEM
10116DC	OEM
10117B	OEM
10117F	OEM
10118B	OEM
10118F	OEM
10119B	OEM
10119F	OEM
10121B	OEM
10121F	OEM
10123DC	OEM
10124B	OEM
10124DC	OEM
10124F	OEM
10124N	OEM
10125B	OEM
10125DC	OEM
10125F	OEM
10125N	OEM
10129F	OEM
10129N	OEM
10130F	OEM
10131F	OEM
10132F	OEM
10132N	OEM
10133F	OEM
10134DC	OEM
10134F	OEM
10134N	OEM
10135F	OEM
10136F	OEM
10137F	OEM
10139F	OEM
10139N	OEM
10140F	OEM
10142F	OEM
10142I	OEM
10145F	OEM
10145I	OEM
10148F	OEM
10149F	OEM
10150BD	OEM
10151F	OEM
10155F	OEM
10155N	OEM
10158F	OEM
10159F	OEM
10159N	OEM
10160F	OEM
10161DC	OEM
10161F	OEM
10162DC	OEM
10162F	OEM
10162N	OEM
10164DC	OEM
10164F	OEM
10164N	OEM
10165DC	OEM
10165F	OEM
10165N	OEM
10170F	OEM
10171DC	OEM
10171F	OEM
10171N	OEM
10172DC	OEM
10172F	OEM
10172N	OEM
10173DC	OEM
10173F	OEM
10173N	OEM
10174DC	OEM
10174F	OEM
10174N	OEM
10175F	OEM
10176F	OEM
10180	0015
10181	0143

DEVICE TYPE	REPL CODE
10181F	OEM
10188F	OEM
10188N	OEM
10189F	OEM
10189N	OEM
10190F	OEM
10191F	OEM
10192DC	OEM
10200BD	OEM
10210B	OEM
10210F	OEM
10211B	OEM
10211F	OEM
10212B	OEM
10212F	OEM
10231B	OEM
10231F	OEM
10250BD	OEM
10300	0037
10300-12	0126
10300BD	OEM
10302-01	0873
10302-02	0697
10302-03	0512
10302-04	0232
10302-05	1168
10302-06	0385
10350	OEM
10350BD	OEM
10400BD	OEM
10410-N0B013	OEM
10410-N0B013-1	OEM
10410-N0B013-2	OEM
10410-N0B013-3	OEM
10410-N0B013-4	OEM
10410-N0B013-5	OEM
10410-N0B013-6	OEM
10410-V0B011	OEM
10410-V0B011-1	OEM
10410-V0B011-2	OEM
10410-V0B011-3	OEM
10410-V0B011-4	OEM
10410-V0B011-5	OEM
10410-V0B011-6	OEM
10410-V0B011-7	OEM
10410-V0B011-8	OEM
10415AF	OEM
10415BF	OEM
10415F	OEM
10422AF	OEM
10422BF	OEM
10422F	OEM
10450BD	OEM
10470AF	OEM
10470F	OEM
10474AF	OEM
10474F	OEM
10508	0150
10514DM	OEM
10514FM	OEM
10515DM	OEM
10515FM	OEM
10516DM	OEM
10516FM	OEM
10523DM	OEM
10523FM	OEM
10524DM	OEM
10524FM	OEM
10525DM	OEM
10525FM	OEM
10534DM	OEM
10534FM	OEM
10541DM	OEM
10561DM	OEM
10562DM	OEM
10562FM	OEM
10564DM	OEM
10564FM	OEM
10565DM	OEM
10565FM	OEM
10571DM	OEM
10571FM	OEM
10572DM	OEM
10572FM	OEM
10573DM	OEM
10573FM	OEM
10574DM	OEM
10574FM	OEM
10592DM	OEM
10592FM	OEM
10650A01	0360
10655A01	1753
10655A03	0856
10655A05	0395
10655A13	0789
10655A15	OEM
10655B01	1753
10655B02	0789
10655B03	0856
10655B13	0167
10655B14	3965
10655C05	0395
10795-5	0081
10795-6	0141
10795-8	0562
10795-9	0358
10795-10	0333
10900-7359	0106
10909	0015

DEVICE TYPE	REPL CODE
10912A	OEM
10912AZ	OEM
10912B	OEM
10912C	OEM
10912D	OEM
10912E	OEM
10912F	OEM
10912Z	OEM
10938	OEM
11005	OEM
11112	0162
11200-1	0512
11202-1	0361
11203-1	0569
11204-1	1926
11205-1	0697
11206-1	1404
11207-1	0314
11208-1	0873
11209-1	0117
11211-1	0492
11213-1	1313
11214-1	1024
11216-1	0232
11229	0418
11233-2	1202
11236-1	0086
11236-3	0103
11252	0143
11252-0	0144
11252-1	0144
11252-2	0144
11252-3	0016
11252-4	0160
106951	0086
11252-5	0086
106952	0630
11252-6	0630
11273-1	0175
11274-1	0335
11276-1	1715
11303-9	0015
11332-1	0015
11339-2	0015
11339-8	0144
11393-8	0144
11399-8	0229
11401-3	0015
11426-7	0144
11457-4829	0532
11503-9	0015
11506-3	0160
11522-5	0016
11522-7	0050
11522-8	0050
11522-9	0050
11526-8	0160
11526-9	0160
11527-5	0136
11555-9	0015
11559-9	0015
11586-7	0229
11587-5	0016
11605-2	0015
11606-8	0975
11607-2	0136
11607-3	0144
11607-4	0016
11607-5	0142
11607-8	0016
11607-9	0144
11608-0	0144
11608-2	0144
11608-3	0144
11608-5	0016
11608-6	0975
11608-7	0975
11608-8	0975
11608-9	0975
11609-1	0004
11609-2	0016
11619-8	0144
11619-9	0144
11620-0	0144
11620-1	0144
11620-2	0050
11620-3	0004
11620-6	0004
11620-7	0050
11620-8	0050
11620-9	0050
11658-8	0016
11668-3	0050
11668-4	0050
11668-5	0004
11668-6	0004
11668-7	0208
11675-6	0136
11675-7	0004
11687-5	0016
11699-7	0004
11746	0015
12061-0013	1198
12110-0	0280
12110-3	0280
12110-4	0280
12110-5	0280
12110-6	0004
12110-7	0004

DEVICE TYPE	REPL CODE
12112-0	0004
12112-8	0280
12112-C	0016
12112-D	0079
12112-E	0016
12112-F	0016
12112C	0079
12112D	0079
12112E	0079
12112F	0079
12113-5	0050
12113-7	0050
12113-8	0050
12113-9	0050
12114-8	0004
12115-0	0050
12115-7	0050
12116-1	0280
12116-2	0280
12116-4	0004
12117-9	0280
12118-0	0280
12118-4	0004
12118-6	0280
12118-9	0136
12119-0	0050
12119-1	0004
12119-2	0004
12122-5	0004
12122-6	0004
12122-7	0004
12122-8	0050
12122-9	0050
12123-0	0050
12123-1	0050
12123-2	0004
12123-3	0050
12123-4	0280
12123-5	0280
12123-6	0280
12124-0	0280
12124-1	0280
12124-2	0136
12124-3	0136
12124-4	0136
12124-6	0004
12125-4	0280
12125-6	0050
12125-7	0136
12125-8	0050
12125-9	0050
12126-0	0050
12126-6	0004
12126-7	0004
12127-0	0160
12127-1	0160
12127-2	0004
12127-3	0280
12127-4	0004
12127-5	0280
12127-6	0016
12127-7	0016
12127-8	0016
12127-9	0016
12128-9	0280
12163	0085
12173	0280
12174	0280
12175	0280
12176	0280
12178	0085
12180	0136
12183	0280
12191	0280
12192	0280
12193	0280
12195	0004
12196	0004
12294B-11	OEM
12296B11	OEM
12429	0774
12536	0103
12536A	0103
12537	0491
12538	0178
12539	0161
12546	0233
12550	0087
12560	0795
12593	0016
12594	0037
12602	0015
12634	0015
12716	OEM
12720	0015
12732	OEM
12736	0015
12746	0015
12764	OEM
12768	0015
12786	0015
12788	0015
12803	0015
12808	0143
12837	0015
12844	0015
12850	0143
12871	0196
12888	0037

DEVICE TYPE	REPL CODE
12901	0188
12951-1	1854
13100BD2	OEM
13100BD4	OEM
13100BD8	OEM
13100BD16	OEM
13150BD2	OEM
13150BD4	OEM
13150BD8	OEM
13150BD16	OEM
13150L	0133
13184	OEM
13200BD2	OEM
13200BD4	OEM
13200BD8	OEM
13200BD16	OEM
13298	0103
13300BD8	OEM
13300BD16	OEM
13400BD	OEM
13579MC	OEM
13782	0015
14027	0015
14126-1	0015
14173I8	0861
14207	2368
14303	0016
14305	0624
14308	1196
14511	1544
14526	OEM
14529	4531
14568	OEM
14573	0440
14588-9	0015
14753A	OEM
14778-5	OEM
15009	0004
15021	OEM
15021Z	OEM
15024	0160
15027	0160
15101GOA	OEM
15101OOA	OEM
15102GOA	OEM
15102OOA	OEM
15103GOA	OEM
15103OOA	OEM
15104GOA	OEM
15104OOA	OEM
15105GOA	OEM
15105OOA	OEM
15106GOA	OEM
15106OOA	OEM
15122	0133
15240	0037
15352R	0030
15354-3	0085
15405-1	0141
15405-2	1844
15405-4	1041
15486	0086
15607GWF	OEM
15607GXF	OEM
15607HWF	OEM
15607HXF	OEM
15608GWF	OEM
15608GXF	OEM
15608HWF	OEM
15608HXF	OEM
15609GWF	OEM
15609GXF	OEM
15609HWF	OEM
15609HXF	OEM
15610GWF	OEM
15610GXF	OEM
15610HWF	OEM
15610HXF	OEM
15611GWF	OEM
15611GXF	OEM
15611HWF	OEM
15611HXF	OEM
15612GWF	OEM
15612GXF	OEM
15612HWF	OEM
15612HXF	OEM
15808-1	0233
15809-1	0198
15810-1	0079
15820-1	0198
15835-1	0198
15840-1	0016
15841-1	0016
15901GPF	OEM
15901GRF	OEM
15901GUF	OEM
15901HPF	OEM
15901HRF	OEM
15901HUF	OEM
15902GPF	OEM
15902GRF	OEM
15902GUF	OEM
15902HPF	OEM
15902HRF	OEM
15902HUF	OEM
15903GPF	OEM
15903GRF	OEM
15903GUF	OEM
15903HPF	OEM

DEVICE TYPE	REPL CODE
15903HRF	OEM
15903HUF	OEM
15904GPF	OEM
15904GRF	OEM
15904GUF	OEM
15904HPF	OEM
15904HRF	OEM
15904HUF	OEM
15905GPF	OEM
15905GRF	OEM
15905GUF	OEM
15905HPF	OEM
15905HRF	OEM
15905HUF	OEM
15906GPF	OEM
15906GRF	OEM
15906GUF	OEM
15906HPF	OEM
15906HRF	OEM
15906HUF	OEM
15907GRF	OEM
15907GUF	OEM
15908GRF	OEM
15908GUE	OEM
15908GUF	OEM
15909GRF	OEM
15909GUF	OEM
15910GRF	OEM
15910GUF	OEM
15911GRF	OEM
15911GUF	OEM
15912GRF	OEM
15912GUF	OEM
15927	0085
15942	0723
15942-1	1876
16002-039	0143
16073	2387
16088	1208
16111	0576
16162	0415
16170	0576
16208	0576
16238	0147
16266	0103
16289	0576
16291	0576
16598	0164
16599	0160
16605	0753
16681	0015
16778	2457
16958	0211
16959	0160
17002	0015
17011-1	0725
17042	0280
17043	0280
17047-1	0211
17142	0004
17143	0004
17144	0198
17246-33	1413
17401GOA	1084
17402GOA	1084
17403GOA	1086
17404GOA	1086
17405GOA	1086
17406GOA	1086
17407GOA	1102
17408GOA	1102
17409GOA	1102
17411GOA	1102
17412-5	0016
17412GOA	1102
17413GOA	OEM
17414GOA	OEM
17443	0015
17444	0076
17607IWA	OEM
17607IXA	OEM
17608IWA	OEM
17608IXA	OEM
17609IWA	OEM
17609IXA	OEM
17610IWA	OEM
17610IXA	OEM
17611IWA	OEM
17611IXA	OEM
17612IWA	OEM
17612IXA	OEM
17887	0160
17901HRA	OEM
17901HUA	OEM
17902HRA	OEM
17902HUA	OEM
17903HRA	OEM
17903HUA	OEM
17904HRA	OEM
17904HUA	OEM
17905HRA	OEM
17905HUA	OEM
17906HRA	OEM
17906HUA	OEM
17945	0085
18031	OEM
18109	0004
18112	0079
18310	0142

DEVICE TYPE	REPL CODE
18410-41	0030
18410-42	0123
18410-43	0165
18410-141	1280
18410-142	0937
18410-143	0076
18410-144	0155
18410-145	0076
18410-146	1419
18410-147	0390
18410-148	2755
18493	0050
18509	0016
18529	0280
18530	0211
18540	0050
18541	0050
18555	0076
18600-51	0030
18600-52	0133
18600-53	0019
18600-54	0165
18600-151	0937
18600-152	0076
18600-153	0076
18600-155	1478
18600-156	2124
18600-I51	0127
18601	0211
18611	0164
18731	0211
18818R3383	OEM
18820S3367	OEM
19042	0015
19278	0142
19420	0144
19645	0079
19865	0015
20013	OEM
20013A	OEM
20014A	OEM
20061	OEM
20082	OEM
20082-0	OEM
20100	OEM
20106	OEM
20112	2870
20125	OEM
20129	OEM
20133	OEM
20139	OEM
20174	OEM
20180	OEM
20187	OEM
20192	OEM
20294	2997
20295	2997
20333	OEM
20452	OEM
20453	OEM
20454	OEM
20505S	1107
20572-307	OEM
20637	OEM
20738	0198
20739	0126
20810-21	0015
20810-22	0015
20810-91	0111
20810-92	0111
20810-93	0086
20810-94	0560
20881	OEM
20957	OEM
20969	OEM
21035H-177	OEM
21123	OEM
21157	OEM
22008	0037
22009	0400
22158	0016
22881	1010
22939	0168
23100BD	OEM
23114-046	0016
23114-050	0126
23114-051	0126
23114-052	0233
23114-053	0016
23114-054	0016
23114-056	0144
23114-057	0144
23114-060	0144
23114-061	0004
23114-070	0103
23114-078	0144
23114-082	0016
23114-095	0016
23114-097	0975
23114-104	0144
23115-042	0071
23115-057	0016
23115-058	0016
23115-072	0071
23115-085	0087
23125-037	0144
23128	OEM
23128-25I	OEM
23128-25N	OEM

If replacement code is OEM, contact original manufacturer for replacement.

DEVICE TYPE	REPL CODE
23128-30I	OEM
23128-30N	OEM
23128-45I	OEM
23128-45N	OEM
23128B	OEM
23128E25I	OEM
23128E25N	OEM
23128E30I	OEM
23128E30N	OEM
23128E45I	OEM
23128E45N	OEM
23128M30I	OEM
23128M45I	OEM
23150BD	OEM
23200BD	OEM
23209-110-760	OEM
23300BD	OEM
23311-006	0160
23350BD	OEM
23400BD	OEM
23402N	OEM
23450BD	OEM
23555	OEM
23606	0325
23754	0042
23762	0103
23785	0136
23785-1	0136
23826	0037
23826(SYLVANIA)	0037
23879	0764
24001GOF	0192
24001IOF	OEM
24002	0127
24002GOF	0096
24002IOF	OEM
24003GOF	0096
24003IOF	OEM
24004GOF	0096
24004IOF	OEM
24005GOF	0159
24005IOF	OEM
24006GOF	0159
24006IOF	OEM
24007GOF	0096
24007IOF	OEM
24008GOF	0096
24008IOF	OEM
24009GOF	0096
24009IOF	OEM
24010GOF	0096
24010IOF	OEM
24011GOF	0096
24011IOF	OEM
24012GOF	0096
24012IOF	OEM
24013GOF	OEM
24013IOF	OEM
24014GOF	OEM
24014IOF	OEM
24198	0136
24198(RECT.)	0015
25114-101	0004
25114-102	0004
25114-103	0004
25114-104	0004
25114-116	0016
25114-121	0144
25114-130	0016
25114-143	0142
25114-161	0079
25115-115	0015
25201-001	0143
25202-002	0015
25211-200	0649
25260-61-067	0133
25642-020	0136
25642-030	0136
25642-031	0136
25642-040	0136
25642-041	0136
25642-110	0136
25642-115	0136
25642-120	0136
25651-020	0004
25651-021	0004
25651-033	0004
25655-055	0004
25655-056	0004
25657-050	0004
25658-120	0222
25658-121	0222
25661-020	0160
25661-022	0085
25671-020	0007
25671-021	0007
25671-023	0007
25672-016	0233
25810-51	0015
25810-52	0133
25810-53	0143
25810-54	0165
25810-55	0143
25810-107	1160
25810-161	0076
25810-162	0076
25810-163	0076
25810-164	0911
25810-165	1478

DEVICE TYPE	REPL CODE
25810-167	1160
25840-51	0015
25840-52	0133
25840-53	0143
25840-54	0244
25840-55	0143
25840-161	0016
25840-162	0111
25840-163	0076
25840-165	1478
25840-167	1160
26011B-575	OEM
26011C-699	OEM
26011D-717	OEM
26011D-718	OEM
26012B-840	OEM
26128A-35F	OEM
26128A-35I	OEM
26128A-35N	OEM
26128A-45F	OEM
26128A-45I	OEM
26128A-45N	OEM
26810-51	0123
26810-52	0064
26810-151	0076
26810-151(1)	0007
26810-152	0006
26810-153	0155
26810-154	0076
26810-155	0018
26810-156	2019
26810-157	0520
26810-158	0433
26810-159	1050
27113-100	0015
27123-050	0015
27123-070	0015
27123-100	0015
27123-120	0015
27125-080	0016
27125-090	0016
27125-110	0086
27125-120	0004
27125-140	0016
27125-150	0004
27125-160	0016
27125-170	0004
27125-270	0016
27125-300	0016
27125-310	0208
27125-330	0004
27125-340	0004
27125-350	0004
27125-360	0280
27125-370	0016
27125-380	0086
27125-460	0111
27125-470	0111
27125-480	0004
27125-490	0208
27125-500	0016
27125-530	0016
27125-540	0004
27125-550	0004
27126-060	0222
27126-090	0160
27126-100	0103
27127-550	0016
27128	1638
27128A	1638
27128G-250	1638
27256	OEM
27256/R09-32015	OEM
27256/R09-32017	OEM
27256-1	OEM
27256-2FXI-A5	OEM
27256-200	OEM
27256-200(JR15C1)	OEM
27256-200(JR15C2)	OEM
27256-200(JR15M)	OEM
27256-250	OEM
27256-XX	OEM
27256D	OEM
27512	OEM
27512-150NS	OEM
27840-41	0015
27840-42	0143
27840-43	0143
27840-161	0079
27840-162	0079
27840-164	1160
27901IXF	OEM
27902IXF	OEM
27903IXF	OEM
27904IXF	OEM
27905IXF	OEM
27906IXF	OEM
27907IWF	OEM
27908IWF	OEM
27909IWF	OEM
27910IWF	OEM
27911IWF	OEM
27912IWF	OEM
27913-14200	0415
28287	0143
28311	0233
28778	OEM
28810-61	0015
28810-62	0030

DEVICE TYPE	REPL CODE
28810-63	0133
28810-64	0123
28810-65	0244
28810-171	0325
28810-172	0127
28810-173	0079
28810-174	0079
28810-175	2261
28810-176	1478
28810-177	0433
28810-178	1160
29611ADC	2389
29611ADM	OEM
29611AFM	OEM
29611DC	2389
29611DM	OEM
29611FM	OEM
29613ADC	OEM
29613ADM	OEM
29613AFM	OEM
29613DC	OEM
29613DM	OEM
29613FM	OEM
29621ADC	2320
29621ADM	OEM
29621AFM	OEM
29621DC	2320
29621DM	OEM
29621FM	OEM
29623ADC	OEM
29623ADM	OEM
29623AFM	OEM
29623DC	OEM
29623DM	OEM
29623FM	OEM
29631ADC	2909
29631ADM	OEM
29631AFM	OEM
29631DC	2909
29631DM	OEM
29631FM	OEM
29633ADC	OEM
29633ADM	OEM
29633AFM	OEM
29633DC	OEM
29633DM	OEM
29633FM	OEM
29641ADC	OEM
29641ADM	OEM
29641AFM	OEM
29641DC	OEM
29641DM	OEM
29641FM	OEM
29643ADC	OEM
29643ADM	OEM
29643AFM	OEM
29643DC	OEM
29643DM	OEM
29643FM	OEM
29651ADC	3166
29651ADM	OEM
29651AFM	OEM
29651DC	3166
29651DM	OEM
29651FM	OEM
29653ADC	OEM
29653ADM	OEM
29653AFM	OEM
29653DC	OEM
29653DM	OEM
29653FM	OEM
29671ADC	OEM
29671ADM	OEM
29671AFM	OEM
29671DC	OEM
29671DM	OEM
29671FM	OEM
29673DC	OEM
29673DM	OEM
29673FM	OEM
29681ADC	OEM
29681ADM	OEM
29681AFM	OEM
29681ASC	OEM
29681ASM	OEM
29681DC	OEM
29681DM	OEM
29681FM	OEM
29681SC	OEM
29681SM	OEM
29683ADC	OEM
29683ADM	OEM
29683AFM	OEM
29683ASC	OEM
29683ASM	OEM
29683DC	OEM
29683DM	OEM
29683FM	OEM
29683SC	OEM
29683SM	OEM
29693DC	OEM
29693DM	OEM
29693FM	OEM
29810-63	0143
29810-173	0016
29810-178	0219
29810-179	2755
29810-180	0520
30001GOF	0192

DEVICE TYPE	REPL CODE
30001IOF	OEM
30002GOF	0192
30002IOF	OEM
30003GOF	4228
30003IOF	OEM
30004GOF	4228
30004IOF	OEM
30005GOF	0159
30005IOF	OEM
30006GOF	0159
30006IOF	OEM
30007GOF	0096
30007IOF	OEM
30008GOF	0096
30008IOF	OEM
30009GOF	0096
30009IOF	OEM
30010GOF	0096
30011GOF	0096
30011IOF	OEM
30012GOF	0096
30012IOF	OEM
30013GOF	OEM
30013IOF	OEM
30014GOF	OEM
30014IOF	OEM
30201	0164
30202	0164
30203	0160
30204	0211
30206	0211
30207	0211
30208	0847
30208-1	0004
30208-2	0211
30210	0076
30211	0160
30213	0050
30214	0050
30215	0136
30215(RCA)	0160
30216	0211
30216(RCA)	0160
30217	0136
30218	0211
30218(RCA)	0050
30219	0198
30221	0136
30222	0136
30223	0136
30224	0198
30226	0198
30227	0016
30228	0016
30229	0016
30230	0136
30231	0280
30234	0142
30235	0111
30236	0178
30238	0050
30239	0050
30240	0050
30241	0079
30242	0016
30243	0079
30244	0211
30245	0142
30246	0160
30246A	0160
30247	0136
30248	0198
30253	0016
30254	0178
30256	0178
30257	0142
30259	0198
30263	0004
30267	2866
30268	0016
30269	0016
30270	0037
30271	0460
30272	0161
30273	0050
30274	0136
30276	0103
30278	0126
30289	0079
30290	0037
30291	4163
30292	0144
30293	0004
30294	0042
30302	0085
30302(RCA)	0004
30498	OEM
30510	0071
30613	0079
30613(RCA)	0079
30710	0167
30901HRF	OEM
30901HUF	OEM
30902HRF	OEM
30902HUF	OEM
30903HRF	OEM
30903HUF	OEM
30904HRF	OEM
30904HUF	OEM

DEVICE TYPE	REPL CODE
30905HRF	OEM
30905HUF	OEM
30906HRF	OEM
30906HUF	OEM
31001	0016
31003	0016
31004	OEM
31013-11	0715
31013E	OEM
31018-277-530	0037
31018-277-560	0471
31450	0242
32040	OEM
32109-101-005	OEM
32109-101-006	6073
32109-101-020	OEM
32109-101-070	OEM
32109-101-100	OEM
32109-101-140	OEM
32109-101-300	OEM
32109-101-310	OEM
32109-101-440	OEM
32109-101-490	OEM
32109-101-550	OEM
32109-110-007	OEM
32109-110-011	OEM
32109-110-084	OEM
32109-301-070	1527
32109-301-610	4310
32109-301-890	0035
32109-410-270	0101
32109-410-280	0101
32109-1001-100	OEM
32117-401-120	OEM
32117-901-110	0334
32119-101-080	3367
32119-101-640	4102
32119-101-870	OEM
32119-101-880	OEM
32119-101-940	0912
32119-102-010	0462
32119-102-020	0486
32119-102-040	0523
32119-102-050	0539
32119-102-060	0557
32119-102-090	0539
32119-102-100	0557
32119-102-120	OEM
32119-102-260	0912
32119-102-300	0976
32119-102-420	OEM
32119-102-570	0976
32119-102-730	1529
32119-102-760	0334
32119-103-100	OEM
32119-103-110	5272
32119-103-420	2178
32119-110-010	OEM
32119-110-016	2178
32119-110-018	OEM
32119-110-050	4360
32119-110-098	OEM
32119-201-180	OEM
32119-201-190	OEM
32119-201-310	0035
32119-201-360	0430
32119-401-010	4189
32119-401-120	1346
32119-401-200	OEM
32119-401-210	OEM
32119-710-019	OEM
32119-801-130	2364
32119-901-020	3957
32119-901-110	0334
32119-901-130	2364
32119-901-220	OEM
32119-910-013	OEM
32137-103-280	0037
32137-103-430	0037
32137-301-010	0261
32137-301-050	0144
32137-301-080	1092
32137-301-090	6657
32137-301-110	3938
32137-301-180	0079
32137-301-530	0261
32137-301-540	0261
32137-301-550	1624
32137-301-560	1624
32137-301-720	0079
32137-401-530	0037
32139-101-420	0148
32139-101-530	0006
32139-301-090	0842
32139-301-140	0261
32139-301-230	0168
32139-301-240	0168
32139-301-350	0168
32139-301-510	0261
32139-302-350	0168
32139-302-810	1284
32139-304-560	0866
32139-304-570	0866
32139-401-100	1541
32139-401-110	1541
32139-401-130	1541
32139-410-006	0643
32147-301-020	1284
32149-101-850	1911

DEVICE TYPE	REPL CODE
32149-101-860	1911
32149-201-450	1648
32149-301-010	3542
32149-301-020	1284
32149-301-040	1541
32149-301-050	0556
32149-301-090	1165
32149-301-140	2797
32149-301-190	0643
32149-401-040	1541
32149-401-050	0556
32149-401-150	OEM
32149-403-020	5330
32149-404-020	0055
32149-404-030	0055
32157-301-170	0079
32159-201-010	OEM
32159-210-002	1541
32159-210-003	0556
32159-210-005	0556
32159-210-006	1541
32159-301-020	0076
32159-301-030	0261
32159-301-170	0079
32159-301-180	0079
32159-301-230	3354
32159-301-290	0261
32159-701-020	0076
32167-101-130	2232
32167-103-080	0023
32167-110-002	0071
32167-201-070	0023
32167-201-110	OEM
32167-201-160	0015
32167-201-170	0023
32167-201-180	0015
32167-201-250	OEM
32167-206-110	0023
32167-206-670	0023
32167-207-120	3087
32167-208-510	0023
32167-208-520	1090
32167-208-550	0023
32167-208-570	0023
32167-208-580	0071
32167-218-130	0023
32167-301-630	0023
32167-301-670	OEM
32167-401-630	0053
32167-401-800	0253
32167-401-860	OEM
32167-401-870	0118
32167-401-880	OEM
32167-402-070	0023
32167-403-560	0471
32167-403-850	0039
32167-405-480	0124
32167-406-040	OEM
32167-406-080	0041
32167-406-110	0023
32167-406-130	0057
32167-406-190	OEM
32167-406-480	0124
32167-406-630	0053
32167-407-480	0124
32167-408-060	6366
32167-408-080	0471
32167-441-011	0062
32169-101-050	0102
32169-101-080	0123
32169-101-090	1127
32169-101-100	1045
32169-101-130	2232
32169-101-180	0023
32169-101-220	0965
32169-201-200	0017
32169-201-230	0031
32169-201-240	0102
32169-202-220	0023
32169-206-110	0023
32169-207-110	0071
32169-208-510	0023
32169-208-530	0071
32169-208-570	0023
32169-208-580	0071
32169-218-130	0023
32169-221-002	3887
32169-301-020	0076
32169-301-090	0102
32169-301-610	0031
32169-301-630	0023
32169-301-670	0017
32169-401-070	0023
32169-403-560	0471
32169-403-850	0039
32169-403-920	0094
32169-404-200	0041
32169-408-080	0471
32169-408-570	0023
32189-106-120	OEM
32199-101-020	OEM
32199-411-090	OEM
32199-500-000	OEM
32199-500-001	OEM
32199-901-200	2289
32309-024-010	1291
32309-024-020	OEM
32309-024-040	OEM
32309-101-090	OEM

DEVICE TYPE	REPL CODE
32309-101-100	OEM
32309-110-001	0600
32309-110-230	OEM
32309-110-420	OEM
32309-110-430	OEM
32309-110-760	OEM
32309-201-003	OEM
32309-202-003	OEM
32310	OEM
32310ZD	OEM
32980	0678
33010	0037
33011	0037
33137-103-430	0037
33324	0030
33511DC	OEM
33512DC	OEM
33563	0016
33571	OEM
33572	OEM
33952	0030
34001PC	0478
34002PC	2055
34011PC	0215
34012PC	0498
34013PC	0413
34014DC	0860
34014FC	OEM
34014FM	OEM
34014PC	0860
34015DC	1014
34015DM	1014
34015FC	OEM
34015FM	OEM
34015PC	1014
34016PC	1143
34017PC	0513
34019PC	1525
34020PC	1661
34021DC	5969
34021DM	OEM
34021FC	OEM
34021FM	OEM
34021PC	1749
34022	0160
34023PC	0519
34025PC	2072
34027PC	1949
34028PC	2228
34029PC	2233
34030PC	0500
34035DC	OEM
34035DM	OEM
34035FC	OEM
34035FM	OEM
34035PC	OEM
34040PC	0056
34042PC	0121
34049PC	0001
34050PC	0398
34051PC	0366
34052PC	0024
34055PC	3304
34066PC	0101
34068PC	2503
34069PC	0119
34070PC	2516
34071PC	0129
34077PC	2558
34078PC	0921
34081PC	0626
34085PC	0303
34086PC	0535
34098	0136
34099	0136
34099PC	3330
34100	0136
34101	0740
34118	0050
34119	0280
34174	0015
34201	0087
34219	0280
34220	0280
34221	0280
34298	0160
34315	0160
34342	0050
34389	0050
34394	0015
34401-1	OEM
34405	0242
34423	0050
34425	0160
34512PC	2120
34518PC	1043
34519-992-010	OEM
34520PC	2675
34527PC	3145
34539PC	3656
34553	0050
34555PC	2938
34556PC	3431
34715	0085
34720DC	OEM
34720DM	OEM
34720FC	OEM
34720FM	OEM
34720PC	OEM

If replacement code is OEM, contact original manufacturer for replacement.

DEVICE TYPE	REPL CODE
34725DC	OEM
34725DM	OEM
34725FC	OEM
34725FM	OEM
34725PC	OEM
34894	0015
34923	0164
34942	0050
35004	0144
35044	0160
35045	0211
35059-2	0167
35070	0050
35084	0085
35086	0211
35144	0160
35168	0136
35169	0050
35170	0050
35201	0085
35218	0847
35219	0137
35231	0160
35242	0079
35259	0127
35260	0085
35287	0015
35287A	0015
35289	0015
35306	0015
35333	0071
35349	0160
35391DC	OEM
35392DC	2642
35405	0178
35449	0144
35452-2	0164
35454	0211
35454-1	0164
35454-2	0164
35454-3	0164
35500	0015
35590	0004
35604	0540
35628	0004
35677	0004
35678	0004
35682-104-620	2289
35683-112-220	2178
35684-107-534	0334
35684-114-010	0643
35684-116-010	1165
35684-117-710	OEM
35684-119-710	0643
35728	0160
35792	0004
35815	0136
35816	0004
35817	0004
35818	0136
35819	0164
35820	0211
35820-1	0164
35820-2	0164
35820-3	0164
35824	0004
35824-	OEM
35860	0015
35885A	0160
35885B	0160
35950	0211
35951	0160
35952	0211
35953	0211
35954	0211
35955	0211
36091R	0816
36147	0242
36201	0015
36203	0160
36212	0144
36212V1	0144
36213	0551
36303	0160
36304	0160
36304-4	0160
36312	0160
36316	OEM
36344	0142
36387	0086
36395	0160
36446-2	1395
36477	0160
36503	0015
36503(AIRCASTLE)	0015
36508	0143
36534	0847
36535	0229
36537	0229
36539	0137
36549	0015
36554	0015
36555	0015
36557	0164
36558	0004
36559	0050
36560	0050
36563	0136
36564	0015
36577	0126
36578	0144
36579	0086
36580	0016
36581	0144
36582	0325
36591	0242
36634	0142
36810-151	OEM
36816	0280
36847	0144
36854A	OEM
36896	0085
36910	0160
36913	0169
36917	0016
36918	0144
36919	0144
36920	0016
36921	0016
37126-1	0015
37278	0050
37279	0208
37334	0007
37383	0144
37384	0144
37464	0086
37510-51	OEM
37510-52	0133
37510-53	0123
37510-54	0244
37510-161	0111
37510-162	0079
37510-163	0079
37510-164	0911
37510-165	4009
37510-166	0433
37510-167	1160
37549	0004
37550	0004
37551	0004
37552	0208
37584	0142
37585	0198
37649	0086
37680	0015
37682	0725
37694	0086
37730	0142
37800	0086
37847	0086
37884	0016
37987	0015
37992	0764
38052	0242
38074	0540
38091	0164
38095	0037
38120	0233
38121	0233
38122	0698
38175	0038
38176	0004
38177	0211
38178	0198
38207	0144
38208	0144
38209	0280
38246	0007
38270	0086
38271	0086
38272	0103
38332	0415
38378	0849
38385	0965
38448	0560
38510-51	0720
38510-52	0123
38510-53	0064
38510-54	0139
38510-161	1280
38510-162	0937
38510-163	0076
38510-164	0155
38510-165	0218
38510-166	0076
38510-167	1419
38510-168	0219
38510-169	2755
38510-170	0520
38510-171	1018
38510-330	0133
38510-331	0157
38510-350	0111
38511	0144
38563	0212
38608	0725
38609	0725
38716	0042
38731(KALOF)	0103
38785	0144
38786	0144
38787	0007
38788	0016
38789	0698
38840	0415
38881-D13	0062
38885	0740
38890	0415
38897	0103
38920	0144
38921	0144
38927	0415
38927-00000	2698
38970	0764
38976	0740
39005	0725
39039	0415
39096	0016
39119	0740
39209-1	1337
39264	2730
39299	0740
39458	0161
39510	0276
39542	1413
39558	2017
39669	0415
39750	0042
39763	2730
39767	0042
39804	0242
39805	0415
39811	1413
39830	0725
39840	0415
39901-0001	0440
39902	2270
39952	2017
40004	0050
40005	0050
40006	0050
40014BDC	OEM
40014BDM	OEM
40014BFC	OEM
40014BFM	OEM
40014BPC	OEM
40022	0160
40024	0015
40024VM	0015
40034	0004
40034-1	0004
40034-2	0004
40034-3	0004
40034VM	0004
40035	0004
40035-1	0004
40035-2	0004
40035-3	0004
40036	0004
40036-1	0004
40036-2	0004
40036-3	0004
40037	0208
40037-1	0208
40037-2	0208
40037-3	0208
40037VM	0208
40038	0004
40038-1	0004
40038-2	0004
40038-3	0004
40038VM	0004
40039T	OEM
40040T	OEM
40041T	OEM
40042	OEM
40043	OEM
40044	OEM
40049	OEM
40050	0160
40051	0160
40051-2	0160
40053	0086
40078	OEM
40080	0086
40081	1411
40082	2700
40082A	0698
40084	0016
40085BDC	3450
40085BDM	3450
40085BFC	3450
40085BFM	3450
40085BPC	3450
40097BDC	6007
40097BDM	OEM
40097BFC	OEM
40097BFM	OEM
40097BPC	6007
40098BDC	0432
40098BDM	OEM
40098BFC	OEM
40098BFM	OEM
40098BPC	0432
40108	0084
40108R	0534
40109	0090
40109R	0748
40110	0097
40110R	0538
40111	0105
40111R	0777
40112	0109
40112R	0538
40113	0116
40113R	0801
40114	1112
40114R	0816
40115	0131
40115R	0545
40143	0015
40160BDC	1359
40160BDM	1359
40160BFC	1359
40160BFM	1359
40160BPC	1359
40161BDC	1373
40161BDM	1373
40161BFC	1373
40161BFM	1373
40161BPC	1373
40162BDC	1388
40162BDM	1388
40162BFC	1388
40162BFM	1388
40162BPC	1388
40163BDC	1407
40163BDM	1407
40163BFC	1407
40163BFM	1407
40163BPC	1407
40192BDC	1764
40192BDM	1764
40192BFC	1764
40192BFM	1764
40192BPC	1764
40193BDC	1776
40193BDM	1776
40193BFC	1776
40193BFM	1776
40193BPC	1776
40194BDC	6609
40194BDM	OEM
40194BFC	OEM
40194BFM	OEM
40194BPC	1769
40195BDC	1784
40195BDM	OEM
40195BFC	OEM
40195BFM	OEM
40195BPC	1784
40208	3189
40208R	1630
40209	0805
40209R	0254
40210	1124
40210R	1107
40211	1126
40211R	1111
40212	0805
40212R	0258
40213	1195
40213R	1644
40214	0319
40214R	0267
40216	0725
40217	0016
40218	0016
40219	0016
40220	0016
40221	0016
40222	0016
40231	0007
40232	0007
40233	0007
40234	0007
40235	0007
40236	0007
40237	0007
40238	0007
40238-3GT	0015
40239	0007
40240	0007
40242	0007
40243	0007
40244	0007
40245	0007
40246	0007
40250	0429
40250V1	0178
40251	0103
40253	0004
40254	0085
40255	3280
40256	0338
40259	0007
40259(RECT)	0122
40260	0007
40261	0050
40262	0050
40263	0004
40263(RCA)	0050
40264	0142
40264V1	0142
40265	0015
40266	0953
40267	0242
40268	0136
40269	0280
40279	0560
40280	2041
40280R	0267
40281	3321
40282	3322
40283	0016
40290	3326
40291	0423
40292	0042
40294	0007
40295	0259
40296	0259
40305	0560
40306	0560
40307	0042
40309	0086
40309L	0630
40309S	0630
40309V1	0086
40309V2	0086
40310	0178
40310V1	0178
40311	0086
40311A	0551
40311B	0551
40311L	0630
40311S	0630
40312	0178
40312V1	0178
40313	0142
40314	0086
40314L	0630
40314S	0630
40314V1	0551
40314V2	0551
40315	0086
40315L	0630
40315S	0630
40315V1	0551
40315V2	0551
40316	0178
40317	0086
40317L	0595
40317S	0595
40317V1	0551
40317V2	0551
40318	0142
40319	0126
40319L	0791
40319S	0126
40319V1	0551
40319V2	0551
40320	0086
40320L	0595
40320S	0595
40320V1	0551
40320V2	0551
40321	0187
40321L	0168
40321V1	0187
40321V2	0187
40322	1027
40323	0086
40323L	0630
40323S	0630
40323V1	0551
40323V2	0551
40324	0178
40325	0103
40326	0086
40326L	0595
40326S	0595
40326V2	0551
40327	0439
40327L	0168
40327V2	0551
40328	0142
40329	0004
40346	0187
40346L	0168
40346S	1978
40346V1	0338
40346V2	3280
40347	1480
40347L	0630
40347S	0018
40347V1	0560
40347V2	3280
40348	0168
40348V1	0423
40348V2	3280
40349	0168
40349V2	3280
40350	0007
40351	0007
40352	0007
40354	0233
40355	0233
40359	0004
40360	0086
40360L	0630
40360S	0630
40360V1	0551
40360V2	0551
40361	0086
40361L	0630
40361V1	3426
40361V2	0551
40361V3	0551
40362	0126
40362S	0126
40362V1	0382
40362V2	0382
40363	0103
40364	0178
40366	0086
40366L	0168
40366S	0630
40367	1480
40368	0042
40369	0103
40372	0178
40373	0178
40374	0142
40375	3362
40378	1487
40379	0740
40383	0015
40385	0187
40385L	0168
40385V1	0551
40385V2	0551
40389	0560
40390	0338
40391	0126
40392	0086
40394	0532
40395	0004
40396	0038
40396/N	0208
40396/P	0164
40396N	0208
40396P	0004
40397	0016
40398	0016
40399	0016
40400	0016
40403	0211
40404	0418
40405	0016
40406	0126
40406L	0791
40406S	0126
40407	0086
40407L	0595
40407S	0595
40408	0126
40408L	0630
40408S	0630
40409	0086
40410	0126
40411	0130
40412	0187
40412L	0168
40412S	OEM
40412V1	0168
40412V2	0187
40413	0144
40414	0144
40414V2	OEM
40421	0160
40422	0142
40423	0142
40424	0178
40425	0142
40426	0142
40427	0142
40428	OEM
40429	0744
40430	3383
40437V1	0551
40437V2	0551
40438V1	0264
40438V2	0264
40439	0338
40439V1	0338
40439V2	0338
40440	0975
40442	0975
40444	2485
40446	0630
40450	0595
40451	0595
40452	0595
40453	0595
40454	0595
40455	0595
40456	0595
40457	0086
40458	0086
40459	0233
40460	OEM
40461	0325
40461-2	0086
40462	0085
40464	0103
40465	0103
40466	0103
40467A	0849
40468	0325
40468A	0849
40469	0493
40470	0144
40471	0493
40472	0007
40473	0016
40474	0016
40475	0007
40476	0007
40477	0016
40478	0007
40479	0007
40480	0144
40481	0007
40482	0144
40485	2394
40487	0050
40488	0136
40489	0050
40490	0004
40491	0142
40500	0016
40501	0086
40502	0411
40503	0415
40504	2270
40505	3402
40506	3404
40507	2270
40508	3402
40509	0411
40510	0415
40511	0411
40512	0415
40513	0103
40514	0103
40517	0198
40518	0198
40519	0016
40525	2352
40526	2353
40527	2356
40528	2352
40529	2353
40530	2356
40531	2383
40532	2353
40533	1413
40534	2352
40535	2394
40536	1413
40537	0126
40537L	0126
40537S	0126
40538	0126
40538L	0126
40538S	0126
40539	0086
40539A	OEM
40539L	0086
40539S	0086
40542	0103
40543	0103
40544	3280
40546	0142
40547	0142
40553	0239
40554	0239
40555	0239
40559	0849
40559A	0849
40561	OEM
40562	OEM
40563	OEM
40564	OEM
40565	OEM
40566	OEM
40567	OEM
40568	OEM
40569	OEM
40570	OEM
40571	OEM
40572	OEM
40573	OEM
40574	OEM
40577	0016
40578	0086
40581	2700
40582	0630
40583	2150
40594	0551
40595	0892
40598	OEM
40598A	OEM
40600	0353
40601	0353
40602	0353
40603	0353
40604	0353
40605	0560
40608	0418
40611	0086
40611S	0630
40612	0160
40613	0042
40616	0086
40616L	0630
40616S	0630
40618	0042
40621	0042
40622	0042
40623	0265
40624	0423
40625	0086
40626	0160
40627	0423
40628	0086
40629	0042
40630	0042
40631	0042
40632	0423
40633	0103
40634	0126
40634L	0382

DEVICE TYPE	REPL CODE
40634S	0382
40635	0086
40635L	0630
40635S	0630
40636	0543
40637	0016
40637A	0018
40638	0411
40639	0415
40640	0239
40641	0061
40642	1217
40643	1217
40644	1217
40650	0285
40654	2095
40655	0710
40662	2015
40663	2017
40664	0178
40665	0042
40666	0560
40667	3473
40668	3474
40669	0576
40672	2018
40673	0212
40675	OEM
40680	1848
40681	1848
40682	1855
40683	3215
40684	2383
40685	2394
40686	1413
40687	2387
40689	2018
40690	2018
40691	2353
40692	2356
40696	2383
40697	2394
40698	1413
40702	2015
40703	2017
40704	2018
40707	2015
40708	2017
40710	2018
40713	0154
40714	0147
40716	0147
40719	0154
40720	0147
40721	0593
40722	0576
40724	OEM
40731	3492
40732	0415
40733	0016
40734	0415
40735	0804
40736R	OEM
40737	1651
40738	1651
40739	1584
40740	1665
40741	2449
40742	0435
40743	1487
40744	0687
40745	3499
40746	1848
40747	1855
40748	3215
40749	1650
40750	1651
40751	1584
40752	1665
40753	0741
40754	0745
40755	0154
40756	0764
40757	1848
40758	1650
40759	2647
40760	0764
40763	0004
40766	2352
40769	2353
40770	2387
40771	2394
40777	0154
40778	0147
40781	0154
40782	0147
40785	0154
40786	0147
40789	3150
40790	2018
40793	2015
40794	2017
40796	0279
40798	0279
40801	0279
40808	0071
40809	0071
40814	0086
40815	0126

DEVICE TYPE	REPL CODE
40816	0561
40819	2458
40820	2458
40821	2458
40822	2458
40823	0212
40828	0004
40829	0905
40833	0863
40836	0264
40837	0264
40841	0353
40842	0577
40843R	OEM
40844R	OEM
40850	0142
40852	0363
40852(VM)	0004
40853	0363
40853(VM)	0164
40854	0642
40867	2089
40868	0505
40869	0710
40871	0236
40872	0681
40874	0324
40875	0561
40877	0561
40878	1199
40880	0239
40881	0060
40882	1308
40884	3560
40885	1708
40886	1708
40887	1708
40888	0239
40889	0061
40890	1217
40891	1217
40892	1217
40894	0259
40895	0259
40896	0259
40897	0259
40899	0061
40900	0571
40901	0593
40902	0576
40910	0178
40911	0178
40912	0142
40913	3553
40915	0861
40934-1	0103
40938	0750
40953	0698
40954	0892
40955	3041
40964	0418
40965	0689
40967	3560
40968	1420
40970	3589
40971	2868
40972	3421
40973	1233
40974	1977
40975	0698
40976	0698
40977	3587
40979	OEM
40980	OEM
41001-4	0015
41004	OEM
41008	0418
41008A	3560
41009	0418
41009A	3560
41010	1420
41012	3554
41013	3554
41014	0593
41015	0576
41020(RCA)	1712
41020-618	0071
41021(RCA)	1712
41023(RCA)	1712
41023-224	0071
41023-225	0071
41051	0016
41052	0126
41053	0086
41178	0161
41179	0460
41201	OEM
41257-P-15	OEM
41342	0460
41344	0161
41464-15	OEM
41500	0042
41501	0854
41502	0086
41503	0126
41504	0042
41505	1708
41506	0074
41508	3822

DEVICE TYPE	REPL CODE
41616	0015
41689	0144
41694	0144
42020	0143
42020-737	0170
42021	0071
42025-850	0137
42065	0178
42221	0015
42302	0136
42304	0164
42305	0004
42311	0136
42321	0847
42322	0004
42323	0142
42324	0211
42342	0178
42396	0325
42464	0079
42942	0042
43021-017	0016
43021-198	0086
43022-860	0144
43044	0016
43046	0160
43054	0111
43055	0111
43060	0103
43062	0133
43074	0604
43095	0103
43104	0543
43200	0232
43200-8	OEM
43201	0361
43202	0467
43205	1172
43256	OEM
43296	0325
43959	0143
43992-2	0600
44001	0071
44002	0124
44003	0071
44004	0071
44005	0071
44006	0071
44007	0071
44011-PC	OEM
44221	OEM
44465-3	0015
44465-4	0015
44465-5	0015
44465-6	0015
44616-1	0211
44763	0551
44764	0382
44765	4579
44766	0325
44933	1217
44934	1217
44935	1217
44936	1217
44936(RCA)	1217
44937	0015
44937(RCA)	1217
44938	0015
44938(RCA)	1217
44967-2	0211
45190	0060
45191	0561
45192	0561
45193	1308
45194	1308
45195	1199
45299	2698
45380	0654
45381	0701
45390	1395
45393	2557
45394	0379
45395	2557
45411	2150
45495-2	0600
45810-51	0015
45810-52	0720
45810-53	0019
45810-54	0064
45810-161	1280
45810-162	0937
45810-163	0076
45810-164	0155
45810-165	0218
45810-166	0076
45810-167	2755
45810-168	0520
45810-169	1018
45810-170	1050
46140-2	0015
46287-4	0123
46490-2	0600
46590-2	0038
46591-2	0038
46592-2	0038
46593-2	0038
46631-2	0600
46774-1	0600
46775-2	0600

DEVICE TYPE	REPL CODE
46914	0242
47126	0015
47126-003	0015
47126-1	0015
47126-1A	0015
47126-2	0015
47126-3	0015
47126-3A	0015
47126-4	0015
47126-4A	0015
47126-6	0015
47126-7	0015
47126-8	0087
47126-9	0015
47126-10	0087
47126-12	0015
47127-3	0015
47394-2	0211
47645-2	0038
47737-2	0211
48106	0302
48159	OEM
48287	0143
48287-4	0123
48385-2	0038
48751-028	0150
48902A0A10	OEM
48937-2	0136
48939-2	0136
49058-2	0038
49138-2	0600
49139-2	0211
49751-163	0222
49939-2	0050
50006G0F	OEM
50006O0F	OEM
50011G0F	OEM
50027	0715
50028	0342
50029	0037
50137-2	0086
50210-2	1041
50210-7	1632
50210-8	0562
50447-4	0160
50464	OEM
50464CP-12	OEM
50477-4	0160
50505-01	0143
50745-01	0015
50745-02	0015
50745-03	0087
50745-04	0087
50745-05	0087
50745-06	0015
50745-07	0015
50745-08	0015
50957-03	0144
51009	OEM
51102-01A	OEM
51194	0086
51194-01	0086
51194-02	0086
51194-03	0086
51202L	OEM
51213	0016
51213-01	0016
51213-02	0016
51213-03	0016
51213-2	0016
51428-01	0016
51429-02	0016
51429-03	0016
51429-3	0016
51441	0016
51441-01	0016
51441-02	0016
51441-03	0016
51442	0016
51442-01	0016
51442-02	0016
51442-03	0016
51545	0016
51547	0016
51650	0160
51750	0410
52000-030	1823
52000-032	1041
52000-033	2097
52000-036	1831
52000-037	1835
52000-044	0033
52000-046	0141
52000-048	3399
52000-049	1844
52000-052	0333
52000-055	1481
52000-057	1433
52000-058	0466
52000-061	1859
52000-062	0562
52000-063	0341
53004-1716	0133
53007-2	0071
53009-1	0030
53016	0133
53016R	0133
53088-4	0071
53092-1	0143

DEVICE TYPE	REPL CODE
53093-1	0196
53099-1	0015
53099-3	0015
53110P-15	OEM
53200-22	0016
53200-23	0016
53200-51	0016
53200-74	0111
53201-01	0004
53201-11	0004
53203-72	0086
53300-31	0071
53300-41	0015
53301-01	0071
53400-01	0016
53704D	0239
54107DM	OEM
54107FM	OEM
54121DM	OEM
54121FM	OEM
54122DM	OEM
54122FM	OEM
54123DM	OEM
54123FM	OEM
54125DM	OEM
54125FM	OEM
54126DM	OEM
54126FM	OEM
54132DM	OEM
54132FM	OEM
54136	OEM
54145DM	OEM
54145FM	OEM
54150DM	OEM
54150FM	OEM
54151ADM	OEM
54151AFM	OEM
54152AFM	OEM
54153DM	OEM
54153FM	OEM
54154DM	OEM
54154FM	OEM
54155DM	OEM
54155FM	OEM
54156DM	OEM
54156FM	OEM
54157DM	OEM
54157FM	OEM
54160DM	OEM
54160FM	OEM
54161DM	OEM
54161FM	OEM
54162FM	OEM
54163DM	OEM
54163FM	OEM
54164DM	OEM
54164FM	OEM
54165DM	OEM
54165FM	OEM
54166DM	OEM
54166FM	OEM
54167FM	OEM
54170DM	OEM
54170FM	OEM
54175DM	OEM
54175FM	OEM
54176DM	OEM
54176FM	OEM
54177DM	OEM
54177FM	OEM
54178DM	OEM
54178FM	OEM
54179DM	OEM
54179FM	OEM
54180DM	OEM
54180FM	OEM
54190FM	OEM
54191DM	OEM
54191FM	OEM
54192DM	OEM
54192FM	OEM
54193DM	OEM
54193FM	OEM
54194DM	OEM
54194FM	OEM
54195DM	OEM
54195FM	OEM
54196DM	OEM
54196FM	OEM
54197DM	OEM
54197FM	OEM
54198DM	OEM
54198FM	OEM
54199DM	OEM
54199FM	OEM
54298DM	OEM
54298FM	OEM
54806P1	0725
55001	0232
55002	0314
55002G0F	OEM
55003	0361
55003G0F	OEM
55004	0385
55005	0512
55006	0697
55007	0873
55008	1052
55009	1108

DEVICE TYPE	REPL CODE
55010	0743
55011	1313
55012	1158
55017	4399
55017H0F	OEM
55021	0492
55022	2242
55023	4603
55024	1349
55027	1441
55029	1202
55032	0268
55034	1024
55035	1339
55036	1206
55107ADM	OEM
55107AFM	OEM
55107BDM	OEM
55107BFM	OEM
55108BDM	OEM
55108BFM	OEM
55110ADM	OEM
55166	0143
55170	OEM
55170-1	0224
55200-500	0015
55325F	OEM
55450ADM	OEM
55450BDM	OEM
55451ARM	OEM
55451BRM	OEM
55452ARM	OEM
55452BRM	OEM
55453ARM	OEM
55453BRM	OEM
55491DM	OEM
55492ADM	OEM
55606	0016
55810-51	0143
55810-52	0143
55810-53	0139
55810-161	0007
55810-162	0079
55810-163	0007
55810-164	0007
55810-165	0079
55810-166	0178
55810-167	0632
55810-168	2393
55810-169	1235
55810-170	2031
55975-1	0081
55976-1	0141
55977	1823
55978-1	0033
55979-1	2078
55980	2097
55982-1	1041
55987-1	0562
56016	OEM
56519	0141
56552	0358
56553	1823
56557	1831
56558	0081
56571	0562
57000-5452	0144
57000-5503	0198
57174	1413
57202	0415
57401AJ	OEM
57401F	OEM
57401J	OEM
57401J16	OEM
57402AF	OEM
57402F	OEM
57402J	OEM
57508	OEM
57516	OEM
57518	OEM
57558	OEM
57558-1	OEM
57558-1F42	OEM
57558-1J40	OEM
57558F42	OEM
57558J40	OEM
58321S	OEM
58810-80	0015
58810-81	0030
58810-82	0133
58810-83	0143
58810-85	0015
58810-86	0170
58810-160	0325
58810-161	0127
58810-162	0079
58810-163	0079
58810-164	0111
58810-165	0079
58810-166	0079
58810-167	0111
58810-168	0079
58810-169	0560
58810-170	0911
58810-171	1478
58819	OEM
58840-111	0015
58840-112	0030
58840-113	0133

DEVICE TYPE	REPL CODE
58840-114	0019
58840-116	0139
58840-117	0170
58840-191	0325
58840-192	0937
58840-193	0079
58840-194	0079
58840-195	0079
58840-196	0079
58840-197	0079
58840-198	0079
58840-199	0079
58840-200	0390
58840-201	0911
58840-203	1018
59395	0143
59395(RCA)	0143
59557-48	0004
59557-49	0015
59625-1	0037
59625-2	0037
59625-3	0037
59625-4	0037
59625-5	0037
59625-6	0037
59625-7	0037
59625-8	0037
59625-9	0037
59625-10	0037
59625-11	0037
59625-12	0037
59840	0133
59840-1	0124
59844-1	0015
59844-2	0015
59844-3	0015
59987-1	0004
59988-1	0086
59989-1	0126
59990-1	0160
59991-1	0137
60034	2017
60045	1413
60059	1413
60108	0725
60132	0560
60133	0042
60136	0415
60182	1413
60202	0415
60213(1C)	2617
60215-1	0143
60245	0415
60246	0415
60256	0240
60259	1413
60270	2353
60304	3291
60306	0411
60317	1413
60319	1413
60322	2015
60323	1413
60327	0155
60337	0353
60341	2356
60360	0576
60373	1217
60401	OEM
60402	OEM
60406	OEM
60407	0560
60408	0042
60409	5628
60410	0593
60413	0560
60415	0576
60416	0423
60424	0147
60451	0415
60466	0593
60478	2387
60496	0147
60510	0348
60533	0147
60550	2387
60572	0593
60579	0415
60582	1413
60594	2015
60613	0415
60615	2387
60650	0576
60663	0415
60667	0147
60673	1413
60694	0415
60732	0415
60733	0415
60741	1413
60793	0849
60806	1413
60816	0576
60817	0576
60833	0576
60879	2387
60908	2730
60916	0576
60968	0147

If replacement code is OEM, contact original manufacturer for replacement.

DEVICE TYPE	REPL CODE	DEVICE TYPE	REPL CODE	DEVICE TYPE	REPL CODE	DEVICE TYPE	REPL CODE	DEVICE TYPE	REPL CODE	DEVICE TYPE	REPL CODE	DEVICE TYPE	REPL CODE
60987	0042	62676	0042	70399-2	2136	72133-1	0623	74156PC	1592	75108BPC	OEM	77272-1	0004
60998	2387	62680	2387	70432-16	0015	72133-2	0623	74157	4445	75110ADC	OEM	77272-5	0004
61003-4	0164	62746	2387	70434	0160	72135	0649	74157DC	1605	75110APC	OEM	77272-7	0004
61008-8	0004	62950	0560	70511	0016	72135-C	0649	74157PC	1605	75150	OEM	77272-9	0004
61008-8-1	0004	63282	0037	70581-1	1303	72137	0102	74158	4446	75150D(SM)	OEM	77273-2	0004
61008-8-2	0004	63722(BOSTON-GEAR)	0109	70622(RCA)	0443	72145	0015	74160DC	1631	75150DC	OEM	77273-3	0004
61009-1	0144	63723(BOSTON-GEAR)	0122	71006-30	0015	72146	0133	74160PC	1631	75150PC	OEM	77273-6	0004
61009-1-1	0144	63728(BOSTON-GEAR)	3868	71037-2	OEM	72147	0133	74161	4447	75150RC	OEM	77273-7	0004
61009-1-2	0144	63729(BOSTON-GEAR)	0740	71037-4	OEM	72148	0196	74161AN	OEM	75150TC	OEM	77489	0917
61009-2	0144	63730(BOSTON-GEAR)	0764	71060-4	OEM	72150	0004	74161DC	1645	75154	OEM	78001	0061
61009-2-1	0144	64071-1	0211	71060-6	OEM	72151	0016	74161N	4447	75154DC	OEM	78005AP	1292
61009-4	0016	64074	0382	71100-4	OEM	72151(RECTIFIER)	0087	74161PC	1645	75154PC	OEM	78009AP	OEM
61009-4-1	0016	64075	0925	71119-1	0133	72152	1149	74162	4448	75170-2	OEM	78010AP	2806
61009-6	0144	64076	0042	71119-2	0124	72158	0015	74162PC	1013	75188	0508	78010BB	OEM
61009-6-1	0144	64090	0239	71193-2	0004	72161-1	0102	74163	4449	75188N	0508	78010BC	OEM
61009-9	0126	64197	0925	71226-1	0155	72162-1	1501	74163DC	1666	75189	0511	78012AP	0334
61009-9-1	0126	65050	OEM	71226-2	0155	72162-2	0023	74163N	1666	75189A	0511	78106C	2301
61009-9-2	0126	65075	OEM	71226-3	0155	72163-1	1149	74163PC	1666	75189AN	0511	78331	0037
61009-9-3	0004	65100	OEM	71226-4	0155	72165-1	0186	74164DC	0734	75212ACW	OEM	78524-39-01	0071
61010-0	0144	65256BLFP-15T	OEM	71226-5	0155	72165-3	0533	74164FC	OEM	75230-9	0015	78527-75-01	0004
61010-0-1	0144	65804-62	0050	71226-6	0155	72165-5	0245	74164PC	0734	75232F	OEM	78527-76-01	0004
61010-2-1	0136	65804-63	0004	71226-10	0016	72165-6	0255	74165	1685	75232N	OEM	78527-78-01	0004
61010-6	0160	66007-4	0086	71226-15	0155	72168-2	0402	74165DC	1685	75234F	OEM	78527-79-01	0004
61010-6-1	0160	66008-2	0004	71266-4	0016	72168-3	0636	74165FC	OEM	75234N	OEM	78606C	2301
61010-7-1	0144	66009-5	0160	71411-1	0794	72171-1	0015	74165PC	1685	75324F	OEM	78894	0015
61010-7-2	0016	66010-3	0160	71411-2	0137	72172-1	0071	74166	4451	75324N	OEM	78896	1030
61011-0	0126	66405B	0286	71411-3	0495	72173-1	0505	74166DC	0231	75325F	OEM	78972	0917
61011-0-1	0126	66526	4606	71411-4	0531	72174-1	0277	74166FC	OEM	75325N	OEM	79408	0012
61011-2	0015	66527	4615	71412-4	0155	72176	0015	74166N	0231	75450ADC	OEM	79855	0144
61011-3-2	0016	66600	0030	71412-5	0155	72180	1067	74166PC	0231	75450APC	1231	79856	0144
61012-4-1	0600	66682-42	0015	71447-1	0218	72181	0175	74167PC	OEM	75450BDC	1231	79922	0178
61012-5-1	0050	67055	0133	71447-2	0218	72185	1158	74170DC	1722	75450BF	OEM	79949	1834
61013-2-1	0016	67193-82	0280	71447-3	0218	72190	0164	74170FC	OEM	75450BN	1231	79985	0143
61013-4-1	0126	67193-85	0211	71448	0160	72191	0208	74170PC	1722	75450BPC	1231	79992	0042
61013-9-1	0144	67401	1087	71448-1	0160	72193	0222	74174	1770	75451ARC	OEM	80011	OEM
61015-0-1	0144	67401AJ	OEM	71448-2	0160	72204	0016	74174DC	1770	75451ATC	6081	80021	OEM
61025	OEM	67401J	OEM	71448-3	0160	72206	0016	74174PC	1770	75451BN	1244	80050	0160
61050	OEM	67401J16	OEM	71448-4	0160	72207	0016	74175	1787	75451BRC	1244	80053	2624
61075	OEM	67402AJ	OEM	71448-5	0160	72784-21	0280	74175DC	1787	75451BT	1244	80070	2624
61088	0953	67402J	OEM	71448-6	0160	72784-22	0004	74175PC	1787	75451BTC	1244	80071	2624
61093	1413	67402J18	OEM	71448-7	0160	72784-23	0004	74176DC	1795	75452	1262	80073	2624
61096	1413	67508	OEM	71449	0242	72797-80	0136	74176PC	1795	75452ARC	OEM	80074	2624
61100	0576	67516	OEM	71449-1	0953	72797-81	0050	74177	1803	75452ATC	4525	80081	2617
61102-0	0136	67518	OEM	71467-1	0133	72799-41	0004	74177DC	1803	75452BN	1262	80083	2624
61104	2387	67544	0157	71488-4	0160	72813-10	0004	74177PC	1803	75452BRC	1262	80090	2624
61122	0764	67558	OEM	71488-5	0160	72847-51	0004	74178DC	1813	75452BT	1262	80094	2624
61134	2387	67558-1	OEM	71488-6	0160	72856-63	0160	74178FC	OEM	75452BTC	1262	80114	2571
61157	1413	67558-1J40	OEM	71588-1	0015	72874-52	0016	74178PC	1813	75453ARC	OEM	80249-910787	0642
61158	1413	67558-F42	OEM	71588-2	0015	72879-39	0136	74179DC	1820	75453ATC	6238	80286	OEM
61189	0147	67558F42	OEM	71588-5	0015	72879-40	0136	74179FC	OEM	75453BN	1272	80287	2571
61190	1413	67558J40	OEM	71588-6	0015	72923-08	0050	74179PC	1820	75453BRC	1272	80287(CO-PROC)	OEM
61224	1413	67586	0079	71667-1	0133	72941-33	0004	74180DC	1829	75453BT	1272	80287(IC)	OEM
61229-1	OEM	67590	0143	71686-4	2950	72949-10	0144	74180PC	1829	75453BTC	1272	80416C	0160
61239	1267	67599	0164	71686-5	3133	72951-95	0144	74181DC	1842	75454BN	1289	80501	0079
61242	0042	67802	0144	71686-6	3133	72951-96	0144	74181PC	1842	75454BT	1289	80540	0016
61244	0126	68001	OEM	71687	0037	72963-14	0016	74182DC	1856	75461TC	OEM	80544	0016
61249-1	0360	68006	OEM	71687-1	0037	72979-80	0144	74182PC	1856	75462TC	OEM	80545	0016
61260	0687	68007	OEM	71687-101	0037	73100-9	0004	74190DC	1912	75468	1134	80551	0037
61291	1413	68177	0015	71748-1	0325	74021-1	5383	74190PC	1912	75471TC	OEM	80710	0167
61292-1	OEM	68177A	0015	71778	0143	74107	4438	74191DC	1917	75472TC	OEM	80813VM	0016
61294-1	OEM	68504-62	0136	71779	0015	74107DC	0942	74191PC	1917	75474P	OEM	80814VM	0016
61295	0147	68504-63	0004	71783	0302	74107PC	0942	74192	1921	75477	OEM	80815VM	0016
61364PD06	OEM	68504-77	0143	71794	0087	74108ADC	OEM	74192DC	1921	75478	OEM	80816VM	0016
61366	0079	68504-78	0015	71818-1	0037	74116	4439	74192PC	1921	75491PC	1729	80817VM	0208
61376	0576	68595-13	0280	71819-1	0016	74120	1116	74193	1926	75492PC	1740	80818VM	0004
61384	2017	68617	0016	71963-1	0111	74121DC	0175	74193DC	1926	75654-1	0259	80829	1843
61399	0415	68645-03	0015	72003	0847	74121N	0175	74193PC	1926	75700-03-01	0160	80910-147	0911
61408	0764	68895-13	0004	72004	0847	74121PC	0175	74194DC	OEM	75700-04	0016	81404-4A	0004
61436	0087	68995-1	1350	72006	0847	74122DC	1139	74194FC	OEM	75700-04-01	0016	81500-3	0004
61452	0576	69011	OEM	72013	0143	74122PC	1139	74194PC	OEM	75700-05	0016	81501-5	0004
61532	0147	69107-42	0136	72017	0196	74123	1157	74195	4267	75700-05-01	0016	81502-0	0280
61566	0701	69107-43	0050	72020	1030	74123DC	1157	74195DC	1943	75700-05-02	0016	81502-0A	0280
61596	0147	69107-44	0136	72025	0015	74123PC	1157	74195FC	OEM	75700-05-03	0016	81502-0B	0280
61612	2018	69107-45	0211	72026	0015	74125	4442	74195N	1943	75700-08	0016	81502-1	0280
61613	2018	69111	0015	72027	0196	74125DC	1182	74195PC	1943	75700-08-02	0016	81502-1A	0280
61652	2018	69213-78	0023	72036	1080	74125N	1182	74196	1950	75700-09-01	0016	81502-1B	0280
61664	1413	70008-0	0103	72041	0015	74125PC	1182	74196DC	1950	75700-09-21	0016	81502-2	0004
61762	0593	70008-3	0103	72053	0196	74126DC	1193	74196PC	1950	75700-13-01	0126	81502-2A	0004
61770	1413	70019-1	0178	72054-1	0400	74126PC	1193	74197	4457	75700-14-02	0015	81502-2B	0004
61807	0242	70019-5	0178	72058	0015	74132DC	1271	74197DC	1956	75700-14-05	0071	81502-3B	0004
61819	0725	70023-0-00	0016	72058A	0015	74132PC	1271	74197PC	1956	75700-22-01	0178	81502-4	0004
61831	0576	70023-1-00	0016	72059	1030	74139	OEM	74198DC	1964	75700-24-02	0071	81502-4A	0004
61832	2018	70064-7	0015	72060	0143	74141DC	1377	74198FC	OEM	75702-15-21	0015	81502-4B	0004
61865	0925	70066-3	0015	72080	0143	74141PC	1377	74198PC	1964	75702-15-24	0229	81502-5	0004
61866	0042	70066-4	0015	72089	0143	74145	0619	74199DC	1971	75702-15-34	0071	81502-5A	0280
61872	0576	70158-9-00	0126	72097	0015	74145DC	0619	74199FC	OEM	75960CH	0211	81502-5B	0280
61926	0423	70167-8-00	0144	72101	0015	74145N	0619	74199PC	1971	76162-2	0190	81502-6	0600
61932	0147	70177A01	1753	72103	0015	74145PC	0619	74200-8	0015	76281A	OEM	81502-6A	0600
61957	0764	70177A02	0789	72104	0143	74148	1464	74200-9	0015	76361A	OEM	81502-6B	0600
62025	OEM	70177A07	0623	72106	0030	74148N	1464	74221	1239	76496	OEM	81502-6C	0600
62075	0764	70177B02	0789	72109	0015	74150	1492	74221N	2142	76600P	0785	81502-6D	0600
62102	3822	70205-8A	0071	72110	0774	74150DC	1492	74273	OEM	76675	0143	81502-7	0280
62104	6819	70231	0144	72111	1303	74150PC	1492	74279	OEM	76675A	0143	81502-7A	0280
62169-403-832	OEM	70260-11	0144	72113	0015	74151ADC	1495	74279DC	OEM	76675B	0143	81502-7B	0280
62191	0576	70260-12	0144	72113A	0015	74151APC	1495	74279PC	OEM	77001	0239	81502-7C	0280
62193	2018	70260-13	0144	72114	0016	74151DC	1495	74283	OEM	77018A	OEM	81502-8	0280
62207	1413	70260-14	0016	72115	0016	74151PC	1495	74283DC	OEM	77027-1	0358	81502-8A	0280
62211	0576	70260-15	0016	72116	0016	74152DC	1517	74283PC	OEM	77027-2	0358	81502-8B	0280
62212	0147	70260-16	0016	72117	0004	74152PC	1517	74290PC	6417	77032-8	OEM	81502-8C	0280
62221	0161	70260-18	0103	72119	0015	74153	4444	74293PC	2644	77052-3	0004	81502-9	0004
62222	0264	70260-19	0126	72123	0015	74153DC	1539	74298DC	OEM	77052-4	0004	81502-9A	0004
62232	1413	70260-20	0016	72123A	0015	74153N	4444	74298PC	OEM	77053-2	0004	81502-9B	0004
62238	0576	70260-29	0126	72128	0123	74153PC	1539	74374	OEM	77068-3170756	0142	81502-9C	0004
62274	1413	70270-05	0071	72128A	0143	74154	1555	74390	3240	77190-7	0015	81503-0	0004
62282	0764	70270-23	0015	72129	0143	74154DC	1555	75107ADC	OEM	77190-8	0071	81503-0A	0004
62334	0147	70270-38	0071	72129A	0143	74154PC	1555	75107APC	OEM	77271-3	0015	81503-0B	0004
62342	0576	70372-2	0015	72130	0015	74155DC	1576	75107BDC	OEM	77271-4	0071	81503-1	0004
62410G	OEM	70398-1	0079	72130-1	0015	74155PC	1576	75107BPC	OEM	77271-8	0050	81503-1A	0004
62457	1413	70399-1	1669	72132-1	1899	74156DC	1592	75108BDC	OEM	77272-0	0004	81503-1B	0004

If replacement code is OEM, contact original manufacturer for replacement.

DEVICE TYPE	REPL CODE
81503-3	0004
81503-3A	0004
81503-4	0004
81503-4A	0004
81503-4B	0004
81503-4C	0004
81503-6	0280
81503-6A	0280
81503-6B	0280
81503-6C	0280
81503-7	0280
81503-7A	0280
81503-7B	0280
81503-7C	0280
81503-8	0280
81503-8B	0004
81503-8C	0004
81504-1	0280
81504-1A	0280
81504-1B	0280
81504-1C	0280
81504-3	0280
81504-3A	0280
81504-3B	0280
81504-3C	0280
81505-8	0004
81505-8A	0004
81505-8B	0004
81505-8C	0004
81505-8X	0004
81506-4	0136
81506-4A	0136
81506-4B	0136
81506-4C	0136
81506-5	0280
81506-5A	0280
81506-5B	0280
81506-5C	0280
81506-6	0280
81506-6A	0280
81506-6B	0280
81506-6C	0280
81506-7	0136
81506-7A	0136
81506-7B	0136
81506-7C	0136
81506-8	0136
81506-8A	0136
81506-8B	0136
81506-8C	0136
81506-9	0004
81506-9A	0004
81506-9B	0004
81506-9C	0004
81507-0	0004
81507-0A	0004
81507-0B	0004
81507-0C	0004
81507-0D	0004
81507-4	0004
81507-5	0208
81507-6	0208
81510-3	0280
81510-4	0004
81510-5	0280
81511-4	0004
81511-5	0280
81511-6	0280
81511-7	0280
81511-8	0004
81512-0	0004
81512-0A	0004
81512-0B	0004
81512-0C	0004
81512-0D	0004
81512-0E	0004
81513-3	0016
81513-6	0004
81513-7	0160
81513-8	0015
81513-9	0004
81514-2	0015
81515-8	0004
81516-0	0004
81516-0A	0004
81516-0B	0004
81516-0C	0004
81516-0D	0004
81516-0E	0004
81516-0F	0004
81516-0G	0004
81516-0H	0004
81516-0J	0004
82716	0144
82965F	0576
83008	0015
83741C01	0710
84001	0126
84001A	0126
84001B	0126
84011	0215
84011U	0215
84323	3061
84324	0562
84353-2	4218
84626	1069
84626-1	1069
84626-2	1069
84626-3	1069
84626-4	1069
84628	3337
84628-1	3337
84628-2	3337
84628-3	3337
84628-4	3337
84630	3418
84630-1	3418
84630-2	3418
84630-3	3418
84631	1810
84631-1	1810
84631-2	1810
84631-3	1810
84631-4	1810
85000-ORG	0042
85549	0016
86345	OEM
86394GL6	1992
86452	0211
86812	0198
86822	0198
86832	0211
86842	0211
86848	0280
87322-04	2472
87532	0086
87756	0133
87757	0079
87758	0037
87759	0037
88060-51	0030
88060-52	0133
88060-53	0123
88060-54	0244
88060-55	0133
88060-141	0325
88060-142	0224
88060-143	0079
88060-144	0111
88060-145	0086
88060-146	0560
88060-147	0911
88510-51	0030
88510-52	0133
88510-53	0123
88510-54	0244
88510-55	0133
88510-171	0325
88510-172	0127
88510-173	0079
88510-174	0111
88510-175	0079
88510-176	0560
88510-177	0911
88510-178	1478
88641	0015
88686	0198
88687	0016
88688	0016
88700	0847
88710S	OEM
88738	OEM
88796-1-1	0764
88832	0160
88834	0086
88862	0016
89028-2	0081
89028-3	0033
89028-4	0141
89028-4-1-4	0141
89028-6	2097
90050	0160
90203-6	0015
91001	0015
91021	0600
91605	0079
93005	0015
93006	0015
93007	0015
93011	0015
93012	0015
93018	0244
93022	0015
93022A	0015
93023	0015
93027	0229
93028	0015
93030	0170
93091-4	1501
93096-1	0190
93141DC	1377
93141PC	1377
93145DC	0619
93145PC	0619
93150DC	1492
93150PC	1492
93151DC	1495
93151PC	1495
93152DC	1517
93152PC	1517
93153DC	1539
93153PC	1539
93154DC	1555
93154PC	1555
93155DC	1576
93155PC	1576
93156DC	1592
93156PC	1592
93164DC	0734
93164PC	0734
93165DC	1685
93165PC	1685
93166DC	0231
93166PC	0231
93170DC	1722
93170PC	1722
93174DC	1770
93174PC	1770
93175DC	1787
93175PC	1787
93176DC	1795
93176PC	1795
93177DC	1803
93177PC	1803
93178DC	1813
93178PC	1813
93179DC	1820
93179PC	1820
93180DC	1829
93180PC	1829
93190DC	1912
93190PC	1912
93191DC	1917
93191PC	1917
93196DC	1950
93196PC	1950
93197DC	1956
93197PC	1956
93198DC	1964
93198PC	1964
93199DC	1971
93199PC	1971
93403DC	OEM
93403DM	OEM
93403FM	OEM
93403PC	OEM
93407ADC	OEM
93407AFC	OEM
93407BDC	OEM
93407BDM	OEM
93407BFC	OEM
93407BFM	OEM
93410ADC	OEM
93415ADC	OEM
93415AFC	OEM
93415APC	OEM
93415DC	OEM
93415DM	OEM
93415FM	OEM
93415PC	OEM
93419ADC	OEM
93419APC	OEM
93419DC	OEM
93419DM	OEM
93419PC	OEM
93422ADC	OEM
93422ADM	OEM
93422AFC	OEM
93422AFM	OEM
93422ALC	OEM
93422ALM	OEM
93422APC	OEM
93422DC	OEM
93422DM	OEM
93422FC	OEM
93422FM	OEM
93422LC	OEM
93422LM	OEM
93422PC	OEM
93425ADC	OEM
93425APC	OEM
93425DC	OEM
93425DM	OEM
93425FM	OEM
93425PC	OEM
93431DC	OEM
93431DM	OEM
93431FC	OEM
93431FM	OEM
93432DC	OEM
93432DM	OEM
93432FC	OEM
93432FM	OEM
93432PC	OEM
93438DM	2324
93438FC	OEM
93438FM	OEM
93438PC	2324
93441DC	OEM
93441DM	OEM
93441FC	OEM
93441FM	OEM
93441PC	OEM
93442DC	OEM
93442DM	OEM
93442FC	OEM
93442FM	OEM
93442PC	OEM
93448DC	2322
93448DM	2322
93448FC	OEM
93448FM	OEM
93448PC	2322
93450DC	2862
93450DM	2862
93450FC	OEM
93450PC	2862
93451DC	2909
93451DM	2909
93451FC	OEM
93451PC	2909
93452DC	2390
93452DM	2390
93452PC	2390
93453DM	2390
93453PC	2392
93454DC	OEM
93454DM	OEM
93454FC	OEM
93454FM	OEM
93454PC	OEM
93458DC	OEM
93458DM	OEM
93459DC	OEM
93459DM	OEM
93470DC	OEM
93470DM	OEM
93471DC	OEM
93471DM	OEM
93471PC	OEM
93475DC	OEM
93475DM	OEM
93475FC	OEM
93475FM	OEM
93475PC	OEM
93475PM	OEM
94000	0004
94001	0004
94002	0004
94003	0164
94004	0085
94005	0164
94006	0004
94007	0050
94008	0004
94009	0164
94010	0604
94013	0050
94014	0004
94015	0004
94016	0004
94017	0164
94018	0164
94019	0164
94020	0164
94021	0164
94022	0164
94023	0208
94024	0160
94025	0160
94026	0160
94027	0016
94028	0050
94029	0208
94030	0004
94032	0160
94033	0050
94034	0160
94035	0050
94036	0050
94037	0037
94037(EICO)	0004
94038	0136
94038(EICO)	0004
94039	0004
94040	0160
94041	0086
94042	0086
94043	0698
94044	0144
94047	0016
94048	0016
94049	0178
94050	0086
94051	0086
94051(EICO)	0233
94052	0086
94062	0595
94063	0126
94064	0126
94065	0103
94065A	0103
94066	0086
94067	0126
94068	0126
94070	0086
94094	0178
94094(EICO)	0103
94132B-2131	OEM
94152	0467
94325	1835
94327	0033
94331	2097
94333	1041
95000	0143
95000DC	OEM
95001	0143
95002	0143
95003	0133
95004	0143
95007	0143
95008	0143
95012	0133
95014	0143
95015	0015
95015(EICO)	0133
95016	0229
95017	0143
95018	0143
95101	0050
95102	0050
95103	0050
95107	0050
95108	0050
95109	0686
95110	0050
95111	0050
95112	0600
95113	0600
95114	0600
95115	0600
95115DC	OEM
95116	0136
95116DC	OEM
95117	0600
95118	0136
95119	0136
95120	0136
95120A	0280
95121	0050
95122	0136
95123	0136
95124	0050
95124DC	OEM
95125	0144
95126	0144
95127	0144
95128	0144
95129	0144
95130	0144
95131	0144
95132	0212
95133	2950
95150	0004
95151	0050
95169-1	OEM
95170-1	0224
95170-2	0224
95170-2(DIO)	0143
95170-2(XSTR)	0224
95171-1	0155
95171-2	0155
95171-3	0076
95171-4	0155
95172-2	0004
95173-1	0004
95173-2	OEM
95176-4	OEM
95180-1	OEM
95181-1	OEM
95182-1	OEM
95186-1	OEM
95187-1	OEM
95191	OEM
95192	OEM
95195	OEM
95196	OEM
95201	0211
95202	0038
95203	0211
95204	0211
95208	0211
95209	0211
95211	0600
95212	0211
95213	0211
95214	0211
95216	0198
95216RED	0198
95216YEL	0198
95217	0211
95218	0211
95219	0211
95220	0086
95221	0037
95222-1	0211
95222-2	0038
95223	0079
95224-1	0004
95224-2	0208
95224-3	0004
95224-4	0208
95225	0198
95226-003	0791
95226-004	0595
95226-1	0086
95226-2	0198
95226-3	0342
95226-4	0324
95226-8	OEM
95227	0037
95227-1	0037
95228	0325
95231	0325
95232	0037
95233	0086
95235	1137
95239-1	0037
95240-1	0037
95241-1	0006
95241-3	0532
95242-1	0144
95246-3	OEM
95250	0160
95250-1	0160
95251	0160
95252	0142
95252-1	0142
95252-2	0142
95252-3	0142
95252-4	0142
95253	0160
95257	3033
95258-1	0551
95258-2	0382
95262-1	0161
95262-2	0460
95263-1	0161
95263-2	0460
95265-2	OEM
95266-1	OEM
95266-2	OEM
95269-1	OEM
95273-1	OEM
95273-2	OEM
95275-1	OEM
95277-1	OEM
95277-3	OEM
95285-2	OEM
95285-4	OEM
95286	1395
95286-3	OEM
95287	2755
95287-2	OEM
95291	0701
95294	0701
95296	0400
95301-1	OEM
95306-3	OEM
95306-9	OEM
95307-1	OEM
95400DC	OEM
95401DC	OEM
96101DC	OEM
96101DM	OEM
96101PC	OEM
96103	OEM
96103DC	OEM
96103DM	OEM
96106DC	OEM
96106DM	OEM
96106FM	OEM
96106PC	OEM
96147-001	0147
96148-001	0147
96702	OEM
97072	OEM
97073	OEM
97074	OEM
97081-001	OEM
97081-016	OEM
97081-116	OEM
97096-001	OEM
97098	OEM
97099	OEM
97100	0424
97101	OEM
97102	OEM
97103-108	OEM
97103-208	OEM
97112	OEM
97114	OEM
97202-222	0495
98496-001	0147
98497-001	0147
99023-3	0015
99101	0136
99102	0136
99103	0050
99104	0050
99105	0050
99106	0050
99107	0050
99108	0050
99109-1	1221
99109-2	0198
99116-1	OEM
99120	0136
99121	0136
99131-2	OEM
99171-1	OEM
99173-1	OEM
99176-1	OEM
99180-1	OEM
99195-1	0024
99195-3	0024
99201	0004
99201-110	0137
99201-128	0166
99201-208	0636
99201-210	0157
99201-211	0636
99201-212	0636
99201-216	0170
99201-219	0137
99201-221	0002
99201-242	0186
99201-312	0025
99201-316	0170
99201-325	0402
99201-331	0931
99202	0004
99202-22	0002
99202-128	0039
99202-220	0365
99202-221	0420
99202-222	0495
99202-226	0441
99202-228	0039
99202-233	0331
99203	0004
99203-005	0015
99203-006	0015
99203-1	6393
99203-3	0554
99203-4	0015
99203-5	0015
99203-6	0015
99203-6(RCA)	0015
99203-7	0071
99203-8	0071
99203-203	0609
99203-206	0795
99204	0004
99205	0004
99206-1	0016
99206-2	0016
99207-2	0016
99210-204	0188
99210-210	0157
99212-1	OEM
99217	0211
99218	0211
99241-2	OEM
99243-1	OEM
99244-1	OEM
99250	0160
99250-1	0160
99251	OEM
99252	0142
99252-1	0142
99252-2	0142
99252-2(METAL)	0142
99252-2(PLASTIC)	0276
99252-3	0142
99252-4	0042
99261-1	OEM
99287-24	OEM
99287-115	OEM
100017	0019
100101D	OEM
100101F	OEM
100102D	OEM
100102F	OEM
100107D	OEM
100107F	OEM
100112D	OEM
100112DC	OEM
100112F	OEM
100112FC	OEM
100114DC	OEM
100114FC	OEM
100117D	OEM
100117F	OEM
100118D	OEM
100118F	OEM
100122D	OEM
100122DC	OEM
100122F	OEM
100122FC	OEM
100123DC	OEM
100123FC	OEM
100130FC	OEM
100136D	OEM
100136F	OEM
100145D	OEM
100145F	OEM
100150FC	OEM
100155DC	OEM
100155FC	OEM
100166D	OEM
100166F	OEM
100180D	OEM
100180F	OEM
100181D	OEM
100412	0015
100415BF	OEM
100415F	OEM
100422AF	OEM
100422BF	OEM
100422F	OEM
100422N	OEM
100449	1030
100451AF	OEM
100470AF	OEM
100470F	OEM
100471	0196
100474AF	OEM
100474F	OEM
100520	0015
100581	0196
100617	0015
100624	0015
100678	0136
100693	0004
100768	2188
100802	0079
100844	0143
101010A	OEM
101010G	OEM
101015A	OEM
101020A	OEM
101025A	OEM

If replacement code is OEM, contact original manufacturer for replacement.

DEVICE TYPE	REPL CODE
101030A	OEM
101040A	OEM
101050A	OEM
101050G	OEM
101078	0050
101087	0050
101089	0136
101100A	OEM
101100G	OEM
101154	0133
101403	0015
101434	0144
101435	0086
101436	0698
101615	0015
101678	0600
101973	0004
101974	0004
102005	1208
102050A	OEM
102050G	OEM
102100A	OEM
102100G	OEM
102249	1030
102510A	OEM
102515A	OEM
102520A	OEM
102525A	OEM
102530A	OEM
102540A	OEM
102989	0143
103318	0015
103443	0600
103508-2	1069
103508-3	3899
103521	0079
103562	0004
103705	0141
103717	0562
103729	1823
103731	1041
103743	1831
103755	0081
103872	0196
104009	0136
104059	0136
104080	0208
104081	0015
104152	0143
104213	0015
104273	0015
104325	0015
104444	0004
104510A	OEM
104510G	OEM
104515A	OEM
104515G	OEM
104520A	OEM
104520G	OEM
104525A	OEM
104525G	OEM
104530A	OEM
104530G	OEM
104540A	OEM
104540G	OEM
104550G	OEM
104615	0015
104648D	2697
104825	2572
104830	1823
104833	2097
104836	1831
104844	0033
104846	0141
104862	0562
104925	2614
105010A	OEM
105010G	OEM
105050A	OEM
105064	0451
105100A	OEM
105100G	OEM
105330	1030
105412(5	3061
105517	0143
106010G	OEM
106015G	OEM
106020G	OEM
106025G	OEM
106030G	OEM
106040G	OEM
106050G	OEM
106379	0229
106525	1714
106625	2781
106719	2305
107225	1540
107255	2019
107268	1080
107510G	OEM
107515G	OEM
107520G	OEM
107525G	OEM
107530G	OEM
107540	0015
107540G	OEM
107550G	OEM
107625	2019
107628	1080

DEVICE TYPE	REPL CODE
107729	0917
107739	0211
108568	OEM
109308	0196
109328	0196
109474	0196
110008-01	OEM
110010-01	OEM
110043	0015
110075	0698
110240-005	3399
110242-003	0562
110242-004	0562
110263	0164
110388	0015
110472-003	0033
110494	0164
110495	0208
110496	0015
110515	0085
110610	0143
110629	0015
110636	0015
110669	0086
110697	0016
110699	0086
110873	0015
110957	0004
110958	0208
110959	0004
111010A	OEM
111015A	OEM
111020A	OEM
111025A	OEM
111030A	OEM
111040A	OEM
111086	0015
111117	0050
111118	0050
111207	0143
111278	0086
111279	0698
111303	0016
111313	0050
111516	0015
111605	0143
111642	0015
111725	0560
111776	0015
111820	0015
111825	2301
111954	0050
111955	0050
111956	0050
111957	0004
111958	0208
111959	0004
112017	0015
112018	0015
112027	OEM
112296	0050
112297	0004
112329	0015
112329(AMPHENOL)	0133
112330	0143
112355	0144
112356	0016
112357	0016
112358	0016
112359	0016
112360	0086
112361	0086
112475	0015
112510A	OEM
112515A	OEM
112520	0016
112520A	OEM
112521	0016
112522	0016
112523	0016
112524	0019
112525	0551
112525A	OEM
112526	0143
112527	0630
112528	0087
112529	0143
112529(AMPHENOL)	0133
112530	0012
112530A	OEM
112531	0015
112540A	OEM
112826	0015
112986	0037
112998	OEM
112999	OEM
113039	0015
113182	0037
113321	0015
113348	0016
113363	0071
113391	1303
113392	2637
113397	0774
113398	0016
113438	0016
113938	0144
113997	0774
113998	0015

DEVICE TYPE	REPL CODE
114013	0015
114267	0212
114401	0764
114510A	OEM
114510G	OEM
114515A	OEM
114515G	OEM
114520A	OEM
114520G	OEM
114525	0144
114525A	OEM
114525G	OEM
114530A	OEM
114530G	OEM
114540A	OEM
114540G	OEM
114550G	OEM
115039	0015
115063	0160
115099	0030
115101	0143
115123	0649
115225	0079
115227	0050
115228	0050
115229	0050
115268	0160
115269	0160
115275	0136
115281	0160
115282	0160
115283	0160
115284	0160
115304	0698
115436	0030
115524	0071
115529	0071
115559	0015
115559A	0015
115599	0015
115792	0925
115867	0920
115875	0016
115910	0144
116010G	OEM
116015G	OEM
116020G	OEM
116021	0133
116021(SEARS)	0133
116025G	OEM
116030G	OEM
116040G	OEM
116048	0143
116050	0123
116050G	OEM
116051	0540
116052	0015
116054	0015
116068	0975
116072	0136
116073	0127
116074	0127
116075	0127
116076	0016
116077	0016
116078	0127
116078(RCA)	0037
116079	0079
116080	0144
116081	0233
116082	0079
116083	0079
116084	0136
116085	0016
116086	0975
116087	0086
116088	0975
116089	0975
116090	0285
116091	0126
116092	0127
116093	0160
116119	0144
116148	0198
116198	0144
116199	0144
116200	0144
116201	0004
116202	0050
116203	0004
116204	0208
116205	0004
116206	0004
116207	0050
116208	0050
116209	0050
116273	0143
116279	0142
116284	0126
116286	0004
116314	0917
116413	2657
116445	0664
116588	0076
116623	1696
116628	0004
116683	0050
116684	0050
116685	0004

DEVICE TYPE	REPL CODE
116686	0004
116687	0208
116756	0136
116757	0004
116761	0861
116875	0016
116988	0164
116997	0004
116998	0004
117145	0015
117145A	0015
117208	0004
117209	0004
117210	0004
117510G	OEM
117515G	OEM
117520G	OEM
117525G	OEM
117530G	OEM
117540G	OEM
117550G	OEM
117616	0280
117617	0280
117618	0050
117658	0050
117659	0143
117724	0136
117725	0050
117726	0050
117727	0004
117728	0004
117729	0720
117730	0143
117760	0143
117823	0144
117824	0136
117866	0050
117867	0164
118125	0144
118200	0016
118244	0015
118279	0142
118284	0037
118358	0649
118361	2624
118686	0142
118713	0016
118822	0127
118825	0015
118873	0015
119013	0050
119035	4788
119119(PAIR)	0123
119199	0143
119412	0127
119414	0144
119526	0136
119554	0007
119555	0007
119556	0007
119557	0007
119594	0170
119596	0133
119597	0133
119609	2624
119635	0079
119636	0079
119650	0142
119661	0030
119662	0917
119719	0924
119721	0160
119722	0975
119723	0975
119724	0016
119725	0016
119726	0016
119727	0164
119728	0086
119730	0037
119822	0086
119823	0007
119824	0007
119825	0007
119919	0143
119956	0133
119982	0016
119983	0126
120073	0016
120074	0016
120075	0004
120143	0004
120144	0004
120481	0016
120482	0016
120483	0016
120503	0015
120504	0137
120544	0015
120545	0004
120546	0004
120617	0143
120818	0293
120909-24.4	0004
121151	0004
121152	0004
121153	0280
121154	0280
121180	0242

DEVICE TYPE	REPL CODE
121243	0160
121244	0604
121467	0037
121468	0015
121655	0016
121658	0016
121659	0037
121660	0016
121661	4163
121661(RCA)	0016
121662	0016
121663	0016
121664	0198
121680	0015
122074	0016
122129	0015
122166	0143
122199	2624
122243	0004
122244	0004
122517	0144
122518	0144
122519	0111
122519(SW)	0079
122520	0143
122524	0143
122664	0016
122665	0016
122725	0136
122781	OEM
122788	0015
122792	0160
122901	0004
122902	0127
122904	0127
123004	0015
123139	0086
123215	0142
123244	0136
123274	0155
123275	0142
123275-14	0142
123276	0030
123296	0015
123375	0142
123379	0004
123429	0144
123430	0144
123431	0144
123511	0136
123555-01	0725
123555-02	0725
123555-03	0725
123702	4423
123703	0086
123726	0030
123791	0004
123792	0160
123804	0015
123804-1	0015
123805	0004
123805(DIODE)	0133
123806	1598
123807	0016
123808	0222
123809	0004
123813	0133
123852	0493
123877	0004
123940	0086
123941	0079
123944	0126
123971	0037
123991	0037
124024	0127
124047	0037
124097	0136
124098	0015
124263	0144
124412	0144
124616	0126
124623	0127
124624	0127
124625	0050
124626	0004
124753	0079
124754	0079
124755	0037
124756	0079
124757	0008
124759	0016
124812	0015
125020	0501
125105	0015
125126	0137
125127	0015
125135	0079
125137	0144
125138	0144
125139	0079
125140	0079
125141	0079
125142	0126
125143	0079
125144	0224
125217	OEM
125261	0917
125263	0144
125264	0144

DEVICE TYPE	REPL CODE
125329	0127
125330	0136
125389	0127
125389-14	0016
125390	0127
125392	0144
125394	0127
125397	0133
125399	0720
125458	0015
125458(HV-23)	0540
125471	0015
125474	0127
125475	0127
125475-14	0144
125488	0102
125499	0317
125519	0155
125528	0133
125529	0102
125549	0015
125588	0133
125703	0160
125707	0126
125757	OEM
125761	0160
125787	0277
125790	0050
125835	0015
125844	0102
125856	0015
125884	0133
125944	0144
125964	0015
125972	0050
125992	0649
125993	0015
125994	0151
125994(RCA)	0127
125994-14	0144
125995	0127
126023	0144
126024	0144
126025	0007
126093	0004
126093-1	0004
126093-2	0004
126093-3	0004
126093-4	0004
126131	0015
126138	0142
126148	0015
126149	0720
126150	0079
126156	0079
126176	0015
126177	0143
126184	0050
126185	0136
126186	0050
126187	0004
126188	0142
126276	0004
126320	0015
126321	0133
126331	0079
126334	0079
126413	0277
126424	0111
126524	0037
126525	0079
126526	0079
126527	0102
126582	0002
126604	2715
126670	0144
126697	0004
126698	0144
126699	0016
126700	0037
126701	0016
126702	0086
126703	0086
126704	0016
126705	0233
126706	0016
126707	0037
126708	0016
126709	0233
126710	0233
126711	0016
126712	0079
126713	0079
126714	0016
126715	0037
126716	0079
126717	0079
126718	0037
126719	0037
126720	0079
126721	0086
126722	0142
126724	0126
126725	0086
126726	0142
126772	0142
126849	0289
126850	OEM
126851	0025

DEVICE TYPE	REPL CODE
126852	0002
126853	0367
126854	0149
126855	0015
126856	0102
126857	1217
126858	1217
126859	1217
126860	0774
126861	0015
126862	2528
126863	0317
126864	0039
126871	0829
126885	0071
126898	0061
126899	0239
126900	0074
126945	0004
127017	0143
127102	0015
127112	0004
127114	0004
127128	1638
127166	2707
127176	0015
127177	0317
127214	0325
127263	0016
127297	0004
127303	0004
127350A	OEM
127350AZ	OEM
127350B	OEM
127350C	OEM
127350D	OEM
127350E	OEM
127350F	OEM
127350Z	OEM
127351A	OEM
127351AZ	OEM
127351B	OEM
127351C	OEM
127351D	OEM
127351E	OEM
127351F	OEM
127351Z	OEM
127354	0079
127355	0111
127376	0086
127379	0071
127382	0002
127393	0079
127396	0015
127397	0164
127474	0133
127529	0079
127532	0917
127533	0917
127589	0004
127590	0847
127693	0144
127695	0015
127712	0233
127767	2763
127784	0143
127792	0144
127793	0144
127794	0144
127798	0042
127828	0604
127835	0087
127845	0086
127899	0079
127905F02	0267
127933	0124
127962	0004
127978	0161
127980	0212
127992	0015
127993	0133
128056	0042
128057	0925
128256	0015
128343	0004
128474	0133
128899	0086
128938	0136
128940	0004
129028	0143
129029	0015
129029(DIO)	0015
129029(XSTR)	0079
129049	0079
129050	0144
129051	0086
129095	0974
129144	0144
129145	0224
129146	0155
129147	0079
129157	0143
129158	0143
129203	0378
129213	0015
129241	0302
129251	0050
129286	0004
129289	0050

DEVICE TYPE	REPL CODE
129334	0015
129347	0050
129348	0071
129348(DIO)	0143
129348(XSTR)	0211
129359	0720
129375	0133
129389	0050
129392	0127
129392-14	0144
129393	0127
129393-14	0144
129394	0127
129394-14	0144
129424	1392
129425	0111
129474	0143
129475	0139
129494	0015
129507	0164
129508	0164
129509	0111
129510	0111
129511	0111
129512	0111
129513	0111
129556	0917
129571	0127
129572	0079
129573	0127
129574	0127
129604	0475
129697	0037
129698	0079
129699	0037
129759	0015
129802	0004
129821	1854
129871	0829
129897	0079
129899	0079
129903	0188
129904	0012
129930	OEM
129938	0636
129940	0365
129946	0002
129979	0127
129980	0212
130013	0079
130013(TRIAC)	2017
130026	1997
130040	0400
130044	0170
130045	0133
130046	2007
130047	0170
130052	0015
130096	1834
130110	0071
130130	0364
130132	0495
130139	0037
130200-00	0004
130200-02	0004
130215	0037
130278	0144
130328	0137
130338-01	0015
130380	0137
130389-00	0015
130400-95	0004
130400-96	0208
130402-36	0004
130403-04	0144
130403-13	0016
130403-17	0016
130403-18	0016
130403-47	0208
130403-52	0004
130403-62	0144
130404-21	0144
130404-29	0126
130404-59	0144
130474	0551
130536	0037
130537	0016
130607	0015
130751	0167
130761	0420
130762	0002
130793	0127
131005	0740
131006	0740
131050	0015
131075	0042
131139	0285
131140	0079
131161	0423
131214	0917
131221	0144
131239	0423
131240	0079
131241	0037
131242	0037
131242-12	0126
131243	0079
131243-12	0079
131245	0015

DEVICE TYPE	REPL CODE
131257	0130
131300	1823
131301	1041
131303	1831
131304	0033
131305	0081
131306	0141
131308	0562
131311	0079
131312	0358
131318	0102
131346	0239
131347	0061
131454	0079
131475	0102
131476	0102
131501	0554
131502	0133
131502(RCA)	0015
131543	0007
131544	0144
131545	0144
131647	0037
131648	0144
131789	0037
131814	0030
131815	0087
131815(RCA)	0917
131844	0144
131848	0224
131848(RCA)	0042
131849	0042
131941	OEM
131950	0015
132148	0102
132149	0015
132175	0086
132176	0037
132180	0086
132313	2743
132314	2549
132315	0328
132325	0239
132326	2327
132327	0086
132328	0086
132329	0079
132416	0137
132418	0102
132445	0551
132446	0551
132447	0382
132448	0854
132478	0693
132488	0382
132492	0783
132495	0161
132498	0037
132499	0042
132500	0086
132501	0102
132509	0102
132547	0102
132548	0102
132549	0102
132558	3508
132571	0460
132573	0042
132574	0854
132616	0025
132634	1997
132642	0155
132643	0155
132645	0133
132650	1893
132776	0161
132813	0042
132814	0015
132815	0087
132823	0079
132824	0039
132830	0037
132865	0157
132912	0143
132915	0143
132966	0288
132995	1199
133026	0288
133171	0007
133176	0595
133177	0086
133178	0079
133182	0037
133218	0079
133249	0079
133253	0037
133265	0285
133266	0015
133275	0079
133276	0007
133390	0133
133390(PWR.-RECT)	0015
133439	0018
133440	0342
133442	OEM
133543	0015
133573	0560
133600	2743
133615	0102

DEVICE TYPE	REPL CODE
133616	0102
133690	0079
133696	0016
133743	0016
133950	0015
134050	OEM
134074	0917
134142	0144
134143	0079
134144	0144
134144-1	0127
134155	0086
134180	0143
134195-001	1823
134196-001	1041
134197-001	0081
134198-001	0141
134254-001	2619
134263	0144
134264	0917
134279	0460
134280	0161
134281	1681
134282	0130
134335	0053
134340	2282
134342	1413
134411	0086
134415	1029
134417	0224
134419	0144
134442	0400
134444	0002
134445	1029
134450	0212
134509	1186
134587	0143
134771	0423
134772	0285
134773	0642
134774	0551
134777	0757
134989	0086
134993	0157
135029	2394
135244	0288
135281	0015
135284	0015
135286	0037
135297	0144
135320	0293
135324	0353
135341	0102
135351	0905
135352	0074
135362	0006
135380	0102
135386	1217
135406	0186
135531	0052
135532	0053
135571	0133
135631	0409
135691	0190
135716	0285
135734	0102
135735	0423
135739	0423
135744	0975
135872	0019
135932	0293
135963	0353
136066	0233
136145	2257
136146	2755
136147	1843
136162	0133
136163	0133
136164	0030
136165	0079
136168	0224
136239	0079
136240	0224
136334	OEM
136355	0725
136423	0126
136424	0086
136430	0079
136601	0460
136605	0071
136634	0365
136635	0015
136648	0042
136688	0133
136696	0086
136721	0039
136766	0050
137031	0409
137065	0382
137066	0551
137075	0102
137093	0004
137127	0224
137155	0037
137161	0517
137241	0086
137245	0364
137276	OEM
137277	OEM

DEVICE TYPE	REPL CODE
137338	0224
137339	0079
137340	0037
137341	0774
137352	0642
137369	0161
137383	0016
137388	0224
137527	0042
137606	0023
137607	0065
137613	0484
137614	0079
137646	0190
137647	0451
137648	0086
137652	0015
137655	0402
137693	0409
137699	OEM
137718	0642
137736	0483
137780	1029
137783	OEM
137833	OEM
137855	0288
137875	0079
137876	1413
137997	0259
138010	0409
138019	0190
138032	2329
138032(RCA)	2329
138033	1712
138033(RCA)	1712
138107	0783
138121	0382
138171	OEM
138172	0015
138173	0023
138174	0441
138175	0015
138191	0079
138192	0042
138193	0460
138194	0103
138195	0543
138196	0133
138218	0155
138310	OEM
138311	0232
138312	0268
138313	0314
138314	0361
138315	0467
138317	0527
138318	0697
138319	1357
138320	1926
138380	1843
138380(IC)	1843
138381	0899
138403	1172
138421	0190
138429	0058
138609	5164
138681	2552
138699	0521
138736	0348
138752	0190
138763	0079
138907	0190
138945	0720
138946	0136
138974	0636
139001	0190
139017	0338
139044	0264
139045	OEM
139061	OEM
139064	0460
139065	0133
139069	0460
139266	0551
139267	0382
139268	3076
139269	0861
139270	0178
139295	0065
139328	0133
139343	0162
139362	0079
139366	5346
139455	0037
139569	0079
139618	0400
139634	0143
139696	0086
139706	0133
139899	OEM
140110	OEM
140129	0551
140130	0382
140259	0285
140290	0037
140365	2333
140501	0086
140503	0102
140506	0561

DEVICE TYPE	REPL CODE
140525	0016
140600-02	OEM
140602-58	OEM
140715	0757
140763	0239
140764	0061
140971	0023
140972	0023
140973	0002
140976	0065
140977	0065
140979	0423
140995	0288
141003	0018
141008	0079
141018	0037
141019	0861
141020	0423
141134	3270
141135	2049
141187	0495
141227	0037
141254	1710
141255	2329
141256	1712
141259	0521
141270	2786
141279	0521
141280	2789
141290	2798
141295	0285
141302	0039
141330	0079
141331	0079
141332	0212
141335	0086
141343	0037
141344	0037
141345	0037
141355	0086
141367	0065
141370	0144
141421	0037
141429	0137
141464	1710
141489	3537
141558	0079
141581	0137
141711	0037
141719	OEM
141767	0079
141783	0079
141849	0015
141872-1	0023
141872-6	0015
141873	2953
141967	0931
142007-2	0753
142137	1329
142190	0037
142251	0551
142251(IC)	1980
142294	0723
142341	3832
142417	0495
142569	2811
142670	0039
142671	0285
142681	0239
142682	0344
142683	0079
142684	0079
142686	0079
142689	0065
142690	0264
142691	0042
142692	0468
142711	0079
142717	0468
142718	2786
142719	3089
142728	1197
142738	0402
142771	2449
142811	0079
142814	0409
142838	0037
142839	0037
142903	3832
143033	2786
143041	5869
143041(IC)	1554
143041(XSTR)	0861
143042	0079
143062	1350
143162	0917
143316	0079
143594	0348
143595	0015
143690	2395
143696	3790
143753	0636
143766	2827
143777	0565
143790	0505
143791	0037
143792	0079
143793	0016
143794	0079

DEVICE TYPE	REPL CODE
143795	0079
143796	0079
143797	0324
143798	0086
143802	0037
143803	0037
143804	0079
143805	0079
143806	0037
143807	0037
143808	3270
143821	2049
143822	3089
143837	0133
143847	0402
143913	1710
143940	0079
143963	0037
143964	0264
144004	0015
144011	4034
144012	2960
144015	4067
144017	4226
144018	4079
144022	4158
144024	4746
144026	0167
144028	2094
144030	0203
144031	0006
144033	0037
144034	0111
144035	1220
144037	0113
144039	0155
144040	1221
144043	0155
144044	0155
144045	0563
144047	0155
144050	0071
144051	0071
144052	0015
144054	0019
144056	0471
144057	0120
144072-2	0334
144076	0925
144182	0563
144581	0133
144582	0151
144585	0143
144612	4759
144691	0133
144761	0006
144848	4744
144849	1580
144850	4758
144851	4752
144852	4747
144853	4751
144854	5071
144855	4162
144856	0563
144857	0925
144858	0693
144859	0325
144860	0023
144861	0471
144863	0693
144920	3862
144921	2259
144922	0456
144923	1329
144966	0218
144967	1599
144968	0215
144969	1143
144970	0398
144971	0119
144973	1603
144974	0133
145113	1329
145172	0382
145173	0079
145174	OEM
145208	1197
145258	0086
145395	0079
145398	0079
145410	0037
145513	0344
145585	OEM
145586	0413
145648	0313
145660	0133
145671	0065
145683	0030
145687	0144
145712	1710
145730	OEM
145776	0037
145803	1328
145817	0317
145818	0215
145833	0079
145838	0155
145839	0023

DEVICE TYPE	REPL CODE
145840	0532
145847	OEM
145966	0203
146027	0124
146052	2786
146081	0925
146136	0023
146137	0023
146138	1329
146139	0042
146141	0079
146142	0037
146143	0264
146149	3976
146150	3784
146151	3832
146152	3089
146164	1491
146173	0123
146260	0057
146261	0210
146264	0420
146273	0157
146301	0079
146316	0348
146320	0162
146359	0037
146364-1	1997
146416	OEM
146460-2	1843
146484	0079
146512	0582
146514	4751
146521	0212
146522	0151
146569	1051
146570	0672
146571	0124
146572	0111
146575	0143
146576	0124
146595	4750
146596	0563
146631	0053
146641	0141
146642	0562
146643	1823
146644	0081
146732	3466
146742	0899
146823	6200
146826	0285
146846	0052
146847	0079
146848	0224
146849	0566
146850	0037
146851	0715
146854	0079
146855	0042
146856	0042
146857	0912
146858	3977
146893	1580
146894	1582
146895	1585
146896	0167
146897	0143
146899	0359
146900	0825
146902	1613
146903	0582
146904	0672
146908	0253
146911	1329
146912	0124
146933-503	0378
147007-3	0015
147015	3537
147111	OEM
147112-7	0126
147171-11	1217
147187-2-14	0071
147213	0133
147214	0143
147215	2542
147216	0015
147218	0133
147219	0165
147220	0181
147221	0062
147232	2656
147234	0215
147242	0582
147243	0148
147245	0191
147245-0-1	0144
147246	0563
147250	0525
147256	0033
147312	0018
147339	5185
147340	5158
147351-4-1	0975
147351-5-1	0160
147352-0-1	0142
147353-0-1	0007
147356-9-1	0144
147357-0-1	0126

If replacement code is OEM, contact original manufacturer for replacement.

DEVICE TYPE	REPL CODE
147357-1-1	0016
147357-2-1	0144
147357-4-1	0126
147357-7-1	0016
147357-9-1	0144
147359-0-1	0126
147360-1	0079
147363-1	0016
147389	0693
147399	OEM
147400	OEM
147406	0053
147437	4782
147438	0167
147464	1329
147477-2	0686
147477-7	0441
147477-7-1	0170
147477-8-2	0071
147477-8-3	0071
147477-8-7	0071
147477-8-8	0071
147477-8-10	0071
147478	0101
147478-8-2	0229
147499	0076
147546	0053
147549-1	0037
147549-2	0037
147555-1	0016
147617-12	0102
147639	OEM
147663	0037
147664	0079
147665	0037
147672	5802
147676-1	0330
147687	0004
147693	0071
147704	0144
147704-6-4	0137
147735	0686
147757	OEM
147886	0144
147887	0413
147888	0679
147922-2	0917
147924	0825
147943	1029
147974	0563
147975	0560
147993	0564
148056	OEM
148061	0079
148062	OEM
148070	1392
148085	0212
148123	0143
148124	3501
148132	2656
148134	3311
148146	1140
148229	0760
148235	0360
148236	0001
148237	0693
148246	0015
148256	0076
148257	0253
148259	0589
148272	0163
148295	0057
148446	OEM
148507	5186
148509	OEM
148529	0062
148531	0057
148533	2223
148664	OEM
148684	0894
148690	0350
148711	2223
148712	1015
148714	0199
148715	5562
148716	1819
148717	3753
148718	5230
148719	5194
148720	1585
148721	5225
148722	5169
148723	5226
148724	0360
148725	5227
148726	3731
148727	5547
148728	0360
148729	5192
148730	0215
148731	0498
148732	2072
148733	0001
148734	0101
148735	0921
148736	2053
148737	3198
148738	2468
148739	5018

DEVICE TYPE	REPL CODE
148740	0176
148741	0683
148742	1349
148743	0867
148744	3738
148745	0624
148746	0859
148747	3165
148748	0648
148749	4348
148750	0018
148751	0423
148751-147	0144
148752	0563
148753	2223
148754	1212
148755	0672
148756	0582
148757	0695
148758	1212
148759	3125
148760	0060
148761	0111
148762	0530
148763	0023
148764	0091
148765	1414
148766	0445
148769	0085
148770	3033
148771	0124
148772	0150
148777	0005
148778	0554
148779	0079
148780	0144
148781	0244
148782	0530
148801	0392
148803	0120
148804	0041
148805	1329
148875	OEM
148907	0079
148919	0392
148952	3528
148953	3529
148954	5888
148955	1212
148956	0923
148957	0062
148958	0700
148959	1603
148964	0313
148968	0212
148970	OEM
148984	0065
148991	1929
148995	0037
148996	0079
148997	0079
148998	0055
148999	0060
149007	1893
149014	OEM
149015	0475
149016	6386
149017	0624
149018	0625
149019	0678
149020	OEM
149033	OEM
149034	6143
149035	6135
149036	6135
149037	1516
149038	1321
149039	OEM
149040	0037
149041	0037
149042	5223
149249	OEM
149251	0239
149252	0167
149253	3997
149254	3977
149255	0026
149433	0343
149434	0120
149488	0672
149539	0080
149540	0053
149544	0224
149546	3383
149585	3731
149586	4258
149594	OEM
149598	0923
149599	4348
149615	0124
149616	0445
149676	0917
149729	OEM
149730	0143
149731	0795
149751	0320
149797	0442
149849	0582
149864	0672

DEVICE TYPE	REPL CODE
149870	3983
149882	5051
149930	0052
149938	1661
149982	1962
150006	OEM
150009	OEM
150038	1351
150068	OEM
150069	OEM
150070	OEM
150117	0076
150156	OEM
150322	0672
150325	0912
150330	0825
150340	2105
150342	4759
150344	0037
150363	0137
150364	0392
150365	1329
150366	0037
150368	5785
150369	4601
150371	1696
150373	1380
150374	2035
150400	OEM
150401	3012
150402	1458
150403	1458
150404	1601
150405	1601
150410	1490
150414	2422
150415	OEM
150416	2424
150417	2424
150418	0176
150419	2938
150420	OEM
150428B	OEM
150430	2105
150431	0870
150440	0427
150447	1402
150448	OEM
150513	0076
150514	0151
150515	3342
150516	1140
150517	2484
150518	0037
150519	0191
150521	0825
150522	1348
150523	0148
150524	0359
150527	0322
150528	0023
150529	0124
150530	0124
150531	0118
150533	0041
150534	0253
150535	0471
150536	1140
150573	0133
150644	OEM
150645	OEM
150646	2744
150651	0023
150671	0071
150703	1333
150704	OEM
150705	0442
150706	OEM
150711	OEM
150764	0532
150767	OEM
151010A	OEM
151010G	OEM
151050A	OEM
151050G	OEM
151056	0631
151100A	OEM
151100G	OEM
151241	0133
151245	OEM
151263	0037
151326	0224
151329	0382
151332	1513
151333	0683
151334	0041
151335	0052
151342	OEM
151343	OEM
151344	OEM
151345	OEM
151476	0102
151544	1041
151548	0333
151579	OEM
151693	0144
151974	0037
152050A	OEM
152050G	OEM

DEVICE TYPE	REPL CODE
152051	3977
152052	0012
152069	0720
152100A	OEM
152100G	OEM
152124	0277
152185	0148
152213	0062
152214	0151
152284	0133
152284(LED)	OEM
152286	0143
152329	0181
152348	0925
152363	OEM
152364	0176
152375	0686
152407	0879
152472	0181
152473K	0133
152500	4511
152544	0151
152545	1085
152546	OEM
152547	OEM
152548	0015
152550	0143
152551	0019
152623	0705
152624	0494
152625	0477
152627	0009
152672	OEM
152682	1674
152750	0253
152766	0050
152793	0445
152805	OEM
152835	0062
152841	0917
152854	0365
152855	0253
152857	0162
152869	0010
152887	0516
152898	2542
152920	1599
152921	0770
152923	0215
152929	3198
152935	0119
152970	0006
152971	1603
152972	2223
152978	0062
152981	OEM
153129	0371
153233	0037
153259	2223
153260	0127
153265	0628
153312	0161
153321	0050
153325	0037
153334	0155
153342	OEM
153350	0925
153351	0144
153399	0079
153433	2484
153443	2484
153444	OEM
153445	0036
153456	0176
153489	0144
153493	0124
153512	OEM
153518	0041
153519	0057
153534	0368
153535	0368
153545	1145
153568	0770
153612	2806
153630	OEM
153632	1458
153633	OEM
153634	OEM
153661	0825
153672	0023
153673	0494
153676	0551
153677	3764
153678	0055
153679	0042
153680	0285
153682	0901
153684	2786
153685	3977
153712	2026
153716	OEM
153826	0052
153844	0071
153875	3997
153882	0203
153937	0215
154002	0686
154027	1143

DEVICE TYPE	REPL CODE
154192	0477
154197	OEM
154201	OEM
154202	OEM
154222	6838
154223	OEM
154224	OEM
154225	OEM
154228	OEM
154230	OEM
154231	OEM
154505	0705
154515	3997
154564	1984
154572	2067
154573	OEM
154574	2068
154575	0037
154576	0699
154577	2066
154578	0917
154579	0009
154580	OEM
154581	0770
154582	0233
154583	0018
154584	0016
154587	0037
154588	0037
154589	0037
154590	0062
154646	0144
154647	0144
154648	2053
154649	0101
154650	0861
154651	0277
154652	2073
154653	0672
154654	0558
154655	0253
154656	0686
154704	0188
154730	2025
154736	OEM
154749	2008
154752	0015
154753	0489
154754	2019
154755	OEM
154756	0253
154765	0203
154766	0582
154767	2055
154768	0498
154770	0129
154771	1014
154772	1957
154773	0056
154774	0478
154782	0624
154783	2923
154784	5350
154785	0928
154807	0191
154812	0016
154822	2026
154831	0079
154832	0928
154856	0140
154988	0825
155001	OEM
155010A	OEM
155010G	OEM
155033	0672
155050A	OEM
155050G	OEM
155100A	OEM
155100G	OEM
155131	0724
155276	0080
155331	1080
155335	OEM
155339	1806
155444	0648
155495	1742
155508	2553
155509	2553
155519	0698
155520	OEM
155522	0077
155523	0445
155525	0936
155527	OEM
155556	0489
155558	0181
155565	0041
155566	0120
155568	0582
155632	2053
155677	0224
155678	0133
155679	2919
155680	0371
155681	2626
155682	0554
155683	1029
155684	0124
155708	1492

DEVICE TYPE	REPL CODE
155709	1555
155710	1921
155711	0969
155712	0973
155713	OEM
155714	0602
155727	OEM
155732	0071
155810	0471
155822	0053
155829	0917
155836	0253
155837	0181
155838	0118
155841	2923
155842	0460
155892	2886
155893	0471
155911	3651
156018	3020
156023	0007
156025	1248
156026	0371
156027	0124
156029	0720
156254	0014
156262	OEM
156297	0455
156313	0309
156317	0023
156369	0794
156372	0739
156468	0002
156477	0167
156545	2913
156546	2186
156548	0006
156549	3314
156567	OEM
156570	OEM
156571	2425
156580	5793
156581	OEM
156582	OEM
156583	OEM
156584	0249
156591	OEM
156592	OEM
156593	OEM
156594	OEM
156616	OEM
156637	0162
156688	0050
156689	0140
156708	OEM
156712	0883
156725	2073
156727	0133
156728	OEM
156729	OEM
156730	2666
156731	OEM
156733	1978
156736	1140
156790	0091
156791	0031
156795	2703
156800	0879
156801	0705
156891	OEM
156892	0004
156893	OEM
156894	5335
156895	2105
156896	0191
156897	0693
156910	0023
156911	3501
156912	5790
156914	OEM
156915	2425
156916	OEM
156917	0693
156918	5793
156919	0006
156920	0883
156928	0280
156931	0144
156932	1699
156933	2366
156934	0006
156935	0161
156936	0210
156937	0053
156938	0057
156939	0589
156942	OEM
156943	0600
156944	0148
156945	0686
156947	0133
156955	0323
156956	0241
156960	0489
156969	0120
156970	OEM
157068	4237
157142	0554
157246	OEM

DEVICE TYPE	REPL CODE
157301	0133
157306	0715
157309	OEM
157424	0133
157425	1513
157427	1083
157437	OEM
157564	1554
157575	1922
157627	1165
157633	0126
157635	0861
157637	1358
157652	OEM
157653	1429
157654	0413
157662	OEM
157663	OEM
157668	OEM
157669	1212
157677	1028
157679	0124
157681	7113
157688	OEM
157689	4406
157694	0323
157805	OEM
157806	2308
157808	0037
157815	OEM
157816	OEM
157832	0253
157833	0079
157835	OEM
157951	0887
157958	0530
157959	0887
157970	0087
157971	4258
157972	4138
157973	OEM
157975	1032
158040	0083
158061	0498
158062	OEM
158063	0498
158064	OEM
158066	0358
158075	0359
158077	0375
158078	OEM
158153	0064
158238	0121
158240	1553
158242	1140
158249	3727
158313	0133
158344	0071
158381	OEM
158387	OEM
158390	1742
158391	OEM
158487	3749
158578	1329
158584	6704
158624	1060
158625	1072
158626	1072
158705	OEM
158768	OEM
158770	OEM
158773	OEM
158836	0502
158837	0505
158869	OEM
159019	OEM
159020	OEM
159021	OEM
159036	OEM
159042	OEM
159044	OEM
159064	0903
159065	0907
159066	OEM
159069	OEM
159216	OEM
159234	OEM
159238	0062
159240	OEM
159242	0151
159249	1167
159250	0133
159277	OEM
159355	1292
159362	0111
159369	0455
159425	1379
159426	0686
159429	0248
159430	1385
159433	1389
159434	0441
159435	0626
159446	OEM
159475	0907
159483	OEM
159490	OEM
159582	OEM
159588	1611

DEVICE TYPE	REPL CODE	DEVICE TYPE	REPL CODE	DEVICE TYPE	REPL CODE	DEVICE TYPE	REPL CODE	DEVICE TYPE	REPL CODE	DEVICE TYPE	REPL CODE	DEVICE TYPE	REPL CODE
159649	0017	161066	0726	162682	0295	164717	0124	168984	0164	171030	0007	171162-252	0015
159890	0052	161077	0079	162706	0854	164719	0015	169113	0015	171030(SEARS)	0144	171162-265	0560
159945	0139	161078	OEM	162735	0124	164733	OEM	169114	0133	171030(TOSHIBA)	0016	171162-269	0143
159946	OEM	161079	0024	162736	0133	164739	OEM	169115	0143	171031	0007	171162-270	0143
159947	0433	161080	0053	162737	OEM	164742	OEM	169116	0015	171031(SEARS)	0144	171162-271	0133
159948	OEM	161081	0883	162739	OEM	164743	OEM	169117	0133	171031(TOSHIBA)	0127	171162-272	0012
159949	OEM	161117	OEM	162740	OEM	164744	OEM	169175	0164	171032	0144	171162-278	1144
159950	0648	161120	0151	162745	0461	164747	OEM	169194	0127	171032-1	2866	171162-279	1144
159951	0151	161123	0151	162772	OEM	164748	OEM	169195	0144	171033	0144	171162-280	0286
159959	1938	161149	0133	162845	OEM	164750	OEM	169196	0127	171033-1	3904	171162-282	0030
159960	1212	161193	0433	162846	OEM	164751	OEM	169197	0016	171033-1(MAGNAVOX)	1657	171162-285	1221
159961	0210	161194	0530	162850	OEM	164752	OEM	169198	0030	171033-2	3732	171162-286	0076
159962	OEM	161195	0648	162851	OEM	164753	OEM	169199	0157	171033-4	3906	171162-287	2046
159963	0052	161199	0636	162871	OEM	164767	OEM	169216	0142	171033-5	OEM	171162-288	0111
159970	1945	161247	0037	162902	OEM	164771	OEM	169359	0004	171034	0127	171162-291	1607
159971	1673	161249	0700	162909	0052	164819	OEM	169360	0004	171038	0086	171162-292	0012
159972	OEM	161265	OEM	162921	1259	164840	OEM	169361	0004	171039	0050	171162-1188	0715
159973	1140	161325	1604	162922	0332	164847	OEM	169362	0143	171039(SEARS)	0050	171174-1	1514
159974	1329	161326	2269	162929	0148	164874	0133	169363	0015	171039(TOSHIBA)	0050	171174-2	3376
160063	OEM	161362	OEM	162931	1396	164959	0148	169403	1313	171040	0079	171174-3	0423
160069	OEM	161392	2956	163033	0636	164964	3133	169501	0143	171040(SEARS)	0079	171175-1	1657
160082	1966	161419	OEM	163060	0898	164981	1176	169505	0127	171040(TOSHIBA)	0079	171175-2	1657
160083	2033	161437	OEM	163064	OEM	164982	1831	169558	0133	171042	OEM	171179-027	2029
160084	OEM	161447	OEM	163198	0951	164984	0215	169565	0015	171044	0184	171179-028	2031
160085	0042	161558	0724	163199	0760	164985	OEM	169570	0015	171044(SEARS)	0086	171179-036	0579
160087	OEM	161571	0016	163202	OEM	165358	0137	169574	0127	171044(TOSHIBA)	0079	171179-038	OEM
160088	OEM	161572	0016	163348	0124	165392	0127	169590	0023	171045	0184	171179-045	1018
160089	OEM	161573	3080	163353	0124	165572	0143	169679	0076	171046	0076	171179-051	1215
160090	0317	161579	0016	163355	OEM	165597	0015	169680	0076	171047	0168	171179-070	2062
160091	0077	161585	OEM	163397	0030	165667	0136	169765	0015	171047(TOSHIBA)	0074	171206-1	0144
160092	0636	161586	OEM	163434	0625	165668	0016	169771	1220	171048	0549	171206-2	0144
160094	2040	161588	0724	163441	0133	165735	0086	169773	0004	171049	0004	171206-4	0144
160175	2105	161669	OEM	163443	OEM	165736	0086	170090-1	OEM	171052	0184	171206-5	0144
160180	OEM	161705	0004	163506	OEM	165737	0693	170128	0693	171053-1	1681	171206-6	0212
160251	OEM	161714	0041	163512	OEM	165738	0178	170132-1	2866	171054	0842	171206-7	0007
160302	1978	161715	3159	163629	OEM	165739	0015	170133	0015	171057	OEM	171206-8	0224
160362	OEM	161717	0208	163755	0101	165740	0100	170287-1	3545	171088-1	OEM	171206-9	0127
160364	0636	161719	OEM	163793	OEM	165827	0016	170294	0079	171090-1	0144	171206-10	0127
160385	OEM	161727	OEM	163794	1292	165828	0016	170297	0015	171091-1	OEM	171206-11	0127
160386	OEM	161728	OEM	163795	OEM	165931	0127	170307-1	0160	171092-1	0002	171206-12	OEM
160387	OEM	161732	0129	163804	OEM	165932	0127	170308	0155	171139-1	0144	171206-13	0127
160412	OEM	161733	3173	163805	OEM	165976	0004	170338	0127	171140-1	0144	171206-14	OEM
160423	OEM	161734	0037	163806	OEM	165995	0144	170370	0133	171141-1	0144	171207-1	0144
160426	OEM	161736	0016	163807	0041	166040	0015	170373	0123	171142-1	OEM	171207-2	0144
160432	OEM	161737	0016	163816	OEM	166272	0144	170376	0160	171143-1	OEM	171207-3	0127
160433	OEM	161738	0016	163818	0030	166273	0143	170376-1	0160	171144-1	0127	171207-4	0325
160434	OEM	161739	OEM	163822	OEM	166400	0164	170388	0127	171149-016	0139	171207-5	OEM
160435	OEM	161743	0360	163823	2385	166593	0023	170407-1	0160	171149-017	0133	171207-6	OEM
160521	OEM	161745	0323	163891	OEM	166726	0023	170479-1	0160	171149-024	0139	171207-7	OEM
160545	OEM	161746	OEM	163892	OEM	166881	0015	170576	0079	171154	0380	171212-2	OEM
160547	0006	161748	OEM	163893	0693	166882	0004	170577	OEM	171162-004	0111	171217-2	OEM
160553	OEM	161749	OEM	163916	OEM	166883	0004	170580	0079	171162-005	0016	171217-3	0224
160554	1063	161752	0016	163917	0061	166906	0127	170666-1	0160	171162-006	0016	171217-4	0133
160555	OEM	161756	3185	163924	OEM	166907	2605	170668-1	0160	171162-008	0016	171268-1	OEM
160556	OEM	161757	3185	163925	OEM	166908	0136	170733-1	1149	171162-009	0016	171304	1023
160557	1513	161869	0365	163926	OEM	166909	0136	170750	0015	171162-026	0595	171305	0015
160562	OEM	161870	0275	163938	OEM	166917	0086	170753-1	0007	171162-027	0144	171318	0475
160568	OEM	161871	0002	164030	OEM	166918	0161	170756-1	0007	171162-039	0005	171319	0475
160569	OEM	161872	0002	164035	OEM	166919	0460	170783-1	2866	171162-042	0143	171320	0127
160574	0015	161877	0495	164049	0015	166920	0015	170783-3	0208	171162-072	0164	171338-1	0127
160575	OEM	161878	0531	164053	OEM	166921	0012	170794	2209	171162-073	0164	171338-2	0127
160576	2339	161879	OEM	164058	3133	166922	0015	170827-1	3545	171162-074	0004	171338-3	0127
160577	OEM	161880	OEM	164061	OEM	166985	0365	170850-1	0265	171162-075	0004	171369-1	OEM
160578	OEM	161896	0931	164062	OEM	166997	0164	170856	0015	171162-076	0004	171416	0015
160589	OEM	161906	0783	164064	OEM	166999	0050	170857	0133	171162-080	0004	171522	0004
160696	0079	161912	OEM	164114	OEM	167034	0015	170890-1	3545	171162-081	0164	171553	0380
160719	OEM	161919	OEM	164115	3133	167235	0222	170891-1	6411	171162-082	0160	171554	0111
160740	OEM	161920	1513	164116	1807	167263	0127	170906	0007	171162-083	0222	171555	0006
160741	2339	161921	OEM	164185	0787	167285	0222	170906-1	0007	171162-086	0160	171556	0436
160747	OEM	161935	OEM	164203	OEM	167413	0015	170953	0111	171162-089	0222	171557	0076
160752	1539	161936	2471	164204	OEM	167540	0155	170954	4571	171162-090	0822	171558	0076
160753	1492	161937	OEM	164214	OEM	167541	0111	170955	0111	171162-095	0552	171559	0151
160763	2489	161976	OEM	164215	OEM	167542	1591	170956	0701	171162-100	0111	171560	0170
160764	1921	161994	0649	164216	OEM	167543	0015	170963	0111	171162-108	0847	171561	0015
160768	OEM	161995	0451	164230	OEM	167544	0015	170964	4561	171162-113	0537	171657	0015
160830	OEM	161998	OEM	164231	OEM	167562	0143	170965	0015	171162-118	0007	171676	0111
160831	OEM	162002-039	0143	164242	OEM	167569	0016	170966-1	0015	171162-119	0016	171677	0111
160832	0770	162002-040	0136	164253	4130	167572	0123	170967	1802	171162-120	0004	171678	0111
160836	0089	162002-041	0136	164323	OEM	167679	0004	170967(TRAN)	1221	171162-121	0004	171814	0005
160837	OEM	162002-042	0136	164324	OEM	167680	0164	170967-1	0079	171162-124	0233	171840	0133
160838	OEM	162002-062	0160	164326	OEM	167688	0016	170968-1	0079	171162-125	0233	171841	0133
160839	0287	162002-062A	0160	164327	OEM	167690	0037	170970-1	0015	171162-126	0233	171842	0157
160844	2539	162002-090	0144	164328	OEM	167691	0042	170993	OEM	171162-128	0144	171843	0133
160851	OEM	162002-095	0160	164343	OEM	167956	0111	170994-1	1657	171162-129	0144	171894	0030
160852	0140	162002-39	0143	164344	OEM	167957	0111	171003	0604	171162-130	0144	171915	0127
160869	2026	162002-40	0136	164345	OEM	167958	0178	171003(SEARS)	0144	171162-131	0144	171916	0004
160872	OEM	162002-41	0136	164351	OEM	167998	0004	171003(TOSHIBA)	0079	171162-132	0016	171917	0004
160879	OEM	162021	0181	164353	0898	167999	0004	171004	0160	171162-143	0016	171982	2591
160890	1348	162037	0360	164354	1221	168339	0015	171005	0208	171162-161	0016	171983	0127
160972	0016	162039	0009	164492	OEM	168405	0111	171005(TOSHIBA)	0208	171162-162	0016	171984	0030
160973	0079	162163	OEM	164493	OEM	168567	0127	171009	0127	171162-163	0086	172201	0133
160979	0079	162201	1030	164494	OEM	168606	0725	171009(SEARS)	0144	171162-164	0566	172252	0579
161001	0123	162259	OEM	164502	OEM	168651	0111	171009(TOSHIBA)	0079	171162-169	0050	172253	0133
161006	0143	162297	OEM	164547	0188	168652	0030	171010-1	2866	171162-172	0649	172254	0030
161015	0123	162394	OEM	164588	4200	168653	0636	171015	0233	171162-180	0076	172272	1215
161016	0143	162446	OEM	164589	0017	168657	0144	171015(TOSHIBA)	0142	171162-186	0224	172318	5383
161016H	0079	162451	2282	164590	0031	168658	0144	171016	0136	171162-187	0224	172336	0126
161021	0133	162505	0018	164591	0162	168659	0144	171016(SEARS)	0004	171162-188	0155	172463	0042
161022	0365	162506	OEM	164593	1276	168660	0086	171017	0004	171162-190	0016	172547	0133
161027	0917	162507	OEM	164594	0451	168692	0120	171018	0004	171162-191	0016	172551	0015
161029	0015	162511	OEM	164595	0181	168692(VARISTOR)	0120	171019	0015	171162-192	OEM	172721	0023
161030	0015	162512	OEM	164599	OEM	168706	0133	171026	0076	171162-193	0126	172722	0133
161031	0015	162514	OEM	164600	OEM	168717	0050	171026(SEARS)	1367	171162-195	1367	172722(AFC)	0030
161032	0102	162665	OEM	164601	OEM	168906	0164	171026(TOSHIBA)	0079	171162-196	0133	172761	0007
161033	0102	162669	OEM	164602	2564	168907	0004	171027	0079	171162-197	0133	172762	0532
161034	0917	162673	OEM	164660	OEM	168910	0143	171028	0144	171162-202	0076	172763	0018
161037	0015	162674	2025	164705	OEM	168953	0208	171029	0007	171162-204	0552	172816	0164
161038	0143	162675	OEM	164707	OEM	168954	0004	171029(SEARS)	0144	171162-235	2016	172856	0628
161046	0244	162677	0151	164708	OEM	168983	0004	171029(TOSHIBA)	0127	171162-247	5470	173078	OEM

DEVICE TYPE	REPL CODE
173082	0164
173083	0164
173115	OEM
173116	3080
173117	OEM
173217	OEM
173219	OEM
173331	3133
173346	OEM
173350	OEM
173357	OEM
173425	OEM
173426	OEM
173427	OEM
173428	OEM
173461	OEM
173590	OEM
173598	0981
173752	4452
173780	0127
173901	OEM
173904	OEM
173906	OEM
173928	OEM
174025	OEM
174098	OEM
174130	OEM
174160	OEM
174169	OEM
174183	OEM
174222	OEM
174223	OEM
174283	0094
174285	0094
174370	0636
174372	1965
174373	0739
174374	OEM
174375	OEM
174376	OEM
174377	OEM
174378	OEM
174381	OEM
174431	0877
174449	OEM
174450	0002
174486	OEM
174489	0133
174576	0124
174631	OEM
174632	OEM
174633	OEM
174640	0724
174641	3478
174643	OEM
174660	OEM
174819	OEM
174858	0124
174863	0124
174994	OEM
175006-181	0050
175006-182	0050
175006-183	0050
175006-184	0050
175006-185	0050
175006-186	0004
175006-187	0144
175006-355	0030
175007-275	0016
175007-276	0016
175007-277	0086
175027-021	0016
175027-022	0085
175037	OEM
175043-023	0160
175043-058	0016
175043-059	0016
175043-060	0016
175043-062	0144
175043-063	0144
175043-064	0144
175043-065	0222
175043-066	0143
175043-068	0143
175043-069	0244
175043-081	0822
175043-81	0160
175043-100	0144
175043-107	0144
175341	0064
175342	OEM
175343	OEM
175344	0002
175382	OEM
175393	OEM
175394	OEM
175401	0144
175424	0700
175663	0016
175704	OEM
175705	OEM
175722	OEM
175886	0160
175910	4130
176144	OEM
176145	OEM
176146	OEM
176147	OEM
176148	OEM
176149	OEM
176150	OEM
176154	OEM
176155	OEM
176156	OEM
176157	OEM
176181	OEM
176222	3997
176223	3365
176224	3884
176225	0001
176226	3772
176258	0212
176296	3537
176363	0224
176383	OEM
176666	0686
176743	0142
176746	0253
176748	OEM
176751	0142
176852	OEM
176853	2957
176854	3236
176873	4383
176911	0137
176921	OEM
176925	0558
176980	0079
177092	0124
177095	OEM
177097	OEM
177098	OEM
177099	OEM
177101	OEM
177102	OEM
177105	0211
177106	6224
177107	OEM
177108	OEM
177109	OEM
177110	5586
177111	OEM
177112	2954
177113	OEM
177114	OEM
177117	OEM
177118	0062
177121	1329
177122	OEM
177189	OEM
177286	OEM
177361-1	0111
177383	OEM
177384	OEM
177388	OEM
177389	OEM
177390	OEM
177523	0077
177547	OEM
177549	OEM
177607	OEM
177612	3250
177613	OEM
177619	2413
177773	OEM
177788	1562
177789	1522
177791	1541
177807	0025
177850	OEM
177857	3997
177901	OEM
177963	OEM
177967	OEM
177972	OEM
177976	OEM
177978	0133
178032	OEM
178063	1032
178064	0887
178200	OEM
178274	OEM
178291	OEM
178295	OEM
178296	OEM
178297	OEM
178387	OEM
178388	OEM
178389	OEM
178392	OEM
178393	OEM
178448	OEM
178452	OEM
178476	1028
178514	OEM
178515	OEM
178519	OEM
178520	OEM
178525	0133
178526	0532
178541	OEM
178576	OEM
178577	OEM
178581	0295
178626	0064
178627	0795
178628	6354
178629	0023
178630	0023
178631	0023
178632	0705
178633	0700
178634	0023
178635	0052
178636	0049
178637	0031
178638	0181
178639	1875
178640	1005
178641	0077
178642	1347
178643	1347
178644	0760
178645	0137
178646	0041
178647	0127
178648	0076
178649	0148
178650	2910
178651	0018
178653	2058
178654	0951
178655	0148
178656	0643
178660	0095
178690	2872
178692	2042
178693	2194
178694	3236
178695	2666
178696	0599
178697	4930
178737	2263
178738	1206
178739	0803
178740	1633
178741	0360
178742	4310
178743	1628
178744	1239
178745	1292
178746	1028
178901	OEM
179047	OEM
179123	OEM
179320	OEM
179473	OEM
179509	OEM
179547	OEM
179603	0151
179627	OEM
179726	3156
179728	OEM
179729	OEM
179730	OEM
179732	OEM
179733	OEM
179734	OEM
179737	OEM
179740	1733
179741	1752
179743	3360
179744	OEM
179751	0922
179862	OEM
179866	OEM
179868	OEM
179890	OEM
179901	OEM
179925	OEM
180010-001	1696
180020	0820
180057	OEM
180060	1807
180113	OEM
180137	1541
180197	0124
180198	0494
180199	0630
180200	3280
180201	0151
180211	OEM
180212	OEM
180213	OEM
180214	0622
180215	5188
180217	1699
180218	0006
180219	0148
180221	0140
180223	OEM
180242	OEM
180323	OEM
180330	OEM
180338	0649
180518	OEM
180519	OEM
180520	OEM
180528	OEM
180530	OEM
180533	OEM
180545	OEM
180546	OEM
180547	OEM
180548	OEM
180549	OEM
180550	OEM
180551	OEM
180552	OEM
180554	OEM
180555	OEM
180556	OEM
180557	OEM
180558	OEM
180559	OEM
180560	OEM
180570	0124
180617	0015
180762	OEM
180763	OEM
180764	OEM
180765	OEM
180766	OEM
180767	OEM
180768	OEM
180769	OEM
180770	OEM
180771	OEM
180772	OEM
180773	OEM
180774	OEM
180775	OEM
180776	OEM
180777	OEM
180778	OEM
180780	OEM
180781	OEM
180783	OEM
180784	OEM
180785	OEM
180786	OEM
180787	OEM
180788	OEM
180789	OEM
180817	OEM
180831	OEM
180851	OEM
180872	OEM
180873	6362
180876	OEM
180877	OEM
180878	OEM
180880	OEM
180881	OEM
180884	OEM
180885	OEM
180887	6562
180889	OEM
180892	OEM
180893	2323
180894	6130
180895	0195
180896	0589
180961	OEM
180968	OEM
181003-7	0144
181003-8	0144
181003-9	0144
181012	0111
181015	0037
181018	OEM
181023	0016
181030	0037
181034	0037
181038	0491
181039	OEM
181039(IC)	OEM
181040	OEM
181073	0133
181151	OEM
181165	OEM
181178	0195
181181	OEM
181183	0631
181214	0016
181239	OEM
181256	OEM
181257	OEM
181281	OEM
181282	OEM
181283	OEM
181284	OEM
181285	OEM
181286	OEM
181287	OEM
181291	OEM
181292	OEM
181295	OEM
181296	OEM
181297	OEM
181298	OEM
181299	OEM
181300	OEM
181311	OEM
181312	OEM
181314	OEM
181315	5201
181316	OEM
181317	OEM
181324	OEM
181325	OEM
181326	OEM
181327	OEM
181328	OEM
181329	OEM
181330	OEM
181331	OEM
181335	OEM
181336	OEM
181337	OEM
181338	OEM
181362	OEM
181363	OEM
181440	OEM
181465	OEM
181466	OEM
181467	OEM
181468	OEM
181475	OEM
181476	OEM
181477	OEM
181478	OEM
181479	OEM
181498	OEM
181503-6	0144
181503-7	0144
181503-9	0144
181504-1	0144
181504-2	0016
181504-7	0144
181506	OEM
181506-7	0144
181507	OEM
181515-4	0086
181515-6	0086
181515-7	0086
181515-9	0086
181516	OEM
181519	1276
181520	0295
181522	0879
181523	OEM
181524	OEM
181526	0133
181530	1742
181533	0703
181553	OEM
181615	OEM
181616	OEM
181619	0124
181654	OEM
181655	OEM
181681	0133
181765	OEM
181772	OEM
181773	OEM
181774	OEM
181775	OEM
181778	OEM
181779	OEM
181780	OEM
181795	OEM
181797	OEM
181826	OEM
181827	OEM
181828	OEM
181829	OEM
181836	OEM
181860	0101
181874	OEM
181876	OEM
181910	OEM
181911	OEM
181912	OEM
181913	OEM
181914	OEM
181915	OEM
181917	OEM
181931	OEM
181940	OEM
181959	OEM
181966	0295
181991	OEM
181993	OEM
182027	OEM
182028	OEM
182029	OEM
182030	OEM
182031	OEM
182032	OEM
182033	OEM
182034	OEM
182035	OEM
182036	OEM
182037	OEM
182038	OEM
182040	OEM
182041	OEM
182042	OEM
182072	OEM
182073	OEM
182074	OEM
182075	OEM
182076	OEM
182100	OEM
182101	OEM
182107	OEM
182115	OEM
182267	OEM
182268	OEM
182301	OEM
182302	OEM
182304	0023
182323	OEM
182324	OEM
182325	OEM
182349	OEM
182350	OEM
182352	OEM
182353	OEM
182354	OEM
182355	OEM
182356	OEM
182357	OEM
182358	OEM
182406	OEM
182407	OEM
182444	OEM
182503	3715
182510	0996
182603	OEM
182604	OEM
182605	OEM
182606	OEM
182607	OEM
182610	OEM
182611	OEM
182612	OEM
182613	OEM
182739	OEM
182753	OEM
182754	OEM
182755	OEM
182756	OEM
182757	1733
182758	OEM
182759	OEM
182760	OEM
182761	OEM
182762	OEM
182763	OEM
182764	OEM
182825	OEM
182827	0041
182838	1396
182898	OEM
182919	OEM
183013	0654
183014	0701
183015	0127
183016	0127
183017	0079
183018	0127
183019	0127
183030	0079
183031	0079
183032	0037
183033	0133
183034	0218
183035	1242
183044	2557
183045	OEM
183046	OEM
183050	OEM
183051	OEM
183052	0165
183053	OEM
183056	OEM
183071	OEM
183072	OEM
183073	OEM
183074	1348
183086	OEM
183087	OEM
183096	OEM
183097	OEM
183098	OEM
183104	OEM
183107	2703
183189	OEM
183405	2557
183532	0410
183896	4011
183898	2094
184073	0645
184074	5376
184078	5354
184086	3749
184087	0903
184093	6832
184106	0041
184574	OEM
184576	OEM
184577	OEM
184578	OEM
184579	OEM
184580	OEM
184581	OEM
184582	OEM
184583	OEM
184585	OEM
184586	OEM
184589	OEM
184591	OEM
184592	OEM
184593	OEM
184615	OEM
184627	OEM
184631	OEM
184632	OEM
184633	OEM
184638	OEM
184640	OEM
184642	OEM
184643	OEM
184644	OEM
184753	OEM
184757	OEM
184779	OEM
184798	0124
184812	OEM
184830	1023
184918	OEM
184920	OEM
184921	OEM
184922	OEM
184929	0242
184985	OEM
185001	OEM
185022	0102
185090	OEM
185104	OEM
185105	OEM
185107	OEM
185118	OEM
185147	OEM
185200	2703
185202	0023
185206	0023
185207	3549
185231	4141
185233	0101
185236	0079
185237	1875
185335	0124
185411	0133
185436	0102
185628	0516
185704	0015
185878	2518
185879	0071
185880	0023
185881	1414
185882	0064
185883	0140
185884	0253
185885	0005
185895	3817
185896	1988
185897	4350
185898	1329
185899	4041
185900	3725
185901	5229
185915	0076
185916	0127
185917	0006
185918	1562
185919	1513
185920	0728
185921	1790
185922	1308
185923	0076
186001	OEM
186050	OEM
186065	OEM
186070	OEM
186112	OEM
186311	OEM
186342A	0086
186383	OEM
186404	0907
186499	5662
186512	0898
186544	0280
186545	0280
186584	0898
186599	OEM
186600	OEM
186610	0037
186611	0280
186642	0079
186645	0079
186648	0041
186650	OEM
186659	0004
186660	4130
186670	OEM
186671	1329
186687	OEM
186703	OEM
186741	0280
186742	0031
186773	0334
186797	3462
186801	0013
187064	OEM
187217	0037
187218	0016
187267	0795
187278	OEM
187553	OEM
187583	2069
187628	OEM
187640	OEM
187659	0887
187660	1032
187749	0832
187774	OEM
187775	OEM
187784	OEM
187901	OEM
187963	OEM

If replacement code is OEM, contact original manufacturer for replacement.

DEVICE TYPE	REPL CODE
188056	0124
188086	0732
188158	OEM
188165	0103
188180	0037
188211	OEM
188226	0126
188232	0002
188233	0062
188234	0023
188235	0091
188236	0023
188237	0158
188238	0071
188240	0032
188241	0257
188242	3927
188243	4422
188244	4346
188245	0050
188246	4129
188271	1162
188272	0276
188273	0168
188274	3619
188400-34	0725
188551	2954
188660-01	0410
188960	0050
188961	0050
188962	0050
188963	0050
188964	0023
188965	0023
188966	0050
188972	0050
188973	0755
189123	OEM
189124	OEM
189148	OEM
189149	OEM
189150	OEM
189151	OEM
189152	1028
189153	OEM
189154	OEM
189155	OEM
189309	OEM
189310	OEM
189311	OEM
189312	OEM
189313	OEM
189314	OEM
189315	OEM
189316	OEM
189319	OEM
189322	OEM
189327	OEM
189328	OEM
189330	OEM
189333	OEM
189424	OEM
189852	OEM
189853	OEM
189972	OEM
189988	OEM
189998	0126
189999	OEM
190206	1090
190208	0064
190224	0112
190227	OEM
190233	OEM
190253	OEM
190277	OEM
190404	0170
190425	0160
190425A	0160
190426	2250
190427	0050
190428	0016
190429	2033
190482	OEM
190483	2129
190484	OEM
190486	0150
190540	OEM
190681	0062
190692	OEM
190714	0007
190715	0016
190716	0143
190718	OEM
190719	OEM
190724	0491
190994	1733
191142	2129
191205	OEM
191208	OEM
191290	OEM
191291	OEM
191292	OEM
191293	OEM
191294	OEM
191295	OEM
191296	0631
191297	0624
191298	OEM
191299	0624
191300	0042
191313	OEM
191316	OEM
191318	OEM
191319	OEM
191328	OEM
191387	OEM
191407	0725
191490	OEM
191491	OEM
191556	OEM
191726	OEM
191729	OEM
191747	OEM
191862	0500
191868	0360
191869	0631
191870	0725
191985	0626
191986	0626
191989	0626
192027	0057
192062	2675
192063	OEM
192064	OEM
192066	OEM
192067	OEM
192068	OEM
192069	OEM
192105	0741
192113	OEM
192135	OEM
192144	OEM
192229	0012
192271	0365
192272	OEM
192276	3389
192277	OEM
192285	0334
192339	OEM
192344	OEM
192351	OEM
192448	OEM
192450	OEM
192544	OEM
192681	1742
192689	0622
192696	OEM
192702	OEM
192703	OEM
192742	2957
192743	OEM
192782	OEM
192789	OEM
192828	OEM
192848	0057
192849	OEM
192902	OEM
192903	OEM
192904	OEM
192909	0286
192963	OEM
192965	0079
193022	0466
193057	1214
193058	1386
193082	3236
193083	OEM
193084	OEM
193159	0495
193207	0410
193228	OEM
193230	OEM
193309	3817
193333	3828
193625	OEM
193626	OEM
193717	0133
193832	0724
194086-3	0847
194168	OEM
194169	OEM
194170	OEM
194171	OEM
194172	OEM
194188	OEM
194189	OEM
194222	OEM
194223	OEM
194225	OEM
194226	OEM
194233	OEM
194234	OEM
194239	OEM
194241	OEM
194243	0086
194244	OEM
194285	0334
194305	0624
194320	OEM
194474-8	0160
194561	0023
194765	OEM
194899	OEM
194907	0925
194917	0023
194929	OEM
195040	OEM
195143	OEM
195288	OEM
195319	0071
195320	0124
195322	OEM
195323	OEM
195336	1790
195478	0077
195492	OEM
195494	OEM
195495	OEM
195496	OEM
195510	0013
195511	0097
195601-6	0004
195606	0770
195617	0123
195648-6	0015
195703	0546
195736	OEM
195768	OEM
195791	4411
195827	OEM
195841	OEM
195857	OEM
195877	OEM
195880	0295
195881	0295
195885	OEM
195889	OEM
195927	2828
195928	4080
195946	OEM
195955	OEM
196000	OEM
196002	OEM
196023-1	0016
196023-2	0016
196058-4	0160
196062	0124
196064-3	0004
196117	0076
196121	1260
196122	3254
196129	OEM
196148-0	0160
196176	2089
196183-5	0160
196183-7	0004
196184-3	0015
196187	4495
196188	4496
196189	OEM
196190	OEM
196192	1682
196193	OEM
196200	2069
196201	3938
196211	OEM
196212	OEM
196220	OEM
196221	0917
196259-4	0015
196296	OEM
196317	OEM
196330	OEM
196331	4524
196333	4130
196337	OEM
196374	1576
196388	OEM
196392	0079
196393	0126
196414	4537
196432	1783
196436	OEM
196496	0738
196501-7	0160
196511	OEM
196512	OEM
196514	OEM
196515	OEM
196516	0738
196517	1212
196582	OEM
196584	OEM
196597	0071
196598	1733
196607-9	0160
196634	0685
196639	1257
196779-9	0086
196779-9-1	0086
196780-1	0086
196961	OEM
197323	1646
197372	0124
197374	0062
197376	0592
197381	3290
197464	0124
197591	1992
197646	OEM
197648	OEM
197666	0081
197713	OEM
197878	OEM
197890	OEM
197935	0832
197936	OEM
198003-1	0016
198003-2	0016
198005-1	0698
198007-3	0016
198010-1	0136
198013-P1	0016
198014-1	0086
198020-1	0086
198020-2	0086
198020-3	0086
198023-1	0016
198023-2	0086
198023-3	0016
198023-4	0016
198023-5	0016
198024	0037
198030	0016
198030-2	0016
198030-3	0016
198030-4	0016
198030-6	0016
198030-7	0016
198031-1	0016
198031-2	0016
198034-1	0103
198034-2	0103
198034-3	0103
198034-4	0103
198034-5	0103
198035-1	0086
198035-3	0086
198036-1	0037
198038-1	0178
198038-3	0178
198038-4	0178
198039-0507	0103
198039-1	0178
198039-3	0178
198039-6	0103
198039-7	0103
198039-501	0178
198039-503	0178
198039-506	0103
198039-507	0103
198042-2	0016
198042-3	0016
198045-4	0086
198047-1	0086
198047-2	0086
198047-3	0086
198047-5	0086
198047-6	0086
198048-1	0086
198048-2	0086
198049-1	0178
198050	0037
198051-1	0016
198051-2	0016
198051-3	0016
198051-4	0016
198063-1	0925
198064-1	0130
198065-1	0126
198065-3	0126
198067-1	0016
198072-1	0086
198074-1	0126
198075-1	0168
198077-1	0086
198078-1	0126
198079-1	0130
198079-2	0103
198400	1301
198409-1	1844
198409-5	0033
198409-6	1041
198409-11	0141
198409-12	0562
198409-13	1823
198409-14	1831
198410-1	1301
198581-1	0016
198581-2	0016
198581-3	0016
198589	0124
198590	5149
198591	OEM
198592	OEM
198593	OEM
198594	OEM
198595	OEM
198596	OEM
198597	0015
198598	OEM
198599	0795
198600	OEM
198601	OEM
198602	OEM
198603	0295
198606	OEM
198607	OEM
198609	4725
198743	0079
198744	OEM
198745	0037
198746	0079
198747	0037
198748	0324
198749	0037
198750	0111
198751	0439
198752	OEM
198763	0715
198763-1	0084
198763-2	0534
198763-3	0097
198763-4	0765
198763-5	0097
198763-6	0765
198763-7	0097
198763-8	0765
198763-9	0105
198763-10	0777
198763-11	0109
198763-12	0538
198763-13	0116
198763-14	0801
198763-15	0122
198763-16	0816
198764-5	1124
198764-6	1107
198764-13	0319
198765-1	3189
198765-2	1630
198765-3	2901
198765-4	0254
198765-5	1124
198765-6	1107
198765-7	1126
198765-8	1111
198765-9	0805
198765-10	0258
198765-11	1195
198765-12	1644
198765-13	0319
198765-14	0267
198766-1	0764
198766-2	0764
198766-3	0764
198766-5	0764
198766-7	0764
198766-8	0764
198773-1	0124
198774	0248
198775-1	0124
198776-1	0124
198779-1	0124
198785	0124
198785-1	0124
198785-2	0124
198794	2490
198794-1	0600
198799-4	0124
198800-1	0415
198801-1	2698
198802	0725
198809-1	0133
198810-1	0124
198813-1	0124
198816	1413
198818	0415
198936-2	0147
199596	0133
199662	0193
199807	0861
199919	0289
199985	0015
200028-7-28	0004
200062-5-31	0004
200062-5-32	0004
200062-5-33	0004
200062-5-34	0004
200062-5-36	0015
200062-6-32	0015
200064-6-103	0144
200064-6-104	0050
200064-6-105	0144
200064-6-106	0050
200064-6-107	0079
200064-6-108	0004
200064-6-109	0975
200064-6-110	0975
200064-6-111	0004
200064-6-115	0015
200064-6-119	0015
200064-6-120	0015
200067	0037
200076	0016
200137	3515
200155	OEM
200156	OEM
200157	0137
200158	OEM
200165	3510
200167	5431
200168	OEM
200200	0111
200200-700	0111
200208	0030
200220	0037
200251-5377	0016
200252	0086
200259-700	0905
200419	OEM
200420	2104
200431	OEM
200433	0037
200471	OEM
200472	OEM
200553	OEM
200554	OEM
200555	OEM
200556	OEM
200557	OEM
200558	OEM
200566	OEM
200567	OEM
200574	OEM
200648-26	0143
200781-702	0142
200838	5276
201034	0142
201133	OEM
201146	OEM
201963	0679
202048	OEM
202055	0217
202066	OEM
202071	5683
202315	0143
202315(THOMAS)	0133
202463	0015
202609-0713	0016
202617	1220
202862-518	0124
202862-947	0016
202907-047P1	0016
202909-577	0037
202909-587	0037
202909-827	0103
202911-737	0037
202913-057	0086
202914-010	1823
202914-010/250-060	0081
202914-010/250-070	1176
202914-010/250-100	1831
202914-020	0141
202914-030	0562
202914-040	1041
202914-050	0033
202914-417	0016
202915-627	0016
202917-137	0126
202920-150	0980
202922-237	0016
202922-280	3722
202925-047	0491
203364	0037
203718	0126
203743	2887
203751	1634
203780	3915
203781	1326
203793	OEM
203811	OEM
204117	0076
204201-001	0086
204210-002	0016
204211-001	0103
204280	OEM
204290	OEM
204292	OEM
204969	0016
205032	0037
205048	0037
205049	0037
205367	0037
205702	0050
205782-97	0079
205782-103	0079
205919	6124
206017	OEM
206038	OEM
206049	OEM
206050	0901
206180	0242
206185	0242
206190	0242
206395	OEM
206617	0133
206730	OEM
206889	0097
207119	0142
207125	0285
207417	0688
207484	1543
207826	OEM
207827	2704
207828	OEM
207831	OEM
207871	OEM
207878	OEM
207887	OEM
207890	0119
208434	3926
209185-962	0160
209417-0714	0144
209741	OEM
209919	0072
210074	0016
210076	0037
210226	OEM
210227	OEM
210236	0079
210533	OEM
210543	OEM
210548	OEM
210646	OEM
210704	OEM
210711	OEM
210736	OEM
210783	OEM
210786	OEM
210911	OEM
211040-1	0130
211083	0124
212465	OEM
212717	1144
213217	1221
214105	0015
214396	0160
215002	0050
215008	0050
215031	0050
215038	0050
215053	0164
215071	0822
215072	0016
215074	0198
215075	0086
215081	0079
215089	0222
215487	OEM
215488	OEM
215489	OEM
215490	OEM
215491	OEM
215495	OEM
215496	OEM
215524	OEM
215525	OEM
215526	OEM
215527	OEM
215528	OEM
215529	OEM
215530	OEM
215531	4385
215532	OEM
215533	OEM
215534	OEM
215669	0015
215861	OEM
216001	0143
216003	0123
216014	0015
216020	0015
216024	0244
216099	OEM
216434	OEM
216435	OEM
216445-2	0016
216449	0015
216467	OEM
216817	0015
216986	0160
217119	0440
217230	0440
217306	OEM
217309	OEM
217320	OEM
217321	OEM
217335	OEM
217654	OEM
217657	OEM
217658	OEM
217892	0160
218012DS	0085
218398	OEM
218401	OEM
218402	OEM
218403	OEM
218502	0004
218503	0004
218511	0086
218537	0126
218612	0015
218817	OEM
218822	OEM
218987	OEM
219016	0600
219025	0807
219245	0229
219301	0160
219361	0160
219412	OEM
219440	0160
219832	OEM
219935	0015
219940	0160
220114	OEM
220536	OEM
220637	OEM
220638	OEM
220639	OEM
220673	0460
220674	0079
220675	2391
221128	0015
221158	0086
221600	0016
221601	0600
221602	0160
221605	0160
221856	0136
221857	0016
221897	0016

If replacement code is OEM, contact original manufacturer for replacement.

DEVICE TYPE	REPL CODE
221918	0016
221924	0600
221940	0160
221941	0160
222131	0016
222509	0004
222611	0015
222867	0071
222915	0160
223083	OEM
223124	0004
223215	0015
223216	0015
223323	0015
223357	0071
223358	0015
223365	0160
223366	0004
223367	0600
223368	0600
223369	0050
223370	0600
223371	0004
223372	0280
223462	0015
223467	0015
223473	0280
223474	0050
223475	0280
223482	0208
223483	0004
223484	0004
223485	0004
223486	0004
223487	0136
223489	0229
223490	0160
223576	0160
223659	OEM
223684	0600
223704	OEM
223720	0229
223724	0015
223753	0015
223806	OEM
223810	0004
223848	4022
223849	4024
223850	OEM
223851	OEM
223852	OEM
223853	4027
223854	OEM
223855	OEM
223856	OEM
223857	OEM
223867	0127
223868	0286
223869	OEM
223870	OEM
223871	0905
223872	OEM
223881	OEM
223882	OEM
223883	4041
223884	OEM
223909	OEM
223942	OEM
223948	OEM
223957	OEM
224159	0015
224503	0160
224506	0016
224584	0280
224586	0050
224587	0050
224597	0015
224696	0004
224774	0015
224780	0137
224820	0208
224857	0004
224873	0160
225200	0015
225265	0319
225267	0229
225300	0208
225301	0157
225311	0050
225316	0157
225410	0143
225592	0015
225593	0004
225594	0050
225594A	0136
225595	0160
225596	0160
225600	0050
225688	OEM
225700	OEM
225702	OEM
225925	0160
225927	0160
226058	0015
226181	0280
226182	0015
226237	0015
226334	0143
226338	0280

DEVICE TYPE	REPL CODE
226344	0143
226441	0600
226517	3046
226546	0015
226634	0160
226788	0015
226789	0440
226791	0208
226922	0071
226924	0004
226999	0160
227000	0144
227015	0229
227348	0015
227517	0368
227565	0015
227566	0160
227567	OEM
227675	0122
227676	0319
227720	0015
227722	OEM
227724	0122
227744	0015
227752	0280
227801	0122
227804	0160
227917	0004
228007	0170
228229	0160
228230	0160
228287	0004
228417	0151
228558	0160
228560	0071
229017	0368
229026	OEM
229042	0122
229045	0440
229088	0319
229133	0050
229220	OEM
229392	0144
229522	0015
229805	0087
230084	0178
230199-4	2977
230208	0160
230209	0600
230214	0086
230218	0015
230225-2	3415
230233	0086
230253	0004
230256	0600
230259	0004
230523	0160
230524	0004
230525	0004
230756	0319
230768	0764
230773	0122
231017	0155
231140-01	0144
231140-04	0975
231140-07	0144
231140-09	0975
231140-11	0160
231140-15	0016
231140-21	0004
231140-23	0144
231140-26	0975
231140-28	0233
231140-31	0144
231140-33	0160
231140-34	0144
231140-36	0007
231140-37	0007
231140-43	0007
231140-44	0144
231140-45	0208
231150	0015
231161	0248
231339	0015
231374	0016
231375	0086
231378	0103
231588	0280
231665	0015
231669	0015
231672	0160
231706	0004
231797	0160
231923	0015
232017	0076
232021	0055
232194	0160
232203	0015
232268	0178
232359	0103
232359A	0103
232519	0122
232520	0319
232674	0160
232675	0160
232676	0050
232678	0016
232680	0050
232681	0136

DEVICE TYPE	REPL CODE
232840	0144
232841	0086
232949	0600
233011	0015
233025	0102
233117	0155
233150	0002
233305	0440
233307	0440
233507	0280
233508	0440
233509	0160
233561	0015
233597	0015
233735	0086
233944	0086
233945	0280
233969	0126
234015	0050
234024	0086
234076	0004
234077	0160
234078	0440
234178	0160
234552	0015
234553	0071
234565	0015
234566	0160
234611	0015
234612	0016
234630	0004
234631	0050
234758	0016
234761	0015
234763	0016
235157	0122
235192	0016
235194	0004
235200	0016
235205	0016
235206	0016
235299	0319
235312	0160
235313	0015
235382	0071
235541	0087
235543	0015
235546	0015
235997	0074
236039	0007
236251	0144
236265	0600
236266	0012
236282	0233
236285	0016
236286	0016
236287	0086
236288	0975
236433	0126
236706	0144
236709	0004
236854	0130
236907	0144
236935	0160
237020	0144
237021	0144
237024	0144
237025	0016
237026	0144
237028	1400
237070	0087
237075	0338
237223	0016
237227	0015
237421	0087
237450	0142
237452	0160
237453	0015
237509	0319
237785	0144
237840	0144
237929	0015
238368	0016
238417	0004
238418	0004
239097	0087
239219	0649
239221	0015
239429	0071
239517	0042
239612	0086
239713	0130
239917	0839
239970	0016
240003	0004
240006	0004
240055	0087
240076	0087
240077	0319
240388	0142
240401	0016
240402	0126
240403	0975
240404	0074
240456	0015
240564	0133
240594	0015
240603	0015
241052	0126

DEVICE TYPE	REPL CODE
241184	0030
241249	0144
241295	0071
241302	0071
241420	0319
241517	0006
241657	0103
241778	0144
241960	0144
242029	0015
242141	0071
242183	0160
242221	0004
242226	0015
242422	0126
242460	0126
242590	0144
242758	0016
242759	0016
242838	0160
242958	0126
242960	0144
243115	0087
243168	0975
243215	0440
243318	0144
243364	0015
243645	0144
243815	0440
243837	0208
243843	1413
243939	0004
244007	0280
244357	0142
244817	0368
245078-3	0144
245117	0165
245217	0057
245489	OEM
245517	0143
245568-2	0847
245917	1946
248017	0018
248717	2710
248817	0133
248917	0080
249217	0080
249508-3	0133
249588	0136
250031	OEM
250032	0299
250062	OEM
250066	OEM
250400	0208
251026	0323
251026-01	1414
251026-02	1203
251026-03	1203
251104-04	OEM
251527-01	OEM
251527-03	OEM
251535-01	OEM
251535-02	OEM
251536-01	OEM
251536-02	OEM
251538-02	OEM
251640-03	OEM
251641-02	OEM
251715-01	OEM
251819-07	0133
251819-16	0124
251828-01	OEM
251828-03	OEM
251829-01	OEM
251853-02	OEM
251871-01	OEM
251913-01	OEM
251968-03	OEM
252034-02	OEM
252308-01	OEM
252333-02	OEM
252371-03	OEM
252372-01	OEM
252817	0023
253704	0440
255728	0004
255903	0319
256068	0160
256071	0160
256122	0319
256126	0280
256127	0600
256217	0151
256317	0006
256319	0006
256417	0151
256480	0440
256517	0224
256617	0113
256717	0030
256728	0319
256729	0122
256730	0319
256817	0076
256917	0076
257017	0151
257242	0440
257243	0440
257340	0004

DEVICE TYPE	REPL CODE
257341	0160
257385	0600
257403	0160
257470	0280
257473	0004
257534	0440
257536	0160
257540	0144
258017	0414
258882	0015
258884	0319
258990	0160
258993	0600
259315	0015
259368	0015
259878	0015
260116-699	OEM
260429	0015
260468	0600
260565	0144
261401	0440
261463	0015
261488	0440
261586	0050
261596	0015
261898	0015
261970	0160
261975	0015
262066	0016
262111	0170
262112	0015
262113	0004
262114	0160
262116	0178
262309	0440
262310	0319
262370	0160
262546	0170
262638	0126
262648	0122
262872	0015
263424	0170
263561	0142
263807	0122
263856	0160
263857	0142
265074	0144
265115	0319
265164	0015
265217	0139
265235	0015
265236	0229
265240	0016
265241	0144
265634	0015
265771	0050
266583	0229
266685	0016
266686	OEM
266702	0280
267272	0441
267791	0103
267797	0144
267878	0130
267898	0016
267899	0016
268717	0693
269367	0208
269374	0004
270642	0071
270744	0160
270745	0160
270746	0160
270779	0122
270780	0160
270781	0600
270785	0160
270786	0764
275131	0016
275612	0160
275831	0071
275845	0975
276097	0015
276160	0086
276331	0016
276413	0086
276415	0086
279317	0208
279417	0079
279517	0042
279617	1591
279717	0936
279817	0558
279917	0151
280017	0124
280117	0139
280217	0071
280317	0012
280417	0005
281917	0649
282217	2793
282317	0155
282601	0015
290458LGD	0079
291509	0130
293118	1685
297065C03	0050
297074C11	0126
297240-1	0004

DEVICE TYPE	REPL CODE
300233	0015
300312	0319
300315	0015
300486	0600
300524	0015
300532	0122
300536	0600
300538	0004
300540	0004
300541	0004
300542	0600
300550	0015
300732	0170
300733	0319
300735	0122
300774	0600
301586	0015
301591	0016
301606	0086
302342	0016
304581B	0016
309004	0723
309316-01	OEM
309412	0160
309421	0004
309504	1413
310017	0004
310030	0050
310035	0004
310132	0050
310157	0633
310158	0633
310159	0004
310160	0004
310162	0050
310201	0164
310204	0050
310221	0050
310223	0136
310224	0004
310225	0004
310254	0467
310389-01	OEM
310651-01	OEM
310654-02	OEM
310654-03	OEM
310654-05	OEM
312558-01	OEM
312558-02	OEM
314849-01	OEM
314858-01	OEM
315009-01	OEM
315012-01	OEM
315014-01	OEM
315014-04	OEM
315020-01	OEM
315092-01	OEM
315093-01	OEM
317053-01	OEM
317054-01	OEM
317093	2884
317208	0137
318004-01	OEM
318004-03	OEM
318005-01	OEM
318005-03	OEM
318005-04	OEM
318006-01	OEM
318011-03	OEM
318012-01	OEM
318013-01	OEM
318018-02	OEM
318018-04	OEM
318019-02	OEM
318019-04	OEM
318020-03	OEM
318020-05	OEM
318022-02	OEM
318023-02	OEM
318087-01	OEM
318091-01	OEM
320007	0143
321119-901-020	OEM
322968-17	0004
322968-140	0160
322968-141	0975
322968-167	0004
325302-01	OEM
325341-08	OEM
325502-01	6181
325502-03	1898
325505-01	0299
325505-02	0376
325506-01	0041
325572-01	OEM
326830	1823
326832	1041
326833	2097
326836	1831
326844	0033
326845	0081
326846	0141
326852	0333
326853	0358
326862	0562
329814	0081
330003	0015
330018	0015
330019	0015

DEVICE TYPE	REPL CODE
331378	1041
336637-20	3418
336638-21	OEM
339002	0682
339003	0685
339009	0559
339300	0232
339300-2	0232
339486	1368
339868	0264
340085DC	OEM
340085DM	OEM
340085FC	OEM
340085FM	OEM
340085PC	3450
340097PC	4441
340098PC	0432
340160PC	1359
340161PC	1373
340162PC	1388
340163PC	1407
340174DC	1551
340174DM	1551
340174FC	OEM
340174FM	OEM
340174PC	1551
340175DC	1528
340175DM	1528
340175FC	OEM
340175FM	OEM
340175PC	1528
340192PC	1764
340193PC	1776
340194DC	1769
340194DM	1769
340194FC	OEM
340194FM	OEM
340194PC	1769
340195DC	1784
340195DM	1784
340195FC	OEM
340195FM	OEM
340195PC	1784
341181	0079
346015-15	0144
346015-16	0144
346015-17	0144
346015-18	0144
346015-19	0144
346015-20	0144
346015-21	0144
346015-22	0144
346015-23	0086
346015-24	0016
346015-25	0144
346015-30	0086
346015-37	0144
346016-1	0004
346016-11	0004
346016-14	0016
346016-16	0086
346016-17	0086
346016-18	0016
346016-19	0016
346016-25	0016
346016-26	0016
346016-27	0086
346016-63	0338
348048-2	0015
348053-3	0015
348054-1	0015
348054-2	0015
348054-5	0015
348054-6	0229
348054-7	0071
348054-9	0015
348054-10	0015
348054-11	0229
348054-14	0229
348054-15	0071
348055-2	0015
348055-3	0015
348057-8	0071
348057-9	0071
348057-17	0157
348058-2	0071
350171-1002	0015
361317-1	0123
373003	0004
373117	0004
373119	0004
373401-1	0232
373404-1	0361
373405-1	0512
373406-1	0697
373407-1	0873
373408-1	1024
373409-1	1313
373410-1	1368
373411-1	1404
373412-1	0117
373413-1	1685
373414-1	1158
373423-1	0492
373424-1	1426
373427-1	1208
373428-1	1715
373429-1	1206
373708-1	1820

If replacement code is OEM, contact original manufacturer for replacement.

DEVICE TYPE	REPL CODE
373712-1	0834
373713-1	1432
373714-1	1185
373714-2	1202
373715-1	1168
373716-1	1574
373718-1	0569
373721-1	1339
374109-1	0467
374110-1	1275
380049	0466
380205-01	OEM
380259-01	OEM
390059-01	OEM
390089-01	OEM
390302-01	OEM
390303-01	OEM
390304-01	OEM
390308-01	OEM
390309-02	OEM
390317-01	OEM
390318-01	OEM
390319-01	OEM
390341-01	OEM
400931	1176
401003-001-0	0037
401113-1	0141
401113-2	1823
401113-3	1041
401113-4	0081
401113-5	0562
401182	2078
405919-35AD	0122
405919-45AD	0816
405965-30A	0604
405965-35A	0126
410221(RCA)	1712
414020	OEM
424863	0124
433836	0016
437476	0042
442077	0879
442079	0053
442085	0041
443070	0532
452077	0325
454549	0280
454760	0050
455947	0101
463984-1	1024
464010	0015
464070	0015
475018	0242
476171-18	0102
480235A57	0137
480235AS7	OEM
485752-090	0917
489043A06	0006
489751-001	0143
489751-020	0917
489751-025	0079
489751-026	0079
489751-027	0144
489751-028	0037
489751-029	0079
489751-030	0079
489751-031	0037
489751-032	0925
489751-033	0042
489751-037	0086
489751-038	0086
489751-039	0007
489751-040	0016
489751-041	0016
489751-042	0037
489751-043	0142
489751-044	0042
489751-045	0004
489751-047	0007
489751-048	0127
489751-049	0127
489751-052	0127
489751-058	0127
489751-097	0037
489751-107	0016
489751-108	0004
489751-109	0004
489751-113	0004
489751-114	0004
489751-115	0975
489751-119	0103
489751-120	0127
489751-121	0007
489751-122	0016
489751-123	0079
489751-124	0037
489751-125	0016
489751-127	0007
489751-128	0007
489751-129	0086
489751-130	0037
489751-131	0144
489751-137	0144
489751-143	0144
489751-144	0086
489751-145	0144
489751-146	0037
489751-147	0144
489751-148	0144

DEVICE TYPE	REPL CODE
489751-162	0144
489751-163	0222
489751-164	0079
489751-165	0144
489751-166	0016
489751-167	0144
489751-168	0144
489751-169	0144
489751-171	0144
489751-172	0079
489751-173	0127
489751-174	0178
489751-175	0528
489751-206	0144
489751-208	0325
489752-001	0143
489752-003	0143
489752-004	0913
489752-005	0015
489752-08	0143
489752-013	0015
489752-014	0015
489752-015	0015
489752-016	0102
489752-017	0484
489752-020	0917
489752-022	0015
489752-025	0015
489752-026	0102
489752-027	0015
489752-028	0015
489752-029	0087
489752-031	0143
489752-035	0015
489752-036	0123
489752-037	0015
489752-038	0277
489752-040	0170
489752-041	0002
489752-042	0143
489752-043	0015
489752-044	0196
489752-045	0137
489752-049	0143
489752-050	0015
489752-051	0015
489752-052	0917
489752-053	OEM
489752-054	0288
489752-066	0015
489752-072	0474
489752-073	0015
489752-088	0030
489752-089	0917
489752-090	0917
489752-091	0317
489752-092	0015
489752-094	0015
489752-095	0007
489752-096	0102
489752-097	0015
489752-108	0015
489752-109	0170
489752-123	0015
489752-124	0162
489752-125	0123
489752-127	0030
489752-169	0030
489765-005	0196
489850-004	0143
500001	0123
500003	0071
500009G	0143
500859	0015
500904	OEM
501010	0015
501152	0015
501170	OEM
501193	OEM
501343	OEM
501697	OEM
501783	OEM
502382	OEM
502394	OEM
502395	OEM
502616	OEM
502617	OEM
502754	OEM
502823	OEM
502824	OEM
503146-1	0913
504720	0133
504833	0580
505254	2622
505287	0086
506911	0242
508590	0361
510007	0004
511534S	0123
511994-4	OEM
512021	OEM
513031S	2556
514023	0079
514045	0127
514067S	0126
514068S	0791
514072S	0126
515039S	0086
515041S	0368

DEVICE TYPE	REPL CODE
515043S	0079
515045S	0079
516009S	0325
518005(0)	OEM
518022S	1478
520120-1001	OEM
530015-2	0143
530019-1002	0124
530042-1	OEM
530043-1	0196
530045-1	0484
530045-2	0484
530045-3	0484
530045-4	0484
530051-1	0015
530051-2	0015
530051-4	OEM
530057-1	0015
530060-1	OEM
530063-1	0123
530063-2	0143
530063-4	0143
530063-6	0015
530063-7	0015
530063-10	0143
530063-11	0015
530063-12	0015
530063-13	0133
530063-14	0123
530065-1	0123
530065-2	0143
530065-3	0143
530065-1002	0143
530065-1002A	0143
530065-1003	0143
530070-1	OEM
530070-4	OEM
530071-1	0242
530071-2	0015
530071-3	0015
530071-1011	0133
530072-1	0015
530072-2	0015
530072-4	0015
530072-5	0015
530072-6	0015
530072-7	0143
530072-8	0133
530072-9	0015
530072-10	0015
530072-11	0015
530072-14	0015
530072-15	0015
530072-18	0133
530072-1001	0123
530072-1002	0133
530072-1006	0124
530072-1008	0133
530072-1009	0133
530072-1010	0133
530072-1011	0133
530072-1014	0087
530072-1015	0133
530072-1017	0015
530072-1019	0015
530073-0030	0256
530073-2	0705
530073-3	0170
530073-4	0636
530073-5	0636
530073-6	0137
530073-8	0071
530073-9	0002
530073-12	0002
530073-13	0137
530073-14	0137
530073-15	0420
530073-16	0421
530073-17	0002
530073-18	0441
530073-20	0820
530073-21	0531
530073-22	0495
530073-23	0100
530073-24	0402
530073-26	0420
530073-28	0531
530073-30	0495
530073-31	0157
530073-32	0234
530073-40	0162
530073-1013	0137
530073-1015	0420
530073-1016	0421
530073-1017	0002
530073-1021	0262
530073-1022	OEM
530073-1023	0100
530073-1024	0402
530073-1025	0245
530073-1028	0531
530073-1029	0274
530073-1030	0256
530073-1031	0636
530073-1034	0025
530073-1035	OEM
530073-1044	0253
530082-1	0015
530082-2	0071

DEVICE TYPE	REPL CODE
530082-3	0133
530082-4	0087
530082-1002	0071
530082-1003	0015
530082-1004	0015
530084-4	0015
530085-2	0143
530086-1	0015
530087-2	0474
530088-1	0071
530088-2	0071
530088-3	0087
530088-4	0015
530088-1002	0277
530088-1003	0015
530088-1004	0015
530089-2	1030
530091-1	3733
530092-1	0143
530092-2	0143
530092-1001	0143
530092-1002	0133
530093-1	0196
530093-2	0484
530093-3	0484
530093-1001	0015
530094-1	0015
530095-1	0015
530096	0774
530096(-1)	0774
530096-1	1303
530096-2	0474
530096-3	1303
530097-1	OEM
530097-2	0474
530097-3	0102
530098-1	0015
530098-1001	0015
530099-1	0015
530099-3	0015
530099-4	0015
530099-5	0196
530099-6	0015
530101-1001	0133
530104-1	0030
530104-2	0030
530104-2(8)	0030
530104-1001	0030
530105-1	0143
530105-1001	0123
530105-1002	OEM
530106-001	0023
530106-1	0920
530106-1001	0920
530109-1	0015
530111-1	0015
530111-1001	0015
530111-1002	0071
530112-1	0015
530113-1	0400
530113-2	0071
530113-1001	0102
530115-1001	0143
530116-1	0288
530116-3	0133
530116-1001	0133
530116-1003	0015
530118-2	0137
530119-1	0288
530119-5	0288
530119-7	0757
530119-8	0378
530119-9	0757
530120-1	0277
530122-1	0015
530122-2	0102
530123-1	0143
530123-3	0102
530123-4	0102
530123-5	0023
530123-7	0102
530124-1	0015
530124-2	0757
530124-3	0015
530126-1	0015
530127-1	0143
530127-4	0015
530127-5	0196
530127-6	0015
530128-1	1030
530129-1	0188
530132	0774
530132-1	0774
530135-1	0071
530135-2	0124
530135-3	0015
530135-1003	0015
530136-1	1303
530140-1	0030
530140-1(8)	0030
530142-7	OEM
530142-8	0030
530144-1	0133
530144-3	0133
530144-1001	0124
530144-1002	0133
530144-1003	0133
530144-1004	0124
530145-100	0170

DEVICE TYPE	REPL CODE
530145-120	0137
530145-130	0053
530145-339	0188
530145-369	OEM
530145-569	0157
530145-689	0025
530145-1100	0170
530145-1470	0999
530145-1569	0157
530146-1	1149
530146-2	1149
530148-3	0102
530148-4	0102
530148-1003	0102
530148-1004	0102
530149-9	0757
530150-1	0133
530151-1	0133
530151-1001	0023
530151-1003	0023
530152-1	0277
530153-1	1501
530154-1	0133
530157-130	0365
530157-569	0157
530157-689	0025
530157-870	3026
530157-950	5696
530157-1100	0170
530157-1101	0170
530157-1120	0052
530157-1131	OEM
530157-1150	0002
530157-1200	0531
530157-1220	0705
530157-1390	0350
530157-1569	0157
530157-1829	0244
530157-1919	0057
530157-1950	0098
530160-1	0280
530162-1	0015
530162-1001	0015
530163-120	0137
530165	2985
530165-1	1501
530165-2	1501
530165-3	1501
530165-4	1501
530165-5	2982
530165-6	2546
530165-7	1054
530165-8	2982
530165-9	1054
530165-10	2982
530165-11	2546
530165-12	1054
530165-13	2546
530165-14	2985
530165-15	2982
530165-16	2982
530165-17	2546
530165-18	2985
530165-20	3667
530165-21	1054
530166-1004	0636
530166-1005	0186
530166-1006	0269
530166-1007	0186
530166-1013	0186
530166-1017	0162
530167-1	0039
530167-1001	0166
530170-1	0133
530171-1	0023
530171-3	0015
530171-1001	0155
530171-1002	0015
530171-1003	0015
530171-1569	0253
530176-1	1329
530179-1	0133
530179-1001	0133
530179-1002	0133
530179-1003	0015
530180-1001	0015
530181-1	0133
530181-3	0124
530181-1001	0133
530181-1002	0124
530181-1003	0124
530181-9001	0133
530181-9003	0133
530182-1	OEM
530184-2	0102
530184-1001	0102
530184-1002	0102
530185-1	0473
530191-5	3280
530192-120	0137
530192-1120	0137
530192-1180	0495
530192-1200	0137
530192-1240	0686
530210-1001	0012
530221-1	0817
530222-1	0071
530224-3	0200
530226-1	0015

DEVICE TYPE	REPL CODE
530226-2	0102
530226-1002	0102
530248-1	OEM
530257-1	0494
530300-1	OEM
530303-1	OEM
530306-3	OEM
530551-6	0015
530972-14	0102
532003	0178
533013	0133
533034	0015
534001H	0133
535007H	OEM
535151-1001	0015
537200	0004
537428	0050
539148-1004	0102
539157-569	0253
540204	0007
540205	0233
543995	0037
551015	0004
551026	0015
551034	0540
551051	0004
552005	0071
552006	0015
552007	0102
552010	0015
552308	0016
552503	0037
558479P	OEM
558875	0232
558876	0314
558877	0512
558878	0697
558879	0873
558880	1024
558881	1172
558882	1313
558883	1208
558885	0569
559507	0268
559509	0743
559510	1275
559613	1185
560004	0265
560020S	0087
560165-20	3667
561103	0349
564171	OEM
567312	0198
568101	0143
570000-5452	0016
570000-5503	0016
570004-503	0016
570005-452	0016
570005-503	0016
570009-01-504	0016
570029	0103
570030	0460
570031	0161
573001	0004
573003	0004
573005	0004
573011	0004
573012	0004
573018	0004
573022	0004
573024	0143
573029	0004
573034	0143
573036	0004
573037	0208
573101	0144
573103	0004
573110	0004
573114	0004
573117	0004
573118	0211
573119	0004
573125	0004
573142	0004
573152	0004
573153	0004
573166	0160
573184	0004
573199	0975
573200	0004
573212	0975
573303	0136
573328	0004
573329	0136
573330	0136
573335	0050
573336	0050
573356	0280
573366	0136
573371	0050
573398	0050
573402	0136
573405	0050
573406	0050
573422	0004
573428	0136
573432	0004
573467	0016
573468	0016

DEVICE TYPE	REPL CODE
573469	0016
573471	0136
573472	0144
573474	0144
573475	0079
573479	0016
573480	0016
573481	0016
573494	0144
573495	0144
573501	0233
573515	0142
573518	0136
573529	0004
573532	0086
574003	0004
575001	0143
575002	0143
575005	0143
575009	0143
575019	0143
575024	0030
575028	0015
575037	0917
575042	0015
575047	0102
575048	6879
575049	0102
575050	0015
575051	0015
575091	0143
575995	0474
576001	0004
576005	0004
576054	0133
576063	0143
580029	0015
581070	0103
581078	2394
581078-1	2394
586343-1	1233
586344-1	1354
599995	1030
600008-001	0064
600080-413-001	0016
601030	0143
601032	0050
601040	0050
601052	0050
601054	0136
601054(SHARP)	0440
601065	0208
601113	0007
601122	0079
602032	0160
602040	0164
602051	0164
602060-76	OEM
602061-73	OEM
602070-05	OEM
602075	0050
602081	0012
602113	0144
602113(SHARP)	0016
602122	0086
602190-01	OEM
602190-03	OEM
602190-06	OEM
602190-34	OEM
602190-38	OEM
602190-39	OEM
602190-40	OEM
602190-44	OEM
602190-45	OEM
602191-58	OEM
602191-59	OEM
602210-03	OEM
602210-05	OEM
602210-06	OEM
602210-07	OEM
602210-08	OEM
602210-09	OEM
602210-10	OEM
602210-18	OEM
602210-26	OEM
602909-2A	0016
602909-3A	0126
602909-7A	0086
603020	0050
603030	0050
603031	0160
603040	0050
603112	0050
603113	0007
603114	0015
603122	0016
603312	0050
604020	OEM
604022-75	OEM
604030	0050
604040	0050
604080	0050
604112	0136
604113	3390
604122	0016
604407	0133
605020	OEM
605030	0004
605112	0136

If replacement code is OEM, contact original manufacturer for replacement.

DEVICE TYPE	REPL CODE
605113	0536
605122	0222
605131	0774
606020	0164
606112	0004
606113	0015
606131	0015
607020	OEM
607030	0208
607101	0196
607112	0975
607113	1303
607122	0142
608020	OEM
608030	0015
608101	0015
608112	0975
608112(SHARP)	0604
608113	0015
608122	0086
609020	0136
609030	0015
609112	0716
609113	0015
610002-011	1667
610002-014	1698
610002-026	1382
610002-038	0996
610002-074	0243
610002-086	0291
610002-123	0979
610010-413	1134
610010-470	2759
610020	0015
610020-917	2318
610024-1	0150
610030	0123
610035	0004
610035-1	0280
610035-2	0211
610036	0004
610036-1	0280
610036-2	0280
610036-3	0280
610036-4	0211
610036-5	0004
610036-6	0211
610036-7	0211
610036-8	0004
610039	0160
610039-1	0160
610040	0004
610040-1	0004
610040-2	0004
610041	0144
610041-1	0144
610041-2	0144
610041-3	0144
610042	0007
610042-1	0144
610043	0004
610043-1	0004
610043-2	0004
610043-3	0004
610043-4	0004
610043-6	0211
610043-7	0211
610044-3	OEM
610044-4	OEM
610045	0144
610045-1	0144
610045-2	0144
610045-3	0016
610045-4	0016
610045-5	0016
610046-7	0144
610050	0050
610050-1	0050
610050-2	0050
610050-3	0050
610051	0050
610051-1	0050
610051-2	0050
610051-4	0050
610052	0136
610052-1	0004
610053	0050
610053-1	0050
610053-2	0050
610055	0050
610055-1	0050
610055-2	0050
610055-3	0050
610056	0136
610056-1	0050
610056-2	0050
610056-3	0050
610056-4	0050
610056-456P4	0143
610061-1	0050
610063-1	0642
610064-1	0074
610067	0160
610067-1	0160
610067-2	0160
610067-3	0265
610068	0160
610068-1	0265
610069	0144
610069-1	0144
610070	0016
610070-1	0016
610070-2	0016
610070-3	0016
610070-4	0079
610071	0142
610071-1	0142
610071-2	0142
610072	0144
610072-1	0144
610072-2	0144
610073	0144
610073-1	0007
610073-13	0137
610073-1013	OEM
610074	0136
610074-1	0037
610074-2	0280
610075-1	0233
610076	0016
610076-1	0016
610076-2	0016
610077	0016
610077-1	0016
610077-2	0016
610077-3	0079
610077-4	0079
610077-5	0079
610077-6	0079
610078	0016
610078-1	0016
610079	0004
610079-1	0004
610079-2	0111
610080	0004
610080-1	0004
610083	0037
610083-1	0693
610083-2	0693
610083-3	0037
610083-4	0693
610083-5	OEM
610083-6	0037
610083-9001	0037
610088	0004
610088-1	0004
610088-2	0004
610091	0144
610091-1	0123
610091-2	0144
610092	0144
610092-1	0144
610092-2	0144
610093-1	0037
610094	0079
610094-1	0079
610094-2	0079
610094-3	0111
610096	0144
610096-1	0144
610099	0126
610099-1	0126
610099-2	0126
610099-3	0004
610099-5	0126
610099-6	0037
610100	0144
610100-1	0007
610100-3	0144
610102-1	0150
610104-2	OEM
610106	0160
610106-1	0160
610107-1	0144
610107-2	0086
610110	0126
610110-1	0037
610110-2	0037
610111	0196
610111-1	0196
610112	0015
610112-1	0925
610113-1	0400
610113-2	0400
610120-1	0037
610121-1	0316
610121-1(PUT)	0316
610121-1(SCR)	0316
610121-2	0316
610121-2(BISW)	4233
610121-2(PUT)	0316
610121-3	0316
610122	0170
610122-1	4527
610122-2	0363
610122-3	0313
610122-6	0313
610123-1	0003
610124-1	0600
610125-1	0037
610126-1	0004
610126-2	0208
610128-2	0111
610128-4	0144
610129-1	0126
610129-D	0224
610131-2	0338
610132	0016
610132-1	0016
610134-1	0693
610134-2	0693
610134-4	0693
610134-6	0693
610135-1	0233
610136-1	0150
610139-1	0224
610139-2	0144
610140-1	0103
610141-2	0016
610142-1	0155
610142-2	0016
610142-3	0079
610142-4	0016
610142-5	0079
610142-6	0144
610142-7	0079
610142-8	0079
610142-10	OEM
610144-1	0285
610144-2	0285
610144-3	0233
610144-4	0233
610144-6	0233
610144-101	0285
610145-1	0330
610146-1	0079
610146-3	0079
610146-5	0079
610147-1	0079
610147-2	0037
610148-1	0086
610148-2	0079
610148-2A	0079
610148-3	0086
610148-4	0086
610149-3	0925
610150-1	0224
610150-2	0079
610150-3	0007
610151-1	0111
610151-2	0079
610151-3	0111
610151-4	0079
610151-5	0111
610153-1	0042
610153-3	0042
610153-5	0042
610153-6	0042
610155-1	0423
610157-3	0551
610157-4	0382
610158-2	0037
610158-3	0037
610161-1	0103
610161-4	0103
610162-3	3327
610162-4	0042
610162-7	0925
610162-8	0042
610164-1	0325
610165-1	0079
610166-1	0212
610167-1	0079
610167-2	0079
610167-4	0016
610168-1	0079
610168-2	0079
610174-1	0144
610180-1	0224
610181-1	0008
610181-2	0007
610181-4	OEM
610186-1	0007
610189-1	0313
610190-1	0264
610190-3	0161
610190-4	0264
610194-1	0065
610194-2	0642
610194-3	0065
610195-1	0423
610195-2	0854
610195-3	0042
610195-4	0925
610195-2002	OEM
610195-2003	OEM
610202-1	0382
610202-2	0382
610203-1	0212
610203-2	0212
610203-3	0212
610203-4	0212
610203-5	0212
610203-6	0212
610209-1	0037
610213-1	0086
610213-2	0086
610216-2	0065
610216-3	0065
610217-1	0177
610217-2	0177
610217-5	0642
610222-5	OEM
610223-1	0037
610223-2	OEM
610224-1	0111
610224-9001	0111
610226-1	0111
610226-2	0111
610226-5	0111
610226-6	0111
610227-1	0382
610227-3	OEM
610228-1	0551
610228-3	1173
610228-4	1173
610228-5	0824
610232-1	0079
610232-2	0076
610232-9001	0076
610232-9002	0079
610233-1	3038
610234-1	OEM
610241-1	0065
610242-1	0065
610245-1	0177
610245-2	0270
610246-1	0037
610249-1	0224
610250-1	0285
610250-3	0285
610291	OEM
610356-5	0111
610358-1	0212
610358-2	0212
610358-3	0212
610360-2	0037
610361-1	1367
610361-5	1367
610370-1	OEM
610386-1	0144
610392-1	0144
610392-2	0313
610395-1	0236
610395-2	0681
610397-1	0925
610397-2	0042
610419-1	0224
610421-3	0042
610421-4	0925
610421-7	OEM
610427-2	0264
610427-3	OEM
610428-2	0824
610433-4	OEM
610435-1	OEM
610435-2	OEM
610517-3	OEM
610525-2	OEM
610540-1	OEM
610559-1	OEM
610569-1	6587
611001-3	2352
611001-4	2353
611001-5	1413
611001-6	1413
611002-3	OEM
611003-3	0058
611003-4	0058
611003-5	0058
611004-3	0717
611004-4	0717
611013-1	4337
611018-1	1222
611019-1	0617
611020	0004
611064	0385
611065	1432
611066	1368
611071	1052
611111	0015
611112	OEM
611132	0196
611563	0232
611564	0314
611565	0361
611566	0512
611567	0697
611568	1024
611569	0743
611570	1185
611571	1313
611572	1208
611573	0834
611730	1926
611731	1921
611844	1368
611845	0385
611870	1158
611872	0942
611878	3158
611900	0942
611901	1829
612003-1	OEM
612004-1	OEM
612005-1	0167
612005-2	0167
612006-1	0701
612006-1M	0701
612006-1Z	0701
612007	0664
612007-1	0664
612007-2	0664
612007-3	0664
612008-2	0692
612009-2	OEM
612010-2	OEM
612011-2	OEM
612012-2	OEM
612013	OEM
612014-2	OEM
612015-2	OEM
612016-2	OEM
612017-2	OEM
612018-2	OEM
612019-2	OEM
612020	0604
612020-1	0632
612021-1	0654
612022	OEM
612024-1	2457
612024-3	2457
612025-1	2743
612025-2	2743
612025-3	2743
612029-1	0349
612029-3	0349
612042-1	2482
612042-2	2472
612044-1	0856
612045-1	2557
612048-4	2283
612049-1	OEM
612054-2	2722
612061-1	0802
612063-1	OEM
612067	0354
612069-1	0354
612070-1	0352
612072-1	0751
612074-1	0413
612075-1	0520
612075-3	0520
612076-1	3735
612076-2	3735
612076-4	3735
612077-2	2755
612080-1	2247
612082-1	0848
612082-2	0848
612082-3	0848
612091-1	2242
612092-1	1313
612100-1	1108
612103-3	0624
612103-4	2239
612105-1	OEM
612106-1	OEM
612107-2	2239
612107-4	0624
612112	0015
612113-1	0360
612114-1	OEM
612120-1	3397
612123-3	OEM
612124-1	OEM
612125-1	OEM
612126	0912
612126-1	0912
612130	0015
612130-1	OEM
612132	0015
612144-1	2557
612159-1	1206
612160-1	OEM
612160-3	OEM
612175-1	OEM
612183-1	0803
612184-1	1846
612185-1	0385
612186-1	1143
612187-1	OEM
612188-1	1082
612189-1	3672
612189-2	3672
612191-1	OEM
612192-1	2086
612194-1	0088
612195-1	3912
612197-1	0943
612199-1	1527
612200-1	0243
612230-1	1654
612236-2	OEM
612247-2	0950
612261-1	1527
612272-1	2049
612273-1	3270
612275-1	1899
612289-1	OEM
612291-1	1888
612294-1	3790
612298-1	OEM
612303-1	0490
612304-1	1022
612305	1057
612305-1	1057
612305-2	1057
612306-3	OEM
612308-1	1867
612310-3	OEM
612318-2	OEM
612319-1	OEM
612330-4	OEM
612331-2	2817
612332-1	3710
612332-6	3711
612334-1	0521
612336-1	OEM
612337-5	OEM
612338-3	2812
612347-1	2831
612351-1	0167
612352-1	3981
612353-1	OEM
612380-1	OEM
612412-2	6910
612438-1	OEM
612448	0624
612449-1	OEM
612472-1	OEM
612479-1	0624
612479-3	OEM
612479-4	OEM
612479-5	1321
612480	OEM
612485-2	OEM
612493-1	0035
612494-1	0360
612496-1	2842
612541-1	OEM
612554-1	OEM
612556-2	OEM
612565-1	OEM
612566-1	OEM
612576-1	OEM
612577-1	OEM
612587-1	OEM
612593-2	OEM
612599-0003	OEM
612626-5	OEM
612658-0003	OEM
612664-2	OEM
612668-0002	OEM
612669-0002	OEM
612731-0001	OEM
612771-1	OEM
612792-1	OEM
613020	0015
613031	OEM
613112	0144
613128P	OEM
613128P(TECEC-0913)	OEM
613130	0015
613132	1303
614010	0015
614012	OEM
614020	0143
614033-3	OEM
614112	0917
615004-8	0133
615010	0143
615130	1030
616010	0196
617020	OEM
617871-1	0160
618020	OEM
618139-1	0160
618580	0325
619009-1	0178
619020	OEM
619094	0415
619130	0015
619256	2018
619256-1	2017
619550-1	1899
630002-001	0076
630003-110	4330
630006-001	4317
630007-001	2033
630008-907	2033
630063	0242
633977	0196
650196	0211
650845	0133
650854	0133
650859-1	0211
650859-2	0211
650859-3	0211
650860	0600
650970	0160
651012	0211
651030	0015
651038	0015
651202	0160
651236	0211
651891	0155
651955-1	0155
651955-2	0155
651955-3	0155
651956	0086
651995-1	0016
651995-2	0016
651995-3	0016
652072	0155
652085	0160
652086	0160
652091	0155
652092	0015
652230	0155
652231	0086
652321	0233
652615	0015
654000	0155
654001	2136
654032	0484
654420	0242
655319	0160
656064	1893
656204	0016
656524	0155
656719	0155
656746	0155
657161	0517
657179	0460
657180	0161
657181	0460
657874	0723
658577	0079
658578	0079
658657	0155
660030	0004
660031	0265
660059	0004
660060	0004
660064	0050
660070	0086
660072	0004
660074	0086
660077	0160
660082	0004
660084	0050
660085	0050
660094	0160
660095	0160
660097	0160
660103	0160
660138	0905
660144	0440
660144A	0440
661010	0196
670850	0265
670850-1	0265
671077-6	0079
699739	0015
700021-00	0774
700043-00	0015
700055-00	0196
700063-00	0015
700080	0103
700080A	0103
700083	0103
700083A	0103
700180-00	0143
700191	0178
700195	0178
700230-00	0016
700231-00	0016
700647	0015
700663	0015
700664	0015
701584-00	0037
701589-00	0037
701662-00	0133
701678-00	0007
702407-00	0086
702415-00	0566
703227-1	OEM
703639-1	3735
703639-2	3735
705269	OEM
717126-505	1907
717136-1	0704
717136-15	1017
717399-22	1350
717399-73	1929
720240	0079
720453	0110
720454	0953
720455	0242
720456	0540
720457	0965
720458	0540
720463	0965
731009	0004
733293	0165
740183	0015
740247	0822
740289	0015
740306	0111
740402	0143
740417	0004
740437	0079
740438	0111
740439	0111
740440	0111
740441	0079
740442	0111
740443	0822
740461	0155
740462	0155
740463	0155
740466	0016
740470	0016
740471	0222
740502	3670
740538	1994
740543	3670
740570	0015
740583	1994
740622	0443
740628	0025
740629	0913

DEVICE TYPE	REPL CODE
740630	0015
740781	4093
740782	5490
740828	0133
740855	1669
740856	0042
740857	0155
740885	0847
740946	0136
740947	0136
740948	0136
740949	0127
740950	0127
740951	0127
741050	0004
741051	0720
741052	0720
741098	1215
741100	0133
741101	0015
741114	0380
741115	0390
741116	0091
741473	2534
741518	2261
741519	0520
741673	3695
741686	1714
741687	1050
741689	0030
741726	0368
741729	0006
741731	0224
741732	0839
741737	1220
741738	0162
741739	0057
741740	0023
741741	0133
741852	2261
741853	1837
741854	2124
741855	0151
741856	0155
741857	0155
741858	0018
741859	0936
741860	0353
741861	0155
741862	0113
741863	0203
741864	0133
741865	0015
741866	0143
741867	0064
741868	0030
741869	0244
741870	0244
742004	0143
742008	0015
742009	0015
742362	0706
742363	1018
742364	1050
742510	1540
742512	0076
742513	1221
742537	1663
742546	0006
742547	0113
742548	0151
742549	0151
742705	0368
742723	1260
742724	OEM
742725	1313
742726	1292
742727	2247
742728	2814
742729	0018
742730	0123
742732	0030
742920	0065
742922	1083
742970	0836
742986	0111
743655	0582
743949	0181
744002	0196
744006	0196
744275	0582
746003	0015
746004	0015
750397	OEM
750746A	1030
751007	OEM
752309	0015
752314	0774
756985	OEM
757004	OEM
757008-02	0016
757024	OEM
757030	OEM
757044	OEM
757080	OEM
757085	OEM
757090	OEM
757095	OEM
757097	OEM

DEVICE TYPE	REPL CODE
757098	OEM
757182	OEM
757302	OEM
757306	OEM
757307	OEM
757308	OEM
757309	OEM
757310	OEM
757311	OEM
757312	OEM
757322	OEM
757326	OEM
757361	OEM
757362	OEM
759500	0071
760005	0212
760006	0030
760007	0133
760011	0232
760013	1208
760014	OEM
760015	1158
760018	0491
760019	OEM
760021	0491
760037	0124
760048	0354
760051	0325
760077	OEM
760101-0005	0143
760101-0006	0143
760104	OEM
760105	1899
760106	1544
760109	OEM
760142	0079
760202-0003	0080
760204-0001	0554
760204-0002	0005
760205-0006	0015
760211-0006	0037
760212-0002	OEM
760212-0114	OEM
760213-0002	0079
760213-0005	0127
760213-0097	0144
760213-0100	0079
760213-0104	OEM
760213-0106	0224
760215-0001	3519
760215-0002	2797
760215-0006	OEM
760251	0155
760253	0144
760269	0006
760275	0560
760276	1267
760284	0493
760298	0162
760304	0137
760309	0442
760522-0002	0701
760522-0004	OEM
760522-0008	1395
760522-0012	4003
761113	0015
762200-14	0358
765722	0143
770339	2100
770523	0004
770524	0004
770525	0164
770730	0004
770768-3170756	0142
771907	0015
771908	0277
771909	0143
771910	0143
771911	0143
772712	0143
772713	0015
772714	0277
772716	0050
772717	0050
772718	0004
772719	0050
772720	0004
772721	0004
772722	0004
772723	0004
772724	0004
772725	0004
772727	0004
772728	0004
772729	0004
772732	0164
772733	0164
772736	0004
772737	0004
772738	0127
772739	0127
772740	0123
772768	0136
785278-01	0143
785278-101	0079
785897-01	0050
793356-1	0103
800020-001	0697
800021-001	0873

DEVICE TYPE	REPL CODE
800022-001	1024
800023-001	0512
800024-001	0232
800025-001	1185
800026-001	0743
800033-1	0764
800073-6	0086
800073-7	0086
800080-001	0314
800382-001	1432
800383-001	0117
800385-001	1052
800386-001	1926
800387-001	0361
800400-001	1313
800491-001	0175
800651-001	1206
800806-001	1339
801500	0211
801501	0211
801507	0280
801509	0211
801510	0211
801511	0211
801512	0155
801513	0155
801514	0155
801515	0155
801516	0155
801517	0155
801518	0160
801519	0085
801520	0004
801522	0160
801523	0604
801524	0155
801525	2136
801527	0218
801529	0155
801530	0155
801531	2136
801532	0155
801533	0400
801534	0155
801535	0316
801536	0155
801537	0103
801538	0604
801540	0037
801541	0285
801543	0155
801545	3620
801707	0015
801711	0015
801712	0133
801714	0015
801715	0071
801716	0015
801722	0143
801723	0071
801724	0124
801726	0012
801728	0133
801729	0155
801730	0242
801731	0137
801800	2867
801804	0723
801805	0268
801806	0361
801807	1052
801808	1208
801809	0723
802008	0124
802032-2	0004
802032-4	0004
802033-3	0004
802054-0	0004
802056-0	0004
802189-7	0004
802189-8	0004
802263-0	0004
802263-1	0004
802389-2	0004
802415-2	0004
802425-0	0975
802439-0	0004
802560	0211
803182-5	0016
803369-6	0016
803372-0	0016
803373-0	0016
803696	0111
803733-0	0016
803733-3	0016
803735-3	0016
810002-269	1823
810002-475	0764
811790	1823
811791	0141
811793	0562
811794	1041
812200	OEM
812201	OEM
814044A	0004
815003	0004
815015	0004
815020	0280
815020A	0280

DEVICE TYPE	REPL CODE
815020B	0280
815021	0280
815021A	0280
815021B	0280
815022	0004
815022A	0004
815022B	0004
815023	0004
815023A	0004
815023B	0004
815024	0004
815024A	0004
815024B	0004
815025	0280
815025A	0280
815025B	0280
815026	0600
815026A	0600
815026B	0600
815026C	0600
815026D	0600
815027	0280
815027A	0280
815027B	0280
815027C	0280
815028	0280
815028A	0280
815028B	0280
815028C	0280
815029	0004
815029A	0004
815029B	0004
815029C	0004
815030	0004
815030A	0004
815030B	0004
815031	0004
815031A	0004
815031B	0004
815033	0211
815034	0211
815034A	0004
815034B	0004
815034C	0004
815036	0280
815036A	0280
815036B	0280
815036C	0280
815037	0280
815037A	0280
815037B	0280
815037C	0280
815038	0004
815038A	0004
815038B	0004
815038C	0004
815041	0280
815041A	0280
815041B	0280
815041C	0280
815043	0280
815043A	0280
815043B	0280
815043C	0280
815055	0280
815056	0280
815057	0280
815058	0004
815058A	0004
815058B	0004
815058C	0004
815058X	0211
815064	0136
815064A	0136
815064B	0136
815064C	0136
815065	0280
815065A	0280
815065B	0280
815065C	0280
815066	0280
815066A	0280
815066B	0280
815066C	0280
815067	0050
815067A	0050
815067B	0050
815067C	0050
815068	0136
815068A	0136
815068B	0136
815068C	0136
815069	0004
815069A	0004
815069B	0004
815069C	0004
815070	0004
815070A	0004
815070B	0004
815070C	0004
815070D	0004
815074	0004
815075	0208
815076	0208
815082	0004
815083	0004
815101	0280
815103	0280
815104	0004

DEVICE TYPE	REPL CODE
815105	0280
815107	0280
815108	0280
815109	0280
815114	0004
815115	0280
815116	0280
815117	0280
815118	0004
815120	0004
815120A	0004
815120B	0004
815120C	0004
815120D	0004
815120E	0004
815122	0004
815133	0079
815134	0079
815135	OEM
815136	0004
815137	0265
815137B	0160
815137Y	0160
815138	0015
815139	0004
815142	0015
815158	0004
815160	0004
815160-C	0211
815160-I	0004
815160-J	0004
815160-K	0004
815160-L	0004
815160-O	0004
815160-P	0004
815160-Q	0004
815160A	0004
815160B	0004
815160C	0004
815160D	0004
815160E	0004
815160F	0004
815160H	0004
815164	0144
815165	0144
815166	0142
815166-4	0142
815167-3	0142
815170	0144
815171	0016
815171D	0016
815172	0144
815172A	0144
815173	0111
815173A	0144
815173C	0144
815173F	0144
815174	0016
815174L	0016
815175	0142
815175H	0142
815177	0004
815178	0004
815179	0004
815180-3	0142
815180-4	0142
815180-7	0142
815181	0004
815181-B	0211
815181A	0004
815181B	0004
815181C	0004
815181D	0004
815182	0016
815183	0016
815184	0016
815184E	0016
815185	0126
815185E	0126
815186	0016
815186C	0016
815186L	0079
815189	0004
815190	0016
815191	0016
815193	0050
815195	0004
815196	0004
815197	0050
815198	0079
815199	0037
815199-6	0037
815201	0079
815202	0079
815203-3	0265
815203-5	0265
815206	0007
815209	0144
815210	0079
815211	0037
815212	0079
815213	0037
815218-3	0004
815218-4	0208
815227	0079
815228	4579
815228A	0004
815228A01	0211
815228A1	0004

DEVICE TYPE	REPL CODE
815228B1	0004
815229	0037
815232	2866
815233	0079
815234	0050
815236	0037
815237	0079
815240	OEM
815243	1381
815247	0037
815308	0280
815308A	0280
816135	0142
817020	0920
817032	0143
817036	1030
817042	0015
817043	0015
817044	0015
817053	0015
817062	1030
817062(CRYSTAL)	0484
817064	0015
817066	0015
817067	0015
817068	0015
817068P	0015
817074	0196
817077	0143
817079	0015
817082	0143
817088	0015
817104	0015
817109	0015
817111	0015
817112	0015
817114	0015
817117	0015
817120	0920
817121	0015
817122	0015
817123	0774
817124	1303
817125	0143
817126	0196
817127	0196
817128	0015
817129	1030
817130	0015
817133	0015
817134	0015
817135	0015
817138	0015
817140	0015
817141	0015
817143	0015
817147	1030
817148	0015
817149	0474
817155	0137
817156	0015
817157	0015
817158	0143
817159	0143
817160	0123
817161	0015
817164	0015
817166	0015
817167	0015
817172	0133
817173	0162
817175	1030
817177	0123
817178	0004
817179	0087
817179A	0087
817180	0015
817190	0133
817193	0071
817194	0143
817195	0015
817197	0012
817199	0143
817208	0137
817209	0087
817962	0484
824960-0	0144
829704-6	4856
829704-7	4856
852158-7-1	0015
852158-7.1	0015
853640	0050
853864-0	0050
860003-99	0033
860003-101	1823
860003-111	3061
860003-121	1041
860003-141	0081
860003-151	0081
860003-161	1831
860011	0133
862200-16	0358
862209-16	0562
870233	1509
870246	1524
871125	0143
880092	0222
881916	1313
889132	4586

DEVICE TYPE	REPL CODE
889302	1421
889304	2557
894876	0079
900750-01	0023
900750-02	0080
900755-01	OEM
900756-01	0277
900757-01	OEM
900850-01	0124
900850-05	0133
900850-16	0124
900927-01	0062
900927-02	0062
900941-01	0077
900948-06	0299
900948-11	0041
901225-01	OEM
901226-01	OEM
901227-02	OEM
901227-03	OEM
901229-03	OEM
901229-05	OEM
901237-01	OEM
901246-01	OEM
901251-30	1698
901337-01	OEM
901342-01	OEM
901343-01	OEM
901401-01	OEM
901435-01	3065
901435-02	OEM
901435-03	OEM
901436-01	OEM
901437-01	OEM
901438-01	OEM
901439-03	OEM
901447-10	OEM
901447-20	OEM
901447-29	OEM
901453-01	2048
901455-01	OEM
901458-01	1973
901460-03	OEM
901465-03	OEM
901465-20	OEM
901465-21	OEM
901465-22	OEM
901465-23	OEM
901466-04	OEM
901468-06	OEM
901468-14	OEM
901468-15	OEM
901468-16	OEM
901472-07	OEM
901474-03	OEM
901479-02	OEM
901482-07	OEM
901483-03	OEM
901483-04	OEM
901484-05	OEM
901486-01	OEM
901486-06	OEM
901486-07	OEM
901491-01	OEM
901499-01	OEM
901502-01	0101
901505-01	2357
901510-01	2286
901521-01	1527
901521-02	1595
901521-03	1633
901521-06	0243
901521-17	1841
901521-18	0153
901521-20	0075
901521-21	1559
901521-22	0183
901521-26	1692
901521-29	0709
901521-30	1698
901521-32	0291
901521-34	1672
901521-54	2469
901521-57	1744
901521-58	1746
901521-68	3175
901522-01	1352
901522-03	1803
901522-04	0232
901522-06	1206
901522-30	1339
901523-01	0973
901523-03	3285
901523-04	1815
901523-08	1118
901527-01	0334
901527-02	0624
901528-03	1922
901528-04	3963
901640-01	OEM
901885-04	OEM
901887-01	OEM
901888-01	OEM
901895-01	OEM
901895-02	OEM
902521	0050
902652-01	0076
902653-01	0061
902658-01	0076

If replacement code is OEM, contact original manufacturer for replacement.

DEVICE TYPE	REPL CODE	DEVICE TYPE	REPL CODE	DEVICE TYPE	REPL CODE	DEVICE TYPE	REPL CODE	DEVICE TYPE	REPL CODE	DEVICE TYPE	REPL CODE	DEVICE TYPE	REPL CODE
902671	0076	922607	OEM	970332	0144	980164	0015	982512	0016	984854	0016	988990	0037
902679	0951	922693	0137	970332-12	0144	980316	0280	982523	0178	984875	0144	988991	0016
902682	0866	922799	0015	970565	OEM	980372	0136	982526	0015	984876	0127	988992	0791
902686-01	0079	922860	0023	970659	0079	980373	0136	982528	0142	984877	0127	988993	0086
902693-01	0076	922873	0133	970660	0016	980374	0136	982531	0004	984878	0127	988994	0143
902694-01	0461	922878	3019	970661	0016	980375	0004	982532	0004	984879	0016	988995	0133
902717	0006	922883	OEM	970662	0079	980376	0004	982601	OEM	984880	0133	988996	0133
902720	0148	922896	0280	970663	0037	980426	0280	982815	0144	984881	0123	988997	0143
902744-01	0148	922903	OEM	970759	0917	980432	0280	982816	0144	984882	0015	988998	0143
906103-02	0760	922950	0317	970762	0037	980434	0280	982817	0144	984932	0142	989171	0440
906107-01	OEM	922969	0015	970762-6	0126	980435	0050	982818	0144	985036	0160	989387	0160
906108-01	OEM	923147	0133	970787	OEM	980437	0160	982819	0144	985087	0079	989615	0440
906108-02	OEM	923233	0181	970905	OEM	980438	0280	982820	0211	985096	0144	989692	0440
906108-08	OEM	924605-3	0015	970911	0144	980439	0280	982822	0143	985097	0144	989693	0440
906109-02	OEM	924801-5	0015	970916	0016	980440	0016	982823	0015	985098	0016	991064	0071
906109-04	OEM	924805-5	0015	970916-6	0016	980441	0050	983011	0649	985099	0016	991129	0071
906111-01	OEM	924805-8	0015	970939	0127	980462	0440	983012	0847	985100	0079	991421	0071
906112-01	OEM	925297	0124	970940	0079	980462A	0440	983036	0160	985101	0079	991422	0071
906114-01	OEM	928510-1	2078	970962	0127	980463	0440	983095	0144	985102	0016	991429	0071
906114-02	OEM	928510-101	2078	970963	0917	980463A	0440	983096	0144	985103	0222	992052	0076
906128-01	0011	928512-101	1823	970964	OEM	980505	0050	983097	0016	985104	0015	992066	0076
906129-01	1875	928514-1	0141	971035	0127	980506	0050	983099	0143	985105	0015	992108	1144
906140-01	OEM	928514-101	0141	971036	0917	980507	0050	983101	0015	985106	0133	992129	0076
906144-01	OEM	928515-101	0081	971059	0037	980508	0004	983233	0127	985175	0325	992143	0019
906145-02	OEM	928517-101	2097	971457	0015	980509	0050	983234	0127	985215	0144	992150	0133
906150-02	OEM	928533-101	0562	971458	0015	980510	0004	983235	0127	985216	0004	992171	0080
910050-2	0004	928560-1	0033	971459	0007	980511	0004	983236	0136	985217	0211	992289	0211
910062-1	0004	928560-101	0033	971460	0111	980514	0143	983237	0004	985218	0015	994634	0144
910070-6	0004	928571-1	0033	971477	0211	980514A	0050	983238	0004	985431	0160	995001	0160
910094-4	0004	930347-1	0697	971526	0127	980540	0015	983239	0143	985432	0440	995002	0004
910799	0144	930347-2	0697	971836	OEM	980545A	0050	983271	0136	985442	0127	995003	0004
910952	0460	930347-3	0232	971904	0007	980626	0136	983272	0136	985442A	0144	995012	0861
912002	0723	930347-4	1426	971905	0037	980636A	0050	983405	0004	985443	0160	995014	0160
913112	OEM	930347-5	1024	971906	OEM	980833	0136	983406	0004	985443A	0144	995015	0160
916002	1479	930347-6	1202	972155	0079	980834	0136	983407	0004	985444	0127	995016	0016
916009	0079	930347-7	1172	972156	0079	980835	0136	983408	0004	985444A	0144	995017	0016
916028	0076	930347-9	0980	972214	0016	980836	0004	983409	0004	985445	0136	995022	0632
916029	0191	930347-10	0697	972215	0016	980837	0004	983411	0004	985445A	0050	995025	0861
916030	0079	930347-11	0314	972216	0015	980958	0136	983413	0015	985446	0136	995029	0861
916031	1221	930347-12	1168	972217	0531	980959	0050	983689	0133	985446A	0050	995030	0142
916031(CARTAPE)	0111	930347-13	0361	972258-1	0143	980960	0004	983742	0127	985447	0160	995042	0829
916031(PENNYS)	0079	930347-15	0385	972258-2	0143	980961	0004	983743	0016	985448	0649	995053	0701
916033	0076	932292-1C	1823	972258-3	0143	980964	0015	983744	0913	985449	0160	995053-1	0701
916034	0086	933044-2D	1041	972258-4	0143	981143	0050	983795	0160	985453	0160	995081-1	0666
916046	0560	934362-3	0285	972258-5	0143	981144	0050	983874	0160	985455	0160	998280-930	1823
916049	0127	936001	1030	972258-6	0143	981145	0050	983945	0440	985468	0004	998280-945	0081
916050	0076	942677-1	0143	972258-8	0143	981146	0050	983975	0160	985468A	0004	998280-962	0562
916051	0037	942677-2	0143	972259-8	0143	981147	0004	983995	0774	985469	0004	1000100	0834
916052	0076	942677-3	0143	972305	0007	981148	0004	984156	0144	985469A	0004	1000100-000	0834
916055	1221	942677-4	0143	972306	0127	981149	0004	984158	0144	985470	0164	1000101	0569
916060	0144	942677-6	0143	972307	0127	981150	0143	984159	0144	985470A	0004	1000101-000	0569
916061	2262	942677-7	0143	972416	OEM	981151	0030	984160	0004	985471	0164	1005191	OEM
916062	0037	944148	2017	972417	0127	981153	0143	984161	0164	985472	0015	1006661REV.C	OEM
916063	2316	945820-4	0015	972418	0127	981203	0136	984162	0133	985543	0016	1043176-1	0015
916064	2393	952851	OEM	972419	0127	981204	0136	984163	0143	985609	0004	1043176-2	0015
916067	1994	962685	0917	972420	0127	981206	0004	984182	0015	985610	0004	1043176-3	0015
916068	0127	964158	0917	972571	1303	981207	0143	984183	0015	985611	0136	1043176-4	0015
916069	0144	964298	0143	972571-1	1303	981248	0774	984184	0015	985619	0127	1043176-5	0015
916070	1478	964634	0144	972571-2	1303	981249	1303	984189	0015	985621	0143	1045013	0015
916072	1215	964688	2866	972571-3	1303	981371	0474	984191	0007	985686	0160	1045154-1	0015
916081	0864	964713	0144	972571-4	0015	981445	0015	984192	0007	985715	0325	1045154-2	0015
916082	0325	964864	OEM	972571-5	0071	981522	0143	984193	0037	985735	0208	1045154-3	0015
916083	1160	964999	0917	972571-6	0015	981672	0004	984194	0144	985735A	0208	1045154-4	0071
916084	0433	965000	0016	972571-7	0015	981673	0004	984195	0144	985957	0030	1045154-5	0087
916091	0079	965073	0917	973918-2	0015	981674	0004	984196	0086	985961	0053	1045154-6	0087
916092	2316	965074	0144	973935-20	0015	981675	0004	984197	0016	986302	0004	1045494-1	0484
916098	1160	965632	0007	973936-1	0015	981676	0143	984198	0016	986305	0164	1107832-6	1080
916100	0325	965633	0144	973936-2	0015	981739	0015	984200	0143	986414	0015	1107832-7	0196
916101	4420	965634	0144	973936-3	0015	981952	0015	984221	0004	986542	0016	1107832-8	0196
916102	1479	965854	OEM	973936-4	0015	981953	0015	984222	0016	986543	0164	1107832-9	0196
916103	OEM	965936	0917	973936-5	0015	981954	0015	984224	0016	986576	0127	1107832-10	0196
916104	OEM	966116	OEM	973936-6	0015	981955	0015	984225	OEM	986577	0133	1107832-11	0196
916105	2057	966117	OEM	973936-7	0015	981956	0015	984226	0143	986578	0162	1107863-1	1030
916106	2534	966368	OEM	973936-8	0015	981959	0050	984227	0233	986634	0144	1107863-2	1030
916108	OEM	970046	0144	973936-9	0015	981969	0440	984228	0004	986635	0144	1107863-3	1030
916109	3366	970046-1	0144	973936-10	0015	981969A	0440	984229	0086	986636	0016	1142415-2	0143
916110	1050	970046-2	0144	973936-11	0015	982065	0143	984252	0649	986637	0015	1195397	0913
916111	0785	970046A	0144	973936-12	0015	982150	0136	984254	0015	986693	0127	1221028	0440
916112	0678	970047	0917	973936-13	0015	982151	0004	984259	0103	986694	0127	1221615	0160
916113	0520	970150-07	OEM	973936-14	0015	982152	0004	984259A	0103	986766	0004	1221625	0085
916114	0161	970153-08	OEM	973936-15	0015	982214	0015	984260	0102	986767	1303	1221648	0280
916118	0936	970244	0144	973936-16	0015	982231	0016	984261	0160	986779	0164	1221649	0280
916119	1591	970245	0144	973936-17	0015	982244	0004	984286	0016	986930	0212	1221900	0244
916121	2019	970246	0037	973936-18	0015	982253	0015	984431	0160	986931	0126	1221962	0016
916125	1816	970247	0079	973936-19	0015	982254	0012	984521	0222	986932	0560	1222123	0016
916138	1645	970248	0037	973936-20	0071	982267	0136	984522	0015	986933	1267	1222133	0016
916149	OEM	970249	0144	973936-21	0015	982268	0144	984577	0144	986934	0133	1222136	0050
916150	1292	970250	0079	973962-1	0015	982269	0144	984590	0016	986935	0015	1222314	0050
916154	OEM	970251	0037	973973-6	OEM	982270	0123	984591	0016	986997	OEM	1222371	0050
916155	0706	970252	0016	980028	OEM	982271	0123	984593	0016	988000	0144	1222424	0016
916157	4428	970253	0325	980052	0440	982275	0143	984594	0015	988001	0144	1222463	0144
921608	0015	970254	0037	980052A	0440	982283	0004	984608	0142	988002	0144	1223770	0143
922021	0143	970255	0142	980132	0160	982284	0004	984666	0143	988003	0016	1223771	0124
922092	0015	970257	OEM	980134	0160	982285	0004	984667	0143	988005	0164	1223772	0015
922183	0015	970309	0144	980135	0160	982289	0136	984685	0136	988048	0030	1223773	0170
922214	0005	970309-1	0144	980136	0280	982290	0143	984686	0016	988049	0229	1223781	0144
922311	0015	970309-2	0144	980138	0144	982300	0086	984687	0016	988050	0170	1223782	0079
922358	0012	970309-3	0144	980139	0144	982307	0222	984688	2866	988051	0015	1223783	2997
922359	0030	970309-4	0144	980140	0050	982321	0144	984690	0277	988080	0160	1223784	0698
922360	0015	970309-5	0144	980142	0050	982322	0050	984713	0015	988336	0160	1223785	1400
922433	0124	970309-12	0144	980143	0015	982324	0030	984743	0144	988413	0160	1223786	0630
922502	OEM	970310	0144	980144	0004	982361	0196	984744	0144	988414	0440	1223909	1714
922503	OEM	970310-1	0144	980146	0050	982363	0196	984745	0016	988468	0160	1223910	2781
922504	OEM	970310-2	0144	980147	0016	982374	0050	984746	0211	988977	0440	1223911	0368
922505	OEM	970310-3	0144	980148	0004	982376	0142	984794	0087	988985	0144	1223912	0151
922524	0057	970310-4	0144	980149	0004	982377	0015	984795	0087	988986	0144	1223913	0155
922545	OEM	970310-5	0144	980150	0975	982497	0050	984851	0144	988987	0144	1223914	0006
922603	0012	970310-12	0144	980153	0004	982510	0079	984852	0144	988988	0144	1223915	0151
922604	0143	970311	0127	980155	0160	982511	0079	984853	0144	988989	0144	1223916	0224

DEVICE TYPE	REPL CODE
1223917	1946
1223918	2495
1223919	0113
1223920	0076
1223921	0019
1223925	0030
1223926	0162
1223927	0057
1223928	0023
1223929	0023
1223930	0133
1223931	0019
1224031	0006
1224076	1540
1224275	3559
1224276	OEM
1224277	OEM
1258444	0015
1303256	0087
1310505	OEM
1317303-1	0037
1320004	OEM
1320031	OEM
1320135	0016
1320135A	0016
1320135BC	0016
1320135C	0016
1321341	OEM
1322966	0288
1330021-0	1854
1330021-1	1854
1331409	2607
1337915	OEM
1340051	OEM
1343041	OEM
1344401	OEM
1353271	OEM
1353331	OEM
1360071	OEM
1360081	OEM
1360271	OEM
1360331	OEM
1360391	OEM
1361192	OEM
1361201	OEM
1371162	OEM
1371174	OEM
1378204	OEM
1401092	OEM
1401100	OEM
1401101(LED)	OEM
1407205-1	0160
1408615-1	0144
1408640-1	0144
1408649-1	0143
1408694-1	0212
1415721-1	0102
1415762-1	0239
1415762-2	0061
1417302-1	0436
1417303-1	0126
1417303-2	0126
1417306-1	0079
1417306-4	0079
1417306-5	0079
1417316-1	0551
1417317-1	0382
1417318-1	0079
1417318-2	0324
1417320-1	0103
1417321	0079
1417321-1	0861
1417322-1	0178
1417324-1	0079
1417325-1	0595
1417330-3	0037
1417330-4	0037
1417331-1	0065
1417335-1	0065
1417338-5	0086
1417339-1	0037
1417340-2	0079
1417342-1	0079
1417344-1	0086
1417345-2	0086
1417347-1	0037
1417349-2	0324
1417352-5	0551
1417354-1	4960
1417356-2	0103
1417358-1	0042
1417359-1	0460
1417362-1	0285
1417362-4	0285
1417363-1	0037
1417364-1	0423
1417366-1	0065
1417370-1	0065
1417372-1	0212
1417380-1	0313
1417381-2	0079
1417399-8	2520
1417400-1	0079
1417872-11	0124
1420427-1	0004
1420427-2	0004
1420427-3	0004
1421207-1	1149
1421729-7	OEM
1421734-1	6363
1421763-2	OEM
1421896-104	OEM
1421896-106	OEM
1421896-107	OEM
1421899-6	OEM
1440614-9	OEM
1440977	0774
1440977-1	0774
1440990	0474
1442415-2	0917
1442415-3	0917
1443200-3	0004
1443223-1	0030
1444875-1	0137
1445470-2	0288
1445470-50	0201
1445470-501	0288
1445470-502	0102
1445470-503	0288
1445470-504	0378
1445740-502	0201
1445829-501	0382
1445829-502	0382
1445829-503	0551
1445829-504	0551
1449098-1	0233
1457061-16	OEM
1460607-9	OEM
1462434-1	0829
1462445-1	2715
1462455-1	OEM
1462506-1	0328
1462516	0167
1462516-001	0167
1462516-1	0167
1462554	2743
1462554-1	2743
1462554-2	2743
1462554-3	2743
1462554-4	2743
1462559-1	2549
1462560-001	0328
1462560-1	0328
1462599-1	OEM
1463641-1	1997
1463641-2	1997
1463677-1	0364
1463677-2	0364
1463677-3	0364
1463681-1	1854
1463686-1	0167
1463695-2	OEM
1464295-1	1186
1464295-2	2282
1464437-2	2257
1464437-3	2257
1464438-2	2755
1464438-3	2755
1464439-2	1843
1464460-2	1843
1464460-3	1843
1464607-1	0409
1464607-2	0409
1464607-3	0409
1464607-4#	OEM
1464607-5	0409
1464607-6	0409
1464607-7	2804
1464607-8	2804
1464607-9	1710
1464607-10	1710
1464609-3	OEM
1464778-9	0920
1464846-1	2552
1464846-2	2552
1464865-1	0409
1464984-1	0409
1464984-2	0409
1465158-1	0521
1465158-2	0521
1465188-1	2549
1465316-1	3270
1465345-1	2049
1465615	2789
1465615-1	2789
1465615-2	2789
1465617-2	2786
1465651-1	OEM
1465819-1	2333
1466701-1	2798
1466860-1	0409
1466860-2	1197
1466862-2	2804
1466865-1	0409
1470872-6	0133
1470990	0474
1470990-1	0474
1471036-14	0160
1471036-20	0160
1471072-4	0133
1471100-1	0004
1471100-8	0004
1471100-9	0004
1471101-2	0004
1471101-3	0004
1471101-4	0004
1471101-15	0004
1471104-5	0136
1471104-6	0136
1471104-7	0136
1471104-8	0136
1471112-7	0037
1471112-8	0037
1471112-8-9	0126
1471112-9	0150
1471112-10	0693
1471112-12	0126
1471113-2	0016
1471114-1	0037
1471115-13	0079
1471115-14	0144
1471117-1	0142
1471120-7	0079
1471120-8	0079
1471120-8-9	0079
1471120-14	0086
1471120-15	0079
1471122-6	0111
1471122-7	0111
1471123-3	0086
1471123-4	0086
1471124-5	0604
1471125-3	0975
1471132-002	0423
1471132-2	0161
1471132-3	0423
1471132-4	0042
1471132-5	0423
1471132-6	0130
1471133-1	0551
1471134-1	0382
1471135-001	0103
1471135-1	0103
1471135-2	0130
1471136-3	0400
1471139-1	1681
1471140-1	0161
1471141-1	0460
1471393-4	0289
1471405-1	0087
1471777-11	0879
1471822-11	0123
1471839-1	0720
1471839-2	OEM
1471839-3	OEM
1471839-5	0720
1471858	2966
1471858-1	0015
1471872-001	0015
1471872-004	0015
1471872-006	0015
1471872-008	0015
1471872-1	0133
1471872-2	0133
1471872-3	0102
1471872-4	0133
1471872-5	0124
1471872-6	0015
1471872-7	0133
1471872-8	0133
1471872-10	0133
1471872-11	0124
1471872-12	0133
1471872-13	0133
1471872-14	0133
1471872-15	0143
1471872-16	0133
1471872-17	0143
1471872-18	0133
1471876-6	1149
1471878-3	0025
1471893-3	OEM
1471893-4	0783
1471898-3	0636
1471898-4	0783
1471898-5	0157
1471908-1	0293
1471908-4	2646
1471922-1	0917
1471922-2	0917
1471922-4	0917
1472164-6	OEM
1472171-32	0015
1472450-1	0144
1472460-1	0229
1472460-2	0133
1472460-3	0133
1472460-4	0229
1472460-5	0229
1472460-6	0133
1472460-7	0133
1472460-8	0133
1472460-9	0924
1472460-13	0102
1472460-14	0087
1472460-16	0015
1472460-18	0484
1472474-2	0233
1472482-1	0004
1472634-1	0144
1472654-1	0224
1472872-4	0133
1473503-1	0142
1473508-1	0086
1473514-1	0975
1473515-1	0160
1473516-1	0126
1473519-1	0079
1473520-1	0142
1473521-1	0127
1473523-2	OEM
1473524-1	0127
1473524-2	0144
1473524-3	0127
1473527-1	0016
1473528-1	2707
1473528-2	2707
1473529-1	0127
1473530-1	0144
1473530-2	0144
1473531-1	0259
1473532-1	0016
1473533-1	0144
1473535-1	0007
1473535-1(RCA)	0007
1473536-001	0079
1473536-1	0079
1473536-2	0007
1473537-1	0144
1473538-1	0016
1473539-1	0079
1473540-1	0037
1473541-1	0233
1473543-1	0007
1473544	0861
1473544-1	0127
1473545-1	0086
1473546-1	0016
1473546-2	0016
1473547-1	0086
1473549-1	0037
1473549-2	0037
1473551-1	0016
1473552-1	0233
1473553-1	0086
1473554-1	0016
1473555-1	0016
1473555-2	0016
1473555-3	0086
1473555-I	0086
1473556-1	0016
1473557-1	0016
1473558-1	0127
1473559-001	0037
1473559-1	0037
1473560	0086
1473560-002	0079
1473560-1	0086
1473560-2	0086
1473561-001	0079
1473561-1	0086
1473561-Y	0079
1473562-1	0037
1473563-1	0037
1473564-1	0074
1473565-001	0086
1473565-1	0086
1473566-1	0086
1473567-1	0142
1473567-2	0161
1473567-4	0042
1473568-1	0144
1473569-1	0016
1473570-1	0037
1473571	0285
1473571-1	0007
1473572-1	0079
1473572-3	0086
1473573-1	0208
1473574	0861
1473574-1	0037
1473576	0007
1473576-1	0330
1473577-1	0007
1473578-1	0004
1473579-1	0007
1473581-1	0037
1473582-1	0016
1473583-7	0239
1473583-8	0239
1473584-1	0142
1473585-1	0061
1473585-5	0239
1473585-6	0061
1473585-7	0239
1473585-8	0061
1473586	0007
1473586-2	0079
1473587-1	OEM
1473588-2	0353
1473588-8	0212
1473589-1	0016
1473590-1	0149
1473591-1	0037
1473592-1	0126
1473593-1	0086
1473595-1	0016
1473597	0150
1473597-1	0037
1473597-2	0037
1473597-7	0150
1473598-1	0136
1473598-2	0004
1473601	0642
1473601-001	0079
1473601-1	0079
1473601-2	0079
1473602	0861
1473603-1	0144
1473604-1	0400
1473604-3	0144
1473606-1	0144
1473608-002	0086
1473608-1	0086
1473608-2	0086
1473608-3	0086
1473610-1	0127
1473611-1	0042
1473612-1	0423
1473612-11	0042
1473613-2	0338
1473613-3	0338
1473613-4	0285
1473614-1	0079
1473614-3	0079
1473615-1	0086
1473616-1	0126
1473617-1	0144
1473617-2	0127
1473618-1	0212
1473619-1	2356
1473619-2	2394
1473619-3	2394
1473620-001	0037
1473620-1	0037
1473620-2	0037
1473621-1	0423
1473622-002	0086
1473622-1	0016
1473622-2	0086
1473623-1	0423
1473624-1	0854
1473624-2	0460
1473625-001	0595
1473625-1	0086
1473626-1	0079
1473627-1	0037
1473628-1	0551
1473628-2	0285
1473628-3	0551
1473629-3	0382
1473631-1	0079
1473632-1	0285
1473632-2	0560
1473632-3	0423
1473633-1	0285
1473634-1	0065
1473635	0212
1473635-1	0212
1473635-2	0353
1473636-1	0142
1473637-1	0074
1473638-1	0905
1473640-1	0423
1473641-2	OEM
1473647-1	0065
1473648-1	0975
1473649-1	0065
1473650-2	OEM
1473651-1	0150
1473652-1	0144
1473656-1	0561
1473656-2	0561
1473656-4	0233
1473657-1	0224
1473657-2	0224
1473665-2	0103
1473666-1	0126
1473669-1	0065
1473669-2	0065
1473676-1	0330
1473676-1(3RD-LF)	0330
1473677-3	0364
1473679-1	0086
1473679-2	0233
1473680-1	0007
1473681-1	0042
1473681-I	0042
1473682-1	0460
1473683-1	0103
1473686-1	0136
1473687-1	0338
1473688-1	0058
1473777-2	0002
1474178-10	0015
1474613-2	0285
1474717-2	0002
1474777-002	0002
1474777-1	0137
1474777-2	0002
1474777-3	0367
1474777-4	0039
1474777-7	0783
1474777-8	0015
1474777-11	0100
1474778	0015
1474778-013	0087
1474778-2	0015
1474778-3	0015
1474778-4	0087
1474778-5	0102
1474778-6	0102
1474778-7	0015
1474778-8	0087
1474778-9	0920
1474778-10	0015
1474778-11	0015
1474778-13	0087
1474778-14	0102
1474778-16	0133
1474778-21	0102
1474788-10	0015
1474872-1	0133
1475721-1	0071
1476049-2	0102
1476161-12	0102
1476161-13	0015
1476171-1	0015
1476171-8	0015
1476171-11	0015
1476171-11(TRANS)	0085
1476171-11(TRANSISTO	0085
1476171-12	1217
1476171-13	0102
1476171-13(TRANSISTO	0085
1476171-14	1217
1476171-15	1217
1476171-17	0102
1476171-18	0102
1476171-19	1217
1476171-20	1217
1476171-21	0087
1476171-22	0102
1476171-24	1217
1476171-25	0102
1476171-26	0102
1476171-27	1217
1476171-28	0015
1476171-29	0015
1476171-30	0023
1476171-31	0015
1476171-32	0102
1476171-33	0133
1476171-34	0102
1476171-35	1217
1476171-36	0348
1476172-12	0102
1476179-001	0143
1476179-1	0143
1476179-3	0143
1476183-2	0102
1476183-3	0288
1476183-4	0757
1476183-5	0102
1476183-6	0102
1476183-8	1303
1476188-1	0016
1476689-1	0030
1476690-1	0133
1476690-2	0133
1476778-5	OEM
1476930	0757
1476933-501	0757
1476933-503	0757
1477022	0162
1477022-1	0163
1477046-1	0002
1477046-4	0137
1477046-5	0820
1477046-10	0495
1477069-1	1029
1477074-1	1029
1477080-501	0002
1477081-501	0002
1477949-1	1029
1477949-2	OEM
1478023-1	0774
1478164-1	0039
1478564-1	2329
1478564-2	1712
1478564-5	2329
1478564-6	1712
1501824	OEM
1503085	OEM
1503135	OEM
1522237-20	0016
1604236	OEM
1642606B1	0162
1642606B4	0636
1642606B6	0157
1662258-6	1107
1700001	0037
1700008	0079
1700019	0079
1700020	0224
1700032	0224
1700033	0224
1700034	0037
1700035	0493
1700036	0042
1700037	0698
1700038	0698
1771872-12	0124
1773599-1	0006
1773631-1	0079
1800002	0143
1800006	0015
1800009	0124
1800012	0157
1800013	0170
1800017	0087
1800018	0015
1800019	2157
1800020	0820
1810037	0144
1810038	0144
1810039	0144
1811119	0015
1815036	0144
1815037	0144
1815039	0144
1815041	0016
1815042	0016
1815043	0016
1815045	0144
1815047	0144
1815054	0016
1815067	0144
1815068	0144
1815139	0038
1815153	0086
1815154	0016
1815154/7	0595
1815154-9	0086
1815156	0086
1815157	0086
1815157-9	0086
1815159	0086
1817004	0144
1817005	0016
1817005-3	0144
1817006	0086
1817006-3	0144
1817007	0016
1817008	0144
1817017	0208
1817045	0144
1817108	0016
1819045	0144
1820657	0774
1826065-1	0409
1826065-2	0409
1826065-3	2804
1826860-1	0409
1827322	0126
1846282-1	0016
1851490	0267
1887048	0474
1895989-1	1632
1895991-1	0141
1895992-1	0562
1895993-1	1823
1895994-1	1041
1895995-1	2097
1914062-1	1831
1944748	0604
1956016	0004
1956197	0244
1956486	0015
1960584	0160
1960632	0604
1960642	0149
1960643	0004
1961479	0160
1961480	0160
1961835	0160
1961837	0004
1961843	0015
1962326	0604
1962594	0015
1965016	0126
1965017	0160
1965019	0012
1966079	0160
1966808	0015
1967799	0086
1967799-1	0086
1967801	0086
1967813	0015
1969497	0015
2000261	OEM
2000287-28	0004
2000391	OEM
2000433-150	0143
2000521	0746
2000625-31	0004
2000625-32	0004
2000625-33	0004
2000625-34	0004
2000625-36	0015
2000626-32	0015
2000646-103	0216
2000646-104	0050
2000646-105	0233
2000646-106	0136
2000646-107	0016
2000646-108	0004
2000646-109	0160
2000646-110	0975
2000646-111	0164
2000646-112	0620
2000646-113	0160
2000646-115	0015
2000646-119	0015
2000646-120	0015
2000648-21	0050
2000648-22	0050
2000648-23	0136
2000648-26	0143
2000648-120	0015
2000752-80	0086

If replacement code is OEM, contact original manufacturer for replacement.

DEVICE TYPE	REPL CODE
2000757	0196
2000757-18	0143
2000757-19	0913
2000757-74	0196
2000757-79	0196
2000757-80	0144
2000804-7	0144
2000804-8	0144
2000804-9	0004
2001221	OEM
2001222	OEM
2001225	OEM
2001372	OEM
2001533	OEM
2001653	0004
2001653-20	0136
2001653-21	0136
2001653-22	0280
2001653-23	0004
2001653-24	0004
2001653-58	0050
2001653-59	0050
2001762	OEM
2001786-134	0143
2001786-139	1080
2001786-141	0015
2001786-142	0071
2001786-169	1030
2001786-207	0015
2001809-47	0004
2001809-48	0004
2001809-48A	0004
2001809-48B	0004
2001812-65	0280
2001822	OEM
2001872	OEM
2002151-020	0143
2002151-18	0050
2002151-18A	0050
2002151-19	0004
2002151-20	0143
2002152-14	0004
2002153-58	0050
2002153-59	0050
2002153-60	0136
2002153-71	0004
2002153-76	0050
2002153-77	0016
2002153-78	0004
2002153-83	0208
2002207	0030
2002207-2	0030
2002209-1	0012
2002209-3	OEM
2002209-4	0636
2002209-5	0170
2002209-7	0495
2002209-9	0170
2002209-10	0137
2002209-11	0244
2002210-110	0004
2002211-24	0004
2002211-25	0004
2002331-46	0365
2002332-53	0144
2002332-54	0144
2002332-55	0144
2002332-56	0144
2002332-57	0015
2002332-58	0015
2002336-12	OEM
2002336-19	0136
2002336-20	0143
2002336-115	0015
2002402	0133
2002402-29	0015
2002403-19	0136
2002620-18	0144
2002620-19	0144
2002621-2	0016
2003069	0071
2003069-1	0071
2003069-2	0133
2003069-4	0143
2003069-5	0139
2003069-6	0133
2003073-0701	0016
2003073-0702	0037
2003073-8	0004
2003073-9	0016
2003073-10	0016
2003073-11	0050
2003073-12	0633
2003073-13	0050
2003073-14	0004
2003073-15	0004
2003073-16	0208
2003073-67	0015
2003073-68	0015
2003073-91	0042
2003073-91A	0103
2003168-135	0016
2003168-136	0016
2003229-25	0016
2003239-65	0016
2003251	0624
2003255	OEM
2003271	OEM
2003311	OEM

DEVICE TYPE	REPL CODE
2003321	3233
2003342-109	0144
2003342-244	0127
2003351	OEM
2003501	OEM
2003522R	OEM
2003779-22	0127
2003779-23	0127
2003779-24	0127
2003779-25	0127
2003779-26	0012
2003811	OEM
2003981	OEM
2004107-40	0196
2004357-106	0143
2004358-123	0004
2004358-142	0015
2004358-168	0004
2004746-87	0142
2004746-114	0144
2004746-115	0144
2004746-116	0126
2004746-117	0126
2004792	OEM
2004851	OEM
2005300	0122
2005779-26	0649
2006226-14	0004
2006227-51	0016
2006334-31	0164
2006334-115	0016
2006334-155	0086
2006422-132	0143
2006422-133	0015
2006431-44	0144
2006431-45	0016
2006431-46	0016
2006431-49	0016
2006431-50	0143
2006436-35	0560
2006436-36	0560
2006436-37	0126
2006436-38	0015
2006436-40	0086
2006436-89	0015
2006441-91	0015
2006441-113	0004
2006441-122	0143
2006441-123	0030
2006441-132	0133
2006463-89	0015
2006512-40	0015
2006512-79	0196
2006512-80	0196
2006513-19	0144
2006513-39	0127
2006513-133	0164
2006514-59	0222
2006514-60	0016
2006514-61	0698
2006582	0015
2006582-20	0143
2006582-21	0071
2006582-22	0015
2006582-23	0015
2006582-24	0015
2006582-25	0016
2006582-101	0007
2006607-59	0160
2006607-60	0016
2006607-61	0698
2006607-63	0649
2006613-77	0016
2006623-47	0325
2006623-48	0133
2006623-49	0015
2006623-88	0161
2006623-128	0698
2006623-145	0016
2006623-148	0086
2006627-54	0133
2006681-93	0127
2006681-94	0127
2006681-95	0127
2006681-96	0016
2006681-97	0030
2006681-120	0164
2006721	OEM
2007831	OEM
2008292-56	0127
2008292-87	0136
2008292-88	0133
2008292-89	0015
2008293-109	0211
2008293-111	0164
2008299-1	0127
2008299-2	0143
2008299-3	0015
2008302-41	0015
2010088-49	0079
2010494-3	0007
2010494-4	0079
2010499-52	0079
2010952-14	0208
2010957-49	0015
2010967-82	0030
2010967-83	0133
2010967-84	0157
2013019-117	0015

DEVICE TYPE	REPL CODE
2020321	OEM
2020341	OEM
2024794-3	OEM
2024794-6	OEM
2024794-8	OEM
2024794-115	OEM
2024794-127	OEM
2024794-140	OEM
2024794-143	OEM
2024794-152	OEM
2024794-187	OEM
2024794-188	OEM
2024794-189	OEM
2024794-190	OEM
2024794-192	OEM
2024794-193	OEM
2024794-194	OEM
2024794-218	OEM
2024794-330	OEM
2024794-364	OEM
2024794-366	OEM
2024794-367	OEM
2026387-1	OEM
2026392-1	OEM
2029655-7	OEM
2057013-0004	0037
2057013-0007	0037
2057013-0008	0037
2057013-0012	0037
2057013-0701	0037
2057013-0702	0037
2057013-0703	0037
2057062-0702	0015
2057100-62	OEM
2057199-070BA	0103
2057199-0700	0103
2057199-0701	0103
2057323-0500	0103
2057323-0501	0103
2060041	0015
2068491-704	0353
2068510-0701	1831
2068510-0702	0141
2068510-0703	0562
2068510-0704	0358
2068510-0705	0333
2076393	0136
2076403	0050
2076403-0703	0050
2076945-0701	0004
2090056-1	0085
2090056-5	0085
2090056-27	0085
2090924-0008	0004
2090924-008	0004
2090924-6	0004
2090924-8	0004
2090924-8A	0004
2091211-0014	0280
2091214-0005	0050
2091217-0014	0050
2091241-0005	0050
2091241-005	0136
2091241-0013	0280
2091241-0014	0280
2091241-0015	0280
2091241-0018	0004
2091241-0719	0136
2091241-1	0280
2091241-2	0280
2091241-3	0280
2091241-4	0280
2091241-5	0136
2091241-5A	0136
2091241-6	0280
2091241-7	0280
2091241-8	0280
2091241-9	0004
2091241-10	0280
2091241-11	0280
2091241-12	0280
2091241-13	0280
2091241-13A	0280
2091241-14	0280
2091241-15	0280
2091241-15A	0280
2091247-005	0050
2091260	0038
2091260-1	2866
2091260-2	2866
2091260-3	0038
2091578-0702	0211
2091578-1	0211
2091858-0712	0160
2091858-0715	0079
2091858-11	0160
2091859-0008	0085
2091859-0011	0160
2091859-0025	0085
2091859-0711	0144
2091859-0712	0160
2091859-0713	0085
2091859-0714	0085
2091859-0715	0160
2091859-0716	0160
2091859-0717	0160
2091859-0718	0160
2091859-0720	0085
2091859-0723	0160

DEVICE TYPE	REPL CODE
2091859-2	0085
2091859-4	0085
2091859-6	0085
2091859-8	0085
2091859-9	0160
2091859-10	0085
2091859-11	0085
2091859-16	0160
2091859-25	0160
2091959-16	0160
2092055-0001	0143
2092055-0002	0244
2092055-0003	0030
2092055-0004	0451
2092055-0005	0649
2092055-0007	0123
2092055-001	0143
2092055-0010	0143
2092055-0016	0157
2092055-0017	0244
2092055-0018	0025
2092055-0024	1606
2092055-0027	0244
2092055-0708	0917
2092055-0710	0025
2092055-0711	0143
2092055-0712	0244
2092055-0713	0143
2092055-0714	0016
2092055-0714(DIODE)	0133
2092055-1	0143
2092055-3	0030
2092055-4	0451
2092055-5	0649
2092055-7	0143
2092117-0018	0127
2092160-3	2866
2092405-7	0143
2092417-005	0050
2092417-0017	0127
2092417-0018	0127
2092417-0019	0127
2092417-0704	0050
2092417-0707	0050
2092417-0708	0050
2092417-0709	0050
2092417-0710	0050
2092417-0711	0007
2092417-0712	0007
2092417-0713	0007
2092417-0714	0007
2092417-0715	0007
2092417-0716	0007
2092417-0717	0050
2092417-0719	0016
2092417-0720	0016
2092417-0721	0016
2092417-0724	0079
2092417-0725	0079
2092417-1	0050
2092417-2	0050
2092417-3	0050
2092417-4	0050
2092417-5	0050
2092417-6	0050
2092417-7	0050
2092417-8	0050
2092417-9	0050
2092417-17	0079
2092417-18	0079
2092417-19	0079
2092418-0022	0007
2092418-0023	0007
2092418-0024	0007
2092418-071	0050
2092418-0708	0050
2092418-0710	0050
2092418-0711	0600
2092418-0712	0050
2092418-0715	0144
2092418-0716	0007
2092418-0717	0007
2092418-0718	0007
2092418-0719	0007
2092418-0720	0007
2092418-0721	0007
2092418-0724	0144
2092418-1	0050
2092418-2	0050
2092418-5	0050
2092418-6	0050
2092418-7	0050
2092418-8	0050
2092418-10	0050
2092418-11	0050
2092605-0705	0079
2092608-22	0079
2092609-0001	0079
2092609-0002	0079
2092609-001	0079
2092609-0022	0079
2092609-0023	0079
2092609-0024	0079
2092609-0025	0037
2092609-0026	0079
2092609-0027	0079
2092609-0028	0079
2092609-0705	0016
2092609-0706	0016

DEVICE TYPE	REPL CODE
2092609-0707	0016
2092609-0713	0016
2092609-0715	0016
2092609-0718	0016
2092609-0720	0079
2092609-0721	0079
2092609-1	0016
2092609-2	0016
2092609-3	0016
2092609-5	0016
2092693-0724	0127
2092693-0725	0127
2092693-1	0004
2092693-2	0050
2092693-3	0050
2092693-4	0050
2092693-8	0004
2092693-9	0050
2093308-070	0144
2093308-0700	0144
2093308-0701	0016
2093308-0702	0016
2093308-0703	0016
2093308-0704	0127
2093308-0704A	0144
2093308-0705	0127
2093308-0706	0127
2093308-0706A	0144
2093308-0708	0016
2093308-0725	0127
2093308-1	0144
2093308-2	0144
2093308-3	0144
2093308-8708	0079
2095083	0143
2097013-0702	0086
2120144	0127
2139103-4001	OEM
2146516	0196
2231154	0052
2234494	3418
2234496	1069
2234497	3924
2243255-1	0004
2295361	0562
2300439-00B	OEM
2300440-00A	OEM
2300441-00A	OEM
2311345	OEM
2312371	3709
2312501	OEM
2312522	OEM
2312532	OEM
2312561	OEM
2312651	OEM
2312771	OEM
2312772	OEM
2312872	OEM
2312873	OEM
2314009	0975
2314381	0477
2314391	1386
2315272	3941
2315411	1165
2315471	OEM
2316177	0127
2318365	0925
2319990	3915
2320011	0004
2320022	0016
2320031	0127
2320041	0151
2320041H	0007
2320042	0127
2320043	0127
2320051	0233
2320051(BRADFORD)	0285
2320051H	0233
2320052	0086
2320062	0144
2320063	0144
2320073	0144
2320083	0042
2320084	0178
2320092	0222
2320111	0016
2320123	0016
2320141	0127
2320141H	0127
2320143	0127
2320154	0280
2320161	0037
2320161(RCA)	0050
2320162	0037
2320191	0716
2320201	0604
2320221	0142
2320222	0142
2320223	0142
2320233	0086
2320242	0126
2320243	0126
2320261	0004
2320271	0074
2320273	0074
2320281	0003
2320291	0065
2320299	0065
2320302	0004

DEVICE TYPE	REPL CODE
2320302H	0004
2320331	0208
2320413	0016
2320415	OEM
2320422	0164
2320422-1	0004
2320423	0004
2320432	0042
2320441	0016
2320471	0127
2320471-1	0127
2320471H	0127
2320482	0042
2320482H	0042
2320483	0042
2320485	0042
2320486	0042
2320492	0004
2320512	0280
2320513	0004
2320514	0050
2320514-1	0280
2320541	0160
2320563	0127
2320591	0076
2320591-1	0079
2320594	0076
2320595	0076
2320596	0076
2320598	0076
2320602	0042
2320631	0148
2320631M	4432
2320632	0148
2320635	0148
2320637	0148
2320643	0191
2320644	0191
2320646	0191
2320646-1	0155
2320647	0191
2320647-1	0155
2320647M	3131
2320651	0042
2320652	0042
2320663	0191
2320664	0191
2320681	0148
2320682	0148
2320696	0016
2320696-1	0016
2320843	0563
2320845	0563
2320846	0563
2320855	0525
2320884	0925
2320913	OEM
2320914	OEM
2320931	0142
2320946	0086
2320961	0642
2320994	0142
2321001	0142
2321095	0168
2321101	0342
2321111	0342
2321112	0342
2321121	2437
2321151	0715
2321154	0052
2321221	1085
2321241	0363
2321264	0940
2321281	0925
2321291	0142
2321301	1284
2321302	0342
2321306	1284
2321321	0693
2321351	0693
2321372	2620
2321374	0074
2321381	1648
2321403	0142
2321411	1085
2321412	1085
2321433	0392
2321438	0392
2321439	0392
2321472	0261
2321511	0224
2321521	0648
2321522	0648
2321541	0151
2321551	2668
2321561	2660
2321563	2660
2321581	0429
2321591	1284
2321621	0151
2321647	0191
2321652	0171
2321693	0392
2321699	0392
2321703	1911
2321751	0006
2321763	0681
2321764	0681
2321791	0392

DEVICE TYPE	REPL CODE
2321803	3354
2321871	1978
2321881	0018
2321882	0018
2321891	0532
2321982	0018
2321984	0018
2321991	0261
2321992	0261
2321993	0261
2322163	0710
2322191	0151
2322223	0728
2322561	0276
2322562	0276
2322851	0055
2322943	0095
2323021	0055
2323102	0111
2323112	0392
2323114	0392
2323123	0236
2323351	2484
2323352	0825
2323411	0392
2323431	6047
2323432	2058
2323434	2058
2323521	2561
2323522	2561
2323523	2561
2323526	2561
2323541	0643
2323642	OEM
2323771	OEM
2323782	0058
2323931	0392
2323941	1648
2324082	1758
2324084	1758
2324171	0563
2324292	0060
2324322	0261
2324411	1541
2324412	1541
2324414	1541
2324471	3728
2325702	1348
2325725	0151
2326022	1513
2326214	0563
2326216	0563
2326218	0563
2326611	0276
2326631	OEM
2326661	0276
2326991	0142
2327022	0086
2327023	0016
2327031	0015
2327041	0015
2327052	0103
2327053	0103
2327061	0142
2327071	0137
2327073	0402
2327074	0053
2327075	0726
2327076	0039
2327077	0420
2327078	0317
2327111	0212
2327122	0016
2327132	0325
2327134	OEM
2327135	OEM
2327142	0325
2327142(JFET)	0325
2327142(MOSFET)	0212
2327152	0560
2327153	0560
2327172	0103
2327182	0142
2327192	0905
2327203	0560
2327206	0042
2327232	0212
2327262	0037
2327282	0126
2327283	0126
2327292	0086
2327293	0086
2327301	OEM
2327302	0632
2327311	1621
2327312	1621
2327332	0086
2327363	0552
2327387	0037
2327393	OEM
2327403	0086
2327411	4570
2327421	0443
2327422	0443
2327431	0212
2327501	2129
2327502	2129
2327504	2129
2327505	2129

DEVICE TYPE	REPL CODE
2327751	0013
2327771	2954
2327772	OEM
2327773	2954
2327783	OEM
2327844	3494
2327883	1522
2327883M	OEM
2327931	0095
2328012	OEM
2328451	OEM
2328454	OEM
2328455	OEM
2330011	0277
2330011H	0277
2330021	0649
2330032	0757
2330034	0378
2330191	0102
2330192	0348
2330201	0015
2330211	0102
2330235	0023
2330241	0137
2330251	0023
2330251(POWER)	0087
2330251H	0015
2330252	0023
2330253	0023
2330254	0023
2330256	0023
2330302	0025
2330302H	0649
2330305	0100
2330307	0244
2330332	0087
2330341	0196
2330351	0133
2330352	0124
2330353	0124
2330354	0023
2330361	0277
2330362	0015
2330381	0378
2330533	0023
2330551	0023
2330553	0023
2330561	0015
2330562	0023
2330564	0023
2330611	0120
2330612	0015
2330631	0062
2330632	0062
2330633	0077
2330634	0062
2330643	0025
2330721	0023
2330741	0322
2330743	0057
2330752	0924
2330771	0087
2330773	0087
2330791	0025
2330921	0201
2331001	0589
2331121	0102
2331141	0015
2331142	0015
2331152	0137
2331154	0052
2331154M	OEM
2331155	0052
2331161	0181
2331164	0248
2331165	0181
2331172	0248
2331173	0317
2331174	0317
2331201	0609
2331351	0124
2331381	0182
2331441	0123
2331481	0015
2331491	0451
2331493	0041
2331495	0253
2331497	0471
2331502	0087
2331546	0376
2331582	0494
2331773	OEM
2331775	0118
2331781	0376
2331782	0376
2331785	0036
2331797	0589
2331799	0589
2331801	0883
2331802	0253
2331804	0883
2331805	0157
2331806	0091
2331807	0091
2331809	0295
2331809M	OEM
2331814	0062
2331815	0062
2331816	0062
2331817	0077
2331819	0077
2331821	0165
2331823	0165
2331826	0322
2331827	0057
2331829	0057
2331831	0064
2331834	0064
2331835	0181
2331841	1005
2331842	1005
2331843	0052
2331844	1005
2331845	0053
2331849	1005
2331851M	OEM
2331857	0376
2331912	0023
2331913	0133
2331941	0102
2331991	0564
2332004	0010
2332031	OEM
2332032	OEM
2332121	0348
2332141	0023
2332142	0023
2332152	0102
2332221	0091
2332242	0023
2332251	0023
2332431	2628
2332483	0760
2332491	4052
2332492	1131
2332501	0010
2332791	0106
2332792	2400
2332794	2400
2332841	0181
2332851	0023
2332891	0516
2332892	0516
2333003	0023
2333011	0023
2333101	0015
2333173	0123
2333201	0023
2333231	0071
2333262	OEM
2333855	0494
2333972	OEM
2334011	OEM
2334021	OEM
2334052	0760
2334061	0118
2334122	0041
2334142	0091
2334243	0445
2334244	0445
2334283	0494
2334303	0195
2334304	OEM
2334305	0195
2334312	0195
2334324M	OEM
2334471	3728
2334581	0023
2334592	0023
2334671	OEM
2334672	OEM
2334881	0124
2335041	0705
2335042	0705
2335181	0700
2335182	0700
2335461	0494
2335462	0494
2335982	0564
2335991	3452
2336291	0023
2336381	OEM
2336612	0283
2336911	OEM
2337011	0133
2337063	0565
2337065	0420
2337071	0015
2337101	0137
2337293	0057
2337372	0106
2338031	1664
2338161	0023
2338162	0023
2338171	OEM
2338321	0133
2338531	0017
2338871	0031
2338902	0023
2338932	OEM
2338941	1127
2338944	OEM
2338971	1570
2339011	0883
2339021	OEM
2339021M	OEM
2339023	OEM
2339031	OEM
2339032	OEM
2339052	OEM
2339053	OEM
2339091	OEM
2339091M	OEM
2339093	OEM
2339102M	OEM
2339121M	OEM
2339142	1005
2339151	OEM
2339152	OEM
2339152M	OEM
2339191M	OEM
2339201	OEM
2339203	0705
2339211	0494
2339212	OEM
2339222	0455
2339223	0714
2339223M	OEM
2339232	OEM
2339242	OEM
2339251	0010
2339252	0010
2339481	0023
2339491	0015
2339531	OEM
2339601M	OEM
2339602	OEM
2339691	OEM
2339692	OEM
2339801M	OEM
2339812M	OEM
2339814	OEM
2339817	OEM
2339822	OEM
2339825	OEM
2339832	OEM
2339833M	OEM
2339834M	OEM
2339835M	OEM
2339837	OEM
2339839	OEM
2339845	3002
2339846	OEM
2339847	OEM
2339848M	OEM
2339849	OEM
2339851	OEM
2339852	OEM
2339854M	OEM
2339857	OEM
2339862M	OEM
2339867	OEM
2339867M	OEM
2339868	OEM
2339868M	OEM
2339872M	OEM
2339873	OEM
2339874	OEM
2339876	OEM
2339882M	OEM
2339885	OEM
2339885M	2421
2339886M	OEM
2339887M	OEM
2339889	2431
2339921	OEM
2339943M	OEM
2339971	OEM
2339972M	OEM
2339981	OEM
2339991	0677
2340211	OEM
2340283	1925
2340331	0120
2340332	0139
2340511	OEM
2340521	OEM
2343961	0023
2344071	4393
2347021	0015
2348062	OEM
2348103	OEM
2348121	OEM
2348141	OEM
2348163	OEM
2348181	OEM
2348183	OEM
2348252	OEM
2348291	OEM
2348391	0241
2348431	1090
2348511	OEM
2360021	2582
2360042	0167
2360141	0878
2360151	4627
2360171	0878
2360181	OEM
2360192	OEM
2360201	0167
2360212	OEM
2360221	0043
2360231	0389
2360241	OEM
2360252	OEM
2360253	OEM
2360254	OEM
2360261	OEM
2360271	4642
2360281	4637
2360291	4641
2360331	0167
2360351	OEM
2360361	0878
2360391	0167
2360401	0878
2360431	1055
2360441	1055
2360501	2122
2360501(SIF)	0167
2360511	0878
2360561	0942
2360611	2184
2360631	4627
2360681	OEM
2360691	1429
2360741	0232
2360751	1931
2360761	OEM
2360782	4782
2360851	0215
2360951	1595
2361251	1206
2361281	0490
2361301	2090
2361461	0334
2361571	4787
2361572	4787
2362051	1633
2362301	0088
2362492	OEM
2362541	5182
2362601	6836
2362605	0360
2362991	4826
2363001	6854
2363191	0101
2363261	4787
2364131	3503
2364134	OEM
2364172	2533
2364181	1201
2365021	3776
2365061	2122
2365062	2122
2365411	0176
2365452	0631
2366151	5786
2366201	3445
2366392	1529
2366411	OEM
2366571	OEM
2366591	1048
2366621	2666
2366631	5215
2366721	5048
2368211	1824
2368271	OEM
2368282	1442
2368283	OEM
2368291	OEM
2368292	OEM
2368451	OEM
2368501	2042
2368671	OEM
2368692	OEM
2368913	1166
2369151	0450
2369582	OEM
2370061	OEM
2370152	OEM
2370873	1641
2371806	OEM
2372161	OEM
2380131	OEM
2380391	OEM
2380426	0334
2380661	OEM
2380662	OEM
2381111	OEM
2381344	OEM
2381471	OEM
2381832	OEM
2381862	OEM
2382952	OEM
2383191	OEM
2383341	OEM
2383342	OEM
2383421	OEM
2390022	0058
2398511	OEM
2398611	0124
2485076-2	0086
2485076-3	0086
2485077-2	0086
2485077-3	0086
2485078-1	0016
2485078-2	0016
2485078-2(PENNCREST)	0079
2485078-2(SEARS)	0079
2485078-3	0016
2485079-1	0016
2485079-2	0016
2485079-3	0016
2485080	0133
2485080(DIODE)	0143
2486836	0015
2487340	0037
2487341	0037
2487424	0595
2495012	0136
2495013	0136
2495014	0004
2495078	0050
2495079	0050
2495080	0004
2495082	0050
2495083	0123
2495083-2	0143
2495084	0143
2495166-1	0144
2495166-2	0016
2495166-4	0016
2495166-8	0016
2495166-9	0016
2495200	0136
2495376	0050
2495377	0050
2495378	0136
2495379	0136
2495380	0143
2495383	0143
2495388	0004
2495388-1	0004
2495388-2	0004
2495488-1	0136
2495488-2	0136
2495520	0007
2495521	0007
2495521-1	0016
2495522-1	0007
2495522-4	0016
2495523-1	0007
2495529	0143
2495529(ARVIN)	0086
2495529(DIODE)	0143
2495567-2	0004
2495567-3	0004
2495568	0164
2495568-2	0004
2496125-2	0016
2496436	0143
2497094-1	0007
2497094-2	0007
2497473	0004
2497473-1	0004
2497496	0004
2497888	0164
2498163	0086
2498456-2	0127
2498457-2	0016
2498482-2	0127
2498507-1	0144
2498507-2	0144
2498507-3	0144
2498508-2	0127
2498508-3	0127
2498512	0037
2498513	0015
2498530	0143
2498837	0050
2498837-4	0004
2498902-1	0127
2498902-2	0127
2498903-1	0127
2498903-2	0224
2498903-3	0127
2498904-3	0016
2498904-4	0016
2498904-6	0016
2499048	0030
2499950	0111
2500389-432	1041
2500389-444	0033
2500390-435	1176
2500390-436	1831
2500390-437	1835
2500390-446	0141
2500390-449	1844
2500390-461	1859
2500390-462	0562
2500391-445	0081
2500747	1041
2501337-433	2097
2501557-430	1823
2501557-436	1831
2501557-437	1835
2501557-446	0141
2501557-449	1844
2501557-461	1859
2501557-462	0562
2520063	0016
2552007	0087
2610783	0314
2610784	1715
2610786	0232
2610788	0175
2650238A	OEM
2652613	0033
2652614	0081
2656211-1	1823
2656212	0141
2656212-1	0141
2656213	0033
2656213-1	0033
2656214	0081
2656214-1	0081
2656747	0033
2656748	0141
2666293	1574
2666294-1	0117
2709759	1922
2711904	0829
2712080	0016
2729789	0847
2780312	1030
2808577	1041
2811925-002	3537
2855296-01	0004
2868536-1	1313
2880661	OEM
2898431-2	0141
2899000-00	0141
2899001-00	0562
2899002-00	1823
2899003-00	1831
2899004-00	0081
2902798-2	1554
2903993-I	0037
2904014	0160
2910011	0557
2910012	0391
2910021	0732
2910932	OEM
2910941	OEM
2911481	3437
2911491	OEM
2911501	0416
2911511	OEM
2912175	3913
2912176	3913
2912177	2364
2912178	3913
2912561	3097
2912961	4102
2913081	4174
2913091	OEM
2913891	0315
2913941	OEM
2913945	OEM
2913981	4129
2914912	OEM
2914921	OEM
2914931	2870
2914941	3083
2914951	4524
2914961	4532
2915191	OEM
2915192	0523
2916242	OEM
2916321	OEM
2916631	OEM
2916633	OEM
2916634	OEM
2916671	OEM
2916681	5054
2917221	OEM
2917222	OEM
2917301	OEM
2917361	5054
2917591	OEM
2917601	3027
2917611	1028
2917621	OEM
2917711	OEM
2917721	OEM
2917731	OEM
2917741	OEM
2917742	OEM
2917802	OEM
2917803	OEM
2970038H05	0280
3004856	0004
3005300	0122
3005861	0016
3007359-00	1835
3007359-01	1844
3007359-02	0341
3007359-03	1041
3007472-00	0081
3007473-00	1041
3007473-01	1041
3007474-00	1823
3007474-01	1823
3007475-00	0562
3007475-01	0562
3007476-00	0141
3007476-01	0141
3007477-00	1831
3007477-01	1831
3007572-00	2097
3007573-00	0358
3007604-00	1619
3008321-00	1922
3020061	0151
3034725-1	2537
3100001	0026
3100002	2247
3100004	0624
3100017	0002
3100021	0565
3100026	0820
3102006	1527
3102007	1559
3102008	1595
3102009	1608
3102010	1667
3102011	0034
3102013	0828
3102014	0088
3102015	0243
3102016	1887
3102017	1888
3102018	1625
3102019	0959
3102020	1161
3102021	4371
3102022	0260
3102023	1672
3102024	0977
3102025	0991
3102026	2112
3102027	2958
3102028	1654
3106001	OEM
3106002	1262
3108001	OEM
3108002	4973
3108003	2307
3108009	0522
3110001	3481
3130057	0178
3130058	0103
3146977	0103
3146977A	0103
3152159	0178
3152170	0103
3152170A	0103
3170717	0233
3170757	0233
3172626	2097
3172627-2	1844
3172629-1	1823
3172629-2	1859
3172630	0033
3172631	3399
3172632-1	0562
3172638	2619
3176135-1	1831
3176135-2	1835
3177200	1041
3230591	OEM
3267390-01	0559
3403787	0016
3404520-601	0353
3412004-1912	0325
3412951	1854
3412986-1	2537
3430063	0015
3430063-1	0540
3430063-I	0540
3457845-1	2041
3457964-1	1233
3457964-2	1233
3457964-3	1233
3457964-4	1233
3459332-1	0007
3460553-2	0160
3460553-4	0160
3460758-2	0953
3462221-1	0160
3462306-1	0160
3463098-1	0084
3463098-2	0534
3463604-1	0178
3463604-2	0178
3463609-2	0007
3468068-1	0007
3468068-2	0007
3468068-3	0007
3468068-4	0007
3520025-001	1831
3520041-001	0232
3520042-001	0512
3520043-001	1172
3520044-001	1024
3520045-001	1168
3520046-001	1313
3520047-001	0873
3520048-001	0361
3520050-001	1426
3539307-001	0007
3539307-002	0007
3596061	0133
3596062	0143
3596063	0037
3596067	0144
3596068	0144
3596069	0144
3596070	0144
3596071	0144
3596072	0144
3596091	0423
3596092	0423
3596100	0854
3596101	0854
3596116	0086
3596117	0079
3596118	0037
3596128	0037
3596242	0030
3596260	0144

If replacement code is OEM, contact original manufacturer for replacement.

DEVICE TYPE	REPL CODE	DEVICE TYPE	REPL CODE	DEVICE TYPE	REPL CODE	DEVICE TYPE	REPL CODE	DEVICE TYPE	REPL CODE	DEVICE TYPE	REPL CODE	DEVICE TYPE	REPL CODE
3596261	0144	4038264-P1	0600	4890210A21	0041	4890478A47	0359	5320031	0849	5330059	0565	5351551	5183
3596316	OEM	4038264-P2	0600	4890210A23	0181	4890478A51	0563	5320032	0325	5330101	0015	5351601	OEM
3596317	OEM	4038359-P1	0050	4890210A24	0064	4890478A52	0582	5320042H	0037	5330101H	0015	5351621	OEM
3596338	0079	4038359-P2	0050	4890210A25	0124	4890478A53	0582	5320043H	0037	5330102	0015	5352631	2105
3596339	0079	4038406-P1	0050	4890210A26	0124	4890478A68	0582	5320051	0127	5330102H	0015	5352791	OEM
3596340	0126	4038406-P2	0050	4890210A28	0062	4890478A69	0672	5320064	0076	5330104	0023	5353002	OEM
3596341	0126	4041200-30110	0133	4890210A29	0253	4890478A83	0203	5320064H	0079	5330104H	0015	5353021	OEM
3596353	1759	4041200-40100	0143	4890210A30	0181	4890488A01	0019	5320067	0076	5330131	0133	5353023	OEM
3596354	1345	4080187-0502	0042	4890210A31	0060	4890488A06	1962	5320069	0076	5330132	0124	5353031	OEM
3596398	0025	4080187-0504	0103	4890210A32	0471	4890488A19	0672	5320074	0016	5330133	0124	5353033	OEM
3596401	1908	4080187-0506	0423	4890210A53	1348	4890488A32	0023	5320101	0004	5330135	0124	5353041	OEM
3596402	0212	4080187-0507	0423	4890210A54	1165	4890488A67	0071	5320111	0126	5330201	0030	5353061	OEM
3596435	0030	4080320-050B	0103	4890210A55	0023	4890488A77	0343	5320141	0004	5330212H	0133	5353081	OEM
3596440	0127	4080320-0501	0103	4890210A56	0735	4890543A15	0158	5320151	1411	5330261	0133	5353091	OEM
3596446	0042	4080320-0504	0103	4890210A57	0041	4890549A20	OEM	5320241	0016	5330261H	0133	5353101	OEM
3596447	0042	4080627-0501	0178	4890210A59	0064	4890549A22	0023	5320295	0004	5330312	0062	5353111	OEM
3596448	0042	4080835-0002	0423	4890210A61	0018	4890549A25	OEM	5320295H	0208	5330315	0077	5353112	OEM
3596449	0042	4080838-0001	0178	4890210A62	0672	4890549A64	0516	5320296	0004	5330322	0012	5353121	OEM
3596451	0925	4080838-0002	0178	4890210A76	0471	4890549A71	0018	5320305	0208	5330325	0165	5353591	2385
3596452	0925	4080838-002	0423	4890210A92	OEM	4890549A99	0077	5320305H	0004	5330328	0057	5354511	OEM
3596453	0925	4080838-2	0178	4890210A96	0077	4890562A02	0015	5320306	0208	5330331	0143	5354523	OEM
3596454	0925	4080838-3	0178	4890223A07	1778	4890562A04	0019	5320326	0016	5330332	0143	5354531	OEM
3596570	0016	4080865-0012	0161	4890223A39	2628	4890562A29	0071	5320326H	0127	5330334	0133	5354541	OEM
3596803	OEM	4080866-0006	0560	4890223A61	1162	4907976	0004	5320328	0144	5330335	0143	5354601	OEM
3596808	2617	4080866-0006(PLASTIC	0042	4890223A62	1212	4914296	4234	5320361	0208	5330336	0015	5354611	OEM
3596809	1843	4080866-0007	0178	4890223A63	3619	4915702	1349	5320372	0016	5330341	0023	5355262	2026
3596810	2755	4080866-000A	0103	4890223A64	0023	4915705	0527	5320372H	0016	5330371	0015	5355611	2666
3597049	1248	4080866-0012	0423	4890223A65	0023	4999774	0160	5320373	0079	5330372	0015	5355751	OEM
3597103	0144	4080866-0013	0042	4890223A66	0166	4999885	0437	5320386	0038	5330381	0015	5356011	OEM
3597104	0144	4080866-1	0178	4890223A67	0091	4999887	0004	5320422H	0042	5330392	0253	5356211	OEM
3597114	0127	4080866-2	0178	4890223A68	0755	5000027	OEM	5320432	5533	5330431	0023	5356251	OEM
3597260	0144	4080866-8006	0103	4890223A88	0170	5073004	0004	5320433	0042	5330521	0005	5356341	OEM
3597261	0144	4080873-0001	0161	4890223A89	0015	5076054	0133	5320475	0208	5330532	0053	5357351	OEM
3597280	1248	4080879-0001	0042	4890223A90	0248	5076204	0704	5320485	0004	5330541	0686	5357531	OEM
3598173	1248	4080879-0006	0423	4890223A92	0071	5076205BZ	1017	5320492H	0042	5330542	0686	5357561	0103
3610001-001	OEM	4080879-0015	0178	4890223A93	1541	5113753A37	5215	5320501	0698	5330551	0064	5359031	0232
3610003	1069	4082501-0001	3033	4890223A94	0359	5113753A47	3916	5320511	1411	5330552	0181	5359251	2616
3610005	3418	4082501-0001A	0160	4890223A95	0582	5113753A68	5208	5320583	0325	5330571	0124	5359261	3815
3650238A	0006	4082501-001	0085	4890223A96	0113	5120134	OEM	5320592	0006	5330572	0133	5359262	3815
3670724H	0076	4082626-0001	1696	4890223A97	0261	5120850	OEM	5320593	0006	5330573	0133	5359271	0569
3671857E	0142	4082664-0001	2617	4890223A98	0261	5165440	1831	5320612	0086	5330661	0628	5359281	0083
3673351K	0191	4082665	1345	4890223A99	0118	5175460	0467	5320613	0191	5330712	0118	5359451	OEM
3673354G	0191	4082665-0001	1345	4890229A10	OEM	5190266A68	3902	5320622	0595	5330714	0275	5359522	4956
3680322-3	1195	4082665-0002	1345	4890229A30	2628	5190458A07	0176	5320632	1267	5330721	0123	5359592	OEM
3720923-1	1974	4082665-0003	1345	4890229A49	0124	5190480A92	0489	5320642	0560	5330731	0019	5359701	2295
3720923-2	1974	4082671-0002	0400	4890229A51	0023	5190480A96	0976	5320643	0161	5330732	0143	5359841	5181
3720923-3	1974	4082748-0002	0157	4890229A53	0276	5190495A88	6040	5320651	0076	5330761	0124	5361071	OEM
3720923-4	1974	4082799-0001	1843	4890229A59	0118	5190495A89	OEM	5320671	0042	5330842	0041	5361671	OEM
3731313-1	0160	4082799-0002	1843	4890229A86	0023	5190495A90	3853	5320702	0325	5330852	0030	5362055	OEM
3939307-002	0007	4082799-1	1843	4890232A40	1220	5190507A20	2019	5320723	0460	5331001	0133	5362121	OEM
4003000	OEM	4082799-2	1843	4890232A67	0124	5190507A39	0770	5320731	0143	5331017	0589	5362291	OEM
4013373-0701	1810	4082802-0001	2747	4890232A78	0023	5190507A40	0770	5320772	0224	5331051	0143	5362382	OEM
4036392-P2	0170	4082802-0002	2747	4890232A94	0071	5190507A46	0699	5320791	0936	5331102	0323	5362741	0215
4036598-P1	0160	4082873-0001	0854	4890232A99	0261	5190507A60	0759	5320813	0076	5331271	0375	5362862	OEM
4036598-P2	0004	4082886-0002	0161	4890233A53	0019	5190507A78	2067	5320851	0079	5331511	3501	5362882	OEM
4036612-P1	0280	4084114-0001	0212	4890233A97	0455	5190507A87	2073	5320861	0127	5331521	0323	5362883	OEM
4036612-P2	0280	4084114-0002	0212	4890234A30	0286	5190518A19	2553	5320872	0004	5331591	0133	5362912	OEM
4036707-P1	0280	4084117-0001	2552	4890234A31	0155	5190528A01	0770	5320921	1790	5331592	0133	5362914	OEM
4036707-P2	0280	4084117-0002	2552	4890234A83	0203	5190528A27	4346	5320942	0325	5331931	0323	5362931	OEM
4036715-P1	0050	4090187-0502	1514	4890235A09	0343	5190528A28	3853	5320943	0325	5332192	0023	5364091	3080
4036715-P2	0050	4090605-1	2557	4890235A14	0023	5190528A29	0334	5321183	0436	5333341	OEM	5364201	3159
4036733-P1	0160	4101685	0015	4890235A24	0023	5190528A30	5961	5321184	0436	5339002	0133	5364306	OEM
4036749-P1	0600	4361620	0103	4890235A25	0023	5190528A30(TNH11304)	OEM	5321204	0532	5339021	0133	5364309	OEM
4036749-P2	0600	4420022-P1	0267	4890235A45	OEM	5190528A41	1641	5321213	0018	5339351	OEM	5364601	1292
4036754-P1	0038	4420022-P2	0267	4890235A51	1992	5190528A43	4194	5321214	0018	5340001	0133	5364602	OEM
4036754-P2	0038	4450028-P1	0725	4890235A52	0825	5190528A44	OEM	5321235	0390	5340001(TRANSISTOR)	0079	5364631	OEM
4036831-P1	0440	4450028-P2	0725	4890235A53	0168	5190528A48	OEM	5321251	0693	5340001H	0133	5364642	OEM
4036832-P1	0440	4450028P4	0725	4890235A54	1162	5190528A50	OEM	5321252	0693	5340021	0015	5364651	OEM
4036887-P8	0016	4450300-1	0147	4890235A55	0023	5190528A53	6257	5321253	0693	5340022	0120	5364881	OEM
4036923-P1	0050	4450300P2	0415	4890235A56	0248	5190528A54	OEM	5321255	0006	5340022H	0133	5364902	OEM
4036923-P2	0050	4450300P3	0415	4890235A57	0137	5190528A55	5260	5321261	0016	5340051	0133	5364911	OEM
4036924-P1	0016	4450303-001	1413	4890235A58	0032	5190528A56	5261	5321291	0151	5340082	0030	5364923	OEM
4036924-P2	0016	4450303-002	1413	4890235A59	0165	5190528A57	5262	5321301	0042	5340111	0133	5364971	OEM
4036937-P1	0280	4511424	0899	4890235A60	0017	5190528A58	4925	5321311	2710	5340113	0139	5366401	OEM
4036937-P2	0280	4613443	4485	4890235A61	0055	5190528A59	5271	5321321	0936	5340202	0120	5366581	OEM
4036962-P1	0050	4663001A905	1313	4890235A62	0609	5190528A60	OEM	5321422	1280	5350051	OEM	5367061	OEM
4036962-P2	0050	4663001A909	0361	4890235A71	0064	5190528A61	5263	5321431	0127	5350081	OEM	5368931	OEM
4036963-P1	0050	4663001A911	1024	4890235A89	0203	5190528A62	OEM	5321501	3342	5350121	2262	5369181	1699
4036963-P2	0050	4663001A912	0512	4890237A60	0695	5190528A63	OEM	5321502	3342	5350132	4561	5369431	1028
4036965-P1	0050	4663001A915	0873	4890237A80	0018	5190528A64	3159	5321662	1140	5350136	4561	5369911	1938
4036965-P2	0050	4663001D907	0232	4890343A49	0155	5190528A65	OEM	5321663	1140	5350141	4562	5369914	OEM
4037145-P1	0004	4800002	0124	4890343A81	0018	5190528A66	OEM	5321901	0224	5350151	4563	5370031	OEM
4037145-P2	0004	4800021	0636	4890343A87	2094	5190538A18	1949	5322216	1513	5350152	4563	5374053	OEM
4037289-P1	0600	4800022	0041	4890420A01	0582	5190538A20	3902	5322733	0006	5350161	4564	5374291	OEM
4037289-P2	0600	4800023	0289	4890420A19	0582	5190538A20(AN5125M)	OEM	5323071	0558	5350182	0470	5374302	OEM
4037325-P1	0015	4800026	0094	4890420A24	0582	5190538A22	0334	5323561	0887	5350211	4390	5376912	OEM
4037410-P1	0050	4800037	0137	4890420A29	0359	5190538A98	4422	5323562	1032	5350231	2124	5377132	OEM
4037410-P2	0050	4800045	1354	4890420A42	0825	5190555A50	1529	5323704	3133	5350251	2656	5377551	OEM
4037413-P1	0636	4800062	2312	4890420A83	0648	5190555A53	3853	5323903	0151	5350321	4401	5377561	OEM
4037586-P1	0016	4804427	2700	4890420A84	0392	5190555A60	0035	5324051	3133	5350481	OEM	5377562	OEM
4037586-P2	0016	4806080	2170	4890420A85	0015	5261001A	OEM	5324111	OEM	5350491	4401	5378641	OEM
4037594-P1	0160	4806095	2170	4890420A86	0015	5261071	OEM	5324631	OEM	5350591	OEM	5380031	OEM
4037607-P1	0160	4813466	0016	4890420A87	0015	5262063	OEM	5326244	OEM	5350604	2913	5380051	0015
4037647-P1	0560	4820004	0042	4890432A02	0563	5262081	OEM	5327011	0151	5350611	1329	5380052	1358
4037647-P2	0560	4822001	0076	4890432A06	0006	5300073-15	0002	5327021	0006	5350702	3196	5380131	OEM
4037764-P1	0050	4822003	0037	4890432A15	0023	5300113-1	0015	5327031	0148	5350961	2656	5380132	OEM
4037764-P2	0050	4824002	0925	4890432A42	0118	5301181	OEM	5327163	OEM	5351021	1621	5381211	OEM
4037800-P1	0016	4824003	0491	4890432A55	0825	5301341	OEM	5328193	OEM	5351031	4571	5381214	OEM
4037800-P2	0016	4890210A03	1162	4890432A56	0825	5301571	OEM	5330001	0015	5351041	0701	5381661	OEM
4037804-P1	0004	4890210A06	0124	4890432A65	1778	5305582	OEM	5330011	0057	5351042	0701	5381662	OEM
4037804-P2	0004	4890210A10	2766	4890445A09	2223	5306567	OEM	5330012	0012	5351051	4570	5381681	5695
4037839-P1	0600	4890210A12	0181	4890445A17	0582	5320003	0178	5330031	0015	5351061	2591	5381682	3185
4037839-P2	0600	4890210A13	0376	4890445A51	0181	5320004	0076	5330041	0023	5351062	2591	5382051	OEM
4037993-P1	0280	4890210A14	0261	4890445A52	0181	5320011	0004	5330041(HR-5A)	0071	5351351	0167	5382092	OEM
4037993-P2	0280	4890210A16	0648	4890448A32	0023	5320022	0076	5330041H	0015	5351361	0878	5390191	OEM
4038256-P1	0208	4890210A17	1165	4890456A24	0124	5320023	0076	5330042	0015	5351411	4805	5391032	0489
4038256-P2	0208	4890210A18	0018	4890456A30	1962	5320023H	0079	5330051	0025	5351501	5180	5406665-P2	0004
4038260-P1	0280	4890210A19	0023	4890456A85	0057	5320024	0079	5330054	0053	5351521	5675	5406665-P5	0004
4038260-P2	0280	4890210A20	0609	4890478A30	0041	5320026	0079	5330054H	0053	5351541	OEM	5490307-P2	0170

DEVICE TYPE	REPL CODE
5490307-P3	0137
5490307-P4	0002
5490415-P4	0015
5490459-P1	0015
5490459-P2	0015
5490459-P3	0015
5490459-P4	0015
5490804-P1	0015
5490804-P2	0015
5490804-P3	0015
5490804-P4	0015
5490804-P5	0015
5490804-P6	0015
5491236-P1	0143
5491236-P2	0143
5491263-P1	0143
5491338-P1	0162
5492153-1	0440
5492153-P1	0440
5492639-P1	0038
5492639-P2	0038
5492653-P1	0600
5492653-P2	0600
5492655-P1	0600
5492655-P2	0600
5492655-P3	0600
5492655-P4	0600
5492655-P5	0600
5492655-P6	0600
5492659-P1	0600
5492659-P2	0600
5493158-1	0160
5493158-P1	0160
5493159-P1	0097
5493957-P1	0050
5493957-P2	0050
5493957-P3	0050
5493957-P4	0050
5493957-P5	0050
5493957-P6	0050
5494922-P1	0015
5494922-P2	0015
5494922-P3	0015
5494922-P4	0015
5494922-P5	0015
5494922-P6	0015
5494922-P7	0015
5494922-P8	0015
5495178-P1	0097
5495912-P2	0441
5495922-P1	0097
5495922-P2	0097
5495922-P3	0097
5495923-P1	0722
5496438-P1	0441
5496663-P1	0160
5496665-P1	0211
5496665-P2	0211
5496665-P3	0211
5496665-P4	0211
5496665-P5	0211
5496665-P6	0211
5496666-P1	0211
5496666-P2	0211
5496666-P3	0211
5496666-P4	0211
5496666-P5	0211
5496666-P6	0211
5496666-P7	0211
5496666-P8	0211
5496667-P1	0211
5496667-P2	0211
5496668-P1	0440
5496774-P1	0211
5496774-P2	0211
5496774-P3	0211
5496774-P4	0211
5496774-P5	0211
5496774-P6	0211
5496839-P1	0211
5496939-P1	0160
5496939-P2	0160
5496947-P3	0441
5518017	0849
5729031	0440
5955748	0004
5955749	0015
5958539	0847
5960904	OEM
5983945A	0440
5985432	0440
5988414	0440
5988977	0440
5989171	0440
5989615	0440
5989692	0440
5989693	0440
6100724-2	0280
6105590K01	OEM
6125070K01	OEM
6126150K01	5391
6127784(SM)	OEM
6127893(SM)	OEM
6142190	0816
6207159	4634
6208839	0076
6212096	0042
6212839	0079
6212922	0144
6218945	0076
6510053-2	0050
6710223-1	0037
6846503	0015
6846812	0474
6879847	OEM
6920802	OEM
6920900	OEM
7011200-02	2097
7011201-02	1844
7011203-02	1432
7011203-03	1432
7012128	1831
7012128-02	1831
7012130	1350
7012131	1929
7012132	1574
7012133-02	0117
7012142-03	1926
7012166	0314
7012167-02	1208
7023416-00	0015
7026011	0007
7026012	0144
7026013	0144
7026014	0016
7026015	0016
7026016	0016
7026018	0177
7026019	0126
7026020	0079
7026024	0042
7026036	0030
7027002	0015
7027005	0015
7027007	0102
7027023	0015
7027031	0139
7027033	0182
7027036	0196
7027038	0015
7027039	0015
7030105	0004
7043130A	OEM
7043216	0015
7043261	0015
7070692	0015
7071021	0079
7071031	0079
7100460	0143
7100628	0015
7100630	0015
7116060	0019
7129386-P2	0143
7200351	0030
7200953	0124
7201090	0244
7201212	0139
7202782	0244
7220132	OEM
7269847	0208
7274653	0160
7276211	0280
7276605	0440
7277066	0208
7278421	0280
7278422	0004
7278423	0004
7279003	0440
7279005	0440
7279007	0440
7279009	0440
7279011	0440
7279013	0440
7279017	0440
7279025	0440
7279027	0440
7279031	0440
7279033	0440
7279035	0440
7279039	0085
7279049	0085
7279069	0604
7279073	0440
7279076	0440
7279281	0847
7279293	0440
7279298	0440
7279379	0280
7279497	0015
7279566	0440
7279779	0136
7279779(G.M.)	0016
7279780	0050
7279781	0050
7279782	0280
7279788	0280
7279789	0280
7279793	0440
7279893	0143
7279940	0280
7279941	0004
7280281	0440
7281307	0280
7281308	0280
7281309	0280
7281310	0004
7281891	0280
7282315	0440
7282358	0143
7283377	0123
7284137	0016
7284513	0050
7284751	0004
7285354	2901
7285663	0160
7285774	0085
7285776	0160
7285778	0085
7286858	0016
7287079	0244
7287107	0604
7287110	0160
7287112	0604
7287117	0604
7287452	0016
7287940	0136
7288019	0030
7288072	0440
7288073	0440
7288076	0440
7288079	0440
7289041	0160
7289047	0085
7289079	0160
7289097	0604
7290593	0160
7290594	0085
7291252	0085
7292308	0050
7292683	0604
7292684	0975
7292689	0604
7292690	0160
7292955	0160
7293818	0074
7293819	0074
7294133	0004
7294796	0160
7294910	0007
7295195	0144
7295196	0144
7295197	0016
7296314	0016
7296476	0244
7297043	0085
7297053	0016
7297054	0016
7297092	0085
7297093	0085
7297347	0604
7297348	0975
7297358	0012
7297980	0007
7298079	0160
7299325	0015
7299720	0440
7299780	0604
7301660	0604
7301661	0604
7301664	0074
7301665	0074
7301666	0074
7302340	0229
7303105	0004
7303120	0016
7303304	0074
7304149	0642
7304971	3620
7305469	0244
7305783	0160
7306982	0103
7311325	0678
7313568	2279
7420683	OEM
7440006	0196
7492377-5	0137
7492377-P1	0162
7492377-P5	0137
7492377-P10	0162
7520136	OEM
7528046-P4	1041
7528048-P4	0141
7528153-P4	3399
7528156-P3	1257
7528158-P4	1257
7528159-P4	0033
7528160-P4	2078
7528374P3	2078
7570003	0160
7570003-01	0160
7570004	0016
7570004-01	0016
7570005	0016
7570005-01	0016
7570005-02	0016
7570005-03	0016
7570008	0016
7570008-02	0016
7570009	0086
7570009-01	0016
7570009-21	0016
7570011-01	0071
7570011-02	0071
7570013-01	0037
7570014	0242
7570014-02	0143
7570014-05	0015
7570014-34	0015
7570015-34	0015
7570016	0143
7570016-02	0143
7570016-03	0143
7570021-12	0441
7570021-21	0441
7570022-01	0178
7570024-02	0015
7570030-01	0460
7570031-01	0042
7570032-01	0126
7570215-21	0133
7570215-24	0015
7570215-34	0015
7571329-01	0277
7576004-01	0037
7576015-01	0016
7582438-01	0019
7582439-01	0143
7777146-P2	0143
7777146-P3	0143
7777146-P4	0143
7777146-P23	0143
7820135	OEM
7835297-01	0280
7851316-01	0004
7851323	0050
7851325	0016
7851326	0016
7851327	0016
7851328-01	0302
7851329-01	0277
7851379-01	0016
7851380-01	0016
7851441-01	0015
7851442-01	0451
7851467-01	0208
7851650-01	0007
7851651-01	0007
7851652-01	0037
7851654-01	0143
7851655-01	0143
7851657-01	0015
7851947-01	0143
7851949-01	0016
7851950-01	0016
7851952-01	0016
7851953-01	0688
7851954-01	0037
7851955-01	3545
7851956-01	0086
7852223-01	0143
7852223-01A	0143
7852264-01	0015
7852438-0	0143
7852438-01	0143
7852439-01	0143
7852452-01	0076
7852454	0016
7852454-01	0111
7852455-01	0016
7852457-01	0191
7852459-01	0016
7852460-01	1147
7852577-01	0277
7852775-01	0004
7852776-01	0004
7852778-01	0004
7852779-01	0004
7852781-01	0076
7852782-01	0143
7852897-01	0050
7852899-01	0136
7852900-01	0050
7852902-01	0123
7852903-01	0030
7852904-01	0162
7852907-01	0015
7853090-01	0688
7853091-01	0191
7853092-01	0191
7853093-01	0286
7853094-01	0076
7853098-01	0143
7853099-01	0277
7853351-01	0004
7853352-01	0004
7853354-01	0004
7853356-01	0004
7853357-01	0143
7853358-01	0015
7853463-01	0076
7853464-01	0016
7853465-01	0016
7855282	0143
7855282-01	0143
7855283-01	0133
7855283-01(AMPEX)	0004
7855284-01	0015
7855291-01	0016
7855292-01	0016
7855293-01	0016
7855294-01	0016
7855295-01	0222
7855296-01	0004
7855297-01	0004
7855298-01	0042
7855352-01	0143
7855357-01	0143
7910070-01	0004
7910071-01	0004
7910072-01	0160
7910073-01	0161
7910076-01	0015
7910108-01	0007
7910111-01	0244
7910134-01	0325
7910267-01	0423
7910268-01	0688
7910269-01	0151
7910270-01	0151
7910271-01	0111
7910272-01	0111
7910273-01	0218
7910274-01	0086
7910275-01	0015
7910276-01	0030
7910278-01	0012
7910584-01	0016
7910585-01	0016
7910586-01	0016
7910587-01	0016
7910588-01	0004
7910589-01	0004
7910590-01	0015
7910780-01	0144
7910781-01	0127
7910783-01	0136
7910784-01	0030
7910801-01	0004
7910803-01	0164
7910804-01	0016
7910805-01	0015
7910872-01	0015
7910873-01	0015
7910875-01	0126
7913605	0074
7914009-01	0127
7914010-01	0127
7914011-01	0015
7914014-01	0133
7914015-01	0030
7932367	1940
7932638	0133
7932980	0678
7938318	0325
8000016	1349
8000017	1352
8000736	0086
8000737	0086
8002494B	0240
8010490	0136
8010520	0280
8010530	0136
8014711	0050
8014712	0280
8015613	0126
8020134	OEM
8020244	0458
8020245	0463
8020322	0004
8020324	0004
8020333	0004
8020334	0004
8020373	0709
8020540	0004
8020560	0004
8021897	0004
8021898	0004
8022630	0004
8022631	0004
8023643	0004
8023892	0004
8024152	0164
8024250	0975
8024390	0050
8024390(AIWA)	0004
8024400	0004
8026002	4910
8026004	3296
8026008	3299
8026014	3300
8026032	3306
8026138	6090
8026195	6100
8026244	6108
8026245	4075
8026373	6123
8030529	4531
8031825	0144
8031836	0144
8031837	0475
8031839	0475
8033690	0016
8033696	0016
8033720	0076
8033730	0016
8033803	0144
8033804	0144
8033943	0144
8033944	0016
8034903	0042
8036683	0127
8037330	0016
8037332	0016
8037333	0016
8037343	0086
8037353	0016
8037722	1144
8037723	1144
8040048	OEM
8040216	OEM
8040253	OEM
8040259	OEM
8040272	OEM
8040316B	OEM
8040328	OEM
8040332	OEM
8040364B	OEM
8040464	OEM
8040496	OEM
8040517	OEM
8040541	OEM
8040542	OEM
8040543	OEM
8040544	OEM
8040545	OEM
8040665	OEM
8040684	OEM
8041016	2307
8041016A	2307
8041088	OEM
8041364	OEM
8042364A	OEM
8044316	OEM
8044364	OEM
8044732	OEM
8045364A	OEM
8046670	OEM
8049008	OEM
8050339	0176
8050358	0770
8050386	3063
8050526	OEM
8050527	OEM
8051060	0019
8052192	0062
8052446	0143
8052805	1292
8075068	OEM
8075069	OEM
8075306	OEM
8075711	OEM
8079015	OEM
8100906	0006
8110014	0325
8110904	0076
8111230	0037
8112023	0004
8112027	0004
8112071	0004
8112090	0004
8112143	0222
8112146	0004
8112162	0004
8113034	0016
8113323	0191
8113327	0380
8114057	0018
8115009	0037
8120126	0133
8120133	0015
8120164	0139
8150148	0124
8150235	0062
8190005	1285
8190104	OEM
8220136	OEM
8249600	0007
8362309	OEM
8362311	OEM
8362312	OEM
8362535N	OEM
8362536	OEM
8362550	OEM
8484771-1	0084
8484771-3	0090
8484771-7	0097
8484771-8	0765
8484771-10	0777
8484894	OEM
8505870-1	1257
8508309	0126
8508331RB	2537
8518237-1	0084
8518237-2	0090
8518237-3	0097
8518237-9	0748
8518237-10	0765
8518382-1	3189
8518382-2	2901
8518382-3	1124
8518382-9	0254
8518382-10	1107
8518383-10	0204
8518383P1	2002
8518383P2	1250
8518383P3	1250
8521587-1	0015
8521587-101	0242
8521587-102	0540
8521692-1	0240
8538640	0050
8620211	OEM
8620815	OEM
8729902XX	OEM
8759602XX	OEM
8975158-1	0004
8981399-1	0015
8981399-2	0015
9001345-02	0569
9001346	0527
9001349-02	1368
9001349-03	1368
9001549-02	1024
9001551-02	0361
9001551-03	0361
9001567	1433
9001567-02	1433
9001570-02	0335
9002097-03	0942
9003091-02	0512
9003091-03	0512
9003096-02	0743
9003096-03	0743
9003097-02	0942
9003148	1823
9003148-01	1823
9003148-02	1823
9003149	0141
9003149-02	0141
9003150	0562
9003150-02	0562
9003150-03	0562
9003151	0232
9003151-03	0232
9003152	1313
9003152-01	1313
9003234-04	0996
9003398-03	0467
9003398-04	0467
9003420-03	1715
9003445-03	0834
9003642-03	0873
9003911	1041
9003911-01	1041
9003911-03	1041
9004075-03	0268
9004076	0697
9004076-03	0697
9004076-04	0697
9004093-03	1172
9004300-03	1158
9004360-03	1202
9004896-04	1185
9004898-04	0527
9100502	0004
9100621	0004
9100706	0004
9100944	0004
9101600	1554
9246055	0015
9248015	0015
9248055	0015
9248058	0015
9320152	OEM
9320170	OEM
9320220	OEM
9330006	0015
9730150	OEM
9739636-20	0015
10112563	3418
10176209	1069
10541284	1257
10541285	3061
10638971	OEM
10658276	1823
10658278	1831
10658279	0033
10658280	0081
10658281	0141
10658282	2078
10674068	1413
11019072	3271
11050925	0006
11056225	0006
11069934A	0103
11113005	1615
11113006	2628
11113007	OEM
11113014	OEM
11113025	OEM
11113035	OEM
11114027	0392
11114035	0042
11114044	0006
11114044T	0006
11114045	0006
11114045T	0006
11114047	0532
11114048	0006
11114048T	0006
11114049	0006
11114049T	0148
11114050	0144
11114055	0286
11114056	0286
11114056T	0286
11114060	0286
11114062	0155
11114063	0261
11114063T	0261
11114064	0261
11114064T	0261
11114066	0286

If replacement code is OEM, contact original manufacturer for replacement.

DEVICE TYPE	REPL CODE
11114066T	0286
11114068	0286
11114073	6186
11114077	0286
11114087	0643
11114088	0643
11114089	0066
11114089T	0066
11114090	0261
11114091	0261
11114091T	0261
11114092	0392
11114093	1911
11114095	0261
11114096	0261
11114103	0224
11114106	1541
11114107	0055
11114112	0392
11114113	3516
11114114	1541
11114119	0016
11114127	0060
11114133	2069
11114134	3548
11114135	1386
11114149T	1522
11114181	3476
11114225T	1562
11114236	OEM
11114237	OEM
11114281T	6241
11114282T	1522
11114284	OEM
11114306	0556
11114307	1541
11114333T	OEM
11115001	0080
11115002	0023
11115002T	0023
11115003	0795
11115003T	0795
11115004	0471
11115014	0023
11115017	0133
11115020	0071
11115024	0124
11115024T	0124
11115025	0133
11115026	0133
11115027	0133
11115029	2232
11115030	0137
11115040	0901
11115043	0023
11115046	0883
11115059	0023
11115060	0182
11115061	0299
11115062	0019
11115063	0023
11115064	0471
11115065	0023
11115066	0087
11115071	1005
11115075	0023
11115089	0471
11115090	0253
11115095	0087
11115096	0166
11115097	0057
11115121	0023
11115155	0023
11115188	0471
11115223	0023
11115223T	0023
11115225T	1562
11115358T	0041
11115364T	0471
11115376T	0057
11115386T	4907
11115446	0540
11116191	0725
11119016	6615
11119023	0167
11119024	4857
11119025	5193
11119026	3609
11119027	2019
11119028	OEM
11119029	OEM
11119030	0362
11119031	3725
11119032	0683
11119033	2090
11119034	2026
11119035	OEM
11119077	OEM
11119112	0732
11119113	3916
11119114	1529
11119117	OEM
11119118	4137
11119120	OEM
11119123	OEM
11119139	2870
11119140	3083
11119144	OEM
11119225	4094

DEVICE TYPE	REPL CODE
11119226	2346
11119256	OEM
11119257	OEM
11119260	OEM
11119399	3515
11119446	OEM
11210016/7825	0143
11220009/7825	0004
11220018/7611	0004
11220022/7611	0004
11220046/7825	0016
11220061/7825	0164
11220076/7825	0004
11220106/7611	0164
11253588	3897
11292312	0682
11292313	0559
11292314	1944
11292315	2172
11369564	3251
11790334	0764
12090924-4	0004
12901503	0037
13020000	0004
13020002	0004
13030189	0133
13030192	0015
13030256	0071
13030266	0133
13030274	0133
13030285	0165
13030301	0143
13030312	0143
13030313	0015
13030401	0365
13033801	0015
13037215	0076
13038115	0113
13038900	0015
13040089	0208
13040095	0004
13040096	0208
13040216	0016
13040229	0143
13040236	0004
13040304	0144
13040313	0016
13040317	0016
13040318	0016
13040329	0208
13040347	2866
13040349	0222
13040352	0004
13040357	0016
13040362	0144
13040421	0144
13040429	0126
13040456	0164
13040459	0144
13050221	0015
13069306	0552
13073525	0191
13083908	0076
13093003	2209
13093004	2209
13093005	2209
13094517	0076
13214051	0267
13411386	0124
13411398	3192
13411409	3501
13419261-00	OEM
13419323	OEM
13421244	0006
13421268	OEM
13428129	4957
13428179	5586
13442206-00	OEM
13442430-00	OEM
13447562	OEM
14001376-01	2018
14001376-02	2018
14001376-03	2018
14388917-1	OEM
14500001-001	0141
14500001-002	1844
14500001-003	0341
14500001-004	1859
14500004-001D	1831
14500004-001P	1831
14500004-002	1835
14500004-003	1176
14500005-001	0033
14636723	0364
14700990	0474
14714729	0267
14769302	0757
14956180	OEM
15105300	1922
15108046	0071
15109200	2263
16103860	3063
16173130	0362
16233010	1215
16271033	1831
16271041	1831
16371420	OEM
16453160	2728
16483610	2295

DEVICE TYPE	REPL CODE
16483611	2295
16906600-01	0037
17100901	0143
17100902	0143
17101881	0143
17101882	0143
17184600	0385
17200000	0133
17200240	0015
17201160	0005
17201350	0023
17210010	0023
17210430	0554
17240010	0023
17241480	0124
18022090	OEM
18080510	OEM
18901105	0079
18901106	0079
18901107	0016
18901402	0111
18901607	0144
18901807	0144
18911202	0037
18955002	0532
19080002	0015
19166123	1859
19901403	0111
19901503	2481
19901606	0079
19901607	0127
19901806	0144
19901807	0144
20001786-134	0143
20001786-139	0484
20001786-141	0015
20001786-142	0015
20025153-77	0536
20030703-0701	0016
20030703-0702	0037
20052600	0016
20080127	0124
20115070	0143
20130109	0564
20502055Z4	OEM
20602065Z4	OEM
20702075Z4	OEM
20912578-1	0847
20918596-2	0160
22114209	OEM
22114210	0595
22114225	0560
22114253	0037
22114254	1267
22114468	0693
22114594	0360
22114626	3716
22114634	0810
22114676	0925
22114677	2276
22115174	0012
22115175	0030
22115181	0143
22115182	0143
22115183	0015
22115184	OEM
22115192	0143
22115304	0628
22115405	0015
22115427	0199
22692012	0133
22901513	0144
22902046	0144
23111006	0160
23113056	0079
23114001	0144
23114004	0975
23114007	0233
23114009	0975
23114011	0160
23114015	0191
23114017	0155
23114021	0004
23114023	0144
23114026	0975
23114028	0233
23114031	0144
23114033	0160
23114034	0144
23114036	0144
23114037	0007
23114038	0178
23114043	0144
23114044	0330
23114045	0208
23114046	0079
23114050	0037
23114051	0037
23114052	0233
23114053	0233
23114054	0233
23114056	0144
23114057	0144
23114058	0086
23114060	0144
23114061	0004
23114070	0103
23114070A	0103
23114071	0103

DEVICE TYPE	REPL CODE
23114072	OEM
23114074	0102
23114078	0842
23114081	0006
23114082	0144
23114083	0178
23114084	0178
23114086	0178
23114087	0079
23114088	0178
23114090	OEM
23114095	0233
23114097	0975
23114100	0103
23114104	0330
23114108	0103
23114109	0144
23114112	0127
23114118	0079
23114119	0079
23114124	0037
23114125	0233
23114126	0127
23114127	0144
23114131	0460
23114132	0161
23114133	0460
23114134	0161
23114136	0037
23114137	0006
23114138	0037
23114148	0127
23114155	0016
23114157	0127
23114163	0184
23114164	0007
23114165	0184
23114171	0144
23114172	0144
23114180	0144
23114181	0144
23114195	0086
23114208	0074
23114211	0004
23114212	0079
23114213	0079
23114214	0079
23114215	0079
23114216	0111
23114217	0111
23114218	0079
23114219	OEM
23114220	0178
23114221	0178
23114226	OEM
23114232	0086
23114238	0144
23114248	0168
23114249	0285
23114250	0285
23114251	0285
23114253	0228
23114254	0228
23114255	0079
23114258	0111
23114259	0155
23114260	0155
23114261	0155
23114262	0006
23114264	OEM
23114265	OEM
23114266	0168
23114267	0079
23114268	0142
23114275	0079
23114276	0076
23114277	0142
23114280	0286
23114282	0286
23114293	0006
23114296	0079
23114297	0191
23114299	0191
23114300	0006
23114301	0006
23114302	0006
23114313	0791
23114314	0155
23114315	0142
23114317	0074
23114321	0065
23114323	0313
23114325	0037
23114326	0065
23114344	0643
23114347	0006
23114349	0076
23114420	3959
23114454	5408
23114457	OEM
23114528	0151
23114530	0148
23114544	5462
23114550	0037
23114552	0286
23114662	0076
23114733	0018
23114735	0434
23114758	0270

DEVICE TYPE	REPL CODE
23114768	0055
23114821	0076
23114864	0620
23114867	0629
23114900	0643
23114915	0076
23114918	1911
23114919	0103
23114923	1284
23114939	0854
23114941	0643
23114944	0220
23114945	0313
23114961	1147
23114962	1150
23114966	0191
23114969	1284
23114974	0168
23114975	0643
23114982	0643
23114983	0264
23114992	0681
23114993	0168
23114994	0155
23114995	0006
23114999	0168
23115012	0071
23115016	0924
23115019	0015
23115022	0774
23115023	0015
23115025	0917
23115038	OEM
23115042	0015
23115044	0015
23115046	0015
23115049	0133
23115050	0133
23115051	0920
23115052	0015
23115057	0016
23115058	0016
23115064	0133
23115065	0133
23115068	0133
23115070	0143
23115071	0133
23115072	0133
23115074	0913
23115076	0012
23115077	OEM
23115078	1080
23115079	0196
23115080	0133
23115085	0015
23115088	0143
23115094	0102
23115098	0071
23115102	0143
23115103	OEM
23115104	OEM
23115108	0133
23115109	0133
23115113	0917
23115115	0143
23115116	0143
23115117	0071
23115118	0071
23115120	0015
23115121	0917
23115129	0143
23115130	0071
23115131	0102
23115140	0015
23115145	0102
23115157	0201
23115168	0636
23115185	0015
23115192	0102
23115194	0133
23115199	0015
23115200	0015
23115201	0015
23115215	0378
23115227	0917
23115238	0402
23115249	0133
23115250	0023
23115263	0015
23115273	0023
23115277	0244
23115285	0133
23115292	0402
23115294	0023
23115296	0023
23115301	0726
23115321	1323
23115322	OEM
23115328	0402
23115334	1323
23115337	0023
23115338	0023
23115366	OEM
23115367	OEM
23115374	0244
23115377	0023
23115455	0041
23115486	0494
23115532	0023

DEVICE TYPE	REPL CODE
23115535	0019
23115549	0700
23115585	0242
23115621	0023
23115622	OEM
23115665	0471
23115690	0023
23115774	0471
23115797	0023
23115800	2628
23115809	0133
23115810	0023
23115822	1329
23115878	1329
23115884	1606
23115888	0023
23115889	0023
23115897	0133
23115903	0316
23115908	0062
23115909	0023
23115910	0023
23115912	0023
23115913	0023
23115914	0102
23115919	0052
23115921	0041
23115922	1329
23115924	OEM
23115931	0471
23115935	OEM
23115938	0378
23115939	OEM
23115940	1005
23115947	0071
23115948	OEM
23115949	0143
23115960	0017
23115965	0348
23115966	0071
23115979	0253
23115981	0102
23115984	0649
23115986	1323
23115987	0244
23115990	0726
23115994	0071
23115999	0102
23118052	0031
23118060	1090
23118094	0023
23118123	OEM
23118173	1045
23118186	0339
23118186(MA700A)	OEM
23118338	0031
23118339	5505
23118450	0071
23118536	0140
23118538	0471
23118634	0118
23118642	OEM
23118708	OEM
23118714	OEM
23118822	0023
23118850	0471
23118859	0133
23118943	1145
23118977	0031
23119003	2715
23119004	0829
23119005	2715
23119007	0829
23119011	2634
23119012	0878
23119013	2639
23119014	1808
23119016	4582
23119017	0403
23119019	0395
23119022	2634
23119023	2639
23119025	0789
23119032	2639
23119033	2639
23119039	OEM
23119060	OEM
23119142	4129
23119144	OEM
23119167	OEM
23119180	OEM
23119182	OEM
23119208	OEM
23119209	OEM
23119210	OEM
23119228	3254
23119259	OEM
23119319	OEM
23119330	OEM
23119336	OEM
23119345	OEM
23119348	OEM
23119355	3236
23119363	OEM
23119450	OEM
23119456	OEM
23119489	3437
23119501	OEM
23119526	OEM

DEVICE TYPE	REPL CODE
23119548	4927
23119566	5256
23119664	OEM
23119702	OEM
23119710	1529
23119723	4310
23119742	1382
23119763	3916
23119796	0663
23119880	6110
23119901	0912
23119905	OEM
23119920	OEM
23119950	0167
23119960	0569
23119961	2062
23119966	1055
23119971	4238
23119978	2284
23119981	3915
23119988	2231
23119989	2949
23119990	3915
23119992	OEM
23119993	1326
23119994	4595
23119995	4594
23119998	OEM
23119999	0789
23124037	0144
23126177	0127
23126183	0007
23126184	0079
23126240	OEM
23126251	OEM
23126252	OEM
23126253	OEM
23126254	OEM
23126289	0127
23126290	0127
23126291	0127
23126577	0917
23126620	0144
23126622	0720
23200596-1	0086
23205955	0076
23205981	0076
23206375	0148
23206471	0191
23206631	0191
23212215	1085
23213515	0693
23219925	0261
23225889	0023
23244121	1541
23284521	OEM
23302565	0023
23303515	0133
23303525	0124
23303531	0124
23311006	0085
23311066	0160
23311545	0052
23311555	0052
23313515	0133
23314055	0203
23314068	3305
23314158	3344
23314162	OEM
23314197	0276
23314356	OEM
23314357	OEM
23314361	OEM
23314511	OEM
23314525	0461
23314699	OEM
23316003	OEM
23316184	1127
23316250	OEM
23316275	OEM
23316288	OEM
23316298	OEM
23316302	0140
23316306	OEM
23316309	0253
23316312	OEM
23316313	0295
23316325	0057
23316332	OEM
23316339	OEM
23316370	OEM
23316380	0182
23316391	0737
23316399	1090
23316406	1179
23316411	3452
23316412	OEM
23316471	OEM
23317975	0041
23318051	OEM
23318075	0091
23318086	2111
23318132	OEM
23318135	OEM
23318137	OEM
23318147	OEM
23318151	0062
23318208	OEM
23318241	OEM

If replacement code is OEM, contact original manufacturer for replacement.

DEVICE TYPE	REPL CODE
23318242	OEM
23318244	OEM
23318271	0057
23318299	OEM
23318361	OEM
23318409	4310
23318466	OEM
23318468	OEM
23318495	OEM
23318627	OEM
23318759	OEM
23318860	OEM
23318861	OEM
23318862	OEM
23318918	OEM
23319002	OEM
23319003	OEM
23319202	4881
23319307	OEM
23319308	OEM
23319364	OEM
23319448	OEM
23319459	OEM
23319497	OEM
23319502	OEM
23319504	OEM
23319592	OEM
23319637	OEM
23319802	0770
23319870	OEM
23319911	0071
23319929	OEM
23319985	OEM
23320315	OEM
23373411	0124
23381625	0023
23383211	0133
23383215	0133
23389311	1664
23390511	OEM
23391411	OEM
23394811	0023
23394815	0023
23394821	0023
23394911	0015
23394915	0015
23399911	0677
23600042	OEM
23683115	OEM
23685015	2042
23687415	OEM
23691515	0450
23904234	OEM
23904303	6322
23904381	OEM
23904391	2922
23904398	OEM
23904727	OEM
23904763	OEM
23996971	OEM
23996981	4290
23996991	OEM
23997001	OEM
23997011	OEM
23997021	OEM
23997031	OEM
23997041	0724
23997051	OEM
23997061	0724
23997071	1733
23997081	1607
23997091	1386
23997101	0124
23997111	5174
23997121	0140
24000097	OEM
24000255	OEM
24561601-E	0267
24561602-E	0267
24561603	0267
24561604-E	0267
24850078-3	OEM
25114101	0004
25114102	0004
25114103	0004
25114104	0004
25114116	0016
25114121	0016
25114130	0233
25114143	0142
25114161	0144
25115102	0143
25115108	0015
25115115	0015
26010006	0688
26010010	0191
26010011	0920
26010016	0006
26010020	0076
26010021	0191
26010022	1411
26010023	0111
26010024	0042
26010026	1144
26010027	0006
26010028	0133
26010029	0636
26010030	0133
26010031	0143
26010032	0139
26010033	0015
26010041	0324
26010044	0244
26010046	0636
26010047	0133
26010050	OEM
26010051	0224
26010052	0324
26010053	0836
26010056	0144
26010057	0133
26010058	0698
26010059	0042
26010062	1029
26411007	OEM
26417402	OEM
26630109	0167
27113100	0015
27123050	0015
27123070	0015
27123100	0015
27123100A	0015
27123120	0015
27123150	0143
27123220	0015
27123240	0143
27123270	0143
27123290	0015
27125080	0016
27125090	0016
27125110	0004
27125120	0004
27125140	0016
27125150	0016
27125160	0016
27125170	0164
27125210	0475
27125220	0475
27125230	0136
27125240	0136
27125250	0475
27125260	0211
27125270	0016
27125300	0016
27125310	0208
27125330	0004
27125340	0004
27125350	0004
27125360	0004
27125370	0016
27125380	0016
27125460	0016
27125470	0016
27125480	0164
27125490	0208
27125500	0016
27125540	0847
27125550	0004
27126060	3545
27126090	0160
27126100	0103
27126130	0160
27126130(TRANSISTOR)	0160
27126130-12	0160
27126220	0144
27420170	OEM
27430040	OEM
27430050	OEM
27430080	OEM
27430090	OEM
27430100	OEM
27430130	OEM
27430150	OEM
27430160	OEM
27430220	OEM
27430240	OEM
27430300	OEM
27430301	OEM
27430360	OEM
27430510	OEM
27430520	OEM
30000010	0136
30000021	0633
30000032	OEM
30200033	0042
30200053	0076
30200061	0151
30200062	0151
30200091	0151
30200101	2059
30400021	0688
30400131	0325
30400152	1758
30400171	0212
30600010	0019
30600020	0123
30600040	0015
30600090	0139
30600140	0137
30601371	0106
30601971	0323
30602521	0277
30602711	0124
30900361	0360
30900460	0651
30900530	4830
30900540	4803
30900720	2252
30900740	0360
30900741	0360
30900900	OEM
30901020	OEM
30901050	1292
30901080	1841
30901090	OEM
30901100	OEM
30901110	OEM
30901120	2252
30901140	OEM
30901150	5306
30901160	OEM
30901170	OEM
30901190	OEM
30901200	OEM
30901350	0489
31210049-00	0050
31210077-44	0050
31210240-33	0050
31210240-44	0050
31210471-11	0050
31210506-11	0050
31210506-22	0050
31220054-00	0004
32119401-200	OEM
32510001	0015
34002412	2537
34002412-3	2537
35002711	0150
35002712	0150
35002713	0150
35002912	0791
35003517	0006
35004112	0342
35004411	0006
35024812	0681
35025012	0532
35025212	1648
35026317	0525
35026712	2897
35044100	0007
35044396	0007
35044397	0007
35044398	0007
35044399	0007
35045402	OEM
35045416	0079
35045502	0079
35045708	0079
35045712	0079
35046506	0003
35046611	0142
35046612	0142
35046613	0142
35046811	0142
35046812	0142
35046813	0142
35046911	0103
35046912	0103
35046913	0103
35047216	0076
35047217	0076
35047218	0079
35047219	0079
35049311	0042
35053012	1327
35053212	0951
35053700	1806
35053811	0276
35053812	0276
35056311	0276
35056312	0276
35056313	0276
35063411	1284
35063412	1284
35063512	0018
35065417	0161
35065700	0055
35065911	0018
35065912	0018
35066100	0055
35066617	0563
35067000	1541
35067212	3703
35067300	0313
35067600	1541
35081400	0168
35120718	1758
35393060-01	0016
35393060-02	0016
35393060-03	0016
36001004	0133
36001008	0170
36001009	0143
36001010	0133
36001015	0133
36001027	0124
36002003	0143
36003007	0098
36003034	0244
36003039	1005
36003049	1329
36003061	0062
36003066	0317
36003096	0012
36003098	0052
36003100	0041
36003106	0471
36003118	0064
36003121	0062
36003122	0253
36003125	0053
36003152	0166
36003154	0052
36003157	0041
36003174	0091
36003175	OEM
36107051	0015
36107062	0015
36107085	0015
36107086	0795
36107094	0023
36107152	0102
36107153	0133
36107154	0102
36107160	0015
36107167	0102
36107169	0023
36107175	0023
36107178	0283
36107253	0102
36107255	0102
36107263	0023
36107286	0071
36107287	0023
36107289	0031
36108003	0106
36108009	0473
36108025	0253
36108026	0041
36108030	OEM
36108031	OEM
36108054	3898
36108055	3898
36186200	1433
36188000	1432
36188700	0682
36501009	OEM
36502004	OEM
36771100	0142
36801061	OEM
36801062	OEM
36801080	OEM
36801081	OEM
36801103	OEM
36801104	OEM
36801118	OEM
36803003	2628
36803009	OEM
37000918	0019
37001003	2882
37001034	2896
37001039	3772
37001040	2872
37001046	OEM
37001047	3916
37001059	1529
37001064	4097
37001065	3365
37001066	3884
37003001	0133
37003001(IC)	0638
37003012	0490
37003020	2623
37003023	0450
37003027	2122
37003029	2122
37005043	OEM
37005064	OEM
37007014	0865
37007021	0539
37009010	1429
37009016	5209
37011020	2026
37011036	4383
37011039	2666
37011041	4496
37011060	OEM
37011063	0107
37051036	0101
37051100	0500
37051246	1949
37051310	0215
37051378	0035
37051523	OEM
37053002	0024
37053003	1359
37058002	OEM
37058026	OEM
37058045	OEM
37246602	0133
37246711	0133
37275350	0133
37568200	0182
37568250	0023
37568800	0023
37582000	0182
38102023	OEM
38112013	OEM
39502017T	OEM
40306312	0789
40306400	0395
40306604	2639
40760211-002	2017
40760211-P2	2017
41013566	0015
41015030AG	3046
41020618	0015
41020687	0087
41021007	0015
41023224	0015
41023225	0015
41025773	0071
41025849	0015
41027063	0015
41027612	0143
41027613	0143
41027991	0023
41027992	0143
41029009	0123
41029290	0123
41029498	0195
41029499	0071
41030618	0087
41514016	OEM
41520419	0015
41522875	0123
41527380	0019
41527570	0120
41622859	0124
41624836	0143
41627989	0139
42020414	0170
42020737	0170
42023425	0365
42024925	0062
42027065	0870
42027092	0420
42029566	0053
43020731	0086
43020735	OEM
43021017	0016
43021067	0127
43021083	0016
43021168	0126
43021198	0086
43021415	0086
43021579	OEM
43021771	OEM
43022055	0050
43022134	0325
43022860	0155
43022861	0016
43023190	0130
43023212	0016
43023221	0016
43023222	0126
43023223	0086
43023843	0103
43023844	0016
43023845	0037
43024216	0103
43024218	0161
43024219	0460
43024225	0325
43024306	0264
43024833	0136
43024834	0633
43024859	0111
43024873	0111
43024878	0079
43024879	0018
43024880	0148
43025055	0076
43025056	0076
43025059	0191
43025538	0111
43025539	0595
43025620	0050
43025762	0325
43025832	0325
43025972	0218
43026284	0086
43026285	0126
43026932	OEM
43027155	OEM
43027213	0236
43027214	0042
43027379	0076
43027571	0786
43027614	0144
43027615	0144
43027616	0144
43027617	0144
43027618	0127
43027619	0150
43027620	0016
43027872	0079
43027987	0042
43029471	0151
43029472	0151
43029483	0552
43029484	0324
43029485	0436
43029486	0236
43100103	0143
43122874	2617
43126551	0701
43126551-A	0701
43126561	OEM
43200103	0143
43300203	0143
43540012	0015
43550009	OEM
43600101	0015
44007301	0144
44008401	0144
44011001	0016
44079004	0111
44089001	0127
44090004	0016
44211001	0164
46861453-3	OEM
46862200	OEM
48134666	0016
48134842	0016
48137195A	0037
48869715	0130
50210600-00-VF	0037
50210700-01-VF	0111
50210710-10-VF	0111
50211500-01VF	0037
50254200	1206
50254400	1554
50254600	0704
50254700	2445
50254900	1017
50255000	1623
51310000	1041
51310001	2097
51310002	1831
51310003	0033
51310004	0141
51310005	3399
51310006	0562
51310007	1632
51320000	0232
51320001	0314
51320002	0361
51320003	0512
51320004	0697
51320005	1024
51320006	1722
51320008	1426
51320009	0335
51320010	0803
51320011	0873
51320012	0232
51320016	1168
51320017	0385
51320018	1368
51330005	0467
51510422A01	0654
51510566A01	0701
51513753A09	0167
51544789J01	2019
51544789J02	2019
51565600-VF	0079
51577500	1823
51577600	1041
51577700	2097
51577800	0033
51577900	0141
51578000	3399
51578100	0562
51624200	0033
51717000	1632
51717100	1481
52335500	1486
52335600	1907
52335600HL	1907
52335700	3715
52335800	1792
52335800HL	1792
52335900	2492
52335900HL	2492
53015201	0087
53020612	0086
53223463	0060
54967774-P5	0211
55330007-001	0920
55335005-001	0084
55919603	OEM
55920301	OEM
57000901-504	0016
57276605	0440
57279005	0440
57279007	0440
57279011	0440
57279017	0440
57279025	0440
57279027	0440
57279033	0440
57279073	0440
57279076	0440
57279293	0440
57279298	0440
57279566	0440
57279793	0440
57280281	0440
57282315	0440
57288072	0440
57288073	0440
57288076	0440
57288079	0440
57299720	0440
60674003	OEM
60674015	OEM
60674016	OEM
60674017	OEM
60674025	OEM
61260039	0136
61260039A	0136
62020911	0540
62045035	2329
62049219	0004
62057957	0004
62113480	0004
62118970	0211
62119292	0280
62154543	0123
62381620	0123
62488131	0922
62525932	OEM
62737654	0789
62771232	0058
62995793	OEM
63000061	1030
63354880	OEM
63551996	OEM
63735248	OEM
63862290	OEM
63962600	OEM
65600500	1823
65600600	1041
65600700	0141
65600800	0358
65600900	0333
65611000	1632
65611100	1481
70114087	1758
70114088	1758
70114113	1946
70114117	3133
70114347	OEM
70115421	OEM
70119023	2960
70119034	4261
70119036	3964
70119037	3966
70119041	4071
70119048	3707
70119049	3970
70119058	OEM
70119060	OEM
70119062	0215
70119063	2228
70119064	0398
70119512	OEM
70119514	OEM
70119517	OEM
70119518	OEM
70119526	2870
70119527	3083
70119561	OEM
70119581	OEM
70119583	OEM
70119627	OEM
70119743	OEM
70119751	OEM
70119767	OEM
70119775	OEM
70119776	OEM
70135052	OEM
70260450	0127
70260490	0030
70260540	0012
70270050	0015
70270250	0133
70270390	0015
70270490	0344
70270500	2270
70270700	OEM
70270710	OEM
70270720	0015
70270740	4485
70270770	5080
70370050	0015
71700901	OEM
71773660	0123
72045600	2899
72729031	0440
76276722	0080
80000900	1030
80001300	0015
80001400	0015
80002400	0015
80050100	0016
80050300	0208
80050400	0050
80050500	0136
80050600	0126
80050700	0160
80051001	0103
80051500	0016
80051600	0126
80052101	0111
80052102	0016
80052201	0086
80052202	0016
80052301	0086
80052302	0016
80052402	0103
80052600	0016
80052700	0126
80052800	0208
80053001	0016
80053300	0042
80053400	0016
80053500	OEM
80053501	0325
80053600	0144
80053702	1514
80104900	0136

If replacement code is OEM, contact original manufacturer for replacement.

DEVICE TYPE	REPL CODE	DEVICE TYPE	REPL CODE	DEVICE TYPE	REPL CODE	DEVICE TYPE	REPL CODE	DEVICE TYPE	REPL CODE	DEVICE TYPE	REPL CODE	DEVICE TYPE	REPL CODE
80105200	0136	91056140	0037	229510033V	0144	570005503	0016	871930213	OEM	872910005	OEM	872960027	OEM
80105300	0136	91190032	0015	231150050	0143	600000060	0019	871930238	OEM	872910075	OEM	872960094	OEM
80146620	0050	92005600	0004	233119336	OEM	600000150	0133	871930243	OEM	872910101	OEM	872960330	OEM
80146630	0050	93037230	0016	233164471	OEM	601100001	0057	871930245	OEM	872910107	OEM	872960632	OEM
80203350	0004	93037240	0016	266910202	OEM	601100002	0471	871930252	OEM	872910162	OEM	872960633	OEM
80205400	0004	93038040	0079	312104732	0050	602000002	0023	871930262	OEM	872910278	OEM	872966347	OEM
80205600	0004	93039440	0079	312104733	0050	603000002	0628	871930267	OEM	872910343	OEM	872967242	OEM
80222631U	0004	93063240	0016	351054000	OEM	610480003	0086	871930274	OEM	872910372	OEM	872967243	OEM
80226310	0004	93063270	0016	370010011	0167	610497001	OEM	871930300	OEM	872910418	OEM	872969792	OEM
80236400	0164	93063280	0016	381079051	OEM	612160030	5398	871930313	OEM	872910426	OEM	872980080	OEM
80236430	0164	93063470	0016	386008040	OEM	612303001	3566	871930333	OEM	872910491	OEM	872980083	OEM
80243900	0004	93064440	0111	387040350	OEM	619000970	OEM	871930359	OEM	872910493	OEM	872980183	OEM
80285400	0847	93064450	0016	390000250	OEM	630000003	0042	871930437	OEM	872910529	OEM	872980193	OEM
80318250	0016	93073260	0111	390000460	OEM	632000001	2710	871930463	OEM	872910553	OEM	872980222	OEM
80337390	0086	93073340	0079	390000850	OEM	632000002	0839	871930507	OEM	872910660	OEM	872980234	OEM
80338030	0144	93073350	0086	390001050	OEM	632037218	0076	871930515	OEM	872910668	OEM	872980250	OEM
80338040	0144	93073440	0233	390001450	OEM	632039418	0155	871931123	OEM	872910725	OEM	872980525	OEM
80339430	0144	93073540	0016	390002750	OEM	632049514	0203	871931187	OEM	872910726	OEM	872980539	OEM
80339440	0144	93078420	0127	390003050	OEM	632049518	0006	871931320	OEM	872910753	OEM	872980541	OEM
80339460	OEM	93082830	0016	390008650	OEM	632073518	0191	871950026	OEM	872910777	OEM	872980543	OEM
80366830	0127	93082840	0016	390013850	OEM	632122617	0219	871950306	OEM	872910798	OEM	872980545	OEM
80366840	0127	93082920	0127	390015450	OEM	632197406	0836	871951001	OEM	872910799	OEM	872980565	OEM
80383840	0016	93097120	0086	390015750	OEM	640000002	1540	871971393	OEM	872910805	OEM	872980567	OEM
80383930	0016	93938040	0016	390016650	OEM	640000003	1048	871975143	OEM	872910806	OEM	872980569	OEM
80383940	0016	93939440	0127	390024450	OEM	640000403	OEM	871976871	OEM	872910814	OEM	872980616	OEM
80389910	0016	94029220	0142	390025950	OEM	640580700	OEM	871980034	OEM	872910953	OEM	872980620	OEM
80414120	0103	94810300	0725	390027350	OEM	661000001	0139	871980076	OEM	872911129	OEM	872980628	OEM
80414130	0103	94825000-05	0358	390037450	OEM	662100200	0178	871980083	OEM	872911155	OEM	872980630	OEM
80421980	0086	94825000-11	1831	391808881	OEM	662400200	OEM	871980148	OEM	872911167	OEM	872980634	OEM
80510600	0143	94846400-00	2017	391974880	OEM	722114299	3729	871980155	OEM	872911297	OEM	872980638	OEM
80521881	0143	94846400-01	2017	394110070	OEM	765232011	OEM	871980170	OEM	872911313	OEM	872980703	OEM
80523520	0030	94846400-02	2017	394110140	OEM	800000029-006	0007	871981585	OEM	872911366	OEM	872980707	OEM
81063405	0681	95114101-00	0004	395008190	OEM	802026310	0211	871990019	OEM	872911512	OEM	872980872	OEM
81073304	0006	95114102-00	0004	399015090	OEM	815992449	OEM	871990023	OEM	872911530	OEM	872980876	OEM
82073206	0111	95114109-00	0004	402003173	0229	871322000	OEM	871990026	OEM	872911577	OEM	872981124	OEM
82409501	0016	95114135-00	0004	410000608F	0006	871900006	OEM	871990057	OEM	872911579	OEM	872983133	OEM
82410300	0126	95114140-00	0160	410000733P	0006	871900069	OEM	871990058	OEM	872911642	OEM	872988391	OEM
83037204	0076	95115115-00	0071	410000733Q	0006	871900090	OEM	871990065	OEM	872911657	OEM	872988392	OEM
83038004	0286	96421866	0071	410001015Y	0148	871901029	OEM	871990133	OEM	872911853	OEM	872988402	OEM
83039404	0155	120001190	0050	410001249R	0525	871901423	OEM	871990198	OEM	872912263	OEM	872990036	OEM
83073205	0016	120001192	0280	410001249S	0525	871902221	OEM	871990279	OEM	872912753	OEM	872990038	OEM
83073206	0016	120001195	0004	410020536G	0537	871910003	OEM	871990290	OEM	872914055	OEM	872990045	OEM
83073304	0016	120001300	0015	410020945K	0076	871910005	OEM	871990309	OEM	872915483	OEM	872990061	OEM
83073305	0016	120002013	0004	410020945P	0076	871910016	OEM	871990328	OEM	872917313	OEM	872990063	OEM
83073306	0111	120002014	0004	410020945Q	0076	871910017	OEM	871990329	OEM	872917723	OEM	872990065	OEM
83073504	0016	120002213	0050	410021568R	1790	871910043	OEM	871990339	OEM	872917732	OEM	872990067	OEM
83078402	1144	120002214	0050	410021568S	1790	871910044	OEM	871990492	OEM	872917742	OEM	872990074	OEM
83093005	0127	120002216	0050	410021674K	1066	871910086	OEM	871990644	OEM	872917743	OEM	872990080	OEM
83094502	0016	120002513	0050	410021674L	1066	871910103	OEM	871990658	OEM	872917863	OEM	872990083	OEM
83167403	0224	120002515	0050	410021730L	0127	871910104	OEM	871990730	OEM	872918092	OEM	872990085	OEM
83167405	0224	120002518	0050	410021815Y	0076	871910108	OEM	871990798	OEM	872918093	OEM	872990089	OEM
83167503	0224	120002520	0050	410022001L	1513	871910123	OEM	871990857	OEM	872919053	OEM	872990103	OEM
83167505	0224	120002521	0004	410022228A	1327	871910138	OEM	871991017	OEM	872919457	OEM	872990105	OEM
84026103	0218	120002656	0050	410022271D	0261	871910158	OEM	871991025	OEM	872919523	OEM	872990210	OEM
86401000	0016	120002748	0004	410022271E	0261	871910260	OEM	871991062	OEM	872919992	OEM	872990239	OEM
87045030	OEM	120004492	0050	410022274E	0155	871910271	OEM	871991065	OEM	872920017	OEM	872990330	OEM
87096010	0023	120004493	0004	410022274F	0155	871910273	OEM	871991068	OEM	872920132	OEM	872990410	OEM
87100600	0143	120004494	0004	410022334L	2268	871910274	OEM	871991072	OEM	872920152	OEM	872990427	OEM
87100605	0019	120004495	0004	410022373K	0728	871910275	OEM	871991106	OEM	872920178	OEM	872990648	OEM
87201881	0019	120004496	0144	410022909R	0261	871910276	OEM	871991155	OEM	872920184	OEM	872992008	OEM
87204260	0123	120004497	0144	410022909S	0261	871910278	OEM	871991165	OEM	872920196	OEM	872992091	OEM
87205530	0030	120004880	0144	410023117R	0563	871910280	OEM	871991302	OEM	872920202	OEM	872996522	OEM
87219410	0015	120004881	0144	410023117S	0563	871910289	OEM	871991355	OEM	872920253	OEM	872996732	OEM
87221394	0005	120004882	0144	410023568K	0728	871910290	OEM	871991359	OEM	872920258	OEM	874110029	OEM
87227880	0133	120004883	0016	410023568L	0728	871910291	OEM	871991552	OEM	872920267	OEM	874113170	OEM
87227881	0124	120004884	1400	410091402C	0016	871910297	OEM	871991845	OEM	872920305	OEM	874113980	OEM
87291320	OEM	120004887	0160	410091502B	0127	871910298	OEM	871991865	OEM	872920419	OEM	874115050	OEM
87291330	OEM	123561011	OEM	410091502C	3807	871910306	OEM	871991866	OEM	872920483	OEM	874901601	OEM
87301620	1925	123578311	OEM	410091502D	0127	871910671	OEM	871991877	OEM	872920502	OEM	874990008	OEM
87302650	0139	123578411	OEM	410091702M	0079	871910782	OEM	871992153	OEM	872920643	OEM	874990015	OEM
87400020	0015	123596211	OEM	410091702N	0286	871910794	OEM	871992223	OEM	872920915	OEM	874990036	OEM
87500620	0123	123596311	OEM	410091802M	0532	871910795	OEM	871992364	OEM	872920960	OEM	874990080	OEM
87500720	0015	123597111	OEM	410149042B	0715	871910804	OEM	871992774	OEM	872921202	OEM	874990135	OEM
87500920	0012	152221011	0004	410149042C	0715	871910968	OEM	871992777	OEM	872921311	OEM	874990149	OEM
87600920	0012	152221483	0050	412050010A	0057	871910974	OEM	871992808	OEM	872921312	OEM	874992003	OEM
87996010	0030	153260500	OEM	412051091M	0057	871910983	OEM	871993105	OEM	872921632	OEM	874992059	OEM
88125010-9	0016	153263700	OEM	412614135	OEM	871910984	OEM	871993130	OEM	872922093	OEM	874992081	OEM
88901010	0812	153267521	OEM	415100574J	1329	871910985	OEM	871993336	OEM	872922462	OEM	874993035	OEM
88902010	0812	153268600	OEM	415101379C	4350	871910989	OEM	871993341	OEM	872923832	OEM	874994121	OEM
88921010	0812	176069200	0368	415200050	2876	871910990	OEM	871993368	OEM	872924583	OEM	874994152	OEM
88922010	2697	178204004	OEM	415207530	3817	871910997	OEM	871993370	OEM	872926693	OEM	874995610	OEM
88922040	OEM	180690121	OEM	415207530N	3817	871911016	OEM	871993377	OEM	872928803	OEM	874995830	OEM
88922060	OEM	180696711	OEM	415207601W	1988	871911017	OEM	871993423	OEM	872930022	OEM	875101870	OEM
88940010	2697	180701811	OEM	415307313P	0362	871911031	OEM	871993683	OEM	872930036	OEM	875200141	OEM
88941010	2697	180730511	OEM	415307368P	OEM	871911036	OEM	871993696	OEM	872930090	OEM	875200610	OEM
88942010	2697	180750911	OEM	415401160	0520	871911048	OEM	871993749	OEM	872930156	OEM	875200730	OEM
88942040	OEM	180751111	OEM	415901220B	OEM	871911053	OEM	871993796	OEM	872930182	OEM	875201080	OEM
88942060	OEM	180768911	OEM	415950445P	OEM	871911078	OEM	871993849	OEM	872930187	OEM	875201090	OEM
88962010	0812	180768931	OEM	415950453P	OEM	871911434	OEM	871993872	OEM	872930429	OEM	875201120	OEM
88962040	OEM	180791321	OEM	430203845	0126	871911829	OEM	871993875	OEM	872930432	OEM	875201290	OEM
88962060	OEM	180839211	OEM	440008401	0015	871919000	OEM	871993884	OEM	872930450	OEM	875201330	OEM
89941008	0086	181515417	0595	440009905	0124	871919104	OEM	871993911	OEM	872930492	OEM	875201390	OEM
89942601	0224	200215358	OEM	440011001	0079	871920002	OEM	871993923	OEM	872930692	OEM	875201870	OEM
89942702	0224	200315428	1085	450010201	0144	871920023	OEM	871994045	OEM	872930906	OEM	875201920	OEM
89946008	0126	211300732	OEM	450010701	0007	871920077	OEM	871994076	OEM	872930908	OEM	875201930	OEM
89960404	0224	213119180	OEM	485134922	0016	871920080	OEM	871994081	OEM	872931142	OEM	875203026	OEM
89962306	0016	226021014	0144	485134923	0016	871920082	OEM	871994134	OEM	872931342	OEM	875203028	OEM
89962307	0079	229005013	0016	485134924	0016	871930033	OEM	871994137	OEM	872931382	OEM	875203093	OEM
89962308	0079	229005014	0016	485134925	0016	871930057	OEM	871994146	OEM	872932278	OEM	875203165	OEM
89962404	0224	229005015	0016	485134926	0016	871930059	OEM	871994214	OEM	872932611	OEM	875203167	OEM
89963008	0016	229018032	0144	485134956	0050	871930065	OEM	871994215	OEM	872935352	OEM	875203180	OEM
89963009	0016	229018033	0144	500000060	0019	871930071	OEM	871994217	OEM	872937884	OEM	875203229	OEM
90200090	1215	229018034	0144	530177001	OEM	871930076	OEM	871994580	OEM	872937891	OEM	875203230	OEM
90200100	2872	229020423	0144	530239038	6356	871930082	OEM	871995062	OEM	872938262	OEM	875203232	OEM
90270020	0240	229021014	0144	530990180	0188	871930118	OEM	871995111	OEM	872938263	OEM	875203233	OEM
90270060	0023	229025010	0144	532390017	2008	871930129	OEM	871996990	OEM	872938303	OEM	875203314	OEM
90270080	0940	229510015V	0144	539100020	OEM	871930138	OEM	871997019	OEM	872938612	OEM	875203334	OEM
90308040	OEM	229510031V	0144	570004503	0016	871930152	OEM	871998101	OEM	872940016	OEM	875203338	OEM
90500160	1395	229510032V	0144	570005452	0016	871930160	OEM	871998204	OEM	872960012	OEM	875203340	OEM

DEVICE TYPE	REPL CODE
875203354	OEM
875203355	OEM
875203361	OEM
875203440	OEM
875203528	OEM
875203552	OEM
875203554	OEM
875232011	OEM
875232044	OEM
875232118	OEM
875232165	OEM
875232232	OEM
875232608	OEM
875232618	OEM
875232872	OEM
875233331	OEM
875233406	OEM
875280016	OEM
875280061	OEM
875280603	OEM
875280878	OEM
875281065	OEM
875794800	OEM
875847000	OEM
875885200	OEM
875900105	OEM
875900868	OEM
875903017	OEM
875903035	OEM
875903148	OEM
875903163	OEM
875903528	OEM
875910060	OEM
875910075	OEM
875910093	OEM
875910155	OEM
875910212	OEM
875910228	OEM
875910233	OEM
875910237	OEM
875910307	OEM
875910325	OEM
875910405	OEM
875910559	OEM
875910582	2042
875910641	OEM
875910645	OEM
875910768	OEM
875910777	OEM
875910785	OEM
875911037	OEM
875911168	OEM
875911188	OEM
875911293	OEM
875912556	OEM
875913240	OEM
875914001	OEM
875914011	OEM
875914017	OEM
875914026	OEM
875914051	OEM
875914053	OEM
875914069	OEM
875914071	OEM
875914337	OEM
875914454	OEM
875914582	OEM
875914634	OEM
875915740	OEM
875917012	OEM
875917812	OEM
875917912	OEM
875920079	OEM
875920123	OEM
875920201	OEM
875920213	OEM
875920245	OEM
875920387	OEM
875920388	OEM
875920430	OEM
875920441	OEM
875920581	OEM
875920582	OEM
875920673	OEM
875920726	OEM
875920776	OEM
875920810	OEM
875920814	OEM
875920882	OEM
875920896	OEM
875920953	OEM
875923026	OEM
875923049	OEM
875923123	OEM
875923426	OEM
875923435	OEM
875924011	OEM
875924012	OEM
875924024	OEM
875924069	OEM
875924071	OEM
875924075	OEM
875924081	OEM
875927630	OEM
875930390	OEM
875931441	OEM
875932040	OEM
875934053	OEM
875940012	OEM
875940088	OEM
875940235	OEM
875940365	OEM
875942001	OEM
875960005	OEM
875960013	OEM
875960035	OEM
875960043	OEM
875960046	OEM
875960102	OEM
875960103	OEM
875960106	OEM
875960141	OEM
875960143	OEM
875960145	OEM
875960148	OEM
875960201	OEM
875960247	OEM
875960254	OEM
875960265	OEM
875960270	OEM
875960271	OEM
875960287	OEM
875960320	OEM
875960328	OEM
875960342	OEM
875960394	OEM
875960400	OEM
875960405	OEM
875960430	OEM
875960431	OEM
875960433	OEM
875960449	OEM
875960463	OEM
875960475	OEM
875960484	OEM
875960486	OEM
875960490	OEM
875960493	OEM
875960539	OEM
875960543	OEM
875960545	OEM
875963021	OEM
875963082	OEM
875963139	OEM
875963140	OEM
875963201	OEM
875963254	OEM
875963261	OEM
875963289	OEM
875964519	OEM
875965142	OEM
875965214	OEM
875970000	OEM
875970006	OEM
875970011	OEM
875970020	OEM
875970024	OEM
875970040	OEM
875970043	OEM
875970050	OEM
875970075	OEM
875970089	OEM
875970112	OEM
875970121	OEM
875970124	OEM
875970192	OEM
875971004	OEM
875974556	OEM
875974564	OEM
875980012	OEM
875980015	OEM
875980018	OEM
875980036	OEM
875980071	OEM
875980101	OEM
875980112	OEM
875980123	OEM
875980173	OEM
875980180	OEM
875980185	OEM
875980198	OEM
875980208	OEM
875980234	OEM
875980239	OEM
875980252	OEM
875980261	OEM
875980274	OEM
875980285	OEM
875980324	OEM
875980329	OEM
875980333	OEM
875980364	OEM
875980472	OEM
875980506	OEM
875980518	OEM
875980534	OEM
875980535	OEM
875980550	OEM
875981290	OEM
875982007	OEM
875982009	OEM
875982010	OEM
875982022	OEM
875982049	OEM
875982051	OEM
875982062	OEM
875982091	OEM
875982145	OEM
875982193	OEM
875982194	OEM
875987802	OEM
875990070	OEM
875990072	OEM
875990239	OEM
875990515	OEM
875990602	OEM
875990716	OEM
875990808	OEM
875990816	OEM
875990817	OEM
875990823	OEM
875990881	OEM
875990889	OEM
875990906	OEM
875990949	OEM
875990950	OEM
875991018	OEM
875991039	OEM
875991252	OEM
875991255	OEM
875991282	OEM
875991311	OEM
875991372	OEM
875991379	OEM
875991380	OEM
875991444	OEM
875991531	OEM
875991554	OEM
875991742	OEM
875991829	OEM
875991876	OEM
875991992	OEM
875992015	OEM
875992027	OEM
875992081	OEM
875992219	OEM
875992341	OEM
875992397	OEM
875992425	OEM
875992426	OEM
875992449	OEM
875992482	OEM
875992564	OEM
875992751	OEM
875992962	OEM
875992965	OEM
875993736	OEM
875993756	OEM
875993795	OEM
875994062	OEM
875994079	OEM
875994088	OEM
875994517	OEM
875994600	OEM
875994662	OEM
875994702	OEM
875994703	OEM
875994761	OEM
875997080	OEM
875997140	OEM
875997152	OEM
875997193	OEM
875997243	OEM
875997249	OEM
875997421	OEM
875997771	OEM
875997834	OEM
875998058	OEM
875998169	OEM
875998210	OEM
875998466	OEM
875999082	OEM
876362300	OEM
876934001	OEM
881250108	0016
881250109	0016
884814411	OEM
885540026-3	0697
885540031-2	4856
910678870	0037
998615901	OEM
1039956300	1329
1522210111	0004
1522210131	0136
1522210300	0004
1522210921	0136
1522211021	0136
1522211200	0004
1522211221	0004
1522211328	0004
1522211921	0050
1522214400	0050
1522214411	0136
1522214435	0136
1522214821	0136
1522214831	0050
1522216500	0004
1522216600	0136
1522217400	0050
1522223720	0016
1522270100	0143
1522270101	0143
1522270208	0143
1523270105	0015
1611190553	0030
1611819064	0016
1716580001	OEM
1760099001	0380
1760579001	0380
1760589001	0380
1760607001	OEM
1760609001	0368
1760609002	0127
1760609003	0380
1760829001	0368
1760979001	0368
1760979002	0368
2002010061	0569
2002010072	1172
2002020066	0035
2002020077	0101
2002030027	OEM
2002100022	0878
2002110206	0167
2002110269	0167
2002110332	3776
2002110345	2896
2002110415	OEM
2002110542	OEM
2002120012	0395
2002120022	0878
2002120033	0855
2002120046	2533
2002120051	0912
2002120067	0069
2002120080	OEM
2002300012	0904
2002300023	1201
2002300029	1429
2002300033	1201
2002300046	2042
2002300067	OEM
2002300078	OEM
2002400617	0520
2002400638	3772
2002400641	OEM
2002400752	4109
2002400779	3635
2002501006	0352
2002501300	1337
2002501708	1055
2002502202	0067
2002502317	0450
2002502338	0497
2002502341	0503
2002502385	0583
2002600107	1329
2002600122	0039
2002600149	1329
2002600183	3503
2002600198	3452
2002600213	1641
2002610039	2269
2002610042	1604
2002610086	3365
2002610090	3884
2002610109	4372
2002620151	OEM
2002700101	5216
2002700130	OEM
2002700158	2026
2002700175	5011
2002700187	OEM
2002800034	0312
2002800112	0360
2002800133	0631
2002800146	0631
2002800178	0631
2002900042	4536
2002900138	OEM
2002900395	0107
2002900408	0969
2002900437	OEM
2002900455	OEM
2002900710	OEM
2003018812	OEM
2003028813	1342
2003028829	0842
2003037227	0076
2003037309	0076
2003038092	0286
2003038208	0184
2003038812	0842
2003038829	0842
2003049501	0786
2003049519	0786
2003049526	0786
2003049614	0786
2003049621	0786
2003055813	4928
2003064314	1150
2003073517	0191
2003082823	1220
2003082837	1220
2003094515	0076
2003094522	0076
2003094536	0076
2003094554	0076
2003094560	0076
2003098316	0066
2003110607	0220
2003117044	1150
2003121216	0563
2003134410	0552
2003134427	0552
2003144795	0643
2003145404	0220
2003150702	0955
2003150710	0955
2003150727	0955
2003151428	1085
2003151432	1085
2003162538	0236
2003162700	0733
2003172208	1085
2003172216	1085
2003172305	1085
2003173009	0127
2003173017	0127
2003174006	0151
2003174014	0151
2003174021	0151
2003174105	1513
2003174113	1513
2003181508	0076
2003181516	0076
2003181523	0076
2003181537	0076
2003184100	0530
2003184502	0530
2003185506	0224
2003189023	OEM
2003189025	0530
2003189109	0223
2003189304	2668
2003189408	0313
2003190604	0127
2003190612	0127
2003194135	1327
2003195906	0286
2003195914	0286
2003200100	1513
2003200118	1513
2003200125	1513
2003200320	0286
2003206800	0643
2003206818	0643
2003207306	1284
2003212007	0866
2003212015	0866
2003212204	0363
2003222907	0066
2003222915	0066
2003223000	0261
2003223018	0261
2003223039	0261
2003223312	1538
2003237331	0728
2003238316	1562
2003238323	1562
2003245819	0076
2003245826	0076
2003245830	0076
2003245843	0076
2003246703	3007
2003248119	0563
2003248126	0563
2003248164	0563
2003248175	0563
2003248217	0261
2003251204	0224
2003251212	0224
2003261019	0261
2003278549	0249
2003278554	0249
2003290117	0392
2003340091	0276
2003340271	0563
2003340283	0563
2004047310	0925
2004049081	1498
2004049514	0006
2004056215	0006
2004056260	0006
2004060858	0006
2004067350	0148
2004073315	0006
2004073322	0006
2004073336	0006
2004073354	0006
2004074304	0525
2004082606	0037
2004082614	0037
2004085415	1348
2004085422	1348
2004085436	1348
2004093316	0148
2004093323	0148
2004094001	1911
2004095005	1348
2004095013	1348
2004095210	0006
2004095227	0006
2004101500	0148
2004101512	0148
2004101518	0148
2004101525	0148
2004104814	0013
2004104821	0013
2004104835	0013
2004120026	0925
2004302210	0164
2004305600	0004
2004305618	0004
2004341519	0164
2004356122	0436
2004356220	0532
2004356234	0532
2004356801	1911
2004503018	0208
2004546819	0018
2004546826	0018
2004547806	0392
2004547814	0392
2004568902	0055
2004581809	2660
2004586902	0055
2004587002	0055
2004588008	0461
2004588016	0461
2004589802	0055
2004590007	0055
2004613904	0392
2004645306	1541
2004645503	1541
2004700108	3108
2007000015	OEM
2007100149	3046
2007100154	3046
2007100171	3133
2007100213	3342
2007110124	0371
2007110138	0371
2007110190	OEM
2007110215	OEM
2007200000	0316
2007200173	0832
2007200236	0832
2008000001	0019
2008000019	0123
2008000026	0019
2008000064	0143
2008000075	0924
2008010028	0133
2008010094	0133
2008010102	0120
2008010110	0133
2008010127	0124
2008010131	0124
2008010165	0133
2008010218	0133
2008010225	0124
2008010263	0133
2008010307	0124
2008100027	0023
2008100093	0348
2008100130	0023
2008110103	0023
2008120009	0023
2008120017	0023
2008120056	0499
2008130084	0023
2008130171	0348
2008130198	0023
2008130234	0023
2008130252	0023
2008140021	0031
2008150011	0031
2008160106	0031
2008209102	OEM
2008220125	0057
2008220174	1005
2008220531	0686
2008220698	0140
2008220703	0140
2008220851	0057
2008220895	0471
2008220970	0041
2008220997	0253
2008221014	0091
2008221048	0062
2008221097	0077
2008221113	0165
2008221168	0064
2008221211	0053
2008221522	0166
2008221772	0471
2008221824	0062
2008221890	0077
2008221975	0057
2008222052	0052
2008222582	0166
2008230039	0162
2008230042	0301
2008230057	0649
2008300017	0720
2008300038	OEM
2008300056	1029
2008300062	OEM
2008300085	OEM
2008310019	0720
2008310092	0720
2008310157	OEM
2008310163	0628
2008310186	OEM
2008310208	0720
2008310216	0720
2008310237	OEM
2008310240	OEM
2008400055	0163
2008400072	0124
2008400084	OEM
2008406103	OEM
2008409101	OEM
2008600011	0917
2008600032	0917
2008607108	OEM
2008609107	OEM
2008700022	2628
2008700036	OEM
2009120101	0023
2009120187	0015
2009120209	0023
2009120224	0023
2009120241	0015
2009120256	0015
2009120262	0015
2009120370	0023
2009130075	0071
2009130100	0564
2009130118	0564
2009700063	0201
2011300055	OEM
2011300183	OEM
2011300469	OEM
2011320047	OEM
2011800017	OEM
2012000092	0019
2012000100	0019
2012000118	0019
2012010028	0133
2012010144	0133
2012010159	0133
2012010165	0133
2012010176	0023
2012010218	0133
2012010225	0124
2012010263	0133
2012100119	0348
2012100126	0023
2012100158	0023
2012100164	0348
2012110103	0071
2012120009	0023
2012120038	0023
2012120056	0023
2012130234	0102
2012220104	0077
2012220139	0064
2012220174	1005
2012220368	0686
2012220851	0057
2012220867	0445
2012220935	0140
2012220970	0041
2012220982	0041
2012221006	0253
2012221014	0077
2012221053	0471
2012221070	0062
2012221120	0165
2012221152	0057
2012221168	0064
2012221179	0064
2012221196	0181
2012221203	0181
2012221228	1005
2012221245	0053
2012221250	0053
2012221294	0686
2012221522	0195
2012221693	0140
2012221723	0041
2012221737	0041
2012221784	0253
2012221817	0471
2012221862	0062
2012221890	0077
2012222005	0057
2012222013	0057
2012222020	0064
2012222052	0064
2012222079	0064
2012222096	0052
2012222129	0052
2012222227	0879
2012222244	0686
2012222570	0195
2012222754	0181
2012222760	0053
2012222861	0053
2012223529	0376
2012230042	0301
2012230109	0137
2012230312	0137
2012400061	0163
2013120208	0023
2013120216	0015
2013120223	0015
2013120255	0015
2013120313	0015
2013120334	0015
2013120347	0609
2013200021	0148
2013200035	0148
2013200048	0148
2013200069	1522
2013200070	1522
2013200082	1522
2013200179	0013
2013200181	0013
2013200196	0013
2013200363	0006

If replacement code is OEM, contact original manufacturer for replacement.

DEVICE TYPE	REPL CODE
2013200779	0013
2013500037	1562
2013500040	1562
2013500084	0076
2013500099	0076
2013500107	0127
2013500115	0261
2013500136	0151
2013500149	0151
2013500154	0151
2013500213	0249
2013500268	0076
2013500279	0076
2013500281	0076
2013500302	0076
2013500310	0076
2013500327	0076
2013500359	2131
2013500365	2131
2013500388	0842
2013500470	2910
2013500482	0261
2013800027	0832
2013800059	4172
2013800088	0832
2013800119	4172
2013800126	4172
2013800143	4172
2018206110	OEM
2018211100	OEM
2018213100	OEM
2028106109	OEM
2028406113	OEM
2028406123	OEM
2029420295	OEM
2058599104	OEM
2058599109	OEM
2058599120	OEM
2058819107	OEM
2102400203	0360
2102400429	0360
2102400451	6498
2103500327	0076
2104000602	0148
2130309103	OEM
2138009101	OEM
2138010101	OEM
2138022101	OEM
2138307122	OEM
2168013105	OEM
2168017131	OEM
2168017134	OEM
2168022105	OEM
2168411105	OEM
2168411106	OEM
2168600106	OEM
2168601105	OEM
2208406113	OEM
2208622106	OEM
2290180119	0144
2295100224	0144
2295100225	0144
2295100226	0144
2295100227	0050
2295100228	0144
2295100232	0229
2295100233	0071
2295100234	0229
2295100235	0229
2302250002	3415
2302550002	3415
2302880023	OEM
2307890127	0120
2308220130	OEM
2308220507	OEM
2308270603	OEM
2309100000	OEM
2309100100	OEM
2309100200	OEM
2309100300	OEM
2309100400	OEM
2338179923	OEM
2505015301	OEM
2723100413	0023
2805015301	0163
2805015302	0163
2805025301	OEM
2805035301	OEM
2805045301	0917
2805055301	OEM
2805066101	OEM
2805076101	0808
2805086121	OEM
2805316101	OEM
2805326101	OEM
2805336101	OEM
2805346101	OEM
3071339004	1540
3071459001	2328
3071511002	OEM
3071517001	OEM
3071519001	4291
3071519002	OEM
3111007300	0143
3112006000	0143
3121004900	0050
3121007744	0050
3121024033	0050
3121024044	0050
3121047111	0050
3121050611	0050
3121050622	0050
3122005400	0004
3539306001	0016
3539306002	0016
3539306003	0016
3618020001	OEM
3630020001	OEM
4004056260	0006
4004085415	1348
4004095210	0006
4100005620	0006
4100006285	0006
4100009870	0006
4100010157	0148
4100104753	0164
4100105440	0532
4100108240	1308
4100204602	0151
4100204603	0151
4100205352	0127
4100206443	0111
4100208271	0007
4100208283	0224
4100208291	0007
4100208292	0224
4100208293	0224
4100210472	0224
4100213421	0127
4100213591	0007
4100217172	0155
4100217177	OEM
4100218157	0076
4100221200	0866
4100222301	0261
4100222710	0261
4100224820	0261
4100230380	3014
4100230420	2185
4100304000	2910
4100308230	1165
4100313980	1541
4100314270	1541
4100900102	0368
4100900103	0368
4100900104	0368
4100900116	0380
4100903554	0018
4100905254	0532
4100907633	0127
4101154010	0342
4101155510	0076
4101490420	0715
4102100220	0164
4104007193	0006
4104007380	0460
4104103242	0164
4104204602	0127
4104204603	0151
4104204612	0151
4104204613	0151
4104205352	0127
4104206440	0111
4104213173	0155
4104213421	0127
4104215684	1790
4104548119	OEM
4104673030	OEM
4108296047	0079
4108296238	0079
4108296255	0079
4108296257	0224
4109208280	1220
4109208284	0016
4109213174	0155
4109213354	0111
4120100090	0143
4120100600	0019
4120100603	0019
4120141480	0124
4120200090	0143
4120200602	0123
4120400260	0120
4120500395	0036
4120500475	0140
4120500565	0253
4120500622	0471
4120500625	0471
4120500685	0062
4120501005	0064
4120501200	0052
4120501205	0052
4120501505	0686
4120501805	0375
4120502405	0494
4120504742	0137
4120505105	0092
4120510513	0248
4120510625	0091
4120511105	0099
4120520513	0041
4120520603	0471
4120521305	0053
4120601550	4052
4120622345	OEM
4120900010	0133
4120900020	0023
4120903100	OEM
4120910010	0023
4120910430	0554
4120922220	0133
4121141480	0124
4121200051	0041
4122510600	0253
4128100020	0023
4129100600	0019
4129100602	0123
4129104002	0080
4129121392	0005
4129200602	0123
4129200857	0030
4129227907	0628
4129301329	0030
4129321292	0030
4129400003	0133
4129501200	0052
4130060010	0023
4130100012	0023
4130100015	0023
4130100021	0023
4130101011	0015
4130104002	0080
4130104006	0072
4130120020	0071
4130151550	0031
4130202040	0071
4130202060	0071
4130508240	1366
4130800060	1042
4131000220	1236
4138104002	0080
4138151520	0071
4138200020	0071
4139000005	0015
4139104002	0080
4150002031	2576
4150003770	2557
4151005753	2872
4151005768	OEM
4151014908	4550
4152032010	2572
4152041000	2614
4152056300	1329
4152075300	3817
4152076200	0557
4152076550	4094
4152078300	0732
4152078350	2346
4152079100	2666
4152085600	3982
4154011370	2755
4154011510	3651
4154011560	0520
4154013890	OEM
4157013102	0520
4158003780	OEM
4159009560	OEM
4159256100	OEM
4159301230	3955
4202003173	0015
4202003500	0143
4202003900	0102
4202005000	0196
4202005600	0123
4202006100	0124
4202006200	0133
4202007600	0102
4202007800	5611
4202007801	0102
4202007806	OEM
4202007860	OEM
4202008000	0102
4202008500	5713
4202008600	0143
4202008700	0023
4202009200	0495
4202011400	0062
4202011500	0181
4202012501	0317
4202012700	0002
4202012800	0495
4202013300	0010
4202013600	0133
4202014400	0080
4202014500	0080
4202014600	0965
4202015100	0182
4202015600	0133
4202016000	0133
4202016200	0133
4202017200	0348
4202017600	0023
4202018500	0023
4202018700	0133
4202019400	0010
4202019500	0010
4202020100	0091
4202020200	0071
4202020300	0564
4202020500	0023
4202020900	0348
4202021000	0023
4202021100	0087
4202021700	0471
4202021900	0133
4202022300	0023
4202022500	0124
4202022700	0124
4202022900	0348
4202023000	0102
4202023100	0023
4202023200	0023
4202023300	0348
4202023400	0015
4202024200	0062
4202024300	0062
4202093599	0123
4202104170	0289
4202104470	0071
4202104570	0102
4202104770	0004
4202104870	0102
4202105000	OEM
4202105070	0924
4202105470	0649
4202106270	0757
4202106970	0102
4202107270	0288
4202107370	0288
4202107470	0133
4202107570	0378
4202107670	0124
4202107970	0102
4202107971	OEM
4202108070	0757
4202108170	0757
4202108270	0015
4202108570	0565
4202109070	0317
4202109370	0023
4202110270	0023
4202110370	0015
4202110470	0102
4202110670	0080
4202110770	0023
4202110870	0317
4202111070	0077
4202115870	0023
4202116370	0077
4202116670	0317
4202970330	0100
4203970101	0076
4206002400	0878
4206002600	0855
4206003900	0855
4206004000	0167
4206004400	0878
4206004600	0167
4206005200	0884
4206007500	0881
4206009100	4196
4206009200	0856
4206009700	0856
4206009800	0886
4206012500	0916
4206042400	0878
4206064400	OEM
4206104970	0167
4206105170	0395
4206105370	2639
4206105470	0167
4206105670	0843
4206105770	0167
4206105870	0167
4206970020	3729
4227920320	OEM
4253401360	OEM
4704020001	OEM
4800155039	0019
4800155047	0124
4800155059	0955
4800155060	0133
4800155073	0286
4800155077	0133
4800155093	0286
4800155094	0286
4800155144	1329
4800155166	0025
4800155172	0825
4800155177	0015
4800155179	0023
4800155258	0124
4800155293	0582
4800155294	0825
4800155296	0023
4800155307	0286
4800155322	4327
4800155330	0002
4800155334	0124
4800155349	0002
4800155364	0018
4800155372	0582
4800155380	0203
4800155400	0582
4800155401	0582
4800155402	0582
4890155343	0825
5023500177	OEM
5147013102	0520
5300731044	0253
5300731045	0062
5300731045,,	0062
5301271120	0052
5301481003	0102
5301511001	0023
5301511003	0023
5301511004	0023
5301570629	0471
5301570689	0062
5301571001	0155
5301571100	0064
5301571100,,	0064
5301571110	0181
5301571120	0052
5301571130	0053
5301571150	0686
5301571160	0445
5301571180	0375
5301571200	0700
5301571220	0705
5301571249	1276
5301571270	0455
5301571300	0195
5301571330	0166
5301571369	0376
5301571399	0036
5301571439	0275
5301571439,,	0275
5301571519	0041
5301571569	0253
5301571629	0471
5301571689	0062
5301571829	0165
5301571919	0057
5301591100	0012
5301670001	OEM
5301710001	0155
5301711001	0023
5301711001,,	0609
5301711002	0609
5301711569	0253
5301791001	0133
5301791002	OEM
5301810001	0133
5301810002	0124
5301811001	0133
5301811002	0124
5301811003	0124
5301819001	0124
5301840001	5987
5301840001,,	0023
5301840002	0102
5301841001	0102
5301880005	OEM
5301881005	OEM
5301890001	OEM
5301890002	OEM
5301920569	OEM
5301920629	0636
5301921120	0137
5301921180	0495
5301921200	0531
5301921240	0402
5301921300	0726
5301921330	0039
5301921919	0012
5301940003	OEM
5301941003	0917
5301950005	OEM
5301951005	OEM
5301990479	0140
5301990689	0062
5301990919	0057
5302050003	0163
5302051003	0163
5302200001	0124
5302261002	0102
5302300002	OEM
5302300003	OEM
5302301003	OEM
5302301004	OEM
5302350001	2628
5302390002	2008
5302390008	1778
5302390011	0071
5302390017	2008
5302390019	0041
5302390020	0023
5302390021	0023
5302390022	0795
5302390023	0023
5302390024	6013
5302390025	0023
5302390026	0023
5302390027	6354
5302390028	6822
5302390029	0271
5302390030	0053
5302390031	0647
5302390032	0705
5302390033	0820
5302390034	OEM
5302390037	6355
5302390038	6356
5302390043	0424
5302390056	0041
5302390057	1993
5302390059	OEM
5302390061	0735
5302390066	1005
5302390086	3452
5302390146	0124
5302390150	OEM
5302390161	0023
5302390168	OEM
5302390182	OEM
5302390187	3501
5302390200	0940
5302390209	0071
5302390210	OEM
5302390252	0023
5302390254	OEM
5302410049	OEM
5302470001	0917
5302471001	0917
5302471001(247-1)	OEM
5302480001	OEM
5302491130	OEM
5302491140	OEM
5302491160	0445
5302491200	OEM
5302491270	4303
5302491300	0195
5302491519	0041
5302491569	0883
5302491629	0295
5302491689	OEM
5302491759	0077
5302491829	OEM
5302511001	0133
5302511002	0124
5302521001	0133
5302521002	0133
5302531005	0531
5302531014	0137
5302531016	OEM
5302541010	0340
5302541011	0052
5302541012	0253
5302541021	0053
5302541027	3946
5302541031	0700
5302541032	0299
5302541034	OEM
5302541036	0375
5302541038	0165
5302541039	0471
5302551001	0015
5302561002	0071
5302570001	0494
5302570001(257-1)	OEM
5302591001	OEM
5302600001	0031
5302600002	0031
5302600004	OEM
5302601001	0031
5302601001,,	0023
5302601002	0023
5302601002,,	0023
5302601003	0102
5302610001	0023
5302611001	0023
5302611001,,	0023
5302620001	0071
5302621001	0071
5302621002	0023
5302641001	1217
5302641002	1217
5302650001	OEM
5302660001	0133
5302661001	0133
5302670002	0271
5302671002	0023
5302680001	OEM
5302681001	0124
5302681002	0124
5302740001	0516
5302740002	5960
5302751001	0348
5302751002	0348
5302751003	0348
5302761003	OEM
5302771001	0143
5302781001	2395
5302781005	0442
5302781006	0870
5302781008	0041
5302801001	0426
5302840001	0473
5302850001	0516
5302860001	2628
5302970001	OEM
5302970002	OEM
5302970003	OEM
5302980001	0315
5302990001	1090
5302990002	1090
5302991001	1090
5302991002	0795
5303000001	0315
5303010002	OEM
5303011001	OEM
5303011002	6027
5303011002,,	OEM
5303011003	OEM
5303020002	0253
5303020003	1005
5303020006	OEM
5303021002	0253
5303021002,,	0253
5303021003	0052
5303021003,,	0052
5303021006	0064
5303030001	OEM
5303040001	OEM
5303040003	OEM
5303046001	OEM
5303047001	OEM
5303050002	0023
5303050003	0023
5303051002	0031
5303051002,,	0023
5303051003	0031
5303051003,,	0023
5303060001	OEM
5303061001	0953
5303100001	0031
5303100003	0031
5303101001	0023
5303101002	0023
5303101003	0031
5303101004	0707
5303121001	2088
5303130001	0163
5303130002	0163
5303131001	OEM
5303131002	OEM
5303150001	OEM
5303150002	OEM
5303150003	OEM
5303176001	OEM
5303177001	OEM
5303200001	OEM
5303200002	OEM
5303211919	0012
5303230001	OEM
5303240001	OEM
5303250001	OEM
5303260003	0546
5303261003	0546
5303277001	OEM
5303287001	OEM
5303291100	6866
5303291130	OEM
5303291160	OEM
5303291200	OEM
5303291330	5880
5303291519	6839
5303291569	3910
5303291629	0295
5303300629	OEM
5303340001	1291
5303371120	0052
5303371220	0705
5303371249	OEM
5303371330	0166
5303371399	0036
5303371479	0140
5303371689	0062
5303371829	0165
5303371919	0057
5303381859	0057
5303390001	OEM
5303410001	0031
5303411001	1191
5310200010	2741
5310200051	0079
5310200053	0006
5310200102	0006
5310200141	0076
5310200142	0076
5310200210	0842
5310200311	0191
5310200482	0643
5310200875	0074
5310200901	0643
5310200902	0643
5310200910	0066
5310200911	0066
5310200920	0261
5310200930	1284
5310200940	0866
5310200950	2668
5310400011	1911
5310400030	0220
5310400050	1284
5310400090	0643
5310400100	2668
5320200050	0167
5320200150	0912
5320200160	2054
5320200170	4134
5320500090	OEM
5320500100	3270
5320500120	2049
5320600070	0521
5320600080	OEM
5329500060	0167
5340100010	0019
5340200040	0133
5340200070	0182
5340200080	0023
5340200430	0124
5340200591	0499
5340200600	0133
5340200601	0133
5340200720	0023
5340200721	0015
5340200722	0071
5340200723	0143
5340200724	0071
5340200741	0071
5340200901	0023

DEVICE TYPE	REPL CODE
5340200902	0023
5340200950	0348
5340300230	0244
5340300350	0053
5340300370	0471
5340300380	0636
5340300430	0533
5340300450	0471
5390100919	0057
5390120949	0057
5390140249	1276
5390150479	0140
5390150519	0041
5390160309	0118
5390190040	0253
5390190060	OEM
5390190090	OEM
5390190230	0077
5390190949	0057
5390620331	0166
5390620399	0036
5390620829	0165
5390750309	0118
5390750519	0041
5390760111	0181
5390760121	0052
5390760131	0053
5390760271	0455
5390760689	0062
5390990150	OEM
5390990180	0188
5390990200	OEM
5390990230	0343
5390990280	0057
5391100060	0124
5391100260	0031
5391100341	0124
5391100460	0124
5391100630	0133
5391100640	0102
5391100650	OEM
5391200020	2542
5391200030	0023
5391200110	0564
5391200302	0023
5391200380	0023
5391200390	0546
5391200420	0546
5391200500	OEM
5391200510	0023
5391200520	0015
5391200530	0071
5391300010	1875
5391300240	0694
5391300380	0199
5391300390	1015
5391400020	3749
5391500170	0242
5391500200	0023
5391500231	0031
5391500270	0133
5391500320	0124
5391500330	0031
5391500390	0023
5391500400	0087
5391500432	0023
5391500440	OEM
5391500460	0348
5391500610	0031
5391510040	0407
5391510050	0031
5391900100	0057
5392000011	OEM
5392000410	OEM
5392100120	OEM
5392200011	OEM
5392400050	OEM
5392400051	OEM
5392400100	2703
5392400110	OEM
5392500860	OEM
5393100131	OEM
5393200010	OEM
5393200040	OEM
5393200050	OEM
5530315001	OEM
5700045452	0198
6020232250	OEM
6020234220	OEM
6040120025	1005
6060102002	OEM
6100830001	0037
6100830005	0037
6100830006	0037
6100839001	0037
6100839005	0037
6100839006	0037
6100940004	0111
6101220003	OEM
6101480003	0079
6101489003	0079
6101490010	0042
6101580003	0037
6102130001	0086
6102230001	0037
6102239001	0037
6102240001	0079
6102249001	0111
6102260001	0079
6102270003	3327
6102280005	0824
6102300003	OEM
6102320001	0079
6102320002	0076
6102329001	0079
6102329002	0076
6102330001	6550
6102340001	2455
6102340001(234-1)	OEM
6102349001	OEM
6102460001	0037
6102460002	OEM
6102500003	0261
6102500004	OEM
6103600001	0342
6103600002	0342
6103609001	0224
6103620001	6466
6103620001(362-1)	OEM
6103680001	0551
6103680002	0079
6103690001	0037
6103699001	0127
6103700001	0079
6103720002	0037
6103900001	0808
6103900001,,	0037
6103902001	0808
6103909001	0808
6103920001	0313
6103920002	0313
6104150001	0261
6104190002	2309
6104190003	2309
6104210001	0042
6104210002	0925
6104210003	0042
6104210004	0925
6104210005	0482
6104210006	4157
6104210008	1369
6104270002	0566
6104270003	0566
6104280002	0824
6104330001	3360
6104330002	1541
6104330002,,	1541
6104330003	3360
6104330004	3514
6104330005	1541
6104340001	0037
6104340001,,	0037
6104340002	OEM
6104340003	OEM
6104340004	0037
6104342001	0037
6104342004	0037
6104349001	0037
6104349002	0037
6104349004	0037
6104350001	0079
6104350001,,	0079
6104350002	0079
6104350002,,	0079
6104350004	0111
6104352001	0079
6104352002	0079
6104359001	0079
6104359002	0079
6104359004	0111
6104360001	3014
6104370001	0261
6104379001	0715
6104380001	1051
6104390001	0276
6104400023	0015
6104400117	0532
6104400118	2923
6104400119	OEM
6104400120	0076
6104400121	1165
6104400122	2954
6104400123	0148
6104410001	0037
6104420001	0008
6104429001	0008
6104430001	3799
6104440001	0055
6104470001	0008
6104490001	0037
6104500001	0079
6104500002	0037
6104509001	0079
6104510002	0037
6104510003	0037
6104570002	0423
6104570003	0423
6104580003	0595
6104600001	0037
6104609001	0037
6104620001	0560
6104620003	0423
6104630002	0313
6104631002	0313
6104640001	3360
6104650001	0079
6104680001	1708
6104710001	0285
6104730001	0313
6104740001	0079
6104750001	0037
6104760001	0439
6104770001	1852
6104790001	0715
6104800001	0861
6104809001	0861
6104820001	0037
6104860001	0079
6104870001	0037
6104880001	0264
6104980001	0037
6104989001	0037
6104990001	0016
6104999001	OEM
6105000001	0079
6105000004	0079
6105009001	0079
6105009004	0079
6105020001	OEM
6105030001	OEM
6105040001	0371
6105070002	1733
6105077001	1733
6105077002	1733
6105080001	5391
6105090001	0313
6105100001	1719
6105107001	1752
6105110001	2860
6105112001	5629
6105119001	2860
6105120001	0482
6105150002	OEM
6105157002	OEM
6105170001	4450
6105170002	4450
6105170003	4450
6105190001	OEM
6105210001	OEM
6105216001	OEM
6105260001	0532
6105269001	0532
6105270001	OEM
6105270002	OEM
6105272001	0477
6105279001	OEM
6105280001	0525
6105292001	0224
6105299001	OEM
6105309001	OEM
6105310001	0168
6105320001	1511
6105320003	1511
6105330001	3209
6105350003	OEM
6105352001	1386
6105359001	OEM
6105360001	0724
6105366001	OEM
6105380002	2860
6105382002	2860
6105389002	0079
6105390001	1165
6105442001	OEM
6105450001	5034
6105459001	OEM
6105460001	1597
6105469001	OEM
6105490001	OEM
6105497001	OEM
6105500001	OEM
6105507001	1435
6105510001	3014
6105520001	1562
6105529001	1562
6105530001	1522
6105539001	1522
6105540001	0261
6105549001	0261
6105550001	1541
6105569001	OEM
6105580001	1526
6105589001	0079
6105590001	OEM
6105590003	OEM
6105607001	OEM
6105610001	1511
6105630001	OEM
6105650001	1165
6105660001	2766
6105670001	0079
6105680001	0037
6105690001	6421
6105767001	OEM
6105807001	OEM
6105847001	OEM
6105869001	0261
6110130001	4337
6110130001(013-1)	OEM
6110130003	6255
6110150001	1137
6110160001	3696
6110180001	1222
6110190001	0617
6110210001	OEM
6120480004	0312
6120720001	OEM
6120750003	0520
6121260001	2826
6121260001(DSD84)	OEM
6121630001	0625
6121860001	1143
6121860001(186-1)	OEM
6122470002	0950
6122500001	0119
6122750001	1899
6122850001	2922
6122850003	2922
6122940001	OEM
6123030001	0490
6123040001	1022
6123050001	3753
6123180001	5875
6123190001	OEM
6123270001	4601
6123280001	5198
6123290001	3713
6123300011	OEM
6123300012	OEM
6123300013	OEM
6123300022	OEM
6123300023	3685
6123300024	5871
6123300025	2042
6123300026	OEM
6123300027	5873
6123300028	OEM
6123300037	OEM
6123300039	OEM
6123300040	OEM
6123300047	OEM
6123300052	OEM
6123300061	OEM
6123300064	OEM
6123300067	OEM
6123300068	OEM
6123300069	OEM
6123300070	OEM
6123300071	OEM
6123300072	OEM
6123300086	OEM
6123300087	OEM
6123300100	OEM
6123300101	4109
6123300102	0583
6123300103	OEM
6123300104	OEM
6123300120	OEM
6123300140	4350
6123300215	3159
6123300218	OEM
6123300224	OEM
6123300227	OEM
6123300229	OEM
6123300230	OEM
6123300231	OEM
6123300232	OEM
6123300293	OEM
6123300320	OEM
6123300332	OEM
6123300334	1988
6123300417	OEM
6123300418	3817
6123300419	OEM
6123300476	OEM
6123300478	OEM
6123300500	OEM
6123300502	OEM
6123300509	OEM
6123300512	OEM
6123300518	OEM
6123300522	OEM
6123300524	OEM
6123310002	2817
6123320001	3710
6123320003	3710
6123320006	3711
6123340001	0521
6123350001	1682
6123360001	3991
6123370003	3994
6123370005	3994
6123380002	2812
6123380003	2812
6123390001	0069
6123410001	OEM
6123420001	5858
6123470002	2831
6123500001	0215
6123510001	0167
6123520001	3981
6123540001	0379
6123590001	3672
6123600001	0624
6123620001	1057
6123620002	1057
6123630001	2022
6123630005	2022
6123700001	0351
6123700002	0351
6123800001	3647
6123940001	0069
6123940002	2826
6123999001	1292
6124120002	3161
6124140001	1122
6124140001(STK4021M)	OEM
6124140002	OEM
6124160001	OEM
6124160002	OEM
6124160003	3590
6124190001	OEM
6124220001	OEM
6124230001	OEM
6124240001	3386
6124260003	OEM
6124260004	OEM
6124340001	OEM
6124400001	OEM
6124410004	OEM
6124410005	OEM
6124420001	OEM
6124440001	2178
6124450001	5198
6124470001	OEM
6124470001(LA7810)	OEM
6124480001	0624
6124490001	3523
6124490001(TDA2546A)	OEM
6124490002	OEM
6124500001	5247
6124500002	5777
6124670001	3495
6124680001	1708
6124680001(468-1)	OEM
6124690001	OEM
6124690001(469-1)	OEM
6124710001	OEM
6124720001	OEM
6124730001	OEM
6124740001	0024
6124760001	OEM
6124770002	OEM
6124780001	OEM
6124790001	0624
6124790003	1196
6124790004	3485
6124790005	1321
6124800001	3391
6124830001	OEM
6124840001	OEM
6124840002	OEM
6124850001	OEM
6124850002	OEM
6124860001	OEM
6124870002	OEM
6124870003	OEM
6124880001	OEM
6124890001	3437
6124890001(AN5291)	OEM
6124920001	1134
6124930001	0035
6124940001	0360
6124960001	2842
6124980001	OEM
6124990001	OEM
6124990002	OEM
6125020004	OEM
6125030001	OEM
6125040001	5741
6125070001	3044
6125080001	5391
6125354053	OEM
6125390001	3391
6125410001	OEM
6125420002	OEM
6125450001	3386
6125510004	OEM
6125520001	OEM
6125530001	OEM
6125540001	OEM
6125560002	OEM
6125570001	OEM
6125580003	OEM
6125590001	OEM
6125600002	OEM
6125610001	OEM
6125610002	OEM
6125610004	OEM
6125620001	OEM
6125630001	3365
6125640001	3884
6125650001	OEM
6125660001	2055
6125670004	OEM
6125680003	OEM
6125710001	3720
6125740001	2957
6125750001	3343
6125760001	OEM
6125770001	OEM
6125770002	OEM
6125780001	OEM
6125790001	OEM
6125830001	OEM
6125840002	OEM
6125840003	OEM
6125840004	OEM
6125840005	OEM
6125840007	OEM
6125860001	3955
6125860002	2364
6125870001	OEM
6125880001	OEM
6125890001	OEM
6125900001	OEM
6125910001	OEM
6125920001	OEM
6125930002	OEM
6125960001	3310
6125990001	OEM
6125990002	OEM
6125990003	OEM
6126010001	OEM
6126010002	OEM
6126050001	OEM
6126060001	3254
6126100002	OEM
6126120001	3044
6126150001	5391
6126190001	OEM
6126210004	OEM
6126210008	OEM
6126210153	OEM
6126210393	OEM
6126210574	OEM
6126210590	OEM
6126214078	OEM
6126217001	OEM
6126250001	0732
6126260003	OEM
6126260005	OEM
6126270001	OEM
6126280001	3097
6126290001	OEM
6126300001	OEM
6126300002	OEM
6126310001	OEM
6126310002	OEM
6126320001	OEM
6126380001	OEM
6126520001	OEM
6126570001	OEM
6126580003	OEM
6126600001	OEM
6126610001	OEM
6126620001	OEM
6126630001	OEM
6126630003	OEM
6126630004	OEM
6126640001	5805
6126670001	OEM
6126680001	OEM
6126680002	OEM
6126690002	OEM
6126710001	0624
6126730001	OEM
6126760001	6643
6126780001	OEM
6126780003	OEM
6126810001	1028
6126820001	OEM
6126830001	OEM
6126830002	OEM
6126860001	OEM
6126880001	5823
6126900001	OEM
6126920001	OEM
6126940001	OEM
6126950001	OEM
6127040001	OEM
6127060001	OEM
6127070001	OEM
6127080001	OEM
6127090001	OEM
6127100001	OEM
6127140001	OEM
6127280001	OEM
6127290001	OEM
6127330001	OEM
6127340002	OEM
6127410001	OEM
6127450001	OEM
6155077002	OEM
6155107001	OEM
6190000080	0582
6190000160	OEM
6190000170	OEM
6190000190	0359
6190000200	0312
6190000210	0416
6190000290	0582
6190000340	0672
6190000433	0832
6190000450	0111
6190000460	0887
6190000462	0887
6190000470	0898
6190000480	OEM
6190000490	0928
6190000500	0014
6190000520	0672
6190000530	0582
6190000540	0842
6190000731	0076
6190000751	OEM
6190001002	0261
6190001230	0076
6190001231	0076
6190001451	1513
6190001470	1140
6190001520	2052
6190001580	0066
6190001620	2131
6190001711	1562
6190001733	0076
6190001740	0261
6190001743	0261
6190001900	0249
6190001902	0249
6190002000	OEM
6190002020	OEM
6190002030	2367
6190002200	1511
6190002420	0842
6190002450	1541
6190002510	OEM
6190002750	0645
6190002960	0014
6190003240	OEM
6190003350	0042
6190003700	OEM
6190003782	1790
6190003900	0161
6190003910	0261
6190003920	3211
6190004020	0728
6190004030	OEM
6190004270	OEM
6190004412	2691
6190004520	2860
6190004530	2766
6190004560	0928
6190004810	OEM
6190004860	OEM
6190004940	1327
6190004950	1790
6190004960	0728
6190005200	4007
6190005440	0127
6190005460	OEM
6190005560	OEM
6190005570	OEM
6190005600	0866
6190005610	0161
6190005920	OEM
6190005950	OEM
6190005960	OEM
6190005970	OEM
6190007060	OEM
6190007070	OEM
6190090770	OEM
6190090810	OEM
6190100010	0808
6190100100	OEM
6190100122	1032
6190100151	4172
6190100402	0006
6190100450	OEM
6190100550	2484
6190100660	0148
6190100661	0148
6190100663	0148
6190100664	OEM
6190100770	OEM
6190100780	OEM
6190100940	5373
6190100950	0825
6190100960	0825
6190101340	4130
6190101480	0006
6190101481	0037
6190101482	OEM
6190101710	0925
6190101720	0006
6190101750	0681
6190101780	0037
6190101860	4130
6190102040	OEM
6190102070	1308
6190102160	4125
6190102180	OEM
6190102200	OEM
6190102320	OEM
6190102330	OEM
6190102420	0148
6190102500	OEM
6190102560	OEM
6190102570	OEM
6190102780	OEM
6190102790	OEM
6190700021	2094
6191400010	0863
6192000040	6259
6192000080	OEM
6192000101	OEM
6192000150	5819
6192000160	6295
6192000170	OEM
6192000180	OEM
6192000910	OEM
6192000920	OEM
6192000930	OEM
6192000940	OEM
6192000950	OEM
6192000960	OEM
6192000980	OEM
6192000990	3254
6192001020	OEM
6192001060	OEM
6192001270	OEM
6192001330	4422

DEVICE TYPE	REPL CODE
6192020160	4041
6192020190	3725
6192020400	0539
6192020620	OEM
6192020630	OEM
6192040031	5262
6192040040	5263
6192040260	0557
6192060090	0976
6192060140	0732
6192060190	OEM
6192080080	4379
6192080200	OEM
6192080250	2042
6192100070	OEM
6192100150	2913
6192100170	OEM
6192100380	OEM
6192120071	OEM
6192120110	4380
6192120120	OEM
6192120250	OEM
6192120300	OEM
6192120302	OEM
6192120360	OEM
6192120370	OEM
6192120380	OEM
6192130070	OEM
6192130130	OEM
6192140020	1329
6192140060	OEM
6192140080	0624
6192140081	1292
6192140170	1641
6192140210	5048
6192140400	3488
6192140620	3452
6192149480	0624
6192160030	3925
6192180010	5324
6192180061	2066
6192180080	3159
6192180110	4117
6192180120	3521
6192180130	4141
6192180150	OEM
6192180160	3925
6192180170	OEM
6192180180	OEM
6192180190	OEM
6192180200	OEM
6192180210	OEM
6192180220	OEM
6192180510	OEM
6192180550	OEM
6192180560	OEM
6192180570	OEM
6192180580	OEM
6192180590	OEM
6192180600	OEM
6192180610	OEM
6192180630	OEM
6192181750	OEM
6192181870	OEM
6192181890	OEM
6192200150	OEM
6192220190	OEM
6192220740	OEM
6192220870	OEM
6192240040	4379
6192240110	OEM
6192240130	OEM
6192240140	5328
6192240150	OEM
6192240151	OEM
6192240160	OEM
6192240580	6335
6192240590	OEM
6192240600	OEM
6192240610	OEM
6192260030	OEM
6192260090	OEM
6192260100	OEM
6192440130	OEM
6193000570	OEM
6193000760	OEM
6193100070	OEM
6193140360	OEM
6193140370	OEM
6193160130	OEM
6193160150	OEM
6193160160	OEM
6193160350	OEM
6193160360	OEM
6193160370	OEM
6193160380	OEM
6193160460	OEM
6193160690	OEM
6193160700	OEM
6193160750	OEM
6193160760	OEM
6193160770	OEM
6193160810	OEM
6193180190	OEM
6193220010	4925
6193240020	OEM
6193300110	OEM
6194000030	0112
6194000031	OEM

DEVICE TYPE	REPL CODE
6194000040	4878
6194030050	4383
6195010740	OEM
6196990020	OEM
6196990030	OEM
6196990040	OEM
6196990051	OEM
6196990060	OEM
6196990080	OEM
6196990160	OEM
6611000900	0182
6611001000	0023
6611001100	0023
6611001200	0102
6611001500	0023
6611001501	0023
6611001700	0023
6611001800	0023
6611002300	0182
6611003000	0201
6611003200	0133
6611003701	0071
6611004601	0023
6611004602	0023
6611004803	0023
6611005100	0106
6611007300	0023
6612002200	0023
6612004000	0019
6612005800	0133
6612005900	0133
6612006000	0133
6612006101	0499
6612009000	0123
6612009002	0143
6613002200	5196
6613003000	0124
6615000100	0091
6615000101	0471
6615000200	0317
6615000600	0649
6615001701	0077
6615002603	0471
6615002788	0275
6615005000	0057
6615005500	1606
6615005600	0445
6621000200	1284
6621000600	2668
6621000700	0643
6621000701	0643
6621001000	0191
6621001100	0142
6621001401	0103
6621001800	0643
6621002000	0178
6621002300	0220
6621002400	1150
6621002800	1284
6621003100	0076
6621003200	0076
6621003400	0076
6621003500	0191
6621004000	0184
6621005000	0842
6621005001	0842
6621005003	0842
6621006200	0066
6621007200	0228
6621007403	0866
6621008100	0191
6621009100	0151
6621009600	0313
6621010001	0629
6621014601	0066
6621015602	0261
6621016000	0155
6621016100	0079
6622000100	0951
6622000300	0079
6622000401	0079
6622005100	0042
6622005400	0392
6623001100	0006
6623001900	0006
6623001902	0006
6623002000	0006
6623002026	0148
6623002100	0006
6623002200	0006
6623002300	0006
6623002304	0006
6623002700	1911
6624001000	0037
6624001100	0436
6624002000	0006
6624003100	0925
6624003200	0076
6631000100	0710
6631000200	0316
6631001200	0316
6640000410	0624
6644000100	0167
6644001100	3915
6644001200	OEM
6644001400	2949
6644001500	2231
6644001700	4594
6644001800	4595

DEVICE TYPE	REPL CODE
6644001900	1326
6644002100	OEM
6644004202	0167
6644004301	1055
6644004400	0878
6644012500	0912
6644012602	2054
6644012700	4134
6644014900	OEM
6645003400	OEM
6647006400	OEM
7044260991	5247
9511410100	0004
9511410200	0004
9511410900	0004
9511413500	0004
9511414000	0222
9511510200	0143
9511511500	0071
12613740490	0211
12618740490	0211
13480085512	1636
13480085513	0070
16000190201	0136
16004191050	0037
16009090545	0791
16100190668	0144
16102190693	0016
16102190929	0144
16102190930	0016
16102190931	0560
16103190668	0144
16103190930	0127
16104190668	0127
16104191168	0016
16104191225	0144
16104191226	0144
16105190536	0016
16105191229	0086
16106190537	0016
16106190772	0079
16108190536	0111
16108290536	0086
16109190536	0111
16109209536	0016
16112190710	0127
16112190772	0079
16114190772	0079
16116190634	0016
16118190634	0079
16147191229	0016
16156197229	0086
16171190693	0016
16171190858	0016
16171191368	0016
16172190693	0016
16172190858	0016
16200190186	0211
16201190022	0004
16201190186	0211
16201190187	0164
16204190457	0164
16207190405	0164
16208190187	0164
16211190022	0164
16212190022	0164
16256197228	0126
16304190031	1984
16304191350	0111
16304197001	0079
16304198000	4463
16304199400	0161
16307190632	0016
16343190142	0560
16356179229	0086
16377190632	0016
16400690060	0143
16401190188	0143
16405990022	0133
16411190188	0143
16411992473	0133
16412190410	0143
16415490000	0143
16419990032	0143
16471190188	OEM
16501090016	0015
16501190005	0015
16501190016	0015
16505090005	0015
16515991002	0015
16516390010	0229
16602990107	0062
16602995856	0064
16611190553	0030
16613190555	0030
16629291210	0015
16669990103	OEM
16740135809	3550
16790036081	OEM
21192010402	2755
21192021701	OEM
21192021706	0520
21192050103	1612
21194010400	OEM
21391018005	0079
21393010508	0224
21393018508	0191
21393026501	0286
21393027409	0079

DEVICE TYPE	REPL CODE
21393033402	2209
21393043405	0127
21393046604	0079
21394015201	0079
21396010204	0688
21396014602	0414
21396015500	OEM
21492019606	1369
21493051102	0728
21494012609	0423
21691010301	0019
21691031603	0143
21692010504	0023
21692010708	0023
21692011305	1747
21692011402	0071
21692011509	0540
21692090100	OEM
21694020505	0137
21694040608	0137
21694040705	0274
21694041700	0100
21695010608	OEM
21695011603	OEM
23091014501	OEM
44202007800	0348
48350947009	OEM
48513047141	OEM
53017101001	0124
53023000003	OEM
53026000001	OEM
57000901504	0016
61250700001	3044
66113002200	0023
66211002300	OEM
66211002800	OEM
66310001000	0710
134800855677	0227
134800863350	0995
134800864205	1153
134800865539	0123
134800868504	1194
134800868825	0691
134800868907	0063
134804204101	0015
134882137801	1402
134882137802	1761
134882137803	0668
134882137804	0547
134882137805	2403
134882137806	0870
134882137807	0524
134882137808	0510
134882137809	0616
134882256333	0760
134882321507	1401
134882321508	1537
134882732304	1250
134883461511	1606
134883461517	2874
134883461528	0309
134883461534	0876
134883461539	0262
134883727201	1756
134883875401	0447
134883875402	0940
134883875403	0469
134883875404	3283
134883875405	3284
134883875406	0940
134884461544	0833
167900306081	0579
260010628057	0006
260010695041	0436
260020176168	0004
260030710037	0368
260030711052	0380
260032878001	3094
260040467037	0951
260040468025	0018
270000600008	0019
270041480007	0124
271026380009	0005
273000068221	0062
273000082221	0165
300003220005	2595
300007313167	0362
300012026007	1260
300012413001	3816
301003450375	3982
301720490201	0143
310720490000	0016
310720490010	0016
310720490020	0016
310720490070	0160
310720490080	0144
310720490100	0016
310720490150	0016
310720490190	0265
310720490201	0143
400100010160	0004
400100020120	0004
404100030180	0164
404100900160	0016
404120010180	0143
404120030100	0143
404120030110	0133
404130010150	0015
482213033019	OEM

DEVICE TYPE	REPL CODE
482213033637	OEM
482213041457	OEM
482213041559	OEM
482213041907	OEM
482213081086	OEM
482213090606	OEM
482220960038	OEM
482220960039	OEM
482220960502	OEM
482220961344	OEM
482220970674	OEM
482220972744	OEM
482220973667	OEM
483513037031	0064
483513037051	0253
483513037084	OEM
483513037094	OEM
483513037134	0495
483513037136	0124
483513037137	2478
483513037138	0495
483513037139	0589
483513037141	0062
483513037142	0062
483513037143	0124
483513037144	2478
483513037145	0124
483513037146	0124
483513037147	1090
483513037148	1875
483513037149	3549
483513037151	0795
483513037153	0424
483513037209	0023
483513037226	0077
483513037229	0023
483513037231	0445
483513037232	5089
483513037234	0023
483513037244	OEM
483513037254	3749
483513037256	0903
483513037326	1962
483513037355	0031
483513037378	OEM
483513047002	OEM
483513047006	OEM
483513047023	OEM
483513047025	3682
483513047035	3723
483513047036	OEM
483513047052	OEM
483513047066	OEM
483513047072	OEM
483513047127	6886
483513047128	3478
483513047129	0724
483513047133	0582
483513047134	0359
483513047135	5376
483513047137	0151
483513047141	0724
483513047142	0724
483513047144	0359
483513047145	1752
483513047146	1742
483513047147	1742
483513047148	2094
483513047226	0558
483513047238	4011
483513047259	0832
483513047261	1140
483513047263	4049
483513047264	OEM
483513047265	0014
483513047266	3674
483513047268	OEM
483513047269	3272
483513047471	0898
483513047481	1892
483513047482	OEM
483513047483	OEM
483513047485	OEM
483513047486	5375
483513047488	1742
483513047493	OEM
483513087016	0516
483520917144	OEM
483520917272	OEM
483520917274	OEM
483520917279	0119
483520917287	OEM
483520947002	OEM
483520947003	OEM
483520947005	OEM
483520947007	OEM
483520947008	OEM
483520947009	OEM
483520947011	OEM
483520947013	OEM
483520947014	OEM
483520947017	OEM
483520947081	OEM
483520947083	0035
483520987033	OEM
483520987057	OEM
483520987058	OEM
483520987063	OEM
483520987111	5681

DEVICE TYPE	REPL CODE
483520987112	4259
483520987113	4259
483520987114	4141
483520987115	OEM
483520987116	OEM
483520987118	OEM
483520987119	OEM
483520987121	0112
483520987171	OEM
483520987189	OEM
483520987196	OEM
483520987197	OEM
483520987198	OEM
483520987199	OEM
483520987202	OEM
483520987342	3521
483520987415	OEM
483520987416	OEM
483520987425	OEM
483520987543	OEM
483520987544	OEM
483520987545	OEM
505130010150	0015
515104221401	0654
933000611112	0015
933001170112	0004
933022960012	0144
933022960112	0016
933022970112	0144
933076760112	0338
1110917004200	OEM
1110917004204	2671
1110917005200	2671
1110991010200	0955
2025911018810	0143
2035510069362	0552
20225230001701	0023
20225230018101	0015
202252300017101	0196
A00001B	OEM
A00005B	OEM
A0000.5B	OEM
A00010B	OEM
A00020B	OEM
A00050B	OEM
A00	0232
A001A0A	OEM
A001A0B	OEM
A001A0C	OEM
A001A0D	OEM
A003	OEM
A007H	0164
A00100B	OEM
A00200B	OEM
A00500A	OEM
A00500B	OEM
A00500C	OEM
A00500D	OEM
A01	0133
A01(MOTOROLA)	0133
A01(TRANSISTOR)	0050
A02	0133
A02(I.C.)	0335
A03	0361
A-04	0015
A04	0015
A04(I.C.)	0385
A05	0015
A05(I.C.)	0512
A05G	0023
A06	0015
A06(I.C.)	0697
A06-1-12	0016
A07	0143
A08	0102
A08(I.C.)	0873
A09	1024
A-025	0302
A025	OEM
A040B250C700	OEM
A054-103	0143
A054-105	0143
A054-108	0016
A054-109	0126
A054-114	0016
A054-115	0016
A054-119	0030
A054-142	0849
A054-148	0127
A054-149	0087
A054-150	0015
A054-151	0137
A054-154	0103
A054-155	0016
A054-156	0086
A054-157	0144
A054-158	0144
A054-159	0144
A054-160	0086
A054-163	0144
A054-164	0144
A054-165	0632
A054-170	0127
A054-173	0079
A054-175	0016
A054-186	0086
A054-187	0143
A054-195	0016
A054-206	0086

DEVICE TYPE	REPL CODE
A054-221	0016
A054-222	0016
A054-223	0037
A054-224	0178
A054-225	0016
A054-226	0123
A054-228	0133
A054-229	0087
A054-230	0015
A054-231	0365
A054-232	0531
A054-233	0016
A054-234	0016
A054-470	0144
A059-100	0050
A059-101	0050
A059-102	0050
A059-103	0050
A059-104	0136
A059-105	0136
A059-106	0004
A059-107	0004
A059-108	0160
A059-109	0016
A059-110	0086
A059-111	0630
A059-114	0015
A059-115	0085
A059-116	0004
A059-117	0649
A059-118	0071
A060-100	0127
A061-105	0127
A061-106	0127
A061-107	0136
A061-108	0127
A061-109	0127
A061-110	0136
A061-111	0136
A061-112	0127
A061-114	0004
A061-115	0004
A061-116	0222
A061-118	0015
A061-119	0012
A065-101	0015
A065-102	0016
A065-103	0016
A065-104	0016
A065-105	0164
A065-106	0004
A065-108	0016
A065-109	0016
A065-110	0016
A065-111	0222
A065-112	0164
A065-113	0016
A066-12	0133
A066-109	0076
A066-110	0688
A066-111	1144
A066-112	0016
A066-113	0037
A066-113(2SC7321)	0079
A066-113A	0037
A066-113AB	0037
A066-114	0228
A066-115	0191
A066-116	1593
A066-117	1411
A066-118	0037
A066-118A	0037
A066-119	0143
A066-119(1S1555)	0133
A066-119(GE)	0143
A066-119(SI)	0133
A066-120	0143
A066-121	0143
A066-124	0015
A066-125	1854
A066-143	0111
A068-100	0015
A068-101	0143
A068-102	0015
A068-103	0012
A068-104	0071
A068-106	0325
A068-107	0325
A068-108	0016
A068-109	0037
A068-109A	0037
A068-111	0144
A068-112	0144
A068-113	0016
A068-114	0178
A069-101	0127
A069-102	0127
A069-102/103	0016
A069-103	0127
A069-104	0016
A069-104/106	0016
A069-105	0004
A069-106	0016
A069-107	0164
A069-109	0143
A069-111	0123
A069-112	0015
A069-114	0127
A069-115	0143

DEVICE TYPE	REPL CODE
A069-116	0127
A069-118	0123
A069-119	0127
A069-120	0016
A069-121	0136
A069-122	0016
A076F	0050
A081C	0164
A090	0143
A098R	0050
A-0205	0250
A0205	0302
A0375	0015
A0377	0015
A0500.5B	OEM
A0503B20C700	OEM
A0506B40C700	OEM
A0512B80C700	OEM
A0526B160C700	OEM
A0540B250C700	OEM
A0553B320C700	OEM
A03004	0415
A04021	1030
A04049	0015
A04049-B	0015
A04049B	0015
A04091A	0015
A04092	0015
A04092-A	0015
A04092-B	0015
A04092A	0015
A04092B	0015
A04093	0015
A04093-A	0102
A04093-X	0102
A04093A2	0102
A04093X	0102
A04099-E	OEM
A04166-2	0002
A04201	0325
A04201(JVC)	0196
A04210-A	0277
A04212-A	0071
A04212-B	0071
A04212A	0015
A04212B	0015
A04226	0015
A04230	0102
A04230-A	0102
A04231-A	0277
A04233	0015
A04234-2	0012
A04241-A	0102
A04242	0015
A04284-A	0277
A04294	0102
A04294-1	0102
A04299-002	OEM
A04299-201	0201
A04299-202	0288
A04299-251	0288
A04331-021	0023
A04331-023	0015
A04332-007	0649
A04334-00A	OEM
A04344-007	0649
A04344-026	1606
A04350-022	0015
A04710	1303
A04716	0277
A04727	0474
A04731	0015
A04735-A	0774
A04901A	0015
A05001B	OEM
A05005B	OEM
A05010B	OEM
A05050B	OEM
A0710500D	OEM
A08010B	OEM
A08010D	OEM
A042313	0015
A072133-001	0623
A0311400	3915
A0354805	2054
A0847905	0055
A1	1718
A1-44	0103
A1A	0233
A1A1	0015
A1A5	0015
A1A9	0015
A1A205A	OEM
A1A210A	OEM
A1A215A	OEM
A1A405A	OEM
A1A410A	OEM
A1A605A	OEM
A1A610A	OEM
A1B	0016
A1B1	0015
A1B5	0015
A1B9	0015
A1B205A	OEM
A1B210A	OEM
A1B215A	OEM
A1B220A	OEM
A1B405A	OEM
A1B410A	OEM
A1B415A	OEM
A1B420A	OEM
A1B425A	OEM
A1B610A	OEM
A1B615A	OEM
A1B620A	OEM
A1B625A	OEM
A1B630A	OEM
A1C	0074
A1C1	0015
A1C5	0015
A1C9	0015
A1D	0642
A1D1	0015
A1D5	0015
A1D9	0015
A1DI	0642
A1DJ	0642
A1E	0007
A1E1	0015
A1E5	0015
A1E9	0015
A1E207A	OEM
A1E210A	OEM
A1E215A	OEM
A1E220A	OEM
A1E250A	OEM
A1E307A	OEM
A1E310A	OEM
A1E315A	OEM
A1E320A	OEM
A1E407A	OEM
A1E410A	OEM
A1E415A	OEM
A1E420A	OEM
A1E507A	OEM
A1E515A	OEM
A1E520A	OEM
A1EK510A	OEM
A1F	0016
A1F1	0015
A1F5	0015
A1F9	0015
A1G	0127
A1G-1	0127
A1G1	0015
A1G-1A	0144
A1G5	0015
A1G9	0015
A1G205A	OEM
A1G205D	OEM
A1G210A	OEM
A1G210D	OEM
A1G215A	OEM
A1G215D	OEM
A1G220D	OEM
A1G225D	OEM
A1G230D	OEM
A1G240D	OEM
A1G250D	OEM
A1G405A	OEM
A1G405D	OEM
A1G410A	OEM
A1G410D	OEM
A1G415A	OEM
A1G415D	OEM
A1G420D	OEM
A1G425D	OEM
A1G430D	OEM
A1G440D	OEM
A1G450D	OEM
A1G605A	OEM
A1G605D	OEM
A1G610A	OEM
A1G610D	OEM
A1G615A	OEM
A1G615D	OEM
A1G620A	OEM
A1G620D	OEM
A1G625A	OEM
A1G625D	OEM
A1G630D	OEM
A1G640D	OEM
A1G650D	OEM
A1H	0016
A1H(MOTOROLA)	0642
A1H001A	OEM
A1H003A	OEM
A1H005A	OEM
A1H010A	OEM
A1H020A	OEM
A1H1	0071
A1H5	0071
A1H9	0071
A1H200A	OEM
A1J	0016
A1K	0144
A1K1	0071
A1K5	0071
A1K9	0071
A1L	0016
A1M	0233
A1M-1	0127
A1M1	0071
A1M5	0071
A1M9	0071
A1N	0142
A1N1	OEM
A1N5	OEM
A1N9	OEM
A1P	0127
A1P/4922	0127
A1P-/4923	0144
A1P/4923	0127
A1P/4923-1	0144
A1P-1	0127
A1P-1A	0127
A1P-5	0144
A1QH3020S	OEM
A1R	0127
A1R/4924	0127
A1R/4925	0127
A1R/4925A	0144
A1R/4926	0127
A1R/4926A	0144
A1R001A	OEM
A1R003A	OEM
A1R005A	OEM
A1R010A	OEM
A1R020A	OEM
A1R050A	OEM
A1R-1	0127
A1R-1A	0144
A1R-2	0127
A1R-2A	0144
A1R-5	0144
A1R100A	OEM
A1R200A	OEM
A1R-24926	0144
A1S	0233
A1T	0016
A1T-1	0016
A1U	0007
A1U(LAST-IF)	0330
A1V	0016
A1VE	0016
A1W	0016
A1Y	0423
A1Z	0330
A-1.5-01	0015
A2	OEM
A2A	0233
A2A1	0015
A2A4	0015
A2A5	0015
A2A9	0015
A2B	0016
A2B0002	OEM
A2B0003	OEM
A2B0005	OEM
A2B1	0015
A2B4	0015
A2B5	0015
A2B9	0015
A2BRN	0016
A2C	0144
A2C1	0015
A2C4	0015
A2C5	0015
A2C9	0015
A2D	0144
A2D1	0015
A2D4	0015
A2D5	0015
A2D9	0015
A2E	0103
A2E1	0015
A2E-2	0103
A2E4	0015
A2E5	0015
A2E9	0015
A2EBLK	0103
A2EBRN	0103
A2EBRN-1	0103
A2F	0144
A2F1	0015
A2F4	0015
A2F5	0015
A2F9	0015
A2FGRN	0016
A2G	0007
A2G1	0015
A2G4	0015
A2G5	0015
A2G9	0015
A2G132	OEM
A2G133	OEM
A2G172	OEM
A2G173	OEM
A2H	0144
A2H1	0071
A2H4	0071
A2H5	0071
A2H9	0071
A2J	0016
A2K	0233
A2K1	0071
A2K4	0071
A2K5	0071
A2K9	0071
A2L	0144
A2M	0127
A2M-1	0127
A2M1	0071
A2M4	0071
A2M5	0071
A2M9	0126
A2MA	0087
A2N	0127
A2N-1	0127
A2N1	OEM
A2N-2	0127
A2N-2A	0144
A2N4	OEM
A2N5	OEM
A2N9	OEM
A2N2156	0437
A2P	0127
A2P-5	0144
A2S	0103
A2S009	OEM
A2S014	OEM
A2S024	OEM
A2S029	OEM
A2S031	OEM
A2S050	OEM
A2S087	OEM
A2S097	OEM
A2S-3	0103
A2S101	OEM
A2S103	OEM
A2S106	OEM
A2S108	OEM
A2S120	OEM
A2S121	OEM
A2S122	OEM
A2S123	OEM
A2S124	OEM
A2S128	OEM
A2S201	OEM
A2S202	OEM
A2S206	OEM
A2S207	OEM
A2S220	OEM
A2S250	OEM
A2S251	OEM
A2S252	OEM
A2S253	OEM
A2S254	OEM
A2S255	OEM
A2S257	OEM
A2S258	OEM
A2S259	OEM
A2S262	OEM
A2S263	OEM
A2S264	OEM
A2S270	OEM
A2S271	OEM
A2S272	OEM
A2S273	OEM
A2S274	OEM
A2S275	OEM
A2S276	OEM
A2S301	OEM
A2S302	OEM
A2S303	OEM
A2S305	OEM
A2S306	OEM
A2S308	OEM
A2S312	OEM
A2S313	OEM
A2S350	OEM
A2S356	OEM
A2S370	OEM
A2S396	OEM
A2S400	OEM
A2S401	OEM
A2S520	OEM
A2S565	OEM
A2S566	OEM
A2S611	OEM
A2S612	OEM
A2S613	OEM
A2S614	OEM
A2S615	OEM
A2S621	OEM
A2S622	OEM
A2S623	OEM
A2S624	OEM
A2S625	OEM
A2S630	OEM
A2S631	OEM
A2S633	OEM
A2S634	OEM
A2S637	OEM
A2S638	OEM
A2S701	OEM
A2S702	OEM
A2S703	OEM
A2S711	OEM
A2S713	OEM
A2S714	OEM
A2S720	OEM
A2S721	OEM
A2S722	OEM
A2S723	OEM
A2S730	OEM
A2S731	OEM
A2S733	OEM
A2S734	OEM
A2S737	OEM
A2S738	OEM
A2S740	OEM
A2S741	OEM
A2S742	OEM
A2S743	OEM
A2S744	OEM
A2S745	OEM
A2S781	OEM
A2S782	OEM
A2S783	OEM
A2S791	OEM
A2S792	OEM
A2S793	OEM
A2S800	OEM
A2S804	OEM
A2S805	OEM
A2S807	OEM
A2S810	OEM
A2S811	OEM
A2S813	OEM
A2S814	OEM
A2S815	OEM
A2S817	OEM
A2S818	OEM
A2S819	OEM
A2S821	OEM
A2S822	OEM
A2S823	OEM
A2S824	OEM
A2S825	OEM
A2S826	OEM
A2S827	OEM
A2S828	OEM
A2S835	OEM
A2S838	OEM
A2S839	OEM
A2S860	OEM
A2S861	OEM
A2S862	OEM
A2S863	OEM
A2S864	OEM
A2S865	OEM
A2S866	OEM
A2S900	OEM
A2S912	OEM
A2S970	OEM
A2SA550P	0037
A2SA564F	0203
A2SA564FR	0203
A2SA666PQR	0006
A2SB240A	0050
A2SB242A	0164
A2SB248A	0085
A2SC538PQR	0079
A2SC538R	0079
A2SD2260P	0178
A2SD226PQ	0178
A2SD2260P	0178
A2T	0144
A2T682	0208
A2T919	0144
A2U	0142
A2V	0144
A2W	0144
A2Y	0127
A2Z	0233
A-3	4917
A3	OEM
A3A	0144
A3A1	0015
A3A3	0015
A3A5	0015
A3A9	0015
A3B1	0015
A3B3	0015
A3B5	0015
A3B9	0015
A3C	0144
A3C1	0015
A3C3	0015
A3C5	0015
A3C9	0015
A3D	0144
A3D1	0015
A3D3	0015
A3D5	0015
A3D9	0015
A3E	0016
A3E1	0015
A3E3	0015
A3E5	0015
A3E9	0015
A3F	0016
A3F1	0015
A3F3	0015
A3F5	0015
A3F9	0015
A3G	0016
A3G1	0015
A3G3	0015
A3G5	0015
A3G9	0015
A3H	0144
A3H-1	0642
A3H1	0071
A3H3	0071
A3H5	0071
A3H9	0071
A3J	0111
A3K	0111
A3K1	0071
A3K3	0071
A3K5	0071
A3K9	0071
A3L	0276
A3M	0233
A3M1	0071
A3M3	0071
A3M5	0071
A3M9	0071
A3MA	0233
A3N	0016
A3N1	OEM
A3N3	OEM
A3N5	OEM
A3N9	OEM
A3N71	0144
A3N72	0144
A3N73	0144
A3P	0144
A3R	0144
A3S	0016
A3SB	0144
A3T	0016
A3T201	0600
A3T203	0600
A3T918	0007
A3T929	0016
A3T930	0016
A3T2221	0016
A3T2221A	0016
A3T2222	0016
A3T2222A	0016
A3T2484	0595
A3T2894	0037
A3T2906	0037
A3T2906A	0037
A3T2907	0037
A3T2907A	0791
A3T3011	0016
A3TE120	0103
A3TE230	0103
A3TE240	0103
A3TX003	0103
A3TX004	0103
A3U	0103
A3U-4	0103
A3W	0016
A3Y	0142
A3Z	0016
A-4	4917
A4A	0016
A4A1	0015
A4A-1-70	0160
A4A-1-705	0160
A4A-1A9G	0160
A4A5	0015
A4A9	0015
A4B	0111
A4B1	0015
A4B5	0015
A4B9	0015
A4C1	0015
A4C5	0015
A4C9	0015
A4D1	0015
A4D5	0015
A4D9	0015
A4E	0127
A4E1	0015
A4E-5	0144
A4E5	0015
A4E9	0015
A4F	0111
A4F1	0015
A4F5	0015
A4F9	0015
A4G	0144
A4G1	0015
A4G5	0015
A4G9	0015
A4H	0233
A4H1	0071
A4H5	0071
A4H9	0071
A4I7014	0130
A4J	0103
A4JBLK	0103
A4JBRN	0103
A4JD3B1	0600
A4JRED-1	0103
A4JX2A822	0208
A4K	0071
A4K1	0071
A4K5	0071
A4K9	0071
A4L	0208
A4M	0016
A4M1	0071
A4M5	0071
A4M9	0071
A4N	0016
A4N1	OEM
A4N5	OEM
A4N9	OEM
A4P	0016
A4P10	OEM
A4R	0016
A4S	0103
A4S008	OEM
A4S015	OEM
A4S017	OEM
A4S020	OEM
A4S021	OEM
A4S-1	0103
A4S112	OEM
A4S151	OEM
A4S153	OEM
A4S180	OEM
A4S312	OEM
A4S316	OEM
A4S317	OEM
A4S318	OEM
A4S360	OEM
A4S360J	OEM
A4S360K	OEM
A4S365	OEM
A4S370	OEM
A4S370I	OEM
A4S370J	OEM
A4S374	OEM
A4S375	OEM
A4S375I	OEM
A4S375J	OEM
A4S376	OEM
A4S386	OEM
A4T	0144
A4U	0016
A4V	0016
A4Y-1	0127
A4Y-1A	0144
A4Y-2	0016
A4Z	0103
A-5	5274
A5	OEM
A5-SAM	0907
A5A	0423
A5A-1	0161
A5A1	0015
A5A-2	0161
A5A2	0015
A5A-3	0161
A5A-4	0161
A5A-5	0161
A5A5	0015
A5A9	0015
A5A-IB	0042
A5B	0086
A5B1	0015
A5B2	0015
A5B5	0015
A5B9	0015
A5C	0016
A5C1	0015
A5C2	0015
A5C5	0015
A5C9	0015
A5D1	0015
A5D2	0015
A5D5	0015
A5D9	0015
A5E	0551
A5E1	0015
A5E2	0015
A5E5	0015
A5E9	0015
A5F	0264
A5F1	0015
A5F2	0015
A5F5	0015
A5F9	0015
A5F15T2	OEM
A5F40R2	OEM
A5F500U2-1	OEM
A5F500U2-2	OEM
A5F500U2-3	OEM
A5F500U2-4	OEM
A5F500U2-5	OEM
A5F500U2-6	OEM
A5F500U2-7	OEM
A5FF	0264
A5G	0161
A5G1	0015
A5G2	0015
A5G5	0015
A5G9	0015
A5H	0016
A5H1	0071
A5H2	0071
A5H5	0071
A5H9	0071
A5J	0144
A5K	0016
A5K1	0071
A5K2	0071
A5K5	0071
A5K9	0071
A5L	0016
A5M	0016
A5M1	0071
A5M2	0071
A5M5	0071
A5M9	0071
A5N	0016
A5N1	OEM
A5N2	OEM
A5N5	OEM
A5N9	OEM
A5P	0016
A5R	0016
A5R4027	OEM
A5S	0016
A5S001	OEM

If replacement code is OEM, contact original manufacturer for replacement.

DEVICE TYPE	REPL CODE
A5S005	OEM
A5S012	OEM
A5S013	OEM
A5S025	OEM
A5S034	OEM
A5S045	OEM
A5S085	OEM
A5S090	OEM
A5S091	OEM
A5S092	OEM
A5S093	OEM
A5S100	OEM
A5S105	OEM
A5S107	OEM
A5S110	OEM
A5S113	OEM
A5S115	OEM
A5S125	OEM
A5S127	OEM
A5S138	OEM
A5S139	OEM
A5S152	OEM
A5S161	OEM
A5S163	OEM
A5S165	OEM
A5S168	OEM
A5S188	OEM
A5S301	OEM
A5S302	OEM
A5S303	OEM
A5S339	OEM
A5S342	OEM
A5S350	OEM
A5S377	OEM
A5S400	OEM
A5S401	OEM
A5S402	OEM
A5S403	OEM
A5S420	OEM
A5T	0338
A5T-1	0338
A5T3707	0086
A5T404	0037
A5T404A	0037
A5T2192	0086
A5T2193	0086
A5T2222	0016
A5T2222A	1655
A5T2243	0622
A5T2604	0037
A5T2605	0037
A5T2907	0037
A5T2907A	2260
A5T3391	0079
A5T3391A	0079
A5T3392	0079
A5T3496	0037
A5T3497	0439
A5T3504	0037
A5T3505	0037
A5T3565	0079
A5T3571	0418
A5T3572	0418
A5T3638	0037
A5T3638A	0037
A5T3644	0037
A5T3645	0037
A5T3707	0079
A5T3708	0079
A5T3709	0079
A5T3710	0079
A5T3711	0079
A5T3751	0086
A5T3821	2950
A5T3824	OEM
A5T3903	0016
A5T3904	0016
A5T3905	0037
A5T3906	0037
A5T4026	0477
A5T4027	0037
A5T4028	0477
A5T4029	0037
A5T4058	0693
A5T4059	0342
A5T4060	0342
A5T4061	0693
A5T4062	0693
A5T4123	0079
A5T4124	0079
A5T4125	0037
A5T4126	0037
A5T4248	0037
A5T4249	0037
A5T4250	0037
A5T4260	0037
A5T4402	0037
A5T4403	0037
A5T4409	0086
A5T4410	0622
A5T4416	OEM
A5T4416A	OEM
A5T4621	0037
A5T5058	0715
A5T5059	0715
A5T5086	0037
A5T5087	0037
A5T5172	0079
A5T5209	0086
A5T5210	0086
A5T5219	0079
A5T5220	0079
A5T5221	0037
A5T5223	0079
A5T5225	0079
A5T5226	0037
A5T5227	0037
A5T5400	0439
A5T5401	0439
A5T5447#	0037
A5T5448#	0037
A5T5450#	0324
A5T5550	0715
A5T5551	0715
A5T6222	0018
A5T6461	0715
A5T6462	0715
A5T6463	0715
A5T6464	0715
A5T6538	0739
A5T6539	0739
A5T6540	1655
A5T6541	1655
A5TA	0338
A5U	0016
A5V	0130
A5W	0016
A5Y	0161
A-6	5274
A6-5002-9	0654
A6A	0642
A6A1	0087
A6A5	0087
A6A9	0087
A6AK	0087
A6B	0144
A6B1	0087
A6B5	0087
A6B9	0087
A6C	0161
A6C-1	0161
A6C1	0087
A6C-1-RED	0161
A6C-2	0161
A6C-2-BLACK	0161
A6C-3	0161
A6C-3-WHITE	0161
A6C-4	0161
A6C5	0087
A6C9	0087
A6C-GREEN	0161
A6D	0551
A6D-1	0551
A6D1	0087
A6D-2	0551
A6D-3	0551
A6D5	0087
A6D9	0087
A6E	0144
A6E1	0087
A6E5	0087
A6E9	0087
A6EA6G	OEM
A6F	0144
A6F1	0087
A6F5	0087
A6F9	0087
A6F10	OEM
A6F60	OEM
A6F80	OEM
A6F100	OEM
A6FK	0087
A6G	0551
A6G1	0087
A6G5	0087
A6G9	0087
A6H	0079
A6H1	0087
A6H5	0087
A6H9	0087
A6HD	0016
A6J	0016
A6K	0016
A6K1	0087
A6K5	0087
A6K9	0087
A6L	0103
A6LBLK	0103
A6LBLK-1	0103
A6LBRN	0103
A6LBRN-1	0103
A6LRED	0103
A6LRED-1	0103
A6M	0642
A6M1	0087
A6M5	0087
A6M9	0087
A6N	0103
A6N1	0102
A6N5	0102
A6N-6	0103
A6N9	0102
A6P	0144
A6R	0016
A6S	0016
A6T	0144
A6T5222	0079
A6U	0144
A6V	0127
A6V-5	0144
A6W	0161
A6Y	0264
A6Z	0642
A6ZH	0642
A7-12	0103
A7-13	0103
A7A	0016
A7A1	0087
A7A5	0087
A7A9	0087
A7A30	0144
A7A31	0144
A7A32	0144
A7B	0142
A7B(MOTOROLA)	0142
A7B1	0087
A7B5	0087
A7B9	0087
A7C	0015
A7C(MOTOROLA)	0161
A7C1	0087
A7C5	0087
A7C7	0087
A7C9	0087
A7D	0015
A7D1	0087
A7D5	0087
A7D9	0087
A7E	0015
A7E(TRANSISTOR)	0016
A7E1	0087
A7E5	0087
A7E9	0087
A7F	0264
A7F1	0087
A7F5	0087
A7F9	0087
A7G	0015
A7G1	0087
A7G5	0087
A7G9	0087
A7H1	0087
A7H5	0087
A7H9	0087
A7K1	0087
A7K5	0087
A7K9	0087
A7M	0103
A7M1	0087
A7M5	0087
A7M-5(TRANSISTOR)	0103
A7M9	0087
A7M42394B39	2027
A7M42894B39	2027
A7N	0144
A7N1	0102
A7N5	0102
A7N9	0102
A7P	0144
A7R	0016
A7S	0016
A7T	0016
A7T3391	0155
A7T3391A	0155
A7T3392	0155
A7T5172	0155
A7T6027	0316
A7T6028	0316
A7U	0144
A7V	0144
A7W	0144
A7Y	0016
A7Z	0161
A8-1	0050
A8-1-70	0050
A8-1-70-1	0050
A8-1-70-12-7	0050
A8-1A	0050
A8-1A0	0050
A8-1A0R	0050
A8-1A1	0050
A8-1A2	0050
A8-1A3	0050
A8-1A3P	0050
A8-1A4	0050
A8-1A4-7	0050
A8-1A4-7B	0050
A8-1A5	0050
A8-1A5L	0050
A8-1A6	0050
A8-1A6-4	0050
A8-1A7	0050
A8-1A7-1	0050
A8-1A8	0050
A8-1A9	0050
A8-1A9G	0050
A8-1A19	0050
A8-1A21	0050
A8-1A82	0050
A8-10	0050
A8-11	0050
A8-12	0050
A8-13	0050
A8-14	0050
A8-15	0050
A8-16	0050
A8-17	0050
A8-18	0050
A8-19	0050
A8A	0861
A8B	0016
A8C	0460
A8D	0086
A8E	0423
A8F	2997
A8G	0016
A8J	0551
A8K	0423
A8L	0016
A8M	0074
A8N	0861
A8P	0103
A8P-2-70	0004
A8P-2-705	0004
A8P-2A9G	0004
A8P404-ORN	0160
A8P404F	0160
A8R	0111
A8S	0144
A8T	0642
A8T404	0037
A8T404A	0037
A8T3391	0079
A8T3391A	0079
A8T3392	0079
A8T3644	2260
A8T3645	2260
A8T3702	0037
A8T3703	0037
A8T3704	0018
A8T3705	0018
A8T3706	0079
A8T3707	0079
A8T3708	0079
A8T3709	0079
A8T3710	0079
A8T3711	0079
A8T4026	0532
A8T4027	0037
A8T4028	0532
A8T4029	0037
A8T4058	0037
A8T4059	0037
A8T4060	0037
A8T4061	0037
A8T4062	0037
A8T5172	0079
A8U	0103
A8V	0233
A8VA	0233
A8W	0103
A8Y	2997
A8Z	0861
A9A	0144
A9B	0016
A9C	0400
A9D	0144
A9E	0016
A9F	0016
A9G	0016
A9G100AN	OEM
A9G100AR	OEM
A9G100BN	OEM
A9G100BR	OEM
A9G200AN	OEM
A9G200AR	OEM
A9G200BN	OEM
A9G200BR	OEM
A9G250AN	OEM
A9G250AR	OEM
A9G250BN	OEM
A9G250BR	OEM
A9G500AN	OEM
A9G500AR	OEM
A9G500BN	OEM
A9G500BR	OEM
A9G1000AN	OEM
A9G1000AR	OEM
A9G1000BN	OEM
A9G1000BR	OEM
A9H	0016
A9J	0016
A9K	0074
A9L-4-70	0004
A9L-4-705	0004
A9L-4A9G	0004
A9M	0338
A9N	0130
A9P	0074
A9R	0642
A9S	0016
A9S161A	OEM
A9S162A	OEM
A9S163A	OEM
A9S164A	OEM
A9S165A	OEM
A9S166A	OEM
A9T	2250
A9U	0016
A9V	0042
A9W	0016
A9Y	0016
A10	0559
A10A	0015
A10B	0015
A10C	0015
A10D	0015
A10E	0015
A10G05-010-A	0016
A10G05-011-A	0016
A10G05-015-D	0016
A10M	0015
A10N	0071
A10P	0071
A11	0913
A11CA	OEM
A11CF	OEM
A11DA	OEM
A12	0004
A12(I.C.)	1168
A12-02	OEM
A12-1-70	0016
A12-1-705	0016
A12-1A9G	0016
A12-2	3280
A12A	0004
A12B	0004
A12C	0004
A12D	0004
A12D36R2-H	OEM
A12F6	OEM
A12F8	OEM
A12F10	OEM
A12F12T2	OEM
A12F36R2	OEM
A12F36R2-H	OEM
A12F60	OEM
A12F80	OEM
A12F100	OEM
A12F500U2-1	OEM
A12F500U2-2	OEM
A12F500U2-3	OEM
A12F500U2-4	OEM
A12F500U2-5	OEM
A12F500U2-6	OEM
A12F500U2-7	OEM
A12H	0004
A12V	0004
A13	0004
A13(TRANS)	0050
A13-0032	0103
A13-17918-1	0103
A13-23594-1	0103
A13-33188-2	0103
A13-86416-1	0208
A13-86420-1	0600
A13-87433-1	0600
A13-SAM	0124
A13A	0050
A13A2	0015
A13AA2	0015
A13B	0050
A13B2	0015
A13C	0050
A13C2	0015
A13D	0050
A13D2	0015
A13E2	0015
A13F2	0015
A13G	0050
A13L	0050
A13M	0050
A13M2	0015
A13N1	0016
A13OR	0050
A13R	0050
A13X	0050
A13Y	0050
A14	0182
A14(TRANS)	0050
A14-586-01	0160
A14-601-10	0103
A14-601-12	0103
A14-601-13	0103
A14-602-63	0144
A14-603-05	0144
A14-603-06	0144
A14-1001	0136
A14-1002	0136
A14-1003	0136
A14-1004	0004
A14-1005	0004
A14-1006	0004
A14-1007	0004
A14-1008	0004
A14-1009	0004
A14-1010	0004
A14A	0071
A14A(TRANS)	0050
A14A-70	0160
A14A-705	0160
A14B	0015
A14B(TRANS)	0050
A14C	0015
A14C(TRANS)	0050
A14D	0015
A14D(TRANS)	0050
A14E	0015
A14E(TRANS)	0050
A14E2	0015
A14F	0015
A14G	0050
A14L	0050
A14M	0015
A14M(TRANS)	0050
A14N	0071
A14OR	0050
A14P	0071
A14PD1	OEM
A14PD2	0102
A14PD3	0102
A14R	0050
A14T	OEM
A14U	1335
A14X	0050
A14Y	0050
A15	1313
A15(TRANS)	0004
A15-2	OEM
A15-6	0004
A15-1001	0136
A15-1002	0136
A15-1003	0136
A15-1004	0004
A15-1005	0004
A15-1007	0143
A15-1008	0015
A15A	0953
A15A(TRANS)	0004
A15B	0242
A15B(TRANS)	0004
A15BK(2S)	0004
A15C	1747
A15C(TRANS)	0004
A15D	0540
A15D(TRANS)	0004
A15D30R2-H	OEM
A15E	1771
A15E(TRANS)	0004
A15F	0110
A15F(TRANS)	0004
A15F15T2	OEM
A15F30R2	OEM
A15F30R2-H	OEM
A15F500U2-1	OEM
A15F500U2-2	OEM
A15F500U2-3	OEM
A15F500U2-4	OEM
A15F500U2-5	OEM
A15F500U2-6	OEM
A15F500U2-7	OEM
A15G	0004
A15H(2S)	0004
A15K(2S)	0004
A15L	0004
A15M	0965
A15M(TRANS)	0004
A15N	0817
A15OR	0004
A15R(2S)	0004
A15RD	0004
A15U	0087
A15U(DIODE)	0110
A15U(RECTIFIER)	0110
A15U(TRANS)	0004
A15V(2S)	0004
A15VR(2S)	0004
A15X	0004
A15Y(2S)	0004
A16(2S)	0004
A16A	0004
A16A1	0208
A16A2	0208
A16B	0004
A16C	0004
A16D	0004
A16E	0004
A16F	0004
A16L	0004
A16M	0004
A16OR	0004
A16R	0004
A16X	0004
A16Y	0004
A17(2S)	0050
A17(I.C.)	1208
A17A	0050
A17B	0050
A17C	0050
A17D	0050
A17E	0050
A17F	0050
A17G	0050
A17H(2S)	0050
A17L	0050
A17OR	0050
A17R	0050
A17X	0050
A17Y	0050
A-18	0103
A18	0050
A18(I.C.)	0569
A18-4	0103
A18A	0050
A18B	0050
A18C	0050
A18D	0050
A18E	0050
A18F	0050
A18F500U2-1	OEM
A18F500U2-2	OEM
A18F500U2-3	OEM
A18F500U2-4	OEM
A18F500U2-5	OEM
A18F500U2-6	OEM
A18F500U2-7	OEM
A18G	0050
A18H(2S)	0050
A18L	0050
A18M	0050
A18OR	0050
A18R	0050
A18X	0050
A18Y	0050
A19(2S)	0164
A19(I.C.)	1486
A19-020-072	0144
A19A	0164
A19B	0164
A19C	0164
A19D	0164
A19E	0164
A19F	0164
A19G	0164
A19L	0164
A19M	0164
A19OR	0164
A19R	0164
A19X	0164
A19Y	0164
A20	0143
A20(2S)	0164
A20A	0164
A20B	0164
A20C	0164
A20D	0164
A20E	0164
A20F	0164
A20G	0633
A20G(2S)	0164
A20H	OEM
A20K	0037
A20KA	0037
A20L	0164
A20M	0164
A20OR	0164
A20R	0164
A20X	0164
A20Y	0164
A21(2S)	0164
A21-02	OEM
A21-04	OEM
A21A	0164
A21B	0164
A21C	0164
A21D	0164
A21E	0164
A21F	0164
A21G	0164
A21L	0164
A21M	0164
A21OR	0164
A21R	0164
A21X	0164
A21Y	0164
A22	0050
A22B	0087
A22B1	0087
A22D	0087
A22D1	0087
A22M	0087
A23	0015
A23(TRANS)	0050
A23-1101-01A	3236
A23B	0580
A23B1	0580
A23D	OEM
A23D1	OEM
A23M	OEM
A24	0050
A24F12T2	OEM
A24F36R2	OEM
A24F36R2-H	OEM
A24F500U2-1	OEM
A24F500U2-2	OEM
A24F500U2-3	OEM
A24F500U2-4	OEM
A24F500U2-5	OEM
A24F500U2-6	OEM
A24F500U2-7	OEM
A24MW594	0144
A24MW595	0144
A24MW596	0144
A24MW597	0144
A24T-016-01L	0144
A24T-016-016	0144
A25	0050
A25-12	3569
A25-1001	0136
A25-1002	0136
A25-1003	0136
A25-1004	0004
A25-1005	0004
A25-1006	0004
A25-1007	0143
A25-1008	0015
A25A	OEM
A25B	OEM
A25D	OEM
A25M	OEM
A25N	OEM
A25P	OEM
A25PB	OEM

If replacement code is OEM, contact original manufacturer for replacement.

DEVICE TYPE	REPL CODE
A25Q	OEM
A26(2S)	0050
A26(W)	0133
A26A	0050
A26B	0050
A26C	0050
A26D	0050
A26E	0050
A26F	0050
A26G	0050
A26L	0050
A26M	0050
A26OR	0050
A26R	0050
A26X	0050
A26Y	0050
A27	0142
A27(2S)	0050
A27BR1200	OEM
A27DR521A	OEM
A27DR521B	OEM
A27DR521M	OEM
A27DR1200	OEM
A27MR1200	OEM
A28	0050
A28A	1567
A28A(TRANS)	0050
A28B	1567
A28B(TRANS)	0050
A28BR1200	OEM
A28BR1201	OEM
A28C	2153
A28C(TRANS)	0050
A28D	2153
A28D(TRANS)	0050
A28DR1201	OEM
A28E	0050
A28F	1567
A28F(TRANS)	0050
A28G	0050
A28L	0050
A28M	0050
A28OR	0050
A28X	0050
A28Y	0050
A29(2S)	0164
A29A	1547
A29A(TRANS)	0164
A29B	1547
A29B(TRANS)	0164
A29C	3139
A29C(TRANS)	0164
A29D	3139
A29D(TRANS)	0164
A29E	0164
A29F	1547
A29F(TRANS)	0164
A29G	0164
A29L	0164
A29M	0164
A29R	0164
A29V082B03	0126
A29X	0164
A29Y	0164
A30	0004
A30(2S)	0136
A30A	0136
A30B	0136
A30C	0136
A30D	0136
A30E	0136
A30F	0136
A30G	0136
A30L	0136
A30M	0136
A30OR	0136
A30X	0136
A30Y	0136
A31	0004
A31(2S)	0280
A31-0206	0144
A31A	0280
A31B	0280
A31C	0280
A31D	0280
A31E	0280
A31F	0280
A31G	0280
A31L	0280
A31M	0280
A31OR	0280
A31X	0280
A31Y	0280
A32(2S)	0050
A32A	0050
A32B	0050
A32C	0050
A32D	0050
A32E	0050
A32F	0050
A32G	0050
A32L	0050
A32M	0050
A32OR	0050
A32X	0050
A32Y	0050
A33	0004
A33(2S)	0004
A34	OEM

DEVICE TYPE	REPL CODE
A34-6001-1	0160
A34-6002-17	0160
A35	0004
A35(2S)	0004
A35A	0004
A35B	0004
A35C	0004
A35D	0004
A35E	0004
A35F	0004
A35G	0004
A35L	0004
A35M	0004
A35OR	0004
A35X	0004
A35Y	0004
A36	0004
A36A	0050
A36B	0050
A36C	0050
A36D	0050
A36E	0050
A36F	0050
A36G	0050
A36L	0050
A36M	0050
A36OR	0050
A36R	0050
A36X	0050
A36Y	0050
A37(2S)	0050
A37A	0050
A37B	0050
A37C	0050
A37D	0050
A37E	0050
A37F	0050
A37G	0050
A37L	0050
A37M	0050
A37OR	0050
A37R	0050
A37X	0050
A37Y	0050
A38	0050
A38A	0050
A38B	0050
A38BR1200	OEM
A38BR1202	OEM
A38C	0050
A38D	0050
A38DR019A	OEM
A38DR019B	OEM
A38DR019M	OEM
A38DR1200	OEM
A38DR1202	OEM
A38E	0050
A38F	0050
A38G	0050
A38L	0050
A38M	0050
A38MR1200	OEM
A38OR	0050
A38R	0050
A38X	0050
A38Y	0050
A39(2S)	0050
A39A	0050
A39B	0050
A39C	0050
A39D	0050
A39E	0050
A39F	0050
A39G	0050
A39L	0050
A39M	0050
A39OR	0050
A39R	0050
A39X	0050
A39Y	0050
A40	0004
A40(2S)	0280
A40-6704	0196
A40-6722	0196
A40A	0590
A40A(TRANS)	0280
A40B	1250
A40B(TRANS)	0280
A40C	5587
A40C(TRANS)	0280
A40D	1581
A40D(TRANS)	0280
A40E	5587
A40E(TRANS)	0280
A40F	2002
A40F(TRANS)	0280
A40G	0280
A40L	0280
A40M	1586
A40M(TRANS)	0280
A40OR	0280
A40R	0280
A40X	0280
A40Y	0280
A41	0050
A41(2S)	0050
A41A	1577
A41A(TRANS)	0050
A41B	1577

DEVICE TYPE	REPL CODE
A41B(TRANS)	0050
A41C	1111
A41C(TRANS)	0050
A41D	3281
A41D(TRANS)	0050
A41E	1195
A41E(TRANS)	0050
A41F	2003
A41F(TRANS)	0050
A41G	0050
A41L	0050
A41M	3294
A41M(TRANS)	0050
A41OR	0050
A41R	0050
A41X	0050
A41Y	0050
A42(2S)	0050
A42X00022-01	0580
A42X00373-01(TRANSIS	0211
A42X00454-01	1402
A42X210	0208
A43(2S)	0136
A44(2S)	0136
A44A	5026
A44A(TRANS)	0136
A44B	5026
A44B(TRANS)	0136
A44C	5026
A44C(TRANS)	0136
A44D	5026
A44D(TRANS)	0136
A44E	2615
A44E(TRANS)	0136
A44F	5026
A44F(TRANS)	0136
A44G	0136
A44L	0136
A44M	2615
A44M(TRANS)	0136
A44OR	0136
A44R	0136
A44X	0136
A44Y	0136
A45	0136
A45(2S)	0136
A45-1(2S)	0136
A45-2(2S)	0136
A45-3(2S)	0136
A45A	2559
A45B	2559
A45C	2559
A45E	2566
A45F	2559
A45J	OEM
A45K	OEM
A46	0280
A46-867-3	0233
A46-8614-3	0600
A46-86101-3	0127
A46-86109-3	0127
A46-86110-3	0127
A46-86133-3	0127
A46-86301-3	0127
A46-86302-3	0127
A46-86303-3	0127
A47	0050
A48	0795
A48(TRANS)	0050
A48-40247G01	0144
A48-43351A02	0144
A48-63078A52	0143
A48-97046A05	0144
A48-97046A06	0144
A48-97046A07	0144
A48-97127A06	0144
A48-97127A12	0144
A48-97127A18	0144
A48-124216	0600
A48-124217	0600
A48-124218	0600
A48-124220	0600
A48-124221	0600
A48-125233	0600
A48-125234	0600
A48-125235	0600
A48-125236	0600
A48-128239	0600
A48-134520	0600
A48-134700	0600
A48-134727	0160
A48-134731	0160
A48-134789	0144
A48-134819	0233
A48-134837	0144
A48-134843	0233
A48-134845	0144
A48-134853	0233
A48-134898	0233
A48-134902	0144
A48-134904	0144
A48-134907	0160
A48-134919	0233
A48-134922	0144
A48-134923	0144
A48-134924	0144
A48-134925	0144
A48-134926	0144

DEVICE TYPE	REPL CODE
A48-134927	0233
A48-134931	0600
A48-134945	0144
A48-134961	0144
A48-134962	0144
A48-134963	0144
A48-134964	0144
A48-134965	0144
A48-134966	0144
A48-134981	0144
A48-137002	0233
A48-137035	0233
A48-137071	0144
A48-137075	0144
A48-137076	0144
A48-137077	0144
A48-137102	0160
A48-137155	0144
A48-137197	0144
A48-137213	0160
A48-137214	0160
A48-137215	0160
A48-137216	0160
A48-137217	0160
A48-137218	0160
A48-137219	0160
A48-137220	0160
A48-869254	0208
A48-869283	0208
A48-869476	0208
A48-869476A	0208
A49(2S)	0633
A49A	0633
A49B	0633
A49C	0633
A49D	0633
A49E	0633
A49F	0633
A49G	0633
A49L	0633
A49M	0633
A49OR	0633
A49R	0633
A49X	0633
A49Y	0633
A50	0015
A50(RECTIFIER)	0015
A50(TRANS)	0050
A50(TRANSISTOR)	0004
A50A	0050
A50B	0050
A50C	0050
A50D	0050
A50E	0050
A50F	0050
A50G	0050
A50L	0050
A50M	0050
A50OR	0050
A50R	0050
A50X	0050
A50Y	0050
A51	0136
A51(2S)	0136
A51A	0136
A51B	0136
A51C	0136
A51D	0136
A51E	0136
A51F	0136
A51G	0136
A51L	0136
A51M	0136
A51OR	0136
A51R	0136
A51X	0136
A51Y	0136
A52(2S)	0136
A52A	0136
A52B	0136
A52C	0136
A52D	0136
A52E	0136
A52F	0136
A52G	0136
A52L	0136
A52M	0136
A52OR	0136
A52R	0136
A52X	0136
A52Y	0136
A53(2S)	0633
A53A	0633
A53B	0633
A53C	0633
A53D	0633
A53E	0633
A53F	0633
A53G	0633
A53L	0633
A53M	0633
A53OR	0633
A53R	0633
A53X	0633
A53Y	0633
A54	0050
A54-3	0042
A54A	0050
A54B	0050

DEVICE TYPE	REPL CODE
A54C	0050
A54D	0050
A54E	0050
A54F	0050
A54G	0050
A54L	0050
A54M	0050
A54OR	0050
A54R	0050
A54X	0050
A54Y	0050
A55(2S)	0136
A55A	0136
A55B	0136
A55C	0136
A55D	0136
A55E	0136
A55F	0136
A55G	0136
A55L	0136
A55M	0136
A55OR	0136
A55R	0136
A55X	0136
A55Y	0136
A56(2S)	0136
A56A	0136
A57(2S)	0136
A57A	0136
A57A144-12	0144
A57A145-12	0144
A57B	0136
A57B124-10	0160
A57C	0136
A57C5	0208
A57C12-1	0233
A57C12-2	0233
A57D	0136
A57D1-122	0233
A57E	0136
A57F	0136
A57G	0136
A57L	0136
A57L5-1	0160
A57M	0136
A57M2-16	0233
A57M2-17	0233
A57M3-7	0160
A57M3-8	0160
A57OR	0136
A57R	0136
A57X	0136
A57Y	0136
A58(2S)	0050
A58A	0050
A58B	0050
A58C	0050
A58D	0050
A58E	0050
A58F	0050
A58G	0050
A58H	0050
A58J	0050
A58K	0050
A58L	0050
A58M	0050
A58OR	0050
A58R	0050
A58X	0050
A58Y	0050
A59(2S)	0050
A59A	0050
A59B	0050
A59C	0050
A59D	0050
A59E	0050
A59F	0050
A59G	0050
A59L	0050
A59M	0050
A59OR	0050
A59R	0050
A59X	0050
A59Y	0050
A60(2S)	0050
A60A	0050
A60B	0050
A60C	0050
A60D	0050
A60E	0050
A60F	0050
A60G	0050
A60H	0050
A60K	0050
A60L	0050
A60M	0050
A60OR	0050
A60R	0050
A60X	0050
A60Y	0050
A61(2S)	0050
A61A	0050
A61C	0050
A61D	0050
A61E	0050
A61G	0050
A61K	0050
A61L	0050
A61OR	0050

DEVICE TYPE	REPL CODE
A61R	0050
A61X	0050
A61Y	0050
A62	0143
A62-18427	0160
A62-19581	0144
A63	0057
A63-18426	0233
A63-18427	0160
A64(2S)	0164
A64A	0164
A64B	0164
A64C	0164
A64D	0164
A64E	0164
A64F	0164
A64G	0164
A64GN	0164
A64H	0164
A64J	0164
A64K	0164
A64L	0164
A64M	0164
A64OR	0164
A64R	0164
A64X	0164
A64Y	0164
A65(2S)	0050
A65/40-0100	OEM
A65/40-1000	OEM
A65/40-7020	OEM
A65/40-7050	OEM
A65-09-220	0006
A65-1-1A9G	0050
A65-1-70	0050
A65-1-705	0050
A65-2-70	0208
A65-2-705	0208
A65-2A9G	0208
A65-4-70	0600
A65-4-705	0600
A65-4A9G	0600
A65-901	OEM
A65-P11305-0001A	0124
A65-P11324-0001	0124
A65A	0050
A65A19G	0050
A65A-70	0050
A65A-705	0050
A65B	0050
A65B19G	0136
A65B-70	0136
A65B-705	0136
A65C	0050
A65C-19G	0160
A65C-70	0160
A65C-705	0160
A65D	0050
A65E	0050
A65F	0050
A65G	0050
A65K	0050
A65L	0050
A65M	0050
A65OR	0050
A65R	0050
A65X	0050
A65Y	0050
A66(2S)	0050
A66-1-70	0004
A66-1-705	0004
A66-1A9G	0004
A66-2-70	0004
A66-2-705	0004
A66-2A9G	0004
A66-3-3A9G	0004
A66-3-70	0004
A66-3-705	0004
A66-3A9G	0004
A66A	0050
A66B	0050
A66C	0050
A66D	0050
A66E	0050
A66F	0050
A66G	0050
A66K	0050
A66L	0050
A66M	0050
A66OR	0050
A66R	0050
A66X	0050
A66Y	0050
A67(2S)	0004
A67-07-244	0076
A67-08-760	0549
A67-15-280	1593
A67-33-340	0111
A67-33-540	0155
A67-37-940	1144
A67-76-200	0836
A67A	0004
A67B	0004
A67C	0004
A67D	0004
A67E	0004
A67F	0004
A67G	0004
A67H	0004

DEVICE TYPE	REPL CODE
A67K	0004
A67L	0004
A67OR	0004
A67R	0004
A67X	0004
A67Y	0004
A68	0050
A68-23-560	0228
A69(2S)	0136
A69A	0136
A69B	0136
A69C	0136
A69D	0136
A69E	0136
A69F	0136
A69G	0136
A69H	0136
A69IM4	0133
A69K	0136
A69L	0136
A69M	0136
A69OR	0136
A69R	0136
A69X	0136
A69Y	0136
A70(2S)	0050
A70-08	0050
A70-0B	0050
A70-0B	0050
A70A	0599
A70A(TRANS)	0050
A70B	0599
A70B(TRANS)	0050
A70C	0599
A70C(TRANS)	0050
A70D	0599
A70D(TRANS)	0050
A70E	1986
A70E(TRANS)	0050
A70F(2S)	0050
A70G	0050
A70H	0050
A70K	0050
A70L(2S)	0050
A70M	1986
A70MA(2S)	0050
A70N	0657
A70OA	0050
A70OR	0050
A70P	0657
A70PA	OEM
A70PB	OEM
A70R	0050
A70S	0657
A70T	0657
A70X	0050
A70Y	0050
A71(2S)	0050
A71A	0050
A71A(2S)	0050
A71A(RECTIFIER)	1347
A71AB(2S)	0050
A71AC(2S)	0050
A71B	0050
A71B(2S)	0050
A71B(RECTIFIER)	1347
A71BS(2S)	0050
A71C	1347
A71C(TRANS)	0050
A71D	0050
A71D(2S)	0050
A71E	1905
A71E(TRANS)	0050
A71F	0050
A71G	0050
A71H	0050
A71K	0050
A71L	0050
A71M	1905
A71M(TRANS)	0050
A71N	0193
A71OR	0050
A71P	0193
A71PA	OEM
A71PB	OEM
A71R	0050
A71S	0193
A71T	0193
A71X	0050
A71Y(2S)	0050
A71YA	0050
A72(2S)	0136
A72-49-600	0133
A72-83-300	2730
A72-86-700	0165
A72A	OEM
A72B	OEM
A72BLU(2S)	0136
A72BRN(2S)	0136
A72D	OEM
A72M	OEM
A72N	OEM
A72ORN(2S)	0136
A72P	OEM
A72PA	OEM
A72PB	OEM
A72S	0136
A72WHT(2S)	0136
A73(2S)	0050

If replacement code is OEM, contact original manufacturer for replacement.

DEVICE TYPE	REPL CODE
A73-16-179	0139
A73A	0050
A73B	0050
A73C	0050
A73D	0050
A73E	0050
A73F	0050
A73G	0050
A73H	0050
A73K	0050
A73L	0050
A73M	0050
A73OR	0050
A73PA	OEM
A73PB	OEM
A73R	0050
A73X	0050
A73Y	0050
A74	0136
A74(2S)	0136
A74-3-70	0280
A74-3-705	0280
A74A	0136
A74B	0136
A74C	0136
A74D	0136
A74E	0136
A74F	0136
A74G	0136
A74H	0136
A74K	0136
A74L	0136
A74M	0136
A74OR	0136
A74R	0136
A74X	0136
A74Y	0136
A75(2S)	0136
A75-68-500	0015
A75A	0136
A75B(2S)	0136
A75C	0136
A75D	0136
A75E	0136
A75F	0136
A75G	0136
A75H	0136
A75K	0136
A75L	0136
A75M	0136
A75OR	0136
A75R	0136
A75X	0136
A75Y	0136
A76(2S)	0050
A76-11770	0160
A76A	0050
A76B	0050
A76C	0050
A76D	0050
A76E	0050
A76F	0050
A76G	0050
A76H	0050
A76K	0050
A76L	0050
A76OR	0050
A76R	0050
A76X	0050
A76Y	0050
A77(2S)	0633
A77A	5636
A77A(2S)	0633
A77A19G	0440
A77A-70	0440
A77A-705	0440
A77B	5636
A77B(2S)	0633
A77C(2S)	0633
A77C19G	0160
A77C-70	0160
A77C-705	0160
A77D	5636
A77D(2S)	0633
A77E	0633
A77F	0633
A77G	0633
A77H	0633
A77K	0633
A77L	0633
A77M	0633
A77M(DIO)	OEM
A77N	OEM
A77OR	0633
A77P	OEM
A77PB	OEM
A77R	0633
A77X	0633
A77Y	0633
A78(2S)	0004
A79(2S)	0004
A80(2S)	0164
A80-12	OEM
A80-12G	OEM
A80A	0164
A80B	0164
A80D	0164
A80E	0164
A80F	0164

DEVICE TYPE	REPL CODE
A80G	0164
A80H	0164
A80K	0164
A80L	0164
A80M	0164
A80OR	0164
A80R	0164
A80X	0164
A80Y	0164
A81(2S)	0050
A81-04F	OEM
A82	0633
A82(2S)	0164
A82A	0164
A82B	0164
A82C	0164
A82D	0164
A82E	0164
A82F	0164
A82G	0164
A82H	0164
A82K	0164
A82M	0164
A82OR	0164
A82R	0164
A82X	0164
A82Y	0164
A83	0633
A83(2S)	0050
A83A	0050
A83B	0050
A83C	0050
A83D	0050
A83E	0050
A83F	0050
A83G	0050
A83H	0050
A83K	0050
A83L	0050
A83M	0050
A83OR	0050
A83P2B	0126
A83R	0050
A83X	0050
A83Y	0050
A84(2S)	0050
A84A	0050
A84A19G	0160
A84A-70	0160
A84A-705	0160
A84B	0050
A84C	0050
A84D	0050
A84E	0050
A84F	0050
A84G	0050
A84H	0050
A84K	0050
A84L	0050
A84M	0050
A84OR	0050
A84R	0050
A84X	0050
A84Y	0050
A85(2S)	0136
A85L	0136
A86(2S)	0050
A-86-3-1	0484
A86-3-1	1080
A-86-4-1	0484
A86-4-1	0484
A-86-9-1	0913
A86-9-1	0196
A86-10-2	0600
A86-44-2	0600
A86-213-2	0233
A86-214-2	0233
A86-215-2	0233
A86-316-2	0233
A87(2S)	0050
A87A	0050
A87B	0050
A87C	0050
A87D	0050
A87E	0050
A87F	0050
A87G	0050
A87H	0050
A87K	0050
A87L	0050
A87M	0050
A87OR	0050
A87R	0050
A87X	0050
A87Y	0050
A88	0050
A88-70	0050
A88-705	0050
A88B19G	0050
A88C19G	0280
A88C-70	0280
A88C-705	0280
A89(2S)	0050
A90	0136
A90A	1023
A90B	1023
A90C	1023
A90D	1023
A90E	1036

DEVICE TYPE	REPL CODE
A90M	1036
A90N	1046
A90P	1046
A90PA	3928
A90PB	3928
A90S	1046
A90T	1046
A90T2	0144
A91A	0501
A91B	0501
A91C	0501
A91D	0501
A91E	1777
A91M	1777
A91N	1789
A91P	1789
A91PA	0501
A91PB	3531
A91S	1789
A91T	1789
A92(2S)	0050
A92-1-70	0440
A92-1-705	0440
A92-1A9G	0440
A92A	0050
A92B	0050
A92C	0050
A92D	0050
A92E	0050
A92F	0050
A92G	0050
A92H	0050
A92K	0050
A92L	0050
A92M	0050
A92OR	0050
A92R	0050
A92X	0050
A92Y	0050
A93	0633
A93(2S)	0050
A93A	0050
A93B	0050
A93C	0050
A93E	0050
A93F	0050
A93G	0050
A93H	0050
A93K	0050
A93L	0050
A93M	0050
A93OR	0050
A93R	0050
A93X	0050
A93Y	0050
A94	0633
A94(2S)	0164
A94A	0164
A94C	0164
A94D	0164
A94E	0164
A94F	0164
A94G	0164
A94H	0164
A94K	0164
A94L	0164
A94M	0164
A94OR	0164
A94R	0164
A94X	0164
A94Y	0164
A95	0633
A95(2S)	0050
A95-5280	0196
A95-5281	0015
A95-5285	0015
A95-5286	1030
A95-5289	0015
A95-5292	1303
A95-5294	0015
A-95-5295	1303
A95-5295	1303
A95-5296	0774
A95-5297	0196
A-95-5314	0190
A95-5314	0190
A96	0050
A96A	OEM
A96B	OEM
A96C	OEM
A96D	OEM
A96E	OEM
A96M	OEM
A96N	OEM
A96P	OEM
A96S	OEM
A96T	OEM
A97	0050
A97A83	0160
A97AA97B	OEM
A97C	OEM
A97D	OEM
A97E	OEM
A97M	OEM
A97N	OEM
A97P	OEM
A97S	OEM
A97T	OEM
A98	0050

DEVICE TYPE	REPL CODE
A99	0633
A99(2S)	0050
A99S07	0600
A99SK5	0600
A99SK7	0600
A100	0015
A100(TRANS)	0136
A100A(2S)	0136
A100B(2S)	0136
A100C(2S)	0136
A100D	0136
A100E	0136
A100F	0136
A100G	0136
A100H	0136
A100J	0136
A100K	0136
A100M	0136
A100OR	0136
A100R	0136
A100U	OEM
A100X	0136
A100Y	0136
A101	0136
A101(2S)	0136
A101(IC)	OEM
A101-A	0015
A101-OR	0136
A101A	0136
A101AA(2S)	0136
A101AY(2S)	0136
A101B(2S)	0136
A101BA(2S)	0136
A101BB(2S)	0136
A101BC(2S)	0136
A101BX(2S)	0136
A101C(2S)	0136
A101CA(2S)	0136
A101CV(2S)	0136
A101CX(2S)	0136
A101D	0136
A101E(2S)	0136
A101F	0136
A101G	0136
A101H	0136
A101K	0136
A101L	0136
A101M	0136
A101OR	0136
A101QA(2S)	0136
A101R	0136
A101X(2S)	0136
A101XBX	0136
A101Y(2S)	0136
A101YA	0136
A101Z(2S)	0136
A102	0136
A102(2S)	0136
A102-OR	0136
A102A(2S)	0136
A102AA(2S)	0136
A102AB(2S)	0136
A102B(2S)	0136
A102BA(2S)	0136
A102BN(2S)	0136
A102C	0136
A102CA(2S)	0136
A102D	0136
A102E	0136
A102F	0136
A102G	0136
A102H	0136
A102K	0136
A102L	0136
A102M	0136
A102OR	0136
A102TV(2S)	0136
A102X	0136
A102Y	0136
A103	0136
A103(2S)	0136
A103A(2S)	0136
A103B(2S)	0136
A103C(2S)	0136
A103CA(2S)	0136
A103CAK(2S)	0136
A103CB	0136
A103CG(2S)	0136
A103D	0136
A103DA(2S)	0136
A103E	0136
A103F	0136
A103G	0136
A103GA	0136
A103K	0136
A103L	0136
A103M	0136
A103OR	0136
A103R	0136
A103X	0136
A103Y	0136
A104	0050
A104(2S)	0050
A104A	0633
A104A(2S)	0050
A104B	0050
A104C	0050
A104D(2S)	0050
A104E	0050

DEVICE TYPE	REPL CODE
A104F	0050
A104G	0050
A104H	0050
A104K	0050
A104L	0050
A104M	0050
A104OR	0050
A104P(2S)	0050
A104R	0050
A104X	0050
A104Y	0050
A105(2S)	0136
A105A	0136
A105B	0136
A105C	0136
A105D	0136
A105E	0136
A105G	0136
A105H	0136
A105K	0136
A105L	0136
A105M	0136
A105OR	0136
A105R	0136
A105X	0136
A105Y	0136
A-106	0079
A106	0136
A106(2S)	0136
A106A	0136
A106B	0136
A106C	0136
A106D	0136
A106E	0136
A106F	0136
A106G	0136
A106H	0136
A106K	0136
A106L	0136
A106M	0136
A106OR	0136
A106R	0136
A106X	0136
A106Y	0136
A107	0136
A107(2S)	0136
A107A	0136
A107B	0136
A107C	0136
A107D	0136
A107E	0136
A107F	0136
A107G	0136
A107H	0136
A107K	0136
A107L	0136
A107M	0136
A107OR	0136
A107R	0136
A107X	0136
A107Y	0136
A108	0016
A108(2S)	0136
A108A	0136
A108B	0136
A108C	0136
A108D	0136
A108E	0136
A108F	0136
A108G	0136
A108H	0136
A108K	0136
A108L	0136
A108M	0136
A108OR	0136
A108R	0136
A108X	0136
A108Y	0136
A109	0136
A109A	0136
A109B	0136
A109C	0136
A109D	0136
A109E	0136
A109F	0136
A109G	0136
A109K	0136
A109L	0136
A109M	0136
A109OR	0136
A109R	0136
A109X	0136
A109Y	0136
A110	0111
A110(2S)	0136
A110A	0136
A110B	0136
A110C	0136
A110D	0136
A110E	0136
A110F	0136
A110G	0136
A110K	0136
A110L	0136
A110M	0136
A110OR	0136
A110R	0136
A110X	0136
A110Y	0136

DEVICE TYPE	REPL CODE
A111	0016
A111(2S)	0136
A111A	0136
A111B	0136
A111C	0136
A111D	0136
A111E	0136
A111F	0136
A111G	0136
A111K	0136
A111L	0136
A111M	0136
A111OR	0136
A111R	0136
A111T2	0233
A111X	0136
A111Y	0136
A112(2S)	0136
A112-000172	0037
A112-000185	0037
A112-000187	0037
A112A	0136
A112B	0136
A112C	0136
A112D	0136
A112E	0136
A112F	0136
A112G	0136
A112GN	0136
A112H	0136
A112K	0136
A112L	0136
A112M	0136
A112OR	0136
A112R	0136
A112X	0136
A112Y	0136
A113(2S)	0136
A113A	0136
A113B	0136
A113C	0136
A113D	0136
A113E	0136
A113F	0136
A113G	0136
A113GN	0136
A113H	0136
A113J	0136
A113L	0136
A113M	0136
A113R	0136
A113X	0136
A113Y	0136
A114	0136
A114A	0023
A114A(TRANS)	0136
A114B	0023
A114B(TRANS)	0136
A114C	0071
A114C(TRANS)	0136
A114D	0023
A114D(TRANS)	0136
A114E	0023
A114E(TRANS)	0136
A114ES	4172
A114F	0920
A114F(TRANS)	0136
A114G	0136
A114H	0136
A114K	0136
A114L	0136
A114M	0023
A114M(TRANS)	0136
A114OR	0136
A114R	0136
A114X	0136
A114Y	0136
A115	0136
A115(2S)	0136
A115A	0031
A115A(TRANS)	0136
A115B	0031
A115B(TRANS)	0136
A115C	0031
A115C(TRANS)	0136
A115D	0031
A115D(TRANS)	0136
A115E	0031
A115E(TRANS)	0136
A115F	0031
A115F(TRANS)	0136
A115G	0136
A115GN	0136
A115H	0136
A115J	0136
A115K	0136
A115L	0136
A115M	0031
A115M(TRANS)	0136
A115OR	0136
A115R	0136
A115X	0136
A115Y	0136
A116	0136
A116(2S)	0136
A116A	0136
A116B	0136
A116C	0136
A116D	0136

DEVICE TYPE	REPL CODE
A116E	0136
A116F	0136
A116G	0136
A116GN	0136
A116H	0136
A116J	0136
A116K	0136
A116L	0136
A116M	0136
A116OR	0136
A116R	0136
A116X	0136
A116Y	0136
A117	0136
A117(2S)	0136
A117A	0136
A117B	0136
A117C	0136
A117D	0136
A117E	0136
A117F	0136
A117G	0136
A117GN	0136
A117H	0136
A117J	0136
A117K	0136
A117L	0136
A117M	0136
A117OR	0136
A117R	0136
A117T	OEM
A117X	0136
A117Y	0136
A118	0136
A118A	0136
A118B	0136
A118C	0136
A118D	0136
A118E	0136
A118F	0136
A118G	0136
A118GN	0136
A118H	0136
A118J	0136
A118K	0136
A118L	0136
A118M	0136
A118OR	0136
A118R	0136
A118X	0136
A118Y	0136
A119	0126
A120	0126
A120A	0126
A120P1	0037
A120UF	OEM
A121(2S)	0136
A121-1	0037
A121-1RED	0037
A121-15	0600
A121-16	0600
A121-17	0600
A121-21	0600
A121-50	0600
A121-361	0233
A121-444	0037
A121-446	0037
A121-480	0144
A121-495	0037
A121-496	0037
A121-497	0037
A121-497WHT	0037
A121-585	0127
A121-585B	0127
A121-602	0037
A121-603	0037
A121-679	0037
A121-687	0127
A121-699	0037
A121-746	0037
A121-762	0600
A121-774	0037
A121-1410	0600
A121A	0136
A121B	0136
A121C	0136
A121D	0136
A121E	0136
A121F	0136
A121G	0136
A121GN	0136
A121H	0136
A121J	0136
A121K	0136
A121L	0136
A121M	0136
A121OR	0136
A121R	0136
A121X	0136
A121Y	0136
A122	0050
A122(2S)	0136
A122-1962	0208
A122A	0136
A122B	0136
A122C	0136
A122D	0136
A122E	0136
A122F	0136

DEVICE TYPE	REPL CODE
A122G	0136
A122GN	0136
A122GRN	0037
A122H	0136
A122K	0136
A122L	0136
A122M	0136
A122OR	0136
A122R	0136
A122X	0136
A122Y	0136
A122YEL	0037
A123	0136
A123(2S)	0136
A123-7	0015
A123A	0136
A123B	0136
A123C	0136
A123D	0136
A123E	0136
A123F	0136
A123G	0136
A123GN	0136
A123H	0136
A123J	0136
A123K	0136
A123L	0136
A123M	0136
A123OR	0136
A123R	0136
A123X	0136
A123Y	0136
A124	0136
A124(2S)	0136
A124A	0136
A124B	0136
A124C	0136
A124D	0136
A124E	0136
A124EF(DTA124EF)	OEM
A124F	0136
A124G	0136
A124GN	0136
A124H	0136
A124J	0136
A124K	0136
A124L	0136
A124M	0136
A124OR	0136
A124R	0136
A124X	0136
A124Y	0136
A125	0136
A125(2S)	0136
A125A	0136
A125B	0136
A125C	0136
A125D	0136
A125E	0136
A125F	0136
A125G	0136
A125GN	0136
A125H	0136
A125J	0136
A125K	0136
A125L	0136
A125M	0136
A125OR	0136
A125R	0136
A125X	0136
A125Y	0136
A126	0136
A126(2S)	0136
A127	0050
A127(2S)	0136
A-128	0079
A128	0016
A128(2S)	0164
A128A	0164
A128B	0164
A128C	0164
A128D	0164
A128E	0164
A128F	0164
A128G	0164
A128GN	0164
A128H	0164
A128J	0164
A128K	0164
A128L	0164
A128M	0164
A128OR	0164
A128R	0164
A128X	0164
A128Y	0164
A129	0164
A129-30	0600
A129-34	0037
A129A	0164
A129B	0164
A129C	0164
A129D	0164
A129E	0164
A129E(2S)	0164
A129F	0164
A129G	0164
A129GN	0164
A129H	0164
A129J	0164
A129K	0164
A129L	0164
A129M	0164
A129M(2S)	0164
A129N	OEM
A129OR	0164
A129P	OEM
A129PB	OEM
A129R	0164
A129X	0164
A129Y	0164
A130	0233
A130(2S)	0050
A130-149	0037
A130-40315	0037
A130-40429	0037
A130-ORN	0233
A130-V10	0233
A130A	0050
A130B	0050
A130C	0050
A130D	0050
A130E	0050
A130F	0050
A130G	0050
A130GN	0050
A130H	0050
A130J	0050
A130K	0050
A130L	0050
A130M	0050
A130OR	0050
A130R	0050
A130X	0050
A130Y	0050
A131	0136
A131(2S)	0136
A131A	0136
A131B	0136
A131C	0136
A131D	0136
A131E	0136
A131F	0136
A131G	0136
A131GN	0136
A131H	0136
A131J	0136
A131K	0136
A131L	0136
A131M	0136
A131OR	0136
A131R	0136
A131X	0136
A131Y	0136
A132	0015
A132(TRANS)	0136
A132-1	0015
A132A	0136
A132B	0136
A132C	0136
A132D	0136
A132E	0136
A132F	0136
A132G	0136
A132H	0136
A132J	0136
A132K	0136
A132L	0136
A132M	0136
A132OR	0136
A132X	0136
A132Y	0136
A133	0861
A133(2S)	0136
A133A	0136
A133B	0136
A133C	0136
A133D	0136
A133E	0136
A133F	0136
A133G	0136
A133GN	0136
A133H	0136
A133J	0136
A133K	0136
A133L	0136
A133M	0136
A133OR	0136
A133R	0136
A133X	0136
A133Y	0136
A134	0136
A134(2S)	0136
A134A	0136
A134B	0136
A134C	0136
A134D	0136
A134E	0136
A134F	0136
A134G	0136
A134H	0136
A134J	0136
A134K	0136
A134L	0136
A134OR	0136
A134R	0136
A134X	0136
A134Y	0136
A135	0076
A135(2S)	0050
A135A	0050
A135B	0050
A135C	0050
A135D	0050
A135E	0050
A135F	0050
A135G	0050
A135GN	0050
A135H	0050
A135J	0050
A135K	0050
A135L	0050
A135M	0050
A135OR	0050
A135R	0050
A135X	0050
A135Y	0050
A136	0633
A136(2S)	0050
A136A	0050
A136B	0050
A136C	0050
A136D	0050
A136E	0050
A136F	0050
A136G	0050
A136GN	0050
A136H	0050
A136J	0050
A136K	0050
A136L	0050
A136M	0050
A136OR	0050
A136R	0050
A136X	0050
A136Y	0050
A137	1381
A137(2S)	0136
A137A	0136
A137B	0136
A137C	0136
A137D	0136
A137E	0136
A137F	0136
A137G	0136
A137GN	0136
A137H	0136
A137J	0136
A137K	0136
A137L	0136
A137M	0136
A137OR	0136
A137R	0136
A137X	0136
A137Y	0136
A138	0111
A138(2S)	0004
A139	0111
A139(2S)	0136
A139A	0136
A139B	0136
A139C	0136
A139D	0136
A139E	1195
A139E(TRANS)	0136
A139F	0136
A139G	0136
A139GN	0136
A139J	0136
A139K	0136
A139L	0136
A139M	0319
A139M(TRANS)	0136
A139N	1132
A139OR	0136
A139P	0044
A139PB	OEM
A139R	0136
A139RE	1644
A139RM	0267
A139RN	1119
A139RP	0281
A139X	0136
A139Y	0136
A141	0004
A141(2S)	0136
A141A	0136
A141B(2S)	0136
A141C(2S)	0136
A141D	0136
A141E	0136
A141F	0136
A141G	0136
A141GN	0136
A141H	0136
A141K	0136
A141L	0136
A141M	0136
A141OR	0136
A141R	0136
A141X	0136
A141Y	0136
A142	0004
A142(2S)	0136
A142A(2S)	0136
A142B(2S)	0136
A142C(2S)	0136
A142D	0136
A142E	0136
A142F	0136
A142G	0136
A142GN	0136
A142H	0136
A142J	0136
A142K	0136
A142L	0136
A142M	0136
A142OR	0136
A142R	0136
A142X	0136
A142Y	0136
A143	0004
A143(2S)	0136
A143A	0136
A143B	0136
A143C	0136
A143D	0136
A143E	0136
A143E(DT)	OEM
A143F	0136
A143G	0136
A143GN	0136
A143H	0136
A143J	0136
A143K	0136
A143L	0136
A143M	0136
A143OR	0136
A143R	0136
A143X	0136
A143XA	OEM
A143Y	0136
A144(2S)	0136
A-144-3078	0725
A144A	0136
A144A-1	0160
A144A-19	0160
A144B	0136
A144C(2S)	0136
A144D	0136
A144E	0136
A144F	0136
A144G	0136
A144GN	0136
A144H	0136
A144J	0136
A144K	0136
A144L	0136
A144M	0136
A144OR	0136
A144R	0136
A144X	0136
A144Y	0136
A145(2S)	0136
A145A(2S)	0136
A145B	0136
A145C(2S)	0136
A145D	0136
A145E	0136
A145G	0136
A145GN	0136
A145K	0136
A145M	0136
A145OR	0136
A145R	0136
A145X	0136
A145Y	0136
A146(2S)	0136
A146A	0136
A146B	0136
A146C	0136
A146D	0136
A146E	0136
A146F	0136
A146G	0136
A146GN	0136
A146H	0136
A146J	0136
A146K	0136
A146L	0136
A146M	0136
A146OR	0136
A146X	0136
A146Y	0136
A147(2S)	0136
A147A	0136
A147B	0136
A147C	0136
A147D	0136
A147E	0136
A147F	0136
A147G	0136
A147H	0136
A147J	0136
A147K	0136
A147L	0136
A147M	0136
A147OR	0136
A147R	0136
A147X	0136
A147Y	0136
A148(2S)	0136
A148A	0136
A148A(2S)	0136
A148B	0136
A148B(2S)	0136
A148C	0136
A148C(2S)	0136
A148D	0136
A148E	0136
A148F	0136
A148G	0136
A148GN	0136
A148H	0136
A148J	0136
A148K	0136
A148L	0136
A148M	0136
A148OR	0136
A148P2	0004
A148P2-29	0004
A148R	0136
A148X	0136
A148Y	0136
A149(2S)	0136
A149A	0136
A149B	0136
A149C	0136
A149D	0136
A149E	0136
A149F	0136
A149G	0136
A149GN	0136
A149H	0136
A149J	0136
A149K	0136
A149L	0136
A149L-4	0004
A149L-49	0004
A149M	0136
A149OR	0136
A149R	0136
A149X	0136
A149Y	0136
A150(2S)	0136
A150A	0136
A150B	0136
A150C	0136
A150D	0136
A150E	0136
A150F	0136
A150G	0136
A150GN	0136
A150H	0136
A150J	0136
A150K	0136
A150L	0136
A150M	0136
A150OR	0136
A150R	0136
A150X	0136
A150Y	0136
A151	0136
A151(2S)	0136
A151A	0136
A151B	0136
A151C	0136
A151D	0136
A151E	0136
A151F	0136
A151G	0136
A151GN	0136
A151H	0136
A151J	0136
A151K	0136
A151L	0136
A151M	0136
A151OR	0136
A151R	0136
A151X	0136
A151Y	0136
A152	0136
A152(2S)	0136
A152A	0136
A152B	0136
A152C	0136
A152D	0136
A152E	0136
A152F	0136
A152G	0136
A152GN	0136
A152H	0136
A152J	0136
A152K	0136
A152L	0136
A152M	0136
A152OR	0136
A152R	0136
A152X	0136
A152Y	0136
A153	0016
A153(2S)	0136
A153(JAPAN)	0136
A153A	0136
A153B	0136
A153C	0136
A153D	0136
A153E	0136
A153F	0136
A153G	0136
A153GN	0136
A153H	0136
A153J	0136
A153K	0136
A153L	0136
A153M	0136
A153OR	0136
A153R	0136
A153X	0136
A153Y	0136
A154	0224
A154(2S)	0136
A154A	0136
A154B	0136
A154C	0136
A154D	0136
A154E	0136
A154F	0136
A154G	0136
A154GN	0136
A154H	0136
A154J	0136
A154K	0136
A154L	0136
A154M	0136
A154OR	0136
A154R	0136
A154X	0136
A154Y	0136
A155	0224
A155(2S)	0136
A155A	0136
A155B	0136
A155C	0136
A155D	0136
A155E	0136
A155F	0136
A155G	0136
A155GN	0136
A155H	0136
A155J	0136
A155K	0136
A155L	0136
A155M	0136
A155OR	0136
A155R	0136
A155X	0136
A155Y	0136
A-156	0016
A156	0016
A156(2S)	0136
A156A	0136
A156B	0136
A156C	0136
A156D	0136
A156E	0136
A156F	0136
A156G	0136
A156GN	0136
A156H	0136
A156J	0136
A156K	0136
A156L	0136
A156M	0136
A156OR	0136
A156R	0136
A156X	0136
A156Y	0136
A157	0016
A157(2S)	0136
A157A	0016
A157A(2S)	0136
A157B	0016
A157B(2S)	0136
A157C	0016
A157C(2S)	0136
A157D	0136
A157E	0136
A157F	0136
A157G	0136
A157GN	0136
A157H	0136
A157J	0136
A157K	0136
A157L	0136
A157M	0136
A157OR	0136
A157R	0136
A157X	0136
A157Y(2S)	0136
A158	0016
A158A	0016
A158B	0079
A158C	0016
A159	0016
A159(2S)	0136
A159A	0136
A159B	0016
A159B(2S)	0136
A159C	0016
A159C(2S)	0136
A159D	0136
A159E	0136
A159F	0136
A159G	0136
A159GN	0136
A159H	0136
A159J	0136
A159K	0136
A159L	0136
A159M	0136
A159OR	0136
A159R	0136
A159X	0136
A159Y	0136
A160	0136
A160(2S)	0136
A160A	0136
A160B	0136
A160C	0136
A160D	0136
A160E	0136
A160F	0136
A160GN	0136
A160H	0136
A160J	0136
A160K	0136
A160L	0136
A160M	0136
A160OR	0136
A160R	0136
A160UF	OEM
A160X	0136
A160Y	0136
A161	0037
A161(2S)	0136
A161A	0136
A161B	0136
A161C	0136
A161D	0136
A161E	0136
A161F	0136
A161G	0136
A161GN	0136
A161H	0136
A161J	0136
A161K	0136
A161L	0136
A161M	0136
A161OR	0136
A161R	0136
A161X	0136
A161Y	0136
A162	0037
A162(2S)	0136
A162A	0136
A162B	0136
A162C	0136
A162D	0136
A162E	0136
A162F	0136
A162G	0136
A162GN	0136
A162H	0136
A162J	0136
A162K	0136
A162L	0136
A162M	0136
A162OR	0136
A162R	0136
A162X	0136
A162Y	0136
A163	0136
A163A	0136
A163B	0136
A163C	0136
A163D	0136
A163E	0136
A163F	0136
A163G	0136
A163GN	0136
A163H	0136
A163J	0136
A163K	0136
A163L	0136
A163M	0136
A163OR	0136
A163R	0136
A163X	0136
A163Y	0136
A164	0224
A164(2S)	0136
A164A	0136
A164B	0136
A164C	0136
A164D	0136
A164E	0136
A164F	0136
A164G	0136
A164GN	0136
A164H	0136
A164J	0136
A164K	0136
A164L	0136
A164M	0136
A164OR	0136
A164R	0136
A164X	0136
A164Y	0136
A165	0224
A165(2S)	0136
A165A	0136
A165B	0136
A165C	0136
A165D	0136
A165E	0136
A165F	0136
A165G	0136
A165GN	0136
A165H	0136
A165J	0136
A165K	0136
A165L	0136
A165M	0136

If replacement code is OEM, contact original manufacturer for replacement.

DEVICE TYPE	REPL CODE
A165OR	0136
A165R	0136
A165X	0136
A165Y	0136
A166(2S)	0136
A166A	0136
A166B	0136
A166C	0136
A166D	0136
A166E	0136
A166F	0136
A166G	0136
A166GN	0136
A166H	0136
A166J	0136
A166K	0136
A166L	0136
A166M	0136
A166OR	0136
A166R	0136
A166X	0136
A166Y	0136
A167(2S)	0280
A167A	0280
A167B	0280
A167C	0280
A167D	0280
A167E	0280
A167F	0280
A167G	0280
A167GN	0280
A167H	0280
A167J	0280
A167K	0280
A167L	0280
A167M	0280
A167OR	0280
A167R	0280
A167X	0280
A167Y	0280
A168	0280
A168A(2S)	0280
A168B	0280
A168C	0280
A168D	0280
A168E	0280
A168F	0280
A168G	0280
A168GN	0280
A168H	0280
A168J	0280
A168K	0280
A168L	0280
A168M	0280
A168OR	0280
A168R	0280
A168X	0280
A168Y	0280
A169(2S)	0280
A169A	0280
A169B	0280
A169C	0280
A169D	0280
A169E	0280
A169F	0280
A169G	0280
A169GN	0280
A169H	0280
A169J	0280
A169K	0280
A169L	0280
A169M	0280
A169OR	0280
A169R	0280
A169X	0280
A169Y	0280
A170	0126
A170(2S)	0280
A170A	0599
A170A(TRANS)	0280
A170AR	1347
A170B	0599
A170B(TRANS)	0280
A170BR	1347
A170C	0599
A170C(TRANS)	0280
A170CR	1347
A170D	0599
A170D(TRANS)	0280
A170DR	1905
A170E	1986
A170E(TRANS)	0280
A170ER	1905
A170F	0280
A170G	0280
A170GN	0280
A170H	0280
A170J	0280
A170K	0280
A170L	0280
A170M	1986
A170M(TRANS)	0280
A170MR	1905
A170N	0657
A170NR	0193
A170OR	0280
A170P	0657
A170PA	3309
A170PB	3309
A170PC	3309
A170PD	3309
A170PE	OEM
A170PR	0193
A170R	0280
A170RA	1347
A170RB	1347
A170RC	1347
A170RD	1347
A170RE	1905
A170RM	1905
A170RN	0193
A170RP	0193
A170RPA	0202
A170RPB	0202
A170RPC	0202
A170RPD	0202
A170RPE	OEM
A170RS	0193
A170RT	0193
A170S	0657
A170SR	0193
A170T	0657
A170TR	0193
A170X	0280
A170Y	0280
A171	0037
A171(2S)	0280
A171A	0280
A171B	0280
A171C	0280
A171D	0280
A171E	0280
A171F	0280
A171G	0280
A171GN	0280
A171H	0280
A171J	0280
A171K	0280
A171L	0280
A171M	0280
A171OR	0280
A171R	0280
A171X	0280
A171Y	0280
A172(2S)	0280
A172A(2S)	0280
A172B	0280
A172C	0280
A172D	0280
A172E	0280
A172F	0280
A172G	0280
A172GN	0280
A172H	0280
A172J	0280
A172K	0280
A172L	0280
A172M	0280
A172OR	0280
A172R	0280
A172X	0280
A172Y	0280
A173(2S)	0004
A173A	0004
A173B(2S)	0004
A173C	0004
A173D	0004
A173E	0004
A173F	0004
A173G	0004
A173GN	0004
A173H	0004
A173J	0004
A173K	0004
A173L	0004
A173M	0004
A173OR	0004
A173R	0004
A173X	0004
A173Y	0004
A174(2S)	0211
A174A	0211
A174B	0211
A174C	0211
A174D	0211
A174E	0211
A174F	0211
A174G	0211
A174GN	0211
A174H	0211
A174J	0211
A174K	0211
A174L	0211
A174M	0211
A174OR	0211
A174R	0211
A174X	0211
A174Y	0211
A175(2S)	0136
A175A	0136
A175B	0136
A175C	0136
A175D	0136
A175E	0136
A175F	0136
A175G	0136
A175GN	0136
A175H	0136
A175J	0136
A175K	0136
A175L	0136
A175M	0136
A175OR	0136
A175R	0136
A175X	0136
A175Y	0136
A176(2S)	0136
A176-025-9-002	0144
A176A	0136
A176B	0136
A176C	0136
A176D	0136
A176E	0136
A176F	0136
A176G	0136
A176GN	0136
A176H	0136
A176J	0136
A176K	0136
A176L	0136
A176M	0136
A176OR	0136
A176R	0136
A176X	0136
A176Y	0136
A177	0037
A177(A)	0037
A177A	0037
A177A(DIO)	OEM
A177AB	0037
A177B	0599
A177C	0599
A177D	0599
A177E	1986
A177M	1986
A177N	0657
A177P	0657
A177PA	3309
A177PB	3309
A177PC	3309
A177PD	3309
A177PE	OEM
A177RA	1347
A177RB	1347
A177RC	1347
A177RD	1347
A177RE	1905
A177RM	1905
A177RN	0193
A177RP	0193
A177RPA	0202
A177RPB	0202
A177RPC	0202
A177RPD	0202
A177RPE	OEM
A177RS	0193
A177RT	0193
A177S	0657
A177T	0657
A178	0050
A178A	0037
A178AB	0037
A178B	0037
A178BA	0037
A179A	0037
A179AC	0037
A179B	0037
A179BB	0037
A180(2S)	0136
A180A	1023
A180A(TRANS)	0136
A180AR	1347
A180B	1023
A180B(TRANS)	0136
A180BR	1347
A180C	1023
A180C(TRANS)	0136
A180CR	1347
A180D	1023
A180D(TRANS)	0136
A180DR	1905
A180E	1036
A180E(TRANS)	0136
A180ER	1905
A180F	0136
A180G	0136
A180GN	0136
A180H	0136
A180J	0136
A180K	0136
A180L	0136
A180M	1036
A180M(TRANS)	0136
A180MR	1905
A180N	1046
A180NR	0193
A180OR	0136
A180P	1046
A180PA	2832
A180PB	2832
A180PC	2832
A180PD	2832
A180PE	OEM
A180PR	0193
A180R	0136
A180RA	0501
A180RB	0501
A180RC	0501
A180RD	0501
A180RE	1777
A180RM	1777
A180RN	1789
A180RP	1789
A180RPA	3167
A180RPB	3167
A180RPC	3167
A180RPD	3167
A180RPE	OEM
A180RS	1789
A180RT	1789
A180S	1046
A180SR	0193
A180T	1046
A180TR	0193
A180X	0136
A180Y	0136
A181(2S)	0136
A181A	0136
A181B	0136
A181C	0136
A181D	0136
A181E	0136
A181F	0136
A181G	0136
A181GN	0136
A181H	0136
A181J	0136
A181K	0136
A181L	0136
A181M	0136
A181OR	0136
A181R	0136
A181X	0136
A181Y	0136
A182(2S)	0050
A182A	0050
A182B	0050
A182C	0050
A182D	0050
A182E	0050
A182F	0050
A182G	0050
A182GN	0050
A182H	0050
A182J	0050
A182K	0050
A182L	0050
A182M	0050
A182OR	0050
A182R	0050
A182X	0050
A182Y	0050
A183(2S)	0136
A183A	0136
A183B	0136
A183C	0136
A183D	0136
A183E	0136
A183F	0136
A183G	0136
A183GN	0136
A183H	0136
A183J	0136
A183K	0136
A183L	0136
A183M	0136
A183OR	0136
A183R	0136
A183X	0136
A183Y	0136
A184	0050
A185	OEM
A186	OEM
A186(2S)	0164
A187A	1023
A187B	1023
A187C	1023
A187D	1023
A187E	1036
A187M	1036
A187N	1046
A187P	1046
A187PA	2832
A187PB	2832
A187PC	2832
A187PD	2832
A187PE	4286
A187RA	0501
A187RB	0501
A187RC	0501
A187RD	0501
A187RE	1777
A187RM	1777
A187RN	1789
A187RP	1789
A187RPA	3167
A187RPB	3167
A187RPC	3167
A187RPD	3167
A187RPE	4285
A187RS	1789
A187RT	1789
A187S	1046
A187T	1046
A187TV(2S)	0004
A188	0136
A188(2S)	0136
A188A	0136
A188B	0136
A188C	0136
A188D	0136
A188E	0136
A188F	0136
A188G	0136
A188GN	0136
A188H	0136
A188J	0136
A188K	0136
A188L	0136
A188M	0136
A188OR	0136
A188R	0136
A188X	0136
A188Y	0136
A189(2S)	0136
A189A	0136
A189B	0136
A189C	0136
A189D	0136
A189E	0136
A189F	0136
A189G	0136
A189GN	0136
A189H	0136
A189J	0136
A189K	0136
A189L	0136
A189M	0136
A189OR	0136
A189R	0136
A189X	0136
A189Y	0136
A190	5213
A190(2S)	0050
A190A	1023
A190AR	0501
A190B	1023
A190BR	0501
A190C	1023
A190CR	0501
A190D	1023
A190DR	1777
A190E	1036
A190ER	1777
A190M	1036
A190MR	1777
A190N	1046
A190NR	1789
A190P	1046
A190PA	3928
A190PB	3928
A190PC	2832
A190PD	2832
A190PE	4286
A190PR	1789
A190RA	0501
A190RB	0501
A190RC	0501
A190RD	0501
A190RE	1777
A190RM	1777
A190RN	1789
A190RP	1789
A190RPA	3167
A190RPB	3167
A190RPC	3167
A190RPD	3167
A190RPE	4285
A190RS	1789
A190RT	1789
A190S	1046
A190SR	1789
A190T	1046
A190TR	1789
A191	5213
A191(2S)	0050
A192	0325
A192(2S)	0050
A193	0164
A193(2S)	0050
A194	0325
A194(2S)	0050
A195	0325
A195(2S)	0050
A-195C	0037
A196	0325
A196(2S)	0050
A197	0136
A197(2S)	0136
A197/197RPB	3167
A197/197RPE	4285
A197A	1023
A197A(TRANS)	0136
A197B	1023
A197B(TRANS)	0136
A197C	1023
A197C(TRANS)	0136
A197D	1023
A197D(TRANS)	0136
A197E	1036
A197E(TRANS)	0136
A197F	0136
A197G	0136
A197GN	0136
A197H	0136
A197J	0136
A197K	0136
A197L	0136
A197M	1036
A197M(TRANS)	0136
A197N	1046
A197OR	0136
A197P	1046
A197PA	2832
A197PB	2832
A197PC	2832
A197PD	2832
A197PE	4286
A197R	0136
A197RA	0501
A197RB	0501
A197RC	0501
A197RD	0501
A197RE	1777
A197RM	1777
A197RN	1789
A197RP	1789
A197RPA	3167
A197RPB	3167
A197RPC	3167
A197RPD	3167
A197RPE	4285
A197RS	1789
A197RT	1789
A197S	1046
A197T	1046
A197X	0136
A197Y	0136
A198	0004
A198(2S)	0050
A198A	1023
A198A(TRANS)	0050
A198B	1023
A198B(TRANS)	0050
A198C	1023
A198C(TRANS)	0050
A198D	1023
A198D(TRANS)	0050
A198E	1036
A198E(TRANS)	0050
A198F	0050
A198G	0050
A198GN	0050
A198H	0050
A198J	0050
A198K	0050
A198L	0050
A198M	1036
A198M(TRANS)	0050
A198N	1046
A198OR	0050
A198P	1046
A198PA	2832
A198PB	2832
A198PC	2832
A198PD	2832
A198PE	4286
A198R	0050
A198RA	0501
A198RB	0501
A198RC	0501
A198RD	0501
A198RE	1777
A198RM	1777
A198RN	1789
A198RP	1789
A198RPA	3167
A198RPB	3167
A198RPC	3167
A198RPD	3167
A198RPE	4285
A198RS	1789
A198RT	1789
A198S	1046
A198T	1046
A198X	0050
A198Y	0050
A199	0633
A199(2S)	0050
A200	0015
A200(TRANS)	0136
A200-052	0037
A200UF	OEM
A201	0136
A201(2S)	0136
A201-0(2S)	0136
A201-N	0136
A201-O	0136
A201-OR	0136
A201A(2S)	0136
A201B(2S)	0136
A201CL	0136
A201D	0136
A201E(2S)	0136
A201F	0136
A201G	0136
A201GN	0136
A201H	0136
A201J	0136
A201K	0136
A201L	0136
A201M	0136
A201N	0136
A201O	0136
A201OR	0136
A201R	0136
A201TV	0136
A201TVO	0136
A201X	0136
A201Y	0136
A202	0136
A202(2S)	0633
A202-OR	0633
A202A(2S)	0633
A202AP	0633
A202B(2S)	0633
A202C(2S)	0633
A202D(2S)	0633
A202D-4	0633
A202E	0633
A202F	0633
A202G	0633
A202GN	0633
A202H	0633
A202J	0633
A202K	0633
A202L	0633
A202M	0633
A202OR	0633
A202R	0633
A202X	0633
A202Y	0633
A203	0004
A203(2S)	0633
A203A(2S)	0633
A203AA(2S)	0633
A203B(2S)	0633
A203C	0633
A203D	0633
A203P(2S)	0633
A204(2S)	0280
A204A	0280
A204B	0280
A204C	0280
A204D	0280
A204E	0280
A204F	0280
A204G	0280
A204GN	0280
A204H	0280
A204J	0280
A204K	0280
A204L	0280
A204M	0280
A204OR	0280
A204R	0280
A204X	0280
A204Y	0280
A205	0280
A205A	0280
A205B	0280
A205C	0280
A205D	0280
A205E	0280
A205F	0280
A205G	0280
A205GN	0280
A205H	0280
A205J	0280
A205K	0280
A205L	0280
A205M	0280
A205OR	0280
A205R	0280
A205X	0280
A205Y	0280
A206	0280
A206A	0280
A206B	0280
A206C	0280
A206D	0280
A206E	0280
A206F	0280
A206G	0280
A206GN	0280
A206H	0280
A206J	0280
A206K	0280
A206L	0280
A206M	0280
A206OR	0280
A206R	0280
A206X	0280
A206Y	0280
A207	0280
A207A	0280
A207B	0280
A207C	0280
A207D	0280
A207E	0280
A207F	0280
A207G	0280
A207GN	0280
A207H	0280
A207J	0280
A207K	0280
A207L	0280
A207M	0280
A207OR	0280
A207R	0280
A207X	0280
A207Y	0280
A208	0847

DEVICE TYPE	REPL CODE
A208(2S)	0847
A208A	0847
A208B	0847
A208C	0847
A208D	0847
A208E	0847
A208F	0847
A208G	0847
A208GN	0847
A208H	0847
A208J	0847
A208K	0847
A208L	0847
A208M	0847
A208OR	0847
A208R	0847
A208X	0847
A208Y	0847
A209	0847
A209(2S)	0847
A209A	0847
A209B	0847
A209C	0847
A209D	0847
A209E	0847
A209F	0847
A209G	0847
A209GN	0847
A209H	0847
A209J	0847
A209K	0847
A209L	0847
A209M	0847
A209OR	0847
A209R	0847
A209X	0847
A209Y	0847
A210	0847
A210(2S)	0847
A210A	0847
A210B	0847
A210C	0847
A210D	0847
A210E	0847
A210F	0847
A210G	0847
A210GN	0847
A210H	0847
A210J	0847
A210K	0847
A210L	0847
A210M	0847
A210OR	0847
A210R	0847
A210X	0847
A210Y	0847
A211	0560
A211(2S)	0280
A211A	0280
A211B	0280
A211C	0280
A211D	0280
A211E	0280
A211F	0280
A211G	0280
A211GN	0280
A211H	0280
A211J	0280
A211K	0280
A211L	0280
A211M	0280
A211OR	0280
A211R	0280
A211X	0280
A211Y	0280
A212	0280
A212(2S)	0050
A212A	0050
A212B	0050
A212C	0050
A212D	0050
A212E	0050
A212F	0050
A212G	0050
A212GN	0050
A212H	0050
A212J	0050
A212K	0050
A212L	0050
A212M	0050
A212OR	0050
A212R	0050
A212X	0050
A212Y	0050
A213	0136
A213(2S)	0136
A213A	0136
A213B	0136
A213C	0136
A213D	0136
A213E	0136
A213F	0136
A213G	0136
A213GN	0136
A213H	0136
A213J	0136
A213K	0136
A213L	0136
A213M	0136
A213OR	0136
A213R	0136
A213X	0136
A213Y	0136
A214	0136
A214(2S)	0136
A214A	0136
A214B	0136
A214C	0136
A214D	0136
A214E	0136
A214F	0136
A214G	0136
A214GN	0136
A214H	0136
A214J	0136
A214K	0136
A214L	0136
A214M	0136
A214OR	0136
A214R	0136
A214X	0136
A214Y	0136
A215	0136
A215(2S)	0136
A215A	0136
A215B	0136
A215C	0136
A215D	0136
A215E	0136
A215F	0136
A215G	0136
A215GN	0136
A215H	0136
A215J	0136
A215K	0136
A215L	0136
A215M	0136
A215OR	0136
A215R	0136
A215X	0136
A215Y	0136
A216(2S)	0136
A216A	0136
A216B	0136
A216C	0136
A216D	0136
A216E	0136
A216F	0136
A216G	0136
A216GN	0136
A216H	0136
A216J	0136
A216K	0136
A216L	0136
A216M	0136
A216OR	0136
A216R	0136
A216X	0136
A216Y	0136
A217(2S)	0280
A217A	0280
A217B	0280
A217C	0280
A217D	0280
A217E	0280
A217F	0280
A217G	0280
A217GN	0280
A217H	0280
A217J	0280
A217K	0280
A217L	0280
A217M	0280
A217OR	0280
A217R	0280
A217X	0280
A217Y	0280
A218(2S)	0050
A218A	0050
A218B	0050
A218C	0050
A218D	0050
A218E	0050
A218F	0050
A218G	0050
A218GN	0050
A218H	0050
A218J	0050
A218K	0050
A218L	0050
A218M	0050
A218OR	0050
A218R	0050
A218X	0050
A218Y	0050
A219(2S)	0136
A219A	0136
A219B	0136
A219C	0136
A219D	0136
A219E	0136
A219F	0136
A219G	0136
A219GN	0136
A219H	0136
A219J	0136
A219K	0136
A219L	0136
A219M	0136
A219OR	0136
A219R	0136
A219X	0136
A219Y	0136
A220	0136
A220A	0136
A220B	0136
A220C	0136
A220D	0136
A220E	0136
A220F	0136
A220G	0136
A220GN	0136
A220H	0136
A220J	0136
A220K	0136
A220L	0136
A220M	0136
A220OR	0136
A220R	0136
A220X	0136
A220Y	0136
A221	0136
A221(2S)	0136
A221-OR	0136
A221A	0136
A221B	0136
A221C	0136
A221D	0136
A221E	0136
A221F	0136
A221G	0136
A221GN	0136
A221H	0136
A221J	0136
A221K	0136
A221L	0136
A221M	0136
A221OR	0136
A221R	0136
A221X	0136
A221Y	0136
A222(2S)	0136
A222A	0136
A222B	0136
A222C	0136
A222D	0136
A222E	0136
A222F	0136
A222G	0136
A222GN	0136
A222H	0136
A222J	0136
A222K	0136
A222L	0136
A222M	0136
A222OR	0136
A222R	0136
A222X	0136
A222Y	0136
A223(2S)	0136
A223A	0136
A223B	0136
A223C	0136
A223D	0136
A223E	0136
A223F	0136
A223G	0136
A223GN	0136
A223H	0136
A223J	0136
A223K	0136
A223L	0136
A223M	0136
A223OR	0136
A223R	0136
A223X	0136
A223Y	0136
A224(2S)	0050
A224A	0050
A224B	0050
A224C	0050
A224D	0050
A224E	0050
A224F	0050
A224G	0050
A224GN	0050
A224H	0050
A224J	0050
A224K	0050
A224L	0050
A224M	0050
A224OR	0050
A224R	0050
A224X	0050
A224Y	0050
A225	0136
A225(2S)	0633
A-225-1	0050
A225A	0633
A225B	0633
A225C	0633
A225D	0633
A225E	0633
A225F	0633
A225G	0633
A225GN	0633
A225H	0633
A225J	0633
A225K	0633
A225L	0633
A225M	0633
A225OR	0633
A225R	0633
A225X	0633
A225Y	0633
A226	0050
A226(2S)	0050
A227(2S)	0050
A227A	0050
A227B	0050
A227C	0050
A227D	0050
A227E	0050
A227F	0050
A227G	0050
A227GN	0050
A227H	0050
A227J	0050
A227K	0050
A227L	0050
A227M	0050
A227OR	0050
A227R	0050
A227X	0050
A227Y	0050
A228(2S)	0050
A229(2S)	0050
A229A	0050
A229B	0050
A229C	0050
A229D	0050
A229E	0050
A229F	0050
A229G	0050
A229GN	0050
A229H	0050
A229J	0050
A229K	0050
A229L	0050
A229OR	0050
A229R	0050
A229X	0050
A229Y	0050
A230	0050
A230(2S)	0050
A230A	0050
A230B	0050
A230C	0050
A230D	0050
A230E	0050
A230F	0050
A230G	0050
A230GN	0050
A230H	0050
A230J	OEM
A230J(2S)	0050
A230K	OEM
A230K(2S)	0050
A230L	OEM
A230L(2S)	0050
A230M	0050
A230OR	0050
A230R	0050
A230X	0050
A230Y	0050
A-231	0023
A231(2S)	0847
A-231-47049	OEM
A232(2S)	0847
A233	0050
A233(2S)	0050
A-233-2032	OEM
A233A(2S)	0050
A233B(2S)	0050
A233C(2S)	0050
A233D	0050
A233E	0050
A233F	0050
A233G	0050
A233GN	0050
A233H	0050
A233J	0050
A233K	0050
A233L	0050
A233M	0050
A233OR	0050
A233R	0050
A233X	0050
A233Y	0050
A234(2S)	0050
A234A(2S)	0050
A234B	0050
A234C	0050
A234D	0050
A234E	0050
A234F	0050
A234G	0050
A234GN	0050
A234H	0050
A234J	0050
A234K	0050
A234L	0050
A234M	0050
A234OR	0050
A234R	0050
A234X	0050
A234Y	0050
A235	0050
A235(2S)	0050
A235A	0050
A235B(2S)	0050
A235C	0050
A235D	0050
A235E	0050
A235F	0050
A235G	0050
A235GN	0050
A235H	0050
A235K	0050
A235M	0050
A235OR	0050
A235R	0050
A235X	0050
A235Y	0050
A236(2S)	0136
A236A	0136
A236B	0136
A236C	0136
A236D	0136
A236E	0136
A236F	0136
A236G	0136
A236GN	0136
A236H	0136
A236J	0136
A236K	0136
A236L	0136
A236M	0136
A236OR	0136
A236R	0136
A236X	0136
A236Y	0136
A237	0136
A237(2S)	0136
A237A	0136
A237B	0136
A237C	0136
A237D	0136
A237E	0136
A237F	0136
A237G	0136
A237GN	0136
A237H	0136
A237J	0136
A237K	0136
A237L	0136
A237M	0136
A237OR	0136
A237R	0136
A237X	0136
A237Y	0136
A238	0136
A238(2S)	0136
A238A	0136
A238B	0136
A238C	0136
A238D	0136
A238E	0136
A238F	0136
A238G	0136
A238GN	0136
A238H	0136
A238J	0136
A238K	0136
A238L	0136
A239(2S)	0050
A239A	0050
A239B	0050
A239C	0050
A239D	0050
A239E	0050
A239F	0050
A239G	0050
A239GN	0050
A239GREEN	0050
A239H	0050
A239J	0050
A239K	0050
A239L	0050
A239M	0050
A239OR	0050
A239R	0050
A239RED	0050
A239X	0050
A239Y	0050
A240	0050
A240(2S)	0050
A240A(2S)	0050
A240B(2S)	0050
A240B2(2S)	0050
A240BL(2S)	0050
A240C	0050
A240D	0050
A240E	0050
A240F	0050
A240G	0050
A240GN	0050
A240GREEN	0050
A240H	0050
A240J	0050
A240K	0050
A240L	0050
A240M	0050
A240OR	0050
A240R	0050
A240RED	0050
A240X	0050
A240Y	0050
A241	0050
A241(2S)	0050
A241A	0050
A241B	0050
A241C	0050
A241D	0050
A241E	0050
A241F	0050
A241G	0050
A241GN	0050
A241H	0050
A241J	0050
A241K	0050
A241L	0050
A241M	0050
A241OR	0050
A241R	0050
A241X	0050
A241Y	0050
A242(2S)	0050
A243(2S)	0050
A243A	0050
A244	0050
A244(AMC)	0007
A244A	0050
A245	0144
A245(2S)	0050
A245(AMC)	0144
A246	0016
A246(2S)	0136
A246(AMC)	0016
A246A	0136
A246B	0136
A246C	0136
A246D	0136
A246E	0136
A246F	0136
A246G	0136
A246GN	0136
A246H	0136
A246J	0136
A246K	0136
A246L	0136
A246M	0136
A246OR	0136
A246R	0136
A246V(2S)	0136
A246X	0136
A246Y	0136
A247	0233
A247(2S)	0136
A247(AMC)	0233
A247A	0136
A247B	0136
A247C	0136
A247D	0136
A247E	0136
A247F	0136
A247G	0136
A247GN	0136
A247H	0136
A247J	0136
A247K	0136
A247L	0136
A247M	0136
A247OR	0136
A247R	0136
A247X	0136
A247Y	0136
A248	0004
A248(AMC)	0079
A248A	0004
A248B	0004
A248C	0004
A248D	0004
A248E	0004
A248F	0004
A248G	0004
A248GN	0004
A248H	0004
A248J	0004
A248K	0004
A248L	0004
A248M	0004
A248OR	0004
A248R	0004
A248X	0004
A248Y	0004
A249	0016
A249(2S)	0164
A249(AMC)	0086
A250(2S)	0136
A250A	0136
A250B	0136
A250C	0136
A250D	0136
A250E	0136
A250F	0136
A250G	0136
A250GN	0136
A250H	0136
A250J	0136
A250L	0136
A250M	0136
A250OR	0136
A250X	0136
A250Y	0136
A251(2S)	0136
A251A	0136
A251B	0136
A251C	0136
A251D	0136
A251E	0136
A251F	0136
A251G	0136
A251H	0136
A251J	0136
A251K	0136
A251L	0136
A251M	0136
A251OR	0136
A251R	0136
A251X	0136
A251Y	0136
A252(2S)	0136
A252A	0136
A252B	0136
A252C	0136
A252D	0136
A252E	0136
A252F	0136
A252G	0136
A252H	0136
A252J	0136
A252K	0136
A252L	0136
A252M	0136
A252OR	0136
A252R	0136
A252X	0136
A252Y	0136
A253	0560
A253(2S)	0136
A253A	0136
A253B	0136
A253C	0136
A253D	0136
A253E	0136
A253F	0136
A253G	0136
A253GN	0136
A253H	0136
A253J	0136
A253K	0136
A253L	0136
A253M	0136
A253OR	0136
A253R	0136
A253X	0136
A253Y	0136
A254(2S)	0136
A254A	0136
A254B	0136
A254C	0136
A254D	0136
A254E	0136
A254F	0136
A254G	0136
A254H	0136
A254J	0136
A254K	0136
A254L	0136
A254M	0136
A254OR	0136
A254R	0136
A254X	0136
A254Y	0136
A255(2S)	0136
A255A	0136
A255B	0136
A255C	0136
A255D	0136
A255E	0136
A255F	0136
A255G	0136
A255GN	0136
A255H	0136
A255J	0136
A255K	0136
A255L	0136
A255M	0136
A255OR	0136
A255R	0136
A255X	0136
A255Y	0136
A256(2S)	0136
A256A	0136
A256B	0136
A256C	0136
A256D	0136
A256E	0136
A256F	0136
A256G	0136
A256GN	0136
A256H	0136
A256J	0136
A256K	0136
A256L	0136
A256M	0136
A256OR	0136
A256R	0136
A256X	0136
A256Y	0136
A257(2S)	0136
A257A	0136

If replacement code is OEM, contact original manufacturer for replacement.

DEVICE TYPE	REPL CODE
A257B	0136
A257C	0136
A257D	0136
A257E	0136
A257F	0136
A257G	0136
A257GN	0136
A257H	0136
A257J	0136
A257K	0136
A257L	0136
A257OR	0136
A257R	0136
A257X	0136
A257Y	0136
A258(2S)	0136
A-258-1	0633
A258A	0136
A258B	0136
A258C	0136
A258D	0136
A258E	0136
A258F	0136
A258G	0136
A258GN	0136
A258H	0136
A258J	0136
A258K	0136
A258L	0136
A258M	0136
A258OR	0136
A258R	0136
A258X	0136
A258Y	0136
A-259	OEM
A259(2S)	0136
A259B	0136
A259C	0136
A259D	0136
A259E	0136
A259F	0136
A259G	0136
A259GN	0136
A259H	0136
A259J	0136
A259L	0136
A259M	0136
A259OR	0136
A259R	0136
A259X	0136
A259Y	0136
A260(2S)	0050
A260A	0050
A260B	0050
A260C	0050
A260D	0050
A260E	0050
A260F	0050
A260G	0050
A260GN	0050
A260H	0050
A260J	0050
A260K	0050
A260L	0050
A260M	0050
A260OR	0050
A260R	0050
A260X	0050
A260Y	0050
A261(2S)	0050
A261A	0050
A261B	0050
A261C	0050
A261D	0050
A261E	0050
A261F	0050
A261G	0050
A261GN	0050
A261H	0050
A261J	OEM
A261K	OEM
A261K(2S)	0050
A261L	0050
A261M	0050
A261OR	0050
A261R	0050
A261X	0050
A261Y	0050
A262(2S)	0050
A262A	0050
A262B	0050
A262C	0050
A262D	0050
A262E	0050
A262F	0050
A262G	0050
A262GN	0050
A262H	0050
A262J	0050
A262K	0050
A262L	0050
A262M	0050
A262OR	0050
A262R	0050
A262X	0050
A262Y	0050
A263(2S)	0050
A263A	0050
A263B	0050

DEVICE TYPE	REPL CODE
A263C	0050
A263D	0050
A263E	0050
A263F	0050
A263G	0050
A263GN	0050
A263H	0050
A263J	0050
A263K	0050
A263L	0050
A263M	0050
A263OR	0050
A263R	0050
A263X	0050
A263Y	0050
A264(2S)	0050
A264A	0050
A264B	0050
A264C	0050
A264D	0050
A264E	0050
A264F	0050
A264G	0050
A264GN	0050
A264H	0050
A264K	0050
A264L	0050
A264M	0050
A264OR	0050
A264R	0050
A264X	0050
A264Y	0050
A265(2S)	0050
A265A	0050
A265B	0050
A265C	0050
A265D	0050
A265E	0050
A265F	0050
A265G	0050
A265GN	0050
A265H	0050
A265J	0050
A265K	0050
A265L	0050
A265M	0050
A265OR	0050
A265R	0050
A265X	0050
A265Y	0050
A266(2S)	0136
A266A	0136
A266B	0136
A266C	0136
A266D	0136
A266E	0136
A266F	0136
A266G	0136
A266GN	0136
A266GREEN	0136
A266H	0136
A266J	0136
A266K	0136
A266L	0136
A266M	0136
A266OR	0136
A266R	0136
A266X	0136
A266Y	0136
A267	0136
A267(2S)	0136
A267A	0136
A267B	0136
A267C	0136
A267D	0136
A267E	0136
A267F	0136
A267G	0136
A267GN	0136
A267H	0136
A267J	0136
A267K	0136
A267L	0136
A267M	0136
A267OR	0136
A267R	0136
A267X	0136
A267Y	0136
A268(2S)	0136
A268A	0136
A268B	0136
A268C	0136
A268D	0136
A268E	0136
A268F	0136
A268G	0136
A268GN	0136
A268H	0136
A268J	0136
A268K	0136
A268L	0136
A268M	0136
A268OR	0136
A268R	0136
A268X	0136
A268Y	0136
A-269	0633
A269(2S)	0136
A269A	0136

DEVICE TYPE	REPL CODE
A269B	0136
A269C	0136
A269D	0136
A269E	0136
A269F	0136
A269GN	0136
A269H	0136
A269J	0136
A269K	0136
A269L	0136
A269M	0136
A269OR	0136
A269X	0136
A269Y	0136
A270	0698
A270(2S)	0136
A270A	0136
A270B	0136
A270C	0136
A270D	0136
A270E	0136
A270F	0136
A270G	0136
A270GN	0136
A270H	0136
A270J	0136
A270K	0136
A270L	0136
A270M	0136
A270OR	0136
A270R	0136
A270X	0136
A270Y	0136
A271	0560
A271(2)	0136
A271(2S)	0136
A271(3)	0136
A271A	0136
A271B	0136
A271C	0136
A271D	0136
A271E	0136
A271F	0136
A271G	0136
A271GN	0136
A271H	0136
A271J	0136
A271K	0136
A271L	0136
A271M	0136
A271OR	0136
A271R	0136
A271X	0136
A271Y	0136
A272	0042
A272(2S)	0136
A272A	0136
A272B	0136
A272C	0136
A272D	0136
A272E	0136
A272F	0136
A272G	0136
A272GN	0136
A272H	0136
A272J	0136
A272K	0136
A272L	0136
A272M	0136
A272OR	0136
A272R	0136
A272X	0136
A272Y	0136
A273	0042
A273(2S)	0136
A273A	0136
A273B	0136
A273C	0136
A273D	0136
A273E	0136
A273F	0136
A273G	0136
A273GN	0136
A273H	0136
A273J	0136
A273K	0136
A273L	0136
A273M	0136
A273OR	0136
A273R	0136
A273X	0136
A273Y	0136
A274	OEM
A274(2S)	0136
A274A	0136
A274B	0136
A274C	0136
A274D	0136
A274E	0136
A274F	0136
A274G	0136
A274GN	0136
A274H	0136
A274J	0136
A274K	0136
A274L	0136
A274M	0136
A274OR	0136
A274R	0136
A274X	0136

DEVICE TYPE	REPL CODE
A274Y	0136
A275	0560
A275(2S)	0136
A275A	0136
A275B	0136
A275C	0136
A275D	0136
A275E	0136
A275F	0136
A275G	0136
A275GN	0136
A275H	0136
A275J	0136
A275L	0136
A275M	0136
A275OR	0136
A275R	0136
A275X	0136
A275Y	0136
A276	0042
A276(2S)	0050
A277	0042
A277(2S)	0280
A277A	0280
A277B	0280
A277C	0280
A277D	0280
A277E	0280
A277F	0280
A277G	0280
A277GN	0280
A277H	0280
A277J	0280
A277K	0280
A277L	0280
A277M	0280
A277OR	0280
A277X	0280
A277Y	0280
A278	0280
A278A	0280
A278B	0280
A278C	0280
A278D	0280
A278E	0280
A278F	0280
A278G	0280
A278GN	0280
A278J	0280
A278K	0280
A278L	0280
A278M	0280
A278OR	0280
A278R	0280
A278X	0280
A278Y	0280
A279	0280
A279A	0280
A279B	0280
A279C	0280
A279D	0280
A279E	0280
A279F	0280
A279G	0280
A279GN	0280
A279H	0280
A279J	0280
A279K	0280
A279L	0280
A279OR	0280
A279R	0280
A279X	0280
A279Y	0280
A280(2S)	0280
A280A	0280
A280B	0280
A280C	0280
A280D	0280
A280E	0280
A280F	0280
A280G	0280
A280GN	0280
A280H	0280
A280J	0280
A280K	0280
A280L	0280
A280M	0280
A280OR	0280
A280R	0280
A280X	0280
A280Y	0280
A281(2S)	0280
A281A	0280
A281B	0280
A281C	0280
A281D	0280
A281E	0280
A281F	0280
A281G	0280
A281GN	0280
A281H	0280
A281J	0280
A281K	0280
A281L	0280
A281M	0280
A281OR	0280
A281R	0280
A281X	0280
A281Y	0280

DEVICE TYPE	REPL CODE
A282(2S)	0004
A282A	0004
A282B	0004
A282C	0004
A282D	0004
A282E	0004
A282F	0004
A282G	0004
A282GN	0004
A282H	0004
A282J	0004
A282K	0004
A282L	0004
A282M	0004
A282OR	0004
A282R	0004
A282X	0004
A282Y	0004
A283(2S)	0280
A283A	0280
A283B	0280
A283C	0280
A283D	0280
A283E	0280
A283F	0280
A283G	0280
A283GN	0280
A283H	0280
A283J	0280
A283K	0280
A283L	0280
A283M	0280
A283OR	0280
A283R	0280
A283X	0280
A283Y	0280
A284(2S)	0280
A284A	0280
A284B	0280
A284C	0280
A284D	0280
A284E	0280
A284F	0280
A284G	0280
A284GN	0280
A284J	0280
A284K	0280
A284L	0280
A284M	0280
A284OR	0280
A284R	0280
A284X	0280
A284Y	0280
A285(2S)	0050
A285B	0050
A285C	0050
A285E	0050
A285F	0050
A285G	0050
A285GN	0050
A285H	0050
A285J	0050
A285L	0050
A285M	0050
A285OR	0050
A285R	0050
A285X	0050
A285Y	0050
A286(2S)	0050
A286A	0050
A286B	0050
A286C	0050
A286D	0050
A286E	0050
A286F	0050
A286G	0050
A286GN	0050
A286H	0050
A286K	0050
A286M	0050
A286OR	0050
A286R	0050
A286X	0050
A286Y	0050
A287(2S)	0050
A287A	0050
A287B	0050
A287C	0050
A287D	0050
A287E	0050
A287F	0050
A287G	0050
A287GN	0050
A287H	0050
A287J	0050
A287K	0050
A287L	0050
A287M	0050
A287OR	0050
A287X	0050
A287Y	0050
A288(2S)	0136
A288A	0136
A288B	0136
A288C	0136
A288D	0136
A288E	0136
A288F	0136
A288G	0136

DEVICE TYPE	REPL CODE
A288GN	0136
A288J	0136
A288K	0136
A288L	0136
A288M	0136
A288OR	0136
A288X	0136
A288Y	0136
A289(2S)	0136
A289A	0136
A289B	0136
A289C	0136
A289D	0136
A289E	0136
A289F	0136
A289G	0136
A289GN	0136
A289H	0136
A289J	0136
A289K	0136
A289L	0136
A289M	0136
A289OR	0136
A289R	0136
A289X	0136
A289Y	0136
A290(2S)	0136
A290A	0136
A290B	0136
A290C	0136
A290D	0136
A290E	0136
A290F	0136
A290G	0136
A290GN	0136
A290H	0136
A290J	0136
A290K	0136
A290L	0136
A290M	0136
A290OR	0136
A290R	0136
A290X	0136
A290Y	0136
A291	0136
A291A	0136
A291B	0136
A291C	0136
A291D	0136
A291E	0136
A291F	0136
A291G	0136
A291GN	0136
A291H	0136
A291J	0136
A291K	0136
A291L	0136
A291M	0136
A291OR	0136
A291PC	OEM
A291PD	OEM
A291PE	OEM
A291PM	OEM
A291PN	OEM
A291PS	OEM
A291R	0136
A291X	0136
A291Y	0136
A292(2S)	0136
A292A	0136
A292B	0136
A292C	0136
A292D	0136
A292E	0136
A292F	0136
A292G	0136
A292GN	0136
A292H	0136
A292J	0136
A292K	0136
A292L	0136
A292M	0136
A292OR	0136
A292PC	OEM
A292PD	OEM
A292PE	OEM
A292PM	OEM
A292PN	OEM
A292PS	OEM
A292R	0136
A292X	0136
A292Y	0136
A293(2S)	0136
A293A	0136
A293B	0136
A293C	0136
A293D	0136
A293E	0136
A293G	0136
A293GN	0136
A293H	0136
A293J	0136
A293K	0136
A293L	0136
A293M	0136
A293OR	0136
A293R	0136
A293X	0136
A293Y	0136

DEVICE TYPE	REPL CODE
A294(2S)	0136
A294A	0136
A294B	0136
A294C	0136
A294D	0136
A294E	0136
A294F	0136
A294GN	0136
A294H	0136
A294J	0136
A294K	0136
A294L	0136
A294M	0136
A294OR	0136
A294R	0136
A294X	0136
A294Y	0136
A295(2S)	0136
A295A	0136
A295B	OEM
A295B(TRANS)	0136
A295C	0136
A295C(TRANS)	0136
A295D	0136
A295D(TRANS)	0136
A295E	0136
A295E(TRANS)	0136
A295F	0136
A295G	0136
A295GN	0136
A295H	0136
A295J	0136
A295K	0136
A295L	0136
A295M	OEM
A295N	OEM
A295OR	0136
A295P	OEM
A295PA	OEM
A295PB	OEM
A295PC	OEM
A295PD	OEM
A295PE	OEM
A295PM	OEM
A295PN	OEM
A295PS	OEM
A295R	0136
A295S	OEM
A295T	OEM
A295Y	0136
A296(2S)	0136
A296A	0136
A296B	OEM
A296B(TRANS)	0136
A296C	OEM
A296C(TRANS)	0136
A296D	0136
A296D(TRANS)	0136
A296E	0136
A296E(TRANS)	0136
A296F	0136
A296G	0136
A296GN	0136
A296H	0136
A296J	0136
A296K	0136
A296L	0136
A296M	OEM
A296M(TRANS)	0136
A296N	OEM
A296OR	0136
A296P	OEM
A296PA	OEM
A296PB	OEM
A296PC	OEM
A296PD	OEM
A296PE	OEM
A296PM	OEM
A296PN	OEM
A296PS	OEM
A296R	0136
A296S	OEM
A296T	OEM
A296X	0136
A296Y	0136
A297(2S)	0136
A297A	0136
A297B	0136
A297C	0136
A297D	0136
A297E	0136
A297F	0136
A297G	0136
A297H	0136
A297J	0136
A297K	0136
A297L	0136
A297L012C01	0037
A297M	0136
A297OR	0136
A297R	0136
A297V073C01	0126
A297V073C02	0126
A297V073C03	0037
A297V073C04	0037
A297X	0136
A297Y	0136
A298	0050
A298(2S)	0050

If replacement code is OEM, contact original manufacturer for replacement.

DEVICE TYPE	REPL CODE
A298A	0050
A298B	0050
A298C	0050
A298D	0050
A298E	0050
A298F	0050
A298G	0050
A298GN	0050
A298H	0050
A298J	0050
A298K	0050
A298L	0050
A298M	0050
A298OR	0050
A298R	0050
A298X	0050
A298Y	0050
A299	0280
A300	0015
A300(TRANS)	0050
A300PA	OEM
A301	0016
A301(2S)	0164
A301(JAPAN)	0136
A301A	0164
A301B	0164
A301C	0164
A301D	0164
A301E	0164
A301F	0164
A301G	0164
A301GN	0164
A301H	0164
A301J	0164
A301K	0164
A301L	0164
A301M	0164
A301OR	0164
A301R	0164
A301X	0164
A301Y	0164
A302	0004
A302(2S)	0004
A303	0004
A303(2S)	0004
A304(2S)	0280
A304-GREEN	0280
A304-RED	0280
A304-YELLOW	0280
A304A	0280
A304B	0280
A304C	0280
A304D	0280
A304F	0280
A304G	0280
A304GN	0280
A304J	0280
A304K	0280
A304L	0280
A304M	0280
A304OR	0280
A304R	0280
A304X	0280
A304Y	0280
A305(2S)	0280
A305-GREEN	0280
A305-RED	0280
A305-YELLOW	0280
A305A	0280
A305B	0280
A305C	0280
A305D	0280
A305E	0280
A305F	0280
A305G	0280
A305GN	0280
A305H	0280
A305J	0280
A305K	0280
A305L	0280
A305M	0280
A305OR	0280
A305R	0280
A305X	0280
A305Y	0280
A306	0016
A306(2S)	0050
A307	0016
A307(2S)	0050
A307A	0050
A307B	0050
A307C	0050
A307D	0050
A307E	0050
A307F	0050
A307G	0050
A307GN	0050
A307J	0050
A307K	0050
A307L	0050
A307M	0050
A307OR	0050
A307R	0050
A307X	0050
A307Y	0050
A308(2S)	0136
A308A	0136
A308B	0136
A308C	0136
A308D	0136
A308E	0136
A308F	0136
A308G	0136
A308GN	0136
A308H	0136
A308J	0136
A308K	0136
A308L	0136
A308M	0136
A308OR	0136
A308R	0136
A308X	0136
A308Y	0136
A309(2S)	1785
A309A	1785
A309B	1785
A309C	1785
A309D	1785
A309E	1785
A309F	1785
A309G	1785
A309GN	1785
A309H	1785
A309J	1785
A309K	1785
A309L	1785
A309M	1785
A309OR	1785
A309R	1785
A309X	1785
A309Y	1785
A310	0233
A310(2S)	0136
A310A	0136
A310B	0136
A310C	0136
A310D	0136
A310E	0136
A310F	0136
A310G	0136
A310GN	0136
A310H	0136
A310J	0136
A310K	0136
A310L	0136
A310M	0136
A310OR	0136
A310R	0136
A310X	0136
A310Y	0136
A311	0004
A311(2S)	0004
A311A	0004
A311B	0004
A311C	0004
A311D	0004
A311E	0004
A311F	0004
A311G	0004
A311GN	0004
A311H	0004
A311J	0004
A311K	0004
A311L	0004
A311M	0004
A311OR	0004
A311R	0004
A311X	0004
A311Y	0004
A312(2S)	0004
A312A	0004
A312B	0004
A312C	0004
A312D	0004
A312E	0004
A312F	0004
A312G	0004
A312GN	0004
A312H	0004
A312J	0004
A312K	0004
A312L	0004
A312M	0004
A312OR	0004
A312R	0004
A312X	0004
A312Y	0004
A313(2S)	0136
A313-BLUE	0136
A313-GREEN	0136
A313-RED	0136
A313-YELLOW	0136
A313A	0136
A313B	0136
A313C	0136
A313D	0136
A313E	0136
A313F	0136
A313G	0136
A313GN	0136
A313H	0136
A313J	0136
A313K	0136
A313L	0136
A313M	0136
A313OR	0136
A313R	0136
A313X	0136
A313Y	0136
A314(2S)	0136
A314-GREEN	0136
A314-RED	0136
A314-YELLOW	0136
A314A	0136
A314B	0136
A314C	0136
A314D	0136
A314E	0136
A314F	0136
A314G	0136
A314GN	0136
A314H	0136
A314J	0136
A314K	0136
A314L	0136
A314M	0136
A314OR	0136
A314R	0136
A314X	0136
A314Y	0136
A315(2S)	0136
A315-GREEN	0136
A315-RED	0136
A315-YELLOW	0136
A315A	0136
A315A(DIO)	OEM
A315B	0136
A315B(DIO)	OEM
A315C	0136
A315D	0136
A315E	0136
A315F	0136
A315F(DIO)	OEM
A315G	0136
A315G(DIO)	OEM
A315GN	0136
A315H	0136
A315J	0136
A315K	0136
A315L	0136
A315M	0136
A315OR	0136
A315R	0136
A315X	0136
A315Y	0136
A316(2S)	0136
A316-GREEN	0136
A316-RED	0136
A316-YELLOW	0136
A316A	0136
A316B	0136
A316C	0136
A316D	0136
A316E	0136
A316F	0136
A316G	0136
A316GN	0136
A316H	0136
A316J	0136
A316K	0136
A316L	0136
A316M	0136
A316OR	0136
A316R	0136
A316X	0136
A316Y	0136
A321	0004
A321(2S)	0004
A321-1	0004
A321A	0004
A321B	0004
A321C	0004
A321D	0004
A321E	0004
A321F	0004
A321G	0004
A321GN	0004
A321H	0004
A321J	0004
A321K	0004
A321L	0004
A321M	0004
A321OR	0004
A321R	0004
A321X	0004
A321Y	0004
A322	0004
A322(2S)	0004
A322A	0004
A322B	0004
A322C	0004
A322D	0004
A322E	0004
A322F	0004
A322G	0004
A322GN	0004
A322H	0004
A322K	0004
A322L	0004
A322M	0004
A322OR	0004
A322R	0004
A322X	0004
A322Y	0004
A323	0016
A323(2S)	0050
A323A	0050
A323B	0050
A323C	0050
A323D	0050
A323E	0050
A323F	0050
A323G	0050
A323GN	0050
A323J	0050
A323K	0050
A323L	0050
A323M	0050
A323R	0050
A323X	0050
A323Y	0050
A324	0136
A324(2S)	0136
A324A	0136
A324B	0136
A324C	0136
A324D	0136
A324E	0136
A324F	0136
A324G	0136
A324GN	0136
A324H	0136
A324K	0136
A324L	0136
A324M	0136
A324OR	0136
A324R	0136
A324X	0136
A324Y	0136
A325(2S)	0136
A325A	0136
A325B	0136
A325D	0136
A325E	0136
A325G	0136
A325GN	0136
A325H	0136
A325J	0136
A325K	0136
A325L	0136
A325M	0136
A325OR	0136
A325R	0136
A325X	0136
A325Y	0136
A326(2S)	0136
A326A	0136
A326B	0136
A326C	0136
A326D	0136
A326E	0136
A326F	0136
A326G	0136
A326GN	0136
A326H	0136
A326J	0136
A326K	0136
A326L	0136
A326M	0136
A326OR	0136
A326R	0136
A326X	0136
A326Y	0136
A327	0050
A327A	OEM
A327A2	OEM
A327B	OEM
A327B2	OEM
A327C	OEM
A327C2	OEM
A327F	OEM
A327F2	OEM
A328	0280
A329(2S)	0136
A329A(2S)	0136
A329B(2S)	0136
A329C	0136
A329D	0136
A329E	0136
A329F	0136
A329G	0136
A329GN	0136
A329H	0136
A329J	0136
A329K	0136
A329L	0136
A329M	0136
A329OR	0136
A329R	0136
A329X	0136
A329Y	0136
A330(2S)	0136
A330A	0136
A330B	0136
A330C	0136
A330D	0136
A330E	OEM
A330E(TRANS)	0136
A330F	0136
A330G	0136
A330GN	0136
A330H	0136
A330J	0136
A330K	0136
A330L	0136
A330M	OEM
A330M(TRANS)	0136
A330N	OEM
A330OR	0136
A330P	OEM
A330PA	OEM
A330PB	OEM
A330PC	OEM
A330PD	OEM
A330PE	OEM
A330S	OEM
A330T	OEM
A330X	0136
A330Y	0136
A331(2S)	0136
A331A	0136
A331B	0136
A331C	0136
A331D	0136
A331E	0136
A331F	0136
A331G	0136
A331GN	0136
A331H	0136
A331J	0136
A331K	0136
A331L	0136
A331M	0136
A331OR	0136
A331R	0136
A331X	0136
A331Y	0136
A332(2S)	0050
A332A	0050
A332B	0050
A332C	0050
A332D	0050
A332E	0050
A332F	0050
A332G	0050
A332GN	0050
A332H	0050
A332J	0050
A332K	0050
A332L	0050
A332M	0050
A332OR	0050
A332R	0050
A332X	0050
A332Y	0050
A333	5213
A334	5213
A335(2S)	0136
A335A	0136
A335B	0136
A335C	0136
A335D	0136
A335E	0136
A335F	0136
A335G	0136
A335GN	0136
A335H	0136
A335J	0136
A335K	0136
A335L	0136
A335M	0136
A335OR	0136
A335R	0136
A335X	0136
A335Y	0136
A336	0050
A337(2S)	0136
A337A	OEM
A337A(TRANS)	0136
A337B	OEM
A337B(TRANS)	0136
A337C	OEM
A337C(TRANS)	0136
A337D	OEM
A337D(TRANS)	0136
A337E	OEM
A337E(TRANS)	0136
A337F	0136
A337G	0136
A337GN	0136
A337H	0136
A337J	0136
A337K	0136
A337L	0136
A337M	OEM
A337M(TRANS)	0136
A337N	OEM
A337OR	0136
A337P	OEM
A337PA	OEM
A337PB	OEM
A337PC	OEM
A337PD	OEM
A337PE	OEM
A337R	0136
A337S	OEM
A337T	OEM
A337X	0136
A337Y	0136
A338(2S)	0136
A338A	0136
A338B	0136
A338C	0136
A338D	0136
A338E	0136
A338F	0136
A338G	0136
A338GN	0136
A338H	0136
A338J	0136
A338K	0136
A338L	0136
A338M	0136
A338OR	0136
A338R	0136
A338X	0136
A338Y	0136
A339(2S)	0136
A339A	0136
A339B	0136
A339C	0136
A339D	0136
A339E	0136
A339F	0136
A339G	0136
A339GN	0136
A339H	0136
A339J	0136
A339K	0136
A339L	0136
A339M	0136
A339OR	0136
A339R	0136
A339Y	0136
A340(2S)	0050
A340A	OEM
A340A2	OEM
A340B	OEM
A340B2	OEM
A340C	OEM
A340C2	OEM
A340F	OEM
A340F2	OEM
A341(2S)	0050
A341-OA	0050
A341-OB(2S)	0050
A341A	0050
A341B	0050
A341C	0050
A341E	0050
A341F	0050
A341G	0050
A341GN	0050
A341H	0050
A341J	0050
A341K	0050
A341L	0050
A341M	0050
A341OR	0050
A341R	0050
A341X	0050
A341Y	0050
A342(2S)	0050
A342A(2S)	0050
A342B	0050
A342C	0050
A342D	0050
A342E	0050
A342F	0050
A342G	0050
A342GN	0050
A342H	0050
A342J	0050
A342K	0050
A342L	0050
A342M	0050
A342OR	0050
A342R	0050
A342X	0050
A342Y	0050
A343(2S)	0050
A343A	0050
A343B	0050
A343C	0050
A343D	0050
A343E	0050
A343F	0050
A343G	0050
A343H	0050
A343J	0050
A343K	0050
A343L	0050
A343M	0050
A343OR	0050
A343R	0050
A343X	0050
A343Y	0050
A344	0016
A344(2S)	0136
A344A	0136
A344B	0136
A344C	0136
A344D	0136
A344E	0136
A344F	0136
A344G	0136
A344GN	0136
A344H	0136
A344J	0136
A344K	0136
A344L	0136
A344M	0136
A344OR	0136
A344R	0136
A344X	0136
A344Y	0136
A345	0016
A345(2S)	0050
A346	0016
A346(2S)	0050
A347(2S)	0050
A348(2S)	0050
A348A	0050
A348B	0050
A348C	0050
A348D	0050
A348E	0050
A348F	0050
A348G	0050
A348GN	0050
A348H	0050
A348J	0050
A348K	0050
A348L	0050
A348M	0050
A348OR	0050
A348R	0050
A348X	0050
A348Y	0050
A349(2S)	0050
A350(2S)	0136
A350A	0136
A350AV	0136
A350B	0136
A350BK	0136
A350C(2S)	0136
A350D	0136
A350E	0136
A350F	0136
A350G	0136
A350GN	0136
A350H(2S)	0136
A350J	0136
A350K	0136
A350L	0136
A350M	0136
A350OR	0136
A350R(2S)	0136
A350T(2S)	0136
A350TY(2S)	0136
A350X	0136
A350Y(2S)	0136
A351(2S)	0136
A351A(2S)	0136
A351A-2	0136
A351B(2S)	0136
A351C	0136
A351D	0136
A351E	0136
A351F	0136
A351G	0136
A351GN	0136
A351GR	0136
A351K	0136
A351L	0136
A351M	0136
A351OR	0136
A351R	0136
A351X	0136
A351Y	0136
A352(2S)	0136
A352A(2S)	0136
A352B(2S)	0136
A352C	0136
A352D	0136
A352E	0136
A352F	0136
A352G	0136
A352GN	0136
A352H	0136
A352J	0136
A352K	0136
A352L	0136
A352M	0136
A352OR	0136
A352R	0136
A352X	0136
A352Y	0136
A353(2S)	0136
A353-9008-001	0126
A353-AC	0136
A353A(2S)	0136
A353AL	0136
A353B	0136
A353C(2S)	0136
A353CL	0136
A353D	0136
A353E	0136
A353F	0136
A353G	0136
A353GN	0136
A353H	0136
A353J	0136
A353K	0136
A353L	0136
A353M	0136
A353OR	0136
A353R	0136
A353X	0136
A353Y	0136
A354(2S)	0136
A354-B	0136
A354A(2S)	0136

If replacement code is OEM, contact original manufacturer for replacement.

DEVICE TYPE	REPL CODE
A354B	0136
A354BK	0136
A354C	0136
A354D	0136
A354E	0136
A354F	0136
A354G	0136
A354GN	0136
A354H	0136
A354J	0136
A354K	0136
A354L	0136
A354M	0136
A354OR	0136
A354R	0136
A354X	0136
A354Y	0136
A355(2S)	0136
A355A(2S)	0136
A355B	0136
A355C	0136
A355D	0136
A355E	0136
A355F	0136
A355G	0136
A355H	0136
A355J	0136
A355K	0136
A355L	0136
A355M	0136
A355OR	0136
A355R	0136
A355X	0136
A355Y	0136
A356(2S)	0136
A356A	0136
A356B	0136
A356C	0136
A356D	0136
A356E	0136
A356F	0136
A356G	0136
A356GN	0136
A356H	0136
A356J	0136
A356K	0136
A356L	0136
A356M	0136
A356OR	0136
A356R	0136
A357(2S)	0136
A357A	0136
A357B	0136
A357C	0136
A357D	0136
A357E	0136
A357F	0136
A357G	0136
A357GN	0136
A357H	0136
A357J	0136
A357K	0136
A357L	0136
A357M	0136
A357OR	0136
A357R	0136
A357X	0136
A357Y	0136
A358(2S)	0136
A358-3	0136
A358A	0136
A358B	0136
A358C	0136
A358D	0136
A358E	0136
A358F	0136
A358G	0136
A358GN	0136
A358H	0136
A358J	0136
A358K	0136
A358L	0136
A358M	0136
A358OR	0136
A358R	0136
A358X	0136
A358Y	0136
A359(2S)	0136
A359A	0136
A359B	0136
A359C	0136
A359D	0136
A359E	0136
A359F	0136
A359G	0136
A359GN	0136
A359H	0136
A359J	0136
A359K	0136
A359L	0136
A359M	0136
A359OR	0136
A359R	0136
A359X	0136
A359Y	0136
A360(2S)	0050
A360A	0050
A360B	0050
A360C	0050
A360E	0050
A360F	0050
A360G	0050
A360GN	0050
A360H	0050
A360J	0050
A360K	0050
A360L	0050
A360M	0050
A360OR	0050
A360R	0050
A360X	0050
A360Y	0050
A361(2S)	0050
A361A	0050
A361B	0050
A361C	0050
A361D	0050
A361E	0050
A361F	0050
A361G	0050
A361GN	0050
A361H	0050
A361J	0050
A361K	0050
A361L	0050
A361M	0050
A361OR	0050
A361R	0050
A361X	0050
A361Y	0050
A362(2S)	0050
A363(2S)	0050
A364(2S)	0136
A364A	0136
A364B	0136
A364C	0136
A364D	0136
A364E	0136
A364F	0136
A364G	0136
A364GN	0136
A364H	0136
A364J	0136
A364K	0136
A364L	0136
A364M	0136
A364OR	0136
A364R	0136
A364X	0136
A364Y	0136
A365(2S)	0136
A365A	0136
A365B	0136
A365C	0136
A365D	0136
A365E	0136
A365F	0136
A365G	0136
A365GN	0136
A365H	0136
A365J	0136
A365K	0136
A365L	0136
A365M	0136
A365OR	0136
A365R	0136
A365X	0136
A365Y	0136
A366(2S)	0136
A366A	0136
A366B	0136
A366C	0136
A366D	0136
A366E	0136
A366F	0136
A366G	0136
A366GN	0136
A366H	0136
A366J	0136
A366K	0136
A366L	0136
A366M	0136
A366OR	0136
A366R	0136
A366X	0136
A366Y	0136
A367(2S)	0136
A367A	0136
A367B	0136
A367C	0136
A367D	0136
A367E	0136
A367F	0136
A367G	0136
A367GN	0136
A367H	0136
A367J	0136
A367K	0136
A367L	0136
A367M	0136
A367OR	0136
A367R	0136
A367X	0136
A367Y	0136
A368(2S)	0136
A368A	0136
A368B	0136
A368C	0136
A368D	0136
A368E	0136
A368F	0136
A368G	0136
A368GN	0136
A368H	0136
A368J	0136
A368K	0136
A368L	0136
A368M	0136
A368OR	0136
A368R	0136
A368X	0136
A368Y	0136
A369(2S)	0136
A369A	0136
A369B	0136
A369C	0136
A369D	0136
A369E	0136
A369F	0136
A369G	0136
A369GN	0136
A369H	0136
A369J	0136
A369K	0136
A369L	0136
A369M	0136
A369OR	0136
A369R	0136
A369X	0136
A369Y	0136
A370	0164
A371	0164
A372(2S)	0050
A373(2S)	0004
A373A	0004
A374(2S)	0004
A375(2S)	0004
A375A	0004
A375B	0004
A375C	0004
A375D	0004
A375E	0004
A375F	0004
A375G	0004
A375GN	0004
A375H	0004
A375J	0004
A375K	0004
A375L	0004
A375M	0004
A375OR	0004
A375R	0004
A375X	0004
A375Y	0004
A376	0136
A376A	0136
A376B	0136
A376C	0136
A376D	0136
A376E	0136
A376F	0136
A376GN	0136
A376H	0136
A376J	0136
A376K	0136
A376L	0136
A376OR	0136
A376R	0136
A376X	0136
A376Y	0136
A377(2S)	0050
A377A	0050
A377B	0050
A377C	0050
A377D	0050
A377E	0050
A377F	0050
A377G	0050
A377GN	0050
A377H	0050
A377J	0050
A377K	0050
A377L	0050
A377M	0050
A377OR	0050
A377R	0050
A377X	0050
A377Y	0050
A378(2S)	0050
A379(2S)	0050
A380(2S)	0136
A380A	0136
A380B	0136
A380C	0136
A380D	0136
A380E	0136
A380F	0136
A380G	0136
A380GN	0136
A380H	0136
A380J	0136
A380K	0136
A380L	0136
A380M	0136
A380OR	0136
A380R	0136
A380X	0136
A380Y	0136
A381(2S)	0136
A381A	0136
A381B	0136
A381D	0136
A381E	0136
A381F	0136
A381G	0136
A381GN	0136
A381H	0136
A381K	0136
A381L	0136
A381M	0136
A381OR	0136
A381R	0136
A381X	0136
A381Y	0136
A382(2S)	0136
A382A	0136
A382B	0136
A382C(2S)	0136
A382D	0136
A382E	0136
A382F	0136
A382G	0136
A382GN	0136
A382H	0136
A382J	0136
A382K	0136
A382L	0136
A382M	0136
A382OR	0136
A382R	0136
A382X	0136
A382Y	0136
A383(2S)	0136
A383A	0136
A383B	0136
A383C	0136
A383D	0136
A383E	0136
A383F	0136
A383G	0136
A383GN	0136
A383H	0136
A383J	0136
A383K	0136
A383L	0136
A383M	0136
A383OR	0136
A383R	0136
A383X	0136
A383Y	0136
A384(2S)	0136
A384A	0136
A384B	0136
A384C	0136
A384D	0136
A384E	0136
A384F	0136
A384G	0136
A384GN	0136
A384H	0136
A384J	0136
A384K	0136
A384L	0136
A384M	0136
A384OR	0136
A384R	0136
A384X	0136
A384Y	0136
A385(2S)	0136
A385A(2S)	0136
A385B	0136
A385C	0136
A385D(2S)	0136
A385E	0136
A385F	0136
A385G	0136
A385GN	0136
A385H	0136
A385J	0136
A385K	0136
A385L	0136
A385M	0136
A385OR	0136
A385R	0136
A385X	0136
A385Y	0136
A386	0050
A387	0050
A387A	OEM
A387B	OEM
A387C	OEM
A387D	OEM
A387E	OEM
A387M	OEM
A387N	OEM
A387P	OEM
A387PA	OEM
A387PB	OEM
A387S	OEM
A387T	OEM
A389	0050
A390	OEM
A390(2S)	0050
A390A	1226
A390B	1226
A390C	1226
A390D	1226
A390E	1226
A390M	1226
A390N	1204
A390P	1204
A390PA	1204
A390PB	1204
A390PC	OEM
A390PD	OEM
A390PE	OEM
A390S	1204
A390T	1204
A391(2S)	0280
A391A	0280
A391B	0280
A391C	0280
A391D	0280
A391E	0280
A391F	0280
A391G	0280
A391GN	0280
A391H	0280
A391J	0280
A391K	0280
A391L	0280
A391M	0280
A391OR	0280
A391R	0280
A391X	0280
A391Y	0280
A392	0280
A392(2S)	0280
A392A	0280
A392B	0280
A392C	0280
A392D	0280
A392E	0280
A392F	0280
A392G	0280
A392GN	0280
A392H	0280
A392J	0280
A392K	0280
A392L	0280
A392M	0280
A392OR	0280
A392R	0280
A392X	0280
A392Y	0280
A393(2S)	0280
A393A(2S)	0280
A394(2S)	0280
A394A	0280
A394B	0280
A394C	0280
A394D	0280
A394E	0280
A394F	0280
A394G	0280
A394GN	0280
A394H	0280
A394J	0280
A394K	0280
A394L	0280
A394M	0280
A394OR	0280
A394R	0280
A394X	0280
A394Y	0280
A395(2S)	0280
A395A	0280
A395B	0280
A395C	0280
A395D	0280
A395E	0280
A395F	0280
A395G	0280
A395GN	0280
A395H	0280
A395J	0280
A395K	0280
A395L	0280
A395M	0280
A395OR	0280
A395R	0280
A395X	0280
A395Y	0280
A396	0211
A396A	OEM
A396B	OEM
A396C	OEM
A396D	OEM
A396E	OEM
A396M	OEM
A396N	OEM
A396P	OEM
A396S	OEM
A396T	OEM
A397(2S)	0211
A397A	1226
A397B	1226
A397C	1226
A397D	1226
A397E	1226
A397M	1226
A397N	1204
A397P	1204
A397PA	1204
A397PB	1204
A397PC	OEM
A397PD	OEM
A397PE	OEM
A397S	1226
A397T	1204
A398(2S)	0211
A398A	1226
A398A(TRANS)	0211
A398B	1226
A398B(TRANS)	0211
A398C	1226
A398C(TRANS)	0211
A398D	1226
A398D(TRANS)	0211
A398E	1226
A398E(TRANS)	0211
A398F	0211
A398G	0211
A398GN	0211
A398H	0211
A398J	0211
A398K	0211
A398L	0211
A398M	1226
A398M(TRANS)	0211
A398N	1204
A398OR	0211
A398P	1204
A398PA	1204
A398PB	1204
A398PC	OEM
A398PD	OEM
A398PE	OEM
A398S	1204
A398T	1204
A398X	0211
A398Y	0211
A399(2S)	0211
A399A	0211
A399B	0211
A399C	0211
A399D	0211
A399E	OEM
A399E(TRANS)	0211
A399F	0211
A399G	0211
A399GN	0211
A399H	0211
A399J	0211
A399K	0211
A399L	0211
A399M	OEM
A399M(TRANS)	0211
A399N	OEM
A399OR	0211
A399P	OEM
A399PA	OEM
A399PB	OEM
A399PC	OEM
A399PD	OEM
A399S	OEM
A399T	OEM
A399X	0211
A399Y	0211
A400	0015
A400(TRANS)	0136
A400A	0136
A400B	0136
A400C	0136
A400D	0136
A400E	0136
A400F	0136
A400G	0136
A400GN	0136
A400H	0136
A400J	0136
A400K	0136
A400L	0136
A400M	0136
A400OR	0136
A400R	0136
A400X	0136
A400Y	0136
A401	0136
A401(2S)	0136
A402(2S)	0037
A402DA	OEM
A402DB	OEM
A402DC	OEM
A402DD	OEM
A402DE	OEM
A402DM	OEM
A402DN	OEM
A402DP	OEM
A402DS	OEM
A402DT	OEM
A402EA	OEM
A402EB	OEM
A402EC	OEM
A402ED	OEM
A402EE	OEM
A402EM	OEM
A402EP	OEM
A403	0050
A403(2S)	0050
A403A	0050
A403B	0050
A403C	0050
A403D	0050
A403E	0050
A403F	0050
A403G	0050
A403GN	0050
A403H	0050
A403J	0050
A403K	0050
A403L	0050
A403M	0050
A403OR	0050
A403R	0050
A403X	0050
A403Y	0050
A404(2S)	0050
A404A	0050
A404B	0050
A404C	0050
A404D	0050
A404E	0050
A404F	0050
A404G	0050
A404GN	0050
A404H	0050
A404K	0050
A404L	0050
A404M	0050
A404OR	0050
A404R	0050
A404X	0050
A404Y	0050
A405(2S)	0050
A405-0	0050
A406	0004
A406(2S)	0004
A406A	0004
A406B	0004
A406C	0004
A406D	0004
A406E	0004
A406G	0004
A406GN	0004
A406H	0004
A406J	0004
A406K	0004
A406L	0004
A406M	0004
A406OR	0004
A406R	0004
A406X	0004
A406Y	0004
A407(2S)	0004
A407A	0004
A407B	0004
A407C	0004
A407D	0004
A407E	0004
A407F	0004
A407G	0004
A407GN	0004
A407H	0004
A407J	0004
A407K	0004
A407L	0004
A407M	0004
A407OR	0004
A407R	0004
A407X	0004
A407Y	0004
A408(2S)	0136
A408A	0136
A408B	0136
A408D	0136
A408E	0136
A408F	0136
A408GN	0136
A408H	0136
A408J	0136
A408K	0136
A408L	0136
A408M	0136
A408OR	0136
A408R	0136
A408X	0136
A408Y	0136
A409(2S)	0136
A409A	0136
A409B	0136
A409C	0136
A409D	0136
A409E	0136
A409F	0136
A409G	0136
A409GN	0136
A409H	0136
A409J	0136
A409K	0136
A409L	0136
A409M	0136
A409OR	0136
A409R	0136
A409X	0136
A409Y	0136
A410(2S)	0050
A411	0050
A411(2S)	0050
A412(2S)	0136
A412A	0136
A412B	0136
A412C	0136

DEVICE TYPE	REPL CODE
A412D	0136
A412E	0136
A412F	0136
A412G	0136
A412GN	0136
A412H	0136
A412J	0136
A412K	0136
A412L	0136
A412M	0136
A412OR	0136
A412R	0136
A412X	0136
A412Y	0136
A413(2S)	0050
A414(2S)	0280
A414A	0280
A414B	0280
A414C	0280
A414D	0280
A414E	0280
A414F	0280
A414G	0280
A414GN	0280
A414H	0280
A414J	0280
A414K	0280
A414L	0280
A414M	0280
A414OR	0280
A414R	0280
A414X	0280
A-415	0079
A415	0016
A415(2S)	0280
A415A	0280
A415B	0280
A415C	0280
A415D	0280
A415E	0280
A415F	0280
A415G	0280
A415H	0280
A415J	0280
A415K	0280
A415L	0280
A415M	0280
A415OR	0280
A415R	0280
A415X	0280
A415Y	0280
A416(2S)	0160
A416A	0160
A416B	0160
A416C	0160
A416D	0160
A416E	0160
A416F	0160
A416G	0160
A416GN	0160
A416H	0160
A416J	0160
A416K	0160
A416L	0160
A416M	0160
A416OR	0160
A416R	0160
A416X	0160
A416Y	0160
A417	0211
A417(2S)	0211
A417-19	0007
A417-43	0126
A417-62	0160
A417-115	0233
A417-116	0037
A417-132	0037
A417-138	0126
A417-153	0037
A417-154	0127
A417-170	0126
A417-176	0037
A417-182	0037
A417-184	0037
A417-190	0127
A417-196	0037
A417-200	0037
A417-201	0037
A417-205	0127
A417-234	0126
A417-235	0037
A418	0144
A419	0144
A419(2S)	0050
A420	0144
A420(2S)	0050
A420A	0050
A420B	0050
A420C	0050
A420D	0050
A420E	0050
A420F	0050
A420G	0050
A420H	0050
A420J	0050
A420K	0050
A420L	0050
A420M	0050
A420OR	0050
A420R	0050
A420X	0050
A420Y	0050
A421(2S)	0050
A422(2S)	0050
A422-A	0015
A423	0280
A424	5584
A424(2S)	0280
A425(2S)	0050
A426(2S)	0050
A426GN	0050
A427	0144
A427(2S)	0136
A427A	0136
A427B	0136
A427C	0136
A427D	0136
A427E	0136
A427F	0136
A427G	0136
A427GN	0136
A427H	0136
A427J	0136
A427K	0136
A427L	0136
A427M	0136
A427OR	0136
A427R	0136
A427X	0136
A427Y	0136
A428(2S)	0136
A428A	0136
A428B	0136
A428C	0136
A428D	0136
A428E	0136
A428F	0136
A428G	0136
A428GN	0136
A428H	0136
A428J	0136
A428K	0136
A428L	0136
A428M	0136
A428OR	0136
A428R	0136
A428X	0136
A428Y	0136
A429(2S)	2741
A429-0981-12	0007
A429-O(2S)	2741
A429A	2741
A429B	2741
A429C	2741
A429D	2741
A429E	2741
A429F	2741
A429G	2741
A429GN	2741
A429G-O	2741
A429G-R	2741
A429G-Y	2741
A429H	2741
A429J	2741
A429K	2741
A429L	2741
A429M	2741
A429OR	2741
A429R(2S)	2741
A429X	2741
A430	0259
A430(2S)	0050
A430A	3768
A430B	3768
A430C	2159
A430D	2159
A430E	2159
A430M	2159
A430N	0479
A430P	0479
A430PA	0479
A430PB	0479
A430PC	OEM
A430PD	OEM
A430PE	OEM
A430PM	OEM
A430PS	OEM
A430S	0479
A430T	0479
A431	0050
A431(2S)	0050
A431A(2S)	0050
A432(2S)	0050
A432A(2S)	0050
A432B	0050
A432C	0050
A432D	0050
A432E	0050
A432F	0050
A432G	0050
A432GN	0050
A432H	0050
A432K	0050
A432L	0050
A432L(DIO)	OEM
A432LA	OEM
A432LB	OEM
A432LC	OEM
A432LD	OEM
A432LE	OEM
A432LM	OEM
A432M	0050
A432OR	0050
A432PM	OEM
A432PN	OEM
A432PT	OEM
A432R	0050
A432X	0050
A432Y	0050
A433(2S)	0050
A433A	0050
A433B	0050
A433C	0050
A433CA	OEM
A433CB	OEM
A433CP	OEM
A433D	0050
A433E	0050
A433F	0050
A433G	0050
A433GN	0050
A433H	0050
A433K	0050
A433L	0050
A433L(DIO)	OEM
A433LB	OEM
A433LD	OEM
A433LM	OEM
A433LN	OEM
A433LS	OEM
A433LT	OEM
A433M	0050
A433OR	0050
A433PM	OEM
A433PN	OEM
A433PS	OEM
A433PT	OEM
A433R	0050
A433X	0050
A433Y	0050
A434(2S)	0050
A434A	0050
A434B	0050
A434C	0050
A434D	0050
A434E	0050
A434F	0050
A434G	0050
A434GN	0050
A434H	0050
A434J	0050
A434K	0050
A434L	0050
A434M	0050
A434OR	0050
A434R	0050
A434X	0050
A434Y	0050
A435	0050
A435(2S)	0050
A435A(2S)	0050
A435B(2S)	0050
A435C	0050
A435D	0050
A435E	0050
A435F	0050
A435G	0050
A435GN	0050
A435H	0050
A435J	0050
A435L	0050
A435M	0050
A435OR	0050
A435R	0050
A435X	0050
A435Y	0050
A436(2S)	0136
A436A	0136
A436B	0136
A436C	0136
A436D	0136
A436E	0136
A436F	0136
A436G	0136
A436GN	0136
A436H	0136
A436J	0136
A436K	0136
A437(2S)	0136
A437A	2159
A437A(TRANS)	0136
A437B	2159
A437B(TRANS)	0136
A437C	2159
A437C(TRANS)	0136
A437D	2159
A437D(TRANS)	0136
A437E	2159
A437E(TRANS)	0136
A437F	0136
A437G	0136
A437GN	0136
A437H	0136
A437J	0136
A437K	0136
A437L	0136
A437M	2159
A437M(TRANS)	0136
A437N	0479
A437OR	0136
A437P	0479
A437PA	0479
A437PB	0479
A437PC	OEM
A437PD	OEM
A437PE	OEM
A437R	0136
A437S	0479
A437T	0479
A437X	0136
A437Y	0136
A438(2S)	0136
A438A	0136
A438B	0136
A438C	0136
A438D	0136
A438E	0136
A438F	0136
A438G	0136
A438GN	0136
A438H	0136
A438J	0136
A438K	0136
A438L	0136
A438M	0136
A438OR	0136
A438R	0136
A438X	0136
A438Y	0136
A440	0050
A440(2S)	0050
A440A(2S)	0050
A440AL	0050
A440B	0050
A440C	0050
A440D	0050
A440E	0050
A440F	0050
A440G	0050
A440GN	0050
A440H	0050
A440J	0050
A440K	0050
A440L	0050
A440M	0050
A440N	OEM
A440OR	0050
A440P	OEM
A440PA	OEM
A440PB	OEM
A440PC	OEM
A440PD	OEM
A440PE	OEM
A440R	0050
A440S	OEM
A440T	OEM
A440X	0050
A440Y	0050
A441	OEM
A441(2S)	0050
A442CA	OEM
A442CB	OEM
A442CP	OEM
A442L	OEM
A442LA	OEM
A442LB	OEM
A442LC	OEM
A442LD	OEM
A442LE	OEM
A442LM	OEM
A442LN	OEM
A442LS	OEM
A442LT	OEM
A442PM	OEM
A442PN	OEM
A442PT	OEM
A443	0004
A444	0004
A445	0004
A446(2S)	0211
A446A	0211
A446B	0211
A446C	0211
A446D	0211
A446E	0211
A446F	0211
A446G	0211
A446GN	0211
A446H	0211
A446J	0211
A446K	0211
A446L	0211
A446M	0211
A446OR	0211
A446R	0211
A446X	0211
A446Y	0211
A447(2S)	0136
A447A	OEM
A447A(2S)	0136
A447B	OEM
A447B(2S)	0136
A447C	OEM
A447C(2S)	0136
A447D	OEM
A447D(2S)	0136
A447E	OEM
A447E(2S)	0136
A447F	0136
A447G	0136
A447GN	0136
A447H	0136
A447J	0136
A447K	0136
A447L	0136
A447M	OEM
A447M(2S)	0136
A447N	OEM
A447OR	0136
A447P	OEM
A447PA	OEM
A447PB	OEM
A447PC	OEM
A447PD	OEM
A447PE	OEM
A447R	0136
A447S	OEM
A447T	OEM
A447X	0136
A447Y	0136
A448(2S)	0050
A449	0136
A450(2S)	0050
A450H(2S)	0050
A451	0050
A451(2S)	0050
A451H(2S)	0050
A452(2S)	0050
A452H(2S)	0050
A453(2S)	0136
A453A	0136
A453B	0136
A453C	0136
A453D	0136
A453E	0136
A453F	0136
A453G	0136
A453GN	0136
A453H	0136
A453J	0136
A453K	0136
A453L	0136
A453M	0136
A453OR	0136
A453R	0136
A453X	0136
A453Y	0136
A454	0016
A454(2S)	0136
A454A	0136
A454B	0136
A454C	0136
A454D	0136
A454E	0136
A454F	0136
A454G	0136
A454GN	0136
A454H	0136
A454J	0136
A454K	0136
A454L	0136
A454OR	0136
A454R	0136
A454X	0136
A454Y	0136
A455	0678
A455(2S)	0136
A455A	0136
A455B	0136
A455C	0136
A455D	0136
A455E	0136
A455F	0136
A455G	0136
A455GN	0136
A455H	0136
A455J	0136
A455K	0136
A455L	0136
A455M	0136
A455OR	0136
A455R	0136
A455X	0136
A455Y	0136
A456(2S)	0136
A456A	0136
A456B	0136
A456C	0136
A456D	0136
A456E	0136
A456F	0136
A456G	0136
A456GN	0136
A456H	0136
A456J	0136
A456K	0136
A456L	0136
A456M	0136
A456OR	0136
A456R	0136
A456X	0136
A456Y	0136
A457(2S)	0136
A457A	0136
A457B	0136
A457C	0136
A457D	0136
A457E	0136
A457F	0136
A457G	0136
A457GN	0136
A457H	0136
A457J	0136
A457K	0136
A457L	0136
A457M	0136
A457OR	0136
A457R	0136
A457X	0136
A457Y	0136
A458	1062
A458C	1062
A459	0004
A459Y	0004
A460(2S)	0050
A461(2S)	0050
A462(2S)	0050
A463(2S)	0050
A463C	0050
A464(2S)	0050
A465	0150
A465-181-19	0007
A466	0086
A466(2S)	0136
A466-2(2S)	0136
A466-3(2S)	0136
A466A	0136
A466B	0136
A466BLK(2S)	0136
A466BLU(2S)	0136
A466C	0136
A466D	0136
A466E	0136
A466F	0136
A466G	0136
A466GN	0136
A466H	0136
A466J	0136
A466K	0136
A466L	0136
A466M	0136
A466OR	0136
A466R	0136
A466X	0136
A466Y	0136
A466YEL(2S)	0136
A467	0006
A467(2S)	0006
A467A	0006
A467B	0006
A467C	0006
A467D	0006
A467E	0006
A467F	0006
A467G(2S)	0006
A467GN	0006
A467G-O(2S)	0006
A467G-R(2S)	0006
A467G-Y(2S)	0006
A467H	0006
A467J	0006
A467K	0006
A467L	0006
A467M	0006
A467OR	0006
A467R	0006
A467X	0006
A467Y	0006
A468(2S)	0136
A468A	0136
A468B	0136
A468C	0136
A468D	0136
A468E	0136
A468F	0136
A468G	0136
A468GN	0136
A468H	0136
A468J	0136
A468K	0136
A468L	0136
A468M	0136
A468OR	0136
A468R	0136
A468X	0136
A468Y	0136
A469(2S)	0136
A469A	0136
A469B	0136
A469C	0136
A469D	0136
A469E	0136
A469F	0136
A469G	0136
A469GN	0136
A469H	0136
A469J	0136
A469K	0136
A469L	0136
A469M	0136
A469OR	0136
A469R	0136
A469X	0136
A469Y	0136
A470(2S)	0136
A470A	0136
A470B	0136
A470C	0136
A470D	0136
A470E	0136
A470F	0136
A470G	0136
A470GN	0136
A470H	0136
A470J	0136
A470K	0136
A470L	0136
A470M	0136
A470OR	0136
A470R	0136
A470X	0136
A470Y	0136
A471(2S)	0136
A471-1(2S)	0136
A471-2(2S)	0136
A471-3(2S)	0136
A471A	0136
A471B	0136
A471C	0136
A471D	0136
A471E	0136
A471F	0136
A471G	0136
A471GN	0136
A471H	0136
A471J	0136
A471K	0136
A471L	0136
A471M	0136
A471OR	0136
A471R	0136
A471X	0136
A471Y	0136
A472	0079
A472(2S)	0136
A472-1	0136
A472-2	0136
A472-3	0136
A472-4	0136
A472-5	0136
A472-6	0136
A472A(2S)	0136
A472B(2S)	0136
A472C(2S)	0136
A472D(2S)	0136
A472E(2S)	0136
A472F	0136
A472G	0136
A472GN	0136
A472H	0136
A472J	0136
A472K	0136
A472L	0136
A472M	0136
A472OR	0136
A472R	0136
A472X	0136
A472Y	0136
A-473	0224
A473	0144
A473(2S)	0925
A473-GR(2S)	0925
A473-O(2S)	0925
A473-O(KOREA)	0925
A473-R(2S)	0925
A473-Y(2S)	0925
A473B	0925
A473D	0925
A473GR	0925
A473R	0925
A473Y(2S)	0925
A474(2S)	0136
A474-G	0136
A474A	0136
A474B	0136
A474C	0136
A474D	0136
A474E	0136
A474F	0136
A474G	0136
A474GN	0136
A474H	0136
A474J	0136
A474K	0136
A474L	0136
A474M	0136
A474OR	0136
A474R	0136
A474X	0136
A474Y	0136
A475	0050
A476(2S)	0136
A476A	0136
A476B	0136
A476C	0136
A476D	0136
A476E	0136
A476F	0136
A476G	0136
A476GN	0136
A476H	0136
A476J	0136
A476K	0136

If replacement code is OEM, contact original manufacturer for replacement.

DEVICE TYPE	REPL CODE
A476L	0136
A476M	0136
A476OR	0136
A476R	0136
A476X	0136
A476Y	0136
A477(2S)	0136
A477A	0136
A477B	0136
A477C	0136
A477D	0136
A477E	0136
A477F	0136
A477G	0136
A477GN	0136
A477H	0136
A477J	0136
A477K	0136
A477L	0136
A477M	0136
A477OR	0136
A477R	0136
A477X	0136
A477Y	0136
A478(2S)	0004
A478-G	0004
A478A	0004
A478B	0004
A478C	0004
A478D	0004
A478E	0004
A478F	0004
A478G	0004
A478GN	0004
A478H	0004
A478J	0004
A478K	0004
A478L	0004
A478M	0004
A478OR	0004
A478R	0004
A478X	0004
A478Y	0004
A479(2S)	0004
A479-G	0004
A479A	0004
A479B	0004
A479C	0004
A479D	0004
A479E	0004
A479F	0004
A479G	0004
A479GN	0004
A479H	0004
A479J	0004
A479K	0004
A479L	0004
A479M	0004
A479OR	0004
A479R	0004
A479X	0004
A479Y	0004
A480	0144
A480(2S)	0037
A480-OR	0037
A480A	0037
A480B	0037
A480C	0037
A480D	0037
A480E	0037
A480F	0037
A480G	0037
A480GN	0037
A480H	0037
A480J	0037
A480K	0037
A480L	0037
A480M	0037
A480OR	0037
A480R	0037
A480X	0037
A480Y	0037
A481	0007
A482	0007
A482(2S)	0037
A482A	0037
A482B	0037
A482C	0037
A482D	0037
A482E	0037
A482F	0037
A482G	0037
A482GN	0037
A482H	0037
A482J	0037
A482K	0037
A482L	0037
A482M	0037
A482OR	0037
A482R	0037
A482X	0037
A482Y	0037
A483	0905
A483(2S)	0905
A483-O(2S)	0905
A483-R	0905
A483-Y	0905
A484	0007
A484(2S)	0892
A484(ADMIRAL)	0007
A484-BL	0892
A484-R	0892
A484-Y	0892
A485	0259
A485(2S)	0892
A485-BL	0892
A485-BLU	0892
A485-R	0892
A485-RED	0892
A485-Y	0892
A485-YEL	0892
A485Y	0892
A486	0007
A486(2S)	0892
A486-BL	0892
A486-BLU	0892
A486-R	0892
A486-RED	0892
A486-Y	0892
A486-YEL	0892
A488-A0060	0124
A489	0007
A489(2S)	0925
A489-O(2S)	0925
A489-R(2S)	0925
A489-Y(2S)	0925
A490	0007
A490(2S)	1498
A490-O	1498
A490-R	1498
A490-Y	1498
A490A	1498
A490B	1498
A490C	1498
A490D	1498
A490E	1498
A490F	1498
A490G	1498
A490GN	1498
A490H	1498
A490J	1498
A490K	1498
A490L	1498
A490M	1498
A490OR	1498
A490R	1498
A490X	1498
A490Y(2S)	1498
A490YA	1498
A492	0007
A493(2S)	0006
A493-GR	0006
A493-O(2S)	0006
A493-Y	0006
A493A	0006
A493B	0006
A493C	0006
A493D	0006
A493E	0006
A493F	0006
A493G	0006
A493G-GR	0006
A493GN	0006
A493G-Y	0006
A493H	0006
A493J	0006
A493K	0006
A493L	0006
A493M	0006
A493OR	0006
A493R	0006
A493X	0006
A493Y(2S)	0006
A-494	0079
A494	0693
A494(2S)	0693
A494-GR(2S)	0693
A494-GR-1	0693
A494-O(2S)	0693
A494-OR	0693
A494-Y(2S)	0693
A494A	0693
A494B	0693
A494C	0693
A494D	0693
A494E	0693
A494F	0693
A494G	0693
A494GN	0693
A494GR	0693
A494H	0693
A494J	0693
A494K	0693
A494L	0693
A494M	0693
A494O	0693
A494OR	0693
A494R	0693
A494X	0693
A494Y	0693
A495	0127
A495(2S)	0006
A495-O	0006
A495-1	0006
A495-GN	0006
A495-O(2S)	0006
A495-OF	0006
A495-OR	0006
A495-ORG	0006
A495-ORG-G	0006
A495-Q	0006
A495-R(2S)	0006
A495-RD	0006
A495-RED	0006
A495-RED-G	0006
A495-Y(2S)	0006
A495-YEL	0006
A495-YEL-G	0006
A495-YL	0006
A495A	0006
A495B	0006
A495C	0006
A495D(2S)	0006
A495E	0006
A495F	0006
A495G	0006
A495G-GR(2S)	0006
A495GN	0006
A495G-O(2S)	0006
A495G-R(2S)	0006
A495G-Y(2S)	0006
A495H	0006
A495J	0006
A495K	0006
A495L	0006
A495M	0006
A495O	0006
A495OF	0006
A495OR	0006
A495R	0006
A495RD	0006
A495RED	0006
A495RED-G	0006
A495RO	0006
A495W(2S)	0006
A495W1	0006
A495WI	0006
A495X	0006
A495Y(2S)	0006
A495YEL	0006
A495YEL-G	0006
A495YL	0006
A496	0006
A496(2S)	0525
A496-O(2S)	0525
A496-ORG	0525
A496-R(2S)	0525
A496-RED	0525
A496-Y(2S)	0525
A496-YEL	0525
A496O	0525
A496ORG	0525
A496R	0525
A496RED	0525
A496Y(2S)	0525
A496YEL	0525
A497	0144
A497(2S)	0126
A497-O	0126
A497-ORG	0126
A497-R	0126
A497-RED	0126
A497-Y	0126
A497-YEL	0126
A497R	0126
A497RED	0126
A497Y	0126
A498	OEM
A498(2S)	0126
A498(F.E.T.)	0212
A498-O	0126
A498-ORG	0126
A498-R	0126
A498-RED	0126
A498-Y	0126
A498-YEL	0126
A498R	0126
A498RED	0126
A498Y(2S)	OEM
A499(2S)	0150
A499-O(2S)	0150
A499-ORG	0150
A499-R(2S)	0150
A499-RED	0150
A499-Y(2S)	0150
A499-YEL	0150
A499A	0150
A499B	0150
A499C	0150
A499D	0150
A499E	0150
A499F	0150
A499G	0150
A499GN	0150
A499H	0150
A499J	0150
A499K	0150
A499L	0150
A499M	0150
A499O	0150
A499OR	0150
A499R	0150
A499X	0150
A499Y	0150
A500	0150
A500(2S)	0150
A500-O(2S)	0150
A500-OR	0150
A500-ORG	0150
A500-R(2S)	0150
A500-RED	0150
A500-Y(2S)	0150
A500-YEL	0150
A500A	0150
A500B	0150
A500C	0150
A500D	0150
A500E	0150
A500F	0150
A500G	0150
A500GN	0150
A500H	0150
A500J	0150
A500K	0150
A500L	OEM
A500L(TRANS)	0150
A500LA	OEM
A500LB	OEM
A500LC	OEM
A500LD	OEM
A500LE	OEM
A500LM	OEM
A500LN	OEM
A500LP	OEM
A500LS	OEM
A500LT	OEM
A500M	0150
A500O	0150
A500OR	0150
A500P	OEM
A500PA	OEM
A500PB	OEM
A500PC	OEM
A500PD	OEM
A500PE	OEM
A500PM	OEM
A500PN	OEM
A500PS	OEM
A500PT	OEM
A500R	0150
A500X	0150
A500Y	0150
A501	0126
A501(2S)	0126
A502(2S)	0006
A502-O	0006
A502-OR	0006
A502-R	0006
A502-Y	0006
A502A	0006
A502B	0006
A502C	0006
A502D	0006
A502E	0006
A502F	0006
A502G	0006
A502GE	OEM
A502GN	0006
A502H	0006
A502J	0006
A502K	0006
A502L	0006
A502M	0006
A502OR	0006
A502R	0006
A502X	0006
A502Y	0006
A503	0126
A503(2S)	0126
A503-GR	0126
A503-GRN	0126
A503-O(2S)	0126
A503-ORG	0126
A503-R(2S)	0126
A503-Y(2S)	0126
A503-YEL	0126
A503G	0126
A503GE	OEM
A503O	0126
A503Y	0126
A504	0126
A504(2S)	0126
A504-GR	0126
A504-GRN	0126
A504-O	0126
A504-ORG	0126
A504-R(2S)	0126
A504-Y(2S)	0126
A504-YEL	0126
A504G	0126
A504GE	OEM
A504O	0126
A504Y	0126
A505(2S)	0525
A505-O(2S)	0525
A505-ORG	0525
A505-R(2S)	0525
A505-RED	0525
A505-Y(2S)	0525
A505-YEL	0525
A505O	0525
A505ORG	0525
A505R	0525
A505RED	0525
A505Y	0525
A505YEL	0525
A506(2S)	0050
A507(2S)	0050
A507A	0050
A507B	0050
A507C	0050
A507D	0050
A507E	0050
A507F	0050
A507G	0050
A507GN	0050
A507H	0050
A507J	0050
A507K	0050
A507L	0050
A507OR	0050
A507R	0050
A507X	0050
A507Y	0050
A508(2S)	0050
A508D	0050
A508F	0050
A509(2S)	0006
A509-O	0006
A509-O(2S)	0006
A509-OR	0006
A509-R	0006
A509-RD	0006
A509-Y(2S)	0006
A509-YE	0006
A509A	0006
A509B	0006
A509BL	0006
A509C	0006
A509D	0006
A509E	0006
A509F	0006
A509G	0006
A509GN	0006
A509G-O	0006
A509GR(2S)	0006
A509GR-1	0006
A509G-Y	0006
A509H	0006
A509J	0006
A509K	0006
A509L	0006
A509M	0006
A509OR	0006
A509Q	0006
A509R(2S)	0006
A509RD	0006
A509T	0006
A509V	0006
A509X	0006
A509Y(2S)	0006
A509YE	0006
A510	0892
A510(2S)	0892
A510-O(2S)	0892
A510-OR	0892
A510-ORG	0892
A510-R(2S)	0892
A510-RD	0892
A510-RED	0892
A510A	0892
A510B	0892
A510C	0892
A510D	0892
A510E	0892
A510F	0892
A510G	0892
A510GN	0892
A510H	0892
A510J	0892
A510K	0892
A510L	0892
A510M	0892
A510OR	0892
A510R	0892
A510X	0892
A510Y	0892
A511	0892
A511(2S)	0892
A511-G	0892
A511-O(2S)	0892
A511-OR	0892
A511-ORG	0892
A511-R(2S)	0892
A511-RD	0892
A511-RED	0892
A511A	0892
A511B	0892
A511C	0892
A511D	0892
A511E	0892
A511F	0892
A511G	0892
A511GN	0892
A511H	0892
A511J	0892
A511K	0892
A511L	0892
A511M	0892
A511OR	0892
A511R	0892
A511X	0892
A511Y	0892
A512(2S)	0892
A512-O(2S)	0892
A512-OR	0892
A512-OR1	0892
A512-ORG	0892
A512-R(2S)	0892
A512-RD	0892
A512-RED	0892
A512A	0892
A512B	0892
A512C	0892
A512D	0892
A512E	0892
A512F	0892
A512G	0892
A512GN	0892
A512H	0892
A512J	0892
A512K	0892
A512L	0892
A512M	0892
A512OR	0892
A512R	0892
A512X	0892
A512Y	0892
A513(2S)	0892
A513-O(2S)	0892
A513-OR	0892
A513-ORG	0892
A513-R(2S)	0892
A513-RD	0892
A513-RED	0892
A513A	0892
A513B	0892
A513C	0892
A513D	0892
A513E	0892
A513F	0892
A513G	0892
A513GN	0892
A513H	0892
A513J	0892
A513K	0892
A513L	0892
A513M	0892
A513OR	0892
A513R	0892
A513X	0892
A513Y	0892
A514-022057	0143
A514-023553	0600
A514-023626	0015
A514-025607	0015
A514-027662	0280
A514-027757	0015
A514-028072	0015
A514-028073	0015
A514-032815	0280
A514-033338	0155
A514-033903	0015
A514-035596	0015
A514-038984	0087
A514-040296	0325
A514-042791	0133
A514-044910	0037
A514-047830	0423
A514-0339903	0015
A515	0103
A516	0892
A516(2S)	0892
A516A(2S)	0892
A517	0136
A517(2S)	0136
A518(2S)	0136
A518-G	0136
A518A	0136
A518B	0136
A518C	0136
A518D	0136
A518E	0136
A518F	0136
A518G	0136
A518GN	0136
A518H	0136
A518J	0136
A518K	0136
A518L	0136
A518M	0136
A518OR	0136
A518R	0136
A518X	0136
A518Y	0136
A520	OEM
A521	OEM
A522	0103
A522(2S)	0037
A522-3	0103
A522A(2S)	0037
A522AL	0037
A522B	0037
A522C	0037
A522D	0037
A522E	0037
A522F	0037
A522G	0037
A522GN	0037
A522H	0037
A522J	0037
A522K	0037
A522L	0037
A522M	0037
A522OR	0037
A522R	0037
A522X	0037
A522Y	0037
A523	0103
A523(2S)	0126
A523A	0126
A524	0037
A525(2S)	0050
A525A(2S)	0050
A525B(2S)	0050
A525C	0050
A525D	0050
A525E	0050
A525F	0050
A525GN	0050
A525H	0050
A525J	0050
A525K	0050
A525L	0050
A525M	0050
A525OR	0050
A525R	0050
A525X	0050
A525Y	0050
A526	0037
A527(2S)	0126
A528(2S)	0126
A529	0126
A530(2S)	0037
A530A	0037
A530B	0037
A530C	0037
A530D	0037
A530E	0037
A530F	0037
A530G	0037
A530GN	0037
A530GR	0037
A530H(2S)	0037
A530H1	0037
A530HA	0037
A530HB	0037
A530HC	0037
A530J	0037
A530K	0037
A530L	0037
A530M	0037
A530OR	0037
A530R	0037
A530X	0037
A530Y	0037
A532(2S)	0126
A532A(2S)	0126
A532B(2S)	0126
A532C(2S)	0126
A532D(2S)	0126
A532E(2S)	0126
A532F(2S)	0126
A535	0037
A536	0050
A537(2S)	0126
A537A(2S)	0126
A537AA(2S)	0126
A537AB(2S)	0126
A537AC(2S)	0126
A537AH(2S)	0126
A537AHA	0126
A537AHB	0126
A537AHC	0126
A537H(2S)	0126
A537HA	0126
A537HB	0126
A537HC	0126
A537K	0126
A537L	0126
A538(2S)	0004
A538-G	0004
A538G	0004
A539(2S)	0006
A539A	0006
A539B	0006
A539C	0006
A539D	0006
A539E	0006
A539F	0006
A539G	0006
A539GN	0006
A539H	0006
A539J	0006
A539K	0006
A539L(2S)	0006
A539M	0006
A539OR	0006
A539R	0006
A539X	0006
A539Y	0006
A539Y(EBC)	2033
A539Y(KOREA)	0037
A540A	OEM
A540B	OEM
A540C	OEM
A540D	OEM
A540E	OEM
A540L	OEM
A540LA	OEM
A540LB	OEM
A540LC	OEM
A540LD	OEM

If replacement code is OEM, contact original manufacturer for replacement.

DEVICE TYPE	REPL CODE
A540LE	OEM
A540LM	OEM
A540M	OEM
A540N	OEM
A540P	OEM
A540PA	OEM
A540PB	OEM
A540PC	OEM
A540PD	OEM
A540PE	OEM
A540PM	OEM
A540PN	OEM
A540PS	OEM
A540PT	OEM
A540S	OEM
A540T	OEM
A542(2S)	0037
A542A	0037
A542B	0037
A542C	0037
A542D	0037
A542E	0037
A542F	0037
A542G	0037
A542GN	0037
A542H	0037
A542J	0037
A542K	0037
A542L	0037
A542M	0037
A542OR	0037
A542R	0037
A542X	0037
A542Y	0037
A543(2S)	0693
A544(2S)	0126
A544A	0126
A544B	0126
A544C	0126
A544D	0126
A544E	0126
A544F	0126
A544G	0126
A544GN	0126
A544H	0126
A544J	0126
A544K	0126
A544L	0126
A544M	0126
A544OR	0126
A544R	0126
A544X	0126
A544Y	0126
A545(2S)	0791
A545A	0791
A545B	0791
A545C	0791
A545D	0791
A545E	0791
A545F	0791
A545G	0791
A545GN	0791
A545H	0791
A545J	0791
A545K(2S)	0791
A545KLM(2S)	0791
A545L(2S)	0791
A545LM(2S)	0791
A545M	0791
A545OR	0791
A545R	0791
A545X	0791
A545Y	0791
A546(2S)	0126
A546A(2S)	0126
A546B(2S)	0126
A546E(2S)	0126
A546H(2S)	0126
A547(2S)	0791
A547A(2S)	0791
A548(2S)	0150
A548G	0150
A548GN	0150
A548H	0150
A548HA	0150
A548HB	0150
A548HC	0150
A548OR	0150
A548R	0150
A548Y	0150
A549	0342
A549A(2S)	0342
A549AH(2S)	0342
A550(2S)	0037
A550A(2S)	0037
A550AB	0037
A550AQ(2S)	0037
A550AR	0037
A550AS	0037
A550B	0037
A550BC	0037
A550BL	0037
A550C	0037
A550D	0037
A550P	0037
A550Q(2S)	0037
A550R(2S)	0037
A550S(2S)	0037
A550Y	0037

DEVICE TYPE	REPL CODE
A551	0126
A551(2S)	0126
A551C(2S)	0126
A551D(2S)	0126
A551E(2S)	0126
A552(2S)	0126
A553	0037
A554	0037
A554A	0037
A555	0006
A556	0006
A556-142	0143
A557	0150
A558	0037
A559	0037
A559A	0037
A560	0126
A560A	0126
A561	0006
A561(2S)	0006
A561-0	0006
A561-GR	0006
A561-GRN	0006
A561-O(2S)	0006
A561-OR	0006
A561-ORG	0006
A561-R(2S)	0006
A561-RD	0006
A561-RED	0006
A561-Y(2S)	0006
A561-YEL	0006
A561-YL	0006
A561GRN	0006
A561R	0006
A561RD	0006
A561RED	0006
A561Y	0006
A561YEL	0006
A561YL	0006
A562	0006
A562-0	0006
A562-GR	0006
A562-GRN	0006
A562-O	0006
A562-O(2S)	0006
A562-O(JAPAN)	0006
A562-O(KOREA)	0006
A562-OR	0006
A562-ORG	0006
A562-R(2S)	0006
A562-RD	0006
A562-RED	0006
A562-Y(2S)	0006
A562-YE	0006
A562-YEL	0006
A562D	0006
A562E	0006
A562G(2S)	0006
A562GR(2S)	0006
A562GRN	0006
A562O	0006
A562Q	0006
A562R(2S)	0006
A562RD	0006
A562RED	0006
A562T	0006
A562TM	0006
A562TM-0	0006
A562TM-O	0006
A562TMO	0006
A562TM-OFA	0006
A562TM-Y	0006
A562TM-YFA	0006
A562TO	0006
A562TQ	0006
A562V	0006
A562VO	0006
A562Y	0006
A562Y(2S)	0006
A562Y(JAPAN)	0006
A562Y(KOREA)	0006
A562YE	0006
A562YEL	0006
A562Y-TM	0006
A564	0203
A564-0(2S)	0203
A564-1	0203
A564-O(2S)	0203
A564-OGD	0203
A564-OR	0203
A564-P	0203
A564-Q	0203
A564-R	0203
A564A(2S)	0203
A564A(JAPAN)	0203
A564ABQ-1	0203
A564AG	0203
A564AK	0203
A564AL	0203
A564AP(2S)	0203
A564A-Q	0203
A564AQ(2S)	0203
A564A-R	0203
A564AR	0203
A564AR(JAPAN)	0203
A564A-S	0203
A564AS(2S)	0203
A564B	0203
A564C	0203

DEVICE TYPE	REPL CODE
A564D	0203
A564E	0203
A564F(2S)	0203
A564FQ(2S)	0203
A564FQ-1	0203
A564FR-1	0203
A564G	0203
A564GN	0203
A564H	0203
A564J(2S)	0203
A564K	0203
A564L	0203
A564M	0203
A564OR	0203
A564P(2S)	0203
A564P.A	0203
A564Q	0203
A564Q(2S)	0203
A564QP	0203
A564QR(2S)	0203
A564QRS	0203
A564R	0203
A564S(2S)	0203
A564T(2S)	0203
A564X	0203
A564XL	0203
A564Y	0203
A565(2S)	0037
A565A(2S)	0037
A565AB	0037
A565B(2S)	0037
A565BA	0037
A565C(2S)	0037
A565D(2S)	0037
A565E	0037
A565F	0037
A565G	0037
A565GN	0037
A565H	0037
A565J	0037
A565K(2S)	0037
A565L	0037
A565M	0037
A565OR	0037
A565R	0037
A565X	0037
A565Y	0037
A566(2S)	0905
A566A(2S)	0905
A566B(2S)	0905
A566C(2S)	0905
A566H	0905
A-567	0038
A567	0037
A567A	0037
A567B	0037
A567C	0037
A567D	0037
A567E	0037
A567F	0037
A567G	0037
A567GN	0037
A567GR	0037
A567H	0037
A567J	0037
A567K	0037
A567L	0037
A567M	0037
A567OR	0037
A567R	0037
A567X	0037
A567Y	0037
A568(2S)	0006
A568-OR	0006
A568A	0006
A568B	0006
A568C	0006
A568D	0006
A568E	0006
A568F	0006
A568G	0006
A568GN	0006
A568H	0006
A568J	0006
A568K	0006
A568L	0006
A568M	0006
A568OR	0006
A568R	0006
A568X	0006
A568Y	0006
A569	0006
A569(2S)	0006
A569A	0006
A569B	0006
A569C	0006
A569D	0006
A569E	0006
A569F	0006
A569G	0006
A569GN	0006
A569H	0006
A569J(2S)	0006
A569K	0006
A569L	0006
A569M	0006
A569OR	0006
A569R	0006
A569X	0006

DEVICE TYPE	REPL CODE
A569Y	0006
A570	5606
A570(2S)	0006
A570A	4922
A570A(TRANS)	0006
A570B	4922
A570B(TRANS)	0006
A570C	4922
A570C(TRANS)	0006
A570D	4922
A570D(TRANS)	0006
A570E	4922
A570E(TRANS)	0006
A570F	0006
A570G	0006
A570GN	0006
A570H	0006
A570J	0006
A570K	0006
A570L	0006
A570M	4922
A570M(TRANS)	0006
A570N	OEM
A570OR	0006
A570P	OEM
A570R	0006
A570S	OEM
A570T	OEM
A570X	0006
A570Y	0006
A571(2S)	0126
A572	0103
A572(2S)	0693
A572-1	0103
A572Y	0693
A573	0006
A574	0436
A575	0006
A576	0150
A576-0001-002	0037
A576-0001-013	0037
A577	0436
A578(2S)	0693
A579(2S)	0693
A580	0126
A580-040215	0130
A580-040315	0130
A580-040515	0130
A580-080215	0130
A580-080315	0130
A580-080515	0130
A581	0126
A590	OEM
A591	OEM
A593	0016
A594(2S)	0126
A594-O(2S)	0126
A594-R(2S)	0126
A594-Y(2S)	0126
A594N	0126
A595	0004
A596M	OEM
A596N	OEM
A596P	OEM
A596PA	OEM
A596PB	OEM
A596PC	OEM
A596PD	OEM
A596PE	OEM
A596S	OEM
A596T	OEM
A597	0126
A599	0791
A599Y	0791
A600	0015
A600(RECTIFIER)	0015
A601	OEM
A602	OEM
A603	0037
A603(2S)	0037
A603A	0037
A603B	0037
A603C	0037
A603D	0037
A603E	0037
A603F	0037
A603G	0037
A603GN	0037
A603H	0037
A603J	0037
A603K	0037
A603L	0037
A603M	0037
A603OR	0037
A603R	0037
A603X	0037
A603Y	0037
A604	0126
A604(2S)	0126
A605	OEM
A605(2S)	0126
A606	0126
A606(2S)	0126
A607	0126
A607(2S)	5008
A607A	1552
A607B	1552
A607C	1552
A607D	1552

DEVICE TYPE	REPL CODE
A607E	1552
A607F	1552
A607G	1552
A607GN	1552
A607H	1552
A607J	1552
A607K	5008
A607L	5008
A607M	5008
A607OR	1552
A607R	1552
A607S	5008
A607SA	1552
A607SB	1552
A607SC	1552
A607SD	1552
A607X	1552
A607Y	1552
A608	0006
A608(2S)	0006
A608-0	0006
A608-C(2S)	0006
A608-E(2S)	0006
A608-F(2S)	0006
A608-O	0006
A608-OR	0006
A608A(2S)	0006
A608AE	0006
A608AF	0006
A608B(2S)	0006
A608BL	0006
A608C(2S)	0006
A608D(2S)	0006
A608E	0006
A608E(2S)	0006
A608E(JAPAN)	0006
A608F	0006
A608F(2S)	0006
A608F(JAPAN)	0006
A608G(2S)	0006
A608GN	0006
A608H	0006
A608J	0006
A608K	0006
A608KE	0006
A608KF	0006
A608KF(JAPAN)	0006
A608L	0006
A608M	0006
A608NP	0006
A608OR	0006
A608P	0006
A608R	0006
A608SP	0006
A608X	0006
A608Y	0006
A609(2S)	0037
A609A(2S)	0037
A609B(2S)	0037
A609C(2S)	0037
A609D(2S)	0037
A609E(2S)	0037
A609F(2S)	0037
A609G(2S)	0037
A609GN	0037
A609J	0037
A609K	0037
A609L	0037
A609M	0037
A609OR	0037
A609R	0037
A609Y	0037
A610	0006
A610(2S)	0006
A610L	OEM
A610S	OEM
A611	0006
A611(2S)	0006
A611L	OEM
A611S	OEM
A612	OEM
A612(2S)	0126
A613(2S)	0905
A614(2S)	0905
A615-1008	0050
A615-1009	0050
A615-1010	0004
A615-1011	0004
A615-1012	0143
A616(1)	0905
A616(2)	0905
A616(2S)	0905
A617	0342
A617K(2S)	0342
A618	0342
A618K(2S)	0342
A620	0006
A620CA	OEM
A620CB	OEM
A620CC	OEM
A620CD	OEM
A620CE	OEM
A620CM	OEM
A620CN	OEM
A620CS	OEM
A620CT	OEM
A620DA	OEM
A620DC	OEM
A620DD	OEM

DEVICE TYPE	REPL CODE
A620DP	OEM
A620LP	OEM
A620PC	OEM
A621	0037
A621A	OEM
A621B	OEM
A621C	OEM
A621D	OEM
A621E	OEM
A621M	OEM
A621N	OEM
A621P	OEM
A621PA	OEM
A621PB	OEM
A621S	OEM
A621T	OEM
A622	3519
A623(2S)	1430
A623-0(2S)	1430
A624(2S)	1430
A625	0126
A626(2S)	3500
A626L	3500
A627(2S)	3500
A628(2S)	0006
A628-0	0006
A628-OR	0006
A628A(2S)	0006
A628A(JAPAN)	0006
A628AA	0006
A628AD	0006
A628A-E	0006
A628AE	0006
A628A-F	0006
A628B	0006
A628C	0006
A628D(2S)	0006
A628E(2S)	0006
A628EF	0006
A628F(2S)	0006
A628G	0006
A628GN	0006
A628H	0006
A628J	0006
A628K	0006
A628L	0006
A628M	0006
A628OR	0006
A628R	0006
A628X	0006
A628Y	0006
A629(2S)	0693
A629A	0693
A629B	0693
A629C	0693
A629D	0693
A629E	0693
A629F	0693
A629G	0693
A629GN	0693
A629H	0693
A629J	0693
A629K	0693
A629L	0693
A629M	0693
A629OR	0693
A629R	0693
A629X	0693
A629Y	0693
A633	1051
A633A	1051
A634(2S)	0681
A634(4)K	0681
A634(4)L	0681
A634-O(KOREA)	1051
A634A(2S)	0681
A634K(2S)	0681
A634L(2S)	0681
A634M(2S)	0681
A635	0382
A636(2S)	2021
A636(4)K	2021
A636(4)L	2021
A636-4K	2021
A636-4L	2021
A636A(2S)	2021
A636B(2S)	2021
A636C(2S)	2021
A636D(2S)	2021
A636K(2S)	2021
A636L(2S)	2021
A636M(2S)	2021
A637(2S)	0342
A637C	0342
A637D	0342
A638(2S)	0342
A638E	0342
A638F	0342
A638S	0342
A639(2S)	0342
A639L	0342
A639S	0342
A640	0693
A640(2S)	0693
A640A(2S)	0693
A640B(2S)	0693
A640C(2S)	0693
A640D(2S)	0693
A640L	0693

DEVICE TYPE	REPL CODE
A640LA	OEM
A640M(2S)	0693
A640P	OEM
A640PA	OEM
A640PB	OEM
A640PC	OEM
A640PD	OEM
A640PE	OEM
A640PM	OEM
A640PN	OEM
A640PS	OEM
A640PT	OEM
A640S	0693
A640T	OEM
A641	0693
A641(2S)	0693
A641(PBP)	0693
A641A(2S)	0693
A641B(2S)	0693
A641BL	0693
A641C(2S)	0693
A641D(2S)	0693
A641G	0693
A641GR	0693
A641K	0693
A641L	0693
A641L(2S)	0693
A641L(DIO)	OEM
A641LA	OEM
A641LB	OEM
A641LC	OEM
A641LD	OEM
A641LE	OEM
A641LM	OEM
A641M(2S)	0693
A641O	0693
A641OR	0693
A641PD	OEM
A641PE	OEM
A641PM	OEM
A641PN	OEM
A641PS	OEM
A641PT	OEM
A641R	0693
A641S	0693
A641Y	0693
A642	0006
A642(2S)	0006
A642-254	0144
A642-260	0144
A642-268	0144
A642-O	0037
A642-O(EBC)	0037
A642A(2S)	0006
A642B(2S)	0006
A642C(2S)	0006
A642D(2S)	0006
A642E(2S)	0006
A642F(2S)	0006
A642G	0006
A642GN	0006
A642H	0006
A642J	0006
A642K	0006
A642L	0006
A642L(2S)	0006
A642M	0006
A642OR	0006
A642R	0006
A642S	0006
A642S(2S)	0006
A642V	0006
A642W(2S)	0006
A642X	0006
A642Y	0006
A643(2S)	1242
A643L	1242
A643R(2S)	1242
A643S	1242
A643V(2S)	1242
A643W(2S)	1242
A644L	0111
A644S	0111
A645(2S)	1367
A645L	0111
A645S	0111
A646	1367
A646(2S)	1367
A647	0126
A648(2S)	2013
A649	2013
A649A	2013
A649L	0016
A649S	0016
A650	2013
A651	2013
A652	0792
A653	0792
A653L	0792
A656(2S)	3500
A656A	3500
A656L(2S)	3500
A656M(2S)	3500
A657	3500
A657A	3500
A658(2S)	3500
A658A	3500
A659(2S)	0006
A659A	0006

DEVICE TYPE	REPL CODE
A659B	0006
A659D(2S)	0006
A659E(2S)	0006
A659F(2S)	0006
A659G	0006
A659L	0006
A659P	0006
A659R	0006
A659Y	0006
A661(2S)	0006
A661-GR	0006
A661-O(2S)	0006
A661-R	0006
A661-Y	0006
A661Y(2S)	0006
A663(2S)	0491
A663-BL	0491
A663-R	0491
A663-Y	0491
A664	0532
A666(2S)	0006
A666A(2S)	0006
A666B	0006
A666BL	0006
A666C	0006
A666D	0006
A666E	0006
A666H(2S)	0006
A666HR(2S)	0006
A666I	0006
A666IQRS(2S)	0006
A666QRS	0006
A666R(2S)	0006
A666S(2S)	0006
A666Y	0006
A667-G	0079
A667-GRN	0111
A667-O	0086
A667-ORG	0111
A667-RED	0111
A667-Y	0079
A667-YEL	0111
A667RED	0144
A668	0126
A668-G	0079
A668-GRN	0111
A668-O	0086
A668-ORG	0111
A668-Y	0079
A668-YEL	0111
A669	0342
A669-G	0079
A669-GRN	0111
A669-Y	0079
A669-YEL	0111
A670(2S)	0925
A670A(2S)	0925
A670B(2S)	0925
A670C(2S)	0925
A671	1308
A671A	1308
A671B	1308
A671C	1308
A671K	1308
A671KA	1308
A671KB	1308
A671KC	1308
A671TD	1308
A672	0037
A672A(2S)	0037
A672B(2S)	0037
A672C(2S)	0037
A673(2S)	0148
A673-C	0148
A673-D	0148
A673-O	0148
A673-OR	0148
A673A	0148
A673A(2S)	0148
A673A(JAPAN)	0148
A673AB(2S)	0148
A673AC(2S)	0148
A673AC(JAPAN)	0148
A673AD(2S)	0148
A673AK	0148
A673AKA	0148
A673AKB	0148
A673AKC	0148
A673AKD	0148
A673AS(2S)	0148
A673B(2S)	0148
A673C	0148
A673C(2S)	0148
A673C(JAPAN)	0148
A673C2(2S)	0148
A673D	0148
A673D(2S)	0148
A673D(JAPAN)	0148
A673E	0148
A673F	0148
A673G	0148
A673GN	0148
A673H	0148
A673J	0148
A673K	0148
A673L	0148
A673M	0148
A673OR	0148
A673R	0148
A673X	0148
A673Y	0148
A675(2S)	0006
A676-60C-1	OEM
A676-60M-1	OEM
A676-90C	OEM
A676-90M	OEM
A676R-60C-1	OEM
A676R-60M1	OEM
A676R-90C	OEM
A676R-90M	OEM
A677(2S)	0006
A677-0	0006
A677-O	0006
A677-OR	0006
A677A	0006
A677B	0006
A677C	0006
A677D	0006
A677E	0006
A677F	0006
A677G	0006
A677GN	0006
A677H	0006
A677HL	0006
A677J	0006
A677K	0006
A677L	0006
A677M	0006
A677OR	0006
A677R	0006
A677X	0006
A677Y	0006
A678(2S)	0006
A678-O	0006
A678-OR	0006
A678A	0006
A678B	0006
A678C	0006
A678D	0006
A678E	0006
A678F	0006
A678G	0006
A678GN	0006
A678H	0006
A678J	0006
A678K	0006
A678L	0006
A678M	0006
A678OR	0006
A678R	0006
A678X	0006
A678Y	0006
A679(2S)	2013
A679-R	2013
A679-Y	2013
A680(2S)	2013
A680-R	2013
A680-Y	2013
A681	0126
A682	0525
A682-O	0525
A682-Y	0525
A6820	0525
A683(2S)	0532
A683NC	0532
A6830	0532
A683P	0532
A683Q(2S)	0532
A683R(2S)	0532
A683S	0532
A683.4	0532
A684(2S)	0532
A684NC	0532
A684Q(2S)	0532
A684R(2S)	0532
A684S	0532
A685	0006
A689C	OEM
A689M	OEM
A690V081H97	0160
A691M4	0133
A691M5	0133
A691M5-2	0133
A692T5	0015
A692T5-0	0015
A692X13-4	0143
A692X16-0	0015
A693	0006
A693C	0006
A694X1	0053
A694X1-0A	0053
A694X1-OA	0365
A695	0436
A695C(2S)	0436
A695D(2S)	0436
A696#	OEM
A696(2S)	0006
A696D(2S)	0006
A696L	OEM
A696PE	OEM
A696PM	OEM
A696PN	OEM
A696PS	OEM
A696PT	OEM
A697(2S)	0006
A697C(2S)	0006
A698	0681
A699	0681
A699(H)	0681
A699A	0681
A699AP(2S)	0681
A699AQ(2S)	0681
A699AR(2S)	0681
A699CF(2S)	0681
A699P(2S)	0681
A699Q(2S)	0681
A699R(2S)	0681
A700(2S)	0925
A700B(2S)	0925
A700Y(2S)	0925
A701	0037
A701F(2S)	0037
A701FO	0037
A701FJ(2S)	0037
A701FO(2S)	0037
A702	0074
A702(2S)	0693
A703(2S)	1430
A704	0006
A704A	0006
A704B	0006
A704C	0006
A704D	0006
A704E	0006
A704F	0006
A704G	0006
A704GN	0006
A704H	0006
A704J	0006
A704K	0006
A704L	0006
A704M	0006
A704OR	0006
A704R	0006
A704X	0006
A704Y	0006
A705	0006
A705(2S)	0006
A705A	0006
A705B	0006
A705C	0006
A705D	0006
A705E	0006
A705F	0006
A705G	0006
A705GN	0006
A705H	0006
A705J	0006
A705K	0006
A705L	0006
A705M	0006
A705OR	0006
A705R	0006
A705X	0006
A705Y	0006
A706	0382
A706-2	0382
A706-3	0382
A706-3A	OEM
A706-3C	OEM
A706-3D	OEM
A706-3E	OEM
A706-3F	OEM
A706-3G	OEM
A706-3H	OEM
A706-3I	OEM
A706-3J	OEM
A706-3K	OEM
A706-3L	OEM
A706-3M	OEM
A706-3N	OEM
A706-4	0382
A706-4A	OEM
A706-4B	OEM
A706-4C	OEM
A706-4D	OEM
A706-4E	OEM
A706-4F	OEM
A706-4G	OEM
A706-4H	OEM
A706-4I	OEM
A706-4J	OEM
A706-4K	OEM
A706-4L	OEM
A706-4M	OEM
A706-5A	OEM
A706-5B	OEM
A706-5C	OEM
A706-5D	OEM
A706-5E	OEM
A706-5F	OEM
A706-5G	OEM
A706-5H	OEM
A706-5I	OEM
A706-5J	OEM
A706-5K	OEM
A706-5L	OEM
A706-5M	OEM
A706-6A	OEM
A706-6B	OEM
A706-6C	OEM
A706-6D	OEM
A706-6E	OEM
A706-6F	OEM
A706-6G	OEM
A706-6H	OEM
A706-6I	OEM
A706-6J	OEM
A706-6K	OEM
A706-6L	OEM
A706-7A	OEM
A706-7B	OEM
A706-7C	OEM
A706-7D	OEM
A706-7E	OEM
A706-7F	OEM
A706-7G	OEM
A706-7H	OEM
A706-7I	OEM
A706-7J	OEM
A706-8A	OEM
A706-8B	OEM
A706-8C	OEM
A706-8D	OEM
A706-8E	OEM
A706-8F	OEM
A706-8G	OEM
A706-8H	OEM
A706H	0382
A706J	0382
A707	OEM
A707(2S)	0436
A707(V)	0436
A707-3A	OEM
A707-3B	OEM
A707-3C	OEM
A707-3D	OEM
A707-3E	OEM
A707-3F	OEM
A707-3G	OEM
A707-3H	OEM
A707-3I	OEM
A707-3J	OEM
A707-3K	OEM
A707-3L	OEM
A707-3M	OEM
A707-4A	OEM
A707-4B	OEM
A707-4C	OEM
A707-4D	OEM
A707-4E	OEM
A707-4F	OEM
A707-4G	OEM
A707-4H	OEM
A707-4I	OEM
A707-4J	OEM
A707-4K	OEM
A707-4L	OEM
A707-5A	OEM
A707-5B	OEM
A707-5C	OEM
A707-5D	OEM
A707-5E	OEM
A707-5F	OEM
A707-5G	OEM
A707-5H	OEM
A707-5I	OEM
A707-5J	OEM
A707-5K	OEM
A707-5L	OEM
A707-6A	OEM
A707-6B	OEM
A707-6C	OEM
A707-6D	OEM
A707-6E	OEM
A707-6F	OEM
A707-6G	OEM
A707-6H	OEM
A707-6I	OEM
A707-6J	OEM
A707-6K	OEM
A707-7A	OEM
A707-7B	OEM
A707-7C	OEM
A707-7D	OEM
A707-7E	OEM
A707-7F	OEM
A707-7G	OEM
A707-7H	OEM
A707-7I	OEM
A707-8A	OEM
A707-8B	OEM
A707-8C	OEM
A707-8D	OEM
A707-8E	OEM
A707-8F	OEM
A707-8G	OEM
A708(2S)	0126
A708-3A	OEM
A708-3B	OEM
A708-3C	OEM
A708-3D	OEM
A708-3E	OEM
A708-3F	OEM
A708-3G	OEM
A708-3H	OEM
A708-3I	OEM
A708-3J	OEM
A708-3K	OEM
A708-4A	OEM
A708-4B	OEM
A708-4C	OEM
A708-4D	OEM
A708-4E	OEM
A708-4F	OEM
A708-4G	OEM
A708-4H	OEM
A708-4I	OEM
A708-5A	OEM
A708-5B	OEM
A708-5C	OEM
A708-5D	OEM
A708-5E	OEM
A708-5F	OEM
A708-5G	OEM
A708-5H	OEM
A708-5I	OEM
A708-5J	OEM
A708-6A	OEM
A708-6B	OEM
A708-6C	OEM
A708-6D	OEM
A708-6E	OEM
A708-6F	OEM
A708-6G	OEM
A708-6H	OEM
A708-6I	OEM
A708-7A	OEM
A708-7B	OEM
A708-7C	OEM
A708-7D	OEM
A708-7E	OEM
A708-7F	OEM
A708-7G	OEM
A708-8A	OEM
A708-8B	OEM
A708-8C	OEM
A708-8D	OEM
A708-8E	OEM
A708-8F	OEM
A708-J	OEM
A708A(2S)	0126
A709	0037
A709-3A	OEM
A709-3B	OEM
A709-3C	OEM
A709-3D	OEM
A709-3E	OEM
A709-3F	OEM
A709-3G	OEM
A709-3H	OEM
A709-3I	OEM
A709-3J	OEM
A709-4A	OEM
A709-4B	OEM
A709-4C	OEM
A709-4D	OEM
A709-4E	OEM
A709-4F	OEM
A709-4G	OEM
A709-4H	OEM
A709-5A	OEM
A709-5B	OEM
A709-5C	OEM
A709-5D	OEM
A709-5E	OEM
A709-5F	OEM
A709-5G	OEM
A709-5H	OEM
A709-6A	OEM
A709-6B	OEM
A709-6C	OEM
A709-6D	OEM
A709-6E	OEM
A709-6F	OEM
A709-6G	OEM
A709-7A	OEM
A709-7B	OEM
A709-7C	OEM
A709-7D	OEM
A709-7E	OEM
A709-7F	OEM
A709-8A	OEM
A709-8B	OEM
A709-8C	OEM
A709-8D	OEM
A709-8E	OEM
A710	0150
A710-3A	OEM
A710-3B	OEM
A710-3C	OEM
A710-3D	OEM
A710-3E	OEM
A710-3F	OEM
A710-3G	OEM
A710-3H	OEM
A710-4A	OEM
A710-4B	OEM
A710-4C	OEM
A710-4D	OEM
A710-4E	OEM
A710-4F	OEM
A710-5A	OEM
A710-5B	OEM
A710-5C	OEM
A710-5D	OEM
A710-5E	OEM
A710-5F	OEM
A710-6A	OEM
A710-6B	OEM
A710-6C	OEM
A710-6D	OEM
A710-6E	OEM
A710-7A	OEM
A710-7B	OEM
A710-7C	OEM
A710-7D	OEM
A710-7E	OEM
A710-8A	OEM
A710-8B	OEM
A710-8C	OEM
A710-8D	OEM
A711	0150
A711-3A	OEM
A711-3B	OEM
A711-3C	OEM
A711-3D	OEM
A711-3E	OEM
A711-4A	OEM
A711-4B	OEM
A711-4C	OEM
A711-4D	OEM
A711-5A	OEM
A711-5B	OEM
A711-5C	OEM
A711-5D	OEM
A711-6A	OEM
A711-6B	OEM
A711-6C	OEM
A711-8A	OEM
A711-8B	OEM
A712	0439
A712-3A	OEM
A712-3B	OEM
A712-3C	OEM
A712-3D	OEM
A712-4A	OEM
A712-4B	OEM
A712-4C	OEM
A712-5A	OEM
A712-5B	OEM
A712-5C	OEM
A712-6A	OEM
A712-6B	OEM
A712-7A	OEM
A712-7B	OEM
A712-8A	OEM
A713	3519
A713-3A	OEM
A713-3B	OEM
A713-3C	OEM
A713-3D	OEM
A713-4A	OEM
A713-4B	OEM
A713-4C	OEM
A713-5A	OEM
A713-5B	OEM
A713-5C	OEM
A713-6A	OEM
A713-6B	OEM
A713-7A	OEM
A713-7B	OEM
A713-8A	OEM
A713A	3519
A714	2013
A714-L	2013
A714L	2013
A715(2S)	0525
A715B(2S)	0525
A715C(2S)	0525
A715D(2S)	0525
A715WBP(2S)	0525
A715WT	0525
A715WTB(2S)	0525
A715WTBC	0525
A715WTC(2S)	0525
A717(2S)	0126
A718(2S)	0037
A719(2S)	0006
A719-Q	0006
A719K	0006
A719NC	0006
A719O	0006
A719PQR	0006
A719Q(2S)	0006
A719R(2S)	0006
A719S(2S)	0006
A720(2S)	0006
A720-G	0006
A720-O	0006
A720-Q	0006
A720-R	0006
A720A	0006
A720B	0006
A720C	0006
A720D	0006
A720L	0006
A720NC	0006
A720Q(2S)	0006
A720QR	0006
A720R	0006
A720R(2S)	0006
A720S(2S)	0006
A720Y	0006
A721(2S)	0693
A721Q	0693
A721R	0693
A721S(2S)	0693
A721T(2S)	0693
A721U(2S)	0693
A722(2S)	0693
A722S(2S)	0693
A723(2S)	0006
A723EH3AB1	OEM
A723EH4AB1	OEM
A723EH5AB1	OEM
A723EH6AB1	OEM
A723EH7AB1	OEM
A723EH8AB1	OEM
A723EH9AB1	OEM
A723EH10AB1	OEM
A723EH11AB1	OEM
A723EH12AB1	OEM
A723R	0006
A724	0148
A725(2S)	0693
A725EH2AB1	0071
A725EH3AB1	OEM
A725EH4AB1	OEM
A725EH6AB1	OEM
A725EH8AB1	OEM
A725EH10AB1	OEM
A725EH12AB1	OEM
A725EH16AB1	OEM
A725EH20AB1	OEM
A725Y	0693
A726(2S)	0693
A726Y	0693
A728	0006
A728A	0006
A730	1242
A730L	OEM
A730LA	OEM
A730LB	OEM
A730LC	OEM
A730LD	OEM
A730P	5299
A730PA	OEM
A730PB	OEM
A730PC	OEM
A730PD	OEM
A730PE	OEM
A730PM	OEM
A730PN	OEM
A730PS	OEM
A730PT	OEM
A731	1242
A731(2S)	1242
A732	0126
A733	0006
A733(2S)	0006
A733-P	0006
A733-Q	0006
A733A(2S)	0006
A733AK	0006
A733AP	0006
A733AP(JAPAN)	0006
A733AQ	0006
A733AQ(JAPAN)	0006
A733H(2S)	0006
A733IO	0006
A733IQ(2S)	0006
A733J	OEM
A733P	0006
A733P(2S)	0006
A733PQ	0006
A733Q	0006
A733Q(2S)	0006
A733Q1P(2S)	0006
A733QP	0006
A733R(2S)	0006
A733Y(KOREA)	0006
A734	OEM
A734(2S)	0126
A735	0037
A736(2S)	0126
A738	0460
A738(B)	0460
A738A	0460
A738C(2S)	0460
A738D(2S)	0460
A739	3606
A740	1648
A740A	1648
A740AB	1648
A740P	OEM
A740PA	OEM
A740PB	OEM
A740PC	OEM
A740PD	OEM
A740PE	OEM
A740PM	OEM
A740PN	OEM
A740PS	OEM
A741	0150
A741H	0150
A742	0126
A742H	0126
A743(2S)	0525
A743A(2S)	0525
A743AB(2S)	0525
A743AC(2S)	0525
A743C(2S)	0525
A744(2S)	0491
A745	2013
A745A	2013
A746(2S)	2013
A747	2013
A747(2S)	2013
A747A	2013
A747A(2S)	2013
A747B	0111
A747C	OEM

If replacement code is OEM, contact original manufacturer for replacement.

DEVICE TYPE	REPL CODE
A748	0925
A748(2S)	0681
A748B	0016
A748C	0111
A748Q(2S)	0681
A749	0016
A749(2S)	0342
A749A	0342
A749B	0016
A749C	0111
A750	0693
A751(2S)	0477
A751(P)	0477
A751Q(2S)	0477
A751R(2S)	0477
A752(2S)	0477
A753(2S)	2013
A753-4004-248	0037
A754(2S)	0925
A755(2S)	0925
A755B(2S)	0925
A755C(2S)	0925
A756(2S)	3500
A757	3500
A757(2S)	3500
A757N	OEM
A758	2013
A758(2S)	2013
A758A	2013
A758B	2013
A759	0861
A759(2S)	0037
A759A	0037
A759B	0037
A760	0693
A761	0126
A761-1	0126
A761-2	0126
A762	0792
A762-1	0792
A762-2	0792
A763	0693
A764(2S)	0905
A765	0905
A765(RECT.)	0922
A766	1648
A766S	1648
A767	0037
A768	0925
A768G	0925
A768O	0925
A768Y	0925
A769	0854
A770	0854
A771	1369
A772(2S)	0532
A772-1	0532
A772-2	0532
A772-23	0532
A772B1	0127
A772EH	0127
A772FE	0127
A772J	0532
A773	2038
A773-1	2038
A773-2	2038
A773H	2038
A774	0693
A774A	0693
A775	0681
A775A	0681
A775AA	0681
A775AB	0681
A775AC	0681
A775B	0681
A775C	0681
A776	0150
A776A	0150
A777	0436
A777(2S)	0436
A777NC	0436
A777Q	0436
A778	0342
A778(2S)	0342
A778A	0342
A778A(2S)	0342
A778AF-02	0342
A778AK	0342
A778AK(2S)	0342
A778AK-02	0342
A778K(2S)	0342
A778K-02	0342
A779	1267
A779(2S)	1051
A779K(2S)	1051
A779KB(2S)	1051
A780	0664
A780(TRANS)	1367
A780A	1367
A780AK(2S)	1367
A780AKB(2S)	1367
A781	0354
A781(TRAN)	0150
A781(TRANS)	0150
A781K	0150
A782	0037
A783	0148
A784	3519
A785	5061
A786	0037
A786-Q	0037
A786Q	0037
A786QL	0037
A786R	0037
A787	0693
A788	0037
A789	0148
A790	0635
A791	3519
A793	0126
A794	0525
A794A	0525
A795	0525
A796	3519
A798	0693
A799	0126
A800	0071
A800(TRANS)	3607
A800-516-00	0126
A800-523-01	0037
A800-523-02	0037
A800-527-00	0037
A807	3500
A808	3500
A808A	3500
A809	0342
A810	0342
A811	0037
A811A	0037
A811C5	0037
A811C6	0037
A811C7	0037
A811C8	0037
A812	1742
A812M3	1742
A812M4	1742
A812M5	1742
A812M6	1742
A812M7	1742
A813	0150
A813S2	0150
A813S3	0150
A813S4	0150
A814	0681
A814-O	0681
A814-Y	0681
A815	0681
A815-O	0681
A815-Y	0681
A816	1648
A816-0	1648
A816-O	1648
A816-Y	1648
A817	0436
A817-Y	0436
A817A	0436
A817AY	0436
A817Y(JAPAN)	0436
A818	0681
A818-O	0681
A818-Y	0681
A825(2S)	0006
A825Q	0006
A825R	0006
A825S	0006
A826	0037
A826P	0037
A826Q	0037
A826R	0037
A826S	0037
A828A	0532
A829	0037
A829A	0037
A829B	0037
A829C	0037
A829D	0037
A829E	0037
A829F	0037
A833	0037
A835	0342
A835H	0342
A836(2S)	0693
A836C	0693
A836D(2S)	0693
A836E(2S)	0693
A836F	0693
A837(2S)	3500
A838	5378
A838B	5378
A838C	5378
A839	1648
A839-O	1648
A839-R	1648
A839-Y	1648
A840(2S)	1648
A840H	1648
A840JC	1648
A840JM	1648
A840K	1648
A841	0693
A841(2S)	0693
A841-BL	0693
A841-GR	0693
A842	0693
A842(2S)	0693
A842-BL	0693
A842-GR	0693
A843	5467
A843(2S)	1648
A844	0693
A844(2S)	0006
A844C	0693
A844C(JAPAN)	0693
A844D	0693
A844D(2S)	0693
A844D(JAPAN)	0693
A844E	0693
A844E(2S)	0693
A844F	0693
A845	0342
A845AH	0342
A845H	0342
A846	OEM
A847	0648
A847(2S)	0648
A847A	0648
A848	0439
A848(2S)	0439
A849	0439
A849(2S)	0439
A850	0006
A854	1348
A854-Q	1348
A854G	1348
A854Q	1348
A854Q(JAPAN)	1348
A854QE	1348
A854R	1348
A855	0693
A856	0693
A856-16	OEM
A856A	0693
A857	0342
A858	0342
A859	0037
A860	0342
A861	1051
A861(3)	1051
A861JK	1051
A866	0037
A866V	0037
A866VG	0037
A867	0150
A867-14-1	OEM
A868	0037
A869	0037
A870	0037
A871	5367
A872	0648
A872A	0648
A872AD	0648
A872AE	0648
A872AF	0648
A872C	0648
A872D	0648
A872E	0648
A872F	0648
A873	0150
A876HA	0037
A876HB	0037
A876HC	0037
A877	3500
A878	3500
A879	0342
A879NC	0342
A879P	0342
A879Q	0342
A880	0693
A880-250-107	0126
A881Q	0006
A881Q(JAPAN)	0006
A882	2013
A883	0532
A884	0532
A885	0525
A886	0525
A886V	0525
A886VQ	0525
A886VR	0525
A887	1267
A887P	1267
A887Q	1267
A888	0037
A889	0037
A890	0037
A891	0037
A892	0602
A893	0648
A893A	0648
A893AC	0648
A893AD	0648
A893AE	0648
A893AF	0648
A893D	0648
A893E	0648
A894	0436
A895	0150
A896	0342
A896-1	0342
A896-2	0342
A897	0648
A898	0525
A899	3703
A900	0525
A900(2S)	0525
A901	0693
A902	0693
A903	0693
A904	0648
A904(2S)	0648
A904A	0648
A905	0338
A905(2S)	0648
A906	0693
A906(2S)	0693
A907	1681
A907(2S)	1681
A908	1598
A908(2S)	1598
A909	1598
A909-1008	0050
A909-1009	0136
A909-1010	0136
A909-1011	0004
A909-1012	0004
A909-1013	0004
A909-1015	0143
A909-1016	0123
A909-1017	0143
A909-1018	0015
A909-1019	0015
A909-27125-160	0144
A912	0342
A912Q	0342
A912R	0342
A912S	0342
A913	1648
A913AQ	1648
A914	0525
A915	0342
A916	0342
A916-31025-58	0144
A916K	0342
A916L	0342
A916M	0342
A917	0477
A918	0150
A920	3758
A921	0648
A921-59B	0144
A921-62B	0144
A921-63B	0144
A921-64B	0144
A921-70B	0037
A921A	0648
A921AS	0648
A922	0126
A923	0342
A923-1	0342
A923-2	0342
A925	0150
A925-1	0150
A925-2	0150
A927	0006
A929	0693
A929G	0693
A929H	0693
A930	0693
A931	5654
A932	5367
A933	0148
A933AQ	0148
A933AR	0148
A933Q	0148
A933Q(JAPAN)	0148
A933QL	0148
A933QY	0148
A933R	0148
A933R(JAPAN)	0148
A933RZ	0148
A933S	0148
A933S(JAPAN)	0148
A933SRR	0148
A933SSJ	0148
A933SSK	0148
A934(JAPAN)	0532
A935	0436
A935R	0436
A935R(JAPAN)	0436
A937	2484
A937-1	0016
A937-3	0016
A937-Q	2484
A937R	2484
A937R(JAPAN)	2484
A939	4479
A940	1911
A940(JAPAN)	1911
A940-Z	1911
A940AD	1911
A940AE(JAPAN)	1911
A940D	1911
A940E	1911
A940GL2	1911
A940LB	1911
A940R	1911
A940RD	1911
A940RE	1911
A941	0006
A942	0006
A945	0164
A945-O	0164
A945-Q	0164
A948	0126
A948E	0126
A949	0477
A949Y	0477
A950	1348
A950-0	1348
A950-O	1348
A950-Y	1348
A950FA	1348
A950Y	1348
A950Y(JAPAN)	1348
A950Y(KOREA)	0532
A951	0681
A951-10	0681
A951KA	0681
A951KD	0681
A951LC	0681
A951M	0681
A951MU	0681
A952	0006
A952(JAPAN)	0006
A952K	0006
A952L	0006
A952L(JAPAN)	0006
A952M(JAPAN)	0006
A953	1348
A954	0006
A955	OEM
A956	0150
A956H3	0150
A956H4	0150
A956H5	0150
A956H6	0150
A957	1648
A957(JAPAN)	1648
A958	1911
A958Y	1911
A962	0382
A963	0525
A964	3758
A964-17887	0160
A964A	3758
A-965	OEM
A965	5329
A966	0532
A966-O(JAPAN)	0532
A966AY(KOREA)	3597
A966Y	0532
A968	1648
A968A	1648
A968B	1648
A970	0006
A971	1598
A973	0693
A-975	OEM
A978	0693
A979	5057
A980	2013
A981	2013
A982	2013
A984	0006
A984(JAPAN)	0006
A984E	0006
A984K	0006
A984KF(JAPAN)	0006
A985	5387
A985A	5387
A987	0006
A988	0648
A990	0693
A991	0693
A991-01-0098	0037
A991-01-1225	0037
A991-01-1316	0144
A991-01-1319	0037
A991-01-3058	0037
A992	0648
A992-00-1192	0160
A993	0037
A995	5059
A999	0006
A999L	0006
A1000	0071
A1001	2013
A1002	2013
A1003	1598
A1003-3	OEM
A1004	0037
A1005	0037
A1006	3543
A1006A	3543
A1007	1598
A1007A	1598
A1008	3544
A1010	1369
A1010-3	OEM
A1010L(JAPAN)	1369
A1010V	1369
A1011	1648
A1011(IC)	3083
A1011(IC-SM)	OEM
A1011(SM)	4524
A1011D	1648
A1011E	1648
A1012	3548
A1012-Y	3548
A1012Y(JAPAN)	3548
A1013	1522
A1013-0	1522
A1013-O	1522
A1013-O(JAPAN)	1522
A1013-R	1522
A1013R	1522
A1013R(JAPAN)	1522
A1015	0148
A1015-0	0148
A1015-O	0148
A1015-O(JAPAN)	0148
A1015-O(KOREA)	0006
A1015-OFA	0148
A1015-Y	0148
A1015-YFA	0148
A1015FA	0148
A1015G	0148
A1015GR	0148
A1015GR(JAPAN)	0148
A1015O	0148
A1015Y	0148
A1015Y(JAPAN)	0148
A1015Y(KOREA)	0148
A1016	0693
A1016K	0693
A1017	0342
A1018	0342
A1019	0648
A1020	0434
A1020-3	OEM
A1020-O	0434
A1020Y	0434
A1020Y(JAPAN)	0434
A1021	0525
A1021(JAPAN)	0525
A1021-O	0525
A1021-R	0525
A1021R	2555
A1021Y	0525
A1023	1603
A1023P	1603
A1023Q	1603
A1024	0917
A1025D	0037
A1025E	0037
A1026	0006
A1026(2S)	0006
A1027	0532
A1027(2S)	0037
A1027R	0037
A1028	2013
A1029	0006
A1029B	0006
A1029C	0006
A1029D	0006
A1030	0006
A1030B	0006
A1030C	0006
A1031B	0037
A1031C	0037
A1032B	0037
A1032C	0037
A1035	OEM
A1040	0006
A1041	1598
A1042	0491
A1043	1681
A1044	1681
A1045	0491
A1046	0491
A1047	0525
A1048	0013
A1048(JAPAN)	0013
A1048-GR	0013
A1048-O	0013
A1048-O(JAPAN)	0013
A1048GR	0013
A1048GR(JAPAN)	0013
A1048O	0013
A1048Y	0013
A1049	0006
A1050	0148
A1060	0859
A1061	0859
A1061P	0859
A1061Q	0859
A1061R	0859
A1062	3571
A1064	1598
A1065	1598
A1066	0477
A1067	1598
A1068	1598
A1069	3577
A1069C	3577
A1072	1598
A1073	1598
A1075	3581
A1076	3581
A1077	0718
A1078	3585
A1080	0681
A1081D	0037
A1081E	0037
A1082D	0037
A1082E	0037
A1083D	0037
A1083E	0037
A1084D	1597
A1084E	1597
A1085D	1597
A1085E	1597
A1086	0079
A1087	0079
A1090	0037
A1091	0342
A1091-O	0342
A1091R	0342
A1092	0693
A1093	3571
A1094	3581
A1095	3581
A1096	0525
A1100	0006
A1100D	0006
A1100F	0006
A1100L	0006
A1101	OEM
A1102	0859
A1103	0859
A1104	3111
A1105	3111
A1106	3111
A1109	0144
A1110	0525
A1111	1648
A1111(IC)	OEM
A1112	1648
A1114-1002	0050
A1115	3314
A1115-E	3314
A1115-F	3314
A1115E	3314
A1116	3606
A1117	3606
A1120	0525
A1123	1522
A1124	1522
A1124C	0160
A1125	1648
A1127NC	0693
A1128	0006
A1133	1911
A1135	0859
A1144	0136
A1145	1522
A1146	3571
A1150	0006
A1150Y	0006
A1151	0532
A1152	0037
A1153	3617
A1154	0532
A1158	0436
A1160	0532
A1164	0037
A1169	3581
A1170	0144
A1170(2S)	3646
A1173	1522
A1174	0693
A1175	3625
A1175(JAPAN)	3625
A1175-F	3625
A1175E	3625
A1175F	3625
A1175F(JAPAN)	3625
A1175FH	3625
A1175H	3625
A1175H(JAPAN)	3625
A1175J	3625
A1180	1598
A1184	0525
A1185	0859
A1186	3571
A1186N	OEM
A1187	3630
A1188D	0477
A1188E	0477
A1189D	0477
A1189E	0477
A1190D	0477
A1190E	0477
A1191D	0477
A1191E	0477
A1193K	3890
A1194K	2897
A1195	1051
A1195-O	1051
A1195O	1051
A1195R	1051
A1198A	OEM
A1201	0579
A1201-H45	2393
A1201-H-75	0167
A-1201B	0579
A1201B	0579
A1205	3111
A1206K	5462
A1207	1522
A1208	1522
A1210	0525
A1214	0037
A1215	3581
A1216	3646
A1217	0525
A1220	0050
A1220(2S)	0525
A1220A	0525
A1221	0525
A1238	0111
A1243	0280
A1246	3654

If replacement code is OEM, contact original manufacturer for replacement.

DEVICE TYPE	REPL CODE
A1246R	3654
A1246S	3654
A1246S(JAPAN)	3654
A1246T	3654
A1246U	3654
A1261L(JAPAN)	1199
A1263	3111
A1264	3111
A1264T	3111
A1265	3111
A1277	1308
A1278	3543
A1301	3673
A1314	0086
A1318S(JAPAN)	OEM
A1320	0342
A1321	0342
A1341	0086
A1357N	0865
A1364N	0878
A1365	0167
A1368	0856
A1369	3705
A1377	0050
A1378	0050
A1379	0016
A1380	0016
A1381	OEM
A1383	0136
A1384	0136
A1390C	0013
A1392	OEM
A1396	0600
A1408	0525
A1409	0233
A1412-1	0016
A1414A	0160
A1414A9	0160
A1418	0144
A1423BC1AB1	OEM
A1423DC1AB1	OEM
A1423MC1AB1	OEM
A1423MC2AB1	OEM
A1423MC3AB1	OEM
A1423MC4AB1	OEM
A1423MC5AB1	OEM
A1423MC6AB1	OEM
A1423MH2AB1	OEM
A1423MH3AB1	OEM
A1423MH4AB1	OEM
A1423MH5AB1	OEM
A1423MH6AB1	OEM
A1423MH7AB1	OEM
A1423MH8AB1	OEM
A1423MH9AB1	OEM
A1423MH10AB1	OEM
A1423MH11AB1	OEM
A1423MH12AB1	OEM
A1423PBC1AB1	OEM
A1423PBC3AB1	OEM
A1423PBC4AB1	OEM
A1423PBC5AB1	OEM
A1423PBC6AB1	OEM
A1423PBH2AB1	OEM
A1423PBH4AB1	OEM
A1423PBH5AB1	OEM
A1423PBH6AB1	OEM
A1423PBH7AB1	OEM
A1423PBH8AB1	OEM
A1423PBH9AB1	OEM
A1423PCB2AB1	OEM
A1425MH8AB1	OEM
A1441L	5485
A1450	0477
A1450S	0477
A1450T	0477
A1460	0111
A1462	0144
A1462-19	0050
A1465-1	0050
A1465-4	0600
A1465-19	0050
A1465-29	0208
A1465-49	0600
A1465A	0050
A1465A9	0050
A1465B	0050
A1465B9	0050
A1465C	0160
A1465C9	0160
A1466-1	0004
A1466-2	0004
A1466-3	0004
A1466-19	0004
A1466-19-	0004
A1466-29	0004
A1466-39	0004
A1469	1308
A1469R	OEM
A1472-19	0198
A1474-3	0280
A1474-39	0280
A1477A	0440
A1477A9	0440
A1477C	0160
A1477C9	0160
A1480	OEM
A1484A	0160
A1484A9	0160
A1488B	0050
A1488B9	0050
A1488C	0280
A1488C9	0280
A1492-1	0440
A1492-19	0440
A1501	0015
A1507S	5901
A1518	0144
A1519	0144
A1520	0144
A1521	0144
A1558-17	0037
A1560P	OEM
A-1567	0016
A1567	0016
A1567-1	0016
A1606E	0718
A1618	OEM
A1618M	OEM
A1624E	0342
A1844-17	0037
A-1854-0492-1	0016
A1854-0533	0861
A1858	0600
A1867-17	0037
A1884-0209	2018
A1884-0218	2018
A1946	0015
A2003-3	OEM
A2010-3	OEM
A2011	OEM
A2020-3	OEM
A2039-2	0600
A2057A2-198	0037
A2057B2-115	0127
A2057B104-8	0233
A2057B106-12	0037
A2057B108-6	0037
A2057B110-9	0126
A2057B112-9	0126
A2057B114-9	0126
A2057B115-9	0126
A2057B116-9	0126
A2057B121-9	0126
A2057B122-9	0126
A2057B145-12	0126
A2057B163-12	0126
A2090	0233
A2202	0023
A2332	0007
A2410	0016
A2411	0016
A2412	0016
A2413	0016
A2414	0164
A2415	0178
A2416	1400
A2417	0642
A2418	0103
A2419	0019
A2420	0143
A2421	0102
A2422	0015
A2423	0015
A2424	0924
A2428	0037
A2434	0016
A2448	0037
A2458	1303
A2459	0774
A2460	0015
A2461	0023
A2462	0102
A2464	0007
A2465	0007
A2466	0016
A2468	0016
A2469	0016
A2470	0016
A2471	0086
A2473	0143
A2474	0137
A2475	0196
A2476	0143
A2479	0007
A2480	0007
A2481	0015
A2485	0015
A2498	0144
A2499	0016
A2511	OEM
A2585	OEM
A2620	0233
A2652-919	0325
A2746	0007
A2798	0037
A3003-3	OEM
A3010-3	OEM
A3101	OEM
A3103BX2	OEM
A3300	2568
A3301	1215
A3350	0416
A3410A	0050
A3410B	0050
A3511	OEM
A3512	OEM
A3513	0037
A3523	0126
A3533	0126
A3533-1	0126
A3540	0037
A3549	0037
A3559	0037
A3562	0037
A3563	0037
A3574	0037
A3581	0037
A3607	0600
A3609	0600
A3616-1	0126
A4000	OEM
A4012-0154	OEM
A4030	3020
A4086	0037
A4087	0037
A4126	0037
A4201	0196
A4212-A	0015
A4212A	0071
A4237M	OEM
A4247	0160
A4310	0037
A4347	0160
A4442	0037
A4478	0126
A4648	0233
A4700	0600
A4745	0037
A4789	0144
A4802-00004	0037
A4815	0037
A4819	0233
A4822-130-40348	0037
A4838	0233
A4843	0233
A4844	0037
A4851	0144
A4853	0233
A5003-3	OEM
A5010-3	OEM
A5027	OEM
A5029	OEM
A5226-1	0037
A5253	0160
A5640R	0693
A6181-1	0233
A6230-AH	1414
A6358S	3233
A6393S	OEM
A6458S	2913
A7011	OEM
A7012	OEM
A7013	OEM
A7014	OEM
A7016	1028
A7205	2457
A7233H2AB1	OEM
A7253	0233
A7520(IC)	0539
A7530N	3817
A8015T	OEM
A8400	OEM
A8402	OEM
A8404	OEM
A8405	0037
A8495	OEM
A8540	0037
A-8865-03	OEM
A-8865-04	OEM
A-8865-05	OEM
A8867	0037
A9218	0124
A9228-3	0124
A-9982	1057
A-9982-02	OEM
A9982-02	4274
A-9982-02A	4274
A-9982-07	1057
A9982-07	4275
A-9982A	1057
A10003A	OEM
A10003D	OEM
A10005D	OEM
A10010D	OEM
A10020D	OEM
A10075A	OEM
A10075D	OEM
A10105	0015
A10113	0015
A10118	0015
A10164	0015
A10165	0015
A-10167	0285
A10169	0015
A11744	0325
A11745	0325
A-11885	OEM
A12010A	OEM
A12010D	OEM
A12163	0160
A12178	0160
A12546	0233
A12594	0037
A-12791	0042
A12888	0037
A-13099	OEM
A-13284	3502
A-13414	OEM
A13658A3	OEM
A13658E3	3612
A-14343-06	OEM
A-14343-09	OEM
A-14343-9	OEM
A14665-2	0208
A14743	0178
A15003A	OEM
A15003D	OEM
A15005D	OEM
A15010D	OEM
A15020D	OEM
A15075A	OEM
A15075D	OEM
A15927	0160
A-16926	OEM
A19040-B	OEM
A19108-A	OEM
A20005D	OEM
A20010D	OEM
A20020D	OEM
A20371	0143
A20372	0016
A21268	0649
A22008	0037
A24015	0133
A24031	OEM
A24060	OEM
A24100	0016
A25005D	OEM
A25010D	OEM
A25020D	OEM
A25762-010	0233
A25762-012	0233
A29035-E	0025
A30005D	OEM
A30010D	OEM
A30020D	OEM
A30270	0037
A30278	0126
A30290	0037
A30302	0160
A34715	0160
A35084	0160
A35201	0160
A35260	0160
A36508	0143
A36539	0137
A36577	0126
A36896	0160
A40005D	OEM
A40010D	OEM
A40020D	OEM
A40410	0126
A41440	0037
A42946	0196
A42946B	0196
A48727	OEM
A49488-B	OEM
A49491-B	OEM
A50020B	OEM
A-54806-A	0725
A59625-1	0037
A59625-2	0037
A59625-3	0037
A59625-4	0037
A59625-5	0037
A59625-6	0037
A59625-7	0037
A59625-8	0037
A59625-9	0037
A59625-10	0037
A59625-11	0037
A59625-12	0037
A71687-1	0037
A75680	0015
A78331	0037
A81004	0735
A82930-6	OEM
A82930-8	OEM
A94004	0160
A94037	0037
A94063	0126
A95115	0600
A95211	0600
A95227	0037
A95232	0037
A112363	0103
A-113110	0037
A-113335	0590
A-113367(DIODE)	0124
A116078	0037
A116081	0233
A116084	0126
A116284	0126
A-118038	0071
A118284	0037
A119730	0037
A119983	0126
A-120018	0086
A-120077	0052
A-120125(PWR.RECT.)	0071
A120150	OEM
A-120278	0016
A-120304	0178
A-120327	0103
A-120407	0870
A-120417	0037
A-120420	0299
A121467	0037
A121659	0037
A124047	0037
A124623	0144
A124624	0144
A124755	0037
A-125278	0471
A125329	0144
A-125332	0016
A126524	0037
A126700	0037
A126705	0233
A126707	0037
A126715	0037
A126718	0037
A126719	0037
A126724	0126
A127712	0233
A129509	0144
A129510	0144
A129511	0144
A129512	0144
A129513	0144
A129571	0144
A129572	0144
A129573	0144
A129574	0144
A129697	0037
A129699	0037
A130139	0037
A-132591	0242
A-134166-2	0137
A190187T	OEM
A190429	0037
A198794-1	0600
A218012DS	0160
A242361	OEM
A251283	OEM
A297074C11	0126
A489751-028	0037
A489751-031	0037
A573501	0233
A610074-1	0037
A610075-1	0233
A610083	0037
A610083-1	0037
A610083-2	0037
A610083-3	0037
A610110-1	0037
A610120-1	0037
A631790L	0313
A650232E	0006
A650235G	0006
A650238A	0006
A650372D	0006
A650923F	0006
A650925H	0006
A660097	0160
A670720K	0076
A670722D	0076
A670729B	0076
A671656A	0786
A671656K	0786
A673351K	0191
A673354K	0191
A673355H	0191
A673355K	0191
A675313F	0066
A675315F	0066
A675318A	0066
A675419H	0111
A678760A	0643
A678760C	0643
A678971C	0643
A678971D	0643
A701584-00	0037
A701589-00	0037
A711002	0041
A720657	0725
A758011	0071
A759500	0071
A772738	0144
A772739	0144
A815185	0126
A815185E	0126
A815199	0037
A815199-6	0037
A815211	0037
A815213	0037
A815229	0037
A815247	0037
A853702	OEM
A970246	0037
A970248	0037
A970251	0037
A970254	0037
A984193	0037
A1471114-1	0037
A1473549-1	0126
A1473563-1	0037
A1473570-1	0037
A1473574-1	0037
A1473590-1	0037
A1473591-1	0037
A1473597-1	0037
A1473616-1	0126
A2006681-95	0144
A2057013-0004	0037
A2057013-0701	0037
A2057013-0702	0037
A2057013-0703	0037
A2090056-1	0160
A2090056-5	0160
A2090056-27	0160
A2091859-0025	0160
A2091859-0720	0160
A2091859-10	0160
A2091859-11	0160
A2092418	0600
A2092418-0711	0600
A2092693-0724	0144
A2092693-0725	0144
A2498512	0037
A3011112	0086
A3115621	0023
A3115690	0023
A3170717	0233
A4037764-2	0126
A5010005	0124
A5320111	0037
A5768475	0023
A6000010	OEM
A6000020	0832
A6000030	0887
A6000040	0898
A6002010	OEM
A6002020	0832
A6002030	0887
A6002040	0898
A6002050	OEM
A6002060	OEM
A6010030	1032
A6010050	OEM
A6010060	OEM
A6012010	OEM
A6012020	7037
A6012030	OEM
A6012040	4130
A6012060	OEM
A6041840	3133
A6041848	3133
A6041876	3133
A6041880	3133
A6048403	3046
A6090130	OEM
A6235070	0261
A6314340	0236
A6314440	0733
A6314446	0733
A6316440	OEM
A6317420	0076
A6317440	0076
A6317460	0076
A6317480	0076
A6317499	0076
A6317520	0076
A6317526	0076
A6317540	0076
A6317547	0076
A6317560	0076
A6317562	0076
A6317567	0076
A6317581	0076
A6317764	0076
A6317800	2668
A6317802	2668
A6317900	0313
A6317901	0313
A6319300	0286
A6319302	0286
A6319400	0643
A6319403	0643
A6319500	1284
A6319550	1284
A6319560	1284
A6319576	1284
A6321241	0866
A6324920	0066
A6324922	0066
A6324942	0066
A6324961	0066
A6325060	0261
A6325070	0261
A6325077	0261
A6328328	1562
A6328333	1562
A6330000	0261
A6330004	0261
A6330059	0261
A6330069	0261
A6333310	0018
A6333320	0018
A6333346	0018
A6333751	0313
A6342200	3094
A6358055	0194
A6359135	OEM
A6360250	0261
A6363200	0276
A6365232	OEM
A6502380	0006
A6502460	0525
A6509120	0006
A6509121	0006
A6509140	0006
A6509141	0006
A6509154	5954
A6531140	0436
A6532300	1348
A6532320	1911
A6532841	0477
A6532853	0477
A6532921	1348
A6532941	1348
A6533240	OEM
A6533730	3548
A6533750	1522
A6533760	1522
A6534020	0148
A6534021	0148
A6534036	4432
A6534040	0148
A6534045	0148
A6534053	0148
A6534060	0148
A6534063	0148
A6534120	0434
A6534125	0434
A6534135	0434
A6534145	0434
A6534430	OEM
A6546025	3111
A6546665	OEM
A6547300	0342
A6610160	0925
A6623870	1369
A6625365	3111
A6633620	3500
A6639320	OEM
A6707244	0076
A6707298	0076
A6708303	0076
A6708850	0842
A6708861	0842
A6708871	0842
A6716560	0786
A6716660	0525
A6733280	0111
A6733340	0191
A6733560	0191
A6734590	0551
A6734592	0551
A6738701	0220
A6754194H	0111
A6771373	1150
A6773802	0103
A6773812	0103
A6789710	0643
A6823565	0228
A6836940	0543
A6841063	3367
A6841420	4247
A6841440	4247
A6841905	OEM
A6842185	4887
A6843401	0178
A6846008	0270
A6846701	2668
A6846801	2051
A6846806	2051
A6846860	2051
A6847905	0055
A6848004	0055
A6848103	0055
A6848520	0461
A6848908	2069
A6852901	3516
A6852961	OEM
A6852981	3516
A6862752	2427
A6862754	2427
A6865553	0313
A6867051	2989
A6867061	2989
A6867970	OEM
A6868654	1541
A6868673	1541
A6868702	1541
A6868706	1541
A6868757	1541
A6868767	1541
A6871242	6089
A6883056	OEM
A6907751	OEM
A6907752	OEM
A6907753	OEM
A7000900	0143
A7001800	0143
A7001957	0143
A7102001	0499
A7109202	OEM
A7109304	0299
A7109395	0036
A7109396	OEM
A7109432	0275
A7109433	0275
A7109473	0140
A7109890	0471
A7110015	0062
A7110016	0041
A7110017	0253
A7110018	0253
A7110023	0041
A7110040	0041
A7110041	0041

If replacement code is OEM, contact original manufacturer for replacement.

DEVICE TYPE	REPL CODE
A7110055	0091
A7110076	0471
A7110077	0471
A7110114	0062
A7110115	0062
A7110116	0062
A7110159	6012
A7110160	0077
A7110207	0244
A7110209	0057
A7110263	0057
A7110359	5953
A7110360	0181
A7110410	1005
A7110411	0137
A7110412	0365
A7110461	0053
A7110462	OEM
A7110508	0879
A7110509	0686
A7110563	0445
A7110604	0495
A7110633	0671
A7110634	0700
A7110645	0705
A7110663	0671
A7110743	0166
A7115315	OEM
A7115414	OEM
A7115415	OEM
A7115605	OEM
A7115615	0760
A7116105	0275
A7116115	0275
A7116205	0589
A7116215	0140
A7116225	OEM
A7116305	0041
A7116315	0589
A7116405	0041
A7116415	0253
A7116505	0295
A7116515	0471
A7116615	0062
A7116625	0062
A7116705	0062
A7116715	OEM
A7116725	0077
A7116805	0077
A7116825	0165
A7116915	0057
A7116925	0057
A7117005	0057
A7117015	OEM
A7117025	OEM
A7117205	1005
A7117215	1005
A7117305	OEM
A7117415	OEM
A7117425	0686
A7117715	0671
A7118115	OEM
A7118305	OEM
A7118425	OEM
A7118925	0057
A7150041	0133
A7150258	0133
A7150351	0124
A7152800	OEM
A7160590	OEM
A7246602	0133
A7246653	0133
A7246700	0133
A7246703	0133
A7246711	0133
A7246727	0133
A7249601	0124
A7272130	1189
A7272143	1189
A7275400	0023
A7279039	0160
A7279049	0160
A7283550	0124
A7285774	0160
A7285778	0160
A7285900	0253
A7286100	0471
A7286107	0471
A7286120	0471
A7286201	0025
A7286500	0077
A7286700	0165
A7287500	0052
A7287513	0052
A7288504	0494
A7289000	0005
A7289047	0160
A7290594	0160
A7291252	0160
A7297043	0160
A7297092	0160
A7297093	0160
A7509201	0015
A7547053	0015
A7568160	0023
A7568250	0023
A7568300	0023
A7568460	0023
A7568475	0023

DEVICE TYPE	REPL CODE
A7568480	0023
A7568487	0023
A7568500	0023
A7568521	0023
A7568700	0023
A7568719	0023
A7568752	0023
A7568754	0795
A7570013-01	0037
A7572100	0023
A7572200	0023
A7576004-01	0037
A7579500	0071
A7580111	0071
A7580310	0564
A7580656	0023
A7580658	OEM
A7580660	0023
A7580910	0017
A7582000	0182
A7671002	0323
A7682012	0196
A7682052	0015
A7801021	0901
A7801166	0407
A7801233	0407
A7803610	0710
A7811055	OEM
A7950540	1282
A7978850	0023
A7978855	0023
A7978950	0023
A7986480	0795
A8015613	0126
A8600606	OEM
A8603110	OEM
A8603140	4767
A8605671	OEM
A8606430	OEM
A8606431	OEM
A8606630	OEM
A8606660	3491
A8608780	OEM
A8608781	OEM
A8612100	OEM
A8612101	OEM
A8612360	OEM
A8620900	OEM
A8626050	OEM
A8626220	OEM
A8632853	0477
A8636410	OEM
A8636690	OEM
A8641063	3367
A8641301	1291
A8641302	1291
A8641303	1291
A8643100	0112
A8643135	0112
A8652035	2628
A11159761	0124
A-11790169	0124
A20030703-0702	0037
A23114050	0037
A23114051	0037
A23114130	0233
A23114550	0037
A25114130	0233
A43023843	0103
A43023845	0037
A67531140	0436
A71909304	0299
A75779500	0071
A80052402	0103
A80414120	0103
A80414130	0103
A86087881	OEM
AA015	0917
AA-07077	0004
AA-07089	0016
AA1	0280
AA1A4M	OEM
AA1L3M	OEM
AA1L3Z	OEM
AA1L4M	OEM
AA2	0600
AA2SB240A	0160
AA3	0050
AA4	0160
AA5	0440
AA8-1-70	0050
AA8-1-705	0050
AA8-1A9G	0050
AA10	0039
AA10-1	0039
AA50	0015
AA100	0015
AA101	0901
AA102	0058
AA103	OEM
AA104	OEM
AA107	0344
AA108	0901
AA109	0901
AA110	OEM
AA111	0143
AA111A	OEM
AA112	0143
AA112A	OEM

DEVICE TYPE	REPL CODE
AA112B	OEM
AA112P	0143
AA113	0143
AA113P	0143
AA114	0143
AA115	0901
AA116	0143
AA116(SRC)	0058
AA117	0143
AA118	0143
AA118P	0143
AA119	0143
AA119P	0143
AA120	0143
AA121	0143
AA123	0143
AA129	OEM
AA130	0143
AA131	0143
AA132	0143
AA133	3280
AA134	0143
AA135	0143
AA136	0143
AA137	0143
AA138	0143
AA139	0143
AA140	0143
AA142	0143
AA143	0143
AA144	0143
AA200	0015
AA210	0143
AA218	0143
AA300	0015
AA310	3280
AA400	0015
AA401	OEM
AA500	0015
AA600	0015
AA779	0143
AA800	0071
AA1000	0071
AA5534S	OEM
AAIA4M	OEM
AAR	OEM
AAV-039	OEM
AAV039	OEM
AAV11A	OEM
AAV1043	OEM
AAV1051	OEM
AAY10-120	0090
AAY11	0124
AAY12	0143
AAY13	OEM
AAY14	0143
AAY15	0019
AAY18	0143
AAY21	OEM
AAY22	0143
AAY27	0143
AAY28	0143
AAY30	0143
AAY32	0143
AAY33	0143
AAY34	OEM
AAY39	OEM
AAY39A	OEM
AAY39R	OEM
AAY40	OEM
AAY42	0143
AAY43	OEM
AAY46	0143
AAY48	0143
AAY49	0143
AAY50	OEM
AAY50R	OEM
AAY51	OEM
AAY51R	OEM
AAY52	OEM
AAY52R	OEM
AAY53	0143
AAY54	0143
AAY55	0143
AAY56	OEM
AAY56R	OEM
AAY59	OEM
AAY60	OEM
AAY139	0143
AAZ10	0143
AAZ13	OEM
AAZ14	OEM
AAZ15	0143
AAZ17	0143
AAZ18	0143
AAZ18D	0015
AB1-9	0012
AB1-11	0181
AB1-35	0039
AB50	0087
AB100	0087
AB107	OEM
AB110	OEM
AB120	OEM
AB200	0087
AB207	OEM
AB210	OEM
AB220	OEM
AB300	0087

DEVICE TYPE	REPL CODE
AB307	OEM
AB310	OEM
AB320	OEM
AB400	0087
AB407	OEM
AB410	OEM
AB420	OEM
AB500	0087
AB507	OEM
AB510	OEM
AB520	OEM
AB600	0087
AB700	0087
AB800	0087
AB800-20	OEM
AB801-20	OEM
AB802-20	OEM
AB803-20	OEM
AB803-51	OEM
AB810-20	OEM
AB810A20	OEM
AB811-20	OEM
AB811A20	OEM
AB812-20	OEM
AB812A20	OEM
AB813A20	OEM
AB813A51	OEM
AB820-20	OEM
AB822-20	OEM
AB822-51	OEM
AB1000	0087
A-B1000-139	0917
A-B1000-162	0917
A-B1002-16	OEM
AB2027	0441
AB4164ANP-12	2357
AB4164ANP-15	2357
AB7543GKN	OEM
ABD100A	OEM
ABD250A	OEM
ABDX25A	OEM
ABDX50A	OEM
ABF	0161
AC02BGM	OEM
AC02BT	OEM
AC02BTR	OEM
AC02DGM	OEM
AC02DT	OEM
AC02DTR	OEM
AC03BGML	2300
AC03BGMR	2300
AC03DGML	2300
AC03DGMR	2300
AC03EGML	OEM
AC03EGMR	OEM
AC03FGML	OEM
AC03FGMR	OEM
AC04BGML	OEM
AC04BGMR	OEM
AC04DGML	OEM
AC04DGMR	OEM
AC04EGML	OEM
AC04EGMR	OEM
AC04FGML	OEM
AC04FGMR	OEM
AC06B	OEM
AC06BS	OEM
AC06BSR	0154
AC06BT	0772
AC06D	OEM
AC06DR	OEM
AC06DS	OEM
AC06DSR	0147
AC06DT	0744
AC08BGML	0593
AC08BGMR	OEM
AC08BIML	OEM
AC08BIMR	OEM
AC08D	OEM
AC08DGML	0576
AC08DGMR	0576
AC08DIML	OEM
AC08DIMR	OEM
AC08EGML	0594
AC08EGMR	0594
AC08EIML	OEM
AC08EIMR	OEM
AC08FGML	0594
AC08FGMR	0594
AC08FIML	OEM
AC08FIMR	OEM
AC031	OEM
AC032	OEM
AC060	OEM
AC061	OEM
AC068BR	OEM
AC0V8BGM	OEM
AC0V8DGM	OEM
AC0V8FGM	OEM
AC10B	OEM
AC10BGML	0593
AC10BGMR	0593
AC10BIML	OEM
AC10BIMR	OEM
AC10BR	OEM
AC10BS	OEM
AC10BSR	0154
AC10BT	0772
AC10D	OEM

DEVICE TYPE	REPL CODE
AC10DGML	OEM
AC10DGMR	0576
AC10DIML	OEM
AC10DIMR	OEM
AC10DR	OEM
AC10DS	OEM
AC10DSR	5143
AC10DT	0744
AC10EGML	0594
AC10EGMR	0594
AC10EIML	OEM
AC10EIMR	OEM
AC10FGML	0594
AC10FGMR	0594
AC10FIML	OEM
AC10FIMR	OEM
AC12BGML	0772
AC12BGMR	0744
AC12BIML	OEM
AC12BIMR	OEM
AC12DGML	0744
AC12DGMR	0744
AC12DIML	OEM
AC12DIMR	OEM
AC12EGML	0744
AC12EGMR	0617
AC12EIML	OEM
AC12EIMR	OEM
AC12FGML	0617
AC12FGMR	0617
AC12FIML	OEM
AC12FIMR	OEM
AC14V	OEM
AC16BGML	OEM
AC16BGMR	OEM
AC16BIFL	OEM
AC16BIFR	OEM
AC16DGML	OEM
AC16DGMR	OEM
AC16DIFL	OEM
AC16DIFR	OEM
AC16EGML	OEM
AC16EGMR	OEM
AC16EIFL	OEM
AC16EIFR	OEM
AC16FGML	OEM
AC16FGMR	OEM
AC16FIFL	OEM
AC16FIFR	OEM
AC18M/U	0208
AC25BIFL	OEM
AC25BIFR	OEM
AC25DIFL	OEM
AC25DIFR	OEM
AC25EIFL	OEM
AC25EIFR	OEM
AC25FIFL	OEM
AC25FIFR	OEM
AC28V	OEM
AC30	0087
AC42V	OEM
AC50	0087
AC50(DIODE)	0133
AC50(RECTIFIER)	0110
AC56V	OEM
AC100	0087
AC105	0004
AC106	0004
AC107	0050
AC107M	0050
AC107N	0050
AC108	0004
AC109	0004
AC110	0004
AC113	0004
AC-113A	0004
AC-114	0004
AC114	0004
AC115	0004
AC116	0004
AC116-G	0280
AC116-GRN	0280
AC116-Y	0280
AC116-YEL	0280
AC117	0004
AC117A	0004
AC117B	0004
AC117P	0004
AC118	0004
AC119	0004
AC120	0004
AC121	0004
AC121-IV	0004
AC121-V	0004
AC121-VI	0004
AC121-VII	0004
AC121IV	0164
AC121V	0004
AC121VI	0004
AC121VII	0004
AC122	0004
AC122/30-GRN	0280
AC122/30-RED	0280
AC122/30-YEL	0280
AC122-30	0004
AC122-G	0280
AC122-GRN	0164
AC122-R	0280
AC122-RED	0164

DEVICE TYPE	REPL CODE
AC122-V	0280
AC122-VIO	0164
AC122-Y	0280
AC122-YEL	0164
AC122GRN	0004
AC122RED	0004
AC122YEL	0004
AC123	0004
AC123-GRN	0280
AC123-YEL	0280
AC124	0004
AC125	0004
AC126	0164
AC127	0208
AC127/AC128-01	OEM
AC127/AC152	OEM
AC127-01	0208
AC127-132	0208
AC128	0164
AC128/01	0164
AC128/AC176	OEM
AC128-01	0164
AC128K	0164
AC129	0050
AC130	0600
AC130V	OEM
AC131	0004
AC131/30	0164
AC131-30	0164
AC132	0004
AC132-01	0004
AC133A	0004
AC134	0004
AC135	0004
AC136	0004
AC137	0004
AC138	0164
AC138H	OEM
AC139	0164
AC139/AC142	OEM
AC139K/AC142K	OEM
AC141	0208
AC-141B	2866
AC141B	0208
AC141H	OEM
AC141H-K	OEM
AC-141K	2866
AC141K	0208
AC142	0164
AC142H	OEM
AC142H-K	OEM
AC142K	0164
AC148	0160
AC150	0004
AC150-G	0280
AC150-GRN	0050
AC150-V	0280
AC150-VIO	0280
AC150-Y	0280
AC150-YEL	0050
AC150GRN	0004
AC150YEL	0004
AC151	0004
AC151-IV	0004
AC151-RIV	0004
AC151-RV	0004
AC151-RVI	0004
AC151-V	0004
AC151-VI	0004
AC151-VII	0004
AC151IV	0004
AC151R	0004
AC151RIV	0004
AC151RV	0004
AC151RVI	0004
AC151V	0004
AC151VI	0004
AC151VII	0004
AC152	0004
AC152-IV	0004
AC152-V	0004
AC152-VI	0004
AC152IV	0004
AC152V	0004
AC152VI	0004
AC153	0164
AC153K	0164
AC154	0164
AC154/AC157	OEM
AC155	0004
AC156	0004
AC157	0038
AC160	0004
AC160-G	0280
AC160-GRN	0050
AC160-R	1303
AC160-RED	0050
AC160-V	0280
AC160-VIO	0050
AC160-Y	1303
AC160-YEL	0050
AC160A	0004
AC160B	0004
AC160GRN	0004
AC160RED	0004
AC160YEL	0004
AC161	0004
AC162	0004
AC163	0004

DEVICE TYPE	REPL CODE
AC164	0136
AC165	0004
AC166	0164
AC166/AC168	OEM
AC167	0164
AC168	0004
AC169	0136
AC170	0004
AC171	0004
AC172	0208
AC173	0004
AC175	0208
AC175A	0208
AC175B	0208
AC175P	0208
AC176	0208
AC176K	0164
AC177	0280
AC178	0164
AC179	0208
AC180	0164
AC180K	0164
AC180KL	OEM
AC181	0208
AC181K	0208
AC181KL	OEM
AC182	0004
AC-183	2866
AC183	0208
AC184	0004
AC185	0208
AC-186	2866
AC186	0208
AC187	0208
AC187/01	0600
AC187/AC188	OEM
AC187-01	4256
AC187K	0208
AC187R	0208
AC188	0164
AC188/01	0164
AC188-01	OEM
AC188-01/AC187-01	OEM
AC188K	0164
AC191	0004
AC192	0004
AC193	0164
AC193K	0164
AC194	0208
AC194K	0208
AC200	0087
AC250V	OEM
AC300	0540
AC400	0540
AC500	0965
AC600	0965
AC800	0817
AC1000	0922
AC4045NL	2048
AC5945	OEM
AC5945N	OEM
AC5947	OEM
AC5947N	OEM
AC5947NF	OEM
AC9082	0037
AC9083	0037
AC9084	0037
AC9085	0037
AC12801	0164
AC18701	0208
AC18801	0164
ACDC9000-1	0037
ACIA-534-200	OEM
ACIA-534-201	OEM
ACIA-534-300	OEM
ACIA-534-301	OEM
ACMB81416-15	OEM
ACR-10	OEM
ACR25U04LG	OEM
ACR25U06LG	OEM
ACR25U08LG	OEM
ACR25U10LG	OEM
ACR35U06LG	OEM
ACR83-1001	0136
ACR83-1002	0136
ACR83-1003	0136
ACR83-1004	0280
ACR83-1005	0280
ACR83-1006	0280
ACR83-1007	0143
ACR83-1008	0015
ACR810-101	0136
ACR810-102	0136
ACR810-103	0136
ACR810-104	0280
ACR810-105	0280
ACR810-106	0280
ACR810-107	0143
ACR810-108	0015
ACS02	OEM
ACS04	OEM
ACS09	OEM
ACS09B	OEM
ACS090EM	OEM
ACS12-PRO	OEM
ACS14PRO	OEM
ACS15	OEM
ACS19	OEM
ACS64	OEM

If replacement code is OEM, contact original manufacturer for replacement.

DEVICE TYPE	REPL CODE
ACS65	OEM
ACS120EM	OEM
ACS140EM	OEM
ACY11	OEM
ACY14	OEM
ACY16	0160
ACY-17	0004
ACY17	0004
ACY-18	0004
ACY18	0004
ACY-19	0004
ACY19	0004
ACY20	0004
ACY20-1	0004
ACY-21	0004
ACY21	0004
ACY21-1	0004
ACY22	0004
ACY22-1	0004
ACY23	0004
ACY23-V	0004
ACY23-VI	0004
ACY23V	0004
ACY23VI	0004
ACY24	0050
ACY27	0004
ACY28	0004
ACY29	0004
ACY30	0004
ACY31	0004
ACY32	0004
ACY32-V	0004
ACY32-VI	0004
ACY32V	0004
ACY32VI	0004
ACY33	0164
ACY33-VI	0164
ACY33-VII	0164
ACY33-VIII	0164
ACY33VI	0164
ACY33VII	0164
ACY34	0004
ACY35	0004
ACY36	0004
ACY38	0004
ACY39	0847
ACY40	0211
ACY41	0211
ACY41-1	0004
ACY44	0004
ACY44-1	0004
ACZ	0050
ACZ21	0050
AD04	OEM
AD-2	0817
AD2	OEM
AD8	0102
AD10	0015
AD27	OEM
AD-29	0102
AD29	0087
AD-29A	0102
AD-29B	0102
AD-29S	0102
AD30	0133
AD50	0015
AD58DTH	OEM
AD100	0015
AD100(DIODE)	0133
AD101	OEM
AD101AH	0093
AD102	OEM
AD103	0604
AD104	0604
AD105	0604
AD107	OEM
AD108	OEM
AD109	OEM
AD110	OEM
AD110A	0133
AD111	OEM
AD114	OEM
AD115	OEM
AD116	OEM
AD117	OEM
AD118	OEM
AD130	0160
AD131	0160
AD132	0160
AD133	0604
AD138	0160
AD138/50	0160
AD-139	0222
AD139	0085
AD-140	0160
AD140	0160
AD142	0604
AD143	0160
AD143B	0160
AD143R	0160
AD145	0160
AD-148	0160
AD148	0160
AD-149	0160
AD149	0085
AD149-01	0160
AD149-02	0160
AD149B	0160
AD149C	0160
AD-150	0160
AD150	0160
AD150(DIODE)	0133
AD-152	0222
AD152	0222
AD153	0160
AD155	0222
AD-156	0222
AD156	0222
AD-157	2763
AD157	2763
AD-159	0160
AD159	0160
AD160	0178
AD161	2763
AD161/AD162	OEM
AD-162	0222
AD162	0222
AD163	0975
AD164	0222
AD165	2763
AD166	0975
AD167	0975
AD169	0222
AD200	0015
AD200(DIODE)	0133
AD201AH	0093
AD201AN	OEM
AD201H	0093
AD262	0222
AD263	0222
AD301AH	0093
AD301ALH	OEM
AD301ALN	OEM
AD301AN	2283
AD346JD	OEM
AD346SD(M)	OEM
AD360SD	OEM
AD380JH	OEM
AD380KH	OEM
AD380LH	OEM
AD381JH	OEM
AD381KH	OEM
AD381LH	OEM
AD382JH	OEM
AD382KH	OEM
AD382LH	OEM
AD411	0133
AD503J	OEM
AD503JH	OEM
AD503K	OEM
AD503KH	OEM
AD503S	OEM
AD503SH	OEM
AD504JH	OEM
AD504KH	OEM
AD504LH	OEM
AD504MH	OEM
AD504SH	OEM
AD506KH	OEM
AD506LH	OEM
AD506SH	OEM
AD507JH	OEM
AD507KH	OEM
AD507SH	OEM
AD509JH	OEM
AD509KH	OEM
AD509SH	OEM
AD510JH	OEM
AD510KH	OEM
AD510LH	OEM
AD515JH	OEM
AD515KH	OEM
AD515LH	OEM
AD516A	OEM
AD516B	OEM
AD517J	OEM
AD517JH	OEM
AD517K	OEM
AD517KH	OEM
AD517LH	OEM
AD517SH	OEM
AD518J	OEM
AD518JH	OEM
AD518K	OEM
AD518KH	OEM
AD521JD	OEM
AD521K	OEM
AD521KD	OEM
AD521LD	OEM
AD522AD	OEM
AD522BD	OEM
AD523JH	OEM
AD523KH	OEM
AD523LH	OEM
AD524A	OEM
AD524B	OEM
AD524C	OEM
AD528J	OEM
AD528K	OEM
AD528S	OEM
AD530JD	OEM
AD530JH	OEM
AD530KD	OEM
AD530KH	OEM
AD530LD	OEM
AD530LH	OEM
AD530SD	OEM
AD530SH	OEM
AD531JD	OEM
AD531KD	OEM
AD531LD	OEM
AD531SD	OEM
AD532J	OEM
AD532JD	OEM
AD532JH	OEM
AD532K	OEM
AD532KD	OEM
AD532KH	OEM
AD532SD	OEM
AD532SH	OEM
AD533JD	OEM
AD533JH	OEM
AD533KD	OEM
AD533KH	OEM
AD533LD	OEM
AD533SD	OEM
AD533SH	OEM
AD534J	OEM
AD534JH	OEM
AD534K	OEM
AD534KH	OEM
AD534LH	OEM
AD534SH	OEM
AD534TH	OEM
AD535JD	OEM
AD535JH	OEM
AD535KD	OEM
AD535KH	OEM
AD536AJD	OEM
AD536AJH	OEM
AD536AKD	OEM
AD536AKH	OEM
AD537JD	OEM
AD537JH	OEM
AD537KD	OEM
AD537KH	OEM
AD537SD	OEM
AD537SH	OEM
AD539JD	OEM
AD539SD	OEM
AD540JH	OEM
AD540KH	OEM
AD542JH	OEM
AD542KH	OEM
AD542LH	OEM
AD544JH	OEM
AD544KH	OEM
AD544LH	OEM
AD545JH	OEM
AD545KH	OEM
AD545LH	OEM
AD545MH	OEM
AD547JH	OEM
AD547KH	OEM
AD547LH	OEM
AD558JH	OEM
AD558JN	OEM
AD558KD	OEM
AD558KN	OEM
AD558SD	OEM
AD558TD	OEM
AD559KD	OEM
AD559SD	OEM
AD561JD	OEM
AD561KD	OEM
AD561SD	OEM
AD561TD	OEM
AD562A	OEM
AD562AD/BCD	OEM
AD562AD/BIN	OEM
AD562KD/BCD	OEM
AD562KD/BIN	OEM
AD562SD/BCD	OEM
AD562SD/BIN	OEM
AD563JD/BCD	OEM
AD563JD/BIN	OEM
AD563KD/BIN	OEM
AD563KD-BCD	OEM
AD563SD/BCD	OEM
AD563SD/BIN	OEM
AD563TD/BCD	OEM
AD563TD/BIN	OEM
AD564KN	OEM
AD565AJD	OEM
AD565AKD	OEM
AD565ASD	OEM
AD565ATD	OEM
AD565JD	OEM
AD565JN	OEM
AD565KD	OEM
AD565KN	OEM
AD565SD	OEM
AD565TD	OEM
AD566AJD	OEM
AD566AKD	OEM
AD566ASD	OEM
AD566ATD	OEM
AD566JD	OEM
AD566JN	OEM
AD566KN	OEM
AD566SD	OEM
AD566TD	OEM
AD567JD	OEM
AD567JN	OEM
AD567KD	OEM
AD567SD	OEM
AD567TD	OEM
AD570JD	OEM
AD570SD	OEM
AD571JD	OEM
AD571KD	OEM
AD571SD	OEM
AD574AJD	OEM
AD574AKD	OEM
AD574ALD	OEM
AD574ASD	OEM
AD574ATD	OEM
AD574AUD	OEM
AD574JD	OEM
AD574KD	OEM
AD574LD	OEM
AD574SD	OEM
AD574TD	OEM
AD574UD	OEM
AD578JD	OEM
AD578KD	OEM
AD578S	OEM
AD580JH	OEM
AD580KH	OEM
AD580LH	OEM
AD580MH	OEM
AD581JH	OEM
AD581KH	OEM
AD581LH	OEM
AD581SH	OEM
AD581TH	OEM
AD581UH(M)	OEM
AD582KD	OEM
AD582KH	OEM
AD582SD	OEM
AD582SH	OEM
AD583KD	OEM
AD584K	OEM
AD584KH	OEM
AD584LH	OEM
AD584SH	OEM
AD585JD	OEM
AD585SD	OEM
AD589JH	OEM
AD589KH	OEM
AD589LH	OEM
AD589MH	OEM
AD589SH	OEM
AD589TH	OEM
AD589UH	OEM
AD590IF	OEM
AD590IH	OEM
AD590JF	OEM
AD590JH	OEM
AD590KF	OEM
AD590KH	OEM
AD590LF	OEM
AD590LH	OEM
AD590MF	OEM
AD590MH	OEM
AD594AD	OEM
AD594CD	OEM
AD611A	OEM
AD611B	OEM
AD611C	OEM
AD630JD	OEM
AD636JH	OEM
AD636KD	OEM
AD636KH	OEM
AD636SD	OEM
AD637J	OEM
AD637K	OEM
AD642JH	OEM
AD642KH	OEM
AD642LH	OEM
AD642SH	OEM
AD647JH	OEM
AD647KH	OEM
AD647LH	OEM
AD647SH	OEM
AD650JN	OEM
AD650JQ	OEM
AD650KN	OEM
AD650KQ	OEM
AD650SQ	OEM
AD741	0410
AD741C	0410
AD741CH	0410
AD741CN	1975
AD741H	0410
AD741JH	OEM
AD741JN	6234
AD741KH	0410
AD741KN	0312
AD741LH	4262
AD741LN	6234
AD741SH	OEM
AD810	OEM
AD811	OEM
AD812	OEM
AD813	OEM
AD814	OEM
AD815	OEM
AD816	OEM
AD818	OEM
AD820	OEM
AD821	OEM
AD822	OEM
AD830	OEM
AD831	OEM
AD832	OEM
AD833	OEM
AD833A	OEM
AD840	OEM
AD841	OEM
AD842	OEM
AD1000	OEM
AD1403AN	OEM
AD1403N	OEM
AD1408-7D	4316
AD1408-8D	OEM
AD1408-9D	OEM
AD1508-8D	OEM
AD1508-9D	OEM
AD1600	OEM
AD2700JD	OEM
AD2700LD	OEM
AD2701JD	OEM
AD2701LD	OEM
AD2701S	OEM
AD2702JD	OEM
AD2702LD	OEM
AD2710KN	OEM
AD2710LN	OEM
AD2712KN	OEM
AD2712LN	OEM
AD2720AH	OEM
AD2720BH	OEM
AD2720SH	OEM
AD2720TH	OEM
AD3554AH	OEM
AD3954	OEM
AD3954A	OEM
AD3954BH	OEM
AD3955	OEM
AD3956	OEM
AD3958	OEM
AD4001	0015
AD4002	0071
AD4003	0071
AD4004	0071
AD4005	0071
AD4006	0071
AD4007	0071
AD5010KD	OEM
AD5201BD	OEM
AD5201TD	OEM
AD5202BD	OEM
AD5202TD	OEM
AD5204BD	OEM
AD5204TD	OEM
AD5205BD	OEM
AD5205TD	OEM
AD6020KD	OEM
AD7420LPE	OEM
AD7501JD	OEM
AD7501JN	OEM
AD7501KD	OEM
AD7501KN	OEM
AD7501SD	OEM
AD7502JD	OEM
AD7502JN	OEM
AD7502KD	OEM
AD7502KN	OEM
AD7502SD	OEM
AD7503JD	OEM
AD7503JN	OEM
AD7503KD	OEM
AD7503KN	OEM
AD7503SD	OEM
AD7506JD	OEM
AD7506KD	OEM
AD7506KN	OEM
AD7506SD	OEM
AD7506TD	OEM
AD7507JD	OEM
AD7507JN	OEM
AD7507KD	OEM
AD7507KN	OEM
AD7507SD	OEM
AD7507TD	OEM
AD7510DIJD	OEM
AD7510DIJN	OEM
AD7510DIKD	OEM
AD7510DIKN	OEM
AD7510DISD	OEM
AD7511DIJD	OEM
AD7511DIJN	OEM
AD7511DIKD	OEM
AD7511DIKN	OEM
AD7511DISD	OEM
AD7511DITD	OEM
AD7512DIJD	OEM
AD7512DIJN	OEM
AD7512DIKD	OEM
AD7512DIKN	OEM
AD7512DISD	OEM
AD7512DITD	OEM
AD7513JH	OEM
AD7513JN	OEM
AD7513KH	OEM
AD7513KN	OEM
AD7513SH	OEM
AD7513TH	OEM
AD7516JN	1143
AD7516SD	OEM
AD7516TD	OEM
AD7519JN	OEM
AD7520JD	OEM
AD7520JJE	OEM
AD7520JN	OEM
AD7520JPE	OEM
AD7520KD	OEM
AD7520KJE	OEM
AD7520KN	OEM
AD7520KPE	OEM
AD7520LD	OEM
AD7520LJE	OEM
AD7520LN	OEM
AD7520SD	OEM
AD7520SJE	OEM
AD7520TD	OEM
AD7520TJE	OEM
AD7520UD	OEM
AD7520UJE	OEM
AD7521JD	OEM
AD7521JJN	OEM
AD7521JN	OEM
AD7521JPN	OEM
AD7521KD	OEM
AD7521KJN	OEM
AD7521KN	OEM
AD7521KPN	OEM
AD7521LD	OEM
AD7521LJN	OEM
AD7521LN	OEM
AD7521LPN	OEM
AD7521SD	OEM
AD7521SJN	OEM
AD7521TD	OEM
AD7521TJN	OEM
AD7521UD	OEM
AD7521UJN	OEM
AD7522JD	OEM
AD7522JN	OEM
AD7522KD	OEM
AD7522KN	OEM
AD7522LD	OEM
AD7522LN	OEM
AD7522SD	OEM
AD7522TD	OEM
AD7522UD	OEM
AD7523AD	OEM
AD7523BD	OEM
AD7523CD	OEM
AD7523JN	OEM
AD7523KN	OEM
AD7523LN	OEM
AD7523SD	OEM
AD7523TD	OEM
AD7523UD	OEM
AD7524AD	OEM
AD7524BD	OEM
AD7524CD	OEM
AD7524JN	OEM
AD7524KN	OEM
AD7524LN	OEM
AD7524SD	OEM
AD7524TD	OEM
AD7524UD	OEM
AD7525BD	OEM
AD7525CD	OEM
AD7525KN	OEM
AD7525LN	OEM
AD7525TD	OEM
AD7525UD	OEM
AD7527BD	OEM
AD7527CD	OEM
AD7527GCD	OEM
AD7527GLN	OEM
AD7527GUD	OEM
AD7527KN	OEM
AD7527LN	OEM
AD7527TD	OEM
AD7527UD	OEM
AD7528AQ	OEM
AD7528BQ	OEM
AD7528CQ	OEM
AD7528JN	OEM
AD7528KN	OEM
AD7528LN	OEM
AD7528SD	OEM
AD7528TD	OEM
AD7528UD	OEM
AD7530JD	OEM
AD7530JN	OEM
AD7530JPE	OEM
AD7530KD	OEM
AD7530KN	OEM
AD7530KPE	OEM
AD7530LD	OEM
AD7530LN	OEM
AD7530LPE	OEM
AD7530SD	OEM
AD7530TD	OEM
AD7530UD	OEM
AD7531JD	OEM
AD7531JN	OEM
AD7531JPN	OEM
AD7531KD	OEM
AD7531KN	OEM
AD7531KPN	OEM
AD7531LD	OEM
AD7531LN	OEM
AD7531LPN	OEM
AD7531SD	OEM
AD7531TD	OEM
AD7531UD	OEM
AD7533AD	OEM
AD7533BD	OEM
AD7533CD	OEM
AD7533JN	OEM
AD7533KN	OEM
AD7533LN	OEM
AD7533SD	OEM
AD7533TD	OEM
AD7533UD	OEM
AD7541AD	OEM
AD7541BD	OEM
AD7541JN	OEM
AD7541KN	OEM
AD7541LN	OEM
AD7541SD	OEM
AD7541TD	OEM
AD7542AD	OEM
AD7542BD	OEM
AD7542JN	OEM
AD7542KN	OEM
AD7542SD	OEM
AD7542TD	OEM
AD7543AD	OEM
AD7543BD	OEM
AD7543GBD	OEM
AD7543GTD	OEM
AD7543JN	OEM
AD7543KN	OEM
AD7543SD	OEM
AD7543TD	OEM
AD7544AD	OEM
AD7544BD	OEM
AD7544GBD	OEM
AD7544GKN	OEM
AD7544GTD	OEM
AD7544JN	OEM
AD7544KN	OEM
AD7544SD	OEM
AD7544TD	OEM
AD7545AQ	OEM
AD7545BQ	OEM
AD7545CQ	OEM
AD7545GCQ	OEM
AD7545GLN	OEM
AD7545GUD	OEM
AD7545JN	OEM
AD7545KN	OEM
AD7545LN	OEM
AD7545SD	OEM
AD7545TD	OEM
AD7545UD	OEM
AD7546AD	OEM
AD7546BD	OEM
AD7546JN	OEM
AD7546KN	OEM
AD7550BD	OEM
AD7552KN	OEM
AD7555BD	OEM
AD7555KN	OEM
AD7570J	OEM
AD7570L	OEM
AD7574AD	OEM
AD7574BD	OEM
AD7574JN	OEM
AD7574KN	OEM
AD7574SD	OEM
AD7574TD	OEM
AD7583KN	OEM
AD8007C	OEM
AD13850	0160
AD51640/10	OEM
ADADC85C-12	OEM
ADB1200PCN	OEM
ADC0800PCD	OEM
ADC0800PCN	OEM
ADC0800PD	OEM
ADC0801	3792
ADC0801LCD	OEM
ADC0801LCN	OEM
ADC0801LD	OEM
ADC0802	3792
ADC0802LCD	3792
ADC0802LCN	3792
ADC0802LD	OEM
ADC0803	OEM
ADC0803LCD	3792
ADC0803LCN	3792
ADC0803LD	OEM
ADC0804	OEM
ADC0804LCD	3792
ADC0804LCN	3792
ADC0805	OEM
ADC0808	OEM
ADC0808CCJ	OEM
ADC0808CCN	OEM
ADC0808N	OEM
ADC0809	OEM
ADC0809CCN	OEM
ADC0809N	OEM
ADC0816	OEM
ADC0816CCJ	OEM
ADC0816CCN	OEM
ADC0816CJ	OEM
ADC0816N	OEM
ADC0817CCN	OEM
ADC0817N	OEM
ADC08041	OEM
ADC11	OEM
ADC12	OEM
ADC-32	OEM

DEVICE TYPE	REPL CODE
ADC-224-200	OEM
ADC-224-300	OEM
ADC-228-200	OEM
ADC-228-201	OEM
ADC-228-300	OEM
ADC-228-301	OEM
ADC830	OEM
ADC833	OEM
ADC833BCD	OEM
ADC833BCN	OEM
ADC833BD	OEM
ADC833CCD	OEM
ADC833CCN	OEM
ADC833CD	OEM
ADC856C	OEM
ADC856M	OEM
ADC1001BCD	OEM
ADC1001BCN	OEM
ADC1001BD	OEM
ADC1001CCD	OEM
ADC1001CCN	OEM
ADC1001CD	OEM
ADC1021BCD	OEM
ADC1021BCN	OEM
ADC1021BD	OEM
ADC1021CCD	OEM
ADC1021CCN	OEM
ADC1021CD	OEM
ADC1080HCD	OEM
ADC1210HCD	OEM
ADC1210HD	OEM
ADC1211HCD	OEM
ADC1211HD	OEM
ADC1280HCD	OEM
ADC3511CC	OEM
ADC3711CC	OEM
ADC7109C	OEM
ADC7109M	OEM
ADC7109R	OEM
ADC-EK8B	OEM
ADC-EK10B	OEM
ADC-EK12B	OEM
ADC-EK12DC	OEM
ADC-EK12DM	OEM
ADC-EK12DR	OEM
ADC-ET8BC	OEM
ADC-ET8BM	OEM
ADC-ET10BC	OEM
ADC-ET10BM	OEM
ADC-ET12BC	OEM
ADC-ET12BM	OEM
ADC-ET12BR	OEM
ADC-MC8BC	OEM
ADC-MC8BM	OEM
ADD3501	5864
ADDAC08AD	OEM
ADDAC08CD	OEM
ADDAC08D	OEM
ADDAC08ED	OEM
ADDAC08HD	OEM
ADDAC80Z-CB1-1	OEM
ADDAC80Z-CBI-I	OEM
ADDAC80Z-CBI-V	OEM
ADDAC80Z-CCD-I	OEM
ADDAC80Z-CCD-V	OEM
ADG200AA	OEM
ADG200AP	OEM
ADG200BA	OEM
ADG200BP	OEM
ADG200CJ	OEM
ADG201AP	OEM
ADG201BP	OEM
ADG201CJ	OEM
ADH03011-8-3	OEM
ADH03011-12-3	OEM
ADLC-540-200	OEM
ADLC-540-201	OEM
ADLC-540-300	OEM
ADLC-540-301	OEM
AD026B	OEM
AD027B	OEM
AD029B	OEM
AD032	OEM
AD044	OEM
AD045	OEM
AD060	OEM
AD072A	OEM
ADOP07AH	OEM
ADOP07C	OEM
ADOP07CH	OEM
ADOP07CN	OEM
ADOP07D	OEM
ADOP07DH	OEM
ADOP07DN	OEM
ADOP07EH	OEM
ADOP07EN	OEM
ADOP07H	OEM
ADP550	OEM
ADP560	OEM
ADP560AC	OEM
ADP560AP	OEM
ADP560AT	OEM
ADP560ATC	OEM
ADP560ATP	OEM
ADP560C	OEM
ADP560P	OEM
ADP560TC	OEM
ADP560TP	OEM
ADP590	OEM
ADP590AC	OEM
ADP590AP	OEM
ADP590ATC	OEM
ADP590ATP	OEM
ADP590C	OEM
ADP590P	OEM
ADP590TC	OEM
ADP590TP	OEM
ADP1010	OEM
ADP1015	OEM
ADP1020	OEM
ADP1400	OEM
ADP1416	OEM
ADP1500	OEM
ADP1520	OEM
ADP1560	OEM
ADP1620/1	OEM
ADP1620/2	OEM
ADP1640/12	OEM
ADP1640/DF	OEM
ADP1640/SH	OEM
ADP1642	OEM
ADSHC85	OEM
ADSHC85ET	OEM
ADV11A	OEM
ADVEFC32BH	OEM
ADVFC32KN	OEM
ADVFC32SH	OEM
ADX0214	0636
ADX1103	0094
ADX1148	0080
ADX1152	0124
ADX1364	0124
ADX1473	0036
ADX1474	0953
ADX1518	0760
ADY10	OEM
ADY11	OEM
ADY12	OEM
ADY13	OEM
ADY18	OEM
ADY20	OEM
ADY22	0160
ADY23	0160
ADY24	0160
ADY25	OEM
ADY-26	0440
ADY26	0040
ADY27	0160
ADY28	0160
ADY30	OEM
ADY32	OEM
ADZ11	0440
ADZ12	0440
AE1A	0087
AE1B	0087
AE1C	0087
AE1D	0087
AE1E	0087
AE1F	0087
AE1G	0087
AE3A	0110
AE3B	0953
AE3C	0242
AE3D	0540
AE3E	0965
AE3F	0817
AE3G	0922
AE10	0133
AE30	0133
AE-50	0004
AE50	0133
AE100	0133
AE150	0133
AE200	0133
AE304	1395
AE623	OEM
AE900	2714
AE904	1395
AE-904-03	1395
AE904-04	1395
AE904-4	1395
AE904-6	1395
AE904-51	1395
AE906	2617
AE-907	1443
AE907	1443
AE907-51	1443
AE907	1345
AE908	1843
AE908-51	1843
AE909	OEM
AE915	3079
AE920	1048
AE921	OEM
AE922	OEM
AEIC50723202	OEM
AEICC4277224	OEM
AEL10	OEM
AEL10H	OEM
AEL13	OEM
AEL14	OEM
AEL16	OEM
AEL1009	OEM
AEL1015	OEM
AEL1024	OEM
AEL1038	OEM
AEL1039	OEM
AEL1044	OEM
AEL1053	OEM
AEL1062	OEM
AEL1067	OEM
AEL1075	OEM
AEL1077	OEM
AEL-A51D	OEM
AEY11	OEM
AEY12	OEM
AEY13	OEM
AEY14	OEM
AEY15	OEM
AEY16	OEM
AEY17	OEM
AEY23	OEM
AEY24	OEM
AEY25	OEM
AEY26	OEM
AEY27	OEM
AEY28	OEM
AEY29	OEM
AEY29R	OEM
AEY30A	OEM
AEY30B	OEM
AEY30C	OEM
AEY30D	OEM
AEY31	OEM
AEY31A	OEM
AEY32	OEM
AEY39S	OEM
AF1(DIODE)	0102
AF4F	OEM
AF63LA	OEM
AF63LB	OEM
AF63LBA	OEM
AF68-3	OEM
AF99	OEM
AF101	0050
AF101CJ	OEM
AF102	0050
AF102CJ	OEM
AF103CJ	OEM
AF104CJ	OEM
AF105	0136
AF105A	0050
AF105CJ	OEM
AF-106	0050
AF106	0050
AF106A	0050
AF107	0050
AF108	0136
AF109	0050
AF109R	0050
AF110	0050
AF110CJ	OEM
AF111	0050
AF111CJ	OEM
AF112	0050
AF112CJ	OEM
AF113	0050
AF113CJ	OEM
AF114	0136
AF114CJ	OEM
AF114N	0050
AF115	0136
AF115N	0050
AF116	0136
AF116N	0050
AF117	0136
AF117C	0050
AF117N	0050
AF118	0050
AF119	0050
AF120	0136
AF120CH	OEM
AF120H	OEM
AF121	0050
AF121-1CJ	OEM
AF121-2CJ	OEM
AF121S	0050
AF122	0050
AF122-1CJ	OEM
AF122-2CJ	OEM
AF124	0136
AF125	0136
AF126	0136
AF127	0136
AF127/01	0050
AF128	0050
AF129	0136
AF130	0136
AF131	0136
AF132	0136
AF133	0136
AF134	0136
AF134-1CJ	OEM
AF134-2CJ	OEM
AF135	0136
AF136	0136
AF-137	0050
AF137	0136
AF137A	0050
AF138	0136
AF138/20	0050
AF139	0050
AF142	0050
AF143	0050
AF144	0136
AF146	0050
AF147	0050
AF148	0050
AF149	0050
AF150	0050
AF150-1CJ	OEM
AF150-2CJ	OEM
AF151-1CJ	OEM
AF151-2CJ	OEM
AF164	0050
AF165	0050
AF166	0050
AF167	0050
AF168	0050
AF169	0050
AF170	0050
AF171	0050
AF172	0050
AF178	0050
AF179	0050
AF180	0050
AF181	0050
AF182	0050
AF185	0050
AF186	0050
AF186G	0050
AF186W	0050
AF187	0004
AF188	0004
AF192	0600
AF193	0050
AF200	0050
AF200U	0050
AF201	0050
AF201C	0050
AF201U	0050
AF202	0050
AF202L	0050
AF202S	0050
AF239	0050
AF239S	0050
AF240	0050
AF251	0050
AF252	0050
AF253	0050
AF254	OEM
AF256	0050
AF267	0050
AF279	0050
AF279S	0050
AF280	0160
AF280S	OEM
AF306	0050
AF367	0050
AF369	0050
AF699	0037
AF3570	0037
AF3590	0037
AF21490	0037
AF16672587	0345
AFC3527	0079
AFCS1170F	0037
AFE428F003X3	OEM
AFE436F003X1	OEM
AFL89WB20000C3	OEM
AFS-160-1017	0143
AFS-160-1020	0123
AFS-160-1021	0015
AFS-760-1017	0143
AFS24226	0037
AFT0019M	0037
AFT052	0037
AFT857P05L	OEM
AFT1341	0037
AFT1746	0037
AFT2031	OEM
AFT2032	OEM
AFT4231	OEM
AFT4232	OEM
AFT6231	OEM
AFT6232	OEM
AFT8431	OEM
AFT8432	OEM
AFT12631	OEM
AFT12632	OEM
AFT18632	OEM
AFT18832	OEM
AFY10	0136
AFY11	0136
AFY12	0050
AFY14	0050
AFY15	0050
AFY16	0050
AFY18	0136
AFY18C	0050
AFY18D	0050
AFY18E	0050
AFY19	0136
AFY34	0050
AFY37	0050
AFY39	0050
AFY40	0050
AFY40K	0050
AFY40R	0050
AFY41	0050
AFY42	0050
AFZ11	0050
AFZ12	0050
AFZ23	0280
AG01	OEM
AG056	OEM
AG6	0551
AG30	0133
AG100	OEM
AG100D	0015
AG100G	0015
AG100J	0015
AG106	OEM
AG134	0136
AG150	OEM
AG206	OEM
AG406	OEM
AG466	OEM
AG606	OEM
AGC330	OEM
AGH-001	4626
AG0512	0319
AG01012	0319
AG01512	0319
AG02012	0319
AG02512	0319
AG03012	0319
AG03512	0319
AG04012	0319
AG05012	0319
AG06012	0319
AGP15-200	0071
AGP15-400	0071
AGP15-600	0071
AGP15-800	0071
AH0010F(A)	OEM
AH0013CA	OEM
AH0013CB	OEM
AH0013MA	OEM
AH0013MB	OEM
AH0014CD	OEM
AH0014CN	OEM
AH0014D	OEM
AH0015CN	OEM
AH0015D	OEM
AH0019CD	OEM
AH0019CN	OEM
AH0019D	OEM
AH0126CD	OEM
AH0126D	OEM
AH0129CD	OEM
AH0129D	OEM
AH0133CD	OEM
AH0133CN	OEM
AH0133D	OEM
AH0134CD	OEM
AH0134CN	OEM
AH0134D	OEM
AH0139CD	OEM
AH0139D	OEM
AH0140CD	OEM
AH0140D	OEM
AH0141CD	OEM
AH0141D	OEM
AH0142CD	OEM
AH0142D	OEM
AH0143CD	OEM
AH0143D	OEM
AH0144CD	OEM
AH0144D	OEM
AH0145CD	OEM
AH0145D	OEM
AH0146CD	OEM
AH0146D	OEM
AH0151CD	OEM
AH0151CN	OEM
AH0151D	OEM
AH0152CD	OEM
AH0152CN	OEM
AH0152D	OEM
AH0153CD	OEM
AH0153D	OEM
AH0154CD	OEM
AH0154D	OEM
AH0161CD	OEM
AH0161D	OEM
AH0162CD	OEM
AH0162D	OEM
AH0163CD	OEM
AH0163D	OEM
AH0164CD	OEM
AH0164D	OEM
AH1	OEM
AH3	OEM
AH5	OEM
AH7	OEM
AH9	OEM
AH11	OEM
AH11-1	OEM
AH11-2	OEM
AH15	OEM
AH17	OEM
AH18	OEM
AH19	OEM
AH23	OEM
AH24	OEM
AH25	OEM
AH27	OEM
AH28	OEM
AH29	OEM
AH31	OEM
AH33	OEM
AH35	OEM
AH37	OEM
AH39	OEM
AH53	OEM
AH54	OEM
AH55	OEM
AH56	OEM
AH57-1	OEM
AH58	OEM
AH59	OEM
AH59-1	OEM
AH60	OEM
AH63	OEM
AH64	OEM
AH65	OEM
AH74	OEM
AH74-1	OEM
AH75	OEM
AH75-2	OEM
AH76	OEM
AH77	OEM
AH77-1	OEM
AH79	OEM
AH401	OEM
AH402	OEM
AH403	OEM
AH404	OEM
AH521	OEM
AH522	OEM
AH591	OEM
AH592	OEM
AH593	OEM
AH594	OEM
AH805	0071
AH810	0071
AH814	0071
AH815	0071
AH1005	0071
AH1010	0071
AH1015	0071
AH1020	0319
AH1510	0319
AH1520	0319
AH2010	0319
AH2020	0319
AH2114CG	OEM
AH2114G	OEM
AH2510	0319
AH2520	0319
AH4013	OEM
AH4077	OEM
AH4229	OEM
AH4230	OEM
AH5009CN	OEM
AH5010CN	OEM
AH5011CN	OEM
AH5012CN	OEM
AH5013CN	OEM
AH5014CN	OEM
AH5015CN	OEM
AH5016CN	OEM
AHB040	OEM
AHD554	OEM
AHD555	OEM
AHD557	OEM
AHD559	OEM
AHD560	OEM
AHD561	OEM
AH0510	0319
AH0520	0319
AHV8401	OEM
AHV8402	OEM
AHV8403	OEM
AHV8501	OEM
AHV8502	OEM
AHV8503	OEM
AHV8601	OEM
AHV8602	OEM
AHV8603	OEM
AHV8701	OEM
AHV8702	OEM
AHV8703	OEM
AHV8801	OEM
AHV8802	OEM
AHV8803	OEM
AHV9001	OEM
AHV9001A	OEM
AHV9002	OEM
AHV9002A	OEM
AHV9004	OEM
AHV9102	OEM
AHV9103A	OEM
AHV9104	OEM
AHV9201	OEM
AHV9201A	OEM
AHV9202	OEM
AHV9202A	OEM
AHV9204	OEM
AHV9301	OEM
AHV9301A	OEM
AHV9302	OEM
AHV9302A	OEM
AHV9304	OEM
AHV9401	OEM
AHV9401A	OEM
AHV9402	OEM
AHV9402A	OEM
AHV9404	OEM
AHV9501	OEM
AHV9501A	OEM
AHV9502	OEM
AHV9502A	OEM
AHV9504	OEM
AHV9801	OEM
AHV9801A	OEM
AHV9802	OEM
AHV9802A	OEM
AHV9803	OEM
AHV9803A	OEM
AHV9804	OEM
AHY10A	OEM
AHY10B	OEM
AHY10C	OEM
AHY10D	OEM
AIE	0007
AIH100A	OEM
AIM-12	OEM
AIM-12B	OEM
AJ5	0015
AJ10	0015
AJ15	0015
AJ20	0015
AJ25	0015
AJ30	0015
AJ35	0015
AJ40	0015
AJ50	0015
AJ60	0015
AJ103	0604
AJ7010	OEM
AJZ61	OEM
AK03	0735
AK1	1414
AK261	0143
AK10182-F20	OEM
AKBP02	0289
AKBP04	0296
AKBP06	0302
AKBP08	0250
AKBPC102	OEM
AKBPC104	OEM
AKBPC106	OEM
AKBPC108	OEM
AKBPC602	OEM
AKBPC604	OEM
AKBPC606	OEM
AKBPC608	OEM
AKBPC802	OEM
AKBPC804	OEM
AKBPC806	OEM
AKBPC808	OEM
AL-0010	OEM
AL01Z	1090
AL1C	0196
AL-10-5	2363
AL-10-10	2363
AL-10-20	2363
AL-10-30	2369
AL-10-40	2369
AL-10-50	2370
AL-10-60	2370
AL-10-70	2372
AL-10-80	2372
AL-10-90	2373
AL-10-100	2373
AL-10-120	OEM
AL-12-5	2363
AL-12-10	2363
AL-12-20	2363
AL-12-30	2369
AL-12-40	2369
AL-12-50	2370
AL-12-60	2370
AL-12-70	2372
AL-12-80	2372
AL-12-90	2373
AL-12-100	2373
AL-12-120	OEM
AL-15-5	2363
AL-15-10	2363
AL-15-20	2363
AL-15-30	2369
AL-15-40	2369
AL-15-50	2370
AL-15-60	2370
AL-15-70	2372
AL-15-80	2372
AL-15-90	2373
AL-15-100	2373
AL-15-120	OEM
AL-20-5	2363
AL-20-10	2363
AL-20-20	2363
AL-20-30	2369
AL-20-40	2369
AL-20-50	2370
AL-20-60	2370
AL-20-70	2372
AL-20-80	2372
AL-20-90	2373
AL-20-100	2373
AL-20-120	OEM
AL-25-5	2363
AL-25-10	2363
AL-25-20	2363
AL-25-30	2369
AL-25-40	2369
AL-25-50	2370
AL-25-60	2370
AL-25-70	2372
AL-25-80	2372

DEVICE TYPE	REPL CODE
AL-25-90	2373
AL-25-100	2373
AL-25-120	OEM
AL-30-5	1643
AL-30-10	1643
AL-30-20	1643
AL-30-30	1673
AL-30-40	1673
AL-30-50	1673
AL-30-60	1673
AL-30-70	1589
AL-30-80	1589
AL-30-90	1589
AL-30-100	1589
AL-30-120	OEM
AL100	0975
AL101	0160
AL102	0975
AL102A	OEM
AL102B	OEM
AL102V	OEM
AL103	0975
AL103A	OEM
AL103B	OEM
AL106A	OEM
AL106B	OEM
AL106V	OEM
AL107A	OEM
AL107B	OEM
AL109A	OEM
AL112	0975
AL112A	OEM
AL112B	OEM
AL112D	OEM
AL112E	OEM
AL112G	OEM
AL112I	OEM
AL112SZ	OEM
AL112V	OEM
AL113	0604
AL113A	OEM
AL113B	OEM
AL113D	OEM
AL113E	OEM
AL113G	OEM
AL113I	OEM
AL113K	OEM
AL113L	OEM
AL113M	OEM
AL113N	OEM
AL113R	OEM
AL113S	OEM
AL113SZ	OEM
AL113V	OEM
AL210	0050
AL304A	OEM
AL304B	OEM
AL304G	OEM
AL304V	OEM
AL307	OEM
AL309A	OEM
AL309B	OEM
AL309D	OEM
AL309E	OEM
AL309G	OEM
AL309I	OEM
AL309K	OEM
AL309SZ	OEM
AL309V	OEM
AL80240	OEM
ALC1	0196
ALC1A	0196
ALCI	0196
ALD980	OEM
ALF1025	OEM
ALF3003	OEM
ALR450	OEM
ALS00A(SM)	OEM
ALS02(SM)	OEM
ALS08(SM)	OEM
ALS27(SM)	OEM
ALS38A(SM)	OEM
ALS74A(SM)	OEM
ALS244A	OEM
ALS244A(SM)	OEM
ALS245A	OEM
ALS257	OEM
ALS311A	OEM
ALS311B	OEM
ALS311G	OEM
ALS311V	OEM
ALS313A5	OEM
ALS374	OEM
ALS573B	OEM
ALS-8922	0016
ALT-512	OEM
ALTR-2480	OEM
ALZ10	0280
AM	0086
AM005	0015
AM01	0015
AM01M	0015
AM01Z	0015
AM010	0015
AM020	0015
AM025	0015
AM030	0015
AM035	0015
AM040	0015
AM050	0015
AM060	0015
AM06200012	0062
AM06200013	0133
AM06200014	0023
AM06200017	0143
AM06200065	0071
AM06200076	0288
AM1	0708
AM1S25610	OEM
AM2	0708
AM3	0015
AM4	0708
AM5	0708
AM9	0233
AM11	0580
AM12	0580
AM13	0015
AM21	0580
AM22	0580
AM23	0015
AM24	0580
AM25L02DC	OEM
AM25L02DM	OEM
AM25L02FM	OEM
AM25L02PC	OEM
AM25L02XC	OEM
AM25L02XM	OEM
AM25L03DC	OEM
AM25L03DM	OEM
AM25L03FM	OEM
AM25L03PC	OEM
AM25L03XC	OEM
AM25L03XM	OEM
AM25L04DC	OEM
AM25L04DM	OEM
AM25L04FM	OEM
AM25L04PC	OEM
AM25L04XC	OEM
AM25L04XM	OEM
AM25L05DC	OEM
AM25L05DM	OEM
AM25L05FM	OEM
AM25L05PC	OEM
AM25L06DC	OEM
AM25L06DM	OEM
AM25L06FM	OEM
AM25L06PC	OEM
AM25LS07DC	OEM
AM25LS07DM	OEM
AM25LS07FM	OEM
AM25LS07PC	OEM
AM25LS08DC	OEM
AM25LS08DM	OEM
AM25LS08FM	OEM
AM25LS08PC	OEM
AM25LS09DC	OEM
AM25LS09DM	OEM
AM25LS09FM	OEM
AM25LS09PC	OEM
AM25LS14DC	OEM
AM25LS14DM	OEM
AM25LS14FM	OEM
AM25LS14PC	OEM
AM25LS15DC	OEM
AM25LS15DM	OEM
AM25LS15FM	OEM
AM25LS15PC	OEM
AM25LS22DC	OEM
AM25LS22DM	OEM
AM25LS22FM	OEM
AM25LS22PC	OEM
AM25LS23DC	OEM
AM25LS23DM	OEM
AM25LS23FM	OEM
AM25LS23PC	OEM
AM25LS138DC	OEM
AM25LS138DM	OEM
AM25LS138FM	OEM
AM25LS138PC	OEM
AM25LS138XC	OEM
AM25LS138XM	OEM
AM25LS139DC	OEM
AM25LS139DM	OEM
AM25LS139FM	OEM
AM25LS139PC	OEM
AM25LS139XC	OEM
AM25LS139XM	OEM
AM25LS151PC	1646
AM25LS160DC	OEM
AM25LS160DM	OEM
AM25LS160FM	OEM
AM25LS160PC	OEM
AM25LS161DC	OEM
AM25LS161DM	OEM
AM25LS161FM	OEM
AM25LS161PC	OEM
AM25LS162DC	OEM
AM25LS162DM	OEM
AM25LS162FM	OEM
AM25LS162PC	OEM
AM25LS163DC	OEM
AM25LS163DM	OEM
AM25LS163FM	OEM
AM25LS163PC	OEM
AM25LS164DC	OEM
AM25LS164DM	OEM
AM25LS164FM	OEM
AM25LS164PC	OEM
AM25LS174DC	OEM
AM25LS174DM	OEM
AM25LS174FM	OEM
AM25LS174PC	OEM
AM25LS175DC	OEM
AM25LS175DM	OEM
AM25LS175FM	OEM
AM25LS175PC	OEM
AM25LS181DC	OEM
AM25LS181DM	OEM
AM25LS181FM	OEM
AM25LS181PC	OEM
AM25LS190DC	OEM
AM25LS190DM	OEM
AM25LS190FM	OEM
AM25LS190PC	OEM
AM25LS191DC	OEM
AM25LS191DM	OEM
AM25LS191FM	OEM
AM25LS191PC	OEM
AM25LS192DC	OEM
AM25LS192DM	OEM
AM25LS192FM	OEM
AM25LS192PC	OEM
AM25LS193DC	OEM
AM25LS193DM	OEM
AM25LS193FM	OEM
AM25LS193PC	OEM
AM25LS194ADC	OEM
AM25LS194ADM	OEM
AM25LS194AFM	OEM
AM25LS194APC	OEM
AM25LS195ADC	OEM
AM25LS195ADM	OEM
AM25LS195AFM	OEM
AM25LS195APC	OEM
AM25LS273DC	OEM
AM25LS273DM	OEM
AM25LS273FM	OEM
AM25LS273PC	OEM
AM25LS299DC	OEM
AM25LS299DM	OEM
AM25LS299FM	OEM
AM25LS299PC	OEM
AM25LS373DC	OEM
AM25LS373DM	OEM
AM25LS373FM	OEM
AM25LS373PC	OEM
AM25LS373XC	OEM
AM25LS373XM	OEM
AM25LS374DC	OEM
AM25LS374DM	OEM
AM25LS374FM	OEM
AM25LS374PC	OEM
AM25LS377BDC	OEM
AM25LS377BDM	OEM
AM25LS377BFM	OEM
AM25LS377BPC	OEM
AM25LS377DC	OEM
AM25LS377DM	OEM
AM25LS377FM	OEM
AM25LS377PC	OEM
AM25LS381DC	OEM
AM25LS381DM	OEM
AM25LS381FM	OEM
AM25LS381PC	OEM
AM25LS533DC	OEM
AM25LS533DM	OEM
AM25LS533FM	OEM
AM25LS533PC	OEM
AM25LS533XC	OEM
AM25LS2513DC	OEM
AM25LS2513DM	OEM
AM25LS2513FM	OEM
AM25LS2513PC	OEM
AM25LS2513XC	OEM
AM25LS2513XM	OEM
AM25LS2516DC	OEM
AM25LS2516DM	OEM
AM25LS2517DC	OEM
AM25LS2517DM	OEM
AM25LS2517FM	OEM
AM25LS2517PC	OEM
AM25LS2519DC	OEM
AM25LS2519DM	OEM
AM25LS2519FM	OEM
AM25LS2519PC	OEM
AM25LS2520DC	OEM
AM25LS2520DM	OEM
AM25LS2520PC	OEM
AM25LS2521PC	OEM
AM25LS2521PC-B	OEM
AM25LS2548DC	OEM
AM25LS2548DM	OEM
AM25LS2548FM	OEM
AM25LS2548PC	OEM
AM25LS2568DC	OEM
AM25LS2568DM	OEM
AM25LS2568FM	OEM
AM25LS2568PC	OEM
AM25LS2569DC	OEM
AM25LS2569DM	OEM
AM25LS2569FM	OEM
AM25LS2569PC	OEM
AM25S05DC	OEM
AM25S05DM	OEM
AM25S05FM	OEM
AM25S05PC	OEM
AM25S07DC	OEM
AM25S07DM	OEM
AM25S07FM	OEM
AM25S07PC	OEM
AM25S08DC	OEM
AM25S08DM	OEM
AM25S08FM	OEM
AM25S08PC	OEM
AM25S09DC	OEM
AM25S09DM	OEM
AM25S09FM	OEM
AM25S09PC	OEM
AM25S10DC	OEM
AM25S10DM	OEM
AM25S10FM	OEM
AM25S10PC	OEM
AM25S18DC	OEM
AM25S18DM	OEM
AM25S18FM	OEM
AM25S18PC	OEM
AM25S557DC	OEM
AM25S557DM	OEM
AM25S558DC	OEM
AM25S558DM	OEM
AM26L02DC	OEM
AM26L02DM	OEM
AM26L02FM	OEM
AM26L02PC	6859
AM26LS30PC	OEM
AM26LS31	OEM
AM26LS31CJ	OEM
AM26LS31CN	OEM
AM26LS31DC	OEM
AM26LS31DM	OEM
AM26LS31FM	OEM
AM26LS31J	OEM
AM26LS31MH	OEM
AM26LS31N	OEM
AM26LS31PC	OEM
AM26LS31XC	OEM
AM26LS31XM	OEM
AM26LS32	OEM
AM26LS32A	OEM
AM26LS32ACJ	OEM
AM26LS32ACN	OEM
AM26LS32AM	OEM
AM26LS32DC	OEM
AM26LS32DM	OEM
AM26LS32FM	OEM
AM26LS32J	OEM
AM26LS32N	OEM
AM26LS32PC	OEM
AM26LS32SM	OEM
AM26LS32XC	OEM
AM26LS33A	OEM
AM26LS33ACJ	OEM
AM26LS33ACN	OEM
AM26LS33AM	OEM
AM26LS33DC	OEM
AM26LS33DM	OEM
AM26LS33FM	OEM
AM26LS33PC	OEM
AM26LS33XC	OEM
AM26LS33XM	OEM
AM26S02DC	OEM
AM26S02DM	OEM
AM26S02FM	OEM
AM26S02PC	OEM
AM26S10C	OEM
AM26S10CJ	OEM
AM26S10CN	OEM
AM26S10DC	OEM
AM26S10DM	OEM
AM26S10FM	OEM
AM26S10M	OEM
AM26S10MJ	OEM
AM26S10MN	OEM
AM26S10PC	OEM
AM26S10XC	OEM
AM26S10XM	OEM
AM26S11C	OEM
AM26S11CJ	OEM
AM26S11CN	OEM
AM26S11DC	OEM
AM26S11DM	OEM
AM26S11FM	OEM
AM26S11MJ	OEM
AM26S11MN	OEM
AM26S11PC	OEM
AM26S11XC	OEM
AM26S11XM	OEM
AM26S12ADC	OEM
AM26S12ADM	OEM
AM26S12AFM	OEM
AM26S12APC	OEM
AM26S12AXC	OEM
AM26S12AXM	OEM
AM26S12DC	OEM
AM26S12DM	OEM
AM26S12FM	OEM
AM26S12PC	OEM
AM26S12XC	OEM
AM26S12XM	OEM
AM27LS00DC	OEM
AM27LS00DM	OEM
AM27LS00FM	OEM
AM27LS00PC	OEM
AM27LS01DC	OEM
AM27LS01DM	OEM
AM27LS01FM	OEM
AM27LS01PC	OEM
AM27LS02ADC	OEM
AM27LS02ADM	OEM
AM27LS02AFM	OEM
AM27LS02APC	OEM
AM27LS02DC	OEM
AM27LS02DM	OEM
AM27LS02FM	OEM
AM27LS02PC	OEM
AM27LS03DC	OEM
AM27LS03DM	OEM
AM27LS03FM	OEM
AM27LS03PC	OEM
AM27LS18DC	OEM
AM27LS18DM	OEM
AM27LS18FM	OEM
AM27LS19DC	OEM
AM27LS19DM	OEM
AM27LS19FM	OEM
AM27S02DC	OEM
AM27S02DM	OEM
AM27S02FM	OEM
AM27S02PC	OEM
AM27S03ADC	OEM
AM27S03ADM	OEM
AM27S03AFM	OEM
AM27S03APC	OEM
AM27S03DC	OEM
AM27S03DM	OEM
AM27S03FM	OEM
AM27S03PC	OEM
AM27S08DC	OEM
AM27S08DM	OEM
AM27S09DC	OEM
AM27S09DM	OEM
AM27S10DC	OEM
AM27S10DM	OEM
AM27S11DC	OEM
AM27S11DM	OEM
AM27S12DC	2388
AM27S12DM	OEM
AM27S12FM	OEM
AM27S13DC	2389
AM27S13DM	OEM
AM27S13FM	OEM
AM27S15DC	OEM
AM27S15DM	OEM
AM27S18DC	3168
AM27S18DM	OEM
AM27S18FM	OEM
AM27S19DC	2175
AM27S19DM	OEM
AM27S19FM	OEM
AM27S20DC	1918
AM27S20DM	OEM
AM27S20FM	OEM
AM27S21DC	2224
AM27S21DM	OEM
AM27S21FM	OEM
AM27S25DC	OEM
AM27S25DM	OEM
AM27S26DC	OEM
AM27S26DM	OEM
AM27S27DC	OEM
AM27S27DM	OEM
AM27S28DC	3288
AM27S28DM	OEM
AM27S29DC	2320
AM27S29DM	OEM
AM27S30DC	2324
AM27S30DM	OEM
AM27S31DC	2322
AM27S31DM	OEM
AM27S32DC	2390
AM27S32DM	OEM
AM27S32PC	OEM
AM27S33DC	2392
AM27S33DM	OEM
AM27S33PC	OEM
AM27S80DC	OEM
AM27S80DM	OEM
AM27S80XX	OEM
AM27S81DC	OEM
AM27S81DM	OEM
AM27S81XX	OEM
AM27S180DC	2862
AM27S180DM	OEM
AM27S181DC	2909
AM27S181DM	OEM
AM27S184DC	OEM
AM27S184DM	OEM
AM27S185DC	3166
AM27S185DM	OEM
AM31	2060
AM31L01DC	OEM
AM31L01E	OEM
AM31L01FM	OEM
AM31L01PC	OEM
AM31L02DM	OEM
AM31L013E	OEM
AM32	2060
AM-33	0015
AM33	0015
AM34	2060
AM41	1000
AM42	1000
AM43	0015
AM44	1000
AM51	2076
AM52	2076
AM53	0015
AM54	2076
AM54LS373J	OEM
AM54LS373W	OEM
AM54LS373X	OEM
AM54LS533J	OEM
AM54LS533W	OEM
AM54LS533X	OEM
AM54S194J	OEM
AM54S194W	OEM
AM54S195J	OEM
AM54S195W	OEM
AM62	0015
AM63	0015
AM64	2081
AM65	0015
AM66	0015
AM74LS373J	0709
AM74LS373N	0709
AM74LS373X	0709
AM74LS533J	OEM
AM74LS533N	OEM
AM74LS533X	OEM
AM74S194J	OEM
AM74S194N	1931
AM74S195J	OEM
AM74S195N	OEM
AM90L44BDC	OEM
AM90L44BDM	OEM
AM90L44BPC	OEM
AM90L44CDC	OEM
AM90L44CDM	OEM
AM90L44CPC	OEM
AM90L44DDC	OEM
AM90L44DPC	OEM
AM90L60CDC	OEM
AM90L60CPC	OEM
AM90L60DDC	OEM
AM90L60DPC	OEM
AM90L60EDC	OEM
AM90L60EPC	OEM
AM91L01ADC	OEM
AM91L01ADM	OEM
AM91L01AFM	OEM
AM91L01APC	OEM
AM91L01BDC	OEM
AM91L01BDM	OEM
AM91L01BFM	OEM
AM91L01BPC	OEM
AM91L01CDC	OEM
AM91L01CDM	OEM
AM91L01CPC	OEM
AM91L02ADC	OEM
AM91L02ADM	OEM
AM91L02AFM	OEM
AM91L02APC	OEM
AM91L02BDC	OEM
AM91L02BDM	OEM
AM91L02BFM	OEM
AM91L02BPC	OEM
AM91L02CDC	OEM
AM91L02CDM	OEM
AM91L02CPC	OEM
AM91L02DC	OEM
AM91L02DM	OEM
AM91L02FM	OEM
AM91L02PC	OEM
AM91L0CDC	OEM
AM91L11ADC	OEM
AM91L11ADM	OEM
AM91L11AFM	OEM
AM91L11APC	OEM
AM91L11BDC	OEM
AM91L11BDM	OEM
AM91L11BFM	OEM
AM91L11BPC	OEM
AM91L11CDC	OEM
AM91L11CDM	OEM
AM91L11CPC	OEM
AM91L12ADC	OEM
AM91L12ADM	OEM
AM91L12AFM	OEM
AM91L12APC	OEM
AM91L12BDC	OEM
AM91L12BDM	OEM
AM91L12BFM	OEM
AM91L12BPC	OEM
AM91L12CDC	OEM
AM91L12CDM	OEM
AM91L12CPC	OEM
AM91L14BDC	OEM
AM91L14BDM	OEM
AM91L14BPC	OEM
AM91L14CDC	OEM
AM91L14CDM	OEM
AM91L14CPC	OEM
AM91L24BDC	OEM
AM91L24BDM	OEM
AM91L24BPC	OEM
AM91L24CDC	OEM
AM91L24CDM	OEM
AM91L24CPC	OEM
AM91L30ADC	OEM
AM91L30ADM	OEM
AM91L30APC	OEM
AM91L30BDC	OEM
AM91L30BDM	OEM
AM91L30BPC	OEM
AM91L30CDC	OEM
AM91L30CDM	OEM
AM91L30CPC	OEM
AM91L30DDC	OEM
AM91L30DPC	OEM
AM91L31ADC	OEM
AM91L31ADM	OEM
AM91L31BDC	OEM
AM91L31BDM	OEM
AM91L31CDC	OEM
AM91L31CDM	OEM
AM91L31DDC	OEM
AM91L40ADC	OEM
AM91L40ADM	OEM
AM91L40APC	OEM
AM91L40BDC	OEM
AM91L40BDM	OEM
AM91L40BPC	OEM
AM91L40CDM	OEM
AM91L40CPC	OEM
AM91L40DDC	OEM
AM91L40DPD	OEM
AM91L41DDC	OEM
AM92L44BDC	OEM
AM92L44BPC	OEM
AM92L44CDC	OEM
AM92L44CDM	OEM
AM92L44CPC	OEM
AM92L44DDC	OEM
AM92L44DPC	OEM
AM93L00DC	OEM
AM93L00FM	OEM
AM93L00PC	OEM
AM93L28DC	OEM
AM93L28DM	OEM
AM93L28FM	OEM
AM93L28PC	OEM
AM93L34DC	OEM
AM93L34DM	OEM
AM93L34FM	OEM
AM93L34PC	OEM
AM93L38DC	OEM
AM93L38DM	OEM
AM93L38FM	OEM
AM93L38PC	OEM
AM95-3310	OEM
AM95-4006	OEM
AM95-4010	OEM
AM95-4620	OEM
AM95-5032	OEM
AM95-5132	OEM
AM95-6011	OEM
AM95-6012	OEM
AM95-6110	OEM
AM95-6120	OEM
AM95-6220	OEM
AM95-6440	OEM
AM95-6450	OEM
AM95-6452	OEM
AM96-1000	OEM
AM96-4016	OEM
AM97C09CN	OEM
AM97C10CN	OEM
AM97C11CN	OEM
AM97C12CN	OEM
AM100A	OEM
AM100B	OEM
AM100C	OEM
AM101A	OEM
AM101B	OEM
AM102A	OEM
AM102B	OEM
AM103A	OEM
AM103B	OEM
AM124D	OEM
AM124F	OEM
AM201A	OEM
AM201B	OEM
AM201C	OEM
AM224D	0625
AM280HM	OEM
AM300	0133
AM300A	0133
AM301	0133
AM301A	0133
AM302	0133
AM302A	0133
AM303	0133
AM303A	0133
AM303B	OEM
AM304	0133
AM304A	0133
AM305	0133
AM305A	0133
AM306	0133
AM306A	0133
AM307	0133
AM307A	0133
AM308	0133
AM308A	0133
AM324D	0625
AM324N	0625
AM405	0015
AM410	0015
AM410-2C	OEM
AM410-2M	OEM
AM411-2C	OEM
AM411-2M	OEM
AM414A	OEM

DEVICE TYPE	REPL CODE
AM414B	OEM
AM414M	OEM
AM415	0015
AM420	0015
AM425	0015
AM430	0015
AM430A	OEM
AM430M	OEM
AM435	0015
AM440	0015
AM445	0015
AM450	0015
AM450-2	OEM
AM450-2M	OEM
AM452-2	OEM
AM452-2M	OEM
AM453-2C	OEM
AM453-2M	OEM
AM460	0015
AM460-2	OEM
AM460-2C	OEM
AM460-2M	OEM
AM462-1	OEM
AM462-1M	OEM
AM462-2	OEM
AM462-2M	OEM
AM464-2	OEM
AM464-2M	OEM
AM470-2C	OEM
AM470-2M	OEM
AM490-2A	OEM
AM490-2B	OEM
AM490-2C	OEM
AM490-2M	OEM
AM500GC	OEM
AM500MC	OEM
AM500MM	OEM
AM500MR	OEM
AM542AMC	OEM
AM542AMM	OEM
AM542AMR	OEM
AM543AMC	OEM
AM543AMM	OEM
AM543AMR	OEM
AM600	OEM
AM620	0133
AM620A	0133
AM626	0133
AM626A	0133
AM632	0133
AM632A	0133
AM670DM	OEM
AM681PC	OEM
AM685DL	OEM
AM685DM	OEM
AM685HL	OEM
AM685HM	OEM
AM686DC	OEM
AM686DM	OEM
AM686HC	OEM
AM686HM	OEM
AM687DL	OEM
AM723HM	1192
AM741HC	0410
AM747HC	2368
AM747HM	2368
AM748DC	2452
AM748DM	2452
AM748HC	2454
AM748HM	2454
AM904EPC	OEM
AM924BDC	OEM
AM1000H	OEM
AM1001H	OEM
AM1002H	OEM
AM1402A	OEM
AM1402A51E	OEM
AM1402A51F	OEM
AM1402A59F	OEM
AM1402ADM	OEM
AM1402APC	OEM
AM1403A	OEM
AM1403A51F	OEM
AM1403A51T	OEM
AM1403A59F	OEM
AM1403AHM	OEM
AM1403APC	OEM
AM1404A	OEM
AM1404A51F	OEM
AM1404A51T	OEM
AM1404A59F	OEM
AM1404AHM	OEM
AM1404APC	OEM
AM1406HM	OEM
AM1407HM	OEM
AM1408L6	OEM
AM1408L7	OEM
AM1408L8	OEM
AM1488XC	OEM
AM1489APC	OEM
AM1489AXC	OEM
AM1489PC	OEM
AM1489XC	OEM
AM1500DC	OEM
AM1500DL	OEM
AM1500DM	OEM
AM1500FM	OEM
AM1506HC	OEM
AM1507HC	OEM

DEVICE TYPE	REPL CODE
AM1508L8	OEM
AM1660HC	OEM
AM1702A	OEM
AM1702A-1	OEM
AM1702A-2	OEM
AM1702AL	OEM
AM1702AL-1	OEM
AM1702AL-2	OEM
AM2009CD	OEM
AM2009CF	OEM
AM2009D	OEM
AM2009F	OEM
AM2501DC	OEM
AM2501DM	OEM
AM2502C/D	OEM
AM2502CJE	OEM
AM2502CPE	OEM
AM2502DC	OEM
AM2502DM	OEM
AM2502FM	OEM
AM2502M/D	OEM
AM2502MJE	OEM
AM2502PC	OEM
AM2502SM	OEM
AM2502XC	OEM
AM2503C/D	OEM
AM2503CJE	OEM
AM2503CPE	OEM
AM2503DC	OEM
AM2503DM	OEM
AM2503FM	OEM
AM2503M/D	OEM
AM2503MJE	OEM
AM2503PC	OEM
AM2503XC	OEM
AM2503XM	OEM
AM2504C/D	OEM
AM2504CJG	OEM
AM2504CPG	OEM
AM2504DC	OEM
AM2504DM	OEM
AM2504FM	OEM
AM2504M/D	OEM
AM2504MJG	OEM
AM2504PC	OEM
AM2504XC	OEM
AM2504XM	OEM
AM2505DC	OEM
AM2505DM	OEM
AM2505FM	OEM
AM2505PC	OEM
AM2506DC	OEM
AM2506DM	OEM
AM2506FM	OEM
AM2506PC	OEM
AM2600DC	OEM
AM2600DM	OEM
AM2600FM	OEM
AM2600PC	OEM
AM2602DC	OEM
AM2602DM	OEM
AM2602FM	OEM
AM2602PC	OEM
AM2708DC	4424
AM2708DM	OEM
AM2716DC	2278
AM2732ADC	5590
AM2764ADC	0812
AM2802DC	OEM
AM2802DM	OEM
AM2802PC	OEM
AM2803HC	OEM
AM2803HM	OEM
AM2803PC	OEM
AM2804HC	OEM
AM2804PC	OEM
AM2805HC	OEM
AM2805HM	OEM
AM2806HC	OEM
AM2806HM	OEM
AM2807PC	OEM
AM2808PC	OEM
AM2809HC	OEM
AM2809HM	OEM
AM2809PC	OEM
AM2810DC	OEM
AM2810DM	OEM
AM2812ADC	OEM
AM2812ADM	OEM
AM2812DC	OEM
AM2812DM	OEM
AM2813ADC	OEM
AM2813ADM	OEM
AM2813DC	OEM
AM2813DM	OEM
AM2814DC	OEM
AM2814DM	OEM
AM2814PC	OEM
AM2833DC	OEM
AM2833DM	OEM
AM2833PC	OEM
AM2841DC	OEM
AM2841DM	OEM
AM2855DC	OEM
AM2855DM	OEM
AM2855PC	OEM
AM2856HC	OEM
AM2856HM	OEM
AM2857DC	OEM

DEVICE TYPE	REPL CODE
AM2857DM	OEM
AM2857PC	OEM
AM2901ADC	OEM
AM2901ADM	OEM
AM2901AFM	OEM
AM2901APC	OEM
AM2901AXC	OEM
AM2901BDC	OEM
AM2901BDM	OEM
AM2901BFM	OEM
AM2901BPC	OEM
AM2901CDC	OEM
AM2901CPC	OEM
AM2901DC	OEM
AM2901PC	OEM
AM2903A	OEM
AM2903DC	OEM
AM2903DM	OEM
AM2903FM	OEM
AM2961DC(M)	OEM
AM2961DC-B	OEM
AM2961DM(M)	OEM
AM2961DM-B(M)	OEM
AM2961FM(M)	OEM
AM2961FM-B(M)	OEM
AM2961XM(M)	OEM
AM2962DC(M)	OEM
AM2962DC-B	OEM
AM2962DM(M)	OEM
AM2962DM-B(M)	OEM
AM2962FM(M)	OEM
AM2962FM-B(M)	OEM
AM2962XC(M)	OEM
AM2962XM(M)	OEM
AM2964BDM	OEM
AM2964BFM	OEM
AM2964BPC	OEM
AM2964BXC	OEM
AM2965DC	OEM
AM2965DM	OEM
AM2965FM	OEM
AM2965PC	OEM
AM2965XC	OEM
AM2965XM	OEM
AM2966DC	OEM
AM2966DM	OEM
AM2966FM	OEM
AM2966PC	OEM
AM2966XC	OEM
AM2966XM	OEM
AM3010	0319
AM3020	0319
AM3101ADC	OEM
AM3101ADM	OEM
AM3101AFM	OEM
AM3101APC	OEM
AM3101DM	OEM
AM3101FM	OEM
AM3208SN	OEM
AM3235	0103
AM3341DC	OEM
AM3341DM	OEM
AM3341PC	OEM
AM3355DC	OEM
AM3355PC	OEM
AM3510	0319
AM3520	0319
AM3705CD	OEM
AM3705CF	OEM
AM3705CN	OEM
AM3705D	OEM
AM3705F	OEM
AM4010	0319
AM4020	0319
AM5005	0319
AM5010	0319
AM6005	0319
AM6010	0319
AM6012ADC	OEM
AM6012ADM	OEM
AM6012DC	OEM
AM6012DM	OEM
AM6070ADC	OEM
AM6070ADM	OEM
AM6070DC	OEM
AM6071ADC	OEM
AM6071ADM	OEM
AM6071DC	OEM
AM6071DM	OEM
AM6072DC	OEM
AM6072DM	OEM
AM6073DC	OEM
AM6073DM	OEM
AM6080ADC	OEM
AM6080ADM	OEM
AM6080APC	OEM
AM6080DC	OEM
AM6080DM	OEM
AM6080PC	OEM
AM6081ADC	OEM
AM6081ADM	OEM
AM6081APC	OEM
AM6081DC	OEM
AM6081DM	OEM
AM6112(A)	OEM
AM7600C	OEM
AM7600M	OEM
AM7600R	OEM
AM7601C	OEM

DEVICE TYPE	REPL CODE
AM7601M	OEM
AM7601R	OEM
AM7605C	OEM
AM7605M	OEM
AM7605R	OEM
AM7606C	OEM
AM7606M	OEM
AM7606R	OEM
AM7611C	OEM
AM7611M	OEM
AM7612C	OEM
AM7612M	OEM
AM7613C	OEM
AM7613M	OEM
AM7614C	OEM
AM7614M	OEM
AM7615C	OEM
AM7615M	OEM
AM7831J	OEM
AM7831W	OEM
AM7831X	OEM
AM7832J	OEM
AM7832W	OEM
AM8085ACC	OEM
AM8085ADC	OEM
AM8085APC	OEM
AM8255A-5PC	0051
AM8510M	OEM
AM8510R	OEM
AM8520M	OEM
AM8520R	OEM
AM8530M	OEM
AM8530R	OEM
AM8831J	OEM
AM8831N	OEM
AM8831X	OEM
AM8832J	OEM
AM8832N	OEM
AM8832X	OEM
AM9016CDC	0522
AM9016CPC	0522
AM9016DDC	0522
AM9016DPC	0522
AM9016EDC	0522
AM9016EPC	0522
AM9016FDC	OEM
AM9016FPC	OEM
AM9044BDC	OEM
AM9044BDM	OEM
AM9044BPC	OEM
AM9044CDC	OEM
AM9044CDM	OEM
AM9044CPC	OEM
AM9044DDC	OEM
AM9044DDM	OEM
AM9044DPC	OEM
AM9044EDC	OEM
AM9050CDC	OEM
AM9050CPC	OEM
AM9050DDC	OEM
AM9050DPC	OEM
AM9050EDC	OEM
AM9050EPC	OEM
AM9060CDC	4902
AM9060CPC	4902
AM9060DDC	4902
AM9060DPC	4902
AM9060EDC	4902
AM9060EPC	4902
AM9080A1DC	OEM
AM9080A1PC	OEM
AM9080A2DC	OEM
AM9080A2DM	OEM
AM9080A2PC	OEM
AM9080A4DC	OEM
AM9080ADC	OEM
AM9080ADM	OEM
AM9080APC	OEM
AM9101	OEM
AM9101ADC	OEM
AM9101ADM	OEM
AM9101AFM	OEM
AM9101APC	OEM
AM9101BDC	OEM
AM9101BDM	OEM
AM9101BFM	OEM
AM9101BPC	OEM
AM9101CDC	OEM
AM9101CDM	OEM
AM9101CPC	OEM
AM9101DDC	OEM
AM9101DPC	OEM
AM9101EDC	OEM
AM9101EPC	OEM
AM9102ADC	4973
AM9102ADM	OEM
AM9102AFM	OEM
AM9102APC	4973
AM9102BDC	4973
AM9102BDM	OEM
AM9102BFM	OEM
AM9102BPC	4973
AM9102CDC	4973
AM9102CDM	OEM
AM9102CPC	4973
AM9102DC	4973
AM9102DDC	OEM
AM9102DM	OEM
AM9102DPC	OEM

DEVICE TYPE	REPL CODE
AM9102EDC	OEM
AM9102EPC	OEM
AM9102FM	OEM
AM9102PC	4973
AM9111	OEM
AM9111ADC	OEM
AM9111ADM	OEM
AM9111AFM	OEM
AM9111APC	OEM
AM9111BDC	OEM
AM9111BDM	OEM
AM9111BFM	OEM
AM9111BPC	OEM
AM9111CDC	OEM
AM9111CDM	OEM
AM9111CPC	OEM
AM9111DDC	OEM
AM9111DPC	OEM
AM9111EDC	OEM
AM9111EPC	OEM
AM9112	OEM
AM9112ADM	OEM
AM9112AFM	OEM
AM9112APC	OEM
AM9112BDC	OEM
AM9112BDM	OEM
AM9112BFM	OEM
AM9112BPC	OEM
AM9112CDC	OEM
AM9112CDM	OEM
AM9112CPC	OEM
AM9112DDC	OEM
AM9112DPC	OEM
AM9112EDC	OEM
AM9112EPC	OEM
AM9114	OEM
AM9114BDC	2048
AM9114BDM	OEM
AM9114BPC	2048
AM9114CDC	2048
AM9114CDM	OEM
AM9114CPC	2048
AM9114EDC	OEM
AM9114EPC	OEM
AM9124BDC	OEM
AM9124BDM	OEM
AM9124BPC	OEM
AM9124CDC	OEM
AM9124CDM	OEM
AM9124CPC	OEM
AM9124EDC	OEM
AM9124EPC	OEM
AM9130ADC	OEM
AM9130ADDM	OEM
AM9130BDC	OEM
AM9130BDM	OEM
AM9130BPC	OEM
AM9130CDC	OEM
AM9130CDM	OEM
AM9130CPC	OEM
AM9130DDC	OEM
AM9130DPC	OEM
AM9130EDC	OEM
AM9130EPC	OEM
AM9131ADC	OEM
AM9131ADM	OEM
AM9131BDC	OEM
AM9131BDM	OEM
AM9131CDC	OEM
AM9131CDM	OEM
AM9131DDC	OEM
AM9131EDC	OEM
AM9140ADC	OEM
AM9140ADM	OEM
AM9140APC	OEM
AM9140BDC	OEM
AM9140BDM	OEM
AM9140BPC	OEM
AM9140CDC	OEM
AM9140CDM	OEM
AM9140CPC	OEM
AM9140DDC	OEM
AM9140DPC	OEM
AM9140EDC	OEM
AM9140EPC	OEM
AM9208BDC	OEM
AM9208BDM	OEM
AM9208CDC	OEM
AM9214DC	OEM
AM9214DM	OEM
AM9216BDC	OEM
AM9216BDM	OEM
AM9216CDC	OEM
AM9217ADC	OEM
AM9217ADM	OEM
AM9217BDC	OEM
AM9217BDM	OEM
AM9218BCC	OEM
AM9218BDC	OEM
AM9218BDM	OEM
AM9218BPC	OEM
AM9218CCC	OEM
AM9218CDC	OEM
AM9218CPC	OEM
AM9232BCC	OEM
AM9232BDC	OEM
AM9232BDM	OEM
AM9232BPC	OEM
AM9232CCC	OEM

DEVICE TYPE	REPL CODE
AM9232CDC	OEM
AM9232CPC	OEM
AM9233BCC	OEM
AM9233BDC	OEM
AM9233BDM	OEM
AM9233BPC	OEM
AM9233CCC	OEM
AM9233CDC	OEM
AM9233CPC	OEM
AM9244BDM	OEM
AM9244BPC	OEM
AM9244CDC	OEM
AM9244CDM	OEM
AM9244CPC	OEM
AM9244DDC	OEM
AM9244DDM	OEM
AM9244DPC	OEM
AM9244EDC	OEM
AM9244EPC	OEM
AM9300DC	OEM
AM9300DM	OEM
AM9300FM	OEM
AM9300PC	4267
AM9328DC	OEM
AM9328DM	OEM
AM9328FM	OEM
AM9328PC	OEM
AM9334DC	OEM
AM9334DM	OEM
AM9334FM	OEM
AM9334PC	OEM
AM9338DC	OEM
AM9338DM	OEM
AM9338FM	OEM
AM9338PC	OEM
AM9401DC	OEM
AM9401DM	OEM
AM9401PC	OEM
AM9404CDC	OEM
AM9404CDM	OEM
AM9404DDC	OEM
AM9404DDM	OEM
AM9404FDC	OEM
AM9517A-5PC	OEM
AM9702A-1HDC	OEM
AM9702AHDC	OEM
AM9702AL-1HDC	OEM
AM9702AL-2HDC	OEM
AM9702ALHDC	OEM
AM9709CN	OEM
AM9710CN	OEM
AM9711CN	OEM
AM9712CN	OEM
AM26123DC	OEM
AM26123DM	OEM
AM26123FM	OEM
AM26123PC	4832
AM27128DC	OEM
AM29203DC	OEM
AM29203DM	OEM
AM29203FM	OEM
AM29720DM	OEM
AM29720FM	OEM
AM29720PC	OEM
AM29721DC	OEM
AM29721DM	OEM
AM29721FM	OEM
AM29721PC	OEM
AM31013	OEM
AM35141DC	OEM
AM35142DC	OEM
AM97021-2HDC	OEM
AM161011H	0286
AM161011J	0079
AM161015H	0006
AM161016H	0079
AM161038H	0079
AM161039H	0079
AM161040H	0342
AM161041H	0715
AM161043H	0925
AM161044E	0392
AM161050H	0037
AM162005A	0015
AM164003A	OEM
AM167001A	0167
AM167013A	0395
AM167014A	1808
AM270059E	OEM
AM270059F	OEM
AM270159E	OEM
AM270159F	OEM
AMB4579	OEM
AMC8013AC	OEM
AMC8013AM	OEM
AMC8013BC	OEM
AMC8013BM	OEM
AMC8013CC	OEM
AMC8013CM	OEM
AMC40800	OEM
AMD-746	0666
AMD746	0349
AMD-780	0352
AMD780	0352
AMD-781	0354
AMD781	0354
AMF101	0561
AMF104	0103
AMF105	0103

DEVICE TYPE	REPL CODE
AMF106	0642
AMF115	0103
AMF116	0103
AMF117	0103
AMF117A	0103
AMF118	0103
AMF118A	0103
AMF119	0103
AMF119A	0103
AMF120	0103
AMF120A	0103
AMF-121	0103
AMF201	0103
AMF201A	OEM
AMF201B	0130
AMF201C	0130
AMF201D	0620
AMF201E	0543
AMF210	0103
AMF210A	0103
AMF210B	0130
AMF210C	0074
AM-G-5	0015
AM-G-5A	0071
AM-G-5C	0071
AM-G-11	0250
AM-G-22	0015
AMI8025DAC	OEM
AMI8239MAQ	OEM
AMI8250MBK	OEM
AMI8314BLU	OEM
AMI8330YX	OEM
AMI8415AAB	OEM
AMI8431CN	OEM
AMI8503MFD	OEM
AMI8503MFX	OEM
AMI8524BT	OEM
AMI8524MAH	OEM
AMI8539MAV	OEM
AMIS25610	OEM
AMLM21D	OEM
AMLM101AH	0093
AMLM101H	0093
AMLM106	OEM
AMLM108	OEM
AMLM108A	OEM
AMLM108AD	OEM
AMLM108AF	OEM
AMLM108D	OEM
AMLM108F	OEM
AMLM111	OEM
AMLM111D	2104
AMLM111F	OEM
AMLM112	OEM
AMLM112D	OEM
AMLM112F	OEM
AMLM118	OEM
AMLM118D	OEM
AMLM118F	OEM
AMLM119D	OEM
AMLM119F	OEM
AMLM119H	OEM
AMLM139AD	OEM
AMLM139AF	OEM
AMLM139D	OEM
AMLM139F	OEM
AMLM148D	OEM
AMLM201	0093
AMLM201A	0093
AMLM201AH	0093
AMLM201H	0093
AMLM206	OEM
AMLM207D	2283
AMLM207F	OEM
AMLM208	OEM
AMLM208A	OEM
AMLM208AD	OEM
AMLM208AF	OEM
AMLM208D	OEM
AMLM208F	OEM
AMLM211	OEM
AMLM216	OEM
AMLM216A	OEM
AMLM216AD	OEM
AMLM216D	OEM
AMLM218	OEM
AMLM218D	OEM
AMLM219D	OEM
AMLM219H	OEM
AMLM239AD	OEM
AMLM239D	OEM
AMLM248D	OEM
AMLM249D	OEM
AMLM301	0093
AMLM301A	0093
AMLM301AH	0093
AMLM306	OEM
AMLM307D	6065
AMLM308	OEM
AMLM308A	0896
AMLM308AD	OEM
AMLM308D	OEM
AMLM311	OEM
AMLM311D	2104
AMLM312	OEM
AMLM312D	OEM
AMLM316	OEM
AMLM316A	OEM
AMLM316AD	OEM

If replacement code is OEM, contact original manufacturer for replacement.

DEVICE TYPE	REPL CODE
AMLM316D	OEM
AMLM318	OEM
AMLM318D	OEM
AMLM319D	OEM
AMLM319H	OEM
AMLM319N	2570
AMLM339AD	OEM
AMLM339AN	OEM
AMLM339D	OEM
AMLM339N	OEM
AMLM348D	OEM
AMLM348N	2823
AMLM349D	OEM
AMLM349N	OEM
AMPAL16L8PC	OEM
AMPAL16R8DC	OEM
AMS-D104	OEM
AMS-D105	OEM
AMS-D106	OEM
AMS-D126-A1	OEM
AMS-D127	OEM
AMS-D128	OEM
AMS-D218	OEM
AMS-D219	OEM
AMS-D220	OEM
AMS-D526	OEM
AMSSS725AJ	OEM
AMSSS725BJ	OEM
AMSSS725EJ	OEM
AMSSS741CJ	4262
AMSSS741J	OEM
AMSSS747CK	OEM
AMSSS747CP	OEM
AMSSS747K	OEM
AMSSS747M	OEM
AMSSS747P	OEM
AMS-SY52	OEM
AMSTRAD40044	OEM
AMU4L93L1851X	OEM
AMU5R7723312	1192
AMU6B930051X	OEM
AMU6B930059X	OEM
AMU6M93L1859X	OEM
AMU7B93L1851X	OEM
AMU7B93L1859X	OEM
AMU7B931851X	OEM
AMU7B931859X	OEM
AMX3548	0026
AMX3550	1527
AMX-3551	1559
AMX3551	1559
AMX3552	1595
AMX3553	1608
AMX3554	1667
AMX3555	0034
AMX3556	0828
AMX3557	0088
AMX3558	0243
AMX3560	1888
AMX3561	1625
AMX3562	0959
AMX3563	1161
AMX3564	4371
AMX3565	0260
AMX3566	1672
AMX3567	0977
AMX3568	0991
AMX3573	1262
AMX3582	0042
AMX3583	4376
AMX3584	0006
AMX3586	3481
AMX3655	0361
AMX-3675	1206
AMX3675	1206
AMX-3683	0996
AMX3683	0996
AMX3684	1339
AMX3688	1118
AMX3698	1633
AMX3701	0291
AMX3706	0819
AMX-3716	1698
AMX3716	1698
AMX3799	0491
AMX3800	0153
AMX3803	0979
AMX3804	1882
AMX3810	1239
AMX3864	0458
AMX3865	OEM
AMX3867	0508
AMX3868	0511
AMX3898	1662
AMX3921	OEM
AMX3955	1157
AMX4178	0076
AMX4181	2318
AMX4188	1838
AMX4200	0176
AMX4225	0452
AMX4227	0894
AMX4258	0312
AMX4260	1285
AMX4261	4516
AMX4263	0016
AMX4321	4525
AMX4321(MC75462P)	OEM
AMX4326	1118
AMX4327	2104
AMX4328	1839
AMX4329	OEM
AMX4330	0037
AMX4331	4330
AMX4470	0463
AMX4471	OEM
AMX4472	OEM
AMX4473	OEM
AMX4574	0814
AMX4575	OEM
AMX4577	0334
AMX4578	0448
AMX4579	OEM
AMX4582	1875
AMX4583	0427
AMX4584	0398
AMX4585	4531
AMX-4591	1206
AMX4642	OEM
AMX4658	0183
AMX4659	1564
AMX4660	0592
AMX4661	0360
AMX4663	3175
AMX4666	0056
AMX-4945	2263
AMX8001DC	OEM
AMZ8002DC	OEM
AMZ8036DM	OEM
AMZ8120	OEM
AMZ8121	OEM
AMZ8133	OEM
AMZ8136DC	OEM
AMZ8136DM	OEM
AMZ8136PC	OEM
AMZ8148DC	OEM
AMZ8148DM	OEM
AMZ8148PC	OEM
AMZ8173	OEM
AN0015CD	OEM
AN0120NA	OEM
AN0120NB	OEM
AN0120ND	OEM
AN0130NA	OEM
AN0130NB	OEM
AN0130ND	OEM
AN0140NA	OEM
AN0140NB	OEM
AN0140ND	OEM
AN-1	0015
AN3	OEM
AN29S	OEM
AN60P	OEM
AN74H72N	OEM
AN74LS107AN	1602
AN78L04	OEM
AN78L05	1292
AN78L05-Y	1292
AN78L05M-E1	OEM
AN78L06	4501
AN78L07	OEM
AN78L08	4502
AN78L09	1786
AN78L09-Y	1786
AN78L10	OEM
AN78L12	1828
AN78L12-Y	1828
AN78L15	4504
AN78L18	4506
AN78L20	OEM
AN78L24	4509
AN78M05	0624
AN78M05F	0624
AN78M05LB	0624
AN78M05R	OEM
AN78M08	1196
AN78M08R	OEM
AN78M09	4522
AN78M09-(LC)	1786
AN78M09F	1786
AN78M09LB	1786
AN78M09R	OEM
AN78M10	2806
AN78M12	0334
AN78M12(LB)	0334
AN78M12-(LB)	0334
AN78M12-LB	0334
AN78M12LB	0334
AN78M15	1321
AN78M15(LB)	1321
AN78M15LB	1321
AN78N05	1292
AN78N05LB	1292
AN78N08	OEM
AN78N09LB	OEM
AN78N12	OEM
AN79L04	OEM
AN79L05	4884
AN79L09	OEM
AN79L09-Y	OEM
AN79L15	4885
AN79L18	4886
AN79L20	OEM
AN79L24	1721
AN79M05	4905
AN79M06	4906
AN79N06	OEM
AN90B00	OEM
AN90B70	OEM
AN90C10	OEM
AN90C20	OEM
AN90C21	4925
AN90C22	5271
AN90C23	OEM
AN91A10S	OEM
AN91A14K	OEM
AN103	1540
AN103-0	1540
AN103G	1540
AN115	0706
AN136	2573
AN155	0039
AN158	0395
AN179	OEM
AN-203	2576
AN203	2576
AN203(C)	2576
AN203AA	2576
AN203BA	2576
AN203BB	2576
AN204D	OEM
AN205	OEM
AN-206	2577
AN206	2577
AN206AB	2577
AN206B	2577
AN206S	2577
AN208	3135
AN-210	2576
AN210	2578
AN210B	2578
AN210C	2578
AN-211	1837
AN211	1837
AN211A	1837
AN211AB	1837
AN211B	1837
AN213	4025
AN213-AB	4025
AN213AB	4025
AN-214	2124
AN214	2124
AN-214-G	2124
AN-214-P	2124
AN214-Q	OEM
AN214-QR	2124
AN214D	2124
AN-214P	2124
AN214P	2124
AN214PQR	2124
AN-214Q	6010
AN214Q	2124
AN214QR	2124
AN214R	2124
AN216	4028
AN217	2581
AN217(BB)	2581
AN217BB	2581
AN217CB	2581
AN217P	2581
AN217PBB	2581
AN220	2582
AN221	0167
AN222	0878
AN225	0403
AN227	2584
AN228V	2586
AN228W	2586
AN229	2585
AN230	2586
AN231	2587
AN234	2588
AN236	4034
AN238	4036
AN238S	0395
AN239	2942
AN239Q	2942
AN239QA	2942
AN239QB	2942
AN240	0167
AN240D	0167
AN240P	0167
AN240PD	0167
AN240PN	0167
AN240S	0167
AN241	4038
AN241A	4038
AN241B	4038
AN241C	4038
AN241D	4038
AN241P	4038
AN241PD	4038
AN242	4039
AN245	3269
AN-246	3269
AN246	3269
AN247	3273
AN247P	3273
AN249	OEM
AN252	2591
AN253	2591
AN253AB	2591
AN-253BB	2591
AN253BB	2591
AN253P	2591
AN255	OEM
AN259	4044
AN260	2592
AN262	2960
AN264	OEM
AN271	2593
AN271B	2593
AN271FB	2593
AN272	4047
AN274	2594
AN277	2596
AN277AB	2596
AN277B	2596
AN277BA	2596
AN278	4051
AN287	0107
AN287NT	0107
AN288	2599
AN289	4055
AN295	4060
AN295N	4060
AN298A	OEM
AN301	4063
AN302	4064
AN303	4065
AN304	1582
AN305	4067
AN313	4068
AN-315	4069
AN315	4069
AN316	4070
AN318	4071
AN320	OEM
AN321	OEM
AN326	4072
AN328	4076
AN331	2602
AN335	OEM
AN335F	OEM
AN337	4079
AN340	3246
AN340P	3246
AN342	2603
AN343	2603
AN345V	OEM
AN352	1260
AN353A	OEM
AN355	OEM
AN360	1524
AN362	1260
AN362L	1260
AN363N	0416
AN366	1509
AN370	OEM
AN374	4086
AN374P	4773
AN380	4087
AN521	OEM
AN605	OEM
AN607	OEM
AN608P	OEM
AN612	4042
AN614	OEM
AN616	OEM
AN640G	5562
AN660	OEM
AN705A	OEM
AN829	3103
AN829P	OEM
AN829S	3103
AN1201BR	0765
AN1339	0176
AN1339S	OEM
AN1358	0770
AN1358S	4259
AN1364	1170
AN1393	5304
AN1393S	OEM
AN1431	5298
AN1458	0360
AN1741	0312
AN2010S	OEM
AN2011S	OEM
AN2065	2577
AN2131	OEM
AN2140	OEM
AN2150S	OEM
AN2153NS	OEM
AN2154FAP	OEM
AN2240	OEM
AN2250S	OEM
AN2254FAP	OEM
AN2330	OEM
AN2340	OEM
AN2341	OEM
AN2350S	OEM
AN2410	OEM
AN2430	OEM
AN2450S	OEM
AN2510S	OEM
AN2512S	OEM
AN2513S	OEM
AN3120	OEM
AN3210K	4379
AN3211K	4379
AN3211NK	4379
AN3211S	OEM
AN3220K	OEM
AN3222	OEM
AN3223K	OEM
AN3224K	OEM
AN3230K	OEM
AN3230NK	OEM
AN3292NS	OEM
AN3292S	OEM
AN3294K	OEM
AN3296	OEM
AN3296S	OEM
AN3310K	5815
AN3311K	4379
AN3311NK	OEM
AN3312	6336
AN3313	5324
AN3313S	OEM
AN3316K	OEM
AN3320	OEM
AN3320K	5258
AN3320S	OEM
AN3321K	OEM
AN3330K	OEM
AN3334S	OEM
AN3346FAS	OEM
AN3394K	OEM
AN3396	OEM
AN3621K	OEM
AN3712K	OEM
AN3720K	OEM
AN3790K	OEM
AN3791	OEM
AN3792	5328
AN3793	OEM
AN3794	OEM
AN3794N	OEM
AN3798NS	OEM
AN3810K	OEM
AN3812K	OEM
AN3813K	OEM
AN3813NK	OEM
AN3815K	OEM
AN3816SCR	OEM
AN3820	OEM
AN3821K	5259
AN3822K	5259
AN3824K	OEM
AN3824NK	OEM
AN3826K	OEM
AN3826NK	OEM
AN3830K	OEM
AN3840SR	OEM
AN3912	OEM
AN3920K	OEM
AN3925K	OEM
AN3928K	OEM
AN3990	OEM
AN3994K	OEM
AN4102	4728
AN5010	3862
AN5011	OEM
AN5013K	OEM
AN5015	OEM
AN5015K	OEM
AN5019	OEM
AN5020	5215
AN5025K	OEM
AN5025KR	OEM
AN5036	OEM
AN5038	OEM
AN5070	5218
AN5071	OEM
AN5072	OEM
AN5079	OEM
AN5111	3738
AN5112	OEM
AN5120N	OEM
AN5125	3902
AN5125M	3902
AN5130	6784
AN5132	3744
AN5135K	3925
AN5135NK	3925
AN5136K	3927
AN5136K-R	3927
AN5136KR	3927
AN5137K	OEM
AN5138K	OEM
AN5138NK	OEM
AN5144	OEM
AN5150	OEM
AN5150N	OEM
AN5151	OEM
AN5151N	OEM
AN5153NK	3925
AN5156K	OEM
AN5156K-N	OEM
AN5156KN	OEM
AN5160NK	6547
AN5160NK)N	OEM
AN5160NK-N	6547
AN5179K	OEM
AN5180K	OEM
AN5180NK	OEM
AN5210	3787
AN5214S	OEM
AN5215	2019
AN5220	OEM
AN5221	OEM
AN5250	4743
AN5255	5208
AN5256	5208
AN5260	OEM
AN5262	4194
AN5265	4422
AN5275	OEM
AN5291	OEM
AN5301K	4241
AN5301NK	4241
AN5302	OEM
AN5302K	4321
AN5304K	OEM
AN5304NK	OEM
AN5307K	OEM
AN5310	2623
AN5310K	2623
AN5310KL	2623
AN5310L	2623
AN5310M	2623
AN5310N	2623
AN5310U	2623
AN5311	2623
AN5311KL	2623
AN5312	OEM
AN5313NK	OEM
AN5313NS	OEM
AN5314K	OEM
AN5315	5234
AN5316	5047
AN5316N	OEM
AN5318	6040
AN5318A	4346
AN5318N	4346
AN5320	4495
AN5321	OEM
AN5322NK	3690
AN5326	OEM
AN5330	4549
AN5332N	OEM
AN5340	3351
AN5341K	OEM
AN5342K	OEM
AN5348K	OEM
AN5352	4310
AN5352N	4310
AN5355	OEM
AN5356	OEM
AN5365	OEM
AN5371S	OEM
AN5372S	OEM
AN5410	2625
AN5411	5095
AN5416	3853
AN5421	OEM
AN5429	OEM
AN5430	OEM
AN5431N	OEM
AN5435	5094
AN5436	3747
AN5436N	3747
AN5437K	OEM
AN5437KR	OEM
AN5440	3359
AN5510	4060
AN5512	0976
AN5515	4927
AN5515X	4927
AN5520	3609
AN-5521	OEM
AN5521	4129
AN5530K	OEM
AN5531	OEM
AN5532	OEM
AN5551	OEM
AN5600K	OEM
AN5601NK	OEM
AN5610N	OEM
AN5612	OEM
AN5613	OEM
AN5615	OEM
AN5622	OEM
AN5625N	OEM
AN5630N	OEM
AN5632K	OEM
AN5633K	OEM
AN5635N	OEM
AN5700	4857
AN5701	OEM
AN5701N	OEM
AN5703	3753
AN5707	OEM
AN5707NS	OEM
AN5710	5193
AN5712	3074
AN5720	3609
AN5722	3084
AN5730	2019
AN5732	2019
AN5733	OEM
AN5740	2787
AN5742	OEM
AN5743	OEM
AN5750	OEM
AN5752	3759
AN5753	3759
AN5760	OEM
AN5762	OEM
AN5763	OEM
AN5817NK	OEM
AN5820	OEM
AN5822	OEM
AN5835	OEM
AN5836	1529
AN5837	OEM
AN5839	OEM
AN5855	OEM
AN5858K	OEM
AN5860	OEM
AN5862	OEM
AN5862K	OEM
AN5900	6287
AN6040	OEM
AN6041	5230
AN6050	OEM
AN6135	OEM
AN6209	2066
AN6209K	2066
AN6209S	OEM
AN6249	OEM
AN6250	OEM
AN6251	OEM
AN6263	OEM
AN6263N	OEM
AN6290K	OEM
AN6291	3437
AN6291K	OEM
AN6291S	OEM
AN6295K	OEM
AN6295NK	OEM
AN6296	OEM
AN6300	4747
AN6306	2067
AN6306S	OEM
AN6307	2553
AN6310	5194
AN6320	4751
AN6320N	4751
AN6321	5071
AN6326	0699
AN6326N	0699
AN6327	OEM
AN6328	2068
AN6328S	OEM
AN6331	1585
AN6332	1585
AN6337	OEM
AN6340	1580
AN6341	4758
AN6341N	4758
AN6342	4759
AN6342N	4759
AN6343	5225
AN6344	1580
AN6345	5169
AN6346	OEM
AN6347	OEM
AN6350	5226
AN6352	0360
AN6354	OEM
AN6356N	5260
AN6357N	OEM
AN6359	5261
AN6359N	5261
AN6360	5227
AN6360S	OEM
AN6361	3731
AN6361N	3731
AN6361NS	OEM
AN6362	5547
AN6366	5262
AN6366K	5262
AN6366NK	5262
AN6366NS	OEM
AN6370	OEM
AN6387	0759
AN6390	5776
AN6391K	OEM
AN6391NK	OEM
AN6517S	OEM
AN6540	OEM
AN6541	OEM
AN6550	OEM
AN6551	2913
AN6552	0360
AN6552(F)	0360
AN6552S	OEM
AN6553	0360
AN6553F	OEM
AN6554	OEM
AN6554NS	OEM
AN6557	OEM
AN6557F	OEM
AN6557FLC	OEM
AN6558	0360
AN6561	3233
AN6562	0770
AN6564	0625
AN6564NS	OEM
AN6570	0312
AN6571	6855
AN6572	0360
AN6609N	OEM
AN6610	OEM
AN6612S	OEM
AN6638	OEM
AN6650-LF	OEM
AN6650S	OEM
AN6651	OEM
AN6652S	OEM
AN6660	OEM
AN6660K	OEM

If replacement code is OEM, contact original manufacturer for replacement.

DEVICE TYPE	REPL CODE
AN6662	OEM
AN6663SP	OEM
AN6671K	OEM
AN6675	OEM
AN6677	5192
AN6683	OEM
AN6683N	OEM
AN6780	5544
AN6811	4752
AN6820	OEM
AN6821	4814
AN6855T	OEM
AN6873	2025
AN6873N	2025
AN6875	4820
AN6880	OEM
AN6884	5162
AN6886	OEM
AN6912	0176
AN6912N	0176
AN6912S	OEM
AN6913	4258
AN6914	0631
AN6914S	OEM
AN6915	OEM
AN7000	OEM
AN7025K	OEM
AN7062	5825
AN7070	OEM
AN7071	OEM
AN7072N	OEM
AN7074P	0395
AN7110	5229
AN7110E	5229
AN7111	OEM
AN7114	OEM
AN7115	4728
AN7115F	4728
AN7120	3196
AN7130	4714
AN7131	4716
AN7140	4716
AN7145	4729
AN7145H	4729
AN7145L	4729
AN7145M	4729
AN7146H	4718
AN7146M	4729
AN7147N	OEM
AN7148	OEM
AN7150	OEM
AN7154	4720
AN7155	OEM
AN7156N	4723
AN7158	3916
AN7158N	3916
AN7161N	6161
AN7163	OEM
AN7166	3916
AN7168	4097
AN7168N	4097
AN7171K	OEM
AN7178	OEM
AN7205	OEM
AN7213	6854
AN7216	OEM
AN7218	2926
AN7218H	2926
AN7220	4290
AN7220A	4290
AN7220B	4290
AN7222	OEM
AN7223	OEM
AN7254	OEM
AN7266	OEM
AN7273	5831
AN7273A	OEM
AN7310	6914
AN7311	4763
AN7312	OEM
AN7351K	OEM
AN7375NS	OEM
AN7384	OEM
AN7410	1318
AN7410N	OEM
AN7410NS	1318
AN7411	OEM
AN7420	OEM
AN7470P	OEM
AN7805	0624
AN7805F	0624
AN7805LB	1812
AN7806	0923
AN7809	1346
AN7809F	OEM
AN7809LB	1346
AN7812	0334
AN7812F	0334
AN7815	6962
AN7818F	2259
AN7905T	1285
AN8003	OEM
AN8083SE2	OEM
AN8370S	OEM
AN8373S	OEM
AN8373SE2	OEM
AN8374S	OEM
AN8374SE2	OEM
AN8377	OEM

DEVICE TYPE	REPL CODE
AN8387SE2	OEM
AN90800	OEM
ANAM1002	OEM
ANC7322	OEM
ANC7332	OEM
ANC-7345	OEM
ANC7345	OEM
ANC-7346	OEM
ANC7346	OEM
ANC-7708	OEM
ANC-7716	OEM
AND700	OEM
AND101R	OEM
AND101RG	3739
AND102G	3455
AND102GKB	OEM
AND102GKW	OEM
AND102R	OEM
AND102RKB	OEM
AND102RKW	OEM
AND103G	3455
AND103GK	OEM
AND103R	OEM
AND103RK	OEM
AND104R	5116
AND104RK	OEM
AND105G	3456
AND105GK	OEM
AND105R	5117
AND105RK	OEM
AND106R	OEM
AND106RK	OEM
AND107G	OEM
AND107GK	OEM
AND107R	OEM
AND107RK	OEM
AND108G	OEM
AND108GKB	OEM
AND108GKW	OEM
AND108R	OEM
AND108RKB	OEM
AND108RKW	OEM
AND109R	OEM
AND109RKB	OEM
AND109RKW	OEM
AND110R	OEM
AND110RK	OEM
AND113G	OEM
AND113R	3182
AND113S	OEM
AND113Y	OEM
AND114G	1778
AND114R	1615
AND114S	OEM
AND114Y	3157
AND115G	OEM
AND115R	OEM
AND116R	OEM
AND120R	OEM
AND121G	OEM
AND121R	1981
AND122G	1983
AND122R	5112
AND123G	OEM
AND123R	3019
AND123S	OEM
AND123Y	OEM
AND124G	0841
AND124R	OEM
AND124S	OEM
AND124Y	3124
AND132G	OEM
AND132O	OEM
AND132Y	OEM
AND202G	OEM
AND202R	OEM
AND205G	OEM
AND205R	OEM
AND205Y	OEM
AND206G	OEM
AND206R	OEM
AND206Y	OEM
AND207G	OEM
AND207R	OEM
AND207Y	OEM
AND208G	OEM
AND208R	OEM
AND208Y	OEM
AND226G	OEM
AND226R	OEM
AND226Y	OEM
AND324G	OEM
AND324R	OEM
AND325G	OEM
AND325R	OEM
AND326G	OEM
AND326R	OEM
AND327G	OEM
AND327R	OEM
AND352G	OEM
AND352R	OEM
AND353G	OEM
AND353R	OEM
AND370G	OEM
AND370R	OEM
AND371G	OEM
AND371R	OEM
AND516	OEM
AND600G	OEM

DEVICE TYPE	REPL CODE
AND600R	OEM
AND622	OEM
AND651	OEM
AND700A	OEM
AND700G	OEM
AND700H	OEM
AND7000	OEM
AND700P	OEM
AND700R	OEM
AND700Y	OEM
AND703A	OEM
AND703G	OEM
AND703H	OEM
AND7030	OEM
AND703P	OEM
AND703R	OEM
AND703Y	OEM
AND704A	OEM
AND704G	OEM
AND704H	OEM
AND704P	OEM
AND704R	OEM
AND704Y	OEM
AND706A	OEM
AND706G	OEM
AND706H	OEM
AND7060	OEM
AND706P	OEM
AND706R	OEM
AND706Y	OEM
AND708H	OEM
AND710A	OEM
AND710G	OEM
AND710H	OEM
AND7100	OEM
AND710P	OEM
AND710R	OEM
AND710Y	OEM
AND713A	OEM
AND713G	OEM
AND713H	OEM
AND7130	OEM
AND713P	OEM
AND713R	OEM
AND713Y	OEM
AND714A	OEM
AND714G	OEM
AND714H	OEM
AND7140	OEM
AND714P	OEM
AND714R	OEM
AND714Y	OEM
AND716A	OEM
AND716G	OEM
AND716H	OEM
AND7160	OEM
AND716P	OEM
AND716R	OEM
AND716Y	OEM
AND718H	OEM
AND720A	OEM
AND720G	OEM
AND720H	OEM
AND7200	OEM
AND720P	OEM
AND720R	OEM
AND720Y	OEM
AND723A	OEM
AND723G	OEM
AND723H	OEM
AND7230	OEM
AND723P	OEM
AND723R	OEM
AND723Y	OEM
AND724A	OEM
AND724G	OEM
AND724H	OEM
AND7240	OEM
AND724P	OEM
AND724R	OEM
AND724Y	OEM
AND726A	OEM
AND726G	OEM
AND726H	OEM
AND7260	OEM
AND726P	OEM
AND726R	OEM
AND726Y	OEM
AND728H	OEM
AND740A	OEM
AND740G	OEM
AND740H	OEM
AND7400	OEM
AND740P	OEM
AND740R	OEM
AND743A	OEM
AND743G	OEM
AND743H	OEM
AND7430	OEM
AND743P	OEM
AND743R	OEM
AND744A	OEM
AND744G	OEM
AND744H	OEM
AND7440	OEM
AND744P	OEM
AND744R	OEM
AND744Y	OEM
AND746A	OEM

DEVICE TYPE	REPL CODE
AND746G	OEM
AND746H	OEM
AND7460	OEM
AND746P	OEM
AND746R	OEM
AND746Y	OEM
AND748H	OEM
AND750A	OEM
AND750G	OEM
AND750H	OEM
AND7500	OEM
AND750P	OEM
AND750R	OEM
AND750Y	OEM
AND753A	OEM
AND753G	OEM
AND753H	OEM
AND7530	OEM
AND753P	OEM
AND753R	OEM
AND753Y	OEM
AND754A	OEM
AND754G	OEM
AND754H	OEM
AND7540	OEM
AND754P	OEM
AND754R	OEM
AND754Y	OEM
AND756A	OEM
AND756G	OEM
AND756H	OEM
AND7560	OEM
AND756P	OEM
AND756R	OEM
AND756Y	OEM
AND758H	OEM
AND1230	OEM
AND1240	OEM
AND1811	OEM
AND1861	OEM
AND3005GAL	OEM
AND3005GCL	OEM
AND3005HAL	OEM
AND3005HCL	OEM
AND30050AL	OEM
AND30050CL	OEM
AND3005RAL	OEM
AND3005RCL	OEM
AND3005YAL	OEM
AND3005YCL	OEM
AND3010GAL	OEM
AND3010GCL	OEM
AND3010HAL	OEM
AND3010HCL	OEM
AND30100AL	OEM
AND30100CL	OEM
AND3010RAL	OEM
AND3010RCL	OEM
AND3010YAL	OEM
AND3010YCL	OEM
AND3020GAL	OEM
AND3020GCL	OEM
AND3020HAL	OEM
AND3020HCL	OEM
AND30200AL	OEM
AND30200CL	OEM
AND3020RAL	OEM
AND3020RCL	OEM
AND3020YAL	OEM
AND3020YCL	OEM
AND4115R	OEM
AND4125R	OEM
AND4135G	OEM
AND4135R	OEM
AND4145G	OEM
AND4145R	OEM
AND4305GAL	OEM
AND4305GCL	OEM
AND4305HAL	OEM
AND4305HCL	OEM
AND43050AL	OEM
AND43050CL	OEM
AND4305RAL	OEM
AND4305RCL	OEM
AND4305YAL	OEM
AND4305YCL	OEM
AND4310GAL	OEM
AND4310GCL	OEM
AND4310HAL	OEM
AND4310HCL	OEM
AND43100AL	OEM
AND43100CL	OEM
AND4310RAL	OEM
AND4310RCL	OEM
AND4310YAL	OEM
AND4310YCL	OEM
AND5005GAL	OEM
AND5005GCL	OEM
AND5005HAL	OEM
AND5005HCL	OEM
AND50050AL	OEM
AND50050CL	OEM
AND5005RAL	OEM
AND5005RCL	OEM
AND5005YAL	OEM
AND5005YCL	OEM
AND5010GAJB	OEM
AND5010GAL	OEM
AND5010GCJB	OEM

DEVICE TYPE	REPL CODE
AND5010GCL	OEM
AND5010HAL	OEM
AND50100AL	OEM
AND50100CL	OEM
AND5010RAJB	OEM
AND5010RAL	OEM
AND5010RCJB	OEM
AND5010RCL	OEM
AND5010YAL	OEM
AND5010YCL	OEM
AND5015GAL	OEM
AND5015GAL1	OEM
AND5015GCL	OEM
AND5015GCL1	OEM
AND5015HAL	OEM
AND5015HAL1	OEM
AND5015HCL	OEM
AND5015HCL1	OEM
AND50150AL	OEM
AND50150AL1	OEM
AND50150CL	OEM
AND50150CL1	OEM
AND5015RAL	OEM
AND5015RAL1	OEM
AND5015RCL	OEM
AND5015RCL1	OEM
AND5015YAL	OEM
AND5015YAL1	OEM
AND5015YCL	OEM
AND5015YCL1	OEM
AND5020GAL	OEM
AND5020GAL1	OEM
AND5020GCL	OEM
AND5020GCL1	OEM
AND5020HAL	OEM
AND5020HAL1	OEM
AND5020HCL	OEM
AND5020HCL1	OEM
AND50200AL	OEM
AND50200AL1	OEM
AND50200CL	OEM
AND50200CL1	OEM
AND5020RAL	OEM
AND5020RAL1	OEM
AND5020RCL	OEM
AND5020RCL1	OEM
AND5020YAL	OEM
AND5020YAL1	OEM
AND5020YCL	OEM
AND5020YCL1	OEM
AND8101G	OEM
AND8101M	OEM
AND8101R	OEM
AND8102M	OEM
AND8152M	OEM
AND8202M	OEM
AND18035RCJ	OEM
AND18040RCJ	OEM
ANDBG05GAJ	OEM
ANDBG05RAJ	OEM
ANDBG10GAJ	OEM
ANDBG10MAJ	OEM
ANDBG10RAJ	OEM
ANDBG20MAJ	OEM
ANDBG30MAJ	OEM
ANDBG40MAJ	OEM
AO1	0136
AO7	0600
AO4091A	0136
AO4092	0015
AO4092A	0015
AO4092B	0015
AO4093	0015
AO4093-A	0015
AO4093-X	0102
AO4093A	0015
AO4166-2	0002
AO4201	0196
AO4210A	0015
AO4212-A	0015
AO4212-B	0015
AO4230-A	0102
AO4233	0015
AO4234-2	0012
AO4241-A	0102
AO4710	1303
AO4716	0015
AO4727	0474
AO4735-A	0774
AO4901A	0015
AOF-1050	OEM
AOF-1052	OEM
AOF-1300	OEM
AOM-12	OEM
AOTMS4416-15	OEM
AP0120NA	OEM
AP0120NB	OEM
AP0120ND	OEM
AP0130NA	OEM
AP0130NB	OEM
AP0130ND	OEM
AP0140NA	OEM
AP0140NB	OEM
AP0140ND	OEM
AP3.3A	0370
AP3.3B	0370
AP3.6A	0188
AP3.6B	0188
AP3.9A	0456

DEVICE TYPE	REPL CODE
AP3.9B	0456
AP4.3A	0533
AP4.3B	0533
AP4.7A	0451
AP4.7B	0451
AP5.1A	0162
AP5.1B	0162
AP5.6A	0157
AP5.6B	0157
AP6.0A	0301
AP6.0B	0301
AP6.2A	0636
AP6.2B	0636
AP6.8A	0025
AP6.8B	0025
AP7.5A	0649
AP7.5B	0649
AP8.2A	0244
AP8.2B	0244
AP9.1A	0012
AP9.1B	0012
AP10.0A	0170
AP10.0B	0170
AP11.0A	0317
AP11.0B	0317
AP12.0A	0137
AP12.0B	0137
AP13.0A	0365
AP13.0B	0365
AP14.0A	0100
AP14.0B	0100
AP15-12	OEM
AP15.0A	0002
AP15.0B	0002
AP16.0A	0420
AP16.0B	0420
AP17.0A	1649
AP17.0B	1649
AP18.0A	0495
AP18.0B	0495
AP19.0A	0949
AP19.0B	0949
AP20.0A	0531
AP20.0B	0531
AP22.0A	0565
AP22.0B	0565
AP24.0A	0402
AP24.0B	0402
AP26.0A	0441
AP26.0B	0441
AP28.0A	1674
AP28.0B	1674
AP30-12	OEM
AP30-12L	OEM
AP30.0A	0726
AP30.0B	0726
AP33.0A	0039
AP33.0B	0039
AP36.0A	0820
AP36.0B	0820
AP39.0A	0350
AP39.0B	0350
AP413	OEM
AP414	OEM
AP433	OEM
AP623	OEM
AP2361	OEM
AP3897	OEM
AP3898	OEM
AP4023S	OEM
AP4156	OEM
AP4157	OEM
AP4829	OEM
AP4830	OEM
AP5179	OEM
AP5317	OEM
APB-11A-1001	0050
APB-11A-1004	0050
APB-11A-1008	0004
APB-11H-1001	0050
APB-11H-1004	0050
APB-11H-1007	0136
APB-11H-1008	0211
APB-11H-1010	0004
APD3.3A	0292
APD3.3B	0292
APD3.6A	0188
APD3.6B	0188
APD3.9A	0456
APD3.9B	0456
APD4C2003I	OEM
APD4C2003M	OEM
APD4C2008I	OEM
APD4C2008M	OEM
APD4C2028I	OEM
APD4C2028M	OEM
APD4C3003I	OEM
APD4C3003M	OEM
APD4C3008I	OEM
APD4C3008M	OEM
APD4C3028I	OEM
APD4C3028M	OEM
APD4C4003I	OEM
APD4C4003M	OEM
APD4C4008M	OEM
APD4C4028I	OEM
APD4C4028M	OEM
APD4C5003I	OEM
APD4C5003M	OEM

DEVICE TYPE	REPL CODE
APD4C5008I	OEM
APD4C5008M	OEM
APD4C5028I	OEM
APD4C5028M	OEM
APD4C20003I	OEM
APD4C20003M	OEM
APD4C20008I	OEM
APD4C20008M	OEM
APD4C20028I	OEM
APD4C20028M	OEM
APD4.3A	0533
APD4.3B	0533
APD4.7A	0451
APD4.7B	0451
APD5.1A	0162
APD5.1B	0162
APD5.6A	0157
APD5.6B	0157
APD6.2A	0636
APD6.2B	0636
APD6.8A	0025
APD6.8B	0025
APD7.5A	0649
APD7.5B	0649
APD8.2A	0244
APD8.2B	0244
APD9.1A	0012
APD9.1B	0057
APD10.0A	0170
APD10.0B	0170
APD11.0A	0317
APD11.0B	0317
APD12.0A	0137
APD12.0B	0137
APD13.0A	0365
APD13.0B	0365
APD15.0A	0002
APD15.0B	0002
APD16.0A	0420
APD16.0B	0420
APD18.0A	0495
APD18.0B	0495
APD20.0A	0531
APD20.0B	0531
APD22.0A	0565
APD22.0B	0565
APD24.0A	0402
APD24.0B	0402
APD27.0A	0441
APD27.0B	0441
APD30.0A	0726
APD30.0B	0726
APD33.0A	0039
APD33.0B	0039
APD36.0A	0820
APD36.0B	0820
APD39.0A	0350
APD39.0B	0350
APD103	OEM
APD200	OEM
APD203	OEM
APD300	OEM
APD303	OEM
APD400	OEM
APD3595	0609
APLA722	OEM
APX43201/202(A)	OEM
APY10/I	OEM
APY10/II	OEM
APY11/I	OEM
APY11/II	OEM
APY12/I	OEM
APY12/II	OEM
APY12/III	OEM
APY13/I	OEM
APY13/II	OEM
APY13/III	OEM
AQ2	0086
AQ2(PHILCO)	0015
AQ3	0086
AQ3(PHILCO)	0015
AQ4	0198
AQ5	0086
AQ6	0198
AR01550	OEM
AR02550	OEM
AR03550	OEM
AR04501	OEM
AR04502	OEM
AR04550	OEM
AR05550	OEM
AR06550	OEM
AR06701	OEM
AR07700	OEM
AR09450	OEM
AR012000	OEM
AR061000	OEM
AR-4	0160
AR4	0085
AR-5	0160
AR5	0085
AR-6	0160
AR6	0085
AR-7	0160
AR7	0085
AR7C	0015
AR-8	0160
AR8	0160
AR8P404R	0160

If replacement code is OEM, contact original manufacturer for replacement.

DEVICE TYPE	REPL CODE
AR-9	0160
AR9	0160
AR-10	0160
AR10	0160
AR-11	0160
AR11	0160
AR-12	0160
AR12	0160
AR-13	0160
AR13	0160
AR-14	0160
AR14	0160
AR-15	0142
AR15	0103
AR16	0015
AR-17	0042
AR17	0178
AR17(PHILCO)	0178
AR17A	0178
AR17B	0178
AR17GREY	0042
AR-18	0142
AR18	0178
AR18(PHILCO)	0178
AR19	0015
AR20	0015
AR21	0015
AR-22	OEM
AR22	0015
AR-22(DIO)	0015
AR22(RECTIFIER)	0015
AR22(TRANSISTOR)	0161
AR-22(XSTR)	0042
AR-23	OEM
AR23	0071
AR-23(DIO)	0071
AR23(RECTIFIER)	0817
AR23(TRANSISTOR)	0460
AR-23(XSTR)	0925
AR-24	0071
AR24	0071
AR24(PHILCO)	0560
AR24(RECTIFIER)	0817
AR24(TRANSISTOR)	0161
AR24(XSTR)	0423
AR24RED	0423
AR-25	0925
AR25	0854
AR25(G)	0925
AR25A	OEM
AR25B	OEM
AR25D	OEM
AR25F	OEM
AR25G	1267
AR25G(DIO)	OEM
AR25H	OEM
AR25J	OEM
AR25K	OEM
AR25M	OEM
AR26	1367
AR27	0854
AR27GREEN	0925
AR28	0423
AR28RED	0423
AR-29	1267
AR29	1367
AR-30	0423
AR30	0423
AR35	0423
AR37	0854
AR37GREEN	0925
AR38	0423
AR38RED	0423
AR44	0925
AR102	0164
AR-103	0004
AR103	0050
AR-104	0004
AR104	0050
AR-105	0004
AR105	0050
AR107	0016
AR108	0016
AR-200	0016
AR200	0144
AR200(W)	0079
AR200W	0144
AR200WHITE	0079
AR-201	0016
AR201	0144
AR201(Y)	0079
AR201Y	0144
AR201YELLOW	0079
AR-202	0016
AR202	0144
AR202G	0144
AR202GREEN	0079
AR203	0086
AR-203(R)	0595
AR203(R)	0595
AR203R	0086
AR203RED	0595
AR204	0016
AR205	0016
AR206	0016
AR207	0595
AR208	0016
AR209	0007
AR210	0007

DEVICE TYPE	REPL CODE
AR211	0007
AR212	0144
AR213	0007
AR213(V)	0111
AR213V	0007
AR213VIOLET	0111
AR218	0224
AR218(RO)	0111
AR218ORANGE	0595
AR218RED	0595
AR218RO	0007
AR219	0007
AR219(YY)	0007
AR219YY	0127
AR220	0007
AR220(GY)	0007
AR220GREEN	0007
AR220GY	0127
AR220YELLOW	0007
AR221	0144
AR222	0007
AR222(BY)	0007
AR222BLUE	0007
AR222BY	0127
AR222YELLOW	0007
AR224	0127
AR224WHITE	0007
AR224YELLOW	0007
AR244	0925
AR303	0079
AR304	0126
AR304GREEN	0791
AR304RED	0791
AR306	0016
AR306BLUE	0086
AR306ORANGE	0086
AR308	0126
AR308VIOLET	0126
AR313	0144
AR343	0079
AR401P11	OEM
AR402P01	OEM
AR402P02	OEM
AR402P03	OEM
AR402P04	OEM
AR402P05	OEM
AR402P06	OEM
AR402P07	OEM
AR402P08	OEM
AR402P09	OEM
AR402P10	OEM
AR402P11	OEM
AR402P12	OEM
AR402S01	OEM
AR402S02	OEM
AR402S03	OEM
AR402S04	OEM
AR402S05	OEM
AR402S06	OEM
AR402S07	OEM
AR402S08	OEM
AR402S09	OEM
AR402S10	OEM
AR402S11	OEM
AR402S12	OEM
AR405P06	OEM
AR501	0212
AR502	0212
AR882	0015
AR904P02	OEM
AR904P03	OEM
AR904P04	OEM
AR904P05	OEM
AR904P06	OEM
AR904P07	OEM
AR904P08	OEM
AR904P09	OEM
AR904P10	OEM
AR904P11	OEM
AR904P12	OEM
AR904P13	OEM
AR904P14	OEM
AR904P15	OEM
AR904P16	OEM
AR904P17	OEM
AR904P18	OEM
AR904P19	OEM
AR904P20	OEM
AR904P21	OEM
AR904P22	OEM
AR904P23	OEM
AR904P24	OEM
AR904P25	OEM
AR904P26	OEM
AR904P27	OEM
AR904P28	OEM
AR904P29	OEM
AR904S02	OEM
AR904S03	OEM
AR904S04	OEM
AR904S05	OEM
AR904S06	OEM
AR904S07	OEM
AR904S08	OEM
AR904S09	OEM
AR904S10	OEM
AR904S11	OEM
AR904S12	OEM
AR904S13	OEM

DEVICE TYPE	REPL CODE
AR904S14	OEM
AR904S15	OEM
AR904S16	OEM
AR904S17	OEM
AR904S18	OEM
AR904S19	OEM
AR904S20	OEM
AR904S21	OEM
AR904S22	OEM
AR904S23	OEM
AR904S24	OEM
AR904S25	OEM
AR904S26	OEM
AR904S27	OEM
AR904S28	OEM
AR904S29	OEM
AR912P25	OEM
AR912P26	OEM
AR912P27	OEM
AR912P28	OEM
AR912P29	OEM
AR912P30	OEM
AR912P31	OEM
AR912P32	OEM
AR912P33	OEM
AR912P34	OEM
AR912P35	OEM
AR912P36	OEM
AR912P37	OEM
AR912P38	OEM
AR912P39	OEM
AR912P40	OEM
AR912P41	OEM
AR912P42	OEM
AR912P43	OEM
AR912P44	OEM
AR912P45	OEM
AR912P46	OEM
AR912P47	OEM
AR912P48	OEM
AR912P49	OEM
AR912P50	OEM
AR912P51	OEM
AR912P52	OEM
AR912P53	OEM
AR912P54	OEM
AR912P55	OEM
AR912P56	OEM
AR912S25	OEM
AR912S26	OEM
AR912S27	OEM
AR912S28	OEM
AR912S29	OEM
AR912S30	OEM
AR912S31	OEM
AR912S32	OEM
AR912S33	OEM
AR912S34	OEM
AR912S35	OEM
AR912S36	OEM
AR912S37	OEM
AR912S38	OEM
AR912S39	OEM
AR912S40	OEM
AR912S41	OEM
AR912S42	OEM
AR912S43	OEM
AR912S44	OEM
AR912S45	OEM
AR912S46	OEM
AR912S47	OEM
AR912S48	OEM
AR912S49	OEM
AR912S50	OEM
AR912S51	OEM
AR912S52	OEM
AR912S53	OEM
AR912S54	OEM
AR912S55	OEM
AR912S56	OEM
AR1101P01	OEM
AR1101P02	OEM
AR1101P03	OEM
AR1101P04	OEM
AR1101P05	OEM
AR1101P06	OEM
AR1101P07	OEM
AR1101P08	OEM
AR1101P09	OEM
AR1101P10	OEM
AR1101S01	OEM
AR1101S02	OEM
AR1101S03	OEM
AR1101S04	OEM
AR1101S05	OEM
AR1101S06	OEM
AR1101S07	OEM
AR1101S08	OEM
AR1101S09	OEM
AR1101S10	OEM
AR1104P10	OEM
AR1104P11	OEM
AR1104P12	OEM
AR1104P13	OEM
AR1104P14	OEM
AR1104P15	OEM
AR1104P16	OEM
AR1104P17	OEM
AR1104P18	OEM

DEVICE TYPE	REPL CODE
AR1104P19	OEM
AR1104P20	OEM
AR1104P21	OEM
AR1104P22	OEM
AR1104P23	OEM
AR1104P24	OEM
AR1104P25	OEM
AR1104P26	OEM
AR1104P27	OEM
AR1104P28	OEM
AR1104P29	OEM
AR1104S10	OEM
AR1104S11	OEM
AR1104S12	OEM
AR1104S13	OEM
AR1104S14	OEM
AR1104S15	OEM
AR1104S16	OEM
AR1104S17	OEM
AR1104S18	OEM
AR1104S19	OEM
AR1104S20	OEM
AR1104S21	OEM
AR1104S22	OEM
AR1104S23	OEM
AR1104S24	OEM
AR1104S25	OEM
AR1104S26	OEM
AR1104S27	OEM
AR1104S28	OEM
AR1104S29	OEM
AR1107P24	OEM
AR1107P25	OEM
AR1107P26	OEM
AR1107P27	OEM
AR1107P28	OEM
AR1107P29	OEM
AR1107P30	OEM
AR1107P31	OEM
AR1107P32	OEM
AR1107P33	OEM
AR1107P34	OEM
AR1107P35	OEM
AR1107P36	OEM
AR1107S24	OEM
AR1107S25	OEM
AR1107S26	OEM
AR1107S27	OEM
AR1107S28	OEM
AR1107S29	OEM
AR1107S30	OEM
AR1107S31	OEM
AR1107S32	OEM
AR1107S33	OEM
AR1107S34	OEM
AR1107S35	OEM
AR1107S36	OEM
AR1109P30	OEM
AR1109P31	OEM
AR1109P32	OEM
AR1109P33	OEM
AR1109P34	OEM
AR1109P35	OEM
AR1109P36	OEM
AR1109P37	OEM
AR1109P38	OEM
AR1109P39	OEM
AR1109P40	OEM
AR1109P41	OEM
AR1109P42	OEM
AR1109P43	OEM
AR1109P44	OEM
AR1109S30	OEM
AR1109S31	OEM
AR1109S32	OEM
AR1109S33	OEM
AR1109S34	OEM
AR1109S35	OEM
AR1109S36	OEM
AR1109S37	OEM
AR1109S38	OEM
AR1109S39	OEM
AR1109S40	OEM
AR1109S41	OEM
AR1109S42	OEM
AR1109S43	OEM
AR1109S44	OEM
AR2004P06	OEM
AR2004P07	OEM
AR2004P08	OEM
AR2004P09	OEM
AR2004P10	OEM
AR2004P11	OEM
AR2004P12	OEM
AR2004P13	OEM
AR2004P14	OEM
AR2004P15	OEM
AR2004P16	OEM
AR2004P17	OEM
AR2004P18	OEM
AR2004P19	OEM
AR2004P20	OEM
AR2004P21	OEM
AR2004P22	OEM
AR2004P23	OEM
AR2004P24	OEM
AR2004S06	OEM
AR2004S07	OEM
AR2004S08	OEM

DEVICE TYPE	REPL CODE
AR2004S09	OEM
AR2004S10	OEM
AR2004S11	OEM
AR2004S12	OEM
AR2004S13	OEM
AR2004S14	OEM
AR2004S15	OEM
AR2004S16	OEM
AR2004S17	OEM
AR2004S18	OEM
AR2004S19	OEM
AR2004S20	OEM
AR2004S21	OEM
AR2004S22	OEM
AR2004S23	OEM
AR2004S24	OEM
AR2009P24	OEM
AR2009P25	OEM
AR2009P26	OEM
AR2009P27	OEM
AR2009P28	OEM
AR2009P29	OEM
AR2009P30	OEM
AR2009P31	OEM
AR2009P32	OEM
AR2009P33	OEM
AR2009P34	OEM
AR2009P35	OEM
AR2009P36	OEM
AR2009P37	OEM
AR2009P38	OEM
AR2009P39	OEM
AR2009P40	OEM
AR2009P41	OEM
AR2009P42	OEM
AR2009P43	OEM
AR2009P44	OEM
AR2009PFS	OEM
AR2009S24	OEM
AR2009S25	OEM
AR2009S26	OEM
AR2009S27	OEM
AR2009S28	OEM
AR2009S29	OEM
AR2009S30	OEM
AR2009S31	OEM
AR2009S32	OEM
AR2009S33	OEM
AR2009S34	OEM
AR2009S35	OEM
AR2009S36	OEM
AR2009S37	OEM
AR2009S38	OEM
AR2009S39	OEM
AR2009S40	OEM
AR2009S41	OEM
AR2009S42	OEM
AR2009S43	OEM
AR2009S44	OEM
AR2400	OEM
AR2401	OEM
AR2402	OEM
AR2404	OEM
AR2406	OEM
AR2432B	OEM
AR3007P16	OEM
AR3007P17	OEM
AR3007P18	OEM
AR3007P19	OEM
AR3007P20	OEM
AR3007P21	OEM
AR3007P22	OEM
AR3007P23	OEM
AR3007P24	OEM
AR3007P25	OEM
AR3007P26	OEM
AR3007P27	OEM
AR3007P28	OEM
AR3007P29	OEM
AR3007P30	OEM
AR3007S16	OEM
AR3007S17	OEM
AR3007S18	OEM
AR3007S19	OEM
AR3007S20	OEM
AR3007S21	OEM
AR3007S22	OEM
AR3007S23	OEM
AR3007S24	OEM
AR3007S25	OEM
AR3007S26	OEM
AR3007S27	OEM
AR3007S28	OEM
AR3007S29	OEM
AR3007S30	OEM
AR3433S	OEM
AR11450	OEM
AR11452	OEM
AR20500	OEM
ARA25N	OEM
ARA25N-H	OEM
ARA25P	OEM
ARA25P-H	OEM
ARA46P	OEM
ARF612P01	OEM
ARF612P02	OEM
ARF612P03	OEM
ARF612P04	OEM
ARF612P05	OEM

DEVICE TYPE	REPL CODE
ARF612P06	OEM
ARF612P07	OEM
ARF612P08	OEM
ARF612P09	OEM
ARF612P10	OEM
ARF612P11	OEM
ARF612P12	OEM
ARF612P13	OEM
ARF612P14	OEM
ARF612P15	OEM
ARF612S01	OEM
ARF612S02	OEM
ARF612S03	OEM
ARF612S04	OEM
ARF612S05	OEM
ARF612S06	OEM
ARF612S07	OEM
ARF612S08	OEM
ARF612S09	OEM
ARF612S10	OEM
ARF612S11	OEM
ARF612S12	OEM
ARF612S13	OEM
ARF612S14	OEM
ARF612S15	OEM
ARF912P01	OEM
ARF912P02	OEM
ARF912P03	OEM
ARF912P04	OEM
ARF912P05	OEM
ARF912P06	OEM
ARF912P07	OEM
ARF912P08	OEM
ARF912P09	OEM
ARF912P10	OEM
ARF912P11	OEM
ARF912P12	OEM
ARF912P13	OEM
ARF912P14	OEM
ARF912P15	OEM
ARF912P16	OEM
ARF912P17	OEM
ARF912P18	OEM
ARF912P19	OEM
ARF912P20	OEM
ARF912P21	OEM
ARF912P22	OEM
ARF912P23	OEM
ARF912P24	OEM
ARF912P25	OEM
ARF912P26	OEM
ARF912S01	OEM
ARF912S02	OEM
ARF912S03	OEM
ARF912S04	OEM
ARF912S05	OEM
ARF912S06	OEM
ARF912S07	OEM
ARF912S08	OEM
ARF912S09	OEM
ARF912S10	OEM
ARF912S11	OEM
ARF912S12	OEM
ARF912S13	OEM
ARF912S14	OEM
ARF912S15	OEM
ARF912S16	OEM
ARF912S17	OEM
ARF912S18	OEM
ARF912S19	OEM
ARF912S20	OEM
ARF912S21	OEM
ARF912S22	OEM
ARF912S23	OEM
ARF912S24	OEM
ARF912S25	OEM
ARF912S26	OEM
ARF2012P01	OEM
ARF2012P02	OEM
ARF2012P03	OEM
ARF2012P04	OEM
ARF2012P05	OEM
ARF2012P06	OEM
ARF2012P07	OEM
ARF2012P08	OEM
ARF2012P09	OEM
ARF2012P10	OEM
ARF2012P11	OEM
ARF2012P12	OEM
ARF2012P13	OEM
ARF2012P14	OEM
ARF2012P15	OEM
ARF2012P16	OEM
ARF2012P17	OEM
ARF2012P18	OEM
ARF2012P19	OEM
ARF2012P20	OEM
ARF2012P21	OEM
ARF2012P22	OEM
ARF2012P23	OEM
ARF2012P24	OEM
ARF2012P25	OEM
ARF2012P26	OEM
ARF2012S01	OEM
ARF2012S02	OEM
ARF2012S03	OEM
ARF2012S04	OEM
ARF2012S05	OEM
ARF2012S06	OEM

DEVICE TYPE	REPL CODE
ARF2012S07	OEM
ARF2012S08	OEM
ARF2012S09	OEM
ARF2012S10	OEM
ARF2012S11	OEM
ARF2012S12	OEM
ARF2012S13	OEM
ARF2012S14	OEM
ARF2012S15	OEM
ARF2012S16	OEM
ARF2012S17	OEM
ARF2012S18	OEM
ARF2012S19	OEM
ARF2012S20	OEM
ARF2012S21	OEM
ARF2012S22	OEM
ARF2012S23	OEM
ARF2012S24	OEM
ARF2012S25	OEM
ARF2012S26	OEM
ARS25A	OEM
ARS25B	OEM
ARS25D	OEM
ARS25G	OEM
ARS25J	OEM
ARS25K	OEM
ARS25M	OEM
ART2739	OEM
ART2895	OEM
AS01	0023
AS01V	0023
AS01V0	0023
AS01Z	0023
AS03250	OEM
AS-2	0015
AS2	0015
AS-3	0015
AS3	0015
AS3A	0110
AS3B	0953
AS3C	0242
AS3D	0540
AS3E	0965
AS3F	0817
AS-4	0015
AS4	0015
AS-5	0015
AS5	0015
AS6	0015
AS11	0015
AS-14	0015
AS-15	0015
AS15	0015
AS80C40	OEM
AS100C40	OEM
AS120C40	OEM
AS215	0160
AS250C40	OEM
AS477	0847
AS3428	0600
AS5040C	OEM
AS5040M	OEM
AS5041C	OEM
AS5041M	OEM
AS5042C	OEM
AS5042M	OEM
AS5043C	OEM
AS5043M	OEM
AS5044C	OEM
AS5044M	OEM
AS5045C	OEM
AS5045M	OEM
AS5046C	OEM
AS5046M	OEM
AS5047C	OEM
AS5047M	OEM
AS5048C	OEM
AS5048M	OEM
AS5049C	OEM
AS5049M	OEM
AS5050C	OEM
AS5050M	OEM
AS5051C	OEM
AS5051M	OEM
AS5140C	OEM
AS5140M	OEM
AS5141C	OEM
AS5141M	OEM
AS5142C	OEM
AS5142M	OEM
AS5143C	OEM
AS5143M	OEM
AS5144C	OEM
AS5144M	OEM
AS5145C	OEM
AS5145M	OEM
AS10400	OEM
AS33867	0004
AS33868	0164
AS34280	0208
ASA2	OEM
ASA31	OEM
ASA51	OEM
ASA1000	OEM
ASA1001	OEM
ASA1003	OEM
ASA1004	OEM
ASBC-65	OEM
ASC/68	OEM

DEVICE TYPE	REPL CODE
ASC/80	OEM
ASC/88	OEM
ASC/Z80	OEM
ASD25	0600
ASEC50	OEM
ASY12	OEM
ASY12-1	0164
ASY12-2	0164
ASY13	OEM
ASY13-1	0164
ASY13-2	0164
ASY14	0164
ASY14-1	0164
ASY14-2	0164
ASY14-3	0164
ASY-24	0050
ASY24	0004
ASY24B	0004
ASY26	0280
ASY26-RT	0050
ASY27	0280
ASY28	0600
ASY28-RT	0600
ASY29	0600
ASY29-RT	0600
ASY30	0136
ASY31	0004
ASY32	0004
ASY48	0004
ASY48-IV	0004
ASY48-V	0004
ASY48-VI	0004
ASY49	0004
ASY50	0004
ASY51	0004
ASY52	0004
ASY53	0600
ASY54	0004
ASY55	0004
ASY56	0004
ASY56N	0280
ASY57	0136
ASY57N	0280
ASY58	0136
ASY58N	0280
ASY59	0136
ASY60	OEM
ASY61	0208
ASY61/TK33	OEM
ASY62	0208
ASY63	0050
ASY63N	0050
ASY64	OEM
ASY66	OEM
ASY67	0050
ASY70	0004
ASY70-IV	0004
ASY70-V	0016
ASY70-VI	0004
ASY70IV	0164
ASY70V	0164
ASY70VI	0164
ASY71	0164
ASY72	0600
ASY73	0600
ASY73-RT	0208
ASY74	0600
ASY74-RT	0208
ASY75	0600
ASY76	0280
ASY76-RT	0847
ASY77	0280
ASY80	0280
ASY80-RT	0847
ASY81	0847
ASY82	0004
ASY83	OEM
ASY84	0004
ASY85	OEM
ASY86	0600
ASY87	0600
ASY88	0600
ASY89	0600
ASY90	0004
ASY91	0004
ASZ10	0136
ASZ10(DIO)	0170
ASZ11	0136
ASZ12	0137
ASZ15	0002
ASZ16	0160
ASZ17	0160
ASZ18	0160
ASZ20	0050
ASZ20N	0050
ASZ21	0136
ASZ30	0136
ASZ1015	OEM
ASZ1016	OEM
ASZ1017	OEM
ASZ1018	OEM
AT/AF1	0211
AT/AF2	0211
AT/RF1	0050
AT/RF2	0050
AT/S13	0050
AT0017	0155
AT0017A	OEM
AT0025	0155
AT0025A	0155
AT0045	0155
AT0801ATI	OEM
AT-1	0050
AT-2	0050
AT-3	0050
AT-4	0050
AT4	0050
AT-5	0280
AT5	0050
AT-6	0050
AT6	0004
AT-6A	0004
AT6A	0004
AT-7	0086
AT7	0595
AT-8	0050
AT-9	0050
AT-10	0103
AT10H	0004
AT10M	0004
AT10N	0004
AT-11	0037
AT11	0150
AT-12	0086
AT12	0086
AT-12(PHILCO)	0086
AT13	0050
AT-14	0050
AT14	0050
AT15	0050
AT16	0050
AT17	0050
AT20H	0004
AT20M	0004
AT20N	0004
AT24C01A	OEM
AT24C04	OEM
AT24C08	OEM
AT30H	0004
AT30M	0004
AT30N	0004
AT50	0004
AT52	0600
AT53	0600
AT71	0600
AT72	0600
AT73R	0600
AT74	0004
AT74S	0004
AT75R	0600
AT76R	0600
AT77	0600
AT100H	0004
AT100M	0004
AT100N	0004
AT110	0975
AT111	0975
AT112	0975
AT113	0975
AT128	OEM
AT200	0975
AT209	OEM
AT210	OEM
AT270	OEM
AT275	OEM
AT310	0127
AT311	0127
AT312	0127
AT313	0127
AT314	0127
AT315	0127
AT316	0127
AT318	0127
AT319	0127
AT321	0127
AT322	0127
AT323	0127
AT324	0127
AT325	0127
AT326	0127
AT327	0127
AT328	0127
AT329	0016
AT330	0127
AT331	0037
AT331A	0037
AT332	0037
AT332A	0037
AT333	0037
AT335	0016
AT335A	0037
AT336	0016
AT337	0016
AT338	0127
AT339	0086
AT340	0007
AT341	0007
AT342	0007
AT343	0007
AT344	0007
AT345	0007
AT346	0007
AT347	0016
AT348	0016
AT349	0016
AT350	0233
AT351	0233
AT353	0016
AT354	0016
AT355	0007
AT356	0007
AT360	OEM
AT361	OEM
AT362	OEM
AT363	OEM
AT364	OEM
AT365	OEM
AT366	OEM
AT367	OEM
AT368	OEM
AT370	0016
AT380	0086
AT381	0086
AT382	0086
AT383	0086
AT384	0086
AT385	0086
AT386	0086
AT387	0086
AT388	0086
AT390	0126
AT391	0126
AT392	0126
AT393	0126
AT394	0126
AT395	0126
AT396	0126
AT397	0126
AT398	0126
AT399	OEM
AT400	0016
AT401	0016
AT402	0016
AT403	0016
AT403P01	OEM
AT403P02	OEM
AT403P03	OEM
AT403P04	OEM
AT403P05	OEM
AT403P06	OEM
AT403S01	OEM
AT403S02	OEM
AT403S03	OEM
AT403S04	OEM
AT403S05	OEM
AT403S06	OEM
AT404	0016
AT405	0016
AT405P01	OEM
AT405P02	OEM
AT405P03	OEM
AT405P04	OEM
AT405P05	OEM
AT405P06	OEM
AT405P07	OEM
AT405P08	OEM
AT405P09	OEM
AT405P10	OEM
AT405P11	OEM
AT405P12	OEM
AT405S01	OEM
AT405S02	OEM
AT405S03	OEM
AT405S04	OEM
AT405S05	OEM
AT405S06	OEM
AT405S07	OEM
AT405S08	OEM
AT405S09	OEM
AT405S10	OEM
AT405S11	OEM
AT405S12	OEM
AT406	0016
AT407	0016
AT410	0037
AT410-1	0037
AT412	0037
AT412-1	0037
AT413	0037
AT413-1	0037
AT414	0037
AT414-1	0037
AT415	0037
AT415-1	0037
AT416	0037
AT416-1	0037
AT417	0037
AT417-1	0037
AT418	0037
AT418-1	0037
AT419	0037
AT419-1	0037
AT420	0016
AT421	0016
AT422	0016
AT423	0016
AT424	0016
AT425	0016
AT426	0016
AT427	0016
AT430	0037
AT430-1	0037
AT431	0037
AT431-1	0037
AT432	0037
AT432-1	0037
AT433	0037
AT433-1	0037
AT434	0037
AT434-1	0037
AT435	0037
AT435-1	0037
AT436	0037
AT436-1	0037
AT437	0037
AT437-1	0037
AT438	0037
AT438-1	0037
AT440	0086
AT441	0086
AT442	0086
AT443	0086
AT444	0086
AT445	0086
AT446	0086
AT447	OEM
AT448	OEM
AT450	OEM
AT451	0037
AT451-1	0037
AT452	0037
AT452-1	0037
AT453	0037
AT453-1	0037
AT454	0037
AT454-1	0037
AT455	0037
AT455-1	0037
AT460	0126
AT461	0126
AT462	0126
AT463	0126
AT464	0126
AT465	0126
AT466	0126
AT467	0126
AT468	0126
AT470	0086
AT471	0086
AT472	0086
AT473	0086
AT474	0086
AT475	0086
AT476	0086
AT477	0086
AT478	0086
AT479	0086
AT480	0126
AT481	0126
AT482	0126
AT483	0126
AT484	0126
AT485	0126
AT490	0016
AT491	0016
AT492	0016
AT493	0016
AT494	0016
AT495	0016
AT504	OEM
AT505P01	OEM
AT505P02	OEM
AT505P03	OEM
AT505P04	OEM
AT505P05	OEM
AT505P06	OEM
AT505P07	OEM
AT505P08	OEM
AT505P09	OEM
AT505P10	OEM
AT505P11	OEM
AT505P12	OEM
AT505P13	OEM
AT505P14	OEM
AT505P15	OEM
AT505S01	OEM
AT505S02	OEM
AT505S03	OEM
AT505S04	OEM
AT505S05	OEM
AT505S06	OEM
AT505S07	OEM
AT505S08	OEM
AT505S09	OEM
AT505S10	OEM
AT505S11	OEM
AT505S12	OEM
AT505S13	OEM
AT505S14	OEM
AT505S15	OEM
AT520	0144
AT521	0600
AT551	0600
AT704P01	OEM
AT704P02	OEM
AT704P03	OEM
AT704P04	OEM
AT704P05	OEM
AT704P06	OEM
AT704P07	OEM
AT704P08	OEM
AT704P09	OEM
AT704P10	OEM
AT704P11	OEM
AT704P12	OEM
AT704P13	OEM
AT704P14	OEM
AT704P15	OEM
AT704P16	OEM
AT704S01	OEM
AT704S02	OEM
AT704S03	OEM
AT704S04	OEM
AT704S05	OEM
AT704S06	OEM
AT704S07	OEM
AT704S08	OEM
AT704S09	OEM
AT704S10	OEM
AT704S11	OEM
AT704S12	OEM
AT704S13	OEM
AT704S14	OEM
AT704S15	OEM
AT704S16	OEM
AT804P01	OEM
AT804P02	OEM
AT804P03	OEM
AT804P04	OEM
AT804P05	OEM
AT804P06	OEM
AT804P07	OEM
AT804P08	OEM
AT804P09	OEM
AT804P10	OEM
AT804P11	OEM
AT804P12	OEM
AT804P13	OEM
AT804P14	OEM
AT804P15	OEM
AT804P16	OEM
AT804S01	OEM
AT804S02	OEM
AT804S03	OEM
AT804S04	OEM
AT804S05	OEM
AT804S06	OEM
AT804S07	OEM
AT804S08	OEM
AT804S09	OEM
AT804S10	OEM
AT804S11	OEM
AT804S12	OEM
AT804S13	OEM
AT804S14	OEM
AT804S15	OEM
AT804S16	OEM
AT807P01	OEM
AT807P02	OEM
AT807P03	OEM
AT807P04	OEM
AT807P05	OEM
AT807P06	OEM
AT807P07	OEM
AT807P08	OEM
AT807P09	OEM
AT807P10	OEM
AT807P11	OEM
AT807P12	OEM
AT807P13	OEM
AT807P14	OEM
AT807P15	OEM
AT807P16	OEM
AT807P17	OEM
AT807P18	OEM
AT807P19	OEM
AT807P20	OEM
AT807P21	OEM
AT807P22	OEM
AT807P23	OEM
AT807P24	OEM
AT807S01	OEM
AT807S02	OEM
AT807S03	OEM
AT807S04	OEM
AT807S05	OEM
AT807S06	OEM
AT807S07	OEM
AT807S08	OEM
AT807S09	OEM
AT807S10	OEM
AT807S11	OEM
AT807S12	OEM
AT807S13	OEM
AT807S14	OEM
AT807S15	OEM
AT807S16	OEM
AT807S17	OEM
AT807S18	OEM
AT807S19	OEM
AT807S20	OEM
AT807S21	OEM
AT807S22	OEM
AT807S23	OEM
AT807S24	OEM
AT874	0004
AT1004P01	OEM
AT1004P02	OEM
AT1004P03	OEM
AT1004P04	OEM
AT1004P05	OEM
AT1004P06	OEM
AT1004P07	OEM
AT1004P08	OEM
AT1004P09	OEM
AT1004P10	OEM
AT1004P11	OEM
AT1004P12	OEM
AT1004P13	OEM
AT1004P14	OEM
AT1004P15	OEM
AT1004S01	OEM
AT1004S02	OEM
AT1004S03	OEM
AT1004S04	OEM
AT1004S05	OEM
AT1004S06	OEM
AT1004S07	OEM
AT1004S08	OEM
AT1004S09	OEM
AT1004S10	OEM
AT1004S11	OEM
AT1004S12	OEM
AT1004S13	OEM
AT1004S14	OEM
AT1004S15	OEM
AT1007P01	OEM
AT1007P02	OEM
AT1007P03	OEM
AT1007P04	OEM
AT1007P05	OEM
AT1007P06	OEM
AT1007P07	OEM
AT1007P08	OEM
AT1007P09	OEM
AT1007P10	OEM
AT1007P11	OEM
AT1007P12	OEM
AT1007P13	OEM
AT1007P14	OEM
AT1007P15	OEM
AT1007P16	OEM
AT1007P17	OEM
AT1007P18	OEM
AT1007P19	OEM
AT1007P20	OEM
AT1007P21	OEM
AT1007P22	OEM
AT1007P23	OEM
AT1007P24	OEM
AT1007S01	OEM
AT1007S02	OEM
AT1007S03	OEM
AT1007S04	OEM
AT1007S05	OEM
AT1007S06	OEM
AT1007S07	OEM
AT1007S08	OEM
AT1007S09	OEM
AT1007S10	OEM
AT1007S11	OEM
AT1007S12	OEM
AT1007S13	OEM
AT1007S14	OEM
AT1007S15	OEM
AT1007S16	OEM
AT1007S17	OEM
AT1007S18	OEM
AT1007S19	OEM
AT1007S20	OEM
AT1007S21	OEM
AT1007S22	OEM
AT1007S23	OEM
AT1007S24	OEM
AT1010S32	OEM
AT1018P20	OEM
AT1018P21	OEM
AT1018P22	OEM
AT1018P23	OEM
AT1018P24	OEM
AT1018P25	OEM
AT1018P26	OEM
AT1018P27	OEM
AT1018P28	OEM
AT1018P29	OEM
AT1018P30	OEM
AT1018P31	OEM
AT1018P32	OEM
AT1018P33	OEM
AT1018P34	OEM
AT1018P35	OEM
AT1018P36	OEM
AT1018S20	OEM
AT1018S21	OEM
AT1018S22	OEM
AT1018S23	OEM
AT1018S24	OEM
AT1018S25	OEM
AT1018S26	OEM
AT1018S27	OEM
AT1018S28	OEM
AT1018S29	OEM
AT1018S30	OEM
AT1018S31	OEM
AT1018S33	OEM
AT1018S34	OEM
AT1018S35	OEM
AT1018S36	OEM
AT1138	0604
AT1138A	0604
AT1138B	0604
AT1825	OEM
AT1833	0604
AT1834	0604
AT1845	OEM
AT1845A	OEM
AT-1856	0103
AT1856	0103
AT2848	0126
AT3006	OEM
AT3007	OEM
AT3008	OEM
AT3009	OEM
AT3010	OEM
AT3011	OEM
AT3012	OEM
AT3013	OEM
AT3014	OEM
AT3015	OEM
AT3016	OEM
AT3017	OEM
AT3018	OEM
AT3019	OEM
AT3020	OEM
AT3021	OEM
AT3260	0103
AT3850	OEM
AT4506	OEM
AT4507	OEM
AT4508	OEM
AT4509	OEM
AT4510	OEM
AT4511	OEM
AT4512	OEM
AT4513	OEM
AT4514	OEM
AT4515	OEM
AT4516	OEM
AT4517	OEM
AT4518	OEM
AT4519	OEM
AT4520	OEM
AT4521	OEM
AT4642	OEM
AT4680	OEM
AT4690	OEM
AT4841	OEM
AT4842	OEM
AT4880	OEM
AT4890	OEM
AT5156	0126
AT6006	OEM
AT6007	OEM
AT6008	OEM
AT6009	OEM
AT6010	OEM
AT6011	OEM
AT6012	OEM
AT6013	OEM
AT6014	OEM
AT6015	OEM
AT6016	OEM
AT6017	OEM
AT6018	OEM
AT6019	OEM
AT6020	OEM
AT6021	OEM
AT7510	OEM
AT8040	OEM
AT8041	OEM
AT8060	OEM
AT8110	OEM
AT8111	OEM
AT8140	OEM
AT8141	OEM
AT8150	OEM
AT8151	OEM
AT8160	OEM
AT8161	OEM
AT8250	OEM
AT8251	OEM
AT9006	OEM
AT9007	OEM
AT9008	OEM
AT9009	OEM
AT9010	OEM
AT9011	OEM
AT9012	OEM
AT9013	OEM
AT9014	OEM
AT9015	OEM
AT9016	OEM
AT9017	OEM
AT9018	OEM
AT9019	OEM
AT9020	OEM
AT9021	OEM
AT10650-1	OEM
AT12570-5	OEM
ATAF1	0004
ATAF2	0004
ATC-SR-3	0087
ATC-TR-4	0086
ATC-TR-5	0160
ATC-TR-6	0160
ATC-TR-7	0086
ATC-TR-13	0086
ATC-TR-14	0160
ATC-TR-15	0103
ATC-TR-19	0042
ATF585P01B	OEM
ATF585P02B	OEM
ATF585P03B	OEM

DEVICE TYPE	REPL CODE
ATF585P04B	OEM
ATF585P05B	OEM
ATF585P06A	OEM
ATF585P06B	OEM
ATF585P06L	OEM
ATF585P06M	OEM
ATF585P07B	OEM
ATF585P08B	OEM
ATF585S01B	OEM
ATF585S02B	OEM
ATF585S03B	OEM
ATF585S04B	OEM
ATF585S05B	OEM
ATF585S06A	OEM
ATF585S06B	OEM
ATF585S06L	OEM
ATF585S06M	OEM
ATF585S07B	OEM
ATF585S08B	OEM
ATF587P01L	OEM
ATF587P02L	OEM
ATF587P03L	OEM
ATF587P04L	OEM
ATF587P05L	OEM
ATF587P06L	OEM
ATF587P07L	OEM
ATF587P08L	OEM
ATF587P09L	OEM
ATF587P10L	OEM
ATF587P11L	OEM
ATF587P12L	OEM
ATF587S01L	OEM
ATF587S02L	OEM
ATF587S03L	OEM
ATF587S04L	OEM
ATF587S05L	OEM
ATF587S06L	OEM
ATF587S07L	OEM
ATF587S08L	OEM
ATF587S09L	OEM
ATF587S10L	OEM
ATF587S11L	OEM
ATF587S12L	OEM
ATF827P01L	OEM
ATF827P02L	OEM
ATF827P03L	OEM
ATF827P04L	OEM
ATF827P05L	OEM
ATF827P06L	OEM
ATF827P07L	OEM
ATF827P08L	OEM
ATF827P09L	OEM
ATF827P09M	OEM
ATF827P09N	OEM
ATF827P09P	OEM
ATF827P10L	OEM
ATF827P11L	OEM
ATF827P12L	OEM
ATF827P13L	OEM
ATF827P14L	OEM
ATF827S01L	OEM
ATF827S02L	OEM
ATF827S03L	OEM
ATF827S04L	OEM
ATF827S05L	OEM
ATF827S06L	OEM
ATF827S07L	OEM
ATF827S08L	OEM
ATF827S09L	OEM
ATF827S09M	OEM
ATF827S09N	OEM
ATF827S09P	OEM
ATF827S10L	OEM
ATF827S11L	OEM
ATF827S12L	OEM
ATF827S13L	OEM
ATF827S14L	OEM
ATF854P01N	OEM
ATF854P02N	OEM
ATF854P03N	OEM
ATF854P04N	OEM
ATF854P05N	OEM
ATF854P06N	OEM
ATF854P07N	OEM
ATF854P08N	OEM
ATF854P09N	OEM
ATF854P10N	OEM
ATF854P11N	OEM
ATF854P12N	OEM
ATF854S01N	OEM
ATF854S02N	OEM
ATF854S03N	OEM
ATF854S04N	OEM
ATF854S05N	OEM
ATF854S06N	OEM
ATF854S07N	OEM
ATF854S08N	OEM
ATF854S09N	OEM
ATF854S10N	OEM
ATF854S11N	OEM
ATF854S12N	OEM
ATF857P01L	OEM
ATF857P02L	OEM
ATF857P03L	OEM
ATF857P06L	OEM
ATF857P07L	OEM
ATF857P08L	OEM
ATF857P09L	OEM
ATF857P0L	OEM
ATF857P10L	OEM
ATF857P11L	OEM
ATF857P12L	OEM
ATF857S01L	OEM
ATF857S02L	OEM
ATF857S03L	OEM
ATF857S04L	OEM
ATF857S05L	OEM
ATF857S06L	OEM
ATF857S07L	OEM
ATF857S08L	OEM
ATF857S09L	OEM
ATF857S10L	OEM
ATF857S11L	OEM
ATF857S12L	OEM
ATF860P01B	OEM
ATF860P02B	OEM
ATF860P03B	OEM
ATF860P04B	OEM
ATF860P05B	OEM
ATF860P06B	OEM
ATF860P07B	OEM
ATF860P08B	OEM
ATF860P09B	OEM
ATF860P10B	OEM
ATF860P11B	OEM
ATF860P12B	OEM
ATF860S01B	OEM
ATF860S02B	OEM
ATF860S03B	OEM
ATF860S04B	OEM
ATF860S05B	OEM
ATF860S06B	OEM
ATF860S07B	OEM
ATF860S08B	OEM
ATF860S09B	OEM
ATF860S10B	OEM
ATF860S11B	OEM
ATF860S12B	OEM
ATF880P01P	OEM
ATF880P02P	OEM
ATF880P03P	OEM
ATF880P04P	OEM
ATF880P05P	OEM
ATF880P06P	OEM
ATF880P07P	OEM
ATF880P08P	OEM
ATF880P09P	OEM
ATF880P10P	OEM
ATF880P11P	OEM
ATF880P12P	OEM
ATF880P13P	OEM
ATF880P14P	OEM
ATF880P15P	OEM
ATF880P16P	OEM
ATF880P17P	OEM
ATF880P18P	OEM
ATF880P19P	OEM
ATF880S01P	OEM
ATF880S02P	OEM
ATF880S03P	OEM
ATF880S04P	OEM
ATF880S05P	OEM
ATF880S06P	OEM
ATF880S07P	OEM
ATF880S08P	OEM
ATF880S09P	OEM
ATF880S10P	OEM
ATF880S11P	OEM
ATF880S12P	OEM
ATF880S13P	OEM
ATF880S14P	OEM
ATF880S15P	OEM
ATF880S16P	OEM
ATF880S17P	OEM
ATF880S18P	OEM
ATF880S19P	OEM
ATF887P01L	OEM
ATF887P02L	OEM
ATF887P03L	OEM
ATF887P04L	OEM
ATF887P05L	OEM
ATF887P06L	OEM
ATF887P07L	OEM
ATF887P08L	OEM
ATF887P09L	OEM
ATF887P10L	OEM
ATF887P11L	OEM
ATF887P12L	OEM
ATF887P13L	OEM
ATF887P14L	OEM
ATF887S01L	OEM
ATF887S02L	OEM
ATF887S03L	OEM
ATF887S04L	OEM
ATF887S05L	OEM
ATF887S06L	OEM
ATF887S07L	OEM
ATF887S08L	OEM
ATF887S09L	OEM
ATF887S10L	OEM
ATF887S11L	OEM
ATF887S12L	OEM
ATF887S13L	OEM
ATF887S14L	OEM
ATF1047P01M	OEM
ATF1047P02M	OEM
ATF1047P03M	OEM
ATF1047P04M	OEM
ATF1047P05M	OEM
ATF1047P06M	OEM
ATF1047P07M	OEM
ATF1047P08M	OEM
ATF1047P09M	OEM
ATF1047P09N	OEM
ATF1047P09P	OEM
ATF1047P09R	OEM
ATF1047P10M	OEM
ATF1047P11M	OEM
ATF1047P12M	OEM
ATF1047P13M	OEM
ATF1047P14M	OEM
ATF1047S01M	OEM
ATF1047S02M	OEM
ATF1047S03M	OEM
ATF1047S04M	OEM
ATF1047S05M	OEM
ATF1047S06M	OEM
ATF1047S07M	OEM
ATF1047S08M	OEM
ATF1047S09M	OEM
ATF1047S09N	OEM
ATF1047S09P	OEM
ATF1047S09R	OEM
ATF1047S10M	OEM
ATF1047S11M	OEM
ATF1047S12M	OEM
ATF1047S13M	OEM
ATF1047S14M	OEM
ATGP	0004
ATM88A	OEM
ATRF1	0136
ATRF2	0136
ATS13	0136
ATS20J	OEM
ATS21H1	OEM
ATS21J	OEM
ATS22H1	OEM
ATS22J	OEM
ATS23J	OEM
ATS24J	OEM
ATS30	OEM
ATS40	OEM
ATS50	OEM
ATS60	OEM
ATT1852ACT	OEM
ATT1853CT	OEM
ATY450	OEM
AU01	0023
AU01-07	0205
AU01-08	0504
AU01-09	0684
AU01-10	0225
AU01-11	0230
AU01-12	0234
AU01-13	0237
AU01-15	0247
AU01-16	0251
AU01-18	0256
AU01-20	0262
AU01-22	0269
AU01-24	0274
AU01-27	0294
AU01-30	0309
AU01-33	0318
AU01V0	OEM
AU01Z	0023
AU02	0023
AU02AU02ZV0	0023
AU02V0	0023
AU02V1	OEM
AU02Z	0023
AU100N	0164
AU101	0975
AU102	0160
AU103	0975
AU104	0975
AU105	0975
AU106	0975
AU107	0975
AU108	0975
AU110	0975
AU111	0975
AU112	0975
AU113	0975
AU206	0975
AU210	0975
AU213	0975
AU2012	0137
A-U2569	1940
AUY10	0160
AUY19	0160
AUY20	0160
AUY-21	0160
AUY21	0085
AUY21A	0160
AUY22	0085
AUY22A	0160
AUY24	OEM
AUY26	OEM
AUY27	OEM
AUY28	0975
AUY29	0604
AUY30	0160
AUY31	0160
AUY32	0160
AUY33	0160
AUY34	0975
AUY37	0604
AUY38	0975
AUZ11	OEM
AV/5R4GY	OEM
AV0000105-0	0133
AV01	OEM
AV03-04	1078
AV03-08	1490
AV03-10	0510
AV03-12	0870
AV03-15	1274
AV03-18	1078
AV03-22	1723
AV03-27	1761
AV2	OEM
AV3B28	OEM
AV4	OEM
AV4B32	OEM
AV5	0137
AV6	OEM
AV7	OEM
AV8	OEM
AV9	OEM
AV10	0039
AV11	OEM
AV12	OEM
AV13	OEM
AV14	OEM
AV15	OEM
AV16	OEM
AV17	OEM
AV18	OEM
AV19	OEM
AV102	OEM
AV104	0510
AV104(ZENER)	0510
AV105	0160
AV105(ZENER)	0870
AV106	OEM
AV107	1078
AV107(ZENER)	1078
AV108	OEM
AV109	OEM
AV110	1788
AV111	OEM
AV112	OEM
AV113	OEM
AV114	OEM
AV115	OEM
AV116	0438
AV117	OEM
AV118	OEM
AV119	OEM
AV2010	OEM
AV2011	OEM
AV2012	0137
AV2013	OEM
AV2014	OEM
AV2015	0002
AV2016	OEM
AV2017	OEM
AV2018	OEM
AV2019	OEM
AV2020	OEM
AV2021	OEM
AV2022	OEM
AV2023	OEM
AV2024	OEM
AV2025	OEM
AV2026	OEM
AV2027	0441
AV2027A	0441
AV2028	OEM
AV2029	OEM
AV2030	OEM
AV2031	OEM
AV2032	OEM
AV2033	0039
AV2034	OEM
AV2035	OEM
AV2036	OEM
AV2037	OEM
AV2038	OEM
AV2039	OEM
AV2040	OEM
AV2041	OEM
AV2042	OEM
AV2043	OEM
AV2044	OEM
AV2045	OEM
AV2046	OEM
AV2047	OEM
AV2048	OEM
AV2049	OEM
AV2050	OEM
AV2051	OEM
AV2052	OEM
AV2053	OEM
AV2054	OEM
AV2055	1834
AV2056	OEM
AV2057	OEM
AV2058	OEM
AV2059	OEM
AV2060	OEM
AV2061	OEM
AV2062	0783
AV2063	OEM
AV2064	OEM
AV2065	OEM
AV2066	OEM
AV2067	OEM
AV2068	OEM
AV2069	OEM
AV2070	OEM
AV2071	OEM
AV2072	OEM
AV2073	OEM
AV2074	OEM
AV2075	OEM
AV2076	OEM
AV2077	OEM
AV2078	OEM
AV2079	OEM
AV2080	OEM
AV2081	OEM
AV2082	0331
AV2083	OEM
AV2084	OEM
AV2085	OEM
AV2086	OEM
AV2087	OEM
AV2088	OEM
AV2089	OEM
AV2090	OEM
AV2091	OEM
AV2092	OEM
AV2093	OEM
AV2094	OEM
AV2095	OEM
AV2096	OEM
AV2097	OEM
AV2098	OEM
AV2099	OEM
AV2100	OEM
AV2105	OEM
AV2110	0149
AV2115	OEM
AV2120	OEM
AV2125	OEM
AV2130	OEM
AV2135	OEM
AV2140	OEM
AV2145	OEM
AV2150	OEM
AV2155	OEM
AV2160	OEM
AV2165	OEM
AV2170	OEM
AV2175	OEM
AV2180	OEM
AV2185	OEM
AV2190	OEM
AV2195	OEM
AV2200	OEM
AV7104	OEM
AV7203	OEM
AV7204	OEM
AV7206	OEM
AV7216	OEM
AV7218	OEM
AV7224	OEM
AV7226	OEM
AV7228	OEM
AV7236	OEM
AV7238	OEM
AV7248	OEM
AV7403	OEM
AV7413	OEM
AV7433	OEM
AV7443	OEM
AV7453	OEM
AV7803	OEM
AV7813	OEM
AV7823	OEM
AV7843	OEM
AV7871	OEM
AV7872	OEM
AV8010	OEM
AV8011	OEM
AV8012	OEM
AV8013	OEM
AV8014	OEM
AV8015	1274
AV8015(ZENER)	1274
AV8016	OEM
AV8017	OEM
AV8018	1078
AV8018(ZENER)	1078
AV8019	OEM
AV8020	OEM
AV8021	OEM
AV8022	1723
AV8022(ZENER)	1723
AV8023	OEM
AV8024	OEM
AV8025	OEM
AV8026	OEM
AV8027	1761
AV8028	OEM
AV8029	OEM
AV8030	OEM
AV8031	OEM
AV8032	OEM
AV8033	OEM
AV8034	OEM
AV8035	OEM
AV8036	OEM
AV8037	OEM
AV8038	OEM
AV8039	OEM
AV8040	OEM
AV8041	OEM
AV8042	OEM
AV8043	OEM
AV8044	OEM
AV8045	OEM
AV8046	OEM
AV8047	OEM
AV8048	OEM
AV8049	OEM
AV8050	OEM
AV8051	OEM
AV8052	OEM
AV8053	OEM
AV8054	OEM
AV8055	OEM
AV8056	0207
AV8057	OEM
AV8058	OEM
AV8059	OEM
AV8060	OEM
AV8061	OEM
AV8062	OEM
AV8063	OEM
AV8064	OEM
AV8065	OEM
AV8066	OEM
AV8067	OEM
AV8068	OEM
AV8069	OEM
AV8070	OEM
AV8071	OEM
AV8072	OEM
AV8073	OEM
AV8074	OEM
AV8075	OEM
AV8076	OEM
AV8077	OEM
AV8078	OEM
AV8079	OEM
AV8080	OEM
AV8081	OEM
AV8082	0356
AV8083	OEM
AV8084	OEM
AV8085	OEM
AV8086	OEM
AV8087	OEM
AV8088	OEM
AV8089	OEM
AV8090	OEM
AV8091	OEM
AV8092	OEM
AV8093	OEM
AV8094	OEM
AV8095	OEM
AV8096	OEM
AV8097	OEM
AV8098	OEM
AV8099	OEM
AV8100	OEM
AV8105	OEM
AV8110	OEM
AV8115	OEM
AV8120	OEM
AV8125	OEM
AV8130	OEM
AV8135	OEM
AV8140	OEM
AV8145	OEM
AV8150	OEM
AV8155	OEM
AV8160	OEM
AV8165	OEM
AV8170	OEM
AV8175	OEM
AV8180	OEM
AV8185	OEM
AV8190	OEM
AV8195	OEM
AV8200	OEM
AV51010	OEM
AV51011	OEM
AV51012	OEM
AV51013	OEM
AV51014	OEM
AV51015	OEM
AV51016	OEM
AV51017	OEM
AV51018	OEM
AV51019	OEM
AV51020	OEM
AV51021	OEM
AV51022	OEM
AV51023	OEM
AV51024	OEM
AV51025	OEM
AV51026	OEM
AV51027	OEM
AV51028	OEM
AV51029	OEM
AV51030	OEM
AV51031	OEM
AV51032	OEM
AV51033	OEM
AV51034	OEM
AV51035	OEM
AV51036	OEM
AV51037	OEM
AV51038	OEM
AV51039	OEM
AV51040	OEM
AV51041	OEM
AV51042	OEM
AV51043	OEM
AV51044	OEM
AV51045	OEM
AV51046	OEM
AV51047	OEM
AV51048	OEM
AV51049	OEM
AV51050	OEM
AV51051	OEM
AV51052	OEM
AV51053	OEM
AV51054	OEM
AV51055	OEM
AV51056	OEM
AV51057	OEM
AV51058	OEM
AV51059	OEM
AV51060	OEM
AV51061	OEM
AV51062	OEM
AV51063	OEM
AV51064	OEM
AV51065	OEM
AV51066	OEM
AV51067	OEM
AV51068	OEM
AV51069	OEM
AV51070	OEM
AV51071	OEM
AV51072	OEM
AV51073	OEM
AV51074	OEM
AV51075	OEM
AV51076	OEM
AV51077	OEM
AV51078	OEM
AV51079	OEM
AV51080	OEM
AV51081	OEM
AV51082	OEM
AV51083	OEM
AV51084	OEM
AV51085	OEM
AV51086	OEM
AV51087	OEM
AV51088	OEM
AV51089	OEM
AV51090	OEM
AV51091	OEM
AV51092	OEM
AV51093	OEM
AV51094	OEM
AV51095	OEM
AV51096	OEM
AV51097	OEM
AV51098	OEM
AV51099	OEM
AV51100	OEM
AV51105	OEM
AV51110	OEM
AV51115	OEM
AV51120	OEM
AV51125	OEM
AV51130	OEM
AV51140	OEM
AV51145	OEM
AV51150	OEM
AV51160	OEM
AV51165	OEM
AV51170	OEM
AV51175	OEM
AV51180	OEM
AV51185	OEM
AV51190	OEM
AV51195	OEM
AV51200	OEM
AV71202	OEM
AV71211	OEM
AV71221	OEM
AV71241	OEM
AV71251	OEM
AV71261	OEM
AV71826	OEM
AV78018	OEM
AV78118	OEM
AV78318	OEM
AW01	0137
AW01-0	0301
AW01-02	OEM
AW01-06	0301
AW-01-07	0025
AW01-07	0025
AW-01-08	0244
AW01-08	0244
AW01-08J	0244
AW-01-09	0012
AW01-09	0012
AW01-7	0025
AW01-9	0012
AW01-9/CP3112030	0012
AW01-9V	0012

If replacement code is OEM, contact original manufacturer for replacement.

DEVICE TYPE	REPL CODE
AW-01-10	0170
AW01-10	0317
AW01-11	0317
AW-01-12	0137
AW01-12	0137
AW-01-12C	0137
AW01-12C	0137
AW01-12V	0137
AW01-13	0053
AW01-15	0002
AW01-16	0420
AW01-18	0495
AW01-20	0531
AW-01-22	0565
AW01-22	0565
AW01-24	0402
AW01-27	0441
AW01-30	0726
AW-01-33	0039
AW01-33	0039
AW02M	0106
AW03-02	OEM
AW03-03	1713
AW03-04	0456
AW03-05	0451
AW04M	0106
AW06M	2010
AW08-7.5A	0649
AW08-7.5B	0649
AW08-8.2A	0244
AW08-8.2B	0244
AW08-9.1A	0012
AW08-9.1B	0012
AW08-10A	0170
AW08-10B	0170
AW08-11A	0317
AW08-11B	0317
AW08-12	0137
AW08-12A	0137
AW08-12B	0137
AW08-13A	0365
AW08-13B	0365
AW08-15A	0002
AW08-15B	0002
AW08-16A	0420
AW08-16B	0420
AW08-18A	0495
AW08-18B	0495
AW08-20A	0531
AW08-20B	0531
AW08-22A	0565
AW08-22B	0565
AW08-24A	0402
AW08-24B	0402
AW08-27A	0441
AW08-27B	0441
AW08-30A	0726
AW08-30B	0726
AW08-33A	0039
AW08-33B	0039
AW08M	2400
AW09	0012
AW0110	0170
AW0113	0137
AW0122	0565
AWOL-13	0365
AWOL-13	0053
AX-7	0649
AX12	0137
AX1600	OEM
AX91770	0144
AXX3021	0522
AXX3037	OEM
AXX3038	2048
AXX3039	OEM
AXX3040	OEM
AXX3041	OEM
AXX3044	4974
AXX3051	2117
AXX3052	OEM
AXX3054	OEM
AXX3055	0522
AY1-0212	4772
AY1-1313	OEM
AY1-1320	OEM
AY1-5050	OEM
AY3-0214	OEM
AY3-0215	OEM
AY3-1015D	OEM
AY3-1270	OEM
AY-3-1350	OEM
AY3-1350	OEM
AY3-4592	OEM
AY-3-8110	OEM
AY3-8211	OEM
AY3-8470	OEM
AY3-8475	OEM
AY3-8500	OEM
AY3-8500-1	OEM
AY-3-8500-1F	OEM
AY3-8550-1	OEM
AY3-8603	OEM
AY3-8603-1	OEM
AY3-8606	OEM
AY3-8606-1	OEM
AY3-8607	OEM
AY3-8607-1	OEM
AY3-8610	OEM
AY3-8610-1	OEM

DEVICE TYPE	REPL CODE
AY3-8765	OEM
AY-3-8900-1	OEM
AY-3-8910	OEM
AY3-8910	OEM
AY-3-8910A	OEM
AY-3-8912	OEM
AY3-8912	OEM
AY-3-8914	OEM
AY-3-8914A	OEM
AY-3-8915	OEM
AY3-8915	OEM
AY-3-8917	OEM
AY3-9400	OEM
AY3-9410	OEM
AY3-9710	OEM
AY3-9725	OEM
AY-3-9735	OEM
AY5-1230	OEM
AY5-1231	OEM
AY5-1232	OEM
AY-5-1317A	OEM
AY5-1317A	OEM
AY-5-3600-PRO	OEM
AY-5-8116	OEM
AY5-8116	OEM
AY5-8116T	OEM
AY5-8136	OEM
AY5-8136T	OEM
AY5-8320	OEM
AY5-8321	OEM
AY5-8450	OEM
AY5-8460	OEM
AY5-9151A	OEM
AY5-9151B	OEM
AY5-9152	OEM
AY5-9152B	OEM
AY5-9153A	OEM
AY5-9153B	OEM
AY5-9154	OEM
AY5-9154A	OEM
AY5-9158	OEM
AY5-9559	OEM
AY102	0975
AY103K	OEM
AY104	OEM
AY104K	OEM
AY105K	OEM
AY106	OEM
AY3427S	OEM
AY3600931PRO	OEM
AYA7104	OEM
AYA7203	OEM
AYA7204	OEM
AYA7206	OEM
AYA7216	OEM
AYA7218	OEM
AYA7226	OEM
AYA7228	OEM
AYA7236	OEM
AYA7238	OEM
AYA7403	OEM
AYA7413	OEM
AYA7433	OEM
AYA7803	OEM
AYA7813	OEM
AYA7823	OEM
AYA71203	OEM
AYA71211	OEM
AYA71221	OEM
AYA78018	OEM
AYA78118	OEM
AYAP2000	OEM
AYY10-120	3733
AZ3.3	0292
AZ3.6	0188
AZ3.9	0456
AZ4.3	0533
AZ4.7	0451
AZ5.1	0162
AZ5.6	0157
AZ6.2	0636
AZ6.8	0062
AZ7	0100
AZ7.5	0649
AZ8.2	0244
AZ9.1	0012
AZ10	0170
AZ11	0317
AZ12	0137
AZ13	0365
AZ15	0002
AZ18	0375
AZ22	0705
AZ27	0441
AZ97B	0054
AZ-110	0317
AZ748	0036
AZ748A	0456
AZ752	0253
AZ752A	0157
AZ754	0062
AZ754A	0025
AZ758	0064
AZ758A	0170
AZ759	0052
AZ759A	0137
AZ957	0062
AZ957A	0062
AZ957B	0025

DEVICE TYPE	REPL CODE
AZ958B	0649
AZ961	0064
AZ961A	0064
AZ961B	0170
AZ963	0052
AZ963A	0052
AZ963B	0137
AZ965	0686
AZ965A	0686
AZ965B	0002
AZ966	0445
AZ966A	0445
AZ966B	0420
AZ967	0375
AZ967A	0375
AZ967B	0495
AZ969	0705
AZ969A	0705
AZ969B	0565
AZ971	0455
AZ971A	0455
AZ971B	0441
AZ973	0166
AZ973A	0166
AZ973B	0039
AZC1028	OEM
AZC1045	OEM
AZC1046	OEM
AZC1048	OEM
AZC1049	OEM
AZC1050	OEM
AZC1051	OEM
AZC1052	OEM
AZC1053	OEM
AZC1054	OEM
AZC1055	OEM
AZC1056	OEM
AZG	0127
AZQQ001GEA	2773
AZY	0127
B	OEM
B00050C	OEM
B00	1052
B004B	0277
B004C	0289
B004E	0296
B006001D	OEM
B006001E	OEM
B006001F	OEM
B006001G	OEM
B006001H	OEM
B006004D	OEM
B006004E	OEM
B006004F	OEM
B006004G	OEM
B006004H	OEM
B006010D	OEM
B006010E	OEM
B006010F	OEM
B006010G	OEM
B0075660	1172
B01	1432
B01-02	0015
B01-2	0133
B02	1158
B03	5302
B071	0649
B090	0012
B094	0170
B0102	0609
B0102V	0609
B01030	OEM
B01031	OEM
B01032	OEM
B01057	OEM
B01058	OEM
B01077	OEM
B01078	OEM
B01097	OEM
B01098	OEM
B03010	OEM
B03011	OEM
B03012	OEM
B04010	OEM
B04011	OEM
B04012	OEM
B04030	OEM
B04031	OEM
B04032	OEM
B05001	OEM
B05007	OEM
B05008	OEM
B05027	OEM
B05028	OEM
B05047	OEM
B05048	OEM
B015001D	OEM
B015001E	OEM
B015001F	OEM
B015001G	OEM
B015001H	OEM
B015004E	OEM
B015004F	OEM
B015004G	OEM
B015004H	OEM
B015010D	OEM
B015010E	OEM
B015010F	OEM

DEVICE TYPE	REPL CODE
B015010G	OEM
B015020C	OEM
B024004D	OEM
B024004E	OEM
B024004F	OEM
B024004G	OEM
B024004H	OEM
B024010C	OEM
B024010D	OEM
B024010E	OEM
B024010F	OEM
B024020C	OEM
B024020D	OEM
B024020E	OEM
B027490	OEM
B030004C	OEM
B030004E	OEM
B030004F	OEM
B030004G	OEM
B030010C	OEM
B030010D	OEM
B030010E	OEM
B030010F	OEM
B030020B	OEM
B030020C	OEM
B030020D	OEM
B030020E	OEM
B038420	OEM
B045004C	OEM
B045004D	OEM
B045004E	OEM
B045004F	OEM
B045010B	OEM
B045010C	OEM
B045010D	OEM
B045010E	OEM
B045010F	OEM
B045020B	OEM
B045020C	OEM
B045020D	OEM
B060010A	OEM
B060010B	OEM
B060010C	OEM
B060010D	OEM
B060010E	OEM
B060020A	OEM
B060020B	OEM
B060020C	OEM
B060020D	OEM
B090010A	OEM
B090010B	OEM
B090010C	OEM
B090010D	OEM
B090020A	OEM
B090020B	OEM
B090020C	OEM
B0100115	OEM
B0100117	OEM
B0272050	5216
B0272054	5216
B0272120	5011
B0272490	5011
B0272632	OEM
B0305401	4633
B0306000	0878
B0306004	0878
B0311000	1018
B0311006	1018
B0311007	1018
B0311008	1018
B0311400	3915
B0311402	3915
B0311405	3915
B0312100	OEM
B0313300	2949
B0313400	2231
B0313500	2949
B0313600	4594
B0313700	4595
B0313800	1326
B0315500	4238
B0316403	2284
B0316451	0167
B0318920	1055
B0319200	1050
B0319960	0799
B0320240	OEM
B0320440	OEM
B0320500	OEM
B0320635	OEM
B0325046	OEM
B0325125	4818
B0325270	4406
B0325290	OEM
B0325350	4410
B0325355	4410
B0325490	OEM
B0325512S	OEM
B0325536	OEM
B0325570	OEM
B0345450	0312
B0345710	0912
B0347230	OEM
B0347500	0770
B0349250	OEM
B0350000	0360
B0350500	0360
B0350510	OEM
B0350602	0360

DEVICE TYPE	REPL CODE
B0351500	0829
B0354700	0912
B0354710	0912
B0354804	2054
B0354805	2054
B0354806	2054
B0354830	2054
B0354832	2054
B0354833	2054
B0354901	4134
B0355431	5205
B0355810	OEM
B0356190	3772
B0356221	6110
B0356227	6110
B0356260	OEM
B0356265	OEM
B0356385	3725
B0356446	0663
B0356448	0663
B0356602	4137
B0356620	6436
B0356640	1676
B0356641	1676
B0356690	4520
B0356711	2284
B0356784	OEM
B0357640	OEM
B0357650	OEM
B0358130	OEM
B0358265	2870
B0358268	3236
B0358272	3236
B0358280	OEM
B0364710	0912
B0372540	0624
B0372900	OEM
B0372960	4480
B0373060	OEM
B0373230	0334
B0376795	OEM
B0376855	OEM
B0376856	OEM
B0376887	OEM
B0377324	OEM
B0377890	OEM
B0379017	OEM
B0379023	OEM
B0379045	OEM
B0379065	OEM
B0379070	OEM
B0379090	OEM
B0379115	OEM
B0379190	OEM
B0379200	OEM
B0379220	OEM
B0379240	OEM
B0379400	3236
B0383100	3515
B0383111	3515
B0383530	OEM
B0383700	OEM
B0383720	OEM
B0383995	OEM
B0383996	OEM
B0384095	OEM
B0384205	OEM
B0384695	OEM
B0384703	OEM
B0402108	OEM
B0403506	OEM
B0403774	OEM
B0406236	OEM
B0406803	OEM
B0406806	OEM
B0410022	OEM
B0428410	OEM
B0455010	OEM
B0470016	3244
B0470116	5891
B0470130	0413
B0470303	OEM
B0470494	0001
B0470522	0024
B0470532	0035
B0470662	0101
B0470932	2384
B0471000	OEM
B0474370	1551
B0487098	OEM
B0487885	OEM
B0526815	OEM
B0589082	OEM
B0589085	OEM
B0589530	OEM
B0589580	OEM
B03544710	0912
B1	1724
B1-8.2	0244
B1-12	OEM
B1A	0561
B1A1	0015
B1A5	0015
B1A9	0015
B1B	0015
B1B1	0015
B1B5	0015
B1B9	0015
B1C	0161

DEVICE TYPE	REPL CODE
B1C1	0015
B1C-2	0161
B1C5	0015
B1C9	0015
B1D	0042
B1D-1	0161
B1D1	0015
B1D5	0015
B1D9	0015
B1E	0338
B1E-1	0338
B1E1	0015
B1E5	0015
B1E9	0015
B1F	0161
B1F1	0015
B1F5	0015
B1F9	0015
B1G	0338
B1G1	0015
B1G5	0015
B1G9	0015
B1H	0007
B1H(DIODE)	0015
B1H(XSTR)	0007
B1H1	0071
B1H5	0071
B1H9	0071
B1J	0037
B1K	0016
B1K1	0071
B1K5	0071
B1K9	0071
B1M	0551
B1M1	0071
B1M5	0071
B1M9	0071
B1N	0016
B1N1	0017
B1N5	0017
B1N9	0017
B1P	0642
B1R	2439
B1T	0363
B1U	2997
B1U148	0161
B1V	1027
B1W	0016
B1Z	0178
B2	0144
B2A	0037
B2A1	0015
B2A5	0015
B2A9	0015
B2B	0178
B2B1	0015
B2B5	0015
B2B9	0015
B2C1	0015
B2C5	0015
B2C9	0015
B2D	0016
B2D1	0015
B2D5	0015
B2D9	0015
B2D746A	OEM
B2D747A	OEM
B2D748A	OEM
B2D749A	OEM
B2D750A	OEM
B2D751A	OEM
B2D752A	OEM
B2D753A	0471
B2D754A	OEM
B2D755A	OEM
B2D756A	OEM
B2D757A	0057
B2D758A	OEM
B2D759A	0052
B2D914	OEM
B2D914F	OEM
B2D914T	OEM
B2D3600	OEM
B2D3600F	OEM
B2D3600T	OEM
B2D4148	OEM
B2D4148F	OEM
B2D4148T	OEM
B2E	0037
B2E1	0015
B2E5	0015
B2E9	0015
B2F1	0015
B2F5	0015
B2F9	0015
B2F10	OEM
B2G	0037
B2G1	0015
B2G5	0015
B2G9	0015
B2H	0074
B2H1	0071
B2H5	0071
B2H9	0071
B2J	0161
B2K	0560
B2K1	0071
B2K5	0071
B2K9	0071

DEVICE TYPE	REPL CODE
B2M	0551
B2M1	0071
B2M5	0071
B2M9	0071
B2N1	0017
B2N5	0017
B2N9	0017
B2P	1027
B2Q90	OEM
B2SB241	0164
B2SB244	0975
B2V	0161
B2Z	0127
B3	0144
B3-12	1974
B3A1	0015
B3A5	0015
B3A9	0015
B3B1	0015
B3B5	0015
B3B9	0015
B3C1	0015
B3C5	0015
B3C9	0015
B3D1	0015
B3D5	0015
B3D9	0015
B3D746A	OEM
B3D747A	OEM
B3D748A	OEM
B3D749A	OEM
B3D750A	OEM
B3D751A	OEM
B3D752A	OEM
B3D753A	0471
B3D754A	OEM
B3D755A	OEM
B3D756A	OEM
B3D757A	0057
B3D758A	OEM
B3D759A	0052
B3D914	OEM
B3D3600	OEM
B3E1	0015
B3E5	0015
B3E9	0015
B3F1	0015
B3F5	0015
B3F9	0015
B3FM1202	OEM
B3G1	0015
B3G5	0015
B3G9	0015
B3H1	0071
B3H5	0071
B3H9	0071
B-3K	OEM
B3K1	0071
B3K5	0071
B3K9	0071
B3M1	0071
B3M5	0071
B3M9	0071
B3N1	0102
B3N5	0102
B3N9	0102
B-3P	0143
B3SD0106	OEM
B4	0144
B4A1	0015
B4A5	0015
B4A9	0015
B4B1	0015
B4B5	0015
B4B9	0015
B4C1	0015
B4C5	0015
B4C9	0015
B4D1	0015
B4D5	0015
B4D9	0015
B4E1	0015
B4E5	0015
B4E9	0015
B4F1	0015
B4F5	0015
B4F9	0015
B4G1	0015
B4G5	0015
B4G9	0015
B4H1	0071
B4H5	0071
B4H9	0071
B4K1	0071
B4K5	0071
B4K9	0071
B4M1	0071
B4M5	0071
B4M9	0071
B4N1	0017
B4N5	0017
B4N9	0017
B4TD0204	OEM
B4Y1-560M	OEM
B4Y2-140M	OEM
B4Y2-280M	OEM
B4Y2-560M	OEM
B4Y5-140M	OEM
B4Y5-280M	OEM

If replacement code is OEM, contact original manufacturer for replacement.

DEVICE TYPE	REPL CODE
B4Y5-420M	OEM
B4Y5-560M	OEM
B4Y10-140M	OEM
B4Y10-280M	OEM
B4Y20-350M	OEM
B4Y20-420M	OEM
B-5	0847
B5	0004
B5-31S	1137
B5-B	OEM
B5A	0004
B5A1	0015
B5A5	0015
B5A9	0015
B5B1	0015
B5B5	0015
B5B9	0015
B5C	0178
B5C1	0015
B5C5	0015
B5C9	0015
B5D	0233
B5D1	0015
B5D5	0015
B5D9	0015
B5E	0423
B5E1	0015
B5E5	0015
B5E9	0015
B5F	0236
B5F1	0015
B5F5	0015
B5F9	0015
B5G	0142
B5G1	0015
B5G5	0015
B5G9	0015
B5H1	0071
B5H5	0071
B5H9	0071
B5J	0016
B5K1	0071
B5K5	0071
B5K9	0071
B5M1	0071
B5M5	0071
B5M9	0071
B5N1	0017
B5N5	0017
B5N9	0017
B5P	0338
B6A1	0087
B6A5	0087
B6A9	0087
B6B1	0087
B6B5	0087
B6B9	0087
B6C1	0087
B6C5	0087
B6C9	0087
B6D1	0087
B6D5	0087
B6D9	0087
B6E1	0087
B6E5	0087
B6E9	0087
B6F1	0087
B6F5	0087
B6F9	0087
B6G1	0087
B6G5	0087
B6G9	0087
B6H1	0087
B6H5	0087
B6H9	0087
B6K1	0087
B6K5	0087
B6K9	0087
B6M1	0087
B6M5	0087
B6M9	0087
B6N1	0102
B6N5	0102
B6N9	0102
B6P	0079
B6RC0911	OEM
B7A1	0087
B7A5	0087
B7A9	0087
B7B1	0087
B7B5	0087
B7B9	0087
B7C1	0087
B7C5	0087
B7C9	0087
B7D1	0087
B7D5	0087
B7D9	0087
B7E1	0087
B7E5	0087
B7E9	0087
B7F1	0087
B7F5	0087
B7F9	0087
B7G1	0087
B7G5	0087
B7G9	0087
B7H1	0087
B7H5	0087
B7H9	0087
B7K	OEM
B7K1	0087
B7K5	0087
B7K9	0087
B7M1	0087
B7M5	0087
B7M9	0087
B7N1	0102
B7N5	0102
B7N9	0102
B8-12	OEM
B8A5	0102
B8B5	0102
B8C5	0102
B8D5	0102
B8E5	0102
B8F5	0102
B8G5	0102
B8H5	0102
B8K5	0102
B8M5	0102
B8N5	0102
B8P-2A21	0004
B9L-4A21	0004
B9TUI	OEM
B10	OEM
B10M	OEM
B12	0150
B12(DIO)	0609
B12(RECT)	0023
B12(RECT),,	0071
B12-02	0015
B12-02R	0023
B12-02RK	0609
B12-1A21	0016
B12-12	1233
B13	0004
B14	0050
B14A-1-21	0160
B15	0164
B15-12	1233
B15D	0164
B15E	0164
B16	0440
B16A(2S)	0440
B16B	0440
B16C	0440
B16D	0440
B16E	0440
B16F	0440
B16G	0440
B16GN	0440
B16H	0440
B16J	0440
B16K	0440
B16L	0440
B16M	0440
B16OR	0440
B16R	0440
B16X	0440
B16Y	0440
B17	0077
B17(TRANS)	0440
B17A(2S)	0440
B17B	0440
B17C	0440
B17D	0440
B17E	0440
B17F	0440
B17G	0440
B17GN	0440
B17H	0440
B17J	0440
B17K	0440
B17L	0440
B17M	0440
B17OR	0440
B17R	0440
B17X	0440
B17Y	0440
B18	0440
B18A(2S)	0440
B18B	0440
B18C	0440
B18D	0440
B18E	0440
B18F	0440
B18G	0440
B18GN	0440
B18H	0440
B18J	0440
B18K	0440
B18L	0440
B18LA	0440
B18M	0440
B18OR	0440
B18R	0440
B18X	0440
B18Y	0440
B19(2S)	0975
B19A	0975
B19B	0975
B19C	0975
B19D	0975
B19E	0975
B19F	0975
B19G	0975
B19GN	0975
B19H	0975
B19J	0975
B19K	0975
B19L	0975
B19M	0975
B19OR	0975
B19R	0975
B19X	0975
B19Y	0975
B20	0085
B20(2S)	0975
B20-001	0160
B20A	0975
B20B	0975
B20C	0975
B20D	0975
B20E	0975
B20F	0975
B20G	0975
B20GN	0975
B20H	0975
B20J	0975
B20K	0975
B20L	0975
B20M	0975
B20OR	0975
B20R	0975
B20S	OEM
B20X	0975
B20Y	0975
B21(2S)	0975
B21A	0975
B21B	0975
B21C	0975
B21D	0975
B21E	0975
B21F	0975
B21G	0975
B21GN	0975
B21H	0975
B21J	0975
B21K	0975
B21L	0975
B21M	0975
B21OR	0975
B21R	0975
B21X	0975
B21Y	0975
B22(2S)	0164
B-22-3	0004
B22-3	0004
B-22-4	0004
B22-4	0004
B22A(2S)	0164
B22B(2S)	0164
B22C	0164
B22D	0164
B22E	0164
B22F	0164
B22G	0164
B22GN	0164
B22H	0164
B22I(2S)	0164
B22J	0164
B22K	0164
B22L	0164
B22M	0164
B22OR	0164
B22P	0164
B22R(2S)	0164
B22X	0164
B22Y(2S)	0164
B-23	0004
B23	0004
B-23-1	0004
B23-1	0004
B23-2	0004
B23A	0004
B23B	0004
B23C	0004
B23D	0004
B23E	0004
B23F	0004
B23G	0004
B23GN	0004
B23H	0004
B23J	0004
B23K	0004
B23L	0004
B23M	0004
B23OR	0004
B23R	0004
B23X	0004
B23Y	0004
B24(2S)	0164
B24-06B	0071
B24-06C	0102
B24-06D	0071
B-24-1	0004
B24-1	0004
B24A	0164
B24B	0164
B24C	0164
B24D	0164
B24E	0164
B24F	0164
B24G	0164
B24GN	0164
B24H	0164
B24J	0164
B24K	0164
B24L	0164
B24M	0164
B24OR	0164
B24R	0164
B24X	0164
B24Y	0164
B25(2S)	0160
B25-12	1977
B25-28	3587
B25-N	0160
B25A	0160
B25B(2S)	0160
B25C	0160
B25CS10	OEM
B25CS20	OEM
B25CS40	OEM
B25CS60	OEM
B25CS80	OEM
B25CS100	OEM
B25CS120	OEM
B25D	0160
B25DA10	OEM
B25DA20	OEM
B25DA40	OEM
B25DA60	OEM
B25DA80	OEM
B25DA100	OEM
B25DA120	OEM
B25DC10	OEM
B25DC20	OEM
B25DC40	OEM
B25DC60	OEM
B25DC80	OEM
B25DC100	OEM
B25DC120	OEM
B25DS10	OEM
B25DS20	OEM
B25DS40	OEM
B25DS60	OEM
B25DS80	OEM
B25DS100	OEM
B25DS120	OEM
B25E	0160
B25F	0160
B25G	0160
B25GN	0160
B25H	0160
B25J	0160
B25JS10	OEM
B25JS20	OEM
B25JS40	OEM
B25JS60	OEM
B25JS80	OEM
B25JS100	OEM
B25JS120	OEM
B25K	0160
B25L	0160
B25M	0160
B25N	0160
B25OR	0160
B25R	0160
B25X	0160
B25Y	0160
B-26	0004
B26	0085
B-26-1	0004
B26A(2S)	0085
B26B	0085
B26C	0085
B26D	0085
B26E	0085
B26F	0085
B26G	0085
B26GN	0085
B26H	0085
B26J	0085
B26K	0085
B26L	0085
B26M	0085
B26OR	0085
B26R	0085
B26X	0085
B26Y	0085
B27(2S)	0085
B27A	0085
B27B	0085
B27C	0085
B27D	0085
B27E	0085
B27F	0085
B27G	0085
B27GN	0085
B27H	0085
B27J	0085
B27K	0085
B27L	0085
B27M	0085
B27OR	0085
B27R	0085
B27X	0085
B27Y	0085
B-28	0143
B28	0085
B28A	0085
B28B	0085
B28C	0085
B28D	0085
B28E	0085
B28F	0085
B28G	0085
B28GN	0085
B28H	0085
B28J	0085
B28K	0085
B28L	0085
B28M	0085
B28OR	0085
B28R	0085
B28X	0085
B28Y	0085
B29(2S)	0085
B29A	0085
B29B	0085
B29C	0085
B29D	0085
B29E	0085
B29F	0085
B29G	0085
B29GN	0085
B29H	0085
B29J	0085
B29K	0085
B29L	0085
B29M	0085
B29OR	0085
B29R	0085
B29X	0085
B29Y	0085
B-30	0143
B30	0143
B30(TRANS)	0085
B30-12	5035
B30A	0085
B30B	0085
B30B250KP	0296
B30C	0085
B30C50KP	0289
B30C250	0015
B30C250-1	0015
B30C250KP	0289
B30C350-1	0015
B30C500	0015
B30C600	0015
B30C600CB	0015
B30C1000	0277
B30D	0085
B30E	0085
B30F	0085
B30G	0085
B30GN	0085
B30H	0085
B30J	0085
B30K	0085
B30L	0085
B30M	0085
B30OR	0085
B-30P	0143
B30R	0085
B30X	0085
B30Y	0085
B-31	0015
B31	0015
B31(TRANS)	0085
B31A	0085
B31B	0085
B31C	0085
B31D	0085
B31E	0085
B31F	0085
B31G	0085
B31GN	0085
B31H	0085
B31J	0085
B31K	0085
B31L	0085
B31M	0085
B31OR	0085
B31R	0085
B31X	0085
B31Y	0085
B32(2S)	0004
B32(3)	0004
B32(N)	0004
B32-0(2S)	0004
B32-1(2S)	0004
B32-2(2S)	0004
B32-4(2S)	0004
B32A	0004
B32B	0004
B32C	0004
B32D	0004
B32E	0004
B32F	0004
B32G	0004
B32GN	0004
B32H	0004
B32J	0004
B32K	0004
B32M	0004
B32OR	0004
B32R	0004
B32X	0004
B32Y	0004
B33	0004
B33(3)	0004
B33(4)	0004
B33(5)	0004
B33-4(2S)	0004
B33-5	0004
B33A	0004
B33B	0004
B33BK	0004
B33C(2S)	0004
B33D(2S)	0004
B33E(2S)	0004
B33F(2S)	0004
B33G	0004
B33GN	0004
B33H	0004
B33J	0004
B33K	0004
B33L	0004
B33M	0004
B33OR	0004
B33R	0004
B33X	0004
B33Y	0004
B34(2S)	0211
B34A	0211
B34B	0211
B34C	0211
B34D	0211
B34E	0211
B34F	0211
B34G	0211
B34GN	0211
B34H	0211
B34J	0211
B34K	0211
B34L	0211
B34M	0211
B34N(2S)	0211
B34OR	0211
B34R	0211
B34X	0211
B34Y	0211
B35	0164
B35-0262	OEM
B35A	0164
B35B	0164
B35C	0164
B35C600	0015
B35D	0164
B35E	0164
B35F	0164
B35G	0164
B35GN	0164
B35H	0164
B35J	0164
B35K	0164
B35L	0164
B35M	0164
B35OR	0164
B35R	0164
B35X	0164
B35Y	0164
B37(2S)	0004
B37A(2S)	0004
B37B(2S)	0004
B37C(2S)	0004
B37D	0004
B37E(2S)	0004
B37F(2S)	0004
B37G	0004
B37GN	0004
B37H	0004
B37J	0004
B37K	0004
B37L	0004
B37M	0004
B37OR	0004
B37R	0004
B37X	0004
B37Y	0004
B38(2S)	0211
B38A	0211
B38B	0211
B38C	0211
B38D	0211
B38E	0211
B38F	0211
B38G	0211
B38GN	0211
B38H	0211
B38J	0211
B38K	0211
B38L	0211
B38M	0211
B38OR	0211
B38R	0211
B38X	0211
B38Y	0211
B39(2S)	0004
B39A	0004
B39B	0004
B39C	0004
B39D	0004
B39E	0004
B39F	0004
B39G	0004
B39GN	0004
B39H	0004
B39J	0004
B39K	0004
B39L	0004
B39M	0004
B39OR	0004
B39R	0004
B39X	0004
B39Y	0004
B40(2S)	0004
B40-12	1974
B40-12A	OEM
B40-28	2506
B40A	0004
B40A10	OEM
B40A20	OEM
B40A40	OEM
B40A60	OEM
B40A80	OEM
B40A100	OEM
B40A120	OEM
B40B	0004
B40C	0004
B40C800	0106
B40C800A	OEM
B40C800D	0241
B40C800D1	0241
B40C800DM	0241
B40C800M	0106
B40C800W	0106
B40C1000	0106
B40C1000A	2785
B40C1000M	0106
B40C1000W	0106
B40C1200S	4073
B40C1200W	0106
B40C1500	0106
B40C1500/1000	OEM
B40C1500/1000SI	OEM
B40C1500/1000SIC	OEM
B40C1500R	0106
B40C1500W	0106
B40C2000/1500	4874
B40C3700/2200	3023
B40C3700/2200SI	0729
B40C3700/2200SIC	0729
B40C5000/3300	1042
B40C5000/3300SI	0729
B40C5000/3300SIC	3023
B40C7500-5000	OEM
B40D	0004
B40E	0004
B40F	0004
B40G	0004
B40GN	0004
B40H	0004
B40J	0004
B40K	0004
B40L	0004
B40M	0004
B40OR	0004
B40R	0004
B40X	0004
B40Y	0004
B41(2S)	0085
B41-04	0023
B41A	0085
B41B	0085
B41C	0085
B41D	0085
B41E	0085
B41F	0085
B41G	0085
B41GN	0085
B41H	0085
B41J	0085
B41K	0085
B41L	0085
B41M	0085
B41OR	0085
B41R	0085
B41X	0085
B41Y	0085
B42(2S)	0085
B42-04	0023
B42A	0085
B42B	0085
B42C	0085
B42D	0085
B42E	0085
B42F	0085
B42G	0085
B42GN	0085
B42H	0085
B42J	0085
B42K	0085
B42L	0085
B42M	0085
B42OR	0085
B42R	0085
B42X	0085
B42Y	0085
B43(2S)	0004
B43-04	0023
B43A(2S)	0004
B43B	0004
B43C	0004
B43D	0004
B43E	0004
B43F	0004
B43G	0004
B43GN	0004
B43H	0004

DEVICE TYPE	REPL CODE
B43J	0004
B43K	0004
B43L	0004
B43M	0004
B430R	0004
B43R	0004
B43X	0004
B43Y	0004
B44(2S)	0004
B44A	0004
B44B	0004
B44C	0004
B44D	0004
B44E	0004
B44F	0004
B44G	0004
B44GN	0004
B44H	0004
B44J	0004
B44K	0004
B44L	0004
B44M	0004
B440R	0004
B44R	0004
B44X	0004
B44Y	0004
B45-12	OEM
B46(2S)	0004
B-46-110	2903
B46-110	0196
B46A	0004
B46B	0004
B46C	0004
B46D	0004
B46E	0004
B46F	0004
B46G	0004
B46GN	0004
B46H	0004
B46J	0004
B46K	0004
B46L	0004
B46M	0004
B460R	0004
B46R	0004
B46X	0004
B46Y	0004
B47(2S)	0004
B47A	0004
B47B	0004
B47C	0004
B47D	0004
B47E	0004
B47F	0004
B47G	0004
B47GN	0004
B47H	0004
B47J	0004
B47K	0004
B47L	0004
B47M	0004
B470R	0004
B47R	0004
B47X	0004
B47Y	0004
B48(2S)	0211
B48A	0211
B48B	0211
B48C	0211
B48D	0211
B48E	0211
B48F	0211
B48G	0211
B48GN	0211
B48H	0211
B48J	0211
B48K	0211
B48L	0211
B48M	0211
B480R	0211
B48R	0211
B48X	0211
B48Y	0211
B49	0211
B49A	0211
B49B	0211
B49C	0211
B49D	0211
B49E	0211
B49F	0211
B49G	0211
B49GN	0211
B49H	0211
B49J	0211
B49K	0211
B49L	0211
B49M	0211
B490R	0211
B49R	0211
B49X	0211
B49Y	0211
B50(2S)	0211
B50A	0211
B50B	0211
B50C	0211
B50C800D	0241
B50C800D1	0241
B50D	0211
B50E	0211

DEVICE TYPE	REPL CODE
B50F	0211
B50G	0211
B50GN	0211
B50H	0211
B50J	0211
B50K	0211
B50L	0211
B50M	0211
B500R	0211
B50R	0211
B50X	0211
B50Y	0211
B51	0015
B51(TRANS)	0211
B51A	0211
B51B	0211
B51C	0211
B51D	0211
B51E	0211
B51F	0211
B51G	0211
B51GN	0211
B51H	0211
B51J	0211
B51K	0211
B51L	0211
B51M	0211
B510R	0211
B51R	0211
B51X	0211
B51Y	0211
B52(2S)	0211
B52A	0211
B52B	0211
B52C	0211
B52D	0211
B52E	0211
B52F	0211
B52G	0211
B52GN	0211
B52H	0211
B52J	0211
B52K	0211
B52L	0211
B52M	0211
B520R	0211
B52R	0211
B52X	0211
B52Y	0211
B53(2S)	0211
B53A	0211
B53B	0211
B53C	0211
B53D	0211
B53E	0211
B53F	0211
B53G	0211
B53GN	0211
B53H	0211
B53J	0211
B53K	0211
B53L	0211
B53M	0211
B530R	0211
B53R	0211
B53X	0211
B53Y	0211
B54(2S)	0004
B54A	0004
B54B(2S)	0004
B54BA	0004
B54C	0004
B54D	0004
B54E(2S)	0004
B54F(2S)	0004
B54G	0004
B54GN	0004
B54H	0004
B54J	0004
B54K	0004
B54L	0004
B54L1	0004
B54M	0004
B540R	0004
B54R	0004
B54X	0004
B54Y(2S)	0004
B55	0004
B55A	0004
B55B	0004
B55C	0004
B55D	0004
B55E	0004
B55F	0004
B55G	0004
B55GN	0004
B55H	0004
B55J	0004
B55K	0004
B55L	0004
B55M	0004
B550R	0004
B55R	0004
B55X	0004
B55Y	0004
B56(2S)	0004
B56-15	0242
B56-33	0953
B56A(2S)	0004

DEVICE TYPE	REPL CODE
B56B(2S)	0004
B56C(2S)	0004
B56CK	0004
B56D	0004
B56E	0004
B56F	0004
B56G	0004
B56GN	0004
B56H	0004
B56J	0004
B56K	0004
B56L	0004
B560R	0004
B56R	0004
B56X	0004
B56Y	0004
B57(2S)	0004
B57A	0004
B57B	0004
B57C	0004
B57D	0004
B57E	0004
B57F	0004
B57G	0004
B57GN	0004
B57H	0004
B57J	0004
B57K	0004
B57L	0004
B57M	0004
B570R	0004
B57R	0004
B57X	0004
B57Y	0004
B58(2S)	0164
B59(2S)	0004
B59A	0004
B59B	0004
B59C	0004
B59D	0004
B59E	0004
B59F	0004
B59G	0004
B59GN	0004
B59J	0004
B59K	0004
B59L	0004
B59M	0004
B590R	0004
B59R	0004
B59X	0004
B59Y	0004
B60(2S)	0004
B60A(2S)	0004
B60B	0004
B60C	0004
B60C300	0015
B60C800W	0106
B60C1000W	0106
B60C1200S	1203
B60C1200W	0106
B60C1500W	0106
B60D	0004
B60E	0004
B60F	0004
B60G	0004
B60GN	0004
B60H	0004
B60J	0004
B60K	0004
B60L	0004
B60M	0004
B600R	0004
B60R	0004
B60X	0004
B60Y	0004
B61(2S)	0004
B61A	0004
B61B	0004
B61C	0004
B61D	0004
B61E	0004
B61F	0004
B61G	0004
B61GN	0004
B61H	0004
B61J	0004
B61K	0004
B61L	0004
B61M	0004
B610R	0004
B61R	0004
B61X	0004
B61Y	0004
B62(2S)	1010
B62A	1010
B62B	1010
B62C	1010
B62D	1010
B62E	1010
B62F	1010
B62G	1010
B62GN	1010
B62H	1010
B62J	1010
B62K	1010
B62L	1010
B62M	1010
B620R	1010

DEVICE TYPE	REPL CODE
B62R	1010
B62X	1010
B62Y	1010
B63(2S)	0222
B63A	0222
B63B	0222
B63C	0222
B63D	0222
B63E	0222
B63F	0222
B63G	0222
B63GN	0222
B63H	0222
B63K	0222
B63L	0222
B63M	0222
B630R	0222
B63R	0222
B63X	0222
B63Y	0222
B64(2S)	0975
B64A	0975
B64B	0975
B64C	0975
B64D	0975
B64E	0975
B64F	0975
B64G	0975
B64H	0975
B64J	0975
B64K	0975
B64L	0975
B64M	0975
B640R	0975
B64R	0975
B64X	0975
B64Y	0975
B65(2S)	0004
B65-1A21	0050
B65-2A21	0208
B65-4A21	0600
B65A	0004
B65A-1-21	0050
B65B	0004
B65B-1-21	0050
B65C	0004
B65C-1-21	0160
B65D	0004
B65F	0004
B65G	0004
B65GN	0004
B65H	0004
B65J	0004
B65K	0004
B65L	0004
B65M	0004
B650R	0004
B65R	0004
B65X	0004
B65Y	0004
B-66	0079
B66(2S)	0004
B66-1A21	0004
B66-2A21	0004
B66-3A21	0004
B66A	0004
B66B	0004
B66C	0004
B66D	0004
B66E	0004
B66F	0004
B66G	0004
B66GN	0004
B66H(2S)	0004
B66J	0004
B66K	0004
B66L	0004
B66M	0004
B660R	0004
B66R	0004
B66X	0004
B66X0033-00	0015
B66X0035-001	0774
B66X0036	1303
B66X0036-001	1303
B66X0041-001	0474
B66Y	0004
B67(2S)	0004
B67A(2S)	0004
B67AH	0004
B67B	0004
B67C	0004
B67D	0004
B67E	0004
B67F	0004
B67G	0004
B67GN	0004
B67H	0004
B67J	0004
B67K	0004
B67L	0004
B67M	0004
B670R	0004
B67R	0004
B67X	0004
B67Y	0004
B68	0004
B68A	0004
B68B	0004

DEVICE TYPE	REPL CODE
B68C	0004
B68D	0004
B68E	0004
B68F	0004
B68G	0004
B68GN	0004
B68H	0004
B68J	0004
B68K	0004
B68L	0004
B68M	0004
B680R	0004
B68R	0004
B68X	0004
B68Y	0004
B69(2S)	0160
B69A	0160
B69B	0160
B69C	0160
B69D	0160
B69E	0160
B69F	0160
B69G	0160
B69GN	0160
B69H	0160
B69J	0160
B69K	0160
B69L	0160
B69M	0160
B690R	0160
B69R	0160
B69X	0160
B69Y	0160
B70	0164
B70-28	OEM
B70H40	OEM
B70H60	OEM
B70H80	OEM
B70H100	OEM
B71(2S)	0004
B71A	0004
B71B	0004
B71C	0004
B71D	0004
B71E	0004
B71F	0004
B71G	0004
B71GN	0004
B71H	0004
B71J	0004
B71K	0004
B71L	0004
B71M	0004
B710R	0004
B71R	0004
B71X	0004
B71Y	0004
B72(2S)	0211
B72A	0211
B72B	0211
B72C	0211
B72D	0211
B72E	0211
B72F	0211
B72G	0211
B72GN	0211
B72H	0211
B72J	0211
B72K	0211
B72L	0211
B72M	0211
B720R	0211
B72R	0211
B72X	0211
B72Y	0211
B73(2S)	0004
B73A(2S)	0004
B73A-1	0004
B73B(2S)	0004
B73C(2S)	0004
B73D	0004
B73E	0004
B73F	0004
B73G	0004
B73GN	0004
B73GR(2S)	0004
B73H	0004
B73J	0004
B73K	0004
B73L	0004
B73M	0004
B730R	0004
B73R	0004
B73S	0004
B73X	0004
B73Y	0004
B74(2S)	0164
B74A	0164
B74B	0164
B74C	0164
B74D	0164
B74E	0164
B74F	0164
B74G	0164
B74GN	0164
B74H	0164
B74J	0164
B74K	0164
B74L	0164

DEVICE TYPE	REPL CODE
B74M	0164
B740R	0164
B74R	0164
B74X	0164
B74Y	0164
B75(2S)	0004
B75A	0004
B75AH(2S)	0004
B75B	0004
B75C(2S)	0004
B75C1	0004
B75C-4	0004
B75D	0004
B75E	0004
B75F(2S)	0004
B75G	0004
B75GN	0004
B75H(2S)	0004
B75J	0004
B75L	0004
B75LB(2S)	0004
B75M	0004
B750R	0004
B75R	0004
B75X	0004
B75Y	0004
B76(2S)	0004
B76A	0004
B76B	0004
B76C	0004
B76D	0004
B76E	0004
B76F	0004
B76G	0004
B76GN	0004
B76H	0004
B76J	0004
B76K	0004
B76L	0004
B76M	0004
B760R	0004
B76R	0004
B76X	0004
B76Y	0004
B77(2S)	0004
B77-A	0004
B77-C	0004
B77-0R	0004
B77A(2S)	0004
B77A-1-21	0440
B77AA(2S)	0004
B77AB(2S)	0004
B77AC(2S)	0004
B77AD(2S)	0004
B77AH(2S)	0004
B77AP(2S)	0004
B77B(2S)	0004
B77B-1-21	0160
B77B2	0004
B77B-11(2S)	0004
B77C	0004
B77D(2S)	0004
B77D0016-1	0147
B77E	0004
B77F	0004
B77G	0004
B77GN	0004
B77H(2S)	0004
B77K	0004
B77L	0004
B77M	0004
B770R	0004
B77P	0004
B77PD	0004
B77R	0004
B77RED	0004
B77T0049	1069
B77V(2S)	0004
B77VRED(2S)	0004
B77X	0004
B77Y	0004
B78(2S)	0004
B78A	0004
B78B	0004
B78C	0004
B78D	0004
B78E	0004
B78F	0004
B78G	0004
B78GN	0004
B78H	0004
B78J	0004
B78K	0004
B78L	0004
B78M	0004
B780R	0004
B78R	0004
B78X	0004
B78Y	0004
B79(2S)	0004
B79A	0004
B79B	0004
B79C	0004
B79D	0004
B79E	0004
B79F	0004
B79G	0004
B79GN	0004
B79H	0004

DEVICE TYPE	REPL CODE
B79J	0004
B79K	0004
B79L	0004
B79M	0004
B790R	0004
B79R	0004
B79X	0004
B79Y	0004
B80(2S)	0222
B80A	0222
B80B	0222
B80C	0222
B80C800	0106
B80C800A	OEM
B80C800D	0241
B80C800D1	0241
B80C800DM	0241
B80C800M	0106
B80C800W	0106
B80C1000	0106
B80C1000A	0106
B80C1000M	0106
B80C1000W	0106
B80C1200S	1203
B80C1200W	0106
B80C1500	0106
B80C1500/1000	OEM
B80C1500/1000SI	OEM
B80C1500/1000SIC	OEM
B80C1500R	0106
B80C1500W	0106
B80C2000/1500	2230
B80C3700/2200	6362
B80C3700/2200SI	0729
B80C3700/2200SIC	3023
B80C5000/3300	3023
B80C5000/3300SI	0729
B80C5000/3300SIC	3023
B80C7500-5000	OEM
B80D	0222
B80E	0222
B80F	0222
B80G	0222
B80GN	0222
B80H	0222
B80J	0222
B80K	0222
B80L	0222
B80M	0222
B800R	0222
B80R	0222
B80X	0222
B80Y	0222
B81(2S)	0975
B81-004	2542
B81A	0975
B81B	0975
B81C	0975
B81D	0975
B81E	0975
B81F	0975
B81GN	0975
B81H	0975
B81J	0975
B81K	0975
B81L	0975
B81M	0975
B810R	0975
B81R	0975
B81X	0975
B81Y	0975
B82(2S)	0975
B82-004	2515
B82A	0975
B82B	0975
B82C	0975
B82D	0975
B82E	0975
B82F	0975
B82G	0975
B82GN	0975
B82H	0975
B82J	0975
B82K	0975
B82L	0975
B82M	0975
B820R	0975
B82R	0975
B82X	0975
B82Y	0975
B83(2S)	0975
B83-7	2018
B83A	0975
B83B	0975
B83C	0975
B83D	0975
B83E	0975
B83F	0975
B83G	0975
B83GN	0975
B83H	0975
B83J	0975
B83K	0975
B83L	0975
B83M	0975
B830R	0975
B83R	0975
B83X	0975
B83Y	0975

If replacement code is OEM, contact original manufacturer for replacement.

DEVICE TYPE	REPL CODE
B84(2S)	0085
B84A	0085
B84A-1-21	0160
B84B	0085
B84C	0085
B84D	0085
B84E	0085
B84F	0085
B84G	0085
B84GN	0085
B84H	0085
B84J	0085
B84K	0085
B84L	0085
B84M	0085
B84OR	0085
B84R	0085
B84X	0085
B84Y	0085
B85(2S)	0164
B85A	0164
B85B	0164
B85C	0164
B85D	0164
B85E	0164
B85F	0164
B85G	0164
B85GN	0164
B85H	0164
B85J	0164
B85K	0164
B85L	0164
B85M	0164
B85OR	0164
B85R	0164
B85X	0164
B85Y	0164
B86	0160
B87(2S)	0164
B87A	0164
B87B	0164
B87C	0164
B87D	0164
B87E	0164
B87F	0164
B87G	0164
B87GN	0164
B87H	0164
B87J	0164
B87K	0164
B87L	0164
B87M	0164
B87OR	0164
B87R	0164
B87X	0164
B87Y	0164
B88	0975
B88B-1-21	0050
B88C-1-21	0280
B89(2S)	0004
B89A(2S)	0004
B89AH(2S)	0004
B89C	0004
B89D	0004
B89E	0004
B89F	0004
B89G	0004
B89GN	0004
B89H(2S)	0004
B89J	0004
B89K	0004
B89L	0004
B89M	0004
B89OR	0004
B89R	0004
B89X	0004
B89Y	0004
B90(2S)	0004
B90A	0004
B90B	0004
B90C	0004
B90D	0004
B90E	0004
B90F	0004
B90G	0004
B90GN	0004
B90H	0004
B90J	0004
B90K	0004
B90L	0004
B90M	0004
B90OR	0004
B90R	0004
B90X	0004
B90Y	0004
B91(2S)	0004
B91A	0004
B91B	0004
B91C	0004
B91D	0004
B91E	0004
B91F	0004
B91G	0004
B91GN	0004
B91H	0004
B91J	0004
B91K	0004
B91L	0004
B91M	0004
B91OR	0004
B91R	0004
B91X	0004
B91Y	0004
B92(2S)	0004
B92-1A21	0440
B92-UA21	0440
B92A	0004
B92B	0004
B92C	0004
B92D	0004
B92E	0004
B92F	0004
B92G	0004
B92GN	0004
B92H	0004
B92J	0004
B92K	0004
B92L	0004
B92M	0004
B92X	0004
B92Y	0004
B93	0004
B94(2S)	0004
B94A	0004
B94B	0004
B94C	0004
B94D	0004
B94E	0004
B94F	0004
B94G	0004
B94GN	0004
B94H	0004
B94J	0004
B94K	0004
B94M	0004
B94OR	0004
B94R	0004
B94X	0004
B94Y	0004
B95(2S)	0004
B95A	0004
B95B	0004
B95C	0004
B95D	0004
B95E	0004
B95F	0004
B95G	0004
B95GN	0004
B95H	0004
B95J	0004
B95L	0004
B95M	0004
B95OR	0004
B95R	0004
B95X	0004
B95Y	0004
B96	0211
B97(2S)	0004
B97A	0004
B97B	0004
B97C	0004
B97D	0004
B97F	0004
B97G	0004
B97GN	0004
B97H	0004
B97K	0004
B97L	0004
B97M	0004
B97OR	0004
B97R	0004
B97X	0004
B97Y	0004
B98(2S)	0211
B98A	0211
B98B	0211
B98C	0211
B98D	0211
B98E	0211
B98F	0211
B98G	0211
B98GN	0211
B98H	0211
B98J	0211
B98K	0211
B98L	0211
B98M	0211
B98OR	0211
B98R	0211
B98X	0211
B98Y	0211
B99	0164
B99A	0164
B99B	0164
B99C	0164
B99D	0164
B99E	0164
B99F	0164
B99G	0164
B99GN	0164
B99H	0164
B99J	0164
B99K	0164
B99L	0164
B99M	0164
B99OR	0164
B99R	0164
B99X	0164
B99Y	0164
B100	0211
B100A	0211
B100B	0211
B100C	0211
B100D	0211
B100E	0211
B100F	0211
B100G	0211
B100GN	0211
B100H	0211
B100J	0211
B100K	0211
B100L	0211
B100M	0211
B100OR	0211
B100R	0211
B100X	0211
B100Y	0211
B101(2S)	0211
B101A	0211
B101B	0211
B101C	0211
B101D	0211
B101E	0211
B101F	0211
B101G	0211
B101GN	0211
B101H	0211
B101J	0211
B101K	0211
B101L	0211
B101M	0211
B101OR	0211
B101R	0211
B101X	0211
B101Y	0211
B102(2S)	0211
B102A	0211
B102B	0211
B102C	0211
B102D	0211
B102E	0211
B102F	0211
B102G	0211
B102GN	0211
B102H	0211
B102J	0211
B102K	0211
B102L	0211
B102M	0211
B102OR	0211
B102R	0211
B102X	0211
B102Y	0211
B103(2S)	0211
B103A	0211
B103B	0211
B103C	0211
B103D	0211
B103E	0211
B103F	0211
B103G	0211
B103GN	0211
B103H	0211
B103J	0211
B103K	0211
B103L	0211
B103M	0211
B103OR	0211
B103R	0211
B103X	0211
B103Y	0211
B104(2S)	0164
B104A	0164
B104B	0164
B104C	0164
B104D	0164
B104E	0164
B104F	0164
B104G	0164
B104GN	0164
B104H	0164
B104J	0164
B104K	0164
B104L	0164
B104M	0164
B104OR	0164
B104R	0164
B104X	0164
B104Y	0164
B105(2S)	0164
B105A	0164
B105B	0164
B105C	0164
B105D	0164
B105E	0164
B105F	0164
B105G	0164
B105GN	0164
B105H	0164
B105K	0164
B105L	0164
B105M	0164
B105OR	0164
B105R	0164
B105X	0164
B105Y	0164
B106(2S)	0164
B107(2S)	0160
B107A(2S)	0160
B107B	0160
B107C	0160
B107D	0160
B107E	0160
B107F	0160
B107G	0160
B107GN	0160
B107H	0160
B107J	0160
B107K	0160
B107L	0160
B107M	0160
B107OR	0160
B107R	0160
B107X	0160
B107Y	0160
B108(2S)	0164
B108A(2S)	0164
B108B(2S)	0164
B108C	0164
B108D	0164
B108E	0164
B108F	0164
B108G	0164
B108GN	0164
B108H	0164
B108J	0164
B108K	0164
B108L	0164
B108M	0164
B108OR	0164
B108X	0164
B108Y	0164
B109(2S)	0164
B109A	0164
B109B	0164
B110	0004
B110A	0004
B110B	0004
B110C	0004
B110D	0004
B110E	0004
B110F	0004
B110G	0004
B110GN	0004
B110H	0004
B110J	0004
B110K	0004
B110L	0004
B110M	0004
B110OR	0004
B110R	0004
B110X	0004
B110Y	0004
B111(2S)	0004
B111A	0004
B111B	0004
B111C	0004
B111D	0004
B111E	0004
B111F	0004
B111G	0004
B111GN	0004
B111H	0004
B111J	0004
B111K(2S)	0004
B111L	0004
B111M	0004
B111OR	0004
B111R	0004
B111X	0004
B111Y	0004
B112(2S)	0004
B112A	0004
B112B	0004
B112C	0004
B112D	0004
B112E	0004
B112F	0004
B112G	0004
B112H	0004
B112J	0004
B112K	0004
B112L	0004
B112M	0004
B112OR	0004
B112R	0004
B112X	0004
B112Y	0004
B113	0004
B113(2S)	0004
B113A	0004
B113B	0004
B113C	0004
B113D	0004
B113E	0004
B113F	0004
B113G	0004
B113GN	0004
B113H	0004
B113J	0004
B113K	0004
B113L	0004
B113M	0004
B113OR	0004
B113R	0004
B113X	0004
B113Y	0004
B114	0004
B114(2S)	0004
B114A	0004
B114B	0004
B114C	0004
B114D	0004
B114E	0004
B114F	0004
B114G	0004
B114GN	0004
B114H	0004
B114J	0004
B114K	0004
B114L	0004
B114M	0004
B114OR	0004
B114R	0004
B114X	0004
B114Y	0004
B115	0004
B115A	0004
B115B	0004
B115C	0004
B115D	0004
B115E	0004
B115F	0004
B115G	0004
B115GN	0004
B115H	0004
B115J	0004
B115K	0004
B115L	0004
B115M	0004
B115OR	0004
B115R	0004
B115X	0004
B115Y	0004
B116(2S)	0004
B116A	0004
B116B	0004
B116C	0004
B116D	0004
B116E	0004
B116F	0004
B116G	0004
B116GN	0004
B116H	0004
B116J	0004
B116K	0004
B116L	0004
B116M	0004
B116OR	0004
B116R	0004
B116X	0004
B116Y	0004
B117	0004
B117A	0004
B117B	0004
B117C	0004
B117D	0004
B117E	0004
B117F	0004
B117G	0004
B117GN	0004
B117H	0004
B117J	0004
B117K(2S)	0004
B117L	0004
B117M	0004
B117OR	0004
B117R	0004
B117X	0004
B117Y	0004
B118	0164
B119(2S)	0160
B119A(2S)	0160
B119B	0160
B119C	0160
B119D	0160
B119E	0160
B119F	0160
B119G	0160
B119GN	0160
B119H	0160
B119J	0160
B119K	0160
B119L	0160
B119M	0160
B119OR	0160
B119R	0160
B119X	0160
B119Y	0160
B120(2S)	0004
B120A	0004
B120B	0004
B120C	0004
B120D	0004
B120E	0004
B120F	0004
B120G	0004
B120GN	0004
B120H	0004
B120J	0004
B120K	0004
B120L	0004
B120M	0004
B120OR	0004
B120R	0004
B120X	0004
B120Y	0004
B121	OEM
B122(2S)	0004
B122A	0160
B122B	0160
B122C	0160
B122D	0160
B122E	0160
B122F	0160
B122G	0160
B122GN	0160
B122H	0160
B122J	0160
B122K	0160
B122L	0160
B122M	0160
B122OR	0160
B122R	0160
B122X	0160
B122Y	0160
B123	0085
B123A(2S)	0085
B123B	0085
B123C	0085
B123D	0085
B123E	0085
B123F	0085
B123G	0085
B123GN	0085
B123H	0085
B123J	0085
B123K	0085
B123L	0085
B123M	0085
B123OR	0085
B123R	0085
B123X	0085
B123Y	0085
B124	0160
B124A	0160
B124B	0160
B124C	0160
B124D	0160
B124E	0160
B124F	0160
B124G	0160
B124H	0160
B124J	0160
B124K	0160
B124L	0160
B124M	0160
B124OR	0160
B124R	0160
B124X	0160
B124Y	0160
B125	0604
B125C800	0106
B125C800A	OEM
B125C800D	4581
B125C800D1	4581
B125C800DM	4581
B125C800M	0106
B125C800W	0106
B125C1000	0106
B125C1000A	2785
B125C1000M	0241
B125C1000W	0106
B125C1200S	5767
B125C1200W	0106
B125C1500	0787
B125C1500/1000	6048
B125C1500/1000SIC	OEM
B125C1500R	2785
B125C1500W	2010
B125C2000/1500	6048
B125C3700/2200	1045
B125C3700/2200SI	0729
B125C3700/2200SIC	3035
B125C5000/3300	1045
B125C5000/3300SI	OEM
B125C5000/3300SIC	3035
B126(2S)	0160
B126A(2S)	0160
B126AA	0160
B126AB	0160
B126AC	0160
B126AD	0160
B126AE	0160
B126AF	0160
B126AG	0160
B126AH	0160
B126B	0160
B126C	0160
B126D	0160
B126E	0160
B126F(2S)	0160
B126FV	0160
B126G	0160
B126GN	0160
B126H	0160
B126J	0160
B126K	0160
B126L	0160
B126M	0160
B126OR	0160
B126P	0160
B126R	0160
B126V(2S)	0160
B126X	0160
B126Y	0160
B127(2S)	0160
B127A(2S)	0160
B127B	0160
B127C	0160
B127D	0160
B127E	0160
B127F	0160
B127G	0160
B127GN	0160
B127H	0160
B127J	0160
B127K	0160
B127L	0160
B127M	0160
B127OR	0160
B127R	0160
B127X	0160
B127Y	0160
B128(2S)	0975
B128A(2S)	0975
B128B	0975
B128C	0975
B128D	0975
B128E	0975
B128F	0975
B128G	0975
B128GN	0975
B128H	0975
B128J	0975
B128K	0975
B128L	0975
B128M	0975
B128OR	0975
B128R	0975
B128V(2S)	0975
B128X	0975
B128Y	0975
B129(2S)	0975
B129A	0975
B129B	0975
B129C	0975
B129D	0975
B129E	0975
B129F	0975
B129G	0975
B129GN	0975
B129H	0975
B129J	0975
B129K	0975
B129L	0975
B129M	0975
B129OR	0975
B129R	0975
B129X	0975
B129Y	0975
B130	0222
B130A(2S)	0222
B130B	0222
B130C	0222
B130D	0222
B130E	0222
B130F	0222
B130G	0222
B130GN	0222
B130H	0222
B130J	0222
B130K	0222
B130L	0222
B130M	0222
B130OR	0222
B130R	0222
B130X	0222
B130Y	0222
B131	0160
B131A(2S)	0160
B131B	0160
B131C	0160
B131D	0160
B131E	0160
B131F	0160
B131G	0160
B131GN	0160
B131H	0160
B131J	0160
B131K	0160
B131L	0160
B131M	0160
B131OR	0160
B131R	0160
B131X	0160
B131Y	0160
B132	0160
B132A(2S)	0160
B132B	0160
B132C	0160
B132D	0160
B132E	0160
B132F	0160
B132G	0160
B132GN	0160
B132H	0160
B132J	0160
B132L	0160
B132M	0160
B132OR	0160
B132R	0160
B132X	0160

If replacement code is OEM, contact original manufacturer for replacement.

DEVICE TYPE	REPL CODE
B132Y	0160
B134	0160
B134(2S)	0004
B134A	0160
B134A(2S)	0004
B134B	OEM
B134B(2S)	0004
B134C	0004
B134D	0004
B134E	0004
B134F	0004
B134G	0004
B134GN	0004
B134H	0004
B134J	0004
B134K	0004
B134L	0004
B134M	0004
B134OR	0004
B134R	0004
B134X	0004
B134Y	0004
B135(2S)	0004
B135A	0004
B135B(2S)	0004
B135C(2S)	0004
B135D	0004
B135E(2S)	0004
B135F	0004
B135G	0004
B135GN	0004
B135H	0004
B135J	0004
B135K	0004
B135L	0004
B135M	0004
B135OR	0004
B135R	0004
B135X	0004
B135Y	0004
B136(2S)	0004
B136A(2S)	0004
B136B(2S)	0004
B136C(2S)	0004
B136D	0004
B136E	0004
B136F	0004
B136G	0004
B136GN	0004
B136H	0004
B136J	0004
B136K	0004
B136L	0004
B136M	0004
B136OR	0004
B136R	0004
B136U(2S)	0004
B136X	0004
B137(2S)	0160
B137A	0160
B137B	0160
B137C	0160
B137D	0160
B137E	0160
B137F	0160
B137G	0160
B137GN	0160
B137H	0160
B137K	0160
B137L	0160
B137OR	0160
B137R	0160
B137X	0160
B137Y	0160
B138(2S)	0160
B138A	0160
B138B	0160
B138C	0160
B138D	0160
B138E	0160
B138G	0160
B138GN	0160
B138H	0160
B138J	0160
B138K	0160
B138L	0160
B138M	0160
B138OR	0160
B138R	0160
B138X	0160
B138Y	0160
B140	0100
B140(TRANS)	0085
B140A	0085
B140B	0085
B140C	0085
B140D	0085
B140E	0085
B140F	0085
B140G	0085
B140GN	0085
B140H	0085
B140J	0085
B140K	0085
B140L	0085
B140M	0085
B140OR	0085
B140X	0085
B140Y	0085
B141(2S)	0085
B142(2S)	0085
B142A	0085
B142B(2S)	0085
B142C(2S)	0085
B142D	0085
B142E	0085
B142F	0085
B142G	0085
B142GN	0085
B142H	0085
B142J	0085
B142K	0085
B142L	0085
B142M	0085
B142OR	0085
B142R	0085
B142X	0085
B142Y	0085
B143(2S)	0085
B143A	0085
B143B	0085
B143C	0085
B143D	0085
B143E	0085
B143F	0085
B143G	0085
B143GN	0085
B143H	0085
B143J	0085
B143K	0085
B143L	0085
B143M	0085
B143OR	0085
B143P(2S)	0085
B143X	0085
B143Y	0085
B144(2S)	0085
B144A	0085
B144B	0085
B144C	0085
B144D	0085
B144E	0085
B144F	0085
B144G	0085
B144GN	0085
B144H	0085
B144J	0085
B144K	0085
B144L	0085
B144M	0085
B144OR	0085
B144P(2S)	0085
B144R	0085
B144X	0085
B144Y	0085
B145(2S)	0085
B145A	0085
B145B	0085
B145C	0085
B145D	0085
B145E	0085
B145F	0085
B145G	0085
B145GN	0085
B145H	0085
B145J	0085
B145K	0085
B145L	0085
B145M	0085
B145OR	0085
B145R	0085
B145X	0085
B145Y	0085
B146(2S)	0085
B146A	0085
B146B	0085
B146C	0085
B146D	0085
B146E	0085
B146F	0085
B146G	0085
B146GN	0085
B146H	0085
B146J	0085
B146K	0085
B146L	0085
B146M	0085
B146OR	0085
B146R	0085
B146X	0085
B146Y	0085
B147(2S)	0085
B147A	0085
B147B	0085
B147C	0085
B147D	0085
B147E	0085
B147F	0085
B147G	0085
B147H	0085
B147J	0085
B147K	0085
B147L	0085
B147M	0085
B147OR	0085
B147X	0085
B147Y	0085
B148	0604
B149	0085
B149-N	0085
B149A	0085
B149B	0085
B149C	0085
B149D	0085
B149E	0085
B149F	0085
B149G	0085
B149GN	0085
B149H	0085
B149J	0085
B149K	0085
B149L	0085
B149M	0085
B149OR	0085
B149R	0085
B149X	0085
B149Y	0085
B150	0002
B150(2S)	0164
B150A	0164
B150B	0164
B150C	0164
B150D	0164
B150E	0164
B150F	0164
B150G	0164
B150GN	0164
B150H	0164
B150J	0164
B150K	0164
B150L	0164
B150M	0164
B150OR	0164
B150R	0164
B150X	0164
B150Y	0164
B151	OEM
B151(2S)	0975
B151A	0975
B151B	0975
B151C	0975
B151D	0975
B151E	0975
B151F	0975
B151G	0975
B151GN	0975
B151H	0975
B151J	0975
B151K	0975
B151L	0975
B151M	0975
B151OR	0975
B151R	0975
B151X	0975
B151Y	0975
B152	5415
B152(2S)	0975
B152A	0975
B152B	0975
B152C	0975
B152D	0975
B152E	0975
B152F	0975
B152G	0975
B152GN	0975
B152H	0975
B152J	0975
B152K	0975
B152L	0975
B152M	0975
B152OR	0975
B152R	0975
B152X	0975
B152Y	0975
B153(2S)	0004
B153A	0004
B153B	0004
B153C	0004
B153D	0004
B153E	0004
B153F	0004
B153G	0004
B153GN	0004
B153H	0004
B153J	0004
B153K	0004
B153L	0004
B153M	0004
B153OR	0004
B153R	0004
B153X	0004
B153Y	0004
B154(2S)	0004
B154A	0004
B154B	0004
B154C	0004
B154D	0004
B154E	0004
B154F	0004
B154G	0004
B154GN	0004
B154H	0004
B154J	0004
B154K	0004
B154L	0004
B154M	0004
B154OR	0004
B154R	0004
B154X	0004
B154Y	0004
B155	0004
B155A(2S)	0004
B155B(2S)	0004
B155C	0004
B155D	0004
B155E	0004
B155F	0004
B155G	0004
B155GN	0004
B155H	0004
B155J	0004
B155K	0004
B155L	0004
B155M	0004
B155OR	0004
B155R	0004
B155X	0004
B155Y	0004
B156(2S)	0004
B156-A	0004
B156A(2S)	0004
B156AA(2S)	0004
B156AB(2S)	0004
B156AC(2S)	0004
B156AD	0004
B156B(2S)	0004
B156B3	0004
B156BK	0004
B156C(2S)	0004
B156D(2S)	0004
B156E	0004
B156F	0004
B156G	0004
B156GN	0004
B156H	0004
B156J	0004
B156K	0004
B156L	0004
B156M	0004
B156OR	0004
B156P(2S)	0004
B156R	0004
B156X	0004
B156Y	0004
B157(2S)	0136
B157A	0136
B157B	0136
B157C	0136
B157D	0136
B157E	0136
B157F	0136
B157G	0136
B157GN	0136
B157H	OEM
B157J	0136
B157K	0136
B157L	0136
B157M	0136
B157OR	0136
B157R	0136
B157X	0136
B157Y	0136
B158(2S)	0136
B158A	0136
B158B	0136
B158C	0136
B158D	0136
B158E	0136
B158F	0136
B158G	0136
B158GN	0136
B158H	0136
B158J	0136
B158K	0136
B158L	0136
B158M	0136
B158OR	0136
B158R	0136
B158X	0136
B158Y	0136
B159(2S)	0136
B159A	0136
B159B	0136
B159C	0136
B159D	0136
B159E	0136
B159F	0136
B159G	0136
B159GN	0136
B159H	0136
B159J	0136
B159K	0136
B159L	0136
B159M	0136
B159OR	0136
B159R	0136
B159X	0136
B159Y	0136
B160(2S)	0136
B160A	0136
B160B	0136
B160C	0136
B160D	0136
B160E	0136
B160F	0136
B160G	0136
B160GN	0136
B160H	0136
B160J	0136
B160K	0136
B160L	0136
B160M	0136
B160OR	0136
B160R	0136
B160X	0136
B160Y	0136
B161(2S)	0211
B161A	0211
B161B	0211
B161C	0211
B161D	0211
B161E	0211
B161F	0211
B161G	0211
B161GN	0211
B161H	0211
B161J	0211
B161K	0211
B161L	0211
B161M	0211
B161OR	0211
B161R	0211
B161X	0211
B161Y	0211
B162(2S)	0004
B162A	0004
B162B	0004
B162C	0004
B162D	0004
B162E	0004
B162F	0004
B162G	0004
B162GN	0004
B162H	0004
B162J	0004
B162K	0004
B162L	0004
B162M	0004
B162OR	0004
B162R	0004
B162X	0004
B162Y	0004
B163(2S)	0211
B163A	0211
B163B	0211
B163C	0211
B163D	0211
B163E	0211
B163F	0211
B163G	0211
B163GN	0211
B163H	0211
B163J	0211
B163K	0211
B163L	0211
B163M	0211
B163OR	0211
B163R	0211
B163X	0211
B163Y	0211
B164(2S)	0211
B164A	0211
B164B	0211
B164C	0211
B164D	0211
B164E	0211
B164F	0211
B164G	0211
B164GN	0211
B164H	0211
B164J	0211
B164L	0211
B164M	0211
B164OR	0211
B164R	0211
B164X	0211
B164Y	0211
B165(2S)	0211
B165A	0211
B165B	0211
B165C	0211
B165D	0211
B165E	0211
B165F	0211
B165G	0211
B165GN	0211
B165H	0211
B165J	0211
B165K	0211
B165M	0211
B165OR	0211
B165R	0211
B165X	0211
B165Y	0211
B166(2S)	0004
B166A	0004
B166B	0004
B166C	0004
B166D	0004
B166E	0004
B166F	0004
B166G	0004
B166GN	0004
B166H	0004
B166J	0004
B166K	0004
B166L	0004
B166M	0004
B166OR	0004
B166R	0004
B166X	0004
B166Y	0004
B167(2S)	0164
B167A	0164
B167B	0164
B167BK	0164
B167C	0164
B167D	0164
B167E	0164
B167F	0164
B167G	0164
B167GN	0164
B167H	0164
B167J	0164
B167K	0164
B167L	0164
B167M	0164
B167OR	0164
B167R	0164
B167X	0164
B167Y	0164
B168(2S)	0004
B168A	0004
B168B	0004
B168C	0004
B168D	0004
B168E	0004
B168F	0004
B168G	0004
B168GN	0004
B168H	0004
B168J	0004
B168K	0004
B168L	0004
B168M	0004
B168OR	0004
B168R	0004
B168X	0004
B168Y	0004
B169	0004
B169A	0004
B169B	0004
B169C	0004
B169D	0004
B169E	0004
B169F	0004
B169G	0004
B169GN	0004
B169H	0004
B169J	0004
B169K	0004
B169L	0004
B169M	0004
B169OR	0004
B169R	0004
B169X	0004
B169Y	0004
B170(2S)	0004
B170A	0004
B170B	0004
B170C	0004
B170D	0004
B170E	0004
B170F	0004
B170G	0004
B170GN	0004
B170H	0004
B170J	0004
B170K	0004
B170L	0004
B170M	0004
B170OR	0004
B170R	0004
B170X	0004
B170Y	0004
B171	0004
B171A(2S)	0004
B171B(2S)	0004
B171C	0004
B171D	0004
B171E	0004
B171F	0004
B171G	0004
B171GN	0004
B171H	0004
B171J	0004
B171K	0004
B171L	0004
B171M	0004
B171OR	0004
B171X	0004
B171Y	0004
B172(2S)	0004
B172-5	0087
B172-10	0087
B172-20	0087
B172-30	0087
B172-40	0087
B172-50	0087
B172-60	0087
B172-70	0087
B172-80	0087
B172-90	0087
B172-100	0087
B172-110	1760
B172-120	1760
B172A(2S)	0004
B172A-1	0004
B172A-F	0004
B172AF(2S)	0004
B172AL	0004
B172B(2S)	0004
B172C(2S)	0004
B172D(2S)	0004
B172E(2S)	0004
B172F(2S)	0004
B172FN	0004
B172G	0004
B172GN	0004
B172H(2S)	0004
B172J	0004
B172K	0004
B172L	0004
B172M	0004
B172OR	0004
B172P(2S)	0004
B172R(2S)	0004
B172X	0004
B172Y	0004
B173(2S)	0004
B173A(2S)	0004
B173B(2S)	0004
B173BL	0004
B173C(2S)	0004
B173CL	0004
B173D	0004
B173E	0004
B173F	0004
B173G	0004
B173GN	0004
B173H	0004
B173J	0004
B173K	0004
B173L(2S)	0004
B173M	0004
B173OR	0004
B173R	0004
B173X	0004
B173Y	0004
B174(2S)	0004
B174A	0004
B174B	0004
B174C	0004
B174D	0004
B174E	0004
B174F	0004
B174G	0004
B174GN	0004
B174H	0004
B174K	0004
B174L	0004
B174M	0004
B174OR	0004
B174R	0004
B174X	0004
B174Y	0004
B175(2S)	0004
B175A(2S)	0004
B175B(2S)	0004
B175B-1	0004
B175BL	0004
B175C(2S)	0004
B175CL	0004
B175D	0004
B175E(2S)	0004
B175F	0004
B175G	0004
B175GN	0004
B175H	0004
B175L	0004
B175M	0004
B175OR	0004
B175R	0004
B175X	0004
B175Y	0004
B176(2S)	0004
B176-O(2S)	0004
B176-P(2S)	0004
B176-PR(2S)	0004
B176A	0004
B176B(2S)	0004
B176C	0004
B176D	0004
B176E	0004
B176F	0004
B176G	0004
B176GN	0004
B176H	0004
B176J	0004
B176K	0004
B176L	0004
B176M(2S)	0004
B176O	0004
B176OR	0004
B176P(2S)	0004
B176PL	0004
B176PR	0004
B176PRC(2S)	0004
B176R(1)	0004
B176R(2S)	0004
B176RG	0004
B176X	0004
B176Y	0004

DEVICE TYPE	REPL CODE
B177	0004
B177(2S)	0004
B177A	0004
B177B	0004
B177C	0004
B177D	0004
B177E	0004
B177F	0004
B177G	0004
B177GN	0004
B177H	0004
B177K	0004
B177L	0004
B177M	0004
B177OR	0004
B177R	0004
B177X	0004
B177Y	0004
B178	0004
B178(2S)	0004
B178-O(2S)	0004
B178-OR	0004
B178-S(2S)	0004
B178A(2S)	0004
B178B	0004
B178C(2S)	0004
B178D(2S)	0004
B178E	0004
B178F	0004
B178G	0004
B178GN	0004
B178H	0004
B178J	0004
B178K	0004
B178L	0004
B178M(2S)	0004
B178N(2S)	0004
B178OR	0004
B178R	0004
B178S	0004
B178T(2S)	0004
B178TC	0004
B178TS	0004
B178U(2S)	0004
B178V(2S)	0004
B178X(2S)	0004
B178Y(2S)	0004
B179	0160
B179(2S)	0050
B180(2S)	0164
B180A(2S)	0164
B181(2S)	0164
B181A(2S)	0164
B182	0050
B183(2S)	0004
B183A	0004
B183B	0004
B183C	0004
B183D	0004
B183E	0004
B183F	0004
B183G	0004
B183GN	0004
B183H	0004
B183J	0004
B183K	0004
B183L	0004
B183M	0004
B183OR	0004
B183R	0004
B183X	0004
B183Y	0004
B184(2S)	0004
B184A	0004
B184B	0004
B184C	0004
B184D	0004
B184E	0004
B184F	0004
B184G	0004
B184GN	0004
B184H	0004
B184J	0004
B184K	0004
B184L	0004
B184M	0004
B184OR	0004
B184R	0004
B184X	0004
B184Y	0004
B185(2S)	0004
B185-O	0004
B185A	0004
B185AA(2S)	0004
B185B	0004
B185C	0004
B185D	0004
B185E	0004
B185F(2S)	0004
B185G	0004
B185GN	0004
B185H	0004
B185I	0004
B185J	0004
B185L	0004
B185M	0004
B185OR	0004
B185P(2S)	0004
B185R	0004
B185X	0004
B185Y	0004
B186	0004
B186(SANYO)	0004
B186-1(2S)	0004
B186-7	0004
B186-K(2S)	0004
B186-O	0004
B186-OR	0004
B186A(2S)	0004
B186AG(2S)	0004
B186B(2S)	0004
B186BY(2S)	0004
B186C	0004
B186D	0004
B186E	0004
B186F	0004
B186G(2S)	0004
B186GN	0004
B186H(2S)	0004
B186J	0004
B186K	0004
B186L(2S)	0004
B186M	0004
B186OR	0004
B186R	0004
B186X	0004
B186Y(2S)	0004
B187	0004
B187(SANYO)	0004
B187-1	0004
B187-OR	0004
B187A	0004
B187AA(2S)	3555
B187B(2S)	3555
B187BK	0004
B187C(2S)	0004
B187D(2S)	0004
B187E	0004
B187F	0004
B187G(2S)	0004
B187GN	0004
B187H	0004
B187K(2S)	0004
B187L	0004
B187M	0004
B187OR	0004
B187R(2S)	0004
B187S	0004
B187TV	0004
B187X	0004
B187Y(2S)	0004
B188(2S)	0004
B188A	0004
B188B	0004
B188C	0004
B188D	0004
B188E	0004
B188F	0004
B188G	0004
B188GN	0004
B188H	0004
B188J	0004
B188K	0004
B188L	0004
B188M	0004
B188OR	0004
B188R	0004
B188X	0004
B188Y	0004
B189(2S)	0004
B189A	0004
B189B	0004
B189C	0004
B189D	0004
B189E	0004
B189F	0004
B189G	0004
B189GN	0004
B189H	0004
B189J	0004
B189K	0004
B189L	0004
B189M	0004
B189OR	0004
B189R	0004
B189X	0004
B189Y	0004
B190	0164
B191	0280
B191-5	2002
B191-10	OEM
B191-20	OEM
B191-30	OEM
B191-40	OEM
B191-50	OEM
B191-60	OEM
B191-70	OEM
B191-80	OEM
B191-90	OEM
B191-100	OEM
B191-120	OEM
B192	0164
B193	0164
B193F	OEM
B194	0164
B195	0164
B196	0164
B197	0164
B198	0164
B199(2S)	0211
B199A	0211
B199B	0211
B199C	0211
B199D	0211
B199E	0211
B199F	0211
B199G	0211
B199GN	0211
B199H	0211
B199J	0211
B199L	0211
B199M	0211
B199OR	0211
B199R	0211
B199X	0211
B199Y	0211
B200	0004
B200(2S)	0004
B200A(2S)	0004
B200B	0004
B200C	0004
B200C40	0015
B200D	0004
B200E	0004
B200F	0004
B200G	0004
B200GN	0004
B200H	0004
B200J	0004
B200K	0004
B200L	0004
B200M	0004
B200OR	0004
B200R	0004
B200X	0004
B200Y	0004
B201	0004
B202(2S)	0004
B202A	0004
B202B	0004
B202C	0004
B202D	0004
B202E	0004
B202F	0004
B202G	0004
B202GN	0004
B202H	0004
B202J	0004
B202L	0004
B202M	0004
B202OR	0004
B202R	0004
B202X	0004
B202Y	0004
B203	0604
B203A	0604
B203AA(2S)	0604
B203B	0604
B203C	0604
B203D	0604
B203E	0604
B203F	0604
B203G	0604
B203GN	0604
B203H	0604
B203J	0604
B203K	0604
B203L	0604
B203M	0604
B203OR	0604
B203R	0604
B203X	0604
B203Y	0604
B204(2S)	0604
B205(2S)	0604
B206(2S)	0604
B207	0604
B207A	0604
B208	0604
B208A	0604
B209	0604
B210	0604
B211	0604
B212	0604
B213	0604
B213A	0604
B214	0604
B214A	0604
B215(2S)	0975
B215A	0975
B215B	0975
B215C	0975
B215D	0975
B215E	0975
B215F	0975
B215G	0975
B215GN	0975
B215H	0975
B215K	0975
B215L	0975
B215M	0975
B215OR	0975
B215R	0975
B215X	0975
B215Y	0975
B216(2S)	0160
B216A(2S)	0160
B216B	0160
B216C	0160
B216D	0160
B216E	0160
B216F	0160
B216G	0160
B216GN	0160
B216H	0160
B216J	0160
B216K	0160
B216L	0160
B216M	0160
B216OR	0160
B216R	0160
B216X	0160
B216Y	0160
B217(2S)	0160
B217A(2S)	0160
B217B	0160
B217C	0160
B217D	0160
B217E	0160
B217F	0160
B217G(2S)	0160
B217GN	0160
B217H	0160
B217K	0160
B217L	0160
B217M	0160
B217OR	0160
B217R	0160
B217U(2S)	0160
B217X	0160
B217Y	0160
B218(2S)	0211
B218A	0211
B218B	0211
B218C	0211
B218D	0211
B218E	0211
B218F	0211
B218G	0211
B218GN	0211
B218H	0211
B218J	0211
B218K	0211
B218L	0211
B218M	0211
B218OR	0211
B218R	0211
B218X	0211
B218Y	0211
B219(2S)	0211
B219A	0211
B219B	0211
B219E	0211
B219F	0211
B219G	0211
B219H	0211
B219J	0211
B219K	0211
B219L	0211
B219R	0211
B219X	0211
B219Y	0211
B220(2S)	0211
B220A(2S)	0211
B220B	0211
B220C	0211
B220D	0211
B220E	0211
B220F	0211
B220GN	0211
B220H	0211
B220J	0211
B220K	0211
B220L	0211
B220M	0211
B220OR	0211
B220R	0211
B220X	0211
B220Y	0211
B221(2S)	0211
B221A(2S)	0211
B221B	0211
B221C	0211
B221D	0211
B221E	0211
B221F	0211
B221G	0211
B221GN	0211
B221H	0211
B221J	0211
B221K	0211
B221L	0211
B221M	0211
B221OR	0211
B221R	0211
B221X	0211
B221Y	0211
B222(2S)	0211
B222A	0211
B222B	0211
B222C	0211
B222D	0211
B222E	0211
B222F	0211
B222G	0211
B222GN	0211
B222H	0211
B222J	0211
B222K	0211
B222L	0211
B222M	0211
B222OR	0211
B222R	0211
B222X	0211
B222Y	0211
B223(2S)	0211
B223-10	0087
B223-20	0348
B223-30	0087
B223-40	0087
B223-50	0087
B223-60	0087
B223A	0211
B223B	0211
B223C	0211
B223D	0211
B223E	0211
B223F	0211
B223G	0211
B223Y	0211
B224	0004
B224A	0004
B224B	0004
B224C	0004
B224D	0004
B224E	0004
B224F	0004
B224G	0004
B224GN	0004
B224H	0004
B224K	0004
B224L	0004
B224M	0004
B224OR	0004
B224R	0004
B224X	0004
B224Y	0004
B225(2S)	0211
B225A	0211
B225B	0211
B225C	0211
B225D	0211
B225E	0211
B225F	0211
B225G	0211
B225GN	0211
B225H	0211
B225K	0211
B225L	0211
B225M	0211
B225OR	0211
B225R	0211
B225X	0211
B225Y	0211
B226(2S)	0211
B226A	0211
B226B	0211
B226C	0211
B226D	0211
B226E	0211
B226F	0211
B226G	0211
B226GN	0211
B226J	0211
B226K	0211
B226L	0211
B226M	0211
B226OR	0211
B226R	0211
B226X	0211
B226Y	0211
B227(2S)	0004
B227A	0004
B227B	0004
B227C	0004
B227D	0004
B227E	0004
B227F	0004
B227G	0004
B227GN	0004
B227H	0004
B227J	0004
B227K	0004
B227L	0004
B227M	0004
B227OR	0004
B227R	0004
B227X	0004
B227Y	0004
B228(2S)	0975
B228A	0975
B228B	0975
B228C	0975
B228D	0975
B228E	0975
B228F	0975
B228G	0975
B228GN	0975
B228H	0975
B228J	0975
B228K	0975
B228L	0975
B228M	0975
B228OR	0975
B228R	0975
B228X	0975
B228Y	0975
B229(2S)	0975
B229A	0975
B229B	0975
B229C	0975
B229D	0975
B229E	0975
B229F	0975
B229G	0975
B229GN	0975
B229J	0975
B229K	0975
B229L	0975
B229M	0975
B229OR	0975
B229R	0975
B229X	0975
B229Y	0975
B230(2S)	0975
B230A	0975
B230B	0975
B230C	0975
B230D	0975
B230E	0975
B230F	0975
B230G	0975
B230GN	0975
B230H	0975
B230J	0975
B230K	0975
B230L	0975
B230M	0975
B230OR	0975
B230R	0975
B230X	0975
B230Y	0975
B231(2S)	0975
B231A	0975
B231B	0975
B231C	0975
B231D	0975
B231E	0975
B231F	0975
B231G	0975
B231GN	0975
B231H	0975
B231J	0975
B231K	0975
B231L	0975
B231M	0975
B231OR	0975
B231R	0975
B231X	0975
B231Y	0975
B232(2S)	0975
B232A	0975
B232B	0975
B232C	0975
B232D	0975
B232E	0975
B232F	0975
B232G	0975
B232GN	0975
B232H	0975
B232K	0975
B232L	0975
B232M	0975
B232OR	0975
B232R	0975
B232X	0975
B232Y	0975
B233(2S)	0975
B233A	0975
B233B	0975
B233C	0975
B233D	0975
B233E	0975
B233F	0975
B233G	0975
B233GN	0975
B233H	0975
B233J	0975
B233K	0975
B233L	0975
B233M	0975
B233OR	0975
B233R	0975
B233X	0975
B233Y	0975
B234(2S)	0975
B234A	0975
B234B	0975
B234C	0975
B234D	0975
B234E	0975
B234F	0975
B234G	0975
B234GN	0975
B234H	0975
B234J	0975
B234K	0975
B234L	0975
B234M	0975
B234N(2S)	0975
B234OR	0975
B234R	0975
B234X	0975
B234Y	0975
B235(2S)	0040
B235A(2S)	0437
B235B	0437
B235C	0437
B235D	0437
B235E	0437
B235F	0437
B235G	0437
B235GN	0437
B235H	0437
B235L	0437
B235M	0437
B235OR	0437
B235R	0437
B235X	0437
B235Y	0437
B236(2S)	0440
B236A	0440
B236B	0440
B236C	0440
B236D	0440
B236E	0440
B236F	0440
B236G	0440
B236GN	0440
B236H	0440
B236J	0440
B236K	0440
B236L	0440
B236M	0440
B236OR	0440
B236R	0440
B236X	0440
B236Y	0440
B237(2S)	0440
B237-12A(2S)	0440
B237-12B(2S)	0440
B237A	0440
B237B	0440
B237C	0440
B237D	0440
B237E	0440
B237F	0440
B237G	0440
B237GN	0440
B237H	0440
B237J	0440
B237K	0440
B237L	0440
B237M	0440
B237OR	0440
B237R	0440
B237X	0440
B237Y	0440
B238(2S)	0164
B238-12A(2S)	0164
B238-12B(2S)	0164
B238-12C(2S)	0164
B238A	0164
B238B	0164
B238C	0164
B238C18	0079
B238D	0164
B238E	0164
B238F	0164
B238G	0164
B238GN	0164
B238H	0164
B238J	0164
B238K	0164
B238L	0164
B238M	0164
B238OR	0164
B238R	0164
B238X	0164
B238Y	0164
B239(2S)	0160
B239A(2S)	0160
B239B	0160
B239C	0160
B239D	0160
B239E	0160
B239F	0160
B239G	0160
B239GN	0160
B239H	0160
B239J	0160
B239K	0160
B239L	0160
B239M	0160
B239OR	0160
B239R	0160
B239X	0160
B239Y	0160
B240	0274
B240(2S)	0050
B240A(2S)	0050
B240B	0050
B240C	0050
B240D	0050
B240E	0050
B240F	0050
B240GN	0050
B240H	0050
B240J	0050
B240K	0050
B240L	0050
B240M	0050

If replacement code is OEM, contact original manufacturer for replacement.

DEVICE TYPE	REPL CODE
B240OR	0050
B240R	0050
B240X	0050
B240Y	0050
B241(2S)	0164
B241A	0164
B241B	0164
B241C	0164
B241D	0164
B241E	0164
B241F	0164
B241G	0164
B241GN	0164
B241H	0164
B241J	0164
B241K	0164
B241L	0164
B241M	0164
B241OR	0164
B241R	0164
B241V	0164
B241X	0164
B241Y	0164
B242	0164
B242A	0164
B242B	0164
B242C	0164
B242D	0164
B242E	0164
B242F	0164
B242G	0164
B242GN	0164
B242H	0164
B242J	0164
B242K	0164
B242L	0164
B242M	0164
B242OR	0164
B242R	0164
B242X	0164
B242Y	0164
B243	0975
B243A	0975
B243B	0975
B243C	0975
B243D	0975
B243E	0975
B243F	0975
B243G	0975
B243GN	0975
B243H	0975
B243J	0975
B243K	0975
B243L	0975
B243M	0975
B243OR	0975
B243R	0975
B243X	0975
B243Y	0975
B244	0975
B244A	0975
B244B	0975
B244C	0975
B244D	0975
B244E	0975
B244F	0975
B244G	0975
B244GN	0975
B244H	0975
B244J	0975
B244K	0975
B244L	0975
B244M	0975
B244OR	0975
B244R	0975
B244X	0975
B244Y	0975
B245	0975
B245A	0975
B245B	0975
B245C	0975
B245D	0975
B245E	0975
B245F	0975
B245G	0975
B245H	0975
B245J	0975
B245L	0975
B245M	0975
B245OR	0975
B245R	0975
B245X	0975
B245Y	0975
B246(2S)	0085
B246A	0085
B246B	0085
B246C	0085
B246D	0085
B246E	0085
B246F	0085
B246G	0085
B246GN	0085
B246H	0085
B246J	0085
B246K	0085
B246L	0085
B246M	0085
B246OR	0085
B246R	0085
B246X	0085
B246Y	0085
B247(2S)	0160
B247A	0160
B247B	0160
B247C	0160
B247D	0160
B247E	0160
B247F	0160
B247G	0160
B247GN	0160
B247H	0160
B247J	0160
B247K	0160
B247L	0160
B247M	0160
B247OR	0160
B247R	0160
B247X	0160
B247Y	0160
B248(2S)	0085
B248A(2S)	0085
B248B	0085
B248C	0085
B248D	0085
B248E	0085
B248F	0085
B248G	0085
B248GN	0085
B248H	0085
B248J	0085
B248K	0085
B248L	0085
B248M	0085
B248OR	0085
B248R	0085
B248X	0085
B248Y	0085
B249(2S)	0604
B249A(2S)	0604
B249B	0604
B249C	0604
B249D	0604
B249E	0604
B249F	0604
B249G	0604
B249GN	0604
B249H	0604
B249J	0604
B249K	0604
B249L	0604
B249M	0604
B249OR	0604
B249X	0604
B249Y	0604
B250(2S)	0085
B250A(2S)	0085
B250B	0085
B250C7K4S	0015
B250C75	0015
B250C75K4	0015
B250C75K5	0015
B250C75K41	0015
B250C75K45	0015
B250C100	0015
B250C100TD	0015
B250C125	0015
B250C125K4	0015
B250C125N2	0015
B250C125X4	0015
B250C150	0015
B250C150K4	0015
B250C800	2010
B250C800A	2189
B250C800D	1875
B250C800D1	1875
B250C800DM	1875
B250C800M	0106
B250C800W	2010
B250C1000	2010
B250C1000A	2189
B250C1000M	0106
B250C1000W	2010
B250C1200S	3414
B250C1200W	2010
B250C1500	2010
B250C1500/1000	6062
B250C1500/1000SIC	OEM
B250C1500R	0302
B250C1500W	2010
B250C2000/1500	OEM
B250C3700/2200	1045
B250C3700/2200SIC	3035
B250C5000/3300	1045
B250C5000/3300SI	OEM
B250C5000/3300SIC	3035
B250D	0085
B250E	0085
B250F	0085
B250G	0085
B250GN	0085
B250H	0085
B250K	0085
B250L	0085
B250M	0085
B250OR	0085
B250X	0085
B250Y	0085
B251(2S)	0975
B251A(2S)	0975
B251B	0975
B251C	0975
B251D	0975
B251E	0975
B251F	0975
B251G	0975
B251GN	0975
B251H	0975
B251J	0975
B251K	0975
B251L	0975
B251M	0975
B251OR	0975
B251R	0975
B251X	0975
B252(2S)	0975
B252A(2S)	0975
B252B	0975
B252C	0975
B252D	0975
B252E	0975
B252F	0975
B252G	0975
B252GN	0975
B252H	0975
B252J	0975
B252L	0975
B252M	0975
B252OR	0975
B252R	0975
B252X	0975
B252Y	0975
B253(2S)	0975
B253A(2S)	0975
B253B	0975
B253C	0975
B253D	0975
B253E	0975
B253F	0975
B253G	0975
B253GN	0975
B253H	0975
B253K	0975
B253L	0975
B253M	0975
B253OR	0975
B253X	0975
B253Y	0975
B254(2S)	1010
B254A	1010
B254B	1010
B254C	1010
B254D	1010
B254E	1010
B254F	1010
B254G	1010
B254GN	1010
B254H	1010
B254J	1010
B254K	1010
B254L	1010
B254M	1010
B254OR	1010
B254R	1010
B254X	1010
B254Y	1010
B255(2S)	1010
B255A	1010
B255B	1010
B255C	1010
B255D	1010
B255E	1010
B255F	1010
B255G	1010
B255GN	1010
B255H	1010
B255J	1010
B255K	1010
B255L	1010
B255M	1010
B255OR	1010
B255R	1010
B255X	1010
B255Y	1010
B256(2S)	1010
B256A	1010
B256B	1010
B256C	1010
B256D	1010
B256E	1010
B256F	1010
B256G	1010
B256GN	1010
B256H	1010
B256J	1010
B256L	1010
B256M	1010
B256OR	1010
B256R	1010
B256X	1010
B256Y	1010
B257(2S)	0004
B257A	0004
B257B	0004
B257C	0004
B257D	0004
B257E	0004
B257F	0004
B257G	0004
B257GN	0004
B257H	0004
B257K	0004
B257L	0004
B257M	0004
B257OR	0004
B257R	0004
B257X	0004
B257Y	0004
B258(2S)	0437
B258A	0437
B258B	0437
B258C	0437
B258D	0437
B258E	0437
B258F	0437
B258G	0437
B258GN	0437
B258H	0437
B258J	0437
B258K	0437
B258L	0437
B258M	0437
B258OR	0437
B258R	0437
B258X	0437
B258Y	0437
B259(2S)	0437
B259A	0437
B259B	0437
B259C	0437
B259D	0437
B259E	0437
B259F	0437
B259GN	0437
B259H	0437
B259J	0437
B259K	0437
B259L	0437
B259OR	0437
B259R	0437
B259X	0437
B259Y	0437
B260(2S)	0040
B260A	0437
B260B	0437
B260C	0437
B260D	0437
B260E	0437
B260F	0437
B260G	0437
B260GN	0437
B260H	0437
B260J	0437
B260K	0437
B260L	0437
B260M	0437
B260OR	0437
B260R	0437
B260X	0437
B260Y	0437
B261(2S)	0004
B261A	0004
B261B	0004
B261C	0004
B261D	0004
B261E	0004
B261F	0004
B261G	0004
B261GN	0004
B261H	0004
B261J	0004
B261K	0004
B261L	0004
B261M	0004
B261N	0004
B261OR	0004
B261R	0004
B261X	0004
B261Y	0004
B262(2S)	0004
B262A	0004
B262B	0004
B262C	0004
B262D	0004
B262E	0004
B262F	0004
B262G	0004
B262GN	0004
B262H	0004
B262J	0004
B262K	0004
B262L	0004
B262M	0004
B262OR	0004
B262R	0004
B262X	0004
B262Y	0004
B263(2S)	0004
B263A	0004
B263B	0004
B263C	0004
B263D	0004
B263E	0004
B263F	0004
B263G	0004
B263GN	0004
B263H	0004
B263J	0004
B263K	0004
B263L	0004
B263M	0004
B263OR	0004
B263R	0004
B263X	0004
B263Y	0004
B264(2S)	0004
B264A	0004
B264B	0004
B264C	0004
B264D	0004
B264E	0004
B264F	0004
B264G	0004
B264GN	0004
B264H	0004
B264J	0004
B264K	0004
B264L	0004
B264M	0004
B264OR	0004
B264R	0004
B264X	0004
B264Y	0004
B265(2S)	0004
B265A	0004
B265B	0004
B265C	0004
B265D	0004
B265E	0004
B265F	0004
B265G	0004
B265GN	0004
B265H	0004
B265J	0004
B265K	0004
B265L	0004
B265M	0004
B265OR	0004
B265X	0004
B265Y	0004
B266(2S)	0004
B266A	0004
B266B	0004
B266C	0004
B266D	0004
B266E	0004
B266F	0004
B266G	0004
B266GN	0004
B266H	0004
B266J	0004
B266K	0004
B266L	0004
B266M	0004
B266OR	0004
B266P(2S)	0004
B266Q(2S)	0004
B266R	0004
B266X	0004
B266Y	0004
B267(2S)	0004
B267A	0004
B267B	0004
B267C	0004
B267D	0004
B267E	0004
B267F	0004
B267G	0004
B267GN	0004
B267H	0004
B267J	0004
B267K	0004
B267L	0004
B267M	0004
B267OR	0004
B267R	0004
B267X	0004
B267Y	0004
B268(2S)	0004
B268A	0004
B268B	0004
B268C	0004
B268D	0004
B268E	0004
B268F	0004
B268G	0004
B268GN	0004
B268H	0004
B268J	0004
B268K	0004
B268L	0004
B268M	0004
B268OR	0004
B268R	0004
B268X	0004
B268Y	0004
B269(2S)	0004
B269A	0004
B269C	0004
B269D	0004
B269E	0004
B269F	0004
B269G	0004
B269GN	0004
B269J	0004
B269K	0004
B269L	0004
B269M	0004
B269OR	0004
B269R	0004
B269X	0004
B269Y	0004
B270(2S)	0004
B270A(2S)	0004
B270B(2S)	0004
B270C(2S)	0004
B270D(2S)	0004
B270E(2S)	0004
B270F	0004
B270G	0004
B270GN	0004
B270H	0004
B270J	0004
B270K	0004
B270L	0004
B270M	0004
B270OR	0004
B270R	0004
B270X	0004
B270Y	0004
B271(2S)	0004
B272(2S)	0004
B272-5	OEM
B272-10	OEM
B272-20	OEM
B272-30	OEM
B272-40	OEM
B272-50	OEM
B272-60	OEM
B272-70	OEM
B272-80	OEM
B272-90	OEM
B272-100	OEM
B272-120	OEM
B272A	0004
B272B	0004
B272C	0004
B272D	0004
B272E	0004
B272F	0004
B272G	0004
B272GN	0004
B272J	0004
B272K	0004
B272L	0004
B272M	0004
B272OR	0004
B272R	0004
B272X	0004
B272Y	0004
B273(2S)	0004
B273A	0004
B273B	0004
B273C	0004
B273D	0004
B273E	0004
B273F	0004
B273G	0004
B273GN	0004
B273H	0004
B273J	0004
B273K	0004
B273L	0004
B273M	0004
B273OR	0004
B273R	0004
B273X	0004
B273Y	0004
B274	0975
B274(SYLVANIA)	0086
B274A	0975
B274B	0975
B274C	0975
B274D	0975
B274E	0975
B274F	0975
B274G	0975
B274H	0975
B274J	0975
B274K	0975
B274L	0975
B274M	0975
B274OR	0975
B274R	0975
B274V	0975
B274X	0975
B274Y	0975
B275(2S)	0975
B275A	0975
B275B	0975
B275C	0975
B275D	0975
B275E	0975
B275F	0975
B275G	0975
B275GN	0975
B275H	0975
B275J	0975
B275K	0975
B275L	0975
B275M	0975
B275OR	0975
B275R	0975
B275X	0975
B275Y	0975
B276(2S)	0975
B276A	0975
B276B	0975
B276C	0975
B276D	0975
B276E	0975
B276F	0975
B276G	0975
B276GN	0975
B276H	0975
B276J	0975
B276K	0975
B276L	0975
B276M	0975
B276OR	0975
B276R	0975
B276X	0975
B276Y	0975
B277	0050
B278	0164
B279	0164
B280	0004
B280C800	OEM
B281	0532
B281C	0532
B282(2S)	0604
B282A	0604
B282B	0604
B282C	0604
B282D	0604
B282E	0604
B282G	0604
B282GN	0604
B282H	0604
B282J	0604
B282K	0604
B282L	0604
B282M	0604
B282OR	0604
B282R	0604
B282X	0604
B282Y	0604
B283(2S)	0160
B283A	0160
B283B	0160
B283C	0160
B283D	0160
B283E	0160
B283F	0160
B283G	0160
B283GN	0160
B283H	0160
B283J	0160
B283K	0160
B283L	0160
B283M	0160
B283OR	0160
B283R	0160
B283X	0160
B283Y	0160
B284(2S)	0160
B284A	0160
B284B	0160
B284C	0160
B284D	0160
B284E	0160
B284F	0160
B284G	0160
B284H	0160
B284J	0160
B284K	0160
B284L	0160
B284M	0160
B284OR	0160
B284R	0160
B284X	0160
B284Y	0160
B285(2S)	0975
B285-5	0277
B285-10	0277
B285-20	0289
B285-30	0296
B285-40	0296
B285-50	0302
B285-60	0302
B285-70	0250
B285-80	0250
B285-90	0250
B285-100	0250
B285-120	OEM
B285A	0975
B285B	0975
B285C	0975
B285D	0975
B285E	0975
B285F	0975
B285G	0975
B285GN	0975
B285H	0975
B285J	0975
B285K	0975
B285L	0975
B285M	0975
B285OR	0975
B285R	0975
B285X	0975
B285Y	0975
B286	0975
B287	0975

If replacement code is OEM, contact original manufacturer for replacement.

DEVICE TYPE	REPL CODE
B288	0050
B289	0050
B290(2S)	0280
B290-GREEN	0280
B290-YELLOW	0280
B290A	0280
B290A-BLUE	0280
B290B	0280
B290C	0280
B290D	0280
B290E	0280
B290F	0280
B290G	0280
B290GN	0280
B290H	0280
B290J	0280
B290K	0280
B290L	0280
B290M	0280
B290OR	0280
B290R	0280
B290X	0280
B290Y	0280
B291(2S)	0280
B291-5	OEM
B291-10	OEM
B291-20	OEM
B291-30	OEM
B291-40	OEM
B291-50	OEM
B291-60	OEM
B291-70	OEM
B291-80	OEM
B291-90	OEM
B291-100	OEM
B291-120	OEM
B291-GREEN	0280
B291-RED	0280
B291-YELLOW	0280
B291A	0280
B291B	0280
B291C	0280
B291D	0280
B291E	0280
B291F	0280
B291G	0280
B291GN	0280
B291H	0280
B291J	0280
B291K	0280
B291L	0280
B291M	0280
B291OR	0280
B291R	0280
B291X	0280
B291Y	0280
B292(2S)	0280
B292-BLUE	0280
B292-GREEN	0280
B292-ORANGE	0280
B292-RED	0280
B292-YELLOW	0280
B292A(2S)	0280
B292A-BLUE	0280
B292A-GREEN	0280
B292A-ORANG	0280
B292A-RED	0280
B292A-YELLO	0280
B292B	0280
B292C	0280
B292D	0280
B292E	0280
B292F	0280
B292G	0280
B292GN	0280
B292H	0280
B292J	0280
B292K	0280
B292L	0280
B292M	0280
B292OR	0280
B292R	0280
B292X	0280
B292Y	0280
B293(2S)	0004
B293A	0004
B293B	0004
B293C	0004
B293D	0004
B293E	0004
B293F	0004
B293G	0004
B293GN	0004
B293H	0004
B293J	0004
B293K	0004
B293L	0004
B293M	0004
B293OR	0004
B293R	0004
B293X	0004
B293Y	0004
B294	0015
B294(TRANS)	0004
B294A	0004
B294B	0004
B294C	0004
B294D	0004
B294E	0004
B294F	0004
B294G	0004
B294GN	0004
B294H	0004
B294J	0004
B294K	0004
B294L	0004
B294M	0004
B294OR	0004
B294R	0004
B294X	0004
B294Y	0004
B295(2S)	0975
B295A	0975
B295B	0975
B295C	0975
B295D	0975
B295E	0975
B295F	0975
B295G	0975
B295GN	0975
B295H	0975
B295J	0975
B295K	0975
B295L	0975
B295M	0975
B295OR	0975
B295R	0975
B295X	0975
B295Y	0975
B296(2S)	0975
B296A	0975
B296B	0975
B296C	0975
B296D	0975
B296E	0975
B296F	0975
B296G	0975
B296GN	0975
B296H	0975
B296J	0975
B296K	0975
B296L	0975
B296M	0975
B296OR	0975
B296R	0975
B296X	0975
B296Y	0975
B299(2S)	0004
B299A	0004
B299B	0004
B299C	0004
B299D	0004
B299E	0004
B299F	0004
B299G	0004
B299GN	0004
B299H	0004
B299J	0004
B299K	0004
B299L	0004
B299M	0004
B299OR	0004
B299R	0004
B299X	0004
B299Y	0004
B300(2S)	0975
B300A	0975
B300B	0975
B300C	0975
B300D	0975
B300E	0975
B300F	0975
B300G	0975
B300GN	0975
B300H	0975
B300J	0975
B300K	0975
B300L	0975
B300M	0975
B300OR	0975
B300R	0975
B300X	0975
B300Y	0975
B301	0975
B301A	0975
B301B	0975
B301C	0975
B301D	0975
B301E	0975
B301F	0975
B301G	0975
B301GN	0975
B301H	0975
B301J	0975
B301L	0975
B301M	0975
B301OR	0975
B301R	0975
B301X	0975
B301Y	0975
B302(2S)	0004
B302A	0004
B302B	0004
B302C	0004
B302D	0004
B302E	0004
B302F	0004
B302G	0004
B302J	0004
B302K	0004
B302L	0004
B302M	0004
B302OR	0004
B302R	0004
B302X	0004
B302Y	0004
B303	0004
B303(0)	0004
B303-0(2S)	0004
B303A(2S)	0004
B303B(2S)	0004
B303BK	0004
B303C(2S)	0004
B303D	0004
B303E	0004
B303F	0004
B303G	0004
B303GN	0004
B303H(2S)	0004
B303J	0004
B303K(2S)	0004
B303L	0004
B303M	0004
B303OR	0004
B303R	0004
B303X	0004
B303Y	0004
B304(2S)	0164
B304A(2S)	0164
B304B	0164
B304C	0164
B304D	0164
B304E	0164
B304F	0164
B304G	0164
B304GN	0164
B304H	0164
B304J	0164
B304K	0164
B304L	0164
B304M	0164
B304OR	0164
B304R	0164
B304X	0164
B304Y	0164
B309(2S)	0975
B309A	0975
B309B	0975
B309C	0975
B309D	0975
B309E	0975
B309F	0975
B309G	0975
B309GN	0975
B309H	0975
B309K	0975
B309L	0975
B309M	0975
B309OR	0975
B309R	0975
B309X	0975
B309Y	0975
B310	0975
B310A	0975
B310B	0975
B310C	0975
B310D	0975
B310E	0975
B310F	0975
B310G	0975
B310GN	0975
B310H	0975
B310J	0975
B310K	0975
B310L	0975
B310M	0975
B310OR	0975
B310R	0975
B310X	0975
B310Y	0975
B311(2S)	0975
B311A	0975
B311B	0975
B311C	0975
B311D	0975
B311E	0975
B311F	0975
B311G	0975
B311GN	0975
B311H	0975
B311J	0975
B311K	0975
B311L	0975
B311M	0975
B311OR	0975
B311R	0975
B311X	0975
B311Y	0975
B312A	0975
B312B	0975
B312C	0975
B312D	0975
B312E	0975
B312F	0975
B312G	0975
B312GN	0975
B312H	0975
B312J	0975
B312K	0975
B312L	0975
B312M	0975
B312X	0975
B312Y	0975
B313(2S)	0975
B313A	0975
B313B	0975
B313C	0975
B313D	0975
B313E	0975
B313F	0975
B313G	0975
B313GN	0975
B313H	0975
B313J	0975
B313K	0975
B313L	0975
B313M	0975
B313OR	0975
B313R	0975
B313X	0975
B313Y	0975
B314(2S)	0280
B314A	0280
B314B	0280
B314C	0280
B314D	0280
B314E	0280
B314F	0280
B314GN	0280
B314H	0280
B314J	0280
B314K	0280
B314L	0280
B314M	0280
B314OR	0280
B314R	0280
B314X	0280
B314Y	0280
B315(2S)	0004
B-315-1	0164
B315A	0004
B315B	0004
B315C	0004
B315D	0004
B315E	0004
B315F	0004
B315G	0004
B315GN	0004
B315H	0004
B315J	0004
B315L	0004
B315M	0004
B315OR	0004
B315R	0004
B315X	0004
B315Y	0004
B316(2S)	0004
B316A	0004
B316B	0004
B316C	0004
B316D	0004
B316E	0004
B316F	0004
B316G	0004
B316GN	0004
B316H	0004
B316OR	0004
B316R	0004
B316X	0004
B316Y	0004
B317(2S)	0004
B318(2S)	0975
B318A	0975
B318B	0975
B318C	0975
B318D	0975
B318E	0975
B318F	0975
B318G	0975
B318GN	0975
B318H	0975
B318J	0975
B318K	0975
B318L	0975
B318M	0975
B318OR	0975
B318R	0975
B318X	0975
B318Y	0975
B319(2S)	0975
B319A	0975
B319B	0975
B319C	0975
B319D	0975
B319E	0975
B319F	0975
B319G	0975
B319GN	0975
B319H	0975
B319J	0975
B319K	0975
B319L	0975
B319M	0975
B319OR	0975
B319R	0975
B319X	0975
B319Y	0975
B320(2S)	0975
B320A	0975
B320B	0975
B320C	0975
B320D	0975
B320E	0975
B320F	0975
B320G	0975
B320GN	0975
B320H	0975
B320J	0975
B320K	0975
B320L	0975
B320M	0975
B320OR	0975
B320R	0975
B320X	0975
B320Y	0975
B321(2S)	0004
B322(2S)	0004
B323(2S)	0004
B-324	0164
B324	1062
B324-OR	1062
B324A(2S)	1062
B324B(2S)	1062
B324C	1062
B324D(2S)	1062
B324E(2S)	1062
B324E-1(2S)	1062
B324E-L	1062
B324F(2S)	1062
B324G(2S)	1062
B324GN	1062
B324H(2S)	1062
B324I(2S)	1062
B324J(2S)	1062
B324K(2S)	1062
B324L	1062
B324M	1062
B324N(2S)	1062
B324OR	1062
B324P(2S)	1062
B324R	1062
B324S(2S)	1062
B324V(2S)	1062
B324X	1062
B324Y	1062
B326(2S)	0004
B326A	0004
B326B	0004
B326C	0004
B326D	0004
B326E	0004
B326F	0004
B326G	0004
B326GN	0004
B326H	0004
B326J	0004
B326K	0004
B326L	0004
B326M	0004
B326OR	0004
B326R	0004
B326X	0004
B326Y	0004
B327(2S)	0004
B327A	0004
B327B	0004
B327C	0004
B327D	0004
B327E	0004
B327F	0004
B327G	0004
B327GN	0004
B327H	0004
B327J	0004
B327K	0004
B327L	0004
B327M	0004
B327OR	0004
B327R	0004
B327X	0004
B327Y	0004
B328(2S)	0211
B328A	0211
B328B	0211
B328C	0211
B328D	0211
B328E	0211
B328F	0211
B328G	0211
B328GN	0211
B328H	0211
B328J	0211
B328K	0211
B328L	0211
B328M	0211
B328OR	0211
B328R	0211
B328X	0211
B328Y	0211
B329(2S)	0004
B329A	0004
B329B	0004
B329C	0004
B329D	0004
B329E	0004
B329F	0004
B329G	0004
B329GN	0004
B329H	0004
B329J	0004
B329K(2S)	0004
B329L	0004
B329M	0004
B329OR	0004
B329R	0004
B329X	0004
B329Y	0004
B331(2S)	0040
B331A	0437
B331B	0437
B331C	0437
B331D	0437
B331E	0437
B331F	0437
B331G	0437
B331GN	0437
B331H(2S)	0040
B331J	0437
B331K	0437
B331L	0437
B331M	0437
B331OR	0437
B331R	0437
B331X	0437
B331Y	0437
B332(2S)	0040
B332A	0437
B332B	0437
B332C	0437
B332D	0437
B332E	0437
B332F	0437
B332G	0437
B332GN	0437
B332H(2S)	0040
B332J	0437
B332K	0437
B332L	0437
B332M	0437
B332OR	0437
B332R	0437
B332X	0437
B332Y	0437
B333(2S)	0040
B333A	0437
B333B	0437
B333D	0437
B333E	0437
B333F	0437
B333G	0437
B333GN	0437
B333H	0437
B333HA	0437
B333HB	0437
B333J	0437
B333K	0437
B333L	0437
B333M	0437
B333OR	0437
B333R	0437
B333X	0437
B333Y	0437
B334	0440
B334-OR	0440
B334A	0440
B334B	0440
B334C	0440
B334D	0440
B334E	0440
B334F	0440
B334G	0440
B334GN	0440
B334H	0440
B334J	0440
B334K	0440
B334L	0440
B334M	0440
B334OR	0440
B334R	0440
B334X	0440
B334Y	0440
B335(2S)	0211
B336(2S)	0004
B337(2S)	0160
B337-B	0160
B337-OR	0160
B337A(2S)	0160
B337B(2S)	0160
B337BK(2S)	0160
B337C	0160
B337D	0160
B337E	0160
B337F	0160
B337G	0160
B337GN	0160
B337H(2S)	0160
B337J	0160
B337K	0160
B337L	0160
B337LB	0160
B337M	0160
B337OR	0160
B337R	0160
B337X	0160
B337Y	0160
B338(2S)	0160
B338A	0160
B338B	0160
B338C	0160
B338D	0160
B338E	0160
B338F	0160
B338G	0160
B338GN	0160
B338H(2S)	0160
B338HA(2S)	0160
B338HB(2S)	0160
B338J	0160
B338K	0160
B338L	0160
B338M	0160
B338OR	0160
B338R	0160
B338X	0160
B338Y	0160
B339(2S)	0604
B339H(2S)	0604
B340(2S)	0604
B340H(2S)	0604
B341(2S)	0975
B341A	0975
B341B	0975
B341C	0975
B341D	0975
B341E	0975
B341F	0975
B341G	0975
B341GN	0975
B341H(2S)	0975
B341J	0975
B341K	0975
B341L	0975
B341M	0975
B341OR	0975
B341R	0975
B341S	0975
B341V(2S)	0975
B341V(S)	0975
B341X	0975
B341Y	0975
B342(2S)	0975
B342A	0975
B342B	0975
B342C	0975
B342D	0975
B342E	0975
B342F	0975
B342G	0975
B342GN	0975
B342H	0975
B342J	0975
B342K	0975
B342L	0975
B342M	0975
B342OR	0975
B342R	0975
B342X	0975
B342Y	0975
B343(2S)	0975
B343A	0975
B343B	0975
B343C	0975
B343D	0975
B343E	0975
B343F	0975
B343G	0975
B343GN	0975
B343H	0975
B343J	0975
B343K	0975
B343L	0975
B343M	0975
B343OR	0975
B343R	0975
B343X	0975
B343Y	0975
B345(2S)	0004
B345A	0004
B345B	0004
B345C	0004
B345D	0004
B345E	0004
B345F	0004
B345G	0004
B345GN	0004
B345H	0004
B345J	0004
B345K	0004
B345L	0004
B345M	0004
B345OR	0004
B345R	0004
B345X	0004
B345Y	0004
B346(2S)	0004
B346A	0004
B346B	0004
B346C	0004
B346D	0004
B346E	0004
B346F	0004
B346G	0004
B346GN	0004
B346H	0004

If replacement code is OEM, contact original manufacturer for replacement.

DEVICE TYPE	REPL CODE
B346J	0004
B346K(2S)	0004
B346L	0004
B346M	0004
B346OR	0004
B346Q(2S)	0004
B346R	0004
B346X	0004
B346Y	0004
B347(2S)	0004
B347A	0004
B347B	0004
B347C	0004
B347D	0004
B347E	0004
B347F	0004
B347G	0004
B347GN	0004
B347H	0004
B347J	0004
B347K	0004
B347L	0004
B347M	0004
B347OR	0004
B347R	0004
B347X	0004
B347Y	0004
B348(2S)	0004
B348A	0004
B348B	0004
B348C	0004
B348D	0004
B348E	0004
B348F	0004
B348G	0004
B348GN	0004
B348H	0004
B348J	0004
B348K	0004
B348L	0004
B348M	0004
B348OR	0004
B348Q(2S)	0004
B348R(2S)	0004
B348X	0004
B348Y	0004
B349(2S)	0004
B350(2S)	0211
B350A	0211
B350B	0211
B350C	0211
B350D	0211
B350E	0211
B350F	0211
B350G	0211
B350GN	0211
B350H	0211
B350J	0211
B350K	0211
B350L	0211
B350M	0211
B350OR	0211
B350R	0211
B350X	0211
B350Y	0211
B351(2S)	0440
B351A	0440
B351B	0440
B351C	0440
B351D	0440
B351E	0440
B351F	0440
B351G	0440
B351GN	0440
B351H	0440
B351J	0440
B351K	0440
B351L	0440
B351M	0440
B351OR	0440
B351R	0440
B351X	0440
B351Y	0440
B352(2S)	0040
B352A	0437
B352B	0437
B352C	0437
B352D(2S)	0040
B352E	0437
B352F	0437
B352G	0437
B352GN	0437
B352H	0437
B352J	0437
B352K	0437
B352L	0437
B352M	0437
B352OR	0437
B352R	0437
B352X	0437
B352Y	0437
B353(2S)	0040
B353A	0437
B353B	0437
B353C	0437
B353D	0437
B353E	0437
B353F	0437
B353G	0437
B353GN	0437
B353H	0437
B353J	0437
B353K	0437
B353L	0437
B353M	0437
B353OR	0437
B353R	0437
B353X	0437
B353Y	0437
B354(2S)	0437
B354A	0437
B354B	0437
B354C	0437
B354D	0437
B354E	0437
B354F	0437
B354G	0437
B354GN	0437
B354H	0437
B354J	0437
B354K	0437
B354L	0437
B354M	0437
B354OR	0437
B354R	0437
B354X	0437
B354Y	0437
B355(2S)	0085
B355A	0085
B355B	0085
B355C	0085
B355D	0085
B355E	0085
B355F	0085
B355G	0085
B355GN	0085
B355H	0085
B355J	0085
B355K	0085
B355L	0085
B355M	0085
B355OR	0085
B355R	0085
B355X	0085
B355Y	0085
B356(2S)	0085
B356A	0085
B356B	0085
B356C	0085
B356D	0085
B356E	0085
B356F	0085
B356G	0085
B356GN	0085
B356H	0085
B356J	0085
B356K	0085
B356L	0085
B356M	0085
B356OR	0085
B356R	0085
B356X	0085
B356Y	0085
B357(2S)	0975
B357A	0975
B357B	0975
B357C	0975
B357D	0975
B357E	0975
B357F	0975
B357G	0975
B357GN	0975
B357H	0975
B357J	0975
B357K	0975
B357L	0975
B357M	0975
B357OR	0975
B357R	0975
B357X	0975
B357Y	0975
B358(2S)	0975
B358A	0975
B358B	0975
B358C	0975
B358D	0975
B358E	0975
B358F	0975
B358G	0975
B358GN	0975
B358H	0975
B358J	0975
B358K	0975
B358L	0975
B358M	0975
B358OR	0975
B358R	0975
B358X	0975
B358Y	0975
B359(2S)	0975
B359A	0975
B359B	0975
B359C	0975
B359D	0975
B359E	0975
B359F	0975
B359G	0975
B359GN	0975
B359H	0975
B359J	0975
B359K	0975
B359L	0975
B359M	0975
B359OR	0975
B359R	0975
B359X	0975
B359Y	0975
B360	0975
B360A	0975
B360B	0975
B360C	0975
B360D	0975
B360E	0975
B360F	0975
B360G	0975
B360GN	0975
B360H	0975
B360J	0975
B360K	0975
B360L	0975
B360M	0975
B360OR	0975
B360R	0975
B360X	0975
B360Y	0975
B361(2S)	0975
B361A	0975
B361B	0975
B361C	0975
B361D	0975
B361E	0975
B361F	0975
B361G	0975
B361GN	0975
B361H	0975
B361J	0975
B361K	0975
B361L	0975
B361M	0975
B361OR	0975
B361R	0975
B361X	0975
B361Y	0975
B362(2S)	0975
B362A	0975
B362B	0975
B362C	0975
B362D	0975
B362E	0975
B362F	0975
B362G	0975
B362GN	0975
B362H	0975
B362J	0975
B362K	0975
B362L	0975
B362M	0975
B362OR	0975
B362R	0975
B362X	0975
B362Y	0975
B363	0854
B364(2S)	0164
B364-OR	0164
B364A	0164
B364B	0164
B364C	0164
B364D	0164
B364E	0164
B364F	0164
B364G	0164
B364GN	0164
B364H	0164
B364J	0164
B364K	0164
B364L	0164
B364M	0164
B364OR	0164
B364R	0164
B364X	0164
B364Y	0164
B365(2S)	0164
B365A	0164
B365B(2S)	0164
B365C	0164
B365D	0164
B365E	0164
B365F	0164
B365G	0164
B365GN	0164
B365H	0164
B365J	0164
B365K	0164
B365L	0164
B365M	0164
B365OR	0164
B365R	0164
B365X	0164
B365Y	0164
B366(2S)	0975
B366A	0975
B366B	0975
B366C	0975
B366D	0975
B366E	0975
B366F	0975
B366G	0975
B366GN	0975
B366H	0975
B366J	0975
B366K	0975
B366L	0975
B366M	0975
B366OR	0975
B366R	0975
B366X	0975
B366Y	0975
B367(2S)	0222
B367-4	0222
B367-5	0222
B367-OR	0222
B367A(2S)	0222
B367AL	0222
B367B(2S)	0222
B367B-2	0222
B367BL	0222
B367BP	0222
B367C(2S)	0222
B367D	0222
B367E	0222
B367F	0222
B367G	0222
B367H(2S)	0222
B367J	0222
B367K	0222
B367L	0222
B367M	0222
B367OR	0222
B367P	0222
B367R	0222
B367X	0222
B367Y	0222
B368(2S)	0222
B368-OR	0222
B368A(2S)	0222
B368B(2S)	0222
B368C	0222
B368D	0222
B368E	0222
B368F	0222
B368G	0222
B368GN	0222
B368H(2S)	0222
B368J	0222
B368K	0222
B368L	0222
B368M	0222
B368OR	0222
B368X	0222
B368Y	0222
B370(2S)	0004
B370-O	0004
B370A(2S)	0004
B370AA(2S)	0004
B370AB(2S)	0004
B370AC(2S)	0004
B370AH	0004
B370AHA(2S)	0004
B370AHB(2S)	0004
B370B(2S)	0004
B370C(2S)	0004
B370D(2S)	0004
B370E	0004
B370F	0004
B370G	0004
B370GN	0004
B370H	0004
B370J	0004
B370K	0004
B370L	0004
B370M	0004
B370OR	0004
B370P(2S)	0004
B370R	0004
B370V(2S)	0004
B370X	0004
B370Y	0004
B371(2S)	0004
B371A	0004
B371B	0004
B371C	0004
B371D(2S)	0004
B371E	0004
B371F	0004
B371G	0004
B371GN	0004
B371H	0004
B371J	0004
B371K	0004
B371L	0004
B371M	0004
B371OR	0004
B371R	0004
B371X	0004
B371Y	0004
B372(2S)	0847
B372-5	OEM
B372-10	OEM
B372-20	OEM
B372-30	OEM
B372-40	OEM
B372-50	OEM
B372-60	OEM
B372-70	OEM
B372-80	OEM
B372-90	OEM
B372-100	OEM
B372-120	OEM
B373(2S)	0847
B373A	0847
B373B	0847
B373C	0847
B373D	0847
B373E	0847
B373F	0847
B373G	0847
B373GN	0847
B373H	0847
B373K	0847
B373L	0847
B373M	0847
B373OR	0847
B373R	0847
B373X	0847
B373Y	0847
B374	0164
B375(2S)	1675
B375-2B(2S)	1675
B375-5B(2S)	1675
B375-OR	1675
B375A(2S)	1675
B375A-2B(2S)	1675
B375A-5B(2S)	1675
B375AL	1675
B375A-NB(2S)	1675
B375ATV	1675
B375B	1675
B375C	1675
B375D	1675
B375E	1675
B375F	1675
B375G	1675
B375GN	1675
B375H	1675
B375J	1675
B375K	1675
B375L	1675
B375M	1675
B375OR	1675
B375R	1675
B375X	1675
B375Y	1675
B376(2S)	0004
B376-5	0106
B376-10	0106
B376-20	0106
B376-30	0106
B376-40	0106
B376-50	2010
B376-60	2010
B376-70	2400
B376-80	2400
B376-90	0787
B376-100	0787
B376-120	OEM
B376A	0004
B376B	0004
B376C	0004
B376D	0004
B376E	0004
B376F	0004
B376G(2S)	0004
B376GN	0004
B376J	0004
B376K	0004
B376L	0004
B376M	0004
B376OR	0004
B376R	0004
B376X	0004
B376Y	0004
B377(2S)	0211
B377B(2S)	0211
B378(2S)	0004
B378A(2S)	0004
B378B	0004
B378C	0004
B378D	0004
B378E	0004
B378F	0004
B378GN	0004
B378J	0004
B378K	0004
B378L	0004
B378M	0004
B378OR	0004
B378R	0004
B378X	0004
B378Y	0004
B379(2S)	0211
B379-2(2S)	0211
B379A(2S)	0211
B379B	0211
B379C	0211
B379D	0211
B379E	0211
B379F	0211
B379G	0211
B379GN	0211
B379H	0211
B379J	0211
B379K	0211
B379L	0211
B379M	0211
B379OR	0211
B379R	0211
B379Y	0211
B380(2S)	0211
B380A(2S)	0211
B380B	0211
B380BX555	OEM
B380C	0211
B380C800	0787
B380C800A	5725
B380C800D	3546
B380C800D1	3546
B380C800M	2010
B380C800W	2010
B380C1000	0787
B380C1000A	5725
B380C1000M	2010
B380C1000W	0787
B380C1200S	6860
B380C1200W	0787
B380C1500	0787
B380C1500/1000	6077
B380C1500/1000SIC	OEM
B380C1500R	0787
B380C1500W	0787
B380C2000/1500	0787
B380C3700/2200	4521
B380C3700/2200SIC	3050
B380C5000/3300	1774
B380D	0211
B380E	0211
B380F	0211
B380GN	0211
B380J	0211
B380K	0211
B380L	0211
B380OR	0211
B380R	0211
B380X	0211
B380Y	0211
B381(2S)	0004
B381A	0004
B381B	0004
B381C	0004
B381D	0004
B381E	0004
B381F	0004
B381G	0004
B381GN	0004
B381H	0004
B381J	0004
B381K	0004
B381L	0004
B381M	0004
B381OR	0004
B381R	0004
B381X	0004
B381Y	0004
B382(2S)	0004
B382A	0004
B382B	0004
B382BK	0004
B382BN	0004
B382C	0004
B382D	0004
B382E	0004
B382F	0004
B382G	0004
B382GN	0004
B382H	0004
B382J	0004
B382K	0004
B382L	0004
B382OR	0004
B382R	0004
B382X	0004
B382Y	0004
B383(2S)	0004
B383-1(2S)	0004
B383-2(2S)	0004
B383A	0004
B383B	0004
B383C	0004
B383D	0004
B383E	0004
B383F	0004
B383G	0004
B383GN	0004
B383H	0004
B383J	0004
B383K	0004
B383L	0004
B383M	0004
B383OR	0004
B383R	0004
B383X	0004
B383Y	0004
B384(2S)	0004
B384A	0004
B384B	0004
B384C	0004
B384D	0004
B384E	0004
B384F	0004
B384G	0004
B384GN	0004
B384H	0004
B384J	0004
B384K	0004
B384L	0004
B384M	0004
B384OR	0004
B384R	0004
B384X	0004
B384Y	0004
B385(2S)	0004
B385A	0004
B385B	0004
B385C	0004
B385D	0004
B385E	0004
B385F	0004
B385G	0004
B385GN	0004
B385H	0004
B385J	0004
B385K	0004
B385L	0004
B385M	0004
B385OR	0004
B385R	0004
B385X	0004
B385Y	0004
B386(2S)	0004
B386A	0004
B386B	0004
B386C	0004
B386D	0004
B386E	0004
B386F	0004
B386G	0004
B386GN	0004
B386H	0004
B386K	0004
B386L	0004
B386M	0004
B386OR	0004
B386R	0004
B386X	0004
B386Y	0004
B387(2S)	0004
B387A	0004
B387B	0004
B387D	0004
B387E	0004
B387F	0004
B387G	0004
B387GN	0004
B387H	0004
B387J	0004
B387K	0004
B387L	0004
B387M	0004
B387OR	0004
B387R	0004
B387X	0004
B387Y	0004
B389(2S)	0004
B389-O	0004
B389A	0004
B389B	0004
B389BK	0004
B389C	0004
B389D	0004
B389E	0004
B389F	0004
B389G	0004
B389GN	0004
B389H	0004
B389J	0004
B389K	0004
B389L	0004
B389M	0004
B389OR	0004
B389R	0004
B389X	0004
B389Y	0004
B390(2S)	0975
B390A	0975
B390B	0975
B390C	0975
B390D	0975
B390E	0975
B390F	0975
B390G	0975
B390GN	0975
B390H	0975
B390J	0975
B390K	0975
B390L	0975
B390M	0975
B390OR	0975
B390R	0975
B390X	0975
B390Y	0975
B391(2S)	0160
B391-5	OEM
B391-10	OEM
B391-20	OEM
B391-30	OEM
B391-40	OEM
B391-50	OEM
B391-60	OEM
B391-70	OEM
B391-80	OEM
B391-90	OEM
B391-100	OEM
B391-120	OEM
B391A	0160

If replacement code is OEM, contact original manufacturer for replacement.

DEVICE TYPE	REPL CODE
B391B	0160
B391C	0160
B391D	0160
B391E	0160
B391F	0160
B391G	0160
B391GN	0160
B391H	0160
B391J	0160
B391K	0160
B391L	0160
B391M	0160
B391OR	0160
B391R	0160
B391X	0160
B391Y	0160
B392(2S)	0280
B392A	0280
B392B	0280
B392C	0280
B392D	0280
B392E	0280
B392F	0280
B392G	0280
B392GN	0280
B392H	0280
B392J	0280
B392K	0280
B392L	0280
B392M	0280
B392OR	0280
B392R	0280
B392X	0280
B392Y	0280
B393(2S)	0280
B393-5	0729
B393-10	0729
B393-20	0729
B393-30	0737
B393-40	0737
B393-50	0737
B393-60	0737
B393-70	0742
B393-80	0742
B393-90	0742
B393-100	0742
B393-120	OEM
B393A	0280
B393B	0280
B393C	0280
B393D	0280
B393E	0280
B393F	0280
B393G	0280
B393GN	0280
B393H	0280
B393J	0280
B393K	0280
B393L	0280
B393M	0280
B393OR	0280
B393R	0280
B393X	0280
B393Y	0280
B394(2S)	0280
B394A	0280
B394B	0280
B394C	0280
B394D	0280
B394E	0280
B394F	0280
B394G	0280
B394GN	0280
B394H	0280
B394K	0280
B394L	0280
B394M	0280
B394OR	0280
B394R	0280
B394X	0280
B394Y	0280
B395(2S)	0280
B395A	0280
B395B	0280
B395C	0280
B395D	0280
B395E	0280
B395F	0280
B395G	0280
B395GN	0280
B395H	0280
B395J	0280
B395K	0280
B395M	0280
B395OR	0280
B395R	0280
B395X	0280
B395Y	0280
B396(2S)	0211
B396A	0211
B396B	0211
B396C	0211
B396D	0211
B396E	0211
B396F	0211
B396G	0211
B396GN	0211
B396H	0211
B396J	0211
B396K	0211
B396L	0211
B396M	0211
B396OR	0211
B396R	0211
B396X	0211
B396Y	0211
B397	0050
B398	0164
B400(2S)	0004
B400A(2S)	0004
B400B(2S)	0004
B400BK	0004
B400C	0004
B400D	0004
B400E	0004
B400F	0004
B400G	0004
B400GN	0004
B400H	0004
B400J	0004
B400K(2S)	0004
B400L	0004
B400M	0004
B400OR	0004
B400R	0004
B400X	0004
B400Y	0004
B401	0280
B401A	0280
B401B	0280
B401C	0280
B401D	0280
B401E	0280
B401F	0280
B401G	0280
B401GN	0280
B401H	0280
B401K	0280
B401L	0280
B401M	0280
B401OR	0280
B401R	0280
B401X	0280
B401Y	0280
B402(2S)	0280
B402A	0280
B402B	0280
B402C	0280
B402D	0280
B402E	0280
B402F	0280
B402G	0280
B402GN	0280
B402H	0280
B402J	0280
B402K	0280
B402L	0280
B402M	0280
B402OR	0280
B402R	0280
B402X	0280
B402Y	0280
B403(2S)	0280
B403A	0280
B403B	0280
B403C	0280
B403D	0280
B403E	0280
B403F	0280
B403G	0280
B403GN	0280
B403H	0280
B403J	0280
B403K	0280
B403L	0280
B403M	0280
B403OR	0280
B403R	0280
B403X	0280
B403Y	0280
B405	0164
B405-0	0164
B405-1	0164
B405-2C(2S)	0164
B405-3C(2S)	0164
B405-4C(2S)	0164
B405-O	0164
B405-OR	0164
B405-R	0164
B405A(2S)	0164
B405AG	0164
B405B(2S)	0164
B405C(2S)	0164
B405D(2S)	0164
B405DK	0164
B405E(2S)	0164
B405EK	0164
B405F	0164
B405G(2S)	0164
B405GN	0164
B405H(2S)	0164
B405J	0164
B405K(2S)	0164
B405L	0164
B405M	0164
B405OR	0164
B405R(2S)	0164
B405RE(2S)	0164
B405X	0164
B405Y	0164
B406	0975
B407(2S)	0160
B407-0(2S)	0160
B407-0	0160
B407-OR	0160
B407A	0160
B407B	0160
B407BK	0160
B407C	0160
B407D	0160
B407E	0160
B407F	0160
B407G	0160
B407GN	0160
B407H	0160
B407J	0160
B407K	0160
B407M	0160
B407OR	0160
B407R	0160
B407TV(2S)	0160
B407TV-2	0160
B407X	0160
B407Y	0160
B408(2S)	0004
B408A	0004
B408B	0004
B408C	0004
B408D	0004
B408E	0004
B408F	0004
B408G	0004
B408GN	0004
B408H	0004
B408J	0004
B408K	0004
B408L	0004
B408M	0004
B408OR	0004
B408R	0004
B408X	0004
B408Y	0004
B409	0280
B410(2S)	0975
B411(2S)	0975
B411-1	OEM
B411-2	OEM
B411-C	OEM
B411F	OEM
B411S	OEM
B411TV	0975
B412	0975
B412-1	OEM
B412-2	OEM
B412-5-5	OEM
B412-5-10	OEM
B412-5-25	OEM
B412-7-5	OEM
B412-7-10	OEM
B412-7-25	OEM
B412-10-5	OEM
B412-10-10	OEM
B412-10-25	OEM
B412-12-5	0201
B412-12-10	OEM
B412-12-25	OEM
B412-15-5	0288
B412-15-10	OEM
B412-15-25	OEM
B412-20-5	0378
B412-20-10	OEM
B412-20-25	OEM
B412-25-5	OEM
B412-25-10	OEM
B412-25-25	OEM
B412-30-5	OEM
B412-30-10	OEM
B412-30-25	OEM
B412-40-5	OEM
B412-40-10	OEM
B412-40-25	OEM
B412-50-5	OEM
B412-50-10	OEM
B412-50-25	OEM
B412-C	OEM
B412F	OEM
B412S	OEM
B413(2S)	0222
B413-1	OEM
B413-2	OEM
B413-C	OEM
B413A	0222
B413B	0222
B413C	0222
B413D	0222
B413E	0222
B413F	OEM
B413F(2S)	0222
B413G	0222
B413GN	0222
B413H	0222
B413J	0222
B413K	0222
B413L	0222
B413M	0222
B413R	0222
B413S	OEM
B413X	0222
B413Y	0222
B414(2S)	1010
B414B	1010
B414C	1010
B414D	1010
B414E	1010
B414F	1010
B414G	1010
B414GN	1010
B414H	1010
B414J	1010
B414K	1010
B414L	1010
B414M	1010
B414OR	1010
B414R	1010
B414X	1010
B414Y	1010
B415(2S)	0164
B415-OR	0164
B415A(2S)	0164
B415B(2S)	0164
B415C	0164
B415D	0164
B415E	0164
B415F	0164
B415G	0164
B415GN	0164
B415H	0164
B415J	0164
B415K	0164
B415L	0164
B415M	0164
B415OR	0164
B415R	0164
B415X	0164
B415Y	0164
B416(2S)	0280
B416A	0280
B416B	0280
B416C	0280
B416D	0280
B416E	0280
B416F	0280
B416G	0280
B416GN	0280
B416H	0280
B416J	0280
B416K	0280
B416L	0280
B416M	0280
B416OR	0280
B416R	0280
B416X	0280
B416Y	0280
B417(2S)	0280
B417A	0280
B417B	0280
B417C	0280
B417D	0280
B417E	0280
B417F	0280
B417G	0280
B417GN	0280
B417H	0280
B417J	0280
B417K	0280
B417L	0280
B417M	0280
B417OR	0280
B417R	0280
B417X	0280
B417Y	0280
B419	0222
B421(2S)	0004
B421-1	OEM
B421-2	OEM
B421-C	OEM
B421F	OEM
B421S	OEM
B422(2S)	0211
B422-1	OEM
B422-2	OEM
B422-C	OEM
B422A	0211
B422B	0211
B422C	0211
B422D	0211
B422E	0211
B422F	OEM
B422F(2S)	0211
B422G	0211
B422GN	0211
B422H	0211
B422J	0211
B422K	0211
B422L	0211
B422M	0211
B422OR	0211
B422R	0211
B422S	OEM
B422X	0211
B422Y	0211
B423(2S)	0004
B423-1	OEM
B423-2	OEM
B423-C	OEM
B423A	0004
B423B	0004
B423C	0004
B423D	0004
B423E	0004
B423F	OEM
B423F(2S)	0004
B423G	0004
B423GN	0004
B423H	0004
B423J	0004
B423K	0004
B423L	0004
B423M	0004
B423OR	0004
B423R	0004
B423S	OEM
B423X	0004
B423Y	0004
B424(2S)	0604
B424A	0604
B424B	0604
B424C	0604
B424D	0604
B424E	0604
B424F	0604
B424G	0604
B424GN	0604
B424H	0604
B424J	0604
B424K	0604
B424L	0604
B424M	0604
B424OR	0604
B424R	0604
B424X	0604
B424Y	0604
B425(2S)	0160
B425-BL	0160
B425-R	0160
B425-Y	0160
B425A	0160
B425B	0160
B425C	0160
B425D	0160
B425E	0160
B425F	0160
B425G	0160
B425G-BL	0160
B425G-R	0160
B425G-Y	0160
B425H	0160
B425J	0160
B425K	0160
B425L	0160
B425M	0160
B425R	0160
B425X	0160
B425Y(2S)	0160
B426(2S)	0160
B426-BL	0160
B426-R	0160
B426-Y	0160
B426A	0160
B426B	0160
B426BL(2S)	0160
B426C	0160
B426D	0160
B426E	0160
B426F	0160
B426G	0160
B426G-BL	0160
B426GN	0160
B426G-R	0160
B426G-Y	0160
B426J	0160
B426K	0160
B426L	0160
B426M	0160
B426OR	0160
B426R(2S)	0160
B426X	0160
B426Y(2S)	0160
B427(2S)	0004
B427-10	0229
B427-15	0102
B427-20	0348
B427-25	OEM
B427-30	OEM
B427-35	OEM
B427-40	OEM
B427-45	OEM
B427-50	OEM
B427-55	OEM
B427-60	OEM
B427-65	OEM
B427-70	OEM
B427A	0004
B427B	0004
B427C	0004
B427D	0004
B427E	0004
B427F	0004
B427G	0004
B427GN	0004
B427H	0004
B427J	0004
B427K	0004
B427L	0004
B427M	0004
B427OR	0004
B427R	0004
B427X	0004
B427Y	0004
B428(2S)	0004
B428A	0004
B428B	0004
B428C	0004
B428D	0004
B428E	0004
B428F	0004
B428G	0004
B428GN	0004
B428H	0004
B428J	0004
B428K	0004
B428L	0004
B428M	0004
B428OR	0004
B428R	0004
B428X	0004
B428Y	0004
B429	0280
B430(2S)	0040
B431(2S)	0164
B431-1	OEM
B431-2	OEM
B431F	OEM
B431S	OEM
B432(2S)	0975
B432-1	OEM
B432-2	OEM
B432-C	OEM
B432A	0975
B432B	0975
B432C	0975
B432D	0975
B432E	0975
B432F	OEM
B432F(2S)	0975
B432G	0975
B432GN	0975
B432H	0975
B432J	0975
B432K	0975
B432L	0975
B432M	0975
B432OR	0975
B432R	0975
B432S	OEM
B432X	0975
B432Y	0975
B433(2S)	5432
B433-1	OEM
B433-2	OEM
B433-C	OEM
B433F	OEM
B433S	OEM
B434(2S)	0925
B434-0	0925
B434-ORG	0925
B434-R(2S)	0925
B434-RED	0925
B434-Y(2S)	0925
B434-YEL	0925
B434G	0925
B434G-0	0925
B434G-R	0925
B434G-Y	0925
B434R	0925
B435(2S)	0925
B435-0(2S)	0925
B435-0	0925
B435-ORG	0925
B435-R(2S)	0925
B435-RED	0925
B435-Y(2S)	0925
B435-YEL	0925
B435G-0	0925
B435G-R	0925
B435G-Y	0925
B4350	0925
B435R	0925
B435RY	0925
B435Y	0925
B436	0050
B437	0050
B438	1062
B439(2S)	0004
B439A(2S)	0004
B439B	0004
B439C	0004
B439D	0004
B439E	0004
B439F	0004
B439G	0004
B439GN	0004
B439H	0004
B439J	0004
B439L	0004
B439M	0004
B439OR	0004
B439R	0004
B439X	0004
B439Y	0004
B440(2S)	0004
B440A	0004
B440B	0004
B440C	0004
B440E	0004
B440F	0004
B440G	0004
B440GN	0004
B440H	0004
B440K	0004
B440L	0004
B440M	0004
B440OR	0004
B440R	0004
B440X	0004
B440Y	0004
B441	0975
B441-1	OEM
B441-2	OEM
B441-C	OEM
B441F	OEM
B441S	OEM
B442	0975
B442-1	OEM
B442-2	OEM
B442-5	0277
B442-10	0277
B442-20	0289
B442-30	0296
B442-40	0296
B442-50	0302
B442-60	0302
B442-70	0250
B442-80	0250
B442-90	0250
B442-100	0250
B442-120	OEM
B442-C	OEM
B442F	OEM
B442S	OEM
B443(2S)	0004
B443-1	OEM
B443-2	OEM
B443-C	OEM
B443A(2S)	0004
B443B(2S)	0004
B443C	0004
B443D	0004
B443E	0004
B443F	OEM
B443F(2S)	0004
B443G	0004
B443GN	0004
B443H	0004
B443J	0004
B443K	0004
B443L	0004
B443M	0004
B443OR	0004
B443R	0004
B443S	OEM
B443X	0004
B443Y	0004
B444(2S)	0136
B444A(2S)	0136
B444B(2S)	0136
B444C	0136
B444D	0136
B444E	0136
B444F	0136
B444G	0136
B444GN	0136
B444H	0136
B444K	0136
B444L	0136
B444M	0136
B444OR	0136
B444R	0136
B444X	0136
B444Y	0136
B445(2S)	1010
B445A	1010
B445B	1010
B445C	1010
B445D	1010
B445E	1010
B445F	1010
B445G	1010
B445GN	1010
B445H	1010
B445J	1010
B445K	1010
B445L	1010
B445M	1010
B445OR	1010
B445R	1010
B445X	1010
B445Y	1010
B446(2S)	1010
B446A	1010
B446B	1010
B446C	1010
B446D	1010
B446E	1010
B446F	1010
B446G	1010
B446GN	1010
B446H	1010
B446J	1010
B446K	1010
B446L	1010
B446M	1010

DEVICE TYPE	REPL CODE
B446OR	1010
B446R	1010
B446X	1010
B446Y	1010
B447(2S)	0975
B447A	0975
B447B	0975
B447C	0975
B447D	0975
B447E	0975
B447F	0975
B447G	0975
B447GN	0975
B447H	0975
B447J	0975
B447K	0975
B447L	0975
B447OR	0975
B447R	0975
B447X	0975
B447Y	0975
B448(2S)	0222
B448A	0222
B448B	0222
B448C	0222
B448D	0222
B448E	0222
B448F	0222
B448G	0222
B448GN	0222
B448H	0222
B448K	0222
B448L	0222
B448M	0222
B448OR	0222
B448R	0222
B448X	0222
B448Y	0222
B449(2S)	0160
B449A	0160
B449B	0160
B449C	0160
B449D	0160
B449E	0160
B449F(2S)	0160
B449G	0160
B449GN	0160
B449H	0160
B449J	0160
B449K	0160
B449L	0160
B449M	0160
B449OR	0160
B449P(2S)	0160
B449PG	0160
B449X	0160
B449Y	0160
B450(2S)	0164
B450A(2S)	0164
B451(2S)	0164
B451-1	OEM
B451-2	OEM
B451-C	OEM
B451F	OEM
B451S	OEM
B452(2S)	0164
B452-1	OEM
B452-2	OEM
B452-C	OEM
B452A(2S)	0164
B452F	OEM
B452S	OEM
B453(2S)	0164
B453-1	OEM
B453-2	OEM
B453-C	OEM
B453F	OEM
B453S	OEM
B454(2S)	0164
B455(2S)	0164
B456(2S)	0164
B457(2S)	0164
B457A	0164
B457AC	0164
B457B	0164
B457C	0164
B457D	0164
B457E	0164
B457F	0164
B457G	0164
B457GN	0164
B457H	0164
B457J	0164
B457K	0164
B457L	0164
B457M	0164
B457OR	0164
B457R	0164
B457X	0164
B457Y	0164
B458(2S)	0222
B458A(2S)	0222
B458B	0222
B458BC	0222
B458BL	0222
B458C	0222
B458D	0222
B458E	0222
B458F	0222
B458G	0222
B458GN	0222
B458H	0222
B458J	0222
B458K	0222
B458L	0222
B458M	0222
B458OR	0222
B458R	0222
B458X	0222
B458Y	0222
B459(2S)	0004
B459-O	0004
B459A(2S)	0004
B459B(2S)	0004
B459C(2S)	0004
B459C-2	0004
B459D(2S)	0004
B459E	0004
B459F	0004
B459G	0004
B459GN	0004
B459H	0004
B459J	0004
B459K	0004
B459L	0004
B459M	0004
B459OR	0004
B459R	0004
B459X	0004
B459Y	0004
B460(2S)	0004
B460A(2S)	0004
B460B(2S)	0004
B460C	0004
B460D	0004
B460E	0004
B460F	0004
B460G	0004
B460GN	0004
B460H	0004
B460J	0004
B460K	0004
B460L	0004
B460M	0004
B460OR	0004
B460R	0004
B460X	0004
B460Y	0004
B461(2S)	0847
B461-1	OEM
B461-2	OEM
B461-C	OEM
B461A	0847
B461B	0847
B461BL	0847
B461C	0847
B461D	0847
B461E	0847
B461F	OEM
B461F(2S)	0847
B461G	0847
B461H	0847
B461J	0847
B461K	0847
B461L	0847
B461M	0847
B461OR	0847
B461R	0847
B461S	OEM
B461X	0847
B461Y	0847
B462(2S)	0222
B462-1	OEM
B462-2	OEM
B462-BL	0222
B462-BLU	0222
B462-BLU-G	0222
B462-C	OEM
B462-R	0222
B462-RED	0222
B462-RED-G	0222
B462-Y	0222
B462-YEL	0222
B462-YEL-G	0222
B462A	0222
B462B	0222
B462C	0222
B462D	0222
B462E	0222
B462F	OEM
B462F(2S)	0222
B462G	0222
B462G-BL	0222
B462GN	0222
B462G-R	0222
B462G-Y	0222
B462H	0222
B462K	0222
B462L	0222
B462M	0222
B462OR	0222
B462R	0222
B462S	OEM
B462X	0222
B462Y	0222
B463(1)	0222
B463(2S)	0222
B463-1	OEM
B463-2	OEM
B463-BL	0222
B463-BLU	0222
B463-BLU-G	0222
B463-C	OEM
B463-O	0222
B463-R	0222
B463-RED	0222
B463-RED-G	0222
B463-Y	0222
B463-YEL	0222
B463-YEL-G	0222
B463A	0222
B463B	0222
B463BL(2S)	0222
B463BLU	0222
B463BLU-G	0222
B463C	0222
B463D	0222
B463E(2S)	0222
B463F	OEM
B463F(2S)	0222
B463G	0222
B463G-BL	0222
B463GN	0222
B463G-R	0222
B463G-Y	0222
B463H	0222
B463J	0222
B463K	0222
B463L	0222
B463M	0222
B463OR	0222
B463R(2S)	5983
B463RED	0222
B463RED-G	0222
B463S	OEM
B463X	0222
B463XL	0222
B463Y(2S)	0222
B463YEL(2S)	0222
B463YEL-G(2S)	0222
B464(2S)	0975
B464A	0975
B464B	0975
B464C	0975
B464D	0975
B464E	0975
B464F	0975
B464G	0975
B464GN	0975
B464H	0975
B464J	0975
B464K	0975
B464M	0975
B464OR	0975
B464X	0975
B464Y	0975
B465(2S)	0975
B465A	0975
B465B	0975
B465C	0975
B465D	0975
B465E	0975
B465F	0975
B465G	0975
B465GN	0975
B465H	0975
B465J	0975
B465K	0975
B465L	0975
B465M	0975
B465R	0975
B465X	0975
B465Y	0975
B466(2S)	1010
B466A	1010
B466B	1010
B466C	1010
B466D	1010
B466E	1010
B466F	1010
B466G	1010
B466GN	1010
B466H	1010
B466J	1010
B466K	1010
B466L	1010
B466M	1010
B466OR	1010
B466R	1010
B466X	1010
B466Y	1010
B467(2S)	1010
B467A	1010
B467C	1010
B467D	1010
B467E	1010
B467F	1010
B467G	1010
B467GN	1010
B467H	1010
B467J	1010
B467K	1010
B467L	1010
B467M	1010
B467OR	1010
B467X	1010
B467Y	1010
B468(2S)	0975
B468-OR	0975
B468A(2S)	0975
B468B(2S)	0975
B468B-5	0975
B468C(2S)	0975
B468D(2S)	0975
B468E	0975
B468F	0975
B468G	0975
B468GN	0975
B468H	0975
B468J	0975
B468K	0975
B468L	0975
B468OR	0975
B468X	0975
B468Y	0975
B470(2S)	0004
B470A	0004
B470B	0004
B470C	0004
B470D	0004
B470E	0004
B470F	0004
B470G	0004
B470GN	0004
B470H	0004
B470J	0004
B470K	0004
B470L	0004
B470M	0004
B470OR	0004
B470R	0004
B470X	0004
B470Y	0004
B471(2S)	0160
B471-1	OEM
B471-2	0160
B471-2(2S)	0160
B471-C	OEM
B471-O	0160
B471A(2S)	0160
B471B(2S)	0160
B471C	0160
B471D	0160
B471E	0160
B471F	OEM
B471F(2S)	0160
B471G	0160
B471GN	0160
B471H	0160
B471J	0160
B471K	0160
B471L	0160
B471M	0160
B471OR	0160
B471R	0160
B471S	OEM
B471X	0160
B471Y	0160
B472(2S)	0604
B472-1	OEM
B472-2	OEM
B472-C	OEM
B472A	0604
B472B	0604
B472C	0604
B472D	0604
B472E	0604
B472F	OEM
B472F(2S)	0604
B472G	0604
B472GN	0604
B472H	0604
B472J	0604
B472K	0604
B472M	0604
B472OR	0604
B472S	OEM
B472X	0604
B472Y	0604
B473(2S)	0222
B473-1	OEM
B473-2	OEM
B473-C	OEM
B473A	0222
B473B	0222
B473C	0222
B473D(2S)	0222
B473E	0222
B473F	0222
B473F(2S)	0222
B473G	0222
B473GN	0222
B473H(2S)	0222
B473K	0222
B473L	0222
B473M	0222
B473S	OEM
B473X	0222
B473Y	0222
B474(2S)	1010
B474-2(2S)	1010
B474-3(2S)	1010
B474-4(2S)	1010
B474-6D(2S)	1010
B474-OR	1010
B474A	1010
B474B	1010
B474C	1010
B474D	1010
B474E	1010
B474F	1010
B474G	1010
B474GN	1010
B474H	1010
B474J	1010
B474K	1010
B474L	1010
B474M	1010
B474MP(2S)	1010
B474OR	1010
B474R	1010
B474S(2S)	1010
B474V4(2S)	1010
B474V10(2S)	1010
B474X	1010
B474Y(2S)	1010
B474YE1	1010
B474YEL	1010
B475(2S)	0004
B475A(2S)	0004
B475B(2S)	0004
B475C	0004
B475D(2S)	0004
B475E(2S)	0004
B475F(2S)	0004
B475G(2S)	0004
B475GN	0004
B475H	0004
B475J	0004
B475K	0004
B475L	0004
B475M	0004
B475OR	0004
B475P(2S)	0004
B475PL	0004
B475Q(2S)	0004
B475R	0004
B475X	0004
B475Y	0004
B476(2S)	0847
B476S	0847
B476W	0847
B477	0440
B477A	0440
B477C	0440
B477D	0440
B477E	0440
B477F	0440
B477G	0440
B477GN	0440
B477H	0440
B477J	0440
B477K	0440
B477L	0440
B477M	0440
B477OR	0440
B477R	0440
B477X	0440
B477Y	0440
B478	0437
B478BS	0437
B479	0437
B480	0437
B481(2S)	0222
B481-1	OEM
B481-2	OEM
B481-C	OEM
B481-OR	0222
B481A	0222
B481A2	OEM
B481B	0222
B481B2	OEM
B481C	0222
B481C2	OEM
B481D(2S)	0222
B481D2	OEM
B481E(2S)	0222
B481E2	OEM
B481F	0222
B481F2	OEM
B481G	0222
B481GN	0222
B481H	0222
B481J	0222
B481K	0222
B481L	0222
B481M	0222
B481OR	0222
B481R	0222
B481X	0222
B482(2S)	0004
B482A	0004
B482A2	OEM
B482B	0004
B482B2	OEM
B482C	0004
B482C2	OEM
B482D	0004
B482D2	OEM
B482E	0004
B482E2	OEM
B482F	OEM
B482F(2S)	0004
B482F2	OEM
B482G	0004
B482GN	0004
B482H	0004
B482J	0004
B482K	0004
B482L	0004
B482M	0004
B482S	OEM
B482V	0004
B482X	0004
B482Y	0004
B483(2S)	0604
B483-1	OEM
B483-2	OEM
B483F	OEM
B483S	OEM
B484(2S)	0604
B485(2S)	0604
B485B2	OEM
B485C2	OEM
B485D2	OEM
B486(2S)	0004
B486A	0004
B486B	0004
B486C	0004
B486D	0004
B486E	0004
B486F	0004
B486G	0004
B486GN	0004
B486H	0004
B486J	0004
B486K	0004
B486L	0004
B486M	0004
B486OR	0004
B486R	0004
B486X	0004
B486Y	0004
B491-1	OEM
B491-2	OEM
B491-C	OEM
B491F	OEM
B491S	OEM
B492(2S)	0847
B492-1	OEM
B492-C	OEM
B492A	0847
B492B(2S)	0847
B492C	0847
B492D	0847
B492E	0847
B492F	OEM
B492F(2S)	0847
B492G	0847
B492GN	0847
B492H	0847
B492J	0847
B492K	0847
B492L	0847
B492M	0847
B492OR	0847
B492R	0847
B492X	0847
B492Y	0847
B493	0164
B493-1	OEM
B493-2	OEM
B493-C	OEM
B493F	OEM
B493S	OEM
B493W	0164
B494(2S)	0164
B495(2S)	0164
B495A(2S)	0164
B495B	0164
B495C(2S)	0164
B495D(2S)	0164
B495E	0164
B495F	0164
B495G	0164
B495GN	0164
B495H	0164
B495J	0164
B495K	0164
B495L	0164
B495M	0164
B495OR	0164
B495R	0164
B495T(2S)	0164
B495X	0164
B495Y	0164
B496(2S)	0004
B496A	0004
B496B	0004
B496C	0004
B496D	0004
B496E	0004
B496F	0004
B496G	0004
B496GN	0004
B496H	0004
B496J	0004
B496K	0004
B496L	0004
B496M	0004
B496OR	0004
B496R	0004
B496X	0004
B496Y	0004
B497(2S)	0004
B497A	0004
B497B	0004
B497C	0004
B497D	0004
B497E	0004
B497F	0004
B497G	0004
B497GN	0004
B497H	0004
B497J	0004
B497K	0004
B497L	0004
B497M	0004
B497OR	0004
B497R	0004
B497X	0004
B497Y	0004
B498(2S)	0004
B498A	0004
B498B	0004
B498C	0004
B498D	0004
B498E	0004
B498F	0004
B498G	0004
B498GN	0004
B498H	0004
B498J	0004
B498K	0004
B498L	0004
B498M	0004
B498OR	0004
B498R	0004
B498X	0004
B498Y	0004
B500	OEM
B502(2S)	0905
B502-O	0905
B502-R	0905
B502-Y	0905
B502A	0905
B502A-O	0905
B502A-R	0905
B502A-Y	0905
B503(2S)	0854
B503-O	0854
B503-R	0854
B503-Y	0854
B503A	0854
B503A-O	0854
B503A-R	0854
B503A-Y	0854
B504	0126
B504A	0126
B505	0681
B506(2S)	3500
B506A(2S)	3500
B507(2S)	0681
B507E	0681
B508(2S)	0925
B509(2S)	0925
B510(2S)	0126
B510S(2S)	0126
B511(2S)	0925
B511-1	OEM
B511-1(2S)	0925
B511-2	OEM
B511-C	OEM
B511C(2S)	0925
B511D(2S)	0925
B511E	0925
B511F	OEM
B511S	OEM
B512(2S)	0925
B512-0	0925
B512-1	OEM
B512-2	OEM
B512-C	OEM
B512-O	0925
B512A(2S)	0925
B512F	OEM
B512P	0925
B512S	OEM
B513(2S)	0925
B513-1	OEM
B513-2	OEM
B513-C	OEM
B513A(2S)	0925
B513F	OEM
B513P	0925
B513Q	0925
B513R	0925
B513S	OEM
B514(2S)	0925
B515(2S)	0925
B516	0004
B516C(2S)	0004
B516CD(2S)	0004
B516D(2S)	0004
B516P(2S)	0004
B518	1681
B519	1681
B520	2013
B520-1	2013
B520-2	2013
B521	0854
B521-1	OEM
B521-2	OEM

DEVICE TYPE	REPL CODE
B521-C	OEM
B521F	OEM
B521S	OEM
B522	0854
B522-1	0854
B522-2	OEM
B522-893	0196
B522-C	OEM
B522F	OEM
B522S	OEM
B523	0925
B523-1	OEM
B523-2	OEM
B523-C	OEM
B523F	OEM
B523S	OEM
B524	0925
B525	0532
B525C	0532
B525D(2S)	0532
B525E(2S)	0532
B526	0681
B526C	0681
B527	0681
B527-062	0196
B528	0681
B528E	0681
B529	0925
B530	2013
B531	2013
B531-1	OEM
B531-2	OEM
B531-C	OEM
B531-O	2013
B531-R	2013
B531-Y	2013
B531F	OEM
B531S	OEM
B532(2S)	2013
B532-1	OEM
B532-2	OEM
B532F	OEM
B532S	2013
B533	0854
B533-O	0854
B533-C	OEM
B533O	0854
B533Y	0854
B534(2S)	0004
B534(A)	0004
B535(2S)	0164
B535G	0164
B536	0681
B536K	0681
B536L	0681
B536M	0681
B537	0681
B537K	0681
B537L	0681
B538	0604
B539(2S)	1598
B539A	1598
B539B	1598
B539C	1598
B540	0164
B541(2S)	3500
B541-1	OEM
B541-2	OEM
B541-C	OEM
B541F	OEM
B541S	OEM
B542	0006
B542-1	OEM
B542-2	OEM
B542-C	OEM
B542D	0006
B542F	OEM
B542S	OEM
B543-1	OEM
B543-2	OEM
B543-C	OEM
B543F	OEM
B543S	OEM
B544(2S)	0532
B544D(2S)	0532
B544E(2S)	0532
B544E(JAPAN)	0532
B544F(2S)	0532
B544P1(2S)	0532
B544P2	0532
B546	1648
B546A	1648
B546AK	1648
B546AK(JAPAN)	1648
B546AL	1648
B546C	1648
B546E	1648
B546EX	1648
B546I	1648
B546K	1648
B546L	1648
B547	1648
B547A	1648
B548	0525
B549	0525
B549P	0525
B549Q	0525
B549R	0525
B550(2S)	3500
B551-1	OEM
B551-2	OEM
B551F	OEM
B551H	0905
B551S	OEM
B552	1598
B552-1	OEM
B552-2	OEM
B552-C	OEM
B552F	OEM
B552S	OEM
B553	0718
B553-1	OEM
B553-2	OEM
B553-C	OEM
B553AC	OEM
B553F	OEM
B553S	OEM
B554(2S)	1598
B555(2S)	2013
B555-O	2013
B555-R	2013
B555R	2013
B556	2013
B556-O	2013
B556-R	2013
B557	2013
B557-O	2013
B557-R	2013
B558	2013
B558-O	2013
B558-R	2013
B559	0460
B560(2S)	0477
B561(2S)	0436
B561B(2S)	0436
B561B(JAPAN)	0436
B561C(2S)	0436
B561C(JAPAN)	0436
B562(2S)	0532
B562-C	0532
B562AC	OEM
B562B(2S)	0532
B562C(2S)	0532
B563	0905
B564	0532
B564(1)K	0532
B564(1)L	0532
B564L	0532
B564M	0532
B564Q	0532
B564R	0532
B564S	0532
B565	0164
B565A	0164
B565B	0164
B565C	0164
B565D	0164
B565E	0164
B565F	0164
B565G	0164
B565GN	0164
B565H	0164
B565J	0164
B565K	0164
B565L	0164
B565M	0164
B565OR	0164
B565R	0164
B565X	0164
B565Y	0164
B566	0681
B566A	0681
B566AKB	0681
B566AKC	0681
B566B	0681
B566C	0681
B566D	0681
B566K	0681
B566KB	0681
B566KC	0681
B567	3543
B568	1911
B568-C	1911
B568B	1911
B568C	1911
B568D	1911
B568HA	OEM
B569	0460
B570	0460
B571	0460
B572	1051
B573	1051
B574	1051
B575	0460
B576	0460
B577	0460
B578	1199
B579	1199
B580	1199
B581	1199
B582	1199
B583	1199
B584	3530
B585	2277
B586	2277
B587	2432
B587-10	0124
B587-15	0102
B587-20	0348
B587-25	OEM
B587-30	OEM
B587-35	OEM
B587-40	OEM
B587-45	OEM
B587-50	OEM
B587-55	OEM
B587-60	OEM
B587-65	OEM
B587-70	OEM
B587-80	OEM
B587-85	OEM
B587-90	OEM
B587-95	OEM
B587-100	OEM
B587-110	OEM
B587-120	OEM
B587-130	OEM
B587-140	OEM
B587-150	OEM
B587-160	OEM
B587-170	OEM
B587-180	OEM
B587-190	OEM
B587-200	OEM
B588	2432
B588-10	OEM
B588-15	OEM
B588-20	0348
B588-25	OEM
B588-30	OEM
B588-35	OEM
B588-40	OEM
B588-45	OEM
B588-50	OEM
B588-55	OEM
B588-60	OEM
B588-65	OEM
B588-70	OEM
B588-75	OEM
B588-80	OEM
B588-85	OEM
B588-90	OEM
B588-95	OEM
B588-96	OEM
B588-100	OEM
B588-110	OEM
B588-120	OEM
B588-130	OEM
B588-140	OEM
B588-150	OEM
B588-160	OEM
B588-170	OEM
B588-180	OEM
B588-190	OEM
B588-200	OEM
B589	2432
B595	1369
B595-Y	1369
B596	0854
B596-0	0854
B596-O	0854
B596R	0854
B596Y	0854
B598	3812
B598E	3812
B598F	3812
B600	3500
B601	0602
B601-1006	0050
B601-1007	0050
B601-1008	0050
B601-1009	0004
B601-1010	0164
B601-1011	0143
B601-1012	0015
B601K(JAPAN)	0602
B604	0854
B605	0532
B605K	0532
B606	0342
B607	0164
B610-5	OEM
B610-10	OEM
B610-20	OEM
B610-30	OEM
B610-40	OEM
B610-50	OEM
B610-60	OEM
B610-70	OEM
B610-80	OEM
B610-90	OEM
B610-100	OEM
B610-120	OEM
B611	2013
B611A	2013
B612	1598
B612A	1598
B613	3606
B616	3571
B617	3571
B617A	3571
B618	3571
B618A	3571
B619	0164
B620	0150
B621	0532
B621A	0532
B621A(JAPAN)	0532
B621C	0532
B621N	0532
B621NC	0532
B621Q	0532
B621R	0532
B621S	0532
B622	0439
B626	1681
B627	3298
B628	1648
B628A	1648
B631	0525
B631K	0525
B631KE	0525
B632	0460
B632K	0460
B632K(JAPAN)	0460
B633	5468
B633E	5468
B633P	5468
B634Y	3500
B635	0004
B636	0004
B637KC	0648
B637KD	0648
B637KE	0648
B638H	2432
B639H	2432
B640	0037
B641	0359
B641-Q	0359
B641-Q(JAPAN)	0359
B641-QR	0359
B641-R	0359
B641P	0359
B641Q	0359
B641R	0359
B641S	0359
B642	0825
B642-O	0825
B642-Q	0825
B642-Q(JAPAN)	0825
B642-QR	0825
B642-R	0825
B642-R(JAPAN)	0825
B642-S	0825
B642P	0825
B642Q	0825
B642R	0825
B642S	0825
B643	0825
B643-O	0825
B643-Q	0825
B643-Q(JAPAN)	0825
B643-R	0825
B643-R(JAPAN)	0825
B643P	0825
B643Q	0825
B643QRS	0825
B643R	0825
B643S	0825
B644	0825
B644-Q	0825
B644-R	0825
B644-R(JAPAN)	0825
B644Q	0825
B644QR	0825
B644R	0825
B644S	0825
B645E	1598
B646	0477
B646A	0477
B646AB	0477
B646B	0477
B646C	0477
B646D	0477
B647	0477
B647A	0477
B647AB	0477
B647AC	0477
B647B	0477
B647C	0477
B647D	0477
B648B	0525
B648C	0525
B649	0525
B649A	0525
B649AB	0525
B649AC	0525
B649AD	0525
B649B	0525
B649C	0525
B649D	0525
B650H	2432
B653	3500
B654	3500
B654-5	0106
B654-10	0106
B654-20	0106
B654-30	0106
B654-40	0106
B654-50	2010
B654-60	2010
B654-70	2400
B654-80	2400
B654-90	0787
B654-100	0787
B654-120	OEM
B655	1598
B656	1598
B668	0602
B669	0602
B673	0602
B674	0602
B675	0602
B676-10C-2	OEM
B676-10M-2	OEM
B676R-10M2	OEM
B679	0602
B683	0854
B683C	0854
B683D	0854
B684	0532
B686	0859
B688	3111
B688-O	3111
B688R	3111
B689	0681
B690	0681
B692X13	0143
B696	3500
B697	1598
B697K	1598
B698	0006
B698F	0006
B699	0925
B699Q	0925
B700	3581
B702	3581
B703	0681
B703A	0681
B703Q	0681
B705	3581
B705A	3581
B705B	3581
B707	0854
B707K(JAPAN)	0854
B708	0854
B709	1742
B709A	1742
B711	0602
B712	0602
B713	5478
B714	0847
B715	0648
B716	0648
B716A	0648
B716AD	0648
B716D	0648
B716E	0648
B717B	5479
B717C	5479
B717D	5479
B719	1648
B721	0037
B723	1598
B726	1603
B726R	1603
B726S	1603
B726T	1603
B727(JAPAN)	OEM
B727K(JAPAN)	OEM
B731	0525
B733	0532
B733K	0532
B733O	0532
B733U	0532
B734	0532
B734K	0532
B734U	0532
B738B	0477
B738C	0477
B739B	0532
B739C	0532
B740	0477
B740-3	0477
B740B	0477
B740C	0477
B741	0477
B741B	0477
B741C	0477
B743Q	0525
B744	1613
B744A	1613
B744Q	1613
B744R	1613
B744S	1613
B745	4139
B745A	4139
B745AS	4139
B745S	4139
B745T	4139
B747	0854
B750	0602
B750A	0602
B751	0602
B751A	0602
B753	1369
B754	5478
B755	3581
B759	0648
B760	0718
B760A	0718
B761	3165
B761A	3165
B761P	3165
B761Q	3165
B761R	3165
B762	0718
B762A	0718
B762P	0718
B762Q	0718
B763	5478
B763A	5478
B764	0477
B765K	1968
B772	0460
B772Q	0460
B774	1613
B774Q	1613
B774R	1613
B774S	1613
B774S(JAPAN)	1613
B774T	1613
B775	5478
B776	5478
B788	0648
B788Q	0648
B788S	0648
B788T	0648
B788U	0648
B790	0148
B791(JAPAN)	OEM
B793	0477
B793A	0477
B793Q	0477
B793R	0477
B794	2897
B794K	2897
B794L	2897
B794L(JAPAN)	2897
B794M	2897
B795L	2897
B796	3606
B808	0006
B808F	0006
B808G1	0006
B810	0532
B811	0477
B812	0859
B812A	0859
B816	3111
B817	3111
B819	4348
B819Q	4348
B819R	4348
B822	4104
B822Q	4104
B823	0854
B824R	1308
B825	0854
B825R	0854
B826	1308
B827	5478
B828	5478
B828S	5478
B833	4104
B833(Q)	4104
B834	0854
B834-O(JAPAN)	0854
B835	0532
B836	3498
B836L	3498
B837	2018
B837B	3498
B837C	3498
B837D	3498
B837LB	3498
B837LC	3498
B837LD	3498
B838B	3498
B839B	3498
B840B	3498
B840C	3498
B840LB	3498
B841C	3498
B841LB	3498
B841LC	3498
B849	5478
B849A	5478
B855A	5492
B855B	5492
B856B	0718
B856C	0718
B857B	2923
B857C	2923
B857D	2923
B858	0718
B858(JAPAN)	2923
B858B	2923
B858C	2923
B858D	2923
B859B	0681
B859C	0681
B860	1369
B861	1648
B861B	1648
B861C	1648
B861D	1648
B863	5478
B865	0635
B867	0718
B868	0681
B869	0718
B870	0718
B882	2237
B882(JAPAN)	2237
B883	0073
B884	0602
B885(JAPAN)	0602
B886	2237
B892S(JAPAN)	0477
B925Q(JAPAN)	OEM
B940P	1648
B941	2923
B941B	2923
B941P	2923
B941Q	2923
B955(K)	1968
B963L	OEM
B994	1308
B995	0718
B996	3165
B1001R	5153
B1012(K)	OEM
B1012/2/16K	OEM
B1012/2/32K	OEM
B1012K	OEM
B1013B	OEM
B1014	OEM
B1015	5475
B1016	4157
B1017	0160
B1018	5680
B1019	1308
B1020/32K	OEM
B1020/64K	OEM
B1020/96K	OEM
B1020/128K	OEM
B1022	0211
B1022-1	0004
B1023	OEM
B1025	OEM
B1026	OEM
B1058	0211
B1058-1	0004
B1085	0160
B1093	OEM
B1093L	OEM
B1093Q	OEM
B1110	0160
B1135R	4125
B1151	0460
B1151A	0160
B1151B	0160
B1152	0160
B1152A	0160
B1152B	0160
B1154	0211
B1154-1	0004
B1165S	OEM
B1178	0975
B1181	0160
B1181MP	OEM
B1191	OEM
B1201	0015
B1202	0609
B1202V	0609
B1204	0071
B1210B60C1000-700	OEM
B1210L	OEM
B1215	0085
B1220B125C1000-700	OEM
B1220L	OEM
B1235	OEM
B1240B250C1000-700	OEM
B1240L	OEM
B1262	OEM
B1272	OEM
B1274	0160
B1274A	0160
B1274B	0160
B1274R	0925
B1280B500C1000-700	OEM
B1280L	OEM
B1310B60C2300-1400	OEM
B1310M	OEM
B1318	OEM
B1318L	OEM
B1320B125C2300-1400	OEM
B1320M	OEM
B1333	OEM
B-1338	0016
B1340B250C2300-1400	OEM
B1340M	OEM
B1359	OEM
B1364	0211
B1368	0160
B1368A	0160
B1368B	0160
B1368C	0160
B1368D	0160
B1368E	0160
B1368F	0160
B1380B500C2300-1400	OEM
B1380M	OEM
B1382	OEM
B1411-1	OEM
B1411-2	OEM
B1411-C	OEM
B1411F	OEM
B1411S	OEM
B1412-1	OEM
B1412-2	OEM
B1412-C	OEM

DEVICE TYPE	REPL CODE
B1412F	OEM
B1412S	OEM
B1413-1	OEM
B1413-2	OEM
B1413-C	OEM
B1413F	OEM
B1413S	OEM
B1420	OEM
B-1421	0016
B1421-1	OEM
B1421-2	OEM
B1421-C	OEM
B1421F	OEM
B1421S	OEM
B1422-1	OEM
B1422-2	OEM
B1422-C	OEM
B1422F	OEM
B1422S	OEM
B1423-1	OEM
B1423-2	OEM
B1423-C	OEM
B1423F	OEM
B1423S	OEM
B-1426	0037
B1431-1	OEM
B1431-2	OEM
B1431-C	OEM
B1431F	OEM
B1431S	OEM
B1432-1	OEM
B1432-2	OEM
B1432-C	OEM
B1432F	OEM
B1432S	OEM
B-1433	0016
B1433-1	OEM
B1433-2	OEM
B1433-C	OEM
B1433F	OEM
B1433S	OEM
B1441-1	OEM
B1441-2	OEM
B1441-C	OEM
B1441F	OEM
B1441S	OEM
B1442-1	OEM
B1442-2	OEM
B1442-C	OEM
B1442F	OEM
B1442S	OEM
B1443-1	OEM
B1443-2	OEM
B1443-C	OEM
B1443F	OEM
B1443S	OEM
B1451-1	OEM
B1451-2	OEM
B1451-C	OEM
B1451F	OEM
B1451S	OEM
B1452-1	OEM
B1452-2	OEM
B1452-C	OEM
B1452S	OEM
B1453-1	OEM
B1453-2	OEM
B1453-C	OEM
B1453F	OEM
B1453S	OEM
B1461-1	OEM
B1461-2	OEM
B1461-C	OEM
B1461F	OEM
B1461S	OEM
B1462-1	OEM
B1462-2	OEM
B1462-C	OEM
B1462F	OEM
B1462S	OEM
B1463-1	OEM
B1463-2	OEM
B1463-C	OEM
B1463F	OEM
B1463S	OEM
B1471-1	OEM
B1471-2	OEM
B1471-C	OEM
B1471F	OEM
B1471S	OEM
B1472-1	OEM
B1472-2	OEM
B1472-C	OEM
B1472F	OEM
B1472S	OEM
B1473-1	OEM
B1473-2	OEM
B1473-C	OEM
B1473F	OEM
B1473S	OEM
B1481-1	OEM
B1481-2	OEM
B1481-C	OEM
B1481F	OEM
B1482-1	OEM
B1482-2	OEM
B1482-C	OEM
B1482F	OEM
B1482S	OEM
B1483-2	OEM
B1483-C	OEM
B1483F	OEM
B1483S	OEM
B1491-1	OEM
B1491-2	OEM
B1491-C	OEM
B1491F	OEM
B1491S	OEM
B1492-1	OEM
B1492-2	OEM
B1492-C	OEM
B1492F	OEM
B1492S	OEM
B1493-1	OEM
B1493-2	OEM
B1493-C	OEM
B1493S	OEM
B-1501U	0015
B-1511	0160
B1511-1	OEM
B1511-2	OEM
B1511-C	OEM
B1511F	OEM
B1511S	OEM
B1512-1	OEM
B1512-2	OEM
B1512-C	OEM
B1512F	OEM
B1512S	OEM
B1513-1	OEM
B1513-2	OEM
B1513-C	OEM
B1513F	OEM
B1513S	OEM
B1521-1	OEM
B1521-2	OEM
B1521-C	OEM
B1521F	OEM
B1521S	OEM
B1522-1	OEM
B1522-2	OEM
B1522-C	OEM
B1522F	OEM
B1522S	OEM
B1523-1	OEM
B1523-2	OEM
B1523-C	OEM
B1523F	OEM
B1523S	OEM
B1531-1	OEM
B1531-2	OEM
B1531-C	OEM
B1531F	OEM
B1531S	OEM
B1532-1	OEM
B1532-2	OEM
B1532F	OEM
B1532S	OEM
B1533-1	OEM
B1533-2	OEM
B1533-C	OEM
B1533F	OEM
B1533S	OEM
B1541-1	OEM
B1541-2	OEM
B1541-C	OEM
B1541S	OEM
B1542-1	OEM
B1542-2	OEM
B1542-C	OEM
B1542F	OEM
B1542S	OEM
B1543-1	OEM
B1543-2	OEM
B1543-C	OEM
B1543F	OEM
B1543S	OEM
B1551-1	OEM
B1551-2	OEM
B1551-C	OEM
B1551F	OEM
B1551S	OEM
B1552-1	OEM
B1552-2	OEM
B1552-C	OEM
B1552F	OEM
B1552S	OEM
B1553-1	OEM
B1553-2	OEM
B1553-C	OEM
B1553F	OEM
B1553S	OEM
B1561-1	OEM
B1561-2	OEM
B1561-C	OEM
B1561F	OEM
B1561S	OEM
B1562-1	OEM
B1562-2	OEM
B1562F	OEM
B1562S	OEM
B1563-1	OEM
B1563-2	OEM
B1563-C	OEM
B1563F	OEM
B1563S	OEM
B1571-1	OEM
B1571-2	OEM
B1571F	OEM
B1571S	OEM
B1572-C	OEM
B1573-1	OEM
B1573-2	OEM
B1573-C	OEM
B1573F	OEM
B1573S	OEM
B1581-1	OEM
B1581-2	OEM
B1581-C	OEM
B1581F	OEM
B1581S	OEM
B1582-1	OEM
B1582-2	OEM
B1582-C	OEM
B1582F	OEM
B1582S	OEM
B1583-1	OEM
B1583-2	OEM
B1583-C	OEM
B1583F	OEM
B1583S	OEM
B1591-1	OEM
B1591-2	OEM
B1591-C	OEM
B1591F	OEM
B1591S	OEM
B1592-1	OEM
B1592-2	OEM
B1592-C	OEM
B1592F	OEM
B1592S	OEM
B1593-1	OEM
B1593-2	OEM
B1593-C	OEM
B1593F	OEM
B-1599	0133
B-1666	0016
B1670C-01	OEM
B1670C-01T-1S	OEM
B1670C-02	OEM
B1670C-02T-1S	OEM
B1670C-03	OEM
B1670C-03T-1S	OEM
B1670C-04	OEM
B1670C-04T-1S	OEM
B1670M-01	OEM
B1670M-01T-1S	OEM
B1670M-02	OEM
B1670M-02T-1S	OEM
B1670M-03	OEM
B1670M-03T-1S	OEM
B1670M-04	OEM
B1670M-04T-1S	OEM
B-1695	0460
B-1702	0133
B1760	0211
B1780	0164
B1780A	0164
B1780B	0164
B1780C	0164
B1780D	0164
B1780E	0164
B1780F	0164
B1780G	0164
B1780GN	0164
B1780H	0164
B1780J	0164
B1780K	0164
B1780L	0164
B1780M	0164
B1780OR	0164
B1780R	0164
B1780X	0164
B1780Y	0164
B1785	0211
B-1790	0161
B-1808	0244
B-1823	0042
B-1842	0016
B1851	0004
B1860	0211
B-1872	0016
B-1881U	0015
B-1882U	0015
B1904	0160
B1906B40C1500-1000	OEM
B1906L	OEM
B-1910	0111
B1912B80C1500-1000	OEM
B1912L	OEM
B1913	OEM
B-1914	0160
B1914	0160
B2000	OEM
B2090	0012
B2167-55(A)	OEM
B2404C	0023
B2404D	0023
B2406C	0023
B2406D	0023
B2411-1	OEM
B2411-2	OEM
B2411-C	OEM
B2411F	OEM
B2411S	OEM
B2412-1	OEM
B2412-2	OEM
B2412-C	OEM
B2412F	OEM
B2412S	OEM
B2413-1	OEM
B2413-2	OEM
B2413-C	OEM
B2413F	OEM
B2413S	OEM
B2421-1	OEM
B2421-2	OEM
B2421-C	OEM
B2421F	OEM
B2421S	OEM
B2422-1	OEM
B2422-2	OEM
B2422-C	OEM
B2422F	OEM
B2422S	OEM
B2423-1	OEM
B2423-2	OEM
B2423-C	OEM
B2423F	OEM
B2431-1	OEM
B2431-2	OEM
B2431F	OEM
B2431S	OEM
B2432-1	OEM
B2432-2	OEM
B2432-C	OEM
B2432F	OEM
B2432S	OEM
B2433-1	OEM
B2433-2	OEM
B2433-C	OEM
B2433F	OEM
B2433S	OEM
B2441-1	OEM
B2441-C	OEM
B2441F	OEM
B2441S	OEM
B2442-1	OEM
B2442-2	OEM
B2442-C	OEM
B2442F	OEM
B2442S	OEM
B2443-1	OEM
B2443-2	OEM
B2443-C	OEM
B2443F	OEM
B2443S	OEM
B2451-1	OEM
B2451-2	OEM
B2451-C	OEM
B2451F	OEM
B2451S	OEM
B2452-1	OEM
B2452-2	OEM
B2452-C	OEM
B2452F	OEM
B2452S	OEM
B2453-1	OEM
B2453-2	OEM
B2453-C	OEM
B2453F	OEM
B2453S	OEM
B2491-1	OEM
B2491-2	OEM
B2491-C	OEM
B2491F	OEM
B2491S	OEM
B2492-1	OEM
B2492-2	OEM
B2492-C	OEM
B2492F	OEM
B2492S	OEM
B2493-1	OEM
B2493-2	OEM
B2493-C	OEM
B2493F	OEM
B2493S	OEM
B2511-1	OEM
B2511-2	OEM
B2511-C	OEM
B2511F	OEM
B2511S	OEM
B2512-1	OEM
B2512-C	OEM
B2512F	OEM
B2512S	OEM
B2513-1	OEM
B2513-2	OEM
B2513-C	OEM
B2513F	OEM
B2513S	OEM
B2521-1	OEM
B2521-2	OEM
B2521-C	OEM
B2521F	OEM
B2521S	OEM
B2522-1	OEM
B2522-2	OEM
B2522-C	OEM
B2522F	OEM
B2522S	OEM
B2523-1	OEM
B2523-2	OEM
B2523-C	OEM
B2523F	OEM
B2523S	OEM
B2531-1	OEM
B2531-2	OEM
B2531-C	OEM
B2531F	OEM
B2531S	OEM
B2532-1	OEM
B2532-2	OEM
B2532-C	OEM
B2532F	OEM
B2532S	OEM
B2533-2	OEM
B2533-C	OEM
B2533F	OEM
B2533S	OEM
B2541-1	OEM
B2541-2	OEM
B2541-C	OEM
B2541F	OEM
B2541S	OEM
B2542-1	OEM
B2542-2	OEM
B2542-C	OEM
B2542F	OEM
B2542S	OEM
B2543-1	OEM
B2543-2	OEM
B2543-C	OEM
B2543F	OEM
B2543S	OEM
B2551-1	OEM
B2551-2	OEM
B2551-C	OEM
B2551F	OEM
B2551S	OEM
B2552-1	OEM
B2552-C	OEM
B2552F	OEM
B2552S	OEM
B2553-1	OEM
B2553-2	OEM
B2553-C	OEM
B2553F	OEM
B2553S	OEM
B2561-1	OEM
B2561-2	OEM
B2561-C	OEM
B2561F	OEM
B2561S	OEM
B2562-1	OEM
B2562-2	OEM
B2562-C	OEM
B2562F	OEM
B2562S	OEM
B2563-1	OEM
B2563-2	OEM
B2563-C	OEM
B2563F	OEM
B2563S	OEM
B2571-1	OEM
B2571-2	OEM
B2571-C	OEM
B2571F	OEM
B2571S	OEM
B2572-1	OEM
B2572-2	OEM
B2572-C	OEM
B2572F	OEM
B2572S	OEM
B2573-1	OEM
B2573-2	OEM
B2573-C	OEM
B2573F	OEM
B2573S	OEM
B2581-1	OEM
B2581-2	OEM
B2581-C	OEM
B2581F	OEM
B2581S	OEM
B2582-1	OEM
B2582-2	OEM
B2582-C	OEM
B2582F	OEM
B2582S	OEM
B2583-1	OEM
B2583-2	OEM
B2583-C	OEM
B2583F	OEM
B2583S	OEM
B2591-1	OEM
B2591-2	OEM
B2591-C	OEM
B2591F	OEM
B2591S	OEM
B2592-1	OEM
B2592-2	OEM
B2592-C	OEM
B2592F	OEM
B2592S	OEM
B2593-1	OEM
B2593-2	OEM
B2593-C	OEM
B2593F	OEM
B2593S	OEM
B2611-1	OEM
B2611-2	OEM
B2611-C	OEM
B2611F	OEM
B2611S	OEM
B2612-1	OEM
B2612-2	OEM
B2612-C	OEM
B2612F	OEM
B2612S	OEM
B2613-1	OEM
B2613-2	OEM
B2613-C	OEM
B2613F	OEM
B2613S	OEM
B2620	0102
B2621-1	OEM
B2621-2	OEM
B2621-C	OEM
B2621F	OEM
B2621S	OEM
B2622-1	OEM
B2622-2	OEM
B2622-C	OEM
B2622F	OEM
B2622S	OEM
B2623-1	OEM
B2623-2	OEM
B2623-C	OEM
B2623F	OEM
B2631-1	OEM
B2631-C	OEM
B2631F	OEM
B2631S	OEM
B2632-1	OEM
B2632-2	OEM
B2632-C	OEM
B2632F	OEM
B2632S	OEM
B2633-1	OEM
B2633-2	OEM
B2633-C	OEM
B2633F	OEM
B2633S	OEM
B2641-1	OEM
B2641-2	OEM
B2641F	OEM
B2641S	OEM
B2804D	0023
B2806B40C900-600	OEM
B2812B80C900-600	OEM
B3015M	0017
B3030(2S)	0004
B3045	0236
B3046	0236
B3141	OEM
B3142	OEM
B3143	OEM
B3144	OEM
B3145	OEM
B3146	OEM
B3147	OEM
B3148	OEM
B3149	OEM
B3161	OEM
B3162	OEM
B3163	OEM
B3224E	0211
B3240	0164
B3240A	0164
B3240B	0164
B3240C	0164
B3240D	0164
B3240E	0164
B3240F	0164
B3240G	0164
B3240GN	0164
B3240H	0164
B3240J	0164
B3240K	0164
B3240L	0164
B3240M	0164
B3240OR	0164
B3240R	0164
B3240X	0164
B3240Y	0164
B3244	0280
B3465	0698
B3466	0423
B3468	0698
B3515	OEM
B3531	0560
B3533	0560
B3537	0560
B3538	0560
B3539	0423
B3540	0560
B3541	0560
B3542	0560
B3543	0423
B3544	0423
B3547	0042
B3548	0042
B3550	0042
B3551	0042
B3570	0560
B3576	0560
B3577	0042
B3578	0042
B3580	0042
B3584	0042
B3585	0042
B3586	0042
B3588	0042
B3589	0042
B3606	0560
B3607	0560
B3608	0560
B3609	0560
B3610	0560
B3611	0560
B3612	0560
B3613	0560
B3614	0560
B3746	0086
B3747	0560
B3748	0560
B3750	0560
B3783A	0164
B3783B	0164
B3783C	0164
B3783D	0164
B3783E	0164
B3783F	0164
B3783G	0164
B3783GN	0164
B3783H	0164
B3783J	0164
B3783K	0164
B3783L	0164
B3783M	0164
B3783OR	0164
B3783R	0164
B3783X	0164
B3783Y	0164
B3812A	0164
B3812B	0164
B3812C	0164
B3812D	0164
B3812F	0164
B3812G	0164
B3812GN	0164
B3812H	0164
B3812J	0164
B3812K	0164
B3812L	0164
B3812M	0164
B3812OR	0164
B3812R	0164
B3812X	0164
B3812Y	0164
B3813A	0164
B3813B	0164
B3813C	0164
B3813D	0164
B3813E	0164
B3813F	0164
B3813G	0164
B3813GN	0164
B3813H	0164
B3813J	0164
B3813K	0164
B3813L	0164
B3813M	0164
B3813OR	0164
B3813R	0164
B3813X	0164
B3813Y	0164
B4002-0289	OEM
B4006R	OEM
B4008R	OEM
B4010R	OEM
B4012-0241	OEM
B4012R	OEM
B4014R	OEM
B4016R	OEM
B4018R	OEM
B4020R	OEM
B4022R	OEM
B4024R	OEM
B4026R	OEM
B4028R	OEM
B4030R	OEM
B4151	0211
B4151M	OEM
B4302	0023
B4304	0023
B4304V	0023
B4310C	OEM
B4340	0854
B4402	0023
B4404	0023
B4406	0023
B4440D	0164
B4631	0222
B5000	0042
B5000-GRN	0042
B5000-RED	0042
B5000-YEL	0042
B5001	0560
B5001-GRN	0042
B5001-ORG	0042
B5001-RED	0042
B5001-YEL	0042
B5002	0042
B5006R	OEM
B5008R	OEM
B5010R0R	OEM
B5012R	OEM
B5014R	OEM
B5016R	OEM
B5018R	OEM
B5020	0103
B5020R	OEM

If replacement code is OEM, contact original manufacturer for replacement.

DEVICE TYPE	REPL CODE
B5021	0042
B5022	0042
B5022R	OEM
B5024R	OEM
B5026R	OEM
B5028R	OEM
B5030R	OEM
B5031	0042
B5032	0042
B-6001	0325
B-6002	0002
B6101C054	OEM
B6101C058	OEM
B6101C084	OEM
B6201/3	OEM
B7513	0280
B8313	OEM
B8507CK	OEM
B8555S	OEM
B8606	OEM
B9096	OEM
B9426	0144
B10060	OEM
B10061	OEM
B10062	OEM
B10063	OEM
B10064	0085
B10065	OEM
B10066	OEM
B10067	OEM
B10068	OEM
B10069	0085
B10142	0975
B10142A	0975
B10142B	0975
B10143	0975
B10143A	0975
B10143B	0975
B10144	OEM
B10144A	OEM
B10144B	OEM
B10162	0160
B10163	0160
B10474	0160
B10475	0160
B10912	0160
B10913	0160
B17307	0103
B35016	0577
B36564	0015
B-75583-I02	0016
B81004	0110
B84009	OEM
B102000	0975
B102001	0975
B102002	0975
B102003	0975
B103000	0975
B103001	0975
B103002	0975
B103003	0975
B103004	0975
B113000	0604
B113000-BRN	0604
B113000-ORG	0604
B113000-RED	0604
B133001	0561
B133002	0561
B133003	0561
B133004	0561
B133005	0561
B133006	0561
B133007	0561
B133008	0561
B143000	0560
B143001	0560
B143003	0560
B143004	0042
B143009	0560
B143010	0560
B143011	0042
B143012	0042
B143015	0560
B143016	0560
B143018	0042
B143019	0042
B143024	0560
B143025	0560
B143026	0042
B143027	0042
B145000	0939
B170000	0103
B170000-B	0065
B170000-BLK	0620
B170000-BRN	0620
B170000-ORG	0103
B170000-ORN	0103
B170000-R	0103
B170000-RED	0103
B170000-Y	0103
B170000-YEL	0130
B170000BLK	0103
B170000BRN	0103
B170001	0103
B170001-B	0103
B170001-BLK	0103
B170001-BRN	0103
B170001-O	0103
B170001-ORG	0103
B170001-R	0103
B170001-RED	0103
B170001-Y	0103
B170001-YEL	0130
B170001BLK	0103
B170001BRN	0103
B170002	0103
B170002-BLK	0103
B170002-BRN	0103
B170002-O	0103
B170002-ORG	0103
B170002-R	0103
B170002-RED	0103
B170002-Y	0103
B170002-YEL	0130
B170003	0103
B170003-B	0065
B170003-BLK	0130
B170003-BRN	0130
B170003-O	0103
B170003-ORG	0130
B170003-R	0103
B170003-RED	0130
B170003-Y	0103
B170003-YEL	0130
B170004	0103
B170004-B	0103
B170004-BLK	0130
B170004-BRN	0130
B170004-O	0130
B170004-ORG	0130
B170004-R	0103
B170004-RED	0130
B170004-Y	0103
B170004-YEL	0130
B170005	0103
B170005-B	0103
B170005-BLK	0130
B170005-BRN	0130
B170005-O	0103
B170005-ORG	0130
B170005-R	0103
B170005-RED	0130
B170005-Y	0103
B170005-YEL	0130
B170006	0103
B170006-B	0065
B170006-BLK	0130
B170006-BRN	0130
B170006-O	0074
B170006-ORG	0130
B170006-R	0074
B170006-RED	0130
B170006-Y	0074
B170006-YEL	0130
B170007	0103
B170007-B	0130
B170007-BLK	0130
B170007-BRN	0130
B170007-O	0074
B170007-ORG	0130
B170007-R	0074
B170007-RED	0130
B170007-Y	0074
B170007-YEL	0130
B170008	0074
B170008-BLK	0130
B170008-BRN	0130
B170008-ORG	0130
B170008-RED	0130
B170008-YEL	0130
B170009	0103
B170010	0103
B170011	0103
B170012	0103
B170013	0103
B170014	0103
B170015	0103
B170016	0103
B170017	0130
B170018	0103
B170019	0103
B170020	0103
B170021	0103
B170022	0103
B170023	0130
B170024	0103
B170025	0103
B170026	0130
B176000	0642
B176001	0074
B176002	0074
B176003	0074
B176004	0642
B176005	0642
B176006	0642
B176007	0642
B176009	0642
B176010	0642
B176011	0642
B176013	0642
B176014	0642
B176015	0642
B176024	0642
B176025	0642
B176026	0642
B176027	0642
B176028	0642
B176029	0642
B177000	0130
B232008	0142
B678760A	0643
B1045494P1	0484
B1760000	0363
B5493957-4	0126
B5493957-5	0126
B5493957-6	0126
B7579500	0071
B7978850	0023
B8400950	OEM
B507138054AQ	OEM
BA-075	0649
BA-07001	0143
BA-07007	0015
BA07018	OEM
BA07021	OEM
BA1A4M	OEM
BA1L4M	OEM
BA5A5	1335
BA5B5	0015
BA5C5	0015
BA5D5	0015
BA5E5	0015
BA5F5	0015
BA5G5	0015
BA5H5	0182
BA5K5	0182
BA5M5	0182
BA5N5	0182
BA5P5	0182
BA6	0164
BA6A	0164
BA8-1A-21	0050
BA10B	OEM
BA20	0473
BA40	0473
BA47	OEM
BA47F	OEM
BA50G	OEM
BA55B	OEM
BA60B	1414
BA60C	OEM
BA65A	OEM
BA67	0016
BA70C	OEM
BA70X	OEM
BA71	0016
BA75C	OEM
BA80A	OEM
BA85A	OEM
BA85X	OEM
BA90A	OEM
BA95A	OEM
BA100	0015
BA100C	OEM
BA102	0030
BA102A	0628
BA102AA	0015
BA102B	OEM
BA102BA	0015
BA102C	0030
BA102CA	0015
BA102D	OEM
BA102DA	0015
BA102EA	0015
BA102FA	0015
BA102GA	0015
BA102HA	0015
BA102KA	0015
BA102MA	0015
BA102PA	0071
BA102RA	0071
BA102TA	0071
BA102VA	0071
BA103	0124
BA104	0080
BA105	0795
BA108	1335
BA109	0030
BA109G	0030
BA110	0030
BA110G	OEM
BA111	0030
BA112	OEM
BA114	0133
BA115	OEM
BA117	OEM
BA119	0015
BA120	0005
BA120A	OEM
BA121	0005
BA122	0133
BA124	0030
BA125	OEM
BA127	0015
BA127D	OEM
BA128	0015
BA129	0609
BA130	1335
BA132AA	0071
BA132BA	0071
BA132CA	0071
BA132DA	0071
BA132EA	0071
BA132FA	0071
BA132GA	0071
BA132HA	0071
BA132KA	0071
BA132MA	0071
BA132PA	0071
BA132RA	0071
BA132TA	0071
BA132VA	0071
BA133	0102
BA133F	0373
BA136	0133
BA136A	0110
BA-138	0030
BA138	OEM
BA139	OEM
BA140	OEM
BA-142	OEM
BA142	OEM
BA-142-01	0015
BA142-01	0015
BA143U	OEM
BA143V	OEM
BA145	0102
BA147	0133
BA148	0102
BA149	OEM
BA150	OEM
BA151	OEM
BA152	OEM
BA152A	0023
BA152PR	OEM
BA153	0015
BA154	0133
BA155	OEM
BA156	OEM
BA157	0017
BA157GP	0271
BA158	0017
BA158GP	0271
BA159	0017
BA159DGP	OEM
BA159GP	0529
BA160	OEM
BA161	OEM
BA162	OEM
BA163	OEM
BA164	1335
BA165	OEM
BA166	0133
BA167	0133
BA168	0133
BA169	OEM
BA170	0133
BA171	OEM
BA172	1335
BA172-5	OEM
BA172-10	OEM
BA172-20	OEM
BA172-30	OEM
BA172-40	OEM
BA172-50	OEM
BA172-60	OEM
BA172-70	OEM
BA172-80	OEM
BA172-90	OEM
BA172-100	OEM
BA172-120	OEM
BA173	0271
BA174	0133
BA175	OEM
BA176	2513
BA177	0133
BA178	0110
BA180	1335
BA181	0133
BA182	2518
BA187	0133
BA188	0015
BA189	0015
BA190	0015
BA191	OEM
BA191-5	0323
BA191-10	0323
BA191-20	1414
BA191-30	0473
BA191-40	0473
BA191-50	0446
BA191-60	0446
BA191-70	1422
BA191-80	1422
BA191-90	2443
BA191-100	2443
BA191-120	OEM
BA192	OEM
BA193	OEM
BA194	OEM
BA195	OEM
BA196	OEM
BA197	OEM
BA198	OEM
BA199	OEM
BA199-250	OEM
BA199-350	OEM
BA199-450	OEM
BA199-550	OEM
BA200	0124
BA201	0124
BA202	0124
BA203	0124
BA204	0124
BA206	OEM
BA207	OEM
BA208	0065
BA209	OEM
BA210	OEM
BA211	OEM
BA212	OEM
BA213	OEM
BA214	OEM
BA215	OEM
BA216	0133
BA217	1335
BA218	1335
BA219	0133
BA220	0124
BA221	0124
BA222	2535
BA223	0124
BA223-10	4503
BA223-15	0102
BA223-20	0348
BA223-25	OEM
BA223-30	OEM
BA223-35	OEM
BA223-40	OEM
BA223-45	OEM
BA223-50	OEM
BA223-55	OEM
BA223-60	OEM
BA223-65	OEM
BA223-70	OEM
BA224	OEM
BA224-150	OEM
BA224-220	OEM
BA224-300	OEM
BA225	OEM
BA226	OEM
BA227	OEM
BA228	OEM
BA232-5	OEM
BA232-10	OEM
BA232-20	OEM
BA232-30	OEM
BA232-40	OEM
BA232-50	OEM
BA232-60	OEM
BA232-70	OEM
BA232-80	OEM
BA232-90	OEM
BA232-100	OEM
BA232-120	OEM
BA235	OEM
BA236	OEM
BA236A	OEM
BA236B	OEM
BA243	0554
BA243A	0124
BA243S	0133
BA244	0163
BA244A	0133
BA244S	0124
BA280	OEM
BA281	OEM
BA282	0163
BA282M3	OEM
BA283	0133
BA284	0124
BA301	2062
BA301B	2062
BA306	OEM
BA308	0433
BA310	OEM
BA311	2631
BA313	2633
BA314	OEM
BA314A	OEM
BA315	1925
BA316	0124
BA317	0133
BA318	2638
BA328	2339
BA328LN	2339
BA328MR	2534
BA333	2063
BA335	OEM
BA336	OEM
BA337	OEM
BA338	2654
BA340	2656
BA343	OEM
BA379	OEM
BA382	OEM
BA389	0124
BA401	2057
BA402	0911
BA403	2693
BA404	2694
BA423	0124
BA430-5	0729
BA430-10	0729
BA430-20	0729
BA430-30	0737
BA430-40	0737
BA430-50	0737
BA430-60	0737
BA430-70	0742
BA430-80	0742
BA430-90	0742
BA430-100	0742
BA430-120	OEM
BA479G	OEM
BA479S	OEM
BA481	2754
BA482	0163
BA483	0163
BA484	0163
BA501	2764
BA501A	2764
BA511	2773
BA511A	2773
BA514	2775
BA514A	2775
BA515	OEM
BA517	0124
BA518	OEM
BA521	2781
BA521A	2781
BA521AX	2781
BA526	2787
BA527	2788
BA532	2794
BA532S	2794
BA534	OEM
BA535	OEM
BA536	2799
BA546	2808
BA547	OEM
BA579A	OEM
BA579C	OEM
BA579S	OEM
BA612	OEM
BA614	2865
BA614A	2865
BA616	OEM
BA618	OEM
BA631	OEM
BA634	OEM
BA656	OEM
BA658	OEM
BA664	OEM
BA668	OEM
BA668A	OEM
BA679	OEM
BA681A	OEM
BA682	OEM
BA682A	OEM
BA683	OEM
BA683A	OEM
BA684A	OEM
BA685	OEM
BA689	OEM
BA695	OEM
BA704	OEM
BA707	OEM
BA707S	OEM
BA715	2913
BA718	2913
BA728	2922
BA728F	OEM
BA751AA	0015
BA751BA	0015
BA751CA	0015
BA751DA	0015
BA751EA	0015
BA751FA	0015
BA751GA	0015
BA751HA	0015
BA751KA	0015
BA751MA	0015
BA751PA	0071
BA751RA	0071
BA751TA	0071
BA751VA	0071
BA806	OEM
BA820	OEM
BA841	OEM
BA842	OEM
BA843	OEM
BA845	OEM
BA847	3012
BA852	OEM
BA857	OEM
BA858	OEM
BA858B	OEM
BA862	OEM
BA867	OEM
BA873	OEM
BA885	OEM
BA1003	0400
BA1102F	OEM
BA1104LS	OEM
BA1310	0706
BA1310F	0706
BA1320	1260
BA1320F	OEM
BA1330	0416
BA1332	OEM
BA1332L	OEM
BA1335	OEM
BA1350	5197
BA1351	OEM
BA1355	OEM
BA1356	OEM
BA1360	OEM
BA1360F	OEM
BA1440	OEM
BA1604	OEM
BA2254F	OEM
BA2266	OEM
BA2435	OEM
BA3112	OEM
BA3302	OEM
BA3304	OEM
BA3304F	OEM
BA3306	OEM
BA3308	OEM
BA3308F	OEM
BA3310N	OEM
BA3312N	OEM
BA3402	OEM
BA3404F	OEM
BA3404L	OEM
BA3416BL	OEM
BA3513	OEM
BA3516	OEM
BA3702	OEM
BA3704	5162
BA3707	OEM
BA3812L	OEM
BA3822LS	OEM
BA3824LS	OEM
BA4110	0465
BA4111	OEM
BA4114	OEM
BA4210	OEM
BA4220	3816
BA4224	OEM
BA4232	OEM
BA4232AL	OEM
BA4233AL	OEM
BA4236L	OEM
BA4403	OEM
BA4404	OEM
BA4410	OEM
BA4412	OEM
BA4424N	OEM
BA4558	0360
BA4558D	OEM
BA4558DX	0360
BA4558F	5531
BA4560F	OEM
BA4561	OEM
BA5101	OEM
BA5102	2105
BA5102A	2105
BA5102AL	OEM
BA5102S	OEM
BA5104A	OEM
BA5112	OEM
BA5112LS	OEM
BA5114F	OEM
BA5114LS	OEM
BA5115	OEM
BA5115L	OEM
BA5116	OEM
BA5119S	OEM
BA5204	OEM
BA5205F	OEM
BA5206BF	OEM
BA5302A	OEM
BA5402A	2799
BA5404	OEM
BA5406	6800
BA6104	OEM
BA6107	OEM
BA6109	OEM
BA6109U	OEM
BA6110	OEM
BA6118	OEM
BA6121	OEM
BA6122	OEM
BA6122A	5232
BA6122AF	OEM
BA6124	5162
BA6125	OEM
BA6133	OEM
BA6135	OEM
BA6137	OEM
BA6138	OEM
BA6139	OEM
BA6144	5826
BA6146	OEM
BA6148	OEM
BA6154	OEM
BA6200AL	OEM
BA6208	OEM
BA6209	3828
BA6209K	OEM
BA6209N	OEM
BA6209U	3828
BA6209U1	3828
BA6209U2	OEM
BA6209U4	OEM
BA6209UI	3828
BA6218	OEM
BA6219	6841
BA6219B	5806
BA6219B-U2	OEM
BA6229	OEM
BA6229W/MNT	OEM
BA6238A	4117
BA6238AU	4117
BA6238AU1	OEM
BA6238AU4	4117
BA6239	OEM
BA6239A	OEM
BA6239AN	OEM
BA6247	OEM
BA6247N	OEM

DEVICE TYPE	REPL CODE
BA6251	OEM
BA6256	OEM
BA6301	OEM
BA6302	OEM
BA6302A	OEM
BA6303	OEM
BA6303F	OEM
BA6304	2425
BA6304A	2425
BA6304AL	2425
BA6304LA	OEM
BA6305	OEM
BA6320L	OEM
BA6321	OEM
BA6328	OEM
BA6334	OEM
BA6340	OEM
BA6352S	OEM
BA6360	OEM
BA6405	OEM
BA6418	OEM
BA6418N	OEM
BA6431S	OEM
BA6455FS	OEM
BA6459P1	OEM
BA6459PI	OEM
BA6562	OEM
BA6565A	OEM
BA6581K	OEM
BA6590S	OEM
BA6805	OEM
BA6822S	OEM
BA6993	0631
BA7001	OEM
BA7004	OEM
BA7007	OEM
BA7021	OEM
BA7050L	OEM
BA7055LS	OEM
BA7058LS	OEM
BA7101	OEM
BA7103	OEM
BA7116	OEM
BA7125L	OEM
BA7212	OEM
BA7212S	OEM
BA7252S	OEM
BA7255S	OEM
BA7266S	OEM
BA7274S	OEM
BA7280AS	OEM
BA7280S	OEM
BA7603F-T1	OEM
BA7603FTI	OEM
BA7604N	OEM
BA7700K1	OEM
BA7703K1	OEM
BA7720S	OEM
BA7743S	OEM
BA7751LS	OEM
BA7755	OEM
BA7765AS	OEM
BA7767AS	OEM
BA7767S	OEM
BA8411	OEM
BA9101	OEM
BA9201	OEM
BA9221	OEM
BA10324	1074
BA10324F	OEM
BA10339	2780
BA10358	6322
BA10358N	OEM
BA10393	OEM
BA12003	1134
BA15218N	1699
BAA6865	OEM
BAC-SHIMI	0124
BACSHIMI	0124
BACSIM1	0124
BAF1360	OEM
BAF5204	OEM
BAL0002-100	OEM
BAL0004-50	OEM
BAL0004-100	OEM
BAL0101-150	OEM
BAL0101-150A	OEM
BAL0101-200	OEM
BAL0105-50	OEM
BAL0105-100	OEM
BAL0204-50	OEM
BAL0204-125	OEM
BAL0305-P300	OEM
BAL0710-50	OEM
BAL0710-75	OEM
BAL74	OEM
BAL99	OEM
BAL6309	OEM
BAM40SR	OEM
BAM80SR	OEM
BAM100SR	OEM
BAM120SR	OEM
BAM150	OEM
BAN62	0133
BAR805	OEM
BAR3S	OEM
BAR10	OEM
BAR11	OEM
BAR12-1	OEM
BAR12-2	OEM
BAR12-3	OEM
BAR12-5	OEM
BAR13-1	OEM
BAR13-2	OEM
BAR13-3	OEM
BAR13-5	OEM
BAR14-1	OEM
BAR14-2	OEM
BAR14-3	OEM
BAR14-5	OEM
BAR15-1	OEM
BAR15-2	OEM
BAR15-3	OEM
BAR15-5	OEM
BAR16-1	OEM
BAR16-2	OEM
BAR16-3	OEM
BAR16-5	OEM
BAR18	OEM
BAR19	OEM
BAR28	OEM
BAR35	OEM
BAR42	OEM
BAR43	OEM
BAR43A	OEM
BAR43C	OEM
BAR43S	OEM
BAR74	OEM
BAR99	OEM
BAR223-10	OEM
BAR223-15	OEM
BAR223-20	OEM
BAR223-25	OEM
BAR223-30	OEM
BAR223-35	OEM
BAR223-40	OEM
BAR223-45	OEM
BAR223-50	OEM
BAR223-55	OEM
BAR223-60	OEM
BAR223-65	OEM
BAR223-70	OEM
BAS11	0023
BAS15	0124
BAS16	3749
BAS17	OEM
BAS18	0023
BAS19	3895
BAS20	3895
BAS21	3895
BAS27	OEM
BAS28	OEM
BAS29	3895
BAS31	OEM
BAS32	OEM
BAS32D	OEM
BAS33	OEM
BAS34	OEM
BAS35	4917
BAS40	OEM
BAS40-01	OEM
BAS40-02	OEM
BAS40-03	OEM
BAS40-04	OEM
BAS40-05	OEM
BAS40-06	OEM
BAS45	OEM
BAS46	OEM
BAS56	OEM
BAS70	OEM
BAS70-01	OEM
BAS70-02	OEM
BAS70-03	OEM
BAS70-04	OEM
BAS70-05	OEM
BAS70-06	OEM
BAS482	OEM
BASIC1	OEM
BASIC2	OEM
BASIC3	OEM
BASIC4	OEM
BAT10	OEM
BAT11	OEM
BAT13	OEM
BAT16-046	OEM
BAT16-096	OEM
BAT17	OEM
BAT17-04	OEM
BAT17-05	OEM
BAT17-06	OEM
BAT17DS	OEM
BAT18	3896
BAT19	OEM
BAT21E	OEM
BAT21J	OEM
BAT22E	OEM
BAT22J	OEM
BAT23G	OEM
BAT23J	OEM
BAT24G	OEM
BAT24H	OEM
BAT25E	OEM
BAT25J	OEM
BAT26E	OEM
BAT26J	OEM
BAT27G	OEM
BAT27H	OEM
BAT28G	OEM
BAT28H	OEM
BAT29	OEM
BAT31	OEM
BAT38	OEM
BAT39	OEM
BAT39A	OEM
BAT41	OEM
BAT42	OEM
BAT43	OEM
BAT45	OEM
BAT46	OEM
BAT47	OEM
BAT48	OEM
BAT49	OEM
BAT50	OEM
BAT51	OEM
BAT52	OEM
BAT54	OEM
BAT59	OEM
BAT74	OEM
BAT81	4945
BAT82	4946
BAT83	4946
BAT85	0735
BAT86	4946
BAV10	OEM
BAV11	OEM
BAV12	OEM
BAV13	OEM
BAV14	OEM
BAV15	OEM
BAV16	OEM
BAV17	0124
BAV18	0124
BAV19	0124
BAV20	0133
BAV21	0133
BAV22	OEM
BAV22R	OEM
BAV23	OEM
BAV24	OEM
BAV25	OEM
BAV26	OEM
BAV27	OEM
BAV28	OEM
BAV29	OEM
BAV30	OEM
BAV31	OEM
BAV32	OEM
BAV33	OEM
BAV34	OEM
BAV35	OEM
BAV36	OEM
BAV37	OEM
BAV38	OEM
BAV39	OEM
BAV40	OEM
BAV41	OEM
BAV42	OEM
BAV44	OEM
BAV45	OEM
BAV45A	OEM
BAV46	OEM
BAV46D	OEM
BAV46E	OEM
BAV46F	OEM
BAV47	OEM
BAV48	OEM
BAV49	OEM
BAV50	OEM
BAV54	OEM
BAV54-30	OEM
BAV54-70	OEM
BAV54-100	OEM
BAV55	OEM
BAV67	OEM
BAV68	OEM
BAV69	OEM
BAV70	0080
BAV71	OEM
BAV72	OEM
BAV74	1335
BAV75	OEM
BAV76	OEM
BAV77	OEM
BAV79	OEM
BAV80	OEM
BAV81	OEM
BAV82	OEM
BAV83	OEM
BAV84	OEM
BAV85	OEM
BAV86	OEM
BAV87	OEM
BAV88	OEM
BAV89	OEM
BAV92	OEM
BAV93	OEM
BAV94	OEM
BAV96A	OEM
BAV96B	OEM
BAV96C	OEM
BAV96D	OEM
BAV97	OEM
BAV99	OEM
BAV100	OEM
BAV101	OEM
BAV102	OEM
BAV103	OEM
BAV205	1000
BAV206	2076
BAV207	2081
BAV208	2088
BAV209	2088
BAV215	1000
BAV216	2076
BAV217	2081
BAV218	2088
BAV219	2088
BAV225	1000
BAV226	2076
BAV227	2081
BAV228	2088
BAV229	2088
BAV305	1000
BAV306	2076
BAV307	2081
BAV308	2088
BAV309	2088
BAV315	1000
BAV316	2076
BAV317	2081
BAV318	2088
BAV319	2088
BAV325	1000
BAV326	2076
BAV327	2081
BAV328	2088
BAV329	2088
BAV405	0197
BAV406	0200
BAV407	0204
BAV408	0206
BAV409	0206
BAV415	0197
BAV416	0200
BAV417	0204
BAV418	0206
BAV419	0206
BAV425	0197
BAV426	0200
BAV427	0204
BAV428	0206
BAV429	0206
BAV505	0109
BAV506	0116
BAV507	0122
BAV508	0131
BAV509	0131
BAV515	0109
BAV516	0116
BAV517	0122
BAV518	0131
BAV519	0131
BAV525	0109
BAV526	0116
BAV527	0122
BAV528	0131
BAV529	0131
BAV705	0197
BAV706	0200
BAV707	0204
BAV708	0206
BAV709	0206
BAV715	0197
BAV716	0200
BAV717	0204
BAV718	0206
BAV719	0206
BAV725	0197
BAV726	0200
BAV727	0204
BAV728	0206
BAV729	0206
BAV805	0109
BAV806	0116
BAV807	0122
BAV808	0131
BAV809	0131
BAV815	0109
BAV816	0116
BAV817	0122
BAV818	0131
BAV819	0131
BAV825	0109
BAV826	0116
BAV827	0122
BAV828	0131
BAV829	0131
BAV3308	2088
BAV3309	2088
BAW10	OEM
BAW10TF20	0133
BAW11	OEM
BAW11TF21	0133
BAW12	OEM
BAW12TF22	0133
BAW13	OEM
BAW13TF23	0133
BAW14	OEM
BAW16	0133
BAW17	0133
BAW18	0133
BAW19	OEM
BAW21A	OEM
BAW21B	OEM
BAW22	OEM
BAW23	OEM
BAW24	0133
BAW25	0133
BAW26	0133
BAW27	0124
BAW28	0080
BAW29	OEM
BAW31	OEM
BAW32A	0133
BAW32B	0133
BAW32C	0080
BAW32D	1335
BAW32E	1335
BAW33	OEM
BAW36	OEM
BAW43	5007
BAW45	0133
BAW46	OEM
BAW47	OEM
BAW48	OEM
BAW49	OEM
BAW50	OEM
BAW51	OEM
BAW52	OEM
BAW53	0133
BAW54	OEM
BAW55	0124
BAW56	0903
BAW56G	OEM
BAW56GT	OEM
BAW57	OEM
BAW58	OEM
BAW59	0133
BAW60	OEM
BAW62	0124
BAW63	OEM
BAW63A	0133
BAW63B	0133
BAW64	OEM
BAW65	OEM
BAW66	OEM
BAW67	OEM
BAW68	OEM
BAW69	OEM
BAW70	OEM
BAW75	0133
BAW76	0124
BAW77	OEM
BAW78A	OEM
BAW78B	OEM
BAW78C	OEM
BAW78D	OEM
BAW79A	OEM
BAW79B	OEM
BAW79C	OEM
BAW79D	OEM
BAW90	OEM
BAW91	OEM
BAW92	OEM
BAW93	OEM
BAW95C	OEM
BAW95D	OEM
BAW95E	OEM
BAW95F	OEM
BAW95G	OEM
BAW96	OEM
BAW99	0133
BAW100	OEM
BAW101	OEM
BAX11	OEM
BAX11IIO	OEM
BAX11IIU	OEM
BAX12	0015
BAX12A	OEM
BAX-13	0124
BAX13	0124
BAX13A	OEM
BAX14	1090
BAX14A	OEM
BAX15	OEM
BAX-16	0133
BAX16	0133
BAX16A	OEM
BAX17	OEM
BAX17A	OEM
BAX18	1090
BAX18A	0080
BAX20	OEM
BAX21	OEM
BAX25	OEM
BAX26	OEM
BAX27	OEM
BAX28	0133
BAX30	0133
BAX32	OEM
BAX33	OEM
BAX34	OEM
BAX35	OEM
BAX36	OEM
BAX37	OEM
BAX38	OEM
BAX39	OEM
BAX40	OEM
BAX41	OEM
BAX42	OEM
BAX43	OEM
BAX44	OEM
BAX45	OEM
BAX46	OEM
BAX47	OEM
BAX48	OEM
BAX49	OEM
BAX50	OEM
BAX51	OEM
BAX52	OEM
BAX53	OEM
BAX54	OEM
BAX55	OEM
BAX56	OEM
BAX57	OEM
BAX58	OEM
BAX59	OEM
BAX60	OEM
BAX61	OEM
BAX62	OEM
BAX63	OEM
BAX64	OEM
BAX65	OEM
BAX66	OEM
BAX67	OEM
BAX68	OEM
BAX69	OEM
BAX70	OEM
BAX71	OEM
BAX72	OEM
BAX73	OEM
BAX74	0133
BAX78	OEM
BAX79C16	0445
BAX79C18	0375
BAX79C20	0700
BAX79C22	0705
BAX79C24	0494
BAX79C27	0455
BAX79C27A	0441
BAX81	OEM
BAX82	OEM
BAX83	OEM
BAX84	OEM
BAX85	OEM
BAX86A	OEM
BAX86B	OEM
BAX87	0133
BAX88	OEM
BAX88TF11	0133
BAX89A	0133
BAX89B	OEM
BAX90A	OEM
BAX90B	OEM
BAX91A	OEM
BAX91B	OEM
BAX91C	OEM
BAX91C/TF102	0124
BAX92	OEM
BAX93	OEM
BAX94	OEM
BAX94C62	0783
BAX95TF600	0124
BAX96C2V7	0760
BAX96C3V3	0299
BAY15	0071
BAY16	0071
BAY17	0133
BAY18	0133
BAY19	0133
BAY20	0133
BAY21	0795
BAY23	0071
BAY24	OEM
BAY25	OEM
BAY26	OEM
BAY31	0133
BAY35	OEM
BAY36	0133
BAY38	0124
BAY41	0133
BAY42	0133
BAY43	0133
BAY44	1335
BAY45	0609
BAY46	0795
BAY52	0133
BAY60	0133
BAY61	0133
BAY63	0133
BAY64	0015
BAY66	OEM
BAY67	0133
BAY68	0133
BAY69	0133
BAY71	5042
BAY72	0133
BAY73	0133
BAY74	0396
BAY78	OEM
BAY79	OEM
BAY79IIIO	OEM
BAY79IIIU	OEM
BAY79IVO	OEM
BAY79IVU	OEM
BAY79V	OEM
BAY80	0182
BAY82	5042
BAY84	OEM
BAY85	OEM
BAY85S	OEM
BAY86	0015
BAY87	0015
BAY88	0424
BAY89	0424
BAY90	0071
BAY91	OEM
BAY91A	OEM
BAY92	0271
BAY93	0124
BAY94	OEM
BAY95	OEM
BAY96	OEM
BB	OEM
BB1	0102
BB1A	0015
BB-2	0015
BB2	0102
BB2A	0023
BB2A184	0015
BB2L	0023
BB2T	0023
BB3	OEM
BB3I	0023
BB3K	0023
BB-4	0023
BB4	0102
BB4-FC	0023
BB4A	OEM
BB4B	OEM
BB4C	5322
BB4T	0023
BB5	OEM
BB5A	OEM
BB5B	OEM
BB5C	OEM
BB-6	0023
BB6	0102
BB6S	0102
BB-8	0023
BB8	0102
BB-10	0102
BB10	0182
BB-10BB-10S	OEM
BB-10S	0102
BB10T	0102
BB12	OEM
BB14	OEM
BB18	OEM
BB22	0030
BB-42-1B	2472
BB-42-2B	OEM
BB71	0016
BB100	1335
BB100GBE	OEM
BB100GRO	OEM
BB100GVE	OEM
BB101	OEM
BB102	OEM
BB102/15	OEM
BB102/16	OEM
BB102/17	OEM
BB102/18	OEM
BB102/19	OEM
BB102/20	OEM
BB103	0030
BB103B	OEM
BB103G	OEM
BB104	0030
BB104B	OEM
BB104G	OEM
BB105	0005
BB-105A	0030
BB105A	0005
BB105B	1029
BB105G	1029
BB106	0628
BB107	0015
BB-109	0015
BB109	1029
BB109G	0030
BB110	OEM
BB110B	OEM
BB110G	OEM
BB112	3171
BB113	OEM
BB117	0015
BB119	OEM
BB121	0720
BB121A	0720
BB121B	1029
BB122	0720
BB122GL	OEM
BB122S	OEM
BB125	0628
BB127	0015
BB130	OEM
BB138	0628
BB139	0030
BB141	OEM
BB141A	OEM
BB141B	OEM
BB142	0720
BB204	OEM
BB204B	2071
BB204G	2071
BB205A	OEM
BB205B	OEM
BB205G	0005
BB209	OEM
BB212	OEM
BB215	OEM
BB219	OEM

If replacement code is OEM, contact original manufacturer for replacement.

DEVICE TYPE	REPL CODE
BB221	1029
BB222	OEM
BB229	OEM
BB304	OEM
BB304I	OEM
BB304II	OEM
BB304III	OEM
BB304IV	OEM
BB304V	OEM
BB305B	OEM
BB305G	OEM
BB312	OEM
BB313	OEM
BB314	OEM
BB329	OEM
BB329A	OEM
BB404	OEM
BB404A	OEM
BB404B	OEM
BB404C	OEM
BB404D	OEM
BB404E	OEM
BB405B	OEM
BB405G	OEM
BB406	OEM
BB409	OEM
BB417	OEM
BB503	OEM
BB504	OEM
BB505B	OEM
BB505G	OEM
BB510	OEM
BB515B	OEM
BB515G	OEM
BB521	OEM
BB523	OEM
BB529	OEM
BB531	OEM
BB601	OEM
BB609	OEM
BB609A	OEM
BB609B	OEM
BB610	OEM
BB619A	OEM
BB619B	OEM
BB620	OEM
BB623	OEM
BB629	OEM
BB631	OEM
BB721	OEM
BB723	OEM
BB729	OEM
BB731	OEM
BB801	OEM
BB804	OEM
BB809	OEM
BB814	OEM
BB909A	OEM
BB909B	OEM
BBCC106	OEM
BBCC106B-1	OEM
BBCC106D-1	OEM
BBCC106E-1	OEM
BBCC106M-1	OEM
BBCS	OEM
BBIA	0015
BBY10	OEM
BBY11	OEM
BBY12	OEM
BBY13	OEM
BBY14	OEM
BBY15	OEM
BBY16	OEM
BBY17	0554
BBY18	OEM
BBY19	OEM
BBY20	OEM
BBY21	OEM
BBY22	OEM
BBY24	OEM
BBY24S1	OEM
BBY25	OEM
BBY25S1	OEM
BBY26	OEM
BBY26S1	OEM
BBY27	OEM
BBY27S1	OEM
BBY28	OEM
BBY29	OEM
BBY30	OEM
BBY31	0133
BBY32CB	OEM
BBY32DA	OEM
BBY32DB	OEM
BBY32EA	OEM
BBY32FA	OEM
BBY33DA2	OEM
BBY34A	OEM
BBY34B	OEM
BBY34C	OEM
BBY34D	OEM
BBY35F	OEM
BBY36	OEM
BBY37	OEM
BBY38	OEM
BBY39	OEM
BBY40	0133
BC10CM-BL	OEM
BC17LB-1	0037
BC45B	0079
BC-71	0016
BC71	0016
BC100	0233
BC103	0086
BC103C	0086
BC107	0016
BC107A	0016
BC107AP	0079
BC107B	0016
BC107BP	0079
BC108	0016
BC108A	0016
BC108AP	0079
BC-108B	0016
BC108B	0016
BC108BP	0079
BC108C	0016
BC108CP	0079
BC109	0016
BC-109B	0016
BC109B	0016
BC109BP	0079
BC109C	0016
BC109CP	0079
BC110	0016
BC111	0144
BC112	0144
BC113	0016
BC113A	OEM
BC114	0016
BC114A	OEM
BC114TR	0016
BC115	0016
BC116	0037
BC116A	0037
BC117	0086
BC118	0016
BC-119	0086
BC119	0086
BC120	0086
BC121	0127
BC122	0127
BC123	0127
BC125	0016
BC125A	0016
BC125B	0016
BC126	0037
BC126A	0150
BC127	0111
BC128	0111
BC129	0016
BC129A	0079
BC129B	0079
BC130	0016
BC130A	0079
BC130B	0079
BC131	0016
BC131B	0079
BC132	0016
BC132A	0079
BC134	0016
BC135	0016
BC135A	0007
BC136	0016
BC137	0037
BC-138	0086
BC138	0086
BC139	0126
BC139A	0861
BC-140	0086
BC140	0086
BC140-06	0324
BC140-6	0086
BC140-10	0086
BC140-16	0086
BC140-25	0086
BC-140A	0086
BC140A	0086
BC-140B	0086
BC140B	0086
BC-140C	0086
BC140C	0086
BC-140D	0086
BC140D	0086
BC-141	0086
BC141	0086
BC141-06	0086
BC141-6	0086
BC141-10	0086
BC141-16	0086
BC141-25	0086
BC141A	OEM
BC141B	OEM
BC141C	OEM
BC-142	0086
BC142	0086
BC142-10	0086
BC142-16	0086
BC143	0126
BC144	0086
BC145	1480
BC146	0111
BC146-01	OEM
BC146-02	OEM
BC146-03	OEM
BC146G	0079
BC146R	0079
BC146Y	0079
BC147	0016
BC147A	0016
BC147B	0016
BC148	0016
BC-148A	0016
BC148A	0016
BC-148B	0016
BC148B	0016
BC-148C	0016
BC148C	0016
BC149	0016
BC149A	0016
BC149B	0016
BC149C	0016
BC149G	0016
BC150	0079
BC151	0079
BC152	0079
BC153	0693
BC154	0693
BC155	0127
BC155A	0144
BC155B	0079
BC155C	0111
BC156	0127
BC156A	0144
BC156B	0079
BC156C	0111
BC157	0037
BC157A	0037
BC157B	0037
BC157V1	OEM
BC157VI	0861
BC158	0037
BC158-1	0037
BC158A	0037
BC158A-1	0037
BC158B	0037
BC158C	0693
BC158V1	OEM
BC158VI	0861
BC159	0037
BC159A	0037
BC159B	0037
BC159C	0693
BC160	0126
BC160-06	0532
BC160-6	0126
BC160-10	0126
BC160-16	0126
BC160-25	0126
BC160A	OEM
BC160B	OEM
BC160C	OEM
BC161	0126
BC161-06	0791
BC161-6	0126
BC161-10	0126
BC161-16	0126
BC161-25	0126
BC161A	OEM
BC161B	OEM
BC161C	OEM
BC167	0079
BC-167-B	0079
BC167A	0155
BC167B	0155
BC168	0079
BC168A	0155
BC168B	0155
BC168C	0155
BC169	0079
BC-169-C	0079
BC169A	0079
BC-169B	0079
BC169B	0079
BC-169C	0079
BC169C	0079
BC170	0111
BC170A	0111
BC170B	0111
BC170C	0111
BC171	0079
BC171A	0079
BC171B	0079
BC172	0079
BC172A	0079
BC172B	0079
BC172C	0079
BC173	0111
BC173,B	0086
BC173A	0111
BC173B	0111
BC173C	0111
BC174	0086
BC174A	0086
BC174B	0086
BC175	0079
BC177	0693
BC177-6	OEM
BC177A	0693
BC177B	0693
BC177C	OEM
BC177PA	OEM
BC177PAK	OEM
BC177PAL	OEM
BC177PAM	OEM
BC177PB	OEM
BC177PBK	OEM
BC177PBL	OEM
BC177PBM	OEM
BC177V	0037
BC177V1	0037
BC177VI	0150
BC178	0693
BC178A	0693
BC178B	0693
BC178C	0693
BC178D	0037
BC178P	OEM
BC178PA	OEM
BC178PAK	OEM
BC178PAL	OEM
BC178PAM	OEM
BC178PB	OEM
BC178PBK	OEM
BC178PBL	OEM
BC178PBM	OEM
BC178PC	OEM
BC178PCK	OEM
BC178PCL	OEM
BC178PCM	OEM
BC178V	0037
BC178V-1	0037
BC178V1	0037
BC178VI	0150
BC179	0693
BC179A	0693
BC179B	0693
BC179C	0693
BC179PB	OEM
BC179PBK	OEM
BC179PBL	OEM
BC179PBM	OEM
BC179PC	OEM
BC179PCK	OEM
BC179PCL	OEM
BC179PCM	OEM
BC179VI	OEM
BC180	0016
BC180B	0016
BC181	0037
BC181A	0037
BC182	0079
BC182A	0155
BC182AP	0079
BC182B	0079
BC182BP	0079
BC182K	0111
BC182KA	0016
BC182KB	0016
BC182L	0155
BC182LA	0155
BC182P	0079
BC183	0079
BC183A	0079
BC183AP	0079
BC183B	0079
BC183BP	0079
BC183C	2860
BC183CP	OEM
BC183K	0111
BC183KA	0111
BC183KB	0016
BC183KC	0016
BC183L	0111
BC183LA	0155
BC183LB	0079
BC183P	0079
BC184	0079
BC184B	0079
BC184BP	0079
BC184C	0079
BC184CP	0079
BC184K	0016
BC184KC	0016
BC184L	0111
BC184P	0079
BC185	0016
BC186	0037
BC187	0037
BC188	0127
BC189	0127
BC190	0079
BC190A	0595
BC190B	0016
BC192	0037
BC194	0144
BC194B	0144
BC195	0144
BC195CD	0144
BC196	0037
BC196A	0037
BC196B	0037
BC196V1	0037
BC196VI	0861
BC197	0016
BC197A	0016
BC197B	0016
BC198	0016
BC198A	0079
BC198B	0111
BC198C	0079
BC199	0016
BC199B	0111
BC199C	0079
BC200	0037
BC200-01	OEM
BC200-02	OEM
BC200-03	OEM
BC200G	0693
BC200R	0037
BC200Y	0037
BC201	0037
BC202	0037
BC203	0037
BC204	0693
BC204-6	0037
BC204A	0693
BC204B	0693
BC204V	0037
BC204V1	0037
BC204VI	0150
BC205	0037
BC205A	0037
BC205B	0693
BC205C	0693
BC205V	0037
BC205V1	0037
BC205VI	0037
BC206	0693
BC206A	0037
BC206B	0693
BC206C	0693
BC-207	0015
BC207	0079
BC207A	0079
BC207B	0079
BC208	0079
BC208A	0079
BC208B	0079
BC208C	0079
BC209	0079
BC209B	0079
BC209C	0079
BC210	0079
BC211	0086
BC211-6	OEM
BC211-10	OEM
BC211-16	OEM
BC211A	0086
BC211A-6	OEM
BC211A-10	OEM
BC211A-16	OEM
BC212	0037
BC212A	0037
BC212AP	0037
BC212B	0037
BC212BP	0037
BC212K	0037
BC212KA	0037
BC212KB	0037
BC212L	0037
BC212LA	0037
BC212LB	0037
BC212P	0037
BC212V1	0037
BC212VI	0037
BC213	0037
BC213A	0037
BC213AP	0037
BC213B	0037
BC213BP	0037
BC213C	0532
BC213CP	OEM
BC213K	0037
BC213KA	0037
BC213KB	0037
BC213KC	0037
BC213L	0037
BC213LA	0037
BC213LB	0037
BC213LC	0037
BC213P	0037
BC214	0037
BC214A	0037
BC214B	0037
BC214BP	0037
BC214C	0532
BC214CP	0037
BC214K	0037
BC214KA	0037
BC214KB	0037
BC214KC	0037
BC214L	0037
BC214LA	0037
BC214LB	0037
BC214LC	0037
BC214P	0037
BC215	OEM
BC215A	0861
BC215B	0861
BC216	0086
BC216A	0086
BC216B	0086
BC220	0016
BC221	0037
BC222	0016
BC223	0191
BC223A	0079
BC223B	0079
BC224	0693
BC225	0693
BC226	0595
BC231	OEM
BC231A	0006
BC231B	0006
BC231M	0126
BC232	OEM
BC232A	0155
BC232B	0155
BC232M	0086
BC233A	0016
BC236	0861
BC237	0079
BC237(PHIN)	0111
BC237-5	OEM
BC237-18	0079
BC237A	0079
BC237A5	0079
BC237A18	0079
BC237B	0079
BC237B5	0079
BC237B18	0079
BC237C	2860
BC237C5	2860
BC237C18	2860
BC237PA	OEM
BC237PAK	OEM
BC237PAL	OEM
BC237PAM	OEM
BC237PB	OEM
BC237PBK	OEM
BC237PBL	OEM
BC237PBM	OEM
BC238	0079
BC238(PHIN)	0111
BC238-5	0079
BC238-16	0037
BC238-18	0079
BC238A	0079
BC238A5	0079
BC238A18	0079
BC238B	0079
BC238B5	0079
BC238B18	0079
BC238C	0111
BC238C5	0079
BC238C18	0079
BC238PA	OEM
BC238PAK	OEM
BC238PAL	OEM
BC238PAM	OEM
BC238PB	OEM
BC238PBK	OEM
BC238PBL	OEM
BC238PBM	OEM
BC238PC	OEM
BC238PCK	OEM
BC238PCL	OEM
BC238PCM	OEM
BC239	0079
BC239(PHIN)	0111
BC239-5	0079
BC239-18	0079
BC239A	0111
BC239B	0111
BC239B5	0079
BC239B18	0079
BC239C	0079
BC239C5	2860
BC239C18	2860
BC239PB	OEM
BC239PBK	OEM
BC239PBL	OEM
BC239PBM	OEM
BC239PC	OEM
BC239PCK	OEM
BC239PCL	OEM
BC239PCM	OEM
BC250	0037
BC250A	0037
BC250B	0037
BC250C	0037
BC251	0037
BC251A	0037
BC251B	0037
BC251C	0037
BC252	0037
BC252A	0037
BC252B	0037
BC252C	0037
BC253	0037
BC253A	0037
BC253B	0037
BC253C	0037
BC254	0086
BC255	0086
BC256	0037
BC256A	0037
BC256B	0037
BC257	0037
BC257A	1242
BC257B	1242
BC257VI	0037
BC258	0037
BC258A	1242
BC258B	1242
BC258C	0693
BC258VI	0037
BC259	0037
BC259A	1242
BC259B	1242
BC259C	3519
BC260	0037
BC260A	0037
BC260B	0037
BC260C	0037
BC-261	0037
BC261	0037
BC261A	0037
BC261B	0037
BC261C	0037
BC262	0037
BC262A	0037
BC262B	0037
BC262C	0037
BC263	0037
BC263A	0037
BC263B	0037
BC263C	0037
BC264	2888
BC264A	2950
BC264B	0325
BC264C	0325
BC264D	OEM
BC264L	OEM
BC264LA	OEM
BC264LB	OEM
BC264LC	OEM
BC264LD	OEM
BC266	0037
BC266A	0037
BC266B	0037
BC267	0016
BC268	0016
BC269	0016
BC270	0016
BC271	0007
BC272	OEM
BC280	0016
BC280A	0016
BC280B	0016
BC280C	0016
BC281	0037
BC281A	0037
BC281B	0037
BC281C	0037
BC282	0079
BC283	0037
BC284	0079
BC284A	0079
BC284B	0079
BC285	0016
BC286	0086
BC287	0126
BC288	0086
BC289	0016
BC289A	0016
BC289B	0016
BC290	0016
BC290B	0016
BC290C	0016
BC291	0037
BC291A	0037
BC291D	0037
BC292	0037
BC292A	0037
BC292B	OEM
BC292D	0037
BC294	0126
BC295	0144
BC297	0037
BC297-7	0037
BC298	0037
BC300	0622
BC300-4	0233
BC300-5	0233
BC300-6	0233
BC301	0086
BC301-4	0086
BC301-5	0086
BC301-6	0086
BC302-4	0086
BC302-5	0086
BC302-6	0086
BC303	0126
BC303-4	0126
BC303-5	0126
BC303-6	0861
BC304	0126
BC304-4	0791
BC304-5	0791
BC304-6	0791
BC-307	0015
BC307	0037
BC307(ALGG)	0037
BC307(PHIN)	0037
BC307-5	0037
BC307-18	0037
BC307A	0037
BC307A5	OEM
BC307A18	OEM
BC307B	0037
BC307B5	OEM
BC307B18	OEM
BC307C	0037
BC307PA	OEM
BC307PAK	OEM
BC307PAL	OEM
BC307PAM	OEM
BC307PB	OEM
BC307PBK	OEM
BC307PBL	OEM
BC307PBM	OEM

DEVICE TYPE	REPL CODE
BC307PC	OEM
BC307V1	0037
BC307VI	0037
BC307VI5	0037
BC307VI18	OEM
BC307VII8	0037
BC308	0037
BC308(ALGG)	0006
BC308(SIEG)	3519
BC308-5	0037
BC308-18	0037
BC308A	0037
BC308A5	OEM
BC308A18	OEM
BC308B	0037
BC308B5	OEM
BC308B18	OEM
BC308C	0037
BC308PA	OEM
BC308PAK	OEM
BC308PAL	OEM
BC308PAM	OEM
BC308PB	OEM
BC308PBK	OEM
BC308PBL	OEM
BC308PBM	OEM
BC308PC	OEM
BC308PCK	OEM
BC308PCL	OEM
BC308PCM	OEM
BC308V1	0037
BC308VI	0861
BC308VI5	0037
BC308VI18	OEM
BC308VII8	0037
BC309	0037
BC309(ALGG)	0037
BC309(SIEG)	0037
BC309-5	0037
BC309-18	0037
BC309A	0037
BC309A5	OEM
BC309A18	OEM
BC309B	0037
BC309B5	OEM
BC309B18	OEM
BC309C	0037
BC309PB	OEM
BC309PBK	OEM
BC309PBL	OEM
BC309PBM	OEM
BC309PC	OEM
BC309PCK	OEM
BC309PCL	OEM
BC309PCM	OEM
BC309VI	OEM
BC310	0086
BC311	0126
BC312	0086
BC313	0126
BC313-6	OEM
BC313-10	OEM
BC313-16	OEM
BC313A	0126
BC313A-6	OEM
BC313A-10	OEM
BC313A-16	OEM
BC315	0037
BC315L	OEM
BC317	0079
BC317A	0079
BC317B	0079
BC318	0079
BC318A	0079
BC318B	0079
BC318C	0079
BC319	0079
BC319B	0079
BC319C	0079
BC320	0037
BC320A	0693
BC320B	OEM
BC321	0693
BC321A	0693
BC321B	0693
BC322	0693
BC322B	0693
BC322C	0693
BC323	0622
BC324	0086
BC325	0037
BC325A	0037
BC326	0037
BC326A	0037
BC327	0037
BC327(25)	0037
BC327-01	OEM
BC327-5	0037
BC327-10	OEM
BC327-16	0436
BC327-18	0037
BC327-25	0436
BC327A	OEM
BC327A(16)	0791
BC327A16	OEM
BC327A25	OEM
BC327AP	0037
BC327BP	OEM
BC327CP	0037
BC328	0006
BC328/B338	OEM
BC328-01	OEM
BC328-5	0037
BC328-10	OEM
BC328-16	0436
BC328-18	0037
BC328-25	0436
BC328-40	OEM
BC328AP	0037
BC328BP	0037
BC328C	0037
BC328CP	0037
BC329B	OEM
BC330	OEM
BC330B	OEM
BC330C	OEM
BC331A	OEM
BC331B	OEM
BC331C	OEM
BC332	OEM
BC332A	OEM
BC332B	OEM
BC332C	OEM
BC333	OEM
BC334	OEM
BC335	OEM
BC336	0018
BC337	0551
BC337(16)	0018
BC337(25)	0018
BC337/BC327	OEM
BC337-01	OEM
BC337-5	0079
BC337-10	OEM
BC337-16	0079
BC337-16-5	OEM
BC337-16-18	OEM
BC337-18	0079
BC-337-25	OEM
BC337-25	0079
BC337-25-5	OEM
BC337-25-18	OEM
BC337-40	0111
BC337A	OEM
BC337A-16	OEM
BC337A-25	OEM
BC337AP	0079
BC337BP	0079
BC337CP	0079
BC338	0079
BC338-01	OEM
BC338-5	0079
BC338-10	OEM
BC338-16	0079
BC338-16-5	OEM
BC338-16-18	OEM
BC338-18	OEM
BC338-20	OEM
BC338-25	0079
BC338-25-5	OEM
BC338-25-28	OEM
BC338-40	0079
BC338AP	OEM
BC338BP	OEM
BC338CP	OEM
BC340	OEM
BC340-06	0018
BC340-6	0086
BC340-10	0086
BC340-16	0086
BC341	OEM
BC341-06	0086
BC341-6	0086
BC341-10	0086
BC342	OEM
BC343	OEM
BC344	OEM
BC345	OEM
BC347	OEM
BC347A	OEM
BC347B	OEM
BC347L	OEM
BC348	OEM
BC348A	OEM
BC348B	OEM
BC348L	OEM
BC349	OEM
BC349A	OEM
BC349B	OEM
BC349L	OEM
BC350	0037
BC350A	0037
BC350B	0037
BC350L	0037
BC351	0037
BC351A	0037
BC351B	0037
BC351L	0037
BC352	0037
BC352A	0037
BC352B	0037
BC352L	0037
BC354	OEM
BC355	OEM
BC355A	OEM
BC355B	OEM
BC355C	OEM
BC357	OEM
BC358	OEM
BC358C	OEM
BC358VI	OEM
BC360	OEM
BC360-06	0532
BC360-6	0126
BC360-10	0126
BC360-16	0126
BC361	OEM
BC361-06	0791
BC361-6	0126
BC361-10	0126
BC362	OEM
BC363	OEM
BC364	OEM
BC365	0551
BC366	OEM
BC367	OEM
BC368	0324
BC368-25	0439
BC369	0532
BC369J	OEM
BC370	0791
BC370VI	0126
BC371	OEM
BC372C	OEM
BC375	3699
BC376	3700
BC377	0016
BC378	0016
BC381	0037
BC382	0016
BC382B	0079
BC382C	0079
BC382L	OEM
BC383	0016
BC383B	0079
BC383C	0016
BC383L	OEM
BC384	0016
BC384B	0079
BC384C	0111
BC384L	OEM
BC385	OEM
BC385A	0111
BC385B	0111
BC386	OEM
BC386A	0111
BC386B	0111
BC387	OEM
BC388	OEM
BC389	OEM
BC390	OEM
BC391	OEM
BC393	0342
BC394	0233
BC395	0861
BC396	0126
BC400	0037
BC404	0037
BC404A	0126
BC404V1	0126
BC404VI	0037
BC405	0037
BC405A	0037
BC405B	0037
BC406	0791
BC406B	0037
BC407	0016
BC407A	0016
BC407B	0076
BC408	0016
BC408A	0111
BC408B	0111
BC408C	0111
BC409	0016
BC409B	0079
BC409C	0079
BC411	OEM
BC412	0698
BC413	0079
BC413B	0079
BC413BP	0079
BC413C	0079
BC414	0016
BC414B	0079
BC414BP	0079
BC414C	0079
BC414CP	2860
BC415	0150
BC415-6	0037
BC415A	0037
BC415AP	0037
BC415B	0037
BC415BP	0037
BC415C	0693
BC415CP	OEM
BC416	0150
BC416-6	0037
BC416A	0037
BC416AP	0037
BC416B	0037
BC416BP	0037
BC416C	0693
BC416CP	0693
BC417	0037
BC418	0037
BC418A	0037
BC418B	0532
BC419	0037
BC419A	0037
BC419B	0532
BC420VI	0342
BC424	OEM
BC425	OEM
BC426	OEM
BC427	OEM
BC429	0086
BC430	0551
BC431	0086
BC431B	0079
BC432	0037
BC437A	OEM
BC437B	OEM
BC438A	OEM
BC438B	OEM
BC438C	OEM
BC440	0622
BC441	0622
BC442	0127
BC445	0037
BC445-5	0037
BC445-18	0037
BC446	0037
BC446-5	0037
BC446-18	0037
BC447	0037
BC447-5	0037
BC447-18	0037
BC448	OEM
BC448-5	0037
BC448-18	OEM
BC449	0079
BC449-5	0037
BC449-18	0861
BC450	0477
BC450-5	0477
BC450-18	0037
BC451	OEM
BC452	0079
BC453	OEM
BC454	OEM
BC455	0079
BC456	0224
BC460	0892
BC461	0892
BC461-4	0126
BC461-5	0126
BC461-6	0126
BC467A	OEM
BC467B	OEM
BC468A	OEM
BC468B	OEM
BC468C	OEM
BC469B	OEM
BC469C	OEM
BC477	0126
BC477A	0126
BC477V1	0126
BC477VI	0861
BC478	0037
BC478A	0037
BC478B	0037
BC479	0037
BC479B	0037
BC485	0037
BC485-5	0037
BC485-18	0037
BC485A	0037
BC485A5	0037
BC485A18	0037
BC485B	0037
BC485B5	0037
BC485B18	0037
BC485L	0037
BC485L5	0037
BC485L18	0037
BC486	0037
BC486-5	0037
BC486-18	0037
BC486A	0037
BC486A5	0037
BC486A18	0037
BC486B	0037
BC486B5	0037
BC486L	0037
BC486L5	0037
BC487	0037
BC487-5	0037
BC487-18	0037
BC487A	0037
BC487A5	0037
BC487A18	0037
BC487B	0037
BC487B5	0037
BC487B18	0037
BC487L	0037
BC487L5	0037
BC487L18	0037
BC488	0037
BC488-5	0037
BC488-18	0037
BC488A	0037
BC488A5	0037
BC488A18	0037
BC488B	0037
BC488B5	0037
BC488B18	0037
BC488L	0037
BC488L5	0037
BC488L18	0037
BC489	0037
BC489-5	0037
BC489-18	0037
BC489A	0037
BC489A5	0037
BC489A18	0037
BC489B	0037
BC489B5	0037
BC489B18	0037
BC489L	0037
BC489L5	0037
BC489L18	0037
BC490	0037
BC490-5	0037
BC490-18	0037
BC490A	0037
BC490A5	0037
BC490A18	0037
BC490B	OEM
BC490B5	0037
BC490B18	0037
BC490L	0037
BC490L5	0037
BC490L18	0037
BC507A	0016
BC507B	0016
BC508	0595
BC508A	0016
BC508B	0016
BC508C	0016
BC509	0595
BC509B	0016
BC509C	0016
BC510	0595
BC510B	0016
BC510C	0127
BC512	0037
BC512A	0037
BC512B	0791
BC513	0037
BC513A	0037
BC513B	0791
BC513C	0532
BC514	0791
BC514A	0037
BC514B	0791
BC514C	0532
BC516	OEM
BC517	OEM
BC517-P	OEM
BC517W	OEM
BC526	OEM
BC527	0126
BC528	0126
BC530	OEM
BC531	OEM
BC532	0861
BC533	0861
BC534	0126
BC535	0086
BC537	0086
BC538	0086
BC546	0111
BC546A	0111
BC546AP	0079
BC546B	0111
BC546BP	0079
BC546VIP	0079
BC547	0079
BC547A	0079
BC547AP	0079
BC547B	0079
BC547BP	0079
BC547C	0079
BC547VI	0079
BC547VIP	0079
BC548	0079
BC548A	0111
BC548AP	0079
BC548B	0079
BC548BC	0079
BC548BP	0079
BC548C	5932
BC548CP	0111
BC548VI	0079
BC548VIP	0079
BC549	0079
BC549B	0079
BC549BP	0079
BC549C	0111
BC550	0079
BC550B	0079
BC550BP	0079
BC550C	0111
BC550CP	0111
BC556	0037
BC556A	0037
BC556B	0037
BC556PA	OEM
BC556PAK	OEM
BC556PAL	OEM
BC556PAM	OEM
BC556PB	OEM
BC556PBK	OEM
BC556PBL	OEM
BC556PBM	OEM
BC556PV1	OEM
BC556PV1K	OEM
BC556PV1L	OEM
BC556PV1M	OEM
BC556VI	0037
BC557	0037
BC557A	0037
BC557B	0037
BC557PA	OEM
BC557PAK	OEM
BC557PAL	OEM
BC557PAM	OEM
BC557PB	OEM
BC557PBK	OEM
BC557PBL	OEM
BC557PBM	OEM
BC557PV1	OEM
BC557PV1K	OEM
BC557PV1L	OEM
BC557PV1M	OEM
BC557VI	0037
BC558	0037
BC558(EUROPE)	2033
BC558A	0037
BC558B	0037
BC558B2	OEM
BC558C	0807
BC558PA	OEM
BC558PAK	OEM
BC558PAL	OEM
BC558PAM	OEM
BC558PB	OEM
BC558PBK	OEM
BC558PBL	OEM
BC558PBM	OEM
BC558PC	OEM
BC558PCK	OEM
BC558PCL	OEM
BC558PCM	OEM
BC558PV1	OEM
BC558PV1K	OEM
BC558PV1L	OEM
BC558PV1M	OEM
BC558VI	0037
BC559	0037
BC559A	0037
BC559B	0037
BC559C	3519
BC559PA	OEM
BC559PAK	OEM
BC559PAL	OEM
BC559PAM	OEM
BC559PB	OEM
BC559PBK	OEM
BC559PBL	OEM
BC559PBM	OEM
BC559PC	OEM
BC559PCK	OEM
BC559PCL	OEM
BC559PCM	OEM
BC560	0037
BC560A	0037
BC560B	0532
BC560C	0693
BC560PA	0037
BC560PAK	0037
BC560PAL	0037
BC560PAM	0037
BC560PB	0037
BC560PBK	0037
BC560PBL	0037
BC560PBM	0037
BC560PC	0693
BC560PCK	0693
BC560PCL	OEM
BC560PCM	0693
BC582	0079
BC582A	0079
BC582B	0079
BC583	0079
BC583A	0079
BC583B	0079
BC583C	0400
BC584	0595
BC584B	0400
BC584C	0400
BC585	OEM
BC586	OEM
BC586L18	OEM
BC612	OEM
BC612L	OEM
BC617	3799
BC618	3799
BC635	1386
BC635/636	OEM
BC636	0477
BC637	0324
BC637/638	OEM
BC638	0477
BC638-BC640	OEM
BC639	0086
BC639/640	OEM
BC640	0477
BC682	0086
BC727	0126
BC728	0126
BC737	0111
BC738	0111
BC807	1752
BC807-16	1499
BC807-25	1499
BC807-40	OEM
BC808	1752
BC808-16	1499
BC808-25	1499
BC808-40	OEM
BC817	1733
BC817-16	1435
BC817-25	1435
BC817-40	OEM
BC818	1733
BC818-16	1435
BC818-25	1435
BC818-40	OEM
BC846	0724
BC846A	0724
BC846B	5385
BC847A	0724
BC847B	0724
BC847C	OEM
BC848A	0724
BC848B	5385
BC848C	OEM
BC849B	5385
BC849C	OEM
BC850	0724
BC850B	5385
BC850C	OEM
BC856	1742
BC856A	1742
BC856B	1742
BC857	1742
BC857A	1742
BC857B	1742
BC857C	OEM
BC858	1742
BC858A	1742
BC858B	1742
BC858C	OEM
BC859	1742
BC859A	3886
BC859B	3886
BC859C	OEM
BC860	1742
BC860A	3886
BC860B	3886
BC860C	OEM
BC868	0667
BC869	3886
BC876	3890
BC877	3892
BC878	3890
BC879	3892
BC880	3890
BC968	OEM
BC-1072	0016
BC1073	0085
BC1073A	0085
BC1073B	OEM
BC-1082	0016
BC-1086	0016
BC-1096	0016
BC1096	0016
BC1274	0085
BC1274A	0085
BC1274B	0085
BC1473	0261
BC1478	0016
BC-1690	0016
BC2290	OEM
BC3337	0018
BC6500	0144
BC486518	OEM
BCD08	5015
BCD10	5015
BCD15	5015
BCD20	OEM
BCD25	OEM
BCD30	OEM
BCD35	OEM
BCD40	OEM
BCD45	OEM
BCD50	OEM
BCD4024BF3	OEM
BCF5	0229
BCF29	1742
BCF29R	OEM
BCF30	1742
BCF30R	OEM
BCF32	0724
BCF32R	OEM
BCF33	0724
BCF33R	OEM
BCF70	1742
BCF70R	OEM
BCF81	0724
BCF81R	OEM
BCK53	OEM
BCM15-60	OEM
BCM15-60E	OEM
BCM15-60J	OEM
BCM15-100	OEM
BCM15-100E	OEM
BCM15-100J	OEM
BCM15-200	OEM
BCM15-200E	OEM
BCM15-200J	OEM
BCM15-300	OEM

If replacement code is OEM, contact original manufacturer for replacement.

DEVICE TYPE	REPL CODE
BCM15-300E	OEM
BCM15-300J	OEM
BCM1002-1	0050
BCM1002-2	0127
BCM1002-3	0164
BCM1002-4	0004
BCM1002-5	0004
BCM1002-6	0222
BCM1002-18	0164
BCM5101-4	OEM
BCP51	OEM
BCP52	OEM
BCP53	OEM
BCP54	OEM
BCP55	OEM
BCP56	OEM
BCP68	OEM
BCP69	OEM
BCR1AM4	OEM
BCR1AM6	OEM
BCR1AM8	OEM
BCR1AM10	OEM
BCR1AM12	OEM
BCR3AM4	OEM
BCR3AM8	OEM
BCR3AM10	OEM
BCR3AM12	OEM
BCR5A2	OEM
BCR5A4	OEM
BCR5A6	OEM
BCR5A8	OEM
BCR5B2	OEM
BCR5B4	OEM
BCR5B6	OEM
BCR5B8	OEM
BCR5C2	OEM
BCR5C4	OEM
BCR5C6	OEM
BCR5C8	OEM
BCR6AM4L	OEM
BCR6AM4R	OEM
BCR6AM6L	OEM
BCR6AM6R	OEM
BCR6AM8L	OEM
BCR6AM8R	OEM
BCR6AM10L	OEM
BCR6AM10R	OEM
BCR6AM12L	OEM
BCR6AM12R	OEM
BCR8A4L	OEM
BCR8A4R	OEM
BCR8A6L	OEM
BCR8A6R	OEM
BCR8A8L	OEM
BCR8A8R	OEM
BCR8A10L	OEM
BCR8A10R	OEM
BCR8CM4L	OEM
BCR8CM4R	OEM
BCR8CM6L	OEM
BCR8CM6R	OEM
BCR8CM8L	OEM
BCR8CM8R	OEM
BCR8CM10L	OEM
BCR8CM10R	OEM
BCR8CM12L	OEM
BCR8CM12R	OEM
BCR8DM	OEM
BCR8DM4L	OEM
BCR8DM4R	OEM
BCR8DM6L	OEM
BCR8DM6R	OEM
BCR8DM8L	OEM
BCR8DM8R	OEM
BCR8DM10L	OEM
BCR8DM10R	OEM
BCR8DM12L	OEM
BCR8DM12R	OEM
BCR10A2	OEM
BCR10A4	OEM
BCR10A6	OEM
BCR10A8	OEM
BCR10AM4L	OEM
BCR10AM4R	OEM
BCR10AM6L	OEM
BCR10AM6R	OEM
BCR10AM8L	0744
BCR10AM8R	0744
BCR10AM10L	0617
BCR10AM10R	0617
BCR10AM12L	OEM
BCR10AM12R	OEM
BCR10B2	OEM
BCR10B4	OEM
BCR10B6	OEM
BCR10B8	OEM
BCR10C2	OEM
BCR10C4	OEM
BCR10C6	OEM
BCR10C8	OEM
BCR10CM4L	OEM
BCR10CM4R	OEM
BCR10CM6L	OEM
BCR10CM6R	OEM
BCR10CM-8L	OEM
BCR10CM8L	OEM
BCR10CM8R	OEM
BCR10CM10L	OEM
BCR10CM10R	OEM
BCR10CM12L	OEM
BCR10CM12R	OEM
BCR10DM4L	OEM
BCR10DM4R	OEM
BCR10DM6L	OEM
BCR10DM6R	OEM
BCR10DM8L	OEM
BCR10DM8R	OEM
BCR10DM10L	OEM
BCR10DM10R	OEM
BCR10DM12L	OEM
BCR10DM12R	OEM
BCR10EM4L	OEM
BCR10EM4R	OEM
BCR10EM6L	OEM
BCR10EM6R	OEM
BCR10EM8L	OEM
BCR10EM-10L	OEM
BCR10EM-10R	OEM
BCR10EM12L	OEM
BCR10EM12R	OEM
BCR10PM	OEM
BCR12AM4L	OEM
BCR12AM4R	OEM
BCR12AM6L	OEM
BCR12AM6R	OEM
BCR12AM8L	OEM
BCR12AM10L	OEM
BCR12AM10R	OEM
BCR12AM12L	OEM
BCR12AM12R	OEM
BCR12CM4L	OEM
BCR12CM4R	OEM
BCR12CM6L	OEM
BCR12CM6R	OEM
BCR12CM8L	OEM
BCR12CM8R	OEM
BCR12CM10L	OEM
BCR12CM10R	OEM
BCR12CM12L	OEM
BCR12CM12R	OEM
BCR12DM4L	OEM
BCR12DM4R	OEM
BCR12DM6L	OEM
BCR12DM6R	OEM
BCR12DM8L	OEM
BCR12DM8R	OEM
BCR12DM10L	OEM
BCR12DM10R	OEM
BCR12DM12L	OEM
BCR12DM12R	OEM
BCR12EM4L	OEM
BCR12EM4R	OEM
BCR12EM6L	OEM
BCR12EM6R	OEM
BCR12EM8L	OEM
BCR12EM8R	OEM
BCR12EM10L	OEM
BCR12EM10R	OEM
BCR12EM12L	OEM
BCR12EM12R	OEM
BCR16A4L	OEM
BCR16A4R	OEM
BCR16A6L	OEM
BCR16A6R	OEM
BCR16A8L	OEM
BCR16A8R	OEM
BCR16A10L	OEM
BCR16A10R	OEM
BCR16AM-4L	OEM
BCR16AM-4R	OEM
BCR16AM-6L	OEM
BCR16AM-6R	OEM
BCR16AM-8L	OEM
BCR16AM-8R	OEM
BCR16AM-10L	OEM
BCR16AM-10R	OEM
BCR16B4L	OEM
BCR16B4R	OEM
BCR16B6L	OEM
BCR16B6R	OEM
BCR16B8L	OEM
BCR16B8R	OEM
BCR16B10L	OEM
BCR16B10R	OEM
BCR16C4L	OEM
BCR16C4R	OEM
BCR16C6L	OEM
BCR16C6R	OEM
BCR16C8L	OEM
BCR16C8R	OEM
BCR16C10L	OEM
BCR16C10R	OEM
BCR16CM4L	OEM
BCR16CM4R	OEM
BCR16CM6L	OEM
BCR16CM6R	OEM
BCR16CM8L	OEM
BCR16CM8R	OEM
BCR16CM10L	OEM
BCR16CM10R	OEM
BCR16CM12L	OEM
BCR16CM12R	OEM
BCR16DM4L	OEM
BCR16DM4R	OEM
BCR16DM6L	OEM
BCR16DM6R	OEM
BCR16DM8L	OEM
BCR16DM8R	OEM
BCR16DM10L	OEM
BCR16DM10R	OEM
BCR16DM12L	OEM
BCR16E4L	OEM
BCR16E4R	OEM
BCR16E6L	OEM
BCR16E6R	OEM
BCR16E8L	OEM
BCR16E8R	OEM
BCR16E10L	OEM
BCR16E10R	OEM
BCR16EM-4L	OEM
BCR16EM-4R	OEM
BCR16EM-6L	OEM
BCR16EM-6R	OEM
BCR16EM-8L	OEM
BCR16EM-8R	OEM
BCR16EM-10L	OEM
BCR16EM-10R	OEM
BCR16FM-4L	OEM
BCR16FM-4R	OEM
BCR16FM-6L	OEM
BCR16FM-6R	OEM
BCR16FM-8L	OEM
BCR16FM-8R	OEM
BCR16FM-10L	OEM
BCR16FM-10R	OEM
BCR16GM-4L	OEM
BCR16GM-4R	OEM
BCR16GM-6L	OEM
BCR16GM-6R	OEM
BCR16GM-8L	OEM
BCR16GM-8R	OEM
BCR16GM-10L	OEM
BCR16GM-10R	OEM
BCR16HM4L	OEM
BCR16HM4R	OEM
BCR16HM6L	OEM
BCR16HM6R	OEM
BCR16HM8L	OEM
BCR16HM8R	OEM
BCR16HM10L	OEM
BCR16M-4L	OEM
BCR16M-4R	OEM
BCR16M-6L	OEM
BCR16M-6R	OEM
BCR16M-8L	OEM
BCR16M-8R	OEM
BCR16M-10L	OEM
BCR16M-10R	OEM
BCR20A4L	OEM
BCR20A4R	OEM
BCR20A6L	OEM
BCR20A6R	OEM
BCR20A8L	OEM
BCR20A8R	OEM
BCR20A10L	OEM
BCR20A10R	OEM
BCR20B4L	OEM
BCR20B4R	OEM
BCR20B6L	OEM
BCR20B6R	OEM
BCR20B8L	OEM
BCR20B8R	OEM
BCR20B10L	OEM
BCR20B10R	OEM
BCR20C4L	OEM
BCR20C4R	OEM
BCR20C6L	OEM
BCR20C6R	OEM
BCR20C8L	OEM
BCR20C8R	OEM
BCR20C10L	OEM
BCR20C10R	OEM
BCR20E4L	OEM
BCR20E4R	OEM
BCR20E6L	OEM
BCR20E6R	OEM
BCR20E8L	OEM
BCR20E8R	OEM
BCR20E10L	OEM
BCR20E10R	OEM
BCR25A4L	OEM
BCR25A4R	OEM
BCR25A6L	OEM
BCR25A6R	OEM
BCR25A8L	OEM
BCR25A8R	OEM
BCR25A10L	OEM
BCR25A10R	OEM
BCR25B4	0908
BCR25B4L	OEM
BCR25B4R	OEM
BCR25B6	0908
BCR25B6L	OEM
BCR25B6R	OEM
BCR25B8	OEM
BCR25B8L	OEM
BCR25B8R	OEM
BCR25B10	OEM
BCR25B10L	OEM
BCR25B10R	OEM
BCR30GM	OEM
BCR30GM4L	OEM
BCR30GM4R	OEM
BCR30GM6L	OEM
BCR30GM6R	OEM
BCR30GM8L	OEM
BCR30GM8R	OEM
BCR30GM10L	OEM
BCR30GM10R	OEM
BCR50A4	OEM
BCR50A6	OEM
BCR50A8	OEM
BCR50A12	OEM
BCR50A16	OEM
BCR70A4	OEM
BCR70A6	OEM
BCR70A8	OEM
BCR70A12	OEM
BCR70A16	OEM
BCR70A20	OEM
BCR70B4	OEM
BCR70B6	OEM
BCR70B8	OEM
BCR70B12	OEM
BCR70B16	OEM
BCR70B20	OEM
BCR70B24	OEM
BCR150A4	OEM
BCR150A6	OEM
BCR150A8	OEM
BCR150A12	OEM
BCR150A16	OEM
BCR150A20	OEM
BCR150A24	OEM
BCR150B4	OEM
BCR150B6	OEM
BCR150B8	OEM
BCR150B12	OEM
BCR150B16	OEM
BCR150B20	OEM
BCR150B24	OEM
BCR300A4	OEM
BCR300A6	OEM
BCR300A8	OEM
BCR300A12	OEM
BCR300A16	OEM
BCR300A20	OEM
BCR300A24	OEM
BCR300B4	OEM
BCR300B6	OEM
BCR300B8	OEM
BCR300B12	OEM
BCR300B16	OEM
BCR300B20	OEM
BCR300B24	OEM
BCS78-25	OEM
BCU75EPAK	OEM
BCV26	5365
BCV27	5364
BCV46	OEM
BCV47	OEM
BCV61	OEM
BCV62	OEM
BCV71	0724
BCV71R	OEM
BCV72	5385
BCV72R	OEM
BCW10	OEM
BCW11	OEM
BCW12	OEM
BCW13	OEM
BCW14	OEM
BCW15	OEM
BCW16	OEM
BCW17	OEM
BCW18	OEM
BCW19	OEM
BCW20	OEM
BCW21	OEM
BCW22	OEM
BCW23	OEM
BCW24	OEM
BCW25	OEM
BCW26	OEM
BCW29	0150
BCW29R	0150
BCW30	0150
BCW30R	0150
BCW31	1733
BCW31R	0144
BCW32	0144
BCW32R	0144
BCW33	0280
BCW33R	0280
BCW34	0016
BCW35	0037
BCW36	0016
BCW37	0037
BCW38C	OEM
BCW44	0086
BCW45	0037
BCW46	0086
BCW46A	0079
BCW46B	0079
BCW47	0086
BCW47A	0079
BCW47B	0079
BCW48	0086
BCW48A	0079
BCW48B	0079
BCW48C	0079
BCW49	0086
BCW49B	0079
BCW49C	0079
BCW50	0285
BCW51	0595
BCW52	0791
BCW56	0037
BCW56A	0037
BCW57	0037
BCW57A	0037
BCW57B	0532
BCW58	0037
BCW58A	0037
BCW58B	0037
BCW59	0037
BCW59A	0037
BCW59B	0693
BCW60	5385
BCW60A	0016
BCW60AA	0016
BCW60AB	0079
BCW60AC	0079
BCW60B	0079
BCW60C	0079
BCW60FF	OEM
BCW60FN	OEM
BCW60R	OEM
BCW61	0037
BCW61A	0791
BCW61B	0791
BCW61BA	0791
BCW61BB	0791
BCW61BC	0693
BCW61BD	0693
BCW61C	0037
BCW61D	0037
BCW61FF	OEM
BCW61FN	OEM
BCW61R	OEM
BCW62	0037
BCW62A	0037
BCW62B	0791
BCW63	0037
BCW63A	0037
BCW63B	0791
BCW63C	0532
BCW64	0791
BCW64A	0037
BCW64B	0791
BCW64C	0532
BCW65	1733
BCW65A	1733
BCW65B	1733
BCW65C	OEM
BCW65EA	0079
BCW65EB	0079
BCW65EC	0018
BCW65R	OEM
BCW66	1733
BCW66EF	0018
BCW66EG	0018
BCW66EH	0018
BCW66EW	0018
BCW66F	1733
BCW66G	1733
BCW66H	1435
BCW66R	OEM
BCW67	1733
BCW67A	1733
BCW67B	1733
BCW67C	1733
BCW67DA	0532
BCW67DB	0532
BCW67DC	0532
BCW67R	OEM
BCW67RA	OEM
BCW68	1733
BCW68DF	0532
BCW68DG	0532
BCW68DH	0532
BCW68F	1733
BCW68G	1733
BCW68H	OEM
BCW68R	OEM
BCW68RH	OEM
BCW69	1742
BCW69R	0126
BCW70	1742
BCW70R	0791
BCW71	0724
BCW71R	0144
BCW72	0144
BCW72R	0144
BCW73	0016
BCW73-16	0595
BCW73-25	0016
BCW74	0016
BCW74-16	0595
BCW74-25	0016
BCW75	OEM
BCW75-10	0791
BCW75-16	0791
BCW75-25	0037
BCW76	OEM
BCW76-10	0791
BCW76-16	0791
BCW77	OEM
BCW77-16	0595
BCW77-25	0086
BCW78	0086
BCW78-16	0595
BCW78-25	0086
BCW79	OEM
BCW79-10	0037
BCW79-16	0037
BCW79-25	0126
BCW80	0126
BCW80-10	0037
BCW80-16	0037
BCW80-25	0126
BCW81	1733
BCW81R	OEM
BCW82	0595
BCW82A	0595
BCW82B	0595
BCW83	0016
BCW83A	0218
BCW83B	0595
BCW84	0595
BCW84B	0286
BCW85	0342
BCW86	0037
BCW87	0111
BCW88	0693
BCW89	2332
BCW89R	OEM
BCW90	0595
BCW90A	0018
BCW90B	0018
BCW90C	0018
BCW90K	0595
BCW90KA	0018
BCW90KB	0018
BCW90KC	0018
BCW91	0595
BCW91A	0595
BCW91B	0595
BCW91K	0595
BCW91KA	0595
BCW91KB	0595
BCW92	0791
BCW92A	0532
BCW92B	0532
BCW92K	0791
BCW92KA	0037
BCW92KB	0037
BCW93	0791
BCW93A	0037
BCW93B	0037
BCW93K	0791
BCW93KA	0037
BCW93KB	0037
BCW94	0079
BCW94A	0079
BCW94B	0079
BCW94C	0079
BCW94K	0595
BCW94KA	0079
BCW94KB	0079
BCW94KC	0079
BCW95	0086
BCW95A	0079
BCW95B	0079
BCW95K	0595
BCW95KA	0079
BCW95KB	0079
BCW96	0037
BCW96A	0037
BCW96B	0037
BCW96K	0791
BCW96KA	0037
BCW96KB	0037
BCW97	0037
BCW97A	0037
BCW97B	0037
BCW97K	0791
BCW97KA	0037
BCW97KB	0037
BCW98A	0111
BCW98B	0111
BCW98C	0111
BCW98D	0111
BCW99A	0532
BCW99B	0532
BCW99C	0532
BCW314	OEM
BCW601	OEM
BCX10	0126
BCX10A	2260
BCX17	0037
BCX17R	0037
BCX18	0342
BCX18R	0342
BCX19	0079
BCX19R	0079
BCX20	0079
BCX20R	0079
BCX21	OEM
BCX22	0233
BCX23	0477
BCX24	0233
BCX25	OEM
BCX26	0037
BCX27	OEM
BCX28	OEM
BCX29	OEM
BCX30	0477
BCX31	0086
BCX32	0324
BCX33	0324
BCX34	0324
BCX35	0037
BCX36	0037
BCX38A	OEM
BCX38B	OEM
BCX38C	OEM
BCX39	OEM
BCX41	OEM
BCX41R	OEM
BCX42	OEM
BCX42R	OEM
BCX45	0037
BCX46	0037
BCX47	1386
BCX48	OEM
BCX49	1386
BCX50	0037
BCX51	3645
BCX51-6	OEM
BCX51-10	OEM
BCX51-16	OEM
BCX52	3645
BCX52-6	OEM
BCX52-10	OEM
BCX53	5406
BCX53-6	OEM
BCX53-10	OEM
BCX54	0667
BCX54-6	OEM
BCX54-10	OEM
BCX54-16	OEM
BCX55	0667
BCX55-6	OEM
BCX55-10	OEM
BCX56	0667
BCX56-6	OEM
BCX56-10	OEM
BCX58	0079
BCX58IX	0018
BCX58VII	0079
BCX58VIII	0079
BCX58X	0018
BCX59	0079
BCX59IX	0018
BCX59VII	0018
BCX59VIII	0018
BCX59X	0018
BCX68	0667
BCX68-10	OEM
BCX68-16	OEM
BCX68-25	OEM
BCX69	3645
BCX69-10	OEM
BCX69-16	OEM
BCX69-25	OEM
BCX69R	OEM
BCX70	0724
BCX70AG	0079
BCX70AH	0079
BCX70AJ	0079
BCX70G	0724
BCX70H	0724
BCX70J	0724
BCX70K	OEM
BCX70R	OEM
BCX71	1742
BCX71BG	0791
BCX71BH	0791
BCX71BJ	0693
BCX71BK	0693
BCX71G	1742
BCX71H	1742
BCX71J	1742
BCX71K	OEM
BCX71R	OEM
BCX73	OEM
BCX73-16	0018
BCX73-25	0018
BCX73-40	0018
BCX74	OEM
BCX74-16	0018
BCX74-25	0018
BCX74-40	0018
BCX75	OEM
BCX75-16	0532
BCX75-25	0532
BCX75-40	0532
BCX76	OEM
BCX76-16	OEM
BCX76-25	0532
BCX76-40	0532
BCX78	2260
BCX78IX	0532
BCX78VII	0150
BCX78VIII	0532
BCX79	2260
BCX79IX	0532
BCX79VII	0150
BCX79VIII	0532
BCX94	OEM
BCY10	0037
BCY11	0037
BCY11S	OEM
BCY12	0037
BCY13	0016
BCY14	OEM
BCY15	0016
BCY16	0016
BCY17	0126
BCY18	0126
BCY19	0126
BCY20	OEM

DEVICE TYPE	REPL CODE
BCY21	0126
BCY22	0126
BCY23	0126
BCY24	0126
BCY25	0126
BCY26	0126
BCY27	0126
BCY28	0126
BCY29	0126
BCY30	0126
BCY30A	OEM
BCY31	0126
BCY31A	OEM
BCY32	0126
BCY32A	OEM
BCY33	0126
BCY33A	OEM
BCY34	0126
BCY34A	0126
BCY35	0037
BCY36	0016
BCY37	0037
BCY38	0037
BCY39	0037
BCY40	0037
BCY42	0016
BCY43	0016
BCY46	0086
BCY47	0086
BCY48	0086
BCY49	0086
BCY50	0016
BCY50I	0050
BCY51	0016
BCY51I	0016
BCY54	0037
BCY55	0111
BCY56	0016
BCY57	0016
BCY58	0037
BCY58A	0016
BCY58B	0016
BCY58C	0016
BCY58D	0016
BCY58IX	0016
BCY58PA	OEM
BCY58PAK	OEM
BCY58PAL	OEM
BCY58PAM	OEM
BCY58PB	OEM
BCY58PBK	OEM
BCY58PBL	OEM
BCY58PBM	OEM
BCY58PC	OEM
BCY58PCK	OEM
BCY58PCL	OEM
BCY58PCM	OEM
BCY58PD	OEM
BCY58PDK	OEM
BCY58PDL	OEM
BCY58VII	0016
BCY58VIII	0016
BCY59	0016
BCY59-7	0016
BCY59-8	0016
BCY59-9	0016
BCY59-10	OEM
BCY59A	0016
BCY59B	0016
BCY59C	0016
BCY59D	0016
BCY59IX	0016
BCY59PA	OEM
BCY59PAK	OEM
BCY59PAL	OEM
BCY59PAM	OEM
BCY59PB	OEM
BCY59PBK	OEM
BCY59PBL	OEM
BCY59PBM	OEM
BCY59PC	OEM
BCY59PCK	OEM
BCY59PCL	OEM
BCY59PCM	OEM
BCY59PD	OEM
BCY59PDK	OEM
BCY59PDL	OEM
BCY59PDM	OEM
BCY59VII	0016
BCY59VIII	0016
BCY65	0086
BCY65E	0016
BCY65EA	OEM
BCY65EB	OEM
BCY65EC	OEM
BCY65EPA	OEM
BCY65EPAL	OEM
BCY65EPAM	OEM
BCY65EPB	OEM
BCY65EPBK	OEM
BCY65EPBL	OEM
BCY65EPBM	OEM
BCY65EPC	OEM
BCY65EPCK	OEM
BCY65EPCL	OEM
BCY65EPCM	OEM
BCY65EPD	OEM
BCY65EPDK	OEM
BCY65EPDL	OEM
BCY65EPDM	OEM
BCY65IX	0016
BCY65VII	0016
BCY65VIII	0016
BCY66	0086
BCY67	0126
BCY69	0016
BCY70	0037
BCY71	0037
BCY71A	0037
BCY72	0037
BCY72A	0037
BCY76	0016
BCY77	OEM
BCY77A	0037
BCY77B	0037
BCY77C	0037
BCY77PA	OEM
BCY77PAK	OEM
BCY77PAL	OEM
BCY77PAM	OEM
BCY77PB	OEM
BCY77PBK	OEM
BCY77PBL	OEM
BCY77PBM	OEM
BCY77PC	OEM
BCY77PCK	OEM
BCY77PCL	OEM
BCY77PCM	OEM
BCY77VII	0037
BCY77VIII	0037
BCY78	0693
BCY78-7	OEM
BCY78-8	OEM
BCY78-9	OEM
BCY78-10	OEM
BCY78A	0126
BCY78B	0126
BCY78C	0532
BCY78D	OEM
BCY78PA	OEM
BCY78PAK	OEM
BCY78PAL	OEM
BCY78PAM	OEM
BCY78PB	OEM
BCY78PBK	OEM
BCY78PBL	OEM
BCY78PBM	OEM
BCY78PC	OEM
BCY78PCK	OEM
BCY78PCL	OEM
BCY78PCM	OEM
BCY78PD	OEM
BCY78PDK	OEM
BCY78PDL	OEM
BCY78PDM	OEM
BCY78VII	0791
BCY78VIII	0791
BCY79	0037
BCY79-7	OEM
BCY79-8	OEM
BCY79-9	OEM
BCY79A	0791
BCY79B	0791
BCY79C	0532
BCY79PA	OEM
BCY79PAK	OEM
BCY79PAL	OEM
BCY79PAM	OEM
BCY79PB	OEM
BCY79PBK	OEM
BCY79PBL	OEM
BCY79PBM	OEM
BCY79PCK	OEM
BCY79PCL	OEM
BCY79PCM	OEM
BCY79VII	0037
BCY79VIII	0791
BCY84A	0016
BCY85	0086
BCY86	0086
BCY87	0144
BCY88	0144
BCY89	0144
BCY90	0037
BCY90B	0037
BCY91	0037
BCY91B	0037
BCY92	0037
BCY92B	0037
BCY93	0037
BCY93B	0037
BCY94	0037
BCY94B	0037
BCY95	0037
BCY95B	0037
BCY96	0126
BCY96B	0126
BCY97	0126
BCY97B	0126
BCY98	0037
BCY98B	0037
BCY443	0086
BCY501	0016
BCY511	0016
BCY771X	0037
BCZ6.8	OEM
BCZ6.8A	OEM
BCZ6.8B	OEM
BCZ7.5	OEM
BCZ7.5A	OEM
BCZ7.5B	OEM
BCZ8.2	OEM
BCZ8.2A	OEM
BCZ8.2B	OEM
BCZ9.1	OEM
BCZ9.1A	OEM
BCZ9.1B	OEM
BCZ10	0037
BCZ10(DIO)	OEM
BCZ10A	3108
BCZ10B	4357
BCZ11	0037
BCZ11(DIO)	OEM
BCZ11A	4357
BCZ11B	OEM
BCZ12	0037
BCZ12(DIO)	OEM
BCZ12A	4357
BCZ12B	4357
BCZ13	0037
BCZ13(DIO)	OEM
BCZ13A	4357
BCZ13B	6683
BCZ14	0037
BCZ15	OEM
BCZ15A	OEM
BCZ15B	OEM
BCZ16	OEM
BCZ16A	OEM
BCZ16B	OEM
BCZ18	OEM
BCZ18A	OEM
BCZ18B	OEM
BCZ20	OEM
BCZ20A	OEM
BCZ20B	OEM
BCZ22	OEM
BCZ22A	OEM
BCZ22B	OEM
BCZ24	OEM
BCZ24A	OEM
BCZ24B	OEM
BCZ27	OEM
BCZ27A	OEM
BCZ27B	OEM
BCZ30	OEM
BCZ30A	OEM
BCZ30B	OEM
BCZ33	OEM
BCZ33A	OEM
BCZ33B	OEM
BCZ36	OEM
BCZ36A	OEM
BCZ36B	OEM
BCZ39	OEM
BCZ39A	OEM
BCZ39B	OEM
BCZ43	OEM
BCZ43A	OEM
BCZ43B	OEM
BCZ47	OEM
BCZ47A	OEM
BCZ47B	OEM
BCZ51	OEM
BCZ51A	OEM
BCZ51B	OEM
BCZ56	OEM
BCZ56A	OEM
BCZ56B	OEM
BCZ62	OEM
BCZ62A	OEM
BCZ62B	OEM
BCZ66	OEM
BCZ66A	OEM
BCZ66B	OEM
BCZ75	OEM
BCZ75A	OEM
BCZ75B	OEM
BCZ82	OEM
BCZ82A	OEM
BCZ82B	OEM
BCZ91	OEM
BCZ91A	OEM
BCZ91B	OEM
BCZ100	OEM
BCZ100A	OEM
BCZ100B	OEM
BCZ110	OEM
BCZ110A	OEM
BCZ110B	OEM
BCZ120	OEM
BCZ120A	OEM
BCZ120B	OEM
BCZ130	OEM
BCZ130A	OEM
BCZ130B	OEM
BCZ150	OEM
BCZ150A	OEM
BCZ150B	OEM
BCZ160	OEM
BCZ160A	OEM
BCZ160B	OEM
BCZ180	OEM
BCZ180A	OEM
BCZ180B	OEM
BCZ200	OEM
BCZ200A	OEM
BCZ200B	OEM
BD-00072	0208
BD0A	0015
BD-1	OEM
BD1	0143
BD1-05	OEM
BD1-06	OEM
BD1-07	OEM
BD1-08	OEM
BD1-09	OEM
BD1-10	OEM
BD-1A	0015
BD1A	OEM
BD1B	OEM
BD-2	OEM
BD2	0143
BD-3	OEM
BD3	0143
BD3A-1B4	0277
BD-3A184	0277
BD3A-184	0229
BD3.6FB	0376
BD-4	OEM
BD4	0143
BD-5	OEM
BD5	0143
BD-6	OEM
BD6	0143
BD-7	OEM
BD7	0143
BD8-05	OEM
BD8-06	OEM
BD8-07	OEM
BD8-08	OEM
BD8-09	OEM
BD8-10	OEM
BD23B	OEM
BD23C	OEM
BD28	3331
BD61YA	OEM
BD71	0144
BD100	OEM
BD100C	OEM
BD101	OEM
BD102	OEM
BD106	0042
BD106A	0042
BD106B	0042
BD-107	0015
BD107	0560
BD107A	0042
BD107B	0042
BD109	0042
BD109-6	0042
BD111	0103
BD111A	0620
BD112	0620
BD113	0224
BD115	0233
BD116	0620
BD117	0543
BD118	0620
BD119	1027
BD120	1027
BD121	0160
BD123	0620
BD124	0042
BD124A	OEM
BD-127	0015
BD127	0276
BD128	0142
BD129	0142
BD130	0103
BD-131	2997
BD131	0161
BD-132	0460
BD132	0460
BD135	0161
BD135-6	0161
BD135-10	0161
BD135-16	0563
BD135-BD-136	OEM
BD135G	0560
BD136	0681
BD136-6	0460
BD136-10	0460
BD136-16	0525
BD136G	1267
BD137	0161
BD137-6	0161
BD137-10	0161
BD137-BD138	OEM
BD137G	0560
BD138	0460
BD138-6	0460
BD138-10	0460
BD138G	1267
BD139	0161
BD139-6	0161
BD139-10	0161
BD139-BD140	OEM
BD139G	0560
BD140	0460
BD140-6	0460
BD140-10	0460
BD140G	1267
BD141	0177
BD142	0103
BD144	0074
BD145	0620
BD148	0561
BD148-6	0561
BD148-10	0561
BD149	0161
BD149-6	0561
BD149-10	0561
BD150	OEM
BD151	0460
BD152	0460
BD153	0161
BD154	0161
BD155	0161
BD156	0460
BD157	0276
BD158	0276
BD160	0223
BD161	OEM
BD162	0042
BD163	0042
BD165	0161
BD166	0460
BD167	0161
BD168	0460
BD169	0161
BD170	0525
BD171	1708
BD172	1708
BD173	1708
BD175	0161
BD176	0460
BD177	0161
BD178	0460
BD179	0161
BD180	0460
BD181	0103
BD182	0103
BD183	0460
BD184	0543
BD185	0161
BD186	0460
BD187	0161
BD188	0460
BD189	0161
BD190	0460
BD195	0561
BD196	1199
BD197	0561
BD198	1199
BD199	0561
BD200	1199
BD201	0482
BD201-BD202	OEM
BD202	1369
BD203	0482
BD203-BD204	OEM
BD204	1369
BD204M	OEM
BD205	0561
BD206	1199
BD207	0561
BD208	1199
BD213	OEM
BD213/45	0561
BD213/60	0561
BD213/80	0561
BD214	OEM
BD214/45	1199
BD214/60	1199
BD214/80	1199
BD214-45	1199
BD215	0276
BD216	0276
BD220	0042
BD221	0042
BD222	0042
BD223	0925
BD224	0925
BD225	0925
BD226	0161
BD227	0460
BD228	0161
BD228-BD229	OEM
BD229	0460
BD230-BD231	OEM
BD232	0276
BD232G	OEM
BD233	0161
BD233G	0236
BD234	0460
BD234G	0681
BD235	0161
BD235-BD236	OEM
BD235G	0236
BD236	0460
BD236G	0681
BD237	0563
BD237-BD238	OEM
BD237G	0236
BD238	0313
BD238G	0681
BD239	0236
BD239A	0236
BD239B	0236
BD240	0925
BD240A	0925
BD240B	0681
BD240C	0681
BD241	0042
BD241A	2997
BD241B	0236
BD241C	0236
BD242	0925
BD242A	0482
BD242B	0925
BD242C	0681
BD243	0042
BD243A	0561
BD243B	0042
BD244	0925
BD244A	0925
BD244B	0925
BD244C	1199
BD245	3081
BD245A	3081
BD245B	3081
BD245C	3081
BD246	3603
BD246A	3603
BD246B	3603
BD246C	3603
BD249	0130
BD249A	0130
BD249B	0130
BD249C	0130
BD250	1681
BD250A	1681
BD250B	1681
BD250C	1681
BD253	OEM
BD253A	OEM
BD253B	OEM
BD253C	OEM
BD260	OEM
BD261	OEM
BD262	2897
BD262(DARLINGTON)	2897
BD262A	2897
BD262A(DARLINGTON)	2897
BD262B	OEM
BD262C	OEM
BD262L	OEM
BD263	0558
BD263A	3529
BD263B	OEM
BD263C	OEM
BD263L	OEM
BD264	2237
BD264A	2237
BD264B	OEM
BD264L	OEM
BD265	2235
BD265A	OEM
BD265B	OEM
BD265L	OEM
BD266	2237
BD266A	2237
BD266B	6554
BD266L	OEM
BD267	2235
BD267A	2235
BD267B	5504
BD267L	OEM
BD271	0042
BD272	0925
BD273	0423
BD274	0681
BD275	0423
BD276	0854
BD277	0854
BD278	0423
BD278A	1361
BD278AE	0423
BD278E	0423
BD279	OEM
BD280	OEM
BD287	OEM
BD288	OEM
BD291	0423
BD292	0681
BD293	0423
BD294	0681
BD295	OEM
BD296	OEM
BD301	0561
BD302	1199
BD303	0561
BD304	1199
BD306A	0161
BD307A	0161
BD311	0177
BD312	0491
BD313	0177
BD314	1681
BD315	2414
BD316	1681
BD317	2414
BD318	1681
BD320A	OEM
BD320B	OEM
BD320C	OEM
BD321A	OEM
BD321B	OEM
BD321C	OEM
BD322A	OEM
BD322B	OEM
BD322C	OEM
BD323A	OEM
BD323B	OEM
BD323C	OEM
BD328	OEM
BD329	0563
BD330	0525
BD331	4077
BD332	4078
BD333	4077
BD334	4078
BD335	4077
BD336	4078
BD337	4077
BD338	4078
BD344	0525
BD345	0563
BD346	OEM
BD347	OEM
BD348	0525
BD349	0563
BD361	OEM
BD361A	OEM
BD362	OEM
BD362A	OEM
BD370A	OEM
BD370A-6	OEM
BD370A-10	OEM
BD370A-16	OEM
BD370B	OEM
BD370B-6	OEM
BD370B-10	OEM
BD370B-16	OEM
BD370C	OEM
BD370C-6	OEM
BD370C-10	OEM
BD370D	OEM
BD371	2525
BD371A	OEM
BD371A-6	OEM
BD371A-10	OEM
BD371A-16	OEM
BD371B	OEM
BD371B-6	OEM
BD371B-10	OEM
BD371B-16	OEM
BD371C	OEM
BD371C-6	OEM
BD371C-10	OEM
BD371D	OEM
BD372A	OEM
BD372A-6	OEM
BD372A-10	OEM
BD372A-16	OEM
BD372B	OEM
BD372B-6	OEM
BD372B-10	OEM
BD372B-16	OEM
BD372C-6	OEM
BD372C-10	OEM
BD372D	OEM
BD373A	OEM
BD373A-6	OEM
BD373A-10	OEM
BD373A-16	OEM
BD373B	OEM
BD373B-6	OEM
BD373B-10	OEM
BD373B-16	OEM
BD373C	OEM
BD373C-6	OEM
BD373C-10	OEM
BD373D	4250
BD375	0161
BD375-6	OEM
BD375-10	OEM
BD375-16	OEM
BD375-25	OEM
BD376	0460
BD376-6	OEM
BD376-10	OEM
BD376-16	OEM
BD376-25	OEM
BD377	0161
BD377-6	OEM
BD377-10	OEM
BD377-16	OEM
BD377-25	OEM
BD378	0460
BD378-6	OEM
BD378-10	OEM
BD378-16	OEM
BD378-25	OEM
BD379-6	OEM
BD379-10	OEM
BD379-16	OEM
BD379-25	OEM
BD380-6	OEM
BD380-10	OEM
BD380-16	OEM
BD380-25	OEM
BD401	OEM
BD-402	OEM
BD402	OEM
BD-403	OEM
BD403	OEM
BD-404	OEM
BD404	OEM
BD-405	OEM
BD405	OEM
BD-406	OEM

DEVICE TYPE	REPL CODE
BD406	OEM
BD-407	OEM
BD407	OEM
BD415	0824
BD416	3327
BD417	0824
BD418	3327
BD424	OEM
BD424A	0130
BD424B	0130
BD427	OEM
BD429	OEM
BD430	OEM
BD-433	0042
BD433	0042
BD433-BD434	OEM
BD433A	0042
BD434	0925
BD434A	0925
BD435	0042
BD435-BD436	OEM
BD435A	0042
BD435C	OEM
BD436	0925
BD436A	0925
BD437	0042
BD437-BD438	OEM
BD438	0925
BD439	0042
BD440	0925
BD441	0423
BD442	0854
BD450	OEM
BD451	OEM
BD457	0080
BD457F	0080
BD457T	0080
BD458	OEM
BD458F	0609
BD458T	0609
BD461	0561
BD462	0460
BD463	0561
BD464	1199
BD466A	OEM
BD466B	OEM
BD477A	OEM
BD477B	OEM
BD483B	0080
BD483BF	0080
BD483BT	0080
BD487	OEM
BD488	OEM
BD497/01	0074
BD500	1199
BD500A	1199
BD500B	1199
BD501	0561
BD501A	0561
BD501B	0561
BD505	OEM
BD505-1	OEM
BD505-5	OEM
BD506	OEM
BD506-1	OEM
BD506-5	OEM
BD507	OEM
BD507-1	OEM
BD507-5	OEM
BD508	OEM
BD508-1	OEM
BD508-5	OEM
BD509	OEM
BD509-1	OEM
BD509-5	OEM
BD510	OEM
BD510-1	OEM
BD510-5	OEM
BD512	OEM
BD515	OEM
BD515-1	OEM
BD515-5	OEM
BD516	OEM
BD517	0551
BD517-1	OEM
BD517-5	OEM
BD518	OEM
BD519	OEM
BD519-1	OEM
BD519-5	OEM
BD520	OEM
BD522	OEM
BD524	0563
BD525	0551
BD525-1	0551
BD525-5	0551
BD526	0382
BD526-1	0382
BD526-5	0382
BD527	0551
BD527-1	0551
BD527-5	0551
BD528	0382
BD528-1	0382
BD528-5	0382
BD529	0551
BD529-1	0551
BD529-5	0551
BD530	0382
BD530-1	OEM
BD530-5	0382
BD533	0042
BD533J	OEM
BD533K	OEM
BD533L	OEM
BD534	0925
BD534J	OEM
BD534K	OEM
BD534L	0925
BD535	0042
BD535J	OEM
BD535K	OEM
BD536	0925
BD536J	OEM
BD536K	OEM
BD537	0423
BD538	0854
BD538J	OEM
BD538K	OEM
BD539	OEM
BD539A	OEM
BD539B	OEM
BD539C	OEM
BD539D	OEM
BD540	OEM
BD540A	OEM
BD540C	OEM
BD540D	OEM
BD543	1369
BD543A	1369
BD543B	1369
BD543C	1369
BD543D	3038
BD544	0482
BD544A	0482
BD544B	0482
BD544C	0482
BD544D	1165
BD545	OEM
BD545A	OEM
BD545B	OEM
BD545C	OEM
BD545D	OEM
BD546	OEM
BD546A	OEM
BD546B	OEM
BD546C	OEM
BD546D	OEM
BD550	4116
BD550A	6450
BD550B	4116
BD561	0161
BD562	0460
BD566	OEM
BD566A	OEM
BD567	OEM
BD567A	OEM
BD575	2997
BD576	3165
BD577	2997
BD578	3165
BD579	0423
BD580	3165
BD581	0561
BD582	1199
BD585	2997
BD586	3165
BD587	2997
BD588	3165
BD589	2997
BD590	3165
BD591	0423
BD592	1199
BD595	0561
BD596	0925
BD597	0561
BD598	1199
BD599	0561
BD600	1199
BD601	0561
BD605	0561
BD606	1199
BD607	0561
BD608	1199
BD609	0561
BD610	1199
BD611	OEM
BD612	OEM
BD613	OEM
BD614	OEM
BD615	OEM
BD616	OEM
BD617	OEM
BD618	OEM
BD619	OEM
BD620	OEM
BD633	0423
BD634	0681
BD635	0423
BD636	0681
BD637	0561
BD643	1212
BD644	2237
BD645	1212
BD646	0602
BD647	2235
BD648	0602
BD649	2235
BD649/BD650	OEM
BD650	2237
BD651	4077
BD652	1968
BD663	OEM
BD663A	0561
BD663B	0561
BD664	OEM
BD675	0558
BD675A	3377
BD676	2897
BD676A	3528
BD677	0558
BD677/678	OEM
BD677A	3377
BD678	2897
BD678A	3528
BD679	0558
BD679/680	OEM
BD679A	3377
BD680	2897
BD680A	3528
BD681	0558
BD681/682	OEM
BD683	OEM
BD683/684	OEM
BD684	OEM
BD695A	3529
BD696A	3530
BD697	0134
BD697A	3529
BD698	0073
BD698A	3530
BD699	3529
BD699A	3529
BD700	3530
BD700A	3530
BD701	3529
BD702	3530
BD705	0482
BD706	1369
BD707	0482
BD708	1369
BD709	0482
BD710	1369
BD711	0482
BD712	1369
BD733	0423
BD734	0681
BD735	0423
BD736	0681
BD737	0423
BD738	0681
BD743	OEM
BD743A	OEM
BD743B	OEM
BD743C	OEM
BD743D	OEM
BD744	OEM
BD744A	OEM
BD744B	OEM
BD744C	OEM
BD744D	OEM
BD745	OEM
BD745A	OEM
BD745B	OEM
BD745C	OEM
BD745D	OEM
BD746	OEM
BD746A	OEM
BD746B	OEM
BD746C	OEM
BD746D	OEM
BD750	4144
BD750A	4144
BD750B	4144
BD750C	4144
BD751	3701
BD751A	3701
BD751B	3701
BD751C	3701
BD795	2235
BD796	2237
BD797	2235
BD798	2237
BD799	2235
BD800	2237
BD801	0236
BD802	0681
BD805	4156
BD806	4157
BD807	4156
BD808	4157
BD813	OEM
BD814	OEM
BD815	OEM
BD816	OEM
BD817	OEM
BD818	OEM
BD825	OEM
BD825-6	OEM
BD825-10	OEM
BD825-16	OEM
BD826	1051
BD826-6	OEM
BD826-10	OEM
BD826-16	OEM
BD827	OEM
BD827-6	OEM
BD827-10	OEM
BD828	OEM
BD828-6	OEM
BD828-10	OEM
BD829	OEM
BD829-6	OEM
BD829-10	OEM
BD830	OEM
BD830-6	OEM
BD830-10	OEM
BD833	0560
BD834	1267
BD835	0560
BD836	1267
BD837	0560
BD838	1267
BD839	0219
BD840	1051
BD841	0824
BD842	3327
BD843	0824
BD844	3327
BD845	OEM
BD846	OEM
BD847	OEM
BD848	OEM
BD849	OEM
BD850	OEM
BD861	OEM
BD863	OEM
BD864	OEM
BD865	OEM
BD866	OEM
BD875	0558
BD876	2897
BD877	0558
BD878	4177
BD879	OEM
BD880	OEM
BD895	2235
BD895A	2235
BD896	2237
BD896A	2237
BD897	2235
BD897A	2235
BD898	2237
BD898A	2237
BD899	2235
BD899A	2235
BD900	2237
BD900A	2237
BD901	2235
BD902	2237
BD905	0482
BD906	1369
BD907	0482
BD908	1369
BD909	0482
BD910	1369
BD911	4184
BD912	1369
BD914	OEM
BD914F	OEM
BD914T	OEM
BD933	1165
BD934	0718
BD935	1165
BD936	0718
BD937	1165
BD938	0718
BD939	1165
BD940	0718
BD941	1165
BD941F	1165
BD942	0718
BD943	0042
BD944	0276
BD945	0042
BD946	0925
BD947	0042
BD948	0925
BD948F	OEM
BD949	0042
BD950	0925
BD951	0060
BD952	1308
BD953	1165
BD954	0482
BD955	1165
BD956	0718
BD975	OEM
BD976	OEM
BD979	OEM
BD980	OEM
BD1530	OEM
BD1540	OEM
BD1550	OEM
BD1560	OEM
BD2530	OEM
BD2540	OEM
BD2550	OEM
BD3595	OEM
BD3595F	0609
BD3595T	0609
BD3600	OEM
BD3600F	OEM
BD3600T	OEM
BD4148	OEM
BD4148F	OEM
BD4148T	OEM
BDF150S01L	OEM
BDF150S02L	OEM
BDF150S03L	OEM
BDF150S04L	OEM
BDF150S05L	OEM
BDF150S06L	OEM
BDF150S07L	OEM
BDF150S08L	OEM
BDF150S09L	OEM
BDF150S10L	OEM
BDF150S11L	OEM
BDF150S12L	OEM
BDF170S01B	OEM
BDF170S02B	OEM
BDF170S03B	OEM
BDF170S04B	OEM
BDF170S05B	OEM
BDF170S07B	OEM
BDF170S08B	OEM
BDM1615500	OEM
BDM1616500	OEM
BDM1617500	OEM
BDM1618500	OEM
BDN23R	OEM
BDOA	0277
BDS40832-200	OEM
BDS40832-201	OEM
BDS40832-206	OEM
BDS40832PD2	OEM
BDS40832S20	OEM
BDT20	OEM
BDT29	0236
BDT29A	0236
BDT29B	0236
BDT29C	1165
BDT30	0681
BDT30A	0681
BDT30B	0681
BDT30C	0718
BDT31	0236
BDT31A	0236
BDT31B	0236
BDT31C	0236
BDT32	0681
BDT32A	0681
BDT32C	0681
BDT41	0482
BDT41A	0482
BDT41B	OEM
BDT41C	OEM
BDT42	1369
BDT42A	1369
BDT42B	OEM
BDT42C	OEM
BDT60	0602
BDT60A	0602
BDT60B	0602
BDT60C	1968
BDT61	1212
BDT61A	1212
BDT61B	1212
BDT61C	1959
BDT62	2237
BDT62A	2237
BDT62B	2237
BDT62C	1968
BDT63	2235
BDT63A	2235
BDT63B	2235
BDT63C	1959
BDT64	1968
BDT64A	1968
BDT64B	1968
BDT64C	1968
BDT65	1959
BDT65A	1959
BDT65B	1959
BDT65C	1959
BDT91	0482
BDT92	1369
BDT93	0482
BDT94	1369
BDT95	0482
BDT96	1369
BDV64	0073
BDV64A	0073
BDV64B	0073
BDV64C	OEM
BDV65	0134
BDV65A	0134
BDV65B	0134
BDV65C	OEM
BDV66	OEM
BDV66A	OEM
BDV66B	OEM
BDV66C	OEM
BDV67	OEM
BDV67A	OEM
BDV67B	OEM
BDV67C	OEM
BDV91	3081
BDV92	0859
BDV93	3081
BDV94	0859
BDV95	3081
BDV96	0859
BDW21	OEM
BDW21A	OEM
BDW21B	OEM
BDW21C	OEM
BDW22	OEM
BDW22A	OEM
BDW22B	OEM
BDW22C	OEM
BDW23	1212
BDW23A	1212
BDW23B	1212
BDW23C	1214
BDW24	0602
BDW24A	0602
BDW24B	0602
BDW24C	0602
BDW26	OEM
BDW32	OEM
BDW38	OEM
BDW51	0543
BDW51A	0543
BDW51B	0543
BDW51C	0543
BDW52	1598
BDW52A	1598
BDW52B	1598
BDW52C	1598
BDW53	OEM
BDW53A	OEM
BDW53B	OEM
BDW53C	OEM
BDW53D	OEM
BDW54	OEM
BDW54C	0602
BDW54D	OEM
BDW55	OEM
BDW56	OEM
BDW57	OEM
BDW58	OEM
BDW59	OEM
BDW60	OEM
BDW63	OEM
BDW63A	OEM
BDW63B	OEM
BDW63C	OEM
BDW63D	OEM
BDW64	OEM
BDW64A	OEM
BDW64B	OEM
BDW64C	OEM
BDW64D	OEM
BDW73	OEM
BDW73A	OEM
BDW73B	OEM
BDW73C	OEM
BDW74	OEM
BDW74B	OEM
BDW74C	OEM
BDW74D	OEM
BDW83	OEM
BDW83A	OEM
BDW83B	OEM
BDW83C	OEM
BDW83D	OEM
BDW84	OEM
BDW84A	OEM
BDW84B	OEM
BDW84C	OEM
BDW84D	OEM
BDW91	OEM
BDW92	OEM
BDW93	1959
BDW93A	1959
BDW93B	1959
BDW93C	1959
BDW94	1968
BDW94A	1968
BDW94B	1968
BDW94C	1968
BDX10	0103
BDX10C	OEM
BDX11	0177
BDX12	0074
BDX13	0130
BDX13/40251	OEM
BDX18	0491
BDX18N	0491
BDX20	1598
BDX23	0543
BDX24	0177
BDX25	0236
BDX27	0925
BDX27-6	0925
BDX27-10	0925
BDX28	0925
BDX28-6	0925
BDX28-10	0925
BDX29	0236
BDX29-6	0236
BDX29-10	0236
BDX30	0236
BDX30-6	0236
BDX30-10	0236
BDX31	OEM
BDX32	0223
BDX33	2235
BDX33A	2235
BDX33B	2235
BDX33C	2235
BDX33D	2235
BDX33E	OEM
BDX34	2237
BDX34A	2237
BDX34B	2237
BDX34C	2240
BDX40	0543
BDX41	0543
BDX42	0558
BDX43	0558
BDX44	OEM
BDX45	2897
BDX46	2897
BDX47	OEM
BDX50	0543
BDX51	0543
BDX53	2235
BDX53A	2235
BDX53B	2235
BDX53C	2235
BDX53D	OEM
BDX53E	OEM
BDX53F	OEM
BDX53S	OEM
BDX54	2237
BDX54A	2237
BDX54B	2237
BDX54C	2237
BDX54D	OEM
BDX54E	OEM
BDX54F	OEM
BDX54S	OEM
BDX60	0543
BDX61	0130
BDX62	2277
BDX62/63	OEM
BDX62/63A	OEM
BDX62A	2277
BDX62B	2277
BDX62B/63B	OEM
BDX62C	2420
BDX62L	OEM
BDX63	2427
BDX63A	2427
BDX63B	2428
BDX63C	2428
BDX63L	OEM
BDX64	2432
BDX64/65	OEM
BDX64A	2432
BDX64A/65	OEM
BDX64B	2432
BDX64B/65B	OEM
BDX64C	2420
BDX64L	OEM
BDX65	2439
BDX65A	2439
BDX65B	2439
BDX65C	2428
BDX65L	OEM
BDX66	2448
BDX66A	2448
BDX66B	2432
BDX66CBDX67	OEM
BDX66L	OEM
BDX67A	1394
BDX67B	1394
BDX67C	1394
BDX67L	OEM
BDX70	0423
BDX71	0423
BDX72	0423
BDX73	0423
BDX74	0042
BDX75	0042
BDX77	0482
BDX78	1369
BDX83	2439
BDX83A	2439
BDX83B	2439
BDX83C	2439
BDX85	2439
BDX85A	2439
BDX85B	2439
BDX85C	2432
BDX86	2432
BDX86A	2432
BDX86B	2432
BDX86C	2432
BDX87	2439
BDX87A	2439
BDX87B	2439
BDX87C	2439
BDX88	2432
BDX88A	2432
BDX88B	2432
BDX88C	2432
BDX91	0177
BDX92	2013
BDX93	0177
BDX94	2013
BDX95	0177
BDX96	1681
BDX331	OEM
BDY10	0103
BDY11	0103
BDY12	0042
BDY13	0042
BDY13-16	OEM
BDY15	2763
BDY15A	2763

DEVICE TYPE	REPL CODE
BDY15B	2763
BDY15C	2763
BDY16	0178
BDY16A	2763
BDY16B	2763
BDY17	0103
BDY18	0543
BDY19	0543
BDY20	0103
BDY21	OEM
BDY22	OEM
BDY23	0177
BDY23A	0177
BDY23B	0177
BDY23C	0177
BDY24	0363
BDY25	0363
BDY26	0363
BDY27	0363
BDY28	0363
BDY29	2201
BDY34	0042
BDY37	0543
BDY37A	0130
BDY38	0103
BDY39	0103
BDY45	1852
BDY46	1852
BDY47	1852
BDY53	0103
BDY54	0177
BDY55	0543
BDY56	0177
BDY57	2485
BDY57A	2433
BDY58	2485
BDY58R	OEM
BDY60	0622
BDY61	0620
BDY62	0160
BDY63	0177
BDY65	0264
BDY66	0264
BDY69	2013
BDY70	0126
BDY71	0178
BDY72	0178
BDY73	0103
BDY74	0543
BDY76	0130
BDY77	0543
BDY78	0429
BDY79	0429
BDY80	0042
BDY80A	OEM
BDY80B	OEM
BDY80C	OEM
BDY81	0236
BDY81A	OEM
BDY81B	OEM
BDY81C	OEM
BDY82	0925
BDY82A	OEM
BDY82B	OEM
BDY82C	OEM
BDY83	0925
BDY83A	OEM
BDY83B	OEM
BDY83C	OEM
BDY90	0620
BDY90A	3259
BDY91	0620
BDY92	0620
BDY93	0065
BDY93/01	0065
BDY94	0620
BDY95	0074
BDY96/01	OEM
BDY97	0074
BDY97/01	0074
BDY98	0074
BDY99	0074
BDY180T2	OEM
BDY181T2	OEM
BDY182T2	OEM
BDY183T2	OEM
BDY184T2	OEM
BDY185T2	OEM
BE6	0280
BE6A	0280
BE-55	0030
BE-66	0016
BE71	0086
BE107	0015
BE117	0015
BE127	0015
BE155	OEM
BE173	0127
BEL100N	OEM
BEL100P	OEM
BEL550A	OEM
BEL550B	OEM
BEL550C	OEM
BEL700	OEM
BEL1044	OEM
BES042	OEM
BES043	OEM
BES044	OEM
BES046	OEM
BES048	OEM
BES050	OEM
BES051	OEM
BES063	OEM
BF7-35	OEM
BF14-35	OEM
BF25-35	OEM
BF36C	OEM
BF40B8H	0729
BF50-35	OEM
BF60B8L	1040
BF60B9A	OEM
BF71	0016
BF100-35	OEM
BF108	0233
BF109	0233
BF110	0233
BF111	0233
BF114	0233
BF115	0079
BF117	0233
BF118	0233
BF119	0233
BF120	0233
BF121	0127
BF123	0127
BF125	0127
BF127	0127
BF137	0285
BF140	0233
BF140A	0233
BF140R	0233
BF140S	0233
BF152	0493
BF153	0493
BF154	0493
BF155	0007
BF155R	0233
BF155S	0233
BF156	0233
BF157	0233
BF157B	0233
BF158	0493
BF159	0493
BF160	0493
BF161	0007
BF162	0144
BF163	0144
BF164	0144
BF165	0144
BF166	0007
BF167	0007
BF168	0007
BF169	0007
BF173	0007
BF173A	0144
BF174	0233
BF175	0007
BF176	0144
BF177	0086
BF178	0233
BF179	0233
BF179A	0233
BF179B	0233
BF179C	0233
BF-180	0079
BF180	0007
BF181	0007
BF182	0007
BF183	0007
BF183A	0016
BF184	0007
BF185	0007
BF186	0233
BF187	0144
BF188	0144
BF189	0016
BF194	0224
BF194A	0144
BF194B	0127
BF195	0224
BF195C	0127
BF195D	0127
BF196	0224
BF196-01	OEM
BF196P	OEM
BF197	0127
BF197-01	OEM
BF197P	OEM
BF-198	0224
BF198	0224
BF199	0224
BF200	0007
BF-200(PENNCREST)	0007
BF200(ZENITH)	0127
BF206	0007
BF207	0007
BF208	0007
BF209	0007
BF212	0007
BF213	0007
BF-214	0016
BF214	0007
BF214A	OEM
BF214B	OEM
BF-215	0016
BF215	0007
BF216	0127
BF217	0127
BF218	0127
BF219	0127
BF220	0127
BF222	0007
BF223	0144
BF224	0224
BF224J	0016
BF225	0224
BF225J	0016
BF-226	0016
BF226	0007
BF227	0007
BF228	0861
BF229	0127
BF230	0127
BF232	0007
BF233	0224
BF233-2	0224
BF233-3	0224
BF233-4	0144
BF233-5	0007
BF233-6	0224
BF234	0224
BF235	0144
BF236	0144
BF237	0079
BF238	0079
BF240	0224
BF240B	0224
BF241	0224
BF241C	0224
BF241D	0224
BF243	0693
BF244	0325
BF244A	0325
BF244B	0325
BF244C	0325
BF245	0325
BF245A	0325
BF245B	0325
BF245C	0325
BF246	0325
BF246A	OEM
BF246B	OEM
BF246C	OEM
BF247	0325
BF247A	OEM
BF247B	OEM
BF247C	OEM
BF248	0016
BF249	0037
BF250	0016
BF251	0007
BF252	0007
BF253	0127
BF253-2	OEM
BF253-3	OEM
BF253-4	OEM
BF253-5	OEM
BF253-6	OEM
BF254	0224
BF254(ALGG)	0191
BF254(PHIN)	0191
BF254(SIEG)	0224
BF254B	0007
BF-255	0079
BF255	0224
BF255(ALGG)	0191
BF255(PHIN)	0191
BF255(SIEG)	0224
BF255C	0007
BF255D	0007
BF256	0325
BF256A	0325
BF256B	0325
BF256C	0325
BF256L	OEM
BF256LA	OEM
BF256LB	OEM
BF256LC	OEM
BF257	0187
BF258	0187
BF259	0233
BF260	0007
BF261	0007
BF262	0144
BF263	0144
BF264	0144
BF270	0007
BF271	0007
BF272	4433
BF273	0007
BF273C	0007
BF273D	0007
BF274	0007
BF274B	0007
BF274C	0007
BF279	0007
BF287	0007
BF288	0007
BF290	0007
BF291	0016
BF291A	0016
BF291B	0016
BF292	0233
BF292A	0233
BF292B	0233
BF292C	0233
BF293	0016
BF293A	0016
BF293D	0016
BF294	0233
BF297	0715
BF297-P	0264
BF297W	OEM
BF297W2	OEM
BF298	0715
BF298-P	0338
BF298W	OEM
BF298W2	OEM
BF299	0342
BF299-P	0338
BF299W	OEM
BF299W2	OEM
BF302	0007
BF303	0493
BF304	0007
BF305	0233
BF306	0007
BF308	0224
BF309	0418
BF310	0224
BF310(ALGG)	0259
BF311	0224
BF314	0224
BF314(ALGG)	0259
BF315	0150
BF316	0150
BF316A	OEM
BF320	OEM
BF320A	OEM
BF320B	OEM
BF320C	OEM
BF321	OEM
BF321A	0198
BF321B	0016
BF321C	0016
BF321D	0016
BF321E	0016
BF321F	0016
BF322	0086
BF323	0126
BF324	0808
BF325	0418
BF327	OEM
BF329	0127
BF330	0418
BF332	0127
BF332B	OEM
BF333	0127
BF333C	0127
BF333D	0127
BF334	0127
BF335	0127
BF336	0233
BF337	0233
BF338	0233
BF339	0150
BF340	0037
BF341	3572
BF342	0037
BF343	0037
BF344	0007
BF345	0007
BF346A	OEM
BF346B	OEM
BF346C	OEM
BF347	OEM
BF348	0325
BF350	OEM
BF351	OEM
BF352	OEM
BF353	OEM
BF354	OEM
BF355	0233
BF357	0144
BF362	0007
BF363	0007
BF364	0224
BF365	0224
BF366	0086
BF367	0224
BF368	0079
BF368K	OEM
BF369	0079
BF369K	OEM
BF370	OEM
BF370R	OEM
BF371	2525
BF372	OEM
BF373	OEM
BF374	2525
BF375	2525
BF375C	2525
BF375D	2525
BF377	0418
BF378	0418
BF379	0037
BF380	OEM
BF380-1	OEM
BF380-5	OEM
BF381	OEM
BF381-5	OEM
BF382	OEM
BF382-1	OEM
BF382-5	OEM
BF384	0144
BF385	0144
BF387	0086
BF388	OEM
BF389B	OEM
BF389C	OEM
BF391	0715
BF391P	OEM
BF391W1	OEM
BF391W2	OEM
BF392	0715
BF392P	OEM
BF392W1	OEM
BF392W2	OEM
BF393	0715
BF393P	OEM
BF393W1	OEM
BF393W2	OEM
BF394	0018
BF394B	0018
BF395	0018
BF395C	0018
BF395D	0018
BF397	0342
BF397A	0342
BF398	0342
BF398A	0342
BF398B	0342
BF400	OEM
BF402	OEM
BF403	OEM
BF404	OEM
BF405	OEM
BF406	OEM
BF410	OEM
BF410A	OEM
BF410B	OEM
BF410C	OEM
BF410D	OEM
BF411	0861
BF412	0861
BF414	0037
BF415	0233
BF416	OEM
BF417	0233
BF418	OEM
BF419	4470
BF420	0261
BF420A	OEM
BF420L	OEM
BF421	0261
BF421A	OEM
BF421L	OEM
BF422	0261
BF422A	OEM
BF422L	OEM
BF422W	OEM
BF422W3	OEM
BF423	0439
BF423A	OEM
BF423L	OEM
BF423W	OEM
BF423W3	OEM
BF424	0808
BF435	OEM
BF436	OEM
BF437	OEM
BF439	OEM
BF440	0037
BF441	0037
BF450	0037
BF451	3617
BF456	0563
BF457	0563
BF458	0276
BF459	0276
BF460	OEM
BF461	OEM
BF462	OEM
BF469	0276
BF470	4479
BF471	0276
BF472	4479
BF479T	OEM
BF480	0259
BF481	OEM
BF482	OEM
BF491	0342
BF492	0342
BF493	0342
BF494	0224
BF494B	OEM
BF495	0224
BF496	OEM
BF497	0007
BF502	OEM
BF503	OEM
BF505	0224
BF506	0037
BF506A	OEM
BF507	0224
BF509S	OEM
BF510	OEM
BF511	OEM
BF512	OEM
BF513	OEM
BF516	0007
BF523	3607
BF536	4489
BF540	0037
BF541	0037
BF542	0037
BF550	0724
BF550R	OEM
BF554	OEM
BF562	OEM
BF568	OEM
BF569	OEM
BF579	4498
BF594	0224
BF595	0224
BF596	0079
BF597	0111
BF597A	OEM
BF597B	OEM
BF599	1435
BF606	3607
BF606A	OEM
BF615	OEM
BF616	OEM
BF617	OEM
BF618	OEM
BF620	4512
BF621	4514
BF622	3471
BF623	4514
BF642	OEM
BF642W	OEM
BF642W3	OEM
BF643	OEM
BF643W	OEM
BF643W3	OEM
BF657	0233
BF658	0233
BF659	0233
BF660	OEM
BF679T	OEM
BF689	0007
BF692	OEM
BF692-P	0439
BF692W	OEM
BF692W2	OEM
BF693	OEM
BF693-P	0439
BF693W	OEM
BF693W2	OEM
BF694	OEM
BF694A	OEM
BF694B	OEM
BF706	0037
BF709	3607
BF715	0285
BF716	OEM
BF717	0285
BF718	OEM
BF720	OEM
BF721	OEM
BF722	OEM
BF723	OEM
BF724	OEM
BF739	OEM
BF740	OEM
BF757	OEM
BF758	OEM
BF759	OEM
BF767	OEM
BF799	OEM
BF800	OEM
BF801	OEM
BF802	OEM
BF805	OEM
BF806	OEM
BF808	OEM
BF810	OEM
BF811	OEM
BF815	OEM
BF816	OEM
BF817	OEM
BF818	OEM
BF819	0168
BF819A	OEM
BF820	3468
BF821	3379
BF822	4542
BF823	4544
BF824	4489
BF-832	0037
BF840	4546
BF841	4546
BF847	OEM
BF848	OEM
BF849	OEM
BF858	OEM
BF859	OEM
BF869	0276
BF870	4479
BF871	0276
BF872	4479
BF881	OEM
BF900	OEM
BF905	OEM
BF907	OEM
BF910	OEM
BF914	OEM
BF926	OEM
BF936	OEM
BF939	3607
BF960	0371
BF961	OEM
BF963	OEM
BF964	0371
BF966	OEM
BF967	OEM
BF968	OEM
BF970	OEM
BF979	OEM
BF979S	OEM
BF980	OEM
BF981	OEM
BF982	OEM
BF989	OEM
BF990AR	OEM
BF994	OEM
BF995	OEM
BF996	OEM
BF998R	OEM
BFAB3/800	OEM
BFC18/500	OEM
BFC91	OEM
BFD150S01L	OEM
BFD150S02L	OEM
BFD150S03L	OEM
BFD150S04L	OEM
BFD150S05L	OEM
BFD150S06L	OEM
BFD150S08L	OEM
BFD150S09L	OEM
BFD150S10L	OEM
BFD150S11L	OEM
BFD150S12L	OEM
BFD170S01B	OEM
BFD170S02B	OEM
BFD170S03B	OEM
BFD170S04B	OEM
BFD170S05B	OEM
BFD170S06B	OEM
BFD170S07B	OEM
BFD170S08B	OEM
BFF150S01L	OEM
BFF150S02B	OEM
BFF150S02L	OEM
BFF150S03B	OEM
BFF150S03L	OEM
BFF150S04B	OEM
BFF150S04L	OEM
BFF150S05B	OEM
BFF150S05L	OEM
BFF150S06B	OEM
BFF150S06L	OEM
BFF150S07B	OEM
BFF150S07L	OEM
BFF150S08B	OEM
BFF150S08L	OEM
BFF150S09L	OEM
BFF150S10L	OEM
BFF150S11L	OEM
BFF150S12L	OEM
BFF170S01B	OEM
BFG16A	OEM
BFG17A	OEM
BFG31	OEM
BFG33	OEM
BFG35	OEM
BFG55	OEM
BFG67	OEM
BFG92A	OEM
BFG93A	OEM
BFG97	OEM
BFG135	OEM
BFG197	OEM
BFG198	OEM
BFN16	OEM
BFN17	OEM
BFN18	OEM
BFN19	OEM
BFN20	OEM
BFN21	OEM
BFN22	OEM
BFN22R	OEM
BFN23	OEM
BFP10	OEM
BFP11	OEM
BFP12	OEM
BFP13	OEM
BFP14	OEM
BFP22	OEM
BFP23	OEM
BFP25	OEM
BFP26	OEM
BFP90	OEM
BFP91	OEM
BFP96	OEM
BFQ10	OEM
BFQ11	OEM
BFQ12	OEM
BFQ13	OEM
BFQ14	OEM
BFQ15	OEM
BFQ16	OEM
BFQ17	OEM
BFQ18	OEM
BFQ18A	OEM
BFQ19	OEM
BFQ22	OEM
BFQ23	OEM
BFQ24	OEM
BFQ28	OEM
BFQ29	OEM
BFQ31	1435

If replacement code is OEM, contact original manufacturer for replacement.

DEVICE TYPE	REPL CODE
BFQ31A	OEM
BFQ31AR	OEM
BFQ31R	OEM
BFQ32	OEM
BFQ33	OEM
BFQ34	OEM
BFQ35	0126
BFQ36	0126
BFQ37	0126
BFQ41	OEM
BFQ42	OEM
BFQ43	OEM
BFQ51	OEM
BFQ52	OEM
BFQ53	OEM
BFQ56	OEM
BFQ57	OEM
BFQ58	OEM
BFQ59	OEM
BFQ60	OEM
BFQ63	OEM
BFQ64	OEM
BFQ67	OEM
BFQ68	OEM
BFQ69	OEM
BFQ70	OEM
BFQ71	OEM
BFQ72	OEM
BFQ73	OEM
BFQ74	OEM
BFQ78	OEM
BFQ86	OEM
BFQ88	OEM
BFQ88A	OEM
BFQ88B	OEM
BFQ89	OEM
BFQ98	OEM
BFQ98B	OEM
BFQ149	OEM
BFR10	0007
BFR11	0016
BFR12	OEM
BFR14B	OEM
BFR14C	OEM
BFR15A	OEM
BFR16	0016
BFR17	0111
BFR18	0086
BFR19	0086
BFR20	0086
BFR21	0730
BFR22	0086
BFR23	0126
BFR24	0126
BFR25	0111
BFR26	0016
BFR27	OEM
BFR28	0144
BFR29	OEM
BFR30	OEM
BFR30R	OEM
BFR31	OEM
BFR31R	OEM
BFR35A	0016
BFR36	0086
BFR37	0259
BFR38	0324
BFR39T05	0187
BFR40	0324
BFR40T05	0086
BFR40T05	0086
BFR41	0324
BFR41T05	0086
BFR41T05	0086
BFR44A	OEM
BFR44B	OEM
BFR44C	OEM
BFR45	OEM
BFR46	OEM
BFR47	OEM
BFR48	OEM
BFR50	0324
BFR51	0086
BFR52	0086
BFR53	2919
BFR54	OEM
BFR56	OEM
BFR57	0233
BFR58	0233
BFR59	0233
BFR60	0436
BFR61	0791
BFR62	0477
BFR63	0418
BFR64	0418
BFR65	2070
BFR77	0086
BFR78	0086
BFR79T05	0477
BFR79T05	0477
BFR80	0791
BFR80T05	0477
BFR81	0532
BFR81T05	0477
BFR81T05	0477
BFR84	OEM
BFR86	0270
BFR86A	OEM
BFR86B	OEM
BFR86T05	2545
BFR86T05	0187
BFR87A	OEM
BFR87B	OEM
BFR87T05	2545
BFR87T05	0187
BFR88	0715
BFR88A	OEM
BFR88B	OEM
BFR88T05	2545
BFR88T05	0187
BFR89	0715
BFR89A	OEM
BFR89B	OEM
BFR89T05	2545
BFR89T05	0187
BFR90	1342
BFR90A	OEM
BFR91A	OEM
BFR92	2919
BFR92A	2919
BFR93	2919
BFR93A	2919
BFR94	2070
BFR96S	OEM
BFR98	2041
BFR99	3607
BFR106	OEM
BFRC90	OEM
BFRC91	OEM
BFRC96	OEM
BFS10	OEM
BFS11	0007
BFS12	OEM
BFS13E	0127
BFS13F	0127
BFS13G	0127
BFS14E	0127
BFS14F	0127
BFS14G	0127
BFS15E	0127
BFS15F	0127
BFS15G	0127
BFS16E	0127
BFS16F	0127
BFS16G	0127
BFS17	0144
BFS17A	3372
BFS17R	0144
BFS18	1733
BFS18CA	0007
BFS18R	0127
BFS19	1733
BFS19CB	0079
BFS19R	0127
BFS20	4489
BFS20R	0127
BFS21	0325
BFS21A	0325
BFS22	0698
BFS22A	0622
BFS22B	OEM
BFS22R	OEM
BFS23	0698
BFS23A	0177
BFS23R	OEM
BFS26E	0037
BFS26F	0037
BFS26G	0037
BFS27E	0007
BFS27F	0007
BFS27G	0007
BFS28	0353
BFS28R	0353
BFS29	0086
BFS29P	0595
BFS30	0595
BFS30P	0595
BFS31	0037
BFS31P	0016
BFS32	0037
BFS32P	0037
BFS33	0037
BFS33P	0037
BFS34	0791
BFS34P	0037
BFS36	0086
BFS36A	0016
BFS37	0037
BFS37A	0037
BFS38	0016
BFS38A	0016
BFS39	0079
BFS40	0037
BFS40A	0037
BFS41	0037
BFS42	0016
BFS43	0016
BFS44	0791
BFS45	0791
BFS46	OEM
BFS46A	OEM
BFS50	0493
BFS51	0560
BFS55A	0111
BFS58	0144
BFS59	0595
BFS59K	OEM
BFS59L	OEM
BFS59M	OEM
BFS60	0595
BFS60K	OEM
BFS60L	OEM
BFS60M	OEM
BFS61K	OEM
BFS61L	OEM
BFS61M	OEM
BFS62	0007
BFS64	OEM
BFS65	OEM
BFS67	OEM
BFS67P	OEM
BFS68	0325
BFS68P	0325
BFS69	0037
BFS70	OEM
BFS71	OEM
BFS72	OEM
BFS73	OEM
BFS74	OEM
BFS75	OEM
BFS76	OEM
BFS77	OEM
BFS78	OEM
BFS79	OEM
BFS80	OEM
BFS85	0418
BFS88	0418
BFS90	OEM
BFS91	OEM
BFS91A	0436
BFS91B	0436
BFS92	0126
BFS93	0126
BFS94	0126
BFS95	0126
BFS96	0037
BFS96K	OEM
BFS96L	OEM
BFS96M	OEM
BFS97	0037
BFS97K	OEM
BFS97L	OEM
BFS97M	OEM
BFS98	0037
BFS98K	OEM
BFS98L	OEM
BFS98M	OEM
BFS99	0285
BFT10	OEM
BFT11	OEM
BFT13	OEM
BFT13A	OEM
BFT14	OEM
BFT14A	OEM
BFT15	OEM
BFT16	OEM
BFT17	OEM
BFT18	OEM
BFT18A	OEM
BFT19	0439
BFT19A	0439
BFT20	0037
BFT21	0037
BFT22	0037
BFT25	2919
BFT27	OEM
BFT28	0439
BFT28A	0439
BFT28B	0439
BFT28C	0439
BFT29	0086
BFT30	0086
BFT31	0086
BFT33	0622
BFT34	0622
BFT39	0086
BFT40	0086
BFT41	0086
BFT42	0622
BFT43	0622
BFT46	OEM
BFT47	OEM
BFT48	OEM
BFT49	OEM
BFT50	OEM
BFT51	OEM
BFT53	0324
BFT54	0086
BFT55	0016
BFT57	0233
BFT59	0233
BFT60	0126
BFT61	0126
BFT62	0126
BFT69	0126
BFT70	0037
BFT71	0037
BFT72	OEM
BFT73	OEM
BFT74	OEM
BFT75	OEM
BFT79	0126
BFT80	0126
BFT81	0126
BFT91	OEM
BFT92	4498
BFT93	4498
BFT95	OEM
BFT96	OEM
BFT97	OEM
BFT98	OEM
BFT99	OEM
BFV10	0079
BFV11	0079
BFV12	0079
BFV13	OEM
BFV14	0018
BFV15	OEM
BFV17	0324
BFV19	OEM
BFV20	0037
BFV21	0037
BFV22	0037
BFV23	OEM
BFV24	OEM
BFV25	0037
BFV26	0037
BFV27	0007
BFV28	0007
BFV29	0037
BFV30	0037
BFV31	0037
BFV32	0037
BFV33	0037
BFV34	0037
BFV35	0342
BFV36	0342
BFV37	0111
BFV38	0111
BFV39	OEM
BFV40	0079
BFV41	0079
BFV42	0079
BFV43	0079
BFV44	0079
BFV45	0079
BFV46	0079
BFV47	0079
BFV48	OEM
BFV49	0079
BFV50	0079
BFV51	0079
BFV52	0018
BFV53	0079
BFV54	0079
BFV55	0079
BFV56	OEM
BFV56A	OEM
BFV57	OEM
BFV57A	OEM
BFV58	OEM
BFV59	0007
BFV60	0007
BFV61	0007
BFV62	0111
BFV63	OEM
BFV63A	OEM
BFV63B	OEM
BFV64	OEM
BFV64A	OEM
BFV64B	OEM
BFV65	OEM
BFV65A	OEM
BFV66	OEM
BFV66A	OEM
BFV67	OEM
BFV68	OEM
BFV68A	OEM
BFV69	OEM
BFV69A	OEM
BFV70	OEM
BFV71	OEM
BFV72	OEM
BFV72N	OEM
BFV73	OEM
BFV73N	OEM
BFV75	OEM
BFV75N	OEM
BFV76	OEM
BFV76N	OEM
BFV77	OEM
BFV78	OEM
BFV79	OEM
BFV80	0007
BFV81	0037
BFV81A	0037
BFV81B	0037
BFV82	0037
BFV82A	0037
BFV82B	0037
BFV82C	0037
BFV83	0144
BFV83A	0144
BFV83B	0016
BFV83C	0016
BFV85	0016
BFV85A	0016
BFV85B	0016
BFV85C	0016
BFV85D	0144
BFV85E	0144
BFV85F	0144
BFV85G	0144
BFV86	0037
BFV86A	0037
BFV86B	0037
BFV86C	0037
BFV87	0016
BFV88	0016
BFV88A	0016
BFV88B	0016
BFV88C	0016
BFV88E	OEM
BFV89	0111
BFV89A	0111
BFV90A	OEM
BFV90B	OEM
BFV91	OEM
BFV91N	OEM
BFV92	OEM
BFV92N	OEM
BFV93	OEM
BFV93A	OEM
BFV93AN	OEM
BFV93N	OEM
BFV94	OEM
BFV94N	OEM
BFV95	OEM
BFV95N	OEM
BFV96	OEM
BFV96N	OEM
BFV97	OEM
BFV97N	OEM
BFV98	OEM
BFV98N	OEM
BFV99	OEM
BFW10	0325
BFW11	0325
BFW12	0325
BFW13	0325
BFW16	0418
BFW16A	0418
BFW17	0418
BFW17A	0418
BFW19	0418
BFW20	0037
BFW21	0037
BFW22	0037
BFW23	0037
BFW24	0622
BFW25	0622
BFW26	0622
BFW29	0016
BFW30	0418
BFW31	0037
BFW32	0016
BFW33	0233
BFW34	OEM
BFW35	OEM
BFW36	0233
BFW37	0233
BFW38	OEM
BFW39	OEM
BFW39A	OEM
BFW40	OEM
BFW41	0007
BFW42	0493
BFW43	0342
BFW44	0126
BFW45	0233
BFW46	0016
BFW47	0418
BFW50	0277
BFW51	OEM
BFW51A	OEM
BFW52	OEM
BFW52A	OEM
BFW54	0325
BFW55	0325
BFW56	0325
BFW57	0861
BFW58	0861
BFW59	0016
BFW60	0016
BFW61	0325
BFW63	0007
BFW64	0144
BFW68	0016
BFW69	OEM
BFW70	1940
BFW71	OEM
BFW73	2525
BFW73A	2525
BFW74	2525
BFW75	OEM
BFW76	2525
BFW76A	2525
BFW77	2525
BFW77A	2525
BFW78	OEM
BFW79	OEM
BFW80	OEM
BFW87	0037
BFW88	0382
BFW89	0037
BFW90	0037
BFW91	0126
BFW92	0259
BFW92A	1514
BFW96	OEM
BFW97	OEM
BFW98	0418
BFW98G	OEM
BFW100	0277
BFW200	0289
BFW300	0296
BFW400	0296
BFW500	0302
BFW600	0302
BFW700	OEM
BFW800	0250
BFW1000	0250
BFX10	OEM
BFX11	2468
BFX12	0150
BFX13	0150
BFX14	OEM
BFX15	OEM
BFX16	OEM
BFX17	0086
BFX18	0144
BFX19	0144
BFX20	0144
BFX21	0144
BFX29	0037
BFX30	0037
BFX31	0007
BFX32	0127
BFX33	0493
BFX34	4225
BFX35	0037
BFX36	2468
BFX37	0693
BFX38	0126
BFX39	0126
BFX40	0126
BFX41	0126
BFX42	2525
BFX43	0016
BFX44	0016
BFX45	0144
BFX47	0007
BFX48	0150
BFX49G	OEM
BFX50	0086
BFX51	0086
BFX52	0086
BFX53	0595
BFX55	0418
BFX59	0008
BFX59F	0016
BFX60	0007
BFX61	0086
BFX62	0007
BFX63	OEM
BFX65	0037
BFX66	0016
BFX67	OEM
BFX68	0086
BFX68A	0086
BFX69	0086
BFX69A	0086
BFX70	OEM
BFX71	OEM
BFX72	OEM
BFX73	0007
BFX74	0086
BFX74A	0126
BFX77	0007
BFX78	OEM
BFX79	OEM
BFX80	OEM
BFX81	OEM
BFX82	OEM
BFX83	OEM
BFX84	0086
BFX85	0086
BFX86	0086
BFX87	0126
BFX88	0126
BFX89	0007
BFX90	0342
BFX91B	OEM
BFX92	0016
BFX92A	0016
BFX93	0016
BFX93A	0111
BFX94	0016
BFX94A	OEM
BFX95	0016
BFX95A	0016
BFX95C	OEM
BFX95C1	OEM
BFX95C2	OEM
BFX96	0198
BFX96A	0086
BFX97	0086
BFX97A	0086
BFX98	0233
BFX99	0086
BFY10	0016
BFY11	0016
BFY12	0016
BFY13	0086
BFY14	0086
BFY15	0086
BFY16	0595
BFY16S	OEM
BFY17	0086
BFY18	0016
BFY19	0016
BFY20	OEM
BFY21	OEM
BFY22	0016
BFY23	0016
BFY23A	0016
BFY24	0016
BFY25	0016
BFY26	0016
BFY27	0086
BFY28	0016
BFY29	0016
BFY30	0016
BFY33	0016
BFY34	0086
BFY37	0016
BFY37I	0016
BFY39	0016
BFY39/1	0016
BFY39/2	0016
BFY39/3	0016
BFY39/11	0861
BFY39/I	0016
BFY39/II	0016
BFY39/III	0016
BFY39-1	0111
BFY39-2	0111
BFY39-3	0111
BFY39I	0016
BFY40	0086
BFY41	0233
BFY43	0233
BFY44	0086
BFY45	0233
BFY46	0086
BFY47	0127
BFY48	0127
BFY49	0127
BFY50	0086
BFY50E	OEM
BFY50I	OEM
BFY51	0086
BFY51(RTCF)	0018
BFY51I	0086
BFY52	0086
BFY53	0086
BFY55	0086
BFY55(RTCF)	0324
BFY56	0086
BFY56A	0086
BFY56B	0086
BFY57	0233
BFY63	0493
BFY64	0126
BFY65	0233
BFY66	0086
BFY67	0086
BFY67A	0086
BFY67C	0086
BFY68	0086
BFY68A	0086
BFY69	0127
BFY69A	0127
BFY69B	0007
BFY70	0086
BFY72	0016
BFY73	0016
BFY74	0016
BFY75	0016
BFY76	0016
BFY77	0016
BFY78	0144
BFY79	0007
BFY80	0086
BFY81	2045
BFY82	2045
BFY83	OEM
BFY84	OEM
BFY85	0016
BFY85A	OEM
BFY85B	OEM
BFY86	0016
BFY86A	OEM
BFY86B	OEM
BFY87	0007
BFY87A	0007
BFY88	0007
BFY90	0493
BFY90(SIEG)	0016
BFY90B	0144
BFY91	OEM
BFY92	OEM
BFY94	0037
BFY99	0086
BFY167	0007
BFY371	0016
BFY391	0016
BFY501	0861
BFZ10	OEM
BG-07077	0012
BG08120K	OEM
BG-66	0016
BG71	0016
BG-94	0016
BG-190	OEM
BG2000-6164-004	OEM
BG2097-641	OEM
BG2097-642	OEM
BG3433S	OEM
BG4524K	OEM
BG5524C	OEM
BG5527S-1	OEM
BG5534S	1778

If replacement code is OEM, contact original manufacturer for replacement.

DEVICE TYPE	REPL CODE
BG5628S	OEM
BG5725S-B3	OEM
BG12201-2	OEM
BG12203-2	OEM
BG12205-2	OEM
BG16101-2	OEM
BGX11-800TT	OEM
BGX11-1200CTT	OEM
BGX11-1200TT	OEM
BGX11-1400CTT	OEM
BGX12-600TT	OEM
BGX12-800TT	OEM
BGX12-1200CTT	OEM
BGX12-1200TT	OEM
BGX12-1400CTT	OEM
BGX13-600	4364
BGX13-600TT	OEM
BGX13-800	4364
BGX13-800TT	OEM
BGX13-1200	4364
BGX13-1200C	OEM
BGX13-1200CTT	OEM
BGX13-1200TT	OEM
BGX13-1400C-TT	OEM
BGX14-600TT	OEM
BGX14-800TT	OEM
BGX14-1200CTT	OEM
BGX14-1200TT	OEM
BGX14-1400CTT	OEM
BGX15-600	2994
BGX15-600TT	OEM
BGX15-800	2994
BGX15-800TT	OEM
BGX15-1200	2994
BGX15-1200C	OEM
BGX15-1200CTT	OEM
BGX15-1200TT	OEM
BGX15-1400C-TT	OEM
BGX17-600	2994
BGX17-600TT	OEM
BGX17-800	2994
BGX17-800TT	OEM
BGX17-1200	2994
BGX17-1200C	OEM
BGX17-1200CTT	OEM
BGX17-1400C-TT	OEM
BGX50A	OEM
BGY12A	OEM
BGY12B	OEM
BGY13A	OEM
BGY13B	OEM
BGY14A	OEM
BGY14B	OEM
BGY15	OEM
BGY16	OEM
BGY17	OEM
BGY18	OEM
BGY19	OEM
BGY21	OEM
BGY22	OEM
BGY22A	OEM
BGY23	OEM
BGY23A	OEM
BGY24	OEM
BGY32	OEM
BGY33	OEM
BGY35	OEM
BGY36	OEM
BGY38	OEM
BH4R1	0015
BH4R2	OEM
BH4R3	OEM
BH4R4	OEM
BH4R5	OEM
BH4R6	0087
BH71	0016
BH200	OEM
BH300	OEM
BH400	OEM
BH500	OEM
BH-2401A	OEM
BH2401A	OEM
BHC0001	OEM
BHC0002	OEM
BHC0003	OEM
BHC0004	OEM
BHC0005	OEM
BHC4038AD	OEM
BHC4054AD	OEM
BH-DX0038CEZZ	0015
B-HQ	OEM
BI-07096	0319
BI71	0016
BI-82	0150
BIP7201	0198
BIP7301A	OEM
BIR	2439
BIW	0016
BJ-155	0030
BK-499	OEM
BK3903	OEM
BK3904	OEM
BKC300	OEM
BKC333	OEM
BKC400	OEM
BKC444	OEM
BKC600	OEM
BKC666	OEM
BKC900	OEM
BKC999	OEM
BKX13A	OEM
BKZ6.8	OEM
BKZ6.8A	OEM
BKZ6.8B	OEM
BKZ7.5	OEM
BKZ7.5A	OEM
BKZ7.5B	OEM
BKZ8.2	OEM
BKZ8.2A	OEM
BKZ8.2B	OEM
BKZ9.1	OEM
BKZ9.1A	OEM
BKZ9.1B	OEM
BKZ10	OEM
BKZ10A	OEM
BKZ10B	OEM
BKZ11	OEM
BKZ11A	OEM
BKZ11B	OEM
BKZ12	OEM
BKZ12A	OEM
BKZ12B	OEM
BKZ13	OEM
BKZ13B	OEM
BKZ15	OEM
BKZ15A	OEM
BKZ15B	OEM
BKZ16	OEM
BKZ16A	OEM
BKZ16B	OEM
BKZ18	OEM
BKZ18A	OEM
BKZ18B	OEM
BKZ20	OEM
BKZ20A	OEM
BKZ20B	OEM
BKZ22	OEM
BKZ22A	OEM
BKZ22B	OEM
BKZ24	OEM
BKZ24A	OEM
BKZ24B	OEM
BKZ27	OEM
BKZ27A	OEM
BKZ27B	OEM
BKZ30	OEM
BKZ30A	OEM
BKZ30B	OEM
BKZ33	OEM
BKZ33A	OEM
BKZ33B	OEM
BKZ36	OEM
BKZ36A	OEM
BKZ36B	OEM
BKZ39	OEM
BKZ39A	OEM
BKZ39B	OEM
BKZ43	OEM
BKZ43A	OEM
BKZ43B	OEM
BKZ47	OEM
BKZ47A	OEM
BKZ47B	OEM
BKZ51	OEM
BKZ51A	OEM
BKZ51B	OEM
BKZ56	OEM
BKZ56A	OEM
BKZ56B	OEM
BKZ62	OEM
BKZ62A	OEM
BKZ62B	OEM
BKZ66	OEM
BKZ66A	OEM
BKZ66B	OEM
BKZ75	OEM
BKZ75A	OEM
BKZ75B	OEM
BKZ82	OEM
BKZ82A	OEM
BKZ82B	OEM
BKZ91	OEM
BKZ91A	OEM
BKZ91B	OEM
BKZ100	OEM
BKZ100A	OEM
BKZ100B	OEM
BKZ110	OEM
BKZ110A	OEM
BKZ110B	OEM
BKZ120	OEM
BKZ120A	OEM
BKZ120B	OEM
BKZ130	OEM
BKZ130A	OEM
BKZ130B	OEM
BKZ150	OEM
BKZ150A	OEM
BKZ150B	OEM
BKZ160	OEM
BKZ160A	OEM
BKZ160B	OEM
BKZ180	OEM
BKZ180A	OEM
BKZ180B	OEM
BKZ200	OEM
BKZ200A	OEM
BKZ200B	OEM
BL221	OEM
BL234C	OEM
BL244	OEM
BL245	OEM
BL246	OEM
BL246A	OEM
BL325	OEM
BL336	OEM
BL350	OEM
BL362	OEM
BL364	OEM
BL366	OEM
BL367A	OEM
BL397A	OEM
BL640	OEM
BL644	OEM
BL696	OEM
BL800A	OEM
BL803	OEM
BL807	OEM
BL814	OEM
BL825	OEM
BL829	OEM
BL843	OEM
BL844	OEM
BL847	OEM
BL849	OEM
BL934A	OEM
BL962	OEM
BL965	OEM
BL967A	OEM
BL970	OEM
BL973	OEM
BL982	OEM
BLA011	OEM
BLA012	OEM
BLA013	OEM
BLA019	OEM
BLA39B	OEM
BLC016	OEM
BLC032	OEM
BLC048	OEM
BLC0128	OEM
BLC0128A	OEM
BLC0512	OEM
BLC80/05	OEM
BLC80/07	OEM
BLC80/10	OEM
BLC80/11	OEM
BLC80/11A	OEM
BLC80/11T	OEM
BLC80/12	OEM
BLC80/12A	OEM
BLC80/12T	OEM
BLC80/14	OEM
BLC80/14A	OEM
BLC80/14T	OEM
BLC80/116	OEM
BLC80/204	OEM
BLC80/316	OEM
BLC86/12B	OEM
BLC104	OEM
BLC116	OEM
BLC300	OEM
BLC406	OEM
BLC501	OEM
BLC508	OEM
BLC517	OEM
BLC519	OEM
BLC556	OEM
BLC8016	OEM
BLC8016A	OEM
BLC8016B	OEM
BLC8032A	OEM
BLC8032B	OEM
BLC8048A	OEM
BLC8048B	OEM
BLC8064A	OEM
BLC8064B	OEM
BLC8201	OEM
BLC8221	OEM
BLC8222	OEM
BLC8228	OEM
BLC8229	OEM
BLC8432	OEM
BLC8488	OEM
BLC8534	OEM
BLC8538	OEM
BLC8824	OEM
BLK001	OEM
BLK002	OEM
BLK003	OEM
BLK004	OEM
BLK005	OEM
BLK006	OEM
BLK007	OEM
BLK008	OEM
BLK010	OEM
BLK011	OEM
BLK012	OEM
BLK020	OEM
BLK022	OEM
BLK024	OEM
BLK829	OEM
BLM003	OEM
BLM006	OEM
BLM014	OEM
BLM017	OEM
BLM071	OEM
BLM076A	OEM
BLM086	OEM
BLM093	OEM
BLM095	OEM
BLM116	OEM
BLM127	OEM
BLM129	OEM
BLM136	OEM
BLM138	OEM
BLM143A	OEM
BLM148	OEM
BLM149A	OEM
BLM153	OEM
BLM154	OEM
BLM155	OEM
BLM162	OEM
BLM167	OEM
BLM176	OEM
BLM179	OEM
BLM181	OEM
BLM183	OEM
BLM184	OEM
BLM186	OEM
BLM188	OEM
BLM190	OEM
BLM192	OEM
BLM195	OEM
BLM196	OEM
BLM198	OEM
BLM199	OEM
BLM201	OEM
BLM203	OEM
BLM204	OEM
BLM205	OEM
BLM207	OEM
BLM211	OEM
BLM213	OEM
BLM256	OEM
BLM304	OEM
BLM306	OEM
BLM309	OEM
BLM310	OEM
BLM311	OEM
BLM312	OEM
BLM313	OEM
BLP005D	OEM
BLP017D	OEM
BLP027D	OEM
BLP136D	OEM
BLP149D	OEM
BLP215	OEM
BLS032	OEM
BLS526	OEM
BLS530	OEM
BLS531	OEM
BLT028	OEM
BLT033	OEM
BLT036	OEM
BLT072	OEM
BLT074	OEM
BLT088	OEM
BLT101	OEM
BLT114A	OEM
BLT119	OEM
BLV10	OEM
BLV11	OEM
BLV20	OEM
BLV21	OEM
BLW10	OEM
BLW11	OEM
BLW12	OEM
BLW13	OEM
BLW14	OEM
BLW15	OEM
BLW16	2041
BLW17	OEM
BLW18	OEM
BLW19	OEM
BLW20	OEM
BLW21	OEM
BLW22	OEM
BLW23	OEM
BLW24	OEM
BLW25	2506
BLW29	OEM
BLW31	OEM
BLW32	OEM
BLW33	OEM
BLW34	OEM
BLW35	OEM
BLW36	OEM
BLW39	OEM
BLW42	OEM
BLW43	OEM
BLW44	OEM
BLW60	1198
BLW60C	OEM
BLW60F	OEM
BLW75	OEM
BLW76	OEM
BLW77	OEM
BLW78	OEM
BLW79	OEM
BLW80	OEM
BLW81	OEM
BLW82	OEM
BLW83	OEM
BLW84	OEM
BLW85	OEM
BLW86	OEM
BLW87	OEM
BLW89	OEM
BLW90	OEM
BLW92	OEM
BLW93	OEM
BLW94	OEM
BLW95	OEM
BLW96	OEM
BLW98	OEM
BLX10	OEM
BLX11	OEM
BLX12	OEM
BLX13C	OEM
BLX17	OEM
BLX18	OEM
BLX34	OEM
BLX35	OEM
BLX36	OEM
BLX37	OEM
BLX38	OEM
BLX39	OEM
BLX40	OEM
BLX41	OEM
BLX47	OEM
BLX48	OEM
BLX57	OEM
BLX65	0418
BLX67	3560
BLX68	2091
BLX69A	OEM
BLX82	OEM
BLX84	OEM
BLX85	OEM
BLX87	OEM
BLX88	0493
BLX91A	OEM
BLX92A	OEM
BLX93A	OEM
BLX94A	OEM
BLX94C	OEM
BLX281	OEM
BLX281A	OEM
BLX281B	OEM
BLX321	OEM
BLX331	OEM
BLX332	OEM
BLX350	OEM
BLX351	OEM
BLY10	0103
BLY11	0103
BLY11S	OEM
BLY12	0620
BLY15	0836
BLY15A	0178
BLY16	OEM
BLY17C	OEM
BLY20	0560
BLY21	0042
BLY25	OEM
BLY26	OEM
BLY27	0079
BLY28	0595
BLY29	OEM
BLY30	OEM
BLY33	0418
BLY34	0418
BLY35	0233
BLY36	0042
BLY37	0560
BLY38	0560
BLY39	OEM
BLY47	0177
BLY47A	0178
BLY48	0177
BLY48A	0178
BLY49	0074
BLY49A	0178
BLY50	0074
BLY53	0560
BLY53AP	OEM
BLY53B	OEM
BLY57	3321
BLY58	3322
BLY59	OEM
BLY60	OEM
BLY61	0086
BLY62	0560
BLY63	0042
BLY64	OEM
BLY65	OEM
BLY66	OEM
BLY68	0236
BLY70	OEM
BLY72	OEM
BLY74	OEM
BLY78	0560
BLY79	0042
BLY82	OEM
BLY86	OEM
BLY87A	1977
BLY87C	OEM
BLY88	0042
BLY88A	1233
BLY88C	OEM
BLY89	0042
BLY89A	1977
BLY89C	OEM
BLY91	0560
BLY91A	3586
BLY91C	OEM
BLY92	0042
BLY92A	2506
BLY92C	OEM
BLY93	3587
BLY93A	2506
BLY93C	OEM
BLY99	0418
BM15-12	OEM
BM25LE	OEM
BM30-12	5614
BM45-12	OEM
BM70-12	3194
BM80-12	3091
BM80-28	OEM
BM100-12	OEM
BM100-28	OEM
BM150-12	OEM
BM1002	OEM
BM1003	OEM
BM1004	OEM
BM1005	OEM
BM1006	OEM
BM1026	OEM
BM1027	OEM
BM1028	OEM
BM1029	OEM
BM1030	OEM
BM1031	OEM
BM1032	OEM
BM1033	OEM
BM1034	OEM
BM1035	OEM
BM1036	OEM
BM1037	OEM
BM1038	OEM
BM1039	OEM
BM1040	OEM
BM4001	OEM
BM4002	OEM
BM4003	OEM
BM4004	OEM
BM4005	OEM
BM4006	OEM
BM4301A	OEM
BM4303	OEM
BM4304	OEM
BM4601A	OEM
BM4603B	OEM
BM4603C	OEM
BM4604	OEM
BM4605	OEM
BM4710	OEM
BM4711	OEM
BM4712	OEM
BM4717	OEM
BM8001	OEM
BM8002	OEM
BM8003	OEM
BM8401A	OEM
BM8403	OEM
BM8404	OEM
BMM-128-200	OEM
BMM-128-300	OEM
BN0335N5	OEM
BN1L4M	OEM
BN12	OEM
BN-66	0016
BN74S301F	OEM
BN5040A	OEM
BN5111	3738
BN5111A	3738
BN5111B	3738
BN5115	OEM
BN5210	3787
BN-5416	3853
BN5416	3853
BN5436N	3747
BN6111	OEM
BN7133	0103
BN7168	0178
BN7214	0103
BN7253	0233
BN7517	0016
BN7518	0016
BN7551	0137
BO-71	0127
BO-190	OEM
BO316415	0167
BO319200	1050
BO351500	0829
BO354710	0912
BO354830	2054
BO356446	0663
BO356620	2054
BO470016	0478
BO470116	0215
BOD1-04	OEM
BOD1-05	OEM
BOD1-06	OEM
BOD1-07	OEM
BOD1-08	OEM
BOD1-09	OEM
BOD1-10	OEM
BOD1-11L	OEM
BOD1-11LD	OEM
BOD1-11R	OEM
BOD1-11RD	OEM
BOD1-12L	OEM
BOD1-12LD	OEM
BOD1-12R	OEM
BOD1-12RD	OEM
BOD1-13L	OEM
BOD1-13LD	OEM
BOD1-13R	OEM
BOD1-13RD	OEM
BOD1-14L	OEM
BOD1-14LD	OEM
BOD1-14R	OEM
BOD1-14RD	OEM
BOD1-15L	OEM
BOD1-15LD	OEM
BOD1-15R	OEM
BOD1-15RD	OEM
BOD1-16L	OEM
BOD1-16LD	OEM
BOD1-16R	OEM
BOD1-16RD	OEM
BOD1-17L	OEM
BOD1-17LD	OEM
BOD1-17R	OEM
BOD1-17RD	OEM
BOD1-18L	OEM
BOD1-18LD	OEM
BOD1-18R	OEM
BOD1-18RD	OEM
BOD1-19L	OEM
BOD1-19LD	OEM
BOD1-19R	OEM
BOD1-19RD	OEM
BOD1-20L	OEM
BOD1-20LD	OEM
BOD1-20R	OEM
BOD1-20RD	OEM
BOD1-21L	OEM
BOD1-21LD	OEM
BOD1-21R	OEM
BOD1-21RD	OEM
BOD1-22L	OEM
BOD1-22LD	OEM
BOD1-22R	OEM
BOD1-22RD	OEM
BOD1-23L	OEM
BOD1-23LD	OEM
BOD1-23R	OEM
BOD1-23RD	OEM
BOD1-24L	OEM
BOD1-24LD	OEM
BOD1-24R	OEM
BOD1-24RD	OEM
BOD1-25L	OEM
BOD1-25LD	OEM
BOD1-25R	OEM
BOD1-25RD	OEM
BOD1-26L	OEM
BOD1-26LD	OEM
BOD1-26R	OEM
BOD1-26RD	OEM
BOD1-28L	OEM
BOD1-28LD	OEM
BOD1-28R	OEM
BOD1-28RD	OEM
BOD1-30L	OEM
BOD1-30LD	OEM
BOD1-30R	OEM
BOD1-30RD	OEM
BOD1-32R	OEM
BOD1-34R	OEM
BOD1-36R	OEM
BOD1-38R	OEM
BOD1-40R	OEM
BODI11LD	OEM
BOT521XXX	OEM
BOT522XXX	OEM
BOV14S10K	OEM
BOV14S14K	OEM
BOV30S10K	OEM
BOV30S14K	OEM
BOV60S05K	OEM
BOV60S07K	OEM
BOV60S10K	OEM
BOV60S14K	OEM
BOV60S20K	OEM
BOV130S10K	OEM
BOV130S14K	OEM
BOV130S20K	OEM
BOV150B32K	OEM
BOV150S05K	OEM
BOV150S07K	OEM
BOV150S10K	OEM
BOV150S14K	OEM
BOV150S20K	OEM
BOV250B25K	OEM
BOV250S05K	OEM
BOV250S07K	OEM
BOV250S10K	OEM
BOV250S14K	OEM
BOV250S20K	OEM
BOV275S05K	OEM
BOV275S07K	OEM
BOV275S10K	OEM
BOV275S14K	OEM
BOV275S20K	OEM
BOV420B32K	OEM
BOV420S10K	OEM
BOV420S14K	OEM
BOV420S20K	OEM

DEVICE TYPE	REPL CODE
BOV550S10K	OEM
BOV550S14K	OEM
BOV550S20K	OEM
BOV625S10K	OEM
BOV625S14K	OEM
BOV625S20K	OEM
BP-0200	OEM
BP-0575	OEM
BP2BC	OEM
BP5BC	OEM
BP8-12	OEM
BP15-12	OEM
BP30-12	OEM
BP30-12L	OEM
BP50	OEM
BP67	0016
BP100P	OEM
BP103BI	OEM
BP103BII	OEM
BP103BIII	OEM
BP103BIV	OEM
BP103I	OEM
BP103II	OEM
BP103III	OEM
BP103IV	OEM
BP-104	OEM
BP104	OEM
BP-2190	OEM
BP3371	OEM
BP5263	0050
BP336933	OEM
BP336940	OEM
BP338978	OEM
BP338979	OEM
BP338986	OEM
BP338987	OEM
BP338995	OEM
BP338998	OEM
BP372882	OEM
BP879849	OEM
BPA07347	OEM
BPA07364	OEM
BPA07374	OEM
BPA07403	OEM
BPA07420	OEM
BPM5-250	OEM
BPM5-250E	OEM
BPM5-250J	OEM
BPM5-500	OEM
BPM5-500E	OEM
BPM5-500J	OEM
BPM12-25D5	OEM
BPM12-25D12	OEM
BPM12-25D28	OEM
BPM12-60	OEM
BPM12-60E	OEM
BPM12-60J	OEM
BPM12-100	OEM
BPM12-100D5	OEM
BPM12-100D12	OEM
BPM12-100D28	OEM
BPM12-100E	OEM
BPM12-200	OEM
BPM12-200E	OEM
BPM12-200J	OEM
BPM12-210D5	OEM
BPM12-210D12	OEM
BPM12-210D28	OEM
BPM12-300	OEM
BPM12-300E	OEM
BPM12-300J	OEM
BPM12-420D5	OEM
BPM12-420D12	OEM
BPM12-420D28	OEM
BPM15-25D5	OEM
BPM15-25D12	OEM
BPM15-25D28	OEM
BPM15-60	OEM
BPM15-60E	OEM
BPM15-60J	OEM
BPM15-100	OEM
BPM15-100D5	OEM
BPM15-100D12	OEM
BPM15-100D28	OEM
BPM15-100E	OEM
BPM15-100J	OEM
BPM15-150D5	OEM
BPM15-150D24	OEM
BPM15-150D28	OEM
BPM15-165D12	OEM
BPM15-165D28	OEM
BPM15-165DS	OEM
BPM15-200	OEM
BPM15-200E	OEM
BPM15-200J	OEM
BPM15-300	OEM
BPM15-300E	OEM
BPM15-300J	OEM
BPM15-330D5	OEM
BPM15-330D12	OEM
BPM15-330D28	OEM
BPM18-25D5	OEM
BPM18-25D12	OEM
BPM18-25D28	OEM
BPM18-100D6	OEM
BPM18-100D12	OEM
BPM18-100D28	OEM
BPM18-140D5	OEM
BPM18-140D12	OEM

DEVICE TYPE	REPL CODE
BPM18-140D28	OEM
BPM18-280D5	OEM
BPM18-280D12	OEM
BPM18-280D28	OEM
BPM120-25	OEM
BPM120-25E	OEM
BPM120-25J	OEM
BPM150-20	OEM
BPM150-20E	OEM
BPM150-20J	OEM
BPM180-16	OEM
BPM180-16E	OEM
BPM180-16J	OEM
BPW10	OEM
BPW13A	OEM
BPW13B	OEM
BPW13C	OEM
BPW14A	OEM
BPW14B	OEM
BPW14C	OEM
BPW20	OEM
BPW21	OEM
BPW21M	OEM
BPW22A	OEM
BPW22A-I	OEM
BPW22A-II	OEM
BPW24	OEM
BPW28	OEM
BPW30	3552
BPW32	OEM
BPW33	OEM
BPW34	OEM
BPW35	OEM
BPW36	OEM
BPW37	OEM
BPW38	OEM
BPW41	2628
BPW41D	OEM
BPW44	OEM
BPW45	OEM
BPW50	OEM
BPW71	OEM
BPX25	OEM
BPX29	OEM
BPX30	OEM
BPX31	OEM
BPX32	OEM
BPX35	OEM
BPX38I	OEM
BPX38II	OEM
BPX38III	OEM
BPX38IV	OEM
BPX40	OEM
BPX41	OEM
BPX42	OEM
BPX43-1	2313
BPX43-2	3878
BPX43-3	3878
BPX43-4	2313
BPX43I	OEM
BPX43II	OEM
BPX43III	OEM
BPX43IV	OEM
BPX47A	OEM
BPX47B-18	OEM
BPX47B-20	OEM
BPX47C36	OEM
BPX48	OEM
BPX58	OEM
BPX59	3552
BPX60	OEM
BPX61	OEM
BPX61P	OEM
BPX63	OEM
BPX65	OEM
BPX66	OEM
BPX66P	OEM
BPX68	OEM
BPX70	OEM
BPX70C	OEM
BPX70D	OEM
BPX70E	OEM
BPX71	OEM
BPX71-201	OEM
BPX71-202	OEM
BPX71-203	OEM
BPX71-204	OEM
BPX72	OEM
BPX72C	OEM
BPX72D	OEM
BPX72E	OEM
BPX72F	OEM
BPX79	OEM
BPX80	OEM
BPX81I	OEM
BPX81II	OEM
BPX81III	OEM
BPX81IV	OEM
BPX82	OEM
BPY11P	OEM
BPY11P/I	OEM
BPY11PIV	OEM
BPY11PV	OEM
BPY11PVI	OEM
BPY12	OEM
BPY15	OEM
BPY20	OEM
BPY21	OEM
BPY22	OEM

DEVICE TYPE	REPL CODE
BPY43	OEM
BPY44	OEM
BPY45	OEM
BPY46	OEM
BPY47	OEM
BPY47P	OEM
BPY48	OEM
BPY48P	OEM
BPY61I	OEM
BPY61II	OEM
BPY61III	OEM
BPY61IV	OEM
BPY62	2313
BPY62-1	2313
BPY62-2	2313
BPY62-3	3878
BPY62I	OEM
BPY62II	OEM
BPY62III	OEM
BPY63	OEM
BPY63P	OEM
BPY64	OEM
BPY64P	OEM
BPY66	OEM
BPY67	OEM
BPY71	OEM
BPY72	OEM
BQ	OEM
BQ67	0016
BQ-94	0016
BQA02-30A	0195
B-R	OEM
BR09250M	OEM
BR09450	OEM
BR-1	0277
BR1	0277
BR5A5	0023
BR5B5	0023
BR5C5	0023
BR5D5	0023
BR5E5	0023
BR5F5	0023
BR5G5	0023
BR5H5	0017
BR5K5	0017
BR5M5	0017
BR5N5	0017
BR5P5	0017
BR31	0323
BR31G	0323
BR32	1414
BR32G	1414
BR32J01	1414
BR34	0473
BR34G	0473
BR36	0446
BR36G	0446
BR38	1422
BR38G	1422
BR42	0015
BR44	0015
BR46	0015
BR47	0015
BR48	0015
BR51	OEM
BR52	0015
BR54	OEM
BR54-600	0308
BR-55	0030
BR56	OEM
BR58	OEM
BR61	0323
BR61G	0323
BR62	1414
BR62G	1414
BR64	0473
BR64G	0473
BR-66	0016
BR66	0446
BR66G	0446
BR67	0016
BR68	1422
BR68G	1422
BR78	0259
BR81	0323
BR81D	0277
BR81DG	0277
BR81G	1422
BR-82	0037
BR82	1414
BR-82D	0289
BR82D	0289
BR82DG	0289
BR82G	1414
BR84	0473
BR84D	0296
BR84DG	0296
BR84G	0473
BR86	0446
BR86D	0302
BR86DG	0302
BR86G	0446
BR88	1422
BR88D	0250
BR88DG	0250
BR88G	1422
BR93LC46A	OEM
BR100	2730
BR-100B	0560

DEVICE TYPE	REPL CODE
BR100B	0560
BR101	OEM
BR-101B	0042
BR101B	0042
BR102	OEM
BR103	1137
BR104	OEM
BR106	OEM
BR108	OEM
BR151	2363
BR151G	2363
BR151W	OEM
BR152	2363
BR152G	2363
BR152W	OEM
BR154	2369
BR154G	2369
BR154W	OEM
BR156	2370
BR156G	2370
BR156W	OEM
BR158	2372
BR158G	2372
BR158W	OEM
BR-190	OEM
BR203	OEM
BR210	OEM
BR220	OEM
BR251	2363
BR251G	2363
BR251W	OEM
BR252	2363
BR252G	2363
BR252W	OEM
BR254	2369
BR254W	OEM
BR256	2370
BR256W	OEM
BR258	2372
BR258W	OEM
BR303	OEM
BR305	0323
BR305G	0323
BR310	2443
BR310G	2443
BR351	1643
BR352	1643
BR354	1673
BR356	1673
BR358	1589
BR358F	1589
BR358G	1589
BR412-5-5	OEM
BR412-5-10	OEM
BR412-5-25	OEM
BR412-7-5	OEM
BR412-7-10	OEM
BR412-7-25	OEM
BR412-10-5	OEM
BR412-10-10	OEM
BR412-10-25	OEM
BR412-12-5	OEM
BR412-12-10	OEM
BR412-12-25	OEM
BR412-15-5	OEM
BR412-15-10	OEM
BR412-15-25	OEM
BR412-20-5	OEM
BR412-20-10	OEM
BR412-20-25	OEM
BR412-25-5	OEM
BR412-25-10	OEM
BR412-25-25	OEM
BR412-30-5	OEM
BR412-30-10	OEM
BR412-30-25	OEM
BR412-40-5	OEM
BR412-40-10	OEM
BR412-40-25	OEM
BR412-50-5	OEM
BR412-50-10	OEM
BR412-50-25	OEM
BR427-10	0017
BR427-15	0017
BR427-20	OEM
BR427-25	OEM
BR427-30	OEM
BR427-35	OEM
BR427-40	OEM
BR427-45	OEM
BR427-50	OEM
BR427-55	OEM
BR427-60	OEM
BR427-65	OEM
BR427-70	OEM
BR577-70	OEM
BR587-10	0017
BR587-15	0017
BR587-20	OEM
BR587-25	OEM
BR587-30	OEM
BR587-35	OEM
BR587-40	OEM
BR587-45	OEM
BR587-50	OEM
BR587-55	OEM
BR587-60	OEM
BR587-65	OEM
BR587-70	OEM

DEVICE TYPE	REPL CODE
BR587-80	OEM
BR587-85	OEM
BR587-90	OEM
BR587-95	OEM
BR587-100	OEM
BR587-110	OEM
BR587-120	OEM
BR587-130	OEM
BR587-140	OEM
BR587-150	OEM
BR587-160	OEM
BR587-170	OEM
BR587-180	OEM
BR587-190	OEM
BR587-200	OEM
BR588-10	0017
BR588-15	0017
BR588-20	OEM
BR588-25	OEM
BR588-30	OEM
BR588-35	OEM
BR588-40	OEM
BR588-45	OEM
BR588-50	OEM
BR588-55	OEM
BR588-60	OEM
BR588-65	OEM
BR588-70	OEM
BR588-80	OEM
BR588-85	OEM
BR588-90	OEM
BR588-95	OEM
BR588-100	OEM
BR588-110	OEM
BR588-120	OEM
BR588-130	OEM
BR588-140	OEM
BR588-150	OEM
BR588-160	OEM
BR588-170	OEM
BR588-180	OEM
BR588-190	OEM
BR588-200	OEM
BR605	0323
BR605G	0323
BR610	2443
BR805	0323
BR805D	0277
BR805DG	0277
BR805G	0323
BR810	2443
BR810D	0250
BR810DG	0250
BR810G	2443
BR-832	0037
BR1005	5295
BR1010	OEM
BR1505	2363
BR1505G	2363
BR1505W	OEM
BR1510	2373
BR1510F	2373
BR1510G	2373
BR1510W	OEM
BR1941L	OEM
BR1941M	OEM
BR2505	2363
BR2505G	2363
BR2505W	OEM
BR2510	2373
BR2510G	2373
BR2510W	OEM
BR2941L	OEM
BR3433S	OEM
BR3505	1643
BR3510	1589
BR5507S	OEM
BR5507S-1	OEM
BR13451	OEM
BR16252	OEM
BR16452	OEM
BR51400-1	0015
BR51401-2	0015
BRA40	0473
BRF151	0277
BRF152	0277
BRF153	0289
BRF154	0296
BRF155	0302
BRF156	0250
BRF157	0250
BRF201	0277
BRF202	0277
BRF203	0289
BRF204	0296
BRF205	0302
BRF206	0250
BRF207	0250
BRF251	0729
BRF252	0729
BRF253	0729
BRF254	0737
BRF255	0737
BRF256	0742
BRF257	0742
BRF401	0729
BRF402	0729
BRF403	0729
BRF404	0737

DEVICE TYPE	REPL CODE
BRF405	0737
BRF406	0742
BRF407	0742
BRH1000	OEM
BRH1001	OEM
BRH1002	OEM
BRH1004	OEM
BRH1006	OEM
BRH2500	OEM
BRH2501	OEM
BRH2502	OEM
BRH2504	OEM
BRH2506	OEM
BRH3500	OEM
BRH3501	OEM
BRH3502	OEM
BRH3504	OEM
BRH3506	OEM
BRHV1502	OEM
BRHV1503	OEM
BRHV1504	OEM
BRHV1505	OEM
BRHV1506	OEM
BRR101	0106
BRR102	0106
BRR103	0106
BRR104	0106
BRR105	2010
BRR106	2400
BRR107	0787
BRR151	0106
BRR152	0106
BRR153	0106
BRR154	0106
BRR155	2010
BRR156	2400
BRR157	0787
BRS10-005	OEM
BRS10-01	OEM
BRS10-02	OEM
BRS10-04	OEM
BRS10-06	OEM
BRS10-08	OEM
BRS10-10	OEM
BRS12-005	OEM
BRS12-01	OEM
BRS12-02	OEM
BRS12-04	OEM
BRS12-06	OEM
BRS12-08	OEM
BRS12-10	OEM
BRS15-005	OEM
BRS15-01	OEM
BRS15-02	OEM
BRS15-04	OEM
BRS15-06	OEM
BRS15-08	OEM
BRS15-10	OEM
BRS25-005	OEM
BRS25-01	OEM
BRS25-02	OEM
BRS25-04	OEM
BRS25-06	OEM
BRS25-08	OEM
BRS25-10	OEM
BRS35-005	OEM
BRS35-01	OEM
BRS35-02	OEM
BRS35-04	OEM
BRS35-06	OEM
BRS35-08	OEM
BRS35-10	OEM
BRS101	1875
BRS102	1875
BRS103	1875
BRS104	1875
BRS105	1875
BRS106	3546
BRS107	3546
BRS301	0323
BRS302	0323
BRS303	1414
BRS304	0473
BRS305	0446
BRS306	1422
BRS307	2443
BRS601	0323
BRS602	0323
BRS603	1414
BRS604	0473
BRS605	0446
BRS606	1422
BRS607	2443
BRT60	OEM
BRT170S01	OEM
BRT170S02	OEM
BRT170S03	OEM
BRT170S04	OEM
BRT170S05	OEM
BRT170S06	OEM
BRT170S07	OEM
BRT170S08	OEM
BRT170S09	OEM
BRT170S10	OEM
BRT170S11	OEM
BRT170S12	OEM
BRT170S13	OEM
BRT170S14	OEM
BRT170S15	OEM

DEVICE TYPE	REPL CODE
BRV615	OEM
BRV615T	OEM
BRX44	1137
BRX45	0344
BRX46	0901
BRX47	0058
BRX48	0407
BRX49	0407
BRX50	OEM
BRX51	OEM
BRX52	OEM
BRX53	OEM
BRX54	OEM
BRX55	OEM
BRX56	OEM
BRX60	0174
BRX61	3854
BRX62	3620
BRX63	OEM
BRX64	3324
BRX65	1502
BRX66	1502
BRY20	OEM
BRY23	OEM
BRY24	OEM
BRY25	OEM
BRY26	OEM
BRY28	OEM
BRY29	OEM
BRY30	OEM
BRY31	OEM
BRY32	OEM
BRY33	OEM
BRY34	OEM
BRY35	OEM
BRY36	OEM
BRY37	OEM
BRY39	4877
BRY39T	OEM
BRY42	OEM
BRY43	OEM
BRY44	OEM
BRY46	OEM
BRY49	OEM
BRY50	OEM
BRY51	OEM
BRY52	OEM
BRY53	OEM
BRY54	OEM
BRY54-100	0179
BRY54-100T	0179
BRY54-200	OEM
BRY54-200T	OEM
BRY54-300	0346
BRY54-300T	OEM
BRY54-400	3349
BRY54-400T	OEM
BRY54-500	3349
BRY54-500T	OEM
BRY54-600	3349
BRY54-600T	OEM
BRY54-1007	0179
BRY55	OEM
BRY55-30	1137
BRY55-60	0344
BRY55-100	0901
BRY55-200	0058
BRY55-300	0407
BRY55-400	OEM
BRY55M-300	OEM
BRY55M-400	OEM
BRY55M-600	OEM
BRY55M-800	OEM
BRY56	OEM
BRY56A	OEM
BRY56B	OEM
BRY56C	OEM
BRY57	OEM
BRY58-30	OEM
BRY58-60	OEM
BRY58-100	OEM
BRY58-200	OEM
BRY58-300	OEM
BRY61	OEM
BS	OEM
BS08A	4233
BS08H	2353
BS-1	0289
BS1	0015
BS2	0015
BS3-04CM1	OEM
BS3-06CM1	OEM
BS4	OEM
BS4-04FM5	OEM
BS4-06FM5	OEM
BS4-08FM5	OEM
BS5-04EM1	OEM
BS5-04FM5	OEM
BS5-06EM1	OEM
BS5-06FM5	OEM
BS5-08EM1	OEM
BS5-08FM5	OEM
BS6	OEM
BS6-01A	0154
BS6-01B	OEM
BS6-01E	1064
BS6-02A	0154
BS6-02B	OEM
BS6-02E	1064

DEVICE TYPE	REPL CODE
BS6-03A	0147
BS6-03B	OEM
BS6-03E	1317
BS6-04A	0147
BS6-04B	OEM
BS6-04E	1317
BS6-05A	0279
BS6-05B	OEM
BS6-05E	1891
BS6-06A	0279
BS6-06B	OEM
BS6-06E	1891
BS6-07A	0875
BS6-07B	OEM
BS7-02A	0772
BS7-04A	0744
BS7-04FM1	OEM
BS7-05A	0617
BS7-06FM1	OEM
BS7-08FM1	OEM
BS8	OEM
BS8-01A	0154
BS8-01B	OEM
BS8-02A	0154
BS8-02B	OEM
BS8-03A	0147
BS8-03B	OEM
BS8-04A	0147
BS8-04B	OEM
BS8-05A	0279
BS8-05B	OEM
BS8-06A	0279
BS8-06B	OEM
BS8-07A	OEM
BS8-07B	OEM
BS9-02A	0772
BS9-04A	0744
BS9-04FM5	OEM
BS9-04HR1	OEM
BS9-05A	0617
BS9-06FM5	OEM
BS9-06HR1	OEM
BS9-08FM5	OEM
BS9-08HR1	OEM
BS10	OEM
BS10-01A	0154
BS10-01B	OEM
BS10-02A	0154
BS10-02B	OEM
BS10-03A	0147
BS10-03B	OEM
BS10-04A	0147
BS10-04B	OEM
BS10-05A	0279
BS10-05B	OEM
BS10-06A	0279
BS10-06B	OEM
BS10-07A	OEM
BS10-07B	OEM
BS12-04HU1	OEM
BS12-04HU5	OEM
BS12-06HU1	OEM
BS12-06HU5	OEM
BS12-08HU1	OEM
BS12-08HU5	OEM
BS16-04HU1	OEM
BS16-06HU1	OEM
BS16-08HU1	OEM
BS29P	OEM
BS48	OEM
BS52	OEM
BS58	OEM
BS60	OEM
BS63	OEM
BS-66	0016
BS67	0016
BS68	0006
BS70	OEM
BS72	OEM
BS76	OEM
BS80	OEM
BS82	OEM
BS84	OEM
BS92	OEM
BS-94	0016
BS100D	OEM
BS102	OEM
BS104	OEM
BS107	OEM
BS107A	OEM
BS107P	OEM
BS110	OEM
BS112	OEM
BS114	OEM
BS116	OEM
BS118	OEM
BS128	OEM
BS138	OEM
BS142	OEM
BS148	OEM
BS156	OEM
BS158	OEM
BS162	OEM
BS166	OEM
BS170P	OEM
BS172	OEM
BS178	OEM
BS188	OEM
BS190	OEM
BS192	OEM
BS194	OEM
BS196	OEM
BS200	OEM
BS202	OEM
BS204	OEM
BS206	OEM
BS216	OEM
BS217	OEM
BS219	OEM
BS220	OEM
BS224	OEM
BS226	OEM
BS228	OEM
BS232	OEM
BS234	OEM
BS246	OEM
BS248	OEM
BS250	OEM
BS250P	OEM
BS256	OEM
BS258	OEM
BS260	OEM
BS262	OEM
BS264	OEM
BS276	OEM
BS277	OEM
BS280	OEM
BS286	OEM
BS310	OEM
BS314	OEM
BS316	OEM
BS320	OEM
BS324	OEM
BS342	OEM
BS344	OEM
BS384	OEM
BS386	OEM
BS390	OEM
BS412	OEM
BS426	OEM
BS430	OEM
BS440	OEM
BS450	OEM
BS452	OEM
BS454	OEM
BS462	OEM
BS466	OEM
BS475	0016
BS500	OEM
BS500A	OEM
BS500B	OEM
BS530UV	OEM
BS608A	OEM
BS608B	OEM
BS616A	OEM
BS616B	OEM
BS624A	OEM
BS624B	OEM
BS624C	OEM
BS624D	OEM
BS624E	OEM
BS624K	OEM
BS624L	OEM
BS624M	OEM
BS624N	OEM
BS624P	OEM
BS630	OEM
BS640	OEM
BS644	OEM
BS646	OEM
BS648	OEM
BS658	OEM
BS660	OEM
BS662	OEM
BS674	OEM
BS676	OEM
BS678	OEM
BS698	OEM
BS702	OEM
BS710	OEM
BS714	OEM
BS716	OEM
BS718	OEM
BS720	OEM
BS724	OEM
BS726	OEM
BS728	OEM
BS732	OEM
BS748	OEM
BS750	OEM
BS752	OEM
BS756	OEM
BS758	OEM
BS760	OEM
BS762	OEM
BS764	OEM
BS766	OEM
BS772	OEM
BS774	OEM
BS776	OEM
BS778	OEM
BS784	OEM
BS798	OEM
BS800	OEM
BS806A	OEM
BS806B	OEM
BS806C	OEM
BS806D	OEM
BS806E	OEM
BS806F	OEM
BS806G	OEM
BS806H	OEM
BS810	OEM
BS814	OEM
BS816	OEM
BS818	OEM
BS822	OEM
BS824	OEM
BS826	OEM
BS828	OEM
BS830	OEM
BS832	OEM
BS834	OEM
BS836	OEM
BS838	OEM
BS842	OEM
BS844	OEM
BS846	OEM
BS848	OEM
BS854	OEM
BS856	OEM
BS858	OEM
BS860	OEM
BS868	OEM
BS870	OEM
BS872	OEM
BS874	OEM
BS876	OEM
BS880	OEM
BS882	OEM
BS894	OEM
BS898	OEM
BS902	OEM
BS904	OEM
BS908	OEM
BS910	OEM
BS912	OEM
BS916	OEM
BS918	OEM
BS928	OEM
BS930	OEM
BS932	OEM
BS940	OEM
BS946	OEM
BS950	OEM
BS952	OEM
BS956	OEM
BS958	OEM
BS958D	OEM
BS959	OEM
BS960	OEM
BS962	OEM
BS966	OEM
BS968	OEM
BS969	OEM
BS970	OEM
BS974	OEM
BS975	OEM
BS976	OEM
BS986	OEM
BS990	OEM
BS994	OEM
BS9011G	0079
BS9301F019-8308	OEM
BS9301F019-8805	OEM
BS74154	1555
BSA01	0133
BSA02	0133
BSA11	0133
BSA12	OEM
BSA21	OEM
BSA31	OEM
BS-B1	0277
BS-B2	0289
BS-B4	0296
BSC01	OEM
BSC02	OEM
BSC03	OEM
BSC04	OEM
BSC05	OEM
BSC11	OEM
BSC12	OEM
BSC13	OEM
BSC14	OEM
BSC21	OEM
BSC22	OEM
BSC23	OEM
BSC24	OEM
BSC25	OEM
BSC26	OEM
BSC51	OEM
BSC52	OEM
BSC-256	OEM
BSC551	OEM
BSC552	OEM
BSC553	OEM
BSC554	OEM
BSC555	OEM
BSC556	OEM
BSC557	OEM
BSC1015	OEM
BSC1015A	OEM
BSC1015B	OEM
BSC1016	OEM
BSC1016A	OEM
BSC1016B	OEM
BSD10	OEM
BSD12	OEM
BSD212	OEM
BSD213	OEM
BSD214	OEM
BSD215	OEM
BSF17	OEM
BSJ111	3842
BSJ112	OEM
BSJ113	3843
BSJ174	OEM
BSJ175	OEM
BSJ176	OEM
BSJ177	OEM
BSM01	OEM
BSM02	OEM
BSM11	OEM
BSM12	OEM
BSM13	OEM
BSM21	OEM
BSM31	OEM
BSM41	OEM
BSM42	OEM
BSM51	OEM
BSM61	OEM
BSM274	OEM
BSP15	OEM
BSP16	OEM
BSP19	OEM
BSP20	OEM
BSP30	OEM
BSP31	OEM
BSP32	OEM
BSP33	OEM
BSP40	OEM
BSP41	OEM
BSP42	OEM
BSP43	OEM
BSP50	OEM
BSP52	OEM
BSP60	OEM
BSP61	OEM
BSP62	OEM
BSR12	OEM
BSR12R	OEM
BSR13	1733
BSR13R	OEM
BSR14	1733
BSR14R	OEM
BSR15	1752
BSR15R	OEM
BSR16	1752
BSR16R	OEM
BSR17	1733
BSR17A	1435
BSR17R	OEM
BSR18	1499
BSR18A	1499
BSR19	6116
BSR19A	6116
BSR20	5366
BSR20A	5366
BSR30	3645
BSR31	3645
BSR32	3645
BSR33	3645
BSR40	0667
BSR41	0667
BSR42	0667
BSR43	0667
BSR50	6911
BSR51	6911
BSR52	3892
BSR56	OEM
BSR57	OEM
BSR58	OEM
BSR60	3890
BSR61	3890
BSR62	3890
BSR64	OEM
BSR65	OEM
BSR66	OEM
BSR67	OEM
BSR70	OEM
BSR72	OEM
BSR76	OEM
BSR78	OEM
BSR80	OEM
BSR81	OEM
BSR82	OEM
BSR89	OEM
BSR92	OEM
BSS10	0016
BSS13	0418
BSS14	0622
BSS15	0622
BSS16	0622
BSS17	0126
BSS18	0126
BSS19	0285
BSS20	0285
BSS21	0016
BSS22	0016
BSS23	0595
BSS24	OEM
BSS25	OEM
BSS26	0016
BSS28	0086
BSS29	0086
BSS30	0630
BSS32	0630
BSS33	0233
BSS34	OEM
BSS35	OEM
BSS36	OEM
BSS37	0342
BSS38	0861
BSS40	0086
BSS41	0086
BSS42	0622
BSS43	0622
BSS44	0126
BSS45	0543
BSS47	OEM
BSS48	OEM
BSS49	OEM
BSS50	OEM
BSS51	OEM
BSS52	OEM
BSS56	OEM
BSS58A	OEM
BSS58B	OEM
BSS60	OEM
BSS61	OEM
BSS62	OEM
BSS63	1503
BSS63R	OEM
BSS64	2332
BSS64R	OEM
BSS65	OEM
BSS65R	OEM
BSS66	OEM
BSS66R	OEM
BSS67	OEM
BSS67R	OEM
BSS68	0342
BSS69	OEM
BSS69R	OEM
BSS70	OEM
BSS70R	OEM
BSS71	OEM
BSS72	OEM
BSS74	OEM
BSS75	OEM
BSS79	OEM
BSS79B	1435
BSS79C	1435
BSS80	OEM
BSS80B	OEM
BSS80C	OEM
BSS81	OEM
BSS81B	OEM
BSS81C	OEM
BSS82	OEM
BSS82B	1499
BSS82C	1499
BSS87	OEM
BSS89	OEM
BSS91	OEM
BSS93	OEM
BSS95	OEM
BSS97	OEM
BSS100	OEM
BSS101	OEM
BSS110	OEM
BSS123	OEM
BSS465	OEM
BST15	5371
BST16	5371
BST39	3471
BST40	3471
BST50	5369
BST51	5369
BST52	5369
BST60	5370
BST61	5370
BST62	5370
BST70A	OEM
BST72A	OEM
BST74A	OEM
BST76A	OEM
BST78	OEM
BST90	OEM
BST97	OEM
BST100	OEM
BST110	OEM
BST250	OEM
BSTA3026	OEM
BSTA3026M	OEM
BSTA3033	OEM
BSTA3033M	OEM
BSTA3040	OEM
BSTA3040J	OEM
BSTA3040K	OEM
BSTA3040L	OEM
BSTA3040M	OEM
BSTA3040MJ	OEM
BSTA3040MK	OEM
BSTA3040ML	OEM
BSTA3046	OEM
BSTA3046J	OEM
BSTA3046K	OEM
BSTA3046L	OEM
BSTA3046M	OEM
BSTA3046MJ	OEM
BSTA3046MK	OEM
BSTA3046ML	OEM
BSTA3053	OEM
BSTA3053M	OEM
BSTB0106	OEM
BSTB0113	OEM
BSTB0126	OEM
BSTB0140	OEM
BSTB0146	OEM
BSTB0206	OEM
BSTB0213	OEM
BSTB0226	OEM
BSTB0240	OEM
BSTB0246	OEM
BSTC0106	2491
BSTC0113	0240
BSTC0126	0676
BSTC0133	0336
BSTC0140	0336
BSTC0146	OEM
BSTC0206	OEM
BSTC0213	OEM
BSTC0226	OEM
BSTC0233	OEM
BSTC0240	OEM
BSTC0246	OEM
BSTC0313	0762
BSTC0313S6	OEM
BSTC0313S6+	0469
BSTC0326	0740
BSTC0326S6	OEM
BSTC0326S6+	0722
BSTC0340	0764
BSTC0340S6	OEM
BSTC0340S6+	0725
BSTC0353	0766
BSTC0353S6	4535
BSTC0353S6+	2036
BSTC0366	4535
BSTC0366S6	4535
BSTC0366S6+	2036
BSTC0380	OEM
BSTC0380S6	OEM
BSTC0380S6+	2036
BSTC0506	OEM
BSTC0506K	OEM
BSTC0506L	OEM
BSTC0513	OEM
BSTC0513K	OEM
BSTC0513L	OEM
BSTC0526	OEM
BSTC0526K	OEM
BSTC0526L	OEM
BSTC0533	OEM
BSTC0533K	OEM
BSTC0533L	OEM
BSTC0540	0061
BSTC0540K	OEM
BSTC0540L	OEM
BSTC0546	OEM
BSTC0546K	OEM
BSTC0546L	OEM
BSTC0606	OEM
BSTC0613	OEM
BSTC0626	OEM
BSTC0626S6	OEM
BSTC0626S9	OEM
BSTC0633	OEM
BSTC0633H	OEM
BSTC0633S6	OEM
BSTC0633S9	OEM
BSTC0640	OEM
BSTC0640H	OEM
BSTC0640S6	OEM
BSTC0640S9	OEM
BSTC0643H	0239
BSTC0646	OEM
BSTC0646H	OEM
BSTC0646S6	OEM
BSTC0646S9	OEM
BSTC0653H	OEM
BSTC0706	OEM
BSTC0713	OEM
BSTC0726	OEM
BSTC0726T	OEM
BSTC0733	OEM
BSTC0733T	OEM
BSTC0740	OEM
BSTC0746	OEM
BSTC0926T92	OEM
BSTC0930T92	OEM
BSTC1026	OEM
BSTC1026M	OEM
BSTC1033	OEM
BSTC1033M	OEM
BSTC1040	OEM
BSTC1040B	OEM
BSTC1040C	OEM
BSTC1040D	OEM
BSTC1040M	OEM
BSTC1040MB	OEM
BSTC1040MC	OEM
BSTC1040MD	OEM
BSTC1040MS	OEM
BSTC1040MS1	OEM
BSTC1040S1	OEM
BSTC1040S2	OEM
BSTC1046	OEM
BSTC1046B	OEM
BSTC1046C	OEM
BSTC1046D	OEM
BSTC1046M	OEM
BSTC1046MB	OEM
BSTC1046MC	OEM
BSTC1046MD	OEM
BSTC1046MS1	OEM
BSTC1046MS2	OEM
BSTC1046S1	OEM
BSTC1046S2	OEM
BSTC1053	OEM
BSTC1053M	OEM
BSTC1226	OEM
BSTC1233	OEM
BSTC1240	OEM
BSTC1246	OEM
BSTC1250	OEM
BSTC3026	OEM
BSTC3033	OEM
BSTC3040	OEM
BSTC3040B	OEM
BSTC3040C	OEM
BSTC3040D	OEM
BSTC3040S1	OEM
BSTC3046	OEM
BSTC3046B	OEM
BSTC3046C	OEM
BSTC3046D	OEM
BSTC3046S1	OEM
BSTC3053	OEM
BSTC3126	OEM
BSTC3126M	OEM
BSTC3133	OEM
BSTC3133M	OEM
BSTC3140	OEM
BSTC3140B	OEM
BSTC3140C	OEM
BSTC3140D	OEM
BSTC3140M	OEM
BSTC3140MB	OEM
BSTC3140MC	OEM
BSTC3140MD	OEM
BSTC3140MS1	OEM
BSTC3140S1	OEM
BSTC3146	OEM
BSTC3146B	OEM
BSTC3146C	OEM
BSTC3146D	OEM
BSTC3146M	OEM
BSTC3146MB	OEM
BSTC3146MC	OEM
BSTC3146MD	OEM
BSTC3146MS1	OEM
BSTC3146S1	OEM
BSTC3153M	OEM
BSTCC0126	OEM
BSTCC0126S6	OEM
BSTCC0126S9	OEM
BSTCC0126T91	OEM
BSTCC0130T91	OEM
BSTCC0133	OEM
BSTCC0133H	OEM
BSTCC0133S6	OEM
BSTCC0133S9	OEM
BSTCC0140	OEM
BSTCC0140H	OEM
BSTCC0140R	OEM
BSTCC0140S6	OEM
BSTCC0140S9	OEM
BSTCC0143	OEM
BSTCC0143H	OEM
BSTCC0143R	OEM
BSTCC0146	OEM
BSTCC0146S6	OEM
BSTCC0146S9	OEM
BSTCC0246H	OEM
BSTCC0246R	OEM
BSTCC0253	OEM
BSTCC0253H	OEM
BSTCC0253R	OEM
BSTD0220	OEM
BSTD0240	OEM
BSTD0260	OEM
BSTD0313	0762
BSTD0313S6	OEM
BSTD0313S6+	0469
BSTD0320	OEM
BSTD0326	0740
BSTD0326S6	OEM
BSTD0326S6+	0722
BSTD0340	0764
BSTD0340S6	OEM
BSTD0340S6+	0725
BSTD0353	0766
BSTD0353S6	4535
BSTD0353S6+	2036
BSTD0366	4535
BSTD0366S6	4535
BSTD0366S6+	0679
BSTD0380	OEM
BSTD0380S6	OEM
BSTD0380S6+	2036
BSTD1026	OEM
BSTD1026M	OEM
BSTD1033	OEM
BSTD1033M	OEM
BSTD1040	OEM
BSTD1040B	OEM
BSTD1040C	OEM
BSTD1040M	OEM
BSTD1040MB	OEM
BSTD1040MC	OEM
BSTD1040MD	OEM

If replacement code is OEM, contact original manufacturer for replacement.

DEVICE TYPE	REPL CODE
BSTD1040MS1	OEM
BSTD1040MS2	OEM
BSTD1040S1	OEM
BSTD1040S2	OEM
BSTD1046	OEM
BSTD1046MC	OEM
BSTD1046MD	OEM
BSTD1046S1	OEM
BSTD1046S2	OEM
BSTD1053	OEM
BSTD1053M	OEM
BSTD1666M	OEM
BSTD1666P	OEM
BSTD1680M	OEM
BSTD1680P	OEM
BSTD3666M	OEM
BSTD3666N	OEM
BSTD3666P	OEM
BSTD3680M	OEM
BSTD3680N	OEM
BSTD3680P	OEM
BSTD4026M	OEM
BSTD4026MM	OEM
BSTD4033M	OEM
BSTD4033MM	OEM
BSTD4040M	OEM
BSTD4040MM	OEM
BSTD4040MN	OEM
BSTD4040N	OEM
BSTD4046M	OEM
BSTD4046MM	OEM
BSTD4046MN	OEM
BSTD4046N	OEM
BSTD4053M	OEM
BSTD4053MM	OEM
BSTD4126M	OEM
BSTD4126MM	OEM
BSTD4133M	OEM
BSTD4133MM	OEM
BSTD4140M	OEM
BSTD4140MM	OEM
BSTD4140MN	OEM
BSTD4140N	OEM
BSTD4146M	OEM
BSTD4146MM	OEM
BSTD4146MN	OEM
BSTD4146N	OEM
BSTD4153M	OEM
BSTD4153MM	OEM
BSTE0240	OEM
BSTE0260	OEM
BSTE0280	OEM
BSTE0326	OEM
BSTE0326T81	OEM
BSTE0326T82	OEM
BSTE0326T83	OEM
BSTE0326T84	OEM
BSTE0326T85	OEM
BSTE0326T86	OEM
BSTE0330T81	OEM
BSTE0330T82	OEM
BSTE0330T83	OEM
BSTE0330T84	OEM
BSTE0330T86	OEM
BSTE0333	OEM
BSTE0340	OEM
BSTE0346	OEM
BSTE0426	OEM
BSTE0426T93	OEM
BSTE0426T94	OEM
BSTE0426T95	OEM
BSTE0426T96	OEM
BSTE0430T93	OEM
BSTE0430T95	OEM
BSTE0430T96	OEM
BSTE0433	OEM
BSTE0440	OEM
BSTE0446	OEM
BSTE0540	OEM
BSTE0580	OEM
BSTE4026MN	OEM
BSTE4026N	OEM
BSTE4033MN	OEM
BSTE4033N	OEM
BSTE4040MN	OEM
BSTE4040MP	OEM
BSTE4046MN	OEM
BSTE4046MP	OEM
BSTE4046N	OEM
BSTE4046P	OEM
BSTE4053MN	OEM
BSTE4053N	OEM
BSTE4126MN	OEM
BSTE4126N	OEM
BSTE4133MN	OEM
BSTE4133N	OEM
BSTE4140MN	OEM
BSTE4140MP	OEM
BSTE4140N	OEM
BSTE4140P	OEM
BSTE4146MN	OEM
BSTE4146MP	OEM
BSTE4146N	OEM
BSTE4146P	OEM
BSTE4153MN	OEM
BSTE4153N	OEM
BSTE05100	OEM
BSTF0320	OEM
BSTF0340	OEM
BSTF0360	OEM
BSTF0420	OEM
BSTF0420S1	OEM
BSTF0426	OEM
BSTF0426S1	OEM
BSTF0433	OEM
BSTF0433S1	OEM
BSTF0440S1	OEM
BSTF0446	OEM
BSTF0446S1	OEM
BSTF0455S7	OEM
BSTF0460S1	OEM
BSTF0460S7	OEM
BSTF0466	OEM
BSTF0466S1	OEM
BSTF0466S7	OEM
BSTF0520	OEM
BSTF0540	OEM
BSTF0560	OEM
BSTF0590	OEM
BSTF05110	OEM
BSTF2520+	4031
BSTF2540	4031
BSTF2540S1	OEM
BSTF2540S7	OEM
BSTF2540S9	OEM
BSTF2540S10	OEM
BSTF2553	OEM
BSTF2553+	0679
BSTF2553S1	OEM
BSTF2553S6	OEM
BSTF2553S7	OEM
BSTF2553S9	OEM
BSTF2553S10	OEM
BSTF2560	OEM
BSTF2560+	0679
BSTF2560S1	OEM
BSTF2560S6	OEM
BSTF2560S7	OEM
BSTF2560S9	OEM
BSTF2560S10	OEM
BSTF2566	OEM
BSTF2566+	0679
BSTF2566S1	OEM
BSTF2566S6	OEM
BSTF2566S7	OEM
BSTF2566S9	OEM
BSTF2566S10	OEM
BSTF2580	OEM
BSTF2580+	0679
BSTF2580S1	OEM
BSTF2580S7	OEM
BSTF2580S9	OEM
BSTF2580S10	OEM
BSTF2590	OEM
BSTF2590S1	OEM
BSTF2590S6	OEM
BSTF2590S7	OEM
BSTF2590S9	OEM
BSTF2590S10	OEM
BSTF25100	OEM
BSTF25100S1	OEM
BSTF25100S6	OEM
BSTF25100S7	OEM
BSTF25100S9	OEM
BSTF25100S10	OEM
BSTF25110	OEM
BSTF25110S1	OEM
BSTF25110S6	OEM
BSTF25110S9	OEM
BSTF25110S10	OEM
BSTF35110	OEM
BSTG0220	OEM
BSTG0240	OEM
BSTG0260	OEM
BSTG0320	OEM
BSTG0340	OEM
BSTG0360	OEM
BSTG3540	OEM
BSTG3560	OEM
BSTG3590	OEM
BSTG35110	OEM
BSTH0420	OEM
BSTH0520	OEM
BSTH0540	OEM
BSTH0540S9	OEM
BSTH0540S10	OEM
BSTH0553	OEM
BSTH0553S9	OEM
BSTH0553S10	OEM
BSTH0560	OEM
BSTH0560S9	OEM
BSTH0560S10	OEM
BSTH0566	OEM
BSTH0566S9	OEM
BSTH0566S10	OEM
BSTH0580	OEM
BSTH0580S9	OEM
BSTH0580S10	OEM
BSTH0590	OEM
BSTH0590S9	OEM
BSTH0590S10	OEM
BSTH05100	OEM
BSTH05100S9	OEM
BSTH05100S10	OEM
BSTH05110	OEM
BSTH05110S9	OEM
BSTH05110S10	OEM
BSTH051120	OEM
BSTH3440I	OEM
BSTH3440IS9	OEM
BSTH3440K	OEM
BSTH3440KS9	OEM
BSTH3453I	OEM
BSTH3453IS9	OEM
BSTH3460K	OEM
BSTH3460KS9	OEM
BSTH3466I	OEM
BSTH3466IS9	OEM
BSTH3466K	OEM
BSTH3480I	OEM
BSTH3480IS9	OEM
BSTH3480K	OEM
BSTH3480KS9	OEM
BSTH3486IS9	OEM
BSTH3486K	OEM
BSTH3486KS9	OEM
BSTH3540	OEM
BSTH3540S9	OEM
BSTH3540S10	OEM
BSTH3553S9	OEM
BSTH3553S10	OEM
BSTH3560	OEM
BSTH3560S9	OEM
BSTH3566S9	OEM
BSTH3566S10	OEM
BSTH3580	OEM
BSTH3580S9	OEM
BSTH3580S10	OEM
BSTH3590	OEM
BSTH3590S9	OEM
BSTH3590S10	OEM
BSTH3733FS9	OEM
BSTH3733G	OEM
BSTH3733GS9	OEM
BSTH3740F	OEM
BSTH3740FS9	OEM
BSTH3740G	OEM
BSTH3740GS9	OEM
BSTH3746F	OEM
BSTH3746FS9	OEM
BSTH3746G	OEM
BSTH3746GS9	OEM
BSTH3753F	OEM
BSTH3753FS9	OEM
BSTH3753G	OEM
BSTH3753GS9	OEM
BSTH3760FS9	OEM
BSTH3760G	OEM
BSTH3766FS9	OEM
BSTH3766G	OEM
BSTH3766GS9	OEM
BSTH3773F	OEM
BSTH4440I	OEM
BSTH4440K	OEM
BSTH4440KS9	OEM
BSTH4453I	OEM
BSTH4453K	OEM
BSTH4453KS9	OEM
BSTH4460I	OEM
BSTH4460K	OEM
BSTH4460KS9	OEM
BSTH4466I	OEM
BSTH4466KS9	OEM
BSTH4480I	OEM
BSTH4486I	OEM
BSTH4486K	OEM
BSTH4540	OEM
BSTH4540S9	OEM
BSTH4540S10	OEM
BSTH4553S9	OEM
BSTH4553S10	OEM
BSTH4560S9	OEM
BSTH4580S9	OEM
BSTH4590S9	OEM
BSTH4760GS9	OEM
BSTH4766F	OEM
BSTH4766GS9	OEM
BSTH6113F	OEM
BSTH6113FS9	OEM
BSTH6113G	OEM
BSTH6113GS9	OEM
BSTH6113Y	OEM
BSTH6120F	OEM
BSTH6120FS9	OEM
BSTH6120G	OEM
BSTH6120GS9	OEM
BSTH6120Y	OEM
BSTH6126F	OEM
BSTH6126FS9	OEM
BSTH6126Y	OEM
BSTH6133F	OEM
BSTH6133FS9	OEM
BSTH6133G	OEM
BSTH6133GS9	OEM
BSTH6133Y	OEM
BSTH6140F	OEM
BSTH6140FS9	OEM
BSTH6140G	OEM
BSTH6140GS9	OEM
BSTH34666KS9	OEM
BSTH35100	OEM
BSTH35100S9	OEM
BSTH35100S10	OEM
BSTH35110	OEM
BSTH35110S9	OEM
BSTH35110S10	OEM
BSTH45100S9	OEM
BSTH45110	OEM
BSTH45110S9	OEM
BSTK4560	OEM
BSTK4560S9	OEM
BSTK4560S10	OEM
BSTK4566	OEM
BSTK4566S9	OEM
BSTK4566S10	OEM
BSTK4580	OEM
BSTK4580S9	OEM
BSTK4580S10	OEM
BSTK4590	OEM
BSTK4590S9	OEM
BSTK4590S10	OEM
BSTK45100	OEM
BSTK45100S9	OEM
BSTK45100S10	OEM
BSTK45120	OEM
BSTK45120S9	OEM
BSTK45120S10	OEM
BSTL0220	OEM
BSTL0240	OEM
BSTL0260	OEM
BSTL0420	OEM
BSTL0440	OEM
BSTL0455	OEM
BSTL0460	OEM
BSTL0520	OEM
BSTL0540	OEM
BSTL0560	OEM
BSTL0590	OEM
BSTL05110	OEM
BSTL06140	OEM
BSTL06160	OEM
BSTL06180	OEM
BSTL2420	OEM
BSTL2440	OEM
BSTL2455	OEM
BSTL2460	OEM
BSTL2520	OEM
BSTL2540	OEM
BSTL2560	OEM
BSTL2580	OEM
BSTL2590	OEM
BSTL3440	OEM
BSTL3440IS9	OEM
BSTL3440K	OEM
BSTL3440KS9	OEM
BSTL3446I	OEM
BSTL3446IS9	OEM
BSTL3446K	OEM
BSTL3446KS9	OEM
BSTL3453I	OEM
BSTL3453IS9	OEM
BSTL3453K	OEM
BSTL3453KS9	OEM
BSTL3460	OEM
BSTL3460I	OEM
BSTL3460IS9	OEM
BSTL3460K	OEM
BSTL3460KS9	OEM
BSTL3466I	OEM
BSTL3466IS9	OEM
BSTL3466K	OEM
BSTL3466KS9	OEM
BSTL3473IS9	OEM
BSTL3473K	OEM
BSTL3473KS9	OEM
BSTL3480	OEM
BSTL3480I	OEM
BSTL3480IS9	OEM
BSTL3480K	OEM
BSTL3480KS9	OEM
BSTL3486I	OEM
BSTL3486IS9	OEM
BSTL3486K	OEM
BSTL3486KS9	OEM
BSTL3520	OEM
BSTL3540	OEM
BSTL3540S9	OEM
BSTL3540S10	OEM
BSTL3553	OEM
BSTL3553S9	OEM
BSTL3560	OEM
BSTL3560S10	OEM
BSTL3566	OEM
BSTL3566S9	OEM
BSTL3566S10	OEM
BSTL3580	OEM
BSTL3580S10	OEM
BSTL3590	OEM
BSTL3590S10	OEM
BSTL3733G	OEM
BSTL3733GS9	OEM
BSTL3740G	OEM
BSTL3740GS9	OEM
BSTL3746F	OEM
BSTL3746G	OEM
BSTL3746GS9	OEM
BSTL3753F	OEM
BSTL3753G	OEM
BSTL3753GS9	OEM
BSTL3760F	OEM
BSTL3760G	OEM
BSTL3760GS9	OEM
BSTL3766F	OEM
BSTL3766G	OEM
BSTL3766GS9	OEM
BSTL4006	OEM
BSTL4020	OEM
BSTL4033	OEM
BSTL4440	OEM
BSTL4440I	OEM
BSTL4440K	OEM
BSTL4440KS9	OEM
BSTL4446I	OEM
BSTL4446K	OEM
BSTL4446KS9	OEM
BSTL4453I	OEM
BSTL4453IS9	OEM
BSTL4453K	OEM
BSTL4453KS9	OEM
BSTL4460	OEM
BSTL4460K	OEM
BSTL4460KS9	OEM
BSTL4460S9	OEM
BSTL4466I	OEM
BSTL4466K	OEM
BSTL4473I	OEM
BSTL4473K	OEM
BSTL4480	OEM
BSTL4480I	OEM
BSTL4480K	OEM
BSTL4486	OEM
BSTL4486I	OEM
BSTL4486K	OEM
BSTL4540	OEM
BSTL4540S9	OEM
BSTL4540S10	OEM
BSTL4553	OEM
BSTL4553S9	OEM
BSTL4553S10	OEM
BSTL4560	OEM
BSTL4560S9	OEM
BSTL4560S10	OEM
BSTL4566	OEM
BSTL4566S9	OEM
BSTL4566S10	OEM
BSTL4580	OEM
BSTL4580S9	OEM
BSTL4580S10	OEM
BSTL4590	OEM
BSTL4590S10	OEM
BSTL4733F	OEM
BSTL4733GS9	OEM
BSTL4740F	OEM
BSTL4740GS9	OEM
BSTL4746F	OEM
BSTL4746GS9	OEM
BSTL4753F	OEM
BSTL4753GS9	OEM
BSTL4760F	OEM
BSTL4760GS9	OEM
BSTL4766F	OEM
BSTL4766GS9	OEM
BSTL6113	OEM
BSTL6113F	OEM
BSTL6113G	OEM
BSTL6113GS9	OEM
BSTL6120	OEM
BSTL6120F	OEM
BSTL6120G	OEM
BSTL6120GS9	OEM
BSTL6126	OEM
BSTL6126F	OEM
BSTL6126G	OEM
BSTL6126GS9	OEM
BSTL6133	OEM
BSTL6133F	OEM
BSTL6133G	OEM
BSTL6133GS9	OEM
BSTL6140F	OEM
BSTL6140G	OEM
BSTL6140GS9	OEM
BSTL25110	OEM
BSTL35100	OEM
BSTL35100S9	OEM
BSTL35100S10	OEM
BSTL35110	OEM
BSTL35110S9	OEM
BSTL35110S10	OEM
BSTL45100	OEM
BSTL45100S9	OEM
BSTL45100S10	OEM
BSTL45110	OEM
BSTL45110S9	OEM
BSTL45110S10	OEM
BSTL45120	OEM
BSTL45120S9	OEM
BSTL45120S10	OEM
BSTM4560	OEM
BSTM4560S9	OEM
BSTM4566S9	OEM
BSTM4580	OEM
BSTM4580S9	OEM
BSTM4590	OEM
BSTM4590S9	OEM
BSTM45100	OEM
BSTM45100S9	OEM
BSTM45110	OEM
BSTM45110S9	OEM
BSTM45120S9	OEM
BSTN0220	OEM
BSTN0240	OEM
BSTN0260	OEM
BSTN0320	OEM
BSTN0340	OEM
BSTN0360	OEM
BSTN0520	OEM
BSTN0540	OEM
BSTN0560	OEM
BSTN0590	OEM
BSTN05100	OEM
BSTN05110	OEM
BSTN4R80K	OEM
BSTN1520	OEM
BSTN1540	OEM
BSTN1560	OEM
BSTN1580	OEM
BSTN1590	OEM
BSTN2320	OEM
BSTN2340	OEM
BSTN2360	OEM
BSTN2520	OEM
BSTN2540	OEM
BSTN2560	OEM
BSTN2590	OEM
BSTN3440I	OEM
BSTN3440K	OEM
BSTN3446I	OEM
BSTN3446K	OEM
BSTN3453I	OEM
BSTN3453K	OEM
BSTN3460I	OEM
BSTN3460K	OEM
BSTN3466+	0738
BSTN3466I	OEM
BSTN3466K	OEM
BSTN3473I	OEM
BSTN3473IS9	OEM
BSTN3480I	OEM
BSTN3480IS9	OEM
BSTN3480K	OEM
BSTN3486I	OEM
BSTN3486IS9	OEM
BSTN3486K	OEM
BSTN3520	OEM
BSTN3540	OEM
BSTN3540S9	OEM
BSTN3540S10	OEM
BSTN3553S10	OEM
BSTN3560	OEM
BSTN3560+	0738
BSTN3560S9	OEM
BSTN3560S10	OEM
BSTN3566S9	OEM
BSTN3566S10	OEM
BSTN3580	OEM
BSTN3580+	0738
BSTN3580S9	OEM
BSTN3580S10	OEM
BSTN3590	OEM
BSTN3590S9	OEM
BSTN3590S10	OEM
BSTN3733G	OEM
BSTN3733GS9	OEM
BSTN3740G	OEM
BSTN3740GS9	OEM
BSTN3746G	OEM
BSTN3746GS9	OEM
BSTN3753F	OEM
BSTN3753G	OEM
BSTN3753GS9	OEM
BSTN3760G	OEM
BSTN3760GS9	OEM
BSTN3766F	OEM
BSTN3766G	OEM
BSTN3766GS9	OEM
BSTN4006	OEM
BSTN4020	OEM
BSTN4033	OEM
BSTN4440K	OEM
BSTN4460K	OEM
BSTN4473IS9	OEM
BSTN4480IS9	OEM
BSTN4486IS9	OEM
BSTN4486K	OEM
BSTN4520	OEM
BSTN4540	OEM
BSTN4540S9	OEM
BSTN4540S10	OEM
BSTN4553	OEM
BSTN4553S9	OEM
BSTN4553S10	OEM
BSTN4560	OEM
BSTN4560S9	OEM
BSTN4560S10	OEM
BSTN4566	OEM
BSTN4566S9	OEM
BSTN4566S10	OEM
BSTN4580	OEM
BSTN4580S9	OEM
BSTN4590	OEM
BSTN4590S9	OEM
BSTN4733F	OEM
BSTN4733G	OEM
BSTN4740F	OEM
BSTN4740G	OEM
BSTN4746F	OEM
BSTN4746G	OEM
BSTN4753	OEM
BSTN4753F	OEM
BSTN4753G	OEM
BSTN4760F	OEM
BSTN4766	OEM
BSTN4766F	OEM
BSTN4766GS9	OEM
BSTN5240S10	OEM
BSTN5253S10	OEM
BSTN5260S9	OEM
BSTN5260S10	OEM
BSTN5266S9	OEM
BSTN5266S10	OEM
BSTN5280S9	OEM
BSTN5280S10	OEM
BSTN5290S9	OEM
BSTN5290S10	OEM
BSTN5540	OEM
BSTN5540S9	OEM
BSTN5540S10	OEM
BSTN5553	OEM
BSTN5553S9	OEM
BSTN5553S10	OEM
BSTN5560S9	OEM
BSTN5560S10	OEM
BSTN5566S9	OEM
BSTN5566S10	OEM
BSTN5580S9	OEM
BSTN5580S10	OEM
BSTN5590S9	OEM
BSTN5590S10	OEM
BSTN6113	OEM
BSTN6113F	OEM
BSTN6113FS9	OEM
BSTN6113G	OEM
BSTN6113Y	OEM
BSTN6120	OEM
BSTN6120F	OEM
BSTN6120FS9	OEM
BSTN6120G	OEM
BSTN6120Y	OEM
BSTN6126	OEM
BSTN6126F	OEM
BSTN6126G	OEM
BSTN6126Y	OEM
BSTN6133	OEM
BSTN6133F	OEM
BSTN6133G	OEM
BSTN6133Y	OEM
BSTN6140F	OEM
BSTN6140G	OEM
BSTN6440IS9	OEM
BSTN6446IS9	OEM
BSTN6453IS9	OEM
BSTN6460IS9	OEM
BSTN6466IS9	OEM
BSTN6473IS9	OEM
BSTN6473K	OEM
BSTN6480IS9	OEM
BSTN6480K	OEM
BSTN6486IS9	OEM
BSTN6486K	OEM
BSTN15110	OEM
BSTN25100	OEM
BSTN25110	OEM
BSTN35100S9	OEM
BSTN35100S10	OEM
BSTN35110	OEM
BSTN35120	OEM
BSTN35120S9	OEM
BSTN35120S10	OEM
BSTN45100	OEM
BSTN45100S9	OEM
BSTN45110	OEM
BSTN45110S9	OEM
BSTN45120	OEM
BSTN45120S9	OEM
BSTN52100S9	OEM
BSTN52100S10	OEM
BSTN52110	OEM
BSTN52110S9	OEM
BSTN52110S10	OEM
BSTN52120	OEM
BSTN52120S9	OEM
BSTN52120S10	OEM
BSTP0220	OEM
BSTP0240	OEM
BSTP0260	OEM
BSTP0280	OEM
BSTP0320	OEM
BSTP0340	OEM
BSTP0360	OEM
BSTP0520	OEM
BSTP0540	OEM
BSTP0560	OEM
BSTP0590	OEM
BSTP05100	OEM
BSTP05110	OEM
BSTP1260	OEM
BSTP1520	OEM
BSTP1560S9	OEM
BSTP1560S10	OEM
BSTP1560T	OEM
BSTP1560TS9	OEM
BSTP1560TS10	OEM
BSTP1566	OEM
BSTP1566S9	OEM
BSTP1566S10	OEM
BSTP1566T	OEM
BSTP1566TS9	OEM
BSTP3520	OEM
BSTP4520	OEM
BSV01	OEM
BSV02	OEM
BSV11	OEM
BSV15	0126

DEVICE TYPE	REPL CODE
BSV15-6	0126
BSV15-10	0126
BSV16	0126
BSV16-6	0126
BSV16-10	0126
BSV17	0126
BSV17-6	0126
BSV17-10	0126
BSV20	OEM
BSV20A	OEM
BSV21	0037
BSV22	OEM
BSV23	OEM
BSV24	OEM
BSV25	OEM
BSV26	OEM
BSV27	OEM
BSV28	OEM
BSV29	OEM
BSV33	OEM
BSV34	OEM
BSV34A	OEM
BSV35	0144
BSV35A	0016
BSV36	0475
BSV38	OEM
BSV38P	OEM
BSV39	OEM
BSV39P	OEM
BSV40	0016
BSV41	0016
BSV42	0037
BSV43A	0037
BSV43B	0791
BSV44A	0037
BSV44B	0791
BSV45A	0037
BSV45B	0791
BSV46	0037
BSV47A	0037
BSV47B	0791
BSV48A	0037
BSV48B	0791
BSV49A	0037
BSV49B	0791
BSV50E	OEM
BSV50F	OEM
BSV50G	OEM
BSV51	0086
BSV52	0144
BSV52R	0144
BSV53	0016
BSV53P	0127
BSV54	0016
BSV54P	0127
BSV55	0016
BSV55A	0037
BSV55AP	0037
BSV55P	0037
BSV56A	OEM
BSV56B	OEM
BSV56C	OEM
BSV57A	OEM
BSV57C	OEM
BSV58A	OEM
BSV58B	OEM
BSV59	0016
BSV60	5533
BSV61	OEM
BSV62	OEM
BSV63	OEM
BSV65FA	0079
BSV65FB	0079
BSV68	0342
BSV69	0086
BSV71	OEM
BSV74	OEM
BSV75	OEM
BSV76	OEM
BSV77	0007
BSV78	1155
BSV79	1155
BSV80	OEM
BSV81	OEM
BSV82	0126
BSV83	0126
BSV84	0086
BSV85	0018
BSV86	0079
BSV87	0079
BSV88	0016
BSV89	0016
BSV90	0016
BSV91	0016
BSV92	0224
BSV95	0418
BSV96	0037
BSV97	0037
BSV98	0037
BSV99	OEM
BSW10	0086
BSW11	0016
BSW12	0016
BSW13	0007
BSW17	OEM
BSW18	OEM
BSW19	0016
BSW19A	0342
BSW19VI	0342
BSW20	0037
BSW20A	0342
BSW20VI	0342
BSW21	0037
BSW21A	0037
BSW22	0037
BSW22A	0037
BSW23	0126
BSW24	0037
BSW25	3607
BSW26	0086
BSW27	0086
BSW28	0086
BSW29	0086
BSW30	OEM
BSW31	OEM
BSW32	0155
BSW33	0016
BSW34	0016
BSW35	0086
BSW36	OEM
BSW37	OEM
BSW38	OEM
BSW39	0016
BSW39-6	0086
BSW39-10	0086
BSW39-16	0086
BSW40	0126
BSW40-6	0126
BSW40-10	0126
BSW40-16	0126
BSW40-25	0126
BSW41	0016
BSW42	0016
BSW42A	0016
BSW42B	OEM
BSW43	0016
BSW43A	0016
BSW44	0037
BSW44A	0037
BSW45	0037
BSW45A	0037
BSW49	0086
BSW51	0016
BSW52	0016
BSW53	0016
BSW54	OEM
BSW58	0079
BSW59	0079
BSW61	OEM
BSW62	OEM
BSW63	OEM
BSW64	OEM
BSW65	0086
BSW66	0086
BSW67	6957
BSW-68	0187
BSW68	6957
BSW68A	OEM
BSW69	0187
BSW70	0233
BSW72	0037
BSW73	0037
BSW74	0037
BSW75	0037
BSW78	OEM
BSW79	OEM
BSW80	2525
BSW81	0037
BSW82	0016
BSW83	0016
BSW84	0016
BSW85	0016
BSW88	0016
BSW88A	0079
BSW88B	0079
BSW89	0016
BSW89A	0079
BSW89B	0079
BSW92	0016
BSW95	OEM
BSW95A	OEM
BSX12	0144
BSX12A	0086
BSX19	0016
BSX20	0493
BSX21	0861
BSX22	1400
BSX23	0086
BSX24	0016
BSX25	0016
BSX26	0144
BSX27	0144
BSX28	0493
BSX29	0150
BSX30	0016
BSX31	OEM
BSX32	0493
BSX33	0086
BSX34	OEM
BSX35	0144
BSX36	0037
BSX38	0016
BSX38A	0016
BSX38B	0016
BSX39	0493
BSX40	0126
BSX41	0126
BSX44	0016
BSX45	0086
BSX45-6	0086
BSX45-10	0086
BSX45-16	0086
BSX46	0086
BSX46-6	0086
BSX46-10	0086
BSX46-16	0086
BSX47	0086
BSX47-6	0086
BSX47-10	0086
BSX48	0016
BSX49	0016
BSX51	0016
BSX51A	0016
BSX51B	OEM
BSX52	0016
BSX52A	0016
BSX52B	OEM
BSX53	0016
BSX53A	OEM
BSX53B	OEM
BSX54	0016
BSX54A	OEM
BSX54B	OEM
BSX58	OEM
BSX59	0086
BSX60	0086
BSX61	0086
BSX62	0086
BSX62-10	0622
BSX62-16	0622
BSX62B	0086
BSX62C	0086
BSX62D	0086
BSX63	0644
BSX63-10	0622
BSX63B	0086
BSX63C	0086
BSX66	0016
BSX67	0016
BSX68	0016
BSX69	0016
BSX70	0086
BSX71	0086
BSX72	0086
BSX75	0016
BSX76	0016
BSX77	0016
BSX78	0016
BSX79	0016
BSX79A	0079
BSX79B	0079
BSX80	0016
BSX81	0016
BSX81A	0079
BSX81B	0079
BSX82	OEM
BSX83	OEM
BSX84	OEM
BSX85	OEM
BSX86	OEM
BSX87	0016
BSX87A	0493
BSX88	0016
BSX88A	0493
BSX89	0016
BSX90	0016
BSX91	0016
BSX92	0144
BSX93	0144
BSX94	OEM
BSX94A	0016
BSX95	0086
BSX96	0086
BSX97	0016
BSY10	0016
BSY11	0016
BSY17	0016
BSY18	0016
BSY19	0079
BSY20	0016
BSY21	0016
BSY22	0144
BSY23	0144
BSY24	0016
BSY25	0016
BSY26	0087
BSY27	0016
BSY28	0016
BSY29	0016
BSY32	OEM
BSY33	OEM
BSY34	0016
BSY36	OEM
BSY38	0016
BSY38A	OEM
BSY39	0016
BSY39A	OEM
BSY40	0037
BSY41	0037
BSY42	OEM
BSY43	OEM
BSY44	0086
BSY45	0086
BSY46	0086
BSY47	OEM
BSY48	0016
BSY49	0016
BSY50	OEM
BSY51	0086
BSY52	0086
BSY53	0086
BSY54	0086
BSY55	0086
BSY56	0086
BSY58	0016
BSY59	0016
BSY61	0016
BSY62	0016
BSY62A	0016
BSY62B	0595
BSY63	0016
BSY68	0861
BSY70	0016
BSY71	0086
BSY72	0016
BSY73	0016
BSY74	0016
BSY75	0016
BSY76	0016
BSY77	0086
BSY78	0086
BSY79	0861
BSY80	0016
BSY81	0086
BSY82	0086
BSY83	0086
BSY84	0086
BSY85	0086
BSY86	0086
BSY87	0086
BSY88	0086
BSY89	0016
BSY90	0086
BSY91	0016
BSY92	0086
BSY93	0016
BSY95	0016
BSY95A	0016
BSY165	0016
BSY168	0016
BT67	0016
BT71	0016
BT82	0037
BT-94	0016
BT100A-300R	1222
BT100A-500R	OEM
BT101-300R	0676
BT101-500R	2809
BT102-300R	2659
BT102-500R	2809
BT106	OEM
BT106A	0447
BT106B	0940
BT106C	0095
BT106D	0095
BT107	2809
BT108	2809
BT109	OEM
BT109-500R	OEM
BT110	OEM
BT112-750R	OEM
BT113-700R	OEM
BT114	OEM
BT115	OEM
BT116	OEM
BT117	OEM
BT118	OEM
BT119	6434
BT120	0061
BT121	0239
BT123	OEM
BT124	OEM
BT126-750R	OEM
BT127-350R	OEM
BT127-750R	OEM
BT138	OEM
BT138-500	0617
BT138-500G	0617
BT138-600G	0617
BT138-800G	0875
BT138F-500	2202
BT138F-600	2202
BT138F-800	2202
BT139-500	1659
BT139-500E	OEM
BT139-500F	6639
BT139-500G	1659
BT139-600	1659
BT139-600E	OEM
BT139-600F	6639
BT139-600G	1659
BT139-800	1659
BT139-800E	OEM
BT139-800F	OEM
BT139-800G	1967
BT139F-500	2202
BT139F-600	2202
BT139F-800	2202
BT143-400R	OEM
BT146	OEM
BT148	OEM
BT149D	OEM
BT151	0863
BT151-500R	0863
BT929	0007
BT930	0259
B-T-1000-139	0127
BT8667	0688
BT151500R	1222
BTA08-200S	5973
BTA08-400S	5973
BTA08-600S	OEM
BTA08-700S	OEM
BTA08-800S	OEM
BTA41-200A	5610
BTA41-400A	5878
BTA41-600A	3708
BTA41-700A	OEM
BTA41-800A	OEM
BTA140-500	1659
BTA140-600	2202
BTA140-800	1967
BTB08-200S	5534
BTB08-400S	5534
BTB08-600S	OEM
BTB08-700S	0059
BTB08-800S	0059
BTB41-200A	3755
BTB41-400A	6151
BTB41-600A	3755
BTB41-700A	OEM
BTB41-800A	OEM
BTB50	OEM
BTB100	OEM
BTB200	OEM
BTB400	OEM
BTB600	OEM
BTB800	OEM
BTB1000	0087
BTB1200	OEM
BTD0105	2383
BTD0110	2383
BTD0120	0481
BTD0140	0485
BTD0305	2383
BTD0310	2383
BTD0320	0481
BTD0340	0485
BTD0360-1	OEM
BTD0360-2	OEM
BTD0360-3	OEM
BTD4	OEM
BTD32B	OEM
BTL0405	OEM
BTL0410	OEM
BTL0420	OEM
BTL0440	OEM
BTL0460	OEM
BTL0605	OEM
BTL0610	OEM
BTL0620	OEM
BTL0640	OEM
BTL0660	OEM
BTL0805	OEM
BTL0810	OEM
BTL0820	OEM
BTL0840	OEM
BTL0860	OEM
BTL1005	OEM
BTL1010	OEM
BTL1020	OEM
BTL1040	OEM
BTL1060	OEM
BTL1605	OEM
BTL1610	OEM
BTL1620	OEM
BTL1640	OEM
BTL1660	OEM
BTM0405	OEM
BTM0410	OEM
BTM0420	OEM
BTM0440	OEM
BTM0460	OEM
BTM0605	OEM
BTM0610	OEM
BTM0620	OEM
BTM0640	OEM
BTM0660	OEM
BTM-50	0110
BTM50	0015
BTM100	OEM
BTM200	OEM
BTM400	OEM
BTM600	OEM
BTM800	OEM
BTM1000	0087
BTM1005	OEM
BTM1010	OEM
BTM1020	OEM
BTM1040	OEM
BTM1060	OEM
BTM1200	OEM
BTM1605	OEM
BTM1610	OEM
BTM1620	OEM
BTM1640	OEM
BTM1660	OEM
BTR0605	0772
BTR0610	0772
BTR0620	0772
BTR0640	0772
BTR0660	OEM
BTR170S15	OEM
BTR171	OEM
BTR174	OEM
BTR198A	OEM
BTR202	OEM
BTR209	OEM
BTR537	OEM
BTR540	OEM
BTR541	OEM
BTR545A	OEM
BTR547	OEM
BTR554	OEM
BTR570A	OEM
BTR574	OEM
BTR576	OEM
BTR583	OEM
BTR608	OEM
BTR614	OEM
BTR629A	OEM
BTR630	OEM
BTR632A	OEM
BTR633	OEM
BTR639	OEM
BTR640	OEM
BTR641	OEM
BTR647	OEM
BTR666	OEM
BTR1005	0593
BTR1010	0593
BTR1020	0593
BTR1040	OEM
BTR1060	OEM
BTS0605	1064
BTS0610	1064
BTS0620	1064
BTS0640	1317
BTS0660	1891
BTS95-500R	OEM
BTS1005	1064
BTS1010	1064
BTS1020	1064
BTS1040	1317
BTS1060	1891
BTS1605	1064
BTS1610	1064
BTS1620	1064
BTS1640	1317
BTS1660	OEM
BTS2505	1064
BTS2510	1064
BTS2520	1064
BTS2540	1317
BTS2560	1891
BTS3500	0133
BTS3600	OEM
BTT24-1200R	OEM
BTT170S01	OEM
BTT170S02	OEM
BTT170S03	OEM
BTT170S04	OEM
BTT170S05	OEM
BTT170S06	OEM
BTT170S07	OEM
BTT170S08	OEM
BTT170S09	OEM
BTT170S10	OEM
BTT170S11	OEM
BTT170S12	OEM
BTT170S13	OEM
BTT170S14	OEM
BTT170S15	OEM
BTU0340	0279
BTU0405	0399
BTU0410	0408
BTU0420	0422
BTU0430	0422
BTU0440	0422
BTU0450	0422
BTU0460	0422
BTU0505	0154
BTU0510	0154
BTU0520	0154
BTU0530	0147
BTU0535	0279
BTU0540	0147
BTU0550	0279
BTU0560	0279
BTU0605	3147
BTU0630	3151
BTU0660	0279
BTU650	3153
BTU1005	0154
BTU1010	0154
BTU1020	0154
BTU1040	0908
BTU1060	0279
BTU1605	0154
BTU1610	0154
BTU1620	0154
BTU1640	0147
BTU1660	0279
BTU2505	2015
BTU2510	2015
BTU2520	2015
BTU2540	2017
BTU2560	2018
BTU3605	2018
BTV0405	3199
BTV0410	3199
BTV0420	3199
BTV0430	3207
BTV0440	3207
BTV0460	3222
BTV24-600R	4031
BTV24-800R	0687
BTV24-1400R	OEM
BTV34-600G	OEM
BTV34-600H	OEM
BTV34-800G	OEM
BTV34-800H	OEM
BTV34-1200G	OEM
BTV34-1200H	OEM
BTV34-1400G	OEM
BTV34-1400H	OEM
BTV58-600R	OEM
BTV58-850R	OEM
BTV58-1000R	OEM
BTW10-50	OEM
BTW10-100	OEM
BTW10-200	OEM
BTW10-300	OEM
BTW10-400	OEM
BTW10-500	OEM
BTW10-600	OEM
BTW11-50	OEM
BTW11-100	OEM
BTW11-200	OEM
BTW11-300	OEM
BTW11-400	OEM
BTW11-500	OEM
BTW11-600	OEM
BTW12-50	OEM
BTW12-100	OEM
BTW12-200	OEM
BTW12-300	OEM
BTW12-400	OEM
BTW12-500	OEM
BTW12-600	OEM
BTW13-50	0154
BTW13-100	0154
BTW13-200	0154
BTW13-300	0147
BTW13-400	0147
BTW13-500	0279
BTW13-600	0279
BTW14-50	0154
BTW14-100	0154
BTW14-200	0154
BTW14-300	0147
BTW14-400	0147
BTW14-500	0279
BTW14-600	0279
BTW15-50	OEM
BTW15-100	OEM
BTW15-200	OEM
BTW15-300	OEM
BTW15-400	OEM
BTW15-500	OEM
BTW15-600	OEM
BTW16-50	0154
BTW16-100	0154
BTW16-200	0154
BTW16-300	0147
BTW16-400	0147
BTW16-500	0279
BTW16-600	0279
BTW17-100	OEM
BTW17-200	OEM
BTW17-300	OEM
BTW17-400	OEM
BTW17-500	OEM
BTW17-600	OEM
BTW18-50	OEM
BTW18-100	OEM
BTW18-200	OEM
BTW18-300	OEM
BTW18-400	OEM
BTW18-500	OEM
BTW18-600	OEM
BTW19-50	0154
BTW19-100	0154
BTW19-200	0154
BTW19-300	0147
BTW19-400	0147
BTW19-500	0279
BTW19-600	0279
BTW20-50	3147
BTW20-100	3148
BTW20-200	3150
BTW20-300	3152
BTW20-400	3152
BTW20-500	4100
BTW20-600	2018
BTW21-50	OEM
BTW21-100	OEM
BTW21-200	OEM
BTW21-300	OEM
BTW21-400	OEM
BTW21-500	OEM
BTW21-600	OEM
BTW23-600	OEM
BTW23-600R	OEM
BTW23-600RM	OEM
BTW23-700R	OEM
BTW23-800	OEM
BTW23-800R	OEM
BTW23-800RM	OEM
BTW23-1000	OEM
BTW23-1000R	OEM
BTW23-1000RM	OEM
BTW23-1200	OEM

If replacement code is OEM, contact original manufacturer for replacement.

DEVICE TYPE	REPL CODE
BTW23-1200R	OEM
BTW23-1200RM	OEM
BTW23-1400	OEM
BTW23-1400R	OEM
BTW23-1400RM	OEM
BTW23-1600	OEM
BTW23-1600R	OEM
BTW23-1600RM	OEM
BTW24-600	OEM
BTW24-600R	OEM
BTW24-600RM	OEM
BTW24-700R	OEM
BTW24-800	OEM
BTW24-800R	OEM
BTW24-800RM	OEM
BTW24-1000	OEM
BTW24-1000R	OEM
BTW24-1000RM	OEM
BTW24-1200	OEM
BTW24-1200R	OEM
BTW24-1200RM	OEM
BTW24-1400	OEM
BTW24-1400R	OEM
BTW24-1400RM	OEM
BTW24-1600	OEM
BTW24-1600R	OEM
BTW24-1600RM	OEM
BTW26-150	OEM
BTW26-300	OEM
BTW26-400	OEM
BTW26-500	OEM
BTW26-600	OEM
BTW27-100R	OEM
BTW27-200R	OEM
BTW27-300R	OEM
BTW27-400R	OEM
BTW27-500R	OEM
BTW27-600R	OEM
BTW27S-200R	OEM
BTW27S-300R	OEM
BTW27S-400R	OEM
BTW27S-500R	OEM
BTW27SA-200R	OEM
BTW27SA-300R	OEM
BTW27SA-400R	OEM
BTW28-500R	OEM
BTW28-600R	0752
BTW28-700R	OEM
BTW28-800R	OEM
BTW28A-500R	OEM
BTW28A500R	OEM
BTW28A-600R	0752
BTW28A600R	0752
BTW28A-700R	OEM
BTW28A700R	OEM
BTW28A-800R	OEM
BTW28A800R	OEM
BTW30-300	OEM
BTW30-300R	OEM
BTW30-300RM	0804
BTW30-400	OEM
BTW30-400R	OEM
BTW30-400RM	0804
BTW30-500	OEM
BTW30-500R	OEM
BTW30-500RM	0804
BTW30-600	0804
BTW30-600R	OEM
BTW30-600RM	0804
BTW30-800	OEM
BTW30-800R	OEM
BTW30-800RM	OEM
BTW30-800RS	OEM
BTW30-1000	OEM
BTW30-1000R	OEM
BTW30-1000RM	OEM
BTW30-1000RS	OEM
BTW30-1200	OEM
BTW30-1200R	OEM
BTW30-1200RM	OEM
BTW30-1200RS	OEM
BTW30-RM	OEM
BTW31-300	OEM
BTW31-300R	OEM
BTW31-300RM	0804
BTW31-400	OEM
BTW31-400R	OEM
BTW31-400RM	0804
BTW31-500	OEM
BTW31-500R	OEM
BTW31-500RM	OEM
BTW31-600	OEM
BTW31-600R	OEM
BTW31-600RM	OEM
BTW31-700R	OEM
BTW31-800	OEM
BTW31-800R	OEM
BTW31-800RM	OEM
BTW31-800RW	OEM
BTW31-1000	OEM
BTW31-1000R	OEM
BTW31-1000RM	OEM
BTW31-1000RW	OEM
BTW31-1200	OEM
BTW31-1200R	OEM
BTW31-1200RM	OEM
BTW31-1200RW	OEM
BTW31-RM	OEM
BTW32-300R	OEM

DEVICE TYPE	REPL CODE
BTW32-400R	OEM
BTW32-500R	OEM
BTW32-600R	OEM
BTW32-600RU	OEM
BTW32-800R	OEM
BTW32-800RU	OEM
BTW32-1000R	OEM
BTW32-1000RU	OEM
BTW32-1200R	OEM
BTW32-1200RU	OEM
BTW32-6004	OEM
BTW32-RM	OEM
BTW33-600R	OEM
BTW33-600RM	OEM
BTW33-600RM10	OEM
BTW33-800R	OEM
BTW33-800RM	OEM
BTW33-800RM10	OEM
BTW33-1000R	OEM
BTW33-1000RM	OEM
BTW33-1200R	OEM
BTW33-1200RM	OEM
BTW33-RU	OEM
BTW34-600	OEM
BTW34-600G	OEM
BTW34-600H	OEM
BTW34-600M	OEM
BTW34-800	OEM
BTW34-800G	OEM
BTW34-800H	OEM
BTW34-1000	OEM
BTW34-1000G	OEM
BTW34-1000H	OEM
BTW34-1200	OEM
BTW34-1200G	OEM
BTW34-1200H	OEM
BTW34-1400	OEM
BTW34-1400G	OEM
BTW34-1400H	OEM
BTW34-1600	OEM
BTW34-1600G	OEM
BTW34-1600H	OEM
BTW35	OEM
BTW36-200RM	OEM
BTW36-400RM	OEM
BTW36-500RM	OEM
BTW36-600RM	OEM
BTW37V600	OEM
BTW37V800	OEM
BTW37V1000	OEM
BTW37V1200	OEM
BTW38-600R	OEM
BTW38-800R	5081
BTW38-1000R	5996
BTW38-1200R	OEM
BTW38V600R	OEM
BTW38V800R	OEM
BTW38V1000R	OEM
BTW38V1200R	OEM
BTW39-50	0740
BTW39-50+	0469
BTW39-100	0567
BTW39-100+	0469
BTW39-200	0762
BTW39-200+	0469
BTW39-300	0740
BTW39-300+	0722
BTW39-400	0740
BTW39-400+	0722
BTW39-500	0764
BTW39-500+	0725
BTW39-600	0764
BTW39-600+	0725
BTW39-700	0766
BTW39-700+	2036
BTW39-800	0766
BTW39-800+	2036
BTW39-900	4535
BTW39-900+	2036
BTW39-1000	4535
BTW39-1000+	2036
BTW39-1100	OEM
BTW39-1100+	2036
BTW39-1200	OEM
BTW39-1200+	2036
BTW40-400R	0722
BTW40-400RM	OEM
BTW40-400RMT	OEM
BTW40-600R	0725
BTW40-600RM	OEM
BTW40-800R	0750
BTW41-500G	OEM
BTW41-500H	OEM
BTW41-600G	OEM
BTW41-600H	OEM
BTW41-800G	OEM
BTW41-800H	OEM
BTW42-600R	OEM
BTW42-800R	OEM
BTW42-1000R	OEM
BTW42-1200R	OEM
BTW43-600G	0422
BTW43-600H	OEM
BTW43-800G	6157
BTW43-800H	OEM
BTW43-1000G	3461
BTW43-1000H	OEM
BTW43-1200G	OEM
BTW43-1200H	OEM

DEVICE TYPE	REPL CODE
BTW44-100	OEM
BTW44-200	OEM
BTW44-300	OEM
BTW44-400	OEM
BTW44-500	OEM
BTW44-600	OEM
BTW44-600R	OEM
BTW44-800R	OEM
BTW44-1000R	OEM
BTW44-1200R	OEM
BTW44-1400R	OEM
BTW44-1600R	OEM
BTW45-400R	0722
BTW45-400RM	OEM
BTW45-600R	0725
BTW45-600RM	OEM
BTW45-800R	0750
BTW45-1000R	OEM
BTW45-1200R	OEM
BTW46-200R	OEM
BTW46-400R	OEM
BTW46-500R	OEM
BTW46-600R	OEM
BTW47-100RM	OEM
BTW47-200RM	OEM
BTW47-200RU	OEM
BTW47-400RM	OEM
BTW47-400RU	OEM
BTW47-600	OEM
BTW47-600RM	OEM
BTW47-600RU	OEM
BTW47-600RU+	2036
BTW47-800	OEM
BTW47-800R	OEM
BTW47-800RM	OEM
BTW47-800RU	6622
BTW47-800RU+	2036
BTW47-1000	OEM
BTW47-1000R	OEM
BTW47-1000RM	OEM
BTW47-1000RU	4535
BTW47-1000RU+	2036
BTW47-1200	OEM
BTW47-1200R	OEM
BTW47-1200RM	OEM
BTW47-1200RU	OEM
BTW47-1200RU+	2036
BTW47-1400	OEM
BTW47-1400R	OEM
BTW47-1400RM	OEM
BTW47-1400RU	OEM
BTW47-1600	OEM
BTW47-1600R	OEM
BTW47-1600RM	OEM
BTW47-1600RU	OEM
BTW47-RU	OEM
BTW48-200	1704
BTW48-200+	4031
BTW48-400	4031
BTW48-400+	4031
BTW48-600	4031
BTW48-600+	4031
BTW48-800	0679
BTW48-800+	0679
BTW48-1200	0679
BTW48-1200+	0679
BTW49-50	0804
BTW49-100	3408
BTW49-200	3408
BTW49-400	3408
BTW49-600	0804
BTW49-800	OEM
BTW50-100	1704
BTW50-200	1704
BTW50-400	4031
BTW50-600	4031
BTW50-800	0679
BTW50-1000	0679
BTW50-1200	0679
BTW58-1000R	OEM
BTW58-1300R	OEM
BTW58-1500R	OEM
BTW63-600RK	OEM
BTW63-600RN	OEM
BTW63-600RP	OEM
BTW63-800RK	OEM
BTW63-800RN	OEM
BTW63-800RP	OEM
BTW66-200N	OEM
BTW66-400	1534
BTW66-400N	OEM
BTW66-600	1534
BTW66-600N	OEM
BTW66-800	OEM
BTW66-800N	OEM
BTW66-1000	OEM
BTW66-1000N	OEM
BTW67-200	1534
BTW67-200N	OEM
BTW67-400	1534
BTW67-400N	OEM
BTW67-600	1534
BTW67-600N	OEM
BTW67-800	OEM
BTW67-800N	OEM
BTW67-1000	OEM
BTW67-1000N	OEM
BTW69-200	2690
BTW69-200N	2696

DEVICE TYPE	REPL CODE
BTW69-400	2690
BTW69-400N	5078
BTW69-600	5974
BTW69-600N	3277
BTW69-800	5974
BTW69-800N	3277
BTW69-1000	OEM
BTW69-1000N	OEM
BTW69-1200	OEM
BTW69-1200N	OEM
BTW92	OEM
BTW92-200RM	OEM
BTW92-200RU	OEM
BTW92-400RM	OEM
BTW92-400RU	OEM
BTW92-600	OEM
BTW92-600R	OEM
BTW92-600RM	OEM
BTW92-600RU	OEM
BTW92-800	OEM
BTW92-800R	OEM
BTW92-800RM	OEM
BTW92-800RU	OEM
BTW92-1000	OEM
BTW92-1000R	OEM
BTW92-1000RM	OEM
BTW92-1000RU	OEM
BTW92-1200	OEM
BTW92-1200R	OEM
BTW92-1200RM	OEM
BTW92-1200RU	OEM
BTW92-1400	OEM
BTW92-1400R	OEM
BTW92-1400RM	OEM
BTW92-1400RU	OEM
BTW92-1600	OEM
BTW92-1600R	OEM
BTW92-1600RM	OEM
BTW92-1600RU	OEM
BTW338-600R	1056
BTX068	0016
BTX070	0004
BTX071	0208
BTX-094	0016
BTX-095	0016
BTX-096	0016
BTX-097	0126
BTX0605	3199
BTX0610	3199
BTX0620	3199
BTX0640	3207
BTX0660	3222
BTX12/100R	OEM
BTX12/200R	OEM
BTX12/300R	OEM
BTX12/400R	OEM
BTX12/450R	OEM
BTX12/500R	OEM
BTX12/600R	OEM
BTX12/700R	OEM
BTX13/300R	OEM
BTX13/700R	OEM
BTX13-100R	OEM
BTX13-200R	OEM
BTX13-400R	OEM
BTX13-450R	OEM
BTX13-500R	OEM
BTX13-600R	OEM
BTX18-100	0179
BTX18-200	0179
BTX18-300	0346
BTX18-400	0346
BTX18-500	3349
BTX20	OEM
BTX21	OEM
BTX22	OEM
BTX23	OEM
BTX24	OEM
BTX25	OEM
BTX26	OEM
BTX29/1000R	OEM
BTX29-200R	OEM
BTX29-300R	OEM
BTX29-400R	OEM
BTX29-500R	OEM
BTX29-600R	OEM
BTX29-700R	OEM
BTX29-800R	OEM
BTX29-900R	OEM
BTX29-1200R	OEM
BTX29-1400R	OEM
BTX29-1600R	OEM
BTX29-1800R	OEM
BTX30-50	OEM
BTX30-100	0179
BTX30-200	0179
BTX30-300	0346
BTX30-400	0346
BTX30-500	3349
BTX30-600	3349
BTX31-50R	2449
BTX31-100	OEM
BTX31-100R	2449
BTX31-200	OEM
BTX31-200R	0435
BTX31-300R	1487
BTX31-400	OEM
BTX31-400R	1487
BTX31-500	OEM

DEVICE TYPE	REPL CODE
BTX31-500R	0687
BTX31-600	OEM
BTX31-600R	0687
BTX31-700	OEM
BTX31-700R	OEM
BTX31-800	OEM
BTX31-800R	OEM
BTX31-900	OEM
BTX31-900R	OEM
BTX31-1000	OEM
BTX31-1000R	OEM
BTX31S50R	OEM
BTX31S100	OEM
BTX31S100R	OEM
BTX31S200	OEM
BTX31S200R	OEM
BTX31S300R	OEM
BTX31S400	OEM
BTX31S400R	OEM
BTX31S500	OEM
BTX31S500R	OEM
BTX31S600	OEM
BTX31S600R	OEM
BTX31S700	OEM
BTX31S700R	OEM
BTX31S800	OEM
BTX31S800R	OEM
BTX31S900	OEM
BTX31S900R	OEM
BTX31S1000	OEM
BTX31S1000R	OEM
BTX32-50R	2519
BTX32-100	OEM
BTX32-100R	0741
BTX32-200	OEM
BTX32-200R	0745
BTX32-300R	0747
BTX32-400	OEM
BTX32-400R	0747
BTX32-500	OEM
BTX32-500R	0752
BTX32-600	OEM
BTX32-600R	OEM
BTX32-700	OEM
BTX32-700R	OEM
BTX32-800	OEM
BTX32-800R	OEM
BTX32-900	OEM
BTX32-900R	OEM
BTX32-1000	OEM
BTX32-1000R	OEM
BTX32S50R	OEM
BTX32S100	OEM
BTX32S100R	OEM
BTX32S200	OEM
BTX32S200R	OEM
BTX32S300R	OEM
BTX32S400	OEM
BTX32S400R	OEM
BTX32S500	OEM
BTX32S500R	OEM
BTX32S600	OEM
BTX32S600R	OEM
BTX32S700	OEM
BTX32S700R	OEM
BTX32S800	OEM
BTX32S800R	OEM
BTX32S900	OEM
BTX32S900R	OEM
BTX32S1000	OEM
BTX32S1000R	OEM
BTX33-50R	0731
BTX33-100	OEM
BTX33-100R	0712
BTX33-200	OEM
BTX33-200R	0469
BTX33-300R	0722
BTX33-400	OEM
BTX33-400R	0722
BTX33-500	OEM
BTX33-500R	0725
BTX33-600	OEM
BTX33-600R	0725
BTX33-700	OEM
BTX33-700R	OEM
BTX33-800	OEM
BTX33-800R	OEM
BTX33-900	OEM
BTX33-900R	OEM
BTX33-1000	OEM
BTX33-1000R	OEM
BTX33S50R	OEM
BTX33S100	OEM
BTX33S100R	OEM
BTX33S200	OEM
BTX33S200R	OEM
BTX33S300R	OEM
BTX33S400	OEM
BTX33S400R	OEM
BTX33S500	OEM
BTX33S500R	OEM
BTX33S600	OEM
BTX33S600R	OEM
BTX33S700	OEM
BTX33S700R	OEM
BTX33S800	OEM
BTX33S800R	OEM
BTX33S900	OEM
BTX33S900R	OEM

DEVICE TYPE	REPL CODE
BTX33S1000	OEM
BTX33S1000R	OEM
BTX35-500R	OEM
BTX35-600R	OEM
BTX35-700R	OEM
BTX35-800R	OEM
BTX36-500R	OEM
BTX36-600R	OEM
BTX36-700R	OEM
BTX36-800R	OEM
BTX37-500R	OEM
BTX37-600R	OEM
BTX37-700R	OEM
BTX37-800R	OEM
BTX38-500R	OEM
BTX38-600R	OEM
BTX38-700R	OEM
BTX38-800R	OEM
BTX41/1800R	OEM
BTX41-200R	OEM
BTX41-300R	OEM
BTX41-400R	OEM
BTX41-500R	OEM
BTX41-600R	OEM
BTX41-700R	OEM
BTX41-800R	OEM
BTX41-900R	OEM
BTX41-1000R	OEM
BTX41-1200R	OEM
BTX41-1400R	OEM
BTX41-1600R	OEM
BTX44-200R	OEM
BTX44-300R	OEM
BTX44-400R	OEM
BTX44-500R	OEM
BTX44-600R	OEM
BTX44-700R	OEM
BTX44-800R	OEM
BTX44-900R	OEM
BTX44-1000R	OEM
BTX44-1200R	OEM
BTX44-1600R	OEM
BTX44-1800R	OEM
BTX45-200R	OEM
BTX45-300R	OEM
BTX45-400R	OEM
BTX45-500R	OEM
BTX45-600R	OEM
BTX45-700R	OEM
BTX45-800R	OEM
BTX45-900R	OEM
BTX45-1000R	OEM
BTX45-1200R	OEM
BTX45-1400R	OEM
BTX45-1600R	OEM
BTX45-1800R	OEM
BTX46-200R	OEM
BTX46-300R	OEM
BTX46-400R	OEM
BTX46-500R	OEM
BTX46-600R	OEM
BTX46-700R	OEM
BTX46-800R	OEM
BTX46-900R	OEM
BTX46-1000R	OEM
BTX46-1200R	OEM
BTX46-1400R	OEM
BTX46-1600R	OEM
BTX46-1800R	OEM
BTX47-1000R	OEM
BTX47-1200R	OEM
BTX47-1400R	OEM
BTX48-1000R	OEM
BTX48-1200R	OEM
BTX48-1400R	OEM
BTX49-600R	OEM
BTX49-700R	OEM
BTX49-800R	OEM
BTX49-900R	OEM
BTX49-1000R	OEM
BTX49-1100R	OEM
BTX49-1200R	OEM
BTX49-1400R	OEM
BTX50-600R	OEM
BTX50-700R	OEM
BTX50-800R	OEM
BTX50-900R	OEM
BTX50-1000R	OEM
BTX50-1100R	OEM
BTX50-1200R	OEM
BTX50-1400R	OEM
BTX51-500R	OEM
BTX51-600R	OEM
BTX51-700R	OEM
BTX51-800R	OEM
BTX52	OEM
BTX53	OEM
BTX54	OEM
BTX55	OEM
BTX57	OEM
BTX58	OEM
BTX59	OEM
BTX60	OEM
BTX64-100R	OEM
BTX64-200R	OEM
BTX64-300R	OEM
BTX64-400R	OEM
BTX64-500R	OEM
BTX64-600R	OEM

DEVICE TYPE	REPL CODE
BTX65-100R	OEM
BTX65-200R	OEM
BTX65-300R	OEM
BTX65-400R	OEM
BTX65-500R	OEM
BTX65-600R	OEM
BTX66-100R	OEM
BTX66-200R	OEM
BTX66-300R	OEM
BTX66-400R	OEM
BTX66-500R	OEM
BTX66-600R	OEM
BTX67-100R	OEM
BTX67-200R	OEM
BTX67-300R	OEM
BTX67-400	OEM
BTX67-400R	OEM
BTX67-500	OEM
BTX67-500R	OEM
BTX67-600	OEM
BTX67-600R	OEM
BTX68-500R	0687
BTX68-600R	0687
BTX68-700R	OEM
BTX68-800R	OEM
BTX68-1000R	OEM
BTX70-100	OEM
BTX70-200	OEM
BTX70-400	OEM
BTX70-500	OEM
BTX70-600	OEM
BTX70-700	OEM
BTX70-800	OEM
BTX70-900	OEM
BTX70-1000	OEM
BTX70S100	OEM
BTX70S200	OEM
BTX70S400	OEM
BTX70S500	OEM
BTX70S600	OEM
BTX70S700	OEM
BTX70S800	OEM
BTX70S900	OEM
BTX70S1000	OEM
BTX71-50R	OEM
BTX71-100	OEM
BTX71-100R	OEM
BTX71-200	OEM
BTX71-200R	OEM
BTX71-300R	OEM
BTX71-400	OEM
BTX71-400R	OEM
BTX71-500	OEM
BTX71-500R	OEM
BTX71-600	OEM
BTX71-600R	OEM
BTX71-700	OEM
BTX71-700R	OEM
BTX71-800	OEM
BTX71-800R	OEM
BTX71-900	OEM
BTX71-900R	OEM
BTX71-1000	OEM
BTX71-1000R	OEM
BTX71S50R	OEM
BTX71S100	OEM
BTX71S100R	OEM
BTX71S200	OEM
BTX71S200R	OEM
BTX71S300R	OEM
BTX71S400	OEM
BTX71S400R	OEM
BTX71S500	OEM
BTX71S500R	OEM
BTX71S600	OEM
BTX71S600R	OEM
BTX71S700	OEM
BTX71S700R	OEM
BTX71S800	OEM
BTX71S800R	OEM
BTX71S900	OEM
BTX71S900R	OEM
BTX71S1000	OEM
BTX71S1000R	OEM
BTX72-50R	OEM
BTX72-100	OEM
BTX72-100R	OEM
BTX72-200	OEM
BTX72-200R	OEM
BTX72-300R	OEM
BTX72-400	OEM
BTX72-400R	OEM
BTX72-500	OEM
BTX72-500R	OEM
BTX72-600	OEM
BTX72-700	OEM
BTX72-700R	OEM
BTX72-800	OEM
BTX72-800R	OEM
BTX72-900	OEM
BTX72-900R	OEM
BTX72-1000	OEM
BTX72-1000R	OEM
BTX72S50R	OEM
BTX72S100	OEM
BTX72S100R	OEM
BTX72S200	OEM
BTX72S200R	OEM
BTX72S300R	OEM

If replacement code is OEM, contact original manufacturer for replacement.

DEVICE TYPE	REPL CODE
BTX72S400	OEM
BTX72S400R	OEM
BTX72S500	OEM
BTX72S500R	OEM
BTX72S600	OEM
BTX72S600R	OEM
BTX72S700	OEM
BTX72S700R	OEM
BTX72S800	OEM
BTX72S800R	OEM
BTX72S900	OEM
BTX72S900R	OEM
BTX72S1000	OEM
BTX72S1000R	OEM
BTX73-50R	OEM
BTX73-100	OEM
BTX73-100R	OEM
BTX73-200	OEM
BTX73-200R	OEM
BTX73-300R	OEM
BTX73-400	OEM
BTX73-500	OEM
BTX73-500R	OEM
BTX73-600	OEM
BTX73-600R	OEM
BTX73-700	OEM
BTX73-700R	OEM
BTX73-800	OEM
BTX73-800R	OEM
BTX73-900	OEM
BTX73-900R	OEM
BTX73-1000	OEM
BTX73-1000R	OEM
BTX73A800	OEM
BTX73A900	OEM
BTX73A1000	OEM
BTX73A1100	OEM
BTX73A1200	OEM
BTX73A1300	OEM
BTX73AS800	OEM
BTX73AS900	OEM
BTX73AS1000	OEM
BTX73AS1100	OEM
BTX73AS1200	OEM
BTX73AS1300	OEM
BTX73B800	OEM
BTX73B900	OEM
BTX73B1000	OEM
BTX73B1100	OEM
BTX73B1200	OEM
BTX73B1300	OEM
BTX73BS800	OEM
BTX73BS900	OEM
BTX73BS1000	OEM
BTX73BS1100	OEM
BTX73BS1200	OEM
BTX73BS1300	OEM
BTX73S50R	OEM
BTX73S100	OEM
BTX73S100R	OEM
BTX73S200	OEM
BTX73S200R	OEM
BTX73S300R	OEM
BTX73S400	OEM
BTX73S500	OEM
BTX73S500R	OEM
BTX73S600	OEM
BTX73S600R	OEM
BTX73S700	OEM
BTX73S800	OEM
BTX73S800R	OEM
BTX73S900	OEM
BTX73S900R	OEM
BTX73S1000	OEM
BTX73S1000R	OEM
BTX74-100	OEM
BTX74-200	OEM
BTX74-400	OEM
BTX74-500	OEM
BTX74-600	OEM
BTX74-700	OEM
BTX74-800	OEM
BTX74-900	OEM
BTX74-1000	OEM
BTX74S100	OEM
BTX74S200	OEM
BTX74S400	OEM
BTX74S500	OEM
BTX74S600	OEM
BTX74S700	OEM
BTX74S800	OEM
BTX74S900	OEM
BTX74S1000	OEM
BTX75-100R	0567
BTX75-200R	0762
BTX75-300R	0740
BTX75-400R	0740
BTX76-100R	0712
BTX76-200R	0469
BTX76-300R	0722
BTX76-400R	0722
BTX81-100R	0712
BTX81-100RM	OEM
BTX81-100RU	OEM
BTX81-200R	0469
BTX81-200RM	OEM
BTX81-200RU	OEM
BTX81-300R	0721
BTX81-300RM	OEM

DEVICE TYPE	REPL CODE
BTX81-300RU	OEM
BTX81-400R	0722
BTX81-400RM	OEM
BTX81-400RU	OEM
BTX81-500R	0778
BTX81-500RM	OEM
BTX81-500RU	OEM
BTX81-600R	0725
BTX81-600RM	OEM
BTX81-600RU	OEM
BTX81-700R	OEM
BTX81-700RM	OEM
BTX81-700RU	OEM
BTX81-800R	OEM
BTX81-800RM	OEM
BTX81-800RU	OEM
BTX82-100R	OEM
BTX82-100RM	OEM
BTX82-100RU	OEM
BTX82-200R	OEM
BTX82-200RM	OEM
BTX82-200RU	OEM
BTX82-300R	OEM
BTX82-300RM	OEM
BTX82-300RU	OEM
BTX82-400R	OEM
BTX82-400RM	OEM
BTX82-400RU	OEM
BTX82-500R	OEM
BTX82-500RM	OEM
BTX82-500RU	OEM
BTX82-600R	OEM
BTX82-600RM	OEM
BTX82-600RU	OEM
BTX82-700R	OEM
BTX82-700RM	OEM
BTX82-700RU	OEM
BTX82-800R	OEM
BTX82-800RM	OEM
BTX82-800RU	OEM
BTX92-600R	OEM
BTX92-800R	OEM
BTX92-900R	OEM
BTX92-1000R	OEM
BTX92-1100R	OEM
BTX92-1200R	OEM
BTX92-1400R	OEM
BTX92-1600R	OEM
BTX94-100	2015
BTX94-200	2015
BTX94-300	2017
BTX94-400	2017
BTX94-400H	2017
BTX94-400J	OEM
BTX94-500	2018
BTX94-600	2018
BTX94-600H	2018
BTX94-600J	OEM
BTX94-700	OEM
BTX94-800	OEM
BTX94-800H	OEM
BTX94-800J	OEM
BTX94-900	OEM
BTX94-1000	OEM
BTX94-1000H	OEM
BTX94-1000J	OEM
BTX94-1100	OEM
BTX94-1200	OEM
BTX94-1200H	OEM
BTX94-1200J	OEM
BTX95-500	OEM
BTX95-500R	OEM
BTX95-600	OEM
BTX95-600R	OEM
BTX95-700	OEM
BTX95-700R	OEM
BTX95-800	OEM
BTX95-800R	OEM
BTX1005	3199
BTX1010	3199
BTX1020	3199
BTX1040	3207
BTX1060	3222
BTX1605	0147
BTX1610	0154
BTX1620	0154
BTX1640	0147
BTX1660	3222
BTX-2367B	0016
BTX2367B	0016
BTX2505	2015
BTX2510	2015
BTX2520	2015
BTX2540	2017
BTX2560	3222
BTY10	OEM
BTY11	OEM
BTY13	OEM
BTY15	OEM
BTY16	OEM
BTY20	OEM
BTY21	OEM
BTY22	OEM
BTY23	OEM
BTY24	OEM
BTY25	OEM
BTY28	OEM
BTY29	OEM
BTY30	OEM

DEVICE TYPE	REPL CODE
BTY31	OEM
BTY34	OEM
BTY34/300	OEM
BTY34/400	OEM
BTY34-100R	2449
BTY34-200R	0435
BTY34-300R	1487
BTY34-400R	1487
BTY35	OEM
BTY36	OEM
BTY37	OEM
BTY38	OEM
BTY39	OEM
BTY41	OEM
BTY43	OEM
BTY44	OEM
BTY46	OEM
BTY47	OEM
BTY50	OEM
BTY51	OEM
BTY52	OEM
BTY53	OEM
BTY54	OEM
BTY57	OEM
BTY58	OEM
BTY59	OEM
BTY60	OEM
BTY61	OEM
BTY62	OEM
BTY64	OEM
BTY65	OEM
BTY66	OEM
BTY67	OEM
BTY68	OEM
BTY69	OEM
BTY70	OEM
BTY79	OEM
BTY79-100	OEM
BTY79-100R	OEM
BTY79-100R05	OEM
BTY79-150R	OEM
BTY79-200	OEM
BTY79-200R	OEM
BTY79-200R05	OEM
BTY79-250R	OEM
BTY79-300	OEM
BTY79-300R	OEM
BTY79-400	OEM
BTY79-400R	0740
BTY79-400R05	OEM
BTY79-500	OEM
BTY79-500R	0764
BTY79-500R05	OEM
BTY79-600	OEM
BTY79-600R	0764
BTY79-700	OEM
BTY79-700R	OEM
BTY79-800	OEM
BTY79-800R	0766
BTY79-1000R	OEM
BTY79A-50	OEM
BTY79A-100	OEM
BTY79A-200	OEM
BTY79A-300	OEM
BTY79A-400	OEM
BTY79A-500	OEM
BTY80	OEM
BTY81	OEM
BTY84	OEM
BTY85	OEM
BTY86	OEM
BTY87	OEM
BTY87-100R	0741
BTY87-100R+	0469
BTY87-100RM	OEM
BTY87-100RU	OEM
BTY87-150R	OEM
BTY87-200R	0745
BTY87-200R+	0469
BTY87-200RM	OEM
BTY87-200RU	OEM
BTY87-250R	OEM
BTY87-300R	2917
BTY87-300R+	0722
BTY87-300RM	OEM
BTY87-300RU	OEM
BTY87-400R	0747
BTY87-400R+	0722
BTY87-400RM	OEM
BTY87-400RU	OEM
BTY87-500R	3243
BTY87-500R+	0725
BTY87-500RM	OEM
BTY87-500RU	OEM
BTY87-600R	0764
BTY87-600R+	0725
BTY87-600RM	OEM
BTY87-600RU	OEM
BTY87-700R	2875
BTY87-700R+	2036
BTY87-700RM	OEM
BTY87-700RU	OEM
BTY87-800R	0766
BTY87-800R+	2036
BTY87-800RM	OEM
BTY87-800RU	OEM
BTY88	OEM
BTY89	OEM
BTY90	OEM

DEVICE TYPE	REPL CODE
BTY91	OEM
BTY91-100R	0567
BTY91-100R+	0469
BTY91-100RM	OEM
BTY91-100RU	OEM
BTY91-150R	OEM
BTY91-200R	0762
BTY91-200R+	0469
BTY91-200RM	OEM
BTY91-200RU	OEM
BTY91-250R	OEM
BTY91-300R	3271
BTY91-300R+	0722
BTY91-300RM	OEM
BTY91-300RU	OEM
BTY91-400R	0740
BTY91-400R+	0722
BTY91-400RM	OEM
BTY91-400RU	OEM
BTY91-500+	0725
BTY91-500R	3291
BTY91-500RM	OEM
BTY91-500RU	OEM
BTY91-600R	0764
BTY91-600R+	0725
BTY91-600RM	OEM
BTY91-600RU	OEM
BTY91-700R	2875
BTY91-700R+	2036
BTY91-800R	0766
BTY91-800R+	2036
BTY91-800RM	OEM
BTY91-800RU	OEM
BTY91-1000RM	OEM
BTY91-1000RU	OEM
BTY91-1200R	OEM
BTY91-1200RM	OEM
BTY91-1200RU	OEM
BTY94	OEM
BTY95	OEM
BTY95-250R	5894
BTY95-800R	0108
BTY98	OEM
BTY99	OEM
BTY99-150R	OEM
BTY99-250R	OEM
BTY99-800R	OEM
BTY-150R	OEM
BTZ10	OEM
BTZ11	OEM
BTZ12	OEM
BTZ13	OEM
BTZ15	OEM
BTZ16	OEM
BTZ18	OEM
BTZ19	OEM
BTZ21	OEM
BTZ35	OEM
BTZ36	OEM
BTZ37	OEM
BTZ38	OEM
BTZ39	OEM
BU029	0229
BU40	6375
BU67	0016
BU71	0016
BU74HC00	OEM
BU102	0363
BU104	0363
BU104D	OEM
BU104DP	OEM
BU104P	0728
BU105	0065
BU105-02	0065
BU106	0363
BU107	0074
BU108	0065
BU109	0642
BU109D	OEM
BU109DP	OEM
BU109P	0728
BU110	0363
BU111	0363
BU113	0363
BU114	0642
BU115	0065
BU118	OEM
BU120	0363
BU121	0363
BU122	0363
BU123	0363
BU124	OEM
BU124A	OEM
BU125	0622
BU125S	OEM
BU126	0642
BU129	0363
BU130	0074
BU131	0074
BU132	0642
BU133	0363
BU137	OEM
BU142	OEM
BU143	OEM
BU157	OEM
BU180	3765
BU180A	3765
BU181	OEM
BU181A	OEM

DEVICE TYPE	REPL CODE
BU184	1511
BU189	1511
BU204	0065
BU205	0065
BU205-01	0065
BU206	0065
BU207	0313
BU207A	0642
BU208	0065
BU208A	0065
BU208B	0065
BU208D	0055
BU209	0065
BU209A	2847
BU210	0363
BU211	0363
BU212	0363
BU225	0065
BU226	2847
BU230F	OEM
BU310	0363
BU311	0363
BU312	0363
BU322	1991
BU322A	OEM
BU323	1991
BU323A	OEM
BU324A	OEM
BU325	OEM
BU326	0313
BU326A	0313
BU326AP	OEM
BU326P	OEM
BU326R	0313
BU326S	0313
BU406	0728
BU406D	0728
BU406H	5637
BU407	0728
BU407D	0728
BU407H	5637
BU408	0728
BU408D	OEM
BU409	0728
BU426	4450
BU426A	4450
BU433	4450
BU500	0313
BU508	3360
BU508A	3360
BU508D	5345
BU508DF	3360
BU508V	3360
BU522	OEM
BU522A	OEM
BU522B	OEM
BU526	0313
BU536	0313
BU546	0313
BU606	OEM
BU606D	OEM
BU607	OEM
BU607D	OEM
BU608	OEM
BU608D	OEM
BU626A	OEM
BU705	3360
BU800	0055
BU801	OEM
BU806	1511
BU807	1511
BU808	1511
BU810	OEM
BU810(A)	OEM
BU826	OEM
BU826A	OEM
BU902	3360
BU903	3360
BU908	3360
BU910	OEM
BU911	OEM
BU912	OEM
BU920	OEM
BU920P	5342
BU920P(A)	OEM
BU920PFI	6856
BU920T	OEM
BU921	OEM
BU921P	5342
BU921P(A)	OEM
BU921PFI	6856
BU921T	OEM
BU922	2626
BU922P	5342
BU922P(A)	OEM
BU922PFI	6856
BU922T	OEM
BU930	2626
BU931	2626
BU931R	2626
BU931RP	4334
BU932	2626
BU932R	2626
BU932RP	4334
BU2305F	OEM
BU2506	OEM
BU2506DF	2129
BU2508DF	5330
BU2508DFLB	OEM

DEVICE TYPE	REPL CODE
BU2702	OEM
BU2742S	OEM
BU2744S	OEM
BU2748S	OEM
BU2763S	OEM
BU3702F	OEM
BU4011B	0215
BU4013B	0413
BU4016B	OEM
BU4030B	OEM
BU4052B	0024
BU4053B	0035
BU4053BF-T1	OEM
BU4066B	0101
BU4066BF	OEM
BU4066BL	0101
BU4066BP	0101
BU4199	OEM
BU4538B	OEM
BU4551B	OEM
BU5702	OEM
BU5777F	OEM
BU8322F	OEM
BUC205	OEM
BUC208	OEM
BUC-97704-2	0079
BUF01AJ	OEM
BUF01AZ	OEM
BUF01AZ(M)	OEM
BUF01BJ	OEM
BUF01BZ	OEM
BUF01BZ(M)	OEM
BUF01EJ	OEM
BUF01EZ	OEM
BUF01FJ	OEM
BUF01FZ	OEM
BUF02AJ	OEM
BUF02AZ(M)	OEM
BUF02BJ	OEM
BUF02BZ(M)	OEM
BUF02EJ	OEM
BUF02EZ	OEM
BUF02FJ	OEM
BUF02FZ	OEM
BUF03AJ	OEM
BUF03BJ	OEM
BUF03EJ	OEM
BUF03FJ	OEM
BUF03G	OEM
BUF03N	OEM
BUK417-500AE	OEM
BUK417-500BE	OEM
BUK426-800A	OEM
BUK426-800B	OEM
BUK426-1000A	OEM
BUK427-400A	OEM
BUK427-400B	OEM
BUK427-500A	OEM
BUK427-500B	OEM
BUK427-600A	OEM
BUK427-600B	OEM
BUK428-500A	OEM
BUK428-500B	OEM
BUK428-800A	OEM
BUK428-800B	OEM
BUK428-1000A	OEM
BUK428-1000B	OEM
BUK436-800A	OEM
BUK436-800B	OEM
BUK436-1000A	OEM
BUK436-1000B	OEM
BUK437-400A	OEM
BUK437-400B	OEM
BUK437-500A	OEM
BUK437-500B	OEM
BUK437-600A	OEM
BUK437-600B	OEM
BUK438-500A	OEM
BUK438-500B	OEM
BUK438-800A	OEM
BUK438-800B	OEM
BUK438-1000A	OEM
BUK438-1000B	OEM
BUK443-60B	OEM
BUK444-400A	OEM
BUK444-400B	OEM
BUK444-500A	OEM
BUK444-500B	OEM
BUK444-600A	OEM
BUK444-600B	OEM
BUK444-800A	OEM
BUK444-800B	OEM
BUK445-400A	OEM
BUK445-400B	OEM
BUK445-500A	OEM
BUK445-500B	OEM
BUK445-600A	OEM
BUK445-600B	OEM
BUK446-800A	3718
BUK446-800B	3718
BUK446-1000A	OEM
BUK446-1000B	OEM
BUK452-60A	OEM
BUK452-60B	OEM
BUK453-60A	OEM
BUK453-60B	OEM
BUK454-400A	OEM
BUK454-400B	OEM
BUK454-500A	OEM

DEVICE TYPE	REPL CODE
BUK454-500B	OEM
BUK454-600A	OEM
BUK454-600B	OEM
BUK454-800A	OEM
BUK454-800B	OEM
BUK455-400A	OEM
BUK455-400B	OEM
BUK455-500A	OEM
BUK455-500B	OEM
BUK455-600A	OEM
BUK455-600B	OEM
BUK456-800A	5096
BUK456-800B	3718
BUK456-1000A	5096
BUK456-1000B	OEM
BUK457-400A	OEM
BUK457-400B	OEM
BUK457-500A	OEM
BUK457-500B	5769
BUK457-600A	OEM
BUK457-600B	OEM
BUK543-50A	OEM
BUK543-50B	OEM
BUK545-50A	OEM
BUK545-50B	OEM
BUK552-60A	OEM
BUK552-60B	OEM
BUK553-50A	OEM
BUK553-50B	OEM
BUK555-50A	OEM
BUK555-50B	OEM
BUKBUK426-1000B	OEM
BUP22A	OEM
BUP22B	3387
BUP22C	3387
BUP23A	OEM
BUP23B	OEM
BUP23C	OEM
BUP60	OEM
BUP61	OEM
BUP62	OEM
BUP63	OEM
BUP64	OEM
BUP65	OEM
BUP66	OEM
BUP67	OEM
BUP68	OEM
BUP69	OEM
BUP70	OEM
BUP71	OEM
BUR10	OEM
BUR11	OEM
BUR12	OEM
BUR13	OEM
BUR20	OEM
BUR21	OEM
BUR22	OEM
BUR23	OEM
BUR24	OEM
BUR50	OEM
BUR51	OEM
BUR52	OEM
BUS11	4450
BUS11A	4450
BUS12	1506
BUS12A	1506
BUS13	7092
BUS13A	OEM
BUS14	OEM
BUS14A	OEM
BUS21A	3388
BUS21B	3388
BUS21C	3388
BUS22A	3388
BUS22B	3388
BUS22C	3388
BUS23A	3388
BUS23B	3388
BUS23C	3388
BUS47P	OEM
BUS133	OEM
BUS133A	OEM
BUS133H	OEM
BUT11	6146
BUT11A	6146
BUT11AF	2766
BUT11F	3339
BUT12	6146
BUT12A	4590
BUT13	OEM
BUT13P	3765
BUT18	6146
BUT18A	6146
BUT18AF	3462
BUT18F	3462
BUT21A	OEM
BUT21B	6146
BUT21C	6146
BUT22A	OEM
BUT22B	6146
BUT22C	2908
BUT54	6146
BUT56	6146
BUT56A	4590
BUT93	0037
BUT131	OEM
BUT131A	OEM
BUT131H	OEM
BUV11	1852

If replacement code is OEM, contact original manufacturer for replacement.

DEVICE TYPE	REPL CODE
BUV18	OEM
BUV19	OEM
BUV20	2433
BUV21	OEM
BUV22	OEM
BUV23	OEM
BUV24	OEM
BUV25	OEM
BUV26	OEM
BUV27	5637
BUV28	5637
BUV30	OEM
BUV37	OEM
BUV46	6146
BUV47	1506
BUV47A	1506
BUV47B	1506
BUV48	1506
BUV48A	1506
BUV82	4450
BUV83	4450
BUV90	OEM
BUV90F	OEM
BUV93	OEM
BUV94	OEM
BUV95	OEM
BUV98	OEM
BUV98A	OEM
BUV98AV	OEM
BUV98V	OEM
BUV298AV	OEM
BUV298V	OEM
BUW11	4450
BUW11A	4450
BUW12	4450
BUW12A	4450
BUW13	1506
BUW13A	1506
BUW16	OEM
BUW17	OEM
BUW18	OEM
BUW19	OEM
BUW22	OEM
BUW23	OEM
BUW24	OEM
BUW25	OEM
BUW26	OEM
BUW32	OEM
BUW34	3038
BUW35	3038
BUW36	7114
BUW38	OEM
BUW39	OEM
BUW40	OEM
BUW40A	6375
BUW40B	6375
BUW41	5637
BUW41A	5637
BUW41B	5637
BUW44	3388
BUW45	3388
BUW46	OEM
BUW48	OEM
BUW49	OEM
BUW57	OEM
BUW58	OEM
BUW64A	5542
BUW64B	5542
BUW64C	OEM
BUW66	OEM
BUW67	OEM
BUW70	OEM
BUW71	OEM
BUW72	OEM
BUW73	OEM
BUW74	1852
BUW75	1852
BUW76	1852
BUW77	OEM
BUW81	OEM
BUW81A	OEM
BUW84	6375
BUW85	6375
BUW86	1852
BUW132	3494
BUW132A	3494
BUW132H	3494
BUW133	OEM
BUW133A	OEM
BUW133H	OEM
BUX10	2485
BUX10A	6454
BUX11	2485
BUX11N	6455
BUX12	2485
BUX13	1852
BUX14	1852
BUX15	1852
BUX16	0363
BUX16A	0363
BUX16B	0363
BUX16C	0363
BUX17	1966
BUX17A	1966
BUX17B	1966
BUX17C	1966
BUX18	0074
BUX18A	0074
BUX18B	0074

DEVICE TYPE	REPL CODE
BUX18C	1966
BUX20A	2433
BUX28	OEM
BUX29	1991
BUX30	OEM
BUX30AV	OEM
BUX30AVA	OEM
BUX30AVB	OEM
BUX30AVC	OEM
BUX31	OEM
BUX31A	OEM
BUX31B	OEM
BUX32	1506
BUX32A	1506
BUX32B	1506
BUX34	OEM
BUX37	OEM
BUX39	2485
BUX40	2485
BUX40A	OEM
BUX41	1341
BUX41N	6517
BUX42	0074
BUX43	1341
BUX44	1852
BUX46	0223
BUX46A	0223
BUX47	1506
BUX47A	1506
BUX47B	1506
BUX48	3388
BUX48A	3388
BUX49	OEM
BUX50	OEM
BUX51	OEM
BUX52	OEM
BUX53	OEM
BUX54	OEM
BUX55	OEM
BUX59	OEM
BUX60	OEM
BUX61	OEM
BUX62	OEM
BUX66	OEM
BUX66A	OEM
BUX66B	OEM
BUX66C	OEM
BUX67	2096
BUX67A	2096
BUX67B	2096
BUX67C	2096
BUX69	OEM
BUX70	OEM
BUX77	OEM
BUX78	OEM
BUX80	1852
BUX81	1852
BUX82	1506
BUX83	1506
BUX84	3014
BUX84F	3547
BUX85	3547
BUX85F	3547
BUX86	3217
BUX87	3217
BUX97	1852
BUX97A	1852
BUX97B	1852
BUX98	OEM
BUX98A	OEM
BUX99	OEM
BUX238	OEM
BUY10	0103
BUY10S	OEM
BUY11	0103
BUY12	0074
BUY13	0074
BUY14	0561
BUY16	OEM
BUY17	OEM
BUY18S	0363
BUY19	OEM
BUY20	0074
BUY21	0074
BUY21A	0074
BUY22	0074
BUY24	0042
BUY26	OEM
BUY27	OEM
BUY28	OEM
BUY29	1341
BUY30	1341
BUY35	0363
BUY38	0429
BUY43	0130
BUY43-6	OEM
BUY43-10	OEM
BUY43-16	OEM
BUY44	0177
BUY46	0130
BUY46-4	OEM
BUY46-6	OEM
BUY50	OEM
BUY51A	2485
BUY52A	2485
BUY53A	2485
BUY54A	2485
BUY55	0074
BUY55-4	0074

DEVICE TYPE	REPL CODE
BUY55-6	0074
BUY55-10	0074
BUY56	0074
BUY56-4	0074
BUY56-6	0074
BUY56-10	0074
BUY59	OEM
BUY60	OEM
BUY61	OEM
BUY62	OEM
BUY63	OEM
BUY64	OEM
BUY65	OEM
BUY66	OEM
BUY67	OEM
BUY69A	1506
BUY69B	0363
BUY69C	0363
BUY70A	1506
BUY70B	0363
BUY70C	0363
BUY71	2847
BUY72	0074
BUY72-4	0074
BUY72-6	0074
BUY72-10	0074
BUY88	OEM
BUY89	0313
BUY90	OEM
BUY91	OEM
BUY92	OEM
BUZ10	OEM
BUZ10A	OEM
BUZ11	OEM
BUZ11A	OEM
BUZ14	OEM
BUZ15	OEM
BUZ20	5060
BUZ21	OEM
BUZ23	OEM
BUZ24	5363
BUZ25	OEM
BUZ30	OEM
BUZ31	2176
BUZ32	5723
BUZ33	OEM
BUZ34	OEM
BUZ35	OEM
BUZ36	OEM
BUZ40	1466
BUZ41	OEM
BUZ41A	1653
BUZ42	1653
BUZ43	OEM
BUZ44	OEM
BUZ44A	OEM
BUZ45	OEM
BUZ45A	OEM
BUZ46	7135
BUZ50	OEM
BUZ50A	OEM
BUZ50B	OEM
BUZ53	OEM
BUZ53A	OEM
BUZ54	OEM
BUZ54A	OEM
BUZ60	1465
BUZ63	OEM
BUZ64	OEM
BUZ71	5060
BUZ71A	5060
BUZ72A	5060
BUZ73A	5060
BUZ74	OEM
BUZ74A	6740
BUZ76	OEM
BUZ76A	OEM
BUZ80	3718
BUZ80A	3718
BUZ83	OEM
BUZ83A	OEM
BUZ84	5899
BUZ84A	5899
BUZ211	OEM
BUZ325	OEM
BV25	0015
BV67	0016
BV69	OEM
BV71	0016
BV100	OEM
BV110	OEM
BV120	OEM
BV130	OEM
BV140	OEM
BV150	OEM
BVDKBPC2502W	OEM
BVDPBM152	OEM
BVIANAM1002	OEM
BVILC7582B	OEM
BVISTK4231-5	OEM
BVX40-1000R	0550
BVX48/600	0122
BVX48-600	0122
BVY27-150	OEM
BW050	OEM
BW080	OEM
BW8T	OEM
BW67	0016
BW71	0016

DEVICE TYPE	REPL CODE
BWALPHA	OEM
BWS1	OEM
BWS2	OEM
BWS66	0168
BWS67	0168
BWX5	OEM
BX090	0012
BX67	0016
BX71	0016
BX213	OEM
BX-324	0164
BX-342	OEM
BX342	OEM
BX-358	OEM
BX358	OEM
BX-360	OEM
BX360	OEM
BX-361	OEM
BX361	OEM
BX-364	OEM
BX364	OEM
BX-369	OEM
BX-376	OEM
BX376	OEM
BX-378	OEM
BX-379	OEM
BX-380	OEM
BX-383	OEM
BX383	OEM
BX-384	OEM
BX384	OEM
BX-384A	OEM
BX-393A	OEM
BX-495	0164
BX495(SONY)	0164
BX909	0012
BX-1013	OEM
BX-1013-1	OEM
BX-1016	OEM
BX-1038	OEM
BX1038	OEM
BX-1039	OEM
BX-1041	OEM
BX-1043	OEM
BX-1044	OEM
BX-1044-1	OEM
BX-1045	OEM
BX-1046	OEM
BX-1047	OEM
BX-1083	OEM
BX-1084	OEM
BX-1086	OEM
BX-1087	OEM
BX-1088	OEM
BX-1090	OEM
BX-1091	OEM
BX-1092	OEM
BX-1093	OEM
BX-1094	OEM
BX-1095	OEM
BX-1153	OEM
BX-1317	OEM
BX1317	OEM
BX-1323	OEM
BX1323	OEM
BX-1357	OEM
BX1357	OEM
BX1387	OEM
BX-1393	OEM
BX1393	OEM
BX-1398	OEM
BX1398	OEM
BX1407	OEM
BX-1435B	OEM
BX1458	OEM
BX-1483	OEM
BX-3901	OEM
BX-3903	OEM
BX3903	OEM
BX-3904	OEM
BX-3912	OEM
BX-3921	OEM
BX-3943	OEM
BX-3945-1	OEM
BX-3946-1	OEM
BX-3947-2	OEM
BX-3948-1	OEM
BX-3949	OEM
BX-3950-1	OEM
BX-3951-2	OEM
BX-3952	OEM
BX-3953	OEM
BX-3954	OEM
BX-3955	OEM
BX-3956-1	OEM
BX-3958-1	OEM
BX-3959-1	OEM
BX-3960	OEM
BX-3969-1	OEM
BX-3969A	OEM
BX-3970-1	OEM
BX-3970A	OEM
BX-3971	OEM
BX-3971A	OEM
BX-3972	OEM
BX-3973	OEM
BX-3976	OEM
BX-3977	OEM
BX-3978	OEM

DEVICE TYPE	REPL CODE
BX-3980	OEM
BX-3981	OEM
BX-3982-1	OEM
BX-3982-2	OEM
BX-3983	OEM
BX-3984	OEM
BX-3988	OEM
BX-3999	OEM
BX-3999-1	OEM
BX6224W	OEM
BX6238	OEM
BX6239	OEM
BX6240	OEM
BX6256	OEM
BX6259	OEM
BX6301	OEM
BX6302	OEM
BX6302A	OEM
BX6335	OEM
BX6410	OEM
BX7197W	OEM
BX7335	OEM
BX7385W	OEM
BX7403	OEM
BX7422F	OEM
BX7426	OEM
BX7426FA	OEM
BX7447	OEM
BX7483	OEM
BX7504A	OEM
BX7521	OEM
BX7580F	OEM
BX7596F	OEM
BX7645	OEM
BX7645A	OEM
BX7645F	OEM
BXB10125	OEM
BXC79C5V6	0253
BXD23/40636	OEM
BXW70-36	OEM
BXY10	0071
BXY10C	OEM
BXY10D	OEM
BXY11E	OEM
BXY11F	OEM
BXY13C	OEM
BXY13CA	OEM
BXY13D	OEM
BXY14E	OEM
BXY14F	OEM
BXY15CA3	OEM
BXY15CA5	OEM
BXY18A2	OEM
BXY18AB	OEM
BXY18AB2	OEM
BXY18AB5	OEM
BXY18AB6	OEM
BXY19E	OEM
BXY19F	OEM
BXY19GB	OEM
BXY22G	OEM
BXY22H	OEM
BXY22J	OEM
BXY23	OEM
BXY27	OEM
BXY28	OEM
BXY29	OEM
BXY30-200	OEM
BXY30-500	OEM
BXY32	OEM
BXY35	OEM
BXY35A	OEM
BXY36	OEM
BXY36B	OEM
BXY36C	OEM
BXY36D	OEM
BXY36E	OEM
BXY37	OEM
BXY37B	OEM
BXY37C	OEM
BXY37D	OEM
BXY37E	OEM
BXY38	2645
BXY38B	OEM
BXY38C	OEM
BXY38D	OEM
BXY38E	OEM
BXY39	OEM
BXY39B	OEM
BXY39C	OEM
BXY39D	OEM
BXY39E	OEM
BXY40	OEM
BXY40B	OEM
BXY40C	OEM
BXY40D	OEM
BXY40E	OEM
BXY41	OEM
BXY41B	OEM
BXY41C	OEM
BXY41D	OEM
BXY41E	OEM
BXY42B	OEM
BXY42BA2	OEM
BXY42BA3	OEM
BXY42BA4	OEM
BXY42BA5	OEM
BXY42BA6	OEM

DEVICE TYPE	REPL CODE
BXY43A	OEM
BXY43B	OEM
BXY43C	OEM
BXY44E	OEM
BXY50	OEM
BXY51	OEM
BXY52	OEM
BXY53	OEM
BXY54	OEM
BXY55	OEM
BXY56	OEM
BXY57	OEM
BXY58EA	OEM
BXY59D	OEM
BXY60	OEM
BXY63	0012
BY12-00	0071
BY54	0071
BY55	0071
BY56	0071
BY67	0016
BY71	0016
BY100	0071
BY100S	0071
BY101	0015
BY102	0015
BY103	0102
BY104	0071
BY105	0071
BY106	0015
BY107	0015
BY108	0071
BY109	0071
BY111	0015
BY112	0015
BY113	0015
BY114	0015
BY115	0015
BY116	0015
BY117	0015
BY118	0071
BY119	0071
BY121	0015
BY122	0277
BY123	0289
BY124	0015
BY125	0015
BY126	0015
BY126/50	0087
BY126/100	0087
BY126/200	0087
BY126/300	0087
BY126/400	0087
BY126MGP	0071
BY-127	0015
BY127	0102
BY127/500	0071
BY127/600	0071
BY127/700	0071
BY127/800	OEM
BY127MGP	0102
BY128	0071
BY129	0071
BY130	0015
BY133	0102
BY133GP	0102
BY134	0015
BY134GP	0071
BY135	0015
BY135GP	0015
BY137-400	OEM
BY137-800	OEM
BY138	OEM
BY139	OEM
BY140A	OEM
BY141	0015
BY142	OEM
BY143	OEM
BY147	OEM
BY151N	OEM
BY152N	OEM
BY153	0015
BY156	0087
BY157	0102
BY158	0102
BY159-50	0106
BY159-200	0106
BY159-400	0106
BY164	0277
BY165	OEM
BY166	OEM
BY167	OEM
BY172	0087
BY173	0087
BY174	OEM
BY176	OEM
BY179	0296
BY182	OEM
BY183-50	OEM
BY183-100	OEM
BY183-200	OEM
BY183-300	OEM
BY183-400	OEM
BY183-500	OEM
BY183-600	OEM
BY184	OEM
BY186	OEM
BY187	OEM
BY187-01	OEM

DEVICE TYPE	REPL CODE
BY188	0071
BY188A	0015
BY188B	0015
BY188G	OEM
BY189	OEM
BY190	OEM
BY191-250	OEM
BY191-400	OEM
BY191P-250	OEM
BY191P-400	OEM
BY192	OEM
BY193	OEM
BY194	OEM
BY195	OEM
BY196	OEM
BY197	OEM
BY198	OEM
BY199	OEM
BY200	OEM
BY201	0540
BY201/2	0580
BY201/4	OEM
BY201/6	OEM
BY201-2	0580
BY201-200	0023
BY201-300	0023
BY201-400	0023
BY201-500	0023
BY201-600	0023
BY202	0580
BY203	0580
BY203/12	0348
BY203/12S	0348
BY203/16	0348
BY203/16S	0348
BY203/20	0348
BY203/20S	0348
BY203-12	0348
BY203-16	0348
BY203-20	0348
BY204	0965
BY204/4	0529
BY204/8	0529
BY204/10	0529
BY205	1000
BY206	0023
BY206GP	0023
BY207	0023
BY207GP	0023
BY-208	0817
BY208	0817
BY208-600	0017
BY208-800	0102
BY208-1000	0017
BY209	2088
BY210-400	0795
BY210-600	0023
BY210-800	OEM
BY211	0708
BY212	4140
BY212-750R	OEM
BY213	0580
BY213-700R	OEM
BY214	1282
BY214-50	1282
BY214-100	1282
BY214-200	1287
BY214-400	1287
BY214-600	1293
BY214-800	1296
BY214-1000	1296
BY215	1000
BY216	2076
BY217	0965
BY217-50	OEM
BY217-100	OEM
BY217-200	OEM
BY217-400	OEM
BY218	2088
BY218-13	OEM
BY218-15	OEM
BY218-100	OEM
BY218-200	OEM
BY218-400	OEM
BY218-600	OEM
BY219	2088
BY221	0708
BY222	4140
BY223	0580
BY224	1296
BY224-200	OEM
BY224-400	OEM
BY224-600	OEM
BY225	1000
BY225-100	1040
BY225-200	1040
BY226	1771
BY226MGP	0071
BY227	0071
BY227MGP	0071
BY228	2088
BY228/13	OEM
BY228/13S	OEM
BY228/15	OEM
BY228/15S	OEM
BY229	2088
BY229-200	1145
BY229-200F	OEM
BY229-200R	OEM

DEVICE TYPE	REPL CODE	DEVICE TYPE	REPL CODE	DEVICE TYPE	REPL CODE	DEVICE TYPE	REPL CODE	DEVICE TYPE	REPL CODE	DEVICE TYPE	REPL CODE	DEVICE TYPE	REPL CODE
BY229-400	1145	BY314	2060	BY550/50	OEM	BY8002	0145	BYQ28-150	1236	BYT30PI800	OEM	BYV21-35A	OEM
BY229-400R	OEM	BY315	1000	BY550/100	OEM	BY8101	0145	BYQ28-200	1236	BYT30PI1000	OEM	BYV21-40	0615
BY229-600	OEM	BY316	2076	BY550/200	OEM	BY8102	0145	BYR28-600	0972	BYT51A	0023	BYV21-40A	OEM
BY229-600F	OEM	BY317	2081	BY550/400	OEM	BY8201	0145	BYR28-800	OEM	BYT51B	0023	BYV21-45	0615
BY229-600R	OEM	BY318	2088	BY550/600	OEM	BY8202	0145	BYR29-600	1664	BYT51D	0023	BYV21-45A	OEM
BY229-800	OEM	BY318-100	0133	BY550/800	OEM	BYD13D	0071	BYR29-700	OEM	BYT51G	0023	BYV22-30	0615
BY229-800R	OEM	BY318-200	OEM	BY584	OEM	BYD13G	0071	BYR29-800	OEM	BYT51J	0023	BYV22-30A	OEM
BY229-1000	OEM	BY318-400	OEM	BY588	0071	BYD13J	0071	BYR34-600	OEM	BYT51K	0017	BYV22-35	1545
BY229-1000R	OEM	BY318-600	OEM	BY609	OEM	BYD13K	0071	BYR34-800	OEM	BYT51M	0017	BYV22-35A	OEM
BY229F200	OEM	BY319	2088	BY610	OEM	BYD13M	0071	BYR79-600	OEM	BYT52A	0023	BYV22-40	1545
BY229F400	OEM	BY321	0708	BY614	OEM	BYD14D	0071	BYR79-800	OEM	BYT52B	0023	BYV22-40A	OEM
BY229F600	OEM	BY322	4140	BY619	OEM	BYD14G	0071	BYS15	OEM	BYT52D	0023	BYV22-45	1545
BY229F800	OEM	BY323	0580	BY620	OEM	BYD14J	0071	BYS21	0080	BYT52G	0023	BYV22-45A	OEM
BY233-200	1127	BY324	2060	BY627	1217	BYD14K	0071	BYS21-45	0080	BYT52J	0023	BYV23-30	OEM
BY233-400	1664	BY325	1000	BY701	0970	BYD14M	0071	BYS21-90	OEM	BYT52K	0017	BYV23-30A	OEM
BY233-600	1664	BY326	2076	BY702	3733	BYD17D	0071	BYS22-45	OEM	BYT52M	0017	BYV23-35	OEM
BY239-200	OEM	BY327	2081	BY703	0989	BYD17G	0071	BYS22-90	OEM	BYT60-200	OEM	BYV23-35A	OEM
BY239-400	OEM	BY328	2088	BY704	3743	BYD17J	0071	BYS24	OEM	BYT60-300	OEM	BYV23-40	OEM
BY239-600	OEM	BY329	2088	BY705	0197	BYD17K	0071	BYS24-45	OEM	BYT60-400	OEM	BYV23-40A	OEM
BY239-800	OEM	BY329-800	OEM	BY706	0200	BYD17M	0071	BYS24-90	OEM	BYT60P200	OEM	BYV23-45	OEM
BY239-1000	OEM	BY329-1000	OEM	BY707	0204	BYD33D	0023	BYS26	OEM	BYT60P300	OEM	BYV23-45A	OEM
BY239-1250	OEM	BY329-1200	OEM	BY708	0206	BYD33DPH	OEM	BYS26-45	OEM	BYT60P400	OEM	BYV24-800	OEM
BY246-600	OEM	BY350-1300	OEM	BY709	0206	BYD33G	0023	BYS26-90	OEM	BYT61-600	OEM	BYV24-800R	OEM
BY246-1000	OEM	BY350-1500	OEM	BY710	OEM	BYD33J	0031	BYS27-45	OEM	BYT61-800	OEM	BYV24-1000	OEM
BY246-1200	OEM	BY359-1000	OEM	BY711	0970	BYD33JPH	OEM	BYS28	OEM	BYT61-900	OEM	BYV24-1000R	OEM
BY249-300	6119	BY359-1300	OEM	BY712	3733	BYD33K	0707	BYS28-45	OEM	BYT61-1000	OEM	BYV25-800	OEM
BY249-300R	OEM	BY359-1500	OEM	BY713	0989	BYD33M	0707	BYS28-90	OEM	BYT61B-600	OEM	BYV25-800R	OEM
BY249-600	1969	BY360-600	OEM	BY714	3743	BYD37D	0031	BYS30	OEM	BYT61B-800	OEM	BYV25-1000	OEM
BY249-600R	OEM	BY396	0102	BY715	0197	BYD37G	0031	BYS31	OEM	BYT61B-900	OEM	BYV25-1000R	OEM
BY250	OEM	BY396P	0102	BY716	0200	BYD37J	0031	BYS32	OEM	BYT61B-1000	OEM	BYV26A	OEM
BY251	0087	BY397	0102	BY717	0204	BYD37K	0707	BYS41	OEM	BYT65-600	OEM	BYV26B	OEM
BY252	0087	BY397P	0031	BY718	0206	BYD37M	0707	BYS42	OEM	BYT65-800	OEM	BYV26C	OEM
BY253	0087	BY398	0031	BY719	0206	BYD73A	OEM	BYS50	OEM	BYT65-900	OEM	BYV26D	OEM
BY254	0087	BY398P	0031	BY721	0970	BYD73B	OEM	BYS51	OEM	BYT65-1000	OEM	BYV26E	OEM
BY255	0087	BY399	0102	BY722	3733	BYD73C	OEM	BYS52-300	OEM	BYT65B-800	OEM	BYV27/50	OEM
BY256	0277	BY399P	1372	BY723	0989	BYD73D	OEM	BYS71	OEM	BYT65B-900	OEM	BYV27/100	OEM
BY257	0302	BY399S	1372	BY724	3743	BYD73E	OEM	BYS72	OEM	BYT65B-1000	OEM	BYV27/150	OEM
BY260-200	OEM	BY401	0015	BY725	0197	BYD73F	OEM	BYS76	OEM	BYT67-800	OEM	BYV27/200	OEM
BY260-400	OEM	BY402	0015	BY726	0200	BYD73G	OEM	BYS79	OEM	BYT67-1000	OEM	BYV27-15	0546
BY260-600	OEM	BY403	0015	BY727	0204	BYD74A	OEM	BYS80	OEM	BYT71-100	OEM	BYV27-50	0546
BY261-200	2363	BY404	0015	BY728	0206	BYD74B	OEM	BYS92-40	OEM	BYT71-200	OEM	BYV27-100	0546
BY261-400	2369	BY405	0015	BY729	0206	BYD74C	OEM	BYS92-45	OEM	BYT71-400	OEM	BYV27-150	0546
BY261-600	2370	BY406	0023	BY801	0084	BYD74D	OEM	BYS92-50	OEM	BYT71-600	OEM	BYV27-200	0546
BY268	0182	BY406A	0200	BY802	0090	BYD74E	OEM	BYS93-40	OEM	BYT71-800	OEM	BYV28/50	OEM
BY269	0182	BY407	0023	BY803	0097	BYD74F	OEM	BYS93-45	OEM	BYT75-200	OEM	BYV28/100	OEM
BY277-600R	OEM	BY407A	0200	BY804	0105	BYD74G	OEM	BYS93-50	OEM	BYT75-400	OEM	BYV28/150	OEM
BY277-750R	OEM	BY408	0206	BY805	0109	BYD77A	OEM	BYS94	OEM	BYT75-500	OEM	BYV28/200	OEM
BY289-150	OEM	BY409	0206	BY806	0116	BYD77B	OEM	BYS95	OEM	BYT75-600	OEM	BYV28-50	0546
BY289-200	OEM	BY409A	OEM	BY807	0122	BYD77C	OEM	BYS95-40	OEM	BYT75-800	OEM	BYV28-100	0546
BY289-300	OEM	BY410	1090	BY808	0131	BYD77D	OEM	BYS95-45	OEM	BYT75-1000	OEM	BYV28-150	0546
BY289-400	OEM	BY411	0970	BY809	0131	BYD77E	OEM	BYS95-50	OEM	BYT77	1928	BYV28-200	0546
BY289-450	OEM	BY412	3733	BY811	0084	BYD77F	OEM	BYS96	OEM	BYT78	OEM	BYV29-300	1969
BY289-600	OEM	BY413	0989	BY812	0090	BYD77G	OEM	BYS97	OEM	BYT79-300	0972	BYV29-400	1664
BY289-900	OEM	BY414	3743	BY813	0097	BYM05-50	OEM	BYS97-40	OEM	BYT79-400	0972	BYV29-500	1664
BY289-1000	OEM	BY415	0197	BY814	0105	BYM05-100	OEM	BYS97-45	OEM	BYT79-500	0972	BYV30-200	1567
BY291-75	OEM	BY416	0200	BY815	0109	BYM05-200	OEM	BYS97-50	OEM	BYT95A	0031	BYV30-300	2153
BY291-150	OEM	BY417	0204	BY816	0116	BYM05-400	OEM	BYS98-40	OEM	BYT95B	0031	BYV30-400	2153
BY291-225	OEM	BY418	0206	BY817	0122	BYM05-600	OEM	BYS98-45	OEM	BYT95C	0031	BYV30-500	OEM
BY291-300	OEM	BY419	0206	BY818	0131	BYM06-40	OEM	BYS98-50	OEM	BYT95D	0017	BYV31-300	OEM
BY291-450	OEM	BY421	0970	BY819	0131	BYM06-50	OEM	BYT01-200	1090	BYT95E	0017	BYV31-400	OEM
BY291-600	OEM	BY422	3733	BY821	0084	BYM06-100	OEM	BYT01-300	OEM	BYT230PI200	OEM	BYV31-500	OEM
BY292-75	OEM	BY423	0989	BY822	0090	BYM06-200	OEM	BYT01-400	OEM	BYT230PI300	OEM	BYV32-50	OEM
BY292-150	OEM	BY424	3743	BY823	0097	BYM06-600	OEM	BYT03-200	0546	BYT230PI400	OEM	BYV32-50F	OEM
BY292-225	OEM	BY425	0197	BY824	0105	BYM10-50	OEM	BYT03-300	1362	BYT230PI600	OEM	BYV32-100	OEM
BY292-300	OEM	BY426	0200	BY825	0109	BYM10-100	OEM	BYT03-400	1362	BYT230PI800	OEM	BYV32-100F	OEM
BY293-75	OEM	BY427	0204	BY826	0116	BYM10-200	OEM	BYT08P200	OEM	BYT230PI1000	OEM	BYV32-150	OEM
BY293-150	OEM	BY428	0206	BY827	0122	BYM10-400	OEM	BYT08P300	OEM	BYT230PIV200	OEM	BYV32-150F	OEM
BY293-225	OEM	BY429	0206	BY828	0131	BYM10-600	OEM	BYT08P400	OEM	BYT230PIV300	OEM	BYV32-200	OEM
BY293-300	OEM	BY438	OEM	BY829	0131	BYM10-800	OEM	BYT08P1200	OEM	BYT230PIV400	OEM	BYV32-200F	OEM
BY294-75	OEM	BY448	OEM	BY1001	0071	BYM10-1000	OEM	BYT08P1300	OEM	BYT230PIV600	OEM	BYV32F50	OEM
BY294-150	OEM	BY458	0102	BY1002	0071	BYM11-50	OEM	BYT08P1400	OEM	BYT230PIV800	OEM	BYV32F100	OEM
BY294-225	OEM	BY476	OEM	BY1101	0071	BYM11-100	OEM	BYT11-600	1372	BYT230PIV1000	OEM	BYV32F150	OEM
BY294-300	OEM	BY476A	OEM	BY1102	0071	BYM11-200	OEM	BYT11-800	1372	BYV10-20	0735	BYV32F200	OEM
BY294-450	OEM	BY477	OEM	BY1200	0071	BYM11-400	OEM	BYT11-1000	1372	BYV10-20A	0735	BYV33-30	2515
BY294-600	OEM	BY478	OEM	BY1201	0071	BYM11-600	OEM	BYT12-200	OEM	BYV10-30	0735	BYV33-30A	OEM
BY295-150	OEM	BY488	0097	BY1202	0071	BYM11-800	OEM	BYT12-300	OEM	BYV10-40	0735	BYV33-30F	OEM
BY295-200	OEM	BY500-100	1362	BY2001	0612	BYM11-1000	OEM	BYT12-400	OEM	BYV10-40A	OEM	BYV33-35	2515
BY295-300	OEM	BY500-200	1362	BY2002	0612	BYM12-50	OEM	BYT12-600	OEM	BYV10-60	OEM	BYV33-35A	OEM
BY295-400	OEM	BY500-400	1362	BY2101	0612	BYM12-100	OEM	BYT12-800	OEM	BYV12	0031	BYV33-35F	OEM
BY295-450	OEM	BY500-600	1372	BY2102	0612	BYM12-150	OEM	BYT12-1000	OEM	BYV13	0031	BYV33-40	2515
BY295-600	OEM	BY500-800	1372	BY2201	0612	BYM12-200	OEM	BYT12P600	OEM	BYV14	0031	BYV33-40A	OEM
BY296	0102	BY501	0084	BY2202	0612	BYM12-300	OEM	BYT12P800	OEM	BYV15	OEM	BYV33-40F	OEM
BY296P	0102	BY502	0090	BY3001	0612	BYM12-400	OEM	BYT12P1000	OEM	BYV16	OEM	BYV33-45	2515
BY297	0102	BY503	0097	BY3002	0612	BYM26A	OEM	BYT12PI600	OEM	BYV18-30	OEM	BYV33-45A	OEM
BY297P	0102	BY504	0105	BY3101	0612	BYM26B	OEM	BYT13-600	1372	BYV18-30A	OEM	BYV33-45F	OEM
BY298	0102	BY505	0109	BY3102	0612	BYM26C	OEM	BYT13-800	1372	BYV18-35	OEM	BYV33F-30	OEM
BY298P	0102	BY506	0116	BY3201	0612	BYM26D	OEM	BYT13-1000	3446	BYV18-35A	OEM	BYV33F-35	OEM
BY299	0102	BY507	0122	BY3202	0612	BYM26E	OEM	BYT16P200	OEM	BYV18-40	OEM	BYV33F-40	OEM
BY299P	0102	BY508	0131	BY4001	0586	BYM36A	0546	BYT16P300	OEM	BYV18-45	OEM	BYV33F-40A	OEM
BY300-500	OEM	BY509	0131	BY4002	0586	BYM36B	OEM	BYT16P400	OEM	BYV18-45A	OEM	BYV33F-45	OEM
BY300-600	OEM	BY510	OEM	BY4101	0586	BYM36C	OEM	BYT28-300	0972	BYV19-30	OEM	BYV33J	0031
BY300-650	OEM	BY511	0084	BY4102	0586	BYM36D	OEM	BYT28-400	0972	BYV19-30A	OEM	BYV34-300	OEM
BY300-700	OEM	BY512	0090	BY4201	0586	BYM36E	OEM	BYT28-500	0972	BYV19-35	OEM	BYV34-400	OEM
BY301	0708	BY513	0097	BY4202	0586	BYM56A	OEM	BYT30-200	OEM	BYV19-35A	OEM	BYV34-500	OEM
BY302	4140	BY514	0105	BY5001	0145	BYM56B	OEM	BYT30-300	OEM	BYV19-40	OEM	BYV36A	OEM
BY302-75	OEM	BY515	0109	BY5002	0145	BYM56C	OEM	BYT30-400	OEM	BYV19-40A	OEM	BYV36B	OEM
BY302-150	OEM	BY516	0116	BY5101	0145	BYM56D	OEM	BYT30-600	OEM	BYV19-45	OEM	BYV36C	OEM
BY302-225	OEM	BY517	0122	BY5102	0145	BYM56E	OEM	BYT30-800	OEM	BYV19-45A	OEM	BYV36D	OEM
BY302-300	OEM	BY518	0131	BY5201	0145	BYO331	0031	BYT30-1000	OEM	BYV20-30	0615	BYV36E	OEM
BY303	0580	BY519	0131	BY5202	0145	BYP21-50	0909	BYT30P200	OEM	BYV20-30A	OEM	BYV37	OEM
BY304	3743	BY521	0084	BY6431	OEM	BYP21-100	0909	BYT30P300	OEM	BYV20-35	0615	BYV38	OEM
BY305	1000	BY522	0090	BY7001	0586	BYP21-150	0909	BYT30P400	OEM	BYV20-35A	OEM	BYV39-30	OEM
BY306	2076	BY523	0097	BY7002	0586	BYP21-200	0909	BYT30P600	OEM	BYV20-40	0615	BYV39-30A	OEM
BY307	2081	BY524	0105	BY7101	0586	BYP22-50	OEM	BYT30P800	OEM	BYV20-40A	OEM	BYV39-35	OEM
BY308	2088	BY525	0109	BY7102	0586	BYP22-100	OEM	BYT30P1000	OEM	BYV20-45	0615	BYV39-35A	OEM
BY309	2088	BY526	0116	BY7201	0586	BYP22-150	OEM	BYT30PI200	OEM	BYV20-45A	OEM	BYV39-40	OEM
BY311	0708	BY527	0122	BY7202	0586	BYP22-200	OEM	BYT30PI300	OEM	BYV21-30	0615	BYV39-40A	OEM
BY312	4140	BY528	0131	BY7637	OEM	BYQ28-50	1236	BYT30PI400	OEM	BYV21-30A	OEM	BYV39-45	OEM
BY313	0580	BY529	0131	BY8001	0145	BYQ28-100	1236	BYT30PI600	OEM	BYV21-35	0615	BYV39-45A	OEM

DEVICE TYPE	REPL CODE
BYV42-50	OEM
BYV42-100	OEM
BYV42-150	OEM
BYV42-200	OEM
BYV43-30	2515
BYV43-30A	OEM
BYV43-35	2515
BYV43-35A	OEM
BYV43-40	2515
BYV43-40A	OEM
BYV43-45	2515
BYV43-45A	OEM
BYV44-300	OEM
BYV44-400	OEM
BYV44-500	OEM
BYV52-50	2234
BYV52-100	2234
BYV52-150	2234
BYV52-200	2234
BYV54-50	OEM
BYV54-100	OEM
BYV54-150	OEM
BYV54-200	OEM
BYV54V50	OEM
BYV54V100	OEM
BYV54V150	OEM
BYV54V200	OEM
BYV61	OEM
BYV62	OEM
BYV63	OEM
BYV71-100	OEM
BYV71-400	OEM
BYV71-600	OEM
BYV71-800	OEM
BYV72-50	2234
BYV72-100	2234
BYV72-150	2234
BYV72-200	2234
BYV73-30	1942
BYV73-30A	OEM
BYV73-35	1942
BYV73-35A	OEM
BYV73-40	1942
BYV73-40A	OEM
BYV73-45	1942
BYV73-45A	OEM
BYV74-300	4488
BYV74-400	4488
BYV74-500	4488
BYV79-50	3551
BYV79-100	3551
BYV79-150	3551
BYV79-200	3551
BYV92-200	1530
BYV92-300	0029
BYV92-400	0029
BYV92-500	OEM
BYV93-300	OEM
BYV93-400	OEM
BYV93-500	OEM
BYV95A	0031
BYV95B	0031
BYV95C	0031
BYV96D	OEM
BYV96E	0707
BYV950	0023
BYW07-50	OEM
BYW07-100	OEM
BYW07-150	OEM
BYW07-150A	OEM
BYW07-200	OEM
BYW08-50	OEM
BYW08-100	OEM
BYW08-150	OEM
BYW08-200	OEM
BYW14-100	OEM
BYW14-200	OEM
BYW14-400	OEM
BYW14-600	OEM
BYW14-800	OEM
BYW15-100	OEM
BYW15-200	OEM
BYW15-400	OEM
BYW15-600	OEM
BYW15-800	OEM
BYW16-100	OEM
BYW16-200	OEM
BYW16-400	OEM
BYW16-600	OEM
BYW16-800	OEM
BYW17-100	OEM
BYW17-200	OEM
BYW17-400	OEM
BYW17-600	OEM
BYW17-800	OEM
BYW17-1000	OEM
BYW17-1200	OEM
BYW18-400	OEM
BYW18-600	OEM
BYW18-800	OEM
BYW18-1000	OEM
BYW19-800	OEM
BYW19-1000	OEM
BYW25	OEM
BYW25-800	OEM
BYW25-800R	OEM
BYW25-1000	OEM
BYW25-1000R	OEM
BYW27-50	1335
BYW27-100	0015
BYW27-200	0015
BYW27-400	0015
BYW27-600	0015
BYW27-800	0071
BYW27-1000	0071
BYW29-50	1127
BYW29-100	1127
BYW29-150	1127
BYW29-200	1127
BYW29F50	OEM
BYW29F100	OEM
BYW29F150	OEM
BYW29F200	OEM
BYW30-50	OEM
BYW30-100	OEM
BYW30-150	OEM
BYW30-200	OEM
BYW31-50	OEM
BYW31-100	OEM
BYW31-150	OEM
BYW31-200	OEM
BYW32	1145
BYW33	1145
BYW34	1145
BYW35	3176
BYW36	3176
BYW51-50	OEM
BYW51-100	1236
BYW51-150	1236
BYW51-200	1236
BYW52	0102
BYW53	0102
BYW54	0102
BYW55	0102
BYW56	0102
BYW72	1145
BYW73	1145
BYW74	1145
BYW75	1191
BYW76	0023
BYW77-50	OEM
BYW77-100	OEM
BYW77-150	OEM
BYW77-150A	OEM
BYW77-200	OEM
BYW77P50	OEM
BYW77P100	OEM
BYW77P150	OEM
BYW77P200	OEM
BYW77PI50	OEM
BYW77PI100	OEM
BYW77PI150	OEM
BYW77PI200	OEM
BYW78-50	OEM
BYW78-100	OEM
BYW78-150	OEM
BYW78-200	OEM
BYW80-50	1127
BYW80-150	1127
BYW80-150A	OEM
BYW80-200	1127
BYW81-50	OEM
BYW81-100	OEM
BYW81-150	OEM
BYW81-150A	OEM
BYW81-200	OEM
BYW81P50	OEM
BYW81P100	OEM
BYW81P150	OEM
BYW81P200	OEM
BYW81PI50	OEM
BYW81PI100	OEM
BYW81PI150	OEM
BYW81PI200	OEM
BYW82	0102
BYW83	0102
BYW84	0102
BYW85	0102
BYW86	0102
BYW88-50	0084
BYW88-100	0097
BYW88-200	0097
BYW88-300	0109
BYW88-400	0109
BYW88-500	0122
BYW88-600	0122
BYW88-800	0131
BYW88-1000	0145
BYW88-1200	OEM
BYW92-50	OEM
BYW92-100	OEM
BYW92-150	OEM
BYW92-150A	OEM
BYW92-200	OEM
BYW93-50	OEM
BYW93-100	OEM
BYW93-150	OEM
BYW93-200	OEM
BYW94-50	OEM
BYW94-50U	OEM
BYW94-100	OEM
BYW94-100U	OEM
BYW94-150	OEM
BYW94-150U	OEM
BYW94-200	OEM
BYW94-200U	OEM
BYW95	0031
BYW95A	0031
BYW95B	0031
BYW95C	0031
BYW96D	1928
BYW96E	1928
BYW98-50	3114
BYW98-100	0546
BYW98-150	0546
BYW98-200	0546
BYW99-50	OEM
BYW99-100	OEM
BYW99-150	OEM
BYW99P50	OEM
BYW99P100	OEM
BYW99P150	OEM
BYW99P200	OEM
BYW100-50	0546
BYW100-100	0546
BYW100-150	0546
BYW100-200	1090
BYW29100	OEM
BYW29200	OEM
BYW81150	OEM
BYX4-1000R	0550
BYX10	0071
BYX10G	1570
BYX10GP	OEM
BYX11	OEM
BYX12/400	0071
BYX13/500	OEM
BYX13/600	0071
BYX13/1000	1119
BYX13/1500	OEM
BYX13-400	0805
BYX13-600	0319
BYX13-800	OEM
BYX13-800R	OEM
BYX13-1000	0044
BYX13-1000R	OEM
BYX13-1200	OEM
BYX13-1200R	OEM
BYX13-1600	OEM
BYX14/500	OEM
BYX14-400	OEM
BYX14-600	OEM
BYX14-800	OEM
BYX14-1000	OEM
BYX14-1200	OEM
BYX15	1119
BYX16	1132
BYX20/200	OEM
BYX21/100	OEM
BYX21/200	OEM
BYX22/200	0015
BYX22/400	0015
BYX22/600	0015
BYX22/800	0071
BYX22-200	0015
BYX22-400	3911
BYX22-600	0015
BYX22-800	0071
BYX22-1200	0102
BYX23-400	OEM
BYX23-600	OEM
BYX23-800	OEM
BYX23-1000	OEM
BYX24	0296
BYX25-600	1609
BYX25-600R	1205
BYX25-800	1610
BYX25-800R	2137
BYX25-1000	1614
BYX25-1000R	2251
BYX25-1200	OEM
BYX25-1400	OEM
BYX26-60	0087
BYX27-400	OEM
BYX27-600	OEM
BYX27-800	OEM
BYX27-1000	OEM
BYX28/200	OEM
BYX28/200A	OEM
BYX28/200B	OEM
BYX28/400	OEM
BYX28/400A	OEM
BYX28/400B	OEM
BYX28/600	2081
BYX29-75	OEM
BYX29-100	OEM
BYX29-125	OEM
BYX29-150	OEM
BYX29-75000	OEM
BYX29-100000	OEM
BYX29-125000	OEM
BYX29-150000	OEM
BYX30/150	0242
BYX30-200	OEM
BYX30-200R	OEM
BYX30-300	OEM
BYX30-300R	OEM
BYX30-400	OEM
BYX30-400R	OEM
BYX30-500	OEM
BYX30-500R	OEM
BYX30-600	OEM
BYX30-600R	1905
BYX32/200	OEM
BYX32-400	OEM
BYX32-400R	OEM
BYX32-600	6076
BYX32-600R	1476
BYX32-800	0657
BYX32-800R	0193
BYX32-1000	0131
BYX32-1000R	0193
BYX32-1200	0145
BYX32-1200R	0550
BYX32-1600	OEM
BYX32-1600R	OEM
BYX33/200	OEM
BYX33-400	OEM
BYX33-600	OEM
BYX33-800	OEM
BYX33-1000	OEM
BYX33-1200	OEM
BYX33-1600	OEM
BYX34-200	OEM
BYX34-300	OEM
BYX34-400	OEM
BYX34-500	OEM
BYX35	OEM
BYX36-150	0087
BYX36-300	0015
BYX36-600	0015
BYX38	0087
BYX38/300	2060
BYX38/300R	1245
BYX38/600	2081
BYX38/600R	1012
BYX38/900	0612
BYX38/900R	1073
BYX38/1200	2088
BYX38/1200R	1138
BYX38-300	5583
BYX38-300+	1112
BYX38-300R	1245
BYX38-300R+	1112
BYX38-600	1609
BYX38-600+	1112
BYX38-600R	1205
BYX38-600R+	0476
BYX38-900	3011
BYX38-900+	3011
BYX38-900R	0449
BYX38-900R+	0449
BYX38-1200	2088
BYX38-1200R	1138
BYX39/600	0122
BYX39/800	0131
BYX39/1000	0145
BYX39-600	0122
BYX39-600+	1112
BYX39-600R	0816
BYX39-600R+	0476
BYX39-800	0131
BYX39-800+	3011
BYX39-800R	0545
BYX39-800R+	0449
BYX39-1000	0145
BYX39-1000+	3011
BYX39-1000R	0550
BYX39-1000R+	0449
BYX39-1200	OEM
BYX39-1400	OEM
BYX40-300+	1112
BYX40-300R+	0476
BYX40-600	0122
BYX40-600+	1112
BYX40-600R	0816
BYX40-600R+	0476
BYX40-800	0131
BYX40-800+	3011
BYX40-800R	0545
BYX40-800R+	0449
BYX40-1000	0145
BYX40-1000+	3011
BYX40-1000R	0550
BYX40-1000R+	0550
BYX40-8000R	0545
BYX42/300	0109
BYX42/600	0122
BYX42/900	0145
BYX42-300	0105
BYX42-300R	0538
BYX42-600	0122
BYX42-600+	1112
BYX42-600R	0816
BYX42-600R+	0476
BYX42-900	0145
BYX42-900+	3011
BYX42-900R	0550
BYX42-900R+	0449
BYX42-1200	OEM
BYX42-1200R	OEM
BYX45-200R	OEM
BYX45-400R	OEM
BYX45-600R	0087
BYX45-800R	0087
BYX45-1000R	0087
BYX45-1200R	OEM
BYX45-1400R	OEM
BYX46-200	0871
BYX46-200R	1635
BYX46-300	0853
BYX46-300R	1251
BYX46-400	0853
BYX46-400R	1251
BYX46-500	1609
BYX46-500R	1205
BYX46-600	1609
BYX46-600R	1205
BYX47	OEM
BYX48/300	0109
BYX48/600	0122
BYX48/900	0145
BYX48-300	0105
BYX48-300+	1112
BYX48-300R	0997
BYX48-300R+	0476
BYX48-600	0122
BYX48-600+	1112
BYX48-600R	0515
BYX48-600R+	0476
BYX48-900	0145
BYX48-900+	3011
BYX48-900R	0948
BYX48-900R+	0449
BYX48-1200	OEM
BYX48-1200R	OEM
BYX49-300	OEM
BYX49-300R	OEM
BYX49-600	OEM
BYX49-600R	OEM
BYX49-900	OEM
BYX49-900R	OEM
BYX49-1200	OEM
BYX49-1200R	OEM
BYX50-200	1567
BYX50-200R	1547
BYX50-300	2153
BYX50-300R	3139
BYX50-400	0197
BYX50-400R	0997
BYX50-500	0204
BYX50-500R	0515
BYX50-600	0204
BYX50-600R	0515
BYX51-100	5582
BYX52-300	1126
BYX52-300R	1111
BYX52-600	0319
BYX52-600R	0267
BYX52-900	0044
BYX52-900R	0281
BYX52-1200	OEM
BYX53	OEM
BYX54	OEM
BYX55	OEM
BYX55-100	OEM
BYX55-200	OEM
BYX55-300	OEM
BYX55-350	0031
BYX55-350P	0031
BYX55-400	OEM
BYX55-500	OEM
BYX55-600	0031
BYX55-600P	0031
BYX56-600	0319
BYX56-600R	0267
BYX56-800	1132
BYX56-800R	1119
BYX56-1000	0044
BYX56-1000R	0281
BYX56-1200	OEM
BYX56-1400	OEM
BYX57-500	0102
BYX57-600	0102
BYX58-50	0133
BYX58-100	0133
BYX58-200	0133
BYX58-300	0023
BYX58-400	0023
BYX60-50	0015
BYX60-100	0015
BYX60-200	0015
BYX60-300	0015
BYX60-400	0015
BYX60-500	0015
BYX60-600	0015
BYX60-700	0071
BYX61-50	1567
BYX61-50U	OEM
BYX61-100	1567
BYX61-100U	OEM
BYX61-200	1567
BYX61-200U	OEM
BYX61-300	2153
BYX61-300U	OEM
BYX61-400	2153
BYX61-400U	OEM
BYX62-600	0711
BYX62-800	OEM
BYX62-1000	OEM
BYX63-600	0601
BYX63-800	OEM
BYX63-1000	OEM
BYX64-600	0601
BYX64-800	OEM
BYX64-1000	OEM
BYX65-50	1530
BYX65-50U	OEM
BYX65-100	1530
BYX65-100U	OEM
BYX65-200	1530
BYX65-200UBYX65-300	OEM
BYX65-300	0029
BYX65-300U	OEM
BYX65-400	0029
BYX65-400U	OEM
BYX66-400	0853
BYX66-500	0711
BYX66-600	0711
BYX66-700	OEM
BYX66-800	3768
BYX66-900	OEM
BYX66-1000	1614
BYX67-400	OEM
BYX67-500	0601
BYX67-600	0601
BYX67-700	OEM
BYX67-800	OEM
BYX67-900	OEM
BYX67-1000	OEM
BYX68	OEM
BYX69	OEM
BYX70-100	OEM
BYX70-300	OEM
BYX70-500	OEM
BYX70-600	OEM
BYX71-350	5915
BYX71-350R	OEM
BYX71-600	OEM
BYX71-600R	OEM
BYX72-150	OEM
BYX72-150R	OEM
BYX72-300	OEM
BYX72-300R	OEM
BYX72-500	OEM
BYX72-500R	OEM
BYX73S700R	OEM
BYX74-100	4874
BYX74-200	2230
BYX74-400	6048
BYX74-600	6062
BYX74-800	6077
BYX75	OEM
BYX76	OEM
BYX77	OEM
BYX78	OEM
BYX79	OEM
BYX80	OEM
BYX81	OEM
BYX82	0102
BYX82-50	OEM
BYX82-100	OEM
BYX82-200	OEM
BYX82-300	OEM
BYX82-400	OEM
BYX83	0102
BYX84	0102
BYX85	0102
BYX86	0102
BYX87	OEM
BYX90	OEM
BYX90G	OEM
BYX91-90K	OEM
BYX91-120K	OEM
BYX91-150K	OEM
BYX91-180K	OEM
BYX91-190K	OEM
BYX92-50	2513
BYX92-100	2513
BYX92-200	2513
BYX92-300	1362
BYX92-400	1362
BYX96-100	OEM
BYX96-200	OEM
BYX96-300	OEM
BYX96-400	OEM
BYX96-600	OEM
BYX96-800	OEM
BYX96-900	OEM
BYX96-1000	OEM
BYX96-1200	OEM
BYX96-1400	OEM
BYX96-1600	OEM
BYX97-300	2006
BYX97-600	2682
BYX97-900	OEM
BYX97-1000	OEM
BYX97-1200	OEM
BYX97-1600	OEM
BYX98-300	0109
BYX98-600	0122
BYX98-900	0145
BYX98-1200	OEM
BYX99-300	3751
BYX99-600	1112
BYX99-900	OEM
BYX99-1200	OEM
BYX38900	0122
BYX39600	0122
BYX40600	0122
BYX42600	0122
BYX48600	0122
BYX55009	0071
BYY10	OEM
BYY15	OEM
BYY16	OEM
BYY20	OEM
BYY20/200	OEM
BYY21	OEM
BYY21/200	OEM
BYY22	OEM
BYY23	OEM
BYY24	OEM
BYY25	OEM
BYY27	OEM
BYY28	OEM
BYY29	OEM
BYY30	OEM
BYY31	0015
BYY32	0015
BYY33	0015
BYY34	0015
BYY35	0071
BYY36	0071
BYY37	0071
BYY38	OEM
BYY39	OEM
BYY39/200	OEM
BYY39/400	OEM
BYY39/600	OEM
BYY39/800	OEM
BYY39/800R	OEM
BYY39/1000R	OEM
BYY39/1200R	OEM
BYY39/1400R	OEM
BYY39/1600R	OEM
BYY39/1800R	OEM
BYY39/2000R	OEM
BYY39/2200R	OEM
BYY39/2400R	OEM
BYY39-800	OEM
BYY56	0529
BYY56/1200	OEM
BYY57/75H	OEM
BYY57/150H	OEM
BYY57/300H	OEM
BYY57/500H	OEM
BYY57/700H	OEM
BYY57-75	OEM
BYY57-150	OEM
BYY57-300	OEM
BYY57-450	OEM
BYY57-600	OEM
BYY58/75H	OEM
BYY58/150H	OEM
BYY58/300H	OEM
BYY58/500H	OEM
BYY58/700H	OEM
BYY58-75	OEM
BYY58-150	OEM
BYY58-300	OEM
BYY58-450	OEM
BYY58-600	OEM
BYY59	OEM
BYY60	0087
BYY61	0087
BYY62	OEM
BYY67	OEM
BYY68	OEM
BYY69	OEM
BYY70	OEM
BYY71	OEM
BYY72	OEM
BYY73	OEM
BYY74	OEM
BYY75	OEM
BYY76	OEM
BYY77	OEM
BYY78	OEM
BYY88	0097
BYY89	0015
BYY90	OEM
BYY91	0071
BYY92	OEM
BYY93	OEM
BYY94	OEM
BYY95	OEM
BYY96	OEM
BYZ10	OEM
BYZ11	OEM
BYZ12	OEM
BYZ13	1250
BYZ14	OEM
BYZ15	OEM
BYZ16	OEM
BYZ17	OEM
BYZ18	OEM
BYZ19	OEM
BZ-050	0162
BZ050	0162
BZ-052	0162
BZ052	0157
BZ-054	0157
BZ054	0157
BZ-056	0157
BZ056	0157
BZ-058	0301
BZ058	0301
BZ-061	0636
BZ061	0636
BZ-063	0636
BZ063	0636
BZ-065	0025
BZ065	0025
BZ-067	0025
BZ067	0025
BZ-069	0025
BZ069	0025
BZ-071	0025
BZ071	0649
BZ-073	0649
BZ073	0649
BZ-075	0649

If replacement code is OEM, contact original manufacturer for replacement.

DEVICE TYPE	REPL CODE
BZ075	0649
BZ-077	0649
BZ077	0649
BZ-079	0244
BZ079	0244
BZ-080	0012
BZ080	0244
BZ-081	0244
BZ081	0244
BZ-083	0244
BZ083	0244
BZ-085	1083
BZ085	0244
BZ-088	1083
BZ088	1083
BZ-090	0012
BZ090	0012
BZ090.1Z9	0012
BZ-092	0012
BZ092	0012
BZ-094	0012
BZ094	0012
BZ-096	0170
BZ096	0170
BZ-098	0170
BZ098	0170
BZ-0800	0244
BZ-0901	0012
BZ1-90	0012
BZ3.6	0299
BZ3.9	0036
BZ4.3	0275
BZ4.7	0140
BZ5.1	0041
BZ5.6	0253
BZ6.2	0471
BZ6.8	0062
BZ7.5	0077
BZ8.2	0165
BZ9.1	0057
BZ11FB	0317
BZ-12	0137
BZ12	0137
BZ55XC20	OEM
BZ67	0016
BZ71	0016
BZ79C10	0170
BZ88/C4V7	0157
BZ-100	0170
BZ100	0170
BZ102	OEM
BZ102(0V7)	OEM
BZ102(1V4)	OEM
BZ102(2V1)	OEM
BZ102(2V8)	0677
BZ102(3V4)	0299
BZ102-0V7	OEM
BZ102-1V4	OEM
BZ102-2V1	OEM
BZ102-2V8	0677
BZ102-3V4	0299
BZ103	0062
BZ103A	0062
BZ104	0165
BZ104A	0165
BZ-105	0317
BZ105	0170
BZ105A	0064
BZ106	0052
BZ106A	0052
BZ107	0686
BZ107A	0686
BZ108	0375
BZ108A	0375
BZ109	0700
BZ109A	0700
BZ-110	0317
BZ110	0317
BZ110A	0494
BZ111	0195
BZ111A	0195
BZ112	0166
BZ112A	0166
BZ113	OEM
BZ-115	0794
BZ115	0794
BZ115-D12	OEM
BZ115-D22	OEM
BZ-120	0137
BZ120	0137
BZ-125	0365
BZ125	0365
BZ-130	0365
BZ130	0365
BZ-135	0100
BZ135	0100
BZ-140	0100
BZ140	0100
BZ-140LF	0100
BZ140LF	0100
BZ-145	0002
BZ145	0002
BZ147	0077
BZ148	0165
BZ-150	0002
BZ150	0002
BZ-157	0420
BZ157	0420
BZ-162	0420
BZ162	0420
BZ-167	1649
BZ167	1649
BZ-172	1649
BZ172	1649
BZ-177	0495
BZ177	0495
BZ-182	0495
BZ182	0495
BZ-187	0949
BZ187	0949
BZ-192	0949
BZ192	0949
BZ-197	0531
BZ197	0531
BZ-210	0531
BZ210	0531
BZ-220	0565
BZ220	0565
BZ-230	0402
BZ230	0565
BZ-240	0402
BZ240	0274
BZ-250	1606
BZ250	1606
BZ-260	0441
BZ260	0441
BZ-270	0441
BZ270	0441
BZ-280	1674
BZ280	1674
BZ-290	0726
BZ290	0726
BZ-300	0726
BZ300	0726
BZ-310	0039
BZ310	0726
BZ-320	0039
BZ320	0039
BZ-320LF	0039
BZ-330	0039
BZ330	0039
BZ-340	0820
BZ340	0039
BZ-350	0820
BZ350	0820
BZ688C5V6	OEM
BZ691C11	OEM
BZ752	OEM
BZ752F	OEM
BZ752T	OEM
BZ758	OEM
BZ758F	OEM
BZ758T	OEM
BZ821T	0471
BZ821TF	0471
BZ821TT	0471
BZ963	0052
BZ963F	0052
BZ963T	0052
BZ965	0686
BZ965F	0686
BZ965T	0686
BZ969	OEM
BZ969F	OEM
BZ969T	OEM
BZ971	OEM
BZ971F	OEM
BZ971T	OEM
BZ1021V4	0133
BZ2090	0012
BZC88C8V2	OEM
BZD10C3V3	0782
BZD10C3V6	0796
BZD10C3V9	0806
BZD10C4V3	0821
BZD10C4V7	0833
BZD10C5V1	0442
BZD10C5V6	0876
BZD10C6V2	0185
BZD10C6V8	0205
BZD10C7V5	0480
BZD10C8V2	0504
BZD10C9V1	0684
BZD10C10	0225
BZD10C11	0230
BZD10C12	0234
BZD10C13	0237
BZD10C15	0247
BZD10C16	0251
BZD10C18	0256
BZD10C20	0262
BZD10C22	0269
BZD10C24	0274
BZD10C27	0294
BZD10C30	0309
BZD10C33	0318
BZD10C36	0320
BZD10C39	0326
BZD10C43	0337
BZD10C47	0347
BZD10C51	0027
BZD10C56	0266
BZD10C62	0386
BZD10C68	0405
BZD10C75	0426
BZD10C82	0444
BZD10C91	0238
BZD10C100	1180
BZD10C110	1190
BZD10C120	1207
BZD10C130	1218
BZD10C150	0647
BZD10C160	1255
BZD10C180	1279
BZD10C200	0605
BZD23C3V9	0456
BZD23C4V3	0533
BZD23C4V7	0451
BZD23C5V1	0162
BZD23C5V6	0157
BZD23C6V2	0636
BZD23C6V8	0025
BZD23C7V5	0649
BZD23C8V2	0244
BZD23C9V1	0012
BZD23C10	0170
BZD23C11	0317
BZD23C12	0137
BZD23C13	0365
BZD23C15	0002
BZD23C16	0420
BZD23C18	0495
BZD23C20	0531
BZD23C22	0565
BZD23C24	0402
BZD23C27	0441
BZD23C30	0726
BZD23C33	0039
BZD23C36	0820
BZD23C39	0350
BZD23C43	0931
BZD23C47	0999
BZD23C51	0502
BZD23C56	1834
BZD23C62	0783
BZD23C68	2157
BZD23C75	1189
BZD23C82	0331
BZD23C91	1311
BZD23C100	0149
BZD23C110	0149
BZD23C120	0186
BZD23C130	0213
BZD23C150	0028
BZD23C160	0255
BZD23C180	0367
BZD23C200	0421
BZD23C220	OEM
BZD23C240	OEM
BZD23C270	OEM
BZD27C3V9	0456
BZD27C4V3	0533
BZD27C4V7	0451
BZD27C5V1	0162
BZD27C5V6	0157
BZD27C6V2	0636
BZD27C6V8	0025
BZD27C7V5	0649
BZD27C8V2	0244
BZD27C9V1	0012
BZD27C10	0170
BZD27C11	0317
BZD27C12	0137
BZD27C13	0365
BZD27C15	0002
BZD27C16	0420
BZD27C18	0495
BZD27C20	0531
BZD27C22	0565
BZD27C24	0402
BZD27C27	0441
BZD27C30	0726
BZD27C36	0820
BZD27C39	0350
BZD27C43	0931
BZD27C47	0999
BZD27C51	0502
BZD27C56	0869
BZD27C62	0783
BZD27C68	2157
BZD27C75	1189
BZD27C82	0331
BZD27C91	1311
BZD27C100	0098
BZD27C110	0149
BZD27C120	0186
BZD27C130	0213
BZD27C150	0028
BZD27C160	0255
BZD27C180	0367
BZD27C200	0421
BZD27C220	OEM
BZD27C240	OEM
BZD27C270	OEM
BZS2.7	0760
BZS3.3	0299
BZT03/D9V1	0684
BZT03/D10	0225
BZT03/D11	0230
BZT03/D12	0234
BZT03/D13	0237
BZT03/D15	0247
BZT03/D16	0251
BZT03/D18	0256
BZT03/D20	0262
BZT03/D22	0269
BZT03/D24	0274
BZT03/D27	0294
BZT03/D30	0309
BZT03/D33	0318
BZT03/D36	0320
BZT03/D39	0326
BZT03/D43	0337
BZT03/D47	0347
BZT03/D51	0027
BZT03/D56	0266
BZT03/D62	0386
BZT03/D68	0405
BZT03/D75	0426
BZT03/D82	0444
BZT03/D91	0238
BZT03/D100	1180
BZT03/D110	1190
BZT03/D120	1207
BZT03/D130	1218
BZT03/D150	0647
BZT03/D160	1255
BZT03/D180	1279
BZT03/D200	0605
BZT03/D220	OEM
BZT03/D240	OEM
BZT03/D270	OEM
BZT03C7V5	OEM
BZT03C8V2	OEM
BZT03C9V1	OEM
BZT03C10	OEM
BZT03C11	OEM
BZT03C12	OEM
BZT03C13	OEM
BZT03C15	OEM
BZT03C16	OEM
BZT03C18	OEM
BZT03C20	OEM
BZT03C22	OEM
BZT03C24	OEM
BZT03C27	OEM
BZT03C30	OEM
BZT03C33	OEM
BZT03C36	OEM
BZT03C39	OEM
BZT03C43	OEM
BZT03C47	OEM
BZT03C51	OEM
BZT03C56	OEM
BZT03C62	OEM
BZT03C68	OEM
BZT03C75	OEM
BZT03C82	OEM
BZT03C91	OEM
BZT03C100	OEM
BZT03C110	OEM
BZT03C120	OEM
BZT03C130	OEM
BZT03C150	OEM
BZT03C160	OEM
BZT03C180	OEM
BZT03C200	OEM
BZT03C220	OEM
BZT03C240	OEM
BZT03C270	OEM
BZV10	OEM
BZV11	OEM
BZV12	OEM
BZV13	OEM
BZV14	OEM
BZV15-C10	OEM
BZV15-C11	OEM
BZV15-C12	OEM
BZV15-C13	OEM
BZV15-C15	OEM
BZV15-C16	OEM
BZV15-C18	OEM
BZV15-C20	OEM
BZV15-C22	OEM
BZV15-C24	OEM
BZV15-C27	OEM
BZV15-C30	OEM
BZV15-C33	OEM
BZV15-C36	OEM
BZV15-C39	OEM
BZV15-C43	OEM
BZV15-C47	OEM
BZV15-C51	OEM
BZV15-C56	OEM
BZV15-C62	OEM
BZV15-C68	OEM
BZV15-C75	OEM
BZV15C10	OEM
BZV15C11	OEM
BZV15C12	OEM
BZV15C13	OEM
BZV15C15	OEM
BZV15C16	OEM
BZV15C18	OEM
BZV15C20	5949
BZV15C22	5950
BZV15C24	OEM
BZV15C27	OEM
BZV15C30	OEM
BZV15C33	OEM
BZV15C36	OEM
BZV15C39	OEM
BZV15C43	OEM
BZV15C47	OEM
BZV15C51	OEM
BZV15C56	OEM
BZV15C62	OEM
BZV15C68	OEM
BZV15C75	OEM
BZV16-C6V8	0205
BZV16-C7V5	0480
BZV16-C8V2	0504
BZV16-C9V1	0684
BZV16-C10	0225
BZV16-C11	0230
BZV16-C12	0234
BZV16-C13	0237
BZV16-C15	0247
BZV16-C16	0251
BZV16-C18	0256
BZV16-C20	0262
BZV16-C22	0269
BZV16-C24	0274
BZV16-C27	0294
BZV16-C30	0309
BZV16-C33	0318
BZV16-C36	0320
BZV16-C39	0326
BZV16-C43	0337
BZV16-C47	0347
BZV16-C51	0027
BZV16-C56	0266
BZV16-C62	0386
BZV16-C68	0405
BZV16-C75	0426
BZV16-C82	0251
BZV16-C91	0238
BZV16-C100	1180
BZV16C3V3	0782
BZV16C3V6	0796
BZV16C3V9	0806
BZV16C4V3	0821
BZV16C4V7	0833
BZV16C5V1	0442
BZV16C5V6	0876
BZV16C6V2	0185
BZV16C6V8	0205
BZV16C7V5	0480
BZV16C8V2	0504
BZV16C9V1	0684
BZV16C10	0225
BZV16C11	0230
BZV16C12	0230
BZV16C13	0237
BZV16C15	0247
BZV16C16	0251
BZV16C18	0256
BZV16C20	0262
BZV16C22	0269
BZV16C24	0274
BZV16C27	0294
BZV16C30	0309
BZV16C33	0318
BZV16C36	0320
BZV16C39	0326
BZV16C43	0337
BZV16C47	0347
BZV16C51	0027
BZV16C56	0266
BZV16C62	0386
BZV16C68	0405
BZV16C75	0426
BZV16C82	0256
BZV16C91	0238
BZV16C100	0225
BZV16C110	1190
BZV16C120	1207
BZV16C130	1218
BZV16C150	0647
BZV16C160	1255
BZV16C180	1279
BZV16C200	0605
BZV18C5V6	0157
BZV18C6V2	0636
BZV18C6V8	0025
BZV18C7V5	0649
BZV18C8V2	0244
BZV18C9V1	0012
BZV18C10	0170
BZV18C11	0317
BZV18C12	0137
BZV27	0471
BZV27A	0471
BZV28	0471
BZV28A	0471
BZV29	0471
BZV29A	0471
BZV30	0471
BZV30A	0471
BZV31	0471
BZV31A	0471
BZV32	0057
BZV32A	0057
BZV32B	0057
BZV33	0057
BZV33A	0057
BZV33B	0057
BZV34	0057
BZV34A	0057
BZV34B	0057
BZV35	0057
BZV35A	0057
BZV35B	0057
BZV36	0057
BZV36A	0057
BZV36B	0057
BZV37	OEM
BZV38	OEM
BZV40C3V3	0782
BZV40C3V6	0796
BZV40C3V9	0806
BZV40C4V3	0821
BZV40C4V7	0833
BZV40C5V1	0442
BZV40C5V6	0876
BZV40C6V2	0185
BZV40C6V8	0205
BZV40C7V5	0480
BZV40C8V2	0504
BZV40C8V7	3317
BZV40C9V1	0684
BZV40C10	0225
BZV40C11	0230
BZV40C12	0234
BZV40C13	0237
BZV40C14	1397
BZV40C15	0247
BZV40C16	0251
BZV40C17	1178
BZV40C18	0256
BZV40C19	2395
BZV40C20	0262
BZV40C22	0269
BZV40C24	0274
BZV40C25	2399
BZV40C27	0294
BZV40C28	1177
BZV40C30	0309
BZV40C33	0318
BZV40C36	0320
BZV40C39	0326
BZV40C43	0337
BZV40C47	0347
BZV40C51	0027
BZV40C56	0266
BZV40C60	2856
BZV40C62	0386
BZV40C68	0405
BZV40C75	0426
BZV40C82	0444
BZV40C87	3028
BZV40C91	0238
BZV40C100	1180
BZV40C110	1190
BZV40C120	1207
BZV40C130	1218
BZV40C140	1881
BZV40C150	0647
BZV40C160	1255
BZV40C170	2102
BZV40C180	1279
BZV40C190	2225
BZV40C200	0605
BZV41C2V7	1312
BZV41C3V0	1713
BZV41C3V3	0292
BZV41C3V6	0188
BZV41C3V9	0456
BZV41C4V3	0533
BZV41C4V7	0451
BZV41C5V1	0162
BZV41C5V6	0157
BZV41C6V2	0636
BZV41C6V8	0025
BZV41C7V5	0649
BZV41C8V2	0244
BZV41C9V1	0012
BZV41C10	0170
BZV41C12	0137
BZV41C13	0365
BZV41C15	0002
BZV41C16	0420
BZV41C18	0495
BZV41C20	0531
BZV41C22	0565
BZV41C24	0402
BZV41C27	0441
BZV41C30	0726
BZV41C33	0039
BZV41C36	0820
BZV41C39	0350
BZV41C43	0931
BZV41C47	0999
BZV43C	0152
BZV44C	0152
BZV45C	0152
BZV46	OEM
BZV46-1V5	OEM
BZV46-2V0	OEM
BZV46-C1V5	OEM
BZV46-C2V0	OEM
BZV46C1V5	OEM
BZV46C2V0	OEM
BZV47C3V3	0782
BZV47C3V6	0796
BZV47C3V9	0806
BZV47C4V3	0821
BZV47C4V7	0833
BZV47C5V1	0442
BZV47C5V6	0876
BZV47C6V2	0185
BZV47C6V8	0205
BZV47C7V5	0480
BZV47C8V2	0504
BZV47C9V1	0684
BZV47C10	0225
BZV47C11	0230
BZV47C12	0234
BZV47C13	0237
BZV47C15	0247
BZV47C16	0251
BZV47C18	0256
BZV47C20	0262
BZV47C22	0269
BZV47C24	0274
BZV47C27	0294
BZV47C30	0309
BZV47C33	0318
BZV47C36	0320
BZV47C39	0326
BZV47C43	0337
BZV47C47	0347
BZV47C51	0027
BZV47C56	0266
BZV47C62	0386
BZV47C68	0405
BZV47C75	0426
BZV47C82	0444
BZV47C91	0238
BZV47C100	1180
BZV47C110	1190
BZV47C120	1207
BZV47C130	1218
BZV47C150	0647
BZV47C160	1255
BZV47C180	1279
BZV47C200	0605
BZV48C3V3	0782
BZV48C3V6	0796
BZV48C3V9	0806
BZV48C4V3	0821
BZV48C4V7	0833
BZV48C5V1	0442
BZV48C5V6	0876
BZV48C6V2	0185
BZV48C6V8	0205
BZV48C7V5	0480
BZV48C8V2	0504
BZV48C8V7	3317
BZV48C9V1	0684
BZV48C10	0225
BZV48C11	0230
BZV48C12	0234
BZV48C13	0237
BZV48C15	0247
BZV48C16	0251
BZV48C18	0256
BZV48C20	0262
BZV48C22	0269
BZV48C24	0274
BZV48C27	0294
BZV48C30	0309
BZV48C33	0318
BZV48C36	0320
BZV48C39	0326
BZV48C43	0337
BZV48C47	0347
BZV48C51	0027
BZV48C56	0266
BZV48C62	0386
BZV48C68	0405
BZV48C75	0426
BZV48C82	0444
BZV48C91	0238
BZV48C100	1180
BZV48C110	1190
BZV48C120	1207
BZV48C130	1218
BZV48C150	0647
BZV48C160	1255
BZV48C180	1279
BZV48C200	0605
BZV49C0V8	OEM
BZV49C2V4	OEM
BZV49C2V7	OEM
BZV49C3V0	OEM
BZV49C3V3	OEM
BZV49C3V6	OEM
BZV49C3V9	OEM
BZV49C4V3	OEM
BZV49C4V7	OEM
BZV49C5V1	OEM
BZV49C5V6	OEM
BZV49C6V2	0471
BZV49C6V8	OEM
BZV49C7V5	OEM
BZV49C8V2	OEM
BZV49C9V1	0057
BZV49C10	OEM
BZV49C11	OEM
BZV49C12	0052
BZV49C13	OEM
BZV49C15	OEM
BZV49C16	OEM
BZV49C18	OEM
BZV49C20	OEM
BZV49C22	OEM
BZV49C24	OEM
BZV49C27	OEM
BZV49C30	OEM
BZV49C33	OEM
BZV49C36	OEM
BZV49C39	OEM

If replacement code is OEM, contact original manufacturer for replacement.

DEVICE TYPE	REPL CODE	DEVICE TYPE	REPL CODE	DEVICE TYPE	REPL CODE	DEVICE TYPE	REPL CODE	DEVICE TYPE	REPL CODE	DEVICE TYPE	REPL CODE	DEVICE TYPE	REPL CODE
BZV49C43	OEM	BZV60C12	0052	BZW03C27	OEM	BZW04-41	0568	BZW04P20B	OEM	BZW06-26B	OEM	BZW06P26	OEM
BZV49C47	OEM	BZV60C13	0053	BZW03C30	OEM	BZW04-41B	OEM	BZW04P23	OEM	BZW06-26V8	OEM	BZW06P26B	OEM
BZV49C51	OEM	BZV60C15	0686	BZW03C33	OEM	BZW04-44	OEM	BZW04P26	OEM	BZW06-28	OEM	BZW06P28	OEM
BZV49C56	OEM	BZV60C16	0445	BZW03C36	OEM	BZW04-44B	OEM	BZW04P26B	OEM	BZW06-28B	OEM	BZW06P28B	OEM
BZV49C62	OEM	BZV60C18	0375	BZW03C39	OEM	BZW04-45	OEM	BZW04P28	OEM	BZW06-29V1	OEM	BZW06P31	OEM
BZV49C68	OEM	BZV60C20	0700	BZW03C43	OEM	BZW04-45B	OEM	BZW04P28B	OEM	BZW06-31	OEM	BZW06P31B	OEM
BZV49C75	OEM	BZV60C22	0705	BZW03C47	OEM	BZW04-47	OEM	BZW04P31	OEM	BZW06-31B	OEM	BZW06P33	OEM
BZV53A	OEM	BZV60C24	0494	BZW03C51	OEM	BZW04-47B	OEM	BZW04P31B	OEM	BZW06-31V6	OEM	BZW06P33B	OEM
BZV53B	OEM	BZV60C27	0455	BZW03C56	OEM	BZW04-48	OEM	BZW04P33	OEM	BZW06-33	OEM	BZW06P37	OEM
BZV54A	OEM	BZV60C30	0195	BZW03C62	OEM	BZW04-48B	OEM	BZW04P37	OEM	BZW06-33B	OEM	BZW06P37B	OEM
BZV54B	OEM	BZV60C33	0166	BZW03C68	OEM	BZW04-50	OEM	BZW04P37B	OEM	BZW06-34V8	OEM	BZW06P40	OEM
BZV55C2V4	OEM	BZV60C36	0010	BZW03C75	OEM	BZW04-50B	OEM	BZW04P40	OEM	BZW06-37	OEM	BZW06P40B	OEM
BZV55C2V7	OEM	BZV60C39	0032	BZW03C82	OEM	BZW04-53	OEM	BZW04P40B	OEM	BZW06-37B	OEM	BZW06P44	OEM
BZV55C3V0	OEM	BZV60C43	0054	BZW03C91	OEM	BZW04-53B	OEM	BZW04P44	OEM	BZW06-38V1	OEM	BZW06P44B	OEM
BZV55C3V3	OEM	BZV60C47	0068	BZW03C100	OEM	BZW04-55	0831	BZW04P44B	OEM	BZW06-40	OEM	BZW06P48	OEM
BZV55C3V6	OEM	BZV60C51	0092	BZW03C110	OEM	BZW04-55B	OEM	BZW04P48	OEM	BZW06-40B	OEM	BZW06P48B	OEM
BZV55C3V9	OEM	BZV60C56	0125	BZW03C120	OEM	BZW04-58	OEM	BZW04P48B	OEM	BZW06-41V3	OEM	BZW06P53	OEM
BZV55C4V3	OEM	BZV60C62	0152	BZW03C130	OEM	BZW04-58B	OEM	BZW04P53	OEM	BZW06-44	OEM	BZW06P53B	OEM
BZV55C4V7	OEM	BZV60C68	0173	BZW03C150	OEM	BZW04-61	OEM	BZW04P53B	OEM	BZW06-44B	OEM	BZW06P58	OEM
BZV55C5V1	OEM	BZV60C75	0094	BZW03C160	OEM	BZW04-61B	OEM	BZW04P58	OEM	BZW06-45V4	OEM	BZW06P58B	OEM
BZV55C5V6	OEM	BZV69C8V2	0244	BZW03C180	OEM	BZW04-64	OEM	BZW04P58B	OEM	BZW06-48	OEM	BZW06P64	OEM
BZV55C6V2	OEM	BZV-80	OEM	BZW03C200	OEM	BZW04-64B	OEM	BZW04P64	OEM	BZW06-48B	OEM	BZW06P64B	OEM
BZV55C6V8	OEM	BZV-81	OEM	BZW03C220	OEM	BZW04-66	OEM	BZW04P64B	OEM	BZW06-50V2	OEM	BZW06P70	OEM
BZV55C7V5	OEM	BZV85C3V6	0188	BZW03C240	OEM	BZW04-66B	OEM	BZW04P70	OEM	BZW06-53	OEM	BZW06P70B	OEM
BZV55C8V2	OEM	BZV85C3V9	0456	BZW03C270	OEM	BZW04-70	OEM	BZW04P70B	OEM	BZW06-53B	OEM	BZW06P78	OEM
BZV55C9V1	OEM	BZV85C4V3	0533	BZW04-5V5	3114	BZW04-70B	OEM	BZW04P78	OEM	BZW06-58	OEM	BZW06P78B	OEM
BZV55C10	OEM	BZV85C4V7	0451	BZW04-5V8	OEM	BZW04-74	OEM	BZW04P78B	OEM	BZW06-58B	OEM	BZW06P85	OEM
BZV55C11	OEM	BZV85C5V1	0162	BZW04-5V8B	OEM	BZW04-74B	OEM	BZW04P85	OEM	BZW06-60V7	OEM	BZW06P85B	OEM
BZV55C12	OEM	BZV85C5V6	0157	BZW04-6V0	OEM	BZW04-78	OEM	BZW04P85B	OEM	BZW06-64	OEM	BZW06P94	OEM
BZV55C13	OEM	BZV85C6V2	0636	BZW04-6V0B	OEM	BZW04-78B	OEM	BZW04P94	OEM	BZW06-64B	OEM	BZW06P94B	OEM
BZV55C15	OEM	BZV85C6V8	0025	BZW04-6V4	OEM	BZW04-81	OEM	BZW04P94B	OEM	BZW06-66V4	OEM	BZW06P102	OEM
BZV55C16	OEM	BZV85C7V5	0649	BZW04-6V4B	OEM	BZW04-81B	OEM	BZW04P102	OEM	BZW06-70	OEM	BZW06P102B	OEM
BZV55C18	OEM	BZV85C8V2	0244	BZW04-6V6	OEM	BZW04-85	OEM	BZW04P102B	OEM	BZW06-70B	OEM	BZW06P111	OEM
BZV55C20	OEM	BZV85C9V1	0012	BZW04-6V6B	OEM	BZW04-85B	OEM	BZW04P111	OEM	BZW06-73V7	OEM	BZW06P111B	OEM
BZV55C22	OEM	BZV85C10	0170	BZW04-7V0	OEM	BZW04-89	OEM	BZW04P111B	OEM	BZW06-78	OEM	BZW06P128	OEM
BZV55C24	OEM	BZV85C11	0317	BZW04-7V0B	OEM	BZW04-89B	OEM	BZW04P128	OEM	BZW06-78B	OEM	BZW06P128B	OEM
BZV55C27	OEM	BZV85C12	0137	BZW04-7V4	OEM	BZW04-94	OEM	BZW04P128B	OEM	BZW06-81	OEM	BZW06P136	OEM
BZV55C30	OEM	BZV85C13	0365	BZW04-7V4B	OEM	BZW04-94B	OEM	BZW04P136	OEM	BZW06-85	OEM	BZW06P136B	OEM
BZV55C33	OEM	BZV85C15	0002	BZW04-7V8	OEM	BZW04-97	OEM	BZW04P136B	OEM	BZW06-85B	OEM	BZW06P145	OEM
BZV55C36	OEM	BZV85C16	0420	BZW04-7V8B	OEM	BZW04-97B	OEM	BZW04P145	OEM	BZW06-89V2	OEM	BZW06P145B	OEM
BZV55C39	OEM	BZV85C18	0495	BZW04-8V1	OEM	BZW04-102	OEM	BZW04P145B	OEM	BZW06-94	OEM	BZW06P154	OEM
BZV55C43	OEM	BZV85C20	0531	BZW04-8V1B	OEM	BZW04-102B	OEM	BZW04P154	OEM	BZW06-94B	OEM	BZW06P154B	OEM
BZV55C47	OEM	BZV85C22	0565	BZW04-8V5	OEM	BZW04-105	OEM	BZW04P154B	OEM	BZW06-97V2	OEM	BZW06P171	OEM
BZV55C51	OEM	BZV85C24	0402	BZW04-8V6	OEM	BZW04-105B	OEM	BZW04P171	OEM	BZW06-102	OEM	BZW06P171B	OEM
BZV55C56	OEM	BZV85C27	0441	BZW04-8V6B	OEM	BZW04-111	OEM	BZW04P171B	OEM	BZW06-102B	OEM	BZW06P188	OEM
BZV55C62	OEM	BZV85C30	0726	BZW04-8V9	OEM	BZW04-111B	OEM	BZW04P188	OEM	BZW06-105	OEM	BZW06P188B	OEM
BZV55C68	OEM	BZV85C33	1083	BZW04-8V9B	OEM	BZW04-120	OEM	BZW04P188B	OEM	BZW06-111	OEM	BZW06P213	OEM
BZV55C75	OEM	BZV85C36	0820	BZW04-9V4	OEM	BZW04-120B	OEM	BZW04P213	OEM	BZW06-111B	OEM	BZW06P213B	OEM
BZV58C3V3	0782	BZV85C39	0350	BZW04-9V4B	OEM	BZW04-121	OEM	BZW04P213B	OEM	BZW06-121	OEM	BZW06P239	OEM
BZV58C3V6	0796	BZV85C43	0931	BZW04-9V7	OEM	BZW04-121B	OEM	BZW04P239	OEM	BZW06-128	OEM	BZW06P239B	OEM
BZV58C3V9	0806	BZV85C47	0999	BZW04-9V7B	OEM	BZW04-128	OEM	BZW04P239B	OEM	BZW06-128B	OEM	BZW06P256	OEM
BZV58C4V3	0821	BZV85C51	0502	BZW04-9V9	OEM	BZW04-128B	OEM	BZW04P256	OEM	BZW06-130	OEM	BZW06P256B	OEM
BZV58C4V7	0833	BZV85C56	0869	BZW04-9V9B	OEM	BZW04-130	OEM	BZW04P256B	OEM	BZW06-136	OEM	BZW06P273	OEM
BZV58C5V1	0442	BZV85C62	0783	BZW04-10	3172	BZW04-130B	OEM	BZW04P273	OEM	BZW06-136B	OEM	BZW06P273B	OEM
BZV58C5V6	0876	BZV85C68	2157	BZW04-10B	OEM	BZW04-136	OEM	BZW04P273B	OEM	BZW06-138	OEM	BZW06P299	OEM
BZV58C6V2	0185	BZV85C75	1189	BZW04-11	OEM	BZW04-136B	OEM	BZW04P299	OEM	BZW06-145	OEM	BZW06P299B	OEM
BZV58C6V8	0205	BZV461V5	OEM	BZW04-11B	OEM	BZW04-138	OEM	BZW04P299B	OEM	BZW06-145B	OEM	BZW06P342	OEM
BZV58C7V5	0480	BZW03/D7V5	OEM	BZW04-12	OEM	BZW04-144	OEM	BZW04P342	OEM	BZW06-146	OEM	BZW06P342B	OEM
BZV58C8V2	0504	BZW03/D8V2	OEM	BZW04-12B	OEM	BZW04-145	OEM	BZW04P342B	OEM	BZW06-154	OEM	BZW06P376	OEM
BZV58C9V1	0684	BZW03/D9V1	OEM	BZW04-12V5	OEM	BZW04-145B	OEM	BZW04P376	OEM	BZW06-154B	OEM	BZW06P376B	OEM
BZV58C10	0225	BZW03/D10	OEM	BZW04-12V5B	OEM	BZW04-146	OEM	BZW04P376B	OEM	BZW06-162	1405	BZW07-10	OEM
BZV58C11	0230	BZW03/D11	OEM	BZW04-13	OEM	BZW04-154	OEM	BZW05-10B	OEM	BZW06-171	OEM	BZW07-10B	OEM
BZV58C12	0234	BZW03/D12	OEM	BZW04-13B	OEM	BZW04-154B	OEM	BZW05-35	OEM	BZW06-171B	OEM	BZW07-27	OEM
BZV58C13	0237	BZW03/D13	OEM	BZW04-13V5	OEM	BZW04-162	1405	BZW06-5V5	3114	BZW06-188	OEM	BZW07-27B	OEM
BZV58C15	0247	BZW03/D15	OEM	BZW04-13V5B	OEM	BZW04-171	OEM	BZW06-5V8	3114	BZW06-188B	OEM	BZW07-43	OEM
BZV58C16	0251	BZW03/D16	OEM	BZW04-14	3191	BZW04-171B	OEM	BZW06-5V8B	OEM	BZW06-213	OEM	BZW07-43B	OEM
BZV58C18	0256	BZW03/D18	OEM	BZW04-14B	OEM	BZW04-188	OEM	BZW06-6V0	OEM	BZW06-213B	OEM	BZW07-110	OEM
BZV58C20	0262	BZW03/D20	OEM	BZW04-15	OEM	BZW04-188B	OEM	BZW06-6V4	OEM	BZW06-239	OEM	BZW07-110B	OEM
BZV58C22	0269	BZW03/D22	OEM	BZW04-15B	OEM	BZW04-213	OEM	BZW06-6V4B	OEM	BZW06-239B	OEM	BZW10-12	OEM
BZV58C24	0274	BZW03/D24	OEM	BZW04-16	3201	BZW04-213B	OEM	BZW06-6V6	OEM	BZW06-256	OEM	BZW10-15	OEM
BZV58C27	0294	BZW03/D27	OEM	BZW04-16B	OEM	BZW04-239	OEM	BZW06-7V0	OEM	BZW06-256B	OEM	BZW11-7V0	OEM
BZV58C30	0309	BZW03/D30	OEM	BZW04-17	OEM	BZW04-239B	OEM	BZW06-7V0B	OEM	BZW06-273	OEM	BZW11-7V8	OEM
BZV58C33	0318	BZW03/D33	OEM	BZW04-17B	OEM	BZW04-256	OEM	BZW06-7V4	OEM	BZW06-273B	OEM	BZW11-8V5	OEM
BZV58C36	0320	BZW03/D36	OEM	BZW04-18	OEM	BZW04-256B	OEM	BZW06-7V8	OEM	BZW06-299	OEM	BZW11-9V4	OEM
BZV58C39	0326	BZW03/D39	OEM	BZW04-18B	OEM	BZW04-273	OEM	BZW06-7V8B	OEM	BZW06-299B	OEM	BZW11-10	OEM
BZV58C43	0337	BZW03/D43	OEM	BZW04-18V5B	OEM	BZW04-273B	OEM	BZW06-8V1	OEM	BZW06-342	OEM	BZW11-10B	OEM
BZV58C47	0347	BZW03/D47	OEM	BZW04-18V15	OEM	BZW04-299	OEM	BZW06-8V5	OEM	BZW06-342B	OEM	BZW11-11	OEM
BZV58C51	0027	BZW03/D51	OEM	BZW04-19	OEM	BZW04-299B	OEM	BZW06-8V5B	OEM	BZW06-376	OEM	BZW11-13	OEM
BZV58C56	0266	BZW03/D56	OEM	BZW04-19B	OEM	BZW04-342	OEM	BZW06-8V9	OEM	BZW06-376B	OEM	BZW11-14	OEM
BZV58C62	0386	BZW03/D62	OEM	BZW04-20	OEM	BZW04-342B	OEM	BZW06-9V4	OEM	BZW06P5V8	OEM	BZW11-15	OEM
BZV58C68	0405	BZW03/D68	OEM	BZW04-20B	OEM	BZW04-376	OEM	BZW06-9V4B	OEM	BZW06P5V8B	OEM	BZW11-17	OEM
BZV58C75	0426	BZW03/D75	OEM	BZW04-22	1915	BZW04-376B	OEM	BZW06-9V7	OEM	BZW06P6V4	OEM	BZW11-19	OEM
BZV58C82	0444	BZW03/D82	OEM	BZW04-22B	OEM	BZW04P5V8	OEM	BZW06-10	3172	BZW06P6V4B	OEM	BZW11-20	OEM
BZV58C91	0238	BZW03/D91	OEM	BZW04-23	OEM	BZW04P5V8B	OEM	BZW06-10B	OEM	BZW06P7V0	OEM	BZW11-23	OEM
BZV58C100	1180	BZW03/D100	OEM	BZW04-23B	OEM	BZW04P6V4	OEM	BZW06-10V5	OEM	BZW06P7V0B	OEM	BZW11-26	OEM
BZV58C110	1190	BZW03/D110	OEM	BZW04-24	OEM	BZW04P6V4B	OEM	BZW06-11	OEM	BZW06P7V8	OEM	BZW11-27	OEM
BZV58C120	1207	BZW03/D120	OEM	BZW04-24B	OEM	BZW04P7V0	OEM	BZW06-11B	OEM	BZW06P7V8B	OEM	BZW11-27B	OEM
BZV58C130	1218	BZW03/D130	OEM	BZW04-26	OEM	BZW04P7V0B	OEM	BZW06-12V1	OEM	BZW06P8V5	OEM	BZW11-28	OEM
BZV58C150	0647	BZW03/D150	OEM	BZW04-26B	OEM	BZW04P7V8	OEM	BZW06-13	OEM	BZW06P8V5B	OEM	BZW11-31	OEM
BZV58C160	1255	BZW03/D160	OEM	BZW04-27	OEM	BZW04P7V8B	OEM	BZW06-13B	OEM	BZW06P9V4	OEM	BZW11-33	OEM
BZV58C180	1279	BZW03/D180	OEM	BZW04-27B	OEM	BZW04P8V5	OEM	BZW06-13V1	OEM	BZW06P9V4B	OEM	BZW11-37	OEM
BZV58C200	0605	BZW03/D200	OEM	BZW04-28	OEM	BZW04P8V5B	OEM	BZW06-14	3191	BZW06P10	OEM	BZW11-40	OEM
BZV60C2V4	1276	BZW03/D220	OEM	BZW04-28B	OEM	BZW04P9V4	OEM	BZW06-14B	OEM	BZW06P10B	OEM	BZW11-43	OEM
BZV60C2V7	0760	BZW03/D240	OEM	BZW04-29	OEM	BZW04P9V4B	OEM	BZW06-14V5	OEM	BZW06P11	OEM	BZW11-43B	OEM
BZV60C3V0	0118	BZW03/D270	OEM	BZW04-29B	OEM	BZW04P10	OEM	BZW06-15	OEM	BZW06P11B	OEM	BZW11-44	OEM
BZV60C3V3	0299	BZW03C6V8	OEM	BZW04-31	OEM	BZW04P10B	OEM	BZW06-15B	OEM	BZW06P13	OEM	BZW11-48	OEM
BZV60C3V6	0376	BZW03C7V5	OEM	BZW04-31B	OEM	BZW04P11	OEM	BZW06-16V2	OEM	BZW06P13B	OEM	BZW11-53	OEM
BZV60C3V9	0036	BZW03C8V2	OEM	BZW04-32	OEM	BZW04P11B	OEM	BZW06-17	OEM	BZW06P14	OEM	BZW11-58	OEM
BZV60C4V3	0275	BZW03C9V1	OEM	BZW04-32B	OEM	BZW04P13	OEM	BZW06-17B	OEM	BZW06P14B	OEM	BZW11-64	OEM
BZV60C4V7	0140	BZW03C10	OEM	BZW04-33	OEM	BZW04P13B	OEM	BZW06-19	OEM	BZW06P15	OEM	BZW11-70	OEM
BZV60C5V1	0041	BZW03C11	OEM	BZW04-33B	OEM	BZW04P14	OEM	BZW06-19B	OEM	BZW06P15B	OEM	BZW11-78	OEM
BZV60C5V6	0253	BZW03C12	OEM	BZW04-35	OEM	BZW04P14B	OEM	BZW06-19V4	OEM	BZW06P17	OEM	BZW11-85	OEM
BZV60C6V2	0471	BZW03C13	OEM	BZW04-35B	OEM	BZW04P15	OEM	BZW06-20	OEM	BZW06P17B	OEM	BZW11-94	OEM
BZV60C6V8	0062	BZW03C15	OEM	BZW04-37	OEM	BZW04P15B	OEM	BZW06-20B	OEM	BZW06P19	OEM	BZW11-102	OEM
BZV60C7V5	0077	BZW03C16	OEM	BZW04-37B	OEM	BZW04P17	OEM	BZW06-21V8	OEM	BZW06P19B	OEM	BZW11-110	OEM
BZV60C8V2	0165	BZW03C18	OEM	BZW04-38	OEM	BZW04P17B	OEM	BZW06-23	OEM	BZW06P20	OEM	BZW11-110B	OEM
BZV60C9V1	0057	BZW03C20	OEM	BZW04-38B	OEM	BZW04P19	OEM	BZW06-23B	OEM	BZW06P20B	OEM	BZW11-128	OEM
BZV60C10	0064	BZW03C22	OEM	BZW04-40	OEM	BZW04P19B	OEM	BZW06-24V3	OEM	BZW06P23	OEM	BZW11-136	OEM
BZV60C11	0181	BZW03C24	OEM	BZW04-40B	OEM	BZW04P20	OEM	BZW06-26	OEM	BZW06P23B	OEM	BZW11-145	OEM

If replacement code is OEM, contact original manufacturer for replacement.

DEVICE TYPE	REPL CODE
BZW11-154	OEM
BZW11-171	OEM
BZW12-11	OEM
BZW12-13	OEM
BZW12-18	OEM
BZW12-22	OEM
BZW12-27	OEM
BZW12-30	OEM
BZW12-36	OEM
BZW12-47	OEM
BZW12-56	OEM
BZW12-68	OEM
BZW12-82	OEM
BZW12-91	OEM
BZW12-110	OEM
BZW12-150	OEM
BZW12-180	OEM
BZW12-220	OEM
BZW12-270	OEM
BZW12-300	OEM
BZW12-360	OEM
BZW12-390	OEM
BZW13-11	OEM
BZW13-13	OEM
BZW13-18	OEM
BZW13-22	OEM
BZW13-27	OEM
BZW13-30	OEM
BZW13-36	OEM
BZW13-47	OEM
BZW13-56	OEM
BZW13-68	OEM
BZW13-82	OEM
BZW13-91	OEM
BZW13-110	OEM
BZW13-150	OEM
BZW13-180	OEM
BZW13-220	OEM
BZW13-270	OEM
BZW13-300	OEM
BZW13-360	OEM
BZW13-390	OEM
BZW14	OEM
BZW20	OEM
BZW22C1	OEM
BZW22C2V7	OEM
BZW22C3V0	OEM
BZW22C3V3	0782
BZW22C3V6	0796
BZW22C3V9	0806
BZW22C4V3	0821
BZW22C4V7	0833
BZW22C5V1	0442
BZW22C5V6	0876
BZW22C6V2	0185
BZW22C6V8	0205
BZW22C7V5	0480
BZW22C8V2	0504
BZW22C9V1	0684
BZW22C10	0225
BZW22C11	0230
BZW22C12	0234
BZW22C13	0237
BZW22C15	0247
BZW22C16	0251
BZW22C18	0256
BZW22C20	0262
BZW22C22	0269
BZW22C24	0274
BZW22C27	0294
BZW22C30	0309
BZW22C33	0318
BZW22C36	0320
BZW22C39	0326
BZW22C43	0337
BZW22C47	0347
BZW22C51	0027
BZW22C56	0266
BZW22C62	0386
BZW22C68	0405
BZW22C75	0426
BZW22C82	0444
BZW22C91	0238
BZW22C100	1180
BZW22C110	1190
BZW22C120	1207
BZW22C130	1218
BZW22C150	0647
BZW22C160	1255
BZW22C180	1279
BZW22C200	0605
BZW25-12	OEM
BZW25-12B	OEM
BZW25-24	OEM
BZW25-24B	OEM
BZW25-47	OEM
BZW25-47B	OEM
BZW25-120	OEM
BZW25-120B	OEM
BZW30-12	OEM
BZW30-13	OEM
BZW30-15	OEM
BZW30-16	OEM
BZW30-18	OEM
BZW30-19	OEM
BZW30-22	OEM
BZW30-24	OEM
BZW30-27	OEM
BZW30-29	OEM

DEVICE TYPE	REPL CODE
BZW30-32	OEM
BZW30-35	OEM
BZW30-38	OEM
BZW30-41	OEM
BZW30-45	OEM
BZW30-47	OEM
BZW30-50	OEM
BZW30-55	OEM
BZW30-61	OEM
BZW30-66	OEM
BZW30-74	OEM
BZW30-81	OEM
BZW30-89	OEM
BZW30-97	OEM
BZW30-105	OEM
BZW30-120	OEM
BZW30-121	OEM
BZW30-130	OEM
BZW30-138	OEM
BZW30-146	OEM
BZW30-162	OEM
BZW30-175	OEM
BZW30-202	OEM
BZW30-243	OEM
BZW30-283	OEM
BZW30-324	OEM
BZW30-405	OEM
BZW30-486	OEM
BZW50-8V2	OEM
BZW50-10	OEM
BZW50-12	OEM
BZW50-12B	OEM
BZW50-15	OEM
BZW50-15B	OEM
BZW50-18	OEM
BZW50-18B	OEM
BZW50-22	OEM
BZW50-22B	OEM
BZW50-27	OEM
BZW50-27B	OEM
BZW50-33	OEM
BZW50-33B	OEM
BZW50-39	OEM
BZW50-39B	OEM
BZW50-47	OEM
BZW50-47B	OEM
BZW50-56	OEM
BZW50-56B	OEM
BZW50-68	OEM
BZW50-68B	OEM
BZW50-82	OEM
BZW50-82B	OEM
BZW50-100	OEM
BZW50-100B	OEM
BZW50-120	OEM
BZW50-120B	OEM
BZW50-150	OEM
BZW50-150B	OEM
BZW50-180	OEM
BZW50-180B	OEM
BZW70-3	OEM
BZW70-5V6	OEM
BZW70-6V2	OEM
BZW70-6V8	3114
BZW70-7V5	OEM
BZW70-8V2	OEM
BZW70-9V1	OEM
BZW70-10	OEM
BZW70-11	OEM
BZW70-12	OEM
BZW70-13	3172
BZW70-15	OEM
BZW70-16	3201
BZW70-18	3191
BZW70-20	3201
BZW70-22	OEM
BZW70-24	OEM
BZW70-27	1915
BZW70-30	OEM
BZW70-33	OEM
BZW70-36	OEM
BZW70-39	OEM
BZW70-43	OEM
BZW70-47	OEM
BZW70-51	0568
BZW70-56	OEM
BZW70-62	OEM
BZW86-7V5	OEM
BZW86-8V2	OEM
BZW86-9V1	OEM
BZW86-10	OEM
BZW86-11	OEM
BZW86-12	OEM
BZW86-13	OEM
BZW86-15	OEM
BZW86-16	OEM
BZW86-18	OEM
BZW86-20	OEM
BZW86-22	OEM
BZW86-24	OEM
BZW86-27	OEM
BZW86-30	OEM
BZW86-33	OEM
BZW86-36	OEM
BZW86-39	OEM
BZW86-43	OEM
BZW86-47	OEM
BZW86-51	OEM
BZW86-56	OEM

DEVICE TYPE	REPL CODE
BZW86-62	OEM
BZW86-R	OEM
BZW86C7V5	OEM
BZW86C8V2	OEM
BZW86C9V1	OEM
BZW86C10	OEM
BZW86C11	OEM
BZW86C12	OEM
BZW86C13	OEM
BZW86C15	OEM
BZW86C16	OEM
BZW86C18	OEM
BZW86C20	OEM
BZW86C22	OEM
BZW86C24	OEM
BZW86C27	OEM
BZW86C30	OEM
BZW86C33	OEM
BZW86C36	OEM
BZW86C39	OEM
BZW86C43	OEM
BZW86C47	OEM
BZW86C51	OEM
BZW86C56	OEM
BZW91	OEM
BZW91-5V6	OEM
BZW91-6V2	OEM
BZW91-6V8	OEM
BZW91-7V5	OEM
BZW91-8V2	OEM
BZW91-9V1	OEM
BZW91-10	OEM
BZW91-11	OEM
BZW91-12	OEM
BZW91-13	OEM
BZW91-15	OEM
BZW91-16	OEM
BZW91-18	OEM
BZW91-20	OEM
BZW91-22	OEM
BZW91-24	OEM
BZW91-27	OEM
BZW91-30	OEM
BZW91-33	OEM
BZW91-36	OEM
BZW91-39	OEM
BZW91-43	OEM
BZW91-47	OEM
BZW91-51	OEM
BZW91-56	OEM
BZW91-62	OEM
BZW91-R	OEM
BZW93-5V6	OEM
BZW93-6V2	OEM
BZW93-6V8	OEM
BZW93-7V5	OEM
BZW93-8V2	OEM
BZW93-9V1	OEM
BZW93-10	OEM
BZW93-11	OEM
BZW93-12	OEM
BZW93-13	OEM
BZW93-15	OEM
BZW93-16	OEM
BZW93-18	OEM
BZW93-20	OEM
BZW93-22	OEM
BZW93-24	OEM
BZW93-27	OEM
BZW93-30	OEM
BZW93-33	OEM
BZW93-36	OEM
BZW93-39	OEM
BZW93-43	OEM
BZW93-47	OEM
BZW93-51	OEM
BZW93-56	OEM
BZW93-62	OEM
BZW93-R	OEM
BZW95-6V2	OEM
BZW95-6V8	OEM
BZW95-7V5	OEM
BZW95-8V2	OEM
BZW95-9V1	OEM
BZW95-10	OEM
BZW95-11	OEM
BZW95-12	OEM
BZW95-13	OEM
BZW95-15	OEM
BZW95-16	OEM
BZW95-18	OEM
BZW95-20	OEM
BZW95-22	OEM
BZW95-24	OEM
BZW95-27	OEM
BZW95-30	OEM
BZW95-33	OEM
BZW95-36	OEM
BZW95-39	OEM
BZW95-43	OEM
BZW95-47	OEM
BZW95-51	OEM
BZW95-56	OEM
BZW95-62	OEM
BZW96-3V9	OEM
BZW96-4V3	OEM
BZW96-4V7	OEM
BZW96-5V1	OEM
BZW96-5V6	OEM

DEVICE TYPE	REPL CODE
BZW96-6V2	OEM
BZW96-6V8	OEM
BZW96-7V5	OEM
BZX9V1	OEM
BZX10	0471
BZX11	0062
BZX12	0077
BZX13	0165
BZX14	0057
BZX15	0064
BZX16	0181
BZX17	0052
BZX18	0053
BZX19	0686
BZX20	0445
BZX21	0375
BZX22	0700
BZX23	0705
BZX24	0705
BZX25	0441
BZX26	0257
BZX27	0039
BZX29-C5V1	0162
BZX29-C5V6	0157
BZX29-C6V2	0636
BZX29-C6V8	0025
BZX29-C7V5	0649
BZX29-C8V2	0244
BZX29-C9V1	0012
BZX29-C10	0170
BZX29-C11	0317
BZX29-C12	0137
BZX29-C15	0002
BZX29-C16	0420
BZX29-C18	0495
BZX29-C20	0531
BZX29-C22	0565
BZX29-C24	0402
BZX29-C27	0441
BZX29-C30	0726
BZX29-C33	0039
BZX29-C36	0820
BZX29-C39	0350
BZX29-C43	0931
BZX29-C47	0999
BZX29-C51	0502
BZX29-C56	0869
BZX29C4V7	0451
BZX29C5V1	0162
BZX29C5V6	0157
BZX29C6V2	0636
BZX29C6V8	0025
BZX29C7V5	0649
BZX29C8V2	0244
BZX29C9V1	0012
BZX29C10	0170
BZX29C11	0317
BZX29C12	0137
BZX29C13	0365
BZX29C15	0002
BZX29C16	0420
BZX29C18	0495
BZX29C20	0531
BZX29C22	0565
BZX29C24	0402
BZX29C27	0441
BZX29C30	0726
BZX29C33	0039
BZX29C36	0820
BZX29C39	0350
BZX29C43	0931
BZX29C47	0999
BZX29C51	0502
BZX29C56	0869
BZX29C82	0331
BZX29C100	0098
BZX30-C3V3	OEM
BZX30-C3V6	OEM
BZX30-C3V9	OEM
BZX30-C4V3	OEM
BZX30-C4V7	OEM
BZX30-C5V1	OEM
BZX30-C5V6	OEM
BZX30-C6V2	OEM
BZX30-C6V8	OEM
BZX30-C7V5	OEM
BZX30-C8V2	OEM
BZX30-C9V1	OEM
BZX30-C10	OEM
BZX30-C11	OEM
BZX30-C12	OEM
BZX30-C13	OEM
BZX30-C15	OEM
BZX30-C16	OEM
BZX30-C18	OEM
BZX30-C20	OEM
BZX30-C22	OEM
BZX30-C24	OEM
BZX30-C27	OEM
BZX30-D15	OEM
BZX30-D18	OEM
BZX30-D22	OEM
BZX30-D27	OEM
BZX31-C3V6	OEM
BZX31-C3V9	OEM
BZX31-C4V3	OEM
BZX33	OEM
BZX33-C4V7	OEM
BZX34	OEM

DEVICE TYPE	REPL CODE
BZX34-C5V1	OEM
BZX35	OEM
BZX35-C5V6	OEM
BZX36	OEM
BZX36-C6V2	OEM
BZX36-C6V8	OEM
BZX36-C7V5	OEM
BZX36-C8V2	OEM
BZX36-C9V1	OEM
BZX43	OEM
BZX44	OEM
BZX45	OEM
BZX46-C2V7	0760
BZX46-C3V0	0118
BZX46-C3V3	0299
BZX46-C3V6	0376
BZX46-C3V9	0036
BZX46-C4V3	0275
BZX46-C4V7	0140
BZX46-C5V1	0041
BZX46-C5V6	0253
BZX46-C6V2	0471
BZX46-C6V8	0062
BZX46-C7V5	0077
BZX46-C8V2	0165
BZX46-C9V1	0057
BZX46-C10	0064
BZX46-C11	0181
BZX46-C12	0052
BZX46-C13	0053
BZX46-C15	0686
BZX46-C16	0445
BZX46-C18	0375
BZX46-C20	0700
BZX46-C22	0705
BZX46-C24	0494
BZX46-C27	0455
BZX46-C30	0195
BZX46-C33	0166
BZX46-C36	0010
BZX46-C39	0032
BZX46-C43	0054
BZX46-C47	0068
BZX46-C51	0092
BZX46-C56	0125
BZX46-C62	0152
BZX46C2V7	0760
BZX46C3V	0118
BZX46C3V0	0118
BZX46C3V3	0299
BZX46C3V6	0376
BZX46C3V9	0036
BZX46C4V3	0275
BZX46C4V7	0140
BZX46C5V1	0041
BZX46C5V6	0253
BZX46C6V2	0471
BZX46C6V8	0062
BZX46C7V5	0077
BZX46C8V2	0165
BZX46C9V1	0057
BZX46C10	0064
BZX46C11	0181
BZX46C12	0052
BZX46C13	0053
BZX46C15	0686
BZX46C16	0445
BZX46C18	0375
BZX46C20	0700
BZX46C22	0705
BZX46C24	0494
BZX46C27	0455
BZX46C30	0195
BZX46C33	0166
BZX46C36	0010
BZX46C39	0032
BZX46C43	0054
BZX46C47	0068
BZX46C51	0092
BZX46C56	0125
BZX46C62	0152
BZX46C68	0173
BZX46C75	0094
BZX46C82	0049
BZX46C91	0156
BZX46C100	0189
BZX46C110	0099
BZX47	0062
BZX48	OEM
BZX49	OEM
BZX50	OEM
BZX51	0165
BZX52	0165
BZX53	0165
BZX54	0165
BZX55-C0V8	OEM
BZX55-C2V4	1276
BZX55-C2V7	0760
BZX55-C3V0	0118
BZX55-C3V3	0299
BZX55-C3V6	0376
BZX55-C3V9	0036
BZX55-C4V3	0275
BZX55-C4V7	0140
BZX55-C5V1	0041
BZX55-C5V6	0253
BZX55-C6V2	0471
BZX55-C6V8	0062
BZX55-C7V5	0077

DEVICE TYPE	REPL CODE
BZX55-C8V2	0165
BZX55-C9V1	0057
BZX55-C10	0064
BZX55-C11	0181
BZX55-C12	0052
BZX55-C13	0053
BZX55-C15	0686
BZX55-C15V	OEM
BZX55-C16	0445
BZX55-C18	0375
BZX55-C20	0700
BZX55-C22	0705
BZX55-C24	0494
BZX55-C27	0455
BZX55-C30	0195
BZX55-C33	0166
BZX55-C36	0010
BZX55-C39	0010
BZX55-C43	0054
BZX55-C47	0068
BZX55-C51	0092
BZX55-C56	0125
BZX55-C62	0152
BZX55C08V	OEM
BZX55C0V8	OEM
BZX55C2V4	1276
BZX55C2V7	0760
BZX55C3V	0118
BZX55C3V0	0118
BZX55C3V3	0299
BZX55C3V6	0376
BZX55C3V9	0036
BZX55C4V3	0275
BZX55C4V7	0140
BZX55C5V1	0041
BZX55C5V6	0253
BZX55C6V2	0471
BZX55C6V8	0062
BZX55C7V5	0077
BZX55C8V2	0165
BZX55C9V1	0057
BZX55C10	0064
BZX55C10V	0064
BZX55C11	0181
BZX55C11V	0181
BZX55C12	0052
BZX55C12V	0052
BZX55C13	0053
BZX55C13V	0053
BZX55C15	0686
BZX55C15V	0686
BZX55C16	0445
BZX55C16V	0445
BZX55C18	0375
BZX55C18V	0375
BZX55C20	0700
BZX55C20V	0700
BZX55C22	0705
BZX55C22V	0705
BZX55C24	0494
BZX55C24V	0494
BZX55C27	0455
BZX55C27V	0455
BZX55C30	0195
BZX55C30V	0195
BZX55C33	0166
BZX55C33V	0166
BZX55C36	0010
BZX55C39	0032
BZX55C43	2317
BZX55C47	0068
BZX55C51	0092
BZX55C56	0125
BZX55C62	0152
BZX55C68	0173
BZX55C75	0094
BZX55C82	0049
BZX55C91	0156
BZX55C100	0189
BZX55C110	0099
BZX55C120	0287
BZX55C130	0287
BZX55C150	0340
BZX55C160	0370
BZX55C180	0425
BZX55C200	1473
BZX55D5V6	0876
BZX55D6V8	0205
BZX55D8V2	0504
BZX55D10	0225
BZX55V36	OEM
BZX57	OEM
BZX58-C6V8	0062
BZX58-C7V5	0077
BZX58-C8V2	0165
BZX58-C9V1	0057
BZX58-C10	0064
BZX58C6V8	0062
BZX58C7V5	0077
BZX58C8V2	0165
BZX58C9V1	0057
BZX58C10	0064
BZX59-C11	0181
BZX59-C13	0053
BZX59-C15	0686
BZX59-C16	0445
BZX59-C18	0375
BZX59-C20	0700
BZX59-C22	0705

DEVICE TYPE	REPL CODE
BZX59-C24	0494
BZX59-C27	0455
BZX59C11	0181
BZX59C12	0052
BZX59C13	0053
BZX59C15	0686
BZX59C16	0445
BZX59C18	0375
BZX59C20	0700
BZX59C22	0705
BZX59C24	0494
BZX59C27	0455
BZX60-C30	0195
BZX60-C33	0166
BZX60-C36	0010
BZX60-C39	0032
BZX60-C43	0054
BZX60-C47	0068
BZX60-C51	0092
BZX60-C56	0125
BZX60C30	0195
BZX60C33	0166
BZX60C36	0010
BZX60C39	0032
BZX60C43	0054
BZX60C47	0068
BZX60C51	0092
BZX60C56	0125
BZX61-C3V6	0188
BZX61-C3V9	0456
BZX61-C4V3	0533
BZX61-C4V7	0451
BZX61-C5V1	0162
BZX61-C5V6	0157
BZX61-C6V2	0636
BZX61-C6V8	0025
BZX61-C7V5	0649
BZX61-C8V2	0244
BZX61-C9V1	0012
BZX61-C10	0170
BZX61-C11	0317
BZX61-C12	0137
BZX61-C13	0365
BZX61-C15	0002
BZX61-C16	0420
BZX61-C18	0495
BZX61-C20	0531
BZX61-C22	0565
BZX61-C24	0402
BZX61-C27	0441
BZX61-C30	0726
BZX61-C33	0039
BZX61-C36	0820
BZX61-C39	0350
BZX61-C43	0931
BZX61-C47	0999
BZX61-C51	0502
BZX61-C56	0869
BZX61-C62	0783
BZX61-C68	2157
BZX61-C75	1189
BZX61-C82	0331
BZX61C3V6	0796
BZX61C3V9	0806
BZX61C4V3	0821
BZX61C4V7	0833
BZX61C5V1	0442
BZX61C5V6	0876
BZX61C6V2	0185
BZX61C6V8	0205
BZX61C7V5	0480
BZX61C8V2	0504
BZX61C9	0225
BZX61C9V1	0684
BZX61C10	0225
BZX61C11	0230
BZX61C12	0234
BZX61C13	0237
BZX61C15	0247
BZX61C16	0251
BZX61C18	0256
BZX61C20	0531
BZX61C22	0269
BZX61C24	0274
BZX61C27	0294
BZX61C30	0309
BZX61C33	0318
BZX61C36	0320
BZX61C39	0326
BZX61C43	0337
BZX61C47	0347
BZX61C51	0027
BZX61C56	0266
BZX61C62	0386
BZX61C68	0405
BZX61C75	0426
BZX61C82	0444
BZX61C91	0238
BZX61C100	1180
BZX61C110	1190
BZX61C120	1207
BZX61C130	1218
BZX61C150	0647
BZX61C160	1255
BZX61C180	1279
BZX61C200	0605
BZX62	OEM
BZX63	OEM
BZX63C8V2	0244

If replacement code is OEM, contact original manufacturer for replacement.

DEVICE TYPE	REPL CODE
BZX63C9V1	0012
BZX64	OEM
BZX64C11	0317
BZX64C12	0137
BZX64C13	0365
BZX64C15	0002
BZX64C16	0420
BZX64C18	0420
BZX64C20	0531
BZX64C22	0565
BZX64C24	0402
BZX64C27	0441
BZX65	OEM
BZX65C30	0726
BZX65C33	0039
BZX65C36	0820
BZX65C39	0350
BZX65C43	0931
BZX65C47	0999
BZX65C51	0502
BZX65C56	0869
BZX66	0025
BZX67-C12	OEM
BZX67-C13	OEM
BZX67-C15	OEM
BZX67-C16	OEM
BZX67-C18	OEM
BZX67-C20	OEM
BZX67-C22	OEM
BZX67-C24	OEM
BZX67-C27	OEM
BZX67-C30	OEM
BZX67-C33	OEM
BZX67-C36	OEM
BZX67-C39	OEM
BZX67-C43	OEM
BZX67-C47	OEM
BZX67-C51	OEM
BZX67-C56	OEM
BZX67-C62	OEM
BZX67-C68	OEM
BZX67-C75	OEM
BZX67-C82	OEM
BZX67-C91	OEM
BZX67-C100	OEM
BZX67-C120	OEM
BZX67-C130	OEM
BZX67-C150	OEM
BZX67-C160	OEM
BZX67-C180	OEM
BZX67-C200	OEM
BZX67C12	OEM
BZX67C13	OEM
BZX67C15	OEM
BZX67C16	OEM
BZX67C18	OEM
BZX67C20	OEM
BZX67C22	OEM
BZX67C24	OEM
BZX67C27	OEM
BZX67C30	OEM
BZX67C33	OEM
BZX67C36	OEM
BZX67C39	OEM
BZX67C43	OEM
BZX67C47	OEM
BZX67C51	OEM
BZX67C56	OEM
BZX67C62	OEM
BZX67C68	OEM
BZX67C75	OEM
BZX67C82	OEM
BZX67C91	OEM
BZX67C100	OEM
BZX67C110	OEM
BZX67C120	OEM
BZX67C130	OEM
BZX67C150	OEM
BZX67C160	OEM
BZX67C180	OEM
BZX67C200	OEM
BZX68-62A	OEM
BZX68-62B	OEM
BZX68-62C	OEM
BZX68C62A	OEM
BZX68C62B	OEM
BZX68C62C	OEM
BZX69-C7V5	0077
BZX69-C8V2	0165
BZX69-C9V2	0057
BZX69-C10	0064
BZX69-C11	0181
BZX69-C12	0052
BZX69C7V5	0649
BZX69C10	0170
BZX69C11	0317
BZX69C12	0137
BZX70-C7V5	0480
BZX70-C8V2	0504
BZX70-C9V2	0684
BZX70-C10	0225
BZX70-C11	0230
BZX70-C12	0234
BZX70-C13	0237
BZX70-C15	0247
BZX70-C16	0251
BZX70-C18	0256
BZX70-C20	0262
BZX70-C22	0269
BZX70-C24	0274
BZX70-C27	0294
BZX70-C30	0309
BZX70-C33	0318
BZX70-C36	0320
BZX70-C39	0326
BZX70-C43	0337
BZX70-C47	0347
BZX70-C51	0027
BZX70-C56	0266
BZX70-C62	0386
BZX70-C68	0405
BZX70-C75	0426
BZX70C7V5	0480
BZX70C8V2	0504
BZX70C9V1	0684
BZX70C10	0225
BZX70C11	0230
BZX70C12	0234
BZX70C13	0237
BZX70C15	0247
BZX70C16	0251
BZX70C18	0256
BZX70C20	0262
BZX70C22	0269
BZX70C24	0274
BZX70C27	0294
BZX70C30	0309
BZX70C33	0318
BZX70C36	0320
BZX70C39	0326
BZX70C43	0337
BZX70C47	0347
BZX70C51	0027
BZX70C56	0266
BZX70C62	0386
BZX70C68	0405
BZX70C75	0426
BZX71-B5V1	0041
BZX71-B5V6	0253
BZX71-B6V2	0471
BZX71-B6V8	0062
BZX71-B7V5	0077
BZX71-B8V2	0165
BZX71-B9V1	0057
BZX71-B10	0064
BZX71-B11	0181
BZX71-B12	0052
BZX71-B13	0053
BZX71-B15	0686
BZX71-B16	0445
BZX71-B18	0375
BZX71-B20	0700
BZX71-B22	0705
BZX71-B24	0494
BZX71-C5V1	0041
BZX71-C5V6	0253
BZX71-C6V2	0471
BZX71-C6V8	0062
BZX71-C7V5	0077
BZX71-C8V2	0165
BZX71-C9V1	0057
BZX71-C10	0064
BZX71-C11	0181
BZX71-C12	0052
BZX71-C13	0053
BZX71-C15	0686
BZX71-C16	0445
BZX71-C18	0375
BZX71-C20	0700
BZX71-C22	0705
BZX71-C24	0494
BZX71B5V1	0087
BZX71B5V6	0253
BZX71B6V2	0471
BZX71B6V8	0062
BZX71B7V5	0077
BZX71B8V2	0165
BZX71B9V1	0057
BZX71B10	0064
BZX71B11	0181
BZX71B12	0052
BZX71B13	0053
BZX71B15	0686
BZX71B16	0445
BZX71B18	0375
BZX71B20	0700
BZX71B22	0705
BZX71B24	0494
BZX71C5V1	0041
BZX71C5V6	0253
BZX71C6V2	0471
BZX71C6V8	0062
BZX71C7V5	0077
BZX71C8V2	0165
BZX71C9V1	0057
BZX71C10	0064
BZX71C11	0181
BZX71C12	0052
BZX71C13	0053
BZX71C15	0686
BZX71C16	0445
BZX71C18	0375
BZX71C20	0700
BZX71C22	0705
BZX71C24	0494
BZX72	0057
BZX72A	0057
BZX72B	0057
BZX72C	0057
BZX74-C5V6	0253
BZX74-C6V2	0471
BZX74-C6V8	0062
BZX74-C7V5	0077
BZX74-C8V2	0165
BZX74-C9V1	0057
BZX74-C10	0064
BZX74-C11	0181
BZX74-C12	0052
BZX74C5V6	0253
BZX74C6V2	0471
BZX74C6V8	0062
BZX74C7V5	0077
BZX74C8V2	0165
BZX74C9V1	0057
BZX74C10	2119
BZX74C11	0181
BZX74C12	0052
BZX75/C2V1	OEM
BZX75-C1V4	OEM
BZX75-C2V1	OEM
BZX75-C2V8	0677
BZX75-C3V6	0376
BZX75C1V4	OEM
BZX75C1V4PN	OEM
BZX75C2V1	OEM
BZX75C2V8	6964
BZX75C3V6	0376
BZX76	0053
BZX77-D5V6	OEM
BZX77-D5V6P	OEM
BZX77-D6V2	OEM
BZX77-D6V2P	OEM
BZX77-D6V8	OEM
BZX77-D6V8P	OEM
BZX77-D7V5	OEM
BZX77-D7V5P	OEM
BZX77-D8V2	OEM
BZX77-D8V2P	OEM
BZX77-D9V1	OEM
BZX77-D9V1P	OEM
BZX78C5V1	OEM
BZX78C5V6	OEM
BZX78C6V2	0636
BZX78C6V8	OEM
BZX78C7V5	OEM
BZX78C8V2	OEM
BZX78C9V1	0012
BZX78C10	OEM
BZX78C11	OEM
BZX78C12	0137
BZX78C13	OEM
BZX78C15	0002
BZX78C16	OEM
BZX78C18	OEM
BZX78C20	OEM
BZX78C22	OEM
BZX78C24	OEM
BZX78C27	OEM
BZX78C30	OEM
BZX78C33	OEM
BZX78C36	OEM
BZX78C39	OEM
BZX78C43	OEM
BZX78C47	OEM
BZX78C51	OEM
BZX78C56	OEM
BZX78C62	OEM
BZX78C68	OEM
BZX78C75	OEM
BZX79/C5V6	0253
BZX79/C6.8	0025
BZX79/C9V1	0057
BZX79/C12	0137
BZX79/C18	0495
BZX79/C24	0402
BZX79-B	OEM
BZX79-C4V7	0140
BZX79-C5V1	0041
BZX79-C5V6	0253
BZX79-C6V2	0471
BZX79-C6V8	0062
BZX79-C7V5	0649
BZX79-C7VS	0077
BZX79-C8V2	0165
BZX79-C9V1	0057
BZX79-C10	0064
BZX79-C11	0181
BZX79-C12	0052
BZX79-C13	0053
BZX79-C15	0686
BZX79-C16	0445
BZX79-C18	0375
BZX79-C20	0700
BZX79-C22	0705
BZX79-C24	0494
BZX79-C27	0455
BZX79-C30	0195
BZX79-C33	0166
BZX79-C36	0010
BZX79-C39	0032
BZX79-C43	0054
BZX79-C47	0068
BZX79-C51	0092
BZX79-C56	0125
BZX79-C62	0152
BZX79-C68	0173
BZX79-C75	0094
BZX79B2V4	OEM
BZX79B2V7	OEM
BZX79B3V0	OEM
BZX79B3V3	OEM
BZX79B3V6	OEM
BZX79B3V9	OEM
BZX79B4V3	OEM
BZX79B4V7	OEM
BZX79B5V1	0589
BZX79B5V6	0883
BZX79B6V2	0295
BZX79B6V8	0062
BZX79B6VB	0062
BZX79B7V5	OEM
BZX79B8V2	3100
BZX79B9V1	0057
BZX79B10	0248
BZX79B11	OEM
BZX79B12	1005
BZX79B12PH	OEM
BZX79B13	OEM
BZX79B15	OEM
BZX79B16	4938
BZX79B18	OEM
BZX79B20	OEM
BZX79B22	4942
BZX79B24	OEM
BZX79B27	OEM
BZX79B30	OEM
BZX79B33	OEM
BZX79B36	OEM
BZX79B39	OEM
BZX79B43	OEM
BZX79B47	OEM
BZX79B51	OEM
BZX79B56	OEM
BZX79B62	OEM
BZX79B68	OEM
BZX79B75	OEM
BZX79C2V4	1276
BZX79C2V5	OEM
BZX79C2V7	0760
BZX79C3V	0118
BZX79C3V0	0118
BZX79C3V3	0299
BZX79C3V6	0376
BZX79C3V9	0036
BZX79C4V3	0275
BZX79C4V7	0140
BZX79C5V1	0041
BZX79C5V6	0253
BZX79C6V2	0471
BZX79C6V8	0062
BZX79C7V5	0077
BZX79C8V2	0165
BZX79C9V1	0057
BZX79C10	0064
BZX79C11	0181
BZX79C12	0052
BZX79C13	0053
BZX79C15	0686
BZX79C15V	0686
BZX79C16	0445
BZX79C18	0375
BZX79C20	0700
BZX79C22	0705
BZX79C24	0494
BZX79C27	0455
BZX79C30	0195
BZX79C33	0166
BZX79C36	0010
BZX79C39	0032
BZX79C43	0054
BZX79C47	0068
BZX79C51	0092
BZX79C56	0125
BZX79C62	0152
BZX79C68	0173
BZX79C75	0094
BZX80-C6V8	0025
BZX80-C7V5	0649
BZX80-C8V2	0244
BZX80-C9V1	0012
BZX80-C10	0170
BZX81-C11	0317
BZX81-C12	0137
BZX81-C13	0365
BZX81-C15	0002
BZX81-C16	0420
BZX81-C18	0495
BZX81-C20	0531
BZX81-C22	0565
BZX81-C24	0402
BZX81-C27	0441
BZX82-C30	0726
BZX82-C33	0039
BZX82-C36	0820
BZX82-C39	0350
BZX82-C43	0931
BZX82-C47	0999
BZX82-C51	0502
BZX82-C56	0869
BZX83(6.2V)	OEM
BZX83-C0V8	OEM
BZX83-C2V4	1276
BZX83-C2V7	0760
BZX83-C3V0	0118
BZX83-C3V3	0299
BZX83-C3V6	0376
BZX83-C3V9	0036
BZX83-C4V3	0275
BZX83-C4V7	0140
BZX83-C5V1	0041
BZX83-C5V6	0253
BZX83-C6V2	0471
BZX83-C6V8	0062
BZX83-C7V5	0062
BZX83-C8V2	0165
BZX83-C9V1	0057
BZX83-C10	0064
BZX83-C11	0181
BZX83-C13	0053
BZX83-C15	0686
BZX83-C16	0445
BZX83-C18	0375
BZX83-C20	0700
BZX83-C22	0705
BZX83-C24	0494
BZX83-C27	0455
BZX83-C30	0195
BZX83-C33	0166
BZX83-C36	0010
BZX83-C39	0032
BZX83-C43	0054
BZX83-C47	0068
BZX83-C51	0092
BZX83-C56	0125
BZX83-C62	0152
BZX83C048	OEM
BZX83C0V8	OEM
BZX83C2V4	1276
BZX83C2V7	0760
BZX83C3V0	0118
BZX83C3V3	0299
BZX83C3V6	0376
BZX83C3V9	0036
BZX83C4V3	0275
BZX83C4V7	0140
BZX83C5V1	0041
BZX83C5V6	0253
BZX83C6V2	0471
BZX83C6V8	0062
BZX83C7V5	0077
BZX83C8V2	0165
BZX83C9V1	0057
BZX83C10	0064
BZX83C10V	0170
BZX83C11	0181
BZX83C11V	0317
BZX83C12	0052
BZX83C12V	0137
BZX83C13	0053
BZX83C13V	0365
BZX83C15	0686
BZX83C15V	0002
BZX83C16	0445
BZX83C16V	0420
BZX83C18	0375
BZX83C18V	0495
BZX83C20	0700
BZX83C20V	0531
BZX83C22	0705
BZX83C22V	0565
BZX83C24	0494
BZX83C24V	0402
BZX83C27	0455
BZX83C27V	0441
BZX83C30	0195
BZX83C30V	0726
BZX83C33	0166
BZX83C33V	0039
BZX83C36	0010
BZX83C39	0032
BZX83C43	0054
BZX83C47	0068
BZX83C51	0092
BZX83C56	0125
BZX83C62	0152
BZX83C68	0173
BZX83C75	0094
BZX84-C4V7	0140
BZX84-C5V1	0041
BZX84-C5V6	0253
BZX84-C6V2	0471
BZX84-C6V8	0062
BZX84-C7V5	0077
BZX84-C8V2	0165
BZX84-C9V1	0057
BZX84-C10	OEM
BZX84-C11	OEM
BZX84-C12	OEM
BZX84C2V4	1276
BZX84C2V7	0760
BZX84C3V0	0118
BZX84C3V3	0299
BZX84C3V6	0376
BZX84C3V9	0036
BZX84C4V3	0275
BZX84C4V7	0140
BZX84C5V1	0041
BZX84C5V6	0253
BZX84C6V2	0471
BZX84C6V8	0062
BZX84C7V5	0077
BZX84C8V2	0165
BZX84C9V1	0057
BZX84C10	OEM
BZX84C11	OEM
BZX84C12	0052
BZX84C13	OEM
BZX84C15	OEM
BZX84C16	OEM
BZX84C18	OEM
BZX84C20	OEM
BZX84C22	OEM
BZX84C24	OEM
BZX84C27	0455
BZX84C30	0195
BZX84C33	0166
BZX84C36	0166
BZX84C39	0032
BZX84C43	0054
BZX84C47	0068
BZX84C51	0092
BZX84C56	0125
BZX84C62	0152
BZX84C68	0173
BZX84C75	0094
BZX85-C2V7	1312
BZX85-C3V0	1713
BZX85-C3V3	0782
BZX85-C3V6	0796
BZX85-C3V9	0806
BZX85-C4V3	0821
BZX85-C4V7	0833
BZX85-C5V1	0442
BZX85-C5V6	0876
BZX85-C6V2	0185
BZX85-C6V8	0205
BZX85-C7V5	0480
BZX85-C8V2	0504
BZX85-C9V1	0684
BZX85-C10	0225
BZX85-C11	0230
BZX85-C12	0234
BZX85-C13	0237
BZX85-C15	0247
BZX85-C16	0251
BZX85-C18	0256
BZX85-C20	0262
BZX85-C22	0269
BZX85-C24	0274
BZX85-C27	0294
BZX85-C30	0309
BZX85-C33	0318
BZX85-C36	0320
BZX85-C39	0326
BZX85-C43	0337
BZX85-C47	0347
BZX85-C51	0027
BZX85-C56	0266
BZX85-C62	0386
BZX85C2V7	1312
BZX85C3V0	1713
BZX85C3V3	0292
BZX85C3V6	0188
BZX85C3V8	0456
BZX85C3V9	0456
BZX85C4V3	0533
BZX85C4V7	0451
BZX85C5V1	0162
BZX85C5V6	0157
BZX85C6V2	0636
BZX85C6V8	0025
BZX85C7V5	0649
BZX85C8V1	0244
BZX85C8V2	0244
BZX85C9V1	0057
BZX85C10	0170
BZX85C11	0317
BZX85C11V	0230
BZX85C12	0137
BZX85C12V	0234
BZX85C13	0365
BZX85C13V	0237
BZX85C15	0002
BZX85C15V	0247
BZX85C16	0420
BZX85C16V	0251
BZX85C18	0495
BZX85C18V	0256
BZX85C20	0531
BZX85C20V	0262
BZX85C22	0565
BZX85C22V	0269
BZX85C24	0402
BZX85C24V	0274
BZX85C27	1606
BZX85C27V	0294
BZX85C30	0726
BZX85C30V	0309
BZX85C33	0039
BZX85C33V	0318
BZX85C36	0820
BZX85C39	0350
BZX85C43	0931
BZX85C47	0999
BZX85C51	0502
BZX85C56	0869
BZX85C62	0783
BZX85C68	2157
BZX85C75	1189
BZX85C82	0331
BZX85C91	1311
BZX85C100	0098
BZX85C110	0149
BZX85C120	0186
BZX85C130	0213
BZX85C150	0028
BZX85C160	0255
BZX85C180	0367
BZX85C200	0605
BZX86-E36	OEM
BZX87-C4V7	0833
BZX87-C5V1	0442
BZX87-C5V6	0876
BZX87-C6V2	0185
BZX87-C6V8	0205
BZX87-C7V5	0480
BZX87-C8V2	0504
BZX87-C9V1	0684
BZX87-C10	0225
BZX87-C11	0230
BZX87-C12	0234
BZX87-C13	0237
BZX87-C15	0247
BZX87-C16	0251
BZX87-C18	0256
BZX87-C20	0262
BZX87-C22	0269
BZX87-C24	0274
BZX87-C27	0294
BZX87-C30	0309
BZX87-C33	0318
BZX87-C36	0320
BZX87-C39	0326
BZX87-C43	0337
BZX87-C47	0347
BZX87-C51	0027
BZX87-C56	0266
BZX87-C62	0386
BZX87-C68	0405
BZX87-C75	0426
BZX87C5V1	0442
BZX87C5V6	0876
BZX87C6V2	0185
BZX87C6V8	0205
BZX87C7V5	0480
BZX87C8V2	0504
BZX87C9V1	0684
BZX87C10	0225
BZX87C11	0230
BZX87C12	0234
BZX87C13	0237
BZX87C15	0247
BZX87C16	0251
BZX87C18	0256
BZX87C20	0262
BZX87C22	0269
BZX87C24	0274
BZX87C27	0294
BZX87C30	0309
BZX87C33	0318
BZX87C36	0320
BZX87C39	0326
BZX87C43	0337
BZX87C47	0347
BZX87C51	0027
BZX87C56	0266
BZX87C62	0386
BZX87C68	0405
BZX87C75	0426
BZX88-C4V7	0140
BZX88-C5V1	0041
BZX88-C5V6	0253
BZX88-C6V2	0471
BZX88-C6V8	0062
BZX88-C7V5	0077
BZX88-C8V2	0165
BZX88-C9V1	0057
BZX88-C10	OEM
BZX88-C11	OEM
BZX88-C12	OEM
BZX88C2V7	0760
BZX88C3V0	0118
BZX88C3V3	0299
BZX88C3V6	0376
BZX88C3V9	0036
BZX88C4V3	0275
BZX88C4V7	0140
BZX88C5V1	0041
BZX88C5V6	0253
BZX88C6V2	0471
BZX88C6V8	0062
BZX88C7V5	0077
BZX88C9V1	0057
BZX88C10	0064
BZX88C11	0181
BZX88C12	0052
BZX88C13	0053
BZX88C15	0686
BZX88C16	0445
BZX88C18	0375
BZX88C20	0700
BZX88C22	0705
BZX88C24	0494
BZX88C27	0455
BZX88C30	0195
BZX88C33	0166
BZX88C36	0010
BZX88C39	0032
BZX88C43	0054
BZX88C47	0068
BZX90	0062
BZX91	0062
BZX92	0062

DEVICE TYPE	REPL CODE
BZX93	0062
BZX94	0062
BZX95-C5V6	0253
BZX95-C6V2	0471
BZX95-C6V8	0062
BZX95-C7V5	0077
BZX95-C8V2	0165
BZX95-C9V1	0057
BZX95-C10	0064
BZX95-C11	0181
BZX95-C12	0052
BZX95-C13	0053
BZX95C-4V7	0140
BZX95C-5V1	0041
BZX95C-5V6	0253
BZX95C-6V2	0471
BZX95C-6V8	0062
BZX95C-7V5	0077
BZX95C-8V2	0165
BZX95C-9V1	0057
BZX95C-10	0064
BZX95C-11	0181
BZX95C-12	0052
BZX95C-13	0053
BZX95C-15	0686
BZX95C-16	0445
BZX95C-18	0375
BZX95C-20	0700
BZX95C-22	0705
BZX95C-24	0494
BZX95C-27	0455
BZX95C-30	0195
BZX95C-33	0166
BZX95C-36	0010
BZX96-C2V7	0760
BZX96-C3	0118
BZX96-C3V3	0299
BZX96-C3V6	0376
BZX96-C3V9	0036
BZX96-C4V3	0275
BZX96-C4V7	0140
BZX96-C5V1	0041
BZX96-C5V6	0253
BZX96-C6V2	0471
BZX96-C6V8	0062
BZX96-C7V5	0077
BZX96-C8V2	0165
BZX96-C9V1	0057
BZX96-C10	0064
BZX96-C11	0181
BZX96-C12	0052
BZX96-C13	0053
BZX96-C15	0686
BZX96-C16	0445
BZX96-C18	0375
BZX96-C20	0700
BZX96-C22	0705
BZX96-C24	0494
BZX96-C27	0455
BZX96-C30	0195
BZX96-C33	0166
BZX96C2V7	0760
BZX96C3	0118
BZX96C3V3	0299
BZX96C3V6	0188
BZX96C3V9	0036
BZX96C4V3	0275
BZX96C4V7	0140
BZX96C5V1	0041
BZX96C5V6	0253
BZX96C6V2	0471
BZX96C6V8	0062
BZX96C7V5	0077
BZX96C8V2	0165
BZX96C9V1	0057
BZX96C10	0064
BZX96C11	0181
BZX96C12	0052
BZX96C13	0053
BZX96C15	0686
BZX96C16	0445
BZX96C18	0375
BZX96C20	0700
BZX96C22	0705
BZX96C24	0494
BZX96C27	0455
BZX96C30	0195
BZX96C33	0166
BZX97-C0V8	OEM
BZX97-C2V4	1276
BZX97-C2V7	0760
BZX97-C3V0	0118
BZX97-C3V3	0299
BZX97-C3V6	0376
BZX97-C3V9	0036
BZX97-C4V3	0275
BZX97-C4V7	0140
BZX97-C5V1	0041
BZX97-C5V6	0253
BZX97-C6V2	0471
BZX97-C6V8	0062
BZX97-C7V5	0077
BZX97-C8V2	0165
BZX97-C9V1	0057
BZX97-C10	0064
BZX97-C11	0181
BZX97-C12	0052
BZX97-C13	0053
BZX97-C15	0686

DEVICE TYPE	REPL CODE
BZX97-C16	0445
BZX97-C18	0375
BZX97-C20	0700
BZX97-C22	0705
BZX97-C24	0494
BZX97-C27	0455
BZX97-C30	0195
BZX97-C33	0166
BZX97C0V8	OEM
BZX97C2V4	1276
BZX97C2V7	0760
BZX97C3V0	0118
BZX97C3V3	0299
BZX97C3V6	0376
BZX97C3V9	0036
BZX97C4V3	0275
BZX97C4V7	0140
BZX97C5V1	0041
BZX97C5V6	0253
BZX97C6V2	0471
BZX97C6V8	0062
BZX97C7V5	0077
BZX97C8V2	0165
BZX97C9V1	0057
BZX97C10	0064
BZX97C11	0181
BZX97C12	0052
BZX97C13	0053
BZX97C15	0686
BZX97C16	0445
BZX97C18	0375
BZX97C20	0700
BZX97C22	0705
BZX97C24	0494
BZX97C27	0455
BZX97C30	0195
BZX97C33	0166
BZX97C36	0010
BZX97C39	0032
BZX97C43	0054
BZX97C47	0068
BZX98C3V9	OEM
BZX98C4V3	OEM
BZX98C4V7	OEM
BZX98C5V1	OEM
BZX98C5V6	OEM
BZX98C6V2	OEM
BZX98C6V8	OEM
BZX98C7V5	OEM
BZX98C8V2	OEM
BZX98C9V1	OEM
BZX98C10	OEM
BZX98C11	OEM
BZX98C12	OEM
BZX98C13	OEM
BZX98C15	OEM
BZX98C16	OEM
BZX98C18	OEM
BZX98C20	1549
BZX98C22	OEM
BZX98C24	OEM
BZX98C27	OEM
BZX98C30	OEM
BZX98C33	OEM
BZX98C36	OEM
BZX98C39	OEM
BZX98C43	OEM
BZX98C47	OEM
BZX98C51	OEM
BZX98C56	OEM
BZX98C62	OEM
BZX98C68	OEM
BZX98C75	OEM
BZX98C82	OEM
BZX98C91	OEM
BZX98C100	OEM
BZX98C110	OEM
BZX98C120	OEM
BZX98C130	OEM
BZX98C150	OEM
BZX98C160	OEM
BZX98C180	OEM
BZX98C200	OEM
BZX79812	OEM
BZY16C6V8	0205
BZY16C7V5	0480
BZY16C8V2	0504
BZY16C9V1	0684
BZY16C10	0225
BZY16C11	0230
BZY16C12	0234
BZY16C13	0237
BZY16C15	0247
BZY16C16	0251
BZY16C18	0256
BZY16C20	0262
BZY16C22	0269
BZY16C24	0274
BZY16C27	0294
BZY16C30	0309
BZY16C33	0318
BZY16C36	0320
BZY16C39	0326
BZY16C43	0337
BZY16C47	0347
BZY16C51	0027
BZY16C56	0266
BZY16C62	0386
BZY16C68	0405

DEVICE TYPE	REPL CODE
BZY16C75	0426
BZY16C82	0444
BZY16C91	0238
BZY16C100	1180
BZY17C5V6	0253
BZY17C6V2	0471
BZY17C6V8	0062
BZY17C7V5	0077
BZY17C8V2	0165
BZY17C9V1	0057
BZY17C10	0064
BZY17C11	0181
BZY17C12	0052
BZY17C13	0053
BZY17C15	0686
BZY17C16	0445
BZY17C18	0375
BZY17C20	0700
BZY17C22	0705
BZY17C24	0494
BZY17C27	0455
BZY17C30	0195
BZY17C33	0166
BZY17C36	0010
BZY17C39	0032
BZY17C43	0054
BZY17C47	0068
BZY17C51	0092
BZY17C56	0125
BZY18	0137
BZY18C5V6	0157
BZY18C6V2	0636
BZY18C6V8	0025
BZY18C7V5	0649
BZY18C8V2	0244
BZY18C9V1	0012
BZY18C10	0170
BZY18C11	0794
BZY18C12	0137
BZY19	0002
BZY22	0165
BZY23	0165
BZY24	0165
BZY25	0165
BZY29C33	0039
BZY56	0451
BZY57	0041
BZY58	0157
BZY59	0636
BZY60	0025
BZY61	0649
BZY62	0244
BZY63	0012
BZY64	0275
BZY65	0041
BZY66	0471
BZY67	0077
BZY68	0057
BZY69	0052
BZY70	0165
BZY71	OEM
BZY74	0696
BZY75	1616
BZY76	0627
BZY78	0253
BZY78P	0253
BZY83/C5V1	0162
BZY83/C5V6	0157
BZY83/C10	0170
BZY83/C15	0002
BZY83/C16	0002
BZY83/D5V6	0157
BZY83/D10	0170
BZY83/D15	0002
BZY83-C4V7	0140
BZY83-C5V1	0041
BZY83-C5V6	0253
BZY83-C6V2	0471
BZY83-C6V8	0062
BZY83-C7V5	0077
BZY83-C8V2	0165
BZY83-C9V1	0057
BZY83-C10	0064
BZY83-C11	0181
BZY83-C12	0052
BZY83-C13	0053
BZY83-C15	0686
BZY83-C16	0445
BZY83-C18	0375
BZY83-C20	0700
BZY83-C22	0705
BZY83-C24	0494
BZY83-D1	OEM
BZY83-D4V7	0140
BZY83-D5V6	0253
BZY83-D6V8	0062
BZY83-D8V2	0165
BZY83-D10	0064
BZY83-D12	0052
BZY83-D15	0686
BZY83-D18	0375
BZY83-D22	0705
BZY83C4V7	0140
BZY83C5V1	0041
BZY83C5V6	0253
BZY83C6V2	0471
BZY83C6V8	0062
BZY83C7V5	0077
BZY83C8V2	0165

DEVICE TYPE	REPL CODE
BZY83C9V1	0057
BZY83C10	0064
BZY83C11	0181
BZY83C12	0052
BZY83C13V5	0053
BZY83C15	0686
BZY83C16	0002
BZY83C16V5	0210
BZY83C18	0375
BZY83C20	0700
BZY83C22	0705
BZY83C24V5	0714
BZY83D1	OEM
BZY83D4V7	0140
BZY83D5V6	0253
BZY83D6V8	0062
BZY83D8V2	0165
BZY83D10	0064
BZY83D12	0052
BZY83D15	0686
BZY83D18	0375
BZY83D22	0705
BZY84D1	OEM
BZY84D5V6	1445
BZY84D6V8	1458
BZY84D8V2	1490
BZY84D10	1618
BZY84D12	0052
BZY85/C3V6	0188
BZY85/C5V1	0162
BZY85/C5V6	0157
BZY85/C10	0170
BZY85/C15	0002
BZY85/C27	0441
BZY85/D5V6	0157
BZY85/D10	0170
BZY85/D15	0002
BZY85-C4V7	0140
BZY85-C5V6	0253
BZY85-C6V1	0471
BZY85-C6V2	0471
BZY85-C6V8	0062
BZY85-C7V5	0077
BZY85-C8V2	0165
BZY85-C9V1	0057
BZY85-C10	0064
BZY85-C11	0181
BZY85-C12	0052
BZY85-C13	0053
BZY85-C15	0686
BZY85-C16	0445
BZY85-C18	0375
BZY85-C20	0700
BZY85-C22	0705
BZY85-C24	0494
BZY85-D1	OEM
BZY85-D4V7	0140
BZY85-D5V6	0253
BZY85-D6V8	0062
BZY85-D8V2	0165
BZY85-D10	0064
BZY85-D12	0052
BZY85-D15	0686
BZY85-D18	0375
BZY85-D22	0705
BZY85B2V7	0760
BZY85B3	OEM
BZY85B3V3	0292
BZY85B3V6	OEM
BZY85B3V9	OEM
BZY85B4V3	OEM
BZY85B4V7	OEM
BZY85B5V6	OEM
BZY85B6V2	0471
BZY85B6V8	6036
BZY85B7V5	6045
BZY85B8V2	6049
BZY85B9V1	4235
BZY85B10	0248
BZY85B11	OEM
BZY85B12	1005
BZY85B13	OEM
BZY85B15	0686
BZY85B16	OEM
BZY85B18	OEM
BZY85B20	OEM
BZY85B22	OEM
BZY85B24	OEM
BZY85B27	OEM
BZY85B30	OEM
BZY85B33	OEM
BZY85C2V7	1312
BZY85C3	1713
BZY85C3V3	0292
BZY85C3V6	0188
BZY85C3V9	0456
BZY85C4V3	0533
BZY85C4V7	0451
BZY85C5V1	0041
BZY85C5V6	0157
BZY85C6V2	0471
BZY85C6V8	0062
BZY85C7V5	0077
BZY85C8V2	0165
BZY85C9V1	0057
BZY85C10	0064
BZY85C11	0181
BZY85C12	0052
BZY85C13	0053

DEVICE TYPE	REPL CODE
BZY85C13V5	0053
BZY85C15	0686
BZY85C16	0445
BZY85C16V5	0445
BZY85C18	0375
BZY85C20	0700
BZY85C22	0705
BZY85C24	0402
BZY85C24V5	0494
BZY85C27	0441
BZY85C30	0726
BZY85C33	0166
BZY85D1	OEM
BZY85D4V7	0140
BZY85D5V6	0253
BZY85D6V8	0062
BZY85D8V2	0165
BZY85D10	0064
BZY85D12	0052
BZY85D15	0686
BZY85D18	0375
BZY85D22	0705
BZY87	OEM
BZY87-0V7	OEM
BZY87-1V4	OEM
BZY87-2V1	OEM
BZY87-2V8	0677
BZY87-3V4	0299
BZY88/C4V7	0157
BZY88/C5V1	0162
BZY88/C5V6	0157
BZY88/C6V2	0636
BZY88/C9V1	0012
BZY88/C9V5	0012
BZY88/C9V7	0162
BZY88-C1V3	OEM
BZY88-C2V7	0760
BZY88-C3V0	0118
BZY88-C3V3	0299
BZY88-C3V6	0376
BZY88-C3V9	0036
BZY88-C4V3	0275
BZY88-C4V7	0140
BZY88-C5V1	0041
BZY88-C5V6	0253
BZY88-C6V2	0471
BZY88-C6V8	0062
BZY88-C7V5	0077
BZY88-C8V2	0165
BZY88-C9V1	0057
BZY88-C10	0064
BZY88-C11	0181
BZY88-C12	0052
BZY88-C13	0053
BZY88-C15	0686
BZY88-C16	0445
BZY88-C18	0375
BZY88-C20	0700
BZY88-C22	0705
BZY88-C24	0494
BZY88-C27	0455
BZY88-C30	0195
BZY88-C33	0166
BZY88C2V7	0760
BZY88C2V9	0677
BZY88C3V0	0118
BZY88C3V1	0118
BZY88C3V3	0299
BZY88C3V6	0376
BZY88C3V9	0036
BZY88C4V3	0275
BZY88C4V7	0140
BZY88C5V1	0041
BZY88C5V6	0253
BZY88C6V2	0471
BZY88C6V8	0062
BZY88C7V5	0077
BZY88C8V2	0165
BZY88C9V1	0057
BZY88C10	0064
BZY88C10V	0064
BZY88C11	0181
BZY88C11V	0181
BZY88C12	0137
BZY88C12V	0052
BZY88C13	0053
BZY88C13V	0053
BZY88C15	0686
BZY88C15V	0686
BZY88C16	0445
BZY88C16V	0445
BZY88C18	0375
BZY88C18V	0375
BZY88C20	0700
BZY88C20V	0700
BZY88C22	0705
BZY88C22V	0705
BZY88C24	0494
BZY88C24V	0494
BZY88C27	0455
BZY88C27V	0455
BZY88C30	0195
BZY88C30V	0195
BZY88C33	0166
BZY88C33V	0166
BZY91-4	OEM
BZY91-C7V5	OEM
BZY91-C8V2	OEM
BZY91-C9V1	OEM

DEVICE TYPE	REPL CODE
BZY91-C10	OEM
BZY91-C11	OEM
BZY91-C12	OEM
BZY91-C13	OEM
BZY91-C15	OEM
BZY91-C16	OEM
BZY91-C18	OEM
BZY91-C20	OEM
BZY91-C22	OEM
BZY91-C24	OEM
BZY91-C27	OEM
BZY91-C30	OEM
BZY91-C33	OEM
BZY91-C36	OEM
BZY91-C39	OEM
BZY91-C43	OEM
BZY91-C47	OEM
BZY91-C51	OEM
BZY91-C56	OEM
BZY91-C62	OEM
BZY91-C68	OEM
BZY91-C75	OEM
BZY91C7V5	OEM
BZY91C8V2	OEM
BZY91C9V1	OEM
BZY91C10	OEM
BZY91C11	OEM
BZY91C12	OEM
BZY91C13	OEM
BZY91C15	OEM
BZY91C16	OEM
BZY91C18	OEM
BZY91C20	1549
BZY91C22	OEM
BZY91C24	OEM
BZY91C27	OEM
BZY91C30	OEM
BZY91C33	OEM
BZY91C36	OEM
BZY91C39	OEM
BZY91C43	OEM
BZY91C47	OEM
BZY91C51	OEM
BZY91C56	OEM
BZY91C62	OEM
BZY91C68	OEM
BZY91C75	OEM
BZY92/C5V1	0162
BZY92/C5V6	0157
BZY92/C15	0002
BZY92/C27	0441
BZY92-C3V9	0806
BZY92-C4V3	0821
BZY92-C4V7	0833
BZY92-C5V1	0442
BZY92-C5V6	0876
BZY92-C6V2	0185
BZY92-C6V8	0205
BZY92-C7V5	0480
BZY92-C8V2	0504
BZY92-C9V1	0684
BZY92-C10	0225
BZY92-C11	0230
BZY92-C12	0234
BZY92-C13	0237
BZY92-C15	0247
BZY92-C16	0251
BZY92-C18	0256
BZY92-C20	0262
BZY92-C22	0269
BZY92-C24	0274
BZY92-C27	0294
BZY92-C30	0309
BZY92-C33	0318
BZY92-C36	0320
BZY92C3V9	0456
BZY92C4V3	0533
BZY92C4V7	0451
BZY92C5V1	0162
BZY92C5V6	0157
BZY92C6V2	0636
BZY92C6V8	0025
BZY92C7V5	0649
BZY92C8V2	0244
BZY92C9V1	0012
BZY92C10	0170
BZY92C11	0317
BZY92C12	0137
BZY92C13	0365
BZY92C15	0002
BZY92C16	0420
BZY92C18	0495
BZY92C20	0531
BZY92C22	0565
BZY92C24	0402
BZY92C27	0441
BZY92C30	0726
BZY92C33	0039
BZY92C36	0820
BZY93-C4	OEM
BZY93-C7V5	OEM
BZY93-C8V2	OEM
BZY93-C9V1	OEM
BZY93-C10	OEM
BZY93-C11	OEM
BZY93-C12	OEM
BZY93-C13	OEM
BZY93-C15	OEM
BZY93-C16	OEM

DEVICE TYPE	REPL CODE
BZY93-C18	OEM
BZY93-C20	OEM
BZY93-C22	OEM
BZY93-C24	OEM
BZY93-C27	OEM
BZY93-C30	OEM
BZY93-C33	OEM
BZY93-C36	OEM
BZY93-C39	OEM
BZY93-C43	OEM
BZY93-C47	OEM
BZY93-C51	OEM
BZY93-C56	OEM
BZY93-C62	OEM
BZY93-C68	OEM
BZY93-C75	OEM
BZY93C6V8	2740
BZY93C7V5	2744
BZY93C8V2	1225
BZY93C9V1	1237
BZY93C10	1252
BZY93C11	1269
BZY93C12	1277
BZY93C13	1294
BZY93C15	1302
BZY93C16	1310
BZY93C18	2746
BZY93C20	1333
BZY93C22	1344
BZY93C24	1356
BZY93C27	1371
BZY93C30	1387
BZY93C33	1406
BZY93C36	1415
BZY93C39	1428
BZY93C43	1440
BZY93C47	2776
BZY93C51	1461
BZY93C56	1474
BZY93C62	0613
BZY93C68	1510
BZY93C75	1523
BZY94/C12	0137
BZY94/C15	0002
BZY94-C10	0064
BZY94-C11	0181
BZY94-C12	0052
BZY94-C13	0053
BZY94-C15	0686
BZY94-C16	0445
BZY94-C18	0375
BZY94-C20	0700
BZY94-C22	0705
BZY94-C24	0494
BZY94-C27	0455
BZY94-C30	0195
BZY94-C33	0166
BZY94-C36	0010
BZY94-C39	0032
BZY94-C43	0054
BZY94-C47	0068
BZY94-C51	0092
BZY94-C56	0125
BZY94-C62	0152
BZY94-C68	0173
BZY94-C75	0094
BZY94C10	0064
BZY94C11	0181
BZY94C12	0052
BZY94C13	0053
BZY94C15	0686
BZY94C16	0445
BZY94C18	0375
BZY94C20	0700
BZY94C22	0705
BZY94C24	0494
BZY94C27	0455
BZY94C30	0195
BZY94C33	0166
BZY94C36	0010
BZY94C39	0032
BZY94C43	0054
BZY94C47	0068
BZY94C51	0099
BZY94C56	0125
BZY94C62	0152
BZY94C68	0173
BZY94C75	0094
BZY95-C10	0225
BZY95-C11	0230
BZY95-C12	0234
BZY95-C13	0237
BZY95-C15	0247
BZY95-C16	0251
BZY95-C18	0256
BZY95-C20	0262
BZY95-C22	0269
BZY95-C24	0274
BZY95-C27	0294
BZY95-C30	0309
BZY95-C33	0318
BZY95-C36	0347
BZY95-C39	0326
BZY95-C43	0337
BZY95-C47	0347
BZY95-C51	0027
BZY95-C56	0266
BZY95-C62	0386
BZY95-C68	0405

If replacement code is OEM, contact original manufacturer for replacement.

DEVICE TYPE	REPL CODE
BZY95-C75	0426
BZY95C10	0225
BZY95C11	0230
BZY95C12	0234
BZY95C13	0237
BZY95C15	0247
BZY95C16	0251
BZY95C18	0256
BZY95C20	0262
BZY95C22	0269
BZY95C24	0274
BZY95C27	0294
BZY95C30	0309
BZY95C33	0318
BZY95C36	0320
BZY95C39	0326
BZY95C43	0337
BZY95C47	0347
BZY95C51	0027
BZY95C56	0266
BZY95C62	0386
BZY95C68	0405
BZY95C75	0426
BZY96-C4V7	0833
BZY96-C5V1	0442
BZY96-C5V6	0876
BZY96-C6V2	0185
BZY96-C6V8	0205
BZY96-C7V5	0480
BZY96-C8V2	0504
BZY96-C9V1	0684
BZY96-C10	0225
BZY96C4V7	0833
BZY96C5V1	0442
BZY96C5V6	0876
BZY96C6V2	0185
BZY96C6V8	0205
BZY96C7V5	0480
BZY96C8V2	0504
BZY96C9V1	0684
BZY96C10	0225
BZY96V9V1	OEM
BZY97C3V3	0782
BZY97C3V6	0796
BZY97C3V9	0806
BZY97C4V3	0821
BZY97C4V7	0833
BZY97C5V1	0442
BZY97C5V6	0876
BZY97C6V2	0185
BZY97C6V8	0205
BZY97C7V5	0480
BZY97C8V2	0504
BZY97C9V1	0684
BZY97C10	0225
BZY97C10V	0225
BZY97C11	0230
BZY97C11V	0230
BZY97C12	0234
BZY97C12V	0234
BZY97C13	0237
BZY97C13V	0237
BZY97C15	0247
BZY97C15V	0247
BZY97C16	0251
BZY97C16V	0251
BZY97C18	0256
BZY97C18V	0256
BZY97C20	0262
BZY97C20V	0262
BZY97C22	0269
BZY97C22V	0269
BZY97C24	0274
BZY97C24V	0274
BZY97C27	0294
BZY97C27V	0294
BZY97C30	0309
BZY97C30V	0309
BZY97C33	0318
BZY97C33V	0318
BZY97C36	0320
BZY97C36V	0320
BZY97C39	0326
BZY97C39V	0326
BZY97C43	0262
BZY97C43V	0262
BZY97C47	0269
BZY97C47V	0269
BZY97C51	0027
BZY97C51V	0027
BZY97C56	0266
BZY97C56V	0266
BZY97C62	0386
BZY97C62V	0386
BZY97C68	0405
BZY97C68V	0405
BZY97C75	0426
BZY97C75V	0426
BZY97C82	0444
BZY97C82V	0444
BZY97C91	0238
BZY97C91V	0238
BZY97C100	1180
BZY97C100V	1180
BZY97C110	1190
BZY97C110V	1190
BZY97C120	1207
BZY97C120V	1207
BZY97C130	1218
BZY97C130V	1218
BZY97C150	0647
BZY97C150V	0647
BZY97C160	1255
BZY97C160V	1255
BZY97C180	1279
BZY97C180V	1279
BZY97C200	0605
BZY97C200V	0605
BZZ10	0091
BZZ11	0062
BZZ12	0077
BZZ13	0165
BZZ14	1445
BZZ15	0696
BZZ16	1601
BZZ17	1616
BZZ18	1622
BZZ19	0627
BZZ20	0992
BZZ21	0995
BZZ22	1263
BZZ23	1249
BZZ24	1639
BZZ25	1703
BZZ26	1716
BZZ27	1731
BZZ28	0727
BZZ29	1756
C006001D	OEM
C006001E	OEM
C006001F	OEM
C006001G	OEM
C006001H	OEM
C006004D	OEM
C006004E	OEM
C006004F	OEM
C006004G	OEM
C006004H	OEM
C006010D	OEM
C006010E	OEM
C006010F	OEM
C006010G	OEM
C0-10750	1973
C0-10816	0398
C02	0244
C03	OEM
C05-03C	0196
C05-06A	OEM
C05-10	0071
C06C	0015
C08P1	0196
C08P1R	1080
C047-52	OEM
C049	0540
C055	OEM
C055P	OEM
C066	OEM
C066P	OEM
C0102	OEM
C0102L7	0071
C0406	0564
C0410	0071
C0410V	0071
C0508	0071
C0510	0071
C0615	0158
C-0799	OEM
C0900P	2563
C0901P	2992
C0902P	2434
C0903P	3015
C0904P	2350
C0905P	2597
C0906P	3032
C0907P	3039
C0908P	3045
C0909P	0078
C0910P	3054
C0911P	3066
C0912P	2992
C03051	OEM
C03069	OEM
C04075	OEM
C010174	6820
C010174-02	1696
C010444	OEM
C010444-22	OEM
C010444D-01	OEM
C010745-03	3070
C010745-12	3070
C010750-03	1973
C010750-12	OEM
C010816	0398
C010816-01	0398
C011298	OEM
C011299	OEM
C011299B-03	OEM
C011464	OEM
C012099	OEM
C012294	OEM
C012294-22	OEM
C012294-31	OEM
C012294B-01	OEM
C012295-03	OEM
C012296D-03	OEM
C012296D-22	OEM
C012298	OEM
C012399	OEM
C012399B-08	OEM
C012499	OEM
C012499B-01	OEM
C014311	OEM
C014312	OEM
C014313	5941
C014329	OEM
C014331-09	0522
C014332	OEM
C014333	OEM
C014336	0366
C014344	5763
C014345	1656
C014347	OEM
C014348	0624
C014349	OEM
C014351	OEM
C014361	OEM
C014362	OEM
C014363	OEM
C014377	7143
C014377-06	OEM
C014599	OEM
C014599B12	OEM
C014776	OEM
C014777	OEM
C014786	OEM
C014795	3065
C014795-12	OEM
C014805	OEM
C014805-01	OEM
C014805-22	OEM
C014805-31	OEM
C014806	3065
C014806-03	3065
C014806-12	OEM
C014806C-29	OEM
C014808-05	OEM
C014809	0037
C015001D	OEM
C015001E	OEM
C015001F	OEM
C015001G	OEM
C015001H	OEM
C015004D	OEM
C015004E	OEM
C015004F	OEM
C015004G	OEM
C015004H	OEM
C015020C	OEM
C015020D	OEM
C015020E	OEM
C015769	OEM
C016364	OEM
C016821	OEM
C017097	1633
C017101	OEM
C017950	OEM
C017951	OEM
C017956	OEM
C018094	OEM
C018991	0006
C019156	OEM
C019156-08	OEM
C021697	OEM
C021697-11	OEM
C024004D	OEM
C024004E	OEM
C024004F	OEM
C024004G	OEM
C024004H	OEM
C024010C	OEM
C024010D	OEM
C024010E	OEM
C024010F	OEM
C024020C	OEM
C024020D	OEM
C024020E	OEM
C024947A	OEM
C025953	OEM
C026028	OEM
C026160-001	OEM
C026162-001	OEM
C026163-001	OEM
C026164-001	OEM
C026165-001	OEM
C030004C	OEM
C030004E	OEM
C030004F	OEM
C030004G	OEM
C030010C	OEM
C030010D	OEM
C030010E	OEM
C030010F	OEM
C030020B	OEM
C030202C	OEM
C030202D	OEM
C030202E	OEM
C045004C	OEM
C045004D	OEM
C045004E	OEM
C045004F	OEM
C045010B	OEM
C045010C	OEM
C045010D	OEM
C045010E	OEM
C045010F	OEM
C045020B	OEM
C045020C	OEM
C045020D	OEM
C060010A	OEM
C060010B	OEM
C060010C	OEM
C060010D	OEM
C060010E	OEM
C060020A	OEM
C060020B	OEM
C060020D	OEM
C060202C	OEM
C060302	OEM
C060302A-29	OEM
C060472	OEM
C060474	1033
C060584	OEM
C060607	0124
C060609	OEM
C060612	2357
C060613	OEM
C060616	OEM
C060616B-05	OEM
C060617	OEM
C060619	OEM
C061428	0427
C061505	OEM
C061598	OEM
C061598B	OEM
C061598B-01	OEM
C061598B-22	OEM
C061618	OEM
C061618-16	OEM
C061702	0770
C061850	1698
C061922	OEM
C061991	OEM
C061991-29	OEM
C070024	OEM
C090010A	OEM
C090010B	OEM
C090010C	OEM
C090010D	OEM
C090020A	OEM
C090020B	OEM
C090020C	OEM
C096509	OEM
C0161991-29	OEM
C0C13000-1C	0595
C1/2-12	0418
C1-12	3560
C1A	1149
C1B	0015
C1D-02A	0196
C1H	0015
C1.0E02	0071
C2-0	OEM
C2AJ102	0015
C2C3.0	OEM
C2C3.3	0782
C2C3.6	0796
C2C3.9	0806
C2C4.3	0821
C2C4.7	0833
C2C5.1	0442
C2C5.6	0876
C2C6.2	0185
C2C6.8	0205
C2C7.5	0480
C2C8.2	0504
C2C9.1	0684
C2C10	0225
C2C11	0230
C2C12	0234
C2C13	0237
C2C15	0247
C2C16	0251
C2C18	0256
C2C20	0262
C2C22	0269
C2C24	0274
C2C27	0294
C2C30	0309
C2C33	0318
C2C36	0320
C2C39	0326
C2C43	0337
C2C47	0347
C2C51	0027
C2C56	0266
C2C62	0386
C2C68	0405
C2C75	0426
C2C82	0444
C2C91	0238
C2C100	1180
C2C110	1190
C2C120	1207
C2C130	1218
C2C150	0647
C2C160	1255
C2C180	1279
C2C200	0605
C2M50-2B	OEM
C2M50-28R	OEM
C2M60-28R	OEM
C2M70-28R	OEM
C2M100-28A	OEM
C3-1	OEM
C3-2B	OEM
C3-4	0725
C3A3.0	OEM
C3A3.3	OEM
C3A3.6	OEM
C3A3.9	OEM
C3A4.3	OEM
C3A4.7	OEM
C3A5.1	OEM
C3A5.6	OEM
C3A6.2	OEM
C3A6.8	OEM
C3A7.5	OEM
C3A8.2	OEM
C3A9.1	OEM
C3C3.0	OEM
C3C3.3	OEM
C3C3.6	OEM
C3C3.9	OEM
C3C4.3	OEM
C3C4.7	OEM
C3C5.1	OEM
C3C5.6	OEM
C3C6.2	OEM
C3C6.8	OEM
C3C7.5	OEM
C3C8.2	OEM
C3C9.1	OEM
C3C10	OEM
C3C11	OEM
C3C12	OEM
C3C13	OEM
C3C15	OEM
C3C16	OEM
C3C18	OEM
C3C20	OEM
C3C22	OEM
C3C24	OEM
C3C27	OEM
C3C30	OEM
C3C33	OEM
C3C36	OEM
C3C39	OEM
C3C43	OEM
C3C47	OEM
C3C51	OEM
C3C56	OEM
C3C62	OEM
C3C68	OEM
C3C75	OEM
C3C82	OEM
C3C91	OEM
C3C100	OEM
C3C110	OEM
C3C120	OEM
C3C130	OEM
C3C150	OEM
C3C160	OEM
C3C180	OEM
C3C200	OEM
C3T225	OEM
C3T225A	OEM
C4C3.0	1713
C4C3.3	0292
C4C3.6	0376
C4C3.9	0456
C4C4.3	0533
C4C4.7	0451
C4C5.1	0162
C4C5.6	0157
C4C6.2	0636
C4C6.8	0025
C4C7.5	0649
C4C8.2	0244
C4C9.1	0012
C4C10	0170
C4C11	0317
C4C12	0137
C4C13	0365
C4C15	0002
C4C16	0420
C4C18	0495
C4C20	0531
C4C22	0565
C4C24	0402
C4C27	0441
C4C30	0726
C4C33	0039
C4C36	0820
C4C39	0350
C4C43	0931
C4C47	0999
C4C51	0502
C4C56	0869
C4C62	0783
C4C68	2157
C4C75	1189
C4C82	0331
C4C91	1311
C4C100	0098
C4C110	0149
C4C120	0186
C4C130	0213
C4C150	0028
C4C160	0255
C4C180	0367
C4C200	0421
C4LS374AP	0713
C4P03Q	1236
C4P04Q	1236
C4P05Q	1236
C4P06Q	1236
C4P09Q	1236
C4P10Q	1236
C5	1718
C5A	0344
C5AR1200	OEM
C5B	0058
C5BR1200	OEM
C5C	0407
C5C11	OEM
C5D	OEM
C5D459	OEM
C5D459CA	OEM
C5D459CC	OEM
C5D485	OEM
C5D485CA	OEM
C5D485CC	OEM
C5D914	OEM
C5D914CA	OEM
C5D914CC	OEM
C5D5226	OEM
C5D5227	OEM
C5D5228	OEM
C5D5229	OEM
C5D5230	OEM
C5D5231	OEM
C5D5232	OEM
C5D5233	OEM
C5D5234	0471
C5D5235	OEM
C5D5236	OEM
C5D5237	OEM
C5D5238	OEM
C5D5239	0057
C5D5240	OEM
C5D5241	OEM
C5D5242	0052
C5D5243	OEM
C5D5244	OEM
C5D5245	0686
C5D5246	OEM
C5D5247	OEM
C5D5248	OEM
C5D5249	OEM
C5D5250	OEM
C5D5251	OEM
C5D5252	OEM
C5D5253	OEM
C5D5254	OEM
C5D5255	OEM
C5D5256	OEM
C5D5257	OEM
C5DR1200	OEM
C5F	1137
C5G	OEM
C5H	OEM
C5U	OEM
C6	0313
C6A	0940
C6B	0940
C6B3	1137
C-6BX212	2136
C6C	0346
C6C3.6	OEM
C6C3.9	OEM
C6C4.3	OEM
C6C4.7	OEM
C6C5.1	OEM
C6C5.6	OEM
C6C6.2	OEM
C6C6.8	OEM
C6C7.5	OEM
C6C8.2	OEM
C6C9.1	OEM
C6C10	OEM
C6C11	OEM
C6C12	OEM
C6C13	OEM
C6C15	OEM
C6C16	OEM
C6C18	OEM
C6C20	OEM
C6C22	OEM
C6C24	OEM
C6C27	OEM
C6C30	OEM
C6C33	OEM
C6C36	OEM
C6C39	OEM
C6C43	OEM
C6C47	OEM
C6C51	OEM
C6C56	OEM
C6C62	OEM
C6C68	OEM
C6C75	OEM
C6C82	OEM
C6C91	OEM
C6C100	OEM
C6C110	OEM
C6C120	OEM
C6C130	OEM
C6C150	OEM
C6C160	OEM
C6C180	OEM
C6C200	OEM
C6D	0346
C6F	0940
C6G	OEM
C6P10FU	1236
C6P10FUR	1727
C6P20FU	1236
C6P20FUR	1727
C6U	0940
C7	0693
C7A	OEM
C7A4	0447
C7B	OEM
C7C	OEM
C7C3.0	OEM
C7C3.3	OEM
C7C3.6	OEM
C7C3.9	OEM
C7C4.3	OEM
C7C4.7	OEM
C7C5.1	OEM
C7C5.6	OEM
C7C6.2	OEM
C7C6.8	OEM
C7C7.5	OEM
C7C8.2	OEM
C7C9.1	OEM
C7C10	OEM
C7C11	OEM
C7C12	OEM
C7C13	OEM
C7C15	OEM
C7C16	OEM
C7C18	OEM
C7C20	OEM
C7C22	OEM
C7C24	OEM
C7C27	OEM
C7C30	OEM
C7C33	OEM
C7C36	OEM
C7C39	OEM
C7C43	OEM
C7C47	OEM
C7C51	OEM
C7C56	OEM
C7C62	OEM
C7C68	OEM
C7C75	OEM
C7C82	OEM
C7C91	OEM
C7C100	OEM
C7C110	OEM
C7C120	OEM
C7C130	OEM
C7C150	OEM
C7C160	OEM
C7C180	OEM
C7C200	OEM
C7C400	OEM
C7F	OEM
C7G	OEM
C7H	OEM
C7U	OEM
C8	0693
C8P03Q	1236
C8P04Q	1236
C8P05Q	1236
C8P06Q	1236
C8P04Q	OEM
C9A	OEM
C9B	OEM
C9C	OEM
C9F	OEM
C9G	OEM
C9H	OEM
C9U	OEM
C-10	0196
C10	0196
C10-04C	0196
C10-1B	0196
C10-2NA	0196
C10-12	OEM
C10-12A	2766
C10-13B	0196
C10-15B	0196
C10-15C	0196
C10-16A	0196
C10-16B	0196
C10-18B	0196
C-10-20A	1149
C10-20A	0913
C10-22C	1149
C10-31A	0196
C10-38C	0196
C10-47B	0196
C10A	2491
C10AR1200	OEM
C10B	0240
C10BR1200	OEM
C10C	2659
C10D	0676
C10D41200	OEM
C10F	1103
C10G	0240
C10H	2659
C10P03Q	2515
C10P04Q	2515
C10P05Q	OEM
C10P06Q	OEM
C10P09Q	OEM
C10P10F	OEM
C10P10FU	1236

DEVICE TYPE	REPL CODE
C10P10FUR	1727
C10P10Q	OEM
C10P20F	OEM
C10P20FU	1236
C10P20FUR	1727
C10P30F	0972
C10P30FR	OEM
C10P40F	0972
C10P40FR	OEM
C10U	3419
C11	0208
C11(2S)	0208
C11A	2491
C11A(TRANS)	0208
C11AR1200	OEM
C11A-U	0240
C11B	0240
C11B(TRANS)	0208
C11BR1200	OEM
C11C	2659
C11C(TRANS)	0208
C11CA	OEM
C11CF	OEM
C11CJ	OEM
C11D	0676
C11D(TRANS)	0208
C11DA	OEM
C11DR700	OEM
C11DR1200	OEM
C11E	2809
C11E(TRANS)	0208
C11F	1103
C11F(TRANS)	0208
C11G	0240
C11G(TRANS)	0208
C11GN	0208
C11H	2659
C11H(TRANS)	0208
C11J	0208
C11K	0208
C11L	0208
C11M	0336
C11M(TRANS)	0208
C110R	0208
C11R	0208
C11S1C1E1C	0015
C11U	1103
C11X	0208
C11Y	0208
C12	0086
C12-12	1420
C12A	OEM
C12A(TRANS)	0086
C12B	OEM
C12B(TRANS)	0086
C12C	OEM
C12C(TRANS)	0086
C12D	OEM
C12D(TRANS)	0086
C12E	0086
C12F	OEM
C12F(TRANS)	0086
C12G	OEM
C12G(TRANS)	0086
C12GN	0086
C12H	OEM
C12H(TRANS)	0086
C12J	0086
C12K	0086
C12L	0086
C12M	0086
C120R	0086
C12R	0086
C12U	OEM
C12X	0086
C12Y	0086
C13	0600
C13A	0600
C13B	0600
C13C	0600
C13D	0600
C13E	0600
C13F	OEM
C13F(2S)	0600
C13G	0600
C13GN	0600
C13H	0600
C13J	0600
C13K	0600
C13L	0600
C13M	0600
C130R	0600
C13R	0600
C13X	0600
C13Y	OEM
C13Y(2S)	0600
C14(2S)	0600
C14A	0600
C14B	0600
C14C	0600
C14D	0600
C14E	0600
C14F	0600
C14G	0600
C14GN	0600
C14H	0600
C14J	0600
C14K	0600
C14L	0600
C14M	0600
C140R	0600
C14R	0600
C14X	0600
C14Y	0600
C15	0016
C15-0	0016
C15-1(2S)	0016
C15-2(2S)	0016
C15-3(2S)	0016
C15A	2491
C15A(TRANS)	0016
C15A-U	0240
C15B	0240
C15B(TRANS)	0016
C15C	2659
C15C(TRANS)	0016
C15D	0676
C15D(TRANS)	0016
C15E	2809
C15E(TRANS)	0016
C15F	1103
C15F(TRANS)	0016
C15G	0240
C15G(TRANS)	0016
C15GN	0016
C15H	0676
C15H(TRANS)	0016
C15J	0016
C15K	0016
C15L	0016
C15M	0336
C15M(TRANS)	0016
C150R	0016
C15R	0016
C15U	1103
C15X	0016
C15Y	0016
C16(2S)	0016
C16A(2S)	0016
C16B	0016
C16C	0016
C16D	0016
C16E	0016
C16F	0016
C16G	0016
C16GN	0016
C16H	0016
C16J	0016
C16K	0016
C16L	0016
C16M	0016
C160R	0016
C16P03Q	1942
C16P04Q	2234
C16P05Q	2234
C16P06Q	2234
C16P09Q	2234
C16P10F	2274
C16P10FU	2234
C16P10FUR	OEM
C16P10Q	2234
C16P20F	2274
C16P20FU	2234
C16P20FUR	OEM
C16P30F	4488
C16P30FR	OEM
C16P40F	4488
C16P40FR	OEM
C16R	0016
C16X	0016
C16Y	0016
C17(2S)	0016
C17A(2S)	0016
C17B	0016
C17C	0016
C17D	0016
C17E	0016
C17F	0016
C17G	0016
C17GN	0016
C17H	0016
C17J	0016
C17K	0016
C17L	0016
C17M	0016
C170R	0016
C17R	0016
C17X	0016
C17Y	0016
C18(2S)	0079
C18A	0079
C18B	0079
C18C	0079
C18D	0079
C18E	0079
C18F	0079
C18G	0079
C18GN	0079
C18H	0079
C18J	0079
C18K	0079
C18L	0079
C18M	0079
C180R	0079
C18R	0079
C18X	0079
C19(2S)	0086
C19A	0086
C19B	0086
C19C	0086
C19D	0086
C19E	0086
C19F	0086
C19G	0086
C19GN	0086
C19H	0086
C19J	0086
C19K	0086
C19L	0086
C19M	0086
C190R	0086
C19R	0086
C19X	0086
C19Y	0086
C20(2S)	0086
C20A	2449
C20A(TRANS)	0086
C20A-U	0240
C20B	0435
C20B(TRANS)	0086
C20C	0722
C20C(TRANS)	0086
C20D	0687
C20D(TRANS)	0086
C20E	0725
C20E(TRANS)	0086
C20F	2449
C20F(TRANS)	0086
C20G	0086
C20GN	0086
C20H	0086
C20J	0086
C20K	0086
C20L	0086
C20M	0725
C20M(TRANS)	0086
C200R	0086
C20R	0086
C20U	2449
C20X	0086
C20Y	0086
C21	0103
C21(2S)	0103
C21A	0103
C21B	0103
C21C	0103
C21D	0103
C21E	0103
C21F	0103
C21G	0103
C21GN	0103
C21H	0103
C21J	0103
C21K	0103
C21L	0103
C21M	0103
C210R	0103
C21R	0103
C21X	0103
C21Y	0103
C22(2S)	0086
C22A	1651
C22A(TRANS)	0086
C22B	1651
C22B(TRANS)	0086
C22C	1584
C22D	1584
C22D(TRANS)	0086
C22E	1665
C22E(TRANS)	0086
C22F	1651
C22F(TRANS)	0086
C22G	0086
C22GN	0086
C22H	0086
C22J	0086
C22K	0086
C22L	0086
C22M	0086
C220R	0086
C22R	0086
C22U	1651
C22X	0086
C22Y	0086
C23(2S)	0086
C23A	0086
C23B	0086
C23D	0086
C23E	0086
C23F	0086
C23G	0086
C23GN	0086
C23H	0086
C23H12B	0015
C23J	0086
C23K	0086
C23L	0086
C23M	0086
C230R	0086
C23R	0086
C23X	0086
C23Y	0086
C24(2S)	0086
C24A	0086
C24B	0086
C24C	0086
C24D	0086
C24E	0086
C24F	0086
C24G	0086
C24GN	0086
C24H	0086
C24H5F	OEM
C24H10F	OEM
C24H15F	OEM
C24H20F	OEM
C24H30F	OEM
C24H40F	OEM
C24J	0086
C24K	0086
C24L	0086
C24M	0086
C240R	0086
C24R	0086
C24X	0086
C24Y	0086
C25	1085
C25-06M	0283
C25P03Q	1942
C25P04Q	1942
C25P05Q	OEM
C25P06Q	OEM
C25P09Q	OEM
C25P10FU	2234
C25P10FUR	OEM
C25P10Q	OEM
C25P20FU	2234
C25P20FUR	OEM
C25P30F	4488
C25P30FR	OEM
C25P40F	4488
C25P40FR	OEM
C26(2S)	0016
C26A	0016
C26B	0016
C26C	0016
C26D	0016
C26E	0016
C26F	0016
C26G	0016
C26GN	0016
C26H	0016
C26J	0016
C26K	0016
C26L	0016
C26M	0016
C260R	0016
C26R	0016
C26X	0016
C26Y	0016
C27(2S)	0016
C27A	0016
C27B	0016
C27D	0016
C27E	0016
C27F	0016
C27G	0016
C27GN	0016
C27H	0016
C27J	0016
C27K	0016
C27L	0016
C27M	0016
C270R	0016
C27R	0016
C27X	0016
C27Y	0016
C28(2S)	0198
C28A	0198
C28B	0198
C28C	0198
C28D	0198
C28E	0198
C28F	0198
C28G	0198
C28GN	0198
C28H	0198
C28J	0198
C28K	0198
C28L	0198
C28M	0198
C280R	0198
C28R	0198
C28X	0198
C28Y	0198
C29(2S)	0198
C29A	0198
C29B	0198
C29C	0198
C29D	0198
C29E	0198
C29F	0198
C29G	0198
C29GN	0198
C29H	0198
C29J	0198
C29K	0198
C29L	0198
C29M	0198
C290R	0198
C29R	0198
C29X	0198
C29Y	0198
C30(2S)	0086
C30-1	0725
C30-2	0725
C30-5	0725
C30-12	OEM
C30-OR	0086
C30A	0567
C30A(TRANS)	0086
C30B	0762
C30B(TRANS)	0086
C30C	3271
C30C(TRANS)	0086
C30D	0740
C30D(TRANS)	0086
C30E	0752
C30E(TRANS)	0086
C30F	2188
C30F(TRANS)	0086
C30G	0086
C30GN	0086
C30H	0086
C30H03Q	OEM
C30H04Q	OEM
C30H05Q	OEM
C30H06Q	OEM
C30J	0086
C30L	0086
C30M	0725
C30M(TRANS)	0086
C300R	0086
C30R	0086
C30U	3307
C30X	0086
C30Y	0086
C31(2S)	0086
C31A	OEM
C31A(TRANS)	0086
C31B	OEM
C31B(TRANS)	0086
C31C	5898
C31C(TRANS)	0086
C31D	OEM
C31D(TRANS)	0086
C31E	OEM
C31E(TRANS)	0086
C31F	OEM
C31F(TRANS)	0086
C31G	0086
C31GN	0086
C31H	0086
C31J	0086
C31K	0086
C31L	0086
C31M	0086
C310R	0086
C31R	0086
C31U	OEM
C31X	0086
C31Y	0086
C32(2S)	0086
C32A	0086
C32A(2S)	0086
C32B	1651
C32B(TRANS)	0086
C32C	1584
C32C(TRANS)	0086
C32D	OEM
C32D(TRANS)	0086
C32E	1665
C32E(TRANS)	0086
C32F	OEM
C32F(TRANS)	0086
C32G	0086
C32GN	0086
C32H	0086
C32J	0086
C32K	0086
C32L	0086
C32M	0086
C320R	0086
C32R	0086
C32U	2519
C32X	0086
C32Y	0086
C33(2S)	0079
C33A	OEM
C33A(TRANS)	0079
C33B	OEM
C33B(TRANS)	0079
C33C	OEM
C33C(TRANS)	0079
C33D	OEM
C33D(TRANS)	0079
C33E	OEM
C33E(TRANS)	0079
C33F	OEM
C33F(TRANS)	0079
C33G	0079
C33GN	0079
C33H	0079
C33J	0079
C33K	0079
C33L	0079
C33M	0079
C330R	0079
C33R	0079
C33U	OEM
C33X	0079
C33Y	0079
C34	0208
C34A	0208
C34A1	1650
C34A2	OEM
C34B1	1650
C34B2	OEM
C34C	0208
C34C1	2647
C34C2	1855
C34D	0208
C34D1	2647
C34D2	0740
C34E	0208
C34E1	2649
C34E2	0764
C34F	0208
C34F1	1650
C34F2	2188
C34G	0208
C34GN	0208
C34H	0208
C34J	0208
C34K	0208
C34L	0208
C34M	0208
C340R	0208
C34R	0208
C34X	0208
C34Y	0208
C35	0208
C35A	0712
C35A(TRANS)	0208
C35AR1200	OEM
C35B	0469
C35B(TRANS)	0208
C35BR1200	OEM
C35C	0721
C35C(TRANS)	0208
C35D	0722
C35DR700	OEM
C35DR1200	OEM
C35E	0778
C35E(TRANS)	0208
C35ER1200	OEM
C35F	0731
C35F(TRANS)	0208
C35G	0469
C35G(TRANS)	0208
C35GN	0208
C35H	0721
C35H(TRANS)	0208
C35J	0208
C35K	0208
C35L	0208
C35M	0725
C35M(TRANS)	0208
C35MR1200	OEM
C35N	0750
C350R	0208
C35P	OEM
C35R	0208
C35S	0750
C35T	OEM
C35U	3277
C35X	0208
C35Y	0208
C36	0208
C36A	0741
C36B	0745
C36B(TRANS)	0208
C36C	2917
C36C(TRANS)	0208
C36D	0747
C36D(TRANS)	0208
C36E	3243
C36E(TRANS)	0208
C36F	2519
C36F(TRANS)	0208
C36G	3105
C36G(TRANS)	0208
C36GN	0208
C36H	3109
C36H(TRANS)	0208
C36J	0208
C36K	0208
C36L	0208
C36M	0725
C36M(TRANS)	0208
C36N	0679
C360R	0208
C36R	0208
C36S	OEM
C36U	3102
C36X	0208
C36Y	0208
C37(2S)	0016
C37A	0567
C37A(TRANS)	0016
C37B	0762
C37B(TRANS)	0016
C37C	3271
C37C(TRANS)	0016
C37D	0740
C37D(TRANS)	0016
C37E	3291
C37E(TRANS)	0016
C37F	2188
C37F(TRANS)	0016
C37G	0722
C37G(TRANS)	0016
C37GN	0016
C37H	0016
C37J	0016
C37K	0016
C37L	0016
C37M	0764
C37M(TRANS)	0016
C37N	0766
C370R	0016
C37R	0016
C37S	2875
C37U	3307
C37X	0016
C37Y	0016
C38	0079
C38(2S)	0016
C38A	0731
C38A(TRANS)	0016
C38B	0469
C38B(TRANS)	0016
C38BR1200	OEM
C38C	0721
C38C(TRANS)	0016
C38D	0722
C38D(TRANS)	0016
C38DR1200	OEM
C38E	0778
C38E(TRANS)	0016
C38F	0731
C38F(TRANS)	0016
C38G	0469
C38G(TRANS)	0016
C38H	0721
C38H(TRANS)	0016
C38HR1200	OEM
C38J	0016
C38K	0016
C38L	0016
C38M	0725
C38M(TRANS)	0016
C380R	0016
C38R	0016
C38U	3277
C38X	0016
C38Y	0016
C39	0144
C39-207	0016
C39A(2S)	0144
C39B	0144
C39C	0144
C39D	0144
C39E	0144
C39F	0144
C39G	0144
C39GN	0144
C39H	0144
C39J	0144
C39K	0144
C39L	0144
C39M	0144
C390R	0144
C39R	0144
C39X	0144
C39Y	0144
C40(2S)	0144
C40A	OEM
C40A(TRANS)	0144
C40B	OEM
C40B(TRANS)	0144
C40C	OEM
C40C(TRANS)	0144
C40D	OEM
C40D(TRANS)	0144
C40E	OEM
C40E(TRANS)	0144
C40F	OEM
C40F(TRANS)	0144
C40G	3408
C40G(TRANS)	0144
C40GN	0144
C40H	OEM
C40H(TRANS)	0144
C40J	0144
C40K	0144
C40L	0144
C40M	OEM
C40N	OEM
C400R	0144
C40R	0144
C40S	OEM
C40U	OEM
C40X	0144
C40Y	0144
C41(2S)	0620
C41B	0620
C41C	0620
C41F	0620
C41GN	0620
C41H	0620
C41J	0620
C41M	0620
C410R	0620
C41R	0620
C41TV(2S)	0620
C41X	0620
C41Y	0620
C42(2S)	0074
C42A(2S)	0074
C42B	0074
C42C	0074
C42E	0074

If replacement code is OEM, contact original manufacturer for replacement.

DEVICE TYPE	REPL CODE
C42F	0074
C42G	0074
C42GN	0074
C42H	0074
C42J	0074
C42K	0074
C42L	0074
C42M	0074
C42OR	0074
C42R	0074
C42X	0074
C42Y	0074
C43(2S)	0074
C44(2S)	0074
C44-04	0023
C45(2S)	0016
C45A	0217
C45B	0608
C45C	0610
C45C10	OEM
C45C12	OEM
C45D	0610
C45D1	OEM
C45D2	OEM
C45D3	OEM
C45D4	OEM
C45D5	OEM
C45D6	OEM
C45E	0610
C45F	0217
C45G	0217
C45H	0610
C45M	0610
C45N	0468
C45P	0468
C45PA	0468
C45PB	0468
C45S	0468
C45T	0468
C45U	0217
C46(2S)	0086
C46A	0217
C46A(TRANS)	0086
C46B	0641
C46B(TRANS)	0086
C46C	0217
C46C(TRANS)	0086
C46D	0217
C46DB	0086
C46E	0217
C46E(TRANS)	0086
C46F	0217
C46F(TRANS)	0086
C46G	0217
C46G(TRANS)	0086
C46GN	0086
C46H	0217
C46H(TRANS)	0086
C46J	0086
C46K	0086
C46L	0086
C46M	0217
C46M(TRANS)	0086
C46N	0658
C46OR	0086
C46P	0658
C46PA	0658
C46PB	0658
C46S	0658
C46T	0658
C46U	0217
C46X	0086
C46Y	0086
C47(2S)	0086
C47A	0086
C47B	0086
C47C	0086
C47D	0086
C47E	0086
C47G	0086
C47GN	0086
C47H	0086
C47J	0086
C47K	0086
C47L	0086
C47M	0086
C47OR	0086
C47R	0086
C47X	0086
C47Y	0086
C48(2S)	0086
C48A	0086
C48B	0086
C48C(2S)	0086
C48D	0086
C48E	0086
C48F	0086
C48G	0086
C48GN	0086
C48H	0086
C48J	0086
C48K	0086
C48L	0086
C48M	OEM
C48M(TRANS)	0086
C48N	OEM
C48OR	0086
C48P	OEM
C48PA	OEM
C48PB	OEM
C48R	0086
C48S	OEM
C48T	OEM
C48X	0086
C48Y	0086
C49(2S)	0086
C49A	OEM
C49A(TRANS)	0086
C49A10	0526
C49A20	0526
C49B	OEM
C49B(TRANS)	0086
C49B10	0526
C49B20	0526
C49C	OEM
C49C(TRANS)	0086
C49C10	0526
C49C20	0526
C49D	OEM
C49D(TRANS)	0086
C49D10	0526
C49D20	0526
C49E	OEM
C49E(TRANS)	0086
C49E10	0526
C49E20	0526
C49F	0086
C49G	OEM
C49G(TRANS)	0086
C49GN	0086
C49H	0086
C49J	0086
C49K	0086
C49L	0086
C49M	OEM
C49M(TRANS)	0086
C49M10	0526
C49M20	0526
C49OR	0086
C49X	0086
C49Y(2S)	0086
C50	0600
C50A	0600
C50A(2S)	0600
C50AX500	OEM
C50B	0217
C50B(TRANS)	0600
C50BX500	OEM
C50C	0217
C50C(TRANS)	0600
C50CX500	OEM
C50D	0217
C50DX500	OEM
C50E	0610
C50E(TRANS)	0600
C50EX500	OEM
C50F	0217
C50F(TRANS)	0600
C50G	0217
C50G(TRANS)	0600
C50GN	0600
C50H	0217
C50H(TRANS)	0600
C50J	0600
C50K	0600
C50L	0600
C50M	0610
C50M(TRANS)	0600
C50MX500	OEM
C50N	0468
C50OR	0600
C50P	0468
C50PA	0468
C50PB	0468
C50R	0600
C50S	0468
C50SX500	OEM
C50T	0468
C50U	0217
C50X	0600
C50Y	0600
C51(2S)	0086
C51A	0086
C51B	0086
C51C	0086
C51D	0086
C51E	0086
C51F	0086
C51G	0086
C51GN	0086
C51H	0086
C51J	0086
C51K	0086
C51L	0086
C51M	0086
C51OR	0086
C51PC	OEM
C51R	0086
C51X	0086
C51Y	0086
C52	0016
C52A	0217
C52A(TRANS)	0016
C52AX500	OEM
C52B	0217
C52B(TRANS)	0016
C52BX500	OEM
C52C	0217
C52C(TRANS)	0016
C52CX500	OEM
C52D	0217
C52D(TRANS)	0016
C52DX500	OEM
C52E	0217
C52E(TRANS)	0016
C52EX500	OEM
C52F	0217
C52F(TRANS)	0016
C52G	0217
C52G(TRANS)	0016
C52GN	0016
C52H	0217
C52H(TRANS)	0016
C52J	0016
C52K	0016
C52L	0016
C52M	0217
C52M(TRANS)	0016
C52MX500	OEM
C52N	0658
C52OR	0016
C52P	0658
C52PA	0658
C52PB	0658
C52S	0658
C52T	0658
C52U	0217
C52X	0016
C52Y	0016
C53	0016
C53A	0016
C53B	0016
C53C	0016
C53D	0016
C53E	0016
C53F	0016
C53G	0016
C53GN	0016
C53H	0016
C53J	0016
C53L	0016
C53OR	0016
C53R	0016
C53X	0016
C53Y	0016
C54(2S)	0016
C54A	0016
C54B	0016
C54C	0016
C54D	0016
C54E	0016
C54F	0016
C54G	0016
C54GN	0016
C54H	0016
C54J	0016
C54K	0016
C54L	0016
C54M	0016
C54OR	0016
C54X	0016
C54Y	0016
C55(2S)	0016
C55A	OEM
C55A(TRANS)	0016
C55B	OEM
C55B(TRANS)	0016
C55C	OEM
C55C(TRANS)	0016
C55D	OEM
C55D(TRANS)	0016
C55E	OEM
C55E(TRANS)	0016
C55F	OEM
C55F(TRANS)	0016
C55G	OEM
C55G(TRANS)	0016
C55GN	0016
C55H	OEM
C55H(TRANS)	0016
C55J	0016
C55K	0016
C55L	0016
C55M	OEM
C55M(TRANS)	0016
C55N	OEM
C55OR	0016
C55R	0016
C55S	OEM
C55U	OEM
C55X	0016
C55Y	0016
C56(2S)	0127
C56A	OEM
C56A(TRANS)	0127
C56B	OEM
C56B(TRANS)	0127
C56C	OEM
C56C(TRANS)	0127
C56D	OEM
C56D(TRANS)	0127
C56E	OEM
C56E(TRANS)	0127
C56F	OEM
C56F(TRANS)	0127
C56G	OEM
C56G(TRANS)	0127
C56GN	0127
C56H	OEM
C56H(TRANS)	0127
C56J	0127
C56K	0127
C56L	0127
C56M	OEM
C56M(TRANS)	0127
C56N	OEM
C56OR	0127
C56R	0127
C56S	OEM
C56U	OEM
C56X	0127
C56Y	0127
C57	0595
C58(2S)	0233
C58A(2S)	0233
C58AC	0233
C58B	0233
C58D	0233
C58E	0233
C58F	0233
C58G	0233
C58GN	0233
C58H	0233
C58J	0233
C58K	0233
C58L	0233
C58M	0233
C58OR	0233
C58R	0233
C58X	0233
C58Y	0233
C59(2S)	0086
C59A	0086
C59B	0086
C59C	0086
C59D	0086
C59E	0086
C59F	0086
C59G	0086
C59GN	0086
C59H	0086
C59J	0086
C59K	0086
C59L	0086
C59M	0086
C59OR	0086
C59R	0086
C59X	0086
C59Y	0086
C60	0208
C60A	OEM
C60A(TRANS)	0208
C60B	OEM
C60B(TRANS)	0208
C60C	OEM
C60C(TRANS)	0208
C60D	OEM
C60D(TRANS)	0208
C60E	OEM
C60E(TRANS)	0208
C60F	OEM
C60F(TRANS)	0208
C60G	OEM
C60G(TRANS)	0208
C60GN	0208
C60H	OEM
C60H(TRANS)	0208
C60H03Q	OEM
C60H04Q	OEM
C60H05Q	OEM
C60H06Q	OEM
C60J	0208
C60K	0208
C60L	0208
C60M	0208
C60OR	0208
C60R	0208
C60U	OEM
C60X	0208
C60Y	0208
C61(2S)	0086
C61A	0086
C61B	0086
C61C	0086
C61D	0086
C61E	0086
C61F	0086
C61G	0086
C61GN	0086
C61H	0086
C61J	0086
C61K	0086
C61L	0086
C61M	0086
C61OR	0086
C61R	0086
C61Y	0086
C62	0016
C62A	OEM
C62A(TRANS)	0016
C62B	OEM
C62B(TRANS)	0016
C62C	OEM
C62C(TRANS)	0016
C62D	OEM
C62D(TRANS)	0016
C62E	OEM
C62E(TRANS)	0016
C62F	OEM
C62F(TRANS)	0016
C62G	OEM
C62G(TRANS)	0016
C62GN	0016
C62H	OEM
C62H(TRANS)	0016
C62J	0016
C62L	0016
C62M	0016
C62OR	0016
C62R	0016
C62U	OEM
C62X	0016
C62Y	0016
C63	0016
C63(2S)	0144
C63A	0144
C63B	0144
C63D	0144
C63E	0144
C63F	0144
C63G	0144
C63GN	0144
C63H	0144
C63J	0144
C63K	0144
C63L	0144
C63M	0144
C63OR	0144
C63R	0144
C63X	0144
C63Y	0144
C64	0086
C64(2S)	0086
C64A	0086
C64B	0086
C64C	0086
C64D	0086
C64E	0086
C64F	0086
C64G	0086
C64GN	0086
C64H	0086
C64K	0086
C64L	0086
C64M	0086
C64OR	0086
C64R	0086
C64X	0086
C64Y	0086
C64Y-RST	0086
C65(2S)	0233
C65-0	0233
C65-OR	0233
C65A	0233
C65B(2S)	0233
C65C	0233
C65D	0233
C65E	0233
C65F	0233
C65G	0233
C65GN	0233
C65H	0233
C65K	0233
C65L	0233
C65M	0233
C65N(2S)	0233
C65OR	0233
C65R	0233
C65X	0233
C65Y(2S)	0233
C65YA(2S)	0233
C65YB(2S)	0233
C65YTV(2S)	0233
C65YTV1	0233
C66(2S)	0233
C66A	0233
C66B	0233
C66C	0233
C66D	0233
C66E	0233
C66EV	0233
C66F	0233
C66G	0233
C66GN	0233
C66H	0233
C66J	0233
C66K	0233
C66L	0233
C66M	0233
C66OR	0233
C66R	0233
C66X	0233
C66Y	0233
C67(2S)	0016
C67A	0016
C67B	0016
C67C	0016
C67D	0016
C67E	0016
C67F	0016
C67G	0016
C67GN	0016
C67H	0016
C67J	0016
C67K	0016
C67L	0016
C67M	0016
C67R	0016
C67X	0016
C67Y	0016
C68(2S)	0016
C68A	0016
C68B	0016
C68C	0016
C68D	0016
C68E	0016
C68F	0016
C68G	0016
C68GN	0016
C68H	0016
C68J	0016
C68K	0016
C68L	0016
C68M	0016
C68OR	0016
C68R	0016
C68X	0016
C68Y	0016
C69(2S)	0086
C69A	0086
C69B	0086
C69C	0086
C69D	0086
C69E	0086
C69F	0086
C69G	0086
C69GN	0086
C69H	0086
C69J	0086
C69K	0086
C69L	0086
C69M	0086
C69OR	0086
C69R	0086
C69X	0086
C69Y	0086
C70(2S)	0233
C70A	0233
C70B	0233
C70C	0233
C70D	0233
C70E	0233
C70F	0233
C70G	0233
C70GN	0233
C70H	0233
C70J	0233
C70K	0233
C70L	0233
C70M	0233
C70OR	0233
C70R	0233
C70X	0233
C70Y	0233
C71(2S)	0600
C71A	0600
C71B	0600
C71C	0600
C71D	0600
C71E	0600
C71F	0600
C71G	0600
C71GN	0600
C71H	0600
C71J	0600
C71M	0600
C71OR	0600
C71R	0600
C71X	0600
C72(2S)	0600
C72A	0600
C72B	0600
C72C	0600
C72D	0600
C72E	0600
C72F	0600
C72G	0600
C72GN	0600
C72H	0600
C72J	0600
C72K	0600
C72L	0600
C72M	0600
C72OR	0600
C72R	0600
C72X	0600
C72Y	0600
C73	0280
C73(2S)	0600
C73A	0600
C73B	0600
C73C	0600
C73D	0600
C73E	0600
C73F	0600
C73G	0600
C73GN	0600
C73H	0600
C73J	0600
C73K	0600
C73L	0600
C73LBGL	0600
C73M	0600
C73OR	0600
C73R	0600
C73X	0600
C73Y	0600
C74(2S)	0198
C74-GR	0198
C74-O	0198
C74-R	0198
C74-Y	0198
C74A	0198
C74B	0198
C74C	0198
C74D	0198
C74E	0198
C74F	0198
C74G	0198
C74GN	0198
C74H	0198
C74J	0198
C74K	0198
C74L	0198
C74LS00P	1527
C74LS02P	1559
C74LS03P	1579
C74LS04P	1595
C74LS05P	1608
C74LS08P	1633
C74LS10P	1662
C74LS11P	1667
C74LS13P	1688
C74LS14P	1698
C74LS20P	0034
C74LS27P	0183
C74LS30P	0828
C74LS32P	0088
C74LS42P	1841
C74LS51P	1033
C74LS54P	1857
C74LS55P	0457
C74LS74P	2504
C74LS86P	0291
C74LS90P	1882
C74LS93P	1888
C74LS109P	1906
C74LS136P	1628
C74LS138P	0427
C74LS151P	1646
C74LS153P	0959
C74LS155P	0209
C74LS157P	1161
C74LS158P	1656
C74LS170P	2629
C74LS174P	0260
C74LS175P	1672
C74LS181P	1678
C74LS190P	1686
C74LS191P	1687
C74LS192P	1689
C74LS193P	1692
C74LS196P	2834
C74LS197P	2469
C74LS251P	1737
C74LS253P	1739
C74LS257P	1744
C74LS258P	1746
C74LS259P	3205
C74LS279P	3290
C74LS298P	3371
C74LS367P	0977
C74LS368P	0991
C74LS670P	1130
C74M	0198
C74R	0198
C74X	0198
C74Y	0198
C75	0280
C75(2S)	0600
C75A	0600
C75B	0600
C75B-1	0600
C75C	0600
C75E	0600
C75F	0600
C75G	0600
C75GN	0600
C75H	0600
C75J	0600
C75K	0600
C75L	0600
C75M	0600
C75OR	0600
C75R	0600
C75X	0600
C75Y	0600
C76	0280
C76(2S)	0600
C76A	0600
C76B	0600
C76C	0600
C76D	0600
C76E	0600
C76F	0600
C76G	0600
C76GN	0600
C76H	0600
C76J	0600
C76K	0600
C76L	0600
C76M	0600
C76OR	0600

DEVICE TYPE	REPL CODE
C76R	0600
C76X	0600
C76Y	0600
C77(2S)	0600
C77A	0600
C77B	0600
C77C(2S)	0600
C77D	0600
C77F	0600
C77G	0600
C77GN	0600
C77H	0600
C77J	0600
C77K	0600
C77L	0600
C77M	0600
C77OR	0600
C77R	0600
C77X	0600
C77Y	0600
C77Z	0600
C78(2S)	0600
C78A	0600
C78B	0600
C78C	0600
C78D	0600
C78E	0600
C78F	0600
C78G	0600
C78GN	0600
C78H	0600
C78J	0600
C78K	0600
C78L	0600
C78M	0600
C78N05	OEM
C78OR	0600
C78R	0600
C78X	0600
C78Y	0600
C79(2S)	0224
C79A	0224
C79B	0224
C79C	0224
C79D	0224
C79E	0224
C79F	0224
C79G	0224
C79GN	0224
C79H	0224
C79J	0224
C79K	0224
C79L	0224
C79M	0224
C79OR	0224
C79R	0224
C79X	0224
C79Y	0224
C80	0007
C80(2S)	0007
C80A	0738
C80A(TRANS)	0007
C80B	0738
C80B(TRANS)	0007
C80C	0738
C80C(TRANS)	0007
C80C39	OEM
C80C49	OEM
C80D	0738
C80D(TRANS)	0007
C80E	0738
C80E(TRANS)	0007
C80F	0738
C80F(TRANS)	0007
C80G	0007
C80G(TRANS)	0007
C80GN	0007
C80H	OEM
C80H(TRANS)	0007
C80H03Q	OEM
C80H04Q	OEM
C80H05Q	OEM
C80H06Q	OEM
C80J	0007
C80K	0007
C80L	0007
C80M	0738
C80M(TRANS)	0007
C80N	0738
C80OR	0007
C80R	0007
C80S	0738
C80U	OEM
C80X	0007
C80Y	0007
C81	0018
C81-004	4299
C81A	0018
C81B	0018
C81C	0018
C81D	0018
C81E	0018
C81F	0018
C81G	0018
C81GN	0018
C81H	0018
C81J	0018
C81K	0018
C81L	0018
C81M	0018
C81OR	0018
C81R	0018
C82	0127
C82-004	1021
C82BN	0127
C82M	3266
C82R	0127
C82S	0144
C83	0127
C83-829	0015
C83-880	0015
C84	OEM
C84(2S)	0600
C84-009	3604
C85	OEM
C85(2S)	0208
C850	0208
C86	0208
C87(2S)	0016
C87A	0016
C87B	0016
C87C	0016
C87D	0016
C87E	0016
C87F	0016
C87G	0016
C87GN	0016
C87H	0016
C87J	0016
C87K	0016
C87L	0016
C87M	0016
C87OR	0016
C87R	0016
C87X	0016
C87Y	0016
C88(2S)	0086
C88A(2S)	0086
C88B	0086
C88C	0086
C88D	0086
C88E	0086
C88F	0086
C88G	0086
C88GN	0086
C88H	0086
C88J	0086
C88K	0086
C88L	0086
C88M	0086
C88OR	0086
C88R	0086
C88X	0086
C88Y	0086
C89(2S)	0600
C89A	0600
C89B	0600
C89C	0600
C89D	0600
C89E	0600
C89F	0600
C89G	0600
C89GN	0600
C89H	0600
C89J	0600
C89K	0600
C89L	0600
C89M	0600
C89OR	0600
C89R	0600
C89X	0600
C89Y	0600
C90(2S)	0600
C90A	0600
C90B	0600
C90C	0600
C90D	0600
C90E	0600
C90F	0600
C90G	0600
C90GN	0600
C90H	0600
C90J	0600
C90K	0600
C90L	0600
C90M	0600
C90OR	0600
C90R	0600
C90X	0600
C90Y	0600
C91	0038
C91(2S)	0600
C91A	0600
C91B	0600
C91C	0600
C91D	0600
C91E	0600
C91F	0600
C91G	0600
C91GN	0600
C91H	0600
C91J	0600
C91K	0600
C91L	0600
C91M	0600
C91OR	0600
C91R	0600
C91X	0600
C91Y	0600
C92	0622
C92(2S)	0042
C92-0025-R0	0823
C93	0622
C93(2S)	0042
C94	0622
C94(2S)	0042
C94A	OEM
C94AP	0042
C94E	2950
C94EG	OEM
C94ER	OEM
C95	0086
C95(2S)	0086
C95A	0086
C95A(2S)	0086
C95B	0086
C95C	0086
C95D	0086
C95E	OEM
C95E(2S)	0086
C95EG	OEM
C95ER	OEM
C95F	0086
C95G	0086
C95GN	0086
C95H	0086
C95J	0086
C95K	0086
C95L	0086
C95M	0086
C95OR	0086
C95R	0086
C95X	0086
C95Y	0086
C96	0079
C96E	OEM
C96EG	OEM
C97	0086
C97(2S)	0086
C97A(2S)	0086
C97B	0086
C97C	0086
C97D	0086
C97E	0086
C97EG	OEM
C97ER	OEM
C97F	0086
C97G	0086
C97GN	0086
C97H	0086
C97J	0086
C97K	0086
C97L	0086
C97M	0086
C97OR	0086
C97R	0086
C97X	0086
C97Y	0086
C98(2S)	0016
C98A	0016
C98B	0016
C98C	0016
C98D	0016
C98E	OEM
C98E(2S)	0016
C98EG	OEM
C98ER	OEM
C98F	0016
C98G	0016
C98GN	0016
C98H	0016
C98J	0016
C98K	0016
C98L	0016
C98M	0016
C98OR	0016
C98R	0016
C98X	0016
C98Y	0016
C99(2S)	0016
C99A	0016
C99B	0016
C99C	0016
C99D	0016
C99E	0016
C99F	0016
C99G	0016
C99GN	0016
C99H	0016
C99J	0016
C99K	0016
C99L	0016
C99M	0016
C99OR	0016
C99R	0016
C99X	0016
C99Y	0016
C100(2S)	0016
C100-OY(2S)	0016
C100A	0016
C100B	0016
C100C	0016
C100D	0016
C100E	0016
C100F	0016
C100G	0016
C100GN	0016
C100J	0016
C100L	0016
C100M	0016
C100OR	0016
C100R	0016
C100X	0016
C100Y	0016
C101	0142
C101(2S)	0074
C101A(2S)	0074
C101B(2S)	0074
C101C	0074
C101D	0074
C101E	0074
C101F	0074
C101G	0074
C101GN	0074
C101H	0074
C101J	0074
C101K	0074
C101L	0074
C101M	0074
C101OR	0074
C101R	0074
C101X(2S)	0074
C101XL	0074
C101Y	0074
C102	0126
C102R472	OEM
C103	0016
C103(2S)	0016
C103A	OEM
C103A(2S)	0016
C103A(SCR)	0901
C103B	0058
C103B(2S)	0016
C103C	0016
C103D	0016
C103E	0016
C103F	0016
C103G	0016
C103GN	0016
C103H	0016
C103J	0016
C103K	0016
C103L	0016
C103M	0016
C103Q	1137
C103X	0016
C103Y	1137
C103Y(2S)	0016
C103YY	0344
C104	0016
C104(2S)	0016
C104A(2S)	0016
C104B	0016
C104C	0016
C104D	0016
C104E	0016
C104F	0016
C104G	0016
C104GN	0016
C104H	0016
C104J	0016
C104K	0016
C104L	0016
C104M	0016
C104OR	0016
C104R	0016
C104X	0016
C104Y	0016
C105(2S)	0016
C105B	0016
C106	0622
C106(2S)	0622
C106A	0447
C106A(2S)	0622
C106A1	0447
C106A2	0447
C106A3	0447
C106A4	0447
C106A11	OEM
C106A12	OEM
C106A21	OEM
C106A32	OEM
C106A41	OEM
C106B	0940
C106B(2S)	0622
C106B1	0940
C106B2	0940
C106B3	0940
C106B4	0940
C106B11	OEM
C106B12	OEM
C106B21	OEM
C106B32	OEM
C106B41	OEM
C106C	1222
C106C(2S)	0622
C106C1	1222
C106C2	1222
C106C3	1222
C106C4	1222
C106C11	OEM
C106C12	OEM
C106C21	OEM
C106C32	OEM
C106C41	OEM
C106D	0095
C106D1	0095
C106D2	0095
C106D3	0095
C106D4	0095
C106D11	OEM
C106D12	OEM
C106D21	OEM
C106D32	OEM
C106D41	OEM
C106E	OEM
C106E1	0308
C106E2	1243
C106E3	1243
C106E4	1243
C106E11	OEM
C106E12	OEM
C106E21	OEM
C106E32	OEM
C106F	1259
C106F1	1259
C106F2	1259
C106F3	1259
C106F4	1259
C106F11	OEM
C106F12	OEM
C106F21	OEM
C106F32	OEM
C106F41	OEM
C106G	0622
C106G1	0940
C106G2	0940
C106G3	0940
C106G4	OEM
C106GN	0622
C106H	0622
C106J	0622
C106K	0622
C106L	0622
C106M	OEM
C106M(TRANS)	0622
C106M1	0308
C106M2	1243
C106M3	1243
C106M4	1243
C106M11	OEM
C106M12	OEM
C106M21	OEM
C106M32	OEM
C106OR	0622
C106Q	1396
C106Q1	1396
C106Q2	1396
C106Q3	1396
C106Q4	1396
C106Q11	OEM
C106Q12	OEM
C106Q21	OEM
C106Q32	OEM
C106Q41	OEM
C106R	0622
C106X	0622
C106Y	1396
C106Y(TRANS)	0622
C106Y1	1396
C106Y2	1396
C106Y3	1396
C106Y4	1396
C106Y11	OEM
C106Y12	OEM
C106Y21	OEM
C106Y32	OEM
C106Y41	OEM
C107	0042
C107A	0447
C107A1	0447
C107A2	0447
C107A3	0447
C107A4	OEM
C107A11	OEM
C107A12	OEM
C107A21	OEM
C107A32	OEM
C107A41	OEM
C107B	0940
C107B1	0940
C107B2	0940
C107B3	0940
C107B4	0940
C107B11	OEM
C107B12	OEM
C107B21	OEM
C107B32	OEM
C107B41	OEM
C107C	1222
C107C1	1222
C107C2	1222
C107C3	1222
C107C4	1222
C107C11	OEM
C107C12	OEM
C107C21	OEM
C107C32	OEM
C107C41	4427
C107D	0095
C107D1	0095
C107D2	0095
C107D3	0095
C107D4	0095
C107D11	OEM
C107D12	OEM
C107D21	OEM
C107D32	OEM
C107D41	OEM
C107E	OEM
C107E1	0308
C107E2	OEM
C107E3	OEM
C107E4	OEM
C107E11	OEM
C107E12	OEM
C107E21	OEM
C107E32	OEM
C107E41	OEM
C107F	1259
C107F1	1259
C107F2	1259
C107F3	1259
C107F4	1259
C107F11	OEM
C107F12	OEM
C107F21	OEM
C107F32	OEM
C107F41	OEM
C107G4	0042
C107G5	0042
C107G6	0042
C107M	OEM
C107M1	0308
C107M2	OEM
C107M3	OEM
C107M4	OEM
C107M11	OEM
C107M12	OEM
C107M21	OEM
C107M32	OEM
C107M41	OEM
C107Q	1396
C107Q1	1396
C107Q2	1396
C107Q3	1396
C107Q4	1396
C107Q11	OEM
C107Q12	OEM
C107Q21	OEM
C107Q32	OEM
C107Q41	OEM
C107Y	1396
C107Y1	1396
C107Y2	1396
C107Y3	1396
C107Y4	1396
C107Y11	OEM
C107Y12	OEM
C107Y21	OEM
C107Y32	OEM
C107Y41	OEM
C108(2S)	0086
C108A	0447
C108A(TRANS)	0086
C108A1	0447
C108A2	0447
C108A3	0447
C108A4	0447
C108A11	OEM
C108A12	OEM
C108A21	OEM
C108A32	OEM
C108A41	OEM
C108A-O	0086
C108A-R	0086
C108B	0940
C108B(TRANS)	0086
C108B1	0940
C108B2	0940
C108B3	0940
C108B4	0940
C108B11	OEM
C108B12	OEM
C108B21	OEM
C108B32	OEM
C108B41	OEM
C108C	0095
C108C(TRANS)	0086
C108C1	0095
C108C2	0095
C108C3	0095
C108C4	0095
C108C11	OEM
C108C12	OEM
C108C21	OEM
C108C32	OEM
C108C41	OEM
C108D	0095
C108D(TRANS)	0086
C108D1	0095
C108D2	0095
C108D3	0095
C108D4	0095
C108D11	OEM
C108D12	OEM
C108D21	OEM
C108D32	OEM
C108D41	OEM
C108E	OEM
C108E(TRANS)	0086
C108E1	OEM
C108E2	1502
C108E3	1502
C108E4	1502
C108E11	OEM
C108E12	OEM
C108E21	OEM
C108E32	OEM
C108E41	OEM
C108F	1259
C108F(TRANS)	0086
C108F1	1259
C108F2	1259
C108F3	1259
C108F4	1259
C108F11	OEM
C108F12	OEM
C108F21	OEM
C108F32	OEM
C108F41	OEM
C108G	0086
C108GN	0086
C108H	0086
C108J	0086
C108K	0086
C108L	0086
C108M	OEM
C108M(TRANS)	0086
C108M1	OEM
C108M2	1502
C108M3	1502
C108M4	1502
C108M11	OEM
C108M12	OEM
C108M21	OEM
C108M32	OEM
C108M41	OEM
C108OR	0086
C108Q	OEM
C108Q1	OEM
C108Q2	1396
C108Q3	1396
C108Q4	1396
C108Q11	OEM
C108Q12	OEM
C108Q21	OEM
C108Q32	OEM
C108Q41	OEM
C108R	0086
C108X	0086
C108Y	OEM
C108Y(TRANS)	0086
C108Y1	OEM
C108Y2	1396
C108Y3	1396
C108Y4	1396
C108Y11	OEM
C108Y12	OEM
C108Y21	OEM
C108Y32	OEM
C108Y41	OEM
C109(2S)	0086
C109A(2S)	0086
C109A-O	0086
C109A-R	0086
C109A-Y	0086
C109B	0086
C109C	0086
C109D	0086
C109E	0086
C109F	0086
C109G	0086
C109G1	0086
C109GN	0086
C109J	0086
C109K	0086
C109L	0086
C109OR	0086
C109R	0086
C109X	0086
C109Y	0086
C110(2S)	0016
C110A	0016
C110B	0016
C110C	0016
C110D	0016
C110E	0016
C110F	0016
C110G	0016
C110GN	0016
C110H	0016
C110J	0016
C110K	0016
C110L	0016
C110M	0016
C110OR	0016
C110R	0016
C110X	0016
C110Y	0016
C111(2S)	0016
C111A	0016
C111B	0144
C111B(TRANS)	0016
C111C	0016
C111D	0016
C111E	0016
C111E(TRANS)	0016
C111F	0016
C111G	0016
C111GN	0016
C111H	0016
C111J	0016

If replacement code is OEM, contact original manufacturer for replacement.

DEVICE TYPE	REPL CODE
C111K	0016
C111L	0016
C111M	0016
C111OR	0016
C111R	0016
C111X	0016
C111Y	0016
C112	2229
C112(2S)	0086
C112A	0086
C112B	0086
C112C	0086
C112D	0086
C112E	0086
C112F	0086
C112G	0086
C112GN	0086
C112H	0086
C112J	0086
C112K	0086
C112L	0086
C112M	0086
C112OR	0086
C112R	0086
C112X	0086
C112Y	0086
C113(2S)	0086
C113A	0086
C113B	0086
C113C	0086
C113D	0086
C113E	0086
C113F	0086
C113G	0086
C113GN	0086
C113H	0086
C113J	0086
C113K	0086
C113L	0086
C113M	0086
C113OR	0086
C113R	0086
C113X	0086
C113Y	0086
C114(2S)	0086
C114A	0086
C114B	0086
C114C	0086
C114D	0086
C114E	0086
C114E(DT)	0832
C114E(SWITCH)	OEM
C114EF	2099
C114ESB	OEM
C114ESC	OEM
C114ESN	0832
C114F	0086
C114G	0086
C114GN	0086
C114H	0086
C114J	0086
C114K	0086
C114L	0086
C114M	0086
C114OR	0086
C114R	0086
C114X	0086
C114Y	0086
C114YSR	OEM
C115(2S)	0086
C115-1	0086
C115-2	0086
C115-3	0086
C115-43	0086
C115A	0086
C115B	0086
C115C	0086
C115D	0086
C115E	0086
C115F	0086
C115G	0086
C115GN	0086
C115H	0086
C115J	0086
C115K	0086
C115L	0086
C115M	0086
C115OR	0086
C115R	0086
C115X	0086
C115Y	0086
C116(2S)	2061
C116-OR	2061
C116A	2089
C116A(TRANS)	2061
C116B	0505
C116B(TRANS)	2061
C116C	0710
C116C(TRANS)	2061
C116D	0710
C116D(TRANS)	2061
C116E	2061
C116F	2095
C116F(TRANS)	2061
C116G	2061
C116GN	2061
C116H	2061
C116J	2061
C116K	2061
C116L	2061
C116M(TRANS)	2061
C116OR	2061
C116R	2061
C116T(2S)	2061
C116X	2061
C116Y	2061
C117(2S)	0086
C117A	0086
C117B	0086
C117C	0086
C117D	0086
C117F	0086
C117G	0086
C117GN	0086
C117H	0086
C117J	0086
C117K	0086
C117L	0086
C117M	0086
C117OR	0086
C117R	0086
C117X	0086
C117Y	0086
C118	0126
C118(2S)	0086
C118A	OEM
C118A(TRANS)	0086
C118B	OEM
C118B(TRANS)	0086
C118C	OEM
C118C(TRANS)	0086
C118D	OEM
C118D(TRANS)	0086
C118E	OEM
C118E(TRANS)	0086
C118F	OEM
C118F(TRANS)	0086
C118G	0086
C118GN	0086
C118H	0086
C118J	0086
C118L	0086
C118M	OEM
C118M(TRANS)	0086
C118OR	0086
C118R	0086
C118X	0086
C118Y	0086
C119	0126
C119(2S)	0086
C119A	0086
C119B	0086
C119C	0086
C119D	0086
C119E	0086
C119F	0086
C119G	0086
C119GN	0086
C119H	0086
C119J	0086
C119K	0086
C119L	0086
C119M	0086
C119OR	0086
C119R	0086
C119X	0086
C119Y	0086
C120(2S)	0016
C120A	0016
C120B	0016
C120C	0016
C120D	0016
C120E	0016
C120F	0016
C120G	0016
C120GN	0016
C120H	0016
C120H03Q	OEM
C120H04Q	OEM
C120H05Q	OEM
C120H06Q	OEM
C120J	0016
C120K	0016
C120M	0016
C120R	0016
C120Y	0016
C121(2S)	0086
C121A	0086
C121B	0086
C121C	0086
C121D	0086
C121E	0086
C121F	0086
C121G	0086
C121GN	0086
C121H	0086
C121J	0086
C121L	0086
C121M	0086
C121OR	0086
C121R	0086
C121X	0086
C121Y	0086
C122	0079
C122(2S)	0086
C122A	2089
C122A(TRANS)	0086
C122A1	2089
C122A2	2089
C122A3	2089
C122A4	2089
C122A5	2089
C122A6	2089
C122B	0505
C122B(TRANS)	0086
C122B1	2095
C122B2	0505
C122B3	0505
C122B4	0505
C122B5	0505
C122B6	2095
C122BB	0505
C122C	0710
C122C(TRANS)	0086
C122C1	0710
C122C2	0710
C122C3	0710
C122C4	0710
C122C5	0710
C122C6	0710
C122D	0710
C122D(TRANS)	0086
C122D1	0710
C122D2	0710
C122D3	0710
C122D4	0710
C122D5	0710
C122D6	0710
C122E	0863
C122E(TRANS)	0086
C122E1	0863
C122E2	OEM
C122E3	OEM
C122E4	OEM
C122E5	OEM
C122E6	OEM
C122F	2095
C122F(TRANS)	0086
C122F1	2095
C122F2	2095
C122F3	2095
C122F4	2095
C122F5	2095
C122F6	2095
C122G	0086
C122GN	0086
C122H	0086
C122J	0086
C122K	0086
C122L	0086
C122M	0863
C122M(TRANS)	0086
C122M1	0863
C122M2	OEM
C122M3	OEM
C122M4	OEM
C122M5	OEM
C122M6	OEM
C122N1	0327
C122OR	0086
C122R	0086
C122S	OEM
C122UX5	2095
C122X	0086
C122Y	0086
C123	0086
C123A	0086
C123B	0086
C123C	0086
C123D	0086
C123E	0086
C123F	0086
C123G	0086
C123GN	0086
C123H	0086
C123IFD	OEM
C123J	0086
C123K	0086
C123L	0086
C123M	0086
C123MDD	OEM
C123MFD	OEM
C123OR	0086
C123R	0086
C123X	0086
C123Y	0086
C124(2S)	0086
C124A	2089
C124A(TRANS)	0086
C124B	0505
C124B(TRANS)	0086
C124C	0710
C124C(TRANS)	0086
C124D	0710
C124D(TRANS)	0086
C124E	0863
C124E(TRANS)	0086
C124EL	OEM
C124ES	0887
C124F	2095
C124F(TRANS)	0086
C124G	0086
C124GN	0086
C124H	0086
C124J	0086
C124K	0086
C124L	0086
C124M	0863
C124M(TRANS)	0086
C124OR	0086
C124R	0086
C124X	0086
C124Y	0086
C125(2S)	0086
C125AL	OEM
C126	0285
C126A	0397
C126A2	OEM
C126A3	OEM
C126A4	OEM
C126A5	OEM
C126A6	OEM
C126B	0397
C126B2	OEM
C126B3	OEM
C126B4	OEM
C126B5	OEM
C126B6	OEM
C126C	0611
C126C2	OEM
C126C3	OEM
C126C4	OEM
C126C5	OEM
C126C6	OEM
C126D	0611
C126D2	OEM
C126D3	OEM
C126D4	OEM
C126D5	OEM
C126D6	OEM
C126E	0952
C126E2	OEM
C126E3	OEM
C126E4	OEM
C126E5	OEM
C126E6	OEM
C126F	2521
C126F2	OEM
C126F3	OEM
C126F4	OEM
C126F5	OEM
C126F6	OEM
C126M	0952
C126M2	OEM
C126M3	OEM
C126M4	OEM
C126M5	OEM
C126M6	OEM
C127(2S)	0016
C127A	0397
C127A(TRANS)	0016
C127B	0397
C127B(TRANS)	0016
C127C	0611
C127C(TRANS)	0016
C127D	0611
C127D(TRANS)	0016
C127E	0952
C127E(TRANS)	0016
C127F	2521
C127F(TRANS)	0016
C127G	0016
C127GN	0016
C127H	0016
C127J	0016
C127K	0016
C127L	0016
C127M	0952
C127M(TRANS)	0016
C127OR	0016
C127R	0016
C127X	0016
C127Y	0016
C128(2S)	0600
C128A	0600
C128B	0600
C128C	0600
C128D	0600
C128D/DOS	OEM
C128E	0600
C128F	0600
C128G	0600
C128GN	0600
C128H	0600
C128J	0600
C128K	0600
C128L	0600
C128M	0600
C128OR	0600
C128R	0600
C128X	0600
C128Y	0600
C129	0600
C129A	0600
C129B	0600
C129C	0600
C129D	0600
C129E	0600
C129F	0600
C129GN	0600
C129H	0600
C129J	0600
C129K	0600
C129L	0600
C129M	0600
C129OR	0600
C129R	0600
C129X	0600
C129Y	0600
C130(2S)	0086
C130A	0086
C130B	0086
C130C	0086
C130D	0086
C130E	0086
C130F	0086
C130GN	0086
C130H	0086
C130J	0086
C130K	0086
C130L	0086
C130M	0086
C130OR	0086
C130R	0086
C130X	0086
C130Y	0086
C131(2S)	0016
C131A	0016
C131B	0016
C131C	0016
C131D	0016
C131E	0016
C131F	0016
C131G	0016
C131GN	0016
C131H	0016
C131J	0016
C131L	0016
C131M	0016
C131OR	0016
C131R	0016
C131T	0016
C131Y	0016
C132(2S)	0016
C132A	0016
C132B	0016
C132C	0016
C132D	0016
C132E	0016
C132F	0016
C132G	0016
C132GN	0016
C132H	0016
C132J	0016
C132K	0016
C132L	0016
C132M	0016
C132OR	0016
C132R	0016
C132X	0016
C132Y	0016
C133(2S)	0016
C133A	0016
C133C	0016
C133D	0016
C133E	0016
C133F	0016
C133G	0016
C133GN	0016
C133H	0016
C133J	0016
C133K	0016
C133L	0016
C133M	0016
C133OR	0016
C133X	0016
C133Y	0016
C134(2S)	0016
C134A	0016
C134B(2S)	0016
C134C	0016
C134D	0016
C134E	0016
C134F	0016
C134G	0016
C134GN	0016
C134H	0016
C134J	0016
C134K	0016
C134L	0016
C134M	0016
C134OR	0016
C134R	0016
C134X	0016
C134Y	0016
C135(2S)	0016
C135A	0712
C135A(TRANS)	0016
C135B	0469
C135B(TRANS)	0016
C135C	0721
C135C(TRANS)	0016
C135D	0722
C135D(TRANS)	0016
C135E	0778
C135E(TRANS)	0016
C135F	0731
C135F(TRANS)	0016
C135G	0016
C135GN	0016
C135H	0016
C135J	0016
C135L	0016
C135M	0725
C135M(TRANS)	0016
C135N	0679
C135OR	0016
C135R	0016
C135S	0679
C135X	0016
C135Y	0016
C136(2S)	0016
C136D	OEM
C136D(2S)	0016
C136E	0725
C136M	0764
C136N	0750
C136S	OEM
C137	0016
C137A	0016
C137B	0016
C137C	0016
C137D	0016
C137E	0778
C137E(TRANS)	0016
C137ER1200	OEM
C137F	0016
C137G	0016
C137GN	0016
C137H	0016
C137J	0016
C137K	0016
C137L	0016
C137M	0725
C137M(TRANS)	0016
C137MR1200	OEM
C137N	0750
C137NR1200	OEM
C137OR	0016
C137P	OEM
C137PA	OEM
C137PB	0679
C137PBR1200	OEM
C137PR1200	OEM
C137R	0016
C137S	0750
C137T	0679
C137X	0016
C137Y	0016
C138	0016
C138A(2S)	0016
C138B	0016
C138C	0016
C138D	0016
C138E	0016
C138E10	OEM
C138E10E	0804
C138E20	OEM
C138E20E	3408
C138F	0016
C138G	0016
C138GN	0016
C138H	0016
C138J	0016
C138L	0016
C138M	0016
C138M10	OEM
C138M10M	0804
C138M20	OEM
C138M20M	0804
C138N10	OEM
C138N10M	OEM
C138N20	OEM
C138N20M	OEM
C138OR	0016
C138R	0016
C138S	0016
C138S10	OEM
C138S10A	OEM
C138S10M	OEM
C138S20	OEM
C138S20M	OEM
C138X	0016
C139	0016
C139A	0016
C139B	0016
C139C	0016
C139D	0016
C139E	0016
C139E10	OEM
C139E10E	0804
C139E20	OEM
C139E20E	3408
C139F	0016
C139G	0016
C139GN	0016
C139H	0016
C139J	0016
C139K	0016
C139M	0016
C139M1M0	OEM
C139M10	OEM
C139M10M	0804
C139M20	OEM
C139M20M	0804
C139N10	OEM
C139N10M	OEM
C139N20	OEM
C139N20M	OEM
C139OR	0016
C139R	0016
C139S10	OEM
C139S10M	OEM
C139S20	OEM
C139S20M	OEM
C139X	0016
C139Y	0016
C140	0086
C140A	0804
C140A(TRANS)	0086
C140B	0804
C140C	0804
C140C(TRANS)	0086
C140D	0804
C140D(TRANS)	0086
C140E	0086
C140F	0804
C140F(TRANS)	0086
C140G	0086
C140GN	0086
C140H	0086
C140J	0086
C140K	0086
C140L	0086
C140M	0086
C140OR	0086
C140R	0086
C140X	0086
C140Y	0086
C141	0079
C141A	0804
C141B	0804
C141C	0804
C141D	0804
C141F	0804
C142	0079
C143	0079
C143E(DT)	OEM
C144	0079
C144A	0079
C144E	0898
C144E(DT)	0898
C144E15	OEM
C144E15E	0804
C144E30	OEM
C144E30E	0804
C144ES	0898
C144ESD	0898
C144M15	OEM
C144M15M	0804
C144M30	OEM
C144M30M	0804
C144N15	OEM
C144N15M	OEM
C144N30	OEM
C144N30M	OEM
C144S15	OEM
C144S15M	OEM
C144S30	OEM
C144S30M	OEM
C145	0079
C145E	OEM
C145M	OEM
C145N	OEM
C145P	OEM
C145PA	OEM
C145PB	OEM
C145S	OEM
C146	0079
C146A	0400
C146A(2S)	0079
C146B	0016
C146C	OEM
C146D	OEM
C146E	OEM
C146F	OEM
C146M	OEM
C146N	OEM
C146P	OEM
C146S	OEM
C146T	OEM
C147(2S)	2354
C147A	1704
C147A(TRANS)	2700
C147B	1704
C147B(TRANS)	2700
C147C	4031
C147C(TRANS)	2700
C147D	4031
C147D(TRANS)	2700
C147E	4031
C147F	1704
C147F(TRANS)	2700
C147G	1704
C147G(TRANS)	2700
C147GN	2700
C147H	4031
C147H(TRANS)	2700
C147J	2700
C147K	2700
C147L	2700
C147M	4031
C147M(TRANS)	2700
C147N	0679
C147P	0679
C147PA	0679
C147PB	0679
C147R	2700
C147S	0679
C147T	0679
C147U	1704
C147X	2700
C147Y	2700

If replacement code is OEM, contact original manufacturer for replacement.

DEVICE TYPE	REPL CODE
C148	0144
C148A	0144
C148B	0144
C148D	0144
C148E	0144
C148F	0144
C148G	0144
C148GN	0144
C148H	0144
C148J	0144
C148K	0144
C148L	0144
C148M	OEM
C148M(TRANS)	0144
C148M30	OEM
C148M40	OEM
C148N	OEM
C148N30	OEM
C148N40	OEM
C148OR	0144
C148P	OEM
C148P30	OEM
C148P40	OEM
C148PA	OEM
C148PA30	OEM
C148PA40	OEM
C148PB	OEM
C148PB30	OEM
C148PB40	OEM
C148R	0144
C148S	OEM
C148S30	OEM
C148S40	OEM
C148T	OEM
C148T30	OEM
C148T40	OEM
C148X	0144
C148Y	0144
C149	0086
C149A	OEM
C149A10	OEM
C149A20	OEM
C149B	OEM
C149B10	OEM
C149B20	OEM
C149C	OEM
C149C10	OEM
C149C20	OEM
C149D	OEM
C149D10	OEM
C149D20	OEM
C149E	OEM
C149E10	OEM
C149E20	OEM
C149G	OEM
C149M	OEM
C149M10	OEM
C149M20	OEM
C150(2S)	2377
C150-OR	2377
C150A	2377
C150B	2377
C150C	2377
C150D	2377
C150E	0610
C150E(TRANS)	2377
C150F	2377
C150G	2377
C150GN	2377
C150H	2377
C150J	2377
C150K	2377
C150L	2377
C150M	0610
C150M(TRANS)	2377
C150N	0468
C150OR	2377
C150P	0468
C150PA	0468
C150PB	0468
C150PC	OEM
C150R	2377
C150S	0468
C150T	0086
C150T(2S)	2377
C150T(SCR)	0468
C150X	2377
C150Y	2377
C151(2S)	0198
C151A	0198
C151B	0198
C151C	0198
C151D	0198
C151E	OEM
C151E(TRANS)	0198
C151G	0198
C151GN	0198
C151H	0198
C151HA	0198
C151HB	0198
C151HC	0198
C151J	0198
C151K	0198
C151L	0198
C151M	OEM
C151M(TRANS)	0198
C151N	OEM
C151OR	0198
C151P	OEM
C151PA	OEM
C151PB	OEM
C151R	0198
C151S	OEM
C151T	OEM
C151X	0198
C151Y	0198
C152	0198
C152A	0198
C152B	0198
C152C	0198
C152D	0198
C152E	0641
C152E(TRANS)	0198
C152F	0198
C152G	0198
C152GN	0198
C152H	0198
C152HA	0198
C152HB	0198
C152HC	0198
C152J	0198
C152K	0198
C152L	0198
C152M	0641
C152M(TRANS)	0198
C152N	0217
C152OR	0198
C152P	0658
C152PA	0658
C152PB	0658
C152PC	OEM
C152R	0198
C152S	0658
C152T	0658
C152X	0198
C152Y	0198
C153	0079
C153E	OEM
C153M	OEM
C153N	OEM
C153P	OEM
C153PA	OEM
C153PB	OEM
C153PC	OEM
C153T	OEM
C154(2S)	0716
C154-OR	0716
C154A	0526
C154A(TRANS)	0716
C154B	0042
C154B(SCR)	0526
C154B(TRANS)	0716
C154C	0233
C154C(SCR)	0526
C154C(TRANS)	0716
C154D	0526
C154D(TRANS)	0716
C154E	0526
C154E(TRANS)	0716
C154F	0716
C154G	0526
C154G(TRANS)	0716
C154GN	0716
C154H(2S)	0716
C154HA	0716
C154HB	0716
C154HC	0716
C154J	0716
C154K	0716
C154L	0716
C154M	0526
C154M(TRANS)	0716
C154OR	0716
C154R	0716
C154X	0716
C154Y	0716
C155	0127
C155(2S)	0127
C155A	0526
C155A(TRANS)	0127
C155B	0526
C155B(TRANS)	0127
C155C	0526
C155C(TRANS)	0127
C155D	0526
C155D(TRANS)	0127
C155E	0526
C155E(TRANS)	0127
C155F	0127
C155G	0526
C155GN	0127
C155H	0127
C155J	0127
C155K	0127
C155L	0127
C155M	0526
C155M(TRANS)	0127
C155OR	0127
C155P	OEM
C155R	0127
C155X	0127
C155Y	0127
C156(2S)	0127
C156A	0655
C156A(TRANS)	0127
C156B	0655
C156B(TRANS)	0127
C156C	0655
C156C(TRANS)	0127
C156D	0655
C156D(TRANS)	0127
C156E	0655
C156E(TRANS)	0127
C156F	0127
C156G	0655
C156G(TRANS)	0127
C156GN	0127
C156H	0127
C156J	0127
C156K	0127
C156L	0127
C156M	0655
C156M(TRANS)	0127
C156OR	0127
C156R	0127
C156X	0127
C156Y	0127
C157(2S)	0198
C157A	0655
C157A(TRANS)	0198
C157B	0655
C157B(TRANS)	0198
C157C	0655
C157C(TRANS)	0198
C157D	0655
C157D(TRANS)	0198
C157E	0655
C157E(TRANS)	0198
C157F	0198
C157G	0264
C157G(TRANS)	0198
C157GN	0198
C157J	0198
C157K	0198
C157L	0198
C157M	0655
C157M(TRANS)	0198
C157OR	0198
C157R	0198
C157Y	0198
C158(2S)	0198
C158A	0198
C158B	0198
C158C	0198
C158D	0198
C158E	0526
C158E(TRANS)	0198
C158F	0198
C158G(TRANS)	0198
C158GN	0198
C158H	0198
C158J	0198
C158K	0198
C158L	0198
C158M	0526
C158M(TRANS)	0198
C158N	0108
C158OR	0198
C158P	0108
C158PA	0108
C158PB	0108
C158R	0198
C158S	0108
C158T	0108
C158X	0198
C158Y	0198
C159(2S)	0016
C159A	0016
C159B	0016
C159C	0016
C159D	0016
C159E	0655
C159E(TRANS)	0016
C159F	0016
C159GN	0016
C159H	0016
C159J	0016
C159K	0016
C159L	0016
C159M	0020
C159M(TRANS)	0016
C159N	0020
C159OR	0016
C159P	0020
C159PA	0020
C159PB	0020
C159R	0016
C159S	0020
C159T	0020
C159X	0016
C159Y	0016
C160(2S)	0016
C160A	0016
C160B	0016
C160C	0016
C160D	0016
C160E	0016
C160F	0016
C160G	0016
C160GN	0016
C160H	0016
C160J	0016
C160K	0016
C160L	0016
C160M	0016
C160OR	0016
C160R	0016
C160X	0016
C160Y	0016
C161	0543
C162	0079
C163(2S)	0493
C163A	0493
C163B	0493
C163C	0493
C163D	0493
C163E	0493
C163F	0493
C163G	0493
C163GN	0493
C163H	0493
C163J	0493
C163K	0493
C163L	0493
C163M	0493
C163OR	0493
C163R	0493
C163X	0493
C163Y	0493
C164	0018
C164A	0526
C164A(TRANS)	0018
C164B	0526
C164B(TRANS)	0018
C164C	0526
C164C(TRANS)	0716
C164D	0526
C164D(TRANS)	0018
C164E	0526
C164E(TRANS)	0018
C164F	0018
C164G	0018
C164GN	0018
C164H	0018
C164J	0018
C164K	0018
C164L	0018
C164M	0526
C164OR	0018
C164R	0018
C164X	0018
C164Y	0018
C165	0595
C165A	0526
C165B	0526
C165C	0526
C165D	0526
C165E	0526
C165M	0526
C165N	0108
C165S	0108
C166	0016
C166(2S)	0016
C166A	0016
C166B	0016
C166C	0016
C166D	0016
C166E	0016
C166F	0016
C166G	0016
C166GN	0016
C166H	0016
C166J	0016
C166K	0016
C166L	0016
C166M	0016
C166OR	0016
C166P	OEM
C166R	0016
C166X	0016
C166Y	0016
C167(2S)	0016
C167A	0016
C167B	0016
C167C	0016
C167D	0016
C167E	0016
C167F	0016
C167G	0016
C167GN	0016
C167H	0016
C167K	0016
C167L	0016
C167M	0016
C167OR	0016
C167X	0016
C167Y	0016
C168	0079
C169	OEM
C169(2S)	0016
C170(2S)	0016
C170A	0016
C170B	0016
C170C	0016
C170D	0016
C170E	0016
C170F	0016
C170G	0016
C170GN	0016
C170H	0016
C170J	0016
C170K	0016
C170L	0016
C170M	0016
C170OR	0016
C170R	0016
C170X	0016
C170Y	0016
C171(2S)	0016
C171A	0016
C171B	0016
C171C	0016
C171D	0016
C171E	0016
C171F	0016
C171G	0016
C171GN	0016
C171H	0016
C171J	0016
C171K	0016
C171L	0016
C171M	0016
C171OR	0016
C171R	0016
C171X	0016
C171Y	0016
C172(2S)	0016
C172A(2S)	0016
C172B	0016
C172C	0016
C172D	0016
C172E	0016
C172F	0016
C172G	0016
C172GN	0016
C172H	0016
C172J	0016
C172K	0016
C172L	0016
C172M	0016
C172OR	0016
C172R	0016
C172Y	0016
C173(2S)	0600
C173A	0600
C173B	0600
C173C	0600
C173D	0600
C173E	0600
C173F	0600
C173G	0600
C173GN	0600
C173H	0600
C173J	0600
C173K	0600
C173L	0600
C173M	0600
C173OR	0600
C173R	0600
C173X	0600
C173Y	0600
C174(2S)	0016
C174A(2S)	0016
C174B	0016
C174C	0016
C174D	0016
C174E	0016
C174F	0016
C174G	0016
C174GN	0016
C174H	0016
C174J	0016
C174K	0016
C174L	0016
C174M	0016
C174OR	0016
C174Q	0016
C174R	0016
C174S	0016
C174X	0016
C174Y	0016
C175(2S)	0600
C175A	0600
C175B(2S)	0600
C175BL	0600
C175C	0600
C175D	0600
C175E	0600
C175F	0600
C175G	0600
C175GN	0600
C175H	0600
C175J	0600
C175K	0600
C175L	0600
C175M	0600
C175OR	0600
C175R	0600
C175X	0600
C175Y	0600
C176(2S)	0600
C176A	0600
C176B	0600
C176C	0600
C176D	0600
C176E	0600
C176F	0600
C176G	0600
C176GN	0600
C176H	0600
C176J	0600
C176K	0600
C176L	0600
C176OR	0600
C176R	0600
C176X	0600
C176Y	0600
C177(2S)	0600
C177A	0600
C177B	0600
C177C	0600
C177D	0600
C177E	0600
C177F	0600
C177G	0600
C177GN	0600
C177H	0600
C177J	0600
C177K	0600
C177L	0600
C177OR	0600
C177R	0600
C177X	0600
C177Y	0600
C178(2S)	0600
C178A	0738
C178A(TRANS)	0600
C178B	0738
C178B(TRANS)	0600
C178C	0738
C178C(TRANS)	0600
C178D	0738
C178D(TRANS)	0600
C178E	0738
C178E(TRANS)	0600
C178F	0600
C178G	0600
C178H	0600
C178J	0600
C178K	0600
C178L	0600
C178M	0738
C178M(TRANS)	0600
C178N	0738
C178OR	0600
C178P	0738
C178PA	0738
C178PB	0738
C178R	0600
C178S	0738
C178T	0738
C178X	0600
C178Y	0600
C179(2S)	0208
C179A	0600
C179B	0208
C179C	0208
C179D	0208
C179E	0208
C179F	0208
C179G	0208
C179GN	0208
C179H	0208
C179J	0208
C179K	0208
C179L	0208
C179M	0208
C179OR	0208
C179R	0208
C179X	0208
C179Y	0208
C180	0208
C180A	0738
C180A(TRANS)	0208
C180AX500	1900
C180B	0738
C180B(TRANS)	0208
C180BX500	1900
C180C	0738
C180C(TRANS)	0208
C180CX500	1900
C180D	0738
C180D(TRANS)	0208
C180DX500	1900
C180E	1086
C180E(TRANS)	0208
C180EX500	1900
C180F	0208
C180G	0208
C180GN	0208
C180H	0208
C180J	0208
C180K	0208
C180L	0208
C180M	1086
C180M(TRANS)	0208
C180MX500	0738
C180N	1102
C180OR	0208
C180P	1102
C180PA	1102
C180PB	1102
C180PC	OEM
C180R	0208
C180S	1102
C180SX500	0738
C180T	1102
C180X	0208
C180Y	0208
C181(2S)	0208
C181A	0738
C181A(TRANS)	0208
C181B	0738
C181B(TRANS)	0208
C181C	0738
C181C(TRANS)	0208
C181D	0738
C181D(TRANS)	0208
C181E	0738
C181E(TRANS)	0208
C181F	0208
C181G	0208
C181GN	0208
C181H	0208
C181J	0208
C181L	0208
C181M	0738
C181M(TRANS)	0208
C181N	0738
C181OR	0208
C181P	0738
C181PA	0738
C181PB	0738
C181R	0208
C181S	0738
C181T	0738
C181X	0208
C181Y	0208
C182(2S)	0475
C182A	0475
C182B	0475
C182C	0475
C182D	0475
C182E	0475
C182F	0475
C182G	0475
C182GN	0475
C182H	0475
C182J	0475
C182K	0475
C182L	0475
C182M	0475
C182Q(2S)	0475
C182R	0475
C182V	0475
C182X	0475
C182Y	0475
C183(2S)	0475
C183-1	0475
C183-OR	0475
C183A	0475
C183AP	0475
C183B	0475
C183BK	0475
C183C	0475
C183D	0475
C183E(2S)	0475
C183F	0475
C183G	0475
C183GN	0475
C183H	0475
C183J(2S)	0475
C183K(2S)	0475
C183L(2S)	0475
C183M(2S)	0475
C183OR	0475
C183P(2S)	0475
C183Q(2S)	0475
C183R(2S)	0475
C183S	0475
C183W(2S)	0475
C183X	0475
C183Y	0475
C184(2S)	0475
C184-OR	0475
C184A	OEM
C184A(TRANS)	0475
C184AP	0475
C184B	OEM
C184B(TRANS)	0475
C184BK	0475
C184C	OEM
C184C(TRANS)	0475
C184D	OEM
C184D(TRANS)	0475
C184E	OEM
C184E(TRANS)	0475
C184F	0475
C184G	0475
C184GN	0475
C184H(2S)	0475
C184J(2S)	0475
C184K	0475
C184L	0475
C184M	OEM
C184M(TRANS)	0475
C184OR	0475
C184P	0475
C184Q	0475
C184R	0475
C184X	0475
C184Y	0475
C185(2S)	0475
C185A	0475
C185A(2S)	0475
C185A(SCR)	OEM
C185B	OEM
C185B(TRANS)	0475
C185C	OEM
C185C(TRANS)	0475
C185D	OEM
C185E	OEM

If replacement code is OEM, contact original manufacturer for replacement.

DEVICE TYPE	REPL CODE
C185E(TRANS)	0475
C185F	0475
C185G	0475
C185GN	0475
C185H	0475
C185J(2S)	0475
C185K	0475
C185L	0475
C185M	0475
C185M(2S)	0475
C185M(SCR)	OEM
C185N	OEM
C185OR	0475
C185Q(2S)	0475
C185R(2S)	0475
C185S	OEM
C185V(2S)	0475
C185X	0475
C185Y	0475
C186(2S)	0127
C186A	0127
C186B	0127
C186C	0127
C186D	0127
C186E	0127
C186F	0127
C186G	0127
C186GN	0127
C186H	0127
C186J	0127
C186K	0127
C186L	0127
C186M	0127
C186OR	0127
C186R	0127
C186X	0127
C186Y	0127
C187(2S)	0674
C187-OR	0674
C187A	0674
C187B	0674
C187C	0674
C187D	0674
C187E	0674
C187F	0674
C187G	0674
C187H	0674
C187I	0674
C187J	0674
C187K	0674
C187L	0674
C187M	0674
C187OR	0674
C187R	0674
C187X	0674
C187Y	0674
C188(2S)	0086
C188A(2S)	0086
C188B	0086
C188C	0086
C188D	0086
C188E	0086
C188F	0086
C188G	0086
C188GN	0086
C188J	0086
C188K	0086
C188L	0086
C188M	0086
C188OR	0086
C188R	0086
C188X	0086
C188Y	0086
C189(2S)	0086
C189A	0086
C189B	0086
C189C	0086
C189D	0086
C189E	0086
C189F	0086
C189G	0086
C189GN	0086
C189H	0086
C189J	0086
C189K	0086
C189L	0086
C189M	0086
C189OR	0086
C189R	0086
C189X	0086
C189Y	0086
C190(2S)	0086
C190A	0086
C190B	0086
C190C	0086
C190D	0086
C190E	0086
C190F	0086
C190GN	0086
C190H	0086
C190J	0086
C190K	0086
C190L	0086
C190M	0086
C190OR	0086
C190R	0086
C190X	0086
C190Y	0086
C191(2S)	0016
C191A	0016
C191B	0016
C191C	0016
C191D	0016
C191E	0016
C191F	0016
C191G	0016
C191GN	0016
C191H	0016
C191J	0016
C191K	0016
C191L	0016
C191M	0016
C191OR	0016
C191R	0016
C191X	0016
C191Y	0016
C192(2S)	0016
C192A	0016
C192B	0016
C192C	0016
C192D	0016
C192E	0016
C192F	0016
C192G	0016
C192GN	0016
C192H	0016
C192J	0016
C192K	0016
C192L	0016
C192M	0016
C192OR	0016
C192R	0016
C192X	0016
C192Y	0016
C193(2S)	0016
C193A	0016
C193B	0016
C193C	0016
C193D	0016
C193E	0016
C193F	0016
C193G	0016
C193GN	0016
C193H	0016
C193J	0016
C193K	0016
C193L	0016
C193M	0016
C193OR	0016
C193R	0016
C193X	0016
C194(2S)	0016
C194A	0016
C194B	0016
C194C	0016
C194D	0016
C194E	0016
C194F	0016
C194G	0016
C194GN	0016
C194H	0016
C194J	0016
C194K	0016
C194L	0016
C194M	0016
C194OR	0016
C194X	0016
C194Y	0016
C195(2S)	0016
C195A	0016
C195B	0016
C195C	0016
C195D	0016
C195E	0016
C195F	0016
C195G	0016
C195GN	0016
C195H	0016
C195J	0016
C195K	0016
C195L	0016
C195M	0016
C195OR	0016
C195R	0016
C195X	0016
C195Y	0016
C196(2S)	0016
C196A	0016
C196B	0016
C196C	0016
C196D	0016
C196E	0016
C196F	0016
C196G	0016
C196GN	0016
C196H	0016
C196J	0016
C196K	0016
C196L	0016
C196M	0016
C196OR	0016
C196R	0016
C196X	0016
C196Y	0016
C197(2S)	0016
C197A	0016
C197B	0016
C197C	0016
C197D	0016
C197E	0016
C197F	0016
C197G	0016
C197GN	0016
C197H	0016
C197J	0016
C197K	0016
C197L	0016
C197M	0016
C197OR	0016
C197R	0016
C197X	0016
C197Y	0016
C198(2S)	0530
C198A	0530
C198H	0530
C198S	0530
C198T	0530
C199(2S)	0016
C199A	0016
C199B	0016
C199C	0016
C199D	0016
C199E	0016
C199F	0016
C199G	0016
C199GN	0016
C199H	0016
C199J	0016
C199K	0016
C199L	0016
C199M	0016
C199OR	0016
C199R	0016
C199X	0016
C199Y	0016
C200	0016
C200(SCR)	1487
C200A	0016
C200B	0016
C200C	0016
C200D	0016
C200E	0016
C200F	0016
C200G	0016
C200GN	0016
C200H	0016
C200J	0016
C200K	0016
C200L	0016
C200M	0016
C200OR	0016
C200R	0016
C200X	0016
C200Y	0016
C201	0126
C201(2S)	0016
C201A	0016
C201B	0016
C201C	0016
C201D	0016
C201E	0016
C201F	0016
C201G	0016
C201GN	0016
C201H	0016
C201J	0016
C201K	0016
C201L	0016
C201M	0016
C201R	0016
C201X	0016
C201Y	0016
C202	0004
C202(2S)	0016
C202A	0016
C202B	0016
C202C	0016
C202D	0016
C202E	0016
C202F	0016
C202G	0016
C202GN	0016
C202H	0016
C202J	0016
C202K	0016
C202L	0016
C202M	0016
C202OR	0016
C202R	0016
C202X	0016
C202Y	0016
C203(2S)	0595
C203A	0901
C203A(TRANS)	0595
C203AA	0595
C203B	0058
C203B(TRANS)	0595
C203C	0407
C203C(TRANS)	0595
C203D	0407
C203D(TRANS)	0595
C203E	0595
C203F	0595
C203G	0595
C203GN	0595
C203H	0595
C203J	0595
C203K	0595
C203L	0595
C203M	0595
C203OR	0595
C203Q	OEM
C203R	0595
C203X	0595
C203Y	1137
C203Y(TRANS)	0595
C203YY	0344
C204(2S)	0016
C204A	0016
C204B	0016
C204C	0016
C204D	0016
C204E	0016
C204F	0016
C204G	0016
C204GN	0016
C204H	0016
C204J	0016
C204K	0016
C204L	0016
C204M	0016
C204OR	0016
C204R	0016
C204X	0016
C204Y	0016
C205(2S)	0016
C205A	OEM
C205A(TRANS)	0016
C205B	OEM
C205B(TRANS)	0016
C205C	OEM
C205C(TRANS)	0016
C205D	OEM
C205D(TRANS)	0016
C205F	0016
C205G	0016
C205GN	0016
C205H	0016
C205J	0016
C205K	0016
C205L	0016
C205M	0016
C205OR	0016
C205Q	OEM
C205R	0016
C205X	0016
C205Y	OEM
C205Y(TRANS)	0016
C205YY	OEM
C206(2S)	0216
C206-OR	0216
C206A	0216
C206B	0216
C206C	0216
C206D	0216
C206E	0216
C206G	0216
C206GN	0216
C206H	0216
C206J	0216
C206K	0216
C206L	0216
C206M	0216
C206OR	0216
C206R	0216
C206RED	0216
C206WHITE	0216
C206X	0216
C206Y	0216
C207	0259
C208(2S)	0007
C208A	0007
C208B	0007
C208C	0007
C208D	0007
C208E	0007
C208F	0007
C208G	0007
C208GN	0007
C208H	0007
C208J	0007
C208K	0007
C208L	0007
C208M	0007
C208OR	0007
C208R	0007
C208X	0007
C208Y	0007
C209	0259
C210	0086
C210(2S)	0086
C210A	0086
C210B	0086
C210C	0086
C210D	0086
C210E	0086
C210F	0086
C210G	0086
C210GN	0086
C210H	0086
C210J	0086
C210K	0086
C210L	0086
C210M	0086
C210OR	0086
C210R	0086
C210X	0086
C210Y	0086
C211(2S)	0086
C211A	0086
C211B	0086
C211C	0086
C211D	0086
C211F	0086
C211G	0086
C211GN	0086
C211H	0086
C211J	0086
C211K	0086
C211L	0086
C211M	0086
C211OR	0086
C211R	0086
C211X	0086
C211Y	0086
C212(2S)	0086
C212A	0086
C212B	0086
C212C	0086
C212D	0086
C212E	0086
C212F	0086
C212G	0086
C212GN	0086
C212H	0086
C212J	0086
C212K	0086
C212L	0086
C212M	0086
C212OR	0086
C212R	0086
C212X	0086
C212Y	0086
C213(2S)	0622
C213A	0622
C213B	0622
C213C	0622
C213D	0622
C213E	0622
C213F	0622
C213G	0622
C213GN	0622
C213H	0622
C213J	0622
C213K	0622
C213L	0622
C213M	0622
C213OR	0622
C213R	0622
C213X	0622
C213Y	0622
C214(2S)	0622
C214A	0622
C214B	0622
C214C	0622
C214D	0622
C214E	0622
C214F	0622
C214G	0622
C214GN	0622
C214H	0622
C214J	0622
C214K	0622
C214L	0622
C214M	0622
C214OR	0622
C214X	0622
C214Y	0622
C215(2S)	0622
C215A	0622
C215B	0622
C215C	0622
C215E	0622
C215F	0622
C215G	0622
C215GN	0622
C215H	0622
C215J	0622
C215K	0622
C215L	0622
C215M	0622
C215OR	0622
C215R	0622
C215X	0622
C215Y	0622
C216(2S)	0086
C216A	0086
C216B	0086
C216C	0086
C216D	0086
C216E	0086
C216F	0086
C216G	0086
C216GN	0086
C216H	0086
C216J	0086
C216K	0086
C216L	0086
C216M	0086
C216OR	0086
C216R	0086
C216X	0086
C216Y	0086
C217(2S)	0086
C217A	0086
C217B	0086
C217C	0086
C217D	0086
C217E	0086
C217F	0086
C217G	0086
C217GN	0086
C217H	0086
C217J	0086
C217K	0086
C217L	0086
C217M	0086
C217OR	0086
C217R	0086
C217X	0086
C217Y	0086
C218(2S)	0086
C218A(2S)	0086
C218B	0086
C218C	0086
C218D	0086
C218E	0086
C218F	0086
C218G	0086
C218GN	0086
C218H	0086
C218J	0086
C218L	0086
C218M	0086
C218OR	0086
C218R	0086
C218T	OEM
C218X	0086
C218Y	0086
C219	0007
C220(2S)	0086
C220A	2449
C220A(TRANS)	0086
C220A2	1848
C220A3	OEM
C220A4	0226
C220A5	0226
C220A6	OEM
C220A7	OEM
C220A8	OEM
C220A9	OEM
C220B	0435
C220B(TRANS)	0086
C220B2	1848
C220B3	1848
C220B4	0226
C220B5	0226
C220B6	OEM
C220B7	OEM
C220B8	OEM
C220B9	OEM
C220C	1487
C220C(TRANS)	0086
C220C2	1855
C220C3	1855
C220C4	0226
C220C5	0226
C220C6	OEM
C220C7	OEM
C220C8	OEM
C220C9	OEM
C220D	1487
C220D(TRANS)	0086
C220D2	1855
C220D3	1855
C220D4	0226
C220D5	0226
C220D6	OEM
C220D7	OEM
C220D8	OEM
C220D9	OEM
C220E	0687
C220E(TRANS)	0086
C220E2	3215
C220E3	3215
C220E4	0226
C220E5	0226
C220E6	OEM
C220E7	OEM
C220E8	OEM
C220E9	OEM
C220F	2449
C220F(TRANS)	0086
C220F2	1848
C220F3	1848
C220F4	0226
C220F5	0226
C220F6	OEM
C220F7	OEM
C220F8	OEM
C220F9	OEM
C220G	0086
C220GN	0086
C220H	0086
C220K	0086
C220L	0086
C220M	0687
C220M(TRANS)	0086
C220M2	3215
C220M3	3215
C220M4	0226
C220M5	0226
C220M6	OEM
C220M7	OEM
C220M8	OEM
C220M9	OEM
C220OR	0086
C220R	0086
C220U	0469
C220U2	1848
C220U3	1848
C220U4	0226
C220U5	0226
C220U6	OEM
C220U7	OEM
C220U8	OEM
C220U9	OEM
C220X	0086
C220Y	0086
C221(2S)	0086
C221A	0086
C221B	0086
C221C	0086
C221D	0086
C221E	0086
C221F	0086
C221G	0086
C221GN	0086
C221H	0086
C221J	0086
C221K	0086
C221L	0086
C221M	0086
C221OR	0086
C221R	0086
C221X	0086
C221Y	0086
C222(2S)	0086
C222A	1651
C222A(TRANS)	0086
C222B	1651
C222B(TRANS)	0086
C222C	1584
C222C(TRANS)	0086
C222D	1584
C222D(TRANS)	0086
C222E	1665
C222E(TRANS)	0086
C222F	1651
C222F(TRANS)	0086
C222G	0086
C222GN	0086
C222H	0086
C222J	0086
C222L	0086
C222M	1665
C222M(TRANS)	0086
C222OR	0086
C222R	0086
C222U	1651
C222X	0086
C222Y	0086
C223	0622
C223A	0622
C223B	0622
C223C	0622
C223D	0622
C223E	0622
C223F	0622
C223G	0622
C223GN	0622
C223H	0622
C223J	0622
C223L	0622
C223M	0622
C223OR	0622
C223R	0622
C223X	0622
C223Y	0622
C224(2S)	0622
C224A	0622
C224B	0622
C224C	0622
C224D	0622
C224E	0622
C224F	0622
C224G	0622
C224GN	0622
C224GR-GL	0622
C224H	0622
C224J	0622
C224K	0622
C224L	0622
C224M	0622
C224OR	0622
C224R	0622
C224X	0622
C224Y	0622
C225(2S)	0086
C225A	0086
C225B	0086
C225C	0086
C225D	0086
C225E	0086
C225F	0086
C225G	0086
C225GN	0086
C225H	0086
C225J	0086
C225L	0086
C225M	0086
C225OR	0086
C225R	0086

DEVICE TYPE	REPL CODE
C225X	0086
C225Y	0086
C226(2S)	0086
C226A	0086
C226B	0086
C226C	0086
C226D	0086
C226E	0086
C226F	0086
C226G	0086
C226GN	0086
C226H	0086
C226J	0086
C226K	0086
C226L	0086
C226M	0086
C226OR	0086
C226R	0086
C226X	0086
C226Y	0086
C227(2S)	0086
C227A	0086
C227B	0086
C227C	0086
C227D	0086
C227E	0086
C227F	0086
C227G	0086
C227GN	0086
C227H	0086
C227J	0086
C227K	0086
C227L	0086
C227M	0086
C227OR	0086
C227R	0086
C227X	0086
C227Y	0086
C228(2S)	0086
C228A	0712
C228A(TRANS)	0086
C228A2	1848
C228A3	1848
C228A4	0226
C228A5	0226
C228A7	OEM
C228A8	OEM
C228A9	OEM
C228B	0469
C228B(TRANS)	0086
C228B2	1848
C228B3	1848
C228B4	1848
C228B5	1848
C228B6	OEM
C228B7	OEM
C228B8	OEM
C228B9	OEM
C228C	0722
C228C(TRANS)	0086
C228C2	1855
C228C3	1855
C228C4	0226
C228C5	0226
C228C6	OEM
C228C7	OEM
C228C8	OEM
C228C9	OEM
C228D	2647
C228D(TRANS)	0086
C228D2	1855
C228D3	1855
C228D4	0226
C228D5	0226
C228D6	OEM
C228D7	OEM
C228D8	OEM
C228D9	OEM
C228E	0725
C228E(TRANS)	0086
C228E2	3215
C228E3	3215
C228E4	0226
C228E5	0226
C228E6	OEM
C228E7	OEM
C228E8	OEM
C228E9	OEM
C228F	0731
C228F(TRANS)	0086
C228F2	1848
C228F3	1848
C228F4	0226
C228F5	OEM
C228F6	OEM
C228F7	OEM
C228F8	OEM
C228F9	OEM
C228G	0086
C228GN	0086
C228H	0086
C228J	0086
C228K	0086
C228L	0086
C228M	0725
C228M(TRANS)	0086
C228M2	3215
C228M3	3215
C228M4	0226
C228M5	0226
C228M6	OEM
C228M7	OEM
C228M8	OEM
C228M9	OEM
C228OR	0086
C228R	0086
C228U	0469
C228U3	1848
C228U4	0226
C228U5	0226
C228U6	OEM
C228U7	OEM
C228U8	OEM
C228U9	OEM
C228X	0086
C228Y	0086
C229(2S)	0622
C229A	1650
C229A(TRANS)	0622
C229B	1650
C229B(TRANS)	0622
C229C	2647
C229C(TRANS)	0622
C229D	2647
C229D(TRANS)	0622
C229E	2649
C229E(TRANS)	0622
C229F	1650
C229F(TRANS)	0622
C229G	0622
C229GN	0622
C229H	0622
C229J	0622
C229K	0622
C229L	0622
C229M	2649
C229M(TRANS)	0622
C229OR	0622
C229R	0622
C229U	1650
C229X	0622
C229Y	0622
C230(2S)	2988
C230A	0226
C230A(TRANS)	0622
C230A2	1848
C230A3	1848
C230A4	0226
C230A5	0226
C230A6	OEM
C230A7	OEM
C230A8	OEM
C230A9	OEM
C230B	2647
C230B(TRANS)	0622
C230B2	1848
C230B3	1848
C230B4	0226
C230B5	0226
C230B6	OEM
C230B7	OEM
C230B8	OEM
C230B9	OEM
C230C	2647
C230C(TRANS)	0622
C230C2	1848
C230C3	1848
C230C4	0226
C230C5	0226
C230C6	OEM
C230C7	OEM
C230C8	OEM
C230C9	OEM
C230D	0740
C230D(TRANS)	0622
C230D2	1855
C230D3	1855
C230D4	0226
C230D5	0226
C230D6	OEM
C230D7	OEM
C230D8	OEM
C230D9	OEM
C230E	0752
C230E(TRANS)	0622
C230E2	3215
C230E3	3215
C230E4	0226
C230E5	0226
C230E6	OEM
C230E7	OEM
C230E8	OEM
C230E9	OEM
C230F	2519
C230F(TRANS)	0622
C230F2	1848
C230F3	1848
C230F4	0226
C230F5	0226
C230F6	OEM
C230F7	OEM
C230F8	OEM
C230F9	OEM
C230G	0622
C230GN	0622
C230H	0622
C230J	0622
C230K	0622
C230L	0622
C230M	0764
C230M(TRANS)	0622
C230M2	3215
C230M3	3215
C230M4	0226
C230M5	0226
C230M6	OEM
C230M7	OEM
C230M8	OEM
C230M9	OEM
C230OR	0622
C230R	0622
C230U	0469
C230U2	1848
C230U3	1848
C230U4	0226
C230U5	0226
C230U6	OEM
C230U7	OEM
C230U8	OEM
C230U9	OEM
C230X	0622
C230Y	0622
C231(2S)	2996
C231A	OEM
C231A(TRANS)	0622
C231A2	OEM
C231A3	OEM
C231A4	OEM
C231A5	OEM
C231A6	OEM
C231A7	OEM
C231A8	OEM
C231A9	OEM
C231B	OEM
C231B(TRANS)	0622
C231B2	OEM
C231B3	OEM
C231B4	OEM
C231B5	OEM
C231B6	OEM
C231B7	OEM
C231B8	OEM
C231B9	OEM
C231C	OEM
C231C(TRANS)	0622
C231C2	OEM
C231C3	OEM
C231C4	OEM
C231C5	OEM
C231C6	OEM
C231C7	OEM
C231C8	OEM
C231C9	OEM
C231D	2647
C231D(TRANS)	0622
C231D2	OEM
C231D3	OEM
C231D4	OEM
C231D5	OEM
C231D6	OEM
C231D7	OEM
C231D8	OEM
C231D9	OEM
C231E	OEM
C231E(TRANS)	0622
C231E2	OEM
C231E3	OEM
C231E4	OEM
C231E5	OEM
C231E6	OEM
C231E7	OEM
C231E8	OEM
C231E9	OEM
C231F	OEM
C231F(TRANS)	0622
C231F2	OEM
C231F3	OEM
C231F4	OEM
C231F5	OEM
C231F6	OEM
C231F7	OEM
C231F8	OEM
C231F9	OEM
C231G	0622
C231GN	0622
C231H	0622
C231J	0622
C231K	0622
C231L	0622
C231M	OEM
C231M(TRANS)	0622
C231M2	OEM
C231M3	OEM
C231OR	0622
C231R	0622
C231U	OEM
C231U2	OEM
C231U3	OEM
C231U4	OEM
C231U5	OEM
C231U6	OEM
C231U7	OEM
C231U8	OEM
C231U9	OEM
C231X	0622
C231Y	0622
C232(2S)	2996
C232A	1650
C232A(TRANS)	0622
C232B	1651
C232B(TRANS)	0622
C232C	1584
C232C(TRANS)	0622
C232D	2647
C232D(TRANS)	0622
C232E	1665
C232E(TRANS)	0622
C232F	1651
C232F(TRANS)	0622
C232G	0622
C232GN	0622
C232H	0622
C232J	0622
C232K	0622
C232L	0622
C232M	1665
C232M(TRANS)	0622
C232OR	0622
C232R	0622
C232U	2519
C232X	0622
C232Y	0622
C233(2S)	0622
C233A	1651
C233A(TRANS)	0622
C233B	OEM
C233B(TRANS)	0622
C233C	OEM
C233C(TRANS)	0622
C233D	1584
C233D(TRANS)	0622
C233E	OEM
C233F	1651
C233F(TRANS)	0622
C233G	0622
C233GN	0622
C233H	0622
C233J	0622
C233L	0622
C233M	1665
C233M(TRANS)	0622
C233OR	0622
C233R	0622
C233U	OEM
C233X	0622
C233Y	0622
C234(2S)	0086
C234A	0567
C234A(TRANS)	0086
C234A2	1848
C234A3	1848
C234A4	0226
C234A5	0226
C234A6	OEM
C234A7	OEM
C234A8	OEM
C234A9	OEM
C234B	0762
C234B(TRANS)	0086
C234B2	1848
C234B3	1848
C234B4	0226
C234B5	0226
C234B6	OEM
C234B7	OEM
C234B8	OEM
C234B9	OEM
C234C	0740
C234C(TRANS)	0086
C234C2	1855
C234C3	1855
C234C4	0226
C234C5	0226
C234C6	OEM
C234C7	OEM
C234C8	OEM
C234C9	OEM
C234D	0740
C234D(TRANS)	0086
C234D2	1855
C234D3	1855
C234D4	0226
C234D5	0226
C234D6	OEM
C234D7	OEM
C234D8	OEM
C234D9	OEM
C234E	0764
C234E(TRANS)	0086
C234E2	3215
C234E3	3215
C234E4	0226
C234E5	0226
C234E6	OEM
C234E7	OEM
C234E8	OEM
C234E9	OEM
C234F	2188
C234F(TRANS)	0086
C234F2	1848
C234F3	1848
C234F4	0226
C234F5	0226
C234F6	OEM
C234F7	OEM
C234F8	OEM
C234F9	OEM
C234G	0086
C234GN	0086
C234H	0086
C234J	0086
C234K	0086
C234L	0086
C234M	2649
C234M(TRANS)	0086
C234M2	3215
C234M3	3215
C234M4	0226
C234M5	0226
C234M6	OEM
C234M7	OEM
C234M8	OEM
C234M9	OEM
C234OR	0086
C234R	0086
C234U	OEM
C234U2	1848
C234U3	1848
C234U4	0226
C234U5	0226
C234U6	OEM
C234U7	OEM
C234U8	OEM
C234U9	OEM
C234X	0086
C234Y	0086
C235(2S)	0018
C235-0(2S)	0018
C235-O	0018
C235A	1650
C235A(TRANS)	0018
C235B	1650
C235B(TRANS)	0018
C235C	2647
C235C(TRANS)	0018
C235D	2647
C235D(TRANS)	0018
C235E	2649
C235E(TRANS)	0018
C235F	1650
C235F(TRANS)	0018
C235G	0018
C235GN	0018
C235H	0018
C235J	0018
C235K	0018
C235L	0018
C235M	OEM
C235M(TRANS)	0018
C235OR	0018
C235R	0018
C235U	OEM
C235X	0018
C235Y	0018
C236(2S)	0086
C236A	0086
C236B	0086
C236C	0086
C236D	0086
C236E	0086
C236F	0086
C236G	0086
C236GN	0086
C236H	0086
C236J	0086
C236K	0086
C236L	0086
C236M	0086
C236OR	0086
C236R	0086
C236X	0086
C236Y	0086
C237(2S)	0016
C237A	0016
C237B	0016
C237C	0016
C237D	0016
C237E	0016
C237F	0016
C237G	0016
C237GN	0016
C237H	0016
C237J	0016
C237K	0016
C237L	0016
C237M	0016
C237OR	0016
C237R	0016
C237X	0016
C237Y	0016
C238(2S)	0016
C238A	0016
C238B	0016
C238C	0016
C238D	0016
C238E	0016
C238F	0016
C238G	0016
C238GN	0016
C238H	0016
C238J	0016
C238K	0016
C238L	0016
C238M	0016
C238OR	0016
C238R	0016
C238X	0016
C238Y	0016
C239(2S)	0016
C239A	0016
C239B	0016
C239C	0016
C239D	0016
C239E	0016
C239F	0016
C239G	0016
C239GN	0016
C239H	0016
C239J	0016
C239K	0016
C239L	0016
C239M	0016
C239OR	0016
C239R	0016
C239X	0016
C239Y	0016
C240(2S)	0103
C240A	0103
C240B	0103
C240C	0103
C240D	0103
C240E	0103
C240F	0103
C240G	0103
C240GN	0103
C240H	0103
C240J	0103
C240K	0103
C240L	0103
C240M	0103
C240OR	0103
C240R	0103
C240X	0103
C240Y	0103
C241(2S)	0103
C241A	0103
C241B	0103
C241C	0103
C241D	0103
C241E	0103
C241F	0103
C241G	0103
C241GN	0103
C241H	0103
C241K	0103
C241L	0103
C241M	0103
C241OR	0103
C241R	0103
C241X	0103
C241Y	0103
C242(2S)	0103
C242A	0103
C242B	0103
C242C	0103
C242D	0103
C242E	0103
C242F	0103
C242G	0103
C242GN	0103
C242H	0103
C242J	0103
C242K	0103
C242L	0103
C242M	0103
C242OR	0103
C242R	0103
C242X	0103
C242Y	0103
C243(2S)	0177
C244(2S)	1966
C244A	1966
C244B	1966
C244C	1966
C244D	1966
C244E	1966
C244F	1966
C244G	1966
C244GN	1966
C244J	1966
C244K	1966
C244L	1966
C244M	1966
C244OR	1966
C244R	1966
C244X	1966
C244Y	1966
C245(2S)	1966
C246(2S)	1966
C247(2S)	0086
C247A	0086
C247B	0086
C247C	0086
C247D	0086
C247E	0086
C247F	0086
C247G	0086
C247GN	0086
C247H	0086
C247J	0086
C247L	0086
C247M	0086
C247OR	0086
C247R	0086
C247X	0086
C247Y	0086
C248(2S)	0016
C248A	0016
C248B	0016
C248C	0016
C248D	0016
C248E	0016
C248F	0016
C248G	0016
C248J	0016
C248K	0016
C248L	0016
C248M	0016
C248OR	0016
C248R	0016
C248X	0016
C248Y	0016
C249(2S)	0086
C249A	0086
C249B	0086
C249C	0086
C249D	0086
C249E	0086
C249F	0086
C249G	0086
C249GN	0086
C249H	0086
C249J	0086
C249K	0086
C249L	0086
C249M	0086
C249OR	0086
C249R	0086
C249X	0086
C249Y	0086
C250(2S)	0016
C250A	0016
C250B	0016
C250C	0016
C250D	0016
C250E	0016
C250F	0016
C250G	0016
C250GN	0016
C250H	0016
C250J	0016
C250K	0016
C250L	0016
C250M	0016
C250OR	0016
C250R	0016
C250X	0016
C250Y	0016
C251(2S)	0259
C251A(2S)	0259
C251B	0259
C251C	0259
C251D	0259
C251E	0259
C251F	0259
C251G	0259
C251GN	0259
C251H	0259
C251J	0259
C251K	0259
C251L	0259
C251M	0259
C251OR	0259
C251R	0259
C251X	0259
C251Y	0259
C252(2S)	0259
C252A	0259
C252B	0259
C252C	0259
C252D	0259
C252E	0259
C252F	0259
C252G	0259
C252GN	0259
C252H	0259
C252J	0259
C252K	0259
C252L	0259
C252M	0259
C252OR	0259
C252R	0259
C252X	0259
C252Y	0259
C253(2S)	0259
C253A	0259
C253B	0259
C253C	0259
C253D	0259
C253E	0259
C253F	0259
C253G	0259
C253GN	0259
C253H	0259
C253J	0259
C253K	0259
C253L	0259
C253M	0259
C253OR	0259
C253R	0259
C253X	0259
C253Y	0259
C254(2S)	0086

If replacement code is OEM, contact original manufacturer for replacement.

DEVICE TYPE	REPL CODE
C255	0086
C256	0086
C257	1984
C258	0264
C259	0264
C260	0595
C260D	0595
C260E	0595
C261(2S)	0698
C262	0595
C263(2S)	0144
C263A	0144
C263B	0144
C263C	0144
C263D	0144
C263E	0144
C263F	0144
C263G	0144
C263GN	0144
C263H	0144
C263J	0144
C263K	0144
C263L	0144
C263M	0144
C263OR	0144
C263R	0144
C263X	0144
C263Y	0144
C264	0079
C265	0079
C266	0475
C266(2S)	0475
C266A	0475
C266B	0475
C266C	0475
C266D	0475
C266E	0475
C266F	0475
C266G	0475
C266GN	0475
C266H	0475
C266J	0475
C266K	0475
C266L	0475
C266M	0475
C266OR	0475
C266P	OEM
C266R	0475
C266X	0475
C266Y	0475
C267(2S)	0016
C267A(2S)	0016
C267B	0016
C267C	0016
C267D	0016
C267E	0016
C267F	0016
C267G	0016
C267GN	0016
C267H	0016
C267J	0016
C267K	0016
C267L	0016
C267M	0016
C267OR	0016
C267R	0016
C267X	0016
C267Y	0016
C268(2S)	0086
C268-OR	0086
C268A(2S)	0086
C268B	0086
C268C	0086
C268D	0086
C268E	0086
C268F	0086
C268G	0086
C268GN	0086
C268H	0086
C268J	0086
C268K	0086
C268L	0086
C268M	0086
C268OR	0086
C268R	0086
C268X	0086
C268Y	0086
C269(2S)	0144
C269A	0144
C269B	0144
C269C	0144
C269D	0144
C269E	0144
C269F	0144
C269G	0144
C269GN	0144
C269H	0144
C269J	0144
C269K	0144
C269L	0144
C269M	0144
C269OR	0144
C269R	0144
C269X	0144
C269Y	0144
C270(2S)	0074
C271(2S)	0144
C271A	0144
C271B	0144
C271C	0144
C271D	0144
C271E	0144
C271F	0144
C271G	0144
C271GN	0144
C271H	0144
C271J	0144
C271K	0144
C271L	0144
C271M	0144
C271OR	0144
C271R	0144
C271X	0144
C271Y	0144
C272(2S)	0144
C272A	0144
C272B	0144
C272C	0144
C272D	0144
C272E	0144
C272F	0144
C272G	0144
C272GN	0144
C272H	0144
C272J	0144
C272K	0144
C272L	0144
C272M	0144
C272OR	0144
C272R	0144
C272X	0144
C272Y	0144
C273(2S)	0233
C277C(2S)	0208
C278-OR	0475
C280	0111
C280AO	0111
C280N	0738
C280P	0738
C280PA	0738
C280PB	0738
C280PC	OEM
C280PD	OEM
C280PE	OEM
C280PM	OEM
C280PS	OEM
C280S	0738
C280T	0738
C281(2S)	0536
C281-OR	0536
C281A(2S)	0536
C281B(2S)	0536
C281BL	0536
C281C(2S)	0536
C281C-EP(2S)	0536
C281D(2S)	0536
C281E	0536
C281EP(2S)	0536
C281F	0536
C281G	0536
C281GN	0536
C281H(2S)	0536
C281HB	0016
C281J	0536
C281K	0536
C281L	0536
C281M	0536
C281N	OEM
C281OR	0536
C281P	OEM
C281PA	OEM
C281PB	OEM
C281PC	OEM
C281PD	OEM
C281PE	OEM
C281PM	OEM
C281PN	OEM
C281PS	OEM
C281R	0536
C281S	OEM
C281X	0536
C281Y	0536
C282(2S)	0016
C282A	0016
C282A(TRANS)	0016
C282B	0016
C282B(TRANS)	0016
C282C	0016
C282C(TRANS)	0016
C282D	OEM
C282D(TRANS)	0016
C282E	OEM
C282E(TRANS)	0016
C282F	0016
C282G	0016
C282GN	0016
C282H(2S)	0016
C282J	0016
C282K	0016
C282L	0016
C282M	OEM
C282M(TRANS)	0016
C282N	OEM
C282OR	0016
C282P	OEM
C282PA	OEM
C282PB	OEM
C282PC	OEM
C282PD	OEM
C282PE	OEM
C282PM	OEM
C282PS	OEM
C282R	0016
C282S	OEM
C282T	OEM
C282X	0016
C282Y	0016
C283(2S)	1562
C283A	0016
C283A(TRANS)	1562
C283B	0016
C283B(TRANS)	1562
C283C	0016
C283C(TRANS)	1562
C283D	OEM
C283D(TRANS)	1562
C283E	OEM
C283E(TRANS)	1562
C283F	1562
C283G	1562
C283GN	1562
C283H	1562
C283J	1562
C283K	1562
C283L	1562
C283M	OEM
C283M(TRANS)	1562
C283N	OEM
C283OR	1562
C283P	OEM
C283PA	OEM
C283PB	OEM
C283PC	OEM
C283PD	OEM
C283PE	OEM
C283PM	OEM
C283PS	OEM
C283R	1562
C283S	OEM
C283T	OEM
C283X	1562
C283Y	1562
C284(2S)	0016
C284A	0016
C284B	0016
C284C	0016
C284D	0016
C284E	0016
C284F	0016
C284G	0016
C284GN	0016
C284H(2S)	0016
C284J	0016
C284K	0016
C284L	0016
C284M	0016
C284OR	0016
C284R	0016
C284X	0016
C284Y	0016
C285(2S)	0493
C285A(2S)	0493
C285B	0493
C285C	0493
C285D	0493
C285E	0493
C285F	0493
C285G	0493
C285GN	0493
C285H	0493
C285J	0493
C285K	0493
C285L	0493
C285M	0493
C285OR	0493
C285R	0493
C285X	0493
C285Y	0493
C286	0144
C286A	OEM
C286A(TRANS)	0144
C286B	0144
C286B(TRANS)	0144
C286C	OEM
C286C(TRANS)	0144
C286D	OEM
C286D(TRANS)	0144
C286E	0144
C286E(TRANS)	0144
C286F	0144
C286G	0144
C286GN	0144
C286H	0144
C286J	0144
C286K	0144
C286L	0144
C286M	OEM
C286M(TRANS)	0144
C286N	OEM
C286OR	0144
C286P	OEM
C286PA	OEM
C286PB	OEM
C286PC	OEM
C286PD	OEM
C286PE	OEM
C286PM	OEM
C286PS	OEM
C286R	0144
C286S	OEM
C286T	OEM
C286X	0144
C286Y	0144
C287	0475
C287-OR	0475
C287A	0475
C287A(2S)	0475
C287B	OEM
C287B(TRANS)	0475
C287C	OEM
C287C(TRANS)	0475
C287D	OEM
C287D(TRANS)	0475
C287E	OEM
C287E(TRANS)	0475
C287F	0475
C287G	0475
C287GN	0475
C287H	0475
C287J	0475
C287K	0475
C287L	0475
C287M	OEM
C287M(TRANS)	0475
C287N	OEM
C287OR	0475
C287P	OEM
C287PA	OEM
C287PB	OEM
C287PC	OEM
C287PD	OEM
C287PE	OEM
C287PM	OEM
C287PS	OEM
C287R	0475
C287S	OEM
C287T	OEM
C287X	0475
C287Y	0475
C288(2S)	1342
C288A(2S)	1342
C288A(5B)	1342
C288A1	1342
C288A1B	1342
C288A5-B	1342
C288AB	1342
C288B	1342
C288C	1342
C288D	1342
C288E	1342
C288F	1342
C288G	1342
C288GN	1342
C288H	1342
C288J	1342
C288K	1342
C288L	1342
C288M	1342
C288OR	1342
C288R	1342
C288X	1342
C288Y	1342
C289(2S)	0144
C289A	0144
C289B	0144
C289C	0144
C289D	0144
C289E	0144
C289F	0144
C289G	0144
C289GN	0144
C289H	0144
C289J	0144
C289K	0144
C289L	0144
C289M	0144
C289OR	0144
C289R	0144
C289X	0144
C289Y	0144
C290(2S)	0698
C290A	1900
C290B	1900
C290C	1900
C290D	0738
C290E	0738
C290F	OEM
C290G	OEM
C290H	OEM
C290J	OEM
C290M	0738
C290N	0738
C290P	0738
C290PA	0738
C290PB	0738
C290S	0738
C290T	0738
C291	0086
C291A	OEM
C291A(TRANS)	0086
C291B	OEM
C291B(TRANS)	0086
C291C	OEM
C291C(TRANS)	0086
C291D	0086
C291D(TRANS)	0086
C291E	OEM
C291E(TRANS)	0086
C291F	OEM
C291F(TRANS)	0086
C291G	OEM
C291G(TRANS)	0086
C291GN	0086
C291H	OEM
C291H(TRANS)	0086
C291J	OEM
C291J(TRANS)	0086
C291K	0086
C291M	OEM
C291M(TRANS)	0086
C291N	OEM
C291OR	0086
C291P	OEM
C291PA	OEM
C291PB	OEM
C291R	0086
C291S	OEM
C291T	OEM
C291X	0086
C291Y	0086
C292(2S)	0622
C292A	0622
C292B	0622
C292C	0622
C292D	0622
C292E	0622
C292F	0622
C292G	0622
C292GN	0622
C292H	0622
C292J	0622
C292K	0622
C292L	0622
C292M	0622
C292OR	0622
C292R	0622
C292X	0622
C292Y	0622
C293(2S)	0622
C293A	0622
C293B	0622
C293C	0622
C293D	0622
C293E	0622
C293F	0622
C293G	0622
C293GN	0622
C293H	0622
C293J	0622
C293K	0622
C293L	0622
C293M	0622
C293OR	0622
C293R	0622
C293X	0622
C293Y	0622
C294	0285
C294X	0285
C295	0079
C296(2S)	0423
C297	0560
C298	0622
C298-4	0622
C299(2S)	3336
C299A	2314
C299B	2314
C299C	2314
C299D	2314
C299E	2314
C299F	2314
C299G	2314
C299GN	2314
C299H	2314
C299J	2314
C299K	2314
C299L	2314
C299M	2314
C299OR	2314
C299R	2314
C299X	2314
C299Y	2314
C300(2S)	0016
C300A	0016
C300B	0016
C300C	0016
C300D	0016
C300E	0016
C300F	0016
C300G	0016
C300GN	0016
C300H	0016
C300J	0016
C300K	0016
C300L	0016
C300M	0016
C300OR	0016
C300R	0016
C300X	0016
C300Y	0016
C301	0016
C301(2S)	0016
C301A	OEM
C301A(2S)	0016
C301B	0016
C301C	0016
C301D	0016
C301E	0016
C301F	0016
C301G	0016
C301GN	0016
C301H	0016
C301J	0016
C301K	0016
C301L	0016
C301M	0016
C301OR	0016
C301R	0016
C301X	0016
C301Y	0016
C302	0150
C302(2S)	0016
C302A	0016
C302B	0016
C302C	0016
C302D	0016
C302E	0016
C302F	0016
C302G	0016
C302GN	0016
C302H	0016
C302J	0016
C302K	0016
C302L	0016
C302M	0016
C302OR	0016
C302R	0016
C302X	0016
C302Y	0016
C303	0595
C304	0595
C305	0595
C305A	0595
C305B	0595
C305C	0595
C305C(IC)	2169
C305D	0595
C305E	0595
C305F	0595
C305G	0595
C305GN	0595
C305H	0595
C305J	0595
C305K	0595
C305L	0595
C305M	0595
C305OR	0595
C305R	0595
C305X	0595
C305Y	0595
C306(2S)	0086
C306A	0086
C306B	0086
C306C	0086
C306D	0086
C306E	0086
C306F	0086
C306G	0086
C306GN	0086
C306H	0086
C306J	0086
C306K	0086
C306L	0086
C306M	0086
C306OR	0086
C306R	0086
C306X	0086
C306Y	0086
C307(2S)	0086
C307-OR	0086
C307A	0086
C307B	0086
C307C	0086
C307D	0086
C307E	0086
C307F	0086
C307G	0086
C307GN	0086
C307H	0086
C307J	0086
C307K	0086
C307L	0086
C307M	0086
C307OR	0086
C307R	0086
C307X	0086
C307Y	0086
C308	0086
C308A	0086
C308B	0086
C308C	0086
C308D	0086
C308E	0086
C308F	0086
C308G	0086
C308GN	0086
C308H	0086
C308J	0086
C308K	0086
C308L	0086
C308M	0086
C308OR	0086
C308R	0086
C308X	0086
C308Y	0086
C309(2S)	0086
C309A	0086
C309B	0086
C309C	0086
C309D	0086
C309E	0086
C309F	0086
C309G	0086
C309GN	0086
C309H	0086
C309J	0086
C309K	0086
C309L	0086
C309M	0086
C309OR	0086
C309R	0086
C309X	0086
C309Y	0086
C310	0622
C310(2S)	0086
C310A	0086
C310B	0086
C310C	0086
C310D	0086
C310E	0086
C310F	0086
C310G	0086
C310GN	0086
C310H	0086
C310J	0086
C310K	0086
C310L	0086
C310M	0086
C310OR	0086
C310R	0086
C310X	0086
C310Y	0086
C311	0698
C311(2S)	0698
C311A	OEM
C311B	0698
C311C	0698
C311D	0698
C311E	0698
C311F	0698
C311G	0698
C311GN	0698
C311H	0698
C311J	0698
C311K	0698
C311L	0698
C311M	0698
C311OR	0698
C311R	0698
C311X	0698
C311Y	0698
C312	0630
C312(2S)	0630
C312A	0630
C312B	0630
C312C	0630
C312D	0630
C312E	0630
C312F	0630
C312G	0630
C312GN	0630
C312H	0630
C312J	0630
C312K	0630
C312L	0630
C312M	0630
C312OR	0630
C312R	0630
C312X	0630
C312Y	0630
C313	3390
C313-OR	3390
C313A	3390
C313B	3390
C313C(2S)	3390
C313D	3390
C313E	3390
C313F	3390
C313G	3390
C313GN	3390
C313H(2S)	3390
C313J	3390
C313K	3390
C313L	3390
C313M	3390
C313OR	3390
C313R	3390
C313X	3390
C313Y	3390
C314	0630
C315(2S)	0016
C315A	0016
C315B	0016
C315C	0016
C315D	0016
C315E	0016
C315F	0016
C315G	0016
C315GN	0016
C315H	0016
C315J	0016
C315K	0016
C315L	0016
C315M	0016

DEVICE TYPE	REPL CODE	DEVICE TYPE	REPL CODE	DEVICE TYPE	REPL CODE	DEVICE TYPE	REPL CODE	DEVICE TYPE	REPL CODE	DEVICE TYPE	REPL CODE	DEVICE TYPE	REPL CODE
C315OR	0016	C321K	0016	C350K	0016	C358M	0459	C366G-0	0286	C371-0(2S)	0191	C374Y	0552
C315R	0016	C321L	0016	C350L	0016	C358N	0587	C366G-R	0286	C371-OR	0191	C375(2S)	0127
C315X	0016	C321M	0016	C350M	0159	C358P	0587	C366G-Y	0286	C371-R(2S)	0191	C375-0(2S)	0127
C315Y	0016	C321OR	0016	C350M(TRANS)	0016	C358PA	0587	C366H	0286	C371-T	0191	C375-O(2S)	0127
C316(2S)	0016	C321R	0016	C350N	0096	C358PB	0587	C366J	0286	C371A	0191	C375-Y(2S)	0127
C316A	0016	C321X	0016	C350OR	0016	C358S	0587	C366K	0286	C371B(2S)	0191	C375A	0127
C316B	0016	C321Y	0016	C350P	0096	C358T	0587	C366L	0286	C371C	0191	C375B	0127
C316C	0016	C322	0079	C350PA	0096	C360(2S)	0016	C366M	0286	C371D	0191	C375C	0127
C316D	0016	C323(2S)	0016	C350PB	0096	C360-OR	0016	C366OR	0286	C371E	0191	C375D	0127
C316E	0016	C323A	0016	C350PC	OEM	C360A	0016	C366X	0286	C371F	0191	C375E	0127
C316F	0016	C323B	0016	C350R	0016	C360B	0016	C367	0286	C371G(2S)	0191	C375F	0127
C316G	0016	C323C	0016	C350S	0096	C360C	0016	C367-0	0286	C371GN	0191	C375G	0127
C316GN	0016	C323D	0016	C350T	0096	C360D(2S)	0016	C367-R	0286	C371G-0	0191	C375GN	0127
C316H	0016	C323E	0016	C350X	0016	C360E	0016	C367-Y	0286	C371G-R	0191	C375H	0127
C316J	0016	C323F	0016	C350Y	0016	C360F	0016	C367A	0286	C371H	0191	C375J	0127
C316K	0016	C323G	0016	C351(2S)	0127	C360G	0016	C367B	0286	C371J	0191	C375K	0127
C316L	0016	C323GN	0016	C351A	0127	C360GN	0016	C367C	0286	C371K	0191	C375L	0127
C316M	0016	C323H	0016	C351B	0127	C360H	0016	C367D	0286	C371L	0191	C375M	0127
C316OR	0016	C323J	0016	C351C	0127	C360J	0016	C367E	0286	C371M	0191	C375OR	0127
C316R	0016	C323K	0016	C351D	0127	C360K	0016	C367F	0286	C371O	0191	C375R	0127
C316X	0016	C323L	0016	C351E	0127	C360L	0016	C367G	0286	C371OR	0191	C375X	0127
C316Y	0016	C323M	0016	C351F	0127	C360M	0016	C367GN	0286	C371R(2S)	0191	C376(2S)	0286
C317(2S)	0016	C323OR	0016	C351FA1	0127	C360OR	0016	C367G-0	0286	C371R-1	0191	C376A	0286
C317A	0016	C323R	0016	C351G	0127	C360R	0016	C367G-R	0286	C371RED-G	0191	C376B	0286
C317B	0016	C323X	0016	C351GN	0127	C360X	0016	C367G-Y	0286	C371T	0191	C376C	0286
C317C(2S)	0016	C323Y	0016	C351H	0127	C360Y	0016	C367H	0286	C371X	0191	C376D	0286
C317D	0016	C324(2S)	0016	C351J	0127	C361(2S)	0155	C367J	0286	C371Y	0191	C376E	OEM
C317E	0016	C324A(2S)	0016	C351K	0127	C361A	0155	C367K	0286	C372	0076	C376E(TRANS)	0286
C317F	0016	C324B	0016	C351L	0127	C361B	0155	C367L	0286	C372(H)	0076	C376E-PD	OEM
C317G	0016	C324C	0016	C351M	0127	C361C	0155	C367M	0286	C372-0(2S)	0076	C376F	0286
C317GN	0016	C324D	0016	C351OR	0127	C361D	0155	C367OR	0286	C372-1(2S)	0076	C376G	0286
C317H	0016	C324E	0016	C351R	0127	C361E	0155	C367R	0286	C372-2(2S)	0076	C376GN	0286
C317J	0016	C324F	0016	C351X	0127	C361F	0155	C367X	0286	C372-O(2S)	0076	C376H	0286
C317K	0016	C324G	0016	C351Y	0127	C361G	0155	C367Y	0286	C372-OR	0076	C376J	0286
C317L	0016	C324GN	0016	C352	0454	C361GN	0155	C368(2S)	0111	C372-ORG	0076	C376K	0286
C317M	0016	C324H(2S)	0016	C352-OR	0454	C361H	0155	C368-BL	0111	C372-Y(2S)	0076	C376L	0286
C317OR	0016	C324HA(2S)	0016	C352A	0454	C361J	0155	C368-GR	0111	C372A	0076	C376M	OEM
C317R	0016	C324J	0016	C352AC	0454	C361K	0155	C368A	0111	C372AR	0076	C376M(TRANS)	0286
C317X	0016	C324K	0016	C352B	0454	C361L	0155	C368B	0111	C372B	0076	C376N	OEM
C317Y	0016	C324L	0016	C352C	0454	C361M	0155	C368C	0111	C372C	0076	C376OR	0286
C318	0016	C324M	0016	C352D	0454	C361OR	0155	C368D	0111	C372D	0076	C376P	OEM
C318A(2S)	0016	C324OR	0016	C352E	0454	C361R	0155	C368E	0111	C372E	0076	C376PA	OEM
C318AB	0016	C324R	0016	C352F	0454	C361X	0155	C368F	0111	C372F	0076	C376PB	OEM
C318B	0016	C324X	0016	C352G	0454	C361Y	0155	C368G	0111	C372G	0076	C376PC	OEM
C318C	0016	C324Y	0016	C352GN	0454	C362(2S)	0155	C368GN	0111	C372GN	0076	C376PD	OEM
C318D	0016	C325(2S)	0259	C352H	0454	C362A	0155	C368H	0111	C372G-O	0076	C376R	0286
C318E	0016	C325C	0259	C352J	0454	C362B	0155	C368J	0111	C372G-Y	0076	C376S	OEM
C318F	0016	C326	0418	C352K	0454	C362C	0155	C368K	0111	C372J	0076	C376T	OEM
C318G	0016	C327	0127	C352L	0454	C362D	0155	C368L	0111	C372K	0076	C376X	0286
C318GN	0016	C327(EUROPE)	0037	C352M	0454	C362E	0155	C368M	0111	C372L	0076	C376Y	0286
C318H	0016	C328	0259	C352OR	0454	C362F	0155	C368OR	0111	C372M	0076	C377(2S)	0076
C318J	0016	C328(EUROPE)	0006	C352R	0454	C362G	0155	C368R	0111	C372O	0076	C377(EUROPE)	OEM
C318K	0016	C329(2S)	0259	C352X	0454	C362GN	0155	C368V(2S)	0111	C372OR	0076	C377-0	0076
C318L	0016	C330	2844	C352Y	0454	C362H	0155	C368X	0111	C372R	0076	C377-BN	0076
C318M	0016	C331	2844	C353(2S)	0086	C362J	0155	C368Y	0111	C372X	0076	C377-BRN	0076
C318OR	0016	C332	0079	C353A(2S)	0086	C362K	0155	C369(2S)	0111	C372Y(2S)	0076	C377-O	0076
C318R	0016	C333	0016	C353AC	0086	C362L	0155	C369-BL	0111	C372Y1	0076	C377-OR	0076
C318X	0016	C334	0016	C353B	0086	C362M	0155	C369-GR	0111	C372YEL	0076	C377-ORG	0076
C318Y	0016	C335(2S)	0536	C353C	0086	C362OR	0155	C369-V	0111	C372YEL-G	0076	C377-R	0076
C319(2S)	0493	C336	0016	C353D	0086	C362R	0155	C369A	0111	C372Z	0076	C377-RED	0076
C319A	0493	C337	0016	C353E	0086	C362X	0155	C369B	0111	C373(2S)	0076	C377A	0076
C319B	0493	C337(EUROPE)	0551	C353F	0086	C362Y	0155	C369BL(2S)	0111	C373-0	0076	C377B	0076
C319C	0493	C338	0076	C353G	0086	C363(2S)	0016	C369C	0111	C373-14	0076	C377BN	0076
C319D	0493	C338(EUROPE)	0079	C353GN	0086	C363-OR	0016	C369D	0111	C373-G	0076	C377BRN	0076
C319E	0493	C339	0079	C353H	0086	C363A	0016	C369E	0111	C373-O	0076	C377C	0076
C319F	0493	C340	0475	C353J	0086	C363B	0016	C369F	0111	C373-OR	0076	C377D	0076
C319G	0493	C341	0324	C353K	0086	C363C	0016	C369G(2S)	0111	C373A	0076	C377E	OEM
C319GN	0493	C342	0079	C353L	0086	C363D	0016	C369G-BL(2S)	0111	C373AL	0076	C377E(TRANS)	0076
C319H	0493	C343	0016	C353M	0086	C363E	0016	C369GBL(2S)	0111	C373B	0076	C377E-PE	OEM
C319J	0493	C344	0127	C353OR	0086	C363F	0016	C369G-GR(2S)	0111	C373C	0076	C377F	0076
C319K	0493	C344Y	0127	C353R	0086	C363G	0016	C369GGR(2S)	0111	C373D	0076	C377G	0076
C319L	0493	C345	0324	C353X	0086	C363GN	0016	C369GN	0111	C373E	0076	C377GN	0076
C319M	0493	C346	0016	C353Y	0086	C363H	0016	C369GR(2S)	0111	C373F	0076	C377H	0076
C319OR	0493	C347	0016	C354	0936	C363J	0016	C369G-V(2S)	0111	C373G	0076	C377J	0076
C319R	0493	C348(2S)	0016	C354A	3349	C363K	0016	C369H	0111	C373GN	0076	C377K	0076
C319X	0493	C348A	0016	C354B	0459	C363L	0016	C369J	0111	C373GR(2S)	0076	C377L	0076
C319Y	0493	C348B	0016	C354C	0459	C363M	0016	C369K	0111	C373H	0076	C377M	OEM
C320(2S)	3421	C348C	0016	C354D	0459	C363OR	0016	C369L	0111	C373J	0076	C377M(TRANS)	0076
C320A	3421	C348D	0016	C354E	0459	C363R	0016	C369M	0111	C373K	0076	C377N	OEM
C320B	3421	C348E	0016	C354G	0459	C363X	0016	C369OR	0111	C373L	0076	C377O	0076
C320C	3421	C348F	0016	C354M	0459	C363Y	0016	C369R	0111	C373M	0076	C377OR	0076
C320D	3421	C348G	0016	C355	0936	C364	0530	C369V(2S)	0111	C373OR	0076	C377P	OEM
C320E	3421	C348GN	0016	C355A	0459	C364A	0459	C369X	0111	C373R	0076	C377PA	OEM
C320F	3421	C348H	0016	C355B	0459	C364B	0459	C369Y	0111	C373W(2S)	0076	C377PB	OEM
C320G	3421	C348J	0016	C355C	0459	C364C	0459	C370(2S)	0155	C373X	0076	C377PC	OEM
C320GN	3421	C348K	0016	C355D	0459	C364D	0459	C370-0	0155	C373Y	0076	C377PD	OEM
C320H	3421	C348L	0016	C355E	0459	C364E	0459	C370-G	0155	C374(2S)	0552	C377R	0076
C320J	3421	C348M	0016	C355G	0459	C364M	0459	C370-O	0155	C374-BL(2S)	0552	C377RED	0076
C320K	3421	C348OR	0016	C355M	0459	C364N	OEM	C370-T	0155	C374-OR	0552	C377S	OEM
C320L	3421	C348X	0016	C356(2S)	0079	C364S	OEM	C370A	0155	C374-V(2S)	0552	C377T	OEM
C320M	3421	C348Y	0016	C356A	0079	C365A	0459	C370B	0155	C374A	0552	C377X	0076
C320OR	3421	C349	0127	C356B	0079	C365B	0459	C370C	0155	C374B	0552	C377Y	0076
C320R	3421	C349R	0127	C356C	0079	C365C	0459	C370D	0155	C374BL	0552	C378(2S)	0155
C320X	3421	C350	0016	C356D	0079	C365D	0459	C370E	0155	C374BLK	0552	C378-0	0155
C320Y	3421	C350A	OEM	C356E	0079	C365E	0459	C370F(2S)	0155	C374C	0552	C378-ORG	0155
C321	0016	C350A(TRANS)	0016	C356F	0079	C365M	0459	C370G(2S)	0155	C374D	0552	C378-R	0155
C321A	0016	C350B	OEM	C356G	0079	C365N	0587	C370GN	0155	C374E	0552	C378-RED	0155
C321B	0016	C350B(TRANS)	0016	C356GN	0079	C365S	0587	C370H(2S)	0155	C374F	0552	C378-Y	0155
C321C	0016	C350C	OEM	C356H	0079	C366	0286	C370I	0155	C374G	0552	C378-YEL	0155
C321D	0016	C350C(TRANS)	0016	C356J	0079	C366(2S)	0286	C370J(2S)	0155	C374GN	0552	C378A	0155
C321E	0016	C350D	OEM	C356K	0079	C366-0	0286	C370K(2S)	0155	C374H	0552	C378B	0155
C321F	0016	C350D(TRANS)	0016	C356L	0079	C366A	0286	C370L	0155	C374J	0552	C378C	0155
C321G	0016	C350E	0159	C356M	0079	C366B	0286	C370M	0155	C374K	0552	C378D	0155
C321GN	0016	C350E(TRANS)	0016	C356OR	0079	C366C	0286	C370OR	0155	C374L	0552	C378E	OEM
C321H(2S)	0016	C350F	0016	C356R	0079	C366D	0286	C370R	0155	C374M	0552	C378E(TRANS)	0155
C321HA(2S)	0016	C350G	0016	C356X	0079	C366E	0286	C370X	0155	C374OR	0552	C378E-PD	OEM
C321HB(2S)	0016	C350GN	0016	C356Y	0079	C366F	0286	C370Y	0155	C374R	0552	C378F	0155
C321HC(2S)	0016	C350H(2S)	0016	C358C	0770	C366G	0286	C371(2S)	0191	C374V	0552	C378G	0155
C321J	0016	C350J	0016	C358E	0459	C366GN	0286	C371-0(2S)	0191	C374X	0552	C378GN	0155

If replacement code is OEM, contact original manufacturer for replacement.

DEVICE TYPE	REPL CODE
C378H	0155
C378J	0155
C378K	0155
C378L	0155
C378M	OEM
C378M(TRANS)	0155
C378N	OEM
C378O	0155
C378OR	0155
C378P	OEM
C378PA	OEM
C378PB	OEM
C378PC	OEM
C378PD	OEM
C378R	0155
C378S	OEM
C378T	OEM
C378X	0155
C378Y	0155
C379(2S)	0286
C379A	0286
C379B	0286
C379C	0286
C379D	0286
C379E	0286
C379F	0286
C379G	0286
C379GN	0286
C379H	0286
C379J	0286
C379K	0286
C379L	0286
C379M	0286
C379OR	0286
C379R	0286
C379X	0286
C379Y	0286
C380	0286
C380-O(2S)	0286
C380-BRN	0286
C380-O(2S)	0286
C380-OR	0286
C380-ORG	0286
C380-R	0286
C380-RED	0286
C380-Y	0286
C380-YEL	0286
C380A	0192
C380A(SCR)	0192
C380A(TRANS)	0286
C380A-O	0286
C380AO	0286
C380A-O	0286
C380AO(2S)	0286
C380AR	0286
C380A-R(2S)	0286
C380ATV(2S)	0286
C380AX500	OEM
C380AX555	OEM
C380A-Y	0286
C380AY(2S)	0286
C380B	0192
C380B(TRANS)	0286
C380BPC	OEM
C380BX500	OEM
C380BX555	OEM
C380B-Y	0286
C380BY	0286
C380C	0159
C380C(TRANS)	0286
C380CX500	OEM
C380CX555	OEM
C380C-Y	0286
C380CY	0286
C380D	0159
C380D(SCR)	0159
C380D(TRANS)	0286
C380DX500	OEM
C380DX555	OEM
C380D-Y	0286
C380DY	0286
C380E	0159
C380E(TRANS)	0286
C380EX500	OEM
C380EX555	OEM
C380E-Y	0286
C380EY	0286
C380F	0286
C380F-Y	0286
C380FY	0286
C380G	0286
C380GN	0286
C380H	0286
C380J	0286
C380K	0286
C380L	0286
C380M	0159
C380M(TRANS)	0286
C380MX500	OEM
C380MX555	OEM
C380N	0096
C380NX500	OEM
C380NX555	OEM
C380O	0286
C380OR	0286
C380P	0096
C380PA	0096
C380PAX555	OEM
C380PB	0096

DEVICE TYPE	REPL CODE
C380PBX555	OEM
C380PC	OEM
C380PX555	OEM
C380R(2S)	0286
C380RED	0286
C380S	0096
C380SX500	OEM
C380SX555	OEM
C380T	0096
C380T(TRANS)	0286
C380TX555	OEM
C380V	0286
C380X	0286
C380Y(2S)	0286
C380YEL	0286
C381(2S)	0113
C381-O(2S)	0113
C381-BN	0113
C381-BRN	0113
C381-O(2S)	0113
C381-OR	0113
C381-ORG	0113
C381-R(2S)	0113
C381-RED	0113
C381A	0113
C381B	0113
C381BN(2S)	0113
C381BN-1	0113
C381C	0113
C381D	0113
C381E	0113
C381F	0113
C381G	0113
C381GN	0113
C381H	0113
C381J	0113
C381K	0113
C381L	0113
C381M	0113
C381O	0113
C381OR	0113
C381R(2S)	0113
C381RED	0113
C381RL	0113
C381X	0113
C381Y	0113
C382(2S)	0184
C382-G	0184
C382-GR(2S)	0184
C382-GY	0184
C382-OR	0184
C382-R	0184
C382-V	0184
C382A	0184
C382B	0184
C382BK(2S)	0184
C382BK1	0184
C382BK2	0184
C382BL(2S)	0184
C382BN(2S)	0184
C382BR(2S)	0184
C382C	0184
C382D	0184
C382E	0184
C382F	0184
C382G(2S)	0184
C382GN	0184
C382GR	0184
C382GY	0184
C382H	0184
C382J	0184
C382K	0184
C382L	0184
C382M	0184
C382OR	0184
C382R(2S)	0184
C382V(2S)	0184
C382W	0184
C382X	0184
C382Y	0184
C383(2S)	0076
C383-OR	0076
C383A	0076
C383B	0076
C383C	0076
C383D	0076
C383E	0076
C383F	0076
C383G(2S)	0076
C383GN	0076
C383H	0076
C383J	0076
C383K	0076
C383L	0076
C383M	0076
C383OR	0076
C383R	0076
C383T(2S)	0076
C383TM	0076
C383W(2S)	0076
C383X	0076
C383Y(2S)	0076
C384(2S)	0127
C384(O)	0127
C384-O	0127
C384A	0459
C384A(TRANS)	0127
C384B	0459
C384B(TRANS)	0127

DEVICE TYPE	REPL CODE
C384C	0459
C384C(TRANS)	0127
C384D	0459
C384D(TRANS)	0127
C384E	0459
C384E(TRANS)	0127
C384F	0127
C384G	0127
C384GN	0127
C384H	0127
C384J	0127
C384K	0127
C384L	0127
C384M	0459
C384M(TRANS)	0127
C384OR	0127
C384R	0127
C384X	0127
C384Y(2S)	0127
C385	0549
C385A	0459
C385A(SCR)	0459
C385A(TRANS)	0549
C385B	0459
C385B(TRANS)	0549
C385C	0459
C385C(TRANS)	0549
C385D	0459
C385D(TRANS)	0549
C385E	0459
C385E(TRANS)	0549
C385F	0549
C385G	0549
C385GN	0549
C385H	0549
C385J	0549
C385K	0549
C385L	0549
C385M	0459
C385M(TRANS)	0549
C385N	0587
C385OR	0549
C385R	0549
C385S	0587
C385X	0549
C385Y	0549
C386	0127
C386-O	0127
C386A	0127
C386AO	0127
C386B	0127
C386C	0127
C386D	0127
C386E	0127
C386F	0127
C386G	0127
C386GN	0127
C386H	0127
C386J	0127
C386K	0127
C386L	0127
C386M	0127
C386OR	0127
C386R	0127
C386X	0127
C386Y	0127
C387(2S)	0549
C387-G	0549
C387-OR	0549
C387A(2S)	0549
C387AG	0549
C387B	0549
C387C	0549
C387D	0549
C387E	0769
C387E(TRANS)	0549
C387F	0549
C387FA3	0549
C387G(2S)	0549
C387GN	0549
C387H	0549
C387J	0549
C387K	0549
C387L	0549
C387M	0769
C387M(TRANS)	0549
C387N	0483
C387OR	0549
C387P	0483
C387PA	0483
C387PB	0483
C387R	0549
C387S	0483
C387T	0483
C387X	0549
C387Y	0549
C388(2S)	0842
C388-A	0842
C388-O	0842
C388-OR	0842
C388A	0842
C388A(2S)	0842
C388A(JAPAN)	0842
C388A(KOREA)	0842
C388A-TM	0842
C388ATM	0842
C388ATMFA	0842
C388B	0842
C388C	0842

DEVICE TYPE	REPL CODE
C388D	0842
C388E	0769
C388E(TRANS)	0842
C388F	0842
C388FA	0842
C388G	0842
C388GN	0842
C388H	0842
C388J	0842
C388K	0842
C388L	0842
C388M	0769
C388M(TRANS)	0842
C388N	0483
C388OR	0842
C388P	0483
C388PA	0483
C388PB	0483
C388R	0842
C388S	0483
C388T	0483
C388TM	0842
C388X	0842
C388Y	0842
C389(2S)	0007
C389-O	0007
C389-OR	0007
C389A	0007
C389AFP	0007
C389B	0007
C389BLB-O	0007
C389C	0007
C389D	0007
C389E	0007
C389F	0007
C389G	0007
C389GN	0007
C389H	0007
C389J	0007
C389K	0007
C389L	0007
C389LP	0007
C389M	0007
C389O	0007
C389OR	0007
C389P	0007
C389R(2S)	0007
C389X	0007
C389Y	0007
C390	0259
C390A	0259
C390AX500	OEM
C390AX550	OEM
C390B	2843
C390B(TRANS)	0259
C390BX500	OEM
C390BX550	OEM
C390C	2843
C390C(TRANS)	0259
C390CX500	OEM
C390CX550	OEM
C390D	2843
C390D(TRANS)	0259
C390DX500	OEM
C390DX550	OEM
C390E	2843
C390E(TRANS)	0259
C390EX500	OEM
C390EX550	OEM
C390EX555	OEM
C390F	0259
C390G	0259
C390GN	0259
C390H	0259
C390J	0259
C390K	0259
C390L	0259
C390M	1765
C390M(TRANS)	0259
C390MX500	OEM
C390MX550	OEM
C390MX555	OEM
C390N	1814
C390NX555	OEM
C390OR	0259
C390P	1814
C390PA	1814
C390PAX555	OEM
C390PB	1814
C390PBX555	OEM
C390PC	OEM
C390PX555	OEM
C390R	0259
C390S	1814
C390S555	OEM
C390SX555	OEM
C390T	1814
C390TX555	OEM
C390X	0259
C390Y	0259
C391	0008
C391A	0008
C391B	0008
C391C	0008
C391D	0008
C391E	0008
C391F	0008
C391G	0008
C391GN	0008

DEVICE TYPE	REPL CODE
C391H	0008
C391J	0008
C391K	0008
C391L	0008
C391M	0008
C391OR	0008
C391PC	OEM
C391PD	OEM
C391PE	OEM
C391PM	OEM
C391PN	OEM
C391PS	OEM
C391R	0008
C391X	0008
C391Y	0008
C392(2S)	0259
C392A	OEM
C392A(TRANS)	0259
C392B	OEM
C392B(TRANS)	0259
C392C	OEM
C392C(TRANS)	0259
C392D	OEM
C392D(TRANS)	0259
C392E	OEM
C392E(TRANS)	0259
C392F(TRANS)	0259
C392G	0259
C392GN	0259
C392H	0259
C392J	0259
C392K	0259
C392L	0259
C392M	OEM
C392M(TRANS)	0259
C392OR	0259
C392R	0259
C392X	0259
C392Y	0259
C393	0079
C393A	OEM
C393B	OEM
C393C	0631
C393C(IC)	0631
C393D	OEM
C393E	OEM
C393M	OEM
C394(2S)	0155
C394-O(2S)	0155
C394-GR	0155
C394-GRN	0155
C394-O(2S)	0155
C394-OR	0155
C394-ORG	0155
C394-R	0155
C394-RED	0155
C394-Y	0155
C394-YEL	0155
C394A	2060
C394A(TRANS)	0155
C394AP	0155
C394B	2060
C394B(TRANS)	0155
C394C	2060
C394C(TRANS)	0155
C394D	2060
C394D(TRANS)	0155
C394E	2060
C394E(TRANS)	0155
C394F	0155
C394G	0155
C394GN	0155
C394GR(2S)	0155
C394GRN	0155
C394H	0155
C394J	0155
C394K	0155
C394L	0155
C394M	2060
C394M(TRANS)	0155
C394O	0155
C394OR	0155
C394R(2S)	0155
C394RED	0155
C394S	OEM
C394W(2S)	0155
C394X	0155
C394Y(2S)	0155
C394YEL	0155
C395(2S)	0016
C395A	4858
C395A(SCR)	OEM
C395A(TRANS)	0016
C395A-O	0016
C395A-ORG	0016
C395A-R	0016
C395A-RED	0016
C395A-Y	0016
C395A-YEL	0016
C395B	OEM
C395B(TRANS)	0016
C395C	OEM
C395C(TRANS)	0016
C395D	OEM
C395D(TRANS)	0016
C395E	OEM
C395E(TRANS)	0016
C395F	0016
C395G	0016

DEVICE TYPE	REPL CODE
C395GN	0016
C395H	0016
C395J	0016
C395K	0016
C395L	0016
C395M	OEM
C395M(TRANS)	0016
C395OR	0016
C395R(2S)	0016
C395S	OEM
C395X	0016
C395Y	0016
C396(2S)	0493
C396A	0493
C396B	0493
C396C	0493
C396D	0493
C396E	0493
C396F	0493
C396G	0493
C396GN	0493
C396GR	0493
C396H	0493
C396J	0493
C396K	0493
C396L	0493
C396M	0493
C396OR	0493
C396R	0493
C396X	0493
C396Y	0493
C397	0007
C397A	0007
C397B	0007
C397C	0007
C397D	0007
C397E	0769
C397E(TRANS)	0007
C397F	0007
C397G	0007
C397GN	0007
C397H	0007
C397J	0007
C397K	0007
C397L	0007
C397M	0769
C397M(TRANS)	0007
C397N	0483
C397OR	0007
C397P	0483
C397PA	0483
C397PB	0483
C397R	0007
C397S	0483
C397T	0483
C397X	0007
C397Y	0007
C398	0008
C398A	0008
C398B	0008
C398C	0008
C398D	0008
C398E	0769
C398E(TRANS)	0008
C398F	0008
C398FA1	0008
C398G	0008
C398GN	0008
C398H	0008
C398J	0008
C398K	0008
C398L	0008
C398M	0769
C398M(TRANS)	0008
C398N	0483
C398OR	0008
C398P	0483
C398PA	0483
C398PB	0483
C398R	0008
C398S	0483
C398T	0483
C398X	0008
C398Y	0008
C399(2S)	0008
C399A	0008
C399B	0008
C399C	0008
C399D	0008
C399E	0008
C399F	0008
C399FA1	0008
C399G	0008
C399GN	0008
C399H	0008
C399J	0008
C399K	0008
C399L	0008
C399M	0008
C399OR	0008
C399R	0008
C399X	0008
C399Y	0008
C400	0016
C400(2S)	0016
C400-O(2S)	0016
C400-GR(2S)	0016
C400-O	0016
C400-R(2S)	0016

DEVICE TYPE	REPL CODE
C400-Y(2S)	0016
C400A	0016
C400B	0016
C400C	0016
C400D	0016
C400E	0016
C400F	0016
C400G	0016
C400GN	0016
C400H	0016
C400J	0016
C400K	0016
C400L	0016
C400M	0016
C400OR	0016
C400R	0016
C400X	0016
C400Y	0016
C401	0155
C401(2S)	0155
C401A	0155
C401B	0155
C401C	0155
C401D	0155
C401E	0155
C401F	0155
C401G	0155
C401GN	0155
C401H	0155
C401J	0155
C401K	0155
C401L	0155
C401M	0155
C401OR	0155
C401R	0155
C401X	0155
C401Y	0155
C402	0155
C402(2S)	0155
C402A(2S)	0155
C402B	0155
C402C	0155
C402D	0155
C402E	0155
C402F	0155
C402G	0155
C402GN	0155
C402H	0155
C402J	0155
C402K	0155
C402L	0155
C402M	0155
C402OR	0155
C402R	0155
C402X	0155
C402Y	0155
C403(2S)	0155
C403-OR	0155
C403A(2S)	0155
C403AL	0155
C403B	0155
C403C(2S)	0155
C403CG	0155
C403D	0155
C403E	0155
C403F	0155
C403G	0155
C403GN	0155
C403H	0155
C403J	0155
C403K	0155
C403L	0155
C403M	0155
C403OR	0155
C403R	0155
C403X	0155
C403Y	0155
C404(2S)	0155
C404A	0155
C404B	0155
C404C	0155
C404D	0155
C404E	0155
C404F	0155
C404G	0155
C404GN	0155
C404H	0155
C404J	0155
C404K	0155
C404L	0155
C404M	0155
C404OR	0155
C404R	0155
C404X	0155
C404Y	0155
C405(2S)	0016
C405A	0016
C405B	0016
C405C	0016
C405D	0016
C405E	0016
C405F	0016
C405G	0016
C405GN	0016
C405H	0016
C405J	0016
C405K	0016
C405L	0016
C405M	0016

DEVICE TYPE	REPL CODE
C405OR	0016
C405R	0016
C405X	0016
C405Y	0016
C406(2S)	0016
C406A	0016
C406B	0016
C406C	0016
C406D	0016
C406E	0016
C406F	0016
C406G	0016
C406GN	0016
C406H	0016
C406J	0016
C406K	0016
C406L	0016
C406M	0016
C406OR	0016
C406R	0016
C406X	0016
C406Y	0016
C407	1966
C407(2S)	1966
C408	1966
C409(2S)	1966
C410	1966
C410(2S)	1966
C410A	1966
C411	1966
C411(2S)	1966
C412(2S)	0642
C413	0630
C413N	OEM
C414	0630
C415	0630
C416	0630
C420	0086
C420(2S)	0630
C423(2S)	0016
C423-0	0016
C423A	0016
C423B(2S)	0016
C423C(2S)	0016
C423D(2S)	0016
C423E(2S)	0016
C423F(2S)	0016
C423G	0016
C423GN	0016
C423H	0016
C423J	0016
C423K	0016
C423L	0016
C423M	0016
C423OR	0016
C423R	0016
C423X	0016
C423Y	0016
C424	0016
C424(2S)	0016
C424D(2S)	0016
C425	0086
C425(2S)	0016
C425A	0016
C425B(2S)	0016
C425C(2S)	0016
C425D(2S)	0016
C425E(2S)	0016
C425F(2S)	0016
C425G	0016
C425GN	0016
C425H	0016
C425K	0016
C425L	0016
C425M	0016
C425OR	0016
C425R	0016
C425X	0016
C425Y	0016
C426	0086
C426(2S)	0079
C426M	OEM
C427	0079
C428	0493
C428(2S)	0079
C429(2S)	0475
C429A	0475
C429B	0475
C429C	0475
C429D	0475
C429E	0475
C429F	0475
C429G	0475
C429GN	0475
C429H	0475
C429J(2S)	0475
C429K	0475
C429L	0475
C429M	0475
C429OR	0475
C429R	0475
C429X(2S)	0475
C429Y	0475
C430(2S)	0475
C430A	0475
C430AX500	OEM
C430AX550	OEM
C430B	0475
C430BX500	OEM
C430BX550	OEM
C430C	0475
C430CX500	OEM
C430CX550	OEM
C430D	0475
C430DX500	OEM
C430DX550	OEM
C430E	OEM
C430E(TRANS)	0475
C430EX500	OEM
C430EX555	OEM
C430F	0475
C430G	0475
C430GN	0475
C430H(2S)	0475
C430J	0475
C430K	0475
C430L	0475
C430M	OEM
C430M(TRANS)	0475
C430MX500	OEM
C430MX555	OEM
C430N	OEM
C430NX555	OEM
C430OR	0475
C430P	OEM
C430PA	OEM
C430PAX555	OEM
C430PB	OEM
C430PBX555	OEM
C430PC	OEM
C430PX555	OEM
C430R	0475
C430S	OEM
C430SX555	OEM
C430T	OEM
C430TX555	OEM
C430W(2S)	0475
C430X	0475
C430Y	0475
C431E1	OEM
C431E2	OEM
C431M1	OEM
C431M2	OEM
C431N1	OEM
C431N2	OEM
C431P1	OEM
C431P2	OEM
C431PA1	OEM
C431PA2	OEM
C431PB	OEM
C431PB1	OEM
C431PB2	OEM
C431PC	OEM
C431PC1	OEM
C431PC2	OEM
C431PD	OEM
C431PD1	OEM
C431PD2	OEM
C431PE	OEM
C431PE1	OEM
C431PE2	OEM
C431PM	OEM
C431PM1	OEM
C431PM2	OEM
C431PN	OEM
C431PN1	OEM
C431PN2	OEM
C431PS	OEM
C431PS1	OEM
C431PS2	OEM
C431S1	OEM
C431S2	OEM
C431T1	OEM
C431T2	OEM
C433	0079
C434	0236
C434A	OEM
C434B	OEM
C434C	OEM
C434D	OEM
C434E	OEM
C434M	OEM
C434S	OEM
C435A	OEM
C435B	OEM
C435C	OEM
C435D	OEM
C435E	OEM
C435M	OEM
C435S	OEM
C436E	OEM
C436M	OEM
C436N	OEM
C436P	OEM
C436PA	OEM
C436PB	OEM
C436S	OEM
C436T	OEM
C437	0630
C437E	OEM
C437M	OEM
C437N	OEM
C437P	OEM
C437PA	OEM
C437PB	OEM
C437S	OEM
C437T	OEM
C438	0698
C438E	OEM
C438M	OEM
C438N	OEM
C438P	OEM
C438PA	OEM
C438PB	OEM
C438S	OEM
C438T	OEM
C439	0018
C440(2S)	0493
C440E	4148
C440M	4148
C440N	1725
C440P	1725
C440PA	1725
C440PB	1725
C440PC	OEM
C440S	1725
C440T	1725
C441	0493
C441(2S)	0493
C441A	0493
C441B	0493
C441C	0493
C441D	0493
C441E	0493
C441F	0493
C441G	0493
C441GN	0493
C441H	0493
C441J	0493
C441K	0493
C441L	0493
C441M	0493
C441OR	0493
C441PC	OEM
C441PD	OEM
C441PE	OEM
C441PM	OEM
C441PN	OEM
C441PS	OEM
C441R	0493
C441X	0493
C441Y	0493
C442	0493
C442(2S)	0493
C442A	0493
C442B	0493
C442C	0493
C442D	0493
C442E	0493
C442F	0493
C442G	0493
C442GN	0493
C442H	0493
C442J	0493
C442K	0493
C442L	0493
C442M	0493
C442OR	0493
C442R	0493
C442X	0493
C442Y	0493
C443(2S)	0086
C443A	0086
C443B	0086
C443C	0086
C443D	0086
C443E	0086
C443F	0086
C443G	0086
C443GN	0086
C443H	0086
C443J	0086
C443K	0086
C443L	0086
C443M	0086
C443OR	0086
C443R	0086
C443X	0086
C443Y	0086
C444	0016
C444(2S)	0595
C444A	OEM
C444B	OEM
C444C	OEM
C444D	OEM
C444E	OEM
C444M	OEM
C444S	OEM
C445	0168
C445A	OEM
C445A(TRANS)	0168
C445B	OEM
C445B(TRANS)	0168
C445C	OEM
C445C(TRANS)	0168
C445D	OEM
C445D(TRANS)	0168
C445E	OEM
C445E(TRANS)	0168
C445F	0168
C445G	0168
C445GN	0168
C445H	0168
C445J	0168
C445K	0168
C445L	0168
C445M	OEM
C445M(TRANS)	0168
C445OR	0168
C445R	0168
C445S	OEM
C445X	0168
C445Y	0168
C446	5568
C447E	OEM
C447M	OEM
C447N	OEM
C447P	OEM
C447PA	OEM
C447PB	OEM
C447S	OEM
C447T	OEM
C448E	OEM
C448M	OEM
C448N	OEM
C448P	OEM
C448PA	OEM
C448PB	OEM
C448S	OEM
C448T	OEM
C449PE	OEM
C450	0016
C450E	OEM
C450E-1	OEM
C450E1	OEM
C450E-2	OEM
C450E2	OEM
C450M	OEM
C450M1	OEM
C450M2	OEM
C450N	OEM
C450N1	OEM
C450N2	OEM
C450P	OEM
C450P1	OEM
C450P2	OEM
C450PA	OEM
C450PA1	OEM
C450PA2	OEM
C450PB	OEM
C450PB1	OEM
C450PB2	OEM
C450PC	OEM
C450PC1	OEM
C450PC2	OEM
C450PD	OEM
C450PD1	OEM
C450PD2	OEM
C450S	OEM
C450S1	OEM
C450S2	OEM
C450T	OEM
C450T1	OEM
C450T2	OEM
C451E1	OEM
C451E2	OEM
C451M1	OEM
C451M2	OEM
C451N-1	OEM
C451N1	OEM
C451N-2	OEM
C451N2	OEM
C451P-1	OEM
C451P1	OEM
C451P-2	OEM
C451P2	OEM
C451PA-1	OEM
C451PA1	OEM
C451PA-2	OEM
C451PA2	OEM
C451PB	OEM
C451PB-1	OEM
C451PB1	OEM
C451PB-2	OEM
C451PB2	OEM
C451PC	OEM
C451PC-1	OEM
C451PC1	OEM
C451PC-2	OEM
C451PC2	OEM
C451PD	OEM
C451PD-1	OEM
C451PD1	OEM
C451PD-2	OEM
C451PD2	OEM
C451PE	OEM
C451PE-1	OEM
C451PE1	OEM
C451PE-2	OEM
C451PE2	OEM
C451PM	OEM
C451PM-1	OEM
C451PM1	OEM
C451PM-2	OEM
C451PM2	OEM
C451PN	OEM
C451PN-1	OEM
C451PN1	OEM
C451PN-2	OEM
C451PN2	OEM
C451PS	OEM
C451PS-1	OEM
C451PS1	OEM
C451PS-2	OEM
C451PS2	OEM
C451S-1	OEM
C451S1	OEM
C451S-2	OEM
C451S2	OEM
C451T-1	OEM
C451T1	OEM
C451T-2	OEM
C451T2	OEM
C454	0076
C454-3	0076
C454-5	0076
C454-C	0076
C454-OR	0076
C454A(2S)	0076
C454B	0076
C454B-6	0076
C454BL	0076
C454C	0076
C454C(JAPAN)	0076
C454D(2S)	0076
C454E	0076
C454F	0076
C454G	0076
C454GN	0076
C454H	0076
C454J	0076
C454K	0076
C454L(2S)	0076
C454LA(2S)	0076
C454M	0076
C454OR	0076
C454R	0076
C454X	0076
C454Y	0076
C455(2S)	0155
C455-OR	0155
C455A	0155
C455B	0155
C455C	0155
C455D	0155
C455E	0155
C455F	0155
C455G	0155
C455GN	0155
C455H	0155
C455J	0155
C455K	0155
C455L	0155
C455M	0155
C455OR	0155
C455R	0155
C455X	0155
C455Y	0155
C456(2S)	0698
C456A(2S)	0698
C456B	0698
C456C	0698
C456D(2S)	0698
C456E	0698
C456F	0698
C456G	0698
C456GN	0698
C456H	0698
C456J	0698
C456K	0698
C456L	0698
C456M	0698
C456OR	0698
C456R	0698
C456X	0698
C456Y	0698
C457	0079
C457E	OEM
C457M	OEM
C457N	OEM
C457P	OEM
C457PA	OEM
C457PB	OEM
C457PC	OEM
C457PD	OEM
C457S	OEM
C457T	OEM
C458	0076
C458(JAPAN)	0076
C458-4	0076
C458-5	0076
C458-B	0076
C458-C	0076
C458-D	0076
C458-GR	0076
C458-O	0076
C458-OR	0076
C458-Y	0076
C458A(2S)	0076
C458AD(2S)	0076
C458AK	0076
C458B	0076
C458B(JAPAN)	0076
C458BC(2S)	0076
C458BD	0076
C458BL(2S)	0076
C458BM	0076
C458C	0076
C458C(JAPAN)	0076
C458CL	0076
C458CLG(2S)	0076
C458CM(2S)	0076
C458D	0076
C458D(2S)	0076
C458E	OEM
C458E(TRANS)	0076
C458F	0076
C458G(2S)	0076
C458GN	0076
C458H	0076
C458J	0076
C458K(2S)	0076
C458KB(2S)	0076
C458L(2S)	0076
C458L6	0076
C458LB(2S)	0076
C458LC	0076
C458LD	0076
C458LG	0076
C458LGA(2S)	0076
C458LGB	0076
C458LGBM(2S)	0076
C458LGC	0076
C458LGC-6	0076
C458LGD	0076
C458LGO(2S)	0076
C458LGR	0076
C458LGS(2S)	0076
C458M	5686
C458M(SCR)	OEM
C458M(TRANS)	0076
C458N	OEM
C458OR	0076
C458P	0286
C458P(SCR)	OEM
C458PA	OEM
C458PB	OEM
C458PC	OEM
C458PD	OEM
C458R	0076
C458RGS(2S)	0076
C458S	OEM
C458T	OEM
C458X	0076
C458Y	0076
C459(2S)	0286
C459A	0286
C459B(2S)	0286
C459C	0286
C459D(2S)	0286
C459E	0286
C459F	0286
C459G	0286
C459GN	0286
C459H	0286
C459J	0286
C459K	0286
C459L	0286
C459M	0286
C459OR	0286
C459R	0286
C459X	0286
C459Y	0286
C460	0151
C460(JAPAN)	0151
C460-5	0151
C460-B	0151
C460-C	0151
C460-OR	0151
C460A(2S)	0151
C460B	0151
C460BL	0151
C460C	0151
C460D(2S)	0151
C460E	0151
C460F	0151
C460G(2S)	0151
C460GB(2S)	0151
C460GN	0151
C460H(2S)	0151
C460J	0151
C460K(2S)	0151
C460L(2S)	0151
C460M	0151
C460OR	0151
C460R	0151
C460X	0151
C460Y	0151
C461	0151
C461-8F	0151
C461-A	0151
C461-B	0151
C461A	0151
C461AL	0151
C461B	0151
C461BF	0151
C461BK	0151
C461BL	0151
C461C	0151
C461E(2S)	0151
C461EP	0151
C461F	0151
C461L(2S)	0151
C462	0079
C463(2S)	0007
C463A	0007
C463B	0007
C463C	0007
C463D	0007
C463E	0007
C463F	0007
C463G	0007
C463GN	0007
C463H(2S)	0007
C463J	0007
C463K	0007
C463L	0007
C463M	0007
C463OR	0007
C463R	0007
C463X	0007
C463Y	0007
C464(2S)	0259
C464A	0259
C464B	0259
C464C(2S)	0259
C464D	0259
C464E	0259
C464F	0259
C464G	0259
C464GN	0259
C464H	0259
C464J	0259
C464K	0259
C464L	0259
C464M	0259
C464OR	0259
C464R	0259
C464X	0259
C464Y	0259
C465(2S)	0259
C465A	0259
C465B	0259
C465C	0259
C465D	0259
C465E	0259
C465F	0259
C465G	0259
C465GN	0259
C465H	0259
C465J	0259
C465K	0259
C465L	0259
C465M	0259
C465OR	0259
C465R	0259
C465X	0259
C465Y	0259
C466(2S)	0259
C466A	0259
C466B	0259
C466C	0259
C466D	0259
C466E	0259
C466F	0259
C466G	0259
C466GN	0259
C466H(2S)	0259
C466J	0259
C466K	0259
C466L	0259
C466M	0259
C466OR	0259
C466R	0259
C466X	0259
C466Y	0259
C467	0144
C468(2S)	0016
C468A(2S)	0016
C468B	0016
C468H	0016
C468LGR	0016
C469(2S)	0475
C469-OR	0475
C469A(2S)	0475
C469B	0475
C469C	0475
C469D	0475
C469E	0475
C469F(2S)	0475
C469G	0475
C469GN	0475
C469H	0475
C469J	0475
C469K(2S)	0475
C469L	0475
C469M	0475
C469OR	0475
C469Q(2S)	0475
C469R(2S)	0475
C469X	0475
C469Y	0475
C470(2S)	0233
C470-3	0233
C470-4	0233
C470-5	0233
C470-6	0233
C470A	0233
C470B	0233
C470C	0233
C470D	0233
C470E	0233
C470F	0233
C470G	0233
C470GN	0233
C470H	0233
C470J	0233
C470K	0233
C470L	0233
C470M	0233
C470OR	0233
C470R	0233
C470X	0233

If replacement code is OEM, contact original manufacturer for replacement.

DEVICE TYPE	REPL CODE
C470Y	0233
C471	0018
C472	0224
C472A	0224
C472B	0224
C472C	0224
C472D	0224
C472E	0224
C472F	0224
C472G	0224
C472GN	0224
C472H	0224
C472J	0224
C472K	0224
C472L	0224
C472M	0224
C472OR	0224
C472R	0224
C472X	0224
C472Y(2S)	0224
C473	0016
C473AC	0016
C474	0079
C475(2S)	0475
C475A	0475
C475B	0475
C475C	0475
C475D	0475
C475E	0475
C475F	0475
C475G	0475
C475GN	0475
C475H	0475
C475J	0475
C475K(2S)	0475
C475L	0475
C475M	0475
C475OR	0475
C475R	0475
C475X	0475
C475Y	0475
C476(2S)	0475
C476A	0475
C476B	0475
C476C	0475
C476D	0475
C476E	OEM
C476E(TRANS)	0475
C476F	0475
C476G	0475
C476GN	0475
C476H	0475
C476J	0475
C476K	0475
C476L	0475
C476M	OEM
C476M(TRANS)	0475
C476N	OEM
C476OR	0475
C476P	OEM
C476PA	OEM
C476PB	OEM
C476R	0475
C476S	OEM
C476T	OEM
C476X	0475
C476Y	0475
C477(2S)	0007
C477A	0007
C477B	0007
C477C	0007
C477D	0007
C477E	OEM
C477E(TRANS)	0007
C477F	0007
C477G	0007
C477GN	0007
C477H	0007
C477J	0007
C477K	0007
C477L	0007
C477M	OEM
C477M(TRANS)	0007
C477N	OEM
C477OR	0007
C477P	OEM
C477PA	OEM
C477PB	OEM
C477PC	OEM
C477PD	OEM
C477R	0007
C477S	OEM
C477T	OEM
C477X	0007
C477Y	0007
C478(2S)	0016
C478-4(2S)	0016
C478-0	0016
C478A	0016
C478B	0016
C478C	0016
C478D(2S)	0016
C478E	OEM
C478E(TRANS)	0016
C478F	0016
C478G	0016
C478GN	0016
C478H	0016
C478J	0016
C478K	0016
C478L	0016
C478M	OEM
C478M(TRANS)	0016
C478N	OEM
C478OR	0016
C478P	OEM
C478PA	OEM
C478PB	OEM
C478PC	OEM
C478PD	OEM
C478R	0016
C478S	OEM
C478T	OEM
C478X	0016
C478Y	0016
C479(2S)	0086
C479A	0086
C479B	0086
C479C	0086
C479D	0086
C479E	0086
C479F	0086
C479G	0086
C479GN	0086
C479H	0086
C479J	0086
C479K	0086
C479L	0086
C479M	0086
C479OR	0086
C479R	0086
C479X	0086
C479Y	0086
C480	0324
C481(2S)	0698
C481-OR	0698
C481A	0698
C481B	0698
C481C	0698
C481D	0698
C481E	0698
C481F	0698
C481G	0698
C481GN	0698
C481H	0698
C481J	0698
C481K	0698
C481L	0698
C481M	0698
C481OR	0698
C481R	0698
C481X(2S)	0698
C481Y	0698
C482(2S)	1593
C482-0	1593
C482-GR(2S)	1593
C482-GRN	1593
C482-O(2S)	1593
C482-OR	1593
C482-ORG	1593
C482-Y(2S)	1593
C482-YEL	1593
C482A	1593
C482B	1593
C482C	1593
C482D	1593
C482E	1593
C482F	1593
C482G	1593
C482GN	1593
C482GR(2S)	1593
C482GRN	1593
C482GRY	1593
C482H	1593
C482J	1593
C482K	1593
C482L	1593
C482M	1593
C482OR	1593
C482R	1593
C482X(2S)	1593
C482Y(2S)	1593
C482YEL	1593
C483	1027
C484(2S)	1480
C484-BL	1480
C484-BLU	1480
C484-R	1480
C484-RED	1480
C484-Y	1480
C484-YEL	1480
C484A	1480
C484B	1480
C484BL(2S)	1480
C484C	1480
C484D	1480
C484E	1480
C484F	1480
C484G	1480
C484GN	1480
C484H	1480
C484J	1480
C484K	1480
C484L	1480
C484M	1480
C484OR	1480
C484R(2S)	1480
C484X	1480
C484Y(2S)	1480
C485(2S)	1480
C485-BL	1480
C485-BLU	1480
C485-OR	1480
C485-R	1480
C485-RED	1480
C485-Y	1480
C485-YEL	1480
C485A	1480
C485B	1480
C485BL(2S)	1480
C485C(2S)	1480
C485D	1480
C485E	1480
C485F	1480
C485G	1480
C485GN	1480
C485H	1480
C485J	1480
C485K	1480
C485L	1480
C485M	1480
C485OR	1480
C485R	1480
C485X	1480
C485Y(2S)	1480
C486(2S)	1480
C486-BL	1480
C486-BLU	1480
C486-R	1480
C486-RED	1480
C486-Y	1480
C486-YEL	1480
C486A	1480
C486B	1480
C486BL(2S)	1480
C486C	1480
C486D	1480
C486E	1480
C486F	1480
C486G	1480
C486GN	1480
C486H	1480
C486J	1480
C486K	1480
C486L	1480
C486M	1480
C486OR	1480
C486R	1480
C486X	1480
C486Y(2S)	1480
C487(2S)	0178
C487A	0178
C488(2S)	0178
C488H	0178
C489(2S)	0178
C489-BL	0178
C489-BLU	0178
C489-R	0178
C489-RED	0178
C489-Y	0178
C489-YEL	0178
C489A	0178
C489B	0178
C489C	0178
C489D	0178
C489E	0178
C489F	0178
C489G	0178
C489GN	0178
C489H	0178
C489J	0178
C489K	0178
C489L	0178
C489M	0178
C489OR	0178
C489R	0178
C489X	0178
C489Y	0178
C490(2S)	0178
C490-BL	0178
C490-BLU	0178
C490-R	0178
C490-RED	0178
C490-Y	0178
C490-YEL	0178
C491(2S)	0178
C491-BL	0178
C491-BLU	0178
C491-R	0178
C491-RED	0178
C491-Y	0178
C491-YEL	0178
C491A	0178
C491B	0178
C491BL(2S)	0178
C491C	0178
C491D	0178
C491E	0178
C491F	0178
C491G	0178
C491GN	0178
C491H	0178
C491J	0178
C491K	0178
C491L	0178
C491M	0178
C491OR	0178
C491R	0178
C491X	0178
C491Y(2S)	0178
C492(2S)	0177
C493(2S)	0867
C493-BL(2S)	0867
C493-R(2S)	0867
C493-Y(2S)	0867
C493A	0867
C493B	0867
C493C	0867
C493D	0867
C493E	0867
C493F	0867
C493G	0867
C493GN	0867
C493H	0867
C493J	0867
C493K	0867
C493L	0867
C493M	0867
C493OR	0867
C493R	0867
C493X	0867
C493Y	0867
C494(2S)	0620
C494-BL(2S)	0620
C494-R(2S)	0620
C494-Y(2S)	0620
C494A	0620
C494B	0620
C494BL(2S)	0620
C494C	0620
C494D	0620
C494E	0620
C494F	0620
C494G	0620
C494GN	0620
C494H	0620
C494J	0620
C494K	0620
C494L	0620
C494M	0620
C494OR	0620
C494R	0620
C494X	0620
C494Y	0620
C495(2S)	0786
C495-0	0786
C495-O(2S)	0786
C495-OR	0786
C495-ORG	0786
C495-R(2S)	0786
C495-RED	0786
C495-Y(2S)	0786
C495-YEL	0786
C495A	0786
C495B	0786
C495C	0786
C495D	0786
C495E	0786
C495F	0786
C495G	0786
C495GN	0786
C495H	0786
C495J	0786
C495K	0786
C495L	0786
C495M	0786
C495OR	0786
C495P	0786
C495R	0786
C495RED	0786
C495T(2S)	0786
C495T(CB)	1591
C495X	0786
C495Y(2S)	0786
C495YEL	0786
C496	0786
C496-0	0786
C496-O(2S)	0786
C496-OR	0786
C496-ORG	0786
C496-R(2S)	0786
C496-RED	0786
C496-Y(2S)	0786
C496-YEL	0786
C496A	0786
C496B	0786
C496C	0786
C496D	0786
C496E	0786
C496F	0786
C496G	0786
C496GN	0786
C496H	0786
C496J	0786
C496K	0786
C496L	0786
C496M	0786
C496OR	0786
C496R	0786
C496RED	0786
C496X	0786
C496Y(2S)	0786
C496YEL	0786
C497(2S)	0086
C497-0	0086
C497-O(2S)	0086
C497-OR	0086
C497-ORG	0086
C497-R(2S)	0086
C497-RED	0086
C497-Y(2S)	0086
C497-YEL	0086
C497A	0086
C497B	0086
C497C	0086
C497D	0086
C497E	0086
C497F	0086
C497G	0086
C497GN	0086
C497H	0086
C497J	0086
C497K	0086
C497L	0086
C497M	0086
C497OR	0086
C497R	0086
C497RED	0086
C497X	0086
C497Y	0086
C498(2S)	0086
C498-0	0086
C498-O(2S)	0086
C498-OR	0086
C498-ORG	0086
C498-R(2S)	0086
C498-RED	0086
C498-Y(2S)	0086
C498-YEL	0086
C498A	0086
C498B	0086
C498C	0086
C498D	0086
C498E	0086
C498F	0086
C498G	0086
C498GN	0086
C498H	0086
C498J	0086
C498K	0086
C498L	0086
C498M	0086
C498OR	0086
C498R	0086
C498RED	0086
C498X	0086
C498Y	0086
C498YEL	0086
C499(2S)	2375
C499-OR	2375
C499-R	2375
C499-RED	2375
C499-RY(2S)	2375
C499-Y	2375
C499-YEL	2375
C499A	2375
C499B	2375
C499C	2375
C499D	2375
C499E	2375
C499F	2375
C499FA1	2375
C499G	2375
C499GN	2375
C499H	2375
C499J	2375
C499K	2375
C499L	2375
C499M	2375
C499OR	2375
C499R(2S)	2375
C499RED	2375
C499RY	2375
C499X	2375
C499Y(2S)	2375
C499YEL	2375
C500(2S)	0233
C500(N)	0233
C500(Y)	0233
C500A	0233
C500B	0233
C500C	0233
C500D	0233
C500E	0233
C500F	0233
C500G	0233
C500GN	0233
C500H	0233
C500J	0233
C500K	0233
C500L	0233
C500M	0233
C500OR	0233
C500R(2S)	0233
C500X	0233
C500Y(2S)	0233
C501	0086
C501-0	0086
C501-ORG	0086
C501-R	0086
C501-RED	0086
C501-Y	0086
C501-YEL	0086
C501A	OEM
C501A(TRANS)	0086
C501B	OEM
C501B(TRANS)	0086
C501C	OEM
C501C(TRANS)	0086
C501D	OEM
C501D(TRANS)	0086
C501E	OEM
C501E(TRANS)	0086
C501F	0086
C501G	0086
C501GN	0086
C501H	0086
C501J	0086
C501K	0086
C501L	OEM
C501L(2S)	0086
C501M	OEM
C501M(TRANS)	0086
C501N	OEM
C501O	0086
C501OR	0086
C501P	OEM
C501PA	OEM
C501PB	OEM
C501PC	OEM
C501PD	OEM
C501PE	OEM
C501PM	OEM
C501PN	OEM
C501PS	OEM
C501PT	OEM
C501R	0086
C501S	OEM
C501T	OEM
C501X	0086
C501Y	0086
C502	0698
C502(2S)	0698
C502A	0698
C502B	0698
C502C	0698
C502D	0698
C502E	0698
C502F	0698
C502G	0698
C502GN	0698
C502H	0698
C502J	0698
C502K	0698
C502L	OEM
C502L(TRANS)	0698
C502LA	OEM
C502M	0698
C502OR	0698
C502PE	OEM
C502PM	OEM
C502PN	OEM
C502PS	OEM
C502PT	OEM
C502R	0698
C502X	0698
C502Y	0698
C503(2S)	0086
C503-GR	0086
C503-O(2S)	0086
C503-Y(2S)	0086
C503A	0086
C503B	0086
C503C	0086
C503D	0086
C503E	0086
C503F	0086
C503G	0086
C503GN	0086
C503GR(2S)	0086
C503H	0086
C503J	0086
C503K	0086
C503L	0086
C503M	0086
C503OR	0086
C503R	0086
C503X	0086
C504(2S)	0086
C504-GR	0086
C504-O(2S)	0086
C504-Y(2S)	0086
C504A	0086
C504B	0086
C504C	0086
C504D	0086
C504E	0086
C504F	0086
C504G	0086
C504GN	0086
C504GR(2S)	0086
C504H	0086
C504J	0086
C504K	0086
C504L	0086
C504M	0086
C504OR	0086
C504R	0086
C504X	0086
C505(2S)	0233
C505-O(2S)	0233
C505-ORG	0233
C505-R(2S)	0233
C505-RED	0233
C505A	0233
C505B	0233
C505C	0233
C505D	0233
C505E	0233
C505F	0233
C505G	0233
C505GN	0233
C505H	0233
C505J	0233
C505K	0233
C505L	0233
C505M	0233
C505O	0233
C505OR	0233
C505R	0233
C505X	0233
C505Y	0233
C506(2S)	0233
C506-O(2S)	0233
C506-ORG	0233
C506-R(2S)	0233
C506-RED	0233
C506A	OEM
C506A(TRANS)	0233
C506B	OEM
C506B(TRANS)	0233
C506C	OEM
C506C(TRANS)	0233
C506D	OEM
C506D(TRANS)	0233
C506E	OEM
C506E(TRANS)	0233
C506F	0233
C506G	0233
C506GN	0233
C506H	0233
C506J	0233
C506K	0233
C506L	0233
C506M	OEM
C506M(TRANS)	0233
C506O	0233
C506OR	0233
C506R	0233
C506X	0233
C506Y	0233
C507(2S)	0233
C507-0	0233
C507-O(2S)	0233
C507-R(2S)	0233
C507-Y(2S)	0233
C507A	OEM
C507A(TRANS)	0233
C507B	OEM
C507B(TRANS)	0233
C507C	OEM
C507C(TRANS)	0233
C507D	OEM
C507D(TRANS)	0233
C507E	OEM
C507E(TRANS)	0233
C507F	0233
C507G	0233
C507GN	0233
C507H	0233
C507J	0233
C507K	0233
C507L	0233
C507M	OEM
C507M(TRANS)	0233
C507N	OEM
C507OR	0233
C507P	OEM
C507R	0233
C507S	OEM
C507T	OEM
C507X	0233
C507Y	0233
C508(2S)	0178
C508A	OEM
C508A(TRANS)	0178
C508B	OEM
C508B(TRANS)	0178
C508C	OEM
C508C(TRANS)	0178
C508D	OEM
C508D(TRANS)	0178
C508E	OEM
C508E(TRANS)	0178
C508F	0178
C508G	0178
C508GN	0178
C508M	OEM
C508N	OEM
C508OR	0178
C508P	OEM
C508PA	OEM
C508PB	OEM
C508S	OEM
C508T	OEM
C508X	0178
C508Y	0178
C509	0155
C509(0)	0155
C509-0	0155
C509-O	0155
C509-Y	0155
C509G(2S)	0155

DEVICE TYPE	REPL CODE
C509G-O	0155
C509G-Y	0155
C509M	OEM
C509N	OEM
C509O	0155
C509P	OEM
C509PA	OEM
C509PB	OEM
C509S	OEM
C509T	OEM
C509Y(2S)	0155
C510	1480
C510-O	1480
C510-ORG	1480
C510-R	1480
C510-RED	1480
C510A	OEM
C510A(TRANS)	1480
C510B	OEM
C510B(TRANS)	1480
C510C	OEM
C510C(TRANS)	1480
C510D	OEM
C510D(TRANS)	1480
C510E	OEM
C510E(TRANS)	1480
C510F	1480
C510G	1480
C510GN	1480
C510H	1480
C510J	1480
C510K	1480
C510L	1480
C510M	OEM
C510M(TRANS)	1480
C510O	1480
C510OR	1480
C510R	1480
C510S	OEM
C510X	1480
C510Y	1480
C511	1480
C511-O	1480
C511-ORG	1480
C511-R	1480
C511-RED	1480
C511A	0447
C511A(TRANS)	1480
C511B	0940
C511B(TRANS)	1480
C511C	1222
C511C(TRANS)	1480
C511D	OEM
C511D(TRANS)	1480
C511E	1480
C511F	0447
C511F(TRANS)	1480
C511G	0447
C511G(TRANS)	1480
C511GN	1480
C511H	1222
C511H(TRANS)	1480
C511J	1480
C511K	1480
C511L	1480
C511M	1480
C511O	1480
C511OR	1480
C511R	1480
C511U	1396
C511X	1480
C511Y	1480
C512(2S)	1480
C512-O	1480
C512-O(2S)	1480
C512-ORG	1480
C512-R(2S)	1480
C512-RED	1480
C512A	1480
C512B	1480
C512C	1480
C512D	1480
C512E	1480
C512F	1480
C512G	1480
C512GN	1480
C512H	1480
C512J	1480
C512K	1480
C512L	1480
C512M	1480
C512O	1480
C512OR	1480
C512R	1480
C512X	1480
C512Y	1480
C513(2S)	1480
C513-O	1480
C513-O(2S)	1480
C513-ORG	1480
C513-R	1480
C513-RED	1480
C513A	1480
C513B	1480
C513C	1480
C513D	1480
C513E	1480
C513F	1480
C513G	1480
C513GN	1480
C513H	1480
C513J	1480
C513K	1480
C513L	1480
C513M	1480
C513O	1480
C513OR	1480
C513R(2S)	1480
C513X	1480
C513Y	1480
C514(2S)	0142
C514A	0142
C514B	0142
C514C	0142
C514D	0142
C514E	0142
C514F	0142
C514G	0142
C514GN	0142
C514H	0142
C514J	0142
C514K	0142
C514L	0142
C514M	0142
C514OR	0142
C514R	0142
C514X	0142
C514Y	0142
C515(2S)	0142
C515-OR	0142
C515A(2S)	0142
C515AM	0142
C515AX(2S)	0142
C515AY	0142
C515B	0142
C515BK	0142
C515C	0142
C515D	0142
C515E	0142
C515F	0142
C515G	0142
C515GN	0142
C515H	0142
C515J	0142
C515K	0142
C515L	0142
C515M	0142
C515OR	0142
C515R	0142
C515X	0142
C515Y	0142
C516(2S)	0086
C516A	0086
C516B	0086
C516C	0086
C516D	0086
C516E	0086
C516F	0086
C516G	0086
C516GN	0086
C516H	0086
C516J	0086
C516K	0086
C516L	0086
C516M	0086
C516OR	0086
C516R	0086
C516X	0086
C516Y	0086
C517(2S)	1411
C517-OR	1411
C517A	1411
C517B	1411
C517C(2S)	1411
C517D	1411
C517E	1411
C517F	1411
C517G	1411
C517GN	1411
C517H	1411
C517J	1411
C517K	1411
C517L	1411
C517M	1411
C517OR	1411
C517R	1411
C517S	1411
C517X	1411
C517Y	1411
C518	0168
C518A	0168
C519(2S)	0867
C519A(2S)	0867
C520	0867
C520A	0867
C520A(2S)	0867
C520B	OEM
C520C	OEM
C520D	OEM
C521(2S)	0867
C521A(2S)	0867
C522(2S)	2314
C522-O	5466
C522-O(2S)	2314
C522-ORG	2314
C522-R(2S)	2314
C522-RED	2314
C522O	2314
C522R	2314
C523(2S)	2314
C523-O	2314
C523-O(2S)	2314
C523-ORG	2314
C523-R(2S)	2314
C523-RED	2314
C523O	2314
C523R	2314
C524(2S)	3336
C524-O	2314
C524-O(2S)	3336
C524-ORG	2314
C524-R(2S)	2314
C524-RED	2314
C524O	2314
C524R	3336
C525(2S)	1411
C525-O	1411
C525-O(2S)	1411
C525-ORG	1411
C525-R(2S)	1411
C525-RED	1411
C525O	1411
C525R	1411
C526(2S)	0233
C526-O	0233
C526A(2S)	0233
C526B	0233
C526C	0233
C526D	0233
C526E	0233
C526F	0233
C526G	0233
C526GN	0233
C526H	0233
C526J	0233
C526K	0233
C526L	0233
C526M	0233
C526OR	0233
C526R	0233
C526X	0233
C526Y	0233
C527	0007
C528	0111
C529	0286
C529A	0286
C529B	0286
C529C	0286
C529D	0286
C529E	0286
C529F	0286
C529G	0286
C529GN	0286
C529H	0286
C529J	0286
C529K	0286
C529L	0286
C529M	0286
C529OR	0286
C529R	0286
C529X	0286
C529Y	0286
C530	0079
C530A	OEM
C530B	OEM
C530C	OEM
C530D	OEM
C530E	OEM
C530M	OEM
C531	0155
C532	0079
C532O	0079
C533	0079
C533A-5	OEM
C533A-6	OEM
C533A-G	OEM
C533A-M	OEM
C533A-X	OEM
C534	0079
C535	0127
C535-OR	0127
C535A	0127
C535ABC	0127
C535AL	0127
C535B	0127
C535B(JAPAN)	0127
C535C	0127
C535D	0127
C535E	0127
C535F	0127
C535G(2S)	0127
C535GN	0127
C535H	0127
C535J	0127
C535K	0127
C535L	0127
C535M	0127
C535OR	0127
C535R	0127
C535X	0127
C535Y	0127
C536	0537
C536(JAPAN)	0537
C536-D	0537
C536-E	0537
C536-F	0537
C536-G	0537
C536-OR	0537
C536A(2S)	0507
C536AF	0537
C536AG(2S)	0537
C536B(2S)	0537
C536C(2S)	0537
C536D(2S)	0537
C536DK(2S)	0537
C536E	0537
C536E(JAPAN)	0537
C536ED(2S)	0537
C536EH(2S)	0537
C536EJ(2S)	0537
C536EN(2S)	0537
C536EP	0537
C536ER(2S)	0537
C536ET(2S)	0537
C536EZ(2S)	0537
C536F	0537
C536F(JAPAN)	0537
C536F1(2S)	0537
C536F2(2S)	0537
C536FC	0537
C536FC(2S)	0537
C536FP(2S)	0537
C536FS(2S)	0537
C536FS6(2S)	0537
C536FZ(2S)	0537
C536G	0537
C536G(2S)	0537
C536G(JAPAN)	0537
C536G-1	0537
C536G1	0537
C536G2	0537
C536GF(2S)	0537
C536GJ	0537
C536GK(2S)	0537
C536GL	0537
C536GM	0537
C536GN	0537
C536GP	0537
C536GT(2S)	0537
C536GV(2S)	0537
C536GY(2S)	0537
C536GZ	0537
C536H(2S)	0537
C536J	0537
C536K	0537
C536K(JAPAN)	0537
C536KE	0537
C536KE(JAPAN)	0537
C536KF	0537
C536KF(JAPAN)	0537
C536KG	0537
C536KNP	0537
C536L	0537
C536M	0537
C536NP	0537
C536NPE	0537
C536NPF	0537
C536NPG	0537
C536OR	0537
C536R	0537
C536SP	0537
C536W(2S)	0537
C536X	0537
C536XL	0537
C536Y	0507
C536YF	0537
C537	0191
C537-C7	0191
C537-EV	0191
C537A	0111
C537ALC	0191
C537B	0191
C537BK	0191
C537C	0191
C537C7	0191
C537D(2S)	0191
C537D1	0191
C537D2(2S)	0191
C537E(2S)	0191
C537EF	0191
C537EH(2S)	0191
C537EJ(2S)	0191
C537EV	0191
C537F	0191
C537F1(2S)	0191
C537F2(2S)	0191
C537FC(2S)	0191
C537F-C7	0191
C537FC7	0191
C537FJ	0191
C537FK	0191
C537FV(2S)	0191
C537G(2S)	0191
C537G1(2S)	0191
C537G2	0191
C537GFL	0191
C537GI(2S)	0191
C537H(2S)	0191
C537HT(2S)	0191
C537WF	0191
C538	0079
C538-Q	0079
C538A(2S)	0079
C538A-P	0079
C538AQ(2S)	0079
C538A-R	0079
C538AR	0079
C538AS	0079
C538K	0079
C538P(2S)	0079
C538Q(2S)	0079
C538R	0079
C538S(2S)	0079
C538T(2S)	0079
C539(2S)	0016
C539R(2S)	0016
C539S(2S)	0016
C539T	0016
C540(2S)	0475
C540A	0475
C540B	0475
C540C	0475
C540D	0475
C540E	0475
C540F	0475
C540G	0475
C540GN	0475
C540H	0475
C540J	0475
C540K	0475
C540L	0475
C540M	0475
C540OR	0475
C540R	0475
C540X	0475
C540Y	0475
C541	0560
C542	0560
C543	0042
C543A	0042
C543B	0042
C543C	0042
C543D	0042
C543E	0042
C543F	0042
C543G	0042
C543GN	0042
C543H	0042
C543J	0042
C543K	0042
C543L	0042
C543M	0042
C543OR	0042
C543R	0042
C543X	0042
C543Y	0042
C544(2S)	2209
C544A	2209
C544AG	2209
C544B	2209
C544C(2S)	2209
C544D(2S)	0127
C544E(2S)	2209
C544F	2209
C544G	2209
C544GN	2209
C544H	2209
C544J	2209
C544K	2209
C544L	2209
C544M	2209
C544OR	2209
C544R	2209
C544X	2209
C544Y	2209
C545(2S)	0127
C545A(2S)	0127
C545B(2S)	0127
C545C(2S)	0127
C545D(2S)	0127
C545E(2S)	0127
C546	0007
C546K	0007
C547	0111
C547C(EUROPE)	0079
C548	0079
C548(EUROPE)	0079
C548A(EUROPE)	0111
C548C	0079
C548C(EUROPE)	0079
C549C(EUROPE)	0111
C550	0560
C550A	0560
C551	0042
C552	0042
C553	0042
C554	0042
C554C	1478
C555	0493
C-555A	0632
C555A	0632
C556	0224
C557(EUROPE)	0037
C558	0037
C558(2S)	4928
C558(EUROPE)	0037
C558B(EUROPE)	0037
C558C(EUROPE)	0648
C559	0018
C559B(EUROPE)	0037
C560(2S)	0086
C560A	0086
C560B	0086
C560C	0086
C560D	0086
C560E	0086
C560F	0086
C560G	0086
C560GN	0086
C560H	0086
C560J	0086
C560K	0086
C560L	0086
C560M	0086
C560OR	0086
C560R	0086
C560X	0086
C560Y	0086
C561(2S)	1431
C561-OR	1431
C561A	1431
C561B	1431
C561C	1431
C561D	1431
C561E	1431
C561F	1431
C561G	1431
C561GN	1431
C561H	1431
C561J	1431
C561K	1431
C561L	1431
C561M	1431
C561OR	1431
C561R	1431
C561X	1431
C561Y	1431
C562(2S)	0216
C562(O)	0216
C562-OR	0216
C562A	0216
C562B	0216
C562C	0216
C562D	0216
C562E	0216
C562F	0216
C562G	0216
C562GN	0216
C562H	0216
C562J	0216
C562K	0216
C562L	0216
C562M	0216
C562O	0007
C562OR	0216
C562R	0216
C562X	0216
C562Y(2S)	0216
C563	0835
C563-F	0835
C563-G	0835
C563-OR	0835
C563A(2S)	0835
C563B	0835
C563C	0835
C563D	0835
C563E	0835
C563F	0835
C563G	0835
C563GN	0835
C563H	0835
C563J	0835
C563K	0835
C563L	0835
C563M	0835
C563OR	0835
C563R	0835
C563X	0835
C563Y	0835
C564(2S)	0016
C564A(2S)	0016
C564AP	0016
C564B	0016
C564C	0016
C564D	0016
C564E	0016
C564F	0016
C564G	0016
C564GN	0016
C564H	0016
C564J	0016
C564K	0016
C564L	0016
C564M	0016
C564OR	0016
C564P(2S)	0016
C564PL	0016
C564Q(2S)	0016
C564QC	0016
C564R(2S)	0016
C564S(2S)	0016
C564T(2S)	0016
C564X	0016
C564Y	0016
C565	0016
C566(2S)	0259
C566A	0259
C566B	0259
C566C	0259
C566D	0259
C566E	0259
C566F	0259
C566G	0259
C566GN	0259
C566H	0259
C566J	0259
C566K	0259
C566L	0259
C566M	0259
C566OR	0259
C566R	0259
C566X	0259
C566Y	0259
C567(2S)	0259
C567A	0259
C567B	0259
C567C	0259
C567D	0259
C567E	0259
C567F	0259
C567G	0259
C567GN	0259
C567H	0259
C567J	0259
C567K	0259
C567L	0259
C567M	0259
C567OR	0259
C567R	0259
C567X	0259
C567Y	0259
C568(2S)	0259
C568A	0259
C568B	0259
C568C	0259
C568D	0259
C568E	0259
C568F	0259
C568G	0259
C568GN	0259
C568H	0259
C568J	0259
C568K	0259
C568L	0259
C568M	0259
C568OR	0259
C568R	0259
C568X	0259
C568Y	0259
C569	0016
C570	0016
C571(2S)	2041
C572	3321
C573	3322
C575	0086
C575C2	2872
C576	0016
C576E	0016
C576F	0016
C576G	0016
C577	0016
C577H	2261
C578	0861
C579	0400
C580(2S)	0086
C580A	0086
C580B	0086
C580C	0086
C580D	0086
C580E	0086
C580F	0086
C580G	0086
C580GN	0086
C580H	0086
C580J	0086
C580K	0086
C580L	0086
C580M	0086
C580OR	0086
C580R	0086
C580T	0086
C580X	0086
C580Y	0086
C581	0007
C582(2S)	2658
C582-OR	2658
C582A(2S)	2658
C582B(2S)	2658
C582BC(2S)	2658
C582BX(2S)	2658
C582BY(2S)	2658
C582C(2S)	2658
C582D	2658
C582E	2658
C582EA	2658
C582EH	2658
C582F	2658
C582G	2658
C582GN	2658
C582H	2658
C582J	2658
C582K	2658
C582L	2658
C582M	2658
C582OR	2658
C582R	2658
C582X	2658
C582Y	2658
C583(2S)	0259
C584	0219
C585	0042
C586(2S)	0177
C587(2S)	0016

If replacement code is OEM, contact original manufacturer for replacement.

DEVICE TYPE	REPL CODE
C587A(2S)	0016
C587B	0016
C587C	0016
C587D	0016
C587E	0016
C587F	0016
C587G	0016
C587GN	0016
C587H	0016
C587J	0016
C587K	0016
C587L	0016
C587M	0016
C587OR	0016
C587R	0016
C587X	0016
C587Y	0016
C588(2S)	0016
C588A	0016
C588B	0016
C588C	0016
C588D	0016
C588E	0016
C588F	0016
C588G	0016
C588GN	0016
C588H	0016
C588J	0016
C588K	0016
C588L	0016
C588M	0016
C588OR	0016
C588R	0016
C588X	0016
C588Y	0016
C589(2S)	0233
C589A	0233
C589B	0233
C589C	0233
C589D	0233
C589E	0233
C589F	0233
C589G	0233
C589GN	0233
C589H	0233
C589J	0233
C589K	0233
C589L	0233
C589M	0233
C589OR	0233
C589R	0233
C589X	0233
C589Y	0233
C590(2S)	0233
C590A	0233
C590B	0233
C590C	0233
C590D	0233
C590E	0233
C590F	0233
C590G	0233
C590GN	0233
C590H	0233
C590J	0233
C590K	0233
C590L	0233
C590M	0233
C590OR	0233
C590X	0233
C590Y(2S)	0233
C591	0168
C591A	0168
C591B	0168
C591C	0168
C591D	0168
C591E	0168
C591F	0168
C591G	0168
C591GN	0168
C591H	0168
C591J	0168
C591K	0168
C591L	0168
C591M	0168
C591OR	0168
C591R	0168
C591X	0168
C591Y	0168
C592	0560
C593(2S)	0016
C594(2S)	0016
C594-0	0016
C594-R	0016
C594-Y	0016
C594A	0016
C594B	0016
C594C	0016
C594D	0016
C594E	0016
C594F	0016
C594G	0016
C594GN	0016
C594H	0016
C594J	0016
C594K	0016
C594L	0016
C594M	0016
C594OR	0016
C594R	0016
C594X	0016
C594Y	0016
C595(2S)	0016
C595A	0016
C595B	0016
C595C	0016
C595D	0016
C595E	0016
C595F	0016
C595G	0016
C595GN	0016
C595H	0016
C595J	0016
C595K	0016
C595L	0016
C595M	0016
C595OR	0016
C595X	0016
C595Y	0016
C596(2S)	0016
C596(IC)	0857
C596A	0016
C596B	0016
C596C	0016
C596C(IC)	0857
C596D	0016
C596E	0016
C596F	0016
C596G	0016
C596GN	0016
C596H	0016
C596J	0016
C596K	0016
C596L	0016
C596M	0016
C596OR	0016
C596R	0016
C596X	0016
C596Y	0016
C597	0936
C598	0560
C599	0042
C600	0042
C600E	OEM
C600M	OEM
C600N	OEM
C600P	OEM
C600PA	OEM
C600PB	OEM
C600S	OEM
C600T	OEM
C601(2S)	0493
C601L	OEM
C601N	OEM
C601N(2S)	0493
C601P	OEM
C601PA	OEM
C601PB	OEM
C601PC	OEM
C601PD	OEM
C601PE	OEM
C601PM	OEM
C601PN	OEM
C601PS	OEM
C601PT	OEM
C601T	OEM
C602	0007
C602L	OEM
C602LA	OEM
C602LB	OEM
C602LC	OEM
C602LD	OEM
C602LE	OEM
C602LM	OEM
C602PM	OEM
C602PN	OEM
C602PS	OEM
C602PT	OEM
C603	0400
C604	0127
C605(2S)	0475
C605-B	0475
C605-OR	0475
C605A	0475
C605B	0475
C605C	0475
C605D	0475
C605E	0475
C605F	0475
C605G	0475
C605GN	0475
C605H	0475
C605J	0475
C605K(2S)	0475
C605L(2S)	0475
C605M(2S)	0475
C605OR	0475
C605Q(2S)	0475
C605R	0475
C605TW(2S)	0475
C605X	0475
C605Y	0475
C606(2S)	0475
C606-B	0475
C606A	0475
C606B	0475
C606C	0475
C606D	0475
C606E	0475
C606F	0475
C606G	0475
C606GN	0475
C606H	0475
C606J	0475
C606K	0475
C606L	0475
C606M	0475
C606N	0475
C606OR	0475
C606R	0475
C606X	0475
C606Y	0475
C607	0086
C608(2S)	1411
C608A	1411
C608AA	1411
C608B	1411
C608C	1411
C608D	1411
C608E(2S)	1411
C608F	1411
C608G	1411
C608GN	1411
C608H	1411
C608J	1411
C608K	1411
C608L	1411
C608M	1411
C608OR	1411
C608R	1411
C608T(2S)	1411
C608X	1411
C608Y	1411
C609(2S)	1411
C609-OR	1411
C609A	1411
C609B	1411
C609C	1411
C609D	1411
C609E	1411
C609F(2S)	1411
C609G	1411
C609GN	1411
C609H	1411
C609J	1411
C609K	1411
C609L	1411
C609M	OEM
C609M(TRANS)	1411
C609N	OEM
C609OR	1411
C609P	OEM
C609PA	OEM
C609PB	OEM
C609R	1411
C609S	OEM
C609T(2S)	1411
C609X	1411
C609Y	1411
C610	0037
C610(2S)	0086
C610A	0086
C610B	0086
C610C	0086
C610D	0086
C610E	0086
C610F	0086
C610G	0086
C610GN	0086
C610H	0086
C610J	0086
C610K	0086
C610L	0086
C610M	0086
C610OR	0086
C610R	0086
C610X	0086
C610Y	0086
C611	0259
C611(2S)	0259
C611A	0447
C611A(TRANS)	0259
C611B	0940
C611B(TRANS)	0259
C611C	OEM
C611C(TRANS)	0259
C611D	OEM
C611D(TRANS)	0259
C611E	0259
C611F	0447
C611F(TRANS)	0259
C611G	0940
C611G(TRANS)	0259
C611GN	0259
C611H	0259
C611J	0259
C611K	0259
C611L	0259
C611M	0259
C611OR	0259
C611R	0259
C611U	1396
C611X	0259
C611Y	0259
C612	0259
C612(2S)	0259
C612A	0259
C612B	0259
C612C	0259
C612D	0259
C612E	0259
C612F	0259
C612G	0259
C612GN	0259
C612H	0259
C612J	0259
C612K	0259
C612L	OEM
C612L(2S)	0259
C612M	0259
C612OR	0259
C612PE	OEM
C612PM	OEM
C612PN	OEM
C612PS	OEM
C612PT	OEM
C612R	0259
C612X	0259
C612Y	0259
C613	0144
C613(2S)	0144
C613A	0144
C613B	0144
C613C	0144
C613D	0144
C613E	0144
C613F	0144
C613G	0144
C613GN	0144
C613H	0144
C613J	0144
C613K	0144
C613L	OEM
C613L(TRANS)	0144
C613M	0144
C613OR	0144
C613PE	OEM
C613PM	OEM
C613PN	OEM
C613PS	OEM
C613PT	OEM
C613R	0144
C613X	0144
C613Y	0144
C614	0698
C614(2S)	0698
C614A	0698
C614B	0698
C614C	0698
C614D	0698
C614E(2S)	0698
C614F(2S)	0698
C614G(2S)	0698
C614GN	0698
C614H	0698
C614J	0698
C614K	0698
C614L	0698
C614M	0698
C614OR	0698
C614R	0698
C614X	0698
C614Y	0698
C615	0698
C615(2S)	0698
C615-OR	0698
C615A	0698
C615B(2S)	0698
C615C(2S)	0698
C615D(2S)	0698
C615E(2S)	0698
C615F	0698
C615G(2S)	0698
C615GN	0698
C615H	0698
C615J	0698
C615K	0698
C615L	0698
C615M	0698
C615OR	0698
C615R	0698
C615X	0698
C615Y	0698
C616	0630
C617	0630
C618	0007
C618A	0007
C619(2S)	0286
C619A	0286
C619B(2S)	0286
C619C(2S)	0286
C619D(2S)	0286
C619E	0286
C619F	0286
C619G	0286
C619GN	0286
C619H	0286
C619J	0286
C619K	0286
C619L	0286
C619M	0286
C619OR	0286
C619R	0286
C619X	0286
C619Y	0286
C620	0079
C620(2S)	0076
C620-D	0076
C620-E	0076
C620-OR	0076
C620A	0076
C620B	0076
C620C(2S)	0076
C620CD(2S)	0076
C620D(2S)	0076
C620DE(2S)	0076
C620E(2S)	0076
C620F	0076
C620G	0076
C620GN	0076
C620H	0076
C620J	0076
C620K	0076
C620L	0076
C620M	0076
C620OR	0076
C620R	0076
C620X	0076
C620Y	0076
C621	0286
C621(2S)	0286
C621A	0286
C621B	0286
C621C	0286
C621D	0286
C621E	0286
C621F	0286
C621G	0286
C621GN	0286
C621H	0286
C621J	0286
C621K	0286
C621L	0286
C621M	0286
C621OR	0286
C621R	0286
C621X	0286
C621Y	0286
C622	0016
C622(2S)	0016
C622A	0016
C622B	0016
C622C	0016
C622D	0016
C622E	0016
C622F	0016
C622G	0016
C622GN	0016
C622H	0016
C622J	0016
C622K	0016
C622L	0016
C622M	0016
C622OR	0016
C622R	0016
C622X	0016
C622Y	0016
C623	OEM
C623(2S)	0016
C624	OEM
C624(2S)	0016
C625	OEM
C626	0016
C627(2S)	0233
C628(2S)	0493
C628E	0493
C628F	0493
C629(2S)	0224
C629-31	0224
C629-41	0224
C629A	0224
C629B	0224
C629C	0224
C629D	0224
C629E	0224
C629F	0224
C629G	0224
C629GN	0224
C629H	0224
C629J	0224
C629K	0224
C629L	0224
C629M	0224
C629OR	0224
C629R	0224
C629X	0224
C629Y	0224
C631	0286
C631(2S)	0286
C631A(2S)	0286
C631B	0286
C631C	0286
C631D	0286
C631E	0286
C631F	0286
C631G	0286
C631GN	0286
C631H	0286
C631J	0286
C631K	0286
C631L	0286
C631M	0286
C631OR	0286
C631R	0286
C631X	0286
C631Y	0286
C632	0286
C632(1)	2075
C632(2S)	2075
C632-OR	2075
C632A(2S)	2075
C632B	2075
C632C	2075
C632D	2075
C632E	2075
C632F	2075
C632G	2075
C632GN	2075
C632H	2075
C632J	2075
C632K	2075
C632L	2075
C632M	2075
C632OR	2075
C632R	2075
C632X	2075
C632Y	2075
C633	0286
C633(2S)	2075
C633-7(2S)	2075
C633-OR	2075
C633A(2S)	2075
C633B	2075
C633C	2075
C633D	2075
C633E	2075
C633F	2075
C633G(2S)	2075
C633GN	2075
C633H(2S)	2075
C633J	2075
C633K	2075
C633L	2075
C633M	2075
C633OR	2075
C633R	2075
C633X	2075
C633Y	2075
C634(2)	2075
C634(2S)	2075
C634-0	2075
C634-OR	2075
C634A(2S)	2075
C634AK	2075
C634AL	2075
C634AXL	2075
C634B	2075
C634C	2075
C634D	2075
C634E	2075
C634F	2075
C634G	2075
C634GN	2075
C634H	2075
C634J	2075
C634K	2075
C634L	2075
C634M	2075
C634OR	2075
C634R	2075
C634SP	2075
C634X	2075
C634Y	2075
C635	0079
C635(EUROPE)	1386
C635A(2S)	0079
C636(2S)	0042
C636(EUROPE)	0477
C637	3321
C637(EUROPE)	0324
C638	3322
C638C	3322
C639	0016
C640	0475
C640(2S)	0475
C640A	0475
C640B(2S)	0475
C640C	0475
C640D	0475
C640E	0475
C640F	0475
C640G	0475
C640GN	0475
C640J	0475
C640K	0475
C640L	0475
C640M	0475
C640R	0475
C640X	0475
C641	0286
C641(2S)	0286
C641A	0286
C641B(2S)	0286
C641C	0286
C641D	0286
C641E	0286
C641F	0286
C641G	0286
C641GN	0286
C641H	0286
C641J	0286
C641K	0286
C641L	0286
C641M	0286
C641OR	0286
C641R	0286
C641X	0286
C641Y	0286
C642	0003
C642(2S)	0003
C642A(2S)	0003
C643	0223
C643(2S)	1150
C643-OR	1150
C643A(2S)	1150
C643B	1150
C643C	1150
C643D	1150
C643E	1150
C643F	1150
C643G	1150
C643GN	1150
C643H	1150
C643J	1150
C643K	1150
C643L	1150
C643M	1150
C643OR	1150
C643R	1150
C643X	1150
C643Y	1150
C644	0111
C644(2S)	0111
C644-OR	0111
C644A	0111
C644B	0111
C644C(2S)	0111
C644D	0111
C644E	0111
C644F	0111
C644F/494	0111
C644FH	0111
C644FHS	0111
C644FR(2S)	0111
C644FS(2S)	0111
C644G	0111
C644GN	0111
C644H(2S)	0111
C644J	0111
C644K	0111
C644L	0111
C644M	0111
C644OR	0111
C644P(2S)	0111
C644PJ(2S)	0111
C644Q(2S)	0111
C644R	0111
C644S	0111
C644S/494	0111
C644ST	0111
C644T(2S)	0111
C644X	0111
C644Y	0111
C645(2S)	0674
C645-OR	0674
C645A(2S)	0674
C645B(2S)	0674
C645B-1	0674
C645C(2S)	0674
C645D	0674
C645E	0674
C645F	0674
C645G(2S)	0674
C645GN	0674
C645GR	0674
C645H	0674
C645J	0674
C645K	0674
C645L	0674
C645M	0674
C645N(2S)	0674
C645OR	0674
C645R	0674
C645V	0674
C645X	0674
C645Y	0674
C646(2S)	0620
C647(2S)	4387
C647Q(2S)	4387
C647R(2S)	4387
C648(2S)	0016
C648A	0016
C648B	0016
C648C	0016
C648D	0016
C648E	OEM
C648E(TRANS)	0016
C648F	0016
C648G	0016
C648H(2S)	0016
C648J	0016
C648K	0016
C648L	0016
C648M	OEM
C648M(TRANS)	0016
C648N	OEM
C648OR	0016
C648P	OEM
C648PA	OEM
C648PB	OEM
C648R	0016
C648S	OEM
C648T	OEM
C648X	0016

If replacement code is OEM, contact original manufacturer for replacement.

DEVICE TYPE	REPL CODE
C648Y	0016
C649(2S)	0016
C649A	0016
C649B	0016
C649C	0016
C649D	0016
C649E	0016
C649F	0016
C649G	0016
C649GN	0016
C649H	0016
C649J	0016
C649K	0016
C649L	0016
C649M	0016
C649OR	0016
C649R	0016
C649X	0016
C649Y	0016
C650	0016
C650(2S)	0536
C650-OR	0536
C650-Y	0536
C650A	0536
C650B(2S)	0536
C650C	0536
C650D	0536
C650E	0536
C650F	0536
C650G	0536
C650GN	0536
C650H	0536
C650J	0536
C650K	0536
C650L	0536
C650M	0536
C650OR	0536
C650R	0536
C650X	0536
C650Y	0536
C651	0418
C651(2S)	0493
C652	0493
C652(2S)	0493
C653	0007
C653(2S)	0259
C654(2S)	0493
C654A	0493
C654B	0493
C654C	0493
C654D	0493
C654E	0493
C654F	0493
C654G	0493
C654GN	0493
C654H	0493
C654J	0493
C654K	0493
C654L	0493
C654M	0493
C654OR	0493
C654R	0493
C654X	0493
C654Y	0493
C655	0079
C655(2S)	0016
C655A	0016
C655B	0016
C655C	0016
C655D	0016
C655E	0016
C655F	0016
C655G	0016
C655GN	0016
C655H	0016
C655J	0016
C655K	0016
C655L	0016
C655M	0016
C655OR	0016
C655R	0016
C655X	0016
C655Y	0016
C656(2S)	0259
C656A	0259
C656B	0259
C656C	0259
C656D	0259
C656E	0259
C656F	0259
C656G	0259
C656GN	0259
C656H	0259
C656J	0259
C656K	0259
C656L	0259
C656M	0259
C656OR	0259
C656R	0259
C656X	0259
C656Y	0259
C657(2S)	0127
C657A	0127
C657B	0127
C657C	0127
C657D	0127
C657E	0127
C657F	0127
C657G	0127
C657GN	0127
C657H	0127
C657J	0127
C657K	0127
C657L	0127
C657M	0127
C657OR	0127
C657R	0127
C657X	0127
C657Y	0127
C658(2S)	0127
C658A(2S)	0127
C658B	0127
C658C	0127
C658D	0127
C658E	0127
C658F	0127
C658G	0127
C658GN	0127
C658H	0127
C658J	0127
C658K	0127
C658L	0127
C658M	0127
C658OR	0127
C658R	0127
C658X	0127
C658Y	0127
C659(2S)	0127
C659A	0127
C659B	0127
C659C	0127
C659D	0127
C659E	0127
C659F	0127
C659G	0127
C659GN	0127
C659H	0127
C659J	0127
C659K	0127
C659L	0127
C659M	0127
C659OR	0127
C659R	0127
C659X	0127
C659Y	0127
C660	0047
C660(2S)	0007
C661	0047
C661(2S)	0007
C661G	OEM
C662(2S)	0144
C662A	0144
C662B	0144
C662C	0144
C662D	0144
C662E	0144
C662F	0144
C662G	0144
C662GN	0144
C662H	0144
C662J	0144
C662K	0144
C662L	0144
C662M	0144
C662OR	0144
C662R	0144
C662X	0144
C662Y	0144
C663(2S)	0007
C663A	0007
C663B	0007
C663C	0007
C663D	0007
C663E	0007
C663F	0007
C663G	0007
C663GN	0007
C663H	0007
C663J	0007
C663K	0007
C663L	0007
C663M	0007
C663OR	0007
C663R	0007
C663X	0007
C663Y	0007
C664	0103
C664-OR	0103
C664A	0103
C664B(2S)	0103
C664C(2S)	0103
C664D	0103
C664E	0103
C664F	0103
C664L	0103
C664OR	0103
C664R	0103
C664Y	0103
C665	0867
C665(2S)	0867
C665H(2S)	0867
C665HA(2S)	0867
C665HB(2S)	0867
C666	OEM
C666(2S)	0177
C666C	0177
C667	OEM
C667(2S)	0007
C668(2S)	0127
C668-OR	0127
C668-SP	0127
C668A(2S)	0127
C668B(2S)	0127
C668B1(2S)	0127
C668BC2(2S)	0127
C668C(2S)	0127
C668C1(2S)	0127
C668C2	0127
C668CD(2S)	0127
C668D(2S)	0127
C668D0(2S)	0127
C668D1(2S)	0127
C668DE(2S)	0127
C668DO(2S)	0127
C668DV(2S)	0127
C668DX(2S)	0127
C668DZ(2S)	0127
C668E(2S)	0127
C668E1(2S)	0127
C668E2(2S)	0127
C668EP(2S)	0127
C668EV(2S)	0127
C668EX(2S)	0127
C668F(2S)	0127
C668G	0127
C668GN	0127
C668H	0127
C668K	0127
C668L	0127
C668M	0127
C668OR	0127
C668R	0127
C668SP	0127
C668X	0127
C668Y	0127
C669	0622
C669A	0622
C669AB	0622
C669C	0622
C669D	0622
C670-090ARC	OEM
C670-090ARCP	OEM
C670-090ARM	OEM
C670-090ARMP	OEM
C670-090ASC	OEM
C670-090ASCP	OEM
C670-090ASM	OEM
C670-090ASMP	OEM
C670-090RC	OEM
C670-090RCP	OEM
C670-090RM	OEM
C670-090RMP	OEM
C670-090SC	OEM
C670-090SCP	OEM
C670-090SM	OEM
C670-090SMP	OEM
C670-180ARC	OEM
C670-180ARCP	OEM
C670-180ARM	OEM
C670-180ARMP	OEM
C670-180ASC	OEM
C670-180ASCP	OEM
C670-180ASM	OEM
C670-180ASMP	OEM
C670-180RC	OEM
C670-180RCP	OEM
C670-180RM	OEM
C670-180RMP	OEM
C670-180SC	OEM
C670-180SCP	OEM
C670-180SM	OEM
C670-180SMP	OEM
C673	0693
C673(2S)	0007
C673B	0007
C673C	0007
C673C(JAPAN)	0007
C673C2(2S)	0007
C673D(2S)	0007
C674	0127
C674(2S)	0127
C674(JAPAN)	0127
C674-B	0127
C674-F	0127
C674B(2S)	0127
C674C(2S)	0127
C674CK	0127
C674CL	0127
C674D(2S)	0127
C674E(2S)	0127
C674F(2S)	0127
C674G(2S)	0127
C675(2S)	0363
C676	0168
C677	0867
C678	0103
C679	1027
C679H	1027
C680	1027
C680(2S)	1504
C680A	1027
C680A(2S)	1504
C680B	1504
C680C	1504
C680G	1504
C680GN	1504
C680H	1504
C680J	1504
C680K	1504
C680L	1504
C680M	1504
C680R(2S)	1504
C680X	1504
C681	0363
C681(2S)	2437
C681(B)	2437
C681A	0363
C681A(2S)	2437
C681AYL	2437
C681L	2437
C681Y	2437
C681YL(2S)	2437
C682	0007
C682(2S)	0047
C682-OR	0047
C682A	0007
C682A(2S)	0047
C682B(2S)	0047
C682C	0047
C682D	0047
C682E	0047
C682F	0047
C682G	0047
C682GN	0047
C682H	0047
C682J	0047
C682K	0047
C682L	0047
C682M	0047
C682OR	0047
C682R	0047
C682X	0047
C682Y	0047
C683	0007
C683(2S)	0047
C683-0	0047
C683-OR	0047
C683A	0007
C683A(2S)	0047
C683B	0047
C683C	0047
C683D	0047
C683E	0047
C683F	0047
C683G	0047
C683GN	0047
C683H	0047
C683J	0047
C683K	0047
C683L	0047
C683M	0047
C683OR	0047
C683P	0047
C683Q	0047
C683R	0047
C683S	0047
C683V(2S)	0047
C683X	0047
C683Y	0047
C684	0127
C684(2S)	0127
C684-OR	0127
C684A	0127
C684A(2S)	0127
C684B(2S)	0127
C684BK(2S)	0127
C684C	0127
C684D	0127
C684E	0127
C684F(2S)	0127
C684G	0127
C684GN	0127
C684H	0127
C684J	0127
C684K	0127
C684L	0127
C684M	0127
C684OR	0127
C684R	0127
C684TM	0127
C684X	0127
C684Y	0127
C685	0142
C685(2S)	0142
C685-0	0142
C685-O	0142
C685A	0142
C685A(2S)	0142
C685ABK	0142
C685AL	0142
C685B(2S)	0142
C685BK	0142
C685C	0142
C685D	0142
C685E	0142
C685F	0142
C685G	0142
C685GN	0142
C685H	0142
C685J	0142
C685K	0142
C685L	0142
C685M	0142
C685OR	0142
C685P(2S)	0142
C685R	0142
C685SY	0142
C685X	0142
C685Y(2S)	0142
C686(2S)	0233
C686A	0233
C686B	0233
C686C	0233
C686D	0233
C686E	0233
C686F	0233
C686G	0233
C686GN	0233
C686H	0233
C686J	0233
C686K	0233
C686L	0233
C686M	0233
C686OR	0233
C686R	0233
C686X	0233
C686Y	0233
C687(2S)	0177
C687A	0177
C687B	0177
C687C	0177
C687D	0177
C687E	0177
C687F	0177
C687G	0177
C687GN	0177
C687H	0177
C687J	0177
C687K	0177
C687L	0177
C687M	0177
C687OR	0177
C687R	0177
C687X	0177
C687Y	0177
C688(2S)	0224
C688-SP	0224
C688A	0224
C688B	0224
C688C	0224
C688D	0224
C688E	0224
C688F	0224
C688G	0224
C688GN	0224
C688H	0224
C688J	0224
C688K	0224
C688L	0224
C688M	0224
C688OR	0224
C688R	0224
C688X	0224
C688Y	0224
C689(2S)	0016
C689A	0016
C689B	0016
C689C	0016
C689D	0016
C689E	0016
C689G	0016
C689GN	0016
C689H(2S)	0016
C689J	0016
C689K	0016
C689L	0016
C689M	0016
C689OR	0016
C689R	0016
C689X	0016
C689Y	0016
C690	3587
C690A	3587
C691	0560
C692	0560
C693	0552
C693-OR	0552
C693A	0552
C693B	0552
C693C	0552
C693D	0552
C693E	0552
C693EB(2S)	0552
C693ET(2S)	0552
C693F(2S)	0552
C693FC(2S)	0552
C693FL(2S)	0552
C693FP	0552
C693FU(2S)	0552
C693G	0552
C693GL(2S)	0552
C693GN	0552
C693GS(2S)	0552
C693GU(2S)	0552
C693GZ(2S)	0552
C693H(2S)	0552
C693J	0552
C693K	0552
C693L	0552
C693M	0552
C693NP	0552
C693OR	0552
C693R	0552
C693U	0552
C693X	0552
C693Y	0552
C694(2S)	0111
C694A	0111
C694B	0111
C694C	0111
C694D	0111
C694E(2S)	0111
C694F(2S)	0111
C694G(2S)	0111
C694GN	0111
C694H	0111
C694J	0111
C694K	0111
C694L	0111
C694M	0111
C694OR	0111
C694R	0111
C694X	0111
C694Y	0111
C694Z(2S)	0111
C695(2S)	0475
C696(2S)	0644
C696-4	0622
C696-OR	0622
C696A(2S)	0644
C696AA	0622
C696AB	0622
C696AD	0644
C696AE	0644
C696AF	0644
C696AG	0622
C696AH	0622
C696AI	0622
C696B(2S)	0622
C696BL	0622
C696C	0622
C696D(2S)	0644
C696E(2S)	0644
C696F(2S)	0644
C696G(2S)	0622
C696GN	0622
C696GU	0644
C696H(2S)	0644
C696I(2S)	0622
C696J	0622
C696K	0622
C696L	0622
C696M	0622
C696OR	0622
C696R	0622
C696U	0622
C696X	0622
C696Y	0622
C697(2S)	1908
C697-OR	1908
C697A(2S)	1908
C697B	1908
C697C	1908
C697D(2S)	1908
C697E	1908
C697F(2S)	1908
C697G	1908
C697GN	1908
C697H	1908
C697J	1908
C697K	1908
C697L	1908
C697M	1908
C697OR	1908
C697R	1908
C697X	1908
C697Y	1908
C698	0630
C699	0595
C700	0595
C701(2S)	0018
C701A	0018
C701B	0018
C701C	0018
C701D	0018
C701E	0018
C701F	0018
C701G	0018
C701GN	0018
C701H	0018
C701J	0018
C701K	0018
C701L	OEM
C701L(TRANS)	0018
C701M	0018
C701OR	0018
C701PA	OEM
C701PB	OEM
C701PC	OEM
C701PCC701PD	OEM
C701PD	OEM
C701PE	OEM
C701PM	OEM
C701PN	OEM
C701PS	OEM
C701PT	OEM
C701R	0018
C701X	0018
C701Y	0018
C702(2S)	0018
C702A	0018
C702B	0018
C702C	0018
C702D	0018
C702E	0018
C702F	0018
C702G	0018
C702GN	0018
C702H	0018
C702J	0018
C702K	0018
C702L	OEM
C702L(TRANS)	0018
C702LA	OEM
C702LB	OEM
C702LC	OEM
C702LD	OEM
C702M	0018
C702OR	0018
C702R	0018
C702X	0018
C702Y	0018
C703	0042
C704	0042
C705(2S)	0127
C705A	0127
C705B(2S)	0127
C705C(2S)	0127
C705D(2S)	0127
C705E(2S)	0127
C705F(2S)	0127
C705G	0127
C705GN	0127
C705J	0127
C705K	0127
C705L	0127
C705M	0127
C705OR	0127
C705R	0127
C705TV(2S)	0127
C705TVV	0127
C705TW	0127
C705X	0127
C705Y	0127
C706	0007
C707(2S)	0259
C707A	0259
C707B	0259
C707C	0259
C707D	0259
C707E	0259
C707F	0259
C707G	0259
C707GN	0259
C707H(2S)	0259
C707K	0259
C707L	0259
C707M	0259
C707OR	0259
C707R	0259
C707X	0259
C707Y	0259
C708(2S)	0086
C708-OR	0086
C708A(2S)	0086
C708AA(2S)	0086
C708AB(2S)	0086
C708AC(2S)	0086
C708AH(2S)	0086
C708B(2S)	0086
C708C(2S)	0086
C708D	0086
C708E	0086
C708F	0086
C708G	0086
C708GN	0086
C708H	0086
C708HA	0086
C708HB	0086
C708L	0086
C708M	0086
C708OR	0086
C708R	0086
C708X	0086
C708Y	0086
C709(2S)	0286
C709A	0286
C709B(2S)	0286
C709C(2S)	0286
C709CD(2S)	0286
C709D(2S)	0286
C709E	0286
C709F	0286
C709G	0286
C709GN	0286
C709H	0286
C709J	0286
C709L	0286
C709M	0286
C709OR	0286
C709X	0286
C709Y	0286
C710(2S)	0368
C710(C)	0079
C710(JAPAN)	0368
C710-0	0368
C710-3	0368
C710-5	0368
C710-13	0368
C710-14	0368
C710-D	0368
C710-E	0368
C710-OR	0368

If replacement code is OEM, contact original manufacturer for replacement.

DEVICE TYPE	REPL CODE
C710AL	0368
C710B(2S)	0368
C710B2	0368
C710BC(2S)	0368
C710C(2S)	0368
C710D(2S)	0368
C710D(JAPAN)	0368
C710DB	0368
C710DE	0368
C710E(2S)	0368
C710F	0368
C710G	0368
C710GN	0368
C710H	0368
C710K	0368
C710L	0368
C710M	0368
C710OR	0368
C710R	0368
C710X	0368
C710XL	0368
C710Y	0368
C711(2S)	0380
C711-OR	0380
C711A(2S)	0380
C711A(JAPAN)	0380
C711AE	0380
C711A-E(2S)	0380
C711AE(JAPAN)	0380
C711A-F	0380
C711AF	0380
C711AG	0380
C711AN	0380
C711B	0380
C711C	0380
C711D(2S)	0380
C711E(2S)	0380
C711EF	0380
C711F(2S)	0380
C711G(2S)	2465
C711GN	0380
C711H	0380
C711J	0380
C711L	0380
C711M	0380
C711OR	0380
C711R	0380
C711X	0380
C711Y	0380
C712(2S)	0076
C712A(2S)	0076
C712B	0076
C712BC	0076
C712C(2S)	0076
C712CD	0076
C712D(2S)	0076
C712DC	0076
C712E(2S)	0076
C712F	0076
C712G	0076
C712GN	0076
C712H	0076
C712J	0076
C712K	0076
C712L	OEM
C712L(TRANS)	0076
C712M	0076
C712OR	0076
C712PE	OEM
C712PM	OEM
C712PN	OEM
C712PS	OEM
C712PT	OEM
C712R	0076
C712W(2S)	0076
C712X	0076
C712Y	0076
C713(2S)	0286
C713A	0286
C713B	0286
C713C	0286
C713D	0286
C713E	0286
C713F	0286
C713G	0286
C713GN	0286
C713H	0286
C713J	0286
C713K	0286
C713L	OEM
C713L(TRANS)	0286
C713M	0286
C713MC	0286
C713MD	0286
C713ME	0286
C713OR	0286
C713PE	OEM
C713PM	OEM
C713PN	OEM
C713PS	OEM
C713PT	OEM
C713R	0286
C713W	0286
C713X	0286
C713Y	0286
C714	OEM
C714(2S)	0155
C714A	0155
C714B	0155
C714C	0155
C714D	0155
C714E	0155
C714F	0155
C714G	0155
C714GN	0155
C714H	0155
C714J	0155
C714K	0155
C714L	0155
C714M	0155
C714OR	0155
C714R	0155
C714X	0155
C714Y	0155
C715	0286
C715-OR	0286
C715A(2S)	0286
C715B(2S)	0286
C715C(2S)	0286
C715D(2S)	0286
C715E(2S)	0286
C715EJ(2S)	0286
C715EV(2S)	0286
C715F(2S)	0286
C715G	0286
C715GN	0286
C715H	0286
C715J	0286
C715K	0286
C715L	0286
C715M	0286
C715OR	0286
C715R	0286
C715X	0286
C715XL(2S)	0286
C715Y	0286
C716(2S)	0155
C716A	0155
C716B(2S)	0155
C716C(2S)	0155
C716D(2S)	0155
C716F(2S)	0155
C716G(2S)	0155
C716GN	0155
C716H	0155
C716J	0155
C716K	0155
C716L	0155
C716M	0155
C716OR	0155
C716R	0155
C716X	0155
C716Y	0155
C717(2S)	0127
C717(JAPAN)	0127
C717-TM	0127
C717B(2S)	0127
C717BK(2S)	0127
C717BLK(2S)	0127
C717F	0127
C717G	0127
C717GN	0127
C717H	0127
C717K	0127
C717L	0127
C717M	0127
C717TM	0127
C717X	0127
C718	0079
C719	0155
C719O	0155
C720	0007
C720(2S)	0007
C721	0079
C722	0127
C722(2S)	0127
C723	0127
C723BL	0127
C724	0155
C725(2S)	0155
C725-06	0155
C726	0079
C727(2S)	0715
C727A	0715
C727B	0715
C727C	0715
C727D	0715
C727E	0715
C727F	0715
C727G	0715
C727GN	0715
C727H	0715
C727J	0715
C727K	0715
C727L	0715
C727M	0715
C727R	0715
C727X	0715
C727Y	0715
C728(2S)	0715
C728A	0715
C728B	0715
C728D	0715
C728E	0715
C728F	0715
C728G	0715
C728GN	0715
C728H	0715
C728J	0715
C728K	0715
C728L	0715
C728M	0715
C728OR	0715
C728R	0715
C728X	0715
C728Y	0715
C729	0418
C730	2041
C731	0493
C731R(2S)	0493
C732(2S)	0111
C732-B	0111
C732-BL	0111
C732-BLU	0111
C732-G	0111
C732-GR	0111
C732-GRN	0111
C732-OR	0111
C732-V	0111
C732-V10	0111
C732-VIO	0111
C732A	0111
C732B	0111
C732BL(2S)	0111
C732BL-1	0111
C732BLU	0111
C732C	0111
C732D	0111
C732E	0111
C732F	0111
C732G	0111
C732GN	0111
C732GR(2S)	0111
C732GRB	0111
C732GRN	0111
C732H	0111
C732J	0111
C732L	0111
C732M	0111
C732OR	0111
C732R	0111
C732S(2S)	0111
C732V(2S)	0111
C732V10	0111
C732VIO	0111
C732X	0111
C732Y(2S)	0111
C733(2S)	0111
C733-0(2S)	0111
C733-B	0111
C733-BL	0111
C733-BLU	0111
C733-G	0111
C733-GR	0111
C733-GRN	0111
C733-O(2S)	0111
C733-OR	0111
C733-ORG	0111
C733-Y	0111
C733-YEL	0111
C733A	OEM
C733B	0111
C733BL(2S)	0111
C733BLK	0111
C733BLU	0111
C733C	0111
C733D	0111
C733E	0111
C733ER	0111
C733F	0111
C733G	0111
C733GN	0111
C733GR(2S)	0111
C733GRN	0111
C733H	0111
C733J	0111
C733K	0111
C733L	0111
C733M	0111
C733O	0111
C733OR	0111
C733Q	0111
C733R(2S)	0111
C733S	0111
C733S-BL	0111
C733V(2S)	0111
C733X	0111
C733Y(2S)	0111
C733YEL	0127
C734	0076
C734-0(2S)	0076
C734-G	0076
C734-GR	0076
C734-GRN	0076
C734-O(2S)	0076
C734-OR	0076
C734-ORG	0076
C734-OY	0076
C734-R(2S)	0076
C734-RED	0076
C734-Y(2S)	0076
C734-YEL	0076
C734A	0076
C734B	0076
C734C	0076
C734D	0076
C734E	0076
C734F	0076
C734G	0076
C734GN	0076
C734GR(2S)	0076
C734GRN	0076
C734H	0076
C734J	0076
C734K	0076
C734L	0076
C734M	0076
C734OR	0076
C734R	0076
C734RED	0076
C734X	0076
C734Y(2S)	0076
C734YEL	0076
C735(2S)	0191
C735-0(2S)	0191
C735-GR	0191
C735-GRN	0191
C735-O(2S)	0191
C735-OR	0191
C735-ORG	0191
C735-ORN	0191
C735-OY	0191
C735-R(2S)	0191
C735-RED	0191
C735-Y	0191
C735-YEL	0191
C735A	0191
C735B(2S)	0191
C735C	0191
C735D	0191
C735E	0191
C735F(2S)	0191
C735FA3(2S)	0191
C735G	0191
C735GN	0191
C735GR(2S)	0191
C735GRN	0191
C735H(2S)	0191
C735J(2S)	0191
C735K(2S)	0191
C735L(2S)	0191
C735M	0191
C735O	0191
C735OR	0191
C735RED	0191
C735X	0191
C735Y(2S)	0191
C735YEL	0191
C735YFA-5	0191
C736(2S)	0867
C736A	0867
C736B	0867
C736C	0867
C736D	0867
C736E	0867
C736F	0867
C736G	0867
C736GN	0867
C736H	0867
C736J	0867
C736K	0867
C736L	0867
C736M	0867
C736OR	0867
C736R	0867
C736X	0867
C736Y	0867
C737	0079
C737Y	0079
C738(2S)	0895
C738A	0895
C738B	0895
C738C(2S)	0895
C738D(2S)	0895
C738E	0895
C738F	0895
C738G	0895
C738GN	0895
C738H	0895
C738J	0895
C738K	0895
C738L	0895
C738M	0895
C738OR	0895
C738R	0895
C738X	0895
C738Y	0895
C739(2S)	0127
C739A	0127
C739B	0127
C739C(2S)	0127
C739D	0127
C739E	0127
C739F	0127
C739G	0127
C739GN	0127
C739H	0127
C739K	0127
C739L	0127
C739M	0127
C739OR	0127
C739R	0127
C739Y	0127
C740	0144
C740(2S)	0144
C740A	0144
C740B	0144
C740C	0144
C740D	0144
C740E	0144
C740F	0144
C740G	0144
C740GN	0144
C740H	0144
C740J	0144
C740K	0144
C740L	0144
C740M	0144
C740OR	0144
C740R	0144
C740X	0144
C740Y	0144
C741(2S)	0689
C741B	0689
C741C	0689
C741D	0689
C741E	0689
C741F	0689
C741GN	0689
C741H	0689
C741J	0689
C741K	0689
C741L	0689
C741M	0689
C741OR	0689
C741R	0689
C741X	0689
C741Y	0689
C742	0079
C742(2S)	0936
C743	0936
C743A(2S)	0936
C744	0086
C744(2S)	0086
C744A	0086
C745	0936
C746	0168
C746A(2S)	0168
C746B	0168
C746C	0168
C746D	0168
C746E	0168
C746F	0168
C746GN	0168
C746H	0168
C746J	0168
C746K	0168
C746L	0168
C746M	0168
C746OR	0168
C746R	0168
C746X	0168
C746Y	0168
C748(2S)	0144
C748A	0144
C748C	0144
C748D	0144
C748E	0144
C748F	0144
C748G	0144
C748GN	0144
C748H	0144
C748J	0144
C748K	0144
C748L	0144
C748M	0144
C748OR	0144
C748R	0144
C748X	0144
C748Y	0144
C749	0936
C751	0111
C752	0551
C752(2S)	0551
C752-ORG-G	0551
C752-RED-G	0551
C752-YEL-G	0551
C752A	0551
C752B	0551
C752C	0551
C752D	0551
C752E	0551
C752F	0551
C752G(2S)	0551
C752GA	0551
C752GN	0551
C752G-O	0551
C752G-R	0551
C752G-TM-Y	0551
C752G-Y	0551
C752GY(JAPAN)	0551
C752H	0551
C752J	0551
C752K	0551
C752L	0551
C752M	0551
C752OR	0551
C752R	0551
C752X	0551
C752Y	0551
C753	0475
C754	0079
C755	0016
C756	0622
C756(4-4)	0622
C756(44)	0622
C756-1(2S)	0622
C756-2(2S)	0622
C756-2-4(2S)	0622
C756-2-5(2S)	0622
C756-3(2S)	0622
C756-4(2S)	0622
C756-OR	0622
C756A(2S)	0622
C756B	0622
C756C(2S)	0622
C756D(2S)	0622
C756E	0622
C756F	0622
C756G	0622
C756GN	0622
C756H	0622
C756J	0622
C756K	0622
C756L	0622
C756M	0622
C756OR	0622
C756R	0622
C756X	0622
C756Y	0622
C757	0475
C758	0074
C758OR	0074
C759	0543
C760	0079
C760(2S)	0103
C761(2S)	1154
C761A	1154
C761B	1154
C761C	1154
C761D	1154
C761E	1154
C761F	1154
C761G	1154
C761GN	1154
C761H	1154
C761J	1154
C761K	1154
C761L	1154
C761M	1154
C761OR	1154
C761P	1154
C761R	1154
C761X	1154
C761Y(2S)	1154
C761Z(2S)	1154
C762	0007
C762(2S)	1154
C762B	1154
C762C	1154
C762D	1154
C762E	1154
C762F	1154
C762G	1154
C762GN	1154
C762H	1154
C762J	1154
C762K	1154
C762L	1154
C762M	1154
C762R	1154
C762X	1154
C762Y	1154
C763(2S)	0127
C763(JAPAN)	0127
C763-C	0127
C763-OR	0127
C763A	0127
C763B(2S)	0127
C763C(2S)	0127
C763D(2S)	0127
C763E	0127
C763F	0127
C763G	0127
C763GN	0127
C763H	0127
C763J	0127
C763K	0127
C763L	0127
C763M	0127
C763OR	0127
C763X	0127
C763Y	0127
C764	0325
C764(2S)	0079
C765(2S)	0130
C765A	0130
C765B	0130
C765C	0130
C765D	0130
C765E	0130
C765F	0130
C765G	0130
C765GN	0130
C765H	0130
C765J	0130
C765K	0130
C765L	0130
C765M	0130
C765OR	0130
C765R	0130
C765X	0130
C765Y	0130
C766	0103
C767	0103
C768(2S)	0620
C768A	0620
C768B	0620
C768C	0620
C768D	0620
C768E	0620
C768F	0620
C768G	0620
C768GN	0620
C768H	0620
C768J	0620
C768K	0620
C768L	0620
C768M	0620
C768OR	0620
C768R	0620
C768X	0620
C768Y	0620
C769	0620
C770	0074
C771	0224
C771A	0224
C771B	0224
C771BX	0224
C771C	0224
C771D	0224
C771E	0224
C771F	0224
C771G	0224
C771GN	0224
C771H	0224
C771J	0224
C771K	0224
C771L	0224
C771M	0224
C771OR	0224
C771R	0224
C771X	0224
C771Y	0224
C772(2S)	1144
C772-OR	1144
C772A	1144
C772B	0224
C772B(2S)	1144
C772BG(2S)	1144
C772BH(2S)	1144
C772BV(2S)	1144
C772BX(2S)	1144
C772BY(2S)	1144
C772C	1144
C772C1(2S)	1144
C772C2(2S)	1144
C772CA	1144
C772CK(2S)	1144
C772CL(2S)	1144
C772CS(2S)	1144
C772CU(2S)	1144
C772CV(2S)	1144
C772CX(2S)	1144
C772D(2S)	1144
C772DJ(2S)	1144
C772DU(2S)	1144
C772DV(2S)	1144
C772DX(2S)	1144
C772DY(2S)	1144
C772E(2S)	1144
C772F(2S)	1144
C772G	1144
C772GN	1144
C772H	1144
C772J	1144
C772K(2S)	1144
C772KB(2S)	1144
C772KC(2S)	1144
C772KD(2S)	1144
C772KD1(2S)	1144
C772KD2(2S)	1144
C772L	1144
C772M	1144
C772OR	1144
C772R	1144
C772RB-D(2S)	1144
C772RD(2S)	1144
C772RS-D	1144
C772X	1144
C772Y	1144
C773(2S)	0788
C773A	0788
C773B	0788
C773C(2S)	0788
C773D(2S)	0788
C773E(2S)	0788
C773F	0788
C773G	0788
C773GN	0788
C773H	0788
C773J	0788
C773K	0788
C773L	0788
C773M	0788
C773OR	0788
C773R	0788
C773X	0788
C773Y	0788
C774(2S)	0698
C774B	0698
C774C	0698
C774D	0698

DEVICE TYPE	REPL CODE
C774E	0698
C774F	0698
C774G	0698
C774GN	0698
C774H	0698
C774J	0698
C774K	0698
C774L	0698
C774M	0698
C774R	0698
C774X	0698
C774Y	0698
C775(2S)	0698
C775A	0698
C775B	0698
C775C	0698
C775D	0698
C775E	0698
C775F	0698
C775G	0698
C775GN	0698
C775H	0698
C775J	0698
C775K	0698
C775L	0698
C775M	0698
C775OR	0698
C775R	0698
C775X	0698
C775Y	0698
C776(2S)	1400
C776A	1400
C776B	1400
C776C	1400
C776D	1400
C776E	1400
C776F	1400
C776G	1400
C776GN	1400
C776H	1400
C776J	1400
C776K	1400
C776L	1400
C776M	1400
C776OR	1400
C776R	1400
C776X	1400
C776Y(2S)	1400
C777(2S)	1411
C777-OR	1411
C777A	1411
C777AP	1411
C777B	1411
C777C	1411
C777D	1411
C777E	1411
C777F	1411
C777G	1411
C777GN	1411
C777H	1411
C777J	1411
C777K	1411
C777L	1411
C777M	1411
C777OR	1411
C777R	1411
C777X	1411
C777Y	1411
C778(2S)	1411
C778A	1411
C778B(2S)	1411
C778C	1411
C778D	1411
C778E	1411
C778F	1411
C778G	1411
C778GN	1411
C778H	1411
C778J	1411
C778K	1411
C778L	1411
C778M	1411
C778OR	1411
C778R	1411
C778X	1411
C778Y	1411
C779(2S)	1027
C779-O	1027
C779-Q	1027
C779-R(2S)	1027
C779-Y(2S)	1027
C779O	1027
C780(2S)	0861
C780-O	0861
C780-ORG-G	0861
C780-R	0861
C780-RED-G	0861
C780-Y	0861
C780-YEL-G	0861
C780A	0861
C780AG(2S)	0861
C780AG-O(2S)	0861
C780AG-R(2S)	0861
C780AG-Y(2S)	0861
C780B	0861
C780C	0861
C780D	0861
C780E	0861
C780F	0861

DEVICE TYPE	REPL CODE
C780G(2S)	0861
C780GA	0861
C780GN	0861
C780G-O	0861
C780G-R	0861
C780G-Y	0861
C780H	0861
C780J	0861
C780K	0861
C780L	0861
C780M	0861
C780O	0861
C780OR	0861
C780R	0861
C780X	0861
C780Y	0861
C781(2S)	0698
C781A	0698
C781AK	0698
C781B	0698
C781C	0698
C781D	0698
C781E	0698
C781F	0698
C781G	0698
C781H	0698
C781J	0698
C781K	0698
C781M	0698
C781OR	0698
C781R	0698
C781X	0698
C781Y	0698
C782(2S)	1469
C782-OR	1469
C782A	1469
C782B	1469
C782C	1469
C782D	1469
C782E	1469
C782F	1469
C782G	1469
C782GN	1469
C782H	1469
C782J	1469
C782K	1469
C782L	OEM
C782L(TRANS)	1469
C782LA	OEM
C782LB	OEM
C782LC	OEM
C782LD	OEM
C782OR	1469
C782R	1469
C782X	1469
C782Y	1469
C783(2S)	1027
C783-O	1027
C783-R	1027
C783-Y	1027
C783CA	OEM
C783CB	OEM
C783CC	OEM
C783CD	OEM
C783CE	OEM
C783LB	OEM
C783LC	OEM
C783LD	OEM
C783LE	OEM
C783LM	OEM
C783LN	OEM
C783LP	OEM
C783LS	OEM
C783LT	OEM
C784(2S)	1144
C784-0(2S)	1144
C784-6	1144
C784-B	1144
C784-BN	1144
C784-BRN	1144
C784-O(2S)	1144
C784-OR	1144
C784-ORG	1144
C784-R	1144
C784-RED	1144
C784-Y	1144
C784A(2S)	1144
C784B	1144
C784BN(2S)	1144
C784BN-1	1144
C784BRN	1144
C784C	1144
C784D	1144
C784E	1144
C784F	1144
C784G	1144
C784GN	1144
C784H	1144
C784J	1144
C784K	1144
C784L	1144
C784M	1144
C784OR	1144
C784P	1144
C784Q	1144
C784R(2S)	1144
C784RA	1144
C784RED	1144
C784X	1144

DEVICE TYPE	REPL CODE
C785(2S)	1144
C785-0	1144
C785-B	1144
C785-BN	1144
C785-BRN	1144
C785-O	1144
C785-ORG	1144
C785-R	1144
C785-RED	1144
C785-Y	1144
C785-YEL	1144
C785A	1144
C785B	1144
C785BL	1144
C785BN(2S)	1144
C785BR	1144
C785BRN	1144
C785C	1144
C785D(2S)	1144
C785E(2S)	1144
C785F	1144
C785G	1144
C785GN	1144
C785GR	1144
C785H	1144
C785JA	1144
C785K	1144
C785L	1144
C785M	1144
C785O	1144
C785R(2S)	1144
C785RA	1144
C785RED	1144
C785V	1144
C785X	1144
C785Y	1144
C785YEL	1144
C786(2S)	0216
C786A	0216
C786B	0216
C786C	0216
C786D	0216
C786E	0216
C786F	0216
C786G	0216
C786GN	0216
C786H	0216
C786J	0216
C786K	0216
C786L	0216
C786M	0216
C786OR	0216
C786R(2S)	0216
C786X	0216
C786Y	0216
C787(2S)	1154
C787A	1154
C787B	1154
C787C	1154
C787D	1154
C787E	1154
C787F	1154
C787G	1154
C787GN	1154
C787H	1154
C787K	1154
C787L	1154
C787M	1154
C787OR	1154
C787R	1154
C787X	1154
C787Y	1154
C788(2S)	0233
C788B	0233
C788C	0233
C788D	0233
C788E	0233
C788F	0233
C788G	0233
C788GN	0233
C788J	0233
C788K	0233
C788L	0233
C788M	0233
C788OR	0233
C788R	0233
C788X	0233
C788Y	0233
C789(2S)	0228
C789-O(2S)	0228
C789-O(2S)	0228
C789-R(2S)	0228
C789-Y(2S)	0228
C789A	0228
C789B	0228
C789C	0228
C789D	0228
C789E	0228
C789F	0228
C789G	0228
C789GN	0228
C789H	0228
C789J	0228
C789K	0228
C789L	0228
C789M	0228
C789OR	0228
C789R	0228
C789X	0228

DEVICE TYPE	REPL CODE
C789Y	0228
C790	1496
C790-O	1496
C790-R	1496
C790-Y	1496
C790Y	1496
C791(2S)	1147
C791-OR	1147
C791A	1147
C791B	1147
C791C	1147
C791D	1147
C791E	1147
C791F	1147
C791FA1	1147
C791G	1147
C791GN	1147
C791H	1147
C791J	1147
C791K	1147
C791M	1147
C791OR	1147
C791R	1147
C791X	1147
C791Y	1147
C792	0220
C792FA-3	0220
C793(2S)	0103
C793-BL	0103
C793-BLU	0103
C793-R	0103
C793-RED	0103
C793-Y	0103
C793-YEL	0103
C793A	0103
C793B	0103
C793BL(2S)	0103
C793C	0103
C793E	0103
C793F	0103
C793G	0103
C793GN	0103
C793H	0103
C793J	0103
C793K	0103
C793L	0103
C793M	0103
C793OR	0103
C793R(2S)	0103
C793X	0103
C793Y(2S)	0103
C794	0103
C794A	0103
C794B	0103
C794C	0103
C794D	0103
C794E	0103
C794F	0103
C794G	0103
C794GN	0103
C794H	0103
C794J	0103
C794K	0103
C794L	0103
C794OR	0103
C794R(2S)	0103
C794RA	0103
C794X	0103
C794Y	0103
C795(2S)	0142
C795A(2S)	0142
C795B	0142
C795BN	0142
C795C	0142
C795D	0142
C795F	0142
C795G	0142
C795GN	0142
C795H	0142
C795J	0142
C795L	0142
C795M	0142
C795OR	0142
C795R	0142
C795X	0142
C795Y	0142
C796(2S)	0016
C796A	0016
C796B	0016
C796C	0016
C796D	0016
C796E	0016
C796F	0016
C796G	0016
C796GN	0016
C796H	0016
C796J	0016
C796K	0016
C796L	0016
C796M	0016
C796OR	0016
C796R	0016
C796X	0016
C796Y	0016
C797(2S)	0079
C797A	0079
C797B	0079
C797C	0079
C797D	0079

DEVICE TYPE	REPL CODE
C797E	0079
C797F	0079
C797G	0079
C797GN	0079
C797H	0079
C797J	0079
C797K	0079
C797L	0079
C797M	0079
C797OR	0079
C797R	0079
C797X	0079
C797Y	0079
C798(2S)	0698
C798A	0698
C798B	0698
C798C	0698
C798D	0698
C798E	0698
C798G	0698
C798GN	0698
C798H	0698
C798J	0698
C798K	0698
C798L	0698
C798M	0698
C798OR	0698
C798R	0698
C798X	0698
C798Y	0698
C799(2S)	1411
C799-4	1411
C799A	1411
C799AP	1411
C799B	1411
C799C	1411
C799D	1411
C799E	1411
C799F	1411
C799G	1411
C799GN	1411
C799H	1411
C799J	1411
C799K(2S)	1411
C799L	1411
C799M	1411
C799OR	1411
C799R	1411
C799X	1411
C799Y	1411
C800(2S)	0630
C800A	0630
C800B	0630
C800C	0630
C800D	0630
C800E	0630
C800F	0630
C800G	0630
C800GN	0630
C800H	0630
C800J	0630
C800K	0630
C800L	0630
C800M	0630
C800OR	0630
C800R	0630
C800X	0630
C800Y	0630
C801	0086
C802	0086
C803(2S)	0698
C803A	0698
C803B	0698
C803C	0698
C803D	0698
C803E	0698
C803F	0698
C803G	0698
C803H	0698
C803J	0698
C803K	0698
C803L	0698
C803M	0698
C803OR	0698
C803R	0698
C803X	0698
C803Y	0698
C804	0475
C804H	0475
C805(2S)	0716
C805A	0716
C805A1	0716
C805A2	0716
C805B	0716
C805C	0716
C805D	0716
C805E	0716
C805F	0716
C805G	0716
C805GN	0716
C805H	0716
C805J	0716
C805L	0716
C805OR	0716
C805R	0716
C805X	0716
C805Y	0716
C806(2S)	0363
C806A	0363

DEVICE TYPE	REPL CODE
C806B	0363
C806C	0363
C806D	0363
C806E	0363
C806F	0363
C806G	0363
C806J	0363
C806L	0363
C806M	0363
C806OR	0363
C806R	0363
C806X	0363
C806Y	0363
C807(2S)	1966
C807A(2S)	1966
C807AK	1966
C808	2414
C809	0259
C810	0493
C811	0007
C812	0630
C812A	0630
C812B	0630
C812BST	OEM
C812C	0630
C812D	0630
C812E	0630
C812F	0630
C812G	0630
C812GN	0630
C812H	0630
C812J	0630
C812K	0630
C812L	0630
C812M	0630
C812OR	0630
C812R	0630
C812Y	0630
C814(2S)	0079
C814A	0079
C814B	0079
C814C	0079
C814D	0079
C814E	0079
C814F	0079
C814G	0079
C814GN	0079
C814H	0079
C814J	0079
C814K	0079
C814M	0079
C814OR	0079
C814R	0079
C814X	0079
C814Y	0079
C815(2S)	0155
C815-1	0155
C815-O	0079
C815A(2S)	0155
C815B(2S)	0155
C815BK	0155
C815C(2S)	0155
C815D	0155
C815E	0155
C815F(2S)	0155
C815G	0155
C815GN	0155
C815H	0155
C815J	0155
C815K(2S)	0155
C815L(2S)	0155
C815LJ	0155
C815M(2S)	0155
C815OR	0155
C815R	0155
C815S(2S)	0155
C815SA(2S)	0155
C815SC(2S)	0155
C815X	0155
C815Y	0155
C815Y(EBC)	1954
C815Y(KOREA)	0079
C816(2S)	0086
C816A	0086
C816B	0086
C816C	0086
C816D	0086
C816E	0086
C816F	0086
C816G	0086
C816GN	0086
C816H	0086
C816HL	0086
C816K(2S)	0086
C816L	0086
C816M	0086
C816OR	0086
C816R	0086
C816X	0086
C816Y	0086
C817	0259
C818(2S)	0168
C818A	0168
C818B	0168
C818C	0168
C818D	0168
C818E	0168
C818F	0168
C818G	0168

DEVICE TYPE	REPL CODE
C818GN	0168
C818H	0168
C818J	0168
C818K	0168
C818L	0168
C818M	0168
C818OR	0168
C818R	0168
C818X	0168
C818Y	0168
C819	2700
C821(2S)	2041
C822(2S)	2041
C823	0418
C824	0418
C825	1027
C825A	1027
C825B	1027
C825C	1027
C825D	1027
C825E	1027
C825F	1027
C825G	1027
C825GN	1027
C825H	1027
C825J	1027
C825K	1027
C825L	1027
C825M	1027
C825OR	1027
C825R	1027
C825X	1027
C825Y	1027
C826(2S)	0086
C826A	0086
C826B	0086
C826C	0086
C826D	0086
C826E	0086
C826F	0086
C826G	0086
C826H	0086
C826J	0086
C826K	0086
C826L	0086
C826M	0086
C826OR	0086
C826R	0086
C826X	0086
C826Y	0086
C827(2S)	0086
C827A	0086
C827B	0086
C827C	0086
C827D	0086
C827E	0086
C827F	0086
C827G	0086
C827GN	0086
C827H	0086
C827J	0086
C827K	0086
C827M	0086
C827OR	0086
C827R	0086
C827X	0086
C827Y	0086
C828	1220
C828(N)	1220
C828-O	1220
C828-O(2S)	1220
C828-OR	1220
C828A(2S)	1220
C828A(JAPAN)	1220
C828AO	1220
C828AQ(2S)	1220
C828AR	1220
C828AR(2S)	1220
C828AS(2S)	1220
C828B	1220
C828C	1220
C828D	1220
C828E(2S)	1220
C828F(2S)	1220
C828G	1220
C828GN	1220
C828H(2S)	1220
C828HR	1220
C828K(2S)	1220
C828L	1220
C828LS(2S)	1220
C828M	1220
C828OR	1220
C828P(2S)	1220
C828PQ	1220
C828Q	1220
C828Q-6	1220
C828QRS(2S)	1220
C828R	1220
C828R(JAPAN)	1220
C828R/494	0111
C828R-1	1220
C828RA	1220
C828RH	1220
C828RS	1220
C828RST	1220
C828RT	1220
C828S(2S)	1220
C828S(JAPAN)	1220

If replacement code is OEM, contact original manufacturer for replacement.

DEVICE TYPE	REPL CODE
C828T(2S)	1220
C828X	1220
C828Y(2S)	1220
C828YL	1220
C829	0151
C829-OR	0151
C829A(2S)	0151
C829AK	0151
C829B	0151
C829BC(2S)	0151
C829BJ	0151
C829BK	0151
C829C	0151
C829CL	0151
C829D(2S)	0151
C829E	0151
C829G	0151
C829GN	0151
C829H	0151
C829K	0151
C829L	0151
C829M	0151
C829OR	0151
C829R(2S)	0151
C829X(2S)	0151
C829Y(2S)	0151
C830(2S)	1147
C830A(2S)	1147
C830B(2S)	1147
C830C(2S)	1147
C830D	1147
C830E	1147
C830F	1147
C830G	1147
C830GN	1147
C830H	1147
C830J	1147
C830K	1147
C830L	1147
C830M	1147
C830OR	1147
C830R	1147
C830X	1147
C830Y	1147
C831	0042
C833	0429
C833BL(2S)	0429
C834L	0429
C835(2S)	0127
C836(2S)	0127
C836M	0127
C837(2S)	0224
C837A	0224
C837B	0224
C837C	0224
C837D	0224
C837E	0224
C837F(2S)	0224
C837G	0224
C837GN	0224
C837H(2S)	0224
C837J	0224
C837K(2S)	0224
C837KL	0224
C837L(2S)	0224
C837M	0224
C837OR	0224
C837R	0224
C837WF(2S)	0224
C837X	0224
C837Y	0224
C838	0191
C838-2	0191
C838-0	0191
C838A(2S)	0191
C838BL	0191
C838C(2S)	0191
C838E(2S)	0191
C838F(2S)	0191
C838H(2S)	0191
C838HF	0191
C838I	0191
C838J	0191
C838K(2S)	0191
C838L(2S)	0191
C838M(2S)	0191
C838R	0155
C839	0076
C839-E	0076
C839-F	0076
C839A(2S)	0076
C839B(2S)	0076
C839C(2S)	0076
C839D(2S)	0076
C839E(2S)	0076
C839F(2S)	0076
C839G	0076
C839GN	0076
C839H(2S)	0076
C839J(2S)	0076
C839J1	0076
C839JH	0076
C839K	0076
C839L(2S)	0076
C839M(2S)	0076
C839OR	0076
C839R	0076
C839X	0076
C839Y	0076
C840(2S)	2028
C840A(2S)	2028
C840AC(2S)	2028
C840B	2028
C840C	2028
C840D	2028
C840E	2028
C840F	2028
C840G	2028
C840GN	2028
C840H(2S)	2028
C840HP	2028
C840J	2028
C840K	2028
C840L	2028
C840M	2028
C840OR	2028
C840P	2028
C840Q	2028
C840R	2028
C840X	2028
C840Y	2028
C841	2041
C841H	2041
C844(2S)	0493
C844D	0493
C845	0493
C847(2S)	0016
C847A	0016
C847B	0016
C847C	0016
C847D	0016
C847E	0016
C847F	0016
C847G	0016
C847GN	0016
C847H	0016
C847J	0016
C847K	0016
C847L	0016
C847M	0016
C847OR	0016
C847R	0016
C847X	0016
C848(2S)	0016
C848A	0016
C848B	0016
C848C	0016
C848D	0016
C848E	0016
C848F	0016
C848G	0016
C848GN	0016
C848H	0016
C848J	0016
C848K	0016
C848L	0016
C848M	0016
C848OR	0016
C848R	0016
C848X	0016
C848Y	0016
C849(2S)	0016
C849A	0016
C849B	0016
C849C	0016
C849D	0016
C849E	0016
C849F	0016
C849G	0016
C849GN	0016
C849H	0016
C849J	0016
C849K	0016
C849L	0016
C849M	0016
C849OR	0016
C849R	0016
C849X	0016
C849Y	0016
C850(2S)	0016
C850A	0016
C850B	0016
C850C	0016
C850D	0016
C850E	0016
C850F	0016
C850G	0016
C850GN	0016
C850H	0016
C850J	0016
C850K	0016
C850L	0016
C850M	0016
C850OR	0016
C850R	0016
C850X	0016
C850Y	0016
C851(2S)	0130
C851A	0130
C851B	0130
C851C	0130
C851D	0130
C851E	0130
C851F	0130
C851G	0130
C851GN	0130
C851H	0130
C851K	0130
C851L	0130
C851M	0130
C851OR	0130
C851R	0130
C851X	0130
C851Y	0130
C852	0493
C852A	0493
C853(2S)	0218
C853-OR	0218
C853A(2S)	0218
C853B(2S)	0218
C853C(2S)	0218
C853D	0218
C853E	0218
C853F	0218
C853G	0218
C853GN	0218
C853H	0218
C853J	0218
C853K	0218
C853KLM(2S)	0218
C853L(2S)	0218
C853M	0218
C853OR	0218
C853R	0218
C853X	0218
C853Y	0218
C854	0418
C855	5962
C855(2S)	0418
C856(2S)	0730
C856-02(2S)	0730
C856-OR	0730
C856A	0730
C856B	0730
C856C(2S)	0730
C856D	0730
C856E	0730
C856F	0730
C856G	0730
C856GN	0730
C856J	0730
C856K	0730
C856L	0730
C856M	0730
C856OR	0730
C856R	0730
C856X	0730
C856Y	0730
C857(2S)	0233
C857H(2S)	0233
C857K	0233
C858(2S)	0111
C858A	0111
C858B	0111
C858C	0111
C858D	0111
C858E	0111
C858F(2S)	0111
C858FG(2S)	0111
C858G(2S)	0111
C858GA	0111
C858GN	0111
C858H	0111
C858J	0111
C858K	0111
C858L	0111
C858M	0111
C858OR	0111
C858R	0111
C858X	0111
C858Y	0111
C859(2S)	0111
C859A	0111
C859B	0111
C859D	0111
C859E(2S)	0111
C859F(2S)	0111
C859FG(2S)	0111
C859G(2S)	0111
C859GK(2S)	0111
C859GL	0111
C859GM	0111
C859GN	0111
C859H	0111
C859J	0111
C859K	0111
C859L	0111
C859M	0111
C859OR	0111
C859R	0111
C859X	0111
C859Y	0111
C860(2S)	0007
C860A	0007
C860B	0007
C860C(2S)	0007
C860D(2S)	0007
C860E(2S)	0007
C860F	0007
C860G	0007
C860GN	0007
C860H	0007
C860J	0007
C860K	0007
C860L	0007
C860M	0007
C860OR	0007
C860R	0007
C860X	0007
C860Y	0007
C861	0642
C862	0363
C863(2S)	0007
C863A	0007
C863B	0007
C863C	0007
C863D	0007
C863E	0007
C863F	0007
C863G	0007
C863GN	0007
C863H	0007
C863J	0007
C863K	0007
C863L	0007
C863M	0007
C863OR	0007
C863R	0007
C863X	0007
C863Y	0007
C864(2S)	0007
C865	0698
C866	5966
C866(2S)	0630
C866A	0630
C866B	0630
C866C	0630
C866D	0630
C866E	0630
C866F	0630
C866G	0630
C866GN	0630
C866H	0630
C866J	0630
C866K	0630
C866L	0630
C866M	0630
C866OR	0630
C866R	0630
C866X	0630
C866Y	0630
C867(2S)	1360
C867-OR	2214
C867-2	2214
C867-3	2214
C867-4	2214
C867-5	2214
C867-OR	2214
C867A	2214
C867B	2214
C867C	2214
C867D	2214
C867E	2214
C867F	2214
C867G	2214
C867GN	2214
C867H	2214
C867J	2214
C867K	2214
C867L	2214
C867M	2214
C867OR	2214
C867R	2214
C867X	2214
C867Y	2214
C868(2S)	0861
C868A	0861
C868B	0861
C868C	0861
C868D	0861
C868E	0861
C868F	0861
C868G	0861
C868GN	0861
C868H	0861
C868J	0861
C868K	0861
C868L	0861
C868M	0861
C868OR	0861
C868R	0861
C868X	0861
C868Y	0861
C869(2S)	0861
C869A	0861
C869B	0861
C869C	0861
C869D	0861
C869E	0861
C869F	0861
C869G	0861
C869GN	0861
C869J	0861
C869K	0861
C869L	0861
C869M	0861
C869OR	0861
C869R	0861
C869X	0861
C869Y	0861
C870(2S)	0076
C870A	0076
C870B	0076
C870C	0076
C870D	0076
C870E(2S)	0076
C870F(2S)	0076
C870FL	0076
C870G	0076
C870GN	0076
C870H	0076
C870J	0076
C870K	0076
C870L	0076
C870M	0076
C870OR	0076
C870R	0076
C870X	0076
C870Y	0076
C871(2S)	0151
C871-G	0151
C871A	0151
C871AM	0151
C871B	0151
C871C	0151
C871D(2S)	0151
C871E(2S)	0151
C871F(2S)	0151
C871G(2S)	0151
C871GN	0151
C871H	0151
C871J	0151
C871K	0151
C871L	0151
C871M	0151
C871OR	0151
C871R	0151
C871X	0151
C871Y	0151
C872	0418
C873	4463
C874	0018
C875(2S)	0086
C875-1(2S)	0086
C875-1C(2S)	0086
C875-1D(2S)	0086
C875-1E(2S)	0086
C875-1F(2S)	0086
C875-2(2S)	0086
C875-2C(2S)	0086
C875-2D(2S)	0086
C875-2E(2S)	0086
C875-2F(2S)	0086
C875-3(2S)	0086
C875-3C(2S)	0086
C875-3D(2S)	0086
C875-3E(2S)	0086
C875-3F(2S)	0086
C875B	0086
C875C(2S)	0086
C875D(2S)	3426
C875DL	0086
C875E(2S)	3426
C875EL	0086
C875F(2S)	0086
C875G	0086
C875GN	0086
C875J	0086
C875K	0086
C875L	0086
C875M	0086
C875OR	0086
C875X	0086
C875Y	0086
C876(2S)	0086
C876-E	0086
C876-F	0086
C876A	0086
C876B	0086
C876C(2S)	0086
C876D(2S)	0086
C876E(2S)	0086
C876F(2S)	0086
C876G	0086
C876GN	0086
C876H	0086
C876J	0086
C876K	0086
C876L	0086
C876M	0086
C876OR	0086
C876R	0086
C876TV(2S)	0086
C876TVD(2S)	0086
C876TV-E	0086
C876TVE(2S)	0086
C876TVEF(2S)	0086
C876TV-F	0086
C876TVF	0086
C876X	0086
C876Y	0086
C877	0016
C878	0016
C879	0127
C880	0324
C880R	0324
C881(2S)	0218
C881A(2S)	0218
C881B(2S)	0218
C881C(2S)	0218
C881D(2S)	0218
C881E	0218
C881F	0218
C881G	0218
C881GN	0218
C881H	0218
C881K(2S)	0218
C881L(2S)	0218
C881M	0218
C881OR	0218
C881R	0218
C881X	0218
C881Y	0218
C882	0620
C883	0423
C884	0178
C885	0074
C885A	OEM
C886	0074
C887	0074
C888	0074
C889(2S)	0103
C890	3846
C891	2091
C892	0042
C893	2314
C894(2S)	0155
C894A	0155
C894B	0155
C894C	0155
C894D	0155
C894E	0155
C894F	0155
C894G	0155
C894GN	0155
C894H	0155
C894J	0155
C894K	0155
C894L	0155
C894M	0155
C894OR	0155
C894R	0155
C894X	0155
C894Y	0155
C895(2S)	0074
C895-2	0074
C895-3	0074
C895-4	0074
C895-5	0074
C896(2S)	0016
C896A	0016
C896B	0016
C896C	0016
C896D	0016
C896E	0016
C896F	0016
C896G	0016
C896GN	0016
C896J	0016
C896K	0016
C896L	0016
C896M	0016
C896OR	0016
C896R	0016
C896X	0016
C896Y	0016
C897(2S)	0177
C897A(2S)	0177
C897B(2S)	0177
C898(2S)	0177
C898A(2S)	0177
C898B(2S)	0177
C898D	0177
C898E	0177
C898F	0177
C898G	0177
C898GN	0177
C898H	0177
C898J	0177
C898K(2S)	0177
C898L	0177
C898M	0177
C898OR	0177
C898X	0177
C898Y	0177
C899	2402
C899A	2402
C899B	2402
C899C	2402
C899D	2402
C899E	2402
C899F	2402
C899G	2402
C899GN	2402
C899H	2402
C899J	2402
C899K(2S)	2402
C899L	2402
C899M	2402
C899OR	2402
C899R	2402
C899X	2402
C899Y	2402
C900(2S)	1221
C900-OR	1221
C900A(2S)	1221
C900AF	1221
C900B(2S)	1221
C900C(2S)	1221
C900D(2S)	1221
C900E(2S)	1221
C900F(2S)	1221
C900G	1221
C900J	1221
C900K	1221
C900L(2S)	1221
C900M(2S)	1221
C900OR	1221
C900R	1221
C900S	1221
C900SA	1221
C900SB	1221
C900SC	1221
C900SD	1221
C900U(2S)	1221
C900UE	1221
C900V	1221
C900VE	1221
C900X	1221
C900Y	1221
C901(2S)	0642
C901A(2S)	0642
C901B	0642
C901C	0642
C901D	0642
C901E	0642
C901F	0642
C901G	0642
C901GN	0642
C901H	0642
C901J	0642
C901K	0642
C901L	0642
C901M	0642
C901OR	0642
C901R	0642
C901X	0642
C901Y	0642
C902	0620
C903	0155
C904	0155
C905	0861
C906	0016
C906F	0016
C907(2S)	0016
C907A(2S)	0016
C907AC(2S)	0016
C907AD(2S)	0016
C907AH(2S)	0016
C907B	0016
C907C(2S)	0016
C907D(2S)	0016
C907E	0016
C907F	0016
C907G	0016
C907GN	0016
C907H(2S)	0016
C907HA(2S)	0016
C907J	0016
C907K	0016
C907L	0016
C907M	0016
C907OR	0016
C907R	0016
C907X	0016
C907Y	0016
C908	1354
C909	0560
C911	0560
C911A	0560
C912(2S)	0144
C912M	0144
C913(2S)	0861
C913A	0861
C913B	0861
C913C	0861
C913D	0861
C913E	0861
C913F	0861
C913G	0861
C913GN	0861
C913H	0861
C913J	0861
C913K	0861
C913L	0861
C913M	0861
C913OR	0861
C913R	0861
C913X	0861
C913Y	0861
C914	0016
C914A	0016
C915	0016
C915A	0016
C916	0042
C917(2S)	0007
C917A	0007
C917B	0007
C917C	0007
C917D	0007
C917E	0007
C917F	0007
C917G	0007
C917GN	0007
C917H	0007
C917J	0007
C917K(2S)	0007
C917L	0007
C917M	0007
C917OR	0007
C917R	0007
C917X	0007
C917Y	0007

DEVICE TYPE	REPL CODE
C918(2S)	0224
C918A	0224
C918AL	0224
C918B	0224
C918C	0224
C918D	0224
C918E	0224
C918F	0224
C918G	0224
C918GN	0224
C918H	0224
C918J	0224
C918K	0224
C918L	0224
C918LF	0224
C918M	0224
C918OR	0224
C918R	0224
C918X	0224
C918XL	0224
C918Y	0224
C920(2S)	1213
C920-0Q	1213
C920-OR	1213
C920A	1213
C920B	1213
C920C	1213
C920CL	1213
C920D	1213
C920E(2S)	1213
C920F	1213
C920G	1213
C920GN	1213
C920H	1213
C920L	1213
C920M	1213
C920OR	1213
C920Q(2S)	1213
C920R(2S)	1213
C920X	1213
C920Y	1213
C921(2S)	2381
C921A	2381
C921B	2381
C921C	2381
C921C1(2S)	2381
C921CL	2381
C921D	2381
C921E	2381
C921F	2381
C921G	2381
C921GN	2381
C921H	2381
C921J	2381
C921K(2S)	2381
C921L(2S)	2381
C921M(2S)	2381
C921OR	2381
C921R	2381
C921W	2381
C921X	2381
C921Y	2381
C922(2S)	0937
C922A(2S)	0937
C922B(2S)	0937
C922C(2S)	0937
C922K(2S)	0937
C922L(2S)	0937
C922M(2S)	0937
C923(2S)	0552
C923A(2S)	0552
C923B(2S)	0552
C923C(2S)	0552
C923D(2S)	0552
C923E(2S)	0552
C923F(2S)	0552
C923G	0552
C923GN	0552
C923H	0552
C923K	0552
C923L	0552
C923M	0552
C923OR	0552
C923R	0552
C923X	0552
C923Y	0552
C924(2S)	0549
C924A	0549
C924B	0549
C924C	0549
C924D	0549
C924E(2S)	0549
C924F(2S)	0549
C924G	0549
C924GN	0549
C924H	0549
C924J	0549
C924K	0549
C924L	0549
C924M(2S)	0549
C924OR	0549
C924R	0549
C924X	0549
C924Y	0549
C925(2S)	0286
C925A	0286
C925B	0286
C925C	0286
C925D	0286

DEVICE TYPE	REPL CODE
C925E	0286
C925F	0286
C925G	0286
C925GN	0286
C925H	0286
C925J	0286
C925K	0286
C925L	0286
C925M	0286
C925OR	0286
C925R	0286
C925X	0286
C925Y	0286
C926(2S)	0261
C926-OR	0261
C926A(2S)	0261
C926B	0261
C926C	0261
C926D	0261
C926E	0261
C926F	0261
C926G	0261
C926GN	0261
C926H	0261
C926J	0261
C926K	0261
C926L	0261
C926M	0261
C926OR	0261
C926R	0261
C926X	0261
C926Y	0261
C927(2S)	0007
C927A(2S)	0007
C927B(2S)	0007
C927C(2S)	0007
C927CJ(2S)	0007
C927CK	0007
C927CT	0007
C927CU(2S)	0007
C927CW(2S)	0007
C927D(2S)	0007
C927E(2S)	0007
C927F	0007
C927G	0007
C927GN	0007
C927H	0007
C927J	0007
C927K	0007
C927L	0007
C927M	0007
C927OR	0007
C927R	0007
C927X	0007
C927XL	0007
C927Y	0007
C927Z	0007
C928(2S)	0007
C928A	0007
C928B(2S)	0007
C928C(2S)	0007
C928D(2S)	0007
C928E(2S)	0007
C928F	0007
C928G	0007
C928GN	0007
C928H	0007
C928J	0007
C928K	0007
C928L	0007
C928M	0007
C928OR	0007
C928R	0007
C928X	0007
C928Y	0007
C929(2S)	1066
C929-0(2S)	1066
C929-O	1066
C929A	1066
C929B(2S)	1066
C929C(2S)	1066
C929C1(2S)	1066
C929D(2S)	1066
C929D1(2S)	1066
C929DE(2S)	1066
C929DP(2S)	1066
C929DU(2S)	1066
C929DV(2S)	1066
C929E(2S)	1497
C929ED(2S)	1066
C929EZ	1066
C929F(2S)	1066
C929FK(2S)	1066
C929G	1066
C929GN	1066
C929H	1066
C929J	1066
C929K	1066
C929L	1066
C929M	1066
C929NP	1066
C929OR	1066
C929R	1066
C929SP	1066
C929X	1066
C929Y	1066
C930(2S)	2209
C930-OR	2209
C930A	2209

DEVICE TYPE	REPL CODE
C930B(2S)	2209
C930BK(2S)	2209
C930C(2S)	2209
C930C-IF	2209
C930CK(2S)	2209
C930CL(2S)	2209
C930CS(2S)	2209
C930D(2S)	2209
C930DB	2209
C930DC	2209
C930DE(2S)	2209
C930DK	2209
C930DS(2S)	2209
C930DT(2S)	2209
C930DT-2(2S)	2209
C930DX(2S)	2209
C930DZ(2S)	2209
C930E(2S)	2209
C930E(JAPAN)	2209
C930EP(2S)	2209
C930ET(2S)	2209
C930EV(2S)	2209
C930EX(2S)	2209
C930F(2S)	2209
C930G	2209
C930GN	2209
C930H	2209
C930J	2209
C930K	2209
C930L	2209
C930M	2209
C930NP	2209
C930OR	2209
C930R	2209
C930SP	2209
C930X	2209
C930Y	2209
C931(2S)	0219
C931C(2S)	0219
C931D(2S)	0219
C931E(2S)	0219
C932(2S)	0560
C932A	0560
C932B	0560
C932BK	0560
C932C	0560
C932D	0560
C932E(2S)	0560
C932F	0560
C932G	0560
C932GN	0560
C932H	0560
C932J	0560
C932K	0560
C932L	0560
C932M	0560
C932OR	0560
C932R	0560
C932X	0560
C932Y	0560
C933(2S)	0191
C933A	0191
C933B	0191
C933BB(2S)	0191
C933C(2S)	0191
C933D(2S)	0191
C933E(2S)	0191
C933F(2S)	0191
C933FB	0191
C933FP(2S)	0191
C933G(2S)	0191
C933GN	0191
C933H	0191
C933J	0191
C933K	0191
C933L	0191
C933M	0191
C933OR	0191
C933R	0191
C933X	0191
C933Y	0191
C934	0016
C934-O	0016
C934A	0016
C934B	0016
C934C(2S)	0016
C934D(2S)	0016
C934E(2S)	0016
C934F(2S)	0016
C934G(2S)	0016
C934GN	0016
C934H	0016
C934J	0016
C934K	0016
C934L	0016
C934M	0016
C934OR	0016
C934Q	0016
C934R	0016
C934X	0016
C934Y	0016
C935(2S)	0074
C935-OR	0074
C935A	0074
C935B	0074
C935C	0074
C935D	0074
C935E	0074
C935F	0074

DEVICE TYPE	REPL CODE
C935G	0074
C935GN	0074
C935H	0074
C935J	0074
C935K	0074
C935M	0074
C935OR	0074
C935R	0074
C935X	0074
C935Y	0074
C936(2S)	0003
C936A	0003
C936B	0003
C936BK	0003
C936C	0003
C936D	0003
C936E	0003
C936F	0003
C936G	0003
C936GN	0003
C936H	0003
C936J	0003
C936K	0003
C936L	0003
C936M	0003
C936OR	0003
C936R	0003
C936X	0003
C936Y	0003
C937(2S)	0103
C937-01(2S)	0103
C937A(2S)	0103
C937B(2S)	0103
C937BK	0103
C937C	0103
C937D	0103
C937E	0103
C937F	0103
C937G	0103
C937GN	0103
C937H	0103
C937J	0103
C937K	0103
C937L	0103
C937M	0103
C937OR	0103
C937R	OEM
C937X	0103
C937Y	0103
C938(2S)	0155
C938-O	0155
C938-Y	1085
C938A(2S)	0155
C938B(2S)	0155
C938C(2S)	0155
C938D	0155
C938E	0155
C938F	0155
C938G	0155
C938GN	0155
C938H	0155
C938J	0155
C938K	0155
C938L	0155
C938M	0155
C938OR	0155
C938R	0155
C938X	0155
C938Y	0155
C939(2S)	0642
C939I	0642
C939L(2S)	0642
C940(2S)	0826
C940L(2S)	0826
C940M(2S)	0826
C940Q	0826
C941(2S)	0155
C941-04	0155
C941-0Y	0155
C941-O(2S)	0155
C941-OY	0155
C941-R(2S)	0155
C941-Y(2S)	0155
C941K	0155
C941O	0155
C941R(2S)	0155
C941TM	0155
C941TM-O	0155
C941Y	0155
C942	0016
C943(2S)	0016
C943A(2S)	0016
C943B(2S)	0016
C943C(2S)	0016
C943D	0016
C943E	0016
C943F	0016
C943G	0016
C943GN	0016
C943H	0016
C943J	0016
C943K	0016
C943L	0016
C943M	0016
C943OR	0016
C943R	0016
C943X	0016
C943Y	0016
C944(2S)	0076

DEVICE TYPE	REPL CODE
C944S	0076
C945	0076
C945(JAPAN)	0076
C945-0	0076
C945-O(2S)	0076
C945-OR	0076
C945-YX	0076
C945A(2S)	0076
C945AK	0076
C945AP(2S)	0076
C945AQ(2S)	0076
C945AR	0076
C945B(2S)	0076
C945C(2S)	0076
C945D(2S)	0076
C945E(2S)	0076
C945F(2S)	0076
C945G(2S)	0076
C945GN	0076
C945H(2S)	0076
C945J	0076
C945K(2S)	0076
C945L(2S)	0076
C945LP	0076
C945LP(JAPAN)	0076
C945LQ	0076
C945M(2S)	0076
C945OR	0076
C945P	0076
C945P(2S)	0076
C945P(JAPAN)	0076
C945P1	0076
C945PJ	0076
C945PQ	0076
C945Q	0076
C945Q(2S)	0076
C945Q(JAPAN)	0076
C945QL(2S)	0076
C945QP(2S)	0076
C945R(2S)	0076
C945RA	0076
C945S(2S)	0076
C945T(2S)	0076
C945TK	0076
C945TP	0076
C945TQ(2S)	0076
C945X(2S)	0076
C945Y	0076
C945Y(KOREA)	0076
C947(2S)	0007
C947A	0007
C947B	0007
C947C	0007
C947D	0007
C947E	0007
C947F	0007
C947G	0007
C947GN	0007
C947H	0007
C947J	0007
C947K	0007
C947L	0007
C947M	0007
C947OR	0007
C947R	0007
C947X	0007
C947Y	0007
C948(2S)	0216
C948A	0216
C948B	0216
C948D	0216
C948E	0216
C948F	0216
C948G	0216
C948GN	0216
C948H	0216
C948J	0216
C948K	0216
C948L	0216
C948M	0216
C948OR	0216
C948R	0216
C948X	0216
C948Y	0216
C949	0216
C950	0016
C951	0079
C952	0079
C953	0016
C954	0016
C955	0016
C956	0016
C957(2S)	0144
C957A	0144
C957AL	0144
C957B	0144
C957C	0144
C957D	0144
C957E	0144
C957F	0144
C957G	0144
C957GN	0144
C957H	0144
C957J	0144
C957K	0144
C957L	0144
C957M	0144
C957OR	0144
C957R	0144

DEVICE TYPE	REPL CODE
C957X	0144
C957XL	0144
C957Y	0144
C959(2S)	0086
C959A(2S)	0086
C959B(2S)	0086
C959C(2S)	0086
C959D(2S)	0086
C959E	0086
C959F	0086
C959G	0086
C959GN	0086
C959H	0086
C959J	0086
C959K	0086
C959L	0086
C959M(2S)	0086
C959OR	0086
C959R	0086
C959S(2S)	0086
C959SA(2S)	0086
C959SB(2S)	0086
C959SC(2S)	0086
C959SD(2S)	0086
C959X	0086
C959Y	0086
C960	5623
C961	0620
C962	0620
C963	0016
C964	0016
C965	0016
C966(2S)	0016
C966A	0016
C966B	0016
C966C	0016
C966D	0016
C966E	0016
C966F	0016
C966G	0016
C966GN	0016
C966H	0016
C966J	0016
C966K	0016
C966L	0016
C966M	0016
C966OR	0016
C966R	0016
C966X	0016
C966Y	0016
C967(2S)	0016
C967A	0016
C967B	0016
C967C	0016
C967D	0016
C967E	0016
C967F	0016
C967G	0016
C967GN	0016
C967H	0016
C967J	0016
C967K	0016
C967L	0016
C967M	0016
C967OR	0016
C967R	0016
C967X	0016
C967Y	0016
C968(2S)	2929
C968A	2929
C968B	2929
C968C	2929
C968D	2929
C968E	2929
C968F	2929
C968G	2929
C968GN	2929
C968H	2929
C968J	2929
C968K	2929
C968L	2929
C968M	2929
C968OR	2929
C968P(2S)	2929
C968R	2929
C968X	2929
C968Y	2929
C969	0016
C970	0016
C971(2S)	2952
C971A	2952
C971B	2952
C971BK	2952
C971C	2952
C971D	2952
C971E	2952
C971F	2952
C971G	2952
C971GN	2952
C971H	2952
C971J	2952
C971K	2952
C971L	2952
C971M	2952
C971OR	2952
C971R	2952
C971X	2952
C971Y	2952
C972(2S)	0086

DEVICE TYPE	REPL CODE
C972A	0086
C972B	0086
C972C(2S)	0086
C972D(2S)	0086
C972E(2S)	0086
C972F	0086
C972G	0086
C972GN	0086
C972H	0086
C972J	0086
C972K	0086
C972L	0086
C972M	0086
C972OR	0086
C972R	0086
C972X	0086
C972Y	0086
C973	2091
C973A	2091
C975	1420
C975A	1420
C976	0086
C976TV	0086
C979	0016
C979-O	0016
C979-R	0016
C979-Y	0016
C979A	0016
C979A-O	0016
C979A-R	0016
C980	0286
C980AG	0286
C980AG-O	0286
C980AG-R	0286
C980G	0286
C980G-O	0286
C980G-R	0286
C980G-Y	0286
C981	0178
C982(2S)	0400
C983(2S)	0066
C983-0(2S)	0066
C983-O	0066
C983-R	0066
C983-Y(2S)	0066
C983FA-1	0066
C983FH-1	0066
C983O	0066
C983R(2S)	0066
C983S(2S)	0066
C983Y(2S)	0066
C983YFA-1	0066
C984(2S)	0536
C984A(2S)	0536
C984B(2S)	0536
C984C(2S)	0536
C984D	0536
C984E	0536
C984F	0536
C984G	0536
C984GN	0536
C984H	0536
C984J	0536
C984L	0536
C984M	0536
C984OR	0536
C984R	0536
C984X	0536
C984Y	0536
C985	0111
C985A	0111
C987	0007
C987A	0007
C988	0007
C988A	0007
C988B	0007
C990	0042
C991(2S)	0493
C991A	0493
C991B	0493
C991C	0493
C991D	0493
C991E	0493
C991F	0493
C991G	0493
C991GN	0493
C991H	0493
C991J	0493
C991K	0493
C991L	0493
C991M	0493
C991OR	0493
C991R	0493
C991X	0493
C991Y	0493
C992(2S)	0493
C992A	0493
C992B	0493
C992C	0493
C992D	0493
C992E	0493
C992F	0493
C992G	0493
C992GN	0493
C992H	0493
C992J	0493
C992K	0493
C992L	0493
C992M	0493

DEVICE TYPE	REPL CODE
C992OR	0493
C992R	0493
C992X	0493
C992Y	0493
C993(2S)	0016
C993D	0016
C993E	0016
C994	0079
C995(2S)	0168
C995A	0168
C995B	0168
C995C	0168
C995E	0168
C995F	0168
C995G	0168
C995GN	0168
C995H	0168
C995J	0168
C995K	0168
C995L	0168
C995M	0168
C995OR	0168
C995R	0168
C995X	0168
C995Y	0168
C996(2S)	3280
C997(2S)	0224
C997A	0224
C997B	0224
C997C	0224
C997D	0224
C997E	0224
C997F	0224
C997G	0224
C997GN	0224
C997H	0224
C997J	0224
C997K	0224
C997L	0224
C997M	0224
C997OR	0224
C997R	0224
C997X	0224
C997Y	0224
C998	3846
C999(2S)	0065
C999A	0065
C999B	0065
C999C	0065
C999D	0065
C999E	0065
C999F	0065
C999G	0065
C999GN	0065
C999H	0065
C999J	0065
C999K	0065
C999L	0065
C999M	0065
C999OR	0065
C999R	0065
C999X	0065
C999Y	0065
C1000	0111
C1000(2S)	0111
C1000-BL(2S)	0111
C1000-GR(2S)	0111
C1000-Y(2S)	0111
C1000A	0111
C1000B	0111
C1000BL	0111
C1000C	0111
C1000D	0111
C1000E	0111
C1000F	0111
C1000G	0111
C1000G-BL	0111
C1000GBL	0111
C1000G-GR	0111
C1000GN	0111
C1000GR	0111
C1000H	0111
C1000J	0111
C1000K	0111
C1000L	0111
C1000M	0111
C1000OR	0111
C1000R	0111
C1000X	0111
C1000Y(2S)	0111
C1001	0418
C1001(2S)	2039
C1002	0111
C1003	0079
C1003A	OEM
C1004	0003
C1004(2S)	0003
C1004(TRANS)	0003
C1004A(2S)	0003
C1004A(TRANS)	0003
C1004B	0003
C1004C	0003
C1004D	0003
C1004E	0003
C1004F	0003
C1004G	0003
C1004GN	0003
C1004H	0003
C1004J	0003
C1004K	0003
C1004L	0003
C1004M	0003
C1004OR	0003
C1004R	0003
C1004X	0003
C1004Y	0003
C1005	0065
C1005(2S)	0065
C1005A	0065
C1005A(2S)	0065
C1005B	0065
C1005C	0065
C1005D	0065
C1005E	0065
C1005F	0065
C1005G	0065
C1005GN	0065
C1005H	0065
C1005J	0065
C1005K	0065
C1005L	0065
C1005M	0065
C1005OR	0065
C1005R	0065
C1005X	0065
C1005Y	0065
C1006(2S)	0111
C1006A	0111
C1006A(2S)	0111
C1006B(2S)	0111
C1006C(2S)	0111
C1007(2S)	0016
C1008(2S)	0086
C1008A	0086
C1008Y	OEM
C1008Y(KOREA)	0079
C1009	0224
C1009A	0224
C1009F1	0224
C1009F2	0224
C1009F3	0224
C1009F4	0224
C1009F5	0224
C1010	0111
C1010(TRANS)	0111
C1010A	0111
C1010A(TRANS)	0111
C1010B	0111
C1010B(TRANS)	0111
C1010C(2S)	0111
C1010D	0111
C1010E	0111
C1010F	0111
C1010G	0111
C1010GN	0111
C1010H	0111
C1010J	0111
C1010K	0111
C1010L	0111
C1010M	0111
C1010OR	0111
C1010R	0111
C1010X	0111
C1010Y	0111
C1011	0698
C1012(2S)	0233
C1012A(2S)	0233
C1012B	0233
C1012C	0233
C1012D	0233
C1012E	0233
C1012F	0233
C1012G	0233
C1012GN	0233
C1012H	0233
C1012J	0233
C1012K	0233
C1012L	0233
C1012M	0233
C1012OR	0233
C1012R	0233
C1012X	0233
C1012Y	0233
C1013(2S)	2046
C1013-OR	2046
C1013A	2046
C1013B	2046
C1013C(2S)	2046
C1013D(2S)	2046
C1013E	2046
C1013F	2046
C1013G	2046
C1013GN	2046
C1013H	2046
C1013J	2046
C1013K	2046
C1013L	2046
C1013LJ	2046
C1013M	2046
C1013OR	2046
C1013PJ	2046
C1013R	2046
C1013X	2046
C1013Y	2046
C1014	2046
C1014(TRANS)	2046
C1014-1	2046
C1014-2	2046
C1014A	OEM
C1014A(TRANS)	2046
C1014B	2046
C1014B(TRANS)	2046
C1014BY	2046
C1014C(2S)	2046
C1014CD(2S)	2046
C1014D(2S)	2046
C1014D1(2S)	2046
C1014E	2046
C1014F	2046
C1014G	2046
C1014GA	2046
C1014GN	2046
C1014H	2046
C1014J	2046
C1014K	2046
C1014L	2046
C1014LG	2046
C1014LR	2046
C1014M	2046
C1014OR	2046
C1014R	2046
C1014W	2046
C1014X	2046
C1014Y	2046
C1015	OEM
C1015A	OEM
C1015B	OEM
C1016	OEM
C1016A	OEM
C1016B	OEM
C1017(2S)	2050
C1018(2S)	2050
C1018A	2050
C1018B(2S)	2050
C1018C	2050
C1018D	2050
C1018E	2050
C1018F	2050
C1018G	2050
C1018GN	2050
C1018H	2050
C1018J	2050
C1018K	2050
C1018L	2050
C1018M	2050
C1018R	2050
C1018X	2050
C1018Y	2050
C1019	OEM
C1019A	OEM
C1019B	OEM
C1019C	0042
C1020	OEM
C1020A	OEM
C1020B	OEM
C1021	OEM
C1021A	OEM
C1021B	OEM
C1022	OEM
C1022A	OEM
C1022B	OEM
C1023	0127
C1023-O(2S)	0127
C1023-Y(2S)	0127
C1023A	0127
C1023B	0127
C1023C	0127
C1023D	0127
C1023E	0127
C1023F	0127
C1023G(2S)	0127
C1023GN	0127
C1023H	0127
C1023J	0127
C1023K	0127
C1023L	0127
C1023M	0127
C1023OR	0127
C1023R	0127
C1023X	0127
C1023Y	0127
C1024(2S)	0178
C1024-E	0178
C1024A	0178
C1024B(2S)	0178
C1024C(2S)	0178
C1024D(2S)	0178
C1024E(2S)	0178
C1024F(2S)	0178
C1024G	0178
C1024L	0178
C1024Y	0178
C1025	0178
C1025(TRANS)	0934
C1025A	OEM
C1025B	OEM
C1025CTV(2S)	0934
C1025D	0934
C1025E	0934
C1025J	0934
C1025MT	0934
C1026	0127
C1026(TRANS)	0127
C1026-0	0127
C1026-R	0127
C1026A	OEM
C1026A(TRANS)	0127
C1026B	OEM
C1026B(TRANS)	0127
C1026BL	0127
C1026C	0127
C1026D	0127
C1026E	0127
C1026F	0127
C1026G(2S)	0127
C1026GN	0127
C1026GR	0127
C1026H	0127
C1026J	0127
C1026K	0127
C1026L	0127
C1026M	0127
C1026OR	0127
C1026R	0127
C1026X	0127
C1026Y(2S)	0127
C1027(2S)	0642
C1029	OEM
C1029A	OEM
C1029B	OEM
C1030	0867
C1030(TRANS)	2059
C1030-OR	2059
C1030A	0867
C1030A(TRANS)	2059
C1030B	0867
C1030B(TRANS)	2059
C1030B2C	2059
C1030C(2S)	2059
C1030D(2S)	2059
C1030E	2059
C1030F	2059
C1030G	2059
C1030H	2059
C1030J	2059
C1030K	2059
C1030L	2059
C1030M	2059
C1030OR	2059
C1030P	0177
C1030R	2059
C1030X	2059
C1030Y	2059
C1031	1027
C1031(TRANS)	1027
C1031A	OEM
C1031B	OEM
C1032	0127
C1032(TRANS)	0127
C1032A	OEM
C1032A(TRANS)	0127
C1032B	OEM
C1032B(TRANS)	0127
C1032BL	0127
C1032C	0127
C1032D	0127
C1032E	0127
C1032F	0127
C1032G(2S)	0127
C1032GN	0127
C1032H	0127
C1032J	0127
C1032K	0127
C1032L	0127
C1032M	0127
C1032OR	0127
C1032P	1041
C1032R	0127
C1032X	0127
C1032Y(2S)	0127
C1033(2S)	0715
C1033A(2S)	0715
C1033B	0715
C1033C	0715
C1033D	0715
C1033E	0715
C1033F	0715
C1033G	0715
C1033GN	0715
C1033H	0715
C1033J	0715
C1033K	0715
C1033L	0715
C1033OR	0715
C1033P	2097
C1033R	0715
C1033X	0715
C1033Y	0715
C1034(2S)	2064
C1034-3	2064
C1034-4	2064
C1034-5	2064
C1035	0007
C1035(TRANS)	0007
C1035A	OEM
C1035A(TRANS)	0007
C1035B	OEM
C1035B(TRANS)	0007
C1035C(2S)	0007
C1035D(2S)	0007
C1035E(2S)	0007
C1035F	0007
C1035GN	0007
C1035H	0007
C1035J	0007
C1035L	0007
C1035M	0007
C1035OR	0007
C1035P	1176
C1035R	0007
C1035X	0007
C1035Y	0007
C1036	0007
C1036(TRANS)	0007
C1036A	OEM
C1036A(TRANS)	0007
C1036B	OEM
C1036B(TRANS)	0007
C1036C	0007
C1036D	0007
C1036E	0007
C1036F	0007
C1036G	0007
C1036GN	0007
C1036H	0007
C1036J	0007
C1036K	0007
C1036L	0007
C1036M	0007
C1036OR	0007
C1036P	1831
C1036R	0007
C1036X	0007
C1036Y	0007
C1038P	OEM
C1039	OEM
C1039A	OEM
C1039B	OEM
C1039P	OEM
C1040	OEM
C1040A	OEM
C1040B	OEM
C1041	OEM
C1041A	OEM
C1041B	OEM
C1042	OEM
C1042(TRANS)	1591
C1042A	OEM
C1042B	OEM
C1043	2070
C1044(2S)	0007
C1044P	0033
C1045	0003
C1045(2S)	0003
C1045A	OEM
C1045B	0003
C1045B(2S)	0003
C1045C(2S)	0003
C1045D(2S)	0003
C1045E(2S)	0003
C1045F	0003
C1045G	0003
C1045GN	0003
C1045H	0003
C1045J	0003
C1045K	0003
C1045L	0003
C1045M	0003
C1045OR	0003
C1045P	0081
C1045R(2S)	0003
C1045X	0003
C1045Y	0003
C1046	0065
C1046(2S)	0065
C1046-OR	0065
C1046A	OEM
C1046A(TRANS)	0065
C1046B	OEM
C1046B(TRANS)	0065
C1046C	0065
C1046D	0065
C1046E	0065
C1046F	0065
C1046G	0065
C1046GN	0065
C1046H	0065
C1046J	0065
C1046K	0065
C1046L	0065
C1046M	0065
C1046N	0065
C1046OR	0065
C1046P	0141
C1046P(HEP)	0141
C1046R	0065
C1046X	0065
C1046Y	0065
C1047(2S)	0113
C1047A	0113
C1047B(2S)	0113
C1047BC	0113
C1047BCD	0113
C1047C(2S)	0113
C1047D(2S)	0113
C1047E(2S)	0113
C1047F	0113
C1047G	0113
C1047GN	0113
C1047GR	0113
C1047H	0113
C1047J	0113
C1047K	0113
C1047L	0113
C1047M	0113
C1047R	0113
C1047X	0113
C1047Y	0113
C1048(2S)	0233
C1048B(2S)	0233
C1048C(2S)	0233
C1048D(2S)	0233
C1048DC	0233
C1048E(2S)	0233
C1048F(2S)	0233
C1049	OEM
C1049A	OEM
C1049B	OEM
C1050	0220
C1050C	0220
C1050D(2S)	0220
C1050E(2S)	0220
C1050F(2S)	0220
C1051(2S)	0177
C1051L	0177
C1052	0086
C1052P	0333
C1053	0086
C1053P	0358
C1054	0007
C1055	0429
C1055H(2S)	0429
C1055U	0429
C1056(2S)	0233
C1056A	0233
C1056B	0233
C1056C	0233
C1056D	0233
C1056E	0233
C1056F	0233
C1056G	0233
C1056GN	0233
C1056H	0233
C1056J	0233
C1056K	0233
C1056L	0233
C1056M	0233
C1056OR	0233
C1056R	0233
C1056X	0233
C1056Y	0233
C1057P	0466
C1058P	0466
C1059(2S)	0142
C1059A	0142
C1059B	0142
C1059C	0142
C1059D	0142
C1059E	0142
C1059F	0142
C1059G	0142
C1059GN	0142
C1059H	0142
C1059J	0142
C1059K	0142
C1059L	0142
C1059M	0142
C1059OR	0142
C1059R	0142
C1059X	0142
C1059Y	0142
C1060	0042
C1060(2S)	0042
C1060A	0042
C1060A(2S)	0042
C1060B	0042
C1060B(2S)	0042
C1060BL	0042
C1060BM(2S)	0042
C1060BY	0042
C1060C(2S)	0042
C1060D(2S)	0042
C1060E	0042
C1060F	0042
C1060G	0042
C1060GN	0042
C1060H	0042
C1060J	0042
C1060K	0042
C1060L	0042
C1060M	0042
C1060OR	0042
C1060R	0042
C1060X	0042
C1060Y	0042
C1061	0042
C1061(2S)	0042
C1061-C	0042
C1061A	0042
C1061A(TRANS)	0042
C1061B	0042
C1061B(TRANS)	0042
C1061BM	0042
C1061C(2S)	0042
C1061D(2S)	0042
C1061K	0042
C1061KA	0042
C1061KB	0042
C1061KC	0042
C1061SC	0042
C1061T(2S)	0042
C1061TB(2S)	0042
C1062	0187
C1062(TRANS)	0233
C1062A	OEM
C1062B	OEM
C1062P	0562
C1063	0018
C1064	0324
C1065	OEM
C1065A	OEM
C1065B	OEM
C1066	0144
C1066(TRANS)	0259
C1066A	OEM
C1066B	OEM
C1067	0007
C1068	0418
C1069	0086
C1069(TRANS)	0630
C1069A	OEM
C1069B	OEM
C1070(2S)	0007
C1070-1	0007
C1070-2	0007
C1070-B	0007
C1071(2S)	0016
C1072(2S)	0086
C1072A(2S)	0086
C1072B	0086
C1072C	0086
C1072D	0086
C1072E	0086
C1072F	0086
C1072G	0086
C1072GN	0086
C1072H	0086
C1072J	0086
C1072K	0086
C1072L	0086
C1072M	0086
C1072OR	0086
C1072R	0086
C1072X	0086
C1072Y	0086
C1074	2091
C1075	1420
C1076	2093
C1078	2096
C1079(2S)	0177
C1079-R	0177
C1079-Y	0177
C1080(2S)	0177
C1080-R	0177
C1080-Y	0177
C1080B	0177
C1080LB	0177
C1083	2041
C1084	0224
C1085	0178
C1086(2S)	0223
C1086A	0223
C1086B	0223
C1086C	0223
C1086D	0223
C1086E	0223
C1086F	0223
C1086G	0223
C1086GN	0223
C1086H	0223
C1086J	0223
C1086K	0223
C1086L	0223
C1086M(2S)	0223
C1086OR	0223
C1086R	0223
C1086X	0223
C1086Y	0223
C1088	0285
C1089	0285
C1089B	0285
C1089C	0285
C1089D	0285
C1090	0233
C1091A	0560
C1095-K	0219
C1096	0390
C1096-4Z-L	0390
C1096-4ZL	0390
C1096-O(KOREA)	0219
C1096-OR	0390
C1096A(2S)	0390
C1096B(2S)	0390
C1096C(2S)	0390
C1096D(2S)	0390
C1096E	0390
C1096F	0390
C1096G	0390
C1096GN	0390
C1096H	0390
C1096J	0390
C1096K(2S)	0390
C1096L(2S)	0390
C1096LM	0390
C1096M(2S)	0390
C1096N(2S)	0390
C1096OR	0390
C1096Q	0390
C1096R	0390
C1096X	0390
C1096Y	0390
C1098(2S)	0560
C1098(4)K	0560

DEVICE TYPE	REPL CODE
C1098(4)L	0560
C1098-4K	0560
C1098-4L	0560
C1098-42M	0560
C1098A(2S)	0560
C1098B(2S)	0560
C1098C(2S)	0560
C1098D(2S)	0560
C1098L(2S)	0560
C1098M(2S)	0560
C1099	1146
C1099K	1146
C1100	0223
C1100(TRANS)	1150
C1100A	OEM
C1100A(TRANS)	1150
C1100B	OEM
C1100B(TRANS)	1150
C1100C	1150
C1100D	1150
C1100E	1150
C1100F	1150
C1100G	1150
C1100GN	1150
C1100H	1150
C1100J	1150
C1100K	1150
C1100L	1150
C1100M	1150
C1100OR	1150
C1100R	1150
C1100X	1150
C1100Y	1150
C1101	0003
C1101(TRANS)	0003
C1101A	0003
C1101A(TRANS)	0003
C1101A1	OEM
C1101A1DM	OEM
C1101A51	OEM
C1101ADM	OEM
C1101B	0003
C1101B(TRANS)	0003
C1101C(2S)	0003
C1101D(2S)	0003
C1101E(2S)	0003
C1101F(2S)	0003
C1101G	0003
C1101GN	0003
C1101H	0003
C1101J	0003
C1101K	0003
C1101L(2S)	0003
C1101M	0003
C1101OR	0003
C1101R	0003
C1101X	0003
C1101Y	0003
C1102	0142
C1102(TRANS)	0142
C1102A	0142
C1102A(TRANS)	0142
C1102B	0142
C1102B(TRANS)	0142
C1102C(2S)	0142
C1102K(2S)	0142
C1102L(2S)	0142
C1102M(2S)	0142
C1103(2S)	0233
C1103A(2S)	0233
C1103B(2S)	0233
C1103C(2S)	0233
C1103L(2S)	0233
C1104(2S)	1209
C1104A(2S)	1209
C1104B(2S)	1209
C1104C(2S)	1209
C1104L	1209
C1105	0142
C1105(2S)	0142
C1105A	0142
C1105A(TRANS)	0142
C1105B	0142
C1105B(TRANS)	0142
C1105C(2S)	0142
C1105D	0142
C1105E	0142
C1105F	0142
C1105G	0142
C1105GN	0142
C1105H	0142
C1105J	0142
C1105K(2S)	0142
C1105L(2S)	0142
C1105M(2S)	0142
C1105OR	0142
C1105R	0142
C1105X	0142
C1105Y	0142
C1106	0270
C1106(TRANS)	0220
C1106A	0270
C1106A(TRANS)	0220
C1106B	0270
C1106B(TRANS)	0220
C1106C(2S)	0220
C1106K(2S)	0220
C1106L(2S)	0220
C1106M(2S)	0220
C1106P	0220
C1106PQ	0220
C1106Q	0220
C1107	0236
C1107Q(2S)	0236
C1107YG	0236
C1108	0236
C1109	0236
C1109(TRANS)	0236
C1109A	OEM
C1109B	OEM
C1110	0236
C1111(2S)	0867
C1112	0867
C1113	0429
C1114(2S)	2125
C1115(2S)	0177
C1116(2S)	0543
C1116-0(2S)	0543
C1116-O	0543
C1116A	0543
C1117	0007
C1117A	0007
C1117B	0007
C1117C	0007
C1117D	0007
C1117E	0007
C1117F	0007
C1117G	0007
C1117GN	0007
C1117H	0007
C1117J	0007
C1117K	0007
C1117L	0007
C1117M	0007
C1117OR	0007
C1117R	0007
C1117X	0007
C1117Y	0007
C1122	1420
C1122A	1420
C1123(2S)	0079
C1123A	0079
C1123B	0079
C1123C	0079
C1123D	0079
C1123E	0079
C1123F	0079
C1123GN	0079
C1123H	0079
C1123J	0079
C1123K	0079
C1123L	0079
C1123M	0079
C1123OR	0079
C1123R	0079
C1123X	0079
C1123Y	0079
C1124(2S)	0264
C1124(13)	0264
C1124-OR	0264
C1124A	0264
C1124B	0264
C1124C	0264
C1124D	0264
C1124E	0264
C1124F	0264
C1124G	0264
C1124GN	0264
C1124H	0264
C1124J	0264
C1124K	0264
C1124L	0264
C1124M	0264
C1124OR	0264
C1124R	0264
C1124X	0264
C1124Y	0264
C1126(2S)	0224
C1126A	0224
C1126B	0224
C1126E	0224
C1126F	0224
C1126GN	0224
C1126J	0224
C1126K	0224
C1126L	0224
C1126M	0224
C1126OR	0224
C1126R	0224
C1126X	0224
C1126Y	0224
C1127(2S)	1068
C1127-1	1068
C1127-2	1068
C1127-O	1068
C1127-OR	1068
C1127A	1068
C1127B	1068
C1127C	1068
C1127D	1068
C1127E	1068
C1127F	1068
C1127G	1068
C1127GA	1068
C1127GN	1068
C1127H	1068
C1127J	1068
C1127JR	1068
C1127K	1068
C1127L	1068
C1127M	1068
C1127OR	1068
C1127R	1068
C1127X	1068
C1127Y	1068
C1128(2S)	0224
C1128-O	0224
C1128A	0224
C1128B	0224
C1128BL	0224
C1128C	0224
C1128D(2S)	0224
C1128G	0224
C1128H	0224
C1128M	0224
C1128R	0224
C1128S	0224
C1128Y	0224
C1129(2S)	0224
C1129-O	0224
C1129A	0224
C1129B	0224
C1129BL	0224
C1129C	0224
C1129G	0224
C1129M	0224
C1129R	0224
C1129Y	0224
C1130	0003
C1131	0642
C1132	0065
C1133	0065
C1133D	0065
C1140	1852
C1141	1852
C1141(TRANS)	1852
C1141A	OEM
C1141B	OEM
C1142	0363
C1143	0363
C1145	OEM
C1145A	OEM
C1145B	OEM
C1146	OEM
C1146A	OEM
C1146B	OEM
C1147	OEM
C1147A	OEM
C1147B	OEM
C1149	OEM
C1149A	OEM
C1149B	OEM
C1150	0086
C1150(TRANS)	0086
C1150A	OEM
C1150B	OEM
C1151	0003
C1151(TRANS)	0003
C1151A	0003
C1151A(TRANS)	0003
C1151B	OEM
C1152	0074
C1152(TRANS)	0074
C1152A	OEM
C1152B	OEM
C1152F(2S)	0074
C1152G(2S)	0074
C1153(2S)	1150
C1153A	1150
C1154(2S)	0223
C1155	0264
C1155(TRANS)	0264
C1155A	OEM
C1155B	OEM
C1156	0264
C1156(TRANS)	0264
C1156A	OEM
C1156A(TRANS)	0264
C1156B	OEM
C1156B(TRANS)	0264
C1156C	0264
C1156D	0264
C1156E	0264
C1156F	0264
C1156G	0264
C1156GN	0264
C1156H	1816
C1156H(TRANS)	0264
C1156K	0264
C1156L	0264
C1156M	0264
C1156OR	0264
C1156R	0264
C1156X	0264
C1156Y	0264
C1157(2S)	0264
C1157A	0264
C1157B	0264
C1157C	0264
C1157E	0264
C1157F	0264
C1157G	0264
C1157GN	0264
C1157H	0264
C1157J	0264
C1157K	0264
C1157L	0264
C1157M	0264
C1157OR	0264
C1157R	0264
C1157X	0264
C1157Y	0264
C1158(2S)	0475
C1159	0475
C1159(TRANS)	0007
C1159A	OEM
C1159B	OEM
C1160(2S)	0178
C1160-O	0178
C1160K(2S)	0178
C1160L(2S)	0178
C1161(2S)	0178
C1162(2S)	0563
C1162A(2S)	0563
C1162B(2S)	0563
C1162C(2S)	0563
C1162C(JAPAN)	0563
C1162CP(2S)	0563
C1162D(2S)	0563
C1162W	0563
C1162WBP(2S)	0563
C1162WT	0563
C1162WTB(2S)	0563
C1162WTC(2S)	0563
C1163	0276
C1164	0418
C1164-O	0418
C1164-R	0418
C1165	2041
C1165(TRANS)	0689
C1165A	OEM
C1165B	OEM
C1166	0155
C1166(TRANS)	0733
C1166-GR	0733
C1166-O	0733
C1166-R	0733
C1166-Y	0733
C1166A	OEM
C1166A(TRANS)	0733
C1166B	OEM
C1166B(TRANS)	0733
C1166C	0733
C1166E	0733
C1166F	0733
C1166G	0733
C1166GN	0733
C1166H	0733
C1166K	0733
C1166L	0733
C1166M	0733
C1166O(2S)	0733
C1166OR	0733
C1166U	0733
C1166Y(2S)	0733
C1167(2S)	0313
C1168(2S)	0142
C1168X(2S)	0142
C1169	6746
C1169(TRANS)	0630
C1169A	OEM
C1169B	OEM
C1170(2S)	1150
C1170-OR	1150
C1170A(2S)	1150
C1170B(2S)	1150
C1170BFA-2	1150
C1170C	1150
C1170D	1150
C1170E	1150
C1170F	1150
C1170FA-2	1150
C1170G	1150
C1170GN	1150
C1170H	1150
C1170J	1150
C1170K	1150
C1170L	1150
C1170M	1150
C1170OR	1150
C1170R	1150
C1170X	1150
C1170Y	1150
C1171(2S)	0065
C1172(2S)	0313
C1172A(2S)	0313
C1172B(2S)	0313
C1172C	0313
C1172D	0313
C1172E	0313
C1172F	0313
C1172G	0313
C1172GN	0313
C1172H	0313
C1172J	0313
C1172K	0313
C1172L	0313
C1172M	0313
C1172R	0313
C1172X	0313
C1172Y	0313
C1173(2S)	0042
C1173-0(2S)	0042
C1173-B	0042
C1173-C	0042
C1173-GR(2S)	0042
C1173-O(2S)	0042
C1173-R(2S)	0042
C1173-Y(2S)	0042
C1173A	0042
C1173B	0042
C1173GR	0042
C1173R(2S)	0042
C1173X(2S)	0042
C1173XO	0042
C1173XY	0042
C1173Y	0042
C1174(2S)	0313
C1174-OR	0313
C1174A	0313
C1174B	0313
C1174C	0313
C1174D	0313
C1174E	0313
C1174F	0313
C1174G	0313
C1174GN	0313
C1174H	0313
C1174J	0313
C1174K	0313
C1174L	0313
C1174M	0313
C1174OR	0313
C1174R	0313
C1174X	0313
C1174Y	0313
C1175(2S)	0191
C1175C(2S)	0191
C1175CTV	0191
C1175D(2S)	0191
C1175E(2S)	0191
C1175F(2S)	0191
C1176	2170
C1177	1198
C1178A	1977
C1180	0007
C1182	0007
C1182B(2S)	0007
C1182C(2S)	0007
C1182D(2S)	0007
C1184(2S)	0003
C1184A(2S)	0003
C1184B(2S)	0003
C1184C(2S)	0003
C1184D(2S)	0003
C1184E(2S)	0003
C1185(2S)	0074
C1185A(2S)	0074
C1185B(2S)	0074
C1185C(2S)	0074
C1185K(2S)	0074
C1185L(2S)	0074
C1185M(2S)	0074
C1187(2S)	0224
C1187R(KOREA)	0224
C1188	0224
C1189	0224
C1189L	0224
C1190	1233
C1195(2S)	0103
C1195A	0103
C1195B	0103
C1195C	0103
C1195D	0103
C1195E	0103
C1195F	0103
C1195FA-1	0103
C1195FA-5	0103
C1195G	0103
C1195GN	0103
C1195GR	0103
C1195H	0103
C1195J	0103
C1195K	0103
C1195L	0103
C1195M	0103
C1195O	0103
C1195OR	0103
C1195R	0103
C1195X	0103
C1195Y	0103
C1199	0418
C1204(2S)	0155
C1204B(2S)	6035
C1204C(2S)	0155
C1204D(2S)	0155
C1205(2S)	2183
C1205A(2S)	2183
C1205B(2S)	2183
C1205C(2S)	2183
C1206	0086
C1206B	0086
C1209(2S)	0076
C1209C(2S)	0076
C1209D	0076
C1209E	0076
C1210(2S)	0951
C1210Y	0951
C1211(2S)	0155
C1212(2S)	0563
C1212A(2S)	0563
C1212AB(2S)	0563
C1212ABWT	0563
C1212AC(2S)	0563
C1212ACWT	0563
C1212B(2S)	0563
C1212C(2S)	0563
C1213	OEM
C1213(2S)	0191
C1213-OR	0191
C1213A	0191
C1213A(2S)	0191
C1213A(JAPAN)	0191
C1213AA(2S)	0191
C1213AB(2S)	0191
C1213AC	0191
C1213AC(2S)	0191
C1213AD(2S)	0191
C1213AK(2S)	0191
C1213AKB(2S)	0191
C1213AKC(2S)	0191
C1213AKD(2S)	0191
C1213B(2S)	0191
C1213BC(2S)	0191
C1213C	0191
C1213C(2S)	0191
C1213C(JAPAN)	0191
C1213CD(2S)	0191
C1213D(2S)	0191
C1213D-24	0191
C1213E	0191
C1213F	0191
C1213G	0191
C1213GN	0191
C1213H	0191
C1213J	0191
C1213K	0191
C1213L	0191
C1213M	0191
C1213OR	0191
C1213Q	0191
C1213X	0191
C1213Y	0191
C1214(2S)	0155
C1214A(2S)	0155
C1214B(2S)	0155
C1214C(2S)	0155
C1214D(2S)	0155
C1215(2S)	0127
C1215C	0127
C1215D	0127
C1215E	0127
C1215F	0127
C1215G	0127
C1215GN	0127
C1215H	0127
C1215J	0127
C1215K	0127
C1215L	0127
C1215M	0127
C1215OR	0127
C1215R	0127
C1215X	0127
C1215Y	0127
C1216	0224
C1217(2S)	0285
C1218(2S)	0086
C1219	0155
C1220	0155
C1220(TRANS)	0155
C1220-003	0155
C1220A	OEM
C1220A(TRANS)	0155
C1220AP	0155
C1220AQ	0155
C1220AR	0155
C1220B	OEM
C1220E(2S)	0155
C1220P	0155
C1220Q	0155
C1220R	0155
C1221	OEM
C1221A	OEM
C1221B	OEM
C1222	0111
C1222(TRANS)	2190
C1222A	0111
C1222A(TRANS)	2190
C1222B	0111
C1222B(TRANS)	2190
C1222C(2S)	2190
C1222D(2S)	2190
C1222E	2190
C1222H	2190
C1222U	2190
C1223	0418
C1224	0338
C1225	OEM
C1225A	OEM
C1225B	OEM
C1226	0219
C1226(TRANS)	0219
C1226-O(2S)	0219
C1226A	0219
C1226A(TRANS)	0219
C1226AC(2S)	0219
C1226ACF	0219
C1226AF	0219
C1226AP(2S)	0219
C1226AQ(2S)	0219
C1226AR	0219
C1226AR(2S)	0219
C1226ARL	0219
C1226B	0219
C1226B(TRANS)	0219
C1226BL	0219
C1226C(2S)	0219
C1226CF(2S)	0219
C1226D	0219
C1226E	0219
C1226F(2S)	0219
C1226G	0219
C1226H	0219
C1226L	0219
C1226OR	0219
C1226P(2S)	0219
C1226Q(2S)	0219
C1226R(2S)	0219
C1226RL	0219
C1226RLP	0219
C1226RLQ	0219
C1226RLR	0219
C1226Y	0219
C1227	0074
C1228	1966
C1228HA	OEM
C1229	1966
C1229(TRANS)	0168
C1229A	OEM
C1229B	OEM
C1229Q	0168
C1229R	0168
C1230	1966
C1231	0861
C1235	0142
C1235A	0142
C1235AL	0142
C1235AM	0142
C1235G	0142
C1236	0224
C1237(2S)	0936
C1237E(2S)	0936
C1239	1908
C1240	0224
C1241	1198
C1241A	1198
C1243(2S)	2016
C1243-24(2S)	2016
C1243C1(2S)	2016
C1243C2(2S)	2016
C1243D1(2S)	2016
C1243D2(2S)	2016
C1244(2S)	0079
C1246	0155
C1246A	0155
C1246R	0155
C1246S	0155
C1246T	0155
C1247	0155
C1247A	0155
C1247AF	0155
C1252	2208
C1252H2	2269
C1253	0418
C1253H2	1604
C1254	0007
C1256	2212
C1257	2170
C1258	1233
C1259	1974
C1260(2S)	0259
C1266	0086
C1267	2170
C1274	0016
C1275	0259
C1276	0111
C1277	0324
C1278(2S)	0861
C1278S(2S)	0861
C1279(2S)	0861
C1279S(2S)	0861
C1280(2S)	0400
C1280A(2S)	0400
C1280AS(2S)	0400
C1280S(2S)	0400
C1281	0079
C1282	0324
C1283	0079
C1285	0155
C1290	2229
C1292	0642
C1293	1066
C1293A	1066
C1293B(2S)	1066
C1293C	1066
C1293D	1066
C1295(2S)	1150
C1295-O	1150
C1295-O(2S)	1150
C1296	0065
C1298	1974
C1300	0065
C1303(2S)	1865
C1304(2S)	1504
C1306(2S)	0839
C1306A	0839
C1306I(2S)	0839
C1306K	0839
C1307(2S)	0836
C1307-1	0836
C1308(2S)	0313
C1308K(2S)	0313
C1308L	0313

DEVICE TYPE	REPL CODE
C1308N	0313
C1309(2S)	0065
C1310	0286
C1311	0286
C1312(2S)	1221
C1312A	1221
C1312BC	1221
C1312C	1221
C1312D	1221
C1312E	1221
C1312F(2S)	1221
C1312G(2S)	1221
C1312GN	1221
C1312H(2S)	1221
C1312J	1221
C1312K	1221
C1312L	1221
C1312M	1221
C1312OR	1221
C1312R	1221
C1312X	1221
C1312Y(2S)	1221
C1313(2S)	0111
C1313B	0111
C1313Y(2S)	0111
C1314	1974
C1315	0233
C1316(2S)	2245
C1316-3	2245
C1317(2S)	0155
C1317-OR	0155
C1317-R	0155
C1317A	0155
C1317BC	0155
C1317C	0155
C1317E	0155
C1317G	0155
C1317GR	0155
C1317L	0155
C1317NC	0155
C1317OR	0155
C1317Q(2S)	0155
C1317R(2S)	0155
C1317S(2S)	0155
C1317T(2S)	0155
C1317V	0155
C1317Y	0155
C1318(2S)	0155
C1318B	0155
C1318C(2S)	0155
C1318E	0155
C1318F	0155
C1318G	0155
C1318GN	0155
C1318H	0155
C1318J	0155
C1318K	0155
C1318L	0155
C1318M	0155
C1318P	0155
C1318PR	0155
C1318Q(2S)	0155
C1318QP	0155
C1318QR	0155
C1318R(2S)	0155
C1318Y	0155
C1319	0007
C1319C	0007
C1320	0224
C1320A	0224
C1320B	0224
C1320C	0224
C1320D	0224
C1320E	0224
C1320F	0224
C1320G	0224
C1320GN	0224
C1320H	0224
C1320J	0224
C1320K	0224
C1320L	0224
C1320M	0224
C1320OR	0224
C1320R	0224
C1320X	0224
C1320Y	0224
C1321	0007
C1321Q2	0007
C1321Q3	0007
C1321Q4	0007
C1321Q5	0007
C1322	0177
C1324	0418
C1324C	0418
C1325(2S)	0891
C1325A(2S)	0891
C1325AK	0891
C1325AL	0891
C1326	0493
C1327(2S)	1221
C1327FS(2S)	1221
C1327R	1221
C1327S	1221
C1327T	1221
C1327TU	1221
C1327TV	1221
C1327U(2S)	1221
C1328(2S)	0111
C1328T(2S)	0111
C1328U(2S)	0111
C1330	0218
C1330(TRANS)	0218
C1330A	0218
C1330B	0218
C1330P	1808
C1331	2525
C1331(TRANS)	0007
C1331A	OEM
C1331B	OEM
C1332	2414
C1332A	OEM
C1332B	OEM
C1335	0155
C1335(TRANS)	1221
C1335-OR	1221
C1335A	0111
C1335A(TRANS)	1221
C1335B	0111
C1335B(TRANS)	1221
C1335C(2S)	1221
C1335D(2S)	1221
C1335E(2S)	1221
C1335F(2S)	1221
C1335G	1221
C1335GN	1221
C1335H	1221
C1335J	1221
C1335K	1221
C1335L	1221
C1335M	1221
C1335OR	1221
C1335R	1221
C1335X	1221
C1335Y	1221
C1336	OEM
C1336(TRANS)	0144
C1336A	OEM
C1336B	OEM
C1336JK	0144
C1337	1420
C1339	OEM
C1339A	OEM
C1339B	OEM
C1340	2170
C1342(2S)	0127
C1342-OR	0127
C1342A(2S)	0127
C1342B(2S)	0127
C1342C(2S)	0127
C1342D	0127
C1342E	0127
C1342F	0127
C1342G	0127
C1342GN	0127
C1342J	0127
C1342K	0127
C1342L	0127
C1342M	0127
C1342OR	0127
C1342R	0127
C1342X	0127
C1342Y	0127
C1343(2S)	0171
C1343-05	0171
C1343B(2S)	0171
C1343BL	0171
C1343D	0171
C1343E	0171
C1343F	0171
C1343G	0171
C1343GN	0171
C1343G-R	0171
C1343GR	0171
C1343H(2S)	0171
C1343J	0171
C1343K	0171
C1343L	0171
C1343M	0171
C1343O	0171
C1343OR	0171
C1343R	0171
C1343X	0171
C1343Y	0171
C1344(2S)	0552
C1344C(2S)	0552
C1344D(2S)	0552
C1344E(2S)	0552
C1344F(2S)	0552
C1345(2S)	0552
C1345C(2S)	0552
C1345D	0552
C1345E	0552
C1345F(2S)	0552
C1345KD	0552
C1345KE	0552
C1345KF	0552
C1346(2S)	0218
C1346R(2S)	0218
C1346S(2S)	0218
C1347(2S)	0218
C1347A	0218
C1347B	0218
C1347C	0218
C1347D	0218
C1347F	0218
C1347G	0218
C1347L	0218
C1347Q(2S)	0218
C1347R(2S)	0218
C1347RQ	0218
C1347X	0218
C1347Y	0218
C1348(2S)	0223
C1348-1	0223
C1348-2	0223
C1348-3	0223
C1349	0861
C1351	0630
C1352	0079
C1353	0630
C1353H2	1604
C1356C	OEM
C1356C2	0865
C1358(2S)	2051
C1358A	2051
C1358K(2S)	2051
C1358K1	2051
C1358K2	2051
C1358K3	2051
C1358L	2051
C1358M	2051
C1359(2S)	0151
C1359A(2S)	0151
C1359B(2S)	0151
C1359BC	0151
C1359C(2S)	0151
C1359Q	0151
C1360	1663
C1360(2S)	1663
C1360A	1663
C1361(2S)	0155
C1362(2S)	0155
C1362(47)	0155
C1363(2S)	0076
C1363CA	0683
C1364(2S)	0076
C1364-6	0076
C1364-8	0076
C1364-OR	0076
C1364A(2S)	0076
C1364B	0076
C1364C	0076
C1364D	0076
C1364E	0076
C1364H	0076
C1364K	0076
C1364L	0076
C1364M	0076
C1364OR	0076
C1364P	0286
C1364R	0076
C1364X	0076
C1364Y	0076
C1365	0418
C1366C2	2533
C1367(2S)	0003
C1367A(2S)	0003
C1368(2S)	1790
C1368B	1790
C1368C(2S)	1790
C1368D(2S)	1790
C1372	0155
C1372Y(2S)	0155
C1373	0111
C1373HA	2026
C1374	2272
C1374H	2272
C1375	0111
C1375H	0111
C1376	0079
C1376H	0079
C1377(2S)	0836
C1380	0016
C1380-BL	0016
C1380-GR	0016
C1380A	0016
C1380A-BL	0016
C1380A-GR	0016
C1381	1591
C1382(2S)	0563
C1382-0(2S)	0563
C1382-O	0563
C1382-Y	0563
C1383	0018
C1383-QR	0018
C1383NC	0018
C1383P(2S)	0018
C1383Q	0018
C1383Q(2S)	0018
C1383R(2S)	0018
C1383R/494	0595
C1383RS	0018
C1383S	0018
C1383S(2S)	0018
C1383X(2S)	0018
C1384	0018
C1384(2S)	0018
C1384-OR	0018
C1384A	0018
C1384B	0018
C1384C	0018
C1384D	0018
C1384E	0018
C1384F	0018
C1384G	0018
C1384GN	0018
C1384H	0018
C1384J	0018
C1384K	0018
C1384L	0018
C1384M	0018
C1384NC	0018
C1384OR	0018
C1384Q(2S)	0018
C1384R	0018
C1384R(2S)	0018
C1384R(JAPAN)	0018
C1384S	0018
C1384S(2S)	0018
C1384X	0018
C1384Y	0018
C1385	0016
C1385H	0016
C1386	0086
C1386H	0086
C1387	0418
C1388	0086
C1390	0597
C1390A	0127
C1390IW	0597
C1390JX	0597
C1390W(2S)	0597
C1390WH(2S)	0597
C1390WX(2S)	0597
C1390X(2S)	0597
C1390XK(2S)	0597
C1390YM(2S)	0597
C1391(2S)	0142
C1391VL(2S)	0142
C1393(2S)	0224
C1393CA	OEM
C1393K	0224
C1393L	0224
C1393M	0224
C1394(2S)	0224
C1395	0007
C1396(2S)	0007
C1398(2S)	0042
C1398P	0042
C1398Q(2S)	0042
C1399	0563
C1399E	0563
C1400	0530
C1402(2S)	0177
C1403	0543
C1403A	0543
C1406	2287
C1406B40C3700-2200	OEM
C1406HA	3254
C1406M	OEM
C1406Q	2287
C1407(2S)	2287
C1407B	2287
C1407Q	2287
C1407X	2287
C1409	0392
C1409A	0392
C1409AA	0392
C1409AB	0392
C1409AC	0392
C1409B	0392
C1409C	0392
C1410	0392
C1410A	0392
C1410AA	0392
C1410AB	0392
C1410AC	0392
C1410B	0392
C1410C	0392
C1411	0016
C1412	0861
C1412B80C3700-2200	OEM
C1412M	OEM
C1413(2S)	0065
C1413A(2S)	0065
C1413AH	0065
C1413B	0065
C1413C	0065
C1413D	0065
C1413E	0065
C1413F	0065
C1413G	0065
C1413GN	0065
C1413K	0065
C1413L	0065
C1413R	0065
C1413X	0065
C1413Y	0065
C1415	1085
C1415-05	1085
C1416(2S)	0016
C1416A	0016
C1417(2S)	0127
C1417C	0127
C1417D(1)	0127
C1417D(U)	0127
C1417F(2S)	0127
C1417V(2S)	0127
C1417VF	0127
C1417VW(2S)	0127
C1417W(2S)	0127
C1418(2S)	0042
C1419(2S)	2292
C1419-C	2292
C1419B(2S)	2292
C1419C(2S)	2292
C1424(2S)	0259
C1426	0630
C1427	0224
C1428	0111
C1429(2S)	0551
C1429-1(2S)	0551
C1429-2(2S)	0551
C1430	0622
C1431	1027
C1431-1	1027
C1431-2	1027
C1433	0642
C1434	1852
C1436	1852
C1437	0280
C1437-NC	0261
C1438	0079
C1439	0079
C1440	1852
C1441	1852
C1444	0423
C1445	0423
C1446(2S)	0643
C1446-0	0643
C1446LB(2S)	0643
C1446P	0643
C1446Q(2S)	0643
C1447	0643
C1447(JAPAN)	0643
C1447-0(2S)	0643
C1447-0	0643
C1447FA-2	0643
C1447GL	0643
C1447GL3	0643
C1448(2S)	1284
C1448A(2S)	1284
C1448C	1284
C1448LB	1284
C1449(2S)	0161
C1449CB	0161
C1449M	0161
C1450	1027
C1450S(2S)	1027
C1451	0715
C1452	0715
C1453	0111
C1454	0220
C1454-0	0220
C1454-O	0220
C1456	0142
C1457	0861
C1463	0642
C1464	0630
C1466	0065
C1467	0065
C1468	1966
C1469	0363
C1469A	0363
C1470	OEM
C1470A	OEM
C1470B	OEM
C1471	OEM
C1471A	OEM
C1471B	OEM
C1472	6777
C1472(TRANS)	0400
C1472A	OEM
C1472B	OEM
C1472K(2S)	0400
C1472KA	0400
C1472KB	0400
C1473	0261
C1473-RNC	0261
C1473A	0261
C1473AH	0261
C1473AQ	0261
C1473AR	0261
C1473CN	0261
C1473NC	0261
C1473NE	0261
C1473P	0261
C1473Q	0261
C1473Q(JAPAN)	0261
C1473QNC	0261
C1473R	0261
C1473R(JAPAN)	0261
C1473RNC	0261
C1474(2S)	0018
C1474-3(2S)	0018
C1474-4(2S)	0018
C1474HA	5256
C1474J	0018
C1474S	0018
C1475	1978
C1475(13)	1978
C1475(TRANS)	1978
C1475-13	1978
C1475A	1978
C1475A(TRANS)	1978
C1475B	OEM
C1475D	1978
C1475H	1978
C1475K	1978
C1476	OEM
C1476A	OEM
C1476B	OEM
C1477(2S)	0363
C1478	0224
C1478A	0224
C1479	0418
C1480CA	3365
C1481CA	3884
C1485	0715
C1490	2170
C1490HA	4550
C1491	1198
C1493	0715
C1493NC	0715
C1493P4	0715
C1494	1974
C1500	OEM
C1501(2S)	0276
C1501P	0276
C1501Q(2S)	0276
C1501R(2S)	0276
C1502L	0168
C1504	2096
C1505(1)	0955
C1505(1)K	0955
C1505(1)L	0955
C1505(2S)	0955
C1505-1	0955
C1505I(2S)	0955
C1505K(2S)	0955
C1505L(2S)	0955
C1505LA(2S)	0955
C1506(2S)	0168
C1506K	0168
C1506L	0168
C1506M	0168
C1507	0955
C1507(2S)	0955
C1507-K	0955
C1507-O	0955
C1507-O(KOREA)	0168
C1507-Y	0955
C1507H	0955
C1507J	0955
C1507K(2S)	0955
C1507L(2S)	0955
C1507L(JAPAN)	0955
C1507M(2S)	0955
C1507Y	0955
C1509(2S)	0324
C1509-O	0324
C1509NC	0324
C1509O	0324
C1509P	0324
C1509Q(2S)	0324
C1509Y	0324
C1510	OEM
C1513	0418
C1513HA	3159
C1514	1085
C1514(2S)	1085
C1514(JAPAN)	1085
C1514-05	1085
C1514-14	1085
C1514-15	1085
C1514BK	1085
C1514BVC	1085
C1514CS-CVC	1085
C1514CVC	1085
C1514VC	1085
C1515	0715
C1515AX	0715
C1515C	0715
C1515K	0715
C1516	1946
C1516K	1946
C1516K-B	1946
C1516KC	1946
C1517	1708
C1517AK	1708
C1517AKB	1708
C1517AKC	1708
C1518	0018
C1518NC	0018
C1519	0168
C1520(1)K(JAPAN)	0168
C1520(2S)	0168
C1520-1	0168
C1520-O(KOREA)	0168
C1520C	0168
C1520I(2S)	0168
C1520K(2S)	0168
C1520K-1	0168
C1520L(2S)	0168
C1520L-1	0168
C1520M(2S)	0168
C1520M-1	0168
C1521	2359
C1521K(2S)	2359
C1521L(2S)	2359
C1522	1085
C1530CA	OEM
C1537(2S)	0111
C1537-O	0111
C1537B(2S)	0111
C1538(2S)	0111
C1538S(2S)	0111
C1538SA(2S)	0111
C1539	0079
C1540	0079
C1541	0155
C1542(2S)	0111
C1543	0224
C1544	0079
C1545	0400
C1546	0400
C1547(2S)	0007
C1550	0276
C1550-1	0276
C1552	2382
C1553	0007
C1553A	0007
C1555A	OEM
C1556	0698
C1565	0563
C1566(2S)	0276
C1566F	0276
C1567	0563
C1567A	0563
C1567R	0563
C1568	1790
C1568-R	1790
C1568R	1790
C1569	0643
C1569(2S)	0643
C1569(KOREA)	1085
C1569-0(2S)	0643
C1569-O	0643
C1569BK	0643
C1569FA-1	0643
C1569FA-3	0643
C1569FA-5	0643
C1569K	0643
C1569LB	0643
C1569LBO	0643
C1569LBR(2S)	0643
C1569LBY(2S)	0643
C1569O	0643
C1569R(2S)	0643
C1569Y	0643
C1570	0111
C1570FL	0111
C1570GL	0111
C1570HL	0111
C1570LH	0111
C1571	0111
C1571G	0111
C1571H	0111
C1571L	0111
C1573(2S)	0261
C1573(TO126)	0261
C1573A	0261
C1573AR	0261
C1573BR	0261
C1573P(2S)	0261
C1573P(TO126)	0261
C1573PQ	0261
C1573PQ(TO126)	0261
C1573Q	0261
C1573Q(TO126)	0261
C1573QNC	0261
C1573R	0261
C1574	0103
C1576	0363
C1577	0363
C1577A	0363
C1578	0363
C1579	0363
C1580	0363
C1581	2312
C1582	2312
C1583	2411
C1583F	2411
C1583G	2411
C1583H	2411
C1584	0543
C1585	2414
C1585F	2414
C1585H	2414
C1586	2414
C1588(2S)	0418
C1589	2050
C1596	0079
C1601	0224
C1605A	1977
C1606	0560
C1607	0330
C1608	0630
C1610	0620
C1617	2437
C1618	2438
C1618B	2438
C1619	0867
C1619A	0867
C1621	0144
C1621-0	0144
C1621B2	0144
C1621B3	0144
C1621B4	0144
C1622	0007
C1622A	0007
C1622D6	0007
C1622D7	0007
C1622D8	0007
C1622O	0007
C1622Y	0007
C1623	0724
C1623L3	0724
C1623L4	0724
C1623L5	0724
C1623L6	0724
C1623L7	0724
C1624	0236

If replacement code is OEM, contact original manufacturer for replacement.

DEVICE TYPE	REPL CODE
C1624-0	0236
C1624-Y	0236
C1625	0236
C1625-0	0236
C1625-O	0236
C1625-Y	0236
C1625Y	0236
C1625YLBGH	0236
C1625YLBGL1	0236
C1626	0236
C1626-0	0236
C1626-Y	0236
C1626Y	0236
C1627	0733
C1627-0	0733
C1627-0(JAPAN)	0733
C1627-Y	0733
C1627-YFA	0733
C1627A	0733
C1627AK	1386
C1627A-Y	0733
C1627AY	0733
C1627AY(KOREA)	1386
C1627Y	0733
C1627Y(JAPAN)	0733
C1628	0264
C1628-0	0264
C1628-Y	0264
C1629	0130
C1629A	0130
C1629AO	0130
C1629AR	0130
C1629M	0130
C1629R	0130
C1630	0233
C1631	0018
C1632	0286
C1633(2S)	2456
C1634	2459
C1635	0086
C1636	2460
C1637	0111
C1638	0418
C1639(2S)	0016
C1641	0016
C1641Q	0016
C1641R	0016
C1647	0111
C1647Q	0111
C1647RY	0111
C1648(2S)	0111
C1648E	0111
C1648EF	0111
C1648SH	0111
C1652	2471
C1653	0861
C1653N2	0861
C1653N3	0861
C1653N4	0861
C1654	0861
C1654N5	0861
C1654N6	0861
C1654N7	0861
C1663	0264
C1663H	0264
C1664	0629
C1664A	0629
C1664O	0629
C1664OR	0629
C1664Q	0629
C1664R	0629
C1665	0259
C1666	0363
C1667(2S)	0103
C1668	1977
C1669	0392
C1669-0	0392
C1669-R	0392
C1669-Y	0392
C1670(2S)	2483
C1670H	2483
C1670J	2483
C1670JA	2483
C1670JD	2483
C1670JW	2483
C1672	2485
C1674	1066
C1674(2S)	1066
C1674-0	0144
C1674AL	1066
C1674K	1066
C1674K(2S)	1066
C1674L	1066
C1674L(2S)	1066
C1674L(JAPAN)	1066
C1674M(2S)	1066
C1674R(KOREA)	0144
C1675(2S)	0076
C1675B	0076
C1675K(2S)	0076
C1675K(JAPAN)	0076
C1675L(2S)	0076
C1675M(2S)	0076
C1676	1974
C1678(2S)	2495
C1678E	2495
C1679(2S)	0836
C1680	OEM
C1680A	OEM
C1680B	OEM
C1681	0530
C1681(TRANS)	1757
C1681-BL	1757
C1681-GR	1757
C1681A	OEM
C1681A(TRANS)	1757
C1681B	0111
C1681B(TRANS)	1757
C1681B6	1757
C1681BL(2S)	1757
C1681C	1757
C1681F	1757
C1681G	1757
C1681GN	1757
C1681GR	1757
C1681H	1757
C1681J	1757
C1681K	1757
C1681L	1757
C1681M	1757
C1681O	1757
C1681OR	1757
C1681R	1757
C1681V	1757
C1682	0111
C1682(TRANS)	0111
C1682-BL	0111
C1682-GR	0111
C1682A	OEM
C1682B	OEM
C1682V	0111
C1683	0392
C1683-0	0392
C1683-Q	0392
C1683A	0392
C1683LA(2S)	0392
C1683P(2S)	0392
C1683Q	0392
C1683R(2S)	0392
C1684(2S)	0155
C1684BL	0155
C1684K	0155
C1684P	0155
C1684Q	0155
C1684R	0155
C1684S	0155
C1684T	0155
C1685	0286
C1685(TRANS)	0286
C1685-0	0286
C1685-P	0286
C1685-Q	0286
C1685-R	0286
C1685A	OEM
C1685B	OEM
C1685C	0286
C1685CR	0286
C1685P(2S)	0286
C1685Q(2S)	0286
C1685Q(JAPAN)	0286
C1685QR	0286
C1685S(2S)	0286
C1685T	0286
C1686(2S)	0224
C1686B	0224
C1686V	0224
C1687(2S)	0224
C1688(2S)	0016
C1689	2506
C1706H	0715
C1707AH	0016
C1707AHB	0016
C1707H	0016
C1707HB	0016
C1707HC	0016
C1707HD	0016
C1708	0530
C1708A	0530
C1709	0079
C1717	0630
C1719	0861
C1720	0861
C1721	0018
C1722	1085
C1722BK	1085
C1722BKS	1085
C1722S	1085
C1723	1085
C1723-02	1085
C1723A	1085
C1724	0630
C1727	2525
C1728(2S)	1173
C1728-3(2S)	1173
C1728D(2S)	1173
C1728F	1173
C1728H	1173
C1729	2526
C1730	0127
C1730K	0127
C1730L	0127
C1734	0016
C1734H	0016
C1735	0286
C1739(2S)	0016
C1740	0151
C1740C	0151
C1740D	0151
C1740E(JAPAN)	0151
C1740L	0151
C1740P	0151
C1740Q	0151
C1740QH	0151
C1740QJ	0151
C1740QR	0151
C1740R	0151
C1740R(JAPAN)	0151
C1740RE	0151
C1740RH	0151
C1740RL	0151
C1740RM	0151
C1740RN(JAPAN)	0151
C1740RS	0151
C1740RZ	0151
C1740S	0151
C1740S(JAPAN)	0151
C1740SEJ	0151
C1740SRJ(JAPAN)	0151
C1740SRS	0151
C1740SSJ	0151
C1740SSK	0151
C1741	1513
C1746A	0079
C1747	2525
C1748	0187
C1749	1085
C1749XD	1085
C1752	0715
C1753	0715
C1755(2S)	1085
C1755C(2S)	1085
C1755D	1085
C1756(2S)	0643
C1756-0	0643
C1756C(2S)	0643
C1756D(2S)	0643
C1756D(JAPAN)	0643
C1756E	0643
C1756K(2S)	0643
C1756M	0643
C1757(2S)	1085
C1758	0285
C1760(2S)	1946
C1760-2	1946
C1760-3(2S)	1946
C1760H	1946
C1760K	1946
C1761	1946
C1761KJ	1946
C1761LE	1946
C1761MC	1946
C1764	2545
C1765	0630
C1766	0380
C1766B	0380
C1768	2548
C1769	2550
C1770	0079
C1771	0016
C1773	2556
C1774	0079
C1775	0530
C1775A	0530
C1775AE	0530
C1775E	0530
C1775F	0530
C1776	0079
C1777	2556
C1778	0224
C1779	0224
C1780	0086
C1781	0016
C1781H	0016
C1781HA	0016
C1781HB	0016
C1781HC	0016
C1782	0867
C1783	0867
C1784	0543
C1785	2414
C1786	2414
C1787	1270
C1788	0155
C1788R	0155
C1789	0007
C1790	0259
C1790JD	0259
C1791	0142
C1800	OEM
C1800A	OEM
C1800B	OEM
C1801	0144
C1801(TRANS)	0259
C1801A	OEM
C1801B	OEM
C1802	0144
C1802(TRANS)	0224
C1802A	OEM
C1802B	OEM
C1807	0493
C1810	2598
C1810HU	2598
C1810J	2598
C1810JL	2598
C1811	0261
C1811-1	0261
C1811-2	0261
C1811KC	0261
C1812	OEM
C1813	0016
C1815	0076
C1815-0	0076
C1815-BL	0076
C1815-GR	0076
C1815-GRFA	0076
C1815-O	0076
C1815-O(JAPAN)	0076
C1815-O(KOREA)	0286
C1815-OFA	0076
C1815-Y	0076
C1815-YFA	0076
C1815BL	0076
C1815BL(KOREA)	0076
C1815FA	0076
C1815G	0076
C1815GR	0076
C1815GR(JAPAN)	0076
C1815GR(KOREA)	0076
C1815O	0076
C1815Y	0076
C1815Y(JAPAN)	0076
C1815Y(KOREA)	0076
C1816(2S)	2495
C1816H	2495
C1816HL	2495
C1817	0836
C1818	0177
C1819	0643
C1819M	0643
C1819ML	0643
C1819MR	0643
C1820	0233
C1826	0423
C1826(JAPAN)	0423
C1826-0	0423
C1826-O	0423
C1826-Y	0423
C1826D	0423
C1826G	0423
C1826O	0423
C1826P	0423
C1826Q	0423
C1826R	0423
C1826Y	0423
C1827	0236
C1828(2S)	2096
C1828B	2096
C1828R	2096
C1829	2620
C1829C	2620
C1830	0543
C1831	0561
C1832	2626
C1833	0286
C1834	0286
C1836	0936
C1840	0111
C1840E	0111
C1840F	0111
C1840P	0111
C1841	0530
C1841E	0530
C1841F	0530
C1841P	0530
C1841U	0530
C1842	0111
C1842E	0111
C1843	0111
C1844	0111
C1845	0530
C1845E	0530
C1845Q	0530
C1845R	0530
C1845S	0530
C1846	0786
C1846-R	0786
C1846-S	0786
C1846B	0786
C1846P	0786
C1846Q	0786
C1846QRS	0786
C1846R	0786
C1846S	0786
C1847	0563
C1847-0	0563
C1847A	0563
C1847B	0563
C1847C	0563
C1847D	0563
C1847E	0563
C1847F	0563
C1847G	0563
C1847GN	0563
C1847LG	0563
C1847P	0563
C1847Q	0563
C1847R	0563
C1847V	0563
C1847VG	0563
C1847VQ	0563
C1847VR	0563
C1847X	0563
C1847Y	0563
C1848	0219
C1848P	0219
C1848Q	0219
C1848R	0219
C1848V	0219
C1849	0079
C1850	0079
C1851	0079
C1852	0079
C1853	1220
C1853B	1220
C1853C	1220
C1854	0286
C1854C	0286
C1854R	0286
C1854S	0286
C1855	0224
C1856	0224
C1856-02	0224
C1856M	0224
C1857	0079
C1859	0630
C1860	2650
C1861	2650
C1862	0187
C1864	0429
C1865	0728
C1866	0620
C1867	1966
C1868	1966
C1869	0620
C1870	2414
C1870CA-001	OEM
C1870CA-002	OEM
C1870CA-003	OEM
C1871A	1966
C1871CU	OEM
C1875	2660
C1875K	2660
C1875L	2660
C1875R	2660
C1881K	1212
C1885	0261
C1886	0259
C1887	0259
C1888	0630
C1890	0530
C1890A	0530
C1890AD	0530
C1890AD(JAPAN)	0530
C1890A-E	0530
C1890AE	0530
C1890AE(JAPAN)	0530
C1890AF	0530
C1890D	0530
C1890E	0530
C1890F	0530
C1891	0223
C1892	2660
C1893	2668
C1893-0	2668
C1893FA-1	2668
C1894	0313
C1894K	0313
C1894N	0313
C1895	0313
C1896	0313
C1897	0086
C1898	0008
C1899	0008
C1900	0530
C1903	0563
C1904	0563
C1905	0955
C1905(1)	0955
C1905H	0955
C1905HLB	0955
C1906	0127
C1906(JAPAN)	0127
C1907	0127
C1908	2679
C1908E(2S)	2679
C1908H	2679
C1909(2S)	0936
C1909K	0936
C1909R	0936
C1913	0392
C1913A	0392
C1913AR	0392
C1914	0530
C1914A	0530
C1915	0530
C1919	0127
C1919C	0127
C1921	0261
C1921(JAPAN)	0261
C1921-03	0261
C1922	2668
C1923	2691
C1923-0	2691
C1923-O	2691
C1923A	2691
C1923BN	2691
C1923Y	2691
C1928	1978
C1929	0168
C1929LB	0168
C1929Q	0168
C1929R	0168
C1940	0079
C1941	1327
C1941-OY	1327
C1941K	1327
C1941L	1327
C1941M	1327
C1941M(JAPAN)	1327
C1942	2660
C1945	0836
C1946	OEM
C1946(TRANS)	2526
C1946A	OEM
C1946A(TRANS)	2526
C1946B	OEM
C1947	2702
C1949	OEM
C1949A	OEM
C1949B	OEM
C1951	1386
C1953	0563
C1954	0127
C1955	2702
C1956	0630
C1957(2S)	2710
C1957K	2710
C1959	0286
C1959-0	0286
C1959-0(KOREA)	0286
C1959-Y	0286
C1959-YFA	0286
C1959GR	0286
C1959Y	0286
C1959Y(JAPAN)	0286
C1959Y(KOREA)	0286
C1960	0286
C1962	0338
C1962-0	0338
C1963	1946
C1963-1	1946
C1963Y	1946
C1964(2S)	2495
C1964D	2495
C1965	2702
C1965A	2702
C1969	0836
C1969B	0836
C1969BH	0836
C1969H	0836
C1970	2719
C1971	2719
C1972	2720
C1973(2S)	0018
C1974(2S)	0836
C1974C	0836
C1975(2S)	0936
C1976	2725
C1978	2719
C1980	1270
C1980S	1270
C1980T	1270
C1982	0264
C1983	2058
C1983-0	2058
C1983Q	2058
C1983R	2058
C1983Y	2058
C1984	1212
C1985	0423
C1985(JAPAN)	0423
C1986	0482
C1987	0429
C1989	0007
C1990	0127
C1990B	0127
C1991	0079
C1992	0016
C1993	0016
C1994	0016
C1995	0016
C1996	0016
C1997	0016
C1998	0079
C1999	0111
C2000	0155
C2000(2S)	0286
C2000L	0286
C2001	1513
C2001-L	1513
C2001A	0286
C2001F	1513
C2001G	1513
C2001I	1513
C2001K	1513
C2001K(JAPAN)	1513
C2001L	1513
C2001L1	1513
C2001M	1513
C2001P	OEM
C2001P(HEP)	2867
C2002	0951
C2002K	0951
C2002L	0951
C2002L(JAPAN)	0951
C2002M	0951
C2002P	OEM
C2003	0286
C2003K	0286
C2003P	OEM
C2004P	OEM
C2005P	3429
C2006G	OEM
C2007G	1364
C2008G	OEM
C2009	0224
C2009G	OEM
C2009G(HEP)	OEM
C2010G	OEM
C2011L	0224
C2012	2757
C2014	1386
C2017	1852
C2018	1852
C2020	0936
C2021	1140
C2021-Q	1140
C2021L	1140
C2021Q	1140
C2021Q(JAPAN)	1140
C2021R	1140
C2021R(JAPAN)	1140
C2021RJ	1140
C2021RL(JAPAN)	1140
C2021RM(JAPAN)	1140
C2021S	1140
C2021ST(JAPAN)	1140
C2021Y	1140
C2022	0955
C2023	2613
C2024	1591
C2025	0259
C2026	1806
C2026K	1806
C2027	0313
C2028	0890
C2028B	0890
C2028B20	0890
C2028D	0890
C2028L	0890
C2029	0936
C2029-1	0936
C2029-3	0936
C2029B	0936
C2029B10	0936
C2029C	0936
C2034	2769
C2035	0079
C2036	0563
C2037	2770
C2043	0836
C2053	0079
C2055	0079
C2056	2702
C2057	0224
C2057-C	0224
C2057C	0224
C2057D	0224
C2057E	0224
C2057E1	0224
C2057F	0224
C2058S	0286
C2060	0018
C2060K	0018
C2060Q	0018
C2060Q(JAPAN)	0018
C2060R	0018
C2060R(JAPAN)	0018
C2061	0324
C2063	0582
C2063P	0582
C2067	1978
C2068	0643
C2068-GS-1	0643
C2068BKLB	0643
C2068FA-1	0643
C2068GS-1	0643
C2068I	0643
C2068LB	0643
C2068LBBK	0643
C2069	0016
C2070	0016
C2071	0276
C2073	1284
C2073(JAPAN)	1284
C2073(KOREA)	0392
C2073-4(JAPAN)	1284
C2073-B	1284
C2073-C	1284
C2073-LBGL2	1284
C2073-Z	1284
C2073A	1284
C2073AD	1284
C2073AD(JAPAN)	1284
C2073AE	1284
C2073B	1284
C2073BGS2	1284
C2073BGSC	1284
C2073C	1284
C2073D	1284
C2073E	1284
C2073FA-4	1284
C2073GL2	1284
C2073J	1284
C2073L	1284
C2073LB	1284
C2073LBGL	1284
C2073LBGL2	1284
C2073LBGS2	1284
C2073R	1284
C2073RD	1284
C2073RE	1284
C2073SO	1284

If replacement code is OEM, contact original manufacturer for replacement.

DEVICE TYPE	REPL CODE
C2073T	1284
C2073Y	1284
C2074	1984
C2074C	1984
C2074Y	1984
C2075	2495
C2076	2814
C2076B	2814
C2076C	2814
C2076CB	2814
C2076CD	2814
C2076D	2814
C2078	2495
C2079	1591
C2079E	1591
C2080	0161
C2081	2821
C2085	0643
C2085P	0643
C2085Q	0643
C2086	2825
C2087	0079
C2088	1562
C2091	1591
C2092	0936
C2093	0836
C2094	2526
C2098	0836
C2099	2835
C2100	2312
C2101	3578
C2101(2S)	1198
C2101-1	OEM
C2101-2	OEM
C2101A	OEM
C2101A-2	OEM
C2101A-4	OEM
C2102	1233
C2102(2S)	1233
C2102-2	OEM
C2102A	4973
C2102A2	OEM
C2102A4	OEM
C2102A6	OEM
C2103	1974
C2103A	1974
C2107	0079
C2107G3	0079
C2109	0079
C2110	0086
C2111	OEM
C2111(2S)	0016
C2111-1	OEM
C2111-2	OEM
C2112	OEM
C2112-2	OEM
C2115	2844
C2117	2702
C2118	2702
C2120	0866
C2120-0	0866
C2120-O	0866
C2120-Y	0866
C2120B125C3000-1800	OEM
C2120FA	0866
C2120M	OEM
C2120Y	0866
C2120Y(JAPAN)	0866
C2120Y(KOREA)	0866
C2120YFA	0866
C2121	0363
C2122	0363
C2122A	0363
C2123	0065
C2124	2847
C2125	2847
C2126	0178
C2127	0543
C2127A	0543
C2129	1978
C2130	1978
C2135	2860
C2137	1966
C2138	0168
C2139	1852
C2140	0074
C2140B250C3000-1800	OEM
C2140M	OEM
C2141	0264
C2141-10	0264
C2141KC	0264
C2141KD	0264
C2141LC	0264
C2141M	0264
C2141NE	0264
C2143	2844
C2145	0698
C2146	0936
C2151	1852
C2153	0079
C2166	2495
C2167	1284
C2167-Y	1284
C2167L	OEM
C2167L-70	OEM
C2167Y	1284
C2168	1284
C2168-0	1284
C2168-Y	1284
C2168Y	1284
C2178	0392
C2178Y	0392
C2180	1974
C2180B500C3000-1800	OEM
C2180M	OEM
C2181	2884
C2182	2506
C2184	0836
C2188	2052
C2189	0620
C2190	0065
C2191	0363
C2193	0715
C2194	0551
C2195	0836
C2196	0630
C2197	1974
C2198	0629
C2199	0620
C2200	0728
C2206	2223
C2206B	2223
C2206B40C3700-2200	OEM
C2206C	2223
C2206N	OEM
C2206Q	2223
C2206R	2223
C2207	0836
C2209	0161
C2209R	0161
C2210	0224
C2212	0144
C2212A	0144
C2212B080C3700-2200	OEM
C2212N	OEM
C2213	0144
C2215	0224
C2216	2902
C2220B125C3700-2200	OEM
C2220N	OEM
C2221	1984
C2222	1984
C2224	0338
C2224A	0338
C2228	1327
C2228A	1327
C2228C	1327
C2228D	1327
C2228D(JAPAN)	1327
C2228E	1327
C2228E(JAPAN)	1327
C2228F	1327
C2228K	1327
C2228M	1327
C2228Y	1327
C2229	0066
C2229(KOREA)	0261
C2229-0	0066
C2229-O	0066
C2229-O(JAPAN)	0066
C2229-O(KOREA)	0261
C2229-Y	0066
C2229D	0066
C2229E	0066
C2229M	0066
C2229O	0066
C2229Y	0066
C2229Y(JAPAN)	0066
C2229Y(KOREA)	0261
C2230	0261
C2230-Y	0261
C2230A	0261
C2230A(JAPAN)	0261
C2230AG	0261
C2230A-GR	0261
C2230AGR	0261
C2230AGR(JAPAN)	0261
C2230A-Y	0261
C2230AY	0261
C2230AY(JAPAN)	0261
C2230AY(KOREA)	0261
C2230AYG	0261
C2230G	0261
C2230GR	0261
C2230Y	0261
C2231	0168
C2231A	0168
C2231Y	0168
C2233	1538
C2234	1974
C2235	1386
C2236	2910
C2236-O	2910
C2236-O(JAPAN)	2910
C2236-Y	2910
C2236AY(KOREA)	1569
C2236Q	2910
C2236Y	2910
C2236Y(JAPAN)	2910
C2237	2526
C2238	2911
C2240	2131
C2240GL	2131
C2240GR	2131
C2241	1085
C2242	1085
C2242BK	1085
C2242GL2	1085
C2242LB	1085
C2242LBGL2	1085
C2243	2912
C2244	0074
C2245	0074
C2246	0363
C2251-O	0261
C2256	2414
C2258	0276
C2259	2924
C2259C	2924
C2259G	2924
C2260	0177
C2261	0177
C2262	0177
C2263	0111
C2264	1978
C2267H	0187
C2271	0261
C2271C	0261
C2271D	0261
C2271D(JAPAN)	0261
C2271E	0261
C2271E(JAPAN)	0261
C2271M	0261
C2271M(JAPAN)	0261
C2271N	0261
C2271Z	0261
C2274	0155
C2274E(JAPAN)	0155
C2274K	0155
C2275	1284
C2275A	1284
C2276	2939
C2277	0079
C2278	0285
C2279	0338
C2281	1198
C2290	2946
C2291	2947
C2292	2928
C2293	2928
C2297	0423
C2298	0161
C2298A	0161
C2298B	0161
C2304	1852
C2306	4386
C2306(2S)	1852
C2307	1506
C2308	0155
C2308B	0155
C2308G	OEM
C2309	0111
C2309D	0111
C2309E	0111
C2309F	0111
C2310	0286
C2310-O	0261
C2311	0161
C2312	0836
C2314	1591
C2314D	1591
C2315	0423
C2315CA	OEM
C2317	0392
C2320	0151
C2320(JAPAN)	0151
C2320L	0151
C2320LE(ECB)	0079
C2321	0620
C2322	0620
C2323	0620
C2324K	2970
C2329	2700
C2330-O	0261
C2330AY(JAPAN)	0261
C2330Y	0261
C2331	0392
C2331-O	1624
C2333	0261
C2334	2268
C2334K	2268
C2334L	2268
C2334L(JAPAN)	2268
C2335	0728
C2336	0392
C2336A	0392
C2336B	0392
C2337	0867
C2337A	0867
C2340N	OEM
C2344	1284
C2344D	1284
C2344E	1284
C2347	1591
C2348	0224
C2349	0127
C2350	0042
C2352	0224
C2354	0429
C2356	1852
C2359	0728
C2362	1227
C2362G	1227
C2362H	1227
C2362K	1227
C2362KF	1227
C2363	0079
C2365	0642
C2368	3007
C2369	3007
C2371	0276
C2371(1)	0276
C2371(1)L	0276
C2371(1)M	0276
C2371(8)L	0276
C2371K	0276
C2371L	0276
C2371M	0276
C2373	0728
C2373K	0728
C2373L	0728
C2373M	0728
C2375	0261
C2377	2223
C2377-C	2223
C2377C	2223
C2377D	2223
C2378	1386
C2382	1974
C2383	1562
C2383-0	1562
C2383-O	1562
C2383-O(JAPAN)	1562
C2383O	1562
C2383R	1562
C2383R(JAPAN)	1562
C2383Y	1562
C2384	0338
C2385	0111
C2388	1852
C2388A	1852
C2393D	0261
C2394D	0836
C2398	0177
C2401	3025
C2402	0543
C2406	0023
C2406M	0023
C2420	3041
C2425	0955
C2427	0728
C2428	0543
C2429	1852
C2429A	1852
C2430	0620
C2431	0620
C2432	0620
C2433	2433
C2434	2433
C2435	1991
C2436	2626
C2441	0563
C2443-24	0551
C2456	0276
C2456(JAPAN)	0276
C2456LB	0276
C2456Z	0276
C2458	0076
C2458-O	0076
C2458BL	0076
C2458GR	0076
C2458Y	0076
C2458Y(JAPAN)	0076
C2459	1270
C2465	3007
C2467	3007
C2470	3007
C2471	0224
C2473	2851
C2481	0563
C2481(JAPAN)	0563
C2481-0	0563
C2481-O	0563
C2481R	0563
C2481Y	0563
C2482	0261
C2482(JAPAN)	0261
C2482(KOREA)	0066
C2482-1	0261
C2482-1(JAPAN)	0261
C2482BK	0261
C2482FA-1	0261
C2482FA1	0261
C2482K	0261
C2482V	0261
C2482Z	0261
C2483	0219
C2483-O	0219
C2483-Y	0219
C2483O	0219
C2483R	0219
C2483R(JAPAN)	0219
C2483Y	0219
C2483Y(JAPAN)	0219
C2484	3081
C2484P	3081
C2485	3081
C2486	3082
C2489	0543
C2491	1165
C2492	0543
C2493	0543
C2497	0563
C2497Q	0563
C2497R	0563
C2498	1806
C2499	2851
C2500	3093
C2500P	OEM
C2501	0728
C2501(JAPAN)	0728
C2501P	OEM
C2502	0728
C2502P	OEM
C2503P	OEM
C2504	0283
C2505	0728
C2506	0283
C2506(TRANS)	0728
C2506B40C1600-1100	OEM
C2507	1852
C2508	2884
C2512	0224
C2512B80C1600-1100	OEM
C2516	1165
C2516AK	1165
C2516AL	1165
C2516AM	1165
C2516K	1165
C2516L	1165
C2516M	1165
C2519	0127
C2520B125C1500-1000	OEM
C2522	0543
C2522A	0543
C2523	0543
C2525	2276
C2526	2276
C2527	1165
C2528	3112
C2530	3115
C2534	0728
C2536	1506
C2540	3119
C2540B250C1500-1000	OEM
C2542(JAPAN)	0728
C2545D	0111
C2545E	0111
C2545F	0111
C2548	1506
C2549	0224
C2550	0016
C2551	0261
C2551-0	0261
C2551-O	0261
C2551-O(JAPAN)	0261
C2551R	0261
C2552	0728
C2553	0728
C2555	2185
C2555Z	2185
C2556	0563
C2556A	0563
C2561	0127
C2562	3125
C2563	3082
C2564	2276
C2565	2276
C2568	0276
C2568(1)	0276
C2568(1)K	0276
C2568(1)KD(JAPAN)	0276
C2568(1)L	0276
C2568(1)LD	0276
C2568(1)M	0276
C2568K	0276
C2568KD	0276
C2568L	0276
C2568LD	0276
C2568LD(JAPAN)	0276
C2568LM	0276
C2568M	0276
C2570	2851
C2570A	2851
C2570AE	2851
C2575	0286
C2575L	0286
C2577	0194
C2578	0194
C2579	3082
C2580	3082
C2581	3082
C2582	0563
C2590	0563
C2591	0392
C2592	0392
C2602	0111
C2603	0518
C2603-E	0518
C2603-F	0518
C2603E	0518
C2603F	0518
C2606	0285
C2607	2414
C2608	2414
C2610	0261
C2610-05	0261
C2610-5	0261
C2610B	0261
C2610B(JAPAN)	0261
C2610BK	0261
C2610K	0261
C2611	0276
C2612	0168
C2613	0728
C2613(JAPAN)	0728
C2613K	0728
C2615A	0017
C2616	2928
C2617	2928
C2621	0276
C2621-C	0276
C2621-D	0276
C2621-E	0276
C2621-RA	0276
C2621-RAC	0276
C2621-RAD	0276
C2621-RAE	0276
C2621C	0276
C2621D	0276
C2621E	0276
C2621E(JAPAN)	0276
C2621RA	0276
C2626	1506
C2630	3119
C2631	0261
C2632	0261
C2633	1085
C2634-RS	0111
C2634NC	0111
C2634R	0111
C2634R(JAPAN)	0111
C2634RS	0111
C2634S	0111
C2634S(JAPAN)	0111
C2636	0127
C2637	1085
C2637L	1085
C2637R	1085
C2637RL	1085
C2646	0127
C2647	0224
C2647C	0224
C2653	0168
C2653(H)	0168
C2653C	0168
C2653CL	0168
C2653H	0168
C2653H(JAPAN)	0168
C2653HCL	0168
C2653HLB	0168
C2653L	0168
C2653LB	0168
C2655	0018
C2655-O	0018
C2655-Y	0018
C2655O(JAPAN)	0018
C2655Y	0018
C2660	0392
C2665	3081
C2668	1144
C2669	0155
C2669-O(JAPAN)	0155
C2670	0155
C2671	2851
C2681	0194
C2687	0261
C2688	0276
C2688(2)	0276
C2688-K	0276
C2688-L	0276
C2688-M	0276
C2688-N	0276
C2688A	0276
C2688K	0276
C2688KA	0276
C2688KA(JAPAN)	0276
C2688L	0276
C2688LA	0276
C2688LA(JAPAN)	0276
C2688M	0276
C2688N	0276
C2690	0563
C2690A	0563
C2694	3194
C2703Y	1562
C2704	0563
C2705	1562
C2706	3082
C2708	4424
C2710	0155
C2710-0	0155
C2710-Y	0155
C2715M	0102
C2717	0313
C2717FA	0313
C2718	0018
C2719	0155
C2719(JAPAN)	0155
C2719K	0155
C2719L	0155
C2719L(JAPAN)	0155
C2720	3203
C2721	0018
C2724	0191
C2724-D	0191
C2724-E	0191
C2727	2844
C2729	2844
C2737	2851
C2738	3209
C2739	0728
C2740	3211
C2751L(JAPAN)	OEM
C2753	1506
C2773	3226
C2774	3226
C2780	1562
C2784	1270
C2785	0249
C2785(JAPAN)	0249
C2785E	0249
C2785F	0249
C2785F(JAPAN)	0249
C2785H	0249
C2785J	0249
C2785K	0249
C2786	0286
C2786L	0286
C2787	0127
C2790	2441
C2791	3229
C2792	3231
C2794	0563
C2794(JAPAN)	0563
C2809	2613
C2809O	2613
C2809Q	2613
C2809V	2613
C2810	0728
C2815	0728
C2816	0728
C2818	0079
C2818H	0079
C2819	1852
C2819H	1852
C2820	1852
C2824	0563
C2826	3014
C2827	0728
C2829	1966
C2830	1852
C2831	2766
C2832	0168
C2832K	0168
C2832L	0168
C2837	0194
C2838	0046
C2839	0151
C2839-D	0151
C2839D	0151
C2839E	0151
C2840	0127
C2840D	0127
C2853D	1386
C2853E	1386
C2854D	1386
C2854E	1386
C2855D	1386
C2855E	1386
C2856E	1386
C2857	0261
C2860	1497
C2877	0161
C2878	3094
C2878A	3094
C2878A(JAPAN)	3094
C2878B	3094
C2878B(JAPAN)	3094
C2884	0127
C2898	0728
C2901	3274
C2901L	3274
C2902	1852
C2903	1852
C2909	0261
C2910	1562
C2911	0563
C2912	0563
C2913	0728
C2914	3267
C2914Z	3267
C2920	1341
C2921	3286
C2922	3226
C2923	0276
C2923RL	0276
C2928	3292
C2939	1506
C2939(JAPAN)	1506
C2958	1386
C2960	3302
C2960E	3302
C2960F	3302
C2960G	3302
C2964	2928
C2965	2928
C2971	1341
C2972	1341
C2976	3060
C2979	2096
C2981	3060
C2995	0530
C3000D	0286
C3000L	OEM
C3000P	0232
C3001	1540
C3001-0	1540
C3001-O	1540
C3001A	1540
C3001L	OEM
C3001M	1540
C3001P	0268

DEVICE TYPE	REPL CODE
C3001T	1540
C3002B	OEM
C3002B-C	OEM
C3002P	0314
C3004L	OEM
C3004P	0361
C3007	3329
C3010	OEM
C3010A	OEM
C3010B	OEM
C3010L	OEM
C3010P	0512
C3015	OEM
C3015A	OEM
C3015B	OEM
C3016	OEM
C3016A	OEM
C3016B	OEM
C3017	0698
C3019	OEM
C3019A	OEM
C3019B	OEM
C3020	OEM
C3020A	OEM
C3020B	OEM
C3020L	OEM
C3020P	0697
C3021	OEM
C3021A	OEM
C3021B	OEM
C3022	OEM
C3022A	OEM
C3022B	OEM
C3025	OEM
C3025A	OEM
C3025B	OEM
C3026	OEM
C3026A	OEM
C3026B	OEM
C3029	OEM
C3029(TRANS)	2844
C3029A	OEM
C3029B	OEM
C3030	OEM
C3030A	OEM
C3030B	OEM
C3030L	OEM
C3030P	0873
C3031	OEM
C3031A	OEM
C3031B	OEM
C3032	OEM
C3032A	OEM
C3032B	OEM
C3035	OEM
C3035A	OEM
C3035B	OEM
C3036	OEM
C3036A	OEM
C3036B	OEM
C3038	3014
C3039	OEM
C3039(JAPAN)	0728
C3039A	OEM
C3039B	OEM
C3040	6858
C3040A	OEM
C3040B	OEM
C3040L	OEM
C3040P	1024
C3041	OEM
C3041A	OEM
C3041B	OEM
C3041L	OEM
C3041P	1038
C3042	6858
C3042A	OEM
C3042B	OEM
C3044	1341
C3044A	1341
C3045	OEM
C3045(TRANS)	1341
C3045A	OEM
C3045B	OEM
C3046	OEM
C3046(TRANS)	3338
C3046A	OEM
C3046B	OEM
C3049	OEM
C3049A	OEM
C3049B	OEM
C3050L	OEM
C3050P	0743
C3055	0728
C3056	0728
C3056A	0728
C3057	0728
C3059	OEM
C3059(TRANS)	0223
C3059A	OEM
C3059B	OEM
C3060	OEM
C3060(TRANS)	0223
C3060A	OEM
C3060B	OEM
C3061	OEM
C3061A	OEM
C3061B	OEM
C3062	OEM
C3062A	OEM
C3062B	OEM
C3063	0276
C3063RL	0276
C3065	OEM
C3065A	OEM
C3065B	OEM
C3066	OEM
C3066A	OEM
C3066B	OEM
C3073L	OEM
C3073P	1172
C3075L	OEM
C3075P	1432
C3076	3350
C3100	2312
C3100A	OEM
C3100B	OEM
C3101	OEM
C3101A	OEM
C3101ADM	OEM
C3101B	OEM
C3102	OEM
C3102A	OEM
C3102B	OEM
C3105	OEM
C3105A	OEM
C3105B	OEM
C3106	OEM
C3106A	OEM
C3106B	OEM
C3109	OEM
C3109A	OEM
C3109B	OEM
C3114T(JAPAN)	0155
C3116T	0563
C3140	OEM
C3140A	OEM
C3140B	OEM
C3141	0133
C3141A	OEM
C3141B	OEM
C3142	0080
C3143	0080
C3144	0609
C3145	OEM
C3145A	OEM
C3145B	OEM
C3146	OEM
C3146A	OEM
C3146B	OEM
C3149	3357
C3149A	OEM
C3149B	OEM
C3150	3357
C3150A	OEM
C3150B	OEM
C3151	3493
C3151A	OEM
C3151B	OEM
C3152	3493
C3152A	OEM
C3152B	OEM
C3155	0223
C3155A	OEM
C3155B	OEM
C3156	3385
C3156A	OEM
C3156B	OEM
C3157(JAPAN)	OEM
C3157L	3060
C3159	0728
C3159A	OEM
C3159B	OEM
C3165	OEM
C3165A	OEM
C3165B	OEM
C3166	OEM
C3166A	OEM
C3166B	OEM
C3174R	0728
C3174S	0728
C3175(JAPAN)	0728
C3180	3082
C3182	0194
C3187	3392
C3209	0261
C3209K	0261
C3209K(JAPAN)	0261
C3209L	0261
C3209L(JAPAN)	0261
C3209M	0261
C3212(JAPAN)	OEM
C3214	0065
C3220	3387
C3220A	OEM
C3220B	OEM
C3221	OEM
C3221A	OEM
C3221B	OEM
C3222	3388
C3222A	OEM
C3222B	OEM
C3225	OEM
C3225A	OEM
C3225B	OEM
C3226	OEM
C3226A	OEM
C3226B	OEM
C3237	3411
C3238	1085
C3258Y(JAPAN)	OEM
C3271N	0276
C3280	3435
C3293	0278
C3298B	OEM
C3299Y	0060
C3300	2367
C3303Y	OEM
C3306(JAPAN)	2185
C3310(JAPAN)	0728
C3311AR	2954
C3315C	OEM
C3330	2954
C3330A	OEM
C3330B	OEM
C3331	0018
C3331A	OEM
C3331B	OEM
C3331T	0018
C3332	OEM
C3332A	OEM
C3332B	OEM
C3334	0261
C3335	OEM
C3335A	OEM
C3335B	OEM
C3387	3476
C3413C	2954
C3450M	OEM
C3460	3360
C3467E	0261
C3470	2954
C3470A	OEM
C3470B	OEM
C3471	OEM
C3471A	OEM
C3471B	OEM
C3472	OEM
C3472A	OEM
C3472B	OEM
C3568(JAPAN)	OEM
C3619	0276
C3621	0563
C3681	OEM
C3681A	OEM
C3681B	OEM
C3682	OEM
C3682A	OEM
C3682B	OEM
C3708S	1386
C3720	0224
C3720A	0224
C3720B	0224
C3720C	0224
C3720D	0224
C3720F	0224
C3720G	0224
C3720GN	0224
C3720H	0224
C3720J	0224
C3720K	0224
C3720L	0224
C3720OR	0224
C3720R	0224
C3720X	0224
C3720Y	0224
C3746	0060
C3746R	0060
C3747	0060
C3748	0060
C3748R	0060
C3789D	0276
C3789E	0276
C3795B(EUROPE)	OEM
C3800	OEM
C3800(TRANS)	0224
C3800A	OEM
C3800A(TRANS)	0224
C3800B	OEM
C3800B(TRANS)	0224
C3800C	0224
C3800D	0224
C3800E	0224
C3800F	0224
C3800G	0224
C3800GN	0224
C3800H	0224
C3800J	0224
C3800K	0224
C3800L	0224
C3800M	0224
C3800OR	0224
C3800P	1208
C3800R	0224
C3800X	0224
C3800Y	0224
C3801L	OEM
C3801P	0834
C3802	OEM
C3802A	OEM
C3802B	OEM
C3802P	OEM
C3803P	OEM
C3804P	OEM
C3805P	OEM
C3806P	1379
C3807	OEM
C3831	OEM
C3832	2908
C3833	1506
C3854	0127
C3890	0728
C3902S	0563
C3940	0224
C3940A	0224
C3940B	0224
C3940C	0224
C3940D	0224
C3940E	0224
C3940F	0224
C3940G	0224
C3940GN	0224
C3940H	0224
C3940J	0224
C3940K	0224
C3940L	0224
C3940M	0224
C3940OR	0224
C3940R	0224
C3940X	0224
C3940Y	0224
C3942	5579
C3987	3723
C4001P	OEM
C4002P	OEM
C4003P	OEM
C4004	3357
C4004P	OEM
C4005P	OEM
C4006P	OEM
C4007P	OEM
C4008P	OEM
C4009P	OEM
C4011	0471
C4012	OEM
C4012A	0018
C4012B	0018
C4012C	0018
C4012D	0018
C4012E	0018
C4012F	0018
C4012G	0018
C4012GN	0018
C4012H	0018
C4012J	0018
C4012L	0018
C4012M	0018
C4012X	0018
C4012Y	0018
C4013	OEM
C4014	3100
C4015	0057
C4016	0170
C4017	OEM
C4018	0052
C4019	OEM
C4020	0686
C4020P	OEM
C4021	OEM
C4021P	OEM
C4022	OEM
C4023	OEM
C4024	OEM
C4025	OEM
C4026	0441
C4027	OEM
C4028	OEM
C4029	OEM
C4030P	OEM
C4030S	OEM
C4031P	OEM
C4032P	OEM
C4033A	0224
C4033B	0224
C4033C	0224
C4033D	0224
C4033E	0224
C4033F	0224
C4033G	0224
C4033GN	0224
C4033H	0224
C4033J	0224
C4033K	0224
C4033L	0224
C4033M	0224
C4033OR	0224
C4033P	OEM
C4033R	0224
C4033X	0224
C4033Y	0224
C4040	OEM
C4040P	OEM
C4041P	OEM
C4042P	OEM
C4050P	OEM
C4051P	OEM
C4052P	OEM
C4053P	OEM
C4054P	OEM
C4055P	OEM
C4056P	OEM
C4057L	OEM
C4057T	OEM
C4058P	OEM
C4059	1506
C4059P	0487
C4108	1506
C4116	0042
C4116A	0042
C4116B	0042
C4116C	0042
C4116D	0042
C4116E	0042
C4116F	0042
C4116G	0042
C4116GN	0042
C4116H	0042
C4116J	0042
C4116K	0042
C4116L	0042
C4116M	0042
C4116OR	0042
C4116R	0042
C4116X	0042
C4116Y	0042
C4141-21816	OEM
C4159E	1165
C4169	1061
C4212(H)	OEM
C4212H	OEM
C4217E	0276
C4274	0728
C-4401	0133
C4404	OEM
C4418	2908
C4418Y	2908
C4423	2185
C4424M	3098
C4434A	OEM
C4547	OEM
C4558C	0360
C4570HA	OEM
C4590E	OEM
C4702	0031
C5001	OEM
C5002	OEM
C5003	OEM
C5004	OEM
C5005	0143
C5100	0168
C5100A	0168
C5100B	0168
C5100C	0168
C5100D	0168
C5100E	0168
C5100F	0168
C5100G	0168
C5100GN	0168
C5100H	0168
C5100J	0168
C5100K	0168
C5100L	0168
C5100M	0168
C5100OR	0168
C5100R	0168
C5100Y	0168
C5110	0168
C5110A	0168
C5110B	0168
C5110C	0168
C5110D	0168
C5110E	0168
C5110F	0168
C5110G	0168
C5110GN	0168
C5110H	0168
C5110J	0168
C5110K	0168
C5110L	0168
C5110M	0168
C5110OR	0168
C5110R	0168
C5110X	0168
C5110Y	0168
C5116#	OEM
C5120	0560
C5130	0560
C5223A4M	OEM
C5335-5	OEM
C5335-6	OEM
C5335-G	OEM
C5335-M	OEM
C5335-X	OEM
C5370	0018
C5370A	0018
C5370B	0018
C5370C	0018
C5370D	0018
C5370E	0018
C5370F	0018
C5370G	0018
C5370GN	0018
C5370H	0018
C5370J	0018
C5370K	0018
C5370L	0018
C5370M	0018
C5370OR	0018
C5370R	0018
C5370X	0018
C5370Y	0018
C5430OR	0224
C6001	2100
C6003	0517
C6003P	OEM
C6004	OEM
C6007	0723
C6009P	1899
C6010	3779
C6010P	3972
C6011	1876
C6012	OEM
C6012A	OEM
C6013	2027
C6013A	OEM
C6014	1186
C6014A	3100
C6015	0057
C6015A	0057
C6016	1980
C6016A	OEM
C6016P	OEM
C6017	1684
C6017A	OEM
C6018	0052
C6018A	0052
C6019	0723
C6019A	OEM
C6020	0686
C6020A	0686
C6021	OEM
C6021A	OEM
C6022	OEM
C6022A	OEM
C6023	OEM
C6023A	OEM
C6024	OEM
C6024A	OEM
C6025	OEM
C6025A	OEM
C6026	OEM
C6026A	OEM
C6027	OEM
C6027A	OEM
C6028	OEM
C6028A	OEM
C6029	OEM
C6029A	OEM
C6030	OEM
C6030A	OEM
C6031	OEM
C6031A	OEM
C6032	OEM
C6032A	OEM
C6041	OEM
C6041M	OEM
C6042	OEM
C6042M	OEM
C6043	OEM
C6043M	OEM
C6044	0677
C6044M	0677
C6045	OEM
C6045M	OEM
C6046	OEM
C6046M	OEM
C6047	OEM
C6047M	OEM
C6048	1276
C6048M	1276
C6049	OEM
C6049G	OEM
C6049M	OEM
C6049R	OEM
C6050	OEM
C6050G	0417
C6050M	OEM
C6051	OEM
C6051M	OEM
C6051P	OEM
C6052	0677
C6052G	OEM
C6052M	0410
C6052M(ZENER)	0677
C6052P	0312
C6052P(HEP)	0312
C6053G	OEM
C6053L	OEM
C6054G	OEM
C6054R	OEM
C6055L(HEP)	0692
C6056P	0701
C6056P(HEP)	0701
C6057P	0349
C6057P(HEP)	0349
C6059P	0785
C6059P(HEP)	0785
C6060P	0789
C6060P(HEP)	0789
C6061P	3219
C6062P	0664
C6062P(HEP)	0664
C6063P	0167
C6063P(HEP)	0167
C6065P	1940
C6065P(HEP)	1940
C6066P	1753
C6067G	0753
C6068P	0443
C6068P(HEP)	0443
C6070P	0352
C6070P(HEP)	0352
C6071P	0354
C6071P(HEP)	0354
C6072P	0751
C6072P(HEP)	0751
C6073P	3006
C6074P	0654
C6074P(HEP)	0654
C6075P	0856
C6076P	0395
C6076P(HEP)	0395
C6078R	OEM
C6079P	1808
C6080P	2279
C6080P(HEP)	2279
C6082P	0664
C6082P(HEP)	0664
C6083P	0167
C6083P(HEP)	0167
C6085P(HEP)	2457
C6087P	OEM
C6089	2161
C6089(HEP)	0753
C6091	1077
C6091G	1077
C6092G	OEM
C6093G	OEM
C6094P	OEM
C6095P	0443
C6096P	0520
C6096P(HEP)	0520
C6099P	3195
C6099T	OEM
C6100	0760
C6100A	0760
C6100P	0802
C6100P(HEP)	0802
C6101	0118
C6101A	0118
C6101P	1345
C6101P(HEP)	1345
C6102	0299
C6102A	0299
C6102P	0360
C6102P(HEP)	0360
C6103	0376
C6103A	0376
C6103P	1301
C6104	0036
C6104A	0036
C6104L	OEM
C6105	2239
C6105A	0275
C6105P	OEM
C6106	0140
C6106A	0140
C6107	0041
C6107A	0041
C6107P	1300
C6108	0253
C6108A	0253
C6109	0471
C6109A	0471
C6110P	OEM
C6111P	0923
C6112	0062
C6112A	0062
C6112P	1196
C6113	0077
C6113A	0077
C6113P	0334
C6114	0165
C6114A	0165
C6114P	1321
C6115	0057
C6115A	0057
C6115P	OEM
C6116	0064
C6116A	0064
C6116P	2239
C6117	0181
C6117A	0181
C6117P	OEM
C6118	0052
C6118A	0052
C6118P	1285
C6119	0053
C6119A	0053
C6119P	OEM
C6120	0686
C6120P	2648
C6121	0445
C6121A	0445
C6121P	OEM
C6122	0375
C6122A	0375
C6122P	1838
C6123	0700
C6123A	0700
C6123P	3829
C6124	0705
C6124A	0705
C6124P	OEM
C6125	0494
C6125A	0494
C6125P	OEM
C6126	0455
C6126A	0455
C6126P	OEM
C6127	0195
C6127A	0195
C6127P	OEM

If replacement code is OEM, contact original manufacturer for replacement.

DEVICE TYPE	REPL CODE	DEVICE TYPE	REPL CODE	DEVICE TYPE	REPL CODE	DEVICE TYPE	REPL CODE	DEVICE TYPE	REPL CODE	DEVICE TYPE	REPL CODE	DEVICE TYPE	REPL CODE
C6128	0166	C7460P	1275	C8283	OEM	C30816	OEM	C74191P	1917	CA0101AT	OEM	CA0555EX	OEM
C6128A	0166	C7470P	1404	C8284	OEM	C30817	OEM	C74192P	1921	CA0101AT3	OEM	CA0555G	OEM
C6128P	OEM	C7472P	1426	C8286	OEM	C30818E	OEM	C74193P	1926	CA0101AT4	OEM	CA0555S	OEM
C6129	0010	C7473P	1172	C8287	OEM	C30822	OEM	C74194P	OEM	CA0101E	OEM	CA0555T	OEM
C6129A	0010	C7473P(HEP)	1172	C8288	OEM	C30831	OEM	C74195P	1943	CA0101S	OEM	CA0555T3W	OEM
C6129P	OEM	C7474P	1313	C8289	OEM	C30843	OEM	C74196P	1950	CA0101T	OEM	CA0723CE	OEM
C6130	0032	C7474P(HEP)	1313	C8290	0079	C30844	OEM	C74197P	1956	CA0107E	OEM	CA0723CEX	OEM
C6130A	0032	C7475P	1432	C8291	OEM	C30845	OEM	C74198P	1964	CA0107G	OEM	CA0723CT	OEM
C6130P	0973	C7476P	1158	C8292	OEM	C30846	OEM	C74199P	1971	CA0107S	OEM	CA0723CTX	OEM
C6131	0054	C7476P(HEP)	1158	C8293	OEM	C30847E	OEM	C74251P	2299	CA0107S3W	OEM	CA0723E	OEM
C6131A	0054	C7483P	0117	C8294	OEM	C30849	OEM	C74259P	OEM	CA0107T	OEM	CA0723EX	OEM
C6131P	0973	C7485P	0374	C8295	OEM	C30850	OEM	C74365P	1459	CA0107T3W	OEM	CA0723T	OEM
C6132	0068	C7486P	1368	C8316A	OEM	C30851	OEM	C74366P	1471	CA0111E	OEM	CA0723T3	OEM
C6132A	0068	C7490AP	1208	C8355	OEM	C30852	OEM	C74367P	1488	CA0111G	OEM	CA0723T3W	OEM
C6132P	1292	C7490P(HEP)	1208	C8355-2	OEM	C30853	OEM	C74368P	1508	CA0111S	OEM	CA0723T4	OEM
C6133P	OEM	C7491AP	0980	C8641A	OEM	C30854	OEM	C75603-2	0622	CA0111T	OEM	CA0741CE	OEM
C6135P	OEM	C7492AP	0834	C8741A	OEM	C30872	OEM	C78010BB014	OEM	CA0124E	OEM	CA0741CEX	OEM
C6136P	OEM	C7493AP	0569	C8748	OEM	C30895	OEM	C78010BD031	OEM	CA0124G	OEM	CA0741CG	OEM
C6139P	OEM	C7495P	1486	C8748-6	OEM	C30900E	OEM	C78020BA	OEM	CA0124GX	OEM	CA0741CGX	OEM
C6140P	OEM	C7496P	1715	C8749H	OEM	C30902E	OEM	C86002E	OEM	CA0139AE	OEM	CA0741CS	OEM
C6143P	OEM	C8000	OEM	C8749H-8	OEM	C30904E	OEM	C86003E-RS	OEM	CA0139AG	OEM	CA0741CSX	OEM
C6144P	OEM	C8000A	OEM	C8751	OEM	C30905E	OEM	C86003E-TS	OEM	CA0139G	OEM	CA0741CT	OEM
C6201	OEM	C8000B	OEM	C8755A	OEM	C30908E	OEM	C86006E	OEM	CA0139GX	OEM	CA0741CTX	OEM
C6202	OEM	C8000C	OEM	C8755A-2	OEM	C30910E	OEM	C86007E	OEM	CA0158AE	OEM	CA0741E	OEM
C6203	OEM	C8000D	OEM	C8783C	OEM	C30914E	OEM	C86008E	OEM	CA0158AG	OEM	CA0741EX	OEM
C6204	OEM	C8001	OEM	C8783CC	OEM	C30916E	OEM	C86009E	OEM	CA0158AS	OEM	CA0741G	OEM
C6205	OEM	C8001A	OEM	C9001	OEM	C30917E	OEM	C86010E	OEM	CA0158AT	OEM	CA0741GX	OEM
C6206	OEM	C8001B	OEM	C9002	OEM	C30919E	OEM	C86011E	OEM	CA0158E	OEM	CA0741S	OEM
C6211	OEM	C8001C	OEM	C9003	OEM	C30920E	OEM	C86012E-RS	OEM	CA0158G	OEM	CA0741S3	OEM
C6212	OEM	C8001D	OEM	C9011E	0079	C30921E	OEM	C86012E-TS	OEM	CA0158S	OEM	CA0741SX	OEM
C6213	OEM	C8002	OEM	C9011F	0079	C30924E	OEM	C86013E	OEM	CA0158T	OEM	CA0741T	OEM
C6214	OEM	C8002A	OEM	C9011H	0079	C30925E	OEM	C86014E	OEM	CA0201AE	OEM	CA0741T1	OEM
C6215	OEM	C8002B	OEM	C9014A	0079	C30950E	OEM	C86022E	OEM	CA0201AG	OEM	CA0741T2	OEM
C6216	OEM	C8002C	OEM	C9014A(KOREA)	0079	C30950EL	OEM	C86023E	OEM	CA0201AS	OEM	CA0741T3	OEM
C6217	OEM	C8002D	OEM	C9080	0126	C30950F	OEM	C86025E	OEM	CA0201AT	OEM	CA0741T4	OEM
C6251	OEM	C8008	OEM	C9081	0126	C30950FL	OEM	C106009	OEM	CA0201G	OEM	CA0741TX	OEM
C6252	OEM	C8008-1	OEM	C9082	0037	C30950G	OEM	C106509	OEM	CA0201S	OEM	CA0747CE	OEM
C6253	OEM	C8020H	OEM	C9083	0037	C30950GL	OEM	C106516	OEM	CA0201T	OEM	CA0747CG	OEM
C6254	OEM	C8021H	OEM	C9084	0037	C30954E	OEM	C107009	OEM	CA0207E	OEM	CA0747CT	OEM
C6311	OEM	C8022	OEM	C9085	0037	C30955E	OEM	C107016	OEM	CA0207G	OEM	CA0747E	OEM
C6330	OEM	C8031	OEM	C9426	0224	C30956E	OEM	C107509	OEM	CA0207S	OEM	CA0747G	OEM
C6330A	OEM	C8035	OEM	C9634	0111	C30971E	OEM	C107516	OEM	CA0207T	OEM	CA0747T	OEM
C6331	OEM	C8035HL	OEM	C10110	0015	C30971EL	OEM	C108016	OEM	CA0211E	OEM	CA0747T1	OEM
C6331A	OEM	C8035HL-1	OEM	C10159	0015	C30979E	OEM	C136009	OEM	CA0211G	OEM	CA0747T2	OEM
C6332	OEM	C8035L	OEM	C10176	0015	C31013	OEM	C136509	OEM	CA0211S	OEM	CA0747T3	OEM
C6332A	OEM	C8039	OEM	C10215-2	0050	C32024S	OEM	C137009	OEM	CA0211T	OEM	CA0748CE	OEM
C6391	1335	C8039-6	OEM	C10227	0004	C32816	OEM	C137509	OEM	CA0224E	OEM	CA0748CEX	OEM
C6391A	1335	C8039H-8	OEM	C10230-3	0004	C33725	0079	C139000	0086	CA0224G	OEM	CA0748CG	OEM
C6392	OEM	C8039HL	OEM	C10258	0050	C36577	0126	C139000A	0086	CA0239AE	OEM	CA0748CS	OEM
C6392A	OEM	C8041A	OEM	C10260	0050	C36578	0007	C206009	OEM	CA0239AG	OEM	CA0748CT	OEM
C6393	OEM	C8048	OEM	C10261	0050	C36579	0086	C206509	OEM	CA0239E	OEM	CA0748E	OEM
C6393A	OEM	C8048H	OEM	C10262	0050	C36580	0016	C206516	OEM	CA0239G	OEM	CA0748EX	OEM
C6394	OEM	C8048H-1	OEM	C10279-1	0111	C-36582	0325	C207009	OEM	CA0239GX	OEM	CA0748S	OEM
C6394A	OEM	C8048L	OEM	C10279-3	0016	C38085	OEM	C207016	OEM	CA0258AE	OEM	CA0748S1	OEM
C6395	OEM	C8049	OEM	C10291	0136	C38121	OEM	C207509	OEM	CA0258AG	OEM	CA0748S2	OEM
C6395A	OEM	C8049H	OEM	C10621	OEM	C38122	OEM	C207516	OEM	CA0258AS	OEM	CA0748S3	OEM
C6396	OEM	C8051	OEM	C11021	0004	C38123	OEM	C208016	OEM	CA0258AT	OEM	CA0748S4	OEM
C6396A	OEM	C8080A	4547	C12712	0086	C38124	OEM	C216509	OEM	CA0258E	OEM	CA0748T	OEM
C6397	OEM	C8080A-1	OEM	C12713	0715	C38129	OEM	C248507	0015	CA0258G	OEM	CA0748T1	OEM
C6397A	OEM	C8080A-2	OEM	C12714	0715	C41010BA	OEM	C251968-03	OEM	CA0258S	OEM	CA0748T2	OEM
C6398	OEM	C8085A	OEM	C12923	0086	C42010EB	OEM	C254187	OEM	CA0258T	OEM	CA0748T3	OEM
C6399	OEM	C8086	OEM	C12924	0086	C42010EC	OEM	C523383	0079	CA0270AE	OEM	CA0748T4	OEM
C6644H(HEP)	0111	C8086-2	OEM	C12925	0079	C51909B	0037	C530041-1	0015	CA0270AW	OEM	CA0758E	OEM
C6690	3384	C8086-4	OEM	C12926	0086	C57401AF	OEM	C642010ED	OEM	CA0270BE	OEM	CA0758Q	OEM
C6691	3384	C8088	OEM	C13390	0079	C57401AJ	OEM	C1107832P6	1080	CA0270BW	OEM	CA0810AQ	OEM
C6692	3384	C8089	OEM	C13391	0079	C57402AF	OEM	C1107832P7	0196	CA0270CE	OEM	CA0810AQM	OEM
C7076	0111	C8146	0018	C13391A	0079	C57402F	OEM	C1107832P9	0196	CA0270CW	OEM	CA0810Q	OEM
C7101	OEM	C8155	OEM	C13392	0086	C57402J	OEM	C2475078-3	0144	CA0301AE	OEM	CA0810QM	OEM
C7161AP	OEM	C8155-2	OEM	C13393	0086	C67401AJ	OEM	C2485076-3	0086	CA0301AEX	OEM	CA0920AE	OEM
C7400P	0232	C8156	OEM	C13394	0086	C67401J	OEM	C2485077-2	0086	CA0301AG	OEM	CA0920AW	OEM
C7400P(HEP)	0232	C8156-2	OEM	C13398	0086	C67402AJ	OEM	C2485078-1	0144	CA0301AS	OEM	CA014803	OEM
C7401P	0268	C8185	OEM	C13402	0715	C67402J	OEM	C2485079-1	0144	CA0301AT	OEM	CA060261-15	OEM
C7401P(HEP)	0268	C8185-2	OEM	C13404	0715	C68000E	OEM	CA078G	OEM	CA0307E	OEM	CA2D2	0004
C7402P	0314	C8202	OEM	C13414	0715	C74107P	4215	CA080AE	OEM	CA0307EX	OEM	CA-6C	OEM
C7402P(HEP)	0314	C8205	OEM	C13415	0715	C74109P	0968	CA080AS	OEM	CA0307G	OEM	CA-7C	OEM
C7403P	0335	C8212	1860	C13416	0715	C74121P	0175	CA080AT	OEM	CA0307GX	OEM	CA-7S	OEM
C7404P	0361	C8216	1863	C13417	0715	C74123P	1157	CA080BE	OEM	CA0307S	OEM	CA-9	OEM
C7405P	0385	C8218	OEM	C13901	0016	C74125P	1182	CA080CS	OEM	CA0307T	OEM	CA10	0015
C7406P	1206	C8219	OEM	C14256	0079	C74126P	1193	CA080CT	OEM	CA0311E	OEM	CA-10X	OEM
C7407P	1339	C8224	1709	C14424	0715	C74132P	1271	CA080E	OEM	CA0311G	OEM	CA-11	OEM
C7408P	0467	C8226	OEM	C14425	0715	C74141P	1377	CA080S	OEM	CA0311GX	OEM	CA-12	OEM
C7408P(HEP)	0467	C8228	1869	C19659	OEM	C74145P	0619	CA080T	OEM	CA0311S	OEM	CA-14	OEM
C7409P	0492	C8231	OEM	C21382	0015	C74147P	1451	CA081AE	OEM	CA0311T	OEM	CA20	0015
C7410P	0512	C8232	OEM	C21383	0012	C74148P	1464	CA081AS	OEM	CA0324E	1074	CA50	0015
C7411P	0527	C8237	OEM	C21480	0143	C74150P	1492	CA081AT	OEM	CA0324EX	OEM	CA-90	0143
C7413P	1441	C8237-2	OEM	C29120E	OEM	C74151AP	1495	CA081BE	OEM	CA0324G	OEM	CA90	0133
C7414P	2243	C8238	OEM	C29120M	OEM	C74153P	1539	CA081CS	OEM	CA0324GX	OEM	CA100	0015
C7416P	1349	C8243	OEM	C29120N	OEM	C74154P	1555	CA081CT	OEM	CA0339AE	OEM	CA101AT	0093
C7417P	1352	C8251A	OEM	C29120P	OEM	C74155P	1576	CA081E	OEM	CA0339AG	OEM	CA101S	0941
C7420P	0697	C8253	OEM	C29120PA	OEM	C74156P	1592	CA081S	OEM	CA0339E	OEM	CA101T	0093
C7422P	OEM	C8253-5	OEM	C29120PB	OEM	C74157P	1605	CA081T	OEM	CA0339EX	OEM	CA102BA	0015
C7423P	3466	C8255A	OEM	C29120S	OEM	C74160AP	1631	CA082AE	OEM	CA0339G	OEM	CA102DA	0015
C7425P	3477	C8255A-5	OEM	C29120T	OEM	C74162AP	1013	CA082AS	OEM	CA0339GX	OEM	CA102FA	0015
C7426P	0803	C8257	OEM	C30012	OEM	C74163AP	1666	CA082AT	OEM	CA0345	1696	CA102HA	0015
C7426P(HEP)	0803	C8257-5	OEM	C30013	OEM	C74164P	0734	CA082BE	OEM	CA0358AE	OEM	CA102KA	1228
C7427P	0818	C8259A	OEM	C30099	OEM	C74165P	1685	CA082CS	OEM	CA0358AG	OEM	CA102MA	0015
C7430P	0873	C8259A-2	OEM	C30116	OEM	C74166P	0231	CA082CT	OEM	CA0358AS	OEM	CA102PA	0071
C7432P	0899	C8259A-8	OEM	C30116F	OEM	C74170P	1722	CA082E	OEM	CA0358AT	OEM	CA102RA	0071
C7437P	3520	C8271	OEM	C30119	OEM	C74173P	1766	CA082S	OEM	CA0358E	OEM	CA102VA	0071
C7440P	1024	C8271-6	OEM	C30122	OEM	C74174P	1770	CA082T	OEM	CA0358EX	OEM	CA124E	0625
C7441AP	1038	C8271-8	OEM	C30123	OEM	C74175P	1787	CA083AE	OEM	CA0358G	OEM	CA139AE	0176
C7445P	1082	C8272	OEM	C30130	OEM	C74176P	1795	CA083BE	OEM	CA0358GX	OEM	CA139F	0176
C7446AP	1098	C8273	OEM	C30133	OEM	C74177P	1803	CA083E	OEM	CA0358S	OEM	CA139G	0176
C7447AP	1108	C8273-4	OEM	C30807	OEM	C74180P	1829	CA-092	0012	CA0358T	OEM	CA150	0015
C7448P	1125	C8273-8	OEM	C30808	OEM	C74181P	1842	CA0101AE	OEM	CA0555CE	OEM	CA152AA	0071
C7450P	0743	C8275	OEM	C30809	OEM	C74182P	1856	CA0101AG	OEM	CA0555CEX	OEM	CA152BA	0071
C7451P	1168	C8279	OEM	C30810	OEM	C74184P	OEM	CA0101AS	OEM	CA0555CS	OEM	CA152CA	0071
C7453P	1185	C8279-5	OEM	C30812	OEM	C74185AP	OEM	CA0101AS3	OEM	CA0555CT	OEM	CA152DA	0071
C7454P	1202	C8282	OEM	C30815	OEM	C74190P	1912	CA0101AS4	OEM	CA0555E	OEM	CA152EA	0071

DEVICE TYPE	REPL CODE
CA152FA	0071
CA152GA	0071
CA152HA	0071
CA152KA	0071
CA152MA	0071
CA152PA	0071
CA152RA	0071
CA152TA	0071
CA152VA	0071
CA158AS	1677
CA158AT	1677
CA158S	1677
CA158T	1677
CA200	0015
CA201T	0093
CA211	1815
CA211AE	0664
CA211AQ	0664
CA211E	2104
CA211G	2104
CA224E	0625
CA224G	0625
CA239	0176
CA239A	0176
CA239AE	0176
CA239AG	0176
CA239E	0176
CA239G	0176
CA250	0015
CA258AG	6018
CA258AS	6334
CA258AT	6334
CA258S	1677
CA258T	1677
CA301AE	1300
CA301AG	1300
CA301AH	OEM
CA301AS	0093
CA301AT	0093
CA301E	1300
CA301G	1300
CA301T	0093
CA307	2283
CA307E	2283
CA307G	2283
CA307H	OEM
CA307S	1781
CA307T	1781
CA311	1815
CA311E	2104
CA311G	2104
CA311S	1815
CA311T	1815
CA314E	2798
CA324E	0625
CA324G	0625
CA324H	OEM
CA339	0176
CA339A	0176
CA339AE	0176
CA339AG	0176
CA339E	0176
CA339G	0176
CA339H	OEM
CA358AS	1677
CA358AT	1677
CA358S	1677
CA358T	1677
CA386M	OEM
CA386N	OEM
CA386P	OEM
CA386PA	OEM
CA386PB	OEM
CA386PC	OEM
CA386PD	OEM
CA386S	OEM
CA386T	OEM
CA555CE	0973
CA555CG	0973
CA555CS	3637
CA555CT	3637
CA555E	0973
CA555G	0973
CA555S	3637
CA555T	3637
CA723	1192
CA723CE	0026
CA723CT	1192
CA723E	0026
CA723T	1192
CA741	0312
CA741C	0312
CA741CE	0312
CA741CG	0312
CA741CS	0312
CA741CT	0410
CA741E	0312
CA741G	0312
CA741S	1301
CA741T	0410
CA747CE	2358
CA747CF	2358
CA747CG	2358
CA747CH	OEM
CA747CT	2368
CA747E	2358
CA747F	2358
CA747G	2358
CA747T	2368
CA748CE	2452
CA748CG	2452
CA748CH	OEM
CA748CJ	2452
CA748CN	2452
CA748CS	OEM
CA748CT	2454
CA748E	2452
CA748G	2452
CA748J	2452
CA748N	2452
CA748S	4369
CA748T	2454
CA758E	1395
CA810Q	3922
CA820C	OEM
CA920	OEM
CA1020A	0015
CA1190	OEM
CA1190G	1170
CA1190GM	1170
CA1190GQ	1170
CA1190GQ/14	OEM
CA1190GQ/M	1170
CA1310	0520
CA1310AE	OEM
CA1310E	0520
CA1352	0395
CA1352E	0395
CA1365	0167
CA1391E	0848
CA1391G	OEM
CA1394E	2493
CA1394G	2493
CA1398E	0856
CA1458	0360
CA1458E	0360
CA1458EX	OEM
CA1458F	0360
CA1458G	0360
CA1458GX	OEM
CA1458S	0360
CA1458T	3137
CA1458TX	OEM
CA1524E	OEM
CA1524EX	OEM
CA1524G	OEM
CA1541D	OEM
CA1558E	0360
CA1558G	OEM
CA1558S	3137
CA1558S3	OEM
CA1558SX	OEM
CA1558T	3137
CA1558T1	OEM
CA1558T2	OEM
CA1558T3	OEM
CA1558T4	OEM
CA1558TX	OEM
CA1724E	OEM
CA1724EX	OEM
CA1724G	OEM
CA1725E	OEM
CA1725EX	OEM
CA1725G	OEM
CA2002	1048
CA2002M	1048
CA2002V	1048
CA2004	1048
CA2004M	OEM
CA2020H	OEM
CA2065	0167
CA2111	0664
CA2111A	0664
CA2111AE	0664
CA2111AQ	0664
CA2136A	1443
CA2136AE	1443
CA2524E	OEM
CA2524EX	OEM
CA2524G	OEM
CA2800B	OEM
CA2800H	OEM
CA2810B	OEM
CA2810H	OEM
CA2812B	OEM
CA2812H	OEM
CA2813B	OEM
CA2813H	OEM
CA2818B	OEM
CA2818H	OEM
CA2820B	OEM
CA2820H	OEM
CA2830B	OEM
CA2830H	OEM
CA2832B	OEM
CA2832H	OEM
CA2840B	OEM
CA2842H	OEM
CA2850RB	OEM
CA2850RH	OEM
CA2870B	OEM
CA2870H	OEM
CA2875RB	OEM
CA2875RH	OEM
CA2876RB	OEM
CA2876RH	OEM
CA2904E	0770
CA2904G	OEM
CA3000	2687
CA3000T	2687
CA3001	2689
CA3001T	2689
CA3002	OEM
CA3004	OEM
CA3005	OEM
CA3006	OEM
CA3007	OEM
CA3008	OEM
CA3008A	OEM
CA3010	2537
CA3010A	2537
CA3010AT	2537
CA3010T	2537
CA3011	2617
CA3011T	2617
CA3012	2617
CA3012H	OEM
CA3012T	2617
CA3013	2624
CA3013S	2624
CA3013T	2624
CA3014	2624
CA3014S	2624
CA3014T	2624
CA3015	2537
CA3015A	2537
CA3015A1	OEM
CA3015A2	OEM
CA3015A3	OEM
CA3015A4	OEM
CA3015AT	2537
CA3015H	OEM
CA3015T	2537
CA3016	OEM
CA3016A	OEM
CA3017	0354
CA3018	1854
CA3018A	1854
CA3018A1	OEM
CA3018A2	OEM
CA3018A3	OEM
CA3018A4	OEM
CA3018AX	OEM
CA3018T	1854
CA3019	2698
CA3019T	2698
CA3020	2699
CA3020A	2699
CA3020A1	OEM
CA3020A2	OEM
CA3020A3	OEM
CA3020A4	OEM
CA3020T	2699
CA3021	OEM
CA3021E	1337
CA3022	OEM
CA3023	OEM
CA3023H	OEM
CA3026	2701
CA3026H	OEM
CA3026T	2701
CA3027EH	OEM
CA3028	0823
CA3028A	0823
CA3028AF	0823
CA3028AH	OEM
CA3028AS	0823
CA3028AT	0823
CA3028B	0823
CA3028B1	OEM
CA3028B2	OEM
CA3028B3	OEM
CA3028B4	OEM
CA3028BF	0823
CA3028BS	0823
CA3028T	0823
CA3029	2705
CA3029A	2705
CA3029E	2705
CA3030	2705
CA3030A	2705
CA3030AE	2705
CA3030E	2705
CA3035	2707
CA3035H	OEM
CA3035T	2707
CA3035V	2707
CA3035V1	2707
CA3036	OEM
CA3037	2705
CA3037A	2705
CA3037AE	2705
CA3037E	2705
CA3038	2705
CA3038A	2705
CA3038AE	2705
CA3038E	2705
CA3039	2711
CA3039T	2711
CA3039X	OEM
CA3040	OEM
CA3041	2571
CA3041P	2571
CA3041T	2571
CA3042	0829
CA3042E	0829
CA3043	2714
CA3043T	2714
CA3044	2715
CA3044T	2715
CA3044V	2715
CA3044V1	2715
CA3044VI	OEM
CA3045	1696
CA3045E	1696
CA3045F	1696
CA3045F3	OEM
CA3045F3W	OEM
CA3045F4	OEM
CA3045FX	OEM
CA3045L	1696
CA3045X	OEM
CA3046	1696
CA3046E	1696
CA3046EX	OEM
CA3047E	3808
CA3048	2529
CA3048H	OEM
CA3049T	2701
CA3049T1	OEM
CA3049T2	OEM
CA3049T3	OEM
CA3049T4	OEM
CA3050	OEM
CA3051	OEM
CA3052	2529
CA3052E	2529
CA3053	0823
CA3053T	0823
CA3054	2722
CA3054E	2722
CA3054H	OEM
CA3054X	OEM
CA3056	0410
CA3056AE	2727
CA3058	2727
CA3058E	2727
CA3059	2727
CA3059E	2727
CA3059EX	OEM
CA3059H	OEM
CA3060	OEM
CA3060AD	OEM
CA3060BD	OEM
CA3060D	OEM
CA3060E	OEM
CA3060H	OEM
CA3062	OEM
CA3064	0364
CA3064E	0802
CA3065	0167
CA3065,RCA	0167
CA3065D	0167
CA3065E	0167
CA3065F	0167
CA3065FC	0167
CA3065F.C.	0167
CA3065N	0167
CA3065PC	0167
CA3065RCA	0167
CA3066	2549
CA3066AE	2549
CA3066E	2549
CA3067	0328
CA3067AE	0328
CA3067E	0328
CA3068	2743
CA3070	0352
CA3070E	0352
CA3070G	0352
CA3071	0354
CA3071E	0354
CA3072	0666
CA3072E	0666
CA3075	1345
CA3075E	1345
CA3076	2747
CA3076T	2747
CA3078AE	1879
CA3078AH	OEM
CA3078AS	1896
CA3078AS3	OEM
CA3078AT	1896
CA3078AT1	OEM
CA3078AT2	OEM
CA3078AT3	OEM
CA3078AT4	OEM
CA3078E	1879
CA3078H	OEM
CA3078S	1896
CA3078T	1896
CA3079	2727
CA3079E	2727
CA3079EX	OEM
CA3080	2748
CA3080A	2748
CA3080A1	OEM
CA3080A2	OEM
CA3080A3	OEM
CA3080A4	OEM
CA3080AS	2032
CA3080ASX	OEM
CA3080AX	OEM
CA3080E	3836
CA3080EX	OEM
CA3080H	OEM
CA3080S	2748
CA3080SX	OEM
CA3080X	OEM
CA3081	2749
CA3081E	2749
CA3081EX	OEM
CA3081F	OEM
CA3081F3W	OEM
CA3081FX	OEM
CA3081N	OEM
CA3082	2750
CA3082F	OEM
CA3082F3W	OEM
CA3082N	OEM
CA3083	2751
CA3083EX	OEM
CA3083F	OEM
CA3083F3W	OEM
CA3084	OEM
CA3085	2333
CA3085A	6753
CA3085A1	OEM
CA3085A2	OEM
CA3085A3	OEM
CA3085A4	OEM
CA3085AE	2333
CA3085AS	OEM
CA3085AS3	OEM
CA3085B	OEM
CA3085B1	OEM
CA3085B2	OEM
CA3085B3	OEM
CA3085B4	OEM
CA3085BS	OEM
CA3085BS3	OEM
CA3085BT3	OEM
CA3085E	2333
CA3085H	OEM
CA3085S	OEM
CA3086	1696
CA3086E	1696
CA3086EX	OEM
CA3086F	OEM
CA-3088E	2257
CA3088E	2257
CA-3089E	2755
CA3089E	2755
CA3089F	2755
CA3089F3W	OEM
CA3089N	OEM
CA3090	2755
CA3090AB	1843
CA3090AE	OEM
CA3090AQ	1843
CA3090E	1843
CA3090Q	1843
CA3091D	OEM
CA3091H	OEM
CA3093E	OEM
CA3094AE	OEM
CA3094AS	3022
CA3094AS3	OEM
CA3094AT	3022
CA3094AT1	OEM
CA3094AT2	OEM
CA3094AT3	OEM
CA3094AT4	OEM
CA3094BS	OEM
CA3094BT	OEM
CA3094E	OEM
CA3094EX	OEM
CA3094H	OEM
CA3094S	3022
CA3094T	3022
CA3094T1	OEM
CA3094T2	OEM
CA3094T3	OEM
CA3094T4	OEM
CA3095E	OEM
CA3096AE	3129
CA3096AEX	OEM
CA3096CE	3129
CA3096E	3129
CA3096EX	OEM
CA3097E	6582
CA3097EX	OEM
CA3098E	OEM
CA3098S	OEM
CA3098T	OEM
CA3099E	OEM
CA3099EX	OEM
CA3100E	OEM
CA3100H	OEM
CA3100S	3394
CA3100S3	OEM
CA3100SX	OEM
CA3100T	3394
CA3100T1	OEM
CA3100T2	OEM
CA3100T3	OEM
CA3100T4	OEM
CA3100TY	3394
CA3102E	OEM
CA3102H	OEM
CA3118AT	OEM
CA3118AT1	OEM
CA3118AT2	OEM
CA3118AT3	OEM
CA3118AT4	OEM
CA3118T	OEM
CA3120E	2457
CA3121	1337
CA3121E	1337
CA3121E-G	1337
CA3121EG	1337
CA3121G	1337
CA3123	1421
CA3123E	1421
CA3123Q	OEM
CA3125	1753
CA3125E	1753
CA3126	0521
CA3126E	OEM
CA3126EM	0521
CA3126EM1	0521
CA3126EMI	0521
CA3126FM1	0521
CA3126P	OEM
CA3126Q	0521
CA3127E	OEM
CA3127EX	OEM
CA3128E	OEM
CA3128Q	OEM
CA3130	2782
CA3130A	2782
CA3130AE	2782
CA3130AS	2782
CA3130AS3	OEM
CA3130AT	2782
CA3130AT1	OEM
CA3130AT2	OEM
CA3130AT3	OEM
CA3130AT4	OEM
CA3130B	2782
CA3130BS	OEM
CA3130BT	OEM
CA3130E	6795
CA3130EX	OEM
CA3130H	OEM
CA3130S	2782
CA3130S3	OEM
CA3130SX	OEM
CA3130T	2782
CA3130T1	OEM
CA3130T2	OEM
CA3130T3	OEM
CA3130T4	OEM
CA3130TX	OEM
CA3131EM	OEM
CA3131PM	OEM
CA3132EM	OEM
CA3132PM	OEM
CA3134	2786
CA3134EM	2786
CA3134G	OEM
CA3134GM	4362
CA3134GQM	2786
CA3134P	OEM
CA3134PM	OEM
CA3134PQM	OEM
CA3134Q	2786
CA3134QM	2786
CA3135E	2501
CA3135G	2501
CA3136E	2502
CA3136G	2502
CA3137	2789
CA3137E	2789
CA3137EM	2789
CA3137EM1	2789
CA3137EMI	2789
CA3137Q	OEM
CA3138AG	OEM
CA3138G	OEM
CA3139	2049
CA3139E	2049
CA3139G	2049
CA3139GM1	2049
CA3139GMI	2049
CA3139GQ	2049
CA3139Q	2049
CA3140AE	4473
CA3140AEX	OEM
CA3140AS	4478
CA3140AS1	OEM
CA3140AS2	OEM
CA3140AS3	OEM
CA3140AS4	OEM
CA3140ASX	OEM
CA3140AT	4478
CA3140AT1	OEM
CA3140AT2	OEM
CA3140AT3	OEM
CA3140AT4	OEM
CA3140ATX	OEM
CA3140BS	OEM
CA3140BT	OEM
CA3140E	4473
CA3140EX	OEM
CA3140H	OEM
CA3140S	4478
CA3140S1	OEM
CA3140S2	OEM
CA3140S3	OEM
CA3140S4	OEM
CA3140SX	OEM
CA3140T	4478
CA3140T1	OEM
CA3140T2	OEM
CA3140T3	OEM
CA3140T4	OEM
CA3140TX	OEM
CA3141E	OEM
CA3142	2457
CA3142E	2457
CA3143	2798
CA3143E	2798
CA3143Q	2798
CA3144E	3832
CA3144G	3832
CA3144GQ	OEM
CA3144Q	3832
CA3145	2801
CA3145G	6803
CA3146AE	2499
CA3146E	2499
CA3151E	5924
CA3151G	3089
CA3151GM1	3089
CA3151GMI	3089
CA3152E	0395
CA3153G	3270
CA3153GM1	3270
CA3153GMI	3270
CA3154G	3720
CA3156	2812
CA3156E	2812
CA3156G	2812
CA3157G	3788
CA3158G	3719
CA3159	2817
CA3159G	2817
CA3160AE	5052
CA3160AS	5055
CA3160AS1	OEM
CA3160AS2	OEM
CA3160AS3	OEM
CA3160AS4	OEM
CA3160AT	5055
CA3160AT1	OEM
CA3160AT2	OEM
CA3160AT3	OEM
CA3160AT4	OEM
CA3160BS	OEM
CA3160BT	OEM
CA3160E	5052
CA3160EX	OEM
CA3160H	OEM
CA3160S	5055
CA3160S1	OEM
CA3160S2	OEM
CA3160S3	OEM
CA3160S4	OEM
CA3160SX	OEM
CA3160T	5055
CA3160T1	OEM
CA3160T2	OEM
CA3160T3	OEM
CA3160T4	OEM
CA3160TX	OEM
CA3161E	4950
CA3161EX	OEM
CA3162E	0521
CA3162EX	OEM
CA3163G	3790
CA3164E	OEM
CA3165E	OEM
CA3165E1	OEM
CA3166E	5609
CA3168	2827
CA3168E	2827
CA3169M	OEM
CA3170	2830
CA3170E	2830
CA3170G	OEM
CA3172	2831
CA3172E	2831
CA3172G	2831
CA3177G	3990
CA3179G	OEM
CA3180Q	OEM
CA3183AE	OEM
CA3183AEX	OEM
CA3183E	OEM
CA3183EX	OEM
CA3189E	6163
CA3190G	OEM
CA3191E	3976
CA3192E	3784
CA3193AE	OEM
CA3193AEX	OEM
CA3193AS	OEM
CA3193ASX	OEM
CA3193AT	OEM
CA3193ATX	OEM
CA3193BE	OEM
CA3193BEX	OEM
CA3193BS	OEM
CA3193BSX	OEM
CA3193BT	OEM
CA3193BTX	OEM
CA3193E	OEM
CA3193EX	OEM
CA3193S	OEM
CA3193SX	OEM
CA3193T	OEM
CA3193TX	OEM

If replacement code is OEM, contact original manufacturer for replacement.

DEVICE TYPE	REPL CODE	DEVICE TYPE	REPL CODE	DEVICE TYPE	REPL CODE	DEVICE TYPE	REPL CODE	DEVICE TYPE	REPL CODE	DEVICE TYPE	REPL CODE	DEVICE TYPE	REPL CODE
CA3194E	6823	CA30262	OEM	CBC800/204TCL	OEM	CCS6168G	0086	CD32-12047	OEM	CD952	OEM	CD3128	OEM
CA3195	2842	CA30263	OEM	CBC800/204TIH	OEM	CCS6225F	0007	CD32-12048	3100	CD962	OEM	CD3129	OEM
CA3195E	2842	CA30264	OEM	CBC800/204TIL	OEM	CCS6226G	0007	CD32-12049	0012	CD962B	OEM	CD3154	OEM
CA3199E	OEM	CA30391	OEM	CBC800/208SCH	OEM	CCS6227F	0007	CD32-12050	OEM	CD963B	0052	CD3155	OEM
CA3207EE	OEM	CA30392	OEM	CBC800/208SCL	OEM	CCS6228F	0037	CD32-12051	OEM	CD964B	OEM	CD3156	OEM
CA3208EE	OEM	CA30393	OEM	CBC800/208SIH	OEM	CCS6229H	0086	CD32-12052	0137	CD965B	0686	CD3168	OEM
CA3208EH	OEM	CA30394	OEM	CBC800/208SIL	OEM	CCS7228	OEM	CD32-12053	OEM	CD966B	OEM	CD3169	OEM
CA3218E	1389	CA30451	OEM	CBC800/208TCH	OEM	CCS9016D	0144	CD32-12054	0002	CD967B	OEM	CD3171	0783
CA3221	1878	CA30452	OEM	CBC800/208TCL	OEM	CCS9016E	0144	CD32-12055	OEM	CD968B	OEM	CD3172	OEM
CA3221G	1878	CA30453	OEM	CBC800/208TIH	OEM	CCS9017	0007	CD32-12056	OEM	CD969B	OEM	CD3173	OEM
CA3240AE	6142	CA30454	OEM	CBC800/208TIL	OEM	CCS9017G925	0007	CD32-12057	OEM	CD970B	OEM	CD3174	0331
CA3240AE1	1883	CA30581	OEM	CBC800/216SCH	OEM	CCS9018E	0016	CD32-12058	OEM	CD971B	OEM	CD3285	OEM
CA3240E	6142	CA30582	OEM	CBC800/216SCL	OEM	CCS9018F	0144	CD32-12059	OEM	CD972	OEM	CD3400	OEM
CA3240E1	1883	CA30583	OEM	CBC800/216SIH	OEM	CCS9018H924	0007	CD32-12060	OEM	CD972B	OEM	CD3401	OEM
CA3240EX	OEM	CA30801	OEM	CBC800/216SIL	OEM	CD-0000	0143	CD32-12061	OEM	CD973B	OEM	CD3403	OEM
CA3240H	OEM	CA30803	OEM	CBC800/216TCH	OEM	CD0000	0133	CD32-12062	OEM	CD974B	OEM	CD3449	OEM
CA3260AE	6319	CA30804	OEM	CBC800/216TCL	OEM	CD-0000N	0143	CD33	OEM	CD975B	OEM	CD3811	OEM
CA3260AS	6321	CA30853	OEM	CBC800/216TIH	OEM	CD000	0143	CD34	OEM	CD976B	OEM	CD3872	OEM
CA3260AT	6321	CA30854	OEM	CBC800/216TIL	OEM	CD005	0015	CD35	OEM	CD977B	OEM	CD4000AD	2024
CA3260BS	OEM	CA30900	1843	CC095509	OEM	CD0014	0133	CD36	OEM	CD978B	OEM	CD4000AD3	OEM
CA3260BT	OEM	CA69001	OEM	CC4	OEM	CD0014(MORSE)	0133	CD-37	0133	CD979B	OEM	CD4000AE	2024
CA3260E	6319	CA69001A	OEM	CC4H	OEM	CD-0014N	0143	CD37	0133	CD980B	OEM	CD4000AEX	OEM
CA3260EX	OEM	CA69002	OEM	CC4V	OEM	CD0014NA	0016	CD-37A	0133	CD981B	OEM	CD4000AF	2024
CA3260S	6321	CA69002A	OEM	CC-8	OEM	CD0014NG	0016	CD37A	0133	CD982	OEM	CD4000AF3	OEM
CA3260SX	OEM	CA69003	OEM	CC8	OEM	CD0015N	0016	CD37A2	0015	CD982B	OEM	CD4000AF3W	OEM
CA3260T	6321	CA69003A	OEM	CC8V	OEM	CD-0021	0133	CD38	6326	CD983B	OEM	CD4000AFB	OEM
CA3260TX	OEM	CA69004	OEM	CC-12	OEM	CD0021	0016	CD39	OEM	CD984B	OEM	CD4000AFX	OEM
CA3280	OEM	CA69004A	OEM	CC16	OEM	CD-0033	0636	CD40A6AK	OEM	CD985B	OEM	CD4000AK	OEM
CA3280AG	OEM	CAC5028A	0127	CC16V	OEM	CD0033	0636	CD43	OEM	CD986B	OEM	CD4000AK3	OEM
CA3280E	OEM	CAG13A	OEM	CC80-C2	OEM	CD0057	0133	CD47	OEM	CD987B	OEM	CD4000AKB	OEM
CA3280G	OEM	CAM601A	OEM	CC80-PP	OEM	CD-0071	0030	CD48	0157	CD988B	OEM	CD4000AZ	2024
CA3289E	2755	CAM604A	OEM	CC80-PU	OEM	CD-05	0015	CD51	OEM	CD989B	OEM	CD4000BD	2024
CA3290AE	0631	CARGENV1.0	OEM	CC102BA	0015	CD05	0015	CD56	OEM	CD990B	OEM	CD4000BD3	OEM
CA3290AE1	OEM	CAT24C01A	OEM	CC102DA	0015	CD-07	0474	CD61	OEM	CD991B	OEM	CD4000BE	2024
CA3290AS	6589	CAT24C01B1	OEM	CC102FA	0015	CD07	0474	CD62	0471	CD992B	OEM	CD4000BEX	OEM
CA3290AT	6589	CAT35C102HP	OEM	CC102HA	0015	CD09	2966	CD63	OEM	CD1111	0015	CD4000BF	2024
CA3290BS	OEM	CAT35C102P	OEM	CC102KA	0015	CD1	OEM	CD68	OEM	CD1112	0015	CD4000BF3	OEM
CA3290BT	OEM	CAT35C202P	OEM	CC102MA	0015	CD-2	0015	CD73/187/72	0569	CD1113	0015	CD4000BF3W	OEM
CA3290E	0631	CAT59C11HP	OEM	CC102PA	0071	CD2	0424	CD73/187/73	5383	CD1114	0015	CD4000BFX	OEM
CA3290E1	OEM	CAT59C11P	OEM	CC102RA	0071	CD2A	5007	CD74AC00E	OEM	CD1115	0015	CD4000BK	OEM
CA3290EX	OEM	CAT93C46AN	OEM	CC102VA	0071	CD-2N	0015	CD74HCT02E	4910	CD1116	0015	CD4000CJ	2024
CA3290S	6589	CAT93C46P	OEM	CC152AA	0071	CD-4	0015	CD74HCT04E	3296	CD1117	0015	CD4000CN	2024
CA3290T	6589	CAX15	OEM	CC152BA	0071	CD5	OEM	CD74HCT08E	3299	CD1121	0015	CD4000MJ	2024
CA3300D	OEM	CAX20	OEM	CC152CA	0071	CD12	OEM	CD74HCT14E	3300	CD1122	0015	CD4000MW	2024
CA3300H	OEM	CAX25	OEM	CC152DA	0071	CD-20	0133	CD74HCT32E	3306	CD1123	0015	CD4000UBD	2024
CA3308D	OEM	CAX30	OEM	CC152EA	0071	CD21	OEM	CD74HCT138E	6990	CD1124	0015	CD4000UBD3	OEM
CA3401E	2247	CAX1124AS	OEM	CC152FA	0071	CD22	OEM	CD74HCT195E	6100	CD1125	0015	CD4000UBE	2024
CA3401H	OEM	CAY10	OEM	CC152GA	0071	CD23	OEM	CD74HCT244E	6108	CD1126	0015	CD4000UBEX	OEM
CA3493AE	OEM	CAY11	OEM	CC152HA	0071	CD24	OEM	CD74HCT245E	4075	CD1127	0015	CD4000UBF	2024
CA3493AS	OEM	CAY12A	OEM	CC152KA	0071	CD26	OEM	CD74HCT373E	6123	CD1141	0015	CD4000UBF3W	OEM
CA3493E	OEM	CAY12B	OEM	CC152MA	0071	CD27	OEM	CD75	OEM	CD1142	0015	CD4000UBFX	OEM
CA3493S	OEM	CAY13	OEM	CC152PA	0071	CD28	OEM	CD82	OEM	CD1143	0015	CD4000UBK	OEM
CA3524E	5233	CAY14	OEM	CC152RA	0071	CD29	OEM	CD88P105	OEM	CD1147	0015	CD4001	0478
CA3524G	OEM	CAY15	OEM	CC152TA	0071	CD31	OEM	CD91	0057	CD1148	0015	CD4001AD	0478
CA3524H	OEM	CAY16	OEM	CC152VA	0071	CD31-00001	OEM	CD92	OEM	CD1149	0015	CD4001AD3	OEM
CA3539	2705	CAY17	OEM	CC280	OEM	CD31-00002	0188	CD93	OEM	CD1151	0015	CD4001AE	0478
CA3545	2711	CB1A	OEM	CC1168F	0016	CD31-00003	OEM	CD94	OEM	CD1224	0133	CD4001AEX	OEM
CA3600E	OEM	CB1F4	0440	CC1600	OEM	CD31-00004	OEM	CD95	OEM	CD1324N	OEM	CD4001AF	0478
CA3741CE	0312	CB2Z-80	OEM	CC4012	OEM	CD31-00005	OEM	CD96	OEM	CD2147	OEM	CD4001AF3	OEM
CA3741E	0312	CB5	0015	CC4013	OEM	CD31-00006	0162	CD97	OEM	CD2155NL	OEM	CD4001AF3W	OEM
CA3748CT	0093	CB5F	OEM	CC4015	OEM	CD31-00007	0157	CD98	OEM	CD2156NL	OEM	CD4001AFB	OEM
CA5470M	OEM	CB10	0015	CC4016	OEM	CD31-00008	0636	CD100	OEM	CD2157NL	OEM	CD4001AFX	OEM
CA6078AH	OEM	CB10F	OEM	CC4018	OEM	CD31-00009	OEM	CD101	0143	CD2300E	1823	CD4001AK	OEM
CA6078AS	OEM	CB20	0015	CC4020	OEM	CD31-00010	OEM	CD200	OEM	CD2300E/830	1823	CD4001AK3	OEM
CA6078AT	OEM	CB20F	OEM	CC4022	OEM	CD31-00011	3100	CD201	OEM	CD2301E	1859	CD4001AKB	OEM
CA6078T	OEM	CB40	OEM	CC4024	OEM	CD31-00012	0012	CD202	OEM	CD2301E/861	1859	CD4001B	OEM
CA6664	0015	CB40F	OEM	CC4027	OEM	CD31-00013	OEM	CD203	OEM	CD2302E	0141	CD4001BCJ	0478
CA6665	0015	CB50	0015	CC4030	OEM	CD31-00014	OEM	CD400EBF	OEM	CD2302E/846	0141	CD4001BCN	0478
CA6741CS	OEM	CB60	OEM	CC4033	OEM	CD31-00015	0137	CD510BE	OEM	CD2303E	1844	CD4001BD	0478
CA6741CT	OEM	CB60F	OEM	CC7228G	OEM	CD31-00016	OEM	CD517BD	OEM	CD2303E/849	1844	CD4001BD3	OEM
CA6741S	OEM	CB80	OEM	CC59018F	0079	CD31-00017	0002	CD610	OEM	CD2304E	0081	CD4001BE	0478
CA6741T	OEM	CB80F	OEM	CC104816	OEM	CD31-00018	OEM	CD643	OEM	CD2304E/845	0081	CD4001BEX	OEM
CA6741T3	OEM	CB100	0087	CC105009	OEM	CD31-00019	OEM	CD746A	OEM	CD2305E	3399	CD4001BF	0478
CA6741TX	OEM	CB100F	OEM	CC105016	OEM	CD31-00020	OEM	CD747A	OEM	CD2305E/848	3399	CD4001BF3	OEM
CA7248	0015	CB101	OEM	CC105209	OEM	CD31-00021	OEM	CD748A	OEM	CD2306E	1041	CD4001BF3W	OEM
CA7520	0539	CB-103	0280	CC105216	OEM	CD31-00022	OEM	CD749A	OEM	CD2306E/832	1041	CD4001BFX	OEM
CA8089F	OEM	CB103	0136	CC105509	OEM	CD31-00023	0441	CD750A	OEM	CD2307E	0033	CD4001BK	OEM
CA8100M	OEM	CB104	0211	CC134816	OEM	CD31-00024	OEM	CD751A	OEM	CD2307E/844	0033	CD4001BK3	OEM
CA8314	0015	CB106	0143	CC135009	OEM	CD31-00025	0039	CD752A	OEM	CD2308E	0562	CD4001BMJ	0478
CA19199	OEM	CB150	0015	CC135016	OEM	CD31-10361	0441	CD753A	0471	CD2308E/862	0562	CD4001BMW	0478
CA19200	OEM	CB156	0050	CC135509	OEM	CD31-10365	0002	CD754A	OEM	CD2309E	0341	CD4001CJ	0478
CA30001	OEM	CB157	0136	CC204816	OEM	CD31-12008	OEM	CD755A	OEM	CD2309E/863	0341	CD4001CN	0478
CA30002	OEM	CB158	0136	CC205009	OEM	CD31-12009	OEM	CD756A	OEM	CD2310E	1831	CD4001MJ	0478
CA30003	OEM	CB161	0004	CC205216	OEM	CD31-12010	OEM	CD758A	OEM	CD2310E/836	1831	CD4001MW	0478
CA30004	OEM	CB163	0143	CC205509	OEM	CD31-12011	OEM	CD759A	0052	CD2311E/837	1835	CD4001UBD	0478
CA30011	OEM	CB200	0015	CC215509	OEM	CD31-12012	OEM	CD821	0471	CD2312E	1176	CD4001UBD3	OEM
CA30012	OEM	CB201	OEM	CCB-1	OEM	CD31-12013	OEM	CD821A	0471	CD2314E	2097	CD4001UBE	0478
CA30013	OEM	CB202	OEM	CCC3500	OEM	CD31-12014	OEM	CD823	0471	CD2314E/833	2097	CD4001UBEX	OEM
CA30014	OEM	CB203	OEM	CCD450ADC	OEM	CD31-12015	0636	CD823A	0471	CD2315E	0358	CD4001UBF	0478
CA30021	OEM	CB244	0050	CCD450DC	OEM	CD31-12016	OEM	CD825	0471	CD2318E	0333	CD4001UBF3	OEM
CA30022	OEM	CB246	0016	CCD3000	OEM	CD31-12017	OEM	CD825A	0471	CD2434	OEM	CD4001UBF3W	OEM
CA30023	OEM	CB248	0004	CCI-2001	OEM	CD31-12018	3100	CD826	OEM	CD2500E	OEM	CD4001UBFX	OEM
CA30024	OEM	CB249	0004	CCI2001	OEM	CD31-12019	0012	CD827	0471	CD2501E	OEM	CD4001UBK	OEM
CA30041	OEM	CB250	0015	CCI2002	OEM	CD31-12020	OEM	CD827A	0471	CD2502E	OEM	CD4002	2055
CA30042	OEM	CB254	0050	CCI3001	OEM	CD31-12021	OEM	CD828	OEM	CD2503E	OEM	CD4002AD	2055
CA30043	OEM	CB-393	0143	CCJ5212	0079	CD31-12022	0137	CD829	0471	CD2505	OEM	CD4002AD3	OEM
CA30044	OEM	CB393	0143	CCS-1025	OEM	CD31-12023	OEM	CD829A	0471	CD2545	2312	CD4002AE	2055
CA30061	OEM	CB5020	OEM	CCS-1025-1	OEM	CD31-12024	0002	CD860	OEM	CD2810	OEM	CD4002AEX	OEM
CA30062	OEM	CB5021	OEM	CCS-1143	OEM	CD31-12025	OEM	CD912	OEM	CD2812	OEM	CD4002AF	2055
CA30063	OEM	CB5022	OEM	CCS1235G	0016	CD31-12026	OEM	CD914	OEM	CD2813	OEM	CD4002AF3	OEM
CA30064	OEM	CB5023	OEM	CCS2001H	0079	CD31-12027	OEM	CD922	OEM	CD3025	OEM	CD4002AF3W	OEM
CA30143	OEM	CBA110	OEM	CCS2004B	0111	CD31-12028	OEM	CD932	OEM	CD3034BFX	OEM	CD4002AFB	OEM
CA30144	OEM	CBA111	OEM	CCS2005B	0037	CD31-12029	OEM	CD935	OEM	CD3080	2312	CD4002AFX	OEM
CA30153	OEM	CBA112	OEM	CCS-2006D	0079	CD31-12030	0441	CD935A	OEM	CD3122	0162	CD4002AK	OEM
CA30191	OEM	CBC800/204SCH	OEM	CCS2006D	0144	CD31-12031	OEM	CD936	OEM	CD3123	OEM	CD4002AK3	OEM
CA30192	OEM	CBC800/204SCL	OEM	CCS2008F015	0144	CD31-12032	0039	CD936A	OEM	CD3124	OEM	CD4002AKB	OEM
CA30193	OEM	CBC800/204SIH	OEM	CCS4004	0016	CD31-12039	0012	CD937	OEM	CD3125	OEM	CD4002BD	2055
CA30194	OEM	CBC800/204SIL	OEM	CCS6168	0016	CD32	OEM	CD937A	OEM	CD3126	OEM	CD4002BD3	OEM
CA30261	OEM	CBC800/204TCH	OEM	CCS6168F	0016	CD32-12046	OEM	CD942	OEM	CD3127	OEM	CD4002BE	2055

DEVICE TYPE	REPL CODE
CD4002BEX	OEM
CD4002BF	2055
CD4002BF3	OEM
CD4002BF3W	OEM
CD4002BFX	OEM
CD4002BK	OEM
CD4002BK3	OEM
CD4002CJ	2055
CD4002CN	2055
CD4002MJ	2055
CD4002MW	2055
CD4002UBD	2055
CD4002UBD3	OEM
CD4002UBE	2055
CD4002UBEX	OEM
CD4002UBF	2055
CD4002UBF3	OEM
CD4002UBF3W	OEM
CD4002UBFX	OEM
CD4002UBK	OEM
CD4004AE	OEM
CD4006	0646
CD4006AD	OEM
CD4006AD3	OEM
CD4006AE	0646
CD4006AEX	OEM
CD4006AF	OEM
CD4006AF3	OEM
CD4006AF3W	OEM
CD4006AFB	OEM
CD4006AFX	OEM
CD4006AH	OEM
CD4006AK	OEM
CD4006AK3	OEM
CD4006AKB	OEM
CD4006BD	OEM
CD4006BD3	OEM
CD4006BE	0646
CD4006BEX	OEM
CD4006BF	OEM
CD4006BF3	OEM
CD4006BF3W	OEM
CD4006BH	OEM
CD4006BK	OEM
CD4006BK3	OEM
CD4007AD	OEM
CD4007AD3	OEM
CD4007AE	2846
CD4007AEX	OEM
CD4007AF	2846
CD4007AF3	OEM
CD4007AF3W	OEM
CD4007AFB	OEM
CD4007AFX	OEM
CD4007AK3	OEM
CD4007AKB	OEM
CD4007C	OEM
CD4007CJ	2846
CD4007CN	2846
CD4007M	OEM
CD4007UBD	4984
CD4007UBE	2846
CD4007UBEX	OEM
CD4007UBF	4984
CD4007UBF3	OEM
CD4007UBF3W	OEM
CD4007UBFX	OEM
CD4007UBK	OEM
CD4007UBK3	OEM
CD4008AD	0988
CD4008AD3	OEM
CD4008AE	0988
CD4008AEX	OEM
CD4008AF	0988
CD4008AF3	OEM
CD4008AF3W	OEM
CD4008AFX	OEM
CD4008AK	OEM
CD4008AK3	OEM
CD4008B	0988
CD4008BCJ	0988
CD4008BCN	0988
CD4008BD	0988
CD4008BD3	OEM
CD4008BE	0988
CD4008BEX	OEM
CD4008BF	0988
CD4008BF3	OEM
CD4008BF3W	OEM
CD4008BJ	0988
CD4008BK	OEM
CD4008BK3	OEM
CD4008BMJ	0988
CD4008BMW	0988
CD4009AD	1999
CD4009AE	0001
CD4009AF	1999
CD4009AH	OEM
CD4009AK	OEM
CD4009AY	OEM
CD4009CJ	0001
CD4009CN	0001
CD4009MJ	3260
CD4009MW	OEM
CD4009UBD	1999
CD4009UBD3	OEM
CD4009UBE	1999
CD4009UBEX	OEM
CD4009UBF	1999
CD4009UBF3	OEM
CD4009UBF3W	OEM
CD4009UBFX	OEM
CD4009UBH	OEM
CD4009UBK	OEM
CD4009UBK3	OEM
CD4010AD	3252
CD4010AE	0398
CD4010AF	3252
CD4010AH	OEM
CD4010AK	OEM
CD4010AY	OEM
CD4010BD	3252
CD4010BD3	OEM
CD4010BE	0398
CD4010BEX	OEM
CD4010BF	3252
CD4010BF3	OEM
CD4010BF3W	OEM
CD4010BFX	OEM
CD4010BH	OEM
CD4010BK	OEM
CD4010BK3	OEM
CD4010CJ	0398
CD4010CN	0398
CD4010MJ	3252
CD4010MW	OEM
CD4011	0215
CD4011AD	0215
CD4011AD3	OEM
CD4011AE	0215
CD4011AEX	OEM
CD4011AF	0215
CD4011AF3	OEM
CD4011AF3W	OEM
CD4011AFB	OEM
CD4011AFX	OEM
CD4011AK	OEM
CD4011AK3	OEM
CD4011AKB	OEM
CD4011BCJ	0215
CD4011BCN	0215
CD4011BD	0215
CD4011BE	0215
CD4011BEX	OEM
CD4011BF	0215
CD4011BF3	OEM
CD4011BF3W	OEM
CD4011BFX	OEM
CD4011BK	OEM
CD4011BK3	OEM
CD4011BMJ	0215
CD4011BMW	0215
CD4011CJ	0215
CD4011CN	0215
CD4011MJ	0215
CD4011MW	0215
CD4011UBD	0215
CD4011UBD3	OEM
CD4011UBE	0215
CD4011UBEX	OEM
CD4011UBF	0215
CD4011UBF3	OEM
CD4011UBF3W	OEM
CD4011UBFX	OEM
CD4011UBK	OEM
CD4011UBK3	OEM
CD4012	0498
CD4012AD	0498
CD4012AD3	OEM
CD4012AE	0498
CD4012AEX	OEM
CD4012AF	0498
CD4012AF3	OEM
CD4012AF3W	OEM
CD4012AFB	OEM
CD4012AFX	OEM
CD4012AK	OEM
CD4012AK3	OEM
CD4012AKB	OEM
CD4012BD	0498
CD4012BD3	OEM
CD4012BE	0498
CD4012BEX	OEM
CD4012BF	0498
CD4012BF3	OEM
CD4012BF3W	OEM
CD4012BFX	OEM
CD4012BK	OEM
CD4012BK3	OEM
CD4012CJ	0498
CD4012CN	0498
CD4012MJ	0498
CD4012MW	0498
CD4012UBD	0498
CD4012UBD3	OEM
CD4012UBE	0498
CD4012UBEX	OEM
CD4012UBF	0498
CD4012UBF3	OEM
CD4012UBF3W	OEM
CD4012UBK	OEM
CD4013	0413
CD4013AD	0413
CD4013AD3	OEM
CD4013AE	0413
CD4013AEX	OEM
CD4013AF	0413
CD4013AF3	OEM
CD4013AF3W	OEM
CD4013AFB	OEM
CD4013AFX	OEM
CD4013AK	OEM
CD4013AK3	OEM
CD4013AKB	OEM
CD4013B	0413
CD4013BCJ	0413
CD4013BCN	0413
CD4013BD	0413
CD4013BD3	OEM
CD4013BE	0413
CD4013BEX	OEM
CD4013BF	0413
CD4013BF3	OEM
CD4013BF3W	OEM
CD4013BFX	OEM
CD4013BK	OEM
CD4013BK3	OEM
CD4013BMJ	0413
CD4013BMW	0413
CD4013BN	0413
CD4014AD	0860
CD4014AD3	OEM
CD4014AE	0860
CD4014AEX	OEM
CD4014AF	0860
CD4014AF3	OEM
CD4014AF3W	OEM
CD4014AFB	OEM
CD4014AFX	OEM
CD4014AH	OEM
CD4014AK	OEM
CD4014AK3	OEM
CD4014AKB	OEM
CD4014BD	0860
CD4014BD3	OEM
CD4014BE	0860
CD4014BEX	OEM
CD4014BF	0860
CD4014BF3	OEM
CD4014BF3W	OEM
CD4014BFX	OEM
CD4014BH	OEM
CD4014BK	OEM
CD4014BK3	OEM
CD4015	1014
CD4015AD	1014
CD4015AD3	OEM
CD4015AE	1014
CD4015AEX	OEM
CD4015AF	1014
CD4015AF3	OEM
CD4015AF3W	OEM
CD4015AFB	OEM
CD4015AFX	OEM
CD4015AH	OEM
CD4015AK	OEM
CD4015AK3	OEM
CD4015AKB	OEM
CD4015BD	1014
CD4015BD3	OEM
CD4015BE	1014
CD4015BEX	OEM
CD4015BF	1014
CD4015BFX	OEM
CD4015BH	OEM
CD4015BK	OEM
CD4015BK3	OEM
CD4016	1143
CD4016AD	1143
CD4016AD3	OEM
CD4016AE	1143
CD4016AEX	OEM
CD4016AF	1143
CD4016AF3	OEM
CD4016AF3W	OEM
CD4016AFX	OEM
CD4016AH	OEM
CD4016AK	OEM
CD4016AK3	OEM
CD4016AY	OEM
CD4016B	1143
CD4016BD	1143
CD4016BD3	OEM
CD4016BE	1143
CD4016BEX	OEM
CD4016BF	1143
CD4016BF3	OEM
CD4016BF3W	OEM
CD4016BFX	OEM
CD4016BH	OEM
CD4016BK	OEM
CD4016CJ	1143
CD4016CN	1143
CD4016MJ	1143
CD4017	0513
CD4017AD	0513
CD4017AD3	OEM
CD4017AE	0513
CD4017AEX	OEM
CD4017AF	0513
CD4017AF3	OEM
CD4017AF3W	OEM
CD4017AFB	OEM
CD4017AFX	OEM
CD4017AK	OEM
CD4017AK3	OEM
CD4017BCJ	0513
CD4017BCN	0513
CD4017BD	0513
CD4017BD3	OEM
CD4017BE	0513
CD4017BEX	OEM
CD4017BF	0513
CD4017BF3	OEM
CD4017BF3W	OEM
CD4017BFX	OEM
CD4017BK	OEM
CD4017BK3	OEM
CD4017BMJ	0513
CD4017BMW	0513
CD4018	1391
CD4018AD	1391
CD4018AD3	OEM
CD4018AE	1391
CD4018AEX	OEM
CD4018AF	1391
CD4018AFB	OEM
CD4018AFX	OEM
CD4018AK	OEM
CD4018AK3	OEM
CD4018AK8	OEM
CD4018AKB	OEM
CD4018BAF3W	OEM
CD4018BCJ	1391
CD4018BCN	1391
CD4018BD	1391
CD4018BD3	OEM
CD4018BE	1391
CD4018BEX	OEM
CD4018BF	1391
CD4018BF3	OEM
CD4018BF3W	OEM
CD4018BFX	OEM
CD4018BK	OEM
CD4018BK3	OEM
CD4018BMJ	1391
CD4018BMW	1391
CD4019	1525
CD4019AD	1525
CD4019AD3	OEM
CD4019AE	1525
CD4019AEX	OEM
CD4019AF	1525
CD4019AF3	OEM
CD4019AF3W	OEM
CD4019AFB	OEM
CD4019AFX	OEM
CD4019AH	OEM
CD4019AK	OEM
CD4019AK3	OEM
CD4019AKB	OEM
CD4019AY	OEM
CD4019BC	1525
CD4019BCJ	1525
CD4019BCN	1525
CD4019BD	1525
CD4019BD3	OEM
CD4019BE	1525
CD4019BEX	OEM
CD4019BF	1525
CD4019BF3	OEM
CD4019BF3W	OEM
CD4019BFX	OEM
CD4019BK	OEM
CD4019BK3	OEM
CD4019BM	OEM
CD4019BMJ	6254
CD4019BMW	OEM
CD4020	1661
CD4020AD	1661
CD4020AD3	OEM
CD4020AE	1661
CD4020AEX	OEM
CD4020AF	1661
CD4020AF3	OEM
CD4020AF3W	OEM
CD4020AFB	OEM
CD4020AFX	OEM
CD4020AK	OEM
CD4020AK3	OEM
CD4020AKB	OEM
CD4020BCJ	1661
CD4020BCN	1661
CD4020BD	1661
CD4020BD3	OEM
CD4020BE	1661
CD4020BEX	OEM
CD4020BF	1661
CD4020BF3	OEM
CD4020BF3W	OEM
CD4020BFX	OEM
CD4020BK3	OEM
CD4020BMJ	1661
CD4020BMW	1661
CD4021	1749
CD4021AD	1749
CD4021AD3	OEM
CD4021AE	1749
CD4021AEX	OEM
CD4021AF	1749
CD4021AF3	OEM
CD4021AF3W	OEM
CD4021AFB	OEM
CD4021AFX	OEM
CD4021AH	OEM
CD4021AK	OEM
CD4021AK3	OEM
CD4021AKB	OEM
CD4021AY	OEM
CD4021BD	1749
CD4021BD3	OEM
CD4021BE	1749
CD4021BEX	OEM
CD4021BF	1749
CD4021BF3	OEM
CD4021BF3W	OEM
CD4021BFX	OEM
CD4021BH	OEM
CD4021BK	OEM
CD4021BK3	OEM
CD4022	1256
CD4022AD	1256
CD4022AD3	OEM
CD4022AE	1256
CD4022AEX	OEM
CD4022AF	1256
CD4022AF3	OEM
CD4022AF3W	OEM
CD4022AFB	OEM
CD4022AFX	OEM
CD4022AK	OEM
CD4022AK3	OEM
CD4022AKB	OEM
CD4022BCJ	1256
CD4022BCN	1256
CD4022BD	1256
CD4022BD3	OEM
CD4022BE	1256
CD4022BEX	OEM
CD4022BF	1256
CD4022BF3	OEM
CD4022BF3W	OEM
CD4022BFX	OEM
CD4022BK	OEM
CD4022BMJ	1256
CD4022BMW	1256
CD4023	0519
CD4023AD	0519
CD4023AD3	OEM
CD4023AE	0519
CD4023AEX	OEM
CD4023AF	0519
CD4023AF3	OEM
CD4023AF3W	OEM
CD4023AFB	OEM
CD4023AFX	OEM
CD4023AK	OEM
CD4023AK3	OEM
CD4023AKB	OEM
CD4023BCJ	0519
CD4023BCN	0519
CD4023BD	0519
CD4023BE	0519
CD4023BEX	OEM
CD4023BF	0519
CD4023BF3	OEM
CD4023BF3W	OEM
CD4023BFX	OEM
CD4023BK	OEM
CD4023BMJ	0519
CD4023BMW	0519
CD4023C	0519
CD4023CJ	0519
CD4023CN	0519
CD4023MJ	0519
CD4023MW	0519
CD4023UBD	0519
CD4023UBD3	OEM
CD4023UBE	0519
CD4023UBEX	OEM
CD4023UBF	0519
CD4023UBF3	OEM
CD4023UBF3W	OEM
CD4023UBFX	OEM
CD4023UBK	OEM
CD4024	1957
CD4024AD	1957
CD4024AD3	OEM
CD4024AE	1957
CD4024AEX	OEM
CD4024AF	1957
CD4024AF3	OEM
CD4024AF3W	OEM
CD4024AFB	OEM
CD4024AFX	OEM
CD4024AK	OEM
CD4024AK3	OEM
CD4024AKB	OEM
CD4024AT	OEM
CD4024BCJ	1957
CD4024BCN	1957
CD4024BD	1957
CD4024BD3	OEM
CD4024BE	1957
CD4024BEX	OEM
CD4024BF	1957
CD4024BF3W	OEM
CD4024BFX	OEM
CD4024BK	OEM
CD4024BK3	OEM
CD4024BMJ	1957
CD4024BMW	1957
CD4025	2072
CD4025AD	2072
CD4025AD3	OEM
CD4025AE	2072
CD4025AEX	OEM
CD4025AF	2072
CD4025AF3	OEM
CD4025AF3W	OEM
CD4025AFB	OEM
CD4025AFX	OEM
CD4025AK	OEM
CD4025AK3	OEM
CD4025AKB	OEM
CD4025BCJ	2072
CD4025BCN	2072
CD4025BD	2072
CD4025BE	2072
CD4025BEX	OEM
CD4025BF	2072
CD4025BF3	OEM
CD4025BF3W	OEM
CD4025BFX	OEM
CD4025BK	OEM
CD4025BK3	OEM
CD4025BMJ	2072
CD4025BMW	2072
CD4025CJ	2072
CD4025CN	2072
CD4025MJ	2072
CD4025MW	2072
CD4025UBD	2072
CD4025UBD3	OEM
CD4025UBE	2072
CD4025UBEX	OEM
CD4025UBF	2072
CD4025UBF3	OEM
CD4025UBF3W	OEM
CD4025UBFX	OEM
CD4025UBK	OEM
CD4026AD	2152
CD4026AD3	OEM
CD4026AE	2152
CD4026AEX	OEM
CD4026AF	2152
CD4026AF3	OEM
CD4026AF3W	OEM
CD4026AFX	OEM
CD4026AK	OEM
CD4026AK3	OEM
CD4026B3W	OEM
CD4026BD	2152
CD4026BE	2152
CD4026BEX	OEM
CD4026BF	2152
CD4026BF3	OEM
CD4026BFX	OEM
CD4026BH	OEM
CD4026BK	OEM
CD4027	1949
CD4027AD	1949
CD4027AD3	OEM
CD4027AE	1949
CD4027AEX	OEM
CD4027AF	1949
CD4027AF3	OEM
CD4027AF3W	OEM
CD4027AFB	OEM
CD4027AFX	OEM
CD4027AK	OEM
CD4027AK3	OEM
CD4027AKB	OEM
CD4027BCJ	1949
CD4027BCN	1949
CD4027BD	1949
CD4027BD3	OEM
CD4027BE	1949
CD4027BEX	OEM
CD4027BF	1949
CD4027BF3	OEM
CD4027BF3W	OEM
CD4027BFX	OEM
CD4027BK	OEM
CD4027BK3	OEM
CD4027BMJ	1949
CD4027BMW	1949
CD4028AD	2228
CD4028AD3	OEM
CD4028AE	2228
CD4028AEX	OEM
CD4028AF	2228
CD4028AF3	OEM
CD4028AF3W	OEM
CD4028AFX	OEM
CD4028AK	OEM
CD4028AK3	OEM
CD4028BCJ	2228
CD4028BCN	2228
CD4028BD	2228
CD4028BD3	OEM
CD4028BE	2228
CD4028BEX	OEM
CD4028BF	2228
CD4028BF3	OEM
CD4028BF3W	OEM
CD4028BFX	OEM
CD4028BK	OEM
CD4028BK3	OEM
CD4028BMJ	2228
CD4028BMW	2228
CD4028F	2228
CD4029AD	2233
CD4029AD3	OEM
CD4029AE	2233
CD4029AEX	OEM
CD4029AF	2233
CD4029AF3	OEM
CD4029AF3W	OEM
CD4029AFX	OEM
CD4029AK	OEM
CD4029AK3	OEM
CD4029BCJ	2233
CD4029BCN	2233
CD4029BD	2233
CD4029BD3	OEM
CD4029BE	2233
CD4029BEX	OEM
CD4029BF	2233
CD4029BF3	OEM
CD4029BF3W	OEM
CD4029BFX	OEM
CD4029BK	OEM
CD4029BMJ	2233
CD4029BMW	2233
CD4030	0500
CD4030AD	0500
CD4030AD3	OEM
CD4030AE	0500
CD4030AEX	OEM
CD4030AF	0500
CD4030AF3	OEM
CD4030AF3W	OEM
CD4030AFX	OEM
CD4030AK	OEM
CD4030AK3	OEM
CD4030BD	0500
CD4030BD3	OEM
CD4030BE	0500
CD4030BEX	OEM
CD4030BF	0500
CD4030BF3	OEM
CD4030BF3W	OEM
CD4030BFX	OEM
CD4030BK	OEM
CD4030BMW	0500
CD4030CJ	0500
CD4030CN	0500
CD4030MJ	0500
CD4030MW	0500
CD4031AD	2971
CD4031AD3	OEM
CD4031AE	2971
CD4031AEX	OEM
CD4031AF	2971
CD4031AF3	OEM
CD4031AF3W	OEM
CD4031AFX	OEM
CD4031AH	OEM
CD4031AK	OEM
CD4031AK3	OEM
CD4031AKB	OEM
CD4031BD	2971
CD4031BE	2971
CD4031BEX	OEM
CD4031BF	2971
CD4031BF3	OEM
CD4031BF3W	OEM
CD4031BFX	OEM
CD4031BH	OEM
CD4031BK	OEM
CD4031BK3	OEM
CD4032AD	2531
CD4032AE	2531
CD4032AEX	OEM
CD4032AF	2531
CD4032AK	OEM
CD4032BD	2531
CD4032BE	2531
CD4032BEX	OEM
CD4032BF	2531
CD4032BFX	OEM
CD4032BK	OEM
CD4033	2635
CD4033AD	2635
CD4033AD3	OEM
CD4033AE	2635
CD4033AEX	OEM
CD4033AF	2635
CD4033AF3	OEM
CD4033AF3W	OEM
CD4033AFX	OEM
CD4033AK	OEM
CD4033AK3	OEM
CD4033BD	2635
CD4033BE	2635
CD4033BEX	OEM
CD4033BF	2635
CD4033BF3	OEM
CD4033BF3W	OEM
CD4033BFX	OEM
CD4033BH	OEM
CD4033BK	OEM
CD4034AD	3615
CD4034AD3	OEM
CD4034AE	3615
CD4034AEX	OEM
CD4034AF	3615
CD4034AF3	OEM
CD4034AF3W	OEM
CD4034AH	OEM
CD4034AK	OEM
CD4034AK3	OEM

If replacement code is OEM, contact original manufacturer for replacement.

DEVICE TYPE	REPL CODE
CD4034AY	3615
CD4034BD	3615
CD4034BE	3615
CD4034BEX	OEM
CD4034BF	3615
CD4034BF3	OEM
CD4034BF3W	OEM
CD4034BH	OEM
CD4034BK	OEM
CD4034BK3	OEM
CD4034FX	OEM
CD4035AD	2777
CD4035AD3	OEM
CD4035AE	2777
CD4035AEX	OEM
CD4035AF	2777
CD4035AF3	OEM
CD4035AF3W	OEM
CD4035AFX	OEM
CD4035AH	OEM
CD4035AK	OEM
CD4035AK3	OEM
CD4035AY	2777
CD4035BD	2777
CD4035BD3	OEM
CD4035BE	2777
CD4035BEX	OEM
CD4035BF	2777
CD4035BF3	OEM
CD4035BF3W	OEM
CD4035BFX	OEM
CD4035BH	OEM
CD4035BK	OEM
CD4035BK3	OEM
CD4036AD	OEM
CD4036AE	OEM
CD4036AEX	OEM
CD4036AF	OEM
CD4037AD	OEM
CD4037AE	OEM
CD4037AEX	OEM
CD4037AF	OEM
CD4037AFX	OEM
CD4037AK	OEM
CD4038AD	2981
CD4038AE	2981
CD4038AEX	OEM
CD4038AF	2981
CD4038AFX	OEM
CD4038BD	2981
CD4038BE	2981
CD4038BEX	OEM
CD4038BF	2981
CD4038BFX	OEM
CD4038BK	OEM
CD4038BK3	OEM
CD4039AD	OEM
CD4039AE	OEM
CD4039AEX	OEM
CD4039AK	OEM
CD4040	0056
CD4040AD	0056
CD4040AD3	OEM
CD4040AE	0056
CD4040AEX	OEM
CD4040AF	0056
CD4040AF3	OEM
CD4040AF3W	OEM
CD4040AFX	OEM
CD4040AK	OEM
CD4040AK3	OEM
CD4040BCJ	0056
CD4040BCN	0056
CD4040BD	0056
CD4040BD3	OEM
CD4040BE	0056
CD4040BEX	OEM
CD4040BF	0056
CD4040BF3	OEM
CD4040BF3W	OEM
CD4040BFX	OEM
CD4040BK	OEM
CD4040BK3	OEM
CD4040BMJ	0056
CD4040BMW	0056
CD4041AD	3174
CD4041AD3	OEM
CD4041AE	3174
CD4041AEX	OEM
CD4041AF	3174
CD4041AF3	OEM
CD4041AF3W	OEM
CD4041AFX	OEM
CD4041AH	OEM
CD4041AK	OEM
CD4041AK3	OEM
CD4041AY	3174
CD4041CJ	3174
CD4041CN	3174
CD4041MJ	3174
CD4041MW	OEM
CD4041UBD	3174
CD4041UBD3	OEM
CD4041UBE	3174
CD4041UBEX	OEM
CD4041UBF	3174
CD4041UBF3	OEM
CD4041UBF3W	OEM
CD4041UBFX	OEM

DEVICE TYPE	REPL CODE
CD4041UBH	OEM
CD4041UBK	OEM
CD4041UBK3	OEM
CD4042	0121
CD4042AD	0121
CD4042AD3	OEM
CD4042AE	0121
CD4042AEX	OEM
CD4042AF	0121
CD4042AF3	OEM
CD4042AF3W	OEM
CD4042AFX	OEM
CD4042AK	OEM
CD4042AK3	OEM
CD4042BCJ	0121
CD4042BCN	0121
CD4042BD	0121
CD4042BD3	OEM
CD4042BE	0121
CD4042BEX	OEM
CD4042BF	0121
CD4042BF3	OEM
CD4042BF3W	OEM
CD4042BFX	OEM
CD4042BMJ	0121
CD4042BMW	0121
CD4043AD	1553
CD4043AD3	OEM
CD4043AE	1553
CD4043AEX	OEM
CD4043AF	1553
CD4043AF3	OEM
CD4043AF3W	OEM
CD4043AFX	OEM
CD4043AK	OEM
CD4043AK3	OEM
CD4043BD	1553
CD4043BD3	OEM
CD4043BE	1553
CD4043BEX	OEM
CD4043BF	1553
CD4043BF3	OEM
CD4043BF3W	OEM
CD4043BFX	OEM
CD4043BK	OEM
CD4043CJ	1553
CD4043CN	1553
CD4043MJ	1553
CD4043MW	1553
CD4044AD	2308
CD4044AD3	OEM
CD4044AE	2308
CD4044AEX	OEM
CD4044AF	2308
CD4044AF3	OEM
CD4044AF3W	OEM
CD4044AFX	OEM
CD4044AK	OEM
CD4044AK3	OEM
CD4044BD	2308
CD4044BD3	OEM
CD4044BE	2308
CD4044BEX	OEM
CD4044BF	2308
CD4044BF3	OEM
CD4044BF3W	OEM
CD4044BFX	OEM
CD4044BK	OEM
CD4044CJ	2308
CD4044CN	2308
CD4044MJ	2308
CD4044MW	2308
CD4045AD	OEM
CD4045AD3	OEM
CD4045AE	3443
CD4045AEX	OEM
CD4045AF	3443
CD4045AFX	OEM
CD4045AK	OEM
CD4045AK3	OEM
CD4045BD	3443
CD4045BE	3443
CD4045BEX	OEM
CD4045BF	3443
CD4045BFX	OEM
CD4045BH	OEM
CD4045BK	OEM
CD4046A	3428
CD4046AD	3428
CD4046AD3	OEM
CD4046AE	3831
CD4046AEX	OEM
CD4046AF	3428
CD4046AF3	OEM
CD4046AF3W	OEM
CD4046AFX	OEM
CD4046AK	OEM
CD4046AK3	OEM
CD4046BD	3428
CD4046BE	3428
CD4046BEX	OEM
CD4046BF	3428
CD4046BFX	OEM
CD4046BH	OEM
CD4046BK	OEM
CD4046BK3	OEM
CD4047AD	2331
CD4047AD3	OEM
CD4047AE	2331

DEVICE TYPE	REPL CODE
CD4047AEX	OEM
CD4047AF	2331
CD4047AF3	OEM
CD4047AF3W	OEM
CD4047AFX	OEM
CD4047AK	OEM
CD4047AK3	OEM
CD4047BCJ	2331
CD4047BCN	2331
CD4047BD	2331
CD4047BD3	OEM
CD4047BE	2331
CD4047BEX	OEM
CD4047BF	2331
CD4047BF3	OEM
CD4047BF3W	OEM
CD4047BFX	OEM
CD4047BK	OEM
CD4047BK3	OEM
CD4047BMJ	2331
CD4047BMW	2331
CD4048AD	3457
CD4048AD3	OEM
CD4048AE	3459
CD4048AEX	OEM
CD4048AF	3459
CD4048AF3	OEM
CD4048AF3W	OEM
CD4048AFX	OEM
CD4048AK	OEM
CD4048AK3	OEM
CD4048BCJ	3459
CD4048BCN	3459
CD4048BD	3457
CD4048BD3	OEM
CD4048BE	3459
CD4048BEX	OEM
CD4048BF	3459
CD4048BF3	OEM
CD4048BF3W	OEM
CD4048BFX	OEM
CD4048BH	OEM
CD4048BK	OEM
CD4048BMJ	3459
CD4048BMW	OEM
CD4049AD	0001
CD4049AD3	OEM
CD4049AE	0001
CD4049AEX	OEM
CD4049AF	0001
CD4049AF3	OEM
CD4049AF3W	OEM
CD4049AFB	OEM
CD4049AFX	OEM
CD4049AH	OEM
CD4049AK	OEM
CD4049AK3	OEM
CD4049AKB	OEM
CD4049CJ	0001
CD4049CN	0001
CD4049MJ	0001
CD4049MW	OEM
CD4049UBD	0001
CD4049UBD3	OEM
CD4049UBE	0001
CD4049UBEX	OEM
CD4049UBF	0001
CD4049UBF3	OEM
CD4049UBF3W	OEM
CD4049UBFX	OEM
CD4049UBH	OEM
CD4049UBK	OEM
CD4049UBK3	OEM
CD4050	0398
CD4050AD	0398
CD4050AD3	OEM
CD4050AE	0398
CD4050AEX	OEM
CD4050AF	0398
CD4050AF3	OEM
CD4050AF3W	OEM
CD4050AFB	OEM
CD4050AFX	OEM
CD4050AH	OEM
CD4050AK	OEM
CD4050AK3	OEM
CD4050AKB	OEM
CD4050B	0398
CD4050BCJ	0398
CD4050BCN	0398
CD4050BD	0398
CD4050BD3	OEM
CD4050BE	0398
CD4050BEX	OEM
CD4050BF	0398
CD4050BF3	OEM
CD4050BF3W	OEM
CD4050BFX	OEM
CD4050BH	OEM
CD4050BK	OEM
CD4050BK3	OEM
CD4050BMJ	6367
CD4050BMW	OEM
CD4051	0366
CD4051AE	0366
CD4051B	0366
CD4051BCJ	0366
CD4051BCN	0366
CD4051BD	0366

DEVICE TYPE	REPL CODE
CD4051BD3	OEM
CD4051BE	0366
CD4051BEX	OEM
CD4051BF	0366
CD4051BF3	OEM
CD4051BF3W	OEM
CD4051BFX	OEM
CD4051BH	OEM
CD4051BK	OEM
CD4051BK3	OEM
CD4051BMJ	0366
CD4051BMW	OEM
CD4052	0024
CD4052AE	0366
CD4052B	3453
CD4052BCJ	0024
CD4052BCN	0024
CD4052BD	0024
CD4052BD3	OEM
CD4052BE	0024
CD4052BEX	OEM
CD4052BF	0024
CD4052BF3	OEM
CD4052BF3W	OEM
CD4052BFX	OEM
CD4052BH	OEM
CD4052BK	OEM
CD4052BK3	OEM
CD4052BMJ	0024
CD4052BMW	OEM
CD4053BC	OEM
CD4053BCJ	0035
CD4053BCN	0035
CD4053BD	0035
CD4053BD3	OEM
CD4053BE	0035
CD4053BEX	OEM
CD4053BF	0035
CD4053BF3	OEM
CD4053BF3W	OEM
CD4053BFX	OEM
CD4053BH	OEM
CD4053BK	OEM
CD4053BK3	OEM
CD4053BMJ	6648
CD4053BMW	OEM
CD4053BP	0035
CD4054BD	4530
CD4054BE	4530
CD4054BEX	OEM
CD4054BF	4530
CD4054BFX	OEM
CD4054BH	OEM
CD4054BK	OEM
CD4054BY	4530
CD4055	3304
CD4055AE	3304
CD4055BD	3304
CD4055BE	3304
CD4055BEX	OEM
CD4055BF	3304
CD4055BFX	OEM
CD4055BH	OEM
CD4055BK	OEM
CD4055BY	3304
CD4056BD	3706
CD4056BE	3706
CD4056BEX	OEM
CD4056BFX	OEM
CD4056BH	OEM
CD4056BK	OEM
CD4056BY	3706
CD4057AD	OEM
CD4057AD3	OEM
CD4057AK	OEM
CD4057AK3	OEM
CD4059	3308
CD4059AD	3308
CD4059AD3	OEM
CD4059AE	3308
CD4059AEX	OEM
CD4059AF	3308
CD4059AFX	OEM
CD4059AK	OEM
CD4059AK3	OEM
CD4060	0146
CD4060AD	0146
CD4060AD3	OEM
CD4060AE	0146
CD4060AEX	OEM
CD4060AF	0146
CD4060AF3	OEM
CD4060AF3W	OEM
CD4060AFX	OEM
CD4060AK	OEM
CD4060AK3	OEM
CD4060BCJ	0146
CD4060BCN	0146
CD4060BD	0146
CD4060BD3	OEM
CD4060BE	0146
CD4060BEX	OEM
CD4060BF	0146
CD4060BF3	OEM
CD4060BF3W	OEM
CD4060BFX	OEM
CD4060BH	0146
CD4060BK	OEM
CD4060BMJ	0146

DEVICE TYPE	REPL CODE
CD4060BMW	0146
CD4061AD	OEM
CD4062AH	OEM
CD4062AK	OEM
CD4062AK3	OEM
CD4062AT	OEM
CD4063BD	3727
CD4063BD3	OEM
CD4063BE	0146
CD4063BEX	OEM
CD4063BF	3727
CD4063BF3	OEM
CD4063BF3W	OEM
CD4063BFX	OEM
CD4063BK	OEM
CD4063BK3	OEM
CD4065BF	OEM
CD4066	0101
CD4066AD	0101
CD4066AD3	OEM
CD4066AE	0101
CD4066AEX	OEM
CD4066AF	0101
CD4066AF3	OEM
CD4066AF3W	OEM
CD4066AFX	OEM
CD4066AH	OEM
CD4066AK	OEM
CD4066AK3	OEM
CD-4066B	0101
CD4066B	0101
CD4066BCJ	0101
CD4066BCN	0101
CD4066BD	0101
CD4066BD3	OEM
CD4066BE	0101
CD4066BEX	OEM
CD4066BEXV	OEM
CD4066BF	0101
CD4066BF3	OEM
CD4066BF3W	OEM
CD4066BFX	OEM
CD4066BH	OEM
CD4066BK	OEM
CD4066BK3	OEM
CD4066BMJ	0101
CD4066BMW	OEM
CD4067BD	3742
CD4067BD3	OEM
CD4067BE	3742
CD4067BEX	OEM
CD4067BF	3742
CD4067BF3	OEM
CD4067BF3W	OEM
CD4067BFX	OEM
CD4067BH	OEM
CD4067BK	OEM
CD4068BD	2503
CD4068BD3	OEM
CD4068BE	2503
CD4068BEX	OEM
CD4068BF	2503
CD4068BF3	OEM
CD4068BF3W	OEM
CD4068BFX	OEM
CD4068BK	OEM
CD4068BK3	OEM
CD4069	3311
CD4069BE	0119
CD4069C	0119
CD4069CJ	0119
CD4069CM(SM)	OEM
CD4069CN	0119
CD4069M	0119
CD4069MJ	0119
CD4069MW	OEM
CD4069N	0119
CD4069UB	3311
CD4069UBD	0119
CD4069UBD3	OEM
CD4069UBE	0119
CD4069UBEX	OEM
CD4069UBF	0119
CD4069UBF3	OEM
CD4069UBF3W	OEM
CD4069UBFX	OEM
CD4069UBH	OEM
CD4069UBK	OEM
CD4069UBK3	OEM
CD4070	0500
CD4070BCJ	2516
CD4070BCN	2516
CD4070BD	2516
CD4070BD3	OEM
CD4070BE	2516
CD4070BEX	OEM
CD4070BF	2516
CD4070BF3	OEM
CD4070BF3W	OEM
CD4070BFB	OEM
CD4070BFX	OEM
CD4070BK	OEM
CD4070BK3	OEM
CD4070BMJ	2516
CD4070BMW	2516
CD4071BCJ	0129
CD4071BCN	0129
CD4071BD	0129
CD4071BD3	OEM

DEVICE TYPE	REPL CODE
CD4071BE	0129
CD4071BEX	OEM
CD4071BF	0129
CD4071BF3	OEM
CD4071BF3W	OEM
CD4071BFB	OEM
CD4071BFX	OEM
CD4071BK	OEM
CD4071BK3	OEM
CD4071BMJ	0129
CD4071BMW	0129
CD4072BD	2524
CD4072BD3	OEM
CD4072BE	2524
CD4072BEX	OEM
CD4072BF	2524
CD4072BF3	OEM
CD4072BF3W	OEM
CD4072BFB	OEM
CD4072BFX	OEM
CD4072BK	OEM
CD4072BK3	OEM
CD4072BSO	OEM
CD4073BCJ	1536
CD4073BCN	1536
CD4073BD	1536
CD4073BD3	OEM
CD4073BE	1536
CD4073BEX	OEM
CD4073BF	1536
CD4073BF3	OEM
CD4073BF3W	OEM
CD4073BFB	OEM
CD4073BFX	OEM
CD4073BK	OEM
CD4073BK3	OEM
CD4073BMJ	1536
CD4073BMW	1536
CD4075BCJ	2540
CD4075BCN	2540
CD4075BD	2540
CD4075BD3	OEM
CD4075BE	2540
CD4075BEX	OEM
CD4075BF	2540
CD4075BF3	OEM
CD4075BF3W	OEM
CD4075BFB	OEM
CD4075BFX	OEM
CD4075BK	OEM
CD4075BK3	OEM
CD4075BMJ	2540
CD4075BMN	2540
CD4075BMW	2540
CD4075BT	2540
CD4076BC	3496
CD4076BD	3496
CD4076BD3	OEM
CD4076BE	3496
CD4076BEX	OEM
CD4076BF	3496
CD4076BF3	OEM
CD4076BF3W	OEM
CD4076BFX	OEM
CD4076BH	OEM
CD4076BK	OEM
CD4076BK3	OEM
CD4076BM	3496
CD4077	3315
CD4077AK	OEM
CD4077BD	2558
CD4077BD3	OEM
CD4077BE	2558
CD4077BEX	OEM
CD4077BF	2558
CD4077BF3	OEM
CD4077BF3W	OEM
CD4077BFB	OEM
CD4077BFX	OEM
CD4077BK	OEM
CD4077BK3	OEM
CD4078BD	0921
CD4078BD3	OEM
CD4078BE	0921
CD4078BEX	OEM
CD4078BF	0921
CD4078BF3	OEM
CD4078BF3W	OEM
CD4078BFX	OEM
CD4078BK	OEM
CD4078BK3	OEM
CD4081	0626
CD4081B	3579
CD4081BCJ	0626
CD4081BCN	0626
CD4081BD	0626
CD4081BD3	OEM
CD4081BE	0626
CD4081BEX	OEM
CD4081BF	0626
CD4081BF3	OEM
CD4081BF3W	OEM
CD4081BFX	OEM
CD4081BK	OEM
CD4081BK3	OEM
CD4081BMJ	0626
CD4081BMW	0626
CD4082	0300
CD4082B	0300

DEVICE TYPE	REPL CODE
CD4082BD	0300
CD4082BD3	OEM
CD4082BE	0300
CD4082BEX	OEM
CD4082BF	0300
CD4082BF3	OEM
CD4082BF3W	OEM
CD4082BFB	OEM
CD4082BFX	OEM
CD4082BK	OEM
CD4082BK3	OEM
CD4085BD	0303
CD4085BD3	OEM
CD4085BE	0303
CD4085BEX	OEM
CD4085BF	0303
CD4085BF3	OEM
CD4085BF3W	OEM
CD4085BFB	OEM
CD4085BFX	OEM
CD4085BK	OEM
CD4085BK3	OEM
CD4086BD	0535
CD4086BD3	OEM
CD4086BE	0535
CD4086BEX	OEM
CD4086BF	0535
CD4086BF3	OEM
CD4086BF3W	OEM
CD4086BFB	OEM
CD4086BFX	OEM
CD4086BK	OEM
CD4086BK3	OEM
CD4089BCJ	3830
CD4089BCN	3830
CD4089BD	3830
CD4089BD3	OEM
CD4089BE	3830
CD4089BEX	OEM
CD4089BF	3830
CD4089BF3	OEM
CD4089BF3W	OEM
CD4089BFX	OEM
CD4089BK	OEM
CD4089BK3	OEM
CD4089BMJ	3830
CD4089BMW	OEM
CD4093BCJ	2384
CD4093BCN	2384
CD4093BD	2384
CD4093BD3	OEM
CD4093BE	2384
CD4093BEX	OEM
CD4093BF	2384
CD4093BF3	OEM
CD4093BF3W	OEM
CD4093BFX	OEM
CD4093BK	OEM
CD4093BK3	OEM
CD4093BMJ	2384
CD4093BMW	OEM
CD4094BD	1682
CD4094BD3	OEM
CD4094BE	1682
CD4094BEX	OEM
CD4094BF	1682
CD4094BF3	OEM
CD4094BF3W	OEM
CD4094BFX	OEM
CD4094BH	OEM
CD4094BK	OEM
CD4094BK3	OEM
CD4095BD	3848
CD4095BD3	OEM
CD4095BE	3848
CD4095BEX	OEM
CD4095BF	3848
CD4095BF3	OEM
CD4095BF3W	OEM
CD4095BFX	OEM
CD4095BK	OEM
CD4095BK3	OEM
CD4096BD	3851
CD4096BD3	OEM
CD4096BE	3851
CD4096BEX	OEM
CD4096BF	3851
CD4096BF3	OEM
CD4096BF3W	OEM
CD4096BFX	OEM
CD4096BK	OEM
CD4096BK3	OEM
CD4097BD	3855
CD4097BD3	OEM
CD4097BE	3855
CD4097BEX	OEM
CD4097BF	3855
CD4097BF3	OEM
CD4097BF3W	OEM
CD4097BFX	OEM
CD4097BH	OEM
CD4097BK	OEM
CD4097BK3	OEM
CD4098BD	3611
CD4098BD3	OEM
CD4098BE	3611
CD4098BEX	OEM
CD4098BF	3611
CD4098BF3	OEM

If replacement code is OEM, contact original manufacturer for replacement.

DEVICE TYPE	REPL CODE
CD4098BF3W	OEM
CD4098BFX	OEM
CD4098BK	OEM
CD4098BK3	OEM
CD4099	3330
CD4099(ZENER)	OEM
CD4099BCJ	3330
CD4099BCN	3330
CD4099BD	3330
CD4099BD3	OEM
CD4099BE	3330
CD4099BEX	OEM
CD4099BF	3330
CD4099BF3	OEM
CD4099BF3W	OEM
CD4099BFX	OEM
CD4099BK	OEM
CD4099BK3	OEM
CD4099BMJ	3330
CD4099BMW	3330
CD4100	OEM
CD4101	OEM
CD4102	OEM
CD4103	0057
CD4104	OEM
CD4105	OEM
CD4106	0052
CD4107	OEM
CD4108	OEM
CD4109	0686
CD4110	OEM
CD4111	OEM
CD4112	OEM
CD4113	OEM
CD4114	OEM
CD4115	OEM
CD4116	0052
CD4117	0052
CD4118	0052
CD4119	OEM
CD4120	OEM
CD4121	0052
CD4122	0052
CD4123	OEM
CD4124	OEM
CD4125	OEM
CD4126	OEM
CD4127	OEM
CD4128	OEM
CD4129	OEM
CD4130	OEM
CD4131	OEM
CD4132	OEM
CD4133	OEM
CD4134	OEM
CD4135	OEM
CD4148	OEM
CD4262	OEM
CD4370A	OEM
CD4371A	OEM
CD4372A	OEM
CD4502BD	1037
CD4502BD3	OEM
CD4502BE	1037
CD4502BEX	OEM
CD4502BF	1037
CD4502BF3	OEM
CD4502BF3W	OEM
CD4502BFX	OEM
CD4502BK	OEM
CD4502BK3	OEM
CD4503B	2053
CD4503BD	2053
CD4503BE	4441
CD4503BEX	OEM
CD4503BF	2053
CD4503BH	OEM
CD4508BD	OEM
CD4508BD3	OEM
CD4508BE	1811
CD4508BEX	OEM
CD4508BF	1811
CD4508BF3	OEM
CD4508BF3W	OEM
CD4508BFX	OEM
CD4508BK	OEM
CD4508BK3	OEM
CD4510BCJ	1963
CD4510BCN	1963
CD4510BD	1963
CD4510BE	1963
CD4510BEX	OEM
CD4510BF	1963
CD4510BF3	OEM
CD4510BF3W	OEM
CD4510BFX	OEM
CD4510BK	OEM
CD4510BMJ	1963
CD4510BMW	1963
CD4511BCJ	1544
CD4511BCN	1544
CD4511BD	1544
CD4511BD3	OEM
CD4511BE	1544
CD4511BEX	OEM
CD4511BF	1544
CD4511BF3	OEM
CD4511BF3W	OEM
CD4511BFX	OEM

DEVICE TYPE	REPL CODE
CD4511BH	OEM
CD4511BK	OEM
CD4511BK3	OEM
CD4511BMJ	1670
CD4511BMW	OEM
CD4512BD	2120
CD4512BE	2120
CD4512BEX	OEM
CD4512BF	2120
CD4512BF3	OEM
CD4512BF3W	OEM
CD4512BFX	OEM
CD4512BH	OEM
CD4512BK	OEM
CD4514BC	1830
CD4514BD	1830
CD4514BD3	OEM
CD4514BE	1830
CD4514BEX	OEM
CD4514BF	1830
CD4514BF3	OEM
CD4514BF3W	OEM
CD4514BFX	OEM
CD4514BH	1830
CD4514BK	1830
CD4514BK3	OEM
CD4514BM	1830
CD4515BC	3600
CD4515BD	3600
CD4515BD3	OEM
CD4515BE	3600
CD4515BEX	OEM
CD4515BF	3600
CD4515BF3	OEM
CD4515BF3W	OEM
CD4515BFX	OEM
CD4515BH	3600
CD4515BK	3600
CD4515BK3	OEM
CD4515BM	3600
CD4516BCJ	2347
CD4516BCN	2347
CD4516BD	2347
CD4516BD3	OEM
CD4516BE	2347
CD4516BEX	OEM
CD4516BF	2347
CD4516BF3	OEM
CD4516BF3W	OEM
CD4516BFX	OEM
CD4516BK	OEM
CD4516BK3	OEM
CD4516BMJ	2347
CD4516BMW	2347
CD4517BE	4284
CD4517BEX	OEM
CD4517BF	4284
CD4517BFX	OEM
CD4517BH	OEM
CD4518B	1043
CD4518BCJ	1043
CD4518BCN	1043
CD4518BD	1043
CD4518BD3	OEM
CD4518BE	1043
CD4518BEX	OEM
CD4518BF	1043
CD4518BF3	OEM
CD4518BF3W	OEM
CD4518BFX	OEM
CD4518BK	OEM
CD4518BK3	OEM
CD4518BMJ	1043
CD4518BMW	1043
CD4519BC	OEM
CD4519BCJ	OEM
CD4519BCN	OEM
CD4519BM	OEM
CD4519BMJ	OEM
CD4519BMW	OEM
CD4520B	2675
CD4520B3W	OEM
CD4520BCJ	2675
CD4520BCN	2675
CD4520BD	2675
CD4520BD3	OEM
CD4520BE	2675
CD4520BEX	2675
CD4520BF	2675
CD4520BF3	OEM
CD4520BFX	OEM
CD4520BK	OEM
CD4520BK3	OEM
CD4520BMJ	2675
CD4520BMW	2675
CD4521BE	4990
CD4522BC	2837
CD4522BE	2837
CD4522BM	2837
CD4526BC	3610
CD4526BCN	3610
CD4526BM	3610
CD4527BCJ	3145
CD4527BCN	3145
CD4527BD	3145
CD4527BD3	OEM
CD4527BE	3145
CD4527BEX	OEM
CD4527BF	3145

DEVICE TYPE	REPL CODE
CD4527BF3	OEM
CD4527BF3W	OEM
CD4527BFX	OEM
CD4527BK	OEM
CD4527BK3	OEM
CD4527BMJ	3145
CD4527BMW	3145
CD4529BCN	4531
CD4532BD	1016
CD4532BD3	OEM
CD4532BE	1016
CD4532BEX	OEM
CD4532BF	1016
CD4532BF3	OEM
CD4532BF3W	OEM
CD4532BFX	OEM
CD4532BH	OEM
CD4532BK	OEM
CD4532BK3	OEM
CD4536BD	3704
CD4536BE	3704
CD4536BF	3704
CD4536BH	OEM
CD4538BCN	4302
CD4538BD	1063
CD4538BE	1063
CD4538BF	1063
CD4541BE	OEM
CD4543BD	5020
CD4543BE	5020
CD4543BF	5020
CD4543BH	OEM
CD4555BD	2938
CD4555BD3	OEM
CD4555BE	2938
CD4555BEX	OEM
CD4555BF	2938
CD4555BF3	OEM
CD4555BF3W	OEM
CD4555BFX	OEM
CD4555BK	OEM
CD4555BK3	OEM
CD4556BD	3431
CD4556BD3	OEM
CD4556BE	3431
CD4556BEX	OEM
CD4556BF	3431
CD4556BF3	OEM
CD4556BF3W	OEM
CD4556BFX	OEM
CD4556BK	OEM
CD4556BK3	OEM
CD4565	OEM
CD4565A	OEM
CD4566	3504
CD4566A	3504
CD4567	OEM
CD4567A	OEM
CD4568	OEM
CD4568A	OEM
CD4569	OEM
CD4569A	OEM
CD4570	OEM
CD4570A	OEM
CD4571	OEM
CD4571A	OEM
CD4572	3480
CD4572A	3480
CD4573	OEM
CD4573A	OEM
CD4574	OEM
CD4574A	OEM
CD4575	OEM
CD4575A	OEM
CD4576	OEM
CD4576A	OEM
CD4577	OEM
CD4577A	OEM
CD4578	OEM
CD4578A	OEM
CD4579	OEM
CD4579A	OEM
CD4580	OEM
CD4580A	OEM
CD4581	OEM
CD4581A	OEM
CD4582	OEM
CD4582A	OEM
CD4583	OEM
CD4583A	OEM
CD4584	OEM
CD4584A	OEM
CD4585BD	1375
CD4585BE	1375
CD4585BEX	OEM
CD4585BF	1375
CD4585BH	OEM
CD4614	OEM
CD4615	OEM
CD4616	OEM
CD4617	OEM
CD4618	OEM
CD4619	OEM
CD4620	OEM
CD4621	OEM
CD4622	OEM
CD4623	OEM
CD4624	OEM
CD4625	OEM

DEVICE TYPE	REPL CODE
CD4626	OEM
CD4627	0471
CD4723BCJ	OEM
CD4723BCN	OEM
CD4723BMJ	OEM
CD4723BMW	OEM
CD4724BCJ	3470
CD4724BCN	3470
CD4724BD	3470
CD4724BE	3470
CD4724BEX	OEM
CD4724BF	3470
CD4724BFX	OEM
CD4724BH	OEM
CD4724BMJ	3470
CD4724BMW	OEM
CD4728	OEM
CD4728A	OEM
CD4729	OEM
CD4729A	OEM
CD4730	OEM
CD4730A	OEM
CD4731	OEM
CD4731A	OEM
CD4732	OEM
CD4732A	OEM
CD4733	OEM
CD4733A	OEM
CD4734	OEM
CD4734A	OEM
CD4735	0636
CD4735A	OEM
CD4736	OEM
CD4736A	OEM
CD4737	OEM
CD4737A	OEM
CD4738	OEM
CD4738A	OEM
CD4739	0012
CD4739A	OEM
CD4740	OEM
CD4740A	OEM
CD4741	OEM
CD4741A	OEM
CD4742	0137
CD4742A	OEM
CD4743	OEM
CD4743A	OEM
CD4744	0002
CD4744A	OEM
CD4745	OEM
CD4745A	OEM
CD4746	OEM
CD4746A	OEM
CD4747	OEM
CD4747A	OEM
CD4748	OEM
CD4748A	OEM
CD4749	OEM
CD4749A	OEM
CD4750	OEM
CD4750A	OEM
CD4751	OEM
CD4751A	OEM
CD4752	OEM
CD4752A	OEM
CD4753	OEM
CD4753A	OEM
CD4754	OEM
CD4754A	OEM
CD4755	OEM
CD4755A	OEM
CD4756	OEM
CD4756A	OEM
CD4757	OEM
CD4757A	OEM
CD4758	OEM
CD4758A	OEM
CD4759	OEM
CD4759A	OEM
CD4760	OEM
CD4760A	OEM
CD4761	OEM
CD4761A	OEM
CD4762	OEM
CD4762A	OEM
CD4763	OEM
CD4763A	OEM
CD4764	OEM
CD4765	0057
CD4765A	0057
CD4766	0057
CD4766A	0057
CD4767	0057
CD4767A	0057
CD4768	0057
CD4768A	OEM
CD4770	0057
CD4770A	0057
CD4771	0057
CD4771A	0057
CD4772	0057
CD4772A	0057
CD4773	0057
CD4773A	0057
CD5221B	OEM
CD5222B	OEM
CD5223B	OEM
CD5224B	OEM

DEVICE TYPE	REPL CODE
CD5225B	OEM
CD5226B	OEM
CD5227B	OEM
CD5228B	OEM
CD5229B	OEM
CD5230B	OEM
CD5231B	OEM
CD5232B	OEM
CD5233B	OEM
CD5234B	0471
CD5235B	0471
CD5236B	OEM
CD5237B	OEM
CD5238B	OEM
CD5239B	0057
CD5240B	OEM
CD5241B	OEM
CD5242B	0052
CD5243B	OEM
CD5244B	OEM
CD5245B	0686
CD5246B	OEM
CD5247B	OEM
CD5248B	OEM
CD5249B	OEM
CD5250B	OEM
CD5251B	OEM
CD5252B	OEM
CD5253B	OEM
CD5254B	OEM
CD5255B	OEM
CD5256B	OEM
CD5257B	OEM
CD5258B	OEM
CD5259B	OEM
CD5260B	OEM
CD5261B	OEM
CD5262B	OEM
CD5263B	OEM
CD5264B	OEM
CD5265B	OEM
CD5266B	OEM
CD5267B	OEM
CD5268B	OEM
CD5269B	OEM
CD5270B	OEM
CD5271B	OEM
CD5272B	OEM
CD5273A	OEM
CD5274A	OEM
CD5275A	OEM
CD5276A	OEM
CD5277B	OEM
CD5278B	OEM
CD5279B	OEM
CD5280B	OEM
CD5281B	OEM
CD5328	1257
CD5333B	OEM
CD5334B	OEM
CD5335B	OEM
CD5336B	OEM
CD5337B	OEM
CD5338B	OEM
CD5339B	OEM
CD5340B	OEM
CD5341B	0471
CD5342B	OEM
CD5343B	OEM
CD5344B	OEM
CD5345B	OEM
CD5346B	0057
CD5347B	OEM
CD5348B	OEM
CD5349B	0052
CD5350B	OEM
CD5351B	OEM
CD5352B	OEM
CD5353B	OEM
CD5354B	OEM
CD5355B	OEM
CD5356B	OEM
CD5357B	OEM
CD5358B	OEM
CD5359B	OEM
CD5360B	OEM
CD5361B	OEM
CD5362B	OEM
CD5363B	OEM
CD5364B	OEM
CD5365B	OEM
CD5366B	OEM
CD5367B	OEM
CD5368B	OEM
CD5369B	OEM
CD5370B	OEM
CD5371B	OEM
CD5372B	OEM
CD5373B	OEM
CD5374B	OEM
CD5375B	OEM
CD5376B	OEM
CD5377B	OEM
CD5378B	OEM
CD5379B	OEM
CD5380B	OEM
CD5381B	OEM
CD5382B	OEM
CD5383B	OEM

DEVICE TYPE	REPL CODE
CD5384B	OEM
CD5385B	OEM
CD5386B	OEM
CD5387B	OEM
CD5388B	OEM
CD5590	1198
CD5645	0133
CD5916	OEM
CD5944	OEM
CD5945	4074
CD6039	1069
CD6111	OEM
CD6112	OEM
CD7012	3565
CD7186D	OEM
CD7211AME	OEM
CD7211E	OEM
CD7211ME	OEM
CD7216AC	OEM
CD7216BC	OEM
CD7216CC	OEM
CD7216DC	OEM
CD7217AC	OEM
CD7217BC	OEM
CD7217C	OEM
CD7217CC	OEM
CD7224AC	OEM
CD7224C	OEM
CD7225AC	OEM
CD7225C	OEM
CD7226AC	OEM
CD7226BC	OEM
CD7227AC	OEM
CD7227BC	OEM
CD7227C	OEM
CD7227CC	OEM
CD8000	0016
CD-8457	0015
CD8457	0133
CD8547	0133
CD12000	0016
CD13332	0015
CD13333	0609
CD13334	0609
CD13335	0015
CD13336	0609
CD13337	0015
CD13338	0609
CD13339	0609
CD22001H	OEM
CD22003	OEM
CD22003E	OEM
CD22003R	OEM
CD22006H	OEM
CD22007V2H	OEM
CD22008V1H	OEM
CD22011E	OEM
CD22012E	OEM
CD22014E	OEM
CD22015E	OEM
CD22018H	OEM
CD22100D	OEM
CD22100E	3042
CD22100EX	OEM
CD22100F	OEM
CD22101D	OEM
CD22101E	3048
CD22101EX	OEM
CD22101F	OEM
CD22101FX	OEM
CD22102D	OEM
CD22102E	3052
CD22102EX	OEM
CD22102F	OEM
CD22102FX	OEM
CD22103D(A)	OEM
CD22103H(A)	OEM
CD22859D	OEM
CD22859E	5310
CD22859H	OEM
CD40061AD	1822
CD40061AE	1822
CD40061AH	OEM
CD40061D	OEM
CD40061E	OEM
CD40100BD	3952
CD40100BE	3952
CD40100BEX	OEM
CD40100BF	3952
CD40100BF3	OEM
CD40100BF3W	OEM
CD40100BH	OEM
CD40100BK	OEM
CD40101BD	4017
CD40101BE	4017
CD40101BEX	OEM
CD40101BF	4017
CD40101BF3	OEM
CD40101BF3W	OEM
CD40101BK	OEM
CD40102BD	4061
CD40102BE	4061
CD40102BEX	OEM
CD40102BF	4061
CD40102BF3	OEM
CD40102BF3W	OEM
CD40102BFX	OEM
CD40102BK	OEM
CD40103BD	4091

DEVICE TYPE	REPL CODE
CD40103BE	4091
CD40103BEX	OEM
CD40103BF	4091
CD40103BF3	OEM
CD40103BF3W	OEM
CD40103BFX	OEM
CD40103BK	OEM
CD40104BD	4108
CD40104BE	4108
CD40104BEX	OEM
CD40104BF	4108
CD40104BF3	OEM
CD40104BFX	OEM
CD40104BH	OEM
CD40104BK	OEM
CD40105BD	4123
CD40105BE	4123
CD40105BEX	OEM
CD40105BF	4123
CD40105BF3	OEM
CD40105BF3W	OEM
CD40105BFX	OEM
CD40105BH	OEM
CD40105BK	OEM
CD40106BCJ	3626
CD40106BCN	3626
CD40106BD	3626
CD40106BE	3626
CD40106BEX	OEM
CD40106BF	3626
CD40106BF3	OEM
CD40106BF3W	OEM
CD40106BFX	OEM
CD40106BH	OEM
CD40106BK	OEM
CD40106BMJ	3626
CD40106BMW	OEM
CD40107BD	4153
CD40107BE	4153
CD40107BEX	OEM
CD40107BF	4153
CD40107BFX	OEM
CD40108BD	4182
CD40108BE	4182
CD40108BEX	OEM
CD40108BF	4182
CD40108BH	OEM
CD40108BK	OEM
CD40109BD	4210
CD40109BE	4210
CD40109BEX	OEM
CD40109BF	4210
CD40109BF3	OEM
CD40109BF3W	OEM
CD40109BFX	OEM
CD40109BH	OEM
CD40109BK	OEM
CD40110BD	4229
CD40110BE	4229
CD40110BEX	OEM
CD40110BF	4229
CD40110BH	OEM
CD40114BD	4298
CD40114BE	4298
CD40114BEX	OEM
CD40114BF	4298
CD40114BH	OEM
CD40115D	OEM
CD40116D	3106
CD40116E	3106
CD40117BD(A)	OEM
CD40117BE(A)	OEM
CD40117BF(A)	OEM
CD40147BD	5102
CD40147BE	5102
CD40147BF	5102
CD40147BH	OEM
CD40160BCJ	1359
CD40160BCN	1359
CD40160BD	1359
CD40160BDCD40160BE	OEM
CD40160BE	1359
CD40160BEX	OEM
CD40160BF	1359
CD40160BF3	OEM
CD40160BF3W	OEM
CD40160BMJ	1359
CD40160BMW	1359
CD40161BCJ	1373
CD40161BCN	1373
CD40161BD	1373
CD40161BE	1373
CD40161BEX	OEM
CD40161BF	1373
CD40161BF3	OEM
CD40161BF3W	OEM
CD40161BFX	OEM
CD40161BMJ	1373
CD40161BMW	1373
CD40162BCJ	1388
CD40162BCN	1388
CD40162BD	1388
CD40162BE	1388
CD40162BEX	OEM
CD40162BF	1388
CD40162BFX	OEM
CD40162BMJ	1388
CD40162BMW	1388
CD40163BCJ	1407

If replacement code is OEM, contact original manufacturer for replacement.

DEVICE TYPE	REPL CODE	DEVICE TYPE	REPL CODE	DEVICE TYPE	REPL CODE	DEVICE TYPE	REPL CODE	DEVICE TYPE	REPL CODE	DEVICE TYPE	REPL CODE	DEVICE TYPE	REPL CODE
CD40163BCN	1407	CD421460	OEM	CDC-9000-1B	0126	CDLL943B	OEM	CDLL4574	OEM	CDLL4744A	OEM	CDLL4909A	OEM
CD40163BD	1407	CD421490	OEM	CDC9000-1B	0126	CDLL944	OEM	CDLL4574A	OEM	CDLL4745	OEM	CDLL4910	OEM
CD40163BE	1407	CD421540	OEM	CDC9000-1C	0791	CDLL944A	OEM	CDLL4575	OEM	CDLL4745A	OEM	CDLL4910A	OEM
CD40163BEX	OEM	CD421560	OEM	CDC9000-1D	0006	CDLL944B	OEM	CDLL4575A	OEM	CDLL4746	OEM	CDLL4911	OEM
CD40163BF	1407	CD421590	OEM	CDC9002	0148	CDLL945	OEM	CDLL4576	OEM	CDLL4746A	OEM	CDLL4911A	OEM
CD40163BF3	OEM	CD430140	0652	CDC9002-1C	0086	CDLL945A	OEM	CDLL4576A	OEM	CDLL4747	OEM	CDLL4912	OEM
CD40163BF3W	OEM	CD430160	2311	CDC9002-18	0086	CDLL945B	OEM	CDLL4577	OEM	CDLL4747A	OEM	CDLL4912A	OEM
CD40163BFX	OEM	CD430190	2311	CDC-9002-IC	0150	CDLL2810	OEM	CDLL4577A	OEM	CDLL4748	OEM	CDLL4913	OEM
CD40163BMJ	1407	CD430240	0652	CDC10000-1E	0037	CDLL3154	OEM	CDLL4578	OEM	CDLL4748A	OEM	CDLL4913A	OEM
CD40163BMW	1407	CD430260	2311	CDC12000	0079	CDLL3154A	OEM	CDLL4578A	OEM	CDLL4749	OEM	CDLL4914	OEM
CD40174BC	1551	CD430290	2311	CDC12000-1C	0016	CDLL3155	OEM	CDLL4579	OEM	CDLL4749A	OEM	CDLL4914A	OEM
CD40174BD	1551	CD430440	0652	CDC12018C	0016	CDLL3155A	OEM	CDLL4579A	OEM	CDLL4750	OEM	CDLL4915	OEM
CD40174BE	1551	CD430460	2311	CDC12030B	0127	CDLL3156	OEM	CDLL4580	OEM	CDLL4750A	OEM	CDLL4915A	OEM
CD40174BEX	OEM	CD430490	2311	CDC12077F	0016	CDLL3156A	OEM	CDLL4580A	OEM	CDLL4751	OEM	CDLL4916	OEM
CD40174BF	1551	CD430640	0652	CDC12108	0079	CDLL3157	OEM	CDLL4581	OEM	CDLL4752	OEM	CDLL4916A	OEM
CD40174BF3	OEM	CD430660	2311	CDC12112C	0144	CDLL3157A	OEM	CDLL4581A	OEM	CDLL4752A	OEM	CDLL4917	OEM
CD40174BF3W	OEM	CD430690	2311	CDC12112D	0144	CDLL4099	OEM	CDLL4582	OEM	CDLL4753	OEM	CDLL4917A	OEM
CD40174BFX	OEM	CD430840	0652	CDC12112E	0144	CDLL4100	OEM	CDLL4582A	OEM	CDLL4753A	OEM	CDLL4918	OEM
CD40174BH	OEM	CD430860	2311	CDC12112F	0144	CDLL4101	OEM	CDLL4583	OEM	CDLL4754	OEM	CDLL4918A	OEM
CD40174BM	OEM	CD430890	2311	CDC13000	0079	CDLL4102	OEM	CDLL4583A	OEM	CDLL4754A	OEM	CDLL4919	OEM
CD40175BC	1528	CD431040	0652	CDC-13000-1	0016	CDLL4103	OEM	CDLL4584	OEM	CDLL4755	OEM	CDLL4919A	OEM
CD40175BM	OEM	CD431060	2311	CDC13000-1	0079	CDLL4104	OEM	CDLL4584A	OEM	CDLL4755A	OEM	CDLL4920	OEM
CD40181BD	5638	CD431090	2311	CDC13000-1B	0016	CDLL4105	OEM	CDLL4614	OEM	CDLL4756	OEM	CDLL4920A	OEM
CD40181BE	5638	CD431240	0652	CDC13000-1C	0016	CDLL4106	OEM	CDLL4615	OEM	CDLL4756A	OEM	CDLL4921	OEM
CD40181BEX	OEM	CD431260	2311	CDC13000-1D	0191	CDLL4107	OEM	CDLL4616	OEM	CDLL4757	OEM	CDLL4921A	OEM
CD40181BF	5638	CD431290	2311	CDC13000-18	0016	CDLL4108	OEM	CDLL4617	OEM	CDLL4757A	OEM	CDLL4922	OEM
CD40181BF3	OEM	CD431440	OEM	CDC13000C	0079	CDLL4109	OEM	CDLL4618	OEM	CDLL4758	OEM	CDLL4922A	OEM
CD40181BF3W	OEM	CD431460	OEM	CDC13016A	0016	CDLL4110	OEM	CDLL4619	OEM	CDLL4758A	OEM	CDLL4923	OEM
CD40181BFX	OEM	CD431490	OEM	CDC13019B	0079	CDLL4111	OEM	CDLL4620	OEM	CDLL4759	OEM	CDLL4923A	OEM
CD40181BH	OEM	CD431540	OEM	CDC13500-1	0016	CDLL4112	OEM	CDLL4621	OEM	CDLL4759A	OEM	CDLL4924	OEM
CD40181BK	OEM	CD431560	OEM	CDC15018	0079	CDLL4113	OEM	CDLL4622	OEM	CDLL4760	OEM	CDLL4924A	OEM
CD40182BD	4660	CD431590	OEM	CDC25100-6	0016	CDLL4114	OEM	CDLL4623	OEM	CDLL4760A	OEM	CDLL4925	OEM
CD40182BE	4660	CD470140	4574	CDC25100-G	0007	CDLL4115	OEM	CDLL4624	OEM	CDLL4761	OEM	CDLL4925A	OEM
CD40182BEX	OEM	CD470160	2878	CDC60132	0079	CDLL4116	OEM	CDLL4625	OEM	CDLL4761A	OEM	CDLL4926	OEM
CD40182BF	4660	CD470190	2878	CDC90001B	0791	CDLL4117	OEM	CDLL4626	OEM	CDLL4762	OEM	CDLL4926A	OEM
CD40182BF3	OEM	CD470240	4574	CDC120700	0086	CDLL4118	OEM	CDLL4627	OEM	CDLL4762A	OEM	CDLL4927	OEM
CD40182BF3W	OEM	CD470260	2878	CDC4306813	0016	CDLL4119	OEM	CDLL4678	OEM	CDLL4763	OEM	CDLL4927A	OEM
CD40182BK	OEM	CD470290	2878	CDCC6V2	OEM	CDLL4121	OEM	CDLL4679	OEM	CDLL4763A	OEM	CDLL4928	OEM
CD40192BCJ	1764	CD470440	4574	CDG-00	0133	CDLL4122	OEM	CDLL4680	OEM	CDLL4764	OEM	CDLL4928A	OEM
CD40192BCN	1764	CD470460	2878	CDG00	0133	CDLL4123	OEM	CDLL4681	OEM	CDLL4764A	OEM	CDLL4929	OEM
CD40192BD	1764	CD470490	2878	CDG005	0015	CDLL4124	OEM	CDLL4682	OEM	CDLL4765	OEM	CDLL4929A	OEM
CD40192BE	1764	CD470640	4574	CDG025	0133	CDLL4125	OEM	CDLL4683	OEM	CDLL4765A	OEM	CDLL4930	OEM
CD40192BEX	OEM	CD470660	2878	CDG-21	0133	CDLL4126	OEM	CDLL4684	OEM	CDLL4766	OEM	CDLL4930A	OEM
CD40192BF	1764	CD470690	2878	CDG21	0133	CDLL4127	OEM	CDLL4685	OEM	CDLL4766A	OEM	CDLL4931	OEM
CD40192BF3	OEM	CD470840	4574	CDG-22	0133	CDLL4128	OEM	CDLL4686	OEM	CDLL4767	OEM	CDLL4931A	OEM
CD40192BF3W	OEM	CD470860	2878	CDG22	0133	CDLL4129	OEM	CDLL4687	OEM	CDLL4767A	OEM	CDLL4932	OEM
CD40192BFX	OEM	CD470890	2878	CDG-24	0133	CDLL4130	OEM	CDLL4688	OEM	CDLL4768	OEM	CDLL4932A	OEM
CD40192BK	OEM	CD471040	4574	CDG24	0015	CDLL4131	OEM	CDLL4689	OEM	CDLL4768A	OEM	CDLL5179	OEM
CD40192BMJ	1764	CD471060	2878	CDG24/3490	0133	CDLL4132	OEM	CDLL4690	OEM	CDLL4769	OEM	CDLL5221	OEM
CD40192BMW	1764	CD471090	2878	CDG27	0133	CDLL4133	OEM	CDLL4691	OEM	CDLL4769A	OEM	CDLL5221A	OEM
CD40192CJ	1764	CD471240	4574	CDJ-00	0133	CDLL4134	OEM	CDLL4692	OEM	CDLL4770	OEM	CDLL5221B	OEM
CD40192CN	1764	CD471260	2878	CDK-4	0015	CDLL4135	OEM	CDLL4693	OEM	CDLL4770A	OEM	CDLL5222	OEM
CD40193BCJ	1776	CD471290	2878	CDLL1W110	OEM	CDLL4156	OEM	CDLL4694	OEM	CDLL4771	OEM	CDLL5222A	OEM
CD40193BCN	1776	CD471440	OEM	CDLL1W110A	OEM	CDLL4157	OEM	CDLL4695	OEM	CDLL4771A	OEM	CDLL5222B	OEM
CD40193BD	1776	CD471460	OEM	CDLL1W120	OEM	CDLL4453	OEM	CDLL4696	OEM	CDLL4772	OEM	CDLL5223	OEM
CD40193BE	1776	CD471490	OEM	CDLL1W120A	OEM	CDLL4460	OEM	CDLL4697	OEM	CDLL4772A	OEM	CDLL5223A	OEM
CD40193BEX	OEM	CD471540	OEM	CDLL1W130	OEM	CDLL4461	OEM	CDLL4698	OEM	CDLL4773	OEM	CDLL5223B	OEM
CD40193BF	1776	CD471560	OEM	CDLL1W130A	OEM	CDLL4462	OEM	CDLL4699	OEM	CDLL4773A	OEM	CDLL5224	OEM
CD40193BF3	OEM	CD471590	OEM	CDLL1W140	OEM	CDLL4463	OEM	CDLL4700	OEM	CDLL4774	OEM	CDLL5224A	OEM
CD40193BF3W	OEM	CD860011	1149	CDLL1W140A	OEM	CDLL4464	OEM	CDLL4701	OEM	CDLL4774A	OEM	CDLL5224B	OEM
CD40193BFX	OEM	CD860037	0133	CDLL1W150	OEM	CDLL4465	OEM	CDLL4702	OEM	CDLL4775	OEM	CDLL5225	OEM
CD40193BK	OEM	CD3112018	0244	CDLL1W150A	OEM	CDLL4466	OEM	CDLL4703	OEM	CDLL4775A	OEM	CDLL5225A	OEM
CD40193BMJ	1776	CD3122055	0012	CDLL1W160	OEM	CDLL4467	OEM	CDLL4704	OEM	CDLL4776	OEM	CDLL5225B	OEM
CD40193BMW	1776	CD3212055	0949	CDLL1W160A	OEM	CDLL4468	OEM	CDLL4705	OEM	CDLL4776A	OEM	CDLL5226	OEM
CD40194BD	1769	CDB-7605	OEM	CDLL1W170	OEM	CDLL4469	OEM	CDLL4706	OEM	CDLL4777	OEM	CDLL5226A	OEM
CD40194BE	1769	CDB-7606	OEM	CDLL1W170A	OEM	CDLL4470	OEM	CDLL4707	OEM	CDLL4777A	OEM	CDLL5226B	OEM
CD40194BEX	OEM	CDC00	OEM	CDLL1W180	OEM	CDLL4471	OEM	CDLL4708	OEM	CDLL4778	OEM	CDLL5227	OEM
CD40194BF	1769	CDC86X7-5	0016	CDLL1W180A	OEM	CDLL4472	OEM	CDLL4709	OEM	CDLL4778A	OEM	CDLL5227A	OEM
CD40194BFX	OEM	CDC430	0016	CDLL1W190	OEM	CDLL4473	OEM	CDLL4710	OEM	CDLL4779	OEM	CDLL5227B	OEM
CD40194BH	OEM	CDC496	0037	CDLL1W190A	OEM	CDLL4474	OEM	CDLL4711	OEM	CDLL4779A	OEM	CDLL5228	OEM
CD40194BK	OEM	CDC587	0086	CDLL1W200	OEM	CDLL4475	OEM	CDLL4712	OEM	CDLL4780	OEM	CDLL5228A	OEM
CD40208BD	6015	CDC731	0325	CDLL1W200A	OEM	CDLL4476	OEM	CDLL4713	OEM	CDLL4780A	OEM	CDLL5228B	OEM
CD40208BE	6015	CDC744	0233	CDLL200	OEM	CDLL4477	OEM	CDLL4714	OEM	CDLL4781	OEM	CDLL5229	OEM
CD40208BF	6015	CDC745	0086	CDLL300	OEM	CDLL4478	OEM	CDLL4715	OEM	CDLL4781A	OEM	CDLL5229A	OEM
CD40208BH	OEM	CDC746	0016	CDLL400	OEM	CDLL4479	OEM	CDLL4716	OEM	CDLL4782	OEM	CDLL5229B	OEM
CD40208BK	OEM	CDC1201BC	0016	CDLL610	OEM	CDLL4480	OEM	CDLL4717	OEM	CDLL4782A	OEM	CDLL5230	OEM
CD40257BD	6415	CDC2010	0016	CDLL821	OEM	CDLL4481	OEM	CDLL4728	OEM	CDLL4783	OEM	CDLL5230A	OEM
CD40257BE	6415	CDC2010D	0016	CDLL821A	OEM	CDLL4482	OEM	CDLL4728A	OEM	CDLL4783A	OEM	CDLL5230B	OEM
CD40257BEX	OEM	CDC2510C-G	0016	CDLL822	OEM	CDLL4483	OEM	CDLL4729	OEM	CDLL4784	OEM	CDLL5231	OEM
CD40257BF	6415	CDC4023A130	0016	CDLL823	OEM	CDLL4484	OEM	CDLL4729A	OEM	CDLL4784A	OEM	CDLL5231A	OEM
CD40257BF3	OEM	CDC5000	0127	CDLL823A	OEM	CDLL4485	OEM	CDLL4730	OEM	CDLL4829	OEM	CDLL5231B	OEM
CD40257BF3W	OEM	CDC5000-1B	0144	CDLL824	OEM	CDLL4486	OEM	CDLL4730A	OEM	CDLL4830	OEM	CDLL5232	OEM
CD40257BFX	OEM	CDC5008	0086	CDLL825	OEM	CDLL4487	OEM	CDLL4731	OEM	CDLL4896	OEM	CDLL5232A	OEM
CD40257BH	OEM	CDC5028A	0086	CDLL825A	OEM	CDLL4488	OEM	CDLL4731A	OEM	CDLL4896A	OEM	CDLL5232B	OEM
CD40257BK	OEM	CDC5038A	0144	CDLL826	OEM	CDLL4489	OEM	CDLL4732	OEM	CDLL4897	OEM	CDLL5233	OEM
CD-84857	0133	CDC5071A	0144	CDLL827	OEM	CDLL4490	OEM	CDLL4732A	OEM	CDLL4897A	OEM	CDLL5233A	OEM
CD86003	0133	CDC5075B	0144	CDLL827A	OEM	CDLL4491	OEM	CDLL4733	OEM	CDLL4898	OEM	CDLL5233B	OEM
CD420140	4364	CDC-7609	OEM	CDLL829	OEM	CDLL4492	OEM	CDLL4733A	OEM	CDLL4898A	OEM	CDLL5234	OEM
CD420160	2994	CDC-7609A	OEM	CDLL829A	OEM	CDLL4493	OEM	CDLL4734	OEM	CDLL4899	OEM	CDLL5234A	OEM
CD420190	2994	CDC-7609B	OEM	CDLL935	OEM	CDLL4494	OEM	CDLL4734A	OEM	CDLL4899A	OEM	CDLL5234B	OEM
CD420240	4364	CDC-7609C	OEM	CDLL935A	OEM	CDLL4495	OEM	CDLL4735	OEM	CDLL4900	OEM	CDLL5235	OEM
CD420260	2994	CDC-7609D	OEM	CDLL935B	OEM	CDLL4496	OEM	CDLL4735A	OEM	CDLL4900A	OEM	CDLL5235A	OEM
CD420290	2994	CDC8000	0076	CDLL936	OEM	CDLL4565	OEM	CDLL4736	OEM	CDLL4901	OEM	CDLL5235B	OEM
CD420440	4364	CDC-8000-1	0079	CDLL936A	OEM	CDLL4565A	OEM	CDLL4736A	OEM	CDLL4901A	OEM	CDLL5236	OEM
CD420460	2994	CDC8000-1	0086	CDLL936B	OEM	CDLL4566	OEM	CDLL4737	OEM	CDLL4902	OEM	CDLL5236A	OEM
CD420490	2994	CDC8000-1B	0016	CDLL937	OEM	CDLL4566A	OEM	CDLL4737A	OEM	CDLL4902A	OEM	CDLL5236B	OEM
CD420640	4364	CDC8000-1C	0086	CDLL937A	OEM	CDLL4567	OEM	CDLL4738	OEM	CDLL4903	OEM	CDLL5237	OEM
CD420660	2994	CDC8000-CM	0086	CDLL937B	OEM	CDLL4567A	OEM	CDLL4738A	OEM	CDLL4903A	OEM	CDLL5237A	OEM
CD420690	2994	CDC-8001	0016	CDLL938	OEM	CDLL4568	OEM	CDLL4739	OEM	CDLL4904	OEM	CDLL5237B	OEM
CD420840	4364	CDC8001	0218	CDLL938A	OEM	CDLL4568A	OEM	CDLL4739A	OEM	CDLL4904A	OEM	CDLL5238	OEM
CD420860	2994	CDC8002	0018	CDLL938B	OEM	CDLL4569	OEM	CDLL4740	OEM	CDLL4905	OEM	CDLL5238A	OEM
CD420890	2994	CDC8002-1	0086	CDLL941	OEM	CDLL4569A	OEM	CDLL4740A	OEM	CDLL4905A	OEM	CDLL5238B	OEM
CD421040	4364	CDC8011B	0016	CDLL941A	OEM	CDLL4570	OEM	CDLL4741	OEM	CDLL4906	OEM	CDLL5239	OEM
CD421060	2994	CDC8021	0016	CDLL941B	OEM	CDLL4570A	OEM	CDLL4741A	OEM	CDLL4906A	OEM	CDLL5239A	OEM
CD421090	2994	CDC8054	0016	CDLL942	OEM	CDLL4571	OEM	CDLL4742	OEM	CDLL4907	OEM	CDLL5239B	OEM
CD421240	4364	CDC8201	0079	CDLL942A	OEM	CDLL4572	OEM	CDLL4742A	OEM	CDLL4907A	OEM	CDLL5240	OEM
CD421260	2994	CDC8457	0133	CDLL942B	OEM	CDLL4572A	OEM	CDLL4743	OEM	CDLL4908	OEM	CDLL5240A	OEM
CD421290	2994	CDC-9000	0136	CDLL943	OEM	CDLL4573	OEM	CDLL4743A	OEM	CDLL4908A	OEM	CDLL5240B	OEM
CD421440	OEM	CDC9000-1	0037	CDLL943A	OEM	CDLL4573A	OEM	CDLL4744	OEM	CDLL4909	OEM	CDLL5241	OEM

DEVICE TYPE	REPL CODE
CDLL5241A	OEM
CDLL5241B	OEM
CDLL5242	OEM
CDLL5242A	OEM
CDLL5242B	OEM
CDLL5243	OEM
CDLL5243A	OEM
CDLL5243B	OEM
CDLL5244	OEM
CDLL5244A	OEM
CDLL5244B	OEM
CDLL5245	OEM
CDLL5245A	OEM
CDLL5245B	OEM
CDLL5246	OEM
CDLL5246A	OEM
CDLL5246B	OEM
CDLL5247	OEM
CDLL5247A	OEM
CDLL5247B	OEM
CDLL5248	OEM
CDLL5248A	OEM
CDLL5248B	OEM
CDLL5249	OEM
CDLL5249A	OEM
CDLL5249B	OEM
CDLL5250	OEM
CDLL5250A	OEM
CDLL5250B	OEM
CDLL5251	OEM
CDLL5251A	OEM
CDLL5251B	OEM
CDLL5252	OEM
CDLL5252A	OEM
CDLL5252B	OEM
CDLL5253	OEM
CDLL5253A	OEM
CDLL5253B	OEM
CDLL5254	OEM
CDLL5254A	OEM
CDLL5254B	OEM
CDLL5255	OEM
CDLL5255A	OEM
CDLL5255B	OEM
CDLL5256	OEM
CDLL5256A	OEM
CDLL5256B	OEM
CDLL5257	OEM
CDLL5257A	OEM
CDLL5257B	OEM
CDLL5258	OEM
CDLL5258A	OEM
CDLL5258B	OEM
CDLL5259	OEM
CDLL5259A	OEM
CDLL5259B	OEM
CDLL5260	OEM
CDLL5260A	OEM
CDLL5260B	OEM
CDLL5261	OEM
CDLL5261A	OEM
CDLL5261B	OEM
CDLL5262	OEM
CDLL5262A	OEM
CDLL5262B	OEM
CDLL5263	OEM
CDLL5263A	OEM
CDLL5263B	OEM
CDLL5264	OEM
CDLL5264A	OEM
CDLL5264B	OEM
CDLL5265	OEM
CDLL5265A	OEM
CDLL5265B	OEM
CDLL5266	OEM
CDLL5266A	OEM
CDLL5266B	OEM
CDLL5267	OEM
CDLL5267A	OEM
CDLL5267B	OEM
CDLL5268	OEM
CDLL5268A	OEM
CDLL5268B	OEM
CDLL5269	OEM
CDLL5269A	OEM
CDLL5269B	OEM
CDLL5270	OEM
CDLL5270A	OEM
CDLL5270B	OEM
CDLL5271	OEM
CDLL5271A	OEM
CDLL5271B	OEM
CDLL5272	OEM
CDLL5272A	OEM
CDLL5272B	OEM
CDLL5273	OEM
CDLL5273A	OEM
CDLL5273B	OEM
CDLL5274	OEM
CDLL5274A	OEM
CDLL5274B	OEM
CDLL5275	OEM
CDLL5275A	OEM
CDLL5275B	OEM
CDLL5276	OEM
CDLL5276A	OEM
CDLL5276B	OEM
CDLL5277	OEM
CDLL5277A	OEM
CDLL5277B	OEM
CDLL5278	OEM
CDLL5278A	OEM
CDLL5278B	OEM
CDLL5279	OEM
CDLL5279A	OEM
CDLL5279B	OEM
CDLL5280	OEM
CDLL5280A	OEM
CDLL5280B	OEM
CDLL5281	OEM
CDLL5281A	OEM
CDLL5281B	OEM
CDLL5283	OEM
CDLL5284	OEM
CDLL5285	OEM
CDLL5286	OEM
CDLL5287	OEM
CDLL5288	OEM
CDLL5289	OEM
CDLL5290	OEM
CDLL5291	OEM
CDLL5292	OEM
CDLL5293	OEM
CDLL5294	OEM
CDLL5295	OEM
CDLL5296	OEM
CDLL5297	OEM
CDLL5298	OEM
CDLL5299	OEM
CDLL5500	OEM
CDLL5501	OEM
CDLL5502	OEM
CDLL5503	OEM
CDLL5504	OEM
CDLL5505	OEM
CDLL5506	OEM
CDLL5507	OEM
CDLL5508	OEM
CDLL5509	OEM
CDLL5510	OEM
CDLL5511	OEM
CDLL5512	OEM
CDLL5513	OEM
CDLL5514	OEM
CDLL5518	OEM
CDLL5518A	OEM
CDLL5518B	OEM
CDLL5518C	OEM
CDLL5518D	OEM
CDLL5519	OEM
CDLL5519A	OEM
CDLL5519B	OEM
CDLL5519C	OEM
CDLL5519D	OEM
CDLL5520	OEM
CDLL5520A	OEM
CDLL5520B	OEM
CDLL5520C	OEM
CDLL5520D	OEM
CDLL5521	OEM
CDLL5521A	OEM
CDLL5521B	OEM
CDLL5521C	OEM
CDLL5521D	OEM
CDLL5522	OEM
CDLL5522A	OEM
CDLL5522B	OEM
CDLL5522C	OEM
CDLL5522D	OEM
CDLL5523	OEM
CDLL5523A	OEM
CDLL5523B	OEM
CDLL5523C	OEM
CDLL5523D	OEM
CDLL5524	OEM
CDLL5524A	OEM
CDLL5524B	OEM
CDLL5524C	OEM
CDLL5524D	OEM
CDLL5525	OEM
CDLL5525A	OEM
CDLL5525B	OEM
CDLL5525C	OEM
CDLL5525D	OEM
CDLL5526	OEM
CDLL5526A	OEM
CDLL5526B	OEM
CDLL5526C	OEM
CDLL5526D	OEM
CDLL5527	OEM
CDLL5527A	OEM
CDLL5527B	OEM
CDLL5527C	OEM
CDLL5527D	OEM
CDLL5528	OEM
CDLL5528A	OEM
CDLL5528B	OEM
CDLL5528C	OEM
CDLL5528D	OEM
CDLL5529	OEM
CDLL5529A	OEM
CDLL5529B	OEM
CDLL5529C	OEM
CDLL5529D	OEM
CDLL5530	OEM
CDLL5530A	OEM
CDLL5530B	OEM
CDLL5530C	OEM
CDLL5530D	OEM
CDLL5531	OEM
CDLL5531A	OEM
CDLL5531B	OEM
CDLL5531C	OEM
CDLL5531D	OEM
CDLL5532	OEM
CDLL5532A	OEM
CDLL5532B	OEM
CDLL5532C	OEM
CDLL5532D	OEM
CDLL5533	OEM
CDLL5533A	OEM
CDLL5533B	OEM
CDLL5533C	OEM
CDLL5533D	OEM
CDLL5534	OEM
CDLL5534A	OEM
CDLL5534B	OEM
CDLL5534C	OEM
CDLL5534D	OEM
CDLL5535	OEM
CDLL5535A	OEM
CDLL5535B	OEM
CDLL5535C	OEM
CDLL5535D	OEM
CDLL5536	OEM
CDLL5536A	OEM
CDLL5536B	OEM
CDLL5536C	OEM
CDLL5536D	OEM
CDLL5537	OEM
CDLL5537A	OEM
CDLL5537B	OEM
CDLL5537C	OEM
CDLL5537D	OEM
CDLL5538	OEM
CDLL5538A	OEM
CDLL5538B	OEM
CDLL5538C	OEM
CDLL5538D	OEM
CDLL5539	OEM
CDLL5539A	OEM
CDLL5539B	OEM
CDLL5539C	OEM
CDLL5539D	OEM
CDLL5540	OEM
CDLL5540A	OEM
CDLL5540B	OEM
CDLL5540C	OEM
CDLL5540D	OEM
CDLL5541	OEM
CDLL5541A	OEM
CDLL5541B	OEM
CDLL5541C	OEM
CDLL5541D	OEM
CDLL5542	OEM
CDLL5542A	OEM
CDLL5542B	OEM
CDLL5542C	OEM
CDLL5542D	OEM
CDLL5543	OEM
CDLL5543A	OEM
CDLL5543B	OEM
CDLL5543C	OEM
CDLL5543D	OEM
CDLL5544	OEM
CDLL5544A	OEM
CDLL5544B	OEM
CDLL5544C	OEM
CDLL5544D	OEM
CDLL5545	OEM
CDLL5545A	OEM
CDLL5545B	OEM
CDLL5545C	OEM
CDLL5545D	OEM
CDLL5546	OEM
CDLL5546A	OEM
CDLL5546B	OEM
CDLL5546C	OEM
CDLL5546D	OEM
CDLL5711	OEM
CDLL5712	OEM
CDLL5817	OEM
CDLL5818	OEM
CDLL5819	OEM
CDLL6309	OEM
CDLL6310	OEM
CDLL6311	OEM
CDLL6312	OEM
CDLL6313	OEM
CDLL6314	OEM
CDLL6315	OEM
CDLL6316	OEM
CDLL6317	OEM
CDLL6318	OEM
CDLL6319	OEM
CDLL6320	OEM
CDLL6321	OEM
CDLL6322	OEM
CDLL6323	OEM
CDLL6324	OEM
CDLL6325	OEM
CDLL6326	OEM
CDLL6327	OEM
CDLL6328	OEM
CDLL6329	OEM
CDLL6330	OEM
CDLL6331	OEM
CDLL6332	OEM
CDLL6333	OEM
CDLL6334	OEM
CDLL6335	OEM
CDLL6336	OEM
CDLL6337	OEM
CDLL6338	OEM
CDLL6339	OEM
CDLL6340	OEM
CDLL6341	OEM
CDLL6342	OEM
CDLL6343	OEM
CDLL6344	OEM
CDLL6345	OEM
CDLL6346	OEM
CDLL6347	OEM
CDLL6348	OEM
CDLL6349	OEM
CDLL6350	OEM
CDLL6351	OEM
CDLL6352	OEM
CDLL6353	OEM
CDLL6354	OEM
CDLL6355	OEM
CDLL6485	OEM
CDLL6486	OEM
CDLL6487	OEM
CDLL6488	OEM
CDLL6490	OEM
CDLL6491	OEM
CDLLL4571A	OEM
CDP18S030	OEM
CDP18S205V1	OEM
CDP18S508	OEM
CDP18S510	OEM
CDP18S601	OEM
CDP18S602	OEM
CDP18S603	OEM
CDP18S604B	OEM
CDP18S605	OEM
CDP18S606	OEM
CDP18S607	OEM
CDP18S608	OEM
CDP18S609	OEM
CDP18S610	OEM
CDP18S620	OEM
CDP18S621	OEM
CDP18S621V1	OEM
CDP18S622	OEM
CDP18S623A	OEM
CDP18S625	OEM
CDP18S626	OEM
CDP18S627	OEM
CDP18S629	OEM
CDP18S640A1	OEM
CDP18S640V1	OEM
CDP18S641	OEM
CDP18S642	OEM
CDP18S643A	OEM
CDP18S644	OEM
CDP18S646	OEM
CDP18S647	OEM
CDP18S648	OEM
CDP18S651	OEM
CDP18S652	OEM
CDP18S653V1	OEM
CDP18S653V2	OEM
CDP18S654	OEM
CDP18S657	OEM
CDP18S658	OEM
CDP18S660	OEM
CDP18S661B	OEM
CDP18S661V3	OEM
CDP18S662	OEM
CDP18S670	OEM
CDP18S675	OEM
CDP18S676	OEM
CDP18S691	OEM
CDP18U42CD	OEM
CDP1801CR	OEM
CDP1801CU	OEM
CDP1801R	OEM
CDP1801U	OEM
CDP1802ACD	3223
CDP1802ACE	3223
CDP1802AD	3223
CDP1802AE	3223
CDP1802BCD	3223
CDP1802BCE	3223
CDP1802CD	3223
CDP1802CD3	OEM
CDP1802CE	3223
CDP1802D	3223
CDP1802D3	OEM
CDP1802E	3223
CDP1804ACD	OEM
CDP1804CE	OEM
CDP1804D	OEM
CDP1804E	OEM
CDP1805CD	3356
CDP1805CE	3356
CDP1805D	3356
CDP1805E	3356
CDP1821CD	OEM
CDP1821CD3	OEM
CDP1821CE	OEM
CDP1821CH	OEM
CDP1821D	OEM
CDP1821E	OEM
CDP1822CD	4043
CDP1822CD3	OEM
CDP1822CE	4043
CDP1822CH	OEM
CDP1822D	4043
CDP1822E	4043
CDP1823CD	4083
CDP1823CD3	OEM
CDP1823CE	4083
CDP1823CH	OEM
CDP1823D	4083
CDP1823E	4083
CDP1824CD	4103
CDP1824CD3	OEM
CDP1824CE	4103
CDP1824CH	OEM
CDP1824D	4103
CDP1824D3	OEM
CDP1824E	4103
CDP1825CD	OEM
CDP1825CE	OEM
CDP1825CH	OEM
CDP1825D	OEM
CDP1825E	OEM
CDP1831CD	OEM
CDP1831CE	OEM
CDP1831CH	OEM
CDP1831D	OEM
CDP1831E	OEM
CDP1832CD	OEM
CDP1832CE	OEM
CDP1832CH	OEM
CDP1832D	OEM
CDP1832E	OEM
CDP1833CD	OEM
CDP1833CE	OEM
CDP1833CH	OEM
CDP1833D	OEM
CDP1833E	OEM
CDP1834CD	OEM
CDP1834CE	OEM
CDP1834CH	OEM
CDP1834D	OEM
CDP1834E	OEM
CDP1851CD	4668
CDP1851CE	4668
CDP1851D	4668
CDP1851E	4668
CDP1852CD	4742
CDP1852CD3	OEM
CDP1852CE	4742
CDP1852D	4742
CDP1852D3	OEM
CDP1852E	4742
CDP1852H	OEM
CDP1853CD	4816
CDP1853CD3	OEM
CDP1853CE	4816
CDP1853D	4816
CDP1853D3	OEM
CDP1853E	4816
CDP1854ACD	4435
CDP1854ACD3	OEM
CDP1854ACE	4435
CDP1854AD	4435
CDP1854AD3	OEM
CDP1854AE	4435
CDP1855CD	4880
CDP1855CE	4880
CDP1855CEX	OEM
CDP1855D	4880
CDP1855DX	OEM
CDP1855E	4880
CDP1855EX	OEM
CDP1856CD	OEM
CDP1856CD3	OEM
CDP1856CDX	OEM
CDP1856CE	OEM
CDP1856CEX	OEM
CDP1856D	OEM
CDP1856D3	OEM
CDP1856DX	OEM
CDP1856E	OEM
CDP1856EX	OEM
CDP1857CD	OEM
CDP1857CD3	OEM
CDP1857CDX	OEM
CDP1857CE	OEM
CDP1857CEX	OEM
CDP1857D	OEM
CDP1857D3	OEM
CDP1857DX	OEM
CDP1857E	OEM
CDP1857EX	OEM
CDP1858CD	OEM
CDP1858CD3	OEM
CDP1858CDX	OEM
CDP1858CE	OEM
CDP1858CEX	OEM
CDP1858D	OEM
CDP1858D3	OEM
CDP1858DX	OEM
CDP1858E	OEM
CDP1858EX	OEM
CDP1859CD	OEM
CDP1859CD3	OEM
CDP1859CDX	OEM
CDP1859CE	OEM
CDP1859CEX	OEM
CDP1859D	OEM
CDP1859D3	OEM
CDP1859DX	OEM
CDP1859E	OEM
CDP1859EX	OEM
CDP1861CD	5009
CDP1861CE	5009
CDP1862CD	5019
CDP1862CE	5019
CDP1863CD	OEM
CDP1863CE	OEM
CDP1863CEX	OEM
CDP1863D	OEM
CDP1863E	OEM
CDP1864CD	5053
CDP1864CE	5053
CDP1866CD	5087
CDP1866CE	5087
CDP1866D	5087
CDP1866E	5087
CDP1867CD	OEM
CDP1867CE	OEM
CDP1867D	OEM
CDP1867E	OEM
CDP1868CD	OEM
CDP1868CE	OEM
CDP1868D	OEM
CDP1868E	OEM
CDP1869CD	5145
CDP1869CE	5145
CDP1869D	5145
CDP1869E	5145
CDP1870CD	OEM
CDP1870CE	OEM
CDP1870D	OEM
CDP1870E	OEM
CDP1871CD	5231
CDP1871CE	5231
CDP1871D	5231
CDP1871E	5231
CDP1872CD	5266
CDP1872CE	5266
CDP1873CD	5288
CDP1873CE	5288
CDP1873CEX	OEM
CDP1874CD	5308
CDP1874CE	5308
CDP1875CD	5317
CDP1875CE	5317
CDP1876CD	5331
CDP1876CDX	OEM
CDP1876CE	5331
CDP1876CEX	OEM
CDP1876D	5331
CDP1876DX	OEM
CDP1876E	5331
CDP1876EX	OEM
CDP1877CD	5343
CDP1877CE	5343
CDP1877E	5343
CDP1888E	OEM
CDP6402CD	OEM
CDP6402CE	5626
CDP6402D	OEM
CDP6402E	OEM
CDPR512CD	OEM
CDPR512D	OEM
CDPR522CD	OEM
CDPR522D	OEM
CDQ1004	0233
CDQ1011	0861
CDQ1012	0861
CDQ1014	0861
CDQ1018	0079
CDQ1021	0079
CDQ1024	0079
CDQ1037	0861
CDQ1044	0861
CDQ1045	0861
CDQ10001	0016
CDQ10002	0016
CDQ10003	0016
CDQ10004	0016
CDQ10005	0016
CDQ10006	0016
CDQ10007	0016
CDQ10008	0016
CDQ10009	0016
CDQ10010	0016
CDQ10011	0086
CDQ10012	0086
CDQ10013	0233
CDQ10014	0086
CDQ10015	0233
CDQ10016	0016
CDQ10017	0016
CDQ10018	0016
CDQ10019	0016
CDQ10020	0016
CDQ10021	0016
CDQ10022	0016
CDQ10023	0016
CDQ10024	0016
CDQ10025	0016
CDQ10026	0016
CDQ10027	0016
CDQ10028	0016
CDQ10032	0016
CDQ10033	0086
CDQ10034	0233
CDQ10035	0198
CDQ10036	0198
CDQ10037	0233
CDQ10044	0595
CDQ10045	0233
CDQ10046	0233
CDQ10047	0233
CDQ10048	0086
CDQ10049	0233
CDQ10051	0086
CDQ10052	0086
CDQ10053	0086
CDQ10054	OEM
CDQ10055	OEM
CDQ10056	OEM
CDQ10057	0086
CDQ10058	0086
CDR-2	0015
CDR-4	0015
CDR125AL	OEM
CDR125AP	OEM
CDS-16B	0015
CDT1309	0160
CDT1310	0160
CDT1311	0160
CDT1312	0975
CDT1313	0975
CDT1315	0975
CDT1319	0160
CDT1320	0160
CDT1321	0160
CDT1322	0975
CDT1349	0160
CDT1349A	0160
CDT1350	0160
CDT1350A	0160
CDT4012	OEM
CDT4013	OEM
CDT4014	OEM
CDT4015	OEM
CDT4016	OEM
CDX34D	OEM
CDX1250M	OEM
CDZ-9V	0012
CDZ-318-75	0649
CDZ318-75	0365
CDZ-C9V	0012
CE0360/7839	0004
CE0361/7839	0004
CE0362/7839	0004
CE0363/7839	0164
CE0378/7839	0030
CE0398/7839	0015
CE0398-7839	0540
CE0495/7839	0143
CE03607839	0211
CE37	0133
CE401	0015
CE402	0015
CE403	0015
CE404	0015
CE405	0015
CE406	0015
CE407	0071
CE408	0071
CE409	0071
CE410	0071
CE501	0015
CE502	0015
CE503	0015
CE504	0015
CE505	0015
CE506	0015
CE507	0071
CE508	0071
CE509	0071
CE510	0071
CE4001B	0079
CE4001C	0079
CE4002D	0126
CE4003D	0086
CE4003E	0079
CE4004C	0079
CE4005C	0037
CE4008B	0224
CE4008C	0224
CE4008D	0224
CE4008E	0224
CE4010D	0144
CE4010E	0144
CE4012D	0037
CE4012UBFX	OEM
CE4013E	0079
CE41009-00F	OEM
CE213811	0004
CEC6050	0015
CEM3310	OEM
CEM3320	OEM
CEM3330	OEM
CEM3335	OEM
CEM3340	OEM
CEM3345	OEM
CEM3350	OEM
CEM3360	OEM
CEN530	OEM
CEN544	OEM
CEN558	OEM
CER6B	0015

If replacement code is OEM, contact original manufacturer for replacement.

DEVICE TYPE	REPL CODE	DEVICE TYPE	REPL CODE	DEVICE TYPE	REPL CODE	DEVICE TYPE	REPL CODE	DEVICE TYPE	REPL CODE	DEVICE TYPE	REPL CODE	DEVICE TYPE	REPL CODE
CER67	0015	CFA6.8D	0062	CFB20D5	0700	CG030B	OEM	CGE-54	2866	CH759	0052	CH4104/25	OEM
CER67A	0015	CFA6.8D5	0062	CFB20D10	0700	CG040A	OEM	CGE-60	0008	CH759/25	OEM	CH4105	OEM
CER67B	0015	CFA6.8D10	0062	CFB22D	0705	CG040C	OEM	CGE-61	0330	CH821	0471	CH4105/25	OEM
CER67C	0015	CFA7.5D	0077	CFB22D5	0705	CG040D	OEM	CGE-62	0111	CH823	0471	CH4106	0052
CER67F	1335	CFA7.5D5	0077	CFB22D10	0705	CG040E	OEM	CGE-64	0400	CH825	0471	CH4107	OEM
CER68	0015	CFA7.5D10	0077	CFB24D	0494	CG-1	OEM	CGE-66	0042	CH827	0471	CH4108	0879
CER68A	0015	CFA8.2D	0165	CFB24D5	0494	CG1	0016	CGE-68	0142	CH829	0471	CH4109	0686
CER68B	0015	CFA8.2D5	0165	CFB24D10	0494	CG1(DIO)	0102	CGE-69	0925	CH962	OEM	CH4110	OEM
CER68C	0015	CFA8.2D10	0165	CFB25D	0714	CG1(DIODE)	0102	CGE-500	0133	CH962/25	OEM	CH4111	OEM
CER-69	0015	CFA9.1D	0057	CFB25D5	0714	CG-1/DG-1	OEM	CGJ-1	0102	CH963	0052	CH4112	OEM
CER69	0015	CFA9.1D5	0057	CFB25D10	0714	CG1/DG1	OEM	CGS-540	OEM	CH963/25	OEM	CH4113	0949
CER69A	0015	CFA9.1D10	0057	CFB27D	0455	CG1-540	OEM	CGY11A	OEM	CH964	OEM	CH4114	OEM
CER69B	0015	CFA10D	0064	CFB27D5	0455	CG2	0102	CGY11B	OEM	CH964/25	OEM	CH4115	OEM
CER69C	0015	CFA10D5	0064	CFB27D10	0455	CG2(DIO)	0102	CGY12A	OEM	CH965	0686	CH4116	OEM
CER70	0015	CFA10D10	0064	CFB30D	0195	CG-2/DG-2	OEM	CGY12B	OEM	CH965/25	OEM	CH4117	1606
CER70A	0015	CFA11D	0181	CFB30D5	0195	CG2/DG2	OEM	CGY13A	OEM	CH966	OEM	CH4118	OEM
CER70B	0015	CFA11D5	0181	CFB30D10	0195	CG3	OEM	CGY13B	OEM	CH966/25	OEM	CH4119	0257
CER70C	0015	CFA11D10	0181	CFB33D	0166	CG3(DIO)	OEM	CGY14A	OEM	CH967	OEM	CH4120	OEM
CER-71	0015	CFA12D	0052	CFB33D5	0166	CG-3/DG-3	OEM	CGY14B	OEM	CH967/25	OEM	CH4121	OEM
CER71	0015	CFA12D5	0052	CFB33D10	0166	CG3/DG3	OEM	CH00031M	OEM	CH968	OEM	CH4122	OEM
CER71A	0015	CFA12D10	0052	CFC4/500	OEM	CG3.22.2FW	OEM	CH10	OEM	CH968/25	OEM	CH4123	OEM
CER71B	0015	CFA13D	0053	CFC6.8D	0062	CG3.22.12FB	OEM	CH16.8	OEM	CH969	OEM	CH4124	OEM
CER71C	0015	CFA13D5	0053	CFC6.8D5	0062	CG11G1	OEM	CH16.8A	OEM	CH969/25	OEM	CH4125	OEM
CER-71CA	0015	CFA13D10	0053	CFC6.8D10	0062	CG12-E	0143	CH17.5	OEM	CH970	OEM	CH4126	OEM
CER71F	OEM	CFA14D	0879	CFC7.5D	0077	CG12E	0143	CH17.5A	OEM	CH970/25	OEM	CH4127	OEM
CER72	0071	CFA14D5	0879	CFC7.5D5	0077	C-G59A	OEM	CH18.2	OEM	CH971	OEM	CH4128	OEM
CER72A	0071	CFA14D10	0879	CFC7.5D10	0077	CG60H	OEM	CH18.2A	OEM	CH971/25	OEM	CH4129	OEM
CER72B	0071	CFA15D	0686	CFC8.2D	0165	CG61H	0143	CH19.1	OEM	CH972	OEM	CH4130	OEM
CER72C	0071	CFA15D5	0686	CFC8.2D5	0165	CG62H	0143	CH19.1A	OEM	CH972/25	OEM	CH4131	OEM
CER72D	0071	CFA15D10	0686	CFC8.2D10	0165	CG63H	0143	CH20	OEM	CH973	OEM	CH4132	OEM
CER72F	0071	CFA16D	0445	CFC9.1D	0057	CG64H	0143	CH30	OEM	CH973/25	OEM	CH4133	3026
CER73	0071	CFA16D5	0445	CFC9.1D5	0057	CG65H	0143	CH40	OEM	CH974	OEM	CH4134	OEM
CER73A	0071	CFA16D10	0445	CFC9.1D10	0057	CG66H	0143	CH50	OEM	CH974/25	OEM	CH4135	OEM
CER73B	0071	CFA17D	0210	CFC10D	0064	CG70H	OEM	CH109A2	OEM	CH975	OEM	CH4370	OEM
CER73C	0071	CFA17D5	0210	CFC10D5	0064	CG71H	OEM	CH109AZ2	OEM	CH975/25	OEM	CH4371	OEM
CER73D	0071	CFA17D10	0210	CFC10D10	0064	CG72H	OEM	CH109B2	OEM	CH976	OEM	CH4372	OEM
CER73F	0071	CFA18D	0375	CFC11D	0181	CG74H	0143	CH109C2	OEM	CH976/25	OEM	CH4565A	OEM
CER500	0015	CFA18D5	0375	CFC11D5	0181	CG80H	OEM	CH109D2	OEM	CH977	OEM	CH4566A	OEM
CER500A	0015	CFA18D10	0375	CFC11D10	0181	CG81H	OEM	CH109E2	OEM	CH977/25	OEM	CH4567A	OEM
CER500B	0015	CFA19D	0671	CFC12D	0052	CG82H	OEM	CH109F2	OEM	CH978	OEM	CH4568A	OEM
CER500C	0015	CFA19D5	0671	CFC12D5	0052	CG83H	OEM	CH109Z2	OEM	CH978/25	OEM	CH4569A	OEM
CER500F	OEM	CFA19D10	0671	CFC12D10	0052	CG84H	OEM	CH110	OEM	CH979	OEM	CH4570A	OEM
CER670	0015	CFA20D	0700	CFC13D	0053	CG85H	OEM	CH110A	OEM	CH979/25	OEM	CH4571A	OEM
CER670A	0015	CFA20D5	0700	CFC13D5	0053	CG86H	0143	CH111	OEM	CH980	OEM	CH4572A	OEM
CER670B	0015	CFA20D10	0700	CFC13D10	0053	CG90H	OEM	CH111A	OEM	CH980/25	OEM	CH4573A	OEM
CER670C	0015	CFA22D	0705	CFC14D	0879	CG91AH	OEM	CH112	OEM	CH981	OEM	CH4574A	OEM
CER680	0015	CFA22D5	0705	CFC14D5	0879	CG92AH	OEM	CH112A	OEM	CH981/25	OEM	CH4575A	OEM
CER680A	0015	CFA22D10	0705	CFC14D10	0879	CG94H	OEM	CH113	OEM	CH982	OEM	CH4576A	OEM
CER680B	0015	CFA24D	0494	CFC15D	0686	CG125	OEM	CH113A	OEM	CH982/25	OEM	CH4577A	OEM
CER680C	0015	CFA24D5	0494	CFC15D5	0686	CG125A	OEM	CH115	OEM	CH983	OEM	CH4578A	OEM
CER690	0015	CFA24D10	0494	CFC15D10	0686	CG125B	OEM	CH115A	OEM	CH983/25	OEM	CH4579A	OEM
CER690A	0015	CFA25D	0714	CFC16D	0445	CG125C	OEM	CH116	OEM	CH984	OEM	CH4580A	OEM
CER690B	0015	CFA25D5	0714	CFC16D5	0445	CG125D	OEM	CH116A	OEM	CH984/25	OEM	CH4581A	OEM
CER690C	0015	CFA25D10	0714	CFC16D10	0445	CG127	OEM	CH118	OEM	CH985	OEM	CH4582A	OEM
CER700	0015	CFA27D	0455	CFC17D	0210	CG127A	OEM	CH118A	OEM	CH985/25	OEM	CH4583A	OEM
CER700A	0015	CFA27D5	0455	CFC17D5	0210	CG127B	OEM	CH119A	OEM	CH986	OEM	CH4584A	OEM
CER700B	0015	CFA27D10	0714	CFC17D10	0210	C-G189A	OEM	CH119AZ	OEM	CH986/25	OEM	CH4614	OEM
CER700C	0015	CFA28D	0257	CFC18D	0375	CG512	OEM	CH119B	OEM	CH987	OEM	CH4614/25	OEM
CER710	0015	CFA28D5	0257	CFC18D5	0375	CG4000	OEM	CH119C	OEM	CH987/25	OEM	CH4615	OEM
CER710A	0015	CFA28D10	0257	CFC18D10	0375	CG4100	OEM	CH119D	0015	CH988	OEM	CH4615/25	OEM
CER710B	0015	CFA30D	0195	CFC19D	0671	CG4103	OEM	CH119E	OEM	CH988/25	OEM	CH4616	OEM
CER710C	0015	CFA30D5	0195	CFC19D5	0671	CG4200	OEM	CH119F	OEM	CH989	OEM	CH4616/25	OEM
CER720	0071	CFA30D10	0195	CFC19D10	0671	CG5500A	OEM	CH119Z	OEM	CH989/25	OEM	CH4617	OEM
CER720A	0071	CFA33D	0166	CFC20D	0700	CG5600A	OEM	CH120	OEM	CH990	OEM	CH4617/25	OEM
CER720B	0071	CFA33D5	0166	CFC20D5	0700	CG5700A	OEM	CH120A	OEM	CH990/25	OEM	CH4618	OEM
CER720C	0071	CFA33D10	0166	CFC20D10	0700	CG5705	OEM	CH122	OEM	CH991	OEM	CH4618/25	OEM
CER730	0071	CFB6.8D	0062	CFC22D	0705	CG24015A	0033	CH122A	OEM	CH991/25	OEM	CH4619	OEM
CER730A	0071	CFB6.8D5	0062	CFC22D5	0705	CG80286-6C	OEM	CH124	OEM	CH992	OEM	CH4619/25	OEM
CER730B	0071	CFB6.8D10	0062	CFC22D10	0705	CG80286-10	OEM	CH124A	OEM	CH992/25	OEM	CH4620	OEM
CER730C	0071	CFB7.5D	OEM	CFC24D	0494	CGB-540	OEM	CH127	OEM	CH1213	OEM	CH4620/25	OEM
CER1200F	OEM	CFB7.5D5	0077	CFC24D5	0494	CGB204001-14	OEM	CH127A	OEM	CH1214	OEM	CH4621	OEM
CF-092	0012	CFB7.5D10	0077	CFC24D10	0494	CGB204002-02	OEM	CH130	OEM	CH1215	OEM	CH4621/25	OEM
CF-07005	0738	CFB8.2D	0165	CFC25D	0714	CGB374008-01	OEM	CH130A	OEM	CH1216	OEM	CH4622	OEM
CF3	0554	CFB8.2D5	0165	CFC25D5	0714	CGB374202-01	OEM	CH133	OEM	CH1253	OEM	CH4622/25	OEM
CF4-28	OEM	CFB8.2D10	0165	CFC25D10	0714	CGB374203-02	OEM	CH133A	OEM	CH1267A	OEM	CH4623	OEM
CF10	OEM	CFB9.1D	0057	CFC27D	0455	CGB374205-01	OEM	CH302BA	OEM	CH1290	OEM	CH4623/25	OEM
CF15	OEM	CFB9.1D5	0057	CFC27D5	0455	CGB394208-01	OEM	CH302BD	OEM	CH1295	OEM	CH4624	OEM
CF20	OEM	CFB9.1D10	0057	CFC27D10	0455	CGB408001-16	OEM	CH302FA	OEM	CH1296	OEM	CH4624/25	OEM
CF24	OEM	CFB10D	0064	CFC30D	0195	CGB408002-03	OEM	CH302HA	OEM	CH1511	OEM	CH4625	OEM
CF25	OEM	CFB10D5	0064	CFC30D5	0195	CGB445005-01	OEM	CH302KA	OEM	CH1513	OEM	CH4626	OEM
CF30	OEM	CFB10D10	0064	CFC30D10	0195	CGB596402-06	OEM	CH302MA	0087	CH1514	OEM	CH4626/25	OEM
CF35	OEM	CFB11D	0181	CFC33D	0166	CGB596402-07	OEM	CH302PA	0087	CH1515	OEM	CH4627	0471
CF40	OEM	CFB11D5	0181	CFC33D5	0166	CGB596403-10	OEM	CH302RA	0087	CH1516	OEM	CH4627/25	OEM
CF45	OEM	CFB11D10	0181	CFC33D10	0166	CGB596403-11	OEM	CH746	OEM	CH1517	OEM	CH4728	0292
CF50	OEM	CFB12D	0052	CFCD/400	OEM	CGB596405-09	OEM	CH746/25	OEM	CH1518	OEM	CH4728/25	OEM
CF102BA	OEM	CFB12D5	0052	CFD8201CP	OEM	CGB596405-10	OEM	CH747	OEM	CH1611	OEM	CH4729	OEM
CF102DA	0015	CFB12D10	0052	CFL-5	OEM	CGB596408-02	OEM	CH747/25	OEM	CH1612	OEM	CH4729/25	OEM
CF102FA	OEM	CFB13D	0053	CFM13026	OEM	CGB596410-03	OEM	CH748	OEM	CH1613	OEM	CH4730	0456
CF102HA	OEM	CFB13D5	0053	CFR-10	0774	CGB647101-09	OEM	CH748/25	OEM	CH1614	OEM	CH4730/25	OEM
CF102KA	OEM	CFB13D10	0053	CFR10	OEM	CGB647103-08	OEM	CH749	OEM	CH1615	OEM	CH4731	0533
CF102MA	OEM	CFB14D	0879	CFR15	OEM	CGB647105-04	OEM	CH749/25	OEM	CH1616	OEM	CH4731/25	OEM
CF102PA	0071	CFB14D5	0879	CFR20	OEM	CGB717701-04	OEM	CH750	OEM	CH1621	OEM	CH4732	OEM
CF102RA	0071	CFB14D10	0879	CFR25	OEM	CGB717702-03	OEM	CH750/25	OEM	CH1622	OEM	CH4732/25	OEM
CF102VA	0087	CFB15D	0686	CFR30	OEM	CGB717704-01	OEM	CH751	OEM	CH1623	OEM	CH4733	OEM
CF152AA	0071	CFB15D5	0686	CFR35	OEM	CGB798405-01	OEM	CH751/25	OEM	CH1624	OEM	CH4733/25	OEM
CF152BA	0071	CFB15D10	0686	CFR40	OEM	CGC081201-07	OEM	CH752	OEM	CH1625	OEM	CH4734	5799
CF152DA	0071	CFB16D	0445	CFR45	OEM	CGC101201-03	OEM	CH752/25	OEM	CH1820(A)	OEM	CH4734/25	OEM
CF152FA	0071	CFB16D5	0445	CFR50	OEM	CGC121301-02	OEM	CH753	0471	CH3065	0167	CH4735	0636
CF152HA	0071	CFB16D10	0445	CFR5185A	OEM	CGD125L	OEM	CH753/25	OEM	CH4099	OEM	CH4735/25	OEM
CF152KA	0071	CFB17D	0210	CFR5186A	OEM	CGD462	0143	CH754	OEM	CH4099/25	OEM	CH4736	OEM
CF152MA	0071	CFB17D5	0210	CFR5187A	OEM	CGD591	0143	CH754/25	OEM	CH4100	OEM	CH4736/25	OEM
CF152PA	0071	CFB17D10	0210	CFR5188A	OEM	CGD685	0143	CH755	OEM	CH4100/25	OEM	CH4737	OEM
CF152RA	0071	CFB18D	0375	CFY11	OEM	CGD1029	0143	CH755/25	OEM	CH4101	OEM	CH4737/25	OEM
CF152VA	0087	CFB18D5	0375	CFY12	OEM	C-GE9A	OEM	CH756	OEM	CH4101/25	OEM	CH4738	OEM
CF-1295H	0086	CFB18D10	0375	CFY20	OEM	C-GE21	OEM	CH756/25	OEM	CH4102	OEM	CH4738/25	OEM
CF2386	4214	CFB19D	0671	CFZ12D	OEM	CGE-50	0050	CH757	0057	CH4102/25	OEM	CH4739	0012
CF37212N	OEM	CFB19D5	0671	CFZ12D5	OEM	CGE-51	0050	CH757/25	OEM	CH4103	0057	CH4739/25	OEM
CF61872N	OEM	CFB19D10	0671	CFZ12D10	OEM	CGE-52	0004	CH758	OEM	CH4103/25	OEM	CH4740	OEM
CF77316AFT	OEM	CFB20D	0700	CG030A	OEM	CGE-53	0164	CH758/25	OEM	CH4104	OEM	CH4740/25	OEM

DEVICE TYPE	REPL CODE
CH4741	OEM
CH4741/25	OEM
CH4742	0137
CH4742/25	OEM
CH4743	OEM
CH4743/25	OEM
CH4744	0002
CH4744/25	OEM
CH4745	OEM
CH4745/25	OEM
CH4746	OEM
CH4746/25	OEM
CH4747	OEM
CH4747/25	OEM
CH4748	OEM
CH4748/25	OEM
CH4749	OEM
CH4749/25	OEM
CH4750	OEM
CH4750/25	OEM
CH4751	OEM
CH4751/25	OEM
CH4752	OEM
CH4752/25	OEM
CH4753	OEM
CH4753/25	OEM
CH4754	OEM
CH4754/25	OEM
CH4755	OEM
CH4755/25	OEM
CH4756	OEM
CH4756/25	OEM
CH4757	OEM
CH4757/25	OEM
CH4758	OEM
CH4758/25	OEM
CH4759	OEM
CH4759/25	OEM
CH4760	OEM
CH4760/25	OEM
CH4761	1189
CH4761/25	OEM
CH4762	OEM
CH4762/25	OEM
CH4763	1311
CH4763/25	OEM
CH4764	0098
CH4764/25	OEM
CH5221	OEM
CH5221/25	OEM
CH5222	OEM
CH5222/25	OEM
CH5223	OEM
CH5223/25	OEM
CH5224	OEM
CH5224/25	OEM
CH5225	OEM
CH5225/25	OEM
CH5226	OEM
CH5226/25	OEM
CH5227	OEM
CH5227/25	OEM
CH5228	OEM
CH5228/25	OEM
CH5229	OEM
CH5229/25	OEM
CH5230	OEM
CH5230/25	OEM
CH5231	OEM
CH5231/25	OEM
CH5232	OEM
CH5232/25	OEM
CH5233	0091
CH5233/25	OEM
CH5234	0471
CH5234/25	OEM
CH5235	OEM
CH5235/25	OEM
CH5236	OEM
CH5236/25	OEM
CH5237	OEM
CH5237/25	OEM
CH5238	OEM
CH5238/25	OEM
CH5239	0057
CH5239/25	OEM
CH5240	OEM
CH5240/25	OEM
CH5241	OEM
CH5241/25	OEM
CH5242	0052
CH5242/25	OEM
CH5243	OEM
CH5243/25	OEM
CH5244	OEM
CH5244/25	OEM
CH5245	0686
CH5245/25	OEM
CH5246	OEM
CH5246/25	OEM
CH5247	1649
CH5247/25	OEM
CH5248	OEM
CH5248/25	OEM
CH5249	OEM
CH5249/25	OEM
CH5250	OEM
CH5250/25	OEM
CH5251	OEM
CH5251/25	OEM
CH5252	OEM
CH5252/25	OEM
CH5253	OEM
CH5253/25	OEM
CH5254	OEM
CH5254/25	OEM
CH5255	OEM
CH5255/25	OEM
CH5256	OEM
CH5256/25	OEM
CH5257	OEM
CH5257/25	OEM
CH5258	OEM
CH5258/25	OEM
CH5259	OEM
CH5259/25	OEM
CH5260	OEM
CH5260/25	OEM
CH5261	OEM
CH5261/25	OEM
CH5262	OEM
CH5262/25	OEM
CH5263	OEM
CH5263/25	OEM
CH5264	1156
CH5264/25	OEM
CH5265	OEM
CH5265/25	OEM
CH5266	OEM
CH5266/25	OEM
CH5267	OEM
CH5267/25	OEM
CH5268	OEM
CH5268/25	OEM
CH5269	OEM
CH5269/25	OEM
CH5270	OEM
CH5270/25	OEM
CH5271	OEM
CH5271/25	OEM
CH5272	OEM
CH5272/25	OEM
CH5273	OEM
CH5273/25	OEM
CH5274	OEM
CH5274/25	OEM
CH5275	0245
CH5275/25	OEM
CH5276	OEM
CH5276/25	OEM
CH5277	OEM
CH5277/25	OEM
CH5278	0877
CH5278/25	OEM
CH5279/25	OEM
CH5280	OEM
CH5280/25	OEM
CH5281	OEM
CH5281/25	OEM
CHA4Z3.6	0188
CHA4Z5.1	0162
CHA4Z5.6	0157
CHAZ6.2	0636
CHAZ6.2A	0636
CHAZ7.5	OEM
CHAZ7.5A	OEM
CHAZ62.A	0636
CHMZ3.6	0188
CHMZ5.1	0162
CHMZ5.6	0157
CHS70ANE11P	OEM
CHS70BGE11P	OEM
CHS70GE11PF	OEM
CHS74GS21PF	OEM
CHV1070A	OEM
CHV1080A	OEM
CHV1090A	OEM
CHV1570A	OEM
CHV1580A	OEM
CHV1590A	OEM
CHV2030-01	OEM
CHV2060A	OEM
CHV2080A	OEM
CHV2090A	OEM
CHZ10	0170
CHZ10A	0170
CHZ210A	0170
CI1	OEM
CI2	OEM
CI3	OEM
CI4	OEM
CI5	OEM
CI-1002	0823
CI-1003	1345
CI-1004	0654
CI1103	OEM
CI1103-2	OEM
CI1123	OEM
CI2711	0016
CI2712	0016
CI2713	0016
CI2714	0016
CI2923	0016
CI2924	0016
CI2925	0016
CI2926	0016
CI3390	0016
CI3391	0016
CI3391A	0016
CI3392	0016
CI3393	0016
CI3394	0016
CI3395	0016
CI3396	0016
CI3397	0016
CI3398	0016
CI3402	0016
CI3403	0016
CI3404	0016
CI3405	0016
CI3414	0079
CI3415	0016
CI3416	0016
CI3417	0016
CI3704	0086
CI3705	0086
CI3706	0086
CI3900	0016
CI3900A	0016
CI3901	0016
CI4256	0016
CI4424	0016
CI4425	0016
CI6800-2	OEM
CI-6800-2/16	OEM
CI-6800-2/32	OEM
CI-6800-2/48	OEM
CI-6800-2/64	OEM
CI8080	OEM
CI8086	OEM
CI68002	OEM
CI-68002A	OEM
CI68002A	OEM
CIJ70645	0123
CIL194	0224
CIL194A	0224
CIL194B	0224
CIL195	0224
CIL195C	OEM
CIL195D	OEM
CIL250	OEM
CIL251	OEM
CIL252	OEM
CIL253	OEM
CIL254	OEM
CIL255	OEM
CIL256	OEM
CIL257	OEM
CIL350	OEM
CIL351	OEM
CIL352	OEM
CIL353	OEM
CIL354	OEM
CIL355	OEM
CIL356	OEM
CIL357	OEM
CIL358	OEM
CIL359	OEM
CIL360	OEM
CIL361	OEM
CIL362	OEM
CIL363	OEM
CIL364	OEM
CIL365	OEM
CIL366	OEM
CIL367	OEM
CIL368	OEM
CIL369	OEM
CIL500/022	OEM
CIL500/024	OEM
CIL500/027	OEM
CIL500/030	OEM
CIL500/033	OEM
CIL500/039	OEM
CIL500/043	OEM
CIL500/047	OEM
CIL500/056	OEM
CIL500/062	OEM
CIL500/068	OEM
CIL500/075	OEM
CIL500/082	OEM
CIL500/091	OEM
CIL500/100	OEM
CIL500/110	OEM
CIL500/120	OEM
CIL500/130	OEM
CIL500/140	OEM
CIL500/150	OEM
CIL500/160	OEM
CIL500/180	OEM
CIL500/200	OEM
CIL500/220	OEM
CIL500/240	OEM
CIL500/270	OEM
CIL500/300	OEM
CIL500/330	OEM
CIL500/360	OEM
CIL500/390	OEM
CIL500/430	OEM
CIL500/470	OEM
CIL500/510	OEM
CIL500/560	OEM
CIL500/620	OEM
CIL500/680	OEM
CIL511	0144
CIL512	0144
CIL513	0144
CIL521	0144
CIL522	0144
CIL523	0144
CIL-531	0079
CIL531	0144
CIL-532	0079
CIL532	0144
CIL533	0144
CIL1300	OEM
CIL1301	OEM
CIL1302	OEM
CIL1303	OEM
CIL1304	OEM
CIL1305	OEM
CINCH	OEM
CI-PCM	OEM
CI-S100/16	OEM
CI-S100/32	OEM
CI-S100/48	OEM
CI-S100/64	OEM
CJ5201	0016
CJ5202	0016
CJ5203	0016
CJ5204	0004
CJ5205	OEM
CJ-5206	0016
CJ5206	0079
CJ5206A	0595
CJ-5207	0016
CJ5207	0079
CJ-5208	0016
CJ-5209	0126
CJ5209	0126
CJ-5210	0086
CJ5210	0086
CJ-5211	0079
CJ5211	0016
CJ-5212	0079
CJ5212	0016
CJ5213	0086
CJ5214	0086
CJ5215	0086
CJCA001	OEM
CJCA002	OEM
CJCA007	OEM
CJCA008	OEM
CJSE001	OEM
CJSE002	OEM
CJSE003	OEM
CJSE004	OEM
CJSE005	OEM
CJSE006	OEM
CJSE009	OEM
CJSE010	OEM
CJSE011	OEM
CJSE012	OEM
CJSE013	OEM
CJSE014	OEM
CJSE017	OEM
CJSE018	OEM
CJSE019	OEM
CJSE020	OEM
CJSE021	OEM
CJSE022	OEM
CJSE033	OEM
CJSE034	OEM
CJSE035	OEM
CJSE036	OEM
CJSE037	OEM
CJSE038	OEM
CJSE039	OEM
CJSE040	OEM
CJSE041	OEM
CJSE042	OEM
CJSE043	OEM
CJSE044	OEM
CJSE045	OEM
CJSE046	OEM
CJSE047	OEM
CJSE048	OEM
CJSE049	OEM
CJSE050	OEM
CJSE052	OEM
CJSE053	OEM
CJSE057	OEM
CJSE058	OEM
CJSE059	OEM
CJSE061	OEM
CJSE062	OEM
CJSE063	OEM
CJSE064	OEM
CJSE065	OEM
CJSE066	OEM
CJSE067	3393
CJSE068	OEM
CJSE069	OEM
CJSE070	OEM
CJSE071	1916
CJSE072	OEM
CK	OEM
CK00121	OEM
CK4	0136
CK4A	0136
CK13	0004
CK13A	0004
CK14	0280
CK14A	0280
CK16	0280
CK16A	0280
CK17	0280
CK17A	0280
CK22	0004
CK22A	0004
CK22B	0004
CK22C	0004
CK25	0280
CK25A	0280
CK26	0280
CK26A	0280
CK27	0280
CK27A	0280
CK28	0136
CK28A	0136
CK64	0004
CK64A	0004
CK64B	0004
CK64C	0004
CK65	0004
CK65A	0004
CK65B	0004
CK65C	0004
CK66	0004
CK66A	0004
CK66B	0004
CK66C	0004
CK67	0004
CK67A	0004
CK67B	0004
CK67C	0004
CK83	0050
CK131	0917
CK-134A	3707
CK261	0600
CK262	0600
CK273	OEM
CK277	OEM
CK302BA	OEM
CK302DA	OEM
CK302FA	OEM
CK302HA	OEM
CK302KA	OEM
CK302MA	2645
CK302PA	2645
CK302RA	2645
CK398	0861
CK419	0016
CK420	0016
CK421	0016
CK422	0016
CK474	0016
CK475	0016
CK476	0016
CK477	0016
CK661	0280
CK662	0280
CK705	0143
CK-706	0143
CK706	0143
CK706-P	0143
CK706A	0143
CK706P	0143
CK710	0917
CK715	0143
CK721	0211
CK722	0211
CK725	0211
CK727	0211
CK731	0917
CK751	0211
CK754	0211
CK759	0280
CK759A	0280
CK760	0280
CK760A	0280
CK761	0280
CK762	0136
CK766	0136
CK766A	0136
CK768	0280
CK776	0280
CK776A	0280
CK790	0211
CK791	0211
CK793	0211
CK794	0211
CK798	OEM
CK799	OEM
CK800	OEM
CK863	OEM
CK863B	OEM
CK870	0211
CK871	0211
CK872	0211
CK875	0211
CK878	0211
CK879	0211
CK882	0211
CK888	0211
CK891	0004
CK892	0004
CK942	0037
CK1101	OEM
CK1101P	OEM
CK1102	OEM
CK1102P	OEM
CK1103	OEM
CK1103P	OEM
CK1104	OEM
CK1104P	OEM
CK1105	OEM
CK1106	OEM
CK1108	OEM
CK1111	OEM
CK1111P	OEM
CK1112	OEM
CK1112P	OEM
CK1113	OEM
CK1113P	OEM
CK1114	OEM
CK1115	OEM
CK1116	OEM
CK1117	OEM
CK1121	OEM
CK1121W	OEM
CK1122	OEM
CK1122B	OEM
CK1122W	OEM
CK1123	OEM
CK1124	OEM
CK1201	OEM
CK1202	OEM
CK1203	OEM
CK1204	OEM
CK1205	OEM
CK1206	OEM
CK1207	OEM
CK1208	OEM
CK1209	OEM
CK1210	OEM
CK1211	OEM
CK1212	OEM
CK1213	OEM
CK1214	OEM
CK1216	OEM
CK1217	OEM
CK1218	OEM
CK1219	OEM
CK1220	OEM
CK1221	OEM
CK1222	OEM
CK1223	OEM
CK1224	OEM
CK1225	OEM
CK1226	OEM
CK1227	OEM
CK1228	OEM
CK1230	OEM
CK1231	OEM
CK1232	OEM
CK1233	OEM
CK1234	OEM
CK1241	OEM
CK1242	OEM
CK1243	OEM
CK1244	OEM
CK1245	OEM
CK1246	OEM
CK1247	OEM
CK1248	OEM
CK1249	OEM
CK1250	OEM
CK1251	OEM
CK1252	OEM
CK1253	OEM
CK1254	OEM
CK1255	OEM
CK1256	OEM
CK1257	OEM
CK1258	OEM
CK1260	OEM
CK1261	OEM
CK1262	OEM
CK1263	OEM
CK1264	OEM
CK1265	OEM
CK1266	OEM
CK1267	OEM
CK1268	OEM
CK1271	OEM
CK1273	OEM
CK1274	OEM
CK1275	OEM
CK1276	OEM
CK1277	OEM
CK1278	OEM
CK1279	OEM
CK1280	OEM
CK1281	OEM
CK1282	OEM
CK1283	OEM
CK1284	OEM
CK1285	OEM
CK1286	OEM
CK1287	OEM
CK1288	OEM
CK1289	OEM
CK1290	OEM
CK1291	OEM
CK1292	OEM
CK1293	OEM
CK2000	OEM
CK2000B	OEM
CK2002	OEM
CK2003	OEM
CK2008P	OEM
CK2013	OEM
CK2015	OEM
CK2018	OEM
CK2019	OEM
CK2024	OEM
CK2025P	OEM
CK2028	OEM
CK2028A	OEM
CK2028B	OEM
CK2029	OEM
CK2030	OEM
CK2032	OEM
CK2033	OEM
CK2034	OEM
CK2035	OEM
CK2037	OEM
CK2038	OEM
CK2039	OEM
CK2040	OEM
CK2041	OEM
CK2042	OEM
CK2043	OEM
CK2046	OEM
CK2051	OEM
CK2052	OEM
CK2053	OEM
CK2060	OEM
CK2065	OEM
CK2066	OEM
CK2067	OEM
CK2070	OEM
CK2071	OEM
CK2072	OEM
CK2078	OEM
CK2080	OEM
CK2086	OEM
CK2095	OEM
CK2101	OEM
CK2102	OEM
CK2103	OEM
CK2104	OEM
CK2105	OEM
CK2106	OEM
CK2112	OEM
CK2114	OEM
CK2117	OEM
CK2118	OEM
CK2124	OEM
CK2125	OEM
CK2127	OEM
CK2129	OEM
CK2132	OEM
CK2137	OEM
CK2139	OEM
CK2140	OEM
CK2141	OEM
CK2142	OEM
CK2143	OEM
CK2145	OEM
CK2146	OEM
CK2147	OEM
CK2148	OEM
CK2149	OEM
CK2150	OEM
CK2151	OEM
CK2153	OEM
CK2154	OEM
CK2155	OEM
CK2157	OEM
CK2160	OEM
CK2161	OEM
CK2162	OEM
CK2170	OEM
CK2171	OEM
CK2172	OEM
CK2173	OEM
CK2177	OEM
CK2178	OEM
CK2190P	OEM
CK2195	OEM
CK-3300	OEM
CK3300	OEM
CL05	0015
CL010	0015
CL025	0015
CL1	0015
CL1.5	0015
CL-2	OEM
CL2	0015
CL-2(F)	OEM
CL3	0015
CL3P7L	OEM
CL4	0015
CL5	0015
CL5M2	OEM
CL5M2L	OEM
CL5M3	OEM
CL5M4	OEM
CL5M4L	OEM
CL5M5	OEM
CL5M5L	OEM
CL5M7	OEM
CL5M9M	OEM
CL5P4	OEM
CL5P4L	OEM
CL5P5	OEM
CL5P5L	OEM
CL5P5M	OEM
CL5P9M	OEM
CL6	0015
CL7	0015

DEVICE TYPE	REPL CODE	DEVICE TYPE	REPL CODE	DEVICE TYPE	REPL CODE	DEVICE TYPE	REPL CODE	DEVICE TYPE	REPL CODE	DEVICE TYPE	REPL CODE	DEVICE TYPE	REPL CODE
CL7P4	OEM	CL905N	OEM	CLA3133-01	OEM	CLR5	OEM	CM4-564	OEM	CM4009AD	1999	CM4500	OEM
CL7P4L	OEM	CL907	OEM	CLA3133-02	OEM	CLR10	OEM	CM4-582B	OEM	CM4009AE	0001	CM4501	OEM
CL7P5	OEM	CL907H	OEM	CLA3133-03	OEM	CLR15	OEM	CM4-582B-2	OEM	CM4010AD	3252	CM4503	OEM
CL7P5H-ID	OEM	CL907HL	OEM	CLA3134-01	OEM	CLR20	OEM	CM4-583B	OEM	CM4010AE	0398	CM4518	1043
CL7P5HL	OEM	CL907HN	OEM	CLA3134-02	OEM	CLR25	OEM	CM4-584B	3157	CM4011	0215	CM4520	2675
CL7P5L	OEM	CL907L	OEM	CLA3135-01	OEM	CLR30	OEM	CM4-584B-2	3157	CM4011AD	0215	CM5770-512	OEM
CL7P7HL	OEM	CL907N	OEM	CLA3135-02	OEM	CLR35	OEM	CM5-5	OEM	CM4011AE	0215	CM6400-3C	OEM
CL7P7HM	OEM	CL909L	OEM	CLA3150-150-806	OEM	CLR40	OEM	CM5-11	OEM	CM4012	0498	CM6400-3P	OEM
CL7P9L	OEM	CL1020	OEM	CLA3150-173-001	OEM	CLR45	OEM	CM5-12	OEM	CM4012AD	0498	CM6400A3C	OEM
CL8	0015	CL1200	OEM	CLA3150-176-001	OEM	CLR50	OEM	CM5-21	OEM	CM4012AE	0498	CM6400A3P	OEM
CL9P4L	OEM	CL1210	OEM	CLA3150-184-802	OEM	CLR2049	OEM	CM5-22	OEM	CM4013	0413	CM6400A-C	OEM
CL9P5	OEM	CL1220	OEM	CLC064	OEM	CLR2050	2313	CM5-23	OEM	CM4013AD	0413	CM6400A-P	OEM
CL9P5HL	OEM	CL1230	OEM	CLD31	OEM	CLR2060	3552	CM5-27	OEM	CM4013AE	0413	CM6400C	OEM
CL9P5L	OEM	CL1305	OEM	CLD41	OEM	CLR2090	OEM	CM5-29	OEM	CM4014AD	0860	CM6400P	OEM
CL9P7HL	OEM	CL1325	OEM	CLD42	OEM	CLR2170	OEM	CM5-31	OEM	CM4014AE	0860	CM7163	0079
CL9P7HLL	OEM	CL1355	OEM	CLD56	OEM	CLR2180	3552	CM5-32	OEM	CM4014AF	OEM	CM8470	0015
CL9P9L	OEM	CL1375	OEM	CLD71	OEM	CLR2191	OEM	CM10-12A	2093	CM4015	1014	CM8640E	0211
CL9P9LL	OEM	CL1395	OEM	CLEAK-02101	OEM	CLR3180	OEM	CM-12	0124	CM4015AD	1014	CMB-7601	OEM
CL10	OEM	CL1506	0273	CLEAK-02200	0323	CLR4180	OEM	CM20-12	OEM	CM4015AE	1014	CMB-7601A	OEM
CL12	0315	CL1506B	OEM	CLEAK-02201	1414	CLR5101	OEM	CM20-12A	2852	CM4015AF	OEM	CMB-7602	OEM
CL13	0315	CL1510	0315	CLEAK-02300	0323	CLR5101-2	OEM	CM25-12	3589	CM4016	1143	CMB-7602A	OEM
CL14	OEM	CL1511	0315	CLEAK-02301	1414	CLT2010	OEM	CM25-28	OEM	CM4016AD	1143	CMC68/04	OEM
CL15	0541	CL1520	OEM	CLEAK-02601	3111	CLT2020	OEM	CM25-28A	OEM	CM4016AE	1143	CMC68/15	OEM
CL16	OEM	CL1800	OEM	CLEAK-02701	1212	CLT2030	OEM	CM30-12	6604	CM4017	0513	CMC68/15B	OEM
CL17	OEM	CL1800M	OEM	CLEAK-02701(2SD560)	OEM	CLT2064	OEM	CM30-12A	2868	CM4017AD	0513	CMC68/15C	OEM
CL18	OEM	CL1800W	OEM	CLEAK-05401	1212	CLT2065	OEM	CM40-12	2868	CM4017AE	0513	CMC68/15G	4013
CL19	OEM	CL1810	OEM	CLEAK-05501	OEM	CLT2130	2313	CM45-12	OEM	CM4018AD	1391	CMC1530A	OEM
CL20	OEM	CL1810M	OEM	CLEAK-05703	OEM	CLT2140	2313	CM45-12A	2868	CM4018AE	1391	CMC1535A	OEM
CL25	OEM	CL1810W	OEM	CLEAK-05704	OEM	CLT2150	3878	CM45-28	OEM	CM4019	1525	CMC1540A	OEM
CL30	OEM	CL1811	OEM	CLEAK-05705	OEM	CLT2160	2313	CM50	OEM	CM4019AD	1525	CMC2510A	OEM
CL35	OEM	CL1811M	OEM	CLEAK-05707	OEM	CLT2164	OEM	CM60/12	OEM	CM4019AE	1525	CMC2512A	OEM
CL40	OEM	CL1811W	OEM	CLEAK-05708	OEM	CLT2165	OEM	CM60/15	OEM	CM4020	1661	CMC2515A	OEM
CL45	OEM	CL1820	OEM	CLEAK-05709	OEM	CLT3020	OEM	CM60-12	OEM	CM4020AD	1661	CMC2530A	OEM
CL50	OEM	CL1820M	OEM	CLEAK-05710	2697	CLT3030	OEM	CM60-12A	3591	CM4020AE	1661	CMC2535A	OEM
CL100	0591	CL1820W	OEM	CLEAK-05903	2697	CLT3160	OEM	CM80-28	OEM	CM4021	1749	CMC2540A	2626
CL110	0591	CL1821	OEM	CLEAK-06002	2697	CLT3170	OEM	CM100/12	OEM	CM4021AD	1749	CMC4010A	OEM
CL111	OEM	CL1821M	OEM	CLEAK-06102	0812	CLT4020	OEM	CM100/15	OEM	CM4021AE	1749	CMC4012A	OEM
CL112	OEM	CL1821W	OEM	CLEAK-06201	0812	CLT4030	OEM	CM106	OEM	CM4021AF	OEM	CMC4015A	OEM
CL113	OEM	CL1830	OEM	CLEAK-09801	OEM	CLT4140	OEM	CM107	OEM	CM4022AD	1256	CMC4030A	OEM
CL114	OEM	CL1830M	OEM	CLEAK-11904	1638	CLT4150	OEM	CM108	OEM	CM4022AE	1256	CMC4035A	OEM
CL116(P)	OEM	CL1830W	OEM	CLEAK-12004	0812	CLT4160	OEM	CM109	OEM	CM4023	0519	CMC4040A	2626
CL116P	6400	CL1831	OEM	CLEAK-14201	1638	CLT4170	OEM	CM110	OEM	CM4023AD	0519	CMC7010A	4013
CL120	0315	CL1831M	OEM	CLEAK-14401	1638	CLT5160	OEM	CM200/12	OEM	CM4023AE	0519	CMC7012A	4013
CL-150R-CD	OEM	CL1831W	OEM	CLED-1	OEM	CLT5170	OEM	CM200/15	OEM	CM4024	1957	CMC7015A	4013
CL155D	OEM	CL1835	OEM	CLED-1F	OEM	CLVA43A	OEM	CM300/12	OEM	CM4024AD	1957	CME30-12	OEM
CL166(D)	OEM	CL1835M	OEM	CLED-5	OEM	CLVA47A	OEM	CM300/15	OEM	CM4024AE	1957	CME50-12	OEM
CL166(P)	OEM	CL1835W	OEM	CLED400	OEM	CLVA51A	OEM	CM301AN	1300	CM4025	2072	CME70-12	OEM
CL166C	0016	CL1836	OEM	CLI5	0315	CLVA56A	OEM	CM500/5	OEM	CM4025AD	2072	CMEM-8K	OEM
CL166C,D	0079	CL1836M	OEM	CLI-6	0315	CLVA62A	0471	CM600	OEM	CM4025AE	2072	CMEM-16K	OEM
CL166D	0016	CL1836W	OEM	CLI-7	0315	CLVA68A	OEM	CM601	OEM	CM4026AD	2152	CMEM-32K	OEM
CL252D57JU	OEM	CL1840	OEM	CLI-8	0315	CLVA75A	OEM	CM602	1155	CM4026AE	2152	CMK-29EHB	OEM
CL254A17JVW	OEM	CL1840M	OEM	CLI-11	1053	CLVA82A	OEM	CM603	1155	CM4027	1949	CM0770	0079
CL254D16EU	OEM	CL1840W	OEM	CLI-12	1053	CLVA91A	0057	CM640	OEM	CM4027AD	1949	CMP01BJ	OEM
CL602	OEM	CL1841	OEM	CLI-13	OEM	CLVA100A	OEM	CM641	OEM	CM4027AE	1949	CMP01BY	OEM
CL603	OEM	CL1841M	OEM	CLI-14	1053	CLZ80-4/2	OEM	CM642	1155	CM4028AD	2228	CMP01BZ	OEM
CL603A	OEM	CL1841W	OEM	CLI25	5425	CLZ80-4/8	OEM	CM643	6792	CM4028AE	2228	CMP01CJ	OEM
CL603AL	OEM	CL1850	OEM	CLI-506	0315	CLZ80-16/2	OEM	CM644	OEM	CM4029AD	2233	CMP01CP	OEM
CL604	OEM	CL1850M	OEM	CLI-800	1417	CLZ80-16/8	OEM	CM645	OEM	CM4029AE	2233	CMP01CY	OEM
CL604L	OEM	CL1850W	OEM	CLI-810	1417	CM	0196	CM646	1155	CM4030	0500	CMP01CZ	OEM
CL604M	OEM	CL1851	OEM	CLI-811	1417	CM05	OEM	CM647	5361	CM4030AD	0500	CMP01EJ	OEM
CL605	OEM	CL1851M	OEM	CLI-820	1417	CM0770	0086	CM697	OEM	CM4030AE	0500	CMP01EP	OEM
CL605L	OEM	CL1851W	OEM	CLI-821	1417	CM0SMM74C00N	2112	CM800	OEM	CM4032AD	2531	CMP01EY	OEM
CL607	OEM	CL1860	OEM	CLI-830	1417	CM4-20	OEM	CM856	OEM	CM4032AE	2531	CMP01EZ	OEM
CL607L	OEM	CL1860M	OEM	CLI-831	1417	CM4-21	OEM	CM860	OEM	CM4033AD	2635	CMP01GR	OEM
CL702	OEM	CL1860W	OEM	CLI-835	1417	CM4-22	OEM	CM1000/5	OEM	CM4033AE	2635	CMP01J	OEM
CL702/2	OEM	CL1861	OEM	CLI-836	1417	CM4-23	OEM	CM1310C	OEM	CM4035AD	2777	CMP01N	OEM
CL702L	OEM	CL1861M	OEM	CLI-840	2636	CM4-24	OEM	CM1310P	OEM	CM4035AE	2777	CMP01Y	OEM
CL702L2	OEM	CL1861W	OEM	CLI-841	2636	CM4-25	OEM	CM1600-1C	OEM	CM4035AF	OEM	CMP01Z	OEM
CL703	OEM	CL1870	OEM	CLI-851	2636	CM4-43	OEM	CM1600-1P	OEM	CM4037AD	OEM	CMP02BJ	OEM
CL703/2	OEM	CL1870M	OEM	CLI-860	2636	CM4-43A	OEM	CM1600-2C	OEM	CM4037AE	OEM	CMP02BY	OEM
CL703A	OEM	CL1870W	OEM	CLI-861	2636	CM4-43B	OEM	CM1600-2P	OEM	CM4038AD	2981	CMP02BZ	OEM
CL703L	OEM	CL1871	OEM	CLI-870	2636	CM4-73A	OEM	CM1600-3C	OEM	CM4038AE	2981	CMP02CJ	OEM
CL703L2	OEM	CL1871M	OEM	CLI-871	2636	CM4-80B	OEM	CM1600-3P	OEM	CM4040	0056	CMP02CP	OEM
CL703M	OEM	CL1871W	OEM	CLK-24	OEM	CM4-81B	OEM	CM1600C	OEM	CM4040AD	0056	CMP02CY	OEM
CL704	OEM	CL2210	OEM	CLK-24C	OEM	CM4-82B	OEM	CM1600P	OEM	CM4040AE	0056	CMP02CZ	OEM
CL704/2	OEM	CL2220	OEM	CLM05	0015	CM4-83	3182	CM2000/5	OEM	CM4041AD	3174	CMP02EJ	OEM
CL704L	OEM	CL3201	OEM	CLM1	0015	CM4-83-1	3182	CM2550	0160	CM4041AE	3174	CMP02EP	OEM
CL704L2	OEM	CL3310	OEM	CLM1.5	OEM	CM4-83-2	3182	CM3200-2C	OEM	CM4042	0121	CMP02EY	OEM
CL704M	OEM	CL3320	OEM	CLM2	OEM	CM4-83-3	3182	CM3200-2D	OEM	CM4042AD	0121	CMP02EZ	OEM
CL705	OEM	CL4710	OEM	CLM3	OEM	CM4-84B	1358	CM3200-2P	OEM	CM4042AE	0121	CMP02GR	OEM
CL705/2	OEM	CL4720	OEM	CLM4	OEM	CM4-84B-0	1358	CM3200-3C	OEM	CM4043AD	1553	CMP02J	OEM
CL705HL	OEM	CL6810	OEM	CLM5	OEM	CM4-84B-1	1358	CM3200-3D	OEM	CM4043AE	1553	CMP02N	OEM
CL705HL2	OEM	CL7500	OEM	CLM5H10A	OEM	CM4-84B-2	OEM	CM3200-3P	OEM	CM4044AD	2308	CMP02Y	OEM
CL705L	OEM	CL7520	OEM	CLM6	OEM	CM4-85B	OEM	CM3200-C	OEM	CM4044AE	2308	CMP02Z	OEM
CL705M	OEM	CL8630	OEM	CLM7	OEM	CM4-86B	OEM	CM3200-D	OEM	CM4047AD	2331	CMP04BY	OEM
CL707	OEM	CL8630S	OEM	CLM7H16A	OEM	CM4-244A	OEM	CM3200-P	OEM	CM4047AE	2331	CMP04FP	OEM
CL707H	OEM	CL8633	OEM	CLM8	OEM	CM4-244B	OEM	CM3200C	OEM	CM4047AF	2331	CMP04FY	OEM
CL707HL	OEM	CL8633S	OEM	CLM50	OEM	CM4-264	OEM	CM3200P	OEM	CM4048AD	3457	CMP05AJ(M)	OEM
CL707HM	OEM	CL8640R	OEM	CLM51	OEM	CM4-282B	OEM	CM3900	2247	CM4048AE	3459	CMP05AZ(M)	OEM
CL707L	OEM	CL8640T	OEM	CLM60	OEM	CM4-282B-2	OEM	CM4000	2024	CM4049	0001	CMP05BJ(M)	OEM
CL707M	OEM	CL8690	OEM	CLM61	OEM	CM4-283B	OEM	CM4000AD	2024	CM4049AD	0001	CMP05BZ(M)	OEM
CL709L	OEM	CLA7	OEM	CLM3000	OEM	CM4-284B	1962	CM4000AE	2024	CM4049AE	0001	CMP05EJ(A)	OEM
CL902	OEM	CLA7AA	OEM	CLM3006A	OEM	CM4-284B-2	1962	CM4001	0478	CM4050	0398	CMP05EP	OEM
CL902L	OEM	CLA7D	OEM	CLM3012A	OEM	CM4-344A	0841	CM4001AD	0478	CM4050AD	0398	CMP05EZ(M)	OEM
CL902N	OEM	CLA7DA	OEM	CLM3120A	OEM	CM4-344B	OEM	CM4001AE	0478	CM4050AE	0398	CMP05FJ(A)	OEM
CL903	OEM	CLA60	OEM	CLM3500	OEM	CM4-382B	OEM	CM4002	2055	CM4051	0366	CMP05FP	OEM
CL903A	OEM	CLA65AA	OEM	CLM4006A	OEM	CM4-382B-2	OEM	CM4002AD	2055	CM4052	0024	CMP05FZ(M)	OEM
CL903AN	OEM	CLA90	OEM	CLM4012A	OEM	CM4-383B	OEM	CM4002AE	2055	CM4066AD	0101	CMP81CY	OEM
CL903L	OEM	CLA90AA	OEM	CLM4120A	OEM	CM4-384B	1778	CM4004AD	OEM	CM4066AE	0101	CMP81EY	OEM
CL903N	OEM	CLA3130-150-806	OEM	CLM6000	OEM	CM4-384B-2	1778	CM4004AE	OEM	CM4066AF	OEM	CMP82CJ	OEM
CL904	OEM	CLA3130-173-001	OEM	CLM6200	OEM	CM4-444A	3096	CM4006AD	0646	CM4066AH	OEM	CMP82CY	OEM
CL904L	OEM	CLA3130-176-001	OEM	CLM6500	OEM	CM4-444B	OEM	CM4006AE	0646	CM4069B	0119	CMP82EJ	OEM
CL904N	OEM	CLA3130-184-802	OEM	CLM8000	OEM	CM4-482B	OEM	CM4006AF	OEM	CM4081	0626	CMP82EY	OEM
CL905	OEM	CLA3131-01	OEM	CLM8200-2	OEM	CM4-482B-2	OEM	CM4007	2846	CM4102AD	OEM	CMP1115	OEM
CL905H	OEM	CLA3131-02	OEM	CLM8500	OEM	CM4-483B	OEM	CM4007AD	2846	CM4102AE	OEM	CMX740	OEM
CL905HL	OEM	CLA3131-03	OEM	CLM8500/2	OEM	CM4-484B	3276	CM4007AE	2846	CM4102D	OEM	CN2484	0086
CL905HLL	OEM	CLA3132-01	OEM	CLM8500HV	OEM	CM4-484B-2	3276	CM4008AD	0988	CM4102E	OEM	CN5310	2623
CL905HN	OEM	CLA3132-02	OEM	CLM8600	OEM	CM4-544A	3124	CM4008AE	0988	CM4116AD	0522	CN5310(K)	2623
CL905L	OEM	CLA3132-03	OEM	CLM9000	OEM	CM4-544B	OEM	CM4009	1999	CM4116AE	0522	CN5310(L)	2623

DEVICE TYPE	REPL CODE
CN5310(M)	2623
CN5310K	2623
CN5310L	2623
CN5310M	2623
CN5311	2623
CN5311CL	2623
CN5311K	2623
CN5311KL	2623
CN5311LM	2623
CN5411	5095
CN5510	4060
CN5520	3609
CN32536N	OEM
CND101A	OEM
CND1001A	OEM
CND1002A	OEM
CND2001A	OEM
CND2002A	OEM
CND3001A	OEM
CND3002A	OEM
CND4001A	OEM
CND6001A	OEM
CND6002A	OEM
CND6003A	OEM
CND6004A	OEM
CND6005A	OEM
CND6006A	OEM
CND6007A	OEM
CND6008A	OEM
CND6009A	OEM
CND6010A	OEM
CND6011A	OEM
CND6012A	OEM
CNM5A	OEM
CNM5B	OEM
CNM5C	OEM
CNM5D	OEM
CNM5E	OEM
CNM5EN	OEM
CNM10A	OEM
CNM10B	OEM
CNM10C	OEM
CNM10D	OEM
CNM10E	OEM
CNM10EN	OEM
CNM20A	OEM
CNM20B	OEM
CNM20C	OEM
CNM20D	OEM
CNM20E	OEM
CNM20EN	OEM
CNM50A	OEM
CNM50B	OEM
CNM50C	OEM
CNM50D	OEM
CNM50E	OEM
CNM50EN	OEM
CNM100A	OEM
CNM100B	OEM
CNM100C	OEM
CNM100D	OEM
CNM100E	OEM
CNM100EN	OEM
CNM1000A	OEM
CNR21	OEM
CNX21	OEM
CNX21T	OEM
CNX35	0315
CNX35U	0315
CNX36	0315
CNX38	OEM
CNX48	OEM
CNY6	OEM
CNY17-1	OEM
CNY17-2	OEM
CNY17-3	OEM
CNY17-4	OEM
CNY17-I	OEM
CNY17-II	OEM
CNY17-III	OEM
CNY17-IV	OEM
CNY17I	OEM
CNY17II	OEM
CNY17III	OEM
CNY17IV	OEM
CNY18-1	OEM
CNY18-2	OEM
CNY18-3	OEM
CNY18-4	OEM
CNY18III	OEM
CNY18IV	OEM
CNY21	OEM
CNY28	2636
CNY29	2636
CNY30	OEM
CNY31	2640
CNY32	2641
CNY33	OEM
CNY34	OEM
CNY35	OEM
CNY36	2640
CNY37	1417
CNY42	OEM
CNY43	OEM
CNY47	OEM
CNY47A	OEM
CNY48	1053
CNY50	OEM
CNY50-1	OEM

DEVICE TYPE	REPL CODE
CNY51	OEM
CNY52	OEM
CNY53	OEM
CNY57	OEM
CNY57A	OEM
CNY62	OEM
CNY63	OEM
CNY64	OEM
CNY65	OEM
CNY65B	OEM
CNY70	OEM
C009	OEM
C02L55A57V	OEM
C06C	0015
C010	OEM
C020A	OEM
C020B	OEM
C040A	OEM
C040B	OEM
C049	0015
C080	OEM
C0231	OEM
C0231-1	OEM
C0231-3	OEM
C0231-4	OEM
C0231H	OEM
C0231H1	OEM
C0231H3	OEM
C0231H4	OEM
C0231HV	OEM
C0231HV1	OEM
C0231HV3	OEM
C0231HV4	OEM
C0231HVY	OEM
C0231HY	OEM
C0231T	OEM
C0231T1	OEM
C0231T3	OEM
C0231T4	OEM
C0231TV	OEM
C0231TV1	OEM
C0231TV3	OEM
C0231TV4	OEM
C0231TVY	OEM
C0231TY	OEM
C0231V	OEM
C0231V1	OEM
C0231V3	OEM
C0231V4	OEM
C0231VH	OEM
C0231VY	OEM
C0231Y	OEM
C0232	OEM
C0232-3	OEM
C0232-5	OEM
C0232T	OEM
C0232T3	OEM
C0232T5	OEM
C0232TV	OEM
C0232TV3	OEM
C0232TV5	OEM
C0232V	OEM
C0232V3	OEM
C0232V5	OEM
C0233	OEM
C0233-1	OEM
C0233-2	OEM
C0233-3	OEM
C0233-4	OEM
C0233-5	OEM
C0233-6	OEM
C0233FU1	OEM
C0233FU2	OEM
C0233FU3	OEM
C0233FU4	OEM
C0233FU5	OEM
C0233FU6	OEM
C0233FUT	OEM
C0233FUT1	OEM
C0233FUT2	OEM
C0233FUT3	OEM
C0233FUT4	OEM
C0233FUT5	OEM
C0233FUT6	OEM
C0233FUTV	OEM
C0233FUTV1	OEM
C0233FUTV2	OEM
C0233FUTV3	OEM
C0233FUTV4	OEM
C0233FUTV5	OEM
C0233FUTV6	OEM
C0233FUV	OEM
C0233FUV1	OEM
C0233FUV2	OEM
C0233FUV3	OEM
C0233FUV4	OEM
C0233FUV5	OEM
C0233FUV6	OEM
C0233FW	OEM
C0233FW1	OEM
C0233FW2	OEM
C0233FW3	OEM
C0233FW4	OEM
C0233FW5	OEM
C0233FW6	OEM
C0233FWT	OEM
C0233FWT1	OEM
C0233FWT2	OEM
C0233FWT3	OEM

DEVICE TYPE	REPL CODE
C0233FWT4	OEM
C0233FWT5	OEM
C0233FWT6	OEM
C0233FWTV	OEM
C0233FWTV1	OEM
C0233FWTV2	OEM
C0233FWTV3	OEM
C0233FWTV4	OEM
C0233FWTV5	OEM
C0233FWTV6	OEM
C0233FWV	OEM
C0233FWV1	OEM
C0233FWV2	OEM
C0233FWV3	OEM
C0233FWV4	OEM
C0233FWV5	OEM
C0233FWV6	OEM
C0233FX	OEM
C0233FX1	OEM
C0233FX2	OEM
C0233FXTV	OEM
C0233FXTV1	OEM
C0233FXTV2	OEM
C0233FXTV3	OEM
C0233FXTV4	OEM
C0233FXTV5	OEM
C0233FXTV6	OEM
C0233FXV	OEM
C0233FXV1	OEM
C0233FXV2	OEM
C0233FXV3	OEM
C0233FXV4	OEM
C0233FXV5	OEM
C0233FXV6	OEM
C0233H	OEM
C0233H1	OEM
C0233H2	OEM
C0233H3	OEM
C0233H4	OEM
C0233H5	OEM
C0233H6	OEM
C0233HR	OEM
C0233HR1	OEM
C0233HR2	OEM
C0233HR3	OEM
C0233HR4	OEM
C0233HR5	OEM
C0233HR6	OEM
C0233HRV	OEM
C0233HRV1	OEM
C0233HRV2	OEM
C0233HRV3	OEM
C0233HRV4	OEM
C0233HRV5	OEM
C0233HRV6	OEM
C0233HTV	OEM
C0233HTV1	OEM
C0233HTV2	OEM
C0233HTV3	OEM
C0233HTV4	OEM
C0233HTV5	OEM
C0233HTV6	OEM
C0233ME	OEM
C0233ME1	OEM
C0233ME2	OEM
C0233ME3	OEM
C0233MEHA	OEM
C0233MEHA1	OEM
C0233MEHAV	OEM
C0233MEHAV1	OEM
C0233MEHB1	OEM
C0233MEHBV	OEM
C0233MEHBV1	OEM
C0233MEHV	OEM
C0233MET	OEM
C0233MET1	OEM
C0233MET2	OEM
C0233MET3	OEM
C0233METV	OEM
C0233METV1	OEM
C0233METV2	OEM
C0233METV3	OEM
C0233MEV	OEM
C0233MEV1	OEM
C0233MEV2	OEM
C0233MEV3	OEM
C0233T	OEM
C0233T1	OEM
C0233T2	OEM
C0233T3	OEM
C0233T4	OEM
C0233T5	OEM
C0233T6	OEM
C0233TV	OEM
C0233TV1	OEM
C0233TV2	OEM
C0233TV3	OEM
C0233TV4	OEM
C0233TV5	OEM
C0233TV6	OEM
C0233U	OEM
C0233U1	OEM
C0233U2	OEM
C0233U3	OEM
C0233U4	OEM
C0233U5	OEM
C0233U6	OEM
C0233UT	OEM
C0233UT1	OEM

DEVICE TYPE	REPL CODE
C0233UT2	OEM
C0233UT3	OEM
C0233UT4	OEM
C0233UT5	OEM
C0233UT6	OEM
C0233UTV	OEM
C0233UTV1	OEM
C0233UTV2	OEM
C0233UTV3	OEM
C0233UTV4	OEM
C0233UTV5	OEM
C0233UTV6	OEM
C0233V1	OEM
C0233V2	OEM
C0233V3	OEM
C0233V4	OEM
C0233V5	OEM
C0233V6	OEM
C0233VFW	OEM
C0233VFWR	OEM
C0233VH	OEM
C0233VHR	OEM
C0233VR	OEM
C0233VW	OEM
C0233VWR	OEM
C0233W	OEM
C0233W1	OEM
C0233W2	OEM
C0233W3	OEM
C0233W4	OEM
C0233W5	OEM
C0233W6	OEM
C0233WT	OEM
C0233WT1	OEM
C0233WT2	OEM
C0233WT3	OEM
C0233WT4	OEM
C0233WT5	OEM
C0233WT6	OEM
C0233WTV	OEM
C0233WTV1	OEM
C0233WTV2	OEM
C0233WTV3	OEM
C0233WTV4	OEM
C0233WTV5	OEM
C0233WTV6	OEM
C0233X	OEM
C0233X1	OEM
C0233X2	OEM
C0233X3	OEM
C0233X4	OEM
C0233X5	OEM
C0233X6	OEM
C0233XT	OEM
C0233XT1	OEM
C0233XT2	OEM
C0233XT3	OEM
C0233XT4	OEM
C0233XT5	OEM
C0233XT6	OEM
C0233XTV	OEM
C0233XTV1	OEM
C0233XTV2	OEM
C0233XTV3	OEM
C0233XTV4	OEM
C0233XTV5	OEM
C0233XTV6	OEM
C0234-3	OEM
C0234A	OEM
C0234A3	OEM
C0234AV	OEM
C0234AV3	OEM
C0234B	OEM
C0234B3	OEM
C0234BV	OEM
C0234BV3	OEM
C0235	OEM
C0235-3	OEM
C0235T	OEM
C0235T3	OEM
C0235T5	OEM
C0235TV	OEM
C0235TV3	OEM
C0235TV5	OEM
C0235V	OEM
C0235V3	OEM
C0235V5	OEM
C0236	OEM
C0236-1	OEM
C0236-3	OEM
C0236-4	OEM
C0236T	OEM
C0236T1	OEM
C0236T3	OEM
C0236T4	OEM
C0236TV	OEM
C0236TV1	OEM
C0236TV3	OEM
C0236TV4	OEM
C0236TVY	OEM
C0236TY	OEM
C0236V	OEM
C0236V1	OEM
C0236V3	OEM
C0236V4	OEM
C0236VY	OEM
C0236Y	OEM
C0238A	OEM
C0238A2	OEM

DEVICE TYPE	REPL CODE
C0238A3	OEM
C0238AV	OEM
C0238AV2	OEM
C0238AV3	OEM
C0238B	OEM
C0238B2	OEM
C0238B3	OEM
C0238BV	OEM
C0238BV2	OEM
C0238BV3	OEM
C0238T2	OEM
C0238T3	OEM
C0238TV	OEM
C0238TV2	OEM
C0238TV3	OEM
C0239A2	OEM
C0239A3	OEM
C0239AV	OEM
C0239AV2	OEM
C0239AV3	OEM
C0239B	OEM
C0239B2	OEM
C0239B3	OEM
C0239BV	OEM
C0239BV2	OEM
C0239BV3	OEM
C0251A27	OEM
C0251A27U	OEM
C0251A27V	OEM
C0251A27VU	OEM
C0251A27VW	OEM
C0251A27VX	OEM
C0251A27W	OEM
C0251A27X	OEM
C0251A57	OEM
C0251A57U	OEM
C0251A57V	OEM
C0251A57VU	OEM
C0251A57VW	OEM
C0251A57VX	OEM
C0251A57W	OEM
C0251A57X	OEM
C0251B16	OEM
C0251B16U	OEM
C0251B16V	OEM
C0251B16VU	OEM
C0251B16VW	OEM
C0251B16VX	OEM
C0251B16W	OEM
C0251B16X	OEM
C0251C36	OEM
C0251C36U	OEM
C0251C36V	OEM
C0251C36VU	OEM
C0251C36VW	OEM
C0251C36VX	OEM
C0251C36W	OEM
C0251C36X	OEM
C0251D56	OEM
C0251D56U	OEM
C0251D56V	OEM
C0251D56VU	OEM
C0251D56VW	OEM
C0251D56VX	OEM
C0251D56W	OEM
C0251D56X	OEM
C0251E15	OEM
C0251E15U	OEM
C0251E15V	OEM
C0251E15VU	OEM
C0251E15VW	OEM
C0251E15VX	OEM
C0251E15W	OEM
C0251E15X	OEM
C0251F25	OEM
C0251F25U	OEM
C0251F25V	OEM
C0251F25VU	OEM
C0251F25VW	OEM
C0251F25VX	OEM
C0251F25W	OEM
C0251F25X	OEM
C0252A17	OEM
C0252A17IVW	OEM
C0252A17IVX	OEM
C0252A17IW	OEM
C0252A17IX	OEM
C0252A17J	OEM
C0252A17JU	OEM
C0252A17JV	OEM
C0252A17JVW	OEM
C0252A17JVX	OEM
C0252A17JW	OEM
C0252A17JX	OEM
C0252A17JXU	OEM
C0252A17K	OEM
C0252A17KU	OEM
C0252A17KV	OEM
C0252A17KVU	OEM
C0252A17U	OEM
C0252A17V	OEM
C0252A17VU	OEM
C0252A17VW	OEM
C0252A17VX	OEM
C0252A17W	OEM
C0252A17X	OEM
C0252A58	OEM
C0252A58J	OEM
C0252A58JU	OEM

DEVICE TYPE	REPL CODE
C0252A58JV	OEM
C0252A58JVU	OEM
C0252A58JVW	OEM
C0252A58JVX	OEM
C0252A58JW	OEM
C0252A58JX	OEM
C0252A58K	OEM
C0252A58KU	OEM
C0252A58KV	OEM
C0252A58KVU	OEM
C0252A58KVW	OEM
C0252A58KVX	OEM
C0252A58KW	OEM
C0252A58KX	OEM
C0252A58U	OEM
C0252A58V	OEM
C0252A58VU	OEM
C0252A58VW	OEM
C0252A58VX	OEM
C0252A58W	OEM
C0252A58X	OEM
C0252B17	OEM
C0252B17J	OEM
C0252B17JU	OEM
C0252B17JV	OEM
C0252B17JVU	OEM
C0252B17JVW	OEM
C0252B17JVX	OEM
C0252B17JW	OEM
C0252B17JX	OEM
C0252B17K	OEM
C0252B17KU	OEM
C0252B17KV	OEM
C0252B17KVU	OEM
C0252B17KVW	OEM
C0252B17KVX	OEM
C0252B17KW	OEM
C0252B17KX	OEM
C0252B17U	OEM
C0252B17V	OEM
C0252B17VU	OEM
C0252B17VW	OEM
C0252B17VX	OEM
C0252B17W	OEM
C0252B17X	OEM
C0252B27	OEM
C0252B27J	OEM
C0252B27JU	OEM
C0252B27JV	OEM
C0252B27JVU	OEM
C0252B27JVW	OEM
C0252B27JVX	OEM
C0252B27JW	OEM
C0252B27JX	OEM
C0252B27K	OEM
C0252B27KU	OEM
C0252B27KV	OEM
C0252B27KVU	OEM
C0252B27KVW	OEM
C0252B27KVX	OEM
C0252B27KW	OEM
C0252B27KX	OEM
C0252B27U	OEM
C0252B27V	OEM
C0252B27VU	OEM
C0252B27VW	OEM
C0252B27VX	OEM
C0252B27W	OEM
C0252B27X	OEM
C0252B57	OEM
C0252B57J	OEM
C0252B57JU	OEM
C0252B57JV	OEM
C0252B57JVU	OEM
C0252B57JVW	OEM
C0252B57JVX	OEM
C0252B57JW	OEM
C0252B57JX	OEM
C0252B57K	OEM
C0252B57KU	OEM
C0252B57KV	OEM
C0252B57KVU	OEM
C0252B57KVW	OEM
C0252B57KVX	OEM
C0252B57KW	OEM
C0252B57KX	OEM
C0252B57U	OEM
C0252B57V	OEM
C0252B57VU	OEM
C0252B57VW	OEM
C0252B57VX	OEM
C0252B57W	OEM
C0252B57X	OEM
C0252C16	OEM
C0252C16J	OEM
C0252C16JU	OEM
C0252C16JV	OEM
C0252C16JVU	OEM
C0252C16JVW	OEM
C0252C16JVX	OEM
C0252C16JW	OEM
C0252C16JX	OEM
C0252C16K	OEM
C0252C16KU	OEM
C0252C16KV	OEM
C0252C16KVU	OEM
C0252C16KVW	OEM
C0252C16KVX	OEM
C0252C16KW	OEM

DEVICE TYPE	REPL CODE
C0252C16KX	OEM
C0252C16U	OEM
C0252C16V	OEM
C0252C16VU	OEM
C0252C16VW	OEM
C0252C16VX	OEM
C0252C16W	OEM
C0252C16X	OEM
C0252C37	OEM
C0252C37J	OEM
C0252C37JU	OEM
C0252C37JV	OEM
C0252C37JVU	OEM
C0252C37JVW	OEM
C0252C37JVX	OEM
C0252C37JW	OEM
C0252C37JX	OEM
C0252C37K	OEM
C0252C37KU	OEM
C0252C37KV	OEM
C0252C37KVU	OEM
C0252C37KVW	OEM
C0252C37KVX	OEM
C0252C37KW	OEM
C0252C37KX	OEM
C0252C37U	OEM
C0252C37V	OEM
C0252C37VU	OEM
C0252C37VW	OEM
C0252C37VX	OEM
C0252C37W	OEM
C0252C37X	OEM
C0252C57	OEM
C0252C57J	OEM
C0252C57JU	OEM
C0252C57JV	OEM
C0252C57JVU	OEM
C0252C57JVW	OEM
C0252C57JVX	OEM
C0252C57JW	OEM
C0252C57JX	OEM
C0252C57K	OEM
C0252C57KU	OEM
C0252C57KV	OEM
C0252C57KVU	OEM
C0252C57KVW	OEM
C0252C57KVX	OEM
C0252C57KW	OEM
C0252C57KX	OEM
C0252C57U	OEM
C0252C57V	OEM
C0252C57VU	OEM
C0252C57VW	OEM
C0252C57VX	OEM
C0252C57W	OEM
C0252C57X	OEM
C0252D16	OEM
C0252D16J	OEM
C0252D16JU	OEM
C0252D16JV	OEM
C0252D16JVU	OEM
C0252D16JVW	OEM
C0252D16JX	OEM
C0252D16K	OEM
C0252D16KU	OEM
C0252D16KV	OEM
C0252D16KVU	OEM
C0252D16KVW	OEM
C0252D16KVX	OEM
C0252D16KW	OEM
C0252D16KX	OEM
C0252D16U	OEM
C0252D16V	OEM
C0252D16VU	OEM
C0252D16VW	OEM
C0252D16VX	OEM
C0252D16W	OEM
C0252D16X	OEM
C0252D57	OEM
C0252D57J	OEM
C0252D57JV	OEM
C0252D57JVU	OEM
C0252D57JVW	OEM
C0252D57JVX	OEM
C0252D57JW	OEM
C0252D57JX	OEM
C0252D57K	OEM
C0252D57KU	OEM
C0252D57KV	OEM
C0252D57KVU	OEM
C0252D57KVW	OEM
C0252D57KVX	OEM
C0252D57KW	OEM
C0252D57KX	OEM
C0252D57U	OEM
C0252D57V	OEM
C0252D57VU	OEM
C0252D57VW	OEM
C0252D57VX	OEM
C0252D57W	OEM
C0252D57X	OEM
C0252E16	OEM
C0252E16J	OEM
C0252E16JU	OEM
C0252E16JV	OEM
C0252E16JVU	OEM
C0252E16JVX	OEM
C0252E16JW	OEM
C0252E16JX	OEM

If replacement code is OEM, contact original manufacturer for replacement.

DEVICE TYPE	REPL CODE
C0252E16K	OEM
C0252E16KU	OEM
C0252E16KV	OEM
C0252E16KVU	OEM
C0252E16KVW	OEM
C0252E16KVX	OEM
C0252E16KW	OEM
C0252E16KX	OEM
C0252E16U	OEM
C0252E16UVW	OEM
C0252E16V	OEM
C0252E16VU	OEM
C0252E16VW	OEM
C0252E16VX	OEM
C0252E16W	OEM
C0252E16X	OEM
C0252E26	OEM
C0252E26J	OEM
C0252E26JU	OEM
C0252E26JV	OEM
C0252E26JVU	OEM
C0252E26JVW	OEM
C0252E26JVX	OEM
C0252E26JW	OEM
C0252E26JX	OEM
C0252E26K	OEM
C0252E26KU	OEM
C0252E26KV	OEM
C0252E26KVU	OEM
C0252E26KVW	OEM
C0252E26KVX	OEM
C0252E26KW	OEM
C0252E26KX	OEM
C0252E26U	OEM
C0252E26V	OEM
C0252E26VU	OEM
C0252E26VW	OEM
C0252E26VX	OEM
C0252E26W	OEM
C0252E26X	OEM
C0252E56	OEM
C0252E56J	OEM
C0252E56JU	OEM
C0252E56JV	OEM
C0252E56JVU	OEM
C0252E56JVW	OEM
C0252E56JVX	OEM
C0252E56JW	OEM
C0252E56JX	OEM
C0252E56K	OEM
C0252E56KU	OEM
C0252E56KV	OEM
C0252E56KVU	OEM
C0252E56KVW	OEM
C0252E56KVX	OEM
C0252E56KW	OEM
C0252E56KX	OEM
C0252E56U	OEM
C0252E56V	OEM
C0252E56VU	OEM
C0252E56VW	OEM
C0252E56VX	OEM
C0252E56W	OEM
C0252E56X	OEM
C0252F16	OEM
C0252F16J	OEM
C0252F16JU	OEM
C0252F16JV	OEM
C0252F16JVU	OEM
C0252F16JVW	OEM
C0252F16JVX	OEM
C0252F16JW	OEM
C0252F16JX	OEM
C0252F16K	OEM
C0252F16KU	OEM
C0252F16KV	OEM
C0252F16KVU	OEM
C0252F16KVW	OEM
C0252F16KVX	OEM
C0252F16KW	OEM
C0252F16KX	OEM
C0252F16U	OEM
C0252F16V	OEM
C0252F16VU	OEM
C0252F16VW	OEM
C0252F16VX	OEM
C0252F16W	OEM
C0252F16X	OEM
C0252F26	OEM
C0252F26J	OEM
C0252F26JU	OEM
C0252F26JV	OEM
C0252F26JVU	OEM
C0252F26JVW	OEM
C0252F26JVX	OEM
C0252F26JW	OEM
C0252F26JX	OEM
C0252F26K	OEM
C0252F26KU	OEM
C0252F26KV	OEM
C0252F26KVU	OEM
C0252F26KVW	OEM
C0252F26KVX	OEM
C0252F26KW	OEM
C0252F26KX	OEM
C0252F26U	OEM
C0252F26V	OEM
C0252F26VU	OEM
C0252F26VW	OEM
C0252F26VX	OEM
C0252F26W	OEM
C0252F26X	OEM
C0252F56	OEM
C0252F56J	OEM
C0252F56JU	OEM
C0252F56JV	OEM
C0252F56JVU	OEM
C0252F56JVW	OEM
C0252F56JVX	OEM
C0252F56JW	OEM
C0252F56JX	OEM
C0252F56K	OEM
C0252F56KU	OEM
C0252F56KV	OEM
C0252F56KVU	OEM
C0252F56KVW	OEM
C0252F56KVX	OEM
C0252F56KW	OEM
C0252F56KX	OEM
C0252F56U	OEM
C0252F56V	OEM
C0252F56VU	OEM
C0252F56VW	OEM
C0252F56VX	OEM
C0252F56W	OEM
C0252F56X	OEM
C0252V16JVX	OEM
C0252V16UW	OEM
C0254A17	OEM
C0254A17E	OEM
C0254A17EU	OEM
C0254A17EV	OEM
C0254A17EVU	OEM
C0254A17EVW	OEM
C0254A17EVX	OEM
C0254A17EW	OEM
C0254A17EX	OEM
C0254A17J	OEM
C0254A17JU	OEM
C0254A17JV	OEM
C0254A17JVU	OEM
C0254A17JVX	OEM
C0254A17JW	OEM
C0254A17JX	OEM
C0254A17K	OEM
C0254A17KU	OEM
C0254A17KV	OEM
C0254A17KVU	OEM
C0254A17KVW	OEM
C0254A17KVX	OEM
C0254A17KW	OEM
C0254A17KX	OEM
C0254A17M	OEM
C0254A17MU	OEM
C0254A17MV	OEM
C0254A17MVU	OEM
C0254A17MVW	OEM
C0254A17MVX	OEM
C0254A17MW	OEM
C0254A17MX	OEM
C0254A17U	OEM
C0254A17V	OEM
C0254A17VU	OEM
C0254A17VW	OEM
C0254A17VX	OEM
C0254A17W	OEM
C0254A17X	OEM
C0254A57	OEM
C0254A57E	OEM
C0254A57EU	OEM
C0254A57EV	OEM
C0254A57EVU	OEM
C0254A57EVW	OEM
C0254A57EVX	OEM
C0254A57EW	OEM
C0254A57EX	OEM
C0254A57J	OEM
C0254A57JU	OEM
C0254A57JV	OEM
C0254A57JVU	OEM
C0254A57JVW	OEM
C0254A57JVX	OEM
C0254A57JW	OEM
C0254A57JX	OEM
C0254A57K	OEM
C0254A57KU	OEM
C0254A57KV	OEM
C0254A57KVU	OEM
C0254A57KVW	OEM
C0254A57KVX	OEM
C0254A57KW	OEM
C0254A57KX	OEM
C0254A57M	OEM
C0254A57MU	OEM
C0254A57MV	OEM
C0254A57MVU	OEM
C0254A57MVW	OEM
C0254A57MVX	OEM
C0254A57MW	OEM
C0254A57MX	OEM
C0254A57U	OEM
C0254A57V	OEM
C0254A57VU	OEM
C0254A57VW	OEM
C0254A57VX	OEM
C0254A57W	OEM
C0254A57X	OEM
C0254B16	OEM
C0254B16E	OEM
C0254B16EU	OEM
C0254B16EV	OEM
C0254B16EVU	OEM
C0254B16EVW	OEM
C0254B16EVX	OEM
C0254B16EW	OEM
C0254B16EX	OEM
C0254B16J	OEM
C0254B16JU	OEM
C0254B16JV	OEM
C0254B16JVU	OEM
C0254B16JVW	OEM
C0254B16JVX	OEM
C0254B16JW	OEM
C0254B16JX	OEM
C0254B16K	OEM
C0254B16KU	OEM
C0254B16KV	OEM
C0254B16KVU	OEM
C0254B16KVW	OEM
C0254B16KVX	OEM
C0254B16KW	OEM
C0254B16KX	OEM
C0254B16M	OEM
C0254B16MU	OEM
C0254B16MV	OEM
C0254B16MVU	OEM
C0254B16MVW	OEM
C0254B16MVX	OEM
C0254B16MW	OEM
C0254B16MX	OEM
C0254B16U	OEM
C0254B16V	OEM
C0254B16VU	OEM
C0254B16VW	OEM
C0254B16VX	OEM
C0254B16W	OEM
C0254B16X	OEM
C0254B27	OEM
C0254B27E	OEM
C0254B27EU	OEM
C0254B27EV	OEM
C0254B27EVU	OEM
C0254B27EVW	OEM
C0254B27EVX	OEM
C0254B27EW	OEM
C0254B27EX	OEM
C0254B27J	OEM
C0254B27JU	OEM
C0254B27JV	OEM
C0254B27JVU	OEM
C0254B27JVW	OEM
C0254B27JVX	OEM
C0254B27JW	OEM
C0254B27JX	OEM
C0254B27K	OEM
C0254B27KU	OEM
C0254B27KV	OEM
C0254B27KVU	OEM
C0254B27KVW	OEM
C0254B27KVX	OEM
C0254B27KW	OEM
C0254B27KX	OEM
C0254B27M	OEM
C0254B27MU	OEM
C0254B27MV	OEM
C0254B27MVU	OEM
C0254B27MVW	OEM
C0254B27MVX	OEM
C0254B27MW	OEM
C0254B27MX	OEM
C0254B27U	OEM
C0254B27V	OEM
C0254B27VU	OEM
C0254B27VW	OEM
C0254B27VX	OEM
C0254B27W	OEM
C0254B27X	OEM
C0254B57	OEM
C0254B57E	OEM
C0254B57EU	OEM
C0254B57EV	OEM
C0254B57EVU	OEM
C0254B57EVW	OEM
C0254B57EVX	OEM
C0254B57EW	OEM
C0254B57EX	OEM
C0254B57J	OEM
C0254B57JU	OEM
C0254B57JV	OEM
C0254B57JVU	OEM
C0254B57JVW	OEM
C0254B57JVX	OEM
C0254B57JW	OEM
C0254B57JX	OEM
C0254B57K	OEM
C0254B57KU	OEM
C0254B57KV	OEM
C0254B57KVU	OEM
C0254B57KVW	OEM
C0254B57KVX	OEM
C0254B57KW	OEM
C0254B57KX	OEM
C0254B57M	OEM
C0254B57MU	OEM
C0254B57MV	OEM
C0254B57MVU	OEM
C0254B57MVW	OEM
C0254B57MVX	OEM
C0254B57MW	OEM
C0254B57MX	OEM
C0254B57U	OEM
C0254B57V	OEM
C0254B57VU	OEM
C0254B57VW	OEM
C0254B57VX	OEM
C0254B57W	OEM
C0254B57X	OEM
C0254C16	OEM
C0254C16E	OEM
C0254C16EU	OEM
C0254C16EV	OEM
C0254C16EVU	OEM
C0254C16EVW	OEM
C0254C16EVX	OEM
C0254C16EW	OEM
C0254C16EX	OEM
C0254C16J	OEM
C0254C16JU	OEM
C0254C16JV	OEM
C0254C16JVU	OEM
C0254C16JVW	OEM
C0254C16JVX	OEM
C0254C16JW	OEM
C0254C16JX	OEM
C0254C16K	OEM
C0254C16KU	OEM
C0254C16KV	OEM
C0254C16KVU	OEM
C0254C16KVW	OEM
C0254C16KVX	OEM
C0254C16KW	OEM
C0254C16KX	OEM
C0254C16M	OEM
C0254C16MU	OEM
C0254C16MV	OEM
C0254C16MVU	OEM
C0254C16MVW	OEM
C0254C16MVX	OEM
C0254C16MW	OEM
C0254C16MX	OEM
C0254C16U	OEM
C0254C16V	OEM
C0254C16VU	OEM
C0254C16VW	OEM
C0254C16VX	OEM
C0254C16W	OEM
C0254C16X	OEM
C0254C36	OEM
C0254C36E	OEM
C0254C36EU	OEM
C0254C36EV	OEM
C0254C36EVU	OEM
C0254C36EVW	OEM
C0254C36EVX	OEM
C0254C36EW	OEM
C0254C36EX	OEM
C0254C36J	OEM
C0254C36JU	OEM
C0254C36JV	OEM
C0254C36JVU	OEM
C0254C36JVW	OEM
C0254C36JVX	OEM
C0254C36JW	OEM
C0254C36JX	OEM
C0254C36K	OEM
C0254C36KU	OEM
C0254C36KV	OEM
C0254C36KVU	OEM
C0254C36KVW	OEM
C0254C36KVX	OEM
C0254C36KW	OEM
C0254C36KX	OEM
C0254C36M	OEM
C0254C36MU	OEM
C0254C36MV	OEM
C0254C36MVU	OEM
C0254C36MVW	OEM
C0254C36MVX	OEM
C0254C36MW	OEM
C0254C36MX	OEM
C0254C36U	OEM
C0254C36V	OEM
C0254C36VU	OEM
C0254C36VW	OEM
C0254C36VX	OEM
C0254C36W	OEM
C0254C36X	OEM
C0254C37	OEM
C0254C37E	OEM
C0254C37EU	OEM
C0254C37EV	OEM
C0254C37EVU	OEM
C0254C37EVW	OEM
C0254C37EVX	OEM
C0254C37EW	OEM
C0254C37EX	OEM
C0254C37J	OEM
C0254C37JU	OEM
C0254C37JV	OEM
C0254C37JVU	OEM
C0254C37JVW	OEM
C0254C37JVX	OEM
C0254C37JW	OEM
C0254C37JX	OEM
C0254C37K	OEM
C0254C37KU	OEM
C0254C37KV	OEM
C0254C37KVU	OEM
C0254C37KVW	OEM
C0254C37KVX	OEM
C0254C37KW	OEM
C0254C37KX	OEM
C0254C37M	OEM
C0254C37MU	OEM
C0254C37MV	OEM
C0254C37MVU	OEM
C0254C37MVW	OEM
C0254C37MVX	OEM
C0254C37MW	OEM
C0254C37MX	OEM
C0254C37U	OEM
C0254C37V	OEM
C0254C37VU	OEM
C0254C37VW	OEM
C0254C37VX	OEM
C0254C37W	OEM
C0254C37X	OEM
C0254C56JX	OEM
C0254D16	OEM
C0254D16E	OEM
C0254D16EV	OEM
C0254D16EVU	OEM
C0254D16EVW	OEM
C0254D16EVX	OEM
C0254D16EW	OEM
C0254D16EX	OEM
C0254D16J	OEM
C0254D16JU	OEM
C0254D16JV	OEM
C0254D16JVU	OEM
C0254D16JVW	OEM
C0254D16JVX	OEM
C0254D16JW	OEM
C0254D16JX	OEM
C0254D16K	OEM
C0254D16KU	OEM
C0254D16KV	OEM
C0254D16KVU	OEM
C0254D16KVW	OEM
C0254D16KVX	OEM
C0254D16KW	OEM
C0254D16KX	OEM
C0254D16M	OEM
C0254D16MU	OEM
C0254D16MV	OEM
C0254D16MVU	OEM
C0254D16MVW	OEM
C0254D16MVX	OEM
C0254D16MW	OEM
C0254D16MX	OEM
C0254D16U	OEM
C0254D16V	OEM
C0254D16VU	OEM
C0254D16VW	OEM
C0254D16VX	OEM
C0254D16W	OEM
C0254D16X	OEM
C0254D56	OEM
C0254D56E	OEM
C0254D56EU	OEM
C0254D56EV	OEM
C0254D56EVU	OEM
C0254D56EVW	OEM
C0254D56EVX	OEM
C0254D56EW	OEM
C0254D56EX	OEM
C0254D56J	OEM
C0254D56JU	OEM
C0254D56JV	OEM
C0254D56JVU	OEM
C0254D56JVW	OEM
C0254D56JVX	OEM
C0254D56JW	OEM
C0254D56JX	OEM
C0254D56K	OEM
C0254D56KU	OEM
C0254D56KV	OEM
C0254D56KVU	OEM
C0254D56KVW	OEM
C0254D56KVX	OEM
C0254D56KW	OEM
C0254D56KX	OEM
C0254D56M	OEM
C0254D56MU	OEM
C0254D56MV	OEM
C0254D56MVU	OEM
C0254D56MVW	OEM
C0254D56MVX	OEM
C0254D56MW	OEM
C0254D56MX	OEM
C0254D56U	OEM
C0254D56V	OEM
C0254D56VU	OEM
C0254D56VW	OEM
C0254D56VX	OEM
C0254D56X	OEM
C0254D57	OEM
C0254D57E	OEM
C0254D57EU	OEM
C0254D57EV	OEM
C0254D57EVU	OEM
C0254D57EVW	OEM
C0254D57EVX	OEM
C0254D57EW	OEM
C0254D57EX	OEM
C0254D57J	OEM
C0254D57JU	OEM
C0254D57JV	OEM
C0254D57JVU	OEM
C0254D57JVW	OEM
C0254D57JVX	OEM
C0254D57JW	OEM
C0254D57JX	OEM
C0254D57K	OEM
C0254D57KU	OEM
C0254D57KV	OEM
C0254D57KVU	OEM
C0254D57KVW	OEM
C0254D57KVX	OEM
C0254D57KW	OEM
C0254D57KX	OEM
C0254D57M	OEM
C0254D57MU	OEM
C0254D57MV	OEM
C0254D57MVU	OEM
C0254D57MVW	OEM
C0254D57MVX	OEM
C0254D57MW	OEM
C0254D57MX	OEM
C0254D57U	OEM
C0254D57V	OEM
C0254D57VU	OEM
C0254D57VW	OEM
C0254D57W	OEM
C0254D57X	OEM
C0254E16	OEM
C0254E16E	OEM
C0254E16EU	OEM
C0254E16EV	OEM
C0254E16EVU	OEM
C0254E16EVW	OEM
C0254E16EVX	OEM
C0254E16EW	OEM
C0254E16EX	OEM
C0254E16J	OEM
C0254E16JU	OEM
C0254E16JV	OEM
C0254E16JVU	OEM
C0254E16JVW	OEM
C0254E16JVX	OEM
C0254E16JW	OEM
C0254E16JX	OEM
C0254E16K	OEM
C0254E16KU	OEM
C0254E16KV	OEM
C0254E16KVU	OEM
C0254E16KVW	OEM
C0254E16KVX	OEM
C0254E16KW	OEM
C0254E16KX	OEM
C0254E16M	OEM
C0254E16MU	OEM
C0254E16MV	OEM
C0254E16MVU	OEM
C0254E16MVW	OEM
C0254E16MVX	OEM
C0254E16MW	OEM
C0254E16MX	OEM
C0254E16U	OEM
C0254E16V	OEM
C0254E16VU	OEM
C0254E16VW	OEM
C0254E16VX	OEM
C0254E16W	OEM
C0254E16X	OEM
C0254E26	OEM
C0254E26E	OEM
C0254E26EU	OEM
C0254E26EV	OEM
C0254E26EVU	OEM
C0254E26EVW	OEM
C0254E26EVX	OEM
C0254E26EW	OEM
C0254E26EX	OEM
C0254E26J	OEM
C0254E26JU	OEM
C0254E26JV	OEM
C0254E26JVU	OEM
C0254E26JVW	OEM
C0254E26JVX	OEM
C0254E26JW	OEM
C0254E26JX	OEM
C0254E26K	OEM
C0254E26KU	OEM
C0254E26KV	OEM
C0254E26KVU	OEM
C0254E26KVW	OEM
C0254E26KVX	OEM
C0254E26KW	OEM
C0254E26KX	OEM
C0254E26M	OEM
C0254E26MU	OEM
C0254E26MV	OEM
C0254E26MVU	OEM
C0254E26MVW	OEM
C0254E26MVX	OEM
C0254E26MW	OEM
C0254E26MX	OEM
C0254E26U	OEM
C0254E26V	OEM
C0254E26VU	OEM
C0254E26VW	OEM
C0254E26VX	OEM
C0254E26W	OEM
C0254E26X	OEM
C0254E56	OEM
C0254E56E	OEM
C0254E56EU	OEM
C0254E56EV	OEM
C0254E56EVU	OEM
C0254E56EVW	OEM
C0254E56EVX	OEM
C0254E56EW	OEM
C0254E56EX	OEM
C0254E56J	OEM
C0254E56JU	OEM
C0254E56JV	OEM
C0254E56JVU	OEM
C0254E56JVW	OEM
C0254E56JVX	OEM
C0254E56JW	OEM
C0254E56K	OEM
C0254E56KU	OEM
C0254E56KV	OEM
C0254E56KVU	OEM
C0254E56KVW	OEM
C0254E56KVX	OEM
C0254E56KW	OEM
C0254E56KX	OEM
C0254E56M	OEM
C0254E56MU	OEM
C0254E56MV	OEM
C0254E56MVU	OEM
C0254E56MVW	OEM
C0254E56MVX	OEM
C0254E56MW	OEM
C0254E56MX	OEM
C0254E56U	OEM
C0254E56V	OEM
C0254E56VU	OEM
C0254E56VW	OEM
C0254E56VX	OEM
C0254E56W	OEM
C0254E56X	OEM
C0254F16E	OEM
C0254F16EU	OEM
C0254F16EV	OEM
C0254F16EVU	OEM
C0254F16EVW	OEM
C0254F16EVX	OEM
C0254F16EW	OEM
C0254F16EX	OEM
C0254F16J	OEM
C0254F16JU	OEM
C0254F16JV	OEM
C0254F16JVU	OEM
C0254F16JVW	OEM
C0254F16JVX	OEM
C0254F16JW	OEM
C0254F16JX	OEM
C0254F16K	OEM
C0254F16KU	OEM
C0254F16KV	OEM
C0254F16KVU	OEM
C0254F16KVW	OEM
C0254F16KVX	OEM
C0254F16KW	OEM
C0254F16KX	OEM
C0254F16M	OEM
C0254F16MU	OEM
C0254F16MV	OEM
C0254F16MVU	OEM
C0254F16MVW	OEM
C0254F16MVX	OEM
C0254F16MW	OEM
C0254F16MX	OEM
C0254F16U	OEM
C0254F16V	OEM
C0254F16VU	OEM
C0254F16VW	OEM
C0254F16VX	OEM
C0254F16W	OEM
C0254F16X	OEM
C0254F26	OEM
C0254F26E	OEM
C0254F26EU	OEM
C0254F26EV	OEM
C0254F26EVU	OEM
C0254F26EVW	OEM
C0254F26EVX	OEM
C0254F26EW	OEM
C0254F26EX	OEM
C0254F26J	OEM
C0254F26JU	OEM
C0254F26JV	OEM
C0254F26JVU	OEM
C0254F26JVW	OEM
C0254F26JVX	OEM
C0254F26JW	OEM
C0254F26JX	OEM
C0254F26K	OEM
C0254F26KU	OEM
C0254F26KV	OEM
C0254F26KVU	OEM
C0254F26KVW	OEM
C0254F26KVX	OEM
C0254F26KW	OEM
C0254F26KX	OEM
C0254F26M	OEM
C0254F26MU	OEM
C0254F26MV	OEM
C0254F26MVU	OEM
C0254F26MVW	OEM
C0254F26MVX	OEM

DEVICE TYPE	REPL CODE
C0254F26MW	OEM
C0254F26MX	OEM
C0254F26U	OEM
C0254F26V	OEM
C0254F26VU	OEM
C0254F26VW	OEM
C0254F26VX	OEM
C0254F26W	OEM
C0254F26X	OEM
C0254F56	OEM
C0254F56E	OEM
C0254F56EU	OEM
C0254F56EV	OEM
C0254F56EVU	OEM
C0254F56EVX	OEM
C0254F56EW	OEM
C0254F56EX	OEM
C0254F56J	OEM
C0254F56JU	OEM
C0254F56JV	OEM
C0254F56JVU	OEM
C0254F56JVW	OEM
C0254F56JVX	OEM
C0254F56JW	OEM
C0254F56JX	OEM
C0254F56K	OEM
C0254F56KU	OEM
C0254F56KV	OEM
C0254F56KVU	OEM
C0254F56KVW	OEM
C0254F56KVX	OEM
C0254F56KW	OEM
C0254F56KX	OEM
C0254F56M	OEM
C0254F56MU	OEM
C0254F56MV	OEM
C0254F56MVU	OEM
C0254F56MVW	OEM
C0254F56MVX	OEM
C0254F56MW	OEM
C0254F56MX	OEM
C0254F56U	OEM
C0254F56V	OEM
C0254F56VU	OEM
C0254F56VW	OEM
C0254F56VX	OEM
C0254F56W	OEM
C0254F56X	OEM
C0255A17	OEM
C0255A17M	OEM
C0255A17MP	OEM
C0255A17MU	OEM
C0255A17MV	OEM
C0255A17MVP	OEM
C0255A17MVU	OEM
C0255A17MVX	OEM
C0255A17MX	OEM
C0255A17P	OEM
C0255A17U	OEM
C0255A17V	OEM
C0255A17VP	OEM
C0255A17VU	OEM
C0255A17VX	OEM
C0255A17X	OEM
C0255A57	OEM
C0255A57M	OEM
C0255A57MP	OEM
C0255A57MU	OEM
C0255A57MV	OEM
C0255A57MVP	OEM
C0255A57MVU	OEM
C0255A57MVX	OEM
C0255A57MX	OEM
C0255A57P	OEM
C0255A57U	OEM
C0255A57VP	OEM
C0255A57VU	OEM
C0255A57VX	OEM
C0255B16	OEM
C0255B16M	OEM
C0255B16MP	OEM
C0255B16MU	OEM
C0255B16MV	OEM
C0255B16MVP	OEM
C0255B16MVU	OEM
C0255B16MVX	OEM
C0255B16MX	OEM
C0255B16P	OEM
C0255B16U	OEM
C0255B16V	OEM
C0255B16VP	OEM
C0255B16VU	OEM
C0255B16VX	OEM
C0255B16X	OEM
C0255B27	OEM
C0255B27M	OEM
C0255B27MP	OEM
C0255B27MU	OEM
C0255B27MV	OEM
C0255B27MVP	OEM
C0255B27MVU	OEM
C0255B27MVX	OEM
C0255B27MX	OEM
C0255B27P	OEM
C0255B27U	OEM
C0255B27V	OEM
C0255B27VP	OEM
C0255B27VU	OEM
C0255B27VX	OEM

DEVICE TYPE	REPL CODE
C0255B27X	OEM
C0255B57	OEM
C0255B57M	OEM
C0255B57MP	OEM
C0255B57MU	OEM
C0255B57MV	OEM
C0255B57MVP	OEM
C0255B57MVU	OEM
C0255B57MVX	OEM
C0255B57MX	OEM
C0255B57P	OEM
C0255B57U	OEM
C0255B57V	OEM
C0255B57VP	OEM
C0255B57VU	OEM
C0255B57VX	OEM
C0255B57X	OEM
C0255C16	OEM
C0255C16M	OEM
C0255C16MP	OEM
C0255C16MU	OEM
C0255C16MV	OEM
C0255C16MVP	OEM
C0255C16MVU	OEM
C0255C16MVX	OEM
C0255C16MX	OEM
C0255C16P	OEM
C0255C16U	OEM
C0255C16V	OEM
C0255C16VP	OEM
C0255C16VU	OEM
C0255C16VX	OEM
C0255C16X	OEM
C0255C36	OEM
C0255C36M	OEM
C0255C36MP	OEM
C0255C36MU	OEM
C0255C36MV	OEM
C0255C36MVP	OEM
C0255C36MVU	OEM
C0255C36MVX	OEM
C0255C36MX	OEM
C0255C36P	OEM
C0255C36U	OEM
C0255C36V	OEM
C0255C36VP	OEM
C0255C36VU	OEM
C0255C36VX	OEM
C0255C36X	OEM
C0255C37	OEM
C0255C37M	OEM
C0255C37MP	OEM
C0255C37MU	OEM
C0255C37MV	OEM
C0255C37MVP	OEM
C0255C37MVU	OEM
C0255C37MVX	OEM
C0255C37MX	OEM
C0255C37P	OEM
C0255C37U	OEM
C0255C37V	OEM
C0255C37VP	OEM
C0255C37VU	OEM
C0255C37VX	OEM
C0255C37X	OEM
C0255D16	OEM
C0255D16M	OEM
C0255D16MP	OEM
C0255D16MU	OEM
C0255D16MV	OEM
C0255D16MVP	OEM
C0255D16MVU	OEM
C0255D16MVX	OEM
C0255D16MX	OEM
C0255D16P	OEM
C0255D16U	OEM
C0255D16V	OEM
C0255D16VP	OEM
C0255D16VU	OEM
C0255D16VX	OEM
C0255D16X	OEM
C0255D56	OEM
C0255D56M	OEM
C0255D56MP	OEM
C0255D56MU	OEM
C0255D56MV	OEM
C0255D56MVP	OEM
C0255D56MVU	OEM
C0255D56MVX	OEM
C0255D56MX	OEM
C0255D56P	OEM
C0255D56U	OEM
C0255D56V	OEM
C0255D56VP	OEM
C0255D56VU	OEM
C0255D56VX	OEM
C0255D56X	OEM
C0255D57	OEM
C0255D57M	OEM
C0255D57MP	OEM
C0255D57MU	OEM
C0255D57MV	OEM
C0255D57MVP	OEM
C0255D57MVU	OEM
C0255D57MVX	OEM
C0255D57MX	OEM
C0255D57P	OEM
C0255D57U	OEM
C0255D57V	OEM

DEVICE TYPE	REPL CODE
C0255D57VP	OEM
C0255D57VU	OEM
C0255D57VX	OEM
C0255D57X	OEM
C0255E16	OEM
C0255E16M	OEM
C0255E16MP	OEM
C0255E16MU	OEM
C0255E16MV	OEM
C0255E16MVP	OEM
C0255E16MVU	OEM
C0255E16MVX	OEM
C0255E16MX	OEM
C0255E16P	OEM
C0255E16U	OEM
C0255E16V	OEM
C0255E16VP	OEM
C0255E16VU	OEM
C0255E16VX	OEM
C0255E16X	OEM
C0255E26	OEM
C0255E26M	OEM
C0255E26MP	OEM
C0255E26MU	OEM
C0255E26MV	OEM
C0255E26MVP	OEM
C0255E26MVU	OEM
C0255E26MVX	OEM
C0255E26MX	OEM
C0255E26P	OEM
C0255E26U	OEM
C0255E26V	OEM
C0255E26VP	OEM
C0255E26VU	OEM
C0255E26VX	OEM
C0255E26X	OEM
C0255E56	OEM
C0255E56MP	OEM
C0255E56MU	OEM
C0255E56MV	OEM
C0255E56MVP	OEM
C0255E56MVU	OEM
C0255E56MVX	OEM
C0255E56MX	OEM
C0255E56P	OEM
C0255E56U	OEM
C0255E56V	OEM
C0255E56VP	OEM
C0255E56VU	OEM
C0255E56VX	OEM
C0255E56X	OEM
C0255F16	OEM
C0255F16M	OEM
C0255F16MP	OEM
C0255F16MU	OEM
C0255F16MV	OEM
C0255F16MVP	OEM
C0255F16MVU	OEM
C0255F16MVX	OEM
C0255F16MX	OEM
C0255F16P	OEM
C0255F16U	OEM
C0255F16V	OEM
C0255F16VP	OEM
C0255F16VU	OEM
C0255F16VX	OEM
C0255F16X	OEM
C0255F26	OEM
C0255F26M	OEM
C0255F26MP	OEM
C0255F26MU	OEM
C0255F26MV	OEM
C0255F26MVP	OEM
C0255F26MVU	OEM
C0255F26MVX	OEM
C0255F26MX	OEM
C0255F26P	OEM
C0255F26U	OEM
C0255F26V	OEM
C0255F26VP	OEM
C0255F26VU	OEM
C0255F26VX	OEM
C0255F26X	OEM
C0255F56	OEM
C0255F56M	OEM
C0255F56MP	OEM
C0255F56MU	OEM
C0255F56MV	OEM
C0255F56MVP	OEM
C0255F56MVU	OEM
C0255F56MVX	OEM
C0255F56MX	OEM
C0255F56P	OEM
C0255F56U	OEM
C0255F56V	OEM
C0255F56VP	OEM
C0255F56VU	OEM
C0255F56VX	OEM
C0255F56X	OEM
C0271VA	OEM
C0271VB	OEM
C0273V	OEM
C0274V	OEM
C0275HA	OEM
C0275HB	OEM
C0275VA	OEM
C0275VB	OEM
C0281	OEM
C0281-1	OEM

DEVICE TYPE	REPL CODE
C0281-2	OEM
C0281-3	OEM
C0281-4	OEM
C0281-5	OEM
C0281-6	OEM
C0281T	OEM
C0281T1	OEM
C0281T2	OEM
C0281T3	OEM
C0281T4	OEM
C0281T5	OEM
C0281T6	OEM
C0281Y	OEM
C0281Y1	OEM
C0281Y2	OEM
C0281Y3	OEM
C0281Y4	OEM
C0281Y5	OEM
C0281Y6	OEM
C0282	OEM
C0282-1	OEM
C0282-2	OEM
C0282-3	OEM
C0282-4	OEM
C0282-5	OEM
C0282-6	OEM
C0282T	OEM
C0282T1	OEM
C0282T2	OEM
C0282T3	OEM
C0282T4	OEM
C0282T5	OEM
C0282T6	OEM
C0284U	OEM
C0284U1	OEM
C0284U2	OEM
C0284U3	OEM
C0284U4	OEM
C0284U5	OEM
C0284U6	OEM
C0284UT	OEM
C0284UT1	OEM
C0284UT2	OEM
C0284UT3	OEM
C0284UT4	OEM
C0284UT5	OEM
C0284UT6	OEM
C0284UTV	OEM
C0284UTV1	OEM
C0284UTV2	OEM
C0284UTV3	OEM
C0284UTV4	OEM
C0284UTV5	OEM
C0284UTV6	OEM
C0284UV	OEM
C0284UV1	OEM
C0284UV2	OEM
C0284UV3	OEM
C0284UV4	OEM
C0284UV5	OEM
C0284UV6	OEM
C0284W	OEM
C0284W1	OEM
C0284W2	OEM
C0284W3	OEM
C0284W4	OEM
C0284W5	OEM
C0284W6	OEM
C0284WT	OEM
C0284WT1	OEM
C0284WT2	OEM
C0284WT3	OEM
C0284WT4	OEM
C0284WT5	OEM
C0284WT6	OEM
C0284WTV	OEM
C0284WTV1	OEM
C0284WTV2	OEM
C0284WTV3	OEM
C0284WTV4	OEM
C0284WTV5	OEM
C0284WTV6	OEM
C0284WV1	OEM
C0284WV2	OEM
C0284WV3	OEM
C0284WV4	OEM
C0284WV5	OEM
C0284WV6	OEM
C0284X	OEM
C0284X1	OEM
C0284X2	OEM
C0284X3	OEM
C0284X4	OEM
C0284X5	OEM
C0284X6	OEM
C0284XT	OEM
C0284XT1	OEM
C0284XT2	OEM
C0284XT3	OEM
C0284XT4	OEM
C0284XT5	OEM
C0284XT6	OEM
C0284XTV	OEM
C0284XTV1	OEM
C0284XTV2	OEM
C0284XTV3	OEM
C0284XTV4	OEM
C0284XTV5	OEM
C0284XTV6	OEM

DEVICE TYPE	REPL CODE
C0284XY	OEM
C0284XY1	OEM
C0284XY2	OEM
C0284XY3	OEM
C0284XY4	OEM
C0284XY5	OEM
C0284XY6	OEM
C0410V	0071
C0510	0071
C0536	OEM
C0536-1	OEM
C0536-3	OEM
C0536T	OEM
C0536T1	OEM
C0536T3	OEM
C0536TV1	OEM
C0536TV3	OEM
C0536V	OEM
C0536V1	OEM
C0536V3	OEM
C0633A3	OEM
C0633AV	OEM
C0633B	OEM
C0633B3	OEM
C0633BV	OEM
C0633BV3	OEM
C0636A	OEM
C0636A1	OEM
C0636A3	OEM
C0636A5	OEM
C0636AV	OEM
C0636AV1	OEM
C0636AV3	OEM
C0636AV5	OEM
C0636B	OEM
C0636B1	OEM
C0636B3	OEM
C0636B5	OEM
C0636BV	OEM
C0636BV1	OEM
C0636BV3	OEM
C0636BV5	OEM
C01433-09	0522
C02138V3	OEM
C02555E56M	OEM
C0-10174-02	1696
C0-10444-22	OEM
C0-10444D-03	OEM
C0-10745-03	3070
C0-10750-03	1973
C0-10816	0398
C010816-XX	0398
C012294-22	OEM
C012294B-01	OEM
C012296D-01	OEM
C012296D-22	OEM
C012399B	OEM
C012499B-01	OEM
C014313	0458
C014336	0366
C014341	1662
C014344	0427
C014345	1656
C0-14348	0624
C014377	3065
C014599B-01	OEM
C014795	OEM
C014805-01	OEM
C014805-22	OEM
C014806-12	3065
C019156-05	OEM
C025912-38	OEM
C025913-38	OEM
C025914-38	OEM
C025915-38	OEM
C026028	OEM
C026034-01	OEM
C026035-01	OEM
COC13000-1C	0595
COC13000-IC	0630
COD1-6045	0015
COD1-6046	0015
COD1-6047	0015
COD1-6048	0015
COD1531	0015
COD1532	0015
COD1533	0015
COD1534	0015
COD1535	0015
COD1536	0015
COD1537	0071
COD1538	0071
COD1551	0015
COD1552	0015
COD1553	0015
COD1554	0015
COD1555	0015
COD1556	0015
COD1575	0015
COD1611	0015
COD1612	0015
COD1613	0015
COD1614	0015
COD1615	0015
COD1616	0015
COD1617	0071
COD1618	0071
COD11556	0015
COD15524	0015

DEVICE TYPE	REPL CODE
COD15534	0015
COD15544	0015
COD15564	0015
COD16047	0015
CODI531	0015
CODI532	0015
CODI533	0015
CODI534	0015
CODI535	0015
CODI536	0015
CODI537	0071
CODI538	0071
CODI551	0015
CODI552	0015
CODI553	0015
CODI554	0015
CODI555	0015
CODI556	0015
CODI617	0071
CODI1531	0015
CODI5314	OEM
CODI5324	OEM
CODI5334	OEM
CODI5344	OEM
CODI5354	OEM
CODI5364	2645
CODI5514	OEM
CODI5524	OEM
CODI5534	OEM
CODI5544	OEM
CODI5554	OEM
CODI5564	OEM
CODI6041	OEM
CODI6042	OEM
CODI6043	OEM
CODI6044	OEM
CODI6045	0015
CODI6046	OEM
CODI6047	0015
CODI6048	OEM
CODI6049	OEM
CODI6050	OEM
CODI6051	OEM
CODI6052	OEM
CODI11556	0015
CODI15524	0015
CODI15534	0015
CODI15544	0015
CODI15564	0015
COM1553A	OEM
COM1671	OEM
COM1863	OEM
COM2017	OEM
COM2017H	OEM
COM2502	OEM
COM2502/H	OEM
COM2601	OEM
COM2651	OEM
COM2661-1	OEM
COM2661-2	OEM
COM2661-3	OEM
COM5016	OEM
COM5016T	OEM
COM5025	OEM
COM5026	OEM
COM5026T	OEM
COM5036	OEM
COM5036T	OEM
COM5046	OEM
COM5046T	OEM
COM8004	OEM
COM8017	OEM
COM8018	OEM
COM8046	OEM
COM8046T	OEM
COM8116	OEM
COM8116T	OEM
COM8126	OEM
COM8126T	OEM
COM8136	OEM
COM8136T	OEM
COM8146	OEM
COM8146T	OEM
COM8251A	OEM
COM8502	OEM
COM9004	OEM
COM9026	OEM
COP302D	OEM
COP302J	OEM
COP302MD	OEM
COP302MJ	OEM
COP302MN	OEM
COP302N	OEM
COP304D	OEM
COP304J	OEM
COP304LD	OEM
COP304LJ	OEM
COP304LN	OEM
COP304N	OEM
COP310CD	OEM
COP310CJ	OEM
COP310CN	OEM
COP310LD	OEM
COP310LJ	OEM
COP310LN	OEM
COP311CD	OEM
COP311CJ	OEM
COP311CN	OEM
COP311LD	OEM

DEVICE TYPE	REPL CODE
COP311LJ	OEM
COP311LN	OEM
COP320CD	OEM
COP320CJ	OEM
COP320CN	OEM
COP320D	OEM
COP320J	OEM
COP320LD	OEM
COP320LJ	OEM
COP320LN	OEM
COP320N	OEM
COP321CD	OEM
COP321CJ	OEM
COP321CN	OEM
COP321D	OEM
COP321J	OEM
COP321LD	OEM
COP321LJ	OEM
COP321LN	OEM
COP321N	OEM
COP322D	OEM
COP322J	OEM
COP322LD	OEM
COP322LJ	OEM
COP322LN	OEM
COP322N	OEM
COP340D	OEM
COP340J	OEM
COP340N	OEM
COP341D	OEM
COP341J	OEM
COP341N	OEM
COP342D	OEM
COP342J	OEM
COP342N	OEM
COP344CD	OEM
COP344CJ	OEM
COP344CN	OEM
COP344LD	OEM
COP344LJ	OEM
COP344LN	OEM
COP345CD	OEM
COP345CJ	OEM
COP345CN	OEM
COP345LD	OEM
COP345LJ	OEM
COP345LN	OEM
COP364D	OEM
COP364J	OEM
COP364N	OEM
COP365D	OEM
COP365J	OEM
COP365N	OEM
COP384D	OEM
COP384J	OEM
COP384N	OEM
COP385D	OEM
COP385J	OEM
COP385N	OEM
COP401LD	OEM
COP401LJ	OEM
COP401LN	OEM
COP402D	OEM
COP402J	OEM
COP402MD	OEM
COP402MJ	OEM
COP402MN	OEM
COP402N	OEM
COP404D	OEM
COP404J	OEM
COP404LD	OEM
COP404LJ	OEM
COP404LN	OEM
COP404N	OEM
COP410CD	OEM
COP410CJ	OEM
COP410CN	OEM
COP410LD	OEM
COP410LJ	OEM
COP410LN	OEM
COP411CD	OEM
COP411CJ	OEM
COP411CN	OEM
COP411LD	OEM
COP411LJ	OEM
COP411LN	OEM
COP420CD	OEM
COP420CJ	OEM
COP420CN	OEM
COP420D	OEM
COP420J	OEM
COP420LD	OEM
COP420LJ	OEM
COP420LN	OEM
COP420N	OEM
COP420RD	OEM
COP420RJ	OEM
COP420RN	OEM
COP421CD	OEM
COP421CJ	OEM
COP421CN	OEM
COP421D	OEM
COP421J	OEM
COP421LD	OEM
COP421LJ	OEM
COP421LN	OEM
COP421N	OEM
COP422D	OEM
COP422J	OEM

If replacement code is OEM, contact original manufacturer for replacement.

DEVICE TYPE	REPL CODE
COP422LD	OEM
COP422LJ	OEM
COP422LN	OEM
COP422N	OEM
COP440D	OEM
COP440J	OEM
COP440N	OEM
COP440RD	OEM
COP440RJ	OEM
COP440RN	OEM
COP441D	OEM
COP441J	OEM
COP441N	OEM
COP442D	OEM
COP442J	OEM
COP442N	OEM
COP444CD	OEM
COP444CJ	OEM
COP444CN	OEM
COP444LJ	OEM
COP444LN	OEM
COP444LRD	OEM
COP444LRJ	OEM
COP444LRN	OEM
COP4440D	OEM
COP445CD	OEM
COP445CJ	OEM
COP445LD	OEM
COP445LJ	OEM
COP445LN	OEM
COP464D	OEM
COP464J	OEM
COP464N	OEM
COP465D	OEM
COP465J	OEM
COP465N	OEM
COP472D	OEM
COP472N	OEM
COP472N-3	OEM
COP484D	OEM
COP484J	OEM
COP484N	OEM
COP485D	OEM
COP485J	OEM
COP485N	OEM
COP2304D	OEM
COP2304J	OEM
COP2304N	OEM
COP2340D	OEM
COP2340N	OEM
COP2341D	OEM
COP2341J	OEM
COP2341N	OEM
COP2342D	OEM
COP2342J	OEM
COP2342N	OEM
COP2404D	OEM
COP2404J	OEM
COP2404N	OEM
COP2440D	OEM
COP2440J	OEM
COP2440N	OEM
COP2440RD	OEM
COP2440RJ	OEM
COP2440RN	OEM
COP2441D	OEM
COP2441N	OEM
COP2442D	OEM
COP2442J	OEM
COP2442N	OEM
CP00202	OEM
CP00811	OEM
CP00821	OEM
CP00851	OEM
CP01132	OEM
CP01771	OEM
CP1	OEM
CP1A	OEM
CP2	OEM
CP2A	OEM
CP10-12	OEM
CP22	OEM
CP82C55A	OEM
CP82C59A	OEM
CP98	OEM
CP100	OEM
CP101	OEM
CP102	0071
CP102BA	0015
CP102DA	0015
CP102FA	0015
CP102HA	0015
CP102KA	0015
CP102MA	0015
CP102PA	0071
CP102RA	0071
CP102VA	0071
CP103	0071
CP104	OEM
CP110	OEM
CP115	OEM
CP119	OEM
CP121	OEM
CP123	OEM
CP124	OEM
CP126	OEM
CP127	OEM
CP128	OEM
CP129	OEM

DEVICE TYPE	REPL CODE
CP130	OEM
CP131	OEM
CP132	OEM
CP133	OEM
CP134	OEM
CP135	OEM
CP136	OEM
CP137	OEM
CP138	OEM
CP139	OEM
CP140	OEM
CP141	OEM
CP142	OEM
CP143	OEM
CP144	OEM
CP145	OEM
CP146	OEM
CP147	OEM
CP148	OEM
CP149	OEM
CP150	OEM
CP151	OEM
CP152	OEM
CP152AA	0071
CP152BA	0071
CP152CA	0071
CP152DA	0071
CP152EA	0071
CP152FA	0071
CP152GA	0071
CP152HA	0071
CP152KA	0071
CP152MA	0071
CP152PA	0071
CP152RA	0071
CP152TA	0071
CP152VA	0071
CP153	OEM
CP154	OEM
CP155	OEM
CP156	OEM
CP157	OEM
CP158	OEM
CP398	OEM
CP400	0103
CP401	0103
CP402	OEM
CP403	0861
CP404	0103
CP405	0103
CP406	0103
CP407	0103
CP408	0103
CP409	0086
CP430	OEM
CP431	OEM
CP432	OEM
CP433	OEM
CP600	OEM
CP601	OEM
CP602	OEM
CP640	OEM
CP643	OEM
CP650	OEM
CP651	OEM
CP652	OEM
CP653	OEM
CP664	OEM
CP665	OEM
CP666	OEM
CP701	0236
CP702	0236
CP703	0861
CP704	0236
CP800	0164
CP801	0164
CP802	0164
CP803	0164
CP1005	OEM
CP1007GL	OEM
CP1009GLR	OEM
CP1600	OEM
CP-1610	OEM
CP-1610K	OEM
CP2006A	OEM
CP2009	OEM
CP2009A	OEM
CP2010	OEM
CP2010A	OEM
CP2011	OEM
CP2011A	OEM
CP2012	OEM
CP2012A	OEM
CP2013	1169
CP2013A	1169
CP2015	OEM
CP2015A	OEM
CP2016	OEM
CP2016A	OEM
CP2018	1767
CP2018A	1767
CP2020	1932
CP2020A	1932
CP2022	OEM
CP2022A	OEM
CP2024	OEM
CP2024A	OEM
CP2025GR	OEM
CP2027	1952

DEVICE TYPE	REPL CODE
CP2027A	1952
CP2030	OEM
CP2030A	OEM
CP2032	OEM
CP2033	OEM
CP2033A	OEM
CP2036	OEM
CP2036A	OEM
CP2039	OEM
CP2039A	OEM
CP2043	OEM
CP2043A	OEM
CP2047	OEM
CP2047A	OEM
CP2051	1972
CP2051A	1972
CP2056	OEM
CP2056A	OEM
CP2062	OEM
CP2062A	OEM
CP2068	1987
CP2068A	1987
CP2075	OEM
CP2075A	OEM
CP2082	OEM
CP2082A	OEM
CP2091	OEM
CP2091A	OEM
CP2100	OEM
CP2100A	OEM
CP2110	OEM
CP2110A	OEM
CP2120	OEM
CP2130	OEM
CP2130A	OEM
CP2140	OEM
CP2143	OEM
CP2150	OEM
CP2150A	OEM
CP2160	OEM
CP2160A	OEM
CP2170	OEM
CP2170A	OEM
CP2180	OEM
CP2180A	OEM
CP2200	1408
CP2200A	1408
CP2220	OEM
CP2220A	OEM
CP2250	OEM
CP2250A	OEM
CP2275	OEM
CP2275A	OEM
CP3009	OEM
CP3009A	OEM
CP3010	OEM
CP3010A	OEM
CP3011	OEM
CP3011A	OEM
CP3012	OEM
CP3012A	OEM
CP3013	OEM
CP3013A	OEM
CP3015	OEM
CP3015A	OEM
CP3016	OEM
CP3016A	OEM
CP3018	OEM
CP3018A	OEM
CP3020	OEM
CP3020A	OEM
CP3022	OEM
CP3022A	OEM
CP3024	OEM
CP3024A	OEM
CP3027	OEM
CP3027A	OEM
CP3030	OEM
CP3030A	OEM
CP3033	OEM
CP3033A	OEM
CP3036	OEM
CP3036A	OEM
CP3039	OEM
CP3039A	OEM
CP3043	OEM
CP3043A	OEM
CP3047	OEM
CP3047A	OEM
CP3051	OEM
CP3051A	OEM
CP3056	OEM
CP3056A	OEM
CP3062	OEM
CP3062A	OEM
CP3068	OEM
CP3068A	OEM
CP3075	OEM
CP3075A	OEM
CP3082	OEM
CP3082A	OEM
CP3091	OEM
CP3091A	OEM
CP3100	OEM
CP3100A	OEM
CP3110	OEM
CP3110A	OEM
CP3120	OEM
CP3120A	OEM

DEVICE TYPE	REPL CODE
CP3120E	2457
CP3130	OEM
CP3130A	OEM
CP3140	OEM
CP3150	OEM
CP3150A	OEM
CP3160	OEM
CP3160A	OEM
CP3170	OEM
CP3170A	OEM
CP3180	OEM
CP3180A	OEM
CP3200	OEM
CP3200A	OEM
CP3220	OEM
CP3220A	OEM
CP3250	OEM
CP3250A	OEM
CP3275	OEM
CP3275A	OEM
CP5011C	OEM
CP3212055	0012
CPS1002	OEM
CPS1113	OEM
CPS1540B	2626
CPS1545B	2626
CPS1550B	2626
CPS2510B	2626
CPS2512B	2626
CPS2515B	2626
CPS2540B	2626
CPS2545B	2626
CPS2550B	2626
CPS4010B	2626
CPS4012B	2626
CPS4015B	2626
CPS4040B	2626
CPS4045B	2626
CPS4050B	2626
CPS7010B	4013
CPS7012B	4013
CPS7015B	4013
CPS15553-101	1041
CPS-16676-1	2078
CPT12035	OEM
CPT12040	OEM
CPT12045	OEM
CPT12050	OEM
CPT20120	OEM
CPT20125	OEM
CPT30035	OEM
CPT30040	OEM
CPT30045	OEM
CPT30050	OEM
CPU/6800	OEM
CPU001	OEM
CPU-001-010	OEM
CPU-001-030	OEM
CPU-001-210	OEM
CPU-001-230	OEM
CPU012	OEM
CPU-012-310	OEM
CPU-012-311	OEM
CPU019	OEM
CPU-019-210	OEM
CPU-019-211	OEM
CPU-019-212	OEM
CPU-019-310	OEM
CPU-019-311	OEM
CPU-019-312	OEM
CPU-800	OEM
CPU-6805	OEM
CPU7811G	OEM
CPU7811HG	OEM
CPX5058H-540Q	OEM
CPZ-48000	OEM
CQ1	0004
CQ4-32	OEM
CQ209S	OEM
CQ216X	OEM
CQ216Y	OEM
CQ327	OEM
CQ327R	OEM
CQ330	OEM
CQ330R	OEM
CQ331	OEM
CQ331R	OEM
CQ332	OEM
CQ332R	OEM
CQ427	OEM
CQ427R	OEM
CQ430	OEM
CQ430R	OEM
CQ431	OEM
CQ431R	OEM
CQ432	OEM
CQ432R	OEM
CQ30148A	OEM
CQB56-2	OEM
CQEAK-00801	OEM
CQEAK-00805	OEM
CQEAK-01011	OEM
CQEAK-01103	0812
CQEAK-03001	OEM
CQEAK-03202	OEM
CQL10	OEM
CQL10A	OEM
CQL10B	OEM
CQT10	OEM

DEVICE TYPE	REPL CODE
CQT940A	0160
CQT940B	0160
CQT940BA	0160
CQT1075	0160
CQT1076	0160
CQT1077	0160
CQT1110	0160
CQT1110A	0160
CQT1111	0160
CQT1111A	0160
CQT1112	0160
CQV10-3	OEM
CQV10-4	OEM
CQV11-4	OEM
CQV11-5	OEM
CQV11-6	OEM
CQV13-4	OEM
CQV13-5	OEM
CQV13-6	OEM
CQV14-3	OEM
CQV14-4	OEM
CQV14-5	OEM
CQV14-6	OEM
CQV15-3	OEM
CQV15-4	OEM
CQV15-6	OEM
CQV16-2	OEM
CQV16-3	OEM
CQV16-4	OEM
CQV18-2	OEM
CQV18-3	OEM
CQV18-4	OEM
CQV19-2	OEM
CQV19-3	OEM
CQV19-4	OEM
CQV20-3	OEM
CQV20-4	OEM
CQV21-4	OEM
CQV21-5	OEM
CQV21-6	OEM
CQV23-4	OEM
CQV23-5	OEM
CQV23-6	OEM
CQV24-3	OEM
CQV24-4	OEM
CQV24-5	OEM
CQV24-6	OEM
CQV25-3	OEM
CQV25-4	OEM
CQV25-5	OEM
CQV25-6	OEM
CQV26-2	OEM
CQV26-3	OEM
CQV26-4	OEM
CQV28-2	OEM
CQV28-3	OEM
CQV28-4	OEM
CQV29-2	OEM
CQV29-3	OEM
CQV29-4	OEM
CQV30A	OEM
CQV30B	OEM
CQV30C	OEM
CQV31D	OEM
CQV31E	OEM
CQV33D	OEM
CQV33E	OEM
CQV35D	OEM
CQV36-3	OEM
CQV36-4	OEM
CQV36-5	OEM
CQV38-3	OEM
CQV38-4	OEM
CQV38-5	OEM
CQV39-3	OEM
CQV39-4	OEM
CQV39-5	OEM
CQV41-3	OEM
CQV41-4	OEM
CQV41-5	OEM
CQV43-3	OEM
CQV43-4	OEM
CQV43-5	OEM
CQV45-3	OEM
CQV45-4	OEM
CQV45-5	OEM
CQV51F	OEM
CQV51G	OEM
CQV51H	OEM
CQV51J	OEM
CQV53F	OEM
CQV53G	OEM
CQV53H	OEM
CQV53J	OEM
CQV55G	OEM
CQV55H	OEM
CQV55J	OEM
CQV55K	OEM
CQV56-3	OEM
CQV56-4	OEM
CQV58-2	OEM
CQV58-3	OEM
CQV58-4	OEM
CQV59-2	OEM
CQV59-4	OEM
CQV60	OEM
CQV61	OEM
CQV62	OEM
CQW10	OEM

DEVICE TYPE	REPL CODE
CQW11	OEM
CQW12	OEM
CQW13	OEM
CQW14	OEM
CQW24	OEM
CQW32	OEM
CQW51	OEM
CQW54	OEM
CQX10	3182
CQX10-I	OEM
CQX10-II	OEM
CQX10-III	OEM
CQX10-IV	OEM
CQX10I	OEM
CQX10II	OEM
CQX10III	OEM
CQX10IV	OEM
CQX11	1778
CQX11-I	OEM
CQX11-II	OEM
CQX11-III	OEM
CQX11-IV	OEM
CQX11I	OEM
CQX11II	OEM
CQX11III	OEM
CQX11IV	OEM
CQX12	3157
CQX12-I	OEM
CQX12-II	OEM
CQX12-III	OEM
CQX12-IV	OEM
CQX12I	OEM
CQX12II	OEM
CQX12III	OEM
CQX12IV	OEM
CQX13	OEM
CQX13-1	OEM
CQX13-2	OEM
CQX13I	OEM
CQX13II	OEM
CQX14	OEM
CQX15	OEM
CQX16	OEM
CQX17	OEM
CQX18	OEM
CQX19	OEM
CQX21	4023
CQX23-1	OEM
CQX23-2	OEM
CQX23I	OEM
CQX23II	OEM
CQX24	OEM
CQX24B	OEM
CQX25N	OEM
CQX26N	OEM
CQX27N	OEM
CQX28	OEM
CQX29	OEM
CQX30	OEM
CQX31	OEM
CQX32	OEM
CQX33-1	OEM
CQX33-2	OEM
CQX33I	OEM
CQX33II	OEM
CQX35	OEM
CQX36	OEM
CQX37	OEM
CQX41NA	OEM
CQX41NB	OEM
CQX42NA	OEM
CQX42NB	OEM
CQX43N	OEM
CQX46	OEM
CQX47	OEM
CQX51	OEM
CQX51-1	OEM
CQX51-2	OEM
CQX51-3	OEM
CQX54	OEM
CQX55	OEM
CQX56	OEM
CQX57	OEM
CQX58	OEM
CQX60	OEM
CQX61	OEM
CQX62	OEM
CQX63	OEM
CQX64	OEM
CQX65	OEM
CQX66	OEM
CQX67	OEM
CQX68	OEM
CQX74	OEM
CQX75	OEM
CQX76	OEM
CQX77	OEM
CQX78	OEM
CQY10	OEM
CQY11A	OEM
CQY11B	OEM
CQY11C	OEM
CQY11D	OEM
CQY12A	OEM
CQY12B	OEM
CQY13	OEM
CQY13B	OEM
CQY14	OEM
CQY15	OEM

DEVICE TYPE	REPL CODE
CQY17IV	OEM
CQY17V	OEM
CQY24B	6601
CQY24B-I	OEM
CQY24B-II	OEM
CQY24B-IV	OEM
CQY27	OEM
CQY31	OEM
CQY32	OEM
CQY33NE	OEM
CQY33NF	OEM
CQY34NE	OEM
CQY34NF	OEM
CQY35NE	OEM
CQY35NR	OEM
CQY36N	OEM
CQY37N	OEM
CQY40	OEM
CQY41N	OEM
CQY49B	OEM
CQY49C	OEM
CQY50	OEM
CQY52	OEM
CQY54	3019
CQY54-I	OEM
CQY54-II	OEM
CQY54-III	OEM
CQY58A	OEM
CQY58A-1	OEM
CQY58A-2	OEM
CQY58A-I	OEM
CQY58A-II	OEM
CQY65	3019
CQY72L	1778
CQY73N	OEM
CQY74	OEM
CQY75N	OEM
CQY77I	OEM
CQY77II	OEM
CQY77III	OEM
CQY78I	OEM
CQY78II	OEM
CQY78III	OEM
CQY80	0315
CQY80N	0541
CQY85NA	OEM
CQY85NB	OEM
CQY86NA	OEM
CQY86NB	OEM
CQY87NQ	OEM
CQY89A	OEM
CQY89A-1	OEM
CQY89A1	OEM
CQY89A-2	OEM
CQY94	1778
CQY94-I	OEM
CQY94-II	OEM
CQY94-III	OEM
CQY94-IV	OEM
CQY95	OEM
CQY95-I	OEM
CQY95-II	OEM
CQY95-III	OEM
CQY96	OEM
CQY96-I	OEM
CQY96-II	OEM
CQY96-III	OEM
CQY96-IV	OEM
CQY97	3124
CQY97-I	OEM
CQY97-II	OEM
CQY97-III	OEM
CQY98	OEM
CQY99	0516
CR	OEM
CR/E	0015
CR0000	0143
CR005-01A	OEM
CR005-01B	OEM
CR005-01C	OEM
CR005-01D	OEM
CR005-01E	OEM
CR005-05A	OEM
CR005-05B	OEM
CR005-05C	OEM
CR005-05D	OEM
CR005-05E	OEM
CR005-10A	OEM
CR005-10B	OEM
CR005-10C	OEM
CR005-10D	OEM
CR005-10E	OEM
CR005-50A	OEM
CR005-50B	OEM
CR005-50C	OEM
CR005-50D	OEM
CR005-50E	OEM
CR02A1	OEM
CR02A2	OEM
CR02A4	OEM
CR02A6	OEM
CR02A8	OEM
CR02AM	0344
CR02AM-1	0344
CR02AM1	0344
CR02AM2	OEM
CR02AM-4	0058
CR02AM4	0058
CR02AM6	OEM

Original Device Types

DEVICE TYPE	REPL CODE
CR02AM-8	0407
CR02AM8	OEM
CR02AM10	OEM
CR02AN4	0058
CR03A-1	OEM
CR03AM1	OEM
CR03AM2	OEM
CR03AM4	OEM
CR03AM6	OEM
CR03AM8	OEM
CR03AM10	OEM
CR03AM12	OEM
CR05	OEM
CR022	OEM
CR024	OEM
CR027	OEM
CR030	OEM
CR033	OEM
CR039	OEM
CR043	OEM
CR047	OEM
CR056	OEM
CR062	OEM
CR068	OEM
CR075	OEM
CR082	OEM
CR091	OEM
CR0121A	1004
CR0121B	1004
CR0121D	1034
CR0121M	1148
CR0122A	2089
CR0122B	0505
CR0122C	OEM
CR0122D	0710
CR0122E	OEM
CR0122F	OEM
CR0122M	0863
CR0122N	OEM
CR0137-14	OEM
CR1-021C	OEM
CR1-051C	0179
CR1-051CA	0174
CR1-051CB	0179
CR1-101C	0179
CR1-101CA	3854
CR1-101CB	0344
CR1-151C	OEM
CR1-201C	0179
CR1-201CA	3620
CR1-201CB	0058
CR1-251C	OEM
CR1-301C	0346
CR1-301CA	0346
CR1-301CB	0407
CR1-401C	0346
CR1-401CA	0346
CR1-401CB	OEM
CR1T01HY1	OEM
CR1T01HY2	OEM
CR1T01HY3	OEM
CR1T02HY1	OEM
CR1T02HY2	OEM
CR1T02HY3	OEM
CR1T03HY1	OEM
CR1T03HY2	OEM
CR1T03HY3	OEM
CR1T04HY1	OEM
CR1T04HY2	OEM
CR1T04HY3	OEM
CR1T05HY1	OEM
CR1T05HY2	OEM
CR1T05HY3	OEM
CR1T06HY1	OEM
CR1T06HY2	OEM
CR1T06HY3	OEM
CR1TX5HY1	OEM
CR1TX5HY2	OEM
CR1TX5HY3	OEM
CR2-160	OEM
CR2AM1	1259
CR2AM2	0447
CR2AM4	0940
CR2AM6	0095
CR2AM-8	0095
CR2AM8	0095
CR2AM10	1243
CR2AM12	1243
CR2T01HY	OEM
CR2T01HY1	OEM
CR2T02HY	OEM
CR2T02HY1	OEM
CR2T03HY	OEM
CR2T03HY1	OEM
CR2T04HY	OEM
CR2T04HY1	OEM
CR2T05HY	OEM
CR2T05HY1	OEM
CR2T06HY	OEM
CR2T06HY1	OEM
CR2TX5HY	OEM
CR2TX5HY1	OEM
CR3-080	OEM
CR3-100	OEM
CR3AM1	1259
CR3AM2	0447
CR3AM4	0940
CR3AM6	0095
CR3AM-8	OEM
CR3AM8	0095
CR3AMZ	0095
CR3AMZ-8	0095
CR3AMZ8	0095
CR3CM	0095
CR3CM-1	1259
CR3CM-2	0447
CR3CM-4	0940
CR3CM-6	0095
CR3CM-8	0095
CR3CM8	0095
CR3CM-8-208	0095
CR3CM10	1243
CR3CM12	1243
CR-3D	OEM
CR3DZ2	OEM
CR3DZ4	OEM
CR3DZ6	OEM
CR3DZ8	OEM
CR3E1	OEM
CR3E2	OEM
CR3E4	OEM
CR3E-6	OEM
CR3E6	OEM
CR3E8	OEM
CR3E10	OEM
CR3E12	OEM
CR3EM1	OEM
CR3EM2	OEM
CR3EM4	OEM
CR3EM6	OEM
CR3EM8	OEM
CR3JM6	OEM
CR3JM8	OEM
CR4-051A	OEM
CR4-051B	OEM
CR4-051C	OEM
CR4-101A	OEM
CR4-101B	OEM
CR4-101C	OEM
CR4-201A	OEM
CR4-201B	OEM
CR4-201C	OEM
CR4-301A	OEM
CR4-301B	OEM
CR4-301C	OEM
CR4-401A	OEM
CR4-401B	OEM
CR4-401C	OEM
CR4-501A	OEM
CR4-501B	OEM
CR4-501C	OEM
CR4-601A	OEM
CR4-601B	OEM
CR4-601C	OEM
CR4-701A	OEM
CR4-701B	OEM
CR4-701C	OEM
CR4-801A	OEM
CR4-801B	OEM
CR4-801C	OEM
CR4-901A	OEM
CR4-901B	OEM
CR4-901C	OEM
CR4-1001A	OEM
CR4-1001B	OEM
CR4-1001C	OEM
CR4U01FY	6103
CR4U01FY1	OEM
CR4U02FY	6112
CR4U02FY1	OEM
CR4U03FY	6127
CR4U03FY1	OEM
CR4U04FY	6127
CR4U04FY1	OEM
CR4U05FY	6147
CR4U05FY1	OEM
CR4U06FY	6147
CR4U06FY1	OEM
CR4U08FY	OEM
CR4U08FY1	OEM
CR4U10FY	OEM
CR4U10FY1	OEM
CR4UX5FY	7138
CR4UX5FY1	OEM
CR4.021A	OEM
CR4.051A	OEM
CR4.071A	OEM
CR4.101A	OEM
CR4.151A	OEM
CR4.201A	OEM
CR4.251A	OEM
CR4.301A	OEM
CR5-021B	OEM
CR5-051B	OEM
CR5-052AB	OEM
CR5-052BB	OEM
CR5-052CB	OEM
CR5-052DB	OEM
CR5-052EB	OEM
CR5-052GB	OEM
CR5-071B	OEM
CR5-101B	OEM
CR5-102AB	OEM
CR5-102B	OEM
CR5-102BB	OEM
CR5-102CCB	OEM
CR5-102DB	OEM
CR5-102EB	OEM
CR5-102GB	OEM
CR5-201B	OEM
CR5-202AB	OEM
CR5-202B	OEM
CR5-202BB	OEM
CR5-202CB	OEM
CR5-202DB	OEM
CR5-202EB	OEM
CR5-202GB	OEM
CR5-301B	OEM
CR5-302AB	OEM
CR5-302B	OEM
CR5-302BB	OEM
CR5-302CB	OEM
CR5-302DB	OEM
CR5-302EB	OEM
CR5-302GB	OEM
CR5-401B	OEM
CR5-402AB	OEM
CR5-402B	OEM
CR5-402BB	OEM
CR5-402CB	OEM
CR5-402DB	OEM
CR5-402EB	OEM
CR5-402GB	OEM
CR5-501B	OEM
CR5-502AB	OEM
CR5-502B	OEM
CR5-502BB	OEM
CR5-502CB	OEM
CR5-502DB	OEM
CR5-502EB	OEM
CR5-502GB	OEM
CR5-601B	OEM
CR5-602AB	OEM
CR5-602B	OEM
CR5-602BB	OEM
CR5-602CB	OEM
CR5-602DB	OEM
CR5-602EB	OEM
CR5-602GB	OEM
CR5-701B	OEM
CR5-801B	OEM
CR5-1029B	OEM
CR5-2029B	OEM
CR5-3029B	OEM
CR5-4029B	OEM
CR5-5029B	OEM
CR5-6029B	OEM
CR5AS-8	OEM
CR5AS8	OEM
CR5B-6	1487
CR5B-8	1487
CR5B-10	0687
CR5B-12	0687
CR6-051B	OEM
CR6-051BA	OEM
CR6-051C	OEM
CR6-051CA	OEM
CR6-052AB	OEM
CR6-052BB	OEM
CR6-052CB	OEM
CR6-101B	OEM
CR6-101BA	OEM
CR6-101C	OEM
CR6-101CA	OEM
CR6-201B	OEM
CR6-201BA	OEM
CR6-201C	OEM
CR6-201CA	OEM
CR6-301B	OEM
CR6-301BA	OEM
CR6-301C	OEM
CR6-301CA	OEM
CR6-401B	OEM
CR6-401BA	OEM
CR6-401C	OEM
CR6-401CA	OEM
CR6-403RB	OEM
CR6-403RBA	OEM
CR6-501B	OEM
CR6-501BA	OEM
CR6-501C	OEM
CR6-501CA	OEM
CR6-503RB	OEM
CR6-503RBA	OEM
CR6-601B	OEM
CR6-601BA	OEM
CR6-601C	OEM
CR6-601CA	OEM
CR6-603RB	OEM
CR6-603RBA	OEM
CR6-703RB	OEM
CR6-703RBA	OEM
CR6-803B	OEM
CR6-803RB	OEM
CR6-803RBA	OEM
CR6-903B	OEM
CR6-903RB	OEM
CR6-903RBA	OEM
CR6-903RC	OEM
CR6-903RCA	OEM
CR6-1003B	OEM
CR6-1003RB	OEM
CR6-1003RBA	OEM
CR6-1003RC	OEM
CR6-1003RCA	OEM
CR6-1103RBA	OEM
CR6-1103RC	OEM
CR6-1103RCA	OEM
CR6-1203B	OEM
CR6-1203RB	OEM
CR6-1203RBA	OEM
CR6-1203RC	OEM
CR6-1203RCA	OEM
CR6-1303RB	OEM
CR6-1303RBA	OEM
CR6-1303RC	OEM
CR6-1303RCA	OEM
CR6-1403RB	OEM
CR6-1403RBA	OEM
CR6-1403RC	OEM
CR6-1403RCA	OEM
CR6-1503RB	OEM
CR6-1503RBA	OEM
CR6-1503RC	OEM
CR6-1503RCA	OEM
CR6AM-2	2089
CR6AM-4	0505
CR6AM-6	0710
CR6AM-8	0710
CR6AM8	OEM
CR6AM10	OEM
CR6AM12	OEM
CR7K103A3	OEM
CR7K103A4	OEM
CR7K103A5	OEM
CR7K103A35	OEM
CR7K103A45	OEM
CR7K203A3	OEM
CR7K203A4	OEM
CR7K203A5	OEM
CR7K203A35	OEM
CR7K203A45	OEM
CR7K303A3	OEM
CR7K303A4	OEM
CR7K303A5	OEM
CR7K303A35	OEM
CR7K303A45	OEM
CR7K403A3	OEM
CR7K403A4	OEM
CR7K403A5	OEM
CR7K403A35	OEM
CR7K403A45	OEM
CR7K503A3	OEM
CR7K503A4	OEM
CR7K503A5	OEM
CR7K503A35	OEM
CR7K503A45	OEM
CR7K603A3	OEM
CR7K603A4	OEM
CR7K603A5	OEM
CR7K603A35	OEM
CR7K603A45	OEM
CR7K703A3	OEM
CR7K703A4	OEM
CR7K703A5	OEM
CR7K703A35	OEM
CR7K703A45	OEM
CR7K803A3	OEM
CR7K803A4	OEM
CR7K803A5	OEM
CR7K803A35	OEM
CR7K803A45	OEM
CR7K903A3	OEM
CR7K903A4	OEM
CR7K903A5	OEM
CR7K903A35	OEM
CR7K903A45	OEM
CR7K1003A3	OEM
CR7K1003A4	OEM
CR7K1003A5	OEM
CR7K1003A35	OEM
CR7K1003A45	OEM
CR7K1103A3	OEM
CR7K1103A4	OEM
CR7K1103A5	OEM
CR7K1103A35	OEM
CR7K1103A45	OEM
CR7K1203A3	OEM
CR7K1203A4	OEM
CR7K1203A5	OEM
CR7K1203A35	OEM
CR7K1203A45	OEM
CR7K1303A3	OEM
CR7K1303A4	OEM
CR7K1303A5	OEM
CR7K1303A35	OEM
CR7K1303A45	OEM
CR7K1403A3	OEM
CR7K1403A4	OEM
CR7K1403A5	OEM
CR7K1403A35	OEM
CR7K1403A45	OEM
CR7K1503A3	OEM
CR7K1503A4	OEM
CR7K1503A5	OEM
CR7K1503A35	OEM
CR7K1503A45	OEM
CR7K1603A4	OEM
CR7K1603A5	OEM
CR7K1603A35	OEM
CR7K1603A45	OEM
CR7K1703A4	OEM
CR7K1703A5	OEM
CR7K1703A45	OEM
CR7K1803A4	OEM
CR7K1803A5	OEM
CR7K1803A45	OEM
CR7K1903A5	OEM
CR7K1903A45	OEM
CR7K2003A5	OEM
CR7K2003A45	OEM
CR7K2103A5	OEM
CR7K2103A45	OEM
CR7K2203A5	OEM
CR7K2203A45	OEM
CR7K2303A5	OEM
CR8-051A	OEM
CR8-051B	OEM
CR8-051C	OEM
CR8-101A	OEM
CR8-101B	OEM
CR8-101C	OEM
CR8-102CC	OEM
CR8-201A	OEM
CR8-201B	OEM
CR8-201C	OEM
CR8-202BC	OEM
CR8-301A	OEM
CR8-301B	OEM
CR8-301C	OEM
CR8-302GC	OEM
CR8-401A	OEM
CR8-401B	OEM
CR8-401C	OEM
CR8-501A	OEM
CR8-501B	OEM
CR8-501C	OEM
CR8-601A	OEM
CR8-601B	OEM
CR8-601C	OEM
CR8-701A	OEM
CR8-701B	OEM
CR8-701C	OEM
CR8-801A	OEM
CR8-801B	OEM
CR8-801C	OEM
CR8-901A	OEM
CR8-901B	OEM
CR8-901C	OEM
CR8-1001A	OEM
CR8-1001B	OEM
CR8-1001C	OEM
CR8AM8	OEM
CR8U01FY	OEM
CR8U02FY	OEM
CR8U02FY1	OEM
CR8U03FY	OEM
CR8U03FY1	OEM
CR8U04FY	OEM
CR8U04FY1	OEM
CR8U05FY	OEM
CR8U05FY1	OEM
CR8U06FY	OEM
CR8U06FY1	OEM
CR8U08FY	OEM
CR8U08FY1	OEM
CR8U10FY	OEM
CR8U10FY1	OEM
CR8UX5FY	OEM
CR8UX5FY1	OEM
CR8.021A	OEM
CR8.051A	OEM
CR8.071A	OEM
CR8.101A	OEM
CR8.151A	OEM
CR8.201A	OEM
CR8.251A	OEM
CR8.301A	OEM
CR9K403A2	OEM
CR9K403A25	OEM
CR9K503A2	OEM
CR9K503A25	OEM
CR9K603A2	OEM
CR9K603A25	OEM
CR9K703A2	OEM
CR9K703A25	OEM
CR9K803A2	OEM
CR9K803A25	OEM
CR9K903A25	OEM
CR9K1003A25	OEM
CR9K1103A25	OEM
CR9K1203A25	OEM
CR9K1303A25	OEM
CR9K1403A25	OEM
CR9K1503A25	OEM
CR10	OEM
CR10-051B	OEM
CR10-052AB	OEM
CR10-052BB	OEM
CR10-052CB	OEM
CR10-052DB	OEM
CR10-052EB	OEM
CR10-052GB	OEM
CR10-101B	OEM
CR10-102AB	OEM
CR10-102B	OEM
CR10-102BB	OEM
CR10-102CB	OEM
CR10-102DB	OEM
CR10-102EB	OEM
CR10-102GB	OEM
CR10-201B	OEM
CR10-202AB	OEM
CR10-202B	OEM
CR10-202BB	OEM
CR10-202CB	OEM
CR10-202DB	OEM
CR10-202EB	OEM
CR10-202GB	OEM
CR10-301B	OEM
CR10-302AB	OEM
CR10-302B	OEM
CR10-302BB	OEM
CR10-302CB	OEM
CR10-302DB	OEM
CR10-302EB	OEM
CR10-302GB	OEM
CR10-401B	OEM
CR10-402AB	OEM
CR10-402B	OEM
CR10-402BB	OEM
CR10-402CB	OEM
CR10-402DB	OEM
CR10-402EB	OEM
CR10-402GB	OEM
CR10-501B	OEM
CR10-502AB	OEM
CR10-502B	OEM
CR10-502BB	OEM
CR10-502CB	OEM
CR10-502DB	OEM
CR10-502EB	OEM
CR10-502GB	OEM
CR10-601B	OEM
CR10-602AB	OEM
CR10-602B	OEM
CR10-602BB	OEM
CR10-602CB	OEM
CR10-602DB	OEM
CR10-602EB	OEM
CR10-701B	OEM
CR10-1001B	OEM
CR10-1029B	OEM
CR10-2029B	OEM
CR10-3029B	OEM
CR10-4029B	OEM
CR10-5029B	OEM
CR10-6029B	OEM
CR10A	OEM
CR10B	OEM
CR10B-1	2449
CR10B-2	2449
CR10B2	OEM
CR10B-4	0435
CR10B4	OEM
CR10B-6	1487
CR10B6	OEM
CR10B-8	1487
CR10B8	OEM
CR10B-10	0687
CR10B10	OEM
CR10B-12	0687
CR10B12	OEM
CR10B16	OEM
CR10C	OEM
CR10C2	0567
CR10C4	0762
CR10C6	0740
CR10C8	0740
CR10C10	0764
CR10C12	0764
CR10C16	0766
CR10CY2	OEM
CR10CY4	OEM
CR10CY6	OEM
CR10CY8	OEM
CR10CY10	OEM
CR10CY12	OEM
CR10D	OEM
CR10E	OEM
CR10EY2	OEM
CR10EY4	OEM
CR10EY6	OEM
CR10EY10	OEM
CR10EY12	OEM
CR10F	OEM
CR10G	OEM
CR10H	OEM
CR12-051B	0469
CR12-051BA	0469
CR12-051C	OEM
CR12-051CA	OEM
CR12-052AB	OEM
CR12-052BB	OEM
CR12-052CB	OEM
CR12-101B	0469
CR12-101BA	0469
CR12-101C	OEM
CR12-101CA	OEM
CR12-201B	0469
CR12-201BA	0469
CR12-201C	OEM
CR12-201CA	OEM
CR12-301B	0722
CR12-301BA	0722
CR12-301C	OEM
CR12-301CA	OEM
CR12-401B	0722
CR12-401BA	0722
CR12-401C	OEM
CR12-401CA	OEM
CR12-403RB	OEM
CR12-403RBA	OEM
CR12-501B	0725
CR12-501BA	0725
CR12-501C	OEM
CR12-501CA	OEM
CR12-503RB	OEM
CR12-503RBA	OEM
CR12-601B	0725
CR12-601BA	0725
CR12-601C	OEM
CR12-601CA	OEM
CR12-603RB	OEM
CR12-603RBA	OEM
CR12-703RB	OEM
CR12-703RBA	OEM
CR12-803B	OEM
CR12-803RB	OEM
CR12-803RBA	OEM
CR12-903B	OEM
CR12-903RB	0679
CR12-903RBA	0679
CR12-903RC	OEM
CR12-903RCA	OEM
CR12-1003B	OEM
CR12-1003RB	0679
CR12-1003RBA	0679
CR12-1003RC	OEM
CR12-1003RCA	OEM
CR12-1103B	OEM
CR12-1103RB	0679
CR12-1103RBA	0679
CR12-1103RC	OEM
CR12-1103RCA	OEM
CR12-1203B	OEM
CR12-1203RB	0679
CR12-1203RBA	0679
CR12-1203RC	OEM
CR12-1203RCA	OEM
CR12-1303RB	OEM
CR12-1303RBA	OEM
CR12-1303RC	0679
CR12-1303RCA	0679
CR12-1403RB	OEM
CR12-1403RBA	OEM
CR12-1403RC	OEM
CR12-1403RCA	OEM
CR12-1503RB	OEM
CR12-1503RBA	OEM
CR12-1503RC	OEM
CR12-1503RCA	OEM
CR12A-6	2647
CR12A-8	2647
CR12A-10	0752
CR12A-12	0764
CR12A-14	0766
CR12A-16	0766
CR12J02JY1	OEM
CR12M01JY	OEM
CR12M01JY1	OEM
CR12M02JY	OEM
CR12M02JY1	OEM
CR12M03JY	OEM
CR12M03JY1	OEM
CR12M04JY	OEM
CR12M04JY1	OEM
CR12M05JY	OEM
CR12M05JY1	OEM
CR12M06JY	OEM
CR12M06JY1	OEM
CR12M08JY	OEM
CR12M08JY1	OEM
CR12M10JY	OEM
CR12M10JY1	OEM
CR12M11JY	OEM
CR12M11JY1	OEM
CR12M12JY	OEM
CR12M12JY1	OEM
CR12M13JY	OEM
CR12M13JY1	OEM
CR12M14JY	OEM
CR12M14JY1	OEM
CR12MX5JY1	OEM
CR12U01JY	OEM
CR12U01JY1	OEM
CR12U02JY	OEM
CR12U03JY	OEM
CR12U03JY1	OEM
CR12U04JY	OEM
CR12U04JY1	OEM
CR12U05JY	OEM
CR12U05JY1	OEM
CR12U06JY	OEM
CR12U06JY1	OEM
CR12U08JY	OEM
CR12U08JY1	OEM
CR12U10JY	OEM
CR12U10JY1	OEM
CR12U11JY	OEM
CR12U11JY1	OEM
CR12U12JY	OEM
CR12U12JY1	OEM
CR12U13JY	OEM
CR12U13JY1	OEM
CR12U14JY	OEM
CR12U14JY1	OEM
CR12UX5JY	OEM
CR12UX5JY1	OEM
CR16-051B	OEM
CR16-052AB	OEM
CR16-052BB	OEM
CR16-052CB	OEM

If replacement code is OEM, contact original manufacturer for replacement.

DEVICE TYPE	REPL CODE
CR16-052DB	OEM
CR16-052EB	OEM
CR16-052GB	OEM
CR16-101B	OEM
CR16-102AB	OEM
CR16-102B	OEM
CR16-102BB	OEM
CR16-102CB	OEM
CR16-102DB	OEM
CR16-102EB	OEM
CR16-102GB	OEM
CR16-201B	OEM
CR16-202AB	OEM
CR16-202B	OEM
CR16-202BB	OEM
CR16-202CB	OEM
CR16-202DB	OEM
CR16-202EB	OEM
CR16-202GB	OEM
CR16-301B	OEM
CR16-302AB	OEM
CR16-302B	OEM
CR16-302BB	OEM
CR16-302CB	OEM
CR16-302DB	OEM
CR16-302EB	OEM
CR16-302GB	OEM
CR16-401B	OEM
CR16-402AB	OEM
CR16-402B	OEM
CR16-402BB	OEM
CR16-402CB	OEM
CR16-402DB	OEM
CR16-402EB	OEM
CR16-402GB	OEM
CR16-501B	OEM
CR16-502AB	OEM
CR16-502B	OEM
CR16-502BB	OEM
CR16-502CB	OEM
CR16-502DB	OEM
CR16-502EB	OEM
CR16-502GB	OEM
CR16-601B	OEM
CR16-602AB	OEM
CR16-602B	OEM
CR16-602BB	OEM
CR16-602CB	OEM
CR16-602DB	OEM
CR16-602EB	OEM
CR16-602GB	OEM
CR16-701B	OEM
CR16-801B	OEM
CR16-1029B	OEM
CR16-2029B	OEM
CR16-3029B	OEM
CR16-4029B	OEM
CR16-5029B	OEM
CR16-6029B	OEM
CR16A	OEM
CR16C	OEM
CR16D	OEM
CR16E	OEM
CR16F	OEM
CR16G	OEM
CR16H	OEM
CR18-051B	0469
CR18-051BA	0469
CR18-051C	OEM
CR18-051CA	OEM
CR18-052AB	OEM
CR18-052BB	OEM
CR18-052CB	OEM
CR18-101B	0469
CR18-101BA	OEM
CR18-101C	OEM
CR18-101CA	OEM
CR18-101RA	0469
CR18-201B	0469
CR18-201BA	0469
CR18-201C	OEM
CR18-201CA	OEM
CR18-301B	0722
CR18-301BA	0722
CR18-301C	OEM
CR18-301CA	OEM
CR18-303RCA	OEM
CR18-401B	0722
CR18-401BA	0722
CR18-401C	OEM
CR18-401CA	OEM
CR18-402AB	OEM
CR18-403RB	OEM
CR18-403RBA	OEM
CR18-501B	0725
CR18-501BA	0725
CR18-501C	OEM
CR18-501CA	OEM
CR18-503RB	OEM
CR18-503RBA	OEM
CR18-601B	0725
CR18-601BA	0725
CR18-601C	OEM
CR18-601CA	OEM
CR18-603RB	OEM
CR18-603RBA	OEM
CR18-703RB	OEM
CR18-703RBA	OEM
CR18-803B	OEM
CR18-803RB	OEM
CR18-803RBA	OEM
CR18-903B	OEM
CR18-903RB	0679
CR18-903RBA	0679
CR18-903RBAC	OEM
CR18-903RBC	OEM
CR18-903RC	OEM
CR18-903RCA	OEM
CR18-903RCAC	OEM
CR18-903RCC	OEM
CR18-1003B	OEM
CR18-1003RB	0679
CR18-1003RBA	0679
CR18-1003RBAC	OEM
CR18-1003RBC	OEM
CR18-1003RC	OEM
CR18-1003RCA	OEM
CR18-1003RCAC	OEM
CR18-1003RCC	OEM
CR18-1103RB	0679
CR18-1103RBA	0679
CR18-1103RBAC	OEM
CR18-1103RBC	OEM
CR18-1103RC	OEM
CR18-1103RCA	OEM
CR18-1103RCAC	OEM
CR18-1103RCC	OEM
CR18-1203B	OEM
CR18-1203RB	0679
CR18-1203RBA	0679
CR18-1203RBAC	OEM
CR18-1203RBC	OEM
CR18-1203RC	OEM
CR18-1203RCA	OEM
CR18-1203RCAC	OEM
CR18-1203RCC	OEM
CR18-1303RB	0679
CR18-1303RBA	0679
CR18-1303RBAC	OEM
CR18-1303RBC	OEM
CR18-1303RC	OEM
CR18-1303RCAC	OEM
CR18-1303RCC	OEM
CR18-1403RB	OEM
CR18-1403RBA	OEM
CR18-1403RBAC	OEM
CR18-1403RBC	OEM
CR18-1403RC	OEM
CR18-1403RCA	OEM
CR18-1403RCAC	OEM
CR18-1403RCC	OEM
CR18-1503RB	OEM
CR18-1503RBA	OEM
CR18-1503RBAC	OEM
CR18-1503RBC	OEM
CR18-1503RC	OEM
CR18-1503RCA	OEM
CR18-1503RCAC	OEM
CR18-1503RCC	OEM
CR20-051B	OEM
CR20-051BA	OEM
CR20-051C	OEM
CR20-051CA	OEM
CR20-101B	OEM
CR20-101BA	OEM
CR20-101C	OEM
CR20-101CA	OEM
CR20-201B	OEM
CR20-201BA	OEM
CR20-201C	OEM
CR20-201CA	OEM
CR20-301B	OEM
CR20-301BA	OEM
CR20-301C	OEM
CR20-301CA	OEM
CR20-401B	OEM
CR20-401BA	OEM
CR20-401C	OEM
CR20-401CA	OEM
CR20-501B	OEM
CR20-501BA	OEM
CR20-501C	OEM
CR20-501CA	OEM
CR20-601B	OEM
CR20-601BA	OEM
CR20-601C	OEM
CR20-601CA	OEM
CR20-903RB	OEM
CR20-903RBA	OEM
CR20-903RBAC	OEM
CR20-903RBC	OEM
CR20-903RC	OEM
CR20-903RCA	OEM
CR20-903RCAC	OEM
CR20-903RCC	OEM
CR20-1003RB	OEM
CR20-1003RBA	OEM
CR20-1003RBAC	OEM
CR20-1003RBC	OEM
CR20-1003RC	OEM
CR20-1003RCA	OEM
CR20-1003RCAC	OEM
CR20-1003RCC	OEM
CR20-1103RB	OEM
CR20-1103RBA	OEM
CR20-1103RBAC	OEM
CR20-1103RBC	OEM
CR20-1103RC	OEM
CR20-1103RCA	OEM
CR20-1103RCAC	OEM
CR20-1103RCC	OEM
CR20-1203RB	OEM
CR20-1203RBA	OEM
CR20-1203RBAC	OEM
CR20-1203RBC	OEM
CR20-1203RC	OEM
CR20-1203RCA	OEM
CR20-1203RCAC	OEM
CR20-1203RCC	OEM
CR20-1303RB	OEM
CR20-1303RBA	OEM
CR20-1303RBAC	OEM
CR20-1303RBC	OEM
CR20-1303RC	OEM
CR20-1303RCA	OEM
CR20-1303RCAC	OEM
CR20-1303RCC	OEM
CR20-1403RB	OEM
CR20-1403RBA	OEM
CR20-1403RBAC	OEM
CR20-1403RBC	OEM
CR20-1403RC	OEM
CR20-1403RCA	OEM
CR20-1403RCAC	OEM
CR20-1403RCC	OEM
CR20-1503RB	OEM
CR20-1503RBA	OEM
CR20-1503RBAC	OEM
CR20-1503RBC	OEM
CR20-1503RC	OEM
CR20-1503RCA	OEM
CR20-1503RCAC	OEM
CR20-1503RCC	OEM
CR20A-1	2519
CR20A-2	0226
CR20A2	OEM
CR20A-4	1650
CR20A4	OEM
CR20A-6	2647
CR20A6	OEM
CR20A-8	2647
CR20A8	OEM
CR20A-10	0752
CR20A10	OEM
CR20A-12	0725
CR20A12	OEM
CR20A-14	0750
CR20A-16	0750
CR20A16	OEM
CR20A20	OEM
CR20AY-2	0804
CR20AY-4	0804
CR20AY-6	0804
CR20AY-8	0804
CR20AY-10	0804
CR20AY-12	0804
CR20EY2	OEM
CR20EY4	OEM
CR20EY6	OEM
CR20EY8	OEM
CR20EY10	OEM
CR20EY12	OEM
CR20EY16	OEM
CR20F2	OEM
CR20F4	OEM
CR20F6	OEM
CR20F8	OEM
CR20F10	OEM
CR20F12	OEM
CR20F16	OEM
CR20F20	OEM
CR20F24	OEM
CR20M01JY	OEM
CR20M01JY1	OEM
CR20M02JY	OEM
CR20M02JY1	OEM
CR20M03JY	OEM
CR20M03JY1	OEM
CR20M04JY	OEM
CR20M04JY1	OEM
CR20M05JY	OEM
CR20M05JY1	OEM
CR20M06JY	OEM
CR20M06JY1	OEM
CR20M08JY	OEM
CR20M08JY1	OEM
CR20M10JY	OEM
CR20M10JY1	OEM
CR20M11JY	OEM
CR20M11JY1	OEM
CR20M12JY	OEM
CR20M12JY1	OEM
CR20M13JY	OEM
CR20M13JY1	OEM
CR20M14JY	OEM
CR20M14JY1	OEM
CR20MX5JY	OEM
CR20MX5JY1	OEM
CR20U01JY	OEM
CR20U01JY1	OEM
CR20U02JY	OEM
CR20U02JY1	OEM
CR20U03JY	OEM
CR20U03JY1	OEM
CR20U04JY	OEM
CR20U04JY1	OEM
CR20U05JY	OEM
CR20U05JY1	OEM
CR20U06JY	OEM
CR20U06JY1	OEM
CR20U08JY	OEM
CR20U08JY1	OEM
CR20U10JY	OEM
CR20U10JY1	OEM
CR20U11JY	OEM
CR20U11JY1	OEM
CR20U12JY	OEM
CR20U12JY1	OEM
CR20U13JY	OEM
CR20U13JY1	OEM
CR20U14JY	OEM
CR20U14JY1	OEM
CR20UX5JY	OEM
CR20UX5JY1	OEM
CR24-051B	OEM
CR24-051BC	OEM
CR24-051C	OEM
CR24-051CC	OEM
CR24-052AB	OEM
CR24-052BB	OEM
CR24-052CB	OEM
CR24-101B	OEM
CR24-101BC	OEM
CR24-101C	OEM
CR24-101CC	OEM
CR24-201B	OEM
CR24-201BC	OEM
CR24-201C	OEM
CR24-201CC	OEM
CR24-301B	OEM
CR24-301BC	OEM
CR24-301C	OEM
CR24-301CC	OEM
CR24-401B	OEM
CR24-401BC	OEM
CR24-401C	OEM
CR24-401CC	OEM
CR24-403RB	OEM
CR24-403RBA	OEM
CR24-501B	OEM
CR24-501BC	OEM
CR24-501C	OEM
CR24-501CC	OEM
CR24-503RB	OEM
CR24-503RBA	OEM
CR24-601B	OEM
CR24-601BC	OEM
CR24-601C	OEM
CR24-601CC	OEM
CR24-603RB	OEM
CR24-603RBA	OEM
CR24-703RB	OEM
CR24-703RBA	OEM
CR24-803B	OEM
CR24-803RB	OEM
CR24-803RBA	OEM
CR24-903B	OEM
CR24-903RB	OEM
CR24-903RBA	OEM
CR24-903RBAC	OEM
CR24-903RBC	OEM
CR24-903RC	OEM
CR24-903RCA	OEM
CR24-903RCAC	OEM
CR24-903RCC	OEM
CR24-1003B	OEM
CR24-1003RB	OEM
CR24-1003RBA	OEM
CR24-1003RBAC	OEM
CR24-1003RBC	OEM
CR24-1003RC	OEM
CR24-1003RCA	OEM
CR24-1003RCAC	OEM
CR24-1003RCC	OEM
CR24-1103RB	OEM
CR24-1103RBA	OEM
CR24-1103RBAC	OEM
CR24-1103RBC	OEM
CR24-1103RC	OEM
CR24-1103RCA	OEM
CR24-1103RCAC	OEM
CR24-1103RCC	OEM
CR24-1203B	OEM
CR24-1203RB	OEM
CR24-1203RBA	OEM
CR24-1203RBAC	OEM
CR24-1203RBC	OEM
CR24-1203RC	OEM
CR24-1203RCA	OEM
CR24-1203RCAC	OEM
CR24-1203RCC	OEM
CR24-1303RB	OEM
CR24-1303RBA	OEM
CR24-1303RBAC	OEM
CR24-1303RBC	OEM
CR24-1303RC	OEM
CR24-1303RCA	OEM
CR24-1303RCAC	OEM
CR24-1303RCC	OEM
CR24-1403RB	OEM
CR24-1403RBA	OEM
CR24-1403RBAC	OEM
CR24-1403RBC	OEM
CR24-1403RC	OEM
CR24-1403RCA	OEM
CR24-1403RCAC	OEM
CR24-1403RCC	OEM
CR24-1503RB	OEM
CR24-1503RBA	OEM
CR24-1503RBAC	OEM
CR24-1503RBC	OEM
CR24-1503RC	OEM
CR24-1503RCA	OEM
CR24-1503RCAC	OEM
CR24-1503RCC	OEM
CR24M01JY	OEM
CR24M01JY1	OEM
CR24M02JY	OEM
CR24M02JY1	OEM
CR24M03JY	OEM
CR24M03JY1	OEM
CR24M04JY	OEM
CR24M04JY1	OEM
CR24M05JY	OEM
CR24M05JY1	OEM
CR24M06JY	OEM
CR24M06JY1	OEM
CR24M08JY	OEM
CR24M08JY1	OEM
CR24M10JY	OEM
CR24M10JY1	OEM
CR24M11JY	OEM
CR24M11JY1	OEM
CR24M12JY	OEM
CR24M12JY1	OEM
CR24M13JY	OEM
CR24M13JY1	OEM
CR24M14JY	OEM
CR24M14JY1	OEM
CR24MX5JY	OEM
CR24MX5JY1	OEM
CR24U01JY	OEM
CR24U01JY1	OEM
CR24U02JY	OEM
CR24U02JY1	OEM
CR24U03JY	OEM
CR24U03JY1	OEM
CR24U04JY	OEM
CR24U04JY1	OEM
CR24U05JY	OEM
CR24U05JY1	OEM
CR24U06JY	OEM
CR24U06JY1	OEM
CR24U08JY	OEM
CR24U08JY1	OEM
CR24U10JY	OEM
CR24U10JY1	OEM
CR24U11JY	OEM
CR24U11JY1	OEM
CR24U12JY	OEM
CR24U12JY1	OEM
CR24U13JY	OEM
CR24U13JY1	OEM
CR24U14JY	OEM
CR24U14JY1	OEM
CR24UX5JY	OEM
CR24UX5JY1	OEM
CR25-051B	OEM
CR25-101B	OEM
CR25-201B	OEM
CR25-301B	OEM
CR25-401B	OEM
CR25-501B	OEM
CR25-601B	OEM
CR25-701B	OEM
CR25-801B	OEM
CR27-051B	OEM
CR27-051BA	OEM
CR27-051C	OEM
CR27-051CA	OEM
CR27-101B	OEM
CR27-101BA	OEM
CR27-101C	OEM
CR27-101CA	OEM
CR27-201B	OEM
CR27-201BA	OEM
CR27-201C	OEM
CR27-201CA	OEM
CR27-301B	OEM
CR27-301BA	OEM
CR27-301C	OEM
CR27-301CA	OEM
CR27-401B	OEM
CR27-401BA	OEM
CR27-401C	OEM
CR27-401CA	OEM
CR27-501B	OEM
CR27-501BA	OEM
CR27-501C	OEM
CR27-501CA	OEM
CR27-601B	OEM
CR27-601BA	OEM
CR27-601C	OEM
CR27-601CA	OEM
CR27-903RB	OEM
CR27-903RBA	OEM
CR27-903RC	OEM
CR27-903RCA	OEM
CR27-1003RB	OEM
CR27-1003RBA	OEM
CR27-1003RC	OEM
CR27-1003RCA	OEM
CR27-1103RB	OEM
CR27-1103RBA	OEM
CR27-1103RC	OEM
CR27-1103RCA	OEM
CR27-1203RB	OEM
CR27-1203RBA	OEM
CR27-1203RC	OEM
CR27-1203RCA	OEM
CR27-1303RB	OEM
CR27-1303RBA	OEM
CR27-1303RC	OEM
CR27-1303RCA	OEM
CR27-1403RB	OEM
CR27-1403RBA	OEM
CR27-1403RC	OEM
CR27-1403RCA	OEM
CR27-1503RB	OEM
CR27-1503RBA	OEM
CR27-1503RC	OEM
CR27-1503RCA	OEM
CR28-051B	OEM
CR28-051BA	OEM
CR28-051C	OEM
CR28-051CA	OEM
CR28-101B	OEM
CR28-101BA	OEM
CR28-101C	OEM
CR28-101CA	OEM
CR28-201B	OEM
CR28-201BA	OEM
CR28-201C	OEM
CR28-201CA	OEM
CR28-301B	OEM
CR28-301BA	OEM
CR28-301C	OEM
CR28-301CA	OEM
CR28-401B	OEM
CR28-401BA	OEM
CR28-401C	OEM
CR28-401CA	OEM
CR28-501B	OEM
CR28-501BA	OEM
CR28-501C	OEM
CR28-501CA	OEM
CR28-601B	OEM
CR28-601BA	OEM
CR28-601C	OEM
CR28-601CA	OEM
CR28-903RB	OEM
CR28-903RBA	OEM
CR28-903RC	OEM
CR28-903RCA	OEM
CR28-1003RB	OEM
CR28-1003RBA	OEM
CR28-1003RC	OEM
CR28-1003RCA	OEM
CR28-1103RB	OEM
CR28-1103RBA	OEM
CR28-1103RC	OEM
CR28-1103RCA	OEM
CR28-1203RB	OEM
CR28-1203RBA	OEM
CR28-1203RC	OEM
CR28-1203RCA	OEM
CR28-1303RB	OEM
CR28-1303RBA	OEM
CR28-1303RC	OEM
CR28-1303RCA	OEM
CR28-1403RB	OEM
CR28-1403RBA	OEM
CR28-1403RC	OEM
CR28-1403RCA	OEM
CR28-1503RB	OEM
CR28-1503RBA	OEM
CR28-1503RC	OEM
CR28-1503RCA	OEM
CR30-01	OEM
CR30-02	OEM
CR30-04	OEM
CR30-06	OEM
CR30-08	OEM
CR30-053A	OEM
CR30-10	OEM
CR30-103A	OEM
CR30-103AA	OEM
CR30-104A	OEM
CR30-104LA	OEM
CR30-203A	OEM
CR30-203AA	OEM
CR30-204A	OEM
CR30-204LA	OEM
CR30-303A	OEM
CR30-303AA	OEM
CR30-304A	OEM
CR30-304LA	OEM
CR30-403A	OEM
CR30-403AA	OEM
CR30-404A	OEM
CR30-503A	OEM
CR30-503AA	OEM
CR30-504A	OEM
CR30-504LA	OEM
CR30-603A	OEM
CR30-603AA	OEM
CR30-604A	OEM
CR30-604LA	OEM
CR30-703A	OEM
CR30-703AA	OEM
CR30-704LA	OEM
CR30-803A	OEM
CR30-803AA	OEM
CR30-804LA	OEM
CR30-903A	OEM
CR30-903AA	OEM
CR30-904LA	OEM
CR30-1003A	OEM
CR30-1003AA	OEM
CR30-1004LA	OEM
CR30-1103A	OEM
CR30-1103AA	OEM
CR30-1104LA	OEM
CR30-1203A	OEM
CR30-1203AA	OEM
CR30-1204LA	OEM
CR30-1303A	OEM
CR30-1303AA	OEM
CR30-1403A	OEM
CR30-1403AA	OEM
CR31-103A	0468
CR31-104LA	OEM
CR31-203A	0738
CR31-204LA	OEM
CR31-303A	0468
CR31-304LA	OEM
CR31-403A	0468
CR31-403AA	OEM
CR31-404LA	OEM
CR31-503A	0468
CR31-503AA	0468
CR31-504LA	OEM
CR31-603A	0468
CR31-603AA	0468
CR31-604LA	OEM
CR31-703A	0468
CR31-703AA	0468
CR31-704LA	OEM
CR31-803A	0468
CR31-803AA	0468
CR31-804LA	OEM
CR31-903A	0468
CR31-903AA	0468
CR31-904LA	OEM
CR31-1003A	0468
CR31-1003AA	0468
CR31-1004LA	OEM
CR31-1103A	0468
CR31-1103AA	0468
CR31-1104LA	OEM
CR31-1203A	0468
CR31-1203AA	0468
CR31-1204LA	OEM
CR31-1303A	0468
CR31-1303AA	0468
CR32M01JY	OEM
CR32M01JY1	OEM
CR32M02JY	OEM
CR32M02JY1	OEM
CR32M03JY	OEM
CR32M03JY1	OEM
CR32M04JY	OEM
CR32M04JY1	OEM
CR32M05JY	OEM
CR32M05JY1	OEM
CR32M06JY	OEM
CR32M06JY1	OEM
CR32M08JY	OEM
CR32M08JY1	OEM
CR32M10JY	OEM
CR32M10JY1	OEM
CR32M11JY	OEM
CR32M11JY1	OEM
CR32M12JY	OEM
CR32M12JY1	OEM
CR32M13JY	OEM
CR32M13JY1	OEM
CR32M14JY	OEM
CR32M14JY1	OEM
CR32MX5JY	OEM
CR32MX5JY1	OEM
CR32U01JY	OEM
CR32U01JY1	OEM
CR32U02JY	OEM
CR32U02JY1	OEM
CR32U03JY	OEM
CR32U03JY1	OEM
CR32U04JY	OEM
CR32U04JY1	OEM
CR32U05JY	OEM
CR32U05JY1	OEM
CR32U06JY	OEM
CR32U06JY1	OEM
CR32U08JY	OEM
CR32U08JY1	OEM
CR32U10JY	OEM
CR32U10JY1	OEM
CR32U11JY	OEM
CR32U11JY1	OEM
CR32U12JY	OEM
CR32U12JY1	OEM
CR32U13JY	OEM
CR32U13JY1	OEM
CR32U14JY	OEM
CR32U14JY1	OEM
CR32UX5JY	OEM
CR32UX5JY1	OEM
CR33J03JY1	OEM
CR33M01JY	OEM
CR33M01JY1	OEM
CR33M02JY	OEM
CR33M02JY1	OEM
CR33M03JY	OEM

If replacement code is OEM, contact original manufacturer for replacement.

DEVICE TYPE	REPL CODE
CR33M03JY1	OEM
CR33M04JY	OEM
CR33M04JY1	OEM
CR33M05JY	OEM
CR33M05JY1	OEM
CR33M06JY	OEM
CR33M06JY1	OEM
CR33M08JY	OEM
CR33M08JY1	OEM
CR33M10JY	OEM
CR33M10JY1	OEM
CR33M11JY	OEM
CR33M11JY1	OEM
CR33M12JY	OEM
CR33M12JY1	OEM
CR33M13JY	OEM
CR33M13JY1	OEM
CR33M14JY	OEM
CR33M14JY1	OEM
CR33MX5JY	OEM
CR33MX5JY1	OEM
CR33U01JY	OEM
CR33U01JY1	OEM
CR33U02JY	OEM
CR33U02JY1	OEM
CR33U03JY	OEM
CR33U04JY	OEM
CR33U04JY1	OEM
CR33U05JY	OEM
CR33U05JY1	OEM
CR33U06JY	OEM
CR33U06JY1	OEM
CR33U08JY	OEM
CR33U08JY1	OEM
CR33U10JY	OEM
CR33U10JY1	OEM
CR33U11JY	OEM
CR33U11JY1	OEM
CR33U12JY	OEM
CR33U12JY1	OEM
CR33U13JY	OEM
CR33U13JY1	OEM
CR33U14JY	OEM
CR33U14JY1	OEM
CR33UX5JY	OEM
CR33UX5JY1	OEM
CR40A4	OEM
CR40A6	OEM
CR40A8	OEM
CR40A10	OEM
CR40A12	OEM
CR40A16	OEM
CR40A20	OEM
CR45U01JY	OEM
CR45U01JY1	OEM
CR45U02JY	OEM
CR45U02JY1	OEM
CR45U03JY	OEM
CR45U03JY1	OEM
CR45U04JY	OEM
CR45U04JY1	OEM
CR45U05JY	OEM
CR45U05JY1	OEM
CR45U06JY	OEM
CR45U06JY1	OEM
CR45U08JY	OEM
CR45U08JY1	OEM
CR45U10JY	OEM
CR45U10JY1	OEM
CR45U11JY	OEM
CR45U11JY1	OEM
CR45U12JY	OEM
CR45U12JY1	OEM
CR45U13JY	OEM
CR45U13JY1	OEM
CR45U14JY	OEM
CR45U14JY1	OEM
CR45UX5JY	OEM
CR45UX5JY1	OEM
CR46U01JY	OEM
CR46U01JY1	OEM
CR46U02JY	OEM
CR46U02JY1	OEM
CR46U03JY	OEM
CR46U03JY1	OEM
CR46U04JY	OEM
CR46U04JY1	OEM
CR46U05JY	OEM
CR46U05JY1	OEM
CR46U06JY	OEM
CR46U06JY1	OEM
CR46UX5JY	OEM
CR46UX5JY1	OEM
CR50	OEM
CR50-01	OEM
CR50-02	OEM
CR50-04	OEM
CR50-06	OEM
CR50-08	OEM
CR50-053A	OEM
CR50-10	OEM
CR50-104A	OEM
CR50-204A	OEM
CR50-304A	OEM
CR50-404A	OEM
CR50-504A	OEM
CR50-604A	OEM
CR50-1004AA	OEM
CR50-1104AA	OEM

DEVICE TYPE	REPL CODE
CR50-1204AA	OEM
CR51-103A	0468
CR51-203A	0468
CR51-303A	0468
CR51-503A	0468
CR51-503AA	0468
CR51-603A	0468
CR51-603AA	0468
CR51-703A	0468
CR51-703AA	0468
CR51-803A	0468
CR51-803AA	0468
CR51-804AA	OEM
CR51-804BA	OEM
CR51-804CA	OEM
CR51-903A	0468
CR51-903AA	0468
CR51-904AA	OEM
CR51-904BA	OEM
CR51-904CA	OEM
CR51-1003A	0468
CR51-1003AA	0468
CR51-1004AA	OEM
CR51-1103A	0468
CR51-1103AA	0468
CR51-1104AA	OEM
CR51-1203A	0468
CR51-1203AA	0468
CR51-1204AA	OEM
CR51-1303A	0468
CR51-1303AA	0468
CR51J06JY	OEM
CR51M01JY	OEM
CR51M01JY1	OEM
CR51M02JY	OEM
CR51M02JY1	OEM
CR51M03JY	OEM
CR51M03JY1	OEM
CR51M04JY	OEM
CR51M04JY1	OEM
CR51M05JY	OEM
CR51M05JY1	OEM
CR51M06JY1	OEM
CR51M08JY	OEM
CR51M08JY1	OEM
CR51M10JY	OEM
CR51M10JY1	OEM
CR51M11JY	OEM
CR51M11JY1	OEM
CR51M12JY	OEM
CR51M12JY1	OEM
CR51M13JY	OEM
CR51M13JY1	OEM
CR51M14JY	OEM
CR51M14JY1	OEM
CR51MX5JY	OEM
CR51MX5JY1	OEM
CR51U01JY	OEM
CR51U01JY1	OEM
CR51U02JY	OEM
CR51U02JY1	OEM
CR51U03JY	OEM
CR51U03JY1	OEM
CR51U04JY	OEM
CR51U04JY1	OEM
CR51U05JY	OEM
CR51U05JY1	OEM
CR51U06JY	OEM
CR51U06JY1	OEM
CR51U08JY	OEM
CR51U08JY1	OEM
CR51U10JY	OEM
CR51U10JY1	OEM
CR51U11JY	OEM
CR51U11JY1	OEM
CR51U12JY	OEM
CR51U12JY1	OEM
CR51U13JY	OEM
CR51U13JY1	OEM
CR51U14JY	OEM
CR51U14JY1	OEM
CR51UX5JY	OEM
CR51UX5JY1	OEM
CR70-051A	0468
CR70-051AA	0468
CR70-051C	0468
CR70-051CA	0468
CR70-101A	0468
CR70-101AA	0468
CR70-101C	0468
CR70-101CA	0468
CR70-201A	0468
CR70-201AA	0468
CR70-201C	0468
CR70-201CA	0468
CR70-301A	0468
CR70-301AA	0468
CR70-301C	0468
CR70-301CA	0468
CR70-401A	0468
CR70-401AA	0468
CR70-401C	0468
CR70-401CA	0468
CR70-501A	0468
CR70-501AA	0468
CR70-501C	0468
CR70-501CA	0468
CR70-601A	0468
CR70-601AA	0468

DEVICE TYPE	REPL CODE
CR70-601C	0468
CR70-601CA	0468
CR70-701A	0468
CR70-701AA	0468
CR70-701C	0468
CR70-701CA	0468
CR70-801A	0468
CR70-801AA	0468
CR70-801C	0468
CR70-801CA	0468
CR70-901A	0468
CR70-901AA	0468
CR70-901C	0468
CR70-901CA	0468
CR70-1001A	0468
CR70-1001AA	0468
CR70-1001C	0468
CR70-1001CA	0468
CR70-1101A	0468
CR70-1101AA	0468
CR70-1101C	0468
CR70-1101CA	0468
CR70-1201A	0468
CR70-1201AA	0468
CR70-1201C	0468
CR70-1201CA	0468
CR70-1301A	0468
CR70-1301AA	0468
CR70-1301C	0468
CR70-1301CA	0468
CR70B2	OEM
CR70B4	OEM
CR70B6	OEM
CR70B8	OEM
CR70B10	OEM
CR70B12	OEM
CR70B14	OEM
CR70B16	OEM
CR70C4	0608
CR70C6	0610
CR70C8	0610
CR70C10	0610
CR70C12	0610
CR70C14	OEM
CR70C16	0468
CR70C20	0468
CR70C24	0468
CR70CX4	OEM
CR70CX6	OEM
CR70CX8	OEM
CR70CX10	OEM
CR70CX12	OEM
CR70CX16	OEM
CR70CX20	OEM
CR70CY4	OEM
CR70CY6	OEM
CR70CY8	OEM
CR70CY10	OEM
CR70CY12	OEM
CR70CY16	OEM
CR71M01JY	OEM
CR71M01JY1	OEM
CR71M02JY	OEM
CR71M02JY1	OEM
CR71M03JY	OEM
CR71M03JY1	OEM
CR71M04JY	OEM
CR71M04JY1	OEM
CR71M05JY	OEM
CR71M05JY1	OEM
CR71M06JY	OEM
CR71M06JY1	OEM
CR71M08JY	OEM
CR71M08JY1	OEM
CR71M10JY	OEM
CR71M10JY1	OEM
CR71M11JY	OEM
CR71M11JY1	OEM
CR71M12JY	OEM
CR71M12JY1	OEM
CR71M13JY	OEM
CR71M13JY1	OEM
CR71M14JY	OEM
CR71M14JY1	OEM
CR71MX5JY	OEM
CR71MX5JY1	OEM
CR71U01JY	OEM
CR71U01JY1	OEM
CR71U02JY	OEM
CR71U02JY1	OEM
CR71U03JY	OEM
CR71U03JY1	OEM
CR71U04JY	OEM
CR71U04JY1	OEM
CR71U05JY	OEM
CR71U05JY1	OEM
CR71U06JY	OEM
CR71U06JY1	OEM
CR71U08JY	OEM
CR71U08JY1	OEM
CR71U10JY	OEM
CR71U10JY1	OEM
CR71U11JY	OEM
CR71U12JY	OEM
CR71U12JY1	OEM
CR71U13JY	OEM
CR71U13JY1	OEM
CR71U14JY	OEM
CR71U14JY1	OEM

DEVICE TYPE	REPL CODE
CR71UX5JY	OEM
CR71UX5JY1	OEM
CR80.021	OEM
CR80.051	OEM
CR80.071	OEM
CR80.101	OEM
CR80.151	OEM
CR80.201	OEM
CR80.251	OEM
CR80.301	OEM
CR81F01JY	OEM
CR81F02JY	OEM
CR81F03JY	OEM
CR81F04JY	OEM
CR81F05JY	OEM
CR81F06JY	OEM
CR81F08JY	OEM
CR81F10JY	OEM
CR81F11JY	OEM
CR81F12JY	OEM
CR81F13JY	OEM
CR81FX5JY	OEM
CR81M01JY	OEM
CR81M02JY	OEM
CR81M02JY1	OEM
CR81M03JY	OEM
CR81M04JY	OEM
CR81M05JY	OEM
CR81M06JY	OEM
CR81M08JY	OEM
CR81M10JY	OEM
CR81M11JY	OEM
CR81M12JY	OEM
CR81M13JY	OEM
CR81M14JY	OEM
CR81MX5JY	OEM
CR81MX5JY1	OEM
CR81U01JY	OEM
CR81U02JY	OEM
CR81U03JY	OEM
CR81U04JY	OEM
CR81U05JY	OEM
CR81U06JY	OEM
CR81U08JY	OEM
CR81U10JY	OEM
CR81U11JY	OEM
CR81U12JY	OEM
CR81U13JY	OEM
CR81U14JY	OEM
CR81UX5JY	OEM
CR100	OEM
CR100-01	OEM
CR100-02	OEM
CR100-04	OEM
CR100-06	OEM
CR100-08	OEM
CR100-10	OEM
CR100AL4	OEM
CR100AL6	OEM
CR100AL8	OEM
CR100AL10	OEM
CR100AL12	OEM
CR100AL16	OEM
CR100AL20	OEM
CR100AL24	OEM
CR100AX4	OEM
CR100AX6	OEM
CR100AX8	OEM
CR100AX10	OEM
CR100AX12	OEM
CR100AX16	OEM
CR100AX20	OEM
CR100AY4	OEM
CR100AY6	OEM
CR100AY8	OEM
CR100AY10	OEM
CR100AY12	OEM
CR100AY16	OEM
CR100CX24	OEM
CR100CX28	OEM
CR100CX30	OEM
CR100CX32	OEM
CR100CX34	OEM
CR100CX36	OEM
CR100DX24	OEM
CR100DX28	OEM
CR100DX32	OEM
CR100DX34	OEM
CR100DX36	OEM
CR100.021	OEM
CR100.051	OEM
CR100.071	OEM
CR100.101	OEM
CR100.151	OEM
CR100.201	OEM
CR100.251	OEM
CR100.301	OEM
CR101F01JY	OEM
CR101F02JY	OEM
CR101F03JY	OEM
CR101F04JY	OEM
CR101F05JY	OEM
CR101F06JY	OEM
CR101F08JY	OEM
CR101F10JY	OEM
CR101F11JY	OEM
CR101F12JY	OEM
CR101F13JY	OEM
CR101F14JY	OEM

DEVICE TYPE	REPL CODE
CR101FX5JY	OEM
CR101M01JY	OEM
CR101M02JY	OEM
CR101M03JY	OEM
CR101M04JY	OEM
CR101M05JY	OEM
CR101M06JY	OEM
CR101M08JY	OEM
CR101M10JY	OEM
CR101M11JY	OEM
CR101M12JY	OEM
CR101M13JY	OEM
CR101M14JY	OEM
CR101MX5JY	OEM
CR101U01JY	OEM
CR101U02JY	OEM
CR101U03JY	OEM
CR101U04JY	OEM
CR101U05JY	OEM
CR101U06JY	OEM
CR101U08JY	OEM
CR101U10JY	OEM
CR101U11JY	OEM
CR101U12JY	OEM
CR101U13JY	OEM
CR101U14JY	OEM
CR101UX5JY	OEM
CR103	0774
CR103A	0901
CR103B	0058
CR103C	0407
CR103D	0407
CR103E	1683
CR103M	1683
CR103N	OEM
CR103S	OEM
CR103Y	1137
CR103YY	0344
CR104	1303
CR105A	2412
CR105B	2412
CR105C	6390
CR105D	6390
CR105E	6244
CR105M	6244
CR110	OEM
CR111F01JY	OEM
CR111F02JY	OEM
CR111F03JY	OEM
CR111F04JY	OEM
CR111F05JY	OEM
CR111F06JY	OEM
CR111F08JY	OEM
CR111F10JY	OEM
CR111F11JY	OEM
CR111F12JY	OEM
CR111F13JY	OEM
CR111F14JY	OEM
CR111FX5JY	OEM
CR111M01JY	OEM
CR111M02JY	OEM
CR111M03JY	OEM
CR111M04JY	OEM
CR111M05JY	OEM
CR111M06JY	OEM
CR111M08JY	OEM
CR111M10JY	OEM
CR111M11JY	OEM
CR111M12JY	OEM
CR111M13JY	OEM
CR111M14JY	OEM
CR111MX5JY	OEM
CR111U01JY	OEM
CR111U02JY	OEM
CR111U03JY	OEM
CR111U04JY	OEM
CR111U05JY	OEM
CR111U06JY	OEM
CR111U08JY	OEM
CR111U10JY	OEM
CR111U11JY	OEM
CR111U12JY	OEM
CR111U13JY	OEM
CR111U14JY	OEM
CR111UX5JY	OEM
CR120	OEM
CR121A	OEM
CR121B	OEM
CR121D	OEM
CR121F01JY	OEM
CR121F02JY	OEM
CR121F03JY	OEM
CR121F04JY	OEM
CR121F05JY	OEM
CR121F06JY	OEM
CR121F08JY	OEM
CR121F10JY	OEM
CR121F11JY	OEM
CR121F12JY	OEM
CR121F13JY	OEM
CR121F14JY	OEM
CR121FX5JY	OEM
CR121M	OEM
CR121M01JY	OEM
CR121M01JY1	OEM
CR121M02JY	OEM
CR121M03JY	OEM
CR121M03JY1	OEM
CR121M04JY	OEM

DEVICE TYPE	REPL CODE
CR121M04JY1	OEM
CR121M05JY	OEM
CR121M05JY1	OEM
CR121M06JY	OEM
CR121M06JY1	OEM
CR121M08JY	OEM
CR121M08JY1	OEM
CR121M10JY	OEM
CR121M10JY1	OEM
CR121M11JY	OEM
CR121M12JY	OEM
CR121M13JY	OEM
CR121M14JY	OEM
CR121MX5JY	OEM
CR121MX5JY1	OEM
CR121N	OEM
CR121U01JY	OEM
CR121U02JY	OEM
CR121U03JY	OEM
CR121U04JY	OEM
CR121U05JY	OEM
CR121U06JY	OEM
CR121U08JY	OEM
CR121U10JY	OEM
CR121U11JY	OEM
CR121U12JY	OEM
CR121U13JY	OEM
CR121U14JY	OEM
CR121UX5JY	OEM
CR130	OEM
CR135F01JY	OEM
CR135F02JY	OEM
CR135F03JY	OEM
CR135F04JY	OEM
CR135F05JY	OEM
CR135F06JY	OEM
CR135F08JY	OEM
CR135F10JY	OEM
CR135FX5JY	OEM
CR135U01JY	OEM
CR135U01JY2	OEM
CR135U02JY	OEM
CR135U02JY2	OEM
CR135U03JY	OEM
CR135U03JY2	OEM
CR135U04JY	OEM
CR135U04JY2	OEM
CR135U05JY	OEM
CR135U05JY2	OEM
CR135U06JY	OEM
CR135U06JY2	OEM
CR135U08JY	OEM
CR135U10JY	OEM
CR135U10JY2	OEM
CR135U11JY	OEM
CR135U11JY2	OEM
CR135U12JY	OEM
CR135U12JY2	OEM
CR135U13JY	OEM
CR135U13JY2	OEM
CR135U14JY	OEM
CR135U14JY2	OEM
CR135UX5JY	OEM
CR135UX5JY2	OEM
CR140	OEM
CR140F01JY	OEM
CR140F02JY	OEM
CR140F03JY	OEM
CR140F04JY	OEM
CR140F05JY	OEM
CR140F06JY	OEM
CR140F08JY	OEM
CR140F10JY	OEM
CR140F11JY	OEM
CR140F12JY	OEM
CR140F13JY	OEM
CR140F14JY	OEM
CR140FX5JY	OEM
CR140U01JY	OEM
CR140U01JY2	OEM
CR140U02JY	OEM
CR140U02JY2	OEM
CR140U03JY	OEM
CR140U03JY2	OEM
CR140U04JY	OEM
CR140U04JY2	OEM
CR140U05JY	OEM
CR140U05JY2	OEM
CR140U06JY	OEM
CR140U06JY2	OEM
CR140U08JY	OEM
CR140U08JY2	OEM
CR140U10JY	OEM
CR140U11JY	OEM
CR140U11JY2	OEM
CR140U12JY	OEM
CR140U12JY2	OEM
CR140U13JY	OEM
CR140U13JY2	OEM
CR140U14JY	OEM
CR140U14JY2	OEM
CR140UX5JY	OEM
CR140UX5JY2	OEM
CR150	OEM
CR150-051A	OEM
CR150-052A3	OEM
CR150-052A4	OEM
CR150-052A5	OEM
CR150-052A6	OEM

DEVICE TYPE	REPL CODE
CR150-101A	OEM
CR150-102A3	OEM
CR150-102A4	OEM
CR150-102A5	OEM
CR150-102A6	OEM
CR150-102B	OEM
CR150-201A	OEM
CR150-202A3	OEM
CR150-202A4	OEM
CR150-202A5	OEM
CR150-202A6	OEM
CR150-202B	OEM
CR150-301A	OEM
CR150-302A3	OEM
CR150-302A4	OEM
CR150-302A5	OEM
CR150-302A6	OEM
CR150-302B	OEM
CR150-401A	OEM
CR150-402A3	OEM
CR150-402A4	OEM
CR150-402A5	OEM
CR150-402A6	OEM
CR150-402B	OEM
CR150-501A	OEM
CR150-502A3	OEM
CR150-502A4	OEM
CR150-502A5	OEM
CR150-502A6	OEM
CR150-502B	OEM
CR150-601A	OEM
CR150-601A3	OEM
CR150-602A3	OEM
CR150-602A4	OEM
CR150-602A5	OEM
CR150-602A6	OEM
CR150-602B	OEM
CR150-701A	OEM
CR150-702A3	OEM
CR150-702A4	OEM
CR150-702A5	OEM
CR150-702A6	OEM
CR150-702B	OEM
CR150-801A	OEM
CR150-801A3	OEM
CR150-802A3	OEM
CR150-802A4	OEM
CR150-802A5	OEM
CR150-802A6	OEM
CR150-802B	OEM
CR150-901A	OEM
CR150-902B	OEM
CR150-1001A	OEM
CR150-1001A3	OEM
CR150-1002B	OEM
CR150-1101A	OEM
CR150-1101A3	OEM
CR150-1102B	OEM
CR150-1201A3	OEM
CR150-1202B	OEM
CR150-1301A3	OEM
CR150B2	OEM
CR150B4	OEM
CR150B6	OEM
CR150B8	OEM
CR150B10	OEM
CR150B12	OEM
CR150B16	OEM
CR150B20	OEM
CR150B24	OEM
CR150BX2	OEM
CR150BX4	OEM
CR150BX6	OEM
CR150BX8	OEM
CR150BX10	OEM
CR150BX12	OEM
CR150BX14	OEM
CR150BX16	OEM
CR150BX18	OEM
CR150BX20	OEM
CR150BY2	OEM
CR150BY4	OEM
CR150BY6	OEM
CR150BY8	OEM
CR150BY10	OEM
CR150BY12	OEM
CR150DM4	OEM
CR150DM6	OEM
CR150DM8	OEM
CR150DM10	OEM
CR150DM12	OEM
CR150DM16	OEM
CR150DM20	OEM
CR150DM24	OEM
CR150DM28	OEM
CR150DM32	OEM
CR150DM36	OEM
CR150DX4	OEM
CR150DX6	OEM
CR150DX8	OEM
CR150DX10	OEM
CR150DX12	OEM
CR150DX16	OEM
CR150DX20	OEM
CR150DY4	OEM
CR150DY6	OEM
CR150DY8	OEM
CR150DY10	OEM

If replacement code is OEM, contact original manufacturer for replacement.

DEVICE TYPE	REPL CODE
CR150DY12	OEM
CR150F2	OEM
CR150F4	1900
CR150F6	1900
CR150F8	1900
CR150F10	1900
CR150F12	1900
CR150F16	0738
CR150F20	0738
CR150F24	OEM
CR150F28	OEM
CR150F32	OEM
CR150F36	OEM
CR151-103A	OEM
CR151-203A	OEM
CR151-303A	OEM
CR151-403A	OEM
CR151-503A	OEM
CR151-603A	OEM
CR151-703A	OEM
CR151-803A	OEM
CR151-903A	OEM
CR151-1003A	OEM
CR151-1103A	OEM
CR151-1203A	OEM
CR151-1303A	OEM
CR152-103A	OEM
CR152-203A	OEM
CR152-303A	OEM
CR152-403A	OEM
CR152-503A	OEM
CR152-603A	OEM
CR152-703A	OEM
CR152-803A	OEM
CR152-903A	OEM
CR152-1003A	OEM
CR152-1103A	OEM
CR152-1203A	OEM
CR152-1303A	OEM
CR152-1403A	OEM
CR152-1503A	OEM
CR153-103A	OEM
CR153-203A	OEM
CR153-303A	OEM
CR153-403A	OEM
CR153-503A	OEM
CR153-603A	OEM
CR153-703A	OEM
CR153-803A	OEM
CR153-903A	OEM
CR153-1003A	OEM
CR153-1103A	OEM
CR153-1203A	OEM
CR153-1303A	OEM
CR153-1403A	OEM
CR153-1503A	OEM
CR155-604NA	OEM
CR155-704NA	OEM
CR155-804NA	OEM
CR155-904NA	OEM
CR155-1004NA	OEM
CR159-901A3	OEM
CR160	OEM
CR180	OEM
CR200	OEM
CR203A	0901
CR203B	0058
CR203C	0407
CR203D	0407
CR203E	1683
CR203M	1683
CR203N	OEM
CR203S	OEM
CR203Y	1137
CR203YY	0344
CR220	OEM
CR240	OEM
CR250AX2	OEM
CR250AX4	OEM
CR250AX6	OEM
CR250AX8	OEM
CR250AX10	OEM
CR250AX16	OEM
CR250AX20	OEM
CR250AXAX12	OEM
CR250AY2	OEM
CR250AY4	OEM
CR250AY6	OEM
CR250AY8	OEM
CR250AY10	OEM
CR250AY12	OEM
CR250B2	OEM
CR250B6	OEM
CR250B8	OEM
CR250B10	OEM
CR250B12	OEM
CR250B14	OEM
CR250B16	OEM
CR250B18	OEM
CR250B20	OEM
CR250B24	OEM
CR250DP4	OEM
CR250DP6	OEM
CR250DP8	OEM
CR250F2	OEM
CR250F4	OEM
CR250F6	OEM
CR250F8	OEM
CR250F10	OEM

DEVICE TYPE	REPL CODE
CR250F12	OEM
CR250F16	OEM
CR250F20	OEM
CR250F24	OEM
CR250F28	OEM
CR250F32	OEM
CR250F36	OEM
CR250GX2	OEM
CR250GX4	OEM
CR250GX6	OEM
CR250GX8	OEM
CR250GX10	OEM
CR250GX12	OEM
CR250GX16	OEM
CR250GX20	OEM
CR250GY2	OEM
CR250GY4	OEM
CR250GY6	OEM
CR250GY8	OEM
CR250GY10	OEM
CR250GY12	OEM
CR250J2	OEM
CR250J4	1900
CR250J6	1900
CR250J8	1900
CR250J10	1900
CR250J12	1900
CR250J16	0738
CR250J20	0738
CR250J24	0738
CR250J28	OEM
CR250J32	OEM
CR250J36	OEM
CR250JX2	OEM
CR250JX4	OEM
CR250JX6	OEM
CR250JX8	OEM
CR250JX10	OEM
CR250JX12	OEM
CR250JX16	OEM
CR250JX20	OEM
CR250JY2	OEM
CR250JY4	OEM
CR250JY6	OEM
CR250JY8	OEM
CR250JY10	OEM
CR250JY12	OEM
CR250JY16	OEM
CR252-103A	OEM
CR252-203A	OEM
CR252-303A	OEM
CR252-403A	OEM
CR252-503A	OEM
CR252-603A	OEM
CR252-703A	OEM
CR252-803A	OEM
CR252-903A	OEM
CR252-1003A	OEM
CR252-1103A	OEM
CR252-1203A	OEM
CR252-1303A	OEM
CR252-1403A	OEM
CR252-1503A	OEM
CR253-103A	OEM
CR253-203A	OEM
CR253-303A	OEM
CR253-403A	OEM
CR253-503A	OEM
CR253-603A	OEM
CR253-605A	OEM
CR253-703A	OEM
CR253-705A	OEM
CR253-803A	OEM
CR253-805A	OEM
CR253-903A	OEM
CR253-905A	OEM
CR253-1003A	OEM
CR253-1005A	OEM
CR253-1103A	OEM
CR253-1105A	OEM
CR253-1203A	OEM
CR253-1205A	OEM
CR253-1303A	OEM
CR253-1305A	OEM
CR253-1403A	OEM
CR253-1503A	OEM
CR270	OEM
CR300	OEM
CR300A2	OEM
CR300A4	OEM
CR300A6	OEM
CR300A8	OEM
CR300A10	OEM
CR300A12	OEM
CR300A14	OEM
CR300A16	OEM
CR300A18	OEM
CR300A20	OEM
CR300A24	OEM
CR300A28	OEM
CR300A32	OEM
CR300A36	OEM
CR300AW2	OEM
CR300AW4	OEM
CR300AW6	OEM
CR300AW8	OEM
CR300AW10	OEM
CR300AW12	OEM
CR300AW16	OEM

DEVICE TYPE	REPL CODE
CR300AW20	OEM
CR300AW24	OEM
CR300AX2	OEM
CR300AX4	OEM
CR300AX6	OEM
CR300AX8	OEM
CR300AX10	OEM
CR300AX12	OEM
CR300AX16	OEM
CR300AX20	OEM
CR300AY2	OEM
CR300AY4	OEM
CR300AY6	OEM
CR300AY8	OEM
CR300AY10	OEM
CR300AY12	OEM
CR300F2	OEM
CR300F4	OEM
CR300F6	OEM
CR300F8	OEM
CR300F10	OEM
CR300F12	OEM
CR300F16	OEM
CR300F20	OEM
CR300F24	OEM
CR300F28	OEM
CR300F32	OEM
CR300F36	OEM
CR300FX40	OEM
CR300FX50	OEM
CR303-404LA	OEM
CR303A	OEM
CR303B	OEM
CR303C	OEM
CR303D	OEM
CR303E	OEM
CR303M	OEM
CR303N	OEM
CR303S	OEM
CR303Y	OEM
CR303YY	OEM
CR330	OEM
CR350A20	OEM
CR350A24	OEM
CR350A28	OEM
CR350A32	OEM
CR350A36	OEM
CR353	0133
CR360	OEM
CR390	OEM
CR400EL4	OEM
CR400EL6	OEM
CR400EL8	OEM
CR400EL10	OEM
CR400EL12	OEM
CR400EL16	OEM
CR400EL20	OEM
CR400EL24	OEM
CR400EX4	OEM
CR400EX6	OEM
CR400EX8	OEM
CR400EX10	OEM
CR400EX12	OEM
CR400EX16	OEM
CR400EX20	OEM
CR400EX24	OEM
CR400EY4	OEM
CR400EY6	OEM
CR400EY8	OEM
CR400EY10	OEM
CR400EY12	OEM
CR400EY16	OEM
CR400FM28	OEM
CR400FM32	OEM
CR400FM36	OEM
CR430	OEM
CR470	OEM
CR530	OEM
CR604SC0101	OEM
CR604SC0202	OEM
CR604SC0303	OEM
CR604SC0404	OEM
CR604SC0505	OEM
CR604SC0606	OEM
CR604SC0707	OEM
CR604SC0808	OEM
CR604SC0909	OEM
CR604SC1010	OEM
CR604SC1111	OEM
CR604SC1212	OEM
CR604SC1313	OEM
CR604SC1414	OEM
CR604SC1415	OEM
CR605SC0101	OEM
CR605SC0202	OEM
CR605SC0303	OEM
CR605SC0404	OEM
CR605SC0505	OEM
CR605SC0606	OEM
CR605SC0707	OEM
CR605SC0808	OEM
CR605SC0909	OEM
CR605SC1010	OEM
CR605SC1111	OEM
CR605SC1212	OEM
CR605SC1313	OEM
CR605SC1414	OEM
CR605SC1515	OEM
CR606SC0101	OEM

DEVICE TYPE	REPL CODE
CR606SC0202	OEM
CR606SC0303	OEM
CR606SC0404	OEM
CR606SC0505	OEM
CR606SC0606	OEM
CR606SC0707	OEM
CR606SC0808	OEM
CR606SC0909	OEM
CR606SC1010	OEM
CR606SC1111	OEM
CR606SC1212	OEM
CR606SC1313	OEM
CR606SC1414	OEM
CR607SC0101	OEM
CR607SC0202	OEM
CR607SC0303	OEM
CR607SC0404	OEM
CR607SC0505	OEM
CR607SC0606	OEM
CR607SC0707	OEM
CR607SC0808	OEM
CR607SC0909	OEM
CR607SC1010	OEM
CR607SC1111	OEM
CR607SC1212	OEM
CR607SC1313	OEM
CR607SC1414	OEM
CR607SC1515	OEM
CR608SA0101	OEM
CR608SA0202	OEM
CR608SA0303	OEM
CR608SA0404	OEM
CR608SA0505	OEM
CR608SA0606	OEM
CR608SA0707	OEM
CR608SA0808	OEM
CR608SA0909	OEM
CR608SA1010	OEM
CR608SA1111	OEM
CR608SA1212	OEM
CR608SA1313	OEM
CR608SA1414	OEM
CR608SA1515	OEM
CR608SC0101	OEM
CR608SC0202	OEM
CR608SC0303	OEM
CR608SC0404	OEM
CR608SC0505	OEM
CR608SC0606	OEM
CR608SC0707	OEM
CR608SC0808	OEM
CR608SC0909	OEM
CR608SC1010	OEM
CR608SC1111	OEM
CR608SC1212	OEM
CR608SC1313	OEM
CR608SC1414	OEM
CR608SC1515	OEM
CR702SC0101	OEM
CR702SC0202	OEM
CR702SC0303	OEM
CR702SC0404	OEM
CR702SC0505	OEM
CR702SC0606	OEM
CR702SC0707	OEM
CR702SC0808	OEM
CR702SC0909	OEM
CR702SC1010	OEM
CR702SC1111	OEM
CR702SC1212	OEM
CR702SC1313	OEM
CR702SC1414	OEM
CR702SC1515	OEM
CR703SC0101	OEM
CR703SC0202	OEM
CR703SC0303	OEM
CR703SC0404	OEM
CR703SC0505	OEM
CR703SC0606	OEM
CR703SC0707	OEM
CR703SC0808	OEM
CR703SC0909	OEM
CR703SC1010	OEM
CR703SC1111	OEM
CR703SC1212	OEM
CR703SC1313	OEM
CR703SC1414	OEM
CR703SC1515	OEM
CR703SW0101	OEM
CR703SW0202	OEM
CR703SW0303	OEM
CR703SW0404	OEM
CR703SW0505	OEM
CR703SW0606	OEM
CR703SW0707	OEM
CR703SW0808	OEM
CR703SW0909	OEM
CR703SW1010	OEM
CR703SW1111	OEM
CR703SW1212	OEM
CR703SW1313	OEM
CR703SW1414	OEM
CR703SW1515	OEM
CR704SC0101	OEM
CR704SC0202	OEM
CR704SC0303	OEM
CR704SC0404	OEM
CR704SC0505	OEM
CR704SC0606	OEM

DEVICE TYPE	REPL CODE
CR704SC0707	OEM
CR704SC0808	OEM
CR704SC0909	OEM
CR704SC1010	OEM
CR704SC1111	OEM
CR704SC1212	OEM
CR704SC1313	OEM
CR704SC1414	OEM
CR704SC1515	OEM
CR704SC1616	OEM
CR705SC0101	OEM
CR705SC0202	OEM
CR705SC0303	OEM
CR705SC0404	OEM
CR705SC0505	OEM
CR705SC0606	OEM
CR705SC0707	OEM
CR705SC0808	OEM
CR705SC0909	OEM
CR705SC1010	OEM
CR705SC1111	OEM
CR705SC1212	OEM
CR705SC1313	OEM
CR705SC1414	OEM
CR705SC1515	OEM
CR705SC1616	OEM
CR705SC1717	OEM
CR705SC1818	OEM
CR705SW0101	OEM
CR705SW0202	OEM
CR705SW0303	OEM
CR705SW0404	OEM
CR705SW0505	OEM
CR705SW0606	OEM
CR705SW0707	OEM
CR705SW0808	OEM
CR705SW1010	OEM
CR705SW1111	OEM
CR705SW1212	OEM
CR705SW1313	OEM
CR705SW1414	OEM
CR705SW1515	OEM
CR705SW1616	OEM
CR705SW1717	OEM
CR705SW1818	OEM
CR705XW0909	OEM
CR706SC0101	OEM
CR706SC0202	OEM
CR706SC0303	OEM
CR706SC0404	OEM
CR706SC0505	OEM
CR706SC0606	OEM
CR706SC0707	OEM
CR706SC0808	OEM
CR706SC0909	OEM
CR706SC1010	OEM
CR706SC1111	OEM
CR706SC1212	OEM
CR706SC1313	OEM
CR706SC1414	OEM
CR706SC1515	OEM
CR706SC1616	OEM
CR706SC1717	OEM
CR706SC1818	OEM
CR706SC1919	OEM
CR706SC2020	OEM
CR707SC0101	OEM
CR707SC0202	OEM
CR707SC0303	OEM
CR707SC0404	OEM
CR707SC0505	OEM
CR707SC0606	OEM
CR707SC0707	OEM
CR707SC0808	OEM
CR707SC0909	OEM
CR707SC1010	OEM
CR707SC1111	OEM
CR707SC1212	OEM
CR707SC1313	OEM
CR707SC1414	OEM
CR707SC1515	OEM
CR707SC1616	OEM
CR707SC1717	OEM
CR707SC1818	OEM
CR707SC1919	OEM
CR707SC2020	OEM
CR707SC2121	OEM
CR707SC2222	OEM
CR707SC2323	OEM
CR707SW0101	OEM
CR707SW0202	OEM
CR707SW0303	OEM
CR707SW0404	OEM
CR707SW0505	OEM
CR707SW0606	OEM
CR707SW0707	OEM
CR707SW0808	OEM
CR707SW0909	OEM
CR707SW1010	OEM
CR707SW1111	OEM
CR707SW1212	OEM
CR707SW1313	OEM
CR707SW1414	OEM
CR707SW1515	OEM
CR707SW1616	OEM
CR707SW1717	OEM
CR707SW1818	OEM
CR707SW1919	OEM
CR707SW2020	OEM

DEVICE TYPE	REPL CODE
CR707SW2121	OEM
CR707SW2222	OEM
CR707SW2323	OEM
CR801	0015
CR804SC0101	OEM
CR804SC0202	OEM
CR804SC0303	OEM
CR804SC0404	OEM
CR804SC0505	OEM
CR804SC0606	OEM
CR804SC0707	OEM
CR804SC0808	OEM
CR804SC0909	OEM
CR804SC1010	OEM
CR804SC1111	OEM
CR804SC1212	OEM
CR804SC1313	OEM
CR804SC1414	OEM
CR804SC1515	OEM
CR804SC1616	OEM
CR804SW0101	OEM
CR804SW0202	OEM
CR804SW0303	OEM
CR804SW0404	OEM
CR804SW0505	OEM
CR804SW0606	OEM
CR804SW0707	OEM
CR804SW0808	OEM
CR804SW0909	OEM
CR804SW1010	OEM
CR804SW1111	OEM
CR804SW1212	OEM
CR804SW1313	OEM
CR804SW1414	OEM
CR804SW1515	OEM
CR804SW1616	OEM
CR807SB0101	OEM
CR807SB0202	OEM
CR807SB0303	OEM
CR807SB0404	OEM
CR807SB0505	OEM
CR807SB0606	OEM
CR807SB0707	OEM
CR807SB0808	OEM
CR807SB0909	OEM
CR807SB1010	OEM
CR807SB1111	OEM
CR807SB1212	OEM
CR807SB1313	OEM
CR807SB1414	OEM
CR807SB1515	OEM
CR807SB1616	OEM
CR807SB1717	OEM
CR807SB1818	OEM
CR807SB1919	OEM
CR807SB2020	OEM
CR807SB2121	OEM
CR807SB2222	OEM
CR807SB2323	OEM
CR807SB2424	OEM
CR807SB2525	OEM
CR807SC0101	OEM
CR807SC0202	OEM
CR807SC0303	OEM
CR807SC0404	OEM
CR807SC0505	OEM
CR807SC0606	OEM
CR807SC0707	OEM
CR807SC0808	OEM
CR807SC0909	OEM
CR807SC1010	OEM
CR807SC1111	OEM
CR807SC1212	OEM
CR807SC1313	OEM
CR807SC1414	OEM
CR807SC1515	OEM
CR807SC1616	OEM
CR807SC1717	OEM
CR807SC1818	OEM
CR807SC1919	OEM
CR807SC2020	OEM
CR807SC2121	OEM
CR807SC2222	OEM
CR807SC2323	OEM
CR807SC2424	OEM
CR807SC2525	OEM
CR807SW0101	OEM
CR807SW0202	OEM
CR807SW0303	OEM
CR807SW0404	OEM
CR807SW0505	OEM
CR807SW0606	OEM
CR807SW0707	OEM
CR807SW0808	OEM
CR807SW0909	OEM
CR807SW1010	OEM
CR807SW1111	OEM
CR807SW1212	OEM
CR807SW1313	OEM
CR807SW1414	OEM
CR807SW1515	OEM
CR807SW1616	OEM
CR807SW1717	OEM
CR807SW1818	OEM
CR807SW1919	OEM
CR807SW2020	OEM
CR807SW2121	OEM
CR807SW2222	OEM
CR807SW2323	OEM

DEVICE TYPE	REPL CODE
CR807SW2424	OEM
CR807SW2525	OEM
CR951	0087
CR952	0087
CR953	0087
CR954	0087
CR955	0087
CR956	0137
CR1034	0015
CR1035	0015
CR1040	OEM
CR1141	OEM
CR1142	OEM
CR1143	OEM
CR1144	OEM
CR1145	OEM
CR1146	OEM
CR1157	0725
CR4120	OEM
CR5000	0293
CR18903RB	OEM
CR32077-1	1030
CR32077R	1030
CR181003RB	OEM
CR181103RB	OEM
CR181203RB	OEM
CR181303RB	OEM
CR181403RB	OEM
CRAM/RAM108	OEM
CRAM-108-230	OEM
CRAM-108-250	OEM
CRAM-108-290	OEM
CRC150AY8	OEM
CRD821	0471
CRD823	0471
CRD825	0471
CRD827	0471
CRD829	0471
CRD921	0471
CRD923	0471
CRD925	0471
CRD927	0471
CRD929	OEM
CRE0004D	OEM
CRE3	OEM
CRG20	OEM
CRG40	OEM
CRG60	OEM
CRG100	OEM
CRG150	OEM
CRG200	OEM
CRHG3	OEM
CRHG4	OEM
CRHG5	OEM
CRHG6	OEM
CRHG7	OEM
CRHG8	OEM
CRHG9	OEM
CRHG10	OEM
CRL-216	OEM
CR02AM	OEM
CRR0240	OEM
CRR0360	OEM
CRR0560	OEM
CRR0800	OEM
CRR1250	OEM
CRR1950	OEM
CRR2900	OEM
CRR4300	OEM
CRS1/35AF	OEM
CRS1-05AF	OEM
CRS1-10AF	OEM
CRS1-20AF	OEM
CRS1-30AF	OEM
CRS1-40AF	OEM
CRS1-50AF	OEM
CRS3/35AF	OEM
CRS3/50	OEM
CRS3/60	OEM
CRS3-05AF	OEM
CRS3-10AF	OEM
CRS3-20AF	OEM
CRS3-30AF	OEM
CRS3-40AF	OEM
CRS3-50AF	OEM
CRS20	OEM
CRS25/05AF	OEM
CRS25/10AF	OEM
CRS25/20AF	OEM
CRS25/30AF	OEM
CRS25/35AF	OEM
CRS25/40AF	OEM
CRS25/50	OEM
CRS25/60	OEM
CRS25-05AF	OEM
CRS25-20AF	OEM
CRS40	OEM
CRS60	OEM
CRS70-05AF	OEM
CRS70-10AF	OEM
CRS70-20AF	OEM
CRS70-30AF	OEM
CRS70-35AF	OEM
CRS70-50	OEM
CRS100	OEM
CRS150	OEM
CRS200	OEM
CRS449-16IY1	OEM
CRS2020	OEM

If replacement code is OEM, contact original manufacturer for replacement.

DEVICE TYPE	REPL CODE
CRT15	OEM
CRT20	OEM
CRT25	OEM
CRT30	OEM
CRT1544	0160
CRT1545	0160
CRT1552	0160
CRT1553	0160
CRT1592	OEM
CRT1602	0160
CRT3602A	0160
CRT5027	OEM
CRT5037	OEM
CRT5047	OEM
CRT5057	OEM
CRT7004A	OEM
CRT7004B	OEM
CRT7004C	OEM
CRT8002A	OEM
CRT8002B	OEM
CRT8002C	OEM
CRT8021	OEM
CRT8021-003	OEM
CRT9006-83	OEM
CRT9006-135	OEM
CRT9007	OEM
CRT9021	OEM
CRT9212	OEM
CRT96364A	OEM
CRT96364B	OEM
CS-0040	OEM
CS08	OEM
CS0.6-02G02	OEM
CS0.6-04G02	OEM
CS0.6-08D02	OEM
CS0.6-08G02	OEM
CS0.6-10D02	OEM
CS0.6-10G02	OEM
CS0.6-12	OEM
CS0.8-02D02	OEM
CS0.8-04D02	OEM
CS0.8-06D02	OEM
CS0.8-07D02	OEM
CS1-02	OEM
CS1-04	OEM
CS1-06	OEM
CS1-08	OEM
CS1-08G03	OEM
CS1-10G03	OEM
CS1-12G03	OEM
CS1.2-02D02	OEM
CS1.2-04D02	OEM
CS1.2-05D02	OEM
CS1.2-06D02	OEM
CS1.2-07D02	OEM
CS2.5-04Z02	OEM
CS2.5-06Z02	OEM
CS3-02	2095
CS3-02D02	OEM
CS3-02G01	OEM
CS3-04	0710
CS3-04D02	OEM
CS3-05G01	OEM
CS3-06	0863
CS3-06D02	OEM
CS3-06G01	OEM
CS3-07D02	OEM
CS3-07G01	OEM
CS3B	OEM
CS3BR	OEM
CS3.5-02D02	OEM
CS3.5-02G02	OEM
CS3.5-04D02	OEM
CS3.5-04G02	OEM
CS3.5-05D02	OEM
CS3.5-05G02	OEM
CS3.5-06D02	OEM
CS3.5-06G02	OEM
CS3.5-07D02	OEM
CS4-100	OEM
CS4-200	OEM
CS4-350	OEM
CS4-450	OEM
CS4B	OEM
CS4.9-02	OEM
CS4.9-04	0740
CS4.9-06	0804
CS5-02+	0469
CS5-02D02	OEM
CS5-02G02	OEM
CS5-02G03	OEM
CS5-04+	0722
CS5-04D02	OEM
CS5-04G01	OEM
CS5-04G02	OEM
CS5-06+	0725
CS5-06D02	OEM
CS5-06G02	OEM
CS5-06G03	OEM
CS5-08+	2036
CS5-08D02	OEM
CS5-10+	2036
CS5-10D02	OEM
CS5-12+	2036
CS5-12D02	OEM
CS5-100	OEM
CS5-200	OEM
CS5-400	OEM
CS5-500	OEM
CS5-600	OEM
CS5-800	OEM
CS5-1000	OEM
CS5B	OEM
CS6-02D01	OEM
CS6-04D01	OEM
CS6-06D01	OEM
CS6-07D01	OEM
CS6-08D01	OEM
CS7	OEM
CS7A	OEM
CS8-02	0745
CS8-02D02	OEM
CS8-02D03	OEM
CS8-02G02	OEM
CS8-02G03	OEM
CS8-02N	2449
CS8-04	0740
CS8-04D02	OEM
CS8-04D03	OEM
CS8-04G02	OEM
CS8-04G03	OEM
CS8-04I03	OEM
CS8-05N	2449
CS8-06	0764
CS8-06D02	OEM
CS8-06D03	OEM
CS8-06G02	OEM
CS8-06I03	OEM
CS8-08D02	OEM
CS8-08D03	OEM
CS8-08I02	OEM
CS8-08I03	OEM
CS8-10D02	OEM
CS8-10D03	OEM
CS8-10I02	OEM
CS8-10I03	OEM
CS8-12D02	OEM
CS8-12D03	OEM
CS8-14I02	OEM
CS8-14I03	OEM
CS8-100	OEM
CS8-200	OEM
CS8-400	OEM
CS8-500	OEM
CS8-600	OEM
CS8-800	OEM
CS8-1000	OEM
CS9B	OEM
CS9BR	OEM
CS10	OEM
CS10-02G02	OEM
CS10-02M	1651
CS10-04G02	OEM
CS10-05M	1651
CS10-05N	2449
CS10-06G02	OEM
CS10-07G02	OEM
CS10-08G02	OEM
CS10-1M	1651
CS10-1N	2449
CS10-2M	1651
CS10-2N	0435
CS10-3M	1584
CS10-3N	1487
CS10-4M	1584
CS10-4N	1487
CS10-6M	1665
CS10-6N	0687
CS10-100	OEM
CS10-200	OEM
CS10-400	OEM
CS10A	OEM
CS10B	OEM
CS10BR	OEM
CS11B	OEM
CS11C	0174
CS11D	0179
CS11E	OEM
CS11G	0179
CS11H	0179
CS11J	OEM
CS11K	0346
CS11L	OEM
CS11M	0346
CS11N	OEM
CS12B	OEM
CS12BR	OEM
CS12C	OEM
CS12D	OEM
CS12E	OEM
CS12G	OEM
CS12J	OEM
CS12L	OEM
CS12M	OEM
CS12N	OEM
CS13-02	0469
CS13-02+	0469
CS13-02G02	OEM
CS13-04	2647
CS13-04+	0722
CS13-04G02	OEM
CS13-04G03	OEM
CS13-06	0725
CS13-06+	0725
CS13-06G02	OEM
CS13-06G03	OEM
CS13-08+	2036
CS13-08G02	OEM
CS13-08G03	OEM
CS13-10+	2036
CS13-10G02	OEM
CS13-10G03	OEM
CS13-12+	2036
CS13-12G02	OEM
CS13-12G03	OEM
CS14	OEM
CS15	OEM
CS15-02G02	OEM
CS15-04G02	OEM
CS15-06G02	OEM
CS15-07G02	OEM
CS15-08G02	OEM
CS15-100	OEM
CS15-100A	OEM
CS15-200	OEM
CS15-200A	OEM
CS15-400	OEM
CS15-400A	OEM
CS15-500	OEM
CS15-500A	OEM
CS15-600	OEM
CS15-600A	OEM
CS15-700	OEM
CS15-700A	OEM
CS15,9-04+	0722
CS15,9-06+	0725
CS15,9-08+	2036
CS15,9-10+	2036
CS15.9-02	OEM
CS15.9-04	0740
CS15.9-04GP2	OEM
CS15.9-04GP3	OEM
CS15.9-05	OEM
CS15.9-06GP2	OEM
CS15.9-06GP3	OEM
CS15.9-08GP2	OEM
CS15.9-08GP3	OEM
CS15.9-08GU2	OEM
CS15.9-08GU3	OEM
CS15.9-10GP2	OEM
CS15.9-10GP3	OEM
CS15.9-10GU2	OEM
CS15.9-10GU3	OEM
CS15.906	0804
CS16-02+	0469
CS16-02D01	OEM
CS16-02D02	OEM
CS16-02D03	OEM
CS16-02G01	OEM
CS16-02G02	OEM
CS16-02G03	OEM
CS16-04+	0722
CS16-04D01	OEM
CS16-04D02	OEM
CS16-04D03	OEM
CS16-04G01	OEM
CS16-04G02	OEM
CS16-04G03	OEM
CS16-06	0725
CS16-06+	0725
CS16-06D01	OEM
CS16-06D02	OEM
CS16-06D03	OEM
CS16-06G01	OEM
CS16-06G02	OEM
CS16-06G03	OEM
CS16-08	0750
CS16-08+	2036
CS16-08D01	OEM
CS16-08D02	OEM
CS16-08D03	OEM
CS16-08I01	OEM
CS16-08I02	OEM
CS16-08I03	OEM
CS16-10	4855
CS16-10+	2036
CS16-10D01	OEM
CS16-10D02	OEM
CS16-10D03	OEM
CS16-10I01	OEM
CS16-10I02	OEM
CS16-10I03	OEM
CS16-12	OEM
CS16-12+	2036
CS16-12D01	OEM
CS16-12D03	OEM
CS16-12I01	OEM
CS16-12I02	OEM
CS16-12I03	OEM
CS16-14I01	OEM
CS16-14I02	OEM
CS16-14I03	OEM
CS16-16I01	OEM
CS16-16I02	OEM
CS16-16I03	OEM
CS16-100	OEM
CS16-200	OEM
CS16-400	OEM
CS16-500	OEM
CS16-600	OEM
CS16-800	OEM
CS16-1000	OEM
CS-16E	0015
CS16E	0015
CS18	OEM
CS20-02R	3307
CS20-05M	3319
CS20-05N	2188
CS20-05R	2188
CS20-1M	1650
CS20-1N	0567
CS20-1R	0567
CS20-1.5R	3258
CS20-2M	1651
CS20-2N	0762
CS20-2R	0762
CS20-2.5R	3268
CS20-3R	3271
CS20-4M	OEM
CS20-4N	0740
CS20-4R	0740
CS20-5R	3291
CS20-6M	1665
CS20-6N	0764
CS20-6R	0764
CS21B	OEM
CS21C	OEM
CS21D	OEM
CS21E	OEM
CS21G	OEM
CS21J	OEM
CS21K	OEM
CS21L	OEM
CS21M	OEM
CS21N	OEM
CS21P	OEM
CS21Q	OEM
CS21R	OEM
CS22-04	OEM
CS22-06	OEM
CS22-08	OEM
CS22-10	OEM
CS22-11	OEM
CS22-12	OEM
CS22-13	OEM
CS22-400	OEM
CS22-500	OEM
CS22-600	OEM
CS22-700	OEM
CS22-800	OEM
CS22-900	OEM
CS22-1000	OEM
CS22-1100	OEM
CS22-1200	OEM
CS22-1300	OEM
CS22B	OEM
CS22C	OEM
CS22D	OEM
CS22E	OEM
CS22G	OEM
CS22J	OEM
CS22K	OEM
CS22L	OEM
CS22M	OEM
CS22N	OEM
CS22P	OEM
CS22Q	OEM
CS22R	OEM
CS23	OEM
CS23-02D03	OEM
CS23-02G03	1704
CS23-02G01	OEM
CS23-02G02	OEM
CS23-02G03	OEM
CS23-04D03	OEM
CS23-04G01	OEM
CS23-04G02	OEM
CS23-04G03	OEM
CS23-06D03	OEM
CS23-06G01	OEM
CS23-06G02	OEM
CS23-06G03	OEM
CS23-08D03	OEM
CS23-08I01	OEM
CS23-08I02	OEM
CS23-08I03	OEM
CS23-10D03	OEM
CS23-10I01	OEM
CS23-10I02	OEM
CS23-10I03	OEM
CS23-12D03	OEM
CS23-12I01	OEM
CS23-12I02	OEM
CS23-12I03	OEM
CS23-14I01	OEM
CS23-14I02	OEM
CS23-14I03	OEM
CS23-16I01	OEM
CS23-16I02	OEM
CS23-16I03	OEM
CS24	OEM
CS25-02M	2519
CS25-02N	3307
CS25-02R	3307
CS25-05M	OEM
CS25-05N	2188
CS25-05R	2188
CS25-1M	1650
CS25-1N	0567
CS25-1R	0567
CS25-2M	1651
CS25-2N	0762
CS25-2R	0762
CS25-3R	3271
CS25-4M	OEM
CS25-4R	0740
CS25-6M	1665
CS25-6N	0764
CS25-6R	0764
CS25-7R	2875
CS25-8N	0766
CS25-8R	0766
CS28	OEM
CS30-100	OEM
CS30-100BX	OEM
CS30-100BY	OEM
CS30-100CX	OEM
CS30-100CY	OEM
CS30-200	OEM
CS30-200BX	OEM
CS30-200BY	OEM
CS30-200CX	OEM
CS30-200CY	OEM
CS30-300BX	OEM
CS30-300BY	OEM
CS30-300CX	OEM
CS30-300CY	OEM
CS30-400	OEM
CS30-400BX	OEM
CS30-400BY	OEM
CS30-400CX	OEM
CS30-400CY	OEM
CS30-500	OEM
CS30-500BX	OEM
CS30-500BY	OEM
CS30-500CX	OEM
CS30-500CY	OEM
CS30-600	OEM
CS30-600BX	OEM
CS30-600BY	OEM
CS30-600CX	OEM
CS30-600CY	OEM
CS30-700BX	OEM
CS30-700BY	OEM
CS30-700CX	OEM
CS30-700CY	OEM
CS31B	OEM
CS31C	OEM
CS31D	OEM
CS31E	OEM
CS31G	OEM
CS31J	OEM
CS31K	OEM
CS31L	OEM
CS31M	OEM
CS31N	OEM
CS31P	OEM
CS31Q	OEM
CS31R	OEM
CS32B	OEM
CS32C	OEM
CS32D	OEM
CS32E	OEM
CS32G	OEM
CS32J	OEM
CS32K	OEM
CS32L	OEM
CS32M	OEM
CS32N	OEM
CS32P	OEM
CS32Q	OEM
CS32R	OEM
CS33-02+	4031
CS33-04+	4031
CS33-06+	4031
CS33-08+	4031
CS33-10+	0679
CS33-12+	0679
CS35-02M	1650
CS35-02N	3277
CS35-02R	3277
CS35-04G03	OEM
CS35-05M	1650
CS35-05N	0731
CS35-05R	0731
CS35-06G03	OEM
CS35-08G03	OEM
CS35-1M	1650
CS35-1N	0712
CS35-1R	0712
CS35-1.5R	0469
CS35-2M	1650
CS35-2N	0469
CS35-2R	0469
CS35-2.5R	0721
CS35-4M	2647
CS35-4N	0722
CS35-4R	0722
CS35-6M	2649
CS35-6N	0725
CS35-6R	0725
CS35-10G03	OEM
CS35-12G03	OEM
CS38	OEM
CS38-08GU2	OEM
CS38-08GV2	OEM
CS38-12GV2	OEM
CS39-02	OEM
CS39-11	OEM
CS39C	OEM
CS40-600	OEM
CS40-800	OEM
CS40-1000	OEM
CS40-1100	OEM
CS40-1200	OEM
CS40-1300	OEM
CS40-1400	OEM
CS40-1500	OEM
CS41B	OEM
CS41C	OEM
CS41D	OEM
CS41E	OEM
CS41G	OEM
CS41J	OEM
CS41K	OEM
CS41L	OEM
CS41M	OEM
CS41N	OEM
CS41P	OEM
CS41Q	OEM
CS41R	OEM
CS42-06+	0468
CS42-08+	0468
CS42-10+	0468
CS42-12+	0468
CS42-12I02	OEM
CS42-14I02	OEM
CS42-16I02	OEM
CS42-600	OEM
CS42-800	OEM
CS42-1000	OEM
CS42-1200	OEM
CS42-1400	OEM
CS42-1600	OEM
CS48	OEM
CS50-06+	0468
CS50-06G01	OEM
CS50-06G02	OEM
CS50-06H01	OEM
CS50-06H02	OEM
CS50-08+	0468
CS50-08G01	OEM
CS50-08H02	OEM
CS50-08H01	OEM
CS50-08I01	OEM
CS50-08I02	OEM
CS50-10+	0468
CS50-10H01	OEM
CS50-10H02	OEM
CS50-10I01	OEM
CS50-10I02	OEM
CS50-11G01	OEM
CS50-11G02	OEM
CS50-11H01	OEM
CS50-11H02	OEM
CS50-11I01	OEM
CS50-11I02	OEM
CS50-12+	0468
CS50-12H01	OEM
CS50-12H02	OEM
CS50-12I01	OEM
CS50-12I02	OEM
CS50-13G01	OEM
CS50-13G02	OEM
CS50-13H01	OEM
CS50-13H02	OEM
CS50-13I01	OEM
CS50-13I02	OEM
CS50-14H01	OEM
CS50-14H02	OEM
CS50-14I01	OEM
CS50-14I02	OEM
CS50-15G01	OEM
CS50-15G02	OEM
CS50-15H01	OEM
CS50-15H02	OEM
CS50-15I01	OEM
CS50-15I02	OEM
CS50-16I01	OEM
CS50-16I02	OEM
CS50-17G01	OEM
CS50-17I01	OEM
CS50-17I02	OEM
CS54-06G01	OEM
CS54-06G04	OEM
CS54-08I01	OEM
CS54-08I04	OEM
CS54-10I01	OEM
CS54-10I04	OEM
CS54-12I01	OEM
CS54-12I04	OEM
CS54-14I01	OEM
CS54-14I04	OEM
CS54-16I01	OEM
CS60	0201
CS65-600	OEM
CS65-800	OEM
CS65-1000	OEM
CS65-1100	OEM
CS65-1200	OEM
CS65-1300	OEM
CS65-1400	OEM
CS65-1500	OEM
CS70-06G01	OEM
CS70-06G02	OEM
CS70-06H01	OEM
CS70-06H02	OEM
CS70-08H01	OEM
CS70-08H02	OEM
CS70-08I01	OEM
CS70-08I02	OEM
CS70-10H01	OEM
CS70-10H02	OEM
CS70-10I01	OEM
CS70-10I02	OEM
CS70-11G01	OEM
CS70-11G02	OEM
CS70-11H01	OEM
CS70-11H02	OEM
CS70-11I01	OEM
CS70-11I02	OEM
CS70-12H01	OEM
CS70-12H02	OEM
CS70-12I01	OEM
CS70-12I02	OEM
CS70-13G01	OEM
CS70-13G02	OEM
CS70-13H01	OEM
CS70-13H02	OEM
CS70-13I01	OEM
CS70-13I02	OEM
CS70-14H01	OEM
CS70-14H02	OEM
CS70-14I01	OEM
CS70-14I02	OEM
CS70-15G01	OEM
CS70-15G02	OEM
CS70-15H01	OEM
CS70-15H02	OEM
CS70-15I01	OEM
CS70-15I02	OEM
CS70-16H01	OEM
CS70-16H02	OEM
CS70-16I01	OEM
CS70-16I02	OEM
CS70-17G01	OEM
CS70-17G02	OEM
CS70-17I01	OEM
CS70-17I02	OEM
CS70-18I01	OEM
CS70-18I02	OEM
CS72-1	OEM
CS72-1M	OEM
CS72-2	OEM
CS72-2M	OEM
CS72-3	OEM
CS72-3M	OEM
CS72-4	OEM
CS72-4M	OEM
CS72-5	OEM
CS72-5M	OEM
CS72-6	OEM
CS72-6M	OEM
CS72-7	OEM
CS72-7M	OEM
CS72-8	OEM
CS72-8M	OEM
CS78-06GU2	OEM
CS78-06GU3	OEM
CS78-06GV2	OEM
CS78-06GV3	OEM
CS78-06H03	OEM
CS78-08GU2	OEM
CS78-08GU3	OEM
CS78-08GV2	OEM
CS78-08H02	OEM
CS78-08H03	OEM
CS78-10GU2	OEM
CS78-10GU3	OEM
CS78-10GV2	OEM
CS78-10GV3	OEM
CS78-10H02	OEM
CS78-10H03	OEM
CS78-12GU2	OEM
CS78-12GU3	OEM
CS78-12GV2	OEM
CS78-12GV3	OEM
CS78-12H02	OEM
CS78-12H03	OEM
CS78-13H02	OEM
CS78-13H03	OEM
CS79-02	OEM
CS79-04	OEM
CS79-06H02	OEM
CS79-06H03	OEM
CS79-08GW2	OEM
CS79-08GW3	OEM
CS79-08H02	OEM
CS79-08H03	OEM
CS79-10GW2	OEM
CS79-10GW3	OEM
CS79-10H02	OEM
CS79-10H03	OEM
CS79-11	OEM
CS79-12GV2	OEM
CS79-12GV3	OEM
CS79-12GW2	OEM
CS79-12GW3	OEM
CS79-12GX2	OEM
CS79-12GX3	OEM
CS79-12H02	OEM
CS79-12H03	OEM
CS79-13GX3	OEM
CS79-13H02	OEM
CS79-13H03	OEM
CS80-06I02	OEM
CS80-08I02	OEM
CS80-10I02	OEM
CS80-11I02	OEM
CS80-12I02	OEM
CS80-15I02	OEM
CS80-16I02	OEM
CS80-17I02	OEM

If replacement code is OEM, contact original manufacturer for replacement.

DEVICE TYPE	REPL CODE
CS80-18I02	OEM
CS80I02	OEM
CS100	OEM
CS100-36I02	OEM
CS101-40I02	OEM
CS-102	OEM
CS-102-1	OEM
CS102-1	OEM
CS110-06G01	OEM
CS110-06G02	OEM
CS110-08I01	OEM
CS110-08I02	OEM
CS110-10I01	OEM
CS110-10I02	OEM
CS110-12I01	OEM
CS110-12I02	OEM
CS110-14I01	OEM
CS110-14I02	OEM
CS110-16I01	OEM
CS110-16I02	OEM
CS110-18G01	OEM
CS110-18G02	OEM
CS110-18I01	OEM
CS110-18I02	OEM
CS120	0201
CS-122	OEM
CS122	OEM
CS122A	OEM
CS122AM	OEM
CS122B	OEM
CS122BM	OEM
CS122C	OEM
CS122CM	OEM
CS122D	OEM
CS122DM	OEM
CS122E	OEM
CS122EM	OEM
CS122M	OEM
CS122MM	OEM
CS122N	OEM
CS122NM	OEM
CS122S	OEM
CS122SM	OEM
CS124E3	OEM
CS124F3	OEM
CS-127	OEM
CS127	OEM
CS-129	OEM
CS129	OEM
CS-130	OEM
CS130	OEM
CS130-11I02	OEM
CS130-12I02	OEM
CS130-13G02	OEM
CS130-13I02	OEM
CS130-14I02	OEM
CS130-16I02	OEM
CS130-17G02	OEM
CS130-17I02	OEM
CS130-18I02	OEM
CS130-600	OEM
CS130-800	OEM
CS130-1000	OEM
CS130-1100	OEM
CS130-1200	OEM
CS130-1300	OEM
CS130-1400	OEM
CS130-1600	OEM
CS130-1800	OEM
CS-131	OEM
CS131A	1023
CS131A+	1023
CS131AZ	1023
CS131AZ+	1023
CS131B	1023
CS131B+	1023
CS131C	1023
CS131C+	1023
CS131D	1228
CS131D(STUD)	1023
CS131D+	1036
CS131E	1036
CS131F	1036
CS131F+	1036
CS131Z	1023
CS131Z+	1023
CS-132	OEM
CS149-04	OEM
CS149-06G02	OEM
CS149-06I02	OEM
CS149-08G02	OEM
CS149-08I02	OEM
CS149-10G02	OEM
CS149-10I02	OEM
CS149-11	OEM
CS149-12G02	OEM
CS149-12I02	OEM
CS149-13G02	OEM
CS149-13I02	OEM
CS160-12I02	OEM
CS160-14I02	OEM
CS160-16I02	OEM
CS166	OEM
CS179	OEM
CS-180	OEM
CS180	0378
CS186-04G01	OEM
CS186-06G01	OEM
CS186-08I01	OEM
CS186-12I01	OEM
CS186-14I01	OEM
CS186-16I01	OEM
CS188	OEM
CS189	OEM
CS189-02G05	OEM
CS189-04G05	OEM
CS189-07	OEM
CS189-11	OEM
CS190-08I01	OEM
CS190-08I04	OEM
CS190-12I01	OEM
CS190-12I04	OEM
CS190-14I01	OEM
CS190-14I04	OEM
CS195-600	OEM
CS195-800	OEM
CS195-1000	OEM
CS195-1100	OEM
CS195-1200	OEM
CS195-1300	OEM
CS195-1400	OEM
CS195-1500	OEM
CS195-1600	OEM
CS210-04G01	OEM
CS210-04G04	OEM
CS210-04I01	OEM
CS210-04I04	OEM
CS210-06G01	OEM
CS210-06G04	OEM
CS210-06I01	OEM
CS210-06I04	OEM
CS210-08G01	OEM
CS210-08G04	OEM
CS210-08I01	OEM
CS210-08I04	OEM
CS210-10G01	OEM
CS210-10G04	OEM
CS210-10I01	OEM
CS210-10I04	OEM
CS210-12G01	OEM
CS210-12G04	OEM
CS210-12I01	OEM
CS210-12I04	OEM
CS220-08I02	OEM
CS220-5	OEM
CS220-5M	OEM
CS220-7	OEM
CS220-7M	OEM
CS220-9	OEM
CS220-9M	OEM
CS220-12I02	OEM
CS220-14I02	OEM
CS220-16I02	OEM
CS221-5	OEM
CS221-5M	OEM
CS221-7	OEM
CS221-7M	OEM
CS221-9	OEM
CS221-9M	OEM
CS226-04G01	OEM
CS226-06G01	OEM
CS226-08I01	OEM
CS226-12I01	OEM
CS226-14I01	OEM
CS226-16I01	OEM
CS239-02G03	OEM
CS239-02I03	OEM
CS239-04G03	OEM
CS239-04I03	OEM
CS239-06G03	OEM
CS239-06I03	OEM
CS250-08I01	OEM
CS250-1	OEM
CS250-10G01	OEM
CS250-10I01	OEM
CS250-12I01	OEM
CS250-14I01	OEM
CS250-15I01	OEM
CS250-16I01	OEM
CS-258	OEM
CS-258A	OEM
CS-263	OEM
CS295-08I01	OEM
CS295-08I04	OEM
CS295-12I01	OEM
CS295-12I04	OEM
CS295-14I01	OEM
CS295-14I04	OEM
CS295-16I01	OEM
CS295-16I04	OEM
CS299DC	OEM
CS299DI	OEM
CS300-06G03	OEM
CS300-06HW2	OEM
CS300-06I03	OEM
CS300-08I03	OEM
CS300-10I01	OEM
CS300-11G03	OEM
CS300-11I03	OEM
CS300-12I03	OEM
CS300-13G03	OEM
CS300-13I03	OEM
CS300-16I03	OEM
CS300-17G03	OEM
CS300-17I03	OEM
CS302BA	OEM
CS302DA	OEM
CS302FA	OEM
CS302HA	OEM
CS302KA	OEM
CS302MA	2645
CS302PA	2645
CS302RA	2645
CS311-27I01	OEM
CS311-30I01	OEM
CS311-32I01	OEM
CS330-04G01	OEM
CS330-04G04	OEM
CS330-04I01	OEM
CS330-04I04	OEM
CS330-06G01	OEM
CS330-06G04	OEM
CS330-06I01	OEM
CS330-06I04	OEM
CS330-08G01	OEM
CS330-08G04	OEM
CS330-08I01	OEM
CS330-08I04	OEM
CS330-10G01	OEM
CS330-10G04	OEM
CS330-10I01	OEM
CS330-10I04	OEM
CS330-12G01	OEM
CS330-12G04	OEM
CS330-12I01	OEM
CS330-12I04	OEM
CS349-06	OEM
CS349-07	OEM
CS349-08	OEM
CS349-10	OEM
CS349-11	OEM
CS349-12	OEM
CS349-13	OEM
CS351-18I01	OEM
CS351-20I01	OEM
CS351-23I01	OEM
CS351-25I01	OEM
CS351-27I01	OEM
CS360	0016
CS369-10I01	OEM
CS370-04G01	OEM
CS370-04G04	OEM
CS370-04I01	OEM
CS370-04I04	OEM
CS370-06G01	OEM
CS370-06G04	OEM
CS370-06I01	OEM
CS370-06I04	OEM
CS399-07	OEM
CS399-08HW2	OEM
CS399-11	OEM
CS399-12I01	OEM
CS399-13I01	OEM
CS400-04G02	OEM
CS400-04H02	OEM
CS400-04I02	OEM
CS400-06G02	OEM
CS400-06H02	OEM
CS400-06I01	OEM
CS400-06I02	OEM
CS400-08G02	OEM
CS400-08H02	OEM
CS400-08I01	OEM
CS400-08I02	OEM
CS400-10H02	OEM
CS400-10I01	OEM
CS400-10I02	OEM
CS400-11G02	OEM
CS400-11H02	OEM
CS400-11I01	OEM
CS400-11I02	OEM
CS400-12H02	OEM
CS400-12I01	OEM
CS400-12I02	OEM
CS400-13G02	OEM
CS400-13H02	OEM
CS400-13I01	OEM
CS400-13I02	OEM
CS400-14H02	OEM
CS400-14I01	OEM
CS400-14I02	OEM
CS400-15H02	OEM
CS400-15I01	OEM
CS400-15I02	OEM
CS400-16H02	OEM
CS400-16I01	OEM
CS400-16I02	OEM
CS-401	OEM
CS401-18H02	OEM
CS401-18I01	OEM
CS401-20H02	OEM
CS401-20I01	OEM
CS401-21H02	OEM
CS401-21I01	OEM
CS401-22G02	OEM
CS401-22H02	OEM
CS401-22I01	OEM
CS401-22I02	OEM
CS401-23H02	OEM
CS401-23I01	OEM
CS401-24H02	OEM
CS401-24I01	OEM
CS401-25H02	OEM
CS401-25I01	OEM
CS-402	OEM
CS411-18I01	OEM
CS411-20I01	OEM
CS411-23I01	OEM
CS411-25I01	OEM
CS415-02G01	OEM
CS415-03G01	OEM
CS415-05G01	OEM
CS-450	OEM
CS-460	OEM
CS460	OEM
CS-460B	0016
CS-461	OEM
CS-461B	0144
CS-462	OEM
CS-463	OEM
CS-464	OEM
CS-480	OEM
CS480	OEM
CS-481	OEM
CS481	OEM
CS-482	OEM
CS482	OEM
CS-483	OEM
CS483	OEM
CS-484	OEM
CS484	OEM
CS-486	OEM
CS486	OEM
CS-487	OEM
CS487	OEM
CS-488	OEM
CS488	OEM
CS-489	OEM
CS489	OEM
CS500-08	OEM
CS500-08I04	OEM
CS500-10	OEM
CS500-10I04	OEM
CS500-12	OEM
CS500-12I04	OEM
CS500-14	OEM
CS500-14I04	OEM
CS500-15	OEM
CS500-15I04	OEM
CS500-16	OEM
CS500-16I04	OEM
CS550-08I01	OEM
CS550-10I01	OEM
CS550-12I01	OEM
CS550-13G01	OEM
CS550-13I01	OEM
CS550-14I01	OEM
CS550-15I01	OEM
CS550-16I01	OEM
CS-560	OEM
CS560	OEM
CS-580	OEM
CS580	OEM
CS601-18I01	OEM
CS601-20I01	OEM
CS601-22I01	OEM
CS601-24I01	OEM
CS601-25I01	OEM
CS661-27I01	OEM
CS661-30I01	OEM
CS661-32I01	OEM
CS661-35I01	OEM
CS696	0016
CS697	0016
CS706	0016
CS718	0016
CS718A	0016
CS720A	0016
CS901	OEM
CS-901ZHF	0037
CS901ZHF	0037
CS902	OEM
CS903	OEM
CS904	OEM
CS905	OEM
CS918	0144
CS925M	0016
CS929	0007
CS930	0007
CS956	0086
CS1000-12I01	OEM
CS1000-14I01	OEM
CS1000-16I01	OEM
CS1000-17I01	OEM
CS1001-20I01	OEM
CS1001-22I01	OEM
CS1001-24I01	OEM
CS1001-25I01	OEM
CS1014	0144
CS1014D	0144
CS1014E	0144
CS1014F	0144
CS1014G	0007
CS1014H	0007
CS1018	0144
CS1021-32I01	OEM
CS1021-38I01	OEM
CS1021-40I01	OEM
CS1068	0079
CS1120C	0007
CS-1120C1	0079
CS-1120C2	0079
CS1120D	0007
CS-1120D1	0079
CS1120E	0007
CS1120F	0007
CS-1120H	0079
CS1120H	0007
CS-1120I	0079
CS1120I	0079
CS1121G	0037
CS-1124G	OEM
CS1124G	0150
CS1129E	0086
CS1166	0016
CS1166D	0079
CS1166D-G	0079
CS1166E	0016
CS1166F	0079
CS1166G	0016
CS1166H	0016
CS1166H/F	0016
CS1168E	0144
CS1168F	0016
CS1168G	0079
CS1168H	0016
CS1170F	0037
CS1221F	0037
CS1225D	0144
CS1225E	0144
CS1225F	0144
CS1225H	0086
CS1226	0144
CS1226E	0144
CS1226F	0144
CS1226G	0144
CS1226H	0144
CS1226N	0079
CS1227	0144
CS1227D	0144
CS1227E	0144
CS1227F	0144
CS1227G	0144
CS1228	0037
CS1228E	0037
CS1229	0016
CS1229A	0016
CS1229B	0016
CS1229C	0016
CS1229D	0016
CS1229E	0016
CS1229F	0016
CS1229G	0016
CS1229H	0016
CS1229K	0086
CS1229N	0086
CS1235C	0016
CS1235E	0144
CS-1235F	0016
CS1235G	0076
CS1236D	0016
CS1236H	0016
CS1237	0126
CS1238	0144
CS-1238F	0079
CS1238F	0016
CS1238G	0144
CS1238H	0144
CS1238I	0144
CS-1238P	0079
CS1238P	0016
CS1243E	0144
CS1243H	0144
CS1244H	0144
CS1244J	0144
CS-1244X	0079
CS1244X	0144
CS1245F	0016
CS1245G	0016
CS1245H	0016
CS1245I	0079
CS1245T	0016
CS1248	0086
CS1248I	0086
CS1248T	0086
CS1250E	0016
CS1250F	0086
CS1251E	0037
CS1251F	0037
CS1252B	0144
CS1252C	0144
CS1255H	0086
CS1255HF	0791
CS1255M	0595
CS1256H	0126
CS1256HG	0595
CS1256MC	0382
CS1257	0016
CS-1258	0079
CS1258	0016
CS-1259	0079
CS1259	0016
CS1283A	0016
CS1284B	0007
CS1284F	0007
CS1284G	0007
CS1284H	0007
CS1286	0079
CS1288	0016
CS1289	0016
CS1293	0144
CS1294E	0037
CS-1294F	0150
CS1294H	0037
CS1295E	0016
CS1295G	0016
CS1295H	0086
CS1300-04G01	OEM
CS1300-08I01	OEM
CS1300-12I01	OEM
CS1300-14I01	OEM
CS1300-16I01	OEM
CS1300-18I01	OEM
CS1303	0037
CS-1305	0079
CS1305	0086
CS1308	0037
CS1312G	0126
CS-1330	0079
CS1330	0007
CS1330A	0144
CS1330B	0144
CS1330C	0144
CS1330D	0079
CS1340D	0144
CS1340E	0144
CS1340F	0144
CS1340G	0144
CS1340H	0144
CS1340I	0016
CS1344	0016
CS1345	0016
CS1347	0233
CS1348	0016
CS1349	0016
CS1350	0144
CS1351	0144
CS-1352	0595
CS1352	0086
CS1353	0016
CS1354	0037
CS-1359	0079
CS1359	0144
CS1360	0144
CS-1361E	0079
CS1361E	0144
CS-1361F	0079
CS1361F	0144
CS-1361G	0079
CS1361G	0016
CS1362	0016
CS1363	0016
CS1368	0016
CS1368A	0016
CS1368B	0016
CS1368C	0016
CS1368D	0016
CS1369	0126
CS1369D	0150
CS1370	0016
CS1371	0016
CS-1372	0086
CS1372	0016
CS1383	0016
CS-1386E	0016
CS-1386H	0079
CS1386H	0144
CS1420	0016
CS1453E	0016
CS1453F	0086
CS1453G	0086
CS1460E	0007
CS1460H	0007
CS1461J	0007
CS1461X	0007
CS1462F	0007
CS1462I	0086
CS1463A	0016
CS1464H	0086
CS1465H	0126
CS1502A	OEM
CS1502B	OEM
CS1504	OEM
CS1508G	0144
CS1509E	0144
CS1509F	0144
CS1518E	0144
CS1555	0144
CS1585	0016
CS1585E/F	0016
CS1585G	0016
CS1585H	0127
CS1589E	0007
CS1589F	0007
CS1589S	0007
CS1591LE	0086
CS1594E	0007
CS1596E	0007
CS1609F	0086
CS1613	0086
CS1661	0144
CS1664	0086
CS1665	0016
CS1702C	0079
CS1711	0086
CS1758	0004
CS1759	0208
CS1834	0144
CS1893	0086
CS1909B	0018
CS1910B	0532
CS1990	0086
CS2000	OEM
CS2001	0016
CS2001H	0079
CS2004	0016
CS-2004C	0144
CS2004C	0079
CS2004D	0016
CS2005	0150
CS-2005B	0037
CS2005B	0150
CS-2005C	0037
CS2006	0016
CS2006G	0144
CS-2007G	0144
CS2007G	0079
CS-2007H	0144
CS2007H	0079
CS2008	0144
CS-2008F	0144
CS-2008G	0144
CS2008G	0144
CS-2008H	0144
CS2008H	0144
CS2008H552	0144
CS2023	0595
CS-2082	OEM
CS2082	0791
CS-2083	0079
CS-2142	0791
CS2142	0791
CS-2143	0595
CS2218	0016
CS2219	0016
CS2221	0016
CS2222	0016
CS2368	OEM
CS2369	0016
CS2481	0016
CS2483	0861
CS2484	0086
CS2639	OEM
CS2640	OEM
CS2641	OEM
CS2642	OEM
CS2643	OEM
CS2644	OEM
CS2657	OEM
CS2711	0016
CS2712	0016
CS2713	0016
CS2714	0016
CS2715	0007
CS2716	0007
CS2907D8	OEM
CS2917	2318
CS2917-1D14	OEM
CS2917D8	OEM
CS2922	0016
CS2923	0016
CS2924	0016
CS2925	0016
CS2941	0037
CS-3001B	0144
CS3001B	0079
CS-3024	0400
CS3390	0016
CS3391	0016
CS3391A	0016
CS3392	0016
CS3393	0016
CS3394	0016
CS3395	0016
CS3396	0016
CS3397	0016
CS3398	0016
CS3402	0016
CS3403	0016
CS3404	0016
CS3405	0016
CS3414	0016
CS3415	0016
CS3416	0016
CS3417	0016
CS3605	0016
CS3606	0016
CS3607	0016
CS3662	0007
CS3663	0007
CS3702	0037
CS3703	0037
CS3704	0086
CS3705	0086
CS3706	0086
CS3707	0007
CS3708	0007
CS3709	0007
CS3710	0007
CS3711	0007
CS3842A	OEM
CS3843	0016
CS3844	0016
CS3845	0016
CS3854	0016
CS3854A	0016
CS3859	0016
CS3860	0016
CS3906	0037
CS4001	0493
CS4003	0016
CS4004D	OEM
CS4005	0086

DEVICE TYPE	REPL CODE	DEVICE TYPE	REPL CODE	DEVICE TYPE	REPL CODE	DEVICE TYPE	REPL CODE	DEVICE TYPE	REPL CODE	DEVICE TYPE	REPL CODE	DEVICE TYPE	REPL CODE
CS4006	0086	CS9013G	0016	CSB156	OEM	CSF369-13IV1	OEM	CSP14	OEM	CT37	0609	CT283	0080
CS4007	0016	CS9013H	0016	CSB158	OEM	CSF369-13IW1	OEM	CSP16.5	OEM	CT38	0080	CT284	0609
CS4012	0016	CS-9013HE	0079	CSB251	OEM	CSF369-14IV1	OEM	CSP28	OEM	CT39	0609	CT285	0609
CS4013	0037	CS9013HE	0016	CSB252	OEM	CSF375-02HP1	OEM	CSR301	OEM	CT40	0124	CT286	0795
CS4021	0016	CS9013HF	0016	CSB254	OEM	CSF375-02HU1	OEM	CSR447-08IU1	OEM	CT41	0124	CT300	0015
CS4060	0016	CS-9013HG	0086	CSB256	OEM	CSF375-02IP1	OEM	CSR447-08IW1	OEM	CT42	0124	CT302-2N	2015
CS4061	0016	CS9013HG	0016	CSB258	OEM	CSF375-02IU1	OEM	CSR447-09IU1	OEM	CT43	0124	CT302-4N	2017
CS4062	0016	CS-9013HH	0150	CSB351	OEM	CSF375-03HP1	OEM	CSR447-09IW1	OEM	CT44	0124	CT302-6N	2018
CS4193	0016	CS9013HH	0016	CSB352	OEM	CSF375-03HU1	OEM	CSR447-10IU1	OEM	CT45	0124	CT401-2N	2015
CS4194	0016	CS-9014	0007	CSB354	OEM	CSF375-03IP1	OEM	CSR447-10IW1	OEM	CT46	0124	CT401-4N	2017
CS5995	0632	CS9014	0016	CSB356	OEM	CSF375-03IU1	OEM	CSR448-11IV1	OEM	CT47	0124	CT401-6N	2018
CS-6168F	0079	CS9014(C)	0079	CSB358	4581	CSF375-04HP1	OEM	CSR448-11IX1	OEM	CT48	OEM	CT402-2N	2015
CS6168F	0016	CS9014A	0016	CSB640J	OEM	CSF375-04HU1	OEM	CSR448-12IV1	OEM	CT48(DIODE)	0124	CT461	0143
CS-6168G	0016	CS9014B	0016	CSB1505	OEM	CSF375-04IP1	OEM	CSR448-12IX1	OEM	CT49	0124	CT600	0015
CS6168G	0086	CS9014B-C	0079	CSB1510	OEM	CSF375-04IU1	OEM	CSR448-13IV1	OEM	CT50	0124	CT760	OEM
CS-6168H	0016	CS9014C	0016	CSB2505	OEM	CSF375-05HP1	OEM	CSR448-13IX1	OEM	CT51	0124	CT762	OEM
CS6168H	0079	CS9014C/3490	0016	CSB2510	OEM	CSF375-05HU1	OEM	CSR449-14IW1	OEM	CT51-403A	0468	CT764	OEM
CS-6225E	0127	CS9014D	0037	CSB3505	OEM	CSF375-05IP1	OEM	CSR449-14IY1	OEM	CT52	0124	CT767	OEM
CS6225E	0079	CS9014G	0079	CSB3510	OEM	CSF375-05IU1	OEM	CSR449-15IW1	OEM	CT53	0124	CT768	OEM
CS-6225F	0079	CS9015	0037	CSB7000B150-802	OEM	CSF399-11IW1	OEM	CSR449-15IY1	OEM	CT54	0124	CT769	OEM
CS6225F	0144	CS9015A	OEM	CSB7000B173-001	OEM	CSF399-12HV1	OEM	CSR449-16IW1	OEM	CT55	0124	CT770	OEM
CS-6225G	0016	CS9015B	0037	CSB7000B176-001	OEM	CSF399-12IW1	OEM	CSR451-18IX1	OEM	CT56	0124	CT1009	0004
CS6225G	0079	CS9015C	0037	CSB7000B184-802	OEM	CSF399-12IX1	OEM	CSR451-18IZ1	OEM	CT57	0124	CT1012	0127
CS6226F	0144	CS9015C2	0037	CSB7002-01	OEM	CSF399-13HV1	OEM	CSR451-20IX1	OEM	CT58	0124	CT1013	0127
CS-6227E	0079	CS9015D	0037	CSB7002-02	OEM	CSF399-13IW1	OEM	CSR451-20IZ1	OEM	CT59	0124	CT1017	0164
CS6227E	0144	CS-9016	0224	CSB7002-03	OEM	CSF399-13IX1	OEM	CSR5700	OEM	CT62-05N	0154	CT1018	0007
CS-6227F	0079	CS9016	0224	CSB7002-04	OEM	CSF495-02HP1	OEM	CSR5701	OEM	CT62-1N	0154	CT1019	0007
CS6227F	0144	CS9016(G)	0079	CSB7002-05	OEM	CSF495-02HU1	OEM	CSR5702	OEM	CT62-2N	0154	CT1020	0007
CS-6227G	0144	CS-9016D	0224	CSB7002-06	OEM	CSF495-02IP1	OEM	CSR5705	OEM	CT62-4N	0147	CT1021	0007
CS6227G	0079	CS9016D	0224	CSB7002-07	OEM	CSF495-02IU1	OEM	CSSD2003	OEM	CT62-5N	0279	CT1022	0211
CS6228F	0037	CS9016E	0224	CSB7003-01	OEM	CSF495-03HP1	OEM	CST30	OEM	CT62-6N	0279	CT1023	0211
CS-6228G	0126	CS9016EF	0224	CSB7003-02	OEM	CSF495-03HU1	OEM	CST1739	0085	CT70	0124	CT1024	0211
CS6228G	0382	CS9016F	0224	CSB7003-03	OEM	CSF495-03IP1	OEM	CST1740	0085	CT71	0124	CT1122	0160
CS6229F	0016	CS9016F(TRUETONE)	0079	CSB7003-04	OEM	CSF495-03IU1	OEM	CST1741	0085	CT72	0124	CT1124	0160
CS6395	OEM	CS9016F(WESTGHSE)	0079	CSB7100B150-802	OEM	CSF495-04HP1	OEM	CST1742	0085	CT73	0124	CT1124A	0160
CS6395M	OEM	CS9016FG	0224	CSB7100B173-001	OEM	CSF495-04HU1	OEM	CST1743	0160	CT74	0124	CT1124B	0160
CS6396	OEM	CS9016G	0224	CSB7100B176-001	OEM	CSF495-04IP1	OEM	CST1744	0160	CT75	0124	CT1300	0144
CS6396M	OEM	CS9016H	0224	CSB7100B184-802	OEM	CSF495-04IU1	OEM	CST1745	0160	CT76	0124	CT1454	OEM
CS6397	OEM	CS9017	0007	CSB7201-01	OEM	CSF495-05HP1	OEM	CST1746	0160	CT77	0124	CT1460	OEM
CS6397M	OEM	CS9017F	0007	CSB7201-02	OEM	CSF495-05HU1	OEM	CT-06C	0239	CT78	0124	CT1462	OEM
CS6398	OEM	CS9017G	0007	CSB7201-03	OEM	CSF495-05IP1	OEM	CT08	OEM	CT79	0124	CT1464	OEM
CS6398M	OEM	CS9017H	0007	CSB7401-01	OEM	CSF495-05IU1	OEM	CT11-1A	2383	CT82-2N	0154	CT1500	0144
CS6399	OEM	CS9018	0144	CSB7401-02	OEM	CSF499-11IW1	OEM	CT11-2A	2394	CT82-4N	0147	CT-2002	0143
CS6399M	OEM	CS-9018D	0079	CSB7401-03	OEM	CSF499-11IX1	OEM	CT11-4A	1413	CT82-6N	0279	CT2002	0143
CS6401	OEM	CS9018D	0144	CSB7500-150-802	OEM	CSF499-12IW1	OEM	CT12-1A	2383	CT100	0015	CT-2005	0143
CS6401M	OEM	CS9018E	0144	CSB7500-173-001	OEM	CSF499-12IX1	OEM	CT12-2A	2394	CT102-05N	0154	CT2005	0123
CS6402	OEM	CS9018EF	0144	CSB7500-176-001	OEM	CSF499-13IW1	OEM	CT12-4A	1413	CT102-1N	0154	CT2006	0030
CS6402M	OEM	CS9018F	0144	CSB7500-184-802	OEM	CSF499-13IX1	OEM	CT15-4	OEM	CT102-2N	0154	CT2007	0143
CS6403	OEM	CS9018FG	0144	CSBC941	OEM	CSF547-08IU1	OEM	CT15-4M	OEM	CT102-4N	0147	CT2008	0123
CS6403M	OEM	CS9018G	0144	CSC12	OEM	CSF547-08IV1	OEM	CT15-5	OEM	CT102-5N	4545	CT2010	OEM
CS6404	OEM	CS9019	0079	CSD2310	OEM	CSF547-08IW1	OEM	CT15-5M	OEM	CT152-05N	0154	CT2012	OEM
CS6404M	OEM	CS9021G-I	0144	CSD2312	OEM	CSF547-09IU1	OEM	CT15-6	OEM	CT152-1N	0154	CT-2017	0030
CS6405	OEM	CS9022LE	0086	CSD2313	OEM	CSF547-09IV1	OEM	CT15-6M	OEM	CT152-2N	0154	CT2017	OEM
CS6405M	OEM	CS9101B	0016	CSD2314	OEM	CSF547-09IW1	OEM	CT15-7	OEM	CT152-4N	0147	CT2200	OEM
CS6505	OEM	CS9102	0037	CSD2315	0124	CSF547-10IU1	OEM	CT15-7M	OEM	CT152-5N	4545	CT-3003	0015
CS6505M	OEM	CS9102B	0126	CSD2316	OEM	CSF547-10IV1	OEM	CT15-8	OEM	CT156	0133	CT3003	0015
CS6506	OEM	CS9103B	0086	CSD2317	OEM	CSF547-10IW1	OEM	CT15-8M	OEM	CT157	0080	CT-3005	0015
CS6506M	OEM	CS9103C	0086	CSD2325	OEM	CSF548-11IV1	OEM	CT15-9	OEM	CT158	0609	CT3005(RATIO-DET.)	0123
CS6507	OEM	CS9104	0111	CSD2359	0124	CSF548-11IW1	OEM	CT15-9M	OEM	CT159	0609	CT-4002	OEM
CS6507M	OEM	CS9123C1	0150	CSF02AM2	OEM	CSF548-11IX1	OEM	CT15-10	OEM	CT161	0133	CT5603	OEM
CS6508	OEM	CS9124-C2	0144	CSF02AM4	OEM	CSF548-12IV1	OEM	CT15-10M	OEM	CT162	0133	CT5604	OEM
CS6508M	OEM	CS9124B1	0144	CSF02AM8	OEM	CSF548-12IW1	OEM	CT15A4	OEM	CT163	0609	CT5606	OEM
CS6509	OEM	CS9124BI	OEM	CSF0.7-04FT1	OEM	CSF548-12IX1	OEM	CT15A4M	OEM	CT164	0609	CT5611	OEM
CS6509M	OEM	CS9125-B1	0144	CSF7.9-04GT2	OEM	CSF548-13IV1	OEM	CT15A5	OEM	CT182	1335	CT5650	OEM
CS7228E	OEM	CS-9125B	0016	CSF7.9-04GT3	OEM	CSF548-13IW1	OEM	CT15A5M	OEM	CT183	0080	CT5651	OEM
CS7228G	0126	CS9125B	0016	CSF7.9-06GT2	OEM	CSF548-13IX1	OEM	CT15A6	OEM	CT184	0609	CT5652	OEM
CS7229F	0086	CS9126	0016	CSF7.9-06GT3	OEM	CSF549-14IV1	OEM	CT15A6M	OEM	CT185	0609	CT6342	OEM
CS7229G	0016	CS9128	0150	CSF7.9-08GT2	OEM	CSF549-14IW1	OEM	CT15A7	OEM	CT186	0795	CT6342A	OEM
CS8050	0324	CS9128-B2	0037	CSF7.9-08GT3	OEM	CSF549-14IX1	OEM	CT15A7M	OEM	CT200	0015	CT6343	OEM
CS9001	0144	CS9128C1	0037	CSF11-02ER1	OEM	CSF549-15IW1	OEM	CT15A8	OEM	CT220-3	OEM	CT6343A	OEM
CS9010	0079	CS9129	0037	CSF11-04ER1	OEM	CSF549-15IX1	OEM	CT15A8M	OEM	CT220-5	OEM	CT6344	OEM
CS-9011	0016	CS9129(B)	0150	CSF11-04ET1	OEM	CSF549-15IY1	OEM	CT15A9	OEM	CT220-7	OEM	CT6344A	OEM
CS9011	0016	CS9129-B1	0150	CSF11-06EP1	OEM	CSF549-16IW1	OEM	CT15A9M	OEM	CT220-9	OEM	CT6345	OEM
CS9011(C)	0016	CS9129B1	0037	CSF11-08EP1	OEM	CSF549-16IX1	OEM	CT15A10	OEM	CT221-3	OEM	CT6345A	OEM
CS9011(E)(F)	0079	CS9129B2	0037	CSF11-08ET1	OEM	CSF549-16IY1	OEM	CT15A10M	OEM	CT221-5	OEM	CT6346	OEM
CS9011(EF)	0079	CS9417	0595	CSF34-08GW3	OEM	CSF551-18IX1	OEM	CT16X7	0913	CT221-7	OEM	CT6346A	OEM
CS9011(F)	0016	CS9600-4	0016	CSF34-10GW3	OEM	CSF551-18IY1	OEM	CT16XT	0196	CT221-9	OEM	CT6347	OEM
CS9011(GH)	0079	CS9600-5	0016	CSF34-12GW3	OEM	CSF551-18IZ1	OEM	CT18	OEM	CT223-4	OEM	CT6347A	OEM
CS9011(I)	0016	CS-12941	0079	CSF37-02GS3	OEM	CSF551-20IX1	OEM	CT21	0133	CT223-4M	OEM	CT6348	OEM
CS9011D	0016	CS13401	0016	CSF37-02GU3	OEM	CSF551-20IY1	OEM	CT22	1335	CT223-5	OEM	CT6348A	OEM
CS9011E	0016	CS20014	OEM	CSF37-04GS3	OEM	CSF551-20IZ1	OEM	CT23	0080	CT223-5M	OEM	CT6349	OEM
CS-9011F	0079	CS23122	OEM	CSF37-04GU3	OEM	CSF595-02HP1	OEM	CT24	0609	CT223-6	OEM	CT6349A	OEM
CS9011F	0016	CS29008	OEM	CSF315-02HP1	OEM	CSF595-02HU1	OEM	CT26	0133	CT223-6M	OEM	CTB-23	1236
CS-9011G	0016	CS29009	OEM	CSF315-02HS1	OEM	CSF595-04HP1	OEM	CT27	1335	CT223-7	OEM	CTB-23L	1236
CS9011G	0016	CS29010	OEM	CSF315-02IP1	OEM	CSF595-05HP1	OEM	CT28	0080	CT223-7M	OEM	CTB-24	1236
CS9011H	0016	CS29011	OEM	CSF315-02IS1	OEM	CSI-06G03	OEM	CT29	0609	CT223-8	OEM	CTB-24L	1236
CS9011I	0016	CS29012	OEM	CSF315-03HP1	OEM	CSKB40C800	OEM	CT30-1A	2352	CT223-8M	OEM	CTB-33	2234
CS9011N	0079	CS29013	OEM	CSF315-03HS1	OEM	CSKB40C1200	OEM	CT30-2A	2353	CT223-9	OEM	CTB-33M	2234
CS9012	0006	CS-90111	0079	CSF315-03IP1	OEM	CSKB80C800	OEM	CT30-4A	2356	CT223-9M	OEM	CTB-33S	2234
CS9012E	0037	CS90111	0016	CSF315-03IS1	OEM	CSKB80C1200	OEM	CT31	0133	CT223-10	OEM	CTB-34	1942
CS9012E-F	0037	CS90144	0917	CSF315-04HP1	OEM	CSKB250C800	OEM	CT31-1A	2383	CT223-10M	OEM	CTB-34H	2234
CS9012F	0037	CSA7200B150-806	OEM	CSF315-04HS1	OEM	CSKB250C1200	OEM	CT31-2A	2394	CT223A4	OEM	CTB-34S	2234
CS9012FG	0037	CSA7200B173-001	OEM	CSF315-04IP1	OEM	CSKB500C800	OEM	CT31-4A	1413	CT223A5	OEM	CTC120A	OEM
CS9012G	0037	CSA7200B176-001	OEM	CSF315-04IS1	OEM	CSKB500C1200	OEM	CT32	1335	CT223A5M	OEM	CTD1322	OEM
CS9012H	0037	CSA7200B184-802	OEM	CSF315-05HP1	OEM	CSL300	OEM	CT32-05N	0154	CT223A6	OEM	CTG-11R	1727
CS9012HE	0037	CSA7205	OEM	CSF315-05HS1	OEM	CSL311	OEM	CT32-1A	2383	CT223A6M	OEM	CTG-11S	1236
CS-9012HF	0037	CSA7205A	OEM	CSF315-05IP1	OEM	CSL311L	OEM	CT32-1N	0154	CT223A7	OEM	CTG12	1030
CS9012HF	0037	CSA7205B	OEM	CSF315-05IS1	OEM	CSM2B2	OEM	CT32-2A	2394	CT223A7M	OEM	CTG-12R	1727
CS9012HG	0037	CSA7205C	OEM	CSF367-06HV1	OEM	CSM3B1	0447	CT32-2N	0154	CT223A8	OEM	CTG-12S	1236
CS-9012HH	0126	CSB05	0241	CSF367-08HU1	OEM	CSM3B2	0940	CT32-4A	1413	CT223A8M	OEM	CTG-14R	OEM
CS9012HH	0037	CSB1	0241	CSF367-08HV1	OEM	CSM7301-01	OEM	CT32-4N	0147	CT223A9	OEM	CTG-14S	0972
CS9012I	0037	CSB2	0241	CSF367-10HU1	OEM	CSM7301-02	OEM	CT32-6A	2387	CT223A9M	OEM	CTG-21R	1727
CS-9013	0079	CSB4	4581	CSF367-10HV1	OEM	CSM7301-03	OEM	CT32-6N	0279	CT223A10	OEM	CTG-21S	1236
CS9013	0016	CSB6	3546	CSF369-11HX1	OEM	CSM7301-04	OEM	CT33	0609	CT223A10M	OEM	CTG-22R	1727
CS9013(HG)	0161	CSB8	3546	CSF369-11IV1	OEM	CSM7301-05	OEM	CT33-1A	2352	CT256	0133	CTG-22S	1236
CS9013B	0016	CSB10	3546	CSF369-12HZ1	OEM	CSM7301-06	OEM	CT33-2A	2353	CT257	0080	CTG-23R	OEM
CS9013E	0016	CSB151	OEM	CSF369-12IV1	OEM	CS068-8	0030	CT33-4A	2356	CT258	0609	CTG-23S	0972
CS9013E-F	0086	CSB152	OEM	CSF369-12IW1	OEM	CSP4.7	OEM	CT34	0609	CT259	0609	CTG-24R	OEM
CS9013F	0016	CSB154	OEM	CSF369-13HX1	OEM	CSP7	OEM	CT36	0133	CT282	0133	CTG-24S	0972

If replacement code is OEM, contact original manufacturer for replacement.

DEVICE TYPE	REPL CODE
CTG-31R	OEM
CTG-31S	2234
CTG-32R	OEM
CTG-32S	2234
CTG-33R	OEM
CTG-33S	4488
CTG-34R	OEM
CTG-34S	4488
CTL-2S	1236
CTL-11S	1236
CTL-12S	1236
CTL-21S	1236
CTL-22S	1236
CTL-31S	2234
CTL-32S	2234
CTM-20	OEM
CTM20	OEM
CTM20B	OEM
CTM-20R	1727
CTM20R	OEM
CTM-20S	1236
CTM20S	OEM
CTM-21	OEM
CTM21	OEM
CTM-21R	1727
CTM21R	OEM
CTM-21S	1236
CTM21S	OEM
CTM-22	OEM
CTM22	OEM
CTM-22R	1727
CTM22R	OEM
CTM-22S	1236
CTM22S	OEM
CTM-22U	OEM
CTM-24	OEM
CTM24	OEM
CTM-24R	OEM
CTM24R	OEM
CTM-24S	OEM
CTM24S	OEM
CTM-24U	OEM
CTM-26	OEM
CTM26	OEM
CTM-26R	OEM
CTM26R	OEM
CTM-26S	OEM
CTM26S	OEM
CTM-30R	OEM
CTM-30S	2234
CTM-31R	OEM
CTM-31S	2234
CTM-32R	OEM
CTM-32S	2234
CTM-34R	OEM
CTM-34S	OEM
CTN50	0071
CTN61	OEM
CTN62	OEM
CTN64	OEM
CTN66	OEM
CTN68	OEM
CTN100	0071
CTN151	OEM
CTN152	OEM
CTN154	OEM
CTN156	OEM
CTN158	OEM
CTN200	0015
CTN251	OEM
CTN252	OEM
CTN254	OEM
CTN256	OEM
CTN258	OEM
CTN300	0071
CTN351	OEM
CTN352	OEM
CTN354	OEM
CTN356	OEM
CTN358	OEM
CTN400	0071
CTN500	0071
CTN600	0071
CTN605	OEM
CTN610	OEM
CTN800	0071
CTN1000	0071
CTN1505	OEM
CTN1510	OEM
CTN2505	OEM
CTN2510	OEM
CTN3505	OEM
CTN3510	OEM
CTNRB151	OEM
CTNRB152	OEM
CTNRB153	OEM
CTNRB154	OEM
CTNRB155	OEM
CTNRB156	OEM
CTNRB157	OEM
CTP50	0071
CTP61	OEM
CTP62	OEM
CTP64	OEM
CTP66	OEM
CTP68	OEM
CTP100	0071
CTP128	OEM
CTP151	OEM
CTP152	OEM
CTP154	OEM
CTP156	OEM
CTP158	OEM
CTP200	0071
CTP251	OEM
CTP252	OEM
CTP254	OEM
CTP256	OEM
CTP258	OEM
CTP300	0071
CTP351	OEM
CTP352	OEM
CTP354	OEM
CTP356	OEM
CTP358	OEM
CTP400	0071
CTP-461	0143
CTP461	3316
CTP500	0071
CTP573	0143
CTP600	0071
CTP605	OEM
CTP610	OEM
CTP800	0071
CTP1000	0071
CTP1002	OEM
CTP1003	OEM
CTP1004	OEM
CTP1005	OEM
CTP1006	OEM
CTP1032	0004
CTP1033	0004
CTP1034	0004
CTP1035	0004
CTP1036	0004
CTP1102	OEM
CTP1104	0085
CTP1106	0085
CTP1108	0085
CTP1109	0085
CTP1111	0160
CTP1117	0085
CTP1119	0085
CTP1124	0160
CTP1127	OEM
CTP1133	0160
CTP1135	0160
CTP1136	0160
CTP1137	0160
CTP1265	0160
CTP1266	0160
CTP1296	0160
CTP1297	0160
CTP1303	OEM
CTP1306	0160
CTP1307	0160
CTP1314	OEM
CTP1320	OEM
CTP1330	OEM
CTP1340	OEM
CTP1350	OEM
CTP1360	OEM
CTP1390	OEM
CTP1400	OEM
CTP1410	OEM
CTP1500	0160
CTP1503	0160
CTP1504	0160
CTP1506	OEM
CTP1507	OEM
CTP1508	0160
CTP1509	0440
CTP1510	OEM
CTP1511	0160
CTP1512	0440
CTP1513	0160
CTP1514	0160
CTP1530	OEM
CTP1544	0604
CTP1545	0604
CTP1550	0160
CTP1551	0160
CTP1552	0604
CTP1553	0604
CTP1728	0160
CTP1729	0160
CTP1730	0160
CTP1731	0160
CTP1732	0160
CTP1733	0160
CTP1735	0160
CTP1736	0160
CTP1739	0160
CTP1740	0085
CTP-2001-1001	0004
CTP-2001-1002	0004
CTP-2001-1003	0004
CTP-2001-1004	0004
CTP-2001-1007	0016
CTP-2001-1008	0016
CTP-2001-1009	0004
CTP-2001-1010	0143
CTP-2001-1011	0015
CTP-2001-1012	0015
CTP-2006-1001	0004
CTP-2006-1002	0004
CTP-2006-1003	0004
CTP-2006-1004	0143
CTP2076-1001	0004
CTP2076-1002	0004
CTP2076-1003	0004
CTP2076-1004	0004
CTP2076-1005	0004
CTP2076-1006	0004
CTP2076-1007	0004
CTP2076-1008	0004
CTP2076-1009	0004
CTP2076-1010	0004
CTP2076-1011	0004
CTP2076-1012	0004
CTP2505	OEM
CTP2510	OEM
CTP3500	0160
CTP3503	0160
CTP3504	0160
CTP3505	OEM
CTP3508	0160
CTP3510	OEM
CTP3544	0604
CTP3545	0604
CTP3550	OEM
CTP3551	OEM
CTP3552	0604
CTP3553	0604
CTPRB151	OEM
CTPRB152	OEM
CTPRB153	OEM
CTPRB154	OEM
CTPRB155	OEM
CTPRB156	OEM
CTPRB157	OEM
CTS0002H/B	OEM
CTS0032	OEM
CTS0033	OEM
CTS0034CB	OEM
CTS-3FU	OEM
CTS11	OEM
CTS101A	OEM
CTS101AH/B	OEM
CTS108A	OEM
CTS108AH/B	OEM
CTS111H/B	OEM
CTS256AL2	OEM
CTS861	OEM
CTS861H/B	OEM
CTU-1S	1236
CTU-2R	4977
CTU-2S	OEM
CTU-3FU	OEM
CTU3FU	OEM
CTU-11R	1727
CTU-11S	1236
CTU-12R	1727
CTU-12S	1236
CTU12S	1236
CTU-14R	OEM
CTU-14S	OEM
CTU-16R	OEM
CTU-16S	OEM
CTU-20	OEM
CTU20	OEM
CTU-20R	1727
CTU20R	1097
CTU-20S	1236
CTU20S	1236
CTU-21	OEM
CTU21	OEM
CTU-21R	1727
CTU21R	1727
CTU-21S	1236
CTU21S	1236
CTU-22	OEM
CTU22	OEM
CTU-22R	1727
CTU22R	1727
CTU-22S	1236
CTU22S	1236
CTU-22U	OEM
CTU-24	OEM
CTU24	OEM
CTU-24R	OEM
CTU24R	OEM
CTU-24S	OEM
CTU24S	OEM
CTU-25	OEM
CTU-26	OEM
CTU26	OEM
CTU-26R	OEM
CTU26R	OEM
CTU-26S	OEM
CTU26S	OEM
CTU-30R	OEM
CTU-30S	2234
CTU-31R	OEM
CTU-31S	2234
CTU-32R	OEM
CTU-32S	2234
CTU-32U	OEM
CTU-34R	OEM
CTU-34S	OEM
CTU-36R	OEM
CTU-36S	OEM
CTU-G3DR	OEM
CTUG26R	OEM
CTUS1R	OEM
CTUS1R6	OEM
CTU-S16R	OEM
CTUS16R	OEM
CTV	1030
CTZ3.0	0118
CTZ3.0A	0118
CTZ3.3	0299
CTZ3.3A	0299
CTZ3.6	0376
CTZ3.6A	0376
CTZ3.9	0036
CTZ3.9A	0036
CTZ4.3	0275
CTZ4.3A	0275
CTZ4.7	0140
CTZ4.7A	0140
CTZ5.1	0041
CTZ5.1A	0041
CTZ5.6	0253
CTZ5.6A	0253
CTZ6.2	0471
CTZ6.2A	0471
CTZ6.8	0062
CTZ6.8A	0062
CTZ7.5	0077
CTZ7.5A	0077
CTZ8.2	0165
CTZ8.2A	0165
CTZ9.1	0057
CTZ9.1A	0057
CTZ10	0064
CTZ10A	0064
CTZ11	0181
CTZ11A	0181
CTZ12	0052
CTZ12A	0052
CTZ13	0053
CTZ13A	0053
CTZ15	0686
CTZ15A	0686
CTZ16	0445
CTZ16A	0445
CTZ18	0375
CTZ18A	0375
CTZ20	0700
CTZ20A	0700
CTZ22	0705
CTZ22A	0705
CTZ24	0494
CTZ24A	0494
CTZ27	0455
CTZ27A	0455
CTZ30	0195
CTZ30A	0195
CTZ33	0166
CTZ33A	0166
CTZ36	0010
CTZ36A	0010
CTZ39	0032
CTZ39A	0032
CTZ43	0054
CTZ43A	0054
CTZ47	0068
CTZ47A	0068
CTZC6.8	2133
CTZC6.8A	2133
CTZC6.8B	2133
CTZC6.8C	OEM
CTZC6.8D	2133
CTZC6.8E	OEM
CTZC7.5	OEM
CTZC7.5A	0480
CTZC7.5B	0480
CTZC7.5C	OEM
CTZC7.5D	0480
CTZC7.5E	OEM
CTZC8.2	OEM
CTZC8.2A	0504
CTZC8.2B	0504
CTZC8.2C	OEM
CTZC8.2D	0504
CTZC8.2E	OEM
CTZC9.1	OEM
CTZC9.1A	0684
CTZC9.1B	0684
CTZC9.1C	OEM
CTZC9.1D	0684
CTZC9.1E	OEM
CTZC10	OEM
CTZC10A	0225
CTZC10B	0225
CTZC10C	OEM
CTZC10D	0225
CTZC10E	OEM
CTZC11	OEM
CTZC11A	0230
CTZC11B	0230
CTZC11C	OEM
CTZC11D	0230
CTZC11E	OEM
CTZC12	OEM
CTZC12A	0234
CTZC12B	0234
CTZC12C	OEM
CTZC12D	0234
CTZC12E	OEM
CTZC13	1169
CTZC13A	1169
CTZC13B	1169
CTZC13C	OEM
CTZC13D	1169
CTZC13E	OEM
CTZC14	OEM
CTZC14A	1397
CTZC14B	1397
CTZC14C	OEM
CTZC14D	1397
CTZC14E	OEM
CTZC15	OEM
CTZC15A	0247
CTZC15B	0247
CTZC15C	OEM
CTZC15D	0247
CTZC15E	OEM
CTZC17	OEM
CTZC17A	1178
CTZC17B	1178
CTZC17C	OEM
CTZC17D	1178
CTZC17E	OEM
CTZC18	1767
CTZC18A	1767
CTZC18B	1767
CTZC18C	OEM
CTZC18D	1767
CTZC18E	OEM
CTZC19	OEM
CTZC19A	2395
CTZC19B	2395
CTZC19C	OEM
CTZC19D	2395
CTZC19E	OEM
CTZC20	1932
CTZC20A	1932
CTZC20B	1932
CTZC20C	OEM
CTZC20D	1932
CTZC20E	OEM
CTZC22	OEM
CTZC22A	0269
CTZC22B	0269
CTZC22C	OEM
CTZC22D	0269
CTZC22E	OEM
CTZC24	OEM
CTZC24A	0274
CTZC24B	0274
CTZC24C	OEM
CTZC24D	0274
CTZC24E	OEM
CTZC25	OEM
CTZC25A	2399
CTZC25B	2399
CTZC25C	OEM
CTZC25D	2399
CTZC25E	OEM
CTZC27	1952
CTZC27A	1952
CTZC27B	1952
CTZC27C	OEM
CTZC27D	1952
CTZC27E	OEM
CTZC30	OEM
CTZC30A	0309
CTZC30B	0309
CTZC30C	OEM
CTZC30D	0309
CTZC30E	OEM
CTZC33	OEM
CTZC33A	0318
CTZC33B	0318
CTZC33C	OEM
CTZC33D	0318
CTZC33E	OEM
CTZC36	OEM
CTZC36A	0320
CTZC36B	0320
CTZC36C	OEM
CTZC36D	0320
CTZC36E	OEM
CTZC39	OEM
CTZC39A	0326
CTZC39B	0326
CTZC39C	OEM
CTZC39D	0326
CTZC39E	OEM
CTZC43	OEM
CTZC43A	0337
CTZC43B	0337
CTZC43C	OEM
CTZC43D	0337
CTZC43E	OEM
CTZC45	OEM
CTZC45A	0347
CTZC45B	0347
CTZC45C	OEM
CTZC45D	OEM
CTZC45E	OEM
CTZC47	OEM
CTZC47A	0347
CTZC47B	0347
CTZC47C	OEM
CTZC47D	0347
CTZC47E	OEM
CTZC50	OEM
CTZC50A	0027
CTZC50B	0027
CTZC50C	OEM
CTZC50D	OEM
CTZC50E	OEM
CTZC51	1972
CTZC51A	1972
CTZC51B	0027
CTZC51C	OEM
CTZC51D	0027
CTZC51E	OEM
CTZC52	OEM
CTZC52A	0027
CTZC52B	0027
CTZC52C	OEM
CTZC52D	OEM
CTZC52E	OEM
CTZC56	OEM
CTZC56A	0266
CTZC56B	0266
CTZC56C	OEM
CTZC56D	0266
CTZC56E	OEM
CTZC62	OEM
CTZC62A	0386
CTZC62B	0386
CTZC62C	OEM
CTZC62D	0386
CTZC62E	OEM
CTZC68	1987
CTZC68A	1987
CTZC68B	1987
CTZC68C	OEM
CTZC68D	1987
CTZC68E	OEM
CTZC75	OEM
CTZC75A	0426
CTZC75B	0426
CTZC75C	OEM
CTZC75D	0426
CTZC75E	OEM
CTZC82	OEM
CTZC82A	0444
CTZC82B	0444
CTZC82C	OEM
CTZC82D	0444
CTZC82E	OEM
CTZC91	OEM
CTZC91A	0238
CTZC91B	0238
CTZC91C	OEM
CTZC91D	0238
CTZC91E	OEM
CTZC100	OEM
CTZC100A	1180
CTZC100B	1180
CTZC100C	OEM
CTZC100D	OEM
CTZC100E	OEM
CTZC105	OEM
CTZC105A	1190
CTZC105B	1190
CTZC105C	OEM
CTZC105D	OEM
CTZC105E	OEM
CTZC110	OEM
CTZC110A	1190
CTZC110B	1190
CTZC110C	OEM
CTZC110D	OEM
CTZC110E	OEM
CTZC120	OEM
CTZC120A	1207
CTZC120B	1207
CTZC120C	OEM
CTZC120D	1207
CTZC120E	OEM
CTZC130	OEM
CTZC130A	1218
CTZC130B	1218
CTZC130C	OEM
CTZC130D	OEM
CTZC130E	OEM
CTZC140	OEM
CTZC140A	1881
CTZC140B	1881
CTZC140C	OEM
CTZC140D	1881
CTZC140E	OEM
CTZC150	OEM
CTZC150A	0647
CTZC150B	0647
CTZC150C	OEM
CTZC150D	0647
CTZC150E	OEM
CTZC160	OEM
CTZC160A	1255
CTZC160B	1255
CTZC160C	OEM
CTZC160D	1255
CTZC160E	OEM
CTZC180	OEM
CTZC180A	1279
CTZC180B	1279
CTZC180C	OEM
CTZC180D	1279
CTZC180E	OEM
CTZC200	1408
CTZC200A	1408
CTZC200B	1408
CTZC200C	OEM
CTZC200D	1408
CTZC200E	OEM
CU-12E	0940
CU1496A	OEM
CV-2	OEM
CV7	OEM
CV8	OEM
CV9	OEM
CV12E	0095
CV71	OEM
CV291	OEM
CV425	0143
CV442	0143
CV448	0143
CV830	OEM
CV831	OEM
CV832	0628
CV833	OEM
CV834	0030
CV835	OEM
CV836	OEM
CV837	OEM
CV838	OEM
CV839	OEM
CV840	OEM
CV1006	OEM
CV1006A	OEM
CV1006B	OEM
CV1006C	OEM
CV1006D	OEM
CV1008	OEM
CV1008A	OEM
CV1008B	OEM
CV1008C	OEM
CV1008D	OEM
CV1010	OEM
CV1010A	OEM
CV1010B	OEM
CV1010C	OEM
CV1010D	OEM
CV1012	OEM
CV1012A	OEM
CV1012B	OEM
CV1012C	OEM
CV1012D	OEM
CV1014	OEM
CV1014A	OEM
CV1014B	OEM
CV1014C	OEM
CV1014D	OEM
CV1016	OEM
CV1016A	OEM
CV1016B	OEM
CV1016C	OEM
CV1016D	OEM
CV1106	OEM
CV1106A	OEM
CV1106B	OEM
CV1106C	OEM
CV1106D	OEM
CV1108	OEM
CV1108A	OEM
CV1108B	OEM
CV1108C	OEM
CV1108D	OEM
CV1110	OEM
CV1110A	OEM
CV1110B	OEM
CV1110C	OEM
CV1110D	OEM
CV1112	OEM
CV1112A	OEM
CV1112B	OEM
CV1112C	OEM
CV1112D	OEM
CV1114	OEM
CV1114A	OEM
CV1114B	OEM
CV1114C	OEM
CV1114D	OEM
CV1116	OEM
CV1116A	OEM
CV1116B	OEM
CV1116C	OEM
CV1116D	OEM
CV1620	0554
CV1622	OEM
CV1624	0005
CV1626	0720
CV1628	OEM
CV1630	OEM
CV1632	OEM
CV1634	0628
CV1636	OEM
CV1638	0030
CV1640	OEM
CV1642	OEM
CV1644	OEM
CV1646	OEM
CV1648	OEM
CV1650	OEM
CV1652	OEM
CV1654	OEM
CV1656	OEM
CV1658	OEM
CV1660	OEM
CV1662	OEM
CV1664	OEM
CV1666	OEM
CV1844	OEM
CV1858B	OEM
CV1858D	OEM

DEVICE TYPE	REPL CODE
CV1860B	OEM
CV1860D	OEM
CV1862B	OEM
CV1862D	OEM
CV1863B	OEM
CV1863D	OEM
CV1864B	OEM
CV1864D	OEM
CV1865B	OEM
CV1865D	OEM
CV1866	OEM
CV1866B	OEM
CV1866D	OEM
CV1868B	OEM
CV1868D	OEM
CV1870B	OEM
CV1870D	OEM
CV1871B	OEM
CV1871D	OEM
CV1872B	OEM
CV1872D	OEM
CV1873B	OEM
CV1873D	OEM
CV2116	OEM
CV2154	OEM
CV2155	OEM
CV2226	OEM
CV2228	OEM
CV2346	OEM
CV2355	OEM
CV2356	OEM
CV2357	OEM
CV3676	OEM
CV5007	0554
CV5008	OEM
CV5010	0005
CV5012	0720
CV5015	OEM
CV5018	OEM
CV5020	OEM
CV5022	0628
CV5027	OEM
CV5033	0030
CV5039	OEM
CV5047	OEM
CV5056	OEM
CV5068	OEM
CV5082	OEM
CV5100	OEM
CV5123	OEM
CV6002	OEM
CV6071	OEM
CV6072	OEM
CV6199	OEM
CV6234	OEM
CV7108	OEM
CV7109	OEM
CV7180	OEM
CV7181	OEM
CV7182	OEM
CV7183	OEM
CV7377	OEM
CV7378	OEM
CV7636	OEM
CV7728	OEM
CV7729	OEM
CV7739	OEM
CV7771	OEM
CV7772	OEM
CV8163	OEM
CV8164	OEM
CV9424	OEM
CV10758	OEM
CVA1115A	OEM
CVA1115B	OEM
CVA1115C	OEM
CVA1116A	OEM
CVA1116B	OEM
CVA1116C	OEM
CVA1116D	OEM
CVA1116E	OEM
CVA1116F	OEM
CVA1116G	OEM
CVA1130A	OEM
CVA1130B	OEM
CVA1130C	OEM
CVA1145B	OEM
CVA1145D	OEM
CVA1145E	OEM
CVB1015A	OEM
CVB1015B	OEM
CVB1015C	OEM
CVB1030A	OEM
CVB1030B	OEM
CVB1030C	OEM
CVB1045B	OEM
CVB1045D	OEM
CVB1045E	OEM
CVH2090-17	OEM
CVH2030-02	OEM
CVH2030-03	OEM
CVH2030-04	OEM
CVH2030-05	OEM
CVH2030-07	OEM
CVH2030-08	OEM
CVH2030-09	OEM
CVH2030-10	OEM
CVH2030-11	OEM
CVH2030-13	OEM

DEVICE TYPE	REPL CODE
CVH2030-14	OEM
CVH2030-15	OEM
CVH2030-16	OEM
CVH2030-17	OEM
CVH2030-19	OEM
CVH2030-20	OEM
CVH2030-21	OEM
CVH2030-22	OEM
CVH2030-23	OEM
CVH2030-25	OEM
CVH2030-26	OEM
CVH2030-27	OEM
CVH2030A	OEM
CVH2030B	OEM
CVH2030C	OEM
CVH2030D	OEM
CVH2030G	OEM
CVH2030H	OEM
CVH2030J	OEM
CVH2045-01	OEM
CVH2045-02	OEM
CVH2045-03	OEM
CVH2045-04	OEM
CVH2045-05	OEM
CVH2045-07	OEM
CVH2045-08	OEM
CVH2045-09	OEM
CVH2045-10	OEM
CVH2045-11	OEM
CVH2045-13	OEM
CVH2045-14	OEM
CVH2045-15	OEM
CVH2045-16	OEM
CVH2045-17	OEM
CVH2045-19	OEM
CVH2045A	OEM
CVH2045B	OEM
CVH2045C	OEM
CVH2045D	OEM
CVH2045E	OEM
CVH2045F	OEM
CVH2045G	OEM
CVH2045H	OEM
CVH2045J	OEM
CVH2060-03	OEM
CVH2060-04	OEM
CVH2060-05	OEM
CVH2060-07	OEM
CVH2060-08	OEM
CVH2060-09	OEM
CVH2060-10	OEM
CVH2060-11	OEM
CVH2060-13	OEM
CVH2060-14	OEM
CVH2060-15	OEM
CVH2060-16	OEM
CVH2060-17	OEM
CVH2060-19	OEM
CVH2060-20	OEM
CVH2060-21	OEM
CVH2060-22	OEM
CVJ2060-23	OEM
CVH2060-25	OEM
CVH2060-26	OEM
CVH2060-27	OEM
CVH2060B	OEM
CVH2060C	OEM
CVH2060D	OEM
CVH2060E	OEM
CVH2060F	OEM
CVH2060G	OEM
CVH2060H	OEM
CVH2060J	OEM
CVH2060K	OEM
CVH2060L	OEM
CVH2060M	OEM
CVH2090-03	OEM
CVH2090-04	OEM
CVH2090-05	OEM
CVH2090-07	OEM
CVH2090-08	OEM
CVH2090-09	OEM
CVH2090-10	OEM
CVH2090-11	OEM
CVH2090-13	OEM
CVH2090-14	OEM
CVH2090-15	OEM
CVH2090-16	OEM
CVH2090-17	OEM
CVH2090-19	OEM
CVH2090-20	OEM
CVH2090-21	OEM
CVH2090B	OEM
CVJ2090C	OEM
CVH2090D	OEM
CVH2090E	OEM
CVH2090F	OEM
CVH2090G	OEM
CVH2090J	OEM
CVH2090K	OEM
CVH2090L	OEM
CVH2090M	OEM
CVH3030E	OEM
CVJ70653	0030
CWW-010	OEM
CX002	OEM
CX003	OEM
CX-0031	0019
CX-0033	0143

DEVICE TYPE	REPL CODE
CX-0035	0015
CX-0036	0143
CX0036	0143
CX-0037	0015
CX0037	0015
CX-0039	0015
CX0039	0015
CX-0040	0015
CX0040	0015
CX-0041	0030
CX0041	0143
CX-0042	0143
CX0042	0123
CX-0045	0143
CX0045	0143
CX-0047	0143
CX0047	0015
CX-0048	0015
CX0048	0015
CX-0049	0531
CX0049	0015
CX-0051	0649
CX0051	0170
CX-0052	0137
CX0052	0137
CX-0054	0015
CX0054	0540
CX-0055	0133
CX-016A	OEM
CX-016B	OEM
CX-025D	OEM
CX-031	OEM
CX-031-10	OEM
CX-031-20	OEM
CX-031-30	OEM
CX032B	OEM
CX064	3932
CX-064-2	3932
CX065	3933
CX065B	3933
CX-075	4669
CX-075B	4669
CX075B	4669
CX-080	OEM
CX085B	OEM
CX089	1634
CX089D	1634
CX-093	0167
CX093D	0167
CX095	3948
CX095A	3948
CX-095C	3948
CX095C	3948
CX-095D	3948
CX095D	3948
CX-095E	3948
CX095E	3948
CX-099	OEM
CX-0411	OEM
CX-0412	OEM
CX-0432	OEM
CX-0441	OEM
CX-0451	OEM
CX-0461	OEM
CX-0462	OEM
CX088535	OEM
CX088535-119S	OEM
CX-100	OEM
CX100B	4434
CX-100D	4434
CX100D	4434
CX-101B	OEM
CX101G	OEM
CX103D	OEM
CX-104A	OEM
CX-104B-1	OEM
CX-130	3961
CX130	3961
CX-131A	4261
CX131A	4261
CX-133A	3697
CX133A	3697
CX-134A	3707
CX134A	3707
CX-135	3964
CX135	3964
CX135A	OEM
CX136	3966
CX-136A	3966
CX136A	3966
CX-137A	4294
CX137A	4294
CX-138	3968
CX138	3968
CX-138A	3968
CX138A	OEM
CX-139A	4242
CX139A	4242
CX-141	3970
CX141	3970
CX-141A	OEM
CX141A	3970
CX-143	3973
CX143	3973
CX-143A	3973
CX143A	3973
CX-145	3975
CX145	3975
CX-150	OEM

DEVICE TYPE	REPL CODE
CX150	OEM
CX-150B	OEM
CX150B	OEM
CX157	3984
CX158	3985
CX161	OEM
CX162	3988
CX170	3992
CX172	3996
CX173	3998
CX177	4000
CX-177B	4000
CX177B	4000
CX-177C	OEM
CX181	4004
CX183	OEM
CX-186	OEM
CX186	OEM
CX-187	OEM
CX187	OEM
CX-187A	OEM
CX187A	OEM
CX-188	OEM
CX188	OEM
CX-194A	OEM
CX-194A-2	OEM
CX-194B-0	OEM
CX-196	OEM
CX196	OEM
CX196A	OEM
CX-501	OEM
CX501	OEM
CX-504	OEM
CX504	OEM
CX-520-102N	OEM
CX520-102N	OEM
CX520-108N	OEM
CX520-111N	OEM
CX522-030	OEM
CX524-216N	OEM
CX-525-1374K	OEM
CX525-1374K	OEM
CX526-017	OEM
CX526-025	OEM
CX530-507N	OEM
CX555	0767
CX-555A	0767
CX555A	0767
CX-556	2623
CX556	2623
CX556A	2623
CX-556B	3566
CX556B	0490
CX-557	2916
CX557	2625
CX-557A	2625
CX557A	2625
CX-557B	6195
CX557B	2916
CX-557S	2625
CX557S	2916
CX701	OEM
CX701A	OEM
CX702	OEM
CX702A	OEM
CX703	OEM
CX703A	OEM
CX703B	OEM
CX704	OEM
CX726A	OEM
CX-728	OEM
CX754	OEM
CX-759-220-04	OEM
CX-760B	OEM
CX-760E	OEM
CX-761A	OEM
CX-766	OEM
CX-775A	OEM
CX775A	OEM
CX-779	OEM
CX-786	OEM
CX-789	OEM
CX789	OEM
CX-804	OEM
CX804	OEM
CX807	OEM
CX-810	OEM
CX-811	OEM
CX-813	OEM
CX-814	OEM
CX-815	OEM
CX-816	OEM
CX822	OEM
CX-825	OEM
CX830A	OEM
CX-832	OEM
CX-841	OEM
CX-841A	OEM
CX841A	OEM
CX842B	3948
CX-843	4165
CX843	4165
CX-843A	4165
CX843A	4165
CX843B	OEM
CX847	OEM
CX-848	4166
CX848	4166
CX855	OEM

DEVICE TYPE	REPL CODE
CX855A	OEM
CX855B	OEM
CX-864	OEM
CX-865	OEM
CX-866	OEM
CX-867	OEM
CX-868	OEM
CX-869	OEM
CX-870A	OEM
CX-874	OEM
CX874	OEM
CX-875A	OEM
CX875A	OEM
CX877	OEM
CX-882	OEM
CX-883	OEM
CX883	OEM
CX885	OEM
CX-885A	OEM
CX885A	OEM
CX885B	OEM
CX-892	OEM
CX-894	OEM
CX894	OEM
CX901	OEM
CX904	OEM
CX906	OEM
CX908	OEM
CX918	OEM
CX954	OEM
CX956	OEM
CX958	OEM
CX-1000	4434
CX1124AS	OEM
CX1534S	OEM
CX2016A	OEM
CX5034H-0805	OEM
CX-7600	OEM
CX-7905A	OEM
CX-7932	OEM
CX7932	OEM
CX7933	OEM
CX7934	OEM
CX7935	OEM
CX-7944	OEM
CX7947	OEM
CX7948	OEM
CX-7958	OEM
CX7958	OEM
CX7959	OEM
CX7977	OEM
CX7978	OEM
CX-9000	1030
CX9000	1030
CX-9001	0015
CX9001	0015
CX-10026	OEM
CX10026	OEM
CX10032A	OEM
CX-20008	OEM
CX20013	OEM
CX20013A	OEM
CX20014	OEM
CX20014A	OEM
CX20017	OEM
CX20029	OEM
CX-20061	OEM
CX20061	OEM
CX20073	OEM
CX20082	OEM
CX20082-0	OEM
CX20083	OEM
CX20085	OEM
CX20100	OEM
CX20106	OEM
CX20106A	OEM
CX20108	OEM
CX20109	OEM
CX20111	OEM
CX20111-T6	OEM
CX-20112	OEM
CX20112	2870
CX20114	OEM
CX20115A	OEM
CX20125	OEM
CX20129	OEM
CX20133	OEM
CX20139	OEM
CX20157	OEM
CX20159	OEM
CX20174	OEM
CX20180	OEM
CX20183	OEM
CX20183F	OEM
CX20185	OEM
CX20187	OEM
CX-20192	OEM
CX20192	OEM
CX-20193	OEM
CX20193	OEM
CX22011	OEM
CX23035	OEM
CX23040	OEM
CX23041-01	OEM
CX23041-09	OEM
CX23042-10	OEM
CX201291	OEM
CX-520102N	OEM
CX520102N	OEM

DEVICE TYPE	REPL CODE
CX-520108N	OEM
CX520108N	OEM
CX-520111N	OEM
CX522030	OEM
CX564171	OEM
CXA153AS	OEM
CXA495D	0126
CXA1011M	4524
CXA1011P	3083
CXA1013	OEM
CXA1013A	OEM
CXA1013AS	OEM
CXA1013S	OEM
CXA1019S	OEM
CXA1072R-T3	OEM
CXA1081	OEM
CXA1081M	OEM
CXA1081Q	OEM
CXA1081S	OEM
CXA1082AQ	OEM
CXA1082AS	OEM
CXA1082BQ	OEM
CXA1082BS	OEM
CXA1082Q	OEM
CXA1100P	OEM
CXA1101	OEM
CXA1101P	OEM
CXA1102P	OEM
CXA1110BS	OEM
CXA1111P	OEM
CXA1114	OEM
CXA1114P	OEM
CXA1115BP	OEM
CXA1124	OEM
CXA1124A	OEM
CXA1124AS	OEM
CXA1124BS	OEM
CXA1124S	OEM
CXA1127M	OEM
CXA1134AS	OEM
CXA1163	OEM
CXA1182	OEM
CXA1182Q-Z	OEM
CXA1182QZ	OEM
CXA1198A	OEM
CXA1200BQ	OEM
CXA1201Q	OEM
CXA1202R	OEM
CXA1204Q	OEM
CXA1228	OEM
CXA1228S	OEM
CXA1238S	OEM
CXA1240	OEM
CXA1264AS	OEM
CXA1264S	OEM
CXA1271Q	OEM
CXA1272Q-Z	OEM
CXA1279S	OEM
CXA1280N	OEM
CXA1291	OEM
CXA1291P	OEM
CXA1313	OEM
CXA1313S	OEM
CXA1314P	OEM
CXA1315M	OEM
CXA1315M-T4	OEM
CXA1315P	OEM
CXA1334S	OEM
CXA1373	OEM
CXA1373Q	OEM
CXA-1387	OEM
CXA1387S	OEM
CXA1414P	OEM
CXA1464AS	OEM
CXA1465AS	OEM
CXA-1526P	OEM
CXA1526P	OEM
CXA1534S	OEM
CXA1545S	OEM
CXA1594L	OEM
CXA1734S	OEM
CXA4559	OEM
CXA20112	OEM
CXA20125	OEM
CXB0026AMR	OEM
CXD1017Q	OEM
CXD1068A	OEM
CXD1095Q	OEM
CXD1120P	OEM
CXD1120P1	OEM
CXD1125Q	OEM
CXD1130Q	OEM
CXD1130QZ	OEM
CXD1135QZ	OEM
CXD1157Q	OEM
CXD1158M	OEM
CXD1159Q-T3	OEM
CXD1161M3	OEM
CXD1161P-2	OEM
CXD1162P	OEM
CXD1204R-T3	OEM
CXD1220AQ	OEM
CXD2500Q	OEM
CXD2550P	OEM
CXD2551P	OEM
CXK1001P	OEM
CXK1004L	OEM
CXK1005P	OEM
CXK1006L	OEM

DEVICE TYPE	REPL CODE
CXK1011	OEM
CXK1011P	OEM
CXK5816M-10L	OEM
CXK5816M-15L	OEM
CXK5816PN-12L	OEM
CXK5816PN-15L	OEM
CXK5864AP-70L	OEM
CXK5864BSP-10L	OEM
CXK5864M-15L	OEM
CXL011	OEM
CXL100	0280
CXL101	0038
CXL102	0211
CXL102A	0136
CXL103	0038
CXL104	0085
CXL104MP	3033
CXL105	0440
CXL106	0150
CXL107	0127
CXL108	0144
CXL109	0143
CXL110	0143
CXL112	0917
CXL113	0196
CXL114	0484
CXL115	1080
CXL116	0015
CXL117	0015
CXL118	0774
CXL119	1303
CXL120	0474
CXL121	0160
CXL121MP	0160
CXL123	1487
CXL123A	0079
CXL124	0142
CXL125	0087
CXL126	0136
CXL127	0975
CXL128	0086
CXL129	0126
CXL130	0103
CXL130MP	1514
CXL131	0222
CXL131MP	0222
CXL132	0325
CXL133	0325
CXL134	0188
CXL135	0162
CXL137	0636
CXL138	0649
CXL139	0012
CXL140	0170
CXL141	0317
CXL142	0137
CXL143	0365
CXL144	0100
CXL145	0002
CXL146	0441
CXL147	0039
CXL148	1834
CXL149	0783
CXL150	0331
CXL151	0149
CXL152	0042
CXL153	0925
CXL154	0233
CXL155	2763
CXL156	0087
CXL157	0276
CXL158	0164
CXL159	0079
CXL160	0050
CXL161	0007
CXL162	0074
CXL163	0642
CXL164	0003
CXL165	0065
CXL166	0277
CXL167	0289
CXL168	0296
CXL169	0302
CXL170	0250
CXL171	0285
CXL172	0400
CXL173BP	0293
CXL175	0178
CXL176	0847
CXL177	0133
CXL178MP	0133
CXL179	0604
CXL180	1681
CXL181	0130
CXL182	0561
CXL183	1199
CXL184	0161
CXL185	0460
CXL186	0560
CXL187	1267
CXL188	0264
CXL189	0382
CXL190	0264
CXL191	0338
CXL192	0595
CXL193	0791
CXL194	0861
CXL195	0698
CXL196	0423

If replacement code is OEM, contact original manufacturer for replacement.

DEVICE TYPE	REPL CODE
CXL197	0854
CXL198	0168
CXL199	0111
CXL210	2050
CXL211	1367
CXL218	0905
CXL219	1681
CXL220	0849
CXL221	0849
CXL222	0212
CXL223	0103
CXL224	0630
CXL225	3280
CXL226	1010
CXL226MP	1010
CXL228	1708
CXL238	0313
CXL309K	1922
CXL370	3810
CXL501	0128
CXL502	0201
CXL503	0288
CXL504	0378
CXL505	0757
CXL506	0102
CXL507	0102
CXL508	1306
CXL509	2083
CXL510	2084
CXL511	1449
CXL512	2085
CXL513	1323
CXL514	3649
CXL516	OEM
CXL518	1791
CXL519	0124
CXL703A	0632
CXL704	2624
CXL705A	0753
CXL706	0579
CXL708	0664
CXL709	0678
CXL710	0829
CXL711	2715
CXL713	0666
CXL714	0352
CXL715	0354
CXL718	0654
CXL722	0701
CXL725	0692
CXL726	2617
CXL727	2529
CXL728	2549
CXL729	0328
CXL730	2743
CXL731	2457
CXL734	1759
CXL738	0856
CXL747	1808
CXL749	0395
CXL760	2100
CXL780	0364
CXL781	2747
CXL783	0802
CXL786	2714
CXL787	2257
CXL788	2755
CXL790	0349
CXL949	2552
CXL1024	1591
CXL1025	1187
CXL1027	2876
CXL1062	0403
CXL1063	2584
CXL1090	2861
CXL1139	3936
CXL1501M	OEM
CXL5000	1276
CXL5001	2874
CXL5002	0760
CXL5003	0677
CXL5004	0118
CXL5005	0292
CXL5005P	OEM
CXL5006	0188
CXL5007	0456
CXL5008	0533
CXL5009	0451
CXL5010	0162
CXL5011	0157
CXL5012	0091
CXL5013	0636
CXL5014	0025
CXL5015	0649
CXL5016	0244
CXL5017	0322
CXL5018	0012
CXL5019	0170
CXL5020	0317
CXL5021	0137
CXL5022	0053
CXL5023	0879
CXL5024	0002
CXL5025	0445
CXL5026	0210
CXL5027	0495
CXL5028	0671
CXL5029	0531
CXL5030	0565
CXL5031	0402
CXL5032	0714
CXL5033	0441
CXL5034	0257
CXL5035	0726
CXL5036	0039
CXL5037	0010
CXL5038	0350
CXL5039	0931
CXL5040	0999
CXL5041	0502
CXL5042	0869
CXL5043	1156
CXL5044	0783
CXL5045	1268
CXL5046	1189
CXL5047	0331
CXL5048	3026
CXL5049	1311
CXL5050	0098
CXL5051	0149
CXL5052	0186
CXL5053	0213
CXL5054	0245
CXL5055	0028
CXL5056	0255
CXL5057	0877
CXL5058	0367
CXL5059	2858
CXL5060	0421
CXL5061	3003
CXL5062	OEM
CXL5063	1312
CXL5064	3010
CXL5065	1713
CXL5066	0292
CXL5067	0456
CXL5068	0451
CXL5069	0451
CXL5070	0301
CXL5071	0025
CXL5072	0244
CXL5073	1083
CXL5074	0317
CXL5075	0420
CXL5076	1649
CXL5077	0495
CXL5078	0949
CXL5079	0531
CXL5080	0565
CXL5081	0402
CXL5082	1606
CXL5083	1674
CXL5084	0726
CXL5085	0350
CXL5086	0350
CXL5087	0931
CXL5088	0999
CXL5089	0502
CXL5090	0869
CXL5091	1156
CXL5092	1268
CXL5093	1189
CXL5094	3026
CXL5095	1311
CXL5096	0098
CXL5097	0186
CXL5098	0028
CXL5099	0245
CXL5100	0028
CXL5101	0255
CXL5102	0877
CXL5103	0367
CXL5104	2858
CXL5105	0421
CXL5111	0782
CXL5112	0796
CXL5113	0806
CXL5114	0821
CXL5115	0833
CXL5116	0442
CXL5117	0876
CXL5118	3127
CXL5119	0185
CXL5120	0205
CXL5121	0480
CXL5122	0504
CXL5123	3317
CXL5124	0684
CXL5125	0225
CXL5126	0230
CXL5127	0234
CXL5128	0237
CXL5129	1397
CXL5130	0251
CXL5131	0251
CXL5132	1178
CXL5133	0256
CXL5134	2395
CXL5135	0262
CXL5136	0269
CXL5137	0274
CXL5138	2399
CXL5139	0294
CXL5140	1177
CXL5141	0309
CXL5142	0318
CXL5143	0320
CXL5144	0326
CXL5145	0337
CXL5146	0347
CXL5147	0027
CXL5148	0266
CXL5149	2856
CXL5150	0386
CXL5151	0405
CXL5152	0426
CXL5153	0444
CXL5154	3028
CXL5155	0238
CXL5156	1180
CXL5157	1190
CXL5158	1207
CXL5159	1218
CXL5160	1881
CXL5161	0647
CXL5162	1255
CXL5163	2102
CXL5164	1279
CXL5165	2225
CXL5166	0605
CXL5172	4231
CXL5173	2397
CXL5174	2035
CXL5175	2401
CXL5176	1438
CXL5177	2407
CXL5178	1445
CXL5179	2415
CXL5180	2221
CXL5181	1458
CXL5182	0221
CXL5183	1490
CXL5184	2422
CXL5185	1618
CXL5186	0510
CXL5187	0691
CXL5188	0870
CXL5189	1020
CXL5190	1153
CXL5191	1274
CXL5192	1402
CXL5193	1532
CXL5194	1078
CXL5195	1711
CXL5196	1717
CXL5197	1723
CXL5198	1736
CXL5199	1748
CXL5200	1761
CXL5201	2450
CXL5202	1772
CXL5203	1788
CXL5204	1796
CXL5205	1804
CXL5206	1194
CXL5207	1821
CXL5208	0022
CXL5209	0070
CXL5210	0132
CXL5211	0172
CXL5212	0207
CXL5213	0227
CXL5214	0263
CXL5215	0310
CXL5216	0329
CXL5217	0356
CXL5218	0381
CXL5220	0438
CXL5221	0464
CXL5222	0488
CXL5223	0509
CXL5224	0524
CXL5225	0542
CXL5226	0063
CXL5227	0401
CXL5228	0598
CXL5229	0616
CXL5230	0634
CXL5231	0650
CXL5232	0668
CXL5240	0815
CXL5241	0827
CXL5242	0846
CXL5243	0868
CXL5244	0885
CXL5245	0897
CXL5246	0914
CXL5247	0926
CXL5248	0944
CXL5249	0958
CXL5250	0978
CXL5251	0994
CXL5252	1009
CXL5253	1019
CXL5254	0889
CXL5255	1049
CXL5256	1058
CXL5257	0932
CXL5258	1079
CXL5259	1096
CXL5260	1106
CXL5261	1123
CXL5262	1135
CXL5263	1152
CXL5264	1164
CXL5265	1174
CXL5266	1184
CXL5267	1200
CXL5268	1210
CXL5269	1223
CXL5270	1232
CXL5271	1246
CXL5272	1265
CXL5273	1273
CXL5274	1290
CXL5275	1299
CXL5276	1307
CXL5277	1322
CXL5278	1331
CXL5279	1340
CXL5280	1353
CXL5281	1365
CXL5282	1384
CXL5283	1401
CXL5284	1412
CXL5285	1423
CXL5286	1436
CXL5287	1444
CXL5290	1484
CXL5291	1505
CXL5292	1521
CXL5293	1531
CXL5294	1548
CXL5295	1568
CXL5296	1587
CXL5400	1137
CXL5401	0344
CXL5402	0901
CXL5403	2342
CXL5404	0058
CXL5452	0940
CXL5453	0940
CXL5454	0940
CXL5455	0940
CXL5456	1222
CXL5457	0095
CXL5461	2095
CXL5462	2089
CXL5463	0505
CXL5465	0710
CXL5470	1487
CXL5471	1487
CXL5472	1487
CXL5473	1487
CXL5474	1487
CXL5475	3671
CXL5476	3439
CXL5480	3419
CXL5481	1103
CXL5482	2491
CXL5483	0240
CXL5484	2659
CXL5485	0676
CXL5486	2809
CXL5487	0336
CXL5491	0722
CXL5492	0722
CXL5494	0722
CXL5496	0687
CXL5500	0722
CXL5501	0722
CXL5502	0722
CXL5503	0722
CXL5504	0745
CXL5505	0722
CXL5506	0722
CXL5507	0722
CXL5508	3243
CXL5511	3696
CXL5512	3383
CXL5513	3624
CXL5514	0722
CXL5515	1584
CXL5516	1665
CXL5517	1650
CXL5518	2647
CXL5519	2649
CXL5520	0722
CXL5521	0722
CXL5522	0722
CXL5523	0722
CXL5524	1487
CXL5525	0722
CXL5526	0722
CXL5527	0722
CXL5528	3291
CXL5529	0764
CXL5530	2875
CXL5531	0766
CXL5540	0722
CXL5541	0722
CXL5542	0722
CXL5543	0469
CXL5544	0722
CXL5545	0722
CXL5546	0778
CXL5547	0725
CXL5640	2383
CXL5641	2394
CXL5642	1413
CXL5643	2387
CXL5650	2352
CXL5651	2353
CXL5652	2356
CXL5661	0404
CXL5662	2356
CXL5663	0422
CXL5664	0422
CXL5665	0422
CXL5666	0422
CXL5667	0422
CXL5673	0154
CXL5675	0147
CXL5676	4545
CXL5677	0279
CXL5680	3146
CXL5681	3147
CXL5682	3148
CXL5683	3150
CXL5684	3151
CXL5686	3153
CXL5693	2015
CXL5695	2017
CXL5697	2018
CXL5800	0087
CXL5801	0087
CXL5802	0087
CXL5803	0087
CXL5804	0087
CXL5805	0087
CXL5806	0087
CXL5808	0087
CXL5809	0087
CXL5830	0097
CXL5831	0933
CXL5832	0097
CXL5833	5517
CXL5834	0097
CXL5835	0947
CXL5836	0805
CXL5837	4523
CXL5839	1012
CXL5840	2076
CXL5841	5567
CXL5842	2081
CXL5843	1073
CXL5846	2088
CXL5847	1138
CXL5848	0612
CXL5849	1188
CXL5850	0097
CXL5851	0985
CXL5852	0097
CXL5853	0910
CXL5854	0097
CXL5855	0990
CXL5856	0097
CXL5857	0993
CXL5858	0805
CXL5859	0997
CXL5860	0200
CXL5861	1001
CXL5862	0204
CXL5863	0515
CXL5866	0206
CXL5867	1008
CXL5868	0586
CXL5869	0948
CXL5870	0097
CXL5871	0534
CXL5872	0097
CXL5873	0748
CXL5874	0097
CXL5875	0765
CXL5876	0805
CXL5877	0777
CXL5878	0805
CXL5879	0538
CXL5880	0116
CXL5881	0801
CXL5882	0122
CXL5883	0816
CXL5886	0131
CXL5887	0545
CXL5890	0145
CXL5891	0550
CXL5940	0805
CXL5941	2003
CXL5942	0805
CXL5943	3265
CXL5944	0805
CXL5945	1577
CXL5946	0805
CXL5947	3275
CXL5948	0805
CXL5949	3281
CXL5950	5589
CXL5951	3287
CXL5952	1586
CXL5953	3294
CXL5980	3189
CXL5981	1630
CXL5982	2901
CXL5986	1124
CXL5987	1107
CXL5988	1126
CXL5989	1111
CXL5990	0805
CXL5991	0258
CXL5992	1195
CXL5993	1644
CXL5994	0319
CXL5995	0267
CXL5998	1132
CXL5999	1119
CXL6002	0044
CXL6003	0281
CXL6020	3762
CXL6021	2665
CXL6022	2653
CXL6023	2695
CXL6026	2657
CXL6027	2768
CXL6030	2664
CXL6031	2855
CXL6034	2006
CXL6035	2907
CXL6038	2677
CXL6039	2974
CXL6040	2682
CXL6041	3051
CXL6042	3900
CXL6043	3117
CXL6044	2655
CXL6045	3208
CXL6050	1560
CXL6051	2179
CXL6054	2840
CXL6055	2182
CXL6058	4308
CXL6059	3601
CXL6060	2850
CXL6061	2191
CXL6064	2871
CXL6065	2197
CXL6068	2833
CXL6069	2216
CXL6072	2473
CXL6073	2340
CXL6400	1669
CXL6401	2136
CXL6402	0316
CXL6406	3648
CXL6407	3331
CXL6408	2150
CXL6409	1175
CXLM1	OEM
CXO-043B	OEM
CXP5016-201	OEM
CXP5024H-003	OEM
CXP5034H-080	OEM
CXP5034H080S	OEM
CXP5048H159S	OEM
CXP5058-051	OEM
CXP5058-088	OEM
CXP5058-133Q	OEM
CXP5058-178Q	OEM
CXP5058H-081Q	OEM
CXP5058H-087Q	OEM
CXP5058H-088	OEM
CXP5058H-114Q	OEM
CXP5068H	OEM
CXP5068H-028S	OEM
CXP5068H-055S	OEM
CXP5068H-057S	OEM
CXP5068H-SV2764	OEM
CXP5086	OEM
CXP5086H	OEM
CXP5086H-004S	OEM
CXP5086H015	OEM
CXP50116	OEM
CXP50116-101Q	OEM
CXP50116-XXX	OEM
CXP50584-540Q	OEM
CXP80116-0900	OEM
CXP80116-615Q	OEM
CXP80424	OEM
CXP80424-065S	OEM
CXP80424-079S	OEM
CXP80424-080S	OEM
CXP80424-SV4782	OEM
CXP85112B	OEM
CXP85112B-638S	OEM
CXQ88501-216	OEM
CXQ88535	OEM
CXQ88535-111	OEM
CXQ88535-111S	OEM
CXQ88535-119	OEM
CXQ-88535-119S	OEM
CXQ88535-119S	OEM
CXQ88551-1300	OEM
CXY10	OEM
CXY11	OEM
CXY11A	OEM
CXY11B	OEM
CXY11C	OEM
CXY11C09	OEM
CXY11C10	OEM
CXY11C11	OEM
CXY11D	OEM
CXY11E	OEM
CXY12	OEM
CXY13D	OEM
CXY13E	OEM
CXY13E09	OEM
CXY13E10	OEM
CXY13E11	OEM
CXY13F	OEM
CXY14A	OEM
CXY14B	OEM
CXY14C	OEM
CXY14D	OEM
CXY14E	OEM
CXY14F	OEM
CXY15A	OEM
CXY15B	OEM
CXY15C	OEM
CXY15D	OEM
CXY15E	OEM
CXY15F	OEM
CXY16A	OEM
CXY16B	OEM
CXY16C	OEM
CXY16D	OEM
CXY16E	OEM
CXY16F	OEM
CXY17A	OEM
CXY17B	OEM
CXY17C	OEM
CXY17D	OEM
CXY17E	OEM
CXY17F	OEM
CXY18A	OEM
CXY18B	OEM
CXY18C	OEM
CXY18D	OEM
CXY18E	OEM
CXY18F	OEM
CXY19	OEM
CXY19A	OEM
CXY19B	OEM
CXY20	OEM
CXY21	OEM
CXY22A	OEM
CXY22B	OEM
CXY23A	OEM
CXY23B	OEM
CXY23C	OEM
CXY23D	OEM
CXY24A	OEM
CXY24B	OEM
CY1	1030
CY2	4465
CY3F	OEM
CY4F	OEM
CY5	OEM
CY5F	OEM
CY6F	OEM
CY7	OEM
CY7F	OEM
CY10	OEM
CY20	OEM
CY40	0015
CY50	0015
CY80	0087
CY100	0071
CY501H	OEM
CY751H	OEM
CY2456	OEM
CY2457	OEM
CYAD471	OEM
CYDAC429	OEM
CYDAC430	OEM
CYN21EXI	OEM
CYN64	OEM
CYN65	OEM
CYN65EXI	OEM
CYN66	OEM
CYN71	OEM
CYN75A	OEM
CYN75B	OEM
CYN75C	OEM
CYT1549	OEM
CYT1550	OEM
CYT1551	OEM
CYT1552	OEM
CYT1553	OEM
CYT1554	OEM
CYT1555	OEM
CYT1556	OEM
CYT1557	OEM
CYT1558	OEM
CYT1559	OEM
CYT1560	OEM
CZ072	0077
CZ078	0649
CZ-092	0012
CZ092	0012
CZ-094	0012
CZ094	0012
CZ3.3	0292
CZ3.6	0188
CZ3.9	0456
CZ4.3	0533
CZ4.7	0451
CZ5	OEM
CZ5.1	0162
CZ5.6	0157
CZ6	OEM
CZ6.2	0636
CZ6.8	0025
CZ6.8A	0025
CZ6.8B	0025
CZ6.8C	OEM
CZ6.8D	0025
CZ7.5	0649
CZ7.5A	0649
CZ7.5B	0649
CZ7.5C	0649
CZ7.5D	0649
CZ8	OEM
CZ8.2	0244
CZ8.2A	0244

If replacement code is OEM, contact original manufacturer for replacement.

DEVICE TYPE	REPL CODE
CZ8.2B	0244
CZ8.2C	3100
CZ8.2D	0244
CZ9	OEM
CZ9.1	0012
CZ9.1A	0012
CZ9.1B	0012
CZ9.1C	0057
CZ9.1D	OEM
CZ10	0170
CZ10A	0170
CZ10B	0170
CZ10C	OEM
CZ10D	0170
CZ11	0317
CZ11A	0317
CZ11B	0317
CZ11C	OEM
CZ11D	0317
CZ12	0137
CZ12A	0137
CZ12B	0137
CZ12C	0137
CZ12D	0137
CZ13	0365
CZ13A	0365
CZ13B	0365
CZ13C	OEM
CZ13D	0365
CZ14	0100
CZ14A	0100
CZ14B	0100
CZ14C	OEM
CZ14D	0100
CZ15	0002
CZ15A	0002
CZ15B	0002
CZ15C	0002
CZ15D	0002
CZ16	0420
CZ16A	0420
CZ16B	0420
CZ16C	OEM
CZ16D	0420
CZ17	1649
CZ17A	1649
CZ17B	1649
CZ17C	OEM
CZ17D	1649
CZ18	0495
CZ18A	0495
CZ18B	0495
CZ18C	OEM
CZ18D	0495
CZ19	0949
CZ19A	0949
CZ19B	0949
CZ19C	OEM
CZ19D	0949
CZ20	0531
CZ20A	0531
CZ20B	0531
CZ20D	0531
CZ22	0565
CZ22A	0565
CZ22B	0565
CZ22C	OEM
CZ22D	0565
CZ24	0402
CZ24A	0402
CZ24B	0402
CZ24C	OEM
CZ24D	0402
CZ25	1606
CZ25A	1606
CZ25B	1606
CZ25C	1606
CZ25D	1606
CZ27	0441
CZ27A	0441
CZ27B	0441
CZ27C	OEM
CZ27D	0441
CZ30	0726
CZ30A	0726
CZ30B	0726
CZ30C	OEM
CZ30D	0726
CZ33	0039
CZ33A	0039
CZ33B	0039
CZ33C	OEM
CZ33D	0039
CZ36	0820
CZ36A	0350
CZ36B	0820
CZ36C	0820
CZ36D	0820
CZ39	0350
CZ39A	0350
CZ39B	0350
CZ39C	0350
CZ39D	0350
CZ43	0931
CZ43A	0931
CZ43B	0931
CZ43C	0931
CZ43D	0931
CZ47	0999

DEVICE TYPE	REPL CODE
CZ47A	0999
CZ47B	0999
CZ47C	0999
CZ47D	0999
CZ50	0502
CZ50A	0502
CZ50B	0502
CZ50C	OEM
CZ50D	OEM
CZ51	0502
CZ51A	0502
CZ51B	0502
CZ51C	0502
CZ51D	0502
CZ52	0502
CZ52A	0502
CZ52B	OEM
CZ52C	OEM
CZ52D	OEM
CZ56	0869
CZ56A	0869
CZ56B	0869
CZ56C	0869
CZ56D	0869
CZ62	0783
CZ62A	0783
CZ62B	0783
CZ62C	0783
CZ62D	0783
CZ68	2157
CZ68A	2157
CZ68B	2157
CZ68C	2157
CZ68D	2157
CZ75	1189
CZ75A	1189
CZ75B	1189
CZ75C	1189
CZ75D	1189
CZ82	0331
CZ82A	0331
CZ82B	0331
CZ82C	0331
CZ82D	0331
CZ91	1311
CZ91A	1311
CZ91B	1311
CZ91C	1311
CZ91D	1311
CZ100	0098
CZ100A	0098
CZ100B	0098
CZ100D	0098
CZ105	0149
CZ105A	0149
CZ105B	0149
CZ105C	OEM
CZ105D	OEM
CZ110	0149
CZ110A	0149
CZ110B	0149
CZ110C	0149
CZ110D	0149
CZ120	0186
CZ120A	0186
CZ120B	0186
CZ120C	0186
CZ120D	0186
CZ130	0213
CZ130A	0213
CZ130B	0213
CZ130C	0213
CZ130D	0213
CZ140	0245
CZ140A	0245
CZ140B	0245
CZ140C	0245
CZ140D	0245
CZ150	0028
CZ150A	0028
CZ150B	0028
CZ150C	0028
CZ150D	0028
CZ160	0255
CZ160A	0255
CZ160B	0255
CZ160C	0255
CZ160D	0255
CZ175	0367
CZ175A	0367
CZ175B	0367
CZ175C	OEM
CZ175D	OEM
CZ180	0367
CZ180A	0367
CZ180B	0367
CZ180C	0367
CZ180D	0367
CZ200	0421
CZ200A	0421
CZ200B	0421
CZ200C	0421
CZ200D	0421
CZ220	OEM
CZ220A	OEM
CZ220B	OEM
CZ220C	OEM
CZ220D	OEM
CZ255	1606

DEVICE TYPE	REPL CODE
CZ821	0471
CZ821A	0471
CZ822	0471
CZ823	0471
CZ823A	0471
CZ824	0471
CZ825	0471
CZ825A	0471
CZ827	0471
CZ827A	0471
CZ829	0471
CZ829A	0471
CZ935	0057
CZ935A	0057
CZ935B	0057
CZ936	0057
CZ936A	0057
CZ936B	0057
CZ937	0057
CZ937A	0057
CZ937B	0057
CZ938	0057
CZ938A	0057
CZ938B	0057
CZ939	0057
CZ939A	0057
CZ939B	0057
CZ4097	1180
CZ4097A	1180
CZ4098	0647
CZ4098A	0647
CZ4565	0471
CZ4565A	0471
CZ4566	0471
CZ4566A	0471
CZ4567	0471
CZ4567A	0471
CZ4568	0471
CZ4568A	0471
CZ4569	0471
CZ4569A	0471
CZ4570	0471
CZ4570A	0471
CZ4571	0471
CZ4571A	0471
CZ4572	0471
CZ4572A	0471
CZ4573	0471
CZ4573A	0471
CZ4574	0471
CZ4574A	0471
CZ4881	0531
CZ4881A	0262
CZ4883	0234
CZ4883A	0234
CZ5063	0205
CZ5063A	0205
CZ5064	0480
CZ5064A	0480
CZ5065	0504
CZ5065A	0504
CZ5066	0684
CZ5066A	0684
CZ5067	0225
CZ5067A	0225
CZ5068	0230
CZ5068A	0230
CZ5069	0237
CZ5069A	0237
CZ5070	1397
CZ5070A	1397
CZ5071	0247
CZ5071A	0247
CZ5072	0251
CZ5072A	0251
CZ5073	0256
CZ5073A	0256
CZ5074	0269
CZ5074A	0269
CZ5075	0274
CZ5075A	0274
CZ5076	0294
CZ5076A	0294
CZ5077	0309
CZ5077A	0309
CZ5078	0318
CZ5078A	0318
CZ5079	0320
CZ5079A	0320
CZ5080	0326
CZ5080A	0326
CZ5081	0326
CZ5081A	0326
CZ5082	0337
CZ5082A	0337
CZ5083	0347
CZ5083A	0347
CZ5084	0347
CZ5084A	0347
CZ5085	0027
CZ5085A	0027
CZ5086	0027
CZ5086A	0027
CZ5087	0266
CZ5087A	0266
CZ5088	2856
CZ5088A	2856
CZ5089	0386
CZ5089A	0386

DEVICE TYPE	REPL CODE
CZ5090	0405
CZ5090A	0405
CZ5091	0405
CZ5091A	0405
CZ5092	0426
CZ5092A	0426
CZ5093	0444
CZ5093A	0444
CZ5094	0444
CZ5094A	0444
CZ5095	0238
CZ5095A	0238
CZ5096	1190
CZ5096A	1190
CZ5097	1207
CZ5097A	1207
CZ5098	1218
CZ5098A	1218
CZ5099	1881
CZ5099A	1881
CZ5100	1255
CZ5100A	1255
CZ5101	2102
CZ5101A	2102
CZ5102	1279
CZ5102A	1279
CZ5103	2225
CZ5103A	2225
CZ5104	0605
CZ5104A	0605
CZA6D5	OEM
CZA6.8D	0062
CZA6.8D5	0062
CZA6.8D10	0062
CZA7.5D	0077
CZA7.5D5	0077
CZA7.5D10	0077
CZA8.2D	0165
CZA8.2D5	0165
CZA8.2D10	0165
CZA9.1D	0057
CZA9.1D5	0057
CZA9.1D10	0057
CZA10D	0064
CZA10D5	0064
CZA10D10	0064
CZA11D	0181
CZA11D5	0181
CZA11D10	0181
CZA12D	0052
CZA12D5	0052
CZA12D10	0052
CZA13D	0053
CZA13D5	0053
CZA13D10	0053
CZA14D	0879
CZA14D5	0879
CZA14D10	0879
CZA15D	0686
CZA15D5	0686
CZA15D10	0686
CZA16D	0445
CZA16D5	0445
CZA16D10	0445
CZA17D	0210
CZA17D5	0210
CZA17D10	0210
CZA18D	0375
CZA18D5	0375
CZA18D10	0375
CZA19D	0671
CZA19D5	0671
CZA19D10	0671
CZA20D	0700
CZA20D5	0700
CZA20D10	0700
CZA22D	0705
CZA22D5	0705
CZA22D10	0705
CZA24D	0494
CZA24D5	0494
CZA24D10	0494
CZA25D	0714
CZA25D5	0714
CZA25D10	0714
CZA27D	0455
CZA27D5	0455
CZA27D10	0455
CZA28D	0257
CZA28D5	0257
CZA28D10	0257
CZA30D	0195
CZA30D5	0195
CZA30D10	0195
CZA33D	0166
CZA33D5	0166
CZA33D10	0166
CZA36D	0010
CZA36D5	0010
CZA36D10	0010
CZA39D	0032
CZA39D5	0032
CZA39D10	0032
CZA43D	0054
CZA43D5	0054
CZA43D10	0054
CZA47D	0068
CZA47D5	0068
CZA47D10	0068

DEVICE TYPE	REPL CODE
CZA51D	0092
CZA51D5	0092
CZA51D10	0092
CZA56D	0125
CZA56D5	0125
CZA56D10	0125
CZA60D	2317
CZA60D5	2317
CZA60D10	2317
CZA62D	0152
CZA62D5	0152
CZA62D10	0152
CZA68D	0173
CZA68D5	0173
CZA68D10	0173
CZA75D	0094
CZA75D5	0094
CZA75D10	0094
CZA82D	0049
CZA82D5	0049
CZA82D10	0049
CZA87D	0104
CZA87D5	0104
CZA87D10	0104
CZA91D	0156
CZA91D5	0156
CZA91D10	0156
CZA100D	0189
CZA100D5	0189
CZA100D10	0189
CZA105D	0099
CZA105D5	0099
CZA105D10	0099
CZA110D	0099
CZA110D5	0099
CZA110D10	0099
CZA120D	0089
CZA120D5	0089
CZA120D10	0089
CZA130D	0287
CZA130D5	0287
CZA130D10	0287
CZA140D	0252
CZA140D5	0252
CZA140D10	0252
CZA150D	0340
CZA150D5	0340
CZA150D10	0340
CZA160D	0370
CZA160D5	0370
CZA160D10	0370
CZA175D	0425
CZA175D5	0425
CZA175D10	0425
CZA180D	0425
CZA180D5	0425
CZA180D10	0425
CZA200D	1473
CZA200D5	1473
CZA200D10	1473
CZAJ102	0015
CZB6.8D	0062
CZB6.8D5	0062
CZB6.8D10	0062
CZB7.5D	0077
CZB7.5D5	0077
CZB7.5D10	0077
CZB8.2D	0165
CZB8.2D5	0165
CZB8.2D10	0165
CZB8.2DD5	OEM
CZB9.1D	0057
CZB9.1D5	0057
CZB9.1D10	0057
CZB10D	0064
CZB10D5	0064
CZB10D10	0064
CZB11D	0181
CZB11D5	0181
CZB11D10	0181
CZB12D	0052
CZB12D5	0052
CZB12D10	0052
CZB13D	0053
CZB13D5	0053
CZB13D10	0053
CZB14D	0879
CZB14D5	0879
CZB14D10	0879
CZB15D	0686
CZB15D5	0686
CZB15D10	0686
CZB16D	0445
CZB16D5	0445
CZB16D10	0445
CZB17D	0210
CZB17D5	0210
CZB17D10	0210
CZB18D	0375
CZB18D5	0375
CZB18D10	0375
CZB19D	0671
CZB19D5	0671
CZB19D10	0671
CZB20D	0700
CZB20D5	0700
CZB20D10	0700
CZB22D	0705
CZB22D5	0705

DEVICE TYPE	REPL CODE
CZB22D10	0705
CZB24D	0494
CZB24D5	0494
CZB24D10	0494
CZB25D	0714
CZB25D5	0714
CZB25D10	0714
CZB30D	0195
CZB30D5	0195
CZB30D10	0195
CZB33D	0166
CZB33D5	0166
CZB33D10	0166
CZB36D	0010
CZB36D5	0010
CZB36D10	0010
CZB39D	0032
CZB39D5	0032
CZB39D10	0032
CZB43D	0054
CZB43D5	0054
CZB43D10	0054
CZB47D	0068
CZB47D5	0068
CZB47D10	0068
CZB51D	0092
CZB51D5	0092
CZB51D10	0092
CZB56D	0125
CZB56D5	0125
CZB56D10	0125
CZB62D	0152
CZB62D5	0152
CZB62D10	0152
CZB68D	0173
CZB68D5	0173
CZB68D10	0173
CZB75D	0094
CZB75D5	0094
CZB75D10	0094
CZB82D	0049
CZB82D5	0049
CZB82D10	0049
CZB91D	0156
CZB91D5	0156
CZB91D10	0156
CZB100D	0189
CZB100D5	0189
CZB100D10	0189
CZB105D	0099
CZB105D5	0099
CZB105D10	0099
CZB110D	0099
CZB110D5	0099
CZB110D10	0099
CZB120D	0089
CZB120D5	0089
CZB120D10	0089
CZB130D	0287
CZB130D5	0287
CZB130D10	0287
CZB140D	0252
CZB140D5	0252
CZB140D10	0252
CZB150D	0340
CZB150D5	0340
CZB150D10	0340
CZB160D	0370
CZB160D5	0370
CZB160D10	0370
CZB175D	0425
CZB175D5	0425
CZB175D10	0425
CZB180D	0425
CZB180D5	0425
CZB180D10	0425
CZB200D	1473
CZB200D5	1473
CZB200D10	1473
CZC6.8D	0062
CZC6.8D5	0062
CZC6.8D10	0062
CZC7.5D	0077
CZC7.5D5	0077
CZC7.5D10	0077
CZC8.2D	0165
CZC8.2D5	0165
CZC8.2D10	0165
CZC9.1D	0057
CZC9.1D5	0057
CZC9.1D10	0057
CZC10D	0064
CZC10D5	0064
CZC10D10	0064
CZC11D	0181
CZC11D5	0181
CZC11D10	0181
CZC12D	0052
CZC12D5	0052
CZC12D10	0052
CZC13D	0053
CZC13D5	0053
CZC13D10	0053
CZC14D	0879
CZC14D5	0879
CZC14D10	0879
CZC15D	0686
CZC15D5	0686
CZC15D10	0686

DEVICE TYPE	REPL CODE
CZC16D	0445
CZC16D5	0445
CZC16D10	0445
CZC17D	0210
CZC17D5	0210
CZC17D10	0210
CZC19D	0671
CZC19D5	0671
CZC19D10	0671
CZC20D	0700
CZC20D5	0700
CZC20D10	0700
CZC22D	0705
CZC22D5	0705
CZC22D10	0705
CZC24D	0494
CZC24D5	0494
CZC24D10	0494
CZC25D	0714
CZC25D5	0714
CZC25D10	0714
CZC27D	0455
CZC27D5	0455
CZC27D10	0455
CZC30D	0195
CZC30D5	0195
CZC30D10	0195
CZC33D	0166
CZC33D5	0166
CZC33D10	0166
CZC36D	0010
CZC36D5	0010
CZC36D10	0010
CZC39D	0032
CZC39D5	0032
CZC39D10	0032
CZC43D	0054
CZC43D5	0054
CZC43D10	0054
CZC47D	0068
CZC47D5	0068
CZC47D10	0068
CZC51D	0092
CZC51D5	0092
CZC51D10	0092
CZC56D	0125
CZC56D5	0125
CZC56D10	0125
CZC62D	0152
CZC62D5	0152
CZC62D10	0152
CZC68D	0173
CZC68D5	0173
CZC68D10	0173
CZC75D	0094
CZC75D5	0094
CZC75D10	0094
CZC82D	0049
CZC82D5	0049
CZC82D10	0049
CZC91D	0156
CZC91D5	0156
CZC91D10	0156
CZC100D	0189
CZC100D5	0189
CZC100D10	0189
CZC105D	0099
CZC105D5	0099
CZC105D10	0099
CZC110D	0099
CZC110D5	0099
CZC110D10	0099
CZC120D	0089
CZC120D5	0089
CZC120D10	0089
CZC130D	0287
CZC130D5	0287
CZC130D10	0287
CZC140D	0252
CZC140D5	0252
CZC140D10	0252
CZC150D	0340
CZC150D5	0340
CZC150D10	0340
CZC160D	0370
CZC160D5	0370
CZC160D10	0370
CZC175D	0425
CZC175D5	0425
CZC175D10	0425
CZC180D	0425
CZC180D5	0425
CZC180D10	0425
CZC200D	1473
CZC200D5	1473
CZC200D10	1473
CZD010	0170
CZD010-5	0170
CZD011	0181
CZD011-5	0181
CZD012	0052
CZD012-5	0052
CZD013	0053
CZD013-5	0053
CZD014	0100
CZD014-5	0100
CZD015	0686
CZD015-5	0686
CZD016	0445

DEVICE TYPE	REPL CODE
CZD016-5	0445
CZD018	0375
CZD018-5	OEM
CZD020	0700
CZD020-5	0700
CZD022	0705
CZD022-5	0705
CZD024	0494
CZD024-5	0494
CZD6.8D	0062
CZD6.8D5	0062
CZD6.8D10	0062
CZD7.5D	0077
CZD7.5D5	0077
CZD7.5D10	0077
CZD8.2D	0165
CZD8.2D5	0165
CZD8.2D10	0165
CZD9.1D	0057
CZD9.1D5	0057
CZD9.1D10	0057
CZD10D	0064
CZD10D5	0064
CZD10D10	0064
CZD11D	0181
CZD11D5	0181
CZD11D10	0181
CZD12D	0052
CZD12D5	0052
CZD12D10	0052
CZD13D	0053
CZD13D5	0053
CZD13D10	0053
CZD14D	0879
CZD14D5	0879
CZD14D10	0879
CZD15D	0686
CZD15D5	0686
CZD15D10	0686
CZD16D	0445
CZD16D5	0445
CZD16D10	0445
CZD17D	0210
CZD17D5	0210
CZD17D10	0210
CZD18D	0375
CZD18D5	0375
CZD18D10	0375
CZD19D	0671
CZD19D5	0671
CZD19D10	0671
CZD20D	0700
CZD20D5	0700
CZD20D10	0700
CZD22D	0705
CZD22D5	0705
CZD22D10	0705
CZD24D	0494
CZD24D5	0494
CZD24D10	0494
CZD25D	0714
CZD25D5	0714
CZD25D10	0714
CZD27D	0455
CZD27D5	0455
CZD27D10	0455
CZD28D	0257
CZD28D5	0257
CZD28D10	0257
CZD30D	0195
CZD30D5	0195
CZD30D10	0195
CZD33D	0166
CZD33D5	0166
CZD33D10	0166
CZD36D	0010
CZD36D5	0010
CZD36D10	0010
CZD39D	0032
CZD39D5	0032
CZD39D10	0032
CZD43D	0054
CZD43D5	0054
CZD43D10	0054
CZD47D	0068
CZD47D5	0068
CZD47D10	0068
CZD51D	0092
CZD51D5	0092
CZD51D10	0092
CZD56D	0125
CZD56D5	0125
CZD56D10	0125
CZD60D	2317
CZD60D5	2317
CZD60D10	2317
CZD62D	0152
CZD62D5	0152
CZD62D10	0152
CZD68D	0173
CZD68D5	0173
CZD68D10	0173
CZD75D	0094
CZD75D5	0094
CZD75D10	0094
CZD82D	0049
CZD82D5	0049
CZD82D10	0049
CZD87D	0104

DEVICE TYPE	REPL CODE
CZD87D5	0104
CZD87D10	0104
CZD91D	0156
CZD91D5	0156
CZD91D10	0156
CZD100D	0189
CZD100D5	0189
CZD100D10	0189
CZD105D	0099
CZD105D5	0099
CZD105D10	0099
CZD110D	0099
CZD110D5	0099
CZD110D10	0099
CZD120D	0089
CZD120D5	0089
CZD120D10	0089
CZD130D	0287
CZD130D5	0287
CZD130D10	0287
CZD140D	0252
CZD140D5	0252
CZD140D10	0252
CZD150D	0340
CZD150D5	0340
CZD150D10	0340
CZD160D	0370
CZD160D5	0370
CZD160D10	0370
CZD175D	0425
CZD175D5	0425
CZD175D10	0425
CZD180D	0425
CZD180D5	0425
CZD180D10	0425
CZD200D	1473
CZD200D5	1473
CZD200D10	1473
CZE6.8D	0062
CZE6.8D5	0062
CZE6.8D10	0062
CZE7.5D	0077
CZE7.5D5	0077
CZE7.5D10	0077
CZE8.2D	0165
CZE8.2D5	0165
CZE8.2D10	0165
CZE9.1D	0057
CZE9.1D5	0057
CZE9.1D10	0057
CZE10D	0064
CZE10D5	0064
CZE10D10	0064
CZE11D	0181
CZE11D5	0181
CZE11D10	0181
CZE12D	0052
CZE12D5	0052
CZE12D10	0052
CZE13D	0053
CZE13D5	0053
CZE13D10	0053
CZE14D	0879
CZE14D5	0879
CZE14D10	0879
CZE15D	0686
CZE15D5	0686
CZE15D10	0686
CZE16D	0445
CZE16D5	0445
CZE16D10	0445
CZE17D	0210
CZE17D5	0210
CZE17D10	0210
CZE18	0375
CZE18D5	0375
CZE18D10	0375
CZE19D	0671
CZE19D5	0671
CZE19D10	0671
CZE20D	0700
CZE20D5	0700
CZE20D10	0700
CZE22D	0705
CZE22D5	0705
CZE22D10	0705
CZE24D	0494
CZE24D5	0494
CZE24D10	0494
CZE25D	0714
CZE25D5	0714
CZE25D10	0714
CZE27D	0455
CZE27D5	0455
CZE27D10	0455
CZE30D	0195
CZE30D5	0195
CZE30D10	0195
CZE33D	0166
CZE33D5	0166
CZE33D10	0166
CZE36D	0010
CZE36D5	0010
CZE36D10	0010
CZE39D	0032
CZE39D5	0032
CZE39D10	0032
CZE43D	0054
CZE43D5	0054

DEVICE TYPE	REPL CODE
CZE43D10	0054
CZE47D	0068
CZE47D5	0068
CZE47D10	0068
CZE51D	0092
CZE51D5	0092
CZE51D10	0092
CZE56D	0125
CZE56D5	0125
CZE56D10	0125
CZE62D	0152
CZE62D5	0152
CZE62D10	0152
CZE68D	0173
CZE68D5	0173
CZE68D10	0173
CZE75D	0094
CZE75D5	0094
CZE75D10	0094
CZE82D	0049
CZE82D5	0049
CZE82D10	0049
CZE91D	0156
CZE91D5	0156
CZE91D10	0156
CZE100D	0189
CZE100D5	0189
CZE100D10	0189
CZE105D	0099
CZE105D5	0099
CZE105D10	0099
CZE110D	0099
CZE110D5	0099
CZE110D10	0099
CZE120D	0089
CZE120D5	0089
CZE120D10	0089
CZE130D	0287
CZE130D5	0287
CZE130D10	0287
CZE140D	0252
CZE140D5	0252
CZE140D10	0252
CZE150D	0340
CZE150D5	0340
CZE150D10	0340
CZE160D	0370
CZE160D5	0370
CZE160D10	0370
CZE175D	0425
CZE175D5	0425
CZE175D10	0425
CZE180D	0425
CZE180D5	0425
CZE180D10	0425
CZE200D	1473
CZE200D5	1473
CZE200D10	1473
CZF1.8D	OEM
CZF1.8D5	OEM
CZF1.8D10	OEM
CZF2.0D	OEM
CZF2.0D5	OEM
CZF2.0D10	OEM
CZF2.2D	OEM
CZF2.2D5	OEM
CZF2.2D10	OEM
CZF2.4D	1276
CZF2.4D5	1276
CZF2.4D10	1276
CZF2.7D	0760
CZF2.7D5	0760
CZF2.7D10	0760
CZF3.0D	0118
CZF3.0D5	0118
CZF3.0D10	0118
CZF3.3D	0299
CZF3.3D5	0299
CZF3.3D10	0299
CZF3.6D	0376
CZF3.6D5	0376
CZF3.6D10	0376
CZF3.9D	0036
CZF3.9D5	0036
CZF3.9D10	0036
CZF4.3D	0275
CZF4.3D5	0275
CZF4.3D10	0275
CZF4.7D	0140
CZF4.7D5	0140
CZF4.7D10	0140
CZF5.1D	0041
CZF5.1D5	0041
CZF5.1D10	0041
CZF5.6D	0253
CZF5.6D5	0253
CZF5.6D10	0253
CZF6.2D	0471
CZF6.2D5	0471
CZF6.2D10	0471
CZF6.8D	0062
CZF6.8D5	0062
CZF6.8D10	0062
CZF7.5D	0077
CZF7.5D5	0077
CZF7.5D10	0077
CZF8.2D	0165
CZF8.2D5	0165
CZF8.2D10	0165

DEVICE TYPE	REPL CODE
CZF9.1D	0057
CZF9.1D5	0057
CZF9.1D10	0057
CZF10D	0064
CZF10D5	0064
CZF10D10	0064
CZF12D	0052
CZF12D5	0052
CZF12D10	0052
CZG2.4D	1276
CZG2.4D5	1276
CZG2.4D10	1276
CZG2.7D	0760
CZG2.7D5	0760
CZG2.7D10	0760
CZG3.0D	0118
CZG3.0D5	0118
CZG3.0D10	0118
CZG3.3D	0299
CZG3.3D5	0299
CZG3.3D10	0299
CZG3.6D	0376
CZG3.6D5	0376
CZG3.6D10	0376
CZG3.9D	0036
CZG3.9D5	0036
CZG3.9D10	0036
CZG4.3D	0275
CZG4.3D5	0275
CZG4.3D10	0275
CZG4.7D	0140
CZG4.7D5	0140
CZG4.7D10	0140
CZG5.1D	0041
CZG5.1D5	0041
CZG5.1D10	0041
CZG5.6D	0253
CZG5.6D5	0253
CZG5.6D10	0253
CZG6.2D	0471
CZG6.2D5	0471
CZG6.2D10	0471
CZG6.8D	0062
CZG6.8D5	0062
CZG6.8D10	0062
CZG7.5D	0077
CZG7.5D5	0077
CZG7.5D10	0077
CZG8.2D	0165
CZG8.2D5	0165
CZG8.2D10	0165
CZG9.1D	0057
CZG9.1D5	0057
CZG9.1D10	0057
CZG10D	0064
CZG10D5	0064
CZG10D10	0064
CZG12D	0052
CZG12D5	0052
CZG12D10	0052
D	OEM
D/7A	OEM
D002	OEM
D004	0015
D006	0127
D008	0004
D009	0224
D-00169C	0143
D-00184R	0133
D-00204R	0143
D-00269C	0143
D-00284R	0143
D-00369C	0649
D-00384R	0015
D-00469C	0636
D-00484R	0143
D-00569C	0012
D-00669C	0143
D01-100	0015
D02S	0005
D02Z6-2W	0471
D02Z62W	0471
D-05	0015
D05	0015
D05G	0023
D010	0123
D018	0004
D019	0136
D020	0633
D021	0633
D030	0004
D031	0004
D031(CHAN.MASTER)	0079
D038	0164
D043	0004
D048	0016
D053	0016
D058	0144
D059	0004
D063	0136
D069	0016
D072	0224
D073	0136
D078	0633
D079	0050
D080	0160
D081	0160
D082	1675
D083	0208

DEVICE TYPE	REPL CODE
D084	0233
D085	0208
D086	0050
D087	0007
D088	0144
D093	0019
D0123WX	2356
D0201YR	3331
D-0501	OEM
D0515L	3837
D0520L	3837
D0535J	OEM
D0535W	OEM
D0550J	OEM
D0555W	OEM
D0565J	OEM
D0570W	OEM
D0J020A90	0123
D0J020A900	0123
D0.2S10	OEM
D0.22/80X41	OEM
D0.22/100X11	OEM
D0.22/125X31	OEM
D0.22/150X31	OEM
D0.24S10	OEM
D0.25S8	OEM
D0.42S5	OEM
D1/2-12	OEM
D1-7E	OEM
D1-8	0012
D1-12B	0794
D1-12E	OEM
D1-26	0015
D1-S2S	0015
D1-S2RB	OEM
D1A	0244
D1AT015550	0133
D1B	0196
D1C	0196
D1CA	OEM
D1CA20R	0015
D1D	0071
D1E	0015
D1F	0102
D1G	0012
D1H	0015
D1I0A243A0	0124
D1J	0015
D1J70542	0143
D1J70543	0143
D1J70544	0015
D1J70545	0133
D1JA	OEM
D1JA20	OEM
D1K	0071
D1K20	0023
D1K20H	1090
D1K40	0023
D1K58	OEM
D1L	0015
D1M	0157
D1M(TRANS)	OEM
D1N20R	5473
D1N20R-TA	OEM
D1N20R-TR	OEM
D1N4002	0080
D1N4148	0124
D1NL20	1090
D1NL20-TR2	1090
D1NS4	OEM
D1NS6	OEM
D1NS6-TA2	OEM
D1R5GZ61	0071
D1R5GZ61FA	0071
D1R20	0182
D1R20Z	0609
D1R38	0016
D1R40Z	0795
D1R60	0182
D1R80	0182
D1R100	0182
D1R150	0182
D1S	0365
D1S34	0143
D1S1553	0133
D1S1554	0133
D1S1555	0133
D1S1887FA	0015
D-1S2473	0133
D1S2775-1	0499
D-1SS133	0133
D1T	0100
D1TH61	0071
D1U	0244
D1V0001320	0124
D1V20	0023
D1V40	0182
D1V60	0182
D1VT001310	0124
D1VT001320	0124
D1VT024720	0133
D1W	0441
D1Y	0127
D1Z	0002
D1ZBLU	0002
D1ZRED	0100
D1ZVIO	0365
D1ZYEL	0002
D1.5/200	OEM

DEVICE TYPE	REPL CODE
D1.5/400	OEM
D1.5/600	OEM
D1.5/800	OEM
D1.5/1000	OEM
D1.5-400	OEM
D1.5-600	OEM
D1.5-800	OEM
D1.5-1000	OEM
D-2	0143
D2-1	0015
D2-12B	OEM
D2-12E	OEM
D2B000R2M0	3883
D2B000RH10	0023
D2B000RU20	0023
D2B000RU30	0283
D2B000SR2M	3883
D2B00EU01A	0023
D2B00RU1P0	0023
D2BF00RU20	0023
D2BF0RU4A0	0102
D2BFMI15MR	1097
D2BFMI15MS	1802
D2BH00EU2Z	0023
D2BHRM11C0	0071
D2BJ00RU4Z	0031
D2BT0EU01A	0023
D2E	0039
D2F	OEM
D2G	0794
D2G-1	0794
D2G-2	0794
D2G-3	0137
D2G-4	0137
D2H	0015
D2J	0015
D2L0040070	0023
D2L0TVR06G	0015
D2LFRGP30G	0017
D2LT040070	0023
D2M	0201
D2M5001	OEM
D2M5002	OEM
D2M5003	OEM
D2N	0288
D2R31	0143
D2R38	0016
D2S4MF	OEM
D2S100	0031
D2S200	0031
D2S400	0031
D2S800	0102
D2SBA	OEM
D2SBA40	0296
D2T918	0259
D2T2904	0037
D2T2904A	0037
D2T2905	0037
D2T2905A	0037
D2U	0143
D2Y	0757
D2Z	0012
D2Z-2	0441
D3-200	0031
D3-400	0031
D3-600	0031
D3-800	0087
D3-1300	OEM
D3A	0920
D3F	0012
D3G	5333
D3H	0636
D3N	0794
D3R	0015
D3R38	0015
D3R39	0015
D3S3M	0031
D3S4M	0031
D3S6M-F	OEM
D3S800	OEM
D3S1000	OEM
D3S1200	OEM
D3SB10	0729
D3SB20	OEM
D3SB60	0737
D3SB60F	0737
D3SB1057	OEM
D3SBA10	6362
D3SBA20	OEM
D3SBA40	1045
D3SBA50	OEM
D3SBA60	1045
D3SBA60-4103	1045
D3SBA601-4103	OEM
D3U	0015
D3V	0015
D3W	0137
D3Y	0636
D3Z	0015
D4	0196
D4-2	0030
D4-28G	OEM
D4BB10	OEM
D4BB20	OEM
D4BB40	OEM
D4C28	0086
D4C29	0086
D4C30	0086
D4C31	0086

DEVICE TYPE	REPL CODE
D4D20	0086
D4D21	0086
D4D22	0007
D4D24	0198
D4D25	0198
D4D26	0198
D4G	0201
D4H	0039
D4J	0039
D4L	0071
D4M	0015
D4N	0794
D4P	0100
D4R	0143
D4R26	0015
D4SB60L	OEM
D4SB60L-F	OEM
D4.5-400	OEM
D4.5-700	OEM
D4.5-1100	OEM
D4.5-1400	OEM
D5	0484
D5B	0002
D5D	OEM
D5E29	OEM
D5E35	OEM
D5E36	OEM
D5E-37	2136
D5E37	2136
D5E-43	1175
D5E43	1175
D5E-44	2136
D5E44	2136
D5E-45	1175
D5E45	1175
D5F840	OEM
D5F860	OEM
D5FB10	OEM
D5FB10F	OEM
D5FB20	OEM
D5FB40	OEM
D5FB60	OEM
D5G	0023
D5H	0023
D5J37	1669
D5J-43	1669
D5J43	1669
D5J-44	1669
D5J44	1669
D5J45	1669
D5K1	OEM
D5K2	OEM
D5KC20	1236
D5KC20H	1236
D5KC20R	1727
D5KC20RH	1727
D5KC40	0972
D5KC40H	OEM
D5KC40R	OEM
D5KD20	OEM
D5KD20H	OEM
D5KD40	OEM
D5LC20	1236
D5LCA20	1236
D5R35	0015
D5R39	0015
D5S3M	OEM
D5S4M	OEM
D5SB10	1040
D5SB20	1040
D5SB20F	OEM
D5SB40	1045
D5SB60	1045
D5SBA10	1040
D5SC4M	OEM
D5SC4MR	OEM
D5V	0133
D5W	0636
D5Z	OEM
D6	1080
D6/400	OEM
D6/400A	OEM
D6/800	OEM
D6/800A	OEM
D6/1200	OEM
D6/1200A	OEM
D6/1400	OEM
D6/1400A	OEM
D6/1600	OEM
D6/1600A	OEM
D6-700	OEM
D6-1100	OEM
D6C	1323
D6HZ	0015
D6K20	1145
D6K20H	1127
D6K20R	OEM
D6K20RH	OEM
D6K40	1145
D6K40R	OEM
D6M	0244
D6SB60L	OEM
D6U	0565
D6.2	0636
D7	1080
D7A30	0086
D7A31	0086
D7A32	0086
D7B2	1966

DEVICE TYPE	REPL CODE
D7C1	OEM
D7C2	OEM
D7C3	OEM
D7D1	OEM
D7D2	OEM
D7D3	OEM
D7D13	OEM
D7E	0133
D7E1	OEM
D7E2	OEM
D7E3	OEM
D7F1	OEM
D7F2	OEM
D7F3	OEM
D7F4	OEM
D7G1	OEM
D7G2	OEM
D7G3	OEM
D7L	1649
D7M	0030
D7N	0030
D7S	0244
D7ST4010(A)	OEM
D7ST4013(A)	OEM
D7ST4015(A)	OEM
D7ST4510(A)	OEM
D7ST4513(A)	OEM
D7ST5010(A)	OEM
D7Z	0133
D8-400B	OEM
D8-400C	OEM
D8-700B	OEM
D8-700C	OEM
D8-800B	OEM
D8-800C	OEM
D8-1100B	OEM
D8-1100C	OEM
D8-1200B	OEM
D8-1200C	OEM
D8-1400B	OEM
D8-1400C	OEM
D8-1600B	OEM
D8-1600C	OEM
D8G	0102
D8HZ	0071
D8L	0102
D8LCA20	1236
D8M	0102
D8N	OEM
D8P	0365
D8U	1083
D9B	0019
D9C	OEM
D9D	0019
D9E	0143
D9F	OEM
D9G	0649
D9I	0019
D9K	0019
D9L	0143
D9M	1083
D9SZ	0143
D9V	0019
D10B551-2,3	OEM
D10B553-2,3	OEM
D10B555-2,3	OEM
D10B556-2,-3	OEM
D10B1051	0127
D10B1055	0127
D10C573-2,-3	OEM
D10G1051	0127
D10G1052	0127
D10H551-2,-3	OEM
D10H553-2,-3	OEM
D10N400B	OEM
D10N400C	OEM
D10N800B	OEM
D10N800C	OEM
D10N1200B	OEM
D10N1200C	OEM
D10N1400B	OEM
D10N1400C	OEM
D10N1600B	OEM
D10N1600C	OEM
D10P	OEM
D10S6CM	OEM
D10SC3M	1021
D10SC3MR	OEM
D10SC4M	1021
D10SC4MR	OEM
D10VD60Z	OEM
D11	0208
D11A	0208
D11B	0208
D11B551-2,-3	OEM
D11B551-2,3	OEM
D11B552-2,3	OEM
D11B555-2,-3	OEM
D11B556-2,-3	OEM
D11B560-2,3	OEM
D11B1052	OEM
D11B1055	OEM
D11C	0208
D11C1B1	0086
D11C1F1	0079
D11C2D1B20	0079
D11C3B1	0086
D11C3F1	0595
D11C5B1	0086
D11C5F1	0079
D11C7B1	0086
D11C7F1	OEM
D11C10B1	0086
D11C10F1	0142
D11C11B1	0086
D11C11F1	0142
D11C201B20	0142
D11C203B20	0142
D11C205B20	0142
D11C207B20	0142
D11C210B20	0142
D11C211B20	0142
D11C551-2,-3	OEM
D11C553-2,-3	OEM
D11C557-2,-3	OEM
D11C702	OEM
D11C704	OEM
D11C710	OEM
D11C1051	OEM
D11C1053	OEM
D11C1057	OEM
D11C1536	0630
D11D	0208
D11E	0208
D11E404	OEM
D11E405	OEM
D11E406	OEM
D11E407	OEM
D11F	0208
D11G	0208
D11GN	0208
D11H	0208
D11J	0208
D11K	0208
D11L	0208
D11M	0208
D110R	0208
D11R	0208
D11X	0208
D11Y	0208
D12	0074
D12-03	OEM
D12-05	OEM
D12-10A	OEM
D12-15A	OEM
D12-20	OEM
D12-28G	OEM
D12-30	OEM
D12-35	OEM
D12-50	OEM
D12A	0074
D12A8	OEM
D12B	0074
D12C	0074
D12E	0074
D12E026	OEM
D12E109	OEM
D12E126	OEM
D12F	0074
D12G	0074
D12GN	0074
D12H	0074
D12J	0074
D12K	0074
D12L	0074
D12M	0074
D120R	0074
D12R	0074
D12X	0074
D12X010	OEM
D12X011	OEM
D12X012	OEM
D12X013	OEM
D12X014	OEM
D12X015	OEM
D12X043	0861
D12X047	1480
D12X070	OEM
D12X084A	OEM
D12Y	0074
D13	0440
D13H1	OEM
D13H2	OEM
D13K1	0316
D13K2	OEM
D13K3	OEM
D13T1	0316
D13T2	0316
D13T3	0316
D13T4	0316
D13TDS442X	0124
D13TGMA010	0124
D14	0150
D14T000880	OEM
D14U001660	OEM
D15(2S)	0130
D15-03	OEM
D15-05	OEM
D15-10A	OEM
D15-15A	OEM
D15-20	OEM
D15-30	OEM
D15-35	OEM
D15-50	OEM
D15-100/5	OEM
D15-100/12	OEM
D15-100/24	OEM
D15-100/28	OEM
D15A	0015
D15A(TRANS)	0130
D15B	0130
D15C	0015
D15C(TRANS)	0130
D15D	0130
D15E	0130
D15F	0130
D15G	0130
D15GN	0130
D15H	0130
D15J	0130
D15K	0130
D15L	0130
D15M	0130
D150R	0130
D15R	0130
D15T015850	0124
D15X	0130
D15Y	0130
D16(2S)	0130
D16A	0130
D16B	0130
D16C	0130
D16D	0130
D16E	0130
D16E7	0016
D16E9	0016
D16EC18	0016
D16F	0130
D16G	0130
D16G6	0127
D16GN	0130
D16H	0130
D16J	0130
D16K	0130
D16K1	0127
D16K2	0127
D16K3	0127
D16K4	0127
D16L	0130
D16M	0130
D160R	0130
D16P1	0400
D16P2	0400
D16P3	OEM
D16P4	0400
D16PC2	OEM
D16R	0130
D16U3	0196
D16U4	0196
D16X	0130
D16Y	0130
D17	0130
D17A	0130
D17C	0130
D17D	0130
D17E	0130
D17F	0130
D17GN	0130
D17H	0130
D17J	0130
D17K	0130
D17L	0130
D17M	0130
D17R	0130
D17S800C	OEM
D17S1000C	OEM
D17S1200C	OEM
D17SR800C	OEM
D17SR1000C	OEM
D17SR1200C	OEM
D17T001320	0124
D17T024720	0133
D17U001310	0124
D17X	0130
D17Y	0130
D18	0015
D18(TRANS)	0867
D18A12	0161
D19	0208
D19A	0208
D19B	0208
D19C	0208
D19E	0208
D19F	0208
D19G	0208
D19GN	0208
D19H	0208
D19J	0208
D19K	0208
D19M	0208
D19Y	0208
D20	0600
D20(2S)	0208
D20A	0208
D20B	0208
D20C	0208
D20D	0208
D20E	0208
D20F	0208
D20G	0208
D20GN	0208
D20H	0208
D20J	0208
D20K	0208
D20L	0208
D20M	0208
D200R	0208
D20R	0208
D20VT60	OEM
D20X	0208
D20Y	0208
D21(2S)	0208
D21A	0208
D21B	0208
D21C	0208
D21D	0208
D21E	0208
D21F	0208
D21G	0208
D21GN	0208
D21H	0208
D21J	0208
D21K	0208
D21L	0208
D21M	0208
D210R	0208
D21R	0208
D21S1000C	OEM
D21S1200C	OEM
D21S1400C	OEM
D21X	0208
D21Y	0208
D22	0208
D22A	0208
D22B	0208
D22C	0208
D22D	0208
D22E	0208
D22F	0208
D22G	0208
D22GN	0208
D22H	0208
D22J	0208
D22K	0208
D22L	0208
D22M	0208
D22N	0087
D220R	0208
D22R	0208
D22X	0208
D22Y	0208
D23(2S)	0208
D23A	0208
D23B	0208
D23C	0208
D23C2000G	OEM
D23D	0208
D23E	0208
D23F	0208
D23F130TAL	0015
D23FDFA05G	0023
D23FDFC10E	0023
D23FDFH10G	0023
D23FFH10TG	0023
D23G	0208
D23GDSF10T	1892
D23GN	0208
D23H	0208
D23HDSA12T	0071
D23J	0208
D23K	0208
D23L	0208
D23M	0208
D230R	0208
D23R	0208
D23TDF05G	0023
D23TDFD05G	0023
D23TDSA12T	0087
D23TDSF10B	0080
D23TDSF10T	0015
D23TFD05TG	0023
D23TFH10TG	0023
D23X	0208
D23Y	0208
D24	0142
D24(TRANS)	0142
D24/400B	OEM
D24/400C	OEM
D24/800B	OEM
D24/800C	OEM
D24/1200B	OEM
D24/1200C	OEM
D24/1400B	OEM
D24/1400C	OEM
D24/1600B	OEM
D24/1600C	OEM
D24/1800B	OEM
D24/1800C	OEM
D24A	0142
D24A3391	0111
D24A3391A	0111
D24A3393	0111
D24A3394	0144
D24A3900	0111
D24A3900A	0111
D24B(2S)	0142
D24C(2S)	0142
D24CK(2S)	0142
D24D(2S)	0142
D24DR	0142
D24E(2S)	0142
D24F(2S)	0142
D24F00V09G	0023
D24G	0142
D24GN	0142
D24H	0142
D24J	0142
D24K(2S)	OEM
D24KC(2S)	0142
D24KD(2S)	0142
D24KE(2S)	0142
D24L	0142
D24M	0142
D24N400B	OEM
D24N400C	OEM
D24N400U	OEM
D24N800B	OEM
D24N800C	OEM
D24N800U	OEM
D24N1200B	OEM
D24N1200C	OEM
D24N1200U	OEM
D24N1400B	OEM
D24N1400C	OEM
D24N1400U	OEM
D24N1600B	OEM
D24N1600C	OEM
D24N1600U	OEM
D24NR400B	OEM
D24NR400C	OEM
D24NR400U	OEM
D24NR800B	OEM
D24NR800C	OEM
D24NR800U	OEM
D24NR1200B	OEM
D24NR1200C	OEM
D24NR1200U	OEM
D24NR1400B	OEM
D24NR1400C	OEM
D24NR1400U	OEM
D24NR1600B	OEM
D24NR1600C	OEM
D24NR1600U	OEM
D240R	0142
D24R	0142
D24X	0142
D24Y(2S)	0142
D24YC	0142
D24YD(2S)	0142
D24YE(2S)	0142
D24YF(2S)	0142
D24YK(2S)	0142
D24YKC	0142
D24YL	0142
D24YLC(2S)	0142
D24YLD(2S)	0142
D24YLE	0142
D24YLF	0142
D24YM	0142
D24YMC	0142
D24YME	0142
D25	0208
D25-02C	2234
D25A	0208
D25B	0208
D25C	0208
D25C(TRANS)	0208
D25D	0208
D25E	0208
D25G	0208
D25GN	0208
D25H	0208
D25J	0208
D25K	0208
D25L	0208
D25M	0208
D250R	0208
D25R	0208
D25T5566B0	0015
D25X	0208
D25Y	0208
D26	0867
D26A(2S)	0867
D26B(2S)	0867
D26B1	0127
D26B2	0127
D26C(2S)	0867
D26C1	0127
D26C2	0127
D26C3	0127
D26C4	0111
D26C5	0111
D26D	0867
D26E	0867
D26E-1	0111
D26E1	OEM
D26E-2	0144
D26E2	0127
D26E-3	0144
D26E3	OEM
D26E-4	0007
D26E4	OEM
D26E-5	0111
D26E5	OEM
D26E-6	0144
D26E6	OEM
D26E-7	0111
D26E7	OEM
D26F	0867
D26G	0867
D26G-1	0007
D26G1	0127
D26GN	0867
D26H	0867
D26J	0867
D26K	0867
D26L	0867
D26M	0867
D260R	0867
D26P1	OEM
D26P2	OEM
D26P3	OEM
D26R	0867
D26X	0867
D26Y	0867
D27	0208
D27C1	0042
D27C2	0042
D27C3	0042
D27C4	0042
D27C256D-20	OEM
D27C512D-15	OEM
D27C5120-15	OEM
D27D1	1267
D27D2	1267
D27D3	1267
D27D4	1267
D28	0015
D28(TRANS)	0042
D28-08	0102
D28A	0042
D28A05	0551
D28A0P04Q0	1021
D28A1	0042
D28A2	0042
D28A3	0042
D28A4	0042
D28A5	0042
D28A5KQ600	OEM
D28A6	0042
D28A7	0042
D28A9	0042
D28A10	0042
D28A12	0042
D28A13	0042
D28B	0042
D28B1	OEM
D28C	0042
D28C1	OEM
D28C2	OEM
D28C3	OEM
D28C4	OEM
D28C5	OEM
D28C7	OEM
D28C8	OEM
D28D	0042
D28D(2S)	0042
D28D07	0595
D28D1	0042
D28D2	0042
D28D3	0042
D28D4	0042
D28D5	0042
D28D7	0042
D28D8	0560
D28D10	0042
D28E	0042
D28E(2S)	0042
D28F	0042
D28F00BB4C	0023
D28F00BB40	0023
D28F00SR2M	3883
D28F30DF20	0087
D28F300F60	0031
D28FMI15MS	0376
D28G	0042
D28G10E100	0071
D28GN	0042
D28H	0042
D28H20E100	0071
D28HRM11C0	0071
D28J	0042
D28K	0042
D28L	0042
D28L30DF20	0031
D28M	0042
D280R	0042
D28R	0042
D28T011E10	0015
D28T011E20	0015
D28T011ES1	0015
D28T0ELS40	0023
D28T10E100	0071
D28T10ELS6	0023
D28X	0042
D28Y	0042
D29	0042
D29-04HZ	OEM
D29-08	1217
D29A	0042
D29A4	0037
D29A5	0037
D29A6	0037
D29A7	0861
D29A8	0861
D29A9	0037
D29A10	0861
D29A11	0861
D29B	0042
D29E	0042
D29E01	0532
D29E01J1	0791
D29E02	0532
D29E02J1	0532
D29E04	0532
D29E04J1	0532
D29E05	0532
D29E05J1	0532
D29E06	0532
D29E06J1	0532
D29E08	0791
D29E08J1	0791
D29E09	0791
D29E09J1	0791
D29E1	0037
D29E1-J1	0791
D29E1J1	0791
D29E2	0037
D29E2-J1	0791
D29E2J1	0791
D29E4	0037
D29E4-J1	0791
D29E4J1	0791
D29E5	0037
D29E5-J1	0791
D29E5J1	0791
D29E6	0037
D29E6-J1	0791
D29E6J1	0791
D29E7	0037
D29E7J1	0791
D29E8	0037
D29E8J1	0791
D29E9	0791
D29E9-J1	0791
D29E9J1	0791
D29E10	0791
D29E10-J1	0791
D29E10J1	0791
D29F	0042
D29F1	0693
D29F2	0693
D29F3	0693
D29F4	0693
D29F5	0037
D29F6	0037
D29F7	0037
D29GN	0042
D29J	0042
D29K	0042
D29L	0042
D29M	0042
D290R	0042
D29R	0042
D29X	0042
D29Y	0042
D30	0208
D30(2S)	0208
D30-0(2S)	0208
D30-0R	0208
D30-N(2S)	0208
D30-O	0208
D30-OR	0208
D30A	0208
D30A1	0037
D30A2	0037
D30A3	0037
D30A4	0037
D30A5	0037
D30B	0208
D30C	0208
D30D	0208
D30E	0208
D30F	0208
D30G	0208
D30GN	0208
D30H	0208
D30J	0208
D30K	0208
D30L	0208
D30M	0208
D30N	0208
D300R	0208
D30P	0208
D30R	0208
D30S3	OEM
D30S4	OEM
D30S1000C	OEM
D30S1200C	OEM
D30S1400C	OEM
D30VT60	OEM
D30X	0208
D30Y	0208
D31	0208
D31-02C	OEM
D31B	0208
D31B(2S)	0208
D31C	0208
D31D(2S)	0208
D31E	0208
D31F	0208
D31G	0208
D31GN	0208
D31H	0208
D31J	0208
D31K	0208
D31L	0208
D31M	0208
D310R	0208
D31R	0208
D31X	0208
D31Y	0208

DEVICE TYPE	REPL CODE
D32	0208
D32A	0208
D32B	0208
D32C	0208
D32D	0208
D32E	0208
D32F	0208
D32G	0208
D32GN	0208
D32H	0208
D32H1	0324
D32H2	0324
D32H4	0324
D32H5	0324
D32H7	0715
D32H8	0715
D32H9	0715
D32J	0208
D32K	0208
D32K1	0079
D32K2	0079
D32L	0208
D32L1	OEM
D32L2	OEM
D32L3	OEM
D32L4	OEM
D32L5	OEM
D32L6	OEM
D32M	0208
D32OR	0208
D32P1	0111
D32P2	0111
D32P3	0111
D32P4	0111
D32R	0208
D32S1	0400
D32S2	0400
D32S3	0400
D32S4	0400
D32S5	0264
D32S6	1212
D32S7	1212
D32S8	0264
D32S9	0264
D32S10	1212
D32W7	0715
D32W12	0079
D32X	0208
D32Y	0208
D33	OEM
D33(2S)	0007
D33A	0007
D33B	0007
D33C(2S)	0007
D33D	0007
D33D1	OEM
D33D2	OEM
D33D3	OEM
D33D4	OEM
D33D5	OEM
D33D6	OEM
D33D21	0324
D33D21-J1	0218
D33D21J1	0218
D33D22	0324
D33D22-J1	0218
D33D22J1	0218
D33D22J2	0018
D33D24	0324
D33D24-J1	0595
D33D24J1	0595
D33D25	0324
D33D25-J1	0218
D33D25J1	0218
D33D26	0324
D33D26-J1	0218
D33D26J1	0218
D33D27	0324
D33D27J1	0218
D33D28	0016
D33D28J1	0595
D33D29	0324
D33D29-J1	0218
D33D29J1	0218
D33D30	0324
D33D30-J1	0218
D33D30J1	0218
D33E	0007
D33E01J1	0532
D33E02J1	0532
D33F	0007
D33G	0007
D33GN	0007
D33H	0007
D33J	0007
D33K	0007
D33K1	2790
D33K2	OEM
D33K3	OEM
D33L	0007
D33OR	0007
D33P2	OEM
D33X	0007
D33Y	0007
D34	0208
D34/400B	OEM
D34/400C	OEM
D34/800B	OEM
D34/800C	OEM
D34/1200B	OEM
D34/1200C	OEM
D34/1400B	OEM
D34/1400C	OEM
D34/1600B	OEM
D34/1600C	OEM
D34/1800B	OEM
D34/1800C	OEM
D34A	0208
D34B	0208
D34C	0208
D34C1	OEM
D34C2	OEM
D34C3	OEM
D34C4	OEM
D34C5	OEM
D34C6	OEM
D34D	0208
D34E	0208
D34F	0208
D34G	0208
D34GN	0208
D34H	0208
D34J	0208
D34J1	0037
D34J2	0037
D34J4	0037
D34J5	0037
D34K	0208
D34L	0208
D34M	0208
D34N400B	OEM
D34N400C	OEM
D34N400U	OEM
D34N800B	OEM
D34N800C	OEM
D34N800U	OEM
D34N1200B	OEM
D34N1200C	OEM
D34N1200U	OEM
D34N1400B	OEM
D34N1400C	OEM
D34N1400U	OEM
D34N1600B	OEM
D34N1600C	OEM
D34N1600U	OEM
D34NR400B	OEM
D34NR400C	OEM
D34NR400U	OEM
D34NR800B	OEM
D34NR800C	OEM
D34NR800U	OEM
D34NR1200B	OEM
D34NR1200C	OEM
D34NR1200U	OEM
D34NR1400B	OEM
D34NR1400C	OEM
D34NR1400U	OEM
D34NR1600B	OEM
D34NR1600C	OEM
D34NR1600U	OEM
D34OR	0208
D34R	0208
D34X	0208
D34Y	0208
D35(2S)	0208
D36	0208
D36A	0208
D36B	0208
D36C	0208
D36D	0208
D36E	0208
D36F	0208
D36G	0208
D36GN	0208
D36H	0208
D36J	0208
D36K	0208
D36L	0208
D36M	0208
D36OR	0208
D36R	0208
D36X	0208
D36Y	0208
D37(2S)	0208
D37A(2S)	0208
D37B(2S)	0208
D37C(2S)	0208
D37D	0208
D37E	0208
D37F	0208
D37G	0208
D37GN	0208
D37H	0208
D37K	0208
D37L	0208
D37M	0208
D37OR	0208
D37R	0208
D37X	0208
D37Y	0208
D38	0208
D38A	0208
D38B	0208
D38C	0208
D38D	0208
D38E	0208
D38F	0208
D38G	0208
D38GN	0208
D38H	0208
D38H1	0079
D38H1-3	OEM
D38H1-6	0079
D38H2	0079
D38H3	0079
D38H4	0861
D38H4-6	0861
D38H5	0861
D38H6	0861
D38H7	0861
D38H8	0861
D38H9	0861
D38J	0208
D38K	0208
D38L	0208
D38L1	2797
D38L1-3	OEM
D38L1-6	OEM
D38L2	2797
D38L3	2797
D38L4	2797
D38L4-6	OEM
D38L5	2797
D38L6	2797
D38M	0208
D38OR	0208
D38R	0208
D38S1	0079
D38S1-4	OEM
D38S1-10	OEM
D38S2	OEM
D38S3	OEM
D38S4	OEM
D38S5	0079
D38S6	OEM
D38S7	OEM
D38S8	0079
D38S8-10	OEM
D38S9	0079
D38S10	OEM
D38V1	0715
D38V2	0715
D38V3	0715
D38W5-11	OEM
D38W7	0861
D38W8	0861
D38W8-10	OEM
D38W9	OEM
D38W10	OEM
D38W12	0861
D38W13	OEM
D38W13-14	OEM
D38W14	OEM
D38X	0208
D38Y	0208
D38Y1-3	0079
D39C1	OEM
D39C1-3	OEM
D39C1-6	OEM
D39C2	OEM
D39C3	OEM
D39C4	OEM
D39C4-6	OEM
D39C5	OEM
D39C6	OEM
D39J1	0037
D39J1-3	OEM
D39J1-6	0079
D39J2	0037
D39J3	0037
D39J4	0037
D39J4-6	OEM
D39J5	0037
D39J6	0037
D39J7	0477
D39J8	0477
D39J9	0477
D40C1	2258
D40C2	3533
D40C3	3534
D40C4	2258
D40C5	3533
D40C6	OEM
D40C7	2258
D40C8	3533
D40D	OEM
D40D1	0566
D40D1F	1984
D40D2	0566
D40D3	0566
D40D4	0566
D40D4F	1984
D40D5	0566
D40D7	0566
D40D8	0566
D40D10	0566
D40D11	0566
D40D13	0566
D40D14	0566
D40E1	0560
D40E5	1690
D40E7	0560
D40H	OEM
D40HL	OEM
D40K1	3535
D40K2	3535
D40K3	3535
D40K4	3535
D40N1	0285
D40N1F	0285
D40N2	0285
D40N2F	0285
D40N3	0285
D40N3F	0285
D40N4	0285
D40N4F	0285
D40N5	0285
D40P1	0338
D40P3	0285
D40P5	0285
D40V1	0285
D40V2	0285
D40V3	0285
D40V4	0285
D40V5	OEM
D40V6	OEM
D41	0130
D41A	0130
D41B	0130
D41C	0130
D41D	0130
D41D1	1367
D41D2	1367
D41D4	1367
D41D5	1367
D41D7	1367
D41D8	1367
D41D10	1367
D41D11	1367
D41D13	1367
D41D14	1367
D41E	0130
D41E1	1173
D41E5	1173
D41E7	1173
D41F	0130
D41G	0130
D41GN	0130
D41H	0130
D41J	0130
D41K	0130
D41K1	3536
D41K2	3536
D41K3	3536
D41K4	3536
D41L	0130
D41OR	0130
D41R	0130
D41X	0130
D41Y	0130
D42-400	OEM
D42-700	OEM
D42-1100	OEM
D42-1400	OEM
D42C1	0560
D42C2	0560
D42C3	0560
D42C4	0560
D42C5	0560
D42C6	0560
D42C7	0560
D42C8	0560
D42C9	0560
D42C10	0423
D42C11	0423
D42C12	0423
D42D1	1212
D42D2	OEM
D42D3	OEM
D42D4	OEM
D42D5	OEM
D42D6	OEM
D42R1	0168
D42R2	0168
D42R3	0168
D42R4	0168
D42T1	OEM
D42T2	OEM
D42T3	OEM
D42T4	OEM
D42T5	OEM
D42T6	OEM
D42T7	OEM
D42T8	OEM
D43	0208
D43A(2S)	0208
D43B	0208
D43C	0208
D43C1	1267
D43C2	1267
D43C3	1267
D43C4	1267
D43C4#	1267
D43C5	1267
D43C6	1267
D43C7	0925
D43C8	1267
D43C9	1267
D43C10	0854
D43C11	0854
D43C12	0854
D43D	0208
D43D1	0602
D43D2	OEM
D43D3	OEM
D43D4	OEM
D43D5	OEM
D43D6	OEM
D43E	0208
D43F	0208
D43G	0208
D43GN	0208
D43H	0208
D43J	0208
D43K	0208
D43L	0208
D43M	0208
D43OR	0208
D43R	0208
D43X	0208
D43Y	0208
D44	0208
D44A	0208
D44A10#	0060
D44B	0208
D44C	0208
D44C1	0060
D44C2	0060
D44C3	0060
D44C4	0236
D44C5	0060
D44C6	0060
D44C7	0060
D44C8	0060
D44C9	0060
D44C10	0060
D44C11	0060
D44C12	0060
D44C88	0560
D44D	0208
D44D1	1212
D44D2	1212
D44D3	1212
D44D4	1212
D44D5	1212
D44D6	1212
D44E	0208
D44E1	2235
D44E2	2235
D44E3	2235
D44F	0208
D44G	0208
D44GN	0208
D44H	0208
D44H1	0561
D44H2	0561
D44H4	0561
D44H5	0561
D44H7	0060
D44H8	0561
D44H10	0060
D44H11	0060
D44J	0208
D44K	0208
D44L	0208
D44M	0208
D44M1	0060
D44M2	0060
D44M4	0060
D44M5	0060
D44M7	0060
D44M8	0060
D44M10	0060
D44M11	0060
D44Q1	3014
D44Q2	3014
D44Q3	3014
D44Q5#	3014
D44R	0208
D44R1	0168
D44R2	0168
D44R3	0168
D44R4	0168
D44R5	0168
D44R6	0168
D44R7	3014
D44R8	0168
D44T1	3014
D44T2	3014
D44T3	3014
D44T4	3449
D44T5	3014
D44T6	3014
D44T7	3014
D44T8	3014
D44TA	OEM
D44TD3	0728
D44TD4	0728
D44TD5	0728
D44TE3	0728
D44TE4	0728
D44TE5	0728
D44VH1	0482
D44VH4	0482
D44VH7	0482
D44VH10	0482
D44VM1	0060
D44VM4	0060
D44VM7	0060
D44VM10	0060
D44X	0208
D44Y	0208
D45	0074
D45C	0015
D45C1	1308
D45C2	1308
D45C3	1308
D45C4	0681
D45C5	1308
D45C6	1308
D45C7	1308
D45C8	1308
D45C9	1308
D45C10	1308
D45C11	1308
D45C12	1308
D45CZ	0015
D45D1	0602
D45D2	0602
D45D3	0602
D45D4	0602
D45D5	0602
D45D6	0602
D45E1	2237
D45E2	2237
D45E3	2237
D45H1	1199
D45H2	1199
D45H4	1199
D45H-5	4430
D45H5	1199
D45H7	1308
D45H8	1369
D45H9	1369
D45H10	1199
D45H11	1308
D45H12	1308
D45HN	0681
D45M1	1308
D45M2	1308
D45M4	1308
D45M5	1308
D45M7	1308
D45M8	1308
D45M9	1308
D45M10	1308
D45M11	1308
D45M12	1308
D45VH1	1369
D45VH4	1369
D45VH7	1369
D45VH10	1369
D45VM1	1308
D45VM4	1308
D45VM7	1308
D45VM10	1308
D46	0074
D47	0620
D47C7	2258
D48	0015
D48(TRANS)	0042
D49(2S)	0042
D49A	0042
D49B	0042
D49C	0042
D49D	0042
D49E	0042
D49F	0042
D49G	0042
D49GN	0042
D49J	0042
D49K	0042
D49L	0042
D49M	0042
D49OR	0042
D49R	0042
D49X	0042
D49Y	0042
D50	1335
D50(TRANS)	0867
D50A	0867
D50B	0867
D50C	0867
D50D	0867
D50E	0867
D50F	0867
D50G	0867
D50GN	0867
D50H	0867
D50J	0867
D50K	0867
D50L	0867
D50M	0867
D50OR	0867
D50R	0867
D50X	0867
D50Y	0867
D51	0867
D51A	0867
D51B	0867
D51C	0867
D51D	0867
D51E	0867
D51F	0867
D51G	0867
D51GN	0867
D51H	0867
D51J	0867
D51K	0867
D51L	0867
D51M	0867
D51OR	0867
D51R	0867
D51X	0867
D51Y	0867
D52	0103
D53(2S)	0130
D53A	0130
D53B	0130
D53C	0130
D53E	0130
D53F	0130
D53G	0130
D53GN	0130
D53H	0130
D53J	0130
D53K	0130
D53L	0130
D53M	0130
D53OR	0130
D53R	0130
D53X	0130
D53Y	0130
D54	0130
D55	0130
D55A(2S)	0130
D56(2S)	4474
D56A	2214
D56B	2214
D56C	2214
D56D	2214
D56E	2214
D56F	2214
D56G	2214
D56GN	2214
D56H	2214
D56J	2214
D56K	2214
D56L	2214
D56M	2214
D56OR	2214
D56R	2214
D56W1	0313
D56W2	0065
D56X	2214
D56Y	2214
D57(2S)	0178
D58(2S)	0178
D59(2S)	0074
D60(2S)	0074
D60/400B	OEM
D60/800B	OEM
D60/1200B	OEM
D60/1400B	OEM
D60/1600B	OEM
D60-700	OEM
D60-1100	OEM
D60H	OEM
D60HL	OEM
D60N400B	OEM
D60N400C	OEM
D60N400U	OEM
D60N800B	OEM
D60N800C	OEM
D60N800U	OEM
D60N1200B	OEM
D60N1200C	OEM
D60N1200U	OEM
D60N1400B	OEM
D60N1400C	OEM
D60N1400U	OEM
D60N1600B	OEM
D60N1600C	OEM
D60N1600U	OEM
D60NR400B	OEM
D60NR400C	OEM
D60NR400U	OEM
D60NR800B	OEM
D60NR800C	OEM
D60NR800U	OEM
D60NR1200B	OEM
D60NR1200C	OEM
D60NR1200U	OEM
D60NR1400B	OEM
D60NR1400C	OEM
D60NR1400U	OEM
D60NR1600B	OEM
D60NR1600C	OEM
D60NR1600U	OEM
D60T257510	OEM
D60T259010	OEM
D60T307510	OEM
D60T309010	OEM
D60T357510	OEM
D60T404010	OEM
D60T405010	OEM
D60T406010	OEM
D60T454010	OEM
D60T455010	OEM
D60T504010	OEM
D60T703005	OEM
D60T704005	OEM
D60T753005	OEM
D60T754005	OEM
D60T803005	OEM
D60T804005	OEM
D61(2S)	0208
D61A	0208
D61B	0208
D61C	0208
D61D	0208
D61E	0208

DEVICE TYPE	REPL CODE
D61F	0208
D61G	0208
D61GN	0208
D61H	0208
D61J	0208
D61K	0208
D61L	0208
D61M	0208
D61OR	0208
D61R	0208
D61X	0208
D61Y	0208
D62(2S)	0208
D62A	0208
D62B	0208
D62C	0208
D62D	0208
D62E	0208
D62G	0208
D62GN	0208
D62H	0208
D62J	0208
D62K	0208
D62L	0208
D62M	0208
D62R	0208
D62T257510	OEM
D62T259010	OEM
D62T307510	OEM
D62T309010	OEM
D62T357510	OEM
D62T404010	OEM
D62T405010	OEM
D62T406010	OEM
D62T454010	OEM
D62T455010	OEM
D62T504010	OEM
D62X	0208
D62Y	0208
D63	0208
D63A	0208
D63B	0208
D63C	0208
D63E	0208
D63F	0208
D63G	0208
D63GN	0208
D63H	0208
D63J	0208
D63K	0208
D63L	0208
D63M	0208
D63OR	0208
D63R	0208
D63X	0208
D63Y	0208
D64	0208
D64A	0208
D64B	0208
D64C	0208
D64D	0208
D64E	0208
D64F	0208
D64G	0208
D64GN	0208
D64H	0208
D64J	0208
D64K	0208
D64M	0208
D64OR	0208
D64R	0208
D64TS3	OEM
D64TS5	OEM
D64VE3	1966
D64VE4	1966
D64VE5	1966
D64VP3	1966
D64VP4	1966
D64VP5	1852
D64VS3	1852
D64VS4	1852
D64VS5	1966
D64X	0208
D64Y	0208
D65	0208
D65-1(2S)	0208
D65A	0208
D65B	0208
D65C	0015
D65C(TRANS)	0208
D65D	0208
D65E	0208
D65F	0208
D65G	0208
D65GN	0208
D65H	0208
D65J	0208
D65K	0208
D65L	0208
D65M	0208
D65OR	0208
D65R	0208
D65X	0208
D65Y	0208
D66	0208
D66A	0208
D66B	0208
D66C	0208
D66D	0208
D66DV5	OEM
D66DV6	OEM
D66DV7	OEM
D66E	0208
D66EV5	OEM
D66EV6	OEM
D66EV7	OEM
D66F	0208
D66G	0208
D66GN	0208
D66GV6	OEM
D66H	0208
D66J	0208
D66K	0208
D66L	0208
D66M	0208
D66OR	0208
D66R	0208
D66X	0208
D66Y	0208
D67(2S)	0177
D67B(2S)	0177
D67C(2S)	0177
D67D(2S)	0177
D67DE5	OEM
D67DE6	OEM
D67DE7	OEM
D67E(2S)	0177
D68	0103
D68(TRANS)	0130
D68A	0130
D68B(2S)	0130
D68C(2S)	0130
D68D(2S)	0130
D68E(2S)	0130
D68F	0130
D68G	0130
D68GN	0130
D68H	0130
D68J	0130
D68L	0130
D68M	0130
D68OR	0130
D68R	0130
D68X	0130
D68Y	0130
D69(2S)	6140
D70(2S)	0130
D70A	0130
D70B	0130
D70C	0130
D70D	0130
D70E	0130
D70F	0130
D70G	0130
D70GN	0130
D70H	0130
D70J	0130
D70K	0130
D70L	0130
D70M	0130
D70OR	0130
D70R	0130
D70X	0130
D70Y	0130
D71(2S)	0178
D72	0208
D72-2C(2S)	0208
D72-3C(2S)	0208
D72-4C(2S)	0208
D72-6	0208
D72-OR	0208
D72A(2S)	0208
D72B(2S)	0208
D72C(2S)	0208
D72D	0208
D72E	0208
D72EJ	0208
D72F	0208
D72G	0208
D72GA	0208
D72H	0208
D72J	0208
D72K	0208
D72L	0208
D72M	0208
D72OR	0208
D72P	0208
D72R	0208
D72RE(2S)	0208
D72X	0208
D72Y	0208
D73(2S)	0867
D73Y	OEM
D73Y(KOREA)	1165
D74(2S)	0286
D75(2S)	0208
D75A(2S)	0208
D75AH(2S)	0208
D75B(2S)	0208
D75C(2S)	0208
D75D	0208
D75E	0208
D75F	0208
D75G	0208
D75GN	0208
D75H(2S)	0208
D75J	0208
D75K	0208
D75L	0208
D75M	0208
D75N400B	OEM
D75N400C	OEM
D75N400U	OEM
D75N800B	OEM
D75N800C	OEM
D75N800U	OEM
D75N1200B	OEM
D75N1200C	OEM
D75N1200U	OEM
D75N1400B	OEM
D75N1400C	OEM
D75N1400U	OEM
D75N1600B	OEM
D75N1600C	OEM
D75N1600U	OEM
D75NR400B	OEM
D75NR400C	OEM
D75NR400U	OEM
D75NR800B	OEM
D75NR800C	OEM
D75NR800U	OEM
D75NR1200B	OEM
D75NR1200C	OEM
D75NR1200U	OEM
D75NR1400B	OEM
D75NR1400C	OEM
D75NR1400U	OEM
D75NR1600B	OEM
D75NR1600C	OEM
D75NR1600U	OEM
D75OR	0208
D76FI3T	OEM
D77(2S)	0208
D77-A	0208
D77A(2S)	0208
D77AH(2S)	0208
D77B(2S)	0208
D77C(2S)	0208
D77D(2S)	0208
D77E	0208
D77F	0208
D77G	0208
D77GN	0208
D77H(2S)	0208
D77J	0208
D77L	0208
D77M	0208
D77OR	0208
D77P(2S)	0208
D77R	0208
D77X	0208
D77Y	0208
D78(2S)	3009
D78A(2S)	3009
D79(2S)	6829
D80(2S)	0130
D80A	0130
D80B	0130
D80C	0130
D80C39	OEM
D80C49	OEM
D80D	0130
D80E	0130
D80F	0130
D80G	0130
D80GN	0130
D80H	0130
D80H(2S)	0130
D80H(DIODE)	OEM
D80HL	OEM
D80J	0130
D80K	0130
D80L	0130
D80OR	0130
D80R	0130
D80X	0130
D80Y	0130
D81(2S)	0867
D81A	0867
D81B	0867
D81C	0867
D81E	0867
D81F	0867
D81G	0867
D81GN	0867
D81H	0867
D81J	0867
D81K	0867
D81L	0867
D81M	0867
D81OR	0867
D81R	0867
D81X	0867
D81Y	0867
D82(2S)	0867
D82-OR	0867
D82A(2S)	0867
D82B	0867
D82C	0867
D82C288-8	OEM
D82D	0867
D82E	0867
D82F	0867
D82G	0867
D82GN	0867
D82H	0867
D82J	0867
D82K	0867
D82L	0867
D82M	0867
D82OR	0867
D82R	0867
D82X	0867
D82Y	0867
D83(2S)	0867
D83-004	OEM
D84(2S)	0867
D85C	0071
D85GR	0728
D88	0867
D88(TRANS)	0867
D88A(2S)	0867
D88OY	0867
D90(2S)	0042
D90A	0042
D90B	0042
D90C	0042
D90D	0042
D90E	0042
D90F	0042
D90G	0042
D90GN	0042
D90H	0042
D90J	0042
D90K	0042
D90L	0042
D90M	0042
D90OR	0042
D90X	0042
D90Y	0042
D91(2S)	0042
D91A	0042
D91B	0042
D91E	0042
D91F(2S)	0042
D91G	0042
D91GN	0042
D91H	0042
D91J	0042
D91L	0042
D91M	0042
D91OR	0042
D91R	0042
D91X	0042
D91Y	0042
D92(2S)	0178
D92D(2S)	0178
D92F13R6B0	0376
D92T03R9B2	0036
D92T05R1B2	0041
D92T05R6B2	0253
D92T06R2B2	0471
D92T06R2B3	0471
D92T06R8B2	0062
D92T09R1B2	0057
D92T0110B2	0181
D92T0110B3	0181
D92T0120B2	1005
D92T0150B1	0879
D92T0200B2	0671
D92T0270B2	0455
D92T0270B3	0441
D92T0300B4	0195
D92T0360B0	0010
D92UA3600B	0010
D93	0074
D93T03R30Y	OEM
D93T04R30Y	0533
D93T04R70X	0140
D93T05R10X	0041
D93T05R10Y	0041
D93T05R60Y	0253
D93T05R100	0041
D93T06R20Y	0471
D93T06R20Z	0295
D93T06R80Y	0062
D93T06R200	0471
D93T07R50Z	0077
D93T08R20X	0165
D93T09R10Y	OEM
D93T01000X	OEM
D93T01100Y	0181
D93T01200Y	0052
D93T01300X	0053
D93T01300Y	0053
D93T01800Y	0375
D93T02000X	0700
D93T02700Y	0455
D93T03000Y	0195
D94(2S)	0392
D94TA5R1B2	0041
D94TA5R6J2	0883
D94TA6R2B2	OEM
D94TA6R8J2	0062
D94TA9R1J2	OEM
D94TA110J3	0181
D94TA120B2	1005
D94TA120J2	1005
D94TA120J3	1005
D94TA130J2	OEM
D94UA6R2J2	0295
D96(2S)	0208
D96A	0208
D96B	0208
D96C	0208
D96D	0208
D96E	0208
D96F	0208
D96G	0208
D96GN	0208
D96H	0208
D96J	0208
D96K	0208
D96L	0208
D96OR	0208
D96R	0208
D96X	0208
D96Y	0208
D97U05R10B	0041
D97U05R60B	0253
D97U05R108	0041
D97U06R20B	0471
D97U06R80B	0471
D97U01100B	0181
D97U01200B	1005
D100	0015
D100(TRANS)	0208
D100-400	OEM
D100-700	OEM
D100-1100	OEM
D100-1400	OEM
D100-1800	OEM
D100A(2S)	0208
D100B	0208
D100C	0208
D100D	0208
D100E	0208
D100F	0208
D100G	0208
D100GN	0208
D100H	OEM
D100H(2S)	0208
D100J	0208
D100K	0208
D100L	0208
D100M	0208
D100OR	0208
D100R	0208
D100X	0208
D100Y	0208
D101	0600
D101AK	0600
D101B	0004
D102	0178
D102-0	0178
D102-O(2S)	0178
D102-R(2S)	0178
D102-Y(2S)	0178
D102A	0178
D102B	0178
D102BS	OEM
D102BS-1	OEM
D102BSM	OEM
D102BSM-1	OEM
D102BSMZS	OEM
D102BSMZS-1	OEM
D102BSQ	OEM
D102BSQ-1	OEM
D102BSQZS	OEM
D102BSQZS-1	OEM
D102BST	OEM
D102BST-1	OEM
D102BSTZS	OEM
D102BSTZS-1	OEM
D102BSZS	OEM
D102BSZS-1	OEM
D102C	0178
D102D	0178
D102E	0178
D102F	0178
D102G	0178
D102GN	0178
D102H	0178
D102J	0178
D102K	0178
D102L	0178
D102M	0178
D102OR	0178
D102R	0178
D102X	0178
D102Y	0178
D103	0178
D103-0	0178
D103-O(2S)	0178
D103-R(2S)	0178
D103-Y(2S)	0178
D104	0208
D104A	0208
D104B	0208
D104BS	OEM
D104BS-1	OEM
D104BSM	OEM
D104BSM-1	OEM
D104BSMZS	OEM
D104BSMZS-1	OEM
D104BSQ	OEM
D104BSQ-1	OEM
D104BSQZS	OEM
D104BSQZS-1	OEM
D104BST	OEM
D104BST-1	OEM
D104BSTZS	OEM
D104BSTZS-1	OEM
D104BSZS	OEM
D104BSZS-1	OEM
D104C	0208
D104D	0208
D104E	0208
D104F	0208
D104G	0208
D104GN	0208
D104H	0208
D104K	0208
D104L	0208
D104M	0208
D104OR	0208
D104R	0208
D104X	0208
D104Y	0208
D105	0208
D105A	0208
D105B	0208
D105C	0071
D105C(TRANS)	0208
D105D	OEM
D105D(2S)	0208
D105DS	OEM
D105DS-1	OEM
D105DSM	OEM
D105DSM-1	OEM
D105DSMZS	OEM
D105DSMZS-1	OEM
D105DSQ	OEM
D105DSQ-1	OEM
D105DSQZS	OEM
D105DSQZS-1	OEM
D105DST	OEM
D105DST-1	OEM
D105DSTZS	OEM
D105DSTZS-1	OEM
D105DSZS	OEM
D105DSZS-1	OEM
D105E	0208
D105F	0208
D105G	0208
D105GN	0208
D105H	0208
D105J	0208
D105K	0208
D105L	0208
D105M	0208
D105OR	0208
D105R	0208
D105X	0208
D105Y	0208
D107	0103
D107E	OEM
D108	0071
D108(TRANS)	0103
D110	0177
D110-0	0177
D110-O(2S)	0177
D110-ORG	0177
D110-R(2S)	0177
D110-RED	0177
D110-Y(2S)	0177
D110-YEL	0177
D110A243A0	0124
D110A243AC	0124
D110DS	OEM
D110DS-1	OEM
D110DSM	OEM
D110DSM-1	OEM
D110DSMZS	OEM
D110DSMZS-1	OEM
D110DSQ	OEM
D110DSQ-1	OEM
D110DSQZS	OEM
D110DSQZS-1	OEM
D110DST	OEM
D110DST-1	OEM
D110DSTZS	OEM
D110DSTZS-1	OEM
D110DSZS	OEM
D110DSZS-1	OEM
D111	0177
D111-0	0177
D111-O(2S)	0177
D111-ORG	0177
D111-R(2S)	0177
D111-RED	0177
D111-Y(2S)	0177
D111-YEL	0177
D111R	0177
D112BS	OEM
D112BS-1	OEM
D112BSM	OEM
D112BSM-1	OEM
D112BSMZS	OEM
D112BSMZS-1	OEM
D112BSQ	OEM
D112BSQ-1	OEM
D112BSQZS	OEM
D112BSQZS-1	OEM
D112BST	OEM
D112BST-1	OEM
D112BSTZS	OEM
D112BSTZS-1	OEM
D112BSZS	OEM
D112BSZS-1	OEM
D113(2S)	0130
D113-0	0130
D113-O(2S)	0130
D113-ORG	0130
D113-R(2S)	0130
D113-RED	0130
D113-Y(2S)	0130
D113-YEL	0130
D113A	0922
D113R	0130
D114(2S)	0130
D114-0	0130
D114-O(2S)	0130
D114-ORG	0130
D114-R(2S)	0130
D114-RED	0130
D114-Y(2S)	0130
D114-YEL	0130
D114R	0130
D116	0103
D117	0620
D117C	0620
D118	0177
D118-B	0177
D118-BL	0177
D118-BLU	0177
D118-R	0177
D118-RED	0177
D118-Y	0177
D118-YEL	0177
D118A	0177
D118BL(2S)	0177
D118C	0177
D118D	0177
D118R(2S)	0177
D118Y(2S)	0177
D119	0177
D119-BL	0177
D119-BLU	0177
D119-R	0177
D119-RED	0177
D119-Y	0177
D119-YEL	0177
D119A	0177
D119B	0177
D119BL(2S)	0177
D119C	0177
D119D	0177
D119R(2S)	0177
D119Y(2S)	0177
D120(2S)	1480
D120/400B	OEM
D120/800B	OEM
D120/1200B	OEM
D120/1400B	OEM
D120/1600B	OEM
D120/1800B	OEM
D120-700	OEM
D120-1100	OEM
D120D	1480
D120E	1480
D120F	1480
D120G	1480
D120GN	1480
D120H(2S)	1480
D120J	1480
D120K	1480
D120L	1480
D120M	1480
D120OR	1480
D120R	1480
D120S800B	OEM
D120S1000B	OEM
D120S1200B	OEM
D120S1400B	OEM
D120SR800B	OEM
D120X	1480
D120Y	1480
D121(2S)	1480
D121C	1480
D121D	1480
D121E	1480
D121F	1480
D121G	1480
D121GN	1480
D121H(2S)	1480
D121J	1480
D121K	1480
D121L	1480
D121M	1480
D121N400B	OEM
D121N400V	OEM
D121N800B	OEM
D121N800V	OEM
D121N1200B	OEM
D121N1200V	OEM
D121N1400B	OEM
D121N1400V	OEM
D121N1800B	OEM
D121N1800V	OEM
D121NR400B	OEM
D121NR400V	OEM
D121NR800B	OEM
D121NR800V	OEM
D121NR1200B	OEM
D121NR1200V	OEM
D121NR1400B	OEM
D121NR1400V	OEM
D121NR1800B	OEM
D121NR1800V	OEM
D121OR	1480
D121R	1480
D121X	1480

If replacement code is OEM, contact original manufacturer for replacement.

DEVICE TYPE	REPL CODE
D121Y	1480
D122	0042
D123	0236
D123AL	OEM
D123AP	OEM
D123BP	OEM
D123CDD	OEM
D123CFD	OEM
D123IDD	OEM
D124	0867
D124(RECT.)	0015
D124A(2S)	0867
D124AH(2S)	0867
D124B	0867
D124C	0867
D124E	0867
D124F	0867
D124G	0867
D124GN	0867
D124H	0867
D124J	0867
D124K	0867
D124L	0867
D124M	0867
D124OR	0867
D124R	0867
D124X	0867
D124Y	0867
D125	0867
D125A(2S)	0867
D125AH(2S)	0867
D125AP	OEM
D125B	0867
D125BP	OEM
D125C	OEM
D125C(TRANS)	0867
D125CDD	OEM
D125CFD	OEM
D125E	0867
D125F	0867
D125G	0867
D125GN	0867
D125H	0867
D125IDD	OEM
D125IFD	OEM
D125J	0867
D125K	0867
D125L	0867
D125M	0867
D125MDD	OEM
D125MFD	OEM
D125OR	0867
D125R	0867
D125X	0867
D125Y	0867
D126	0867
D126H(2S)	0867
D126HA(2S)	0867
D126HB(2S)	0867
D127	0208
D127A(2S)	0208
D127B	0208
D127C	0208
D127D	0208
D127E	0208
D127F	0208
D127G	0208
D127GN	0208
D127H	0208
D127J	0208
D127K	0208
D127L	0208
D127M	0208
D127OR	0208
D127R	0208
D127X	0208
D127Y	0208
D128	0208
D128A(2S)	0208
D128B	0208
D128C	0208
D128D	0208
D128E	0208
D128F	0208
D128G	0208
D128GN	0208
D128H	0208
D128J	0208
D128K	0208
D128L	0208
D128M	0208
D128OR	0208
D128R	0208
D128X	0208
D128Y	0208
D129	0178
D129-BL(2S)	0178
D129-BLU	0178
D129-R(2S)	0178
D129-RED	0178
D129-Y(2S)	0178
D129-YEL	0178
D129AL	OEM
D129AP	OEM
D129BP	OEM
D129IDD	OEM
D129IFD	OEM
D129MDD	OEM
D129MFD	OEM

DEVICE TYPE	REPL CODE
D130	1147
D130-BL	1147
D130-BLU	1147
D130-R(2S)	1147
D130-RED	1147
D130-Y(2S)	1147
D130-YEL	1147
D130A	1147
D130B	1147
D130BL(2S)	1147
D130C	1147
D130D	1147
D130DS442X	0124
D130E	1147
D130F	1147
D130G	1147
D130GMA010	0124
D130GN	1147
D130H	1147
D130J	1147
D130K	1147
D130OR	1147
D130R	1147
D130X	1147
D130Y	1147
D131	0620
D132	0130
D133	0079
D133(CHAN.MASTER)	0144
D134	0018
D134A	0018
D134C	0018
D134D	0018
D134F	0018
D134G	0018
D134GN	0018
D134H	0018
D134J	0018
D134K	0018
D134L	0018
D134M	0018
D134OR	0018
D134R	0018
D134X	0018
D134Y	0018
D135	0004
D136	0142
D136D	0142
D136E	0142
D136F	0142
D136G	0142
D136GN	0142
D136J	0142
D136K	0142
D136L	0142
D136M	0142
D136OR	0142
D136R	0142
D136X	0142
D136Y	0142
D137	0142
D137A	0142
D137B	0142
D137C	0142
D137D	0142
D137F	0142
D137G	0142
D137H	0142
D137J	0142
D137K	0142
D137L	0142
D137M	0142
D137OR	0142
D137R	0142
D137X	0142
D137Y	0142
D138	0168
D139	0168
D139AA	OEM
D139AL	OEM
D139BA	OEM
D139BP	OEM
D139CJ	OEM
D141	0016
D141(2S)	0042
D141(CHAN.MASTER)	0144
D141A	0042
D141B	0042
D141C	0042
D141E	0042
D141F	0042
D141G	0042
D141GN	0042
D141H	0042
D141H01(2S)	0042
D141J	0042
D141K	0042
D141L	0042
D141M	0042
D141OR	0042
D141R	0042
D141X	0042
D141Y	0042
D142(2S)	0178
D142A	0178
D142B	0178
D142C	0178
D142D	0178
D142E	0178

DEVICE TYPE	REPL CODE
D142F	0178
D142G	0178
D142GN	0178
D142H	0178
D142J	0178
D142K	0178
D142L	0178
D142M(2S)	0178
D142OR	0178
D142R	0178
D142X	0178
D142Y	0178
D143(2S)	0178
D144(2S)	0178
D144S	0015
D145(2S)	0178
D146(2S)	0042
D146A	0042
D146B	0042
D146C	0042
D146D	0042
D146E	0042
D146F	0042
D146G	0042
D146GN	0042
D146H	0042
D146J	0042
D146K	0042
D146L	0042
D146OR	0042
D146R	0042
D146UK(2S)	0042
D146VK	0042
D146X	0042
D146Y	0042
D147(2S)	0042
D147A	0042
D147B	0042
D147C	0042
D147D	0042
D147E	0042
D147F	0042
D147GN	0042
D147H	0042
D147J	0042
D147K	0042
D147L	0042
D147M	0042
D147OR	0042
D147R	0042
D147X	0042
D147Y	0042
D148	0042
D149	0086
D150(2S)	0178
D150/2800B	OEM
D150/3200B	OEM
D150/3600B	OEM
D150A	0178
D150B	0178
D150C	0178
D150E	0178
D150F	0178
D150G	0178
D150GN	0178
D150H	OEM
D150H(2S)	0178
D150J	0178
D150K	0178
D150L	0178
D150M	0178
D150OR	0178
D150R	0178
D150R3200B	OEM
D150R3600B	OEM
D150X	0178
D150Y	0178
D151(2S)	0620
D151A	0620
D151B	0620
D151C	0620
D151E	0620
D151F	0620
D151G	0620
D151GN	0620
D151H	0620
D151J	0620
D151K	0620
D151L	0620
D151M	0620
D151OR	0620
D151R	0620
D151Y	0620
D152(2S)	0392
D152(CHAN.MASTER)	0698
D152A	0392
D152B	0392
D152C	0392
D152D	0392
D152E	0392
D152F	0392
D152G	0392
D152GN	0392
D152H	0392
D152J	0392
D152K	0392
D152L	0392
D152M	0392
D152OR	0392

DEVICE TYPE	REPL CODE
D152R	0392
D152X	0392
D152Y	0392
D154(2S)	0236
D154A	0236
D154B	0236
D154C	0236
D154E	0236
D154F	0236
D154G	0236
D154GN	0236
D154H	0236
D154J	0236
D154K	0236
D154L	0236
D154M	0236
D154R	0236
D154X	0236
D154Y	0236
D155	0178
D155-H	0178
D155-K	0178
D155H(2S)	0178
D155K(2S)	0178
D155L(2S)	0178
D156	0142
D156B	0142
D156C	0142
D156D	0142
D156E	0142
D156F	0142
D156G	0142
D156GN	0142
D156H	0142
D156K	0142
D156L	0142
D156M	0142
D156OR	0142
D156R	0142
D156X	0142
D156Y	0142
D157(2S)	0142
D157A	0142
D157B(2S)	0142
D157C	0142
D157D	0142
D157G	0142
D157GN	0142
D157H	0142
D157J	0142
D157K	0142
D157L	0142
D157OR	0142
D157R	0142
D157X	0142
D157Y	0142
D158(2S)	1027
D158A	1027
D158B	1027
D158C	1027
D158D	1027
D158E	1027
D158F	1027
D158GN	1027
D158H	1027
D158J	1027
D158K	1027
D158L	1027
D158M	1027
D158OR	1027
D158R	1027
D158X	1027
D158Y	1027
D159(2S)	1027
D159A	1027
D159B	1027
D159C	1027
D159D	1027
D159E	1027
D159F	1027
D159G	1027
D159GN	1027
D159H	1027
D159J	1027
D159K	1027
D159L	1027
D159M	1027
D159OR	1027
D159R	1027
D159X	1027
D159Y	1027
D160(2S)	0630
D161(2S)	0620
D161A	0620
D161B	0620
D161C	0620
D161D	0620
D161E	0620
D161F	0620
D161G	0620
D161GN	0620
D161H	0620
D161J	0620
D161K	0620
D161L	0620
D161M	0620
D161OR	0620
D161R	0620
D161X	0620

DEVICE TYPE	REPL CODE
D161Y	0620
D162(2S)	0208
D162A	0208
D162B	0208
D162C	0208
D162D	0208
D162E	0208
D162F	0208
D162G	0208
D162GN	0208
D162H	0208
D162J	0208
D162K	0208
D162L	0208
D162M	0208
D162OR	0208
D162R	0208
D162X	0208
D162Y	0208
D163(2S)	0103
D163A	0103
D163B	0103
D163C	0103
D163D	0103
D163E	0103
D163F	0103
D163G	0103
D163GN	0103
D163H	0103
D163J	0103
D163K	0103
D163L	0103
D163M	0103
D163OR	0103
D163R	0103
D163X	0103
D163Y	0103
D164(2S)	0103
D164A	0103
D164B	0103
D164C	0103
D164D	0103
D164E	0103
D164F	0103
D164G	0103
D164GN	0103
D164H	0103
D164J	0103
D164K	0103
D164L	0103
D164M	0103
D164OR	0103
D164R	0103
D164X	0103
D164Y	0103
D165	1966
D166(2S)	1966
D167(2S)	0208
D167A	0208
D167B	0208
D167C	0208
D167E	0208
D167F	0208
D167G	0208
D167GN	0208
D167H	0208
D167J	0208
D167K	0208
D167L	0208
D167M	0208
D167OR	0208
D167R	0208
D167X	0208
D167Y	0208
D168	3373
D169AK	OEM
D169AP	OEM
D169CJ	OEM
D169CK	OEM
D170(2S)	0208
D170/2800T	OEM
D170A(2S)	0208
D170AA(2S)	0208
D170AB(2S)	0208
D170AC(2S)	0208
D170B(2S)	0208
D170BC(2S)	0208
D170C(2S)	0208
D171	0642
D172	0103
D172(IR)	0102
D172-F	0102
D172A	0103
D172B	0103
D172C	0103
D172D	0103
D172E	0103
D172F	0103
D172G	0103
D172GN	0103
D172H	0103
D172J	0103
D172K	0103
D172L	0103
D172M	0103
D172OR	0103
D172R	0103
D172X	0103
D172Y	0103

DEVICE TYPE	REPL CODE
D173	0130
D173A	0130
D173B	0130
D173C	0130
D173D	0130
D173E	0130
D173F	0130
D173G	0130
D173GN	0130
D173H	0130
D173J	0130
D173K	0130
D173M	0130
D173OR	0130
D173X	0130
D173Y	0130
D174	0103
D174A	0103
D174B	0103
D174C	0103
D174D	0103
D174E	0103
D174F	0103
D174G	0103
D174GN	0103
D174H	0103
D174J	0103
D174K	0103
D174L	0103
D174M	0103
D174R	0103
D174X	0103
D174Y	0103
D175(2S)	0103
D175A	0103
D175B	0103
D175C	0103
D175D	0103
D175E	0103
D175F	0103
D175G	0103
D175GN	0103
D175H	0103
D175J	0103
D175K	0103
D175L	0103
D175M	0103
D175OR	0103
D175R	0103
D175X	0103
D175Y	0103
D176	0012
D176(TRANS)	0130
D176A	0130
D176B	0130
D176C	0130
D176D	0130
D176E	0130
D176F	0130
D176G	0130
D176GN	0130
D176H	0130
D176J	0130
D176K	0130
D176L	0130
D176M	0130
D176OR	0130
D176X	0130
D176Y	0130
D177(2S)	0620
D178(2S)	0208
D178A(2S)	0208
D178B	0208
D178C	0208
D178D	0208
D178E	0208
D178F	0208
D178G	0208
D178GN	0208
D178H	0208
D178J	0208
D178K	0208
D178L	0208
D178M	0208
D178Q(2S)	0208
D178T(2S)	0208
D178X	0208
D178Y	0208
D180	0867
D180(2S)	2793
D180A(2S)	2793
D180B(2S)	2793
D180C(2S)	2793
D180D(2S)	2793
D180F	2793
D180G	2793
D180GN	2793
D180H	2793
D180J	2793
D180K	2793
D180L	2793
D180M(2S)	2793
D180OR	2793
D180R	2793
D180X	2793
D180Y	2793
D181(2S)	1966
D181A	1966
D182(2S)	0086

DEVICE TYPE	REPL CODE
D182A	0086
D182B	0086
D182C	0086
D182D	0086
D182E	0086
D182F	0086
D182G	0086
D182GN	0086
D182H	0086
D182J	0086
D182K	0086
D182L	0086
D182OR	0086
D182R	0086
D182X	0086
D182Y	0086
D183(2S)	0086
D183B	0086
D183E	0086
D183GN	0086
D183K	0086
D183L	0086
D183OR	0086
D183R	0086
D184(2S)	0042
D184A	0042
D184B	0042
D184C	0042
D184D	0042
D184E	0042
D184F	0042
D184H	0042
D184J	0042
D184K	0042
D184L	0042
D184M	0042
D184OR	0042
D184Y	0042
D185	0042
D186(2S)	0208
D186A(2S)	0208
D186B(2S)	0208
D186C	0208
D186D	0208
D186E	0208
D186F	0208
D186G	0208
D186GN	0208
D186H	0208
D186J	0208
D186K	0208
D186L	0208
D186M	0208
D186OR	0208
D186R	0208
D186X	0208
D186Y	0208
D187	0208
D187-OR	0208
D187A(2S)	0208
D187B	0208
D187C	0208
D187D	0208
D187E	0208
D187F	0208
D187G	0208
D187GN	0208
D187H	0208
D187J	0208
D187K	0208
D187L	0208
D187M	0208
D187OR	0208
D187R(2S)	0208
D187X	0208
D187Y(2S)	0208
D188	2556
D188A(2S)	2556
D188B(2S)	2556
D188C(2S)	2556
D188D	2556
D188E	2556
D188F	2556
D188G	2556
D188GN	2556
D188H	2556
D188J	2556
D188L	2556
D188M	2556
D188OR	2556
D188R	2556
D188X	2556
D188Y	2556
D189(2S)	0177
D189A(2S)	0177
D189B	0177
D189C	0177
D189D	0177
D189E	0177
D189F	0177
D189G	0177
D189GN	0177
D189H	0177
D189J	0177
D189K	0177
D189L	0177
D189M	0177
D189OR	0177
D189R	0177

DEVICE TYPE	REPL CODE
D189X	0177
D189Y	0177
D190	0142
D190B	0142
D190C	0142
D190D	0142
D190E	0142
D190F	0142
D190GN	0142
D190H	0142
D190J	0142
D190K	0142
D190L	0142
D190M	0142
D190OR	0142
D190R	0142
D190X	0142
D190Y	0142
D191	0208
D192(2S)	0208
D193	0208
D194(2S)	0208
D195	0208
D195A(2S)	0208
D195B	0208
D195C	0208
D195D	0208
D195E	0208
D195F	0208
D195G	0208
D195GN	0208
D195H	0208
D195J	0208
D195K	0208
D195L	0208
D195M	0208
D195OR	0208
D195R	0208
D195X	0208
D195Y	0208
D198(2S)	0042
D198A	0042
D198AP	0042
D198AQ	0042
D198AR	0042
D198B	0042
D198C	0042
D198D	0042
D198F	0042
D198G	0042
D198H(2S)	0042
D198HQ(2S)	0042
D198HR(2S)	0042
D198J	0042
D198K	0042
D198OR	0042
D198P(2S)	0042
D198Q(2S)	0042
D198R(2S)	0042
D198S(2S)	0042
D198V(2S)	0042
D198X	0042
D198Y	0042
D199(2S)	0003
D200	0223
D200(2S)	0223
D200(IR)	0133
D200-OR	0223
D200A(2S)	0223
D200B	0223
D200C	0223
D200D	0223
D200E	0223
D200F	0223
D200G	0223
D200GN	0223
D200H	0223
D200H(TRANS)	0223
D200HL	OEM
D200J	0223
D200K	0223
D200L	0223
D200M	0223
D200MP	0133
D200MP(IR)	0133
D200OR	0223
D200X	0223
D200Y	0223
D201(0)	0867
D201(2S)	0867
D201-O	0867
D201A	0867
D201B	0867
D201C	0867
D201F	0867
D201G	0867
D201GN	0867
D201H	0867
D201J	0867
D201K	0867
D201L	0867
D201M(2S)	0867
D201MO	0867
D201MY	0867
D201Q	0867
D201X	0867
D201Y(2S)	0867
D202(2S)	0867
D203(2S)	0867
D204(2S)	0086
D204A	0086
D204B	0086
D204BL	0086
D204BS	OEM
D204BS-1	OEM
D204BSM	OEM
D204BSM-1	OEM
D204BSMZS	OEM
D204BSMZS-1	OEM
D204BSQ	OEM
D204BSQ-1	OEM
D204BSQZS	OEM
D204BSQZS-1	OEM
D204BST	OEM
D204BST-1	OEM
D204BSTZS	OEM
D204BSTZS-1	OEM
D204BSZS	OEM
D204BSZS-1	OEM
D204C	0086
D204D	0086
D204E	0086
D204F	0086
D204G	0086
D204GA	0086
D204GN	0086
D204H	0086
D204J	0086
D204K	0086
D204L(2S)	0086
D204M	0086
D204R	0086
D204Y	0086
D205(2S)	0086
D205A	0086
D205B	0086
D205C	0086
D205D	0086
D205E	0086
D205F	0086
D205G	0086
D205GN	0086
D205H	0086
D205J	0086
D205L	0086
D205M	0086
D205OR	0086
D205R	0086
D205X	0086
D205Y	0086
D206	0133
D206(2S)	0543
D206A	0543
D207	0074
D207(2S)	0103
D207(DIODE)	OEM
D207A	0103
D208	0543
D208(2S)	0620
D208(DIODE)	OEM
D208A	0620
D209	OEM
D210	OEM
D211	0103
D211(2S)	0103
D211(DIODE)	OEM
D211-OR	0103
D211A	0103
D211B	0103
D211C	0103
D211D	0103
D211E	0103
D211F	0103
D211G	0103
D211GN	0103
D211H	0103
D211J	0103
D211K	0103
D211L	0103
D211M	0103
D211OR	0103
D211R	0103
D211X	0103
D211Y	0103
D212(2S)	0103
D212A	0103
D212B	0103
D212BS	OEM
D212BS-1	OEM
D212BSM	OEM
D212BSM-1	OEM
D212BSMZS	OEM
D212BSMZS-1	OEM
D212BSQ	OEM
D212BSQ-1	OEM
D212BSQZS	OEM
D212BSQZS-1	OEM
D212BST	OEM
D212BST-1	OEM
D212BSTZS	OEM
D212BSTZS-1	OEM
D212BSZS	OEM
D212BSZS-1	OEM
D212C	0103
D212D	0103
D212E	0103
D212F	0103
D212G	0103
D212GN	0103
D212H	0103
D212J	0103
D212K	0103
D212L	0103
D212M	0103
D212OR	0103
D212R	0103
D212X	0103
D212Y	0103
D213(2S)	0177
D214	0177
D215(2S)	0086
D216	0086
D217(2S)	0867
D218(2S)	0867
D219(2S)	0086
D219A	0086
D219A(2S)	0086
D219A(DIODE)	OEM
D219B	0086
D219C	0086
D219D	0086
D219E	0086
D219F	0086
D219G	0086
D219GN	0086
D219H	0086
D219J	0086
D219K	0086
D219L	0086
D219M	0086
D219OR	0086
D219R	0086
D219X	0086
D219Y	0086
D220	0023
D220(TRANS)	0086
D220-1M	1217
D220-1N	1217
D220A	0086
D220A(2S)	0086
D220A(DIODE)	OEM
D220B	0086
D220B(2S)	0086
D220B(DIODE)	OEM
D220C	0086
D220D	0086
D220E	0086
D220F	0086
D220G	0086
D220GN	0086
D220H	0086
D220J	0086
D220K	0086
D220L	0086
D220M(2S)	0086
D220OR	0086
D220R	0086
D220X	0086
D220Y	0086
D221(2S)	0086
D221F	0086
D222	1480
D222A	1480
D222B	1480
D222C	1480
D222D	1480
D222E	1480
D222F	1480
D222G	1480
D222GN	1480
D222H	1480
D222J	1480
D222K	1480
D222L	1480
D222M	1480
D222OR	1480
D222R	1480
D222X	1480
D222Y	1480
D223(2S)	1480
D223A	1480
D223B	1480
D223B(2S)	1480
D223B(DIODE)	OEM
D223C	1480
D223D	1480
D223E	1480
D223F	1480
D223G	1480
D223GN	1480
D223H	1480
D223J	1480
D223K	1480
D223L	1480
D223M	1480
D223OR	1480
D223R	1480
D223X	1480
D223Y	1480
D224	1480
D226(2S)	0178
D226-0(2S)	0178
D226-0	0178
D226A(2S)	0178
D226A(0)	0178
D226AO	0178
D226AP(2S)	0178
D226B	0178
D226B(2S)	0178
D226B(DIODE)	0015
D226BP(2S)	0178
D226C	0178
D226D	0015
D226E	0178
D226F	0178
D226G	0015
D226G(2S)	0178
D226GN	0178
D226H	0178
D226J	0178
D226K	0178
D226L	0178
D226M	0178
D226OR	0178
D226P(2S)	0178
D226Q(2S)	0178
D226R	0178
D226V	0015
D226X	0178
D226Y	0178
D227	1419
D227-175	1419
D227-OR	1419
D227A(2S)	1419
D227B(2S)	1419
D227C(2S)	1419
D227D(2S)	1419
D227E(2S)	1419
D227G	1419
D227GN	1419
D227H	1419
D227J	1419
D227K	1419
D227L(2S)	1419
D227LF	1419
D227M	1419
D227OR	1419
D227R(2S)	1419
D227V	1419
D227X	1419
D227Y	1419
D228(2S)	0155
D228F	0155
D228G	0155
D228GN	0155
D228H	0155
D228J	0155
D228K	0155
D228L	0155
D228M	0155
D228OR	0155
D228R	0155
D228X	0155
D228Y	0155
D229D	OEM
D229E	OEM
D229G	0223
D229G(DIODE)	OEM
D229V	4085
D230DFA05G	0023
D230HI	0102
D231	0130
D231DFC15L	0102
D232(2S)	0130
D232A	0130
D234(2S)	0228
D234-0(2S)	0228
D234-O(2S)	0228
D234-ORG	0228
D234-R(2S)	0228
D234-RED	0228
D234-Y(2S)	0228
D234-YEL	0228
D234A	0228
D234B	0228
D234C	0228
D234D	0228
D234E	0228
D234F	0228
D234G	0228
D234GA	0228
D234GN	0228
D234G-O	0228
D234G-R	0228
D234GR	0228
D234G-Y	0228
D234H	0228
D234J	0228
D234K	0228
D234L	0228
D234M	0228
D234N	0228
D234O	0228
D234OR	0228
D234R	0228
D234X	0228
D234Y	0228
D235(2S)	0228
D235-0(2S)	0228
D235-O(2S)	0228
D235-OR	0228
D235-ORG	0228
D235-R(2S)	0228
D235-RED	0228
D235-Y(2S)	0228
D235-YEL	0228
D235A	0228
D235B	0228
D235C	0228
D235D(2S)	0228
D235E	0228
D235F	0228
D235G(2S)	0228
D235GN	0228
D235G-O	0228
D235G-R	0228
D235G-Y	0228
D235H	0228
D235J	0228
D235L	0228
D235LBY	0228
D235M	0228
D235O	0228
D235OR	0228
D235R(2S)	0228
D235RED	0228
D235RY	0228
D235X	0228
D235Y(2S)	0228
D236	4098
D236(RECT.)	0277
D236A	4098
D236B	4098
D236C	4098
D236D	4098
D236E	4098
D236F	4098
D236G	4098
D236GN	4098
D236H	4098
D236J	4098
D236K	4098
D236L	4098
D236M	4098
D236R	4098
D236X	4098
D236Y	4098
D237	0178
D238(2S)	0178
D238F(2S)	0178
D240/200B	OEM
D240/400B	OEM
D240/600B	OEM
D240/700B	OEM
D240-100	OEM
D240-300	OEM
D240-500	OEM
D240R200B	OEM
D240R200B#	1023
D240R400B	OEM
D240R400B#	1036
D240R600B	OEM
D240R600B#	1036
D240R700B	OEM
D240R700B#	1046
D240S800B	OEM
D240S1000B	OEM
D240S1200B	OEM
D240S1400B	OEM
D241	0042
D242	OEM
D242(2S)	0423
D242A	OEM
D242B	OEM
D242YLC	0423
D243	5991
D243(2S)	0236
D243A	OEM
D243B	OEM
D244	0236
D245	OEM
D245A	OEM
D245B	OEM
D246	0223
D246(2S)	0223
D246(DIODE)	OEM
D246A	OEM
D246A(2S)	0223
D246B	OEM
D246B(2S)	0223
D246C	0223
D246D	0223
D246E	0223
D246F	0223
D246G	0223
D246GN	0223
D246H	0223
D246J	0223
D246K	0223
D246L	0223
D246M	0223
D246OR	0223
D246R	0223
D246X	0223
D246Y	0223
D247	OEM
D247(2S)	0620
D247B	OEM
D247K	0620
D248B	OEM
D249	0130
D250	0130
D250/400E	OEM
D250/800E	OEM
D250/1200E	OEM
D250/1400E	OEM
D250/1600E	OEM
D250/1800E	OEM
D250-700	OEM
D250-1100	OEM
D250H	OEM
D250HL	OEM
D250R400E	OEM
D250R800E	OEM
D250R1200E	OEM
D250R1400E	OEM
D250R1600E	OEM
D250R1800E	OEM
D250S1000E	OEM
D250S1200E	OEM
D250S1400E	OEM
D250S1600E	OEM
D251	1027
D251N800B	OEM
D251N800E	OEM
D251N800V	OEM
D251N1200B	OEM
D251N1200E	OEM
D251N1200V	OEM
D251N1400B	OEM
D251N1400E	OEM
D251N1400V	OEM
D251NR800B	OEM
D251NR800V	OEM
D251NR1200B	OEM
D251NR1200E	OEM
D251NR1200V	OEM
D251NR1400B	OEM
D251NR1400E	OEM
D251NR1800B	OEM
D251NR1800E	OEM
D251NR1800V	OEM
D254(2S)	0178
D255(2S)	0178
D255N200B	OEM
D255N200V	OEM
D255N400B	OEM
D255N400V	OEM
D255N600B	OEM
D255N600V	OEM
D255N700B	OEM
D255N700V	OEM
D255N800B	OEM
D255N800V	OEM
D255NR200B	OEM
D255NR200V	OEM
D255NR400B	OEM
D255NR400V	OEM
D255NR600B	OEM
D255NR600V	OEM
D255NR700B	OEM
D255NR700V	OEM
D255NR800B	OEM
D255NR800V	OEM
D256	0042
D257(2S)	0423
D258	0236
D258S800T	OEM
D258S1000T	OEM
D258S1200T	OEM
D258S1400T	OEM
D258S1600T	OEM
D259	0236
D260	0620
D261(2S)	0218
D261-0	0218
D261-O	0218
D261A(2S)	0218
D261B(2S)	0218
D261C(2S)	0218
D261D(2S)	0218
D261E(2S)	0218
D261F(2S)	0218
D261G	0218
D261GN	0218
D261H	0218
D261J	0218
D261K	0218
D261L(2S)	0218
D261M	0218
D261OR	0218
D261P(2S)	0218
D261Q	0218
D261R(2S)	0218
D261S	0218
D261U	0218
D261V(2S)	0218
D261W(2S)	0218
D261X	0218
D261Y	0218
D262	0363
D265	0363
D266	0363
D269/2800	OEM
D269/3200	OEM
D269/3600	OEM
D270/3200T	OEM
D270/3600T	OEM
D272	0363
D273	0363
D274	0363
D280-400	OEM
D280-700	OEM
D280-1100	OEM
D280-1400	OEM
D280-1800	OEM
D280F5KF20	1127
D283(2S)	0074
D284(2S)	0429
D285(2S)	0074
D286	0543
D287(2S)	0543
D287A	0543
D287B	0543
D287C	0543
D288(2S)	0423
D288A(2S)	0423
D288B(2S)	0423
D288C(2S)	0423
D288K	0423
D288L(2S)	0423
D289(2S)	0423
D289A(2S)	0423
D289B(2S)	0423
D289C(2S)	0423
D290(2S)	0178
D291(2S)	4098
D291-0	4098
D291B	4098
D291BL	4098
D291C	4098
D291D	4098
D291E	4098
D291F	4098
D291G	4098
D291GA	4098
D291GN	4098
D291H	4098
D291J	4098
D291K	4098
D291L	4098
D291M	4098
D291OR	4098
D291R	4098
D291X	4098
D291Y	4098
D292	0042
D292(CHAN.MASTER)	0127
D292-0	0042
D292A	0042
D292B	0042
D292BL	0042
D292C	0042
D292D	0042
D292E	0042
D292F	0042
D292G	0042
D292GA	0042
D292GN	0042
D292H	0042
D292J	0042
D292K	0042
D292L	0042
D292M	0042
D292OR	0042
D292R	0042
D292X	0042
D292Y	0042
D293	0363
D294(2S)	0363
D297(2S)	0178
D299(2S)	0223
D299A	0223
D299B	0223
D299C	0223
D299E	0223
D299F	0223
D299G	0223
D299GN	0223
D299H	0223
D299J	0223
D299K	0223
D299M	0223
D299OR	0223
D299R	0223
D299X	0223
D299Y	0223
D300	0223
D300(2S)	0223
D300-700	OEM
D300-1100	OEM
D300A	0223
D300B(2S)	0223
D300C	0223
D300D	0223
D300G	0223
D300H	0223
D300J	0223
D300N400E	OEM
D300N800E	OEM
D300N1200E	OEM
D300N1400E	OEM
D300N1600E	OEM
D300N1800E	OEM
D300R400E	OEM
D300R800E	OEM
D300R1200E	OEM
D300R1400E	OEM
D300R1600E	OEM
D300R1800E	OEM
D301	4013
D302	OEM
D303	OEM

If replacement code is OEM, contact original manufacturer for replacement.

DEVICE TYPE	REPL CODE
D304	OEM
D305	OEM
D310	1852
D311	1852
D312(2S)	0003
D312A	0003
D312B	0003
D312C	0003
D312D	0003
D312E	0003
D312F	0003
D312G	0003
D312GN	0003
D312H	0003
D312J	0003
D312K	0003
D312L	0003
D312M	0003
D312X	0003
D312Y	0003
D313	0042
D313(JAPAN)	0042
D313A	0042
D313B	0042
D313C(2S)	0042
D313D(2S)	0042
D313DE	0042
D313E	0042
D313F(2S)	0042
D313G	0042
D313GN	0042
D313H	0042
D313L	0042
D313M	0042
D313N	0042
D313R	0042
D313Y	0042
D314	0042
D314A	0042
D314B	0042
D314C	0042
D314D	0042
D314E	0042
D314GN	0042
D314H	0042
D314L	0042
D314M	0042
D314N	0042
D314R	0042
D314Y	0042
D315(2S)	4251
D315C	4251
D315D	4251
D315E	4251
D316	0177
D316-1	0177
D316-2	0177
D317(2S)	0423
D317A(2S)	0423
D317AF	0423
D317AP	0423
D317F(2S)	0423
D317P(2S)	0423
D318(2S)	0423
D318-O	0423
D318A(2S)	0423
D318B	0423
D318P	0423
D318Q	0423
D319(2S)	0543
D320(2S)	0074
D320(DIODE)	0293
D320(IR)	0293
D320F	0293
D321(2S)	0363
D322(2S)	0177
D322A(2S)	0177
D322B(2S)	0177
D322C(2S)	0177
D323(2S)	0177
D323A(2S)	0177
D323B(2S)	0177
D323C(2S)	0177
D324(2S)	0142
D324E	0142
D325(2S)	0560
D325-1	0560
D325-OR	0560
D325A	0560
D325B	0560
D325C(2S)	0560
D325D(2S)	0560
D325E(2S)	0560
D325F(2S)	0560
D325G	0560
D325GN	0560
D325H	0560
D325J	0560
D325K	0560
D325L	0560
D325M	0560
D325OR	0560
D325R	OEM
D325X	0560
D325Y	0560
D326	0142
D326A	0142
D326B	0142
D326C	0142
D326D	0142
D326E	0142
D326L	0142
D326M	0142
D326OR	0142
D326X	0142
D326Y	0142
D327(2S)	0286
D327-V	0286
D327A(2S)	0286
D327B(2S)	0286
D327C(2S)	0286
D327D(2S)	0286
D327E(2S)	0286
D327F(2S)	0286
D327L	0286
D327OR	0286
D327R	0286
D328(2S)	0086
D328S(2S)	0086
D329	0187
D330	1607
D330D(2S)	1607
D330E	1607
D331	0042
D332	0867
D334(2S)	0177
D334A(2S)	0177
D334R(2S)	0177
D335	0130
D336(2S)	0324
D338	0130
D339	0177
D339-1	0177
D339-2	0177
D340	0177
D340-1	0177
D340-2	0177
D341(2S)	0103
D341H(2S)	0103
D342(2S)	0423
D343(2S)	0228
D343A	0228
D343B	0228
D343C	0228
D343D	0228
D343H	0228
D343J	0228
D343K	0228
D344	0423
D345	0423
D346	0423
D347	0423
D348	0313
D349	0400
D350	2051
D350A	2051
D350Q(2S)	2051
D350T	2051
D351(2S)	0363
D352	0208
D352D(2S)	0208
D352H	0208
D353(2S)	0220
D354	0628
D355(2S)	0018
D355C(2S)	0018
D355D	0018
D356	0236
D356C	0236
D356D	0236
D356E	0236
D357	0236
D357C	0236
D357D	0236
D357E	0236
D358	0236
D358C	0236
D358D	0236
D358E	0236
D358N400T	OEM
D358N800T	OEM
D358N1200T	OEM
D358N1400T	OEM
D358N1600T	OEM
D358N1800T	OEM
D359(2S)	1607
D359C(2S)	1607
D359C2(2S)	1607
D359D(2S)	1607
D359D1(2S)	1607
D359D2(2S)	1607
D360(2S)	0042
D360D(2S)	0042
D360E(2S)	0042
D361	0042
D361C	0042
D361D	0042
D362	1165
D365	0042
D365A	0042
D365B	0042
D365H(2S)	0042
D365P	0042
D366	0042
D366A	0042
D367(2S)	0208
D367A(2S)	0208
D367B(2S)	0208
D367C(2S)	0208
D367E(2S)	0208
D367F(2S)	0208
D367H	0208
D367J	0208
D367K	0208
D367L	0208
D367M	0208
D367OR	0208
D367R	0208
D367X	0208
D367Y	0208
D368(2S)	0313
D369	1160
D369(2S)	0543
D369-0	0543
D369-O	0543
D369C	0543
D369Y	0543
D370	0177
D371	0867
D371-O	0867
D371-R	0867
D371-Y	0867
D375	0620
D376	1966
D376A	1966
D377	1966
D378	0630
D379(2S)	0177
D380(2S)	0313
D380A	0313
D381	0236
D381K	0236
D381L	0236
D381M	0236
D382(2S)	0236
D383(2S)	0363
D384	3370
D385	1212
D386	0392
D386A(2S)	0392
D386A-D	0392
D386AD	0392
D386A-E	0392
D386AE	0392
D386E	0392
D386Y(2S)	0392
D387	0392
D387A	0392
D387AC	0392
D387AD	0392
D387AE	0392
D387AF(2S)	0392
D387AS	0392
D387D	0392
D387E	0392
D387S	0392
D388	0867
D389(2S)	0042
D389-0	0042
D389-O(2S)	0042
D389A(2S)	0042
D389AF	0042
D389AF0	0042
D389AFO	0042
D389AFP(2S)	0042
D389AO	0042
D389AP	0042
D389AQ	0042
D389BLB	0042
D389BLB-O	0042
D389BLB-P	0042
D389BP	0042
D389L	0042
D389LB(2S)	0042
D389LBP	0042
D389LP	0042
D389P	0042
D389Q	0042
D389R	0042
D390	0042
D390A	0042
D390P	0042
D390Q	0042
D392	0286
D393	0363
D394	0313
D395	0363
D396	0363
D400	0018
D400(2S)	2910
D400-1400	OEM
D400E	2910
D400E(2S)	2910
D400F(2S)	2910
D400N1600E	OEM
D400N1800E	OEM
D400N2000E	OEM
D400N2200E	OEM
D400N2400E	OEM
D400N2600E	OEM
D400N2800E	OEM
D400P1(2S)	2910
D400P2	2910
D400R1600E	OEM
D400R1800E	OEM
D400R2000E	OEM
D400R2200E	OEM
D400R2400E	OEM
D400R2600E	OEM
D400R2800E	OEM
D401	1284
D401-3800	OEM
D401-4200	OEM
D401-4600	OEM
D401-O	1284
D401-Y	1284
D401A	1284
D401A(JAPAN)	1284
D401A-K	1284
D401AK	1284
D401AK(JAPAN)	1284
D401A-L	1284
D401AL	1284
D401AL(JAPAN)	1284
D401AM	1284
D401E	1284
D401EK	1284
D401H	1284
D401I	1284
D401IJ	1284
D401K	1284
D401L	1284
D401LA	1284
D401M	1284
D401N3600E	OEM
D401N3800E	OEM
D401N4000E	OEM
D401N4400E	OEM
D401N4800E	OEM
D401R	1284
D401R3600E	OEM
D401R4000E	OEM
D401R4400E	OEM
D401R4800E	OEM
D401Y	0392
D401Y(JAPAN)	1284
D402	0392
D402A	0392
D402AK	0392
D402AL	0392
D402AM	0392
D402K	0392
D402L	0392
D402M	0392
D404	0423
D404G	0423
D405	OEM
D405(2S)	1212
D405NR2000	OEM
D405NR2200	OEM
D405NR2400	OEM
D405NR2600	OEM
D405NR2800	OEM
D408BS	OEM
D408BS-1	OEM
D408BSM	OEM
D408BSM-1	OEM
D408BSMZS	OEM
D408BSMZS-1	OEM
D408BSQ	OEM
D408BSQ-1	OEM
D408BSQZS	OEM
D408BSQZS-1	OEM
D408BST	OEM
D408BST-1	OEM
D408BSTZS	OEM
D408BSTZS-1	OEM
D408BSZS	OEM
D408BSZS-1	OEM
D409	1212
D411	2439
D413	0233
D414	0563
D414Q	0563
D415	0563
D415P	0563
D415Q	0563
D415R	0563
D415Y	0563
D416	0522
D416(TRANS)	0313
D417	0867
D418(2S)	0065
D419	1212
D422	0429
D423	0429
D424(2S)	0543
D425(2S)	0177
D425-O	0177
D425-R	0177
D425O	0177
D426	0177
D426-O	0177
D426-R	0177
D427	0177
D427-O	0177
D427-R	0177
D428	0867
D428-O	0867
D428-R	0867
D428N800	OEM
D428N1200	OEM
D428N1400	OEM
D428N1800	OEM
D429	0363
D430	0620
D431	0620
D431(IR)	0023
D431-F	0102
D431S2400T	OEM
D431S2800T	OEM
D431S3200T	OEM
D431S3600T	OEM
D432	0620
D432C1	1267
D437	0363
D437W	0363
D438	0191
D438/200T	OEM
D438/400T	OEM
D438/600T	OEM
D438/700T	OEM
D438/800T	OEM
D438E	0191
D439	0563
D439E	0563
D448N400	OEM
D448N600	OEM
D448N700	OEM
D448N800	OEM
D450-700	OEM
D450-1100	OEM
D450N400E	OEM
D450N800E	OEM
D450N1200E	OEM
D450N1400E	OEM
D450R400E	OEM
D450R800E	OEM
D450R1200E	OEM
D450R1400E	OEM
D452N400	OEM
D452N800	OEM
D452N1200	OEM
D452N1400	OEM
D452N1800	OEM
D452NR400	OEM
D452NR800	OEM
D452NR1200	OEM
D452NR1400	OEM
D452NR1800	OEM
D458(2S)	0363
D458C	0363
D459	2235
D460	1212
D461	0642
D463	2427
D464	2439
D467	0951
D467-C	0951
D467B	0951
D467C	0951
D467C(JAPAN)	0951
D468	0018
D468A	0018
D468AC	0018
D468B(2S)	0018
D468C(2S)	0018
D468D	0018
D468E	0018
D468F	0018
D468G	0018
D468GN	0018
D468H	0018
D468L	0018
D468LN	0018
D468Y	0018
D469	0177
D470	1751
D470-B	1751
D470A	1751
D470B	1751
D471	0018
D471(2S)	0018
D471(JAPAN)	0018
D471-L	0018
D471A	0018
D471B	0018
D471C	0018
D471D	0018
D471E	0018
D471G	0018
D471GN	0018
D471K	0018
D471L(2S)	0018
D471M	0018
D471N	0018
D471OR	0018
D471R	0018
D471X	0018
D471Y	0018
D473H	0018
D474K	0007
D475	0042
D475A	0042
D476	0423
D476A	0423
D476AKB	0423
D476AKC	0423
D476B	0423
D476C	0423
D476D	0423
D476G	0423
D476KB	0423
D476KC	0423
D4760	0423
D476YL	0423
D477	0042
D478	0392
D478-C	0392
D478A-01	0392
D478B	0392
D478BS	0392
D478C	0392
D478CS	0392
D478D	0392
D478Y	0392
D478YL	0392
D479	0161
D480	0161
D480-1600	OEM
D480-1800	OEM
D480-2000	OEM
D480-2200	OEM
D480-2400	OEM
D480-2600	OEM
D480-2800	OEM
D481	0161
D482	0276
D483	0276
D484	0168
D485	0161
D486	0161
D487	0161
D488	0161
D489	0161
D490	0161
D491	0561
D492	0103
D493	0561
D494	0561
D495	0561
D495S2000	OEM
D495S2200T	OEM
D495S2400T	OEM
D495S2600T	OEM
D496	3529
D497	3529
D498	3529
D499	0561
D500	0015
D500(TRANS)	0561
D501	0561
D502	2427
D503	2427
D504	2439
D505	2439
D506	2439
D510DS	OEM
D510DS-1	OEM
D510DSM	OEM
D510DSM-1	OEM
D510DSMZS	OEM
D510DSMZS-1	OEM
D510DSQ	OEM
D510DSQ-1	OEM
D510DSQZS	OEM
D510DST	OEM
D510DST-1	OEM
D510DSTZS	OEM
D510DSTZS-1	OEM
D510DSZS	OEM
D510DSZS-1	OEM
D513	0042
D517	0313
D518	0178
D519	0429
D520	0130
D520DS	OEM
D520DS-1	OEM
D520DSM	OEM
D520DSM-1	OEM
D520DSMZS	OEM
D520DSMZS-1	OEM
D520DSQ	OEM
D520DSQ-1	OEM
D520DSQZS	OEM
D520DSQZS-1	OEM
D520DST	OEM
D520DST-1	OEM
D520DSTZS	OEM
D520DSTZS-1	OEM
D520DSZS	OEM
D520DSZS-1	OEM
D522	0103
D523	2427
D524	1394
D525	4247
D525-O	4247
D525-Y	4247
D525R	4247
D525Y	4247
D526	2997
D526-O	2997
D528/400	OEM
D528/800	OEM
D528/1200	OEM
D528/1400	OEM
D528N1600T	OEM
D528N1800T	OEM
D528N2000T	OEM
D528N2200T	OEM
D528N2400T	OEM
D528N2600T	OEM
D528N2800T	OEM
D531	0423
D532	5709
D533	0363
D535	2414
D536	1966
D537	1966
D538	1852
D538A	1852
D539	0363
D539A	0363
D544	0423
D545	0018
D545E	0018
D545F	0018
D545G	0018
D546	2096
D546K	2096
D549	0558
D552	2414
D553	4887
D553-O(JAPAN)	4887
D553Y	4887
D554	1027
D554C-118	OEM
D554C-188	OEM
D555	0867
D556	0543
D557	0543
D560	1212
D562(2S)	0836
D562A	0836
D562B	0836
D562C	0836
D562D	0836
D562E	0836
D562K	0836
D562L	0836
D562OR	0836
D562R	0836
D562X	0836
D562Y	0836
D564	4246
D570	0423
D571	1978
D571K	1978
D571K(JAPAN)	1978
D571L	1978
D572	0363
D573	0363
D575	4928
D575L	4928
D576	0233
D577	0065
D579	0178
D580	0747
D581	0177
D581A	0177
D582	4944
D582A	4944
D586	2939
D586A	2939
D586R	2939
D587	2939
D587A	2939
D588	2939
D588A	2939
D589	0065
D590	0086
D591	0111
D591R	0111
D592	0286
D592ANC	0286
D592NC	0286
D592R	0286
D593	0168
D597	0867
D598	0867
D598N400T	OEM
D598N800T	OEM
D598N1200T	OEM
D598N1400T	OEM
D599	0111
D600	0563
D600K	0563
D603	0155
D604	0074
D605	0363
D606	0363
D608	0392
D608A	0392
D611	0429
D612	1790
D612E	1790
D612K	1790
D613	0060
D613P	0060
D617	2439
D619	1946
D620	0161
D620D12200	0120
D621	2847
D621EB	OEM
D621EC	OEM
D621ED	OEM
D621EE	OEM
D621EF	OEM
D621EG	OEM
D622	0429
D624	0233

DEVICE TYPE	REPL CODE
D625	5727
D627	2660
D628EB	OEM
D628EC	OEM
D628ED	OEM
D628EE	OEM
D628EF	OEM
D628EG	OEM
D628H	2439
D629/3600	OEM
D629/4000	OEM
D629/4400	OEM
D629/4800	OEM
D629H	2439
D630	0130
D631	2433
D631N3600T	OEM
D631N4000T	OEM
D631N4400T	OEM
D631N4800T	OEM
D632	0270
D632P	0270
D632Q	0270
D632R	0270
D633	1212
D634	1212
D635	1212
D636	0582
D636-0	0582
D636-O	0582
D636-Q	0582
D636-Q(JAPAN)	0582
D636-R	0582
D636-R(JAPAN)	0582
D636-S	0582
D636P	0582
D636Q	0582
D636R	0582
D636S	0582
D636T	0582
D637	0582
D637-0	0582
D637-P	0582
D637-Q	0582
D637-Q(JAPAN)	0582
D637-QR	0582
D637-R	0582
D637-R(JAPAN)	0582
D637-S	0582
D637C	0582
D637CR	0582
D637O	0582
D637P	0582
D637PQR	0582
D637Q	0582
D637Q(JAPAN)	0582
D637QR	0582
D637R	0582
D637S	0582
D637T	0582
D638	0672
D638-R	0672
D638P	0672
D638Q	0672
D638R	0672
D638S	0672
D639	0672
D639P	0672
D639Q	0672
D639R	0672
D639S	0672
D639T	0672
D640	0363
D641	1852
D647A	1386
D649	0065
D649Q	0065
D654	0086
D655D	1978
D655D(JAPAN)	0016
D655E	1978
D655F	1978
D656	0178
D657	0074
D660	0630
D661	0582
D661-R(JAPAN)	0582
D661S	0582
D661T	0582
D662	0715
D663	0363
D664	1212
D665	0543
D666	1386
D666A	1386
D666AB	1386
D666AC	1386
D666B	1386
D666C	1386
D666D	1386
D667	1386
D667A	1386
D667AB	1386
D667AC	1386
D667B	1386
D667C	1386
D667D	1386
D668AB	0563
D668AC	0563
D668B	0563
D668D	0563
D668N2000	OEM
D668N2200	OEM
D668N2400	OEM
D668N2600	OEM
D668N2800	OEM
D669	0563
D669AB	0563
D669AC	0563
D669B	0563
D669C	0563
D669D	0563
D670H	1394
D671	0155
D673	0867
D674	0867
D675	0543
D676	0543
D677	0074
D683	1852
D683A	1852
D683S	1852
D685	1852
D686	1212
D687	1212
D688	1480
D689	0055
D691	1212
D692	2427
D692M	2427
D692Q	2427
D693	0074
D704	0060
D704D	0060
D704E	0060
D705	0363
D706	0074
D707	0363
D711BS	OEM
D711BS-1	OEM
D711BSM	OEM
D711BSM-1	OEM
D711BSMZS	OEM
D711BSMZS-1	OEM
D711BSQ	OEM
D711BSQ-1	OEM
D711BSQZS	OEM
D711BSQZS-1	OEM
D711BST	OEM
D711BST-1	OEM
D711BSTZS	OEM
D711BSTZS-1	OEM
D716	3081
D717	3081
D718	0194
D720D(2S)	0208
D720E(2S)	0208
D720PJ	0208
D721	1212
D722	5748
D723	0236
D724	0392
D725	0065
D725-06	0065
D726	0236
D727	3602
D729H	3525
D730H	1394
D732	0867
D732K	0867
D733	0543
D733K	0543
D733U	0543
D734	1513
D734E	1513
D734F	1513
D734G	1513
D735	2276
D735-O	2276
D736	2276
D738	2276
D743	1284
D743A	1284
D743R	1284
D745	2276
D745A	2276
D745B	2276
D748	3227
D748A	3227
D748A-01	3227
D748A-01A	3227
D748B	3227
D749	0363
D749P	0363
D750	0543
D751	0194
D753	0543
D754	0324
D755	0530
D755D	0530
D755E	0530
D755F	0530
D756	0530
D756A	0530
D756AF	0530
D756D	0530
D756E	0530
D757	0168
D757C	0168
D757D	0168
D758	1085
D758C	1085
D758D	1085
D761	0392
D761D	0392
D761E	0392
D761V	0392
D761VL	0392
D761VR	0392
D761VRD	0392
D761VRE	0392
D762	0695
D762-O(JAPAN)	0695
D762P	0695
D762P(JAPAN)	0695
D762Q	0695
D763	1386
D764	0223
D765	0223
D765AC	OEM
D766	2096
D766D	2096
D766P	2096
D766Q	2096
D766R	2096
D767R	0155
D768(JAPAN)	1959
D768K	1959
D772	0392
D772B	0392
D772BLB	0392
D773	0018
D773K	0018
D773K(JAPAN)	0018
D773L	0018
D773U	0018
D774	2030
D774-4	2030
D774-K	2030
D774K	2030
D774K(JAPAN)	2030
D774U2	2030
D776	2548
D777	0270
D777-1	0270
D777FA-1	0270
D777FA-5	0270
D777FA5	0270
D778	0286
D778Q	0286
D778R	0286
D780C-1	3481
D781	0563
D783	0313
D784	2847
D785	2847
D787B	0715
D787C	0715
D787D	0715
D787E	0715
D788	0018
D788B	0018
D788C	0018
D788D	0018
D788E	0018
D789	2561
D789B	2561
D789C	2561
D789C(JAPAN)	2561
D789D	2561
D789D(JAPAN)	2561
D789E	2561
D789E(JAPAN)	2561
D789M	2561
D789P	2561
D790B	1386
D790C	1386
D792	0313
D792S	0313
D792T	0313
D793	0563
D793Q	0563
D794	0161
D794(2)	0161
D794A	0161
D794AD	0161
D794AQ	0161
D794AQ(JAPAN)	0161
D794D	0161
D794P	0161
D794Q	0161
D794R	0161
D795	0836
D795A	0836
D795AQ	0836
D795P	0836
D795Q	0836
D797	2433
D798N400	OEM
D798N800	OEM
D798N1200	OEM
D798N1400	OEM
D798N1800	OEM
D800	0071
D801	1852
D802	3060
D803	2439
D803Q	2439
D804	0042
D804H	0042
D804HLB	0042
D804HP	0042
D807	1493
D807M	1493
D809	0563
D810	4450
D811	5758
D812	1165
D812BS	OEM
D812BS-1	OEM
D812BSM-1	OEM
D812BSMZS	OEM
D812BSMZS-1	OEM
D812BSQ	OEM
D812BSQZS	OEM
D812BSQZS-1	OEM
D812BSTZS	OEM
D812BSTZS-1	OEM
D812BSZS	OEM
D812BSZS-1	OEM
D814A	0077
D814B	0322
D814D	0053
D814G	0181
D814V	0064
D815A	OEM
D815B	OEM
D815D	OEM
D815E	OEM
D815G	OEM
D815SZ	OEM
D815V	OEM
D816A	OEM
D816B	OEM
D816BS	OEM
D816BS-1	OEM
D816BSM	OEM
D816BSM-1	OEM
D816BSMZS	OEM
D816BSMZS-1	OEM
D816BSQ	OEM
D816BSQ-1	OEM
D816BSQZS	OEM
D816BSQZS-1	OEM
D816BST	OEM
D816BST-1	OEM
D816BSTZS	OEM
D816BSTZS-1	OEM
D816BSZS	OEM
D816BSZS-1	OEM
D816D	OEM
D816G	OEM
D816V	OEM
D817A	OEM
D817B	OEM
D817G	OEM
D817V	OEM
D818	2660
D818(IR)	0293
D819	2668
D820	2051
D820-1(JAPAN)	2051
D821	0313
D822	5407
D822P	5407
D822Q	5407
D823	1165
D829	OEM
D835	5307
D836	1212
D836A	1212
D836Q	1212
D836R	1212
D837	1212
D837A	1212
D837AB	1212
D837AD	1212
D837AQ	1212
D837AR	1212
D837B	1212
D837LB	1212
D837O	1212
D837P	1212
D837Q	1212
D837R	1212
D838	2847
D838B	2847
D841	3014
D843	0060
D844	0194
D844E	0194
D845	2276
D8450	2276
D849	0065
D850	2668
D855	1165
D855A	1165
D855B	1165
D856	0060
D856A	0060
D856B	0060
D856L	0060
D856LB	0060
D856M	0060
D856MLB	0060
D856O	0060
D856P	0060
D856Q	0060
D856R	0060
D857	0060
D857A	0060
D857B	0060
D858	5451
D858(IC)	2253
D858A	5451
D858B	5451
D859	0168
D859A	0168
D859B	0168
D860	0168
D860A	0168
D860B	0168
D863	1386
D864K	1959
D866	3125
D866A	3125
D866B	3125
D866D	3125
D866P	3125
D866Q	3125
D866R	3125
D867	0177
D868	0055
D869	0055
D869-L	0055
D869B	0055
D869FA	0055
D869L	0055
D870	0055
D870A	0055
D870FA	0055
D870Z	0055
D871	0055
D871Z	0055
D873	4786
D876	0392
D877	0178
D878	0103
D880	0461
D880-O	0461
D880-Y	0461
D880GR	0461
D880Y	0461
D880Y(JAPAN)	0461
D880Y(KOREA)	0461
D882	0161
D882P	0161
D882Q	0161
D882R	0161
D884	0728
D886	0060
D886A	0060
D886F	0060
D886P	0060
D886Q	0060
D889	0111
D889Q	0111
D889R	0111
D889T	0111
D890	0530
D892	2797
D892A	2797
D892AR	2797
D893	0400
D893A	0400
D894	0558
D895	3082
D896	3082
D897A	0055
D898	0055
D898-B13	0055
D898A	0055
D898B	0055
D900A	0055
D900B	0055
D900B-06	0055
D900D	0055
D903	0055
D904	0055
D904P	0055
D904P(JAPAN)	OEM
D906	0055
D907	3081
D917	0728
D918	0482
D919	0155
D920	0074
D920B	0074
D929	2548
D933H	5419
D940W01120	0137
D946	0558
D946A	0558
D950	2668
D950Q	2668
D951	0055
D952	0223
D953	0223
D957A	0055
D957A-03	0055
D958	0530
D958R	0530
D958S	0530
D958T	0530
D958U	0530
D959LB	1165
D959Q	1165
D960	0236
D960Q(JAPAN)	5623
D961	1165
D969	0672
D973	0009
D973A	0009
D973AR	1386
D973Q	0009
D973R	0009
D973S	0009
D974	1386
D975	0917
D975(TRANS)	0563
D975C	0563
D976	0917
D976(TRANS)	0728
D978	5307
D985	3703
D985-L	3703
D985K	3703
D985L	3703
D985M	3703
D985T	3703
D986	2970
D986(1)KA(JAPAN)	2970
D986(JAPAN)	2970
D986K	2970
D986L(JAPAN)	2970
D987	0728
D991K	5307
D992M	3350
D993	0055
D993A	0055
D994	0055
D995	2847
D997	2414
D1000	0071
D1000A60	OEM
D1010	0111
D1010S	0111
D1011	0530
D1012	0286
D1012F	0286
D1012G1	0286
D1016	0055
D1017	2613
D1018	2613
D1020	0018
D1020H	0018
D1021	0018
D1022	1212
D1024	1212
D1025	2096
D1032	3081
D1032A	3081
D1035	1165
D1036	2276
D1039	3553
D1040	0543
D1041	3554
D1044	0134
D1044A	0134
D1046	0194
D1047	0194
D1051	0018
D1051Q	0018
D1051R	0018
D1052	2069
D1052A	2069
D1052AFA	2069
D1052FA	2069
D1055	0009
D1055Q	0009
D1055R	0009
D1059	2201
D1060(JAPAN)	0060
D1060K	0060
D1061	0060
D1062	1966
D1063	3081
D1064	3081
D1064(JAPAN)	3081
D1069	0728
D1076	3350
D1076L	3350
D1078B	3350
D1079C	3350
D1079LB	3350
D1079LC	3350
D1080C	3350
D1080LC	3350
D1081B	3350
D1081C	3350
D1081LB	3350
D1081LC	3350
D1083	3350
D1083L	3350
D1085K	5307
D1092	3516
D1092FA	3516
D1093	2993
D1094	3588
D1098-42M	0560
D1101	0325
D1102	0325
D1103	0065
D1110	0130
D1110A	0130
D1115K	5307
D1126K	1959
D1127K	1959
D1130	0130
D1130K	0130
D1131B	0236
D1131C	0236
D1131D	0236
D1132B	0236
D1132C	0236
D1132D	0236
D1133B	0236
D1133C	0236
D1133D	0236
D1135(JAPAN)	0236
D1135-5(JAPAN)	0236
D1135C(JAPAN)	0236
D1136	0392
D1138	0392
D1138-C	0392
D1138-D	0392
D1138-O	0392
D1138-Y	0392
D1138C	0392
D1138C(JAPAN)	0392
D1138Y	0392
D1139B	0392
D1140	0130
D1142	1541
D1148	3082
D1153	0400
D1161-1	OEM
D1161-2	OEM
D1161-3	OEM
D1161-4	OEM
D1161-5	OEM
D1161-6	OEM
D1161-7	OEM
D1161-8	OEM
D1161-9	OEM
D1161-10	OEM
D1161-11	OEM
D1161-12	OEM
D1161C1	OEM
D1161C2	OEM
D1161C3	OEM
D1161C4	OEM
D1161C5	OEM
D1161C6	OEM
D1161C7	OEM
D1161C8	OEM
D1161C9	OEM
D1161C10	OEM
D1161C11	OEM
D1161C12	OEM
D1162K	5307
D1162L	5307
D1162M	5307
D1164K	3225
D1164L	3225
D1164M	3225
D1171	0055
D1172S	0015
D1173	0055
D1175	3619
D1177	0325
D1177C	0563
D1178	0325
D1178C	0563
D1179	0563
D1180	3622
D1181	0325
D1182	OEM
D1183	0313
D1184	0313
D1184T41	OEM
D1184T42	OEM
D1184T43	OEM
D1184T44	OEM
D1184T46	OEM
D1184T47	OEM
D1184T48	OEM
D1184T49	OEM
D1184T51	OEM
D1184T52	OEM
D1184T53	OEM
D1184T54	OEM
D1185	0313
D1186	0313
D1187	3279
D1190(JAPAN)	5457
D1193	3627
D1195	1212
D1197	1573
D1201	3038
D1201(2S)	1852
D1202	1852
D1203	3038
D1203(2S)	3038
D1206Q	0672
D1206R	0672
D1207S	3938
D1207T(JAPAN)	3938
D1208	2427
D1208FA	2427
D1218	2366
D1226	2471
D1226Q	2471

If replacement code is OEM, contact original manufacturer for replacement.

DEVICE TYPE	REPL CODE
D1246T	1500
D1264P	0392
D1265	1162
D1265-LB	1162
D1265-O	1162
D1265-O(JAPAN)	1162
D1265-OP	1162
D1265-P	1162
D1265B	1162
D1265L	1162
D1265LB	1162
D12650	1162
D12650P	1162
D1265P	1162
D1265P(JAPAN)	1162
D1266AQ	0060
D1266P	0060
D1266P(JAPAN)	0060
D1266Q	0060
D1271BP	1165
D1273	0928
D1273P	0928
D1273Q	0928
D1274B	4007
D1276-AP	1212
D1276-LB	1212
D1276A	1212
D1276AP	1212
D1276B	1212
D1276LB	1212
D1276P	1212
D1277B	1212
D1277B(JAPAN)	1212
D1283	0208
D1294	2989
D1294FA	2989
D1301	0325
D1301(2S)	0055
D1302	0325
D1303	0325
D1308K(JAPAN)	3136
D1308L(JAPAN)	3136
D1314	1506
D1318	0086
D1344E	0055
D1347	0018
D1353	3691
D1354	3692
D1355	3693
D1356	3694
D1376	3703
D1376K	3703
D1390	2766
D1392(JAPAN)	2366
D1395	2366
D1396	1541
D1397	1541
D1397B	1541
D1398	1541
D1398S	1541
D1398S-CA	1541
D1406	3125
D1406-Y	3125
D1406GR	3125
D1406Y	3125
D1406Y(JAPAN)	3125
D1420	0325
D1421	0325
D1422	OEM
D1423AS	2954
D1426	1541
D1426(JAPAN)	1541
D1427	1541
D1427-LB	1541
D1428	1541
D1438	3703
D1439	1541
D1439(JAPAN)	1541
D1439P	1541
D1439Q	1541
D1439RL	1541
D1441	1541
D1445	0015
D1453	1541
D1455	1541
D1459	3728
D1459Q	3728
D1459R	3728
D1491L(JAPAN)	1061
D1497	3360
D1511C-072	OEM
D1511C-073	OEM
D1511C-074	OEM
D1554	1541
D1555	0556
D1555-LB	0556
D1556	1541
D1556-LB	1541
D1579K	OEM
D1585K	0236
D1593L	OEM
D1593M	OEM
D1617	OEM
D1626	2828
D1627	3804
D1637	1061
D1637Y	1061
D1645	OEM
D1647	0278
D1649	1541
D1650	1541
D1651	0556
D1652	1541
D1666	0144
D1709ACT-128	OEM
D1709ACT-746	OEM
D1709C	OEM
D1709C-011	OEM
D1709C-515	OEM
D1709C-520	OEM
D1709C521	OEM
D1709C-538	OEM
D1709CT	OEM
D1709CT-113	OEM
D1709CT-118	OEM
D1709CT-121	OEM
D1709CT-713	OEM
D1709CT-715	OEM
D1709CT715	OEM
D1711CU517	OEM
D1720B	OEM
D1720C	OEM
D1722B	OEM
D1722C	OEM
D1724B	OEM
D1724BS	OEM
D1724BS-1	OEM
D1724BSM	OEM
D1724BSM-1	OEM
D1724BSMZS	OEM
D1724BSMZS-1	OEM
D1724BSQ	OEM
D1724BSQ-1	OEM
D1724BSQZS	OEM
D1724BSQZS-1	OEM
D1724BST	OEM
D1724BST-1	OEM
D1724BSTZS	OEM
D1724BSTZS-1	OEM
D1724BSZS	OEM
D1724BSZS-1	OEM
D1724C	OEM
D1726	0720
D1726B	OEM
D1726C	OEM
D1728	1541
D1728B	OEM
D1728C	OEM
D1730	1541
D1730B	OEM
D1730C	OEM
D1732	OEM
D1732B	OEM
D1732C	OEM
D1734	0628
D1734B	OEM
D1734C	OEM
D1736	OEM
D1736B	OEM
D1736C	OEM
D1738	0030
D1738B	OEM
D1738C	OEM
D1740	OEM
D1740B	OEM
D1740C	OEM
D1742	OEM
D1742B	OEM
D1742C	OEM
D1744	OEM
D1744B	OEM
D1744C	OEM
D1746	OEM
D1746B	OEM
D1746C	OEM
D1748	OEM
D1748B	OEM
D1748C	OEM
D1750	OEM
D1750B	OEM
D1750C	OEM
D1783L(JAPAN)	2366
D1785	OEM
D1788	3136
D1800/2000	OEM
D1800/2200	OEM
D1800/2600	OEM
D1800/3000	OEM
D1800N2400	OEM
D1800N2400T	OEM
D1800N2800	OEM
D1800N2800T	OEM
D1800N3200	OEM
D1800N3200T	OEM
D1800N3600	OEM
D1809N2400	OEM
D1809N2800	OEM
D1809N3200	OEM
D1809N3600	OEM
D1832	0042
D1833	0042
D1833F	0042
D1843	3874
D1843L	3874
D1846	2129
D1847	OEM
D1876	2129
D1877	2129
D1878	2129
D1879	2129
D1880	0392
D1881	OEM
D1888	1959
D1894	0065
D1900	OEM
D1910	1541
D1911	1541
D1913-O	0042
D1913R	0042
D1929	1061
D1929Q	1061
D1941	3360
D1943G	OEM
D1959-02	3494
D1978	OEM
D2010	OEM
D2010P	OEM
D2012	1165
D2015L	3837
D2020L	3837
D2035J	OEM
D2035W	OEM
D2046	OEM
D2050J	OEM
D2055W	OEM
D2065J	OEM
D2070W	OEM
D2101S	1217
D2102S	1217
D2103S	1217
D2103SF	1217
D2112A	OEM
D2112A-2	OEM
D2112A-4	OEM
D2114	2048
D2114-2	2048
D2114A4	2048
D2114A5	2048
D2114AL1	OEM
D2114AL2	OEM
D2114AL3	2048
D2114AL4	2048
D2114L	2048
D2114L2	2048
D2114L3	2048
D2115A	OEM
D2115A-2	OEM
D2115AL	OEM
D2115AL-2	OEM
D2115H2	OEM
D2115H3	OEM
D2115H4	OEM
D2125A	OEM
D2125A-2	OEM
D2125AL	OEM
D2125AL-2	OEM
D2125H1	OEM
D2125H2	OEM
D2125H3	OEM
D2125H4	OEM
D2141-2	OEM
D2141-3	OEM
D2141-4	OEM
D2141-5	OEM
D2141L-3	OEM
D2141L-4	OEM
D2141L-5	OEM
D2142	OEM
D2142-2	OEM
D2142-3	OEM
D2142L	OEM
D2142L3	OEM
D2144-3	OEM
D2147	OEM
D2147-3	OEM
D2147H	OEM
D2147H1	OEM
D2147H2	OEM
D2147H3	OEM
D2147HL	OEM
D2147L	OEM
D2148	OEM
D2148-3	OEM
D2148H	OEM
D2148H3	OEM
D2148HL	OEM
D2148HL3	OEM
D2148L	OEM
D2148L-3	OEM
D2149H	OEM
D2149H2	OEM
D2149H3	OEM
D2149HL	OEM
D2149HL3	OEM
D2187A-25	OEM
D2200N2000	OEM
D2200N2200	OEM
D2200N2400	OEM
D2200N2600	OEM
D2200N2800	OEM
D2201A	1217
D2201B	1217
D2201D	1217
D2201F	1217
D2201M	1217
D2201N	1217
D2209N2000	OEM
D2209N2200	OEM
D2209N2400	OEM
D2209N2600	OEM
D2209N2800	OEM
D2223BS	OEM
D2223BS-1	OEM
D2223BSM	OEM
D2223BSM-1	OEM
D2223BSMZS	OEM
D2223BSMZS-1	OEM
D2223BSQ	OEM
D2223BSQ-1	OEM
D2223BSQZS	OEM
D2223BSQZS-1	OEM
D2223BST	OEM
D2223BST-1	OEM
D2223BSTZS	OEM
D2223BSTZS-1	OEM
D2223BSZS	OEM
D2223BSZS-1	OEM
D2228N200	OEM
D2228N400	OEM
D2228N600	OEM
D2228N800	OEM
D2340	0042
D2350	0042
D2364C	OEM
D2600EF	0015
D2600F	1217
D2600M	1217
D2601DF	1217
D2601E	1217
D2601EF	1217
D2601M	1217
D2716D	2278
D2716D-Q1P-4	2278
D2716D-Q1X-0	2278
D2732A	2697
D2732D	2697
D2764	0812
D2764A-2	0812
D2764D	0812
D2801C	OEM
D2803C	OEM
D2812C	OEM
D2814C	OEM
D2816C	OEM
D3202Y	2150
D3205	OEM
D3207A	OEM
D3207A-1	OEM
D3222	OEM
D3232	OEM
D3242	OEM
D3245	OEM
D3313J	OEM
D3313K	OEM
D3404	OEM
D3436	0015
D3508	OEM
D3530	OEM
D3628A	2909
D3628A-1	2909
D3858-10	OEM
D4001-1400	OEM
D4001-1600	OEM
D4001-1800	OEM
D4001-2000	OEM
D4001-2200	OEM
D4002BC	2055
D4003	OEM
D4004	OEM
D4011BC	0215
D4012BC	0498
D4013BC	0413
D4015L	3837
D4016C-3	OEM
D4016CX-15	1898
D4017BC	0513
D4020L	3837
D4027BC	1949
D4030BC	0500
D4035J	OEM
D4035W	OEM
D4040	OEM
D4040BC	0056
D4049UBC	0001
D4050J	OEM
D4052BC	0024
D4053BC	0035
D4053G	OEM
D4055W	OEM
D4065J	OEM
D4066BC	0101
D4066SIP	OEM
D4069C	0119
D4069UBC	0119
D4070W	OEM
D4074	OEM
D4074A	OEM
D4075	OEM
D4075A	OEM
D4075B	OEM
D4075C	OEM
D4075D	OEM
D4075E	OEM
D4075F	OEM
D4075G	OEM
D4075H	OEM
D4081	0626
D4081A	OEM
D4081BC	0626
D4082	OEM
D4082A	OEM
D4082B	OEM
D4082D	OEM
D4084	OEM
D4084A	OEM
D4084B	OEM
D4089	OEM
D4092	OEM
D4121	OEM
D4127	OEM
D4128	OEM
D4136	OEM
D4136A	OEM
D4136B	OEM
D4140	OEM
D4140A	OEM
D4140B	OEM
D4140C	OEM
D4140D	OEM
D4140E	OEM
D4141	OEM
D4141A	OEM
D4141B	OEM
D4141C	OEM
D4141D	OEM
D4141E	OEM
D4148C	OEM
D4148E	OEM
D4151	OEM
D4151A	OEM
D4155	OEM
D4160	OEM
D4160BC	1359
D4164C-2	2357
D4164C-3	2357
D4164C-15	2357
D4168	OEM
D4168A	OEM
D4168B	OEM
D4168C	OEM
D4168C-15	OEM
D4168C-20	OEM
D4168D	OEM
D4175	OEM
D4175A	OEM
D4180	OEM
D4180C	OEM
D4180E	OEM
D4181C	OEM
D4181D	OEM
D4181E	OEM
D4186C	OEM
D4186D	OEM
D4186E	OEM
D4186F	OEM
D4187C	OEM
D4187D	OEM
D4187E	OEM
D4188B	OEM
D4188C	OEM
D4188D	OEM
D4188E	OEM
D4188F	OEM
D4189B	OEM
D4189C	OEM
D4189D	OEM
D4189E	OEM
D4189F	OEM
D4194	OEM
D4194A	OEM
D4195	OEM
D4195A	OEM
D4196	OEM
D4196A	OEM
D4200	OEM
D4200-1	OEM
D4200-2	OEM
D4200-3	OEM
D4200A-E	OEM
D4200F	OEM
D4200G	OEM
D4200H	OEM
D4200J-L	OEM
D4201	OEM
D4201-1	OEM
D4201-2	OEM
D4201-3	OEM
D4201A-E	OEM
D4201F	OEM
D4201G	OEM
D4201H	OEM
D4201J	OEM
D4201K	OEM
D4202	OEM
D4202-1	OEM
D4202-2	OEM
D4202-3	OEM
D4202A-E	OEM
D4202F	OEM
D4202G	OEM
D4202H	OEM
D4203	OEM
D4203-1	OEM
D4203-2	OEM
D4203-3	OEM
D4203A-E	OEM
D4203F	OEM
D4203G	OEM
D4204	OEM
D4204-1	OEM
D4204-2	OEM
D4204-3	OEM
D4204A-E	OEM
D4205	OEM
D4205-1	OEM
D4205-2	OEM
D4205-3	OEM
D4205A	OEM
D4205B	OEM
D4206	OEM
D4206-1	OEM
D4206-2	OEM
D4206-3	OEM
D4206A	OEM
D4207	OEM
D4207-1	OEM
D4207-2	OEM
D4207-3	OEM
D4210	OEM
D4210-1	OEM
D4210-2	OEM
D4210-3	OEM
D4210A-E	OEM
D4210F	OEM
D4210G	OEM
D4210H	OEM
D4210J-L	OEM
D4211	OEM
D4211-1	OEM
D4211-2	OEM
D4211-3	OEM
D4211A-E	OEM
D4211F	OEM
D4211G	OEM
D4211H	OEM
D4211J	OEM
D4211K	OEM
D4212	OEM
D4212-1	OEM
D4212-2	OEM
D4212-3	OEM
D4212A-E	OEM
D4212F	OEM
D4212G	OEM
D4212H	OEM
D4213	OEM
D4213-1	OEM
D4213-2	OEM
D4213-3	OEM
D4213A-E	OEM
D4213F	OEM
D4213G	OEM
D4214	OEM
D4214-1	OEM
D4214-2	OEM
D4214-3	OEM
D4214A-E	OEM
D4215	OEM
D4215-1	OEM
D4215-2	OEM
D4215-3	OEM
D4215A	OEM
D4215B	OEM
D4216	OEM
D4216-1	OEM
D4216-2	OEM
D4216-3	OEM
D4216A	OEM
D4217-1	OEM
D4217-2	OEM
D4217-3	OEM
D4220	OEM
D4220-1	OEM
D4220-2	OEM
D4220-3	OEM
D4220A-E	OEM
D4220F	OEM
D4220G	OEM
D4220H	OEM
D4220J	OEM
D4220K	OEM
D4221	OEM
D4221-1	OEM
D4221-2	OEM
D4221-3	OEM
D4221A-E	OEM
D4221F	OEM
D4221G	OEM
D4221H	OEM
D4221J	OEM
D4222	OEM
D4222-1	OEM
D4222-2	OEM
D4222-3	OEM
D4222A-E	OEM
D4222F	OEM
D4222G	OEM
D4223	OEM
D4223-1	OEM
D4223-2	OEM
D4223-3	OEM
D4223A-E	OEM
D4223F	OEM
D4224	OEM
D4224-1	OEM
D4224-2	OEM
D4224-3	OEM
D4224A-D	OEM
D4225	OEM
D4225-1	OEM
D4225-2	OEM
D4225-3	OEM
D4225A	OEM
D4226	OEM
D4226-1	OEM
D4226-2	OEM
D4226-3	OEM
D4227	OEM
D4227-1	OEM
D4227-2	OEM
D4227-3	OEM
D4230	OEM
D4230-1	OEM
D4230-2	OEM
D4230-3	OEM
D4230A-E	OEM
D4230F	OEM
D4230G	OEM
D4230H	OEM
D4230J	OEM
D4231	OEM
D4231-1	OEM
D4231-2	OEM
D4231-3	OEM
D4231A-E	OEM
D4231F	OEM
D4231G	OEM
D4231H	OEM
D4232	OEM
D4232-1	OEM
D4232-2	OEM
D4232-3	OEM
D4232A-E	OEM
D4232F	OEM
D4233	OEM
D4233-1	OEM
D4233-2	OEM
D4233-3	OEM
D4233A-D	OEM
D4234	OEM
D4234-1	OEM
D4234-2	OEM
D4234-3	OEM
D4234A	OEM
D4234B	OEM
D4235	OEM
D4235-1	OEM
D4235-2	OEM
D4235-3	OEM
D4236-1	OEM
D4236-2	OEM
D4236-3	OEM
D4237-1	OEM
D4237-2	OEM
D4237-3	OEM
D4240	OEM
D4240-1	OEM
D4240-2	OEM
D4240-3	OEM
D4240A-E	OEM
D4240F	OEM
D4240G	OEM
D4240H	OEM
D4240J	OEM
D4241	OEM
D4241-1	OEM
D4241-2	OEM
D4241-3	OEM
D4241A-E	OEM
D4241F	OEM
D4241G	OEM
D4241H	OEM
D4242	OEM
D4242-1	OEM
D4242-2	OEM
D4242-3	OEM
D4242A-E	OEM
D4242F	OEM
D4243	OEM
D4243-1	OEM
D4243-2	OEM
D4243-3	OEM
D4243A-D	OEM
D4244	OEM
D4244-1	OEM
D4244-2	OEM
D4244-3	OEM
D4244A	OEM
D4244B	OEM
D4245	OEM
D4245-1	OEM
D4245-2	OEM
D4245-3	OEM
D4246-1	OEM
D4246-2	OEM
D4246-3	OEM
D4247-1	OEM
D4247-2	OEM
D4247-3	OEM
D4250	0177
D4250-1	OEM
D4250-2	OEM
D4250-3	OEM

DEVICE TYPE	REPL CODE
D4250A-E	OEM
D4250F	OEM
D4251	OEM
D4251-1	OEM
D4251-2	OEM
D4251-3	OEM
D4251A-E	OEM
D4252	OEM
D4252-1	OEM
D4252-2	OEM
D4252-3	OEM
D4252A-D	OEM
D4253	OEM
D4253-1	OEM
D4253-2	OEM
D4253-3	OEM
D4253A-C	OEM
D4254	OEM
D4254-1	OEM
D4254-2	OEM
D4254-3	OEM
D4254A	OEM
D4255-1	OEM
D4255-2	OEM
D4255-3	OEM
D4256-2	OEM
D4256-3	OEM
D4257-2	OEM
D4257-3	OEM
D4260	OEM
D4260-1	OEM
D4260-2	OEM
D4260-3	OEM
D4260A	OEM
D4260B	OEM
D4260C	OEM
D4261	OEM
D4261-1	OEM
D4261-2	OEM
D4261-3	OEM
D4261A	OEM
D4261B	OEM
D4262	OEM
D4262-1	OEM
D4262-2	OEM
D4262-3	OEM
D4262A	OEM
D4263	OEM
D4263-1	OEM
D4263-2	OEM
D4263-3	OEM
D4263A	OEM
D4264	OEM
D4264-1	OEM
D4264-2	OEM
D4264-3	OEM
D4265-1	OEM
D4265-2	OEM
D4265-3	OEM
D4266-3	OEM
D4267-3	OEM
D4270	OEM
D4270-1	OEM
D4270-2	OEM
D4270-3	OEM
D4271	OEM
D4271-1	OEM
D4271-2	OEM
D4271-3	OEM
D4272-1	OEM
D4272-2	OEM
D4272-3	OEM
D4274-2	OEM
D4274-3	OEM
D4275-2	OEM
D4275-3	OEM
D4276-3	OEM
D4277-3	OEM
D4300	OEM
D4300A	OEM
D4300B-E	OEM
D4300F	OEM
D4300G	OEM
D4300H	OEM
D4300J	OEM
D4300K	OEM
D4301	OEM
D4301A	OEM
D4301B-E	OEM
D4301F	OEM
D4301G	OEM
D4301H	OEM
D4301J	OEM
D4301K	OEM
D4302	OEM
D4302A	OEM
D4302B-E	OEM
D4302F	OEM
D4302G	OEM
D4302H	OEM
D4302J	OEM
D4302K	OEM
D4303	OEM
D4303A	OEM
D4303B-E	OEM
D4303F	OEM
D4303G	OEM
D4303H	OEM
D4303J	OEM
D4303K	OEM
D4310	OEM
D4310A	OEM
D4310B-E	OEM
D4310F	OEM
D4310G	OEM
D4310H	OEM
D4310J	OEM
D4310K	OEM
D4311	OEM
D4311A	OEM
D4311B-E	OEM
D4311F	OEM
D4311G	OEM
D4311H	OEM
D4311J	OEM
D4311K	OEM
D4312	OEM
D4312A	OEM
D4312B-E	OEM
D4312F	OEM
D4312G	OEM
D4312H	OEM
D4312J	OEM
D4312K	OEM
D4313	OEM
D4313A	OEM
D4313B-E	OEM
D4313F	OEM
D4313G	OEM
D4313H	OEM
D4313J	OEM
D4313K	OEM
D4320	OEM
D4320A	OEM
D4320B-E	OEM
D4320F	OEM
D4320G	OEM
D4320H	OEM
D4320J	OEM
D4321	OEM
D4321A	OEM
D4321B-E	OEM
D4321F	OEM
D4321G	OEM
D4321H	OEM
D4321J	OEM
D4322	OEM
D4322A	OEM
D4322B-E	OEM
D4322F	OEM
D4322G	OEM
D4322H	OEM
D4322J	OEM
D4323	OEM
D4323A	OEM
D4323B-E	OEM
D4323F	OEM
D4323G	OEM
D4323H	OEM
D4323J	OEM
D4330	OEM
D4330A	OEM
D4330B-E	OEM
D4330F	OEM
D4330G	OEM
D4330H	OEM
D4330J	OEM
D4331	OEM
D4331A	OEM
D4331B-E	OEM
D4331F	OEM
D4331G	OEM
D4331H	OEM
D4331J	OEM
D4332	OEM
D4332A	OEM
D4332B-E	OEM
D4332F	OEM
D4332G	OEM
D4332H	OEM
D4332J	OEM
D4333	OEM
D4333A	OEM
D4333B-E	OEM
D4333F	OEM
D4333G	OEM
D4333H	OEM
D4333J	OEM
D4340	OEM
D4340A	OEM
D4340B-E	OEM
D4340F	OEM
D4341	OEM
D4341A	OEM
D4341B-E	OEM
D4341F	OEM
D4342	OEM
D4342A	OEM
D4342B-E	OEM
D4342F	OEM
D4343	OEM
D4343A	OEM
D4343B-E	OEM
D4343F	OEM
D4350	OEM
D4350A	OEM
D4350B-D	OEM
D4351	OEM
D4351A	OEM
D4351B-D	OEM
D4352	OEM
D4352A	OEM
D4352B-D	OEM
D4353	OEM
D4353A	OEM
D4353B-D	OEM
D4360	OEM
D4360A	OEM
D4361	OEM
D4361A	OEM
D4362	OEM
D4362A	OEM
D4363	OEM
D4363A	OEM
D4364C-15L	OEM
D4364C-20L	OEM
D4400	OEM
D4400A	OEM
D4400BE	OEM
D4400F	OEM
D4400G	OEM
D4400H	OEM
D4400J	OEM
D4400K	OEM
D4400N200	OEM
D4400N400	OEM
D4400N600	OEM
D4400N800	OEM
D4401	OEM
D4401A	OEM
D4401B-E	OEM
D4401F	OEM
D4401G	OEM
D4401H	OEM
D4401J	OEM
D4401K	OEM
D4402	OEM
D4402A-E	OEM
D4402F	OEM
D4402G	OEM
D4402H	OEM
D4402J	OEM
D4402K	OEM
D4403	OEM
D4403A-E	OEM
D4403F	OEM
D4403G	OEM
D4403H	OEM
D4403J	OEM
D4403K	OEM
D4409N200	OEM
D4409N400	OEM
D4409N600	OEM
D4409N800	OEM
D4410	OEM
D4410A	OEM
D4410B-E	OEM
D4410F	OEM
D4410G	OEM
D4410H	OEM
D4410J	OEM
D4410K	OEM
D4411	OEM
D4411A	OEM
D4411B-E	OEM
D4411F	OEM
D4411G	OEM
D4411H	OEM
D4411J	OEM
D4411K	OEM
D4412	OEM
D4412A-E	OEM
D4412F	OEM
D4412G	OEM
D4412H	OEM
D4412J	OEM
D4412K	OEM
D4413	OEM
D4413A-E	OEM
D4413F	OEM
D4413G	OEM
D4413H	OEM
D4420	OEM
D4420A	OEM
D4420B-E	OEM
D4420F	OEM
D4420G	OEM
D4420H	OEM
D4420J	OEM
D4421A	OEM
D4421B-E	OEM
D4421F	OEM
D4421G	OEM
D4421H	OEM
D4421J	OEM
D4422	OEM
D4422A-E	OEM
D4422F	OEM
D4422G	OEM
D4422H	OEM
D4422J	OEM
D4423	OEM
D4423A-E	OEM
D4423F	OEM
D4423G	OEM
D4423H	OEM
D4423J	OEM
D4430	OEM
D4430A	OEM
D4430B-E	OEM
D4430F	OEM
D4430G	OEM
D4430H	OEM
D4430J	OEM
D4431	OEM
D4431A	OEM
D4431A-B	OEM
D4431F	OEM
D4431G	OEM
D4431H	OEM
D4431J	OEM
D4432	OEM
D4432A-E	OEM
D4432F	OEM
D4432G	OEM
D4432H	OEM
D4432J	OEM
D4433	OEM
D4433A-E	OEM
D4433F	OEM
D4433G	OEM
D4433H	OEM
D4433J	OEM
D4440	OEM
D4440A	OEM
D4440B-E	OEM
D4440F	OEM
D4441	OEM
D4441A	OEM
D4441B-E	OEM
D4441F	OEM
D4442	OEM
D4442A-E	OEM
D4442F	OEM
D4443	OEM
D4443A-E	OEM
D4443F	OEM
D4450	OEM
D4450A	OEM
D4450B-D	OEM
D4451	OEM
D4451A	OEM
D4451B-D	OEM
D4452	OEM
D4452A-D	OEM
D4453	OEM
D4453A-D	OEM
D4460	OEM
D4460A	OEM
D4462	OEM
D4462A	OEM
D4463	OEM
D4463A	OEM
D4500X	OEM
D4500Y	OEM
D4500Z	OEM
D4501X	OEM
D4501Y	OEM
D4501Z	OEM
D4502X	OEM
D4502Y	OEM
D4502Z	OEM
D4503X	OEM
D4503Y	OEM
D4503Z	OEM
D4504X	OEM
D4504Y	OEM
D4504Z	OEM
D4505X	OEM
D4505Y	OEM
D4505Z	OEM
D4506X	OEM
D4506Y	OEM
D4506Z	OEM
D4510X	OEM
D4510Y	OEM
D4510Z	OEM
D4511X	OEM
D4511Y	OEM
D4511Z	OEM
D4512X	OEM
D4512Y	OEM
D4512Z	OEM
D4513X	OEM
D4513Y	OEM
D4513Z	OEM
D4514X	OEM
D4514Y	OEM
D4514Z	OEM
D4515X	OEM
D4515Y	OEM
D4515Z	OEM
D4516X	OEM
D4516Y	OEM
D4516Z	OEM
D4520Y	OEM
D4520Z	OEM
D4521X	OEM
D4521Y	OEM
D4521Z	OEM
D4522X	OEM
D4522Y	OEM
D4522Z	OEM
D4523X	OEM
D4523Y	OEM
D4523Z	OEM
D4524X	OEM
D4524Y	OEM
D4524Z	OEM
D4525X	OEM
D4525Y	OEM
D4525Z	OEM
D4526X	OEM
D4526Y	OEM
D4526Z	OEM
D4530X	OEM
D4530Y	OEM
D4530Z	OEM
D4531X	OEM
D4531Y	OEM
D4531Z	OEM
D4532X	OEM
D4532Y	OEM
D4532Z	OEM
D4533X	OEM
D4533Y	OEM
D4533Z	OEM
D4534X	OEM
D4534Y	OEM
D4534Z	OEM
D4535X	OEM
D4535Y	OEM
D4535Z	OEM
D4536X	OEM
D4536Y	OEM
D4536Z	OEM
D4540X	OEM
D4540Y	OEM
D4540Z	OEM
D4541X	OEM
D4541Y	OEM
D4541Z	OEM
D4542X	OEM
D4542Y	OEM
D4542Z	OEM
D4543X	OEM
D4543Y	OEM
D4543Z	OEM
D4544X	OEM
D4544Y	OEM
D4544Z	OEM
D4545X	OEM
D4545Y	OEM
D4545Z	OEM
D4546X	OEM
D4546Y	OEM
D4546Z	OEM
D4550X	OEM
D4550Y	OEM
D4550Z	OEM
D4551X	OEM
D4551Y	OEM
D4551Z	OEM
D4552X	OEM
D4552Y	OEM
D4552Z	OEM
D4553X	OEM
D4553Y	OEM
D4553Z	OEM
D4554X	OEM
D4554Y	OEM
D4554Z	OEM
D4555X	OEM
D4555Y	OEM
D4555Z	OEM
D4556X	OEM
D4556Y	OEM
D4556Z	OEM
D4560X	OEM
D4560Y	OEM
D4560Z	OEM
D4561X	OEM
D4561Y	OEM
D4561Z	OEM
D4562X	OEM
D4562Y	OEM
D4562Z	OEM
D4563X	OEM
D4563Y	OEM
D4563Z	OEM
D4564X	OEM
D4564Y	OEM
D4564Z	OEM
D4565X	OEM
D4565Y	OEM
D4565Z	OEM
D4566X	OEM
D4566Y	OEM
D4566Z	OEM
D4570X	OEM
D4570Y	OEM
D4570Z	OEM
D4571X	OEM
D4571Y	OEM
D4571Z	OEM
D4572X	OEM
D4572Y	OEM
D4572Z	OEM
D4573X	OEM
D4573Y	OEM
D4573Z	OEM
D4574X	OEM
D4574Y	OEM
D4574Z	OEM
D4575X	OEM
D4575Y	OEM
D4575Z	OEM
D4576X	OEM
D4576Y	OEM
D4576Z	OEM
D4600	OEM
D4600-1	OEM
D4600-2	OEM
D4600-3	OEM
D4600A-E	OEM
D4600FH	OEM
D4600J-L	OEM
D4600N	OEM
D4600P	OEM
D4601	OEM
D4601-1	OEM
D4601-2	OEM
D4601-3	OEM
D4601A-E	OEM
D4601F-H	OEM
D4601J-L	OEM
D4602	OEM
D4602-1	OEM
D4602-2	OEM
D4602-3	OEM
D4602A-E	OEM
D4602F	OEM
D4602G	OEM
D4602H	OEM
D4603	OEM
D4603-1	OEM
D4603-2	OEM
D4603-3	OEM
D4603A-E	OEM
D4603F	OEM
D4603G	OEM
D4604	OEM
D4604-1	OEM
D4604-2	OEM
D4604-3	OEM
D4604A-E	OEM
D4605	OEM
D4605-1	OEM
D4605-2	OEM
D4605-3	OEM
D4605A	OEM
D4605B	OEM
D4606	OEM
D4606-1	OEM
D4606-2	OEM
D4606-3	OEM
D4606A	OEM
D4607	OEM
D4607-1	OEM
D4607-2	OEM
D4607-3	OEM
D4610	OEM
D4610-1	OEM
D4610-2	OEM
D4610-3	OEM
D4610A-E	OEM
D4610F	OEM
D4610G	OEM
D4610H	OEM
D4610J-L	OEM
D4611	OEM
D4611-1	OEM
D4611-2	OEM
D4611-3	OEM
D4611A-E	OEM
D4611F-H	OEM
D4611J	OEM
D4611K	OEM
D4612	OEM
D4612-1	OEM
D4612-2	OEM
D4612-3	OEM
D4612A-E	OEM
D4612F	OEM
D4612G	OEM
D4612H	OEM
D4613	OEM
D4613-1	OEM
D4613-2	OEM
D4613-3	OEM
D4613A-E	OEM
D4613F	OEM
D4613G	OEM
D4614	OEM
D4614-1	OEM
D4614-2	OEM
D4614-3	OEM
D4614A-E	OEM
D4615	OEM
D4615-1	OEM
D4615-2	OEM
D4615-3	OEM
D4615A	OEM
D4615B	OEM
D4616	OEM
D4616-1	OEM
D4616-2	OEM
D4616-3	OEM
D4616A	OEM
D4617	OEM
D4617-1	OEM
D4617-2	OEM
D4617-3	OEM
D4620	OEM
D4620-1	OEM
D4620-2	OEM
D4620-3	OEM
D4620A-E	OEM
D4620F-H	OEM
D4620J-L	OEM
D4620X	OEM
D4621	OEM
D4621-1	OEM
D4621-2	OEM
D4621-3	OEM
D4621A-E	OEM
D4621F-H	OEM
D4621J	OEM
D4622	OEM
D4622-1	OEM
D4622-2	OEM
D4622-3	OEM
D4622A-E	OEM
D4622F	OEM
D4622G	OEM
D4623	OEM
D4623-1	OEM
D4623-2	OEM
D4623-3	OEM
D4623A-E	OEM
D4623F	OEM
D4624	OEM
D4624-1	OEM
D4624-2	OEM
D4624-3	OEM
D4624A-D	OEM
D4625	OEM
D4625-1	OEM
D4625-2	OEM
D4625-3	OEM
D4625A	OEM
D4626	OEM
D4626-1	OEM
D4626-2	OEM
D4626-3	OEM
D4627	OEM
D4627-1	OEM
D4627-2	OEM
D4627-3	OEM
D4630	OEM
D4630-1	OEM
D4630-2	OEM
D4630-3	OEM
D4630A-E	OEM
D4630F-H	OEM
D4630J	OEM
D4630K	OEM
D4631	OEM
D4631-2	OEM
D4631-3	OEM
D4631A-E	OEM
D4631F-H	OEM
D4632	OEM
D4632-1	OEM
D4632-2	OEM
D4632-3	OEM
D4632A-E	OEM
D4632AMR	OEM
D4632F	OEM
D4633	OEM
D4633-1	OEM
D4633-2	OEM
D4633-3	OEM
D4633A-D	OEM
D4634	OEM
D4634-1	OEM
D4634-2	OEM
D4634-3	OEM
D4634A	OEM
D4634B	OEM
D4635	OEM
D4635-1	OEM
D4635-2	OEM
D4635-3	OEM
D4636-1	OEM
D4636-2	OEM
D4636-3	OEM
D4637-1	OEM
D4637-2	OEM
D4637-3	OEM
D4640	OEM
D4640-1	OEM
D4640-2	OEM
D4640-3	OEM
D4640A-E	OEM
D4640F	OEM
D4640G	OEM
D4640H	OEM
D4640J	OEM
D4641	OEM
D4641-1	OEM
D4641-2	OEM
D4641-3	OEM
D4641A-E	OEM
D4641F-H	OEM
D4642	OEM
D4642-1	OEM
D4642-2	OEM
D4642-3	OEM
D4642A-E	OEM
D4642F	OEM
D4643	OEM
D4643-2	OEM

If replacement code is OEM, contact original manufacturer for replacement.

DEVICE TYPE	REPL CODE	DEVICE TYPE	REPL CODE	DEVICE TYPE	REPL CODE	DEVICE TYPE	REPL CODE	DEVICE TYPE	REPL CODE	DEVICE TYPE	REPL CODE	DEVICE TYPE	REPL CODE
D4643-3	OEM	D4711A	OEM	D4761	OEM	D4822A-E	OEM	D4854-2	OEM	D4974	OEM	D5065A	OEM
D4643A-D	OEM	D4711B-E	OEM	D4761A	OEM	D4822F	OEM	D4854-3	OEM	D4974A	OEM	D5065B	OEM
D4644	OEM	D4711F	OEM	D4762	OEM	D4822G	OEM	D4854A	OEM	D4974B	OEM	D5065C	OEM
D4644-1	OEM	D4711G	OEM	D4762A	OEM	D4823	OEM	D4855-1	OEM	D4974C	OEM	D5066	OEM
D4644-2	OEM	D4711H	OEM	D4763	OEM	D4823-1	OEM	D4855-2	OEM	D4975	OEM	D5066A	OEM
D4644-3	OEM	D4711J	OEM	D4763A	OEM	D4823-2	OEM	D4855-3	OEM	D4975A	OEM	D5066B	OEM
D4644A	OEM	D4711K	OEM	D4767	2258	D4823-3	OEM	D4856-2	OEM	D4975B	OEM	D5066C	OEM
D4644B	OEM	D4712	OEM	D4770	OEM	D4823A-E	OEM	D4856-3	OEM	D4975C	OEM	D5071	OEM
D4645	OEM	D4712A	OEM	D4771	OEM	D4823F	OEM	D4857-2	OEM	D4976	OEM	D5071A	OEM
D4645-1	OEM	D4712B-E	OEM	D4772	OEM	D4824	OEM	D4857-3	OEM	D4976A	OEM	D5071B	OEM
D4645-2	OEM	D4712F	OEM	D4773	OEM	D4824-1	OEM	D4861	OEM	D4976B	OEM	D5071C	OEM
D4645-3	OEM	D4712G	OEM	D4782	OEM	D4824-2	OEM	D4861-1	OEM	D4976C	OEM	D5072	OEM
D4646-1	OEM	D4712H	OEM	D4782A	OEM	D4824-3	OEM	D4861-2	OEM	D5002-06	OEM	D5072A	OEM
D4646-2	OEM	D4712J	OEM	D4801	OEM	D4824A-D	OEM	D4861-3	OEM	D5002-12	OEM	D5072B	OEM
D4646-3	OEM	D4712K	OEM	D4801-1	OEM	D4825	OEM	D4861A	OEM	D5002-18	OEM	D5072C	OEM
D4647-1	OEM	D4713	OEM	D4801-2	OEM	D4825-1	OEM	D4861B	OEM	D5002-24	OEM	D5073	OEM
D4647-2	OEM	D4713A	OEM	D4801-3	OEM	D4825-2	OEM	D4862	OEM	D5002-30	OEM	D5073A	OEM
D4647-3	OEM	D4713B-E	OEM	D4801A-E	OEM	D4825-3	OEM	D4862-1	OEM	D5002-36	OEM	D5073B	OEM
D4650	OEM	D4713F	OEM	D4801F	OEM	D4825A	OEM	D4862-2	OEM	D5002-42	OEM	D5073C	OEM
D4650-1	OEM	D4713G	OEM	D4801G	OEM	D4826	OEM	D4862-3	OEM	D5002-48	OEM	D5074	OEM
D4650-2	OEM	D4713H	OEM	D4801H	OEM	D4826-1	OEM	D4862A	OEM	D5005-06	OEM	D5074A	OEM
D4650-3	OEM	D4713J	OEM	D4801J	OEM	D4826-2	OEM	D4863	OEM	D5005-12	OEM	D5074B	OEM
D4650A-E	OEM	D4713K	OEM	D4801K	OEM	D4826-3	OEM	D4863-1	OEM	D5005-18	OEM	D5074C	OEM
D4650F	OEM	D4720	OEM	D4802	OEM	D4827	OEM	D4863-2	OEM	D5006-06	OEM	D5075	OEM
D4651	OEM	D4720A	OEM	D4802-1	OEM	D4827-1	OEM	D4863-3	OEM	D5006-12	OEM	D5075A	OEM
D4651-1	OEM	D4720B-E	OEM	D4802-2	OEM	D4827-2	OEM	D4863A	OEM	D5006-18	OEM	D5075B	OEM
D4651-2	OEM	D4720F	OEM	D4802-3	OEM	D4827-3	OEM	D4864	OEM	D5006-24	OEM	D5075C	OEM
D4651-3	OEM	D4720G	OEM	D4802A-E	OEM	D4831	OEM	D4864-1	OEM	D5006-30	OEM	D5076	OEM
D4651A-E	OEM	D4720H	OEM	D4802F-H	OEM	D4831-1	OEM	D4864-2	OEM	D5006-36	OEM	D5076A	OEM
D4652	OEM	D4720J	OEM	D4803	OEM	D4831-2	OEM	D4864-3	OEM	D5007-06	OEM	D5076B	OEM
D4652-1	OEM	D4721	OEM	D4803-1	OEM	D4831-3	OEM	D4865-1	OEM	D5007-12	OEM	D5076C	OEM
D4652-2	OEM	D4721A	OEM	D4803-2	OEM	D4831A-E	OEM	D4865-2	OEM	D5007-18	OEM	D5082B	OEM
D4652-3	OEM	D4721B-E	OEM	D4803-3	OEM	D4831F-H	OEM	D4865-3	OEM	D5008-06	OEM	D5082C	OEM
D4652A-D	OEM	D4721F	OEM	D4803A-E	OEM	D4832	OEM	D4866-3	OEM	D5008-12	OEM	D5082D	OEM
D4653	OEM	D4721G	OEM	D4803F	OEM	D4832-1	OEM	D4867-3	OEM	D5008-18	OEM	D5082E	OEM
D4653-1	OEM	D4721H	OEM	D4803G	OEM	D4832-2	OEM	D4871	OEM	D5008-24	OEM	D5082F	OEM
D4653-2	OEM	D4721J	OEM	D4804	OEM	D4832-3	OEM	D4871-1	OEM	D5008-30	OEM	D5090	OEM
D4653-3	OEM	D4722	OEM	D4804-1	OEM	D4832A-E	OEM	D4871-2	OEM	D5009-06	OEM	D5090A	OEM
D4653A-C	OEM	D4722A	OEM	D4804-2	OEM	D4832F	OEM	D4871-3	OEM	D5009-12	OEM	D5091AM	OEM
D4654	OEM	D4722B-E	OEM	D4804-3	OEM	D4833	OEM	D4872	OEM	D5018-06	OEM	D5091M	OEM
D4654-1	OEM	D4722F	OEM	D4804A-E	OEM	D4833-1	OEM	D4872-1	OEM	D5020	OEM	D5092AM	OEM
D4654-2	OEM	D4722G	OEM	D4805	OEM	D4833-2	OEM	D4872-2	OEM	D5020A	OEM	D5092BM	OEM
D4654-3	OEM	D4722H	OEM	D4805-1	OEM	D4833-3	OEM	D4872-3	OEM	D5020B	OEM	D5092M	OEM
D4654A	OEM	D4722J	OEM	D4805-2	OEM	D4833A-D	OEM	D4873-1	OEM	D5020C	OEM	D5104-10	OEM
D4655-1	OEM	D4723	OEM	D4805-3	OEM	D4834	OEM	D4873-2	OEM	D5020D	OEM	D5104-15	OEM
D4655-2	OEM	D4723A	OEM	D4805A	OEM	D4834-1	OEM	D4873-3	OEM	D5021	OEM	D5104-20	OEM
D4655-3	OEM	D4723B-E	OEM	D4805B	OEM	D4834-2	OEM	D4874-1	OEM	D5021A	OEM	D5109-2	OEM
D4656-2	OEM	D4723F	OEM	D4806	OEM	D4834-3	OEM	D4874-2	OEM	D5021B	OEM	D5109-4	OEM
D4656-3	OEM	D4723G	OEM	D4806-1	OEM	D4834A	OEM	D4874-3	OEM	D5021C	OEM	D5109-6	OEM
D4657-2	OEM	D4723H	OEM	D4806-2	OEM	D4834B	OEM	D4875-2	OEM	D5021D	OEM	D5109-8	OEM
D4657-3	OEM	D4723J	OEM	D4806-3	OEM	D4835	OEM	D4875-3	OEM	D5022	OEM	D5146	OEM
D4660	OEM	D4730	OEM	D4806A	OEM	D4835-1	OEM	D4876-3	OEM	D5022A	OEM	D5146A	OEM
D4660-1	OEM	D4730A	OEM	D4807	OEM	D4835-2	OEM	D4877-3	OEM	D5022B	OEM	D5146B	OEM
D4660-2	OEM	D4730B-E	OEM	D4807-1	OEM	D4835-3	OEM	D4912	OEM	D5022C	OEM	D5146C	OEM
D4660-3	OEM	D4730F	OEM	D4807-2	OEM	D4836-1	OEM	D4912A	OEM	D5022D	OEM	D5146D	OEM
D4660A-C	OEM	D4730G	OEM	D4807-3	OEM	D4836-2	OEM	D4912B	OEM	D5023	OEM	D5146E	OEM
D4661	OEM	D4730H	OEM	D4811	OEM	D4836-3	OEM	D4920F	OEM	D5023A	OEM	D5146F	OEM
D4661-1	OEM	D4730J	OEM	D4811-1	OEM	D4837-1	OEM	D4920G	OEM	D5023B	OEM	D5146G	OEM
D4661-2	OEM	D4731	OEM	D4811-2	OEM	D4837-2	OEM	D4920H	OEM	D5023C	OEM	D5146H	OEM
D4661-3	OEM	D4731A	OEM	D4811-3	OEM	D4837-3	OEM	D4920J	OEM	D5023D	OEM	D5147	OEM
D4661A	OEM	D4731B-E	OEM	D4811A-3	OEM	D4841	OEM	D4926A	OEM	D5024	OEM	D5147A	OEM
D4661B	OEM	D4731F	OEM	D4811A-E	OEM	D4841-1	OEM	D4927	OEM	D5024A	OEM	D5147B	OEM
D4662	OEM	D4731G	OEM	D4811F-H	OEM	D4841-2	OEM	D4938	OEM	D5024B	OEM	D5147C	OEM
D4662-1	OEM	D4731H	OEM	D4811F-J	OEM	D4841-3	OEM	D4938A	OEM	D5025	OEM	D5147D	OEM
D4662-2	OEM	D4731J	OEM	D4811F-K	OEM	D4841A-E	OEM	D4939	OEM	D5025A	OEM	D5147E	OEM
D4662-3	OEM	D4732	OEM	D4812	OEM	D4841F-H	OEM	D4940	OEM	D5025B	OEM	D5147F	OEM
D4662A	OEM	D4732A	OEM	D4812-1	OEM	D4842	OEM	D4941	OEM	D5025C	OEM	D5147G	OEM
D4663	OEM	D4732B-E	OEM	D4812-2	OEM	D4842-1	OEM	D4942	OEM	D5026	OEM	D5147H	OEM
D4663-1	OEM	D4732F	OEM	D4812-3	OEM	D4842-3	OEM	D4957	OEM	D5026A	OEM	D5151	OEM
D4663-2	OEM	D4732G	OEM	D4812A-E	OEM	D4842A-E	OEM	D4957A	OEM	D5026B	OEM	D5151A	OEM
D4663-3	OEM	D4732H	OEM	D4812F-H	OEM	D4842F	OEM	D4957B	OEM	D5030A	OEM	D5151B	OEM
D4663A	OEM	D4732J	OEM	D4813	OEM	D4843	OEM	D4957H	OEM	D5030B	OEM	D5167	OEM
D4664	OEM	D4733	OEM	D4813-1	OEM	D4843-1	OEM	D4957J-L	OEM	D5030C	OEM	D5221CM	OEM
D4664-1	OEM	D4733A	OEM	D4813-2	OEM	D4843-2	OEM	D4957J-N	OEM	D5030D	OEM	D5221EM	OEM
D4664-2	OEM	D4733B-E	OEM	D4813-3	OEM	D4843-3	OEM	D4961	OEM	D5031A	OEM	D5221FM	OEM
D4664-3	OEM	D4733F	OEM	D4813A-3	OEM	D4843A-D	OEM	D4961A	OEM	D5031B	OEM	D5221GM	OEM
D4665-2	OEM	D4733G	OEM	D4813F	OEM	D4844	OEM	D4961B	OEM	D5031C	OEM	D5223CM	OEM
D4665-3	OEM	D4733H	OEM	D4813G	OEM	D4844-1	OEM	D4961C	OEM	D5031D	OEM	D5223EM	OEM
D4666-3	OEM	D4733J	OEM	D4814	OEM	D4844-2	OEM	D4962	OEM	D5036A	OEM	D5223F	OEM
D4667-3	OEM	D4740	OEM	D4814-1	OEM	D4844-3	OEM	D4962B	OEM	D5046	OEM	D5223FM	OEM
D4670	OEM	D4740A	OEM	D4814-2	OEM	D4844A	OEM	D4962C	OEM	D5046A	OEM	D5223G	OEM
D4670-1	OEM	D4740B-E	OEM	D4814-3	OEM	D4844B	OEM	D4963	OEM	D5046B	OEM	D5233	OEM
D4670-2	OEM	D4740F	OEM	D4814A-E	OEM	D4845	OEM	D4963A	OEM	D5046C	OEM	D5235	OEM
D4670-3	OEM	D4741	OEM	D4815	OEM	D4845-1	OEM	D4963B	OEM	D5047	OEM	D5235-1	OEM
D4670A	OEM	D4741A	OEM	D4815-1	OEM	D4845-2	OEM	D4963C	OEM	D5047A	OEM	D5236	OEM
D4671	OEM	D4741B-E	OEM	D4815-2	OEM	D4845-3	OEM	D4964	OEM	D5047B	OEM	D5244	OEM
D4671-1	OEM	D4741F	OEM	D4815-3	OEM	D4846-1	OEM	D4964A	OEM	D5047C	OEM	D5244-06	OEM
D4671-2	OEM	D4742	OEM	D4815A	OEM	D4846-2	OEM	D4964B	OEM	D5047D	OEM	D5244-12	OEM
D4671-3	OEM	D4742A	OEM	D4815B	OEM	D4846-3	OEM	D4964C	OEM	D5047E	OEM	D5244-18	OEM
D4672	OEM	D4742B-E	OEM	D4816	OEM	D4847-1	OEM	D4965	OEM	D5047F	OEM	D5244-24	OEM
D4672-1	OEM	D4742F	OEM	D4816-1	OEM	D4847-2	OEM	D4965A	OEM	D5047G	OEM	D5244-30	OEM
D4672-2	OEM	D4743	OEM	D4816-2	OEM	D4847-3	OEM	D4965B	OEM	D5047H	OEM	D5244-36	OEM
D4672-3	OEM	D4743A	OEM	D4816-3	OEM	D4851	OEM	D4965C	OEM	D5061	OEM	D5244-42	OEM
D4674-1	OEM	D4743B-E	OEM	D4816A	OEM	D4851-1	OEM	D4966	OEM	D5061A	OEM	D5244A	OEM
D4674-2	OEM	D4743F	OEM	D4817	OEM	D4851-2	OEM	D4966A	OEM	D5061B	OEM	D5244B	OEM
D4674-3	OEM	D4750	OEM	D4817-1	OEM	D4851-3	OEM	D4966B	OEM	D5061C	OEM	D5244C	OEM
D4675-2	OEM	D4750A	OEM	D4817-2	OEM	D4851A-E	OEM	D4966C	OEM	D5062	OEM	D5244D	OEM
D4675-3	OEM	D4750B-D	OEM	D4817-3	OEM	D4852	OEM	D4971	OEM	D5062A	OEM	D5244E	OEM
D4676-3	OEM	D4751	OEM	D4821	OEM	D4852-1	OEM	D4971A	OEM	D5062B	OEM	D5244F	OEM
D4677-3	OEM	D4751A	OEM	D4821-1	OEM	D4852-2	OEM	D4971B	OEM	D5062C	OEM	D5244G	OEM
D4710	OEM	D4751B-D	OEM	D4821-2	OEM	D4852-3	OEM	D4971C	OEM	D5063	OEM	D5245-06	OEM
D4710A	OEM	D4752	OEM	D4821-3	OEM	D4852A-D	OEM	D4972	OEM	D5063A	OEM	D5245-12	OEM
D4710B-E	OEM	D4752A	OEM	D4821A-E	OEM	D4853	OEM	D4972A	OEM	D5063B	OEM	D5245-18	OEM
D4710F	OEM	D4752B-D	OEM	D4821F-H	OEM	D4853-1	OEM	D4972B	OEM	D5063C	OEM	D5245-24	OEM
D4710G	OEM	D4753	OEM	D4821F-J	OEM	D4853-2	OEM	D4972C	OEM	D5064	OEM	D5245-30	OEM
D4710H	OEM	D4753A	OEM	D4822	OEM	D4853-3	OEM	D4973	OEM	D5064A	OEM	D5245-36	OEM
D4710J	OEM	D4753B-D	OEM	D4822-1	OEM	D4853A-C	OEM	D4973A	OEM	D5064B	OEM	D5245A	OEM
D4710K	OEM	D4760	OEM	D4822-2	OEM	D4854	OEM	D4973B	OEM	D5064C	OEM	D5245B	OEM
D4711	OEM	D4760A	OEM	D4822-3	OEM	D4854-1	OEM	D4973C	OEM	D5065	OEM	D5245C	OEM

DEVICE TYPE	REPL CODE
D5245D	OEM
D5245E	OEM
D5245F	OEM
D5246-06	OEM
D5246-12	OEM
D5246-18	OEM
D5246-24	OEM
D5246-30	OEM
D5246A	OEM
D5246B	OEM
D5246C	OEM
D5246D	OEM
D5246E	OEM
D5252	OEM
D5253B	OEM
D5253BM	OEM
D5253C	OEM
D5253CM	OEM
D5253D	OEM
D5253DM	OEM
D5253E	OEM
D5253EM	OEM
D5254-06	OEM
D5254-12	OEM
D5254-18	OEM
D5254A	OEM
D5254A-D	OEM
D5254B	OEM
D5254C	OEM
D5255-06	OEM
D5255-12	OEM
D5255-18	OEM
D5255A	OEM
D5255B	OEM
D5255C	OEM
D5256-06	OEM
D5256-12	OEM
D5256A	OEM
D5256B	OEM
D5258	OEM
D5264	OEM
D5264A	OEM
D5264B	OEM
D5265	OEM
D5265A	OEM
D5277D	0015
D5278B	OEM
D5278BM	OEM
D5278C	OEM
D5278CM	OEM
D5278D	OEM
D5278DM	OEM
D5278E	OEM
D5278EM	OEM
D5282C	OEM
D5282D	OEM
D5282E	OEM
D5282F	OEM
D5282FM	OEM
D5282FMR	OEM
D5282G	OEM
D5282GM	OEM
D5282GMR	OEM
D5298	OEM
D5320	OEM
D5320A	OEM
D5320B	OEM
D5321	OEM
D5321A	OEM
D5321B	OEM
D5322	OEM
D5322A	OEM
D5322B	OEM
D5323	OEM
D5323A	OEM
D5323B	OEM
D5326	OEM
D5326A	OEM
D5326AM	OEM
D5326AMR	OEM
D5326B	OEM
D5326BM	OEM
D5326BMR	OEM
D5326M	OEM
D5326MR	OEM
D5337	OEM
D5337-00	OEM
D5337-06	OEM
D5337-12	OEM
D5337A	OEM
D5337B	OEM
D5338	OEM
D5339	OEM
D5340-5	OEM
D5340-6	OEM
D5340-G	OEM
D5340-M	OEM
D5340-X	OEM
D5345-5	OEM
D5345-6	OEM
D5345-G	OEM
D5345-M	OEM
D5345-X	OEM
D5347	OEM
D5347A	OEM
D5347B	OEM
D5353B	OEM
D5353BM	OEM
D5353BMR	OEM
D5353C	OEM
D5353CM	OEM
D5353CMR	OEM
D5353D	OEM
D5353DM	OEM
D5353DMR	OEM
D5353E	OEM
D5353EM	OEM
D5353EMR	OEM
D5353F	OEM
D5353FM	OEM
D5353MR	OEM
D5357	OEM
D5357A	OEM
D5357B	OEM
D5360	OEM
D5360-5	OEM
D5360-6	OEM
D5360-G	OEM
D5360-M	OEM
D5360-X	OEM
D5360A	OEM
D5361	OEM
D5361A	OEM
D5362	OEM
D5362A	OEM
D5363	OEM
D5363A	OEM
D5364	OEM
D5364A	OEM
D5365	OEM
D5365A	OEM
D5371	OEM
D5371A	OEM
D5371B	OEM
D5371C	OEM
D5371D	OEM
D5391C	OEM
D5392	OEM
D5392A	OEM
D5392AM	OEM
D5392AMR	OEM
D5392B	OEM
D5392BM	OEM
D5392BMR	OEM
D5392M	OEM
D5392MR	OEM
D5400	OEM
D5400-1	OEM
D5400-2	OEM
D5400-3	OEM
D5400A-E	OEM
D5400F-H	OEM
D5400J-L	OEM
D5401	OEM
D5401-1	OEM
D5401-2	OEM
D5401-3	OEM
D5401A-E	OEM
D5401F-H	OEM
D5401F-J	OEM
D5401F-K	OEM
D5402	OEM
D5402-1	OEM
D5402-2	OEM
D5402-3	OEM
D5402A-E	OEM
D5402F-H	OEM
D5403	2977
D5403-1	OEM
D5403-2	OEM
D5403-3	OEM
D5403A-E	OEM
D5403F	OEM
D5403G	OEM
D5404	2978
D5404-1	OEM
D5404-2	OEM
D5404-3	OEM
D5404A-E	OEM
D5405	OEM
D5405-1	OEM
D5405-2	OEM
D5405-3	OEM
D5405A	OEM
D5405B	OEM
D5410	2979
D5410-1	OEM
D5410-2	OEM
D5410-3	OEM
D5410A-E	OEM
D5410F-H	OEM
D5410J-L	OEM
D5411	OEM
D5411-1	OEM
D5411-2	OEM
D5411-3	OEM
D5411A-E	OEM
D5411F-H	OEM
D5411F-J	OEM
D5411F-K	OEM
D5412	OEM
D5412-1	OEM
D5412-2	OEM
D5412-3	OEM
D5412A-E	OEM
D5412F-H	OEM
D5413	OEM
D5413-1	OEM
D5413-2	OEM
D5413-3	OEM
D5413A-E	OEM
D5413F	OEM
D5413G	OEM
D5414	OEM
D5414-1	OEM
D5414-2	OEM
D5414-3	OEM
D5414A-E	OEM
D5415	OEM
D5415-1	OEM
D5415-2	OEM
D5415-3	OEM
D5415A	OEM
D5415B	OEM
D5420	OEM
D5420-1	OEM
D5420-2	OEM
D5420-3	OEM
D5420A-E	OEM
D5420F-H	OEM
D5420F-J	OEM
D5420F-K	OEM
D5421	OEM
D5421-1	OEM
D5421-2	OEM
D5421-3	OEM
D5421A-E	OEM
D5421F-H	OEM
D5421F-J	OEM
D5422	OEM
D5422-1	OEM
D5422-2	OEM
D5422-3	OEM
D5422A-E	OEM
D5422F	OEM
D5422G	OEM
D5423	OEM
D5423-1	OEM
D5423-2	OEM
D5423-3	OEM
D5423A-E	OEM
D5423F	OEM
D5424	OEM
D5424-1	OEM
D5424-2	OEM
D5424-3	OEM
D5424A-D	OEM
D5425	OEM
D5425-1	OEM
D5425-2	OEM
D5425-3	OEM
D5425A	OEM
D5430	OEM
D5430-1	OEM
D5430-2	OEM
D5430-3	OEM
D5430A-E	OEM
D5430F-H	OEM
D5430F-J	OEM
D5431	OEM
D5431-2	OEM
D5431-3	OEM
D5431A-E	OEM
D5431F-H	OEM
D5432	OEM
D5432-1	OEM
D5432-2	OEM
D5432-3	OEM
D5432A-E	OEM
D5432F	OEM
D5433	OEM
D5433-1	OEM
D5433-2	OEM
D5433-3	OEM
D5433A-D	OEM
D5434	OEM
D5434-1	OEM
D5434-2	OEM
D5434-3	OEM
D5434A	OEM
D5434B	OEM
D5435	OEM
D5435-1	OEM
D5435-2	OEM
D5435-3	OEM
D5440	OEM
D5440-1	OEM
D5440-2	OEM
D5440-3	OEM
D5440A-E	OEM
D5440F-H	OEM
D5440F-J	OEM
D5441	OEM
D5441-1	OEM
D5441-2	OEM
D5441-3	OEM
D5441A-E	OEM
D5441F-H	OEM
D5442	OEM
D5442-1	OEM
D5442-2	OEM
D5442-3	OEM
D5442A-F	OEM
D5442F	OEM
D5443	OEM
D5443-1	OEM
D5443-2	OEM
D5443-3	OEM
D5443A-D	OEM
D5444	OEM
D5444-1	OEM
D5444-2	OEM
D5444-3	OEM
D5444-48	OEM
D5444A	OEM
D5444B	OEM
D5450	OEM
D5450-1	OEM
D5450-2	OEM
D5450-3	OEM
D5450A-E	OEM
D5450F	OEM
D5451	OEM
D5451-1	OEM
D5451-2	OEM
D5451-3	OEM
D5451A-E	OEM
D5452	OEM
D5452-1	OEM
D5452-2	OEM
D5452-3	OEM
D5452A-D	OEM
D5453	OEM
D5453-1	OEM
D5453-2	OEM
D5453-3	OEM
D5453A-C	OEM
D5459BSQ	OEM
D5459BSQ-1	OEM
D5459BSQZS	OEM
D5459BSQZS-1	OEM
D5459BST	OEM
D5459BST-1	OEM
D5459BSTZS	OEM
D5459BSTZS-1	OEM
D5459BSZS	OEM
D5459BSZS-1	OEM
D5495BS	OEM
D5495BS-1	OEM
D5495BSM	OEM
D5495BSM-1	OEM
D5495BSMZS	OEM
D5495BSMZS-1	OEM
D5503	OEM
D5503A	OEM
D5503B	OEM
D5503C	OEM
D5506A	OEM
D5506B	OEM
D5506C	OEM
D5507	OEM
D5507A	OEM
D5507B	OEM
D5507C	OEM
D5509	OEM
D5509A	OEM
D5509B	OEM
D5509C	OEM
D5509D	OEM
D5520	OEM
D5520A	OEM
D5520B	OEM
D5521	OEM
D5521A	OEM
D5521B	OEM
D5522	OEM
D5522A	OEM
D5522B	OEM
D5523	OEM
D5523A	OEM
D5523B	OEM
D5540	OEM
D5560	OEM
D5560A	OEM
D5560B	OEM
D5561	OEM
D5561A	OEM
D5561B	OEM
D5562	OEM
D5562A	OEM
D5562B	OEM
D5563	OEM
D5563A	OEM
D5563B	OEM
D5564	OEM
D5564A	OEM
D5564B	OEM
D5565	OEM
D5565A	OEM
D5565B	OEM
D5570	OEM
D5570A	OEM
D5570B	OEM
D5571	OEM
D5571A	OEM
D5571B	OEM
D5572	OEM
D5572A	OEM
D5572B	OEM
D5573	OEM
D5573A	OEM
D5573B	OEM
D5574	OEM
D5574A	OEM
D5574B	OEM
D5575	OEM
D5575A	OEM
D5575B	OEM
D5600	OEM
D5600A	OEM
D5610	OEM
D5611	OEM
D5612	OEM
D5615H	OEM
D5616H	OEM
D5621G	OEM
D5621H	OEM
D5623G	OEM
D5623H	OEM
D5632	OEM
D5632A	OEM
D5632AM	OEM
D5632M	OEM
D5632MR	OEM
D5634	OEM
D5638	OEM
D5650A	OEM
D5650B	OEM
D5650C	OEM
D5650D	OEM
D5650E	OEM
D5668A	OEM
D5668B	OEM
D5668C	OEM
D5668D	OEM
D5668E	OEM
D5669A	OEM
D5669B	OEM
D5669C	OEM
D5669D	OEM
D5669E	OEM
D5690B	OEM
D5720	OEM
D5720A	OEM
D5720B	OEM
D5721	OEM
D5722	OEM
D5723	OEM
D5723A	OEM
D5723B	OEM
D5733	OEM
D5754	OEM
D5754A	OEM
D5760A	OEM
D5760B	OEM
D5760C	OEM
D5760D	OEM
D5761A	OEM
D5761B	OEM
D5761C	OEM
D5761D	OEM
D5762A	OEM
D5762B	OEM
D5762C	OEM
D5762D	OEM
D5763A	OEM
D5763B	OEM
D5763C	OEM
D5763D	OEM
D5807A	OEM
D5807B	OEM
D5807C	OEM
D5807D	OEM
D5807E	OEM
D5817	OEM
D5817A	OEM
D5818	OEM
D5818A	OEM
D5820	OEM
D5820A	OEM
D5820B	OEM
D5820C	OEM
D5821	OEM
D5821A	OEM
D5821B	OEM
D5821C	OEM
D5822	OEM
D5822A	OEM
D5822B	OEM
D5822C	OEM
D5823	OEM
D5823A	OEM
D5823B	OEM
D5823C	OEM
D5824	OEM
D5824A	OEM
D5824B	OEM
D5824C	OEM
D5825A	OEM
D5825B	OEM
D5825C	OEM
D5826	OEM
D5826A	OEM
D5826B	OEM
D5826C	OEM
D5827	OEM
D5827A	OEM
D5828	OEM
D5828A	OEM
D5829	OEM
D5829A	OEM
D5840	OEM
D5840A	OEM
D5840B	OEM
D5840C	OEM
D5840E	OEM
D5841	OEM
D5841A	OEM
D5842	OEM
D5842A	OEM
D5843	OEM
D5843A	OEM
D5844	OEM
D5844A	OEM
D5845	OEM
D5845A	OEM
D5846	OEM
D5846A	OEM
D5847	OEM
D5847A	OEM
D5848	OEM
D5848A	OEM
D5855	OEM
D5880C	OEM
D5890A	OEM
D5890B	OEM
D5890C	OEM
D5890D	OEM
D5890E	OEM
D5905A	OEM
D5905B	OEM
D5905C	OEM
D5905D	OEM
D5910	OEM
D5910A	OEM
D5910B	OEM
D5910C	OEM
D5940	OEM
D5940A	OEM
D5940B	OEM
D5940C	OEM
D5940D	OEM
D5940E	OEM
D5940F	OEM
D5940G	OEM
D5940H	OEM
D5940J	OEM
D5950	OEM
D5950A	OEM
D5950B	OEM
D5950C	OEM
D5950D	OEM
D5950E	OEM
D5950F	OEM
D5950G	OEM
D5950H	OEM
D5950J	OEM
D5964A	OEM
D5964B	OEM
D5964C	OEM
D5964D	OEM
D5964E	OEM
D5968	OEM
D5968A	OEM
D5968B	OEM
D5968C	OEM
D5968D	OEM
D5968E	OEM
D5968F	OEM
D5982A	OEM
D5982B	OEM
D5982C	OEM
D5982D	OEM
D5982E	OEM
D5994A	OEM
D5994B	OEM
D5994C	OEM
D5994D	OEM
D5994E	OEM
D6015L	3837
D6020L	3837
D6035J	OEM
D6035W	OEM
D6050J	OEM
D6055W	OEM
D6065J	OEM
D6070W	OEM
D6104C	OEM
D6104C001	OEM
D6105C	OEM
D6105C(UPD6105C)	OEM
D6105C02	OEM
D6117C	OEM
D6121G001	OEM
D6124CA-612	OEM
D6125A-712	OEM
D6140C001	OEM
D6145C	OEM
D6145C-001	OEM
D6325C	1880
D6326C	OEM
D6330	OEM
D6330A	OEM
D6330B	OEM
D6330C	OEM
D6330D	OEM
D6331	OEM
D6331A	OEM
D6331B	OEM
D6331C	OEM
D6331D	OEM
D6332	OEM
D6332A	OEM
D6332B	OEM
D6332C	OEM
D6332D	OEM
D6333	OEM
D6333A	OEM
D6333B	OEM
D6333C	OEM
D6333D	OEM
D6334	OEM
D6334A	OEM
D6334B	OEM
D6334C	OEM
D6334D	OEM
D6335	OEM
D6335A	OEM
D6335B	OEM
D6335C	OEM
D6335D	OEM
D6336	OEM
D6336A	OEM
D6336B	OEM
D6336C	OEM
D6336D	OEM
D6337	OEM
D6337A	OEM
D6337B	OEM
D6337C	OEM
D6337D	OEM
D6338	OEM
D6338A	OEM
D6338B	OEM
D6338C	OEM
D6338D	OEM
D6462	0133
D6599	OEM
D6623	0015
D6623A	0015
D6624	0015
D6624A	0015
D6625	0015
D6625A	0015
D6645	OEM
D6701	OEM
D6718	OEM
D6719	OEM
D6721	OEM
D6722	OEM
D6726	0133
D6743	OEM
D6900	OEM
D6901	OEM
D6901G	OEM
D6902	OEM
D6903	OEM
D6904	OEM
D6905	OEM
D6906	OEM
D6907	OEM
D6908	OEM
D6909	OEM
D6910	OEM
D6911	OEM
D7132	OEM
D7201C	OEM
D7220D-1	OEM
D7486J	1368
D7503G-186	OEM
D7507CU-246	OEM
D7516AGF558	OEM
D7519HG-555	OEM
D7537AC-026	OEM
D7585BS	OEM
D7585BS-1	OEM
D7585BSM	OEM
D7585BSM-1	OEM
D7585BSMZS	OEM
D7585BSMZS-1	OEM
D7585BSQ	OEM
D7585BSQ-1	OEM
D7585BSQZS	OEM
D7585BSQZS-1	OEM
D7585BST	OEM
D7585BST-1	OEM
D7585BSTZS	OEM
D7585BSTZS-1	OEM
D7585BSZS	OEM
D7585BSZS-1	OEM
D7800G	OEM
D7801G-176	OEM
D7810HCW	OEM
D7810HG	OEM
D7811G	OEM
D8015L	OEM
D8020H	OEM
D8020L	OEM
D8021	OEM
D8021H	OEM
D8022	OEM
D8031	OEM
D8035	OEM
D8035HL	OEM
D8035HL-1	OEM
D8035J	OEM
D8035L	OEM
D8035W	OEM
D8039	OEM
D8039-6	OEM
D8039H-8	OEM
D8039HL	OEM
D8039LC	OEM

If replacement code is OEM, contact original manufacturer for replacement.

DEVICE TYPE	REPL CODE
D8041A	OEM
D8041AC	OEM
D8048	OEM
D8048H-1	OEM
D8048HC	OEM
D8048HC545	OEM
D8048L	OEM
D8049	OEM
D8049H	OEM
D8049HC	OEM
D8049HC-029	OEM
D8050J	OEM
D8051	OEM
D8055W	OEM
D8065J	OEM
D8070W	OEM
D8080A	OEM
D8080A-1	OEM
D8080A-2	OEM
D8085A	OEM
D8085AC	OEM
D8085AC-2	OEM
D8085AH-2	OEM
D8086	OEM
D8086-2	OEM
D8086-4	OEM
D8088	OEM
D8088-2	OEM
D8088D	OEM
D8089	OEM
D8155	OEM
D8155-2	OEM
D8155C	OEM
D8155H	OEM
D8156	OEM
D8156-2	OEM
D8185	OEM
D8185-2	OEM
D8202	OEM
D8205	OEM
D8212	OEM
D8216	OEM
D8218	OEM
D8219	OEM
D8224	OEM
D8226	OEM
D8228	OEM
D8231	OEM
D8232	OEM
D8237	OEM
D8237-2	OEM
D8237A-5	OEM
D8237AC-5	OEM
D8237AS	OEM
D8238	OEM
D8243	OEM
D8251A	OEM
D8251AC	OEM
D8251AFC	OEM
D8253	OEM
D8253-5	OEM
D8253C-2	OEM
D8253C-5	OEM
D8253C5	OEM
D8255A	OEM
D8255A-5	0051
D8255AC-2	0051
D8255AC-5	0051
D8257	OEM
D8257-5	OEM
D8259A	OEM
D8259A-2	OEM
D8259A-8	OEM
D8259AC	OEM
D8259AC-2	OEM
D8271-6	OEM
D8271-8	OEM
D8272	OEM
D8273	OEM
D8273-	OEM
D8273-8	OEM
D8275	OEM
D8279	OEM
D8279-5	OEM
D8282	OEM
D8283	OEM
D8284	OEM
D8284A	OEM
D8286	OEM
D8287	OEM
D8288	OEM
D8289	OEM
D8291	OEM
D8292	OEM
D8293	OEM
D8295	OEM
D8355	OEM
D8355-2	OEM
D8410	0143
D8596BS	OEM
D8596BS-1	OEM
D8596BSM	OEM
D8596BSM-1	OEM
D8596BSMZS	OEM
D8596BSMZS-1	OEM
D8596BSQ	OEM
D8596BSQ-1	OEM
D8596BSQZS	OEM
D8596BSQZS-1	OEM

DEVICE TYPE	REPL CODE
D8596BST	OEM
D8596BST-1	OEM
D8596BSTZS	OEM
D8596BSTZS-1	OEM
D8596BSZS	OEM
D8596BSZS-1	OEM
D8641A	OEM
D8741A	OEM
D8742	OEM
D8748	OEM
D8748-6	OEM
D8748HD	OEM
D8749H	OEM
D8749H-8	OEM
D8749HC	OEM
D8751	OEM
D8755A	OEM
D8755A-2	OEM
D9011	OEM
D9011A	OEM
D9011B	OEM
D9011C	OEM
D9011D	OEM
D9011E	OEM
D9012	OEM
D9012A	OEM
D9012B	OEM
D9012C	OEM
D9012D	OEM
D9012E	OEM
D9013	OEM
D9013A	OEM
D9013B	OEM
D9013C	OEM
D9013D	OEM
D9013E	OEM
D9014	OEM
D9014A	OEM
D9014B	OEM
D9014C	OEM
D9014D	OEM
D9014E	OEM
D9634	0111
D10167	0015
D10168	0015
D10611DM	OEM
D20000BB40	0102
D21489	0015
D22400C#	0319
D22800C#	0044
D23128EC	OEM
D24000V09G	0023
D27128	1638
D27128A	1638
D27128A-3	1638
D27128A20	1638
D27128D	1638
D27128D-2	OEM
D27128D-482US	OEM
D27256	OEM
D27256-1	OEM
D27256-2	OEM
D27256-25	OEM
D27256-XX	OEM
D27256AD-2	OEM
D27256D	OEM
D27256J-2	OEM
D27512	OEM
D27512J	OEM
D27512J-1	OEM
D27512J-2	OEM
D28000BB40	0023
D28010ELS6	0023
D33400C#	0319
D33800C#	0044
D41256C	OEM
D41256C-12	OEM
D41256C-15	OEM
D41264C-12	OEM
D41264C-15	OEM
D41416C-15	OEM
D41464C-15	OEM
D42272AGF	OEM
D42832C-12L	OEM
D42832C-15L	OEM
D60400C#	2813
D60800C#	2845
D65006CW015	OEM
D65006CW-LC	OEM
D65006CW-LC2	OEM
D65006CW-LCII	OEM
D65006CW-NL	OEM
D65013GF216	OEM
D65022GF116	OEM
D71054GB	OEM
D75104G-576	OEM
D75104GF-752	OEM
D75108GF-720	OEM
D75116CW-091	OEM
D75208CW-058	OEM
D75208CW-069	OEM
D75208CW-287	OEM
D75216ACW095	OEM
D75216AGF558	OEM
D75216AGF568	OEM
D75216AGF609	OEM
D75516GF-019	OEM
D75516GF-025	OEM
D75516GF-045	OEM

DEVICE TYPE	REPL CODE
D82284-6	OEM
D82288-6	OEM
D82871	OEM
D92004R7B0	0140
D92004R7B1	0140
D92005R1B0	0041
D92007R5B0	0077
D92007R5B1	0077
D92013R6B0	0376
D93004R70X	0140
D93004R70Z	0041
D93005R10X	0041
D93005R600	0253
D93006R80Y	0062
D93007R50Y	0077
D93019R11B	0012
D97006R20B	0471
D101167	0015
D118441-JH00	OEM
D118441-LE00	OEM
D118441-LH00	OEM
D118442-JH00	OEM
D118442-LE00	OEM
D118442-LH00	OEM
D118443-JH00	OEM
D118443-LE00	OEM
D118443-LH00	OEM
D118444-JH00	OEM
D118444-LE00	OEM
D118444-LH00	OEM
D118445-JH00	OEM
D118445-LE00	OEM
D118445-LH00	OEM
D118446-JH00	OEM
D118446-LE00	OEM
D118446-LH00	OEM
D118447-JH00	OEM
D118447-LE00	OEM
D118447-LH00	OEM
D118448-JH00	OEM
D118448-LE00	OEM
D118448-LH00	OEM
D118449-JH00	OEM
D118449-LE00	OEM
D118449-LH00	OEM
D118450-JH00	OEM
D118450-LE00	OEM
D118450-LH00	OEM
D118451-JH00	OEM
D118451-LE00	OEM
D118451-LH00	OEM
D118452-JH00	OEM
D118452-LE00	OEM
D118452-LH00	OEM
D118453-JH00	OEM
D118453-LE00	OEM
D118453-LH00	OEM
D118454-JH00	OEM
D118454-LE00	OEM
D118454-LH00	OEM
D118455-JH00	OEM
D118455-LE00	OEM
D118455-LH00	OEM
D120400C#	1986
D120800C#	0657
D150400C#	1986
D150800C#	0657
D230130TAL	0015
D240200B#	1023
D240400B#	1036
D240600B#	1036
D240700B#	1046
D280010D10	0015
D280010D40	0071
D481111	OEM
D484242-2	OEM
D756012	OEM
D919695	3418
D919698	3861
D919699	3833
D919700	3714
D6400023GB	1925
D9200110B0	0181
D9200110B2	0181
D9200110B3	0181
D9200120B1	0052
D9200180B0	0375
D9200300B4	0195
D34002410-001	1069
D34013890-002	1907
D92015600B	0869
D93001000Y	0064
D93001100Y	0181
D93001200X	0052
D93001200Y	1005
D93001300Y	0053
D93001300Z	0053
D93001600Y	0445
D93002700X	0455
D93002700Y	0455
D93003000Y	0195
D93011201B	0137
D93011801B	0495
D93012701C	1674
D93013001B	0726
D170001310	0124
D920011082	0181
D930013000	0053
DA000	0015

DEVICE TYPE	REPL CODE
DA001	0015
DA002	0015
DA058	0071
DA0601	OEM
DA0602	OEM
DA0D	OEM
DA1Z-8	OEM
DA1.5/200	0087
DA1.5/400	0087
DA1.5/600	0087
DA1.5/800	0087
DA1.5/1000	0087
DA3	OEM
DA3/200	0087
DA3/400	0087
DA3/600	0087
DA3/800	0087
DA3/1000	0087
DA3F3	0437
DA-3N	OEM
DA-3P	OEM
DA4	OEM
DA4/600	OEM
DA4/800	OEM
DA-4N	OEM
DA4N	OEM
DA-4P	OEM
DA4P	OEM
DA6/400	OEM
DA6/800	OEM
DA6/1200	OEM
DA6/1400	OEM
DA6/1600	OEM
DA6-1100	OEM
DA6-1400	OEM
DA6-1600	OEM
DA-6NC	OEM
DA-6PC	OEM
DA-8N	OEM
DA8N	OEM
DA-8P	OEM
DA8P	OEM
DA90	0143
DA90M	OEM
DA101	0124
DA101A	0124
DA101B	0124
DA101G	OEM
DA101GR	OEM
DA101N	OEM
DA102	0124
DA102A	0124
DA102B	0124
DA103	OEM
DA106	OEM
DA106K	OEM
DA203	1091
DA-203X	1091
DA203X	OEM
DA-203Y	1091
DA203Y	OEM
DA204	OEM
DA204K	OEM
DA-204X	1091
DA204X	OEM
DA-204Y	1091
DA204Y	OEM
DA205	0133
DA210S	OEM
DA402	OEM
DA711	OEM
DA716	OEM
DA776	OEM
DA777	OEM
DA778	OEM
DA1000	0080
DA1001	0015
DA1002	0015
DA1006	0015
DA1058	0071
DA1200CN	OEM
DA1201CN	OEM
DA1202CN	OEM
DA1203CN	OEM
DA1600	OEM
DA1701	0102
DA1702	0102
DA1703	0920
DA1704	0920
DA2000	OEM
DA2010	OEM
DA2020	OEM
DA2068	0071
DA2080	OEM
DA2100	OEM
DA2100V1	OEM
DA2110	OEM
DA2110V1	OEM
DA2120	OEM
DA2120V1	OEM
DA2500	OEM
DAAY001002	0123
DAAY002001	0015
DAAY003002	0133
DAAY004001	0133
DAAY010092	0012
DAAY032001	OEM
DAC01AY	OEM
DAC01BY	OEM

DEVICE TYPE	REPL CODE
DAC01CY	OEM
DAC01DY	OEM
DAC01FY	OEM
DAC01G	OEM
DAC01GR	OEM
DAC01HY	OEM
DAC01N	OEM
DAC01Y	OEM
DAC02ACX	OEM
DAC02BCX	OEM
DAC02CCX	OEM
DAC02DDX	OEM
DAC02GR	OEM
DAC02N	OEM
DAC03ADX1	OEM
DAC03BDX1	OEM
DAC03CDX1	OEM
DAC03DDX1	OEM
DAC04BCX	OEM
DAC04CCX	OEM
DAC04DDX	OEM
DAC04G	OEM
DAC04GR	OEM
DAC04GT	OEM
DAC04N	OEM
DAC04NT	OEM
DAC05AX	OEM
DAC05BX	OEM
DAC05CX	OEM
DAC05EX	OEM
DAC05FX	OEM
DAC05GX	OEM
DAC06BX	OEM
DAC06CX	OEM
DAC06EX	OEM
DAC06FX	OEM
DAC06GX	OEM
DAC-08ADM	OEM
DAC-08ADM/883B	OEM
DAC08AF	OEM
DAC-08AQ	OEM
DAC08AQ	OEM
DAC-08BC	OEM
DAC-08BM	OEM
DAC-08CDC	OEM
DAC-08CDM	OEM
DAC08CF	OEM
DAC08CN	OEM
DAC-08CP	OEM
DAC08CP	OEM
DAC-08CQ	OEM
DAC08CQ	OEM
DAC-08DM	OEM
DAC-08DM/883B	OEM
DAC08ED	OEM
DAC-08EDC	OEM
DAC-08EDM	OEM
DAC08EN	OEM
DAC-08EP	OEM
DAC08EP	OEM
DAC-08EQ	OEM
DAC08EY	OEM
DAC08F	OEM
DAC08G	OEM
DAC08GR	OEM
DAC08GT	OEM
DAC-08HDC	OEM
DAC-08HDM	OEM
DAC08HF	OEM
DAC08HN	OEM
DAC-08HP	OEM
DAC08HP	OEM
DAC-08HQ	OEM
DAC08HQ	OEM
DAC08N	OEM
DAC08NT	OEM
DAC08Q	OEM
DAC0800LAJ	OEM
DAC0800LCJ	OEM
DAC0800LCN	OEM
DAC0800LD	OEM
DAC0800LJ	OEM
DAC0801LCJ	OEM
DAC0801LCN	OEM
DAC0802LCJ	OEM
DAC0802LCN	OEM
DAC0802LD	OEM
DAC0806LCJ	OEM
DAC0806LCN	5312
DAC0807LCJ	OEM
DAC0807LCN	5312
DAC0808LCJ	OEM
DAC0808LCN	OEM
DAC0808LD	OEM
DAC0808LJ	OEM
DAC0830LCD	OEM
DAC0830LCN	OEM
DAC0830LD	OEM
DAC0831LCD	OEM
DAC0831LCN	OEM
DAC0831LD	OEM
DAC0832LCD	OEM
DAC0832LCN	OEM
DAC0832LD	OEM
DAC02701-102	OEM
DAC02701-103	OEM
DAC02701-112(M)	OEM
DAC02701-113(M)	OEM
DAC02701-202	OEM

DEVICE TYPE	REPL CODE
DAC02701-203	OEM
DAC02701-212(M)	OEM
DAC02701-213(M)	OEM
DAC1	OEM
DAC2	OEM
DAC3	OEM
DAC-10BDM	OEM
DAC-10BDM/883B	OEM
DAC10BX	OEM
DAC-10CDM	OEM
DAC-10CDM/883B	OEM
DAC10CX	OEM
DAC-10FDC	OEM
DAC10FX	OEM
DAC10G	OEM
DAC-10GDC	OEM
DAC10GX	OEM
DAC10HT	OEM
DAC10HT-1	OEM
DAC10N	OEM
DAC11	OEM
DAC12	OEM
DAC12AV	OEM
DAC12BV	OEM
DAC12EV	OEM
DAC12FV	OEM
DAC12GV	OEM
DAC20CP	OEM
DAC20CQ	OEM
DAC20EP	OEM
DAC20EQ	OEM
DAC20G	OEM
DAC20N	OEM
DAC76CX	OEM
DAC76DX	OEM
DAC76EX	OEM
DAC76G	OEM
DAC76N	OEM
DAC76X	OEM
DAC86CX	OEM
DAC86EX	OEM
DAC87CX	OEM
DAC87EX	OEM
DAC88CX	OEM
DAC88EX	OEM
DAC89CX	OEM
DAC89EX	OEM
DAC90BG	OEM
DAC90BGQ(M)	OEM
DAC90SG	OEM
DAC90SGQ(M)	OEM
DAC100AAQ7	OEM
DAC100AAQ8	OEM
DAC100ABQ7	OEM
DAC100ABQ8	OEM
DAC100ACQ3	OEM
DAC100ACQ4	OEM
DAC100ACQ5	OEM
DAC100ACQ6	OEM
DAC100ACQ7	OEM
DAC100ACQ8	OEM
DAC100BBQ5	OEM
DAC100BBQ6	OEM
DAC100BBQ7	OEM
DAC100BBQ8	OEM
DAC100BCQ3	OEM
DAC100BCQ4	OEM
DAC100BCQ5	OEM
DAC100BCQ6	OEM
DAC100BCQ7	OEM
DAC100BCQ8	OEM
DAC100CCQ3	OEM
DAC100CCQ4	OEM
DAC100CCQ5	OEM
DAC100CCQ6	OEM
DAC100CCQ7	OEM
DAC100CCQ8	OEM
DAC100DDQ3	OEM
DAC100DDQ4	OEM
DAC100DDQ7	OEM
DAC100DDQ8	OEM
DAC101EQ	OEM
DAC101FQ	OEM
DAC101GQ	OEM
DAC206AY	OEM
DAC206BY	OEM
DAC206EY	OEM
DAC206FY	OEM
DAC206G	OEM
DAC206N	OEM
DAC208AX	OEM
DAC208BX	OEM
DAC208EX	OEM
DAC208FX	OEM
DAC210ASX	OEM
DAC210AX	OEM
DAC210BSX	OEM
DAC210BX	OEM
DAC210ESX	OEM
DAC210EX	OEM
DAC210FSX	OEM
DAC210FX	OEM
DAC210G	OEM
DAC210GR	OEM
DAC210GX	OEM
DAC210N	OEM
DAC230/E-6	OEM
DAC-230-200	OEM
DAC-230-300	OEM

DEVICE TYPE	REPL CODE
DAC331B-14-1	OEM
DAC331B-14-2	OEM
DAC331C-14-1	OEM
DAC331C-14-2	OEM
DAC562C	OEM
DAC562M	OEM
DAC-681C	OEM
DAC-681M	OEM
DAC685V-BC	OEM
DAC685V-BR	OEM
DAC685V-DC	OEM
DAC685V-DR	OEM
DAC687I-BM	OEM
DAC687I-DM	OEM
DAC736K	OEM
DAC800CBI-I	OEM
DAC800CBI-V	OEM
DAC808A	OEM
DAC808AX	OEM
DAC808BX	OEM
DAC808EX	OEM
DAC808FX	OEM
DAC808G	OEM
DAC808GR	OEM
DAC808GX	OEM
DAC808N	OEM
DAC850CBI-I	OEM
DAC850CBI-V	OEM
DAC851CBI-I	OEM
DAC851CBI-V	OEM
DAC888AX	OEM
DAC888BX	OEM
DAC888EX	OEM
DAC888FX	OEM
DAC888G	OEM
DAC888N	OEM
DAC1000LCD	OEM
DAC1000LCN	OEM
DAC1000LD	OEM
DAC1001LCD	OEM
DAC1001LCN	OEM
DAC1001LD	OEM
DAC1002LCD	OEM
DAC1002LCN	OEM
DAC1002LD	OEM
DAC1003LCD	OEM
DAC1003LCN	OEM
DAC1003LD	OEM
DAC1004LCD	OEM
DAC1004LCN	OEM
DAC1004LD	OEM
DAC1005LCD	OEM
DAC1005LCN	OEM
DAC1005LD	OEM
DAC1006LCD	OEM
DAC1006LCN	OEM
DAC1006LD	OEM
DAC1007LCD	OEM
DAC1007LCN	OEM
DAC1007LD	OEM
DAC1008LCD	OEM
DAC1008LCN	OEM
DAC1008LD	OEM
DAC1010LD	OEM
DAC1020LCD	OEM
DAC1021LCD	OEM
DAC1021LD	OEM
DAC1022LCD	OEM
DAC1022LD	OEM
DAC1200HCD	OEM
DAC1200HD	OEM
DAC1201HCD	OEM
DAC1201HD	OEM
DAC1202HCD	OEM
DAC1202HD	OEM
DAC1203HCD	OEM
DAC1203HD	OEM
DAC1208LCD	OEM
DAC1209LCD	OEM
DAC1210LCD	OEM
DAC1218LD	OEM
DAC1219LD	OEM
DAC1220LCD	OEM
DAC1220LD	OEM
DAC1221LCD	OEM
DAC1221LD	OEM
DAC1222LCD	OEM
DAC1222LD	OEM
DAC1230LCD	OEM
DAC1231LCD	OEM
DAC1232LCD	OEM
DAC1280A	OEM
DAC1280ACD	OEM
DAC1280HCD	OEM
DAC1285ACD	OEM
DAC1285AD	OEM
DAC1285HCD	OEM
DAC1285HD	OEM
DAC1286HCD	OEM
DAC1286HD	OEM
DAC1287HCD	OEM
DAC1408A6P	OEM
DAC1408A6Q	OEM
DAC1408A7P	OEM
DAC1408A7Q	OEM
DAC1408A8P	OEM
DAC1408A8Q	OEM
DAC1408AG	OEM
DAC1408G	OEM

If replacement code is OEM, contact original manufacturer for replacement.

DEVICE TYPE	REPL CODE
DAC1508ABQ	OEM
DAC-4565DDC	OEM
DAC-4565JDC	OEM
DAC-4565SDM	OEM
DAC-4565SDM/883B	OEM
DAC-4881B	OEM
DAC-4881B/883B	OEM
DAC-4881D	OEM
DAC-4881F	OEM
DAC-6012ADC	OEM
DAC-6012ADM	OEM
DAC-6012DC	OEM
DAC-6012DM/883B	OEM
DAC6851-BC	OEM
DAC6851-BR	OEM
DAC6851-DC	OEM
DAC6851-DR	OEM
DAC7520C	OEM
DAC7520M	OEM
DAC7520R	OEM
DAC7521C	OEM
DAC7521M	OEM
DAC7521R	OEM
DAC7523C	OEM
DAC7523M	OEM
DAC7523R	OEM
DAC7533C	OEM
DAC7533M	OEM
DAC7533R	OEM
DAC7541C	OEM
DAC7541M	OEM
DAC7541R	OEM
DAC-HA10BC	OEM
DAC-HA10BC1	OEM
DAC-HA10BM	OEM
DAC-HA10BM1	OEM
DAC-HA10BR	OEM
DAC-HA10BR1	OEM
DAC-HA12BC	OEM
DAC-HA12BC1	OEM
DAC-HA12BM	OEM
DAC-HA12BM1	OEM
DAC-HA12BR	OEM
DAC-HA12BR1	OEM
DAC-HA12DC	OEM
DAC-HA12DC1	OEM
DAC-HA12DM	OEM
DAC-HA12DM1	OEM
DAC-HA12DR	OEM
DAC-HA12DR1	OEM
DAC-HA14BC	OEM
DAC-HA14BC1	OEM
DAC-HA14BM	OEM
DAC-HA14BM1	OEM
DAC-HA14BR	OEM
DAC-HA14BR1	OEM
DAC-IC8BC	OEM
DAC-IC8BM	OEM
DAC-IC10B	OEM
DAC-IC10BC	OEM
DAC-IC10BM	OEM
DAC-UP8BC	OEM
DAC-UP8BM	OEM
DAD02G	OEM
DAG1A05	0031
DAG1A1	0031
DAG1A2	0031
DAG1A3	0031
DAG1A4	0031
DAJ201	0694
DAN-4	OEM
DAN-4A	OEM
DAN-4B	OEM
DAN-10	OEM
DAN-110	OEM
DAN-201	OEM
DAN201	3501
DAN201K	OEM
DAN201V	3501
DAN201VA	3501
DAN201W	0196
DAN201X	3501
DAN202	0907
DAN202K	0907
DAN202K-N	OEM
DAN202KVA	0907
DAN202VA	0907
DAN202VAK	OEM
DAN205	OEM
DAN206	OEM
DAN207-100	4721
DAN207-200	4721
DAN207-400	4721
DAN-208	OEM
DAN208	OEM
DAN208-100	1802
DAN208-200	0199
DAN209	3192
DAN209S	3192
DAN211	1802
DAN217	OEM
DAN217-T147	OEM
DAN235K	OEM
DAN401	OEM
DAN401VA	OEM
DAN403	OEM
DAN404	OEM
DAN601	OEM
DAN601VA	OEM

DEVICE TYPE	REPL CODE
DAN801	OEM
DAN801VA	OEM
DAN802	OEM
DAN802VA	OEM
DANZ00600	0143
DANZ006000	0019
DANZ0060000	0143
DAP-4	0396
DAP-4A	0396
DAP-4B	0396
DAP-10	0396
DAP-110	0396
DAP201	2203
DAP201N	3894
DAP201VA	0694
DAP202	0903
DAP202K	0903
DAP202K-T96	OEM
DAP202KVA	OEM
DAP202VA	OEM
DAP202VAK	OEM
DAP204K	OEM
DAP207-100	4783
DAP207-200	4783
DAP207-400	4783
DAP208	OEM
DAP208-100	1097
DAP208-200	1015
DAP209	3894
DAP209S	3894
DAP211	1097
DAP401	OEM
DAP401VA	OEM
DAP404	OEM
DAP601	3126
DAP601VA	OEM
DAP801	OEM
DAP801VA	OEM
DAP802	OEM
DAP802VA	OEM
DAR01G	OEM
DAR01GR	OEM
DAR01N	OEM
DAS-250A	OEM
DAS-250B	OEM
DAS400	OEM
DAS-952R	OEM
DASZ158800	0133
DAT1	OEM
DAT1A	0136
DAT2	0136
DB/65	OEM
DB08	OEM
DB1A	6388
DB1B	OEM
DB1D	6388
DB1F	6388
DB1M	3546
DB1N	3546
DB1P	3546
DB3	5388
DB4	5389
DB6	5390
DB7T2MR150	OEM
DB16	OEM
DB101	0241
DB101G	0241
DB102	0241
DB102G	0241
DB103	0241
DB103G	0241
DB104	4581
DB104BSXXXX	OEM
DB104G	4581
DB105	1875
DB105DXXXX	OEM
DB105G	1875
DB106	3546
DB106G	3546
DB107	3546
DB107G	3546
DB110DSXXXX	OEM
DB112BSXXXX	OEM
DB202	0015
DB204	0795
DB204BSXXXX	OEM
DB206	0015
DB208	0072
DB210	0071
DB212	0102
DB212BSXXXX	OEM
DB510DSXXXX	OEM
DB600T	OEM
DB600W	OEM
DB601T	OEM
DB601W	OEM
DB602T	OEM
DB602W	OEM
DB604T	OEM
DB604W	OEM
DB606T	OEM
DB606W	OEM
DB608T	OEM
DB608W	OEM
DB610T	OEM
DB610W	OEM
DB711BSXXXX	OEM
DB776	OEM
DB777	OEM

DEVICE TYPE	REPL CODE
DB778	OEM
DB800T	OEM
DB800W	OEM
DB801T	OEM
DB801W	OEM
DB802T	OEM
DB802W	OEM
DB804T	OEM
DB804W	OEM
DB806T	OEM
DB806W	OEM
DB808T	OEM
DB808W	OEM
DB810T	OEM
DB810W	OEM
DB812BSXXXX	OEM
DB816BSXXXX	OEM
DB826	OEM
DB1200T	OEM
DB1200W	OEM
DB1201T	OEM
DB1201W	OEM
DB1202T	OEM
DB1202W	OEM
DB1204T	OEM
DB1204W	OEM
DB1206T	OEM
DB1206W	OEM
DB1208T	OEM
DB1208W	OEM
DB1210T	OEM
DB1210W	OEM
DB1500T	OEM
DB1500W	OEM
DB1501T	OEM
DB1501W	OEM
DB1502T	OEM
DB1502W	OEM
DB1504T	OEM
DB1504W	OEM
DB1506T	OEM
DB1506W	OEM
DB1508T	OEM
DB1508W	OEM
DB1510T	OEM
DB1510W	OEM
DB1724BSXXXX	OEM
DB2500T	OEM
DB2500W	OEM
DB2501T	OEM
DB2501W	OEM
DB2502T	OEM
DB2502W	OEM
DB2504T	OEM
DB2504W	OEM
DB2506T	OEM
DB2506W	OEM
DB2508T	OEM
DB2508W	OEM
DB2510T	OEM
DB2510W	OEM
DB3500T	OEM
DB3500W	OEM
DB3501T	OEM
DB3501W	OEM
DB3502T	OEM
DB3502W	OEM
DB3504T	OEM
DB3504W	OEM
DB3506T	OEM
DB3506W	OEM
DB3508T	OEM
DB3508W	OEM
DB3510T	OEM
DB3510W	OEM
DB5459BSXXXX	OEM
DB8596BSXXXX	OEM
DBA10B	0277
DBA10C	0289
DBA10E	0296
DBA10G	0302
DBA20B	0323
DBA20C	1414
DBA20CK15	OEM
DBA20E	0473
DBA30B	0323
DBA30C	1414
DBA30CK12	OEM
DBA30E	0473
DBA30G	0446
DBA40B	0323
DBA40C	1414
DBA40E	0473
DBA60B	1414
DBA60C	0473
DBA60E	0473
DBA100B	0323
DBA100C	1414
DBA100E	0473
DBA408BSXXXX	OEM
DBAZ073304	0532
DBB10	OEM
DBB10B	1875
DBB10C	1875
DBB10E	1875
DBB10G	1875
DBBY001003	0622
DBBY003001	0219
DBBY003002	0219

DEVICE TYPE	REPL CODE
DBBY005001	0936
DBBY407004	0423
DBC1	OEM
DBC2	OEM
DBC3	OEM
DBC4	OEM
DBC5	OEM
DBC6	OEM
DBC7	OEM
DBC8	OEM
DBC9	OEM
DBCZ037300	0111
DBCZ039404	0224
DBCZ073304	0111
DBCZ073503	0155
DBCZ073504	0191
DBCZ083905	0076
DBCZ083906	0016
DBCZ094504	0111
DBCZ0373000	1303
DBCZ101800	2050
DBCZ136406	0016
DBCZ176003	1946
DBCZ373000	0111
DBF60B	1040
DBL45-12	OEM
DBL60-12	OEM
DBL567	OEM
DBX72L	OEM
DBX158L	OEM
DBX158M	OEM
DBX167L	OEM
DBX167M	OEM
DBX184LS	OEM
DBX184MX	OEM
DBX185L	OEM
DBX185M	OEM
DBXAN6291	3437
DC	1802
DC3B24	OEM
DC3B28	OEM
DC3B29	OEM
DC5R4	OEM
DC5U4	OEM
DC5V4	OEM
DC5Y3	OEM
DC5Z3	OEM
DC5Z4	OEM
DC6AX5	OEM
DC6X4	OEM
DC7	OEM
DC7A	OEM
DC7B	OEM
DC7C	OEM
DC7D	OEM
DC-9	0004
DC-10	0004
DC10.5	OEM
DC-12	0164
DC-13	0143
DC14V	OEM
DC25Z6	OEM
DC28V	OEM
DC42V	OEM
DC56V	OEM
DC80	OEM
DC95A2	OEM
DC101	OEM
DC102	OEM
DC104	OEM
DC106	OEM
DC108	OEM
DC110	OEM
DC165A2	OEM
DC169A2	OEM
DC169A2A	OEM
DC205A2	OEM
DC209A2	OEM
DC209B2	OEM
DC209B2A	OEM
DC250R	OEM
DC269A2	OEM
DC300.5	OEM
DC301	OEM
DC302	OEM
DC304	OEM
DC306	OEM
DC308	OEM
DC310	OEM
DC326A2	OEM
DC329A2	OEM
DC371A	OEM
DC405A2	OEM
DC406A2	OEM
DC442	OEM
DC575	OEM
DC576	OEM
DC600.5	OEM
DC601	OEM
DC602	OEM
DC604	OEM
DC606	OEM
DC608	OEM
DC610	OEM
DC610A2	OEM
DC673	OEM
DC806A2	OEM
DC809B2	OEM
DC866	OEM

DEVICE TYPE	REPL CODE
DC872	OEM
DC1015A2	OEM
DC1016	OEM
DC1017	OEM
DC1018	OEM
DC1019	OEM
DC1020	OEM
DC1020A	OEM
DC1022	OEM
DC1028A	OEM
DC1100-01	OEM
DC1100-03	OEM
DC1100-05	OEM
DC1100-10	OEM
DC1100-20	OEM
DC1151	OEM
DC1152	OEM
DC1181G	OEM
DC1181H	OEM
DC1181J	OEM
DC1186F	OEM
DC1186G	OEM
DC1201A	OEM
DC1201B	OEM
DC1201C	OEM
DC1201D	OEM
DC1201E	OEM
DC1201F	OEM
DC1202A	OEM
DC1202B	OEM
DC1202C	OEM
DC1202E	OEM
DC1203A	OEM
DC1203B	OEM
DC1203C	OEM
DC1203D	OEM
DC1203E	OEM
DC1204A	OEM
DC1204B	OEM
DC1204C	OEM
DC1204D	OEM
DC1204E	OEM
DC1205A	OEM
DC1205B	OEM
DC1205C	OEM
DC1205D	OEM
DC1222A	OEM
DC1222B	OEM
DC1222C	OEM
DC1222D	OEM
DC1222E	OEM
DC1231A	OEM
DC1231B	OEM
DC1231C	OEM
DC1231D	OEM
DC1231E	OEM
DC1232A	OEM
DC1232B	OEM
DC1232C	OEM
DC1232D	OEM
DC1232E	OEM
DC1233A	OEM
DC1233B	OEM
DC1233C	OEM
DC1233D	OEM
DC1233E	OEM
DC1234A	OEM
DC1234B	OEM
DC1234C	OEM
DC1234D	OEM
DC1234E	OEM
DC1251F	OEM
DC1251G	OEM
DC1251H	OEM
DC1251J	OEM
DC1252F	OEM
DC1252G	OEM
DC1252H	OEM
DC1253F	OEM
DC1253G	OEM
DC1253H	OEM
DC1254E	OEM
DC1254F	OEM
DC1254G	OEM
DC1255F	OEM
DC1255G	OEM
DC1255H	OEM
DC1256F	OEM
DC1256G	OEM
DC1272F	OEM
DC1272G	OEM
DC1272H	OEM
DC1281F	OEM
DC1281G	OEM
DC1281H	OEM
DC1281J	OEM
DC1282F	OEM
DC1282G	OEM
DC1282H	OEM
DC1283F	OEM
DC1283G	OEM
DC1283H	OEM
DC1284E	OEM
DC1284F	OEM
DC1284G	OEM
DC1301	OEM
DC1301A	OEM
DC1301C	OEM
DC1302	OEM

DEVICE TYPE	REPL CODE
DC1302C	OEM
DC1303	OEM
DC1304	OEM
DC1304/3	OEM
DC1304C	OEM
DC1306	OEM
DC1307	OEM
DC1308	OEM
DC1309	OEM
DC1311	OEM
DC1312	OEM
DC1313	OEM
DC1314	OEM
DC1316	OEM
DC1320	OEM
DC1321	OEM
DC1322	OEM
DC1323	OEM
DC1324	OEM
DC1325	OEM
DC1332	OEM
DC1333	OEM
DC1333C	OEM
DC1334	OEM
DC1335	OEM
DC1336	OEM
DC1340	OEM
DC1346	OEM
DC1501	OEM
DC1501/1	OEM
DC1501E	OEM
DC1501F	OEM
DC1501G	OEM
DC1502	OEM
DC1502E	OEM
DC1502F	OEM
DC1502G	OEM
DC1503	OEM
DC1504	OEM
DC1504E	OEM
DC1504F	OEM
DC1504G	OEM
DC1505	OEM
DC1505F	OEM
DC1505G	OEM
DC1505H	OEM
DC1506	OEM
DC1506E	OEM
DC1506F	OEM
DC1506G	OEM
DC1507	OEM
DC1507F	OEM
DC1507G	OEM
DC1507H	OEM
DC1508F	OEM
DC1508G	OEM
DC1508H	OEM
DC1509	OEM
DC1510	OEM
DC1511F	OEM
DC1511G	OEM
DC1511H	OEM
DC1512	OEM
DC1512/1	OEM
DC1513	OEM
DC1514	OEM
DC1515	OEM
DC1516	OEM
DC1517	OEM
DC1518	OEM
DC1519	OEM
DC1520	OEM
DC1521	OEM
DC1523	OEM
DC1526	OEM
DC1527	OEM
DC1528	OEM
DC1529	OEM
DC1533E	OEM
DC1533F	OEM
DC1533G	OEM
DC1534E	OEM
DC1534F	OEM
DC1534G	OEM
DC1535	OEM
DC1536F	OEM
DC1536G	OEM
DC1539/1	OEM
DC1542	OEM
DC1544/1	OEM
DC1544/3	OEM
DC1544/5	OEM
DC1544/7	OEM
DC1544/8	OEM
DC1546	OEM
DC1547	OEM
DC1551	OEM
DC1553	OEM
DC1554	OEM
DC1557	OEM
DC1558	OEM
DC1560	OEM
DC1570E	OEM
DC1570F	OEM
DC1570G	OEM
DC1571E	OEM
DC1571F	OEM
DC1571G	OEM
DC1572E	OEM

DEVICE TYPE	REPL CODE
DC1572F	OEM
DC1572G	OEM
DC1573E	OEM
DC1573F	OEM
DC1573G	OEM
DC1574E	OEM
DC1574F	OEM
DC1574G	OEM
DC1575E	OEM
DC1575F	OEM
DC1575G	OEM
DC1576E	OEM
DC1576F	OEM
DC1576G	OEM
DC1577E	OEM
DC1577F	OEM
DC1577G	OEM
DC1578E	OEM
DC1578F	OEM
DC1578G	OEM
DC1579E	OEM
DC1585	OEM
DC1590/1	OEM
DC1590/2	OEM
DC1590/3	OEM
DC1591	OEM
DC1593	OEM
DC1595	OEM
DC1596	OEM
DC1597	OEM
DC1612B2	OEM
DC1612B2A	OEM
DC1612C2	OEM
DC1612C2A	OEM
DC1612D2	OEM
DC2010	OEM
DC2011	OEM
DC2011A	OEM
DC2011AR	OEM
DC2011B	OEM
DC2011BR	OEM
DC2011C	OEM
DC2011CR	OEM
DC2011D	OEM
DC2011DR	OEM
DC2011G	OEM
DC2011GR	OEM
DC2012	OEM
DC2013	OEM
DC2013A	OEM
DC2013AR	OEM
DC2013B	OEM
DC2013BR	OEM
DC2013C	OEM
DC2013CR	OEM
DC2013D	OEM
DC2013DR	OEM
DC2013G	OEM
DC2013GR	OEM
DC2014	OEM
DC2014A	OEM
DC2014B	OEM
DC2014C	OEM
DC2015	OEM
DC2015B2	OEM
DC2016	OEM
DC2017	OEM
DC2018	OEM
DC2020	OEM
DC2023	OEM
DC2101A	OEM
DC2101B	OEM
DC2101C	OEM
DC2101G	OEM
DC2103A	OEM
DC2103B	OEM
DC2103C	OEM
DC2103G	OEM
DC2104A	OEM
DC2104B	OEM
DC2104C	OEM
DC2104F	OEM
DC2104G	OEM
DC2110A	OEM
DC2110B	OEM
DC2110C	OEM
DC2110G	OEM
DC2118B	OEM
DC2118C	OEM
DC2118G	OEM
DC2130A	OEM
DC2130G	OEM
DC2133A	OEM
DC2133G	OEM
DC2135A	OEM
DC2135G	OEM
DC2140A	OEM
DC2140G	OEM
DC2171H	OEM
DC2172H	OEM
DC2403A	OEM
DC2410A	OEM
DC2412A	OEM
DC2418A	OEM
DC2418A1	OEM
DC2419A	OEM
DC2419A1	OEM
DC2443A	OEM
DC2460A	OEM

Original Device Types

DEVICE TYPE	REPL CODE
DC2510A	OEM
DC2510B	OEM
DC2510C	OEM
DC2510D	OEM
DC2510G	OEM
DC2512A	OEM
DC2512B	OEM
DC2512C	OEM
DC2512D	OEM
DC2512G	OEM
DC2518A	OEM
DC2518B	OEM
DC2518C	OEM
DC2518D	OEM
DC2518G	OEM
DC2519A	OEM
DC2519B	OEM
DC2519C	OEM
DC2519D	OEM
DC2519G	OEM
DC2521A	OEM
DC2521B	OEM
DC2521C	OEM
DC2521D	OEM
DC2521G	OEM
DC2552A	OEM
DC2552B	OEM
DC2552C	OEM
DC2552D	OEM
DC2552G	OEM
DC2610A	OEM
DC2611	OEM
DC2612A	OEM
DC2613	OEM
DC2614	OEM
DC2615	OEM
DC2616	OEM
DC2618A	OEM
DC2619A	OEM
DC2624D	OEM
DC2652A	OEM
DC2801A	OEM
DC2802G	OEM
DC2825E	OEM
DC2840E	OEM
DC2841E	OEM
DC2842E	OEM
DC2843E	OEM
DC2844E	OEM
DC2845E	OEM
DC2846E	OEM
DC2847E	OEM
DC2848E	OEM
DC2849E	OEM
DC2850E	OEM
DC2870C	OEM
DC2906A	OEM
DC2907A	OEM
DC3010	OEM
DC3011	OEM
DC3012	OEM
DC3012S	OEM
DC3013	OEM
DC3015	OEM
DC3016	OEM
DC3018	OEM
DC3021	OEM
DC3022	OEM
DC3024	OEM
DC3439	OEM
DC3440	OEM
DC4021A00	OEM
DC4021A01	OEM
DC4021A02	OEM
DC4021B00	OEM
DC4021B01	OEM
DC4021B02	OEM
DC4021C00	OEM
DC4021C01	OEM
DC4021C02	OEM
DC4021D00	OEM
DC4021D01	OEM
DC4021D02	OEM
DC4021E00	OEM
DC4021E01	OEM
DC4021E02	OEM
DC4021F00	OEM
DC4021F01	OEM
DC4021F02	OEM
DC4021H00	OEM
DC4021H01	OEM
DC4021H02	OEM
DC4021J00	OEM
DC4021J01	OEM
DC4021J02	OEM
DC4021K00	OEM
DC4021K01	OEM
DC4021K02	OEM
DC4021L00	OEM
DC4021L01	OEM
DC4021L02	OEM
DC4021M00	OEM
DC4021M01	OEM
DC4021M02	OEM
DC4021N00	OEM
DC4021N01	OEM
DC4021N02	OEM
DC4021P00	OEM
DC4021P01	OEM
DC4021P02	OEM
DC4022A00	OEM
DC4022A01	OEM
DC4022A02	OEM
DC4022B00	OEM
DC4022B01	OEM
DC4022B02	OEM
DC4022C00	OEM
DC4022C01	OEM
DC4022C02	OEM
DC4022D00	OEM
DC4022D01	OEM
DC4022D02	OEM
DC4022E00	OEM
DC4022E01	OEM
DC4022E02	OEM
DC4022F00	OEM
DC4022F01	OEM
DC4022F02	OEM
DC4022G00	OEM
DC4022G01	OEM
DC4022G02	OEM
DC4022H00	OEM
DC4022H01	OEM
DC4022H02	OEM
DC4022J00	OEM
DC4022J01	OEM
DC4022J02	OEM
DC4022K00	OEM
DC4022K01	OEM
DC4022K02	OEM
DC4022L00	OEM
DC4022L01	OEM
DC4022L02	OEM
DC4023A00	OEM
DC4023A01	OEM
DC4023A02	OEM
DC4023B00	OEM
DC4023B01	OEM
DC4023B02	OEM
DC4023C00	OEM
DC4023C01	OEM
DC4023C02	OEM
DC4023D00	OEM
DC4023D01	OEM
DC4023D02	OEM
DC4023E00	OEM
DC4023E01	OEM
DC4023E02	OEM
DC4023F00	OEM
DC4023F01	OEM
DC4023F02	OEM
DC4023G00	OEM
DC4023G01	OEM
DC4023G02	OEM
DC4023H00	OEM
DC4023H01	OEM
DC4023H02	OEM
DC4023J00	OEM
DC4023J01	OEM
DC4023J02	OEM
DC4031A00	OEM
DC4031A01	OEM
DC4031A02	OEM
DC4031B00	OEM
DC4031B01	OEM
DC4031B02	OEM
DC4031C00	OEM
DC4031C01	OEM
DC4031C02	OEM
DC4031D00	OEM
DC4031D01	OEM
DC4031D02	OEM
DC4031E00	OEM
DC4031E01	OEM
DC4031E02	OEM
DC4031F00	OEM
DC4031F01	OEM
DC4031F02	OEM
DC4031G00	OEM
DC4031G01	OEM
DC4031G02	OEM
DC4031H00	OEM
DC4031H01	OEM
DC4031H02	OEM
DC4031J00	OEM
DC4031J01	OEM
DC4031J02	OEM
DC4031K00	OEM
DC4031K01	OEM
DC4031K02	OEM
DC4031L00	OEM
DC4031L01	OEM
DC4031L02	OEM
DC4031M00	OEM
DC4031M01	OEM
DC4031M02	OEM
DC4032A00	OEM
DC4032A01	OEM
DC4032A02	OEM
DC4032B00	OEM
DC4032B01	OEM
DC4032B02	OEM
DC4032C00	OEM
DC4032C01	OEM
DC4032C02	OEM
DC4032D00	OEM
DC4032D01	OEM
DC4032D02	OEM
DC4032E00	OEM
DC4032E01	OEM
DC4032E02	OEM
DC4032F00	OEM
DC4032F01	OEM
DC4032F02	OEM
DC4032G00	OEM
DC4032G01	OEM
DC4032G02	OEM
DC4032H00	OEM
DC4032H01	OEM
DC4032H02	OEM
DC4032J00	OEM
DC4032J01	OEM
DC4032J02	OEM
DC4032K00	OEM
DC4032K01	OEM
DC4032K02	OEM
DC4032L00	OEM
DC4032L01	OEM
DC4032L02	OEM
DC4033A00	OEM
DC4033A01	OEM
DC4033A02	OEM
DC4033B00	OEM
DC4033B01	OEM
DC4033B02	OEM
DC4033C00	OEM
DC4033C01	OEM
DC4033C02	OEM
DC4033D00	OEM
DC4033D01	OEM
DC4033D02	OEM
DC4033E00	OEM
DC4033E01	OEM
DC4033E02	OEM
DC4033F00	OEM
DC4033F01	OEM
DC4033F02	OEM
DC4033G00	OEM
DC4033G01	OEM
DC4033G02	OEM
DC4033H00	OEM
DC4033H01	OEM
DC4033H02	OEM
DC4033J00	OEM
DC4033J01	OEM
DC4033J02	OEM
DC4034A00	OEM
DC4034A01	OEM
DC4034A02	OEM
DC4034B00	OEM
DC4034B01	OEM
DC4034B02	OEM
DC4034C00	OEM
DC4034C01	OEM
DC4034C02	OEM
DC4034D00	OEM
DC4034D01	OEM
DC4034D02	OEM
DC4034E00	OEM
DC4034E01	OEM
DC4034E02	OEM
DC4041A00	OEM
DC4041A01	OEM
DC4041A02	OEM
DC4041B00	OEM
DC4041B01	OEM
DC4041B02	OEM
DC4041C00	OEM
DC4041C01	OEM
DC4041C02	OEM
DC4041D00	OEM
DC4041D01	OEM
DC4041D02	OEM
DC4041E00	OEM
DC4041E01	OEM
DC4041E02	OEM
DC4041F00	OEM
DC4041F01	OEM
DC4041F02	OEM
DC4041G00	OEM
DC4041G01	OEM
DC4041G02	OEM
DC4041H00	OEM
DC4041H01	OEM
DC4041H02	OEM
DC4041J00	OEM
DC4041J01	OEM
DC4041J02	OEM
DC4041K00	OEM
DC4041K01	OEM
DC4041K02	OEM
DC4041L00	OEM
DC4041L01	OEM
DC4041L02	OEM
DC4042A00	OEM
DC4042A01	OEM
DC4042A02	OEM
DC4042B00	OEM
DC4042B01	OEM
DC4042B02	OEM
DC4042C00	OEM
DC4042C01	OEM
DC4042C02	OEM
DC4042D00	OEM
DC4042D01	OEM
DC4042D02	OEM
DC4042E00	OEM
DC4042E01	OEM
DC4042E02	OEM
DC4042F00	OEM
DC4042F01	OEM
DC4042F02	OEM
DC4042G00	OEM
DC4042G01	OEM
DC4042G02	OEM
DC4042H00	OEM
DC4042H01	OEM
DC4042H02	OEM
DC4042J00	OEM
DC4042J01	OEM
DC4042J02	OEM
DC4042K00	OEM
DC4042K01	OEM
DC4042K02	OEM
DC4042L00	OEM
DC4042L01	OEM
DC4042L02	OEM
DC4043A00	OEM
DC4043A01	OEM
DC4043A02	OEM
DC4043B00	OEM
DC4043B01	OEM
DC4043B02	OEM
DC4043C00	OEM
DC4043C01	OEM
DC4043C02	OEM
DC4043D00	OEM
DC4043D01	OEM
DC4043D02	OEM
DC4043E00	OEM
DC4043E01	OEM
DC4043E02	OEM
DC4043F00	OEM
DC4043F01	OEM
DC4043F02	OEM
DC4043G00	OEM
DC4043G01	OEM
DC4043G02	OEM
DC4043H00	OEM
DC4043H01	OEM
DC4043H02	OEM
DC4043J00	OEM
DC4043J01	OEM
DC4043J02	OEM
DC4044A00	OEM
DC4044A01	OEM
DC4044A02	OEM
DC4044B00	OEM
DC4044B01	OEM
DC4044B02	OEM
DC4044C00	OEM
DC4044C01	OEM
DC4044C02	OEM
DC4044D00	OEM
DC4044D01	OEM
DC4044D02	OEM
DC4045A00	OEM
DC4045A01	OEM
DC4045A02	OEM
DC4045B00	OEM
DC4045B01	OEM
DC4045B02	OEM
DC4051A00	OEM
DC4051A01	OEM
DC4051A02	OEM
DC4051B00	OEM
DC4051B01	OEM
DC4051B02	OEM
DC4051C00	OEM
DC4051C01	OEM
DC4051C02	OEM
DC4051D00	OEM
DC4051D01	OEM
DC4051D02	OEM
DC4051E00	OEM
DC4051E01	OEM
DC4051E02	OEM
DC4051F00	OEM
DC4051F01	OEM
DC4051F02	OEM
DC4051G00	OEM
DC4051G01	OEM
DC4051G02	OEM
DC4051H00	OEM
DC4051H01	OEM
DC4051H02	OEM
DC4051J00	OEM
DC4051J01	OEM
DC4051J02	OEM
DC4051K00	OEM
DC4051K01	OEM
DC4051K02	OEM
DC4051L00	OEM
DC4051L01	OEM
DC4051L02	OEM
DC4052A00	OEM
DC4052A01	OEM
DC4052A02	OEM
DC4052B00	OEM
DC4052B01	OEM
DC4052B02	OEM
DC4052C00	OEM
DC4052C01	OEM
DC4052C02	OEM
DC4052D00	OEM
DC4052D01	OEM
DC4052D02	OEM
DC4052E00	OEM
DC4052E01	OEM
DC4052E02	OEM
DC4052F00	OEM
DC4052F01	OEM
DC4052F02	OEM
DC4052G00	OEM
DC4052G01	OEM
DC4052G02	OEM
DC4052H00	OEM
DC4052H01	OEM
DC4052H02	OEM
DC4052J00	OEM
DC4052J01	OEM
DC4052J02	OEM
DC4053A00	OEM
DC4053A01	OEM
DC4053A02	OEM
DC4053B00	OEM
DC4053B01	OEM
DC4053B02	OEM
DC4053C00	OEM
DC4053C01	OEM
DC4053C02	OEM
DC4053D00	OEM
DC4053D01	OEM
DC4053D02	OEM
DC4053E00	OEM
DC4053E01	OEM
DC4053E02	OEM
DC4053F00	OEM
DC4053F01	OEM
DC4053F02	OEM
DC4054A00	OEM
DC4054A01	OEM
DC4054A02	OEM
DC4054B00	OEM
DC4054B01	OEM
DC4054B02	OEM
DC4054C00	OEM
DC4054C01	OEM
DC4054C02	OEM
DC4054D00	OEM
DC4054D01	OEM
DC4054D02	OEM
DC4061A00	OEM
DC4061A01	OEM
DC4061A02	OEM
DC4061B00	OEM
DC4061B01	OEM
DC4061B02	OEM
DC4061C00	OEM
DC4061C01	OEM
DC4061C02	OEM
DC4061D00	OEM
DC4061D01	OEM
DC4061D02	OEM
DC4061E00	OEM
DC4061E01	OEM
DC4061E02	OEM
DC4061F00	OEM
DC4061F01	OEM
DC4061F02	OEM
DC4061G00	OEM
DC4061G01	OEM
DC4061G02	OEM
DC4061H00	OEM
DC4061H01	OEM
DC4061H02	OEM
DC4061J00	OEM
DC4061J01	OEM
DC4061J02	OEM
DC4062A00	OEM
DC4062A01	OEM
DC4062A02	OEM
DC4062B00	OEM
DC4062B01	OEM
DC4062B02	OEM
DC4062C00	OEM
DC4062C01	OEM
DC4062C02	OEM
DC4062D00	OEM
DC4062D01	OEM
DC4062D02	OEM
DC4062E00	OEM
DC4062E01	OEM
DC4062E02	OEM
DC4062F00	OEM
DC4062F01	OEM
DC4062F02	OEM
DC4063A00	OEM
DC4063A01	OEM
DC4063A02	OEM
DC4063B00	OEM
DC4063B01	OEM
DC4063B02	OEM
DC4063C00	OEM
DC4063C01	OEM
DC4063C02	OEM
DC4063D00	OEM
DC4063D01	OEM
DC4063D02	OEM
DC4063E00	OEM
DC4063E01	OEM
DC4063E02	OEM
DC4064A00	OEM
DC4064A01	OEM
DC4064A02	OEM
DC4064B00	OEM
DC4064B01	OEM
DC4064B02	OEM
DC4064C00	OEM
DC4064C01	OEM
DC4064C02	OEM
DC4121A01	OEM
DC4121A02	OEM
DC4121B01	OEM
DC4121B02	OEM
DC4121C01	OEM
DC4121C02	OEM
DC4121D01	OEM
DC4121D02	OEM
DC4121E01	OEM
DC4121E02	OEM
DC4121F01	OEM
DC4121F02	OEM
DC4121G01	OEM
DC4121G02	OEM
DC4121H01	OEM
DC4121H02	OEM
DC4121J01	OEM
DC4121J02	OEM
DC4121K01	OEM
DC4121K02	OEM
DC4121L01	OEM
DC4121L02	OEM
DC4121M01	OEM
DC4121M02	OEM
DC4121N01	OEM
DC4121N02	OEM
DC4121P01	OEM
DC4121P02	OEM
DC4122A01	OEM
DC4122A02	OEM
DC4122B01	OEM
DC4122B02	OEM
DC4122C01	OEM
DC4122C02	OEM
DC4122D01	OEM
DC4122D02	OEM
DC4122E01	OEM
DC4122E02	OEM
DC4122F01	OEM
DC4122F02	OEM
DC4122G01	OEM
DC4122G02	OEM
DC4122H01	OEM
DC4122H02	OEM
DC4122J01	OEM
DC4122J02	OEM
DC4122K01	OEM
DC4122K02	OEM
DC4122L01	OEM
DC4122L02	OEM
DC4122M01	OEM
DC4122M02	OEM
DC4131A01	OEM
DC4131A02	OEM
DC4131B01	OEM
DC4131B02	OEM
DC4131C01	OEM
DC4131C02	OEM
DC4131D01	OEM
DC4131D02	OEM
DC4131E01	OEM
DC4131E02	OEM
DC4131F01	OEM
DC4131F02	OEM
DC4131G01	OEM
DC4131G02	OEM
DC4131H01	OEM
DC4131H02	OEM
DC4131J01	OEM
DC4131J02	OEM
DC4131K01	OEM
DC4131K02	OEM
DC4131L01	OEM
DC4131L02	OEM
DC4131M01	OEM
DC4131M02	OEM
DC4132A01	OEM
DC4132A02	OEM
DC4132B01	OEM
DC4132B02	OEM
DC4132C01	OEM
DC4132C02	OEM
DC4132D01	OEM
DC4132D02	OEM
DC4132E01	OEM
DC4132E02	OEM
DC4132F01	OEM
DC4132F02	OEM
DC4132G01	OEM
DC4132G02	OEM
DC4132H01	OEM
DC4132H02	OEM
DC4132J01	OEM
DC4132J02	OEM
DC4132K01	OEM
DC4132K02	OEM
DC4132L01	OEM
DC4132L02	OEM
DC4132M01	OEM
DC4132M02	OEM
DC4141A01	OEM
DC4141A02	OEM
DC4141B01	OEM
DC4141B02	OEM
DC4141C01	OEM
DC4141C02	OEM
DC4141D01	OEM
DC4141D02	OEM
DC4141E01	OEM
DC4141E02	OEM
DC4141F01	OEM
DC4141F02	OEM
DC4141G01	OEM
DC4141G02	OEM
DC4141H01	OEM
DC4141H02	OEM
DC4141J01	OEM
DC4141J02	OEM
DC4141K01	OEM
DC4141K02	OEM
DC4141L01	OEM
DC4141L02	OEM
DC4141M01	OEM
DC4141M02	OEM
DC4142A01	OEM
DC4142A02	OEM
DC4142B01	OEM
DC4142B02	OEM
DC4142C01	OEM
DC4142C02	OEM
DC4142D01	OEM
DC4142D02	OEM
DC4142E01	OEM
DC4142E02	OEM
DC4142F01	OEM
DC4142F02	OEM
DC4142G01	OEM
DC4142G02	OEM
DC4142H01	OEM
DC4142H02	OEM
DC4142J01	OEM
DC4142J02	OEM
DC4142K01	OEM
DC4142K02	OEM
DC4142L01	OEM
DC4142L02	OEM
DC4151A01	OEM
DC4151A02	OEM
DC4151B01	OEM
DC4151B02	OEM
DC4151C01	OEM
DC4151C02	OEM
DC4151D01	OEM
DC4151D02	OEM
DC4151E01	OEM
DC4151E02	OEM
DC4151F01	OEM
DC4151F02	OEM
DC4151G01	OEM
DC4151G02	OEM
DC4151H01	OEM
DC4151H02	OEM
DC4151J01	OEM
DC4151J02	OEM
DC4151K01	OEM
DC4151K02	OEM
DC4151L01	OEM
DC4151L02	OEM
DC4152A01	OEM
DC4152A02	OEM
DC4152B01	OEM
DC4152B02	OEM
DC4152C01	OEM
DC4152C02	OEM
DC4152D01	OEM
DC4152D02	OEM
DC4152E01	OEM
DC4152E02	OEM
DC4152F01	OEM
DC4152F02	OEM
DC4152G01	OEM
DC4152G02	OEM
DC4152H01	OEM
DC4152H02	OEM
DC4152J01	OEM
DC4152J02	OEM
DC4161A01	OEM
DC4161A02	OEM
DC4161B01	OEM
DC4161B02	OEM
DC4161C01	OEM
DC4161C02	OEM
DC4161D01	OEM
DC4161D02	OEM
DC4161E01	OEM
DC4161E02	OEM
DC4161F01	OEM
DC4161F02	OEM
DC4161G01	OEM
DC4161G02	OEM
DC4161H01	OEM
DC4161H02	OEM
DC4161J01	OEM
DC4161J02	OEM
DC4162A01	OEM
DC4162A02	OEM
DC4162B01	OEM
DC4162B02	OEM
DC4162C01	OEM
DC4162C02	OEM
DC4162D01	OEM
DC4162D02	OEM

If replacement code is OEM, contact original manufacturer for replacement.

DEVICE TYPE	REPL CODE
DC4162E01	OEM
DC4162E02	OEM
DC4162F01	OEM
DC4162F02	OEM
DC4201	OEM
DC4202	OEM
DC4210	OEM
DC4210B	3934
DC4211	OEM
DC4211B	OEM
DC4212	OEM
DC4212B	OEM
DC4213	OEM
DC4213B	OEM
DC4214	OEM
DC4214B	OEM
DC4215	OEM
DC4215B	OEM
DC4216	OEM
DC4216B	OEM
DC4217	OEM
DC4217B	OEM
DC4218B	OEM
DC4220	OEM
DC4221	OEM
DC4222	OEM
DC4223	OEM
DC4224	OEM
DC4224B	OEM
DC4225	OEM
DC4225B	OEM
DC4226	OEM
DC4226B	OEM
DC4227B	OEM
DC4228B	OEM
DC4229D	OEM
DC4229F	OEM
DC4230	OEM
DC4231	OEM
DC4232	OEM
DC4232B	OEM
DC4233	OEM
DC4233B	OEM
DC4234	OEM
DC4234B	OEM
DC4235	OEM
DC4240	OEM
DC4241	OEM
DC4242	OEM
DC4243	OEM
DC4243C	OEM
DC4244	OEM
DC4244C	OEM
DC4245	OEM
DC4253	OEM
DC4254	OEM
DC4255	OEM
DC4255B	OEM
DC4256	OEM
DC4256B	OEM
DC4257	OEM
DC4257B	OEM
DC4265B	OEM
DC4266B	OEM
DC4267B	OEM
DC4285B	OEM
DC4286B	OEM
DC4287B	OEM
DC4298	OEM
DC4299	OEM
DC4301A	OEM
DC4301B	OEM
DC4301C	OEM
DC4302A	OEM
DC4302B	OEM
DC4303A	OEM
DC4303B	OEM
DC4304A	OEM
DC4304B	OEM
DC4305A	OEM
DC4305B	OEM
DC4352	OEM
DC4371A	OEM
DC4371B	OEM
DC4372A	OEM
DC4372B	OEM
DC4373A	OEM
DC4373B	OEM
DC4374A	OEM
DC4374B	OEM
DC4375A	OEM
DC4375B	OEM
DC5021	OEM
DC5022	OEM
DC5023	OEM
DC5108	OEM
DC5125	OEM
DC5141	OEM
DC5142	OEM
DC5143	OEM
DC5404	OEM
DC5405	OEM
DC5412	OEM
DC5414	OEM
DC5415	OEM
DC5416	OEM
DC5423	OEM
DC5424	OEM
DC5425	OEM
DC5441	OEM
DC5442	OEM
DC5443	OEM
DC5445	OEM
DC5623	OEM
DC5631	OEM
DC5632	OEM
DC5651	OEM
DC5652	OEM
DC5653	OEM
DC8008	OEM
DC8020	OEM
DC8457	0133
DC16055A	OEM
DC20026A2	OEM
DC40026B2	OEM
DC40046A	OEM
DC40066A	OEM
DC40125A	OEM
DC80025B	OEM
DCA010	0903
DCA010W5-TA	OEM
DCA015	4917
DCA015A4-TA	OEM
DCA25B	0199
DCA25C	0199
DCA25E	0199
DCB010	5274
DCB010W6-TA	OEM
DCB015	0907
DCB015-TB	OEM
DCB015A6-TA	OEM
DCB015TA	OEM
DCB015TB	0907
DCB25B	1015
DCB25C	1015
DCB25E	1015
DCB100	OEM
DCB100F	OEM
DCB101	OEM
DCB101F	OEM
DCB102	OEM
DCB102F	OEM
DCB104	OEM
DCB104F	OEM
DCB106	OEM
DCB106F	OEM
DCB108	OEM
DCB110	OEM
DCBFTS3FU0	OEM
DCC010	OEM
DCC010W7-TA	OEM
DCC016	OEM
DCC025	OEM
DCC050GACC	OEM
DCC050GBCA	OEM
DCC050RACC	OEM
DCC050RBCA	OEM
DCC050RBCC	OEM
DCC050YACC	OEM
DCC050YBCA	OEM
DCC0501RACC	OEM
DCC0501RBCA	OEM
DCC100GCA	OEM
DCC100GCC	OEM
DCC100GH.E.CC	OEM
DCC100YCA	OEM
DCC100YCC	OEM
DCC100YH.E.CC	OEM
DCC1001RCA	OEM
DCC1001RCC	OEM
DCC1001RH.E.CC	OEM
DCC1630	OEM
DCC10018	OEM
DCC10018AG	OEM
DCC10018AIR	OEM
DCC10018AR	OEM
DCC10018AY	OEM
DCC10018LG	OEM
DCC10018LIR	OEM
DCC10018LR	OEM
DCC10018LY	OEM
DCC10030	OEM
DCD015	2203
DCE015	0694
DCE5/15/150	OEM
DCE12/15/150	OEM
DCE24/15/150	OEM
DCE28/15/150	OEM
DCERB12-01	0023
DCESAC82-004	1021
DCI5/15/60	OEM
DCI12/15/60	OEM
DCI28/15/60	OEM
DCI177	OEM
DCI178	OEM
DCP102	OEM
DCR5-5	OEM
DCR5-5-5	OEM
DCR5-12	OEM
DCR5-12-12	OEM
DCR5-15	OEM
DCR5-15-15	OEM
DCR12-5	OEM
DCR12-12	OEM
DCR12-12-12	OEM
DCR12-15	OEM
DCR12-15-15	OEM
DCR403ST0101	OEM
DCR403ST0102	OEM
DCR403ST0202	OEM
DCR403ST0203	OEM
DCR403ST0303	OEM
DCR403ST0304	OEM
DCR403ST0404	OEM
DCR403ST0405	OEM
DCR403ST0505	OEM
DCR403ST0506	OEM
DCR403ST0606	OEM
DCR403ST0607	OEM
DCR403ST0707	OEM
DCR403ST0709	OEM
DCR405ST0101	OEM
DCR405ST0102	OEM
DCR405ST0202	OEM
DCR405ST0203	OEM
DCR405ST0303	OEM
DCR405ST0304	OEM
DCR405ST0404	OEM
DCR405ST0405	OEM
DCR405ST0505	OEM
DCR405ST0506	OEM
DCR405ST0606	OEM
DCR405ST0607	OEM
DCR405ST0707	OEM
DCR405ST0709	OEM
DCR405ST0808	OEM
DCR405ST0810	OEM
DCR405ST0909	OEM
DCR405ST1010	OEM
DCR405ST1012	OEM
DCR405ST1111	OEM
DCR405ST1113	OEM
DCR405ST1212	OEM
DCR405ST1215	OEM
DCR405ST1313	OEM
DCR405ST1316	OEM
DCR424ST0101T	OEM
DCR424ST0101V	OEM
DCR424ST0101W	OEM
DCR424ST0101X	OEM
DCR424ST0101Z	OEM
DCR424ST0202T	OEM
DCR424ST0202V	OEM
DCR424ST0202W	OEM
DCR424ST0202X	OEM
DCR424ST0202Z	OEM
DCR424ST0303T	OEM
DCR424ST0303V	OEM
DCR424ST0303W	OEM
DCR424ST0303X	OEM
DCR424ST0303Z	OEM
DCR424ST0404T	OEM
DCR424ST0404V	OEM
DCR424ST0404W	OEM
DCR424ST0404X	OEM
DCR424ST0404Z	OEM
DCR424ST0505T	OEM
DCR424ST0505V	OEM
DCR424ST0505X	OEM
DCR424ST0505Z	OEM
DCR424ST0606	OEM
DCR424ST0606T	OEM
DCR424ST0606V	OEM
DCR424ST0606W	OEM
DCR424ST0606X	OEM
DCR424ST0606Z	OEM
DCR425ST0101A	OEM
DCR425ST0101B	OEM
DCR425ST0101M	OEM
DCR425ST0101N	OEM
DCR425ST0101P	OEM
DCR425ST0101R	OEM
DCR425ST0101S	OEM
DCR425ST0101T	OEM
DCR425ST0101W	OEM
DCR425ST0101X	OEM
DCR425ST0101Z	OEM
DCR425ST0202A	OEM
DCR425ST0202B	OEM
DCR425ST0202L	OEM
DCR425ST0202M	OEM
DCR425ST0202N	OEM
DCR425ST0202P	OEM
DCR425ST0202R	OEM
DCR425ST0202T	OEM
DCR425ST0202W	OEM
DCR425ST0202Z	OEM
DCR425ST0303A	OEM
DCR425ST0303B	OEM
DCR425ST0303L	OEM
DCR425ST0303M	OEM
DCR425ST0303N	OEM
DCR425ST0303P	OEM
DCR425ST0303R	OEM
DCR425ST0303S	OEM
DCR425ST0303T	OEM
DCR425ST0303V	OEM
DCR425ST0303W	OEM
DCR425ST0303X	OEM
DCR425ST0303Z	OEM
DCR425ST0404A	OEM
DCR425ST0404L	OEM
DCR425ST0404M	OEM
DCR425ST0404N	OEM
DCR425ST0404P	OEM
DCR425ST0404R	OEM
DCR425ST0404S	OEM
DCR425ST0404V	OEM
DCR425ST0404W	OEM
DCR425ST0404X	OEM
DCR425ST0404Z	OEM
DCR425ST0505A	OEM
DCR425ST0505B	OEM
DCR425ST0505L	OEM
DCR425ST0505M	OEM
DCR425ST0505N	OEM
DCR425ST0505P	OEM
DCR425ST0505R	OEM
DCR425ST0505S	OEM
DCR425ST0505V	OEM
DCR425ST0505W	OEM
DCR425ST0505X	OEM
DCR425ST0505Z	OEM
DCR425ST0606A	OEM
DCR425ST0606B	OEM
DCR425ST0606L	OEM
DCR425ST0606M	OEM
DCR425ST0606N	OEM
DCR425ST0606P	OEM
DCR425ST0606R	OEM
DCR425ST0606S	OEM
DCR425ST0606V	OEM
DCR425ST0606W	OEM
DCR425ST0606X	OEM
DCR425ST0606Z	OEM
DCR485ST0101Z	OEM
DCR503ST0202	OEM
DCR504ST0101	OEM
DCR504ST0202	OEM
DCR504ST0303	OEM
DCR504ST0404	OEM
DCR504ST0505	OEM
DCR504ST0606	OEM
DCR504ST0707	OEM
DCR504ST0808	OEM
DCR504ST0909	OEM
DCR504ST1010	OEM
DCR504ST1111	OEM
DCR504ST1212	OEM
DCR504ST1313	OEM
DCR504ST1414	OEM
DCR504ST1515	OEM
DCR508ST0101	OEM
DCR508ST0911	OEM
DCR525SM0101M	OEM
DCR525SM0101N	OEM
DCR525SM0101P	OEM
DCR525SM0101R	OEM
DCR525SM0101S	OEM
DCR525SM0101T	OEM
DCR525SM0101V	OEM
DCR525SM0101W	OEM
DCR525SM0101X	OEM
DCR525SM0101Z	OEM
DCR525SM0202M	OEM
DCR525SM0202N	OEM
DCR525SM0202P	OEM
DCR525SM0202R	OEM
DCR525SM0202S	OEM
DCR525SM0202T	OEM
DCR525SM0202V	OEM
DCR525SM0202W	OEM
DCR525SM0202X	OEM
DCR525SM0202Z	OEM
DCR525SM0303M	OEM
DCR525SM0303N	OEM
DCR525SM0303P	OEM
DCR525SM0303R	OEM
DCR525SM0303S	OEM
DCR525SM0303T	OEM
DCR525SM0303V	OEM
DCR525SM0303W	OEM
DCR525SM0303Z	OEM
DCR525SM0404M	OEM
DCR525SM0404N	OEM
DCR525SM0404P	OEM
DCR525SM0404R	OEM
DCR525SM0404S	OEM
DCR525SM0404T	OEM
DCR525SM0404V	OEM
DCR525SM0404W	OEM
DCR525SM0404X	OEM
DCR525SM0404Z	OEM
DCR525SM0505M	OEM
DCR525SM0505N	OEM
DCR525SM0505P	OEM
DCR525SM0505R	OEM
DCR525SM0505S	OEM
DCR525SM0505T	OEM
DCR525SM0505V	OEM
DCR525SM0505W	OEM
DCR525SM0505X	OEM
DCR525SM0505Z	OEM
DCR525SM0606M	OEM
DCR525SM0606N	OEM
DCR525SM0606P	OEM
DCR525SM0606R	OEM
DCR525SM0606S	OEM
DCR525SM0606T	OEM
DCR525SM0606V	OEM
DCR525SM0606W	OEM
DCR525SM0606X	OEM
DCR525SM0606Z	OEM
DCR604SE0101	OEM
DCR604SE0202	OEM
DCR604SE0303	OEM
DCR604SE0404	OEM
DCR604SE0505	OEM
DCR604SE0606	OEM
DCR604SE0707	OEM
DCR604SE0808	OEM
DCR604SE0909	OEM
DCR604SE1010	OEM
DCR604SE1111	OEM
DCR604SE1212	OEM
DCR604SE1313	OEM
DCR604SE1414	OEM
DCR604SK0101	OEM
DCR604SK0202	OEM
DCR604SK0303	OEM
DCR604SK0404	OEM
DCR604SK0505	OEM
DCR604SK0606	OEM
DCR604SK0707	OEM
DCR604SK0808	OEM
DCR604SK0909	OEM
DCR604SK1010	OEM
DCR604SK1111	OEM
DCR604SK1212	OEM
DCR604SK1313	OEM
DCR604SM0101	OEM
DCR604SM0202	OEM
DCR604SM0303	OEM
DCR604SM0404	OEM
DCR604SM0505	OEM
DCR604SM0606	OEM
DCR604SM0707	OEM
DCR604SM0808	OEM
DCR604SM0909	OEM
DCR604SM1010	OEM
DCR604SM1111	OEM
DCR604SM1212	OEM
DCR604SM1313	OEM
DCR604SM1414	OEM
DCR604SM1515	OEM
DCR605SK0101	OEM
DCR605SK0202	OEM
DCR605SK0303	OEM
DCR605SK0404	OEM
DCR605SK0505	OEM
DCR605SK0606	OEM
DCR605SK0707	OEM
DCR605SK0808	OEM
DCR605SK0909	OEM
DCR605SK1010	OEM
DCR605SK1111	OEM
DCR605SK1212	OEM
DCR605SK1313	OEM
DCR605SM0101	OEM
DCR605SM0202	OEM
DCR605SM0303	OEM
DCR605SM0404	OEM
DCR605SM0505	OEM
DCR605SM0606	OEM
DCR605SM0707	OEM
DCR605SM0808	OEM
DCR605SM0909	OEM
DCR605SM1010	OEM
DCR605SM1111	OEM
DCR605SM1212	OEM
DCR605SM1313	OEM
DCR605SM1414	OEM
DCR605SM1515	OEM
DCR703SL0101	OEM
DCR703SL0202	OEM
DCR703SL0303	OEM
DCR703SL0404	OEM
DCR703SL0505	OEM
DCR703SL0606	OEM
DCR703SL0707	OEM
DCR703SL0808	OEM
DCR703SL0909	OEM
DCR703SL1010	OEM
DCR703SL1111	OEM
DCR703SL1212	OEM
DCR703SL1313	OEM
DCR703SL1414	OEM
DCR703SL1515	OEM
DCR703SM0101	OEM
DCR703SM0202	OEM
DCR703SM0203	OEM
DCR703SM0303	OEM
DCR703SM0404	OEM
DCR703SM0505	OEM
DCR703SM0606	OEM
DCR703SM0707	OEM
DCR703SM0808	OEM
DCR703SM0909	OEM
DCR703SM1010	OEM
DCR703SM1111	OEM
DCR703SM1212	OEM
DCR703SM1313	OEM
DCR703SM1414	OEM
DCR703SM1515	OEM
DCR704SL0101	OEM
DCR704SL0202	OEM
DCR704SL0303	OEM
DCR704SL0404	OEM
DCR704SL0505	OEM
DCR704SL0606	OEM
DCR704SL0707	OEM
DCR704SL0808	OEM
DCR704SL0909	OEM
DCR704SL1010	OEM
DCR704SL1111	OEM
DCR704SL1212	OEM
DCR704SL1313	OEM
DCR704SL1414	OEM
DCR704SL1515	OEM
DCR704SL1616	OEM
DCR704SM0101	OEM
DCR704SM0202	OEM
DCR704SM0303	OEM
DCR704SM0404	OEM
DCR704SM0505	OEM
DCR704SM0606	OEM
DCR704SM0707	OEM
DCR704SM0808	OEM
DCR704SM0909	OEM
DCR704SM1010	OEM
DCR704SM1111	OEM
DCR704SM1212	OEM
DCR704SM1313	OEM
DCR704SM1414	OEM
DCR704SM1515	OEM
DCR704SM1616	OEM
DCR705SM0101	OEM
DCR705SM0202	OEM
DCR705SM0303	OEM
DCR705SM0404	OEM
DCR705SM0505	OEM
DCR705SM0606	OEM
DCR705SM0707	OEM
DCR705SM0808	OEM
DCR705SM0909	OEM
DCR705SM1010	OEM
DCR705SM1111	OEM
DCR705SM1212	OEM
DCR705SM1313	OEM
DCR705SM1414	OEM
DCR705SM1515	OEM
DCR705SM1616	OEM
DCR705SM1717	OEM
DCR705SM1818	OEM
DCR706SL0202	OEM
DCR706SL0303	OEM
DCR706SL0404	OEM
DCR706SL0505	OEM
DCR706SL0606	OEM
DCR706SL0707	OEM
DCR706SL0808	OEM
DCR706SL0909	OEM
DCR706SL1010	OEM
DCR706SL1111	OEM
DCR706SL1212	OEM
DCR706SL1313	OEM
DCR706SL1414	OEM
DCR706SL1515	OEM
DCR706SL1616	OEM
DCR706SL1717	OEM
DCR706SM0101	OEM
DCR706SM0202	OEM
DCR706SM0303	OEM
DCR706SM0404	OEM
DCR706SM0505	OEM
DCR706SM0606	OEM
DCR706SM0707	OEM
DCR706SM0808	OEM
DCR706SM0909	OEM
DCR706SM1010	OEM
DCR706SM1111	OEM
DCR706SM1212	OEM
DCR706SM1313	OEM
DCR706SM1414	OEM
DCR706SM1515	OEM
DCR706SM1616	OEM
DCR706SM1717	OEM
DCR706SM1818	OEM
DCR706SM1919	OEM
DCR706SM2020	OEM
DCR707SL0101	OEM
DCR707SL0202	OEM
DCR707SL0303	OEM
DCR707SL0404	OEM
DCR707SL0505	OEM
DCR707SL0606	OEM
DCR707SL0707	OEM
DCR707SL0808	OEM
DCR707SL0909	OEM
DCR707SL1010	OEM
DCR707SL1111	OEM
DCR707SL1212	OEM
DCR707SL1313	OEM
DCR707SL1414	OEM
DCR707SL1515	OEM
DCR707SL1616	OEM
DCR707SL1717	OEM
DCR707SL1818	OEM
DCR707SL1919	OEM
DCR707SL2020	OEM
DCR707SL2121	OEM
DCR707SL2222	OEM
DCR707SL2323	OEM
DCR707SM0101	OEM
DCR707SM0202	OEM
DCR707SM0303	OEM
DCR707SM0404	OEM
DCR707SM0505	OEM
DCR707SM0606	OEM
DCR707SM0707	OEM
DCR707SM0808	OEM
DCR707SM0909	OEM
DCR707SM1010	OEM
DCR707SM1111	OEM
DCR707SM1212	OEM
DCR707SM1313	OEM
DCR707SM1414	OEM
DCR707SM1417	OEM
DCR707SM1515	OEM
DCR707SM1616	OEM
DCR707SM1717	OEM
DCR707SM1818	OEM
DCR707SM1919	OEM
DCR707SM2020	OEM
DCR707SM2121	OEM
DCR707SM2222	OEM
DCR707SM2323	OEM
DCR707SM2424	OEM
DCR707SM2525	OEM
DCR802SM0101	OEM
DCR802SM0202	OEM
DCR802SM0303	OEM
DCR802SM0404	OEM
DCR802SM0505	OEM
DCR802SM0606	OEM
DCR803SM0101	OEM
DCR803SM0202	OEM
DCR803SM0303	OEM
DCR803SM0404	OEM
DCR803SM0505	OEM
DCR803SM0606	OEM
DCR803SM0707	OEM
DCR803SM0808	OEM
DCR803SM0909	OEM
DCR803SM1010	OEM
DCR803SM1111	OEM
DCR803SM1212	OEM
DCR803SM1313	OEM
DCR803SM1414	OEM
DCR803SM1515	OEM
DCR803SM1616	OEM
DCR803SM1717	OEM
DCR803WM1818	OEM
DCR804SM0101	OEM
DCR804SM0102	OEM
DCR804SM0203	OEM
DCR804SM0303	OEM
DCR804SM0404	OEM
DCR804SM0505	OEM
DCR804SM0606	OEM
DCR804SM0707	OEM
DCR804SM0808	OEM
DCR804SM0909	OEM
DCR804SM1010	OEM
DCR804SM1111	OEM
DCR804SM1212	OEM
DCR804SM1313	OEM
DCR804SM1515	OEM
DCR804SM1616	OEM
DCR806SM1515	OEM
DCR806SM1616	OEM
DCR806SM1717	OEM
DCR806SM1818	OEM
DCR806SM1919	OEM
DCR806SM2020	OEM
DCR806SM2121	OEM
DCR806SM2222	OEM
DCR806SM2323	OEM
DCR806SM2424	OEM
DCR806SM2525	OEM
DCR806SM2626	OEM
DCR806SM2727	OEM
DCR806SM2828	OEM
DCR806SM2929	OEM
DCR806SM3030	OEM
DCR807SM0101	OEM
DCR807SM0202	OEM
DCR807SM0303	OEM
DCR807SM0404	OEM
DCR807SM0505	OEM
DCR807SM0606	OEM
DCR807SM0707	OEM
DCR807SM0808	OEM
DCR807SM0909	OEM
DCR807SM1010	OEM
DCR807SM1111	OEM
DCR807SM1212	OEM
DCR807SM1313	OEM
DCR807SM1414	OEM
DCR807SM1515	OEM
DCR807SM1616	OEM
DCR807SM1717	OEM
DCR807SM1818	OEM
DCR807SM1919	OEM
DCR807SM2020	OEM
DCR807SM2121	OEM
DCR807SM2222	OEM
DCR807SM2323	OEM
DCR807SM2424	OEM
DCR807SM2525	OEM
DCR807SM2626	OEM
DCR818SM2525	OEM
DCR818SM2626	OEM
DCR818SM2727	OEM
DCR818SM2828	OEM
DCR818SM2929	OEM
DCR818SM3030	OEM
DCR818SM3131	OEM
DCR818SM3232	OEM
DCR818SM3333	OEM
DCR818SM3434	OEM
DCR818SM3535	OEM
DCR818SM3636	OEM

If replacement code is OEM, contact original manufacturer for replacement.

DEVICE TYPE	REPL CODE
DCR818SM3737	OEM
DCR818SM3838	OEM
DCR818SM3939	OEM
DCR818SM4040	OEM
DCR854SM0101	OEM
DCR854SM0202	OEM
DCR854SM0303	OEM
DCR854SM0404	OEM
DCR854SM0505	OEM
DCR854SM0606	OEM
DCR854SM0707	OEM
DCR854SM0808	OEM
DCR854SM0909	OEM
DCR854SM1010	OEM
DCR854SM1111	OEM
DCR854SM1212	OEM
DCR854SM1313	OEM
DCR857SM0101	OEM
DCR857SM0202	OEM
DCR857SM0303	OEM
DCR857SM0404	OEM
DCR857SM0505	OEM
DCR857SM0606	OEM
DCR857SM0707	OEM
DCR857SM0808	OEM
DCR857SM0909	OEM
DCR857SM1010	OEM
DCR857SM1111	OEM
DCR857SM1212	OEM
DCR860SM0101	OEM
DCR860SM0202	OEM
DCR860SM0303	OEM
DCR860SM0404	OEM
DCR860SM0505	OEM
DCR860SM0606	OEM
DCR860SM0707	OEM
DCR860SM0808	OEM
DCR860SM0909	OEM
DCR860SM1010	OEM
DCR860SM1111	OEM
DCR860SM1212	OEM
DCR860SM1313	OEM
DCR880SM0101	OEM
DCR880SM0202	OEM
DCR880SM0303	OEM
DCR880SM0404	OEM
DCR880SM0505	OEM
DCR880SM0606	OEM
DCR880SM0707	OEM
DCR880SM0808	OEM
DCR880SM0909	OEM
DCR880SM1010	OEM
DCR880SM1111	OEM
DCR880SM1212	OEM
DCR880SM1313	OEM
DCR880SM1414	OEM
DCR880SM1515	OEM
DCR880SM1616	OEM
DCR880SM1717	OEM
DCR880SM1818	OEM
DCR880SM1919	OEM
DCR880SM2020	OEM
DCR884SM0101	OEM
DCR884SM0202	OEM
DCR884SM0303	OEM
DCR884SM0404	OEM
DCR884SM0505	OEM
DCR884SM0606	OEM
DCR884SM0707	OEM
DCR884SM0808	OEM
DCR884SM0909	OEM
DCR884SM1010	OEM
DCR884SM1111	OEM
DCR884SM1212	OEM
DCR884SM1313	OEM
DCR884SM1414	OEM
DCR884SM1515	OEM
DCR887SM0101	OEM
DCR887SM0202	OEM
DCR887SM0303	OEM
DCR887SM0404	OEM
DCR887SM0505	OEM
DCR887SM0606	OEM
DCR887SM0707	OEM
DCR887SM0808	OEM
DCR887SM0909	OEM
DCR887SM1010	OEM
DCR887SM1111	OEM
DCR887SM1212	OEM
DCR887SM1313	OEM
DCR1003SN0101	OEM
DCR1003SN0202	OEM
DCR1003SN0303	OEM
DCR1003SN0404	OEM
DCR1003SN0505	OEM
DCR1003SN0606	OEM
DCR1003SN0707	OEM
DCR1003SN0808	OEM
DCR1003SN0909	OEM
DCR1003SN1010	OEM
DCR1003SN1111	OEM
DCR1003SN1212	OEM
DCR1003SN1313	OEM
DCR1003SN1414	OEM
DCR1003SN1515	OEM
DCR1003SN1616	OEM
DCR1003SN1717	OEM
DCR1003SN1818	OEM
DCR1004SM0101	OEM

DEVICE TYPE	REPL CODE
DCR1004SM0202	OEM
DCR1004SM0303	OEM
DCR1004SM0404	OEM
DCR1004SM0505	OEM
DCR1004SM0606	OEM
DCR1004SM0707	OEM
DCR1004SM0808	OEM
DCR1004SM0909	OEM
DCR1004SM1010	OEM
DCR1004SM1111	OEM
DCR1004SM1212	OEM
DCR1004SM1313	OEM
DCR1004SM1414	OEM
DCR1004SM1515	OEM
DCR1004SN0101	OEM
DCR1004SN0202	OEM
DCR1004SN0303	OEM
DCR1004SN0404	OEM
DCR1004SN0505	OEM
DCR1004SN0606	OEM
DCR1004SN0707	OEM
DCR1004SN0808	OEM
DCR1004SN0909	OEM
DCR1004SN1010	OEM
DCR1004SN1111	OEM
DCR1004SN1212	OEM
DCR1004SN1313	OEM
DCR1004SN1414	OEM
DCR1004SN1515	OEM
DCR1004SN1616	OEM
DCR1004SN1717	OEM
DCR1004SN1818	OEM
DCR1006SN1515	OEM
DCR1006SN1616	OEM
DCR1006SN1717	OEM
DCR1006SN1818	OEM
DCR1006SN1919	OEM
DCR1006SN2020	OEM
DCR1006SN2121	OEM
DCR1006SN2222	OEM
DCR1006SN2323	OEM
DCR1006SN2424	OEM
DCR1006SN2525	OEM
DCR1006SN2626	OEM
DCR1006SN2727	OEM
DCR1006SN2828	OEM
DCR1006SN2929	OEM
DCR1006SN3030	OEM
DCR1007SN0101	OEM
DCR1007SN0202	OEM
DCR1007SN0303	OEM
DCR1007SN0404	OEM
DCR1007SN0505	OEM
DCR1007SN0606	OEM
DCR1007SN0707	OEM
DCR1007SN0808	OEM
DCR1007SN0909	OEM
DCR1007SN1010	OEM
DCR1007SN1111	OEM
DCR1007SN1212	OEM
DCR1007SN1313	OEM
DCR1007SN1414	OEM
DCR1007SN1515	OEM
DCR1007SN1616	OEM
DCR1007SN1717	OEM
DCR1007SN1818	OEM
DCR1007SN1919	OEM
DCR1007SN2020	OEM
DCR1007SN2121	OEM
DCR1007SN2222	OEM
DCR1007SN2323	OEM
DCR1007SN2424	OEM
DCR1007SN2525	OEM
DCR1007SN2626	OEM
DCR1007SN2727	OEM
DCR1007SN2828	OEM
DCR1007SN2929	OEM
DCR1007SN3030	OEM
DCR1018SN3131	OEM
DCR1018SN4141	OEM
DCR1018SN4242	OEM
DCR1018SN4343	OEM
DCR1018SN4444	OEM
DCR1018SN4545	OEM
DCR1054SD0606N	OEM
DCR1054SD0606P	OEM
DCR1054SD0606R	OEM
DCR1054SD0606S	OEM
DCR1054SD0707N	OEM
DCR1054SD0707P	OEM
DCR1054SD0707R	OEM
DCR1054SD0707S	OEM
DCR1054SD0808N	OEM
DCR1054SD0808P	OEM
DCR1054SD0808R	OEM
DCR1054SD0808S	OEM
DCR1054SD0909N	OEM
DCR1054SD0909P	OEM
DCR1054SD0909R	OEM
DCR1054SD0909S	OEM
DCR1054SD1010N	OEM
DCR1054SD1010P	OEM
DCR1054SD1010R	OEM
DCR1054SD1010S	OEM
DCR1054SD1111N	OEM
DCR1054SD1111P	OEM
DCR1054SD1111R	OEM
DCR1054SD1111S	OEM
DCR1054SD1212N	OEM

DEVICE TYPE	REPL CODE
DCR1054SD1212P	OEM
DCR1054SD1212R	OEM
DCR1054SD1212S	OEM
DCR1054SD1313N	OEM
DCR1054SD1313P	OEM
DCR1054SD1313R	OEM
DCR1054SD1313S	OEM
DCR1054SD1414N	OEM
DCR1054SD1414P	OEM
DCR1054SD1414R	OEM
DCR1054SD1414S	OEM
DCR1054SD1415N	OEM
DCR1054SD1415P	OEM
DCR1054SD1415R	OEM
DCR1054SD1415S	OEM
DCR1278SD2424	OEM
DCR1278SD2525	OEM
DCR1278SD2626	OEM
DCR1278SD2727	OEM
DCR1278SD2828	OEM
DCR1278SD2929	OEM
DCR1278SD3030	OEM
DCR1278SD3131	OEM
DCR1278SD3232	OEM
DCR1278SD3333	OEM
DCR1278SD3535	OEM
DCR1278SD3636	OEM
DCR1278SD3737	OEM
DCR1284SD2121	OEM
DCR1284SD2222	OEM
DCR1284SD2323	OEM
DCR1284SD2424	OEM
DCR1284SD2525	OEM
DCR1284SD2626	OEM
DCR1284SD2727	OEM
DCR1284SD2828	OEM
DCR1284SD2929	OEM
DCR1284SD3030	OEM
DCR1284SD3131	OEM
DCR1284SD3232	OEM
DCR1284SD3333	OEM
DCR1284SD3434	OEM
DCR1284SD3535	OEM
DCR1284SD3636	OEM
DCR1284SD3737	OEM
DCR1476SU2424	OEM
DCR1476SU2525	OEM
DCR1476SU2626	OEM
DCR1476SU2727	OEM
DCR1476SU2828	OEM
DCR1476SU2929	OEM
DCR1476SU3030	OEM
DCR1476SU3131	OEM
DCR1476SU3232	OEM
DCR1476SU3333	OEM
DCR1476SU3434	OEM
DCR1476SU3535	OEM
DCR1476SU3636	OEM
DCR1476SU3737	OEM
DCR1476SU3838	OEM
DCR1476SU3939	OEM
DCR1478SU3131	OEM
DCR1478SU3232	OEM
DCR1478SU3333	OEM
DCR1478SU3434	OEM
DCR1478SU3535	OEM
DCR1478SU3737	OEM
DCR1478SU3838	OEM
DCR1478SU3939	OEM
DCR1478SU4040	OEM
DCR1478SU4141	OEM
DCR1478SU4242	OEM
DCR1478SU4343	OEM
DCS025	OEM
DCS026-014	OEM
DCS026-014-25	OEM
DCS050	OEM
DCS86/16	OEM
DCU5-5	OEM
DCU5-12-12	OEM
DCU5-15-15	OEM
DCU12-5	OEM
DCU12-12-12	OEM
DCU12-15-15	OEM
DCV1	OEM
DCV3	OEM
DCV6	0087
DCV12	OEM
DCV15	OEM
DCW12/5/3000	OEM
DCW12/5/4000	OEM
DCW12/12/1700	OEM
DCW12/15/1500	OEM
DCW14/15/1500	OEM
DCW24/5/3000	OEM
DCW24/5/4000	OEM
DCW24/12/1700	OEM
DCW48/5/3000	OEM
DCW48/5/4000	OEM
DCW48/12/1700	OEM
DCW48/15/1500	OEM
DD-000	0015
DD000	1335
DD-003	0015
DD003	0015
DD-006	0015
DD006	0015
DD-007	0015
DD007	0015

DEVICE TYPE	REPL CODE
DD04	0196
DD04(IR)	0196
DD05	0484
DD05(IR)	0484
DD06	1080
DD06(IR)	1080
DD07(IR)	1080
DD07K	OEM
DD08K	OEM
DD09K	OEM
DD056	0015
DD058	0071
DD0.22/80X41	OEM
DD1S954	0133
DD3RLFB01L	OEM
DD10	OEM
DD10A	0373
DD10F-100	OEM
DD10K	OEM
DD11K	OEM
DD12J	OEM
DD13K	OEM
DD15	OEM
DD15K	OEM
DD20	OEM
DD20A	OEM
DD21S100	OEM
DD21S800	OEM
DD21S1200	OEM
DD22S800	OEM
DD22S1000	OEM
DD22S1200	OEM
DD22S1400	OEM
DD25	OEM
DD25F-20	2255
DD25F-40	2255
DD25F-60	2255
DD25F-80	2255
DD25F-100	OEM
DD25F-120	OEM
DD25F-140	OEM
DD25F-160	OEM
DD30	OEM
DD30A	OEM
DD31N400	1798
DD31N800	1798
DD31N1200	1798
DD31N1400	OEM
DD31S800	OEM
DD31S1000	OEM
DD31S1200	OEM
DD31S1400	OEM
DD35	OEM
DD40	OEM
DD40A	OEM
DD40F-20	2255
DD40F-40	2255
DD40F-60	2255
DD40F-80	2255
DD40F-100	OEM
DD40F-120	OEM
DD40F-140	OEM
DD40F-160	OEM
DD45S800	OEM
DD45S1000	OEM
DD45S1200	OEM
DD53S1000	OEM
DD53S1200	OEM
DD53S1400	OEM
DD55F-100	OEM
DD55FF-20	OEM
DD55FF-40	OEM
DD55FF-60	OEM
DD55FF-80	OEM
DD55FF-120	OEM
DD55FF-140	OEM
DD55FF-160	OEM
DD61N400K	OEM
DD61N800K	OEM
DD61N1200K	OEM
DD61N1400K	OEM
DD61N1600K	OEM
DD61S1000	OEM
DD61S1200	OEM
DD61S1400	OEM
DD62S400	OEM
DD62S600	OEM
DD62S800	OEM
DD62S1000	OEM
DD65N400	2192
DD65N800	2192
DD65N1200	2192
DD65N1400	OEM
DD65N1600	OEM
DD68S1000	OEM
DD68S1200	OEM
DD68S1400	OEM
DD70F-20	2192
DD70F-40	2192
DD70F-60	2192
DD70F-80	2192
DD70F-100	OEM
DD70F-120	OEM
DD70F-140	OEM
DD70F-160	OEM
DD76N400K	OEM
DD76N800K	OEM
DD76N1200K	OEM
DD76N1400K	OEM

DEVICE TYPE	REPL CODE
DD76N1600K	OEM
DD81S1000	OEM
DD81S1200	OEM
DD81S1400	OEM
DD82S400	OEM
DD82S600	OEM
DD82S800	OEM
DD85N400	2192
DD85N800	2192
DD85N1200	2192
DD85N1400	OEM
DD85N1600	OEM
DD85S800	OEM
DD85S1000	OEM
DD85S1200	OEM
DD85S1400	OEM
DD86N400K	OEM
DD86N800K	OEM
DD86N1200K	OEM
DD86N1400K	OEM
DD86N1600K	OEM
DD86N1800K	OEM
DD90F-20	6465
DD90F-40	6465
DD90F-60	6465
DD90F-80	6465
DD90F-100	OEM
DD90F-120	OEM
DD90F-140	OEM
DD90F-160	OEM
DD106N400K	OEM
DD106N800K	OEM
DD106N1200K	OEM
DD106N1400K	OEM
DD106N1600K	OEM
DD106N1800K	OEM
DD110F-20	OEM
DD110F-40	OEM
DD110F-60	OEM
DD110F-80	OEM
DD110F-120	OEM
DD110F-140	OEM
DD110F-160	OEM
DD130F-20	OEM
DD130F-40	OEM
DD130F-60	OEM
DD130F-80	OEM
DD130F-100	OEM
DD130F-120	OEM
DD130F-140	OEM
DD130F-160	OEM
DD151N400K	OEM
DD151N800K	OEM
DD151N1200K	OEM
DD151N1400K	OEM
DD151N1600K	OEM
DD151N1800K	OEM
DD160F-20	OEM
DD160F-40	OEM
DD160F-60	OEM
DD160F-80	OEM
DD160F-100	OEM
DD160F-120	OEM
DD160F-140	OEM
DD160F-160	OEM
DD175C	0015
DD176C	0015
DD177C	0015
DD236	0015
DD250N2000K	OEM
DD250N2200K	OEM
DD250N2400K	OEM
DD250N2500K	OEM
DD260N400K	OEM
DD260N800K	OEM
DD260N1200K	OEM
DD260N1400K	OEM
DD260N1600K	OEM
DD260N1800K	OEM
DD261N2000K	OEM
DD261N2200K	OEM
DD261N2400K	OEM
DD261N2500K	OEM
DD266	0015
DD268	0071
DD700	OEM
DD710	OEM
DD711	OEM
DD713	OEM
DD716	OEM
DD810A(A)	OEM
DD811A(A)	OEM
DD813A(A)	OEM
DD816A(A)	OEM
DD1000	1335
DD1001	0015
DD1003	0015
DD1006	0015
DD1056	0015
DD1057	0015
DD1058	0071
DD1712A	OEM
DD2066	0015
DD2068	0071
DD2320	0015
DD2321	0015
DD2326	OEM
DD2327	0087
DD2520	OEM

DEVICE TYPE	REPL CODE
DD2520A	OEM
DD2573	OEM
DD2573A	OEM
DD3020	OEM
DD3026	OEM
DD3076	OEM
DD3078	0087
DD3520	OEM
DD3520A	OEM
DD3573	OEM
DD3573A	OEM
DD3750	OEM
DD3751	OEM
DD3752	OEM
DD3753	OEM
DD3754	OEM
DD3755	OEM
DD3756	OEM
DD4020	OEM
DD4026	OEM
DD4066	OEM
DD4067	OEM
DD4068	OEM
DD4520	OEM
DD4521	OEM
DD4523	OEM
DD4526	OEM
DD5620H	OEM
DD5620S	OEM
DD5620V	OEM
DD5621H	OEM
DD5621S	OEM
DD5621V	OEM
DD5623H	OEM
DD5623S	OEM
DD5623V	OEM
DD5626H	OEM
DD5626S	OEM
DD5626V	OEM
DD6120H	OEM
DD6120S	OEM
DD6120V	OEM
DD6121H	OEM
DD6121S	OEM
DD6121V	OEM
DD6123H	OEM
DD6123S	OEM
DD6123V	OEM
DD6126H	OEM
DD6126S	OEM
DD6126V	OEM
DD6202	OEM
DD6205	OEM
DD6210	OEM
DD6220	OEM
DD6230	OEM
DD6240	OEM
DD6520(A)	OEM
DD6521(A)	OEM
DD6523(A)	OEM
DD6526(A)	OEM
DD6620(A)	OEM
DD6621(A)	OEM
DD6623(A)	OEM
DD6626(A)	OEM
DD6720(A)	OEM
DD6721(A)	OEM
DD6723(A)	OEM
DD6726(A)	OEM
DD6727(A)	OEM
DD6802	OEM
DD6805	OEM
DD6810	OEM
DD6820	OEM
DD6830	OEM
DD6840	OEM
DD7120	OEM
DD7121	OEM
DD7123	OEM
DD7126	OEM
DD7211AC	OEM
DD7211AMC	OEM
DD7211C	OEM
DD7211MC	OEM
DD7212AC	OEM
DD7212AMC	OEM
DD7212C	OEM
DD7212MC	OEM
DD7218BAC	OEM
DD7218BBC	OEM
DD7218BCC	OEM
DD7218BDC	OEM
DD7218EC	OEM
DD75516GF-045	OEM
DD232323	OEM
DD10000019	OEM
DD10000020	OEM
DD20000016	OEM
DD20000028	OEM
DD20000041	OEM
DD20000077	OEM
DD20000107	OEM
DD20000181	OEM
DD40000460	OEM
DD40000563	OEM
DD40000605	OEM
DD40000617	OEM
DD40001002	OEM
DD40001026	OEM

DEVICE TYPE	REPL CODE
DD50000020	OEM
DDAC-10-1	OEM
DDAC-10-3	OEM
DDAC-11-1	OEM
DDAC-11-3	OEM
DDAC-12-1	OEM
DDAC-12-3	OEM
DDAY001001	0019
DDAY001010	0019
DDAY001022	0019
DDAY002001	0023
DDAY002002	0015
DDAY003001	OEM
DDAY004001	0133
DDAY006002	0030
DDAY007001	5116
DDAY008001	0471
DDAY008003	0244
DDAY008020	0062
DDAY009001	0157
DDAY009003	0420
DDAY009007	0420
DDAY09001	OEM
DDAY010002	0012
DDAY010005	0012
DDAY010007	0077
DDAY042001	0015
DDAY047001	0133
DDAY047005	0133
DDAY-048001	0124
DDAY048001	0133
DDAY048007	0124
DDAY048008	0124
DDAY048012	0133
DDAY048013	0124
DDAY048014	0133
DDAY067001	0720
DDAY069001	0133
DDAY090001	0163
DDAY101001	0019
DDAY103001	0015
DDAY108001	0649
DDAY126001	0057
DDAY1003001	OEM
DDB-4009	OEM
DDB-4037	OEM
DDB-4503	OEM
DDB-4504	OEM
DDB-4517	OEM
DDB-4517A	OEM
DDB5098	OEM
DDB-6224	OEM
DDB6224	OEM
DDB-6672	OEM
DDB-6672A	OEM
DDB-6673	OEM
DDB6673Y	OEM
DDB6783	OEM
DDBY002001	0023
DDBY003001	0006
DDBY003002	0006
DDBY003003	0006
DDBY004001	0006
DDBY007002	0148
DDBY008001	0693
DDBY104002	0532
DDBY209003	0368
DDBY216002	0224
DDBY219001	1144
DDBY222002	0076
DDBY224001	0076
DDBY224003	0076
DDBY224004	0076
DDBY224006	0076
DDBY227001	0219
DDBY227004	0390
DDBY228001	0042
DDBY230001	0839
DDBY231002	0836
DDBY233001	0076
DDBY246001	0786
DDBY256001	0890
DDBY257001	0936
DDBY259001	0076
DDBY259002	0076
DDBY261002	2209
DDBY262001	1220
DDBY269001	0127
DDBY270001	2814
DDBY272001	0018
DDBY273001	0076
DDBY277002	0151
DDBY278001	2292
DDBY278002	2292
DDBY283001	0127
DDBY287001	0127
DDBY288001	1591
DDBY289001	0936
DDBY295002	1066
DDBY299001	0076
DDBY301001	0155
DDBY307003	0836
DDBY407001	0042
DDBY407004	0423
DDBY410001	0951
DDBY410002	0951
DDBY4233001	0111
DDC-4561	OEM
DDC-4561A	OEM

DEVICE TYPE	REPL CODE
DDC-4561B	OEM
DDC-4561C	OEM
DDC-4561D	OEM
DDC-4562	OEM
DDC-4562A	OEM
DDC-4562B	OEM
DDC-4562C	OEM
DDC-4562D	OEM
DDC-4563	OEM
DDC-4563A	OEM
DDC-4563B	OEM
DDC-4563C	OEM
DDC-4563D	OEM
DDC-4564	OEM
DDC-4564A	OEM
DDC-4564B	OEM
DDC-4564C	OEM
DDC-4564D	OEM
DDC-4565	OEM
DDC-4565A	OEM
DDC-4565B	OEM
DDC-4565C	OEM
DDC-4565D	OEM
DDC6757	OEM
DDC6757A	OEM
DDC6757B	OEM
DDC6757C	OEM
DDC6757D	OEM
DDCY001002	0688
DDCY002002	0325
DDCY006001	0325
DDCY007002	0148
DDCY103001	0849
DDCY104001	0414
DDCY104002	0212
DDCY104003	0414
DDE-201	0015
DDE6889	OEM
DDE6890	OEM
DDEY001001	0417
DDEY002001	0823
DDEY004001	0911
DDEY007001	OEM
DDEY015001	1013
DDEY017001	1379
DDEY019001	0011
DDEY020001	3801
DDEY026001	1300
DDEY027001	1160
DDEY028001	1196
DDEY029001	1208
DDEY030001	0232
DDEY031001	OEM
DDEY032001	1379
DDEY033001	OEM
DDEY037001	3805
DDEY042001	1292
DDEY046001	1292
DDEY055001	2253
DDEY058001	0215
DDEY061001	1816
DDEY064001	2261
DDEY070001	OEM
DDEY082001	4484
DDEY084001	0215
DDEY087001	2306
DDEY088001	1292
DDEY089001	0215
DDEY091001	1816
DDEY093001	1292
DDEY097001	4209
DDEY101001	OEM
DDEY109001	1540
DDEY123001	2534
DDEY130001	4042
DDEY131001	4648
DDEY132001	6975
DDEY133001	4645
DDEY146001	4652
DDEY147001	2328
DDEY148001	OEM
DDEY149001	0756
DDEY155001	OEM
DDEY157001	4291
DDEY171001	4291
DDEY172001	4476
DDEY173001	OEM
DDEY179001	0706
DDEY196001	OEM
DDEY1089001	OEM
DDF4021001	0120
DDF4091001	1816
DDFY007001	0139
DDFY017001	0316
DDFY020001	0139
DDFY021001	0120
DDFY021001#	0343
DDFY055001	2253
DDFY091001	1816
DDMV-1	0143
DDMV-2	0143
DDT12	OEM
DDV325	OEM
DE	OEM
DE-087	2306
DE2A	0124
DE3	0124
DE7RA5R10B	OEM
DE7RA5R60B	OEM

DEVICE TYPE	REPL CODE
DE7RA6R80B	OEM
DE7RA8R20A	OEM
DE14	0015
DE14A	0015
DE16	0015
DE16A	0015
DE100	OEM
DE101	OEM
DE102	OEM
DE103	OEM
DE104	0124
DE105	OEM
DE106	OEM
DE110	0124
DE111	0124
DE112	0124
DE113	0124
DE114	0124
DE115	0124
DE125	OEM
DE-201	0015
DE201	0015
DE210	OEM
DE220	OEM
DE232	OEM
DE240	OEM
DE420A	OEM
DE432A	OEM
DE1000	OEM
DE1714A	OEM
DE1716A	OEM
DE1717A	OEM
DE1718A	OEM
DE1719A	OEM
DE1720A	OEM
DE1722A	OEM
DE1724A	OEM
DE1726A	OEM
DE3101/2	OEM
DE3101/5	OEM
DEC5651	OEM
DEEY004001	0911
DEP1712A	OEM
DEP1714A	OEM
DEP1716A	OEM
DEP1717A	OEM
DEP1718A	OEM
DEP1719A	OEM
DEP1720A	OEM
DEP1722A	OEM
DEP1724A	OEM
DEP1726A	OEM
DER1	0071
DER2	OEM
DER3	OEM
DER4	OEM
DER5	OEM
DER6	OEM
DER7	OEM
DER8	OEM
DER9	OEM
DER10	OEM
DER11	OEM
DER12	OEM
DER13	OEM
DER14	OEM
DER15	OEM
DERB26-20	0348
DESFH205	OEM
DF005	0241
DF005M	0241
DF005S	OEM
DF01	0241
DF01M	0241
DF01S	OEM
DF02	0241
DF02M	0241
DF02S	OEM
DF04	4581
DF04M	4581
DF04S	OEM
DF06	1875
DF06M	1875
DF06S	OEM
DF08	3546
DF08M	3546
DF08S	OEM
DF1H	0102
DF-2	0103
DF2	0720
DF10	3546
DF10M	3546
DF10S	OEM
DF15	OEM
DF20	OEM
DF25	OEM
DF30	OEM
DF35	OEM
DF40	OEM
DF64D	0071
DF100	OEM
DF252-02	OEM
DF252-04	OEM
DF252-06	OEM
DF252-08	OEM
DF252-10	OEM
DF252-12	OEM
DF252-14	OEM
DF320	6631

DEVICE TYPE	REPL CODE
DF320ADJ	OEM
DF320ADK	OEM
DF320DJ	OEM
DF320DK	OEM
DF320DP	OEM
DF321	OEM
DF321DP	OEM
DF322	OEM
DF322DJ	OEM
DF322DK	OEM
DF322DP	OEM
DF323DJ	OEM
DF323DP	OEM
DF328DJ	OEM
DF331/332	OEM
DF412CJ	OEM
DF451-01	OEM
DF451-02	OEM
DF451-04	OEM
DF451-06	OEM
DF451-08	OEM
DF451-10	OEM
DF451-12	OEM
DF451-14	OEM
DF452-02	OEM
DF452-04	OEM
DF452-06	OEM
DF452-08	OEM
DF452-10	OEM
DF452-12	OEM
DF452-14	OEM
DF452-15	OEM
DF652-02	OEM
DF652-04	OEM
DF652-06	OEM
DF652-08	OEM
DF652-10	OEM
DF652-12	OEM
DF652-14	OEM
DF652-16	OEM
DF652-18	OEM
DF652-20	OEM
DF652-22	OEM
DF652-24	OEM
DF652-25	OEM
DF752-02	OEM
DF752-04	OEM
DF752-06	OEM
DF752-08	OEM
DF752-11	OEM
DF752-12	OEM
DF752-14	OEM
DF752-16	OEM
DF752-18	OEM
DF752-20	OEM
DFA01C	0133
DFA01G	0023
DFA01J	0017
DFA01L	0017
DFA01R	0017
DFA03L	OEM
DFA03R	OEM
DFA05B	0023
DFA05C	0023
DFA05E	0023
DFA05G	0023
DFA05G-KB4	0023
DFA05G-KBG	0023
DFA1A1	1090
DFA1A2	0023
DFA1A4	0023
DFA1A4-T3	0023
DFA1A4-Y	0023
DFA1A4-Z	0023
DFB03R	0348
DFB03T	0348
DFB03W	0348
DFB05B	0023
DFB05C	0023
DFB05E	0023
DFB05G	0023
DFC05J	0017
DFC05L	0017
DFC05N	0017
DFC05R	0017
DFC10E	0023
DFC10E-KB4	0023
DFC10G	0023
DFC15L	0102
DFC15L-KC5	0102
DFD05	OEM
DFD05B	0015
DFD05C	0015
DFD05E	0015
DFD05G	0023
DFD05T	0023
DFD05T-BT	OEM
DFD05TE	OEM
DFD05TE-BT	OEM
DFD05TG	0023
DFD05TG-BT	0023
DFEQA02-05C	0041
DFEQA02-06	0091
DFEQA02-06A	0253
DFEQA02-06C	0091
DFEQA02-06CDE	0091
DFEQA02-06D	0471
DFEQA02-08A	0077
DFEQA02-08D	0077

DEVICE TYPE	REPL CODE
DFEQA02-09D	0057
DFEQA02-10B	0064
DFEQA02-10C	0064
DFEQA02-12B	0053
DFEQA02-16A	0445
DFEQA02-23A	0494
DFEQA02-30C	0166
DFEQA02-35B	0010
DFERA81-004	0023
DFERB12-01	0023
DFERB12-02	0023
DFERC04-02	0071
DFERC04-02F	0564
DFF20B10	OEM
DFF20B12	OEM
DFF50B10	OEM
DFF50B12	OEM
DFG1A2	0609
DFG1A4	0795
DFG1A6	0015
DFG1A8	0072
DFG2A2	0087
DFG2A4	0087
DFG2A6	0087
DFG2A8	0087
DFH10B	0015
DFH10C	0015
DFH10E	0015
DFH10G	0015
DFH10G-KB4	0023
DFH10GKB4	0023
DFH10J	0071
DFH10TG	0023
DFH10TG-AT1	0023
DFH10TG-KB4	0023
DFM1A2	0023
DFM1A4	0023
DFM1SA4	0023
DFNA3-50	OEM
DFNA3-100	OEM
DFP027	OEM
DFP068	OEM
DFP140	OEM
DFP180	OEM
DFP240	OEM
DFS80A8	OEM
DFS80A8R	OEM
DFS80A10	OEM
DFS80A10R	OEM
DFS80A13	OEM
DFS80A13R	OEM
DFS80A15	OEM
DFS80A15R	OEM
DFS250A8	OEM
DFS250A10	OEM
DFS250A13	OEM
DFS250A15	OEM
DFS250AR8	OEM
DFS250AR10	OEM
DFS250AR13	OEM
DFS250AR15	OEM
DG1	0102
DG1-005	0023
DG1K	0071
DG1M	0015
DG1MR	0015
DG-1N	0102
DG1N	0102
DG1N60	0019
DG1NK	OEM
DG-1NR	0102
DG1NR	0182
DG1PR	0015
DG1S34	0143
DG1S188	0143
DG1S188-FM	2232
DG2	0102
DG2-006	0023
DG3	OEM
DG11	OEM
DG-13	0924
DG13	5415
DG13TV	5415
DG-14	0975
DG14	5415
DG20ABK	OEM
DG100J	OEM
DG100K	OEM
DG100M	OEM
DG106AAR	OEM
DG111CDD	OEM
DG111CFD	OEM
DG111IDD	OEM
DG111IFD	OEM
DG111MDD	OEM
DG111MFD	OEM
DG112CDD	OEM
DG112CFD	OEM
DG112IDD	OEM
DG112IFD	OEM
DG112MDD	OEM
DG112MFD	OEM
DG116CDD	OEM
DG116CFD	OEM
DG116IDD	OEM
DG116IFD	OEM
DG116MDD	OEM
DG116MFD	OEM
DG118CDD	OEM

DEVICE TYPE	REPL CODE
DG118CFD	OEM
DG118IDD	OEM
DG118IFD	OEM
DG118MDD	OEM
DG118MFD	OEM
DG120CDD	OEM
DG120CFD	OEM
DG120IDD	OEM
DG120IFD	OEM
DG120MDD	OEM
DG120MFD	OEM
DG121CDD	OEM
DG121CFD	OEM
DG121IDD	OEM
DG121IFD	OEM
DG121MDD	OEM
DG121MFD	OEM
DG123AL	OEM
DG123AP	OEM
DG123BP	OEM
DG123CDD	OEM
DG123CFD	OEM
DG123IDD	OEM
DG123IFD	OEM
DG123MDD	OEM
DG123MFD	OEM
DG125AL	OEM
DG125AP	OEM
DG125BP	OEM
DG125CDD	OEM
DG125CFD	OEM
DG125IDD	OEM
DG125IFD	OEM
DG125MDD	OEM
DG125MFD	OEM
DG126ADD	OEM
DG126AFD	OEM
DG126AL	OEM
DG126AP	OEM
DG126BDD	OEM
DG126BFD	OEM
DG126BP	OEM
DG129ADD	OEM
DG129AFD	OEM
DG129AL	OEM
DG129AP	OEM
DG129BDD	OEM
DG129BFD	OEM
DG129BP	OEM
DG-130	OEM
DG130	5415
DG133ADD	OEM
DG133AFD	OEM
DG133AL	OEM
DG133AP	OEM
DG133BDD	OEM
DG133BFD	OEM
DG133BP	OEM
DG134ADD	OEM
DG134AFD	OEM
DG134AL	OEM
DG134AP	OEM
DG134BDD	OEM
DG134BFD	OEM
DG134BP	OEM
DG139ADD	OEM
DG139AFD	OEM
DG139AL	OEM
DG139AP	OEM
DG139BDD	OEM
DG139BFD	OEM
DG139BP	OEM
DG140ADD	OEM
DG140AFD	OEM
DG140AL	OEM
DG140BDD	OEM
DG140BFD	OEM
DG141ADD	OEM
DG141AFD	OEM
DG141AL	OEM
DG141AP	OEM
DG141BDD	OEM
DG141BFD	OEM
DG141BP	OEM
DG142ADD	OEM
DG142AFD	OEM
DG142AL	OEM
DG142AP	OEM
DG142BDD	OEM
DG142BFD	OEM
DG142BP	OEM
DG143ADD	OEM
DG143AFD	OEM
DG143AL	OEM
DG143AP	OEM
DG143BDD	OEM
DG143BFD	OEM
DG143BP	OEM
DG144ADD	OEM
DG144AFD	OEM
DG144AL	OEM
DG144AP	OEM
DG144BDD	OEM
DG144BFD	OEM
DG144BP	OEM
DG145ADD	OEM
DG145AFD	OEM
DG145AL	OEM
DG145BDD	OEM

DEVICE TYPE	REPL CODE
DG145BFD	OEM
DG146ADD	OEM
DG146AFD	OEM
DG146AL	OEM
DG146AP	OEM
DG146BDD	OEM
DG146BFD	OEM
DG146BP	OEM
DG151ADD	OEM
DG151AFD	OEM
DG151AL	OEM
DG151AP	OEM
DG151BDD	OEM
DG151BFD	OEM
DG151BP	OEM
DG152ADD	OEM
DG152AFD	OEM
DG152AL	OEM
DG152AP	OEM
DG152BDD	OEM
DG152BFD	OEM
DG152BP	OEM
DG153ADD	OEM
DG153AFD	OEM
DG153AL	OEM
DG153BDD	OEM
DG153BFD	OEM
DG154ADD	OEM
DG154AFD	OEM
DG154AL	OEM
DG154AP	OEM
DG154BDD	OEM
DG154BFD	OEM
DG154BP	OEM
DG161ADD	OEM
DG161AFD	OEM
DG161AL	OEM
DG161AP	OEM
DG161BDD	OEM
DG161BFD	OEM
DG161BP	OEM
DG162ADD	OEM
DG162AFD	OEM
DG162AL	OEM
DG162AP	OEM
DG162BDD	OEM
DG162BFD	OEM
DG162BP	OEM
DG163ADD	OEM
DG163AFD	OEM
DG163AL	OEM
DG163BDD	OEM
DG163BFD	OEM
DG164ADD	OEM
DG164AFD	OEM
DG164AL	OEM
DG164AP	OEM
DG164BDD	OEM
DG164BFD	OEM
DG164BP	OEM
DG172AL	OEM
DG172AP	OEM
DG172BP	OEM
DG172CJ	OEM
DG180AA	OEM
DG180AL	OEM
DG180AP	OEM
DG180APDD	OEM
DG180APDE	OEM
DG180APJD	OEM
DG180BA	OEM
DG180BL	OEM
DG180BP	OEM
DG180BPDD	OEM
DG180BPJD	OEM
DG181AA	OEM
DG181AL	OEM
DG181AP	OEM
DG181APDD	OEM
DG181APJD	OEM
DG181BA	OEM
DG181BL	OEM
DG181BP	OEM
DG181BPDD	OEM
DG181BPJD	OEM
DG182AA	OEM
DG182AL	OEM
DG182AP	OEM
DG182APDD	OEM
DG182APJD	OEM
DG182BA	OEM
DG182BL	OEM
DG182BP	OEM
DG182BPDD	OEM
DG182BPDE	OEM
DG182BPJD	OEM
DG182BPJE	OEM
DG183AL	OEM
DG183AP	OEM
DG183APDE	OEM
DG183APJE	OEM
DG183BL	OEM
DG183BP	OEM
DG184AL	OEM
DG184AP	OEM
DG184APDE	OEM
DG184APJE	OEM
DG184BL	OEM
DG184BP	OEM

DEVICE TYPE	REPL CODE
DG185AL	OEM
DG185AP	OEM
DG185APDE	OEM
DG185APJE	OEM
DG185BL	OEM
DG185BP	OEM
DG185BPDE	OEM
DG185BPJE	OEM
DG186AA	OEM
DG186AL	OEM
DG186AP	OEM
DG186APDD	OEM
DG186APJD	OEM
DG186BA	OEM
DG186BL	OEM
DG186BP	OEM
DG186BPDD	OEM
DG186BPJD	OEM
DG187AL	OEM
DG187AP	OEM
DG187APDD	OEM
DG187APJD	OEM
DG187BA	OEM
DG187BL	OEM
DG187BP	OEM
DG187BPDD	OEM
DG187BPJD	OEM
DG188AA	OEM
DG188AL	OEM
DG188AP	OEM
DG188APDD	OEM
DG188APJD	OEM
DG188BA	OEM
DG188BL	OEM
DG188BP	OEM
DG188BPDD	OEM
DG188BPJD	OEM
DG189AL	OEM
DG189AP	OEM
DG189APDE	OEM
DG189APJE	OEM
DG189BL	OEM
DG189BP	OEM
DG189BPDE	OEM
DG189BPJE	OEM
DG190AL	OEM
DG190AP	OEM
DG190APDE	OEM
DG190APJE	OEM
DG190BL	OEM
DG190BP	OEM
DG190BPDE	OEM
DG190BPJE	OEM
DG191AL	OEM
DG191AP	OEM
DG191APDE	OEM
DG191APJE	OEM
DG191BL	OEM
DG191BP	OEM
DG191BPDE	OEM
DG191BPJE	OEM
DG200AA	OEM
DG200AAA	OEM
DG200AAK	OEM
DG200ABA	OEM
DG200ABK	OEM
DG200ACJ	OEM
DG200ACK	OEM
DG200AK	OEM
DG200AL	OEM
DG200AP	OEM
DG200BA	OEM
DG200BK	OEM
DG200BL	OEM
DG200BP	OEM
DG200CJ	OEM
DG201AAK	OEM
DG201ABK	OEM
DG201ACJ	OEM
DG201ACK	OEM
DG201AK	OEM
DG201CJ	OEM
DG202AK(A)	OEM
DG202BK(A)	OEM
DG202CJ(A)	OEM
DG202CK(A)	OEM
DG211CJ	OEM
DG212CJ(A)	OEM
DG243AK(A)	OEM
DG243CJ(A)	OEM
DG243CK(A)	OEM
DG281AA	OEM
DG281AP	OEM
DG281BA	OEM
DG281BP	OEM
DG284AP	OEM
DG284BP	OEM
DG287AA	OEM
DG287AP	OEM
DG287BA	OEM
DG287BP	OEM
DG290AP	OEM
DG290BP	OEM
DG300AA	OEM
DG300AAA	OEM
DG300AAK	OEM
DG300AAL	OEM
DG300ABK	OEM
DG300ACJ	OEM

If replacement code is OEM, contact original manufacturer for replacement.

DEVICE TYPE	REPL CODE
DG300ACK	OEM
DG300AP	OEM
DG300BA	OEM
DG300BP	OEM
DG300CJ	OEM
DG301AA	OEM
DG301AAA	OEM
DG301AAK	OEM
DG301ABA	OEM
DG301ABK	OEM
DG301ACJ	OEM
DG301ACK	OEM
DG301AP	OEM
DG301BA	OEM
DG301BP	OEM
DG301CJ	OEM
DG302AAK	OEM
DG302AAL	OEM
DG302ACJ	OEM
DG302ACK	OEM
DG302AP	OEM
DG302BP	OEM
DG302CJ	OEM
DG303AAK	OEM
DG303AAL	OEM
DG303ABK	OEM
DG303ACJ	OEM
DG303ACK	OEM
DG303AP	OEM
DG303BP	OEM
DG303CJ	OEM
DG304AA	OEM
DG304AAK	OEM
DG304AAL	OEM
DG304ABA	OEM
DG304ABK	OEM
DG304ACJ	OEM
DG304ACK	OEM
DG304AP	OEM
DG304BA	OEM
DG304BP	OEM
DG304CJ	OEM
DG305AA	OEM
DG305AAA	OEM
DG305AAK	OEM
DG305AAL	OEM
DG305ABA	OEM
DG305ABK	OEM
DG305ACJ	OEM
DG305ACK	OEM
DG305AP	OEM
DG305BA	OEM
DG305BP	OEM
DG305CJ	OEM
DG306AAK	OEM
DG306AAL	OEM
DG306ABK	OEM
DG306ACJ	OEM
DG306ACK	OEM
DG306AP	OEM
DG306BP	OEM
DG306CJ	OEM
DG307AAK	OEM
DG307AAL	OEM
DG307ABK	OEM
DG307ACJ	OEM
DG307ACK	OEM
DG307AP	OEM
DG307BP	OEM
DG307CJ	OEM
DG308CJ	OEM
DG309CJ	OEM
DG381AA	OEM
DG381AAA	OEM
DG381AAK	OEM
DG381ABA	OEM
DG381ABK	OEM
DG381ACJ	OEM
DG381ACK	OEM
DG381AP	OEM
DG381BA	OEM
DG381BP	OEM
DG381CJ	OEM
DG384AAK	OEM
DG384ABK	OEM
DG384ACJ	OEM
DG384ACK	OEM
DG384AP	OEM
DG384BP	OEM
DG384CJ	OEM
DG387AA	OEM
DG387AAA	OEM
DG387AAK	OEM
DG387ABA	OEM
DG387ABK	OEM
DG387ACJ	OEM
DG387ACK	OEM
DG387AP	OEM
DG387BA	OEM
DG387BP	OEM
DG387CJ	OEM
DG390AAK	OEM
DG390ABK	OEM
DG390ACJ	OEM
DG390ACK	OEM
DG390AP	OEM
DG390BP	OEM
DG390CJ	OEM
DG501AP	OEM

DEVICE TYPE	REPL CODE
DG501BP	OEM
DG501CJ	OEM
DG503AP	OEM
DG503BP	OEM
DG506ABR	OEM
DG506ACJ	OEM
DG506AR	OEM
DG506BR	OEM
DG506CJ	OEM
DG507AAR	OEM
DG507ABR	OEM
DG507ACJ	OEM
DG507AR	OEM
DG507BR	OEM
DG507CJ	OEM
DG508AAK	OEM
DG508ABK	OEM
DG508ACJ	OEM
DG508ACK	OEM
DG508AL	OEM
DG508AP	OEM
DG508BP	OEM
DG508CJ	OEM
DG509AAK	OEM
DG509ABK	OEM
DG509ACJ	OEM
DG509ACK	OEM
DG509AL	OEM
DG509AP	OEM
DG509BP	OEM
DG509CJ	OEM
DG515ADICE	OEM
DG515AP	OEM
DG515BP	OEM
DG515CJ	OEM
DG516ADICE	OEM
DG516AR	OEM
DG516BR	OEM
DG516CJ	OEM
DG528AK	OEM
DG528BK(A)	OEM
DG528CJ(A)	OEM
DG529AK(A)	OEM
DG529BK(A)	OEM
DG529CJ(A)	OEM
DG5040AK(A)	OEM
DG5040AL(A)	OEM
DG5040CJ(A)	OEM
DG5040CK(A)	OEM
DG5041AA(A)	OEM
DG5041AK(A)	OEM
DG5041AL(A)	OEM
DG5041CJ(A)	OEM
DG5041CK(A)	OEM
DG5042AA(A)	OEM
DG5042AK(A)	OEM
DG5042AL(A)	OEM
DG5042CJ(A)	OEM
DG5042CK(A)	OEM
DG5043AK(A)	OEM
DG5043AL(A)	OEM
DG5043CJ(A)	OEM
DG5043CK(A)	OEM
DG5044AA(A)	OEM
DG5044AK(A)	OEM
DG5044AL(A)	OEM
DG5044CK(A)	OEM
DG5045AK(A)	OEM
DG5045AL(A)	OEM
DG5045CJ(A)	OEM
DG5045CK(A)	OEM
DGA5	0133
DGA187AA	OEM
DGA5768	OEM
DGA5768A	OEM
DGA5768B	OEM
DGA5768C	OEM
DGA5768D	OEM
DGA6036	OEM
DGA6036A	OEM
DGA6036B	OEM
DGA6238	OEM
DGA6238A	OEM
DGA6238B	OEM
DGB6126	OEM
DGB6126A	OEM
DGB6200	OEM
DGB6200A	OEM
DGB6201	OEM
DGB6201A	OEM
DGB6202	OEM
DGB6202A	OEM
DGB6203	OEM
DGB6203A	OEM
DGB6204	OEM
DGB6204A	OEM
DGB8081	OEM
DGB8091	OEM
DGB8094	OEM
DGB8095	OEM
DGB8121	OEM
DGB8122	OEM
DGB8123	OEM
DGB8125	OEM
DGB8131	OEM
DGB8132	OEM
DGB8133	OEM
DGB8134	OEM
DGB8135	OEM

DEVICE TYPE	REPL CODE
DGB8141	OEM
DGB8144	OEM
DGB8145	OEM
DGB8154	OEM
DGB8155	OEM
DGB8156	OEM
DGB8181	OEM
DGB8191	OEM
DGB8194	OEM
DGB8195	OEM
DGB8211	OEM
DGB8212	OEM
DGB8213	OEM
DGB8214	OEM
DGB8215	OEM
DGB8216	OEM
DGB8221	OEM
DGB8222	OEM
DGB8223	OEM
DGB8224	OEM
DGB8225	OEM
DGB8231	OEM
DGB8232	OEM
DGB8233	OEM
DGB8234	OEM
DGB8235	OEM
DGB8241	OEM
DGB8244	OEM
DGB8245	OEM
DGB8254	OEM
DGB8255	OEM
DGB8256	OEM
DGB8281	OEM
DGB8291	OEM
DGB8294	OEM
DGB8295	OEM
DGB8311	OEM
DGB8312	OEM
DGB8313	OEM
DGB8314	OEM
DGB8315	OEM
DGB8321	OEM
DGB8322	OEM
DGB8323	OEM
DGB8324	OEM
DGB8325	OEM
DGB8332	OEM
DGB8333	OEM
DGB8334	OEM
DGB8335	OEM
DGB8344	OEM
DGB8345	OEM
DGB8354	OEM
DGB8355	OEM
DGB8356	OEM
DGB8411	OEM
DGB8412	OEM
DGB8413	OEM
DGB8414	OEM
DGB8415	OEM
DGB8421	OEM
DGB8422	OEM
DGB8423	OEM
DGB8424	OEM
DGB8425	OEM
DGB8431	OEM
DGB8432	OEM
DGB8433	OEM
DGB8434	OEM
DGB8435	OEM
DGB8444	OEM
DGB8445	OEM
DGB8454	OEM
DGB8455	OEM
DGB8456	OEM
DGB8511	OEM
DGB8512	OEM
DGB8513	OEM
DGB8515	OEM
DGB8521	OEM
DGB8522	OEM
DGB8523	OEM
DGB8524	OEM
DGB8525	OEM
DGB8531	OEM
DGB8532	OEM
DGB8533	OEM
DGB8534	OEM
DGB8535	OEM
DGB8544	OEM
DGB8545	OEM
DGB8554	OEM
DGB8555	OEM
DGB8556	OEM
DGB8612	OEM
DGB8613	OEM
DGB8615	OEM
DGB8622	OEM
DGB8623	OEM
DGB8624	OEM
DGB8625	OEM
DGB8632	OEM
DGB8633	OEM
DGB8634	OEM
DGB8635	OEM
DGB8712	OEM
DGB8713	OEM
DGB8714	OEM
DGB8715	OEM

DEVICE TYPE	REPL CODE
DGB8722	OEM
DGB8724	OEM
DGB8725	OEM
DGB8732	OEM
DGB8733	OEM
DGB8734	OEM
DGB8735	OEM
DGB8812	OEM
DGB8813	OEM
DGB8822	OEM
DGB8823	OEM
DGB8912	OEM
DGB8913	OEM
DGB8922	OEM
DGB8923	OEM
DGB9211	OEM
DGB9212	OEM
DGB9213	OEM
DGB9214	OEM
DGB9215	OEM
DGB9221	OEM
DGB9222	OEM
DGB9223	OEM
DGB9224	OEM
DGB9225	OEM
DGB9234	OEM
DGB9235	OEM
DGB9244	OEM
DGB9245	OEM
DGB9246	OEM
DGB9311	OEM
DGB9312	OEM
DGB9313	OEM
DGB9314	OEM
DGB9315	OEM
DGB9321	OEM
DGB9322	OEM
DGB9323	OEM
DGB9324	OEM
DGB9325	OEM
DGB9334	OEM
DGB9335	OEM
DGB9344	OEM
DGB9346	OEM
DGB9411	OEM
DGB9412	OEM
DGB9413	OEM
DGB9414	OEM
DGB9415	OEM
DGB9421	OEM
DGB9422	OEM
DGB9423	OEM
DGB9424	OEM
DGB9425	OEM
DGB9434	OEM
DGB9435	OEM
DGDS135D	0023
DGDS135E	0023
DGDS442	0133
DGDS442X	0124
DGDSB15C	0071
DGDSF10C	0023
DGGMA01	0124
DGGMA01-L	0133
DGGZA3.9-X	0036
DGGZA3.9X	0036
DGGZA4.3-Y	0275
DGGZA4.3Y	0275
DGGZA5.1-X	0041
DGGZA5.1-Y	0589
DGGZA5.1-Z	0041
DGGZA5.1X	0041
DGGZA5.1Y	0041
DGGZA5.6-Y	0253
DGGZA6.2	0471
DGGZA6.2-X	0471
DGGZA6.2-XYZ	0091
DGGZA6.2-Y	0471
DGGZA6.2X	0471
DGGZA7.5-Y	0077
DGGZA8.2-Y	0165
DGGZA9.1-Z	0057
DGGZA9.1Z	0057
DGGZA10-T	0064
DGGZA10-Y	0064
DGGZA10-Z	0064
DGGZA10Y	0064
DGGZA13-X	0053
DGGZA13X	0053
DGGZA16-Y	0445
DGGZA22-Z	0705
DGGZA24-Y	0494
DGGZA27-Y	0455
DGGZA30	0195
DGGZA30-Y	0195
DGGZA36-Y	0010
DGGZA36Y	0010
DGGZA39-Y	0032
DGGZM6.2-B2	0471
DGGZM6.8-B2	OEM
DGGZM7.5-B2	0077
DGGZM8.2-B2	0165
DGGZM9.1-B2	0057
DGGZM20-B2	0700
DGGZM24-B2	0494
DGIN	0102
DGL5630	1329
DGM	0244
D-GM-2	0143

DEVICE TYPE	REPL CODE
DGM-2	0143
DGM-3	0143
DGM3	0143
DGM111AL	OEM
DGM111AP	OEM
DGM111BP	OEM
DGM122AL	OEM
DGM122AP	OEM
DGM122BP	OEM
DGM181BA	OEM
DGM181BJ	OEM
DGM181BK	OEM
DGM181BL	OEM
DGM182AA	OEM
DGM182AJ	OEM
DGM182AK	OEM
DGM182AL	OEM
DGM182BA	OEM
DGM182BJ	OEM
DGM182BK	OEM
DGM184BJ	OEM
DGM184BK	OEM
DGM184BL	OEM
DGM185AJ	OEM
DGM185AL	OEM
DGM185BJ	OEM
DGM185BK	OEM
DGM185BL	OEM
DGM187BA	OEM
DGM187BJ	OEM
DGM188AA	OEM
DGM188AJ	OEM
DGM188AK	OEM
DGM188AL	OEM
DGM188BA	OEM
DGM188BK	OEM
DGM188BL	OEM
DGM190BJ	OEM
DGM190BK	OEM
DGM190BL	OEM
DGM191AJ	OEM
DGM191AK	OEM
DGM191AL	OEM
DGM191BJ	OEM
DGM191BK	OEM
DGM191BL	OEM
DGMS181BA	OEM
DGMS181BK	OEM
DGMS181BL	OEM
DGMS182AA	OEM
DGMS182AJ	OEM
DGMS182AK	OEM
DGMS182AL	OEM
DGMS182BA	OEM
DGMS182BJ	OEM
DGMS182BK	OEM
DGMS182BL	OEM
DGMS184BJ	OEM
DGMS184BK	OEM
DGMS184BL	OEM
DGMS185AJ	OEM
DGMS185AK	OEM
DGMS185AL	OEM
DGMS185BJ	OEM
DGMS185BK	OEM
DGMS185BL	OEM
DGMS187BA	OEM
DGMS187BJ	OEM
DGMS187BK	OEM
DGMS187BL	OEM
DGMS188AA	OEM
DGMS188AJ	OEM
DGMS188AK	OEM
DGMS188AL	OEM
DGMS188BA	OEM
DGMS188BJ	OEM
DGMS188BK	OEM
DGMS188BL	OEM
DGMS191BJ	OEM
DGMS191BK	OEM
DGS4B10	0323
DH0006CH	OEM
DH0006CN	OEM
DH0006H	OEM
DH0008CH	OEM
DH0008CN	OEM
DH0008H	OEM
DH-001	0015
DH-002	0015
DH0011AH	OEM
DH0011CH	OEM
DH0011CN	OEM
DH0011H	OEM
DH0016CN	OEM
DH0017CN	OEM
DH0018CN	OEM
DH0028CH	OEM
DH0028CN	OEM
DH0034CD	OEM
DH0034CH	OEM
DH0034D	OEM
DH0034H	OEM
DH0069	OEM
DH1S2076	0133
DH1S2076A	0124
DH4R2	0015
DH4R4	OEM
DH4R6	OEM
DH14	0015

DEVICE TYPE	REPL CODE
DH14A	0015
DH16	0015
DH16A	0015
DH3467CD	OEM
DH3467CN	OEM
DH3468CD	OEM
DH3468CN	OEM
DH3724CD	OEM
DH3724CN	OEM
DH3725CD	OEM
DH3725CN	OEM
DHA16	OEM
DHA20	OEM
DHA6520	OEM
DHA6520B	OEM
DHA6520C	OEM
DHA6520D	OEM
DHA6522	OEM
DHA6522A	OEM
DHA6522B	OEM
DHA6522C	OEM
DHA6522D	OEM
DHA6523	OEM
DHA6523A	OEM
DHA6523B	OEM
DHA6523C	OEM
DHA6523D	OEM
DHA6524	OEM
DHA6524A	OEM
DHA6524B	OEM
DHA6524C	OEM
DHA6524D	OEM
DHA6525	OEM
DHA6525A	OEM
DHA6525B	OEM
DHA6525C	OEM
DHA6525D	OEM
DHB08	OEM
DHB10	OEM
DHB12	OEM
DHB14	OEM
DHB16	OEM
DHB18	OEM
DHD800	0133
DHD805	0133
DHD806	0133
DHD8001	0133
DHG3A120	OEM
DHG3B140	OEM
DHG3C160	OEM
DHG3D200	OEM
DHG3F250	OEM
DHG3G80	OEM
DHHZT33	3452
DHR21	OEM
DI-1	0143
DI-2	0143
DI-5	0760
DI-7	0015
DI7	0015
DI-8	0057
DI-8(COURIER)	0012
DI-8(DIODE)	0143
DI-20	0124
DI-42S	0015
DI42S	0015
DI-46	0015
DI46	0015
DI-52S	0015
DI52S	0015
DI-55	0015
DI-56	0015
DI56	0015
DI-71	0015
DI-72S	0015
DI72S	0015
DI-172S	0015
DI210	OEM
DI220	OEM
DI230	OEM
DI232	OEM
DI240	OEM
DI242	OEM
DI300	OEM
DI302	OEM
DI445	OEM
DI446	OEM
DI500B	OEM
DI502B	OEM
DI503B	OEM
DI504B	OEM
DI505B	OEM
DI507B	OEM
DI508B	OEM
DI509B	OEM
DI510B	OEM
DI512B	OEM
DI513B	OEM
DI514B	OEM
DI603A	OEM
DI604A	OEM
DI605A	OEM
DI-645	0015
DI645	0015
DI-646	0015
DI646	0015
DI-647	0015
DI-648	0015

DEVICE TYPE	REPL CODE
DI-649	0015
DI-650	0071
DI650	0071
DI-705	0015
DI803A	OEM
DI804A	OEM
DI805A	OEM
DI914-1B	OEM
DI914-1Q	OEM
DI914-1QM	OEM
DI914-2B	OEM
DI914-2Q	OEM
DI914-2QM	OEM
DI914-3B	OEM
DI914-3Q	OEM
DI914-3QM	OEM
DI-1649	0015
DI1649	0015
DI5117/5120	OEM
DI5118/5121	OEM
DI5119/5122	OEM
DI5119-1	OEM
DI5140	OEM
DI5180	OEM
DI5240	OEM
DI5280	OEM
DI-8020(A)	OEM
DIA	0244
DIAM01Z	0015
DIB	0196
DIB005	OEM
DIB01	OEM
DIB02	OEM
DIB04	OEM
DIB06	OEM
DIB08	OEM
DIB10	OEM
DIC	0196
DICA1	OEM
DICR1	0071
DICR2	OEM
DICR3	OEM
DICR4	OEM
DICR5	OEM
DICR6	OEM
DICR7	OEM
DICR8	OEM
DICR9	OEM
DICR10	OEM
DICR11	OEM
DICR12	OEM
DICR13	OEM
DICR14	OEM
DICR15	OEM
DID	0071
DIDS442X	0124
DIDS448-FA1	0124
DIE	0015
DII-203-200	OEM
DII-203-201	OEM
DII-203-300	OEM
DII-203-301	OEM
DIJ	0015
DIJ61224	0143
DIJ70485	0030
DIJ70486	0133
DIJ70488	0015
DIJ70542	0143
DIJ70543	0123
DIJ70544	0015
DIJ70545	0133
DIJ70644	0123
DIJ70645	0143
DIJ70646	0143
DIJ70695	0015
DIJ71273	0133
DIJ71711	0133
DIJ71776	0143
DIJ71777	0030
DIJ71778	0123
DIJ71779	0396
DIJ71895-1	1149
DIJ71958	0015
DIJ71959	0015
DIJ71960	0133
DIJ71961	0757
DIJ72163	0102
DIJ72164	0133
DIJ72165	0012
DIJ72166	0143
DIJ72167	0102
DIJ72168	0015
DIJ72169	0098
DIJ72170	0102
DIJ72171	0102
DIJ72172	1083
DIJ72174	0102
DIJ72290	0102
DIJ72291	0102
DIJ72292	0087
DIJ72293	0636
DIJ72294	0143
DIJ72653	OEM
DIJ72702	0554
DIK	0071
DIK40	OEM
DIL	0015
DIM	0157
DINS4	OEM

DEVICE TYPE	REPL CODE
DIO11	OEM
DIO-204-200	OEM
DIO-204-300	OEM
DIP630	OEM
DIP630PM	OEM
DIP631L	OEM
DIP631R	OEM
DIP632	OEM
DIP640	OEM
DIP640PM	OEM
DIP641L	OEM
DIP641R	OEM
DIP642	OEM
DIP650	OEM
DIP650PM	OEM
DIP651L	OEM
DIP651R	OEM
DIP652	OEM
DIP850	OEM
DIP850PM	OEM
DIP851L	OEM
DIP851R	OEM
DIP852	OEM
DIP1050	OEM
DIP1050A	OEM
DIP1050PM-1	OEM
DIP1051L	OEM
DIP1051R	OEM
DIP1052	OEM
DIPLINK-1A	OEM
DIPLINK-1B	OEM
DIR100	OEM
DIR170	OEM
DIS-1S	0015
DIS1S	0015
DIS990-060	OEM
DIV100HP	OEM
DIV100JP	OEM
DIV100KP	OEM
DJ74S472J	OEM
DK19	0143
DK20	0143
DK21	0143
DK48	OEM
DK130FA	OEM
DK130FB	OEM
DK130FX	OEM
DK130FY	OEM
DK150FA	OEM
DK150FB	OEM
DK150FX	OEM
DK150FY	OEM
DK170FA	OEM
DK170FB	OEM
DK170FX	OEM
DK170FY	OEM
DK750	OEM
DK751	OEM
DK752	OEM
DK753	OEM
DK754	OEM
DK755	OEM
DK756	OEM
DK757	OEM
DK758	OEM
DK1002FA	OEM
DK1002FQ	OEM
DK1002FW	OEM
DK1002FX	OEM
DK1002FZ	OEM
DK1004FA	OEM
DK1004FW	OEM
DK1004FX	OEM
DK1004FZ	OEM
DK1006FA	OEM
DK1006FQ	OEM
DK1006FW	OEM
DK1006FX	OEM
DK1006FZ	OEM
DK1008FA	OEM
DK1008FW	OEM
DK1008FX	OEM
DK1008FZ	OEM
DK1010FA	OEM
DK1010FW	OEM
DK1010FX	OEM
DK1010FZ	OEM
DK1012FA	OEM
DK1012FX	OEM
DK1012FZ	OEM
DK1014FZ	OEM
DK1102FA	OEM
DK1102FQ	OEM
DK1102FW	OEM
DK1102FX	OEM
DK1102FZ	OEM
DK1104FA	OEM
DK1104FQ	OEM
DK1104FW	OEM
DK1104FX	OEM
DK1104FZ	OEM
DK1106FA	OEM
DK1106FB	OEM
DK1106FQ	OEM
DK1106FW	OEM
DK1106FX	OEM
DK1106FZ	OEM
DK1108FA	OEM
DK1108FW	OEM
DK1108FX	OEM
DK1108FZ	OEM
DK1110FA	OEM
DK1110FX	OEM
DK1110FZ	OEM
DK1112FZ	OEM
DK1301FA	OEM
DK1301FQ	OEM
DK1302FA	OEM
DK1302FQ	OEM
DK1302FW	OEM
DK1302FX	OEM
DK1302FZ	OEM
DK1304FA	OEM
DK1304FQ	OEM
DK1304FW	OEM
DK1304FX	OEM
DK1304FZ	OEM
DK1306FA	OEM
DK1306FQ	OEM
DK1306FW	OEM
DK1306FX	OEM
DK1306FZ	OEM
DK1308FA	OEM
DK1308FW	OEM
DK1308FX	OEM
DK1308FZ	OEM
DK1310FA	OEM
DK1310FX	OEM
DK1310FZ	OEM
DK1312FZ	OEM
DK1408FX	OEM
DK2401FW	OEM
DK2401FX	OEM
DK2401FY	OEM
DK2401FZ	OEM
DK2402FA	OEM
DK2402FB	OEM
DK2402FQ	OEM
DK2402FW	OEM
DK2402FX	OEM
DK2402FY	OEM
DK2402FZ	OEM
DK2404FA	OEM
DK2404FB	OEM
DK2404FQ	OEM
DK2404FW	OEM
DK2404FX	OEM
DK2404FY	OEM
DK2404FZ	OEM
DK2406FA	OEM
DK2406FB	OEM
DK2406FQ	OEM
DK2406FW	OEM
DK2406FX	OEM
DK2406FY	OEM
DK2408FA	OEM
DK2408FB	OEM
DK2408FW	OEM
DK2408FX	OEM
DK2408FY	OEM
DK2408FZ	OEM
DK2410FA	OEM
DK2410FB	OEM
DK2410FX	OEM
DK2410FY	OEM
DK2412FB	OEM
DK2502FA	OEM
DK2502FB	OEM
DK2502FQ	OEM
DK2502FW	OEM
DK2502FX	OEM
DK2502FY	OEM
DK2504FA	OEM
DK2504FB	OEM
DK2504FQ	OEM
DK2504FW	OEM
DK2504FX	OEM
DK2504FY	OEM
DK2506FA	OEM
DK2506FB	OEM
DK2506FQ	OEM
DK2506FW	OEM
DK2506FX	OEM
DK2506FY	OEM
DK2508FA	OEM
DK2508FB	OEM
DK2508FW	OEM
DK2508FX	OEM
DK2508FY	OEM
DK2510FA	OEM
DK2510FB	OEM
DK2510FX	OEM
DK2510FY	OEM
DK2512FB	OEM
DK2702FA	OEM
DK2702FQ	OEM
DK2702FX	OEM
DK2702FZ	OEM
DK2704FA	OEM
DK2704FQ	OEM
DK2704FX	OEM
DK2704FZ	OEM
DK2706FA	OEM
DK2706FQ	OEM
DK2706FW	OEM
DK2706FX	OEM
DK2706FY	OEM
DK2706FZ	OEM
DK2708FA	OEM
DK2708FB	OEM
DK2708FX	OEM
DK2708FY	OEM
DK2708FZ	OEM
DK2710FB	OEM
DK2710FZ	OEM
DK7510	OEM
DK7512	OEM
DK7515	OEM
DKDS1555	0133
DKEK-04	1743
DKLR114E	1615
DKLR124E	OEM
DKV6510	OEM
DKV6510A	OEM
DKV6510B	OEM
DKV6515	OEM
DKV6515A	OEM
DKV6515B	OEM
DKV6516	OEM
DKV6516A	OEM
DKV6516B	OEM
DKV6517	OEM
DKV6517A	OEM
DKV6517B	OEM
DKV6518	OEM
DKV6518A	OEM
DKV6518B	OEM
DKV6520	OEM
DKV6520A	OEM
DKV6520B	OEM
DKV6520C	OEM
DKV6520D	OEM
DKV6522	OEM
DKV6522A	OEM
DKV6522B	OEM
DKV6522C	OEM
DKV6522D	OEM
DKV6523	OEM
DKV6523A	OEM
DKV6523B	OEM
DKV6523C	OEM
DKV6523D	OEM
DKV6524	OEM
DKV6524A	OEM
DKV6524B	OEM
DKV6524C	OEM
DKV6524D	OEM
DKV6525	OEM
DKV6525A	OEM
DKV6525B	OEM
DKV6525C	OEM
DKV6525D	OEM
DKV6526	OEM
DKV6526A	OEM
DKV6526B	OEM
DKV6526C	OEM
DKV6526D	OEM
DKV6527	OEM
DKV6527A	OEM
DKV6527B	OEM
DKV6527C	OEM
DKV6527D	OEM
DKV6533	OEM
DKV6533A	0005
DKV6533B	OEM
DKV6533C	OEM
DKV6533D	OEM
DKV6533E	OEM
DKV6533F	OEM
DKV6534	OEM
DKV6534A	OEM
DKV6534B	OEM
DKV6534C	OEM
DKV6534D	OEM
DKV6534E	OEM
DKV6534F	OEM
DKV6550A	OEM
DKV6550B	OEM
DKV6550C	OEM
DKV6550D	OEM
DKV6550E	OEM
DL1CL3	OEM
DL1CL3G	OEM
DL1F	OEM
DL1FG	OEM
DL1KL	OEM
DL1KLG	OEM
DL1L	OEM
DL-1L3	OEM
DL-1LGN	OEM
DL-1LRN	OEM
DL1ML	OEM
DL-11S2	OEM
DL11S2	OEM
DL11S2G	OEM
DL-11S2R1	OEM
DL11S2R1	OEM
DL11S2RS	OEM
DL-25G	OEM
DL25G	OEM
DL-25R	OEM
DL25R	OEM
DL57	OEM
DL59	OEM
DL201	OEM
DL202	OEM
DL203	OEM
DL204	OEM
DL330M	OEM
DL340M	OEM
DL401	OEM
DL402	OEM
DL403	OEM
DL404	OEM
DL416	OEM
DL430M	OEM
DL440M	OEM
DL601	OEM
DL602	OEM
DL603	OEM
DL604	OEM
DL801	OEM
DL802	OEM
DL803	OEM
DL804	OEM
DL-1152R1	OEM
DL1414	OEM
DL1416	OEM
DL2416	OEM
DL3400	OEM
DL3401	OEM
DL3403	OEM
DL3405	OEM
DL3406	OEM
DL3416	OEM
DL3416H	OEM
DL3422	OEM
DL3531	OEM
DL4509	OEM
DL4530	OEM
DL4770	OEM
DL5735	OEM
DL7750	OEM
DL7751	OEM
DL7756	OEM
DL7760	OEM
DLA5544	OEM
DLA5544A	OEM
DLA5544B	OEM
DLA5544C	OEM
DLA5544D	OEM
DLA5544E	OEM
DLA5544F	OEM
DLA6929	OEM
DLA6929A	OEM
DLA6929B	OEM
DLB156	2189
DLD271	0516
DLG7670	OEM
DLG7671	OEM
DLG7673	OEM
DLG7676	OEM
DL-IIS2R1	OEM
DL-ILR	OEM
DL03900	OEM
DL03901	OEM
DL03903	OEM
DL03905	OEM
DL03906	OEM
DL04770	OEM
DL07610	OEM
DL07611	OEM
DL07613	OEM
DL07614	OEM
DL07616	OEM
DL07650	OEM
DL07651	OEM
DL07653	OEM
DL07656	OEM
DLP238	OEM
DLS1585	OEM
DLS1586	OEM
DLV11E	OEM
DLV11F	OEM
DLV11J	OEM
DLV11KA	OEM
DLXAAEL0196	OEM
DLY7660	OEM
DLY7661	OEM
DLY7663	OEM
DLY7666	OEM
DM01B	OEM
DM02B	OEM
DM03B	OEM
DM05A	OEM
DM06A	OEM
DM0700	OEM
DM0A90A-M	0019
DM-2	OEM
DM-3	OEM
DM3-7E	OEM
DM-4	OEM
DM4	OEM
DM-7	OEM
DM7S65N	OEM
DM-8	OEM
DM-9	1759
DM-10	OEM
DM10	OEM
DM10-28B1	OEM
DM-11	0678
DM11	0678
DM-11A	0678
DM-12	OEM
DM-14	0654
DM14	0654
DM-19	3410
DM-20	1421
DM20-28B1	OEM
DM-22	OEM
DM22	OEM
DM-24	1940
DM24	1940
DM-25	OEM
DM2503W	OEM
DM-26	2279
DM28	OEM
DM-29	OEM
DM-30	2279
DM-31	0678
DM31	0678
DM-32	1421
DM32	1421
DM-33	1196
DM-35	3422
DM-36	0520
DM36	OEM
DM-37	3423
DM-39	OEM
DM-40	0360
DM40-12BA	OEM
DM40-28BY	OEM
DM-41	1443
DM41X40B	OEM
DM41X40C	OEM
DM-43	OEM
DM-44	1395
DM44	1395
DM-45	OEM
DM-46	OEM
DM-47	OEM
DM-48	OEM
DM-50	3425
DM-51	2755
DM-54	1940
DM54	1940
DM54H00J	OEM
DM54H01J	OEM
DM54H08J	OEM
DM54H10J	OEM
DM54H11J	OEM
DM54H20J	OEM
DM54H21J	OEM
DM54H30J	OEM
DM54H40J	OEM
DM54H50J	OEM
DM54H62J	OEM
DM54H74J	OEM
DM54L002	OEM
DM54L00J	OEM
DM54L02J	OEM
DM54L02W	OEM
DM54L03J	OEM
DM54L10W	OEM
DM54L20J	OEM
DM54L20W	OEM
DM54L30J	OEM
DM54L42AJ	OEM
DM54L42AW	OEM
DM54L51J	OEM
DM54L54J	OEM
DM54L54W	OEM
DM54L55J	OEM
DM54L55W	OEM
DM54L71J	OEM
DM54L71W	OEM
DM54L72J	OEM
DM54L72W	OEM
DM54L73J	OEM
DM54L73W	OEM
DM54L74J	OEM
DM54L74W	OEM
DM54L75AJ	OEM
DM54L75AW	OEM
DM54L78J	OEM
DM54L78W	OEM
DM54L85J	OEM
DM54L85W	OEM
DM54L86J	OEM
DM54L86W	OEM
DM54L89AJ	OEM
DM54L89AW	OEM
DM54L90J	OEM
DM54L90W	OEM
DM54L91J	OEM
DM54L91W	OEM
DM54L93J	OEM
DM54L93W	OEM
DM54L95J	OEM
DM54L95W	OEM
DM54L98J	OEM
DM54L98W	OEM
DM54L154AF	OEM
DM54L154AJ	OEM
DM54L164AJ	OEM
DM54L164AW	OEM
DM54L165AJ	OEM
DM54L165AW	OEM
DM54L192J	OEM
DM54L192W	OEM
DM54L193J	OEM
DM54L193W	OEM
DM54LS13J	OEM
DM54LS13N	OEM
DM54LS13W	OEM
DM54LS14J	OEM
DM54LS14N	OEM
DM54LS14W	OEM
DM54LS77W	OEM
DM54LS83AJ	OEM
DM54LS83AW	OEM
DM54LS85J	OEM
DM54LS85W	OEM
DM54LS90J	OEM
DM54LS90W	OEM
DM54LS92J	OEM
DM54LS92W	OEM
DM54LS93J	OEM
DM54LS93W	OEM
DM54LS125AJ	OEM
DM54LS125AN	OEM
DM54LS125AW	OEM
DM54LS126AJ	OEM
DM54LS126AW	OEM
DM54LS132J	OEM
DM54LS132N	OEM
DM54LS132W	OEM
DM54LS154F	OEM
DM54LS154J	OEM
DM54LS155J	OEM
DM54LS155W	OEM
DM54LS156J	OEM
DM54LS156W	OEM
DM54LS160AJ	OEM
DM54LS160AW	OEM
DM54LS161AJ	OEM
DM54LS161AW	OEM
DM54LS162AJ	OEM
DM54LS162AW	OEM
DM54LS163AJ	OEM
DM54LS163AW	OEM
DM54LS164J	OEM
DM54LS164W	OEM
DM54LS165J	OEM
DM54LS165W	OEM
DM54LS166J	OEM
DM54LS166W	OEM
DM54LS170J	OEM
DM54LS170W	OEM
DM54LS173J	OEM
DM54LS173W	OEM
DM54LS174J	OEM
DM54LS174W	OEM
DM54LS175J	OEM
DM54LS175W	OEM
DM54LS190J	OEM
DM54LS190W	OEM
DM54LS191J	OEM
DM54LS191W	OEM
DM54LS192J	OEM
DM54LS192W	OEM
DM54LS193J	OEM
DM54LS193W	OEM
DM54LS194AJ	OEM
DM54LS194AW	OEM
DM54LS195AJ	OEM
DM54LS195AW	OEM
DM54LS196J	OEM
DM54LS196W	OEM
DM54LS197J	OEM
DM54LS197W	OEM
DM54LS240J	OEM
DM54LS241J	OEM
DM54LS242J	OEM
DM54LS243J	OEM
DM54LS244J	OEM
DM54LS245J	OEM
DM54LS245N	OEM
DM54LS251J	OEM
DM54LS251W	OEM
DM54LS253J	OEM
DM54LS253W	OEM
DM54LS255BJ	OEM
DM54LS255BW	OEM
DM54LS257BJ	OEM
DM54LS257BW	OEM
DM54LS259J	OEM
DM54LS259W	OEM
DM54LS283J	OEM
DM54LS283W	OEM
DM54LS290J	OEM
DM54LS290W	OEM
DM54LS293J	OEM
DM54LS293W	OEM
DM54LS352J	OEM
DM54LS352N	OEM
DM54LS352W	OEM
DM54LS353J	OEM
DM54LS353N	OEM
DM54LS353W	OEM
DM54LS365AJ	OEM
DM54LS365AN	OEM
DM54LS365AW	OEM
DM54LS366AJ	OEM
DM54LS366AN	OEM
DM54LS366AW	OEM
DM54LS367AJ	OEM
DM54LS367AN	OEM
DM54LS367AW	OEM
DM54LS368AJ	OEM
DM54LS368AN	OEM
DM54LS368AW	OEM
DM54LS373J	OEM
DM54LS373W	OEM
DM54LS374J	OEM
DM54LS374W	OEM
DM54LS390J	OEM
DM54LS390W	OEM
DM54LS393J	OEM
DM54LS393W	OEM
DM54LS670J	OEM
DM54LS670N	OEM
DM54LS670W	OEM
DM54LS952	OEM
DM54LS962	OEM
DM54S160J	OEM
DM54S160W	OEM
DM54S162J	OEM
DM54S162W	OEM
DM54S163J	OEM
DM54S163W	OEM
DM54S174J	OEM
DM54S174W	OEM
DM54S175J	OEM
DM54S175W	OEM
DM54S181J	OEM
DM54S182J	OEM
DM54S188J	OEM
DM54S194J	OEM
DM54S194W	OEM
DM54S195J	OEM
DM54S195W	OEM
DM54S196J	OEM
DM54S196W	OEM
DM54S197J	OEM
DM54S197W	OEM
DM54S240J	OEM
DM54S241J	OEM
DM54S242J	OEM
DM54S243J	OEM
DM54S244J	OEM
DM54S251J	OEM
DM54S251W	OEM
DM54S253J	OEM
DM54S253W	OEM
DM54S257J	OEM
DM54S257W	OEM
DM54S258J	OEM
DM54S258W	OEM
DM54S280J	OEM
DM54S280W	OEM
DM54S283J	OEM
DM54S283W	OEM
DM54S287J	OEM
DM54S288J	OEM
DM54S299J	OEM
DM54S373J	OEM
DM54S373W	OEM
DM54S381J	OEM
DM54S387J	OEM
DM54S472AJ	OEM
DM54S472J	OEM
DM54S473AJ	OEM
DM54S473J	OEM
DM54S474AJ	OEM
DM54S474J	OEM
DM54S475AJ	OEM
DM54S475J	OEM
DM54S570J	OEM
DM54S571J	OEM
DM54S572J	OEM
DM54S573AJ	OEM
DM54S573J	OEM
DM54S940J	OEM
DM54S941J	OEM
DM-56	OEM
DM-59	3432
DM64X64A	OEM
DM64X64B	OEM
DM-65	0845
DM-68	OEM
DM-70	OEM
DM-71	OEM
DM71L22J	OEM
DM71L22W	OEM
DM71L23J	OEM
DM71L23W	OEM
DM-72	OEM
DM-73	OEM
DM-74	OEM
DM74ALS	OEM
DM74ALS00N	3774
DM74ALS02N	OEM
DM74ALS04AN	OEM
DM74ALS08N	OEM
DM74ALS10N	OEM
DM74ALS27N	OEM
DM74ALS32N	3775
DM74ALS74AN	OEM
DM74ALS138N	5763
DM74ALS175N	OEM
DM74ALS244N	OEM
DM74ALS245AN	OEM
DM74ALS573N	OEM
DM74H00J	0682
DM74H00N	0682
DM74H01J	5338
DM74H01N	5338
DM74H04J	1907
DM74H04N	1907
DM74H05	3251
DM74H05J	3251
DM74H05L	3251

If replacement code is OEM, contact original manufacturer for replacement.

DEVICE TYPE	REPL CODE
DM74H05N	3251
DM74H08J	5355
DM74H08N	5355
DM74H10J	0685
DM74H10N	0685
DM74H11J	2398
DM74H11N	2398
DM74H20J	3715
DM74H20N	3715
DM74H21J	4856
DM74H21N	4856
DM74H22J	4596
DM74H22N	4596
DM74H30J	5383
DM74H30N	5383
DM74H40J	0559
DM74H40N	0559
DM74H50J	1792
DM74H50N	1792
DM74H51J	1944
DM74H51N	1944
DM74H52J	2020
DM74H52N	2020
DM74H53J	2101
DM74H53N	2101
DM74H54J	2172
DM74H54N	2172
DM74H55J	3158
DM74H55N	3158
DM74H60J	5410
DM74H60N	5410
DM74H61J	2663
DM74H61N	2663
DM74H62J	2731
DM74H62N	2731
DM74H71J	3264
DM74H71N	3264
DM74H72J	3313
DM74H72N	3313
DM74H73J	2463
DM74H73N	2463
DM74H74J	2492
DM74H74N	2492
DM74H76J	5302
DM74H76N	5302
DM74H78J	5418
DM74H78N	5418
DM74L00N	OEM
DM74L02N	OEM
DM74L03N	OEM
DM74L04	OEM
DM74L10N	OEM
DM74L20N	OEM
DM74L30N	OEM
DM74L42AN	OEM
DM74L51N	OEM
DM74L54N	OEM
DM74L55N	OEM
DM74L71N	OEM
DM74L72N	OEM
DM74L73N	OEM
DM74L74N	OEM
DM74L78N	OEM
DM74L85N	OEM
DM74L85W	OEM
DM74L86N	OEM
DM74L89AN	OEM
DM74L90N	OEM
DM74L91N	OEM
DM74L91W	OEM
DM74L93F	0656
DM74L93J	0656
DM74L93N	0656
DM74L95N	OEM
DM74L98N	OEM
DM74L123N	OEM
DM74L154AN	OEM
DM74L164AN	OEM
DM74L165AN	OEM
DM74L192N	OEM
DM74L193N	OEM
DM74LS00N	1527
DM74LS01N	1546
DM74LS02N	1559
DM74LS03N	1579
DM74LS04M	OEM
DM74LS04N	1595
DM74LS05N	1608
DM74LS08N	1633
DM74LS09N	1642
DM74LS10N	1662
DM74LS11N	1667
DM74LS12N	1679
DM74LS13J	1688
DM74LS13N	1688
DM74LS13W	1688
DM74LS14J	1698
DM74LS14N	1698
DM74LS14W	1698
DM74LS20N	0034
DM74LS26N	1382
DM74LS27N	0183
DM74LS32N	0088
DM74LS38N	1839
DM74LS40N	0135
DM74LS42N	1841
DM74LS47N	1845
DM74LS48N	1849
DM74LS51	1033
DM74LS51N	1033
DM74LS55N	0457
DM74LS74AN	0243
DM74LS75N	1870
DM74LS83AN	2218
DM74LS85N	0431
DM74LS86N	0291
DM74LS92N	1887
DM74LS93N	1888
DM74LS109AN	1906
DM74LS112AN	2128
DM74LS123N	7068
DM74LS125AN	0075
DM74LS126AJ	2877
DM74LS126AN	2877
DM74LS126AW	2877
DM74LS132J	1625
DM74LS132N	1625
DM74LS132W	1625
DM74LS138N	0427
DM74LS139N	0153
DM74LS151	5136
DM74LS151N	1646
DM74LS153N	0959
DM74LS154N	5045
DM74LS155N	0209
DM74LS156N	1654
DM74LS157N	5801
DM74LS158N	OEM
DM74LS160AN	0837
DM74LS161AN	0858
DM74LS161N	5166
DM74LS162AN	0880
DM74LS163AN	0893
DM74LS163N	0893
DM74LS164N	4339
DM74LS165N	4356
DM74LS166N	4371
DM74LS170N	2629
DM74LS173N	5217
DM74LS174M	OEM
DM74LS174N	0260
DM74LS175N	1672
DM74LS190N	1686
DM74LS191N	1687
DM74LS191NB	OEM
DM74LS192N	1689
DM74LS193N	1692
DM74LS194AN	1304
DM74LS195AN	1315
DM74LS196	2834
DM74LS196N	2834
DM74LS197N	2469
DM74LS221N	7130
DM74LS240N	0452
DM74LS240WM	OEM
DM74LS241N	1726
DM74LS242N	1728
DM74LS243N	0906
DM74LS244N	0458
DM74LS244WM	OEM
DM74LS245N	0463
DM74LS247N	1732
DM74LS251N	1737
DM74LS253N	1739
DM74LS257BN	1744
DM74LS258BN	1746
DM74LS259N	3205
DM74LS279N	3290
DM74LS283N	1779
DM74LS290N	4425
DM74LS293N	0082
DM74LS352J	0773
DM74LS352N	0761
DM74LS352W	0761
DM74LS353J	0773
DM74LS353N	0773
DM74LS353W	0773
DM74LS365AJ	0943
DM74LS365AN	0943
DM74LS365AW	0943
DM74LS366AJ	0956
DM74LS366AN	0956
DM74LS366AW	0956
DM74LS367AJ	0977
DM74LS367AN	0977
DM74LS367AW	0977
DM74LS367N	0977
DM74LS368AJ	0991
DM74LS368AN	0991
DM74LS368AW	0991
DM74LS373N	0709
DM74LS374N	0713
DM74LS390N	1288
DM74LS393N	0819
DM74LS670J	1130
DM74LS670N	1130
DM74LS670W	1130
DM74LS952	OEM
DM74LS962	OEM
DM74S00J	0704
DM74S00N	0704
DM74S03J	2217
DM74S03N	2217
DM74S04J	2263
DM74S04N	2263
DM74S05J	2321
DM74S05N	2321
DM74S08N	OEM
DM74S10J	2445
DM74S10N	2445
DM74S11J	2447
DM74S11N	2447
DM74S15N	2451
DM74S20J	1017
DM74S20N	1017
DM74S22J	2461
DM74S22N	2461
DM74S40N	2475
DM74S51N	4305
DM74S64N	2496
DM74S65N	2497
DM74S74N	2504
DM74S86N	2511
DM74S112N	1617
DM74S113N	1623
DM74S114N	1629
DM74S133N	OEM
DM74S135N	OEM
DM74S138N	OEM
DM74S139N	OEM
DM74S140N	1886
DM74S151N	1955
DM74S153N	2151
DM74S157N	1695
DM74S158N	OEM
DM74S160N	OEM
DM74S161N	OEM
DM74S162N	OEM
DM74S163N	0893
DM74S174N	2132
DM74S175N	2141
DM74S181N	2165
DM74S182N	2166
DM74S188J	3168
DM74S188N	3168
DM74S194N	1931
DM74S195N	OEM
DM74S196N	OEM
DM74S197N	OEM
DM74S240J	OEM
DM74S240N	OEM
DM74S241J	OEM
DM74S241N	OEM
DM74S242N	OEM
DM74S243N	OEM
DM74S244N	OEM
DM74S251N	2198
DM74S253N	OEM
DM74S257N	OEM
DM74S258N	2205
DM74S280N	2220
DM74S283N	OEM
DM74S287J	2224
DM74S287N	2224
DM74S288J	2175
DM74S288N	2175
DM74S299N	OEM
DM74S373N	2264
DM74S374N	2266
DM74S381N	OEM
DM74S387J	1918
DM74S387N	1918
DM74S472AJ	2320
DM74S472AN	2320
DM74S472N	2320
DM74S473AJ	3288
DM74S473AN	3288
DM74S473J	3288
DM74S473N	3288
DM74S474AJ	2322
DM74S474AN	2322
DM74S474J	2322
DM74S474N	2322
DM74S475AJ	2324
DM74S475AN	2324
DM74S475J	2324
DM74S475N	2324
DM74S570J	2388
DM74S570N	2388
DM74S571J	2389
DM74S571N	2389
DM74S572AJ	2390
DM74S572AN	2390
DM74S572J	2390
DM74S572N	2390
DM74S573AJ	2392
DM74S573AN	2392
DM74S573J	2392
DM74S573N	2392
DM74S940J	OEM
DM74S940N	OEM
DM74S941N	OEM
DM-75	OEM
DM75L11D	OEM
DM75L11F	OEM
DM75L11J	OEM
DM75L11N	OEM
DM75L11W	OEM
DM75L12F	OEM
DM75L12J	OEM
DM75L12N	OEM
DM75L12W	OEM
DM75L51J	OEM
DM75L51W	OEM
DM75L52F	OEM
DM75L52J	OEM
DM75L52N	OEM
DM75L52W	OEM
DM75L54F	OEM
DM75L54J	OEM
DM75L54N	OEM
DM75L54W	OEM
DM75L60J	OEM
DM75L60W	OEM
DM75L63J	OEM
DM75L63W	OEM
DM-76	OEM
DM76L13J	OEM
DM76L13W	OEM
DM76L24J	OEM
DM76L24N	OEM
DM76L75F	OEM
DM76L75J	OEM
DM76L75N	OEM
DM76L75W	OEM
DM76L76F	OEM
DM76L76J	OEM
DM76L76N	OEM
DM76L76W	OEM
DM76L90J	OEM
DM76L90W	OEM
DM76L93F	OEM
DM76L93J	OEM
DM76L93N	OEM
DM76L97J	OEM
DM76L97W	OEM
DM76L99J	OEM
DM76L99W	OEM
DM76LS52	OEM
DM76LS62	OEM
DM76S64J	OEM
DM76S128J	OEM
DM-77	0119
DM77S180J	OEM
DM77S181J	OEM
DM77S184J	OEM
DM77S185J	OEM
DM77S188J	OEM
DM77S190J	OEM
DM77S191J	OEM
DM77S288J	OEM
DM-78	OEM
DM-79	OEM
DM-80	OEM
DM80L06N	OEM
DM80X80C	OEM
DM-81	OEM
DM81L22N	OEM
DM81L23N	OEM
DM-82	OEM
DM-83	OEM
DM-84	3432
DM-85	3444
DM85L11D	OEM
DM85L11F	OEM
DM85L11N	OEM
DM85L12F	OEM
DM85L12J	OEM
DM85L12N	OEM
DM85L51N	OEM
DM85L52F	OEM
DM85L52J	OEM
DM85L52N	OEM
DM85L54F	OEM
DM85L54J	OEM
DM85L54N	OEM
DM85L60N	OEM
DM85L63N	OEM
DM85S68D	OEM
DM85S68N	OEM
DM-86	1143
DM86L13N	OEM
DM86L24J	OEM
DM86L75F	OEM
DM86L75J	OEM
DM86L75N	OEM
DM86L76F	OEM
DM86L76J	OEM
DM86L76N	OEM
DM86L90N	OEM
DM86L93F	OEM
DM86L93J	OEM
DM86L93N	OEM
DM86L97N	OEM
DM86L99N	OEM
DM86LS52	OEM
DM86LS62	OEM
DM86S64J	OEM
DM86S128J	OEM
DM86S128N	OEM
DM-87	0176
DM87	0176
DM87S180J	OEM
DM87S180N	OEM
DM87S181J	OEM
DM87S181N	OEM
DM87S184J	OEM
DM87S184N	OEM
DM87S185J	3166
DM87S185N	3166
DM87S188J	OEM
DM87S188N	OEM
DM87S190J	OEM
DM87S190N	OEM
DM87S191J	OEM
DM87S191N	OEM
DM87S288J	OEM
DM87S288N	OEM
DM-88	2257
DM88	2257
DM-89	OEM
DM-90	2782
DM90	2782
DM-91	1196
DM91	1196
DM-92	1196
DM-94	0624
DM-96	OEM
DM-97	OEM
DM-98	3432
DM101A	OEM
DM102BSXXXX	OEM
DM-103	0626
DM-104	1292
DM104BSXXXX	OEM
DM105DXXXX	OEM
DM-106	0083
DM106	0083
DM106A	OEM
DM-107	2310
DM-108	OEM
DM-110	4284
DM110DSXXXX	OEM
DM112BSXXXX	OEM
DM128X24A	OEM
DM128X128C	OEM
DM150P	OEM
DM204BSXXXX	OEM
DM208FS	OEM
DM208FS-1	OEM
DM208FSM	OEM
DM208FSM-1	OEM
DM208FSMZS	OEM
DM208FSMZS-1	OEM
DM208FSQ	OEM
DM208FSQ-1	OEM
DM208FSQZS	OEM
DM208FSQZS-1	OEM
DM208FST	OEM
DM208FST-1	OEM
DM208FSTZS	OEM
DM208FSTZS-1	OEM
DM212BSXXXX	OEM
DM218BS	OEM
DM218BS-1	OEM
DM218BSM	OEM
DM218BSM-1	OEM
DM218BSMZS	OEM
DM218BSMZS-1	OEM
DM218BSQ	OEM
DM218BSQZS	OEM
DM218BSQZS-2	OEM
DM218BST	OEM
DM218BST-1	OEM
DM218BSTZS	OEM
DM218BSTZS-1	OEM
DM218BSZS	OEM
DM218BSZS-1	OEM
DM250	OEM
DM256X64A	OEM
DM256X256B	OEM
DM258X26B	OEM
DM350	OEM
DM400B	OEM
DM408BSXXXX	OEM
DM450	OEM
DM510DSXXXX	OEM
DM520DSXXXX	OEM
DM550	OEM
DM650	OEM
DM711BSXXXX	OEM
DM750	OEM
DM812BSXXXX	OEM
DM816BSXXXX	OEM
DM852N	OEM
DM930N	1823
DM932N	1041
DM933N	2097
DM935N	1176
DM936N	1831
DM937N	1835
DM944N	0033
DM945N	0081
DM946N	0141
DM948N	3399
DM949N	1844
DM950	OEM
DM958N	0466
DM961N	1859
DM962N	0562
DM963N	0341
DM1724BSXXXX	OEM
DM1800N	OEM
DM1801N	OEM
DM1883A	OEM
DM1883B	OEM
DM2223BSXXXX	OEM
DM2502CJ	OEM
DM2502CN	OEM
DM2502CW	OEM
DM2502J	OEM
DM2502W	OEM
DM2503CJ	OEM
DM2503CN	OEM
DM2503CW	OEM
DM2503J	OEM
DM2504CF	OEM
DM2504CN	OEM
DM2504F	OEM
DM2504J	OEM
DM3000	OEM
DM3001	OEM
DM3510	OEM
DM3520	OEM
DM3530	OEM
DM3540	OEM
DM5400J	OEM
DM5400W	OEM
DM5401J	OEM
DM5401W	OEM
DM5402J	OEM
DM5402W	OEM
DM5403J	OEM
DM5408J	OEM
DM5408W	OEM
DM5409J	OEM
DM5409W	OEM
DM5410J	OEM
DM5410W	OEM
DM5411J	OEM
DM5413J	OEM
DM5413N	OEM
DM5413W	OEM
DM5414J	OEM
DM5414N	OEM
DM5414W	OEM
DM5416J	OEM
DM5416W	OEM
DM5417J	OEM
DM5417W	OEM
DM5420J	OEM
DM5420W	OEM
DM5423J	OEM
DM5423W	OEM
DM5425J	OEM
DM5425W	OEM
DM5427J	OEM
DM5427W	OEM
DM5430J	OEM
DM5430W	OEM
DM5432J	OEM
DM5432W	OEM
DM5437J	OEM
DM5437W	OEM
DM5438J	OEM
DM5438W	OEM
DM5440J	OEM
DM5440W	OEM
DM5441AJ	OEM
DM5441AW	OEM
DM5442J	OEM
DM5442W	OEM
DM5445J	OEM
DM5445W	OEM
DM5446AJ	OEM
DM5446AN	OEM
DM5446AW	OEM
DM5447AJ	OEM
DM5447AN	OEM
DM5447AW	OEM
DM5448J	OEM
DM5448N	OEM
DM5448W	OEM
DM5450J	OEM
DM5450W	OEM
DM5451J	OEM
DM5451W	OEM
DM5453J	OEM
DM5453W	OEM
DM5454J	OEM
DM5454W	OEM
DM5459BSXXXX	OEM
DM5460J	OEM
DM5460W	OEM
DM5470J	OEM
DM5470W	OEM
DM5472J	OEM
DM5472W	OEM
DM5473J	OEM
DM5473W	OEM
DM5474J	OEM
DM5474W	OEM
DM5475J	OEM
DM5475W	OEM
DM5476J	OEM
DM5476W	OEM
DM5483J	OEM
DM5483W	OEM
DM5485J	OEM
DM5485W	OEM
DM5486J	OEM
DM5486W	OEM
DM5489BJ	OEM
DM5489BN	OEM
DM5490AJ	OEM
DM5490J	OEM
DM5490W	OEM
DM5492AJ	OEM
DM5492AW	OEM
DM5492J	OEM
DM5492W	OEM
DM5493AJ	OEM
DM5493AW	OEM
DM5493J	OEM
DM5493W	OEM
DM5495J	OEM
DM5495W	OEM
DM5496J	OEM
DM5496W	OEM
DM6003	OEM
DM6004	OEM
DM6009	OEM
DM6010	OEM
DM6015	OEM
DM6016	OEM
DM6020	OEM
DM6021	OEM
DM6022	OEM
DM6030	OEM
DM6040	OEM
DM6050	OEM
DM7091J	OEM
DM7091N	OEM
DM7091W	OEM
DM7092J	OEM
DM7092N	OEM
DM7092W	OEM
DM7121J	OEM
DM7121W	OEM
DM7123J	OEM
DM7123W	OEM
DM7130D	OEM
DM7130F	OEM
DM7130J	OEM
DM7131J	OEM
DM7131N	OEM
DM7131W	OEM
DM7136J	OEM
DM7136W	OEM
DM7145J	OEM
DM7160D	OEM
DM7160J	OEM
DM7160W	OEM
DM7200J	OEM
DM7200N	OEM
DM7200W	OEM
DM7214J	OEM
DM7214W	OEM
DM7219F	OEM
DM7219J	OEM
DM7220J	OEM
DM7220N	OEM
DM7220W	OEM
DM7280J	OEM
DM7280W	OEM
DM7281J	OEM
DM7281W	OEM
DM7373N	1172
DM7400	0232
DM7400J	0232
DM7400N	0232
DM7401J	0268
DM7401N	0268
DM7402J	0314
DM7402N	0314
DM7403J	0335
DM7403N	0335
DM7404N	0361
DM7405N	0385
DM7406N	1206
DM7407	6069
DM7407N	1339
DM7408J	0467
DM7408N	0467
DM7409J	0492
DM7409N	0492
DM7410J	0512
DM7410N	0512
DM7411J	0527
DM7411N	0527
DM7413J	1441
DM7413N	1441
DM7413W	OEM
DM7414J	2243
DM7414N	2243
DM7414W	OEM
DM74150	1492
DM7416N	1349
DM7417N	1352
DM7420J	0697
DM7420N	0697
DM7423J	3466
DM7423N	3466
DM7423W	3466
DM7425J	3477
DM7425N	3477
DM7425W	3477
DM7426N	0803
DM7427J	0818
DM7427N	0818
DM7427W	0818
DM7430J	0873
DM7430N	0873
DM7432J	0899
DM7432N	0899
DM7437J	3520
DM7437N	3520
DM7437W	3520
DM7438J	0996
DM7438N	0996
DM7438W	0996
DM7440J	1024
DM7440N	1024

DEVICE TYPE	REPL CODE
DM7441AJ	1038
DM7441AN	1038
DM7441N	1038
DM7442J	1052
DM7442N	1052
DM7445J	1082
DM7445N	1082
DM7445W	1082
DM7446AJ	1098
DM7446AN	1098
DM7446AW	OEM
DM7447AJ	1108
DM7447AN	1108
DM7447AW	OEM
DM7447N	1108
DM7448J	1125
DM7448N	1125
DM7448W	1125
DM7450J	0743
DM7450N	0743
DM7451J	1168
DM7451N	1168
DM7453J	1185
DM7453N	1185
DM7454J	1202
DM7454N	1202
DM7460J	1275
DM7460N	1275
DM7470J	1404
DM7470N	1404
DM7472J	1426
DM7472N	1426
DM7473	6074
DM7473J	1172
DM7473N	1172
DM7474J	1313
DM7474M	1313
DM7474N	1313
DM7475J	1432
DM7475N	1432
DM7476J	1158
DM7476N	1158
DM7483J	3121
DM7483N	0117
DM7485N	1432
DM7486J	1368
DM7486N	1368
DM7488N	4904
DM7489N	5455
DM7490	1208
DM7490AN	1208
DM7490J	1208
DM7490N	1208
DM7491AN	0980
DM7492AN	0834
DM7492J	0834
DM7492N	0834
DM7493AN	0569
DM7493J	0569
DM7493N	0569
DM7495N	1486
DM7496J	1715
DM7496N	1715
DM7496W	OEM
DM7497J	2940
DM7511J	OEM
DM7511N	OEM
DM7511W	OEM
DM7512J	OEM
DM7512N	OEM
DM7512W	OEM
DM7542J	OEM
DM7542W	OEM
DM7544J	OEM
DM7544W	OEM
DM7546J	OEM
DM7546W	OEM
DM7551J	OEM
DM7551W	OEM
DM7552J	OEM
DM7552W	OEM
DM7553J	OEM
DM7553W	OEM
DM7554J	OEM
DM7554W	OEM
DM7555J	OEM
DM7555N	OEM
DM7555W	OEM
DM7556J	OEM
DM7556N	OEM
DM7556W	OEM
DM7560J	OEM
DM7560W	OEM
DM7563J	OEM
DM7563W	OEM
DM7570J	OEM
DM7570W	OEM
DM7585BSXXXX	OEM
DM7590J	OEM
DM7590W	OEM
DM7599J	OEM
DM7599N	OEM
DM7613J	OEM
DM7613W	OEM
DM7819J	OEM
DM7819N	OEM
DM7819W	OEM
DM7853J	OEM
DM7853W	OEM
DM7875AD	OEM
DM7875AJ	OEM
DM7875AW	OEM
DM7875BD	OEM
DM7875BJ	OEM
DM7875BW	OEM
DM8091J	OEM
DM8091N	OEM
DM8091W	OEM
DM8092J	OEM
DM8092N	OEM
DM8092W	OEM
DM8121N	2299
DM8123N	5939
DM8130D	OEM
DM8130F	OEM
DM8130N	OEM
DM8131J	OEM
DM8131N	OEM
DM8136N	OEM
DM8160	OEM
DM8160J	OEM
DM8160N	OEM
DM8160W	OEM
DM8200J	OEM
DM8200N	OEM
DM8214N	6798
DM8219N	6067
DM8220J	OEM
DM8220N	OEM
DM8280J	1795
DM8280N	1795
DM8280W	OEM
DM8281J	1803
DM8281N	1803
DM8281W	OEM
DM8300N	OEM
DM8301J	2196
DM8301N	2196
DM8301W	OEM
DM8310N	OEM
DM8316N	5957
DM8318N	4145
DM8334J	OEM
DM8334N	OEM
DM8334W	OEM
DM8511J	OEM
DM8511N	OEM
DM8511W	OEM
DM8512W	OEM
DM8542N	5976
DM8544N	OEM
DM8546N	7015
DM8551N	1926
DM8552J	OEM
DM8552W	OEM
DM8553J	OEM
DM8553N	6425
DM8553W	OEM
DM8554J	OEM
DM8554N	6426
DM8554W	OEM
DM8555N	6427
DM8556N	5978
DM8560	1921
DM8560J	1921
DM8560N	1921
DM8563J	1926
DM8563N	1926
DM8570N	0734
DM8590N	1685
DM8596BSXXXX	OEM
DM8601J	2267
DM8601N	2267
DM8601W	OEM
DM8602J	OEM
DM8602N	2286
DM8602W	OEM
DM8613N	6091
DM8678J	OEM
DM8678N	OEM
DM8810N	0803
DM8811N	OEM
DM8819J	OEM
DM8819N	OEM
DM8819W	OEM
DM8853J	OEM
DM8853N	1741
DM8853W	OEM
DM8875AJ	OEM
DM8875AN	OEM
DM8875BJ	OEM
DM8875BN	OEM
DM8890N	2043
DM8898N	OEM
DM8899N	OEM
DM9002CJ	OEM
DM9002CN	OEM
DM9003CJ	OEM
DM9003CN	OEM
DM9004CJ	OEM
DM9004CN	OEM
DM9005CJ	OEM
DM9005CN	OEM
DM9006CJ	OEM
DM9006CN	OEM
DM9008CJ	OEM
DM9008CN	OEM
DM9009CJ	OEM
DM9009CN	OEM
DM9012CJ	OEM
DM9012CN	OEM
DM9016CJ	OEM
DM9016CN	OEM
DM9093N	0358
DM9094N	1632
DM9097N	1481
DM9099N	0333
DM9300J	OEM
DM9300W	OEM
DM9301J	2196
DM9301N	2196
DM9301W	OEM
DM9310J	OEM
DM9310W	OEM
DM9316J	OEM
DM9316W	OEM
DM9318J	4145
DM9334J	OEM
DM9334W	OEM
DM9601J	2267
DM9601N	2267
DM9601W	OEM
DM9602J	OEM
DM9602N	OEM
DM9602W	OEM
DM10101J	OEM
DM10102J	OEM
DM10105J	OEM
DM10106J	OEM
DM10107J	OEM
DM10109J	OEM
DM10110J	OEM
DM10111J	OEM
DM10112J	OEM
DM10117J	OEM
DM10118J	OEM
DM10119J	OEM
DM10121J	OEM
DM10348-03	OEM
DM10348-04	OEM
DM10414J	OEM
DM10414N	OEM
DM10415AJ	OEM
DM10415AN	OEM
DM10415J	OEM
DM10415N	OEM
DM10470A	OEM
DM10470J	OEM
DM10470L	OEM
DM54107J	OEM
DM54121J	OEM
DM54121W	OEM
DM54125J	OEM
DM54125N	OEM
DM54125W	OEM
DM54126J	OEM
DM54126W	OEM
DM54132J	OEM
DM54132N	OEM
DM54132W	OEM
DM54141J	OEM
DM54141W	OEM
DM54147J	OEM
DM54147W	OEM
DM54148J	OEM
DM54148W	OEM
DM54155J	OEM
DM54155W	OEM
DM54156J	OEM
DM54156W	OEM
DM54160AJ	OEM
DM54160AW	OEM
DM54161AJ	OEM
DM54161AW	OEM
DM54162AJ	OEM
DM54162AW	OEM
DM54163AJ	OEM
DM54163AW	OEM
DM54164J	OEM
DM54164W	OEM
DM54165J	OEM
DM54165W	OEM
DM54166J	OEM
DM54166W	OEM
DM54170N	OEM
DM54173J	OEM
DM54173W	OEM
DM54174J	OEM
DM54174W	OEM
DM54175J	OEM
DM54175W	OEM
DM54182J	OEM
DM54184J	OEM
DM54184W	OEM
DM54185AJ	OEM
DM54185AW	OEM
DM54191J	OEM
DM54191W	OEM
DM54192W	OEM
DM54193J	OEM
DM54193W	OEM
DM54194J	OEM
DM54194W	OEM
DM54195J	OEM
DM54195W	OEM
DM54196J	OEM
DM54197J	OEM
DM54198J	OEM
DM54198W	OEM
DM54199J	OEM
DM54199W	OEM
DM54251J	OEM
DM54251W	OEM
DM54259J	OEM
DM54259W	OEM
DM54365J	OEM
DM54365N	OEM
DM54365W	OEM
DM54366J	OEM
DM54366N	OEM
DM54366W	OEM
DM54367J	OEM
DM54367N	OEM
DM54367W	OEM
DM54368J	OEM
DM54368N	OEM
DM54368W	OEM
DM74107J	0942
DM74107N	0942
DM74121J	0175
DM74121N	0175
DM74123	1157
DM74123N	1157
DM74125J	1182
DM74125N	1182
DM74125W	OEM
DM74126N	1193
DM74132J	1271
DM74132N	1271
DM74132W	OEM
DM74141J	1377
DM74141N	1377
DM74141W	OEM
DM74145	0619
DM74145N	0619
DM74145W	OEM
DM74147J	1451
DM74147N	1451
DM74147W	OEM
DM74148G	OEM
DM74148GM	OEM
DM74148GMR	OEM
DM74148J	OEM
DM74148N	OEM
DM74148W	OEM
DM74150	1492
DM74150N	1492
DM74151	1495
DM74151N	1495
DM74153	1539
DM74153N	1539
DM74154	1555
DM74154F	OEM
DM74154J	1555
DM74154N	1555
DM74155N	1576
DM74156N	1592
DM74157N	1605
DM74160AN	1631
DM74160N	1631
DM74161AN	1645
DM74161N	1645
DM74162	1013
DM74162AN	1013
DM74162N	1013
DM74163AN	1666
DM74163N	1666
DM74164N	0734
DM74165N	1685
DM74166N	0231
DM74170N	1722
DM74173N	1766
DM74174	1770
DM74174N	1770
DM74175N	1787
DM74176J	1795
DM74176N	1795
DM74176W	OEM
DM74177J	1803
DM74177N	1803
DM74177W	OEM
DM74180J	1829
DM74180N	1829
DM74180W	OEM
DM74181J	1842
DM74181N	1842
DM74182N	1856
DM74184N	OEM
DM74185AN	1868
DM74190	1912
DM74190J	1912
DM74190N	1912
DM74190W	OEM
DM74191J	1917
DM74191N	1917
DM74191W	OEM
DM74192J	1921
DM74192N	1921
DM74193	1926
DM74193J	1926
DM74193N	1926
DM74194N	OEM
DM74195N	1943
DM74196N	1950
DM74197N	1956
DM74198N	1964
DM74199N	1971
DM74251N	2299
DM74259N	OEM
DM74365J	1459
DM74365N	1459
DM74365W	OEM
DM74366J	1471
DM74366N	1471
DM74366W	OEM
DM74367J	1488
DM74367N	1488
DM74367W	OEM
DM74368J	1508
DM74368N	1508
DM74368W	OEM
DMA6497	OEM
DMA6497A	OEM
DMA6497AM	OEM
DMA6497AMR	OEM
DMA6497B	OEM
DMA6497BM	OEM
DMA6497BMR	OEM
DMA6497C	OEM
DMA6497CM	OEM
DMA6497CMR	OEM
DMA6497M	OEM
DMA6497MR	OEM
DMA6498AM	OEM
DMA6498AMR	OEM
DMA6498CM	OEM
DMA6498CMR	OEM
DMA6499	OEM
DMA6499A	OEM
DMA6499AMR	OEM
DMA6499B	OEM
DMA6499BM	OEM
DMA6499BMR	OEM
DMA6499MR	OEM
DMB5-12	OEM
DMB15-12	OEM
DMB20-12	OEM
DMB30-12	OEM
DMB-4500	OEM
DMB-4500A	OEM
DMB-4501	OEM
DMB-4501A	OEM
DMB6411	OEM
DMB6411A	OEM
DMB6411B	OEM
DMB6411C	OEM
DMB-6780	OEM
DMB-6780A	OEM
DMB-6781	OEM
DMB-6781A	OEM
DMB-6782	OEM
DMB-6782A	OEM
DMD5600	OEM
DMD5600A	OEM
DMD5817	OEM
DMD5817A	OEM
DMD5818	OEM
DMD5818A	OEM
DMD5819	OEM
DMD5827	OEM
DMD5827A	OEM
DMD5828	OEM
DMD5828A	OEM
DMD5835	OEM
DMD5841	OEM
DMD5841A	OEM
DMD5842	OEM
DMD5845	OEM
DMD5845A	OEM
DMD5846	OEM
DMD-6022	OEM
DMD6022	OEM
DMD-6022A	OEM
DMD6022A	OEM
DMD-6023	OEM
DMD6023	OEM
DMD-6023A	OEM
DMD6023A	OEM
DMD-6288	OEM
DMD-6288A	OEM
DMD-6298	OEM
DMD-6298A	OEM
DMD-6395	OEM
DMD-6395A	OEM
DMD6459	OEM
DMD6459A	OEM
DMD-6460	OEM
DMD-6460A	OEM
DMD-6829	OEM
DMD-6829A	OEM
DMD-6884	OEM
DMD-6884A	OEM
DMD-6885	OEM
DMD-6885A	OEM
DMD-6886	OEM
DMD-6886A	OEM
DME6L	OEM
DME7	OEM
DME25	OEM
DME30L	OEM
DME50	OEM
DME120L	OEM
DME150	OEM
DME250	OEM
DME300	OEM
DME375	OEM
DME600	OEM
DME800	OEM
DME6507	OEM
DME6561	OEM
DME6562	OEM
DME6563	OEM
DME6957	OEM
DMF-4000	OEM
DMF4000	OEM
DMF-4011	OEM
DMF-4011A	OEM
DMF-4012	OEM
DMF-4012A	OEM
DMF-4035	OEM
DMF-4035A	OEM
DMF-4040	OEM
DMF-4040A	OEM
DMF-4052	OEM
DMF-4052A	OEM
DMF-4059	OEM
DMF-4059A	OEM
DMF4788	OEM
DMF4788A	OEM
DMF4792	OEM
DMF-5079	OEM
DMF-5079A	OEM
DMF-5080	OEM
DMF-5080A	OEM
DMF6034	OEM
DMF6034A	OEM
DMF6034B	OEM
DMF6035	OEM
DMF6035A	OEM
DMF6035B	OEM
DMF6037	OEM
DMF6037A	OEM
DMF6037B	OEM
DMF6106	OEM
DMF6106A	OEM
DMF6106AM	OEM
DMF6106B	OEM
DMF6106CM	OEM
DMF6106M	OEM
DMF6107	OEM
DMF6107A	OEM
DMF6107AM	OEM
DMF6107B	OEM
DMF6107BM	OEM
DMF6107CM	OEM
DMF6107M	OEM
DMF6130	OEM
DMF6130A	OEM
DMF6131	OEM
DMF6131A	OEM
DMF6131B	OEM
DMF6132	OEM
DMF6132A	OEM
DMF6132B	OEM
DMF-6426	OEM
DMF-6426A	OEM
DMG-6412	OEM
DMG6412	OEM
DMG-6412A	OEM
DMG-6413	OEM
DMG6413	OEM
DMG-6413A	OEM
DMG-6414	OEM
DMG6414	OEM
DMG-6414A	OEM
DMG-6415	OEM
DMJ-4007	OEM
DMJ-4014	OEM
DMJ-4015	OEM
DMJ-4502	OEM
DMJ4759	OEM
DMJ4771	OEM
DMJ4798	OEM
DMJ-5034	OEM
DMJ6564	OEM
DMJ-6665	OEM
DMJ-6666	OEM
DMJ-6668	OEM
DMJ-6669	OEM
DMJ-6670	OEM
DMJ-6777	OEM
DMJ-6778	OEM
DMJ-6784	OEM
DMJ-6785	OEM
DMJ-6786	OEM
DMJ-6788	OEM
DMJ-6789	OEM
DMJ-6990	OEM
DMK4791	OEM
DMK-5068	OEM
DMK-5068A	OEM
DMM102BSXXXX	OEM
DMM104BSXXXX	OEM
DMM105DXXXX	OEM
DMM110DSXXXX	OEM
DMM112BSXXXX	OEM
DMM204BSXXXX	OEM
DMM208FSXXXX	OEM
DMM212BSXXXX	OEM
DMM218BSXXXX	OEM
DMM408BSXXXX	OEM
DMM510DSXXXX	OEM
DMM520DSXXXX	OEM
DMM711BSXXXX	OEM
DMM812BSXXXX	OEM
DMM816BSXXXX	OEM
DMM1724BSXXXX	OEM
DMM5459BSXXXX	OEM
DMM7585BSXXXX	OEM
DMM8596BSXXXX	OEM
DMMA700	0339
DMMA4043-H	0275
DMMA4051-H	0041
DMMA4051-M	0589
DMMA4075-M	0077
DMMA4082-M	0165
DMMA4091-M	0057
DMMA4100-M	0064
DMMA4200-M	0700
DMMA4270-M	0441
DMOA90A-M	0143
DMP4025DB	OEM
DMPAL10H8JM	OEM
DMPAL10H8NC	OEM
DMPAL10L8JM	OEM
DMPAL10L8NC	OEM
DMPAL12H6JM	OEM
DMPAL12H6NC	OEM
DMPAL12L6JM	OEM
DMPAL12L6NC	OEM
DMPAL14H4JM	OEM
DMPAL14H4NC	OEM
DMPAL14L4JM	OEM
DMPAL14L4NC	OEM
DMPAL16A4JM	OEM
DMPAL16A4NC	OEM
DMPAL16C1JM	OEM
DMPAL16C1NC	OEM
DMPAL16H2JM	OEM
DMPAL16H2NC	OEM
DMPAL16L2JM	OEM
DMPAL16L2NC	OEM
DMPAL16L8JM	OEM
DMPAL16R4JM	OEM
DMPAL16R6JM	OEM
DMPAL16R6NC	OEM
DMPAL16R8JM	OEM
DMPAL16R8NC	OEM
DMPAL16X4JM	OEM
DMPAL16X4NC	OEM
DMX88EQ	OEM
DMX88FQ	OEM
DN0700A	OEM
DN0701	OEM
DN0702	OEM
DN74LS00	OEM
DN74LS02	1559
DN74LS04	OEM
DN74LS05	OEM
DN74LS08	1633
DN74LS09	1642
DN74LS14P	1698
DN74LS32	0088
DN74LS74	OEM
DN74LS125P	0075
DN74LS132	1625
DN74LS138	OEM
DN74LS156	1654
DN74LS174P	0260
DN74LS240P	0452
DN74LS244	0458
DN74LS245P	0463
DN74LS373	0709
DN262-02	OEM
DN262-04	OEM
DN262-06	OEM
DN262-08	OEM
DN262-10	OEM
DN262-12	OEM
DN262-14	OEM
DN262-16	OEM
DN262-18	OEM
DN262-20	OEM
DN304B	OEM
DN462-02	OEM
DN462-04	OEM
DN462-06	OEM
DN462-08	OEM
DN462-10	OEM
DN462-12	OEM
DN462-14	OEM
DN462-16	OEM
DN462-18	OEM
DN462-20	OEM
DN662-02	OEM
DN662-04	OEM
DN662-06	OEM
DN662-08	OEM
DN662-10	OEM
DN662-12	OEM
DN662-14	OEM
DN662-16	OEM
DN662-18	OEM
DN662-20	OEM
DN662-22	OEM
DN662-24	OEM
DN662-25	OEM
DN681L	OEM
DN761-20	OEM
DN761-22	OEM
DN761-24	OEM
DN761-26	OEM
DN761-28	OEM

If replacement code is OEM, contact original manufacturer for replacement.

DEVICE TYPE	REPL CODE
DN761-30	OEM
DN761-32	OEM
DN761-34	OEM
DN762-02	OEM
DN762-04	OEM
DN762-06	OEM
DN762-08	OEM
DN762-10	OEM
DN762-12	OEM
DN762-14	OEM
DN762-16	OEM
DN762-18	OEM
DN762-20	OEM
DN811	4154
DN819	4158
DN834	OEM
DN835	OEM
DN837	OEM
DN838	4162
DN839	OEM
DN850	4168
DN851	OEM
DN852	1599
DN852P	1599
DN918	OEM
DN961-02	OEM
DN961-04	OEM
DN961-06	OEM
DN961-08	OEM
DN961-10	OEM
DN961-12	OEM
DN961-14	OEM
DN961-16	OEM
DN961-18	OEM
DN961-20	OEM
DN961-22	OEM
DN961-24	OEM
DN961-25	OEM
DN1000	OEM
DN1000A	OEM
DN1001	OEM
DN1002	OEM
DN1005AM	OEM
DN1005BM	OEM
DN1005CM	OEM
DN-1006-2	OEM
DN1946	0141
DN2000	OEM
DN2001	OEM
DN2002	OEM
DN2175	OEM
DN2219A	OEM
DN2222A	OEM
DN2484	OEM
DN3000	OEM
DN3000A	OEM
DN3001	OEM
DN3002	OEM
DN3066A	OEM
DN3067A	OEM
DN3068A	OEM
DN3069A	OEM
DN3070A	OEM
DN3071A	OEM
DN3365A	OEM
DN3366A	OEM
DN3367A	OEM
DN3368A	OEM
DN3369A	OEM
DN3370A	OEM
DN3436A	OEM
DN3437A	OEM
DN3438A	OEM
DN3439	OEM
DN3458A	OEM
DN3459A	OEM
DN3460A	OEM
DN5000	OEM
DN5001	OEM
DN5002	OEM
DN5564	OEM
DN5565	OEM
DN5566	OEM
DN5567	OEM
DN6338-A	0489
DN6368	OEM
DN6811	4752
DN6835	OEM
DN6837	OEM
DN6838	0489
DN6838A	0489
DN6839	OEM
DN6851	OEM
DN6851H1	2557
DN6851HI	2557
DN7432	OEM
DN8000	OEM
DN8001	OEM
DN8002	OEM
DN9000	OEM
DNA5856	OEM
DND800	0133
DNMPC574J	1329
DNN03602D9	OEM
DNN03604D9	OEM
DNN03606D9	OEM
DNN03608D9	OEM
DNN03610D9	OEM
DNN03612D9	OEM
DNN03614D9	OEM
DNN03616D9	OEM
DNN03618D9	OEM
DNN03620D9	OEM
DNN05002D13	OEM
DNN05004D13	OEM
DNN05006D13	OEM
DNN05008D13	OEM
DNN05010D13	OEM
DNN05012D13	OEM
DNN05014D13	OEM
DNN05016D13	OEM
DNN05018D13	OEM
DNN05020D13	OEM
DNN07802D20	OEM
DNN07804D20	OEM
DNN07806D20	OEM
DNN07808D20	OEM
DNN07810D20	OEM
DNN07812D20	OEM
DNN07814D20	OEM
DNN07816D20	OEM
DNN07818D20	OEM
DNN07820D20	OEM
DNN07822D20	OEM
DNN07824D20	OEM
DNN07826D20	OEM
DNN07828D20	OEM
DNN07830D20	OEM
DNN07832D20	OEM
DNN07834D20	OEM
DNN08502D20	OEM
DNN08504D20	OEM
DNN08506D20	OEM
DNN08508D20	OEM
DNN08510D20	OEM
DNN08512D20	OEM
DNN08514D20	OEM
DNN08516D20	OEM
DNN08518D20	OEM
DNN08520D20	OEM
DNRD3.9E-B1	0036
DNRD5.1E-B3	0041
DNRD5.6E-B2	0253
DNRD6.2E	0471
DNRD6.2E-B1-B3	0471
DNRD6.2E-B2	0471
DNRD6.2E-B3	0471
DNRD6.2L	0471
DNRD7.5E	0077
DNRD7.5E-B3	0077
DNRD8.2E-B3	0165
DNRD10E-B1	0064
DNRD10E-B3	0064
DNRD10F-B1	0064
DNRD11E-B1	0181
DNRD13E-B2	0053
DNRD16E-B3	0445
DNRD24E-B2	0494
DNRD27E-B4	0441
DNRD33E-B1	0166
DNRD39E-B1	0032
DNRD39E-C2	0032
DNUPC574J	1329
DNX1	OEM
DNX1A	OEM
DNX2	OEM
DNX2A	OEM
DNX3	OEM
DNX4	3133
DNX4A	3133
DNX5	OEM
DNX5A	OEM
DNX6	OEM
DNX6A	OEM
DNX7	OEM
DNX7A	OEM
DNX8	OEM
DNX8A	OEM
DNX9	OEM
DNX9A	OEM
DNX31	OEM
DO-7	0292
DO19	0211
DO31	0079
DO38	0211
DO43	0211
DO48	OEM
DO63	0050
DO78	0211
DO79	0050
DO80	0085
DO81	0160
DO83	0038
DO85	0038
DO86	0050
DO87	0127
DO88	0127
DO104BSXXXX	OEM
DO105DXXXX	OEM
DO110DSXXXX	OEM
DO112BSXXXX	OEM
DO204BSXXXX	OEM
DO208FSXXXX	OEM
DO212BSXXXX	OEM
DO218BSXXXX	OEM
DO408BSXXXX	OEM
DO510DSXXXX	OEM
DO520DSXXXX	OEM
DO711BSXXXX	OEM
DO812BSXXXX	OEM
DO816BSXXXX	OEM
DO1724BSXXXX	OEM
DO5459BSXXXX	OEM
DO7585BSXXXX	OEM
DO8596BSXXXX	OEM
DOD-202-200	OEM
DOD-202-201	OEM
DOD-202-300	OEM
DOD-202-301	OEM
DOF4021001	0120
DOM102BSXXXX	OEM
DOM104BSXXXX	OEM
DOM105DXXXX	OEM
DOM110DSXXXX	OEM
DOM112BSXXXX	OEM
DOM204BSXXXX	OEM
DOM208FSXXXX	OEM
DOM212BSXXXX	OEM
DOM218BSXXXX	OEM
DOM408BSXXXX	OEM
DOM510DSXXXX	OEM
DOM520DSXXXX	OEM
DOM711BSXXXX	OEM
DOM812BSXXXX	OEM
DOM816BSXXXX	OEM
DOM1724BSXXXX	OEM
DOM5495BSXXXX	OEM
DOM7585BSXXXX	OEM
DOM8596BSXXXX	OEM
DOS118B	OEM
DOS118B-1	OEM
DOS118BM	OEM
DOS118BM-1	OEM
DOS118BMZS	OEM
DOS118BMZS-1	OEM
DOS118BQ	OEM
DOS118BQ-1	OEM
DOS118BQZS	OEM
DOS118BQZS-1	OEM
DOS118BT	OEM
DOS118BT-1	OEM
DOS118BTZS	OEM
DOS118BTZS-1	OEM
DOS118BZS	OEM
DOS118BZS-1	OEM
DOUBLER	OEM
DOZ105DXXXX	OEM
DP0300	OEM
DP0301	OEM
DP0302	OEM
DP0700	OEM
DP0701	OEM
DP0702	OEM
DP2	OEM
DP3	OEM
DP4	OEM
DP13B00-2	0277
DP13B01-2	0277
DP13B02-2	0289
DP13B04-2	0296
DP100	0071
DP146	OEM
DP147	OEM
DP1000	OEM
DP1000A	OEM
DP1001	OEM
DP1002	OEM
DP1003	OEM
DP1004	OEM
DP1005	OEM
DP1005A000	OEM
DP1005A003	OEM
DP1005AM	OEM
DP1005B000	OEM
DP1005B003	OEM
DP1005BM	OEM
DP1005C000	OEM
DP1005C003	OEM
DP1005CM	OEM
DP1006	OEM
DP1007	OEM
DP1008	OEM
DP1009	OEM
DP1010	OEM
DP1501	OEM
DP1502	OEM
DP1503	OEM
DP1504	OEM
DP1505	OEM
DP1506	OEM
DP2000	OEM
DP2001	OEM
DP2002	OEM
DP2605	OEM
DP2907A	OEM
DP3000	OEM
DP3000A	OEM
DP3001	OEM
DP3002	OEM
DP5000	OEM
DP5001	OEM
DP5002	OEM
DP7303J	OEM
DP7304BJ	OEM
DP7307J	OEM
DP7308J	OEM
DP7310J	OEM
DP7311J	OEM
DP8000	OEM
DP8001	OEM
DP8002	OEM
DP8216	1863
DP8216J	1863
DP8216M	OEM
DP8216MJ	OEM
DP8216N	1863
DP8226	OEM
DP8226J	OEM
DP8226M	OEM
DP8226MJ	OEM
DP8226N	OEM
DP8303J	OEM
DP8303N	OEM
DP8304BJ	OEM
DP8304BN	OEM
DP8307J	OEM
DP8307N	OEM
DP8308J	OEM
DP8308N	OEM
DP8310J	OEM
DP8310N	OEM
DP8311J	OEM
DP8311N	OEM
DP8340J	OEM
DP8340N	OEM
DP8341J	OEM
DP8341N	OEM
DP8342J	OEM
DP8342N	OEM
DP8343N	OEM
DP8350N	OEM
DP8352N	OEM
DP8353N	OEM
DP8408	OEM
DP8409	OEM
DPA68/1M	OEM
DPB101-00S	OEM
DPB102-00S	OEM
DPB103-00S	OEM
DPB104-00S	OEM
DPB105-00S	OEM
DPH302	2628
DPT121	OEM
DPT122	OEM
DPT123	OEM
DPT124	OEM
DPT200	OEM
DPT2600	OEM
DPZ15-30R	OEM
DPZ15-33R	OEM
DPZ15-36R	OEM
DPZ15-45R	OEM
DPZ15-47R	OEM
DPZ15-50R	OEM
DQA05	OEM
DQA08	OEM
DQA0115V	OEM
DQA10	OEM
DQA12	OEM
DQA15	OEM
DQA20	OEM
DQA25	OEM
DQA30	OEM
DQA50	OEM
DQB05	OEM
DQB08	OEM
DQB10	OEM
DQB12	OEM
DQB15	OEM
DQB20	OEM
DQB25	OEM
DQB30	OEM
DQB50	OEM
DQN1006	OEM
DR1	0015
DR1PR	0015
DR2	4240
DR3	0015
DR4	0015
DR5	0015
DR50	0080
DR100	0015
DR120GB2	OEM
DR128	0143
DR150	OEM
DR200	0015
DR202	0015
DR204	0795
DR206	0015
DR207	0143
DR208	0072
DR209	0133
DR210	0071
DR211	OEM
DR212	0102
DR213	0143
DR272	OEM
DR283	0143
DR291	0143
DR292	OEM
DR295	0143
DR300	0015
DR301	0133
DR302	0143
DR303	0143
DR304	OEM
DR305	OEM
DR306	0143
DR307	0143
DR308	0143
DR309	0143
DR310	OEM
DR311	OEM
DR312	OEM
DR313	0143
DR314	0143
DR315	OEM
DR316	OEM
DR317	0143
DR318	0143
DR319	0143
DR321	OEM
DR322	OEM
DR323	0143
DR324	0143
DR325	OEM
DR326	0143
DR327	0133
DR328	0143
DR329	0143
DR330	0143
DR336	0143
DR337	0143
DR338	0143
DR351	0143
DR352	0143
DR362	OEM
DR365	0143
DR366	0143
DR379	OEM
DR385	0143
DR389	0143
DR400	0015
DR401	0143
DR402	OEM
DR403	OEM
DR404	OEM
DR407	OEM
DR408	OEM
DR412-12-10	OEM
DR418	0143
DR419	OEM
DR422	OEM
DR426	0143
DR427	0015
DR434	0143
DR435	1335
DR437	OEM
DR449	0143
DR450	0071
DR459	OEM
DR463	0143
DR464	0143
DR481	OEM
DR482	0143
DR498	OEM
DR500	0015
DR521	0023
DR562	OEM
DR600	0023
DR-630	OEM
DR-630EW	OEM
DR-630NS	OEM
DR-640	OEM
DR-640EW	OEM
DR-640NS	OEM
DR650	OEM
DR650EW	OEM
DR650NS	OEM
DR661	OEM
DR664	0124
DR667	OEM
DR668	0080
DR669	0015
DR670	0795
DR671	0015
DR672	OEM
DR673	OEM
DR674	OEM
DR675	OEM
DR677	0124
DR688	OEM
DR694	0023
DR695	0015
DR698	0015
DR699	0795
DR700	0071
DR800	0071
DR826	0795
DR827	OEM
DR833	0023
DR848	0015
DR852	0124
DR863	0015
DR900	0071
DR999	OEM
DR1000	0071
DR1100	0015
DR1200	0984
DR5101	0015
DR5102	0015
DR25387T1	OEM
DR-104532	1413
DRA01B	0901
DRA01C	0058
DRA01E	0407
DRA03A	0344
DRA03B	0901
DRA03C	0058
DRA03D	0407
DRA03E	0407
DRA03F	1683
DRA2A	1259
DRA2B	0447
DRA2C	0940
DRA2E	0095
DRA2F	OEM
DRA2G	OEM
DRA3B	OEM
DRA3C	OEM
DRA3E	OEM
DRA3F	OEM
DRA3GB	OEM
DRA5B	2089
DRA5C	0505
DRA5E	0710
DRA5G	0863
DRA8B	1683
DRA8C	1683
DRA8E	0179
DRA8G	0952
DRA40	5767
DRA402	OEM
DRA404	OEM
DRA406	OEM
DRA408	OEM
DRA410	OEM
DRB2B	OEM
DRB2C	OEM
DRB2E	OEM
DRB3B	0447
DRB3C	0940
DRB3E	0095
DRB100-02R	OEM
DRB100-04R	OEM
DRB100-06R	OEM
DRB100-08R	OEM
DRB100-10R	OEM
DRB100-12R	OEM
DRBT100-02R	OEM
DRBT100-04R	OEM
DRBT100-06R	OEM
DRBT100-08R	OEM
DRBT100-10R	OEM
DRBT100-12R	OEM
DRC2E	OEM
DRC3E	0710
DRN02702D9	OEM
DRN02704D9	OEM
DRN02706D9	OEM
DRN02708D9	OEM
DRN02710D9	OEM
DRN02712D9	OEM
DRN02714D9	OEM
DRN03902D13	OEM
DRN03904D13	OEM
DRN03906D13	OEM
DRN03908D13	OEM
DRN03910D13	OEM
DRN03912D13	OEM
DRN03914D13	OEM
DRN03915D13	OEM
DRN05502D17	OEM
DRN05504D17	OEM
DRN05506D17	OEM
DRN05508D17	OEM
DRN05510D17	OEM
DRN05512D17	OEM
DRN05514D17	OEM
DRN05516D17	OEM
DRN05518D17	OEM
DRN05520D17	OEM
DRN05522D17	OEM
DRN05524D17	OEM
DRN05525D17	OEM
DRN07002D20	OEM
DRN07004D20	OEM
DRN07006D20	OEM
DRN07008D20	OEM
DRN07010D20	OEM
DRN07012D20	OEM
DRN07014D20	OEM
DRN07016D20	OEM
DRN07018D20	OEM
DRN07020D20	OEM
DRS1	0015
DRS2	0087
DRS5E	0710
DRS101	0015
DRS102	0015
DRS104	0015
DRS106	0015
DRS107	0071
DRS108	0071
DRS110	0071
DRS112	OEM
DRS114	OEM
DRS114C	OEM
DRS114D	OEM
DRS163	1023
DRS164	1036
DRS165	1036
DRS166	1036
DRS167	4287
DRS168	4287
DRS169	4287
DRS185R	OEM
DRS203	1023
DRS204	1036
DRS205	1036
DRS206	1036
DRS207	4287
DRS208	4287
DRS209	4287
DRS243	1023
DRS244	1036
DRS245	1036
DRS246	1036
DRS247	4287
DRS248	4287
DRS249	4287
DRS250	3928
DRS251	3928
DRS252	3928
DRS253	OEM
DRS254	OEM
DRS1851	OEM
DRS1852	OEM
DRS1852R	OEM
DRS1853	OEM
DRS1853R	OEM
DRS1854	OEM
DRS1854R	OEM
DRS1855	OEM
DRS1855R	OEM
DRV11	OEM
DRV11B	OEM
DRX10	OEM
DRZ4300R	OEM
DS(I)35-02A#	2813
DS(I)35-04A#	2813
DS(I)35-07A#	2845
DS0025CH	OEM
DS0025CN	OEM
DS0025CN8	OEM
DS0025H	OEM
DS0026CG	OEM
DS0026CH	OEM
DS0026CJ	OEM
DS0026CJ-8	OEM
DS0026CJ8	OEM
DS0026CN	OEM
DS0026CN8	OEM
DS0026F	OEM
DS0026G	OEM
DS0026H	OEM
DS0026J	OEM
DS0026J-8	OEM
DS0026J8	OEM
DS0026W	OEM
DS0056CG	OEM
DS0056CH	OEM
DS0056CJ	OEM
DS0056CJ-8	OEM
DS0056CJ8	OEM
DS0056CN	OEM
DS0056CN8	OEM
DS0056G	OEM
DS0056H	OEM
DS0056J	OEM
DS0056J-8	OEM
DS0056J8	OEM
DS-0065	0015
DS0065	0015
DS0.5-200A	OEM
DS0.5-400A	OEM
DS0.5-600A	OEM
DS0.5-800A	OEM
DS0.5-1000A	OEM
DS0.5A	OEM
DS0.5A1	OEM
DS0.5A2	OEM
DS0.5A3	OEM
DS0.5A4	OEM
DS0.9-04A	OEM
DS0.9-08A	OEM
DS0.9-12A	OEM
DS0.9-14A	OEM
DS0.9-16A	OEM
DS0.9-18A	OEM
DS-1	0208
DS1	0015
DS-1(DELCO)	0208
DS1-200B	OEM
DS1-400B	2645
DS1-600B	OEM
DS1-800B	OEM
DS1-1000B	OEM
DS1A	OEM
DS1B	0016
DS1B1	2645
DS1B2	2645
DS1B3	OEM
DS1B4	OEM
DS-1K	0071
DS1K	0023
DS1K7	0015
DS-1M	0015
DS1M	0023
DS-1N	0222
DS1N	OEM
DS-1P	0015
DS1P	0015

DEVICE TYPE	REPL CODE
DS-1U	0920
DS1U	OEM
DS1.2-04A	OEM
DS1.2-08A	OEM
DS1.2-12A	OEM
DS1.2-14A	OEM
DS1.2-16A	OEM
DS1.2-18A	OEM
DS1.2A6	OEM
DS1.2A9	OEM
DS1.2A12	OEM
DS1.2A15	OEM
DS-1.5-2	0087
DS1.8-18A	OEM
DS2	0600
DS2(LEAD)	OEM
DS2(LED)	OEM
DS2-04A	OEM
DS2-08A	OEM
DS2-12A	OEM
DS2-14A	OEM
DS2-16A	OEM
DS2-18A	OEM
DS2B6	2645
DS2B9	2645
DS2B12	OEM
DS2B15	OEM
DS2CDL-25	OEM
DS2DL-1LGN	OEM
DS2K	0015
DS2L	OEM
DS-2M	0102
DS2M	0023
DS-2N	0087
DS2N	0023
DS2N22	0015
DS2P	0540
DS3	0004
DS3-200A	OEM
DS3-400A	OEM
DS3-600A	OEM
DS3-800A	OEM
DS3-1000A	OEM
DS3A	OEM
DS3A1	OEM
DS3A2	OEM
DS3A3	OEM
DS3A4	OEM
DS4	0600
DS5	0600
DS5-200B	OEM
DS5-400B	OEM
DS5-600B	OEM
DS5-800B	OEM
DS5-1000B	OEM
DS5A	OEM
DS5B1	OEM
DS5B2	OEM
DS5B3	OEM
DS5B4	OEM
DS5BN	0323
DS5BN-L	0015
DS6	0600
DS6-08F	OEM
DS6-12F	OEM
DS6-14F	OEM
DS6-16F	OEM
DS6-18A	OEM
DS6A6	OEM
DS6A9	OEM
DS6A12	OEM
DS6A15	OEM
DS7	0600
DS-8	0004
DS8	0600
DS8T26AJ	4084
DS8T26AN	4084
DS8T28J	0581
DS8T28MJ	0581
DS8T28N	0581
DS9	0600
DS9-04F	OEM
DS9-08F	OEM
DS9-12F	OEM
DS9-14F	OEM
DS9-16F	OEM
DS9-18A	OEM
DS9B6	OEM
DS9B9	OEM
DS9B12	OEM
DS9B15	OEM
DS-10	0015
DS10-200C	OEM
DS10-400C	OEM
DS10-600C	OEM
DS10-800C	OEM
DS10-1000C	OEM
DS10C	OEM
DS10C1	OEM
DS10C2	OEM
DS10C3	OEM
DS10C4	OEM
DS10S1	OEM
DS10S2	OEM
DS10S3	OEM
DS11	0600
DS12	0600
DS-13	0071
DS13	0004

DEVICE TYPE	REPL CODE
DS-13(COURIER)	0015
DS-13A(SANYO)	0015
DS-13B(SANYO)	0015
DS13C	0871
DS13E	0853
DS-14	0004
DS14	0136
DS-14(DIODE)	0015
DS-14(SANYO)	0920
DS14C	0871
DS14E	0853
DS15A	0071
DS15B	0072
DS15C	0023
DS15E	0023
DS15F	OEM
DS15G	0017
DS16	0004
DS-16(SEARS-DIODE)	0015
DS-16A	0015
DS16A	0015
DS-16A(SANYO)	0071
DS-16B	0015
DS16B	0015
DS-16B(SANYO)	0015
DS-16B(SYLVANIA)	0015
DS16C	0015
DS-16C(SANYO)	0015
DS-16D	0015
DS16D	0023
DS-16D(SANYO)	0015
DS-16E	0015
DS16E	0015
DS-16E(SANYO)	0015
DS16N	0015
DS16NA	0071
DS16NB	0015
DS16NC	0015
DS-16ND	0015
DS16ND	0023
DS-16NE	0015
DS16NE	0015
DS-16NY	0015
DS-16YA	0015
DS-17	0196
DS17	0023
DS17(ADMIRAL)	0015
DS-17(SANYO)	0015
DS17-02A	OEM
DS17-04A	OEM
DS17-06A	OEM
DS17-08A	OEM
DS-17-6A	0015
DS17-12A	OEM
DS17-14A	OEM
DS17-16A	OEM
DS17-18A	OEM
DS17N	0015
DS17S2	OEM
DS17S3	OEM
DS17S6	OEM
DS-18	0015
DS18	0023
DS-18(DELCO)	0150
DS-18(GE)	0015
DS-18(SANYO)	0015
DS18(SEARS)	0015
DS18N	0015
DS-19	0004
DS19	0609
DS-19(RECTIFIER)	0015
DS-19(SANYO)	0015
DS19C	OEM
DS19E	OEM
DS19G	OEM
DS19J	OEM
DS20-200C	OEM
DS20-400C	OEM
DS20-600C	OEM
DS20-800	OEM
DS20-800C	OEM
DS20-1000C	OEM
DS20B	OEM
DS20C	OEM
DS20C1	OEM
DS20C2	OEM
DS20C3	OEM
DS20C4	OEM
DS20E	OEM
DS20G	OEM
DS20J	OEM
DS20S1	OEM
DS20S2	OEM
DS20S3	OEM
DS21	0280
DS-22	0280
DS22	0280
DS22-04A	0258
DS22-08A	OEM
DS22-12A	OEM
DS22-14A	OEM
DS22-16A	OEM
DS22-18A	OEM
DS22C6	OEM
DS22C9	OEM
DS22C12	OEM
DS22C15	OEM
DS23	0280
DS-24	0136

DEVICE TYPE	REPL CODE
DS24	0136
DS-25	0136
DS25	0136
DS-26	0164
DS26	0004
DS26LS31CJ	OEM
DS26LS31CN	OEM
DS26LS31MJ	OEM
DS26LS32CJ	OEM
DS26LS32CN	OEM
DS26LS32MJ	OEM
DS26LS33CJ	OEM
DS26LS33CN	OEM
DS26LS33MJ	OEM
DS26S10J	OEM
DS26S10MJ	OEM
DS26S10N	OEM
DS26S11J	OEM
DS26S11MJ	OEM
DS26S11N	OEM
DS-27	0143
DS27	0143
DS-28(DELCO)	0280
DS29	0004
DS-31	0025
DS31	0133
DS31(DELCO)	0143
DS-32	0143
DS32	0143
DS-33	0143
DS33	0143
DS33(DELCO)	0143
DS-34	0050
DS34	0050
DS-35	0050
DS35	0050
DS35-02A	2768
DS35-04A	2907
DS35-08A	OEM
DS35-12A	OEM
DS35-14A	OEM
DS35-16A	OEM
DS35-18A	OEM
DS35S2	OEM
DS35S3	OEM
DS35S6	OEM
DS-36	0050
DS36	0136
DS-37	0050
DS37	0050
DS-38	0133
DS38	0136
DS38(DELCO)	0050
DS38(SANYO)	0015
DS-39	0143
DS39	0123
DS40Q-04F	OEM
DS-41	0050
DS41	0050
DS-42	0050
DS42	0050
DS42-04#	2813
DS42-04A	OEM
DS42-07#	2845
DS42-08A	OEM
DS42-12A	OEM
DS42-14A	OEM
DS42-16A	OEM
DS42-18A	OEM
DS42C6	OEM
DS42C9	OEM
DS42C12	OEM
DS42C15	OEM
DS-43	0012
DS43	0012
DS-44	0079
DS44	0016
DS-45	0079
DS45	0016
DS45-200C	OEM
DS45-400	OEM
DS45-400C	OEM
DS45-600	OEM
DS45-600C	OEM
DS45-800	OEM
DS45-800C	OEM
DS45-1000C	OEM
DS45B	OEM
DS45C1	OEM
DS45C2	OEM
DS45C3	OEM
DS45C4	OEM
DS-46	0079
DS46	0016
DS-47	0016
DS47	0016
DS48	0012
DS-49	0244
DS50	0157
DS-51	0050
DS51	0136
DS-52	0050
DS52	0136
DS-53	0211
DS53	0280
DS-55	0030
DS55	0030
DS-56	0050
DS56	0050

DEVICE TYPE	REPL CODE
DS58(SANYO)	0015
DS-60	0400
DS-62	0050
DS62	0050
DS-63	0050
DS63	0050
DS-64	0050
DS64	0050
DS-65	0050
DS65	0050
DS-66	0016
DS66	0016
DS-66W	0079
DS-67	0016
DS67	0016
DS-67W	0079
DS67W	0016
DS-68	0150
DS68	0150
DS-70	2235
DS-71	0127
DS71	0127
DS-72	0127
DS72	0127
DS-73	0127
DS73	0127
DS-74	0127
DS74	0127
DS-75	0079
DS75	0127
DS75-02B	OEM
DS75-04B	OEM
DS75-06B	OEM
DS75-08B	OEM
DS75-12B	OEM
DS75-14B	OEM
DS75-16B	OEM
DS-76	0016
DS76	0016
DS-77	0079
DS77	0016
DS-78	0007
DS78	0007
DS78C20J	OEM
DS78C120J	OEM
DS78L12J	OEM
DS78LS120J	OEM
DS-79	0015
DS79	0015
DS-79(DELCO)	0015
DS-80	0015
DS80	0015
DS80-04A	1347
DS80-08A	OEM
DS80-12A	OEM
DS80-14A	OEM
DS80-16A	OEM
DS80-18A	OEM
DS-81	0224
DS81	0007
DS-82	0037
DS82	0037
DS-83	0037
DS83	0037
DS-85	0007
DS85	0007
DS-86	0126
DS86	0150
DS-87	0908
DS-88	0325
DS88	0325
DS88C20J	OEM
DS88C20N	OEM
DS88C120J	OEM
DS88C120N	OEM
DS88L12J	OEM
DS88L12N	OEM
DS88LS120J	OEM
DS88LS120N	OEM
DS89	0229
DS90SM19	OEM
DS-94	0155
DS96	0150
DS-97	0133
DS97	0133
DS97(DELCO)	0133
DS99	0012
DS-102	0212
DS-103	OEM
DS104	0133
DS-105	0212
DS-106	0212
DS-107	0162
DS-108	0137
DS-110	0100
DS110-04A	OEM
DS110-08A	OEM
DS110-12A	OEM
DS110-14A	OEM
DS110-16A	OEM
DS110-18A	OEM
DS110C6	OEM
DS110C9	OEM
DS110C12	OEM
DS110C15	OEM
DS111	5062
DS-113	0124
DS-113A	0102
DS113A	OEM

DEVICE TYPE	REPL CODE
DS-113B	0102
DS113B	5309
DS113C	OEM
DS-114	0030
DS114	0030
DS-117	0133
DS117	0133
DS118	OEM
DS118A	0242
DS118B	0540
DS118C	0965
DS118D	0817
DS118E	0922
DS-130	0015
DS130A	0072
DS-130B	0015
DS130B	0015
DS-130C	0015
DS130C	0015
DS130D	0015
DS-130E	0015
DS130E	0015
DS130NA	0071
DS130NB	0015
DS130NC	0015
DS130ND	0015
DS130N-E	OEM
DS130NE	0015
DS130T	0071
DS130TA	0071
DS-130TA-L	0015
DS130TA-L	0015
DS130TAL-FB2	0015
DS-130TA-S	0015
DS130TB	0071
DS130TC	0071
DS130TD	0071
DS130TE	0071
DS-130YB	0015
DS130YC	0015
DS130YE	0015
DS-131	0015
DS131	0023
DS-131A	0015
DS131A	0015
DS-131B	0015
DS131B	0015
DS132	1097
DS132A	0015
DS-132B	0015
DS132B	0015
DS133	OEM
DS133A	0015
DS133B	0102
DS135	0023
DS135AT	OEM
DS135B	OEM
DS135C	0023
DS135C-AT	0023
DS135D	0023
DS135D-FA3	0023
DS135D-KB1	OEM
DS135D-KB3	OEM
DS135E	0023
DS135F	0023
DS-149	0244
DS149	0244
DS150	0071
DS150A	0071
DS150B	0071
DS150C	0071
DS151	OEM
DS152	OEM
DS-159	0025
DS159	0636
DS160(G.E.)	0015
DS-170	OEM
DS170-18L	OEM
DS170-20L	OEM
DS170-25L	OEM
DS175-02L	OEM
DS175-04L	OEM
DS175-06L	OEM
DS175-08L	OEM
DS175-10L	OEM
DS175-12L	OEM
DS175-14L	OEM
DS175-16L	OEM
DS183AA	3189
DS183BA	2901
DS183CA	1124
DS183DA	1124
DS183EA	1126
DS183FA	1126
DS183GA	0805
DS183HA	0805
DS183KA	1195
DS183MA	0319
DS183PA	1132
DS183RA	1132
DS183TA	0044
DS183VA	0044
DS184	OEM
DS185D	OEM
DS185E	OEM
DS185F	OEM
DS-189	0317
DS189	0317
DS-190	2071

DEVICE TYPE	REPL CODE
DS190	OEM
DS200	OEM
DS201	OEM
DS202	OEM
DS203AA	3189
DS203BA	2901
DS203CA	1124
DS203DA	1124
DS203EA	1126
DS203FA	1126
DS203GA	0805
DS203HA	0805
DS203KA	1195
DS203MA	0319
DS203PA	1132
DS203RA	1132
DS203TA	0044
DS203VA	1630
DS230A	0102
DS230B	0023
DS230C	0017
DS230D	0017
DS240-02A	OEM
DS240-02A#	1023
DS240-04A	OEM
DS240-04A#	1036
DS240-06A	OEM
DS240-06A#	1036
DS240-07A	OEM
DS240-07A#	1046
DS250-04F	OEM
DS250-04L	OEM
DS250-07F	OEM
DS250-07L	OEM
DS250-11F	OEM
DS250-11L	OEM
DS250-14F	OEM
DS250-14L	OEM
DS250-17F	OEM
DS250-17L	OEM
DS250-20F	OEM
DS250-20L	OEM
DS250-23F	OEM
DS250-23L	OEM
DS253AA	3189
DS253BA	2901
DS253CA	1124
DS253DA	1124
DS253EA	1126
DS253FA	1126
DS253GA	0805
DS253HA	0805
DS253KA	1195
DS253MA	0319
DS253PA	1132
DS253RA	1132
DS253TA	0044
DS253VA	0044
DS290-08J	OEM
DS290-12J	OEM
DS290-14J	OEM
DS290-16J	OEM
DS300	OEM
DS301	OEM
DS302	OEM
DS303AA	3189
DS303BA	2901
DS303CA	1124
DS303DA	1124
DS303EA	1126
DS303FA	1126
DS303GA	0805
DS303HA	0805
DS303KA	1195
DS303MA	0319
DS303PA	1132
DS303RA	1132
DS303TA	0044
DS303VA	0044
DS315N	OEM
DS316N	OEM
DS330A	0023
DS330B	0023
DS330C	0017
DS330D	0017
DS353AA	3189
DS353BA	2901
DS353CA	1124
DS353DA	1124
DS353EA	1126
DS353FA	1126
DS353GA	0805
DS353HA	0805
DS353KA	1195
DS353MA	0319
DS353PA	1132
DS353RA	1132
DS353TA	0044
DS353VA	0044
DS390-04	OEM
DS390-04J	OEM
DS390-06J	OEM
DS390-08J	OEM
DS400	OEM
DS400-07F	OEM
DS400-11F	OEM
DS400-14F	OEM
DS400-17F	OEM
DS400-20F	OEM

DEVICE TYPE	REPL CODE
DS400-23F	OEM
DS401	OEM
DS402	OEM
DS402-ST01	OEM
DS402-ST02	OEM
DS402-ST04	OEM
DS402-ST05	OEM
DS402-ST06	OEM
DS402ST01	OEM
DS402ST02	OEM
DS402ST03	OEM
DS402ST04	OEM
DS402ST05	OEM
DS402ST06	OEM
DS402ST07	OEM
DS402ST08	OEM
DS402ST09	OEM
DS402ST10	OEM
DS402ST11	OEM
DS402ST12	OEM
DS403AA	3189
DS403BA	2901
DS403CA	1124
DS403DA	1124
DS403EA	1126
DS403FA	1126
DS403GA	0805
DS403HA	0805
DS403KA	1195
DS403MA	0319
DS403PA	1132
DS403RA	1132
DS403TA	0044
DS403VA	0044
DS-410	0133
DS410	0133
DS-410(AMPEX)	0133
DS410(COURIER)	0133
DS410(EMERSON)	0133
DS410(FANON)	0133
DS410(G.E.)	0143
DS-410(MOTOROLA)	0016
DS410(OLYMPIC)	0133
DS410(WESTINGHOUSE)	0133
DS410R	0133
DS410R(G.E.)	0143
DS412SE40	OEM
DS412SE41	OEM
DS412SE42	OEM
DS412SE43	OEM
DS412SE44	OEM
DS422S2	OEM
DS-430	0015
DS430	0015
DS430(VARISTOR)	0139
DS430-04J	OEM
DS430-06J	OEM
DS430-08J	OEM
DS430-10J	OEM
DS430-12J	OEM
DS430-14J	OEM
DS430-16J	OEM
DS441	0124
DS-442	0133
DS442	0133
DS442AT	OEM
DS442BT	OEM
DS442FM	0133
DS442S2	0133
DS442VB5	OEM
DS442VG1	OEM
DS442X	0124
DS442X-BT	0124
DS443	0133
DS446	0124
DS448	0133
DS448-FA1	0124
DS448FA1	0124
DS452	0023
DS454	0023
DS456	OEM
DS500	OEM
DS-501	0440
DS501	0440
DS502	0440
DS-503	0085
DS503	0085
DS504	0440
DS505	0440
DS506	0440
DS-509	0103
DS509	0103
DS-512	0086
DS512	0086
DS-513	0042
DS513	0042
DS-514	0103
DS514	0103
DS-515	0160
DS515	0160
DS-519	0103
DS519	0103
DS-520	0160
DS520	0160
DS525	0440
DS570	0440
DS677	OEM
DS-781	0127
DS804SE07	OEM

If replacement code is OEM, contact original manufacturer for replacement.

DEVICE TYPE	REPL CODE
DS804SE08	OEM
DS804SE09	OEM
DS804SE10	OEM
DS804SE11	OEM
DS804SE12	OEM
DS804SE13	OEM
DS804SE14	OEM
DS804SE15	OEM
DS804SE16	OEM
DS804SE17	OEM
DS804SE18	OEM
DS804SE19	OEM
DS804SE20	OEM
DS804SE21	OEM
DS820A	OEM
DS904LS18	OEM
DS904SL10	OEM
DS904SL11	OEM
DS904SL12	OEM
DS904SL13	OEM
DS904SL14	OEM
DS904SL15	OEM
DS904SL16	OEM
DS904SL17	OEM
DS904SL19	OEM
DS904SL20	OEM
DS904SL21	OEM
DS904SL22	OEM
DS904SL23	OEM
DS904SL24	OEM
DS904SL25	OEM
DS904SL26	OEM
DS904SL27	OEM
DS904SL28	OEM
DS904SL29	OEM
DS904SM10	OEM
DS904SM11	OEM
DS904SM12	OEM
DS904SM13	OEM
DS904SM14	OEM
DS904SM15	OEM
DS904SM16	OEM
DS904SM17	OEM
DS904SM18	OEM
DS904SM19	OEM
DS904SM20	OEM
DS904SM21	OEM
DS904SM22	OEM
DS904SM23	OEM
DS904SM24	OEM
DS904SM25	OEM
DS904SM26	OEM
DS904SM27	OEM
DS904SM28	OEM
DS904SM29	OEM
DS912SM25	OEM
DS912SM26	OEM
DS912SM27	OEM
DS912SM28	OEM
DS912SM29	OEM
DS912SM30	OEM
DS912SM31	OEM
DS912SM32	OEM
DS912SM33	OEM
DS912SM34	OEM
DS912SM35	OEM
DS912SM36	OEM
DS912SM37	OEM
DS912SM38	OEM
DS912SM39	OEM
DS912SM40	OEM
DS912SM41	OEM
DS912SM42	OEM
DS912SM43	OEM
DS912SM44	OEM
DS912SM45	OEM
DS912SM46	OEM
DS912SM47	OEM
DS912SM48	OEM
DS912SM49	OEM
DS912SM50	OEM
DS912SM51	OEM
DS912SM52	OEM
DS912SM53	OEM
DS912SM54	OEM
DS912SM55	OEM
DS912SM56	OEM
DS923SM47	OEM
DS1101SM01	OEM
DS1101SM02	OEM
DS1101SM03	OEM
DS1101SM04	OEM
DS1101SM05	OEM
DS1101SM06	OEM
DS1101SM07	OEM
DS1101SM08	OEM
DS1101SM09	OEM
DS1101SM10	OEM
DS1101SM11	OEM
DS1101SM12	OEM
DS1104SM01	OEM
DS1104SM02	OEM
DS1104SM03	OEM
DS1104SM04	OEM
DS1104SM05	OEM
DS1104SM06	OEM
DS1104SM07	OEM
DS1104SM08	OEM
DS1104SM09	OEM
DS1104SM10	OEM
DS1104SM11	OEM
DS1104SM12	OEM
DS1104SM13	OEM
DS1104SM14	OEM
DS1104SM15	OEM
DS1104SM16	OEM
DS1104SM17	OEM
DS1104SM18	OEM
DS1104SM19	OEM
DS1104SM20	OEM
DS1104SM21	OEM
DS1104SM22	OEM
DS1104SM23	OEM
DS1104SM24	OEM
DS1104SM25	OEM
DS1104SM26	OEM
DS1104SM27	OEM
DS1104SM28	OEM
DS1104SM29	OEM
DS1107SM24	OEM
DS1107SM25	OEM
DS1107SM26	OEM
DS1107SM27	OEM
DS1107SM28	OEM
DS1107SM29	OEM
DS1107SM30	OEM
DS1107SM31	OEM
DS1107SM32	OEM
DS1107SM33	OEM
DS1107SM34	OEM
DS1107SM35	OEM
DS1107SM36	OEM
DS1109SM30	OEM
DS1109SM31	OEM
DS1109SM32	OEM
DS1109SM33	OEM
DS1109SM34	OEM
DS1109SM35	OEM
DS1109SM36	OEM
DS1109SM37	OEM
DS1109SM38	OEM
DS1109SM39	OEM
DS1109SM40	OEM
DS1109SM41	OEM
DS1109SM42	OEM
DS1109SM43	OEM
DS1109SM44	OEM
DS1112SM47-18	OEM
DS1112SM48-18	OEM
DS1112SM49-18	OEM
DS1112SM50-18	OEM
DS1112SM51-18	OEM
DS1112SM52-18	OEM
DS1112SM53-18	OEM
DS1112SM54-18	OEM
DS1112SM55-18	OEM
DS1112SM56-18	OEM
DS1488J	0508
DS1488N	0508
DS1489AJ	0511
DS1489AN	0511
DS1489J	0511
DS1489N	0511
DS1603J	OEM
DS1611H	OEM
DS1611J-8	OEM
DS1612H	OEM
DS1612J-8	OEM
DS1613H	OEM
DS1613J-8	OEM
DS1614H	OEM
DS1614J-8	OEM
DS1628J	OEM
DS1630J	OEM
DS1631H	OEM
DS1631J-8	OEM
DS1632H	OEM
DS1632J-8	OEM
DS1633H	OEM
DS1633J-8	OEM
DS1634H	OEM
DS1634J-8	OEM
DS1644J	OEM
DS1645J	OEM
DS1647D	OEM
DS1648J	OEM
DS1649J	OEM
DS1650J	OEM
DS1651J	OEM
DS1652J	OEM
DS1653J	OEM
DS1671H	OEM
DS1671J	OEM
DS1671J-8	OEM
DS1674J	OEM
DS1675J	OEM
DS1677D	OEM
DS1678J	OEM
DS1679J	OEM
DS1686H	OEM
DS1686J	OEM
DS1686J-8	OEM
DS1687H	OEM
DS1687J	OEM
DS1687J-8	OEM
DS1691J	OEM
DS1692J	OEM
DS2004SN10	OEM
DS2004SN11	OEM
DS2004SN12	OEM
DS2004SN13	OEM
DS2004SN14	OEM
DS2004SN15	OEM
DS2004SN16	OEM
DS2004SN17	OEM
DS2004SN18	OEM
DS2004SN19	OEM
DS2004SN20	OEM
DS2004SN21	OEM
DS2004SN22	OEM
DS2004SN23	OEM
DS2004SN24	OEM
DS2007SN24	OEM
DS2007SN25	OEM
DS2007SN26	OEM
DS2007SN29	OEM
DS2007SN30	OEM
DS2007SN31	OEM
DS2007SN32	OEM
DS2007SN33	OEM
DS2007SN34	OEM
DS2007SN35	OEM
DS2007SN36	OEM
DS2007SN37	OEM
DS2007SN38	OEM
DS2007SN39	OEM
DS2007SN40	OEM
DS2007SN2728	OEM
DS2009SN30	OEM
DS2009SN31	OEM
DS2009SN32	OEM
DS2009SN33	OEM
DS2009SN34	OEM
DS2009SN35	OEM
DS2009SN36	OEM
DS2009SN37	OEM
DS2009SN38	OEM
DS2009SN39	OEM
DS2009SN40	OEM
DS2009SN41	OEM
DS2009SN42	OEM
DS2009SN43	OEM
DS2009SN44	OEM
DS2012SM42	OEM
DS2012SM43	OEM
DS2012SM44	OEM
DS2012SM45	OEM
DS2012SM46	OEM
DS2012SM47	OEM
DS2012SM48	OEM
DS2012SM49	OEM
DS2012SM50	OEM
DS2012SM51	OEM
DS2012SM52	OEM
DS2012SM53	OEM
DS2012SM54	OEM
DS2012SM55	OEM
DS2012SM56	OEM
DS2016	OEM
DS2020	OEM
DS2030	OEM
DS2040	OEM
DS2050	OEM
DS2060	OEM
DS2070	OEM
DS2080	OEM
DS2106SU20	OEM
DS2106SU21	OEM
DS2106SU22	OEM
DS2106SU23	OEM
DS2106SU24	OEM
DS2106SU25	OEM
DS2106SU26	OEM
DS2106SU27	OEM
DS2106SU28	OEM
DS2106SU29	OEM
DS2106SU30	OEM
DS2106SU31	OEM
DS2106SU32	OEM
DS2106SU33	OEM
DS2106SU34	OEM
DS2106SU35	OEM
DS2106SU36	OEM
DS2657	OEM
DS3010	OEM
DS3020	OEM
DS3030	OEM
DS3040	OEM
DS3245J	OEM
DS3245N	OEM
DS3486J	OEM
DS3486N	OEM
DS3487J	OEM
DS3487N	OEM
DS3587J	OEM
DS3603J	OEM
DS3603N	OEM
DS3604J	OEM
DS3604N	OEM
DS3611H	OEM
DS3611J-8	OEM
DS3611N	OEM
DS3612H	OEM
DS3612J-8	OEM
DS3612N	OEM
DS3613H	OEM
DS3613J-8	OEM
DS3613N	OEM
DS3614H	OEM
DS3614J-8	OEM
DS3614N	OEM
DS3617N	OEM
DS3628J	OEM
DS3628N	OEM
DS3630J	OEM
DS3630N	OEM
DS3631H	OEM
DS3631J-8	OEM
DS3631N	OEM
DS3632H	OEM
DS3632J-8	OEM
DS3632N	OEM
DS3633H	OEM
DS3633J	OEM
DS3633J-8	OEM
DS3633N	OEM
DS3634H	OEM
DS3634N	OEM
DS3642H	OEM
DS3642J-8	OEM
DS3644J	OEM
DS3644N	OEM
DS3645J	OEM
DS3647D	OEM
DS3647N	OEM
DS3648N	OEM
DS3649J	OEM
DS3649N	OEM
DS3650J	OEM
DS3650N	OEM
DS3651J	OEM
DS3651N	OEM
DS3652J	OEM
DS3652N	OEM
DS3653J	OEM
DS3653N	OEM
DS3654N	OEM
DS3662N	OEM
DS3671H	OEM
DS3671J	OEM
DS3671J-8	OEM
DS3671N	OEM
DS3674J	OEM
DS3674N	OEM
DS3675J	OEM
DS3675N	OEM
DS3677D	OEM
DS3677N	OEM
DS3678J	OEM
DS3678N	OEM
DS3679J	OEM
DS3679N	OEM
DS3680J	OEM
DS3680N	OEM
DS3686H	OEM
DS3686J	OEM
DS3686J-8	OEM
DS3686N	OEM
DS3687H	OEM
DS3687J	OEM
DS3687J-8	OEM
DS3687N	OEM
DS3691J	OEM
DS3691N	OEM
DS3692J	OEM
DS3692N	OEM
DS4001	OEM
DS4003	0023
DS4148	0133
DS4510	OEM
DS4520	OEM
DS4530	OEM
DS4540	OEM
DS4550	OEM
DS4560	OEM
DS4570	OEM
DS4638J	OEM
DS4645N	OEM
DS5295G	0023
DS5295J	0023
DS5534J	OEM
DS5538J	OEM
DS7640J	OEM
DS7641J	OEM
DS7800H	OEM
DS7810J	OEM
DS7811J	OEM
DS7812J	OEM
DS7819J	OEM
DS7819W	OEM
DS7820AF	OEM
DS7820AJ	OEM
DS7820AN	OEM
DS7820AW	OEM
DS7820F	OEM
DS7820J	OEM
DS7820N	OEM
DS7830F	OEM
DS7830J	OEM
DS7830N	OEM
DS7831	OEM
DS7831J	OEM
DS7832	OEM
DS7832J	OEM
DS7833J	OEM
DS7834J	OEM
DS7835J	OEM
DS7836J	OEM
DS7837J	OEM
DS7838J	OEM
DS7839J	OEM
DS7856J	OEM
DS7858J	OEM
DS7880J	OEM
DS7889J	OEM
DS7897AJ	OEM
DS8626N	OEM
DS8629N	OEM
DS8640J	OEM
DS8640N	OEM
DS8641J	OEM
DS8641N	OEM
DS8647N	OEM
DS8648N	OEM
DS8654N	OEM
DS8656N	OEM
DS8664N	OEM
DS8665N	OEM
DS8666N	OEM
DS8667N	OEM
DS8668	OEM
DS8669N	OEM
DS8692N	OEM
DS8800H	OEM
DS8810J	OEM
DS8810N	OEM
DS8811J	OEM
DS8811N	OEM
DS8812J	OEM
DS8812N	OEM
DS8819J	OEM
DS8819N	OEM
DS8820AF	OEM
DS8820AJ	OEM
DS8820AN	OEM
DS8820F	OEM
DS8820J	OEM
DS8820N	OEM
DS8822N	OEM
DS8830	OEM
DS8830F	OEM
DS8830J	OEM
DS8830N	OEM
DS8831	OEM
DS8831J	OEM
DS8831N	OEM
DS8832	OEM
DS8832J	OEM
DS8832N	OEM
DS8833J	OEM
DS8833N	OEM
DS8834J	OEM
DS8834N	OEM
DS8835J	OEM
DS8835N	OEM
DS8836J	OEM
DS8836N	OEM
DS8837J	OEM
DS8837N	OEM
DS8838	OEM
DS8838J	OEM
DS8838N	OEM
DS8839J	OEM
DS8839N	OEM
DS8856J	OEM
DS8856N	OEM
DS8857J	OEM
DS8858J	OEM
DS8858N	OEM
DS8859AJ	OEM
DS8859AN	OEM
DS8861N	OEM
DS8863N	OEM
DS8867N	OEM
DS8869AJ	OEM
DS8869AN	OEM
DS8870	1701
DS8870J	1697
DS8870N	1701
DS8871N	OEM
DS8872N	OEM
DS8873N	OEM
DS8874N	OEM
DS8877N	OEM
DS8880F	1793
DS8880J	1793
DS8880N	1793
DS8881N	OEM
DS8884AN	OEM
DS8885J	OEM
DS8885N	OEM
DS8887J	OEM
DS8887N	OEM
DS8889J	1877
DS8889N	1877
DS8891J	OEM
DS8891N	OEM
DS8897AJ	OEM
DS8897AN	OEM
DS8906	OEM
DS8907	OEM
DS8908	OEM
DS8963N	OEM
DS8973N	OEM
DS8975N	OEM
DS8980N	OEM
DS8981J	OEM
DS8981N	OEM
DS16149J	OEM
DS16177D	OEM
DS16179J	OEM
DS36147D	OEM
DS36147N	OEM
DS36149J	OEM
DS36149N	OEM
DS36177D	OEM
DS36177N	OEM
DS36179J	OEM
DS36179N	OEM
DS54552J-8	OEM
DS54553J-8	OEM
DS54554J-8	OEM
DS55107J	OEM
DS55108J	OEM
DS55109J	OEM
DS55113J	OEM
DS55114J	OEM
DS55115J	OEM
DS55121J	OEM
DS55122J	OEM
DS55325J	OEM
DS55451H	OEM
DS55451J-8	OEM
DS55452H	OEM
DS55452J-8	OEM
DS55453H	OEM
DS55453J-8	OEM
DS55454H	OEM
DS55454J-8	OEM
DS55461H	OEM
DS55461J-8	OEM
DS55462H	OEM
DS55462J-8	OEM
DS55463H	OEM
DS55463J-8	OEM
DS55464H	OEM
DS55464J-8	OEM
DS55493J	OEM
DS55494J	OEM
DS74592J	OEM
DS75107J	OEM
DS75107N	OEM
DS75108J	OEM
DS75108N	OEM
DS75113J	OEM
DS75113N	OEM
DS75114J	OEM
DS75114N	OEM
DS75115J	OEM
DS75115N	OEM
DS75121J	OEM
DS75121N	OEM
DS75122J	OEM
DS75122N	OEM
DS75123J	OEM
DS75123N	OEM
DS75124J	OEM
DS75124N	OEM
DS75125J	OEM
DS75125N	OEM
DS75127J	OEM
DS75127N	OEM
DS75128J	OEM
DS75128N	OEM
DS75129J	OEM
DS75129N	OEM
DS75150J	OEM
DS75150J-8	OEM
DS75150N	OEM
DS75154J	OEM
DS75154N	OEM
DS75207J	OEM
DS75207N	OEM
DS75208J	OEM
DS75208N	OEM
DS75325J	OEM
DS75325N	OEM
DS75361J-8	OEM
DS75361N	OEM
DS75362J-8	OEM
DS75362N	OEM
DS75364J-8	OEM
DS75365J	OEM
DS75365N	OEM
DS75450J	OEM
DS75450N	1231
DS75451H	OEM
DS75451J-8	OEM
DS75451M	OEM
DS75451N	1244
DS75452H	OEM
DS75452J-8	OEM
DS75452M	OEM
DS75452N	1262
DS75453H	OEM
DS75453J-8	OEM
DS75453M	OEM
DS75453N	1272
DS75454H	OEM
DS75454J-8	OEM
DS75454N	1289
DS75461H	OEM
DS75461J-8	OEM
DS75461N	OEM
DS75462H	OEM
DS75462J-8	OEM
DS75462N	OEM
DS75463H	OEM
DS75463J-8	OEM
DS75463N	OEM
DS75464H	OEM
DS75464J-8	OEM
DS75464N	OEM
DS75491J	OEM
DS75491N	1729
DS75492J	OEM
DS75492N	1740
DS75493J	OEM
DS75493N	5021
DS75494J	OEM
DS75494N	5022
DSA010	0124
DSA015	0124
DSA0.5A1100	OEM
DSA0.5A1400	OEM
DSA0.5A1700	OEM
DSA0.9-12A	OEM
DSA0.9-14A	OEM
DSA0.9-16A	OEM
DSA0.9-18A	OEM
DSA1A1	0080
DSA1A2	0609
DSA1A4	0795
DSA1A1100	OEM
DSA1A1400	OEM
DSA1A1700	OEM
DSA1C1	0023
DSA1.2-12A	OEM
DSA1.2-14A	OEM
DSA1.2-16A	OEM
DSA1.2-18A	OEM
DSA1.8-18A	OEM
DSA2-12A	OEM
DSA2-14A	OEM
DSA2-16A	OEM
DSA2-18A	OEM
DSA3A1	OEM
DSA3A1100	OEM
DSA3A1400	OEM
DSA3A1700	OEM
DSA5B1100	OEM
DSA5B1400	OEM
DSA5B1700	OEM
DSA6-12F	OEM
DSA6-14F	OEM
DSA6-16F	OEM
DSA6-18A	OEM
DSA9-12F	OEM
DSA9-14F	OEM
DSA9-16F	OEM
DSA9-18A	OEM
DSA10C1100	OEM
DSA10C1400	OEM
DSA10C1700	OEM
DSA10G	0071
DSA10J	0071
DSA10L	0071
DSA12B	0087
DSA12C	0087
DSA12E	0087
DSA12EQ	0087
DSA12G	0087
DSA12J	0087
DSA12L	0087
DSA12T	0071
DSA12TL	0087
DSA12TL-AT1	0087
DSA12TL-FB2	0071
DSA17-12A	OEM
DSA17-14A	OEM
DSA17-16A	OEM
DSA17-18A	OEM
DSA17B	0071
DSA17C	0564
DSA17E	0071
DSA17KC-5	OEM
DSA20B	0102
DSA20C	0102
DSA20E	0102
DSA20G	0102
DSA20J	0102
DSA22-12A	OEM
DSA22-14A	OEM
DSA22-16A	OEM
DSA22-18A	OEM
DSA26B	0087
DSA26C	0087
DSA26E	0087
DSA35-12A	OEM
DSA35-14A	OEM
DSA35-16A	OEM
DSA35-18A	OEM
DSA-40SN110	OEM
DSA42-12A	OEM
DSA42-14A	OEM
DSA42-16A	OEM
DSA42-18A	OEM
DSA-64SN110	OEM
DSA75-12B	OEM
DSA75-14B	OEM
DSA75-16B	OEM
DSA80-12A	OEM
DSA80-14A	OEM

DEVICE TYPE	REPL CODE
DSA80-16A	OEM
DSA80-18A	OEM
DSA110-12A	OEM
DSA110-14A	OEM
DSA110-16A	OEM
DSA110-18A	OEM
DSA150	0133
DSA175-11B	OEM
DSA175-14B	OEM
DSA200B1100	OEM
DSA200B1400	OEM
DSA200B1700	OEM
DSA250-11F	OEM
DSA250-11L	OEM
DSA250-14F	OEM
DSA250-14L	OEM
DSA250-17F	OEM
DSA250-17L	OEM
DSA250-20F	OEM
DSA250-20L	OEM
DSA250-23F	OEM
DSA250-23L	OEM
DSA251-54F	OEM
DSA301LS-C	OEM
DSA304-11A	OEM
DSA304-14A	OEM
DSA304-17A	OEM
DSA304-20A	OEM
DSA304-23A	OEM
DSA400-11F	OEM
DSA400-11KC	OEM
DSA400-11NC	OEM
DSA400-14F	OEM
DSA400-14KC	OEM
DSA400-14NC	OEM
DSA400-17F	OEM
DSA400-17KC	OEM
DSA400-17NC	OEM
DSA400-20F	OEM
DSA400-20KC	OEM
DSA400-20NC	OEM
DSA400-23F	OEM
DSA400-23KC	OEM
DSA400-23NC	OEM
DSA401-26GE	OEM
DSA401-26LE	OEM
DSA401-29GE	OEM
DSA401-29LE	OEM
DSA401-32GE	OEM
DSA401-32LE	OEM
DSA401-38GG	OEM
DSA401-38LG	OEM
DSA401-44GG	OEM
DSA401-44LG	OEM
DSA401-50GG	OEM
DSA401-50LG	OEM
DSA401-54F	OEM
DSA403-38GG	OEM
DSA403-38L1G	OEM
DSA403-44GG	OEM
DSA403-44L1G	OEM
DSA403-50GG	OEM
DSA403-50L1G	OEM
DSA405-38AG	OEM
DSA405-44AG	OEM
DSA405-50AG	OEM
DSA503-23GE	OEM
DSA503-23L1E	OEM
DSA503-26GE	OEM
DSA503-26L1E	OEM
DSA503-29G	OEM
DSA503-29GE	OEM
DSA503-29L1E	OEM
DSA503-32GE	OEM
DSA503-32L1E	OEM
DSA603-11GC	OEM
DSA603-11L1C	OEM
DSA603-14GC	OEM
DSA603-14L1C	OEM
DSA603-17GC	OEM
DSA603-17L1C	OEM
DSA603-20GC	OEM
DSA603-20L1C	OEM
DSA605-23AE	OEM
DSA605-26AE	OEM
DSA605-29AE	OEM
DSA605-32AE	OEM
DSA607-38A	OEM
DSA705-11A	OEM
DSA705-11AC	OEM
DSA705-14A	OEM
DSA705-14AC	OEM
DSA705-17A	OEM
DSA705-17AC	OEM
DSA705-20A	OEM
DSA705-20AC	OEM
DSA705-23B	OEM
DSA707-23A	OEM
DSA707-26A	OEM
DSA707-29A	OEM
DSA707-32A	OEM
DSA807-11A	OEM
DSA807-14A	OEM
DSA807-17A	OEM
DSA807-20A	OEM
DSA908-38AG	OEM
DSA908-44AG	OEM
DSA908-50AG	OEM
DSA1208-23AE	OEM
DSA1208-26AE	OEM
DSA1208-29AE	OEM
DSA1208-32AE	OEM
DSA1508-11AC	OEM
DSA1508-14AC	OEM
DSA1508-17AC	OEM
DSA1508-20AC	OEM
DSA1508-23A	OEM
DSA6920	OEM
DSA6920A	OEM
DSA6920B	OEM
DSA6920C	OEM
DSA6921	OEM
DSA6921A	OEM
DSA6922	OEM
DSA6922A	OEM
DSA6922B	OEM
DSA6922C	OEM
DSA6923	OEM
DSA6923A	OEM
DSA6923B	OEM
DSA6923C	OEM
DSA-6925	OEM
DSA6925	OEM
DSA-6925A	OEM
DSA6925A	OEM
DSA-6925B	OEM
DSA6925B	OEM
DSA-6925C	OEM
DSA6925C	OEM
DSA6926	OEM
DSA6926A	OEM
DSA6927	OEM
DSA6927A	OEM
DSA6927B	OEM
DSA6927C	OEM
DSA-6928	OEM
DSA6928	OEM
DSA-6928A	OEM
DSA6928A	OEM
DSA-6928B	OEM
DSA6928B	OEM
DSA-6928C	OEM
DSA6928C	OEM
DSAI17-16A	OEM
DSAI17-18A	OEM
DSAI35-18A	OEM
DSAI75-11B	OEM
DSAI75-14B	OEM
DSAI75-16A	OEM
DSAI80-16A	OEM
DSAI80-18A	OEM
DSAI110-16A	OEM
DSAI110-18A	OEM
DSAI250-11F	OEM
DSAI250-11L	OEM
DSAI250-14F	OEM
DSAI250-14L	OEM
DSAI250-17F	OEM
DSAI250-17G	OEM
DSAI250-17L	OEM
DSAI250-17M	OEM
DSAI250-20F	OEM
DSAI250-20G	OEM
DSAI250-20L0L	OEM
DSAI250-20M	OEM
DSAI250-23F	OEM
DSAI250-23G	OEM
DSAI250-23L	OEM
DSAI250-23M	OEM
DSAS05-0	OEM
DSAS05-2	OEM
DSAS05-4	OEM
DSAS05-6	OEM
DSAS07-0	OEM
DSAS07-2	OEM
DSAS07-4	OEM
DSAS07-6	OEM
DSAS09-0	OEM
DSAS09-2	OEM
DSAS09-4	OEM
DSAS09-6	OEM
DSAS5U	OEM
DSAS6U	OEM
DSAS7U	OEM
DSAS8U	OEM
DSAS9U	OEM
DSAS10U	OEM
DSAS11-0	OEM
DSAS11-2	OEM
DSAS11-4	OEM
DSAS11-6	OEM
DSAS11U	OEM
DSAS12U	OEM
DSAS13-0	OEM
DSAS13-2	OEM
DSAS13-4	OEM
DSAS13-6	OEM
DSAS13U	OEM
DSAS14U	OEM
DSAS15-0	OEM
DSAS15-1	OEM
DSAS15-2	OEM
DSAS15-3	OEM
DSAS15-4	OEM
DSAS15-5	OEM
DSAS15-6	OEM
DSAS15-7	OEM
DSAS15-8	OEM
DSAS15-9	OEM
DSAS15U	OEM
DSAS16U	OEM
DSAS17-0	OEM
DSAS17-4	OEM
DSAS19-0	OEM
DSAS19-4	OEM
DSAS21-0	OEM
DSAS21-4	OEM
DSAS23-0	OEM
DSAS23-4	OEM
DSAS25-0	OEM
DSAS25-4	OEM
DSAS27-0	OEM
DSAS27-4	OEM
DSAS29-0	OEM
DSAS29-4	OEM
DSAS31-0	OEM
DSAS31-4	OEM
DSB010	0124
DSB015	0124
DSB015A2-TA	OEM
DSB15B	0102
DSB15C	0071
DSB15E	0102
DSB15G	0102
DSB15TC	OEM
DSB6477A	OEM
DSB6477B	OEM
DSB6477C	OEM
DSB6477D	OEM
DSB6477E	OEM
DSB6478A	OEM
DSB6478B	OEM
DSB6478C	OEM
DSB6478D	OEM
DSB6478E	OEM
DSB6479A	OEM
DSB6479B	OEM
DSB6479C	OEM
DSB6479D	OEM
DSB6479E	OEM
DSB6480A	OEM
DSB6480B	OEM
DSB6480C	OEM
DSB6480D	OEM
DSB6480E	OEM
DSB6481A	OEM
DSB6481B	OEM
DSB6481C	OEM
DSB6481D	OEM
DSB6481E	OEM
DSB6482A	OEM
DSB6482B	OEM
DSB6482C	OEM
DSB6482D	OEM
DSB6482E	OEM
DSB6483A	OEM
DSB6483B	OEM
DSB6483C	OEM
DSB6483D	OEM
DSB6483E	OEM
DSB6484A	OEM
DSB6484B	OEM
DSB6484C	OEM
DSB6484D	OEM
DSB6484E	OEM
DSB6485A	OEM
DSB6485B	OEM
DSB6485C	OEM
DSB6485D	OEM
DSB6485E	OEM
DSB6486A	OEM
DSB6486B	OEM
DSB6486C	OEM
DSB6486D	OEM
DSB6486E	OEM
DSB6487A	OEM
DSB6487B	OEM
DSB6487C	OEM
DSB6487D	OEM
DSB6487E	OEM
DSB6488A	OEM
DSB6488B	OEM
DSB6488C	OEM
DSB6488D	OEM
DSB6488E	OEM
DSB6489A	OEM
DSB6489B	OEM
DSB6489C	OEM
DSB6489D	OEM
DSB6489E	OEM
DSB6490A	OEM
DSB6490B	OEM
DSB6490C	OEM
DSB6490D	OEM
DSB6490E	OEM
DSB6491A	OEM
DSB6491B	OEM
DSB6491C	OEM
DSB6491D	OEM
DSB6491E	OEM
DSB6492A	OEM
DSB6492B	OEM
DSB6492C	OEM
DSB6492D	OEM
DSB6492E	OEM
DSBC941	OEM
DSBN-1	0015
DSC010	0124
DSC30TC-KD2	OEM
DSC5012A115-60	OEM
DSC5012A115-60M	OEM
DSC5012B26-400	OEM
DSC5012B26-400M	OEM
DSC5012B115-60	OEM
DSC5012B115-60M	OEM
DSC5012B115-400	OEM
DSC5012B115-400M	OEM
DSC5012C115-60M	OEM
DSC5112A115-400M	OEM
DSD010	0124
DSD0.8-02B	0031
DSD0.9-02A	OEM
DSD1.2-01A	OEM
DSD1.2-06A	OEM
DSD2-01A	OEM
DSD2-02A	OEM
DSD2-04A	OEM
DSD2-06A	OEM
DSD2-08A	OEM
DSD2-10A	OEM
DSD12-02A	OEM
DSD12-04A	OEM
DSD12-06A	OEM
DSD12-08A	OEM
DSD12-12A	OEM
DSD12-14A	OEM
DSD12-16A	OEM
DSD17-01A	OEM
DSD17-01B	OEM
DSD17-02A	OEM
DSD17-02B	OEM
DSD17-04A	OEM
DSD17-04B	OEM
DSD17-06B	OEM
DSD17-07A	OEM
DSD17-08B	OEM
DSD17-015B	OEM
DSD17-11A	OEM
DSD17-12B	OEM
DSD17-14A	OEM
DSD17-14B	OEM
DSD17-16B	OEM
DSD35-01A	OEM
DSD35-02A	OEM
DSD35-04A	OEM
DSD35-06A	OEM
DSD35-07A	OEM
DSD35-08A	OEM
DSD35-12A	OEM
DSD35-14A	OEM
DSD35-16A	OEM
DSD84	0912
DSD110-12A	OEM
DSD110-14A	OEM
DSD110-16A	OEM
DSD113-01B	OEM
DSD113-015B	OEM
DSD120-01B	OEM
DSD120-015B	OEM
DSD136-01B	OEM
DSD136-015B	OEM
DSD304-06A	OEM
DSD304-06C	OEM
DSD304-08A	OEM
DSD304-08C	OEM
DSD304-10A	OEM
DSD304-10C	OEM
DSD304-11A	OEM
DSD304-11C	OEM
DSD304-12A	OEM
DSD304-12C	OEM
DSD304-14A	OEM
DSD304-16A	OEM
DSD605-11A	OEM
DSD605-14A	OEM
DSD605-16A	OEM
DSD605-18A	OEM
DSD605-20A	OEM
DSD625-25A	OEM
DSD625-25D	OEM
DSD625-30A	OEM
DSD635-22A	OEM
DSD635-22D	OEM
DSD635-25A	OEM
DSD1028-25A	OEM
DSD1028-25D	OEM
DSD1028-30A	OEM
DSD1038-22A	OEM
DSD1038-22D	OEM
DSD8649V3FY	OEM
DSD8744	OEM
DSD8923	OEM
DSDI7-01B	OEM
DSDI7-015B	OEM
DSDI13-01B	OEM
DSDI13-015B	OEM
DSDI17-16B	OEM
DSDI20-01B	OEM
DSDI20-015B	OEM
DSDI35-16A	OEM
DSDI36-01B	OEM
DSDI36-015B	OEM
DSF10	OEM
DSF10A	1335
DSF10B	0023
DSF-10B-BT	0080
DSF10BT	OEM
DSF10C	0023
DSF10CBT	OEM
DSF10C-KB1	OEM
DSF10E	0023
DSF10G	0015
DSF10TB	0015
DSF10TB-BT-A	0015
DSF10TB-KB1	1892
DSF10TC	OEM
DSF402ST01	OEM
DSF402ST02	OEM
DSF402ST03	OEM
DSF402ST04	OEM
DSF402ST05	OEM
DSF402ST06	OEM
DSF402ST07	OEM
DSF402ST08	OEM
DSF402ST09	OEM
DSF402ST10	OEM
DSF402ST11	OEM
DSF402ST12	OEM
DSF502ST06	OEM
DSF502ST07	OEM
DSF502ST08	OEM
DSF502ST09	OEM
DSF502ST10	OEM
DSF502ST11	OEM
DSF502ST12	OEM
DSF805SE01	OEM
DSF805SE02	OEM
DSF805SE03	OEM
DSF805SE04	OEM
DSF805SE05	OEM
DSF805SE06	OEM
DSF805SE07	OEM
DSF805SE08	OEM
DSF805SE09	OEM
DSF805SE10	OEM
DSF805SE11	OEM
DSF805SE12	OEM
DSF808SE06	OEM
DSF808SE07	OEM
DSF808SE08	OEM
DSF808SE09	OEM
DSF808SE10	OEM
DSF808SE11	OEM
DSF808SE12	OEM
DSF912SM01	OEM
DSF912SM02	OEM
DSF912SM03	OEM
DSF912SM04	OEM
DSF912SM05	OEM
DSF912SM06	OEM
DSF912SM07	OEM
DSF912SM08	OEM
DSF912SM09	OEM
DSF912SM10	OEM
DSF912SM11	OEM
DSF912SM12	OEM
DSF912SM13	OEM
DSF912SM14	OEM
DSF912SM15	OEM
DSF912SM16	OEM
DSF912SM17	OEM
DSF912SM18	OEM
DSF912SM19	OEM
DSF912SM20	OEM
DSF912SM21	OEM
DSF912SM22	OEM
DSF912SM23	OEM
DSF912SM24	OEM
DSF912SM25	OEM
DSF912SM26	OEM
DSF1108SM01	OEM
DSF1108SM02	OEM
DSF1108SM03	OEM
DSF1108SM04	OEM
DSF1108SM05	OEM
DSF1108SM06	OEM
DSF1108SM07	OEM
DSF1108SM08	OEM
DSF1108SM09	OEM
DSF1108SM10	OEM
DSF1108SM11	OEM
DSF1108SM12	OEM
DSF1112SG09	OEM
DSF1112SG10	OEM
DSF1112SG11	OEM
DSF1112SG12	OEM
DSF1112SG13	OEM
DSF1112SG14	OEM
DSF1112SG15	OEM
DSF1112SG16	OEM
DSF1112SG17	OEM
DSF1112SG18	OEM
DSF1112SG19	OEM
DSF1112SG20	OEM
DSF1112SG21	OEM
DSF1112SG22	OEM
DSF1112SG23	OEM
DSF1112SG24	OEM
DSF1112SG25	OEM
DSF1112SM20	OEM
DSF1112SM21	OEM
DSF1112SM22	OEM
DSF1112SM23	OEM
DSF1112SM24	OEM
DSF1112SM25	OEM
DSF1114SM21	OEM
DSF1114SM22	OEM
DSF1114SM23	OEM
DSF1114SM24	OEM
DSF1114SM25	OEM
DSF1114SM26	OEM
DSF1114SM27	OEM
DSF1114SM28	OEM
DSF1114SM29	OEM
DSF1114SM30	OEM
DSF1114SM31	OEM
DSF1114SM32	OEM
DSF1114SM33	OEM
DSF1114SM34	OEM
DSF1114SM35	OEM
DSF1114SM36	OEM
DSF2013SD13	OEM
DSF2013SD14	OEM
DSF2013SD15	OEM
DSF2013SD16	OEM
DSF2013SD17	OEM
DSF2013SD18	OEM
DSF2013SD19	OEM
DSF2013SD20	OEM
DSF2013SD21	OEM
DSF2013SD22	OEM
DSF2013SD23	OEM
DSF2013SD24	OEM
DSF2013SD25	OEM
DSF2013SD26	OEM
DSF2030SD21	OEM
DSF2030SD22	OEM
DSF2030SD23	OEM
DSF2030SD25	OEM
DSF2030SD26	OEM
DSF2030SD27	OEM
DSF2030SD28	OEM
DSF2030SD29	OEM
DSF2030SD30	OEM
DSF2030SD31	OEM
DSF2030SD32	OEM
DSF2030SD33	OEM
DSF2030SD34	OEM
DSF2030SD35	OEM
DSF2030SD36	OEM
DSF2030SDD24	OEM
DSG6358A	OEM
DSG6358B	OEM
DSG6358C	OEM
DSG6358D	OEM
DSG6358E	OEM
DSG6470A	OEM
DSG6470B	OEM
DSG6470C	OEM
DSG6470D	OEM
DSG6470E	OEM
DSG6474A	OEM
DSG6474B	OEM
DSG6474C	OEM
DSG6474D	OEM
DSG6474E	OEM
DSH4785-10	OEM
DSH4785-40	OEM
DSH4787-10	OEM
DSH4787-15	OEM
DSH4787-20	OEM
DSH4787-30	OEM
DSH4787-40	OEM
DSH6172	OEM
DSH6173	OEM
DSH6174	OEM
DSI17-16A	OEM
DSI17-18A	OEM
DSI35-16A	OEM
DSI35-18A	OEM
DSI75-16A	OEM
DSI80-16A	OEM
DSI80-18A	OEM
DSI110-16A	OEM
DSI110-18A	OEM
DSI250-04F	OEM
DSI250-04L	OEM
DSI250-07F	OEM
DSI250-07L	OEM
DSI250-11F	OEM
DSI250-11L	OEM
DSI250-14F	OEM
DSI250-14L	OEM
DSI250-17F	OEM
DSI250-17G	OEM
DSI250-17L	OEM
DSI250-17M	OEM
DSI250-20F	OEM
DSI250-20G	OEM
DSI250-20L	OEM
DSI250-20M	OEM
DSI250-23F	OEM
DSI250-23G	OEM
DSI250-23L	OEM
DSI250-23M	OEM
DS-IM	0015
DS-IN	0208
DSJ6110	OEM
DSJ6111	OEM
DSJ6112	OEM
DSJ6811A	OEM
DSJ6811B	OEM
DSJ6812A	OEM
DSK10B	OEM
DSK10C	OEM
DSK10C-BT	OEM
DSM1A1	1090
DSM1D1	OEM
DSM30A4	OEM
DSM30A6	OEM
DSM4380A	OEM
DSM4380B	OEM
DSM4380C	OEM
DSM4380D	OEM
DSM4380E	OEM
DSM4381A	OEM
DSM4381B	OEM
DSM4381C	OEM
DSM4381D	OEM
DSM4381E	OEM
DSN6560-50	OEM
DSN6566-50	OEM
DSP2500A20	OEM
DSP2500A25	OEM
DSP2500A30	OEM
DSR51	OEM
DSR54	OEM
DSR55	OEM
DSR56	OEM
DSR3050	1090
DSR3051	1090
DSR3100	1090
DSR3101	1090
DSR3150	OEM
DSR3151	1090
DSR3200	OEM
DSR3201	1090
DSR3400X	OEM
DSR3500X	OEM
DSR3600X	OEM
DSR3700X	OEM
DSR3800X	OEM
DSR5050	OEM
DSR5051	OEM
DSR5052	OEM
DSR5100	OEM
DSR5101	OEM
DSR5102	OEM
DSR5150	OEM
DSR5151	OEM
DSR5200	OEM
DSR5201	OEM
DSR5202	OEM
DSR5400X	OEM
DSR5500X	OEM
DSR5600X	OEM
DSR5700X	OEM
DSR5800X	OEM
DSR31000X	OEM
DSR34000X	OEM
DSR35000X	OEM
DSR36000X	OEM
DSR37000X	OEM
DSR38000X	OEM
DSR51000X	OEM
DSR54000X	OEM
DSR55000X	OEM
DSR56000X	OEM
DSR57000X	OEM
DSR58000X	OEM
DSS100A8	OEM
DSS100A8R	OEM
DSS100A10	OEM
DSS100A10R	OEM
DSS100A13	OEM
DSS100A13R	OEM
DSS100A15	OEM
DSS100A15R	OEM
DSS300A8	OEM
DSS300A8R	OEM
DSS300A10	OEM
DSS300A10R	OEM
DSS300A13	OEM
DSS300A13R	OEM
DSS300A15	OEM
DSS300A15R	OEM
DSS953	0133
DSS16685	0143
DST1	OEM
DSZ412SE40	OEM
DSZ412SE41	OEM
DSZ412SE42	OEM
DSZ412SE43	OEM
DSZ412SE44	OEM
DSZ3006	OEM
DSZ3007	OEM
DSZ3008	3100
DSZ3009	0057
DSZ3010	OEM
DSZ3011	OEM
DSZ3012	0052
DSZ3013	OEM
DSZ3014	OEM
DSZ3015	OEM
DSZ3016	OEM
DSZ3018	OEM
DSZ3020	OEM
DSZ3022	OEM
DSZ3024	OEM
DSZ3027	OEM
DSZ3030	OEM
DSZ3033	OEM
DSZ3036	OEM

If replacement code is OEM, contact original manufacturer for replacement.

DEVICE TYPE	REPL CODE
DSZ3040	OEM
DSZ3045	OEM
DSZ3050	OEM
DSZ3056	OEM
DSZ3060	1156
DSZ3062	OEM
DSZ3070	OEM
DSZ3075	OEM
DSZ3080	OEM
DSZ3090	OEM
DSZ3100	OEM
DSZ4110	1180
DSZ4210	1180
DSZ4706	0205
DSZ4707	0480
DSZ4708	0504
DSZ4709	0684
DSZ4710	0225
DSZ4712	0234
DSZ4713	0237
DSZ4715	0247
DSZ4716	0251
DSZ4718	0256
DSZ4720	0262
DSZ4722	0269
DSZ4724	0274
DSZ4727	0294
DSZ4730	0309
DSZ4733	0318
DSZ4736	0010
DSZ4739	0320
DSZ4743	0337
DSZ4747	0347
DSZ4751	0027
DSZ4756	0266
DSZ4762	0386
DSZ4768	0405
DSZ4775	0426
DSZ4782	0444
DSZ4791	0238
DSZ4806	0205
DSZ4807	0480
DSZ4808	0504
DSZ4809	0684
DSZ4810	0225
DSZ4812	0234
DSZ4813	0237
DSZ4815	0247
DSZ4816	0256
DSZ4818	0256
DSZ4820	0262
DSZ4822	0269
DSZ4824	0274
DSZ4827	0294
DSZ4830	0309
DSZ4833	0318
DSZ4836	0320
DSZ4839	0326
DSZ4843	0337
DSZ4847	0347
DSZ4851	0027
DSZ4856	0266
DSZ4862	0386
DSZ4868	0405
DSZ4875	0426
DSZ4882	0444
DSZ4891	0238
DSZ5110	1180
DSZ5210	1180
DSZ5706	0205
DSZ5707	0480
DSZ5708	0504
DSZ5709	0684
DSZ5710	0225
DSZ5712	0234
DSZ5713	0237
DSZ5714	1397
DSZ5715	0247
DSZ5716	0251
DSZ5718	0256
DSZ5720	0262
DSZ5722	0269
DSZ5724	0274
DSZ5727	0294
DSZ5730	0309
DSZ5733	0318
DSZ5736	0320
DSZ5740	0326
DSZ5745	0347
DSZ5750	0027
DSZ5756	0266
DSZ5760	2856
DSZ5766	0405
DSZ5770	0405
DSZ5775	0426
DSZ5780	0444
DSZ5790	0238
DSZ5806	0205
DSZ5807	0480
DSZ5808	0504
DSZ5809	0684
DSZ5810	0225
DSZ5812	0234
DSZ5813	0237
DSZ5814	1397
DSZ5815	0247
DSZ5816	0251
DSZ5818	0256
DSZ5820	0262
DSZ5822	0269
DSZ5824	0274
DSZ5827	0294
DSZ5830	0309
DSZ5833	0318
DSZ5836	0320
DSZ5840	0326
DSZ5845	0347
DSZ5850	0027
DSZ5856	0266
DSZ5860	2856
DSZ5870	0405
DSZ5875	0426
DSZ5880	0444
DSZ5890	0238
DT2C14	OEM
DT2C15	OEM
DT10E	OEM
DT10F	OEM
DT25	OEM
DT41	0160
DT57C01	OEM
DT57C01DI	OEM
DT57C01SE	OEM
DT57C02	OEM
DT57C03	OEM
DT80	0440
DT88C	OEM
DT88E	OEM
DT88F	OEM
DT88G	OEM
DT100	0440
DT104BSXXXX	OEM
DT105DXXXX	OEM
DT110	OEM
DT110DSXXXX	OEM
DT112BSXXXX	OEM
DT161	0016
DT180FA	OEM
DT183AA	1630
DT183BA	0254
DT183CA	1107
DT183DA	1107
DT183EA	1111
DT183FA	1111
DT183GA	0258
DT183HA	0258
DT183KA	1644
DT183MA	0267
DT183PA	1119
DT183RA	1119
DT183TA	0281
DT183VA	0281
DT203AA	1630
DT203BA	0254
DT203CA	1107
DT203DA	1107
DT203EA	1111
DT203FA	1111
DT203GA	0258
DT203HA	0258
DT203KA	1644
DT203MA	0267
DT203PA	1119
DT203RA	1119
DT203TA	0281
DT203VA	0281
DT204BSXXXX	OEM
DT210FA	OEM
DT210FB	OEM
DT210FQFQ	OEM
DT210FW	OEM
DT210FX	OEM
DT210FY	OEM
DT212	OEM
DT212BSXXXX	OEM
DT214	OEM
DT215	OEM
DT220FA	OEM
DT220FB	OEM
DT220FQ	OEM
DT220FW	OEM
DT220FX	OEM
DT220FY	OEM
DT230A	0133
DT230B	0015
DT230C	OEM
DT230C#	OEM
DT230F	0133
DT230FA	OEM
DT230FB	OEM
DT230FQ	OEM
DT230FX	OEM
DT230FY	OEM
DT230G	0133
DT230H	0102
DT230H1	0015
DT230HI	OEM
DT230HI#	OEM
DT240FA	OEM
DT240FB	OEM
DT240FQ	OEM
DT240FW	OEM
DT240FX	OEM
DT240FY	OEM
DT250FA	OEM
DT250FB	OEM
DT250FW	OEM
DT250FX	OEM
DT250FY	OEM
DT253AA	1630
DT253BA	0254
DT253CA	1107
DT253DA	1107
DT253EA	1111
DT253FA	1111
DT253GA	0258
DT253HA	0258
DT253KA	1644
DT253MA	0267
DT253PA	1119
DT253RA	1119
DT253TA	0281
DT253VA	0281
DT260FA	OEM
DT260FB	OEM
DT260FQ	OEM
DT260FW	OEM
DT260FX	OEM
DT260FY	OEM
DT270FA	OEM
DT270FB	OEM
DT270FW	OEM
DT270FX	OEM
DT270FY	OEM
DT280FB	OEM
DT280FW	OEM
DT280FX	OEM
DT280FY	OEM
DT300	OEM
DT300A	OEM
DT301	OEM
DT302	OEM
DT303	OEM
DT303AA	1630
DT303BA	0254
DT303CA	1107
DT303DA	1107
DT303EA	1111
DT303FA	1111
DT303GA	0258
DT303HA	0258
DT303KA	1644
DT303MA	0267
DT303PA	1119
DT303RA	1119
DT303TA	0281
DT303VA	0281
DT304	OEM
DT305	OEM
DT306	OEM
DT306-300	OEM
DT307	OEM
DT307-300	OEM
DT308	OEM
DT308-300	OEM
DT309	OEM
DT309-300	OEM
DT310	OEM
DT310-300	OEM
DT310FA	OEM
DT310FB	OEM
DT310FQ	OEM
DT310FW	OEM
DT310FX	OEM
DT310FY	OEM
DT311	OEM
DT311-300	OEM
DT312	OEM
DT312-300	OEM
DT313	OEM
DT313-300	OEM
DT314	OEM
DT314-300	OEM
DT315	OEM
DT315-300	OEM
DT316	OEM
DT320FA	OEM
DT320FB	OEM
DT320FQ	OEM
DT320FW	OEM
DT320FX	OEM
DT320FY	OEM
DT330FA	OEM
DT330FB	OEM
DT330FX	OEM
DT330FY	OEM
DT340FA	OEM
DT340FB	OEM
DT340FQ	OEM
DT340FW	OEM
DT340FX	OEM
DT340FY	OEM
DT350FA	OEM
DT350FB	OEM
DT350FX	OEM
DT350FY	OEM
DT353AA	1630
DT353BA	0254
DT353CA	1107
DT353DA	1107
DT353EA	1111
DT353FA	1111
DT353GA	0258
DT353HA	0258
DT353KA	1644
DT353MA	0267
DT353PA	1119
DT353RA	1119
DT353TA	0281
DT353VA	0281
DT360FA	OEM
DT360FB	OEM
DT360FQ	OEM
DT360FW	OEM
DT360FX	OEM
DT360FY	OEM
DT380FA	OEM
DT380FB	OEM
DT380FW	OEM
DT380FX	OEM
DT380FY	OEM
DT390FA	OEM
DT390FB	OEM
DT390FX	OEM
DT390FY	OEM
DT401	0085
DT403AA	1630
DT403BA	0254
DT403CA	1107
DT403DA	1107
DT403EA	1111
DT403FA	1111
DT403GA	0258
DT403HA	0258
DT403KA	1644
DT403MA	0267
DT403PA	1119
DT403RA	1119
DT403TA	0281
DT403VA	0281
DT408BSXXXX	OEM
DT510DSXXXX	OEM
DT520DSXXXX	OEM
DT600	OEM
DT600A	OEM
DT601	OEM
DT602	OEM
DT603	OEM
DT604	OEM
DT605	OEM
DT606	OEM
DT606-300	OEM
DT607	OEM
DT607-300	OEM
DT608	OEM
DT608-300	OEM
DT609	OEM
DT609-300	OEM
DT610	OEM
DT610-300	OEM
DT611	OEM
DT611-300	OEM
DT612	OEM
DT612-300	OEM
DT613	OEM
DT613-300	OEM
DT614	OEM
DT614-300	OEM
DT615	OEM
DT615-300	OEM
DT616	OEM
DT711BSXXXX	OEM
DT812	0286
DT816BSXXXX	OEM
DT820	OEM
DT825	OEM
DT830	OEM
DT835	OEM
DT900	OEM
DT900A	OEM
DT901	OEM
DT902	OEM
DT903	OEM
DT904	OEM
DT905	OEM
DT906	OEM
DT906-300	OEM
DT907	OEM
DT907-300	OEM
DT908	OEM
DT908-300	OEM
DT909	OEM
DT909-300	OEM
DT910	OEM
DT910-300	OEM
DT911	OEM
DT911-300	OEM
DT912	OEM
DT912-300	OEM
DT913	OEM
DT913-300	OEM
DT914	OEM
DT914-300	OEM
DT915	OEM
DT915-300	OEM
DT916	OEM
DT1003	0233
DT1040	0160
DT1110	0086
DT1111	0086
DT1112	0086
DT1120	0086
DT1121	0086
DT1122	0086
DT1311	0086
DT1321	0086
DT1510	0086
DT1511	0086
DT1512	0086
DT1520	0086
DT1521	0086
DT1522	0086
DT1602	0233
DT1603	0233
DT1610	0198
DT1612	0233
DT1613	0233
DT1621	0086
DT1711D114	OEM
DT1711D116	OEM
DT1711D132	OEM
DT1711DI	OEM
DT1711DIC	OEM
DT1711DIDMA	OEM
DT1711DIPG	OEM
DT1711SE	OEM
DT1711SE14	OEM
DT1711SE32	OEM
DT1711SE64	OEM
DT1711SEC	OEM
DT1711SEDMA	OEM
DT1711SEPG	OEM
DT1712DI	OEM
DT1712DI14	OEM
DT1712DI16	OEM
DT1712DI32	OEM
DT1712DIC	OEM
DT1712DIDMA	OEM
DT1712DIPG	OEM
DT1712SE	OEM
DT1712SE14	OEM
DT1712SE32	OEM
DT1712SE64	OEM
DT1712SEC	OEM
DT1712SEDMA	OEM
DT1712SEPG	OEM
DT1715	OEM
DT1715DMA	OEM
DT1715R	OEM
DT1719	OEM
DT1719DMA	OEM
DT1719PG	OEM
DT1719R	OEM
DT1722	OEM
DT1723	OEM
DT1724BSXXXX	OEM
DT1735	OEM
DT1738	OEM
DT1739	OEM
DT1741	OEM
DT1742	OEM
DT1744	OEM
DT1748	OEM
DT1751	OEM
DT1755	OEM
DT1759	OEM
DT1761	OEM
DT1762	OEM
DT1764	OEM
DT1765	OEM
DT1768	OEM
DT1769	OEM
DT1781	OEM
DT1782	OEM
DT1784	OEM
DT1785	OEM
DT1788	OEM
DT1789	OEM
DT1841	OEM
DT1842	OEM
DT1843	OEM
DT2009	OEM
DT2010	OEM
DT2100FA	OEM
DT2100FB	OEM
DT2100FX	OEM
DT2100FY	OEM
DT2120FB	OEM
DT2722	OEM
DT2724	OEM
DT2725	OEM
DT2726	OEM
DT2727	OEM
DT2762	OEM
DT2764	OEM
DT2765	OEM
DT2766	OEM
DT2767	OEM
DT2768	OEM
DT2769	OEM
DT2771	OEM
DT2781	OEM
DT2782	OEM
DT2784	OEM
DT2785	OEM
DT3100FA	OEM
DT3100FB	OEM
DT3100FX	OEM
DT3100FY	OEM
DT3120FB	OEM
DT3301	0178
DT3302	0178
DT3752	OEM
DT3754	OEM
DT3755	OEM
DT3762	OEM
DT3772D15710-0M	OEM
DT3772D15710-0V	OEM
DT3772D15710-4M	OEM
DT3772D15710-4V	OEM
DT3772D15710-8M	OEM
DT3772D15710-8V	OEM
DT3772D15712-0M	OEM
DT3772D15712-0V	OEM
DT3772D15712-4M	OEM
DT3772D15712-4V	OEM
DT3772D15712-8M	OEM
DT3772D15712-8V	OEM
DT3772D15714-0M	OEM
DT3772D15714-0V	OEM
DT3772D15714-4M	OEM
DT3772D15714-4V	OEM
DT3772D15714-8M	OEM
DT3772D15714-8V	OEM
DT3772D15716-0M	OEM
DT3772D15716-0V	OEM
DT3772D15716-4M	OEM
DT3772D15716-4V	OEM
DT3772D15716-8M	OEM
DT3772D15716-8V	OEM
DT3772SE5710-0M	OEM
DT3772SE5710-0V	OEM
DT3772SE5710-4M	OEM
DT3772SE5710-4V	OEM
DT3772SE5710-8M	OEM
DT3772SE5710-8V	OEM
DT3772SE5712-0M	OEM
DT3772SE5712-0V	OEM
DT3772SE5712-4M	OEM
DT3772SE5712-4V	OEM
DT3772SE5712-8M	OEM
DT3772SE5712-8V	OEM
DT3772SE5714-0M	OEM
DT3772SE5714-0V	OEM
DT3772SE5714-4M	OEM
DT3772SE5714-4V	OEM
DT3772SE5714-8M	OEM
DT3772SE5714-8V	OEM
DT3772SE5716-0M	OEM
DT3772SE5716-0V	OEM
DT3772SE5716-4M	OEM
DT3772SE5716-4V	OEM
DT3772SE5716-8M	OEM
DT3772SE5716-8V	OEM
DT3774D15710-0M	OEM
DT3774D15710-0V	OEM
DT3774D15710-4M	OEM
DT3774D15710-4V	OEM
DT3774D15710-8M	OEM
DT3774D15710-8V	OEM
DT3774D15712-0M	OEM
DT3774D15712-0V	OEM
DT3774D15712-4M	OEM
DT3774D15712-4V	OEM
DT3774D15712-8V	OEM
DT3774D15714-0M	OEM
DT3774D15714-0V	OEM
DT3774D15714-4M	OEM
DT3774D15714-4V	OEM
DT3774D15714-8M	OEM
DT3774D15714-8V	OEM
DT3774D15716-0M	OEM
DT3774D15716-0V	OEM
DT3774D15716-4M	OEM
DT3774D15716-4V	OEM
DT3774D15716-8M	OEM
DT3774D15716-8V	OEM
DT3774SE5710-0M	OEM
DT3774SE5710-0V	OEM
DT3774SE5710-4M	OEM
DT3774SE5710-4V	OEM
DT3774SE5710-8M	OEM
DT3774SE5710-8V	OEM
DT3774SE5712-0M	OEM
DT3774SE5712-0V	OEM
DT3774SE5712-4M	OEM
DT3774SE5712-4V	OEM
DT3774SE5712-8M	OEM
DT3774SE5712-8V	OEM
DT3774SE5714-0M	OEM
DT3774SE5714-0V	OEM
DT3774SE5714-4M	OEM
DT3774SE5714-4V	OEM
DT3774SE5714-8M	OEM
DT3774SE5714-8V	OEM
DT3774SE5716-0M	OEM
DT3774SE5716-0V	OEM
DT3774SE5716-4M	OEM
DT3774SE5716-4V	OEM
DT3774SE5716-8M	OEM
DT3774SE5716-8V	OEM
DT4011	0103
DT4110	0103
DT4111	0103
DT4112	0130
DT4120	0103
DT4121	0103
DT4303	0363
DT4304	0363
DT4305	0363
DT4306	0363
DT4335	OEM
DT4336	OEM
DT4500	OEM
DT4500A	OEM
DT4501	OEM
DT4502	OEM
DT4503	OEM
DT4504	OEM
DT4505	OEM
DT4506	OEM
DT4506-300	OEM
DT4507-300	OEM
DT4508	OEM
DT4508-300	OEM
DT4509	OEM
DT4509-300	OEM
DT4510	OEM
DT4510-300	OEM
DT4511	OEM
DT4511-300	OEM
DT4512	OEM
DT4512-300	OEM
DT4513	OEM
DT4513-300	OEM
DT4514	OEM
DT4514-300	OEM
DT4515	OEM
DT4515-300	OEM
DT4516	OEM
DT5335	OEM
DT5336	OEM
DT5459BSXXXX	OEM
DT5701	OEM
DT5702	OEM
DT5703	OEM
DT5710	OEM
DT5710A	OEM
DT5714	OEM
DT5716	OEM
DT5720	OEM
DT5750	OEM
DT6103	OEM
DT6104	OEM
DT6110	0160
DT6812	OEM
DT7585BSXXXX	OEM
DT8596BSXXXX	OEM
DT15060	OEM
DT15150	OEM
DT171514	OEM
DT171514R4	OEM
DTA05B	4455
DTA05C	4455
DTA05E	4456
DTA2B	2300
DTA2C	2300
DTA2E	2300
DTA3C	2300
DTA3E	2300
DTA3F	OEM
DTA3G	OEM
DTA8C	0593
DTA8E	0576
DTA8F	0594
DTA8G	0594
DTA10C	0593
DTA10E	0576
DTA10F	0594
DTA10G	0594
DTA12C	0772
DTA12E	0744
DTA12F	0617
DTA12G	0617
DTA16C	0772
DTA16E	0744
DTA16F	0617
DTA16G	0617
DTA113ZS	OEM
DTA114	OEM
DTA114A	4172
DTA114EA	4172
DTA114EF	6600
DTA114EK	1892
DTA114EKT97	OEM
DTA114ES	4172
DTA114F	OEM
DTA114K	1892
DTA114S	4172
DTA114STP	4172
DTA114T	OEM
DTA114TF	OEM
DTA114TK	OEM
DTA114Y	OEM
DTA114YK	OEM
DTA114YS	OEM
DTA123YS	OEM
DTA124	4852
DTA124(N)	1032
DTA124A	1032
DTA124EA	1032
DTA124EF	6684
DTA124EFT	OEM
DTA124EK	0703
DTA124EKTP	0703
DTA124EL	OEM
DTA124ES	1032
DTA124ES-T	1032
DTA124ESZ	1032
DTA124EU	OEM
DTA124EV	OEM
DTA124F	1032
DTA124FS	OEM

DEVICE TYPE	REPL CODE
DTA124K	0703
DTA124K(15)	0703
DTA124L	OEM
DTA124N	1032
DTA124S	1032
DTA124TA	OEM
DTA124TF	OEM
DTA124TK	OEM
DTA124TKT97	OEM
DTA124TL	OEM
DTA124TS	OEM
DTA124TU	OEM
DTA124TV	OEM
DTA124X	OEM
DTA124XA	OEM
DTA124XF	OEM
DTA124XK	OEM
DTA124XL	OEM
DTA124XS	OEM
DTA124XU	OEM
DTA124XV	OEM
DTA143	OEM
DTA143A	OEM
DTA143E	OEM
DTA143EA	OEM
DTA143EK	OEM
DTA143ES	OEM
DTA143ESA	OEM
DTA143X	OEM
DTA143XF	OEM
DTA143XK	OEM
DTA143XS	OEM
DTA143XSZ	OEM
DTA144	4130
DTA144(N)	4130
DTA144A	4130
DTA144E	OEM
DTA144EA	4130
DTA144EF	6835
DTA144EK	3272
DTA144EK-16	OEM
DTA144ES	4130
DTA144ES-T	OEM
DTA144F	5586
DTA144KK(16)	OEM
DTA144N	5586
DTA144S	4130
DTA144STP	4130
DTA144TKT97	OEM
DTA144TK-W	OEM
DTA144TL	OEM
DTA144TS	4130
DTA144TSTP	4130
DTA144W	OEM
DTA144WS	OEM
DTA363TK-W	OEM
DTA1011	0160
DTA6001	OEM
DTA6001A	OEM
DTA6002	OEM
DTA6002A	OEM
DTA6003	OEM
DTA6003A	OEM
DTA6004	OEM
DTA6004A	OEM
DTA6005	OEM
DTA6005A	OEM
DTA6007	OEM
DTA6007A	OEM
DTA6008	OEM
DTA6008A	OEM
DTA6009	OEM
DTA6009A	OEM
DTA6010	OEM
DTA6010A	OEM
DTA6011	OEM
DTA6011A	OEM
DTA6013	OEM
DTA6013A	OEM
DTA6014	OEM
DTA6014A	OEM
DTA6015	OEM
DTA6015A	OEM
DTA6016	OEM
DTA6016A	OEM
DTA6017	OEM
DTA6017A	OEM
DTA6019	OEM
DTA6019A	OEM
DTA6027	OEM
DTA6027A	OEM
DTA6028	OEM
DTA6028A	OEM
DTA6029	OEM
DTA6029A	OEM
DTA6030	OEM
DTA6030A	OEM
DTB10C	0555
DTB10E	0555
DTB10F	0555
DTB10G	0555
DTB12C	OEM
DTB12E	OEM
DTB12F	OEM
DTB12G	OEM
DTB16C	0908
DTB16E	0908
DTB16F	0908
DTB16G	0908
DTB5724	OEM
DTB5725	OEM
DTC16C	0908
DTC16E	0908
DTC16F	0908
DTC16G	0908
DTC-114	0832
DTC114	0832
DTC114A	0832
DTC114E	0832
DTC114EA	0832
DTC114EF	0832
DTC114EK	3482
DTC114EKA	3482
DTC114EKT96	OEM
DTC114ES	0832
DTC114ESB	2099
DTC114ESC	2099
DTC114ESE	0832
DTC114ESN	0832
DTC114E-S-T	0832
DTC114ES-T	0832
DTC114ESZ	0832
DTC114F	OEM
DTC114FS	OEM
DTC114N	OEM
DTC114S	0832
DTC114TK	OEM
DTC114TKT97	OEM
DTC114Y	OEM
DTC114YK	OEM
DTC114YS	0832
DTC114YSK	OEM
DTC114YSR	OEM
DTC124	5301
DTC124(N)	0887
DTC124A	0887
DTC124EA	0887
DTC124EF	5687
DTC124EFM	OEM
DTC124EFT	OEM
DTC124EK	5376
DTC124EKTP	3482
DTC124EL	OEM
DTC124ES	0887
DTC124ES-DCTP	0887
DTC124ES-T	0887
DTC124EST	0887
DTC124ESZ	0887
DTC124EU	OEM
DTC124EV	OEM
DTC124F	0887
DTC124K	0981
DTC124K(25)	0981
DTC124L	OEM
DTC124N	0887
DTC124S	0887
DTC124TA	OEM
DTC124TF	OEM
DTC124TK	OEM
DTC124TL	OEM
DTC124TS	OEM
DTC124TU	OEM
DTC124TV	OEM
DTC124X	OEM
DTC124XA	OEM
DTC124XF	OEM
DTC124XK	OEM
DTC124XL	OEM
DTC124XS	OEM
DTC124XU	OEM
DTC124XV	OEM
DTC143	OEM
DTC143A	OEM
DTC143E	OEM
DTC143EA	OEM
DTC143EF	OEM
DTC143EK	OEM
DTC143ES	OEM
DTC143T	OEM
DTC143TA	OEM
DTC143TS	OEM
DTC143X	OEM
DTC143XS	OEM
DTC143XSZ	OEM
DTC143Z	OEM
DTC143ZS	OEM
DTC144	0898
DTC144A	0898
DTC144E	0898
DTC144EA	0898
DTC144EF	5408
DTC144EK	3478
DTC144EL-TL2	OEM
DTC144E-S	5408
DTC144ES	0898
DTC144ESD	0898
DTC144ES-DCTP	0898
DTC144ES-T	OEM
DTC144ES-TP	0898
DTC144ESTP	0898
DTC144ESZ	0898
DTC144F	0898
DTC144K	3478
DTC144K(26)	OEM
DTC144K-26	OEM
DTC144N	0898
DTC144N-DCTP	OEM
DTC144S	0898
DTC144SD	0898
DTC144TK	OEM
DTC144W	OEM
DTC144WF	OEM
DTC144WK	OEM
DTC314TA	OEM
DTC314TS	OEM
DTC323TK-W	OEM
DTC323TS-T	OEM
DTC363EK	OEM
DTC363ES	OEM
DTC363TK	OEM
DTC363TK-W	OEM
DTC363TS	OEM
DTC2400	0604
D-TEX-8	1854
DTG-110	0085
DTG110	0160
DTG110A	0604
DTG110B	0604
DTG400M	0604
DTG-600	0604
DTG600	0604
DTG-601	0604
DTG601	0604
DTG-602	0604
DTG602	0604
DTG-603	0604
DTG603	0604
DTG-603M	0604
DTG603M	0604
DTG1000	OEM
DTG-1010	0975
DTG1010	0975
DTG1011	0160
DTG1040	0160
DTG1110	0975
DTG1110B	OEM
DTG1200	0604
DTG1210A	OEM
DTG-2000	0604
DTG2000	0604
DTG2000A	0604
DTG2000H	0604
DTG-2100	0604
DTG2100	0604
DTG2100A	0604
DTG2200	0604
DTG2400	0604
DTH2411	OEM
DTH2412	OEM
DTH2413	OEM
DTH2511	OEM
DTH2512	OEM
DTH2513	OEM
DTH2611	OEM
DTH2612	OEM
DTH2613	OEM
DTH2711	OEM
DTH2712	OEM
DTH2713	OEM
DTH2811	OEM
DTH2812	OEM
DTH2813	OEM
DTH3011	OEM
DTH3011A	OEM
DTH3012	OEM
DTH3012A	OEM
DTH3014	OEM
DTH3111	OEM
DTH3111A	OEM
DTH3112	OEM
DTH3112A	OEM
DTH3113	OEM
DTH3113A	OEM
DTH3114	OEM
DTH3211	OEM
DTH3211A	OEM
DTH3212	OEM
DTH3212A	OEM
DTH3214	OEM
DTH3311	OEM
DTH3311A	OEM
DTH3312	OEM
DTH3312A	OEM
DTH3314	OEM
DTH3411	OEM
DTH3411A	OEM
DTH3412	OEM
DTH3412A	OEM
DTH3414	OEM
DTH3511	OEM
DTH3511A	OEM
DTH3512	OEM
DTH3512A	OEM
DTH3514	OEM
DTH3611	OEM
DTH3611A	OEM
DTH3612	OEM
DTH3612A	OEM
DTH3614	OEM
DTH3711	OEM
DTH3711A	OEM
DTH3712	OEM
DTH3712A	OEM
DTH3714	OEM
DTH3811	OEM
DTH3811A	OEM
DTH3812	OEM
DTH3812A	OEM
DTH3813	OEM
DTH3813A	OEM
DTH3814	OEM
DTML9093	0358
DTML9099	0333
DTML9930	1823
DTML9932	1041
DTML9933	2097
DTML9935	1176
DTML9936	1831
DTML9944	0033
DTML9945	0081
DTML9946	0141
DTML9948	3399
DTML9949	1844
DTML9961	1859
DTML9962	0562
DTML9963	0341
DTN1006	OEM
DTN9000	OEM
DTN9000T	OEM
DTN9001	OEM
DTN9001T	OEM
DTS013	0003
DTS-0710	0074
DTS0710	0074
DTS-0713	0003
DTS0713	0003
DTS06506	OEM
DTS-103	0543
DTS103	0543
DTS-104	0543
DTS104	0543
DTS-105	0543
DTS105	0543
DTS-106	0543
DTS106	0543
DTS-107	0543
DTS107	0436
DTS108	0543
DTS-310	0363
DTS310	0363
DTS-311	0363
DTS311	0363
DTS400	OEM
DTS-401	0074
DTS401	0074
DTS-402	0074
DTS402	0363
DTS-403	0065
DTS403	0363
DTS-409	0065
DTS409	2414
DTS-410	0074
DTS410	0103
DTS-411	0074
DTS411	0270
DTS-413	0074
DTS413	0313
DTS-423	1966
DTS423	0313
DTS-423M	3038
DTS423M	0363
DTS-424	0363
DTS424	0363
DTS-425	0363
DTS425	0363
DTS-430	4591
DTS430	1966
DTS-431	0065
DTS431	0363
DTS-431M	0065
DTS431M	0363
DTS-515	3038
DTS515	0074
DTS-516	3038
DTS516	3038
DTS-517	3038
DTS517	3038
DTS-518	3038
DTS518	3038
DTS-519	3038
DTS519	1966
DTS520	OEM
DTS520MX	OEM
DTS-701	0003
DTS701	0003
DTS-702	0065
DTS702	0065
DTS-704	0003
DTS704	0065
DTS-708	0065
DTS-709	0065
DTS-710	0065
DTS-712	0065
DTS-714	0065
DTS720	OEM
DTS-723	0065
DTS-801	0223
DTS801	0223
DTS-802	0065
DTS802	0065
DTS-804	0003
DTS804	0313
DTS812MX	OEM
DTS2000	OEM
DTS2001	OEM
DTS2002	OEM
DTS2003	OEM
DTS3704	0065
DTS3704A	0065
DTS3704B	0065
DTS3705	0074
DTS3705A	0074
DTS3705B	0074
DTS4010	OEM
DTS4025	OEM
DTS4026	OEM
DTS4039	OEM
DTS4041	OEM
DTS4059	OEM
DTS4061	OEM
DTS4062	2626
DTS4066	2626
DTS4067	2626
DTS4072	2626
DTS4074	2626
DTS4075	2626
DTZ6V8A	OEM
DTZ6V8P	OEM
DTZ7V5	OEM
DTZ7V5A	OEM
DTZ7V5P	OEM
DTZ8V2	OEM
DTZ8V2A	OEM
DTZ8V2P	OEM
DTZ9V1	OEM
DTZ9V1A	OEM
DTZ9V1P	OEM
DTZ10	OEM
DTZ10A	OEM
DTZ10P	OEM
DTZ11	OEM
DTZ11A	OEM
DTZ11P	OEM
DTZ12	OEM
DTZ12A	OEM
DTZ12P	OEM
DTZ13	1169
DTZ13A	1169
DTZ13P	OEM
DTZ15	OEM
DTZ15A	OEM
DTZ15P	OEM
DTZ16	OEM
DTZ16A	OEM
DTZ16P	OEM
DTZ18	1767
DTZ18A	1767
DTZ18P	OEM
DTZ20	1932
DTZ20A	1932
DTZ20P	OEM
DTZ22	OEM
DTZ22A	OEM
DTZ22P	OEM
DTZ24	OEM
DTZ24A	OEM
DTZ24P	OEM
DTZ27	1952
DTZ27A	1952
DTZ27P	OEM
DTZ30	OEM
DTZ30A	OEM
DTZ30P	OEM
DTZ33	OEM
DTZ33A	OEM
DTZ33P	OEM
DTZ36	OEM
DTZ36A	OEM
DTZ36P	OEM
DTZ39	OEM
DTZ39A	OEM
DTZ39P	OEM
DTZ43	OEM
DTZ43A	OEM
DTZ43P	OEM
DTZ47	OEM
DTZ47A	OEM
DTZ47P	OEM
DTZ51	1972
DTZ51A	1972
DTZ51P	OEM
DTZ56	OEM
DTZ56A	OEM
DTZ56P	OEM
DTZ62	OEM
DTZ62A	OEM
DTZ62P	OEM
DTZ68	1987
DTZ68A	1987
DTZ68P	OEM
DTZ75	OEM
DTZ75A	OEM
DTZ75P	OEM
DTZ82	OEM
DTZ82A	OEM
DTZ82P	OEM
DTZ91	OEM
DTZ91A	OEM
DTZ91P	OEM
DTZ100	OEM
DTZ100A	OEM
DTZ100P	OEM
DTZ110	OEM
DTZ110A	OEM
DTZ110P	OEM
DTZ120	OEM
DTZ120A	OEM
DTZ120P	OEM
DTZ130	OEM
DTZ130A	OEM
DTZ130P	OEM
DTZ150	OEM
DTZ150A	OEM
DTZ150P	OEM
DTZ160	OEM
DTZ160A	OEM
DTZ160P	OEM
DTZ170	OEM
DTZ170A	OEM
DTZ170P	OEM
DTZ180	OEM
DTZ180A	OEM
DTZ180P	OEM
DTZ200	1408
DTZ200A	1408
DTZ200P	OEM
DTZ220	OEM
DTZ220A	OEM
DTZ220P	OEM
DTZ250A	OEM
DTZ250P	OEM
DTZ280A	OEM
DTZ280P	OEM
DTZ300A	OEM
DTZ300P	OEM
DTZ320A	OEM
DTZ320P	OEM
DTZ350A	OEM
DTZ350P	OEM
DTZ400A	OEM
DTZ400P	OEM
DTZ440A	OEM
DTZ440P	OEM
DU1	0050
DU2	0050
DU3	0004
DU4	0164
DU5	0004
DU6	0085
DU7	0440
DU12	0050
DU47	0440
DU400	0015
DU600	0015
DU800	0071
DU1000	0071
DU4339	OEM
DU4340	3384
DUNTH0002CE00	OEM
DUNTH0017DE00	OEM
DUNTH0032DE00	OEM
DUV11	OEM
DV13C	0871
DV13E	0853
DV14C	0871
DV14E	0853
DV45	OEM
DV1006	OEM
DV1007	OEM
DV1110	OEM
DV1111	OEM
DV1112	OEM
DV1202S	OEM
DV1202W	OEM
DV1205S	OEM
DV1205W	OEM
DV1210S	OEM
DV1210W	OEM
DV1220S	OEM
DV1220W	OEM
DV1230T	OEM
DV1230W	OEM
DV1240T	OEM
DV1240W	OEM
DV2209	0030
DV2805S	OEM
DV2810S	OEM
DV2810X	OEM
DV2820S	OEM
DV2820V	OEM
DV2820W	OEM
DV2820X	OEM
DV2840V	OEM
DV2880T	OEM
DV2880U	OEM
DV2880V	OEM
DV2880W	OEM
DV2880X	OEM
DV3805W	OEM
DV3810W	OEM
DV3840X	OEM
DV28120T	OEM
DV28120U	OEM
DV28120V	OEM
DV28120X	OEM
DVA4580A	OEM
DVA4580B	OEM
DVA4580C	OEM
DVA4580D	OEM
DVA4580E	OEM
DVA4580F	OEM
DVA6180	OEM
DVA6181	OEM
DVA6181A	OEM
DVA6181B	OEM
DVA6182	OEM
DVA6182A	OEM
DVA6182B	OEM
DVA6183	OEM
DVA6183A	OEM
DVA6183B	OEM
DVA6184	OEM
DVA6184A	OEM
DVA6185	OEM
DVA6185A	OEM
DVA6185B	OEM
DVA6186	OEM
DVA6186A	OEM
DVA6186B	OEM
DVA6187	OEM
DVA6187A	OEM
DVA6187B	OEM
DVA6188	OEM
DVA6188A	OEM
DVA6188B	OEM
DVA6735A	OEM
DVA6735B	OEM
DVA6735C	OEM
DVA6735D	OEM
DVA6736A	OEM
DVA6736B	OEM
DVA6736C	OEM
DVA6736D	OEM
DVA6737A	OEM
DVA6737B	OEM
DVA6737C	OEM
DVA6737D	OEM
DVA6737E	OEM
DVA6738A	OEM
DVA6738B	OEM
DVA6738C	OEM
DVA6738D	OEM
DVA6738E	OEM
DVA6738F	OEM
DVA6738G	OEM
DVA6738H	OEM
DVA6739A	OEM
DVA6739B	OEM
DVA6739C	OEM
DVA6739D	OEM
DVA6739E	OEM
DVA6739F	OEM
DVB6100A	OEM
DVB6100B	OEM
DVB6100C	OEM
DVB6101A	OEM
DVB6101B	OEM
DVB6101C	OEM
DVB6101D	OEM
DVB6101E	OEM
DVB6102A	OEM
DVB6102B	OEM
DVB6102C	OEM
DVB6102D	OEM
DVB6103A	OEM
DVB6103B	OEM
DVB6103C	OEM
DVB6103D	OEM
DVB6104A	OEM
DVB6104B	OEM
DVB6104C	OEM
DVB6104D	OEM
DVB6850A	OEM
DVB6850B	OEM
DVB6850C	OEM
DVB6851A	OEM
DVB6851B	OEM
DVB6851C	OEM
DVB6852A	OEM
DVB6852B	OEM
DVB6852C	OEM
DVB6860A	OEM
DVB6860B	OEM
DVB6860C	OEM
DVB6861A	OEM
DVB6861B	OEM
DVB6861C	OEM
DVB6862A	OEM
DVB6862B	OEM
DVB6870A	OEM
DVB6870B	OEM
DVB6870C	OEM
DVB6871A	OEM
DVB6871B	OEM
DVB6871C	OEM
DVB6872A	OEM
DVB6872B	OEM
DVD5001	OEM
DVD5001A	OEM
DVE955B	OEM
DVE4550A	OEM
DVE4550B	OEM
DVE4550C	OEM
DVE4550D	OEM
DVE4550E	OEM
DVE4550F	OEM
DVE4550G	OEM
DVE4550H	OEM
DVE4551A	OEM
DVE4551B	OEM
DVE4551C	OEM
DVE4551D	OEM
DVE4551E	OEM

If replacement code is OEM, contact original manufacturer for replacement.

DEVICE TYPE	REPL CODE	DEVICE TYPE	REPL CODE	DEVICE TYPE	REPL CODE	DEVICE TYPE	REPL CODE	DEVICE TYPE	REPL CODE	DEVICE TYPE	REPL CODE	DEVICE TYPE	REPL CODE
DVE4551F	OEM	DVE4558-22	OEM	DVE6954F	OEM	DVH6641J	OEM	DVH6694F	OEM	DVH6734-13	OEM	DVH6760-22	OEM
DVE4551G	OEM	DVE4558-23	OEM	DVE6954G	OEM	DVH6642A	OEM	DVH6694G	OEM	DVH6734-14	OEM	DVH6760-23	OEM
DVE4551H	OEM	DVE4558-24	OEM	DVE6954H	OEM	DVH6642B	OEM	DVH6694H	OEM	DVH6734-15	OEM	DVH6760-25	OEM
DVE4552A	OEM	DVE4558-25	OEM	DVE6954J	OEM	DVH6642C	OEM	DVH6694J	OEM	DVH6734-16	OEM	DVH6760-26	OEM
DVE4552B	OEM	DVE4558-26	OEM	DVE6954K	OEM	DVH6642D	OEM	DVH6694K	OEM	DVH6734-17	OEM	DVH6760-27	OEM
DVE4552C	OEM	DVE4558-31	OEM	DVE6955A	OEM	DVH6642E	OEM	DVH6694L	OEM	DVH6734-19	OEM	DVH6761-03	OEM
DVE4552D	OEM	DVE4558-32	OEM	DVE6955C	OEM	DVH6642F	OEM	DVH6694M	OEM	DVH6734-20	OEM	DVH6761-04	OEM
DVE4552E	OEM	DVE4558-33	OEM	DVE6955D	OEM	DVH6642G	OEM	DVH6730-01	OEM	DVH6734-21	OEM	DVH6761-05	OEM
DVE4552F	OEM	DVE4558-34	OEM	DVE6955E	OEM	DVH6642H	OEM	DVH6730-02	OEM	DVH6734-22	OEM	DVH6761-07	OEM
DVE4552G	OEM	DVE4558-35	OEM	DVE6955F	OEM	DVH6642J	OEM	DVH6730-03	OEM	DVH6734-23	OEM	DVH6761-08	OEM
DVE4552H	OEM	DVE4558-41	OEM	DVE6955G	OEM	DVH6643A	OEM	DVH6730-04	OEM	DVH6734-25	OEM	DVH6761-09	OEM
DVE4555A	OEM	DVE4558-42	OEM	DVE6955H	OEM	DVH6643B	OEM	DVH6730-05	OEM	DVH6734-26	OEM	DVH6761-10	OEM
DVE4555B	OEM	DVE4558-43	OEM	DVE6955J	OEM	DVH6643C	OEM	DVH6730-07	OEM	DVH6734-27	OEM	DVH6761-11	OEM
DVE4555C	OEM	DVE4558-44	OEM	DVE6955K	OEM	DVH6643D	OEM	DVH6730-08	OEM	DVH6740-01	OEM	DVH6761-13	OEM
DVE4555D	OEM	DVE4558-51	OEM	DVE6956A	OEM	DVH6643E	OEM	DVH6730-09	OEM	DVH6740-02	OEM	DVH6761-14	OEM
DVE4555E	OEM	DVE4558-52	OEM	DVE6956B	OEM	DVH6643F	OEM	DVH6730-10	OEM	DVH6740-03	OEM	DVH6761-15	OEM
DVE4555F	OEM	DVE4558-53	OEM	DVE6956C	OEM	DVH6643G	OEM	DVH6730-11	OEM	DVH6740-04	OEM	DVH6761-16	OEM
DVE4555G	OEM	DVE4558-61	OEM	DVE6956D	OEM	DVH6643H	OEM	DVH6730-13	OEM	DVH6740-05	OEM	DVH6761-17	OEM
DVE4555H	OEM	DVE4558-62	OEM	DVE6956E	OEM	DVH6643J	OEM	DVH6730-14	OEM	DVH6740-07	OEM	DVH6761-19	OEM
DVE4556-01	OEM	DVE4558-71	OEM	DVE6956F	OEM	DVH6644A	OEM	DVH6730-15	OEM	DVH6740-08	OEM	DVH6761-20	OEM
DVE4556-02	OEM	DVE4560-00	OEM	DVE6956G	OEM	DVH6644B	OEM	DVH6730-16	OEM	DVH6740-09	OEM	DVH6761-21	OEM
DVE4556-03	OEM	DVE4560-06	OEM	DVE6956H	OEM	DVH6644C	OEM	DVH6730-17	OEM	DVH6740-10	OEM	DVH6761-22	OEM
DVE4556-04	OEM	DVE4560-12	OEM	DVE6956J	OEM	DVH6644D	OEM	DVH6730-19	OEM	DVH6740-11	OEM	DVH6761-23	OEM
DVE4556-05	OEM	DVE4560-18	OEM	DVE6956K	OEM	DVH6644E	OEM	DVH6730-20	OEM	DVH6740-13	OEM	DVH6761-25	OEM
DVE4556-06	OEM	DVE4560-24	OEM	DVF4559-01	OEM	DVH6644F	OEM	DVH6730-21	OEM	DVH6740-14	OEM	DVH6761-26	OEM
DVE4556-07	OEM	DVE4576A	OEM	DVF4559-02	OEM	DVH6644G	OEM	DVH6730-22	OEM	DVH6740-15	OEM	DVH6761-27	OEM
DVE4556-08	OEM	DVE4576B	OEM	DVF4559-03	OEM	DVH6644H	OEM	DVH6730-23	OEM	DVH6740-16	OEM	DVH6761B	OEM
DVE4556-11	OEM	DVE4576C	OEM	DVF4559-04	OEM	DVH6644J	OEM	DVH6730-25	OEM	DVH6740-17	OEM	DVH6761C	OEM
DVE4556-12	OEM	DVE4576D	OEM	DVF4559-11	OEM	DVH6661B	OEM	DVH6730-26	OEM	DVH6740-19	OEM	DVH6761D	OEM
DVE4556-13	OEM	DVE4576E	OEM	DVF4559-12	OEM	DVH6661C	OEM	DVH6730-27	OEM	DVH6741-01	OEM	DVH6761E	OEM
DVE4556-14	OEM	DVE4576F	OEM	DVF4559-21	OEM	DVH6661D	OEM	DVH6731-01	OEM	DVH6741-02	OEM	DVH6761F	OEM
DVE4556-15	OEM	DVE4576G	OEM	DVF4559-22	OEM	DVH6661E	OEM	DVH6731-02	OEM	DVH6741-03	OEM	DVH6761G	OEM
DVE4556-16	OEM	DVE4576H	OEM	DVF4559-23	OEM	DVH6661F	OEM	DVH6731-03	OEM	DVH6741-04	OEM	DVH6761H	OEM
DVE4556-17	OEM	DVE4588-01	OEM	DVF4559-24	OEM	DVH6661G	OEM	DVH6731-04	OEM	DVH6741-05	OEM	DVH6761J	OEM
DVE4556-21	OEM	DVE4588-02	OEM	DVF4559-31	OEM	DVH6661H	OEM	DVH6731-05	OEM	DVH6741-07	OEM	DVH6761K	OEM
DVE4556-22	OEM	DVE4588-03	OEM	DVF4559-32	OEM	DVH6661J	OEM	DVH6731-07	OEM	DVH6741-08	OEM	DVH6761L	OEM
DVE4556-23	OEM	DVE4588-11	OEM	DVF4559-41	OEM	DVH6661K	OEM	DVH6731-08	OEM	DVH6741-09	OEM	DVH6761M	OEM
DVE4556-24	OEM	DVE4588-12	OEM	DVF4559-42	OEM	DVH6661L	OEM	DVH6731-09	OEM	DVH6741-10	OEM	DVH6762-03	OEM
DVE4556-25	OEM	DVE4588-13	OEM	DVF4559-43	OEM	DVH6661M	OEM	DVH6731-10	OEM	DVH6741-11	OEM	DVH6762-04	OEM
DVE4556-26	OEM	DVE5003-04	OEM	DVH6150	OEM	DVH6662B	OEM	DVH6731-11	OEM	DVH6741-13	OEM	DVH6762-05	OEM
DVE4556-31	OEM	DVE5033-01	OEM	DVH6150A	OEM	DVH6662C	OEM	DVH6731-13	OEM	DVH6741-14	OEM	DVH6762-07	OEM
DVE4556-32	OEM	DVE5033-02	OEM	DVH6150B	OEM	DVH6662D	OEM	DVH6731-14	OEM	DVH6741-15	OEM	DVH6762-08	OEM
DVE4556-33	OEM	DVE5033-03	OEM	DVH6150C	OEM	DVH6662E	OEM	DVH6731-15	OEM	DVH6741-16	OEM	DVH6762-09	OEM
DVE4556-34	OEM	DVE5033-05	OEM	DVH6150D	OEM	DVH6662F	OEM	DVH6731-16	OEM	DVH6741-17	OEM	DVH6762-10	OEM
DVE4556-35	OEM	DVE5033-06	OEM	DVH6151	OEM	DVH6662G	OEM	DVH6731-17	OEM	DVH6741-19	OEM	DVH6762-11	OEM
DVE4556-41	OEM	DVE5033-07	OEM	DVH6151A	OEM	DVH6662H	OEM	DVH6731-19	OEM	DVH6742-01	OEM	DVH6762-13	OEM
DVE4556-42	OEM	DVE5033-08	OEM	DVH6151B	OEM	DVH6662J	OEM	DVH6731-20	OEM	DVH6742-02	OEM	DVH6762-14	OEM
DVE4556-43	OEM	DVE5033-09	OEM	DVH6151C	OEM	DVH6662K	OEM	DVH6731-21	OEM	DVH6742-03	OEM	DVH6762-15	OEM
DVE4556-44	OEM	DVE5033-10	OEM	DVH6151D	OEM	DVH6662L	OEM	DVH6731-22	OEM	DVH6742-04	OEM	DVH6762-16	OEM
DVE4556-51	OEM	DVE5033-11	OEM	DVH6151E	OEM	DVH6662M	OEM	DVH6731-23	OEM	DVH6742-05	OEM	DVH6762-17	OEM
DVE4556-52	OEM	DVE5033-12	OEM	DVH6151F	OEM	DVH6663B	OEM	DVH6731-25	OEM	DVH6742-07	OEM	DVH6762-19	OEM
DVE4556-53	OEM	DVE5033-13	OEM	DVH6151G	OEM	DVH6663C	OEM	DVH6731-26	OEM	DVH6742-08	OEM	DVH6762-20	OEM
DVE4556-61	OEM	DVE5033-14	OEM	DVH6152	OEM	DVH6663D	OEM	DVH6731-27	OEM	DVH6742-09	OEM	DVH6762-21	OEM
DVE4556-62	OEM	DVE6347	OEM	DVH6152A	OEM	DVH6663E	OEM	DVH6732-01	OEM	DVH6742-10	OEM	DVH6762-22	OEM
DVE4556-71	OEM	DVE6347A	OEM	DVH6152B	OEM	DVH6663F	OEM	DVH6732-02	OEM	DVH6742-11	OEM	DVH6762-23	OEM
DVE4557-01	OEM	DVE6347B	OEM	DVH6152C	OEM	DVH6663G	OEM	DVH6732-03	OEM	DVH6742-13	OEM	DVH6762-25	OEM
DVE4557-02	OEM	DVE6347C	OEM	DVH6160	OEM	DVH6663H	OEM	DVH6732-04	OEM	DVH6742-14	OEM	DVH6762-26	OEM
DVE4557-03	OEM	DVE6347D	OEM	DVH6160A	OEM	DVH6663J	OEM	DVH6732-05	OEM	DVH6742-15	OEM	DVH6762-27	OEM
DVE4557-04	OEM	DVE6347E	OEM	DVH6161	OEM	DVH6663K	OEM	DVH6732-07	OEM	DVH6742-16	OEM	DVH6762B	OEM
DVE4557-05	OEM	DVE6347F	OEM	DVH6161A	OEM	DVH6663L	OEM	DVH6732-08	OEM	DVH6742-17	OEM	DVH6762C	OEM
DVE4557-06	OEM	DVE6347G	OEM	DVH6161B	OEM	DVH6663M	OEM	DVH6732-09	OEM	DVH6742-19	OEM	DVH6762D	OEM
DVE4557-07	OEM	DVE6347H	OEM	DVH6161C	OEM	DVH6664B	OEM	DVH6732-10	OEM	DVH6743-01	OEM	DVH6762E	OEM
DVE4557-08	OEM	DVE6722A	OEM	DVH6162	OEM	DVH6664C	OEM	DVH6732-11	OEM	DVH6743-02	OEM	DVH6762F	OEM
DVE4557-11	OEM	DVE6722B	OEM	DVH6162A	OEM	DVH6664D	OEM	DVH6732-13	OEM	DVH6743-03	OEM	DVH6762G	OEM
DVE4557-12	OEM	DVE6722C	OEM	DVH6162B	OEM	DVH6664E	OEM	DVH6732-14	OEM	DVH6743-04	OEM	DVH6762H	OEM
DVE4557-13	OEM	DVE6722D	OEM	DVH6631A	OEM	DVH6664F	OEM	DVH6732-15	OEM	DVH6743-05	OEM	DVH6762J	OEM
DVE4557-14	OEM	DVE6722E	OEM	DVH6631B	OEM	DVH6664G	OEM	DVH6732-16	OEM	DVH6743-07	OEM	DVH6762K	OEM
DVE4557-15	OEM	DVE6722F	OEM	DVH6631C	OEM	DVH6664H	OEM	DVH6732-17	OEM	DVH6743-08	OEM	DVH6762L	OEM
DVE4557-16	OEM	DVE6810A	OEM	DVH6631D	OEM	DVH6664J	OEM	DVH6732-19	OEM	DVH6743-09	OEM	DVH6762M	OEM
DVE4557-17	OEM	DVE6810B	OEM	DVH6631E	OEM	DVH6664K	OEM	DVH6732-20	OEM	DVH6743-10	OEM	DVH6763-03	OEM
DVE4557-21	OEM	DVE6810C	OEM	DVH6631F	OEM	DVH6664L	OEM	DVH6732-21	OEM	DVH6743-11	OEM	DVH6763-04	OEM
DVE4557-22	OEM	DVE6810D	OEM	DVH6631G	OEM	DVH6664M	OEM	DVH6732-22	OEM	DVH6743-13	OEM	DVH6763-05	OEM
DVE4557-23	OEM	DVE6810E	OEM	DVH6631H	OEM	DVH6691B	OEM	DVH6732-23	OEM	DVH6743-14	OEM	DVH6763-07	OEM
DVE4557-24	OEM	DVE6810F	OEM	DVH6631J	OEM	DVH6691C	OEM	DVH6732-25	OEM	DVH6743-15	OEM	DVH6763-08	OEM
DVE4557-25	OEM	DVE6951A	OEM	DVH6632A	OEM	DVH6691D	OEM	DVH6732-26	OEM	DVH6743-16	OEM	DVH6763-09	OEM
DVE4557-26	OEM	DVE6951B	OEM	DVH6632B	OEM	DVH6691E	OEM	DVH6732-27	OEM	DVH6743-17	OEM	DVH6763-10	OEM
DVE4557-31	OEM	DVE6951C	OEM	DVH6632C	OEM	DVH6691F	OEM	DVH6733-01	OEM	DVH6743-19	OEM	DVH6763-11	OEM
DVE4557-32	OEM	DVE6951D	OEM	DVH6632D	OEM	DVH6691G	OEM	DVH6733-02	OEM	DVH6744-01	OEM	DVH6763-13	OEM
DVE4557-33	OEM	DVE6951E	OEM	DVH6632E	OEM	DVH6691H	OEM	DVH6733-03	OEM	DVH6744-02	OEM	DVH6763-14	OEM
DVE4557-34	OEM	DVE6951F	OEM	DVH6632F	OEM	DVH6691J	OEM	DVH6733-04	OEM	DVH6744-03	OEM	DVH6763-15	OEM
DVE4557-35	OEM	DVE6951G	OEM	DVH6632G	OEM	DVH6691K	OEM	DVH6733-05	OEM	DVH6744-04	OEM	DVH6763-16	OEM
DVE4557-41	OEM	DVE6951H	OEM	DVH6632H	OEM	DVH6691L	OEM	DVH6733-07	OEM	DVH6744-05	OEM	DVH6763-17	OEM
DVE4557-42	OEM	DVE6951J	OEM	DVH6632J	OEM	DVH6691M	OEM	DVH6733-08	OEM	DVH6744-07	OEM	DVH6763-19	OEM
DVE4557-43	OEM	DVE6951K	OEM	DVH6633A	OEM	DVH6692B	OEM	DVH6733-09	OEM	DVH6744-08	OEM	DVH6763-20	OEM
DVE4557-44	OEM	DVE6952A	OEM	DVH6633B	OEM	DVH6692C	OEM	DVH6733-10	OEM	DVH6744-09	OEM	DVH6763-21	OEM
DVE4557-51	OEM	DVE6952B	OEM	DVH6633C	OEM	DVH6692D	OEM	DVH6733-11	OEM	DVH6744-10	OEM	DVH6763-22	OEM
DVE4557-52	OEM	DVE6952C	OEM	DVH6633D	OEM	DVH6692E	OEM	DVH6733-13	OEM	DVH6744-11	OEM	DVH6763-23	OEM
DVE4557-53	OEM	DVE6952D	OEM	DVH6633E	OEM	DVH6692F	OEM	DVH6733-14	OEM	DVH6744-13	OEM	DVH6763-25	OEM
DVE4557-61	OEM	DVE6952E	OEM	DVH6633F	OEM	DVH6692G	OEM	DVH6733-15	OEM	DVH6744-14	OEM	DVH6763-26	OEM
DVE4557-62	OEM	DVE6952F	OEM	DVH6633G	OEM	DVH6692H	OEM	DVH6733-16	OEM	DVH6744-15	OEM	DVH6763-27	OEM
DVE4557-71	OEM	DVE6952G	OEM	DVH6633H	OEM	DVH6692J	OEM	DVH6733-17	OEM	DVH6744-16	OEM	DVH6763B	OEM
DVE4557G	OEM	DVE6952H	OEM	DVH6633J	OEM	DVH6692K	OEM	DVH6733-19	OEM	DVH6744-17	OEM	DVH6763C	OEM
DVE4557H	OEM	DVE6952J	OEM	DVH6634A	OEM	DVH6692L	OEM	DVH6733-20	OEM	DVH6744-19	OEM	DVH6763D	OEM
DVE4558-01	OEM	DVE6952K	OEM	DVH6634B	OEM	DVH6692M	OEM	DVH6733-21	OEM	DVH6760-03	OEM	DVH6763E	OEM
DVE4558-02	OEM	DVE6953A	OEM	DVH6634C	OEM	DVH6693B	OEM	DVH6733-22	OEM	DVH6760-04	OEM	DVH6763F	OEM
DVE4558-03	OEM	DVE6953B	OEM	DVH6634D	OEM	DVH6693C	OEM	DVH6733-23	OEM	DVH6760-05	OEM	DVH6763G	OEM
DVE4558-04	OEM	DVE6953C	OEM	DVH6634E	OEM	DVH6693D	OEM	DVH6733-25	OEM	DVH6760-07	OEM	DVH6763H	OEM
DVE4558-05	OEM	DVE6953D	OEM	DVH6634F	OEM	DVH6693E	OEM	DVH6733-26	OEM	DVH6760-08	OEM	DVH6763J	OEM
DVE4558-06	OEM	DVE6953E	OEM	DVH6634G	OEM	DVH6693F	OEM	DVH6733-27	OEM	DVH6760-09	OEM	DVH6763K	OEM
DVE4558-07	OEM	DVE6953F	OEM	DVH6634H	OEM	DVH6693G	OEM	DVH6734-01	OEM	DVH6760-10	OEM	DVH6763L	OEM
DVE4558-08	OEM	DVE6953G	OEM	DVH6634J	OEM	DVH6693H	OEM	DVH6734-02	OEM	DVH6760-11	OEM	DVH6763M	OEM
DVE4558-11	OEM	DVE6953H	OEM	DVH6641A	OEM	DVH6693J	OEM	DVH6734-03	OEM	DVH6760-13	OEM	DVH6764-03	OEM
DVE4558-12	OEM	DVE6953J	OEM	DVH6641B	OEM	DVH6693K	OEM	DVH6734-04	OEM	DVH6760-14	OEM	DVH6764-04	OEM
DVE4558-13	OEM	DVE6953K	OEM	DVH6641C	OEM	DVH6693L	OEM	DVH6734-05	OEM	DVH6760-15	OEM	DVH6764-05	OEM
DVE4558-14	OEM	DVE6954A	OEM	DVH6641D	OEM	DVH6693M	OEM	DVH6734-07	OEM	DVH6760-16	OEM	DVH6764-07	OEM
DVE4558-15	OEM	DVE6954B	OEM	DVH6641E	OEM	DVH6694B	OEM	DVH6734-08	OEM	DVH6760-17	OEM	DVH6764-08	OEM
DVE4558-16	OEM	DVE6954C	OEM	DVH6641F	OEM	DVH6694C	OEM	DVH6734-09	OEM	DVH6760-19	OEM	DVH6764-09	OEM
DVE4558-17	OEM	DVE6954D	OEM	DVH6641G	OEM	DVH6694D	OEM	DVH6734-10	OEM	DVH6760-20	OEM	DVH6764-10	OEM
DVE4558-21	OEM	DVE6954E	OEM	DVH6641H	OEM	DVH6694E	OEM	DVH6734-11	OEM	DVH6760-21	OEM	DVH6764-11	OEM

DEVICE TYPE	REPL CODE
DVH6764-13	OEM
DVH6764-14	OEM
DVH6764-15	OEM
DVH6764-16	OEM
DVH6764-17	OEM
DVH6764-19	OEM
DVH6764-20	OEM
DVH6764-21	OEM
DVH6764-22	OEM
DVH6764-23	OEM
DVH6764-25	OEM
DVH6764-26	OEM
DVH6764-27	OEM
DVH6764B	OEM
DVH6764C	OEM
DVH6764D	OEM
DVH6764E	OEM
DVH6764F	OEM
DVH6764G	OEM
DVH6764H	OEM
DVH6764J	OEM
DVH6764K	OEM
DVH6764L	OEM
DVH6764M	OEM
DVH6790-03	OEM
DVH6790-04	OEM
DVH6790-05	OEM
DVH6790-07	OEM
DVH6790-08	OEM
DVH6790-09	OEM
DVH6790-10	OEM
DVH6790-11	OEM
DVH6790-13	OEM
DVH6790-14	OEM
DVH6790-15	OEM
DVH6790-16	OEM
DVH6790-17	OEM
DVH6790-19	OEM
DVH6790-20	OEM
DVH6790-21	OEM
DVH6791-03	OEM
DVH6791-04	OEM
DVH6791-05	OEM
DVH6791-07	OEM
DVH6791-08	OEM
DVH6791-09	OEM
DVH6791-10	OEM
DVH6791-11	OEM
DVH6791-13	OEM
DVH6791-14	OEM
DVH6791-15	OEM
DVH6791-16	OEM
DVH6791-17	OEM
DVH6791-19	OEM
DVH6791-20	OEM
DVH6791-21	OEM
DVH6792-03	OEM
DVH6792-04	OEM
DVH6792-05	OEM
DVH6792-07	OEM
DVH6792-08	OEM
DVH6792-09	OEM
DVH6792-10	OEM
DVH6792-11	OEM
DVH6792-13	OEM
DVH6792-14	OEM
DVH6792-15	OEM
DVH6792-16	OEM
DVH6792-17	OEM
DVH6792-19	OEM
DVH6792-20	OEM
DVH6792-21	OEM
DVH6793-03	OEM
DVH6793-04	OEM
DVH6793-05	OEM
DVH6793-07	OEM
DVH6793-08	OEM
DVH6793-09	OEM
DVH6793-10	OEM
DVH6793-11	OEM
DVH6793-13	OEM
DVH6793-14	OEM
DVH6793-15	OEM
DVH6793-16	OEM
DVH6793-17	OEM
DVH6793-19	OEM
DVH6793-20	OEM
DVH6793-21	OEM
DVH6794-04	OEM
DVH6794-05	OEM
DVH6794-07	OEM
DVH6794-08	OEM
DVH6794-09	OEM
DVH6794-10	OEM
DVH6794-11	OEM
DVH6794-13	OEM
DVH6794-14	OEM
DVH6794-15	OEM
DVH6794-16	OEM
DVH6794-17	OEM
DVH6794-19	OEM
DVH6794-20	OEM
DVH6794-21	OEM
DVV004	0030
DW144	OEM
DWA010	OEM
DWMA154WA	2203
DWMA154WK	0694
DWP-300A	OEM
DWP400	OEM
DX-0007	0023
DX-0022	0124
DX0022	0124
DX0033TA	0133
DX0038CE	0023
DX0045GE	0124
DX0049GE	OEM
DX0053GE	0124
DX0055CE	0015
DX-0061	0162
DX-0079	0133
DX-0080	0133
DX-0081	0133
DX0085TA	0023
DX0086CE	0023
DX0086TA	0023
DX-0087	0012
DX-0099	0015
DX0101CE	0023
DX0105TA	0023
DX0110CE	0795
DX0113TA	2518
DX0115CE	0023
DX0117TA	0023
DX0123CE	0023
DX0124TA	0071
DX0125CE	0023
DX-0126	0554
DX0126CE	0023
DX0128CE	0087
DX0130CE	0023
DX0131CE	0023
DX0132CE	6354
DX0135TA	0023
DX0142CE	0124
DX-0143	0031
DX-0147	OEM
DX-0150	0139
DX0152CE	0473
DX0153CE	0023
DX-0154	OEM
DX0154CE	0023
DX0155CE	1090
DX-0156	0123
DX-0159	0253
DX0159CE	0023
DX0159DE	0023
DX-0161	0019
DX0161CE	0023
DX-0162	0123
DX-0164	0760
DX0164CE	0348
DX0179CE	0124
DX0181CE	0031
DX0200CE	1875
DX0202CE	0023
DX0203CE	0031
DX0204CE	1091
DX0206	0080
DX0208CE	1875
DX0220CE	0609
DX0226CE	0023
DX0229CE	0102
DX0234CE	0031
DX-0235	0705
DX0235CE	0023
DX0236CE	0133
DX0237CE	0124
DX-0241	0143
DX0247CE	0283
DX0248CE	0023
DX-0253	0019
DX-0255	0133
DX-0259	0133
DX-0270	0133
DX-0273	0133
DX-0287	0133
DX-0289	OEM
DX-0299	0133
DX-0301	0005
DX-0307	0030
DX-0313	0139
DX-0314	0139
DX-0316	0139
DX-0349	0015
DX-0365	0139
DX0382CE	OEM
DX-0398	0041
DX-0402	0057
DX-0403	1005
DX-0405	0053
DX-0408	0471
DX-0445	0080
DX-0475	0015
DX0475	0015
DX-0516	0120
DX-0529	0188
DX-0530	0471
DX-0543	0133
DX-0576	0143
DX-0587	0023
DX-0611	0123
DX-0650	1925
DX-0685	0133
DX-0697	0120
DX-0713	0100
DX-0721	0023
DX-0723	0554
DX-0725	0143
DX-0726	0323
DX-0727	0253
DX-0728	0057
DX-0729	0165
DX-0735	0253
DX-0749	0720
DX-0750	0139
DX-0752	0139
DX-0856	0213
DX-0862	0181
DX-0872	0139
DX-0983	0062
DX-0987	0162
DX-0999	0140
DX57	OEM
DX58	OEM
DX206	OEM
DX260	OEM
DX267	OEM
DX276	OEM
DX285	OEM
DX286	OEM
DX290	OEM
DX332	OEM
DX342	OEM
DX343	OEM
DX368	OEM
DX373	OEM
DX423	OEM
DX450	OEM
DX453	OEM
DX453B	OEM
DX481	OEM
DX482	OEM
DX520	0015
DX1007	0030
DX-1034	0170
DX-1039	0015
DX1039	0015
DX-1043	0062
DX-1044	0077
DX-1080	0030
DX1080	0628
DX-1082	0102
DX-1126	0253
DX-1131	0030
DX-1132	0253
DX-1139	0133
DX-1147	4327
DX-1161	0023
DX-1162	1875
DX-1163	0137
DX-1165	0095
DX1193	0041
DX1194	0091
DX-1195	0039
DX-1196	0005
DX1196	0005
DX-1221	OEM
DX-1222	0062
DX-1228	0275
DX-1229	0564
DX-1256	0396
DX-1258	0299
DX-1325	0143
DX-1346	0057
DX-1392	0005
DX-1430	0023
DX-1535	0879
DX-1662	0071
DX-1663	0064
DX-1691	0649
DX-1692	0106
DX-1753	0062
DX-1772	0376
DX-1773	0023
DX-1829	0165
DX-1834	0057
DX-1912	0023
DX-1968	OEM
DX-1969	OEM
DX-1991	0064
DX-1993	0041
DX-2060	0062
DX-2061	0165
DX-2137	1875
DX-2149	0253
DX-2150	0181
DX-2230	0133
DX-2231	0041
DX-2232	0609
DX-2236	OEM
DX-2253	0636
DX-2257	0283
DX-2272	0124
DX-2273	0023
DX-2274	0023
DX-2275	0023
DX-2276	0795
DX-2277	0023
DX-2278	0565
DX-2279	0700
DX-2280	0049
DX-2281	0095
DX-2285	OEM
DX-2306	0052
DX2321	0087
DX2322	0143
DX2323	0036
DX2327	0133
DX-2443	0023
DX-2446	0031
DX-2447	0124
DX-2451	0137
DX-2510	OEM
DX-2511	OEM
DX-2517	0064
DX-2555	0165
DX-2556	0023
DX-2558	OEM
DX-2617	0023
DX-3975	0150
DX6873	0143
DXA08A818	OEM
DXA020401	OEM
DXA020402	OEM
DXA020403	OEM
DXA020404	OEM
DXA020405	OEM
DXA020406	OEM
DXA020407	OEM
DXA020408	OEM
DXA020409	OEM
DXA020410	OEM
DXA020411	OEM
DXA020412	OEM
DXA020413	OEM
DXA020414	OEM
DXA020415	OEM
DXA020416	OEM
DXA020417	OEM
DXA020418	OEM
DXA020419	OEM
DXA020420	OEM
DXA020601	OEM
DXA020602	OEM
DXA020603	OEM
DXA020604	OEM
DXA020605	OEM
DXA020606	OEM
DXA020607	OEM
DXA020608	OEM
DXA020609	OEM
DXA020610	OEM
DXA020611	OEM
DXA020612	OEM
DXA020613	OEM
DXA020614	OEM
DXA020615	OEM
DXA020801	OEM
DXA020802	OEM
DXA020803	OEM
DXA020804	OEM
DXA020805	OEM
DXA020806	OEM
DXA020807	OEM
DXA020808	OEM
DXA020809	OEM
DXA020810	OEM
DXA020811	OEM
DXA020812	OEM
DXA020813	OEM
DXA020814	OEM
DXA020815	OEM
DXA040802	OEM
DXA040803	OEM
DXA040804	OEM
DXA040805	OEM
DXA040806	OEM
DXA040807	OEM
DXA040808	OEM
DXA040809	OEM
DXA040810	OEM
DXA040811	OEM
DXA040812	OEM
DXA040813	OEM
DXA040814	OEM
DXA040815	OEM
DXA040816	OEM
DXA040817	OEM
DXA040818	OEM
DXA040819	OEM
DXA040820	OEM
DXA040821	OEM
DXA040822	OEM
DXA040823	OEM
DXA040824	OEM
DXA051001	OEM
DXA051002	OEM
DXA051003	OEM
DXA051004	OEM
DXA051005	OEM
DXA051006	OEM
DXA051007	OEM
DXA051008	OEM
DXA051009	OEM
DXA051010	OEM
DXA051011	OEM
DXA051012	OEM
DXA061801	OEM
DXA061802	OEM
DXA061803	OEM
DXA061804	OEM
DXA061805	OEM
DXA061806	OEM
DXA061807	OEM
DXA061808	OEM
DXA061809	OEM
DXA061810	OEM
DXA061811	OEM
DXA061812	OEM
DXA061813	OEM
DXA061814	OEM
DXA061815	OEM
DXA081201	OEM
DXA081202	OEM
DXA081203	OEM
DXA081204	OEM
DXA081205	OEM
DXA081206	OEM
DXA081207	OEM
DXA081208	OEM
DXA081209	OEM
DXA081210	OEM
DXA081211	OEM
DXA081212	OEM
DXA081801	OEM
DXA081802	OEM
DXA081803	OEM
DXA081804	OEM
DXA081805	OEM
DXA081806	OEM
DXA081807	OEM
DXA081808	OEM
DXA081809	OEM
DXA081810	OEM
DXA081811	OEM
DXA081812	OEM
DXA081813	OEM
DXA081814	OEM
DXA081815	OEM
DXA081816	OEM
DXA081817	OEM
DXA081819	OEM
DXA081820	OEM
DXA3115-01	OEM
DXA3115-02	OEM
DXA3115-03	OEM
DXA3157-01	OEM
DXA3157-02	OEM
DXA3157-03	OEM
DXL2501A	OEM
DXL2501A-P70	OEM
DXL2502A	OEM
DXL2502A-P70	OEM
DXL2503A	OEM
DXL2503A-CR	OEM
DXL2503A-CRES	OEM
DXL3501A	OEM
DXL3501A-P100F	OEM
DXL3503A	OEM
DXL3503A-CR	OEM
DXL3503A-CRES	OEM
DXL3504A	OEM
DXL3504A-CR	OEM
DXL3504A-CRES	OEM
DXL3615A	OEM
DXL3615A-P100F	OEM
DXL3630A	OEM
DXL3630A-P100F	OEM
DY1SS131	0124
DY1SS133	0133
DYC124XS	OEM
DYDA203	1091
DYDAN201VA	3501
DYDAP201VA	0694
DYLTZ-G15	OEM
DYLTZ-R15	OEM
DYMTZ5.1-B	0589
DYMTZ5.1-C	0041
DYMTZ6.2-C	0471
DYMTZ6.8-C	OEM
DYMTZ7.5-C	0077
DYMTZ7.5C	0077
DYMTZ8.2-C	0165
DYMTZ9.1-C	0057
DYMTZ10-C	0064
DYMTZ20-D	0700
DYMTZ24-C	0494
DYSLR-34UR3F	OEM
DYSLR-34UR5F	OEM
DYSLR-40MC3F	OEM
DYSLR-40UR3F	OEM
DYSLR-40UR5F	OEM
DYSLR-54UR	OEM
DYSLR-54UR-H	OEM
DYSLR-55MC	OEM
DYSLR-55MC-H	OEM
DZ-081	0244
DZ0820	0244
DZ5.1B	OEM
DZ5.6B	OEM
DZ5.6BH	OEM
DZ6.8B	OEM
DZ6.8BM	OEM
DZ9A4	0037
DZ9A5	OEM
DZ10A	0170
DZ10B	0189
DZ10BM	OEM
DZ-12	0137
DZ12	0365
DZ-12A	0137
DZ12A	0052
DZ12B	0089
DZ13BM	OEM
DZ-15A	0002
DZ15A	0686
DZ15B	0340
DZ15BH	OEM
DZ15BM	OEM
DZ16BM	OEM
DZ18A	0375
DZ18B	0425
DZ22A	0705
DZ27A	0441
DZ33A	0039
DZ39A	0032
DZ47A	0068
DZ56A	0125
DZ68A	0173
DZ82A	0331
DZ120	OEM
DZ120A	1005
DZ800	OEM
DZ805	OEM
DZ806	OEM
DZ78103C	OEM
DZB6.2	0636
DZB6.8	0025
DZB6.8C	0025
DZB7.5	0649
DZB7.5C	0649
DZB8.2	0244
DZB8.2C	0244
DZB9.1	0012
DZB9.1C	0012
DZB10	0170
DZB10C	0170
DZB11	0317
DZB11C	0317
DZB12	0137
DZB12C	0137
DZB13	0365
DZB13C	0365
DZB15	0002
DZB15C	0002
DZB16	0420
DZB16C	0420
DZB18	0495
DZB18C	0495
DZB20	0531
DZB20C	0531
DZB22	0565
DZB22C	0565
DZB24	0402
DZB24C	0402
DZB27	0441
DZB27C	0441
DZB30	0726
DZC30	0726
DZC50	0502
DZDP12	OEM
DZJ.6BM	OEM
DZPD9.1	0057
DZPD22	5248
DZSLP-138C-51	OEM
DZSLP-982A-50	OEM
DZ-TE-17-6.2B	OEM
DZV11B	OEM
E	OEM
E0011	0030
E0018	0143
E01A05	OEM
E01A05KA	OEM
E01A05KB	OEM
E01A05LC	OEM
E05	OEM
E05A01LA	OEM
E05A02	OEM
E05A02LA	OEM
E05A03	OEM
E05A03BA	OEM
E05A03CA	OEM
E05A03EA	OEM
E05A09BA	OEM
E05A09KA0123	OEM
E05A10AA	OEM
E05A15HA	OEM
E05A16GA	OEM
E05A29YA	OEM
E05A30	OEM
E05A30YA	OEM
E09-306112	0133
E-044A	0004
E044A	0211
E-065	0136
E-066	0136
E066	0211
E-067	0136
E067	0211
E-068	0136
E068	0211
E-070	0004
E070	0050
E-075L	0015
E075L	0015
E0100AD	0901
E0100FD	0344
E0100YD	1137
E0102AA	0901
E0102AB	0901
E0102AD	0901
E0102FA	0344
E0102FB	0344
E0102FD	0344
E0102YA	1137
E0102YB	1137
E0102YD	1137
E0105	0211
E0350-002	OEM
E0350-003	OEM
E0704-W	0015
E-0704W	0015
E0704W	0540
E0771-3	0012
E0771-6	0100
E0771-7	0137
E0788-C	0015
E0788C	0015
E01011BA	OEM
E02041HA	OEM
E03090-002	0289
E03155-001	0015
E03155-002	0015
E03155-091	OEM
E03450-00	OEM
E05020AA	OEM
E05203BA	OEM
E1	0015
E1A	0127
E1A3	1335
E1B	1733
E1B3	OEM
E1B5	OEM
E1B6	OEM
E1B7	OEM
E1B8	OEM
E1B9	OEM
E1C3	OEM
E1CM-0060	1050
E1D3	OEM
E1E3	OEM
E1F3	OEM
E1G3	OEM
E1H3	1637
E1K3	1637
E1M3	0071
E1N3	0071
E1P	OEM
E2	0015
E2A3	OEM
E2B3	OEM
E2C3	OEM
E2D3	OEM
E2E3	OEM
E2F3	OEM
E2G3	OEM
E2H3	1637
E2K3	2645
E2M3	2645
E2N3	2645
E-2SC536AG-SP	0537
E-2SC536E-NP	0537
E-2SC930D-NP	2209
E-2SC930E-NP	2209
E-2SC1923-O	2691
E-2SD734G	1513
E3	0015
E3A3	OEM
E3B3	OEM
E3C3	OEM
E3D3	OEM
E3E3	OEM
E3F3	OEM
E3G3	OEM
E3H3	1637
E3K3	1637
E3M3	OEM
E3N3	OEM
E4	0164
E4A3	OEM
E4B3	OEM
E4C	OEM
E4C3	OEM
E4D3	OEM
E4E3	OEM
E4F3	OEM
E4G3	OEM
E4H3	1637
E4HZ	OEM
E4K	OEM
E4K3	1637
E4M3	OEM
E4N3	OEM
E5	0015
E5A3	OEM
E5B3	OEM
E5C3	OEM
E5D3	OEM
E5E3	OEM
E5F3	OEM
E5G3	OEM
E5H3	1637
E5K3	1637
E5M3	OEM
E5N3	OEM
E5T20A12.4	0234
E5T20A14.6	0247
E5T20A16.8	1178
E5T20A18.5	2395
E5T20A21	0269
E5T20A23	0274
E5T20A27	0294

DEVICE TYPE	REPL CODE
E5T20A30	0309
E5T20A33	0318
E5T20A37	0320
E5T20A43	0337
E5T20A47	0347
E5T20A51	0027
E5T20A56	0266
E5T20A62	0386
E5T20A68	0405
E5T20A75	0426
E5T20A82	0444
E5T20A87	3028
E5T20A91	0238
E5T20A100	1180
E5T20A105	1190
E5T20A110	1190
E5T20A120	1207
E5T20A130	1218
E5T20A140	1881
E5T20A150	0647
E5T20A175	1279
E5T20A200	0605
E5T50A12.4	0234
E5T50A14.6	0247
E5T50A16.8	1178
E5T50A18.5	2395
E5T50A21	0269
E5T50A23	0274
E5T50A27	0294
E5T50A30	0309
E5T50A33	0318
E5T50A37	0320
E5T50A43	0337
E5T50A47	0347
E5T50A51	0027
E5T50A56	0266
E5T50A62	0386
E5T50A68	0405
E5T50A75	0426
E5T50A82	0444
E5T50A87	3028
E5T50A91	0238
E5T50A100	1180
E5T50A105	1190
E5T50A110	1190
E5T50A120	1207
E5T50A130	1218
E5T50A140	1881
E5T50A150	0647
E5T50A175	1279
E5T50A200	0605
E6	0015
E6A3	OEM
E6B3	OEM
E6C3	OEM
E6D3	OEM
E6E3	OEM
E6F3	OEM
E6G3	OEM
E6H3	1637
E6HZS	OEM
E6K3	1637
E6M3	OEM
E6N3	OEM
E7	OEM
E7A3	OEM
E7B3	OEM
E7C3	OEM
E7D3	OEM
E7E3	6158
E7F3	OEM
E7G3	OEM
E7H3	1637
E7K3	1637
E7M3	OEM
E7N3	OEM
E8	0071
E8HZ	OEM
E9	0071
E10	0071
E10DS1	OEM
E10DS2	OEM
E10DS4	OEM
E10FS1	OEM
E10FS2	OEM
E10HZ	OEM
E10QS03	OEM
E10QS04	OEM
E10QS05	OEM
E10QS06	OEM
E10QS09	OEM
E10QS10	OEM
E11	OEM
E11FS1	OEM
E11FS2	OEM
E11FS3	OEM
E11FS4	OEM
E12	OEM
E13-000-03	0016
E13-000-04	0016
E13-001-02	0037
E13-001-03	0037
E13-001-04	0037
E13-002-03	0016
E13-003-00	0016
E13-003-01	0016
E13-004-00	0330
E13-005-02	0016
E13-006-02	0037
E13-007-00	0285
E13-008-00	0074
E13-009-00	0065
E13-0-1-01	0757
E13-010-00	0338
E13-011-00	0338
E13-012-00	0561
E13-013-03	0474
E13-013-04	0474
E13-017-01	0133
E13-020-00	0015
E13-021-01	0757
E13-021-03	0474
E13-20-00	0015
E14C350	0015
E20C30	0143
E21	0015
E25C5	0015
E30C60	0143
E30Hs	OEM
E39ED	OEM
E-41	0015
E41	0015
E49A-B380-340-30	OEM
E49A-B500-450-30	OEM
E49B30-25-30	OEM
E49B60-52-30	OEM
E49B125-110-30	OEM
E49B250-225-30	OEM
E49B380-340-30	OEM
E49B500-450-30	OEM
E50A-DB380-500-40	OEM
E50A-DB500-670-40	OEM
E50DB30-38-40	OEM
E50DB60-78-40	OEM
E50DB125-165-40	OEM
E50DB250-330-40	OEM
E50DB380-500-40	OEM
E50DB500-670-40	OEM
E50HR	OEM
E50HS	OEM
E61	OEM
E-80	OEM
E81	OEM
E84	0170
E100	4214
E100HR	OEM
E100HS	OEM
E-101	OEM
E101	2950
E-101L	OEM
E-102	5213
E102	2950
E102(ELCOM)	0015
E103	0325
E105	0136
E106(ELCOM)	0015
E108	0071
E108(ELCOM)	0071
E109	OEM
E110	OEM
E111	1155
E112	3130
E113	1155
E114	3130
E125C200	0015
E132	0004
E135	0015
E143	0015
E146	0015
E150HR	OEM
E150L	0015
E151A	0469
E151B	0469
E151C	0469
E151D	0469
E151E	0722
E151F	0722
E151H	0722
E151K	0725
E151M	0725
E151P	0679
E151S	0679
E151V	0679
E-158	0211
E158	0004
E161A	0469
E161B	0469
E161C	0469
E161D	0469
E161E	0722
E174	OEM
E175	OEM
E176	OEM
E177	OEM
E181	0004
E181A	0004
E181B	0004
E181C	0004
E181D	0004
E-185B121712	0086
E185B121712	0276
E201	3384
E-202	OEM
E202	0087
E202(ELCOM)	0087
E203	3384
E210	0079
E211	1392
E212	0079
E213	0037
E214B	0211
E230	3384
E231	3384
E232	1155
E241	0004
E241A	0004
E241A(SCR)	0469
E241B	0004
E241B(SCR)	0469
E241C	0469
E241D	0469
E241E	0722
E241F	0722
E241H	0722
E241K	0725
E241M	0725
E241P	0679
E241S	0679
E241V	0679
E241Z	0679
E241ZB	0679
E241ZD	0679
E262	0137
E270	OEM
E271	OEM
E-272	OEM
E295ZZ01	0143
E300	0325
E300(ELCOM)	0922
E300D2	OEM
E300D7	OEM
E300F2	OEM
E300F7	OEM
E300L	0015
E-301	OEM
E301D2	OEM
E301D7	OEM
E301F2	OEM
E301F7	OEM
E302(ELCOM)	0242
E302D2	OEM
E302D7	OEM
E302F2	OEM
E302F7	OEM
E303D2	OEM
E303D7	OEM
E303F2	OEM
E303F7	OEM
E304	3384
E304(ELCOM)	0540
E304D2	OEM
E304D7	OEM
E304F2	OEM
E304F7	OEM
E305	3384
E305D2	OEM
E305D7	OEM
E305F2	OEM
E305F7	OEM
E306(ELCOM)	0965
E306D2	OEM
E306D7	OEM
E306F2	OEM
E306F7	OEM
E308	0817
E308(ELCOM)	0817
E309	OEM
E310	OEM
E311	OEM
E312	OEM
E-352	OEM
E413	OEM
E415	OEM
E-452	OEM
E-452-2	OEM
E500	OEM
E500L	0015
E-501	OEM
E501	OEM
E502	OEM
E503	OEM
E504	OEM
E505	OEM
E506	OEM
E507	OEM
E-562	OEM
E602	OEM
E603	OEM
E605	OEM
E607	OEM
E610	OEM
E650L	0015
E656	0133
E660	0196
E700	OEM
E-701	OEM
E701	OEM
E702	OEM
E703	OEM
E705	OEM
E707	OEM
E710	OEM
E715	OEM
E722	OEM
E744B	OEM
E750(ELCOM)	0015
E752(ELCOM)	0015
E756(ELCOM)	0015
E758(ELCOM)	0071
E760(ELCOM)	0071
E814-0	OEM
E814-1	OEM
E814-2	OEM
E814-3	OEM
E814-4	OEM
E814-6	OEM
E814-8	OEM
E815-0	OEM
E815-1	OEM
E815-2	OEM
E815-3	OEM
E815-4	OEM
E815-6	OEM
E815-8	OEM
E842	0144
E843	0144
E844	0144
E1000	OEM
E1010	0015
E1011	0015
E1018N	0015
E1020	OEM
E1020F	OEM
E1031RT	0143
E1031RXT	0143
E1042	0289
E1100	OEM
E1121R	0133
E1124	0015
E1138R	0133
E1143RD	0087
E1145R	0102
E1145RED	0102
E1145RJH	0102
E1146J	0015
E1146R	0015
E1153RB	0015
E1156RD	0015
E1157RNA	0015
E1158ZA	0301
E1174R	1303
E1176	0196
E1176R	1149
E1184	OEM
E1200	OEM
E1221R	0133
E1229R	0133
E1237RF	0023
E1256R	0133
E1266R	0348
E1270R	0023
E1286R	0133
E1293Z	0010
E1294Z	0010
E1300	OEM
E1303Z	0091
E1304R	0071
E1308R	0023
E1314R	2280
E1315R	0087
E1322Z	0471
E1324R	0133
E1332R	0124
E1336R	0348
E1339R	0023
E1340R	0348
E1410	0015
E1411	0015
E1412	0015
E1412R	0023
E1413	0015
E1413R	0023
E1415	0015
E1428R	0023
E1436ZB	6906
E1440	0015
E1442R	0023
E1466Z	0166
E1490R	0283
E1500	OEM
E1500-1	OEM
E1500F	OEM
E1510	OEM
E1510F	OEM
E1512L	5174
E1516R	0023
E1518R	0023
E1520	OEM
E1520F	OEM
E1533R	0023
E1534R	0023
E1577P	OEM
E1598R	0133
E1600	OEM
E1614L	OEM
E1616R	0283
E1618R	0023
E1634ZXU	0471
E1639Z	0057
E1645Z	0137
E1650Z	0181
E1660Z	0041
E1670Z	0376
E1672Z	0166
E1686ZXU	OEM
E1704R	0564
E1705R	0348
E1712A	OEM
E1714A	OEM
E1716A	OEM
E1717A	OEM
E1717R	0564
E1718A	OEM
E1719A	OEM
E1720A	OEM
E1720R	0071
E1722A	OEM
E1724A	OEM
E1725RM	0283
E1726A	OEM
E1750Z	0301
E1756ZB	0057
E1756ZC	0057
E1775R	0133
E1779R	0283
E1786R	0023
E1789R	0023
E1791R	0023
E1794R	0023
E1810ZA	0195
E1810ZB	0195
E1821ZA	0062
E1821ZB	0062
E1821ZC	0062
E1821ZD	0077
E1823Z	0077
E1823ZA	0077
E1823ZB	0077
E1828ZA	0053
E1828ZB	1005
E1829R	0564
E1852	0137
E1860Z	OEM
E2000	OEM
E2004	OEM
E2070	3011
E2106B40C5000-3300	OEM
E2112B80C5000-3300	OEM
E2220B125C5000-3300	OEM
E2240B250C5000-3300	OEM
E2412	0280
E2427	0600
E-2428	5486
E2428	0600
E-2429	5486
E2429	0600
E2430	0016
E2431	0016
E2434	0144
E-2435	5487
E2435	0144
E-2436	2536
E2436	0016
E2438	0050
E2439	0050
E2440	0050
E2441	0086
E2444	0016
E2445	0164
E2447	0208
E2448	0004
E2449	0086
E2450	0050
E2451	0136
E2452	0016
E2453	0004
E2454	0016
E2455	0016
E2456	OEM
E2457	OEM
E2459	0016
E2460	0142
E2461	0016
E-2462	0050
E2462	0050
E2464	2866
E2465	0004
E2466	0208
E2467	0004
E2474	0050
E2475	0050
E2476	0004
E2477	0050
E2478	0050
E2479	0050
E2480	0004
E2481	0004
E2482	0004
E2484	0123
E2486	0636
E2495	2529
E2496	0042
E2497	0016
E2498	0126
E2499	0016
E2500	OEM
E2502	OEM
E2506B40C5000-3300	OEM
E2506N	OEM
E2512B80C5000-3300	OEM
E2512N	OEM
E2525	OEM
E2620B125C5000-3300	OEM
E2620N	OEM
E2640B250C5000-3300	OEM
E2640N	OEM
E3006	0015
E3274A	OEM
E3274B	OEM
E3389	OEM
E3469	OEM
E3509A	OEM
E3509A-H	OEM
E3509H	OEM
E3509L	OEM
E3513	OEM
E3513H	OEM
E3513L	OEM
E3514	OEM
E3518	OEM
E3521	OEM
E3521H	OEM
E3521L	OEM
E3522	OEM
E3523	OEM
E3524	OEM
E3526	OEM
E3753	OEM
E3774	OEM
E3776	OEM
E3803	OEM
E3806B40C5000-3300	OEM
E3812B80C5000-3300	OEM
E3820B125C5000-3300	OEM
E3824	OEM
E3824A	OEM
E3825	OEM
E3832	OEM
E3840B250C5000-3300	OEM
E3902	OEM
E4676B	0015
E5100	OEM
E5102	OEM
E5103	OEM
E5104	OEM
E5106	OEM
E5108	OEM
E7441	0015
E7460	OEM
E7462	OEM
E7464	OEM
E7466	OEM
E7468	OEM
E8327B	OEM
E8343	OEM
E9625	0224
E10116	0015
E10157	0015
E10171	0015
E10172	0015
E10416A	OEM
E11401	OEM
E17336	2882
E21135	0123
E21430	0143
E21431	0636
E23036	0930
E24100	0015
E24101	0087
E24103	0016
E24104	0004
E24105	0208
E24106	0004
E24107	0222
E25056	0930
E570022-01	0178
EA000-13000	0215
EA000-17200	1698
EA000-22600	0709
EA000-36400	OEM
EA000-44600	0051
EA000-75500	OEM
EA000-95300	OEM
EA0002	0050
EA0007	0050
EA0009	0004
EA001-02200	0051
EA001-02700	1898
EA001-02703	1898
EA001-06300	OEM
EA001-06700	OEM
EA001-07100	OEM
EA001-07200	OEM
EA001-08700	OEM
EA001-21400	OEM
EA005	0015
EA006-41400	OEM
EA006-41500	OEM
EA0013	0144
EA0015	0015
EA0016	0015
EA0031	0015
EA0053	0050
EA0081	0086
EA0083	OEM
EA0085	OEM
EA0086	0126
EA0087	0126
EA0088	0150
EA0090	0086
EA0091	0144
EA0092	0016
EA0093	0144
EA0094	0144
EA0095	0144
EA01-07R	0002
EA010	0015
EA020	0015
EA030	0015
EA040	0015
EA050	0015
EA060	0015
EA080	0071
EA0X12	3732
EA1-380	0161
EA1SY11	0050
EA15X1	0016
EA15X2	0004
EA15X3	0004
EA15X4	0004
EA15X5	0050
EA15X6	0440
EA15X7	0004
EA15X8	0004
EA15X9	0016
EA15X10	0160
EA15X11	0050
EA15X11CL	0050
EA15X12	0160
EA15X13	0050
EA15X14	1015
EA15X15	0160
EA15X15CL	0160
EA15X18	0016
EA15X18CL	0155
EA15X19	0004
EA15X19CL	0004
EA15X19P	0004
EA15X20	0016
EA15X22	0016
EA15X23	0004
EA15X24	0016
EA15X25	0004
EA15X26	0160
EA15X27	0050
EA15X28	0004
EA15X29	0050
EA15X30	0050
EA15X31	0016
EA15X33	0160
EA15X35	0160
EA15X36	0004
EA15X37	0016
EA15X38	0160
EA15X40	0050
EA15X41	0050
EA15X43	0050
EA15X44	0016
EA15X44P	0144
EA15X45	0016
EA15X45CL	0144
EA15X45P	0144
EA15X48	0144
EA15X49	0144
EA15X50	0144
EA15X51	0144
EA15X52	0016
EA15X53	0160
EA15X54	0144
EA15X55	0144
EA15X55P	0144
EA15X56	0016
EA15X56CL	0042
EA15X57	0086
EA15X58	0016
EA15X59	0016
EA15X60	0079
EA15X63	0016
EA15X63CL	0155
EA15X66	0050
EA15X67	0004
EA15X67CL	0004
EA15X68	0016
EA15X69	0150
EA15X70	0150
EA15X71	0150
EA15X72	0016
EA15X73	0016
EA15X75	0016
EA15X75CL	0155
EA15X76	0016
EA15X76CL	0155
EA15X77	0016
EA15X77CL	0155
EA15X83	0016
EA15X84	0016
EA15X84CL	0155
EA15X85	0016
EA15X85CL	0155
EA15X86	0016
EA15X88	0160
EA15X90	0079
EA15X91	0016
EA15X94	0144
EA15X94CL	0144
EA15X94P	0144
EA15X96	0016
EA15X101	0016
EA15X102	0086
EA15X103	0016
EA15X103CL	0155
EA15X111	0016
EA15X112	0016

DEVICE TYPE	REPL CODE
EA15X113	0144
EA15X121	0178
EA15X121CL	0178
EA15X122	0124
EA15X123	0130
EA15X124	1681
EA15X126	0451
EA15X130	0224
EA15X130A	0144
EA15X131	0144
EA15X131A	0144
EA15X132	0144
EA15X132L	0144
EA15X133	0136
EA15X133P	0050
EA15X134	0127
EA15X134L	0144
EA15X135	0127
EA15X135L	0144
EA15X136	0016
EA15X137	0076
EA15X139	0222
EA15X140	0136
EA15X141	0136
EA15X142	0016
EA15X143	0016
EA15X144	0086
EA15X152	0111
EA15X153	0016
EA15X154	0222
EA15X157	0076
EA15X160	0560
EA15X161	0111
EA15X162	0016
EA15X163	0016
EA15X164	0004
EA15X165	0325
EA15X167	0016
EA15X168	0086
EA15X169	0325
EA15X171	0133
EA15X172	0133
EA15X173	0160
EA15X180	0076
EA15X185	0006
EA15X189	0597
EA15X190	0151
EA15X192	1211
EA15X193	3155
EA15X194	0006
EA15X203	0004
EA15X207	0004
EA15X212	0004
EA15X213	0111
EA15X233	0006
EA15X237	0016
EA15X238	0325
EA15X239	0127
EA15X240	0597
EA15X241	1221
EA15X242	0148
EA15X243	0460
EA15X244	1790
EA15X245	0111
EA15X246	1480
EA15X247	0400
EA15X248	0560
EA15X249	0155
EA15X250	0018
EA15X251	0151
EA15X256	0155
EA15X257	0164
EA15X258	0111
EA15X259	0111
EA15X264	0111
EA15X266	0400
EA15X267	0155
EA15X268	0532
EA15X269	0563
EA15X270	0525
EA15X272	0016
EA15X273	0791
EA15X274	0595
EA15X288	0111
EA15X325	1220
EA15X326	1062
EA15X327	0042
EA15X328	0925
EA15X329	3533
EA15X330	0016
EA15X331	0016
EA15X332	2258
EA15X333	0042
EA15X334	0925
EA15X335	0076
EA15X336	0127
EA15X349	0595
EA15X350	0076
EA15X351	1144
EA15X352	1221
EA15X353	0380
EA15X354	0380
EA15X355	0111
EA15X356	1865
EA15X359	2910
EA15X360	0191
EA15X361	0111
EA15X362	1591
EA15X363	0936
EA15X364	0368
EA15X365	0368
EA15X367	0368
EA15X370	0016
EA15X371	0079
EA15X372	2710
EA15X373	0155
EA15X374	0076
EA15X376	1066
EA15X378	0155
EA15X379	0224
EA15X380	1946
EA15X381	0839
EA15X383	0786
EA15X385	0006
EA15X386	1221
EA15X393	0127
EA15X394	3155
EA15X395	0006
EA15X396	0532
EA15X397	0018
EA15X400	1280
EA15X401	1280
EA15X402	0212
EA15X404	0076
EA15X405	0212
EA15X408	0733
EA15X412	1400
EA15X413	2825
EA15X414	0936
EA15X415	0549
EA15X421	0006
EA15X437	0224
EA15X441	0286
EA15X446	3046
EA15X456	0018
EA15X461	1066
EA15X483	0224
EA15X488	1513
EA15X524	0111
EA15X525	0042
EA15X534	5954
EA15X625	1386
EA15X945	0076
EA15X947	4432
EA15X1014	0079
EA15X1015	OEM
EA15X1061	OEM
EA15X1117	0079
EA15X1150	OEM
EA15X1422	0127
EA15X1710	0042
EA15X2015	0144
EA15X2024	0127
EA15X2025	0127
EA15X2110	0079
EA15X2112	0079
EA15X2522	0018
EA15X3118	0925
EA15X4064	0127
EA15X4152	0144
EA15X4335	0286
EA15X4531	0595
EA15X4532	0016
EA15X4631	0342
EA15X4662	OEM
EA15X4810	0286
EA15X4812	0286
EA15X5939	0079
EA15X6840	0004
EA15X7112	0016
EA15X7113	0144
EA15X7114	0144
EA15X7115	0016
EA15X7117	0127
EA15X7118	0016
EA15X7119	0016
EA15X7120	0016
EA15X7121	0042
EA15X7125	0127
EA15X7136	0224
EA15X7137	0037
EA15X7140	0144
EA15X7141	0127
EA15X7173	0008
EA15X7174	0008
EA15X7175	0016
EA15X7176	0016
EA15X7177	0144
EA15X7178	0008
EA15X7179	0008
EA15X7215	0127
EA15X7228	0144
EA15X7231	0127
EA15X7232	0016
EA15X7233	0224
EA15X7234	0224
EA15X7235	0224
EA15X7236	0224
EA15X7242	0224
EA15X7243	0144
EA15X7244	0224
EA15X7245	0111
EA15X7262	0079
EA15X7263	0127
EA15X7264	0127
EA15X7514	0016
EA15X7517	0016
EA15X7519	0595
EA15X7583	0111
EA15X7586	0016
EA15X7587	0144
EA15X7588	0079
EA15X7589	0079
EA15X7590	0016
EA15X7592	0532
EA15X7635	0086
EA15X7638	0016
EA15X7639	0532
EA15X7643	0016
EA15X7722	0144
EA15X7732	0037
EA15X7737	0037
EA15X8118	0925
EA15X8119	0042
EA15X8122	0382
EA15X8124	0042
EA15X8126	0491
EA15X8130	0925
EA15X8180	OEM
EA15X8442	0004
EA15X8443	0208
EA15X8444	0004
EA15X8502	0016
EA15X8511	0016
EA15X8514	0018
EA15X8515	0016
EA-15X8517	0086
EA15X8517	0018
EA15X8518	0079
EA15X8519	0079
EA15X8520	0079
EA15X8521	0324
EA15X8522	0436
EA15X8523	0079
EA15X8524	0436
EA15X8525	0079
EA15X8529	0016
EA15X8576	0079
EA15X8589	0144
EA15X8601	0079
EA15X8602	0079
EA15X8605	0460
EA15X8608	0144
EA15X8609	0144
EA15X8610	0144
EA16X1	0143
EA16X2	0015
EA16X4	0794
EA16X5	0143
EA16X6	0794
EA16X8	0015
EA16X9	0143
EA16X11	0143
EA16X13	0015
EA16X16	0794
EA16X19	0030
EA16X20	0133
EA16X21	0015
EA16X22	0143
EA16X27	0143
EA16X28	0030
EA16X29	0137
EA16X30	0015
EA16X31	0030
EA16X33	0015
EA16X34	0015
EA16X35	0579
EA16X38	0402
EA16X39	0133
EA16X48	0143
EA16X49	0133
EA16X54	0080
EA16X55	0015
EA16X59	0133
EA16X60	0133
EA16X61	0133
EA16X62	0012
EA16X68	0139
EA16X69	0139
EA16X71	0015
EA16X73	0120
EA16X75	0133
EA16X77	0137
EA16X80	0636
EA16X81	0365
EA16X82	0012
EA16X84	0133
EA16X88	0649
EA16X89	0036
EA16X91	0124
EA16X92	0023
EA16X94	0133
EA16X95	0124
EA16X97	0143
EA16X101	0133
EA16X105	0133
EA16X106	0030
EA16X107	0139
EA16X110	0133
EA16X116	0120
EA16X117	0628
EA16X118	0157
EA16X122	0133
EA16X123	0301
EA16X124	0002
EA16X125	0649
EA16X126	0041
EA16X127	0030
EA16X134	0133
EA16X135	0124
EA16X136	0041
EA16X140	0143
EA16X144	0140
EA16X146	0133
EA16X147	0628
EA16X148	0143
EA16X149	0540
EA16X150	0064
EA16X152	0133
EA16X154	0253
EA16X157	0170
EA16X162	0137
EA16X166	0554
EA16X171	0133
EA16X177	0030
EA16X192	0005
EA16X193	1276
EA16X203	1778
EA16X402	1083
EA16X421	0471
EA16X422	OEM
EA16X424	0064
EA16X426	0165
EA16X427	0005
EA16X438	0165
EA16X441	0143
EA16X442	0133
EA16X443	0317
EA16X445	0064
EA16X446	0133
EA16X447	0253
EA16X451	0005
EA16X455	0064
EA16X456	OEM
EA16X462	0140
EA16X463	0005
EA16X464	0015
EA16X474	0686
EA16X475	0253
EA16X476	0106
EA16X477	0133
EA16X483	0062
EA16X494	OEM
EA16X504	0123
EA16X505	0064
EA16X506	0080
EA16X939	0124
EA16X1049	5733
EA16X4532	0079
EA33C8390	2063
EA33X8333	4669
EA33X8341	OEM
EA33X8348	OEM
EA33X8351	2876
EA33X8352	2509
EA33X8356	2568
EA33X8363	2614
EA33X8364	1611
EA33X8367	0579
EA33X8368	4566
EA33X8369	OEM
EA33X8371	4543
EA33X8372	1478
EA33X8373	1843
EA33X8374	2572
EA33X8375	0520
EA33X8382	OEM
EA33X8383	4669
EA33X8384	3318
EA33X8385	1172
EA33X8386	1172
EA33X8388	1714
EA33X8389	1050
EA33X8392	1540
EA33X8393	OEM
EA33X8394	1540
EA33X8395	4042
EA33X8396	1050
EA33X8399	0141
EA33X8450	2295
EA33X8455	OEM
EA33X8456	OEM
EA33X8465	2728
EA33X8467	OEM
EA33X8468	0973
EA33X8469	0026
EA33X8470	OEM
EA33X8471	OEM
EA33X8475	OEM
EA33X8476	OEM
EA33X8477	OEM
EA33X8478	OEM
EA33X8479	0312
EA33X8487	OEM
EA33X8488	OEM
EA33X8489	OEM
EA33X8492	OEM
EA33X8493	OEM
EA33X8497	5337
EA33X8500	0911
EA33X8501	3559
EA33X8506	1910
EA33X8508	4042
EA33X8509	1540
EA33X8511	0520
EA33X8520	OEM
EA33X8521	4830
EA33X8522	5344
EA33X8523	2828
EA33X8524	3804
EA33X8532	5195
EA33X8535	0706
EA33X8536	OEM
EA33X8537	5195
EA33X8543	OEM
EA33X8544	OEM
EA33X8545	0756
EA33X8548	3063
EA33X8559	0362
EA33X8562	OEM
EA33X8564	5195
EA33X8569	5205
EA33X8570	OEM
EA33X8571	OEM
EA33X8572	2304
EA33X8583	5357
EA33X8584	2728
EA33X8585	0433
EA33X8593	OEM
EA33X8600	2728
EA33X8601	OEM
EA33X8605	1527
EA33X8607	1260
EA33X8609	OEM
EA33X8626	OEM
EA33X8642	OEM
EA33X8675	OEM
EA33X8705	1260
EA33X8706	2595
EA33X8707	4726
EA33X8717	3816
EA33X8956	OEM
EA33X9054	OEM
EA33X9213	OEM
EA41X556	4629
EA50X12	3732
EA57X1	0015
EA57X3	0015
EA57X4	0080
EA57X8	0015
EA57X10	0015
EA57X11	0015
EA57X14	0023
EA75X1	0015
EA85X8476	OEM
EA93X8542	OEM
EA100	0071
EA330E	OEM
EA330M	OEM
EA330N	OEM
EA330P	OEM
EA330PB	OEM
EA330PD	OEM
EA330PE	OEM
EA390A	OEM
EA390B	OEM
EA390D	OEM
EA390M	OEM
EA390N	OEM
EA390P	OEM
EA390PB	OEM
EA390PD	OEM
EA390PM	OEM
EA397A	OEM
EA397B	OEM
EA397D	OEM
EA397M	OEM
EA397N	OEM
EA397P	OEM
EA397PB	OEM
EA397PD	OEM
EA397PM	OEM
EA398A	OEM
EA398B	OEM
EA398D	OEM
EA398M	OEM
EA398N	OEM
EA398P	OEM
EA398PB	OEM
EA398PD	OEM
EA398PM	OEM
EA403	0396
EA430E	OEM
EA430M	OEM
EA430N	OEM
EA430P	OEM
EA430PB	OEM
EA430PD	OEM
EA430PM	OEM
EA437A	OEM
EA437B	OEM
EA437D	OEM
EA437M	OEM
EA437N	OEM
EA437P	OEM
EA437PB	OEM
EA437PD	OEM
EA437PM	OEM
EA540L	OEM
EA540LA	OEM
EA540LB	OEM
EA540LC	OEM
EA540LD	OEM
EA596N	OEM
EA596P	OEM
EA596PA	OEM
EA596PB	OEM
EA596PC	OEM
EA596PD	OEM
EA596T	OEM
EA640L	OEM
EA640LQ	OEM
EA640P	OEM
EA640PB	OEM
EA640PD	OEM
EA640PM	OEM
EA640PN	OEM
EA641L	OEM
EA641LA	OEM
EA641LB	OEM
EA641LC	OEM
EA641LD	OEM
EA641LE	OEM
EA641LM	OEM
EA696L	OEM
EA696PE	OEM
EA696PM	OEM
EA696PN	OEM
EA696PS	OEM
EA696PT	OEM
EA740L	OEM
EA740PD	OEM
EA740PE	OEM
EA740PM	OEM
EA740PN	OEM
EA1072	0015
EA1080	0016
EA1081	0004
EA1082	0160
EA1085	OEM
EA1123	0143
EA1128	0016
EA1129	0016
EA1135	0016
EA1145	0016
EA1318	0137
EA1337	0050
EA1338	0050
EA1339	0050
EA1340	0050
EA1341	0160
EA1342	0050
EA1343	0144
EA1344	0016
EA1345	0016
EA1346	0004
EA1385	0030
EA1405	0133
EA1406	0016
EA1407	0016
EA1408	0016
EA1448	0015
EA1451	0016
EA1452	0016
EA1499	0016
EA1549	0086
EA1562	0144
EA1563	0144
EA1564	0016
EA1578	0016
EA1581	0016
EA1628	0016
EA1629	0016
EA1630	0016
EA1638	0016
EA1661	0133
EA1672	0015
EA1684	0086
EA1695	0016
EA1696	0016
EA1697	0016
EA1698	0086
EA1700	0160
EA1703	0016
EA1716	0016
EA1718	0016
EA1733	0144
EA1735	0016
EA1740	0103
EA1760	3724
EA1778	0079
EA1793	0144
EA1872	0016
EA1873	0086
EA2114L-15PC	OEM
EA2114L-20PC	OEM
EA2114L-25PC	OEM
EA2114L-30PC	OEM
EA2114LPC	OEM
EA2131	0127
EA2132	0127
EA2133	0136
EA2134	0004
EA2135	0004
EA2136	0164
EA2137	0123
EA2138	0030
EA2140	0015
EA2176	0004
EA2271	0016
EA2429	0111
EA2488	0560
EA2489	0016
EA2490	0016
EA2491	0136
EA2493	0151
EA2494	0151
EA2495	0151
EA2496	0127
EA2497	0136
EA2498	0136
EA2499	0015
EA2500	0137
EA2501	0015
EA2502	0143
EA2503	0030
EA2600	0144
EA2601	0144
EA2602	0144
EA2603	0144
EA2604	0144
EA2605	0144
EA2606	0123
EA2607	0133
EA2608	0157
EA2738	0111
EA2739	0016
EA2740	0537
EA2741	0015
EA2770	0016
EA2770(N)	0079
EA2771	0111
EA2812	0127
EA3127	0143
EA3149	0016
EA3211	0111
EA3278	0325
EA3280	OEM
EA3281	OEM
EA3282	0579
EA3406	0144
EA3447	0133
EA3674	0560
EA3713	1144
EA3714	0006
EA3715	0925
EA3716	0042
EA3717	0005
EA3718	0143
EA3719	0137
EA3763	0111
EA3827	0015
EA3866	0012
EA3989	0015
EA3990	0218
EA4025	1221
EA4055	0560
EA4085	0560
EA4104L	OEM
EA4104P	OEM
EA4104PA	OEM
EA4104PB	OEM
EA4104PD	OEM
EA4104PE	OEM
EA4104PM	OEM
EA4104PN	OEM
EA4112	0016
EA4156N	OEM
EA4156P	OEM
EA4156PA	OEM
EA4156PB	OEM
EA4156PD	OEM
EA4156PE	OEM
EA4157L	OEM
EA4157LB	OEM
EA4157LD	OEM
EA4157LM	OEM
EA4157PN	OEM
EA4157PS	OEM
EA4158L	OEM
EA4158LB	OEM
EA4158LD	OEM
EA4158LM	OEM
EA4158LN	OEM
EA4158LP	OEM
EA4158PN	OEM
EA4158PS	OEM
EA5316	OEM
EA5711	0015
EA-6801	OEM
EA6801	1235
EA7316A-1	2295
EA-7316B	2295
EA7316B	2295
EA7316D	OEM
EA7317B	2295
EA7317C	OEM
EA7317D	OEM
EA8308A-5DC	OEM
EA8308A-5PC	OEM
EA8308AC	OEM
EA8308ADC	OEM
EA8308AL	OEM
EA8308AP	OEM
EA8308APC	OEM
EA8316ADC	OEM
EA8316ADM	OEM
EA8316E-5DC	OEM
EA8316E-5PC	OEM
EA8316EDC	OEM
EA8316EDL	OEM
EA8316EPC	OEM
EA8332A1DC	OEM

If replacement code is OEM, contact original manufacturer for replacement.

DEVICE TYPE	REPL CODE
EA8332A1PC	OEM
EA8332ADC	OEM
EA8332APC	OEM
EA8332B1DC	OEM
EA8332B1PC	OEM
EA8332BDC	OEM
EA8332BPC	OEM
EA8364DC	OEM
EA8364PC	OEM
EAA00-02700	0786
EAA00-11300	1212
EAA00-12800	1212
EAA00-13400	0076
EAA00-13900	2268
EAA00-16800	1212
EAA00-17400	1212
EAA00-17500	0076
EAA00-19700	OEM
EAA00-19800	OEM
EAA00-19900	OEM
EAB00-04000	0525
EAB00-06500	0602
EAB00-06600	0148
EAB00-07100	0148
EAB00-08300	0681
EAB00-08600	0602
EAB00-08900	0434
EAB00-09500	1348
EAB00-09900	0434
EAB00-Q6600	0148
EAC00-00400	0133
EAC00-00600	0023
EAC00-05400	0023
EAC00-09300	3501
EAC00-09500	3604
EAC00-10000	OEM
EAC00-13100	0015
EAC00-13200	OEM
EACT0-00300	0124
EACT0-00400	0133
EACT0-08100	2542
EAD00-08200	0052
EAD00-08400	0053
EAD00-08700	0441
EAD00-09200	0471
EAD00-10100	0140
EAD00-10400	0140
EAD00-15000	0062
EAD00-15100	0002
EAD00-16700	0636
EAD00-17200	0999
EAD00-20200	0471
EAD00-20700	0318
EAD00-20800	0820
EAD00-21700	0170
EAD00-25300	0441
EADT0-16700	0636
EADT0-25300	0441
EAE00-12600	1414
EAE00-14700	1040
EAF00-01600	0273
EAI-380	0161
E-AN7110E	5229
EA000-36400	OEM
EA000-39500	OEM
EA000-75500	OEM
EA000-95300	OEM
EA001-00600	1898
EA001-21400	OEM
EA001-23300	OEM
EA006-30000	OEM
EA006-30100	OEM
EA006-56300	OEM
EA006-64300	OEM
EAQ00-02000	0361
EAQ00-05000	1262
EAQ00-07500	1206
EAQ00-07514	1206
EAQ00-09600	OEM
EAQ00-09800	OEM
EAQ00-1200	1595
EAQ00-11100	0973
EAQ00-11300	3520
EAQ00-12100	1527
EAQ00-12200	1595
EAQ00-12300	1633
EAQ00-12400	1662
EAQ00-12700	0243
EAQ00-12900	0979
EAQ00-12914	0979
EAQ00-13000	0209
EAQ00-13100	0260
EAQ00-15200	1339
EAQ00-17200	1698
EAQ00-17600	0088
EAQ00-18700	0427
EAQ00-19000	0943
EAQ00-19300	0713
EAQ00-19400	0458
EAQ00-20300	0893
EAQ00-21500	0153
EAQ00-22100	0463
EAQ00-22600	0709
EAQ00-22614	0709
EAQ00-22900	0209
EAQ00-23600	1744
EAQ00-24000	1832
EAQ00-24700	1654
EAQ00-30000	0508

DEVICE TYPE	REPL CODE
EAQ00-30200	0511
EAQ00-30300	1133
EAQ00-75500	OEM
EAS00-00700	0624
EAS00-05800	0176
EAS00-06100	0360
EAS00-09000	0312
EAS00-09700	1838
EAS00-12014	OEM
EAS00-12800	OEM
EAS00-12900	0631
EAS00-13500	0334
EAS00-16800	OEM
EAS00-17800	OEM
EAT00-04200	OEM
EAT00-04300	OEM
EAT00-04500	OEM
EAT00-08900	OEM
EAT00-09000	OEM
EAT00-09100	1596
EAW00-00700	0325
EAX5X249	0155
EAY00-01100	OEM
EAY00-01300	OEM
EAY00-02600	OEM
EB0001	0004
EB0003	0004
EB1(ELCOM)	0277
EB2(ELCOM)	0296
EB3(ELCOM)	0277
EB4(ELCOM)	0289
EB6(ELCOM)	0296
EB7(ELCOM)	0302
EB9(ELCOM)	0277
EB10(ELCOM)	0289
EB11(ELCOM)	0296
EB12(ELCOM)	0302
EB13(ELCOM)	0250
EB383	OEM
EBR11-01	4240
EBR3432S	OEM
EBR5504S	OEM
EC080	0087
EC-0677	OEM
EC-0678	OEM
EC-0726	OEM
EC-0751	OEM
EC-0776	OEM
EC-0777	OEM
EC-0856	OEM
EC-0877	OEM
EC-0884	OEM
EC-0885	OEM
EC-0973	OEM
EC60L	0201
EC68A21P	OEM
EC72L	0201
EC100	0071
EC103A	0901
EC103B	0058
EC103C	OEM
EC103D	0407
EC103Y	1137
EC106A1	0447
EC106A2	0447
EC106B1	0940
EC106B2	0940
EC106D1	0095
EC106D2	0095
EC106F1	1259
EC106F2	1259
EC106M1	5732
EC106M2	3620
EC106Y1	1396
EC106Y2	1396
EC107A1	3854
EC107A2	3854
EC107B1	3620
EC107B2	3620
EC107D1	0095
EC107D2	0095
EC107F1	0174
EC107F2	0174
EC107M1	OEM
EC107M2	0174
EC107Y1	4458
EC107Y2	4458
EC120	OEM
EC350E	OEM
EC350M	OEM
EC350N	OEM
EC350P	OEM
EC350PA	OEM
EC350PB	OEM
EC350PD	OEM
EC350PM	OEM
EC354A	OEM
EC354B	OEM
EC354C	OEM
EC354D	OEM
EC354E	OEM
EC354M	OEM
EC355A	OEM
EC355B	OEM
EC355C	OEM
EC355D	OEM
EC355E	OEM
EC355M	OEM
EC358E	OEM

DEVICE TYPE	REPL CODE
EC358M	OEM
EC358N	OEM
EC358P	OEM
EC358PB	OEM
EC364A	OEM
EC364B	OEM
EC364C	OEM
EC364D	OEM
EC364E	OEM
EC364M	OEM
EC364N	OEM
EC365A	OEM
EC365B	OEM
EC365C	OEM
EC365D	OEM
EC365E	OEM
EC365M	OEM
EC365N	OEM
EC376E	OEM
EC376M	OEM
EC376N	OEM
EC376P	OEM
EC376PB	OEM
EC376PD	OEM
EC377E	OEM
EC377M	OEM
EC377N	OEM
EC377P	OEM
EC377PB	OEM
EC377PD	OEM
EC378E	OEM
EC378M	OEM
EC378N	OEM
EC378P	OEM
EC378PB	OEM
EC378PD	OEM
EC380A	OEM
EC380AX500	OEM
EC380AX555	OEM
EC380B	OEM
EC380BX500	OEM
EC380BX555	OEM
EC380CX500	OEM
EC380CX555	OEM
EC380D	OEM
EC380DX500	OEM
EC380DX555	OEM
EC380M	OEM
EC380MX500	OEM
EC380MX555	OEM
EC380N	OEM
EC380NX500	OEM
EC380NX555	OEM
EC380P	OEM
EC380PB	OEM
EC380PBX500	OEM
EC380PBX555	OEM
EC380PD	OEM
EC380PM	OEM
EC380PX500	OEM
EC380PX555	OEM
EC384A	OEM
EC384B	OEM
EC384C	OEM
EC384D	OEM
EC384E	OEM
EC384M	OEM
EC384N	OEM
EC384S	OEM
EC385A	OEM
EC385C	OEM
EC385D	OEM
EC385E	OEM
EC385M	OEM
EC385N	OEM
EC385S	OEM
EC387E	OEM
EC387M	OEM
EC387N	OEM
EC387P	OEM
EC387PB	OEM
EC388E	OEM
EC388M	OEM
EC388N	OEM
EC388P	OEM
EC388PB	OEM
EC390AX500	OEM
EC390AX550	OEM
EC390AX555	OEM
EC390BX500	OEM
EC390BX550	OEM
EC390CX500	OEM
EC390CX550	OEM
EC390DX500	OEM
EC390DX550	OEM
EC390DX555	OEM
EC390EX550	OEM
EC390MX500	OEM
EC390MX555	OEM
EC390NX500	OEM
EC390NX555	OEM
EC390PBX555	OEM
EC390PX555	OEM
EC391L	OEM
EC391P	OEM
EC391PB	OEM
EC391PD	OEM
EC391PM	OEM
EC391PN	OEM

DEVICE TYPE	REPL CODE
EC392A	OEM
EC392B	OEM
EC392C	OEM
EC392D	OEM
EC392E	OEM
EC392M	OEM
EC393A	OEM
EC393B	OEM
EC393C	OEM
EC393D	OEM
EC393E	OEM
EC393M	OEM
EC394A	OEM
EC394B	OEM
EC394C	OEM
EC394D	OEM
EC394E	OEM
EC394M	OEM
EC395A	OEM
EC395B	OEM
EC395C	OEM
EC395D	OEM
EC395E	OEM
EC395M	OEM
EC397E	OEM
EC397M	OEM
EC397N	OEM
EC397P	OEM
EC397PB	OEM
EC398E	OEM
EC398M	OEM
EC398N	OEM
EC398P	OEM
EC398PB	OEM
EC401	0015
EC402	0015
EC430AX500	OEM
EC430AX550	OEM
EC430AX555	OEM
EC430BX500	OEM
EC430BX555	OEM
EC430CX500	OEM
EC430CX550	OEM
EC430DX500	OEM
EC430DX550	OEM
EC430DX555	OEM
EC430EX550	OEM
EC430MX500	OEM
EC430MX555	OEM
EC430NX500	OEM
EC430NX555	OEM
EC430PBX555	OEM
EC430PX555	OEM
EC431L	OEM
EC431P	OEM
EC431PB	OEM
EC431PD	OEM
EC431PM	OEM
EC431PN	OEM
EC434A	OEM
EC434B	OEM
EC434C	OEM
EC434D	OEM
EC434E	OEM
EC434M	OEM
EC434S	OEM
EC435A	OEM
EC435B	OEM
EC435C	OEM
EC435D	OEM
EC435E	OEM
EC435M	OEM
EC435S	OEM
EC436E	OEM
EC436M	OEM
EC436N	OEM
EC436P	OEM
EC436PB	OEM
EC437E	OEM
EC437M	OEM
EC437N	OEM
EC437P	OEM
EC437PB	OEM
EC438E	OEM
EC438M	OEM
EC438N	OEM
EC438P	OEM
EC438PB	OEM
EC441L	OEM
EC441P	OEM
EC441PB	OEM
EC441PD	OEM
EC441PM	OEM
EC441PN	OEM
EC444A	OEM
EC444B	OEM
EC444C	OEM
EC444D	OEM
EC444E	OEM
EC444M	OEM
EC444S	OEM
EC445A	OEM
EC445B	OEM
EC445C	OEM
EC445D	OEM
EC445E	OEM
EC445M	OEM
EC445S	OEM

DEVICE TYPE	REPL CODE
EC447E	OEM
EC447M	OEM
EC447N	OEM
EC447P	OEM
EC447PB	OEM
EC448E	OEM
EC448M	OEM
EC448N	OEM
EC448P	OEM
EC448PB	OEM
EC449PE	OEM
EC449PM	OEM
EC449PN	OEM
EC449PS	OEM
EC450E	OEM
EC450M	OEM
EC450N	OEM
EC450P	OEM
EC450PA	OEM
EC450PB	OEM
EC450PD	OEM
EC451E1	OEM
EC451M1	OEM
EC451N1	OEM
EC451PB1	OEM
EC451PD1	OEM
EC451PM1	OEM
EC451PN1	OEM
EC457E	OEM
EC457M	OEM
EC457N	OEM
EC457P	OEM
EC457PB	OEM
EC457PD	OEM
EC458E	OEM
EC458M	OEM
EC458N	OEM
EC458P	OEM
EC458PB	OEM
EC458PD	OEM
EC476E	OEM
EC476M	OEM
EC476N	OEM
EC476P	OEM
EC476PB	OEM
EC476PD	OEM
EC477E	OEM
EC477M	OEM
EC477N	OEM
EC477P	OEM
EC477PB	OEM
EC477PD	OEM
EC478E	OEM
EC478M	OEM
EC478N	OEM
EC478P	OEM
EC478PB	OEM
EC478PD	OEM
EC590BX555	OEM
EC601L	OEM
EC601P	OEM
EC601PB	OEM
EC601PD	OEM
EC601PM	OEM
EC601PN	OEM
EC602L	OEM
EC602LB	OEM
EC602LD	OEM
EC602LM	OEM
EC602PN	OEM
EC612L	OEM
EC612PE	OEM
EC612PM	OEM
EC612PN	OEM
EC612PS	OEM
EC612PT	OEM
EC613L	OEM
EC613PE	OEM
EC613PM	OEM
EC613PN	OEM
EC613PS	OEM
EC613PT	OEM
EC648E	OEM
EC648M	OEM
EC648N	OEM
EC648P	OEM
EC648PB	OEM
EC702L	OEM
EC702LB	OEM
EC702LD	OEM
EC702LM	OEM
EC702PN	OEM
EC712L	OEM
EC712PA	OEM
EC712PB	OEM
EC712PD	OEM
EC712PN	OEM
EC713L	OEM
EC713PA	OEM
EC713PB	OEM
EC713PD	OEM
EC713PM	OEM
EC713PN	OEM
EC-829	OEM
EC-877	OEM
EC-877-JPN	OEM
EC-884	OEM
EC-885	OEM
EC-936	OEM

DEVICE TYPE	REPL CODE
EC961	0867
EC-973	OEM
EC973	OEM
EC4101-1M	OEM
EC4101-1N	OEM
EC4101-1P	OEM
EC4101-1PB	OEM
EC4105-1A	OEM
EC4105-1B	OEM
EC4105-1C	OEM
EC4105-1D	OEM
EC4105-1E	OEM
EC4131-1M	OEM
EC4131-1N	OEM
EC4131-PB	OEM
EC4135-1A	OEM
EC4135-1B	OEM
EC4135-1C	OEM
EC4135-1D	OEM
EC4135-1E	OEM
EC4158L	OEM
EC4158LB	OEM
EC4158LD	OEM
EC4158LM	OEM
EC4158LN	OEM
EC4158LR	OEM
EC4158PN	OEM
EC-A041	OEM
EC-A042	OEM
EC-A043	OEM
EC-A051	OEM
EC-A056	OEM
EC-A063	1596
EC-A078	OEM
EC-A104	OEM
EC-A104J	OEM
EC-A137	OEM
ECC-01262	1041
ECC-01263	0081
ECC-01264	0141
ECC-01265	0562
ECC-01266	2078
ECC-01267	1823
ECCG5220AK	OEM
ECG1V010	1691
ECG1V014	2008
ECG1V015	1314
ECG1V017	2325
ECG1V020	2790
ECG1V025	4367
ECG1V030	5744
ECG1V035	2935
ECG1V040	4263
ECG1V050	4472
ECG1V060	4222
ECG1V075	4921
ECG1V095	5157
ECG1V115	4135
ECG1V130	1374
ECG1V150	0246
ECG1V175	6630
ECG1V250	1332
ECG1V275	1658
ECG1V300	4471
ECG2V010	1694
ECG2V014	2009
ECG2V015	1319
ECG2V017	2326
ECG2V020	2796
ECG2V025	3073
ECG2V030	3075
ECG2V035	2941
ECG2V040	3078
ECG2V050	3085
ECG2V060	3090
ECG2V075	3104
ECG2V095	3116
ECG2V115	3123
ECG2V130	0838
ECG2V150	1993
ECG2V250	1334
ECG2V275	1660
ECG2V300	2034
ECG2V420	3257
ECG2V480	1101
ECG10	2851
ECG11	1500
ECG12	4011
ECG13	0014
ECG14	5061
ECG15	2223
ECG16	0582
ECG17	2484
ECG18	2471
ECG19	0359
ECG20	0009
ECG21	4104
ECG22	3608
ECG23	1806
ECG24	3938
ECG25	4548
ECG26	1270
ECG27	4099
ECG28	4012
ECG29	4345
ECG30	4344
ECG31	1562
ECG32	1522
ECG33	0046

DEVICE TYPE	REPL CODE
ECG34	3630
ECG35	5056
ECG36	0194
ECG36MP	5393
ECG37	3111
ECG37MCP	4475
ECG38	0792
ECG39	4479
ECG40	2924
ECG41	5057
ECG42	2411
ECG43	5058
ECG44	2947
ECG45	5059
ECG46	2797
ECG47	2860
ECG48	3799
ECG49	0824
ECG50	3327
ECG51	3014
ECG52	5066
ECG53	3038
ECG54	1165
ECG54MP	5603
ECG55	0718
ECG55MCP	3789
ECG56	2069
ECG57	0629
ECG58	3226
ECG59	3646
ECG60	3701
ECG60MP	5693
ECG61	4144
ECG61MCP	2011
ECG61MP	5704
ECG62	2847
ECG63	2844
ECG64	3007
ECG65	1342
ECG66	5060
ECG67	1465
ECG68	3606
ECG68MCP	4864
ECG69	2309
ECG70	3489
ECG71	3512
ECG72	0939
ECG73	2662
ECG74	2873
ECG74C00	2112
ECG74C02	2114
ECG74C04	2958
ECG74C08	2118
ECG74C10	2119
ECG74C14	3031
ECG74C20	2123
ECG74C30	3128
ECG74C32	3134
ECG74C42	2143
ECG74C48	2148
ECG74C73	2162
ECG74C74	2163
ECG74C76	2164
ECG74C85	2171
ECG74C90	2174
ECG74C93	2177
ECG74C95	3401
ECG74C107	3834
ECG74C151	4001
ECG74C154	4014
ECG74C157	4033
ECG74C161	4046
ECG74C164	4062
ECG74C173	4088
ECG74C174	4092
ECG74C175	4096
ECG74C192	4119
ECG74C193	4122
ECG74C221	3047
ECG74C240	2956
ECG74C244	4230
ECG74C373	4462
ECG74C374	4467
ECG74C901	4607
ECG74C902	4608
ECG74C903	4609
ECG74C904	4610
ECG74C922	5597
ECG74C923	5599
ECG74C925	4621
ECG74H00	0682
ECG74H01	5338
ECG74H04	1907
ECG74H05	3251
ECG74H08	5355
ECG74H10	0685
ECG74H11	2398
ECG74H20	3715
ECG74H21	4856
ECG74H22	4596
ECG74H30	5383
ECG74H40	0559
ECG74H50	1792
ECG74H51	1944
ECG74H52	2020
ECG74H53	2101
ECG74H54	2172
ECG74H55	3158
ECG74H60	5410

DEVICE TYPE	REPL CODE
ECG74H61	2663
ECG74H62	2731
ECG74H71	3264
ECG74H72	3313
ECG74H73	2463
ECG74H74	2492
ECG74H76	5302
ECG74H78	5418
ECG74H86	5791
ECG74H87	2579
ECG74H101	5521
ECG74H102	5523
ECG74H103	2969
ECG74H106	5253
ECG74H108	0180
ECG74H183	4399
ECG74HC00	6625
ECG74HC02	1452
ECG74HC04	1455
ECG74HC08	4289
ECG74HC10	6951
ECG74HC11	5530
ECG74HC14	6650
ECG74HC32	6672
ECG74HC86	5535
ECG74HC109	6642
ECG74HC123	3905
ECG74HC125	6915
ECG74HC126	6916
ECG74HC132	6917
ECG74HC138	6918
ECG74HC139	6919
ECG74HC151	6920
ECG74HC153	6921
ECG74HC154	6922
ECG74HC161	5413
ECG74HC163	6923
ECG74HC164	6924
ECG74HC165	6925
ECG74HC173	6926
ECG74HC174	6928
ECG74HC175	6929
ECG74HC240	1979
ECG74HC244	6932
ECG74HC257	6933
ECG74HC259	6934
ECG74HC273	6935
ECG74HC299	6938
ECG74HC373	6943
ECG74HC374	6944
ECG74HC377	6945
ECG74HC390	6946
ECG74HC393	6947
ECG74HC573	6958
ECG74HC574	6959
ECG74HC4020	5613
ECG74HC4040	5615
ECG74HC4053	5617
ECG74HC4060	5619
ECG74HC4067	5620
ECG74HC40105	6055
ECG74HCT00	3295
ECG74HCT04	3296
ECG74HCT08	3299
ECG74HCT14	3300
ECG74HCT32	3306
ECG74HCT138	6095
ECG74HCT161	6099
ECG74HCT163	6101
ECG74HCT174	6105
ECG74HCT240	6111
ECG74HCT244	6108
ECG74HCT273	6113
ECG74HCT373	6123
ECG74HCT374	6131
ECG74HCT573	6148
ECG74HCT574	6149
ECG74L93	0656
ECG74LS00	1527
ECG74LS01	1546
ECG74LS02	1559
ECG74LS03	1579
ECG74LS04	1595
ECG74LS05	1608
ECG74LS08	1633
ECG74LS09	1642
ECG74LS10	1662
ECG74LS11	1667
ECG74LS12	1679
ECG74LS13	1688
ECG74LS14	1698
ECG74LS15	1707
ECG74LS20	0034
ECG74LS21	1763
ECG74LS22	1775
ECG74LS26	1382
ECG74LS27	0183
ECG74LS28	0472
ECG74LS30	0828
ECG74LS32	0088
ECG74LS33	1832
ECG74LS37	1730
ECG74LS38	1839
ECG74LS40	0135
ECG74LS42	1841
ECG74LS47	1845
ECG74LS48	1849
ECG74LS49	1850
ECG74LS51	1033

DEVICE TYPE	REPL CODE
ECG74LS54	1857
ECG74LS55	0457
ECG74LS73	1867
ECG74LS74A	0243
ECG74LS75	1870
ECG74LS76A	2180
ECG74LS77	1872
ECG74LS78	1873
ECG74LS83A	2218
ECG74LS85	0431
ECG74LS86	0291
ECG74LS90	1882
ECG74LS91	1885
ECG74LS92	1887
ECG74LS93	1888
ECG74LS95	2915
ECG74LS95B	0771
ECG74LS107	1602
ECG74LS109A	1906
ECG74LS112A	7157
ECG74LS113	2256
ECG74LS114	2302
ECG74LS122	1620
ECG74LS123	0979
ECG74LS125A	0075
ECG74LS126	2877
ECG74LS132	1625
ECG74LS133	3400
ECG74LS136	1628
ECG74LS138	0427
ECG74LS139	0153
ECG74LS145	1564
ECG74LS147	4400
ECG74LS148	3912
ECG74LS151	1646
ECG74LS153	0959
ECG74LS155	0209
ECG74LS156	1654
ECG74LS157	1161
ECG74LS158	1656
ECG74LS160A	0837
ECG74LS161A	0858
ECG74LS162A	0880
ECG74LS163A	0893
ECG74LS164	4339
ECG74LS164A	OEM
ECG74LS165	4356
ECG74LS166	4371
ECG74LS168A	0967
ECG74LS169A	0986
ECG74LS170	2629
ECG74LS173	5217
ECG74LS174	0260
ECG74LS175	1672
ECG74LS181	1678
ECG74LS190	1686
ECG74LS191	1687
ECG74LS192	1689
ECG74LS193	1692
ECG74LS194A	1304
ECG74LS195A	1315
ECG74LS196	2834
ECG74LS197	2469
ECG74LS221	1239
ECG74LS240	0452
ECG74LS241	1726
ECG74LS242	1728
ECG74LS243	0906
ECG74LS244	0458
ECG74LS245	0463
ECG74LS246A	OEM
ECG74LS247	1732
ECG74LS248	1734
ECG74LS249	1735
ECG74LS251	1737
ECG74LS253	1739
ECG74LS257	1744
ECG74LS258	1746
ECG74LS259	3205
ECG74LS260	5995
ECG74LS266	0592
ECG74LS273	0894
ECG74LS279	3290
ECG74LS280	1773
ECG74LS283	5997
ECG74LS290	4425
ECG74LS293	0082
ECG74LS295A	2227
ECG74LS298	3371
ECG74LS299	4426
ECG74LS348	0665
ECG74LS352	0761
ECG74LS353	0773
ECG74LS363	6005
ECG74LS364	6006
ECG74LS365A	0943
ECG74LS366A	0956
ECG74LS367	0977
ECG74LS368	0991
ECG74LS373	0709
ECG74LS374	0713
ECG74LS377	1120
ECG74LS378	1133
ECG74LS379	1151
ECG74LS386	1230
ECG74LS390	1288
ECG74LS393	0819
ECG74LS395A	1330
ECG74LS398	1383

DEVICE TYPE	REPL CODE
ECG74LS399	1398
ECG74LS490	2213
ECG74LS540	2541
ECG74LS541	2547
ECG74LS624	3141
ECG74LS625	3149
ECG74LS626	3154
ECG74LS627	3162
ECG74LS629	3175
ECG74LS640	0669
ECG74LS641	0690
ECG74LS642	0719
ECG74LS643	2056
ECG74LS645	0775
ECG74LS670	1130
ECG74S00	0704
ECG74S02	2238
ECG74S03	2217
ECG74S04	2263
ECG74S05	2321
ECG74S08	2569
ECG74S09	2667
ECG74S10	2445
ECG74S11	2447
ECG74S15	2451
ECG74S20	1017
ECG74S22	2461
ECG74S30	3726
ECG74S40	2475
ECG74S51	4305
ECG74S64	2496
ECG74S65	2497
ECG74S74	2504
ECG74S86	2511
ECG74S112	1617
ECG74S113	1623
ECG74S114	1629
ECG74S124	2126
ECG74S133	1819
ECG74S134	1827
ECG74S138	2138
ECG74S140	1886
ECG74S151	1955
ECG74S153	2151
ECG74S157	1695
ECG74S158	2154
ECG74S174	2132
ECG74S175	2141
ECG74S181	2165
ECG74S188	3168
ECG74S194	1931
ECG74S251	2198
ECG74S258	2205
ECG74S287	2224
ECG74S288	2175
ECG74S387	1918
ECG74S454	3166
ECG74S472	2320
ECG74S473	3288
ECG74S474	2322
ECG74S475	2324
ECG74S478	2909
ECG74S479	2862
ECG74S570	2388
ECG74S571	2389
ECG74S572	2390
ECG74S573	2392
ECG75	2964
ECG76	2070
ECG76MP	5956
ECG77	2208
ECG78	5062
ECG79	3797
ECG80	2598
ECG80C95	3177
ECG80C96	3179
ECG80C97	3180
ECG81	2045
ECG82	2468
ECG83	5018
ECG84	5064
ECG85	0286
ECG86	2548
ECG87	0867
ECG87MP	6034
ECG88	3500
ECG88MCP	4500
ECG88MP	6041
ECG89	0055
ECG90	0530
ECG91	0648
ECG92	2276
ECG93	3581
ECG93L08	2127
ECG93L16	2685
ECG93MCP	3303
ECG94	0270
ECG95	4252
ECG96	4029
ECG96L02	1468
ECG96LS02	2934
ECG96S02	4292
ECG97	1991
ECG98	2626
ECG99	4013
ECG100	0280
ECG101	0600
ECG102	0211
ECG102A	0004

DEVICE TYPE	REPL CODE
ECG103	0038
ECG103A	0208
ECG104	0085
ECG104MP	3033
ECG105	0440
ECG106	0150
ECG107	0127
ECG108	0144
ECG109	0143
ECG110	1114
ECG110A	1114
ECG110MP	0123
ECG111	0917
ECG112	0917
ECG113	0196
ECG113A	0196
ECG114	0484
ECG115	1080
ECG116	0015
ECG116(4)	OEM
ECG117	0229
ECG117A	4630
ECG118	0774
ECG119	1303
ECG120	0474
ECG121	0160
ECG121MP	0265
ECG122	0240
ECG123	0198
ECG123A	0016
ECG123AP	0079
ECG124	0142
ECG125	0071
ECG125(2)	OEM
ECG126	0136
ECG126A	0633
ECG127	0975
ECG128	0086
ECG128P	5034
ECG129	0126
ECG129MCP	4163
ECG129P	1597
ECG130	0103
ECG130MP	1514
ECG131	0222
ECG131MP	0822
ECG132	3712
ECG133	3483
ECG134	0188
ECG134A	0188
ECG135	0162
ECG135A	0162
ECG136	0157
ECG136A	0157
ECG137	0636
ECG137A	0636
ECG138	0649
ECG138A	0649
ECG139	0012
ECG139A	0012
ECG140	0170
ECG140A	0170
ECG141	0794
ECG141A	0794
ECG142	0137
ECG142A	0137
ECG143	0365
ECG143A	0365
ECG144	0100
ECG144A	0100
ECG145	0002
ECG145A	0002
ECG146	0441
ECG146A	0441
ECG147	0039
ECG147A	0039
ECG148	1834
ECG148A	1834
ECG149	0783
ECG149A	0783
ECG150	0331
ECG150A	0331
ECG151	0149
ECG151A	0149
ECG152	0042
ECG152MP	5088
ECG153	0925
ECG153MCP	4220
ECG154	0233
ECG155	2763
ECG156	0087
ECG157	0276
ECG158	0164
ECG159	0037
ECG159MCP	1381
ECG160	0050
ECG161	0007
ECG162	0074
ECG163	0642
ECG163A	0642
ECG164	0003
ECG165	0065
ECG166	0277
ECG167	0289
ECG168	0296
ECG169	0302
ECG170	0250
ECG171	0285
ECG172	0400

DEVICE TYPE	REPL CODE
ECG172A	0400
ECG173BP	0293
ECG174	0924
ECG175	0178
ECG176	0847
ECG177	0133
ECG178MP	0913
ECG179	0604
ECG179MP	5536
ECG180	1681
ECG180MCP	3122
ECG181	0130
ECG181MP	5474
ECG182	0561
ECG183	1199
ECG184	0161
ECG184MP	5595
ECG185	0460
ECG185MCP	4605
ECG186	0560
ECG186A	0219
ECG187	1267
ECG187A	1051
ECG188	0551
ECG189	0382
ECG190	0264
ECG191	0338
ECG192	0595
ECG192-1	0218
ECG192A	0218
ECG193	0791
ECG193-1	1242
ECG193A	1242
ECG194	0861
ECG195	0698
ECG195A	0698
ECG196	0423
ECG197	0854
ECG198	0168
ECG199	0111
ECG210	0566
ECG211	1367
ECG213	0437
ECG214	3746
ECG215	3748
ECG216	1526
ECG217	3750
ECG218	0905
ECG219	0491
ECG219MCP	5498
ECG220	0849
ECG221	0353
ECG222	0212
ECG223	0103
ECG224	0630
ECG225	3280
ECG226	1010
ECG226MP	1862
ECG227	3754
ECG228	1708
ECG228A	1708
ECG229	0224
ECG230	0239
ECG231	0061
ECG232	3519
ECG233	0330
ECG234	0693
ECG235	0936
ECG236	0836
ECG237	1411
ECG238	0313
ECG239	2336
ECG240	3758
ECG241	2997
ECG242	3165
ECG243	2427
ECG244	2277
ECG245	3373
ECG246	3374
ECG247	2439
ECG248	2432
ECG249	1394
ECG250	2448
ECG251	3525
ECG252	3526
ECG253	0558
ECG254	2897
ECG255	3764
ECG256	3765
ECG257	3377
ECG258	3528
ECG259	3529
ECG260	3530
ECG261	1212
ECG262	0602
ECG263	2235
ECG264	2237
ECG265	2258
ECG266	3533
ECG267	3534
ECG268	3535
ECG269	3536
ECG270	0134
ECG271	0073
ECG272	3538
ECG273	3539
ECG274	3370
ECG275	3540
ECG276	1652

DEVICE TYPE	REPL CODE
ECG277	2214
ECG278	0418
ECG279	2337
ECG279A	2337
ECG280	0177
ECG280MP	2140
ECG281	2013
ECG281MCP	5778
ECG282	0622
ECG283	0363
ECG284	0543
ECG284MP	2371
ECG285	1598
ECG285MCP	6141
ECG286	1027
ECG287	0715
ECG288	0342
ECG289	0155
ECG289A	0155
ECG289AMP	1224
ECG289MP	2767
ECG290	0006
ECG290A	0006
ECG290AMCP	3689
ECG291	0236
ECG292	0681
ECG292MCP	3255
ECG293	0018
ECG293MP	3049
ECG294	0532
ECG295	1591
ECG296	2200
ECG297	0324
ECG297MP	3247
ECG298	0436
ECG299	2050
ECG300	2046
ECG300MP	3376
ECG302	1173
ECG306	1946
ECG307	1430
ECG308	1712
ECG308P	4615
ECG309K	1922
ECG310	2329
ECG310P	4606
ECG311	0493
ECG312	0325
ECG313	0475
ECG314	1825
ECG315	1978
ECG316	0259
ECG317	2946
ECG318	2312
ECG319	0008
ECG319P	0008
ECG320	1974
ECG320F	2884
ECG321	1751
ECG322	1984
ECG323	0892
ECG324	1480
ECG325	1198
ECG326	2987
ECG327	2485
ECG328	0620
ECG329	2700
ECG330	3561
ECG330W	6003
ECG331	0482
ECG331MP	4184
ECG332	1369
ECG332MCP	3949
ECG333	2835
ECG334	3563
ECG335	3565
ECG336	3567
ECG337	3568
ECG338	3569
ECG338F	5202
ECG339	3570
ECG340	2725
ECG341	3572
ECG342	2719
ECG343	2720
ECG344	2526
ECG345	3041
ECG346	2041
ECG347	3576
ECG348	3578
ECG349	2170
ECG350	1233
ECG350F	5303
ECG351	1977
ECG352	3194
ECG353	3580
ECG354	3582
ECG355	3583
ECG356	3584
ECG357	3586
ECG358	1030
ECG358A	4465
ECG358B	5036
ECG358C	5503
ECG359	3587
ECG360	2506
ECG361	2039
ECG362	3560
ECG363	2091

DEVICE TYPE	REPL CODE
ECG364	1420
ECG365	2093
ECG366	3589
ECG367	2868
ECG368	3591
ECG369	2096
ECG370	3810
ECG370A	3810
ECG371	OEM
ECG372	OEM
ECG373	0563
ECG374	0525
ECG375	0392
ECG376	1085
ECG377	0060
ECG378	1308
ECG379	0728
ECG380	2939
ECG380MP	2735
ECG381	3571
ECG382	1386
ECG383	0477
ECG384	0429
ECG385	1966
ECG386	1852
ECG387	2433
ECG387MP	3187
ECG388	2414
ECG389	0223
ECG390	3081
ECG391	0859
ECG392	3602
ECG393	3603
ECG394	2613
ECG395	3607
ECG396	0187
ECG397	0439
ECG398	1648
ECG399	0261
ECG425A	OEM
ECG425B	OEM
ECG425C	OEM
ECG425D	OEM
ECG425E	OEM
ECG425F	OEM
ECG425G	OEM
ECG425H	OEM
ECG425J	OEM
ECG425K	OEM
ECG425L	OEM
ECG435K28	OEM
ECG435K30	OEM
ECG435K42	OEM
ECG435K64	OEM
ECG451	1392
ECG452	2888
ECG453	3072
ECG454	2458
ECG455	0371
ECG456	3622
ECG457	2950
ECG458	1758
ECG459	3133
ECG460	1141
ECG461	2945
ECG462	3628
ECG463	3839
ECG464	0844
ECG465	0983
ECG466	1155
ECG467	3130
ECG468	3842
ECG469	3843
ECG470	3631
ECG471	2545
ECG472	3421
ECG473	3326
ECG474	3632
ECG475	3321
ECG476	3322
ECG477	3119
ECG478	3636
ECG479	1354
ECG480	3638
ECG481	3640
ECG482	3641
ECG483	3642
ECG484	3643
ECG485	3644
ECG486	0689
ECG487	3846
ECG488	2702
ECG489	2455
ECG490	3849
ECG500	0190
ECG500A	0190
ECG501	0128
ECG501A	0128
ECG501B	0128
ECG502	0201
ECG503	0288
ECG504	0378
ECG505	0757
ECG506	0102
ECG507	0920
ECG508	1306
ECG508/R-3A3	1306
ECG509	2083
ECG509/R-3AT2	2083

If replacement code is OEM, contact original manufacturer for replacement.

DEVICE TYPE	REPL CODE
ECG510	2084
ECG510/R-3DB3	2084
ECG511	1449
ECG511/R-2AV2	1449
ECG512	2085
ECG512/R-6DW4	2085
ECG513	1323
ECG514	3649
ECG514/R-3DS3	3649
ECG515	1217
ECG516	3856
ECG517	0409
ECG518	1791
ECG519	0124
ECG520	2984
ECG521	3858
ECG522	1501
ECG523	1067
ECG523/3306	1067
ECG524	3652
ECG524V13	0830
ECG524V15	3415
ECG524V17	6952
ECG524V25	1336
ECG524V27	4867
ECG524V30	4872
ECG524V42	3859
ECG524V48	0781
ECG525	0348
ECG526	1706
ECG526A	1706
ECG527	3860
ECG527A	5731
ECG528	2109
ECG529	2546
ECG530	3655
ECG531	2804
ECG532	0409
ECG533	1710
ECG534	2982
ECG535	2985
ECG536	1997
ECG536A	1997
ECG537	1197
ECG538	2984
ECG539	1054
ECG540	3863
ECG541	3658
ECG542	3659
ECG543	3864
ECG544	3660
ECG545	3865
ECG546	3661
ECG547	3866
ECG548	3662
ECG549	3867
ECG550	3868
ECG551	0182
ECG552	0023
ECG553	0163
ECG554	3867
ECG555	2518
ECG556	3666
ECG556A	6189
ECG557	3667
ECG558	0017
ECG559	3871
ECG560	3358
ECG561	3872
ECG562	3875
ECG563	3877
ECG564	3879
ECG565	3669
ECG568	2671
ECG568A	6194
ECG569	3882
ECG570	3883
ECG576	1366
ECG577	3887
ECG578	3888
ECG579	3604
ECG580	0031
ECG581	1145
ECG582	3889
ECG583	3891
ECG584	0339
ECG585	0735
ECG586	2542
ECG587	1090
ECG588	0546
ECG590	3192
ECG591	3894
ECG592	3895
ECG593	3749
ECG594	3896
ECG595	0907
ECG596	0903
ECG597	1127
ECG598	1664
ECG599	3551
ECG600	1925
ECG601	0139
ECG604	3683
ECG605	0120
ECG605A	0343
ECG606	3684
ECG607	3442
ECG610	0554
ECG611	0005

DEVICE TYPE	REPL CODE
ECG612	0720
ECG613	0628
ECG614	0030
ECG615	3452
ECG615A	1329
ECG616	1029
ECG617	3687
ECG618	3688
ECG700	3719
ECG701	3720
ECG702	2801
ECG703	0632
ECG703A	5986
ECG704	2624
ECG705	2161
ECG705A	2161
ECG706	2571
ECG707	0753
ECG708	0664
ECG709	0678
ECG710	0829
ECG711	2715
ECG712	0167
ECG713	0666
ECG714	0352
ECG715	0354
ECG716	3722
ECG717	3724
ECG718	0654
ECG719	1940
ECG720	0443
ECG721	2279
ECG722	0701
ECG723	1345
ECG724	0823
ECG725	0692
ECG726	2617
ECG727	2529
ECG728	2549
ECG729	0328
ECG730	2743
ECG731	2457
ECG732	2557
ECG733	2557
ECG734	1759
ECG735	3008
ECG736	1759
ECG737	1443
ECG738	0856
ECG739	0751
ECG740	0379
ECG740A	0379
ECG741	3734
ECG742	3735
ECG743	1395
ECG744	1421
ECG745	2930
ECG746	0785
ECG747	1808
ECG748	0789
ECG748A	0789
ECG749	0395
ECG750	3219
ECG752	0517
ECG753	3779
ECG754	0723
ECG755	2027
ECG756	1186
ECG757	1186
ECG758	3931
ECG759	1671
ECG760	2100
ECG761	1876
ECG762	OEM
ECG763	2282
ECG764	3935
ECG765	3935
ECG766	1899
ECG766A	1899
ECG767	0623
ECG768	3940
ECG769	OEM
ECG770	2906
ECG771	OEM
ECG772	1980
ECG772A	1980
ECG773	1684
ECG774	OEM
ECG775	3943
ECG776	3944
ECG777	OEM
ECG778	0360
ECG778A	0360
ECG778S	1699
ECG778SM	5531
ECG779	0372
ECG779-1	0372
ECG779A	0372
ECG780	0364
ECG781	2747
ECG782	0377
ECG783	0802
ECG784	2699
ECG785	2707
ECG786	2714
ECG787	2257
ECG788	2755
ECG789	1843
ECG790	0349

DEVICE TYPE	REPL CODE
ECG791	1337
ECG792	2167
ECG793	2043
ECG794	3380
ECG795	3195
ECG796	1980
ECG797	0521
ECG798	1753
ECG799	1611
ECG800	2975
ECG801	0520
ECG802	3761
ECG803	3763
ECG804	2557
ECG805	3410
ECG806	3422
ECG807	3766
ECG808	2608
ECG809	2688
ECG810	2480
ECG810A	2480
ECG811	OEM
ECG812	3770
ECG813	3954
ECG814	3006
ECG814A	3006
ECG815	0848
ECG816	1077
ECG817	1913
ECG818	2501
ECG819	1878
ECG820	2817
ECG821	2831
ECG822	2812
ECG823	3063
ECG824	3079
ECG825	3099
ECG826	3962
ECG828	3086
ECG829	1899
ECG830	1328
ECG831	3965
ECG832	3781
ECG832SM	6080
ECG833	1543
ECG834	0176
ECG834SM	2742
ECG835	3397
ECG836	2493
ECG837	3971
ECG838	3972
ECG839	3974
ECG840	2144
ECG841	3784
ECG842	3976
ECG843	2502
ECG844	3977
ECG845	3978
ECG846	1516
ECG847	3979
ECG848	3981
ECG849	3788
ECG850	3983
ECG851	3790
ECG852	2022
ECG853	3791
ECG854	3986
ECG855	0814
ECG856	3793
ECG857M	3672
ECG857SM	6365
ECG858M	3741
ECG858SM	4105
ECG859	3391
ECG859SM	4136
ECG860	3795
ECG861	3989
ECG862	3056
ECG863	3990
ECG864	0304
ECG865	3991
ECG866	3994
ECG867	3995
ECG868	3997
ECG869	2893
ECG870	2168
ECG871	3394
ECG872	3999
ECG873	2107
ECG874	4002
ECG875	1104
ECG876	3062
ECG877	4005
ECG878	1425
ECG879	3312
ECG880	4008
ECG887M	3181
ECG888M	3942
ECG889M	3320
ECG900	2687
ECG901	2689
ECG902	2748
ECG903	2537
ECG904	1854
ECG905	2698
ECG906	2701
ECG907	2711
ECG908	2705
ECG909	1301

DEVICE TYPE	REPL CODE
ECG909D	1705
ECG910	1797
ECG910D	1800
ECG911	1890
ECG911D	1897
ECG912	1696
ECG913	3808
ECG914	2727
ECG915	1554
ECG916	2749
ECG917	2722
ECG918	2562
ECG918M	2567
ECG918SM	2565
ECG919	2574
ECG919D	2570
ECG920	4026
ECG921	3686
ECG922	1815
ECG922M	2104
ECG922SM	4169
ECG923	1192
ECG923D	0026
ECG924	2222
ECG924M	2517
ECG925	2890
ECG926	3798
ECG927	1121
ECG927D	1118
ECG927SM	4273
ECG928	1677
ECG928M	0770
ECG928S	3233
ECG928SM	2881
ECG929	2751
ECG930	2782
ECG931	1916
ECG932	2863
ECG933	3393
ECG934	3813
ECG935	2737
ECG936	3814
ECG937	3621
ECG937M	6039
ECG938	0896
ECG938M	2246
ECG939	3818
ECG940	1919
ECG941	0410
ECG941D	1975
ECG941M	0312
ECG941S	6522
ECG941SM	1976
ECG942	3029
ECG943	3819
ECG943M	0631
ECG943SM	3118
ECG944	3820
ECG944M	3434
ECG945	4037
ECG946	3663
ECG947	2368
ECG947D	2358
ECG948	2823
ECG948SM	2816
ECG949	2552
ECG950	1828
ECG951	3395
ECG952	3821
ECG953	3823
ECG954	3824
ECG955M	0973
ECG955MC	0585
ECG955S	3956
ECG955SM	2771
ECG956	2564
ECG957	2736
ECG958	2259
ECG959	3826
ECG960	0624
ECG961	1285
ECG962	0923
ECG963	2648
ECG964	1196
ECG965	2791
ECG966	0334
ECG967	1838
ECG968	1321
ECG969	3829
ECG970	2838
ECG971	3575
ECG972	2239
ECG973	0417
ECG973D	3801
ECG974	0011
ECG975	1300
ECG975SM	2355
ECG976	2283
ECG977	1292
ECG978	3285
ECG978C	2869
ECG978SM	2606
ECG979	4048
ECG980	3831
ECG981	0083
ECG982	2830
ECG983	3425
ECG984	3270
ECG985	3832

DEVICE TYPE	REPL CODE
ECG986	3089
ECG987	0625
ECG987SM	3202
ECG988	2301
ECG989	3381
ECG990	1491
ECG991	3594
ECG992	2247
ECG993	3835
ECG994	3825
ECG994M	3827
ECG995	2318
ECG995M	4551
ECG996	3836
ECG997	3024
ECG998	4058
ECG999	4059
ECG999M	5298
ECG999SM	5294
ECG1000	1235
ECG1001	4437
ECG1002	0387
ECG1003	0579
ECG1004	0878
ECG1005	2568
ECG1006	1215
ECG1008	3020
ECG1009	0843
ECG1010	0864
ECG1011	0882
ECG1012	0021
ECG1013	0138
ECG1014	0355
ECG1015	2404
ECG1016	2406
ECG1017	2440
ECG1018	2476
ECG1019	2509
ECG1020	OEM
ECG1021	2155
ECG1022	4556
ECG1023	4557
ECG1024	3550
ECG1025	1187
ECG1026	2464
ECG1027	2876
ECG1028	4559
ECG1029	2262
ECG1030	4561
ECG1031	4562
ECG1032	4563
ECG1033	4564
ECG1034	2889
ECG1035	4565
ECG1036	4566
ECG1037	4093
ECG1038	4409
ECG1039	1621
ECG1040	4569
ECG1041	4570
ECG1042	4571
ECG1043	4390
ECG1045	2393
ECG1046	0528
ECG1047	2882
ECG1048	0673
ECG1049	0653
ECG1050	2891
ECG1051	3785
ECG1052	0433
ECG1053	2573
ECG1054	2576
ECG1055	2578
ECG1056	1837
ECG1057	2577
ECG1058	2124
ECG1059	4025
ECG1060	2581
ECG1061	2582
ECG1062	0403
ECG1063	2584
ECG1064	2585
ECG1065	2586
ECG1066	2587
ECG1067	2588
ECG1068	2590
ECG1069	2599
ECG1070	4076
ECG1071	2603
ECG1072	2591
ECG1073	2596
ECG1074	2592
ECG1075	0470
ECG1075A	0470
ECG1076	0946
ECG1077	0930
ECG1078	3352
ECG1079	4575
ECG1080	0855
ECG1081	1160
ECG1081A	1160
ECG1082	2261
ECG1083	2854
ECG1084	2886
ECG1085	2631
ECG1086	2887
ECG1087	1018
ECG1088	5682
ECG1089	0638

DEVICE TYPE	REPL CODE
ECG1090	2861
ECG1091	0964
ECG1092	2316
ECG1093	4578
ECG1094	0043
ECG1095	4485
ECG1096	1634
ECG1097	3263
ECG1098	3958
ECG1099	4260
ECG1100	0911
ECG1101	2634
ECG1102	2632
ECG1103	1994
ECG1104	2057
ECG1105	4582
ECG1106	2630
ECG1107	3670
ECG1108	1393
ECG1109	2639
ECG1110	2305
ECG1111	2472
ECG1112	2482
ECG1113	2158
ECG1114	4586
ECG1115	1248
ECG1115A	3922
ECG1116	4147
ECG1117	4003
ECG1118	1920
ECG1119	4589
ECG1120	2031
ECG1121	0389
ECG1122	0607
ECG1123	4034
ECG1124	4543
ECG1125	4028
ECG1126	4593
ECG1127	4420
ECG1128	3915
ECG1130	1326
ECG1131	4594
ECG1132	4595
ECG1133	2231
ECG1134	2949
ECG1135	2062
ECG1136	3135
ECG1137	2594
ECG1139	3936
ECG1140	2872
ECG1141	3840
ECG1142	1478
ECG1148	2864
ECG1149	4597
ECG1150	4598
ECG1151	5080
ECG1152	2505
ECG1153	3366
ECG1154	4600
ECG1155	1050
ECG1156	3017
ECG1158	0881
ECG1159	0884
ECG1160	3695
ECG1161	2942
ECG1162	4038
ECG1162A	OEM
ECG1163	2604
ECG1164	3269
ECG1165	2773
ECG1166	2781
ECG1167	1714
ECG1168	2602
ECG1169	4401
ECG1170	2534
ECG1171	0093
ECG1172	1379
ECG1173	3273
ECG1174	2049
ECG1175	2786
ECG1176	2789
ECG1177	2798
ECG1178	3705
ECG1179	2572
ECG1180	2614
ECG1181	2593
ECG1182	4055
ECG1183	4196
ECG1184	4612
ECG1185	1479
ECG1186	0852
ECG1187	0857
ECG1188	2618
ECG1189	4395
ECG1190	4616
ECG1191	4087
ECG1192	1540
ECG1193	3318
ECG1194	1816
ECG1195	4209
ECG1196	1055
ECG1197	4620
ECG1198	2253
ECG1199	3805
ECG1200	4238
ECG1201	4622
ECG1202	2380
ECG1203	2306
ECG1204	4484

DEVICE TYPE	REPL CODE
ECG1205	2931
ECG1206	4623
ECG1207	4700
ECG1208	4624
ECG1209	4625
ECG1210	2633
ECG1211	2193
ECG1212	4626
ECG1213	4627
ECG1214	0651
ECG1215	2532
ECG1216	0916
ECG1217	0416
ECG1218	1571
ECG1219	1612
ECG1220	5697
ECG1221	4631
ECG1222	5698
ECG1223	1524
ECG1224	4086
ECG1225	2670
ECG1226	0706
ECG1227	0621
ECG1228	3190
ECG1229	4632
ECG1230	2669
ECG1231	1170
ECG1231A	1057
ECG1232	1048
ECG1233	4224
ECG1234	2019
ECG1235	4633
ECG1236	2284
ECG1237	3651
ECG1238	4018
ECG1239	3262
ECG1240	4069
ECG1241	2638
ECG1242	1509
ECG1243	4634
ECG1244	2764
ECG1245	0904
ECG1246	4635
ECG1247	4636
ECG1248	1260
ECG1249	4042
ECG1250	4072
ECG1251	4637
ECG1252	4638
ECG1253	4639
ECG1254	3815
ECG1255	2616
ECG1256	4640
ECG1257	4641
ECG1258	4642
ECG1259	5618
ECG1260	4644
ECG1261	1128
ECG1262	4645
ECG1263	2960
ECG1264	4064
ECG1265	4065
ECG1266	4067
ECG1267	4070
ECG1268	4071
ECG1269	4079
ECG1270	5562
ECG1271	4648
ECG1272	3559
ECG1273	4649
ECG1274	5705
ECG1275	5706
ECG1276	3769
ECG1277	3432
ECG1278	4652
ECG1279	5707
ECG1280	5708
ECG1281	4653
ECG1282	4654
ECG1283	4656
ECG1284	4657
ECG1285	0754
ECG1286	0756
ECG1287	4659
ECG1288	4573
ECG1289	4661
ECG1290	4089
ECG1291	4663
ECG1292	3451
ECG1293	4665
ECG1294	4006
ECG1295	5094
ECG1296	0067
ECG1297	4389
ECG1298	4672
ECG1299	5095
ECG1300	3933
ECG1301	4669
ECG1302	3992
ECG1303	4670
ECG1304	3996
ECG1305	3998
ECG1306	4000
ECG1307	4004
ECG1308	3948
ECG1309	4671
ECG1310	3988
ECG1311	3932
ECG1312	3984

DEVICE TYPE	REPL CODE
ECG1313	3985
ECG1314	5547
ECG1315	5714
ECG1316	5715
ECG1317	5716
ECG1318	2297
ECG1319	1122
ECG1320	4673
ECG1321	4674
ECG1322	4675
ECG1323	4676
ECG1324	4677
ECG1325	4678
ECG1326	4680
ECG1327	4681
ECG1328	4682
ECG1329	4683
ECG1330	4684
ECG1331	4685
ECG1332	5721
ECG1333	4687
ECG1334	4688
ECG1335	4689
ECG1336	4690
ECG1337	4691
ECG1338	1437
ECG1339	4693
ECG1340	4694
ECG1341	4695
ECG1342	4696
ECG1343	4698
ECG1344	4699
ECG1345	1903
ECG1346	4701
ECG1347	4702
ECG1348	4703
ECG1349	4704
ECG1350	4705
ECG1351	4706
ECG1352	4707
ECG1353	5722
ECG1354	4708
ECG1355	4710
ECG1356	5099
ECG1357	5100
ECG1358	4151
ECG1359	5101
ECG1360	5724
ECG1361	4713
ECG1362	5104
ECG1363	4714
ECG1364	4715
ECG1365	4716
ECG1366	4717
ECG1367	4718
ECG1368	4719
ECG1369	4720
ECG1370	4722
ECG1371	4723
ECG1372	4724
ECG1373	3916
ECG1374	3710
ECG1375	2794
ECG1376	3711
ECG1377	3333
ECG1378	4725
ECG1379	4726
ECG1380	4727
ECG1381	4728
ECG1382	1951
ECG1383	4729
ECG1384	4730
ECG1385	4732
ECG1386	3239
ECG1387	4372
ECG1388	4734
ECG1389	4735
ECG1390	5107
ECG1391	4736
ECG1392	4737
ECG1393	4739
ECG1394	4740
ECG1395	4097
ECG1396	0496
ECG1397	5108
ECG1398	5110
ECG1399	5111
ECG1400	1582
ECG1401	3351
ECG1402	3862
ECG1403	3359
ECG1404	4743
ECG1405	4744
ECG1406	4060
ECG1407	4745
ECG1408	4746
ECG1409	0683
ECG1409C	6096
ECG1410	2623
ECG1411	0767
ECG1412	4747
ECG1413	0069
ECG1414	4750
ECG1415	4495
ECG1416	0490
ECG1417	2625
ECG1418	4165
ECG1419	4751
ECG1420	0553
ECG1421	4752
ECG1422	1585
ECG1423	4291
ECG1424	4476
ECG1425	4758
ECG1426	0675
ECG1427	4759
ECG1428	1580
ECG1429	4760
ECG1430	0918
ECG1431	4761
ECG1432	5118
ECG1433	4762
ECG1434	0872
ECG1435	4763
ECG1436	4428
ECG1437	4549
ECG1438	4764
ECG1439	3103
ECG1440	3738
ECG1441	4765
ECG1442	4766
ECG1443	5121
ECG1444	4068
ECG1445	1022
ECG1446	4063
ECG1447	4768
ECG1448	4051
ECG1449	4770
ECG1450	4047
ECG1451	4771
ECG1452	4773
ECG1453	2514
ECG1454	3246
ECG1455	4776
ECG1456	5746
ECG1457	4777
ECG1458	4778
ECG1459	2530
ECG1460	4779
ECG1461	2712
ECG1462	4780
ECG1463	3196
ECG1464	4781
ECG1465	0362
ECG1466	2063
ECG1467	4757
ECG1468	3980
ECG1469	4782
ECG1470	3228
ECG1471	1201
ECG1472	4785
ECG1473	4787
ECG1474	4789
ECG1475	4791
ECG1476	4793
ECG1477	4794
ECG1478	4795
ECG1479	1446
ECG1480	4560
ECG1481	4797
ECG1482	4798
ECG1483	0865
ECG1484	4799
ECG1485	3609
ECG1486	3716
ECG1487	4044
ECG1488	2252
ECG1489	4803
ECG1490	4805
ECG1491	4807
ECG1492	4808
ECG1493	4809
ECG1494	4810
ECG1495	4811
ECG1496	4134
ECG1497	4812
ECG1498	4813
ECG1499	4814
ECG1500	4815
ECG1501	4817
ECG1502	4819
ECG1503	4820
ECG1504	4821
ECG1505	4823
ECG1506	5720
ECG1507	4824
ECG1508	3068
ECG1509	3069
ECG1510	0810
ECG1511	4825
ECG1512	4280
ECG1513	4826
ECG1514	4827
ECG1515	4830
ECG1516	4831
ECG1517	4834
ECG1518	4835
ECG1519	4836
ECG1520	1627
ECG1521	3945
ECG1522	2533
ECG1523	4838
ECG1525	2694
ECG1526	4841
ECG1527	4842
ECG1528	2328
ECG1529	2913
ECG1530	4844
ECG1531	4846
ECG1532	2054
ECG1533	4847
ECG1534	4849
ECG1535	5156
ECG1536	4850
ECG1537	4851
ECG1538	0332
ECG1539	1590
ECG1540	4351
ECG1541	2693
ECG1542	4307
ECG1543	4853
ECG1544	4854
ECG1545	0912
ECG1546	3503
ECG1547	0663
ECG1548	3502
ECG1549	3071
ECG1550	1429
ECG1551	3744
ECG1552	2916
ECG1553	3518
ECG1554	5158
ECG1555	5755
ECG1556	5159
ECG1557	3816
ECG1558	2105
ECG1559	5160
ECG1560	1318
ECG1561	5162
ECG1562	5163
ECG1563	2926
ECG1564	5759
ECG1565	3057
ECG1566	4601
ECG1567	3713
ECG1568	5169
ECG1569	0107
ECG1570	4986
ECG1571	5172
ECG1572	4109
ECG1573	1676
ECG1574	3921
ECG1575	2184
ECG1576	3772
ECG1577	5175
ECG1578	3423
ECG1579	5177
ECG1580	0351
ECG1581	4112
ECG1582	3973
ECG1583	5770
ECG1584	3975
ECG1585	5180
ECG1586	3970
ECG1587	5771
ECG1588	4242
ECG1589	0450
ECG1590	3968
ECG1591	5181
ECG1592	4294
ECG1593	5182
ECG1594	3966
ECG1595	5655
ECG1596	3964
ECG1597	5183
ECG1598	3707
ECG1599	3697
ECG1600	5185
ECG1601	4261
ECG1602	5187
ECG1603	3961
ECG1604	0465
ECG1605	4434
ECG1606	5188
ECG1607	5189
ECG1608	0662
ECG1609	2535
ECG1610	3084
ECG1611	5191
ECG1612	2787
ECG1613	5192
ECG1614	5774
ECG1615	4857
ECG1616	2896
ECG1617	5775
ECG1618	3074
ECG1619	2799
ECG1620	5193
ECG1621	5194
ECG1622	5071
ECG1623	5776
ECG1624	5195
ECG1625	3753
ECG1626	4840
ECG1627	2808
ECG1628	2073
ECG1629	3759
ECG1630	0759
ECG1631	5197
ECG1632	5198
ECG1633	4274
ECG1634	2595
ECG1635	5199
ECG1636	3012
ECG1637	5200
ECG1638	5779
ECG1639	4278
ECG1640	5780
ECG1641	5201
ECG1642	5781
ECG1644	5782
ECG1645	5783
ECG1646	5784
ECG1647	4355
ECG1648	5203
ECG1649	5204
ECG1650	2122
ECG1651	4520
ECG1652	5785
ECG1653	5786
ECG1654	5205
ECG1655	2842
ECG1656	1442
ECG1657	4410
ECG1658	2666
ECG1659	5206
ECG1660	2656
ECG1661	4350
ECG1662	5207
ECG1663	5208
ECG1664	5209
ECG1665	4166
ECG1666	5210
ECG1667	2386
ECG1668	3776
ECG1669	3282
ECG1670	4553
ECG1671	3685
ECG1672	5212
ECG1673	1988
ECG1674	0976
ECG1675	3363
ECG1676	2042
ECG1677	5011
ECG1678	5048
ECG1679	2425
ECG1680	3126
ECG1681	5214
ECG1682	5215
ECG1683	3853
ECG1684	4927
ECG1685	3253
ECG1686	5790
ECG1687	2304
ECG1688	5218
ECG1689	0699
ECG1690	5219
ECG1691	5220
ECG1692	5221
ECG1693	5222
ECG1700	5225
ECG1701	5226
ECG1702	5227
ECG1703	3731
ECG1704	5229
ECG1705	0489
ECG1706	5793
ECG1707	3335
ECG1708	4384
ECG1709	5230
ECG1710	2066
ECG1711	2553
ECG1712	2067
ECG1713	2068
ECG1714M	2686
ECG1714S	2026
ECG1715	2025
ECG1716	3828
ECG1717	5232
ECG1718	4258
ECG1719	2103
ECG1720	5233
ECG1721	2829
ECG1722	2841
ECG1723	2839
ECG1724	4796
ECG1725	2116
ECG1726	5234
ECG1727	3747
ECG1728	0539
ECG1729	1824
ECG1730	5236
ECG1731	5237
ECG1732	1572
ECG1733	2374
ECG1734	5238
ECG1735	2385
ECG1736	3361
ECG1737	1596
ECG1738	2879
ECG1739	1389
ECG1740	4146
ECG1741	4650
ECG1742	1641
ECG1743	3445
ECG1744	3437
ECG1745	5239
ECG1746	5216
ECG1747	4346
ECG1748	3348
ECG1749	5241
ECG1750	5242
ECG1751	2194
ECG1752	3523
ECG1753	5243
ECG1754	2178
ECG1755	1094
ECG1756	5800
ECG1757	5244
ECG1758	4788
ECG1759	2090
ECG1760	0950
ECG1761	5246
ECG1762	5247
ECG1763	5804
ECG1764	5250
ECG1765	5805
ECG1766	5806
ECG1767	5054
ECG1768	4613
ECG1769	4618
ECG1770	4666
ECG1771	0391
ECG1772	4353
ECG1773	0732
ECG1774	5251
ECG1775	4137
ECG1776	3955
ECG1777	2364
ECG1778	3913
ECG1779	3902
ECG1780	1529
ECG1781	1028
ECG1782	4129
ECG1783	4194
ECG1784	5809
ECG1785	5810
ECG1786	3386
ECG1787	3159
ECG1788	4385
ECG1789	4422
ECG1790	3236
ECG1791	5256
ECG1792	3254
ECG1793	0583
ECG1794	2269
ECG1795	1604
ECG1796	5257
ECG1797	2957
ECG1798	3097
ECG1799	3027
ECG1800	3365
ECG1801	3884
ECG1802	5814
ECG1803	3161
ECG1804	1254
ECG1805	4379
ECG1806	5815
ECG1807	5324
ECG1808	5258
ECG1809	5259
ECG1810	3925
ECG1811	5260
ECG1812	5261
ECG1813	5262
ECG1814	5263
ECG1815	2512
ECG1816	5265
ECG1817	5267
ECG1818	5268
ECG1819	5269
ECG1820	5270
ECG1821	5817
ECG1822	2405
ECG1823	2418
ECG1824	3340
ECG1825	3355
ECG1826	4383
ECG1827	3817
ECG1828	4925
ECG1829	5271
ECG1830	5272
ECG1831	5273
ECG1832	5818
ECG1833	5819
ECG1834	4117
ECG1835	4310
ECG1836	4381
ECG1837	4380
ECG1838	0462
ECG1839	3953
ECG1840	3574
ECG1841	3305
ECG1842	3725
ECG1843	4041
ECG1844	3250
ECG1845	0557
ECG1846	3343
ECG1847	2111
ECG1848	2921
ECG1849	5275
ECG1850	3156
ECG1851	5820
ECG1852	3495
ECG1853	5821
ECG1854D	6122
ECG1854M	6098
ECG1855	2346
ECG1856	0486
ECG1857	5823
ECG1858	5824
ECG1859	3083
ECG1860	2870
ECG1861	3927
ECG1862	3773
ECG1863	4094
ECG1864	1482
ECG1865	5825
ECG1866	5826
ECG1867	1167
ECG1868	4882
ECG1869	5827
ECG1869SM	4118
ECG1870	3488
ECG1871	5276
ECG1872	2419
ECG1873	4496
ECG1874	5277
ECG1875	5278
ECG1876	5279
ECG1877	1483
ECG1878	5828
ECG1879	1439
ECG1880	2303
ECG1881	2413
ECG1882	4143
ECG1883	2334
ECG1884	2417
ECG1885	2765
ECG1886	4869
ECG1887	5829
ECG1888	5830
ECG1889	5831
ECG1890	5391
ECG1891	1938
ECG1892	3080
ECG1893	2186
ECG1894	3344
ECG1895	2289
ECG1896	3951
ECG1897	3957
ECG1898	5833
ECG1899	5834
ECG1900	5280
ECG1901	5281
ECG1902	1786
ECG1903	1836
ECG1905	5041
ECG1906	4506
ECG1907	4886
ECG1908	4509
ECG1909	1721
ECG1910	4881
ECG1911	2723
ECG1912	2254
ECG1913	2004
ECG1914	3963
ECG1915	5287
ECG1916	2000
ECG1917	4508
ECG1918	2462
ECG1919	5289
ECG1920	3650
ECG1923	5290
ECG1924	3881
ECG1925	3573
ECG1926	5291
ECG1927	5292
ECG1928	2470
ECG1930	2169
ECG1932	2806
ECG1934	5293
ECG1934X	2539
ECG1936	4329
ECG1938	5296
ECG1940	5297
ECG1941	5837
ECG1942	5838
ECG1960	5839
ECG1961	5840
ECG1962	5841
ECG1963	5842
ECG1964	5843
ECG1965	5845
ECG1966	5846
ECG1967	5847
ECG1968	5848
ECG1970	5849
ECG1971	5850
ECG1972	5851
ECG1973	5852
ECG1974	5853
ECG1975	5854
ECG1976	5855
ECG1977	5856
ECG2000	5306
ECG2001	3729
ECG2002	4219
ECG2003	4933
ECG2004	4106
ECG2011	0845
ECG2012	1007
ECG2013	1134
ECG2014	1261
ECG2015	4935
ECG2016	4936
ECG2017	4937
ECG2018	4939
ECG2019	4940
ECG2020	4941
ECG2021	3893
ECG2022	3920
ECG2023	2750
ECG2024	2827
ECG2025	4943
ECG2026	4081
ECG2027	1701
ECG2028	1793
ECG2029	4947
ECG2030	1877
ECG2031	4949
ECG2032	4950
ECG2033	5859
ECG2040	4358
ECG2041	5860
ECG2042	4599
ECG2043	4772
ECG2045	5862
ECG2046	5863
ECG2047	5311
ECG2049	0506
ECG2050	4951
ECG2051	4952
ECG2052	5864
ECG2053	3792
ECG2054	4954
ECG2055	0487
ECG2056	5312
ECG2057	5865
ECG2060	2295
ECG2061	2728
ECG2062	3982
ECG2063	5866
ECG2064	5867
ECG2065	5868
ECG2070	4955
ECG2071	4956
ECG2072	4957
ECG2073	4959
ECG2074	4536
ECG2075	4961
ECG2076	4962
ECG2077	4963
ECG2078	4965
ECG2079	0969
ECG2080	4966
ECG2081	4967
ECG2082	4968
ECG2083	4969
ECG2084	4138
ECG2085	4868
ECG2086	5314
ECG2087	4970
ECG2088	5315
ECG2090	5870
ECG2102	4973
ECG2104	4974
ECG2107	4902
ECG2114	2048
ECG2117	0522
ECG2128	1898
ECG2147	1693
ECG2164	2357
ECG2200	5858
ECG2201	5884
ECG2202	3590
ECG2203	5875
ECG2204	5885
ECG2205	5886
ECG2206	5887
ECG2207	5888
ECG2300	3360
ECG2301	3476
ECG2302	1541
ECG2303	2766
ECG2304	2367
ECG2305	3435
ECG2306	3673
ECG2307	3001
ECG2308	2185
ECG2309	2784
ECG2310	4450
ECG2311	1506
ECG2312	2908
ECG2313	3547
ECG2314	5341
ECG2315	1511
ECG2316	5342
ECG2317	4334
ECG2318	5345
ECG2319	3388
ECG2320	3005
ECG2321	0544
ECG2322	0282
ECG2323	5348
ECG2324	3494
ECG2325	3357
ECG2326	3136
ECG2327	3217
ECG2328	3436
ECG2329	5349
ECG2330	3516
ECG2331	2129
ECG2332	0278
ECG2333	4590
ECG2334	2366
ECG2335	2989
ECG2336	3723
ECG2337	3462
ECG2338	1061
ECG2339	3460
ECG2340	5350
ECG2341	3892

If replacement code is OEM, contact original manufacturer for replacement.

DEVICE TYPE	REPL CODE
ECG2342	3890
ECG2343	1959
ECG2344	1968
ECG2345	4077
ECG2346	4078
ECG2348	3542
ECG2349	2428
ECG2350	2420
ECG2351	5352
ECG2352	5353
ECG2353	5330
ECG2354	5896
ECG2355	0832
ECG2356	4172
ECG2357	2323
ECG2358	3143
ECG2359	0898
ECG2360	4130
ECG2361	2954
ECG2362	0013
ECG2369	3796
ECG2380	1466
ECG2381	5360
ECG2382	3088
ECG2383	2271
ECG2384	5899
ECG2385	5769
ECG2386	1668
ECG2387	5096
ECG2388	2176
ECG2390	5361
ECG2392	5363
ECG2400	5900
ECG2401	4489
ECG2402	2919
ECG2403	4498
ECG2404	5364
ECG2405	5365
ECG2406	1733
ECG2407	1752
ECG2408	0724
ECG2409	1742
ECG2410	2332
ECG2411	5454
ECG2412	3379
ECG2413	5368
ECG2414	3482
ECG2415	1892
ECG2416	0981
ECG2417	0703
ECG2418	3478
ECG2419	3272
ECG2426	5369
ECG2427	5370
ECG2428	0667
ECG2429	3645
ECG2430	3471
ECG2431	5371
ECG2532	5098
ECG2708	4424
ECG2716	2278
ECG2732	2697
ECG2764	0812
ECG2800	1846
ECG2909	OEM
ECG3000	1981
ECG3001	1989
ECG3002	1985
ECG3003	1983
ECG3004	5112
ECG3005	5113
ECG3006	5114
ECG3007	3019
ECG3008	3096
ECG3009	5115
ECG3010	0841
ECG3011	3124
ECG3012	5116
ECG3012A	3163
ECG3013	5117
ECG3013A	3169
ECG3014	3455
ECG3015	3456
ECG3016	3739
ECG3017	0516
ECG3018	1615
ECG3019	5119
ECG3020	1358
ECG3021	3157
ECG3022	1962
ECG3023	3276
ECG3024	1778
ECG3025	3182
ECG3026	2195
ECG3027	4183
ECG3028	0591
ECG3029	3232
ECG3029A	3232
ECG3030	4023
ECG3031	5122
ECG3032	2313
ECG3033	2628
ECG3034	3878
ECG3035	3185
ECG3035A	3185
ECG3036	3552
ECG3037	5123
ECG3038	5124
ECG3039	5125
ECG3040	0541
ECG3041	0315
ECG3042	0273
ECG3043	4431
ECG3044	4932
ECG3045	2108
ECG3046	2236
ECG3047	5051
ECG3048	5065
ECG3049	5126
ECG3050	4987
ECG3051	4988
ECG3052	4870
ECG3053	5127
ECG3054	5128
ECG3055	5129
ECG3056	4873
ECG3057	4190
ECG3058	5130
ECG3059	5131
ECG3060	5132
ECG3061	4989
ECG3062	5134
ECG3063	5135
ECG3064	5137
ECG3065	4871
ECG3068	5138
ECG3069	5139
ECG3070	5140
ECG3071	5141
ECG3074	4327
ECG3075	4332
ECG3076	4331
ECG3077	4333
ECG3078	2206
ECG3079	2199
ECG3080	2378
ECG3081	3367
ECG3082	5142
ECG3083	1109
ECG3084	1053
ECG3085	5421
ECG3086	1403
ECG3087	5423
ECG3088	5424
ECG3089	5425
ECG3090	5426
ECG3091	0235
ECG3092	1291
ECG3093	5427
ECG3094	5674
ECG3095	5428
ECG3096	5429
ECG3097	5430
ECG3098	0112
ECG3099	2703
ECG3100	1417
ECG3101	2636
ECG3102	2640
ECG3103	5433
ECG3104	5662
ECG3105	4878
ECG3111	5434
ECG3112	5435
ECG3113	5436
ECG3114	5151
ECG3115	5153
ECG3116	5154
ECG3117	5155
ECG3120	2094
ECG3121	3987
ECG3130	5438
ECG3131	5016
ECG3150	5165
ECG3151	5167
ECG3152	5168
ECG3153	5170
ECG3154	5211
ECG3155	5171
ECG3156	4443
ECG3157	2047
ECG3158	5037
ECG3160	5174
ECG3161	1076
ECG3162	5176
ECG3163	4767
ECG3164	3491
ECG3165	5178
ECG3166	4052
ECG3167	1131
ECG3168	5179
ECG3169	5443
ECG3170	5444
ECG3171	5063
ECG3172	4859
ECG3173	1970
ECG3174	4861
ECG3180	5446
ECG3181	5447
ECG3182	5448
ECG3200	5990
ECG3201	2708
ECG3202	2753
ECG3470	2759
ECG3539	2642
ECG3880	3481
ECG3881	1904
ECG3882	5339
ECG4000	2024
ECG4001	0478
ECG4001B	0478
ECG4002	2055
ECG4002B	2055
ECG4006B	0646
ECG4007	2846
ECG4007B	2846
ECG4008B	0988
ECG4011	0215
ECG4011B	0215
ECG4012	0498
ECG4012B	0498
ECG4013	0413
ECG4013B	0413
ECG4014	0860
ECG4014B	0860
ECG4015	1014
ECG4015B	1014
ECG4016	1143
ECG4016B	1143
ECG4017	0513
ECG4017B	0513
ECG4018B	1391
ECG4019	1525
ECG4019B	1525
ECG4020	1661
ECG4020B	1661
ECG4021	1749
ECG4021B	1749
ECG4022B	1256
ECG4023	0519
ECG4023B	0519
ECG4024	1957
ECG4024B	1957
ECG4025	2072
ECG4025B	2072
ECG4026B	2152
ECG4027	1949
ECG4027B	1949
ECG4028	2228
ECG4028B	2228
ECG4029	2233
ECG4029B	2233
ECG4030	0500
ECG4030B	0500
ECG4031B	2971
ECG4032B	2531
ECG4033B	2635
ECG4034B	3615
ECG4035B	2777
ECG4038B	2981
ECG4040	0056
ECG4040B	0056
ECG4041	3174
ECG4042	0121
ECG4042B	0121
ECG4043B	1553
ECG4044B	2308
ECG4045B	3443
ECG4046B	3428
ECG4047B	2331
ECG4048B	3459
ECG4049	0001
ECG4049B	0001
ECG4050	0398
ECG4050B	0398
ECG4051	0366
ECG4051B	0366
ECG4052	0024
ECG4052B	0024
ECG4053B	0035
ECG4055	3304
ECG4055B	3304
ECG4056B	3706
ECG4060B	0146
ECG4063B	3727
ECG4066B	0101
ECG4067B	3742
ECG4068	2503
ECG4068B	2503
ECG4069	0119
ECG4070B	2516
ECG4071	0129
ECG4071B	0129
ECG4072B	2524
ECG4073B	1536
ECG4075B	2540
ECG4076B	3496
ECG4077	2558
ECG4077B	2558
ECG4078	0921
ECG4078B	0921
ECG4081	0626
ECG4081B	0626
ECG4082B	0300
ECG4085	0303
ECG4085B	0303
ECG4086	0535
ECG4086B	0535
ECG4089B	3830
ECG4093B	2384
ECG4094B	1682
ECG4095B	3848
ECG4096B	3851
ECG4097B	3855
ECG4098B	3611
ECG4099	3330
ECG4099B	3330
ECG4256	1472
ECG4501	3809
ECG4502	1037
ECG4502B	1037
ECG4506	6078
ECG4506B	3767
ECG4508B	1811
ECG4510B	1963
ECG4511B	1544
ECG4512	2120
ECG4512B	2120
ECG4513B	4976
ECG4514B	1830
ECG4515B	3600
ECG4516B	2347
ECG4517B	4284
ECG4518	1043
ECG4518B	1043
ECG4520	2675
ECG4520B	2675
ECG4521B	4990
ECG4522	2837
ECG4522B	2837
ECG4526B	3610
ECG4527	3145
ECG4527B	3145
ECG4528B	3198
ECG4529B	4531
ECG4531B	3325
ECG4532B	1016
ECG4536B	3704
ECG4538B	1063
ECG4539	3656
ECG4539B	3656
ECG4541B	5017
ECG4543B	5020
ECG4547B	5032
ECG4551B	5039
ECG4553B	5040
ECG4555	2938
ECG4555B	2938
ECG4556	3431
ECG4556B	3431
ECG4558B	5049
ECG4562B	1035
ECG4566B	3504
ECG4568B	3618
ECG4569B	4487
ECG4582B	4660
ECG4583B	1297
ECG4585B	1375
ECG4597B	4110
ECG4598B	4301
ECG4900	2396
ECG4901	6118
ECG4902	3114
ECG4903	2133
ECG4918	3172
ECG4919	1169
ECG4926	3191
ECG4927	1767
ECG4928	3201
ECG4929	1932
ECG4934	1915
ECG4935	1952
ECG4950	0568
ECG4951	1972
ECG4958	0831
ECG4959	1987
ECG4988	1405
ECG4989	1408
ECG5000	1276
ECG5000A	1276
ECG5001	2874
ECG5001A	2874
ECG5002	0760
ECG5002A	0760
ECG5003	0677
ECG5003A	0677
ECG5004	0118
ECG5004A	0118
ECG5005	0299
ECG5005A	0299
ECG5006	0376
ECG5006A	0376
ECG5007	0036
ECG5007A	0036
ECG5008	0275
ECG5008A	0275
ECG5009	0140
ECG5009A	0140
ECG5010	0041
ECG5010A	0041
ECG5010T1	0589
ECG5011	0253
ECG5011A	0253
ECG5011T1	0883
ECG5012	0091
ECG5012A	0091
ECG5013	0471
ECG5013A	0471
ECG5013T1	0295
ECG5014	0062
ECG5014A	0062
ECG5015	0077
ECG5015A	0077
ECG5016	0165
ECG5016A	0165
ECG5017	0322
ECG5017A	0322
ECG5018	0057
ECG5018A	0057
ECG5019	0064
ECG5019A	0064
ECG5019T1	0248
ECG5020	0181
ECG5020A	0181
ECG5021	0052
ECG5021A	0052
ECG5021T1	1005
ECG5022	0053
ECG5022A	0053
ECG5023	0879
ECG5023A	0879
ECG5024	0686
ECG5024A	0686
ECG5025	0445
ECG5025A	0445
ECG5026	0210
ECG5026A	0210
ECG5027	0375
ECG5027A	0375
ECG5028	0671
ECG5028A	0671
ECG5029	0700
ECG5029A	0700
ECG5030	0705
ECG5030A	0705
ECG5031	0494
ECG5031A	0494
ECG5032	0714
ECG5032A	0714
ECG5033	0455
ECG5033A	0455
ECG5034	0257
ECG5034A	0257
ECG5034C	OEM
ECG5035	0195
ECG5035A	0195
ECG5036	0166
ECG5036A	0166
ECG5037	0010
ECG5037A	0010
ECG5038	0032
ECG5038A	0032
ECG5039	0054
ECG5039A	0054
ECG5040	0068
ECG5040A	0068
ECG5041	0092
ECG5041A	0092
ECG5042	0125
ECG5042A	0125
ECG5043	2317
ECG5043A	2317
ECG5044	0152
ECG5044A	0152
ECG5045	0173
ECG5045A	0173
ECG5046	0094
ECG5046A	0094
ECG5047	0049
ECG5047A	0049
ECG5048	0104
ECG5048A	0104
ECG5049	0156
ECG5049A	0156
ECG5050	0189
ECG5050A	0189
ECG5051	0099
ECG5051A	0099
ECG5052	0089
ECG5052A	0089
ECG5053	0287
ECG5053A	0287
ECG5054	0252
ECG5054A	0252
ECG5055	0340
ECG5055A	0340
ECG5056	0370
ECG5056A	0370
ECG5057	0394
ECG5057A	0394
ECG5058	0425
ECG5058A	0425
ECG5059	0453
ECG5059A	0453
ECG5060	1473
ECG5060A	1473
ECG5061	3003
ECG5061A	3003
ECG5063	1312
ECG5063A	1312
ECG5064	3010
ECG5064A	3010
ECG5065	1713
ECG5065A	1713
ECG5066	0292
ECG5066A	0292
ECG5067	0456
ECG5067A	0456
ECG5068	0533
ECG5068A	0533
ECG5069	0451
ECG5069A	0451
ECG5070	0301
ECG5070A	0301
ECG5071	0025
ECG5071A	0025
ECG5072	0244
ECG5072A	0244
ECG5073	1083
ECG5073A	1083
ECG5074	0317
ECG5074A	0317
ECG5075	0420
ECG5075A	0420
ECG5076	1649
ECG5076A	1649
ECG5077	0495
ECG5077A	0495
ECG5078	0949
ECG5078A	0949
ECG5079	0531
ECG5079A	0531
ECG5080	0565
ECG5080A	0565
ECG5081	0402
ECG5081A	0402
ECG5082	1606
ECG5082A	1606
ECG5083	1674
ECG5083A	1674
ECG5084	0726
ECG5084A	0726
ECG5085	0820
ECG5085A	0820
ECG5086	0350
ECG5086A	0350
ECG5087	0931
ECG5087A	0931
ECG5088	0999
ECG5088A	0999
ECG5089	0502
ECG5089A	0502
ECG5090	0869
ECG5090A	0869
ECG5091	1156
ECG5091A	1156
ECG5092	1268
ECG5092A	1268
ECG5093	1189
ECG5093A	1189
ECG5093T	OEM
ECG5094	3026
ECG5094A	3026
ECG5095	1311
ECG5095A	1311
ECG5096	0098
ECG5096A	0098
ECG5097	0186
ECG5097A	0186
ECG5098	0213
ECG5098A	0213
ECG5099	0245
ECG5099A	0245
ECG5100	0028
ECG5100A	0028
ECG5101	0255
ECG5101A	0255
ECG5102	0877
ECG5102A	0877
ECG5103	0367
ECG5103A	0367
ECG5104	2858
ECG5104A	2858
ECG5105	0421
ECG5105A	0421
ECG5107T2	3946
ECG5111	0782
ECG5111A	0782
ECG5112	0796
ECG5112A	0796
ECG5113	0806
ECG5113A	0806
ECG5114	0821
ECG5114A	0821
ECG5115	0833
ECG5115A	0833
ECG5116	0442
ECG5116A	0442
ECG5117	0876
ECG5117A	0876
ECG5118	3127
ECG5118A	3127
ECG5119	0185
ECG5119A	0185
ECG5120	0205
ECG5120A	0205
ECG5121	0480
ECG5121A	0480
ECG5122	0504
ECG5122A	0504
ECG5123	3317
ECG5123A	3317
ECG5124	0684
ECG5124A	0684
ECG5125	0225
ECG5125A	0225
ECG5126	0230
ECG5126A	0230
ECG5127	0234
ECG5127A	0234
ECG5128	0237
ECG5128A	0237
ECG5129	1397
ECG5129A	1397
ECG5130	0247
ECG5130A	0247
ECG5131	0251
ECG5131A	0251
ECG5132	1178
ECG5132A	1178
ECG5133	0256
ECG5133A	0256
ECG5134	2395
ECG5134A	2395
ECG5135	0262
ECG5135A	0262
ECG5136	0269
ECG5136A	0269
ECG5137	0274
ECG5137A	0274
ECG5138	2399
ECG5138A	2399
ECG5139	0294
ECG5139A	0294
ECG5140	1177
ECG5140A	1177
ECG5141	0309
ECG5141A	0309
ECG5142	0318
ECG5142A	0318
ECG5143	0320
ECG5143A	0320
ECG5144	0326
ECG5144A	0326
ECG5145	0337
ECG5145A	0337
ECG5146	0347
ECG5146A	0347
ECG5147	0027
ECG5147A	0027
ECG5148	0266
ECG5148A	0266
ECG5149	2856
ECG5149A	2856
ECG5150	0386
ECG5150A	0386
ECG5151	0405
ECG5151A	0405
ECG5152	0426
ECG5152A	0426
ECG5153	0444
ECG5153A	0444
ECG5154	3028
ECG5154A	3028
ECG5155	0238
ECG5155A	0238
ECG5156	1180
ECG5156A	1180
ECG5157	1190
ECG5157A	1190
ECG5158	1207
ECG5158A	1207
ECG5159	1218
ECG5159A	1218
ECG5160	1881
ECG5160A	1881
ECG5161	0647
ECG5161A	0647
ECG5162	1255
ECG5162A	1255
ECG5163	2102
ECG5163A	2102
ECG5164	1279
ECG5164A	1279
ECG5165	2225
ECG5165A	2225
ECG5166	0605
ECG5166A	0605
ECG5172	4231
ECG5172A	4231
ECG5172AK	1006
ECG5172K	1006
ECG5173	2397
ECG5173A	2397
ECG5173AK	1380
ECG5173K	1380
ECG5174	2035
ECG5174A	2035
ECG5174AK	0547
ECG5174K	0547
ECG5175	2401
ECG5175A	2401
ECG5175AK	2403
ECG5175K	2403
ECG5176	1438
ECG5176A	1438
ECG5176AK	2113
ECG5176K	2113
ECG5177	2407
ECG5177A	2407
ECG5177AK	2410
ECG5177K	2410
ECG5178	1445
ECG5178A	1445
ECG5178AK	1901
ECG5178K	1901
ECG5179	2415
ECG5179A	2415
ECG5179AK	2416
ECG5179K	2416
ECG5180	2221
ECG5180A	2221
ECG5180AK	0696
ECG5180K	0696

If replacement code is OEM, contact original manufacturer for replacement.

DEVICE TYPE	REPL CODE	DEVICE TYPE	REPL CODE	DEVICE TYPE	REPL CODE	DEVICE TYPE	REPL CODE	DEVICE TYPE	REPL CODE	DEVICE TYPE	REPL CODE	DEVICE TYPE	REPL CODE
ECG5181	1458	ECG5208A	0022	ECG5242AK	2726	ECG5269K	1406	ECG5304	0106	ECG5482	2491	ECG5625	1095
ECG5181A	1458	ECG5208AK	1853	ECG5242K	2726	ECG5270	1232	ECG5305	2010	ECG5483	0240	ECG5626	0415
ECG5181AK	1601	ECG5208K	1853	ECG5243	0868	ECG5270A	1232	ECG5306	2400	ECG5484	2659	ECG5627	1099
ECG5181K	1601	ECG5209	0070	ECG5243A	0868	ECG5270AK	1415	ECG5307	0787	ECG5485	0676	ECG5628	1100
ECG5182	0221	ECG5209A	0070	ECG5243AK	2729	ECG5270K	1415	ECG5309	0729	ECG5486	2809	ECG5629	2300
ECG5182A	0221	ECG5209AK	1861	ECG5243K	2729	ECG5271	1246	ECG5310	0737	ECG5487	0336	ECG5631	0578
ECG5182AK	1616	ECG5209K	1861	ECG5244	0885	ECG5271A	1246	ECG5311	0742	ECG5491	2449	ECG5632	0571
ECG5182K	1616	ECG5210	0132	ECG5244A	0885	ECG5271AK	1428	ECG5312	0323	ECG5492	0435	ECG5633	0593
ECG5183	1490	ECG5210A	0132	ECG5244AK	2732	ECG5271K	1428	ECG5313	1414	ECG5494	1487	ECG5634	0574
ECG5183A	1490	ECG5210AK	1866	ECG5244K	2732	ECG5272	1265	ECG5314	0473	ECG5496	0687	ECG5635	0576
ECG5183AK	1622	ECG5210K	1866	ECG5245	0897	ECG5272A	1265	ECG5315	0446	ECG5500	3102	ECG5636	0577
ECG5183K	1622	ECG5211	0172	ECG5245A	0897	ECG5272AK	1440	ECG5316	1422	ECG5501	2519	ECG5637	0594
ECG5184	2422	ECG5211A	0172	ECG5245AK	2734	ECG5272K	1440	ECG5317	2443	ECG5502	0741	ECG5638	5322
ECG5184A	2422	ECG5211AK	1874	ECG5245K	2734	ECG5273	1273	ECG5318	3023	ECG5503	3105	ECG5640	2383
ECG5184AK	2424	ECG5211K	1874	ECG5246	0914	ECG5273A	1273	ECG5319	3035	ECG5504	0745	ECG5641	2394
ECG5184K	2424	ECG5212	0207	ECG5246A	0914	ECG5273AK	2772	ECG5320	3050	ECG5505	3109	ECG5642	1413
ECG5185	1618	ECG5212A	0207	ECG5246AK	2738	ECG5273K	2772	ECG5322	2363	ECG5506	2917	ECG5643	2387
ECG5185A	1618	ECG5212AK	1884	ECG5246K	2738	ECG5274	1290	ECG5324	2369	ECG5507	0747	ECG5645	0555
ECG5185AK	0627	ECG5212K	1884	ECG5247	0926	ECG5274A	1290	ECG5326	2370	ECG5508	3243	ECG5646	1456
ECG5185K	0627	ECG5213	0227	ECG5247A	0926	ECG5274AK	1447	ECG5327	2372	ECG5509	0752	ECG5649	5553
ECG5186	0510	ECG5213A	0227	ECG5247AK	2740	ECG5274K	1447	ECG5328	2373	ECG5511	3696	ECG5650	2352
ECG5186A	0510	ECG5213AK	2474	ECG5247K	2740	ECG5275	1299	ECG5329	1042	ECG5512	3383	ECG5651	2353
ECG5186AK	0992	ECG5213K	2474	ECG5248	0944	ECG5275A	1299	ECG5330	1045	ECG5513	3624	ECG5652	2356
ECG5186K	0992	ECG5214	0263	ECG5248A	0944	ECG5275AK	2776	ECG5331	1047	ECG5514	1651	ECG5653	5555
ECG5187	0691	ECG5214A	0263	ECG5248AK	2744	ECG5275K	2776	ECG5332	1875	ECG5515	1584	ECG5655	4455
ECG5187A	0691	ECG5214AK	1895	ECG5248K	2744	ECG5276	1307	ECG5334	3546	ECG5516	1665	ECG5656	4456
ECG5187AK	0995	ECG5214K	1895	ECG5249	0958	ECG5276A	1307	ECG5335	5734	ECG5517	1650	ECG5657	4482
ECG5187K	0995	ECG5215	0310	ECG5249A	0958	ECG5276AK	1461	ECG5338	5736	ECG5518	2647	ECG5660	0399
ECG5188	0870	ECG5215A	0310	ECG5249AK	1225	ECG5276K	1461	ECG5340	1643	ECG5519	2649	ECG5661	0404
ECG5188A	0870	ECG5215AK	1902	ECG5249K	1225	ECG5277	1322	ECG5342	1673	ECG5520	3307	ECG5662	0408
ECG5188AK	1263	ECG5215K	1902	ECG5250	0978	ECG5277A	1322	ECG5344	1589	ECG5521	2188	ECG5663	0422
ECG5188K	1263	ECG5216	0329	ECG5250A	0978	ECG5277AK	2779	ECG5346	5738	ECG5522	0567	ECG5664	0422
ECG5189	1020	ECG5216A	0329	ECG5250AK	2746	ECG5277K	2779	ECG5348	5739	ECG5523	3258	ECG5665	0422
ECG5189A	1020	ECG5216AK	0736	ECG5250K	2746	ECG5278	1331	ECG5351	4337	ECG5524	0762	ECG5666	0422
ECG5189AK	1249	ECG5216K	0736	ECG5251	0994	ECG5278A	1331	ECG5360	0804	ECG5525	3268	ECG5667	0422
ECG5189K	1249	ECG5217	0356	ECG5251A	0994	ECG5278AK	1474	ECG5368	0655	ECG5526	3271	ECG5667A	0422
ECG5190	1153	ECG5217A	0356	ECG5251AK	1237	ECG5278K	1474	ECG5369	0020	ECG5527	0740	ECG5668A	6157
ECG5190A	1153	ECG5217AK	1909	ECG5251K	1237	ECG5279	1340	ECG5371	0526	ECG5528	3291	ECG5669A	3461
ECG5190AK	1636	ECG5217K	1909	ECG5252	1009	ECG5279A	1340	ECG5372	0108	ECG5529	0764	ECG5671	2202
ECG5190K	1636	ECG5218	0381	ECG5252A	1009	ECG5279AK	2783	ECG5374	0428	ECG5530	2875	ECG5672	5560
ECG5191	1274	ECG5218A	0381	ECG5252AK	1252	ECG5279K	2783	ECG5375	1871	ECG5531	0766	ECG5673	0154
ECG5191A	1274	ECG5218AK	2479	ECG5252K	1252	ECG5280	1353	ECG5377	1700	ECG5534	0226	ECG5675	0147
ECG5191AK	1639	ECG5218K	2479	ECG5253	1019	ECG5280A	1353	ECG5378	0345	ECG5534A	1534	ECG5676	4545
ECG5191K	1639	ECG5219	0412	ECG5253A	1019	ECG5280AK	0613	ECG5380	0459	ECG5536	5079	ECG5677	0279
ECG5192	1402	ECG5219A	0412	ECG5253AK	1269	ECG5280K	0613	ECG5381	0587	ECG5538	5974	ECG5679	0908
ECG5192A	1402	ECG5219AK	1914	ECG5253K	1269	ECG5281	1365	ECG5386	0769	ECG5539	5078	ECG5680	3146
ECG5192AK	1703	ECG5219K	1914	ECG5254	0889	ECG5281A	1365	ECG5387	0483	ECG5540	3277	ECG5681	3147
ECG5192K	1703	ECG5220	0438	ECG5254A	0889	ECG5281AK	1510	ECG5400	1137	ECG5541	0731	ECG5682	3148
ECG5193	1532	ECG5220A	0438	ECG5254AK	1277	ECG5281K	1510	ECG5401	0344	ECG5542	0712	ECG5683	3150
ECG5193A	1532	ECG5220AK	1163	ECG5254K	1277	ECG5282	1384	ECG5402	0901	ECG5543	0469	ECG5684	3151
ECG5193AK	1640	ECG5220K	1163	ECG5255	1049	ECG5282A	1384	ECG5403	2342	ECG5544	0721	ECG5685	3152
ECG5193K	1640	ECG5221	0464	ECG5255A	1049	ECG5282AK	1523	ECG5404	0058	ECG5545	0722	ECG5686	3153
ECG5194	1078	ECG5221A	0464	ECG5255AK	1294	ECG5282K	1523	ECG5405	0407	ECG5546	0778	ECG5687	4100
ECG5194A	1078	ECG5221AK	1924	ECG5255K	1294	ECG5283	1401	ECG5406	1683	ECG5547	0725	ECG5688	1064
ECG5194AK	1716	ECG5221K	1924	ECG5256	1058	ECG5283A	1401	ECG5408	0179	ECG5548	0750	ECG5689	1317
ECG5194K	1716	ECG5222	0488	ECG5256A	1058	ECG5283AK	1537	ECG5409	0346	ECG5550	2521	ECG5690	1891
ECG5195	1711	ECG5222A	0488	ECG5256AK	2752	ECG5283K	1537	ECG5410	3349	ECG5552	0397	ECG5693	2015
ECG5195A	1711	ECG5222AK	1933	ECG5256K	2752	ECG5284	1412	ECG5411	4458	ECG5554	0611	ECG5695	2017
ECG5195AK	1720	ECG5222K	1933	ECG5257	0932	ECG5284A	1412	ECG5412	0174	ECG5556	0952	ECG5697	2018
ECG5195K	1720	ECG5223	0509	ECG5257A	0932	ECG5284AK	1550	ECG5413	3854	ECG5558	1512	ECG5699	0793
ECG5196	1717	ECG5223A	0509	ECG5257AK	1302	ECG5284K	1550	ECG5414	3620	ECG5562	1848	ECG5700	1434
ECG5196A	1717	ECG5223AK	1941	ECG5257K	1302	ECG5285	1423	ECG5415	3324	ECG5564	1855	ECG5701	1680
ECG5196AK	1731	ECG5223K	1941	ECG5258	1079	ECG5285A	1423	ECG5416	1502	ECG5566	3215	ECG5702	2294
ECG5196K	1731	ECG5224	0524	ECG5258A	1079	ECG5285AK	1575	ECG5417	1004	ECG5567	1704	ECG5703	2298
ECG5197	1723	ECG5224A	0524	ECG5258AK	1310	ECG5285K	1575	ECG5418	1034	ECG5568	4031	ECG5704	5812
ECG5197A	1723	ECG5224AK	1947	ECG5258K	1310	ECG5286	1436	ECG5419	1148	ECG5569	0679	ECG5705	5813
ECG5197AK	0727	ECG5224K	1947	ECG5259	1096	ECG5286A	1436	ECG5421	6169	ECG5570	0608	ECG5710	0652
ECG5197K	0727	ECG5225	0542	ECG5259A	1096	ECG5286AK	2802	ECG5422	3284	ECG5572	0610	ECG5711	4364
ECG5198	1736	ECG5225A	0542	ECG5259AK	2756	ECG5286K	2802	ECG5423	1923	ECG5574	0468	ECG5712	4574
ECG5198A	1736	ECG5225AK	1953	ECG5259K	2756	ECG5287	1444	ECG5424	1992	ECG5575	0641	ECG5720	2311
ECG5198AK	1756	ECG5225K	1953	ECG5260	1106	ECG5287A	1444	ECG5425	5527	ECG5577	0217	ECG5721	2994
ECG5198K	1756	ECG5226	0063	ECG5260A	1106	ECG5287AK	2803	ECG5426	1044	ECG5579	0658	ECG5722	2878
ECG5199	1748	ECG5226A	0063	ECG5260AK	1324	ECG5287K	2803	ECG5427	5750	ECG5580	1084	ECG5800	0110
ECG5199A	1748	ECG5226AK	1961	ECG5260K	1324	ECG5288	1457	ECG5428	5751	ECG5582	1086	ECG5801	0953
ECG5199AK	1768	ECG5226K	1961	ECG5261	1123	ECG5288A	1457	ECG5429	5469	ECG5584	1102	ECG5802	0242
ECG5199K	1768	ECG5227	0401	ECG5261A	1123	ECG5288AK	1507	ECG5431	3283	ECG5587	1900	ECG5803	1747
ECG5200	1761	ECG5227A	0401	ECG5261AK	2758	ECG5288K	1507	ECG5432	6171	ECG5589	0738	ECG5804	0540
ECG5200A	1761	ECG5227AK	0357	ECG5261K	2758	ECG5289	1470	ECG5433	6172	ECG5590	0192	ECG5805	1771
ECG5200AK	1782	ECG5227K	0357	ECG5262	1135	ECG5289A	1470	ECG5434	2270	ECG5591	0159	ECG5806	0965
ECG5200K	1782	ECG5228	0598	ECG5262A	1135	ECG5289AK	2805	ECG5437	0717	ECG5592	0096	ECG5808	0817
ECG5201	2450	ECG5228A	0598	ECG5262AK	1333	ECG5289K	2805	ECG5438	0308	ECG5594	2843	ECG5809	0922
ECG5201A	2450	ECG5228AK	2500	ECG5262K	1333	ECG5290	1484	ECG5440	3908	ECG5595	1765	ECG5812	1282
ECG5201AK	2453	ECG5228K	2500	ECG5263	1152	ECG5290A	1484	ECG5442	4588	ECG5596	1814	ECG5814	1287
ECG5201K	2453	ECG5229	0616	ECG5263A	1152	ECG5290AK	2807	ECG5444	2270	ECG5598	4148	ECG5815	1293
ECG5202	1772	ECG5229A	0616	ECG5263AK	1344	ECG5290K	2807	ECG5446	3402	ECG5599	1725	ECG5817	1296
ECG5202A	1772	ECG5229AK	0670	ECG5263K	1344	ECG5291	1505	ECG5448	3404	ECG5600	4412	ECG5818	1567
ECG5202AK	1794	ECG5229K	0670	ECG5264	1164	ECG5291A	1505	ECG5452	1396	ECG5601	1755	ECG5819	1547
ECG5202K	1794	ECG5230	0634	ECG5264A	1164	ECG5291AK	2810	ECG5453	1259	ECG5602	4414	ECG5820	2153
ECG5203	1788	ECG5230A	0634	ECG5264AK	1356	ECG5291K	2810	ECG5454	0447	ECG5603	3499	ECG5821	3139
ECG5203A	1788	ECG5230AK	0776	ECG5264K	1356	ECG5292	1521	ECG5455	0940	ECG5604	4121	ECG5822	0711
ECG5203AK	1799	ECG5230K	0776	ECG5265	1174	ECG5292A	1521	ECG5456	1222	ECG5605	0485	ECG5823	2967
ECG5203K	1799	ECG5231	0650	ECG5265A	1174	ECG5292AK	2815	ECG5457	0095	ECG5606	4416	ECG5826	3517
ECG5204	1796	ECG5231A	0650	ECG5265AK	2760	ECG5292K	2815	ECG5458	1243	ECG5607	4417	ECG5827	5012
ECG5204A	1796	ECG5231AK	2507	ECG5265K	2760	ECG5293	1531	ECG5460	4576	ECG5608	0572	ECG5828	3541
ECG5204AK	1809	ECG5231K	2507	ECG5266	1184	ECG5293A	1531	ECG5461	2095	ECG5609	0935	ECG5829	5577
ECG5204K	1809	ECG5232	0668	ECG5266A	1184	ECG5293AK	2818	ECG5462	2089	ECG5610	0059	ECG5830	0708
ECG5205	1804	ECG5232A	0668	ECG5266AK	1371	ECG5293K	2818	ECG5463	0505	ECG5611	5549	ECG5831	0933
ECG5205A	1804	ECG5232AK	1071	ECG5266K	1371	ECG5294	1548	ECG5465	0710	ECG5612	5550	ECG5832	4140
ECG5205AK	1817	ECG5232K	1071	ECG5267	1200	ECG5294A	1548	ECG5466	0863	ECG5613	0954	ECG5833	5517
ECG5205K	1817	ECG5240	0815	ECG5267A	1200	ECG5294AK	2820	ECG5468	0327	ECG5614	4440	ECG5834	0580
ECG5206	1194	ECG5240A	0815	ECG5267AK	2762	ECG5294K	2820	ECG5470	1110	ECG5615	0957	ECG5835	0947
ECG5206A	1194	ECG5240AK	2721	ECG5267K	2762	ECG5295	1568	ECG5471	2932	ECG5616	0960	ECG5836	2060
ECG5206AK	1826	ECG5240K	2721	ECG5268	1210	ECG5295A	1568	ECG5472	0963	ECG5617	0961	ECG5837	4523
ECG5206K	1826	ECG5241	0827	ECG5268A	1210	ECG5295AK	2822	ECG5473	2933	ECG5618	0966	ECG5838	1000
ECG5207	1821	ECG5241A	0827	ECG5268AK	1387	ECG5295K	2822	ECG5474	2936	ECG5620	3593	ECG5839	1012
ECG5207A	1821	ECG5241AK	2724	ECG5268K	1387	ECG5296	1587	ECG5475	3671	ECG5621	5551	ECG5840	2076
ECG5207AK	1840	ECG5241K	2724	ECG5269	1223	ECG5296A	1587	ECG5476	3439	ECG5622	1089	ECG5841	5567
ECG5207K	1840	ECG5242	0846	ECG5269A	1223	ECG5296AK	2824	ECG5480	3419	ECG5623	1093	ECG5842	2081
ECG5208	0022	ECG5242A	0846	ECG5269AK	1406	ECG5296K	2824	ECG5481	1103	ECG5624	0411	ECG5843	1073

If replacement code is OEM, contact original manufacturer for replacement.

DEVICE TYPE	REPL CODE
ECG5846	2088
ECG5847	1138
ECG5848	0612
ECG5849	1188
ECG5850	0970
ECG5851	0985
ECG5852	3733
ECG5853	0910
ECG5854	0989
ECG5855	0990
ECG5856	3743
ECG5857	0993
ECG5858	0197
ECG5859	0997
ECG5860	0200
ECG5861	1001
ECG5862	0204
ECG5863	0515
ECG5866	0206
ECG5867	1008
ECG5868	0586
ECG5869	0948
ECG5870	0084
ECG5871	0534
ECG5872	0090
ECG5873	0748
ECG5874	0097
ECG5875	0765
ECG5876	0105
ECG5877	0777
ECG5878	0109
ECG5879	0538
ECG5880	0116
ECG5881	0801
ECG5882	0122
ECG5883	0816
ECG5886	0131
ECG5887	0545
ECG5890	0145
ECG5891	0550
ECG5892	5514
ECG5893	1238
ECG5894	5582
ECG5895	1240
ECG5896	2900
ECG5897	1241
ECG5898	5583
ECG5899	1245
ECG5900	3751
ECG5901	2365
ECG5902	3757
ECG5903	1253
ECG5904	1112
ECG5905	2376
ECG5908	3768
ECG5909	1264
ECG5910	3011
ECG5911	0449
ECG5912	1600
ECG5913	5005
ECG5914	5282
ECG5915	5491
ECG5916	0871
ECG5917	1635
ECG5918	5283
ECG5919	5495
ECG5920	0853
ECG5921	1251
ECG5922	5286
ECG5923	5496
ECG5924	1609
ECG5925	1205
ECG5928	1610
ECG5929	2137
ECG5932	1614
ECG5933	2251
ECG5940	2002
ECG5941	2003
ECG5942	1250
ECG5943	3265
ECG5944	1250
ECG5945	1577
ECG5946	5587
ECG5947	3275
ECG5948	1581
ECG5949	3281
ECG5950	5589
ECG5951	3287
ECG5952	1586
ECG5953	3294
ECG5962	5026
ECG5963	2559
ECG5966	2615
ECG5967	2566
ECG5980	3189
ECG5981	1630
ECG5982	2901
ECG5983	0254
ECG5986	1124
ECG5987	1107
ECG5988	1126
ECG5989	1111
ECG5990	0805
ECG5991	0258
ECG5992	1195
ECG5993	1644
ECG5994	0319
ECG5995	0267
ECG5998	1132

DEVICE TYPE	REPL CODE
ECG5999	1119
ECG6002	0044
ECG6003	0281
ECG6006	1530
ECG6007	1520
ECG6008	0029
ECG6009	1847
ECG6010	0601
ECG6011	1851
ECG6013	3837
ECG6020	3762
ECG6021	2665
ECG6022	2653
ECG6023	2695
ECG6026	2657
ECG6027	2768
ECG6030	2664
ECG6031	2855
ECG6034	2006
ECG6035	2907
ECG6038	2677
ECG6039	2974
ECG6040	2682
ECG6041	3051
ECG6042	3900
ECG6043	3117
ECG6044	2655
ECG6045	3208
ECG6048	4306
ECG6049	3771
ECG6050	1560
ECG6051	2179
ECG6054	2840
ECG6055	2182
ECG6058	4308
ECG6059	3601
ECG6060	2850
ECG6061	2191
ECG6064	2871
ECG6065	2197
ECG6068	2833
ECG6069	2216
ECG6072	2473
ECG6073	2340
ECG6074	2778
ECG6075	1745
ECG6076	2813
ECG6077	1783
ECG6078	2845
ECG6079	1818
ECG6084	0615
ECG6085	1021
ECG6087	2515
ECG6090	1942
ECG6094	1545
ECG6102	5046
ECG6103	2315
ECG6104	0749
ECG6105	2792
ECG6106	4286
ECG6107	4285
ECG6110	1226
ECG6112	1204
ECG6113	2159
ECG6115	0479
ECG6116	4922
ECG6118	2139
ECG6154	0599
ECG6155	1347
ECG6156	1986
ECG6157	1905
ECG6158	0657
ECG6159	0193
ECG6162	3309
ECG6163	0202
ECG6200	3676
ECG6202	3679
ECG6204	3680
ECG6206	2836
ECG6208	4188
ECG6210	4197
ECG6220	1798
ECG6230	2192
ECG6240	1236
ECG6241	5796
ECG6244	1727
ECG6246	2234
ECG6247	5992
ECG6354	1023
ECG6355	0501
ECG6356	1036
ECG6357	1777
ECG6358	1046
ECG6359	1789
ECG6362	2832
ECG6363	3167
ECG6400	1669
ECG6400A	1669
ECG6400B	4374
ECG6401	2136
ECG6402	0316
ECG6403	4233
ECG6404	3407
ECG6405	4405
ECG6406	4585
ECG6407	3331
ECG6408	5911
ECG6409	1175
ECG6410	1893

DEVICE TYPE	REPL CODE
ECG6411	5389
ECG6412	5643
ECG6415	5645
ECG6416	5646
ECG6417	5649
ECG6418	5650
ECG6419	5651
ECG6502	3065
ECG6507	3070
ECG6508	3193
ECG6532	1973
ECG6664	4315
ECG6800	0388
ECG6802	0393
ECG6809	5967
ECG6809E	2117
ECG6810	2086
ECG6820	OEM
ECG6821	0448
ECG6850	0514
ECG6860	2692
ECG6875	4181
ECG6880	4516
ECG6885	4647
ECG6886	4651
ECG6887	4658
ECG6888	4667
ECG6889	0581
ECG7000	6159
ECG7001	5745
ECG7002	6262
ECG7003	4241
ECG7004	6028
ECG7005	4953
ECG7006	6160
ECG7007	4532
ECG7008	0523
ECG7009	3690
ECG7010	3515
ECG7011	6161
ECG7012	3378
ECG7013	4550
ECG7014	6162
ECG7015	3635
ECG7016	6163
ECG7017	6164
ECG7018	3044
ECG7019	4930
ECG7020	6165
ECG7021	6263
ECG7022	6264
ECG7023	6266
ECG7024	6267
ECG7025	6268
ECG7026	6269
ECG7027	6270
ECG7028	6271
ECG7029	6272
ECG7030	6273
ECG7031	6274
ECG7032	6275
ECG7033	6276
ECG7034	6277
ECG7035	6278
ECG7036	6279
ECG7037	6280
ECG7038	6281
ECG7039	6282
ECG7040	6283
ECG7041	6284
ECG7042	6285
ECG7043	6286
ECG7044	6287
ECG7045	6288
ECG7046	6289
ECG7047	6290
ECG7048	6291
ECG7049	6292
ECG7050	6293
ECG7051	6294
ECG7052	6296
ECG7053	6297
ECG7054	6298
ECG7055	6299
ECG7056	6300
ECG7057	6301
ECG7058	6302
ECG7059	6303
ECG7060	6304
ECG7061	6305
ECG7062	6306
ECG7063	6307
ECG7064	6308
ECG7065	6309
ECG7066	6310
ECG7067	6311
ECG7068	6312
ECG7069	6313
ECG7070	6314
ECG7071	6315
ECG7072	6316
ECG7073	6317
ECG7074	6318
ECG7085	1266
ECG7104	OEM
ECG7105	6320
ECG7111	4321
ECG7116	OEM
ECG7214	5994

DEVICE TYPE	REPL CODE
ECG7400	0232
ECG7401	6337
ECG7402	0314
ECG7403	0335
ECG7404	0361
ECG7405	0385
ECG7406	1206
ECG7407	1339
ECG7408	0467
ECG7409	0492
ECG7410	0512
ECG7411	0527
ECG7412	2242
ECG7413	1441
ECG7414	2243
ECG7416	1349
ECG7417	1352
ECG7420	0697
ECG7421	1357
ECG7422	4603
ECG7423	3466
ECG7425	3477
ECG7426	0803
ECG7427	0818
ECG7428	4180
ECG7430	0873
ECG7432	0899
ECG7433	4193
ECG7437	3520
ECG7438	0996
ECG7439	5836
ECG7440	1024
ECG7441	1038
ECG7442	1052
ECG7443	1060
ECG7444	1072
ECG7445	1082
ECG7446	1098
ECG7447	1108
ECG7448	1125
ECG7450	0743
ECG7451	1168
ECG7453	1185
ECG7454	1202
ECG7460	1275
ECG7470	1404
ECG7472	1426
ECG7473	1172
ECG7474	1313
ECG7475	1432
ECG7476	1158
ECG7480	1535
ECG7481	3121
ECG7482	1574
ECG7483	0117
ECG7485	0374
ECG7486	1368
ECG7488A	4904
ECG7489	5455
ECG7490	1208
ECG7491	0980
ECG7492	0834
ECG7493A	0569
ECG7494	1702
ECG7495	1486
ECG7496	1715
ECG7497	2940
ECG8063	2937
ECG8070	2943
ECG8076	2944
ECG8080A	4547
ECG8081	2948
ECG8085	2951
ECG8090	2955
ECG8092	6391
ECG8096	2961
ECG8098	2379
ECG8103	2963
ECG8108	2968
ECG8115	2972
ECG8118	2973
ECG8123	5939
ECG8125	2976
ECG8139	2990
ECG8149	2998
ECG8167	3034
ECG8181	3036
ECG8212	1860
ECG8213	3000
ECG8214	6064
ECG8216	1863
ECG8219	6067
ECG8223	5951
ECG8224	1709
ECG8226	3013
ECG8228	1869
ECG8233	2074
ECG8234	2080
ECG8235	2087
ECG8242	3021
ECG8250	2187
ECG8255	0051
ECG8266	2275
ECG8301	2196
ECG8308	1070
ECG8309	5528
ECG8314	2488
ECG8316	5957
ECG8318	4145

DEVICE TYPE	REPL CODE
ECG8321	5958
ECG8328	5959
ECG8345M	OEM
ECG8368	5963
ECG8370	5964
ECG8374	5965
ECG8520	5975
ECG8542	5976
ECG8546	5977
ECG8552	6423
ECG8553	6425
ECG8554	6426
ECG8555	6427
ECG8556	5978
ECG8613	6091
ECG8853	1741
ECG9093	0358
ECG9094	1632
ECG9097	1481
ECG9099	0333
ECG9109	1350
ECG9110	2899
ECG9111	2959
ECG9112	1929
ECG9135	1619
ECG9157	1433
ECG9158	0466
ECG9200	4158
ECG9221	OEM
ECG9285	OEM
ECG9301	6467
ECG9302	6468
ECG9303	6469
ECG9304	6470
ECG9306	6471
ECG9307	6472
ECG9311	6474
ECG9312	6475
ECG9321	6478
ECG9322	6479
ECG9323	6480
ECG9324	6481
ECG9325	6482
ECG9326	6483
ECG9331	6484
ECG9332	6485
ECG9333	6486
ECG9334	6487
ECG9335	6488
ECG9342	6489
ECG9343	6490
ECG9347	6492
ECG9361	6495
ECG9362	6496
ECG9363	6497
ECG9367	6499
ECG9368	6500
ECG9370	1295
ECG9371	1416
ECG9372	1556
ECG9375	6501
ECG9380	6502
ECG9381	6503
ECG9382	6504
ECG9383	5028
ECG9390	6505
ECG9391	6506
ECG9392	6507
ECG9393	6508
ECG9394	6509
ECG9400	6510
ECG9401	4168
ECG9402	4162
ECG9403	1599
ECG9600	6476
ECG9601	2267
ECG9602	2286
ECG9615	5316
ECG9660	0618
ECG9661	2597
ECG9662	2563
ECG9663	2992
ECG9664	3032
ECG9665	3054
ECG9666	4142
ECG9667	3066
ECG9668	2434
ECG9669	3015
ECG9670	2350
ECG9671	4149
ECG9672	0078
ECG9673	0800
ECG9674	4161
ECG9675	4164
ECG9676	4170
ECG9677	4171
ECG9678	4173
ECG9679	4175
ECG9680	4178
ECG9681	3045
ECG9682	3039
ECG9683	4185
ECG9684	4187
ECG9685	4191
ECG9686	4195
ECG9688	4199
ECG9689	4201
ECG9690	4204
ECG9691	4207

DEVICE TYPE	REPL CODE
ECG9696	4221
ECG9800	0850
ECG9801	0874
ECG9802	0888
ECG9803	0900
ECG9804	0919
ECG9805	0929
ECG9806	0945
ECG9807	0962
ECG9808	0982
ECG9809	0998
ECG9810	1011
ECG9811	1025
ECG9812	1039
ECG9813	4538
ECG9814	4541
ECG9900	3418
ECG9903	3861
ECG9904	3833
ECG9905	3714
ECG9906	4234
ECG9907	4236
ECG9908	3845
ECG9909	3850
ECG9910	1257
ECG9911	1364
ECG9912	3897
ECG9913	3901
ECG9914	1069
ECG9915	3899
ECG9917	OEM
ECG9921	3337
ECG9923	OEM
ECG9924	2867
ECG9926	3924
ECG9927	4218
ECG9930	1823
ECG9931	3061
ECG9932	1041
ECG9933	2097
ECG9935	1176
ECG9936	1831
ECG9937	1835
ECG9941	2622
ECG9944	0033
ECG9945	0081
ECG9946	0141
ECG9948	3399
ECG9949	1844
ECG9950	2619
ECG9951	2078
ECG9961	1859
ECG9962	0562
ECG9963	0341
ECG9974	1810
ECG9976	3427
ECG9982	6572
ECG9989	3429
ECG9990	6575
ECG15005	6128
ECG15006	5356
ECG15007	5000
ECG15008	4629
ECG15009	5681
ECG15019	1807
ECG15020	3914
ECG15021	6129
ECG15022	6130
ECG15023	6132
ECG40085	3450
ECG40085B	3450
ECG40097	4441
ECG40097B	4441
ECG40098	0432
ECG40098B	0432
ECG40100B	3952
ECG40106B	3626
ECG40160	1359
ECG40160B	1359
ECG40161	1373
ECG40161B	1373
ECG40162	1388
ECG40162B	1388
ECG40163	1407
ECG40163B	1407
ECG40174	1551
ECG40174B	1551
ECG40175B	1528
ECG40182B	4660
ECG40192	1764
ECG40192B	1764
ECG40193	1776
ECG40193B	1776
ECG40194	1769
ECG40194B	1769
ECG40195	1784
ECG40195B	1784
ECG56004	0772
ECG56006	0744
ECG56008	0617
ECG56010	0875
ECG56014	6340
ECG56015	5323
ECG56016	5326
ECG56017	1659
ECG56018	1967
ECG56019	6139
ECG56020	5612
ECG56022	3199

DEVICE TYPE	REPL CODE
ECG56024	3207
ECG56026	3222
ECG56028	6152
ECG56030	5878
ECG56031	3708
ECG56033	6151
ECG65101	4604
ECG74107	0942
ECG74109	0968
ECG74110	0987
ECG74111	1002
ECG74121	0175
ECG74122	1139
ECG74123	1157
ECG74125	1182
ECG74126	1193
ECG74128	1219
ECG74132	1271
ECG74136	1316
ECG74141	1377
ECG74142	1390
ECG74143	1409
ECG74144	1418
ECG74145	0619
ECG74147	1451
ECG74150	1492
ECG74151	1495
ECG74152	1517
ECG74153	1539
ECG74154	1555
ECG74155	1576
ECG74156	1592
ECG74157	1605
ECG74158	4446
ECG74160	1631
ECG74161	1645
ECG74162	1013
ECG74163	1666
ECG74164	0734
ECG74165	1685
ECG74166	0231
ECG74170	1722
ECG74173	1766
ECG74174	1770
ECG74175	1787
ECG74176	1795
ECG74177	1803
ECG74178	1813
ECG74179	1820
ECG74180	1829
ECG74181	1842
ECG74182	1856
ECG74190	1912
ECG74191	1917
ECG74192	1921
ECG74193	1926
ECG74195	1943
ECG74196	1950
ECG74197	1956
ECG74198	1964
ECG74199	1971
ECG74221	2142
ECG74249	2290
ECG74251	2299
ECG74290	2612
ECG74293	2644
ECG74365	1459
ECG74366	1471
ECG74367	1488
ECG74368	1508
ECG74390	3240
ECG74393	3256
ECG74490	3664
ECG75188	0508
ECG75189	0511
ECG75450B	1231
ECG75451B	1244
ECG75452B	1262
ECG75453B	1272
ECG75454B	1289
ECG75491B	1729
ECG75492B	1740
ECG75493	5021
ECG75494	5022
ECG75497	5857
ECG75498	5024
ECG-HIDIV-1	2977
ECG-HIDIV-2	2978
ECG-HIDIV-3	2980
ECG-HIDIV-4	2979
ECG-HIDIV-12	5362
ECN2511SPV	OEM
ECR0510	0071
ECR405ST0911	OEM
ECR-600-2	0015
ECR600-2	0015
ECXB0061	OEM
ECX-B0063	OEM
ECX-B0085	OEM
ECX-B0085C	OEM
ECXB0087	OEM
ECX-B0095	OEM
ECXB0095	OEM
ECXB1026	OEM
ECXB1027	OEM
ECXB1028	OEM
ECXB1028B	OEM
ECX-B1040	OEM
ECXB1050C	OEM

DEVICE TYPE	REPL CODE
ECXB1060	OEM
ED0011	0057
ED0012	0057
ED0031	0283
ED0032	0133
ED0036	0091
ED0039	0686
ED0056	0283
ED0059	0133
ED0067	0485
ED0084A	0253
ED0084B	0253
ED0084C	0091
ED0084D	0471
ED01	OEM
ED01-06	OEM
ED03	OEM
ED05	OEM
ED0109E	0471
ED0111	0118
ED0114XT	0907
ED0116XT	0907
ED0120	0023
ED0147	0677
ED0152	OEM
ED0173	0071
ED0202	OEM
ED0214F	0471
ED0226A	0879
ED0278	0023
ED0328	0062
ED0353	0023
ED0366	0023
ED0372	0031
ED2(ELCOM)	0917
ED3(ELCOM)	0917
ED-3TV	OEM
ED-3TV1	OEM
ED-4	0015
ED4	0015
ED4(ELCOM)	0143
ED-5	0015
ED5	0015
ED-6	0015
ED6	0015
ED6(ELCOM)	0143
ED-6C3	OEM
ED6C3	OEM
ED-6C3R	OEM
ED6.2EB	0636
ED7	0015
ED7(ELCOM)	0917
ED-7TV	OEM
ED-7TV1	0201
ED8(ELCOM)	0087
ED-8H1	OEM
ED8H1	OEM
ED-8N1	OEM
ED8N1	OEM
ED-9	OEM
ED9	0143
ED9(ELCOM)	0143
ED10(ELCOM)	0123
ED10L	0102
ED-11	OEM
ED11	6460
ED11(ELCOM)	0123
ED11EB	0181
ED12(ELCOM)	0143
ED-13TV1	0288
ED-15	OEM
ED15	OEM
ED-15TV1	0288
ED-15TV3	0288
ED-16H1	0378
ED16H1	OEM
ED-16N1	OEM
ED16N1	OEM
ED20	OEM
ED-22TV1	OEM
ED24(ELCOM)	0293
ED-24H1	OEM
ED24H1	OEM
ED29(ELCOM)	0087
ED31(ELCOM)	0133
ED32(ELCOM)	0133
ED33(ELCOM)	0924
ED-45TV1	OEM
ED46	0143
ED50	OEM
ED50-01	OEM
ED50-02	OEM
ED50-03	OEM
ED51	0050
ED52	0280
ED53	0280
ED54B	0280
ED55	0004
ED56	0004
ED57	0004
ED-60	0143
ED60	0143
ED-75X1	OEM
ED75X1	OEM
ED95	OEM
ED-100X1	OEM
ED100X1	OEM
ED101R	OEM
ED101RG	OEM
ED102G	OEM
ED102GKB	OEM
ED102R	OEM
ED102RKB	OEM
ED102RKW	OEM
ED103G	OEM
ED103GK	OEM
ED103R	OEM
ED103RK	OEM
ED104R	OEM
ED104RK	OEM
ED105G	OEM
ED105GK	OEM
ED105R	OEM
ED105RK	OEM
ED106R	OEM
ED106RK	OEM
ED107G	OEM
ED107GK	OEM
ED107R	OEM
ED107RK	OEM
ED108G	OEM
ED108GKB	OEM
ED108GKW	OEM
ED108R	OEM
ED108RKB	OEM
ED108RKW	OEM
ED109R	OEM
ED109RKB	OEM
ED109RKW	OEM
ED110R	OEM
ED113G	OEM
ED113R	OEM
ED113Y	OEM
ED114G	OEM
ED114R	OEM
ED114Y	OEM
ED115G	OEM
ED115R	OEM
ED116R	OEM
ED120L	0201
ED120R	OEM
ED121G	OEM
ED121R	OEM
ED122G	OEM
ED122R	OEM
ED123G	OEM
ED123R	OEM
ED123Y	OEM
ED124G	OEM
ED124R	OEM
ED124Y	OEM
ED-125X1	OEM
ED125X1	OEM
ED-150X1	OEM
ED150X1	OEM
ED150Z	0016
ED202G	OEM
ED202R	OEM
ED209G	OEM
ED209R	OEM
ED209Y	OEM
ED216G	OEM
ED216R	OEM
ED216Y	OEM
ED220G	OEM
ED220R	OEM
ED220Y	OEM
ED226A	OEM
ED226G	OEM
ED226R	OEM
ED226Y	OEM
ED322	OEM
ED402	OEM
ED553G	OEM
ED553R	OEM
ED553Y	OEM
ED592K	0224
ED592M	0144
ED700A	OEM
ED700G	OEM
ED700H	OEM
ED7000	OEM
ED700P	OEM
ED700R	OEM
ED700Y	OEM
ED703A	OEM
ED703G	OEM
ED703H	OEM
ED7030	OEM
ED703P	OEM
ED703R	OEM
ED703Y	OEM
ED704A	OEM
ED704G	OEM
ED704H	OEM
ED7040	OEM
ED704P	OEM
ED704R	OEM
ED704Y	OEM
ED706A	OEM
ED706G	OEM
ED706H	OEM
ED7060	OEM
ED706P	OEM
ED706R	OEM
ED706Y	OEM
ED708H	OEM
ED710A	OEM
ED710G	OEM
ED710H	OEM
ED7100	OEM
ED710P	OEM
ED710R	OEM
ED710Y	OEM
ED713A	OEM
ED713G	OEM
ED713J	OEM
ED7130	OEM
ED713P	OEM
ED713R	OEM
ED713Y	OEM
ED714A	OEM
ED714G	OEM
ED714H	OEM
ED7140	OEM
ED714P	OEM
ED714R	OEM
ED714Y	OEM
ED716A	OEM
ED716G	OEM
ED716H	OEM
ED7160	OEM
ED716P	OEM
ED716R	OEM
ED716Y	OEM
ED718H	OEM
ED720A	OEM
ED720G	OEM
ED720H	OEM
ED7200	OEM
ED720P	OEM
ED720R	OEM
ED720Y	OEM
ED723A	OEM
ED723G	OEM
ED723H	OEM
ED7230	OEM
ED723P	OEM
ED723R	OEM
ED723Y	OEM
ED724A	OEM
ED724G	OEM
ED724H	OEM
ED7240	OEM
ED724P	OEM
ED724R	OEM
ED724Y	OEM
ED726A	OEM
ED726G	OEM
ED726H	OEM
ED7260	OEM
ED726P	OEM
ED726R	OEM
ED726Y	OEM
ED728H	OEM
ED740A	OEM
ED740G	OEM
ED740H	OEM
ED7400	OEM
ED740P	OEM
ED740R	OEM
ED740Y	OEM
ED743A	OEM
ED743G	OEM
ED743H	OEM
ED7430	OEM
ED743P	OEM
ED743R	OEM
ED743Y	OEM
ED744A	OEM
ED744G	OEM
ED744H	OEM
ED7440	OEM
ED744P	OEM
ED744R	OEM
ED744Y	OEM
ED746A	OEM
ED746G	OEM
ED746H	OEM
ED7460	OEM
ED746P	OEM
ED746R	OEM
ED746Y	OEM
ED748H	OEM
ED750A	OEM
ED750G	OEM
ED750H	OEM
ED7500	OEM
ED750P	OEM
ED750R	OEM
ED750Y	OEM
ED753A	OEM
ED753G	OEM
ED753H	OEM
ED7530	OEM
ED753P	OEM
ED753R	OEM
ED753Y	OEM
ED754A	OEM
ED754G	OEM
ED754H	OEM
ED7540	OEM
ED754P	OEM
ED754R	OEM
ED754Y	OEM
ED756A	OEM
ED756G	OEM
ED756H	OEM
ED7560	OEM
ED756P	OEM
ED756R	OEM
ED756Y	OEM
ED758H	OEM
ED1052	OEM
ED1104K	OEM
ED-1402	0016
ED1402	1954
ED1402A	0016
ED1402A/09-30506	0079
ED1402A/09-305066	0016
ED1402B	0016
ED1402C	0076
ED1402D	0079
ED1402E	0076
ED1502	2209
ED1502A	0224
ED1502B	0079
ED-1502C	0224
ED1502C	0127
ED1502D	0079
ED1502E	0127
ED1502F	0224
ED1502P	0224
ED1502R	0330
ED-1602	OEM
ED1602(C)	0037
ED1602A	2260
ED1602B	2260
ED1602C	0037
ED1602D	0037
ED1602E	0037
ED1702(L)	OEM
ED1702K	0079
ED1702L	0016
ED1702L/09-30506	0079
ED1702L/09-305068	0016
ED1702M	0076
ED1702N	0018
ED1704L	0079
ED1802-0	0037
ED1802-O	0150
ED1802C	0532
ED1802K	OEM
ED1802L	0342
ED1802M	0037
ED1802N	0532
ED1802N,M	0150
ED1804	0015
ED1892	0015
ED2106	0015
ED2107	0015
ED2108	0015
ED2109	0015
ED2110	0015
ED2200	OEM
ED2842	0015
ED2843	0015
ED2844	0015
ED2845	0015
ED2846	0015
ED2847	0071
ED2848	0071
ED2849	0071
ED2910	0071
ED2911	0071
ED2912	0071
ED2913	0071
ED2914	0015
ED2915	0015
ED2916	0015
ED2917	0015
ED2918	0015
ED2919	0015
ED2920	0015
ED2921	0015
ED2922	0071
ED2923	0071
ED2924	0071
ED3000	0015
ED3000A	0015
ED3000B	0015
ED3001	0015
ED3001A	0015
ED3001B	0015
ED3002	0015
ED3002A	0015
ED3002B	0015
ED3003	0015
ED3003A	0015
ED3003B	0015
ED3003S	0015
ED3004	0015
ED3004A	0015
ED3004B	0015
ED3005	0023
ED3005A	0015
ED3005B	0015
ED3006	0015
ED3006A	0015
ED3006B	0015
ED3007	0071
ED3007A	0071
ED3007B	0071
ED3008	0071
ED3008A	0071
ED3008B	0071
ED3009	0071
ED3009A	0071
ED3009B	0071
ED3010	0071
ED3010A	0071
ED3010B	0071
ED5053	OEM
ED5053G	OEM
ED5053R	OEM
ED5053Y	OEM
ED5290	OEM
ED8307	OEM
ED8308	OEM
ED8309	OEM
ED8310	OEM
ED8311	OEM
ED8312	OEM
ED219464	0143
ED224548	0015
ED224550	0015
ED269464	OEM
ED302379	0133
ED329128	0015
ED329130	0015
ED491130	1083
ED494583	0015
ED498150	0949
ED511097	0015
ED511918	0402
ED514721	0133
ED515790	0133
ED516420	0133
ED520762	OEM
ED536062	0133
ED560913	0133
ED703876	0124
ED703883	0210
ED703904	0023
EDF1A	1875
EDF1AM	OEM
EDF1B	1875
EDF1BM	OEM
EDF1C	1875
EDF1CM	OEM
EDF1D	1875
EDF1DM	OEM
EDF703882	0053
EDG-0001	0143
EDG-0003	0019
EDG-0006	0123
EDG-0014	0133
EDG-1	0143
EDG-3	0019
EDH2-30	OEM
EDH4-75	OEM
EDH4-100	OEM
EDH4-125	OEM
EDH4-150	OEM
EDH12	OEM
EDH18	OEM
EDH36	OEM
EDH36-1	OEM
EDH36-2	OEM
EDH36-3	OEM
EDJ-363	0015
EDMF-15B	0288
EDMF-25B	0378
EDMF25B	0378
EDS-0001	0133
EDS-0002	0023
EDS-0004	0023
EDS-0014	0133
EDS-0017	0015
EDS-0024	0023
EDS-0038	0139
EDS-0042	0030
EDS-1	0133
EDS-4	0023
EDS-11	0015
E-DS17	0023
EDS-17	0015
E-DS18	0023
EDS-31	0030
E-DS135D	0023
E-DS442	0133
E-DS442-VG1	0133
E-DS442VG1	0133
E-DS442YG1	0133
EDZ-0045	0057
EDZ-2	0649
EDZ-11	0420
EDZ-14	0244
EDZ-19	0091
EDZ-20	0025
EDZ-23	0012
EDZ-24	0244
EE5A-3.3	0292
EE5A-3.6	0188
EE5A-3.9	0456
EE5A-4.3	0533
EE5A-4.7	0451
EE5A-5.1	0162
EE5A-5.6	0157
EE5A-6.2	0636
EE5A-6.8	0025
EE5A-7.5	0649
EE5A-8.2	0244
EE5A-9.1	0012
EE5A-10	0170
EE5A-11	0317
EE5A-12G	0137
EE5A-14	0100
EE5A-15	0002
EE5A-20G	0531
EE5A-22	0565
EE5A-24	0402
EE5A-26	0441
EE5A-27	0441
EE5A-29	0726
EE5A-33G	0039
EE5A-36	0820
EE5A-39	0350
EE5A-43	0931
EE5A-47	0999
EE5A-51	0502
EE5A-55	1834
EE5A-56	0869
EE5A-62A	0783
EE5A-68	1268
EE5A-82P	0331
EE5A-91	1311
EE5A-110A	0149
EE5A-115	0794
EE5A-120	0186
EE5A-129	0365
EE5A-130	0213
EE5A-140	0245
EE5A-150	0028
EE5A-160	0255
EE5A-162	0420
EE5A-170	0877
EE5A-177	0495
EE5A-180	0367
EE5A-190	2858
EE5A-200	0421
EE5A-200G	0421
EE301	0259
EECG-1234	0155
EER600-2	0229
EE-SJ3	OEM
EE-SJ5	OEM
EE-SX315	OEM
EE-TP109	OEM
EF1(ELCOM)	0325
EF2(ELCOM)	0325
EF3	0325
EF4(ELCOM)	0212
EF5(ELCOM)	0212
EF68A00C	OEM
EF68A00CV	OEM
EF68A00P	OEM
EF68A00PV	OEM
EF68A10C	OEM
EF68A10CP	OEM
EF68A10CV	OEM
EF68A10P	OEM
EF68A21C	OEM
EF68A21CV	OEM
EF68A21PV	OEM
EF68A50C	OEM
EF68A50CP	OEM
EF68A50CV	OEM
EF68A50P	OEM
EF68A50PV	OEM
EF68A52C	OEM
EF68A52CV	OEM
EF68A52P	OEM
EF68A52PV	OEM
EF68B00C	OEM
EF68B00P	OEM
EF68B10C	OEM
EF68B10P	OEM
EF68B21C	OEM
EF68B21P	OEM
EF68B50C	OEM
EF68B50P	OEM
EF68B52C	OEM
EF68B52P	OEM
EF75CN1	OEM
EF75CN3	OEM
EF75M1	OEM
EF75M3	OEM
EF75N6	OEM
EF75N12	OEM
EF100	0071
EF125CN1	OEM
EF125CN3	OEM
EF125M1	OEM
EF125M3	OEM
EF125N6	OEM
EF125N12	OEM
EF150	OEM
EF175CN1	OEM
EF175CN3	OEM
EF175M1	OEM
EF175M3	OEM
EF175N6	OEM
EF175N12	OEM
EF200	0348
EF4116BC-15	OEM
EF4116BC-20	OEM
EF4116BC-25	OEM
EF4116BC-30	OEM
EF4116BJ-15	OEM
EF4116BJ-25	OEM
EF4116BJ-30	OEM
EF6800C	OEM
EF6800CMB	OEM
EF6800CV	OEM
EF6800P	OEM
EF6800PV	OEM
EF6802C	OEM
EF6802P	OEM
EF6805P2C	OEM
EF6805P2J	OEM
EF6805P2P	OEM
EF6810C	2086
EF6810CMB	OEM
EF6810CP	OEM
EF6810CV	OEM
EF6810P	2086
EF6821C	OEM
EF6821CMB	OEM
EF6821CV	OEM
EF6821P	OEM
EF6821PV	OEM
EF6850C	OEM
EF6850CMB	OEM
EF6850CP	OEM
EF6850CV	OEM
EF6850P	OEM
EF6850PV	OEM
EF6852C	OEM
EF6852CMB	OEM
EF6852CV	OEM
EF6852P	OEM
EF6852PV	OEM
EF9364AC	OEM
EF9364AP	OEM
EF9364BC	OEM
EF9364BP	OEM
EF68000	OEM
EFD13A	OEM
EFD13B	OEM
EFD13C	OEM
EFD13D	OEM
EFD13E	OEM
EFD13F	OEM
EFD13G	OEM
EFE13A	OEM
EFE13B	OEM
EFE13C	OEM
EFE13D	OEM
EFE13E	OEM
EFE13F	OEM
EFE13G	OEM
EFE14A	OEM
EFE14B	OEM
EFE14C	OEM
EFE14D	OEM
EFE14E	OEM
EFE14F	OEM
EFE14G	OEM
EFE15A	OEM
EFE15B	OEM
EFE15C	OEM
EFE15D	OEM
EFE15E	OEM
EFE15F	OEM
EFE15G	OEM
EFF13A	OEM
EFF13B	OEM
EFF13C	OEM
EFF13D	OEM
EFF13E	OEM
EFF13F	OEM
EFF13G	OEM
EFF14A	OEM
EFF14B	OEM
EFF14C	OEM
EFF14D	OEM
EFF14E	OEM
EFF14F	OEM
EFF14G	OEM
EFF15A	OEM
EFF15B	OEM
EFF15C	OEM
EFF15D	OEM
EFF15E	OEM
EFF15F	OEM
EFF15G	OEM
EFG13A	OEM
EFG13B	OEM
EFG13C	OEM
EFG13D	OEM
EFG13E	OEM
EFG13F	OEM
EFG13G	OEM
EFG14A	OEM
EFG14B	OEM
EFG14C	OEM
EFG14D	OEM
EFG14E	OEM
EFG14F	OEM
EFG14G	OEM
EFG15A	OEM
EFG15B	OEM
EFG15C	OEM
EFG15D	OEM
EFG15E	OEM
EFG15F	OEM
EFG15G	OEM
EFM-VE10	OEM
EFR1A	OEM
EFR1B	OEM
EFR1C	OEM
EFR1D	OEM

DEVICE TYPE	REPL CODE
EFR3A	OEM
EFR3B	OEM
EFR3C	OEM
EFR3D	OEM
EFR5A	OEM
EFR5B	OEM
EFR5C	OEM
EFR5D	OEM
EFR8A	OEM
EFR8B	OEM
EFR8C	OEM
EFR8D	OEM
EFS-16S1	OEM
EFS-32D1/D2	OEM
EFS-32U2	OEM
EFS-ADC1	OEM
EFS-DAC1	OEM
EFS-MPU2	OEM
EFS-MPU9	OEM
EFS-P8I0	OEM
EFS-P160	OEM
EFS-P161	OEM
EFS-PTIM	OEM
EFS-VIE1	OEM
EFS-VIG1	OEM
EFT-P064	OEM
EG-01	OEM
EG01	0795
EG01A	0015
EG01C	0017
EG01Y	1090
EG01Z	1090
EG1	0795
EG1A	0015
EG1Y	1090
EG1ZV	OEM
EG100	0015
EG100H	0015
EG200	OEM
EG250	OEM
EG300	OEM
EGL27A	OEM
EGL27B	OEM
EGL27C	OEM
EGL27D	OEM
EGL27F	OEM
EGL27G	OEM
EGL41A	OEM
EGL41B	OEM
EGL41C	OEM
EGL41D	OEM
EGL41F	OEM
EGL41G	OEM
EGP10A	0499
EGP10B	0499
EGP10C	0499
EGP10D	0499
EGP10F	0499
EGP10G	0499
EGP20A	0031
EGP20B	0031
EGP20C	0031
EGP20D	0031
EGP20F	0031
EGP20G	0031
EGP30A	0031
EGP30B	0031
EGP30C	0031
EGP30D	0031
EGP30F	0031
EGP30G	0031
EGP30GL	OEM
EGP30GL-6072	OEM
EGP50A	1362
EGP50B	1362
EGP50C	1362
EGP50D	1362
EGP50F	1362
EGP50G	1362
E-GZA5R1Z	0041
EH1	0023
EH-1A	0023
EH1A	0023
EH1Y1	OEM
EH-1Z	0023
EH1Z	0023
EH-1ZV	0023
EH2C11	0015
EH2C11/7+12/1	0015
EH250	OEM
EH300	OEM
EH350	OEM
EH400	OEM
E-HA12413	3816
EHD-AP4023S	OEM
EHD-HA6475	OEM
EHF4047B	OEM
EHF40163B	OEM
EHM312A12	OEM
EHM315S02	OEM
EHM-822A29	OEM
EHM-J12E94	OEM
EHM-K046W66(K)	OEM
EHM-K320E69	0679
EHM-K320F69	OEM
EHM-K720A72FB	OEM
EHM-K720A73FB	OEM
EHMK722P	OEM
EHM-K724F13S	OEM

DEVICE TYPE	REPL CODE
EHMK3342W92K	OEM
EHM-VCR0019	OEM
EHT1500B	OEM
EHT1700B	OEM
EHT1750B	OEM
EI1001	OEM
EI299700	3475
EI307200	4803
EI703843	2252
EI703907	4280
EICM-0060	1050
EICM-14	1994
EICM-19	2618
EIP	OEM
EJ350	OEM
EJ400	OEM
EJ450	OEM
EJ500	OEM
EJA17-25	0378
EK-03	0735
EK03	0735
EK03W	OEM
EK-04	1743
EK04	0735
EK04LF	0735
EK06W	OEM
EK13	0499
EK14	0499
EK14-F5	4299
EK136	0004
EK159	0280
EK450	OEM
EK500	OEM
EK600	OEM
EKB32-02	OEM
EL0018	OEM
EL0069	OEM
EL-1	OEM
EL1	OEM
EL1-Z	1090
EL1CL3	OEM
EL1F	OEM
EL1K3	OEM
EL1KL2	OEM
EL1KL3	OEM
EL1L1	OEM
EL-1L2	OEM
EL1L2	OEM
EL-1L3	3334
EL1ML2	OEM
EL1Z	1090
EL1Z-V1	0023
EL2	OEM
EL7L	OEM
EL23F	OEM
EL45F	OEM
EL74A	OEM
EL74HCT244	OEM
EL74HCT374	OEM
EL75	0224
EL119	OEM
EL131	0325
EL214	0086
EL221	OEM
EL231	0127
EL232	0016
EL238	0016
EL264	0126
EL270	OEM
EL271	OEM
EL401	0400
EL403	0076
EL434	0127
EL453	OEM
EL603	OEM
EL613	5701
EL627	0127
EL642	0079
EL645	OEM
EL-724	0402
EL850	OEM
E-LA1492	OEM
E-LA4192	2386
E-LB1423	OEM
ELI	OEM
ELIZ-V1	0124
ELM100	0280
ELM102	0211
ELM102A	0004
ELM103A	0208
ELM104	0085
ELM104MP	3033
ELM105	0440
ELM106	0150
ELM107	0127
ELM108	0144
ELM109	0143
ELM110MP	0123
ELM112	0917
ELM113	0196
ELM114	0484
ELM116	0015
ELM117	0015
ELM118	0774
ELM120	0474
ELM121	0160
ELM121MP	0265
ELM123	0198
ELM123A	0016

DEVICE TYPE	REPL CODE
ELM124	0142
ELM125	0071
ELM127	0975
ELM128	0086
ELM129	0126
ELM130	0103
ELM130MP	1514
ELM131	0222
ELM135	0162
ELM136	0157
ELM137	0636
ELM138	0649
ELM139	0012
ELM140	0170
ELM142	0137
ELM144	0100
ELM152	0042
ELM153	0854
ELM154	0233
ELM156	0087
ELM157	0276
ELM158	0164
ELM159	0037
ELM160	0050
ELM161	0007
ELM162	0074
ELM163	0642
ELM164	0003
ELM165	0065
ELM166	0277
ELM167	0289
ELM168	0106
ELM169	0787
ELM170	0787
ELM171	0285
ELM172A	0400
ELM175	0178
ELM177	0133
ELM180	1681
ELM181	0130
ELM182	0561
ELM183	1199
ELM184	0161
ELM185	0460
ELM186	0219
ELM187	1051
ELM188	0551
ELM189	0382
ELM190	0338
ELM191	0338
ELM192	0218
ELM193	0791
ELM194	0861
ELM195A	0698
ELM196	0423
ELM197	0854
ELM198	0168
ELM199	0111
ELM210	0566
ELM211	1367
ELM218	0905
ELM219	0491
ELM221	0353
ELM222	0212
ELM224	0630
ELM229	0224
ELM234	0693
ELM235	0936
ELM236	0836
ELM237	1411
ELM241	0561
ELM242	1199
ELM243	2427
ELM263	2235
ELM281	2013
ELM287	0715
ELM288	0342
ELM500A	0190
ELM502	0201
ELM503	0288
ELM504	0757
ELM505	0757
ELM506	0102
ELM513	1323
ELM519	0124
ELM5400	0901
ELM5404	0058
ELM5800	0965
ELM5804	0965
ELM5809	0087
ELM5830	0970
ELM5831	0816
ELM5842	0204
ELM5843	0816
ELM5850	0970
ELM5851	0816
ELM5862	0122
ELM5863	0816
ELM6408	2150
ELP3.3	0292
ELP3.6	0188
ELP3.9	0456
ELP4.3	0533
ELP4.7	0451
ELP5.6	0157
ELP6.2	0636
ELP6.8	0025
ELP7.5	0649
ELP8.2	0244

DEVICE TYPE	REPL CODE
ELP9.1	0012
ELP10	0170
ELP11	0317
ELP12G	0137
ELP14	0002
ELP15	0002
ELP20G	0531
ELP22	0565
ELP24	0402
ELP26	0441
ELP27	0441
ELP29	0726
ELP33G	0039
ELP36	0820
ELP39G	0350
ELP43	0931
ELP47	0999
ELP51	0502
ELP55	1834
ELP56	0869
ELP62C	0783
ELP68	1268
ELP82B	0331
ELP91	1311
ELP110	0149
ELP115G	0794
ELP120	0186
ELP129	0365
ELP130	0213
ELP140	0245
ELP150	0028
ELP160	0255
ELP170	0877
ELP178	0495
ELP180	0367
ELP190	2858
ELP200	0421
ELP200G	0421
EL-SKL	OEM
EL-SKLT	OEM
EM01	0015
EM01A	0072
EM01Z	0133
EM01ZW	0015
EM02B	4415
EM02BM	0071
EM02BMV	0071
EM02BMV0	0071
EM-1	0015
EM1	0015
EM1A	0072
EM1B	0071
EM1C	0017
EM1J2	0015
EM1Y	0609
EM-1Z	0609
EM1Z	0023
EM1Z1	0015
EM-1ZS	0609
EM1ZV	0609
EM2	0071
EM2A	0071
EM2B	0071
EM51	OEM
EM401	0015
EM402	0015
EM403	0015
EM404	0015
EM405	0015
EM406	0015
EM407	0071
EM408	0071
EM409	0071
EM410	0071
EM500	0015
EM501	0015
EM502	0015
EM503	0015
EM504	0015
EM505	0015
EM506	0015
EM507	0071
EM508	0071
EM509	0071
EM510	0071
EM513	0348
EM600	OEM
EM1021	0015
EM1095	OEM
EM1114	OEM
EM1114A	OEM
EM1114B	OEM
EM1114C	OEM
EM1137L	OEM
EM1138	OEM
EM1138M	OEM
EM1149	OEM
EM-1171	0015
EM1188	OEM
EM1502	OEM
EM1507	OEM
EM1508	OEM
EMI04700	OEM
EMK3342W98K	OEM
EMM-VCR0019	OEM
EMRL01-06	OEM
EMRL01-08	OEM
EMS-72272	0124
EN10	0079

DEVICE TYPE	REPL CODE
EN30	0079
EN40	0079
EN697	0016
EN706	0016
EN708	0016
EN718A	0144
EN722	0037
EN744	0016
EN870	0086
EN871	0861
EN914	0016
EN915	0079
EN916	0144
EN918	0144
EN930	0016
EN956	0016
EN1132	0037
EN1364	1170
EN1364/1231	1170
EN1613	0086
EN1711	0086
EN2219	0016
EN2222	0016
EN2369	OEM
EN2369A	0016
EN2484	0111
EN2894	0037
EN2894A	0037
EN2905	0126
EN2907	0037
EN3009	0016
EN3011	0016
EN3013	0016
EN3014	0016
EN3250	0037
EN3502	0126
EN3504	0037
EN3903	0076
EN3904	0016
EN3905	0037
EN3906	0006
EN3962	0037
EN11235	1429
EN11238	3282
EN11301	1170
EN11401	3351
EN11414	3359
EN11436	2122
EN11438	3089
ENC241D-10AS	OEM
ENC431D-10A	OEM
E0-44A	0211
E044A	0004
E065	0280
E066	0280
E067	0280
E068	0280
E070	0136
E0105	0280
E0704	0015
E0T521XXX	OEM
E0T522XXX	OEM
EP01C	0546
EP6X10	0133
EP6X11	0102
EP14X226	OEM
EP14X326	OEM
EP15X0045	OEM
EP15X0217	OEM
EP15X1	0079
EP15X2	0079
EP15X3	0111
EP15X4	0126
EP15X5	0086
EP15X6	0155
EP15X6(PNP)	2484
EP15X7	0016
EP15X8	0079
EP15X9	0079
EP15X10	0058
EP15X11	0423
EP15X12	0313
EP15X13	0791
EP15X14	0423
EP15X15	0925
EP15X16	0142
EP15X17	0037
EP15X18	0233
EP15X19	0392
EP15X20	0330
EP15X21	0791
EP15X22	0042
EP15X23	0925
EP15X24	1367
EP15X25	0566
EP15X26	0037
EP15X27	0285
EP15X28	0065
EP15X29	0130
EP15X30	0423
EP15X31	0095
EP15X32	0563
EP15X33	0086
EP15X34	1027
EP15X35	0142
EP15X36	0212
EP15X37	0127
EP15X38	0127
EP15X39	0079

DEVICE TYPE	REPL CODE
EP15X40	0212
EP15X41	0224
EP15X42	0224
EP15X43	0423
EP15X44	0854
EP15X45	0065
EP15X48	0203
EP15X50	0168
EP15X51	0791
EP15X52	0006
EP15X53	0037
EP15X54	0330
EP15X55	0144
EP15X57	0006
EP15X58	0210
EP15X60	0037
EP15X61	0285
EP15X64	0212
EP15X68	0042
EP15X83	0079
EP15X85	0079
EP15X86	0079
EP15X87	0079
EP15X88	0079
EP15X89	0079
EP15X90	0037
EP15X90(IC)	0624
EP15X91	0264
EP15X92	0076
EP15X93	0191
EP15X94	3689
EP15X94(PNP)	OEM
EP15X95	3342
EP15X96	0390
EP15X97	0042
EP15X98	0224
EP15X99	0127
EP15X101	0191
EP15X102	0423
EP15X104	3937
EP15X105	0392
EP15X106	0058
EP15X107	0037
EP15X108	2620
EP15X109	2797
EP15X110	0076
EP15X123	0392
EP15X124	3342
EP15X125	0127
EP15X126	0006
EP15X128	0006
EP15X141	0224
EP15X142	0236
EP15X143	0079
EP15X217	1992
EP15X218	0055
EP15X221	0824
EP15X222	3327
EP15X223	4591
EP15X224	5534
EP15X225	0006
EP15X226	0079
EP15X227	OEM
EP15X228	0824
EP15X231	OEM
EP15X232	OEM
EP15X233	OEM
EP15X244	OEM
EP15X288	0824
EP15X400	OEM
EP16X1	1149
EP16X2	0269
EP16X3	0019
EP16X4	0133
EP16X5	0002
EP16X6	0196
EP16X7	0086
EP16X8	0030
EP16X10	0102
EP16X11	0102
EP16X12	0649
EP16X13	0015
EP16X14	OEM
EP16X15	0133
EP16X16	0402
EP16X17	0028
EP16X18	OEM
EP16X19	0181
EP16X20	0133
EP16X21	0123
EP16X22	0133
EP16X23	0133
EP16X24	0017
EP16X25	0162
EP16X27	0124
EP16X29	0292
EP16X30	0062
EP16X32	0726
EP16X33	OEM
EP16X34	2518
EP16X36	0137
EP16X42	0917
EP16X49	0124
EP16X50	0124
EP16X51	0133
EP16X52	0686
EP16X53	0123
EP16X54	1329
EP16X58	0210

DEVICE TYPE	REPL CODE
EP16X59	0124
EP16X61	0471
EP16X66	0124
EP16X68	0002
EP16X69	0181
EP16X70	0077
EP16X71	0023
EP16X73	0451
EP16X74	0533
EP16X135	OEM
EP16X145	0023
EP16X147	0133
EP16X148	0820
EP16X149	0023
EP16X152	0041
EP16X153	0471
EP16X154	0023
EP16X155	0064
EP16X156	0441
EP16X157	1414
EP16X158	OEM
EP16X161	OEM
EP16X165	OEM
EP16X166	OEM
EP16X167	0140
EP16X217	OEM
EP20	0342
EP21G	OEM
EP21R	OEM
EP21Y	OEM
EP25	0037
EP35	0791
EP36X12	0649
EP41X1	OEM
EP41X14	OEM
EP41X68	OEM
EP41X70	OEM
EP57X1	0023
EP57X2	0757
EP57X3	1323
EP57X4	0023
EP57X5	0017
EP57X7	OEM
EP57X9	0102
EP57X10	0023
EP57X11	1145
EP57X12	0015
EP57X54	0015
EP57X156	OEM
EP62G	OEM
EP62R	OEM
EP62X41	1501
EP62X57	3868
EP62X61	1501
EP62X76	6446
EP62X84	6447
EP62X167	OEM
EP62Y	OEM
EP64G	OEM
EP64R	OEM
EP64Y	OEM
EP84X0035	OEM
EP84X0222	OEM
EP84X1	0852
EP84X2	0167
EP84X3	0666
EP84X4	0802
EP84X5	0802
EP84X6	0167
EP84X7	2549
EP84X8	0395
EP84X9	0666
EP84X10	0852
EP84X11	0697
EP84X12	0856
EP84X13	0232
EP84X14	OEM
EP84X15	OEM
EP84X16	OEM
EP84X17	OEM
EP84X18	0569
EP84X19	1157
EP84X20	2247
EP84X25	3921
EP84X30	1143
EP84X32	OEM
EP84X34	0069
EP84X35	1055
EP84X36	3735
EP84X37	0848
EP84X50	OEM
EP84X59	OEM
EP84X60	4549
EP84X62	5180
EP84X63	5655
EP84X64	OEM
EP84X65	OEM
EP84X66	OEM
EP84X67	5183
EP84X69	0360
EP84X71	2656
EP84X72	4390
EP84X73	0167
EP84X74	0878
EP84X76	OEM
EP84X77	OEM
EP84X78	1627
EP84X80	0918
EP84X81	1170

DEVICE TYPE	REPL CODE
EP84X82	2022
EP84X100	5181
EP84X101	OEM
EP84X102	OEM
EP84X103	OEM
EP84X104	OEM
EP84X105	OEM
EP84X106	OEM
EP84X107	OEM
EP84X108	OEM
EP84X109	3711
EP84X119	0625
EP84X143	OEM
EP84X154	0865
EP84X213	0865
EP84X214	2122
EP84X215	1201
EP84X216	2896
EP84X217	0334
EP84X220	OEM
EP84X221	2122
EP84X222	OEM
EP84X224	OEM
EP84X226	OEM
EP84X227	5509
EP84X228	OEM
EP84X229	3772
EP84X230	OEM
EP84X232	OEM
EP84X233	OEM
EP84X234	OEM
EP84X235	OEM
EP84X236	OEM
EP84X237	OEM
EP84X238	OEM
EP84X239	OEM
EP84X240	3626
EP84X241	3198
EP84X242	0513
EP84X243	OEM
EP84X244	OEM
EP84X245	OEM
EP84X246	OEM
EP84X250	4496
EP84X251	OEM
EP84X252	OEM
EP84X253	OEM
EP84X254	OEM
EP84X255	OEM
EP84X256	OEM
EP84X257	OEM
EP84X258	OEM
EP84X261	OEM
EP84X288	OEM
EP-100	0042
EP100	1367
EP-101A	0149
EP101A	0566
EP120	OEM
EP121	OEM
EP122	OEM
EP123	OEM
EP162	0420
EP200	0015
EP-276	0423
EP276	0423
EP400	0015
EP-421	0925
EP-422	0042
EP600	0015
EP-637	0264
EP-797	0042
EP800	0071
EP-801	0042
EP-802	0925
EP-943	0925
EP-944	0042
EP-976	0382
EP1000	0071
EP1014A	OEM
EP1014B	OEM
EP1014D	OEM
EP1014E	OEM
EP1014H	OEM
EP1016A	OEM
EP1016B	OEM
EP1016D	OEM
EP1016E	OEM
EP1016H	OEM
EP1017A	OEM
EP1017B	OEM
EP1017D	OEM
EP1017E	OEM
EP1017H	OEM
EP1018A	OEM
EP1018B	OEM
EP1018D	OEM
EP1018E	OEM
EP1018H	OEM
EP1019A	OEM
EP1019B	OEM
EP1019D	OEM
EP1019E	OEM
EP1019H	OEM
EP1020A	OEM
EP1020B	OEM
EP1020D	OEM
EP1020E	OEM
EP1020H	OEM
EP1022A	OEM
EP1022B	OEM
EP1022D	OEM
EP1022E	OEM
EP1022H	OEM
EP1024A	OEM
EP1024B	OEM
EP1024D	OEM
EP1024E	OEM
EP1024H	OEM
EP1026A	OEM
EP1026B	OEM
EP1026D	OEM
EP1026E	OEM
EP1026H	OEM
EP1028A	OEM
EP1028B	OEM
EP1028D	OEM
EP1028E	OEM
EP1028H	OEM
EP1030A	OEM
EP1030B	OEM
EP1030D	OEM
EP1030E	OEM
EP1030H	OEM
EP1032A	OEM
EP1032B	OEM
EP1032D	OEM
EP1032E	OEM
EP1032H	OEM
EP1034A	OEM
EP1034B	OEM
EP1034D	OEM
EP1034E	OEM
EP1034H	OEM
EP1036A	OEM
EP1036B	OEM
EP1036D	OEM
EP1036E	OEM
EP1036H	OEM
EP1038A	OEM
EP1038B	OEM
EP1038D	OEM
EP1038E	OEM
EP1038H	OEM
EP1259	0015
EP1259-2	0015
EP-1428-2H	0015
EP1428-2H	0015
EP1712A	OEM
EP1714A	OEM
EP1716A	OEM
EP1717A	OEM
EP1718A	OEM
EP1719A	OEM
EP1720A	OEM
EP1722A	OEM
EP1724A	OEM
EP1726A	OEM
EP1736A	OEM
EP-2798	0196
EP2798	0133
EP-3053	0042
EP-3149	0540
EP3149	0015
EP-3572	0196
EP-5219-1	1080
EP-5619-2/7628	0015
EP-5641-1	0288
EP-5641H-2	0378
EP5641H-2	0378
EP5641H2	0378
EPB22-15	0102
EPG20G	0031
EPG30GL-6072	OEM
EPM60/12	OEM
EPM60/15	OEM
EPM100/12	OEM
EPM100/15	OEM
EPM200/12	OEM
EPM200/15	OEM
EPM300/12	OEM
EPM300/15	OEM
EPR9	0012
EPROM-32	OEM
EPROM-64	OEM
EPROM306/E-6	OEM
EPROM332/E-6	OEM
EPROM-332-200	OEM
EPROM-332-300	OEM
EPW29	OEM
EPX2	0079
EPX15X17	0037
EPX59	0124
EPX68	0002
EQ01-07RE	0025
EQ-09R	0012
EQ1	2950
EQA01-01	0012
EQA01-05	0041
EQA01-05R	0140
EQA01-05S	0041
EQA01-05T	0041
EQA01-06	0091
EQA01-06D	OEM
EQA01-06R	0253
EQA01-06S	0091
EQA01-06S13	0091
EQA01-06SB	0091
EQA01-06SC	0091
EQA01-06SD	0301
EQA01-06T	0471
EQA01-07	0062
EQA01-07R	0062
EQA01-07R2	3465
EQA01-07RE	0025
EQA01-07S	0062
EQA01-08	0077
EQA01-08A	0077
EQA01-08D	OEM
EQA01-08R	0077
EQA01-08RG	OEM
EQA01-08S	0165
EQA01-08SF	OEM
EQA01-09	0322
EQA01-09(R)	0012
EQA01-09C	0057
EQA01-09R	0322
EQA01-09S	0057
EQA01-09S1	0057
EQA01-067	0636
EQA01-072	0649
EQA01-6S	0091
EQA01-10	0064
EQA01-10B	OEM
EQA01-10R	0537
EQA01-10R2	6243
EQA01-10S	0064
EQA01-11	0181
EQA01-11R	0317
EQA01-11S	0181
EQA01-11SB	OEM
EQA01-11SE	0317
EQA01-11SV	0181
EQA01-11Z	5186
EQA01-12	0052
EQA01-12M	OEM
EQA01-12R	1005
EQA01-12R1A	OEM
EQA01-12S	0052
EQA01-12S1	0052
EQA01-12Z	0137
EQA01-13	0053
EQA01-13R	0053
EQA01-13R2	OEM
EQA01-14	0879
EQA01-14R	0879
EQA01-14RD	0100
EQA01-15	0686
EQA01-15R	0686
EQA01-15RL	0686
EQA01-15RLV	0686
EQA01-16	0445
EQA01-16R	0445
EQA01-17	0210
EQA01-17R	0210
EQA01-18	0375
EQA01-18A	0375
EQA01-18R	0375
EQA01-19	0671
EQA01-19R	0671
EQA01-20	0700
EQA01-20R	0700
EQA01-20RB	0531
EQA01-21	0705
EQA01-21R	0705
EQA01-22	0705
EQA01-22R	0705
EQA01-23AC	OEM
EQA01-24	0494
EQA01-24R	0494
EQA01-24RR	0402
EQA01-25	0714
EQA01-25R	0714
EQA01-26	0455
EQA01-26R	0455
EQA01-28	0257
EQA01-28R	0257
EQA01-28R2	0257
EQA01-30	0195
EQA01-30R	0195
EQA01-32	0195
EQA01-32R	0195
EQA01-33	0166
EQA01-33F	0166
EQA01-33R	0166
EQA01-33RF	0166
EQA01-35	0010
EQA01-35R	0010
EQA01-35RC	0010
EQA01-115	0317
EQA01-Z4RA	0726
EQA02-03M	OEM
EQA02-05	0041
EQA02-05(C)	0041
EQA02-05(D)	0041
EQA02-05-A	0041
EQA02-05-B	0041
EQA02-05-D	0041
EQA02-05A	0140
EQA02-05AB	0140
EQA02-05B	0140
EQA02-05C	0041
EQA02-05CD	0041
EQA02-05CDB	0041
EQA02-05D	0041
EQA02-05E	0041
EQA02-05EF	0041
EQA02-05F	0041
EQA02-05M1	0041
EQA02-06	0091
EQA02-06(A)	0253
EQA02-06(B)	0253
EQA02-06(C)	0091
EQA02-06(CDE)	0091
EQA02-06(D)	0091
EQA02-06-A	0253
EQA02-06-B	0253
EQA02-06A	0253
EQA02-06AB	0253
EQA02-06B	0253
EQA02-06C	0091
EQA02-06CD	0091
EQA-02-06D	0091
EQA02-06D	0471
EQA02-06D-TAPG	0091
EQA02-06E	0471
EQA02-06EF	0471
EQA02-06F	0471
EQA02-06M2	OEM
EQA02-06M4	OEM
EQA02-06M6	0091
EQA02-06S	0091
EQA02-07	0062
EQA02-07A	0062
EQA02-07AB	0062
EQA02-07ABD	0062
EQA02-07B	0062
EQA02-07C	0062
EQA02-07CD	0062
EQA02-07CDA	0077
EQA02-07D	0077
EQA02-07M6	0062
EQA02-07Z	OEM
EQA02-08	0165
EQA02-08(D)	0077
EQA02-08A	0077
EQA02-08AB	0077
EQA02-08B	0077
EQA02-08C	0165
EQA02-08CD	0165
EQA02-08D	0077
EQA02-08M3	0077
EQA02-08S	OEM
EQA02-08S3	0165
EQA02-09	0057
EQA02-09(C)	0057
EQA02-09(D)	0057
EQA02-09-C0	0057
EQA02-09-CD	0057
EQA02-09A	0057
EQA02-09AB	0322
EQA02-09AV	0057
EQA02-09B	0057
EQA02-09B(1%)	OEM
EQA02-09C	0057
EQA02-09CD	0057
EQA02-09D	0057
EQA02-10	0064
EQA02-10(A)	0064
EQA02-10(B)	0064
EQA02-10(C)	0064
EQA02-10-A	0064
EQA02-10-B	0064
EQA02-10-C	0064
EQA02-10-CD	0064
EQA02-10-D	0064
EQA02-10A	0064
EQA02-10AB	0064
EQA02-10B	0064
EQA02-10C	0064
EQA02-10CD	0064
EQA02-10CDA	0064
EQA02-10D	0064
EQA02-11	0181
EQA02-11A	0181
EQA02-11AB	0181
EQA02-11B	0052
EQA02-11C	0052
EQA02-11CD	0052
EQA02-11CDB	0052
EQA02-11D	0181
EQA02-11N4N	OEM
EQA02-11Z	OEM
EQA02-12	0052
EQA02-12A	0053
EQA02-12AB	0053
EQA02-12B	0053
EQA02-12M	OEM
EQA02-12M1	0052
EQA02-12Z	OEM
EQA02-13	0053
EQA02-13-A	0053
EQA02-13-B	0053
EQA02-13A	0053
EQA02-13AB	0053
EQA02-13B	6756
EQA02-13M2	OEM
EQA02-14	0879
EQA02-14-A	0879
EQA02-14-AB	0686
EQA02-14-B	0879
EQA02-14A	0879
EQA02-14AB	0879
EQA02-14B	0879
EQA02-15	0686
EQA02-15A	0686
EQA02-15AB	0686
EQA02-15M	0686
EQA02-16	0445
EQA02-16A	0445
EQA02-16AB	0445
EQA02-16B	1229
EQA02-17	0210
EQA02-17AB	0210
EQA02-17B	4848
EQA02-17M2	OEM
EQA02-18	0375
EQA02-18A	0375
EQA02-18AB	0671
EQA02-18B	5252
EQA02-20	0700
EQA02-20AB	0700
EQA02-20B	0700
EQA02-21	0705
EQA02-21AB	0705
EQA02-22	0705
EQA02-22AB	0705
EQA02-23	0494
EQA02-23(A)	0494
EQA02-23AC	OEM
EQA02-23ACA	0494
EQA02-25	0714
EQA02-25AC	0714
EQA02-25C	0714
EQA02-28	0257
EQA02-28AC	0257
EQA02-28B	OEM
EQA02-28C	0257
EQA02-30	0195
EQA02-30A	0195
EQA02-30AC	0195
EQA02-30B	0195
EQA02-30C	0166
EQA02-32	0166
EQA02-32AC	0166
EQA02-32B	0166
EQA02-33	0166
EQA02-33A	0166
EQA02-33AC	0166
EQA02-33B	0166
EQA02-35	0010
EQA02-35B	0010
EQA02-35R3	0010
EQA03-05A	OEM
EQA03-05B	0041
EQA03-05C	OEM
EQA03-05D	OEM
EQA03-05E	OEM
EQA03-05F	OEM
EQA03-06A	OEM
EQA03-06B	OEM
EQA03-06C	0091
EQA03-06D	OEM
EQA03-06E	OEM
EQA03-06F	OEM
EQA03-07A	OEM
EQA03-07B	OEM
EQA03-07C	OEM
EQA03-07D	OEM
EQA03-08A	OEM
EQA03-08B	OEM
EQA03-08C	OEM
EQA03-08D	OEM
EQA03-09A	0322
EQA03-09B	OEM
EQA03-09C	OEM
EQA03-09D	OEM
EQA03-10A	OEM
EQA03-10B	OEM
EQA03-10C	OEM
EQA03-10D	OEM
EQA03-11A	0181
EQA03-11B	OEM
EQA03-11C	OEM
EQA03-11D	OEM
EQA03-12A	OEM
EQA03-12B	OEM
EQA03-13A	OEM
EQA03-13B	OEM
EQA03-14A	OEM
EQA03-14B	OEM
EQA03-15A	OEM
EQA03-15B	OEM
EQA03-16A	0445
EQA03-16B	OEM
EQA03-17A	OEM
EQA03-17B	OEM
EQA03-18A	OEM
EQA03-18B	OEM
EQA03-20A	OEM
EQA03-20B	OEM
EQA03-21A	OEM
EQA03-21B	OEM
EQA03-22A	OEM
EQA03-22B	OEM
EQA03-23A	OEM
EQA03-23B	OEM
EQA03-23C	OEM
EQA03-25A	OEM
EQA03-25B	OEM
EQA03-25C	OEM
EQA03-28A	OEM
EQA03-28B	OEM
EQA03-28C	OEM
EQA03-30A	OEM
EQA03-30B	OEM
EQA03-30C	OEM
EQA03-32A	OEM
EQA03-32B	OEM
EQA03-32C	OEM
EQA03-33A	OEM
EQA03-33B	OEM
EQA03-33C	OEM
EQA03-35A	OEM
EQA03-35B	OEM
EQA03-35C	OEM
EQA09R	0012
EQA010ST	0157
EQA0105T	0157
EQA0106R	0253
EQA0106T	0471
EQA0107	0062
EQA0108	0077
EQA0108R	0077
EQA-0109S	0012
EQA0111	0181
EQA0116	0445
EQA0205	0041
EQA0205A	0041
EQA0205B	0041
EQA0206	0091
EQA0206C	0471
EQA0207	0062
EQA0207D	0077
EQA0209CD	0057
EQA0210(B)	0064
EQA0210A	0064
EQA0210B	0064
EQA0210C	0064
EQA0210D	0064
EQA0212AB	0052
EQA0213	0053
EQA0213A	0053
EQA0213B	0053
EQA0214	0879
EQA0214A	0879
EQA0214B	0879
EQA0220	0700
EQA0305B	0041
EQA0305C	OEM
EQA0306C	0091
EQA0308AB	4860
EQA0309A	0322
EQA0311A	0181
EQA0313AB	4964
EQA0313B	0471
EQA0313BV3	4964
EQA0313BV5	4964
EQA0333A	OEM
EQA-9	0012
EQA107R	0025
EQA107RE	0025
EQA120R	0531
EQA01-06	OEM
EQA01-06R	OEM
EQA01-24RR	OEM
EQAQ1-35RC	0320
EQA.9	0012
EQB0-06V	0301
EQB01	0137
EQB01-05	0162
EQB01-05SA	0162
EQB01-05TV	OEM
EQB01-06	0301
EQB01-06V	0301
EQB01-06Y	0301
EQB01-07	0649
EQB01-07S	OEM
EQB01-08	0244
EQB01-08S1	0244
EQB01-09	1083
EQB01-09B	1083
EQB01-011Z	0317
EQB01-095	OEM
EQB01-1R50	OEM
EQB01-2R25	OEM
EQB01-3R00	1713
EQB01-3R75	OEM
EQB01-4R50	OEM
EQB01-10	0170
EQB01-11	0317
EQB01-11Q	0317
EQB01-11Q(ZD)	0317
EQB01-11QV	0317
EQB01-11Z	0317
EQB01-12	0137
EQB01-12A	0137
EQB01-12B	0137
EQB01-12BV	0137
EQB01-12M	0137
EQB01-12R	0137
EQB01-12R1V	OEM
EQB01-12Z	0137
EQB01-13	0365
EQB01-14	0100
EQB01-15	0002
EQB01-15Z	0002
EQB01-15ZB	0002
EQB01-16	0420
EQB01-17	1649
EQB01-18	0495
EQB01-19	0949
EQB01-20	0531
EQB01-21	0565
EQB01-22	0565
EQB01-24	0402
EQB01-24RR	0402
EQB01-25	1606
EQB01-26	0441
EQB01-28	1674
EQB01-30	0726
EQB01-32	0039
EQB01-33	0039
EQB01-35	0820
EQB01-35E	OEM
EQB01-50	0502
EQB01-90S	0012
EQB01-100	0098
EQB01-112	3930
EQB01-150	0028
EQB-0108	0244
EQB11-50	0092
EQB35E	OEM
EQB01-06V	OEM
EQB01-07	OEM
EQB01-10	0170
EQC01-10	0087
EQC04-02	0071
EQD01-09	0012
EQF-0004	3717
EQF-0009	3717
EQF-3	0325
EQF-4	0688
EQG01-12A	0137
EQG-6	0160
EQG-8	0160
EQG-9	0004
EQG-12A	0137
EQG-15	0004
EQG-20	0280
EQH-1	0208
EQH-20	0600
EQR-0016	0006
EQR-0038	0006
EQR-1	0150
EQS-0018	0151
EQS-0061	0076
EQS-0100	0076
EQS-0102	1221
EQS-0159	0836
EQS-0160	0839
EQS-0165	0155
EQS-0184	1946
EQS-0192	0191
EQS-0195	0368
EQS-0196	0368
EQS-0198	0224
EQS-5	0151
EQS-9	0191
EQS-10	0111
EQS-11	0111
EQS-18	0151
EQS-19	0475
EQS-20	0475
EQS-21	0144
EQS-22	0155
EQS-56	4208
EQS-57	0698
EQS-60	1411
EQS-61	0076
EQS-62	0079
EQS-64	0079
EQS-66	0560
EQS-67	0560
EQS-78	1221
EQS-86	1908
EQS-89	0390
EQS-100	0076
EQS131	0076
EQS-139	2209
EQS-141	0836
EQZ02-06D	6260
ER007-15-L	OEM
ER0082	OEM
ER1	0023
ER2	0023
ER2-06	OEM
ER4	0023
ER5(ELCOM)	0474
ER6	2527
ER7(ELCOM)	0190
ER8	OEM
ER10(ELCOM)	5514
ER10R(ELCOM)	1238
ER11	0015
ER11(ELCOM)	5582
ER11R(ELCOM)	1240
ER12	0015
ER12(ELCOM)	2900
ER12R(ELCOM)	1241
ER13(ELCOM)	1245
ER13R(ELCOM)	1245
ER14(ELCOM)	3751
ER14R(ELCOM)	2365
ER15(ELCOM)	3757
ER15R(ELCOM)	1253
ER15X4	0050
ER15X5	0050
ER15X6	0050
ER15X7	0004
ER15X8	0016
ER15X9	0004
ER15X10	0160

DEVICE TYPE	REPL CODE
ER15X11	0050
ER15X12	0050
ER15X13	0050
ER15X14	0050
ER15X15	0050
ER15X16	0050
ER15X17	0004
ER15X18	0016
ER15X19	0050
ER15X20	0050
ER15X21	0050
ER15X22	0004
ER15X23	0004
ER15X24	0136
ER15X25	0136
ER15X26	0280
ER16(ELCOM)	1112
ER17(ELCOM)	3768
ER17R(ELCOM)	1264
ER18(ELCOM)	3011
ER18R(ELCOM)	0449
ER20(ELCOM)	3189
ER20R(ELCOM)	1630
ER21	0015
ER21(ELCOM)	1124
ER21R(ELCOM)	0254
ER22	0015
ER22(ELCOM)	1124
ER22R(ELCOM)	1107
ER23(ELCOM)	1126
ER23R(ELCOM)	1111
ER24(ELCOM)	0805
ER24R(ELCOM)	0258
ER25(ELCOM)	1195
ER25R(ELCOM)	0267
ER26(ELCOM)	0319
ER26R(ELCOM)	0267
ER27(ELCOM)	1132
ER28(ELCOM)	0044
ER28R(ELCOM)	0281
ER31	0015
ER32(ELCOM)	0190
ER41	0015
ER42	0015
ER51	0015
ER57X2	0015
ER57X3	0015
ER57X4	0015
ER61	0015
ER62	0015
ER81	0071
ER101	0015
ER102	0229
ER102D	0015
ER103D	0015
ER103E	0015
ER104D	0015
ER105D	0015
ER106D	0015
ER107D	0071
ER108D	0071
ER181	0015
ER182	0015
ER183	0015
ER184	0015
ER185	0015
ER186	0071
ER187	0071
ER201	0015
ER-300	4304
ER301	0015
ER308	0071
ER310	0071
ER312	OEM
ER381	0015
ER401	0015
ER402	0015
ER403	0015
ER404	0015
ER405	0015
ER406	0071
ER407	0071
ER410	0015
ER442	0348
ER443	0348
ER444	0348
ER501	0015
ER510	0015
ER601	0015
ER801	0071
ER881-004	OEM
ER1001	0071
ER1075	OEM
ER1400	1846
ER2000	0031
ER2001	0031
ER2002	0031
ER2003	0031
ER2004	0031
ER2005	1217
ER2006	OEM
ER2051	OEM
ER2051HR	OEM
ER2051IR	OEM
ER2055	OEM
ER2055HR	OEM
ER2055IR	OEM
ER2810HR	OEM
ER2810IR	OEM
ER3400	OEM
ER3400HR	OEM
ER3400I	OEM
ER3400IR	OEM
ER5304(A)	OEM
ER5716	OEM
ER5716HR(A)	OEM
ER5816(A)	OEM
ER5816HR(A)	OEM
ER5816IR(A)	OEM
ER5901(A)	OEM
ER5901HR(A)	OEM
ER5901IR(A)	OEM
ER5916(A)	OEM
ER5916HR(A)	OEM
ER5916IR(A)	OEM
ER59256	OEM
ERA01-35RC	0365
ERA0106R	0253
ERA15-01	0023
ERA15-01AVRB	OEM
ERA15-01V3	0023
ERA15-01V5	OEM
ERA15-02	0023
ERA15-02A	OEM
ERA15-02KFRB	0023
ERA15-02V3	0015
ERA15-02VH	OEM
ERA15-02VH-T	0023
ERA15-02Y	OEM
ERA15-04	0023
ERA15-04G1	0023
ERA15-06	0023
ERA15-08	0017
ERA15-10	0017
ERA18-02	0023
ERA18-04	0023
ERA18-04V3	0023
ERA21-02	0609
ERA21-04	0795
ERA21-06	0015
ERA22-02	0609
ERA22-04	0795
ERA22-04FRB	0795
ERA22-04V	0795
ERA22-04V3	0795
ERA22-06	0015
ERA22-08	0072
ERA22-10	0017
ERA32-01	1090
ERA34-10	OEM
ERA38-04	0015
ERA38-05	0015
ERA38-06	OEM
ERA48-02	0023
ERA48-04	0023
ERA81-004	0023
ERA81-004G	0735
ERA81-005	OEM
ERA82-004	0499
ERA82-005	OEM
ERA83-006	OEM
ERA84-009	0023
ERA85-009	OEM
ERA91-02	1090
ERA1501	0023
ERA1501-01V3	0023
ERA1502	5473
ERA1804	0023
ERA2202	0023
ERA2204	0023
ERA2206	0271
ERA4804	OEM
ERA81004	0735
ERA81004G	0735
ERB01-01	0015
ERB01-05	0162
ERB01-07	0025
ERB01-10	0071
ERB02-10	0102
ERB02-12	0102
ERB06-13	0017
ERB06-15	0017
ERB06-15V	0017
ERB-06B	0015
ERB-07RE	0025
ERB0615V	0017
ERB2P-04	0023
ERB6.2B2	OEM
ERB-6.2EB2	0471
ERB11-01	0015
ERB-12-01	0080
ERB12-01	0023
ERB-12-01G	0015
ERB12-01G	0015
ERB12-01R	0023
ERB-12-01RK	0609
ERB12-01RK	0023
ERB12-01U4	0023
ERB12-01V	0080
ERB-12-02	0124
ERB12-02	0023
ERB-12-02R	0015
ERB12-02RK	0023
ERB12-02S	OEM
ERB12-02V	0015
ERB12-04	0071
ERB12-04RK	0023
ERB12-06	0071
ERB12-10	0071
ERB12-11	0015
ERB15	4240
ERB16-08	0102
ERB16-12	0102
ERB21-10	0102
ERB21-12	0102
ERB21-15	0102
ERB22-10	0102
ERB22-12	0102
ERB22-15	0015
ERB24-02B	0102
ERB24-02D	0023
ERB24-04	0023
ERB24-04A	0023
ERB24-04B	0023
ERB-24-04C	0023
ERB24-04C	0023
ERB24-04CV	0023
ERB24-04D	0023
ERB24-06	0023
ERB24-06A	0015
ERB24-06B	0023
ERB24-06B-D	OEM
ERB24-06C	0023
ERB24-06D	0102
ERB26-15	0348
ERB26-20	0348
ERB26-20L	0348
ERB26-20M	0348
ERB26-20MV	0348
ERB28-02	0023
ERB28-04	0023
ERB28-04B	0023
ERB28-04C	0023
ERB-28-04D	0023
ERB28-04D	0023
ERB28-04DV	0071
ERB28-06	2527
ERB28-06D	0023
ERB28-06DV	0023
ERB29-02	0023
ERB29-04	0023
ERB30-13	0102
ERB30-15	0182
ERB32-01	0546
ERB32-01L3	0546
ERB35-02	0609
ERB38-05	OEM
ERB41-04	OEM
ERB42-04	OEM
ERB43	0023
ERB43-02	0023
ERB43-04	0023
ERB43-04G	0023
ERB43-04V	0023
ERB43-08	0017
ERB44-02	0023
ERB44-04	0023
ERB44-04L3	0023
ERB44-04V	0023
ERB44-06	0023
ERB44-08	0017
ERB44-6	0023
ERB44-10	0071
ERB44-D6E	OEM
ERB47-02	0609
ERB-81	2542
ERB81	0071
ERB81-004	2542
ERB83-004	0031
ERB84-009	0031
ERB91-02	1090
ERB93-02	0546
ERB1201	6624
ERB1201U4	0015
ERB1202	0071
ERB2404C	0023
ERB4302	0023
ERB4304	0023
ERB4304G	0023
ERB4304V	0023
ERB4404V	0023
ERB4702	0609
ERB5104	OEM
ERB81004	2542
ERC01-005	OEM
ERC-01-02	0071
ERC01-02	0564
ERC01-02E	OEM
ERC01-02F	0071
ERC01-02FL	0031
ERC01-02FL12	0071
ERC01-02L	0564
ERC01-02L-12	0071
ERC01-02L12	0071
ERC01-04	0071
ERC01-04F	0087
ERC01-06	0564
ERC01-06C	OEM
ERC01-06E	OEM
ERC01-10	0087
ERC03-01	0871
ERC03-03	0853
ERC04-02	0071
ERC04-02E3	0071
ERC04-02F	0564
ERC04-02L	OEM
ERC04-04	0564
ERC04-04C	0071
ERC04-04F	0071
ERC04-06	0564
ERC04-10	0071
ERC05-06	0564
ERC05-08	0071
ERC05-08V	0071
ERC05-10	0071
ERC05-10B	OEM
ERC06-13	0102
ERC06-15	0017
ERC06-15L	0017
ERC06-15S	0348
ERC0102L	0071
ERC0402	0071
ERC0402E3	0071
ERC0508	0071
ERC0508V	0071
ERC0510	0071
ERC0510V	0071
ERC0615	0348
ERC10ZK241U	OEM
ERC12-08	0071
ERC13-08	0071
ERC13-08V3	OEM
ERC16-06	0023
ERC16-10	0087
ERC16-12	3241
ERC20-02	1145
ERC20-04	1145
ERC20-06	1664
ERC20-08	OEM
ERC20M-04	1179
ERC24-04	0023
ERC24-06	0023
ERC24-06S	0023
ERC25-04	0283
ERC25-06	0283
ERC25-06M	OEM
ERC25-06S	0023
ERC26-13	0102
ERC-26-13L	0348
ERC26-13L	0348
ERC26-15	0182
ERC-26-15S	0102
ERC26-15S	0158
ERC26-15T	0102
ERC26-131	0102
ERC27-13	0102
ERC27-15	0182
ERC29-02	OEM
ERC29-04	OEM
ERC30-01	0546
ERC30-02K	OEM
ERC33-02	1090
ERC35-02	0031
ERC38-04	1362
ERC38-05	1372
ERC38-06	3887
ERC44-04	0023
ERC46-02	0031
ERC46-04	0031
ERC47-02	0031
ERC47-04	0031
ERC48-02	OEM
ERC62-004	OEM
ERC80-004	OEM
ERC81-004	0031
ERC84-009	3604
ERC88-009	OEM
ERC90-02	6119
ERC91-02	0546
ERC510	0071
ERC1208	0071
ERC1308	0071
ERC2902	OEM
ERC4702	0031
ERC81004	0087
ERCC4-06	0071
ERD02-06	OEM
ERD03-02	0564
ERD03-02J	OEM
ERD03-04	0540
ERD07-13	0182
ERD07-15	0182
ERD07-15-L	OEM
ERD07-15L	0102
ERD08M-15	OEM
ERD0715	0182
ERD18-08S	OEM
ERD24-005	1600
ERD24-01	0871
ERD24-02	0871
ERD24-04	3139
ERD24-06	2967
ERD27-10	0550
ERD28-04	0031
ERD28-04S	0031
ERD28-06	0031
ERD28-08	0102
ERD28-08S	0102
ERD29-02	0031
ERD29-02H3	0031
ERD29-04	0031
ERD29-04H2	0564
ERD29-06	0031
ERD29-08J	1217
ERD32-01	0546
ERD32-01L22	OEM
ERD33-02	0546
ERD47-02	0031
ERD51-01	0097
ERD51-03	0109
ERD51-06	0122
ERD51-09	0145
ERD51-12	OEM
ERD74-005	1600
ERD74-01	0871
ERD74-02	0871
ERD74-04	2153
ERD74-06	0711
ERD75-005	4998
ERD75-01	4998
ERD75-02	4998
ERD77-10	OEM
ERD80-004	OEM
ERD81-004	0615
ERD81B004	OEM
ERD100	0015
ERD200	0015
ERD300	0015
ERD400	0015
ERD500	0015
ERD600	0015
ERD700	0071
ERD800	0071
ERD900	0071
ERD1000	0071
ERD2902H3	0031
ERD2904	0023
ERE24-005	1600
ERE24-01	0871
ERE24-02	0871
ERE24-04	1847
ERE24-06	1851
ERE51-01	2900
ERE51-03	1112
ERE51-06	3011
ERE51-09	7046
ERE51-12	OEM
ERE74-005	1600
ERE74-01	0871
ERE74-02	0871
ERE74-04	0853
ERE74-06	0601
ERE75-005	1530
ERE75-01	1530
ERE75-02	1530
ERE81-004	1545
ERE81B004	OEM
ERG24-005	1520
ERG24-01	1520
ERG24-02	1520
ERG24-04	1847
ERG27-10	OEM
ERG51-01	1530
ERG51-03	0029
ERG51-06	0319
ERG51-09	0044
ERG51-12	OEM
ERG74-005	1530
ERG74-01	1530
ERG74-02	1530
ERG74-04	0029
ERG75-005	1530
ERG75-01	1530
ERG75-02	1530
ERG77-10	OEM
ERG81-004	1545
ERG81A-004	1545
ERN04-20	OEM
ERN26-08	OEM
ERP04-25	OEM
ERP04-30	OEM
ERP15-16	OEM
ERPF5B0M050F	OEM
ERPF5B0M050K	OEM
ERPF5B0M080F	OEM
ERP-F5B0M080H	OEM
ERPF5B0M080K	OEM
ERPW5B0M030D	OEM
ERPW5B0M050D	OEM
ERPZ4B0M080B	OEM
ERPZ5B0M080A	OEM
ERR03-25	OEM
ERR03-30	OEM
ERS37	OEM
ERS100	0086
ERS120	0187
ERS123AP	OEM
ERS128	OEM
ERS140	0187
ERS152	OEM
ERS156	OEM
ERS160	0187
ERS180	0187
ERS200	0187
ERS210	OEM
ERS225	0187
ERS229	OEM
ERS230	OEM
ERS234	OEM
ERS244	OEM
ERS246	OEM
ERS250	0187
ERS261	OEM
ERS275	0187
ERS301	0187
ERS325	0187
ERS350	0187
ERS375	0187
ERS385	OEM
ERS401	0187
ERS425	0187
ERS450	0187
ERS475	0168
ERS519	OEM
ERS552	OEM
ERS856	OEM
ERS960	OEM
ERS961	OEM
ERS962	OEM
ERS966	OEM
ERS967	OEM
ERS972	OEM
ERTD2ZFL351S	OEM
ERTD2ZFL601S	OEM
ERTD2ZGL351S	OEM
ERV-02F2150	0015
ERZ08A3K11	0139
ERZ-08A3K101	0139
ERZ08A3M	OEM
ERZ08D3J301	OEM
ERZ08D3K101	OEM
ERZ-08D3K121	0186
ERZ1SA32S251	OEM
ERZ15A32S251	0139
ERZ15D3K471	0139
ERZ-B14DK510Z	OEM
ERZ-C03DK220	2008
ERZC03DK220	2008
ERZC03DK220S	2008
ERZ-C07DK101	4222
ERZ-C07DK121	4921
ERZ-C07DK201	1374
ERZ-C07DK241	0246
ERZ-C07DK391	1332
ERZ-C07DK431	1658
ERZ-C07DK471	4471
ERZ-C10DK241	1993
ERZC10DK241U	1993
ERZC10ZK241U	1993
ERZC10ZK24166	OEM
ERZ-C14DK101	3090
ERZ-C14DK102	OEM
ERZ-C14DK151	3116
ERZ-C14DK182	OEM
ERZ-C14DK201	0838
ERZ-C14DK201U	0838
ERZ-C14DK241	1993
ERZ-C14DK330	2796
ERZ-C14DK390	3073
ERZ-C14DK391	1334
ERZ-C14DK431	1660
ERZ-C14DK470	3075
ERZ-C14DK560	2941
ERZ-C14DK680	3078
ERZ-C14DK681	3257
ERZ-C14DK751	OEM
ERZ-C14DK781	1101
ERZ-C14DK820	3085
ERZ-C14DK821	OEM
ERZ-C14DK911	OEM
ERZ-C20DK102	OEM
ERZ-C20DK182	OEM
ERZ-C20DK201	0830
ERZ-C20DK201U	0830
ERZ-C20DK241	3415
ERZ-C20DK391	1336
ERZ-C20DK431	4867
ERZ-C20DK681	3859
ERZ-C20DK751	OEM
ERZ-C20DK781	0781
ERZ-C20DK821	OEM
ERZ-C20DK911	OEM
ERZC100K241	1993
ERZC100K241U	1993
ERZ-M10DK220	OEM
ERZV07D361	OEM
ES01	0182
ES01A	0182
ES01F	0182
ES01V	0182
ES01Z	0182
ES-1	0023
ES1	0023
ES1(ELCOM)	0050
ES-1A	0182
ES1A	0023
ES-1C	OEM
ES1C	0023
ES1F	0017
ES1F-LF-G2	OEM
ES1F-LFG2	OEM
ES1F-V1	0017
ES1KX122	0224
ES-1V	0023
ES1V	0023
ES1Z	0023
ES2(ELCOM)	0004
ES3	0136
ES3(ELCOM)	0004
ES4(ELCOM)	0164
ES5	0600
ES5(ELCOM)	0208
ES6(ELCOM)	0208
ES7	0085
ES7(ELCOM)	0222
ES8(ELCOM)	0050
ES9	0085
ES9(ELCOM)	0085
ES10	0440
ES10(ELCOM)	0440
ES11	0127
ES11(ELCOM)	0050
ES13	0085
ES13(ELCOM)	0085
ES14	0136
ES14(ELCOM)	0050
ES15	0127
ES15(ELCOM)	0050
ES15X1	0016
ES15X2	0144
ES15X3	0144
ES15X4	0004
ES15X6	0144
ES15X7	0016
ES15X8	0004
ES15X9	0037
ES15X10	0144
ES15X11	0016
ES15X12	0042
ES15X14	0016
ES15X16	0016
ES15X17	0160
ES15X18	0144
ES15X19	0144
ES15X20	0016
ES15X21	OEM
ES15X22	0330
ES15X23	0016
ES15X24	0016
ES15X29	OEM
ES15X30	0144
ES15X31	0004
ES15X32	0004
ES15X33	OEM
ES15X35	OEM
ES15X36	OEM
ES15X37	0016
ES15X38	OEM
ES15X39	OEM
ES15X40	OEM
ES15X41	OEM
ES15X42	0016
ES15X43	0160
ES15X44	OEM
ES15X45	0085
ES15X46	OEM
ES15X48	0208
ES15X49	0004
ES15X50	0004
ES15X51	0085
ES15X52	0975
ES15X53	0004
ES15X54	0975
ES15X55	0004
ES15X56	0007
ES15X57	0007
ES15X58	0016
ES15X59	0233
ES15X60	0144
ES15X61	0136
ES15X62	0016
ES15X63	0004
ES15X64	0016
ES15X65	0007
ES15X66	0127
ES15X67	0286
ES15X68	0016
ES15X69	0233
ES15X70	0016
ES15X71	0208
ES15X72	0208
ES15X73	0136
ES15X74	0208
ES15X75	0164
ES15X76	0016
ES15X77	1675
ES15X78	0085
ES15X79	0007
ES15X80	0127
ES15X81	0127
ES15X82	0127
ES15X83	0016
ES15X84	0016
ES15X85	0016
ES15X86	0042
ES15X87	0007
ES15X88	0007
ES15X89	0233
ES15X90	0203
ES15X91	1708
ES15X92	0325
ES15X93	0086
ES15X94	0065
ES15X95	0142
ES15X96	0007
ES15X97	2209
ES15X98	0178
ES15X99	1010
ES15X100	0004
ES15X101	0037
ES15X102	0127
ES15X103	OEM
ES15X104	0007

DEVICE TYPE	REPL CODE
ES15X105	0007
ES15X106	0330
ES15X107	0037
ES15X119	0007
ES15X120	0127
ES15X121	0127
ES15X122	0007
ES15X123	0007
ES15X125	0392
ES15X126	0642
ES15X127	0330
ES15X128	0037
ES15X502SB77C	0004
ES15X532SB155A	0004
ES15X552SA15V	0004
ES15X612SA351B	0136
ES16(ELCOM)	0103
ES16X2	0143
ES16X3	0143
ES16X4	0133
ES16X5	0143
ES16X6	0143
ES16X7	0143
ES16X8	0137
ES16X9	0015
ES16X10	0229
ES16X11	OEM
ES16X12	0143
ES16X13	0015
ES16X14	0143
ES16X15	0924
ES-16X16	0102
ES16X16	0102
ES16X17	OEM
ES16X20	0102
ES16X21	0123
ES16X24	0133
ES16X25	0015
ES16X26	OEM
ES16X27	0124
ES16X28	0102
ES16X29	0565
ES16X30	0133
ES16X31	0102
ES16X32	0133
ES16X33	0087
ES16X38	0015
ES16X40	0133
ES16X41	0531
ES16X61	OEM
ES16X70	0123
ES16X103	0143
ES17(ELCOM)	0004
ES18	0160
ES18(ELCOM)	0085
ES19	0004
ES19(ELCOM)	0050
ES20(ELCOM)	0016
ES21	0440
ES21(ELCOM)	0160
ES22	0161
ES22(ELCOM)	0086
ES23	0004
ES23(ELCOM)	0050
ES24(ELCOM)	0142
ES25	0280
ES25(ELCOM)	0050
ES26	0280
ES26(ELCOM)	0004
ES27(ELCOM)	0604
ES28(ELCOM)	2763
ES29	0222
ES29(ELCOM)	0222
ES30(ELCOM)	2866
ES31(ELCOM)	0103
ES32(ELCOM)	0233
ES33(ELCOM)	0074
ES34(ELCOM)	0037
ES35(ELCOM)	1514
ES36(ELCOM)	0178
ES36X103	0143
ES37(ELCOM)	0164
ES38	0161
ES38(ELCOM)	0208
ES39(ELCOM)	0276
ES41	0136
ES41(ELCOM)	0050
ES42(ELCOM)	0975
ES43(ELCOM)	0130
ES44(ELCOM)	0178
ES45(ELCOM)	0178
ES46	0198
ES46(ELCOM)	0016
ES47(ELCOM)	1669
ES47X1	0015
ES47X4	0015
ES49(ELCOM)	0905
ES50	0222
ES50(ELCOM)	0222
ES51	0015
ES51(ELCOM)	0126
ES51X65	0086
ES53(ELCOM)	0016
ES54(ELCOM)	0007
ES55(ELCOM)	0074
ES56(ELCOM)	0086
ES57(ELCOM)	0086
ES57X1	0015
ES57X2	0015

DEVICE TYPE	REPL CODE
ES57X4	0015
ES57X5	0015
ES57X6	0196
ES57X7	0649
ES57X8	0102
ES57X9	0102
ES57X11	0757
ES57X12	0133
ES57X13	0288
ES57X14	0289
ES57X20	0288
ES57X21	0288
ES57X22	OEM
ES57X23	OEM
ES57X35	OEM
ES57X39	OEM
ES58	0161
ES59(ELCOM)	0604
ES61(ELCOM)	1010
ES62(ELCOM)	0086
ES63	0222
ES63(ELCOM)	0847
ES65(ELCOM)	0037
ES66(ELCOM)	0561
ES68	0460
ES69(ELCOM)	0103
ES70	0715
ES71(ELCOM)	0074
ES73(ELCOM)	0007
ES75(ELCOM)	0400
ES75X1	0167
ES76(ELCOM)	0285
ES80(ELCOM)	0042
ES81	0382
ES81(ELCOM)	0925
ES82	0161
ES82(ELCOM)	0551
ES83	0460
ES83(ELCOM)	0382
ES84	0642
ES84(ELCOM)	0642
ES84X1	3020
ES84X2	2476
ES84X3	0167
ES84X6	0167
ES85(ELCOM)	0016
ES86(ELCOM)	0007
ES88	0161
ES89	0285
ES89(ELCOM)	0285
ES90	0491
ES91(ELCOM)	0847
ES92	3377
ES93	3528
ES94	3280
ES95	0065
ES95(ELCOM)	0065
ES96(ELCOM)	0338
ES97	0861
ES105	0558
ES106	2897
ES108	2427
ES109	2277
ES481A	OEM
ES481B	OEM
ES481C	OEM
ES481D	OEM
ES481E	OEM
ES481F	OEM
ES482A	OEM
ES482B	OEM
ES482C	OEM
ES482D	OEM
ES482E	OEM
ES482F	OEM
ES483A	OEM
ES483B	OEM
ES483C	OEM
ES483D	OEM
ES483E	OEM
ES483F	OEM
ES484A	OEM
ES484B	OEM
ES484C	OEM
ES484D	OEM
ES484E	OEM
ES484F	OEM
ES485A	OEM
ES485B	OEM
ES485C	OEM
ES485D	OEM
ES485E	OEM
ES485F	OEM
ES501	0440
ES501(ELCOM)	0440
ES503	0160
ES503(ELCOM)	0085
ES581A	OEM
ES581B	OEM
ES581C	OEM
ES581D	OEM
ES581E	OEM
ES581F	OEM
ES582A	OEM
ES582B	OEM
ES582C	OEM
ES582D	OEM
ES582E	OEM
ES582F	OEM

DEVICE TYPE	REPL CODE
ES583A	OEM
ES583B	OEM
ES583C	OEM
ES583D	OEM
ES583E	OEM
ES583F	OEM
ES584A	OEM
ES584B	OEM
ES584C	OEM
ES584D	OEM
ES584E	OEM
ES584F	OEM
ES585A	OEM
ES585B	OEM
ES585C	OEM
ES585D	OEM
ES585E	OEM
ES585F	OEM
ES1010	OEM
ES1015	OEM
ES1025	OEM
ES1035	OEM
ES1050	OEM
ES1063	OEM
ES1075	OEM
ES1513	0288
ES1627	0133
ES2010	OEM
ES2015	OEM
ES2025	OEM
ES2035	OEM
ES2050	OEM
ES2063	OEM
ES2075	OEM
ES3010	OEM
ES3015	OEM
ES3025	OEM
ES3035	OEM
ES3050	OEM
ES3063	OEM
ES3075	OEM
ES3110	0136
ES3111	0136
ES3112	0136
ES3113	0136
ES3114	0136
ES3115	0136
ES3116	0136
ES3120	0004
ES3121	0004
ES3122	0004
ES3123	0004
ES3124	0004
ES3125	0004
ES3126	0004
ES3266	0144
ES3501	OEM
ES3511	OEM
ES4010	OEM
ES4015	OEM
ES4025	OEM
ES4050	OEM
ES4063	OEM
ES4075	OEM
ES10100	OEM
ES10110	0222
ES10125	OEM
ES10155	OEM
ES10186	0127
ES10187	0151
ES10188	0151
ES10189	0143
ES10190	0030
ES10222	1220
ES10223	1220
ES10224	0143
ES10225	0143
ES10231	0111
ES10232	0079
ES10233	0015
ES10234	0137
ES15046	0127
ES15047	0127
ES15048	0079
ES15049	0111
ES15050	0016
ES15051	1267
ES15052	0111
ES15054	0143
ES15055	0030
ES15056	0015
ES15057	0133
ES15102	0127
ES15226	0086
ES15227	0560
ES20100	OEM
ES20125	OEM
ES20155	OEM
ES30100	OEM
ESA-06	0296
ESA06	0296
ESA10-98	0015
ESA10-99	0015
ESA-10C	0015
ESA-10N	0015
ESA213	0136
ESA233	0136
ESAB03-01	OEM
ESAB03-02	OEM

DEVICE TYPE	REPL CODE
ESAB03-04	OEM
ESAB33-02CS	1236
ESAB82-004	2515
ESAB85-009	1236
ESAB92-02	1236
ESAB92M-02	1021
ESAC02-C	OEM
ESAC02-N	OEM
ESAC06-06	0446
ESAC25-02C	1236
ESAC25-02D	OEM
ESAC25-02N	1727
ESAC25-04C	OEM
ESAC25-04D	OEM
ESAC25-04N	OEM
ESAC25C/N02	OEM
ESAC25C/N04	OEM
ESAC31-01C	1236
ESAC31-01D	OEM
ESAC31-01N	1727
ESAC31-02C	1236
ESAC31-02D	OEM
ESAC31-02N	1727
ESAC33-02C	1236
ESAC33-02CS	1236
ESAC33-02D	OEM
ESAC33-02N	1727
ESAC75-005	6551
ESAC75-01	2836
ESAC75-02	2836
ESAC81-004	0584
ESAC82-004	1021
ESAC83-004	1942
ESAC85-009	1236
ESAC87-009	2234
ESAC92-02	1236
ESAC93-02	2234
ESAC2504C	OEM
ESAD25-02C	2234
ESAD25-02D	OEM
ESAD25-02N	OEM
ESAD25-04C	4488
ESAD25-04D	OEM
ESAD25-04N	OEM
ESAD25-06	OEM
ESAD31-02C	OEM
ESAD33-02C	2345
ESAD33-02CS	2234
ESAD33-02D	OEM
ESAD33-02N	OEM
ESAD75-005	2226
ESAD75-01	2836
ESAD75-02	2836
ESAD81-004	0584
ESAD81-004A	OEM
ESAD82-004	OEM
ESAD85-009	2234
ESAD92-02	2234
ESAE31-06T	OEM
ESAE31-08T	OEM
ESAG31-06T	OEM
ESAG31-08T	OEM
ESAG32-06T	OEM
ESAG32-08T	OEM
ESAG73-03C	OEM
ESAG73-03D	OEM
ESAG73-03N	OEM
ESAG73-06C	OEM
ESAG73-06D	OEM
ESAG73-06N	OEM
ESAG84-004	OEM
ESAH73-03C	OEM
ESAH73-03D	OEM
ESAH73-03N	OEM
ESAH73-06C	OEM
ESAH73-06D	OEM
ESAH73-06N	OEM
ESBB01-1R50	OEM
ESBB01-2R25	OEM
ESBB01-3R00	1713
ESBB01-3R75	OEM
ESBB01-4R50	OEM
ESBC604	OEM
ESBC604H	OEM
ESBC614G	OEM
ESCA25-02C	OEM
ESCA83-004	OEM
ESDA-5	OEM
ESDA5	OEM
ESDA5H	OEM
ESDA5J	OEM
ESDA5K	OEM
ESDA-8	OEM
ESDA8	OEM
ESDA-10	OEM
ESDA10	OEM
ESDA-12	OEM
ESDA12	OEM
ESDA-15	OEM
ESDA15	OEM
ESDA-20	OEM
ESDA20	OEM
ESDA-25	OEM
ESDA25	OEM
ESDA-30	OEM
ESDA30	OEM
ESDA-50	OEM
ESDA50	OEM
ESI-LA1111P	0387

DEVICE TYPE	REPL CODE
ESI-MSM5807	OEM
ESI-UPC30C	0653
ESI-UPC1020H	3695
ESJA03-13	OEM
ESJA17-16	OEM
ESJA17-20	OEM
ESJA17-25	OEM
ESJA18-16	OEM
ESJA18-20	OEM
ESJA18-25	OEM
ESJA18-30	OEM
ESJA19-30	0288
ESJA25-04	OEM
ESJA25-06	OEM
ESJA25-08	OEM
ESJA25-10	OEM
ESJA25-12	OEM
ESJA25-12S	OEM
ESJA25-16	OEM
ESJA25-20	OEM
ESJA31-16	OEM
ESJA31-20	0378
ESJA31-24	OEM
ESJA34-06	OEM
ESJA35-10	OEM
ESJA35-12	OEM
ESJA37-16	OEM
ESJA37-20	OEM
ESJA37-24	OEM
ESJA52-10	OEM
ESJA52-12	OEM
ESJA52-14	OEM
ESJA53-16	OEM
ESJA53-18	OEM
ESJA53-20	OEM
ESJA54-06	OEM
ESJA54-08	OEM
ESJA56-20	OEM
ESJA56-24	OEM
ESJA57-04	OEM
ESK-1	0015
ESK1	0015
ESK1/02	0015
ESK1/06	0015
ESK1/10	OEM
ESK1/12	OEM
ESK100/08	OEM
ESK100/10	OEM
ESK100/12	OEM
ESK100/14	OEM
ESK100/16	OEM
ESKA1/10	OEM
ESKA1/12	OEM
ESKE40C500	0015
ESKE125C500	0071
ESM16	1852
ESM22-100	OEM
ESM22-100N	OEM
ESM22-200	OEM
ESM22-200N	OEM
ESM22-300	OEM
ESM22-300N	OEM
ESM22-400	OEM
ESM22-400N	OEM
ESM22-500	OEM
ESM22-500N	OEM
ESM22-600	OEM
ESM22-600N	OEM
ESM23-100	OEM
ESM23-100N	OEM
ESM23-200	OEM
ESM23-200N	OEM
ESM23-300	OEM
ESM23-300N	OEM
ESM23-400	OEM
ESM23-400N	OEM
ESM23-500	OEM
ESM23-500N	OEM
ESM23-600	OEM
ESM23-600N	OEM
ESM25	OEM
ESM25A	OEM
ESM100	OEM
ESM113	2427
ESM114	2427
ESM117	3373
ESM118	3373
ESM1350P	OEM
ESM181-200R	OEM
ESM181-300R	OEM
ESM181-400R	OEM
ESM181-500R	OEM
ESM181-600R	OEM
ESM193-400	OEM
ESM193-600	OEM
ESM193-800	OEM
ESM193-900	OEM
ESM193-1000	OEM
ESM194-600	OEM
ESM194-800	OEM
ESM194-1000	OEM
ESM194-1200	OEM
ESM213	OEM
ESM214	OEM
ESM217	OEM
ESM218	OEM
ESM222R	OEM
ESM227	OEM
ESM227N	OEM

DEVICE TYPE	REPL CODE
ESM231	OEM
ESM243-50	OEM
ESM243-100	OEM
ESM243-200	OEM
ESM243-300	OEM
ESM243-300R	OEM
ESM243-400	OEM
ESM244-50	OEM
ESM244-100	OEM
ESM244-200	OEM
ESM244-300	OEM
ESM244-400	OEM
ESM244-500	OEM
ESM244-600	OEM
ESM244-800	OEM
ESM244-1000	OEM
ESM245-50	OEM
ESM245-200	OEM
ESM245-400	OEM
ESM245-600	OEM
ESM245-800	OEM
ESM245-1000	OEM
ESM245-1000R	OEM
ESM255-50	OEM
ESM255-100	OEM
ESM255-200	OEM
ESM255-300	OEM
ESM255-400	OEM
ESM269	OEM
ESM282	OEM
ESM282A	OEM
ESM283	OEM
ESM310	OEM
ESM381	OEM
ESM382	OEM
ESM383	OEM
ESM432	OEM
ESM432C	OEM
ESM532C	OEM
ESM579	OEM
ESM582	0133
ESM622	OEM
ESM623	OEM
ESM632C	OEM
ESM635	OEM
ESM636	OEM
ESM637	OEM
ESM638	OEM
ESM639	OEM
ESM640	OEM
ESM642	OEM
ESM643	OEM
ESM691	OEM
ESM692	OEM
ESM693	OEM
ESM700	OEM
ESM707	OEM
ESM732C	OEM
ESM737	OEM
ESM738	OEM
ESM765-100	OEM
ESM765-200	OEM
ESM765-400	OEM
ESM765-600	OEM
ESM765-800	OEM
ESM837	OEM
ESM855	OEM
ESM856	OEM
ESM870	OEM
ESM871	OEM
ESM900	OEM
ESM910	OEM
ESM990-200	OEM
ESM990-300	OEM
ESM990-400	OEM
ESM1000	OEM
ESM1406	OEM
ESM1410	OEM
ESM1503	OEM
ESM1600B	OEM
ESM1602	OEM
ESM2060	OEM
ESM3045AV	OEM
ESM3045DV	OEM
ESM4014	OEM
ESM4015	OEM
ESM4016	OEM
ESM4017	OEM
ESM4045AV	OEM
ESM4045DV	OEM
ESM4091	OEM
ESM4092	OEM
ESM4093	OEM
ESM4116-200	OEM
ESM4116-400	OEM
ESM4116-500	OEM
ESM4116-600	OEM
ESM4120-600	OEM
ESM4120-800	OEM
ESM4120-900	OEM
ESM4120-1000	OEM
ESM4391	OEM
ESM4392	OEM
ESM4393	OEM
ESM4446	OEM
ESM4448	OEM
ESM5045DV	OEM
ESM6045AV	OEM
ESM6045DV	OEM

DEVICE TYPE	REPL CODE
ESP43	OEM
ESP550	OEM
ESP5100	OEM
ESP5200	OEM
ESP5300	OEM
ESP5400	OEM
ESP5500	OEM
ESP5600	OEM
ESP5800	OEM
ESP51000	OEM
ESPA02-30A	OEM
ESPA03	OEM
ESPA03-40	OEM
ESPA05-30	3860
ESPA11-40	1323
ESPA91	2671
ESPA94	2671
ESZ5.1	0041
ESZ5.6	0253
ESZ6.2	0471
ESZ6.8	0062
ESZ7.5	0077
ESZ8.2	0165
ESZ9.1	0057
ESZ10	0064
ESZ11	0181
ESZ12	0052
ET1	0050
ET1X1	OEM
ET2	0050
ET3	0211
ET4	0211
ET5	0211
ET6	0085
ET7	0440
ET8	0600
ET9	0600
ET10	0038
ET11	0038
ET12	0136
ET15X1	0004
ET15X2	0127
ET15X3	0144
ET15X4	0160
ET15X5	0975
ET15X7	0144
ET15X8	0233
ET15X9	0127
ET15X10	0155
ET15X11	0155
ET15X12	0086
ET15X13	0016
ET15X14	0016
ET15X15	0016
ET15X16	0016
ET15X17	0604
ET15X18	0144
ET15X19	0155
ET15X20	0155
ET15X21	0144
ET15X23	0079
ET15X24	0016
ET15X25	0004
ET15X26	0975
ET15X27	0079
ET15X29	0050
ET15X30	0127
ET15X31	0004
ET15X32	0280
ET15X33	0037
ET15X34	0233
ET15X35	OEM
ET15X36	0086
ET15X37	0016
ET15X38	0126
ET15X39	0126
ET15X40	0975
ET15X41	0016
ET15X42	0016
ET15X43	0160
ET15X44	OEM
ET15X45	0016
ET15X54	0079
ET16X1	0143
ET16X6	0917
ET16X7	0196
ET16X10	0196
ET16X11	0196
ET16X13	0917
ET16X14	0917
ET16X15	0137
ET16X16	0229
ET16X16X	0229
ET16X17	0157
ET16X19	0143
ET16X20	0143
ET16X21	0143
ET16X22	OEM
ET16X23	OEM
ET30X85	OEM
ET30X87	OEM
ET41X27	4465
ET41X37	0143
ET41X47	4465
ET47X27	4465
ET51X3	0016
ET51X25	0015
ET52X25	0015
ET55-25	0015

If replacement code is OEM, contact original manufacturer for replacement.

Original Device Types

DEVICE TYPE	REPL CODE
ET55X25	0015
ET55X29	0015
ET57X25	0015
ET57X26	0474
ET57X29	0015
ET57X30	0015
ET57X31	1303
ET57X32	0774
ET57X33	0015
ET57X34	OEM
ET57X35	0015
ET57X36	OEM
ET57X37	OEM
ET57X38	0015
ET57X39	0102
ET57X40	0102
ET57X41	1303
ET57X44	OEM
ET58X32	0774
ET161X1	OEM
ET192	OEM
ET-192-01	OEM
ET200	0015
ET400	0015
ET401	OEM
ET402	OEM
ET403	2784
ET409	OEM
ET410	OEM
ET411	OEM
ET413	OEM
ET423	OEM
ET424	OEM
ET425	OEM
ET430	OEM
ET431	OEM
ET515	OEM
ET516	OEM
ET517	OEM
ET518	OEM
ET519	OEM
ET600	0015
ET670	0847
ET701	OEM
ET708	OEM
ET709	OEM
ET710	OEM
ET721	OEM
ET801	OEM
ET1002	OEM
ET1004	OEM
ET1006	OEM
ET1008	OEM
ET1511	0004
ET2002	OEM
ET2004	OEM
ET2006	OEM
ET2008	OEM
ET3002	OEM
ET3004	OEM
ET3006	OEM
ET3008	OEM
ET4002	OEM
ET4004	OEM
ET4006	OEM
ET4008	OEM
ET4039	OEM
ET4040	OEM
ET4041	OEM
ET4045	OEM
ET4050	OEM
ET4055	OEM
ET4059	OEM
ET4060	OEM
ET4061	OEM
ET4065	OEM
ET5000	OEM
ET5001	OEM
ET5002	OEM
ET5004	OEM
ET5006	OEM
ET5008	OEM
ET5060	OEM
ET5061	OEM
ET5062	OEM
ET5063	2626
ET5064	OEM
ET5065	OEM
ET5066	OEM
ET5251	1991
ET5252	1991
ET5253	1991
ET6000	OEM
ET6001	OEM
ET6002	OEM
ET6060	OEM
ET6061	OEM
ET6062	OEM
ET6251	OEM
ET6252	OEM
ET6253	OEM
ET6542	OEM
ET6543	OEM
ET6544	OEM
ET6545	OEM
ET6546	OEM
ET6547	OEM
ET10015	OEM
ET10016	OEM
ET10020	OEM
ET10021	OEM
ET35035	0693
ET234854	0155
ET238894	0016
ET307195	2131
ET329218	0155
ET338894	1480
ET350335	0693
ET368021	0016
ET379462	0111
ET398711	0111
ET398777	0111
ET412626	0016
ET453611	0219
ET491051	0325
ET495371	0560
ET517263	0111
ET517375	0936
ET539122	0532
ET703844	0076
ET703854	0076
ET703865	0693
ET703867	3081
ET703872	0194
ET703873	0859
ET703874	0733
ET703875	0693
ET703878	0436
ET703879	3111
ET703881	0066
ET703887	5329
ET703894	0461
ET704334	0127
E-TA7328AP	4818
E-TA7335P	4406
E-TA7343P	4410
E-TA7343PS	4410
ETD-1N60	0019
ETD1S2788	0124
ETD-10D1	0015
ETD-10D2	0015
ETD-5046	0143
ETDCDG21	0133
ETD-HZ7C	0077
ETD-RD9.1FB	0012
ETD-RD75	0077
ETD-SD46	0143
ETD-SG9150	0133
ETD-V06C	0015
ETI-13	5746
ETI-21	0387
ETI-22	0520
ETI-23	1050
ETP2008	0233
ETP3114	0233
ETP3923	0233
ETP5092	0187
ETP5095	0168
ETS-003	0103
ETS-068	0016
ETS-069	0126
ETS-070	0086
ETS-071	0126
ETTB-2SB176	0004
ETTB-2SB176A	0004
ETTB-2SB176B	0004
ETTB-2SB176R	0004
ETTB-75LB	0004
ETTB-367B	0222
ETTC-2SC490	0178
ETTC-458LG	0016
ETTC-930D	2209
ETTC-945	0076
ETTC-CD8000	0016
ETTC-CD12000	0016
ETTC-CD13000	0016
ETT-CDC-12000	0016
ETTD-235	0228
ETX18	0155
EU01	0023
EU01A	0023
EU01Z	0023
EU02	0182
EU02A	0023
EU02Z	0023
EU-1	0023
EU1	0023
EU-1A	0023
EU1A	0023
EU-1AV	0023
EU-1V	0023
EU1V	6027
EU-1Z	0023
EU1Z	0023
EU1ZV	OEM
EU1ZV1	0023
EU-2	0023
EU2	0023
EU2A	0023
EU-2V	0023
EU2V	0023
EU2YX	0023
EU2YXV1	0023
EU2Z	0023
EU2ZV	0023
EU15X1	0144
EU15X2	0144
EU15X3	0144
EU15X4	0917
EU15X5	OEM
EU15X6	0144
EU15X27	0086
EU15X34	0086
EU16X1	0143
EU16X2	0143
EU16X3	OEM
EU16X4	0917
EU16X5	OEM
EU16X7	0015
EU16X8	0484
EU16X11	0133
EU16X14	0917
EU16X19	0123
EU16X20	0015
EU16X27	4179
EU16X40	OEM
EU30X87	0030
EU57X30	0015
EU57X31	1303
EU57X32	0774
EU57X38	0474
EU57X40	0015
EU-Z	0023
EUZ	0023
EV0004	OEM
EV12R	OEM
EV15	0002
EV16X20	0102
EVD-3	0133
EVM511	0015
EVM711	OEM
EVMEH2AC11/1	OEM
EVMEH2C12/1	OEM
EVR1	OEM
EVR1A	OEM
EVR1B	OEM
EVR4	0157
EVR4A	0157
EVR4B	0157
EVR5	0636
EVR5A	0636
EVR5B	0636
EVR6	0301
EVR6A	0301
EVR6B	0301
EVR7	0025
EVR7A	0025
EVR7B	0025
EVR8	0244
EVR8A	0244
EVR8B	0244
EVR9	0012
EVR9A	0012
EVR9B	0012
EVR10	0170
EVR10A	0170
EVR10B	0170
EVR11	0794
EVR11A	0794
EVR11B	0794
EVR12	0137
EVR12A	0137
EVR12B	0137
EVR13	0365
EVR13A	0053
EVR13B	0365
EVR14	0100
EVR14A	0100
EVR14B	0100
EVR15	0002
EVR15A	0002
EVR15B	0002
EVR16	0420
EVR16A	0420
EVR16B	0420
EVR17	1649
EVR17A	1649
EVR17B	1649
EVR18	0495
EVR18A	0495
EVR18B	0495
EVR19	0949
EVR19A	0949
EVR19B	0949
EVR20	0531
EVR20A	0531
EVR20B	0531
EVR21	0565
EVR21A	0565
EVR21B	0565
EVR22	0565
EVR22A	0565
EVR22B	0565
EVR23	0402
EVR23A	0402
EVR23B	0402
EVR24	0402
EVR24A	0402
EVR24B	0402
EVR25	1606
EVR25A	1606
EVR25B	1606
EVR26	0441
EVR26A	0441
EVR26B	0441
EVR27	0441
EVR27A	0441
EVR27B	0441
EVR28	1674
EVR28A	1674
EVR28B	1674
EVR30	0726
EVR30A	0726
EVR30B	0726
EVR32	0039
EVR32A	0039
EVR32B	0039
EVR36	0820
EVR36A	0820
EVR36B	0820
EVR39	0350
EVR39A	0350
EVR39B	0350
EVR43	0350
EVR43A	0350
EVR43B	0350
EVR47	0999
EVR47A	0999
EVR47B	0999
EVR50	0502
EVR50A	0502
EVR50B	0502
EVR56	1834
EVR56A	1834
EVR56B	1834
EVR60	1156
EVR60A	1156
EVR60B	1156
EVR68	2157
EVR68A	2157
EVR68B	2157
EVR75	1189
EVR75A	1189
EVR75B	1189
EVR82	0331
EVR82A	0331
EVR82B	0331
EVR91	1311
EVR91A	1311
EVR91B	1311
EVR100	0098
EVR100A	0098
EVR100B	0098
EVR110	0149
EVR110A	0149
EVR110B	0149
EVR150	0028
EVR150A	0028
EVR150B	0028
EVS-K112C	0196
EW02-05A	OEM
EW10X76	5000
EW10X103	6128
EW14X18	0120
EW14X232	1993
EW14X258	OEM
EW14X486	1993
EW15R	OEM
EW15X0399	OEM
EW15X6	0155
EW15X19	0127
EW15X20	1386
EW15X21	0076
EW15X22	0006
EW15X23	1513
EW15X24	0728
EW15X25	0076
EW15X26	0925
EW15X27	0006
EW15X28	OEM
EW15X29	0037
EW15X30	0076
EW15X31	0563
EW15X32	0076
EW15X33	0563
EW15X42	0076
EW15X44	0203
EW15X51	0261
EW15X55	0037
EW15X57	0079
EW15X58	0037
EW15X59	0925
EW15X63	0079
EW15X64	0168
EW15X66	0076
EW15X67	0055
EW15X70	0155
EW15X71	0563
EW15X78	0113
EW15X106	0400
EW15X108	0928
EW15X110	0076
EW15X130	1140
EW15X135	2484
EW15X139	0161
EW15X144	0825
EW15X148	1348
EW15X151	0477
EW15X152	4348
EW15X154	2223
EW15X156	0887
EW15X158	0582
EW15X159	0165
EW15X160	0672
EW15X165	0018
EW15X168	0825
EW15X171	0582
EW15X172	0582
EW15X178	0825
EW15X183	0582
EW15X187	2223
EW15X188	1978
EW15X190	0359
EW15X199	0672
EW15X204	0009
EW15X206	0582
EW15X215	0825
EW15X219	0672
EW15X223	0006
EW15X224	2209
EW15X225	0042
EW15X226	1513
EW15X231	0127
EW15X232	2910
EW15X233	0532
EW15X234	0191
EW15X235	3125
EW15X236	0148
EW15X237	0925
EW15X238	0563
EW15X240	0148
EW15X241	0423
EW15X242	0066
EW15X243	2902
EW15X244	3046
EW15X254	0076
EW15X258	0648
EW15X260	0825
EW15X262	0060
EW15X267	0887
EW15X268	0014
EW15X269	1032
EW15X272	6342
EW15X276	2484
EW15X278	0060
EW15X279	0825
EW15X285	4348
EW15X286	0672
EW15X287	5464
EW15X291	5450
EW15X292	0645
EW15X310	0168
EW15X315	0582
EW15X316	0582
EW15X317	0672
EW15X320	0060
EW15X322	0532
EW15X327	0261
EW15X328	0695
EW15X329	0825
EW15X330	2668
EW15X341	0648
EW15X342	1140
EW15X343	2052
EW15X344	0728
EW15X345	0728
EW15X346	0060
EW15X359	0261
EW15X360	0582
EW15X363	0018
EW15X366	4172
EW15X367	4130
EW15X368	0832
EW15X369	0898
EW15X370	4172
EW15X372	0832
EW15X373	0898
EW15X374	0825
EW15X375	0825
EW15X376	0460
EW15X377	1308
EW15X378	3014
EW15X379	0009
EW15X380	OEM
EW15X383	0006
EW15X386	OEM
EW15X406	0060
EW15X413	1162
EW15X415	1162
EW15X421	1032
EW15X427	OEM
EW15X428	1140
EW15X433	0009
EW15X450	0076
EW15X457	0825
EW15X467	3133
EW15X487	3314
EW15X520	0392
EW15X521	0842
EW15X522	0286
EW15X523	0066
EW15X524	1541
EW15X525	0887
EW15X526	4172
EW15X529	0898
EW15X531	OEM
EW15X549	0111
EW15X550	OEM
EW15X564	1308
EW15X566	1513
EW15X567	0728
EW15X568	0127
EW15X570	4650
EW15X571	0261
EW15X572	0006
EW15X573	0261
EW15X574	0866
EW15X575	2367
EW15X576	0191
EW15X577	0518
EW15X578	0006
EW15X579	1541
EW15X580	1284
EW15X581	0155
EW15X582	0191
EW15X583	0155
EW15X584	2910
EW15X585	0518
EW15X588	0007
EW15X589	1327
EW15X590	1790
EW15X591	1212
EW15X592	0276
EW15X593	3619
EW15X594	0755
EW15X595	0261
EW15X596	0018
EW15X597	0286
EW15X598	0111
EW15X619	0695
EW15X620	0261
EW15X628	0261
EW15X629	2795
EW15X634	0276
EW15X635	1541
EW16X9	0133
EW16X10	0124
EW16X11	0023
EW16X12	0062
EW16X13	0471
EW16X14	0140
EW16X15	0133
EW16X17	0030
EW16X19	0139
EW16X21	0077
EW16X22	0015
EW16X26	1414
EW16X36	0023
EW16X39	0471
EW16X59	0023
EW16X76	3501
EW16X82	0760
EW16X92	0735
EW16X93	2542
EW16X97	0253
EW16X98	0023
EW16X102	0124
EW16X105	0015
EW16X106	0133
EW16X109	0133
EW16X110	0124
EW16X111	0124
EW16X114	0077
EW16X116	0181
EW16X118	0700
EW16X121	0023
EW16X124	0636
EW16X135	0015
EW16X159	0165
EW16X160	0879
EW16X165	0120
EW16X167	0140
EW16X169	0023
EW16X171	0299
EW16X175	0023
EW16X176	0343
EW16X189	0019
EW16X190	0677
EW16X193	0124
EW16X194	0143
EW16X196	0023
EW16X197	0036
EW16X200	0023
EW16X201	OEM
EW16X202	0071
EW16X204	0253
EW16X205	0057
EW16X206	0445
EW16X207	0705
EW16X209	5499
EW16X218	2232
EW16X221	0023
EW16X223	0041
EW16X224	0253
EW16X226	0124
EW16X227	0124
EW16X231	0694
EW16X232	2203
EW16X234	0471
EW16X235	2008
EW16X237	0053
EW16X238	0455
EW16X242	0181
EW16X243	0053
EW16X245	0565
EW16X248	0041
EW16X249	1005
EW16X251	0062
EW16X253	0375
EW16X254	0195
EW16X255	0140
EW16X256	1414
EW16X257	0124
EW16X258	0495
EW16X259	0244
EW16X260	0165
EW16X261	0077
EW16X262	OEM
EW16X264	0165
EW16X265	0455
EW16X266	0195
EW16X270	0071
EW16X292	0124
EW16X296	0140
EW16X297	0726
EW16X298	0062
EW16X300	0053
EW16X301	0686
EW16X302	0077
EW16X303	0133
EW16X304	0120
EW16X305	0133
EW16X306	0015
EW16X307	0025
EW16X308	0023
EW16X309	0023
EW16X310	0181
EW16X311	0019
EW16X312	0023
EW16X313	0071
EW16X314	1091
EW16X321	0182
EW16X323	0031
EW16X324	0451
EW16X325	0124
EW16X326	0023
EW16X327	0343
EW16X328	0062
EW16X329	0064
EW16X330	0053
EW16X331	0053
EW16X332	0640
EW16X334	0296
EW16X336	0124
EW16X337	0023
EW16X338	0071
EW16X343	0140
EW16X352	0023
EW16X354	0023
EW16X358	0019
EW16X363	0023
EW16X365	1091
EW16X366	0253
EW16X367	0062
EW16X368	0879
EW16X370	0052
EW16X371	0694
EW16X372	0445
EW16X373	1236
EW16X374	0057
EW16X375	0064
EW16X376	0057
EW16X377	0473
EW16X378	0077
EW16X381	3452
EW16X391	0471
EW16X395	0057
EW16X397	0057
EW16X405	1005
EW16X407	0181
EW16X408	0031
EW16X409	0041
EW16X415	0053
EW16X437	0343
EW16X478	0396
EW16X480	OEM
EW16X481	OEM
EW16X482	0124
EW16X483	4403
EW16X484	0023
EW16X485	0023
EW16X486	0795
EW16X487	0137
EW16X488	0564
EW16X489	2234
EW16X490	0124
EW16X491	0705
EW16X492	0546
EW16X494	1875
EW16X501	0077
EW16X518	OEM
EW16X526	0124
EW16X527	0005
EW16X529	0017
EW16X530	0023
EW16X531	0031
EW16X532	0036
EW16X533	0041
EW16X534	OEM
EW16X535	0052
EW16X536	0057
EW16X537	0062
EW16X538	0064
EW16X539	OEM
EW16X540	0072
EW16X541	0023
EW16X542	0015
EW16X543	0091
EW16X544	0100
EW16X545	0106
EW16X546	0112
EW16X547	0118
EW16X548	0124

DEVICE TYPE	REPL CODE
EW16X549	0071
EW16X550	0133
EW16X551	0140
EW16X552	0023
EW16X553	0064
EW16X554	0023
EW16X555	0158
EW16X556	0091
EW16X557	0165
EW16X558	0166
EW16X559	0023
EW16X560	0023
EW16X561	0002
EW16X562	0181
EW16X563	0071
EW16X564	0041
EW16X565	OEM
EW16X566	0195
EW16X567	0052
EW16X572	OEM
EW16X583	0124
EW16X584	0241
EW16X585	0248
EW16X586	0062
EW16X587	0077
EW16X588	0257
EW16X589	0158
EW16X590	0091
EW16X594	0275
EW16X595	0283
EW16X596	0053
EW16X597	0295
EW16X598	0301
EW16X599	0057
EW16X600	0023
EW16X601	0071
EW16X603	OEM
EW16X605	0023
EW16X606	0091
EW16X607	0322
EW16X608	0181
EW16X609	OEM
EW16X610	0017
EW16X611	0041
EW16X612	0023
EW16X613	OEM
EW16X614	0339
EW16X615	0057
EW16X616	0343
EW16X617	OEM
EW16X618	0124
EW16X619	0077
EW16X620	0023
EW16X622	0071
EW16X630	OEM
EW16X651	0441
EW16X652	0077
EW16X656	0455
EW16X657	0166
EW16X658	0071
EW16X712	0017
EW20	OEM
EW41X7	OEM
EW41X29	2094
EW41X59	2628
EW41X97	1778
EW41X98	1962
EW41X129	OEM
EW41X146	OEM
EW41X159	OEM
EW41X184	1778
EW41X216	OEM
EW41X224	OEM
EW41X257	OEM
EW41X258	1778
EW41X259	OEM
EW41X397	2703
EW41X530	OEM
EW41X561	OEM
EW41X595	OEM
EW41X609	OEM
EW41X619	0241
EW41X620	OEM
EW41X650	OEM
EW41X652	OEM
EW41X685	OEM
EW41X723	OEM
EW41X724	OEM
EW51	OEM
EW53/1	OEM
EW53/2	OEM
EW57X1	0080
EW57X2	0023
EW58/1	OEM
EW58/2	OEM
EW59	OEM
EW69	OEM
EW84X0425	5234
EW84X0594	OEM
EW84X0622	OEM
EW84X0783	OEM
EW84X0898	OEM
EW84X12	1329
EW84X13	OEM
EW84X14	0603
EW84X15	4410
EW84X16	OEM
EW84X17	OEM
EW84X18	4830
EW84X30	0513
EW84X60	2026
EW84X114	0167
EW84X115	2533
EW84X156	0176
EW84X172	2053
EW84X192	0360
EW84X196	0631
EW84X217	2019
EW84X224	3233
EW84X230	OEM
EW84X232	0770
EW84X233	0770
EW84X239	0699
EW84X252	0759
EW84X254	OEM
EW84X255	6324
EW84X267	4137
EW84X268	0167
EW84X269	1590
EW84X270	0490
EW84X271	OEM
EW84X272	1846
EW84X273	1329
EW84X296	2073
EW84X310	1201
EW84X312	2067
EW84X313	OEM
EW84X314	2553
EW84X315	5218
EW84X317	5261
EW84X318	OEM
EW84X319	4925
EW84X320	5260
EW84X322	5271
EW84X323	3159
EW84X324	OEM
EW84X326	OEM
EW84X327	3902
EW84X329	6328
EW84X352	3159
EW84X355	5819
EW84X365	OEM
EW84X370	1824
EW84X372	0176
EW84X375	2068
EW84X376	OEM
EW84X377	5236
EW84X378	OEM
EW84X379	OEM
EW84X380	OEM
EW84X381	5259
EW84X382	OEM
EW84X383	OEM
EW84X384	OEM
EW84X385	2025
EW84X386	OEM
EW84X387	OEM
EW84X388	OEM
EW84X389	1065
EW84X390	OEM
EW84X391	OEM
EW84X394	4495
EW84X416	1945
EW84X422	3744
EW84X423	5208
EW84X424	5094
EW84X425	5234
EW84X426	0976
EW84X427	3503
EW84X431	1329
EW84X439	OEM
EW84X440	4379
EW84X441	6336
EW84X442	5258
EW84X443	5259
EW84X444	3925
EW84X445	5262
EW84X446	5263
EW84X447	OEM
EW84X448	0112
EW84X451	3902
EW84X452	4346
EW84X453	0624
EW84X464	1949
EW84X470	0024
EW84X473	OEM
EW84X474	OEM
EW84X475	OEM
EW84X476	OEM
EW84X477	OEM
EW84X478	OEM
EW84X479	2066
EW84X480	3437
EW84X481	0699
EW84X482	OEM
EW84X483	0360
EW84X484	6339
EW84X485	OEM
EW84X486	2535
EW84X487	2062
EW84X488	0360
EW84X489	OEM
EW84X490	3828
EW84X491	OEM
EW84X492	OEM
EW84X493	OEM
EW84X494	OEM
EW84X495	OEM
EW84X496	0101
EW84X497	OEM
EW84X498	4383
EW84X499	2913
EW84X502	OEM
EW84X503	OEM
EW84X504	OEM
EW84X505	OEM
EW84X506	OEM
EW84X526	1321
EW84X536	5258
EW84X537	OEM
EW84X538	OEM
EW84X539	OEM
EW84X589	5815
EW84X594	OEM
EW84X595	4496
EW84X596	OEM
EW84X598	OEM
EW84X624	3159
EW84X625	6165
EW84X635	OEM
EW84X738	5229
EW84X744	5324
EW84X747	0334
EW84X756	0119
EW84X761	4379
EW84X762	OEM
EW84X763	4117
EW84X764	OEM
EW84X765	OEM
EW84X766	1065
EW84X767	3925
EW84X768	OEM
EW84X769	OEM
EW84X770	4141
EW84X771	5328
EW84X772	OEM
EW84X774	4379
EW84X776	OEM
EW84X777	OEM
EW84X782	2122
EW84X783	4146
EW84X784	2116
EW84X785	0312
EW84X786	OEM
EW84X787	OEM
EW84X788	0360
EW84X789	OEM
EW84X790	OEM
EW84X791	OEM
EW84X792	6295
EW84X793	0416
EW84X794	3521
EW84X795	OEM
EW84X796	OEM
EW84X797	OEM
EW84X798	OEM
EW84X799	OEM
EW84X801	OEM
EW84X802	OEM
EW84X803	OEM
EW84X804	OEM
EW84X805	OEM
EW84X806	OEM
EW84X807	OEM
EW84X808	OEM
EW84X810	4380
EW84X811	OEM
EW84X812	OEM
EW84X813	OEM
EW84X822	OEM
EW84X863	0413
EW84X864	OEM
EW84X865	0360
EW84X866	OEM
EW84X868	OEM
EW84X875	OEM
EW84X888	2666
EW84X894	3725
EW84X895	4041
EW84X897	4350
EW84X898	OEM
EW84X899	0912
EW84X900	0663
EW84X901	0167
EW84X902	0841
EW84X903	OEM
EW84X904	OEM
EW84X905	1028
EW84X906	0539
EW84X907	OEM
EW84X908	OEM
EW84X909	3955
EW84X910	1988
EW84X911	3817
EW84X912	5961
EW84X913	5208
EW84X914	3853
EW84X915	OEM
EW84X920	4266
EW84X976	1529
EW84X977	6162
EW84X978	3916
EW84X979	0335
EW84X980	OEM
EW84X981	OEM
EW84X1002	OEM
EW84X1004	OEM
EW84X1006	OEM
EW84X1007	1579
EW84X1008	4422
EW84X1009	3747
EW84X1010	2364
EW84X1072	OEM
EW84Z316	5263
EW162	0144
EW163	0127
EW164	0127
EW165	0127
EW165V	0016
EW166	0143
EW167	0143
EW168	0123
EW169	0087
EW169B	0015
EW181	0016
EW182	0016
EW183	0925
EW183B	0178
EW212	0007
EW721	OEM
EW722	OEM
EW723	OEM
EWA127	OEM
EWC108	OEM
EWQ202	0037
EX0010GE	0253
EX0011GE	0686
EX0019TA	0053
EX0020CE	OEM
EX0020GE	0041
EX0021TA	0181
EX0023GE	0170
EX0024CE	0471
EX0038CE	0137
EX0041TA	0062
EX0042TA	0140
EX0044CE	0064
EX0045TA	0883
EX0046TA	0077
EX0047CE	1005
EX0047TA	0064
EX0048CE	0471
EX0051CE22.5V	0705
EX0061CE	0137
EX0065CE	0445
EX0074CE	0647
EX0084CE	1189
EX0085CE	0062
EX0088CE	0062
EX0089CE	0165
EX0091CE	0705
EX0092CE	0376
EX0092CE3.9V	0036
EX0101CE	0760
EX0102CE	OEM
EX0103CE	0253
EX0106CE	0049
EX0111-XS	OEM
EX0112CE	0375
EX0113CE15V	0445
EX0114CE9.1V	0057
EX0116CE	OEM
EX0116GE	0062
EX0130CE	0700
EX0131CE	0041
EX0136CE	0820
EX0137CE	0041
EX0137TA	0140
EX0138CE	0062
EX0140CE	0062
EX0146-XS	OEM
EX0147CE	0077
EX0152CE	0213
EX0155CE	0041
EX0161CE	0077
EX0161CEZZ	0077
EX0163CE	0156
EX0166CE	0686
EX0192CE	0636
EX0193CE	0181
EX0200CE	0052
EX0216-XS	OEM
EX0217CE	0879
EX0219CE	0166
EX0221-XS	OEM
EX0222-XS	OEM
EX0225-XS	OEM
EX0375GEZZ	0883
EX0443CE	0036
EX0444CE	0376
EX0598CE	OEM
EX0D43TA5.6V	0253
EX15X25	0004
EX15X336	0127
EX16X10	0133
EX16X27	0133
EX33X8392	1540
EX39	2161
EX39-X	2161
EX42	2830
EX42-X	0352
EX42X	0352
EX46-X	0666
EX48	0167
EX48-X	0354
EX62-X	0349
EX76-X	0015
EX85B-X	0774
EX89/211-X	0143
EX142-X	0133
EX215-X	1323
EX499-X	0079
EX500-X	0079
EX524-X	3081
EX695-X	0079
EX699-X	0150
EX743-X	0233
EX744-X	0086
EX746-X	0037
EX748-X	0079
EX831-X	0642
EX868-X	0285
EX888-X	0079
EX911-X	0276
EX4035	0701
EX4053	0701
EX795427C	OEM
EXB341	OEM
EXB341-0A	OEM
EXB341-0A	OEM
EXC	OEM
EXD019TA	0053
EXDA415	OEM
EXDA420	OEM
EXDA435	OEM
EXDA435C	OEM
EXDA635	OEM
EXO-512	OEM
EXO-2480	OEM
EXPANDOPROM	OEM
EXPANDORAM	OEM
EXPANDORAMII	OEM
EXT954-20C	2103
EXT954-25C	3518
EXT95247C	3518
EXT95423C	3502
EXT95425C	OEM
EXT95427C	3518
EXV420DIR5JB	0143
EYV320D010JB	0139
EYV320D1R2J3	0139
EYV320DIR2J2	OEM
EYV320DIR213	0139
EYV-420D1R5JA	0015
EYV420D1R5JB	0143
EYV420DIR5JB	0143
EYV3200010IB	OEM
EYZP-307-900	0764
EYZP623	0037
EYZP632	0079
EYZP791	0079
EYZP808	0693
EZ01	0023
EZ01Z	0023
EZ055	OEM
EZ056	OEM
EZ057	OEM
EZ058	OEM
EZ059	OEM
EZ060	OEM
EZ062	OEM
EZ063	OEM
EZ065	OEM
EZ066	OEM
EZ068	OEM
EZ069	OEM
EZ070	OEM
EZ072	OEM
EZ074	OEM
EZ075	OEM
EZ077	OEM
EZ079	OEM
EZ081	OEM
EZ083	OEM
EZ085	OEM
EZ087	OEM
EZ091	0057
EZ093	OEM
EZ095	OEM
EZ097	OEM
EZ099	OEM
EZ3-7(ELCOM)	0205
EZ3-12(ELCOM)	0234
EZ3-15(ELCOM)	0247
EZ3-20(ELCOM)	0262
EZ3A	OEM
EZ4(ELCOM)	0456
EZ5(ELCOM)	0451
EZ5R6(ELCOM)	0157
EZ6(ELCOM)	0025
EZ6A	OEM
EZ7(ELCOM)	0649
EZ8A	3100
EZ9(ELCOM)	0012
EZ10(ELCOM)	0170
EZ10-7(ELCOM)	1458
EZ10-8(ELCOM)	1490
EZ10-10(ELCOM)	0510
EZ10-12(ELCOM)	0870
EZ10-24(ELCOM)	1736
EZ10-28(ELCOM)	2450
EZ10A	OEM
EZ10B	OEM
EZ12(ELCOM)	0137
EZ12A	0137
EZ12B	OEM
EZ15(ELCOM)	0002
EZ15A	0002
EZ15B	OEM
EZ18(ELCOM)	0495
EZ18A	OEM
EZ18B	OEM
EZ19G	OEM
EZ19R	OEM
EZ19Y	OEM
EZ20G	OEM
EZ20R	OEM
EZ20Y	OEM
EZ22(ELCOM)	0565
EZ22A	OEM
EZ27A	OEM
EZ33(ELCOM)	0039
EZ33A	OEM
EZ39(ELCOM)	0350
EZ47(ELCOM)	0999
EZ47A	OEM
EZ56(ELCOM)	0869
EZ56A	OEM
EZ68(ELCOM)	2157
EZ68A	OEM
EZ82(ELCOM)	0331
EZ82A	OEM
EZ100	OEM
EZ101	OEM
EZ104	OEM
EZ106	OEM
EZ109	OEM
EZ111	OEM
EZ114	OEM
EZ117	OEM
EZ119	OEM
EZ122	OEM
EZ125	OEM
EZ127	OEM
EZ129	OEM
EZ132	OEM
EZ135	OEM
EZ138	OEM
EZ141	OEM
EZ144	OEM
EZ148	OEM
EZ150	0002
EZ152	OEM
EZ156	OEM
EZ161	OEM
EZ167	OEM
EZ173	OEM
EZ179	OEM
EZ186	OEM
EZ193	OEM
EZ200	OEM
EZ207	OEM
EZ214	OEM
EZ221	OEM
EZ228	OEM
EZ235	OEM
EZ242	OEM
EZ249	OEM
EZ250	OEM
EZ259	OEM
EZ269	OEM
EZ279	OEM
EZ289	OEM
EZ299	OEM
EZ300	OEM
EZ309	OEM
EZ319	OEM
EZ329	OEM
EZ339	OEM
EZ350	OEM
EZ361	OEM
EZ372	OEM
EZ400	OEM
EZRC07DK201	OEM
EZT308	1854
E.S57X9	0102
F0010	2888
F0015	0325
F0021	0325
F0-02-200D	OEM
F0-02-400D	OEM
F0-02ED	OEM
F0-04ED	OEM
F0-30EDP	OEM
F-05	0015
F05	0015
F031	0133
F046	0133
F060	0133
F084	0133
F0111	OEM
F0810BH	0717
F0810DH	0717
F0810MH	0308
F0810NH	0327
F0R1B	0901
F0R3G	0407
F0R3G6.J	0407
F1	0079
F1(DIODE)	0023
F1-02	0023
F1-04	0023
F1-06	0023
F1-08	0023
F1A3	OEM
F1B3	OEM
F1B5	OEM
F1B6	OEM
F1B7	OEM
F1B8	OEM
F1B9	OEM
F1C3	OEM
F1D3	OEM
F1E3	OEM
F1F3	OEM
F1G3	OEM
F1H3	6192
F1K3	6192
F1M3	OEM
F1N3	OEM
F1V	OEM
F2	0015
F2A3	OEM
F2B3	OEM
F2C3	OEM
F2D	0940
F2D3	OEM
F2E3	OEM
F2F3	OEM
F2G3	OEM
F2H3	6192
F2K3	6192
F2M3	OEM
F2N3	OEM
F3	0015
F3A3	OEM
F3B3	OEM
F3C3	OEM
F3D3	OEM
F3E3	OEM
F3F3	OEM
F3G3	OEM
F3H3	OEM
F3K3	OEM
F3M3	OEM
F3N3	OEM
F3P3	OEM
F3R3	OEM
F3S3	OEM
F4	0015
F4-005	0692
F4A3	OEM
F4B3	OEM
F4C3	OEM
F4D3	OEM
F4E3	OEM
F4F3	OEM
F4G3	OEM
F4H3	OEM
F4K3	OEM
F4M3	OEM
F4N3	OEM
F4P3	OEM
F4R3	OEM
F4S3	OEM
F5	0015
F5A3	OEM
F5B3	OEM
F5C3	OEM
F5D1	OEM
F5D2	OEM
F5D3	OEM
F5E1	OEM
F5E2	OEM
F5E3	OEM
F5F1	3232
F5F3	OEM
F5G3	6220
F5H3	OEM
F5K3	OEM
F5KF20	1127
F5KQ60	OEM
F5M3	OEM
F5N3	OEM
F5P3	OEM
F5R3	OEM
F5S3	OEM
F6	0015
F6A3	OEM
F6B3	OEM
F6C3	OEM
F6D3	OEM
F6E3	OEM
F6F3	OEM
F6G3	OEM
F6H3	OEM
F6K3	OEM
F6M3	OEM
F6N3	OEM
F6P3	OEM
F6R3	OEM
F6S3	OEM
F7A3	OEM
F7B3	OEM
F7C3	OEM
F7D3	OEM
F7E3	OEM
F7F3	OEM
F7G3	OEM
F7H3	OEM
F7K3	OEM
F7M3	OEM
F7N3	OEM

If replacement code is OEM, contact original manufacturer for replacement.

Original Device Types

DEVICE TYPE	REPL CODE
F7P3	OEM
F7R3	OEM
F7S3	OEM
F8	0071
F10	0071
F10P04Q	1021
F11	2336
F11C01FC	OEM
F11C05DC	OEM
F11C05DM	OEM
F11C06DC	OEM
F11C24DC	OEM
F11C24DM	OEM
F11C58DC	OEM
F11C70DC	OEM
F11C70DM	OEM
F11C83DC	OEM
F11C90DC	OEM
F11C90DM	OEM
F11C91DC	OEM
F11C91DM	OEM
F11Q	2336
F12	OEM
F12(DIODE)	0031
F14	0015
F14-C	0015
F-14A	0015
F14A	0015
F14A-P	0015
F14AP	0015
F14B	0023
F14BP	0015
F-14C	0015
F14C	0023
F14CP	0015
F14D	0071
F14DP	0023
F14E	0071
F14F	0071
F14H	0071
F14J	0071
F15	5356
F15X	OEM
F16H1	0471
F16K2DC	OEM
F16K3DC	0522
F16K4DC	0522
F16K5DC	0522
F20	0160
F20-1001	0144
F20-1002	0144
F20-1003	0144
F20-1004	0144
F20-1005	0144
F20-1006	0004
F20-1007	0004
F20-1008	0004
F20-1009	0004
F20-1010	0143
F20-1012	0143
F20-1013	0143
F20-1014	0143
F20-1015	0015
F20-1016	0015
F21	2645
F22	0741
F24E-001	0030
F24T-011-013	0144
F24T-011-015	0224
F24T-016-024	0127
F24T-029	OEM
F24T-029-001	0016
F25	0747
F30	2519
F32	0567
F34	0762
F36	3271
F38	OEM
F38C70DC	OEM
F38C70DL	OEM
F38C70DM	OEM
F38C70PC	OEM
F38C70PM	OEM
F38E70DC	OEM
F38E70DL	OEM
F38E70DM	OEM
F38E70PC	OEM
F38E70PL	OEM
F38E70PM	OEM
F38L72DC	OEM
F38L72DL	OEM
F38L72DM	OEM
F38L72PC	OEM
F38L72PL	OEM
F38L72PM	OEM
F38T56DC	OEM
F38T56DL	OEM
F38T56DM	OEM
F38T56PC	OEM
F38T56PL	OEM
F38T56PM	OEM
F38Z	OEM
F40A	0731
F40H	OEM
F41	OEM
F41Z	OEM
F42	OEM
F42Z	OEM
F50A	OEM
F60A	OEM
F61	2645
F61Z	2645
F62	2645
F62Z	2645
F-67-E	0085
F67E	0160
F68A10CS	OEM
F68A10P	OEM
F68A10S	OEM
F68A21CP	OEM
F68A21CS	OEM
F68A21P	OEM
F68A21S	OEM
F68A40CP	OEM
F68A40CS	OEM
F68A40P	OEM
F68A40S	OEM
F68A44P	OEM
F68A45ACP	OEM
F68A45ACS	OEM
F68A45AP	OEM
F68A45AS	OEM
F68A45CP	OEM
F68A45CS	OEM
F68A45P	OEM
F68A45S	OEM
F68A50P	OEM
F68A50S	OEM
F68A52CP	OEM
F68A52CS	OEM
F68A52P	OEM
F68A52S	OEM
F68A54CP	OEM
F68A54CS	OEM
F68A54P	OEM
F68A54S	OEM
F68A488CP	OEM
F68A488CS	OEM
F68A488P	OEM
F68A488S	OEM
F68B21P	OEM
F68B21S	OEM
F68B40P	OEM
F68B40S	OEM
F68B44P	OEM
F68B45AP	OEM
F68B45AS	OEM
F68B45P	OEM
F68B45S	OEM
F68B50P	OEM
F68B50S	OEM
F68B52P	OEM
F68B52S	OEM
F68B54P	OEM
F68B54S	OEM
F68B488P	OEM
F68B488S	OEM
F-74LS04PC	1595
F-74LS86PC	0291
F74LS174PC	OEM
F-74LS390PC	1288
F74S138	2138
F-74S138PC	2138
F75F	OEM
F78L05AC	1292
F78L05AV	1292
F78L05AWC	1292
F78L06	2301
F78L06AC	2853
F78L06C	2301
F78L062AC	2853
F78L62AC	2301
F78L62AWC	2853
F78L62WV	2301
F80H	OEM
F81	0747
F81Z	2645
F82	0747
F82Z	0087
F96N	0144
F100	OEM
F100H	OEM
F101	0087
F102	0087
F102Z	OEM
F103	OEM
F103P	0150
F104	0065
F105	0065
F106	0065
F107	OEM
F108	OEM
F109	OEM
F110	OEM
F111	OEM
F112	OEM
F112C1	OEM
F113	OEM
F113A	0023
F113B	0023
F113D	0023
F113E	0023
F113F	0182
F114	0023
F114B	0102
F114D	0102
F114E	0023
F114F	0102
F115	0103
F116	0103
F117	0065
F117A	0065
F118	0065
F118A	0065
F119	0065
F119A	0065
F120	0065
F120A	0065
F121	OEM
F121-546	0079
F121-603	0150
F121-60216	0150
F121-433804	0016
F121A	OEM
F122	OEM
F122A	OEM
F123	OEM
F123A	OEM
F124	OEM
F124A	OEM
F136	0143
F150H	OEM
F150HL	OEM
F200	OEM
F200H	OEM
F200HL	OEM
F204	0382
F209	0037
F211	0725
F211A	0725
F215-1001	0144
F215-1002	0144
F215-1003	0144
F215-1004	0144
F215-1005	0144
F215-1006	0004
F215-1007	0004
F215-1008	0004
F215-1009	0004
F215-1010	0143
F215-1012	0143
F215-1013	0143
F215-1014	0143
F215-1016	0015
F215-1017	0015
F222	0016
F230GG0A	OEM
F230HH0A	OEM
F248A	1825
F250H	OEM
F250HL	OEM
F266	2659
F266A	2933
F267	0676
F267A	2936
F268	0676
F268A	2936
F274	0354
F300HL	OEM
F-302-1	0079
F302-1	0016
F302-2	0016
F-302-1532	0079
F302-1532	0016
F-302-2532	0079
F302-2532	0016
F306-001	0016
F306-022	0016
F324	0808
F324(EUROPE)	0808
F336GD	5016
F336HD	4023
F366	0016
F400	OEM
F422	0261
F422(EUROPE)	0261
F422P	0261
F423	OEM
F450	OEM
F495	OEM
F501	0007
F501(ZENITH)	0007
F501-16	0007
F502	0007
F502(ZENITH)	0007
F523	0007
F531	0016
F549-1	0037
F572-1	0016
F587	0016
F600	OEM
F610	0030
F625-1	0086
F-677	0100
F699	0037
F757PC	0701
F767PC	0701
F927-11G	OEM
F927-21G	OEM
F927-31G	OEM
F927-41G	OEM
F927-51G	OEM
F927-61G	OEM
F927-71G	OEM
F1005	OEM
F1016DC	OEM
F1016DM	OEM
F1016FM	OEM
F1022	OEM
F1025	OEM
F1030	OEM
F1031	OEM
F1032	OEM
F1032B	OEM
F1032L	OEM
F1032U	OEM
F1032Z	OEM
F1033	OEM
F1034	OEM
F1035	2987
F1036	1141
F1037	OEM
F1038	OEM
F1039	OEM
F1040	OEM
F1041	OEM
F1042	OEM
F1043	OEM
F1044	OEM
F1050	OEM
F1054	OEM
F1055	OEM
F1056	OEM
F1088	0468
F1089	OEM
F1090	OEM
F1091	OEM
F1092	OEM
F1093	1102
F1094	0963
F1095	2936
F1096	3439
F1103	1086
F1119	OEM
F1120	OEM
F1121	OEM
F1123	2671
F1127	OEM
F1128	OEM
F1168A	0239
F1203M	OEM
F1212M	OEM
F1220M	OEM
F1227L	OEM
F1228L	OEM
F1322	OEM
F1462	0325
F1463	0325
F1612BH	0397
F1612DH	0611
F1612MH	0952
F1612NH	1512
F1802H	OEM
F1802R	OEM
F2004	2458
F2005(HEP)	0849
F2006	OEM
F2007A(HEP)	0212
F2008	OEM
F2009	OEM
F2016AP12	OEM
F2040	OEM
F2041	0791
F2042	OEM
F2043	OEM
F2055	OEM
F2084	OEM
F2085	OEM
F2098	OEM
F2099	OEM
F2101	OEM
F2104	OEM
F2106	OEM
F2107	OEM
F2114-2DC	OEM
F2114-2PC	OEM
F2114-3DC	OEM
F2114-3PC	OEM
F2114DC	OEM
F2114L2DC	OEM
F2114L2PC	OEM
F2114L3DC	OEM
F2114L3PC	OEM
F2114LDC	OEM
F2114LPC	OEM
F2114PC	OEM
F2120	OEM
F2124	OEM
F2125	OEM
F2128	OEM
F2129	OEM
F2130	OEM
F2131	OEM
F2137	OEM
F2147	OEM
F2153	OEM
F2154	OEM
F2161	OEM
F2169	OEM
F2180	OEM
F2184	OEM
F2191	OEM
F2194	OEM
F2195	OEM
F2196	OEM
F2197	OEM
F2427	0144
F2443	0016
F2448	0016
F2450	0144
F2480	0211
F2547	OEM
F2548	OEM
F2552	OEM
F2576	OEM
F2577	OEM
F2578	OEM
F2584	0016
F2633	0144
F2634	0144
F2636	0144
F2708-1DC	OEM
F2708DC	OEM
F2708DL	OEM
F2708DM	OEM
F2716DC	OEM
F2901ADC	OEM
F2901ADM	OEM
F2901APC	OEM
F2909	OEM
F2910	OEM
F2911	OEM
F2912	OEM
F2914	OEM
F2917/8315	OEM
F2919	OEM
F2920	OEM
F2922	OEM
F2923	OEM
F2924	OEM
F2932	OEM
F2933/DOD034	OEM
F2934	OEM
F2945	OEM
F2946	OEM
F2947	OEM
F2948	OEM
F3003	OEM
F3004	OEM
F3012	OEM
F3519	0016
F3530	0144
F3532	0016
F3532-30P	OEM
F3532-30S	OEM
F3532-35P	OEM
F3532-35S	OEM
F3532-45P	OEM
F3532-45S	OEM
F3533-30P	OEM
F3533-30S	OEM
F3533-35P	OEM
F3533-35S	OEM
F3533-45P	OEM
F3533-45S	OEM
F3535	0007
F3549	0126
F3559	0037
F3560	0086
F3561	0086
F3564-25(A)	OEM
F3564-35(A)	OEM
F3565	0086
F3565-25(A)	OEM
F3565-35(A)	OEM
F3566-25(A)	OEM
F3566-35(A)	OEM
F3568-25(A)	OEM
F3568-35(A)	OEM
F3569	0016
F3569-25(A)	OEM
F3569-35(A)	OEM
F3570	0037
F3570-25(A)	OEM
F3570-35(A)	OEM
F3571	0016
F3574	0007
F3589	0086
F3590	0037
F3597	0037
F3846CP	OEM
F3846CS	OEM
F3846DM	OEM
F3846P	OEM
F3846S	OEM
F3850DC	OEM
F3850DL	OEM
F3850DM	OEM
F3851DC	OEM
F3851DL	OEM
F3851DM	OEM
F3851PC	OEM
F3851PL	OEM
F3851PM	OEM
F3854DC	OEM
F3854DL	OEM
F3854DM	OEM
F3854PC	OEM
F3854PL	OEM
F3856DC	OEM
F3856DL	OEM
F3856DM	OEM
F3856PC	OEM
F3856PL	OEM
F3856PM	OEM
F3861DC	OEM
F3861DL	OEM
F3861DM	OEM
F3861PC	OEM
F3861PL	OEM
F3861PM	OEM
F3870DC	OEM
F3870DL	OEM
F3870DM	OEM
F3870PC	OEM
F3870PL	OEM
F3870PM	OEM
F3871DC	OEM
F3871DL	OEM
F3871DM	OEM
F3871PC	OEM
F3871PL	OEM
F3871PM	OEM
F3872DC	OEM
F3872DL	OEM
F3872DM	OEM
F3872PC	OEM
F3872PL	OEM
F3872PM	OEM
F3876DC	OEM
F3876DL	OEM
F3876DM	OEM
F3876PC	OEM
F3876PL	OEM
F3876PM	OEM
F4001	0478
F4002	2055
F4007	2846
F4011	0215
F4011P	OEM
F4011PC	0215
F4012	0498
F4013	0413
F4015	1014
F4016	1143
F4017	0513
F4017D	OEM
F4019	1525
F4020	1661
F4021	1749
F4023	0519
F4024	1957
F4024B	1957
F4025	2072
F4025B	2072
F4026B	2152
F4027	1949
F4029	2233
F4029E	OEM
F4029PC	2233
F4030	0500
F4032B	2531
F4040	0056
F4042	0121
F4049	0001
F4049PC	0001
F4050	0398
F4051	0366
F4052	0024
F4074	OEM
F4075	2540
F4076	3496
F4077BPC	2558
F4081	0626
F4084	OEM
F4123A	OEM
F4129	OEM
F4146	OEM
F4150	OEM
F4156	OEM
F4171A	OEM
F4185	OEM
F4186	OEM
F4187F	OEM
F4511	1544
F4511PC	1544
F4512BPC	4217
F4518	1043
F4528B	3198
F4706	0144
F4709	0086
F4846-1	2552
F5380-1	OEM
F5380-5	OEM
F5380-6	OEM
F5380-G	OEM
F5380-M	OEM
F5380-X	OEM
F-5404DM	OEM
F5404DM	0361
F6446P	OEM
F6810CP	2086
F6810CS	2086
F6810DM	2086
F6810P	2086
F6810S	2086
F6820P	OEM
F6820S	OEM
F6821CP	0448
F6821DM	0448
F6821P	0448
F6821S	0448
F6840CP	OEM
F6840CS	OEM
F6840DL	OEM
F6840DM	OEM
F6840P	OEM
F6840S	OEM
F6844P	OEM
F6845ACP	OEM
F6845ACS	OEM
F6845AP	OEM
F6845AS	OEM
F6845CS	OEM
F6845P	OEM
F6845S	OEM
F6846S	OEM
F6847P	OEM
F6847S	OEM
F6850CP	0514
F6850CS	0514
F6850DL	0514
F6850DM	0514
F6850P	0514
F6850S	0514
F6852CP	OEM
F6852CS	OEM
F6852DLQB	OEM
F6852DMQB	OEM
F6852P	OEM
F6852S	OEM
F6854CP	OEM
F6854CS	OEM
F6854DL	OEM
F6854DM	OEM
F6854P	OEM
F6854S	OEM
F6856CP	OEM
F6856CS	OEM
F6856DM	OEM
F6856P	OEM
F6856S	OEM
F7316	0079
F-7402PC	0314
F7403PC	OEM
F-7404PC	0361
F-7417PC	1352
F7417PC	1352
F7430PC	0873
F7473PC	OEM
F7474PC	1313
F7476PC	1158
F8114	0740
F9014D	0155
F9316PC	1645
F9423DC	OEM
F9423PC	OEM
F9443	OEM
F9444	OEM
F9445-16DC	OEM
F9445-16DM	OEM
F9445-16DMQB	OEM
F9445-20DC	OEM
F9445-20DM	OEM
F9445-20DMQB	OEM
F9445-24DC	OEM
F9445-24DM	OEM
F9445-24DMQB	OEM
F9446	OEM
F9447	OEM
F9448	OEM
F9449DC	OEM
F9449DM	OEM
F9470DC	OEM
F9470DL	OEM
F9470DM	OEM
F9600	0016
F9623	0016
F9623(G.E.)	0144
F9623F	0079
F9625	0144
F10000DC	OEM
F10000DM	OEM
F10000FC	OEM
F10000FM	OEM
F10010DC	OEM
F10010DM	OEM
F10010FM	OEM
F10100DC	OEM
F10101DC	OEM
F10103DC	OEM
F10104DC	OEM
F10105DC	OEM
F10106DC	OEM
F10107DC	OEM
F10109DC	OEM
F10110DC	OEM
F10111DC	OEM
F10113DC	OEM
F10117DC	OEM
F10118DC	OEM
F10119DC	OEM
F10121DC	OEM
F10124	0015
F10130DC	OEM
F10131DC	OEM
F10133DC	OEM
F10135DC	OEM
F10136DC	OEM
F10137DC	OEM
F10141DC	OEM
F10141FC	OEM
F10145ADC	OEM

If replacement code is OEM, contact original manufacturer for replacement.

Original Device Types

DEVICE TYPE	REPL CODE
F10145AFC	OEM
F10148	0015
F10153DC	OEM
F10160DC	OEM
F10166DC	OEM
F10168DC	OEM
F10170DC	OEM
F10175DC	OEM
F10176DC	OEM
F10176FC	OEM
F10176PC	OEM
F10179DC	OEM
F10180	0015
F10180DC	OEM
F10181DC	OEM
F10186DC	OEM
F10186FC	OEM
F10186PC	OEM
F10210DC	OEM
F10211DC	OEM
F10231DC	OEM
F10405DC	OEM
F10405FC	OEM
F10410DC	OEM
F10410FC	OEM
F10410PC	OEM
F10414DC	OEM
F10414FC	OEM
F10415ADC	OEM
F10415AFC	OEM
F10415DC	OEM
F10415FC	OEM
F10416DC	OEM
F10416FC	OEM
F10422DC	OEM
F10422FC	OEM
F10470DC	OEM
F10500DM	OEM
F10501DM	OEM
F10501FM	OEM
F10502DM	OEM
F10502FM	OEM
F10503DM	OEM
F10503FM	OEM
F10504DM	OEM
F10504FM	OEM
F10505DM	OEM
F10505FM	OEM
F10506DM	OEM
F10506FM	OEM
F10507DM	OEM
F10507FM	OEM
F10509DM	OEM
F10509FM	OEM
F10510DM	OEM
F10510FM	OEM
F10511DM	OEM
F10511FM	OEM
F10513DM	OEM
F10513FM	OEM
F10517DM	OEM
F10517FM	OEM
F10518DM	OEM
F10518FM	OEM
F10519DM	OEM
F10519FM	OEM
F10521DM	OEM
F10521FM	OEM
F10530DM	OEM
F10530FM	OEM
F10531DM	OEM
F10531FM	OEM
F10533DM	OEM
F10533FM	OEM
F10535DM	OEM
F10535FM	OEM
F10536DM	OEM
F10536FM	OEM
F10537DM	OEM
F10537FM	OEM
F10541DM	OEM
F10541FM	OEM
F10545ADM	OEM
F10545AFM	OEM
F10553DM	OEM
F10553FM	OEM
F10560DM	OEM
F10560FM	OEM
F10566DM	OEM
F10566FM	OEM
F10568DM	OEM
F10568FM	OEM
F10570DM	OEM
F10570FM	OEM
F10575DM	OEM
F10575FM	OEM
F10576DM	OEM
F10576FM	OEM
F10579DM	OEM
F10579FM	OEM
F10580DM	OEM
F10580FM	OEM
F10581DM	OEM
F10581FM	OEM
F10586DM	OEM
F10586FM	OEM
F10610DM	OEM
F10610FM	OEM
F10611FM	OEM
F10631DM	OEM
F10631FM	OEM
F-10896	6170
F-12227	0023
F-12737	0042
F12737	2520
F-13316	0042
F-13317	0236
F-14168	0133
F-14201	OEM
F-14753	2520
F-14959	0133
F-15023	0133
F-15024	0133
F-15684	3016
F15810	0144
F15835	0144
F15840	0016
F15840-1	0016
F15841	0144
F16032	OEM
F16081	OEM
F16105	OEM
F-16147	0133
F16201	OEM
F16202	OEM
F16203	OEM
F16413	OEM
F16456	OEM
F16488	OEM
F20303	0143
F21490	0037
F27002	0015
F-28309	OEM
F-28908	OEM
F-28909	OEM
F-28910	OEM
F-28911	OEM
F-28912	OEM
F-28936	OEM
F35316-25P	OEM
F35316-25S	OEM
F35316-30P	OEM
F35316-30S	OEM
F50116	0007
F68316-35P	OEM
F68316-35S	OEM
F68316-45P	OEM
F68316-45S	OEM
F68332-30P	OEM
F68332-30S	OEM
F68332-35P	OEM
F68332-45P	OEM
F68332-45S	OEM
F68488CP	OEM
F68488CS	OEM
F68488DL	OEM
F68488DM	OEM
F68488P	OEM
F68488S	OEM
F69916	0086
F73216	0050
F-74390PC	OEM
F75116	0016
F78105AV	1292
F95000DC	OEM
F95002DC	OEM
F95003DC	OEM
F95004DC	OEM
F95010DC	OEM
F95016DC	OEM
F95029DC	OEM
F95101DC	OEM
F95102DC	OEM
F95103DC	OEM
F95105DC	OEM
F95106DC	OEM
F95107DC	OEM
F95109DC	OEM
F95110DC	OEM
F95111DC	OEM
F95130DC	OEM
F95231DC	OEM
F100101DC	OEM
F100101FC	OEM
F100102DC	OEM
F100102FC	OEM
F100107DC	OEM
F100107FC	OEM
F100112DC	OEM
F100112FC	OEM
F100113DC	OEM
F100113FC	OEM
F100114DC	OEM
F100114FC	OEM
F100117DC	OEM
F100117FC	OEM
F100118DC	OEM
F100118FC	OEM
F100125DC	OEM
F100125FC	OEM
F100130DC	OEM
F100130FC	OEM
F100131DC	OEM
F100131FC	OEM
F100141DC	OEM
F100141FC	OEM
F100142DC	OEM
F100142FC	OEM
F100145DC	OEM
F100145FC	OEM
F100150DC	OEM
F100150FC	OEM
F100151DC	OEM
F100151FC	OEM
F100155DC	OEM
F100155FC	OEM
F100156DC	OEM
F100156FC	OEM
F100160DC	OEM
F100160FC	OEM
F100163DC	OEM
F100163FC	OEM
F100164DC	OEM
F100164FC	OEM
F100165DC	OEM
F100165FC	OEM
F100166DC	OEM
F100166FC	OEM
F100170DC	OEM
F100170FC	OEM
F100171DC	OEM
F100171FC	OEM
F100179DC	OEM
F100179FC	OEM
F100180DC	OEM
F100180FC	OEM
F100181DC	OEM
F100181FC	OEM
F100182DC	OEM
F100182FC	OEM
F100183DC	OEM
F100183FC	OEM
F100194DC	OEM
F100194FC	OEM
F100220	OEM
F100414DC	OEM
F100414FC	OEM
F100415DC	OEM
F100415FC	OEM
F100416DC	OEM
F100416FC	OEM
F100422DC	OEM
F100422FC	OEM
F100470DC	OEM
F100470FC	OEM
F814997G8503	OEM
F303030000	OEM
F304059000	OEM
F315056000	OEM
F315060000	OEM
F317055000	OEM
FA002	OEM
FA-1	0143
FA1A4M-L33	OEM
FA1A4P	OEM
FA1L3Z	OEM
FA1L4M-L31	OEM
FA-2	0030
FA-3	OEM
FA-4	0071
FA4	0015
FA6	0015
FA-7	OEM
FA8	0071
FA111	0133
FA2000	OEM
FA2001	OEM
FA2001T	OEM
FA2002	OEM
FA2002T	OEM
FA2003	OEM
FA2004	OEM
FA2005	OEM
FA2006	OEM
FA2007	OEM
FA2008	OEM
FA2009	OEM
FA2010	OEM
FA2011	OEM
FA2310	OEM
FA2310E	0133
FA2310U	0133
FA2320	OEM
FA2320E	0133
FA2320U	0133
FA2330	OEM
FA2330E	0133
FA2330U	0133
FA-2711	OEM
FA4000	OEM
FA4001	OEM
FA4002	OEM
FA4003	OEM
FA4004	OEM
FA4005	OEM
FA4006	OEM
FA4007	OEM
FA4008	OEM
FA4009	OEM
FA4010	OEM
FA4011	OEM
FA4071A	OEM
FA4075	OEM
FA4310	OEM
FA4320	OEM
FA4330	OEM
FA5301P	OEM
FA6001	OEM
FA6001D	0632
FA6001T	OEM
FA6004	OEM
FA6005	OEM
FA6005T	OEM
FA6007	OEM
FA6007M	OEM
FA6007N	OEM
FA6007T	OEM
FA6008(M)	OEM
FA-6008D	OEM
FA-6008T	4570
FA6008T	6583
FA6013	OEM
FA6013D	OEM
FA6016	OEM
FA6017	OEM
FA6018	OEM
FA6018(D)	OEM
FA-6018(M)	OEM
FA6019	OEM
FA-6019(M)	OEM
FA8001	OEM
FA8002	OEM
FA8003	OEM
FA8004	OEM
FA8005	0794
FA8006	0794
FA8007	0794
FA8008	0794
FA8009	0002
FA8010	0002
FA8011	0002
FA8012	0002
FA100252	OEM
FA100443	OEM
FA5353121	OEM
FAA237D	OEM
FAG013D	OEM
FAN3132R	OEM
FAN3132T	OEM
FAN4050D	OEM
FAN4050R	OEM
FAN4050T	OEM
FAN4071D	OEM
FAN4071R	OEM
FAN4071T	OEM
FAN4137R	OEM
FAN4137T	OEM
FAN5132D	OEM
FAN5132R	OEM
FAN5132T	OEM
FAN6102D	OEM
FAN6102R	OEM
FAN6102T	OEM
FAN6133R	OEM
FAN31861R	OEM
FAN31861T	OEM
FAN40551D	OEM
FAN40551R	OEM
FAN40551T	OEM
FAN41041D	OEM
FAN41041R	OEM
FAN41041T	OEM
FAN41052D	OEM
FAN41052R	OEM
FAN41052T	OEM
FAN41851D	OEM
FAN41851R	OEM
FAN41851T	OEM
FAN41860R	OEM
FAN41860T	OEM
FAN60584D	OEM
FAN60584R	OEM
FAN60584T	OEM
FAN60585D	OEM
FAN60585R	OEM
FAN60585T	OEM
FAN61821D	OEM
FAN61821R	OEM
FAN61821T	OEM
FAN605831D	OEM
FAN605831R	OEM
FAN605831T	OEM
FAR-3HS	OEM
FAR-4HS	OEM
FAS16061R	OEM
FAT-3HS	OEM
FAT-4HS	OEM
FB30G	0031
FB100D4	OEM
FB100D-6	OEM
FB100D-8	OEM
FB100D-12	OEM
FB100D-16	OEM
FB100D-20	OEM
FB100D-24	OEM
FB150D-4	OEM
FB150D-6	OEM
FB150D-8	OEM
FB150D-12	OEM
FB150D-16	OEM
FB150D-20	OEM
FB150D-24	OEM
FB-200	0296
FB200	0296
FB-274	OEM
FB274	2393
FB300D-4	OEM
FB300D-6	OEM
FB300D-8	OEM
FB300D-12	OEM
FB300D-16	OEM
FB300D-20	OEM
FB300D-24	OEM
FB401	0050
FB402	0050
FB403	0050
FB420	0004
FB421	0004
FB440	0050
FB1001	2363
FB1001L	0323
FB1002	0323
FB1002L	1414
FB1003	OEM
FB1004	2369
FB1004L	0473
FB1006	2370
FB1006L	0446
FB1043	0143
FB2500	2363
FB2501	2363
FB2502	2363
FB2504	2363
FB2506	2370
FB3500	1643
FB3501	1643
FB3502	1643
FB3504	2925
FB3506	2925
FB6853	0016
FB8056	6134
FB8550	6174
FBC404M3L	OEM
FBM31DB	OEM
FBM43DA	OEM
FBN5175D	OEM
FBN5185R	OEM
FBN5185T	OEM
FBP-00-024	0725
FBS3425F	3152
FBS3515F	0279
FBS3525F	3153
FBS3610F	0422
FBTX070	0086
FC10D	OEM
FC10E	OEM
FC11D	OEM
FC11E	OEM
FC12D	OEM
FC12E	OEM
FC13D	OEM
FC13E	OEM
FC14D	OEM
FC14E	OEM
FC31A	OEM
FC31B	OEM
FC33B	OEM
FC34B	OEM
FC42	OEM
FC44	OEM
FC51	OEM
FC51A	OEM
FC51H	OEM
FC51M	OEM
FC-52	0030
FC52	0030
FC52A	OEM
FC-52M	0030
FC52M	0030
FC53	OEM
FC53A	OEM
FC53M	OEM
FC54	OEM
FC54A	OEM
FC54M	OEM
FC55	0005
FC56	1029
FC57	1029
FC62	OEM
FC63	OEM
FC64M	OEM
FC65M	OEM
FC66M	OEM
FC74LS90	1882
FC74LS93	1888
FC74LS196N	2834
FC81	0005
FC82	0554
FC91	0133
FC210R	OEM
FC210S	OEM
FC225R	OEM
FC225S	OEM
FC240R	OEM
FC240S	OEM
FC5006	0144
FC100240	OEM
FC100347	OEM
FC100348	OEM
FC100350	OEM
FC100535	OEM
FC100541	OEM
FC100542	OEM
FC100611	OEM
FC100628	OEM
FCB61C65L-70P	OEM
FCD0003PC	0133
FCD0014NCS	0133
FCD0070ANC	0030
FCD0070PC	0030
FCD0074	0030
FCD810	0541
FCD810C	0315
FCD810D	0315
FCD811	0315
FCD820	0541
FCD820A	0541
FCD820B	0541
FCD820C	0315
FCD820D	0315
FCD825	0315
FCD825A	0315
FCD825C	0315
FCD825D	0315
FCD830	0541
FCD830C	0315
FCD830D	0315
FCD831C	0315
FCD831D	0315
FCD836C	0315
FCD836D	0315
FCD850	1053
FCD850C	1053
FCD855	1109
FCD855C	1109
FCD860	1053
FCD860C	1053
FCD865	1053
FCD865C	1053
FCD880	1403
FCD885	1403
FCH101	OEM
FCH111	OEM
FCH121	OEM
FCH131	OEM
FCH141	OEM
FCH151	OEM
FCH161	OEM
FCH171	OEM
FCH181	OEM
FCH191	OEM
FCH221	OEM
FCH231	OEM
FCH301	OEM
FCJ101	OEM
FCJ111	OEM
FCJ121	OEM
FCJ131	OEM
FCJ141	OEM
FCJ191	OEM
FCJ201	OEM
FCJ211	OEM
FCJ221	OEM
FCK111	OEM
FCN16062R	OEM
FCR8U01HA	OEM
FCR8U01HC	OEM
FCR8U01HD	OEM
FCR8U02HA	OEM
FCR8U02HC	OEM
FCR8U02HD	OEM
FCR8U03HA	OEM
FCR8U03HC	OEM
FCR8U03HD	OEM
FCR8U04HA	OEM
FCR8U04HC	OEM
FCR8U04HD	OEM
FCR8U05HA	OEM
FCR8U05HC	OEM
FCR8U05HD	OEM
FCR8U06HA	OEM
FCR8U06HC	OEM
FCR8U06HD	OEM
FCR9U01JC	OEM
FCR9U01JD	OEM
FCR9U01JG	OEM
FCR9U02JC	OEM
FCR9U02JD	OEM
FCR9U02JG	OEM
FCR9U03JC	OEM
FCR9U03JD	OEM
FCR9U03JG	OEM
FCR9U04JC	OEM
FCR9U04JD	OEM
FCR9U04JG	OEM
FCR9U05JC	OEM
FCR9U05JD	OEM
FCR9U05JG	OEM
FCR9U06JD	OEM
FCR9U06JG	OEM
FCR9U07JA	OEM
FCR9U07JC	OEM
FCR9U08JA	OEM
FCR9U08JC	OEM
FCR9U09JA	OEM
FCR9U09JC	OEM
FCR9U10JA	OEM
FCR9U10JC	OEM
FCR9U11JA	OEM
FCR9U11JC	OEM
FCR9U12JA	OEM
FCR9U12JC	OEM
FCR12U01JC	OEM
FCR12U01JD	OEM
FCR12U01JG	OEM
FCR12U02JC	OEM
FCR12U02JD	OEM
FCR12U02JG	OEM
FCR12U03JC	OEM
FCR12U03JD	OEM
FCR12U03JG	OEM
FCR12U04JC	OEM
FCR12U04JD	OEM
FCR12U04JG	OEM
FCR12U05JC	OEM
FCR12U05JD	OEM
FCR12U05JG	OEM
FCR12U06JC	OEM
FCR12U06JD	OEM
FCR12U06JG	OEM
FCR12U07JA	OEM
FCR12U07JC	OEM
FCR12U08JA	OEM
FCR12U08JC	OEM
FCR12U09JA	OEM
FCR12U09JC	OEM
FCR12U10JA	OEM
FCR12U10JC	OEM
FCR12U11JA	OEM
FCR12U11JC	OEM
FCR12U12JA	OEM
FCR12U12JC	OEM
FCR20U01JC	OEM
FCR20U01JD	OEM
FCR20U01JG	OEM
FCR20U02JC	OEM
FCR20U02JD	OEM
FCR20U02JG	OEM
FCR20U03JC	OEM
FCR20U03JD	OEM
FCR20U03JG	OEM
FCR20U04JC	OEM
FCR20U04JD	OEM
FCR20U04JG	OEM
FCR20U05JC	OEM
FCR20U05JD	OEM
FCR20U05JG	OEM
FCR20U06JC	OEM
FCR20U06JD	OEM
FCR20U06JG	OEM
FCR20U07JA	OEM
FCR20U07JC	OEM
FCR20U08JA	OEM
FCR20U08JC	OEM
FCR20U09JA	OEM
FCR20U09JC	OEM
FCR20U10JA	OEM
FCR20U10JC	OEM
FCR20U11JA	OEM
FCR20U11JC	OEM
FCR20U12JA	OEM
FCR20U12JC	OEM
FCR24U01JC	OEM
FCR24U01JD	OEM
FCR24U01JG	OEM
FCR24U02JC	OEM
FCR24U02JD	OEM
FCR24U02JG	OEM
FCR24U03JC	OEM
FCR24U03JD	OEM
FCR24U03JG	OEM
FCR24U04JC	OEM
FCR24U04JD	OEM
FCR24U04JG	OEM
FCR24U05JC	OEM
FCR24U05JD	OEM
FCR24U05JG	OEM
FCR24U06JC	OEM
FCR24U06JD	OEM
FCR24U06JG	OEM
FCR24U07JA	OEM
FCR24U07JC	OEM
FCR24U08JA	OEM
FCR24U08JC	OEM
FCR24U09JA	OEM
FCR24U09JC	OEM
FCR24U10JA	OEM
FCR24U10JC	OEM
FCR24U11JA	OEM
FCR24U11JC	OEM
FCR24U12JA	OEM
FCR24U12JC	OEM
FCR31U01JB	OEM
FCR31U01JC	OEM
FCR31U01JD	OEM
FCR31U02JB	OEM
FCR31U02JC	OEM
FCR31U02JD	OEM
FCR31U03JB	OEM
FCR31U03JC	OEM
FCR31U03JD	OEM
FCR31U04JB	OEM
FCR31U04JC	OEM
FCR31U04JD	OEM
FCR31U05JB	OEM
FCR31U05JC	OEM
FCR31U05JD	OEM
FCR31U06JB	OEM
FCR31U06JC	OEM
FCR31U06JD	OEM
FCR31U07JA	OEM
FCR31U07JB	OEM

If replacement code is OEM, contact original manufacturer for replacement.

DEVICE TYPE	REPL CODE
FCR31U08JA	OEM
FCR31U08JB	OEM
FCR31U09JA	OEM
FCR31U09JB	OEM
FCR31U10JA	OEM
FCR31U10JB	OEM
FCR31U11JA	OEM
FCR31U11JB	OEM
FCR43U01JA	OEM
FCR43U01JB	OEM
FCR43U01JM	OEM
FCR43U02JA	OEM
FCR43U02JB	OEM
FCR43U02JM	OEM
FCR43U03JA	OEM
FCR43U03JB	OEM
FCR43U03JM	OEM
FCR43U04JA	OEM
FCR43U04JB	OEM
FCR43U04JM	OEM
FCR43U05JA	OEM
FCR43U05JB	OEM
FCR43U06JA	OEM
FCR43U06JB	OEM
FCR43U06JM	OEM
FCR43U07JA	OEM
FCR43U07JB	OEM
FCR43U07JM	OEM
FCR43U08JA	OEM
FCR43U08JB	OEM
FCR43U08JM	OEM
FCR43U09J,	OEM
FCR43U09JA	OEM
FCR43U09JB	OEM
FCR43U10JA	OEM
FCR43U10JB	OEM
FCR43U10JM	OEM
FCR43U11JA	OEM
FCR43U11JB	OEM
FCR43U11JM	OEM
FCR43U12JA	OEM
FCR43U12JB	OEM
FCR44U01JC	OEM
FCR44U01UD	OEM
FCR44U02JC	OEM
FCR44U02JD	OEM
FCR44U03JC	OEM
FCR44U03JD	OEM
FCR44U04JC	OEM
FCR44U04JD	OEM
FCR44U05JC	OEM
FCR44U05JD	OEM
FCR44U06JC	OEM
FCR44U06JD	OEM
FCR50U01JC	OEM
FCR70U05JD	OEM
FCR80U03JB	OEM
FCR80U03JC	OEM
FCR80U03JD	OEM
FCR80U04JC	OEM
FCR80U04JD	OEM
FCR80U05JC	OEM
FCR80U05JD	OEM
FCR80U06JC	OEM
FCR80U06JD	OEM
FCR80U07JB	OEM
FCR80U08JB	OEM
FCR80U09JB	OEM
FCR80U10JB	OEM
FCR80U11JB	OEM
FCR100U09JB	OEM
FCS916F	0224
FCS1168E	0144
FCS1168E641	0144
FCS1168F813	0016
FCS1168G	0016
FCS1168G704	0016
FCS1170F	0037
FCS1225E	0144
FCS1227E	0144
FCS1227E814	0144
FCS1227F	0144
FCS1227F743	0144
FCS1227G	0144
FCS1227G810	0144
FCS1229	0086
FCS1229F	0016
FCS1229G	0016
FCS1795D	0126
FCS8013HG	OEM
FCS8050C	0155
FCS8550	0006
FCS8550C	0006
FCS9011E	0079
FCS9011F	0079
FCS9011G	0286
FCS9011H	0079
FCS9012	0037
FCS-9012F	0037
FCS-9012G	0037
FCS9012G	0037
FCS9012H	0037
FCS9012HE	0037
FCS9012HG	0037
FCS9012HH	0150
FCS9013	0286
FCS-9013F	0079
FCS9013F	0079
FCS9013F,G	0016
FCS9013FG	0016
FCS-9013G	0079
FCS9013G	0286
FCS9013H	0286
FCS9013HG	0079
FCS9013HH	0079
FCS9014	0079
FCS9014(B)	0079
FCS9014B	0079
FCS9014C	0111
FCS9014D	0144
FCS9015B	0037
FCS-9015C	0150
FCS9015C	2481
FCS9015D	0037
FCS9016	0127
FCS9016D	0127
FCS9016E	0127
FCS-9016F	0368
FCS9016F	0079
FCS-9016G	0079
FCS9016G	0127
FCS9016H	0144
FCS9018	0144
FCS9018D	0144
FCS9018E	0144
FCS-9018F	0016
FCS9018F	0144
FCS9018G	0144
FCS9018H	0144
FCS9066	0144
FCT1035	OEM
FCT1135	OEM
FCY101	OEM
FD3	0015
FD6	0015
FD8	OEM
FD13	OEM
FD13A	OEM
FD13B	OEM
FD13C	OEM
FD13D	OEM
FD13E	OEM
FD13F	OEM
FD13G	OEM
FD23	OEM
FD43	OEM
FD63	0087
FD80A4	OEM
FD80A6	OEM
FD80A8	OEM
FD80A10	OEM
FD83	2645
FD100	0124
FD103	0087
FD111	0133
FD150B4	OEM
FD150B6	OEM
FD150B8FD150B10	OEM
FD160L	0288
FD200	0133
FD200B-2	OEM
FD200B2	OEM
FD200B-4	OEM
FD200B4	OEM
FD200B-8	OEM
FD200B8	OEM
FD200B-16	OEM
FD200B16	OEM
FD200E-2	OEM
FD200E2	OEM
FD200E-4	OEM
FD200E4	OEM
FD200E-8	OEM
FD200E8	OEM
FD200E-16	OEM
FD200E16	OEM
FD222	0133
FD250A12	OEM
FD250A16	OEM
FD250A-20	OEM
FD250A-24	OEM
FD250A-28	OEM
FD250A30	OEM
FD250A-32	OEM
FD250A-36	OEM
FD250A-40	OEM
FD250B-12	OEM
FD250B-16	OEM
FD250B-20	OEM
FD250DM28	OEM
FD250DM32	OEM
FD250DM36	OEM
FD250DM40	OEM
FD300	0609
FD333	0133
FD400	0609
FD400DL12	OEM
FD400DL20	OEM
FD400DL24	OEM
FD500C32	OEM
FD500C36	OEM
FD500C-40	OEM
FD500C40	OEM
FD500C-50	OEM
FD500C50	OEM
FD500C-56	OEM
FD500C56	OEM
FD500C-60	OEM
FD500C60	OEM
FD500D10	OEM
FD500D12	OEM
FD500D16	OEM
FD500D20	OEM
FD500E12	OEM
FD500E-16	OEM
FD500E16	OEM
FD500E-20	OEM
FD500E20	OEM
FD500E-24	OEM
FD500E24	OEM
FD500E-28	OEM
FD500E28	OEM
FD500E30	OEM
FD500E-32	OEM
FD500E32	OEM
FD500E-36	OEM
FD500E36	OEM
FD500E-40	OEM
FD500E40	OEM
FD500E-50	OEM
FD500E50	OEM
FD500E-60	OEM
FD500E60	OEM
FD500E70	OEM
FD500E80	OEM
FD500EV-70	OEM
FD500EV70	OEM
FD500EV-80	OEM
FD500EV80	OEM
FD500d24	OEM
FD600	0124
FD600-1	OEM
FD600-2	OEM
FD600-3	OEM
FD600-4	OEM
FD600-5	OEM
FD600-6	OEM
FD600-7	OEM
FD600-8	OEM
FD600-9	OEM
FD600-10	OEM
FD600-12	OEM
FD600-14	OEM
FD600-16	OEM
FD600-18	OEM
FD600-20	OEM
FD600-22	OEM
FD600-24	OEM
FD600-26	OEM
FD600-28	OEM
FD600-30	OEM
FD700	5042
FD750-2	OEM
FD750-4	OEM
FD750-6	OEM
FD750-12	OEM
FD750-16	OEM
FD750-20	OEM
FD750-24	OEM
FD777	0124
FD900-1	OEM
FD900-2	OEM
FD900-3	OEM
FD900-4	OEM
FD900-5	OEM
FD900-6	OEM
FD900-7	OEM
FD900-8	OEM
FD900-9	OEM
FD900-10	OEM
FD900-12	OEM
FD900-14	OEM
FD900-16	OEM
FD900-18	OEM
FD900-20	OEM
FD1000-2	OEM
FD1000-4	OEM
FD1000-6	OEM
FD1000-12	OEM
FD1000-16	OEM
FD1000-20	OEM
FD1000-24	OEM
FD1000A12	OEM
FD1000A-16	OEM
FD1000A16	OEM
FD1000A-20	OEM
FD1000A20	OEM
FD1000A-24	OEM
FD1000A24	OEM
FD1000A-28	OEM
FD1000A28	OEM
FD1000A-32	OEM
FD1000A32	OEM
FD1000A-36	OEM
FD1000A36	OEM
FD1000A-40	OEM
FD1000A40	OEM
FD1000A-50	OEM
FD1000A50	OEM
FD1000A-56	OEM
FD1000A56	OEM
FD1000A70	OEM
FD1000A80	OEM
FD1000D16	OEM
FD1000D20	OEM
FD1000D24	OEM
FD1000D28	OEM
FD1000D32	OEM
FD1000D36	OEM
FD1000D40	OEM
FD1000D50	OEM
FD1000D56	OEM
FD-1029-4	0037
FD-1029-4&	0037
FD-1029-EE	0211
FD-1029-JE	0122
FD-1073-BF	0232
FD-1073-BG	0314
FD-1073-BH	0335
FD-1073-BJ	0361
FD-1073-BM	0467
FD-1073-BN	0512
FD-1073-BR	0697
FD-1073-BS	0873
FD-1073-BU	1024
FD-1073-BW	1168
FD-1073-CA	1368
FD1389	OEM
FD1559	0540
FD1599	0015
FD1600A-16	OEM
FD1600A16	OEM
FD1600A-20	OEM
FD1600A20	OEM
FD1600A-24	OEM
FD1600A24	OEM
FD1600A-28	OEM
FD1600A28	OEM
FD1600A-32	OEM
FD1600A32	OEM
FD1600A-36	OEM
FD1600A36	OEM
FD1600A-40	OEM
FD1600A40	OEM
FD1600A-50	OEM
FD1600A50	OEM
FD1600BP-2	OEM
FD1600BP2	OEM
FD1600BP-4	OEM
FD1600BP4	OEM
FD1600BP-6	OEM
FD1600BP6	OEM
FD1600BP-8	OEM
FD1600BP8	OEM
FD1600BP-10	OEM
FD1600BP10	OEM
FD1708	0133
FD1761A-02	OEM
FD1761B-02	OEM
FD1763A-02	OEM
FD1765A-02	OEM
FD1765B-02	OEM
FD1767A-02	OEM
FD1767B-02	OEM
FD1771-B01	OEM
FD1771A01	OEM
FD1771B01	OEM
FD1791A02	OEM
FD1791B02	OEM
FD1792A02	OEM
FD1792B02	OEM
FD1793A02	OEM
FD1793B-02	OEM
FD1793B02	OEM
FD1794A02	OEM
FD1794B02	OEM
FD1795A02	OEM
FD1795B02	OEM
FD1797A02	OEM
FD1797B02	OEM
FD1843	0133
FD1873B-02	OEM
FD1980	0143
FD2055	OEM
FD2389	OEM
FD3389	0015
FD3405	0920
FD3500AH-16	OEM
FD3500AH-20	OEM
FD3500AH-24	OEM
FD3500AH-28	OEM
FD3500AH-32	OEM
FD3500AH-36	OEM
FD3500AH-40	OEM
FD3500AH-50	OEM
FD3500AH56	OEM
FD6389	OEM
FD6451	0133
FD6489	0133
FD6666	OEM
FD17718-01	OEM
FDA548	OEM
FDA555	OEM
FDC11	OEM
FDC91C36	OEM
FDC765AC	OEM
FDC1761-02	OEM
FDC1763-02	OEM
FDC1765-02	OEM
FDC1767-02	OEM
FDC1791-02	OEM
FDC1792-02	OEM
FDC1793-02	OEM
FDC1794-02	OEM
FDC1795-02	OEM
FDC1797-02	OEM
FDC3070	OEM
FDC3400	OEM
FDC3400-1	OEM
FDC3600	OEM
FDC4376	OEM
FDC4558	OEM
FDC9216	OEM
FDC9216B	OEM
FDC9229	OEM
FDC9229B	OEM
FDD-1N	OEM
FDD-2N	OEM
FDD-3N	OEM
FDD-4N	OEM
FDG559	OEM
FDH-9	0133
FDH300	0609
FDH333	0609
FDH400	0015
FDH-444	0124
FDH444	0124
FDH600	0102
FDH666	0133
FDH694	0133
FDH900	0133
FDH999	0124
FDH1000	1335
FDH6072	OEM
FDH6229	0133
FDM1006	0133
FDM1007	0133
FDN400	0609
FDN444	0609
FDN600	0133
FDN666	0133
FDR-2L	OEM
FDR-5AL	OEM
FDR-5L	OEM
FDR116Z	OEM
FDR126Z1	OEM
FDR131Z2	OEM
FDR300	OEM
FDR600	OEM
FDR700	OEM
FDT-2L	OEM
FDT-5L	OEM
FDU6603	OEM
FDZ10A	OEM
FDZ10B	OEM
FDZ12A	OEM
FDZ12B	OEM
FDZ15A	OEM
FDZ15B	OEM
FDZ18A	OEM
FDZ18B	OEM
FDZ22A	OEM
FDZ27A	OEM
FDZ33A	OEM
FDZ39A	OEM
FDZ47A	OEM
FDZ56A	OEM
FDZ68A	OEM
FDZ82A	OEM
FE00202ATF	OEM
FE040EABR	OEM
FE071AAB	OEM
FE0101	OEM
FE0101ATF	OEM
FE0101ATFB	OEM
FE0101ATFG	OEM
FE0101ATFQ	OEM
FE0101ATFR	OEM
FE0101ATG	OEM
FE0101ATGG	OEM
FE0101ATGO	OEM
FE0101ATGR	OEM
FE0101ATH	OEM
FE0101ATHB	OEM
FE0101ATHG	OEM
FE0101ATHR	OEM
FE0101ATI	OEM
FE0101ATIB	OEM
FE0101ATIG	OEM
FE0101ATIQ	OEM
FE0101ATIR	OEM
FE0101ATJ	OEM
FE0101ATJB	OEM
FE0101ATJG	OEM
FE0101ATJQ	OEM
FE0101ATJR	OEM
FE0101TFB	OEM
FE0101TFG	OEM
FE0101TFQ	OEM
FE0101TFR	OEM
FE0101TG	OEM
FE0101TGB	OEM
FE0101TGG	OEM
FE0101TGQ	OEM
FE0101TGR	OEM
FE0101TH	OEM
FE0101THB	OEM
FE0101THG	OEM
FE0101THR	OEM
FE0101TI	OEM
FE0101TIB	OEM
FE0101TIG	OEM
FE0101TIQ	OEM
FE0101TIR	OEM
FE0101TJ	OEM
FE0101TJB	OEM
FE0101TJG	OEM
FE0101TJQ	OEM
FE0101TJR	OEM
FE0201	OEM
FE0201A	OEM
FE0201AA	OEM
FE0201AAB	OEM
FE0201AAG	OEM
FE0201AAQ	OEM
FE0201AAR	OEM
FE0201AB	OEM
FE0201ABB	OEM
FE0201ABG	OEM
FE0201ABQ	OEM
FE0201ABR	OEM
FE0201AC	OEM
FE0201ACB	OEM
FE0201ACG	OEM
FE0201ACR	OEM
FE0201AD	OEM
FE0201ADB	OEM
FE0201ADG	OEM
FE0201ADQ	OEM
FE0201ADR	OEM
FE0201AE	OEM
FE0201AEG	OEM
FE0201AEQ	OEM
FE0201AER	OEM
FE0201AG	OEM
FE0201AQ	OEM
FE0201AR	OEM
FE0201ATF	OEM
FE0201ATFB	OEM
FE0201ATFG	OEM
FE0201ATFQ	OEM
FE0201ATFR	OEM
FE0201ATG	OEM
FE0201ATGB	OEM
FE0201ATGG	OEM
FE0201ATGQ	OEM
FE0201ATGR	OEM
FE0201ATH	OEM
FE0201ATHB	OEM
FE0201ATHG	OEM
FE0201ATHR	OEM
FE0201ATI	OEM
FE0201ATIB	OEM
FE0201ATIG	OEM
FE0201ATIQ	OEM
FE0201ATIR	OEM
FE0201ATJ	OEM
FE0201ATJB	OEM
FE0201ATJG	OEM
FE0201ATJQ	OEM
FE0201ATJR	OEM
FE0201B	OEM
FE0201BB	OEM
FE0201BG	OEM
FE0201BQ	OEM
FE0201BR	OEM
FE0201C	OEM
FE0201CB	OEM
FE0201CG	OEM
FE0201CR	OEM
FE0201D	OEM
FE0201DB	OEM
FE0201DG	OEM
FE0201DQ	OEM
FE0201DR	OEM
FE0201E	OEM
FE0201EB	OEM
FE0201EG	OEM
FE0201EQ	OEM
FE0201ER	OEM
FE0201TF	OEM
FE0201TFB	OEM
FE0201TFG	OEM
FE0201TFQ	OEM
FE0201TFR	OEM
FE0201TG	OEM
FE0201TGB	OEM
FE0201TGG	OEM
FE0201TGQ	OEM
FE0201TGR	OEM
FE0201TH	OEM
FE0201THB	OEM
FE0201THG	OEM
FE0201THR	OEM
FE0201TI	OEM
FE0201TIB	OEM
FE0201TIG	OEM
FE0201TIQ	OEM
FE0201TIR	OEM
FE0201TJ	OEM
FE0201TJB	OEM
FE0201TJG	OEM
FE0201TJQ	OEM
FE0201TJR	OEM
FE0202	OEM
FE0202A	OEM
FE0202AA	OEM
FE0202AAB	OEM
FE0202AAG	OEM
FE0202AAQ	OEM
FE0202AAR	OEM
FE0202AB	OEM
FE0202ABB	OEM
FE0202ABG	OEM
FE0202ABQ	OEM
FE0202ABR	OEM
FE0202AC	OEM
FE0202ACB	OEM
FE0202ACG	OEM
FE0202AD	OEM
FE0202ADB	OEM
FE0202ADG	OEM
FE0202ADQ	OEM
FE0202ADR	OEM
FE0202AE	OEM
FE0202AEB	OEM
FE0202AEQ	OEM
FE0202AER	OEM
FE0202AG	OEM
FE0202AQ	OEM
FE0202AR	OEM
FE0202ATFB	OEM
FE0202ATFG	OEM
FE0202ATFQ	OEM
FE0202ATFR	OEM
FE0202ATG	OEM
FE0202ATGB	OEM
FE0202ATGG	OEM
FE0202ATGQ	OEM
FE0202ATGR	OEM
FE0202ATH	OEM
FE0202ATHB	OEM
FE0202ATHG	OEM
FE0202ATI	OEM
FE0202ATIB	OEM
FE0202ATIG	OEM
FE0202ATIQ	OEM
FE0202ATIR	OEM
FE0202ATJ	OEM
FE0202ATJB	OEM
FE0202ATJG	OEM
FE0202ATJQ	OEM
FE0202ATJR	OEM
FE0202B	OEM
FE0202BB	OEM
FE0202BG	OEM
FE0202BQ	OEM
FE0202BR	OEM
FE0202C	OEM
FE0202CB	OEM
FE0202CG	OEM
FE0202CR	OEM
FE0202D	OEM
FE0202DB	OEM
FE0202DG	OEM
FE0202DQ	OEM
FE0202DR	OEM
FE0202E	OEM
FE0202EB	OEM
FE0202EG	OEM
FE0202EQ	OEM
FE0202ER	OEM
FE0202TFB	OEM
FE0202TFG	OEM
FE0202TFQ	OEM
FE0202TFR	OEM
FE0202TGB	OEM
FE0202TGG	OEM
FE0202TGQ	OEM
FE0202TGR	OEM
FE0202THB	OEM
FE0202THG	OEM
FE0202THR	OEM
FE0202TIB	OEM
FE0202TIG	OEM
FE0202TIQ	OEM
FE0202TIR	OEM
FE0202TJB	OEM
FE0202TJG	OEM
FE0202TJQ	OEM
FE0202TJR	OEM
FE0203	OEM
FE0203A	OEM
FE0203AA	OEM
FE0203AAB	OEM
FE0203AAG	OEM
FE0203AAQ	OEM
FE0203AAR	OEM
FE0203AB	OEM
FE0203ABB	OEM
FE0203ABG	OEM
FE0203ABQ	OEM
FE0203ABR	OEM
FE0203AC	OEM
FE0203ACB	OEM
FE0203ACR	OEM
FE0203AD	OEM
FE0203ADB	OEM
FE0203ADG	OEM
FE0203ADQ	OEM
FE0203ADR	OEM
FE0203AE	OEM
FE0203AEB	OEM
FE0203AEG	OEM
FE0203AEQ	OEM
FE0203AER	OEM
FE0203AG	OEM
FE0203AQ	OEM
FE0203AR	OEM
FE0203ATF	OEM
FE0203ATFB	OEM
FE0203ATFG	OEM
FE0203ATFQ	OEM

DEVICE TYPE	REPL CODE
FE0203ATFR	OEM
FE0203ATGB	OEM
FE0203ATGG	OEM
FE0203ATGQ	OEM
FE0203ATGR	OEM
FE0203ATH	OEM
FE0203ATHB	OEM
FE0203ATHG	OEM
FE0203ATHR	OEM
FE0203ATI	OEM
FE0203ATIB	OEM
FE0203ATIG	OEM
FE0203ATIQ	OEM
FE0203ATIR	OEM
FE0203ATJB	OEM
FE0203ATJG	OEM
FE0203ATJQ	OEM
FE0203ATJR	OEM
FE0203B	OEM
FE0203BB	OEM
FE0203BG	OEM
FE0203BQ	OEM
FE0203BR	OEM
FE0203C	OEM
FE0203CB	OEM
FE0203CG	OEM
FE0203CR	OEM
FE0203D	OEM
FE0203DG	OEM
FE0203DQ	OEM
FE0203DR	OEM
FE0203E	OEM
FE0203EB	OEM
FE0203EG	OEM
FE0203EQ	OEM
FE0203ER	OEM
FE0203TF	OEM
FE0203TFB	OEM
FE0203TFQ	OEM
FE0203TFR	OEM
FE0203TG	OEM
FE0203TGB	OEM
FE0203TGG	OEM
FE0203TGQ	OEM
FE0203TGR	OEM
FE0203TH	OEM
FE0203THB	OEM
FE0203THG	OEM
FE0203THR	OEM
FE0203TI	OEM
FE0203TIB	OEM
FE0203TIG	OEM
FE0203TIQ	OEM
FE0203TIR	OEM
FE0203TJ	OEM
FE0203TJB	OEM
FE0203TJQ	OEM
FE0203TJR	OEM
FE0204A	OEM
FE0204B	OEM
FE0204C	OEM
FE0204D	OEM
FE0204E	OEM
FE0204F	OEM
FE0204G	OEM
FE0204H	OEM
FE0204I	OEM
FE0204J	OEM
FE0205	OEM
FE0205A	OEM
FE0205B	OEM
FE0205C	OEM
FE0205D	OEM
FE0205E	OEM
FE0205F	OEM
FE0205G	OEM
FE0205H	OEM
FE0205I	OEM
FE0205J	OEM
FE0206	OEM
FE0206A	OEM
FE0206AA	OEM
FE0206AAB	OEM
FE0206AAG	OEM
FE0206AAQ	OEM
FE0206AAR	OEM
FE0206AB	OEM
FE0206ABB	OEM
FE0206ABG	OEM
FE0206ABQ	OEM
FE0206ABR	OEM
FE0206AC	OEM
FE0206ACB	OEM
FE0206ACG	OEM
FE0206ACR	OEM
FE0206AD	OEM
FE0206ADB	OEM
FE0206ADG	OEM
FE0206ADQ	OEM
FE0206ADR	OEM
FE0206AE	OEM
FE0206AEB	OEM
FE0206AEG	OEM
FE0206AEQ	OEM
FE0206AER	OEM
FE0206AG	OEM
FE0206AQ	OEM
FE0206AR	OEM
FE0206ATF	OEM
FE0206ATFB	OEM
FE0206ATFG	OEM
FE0206ATFQ	OEM
FE0206ATFR	OEM
FE0206ATG	OEM
FE0206ATGB	OEM
FE0206ATGG	OEM
FE0206ATGQ	OEM
FE0206ATGR	OEM
FE0206ATH	OEM
FE0206ATHB	OEM
FE0206ATHG	OEM
FE0206ATHR	OEM
FE0206ATI	OEM
FE0206ATIB	OEM
FE0206ATIG	OEM
FE0206ATIQ	OEM
FE0206ATIR	OEM
FE0206ATJ	OEM
FE0206ATJB	OEM
FE0206ATJG	OEM
FE0206ATJQ	OEM
FE0206ATJR	OEM
FE0206B	OEM
FE0206BB	OEM
FE0206BG	OEM
FE0206BQ	OEM
FE0206BR	OEM
FE0206C	OEM
FE0206CB	OEM
FE0206CG	OEM
FE0206CR	OEM
FE0206D	OEM
FE0206DB	OEM
FE0206DG	OEM
FE0206DQ	OEM
FE0206DR	OEM
FE0206E	OEM
FE0206EB	OEM
FE0206EG	OEM
FE0206EQ	OEM
FE0206ER	OEM
FE0206TF	OEM
FE0206TFB	OEM
FE0206TFG	OEM
FE0206TFQ	OEM
FE0206TFR	OEM
FE0206TG	OEM
FE0206TGB	OEM
FE0206TGG	OEM
FE0206TGQ	OEM
FE0206TGR	OEM
FE0206TH	OEM
FE0206THB	OEM
FE0206THG	OEM
FE0206THR	OEM
FE0206TI	OEM
FE0206TIB	OEM
FE0206TIG	OEM
FE0206TIQ	OEM
FE0206TIR	OEM
FE0206TJ	OEM
FE0206TJG	OEM
FE0206TJQ	OEM
FE0206TJR	OEM
FE0208	OEM
FE0208A	OEM
FE0208AA	OEM
FE0208AAB	OEM
FE0208AAG	OEM
FE0208AAR	OEM
FE0208AB	OEM
FE0208ABB	OEM
FE0208ABG	OEM
FE0208ABR	OEM
FE0208AC	OEM
FE0208ACB	OEM
FE0208ACG	OEM
FE0208ACR	OEM
FE0208AD	OEM
FE0208ADB	OEM
FE0208ADG	OEM
FE0208ADR	OEM
FE0208AE	OEM
FE0208AEB	OEM
FE0208AEG	OEM
FE0208AER	OEM
FE0208AG	OEM
FE0208AQ	OEM
FE0208AR	OEM
FE0208ATF	OEM
FE0208ATFB	OEM
FE0208ATFG	OEM
FE0208ATFR	OEM
FE0208ATG	OEM
FE0208ATGB	OEM
FE0208ATGG	OEM
FE0208ATGR	OEM
FE0208ATH	OEM
FE0208ATHB	OEM
FE0208ATHG	OEM
FE0208ATHR	OEM
FE0208ATI	OEM
FE0208ATIB	OEM
FE0208ATIG	OEM
FE0208ATIR	OEM
FE0208ATJ	OEM
FE0208ATJB	OEM
FE0208ATJG	OEM
FE0208B	OEM
FE0208BB	OEM
FE0208BG	OEM
FE0208BQ	OEM
FE0208BR	OEM
FE0208C	OEM
FE0208CB	OEM
FE0208CG	OEM
FE0208CR	OEM
FE0208D	OEM
FE0208DB	OEM
FE0208DG	OEM
FE0208DQ	OEM
FE0208DR	OEM
FE0208E	OEM
FE0208EB	OEM
FE0208EG	OEM
FE0208EQ	OEM
FE0208ER	OEM
FE0208TF	OEM
FE0208TFB	OEM
FE0208TFG	OEM
FE0208TFQ	OEM
FE0208TFR	OEM
FE0208TG	OEM
FE0208TGB	OEM
FE0208TGG	OEM
FE0208TGQ	OEM
FE0208TGR	OEM
FE0208TH	OEM
FE0208THB	OEM
FE0208THG	OEM
FE0208THR	OEM
FE0208TI	OEM
FE0208TIB	OEM
FE0208TIG	OEM
FE0208TIQ	OEM
FE0208TIR	OEM
FE0208TJ	OEM
FE0208TJB	OEM
FE0208TJG	OEM
FE0208TJQ	OEM
FE0208TJR	OEM
FE0220ACR	OEM
FE0401	OEM
FE0401A	OEM
FE0401AA	OEM
FE0401AAB	OEM
FE0401AAG	OEM
FE0401AAR	OEM
FE0401AB	OEM
FE0401ABB	OEM
FE0401ABG	OEM
FE0401ABQ	OEM
FE0401ABR	OEM
FE0401AC	OEM
FE0401ACB	OEM
FE0401ACG	OEM
FE0401ACR	OEM
FE0401AD	OEM
FE0401ADB	OEM
FE0401ADG	OEM
FE0401ADQ	OEM
FE0401ADR	OEM
FE0401AE	OEM
FE0401AEB	OEM
FE0401AEG	OEM
FE0401AEQ	OEM
FE0401AER	OEM
FE0401AG	OEM
FE0401AQ	OEM
FE0401AR	OEM
FE0401ATF	OEM
FE0401ATFB	OEM
FE0401ATFG	OEM
FE0401ATFQ	OEM
FE0401ATFR	OEM
FE0401ATG	OEM
FE0401ATGB	OEM
FE0401ATGG	OEM
FE0401ATGQ	OEM
FE0401ATGR	OEM
FE0401ATH	OEM
FE0401ATHB	OEM
FE0401ATHG	OEM
FE0401ATHR	OEM
FE0401ATI	OEM
FE0401ATIG	OEM
FE0401ATIQ	OEM
FE0401ATIR	OEM
FE0401ATJ	OEM
FE0401ATJG	OEM
FE0401ATJQ	OEM
FE0401ATJR	OEM
FE0401B	OEM
FE0401BB	OEM
FE0401BG	OEM
FE0401BQ	OEM
FE0401BR	OEM
FE0401C	OEM
FE0401CB	OEM
FE0401CG	OEM
FE0401CR	OEM
FE0401D	OEM
FE0401DB	OEM
FE0401DG	OEM
FE0401DQ	OEM
FE0401DR	OEM
FE0401E	OEM
FE0401EB	OEM
FE0401EG	OEM
FE0401EQ	OEM
FE0401ER	OEM
FE0401TF	OEM
FE0401TFB	OEM
FE0401TFG	OEM
FE0401TFQ	OEM
FE0401TFR	OEM
FE0401TG	OEM
FE0401TGB	OEM
FE0401TGG	OEM
FE0401TGQ	OEM
FE0401TGR	OEM
FE0401TH	OEM
FE0401THB	OEM
FE0401THG	OEM
FE0401THR	OEM
FE0401TI	OEM
FE0401TIB	OEM
FE0401TIG	OEM
FE0401TIQ	OEM
FE0401TIR	OEM
FE0401TJ	OEM
FE0401TJB	OEM
FE0401TJG	OEM
FE0401TJQ	OEM
FE0401TJR	OEM
FE0402	OEM
FE0402A	OEM
FE0402AA	OEM
FE0402AAB	OEM
FE0402AAG	OEM
FE0402AAQ	OEM
FE0402AAR	OEM
FE0402AB	OEM
FE0402ABB	OEM
FE0402ABG	OEM
FE0402ABQ	OEM
FE0402ABR	OEM
FE0402AC	OEM
FE0402ACB	OEM
FE0402ACG	OEM
FE0402ACR	OEM
FE0402AD	OEM
FE0402ADB	OEM
FE0402ADG	OEM
FE0402ADQ	OEM
FE0402ADR	OEM
FE0402AE	OEM
FE0402AEB	OEM
FE0402AEG	OEM
FE0402AEQ	OEM
FE0402AER	OEM
FE0402AG	OEM
FE0402AQ	OEM
FE0402AR	OEM
FE0402ATF	OEM
FE0402ATFB	OEM
FE0402ATFG	OEM
FE0402ATFQ	OEM
FE0402ATFR	OEM
FE0402ATG	OEM
FE0402ATGB	OEM
FE0402ATGG	OEM
FE0402ATGQ	OEM
FE0402ATGR	OEM
FE0402ATH	OEM
FE0402ATHB	OEM
FE0402ATHG	OEM
FE0402ATHR	OEM
FE0402ATI	OEM
FE0402ATIB	OEM
FE0402ATIG	OEM
FE0402ATIQ	OEM
FE0402ATIR	OEM
FE0402ATJ	OEM
FE0402ATJB	OEM
FE0402ATJG	OEM
FE0402ATJR	OEM
FE0402B	OEM
FE0402BB	OEM
FE0402BG	OEM
FE0402BQ	OEM
FE0402BR	OEM
FE0402C	OEM
FE0402CB	OEM
FE0402CG	OEM
FE0402CR	OEM
FE0402D	OEM
FE0402DB	OEM
FE0402DG	OEM
FE0402DQ	OEM
FE0402DR	OEM
FE0402E	OEM
FE0402EB	OEM
FE0402EG	OEM
FE0402EQ	OEM
FE0402ER	OEM
FE0402TF	OEM
FE0402TFB	OEM
FE0402TFG	OEM
FE0402TFQ	OEM
FE0402TFR	OEM
FE0402TG	OEM
FE0402TGB	OEM
FE0402TGG	OEM
FE0402TGQ	OEM
FE0402TGR	OEM
FE0402TH	OEM
FE0402THB	OEM
FE0402THG	OEM
FE0402THR	OEM
FE0402TI	OEM
FE0402TIB	OEM
FE0402TIG	OEM
FE0402TIQ	OEM
FE0402TIR	OEM
FE0402TJ	OEM
FE0402TJB	OEM
FE0402TJG	OEM
FE0402TJQ	OEM
FE0402TJR	OEM
FE0403	OEM
FE0403A	OEM
FE0403AA	OEM
FE0403AAB	OEM
FE0403AAG	OEM
FE0403AAQ	OEM
FE0403AAR	OEM
FE0403AB	OEM
FE0403ABB	OEM
FE0403ABG	OEM
FE0403ABQ	OEM
FE0403AC	OEM
FE0403ACB	OEM
FE0403ACG	OEM
FE0403ACR	OEM
FE0403AD	OEM
FE0403ADB	OEM
FE0403ADG	OEM
FE0403ADQ	OEM
FE0403AE	OEM
FE0403AEB	OEM
FE0403AEG	OEM
FE0403AEQ	OEM
FE0403AER	OEM
FE0403AG	OEM
FE0403AQ	OEM
FE0403AR	OEM
FE0403ATF	OEM
FE0403ATFB	OEM
FE0403ATFG	OEM
FE0403ATFQ	OEM
FE0403ATFR	OEM
FE0403ATG	OEM
FE0403ATGB	OEM
FE0403ATGG	OEM
FE0403ATGQ	OEM
FE0403ATGR	OEM
FE0403ATH	OEM
FE0403ATHB	OEM
FE0403ATHG	OEM
FE0403ATHR	OEM
FE0403ATI	OEM
FE0403ATIB	OEM
FE0403ATIG	OEM
FE0403ATIQ	OEM
FE0403ATIR	OEM
FE0403ATJ	OEM
FE0403ATJB	OEM
FE0403ATJG	OEM
FE0403ATJQ	OEM
FE0403ATJR	OEM
FE0403BB	OEM
FE0403BG	OEM
FE0403BQ	OEM
FE0403BR	OEM
FE0403C	OEM
FE0403CB	OEM
FE0403CG	OEM
FE0403CR	OEM
FE0403D	OEM
FE0403DB	OEM
FE0403DG	OEM
FE0403DQ	OEM
FE0403DR	OEM
FE0403E	OEM
FE0403EB	OEM
FE0403EG	OEM
FE0403EQ	OEM
FE0403ER	OEM
FE0403TF	OEM
FE0403TFB	OEM
FE0403TFG	OEM
FE0403TFQ	OEM
FE0403TFR	OEM
FE0403TG	OEM
FE0403TGB	OEM
FE0403TGG	OEM
FE0403TGQ	OEM
FE0403TGR	OEM
FE0403TH	OEM
FE0403THB	OEM
FE0403THG	OEM
FE0403THR	OEM
FE0403TI	OEM
FE0403TIB	OEM
FE0403TIG	OEM
FE0403TIQ	OEM
FE0403TIR	OEM
FE0403TJ	OEM
FE0403TJB	OEM
FE0403TJQ	OEM
FE0403TJR	OEM
FE0404AB	OEM
FE0404F	OEM
FE0404G	OEM
FE0404H	OEM
FE0404I	OEM
FE0404J	OEM
FE0404TF	OEM
FE0404TFB	OEM
FE0404TFG	OEM
FE0404TFQ	OEM
FE0404TFR	OEM
FE0404TG	OEM
FE0404TGB	OEM
FE0404TGG	OEM
FE0404TGQ	OEM
FE0404TGR	OEM
FE0404TH	OEM
FE0404THB	OEM
FE0404THG	OEM
FE0404THR	OEM
FE0404TI	OEM
FE0404TIB	OEM
FE0404TIG	OEM
FE0404TIQ	OEM
FE0404TIR	OEM
FE0404TJ	OEM
FE0404TJB	OEM
FE0404TJG	OEM
FE0404TJQ	OEM
FE0404TJR	OEM
FE0405	OEM
FE0405A	OEM
FE0405AA	OEM
FE0405AAB	OEM
FE0405AAG	OEM
FE0405AAQ	OEM
FE0405AAR	OEM
FE0405ABB	OEM
FE0405ABG	OEM
FE0405ABQ	OEM
FE0405ABR	OEM
FE0405AC	OEM
FE0405ACB	OEM
FE0405ACG	OEM
FE0405ACR	OEM
FE0405AD	OEM
FE0405ADB	OEM
FE0405ADG	OEM
FE0405ADQ	OEM
FE0405ADR	OEM
FE0405AE	OEM
FE0405AEB	OEM
FE0405AEG	OEM
FE0405AEQ	OEM
FE0405AER	OEM
FE0405AG	OEM
FE0405AQ	OEM
FE0405AR	OEM
FE0405ATF	OEM
FE0405ATFB	OEM
FE0405ATFG	OEM
FE0405ATFQ	OEM
FE0405ATFR	OEM
FE0405ATG	OEM
FE0405ATGB	OEM
FE0405ATGG	OEM
FE0405ATGQ	OEM
FE0405ATGR	OEM
FE0405ATH	OEM
FE0405ATHB	OEM
FE0405ATHG	OEM
FE0405ATHR	OEM
FE0405ATI	OEM
FE0405ATIB	OEM
FE0405ATIG	OEM
FE0405ATIQ	OEM
FE0405ATIR	OEM
FE0405ATJ	OEM
FE0405ATJB	OEM
FE0405ATJG	OEM
FE0405ATJQ	OEM
FE0405ATJR	OEM
FE0405B	OEM
FE0405BB	OEM
FE0405BG	OEM
FE0405BQ	OEM
FE0405BR	OEM
FE0405C	OEM
FE0405CB	OEM
FE0405CG	OEM
FE0405CR	OEM
FE0405DB	OEM
FE0405DG	OEM
FE0405DQ	OEM
FE0405DR	OEM
FE0405E	OEM
FE0405EB	OEM
FE0405EG	OEM
FE0405EQ	OEM
FE0405ER	OEM
FE0405TF	OEM
FE0405TFB	OEM
FE0405TFG	OEM
FE0405TFQ	OEM
FE0405TFR	OEM
FE0405TG	OEM
FE0405TGB	OEM
FE0405TGG	OEM
FE0405TGQ	OEM
FE0405TGR	OEM
FE0405TH	OEM
FE0405THB	OEM
FE0405THG	OEM
FE0405THR	OEM
FE0405TI	OEM
FE0405TIB	OEM
FE0405TIG	OEM
FE0405TIQ	OEM
FE0405TIR	OEM
FE0405TJ	OEM
FE0405TJB	OEM
FE0405TJG	OEM
FE0405TJQ	OEM
FE0405TJR	OEM
FE0501	OEM
FE0501A	OEM
FE0501AA	OEM
FE0501AAB	OEM
FE0501AAG	OEM
FE0501AAQ	OEM
FE0501AAR	OEM
FE0501AB	OEM
FE0501ABB	OEM
FE0501ABG	OEM
FE0501ABQ	OEM
FE0501ABR	OEM
FE0501AC	OEM
FE0501ACB	OEM
FE0501ACG	OEM
FE0501ACR	OEM
FE0501AD	OEM
FE0501ADB	OEM
FE0501ADG	OEM
FE0501ADQ	OEM
FE0501ADR	OEM
FE0501AE	OEM
FE0501AEB	OEM
FE0501AEG	OEM
FE0501AEQ	OEM
FE0501AER	OEM
FE0501AG	OEM
FE0501AQ	OEM
FE0501AR	OEM
FE0501ATF	OEM
FE0501ATFB	OEM
FE0501ATFG	OEM
FE0501ATFQ	OEM
FE0501ATFR	OEM
FE0501ATG	OEM
FE0501ATGB	OEM
FE0501ATGQ	OEM
FE0501ATGR	OEM
FE0501ATH	OEM
FE0501ATHB	OEM
FE0501ATHG	OEM
FE0501ATHR	OEM
FE0501ATI	OEM
FE0501ATIB	OEM
FE0501ATIG	OEM
FE0501ATIQ	OEM
FE0501ATIR	OEM
FE0501ATJ	OEM
FE0501ATJB	OEM
FE0501ATJG	OEM
FE0501ATJQ	OEM
FE0501ATJR	OEM
FE0501B	OEM
FE0501BB	OEM
FE0501BG	OEM
FE0501BQ	OEM
FE0501BR	OEM
FE0501C	OEM
FE0501CB	OEM
FE0501CG	OEM
FE0501D	OEM
FE0501DB	OEM
FE0501DG	OEM
FE0501DQ	OEM
FE0501DR	OEM
FE0501E	OEM
FE0501EB	OEM
FE0501EG	OEM
FE0501EQ	OEM
FE0501ER	OEM
FE0501TFB	OEM
FE0501TFG	OEM
FE0501TFQ	OEM
FE0501TFR	OEM
FE0501TG	OEM
FE0501TGB	OEM
FE0501TGG	OEM
FE0501TGQ	OEM
FE0501TGR	OEM
FE0501TH	OEM
FE0501THB	OEM
FE0501THG	OEM
FE0501THR	OEM
FE0501TI	OEM
FE0501TIB	OEM
FE0501TIG	OEM
FE0501TIQ	OEM
FE0501TIR	OEM
FE0501TJ	OEM
FE0501TJB	OEM
FE0501TJG	OEM
FE0501TJQ	OEM
FE0501TJR	OEM
FE0502	OEM
FE0502A	OEM
FE0502AA	OEM
FE0502AAB	OEM

If replacement code is OEM, contact original manufacturer for replacement.

DEVICE TYPE	REPL CODE
FE0502AAG	OEM
FE0502AAQ	OEM
FE0502AAR	OEM
FE0502AB	OEM
FE0502ABB	OEM
FE0502ABG	OEM
FE0502ABQ	OEM
FE0502ABR	OEM
FE0502AC	OEM
FE0502ACB	OEM
FE0502ACG	OEM
FE0502ACR	OEM
FE0502AD	OEM
FE0502ADB	OEM
FE0502ADG	OEM
FE0502ADQ	OEM
FE0502ADR	OEM
FE0502AE	OEM
FE0502AEB	OEM
FE0502AEG	OEM
FE0502AEQ	OEM
FE0502AER	OEM
FE0502AG	OEM
FE0502AQ	OEM
FE0502AR	OEM
FE0502ATF	OEM
FE0502ATFB	OEM
FE0502ATFG	OEM
FE0502ATFQ	OEM
FE0502ATFR	OEM
FE0502ATG	OEM
FE0502ATGB	OEM
FE0502ATGG	OEM
FE0502ATGQ	OEM
FE0502ATGR	OEM
FE0502ATH	OEM
FE0502ATHB	OEM
FE0502ATHG	OEM
FE0502ATHR	OEM
FE0502ATI	OEM
FE0502ATIB	OEM
FE0502ATIG	OEM
FE0502ATIQ	OEM
FE0502ATIR	OEM
FE0502ATJ	OEM
FE0502ATJB	OEM
FE0502ATJG	OEM
FE0502ATJQ	OEM
FE0502ATJR	OEM
FE0502B	OEM
FE0502BB	OEM
FE0502BG	OEM
FE0502BQ	OEM
FE0502BR	OEM
FE0502C	OEM
FE0502CB	OEM
FE0502CG	OEM
FE0502CR	OEM
FE0502D	OEM
FE0502DB	OEM
FE0502DG	OEM
FE0502DQ	OEM
FE0502DR	OEM
FE0502E	OEM
FE0502EB	OEM
FE0502EG	OEM
FE0502EQ	OEM
FE0502ER	OEM
FE0502TF	OEM
FE0502TFB	OEM
FE0502TFG	OEM
FE0502TFQ	OEM
FE0502TFR	OEM
FE0502TG	OEM
FE0502TGB	OEM
FE0502TGG	OEM
FE0502TGQ	OEM
FE0502TGR	OEM
FE0502TH	OEM
FE0502THB	OEM
FE0502THG	OEM
FE0502THR	OEM
FE0502TI	OEM
FE0502TIB	OEM
FE0502TIG	OEM
FE0502TIQ	OEM
FE0502TIR	OEM
FE0502TJ	OEM
FE0502TJB	OEM
FE0502TJG	OEM
FE0502TJQ	OEM
FE0502TJR	OEM
FE0505CR	OEM
FE0601	OEM
FE0601A	OEM
FE0601AA	OEM
FE0601AAB	OEM
FE0601AAG	OEM
FE0601AAQ	OEM
FE0601AAR	OEM
FE0601AB	OEM
FE0601ABB	OEM
FE0601ABG	OEM
FE0601ABQ	OEM
FE0601ABR	OEM
FE0601AC	OEM
FE0601ACB	OEM
FE0601ACG	OEM
FE0601ACR	OEM
FE0601AD	OEM
FE0601ADB	OEM
FE0601ADG	OEM
FE0601ADQ	OEM
FE0601ADR	OEM
FE0601AE	OEM
FE0601AEG	OEM
FE0601AEQ	OEM
FE0601AER	OEM
FE0601AQ	OEM
FE0601AR	OEM
FE0601ATF	OEM
FE0601ATFB	OEM
FE0601ATFG	OEM
FE0601ATFQ	OEM
FE0601ATFR	OEM
FE0601ATG	OEM
FE0601ATGB	OEM
FE0601ATGG	OEM
FE0601ATGQ	OEM
FE0601ATGR	OEM
FE0601ATH	OEM
FE0601ATHB	OEM
FE0601ATHG	OEM
FE0601ATHR	OEM
FE0601ATI	OEM
FE0601ATIB	OEM
FE0601ATIG	OEM
FE0601ATIQ	OEM
FE0601ATIR	OEM
FE0601ATJ	OEM
FE0601ATJB	OEM
FE0601ATJG	OEM
FE0601ATJR	OEM
FE0601B	OEM
FE0601BB	OEM
FE0601BG	OEM
FE0601BQ	OEM
FE0601BR	OEM
FE0601C	OEM
FE0601CB	OEM
FE0601CG	OEM
FE0601CR	OEM
FE0601D	OEM
FE0601DB	OEM
FE0601DG	OEM
FE0601DQ	OEM
FE0601DR	OEM
FE0601E	OEM
FE0601EB	OEM
FE0601EG	OEM
FE0601EQ	OEM
FE0601ER	OEM
FE0601TF	OEM
FE0601TFB	OEM
FE0601TFG	OEM
FE0601TFQ	OEM
FE0601TFR	OEM
FE0601TG	OEM
FE0601TGB	OEM
FE0601TGG	OEM
FE0601TGQ	OEM
FE0601TGR	OEM
FE0601TH	OEM
FE0601THB	OEM
FE0601THG	OEM
FE0601THR	OEM
FE0601TI	OEM
FE0601TIB	OEM
FE0601TIG	OEM
FE0601TIQ	OEM
FE0601TIR	OEM
FE0601TJ	OEM
FE0601TJB	OEM
FE0601TJG	OEM
FE0601TJQ	OEM
FE0601TJR	OEM
FE0602F	OEM
FE0602G	OEM
FE0602H	OEM
FE0602I	OEM
FE0602J	OEM
FE0602TF	OEM
FE0602TFB	OEM
FE0602TFG	OEM
FE0602TFR	OEM
FE0602TG	OEM
FE0602TGB	OEM
FE0602TGG	OEM
FE0602TGR	OEM
FE0602TH	OEM
FE0602THB	OEM
FE0602THG	OEM
FE0602THR	OEM
FE0602TI	OEM
FE0602TIB	OEM
FE0602TIF	OEM
FE0602TIG	OEM
FE0602TJ	OEM
FE0602TJB	OEM
FE0602TJG	OEM
FE0602TJR	OEM
FE0701	OEM
FE0701A	OEM
FE0701AA	OEM
FE0701AAG	OEM
FE0701AAQ	OEM
FE0701AAR	OEM
FE0701AB	OEM
FE0701ABB	OEM
FE0701ABG	OEM
FE0701ABQ	OEM
FE0701ABR	OEM
FE0701AC	OEM
FE0701ACB	OEM
FE0701ACG	OEM
FE0701ACR	OEM
FE0701AD	OEM
FE0701ADB	OEM
FE0701ADG	OEM
FE0701ADQ	OEM
FE0701ADR	OEM
FE0701AE	OEM
FE0701AEB	OEM
FE0701AEG	OEM
FE0701AEQ	OEM
FE0701AER	OEM
FE0701AG	OEM
FE0701AQ	OEM
FE0701AR	OEM
FE0701ATF	OEM
FE0701ATFB	OEM
FE0701ATFG	OEM
FE0701ATFQ	OEM
FE0701ATFR	OEM
FE0701ATG	OEM
FE0701ATGB	OEM
FE0701ATGG	OEM
FE0701ATGQ	OEM
FE0701ATH	OEM
FE0701ATHB	OEM
FE0701ATHG	OEM
FE0701ATHR	OEM
FE0701ATI	OEM
FE0701ATIB	OEM
FE0701ATIG	OEM
FE0701ATIQ	OEM
FE0701ATIR	OEM
FE0701ATJ	OEM
FE0701ATJB	OEM
FE0701ATJG	OEM
FE0701ATJQ	OEM
FE0701ATJR	OEM
FE0701B	OEM
FE0701BB	OEM
FE0701BG	OEM
FE0701BQ	OEM
FE0701BR	OEM
FE0701C	OEM
FE0701CB	OEM
FE0701CG	OEM
FE0701CR	OEM
FE0701D	OEM
FE0701DB	OEM
FE0701DG	OEM
FE0701DQ	OEM
FE0701DR	OEM
FE0701E	OEM
FE0701EB	OEM
FE0701EG	OEM
FE0701EQ	OEM
FE0701ER	OEM
FE0701TF	OEM
FE0701TFB	OEM
FE0701TFG	OEM
FE0701TFQ	OEM
FE0701TFR	OEM
FE0701TG	OEM
FE0701TGB	OEM
FE0701TGG	OEM
FE0701TGQ	OEM
FE0701TGR	OEM
FE0701TH	OEM
FE0701THB	OEM
FE0701THG	OEM
FE0701THR	OEM
FE0701TI	OEM
FE0701TIB	OEM
FE0701TIG	OEM
FE0701TIQ	OEM
FE0701TIR	OEM
FE0701TJ	OEM
FE0701TJG	OEM
FE0701TJQ	OEM
FE0701TJR	OEM
FE0801	OEM
FE0801A	OEM
FE0801AA	OEM
FE0801AAB	OEM
FE0801AAG	OEM
FE0801AAQ	OEM
FE0801AAR	OEM
FE0801AB	OEM
FE0801ABB	OEM
FE0801ABG	OEM
FE0801ABQ	OEM
FE0801ABR	OEM
FE0801AC	OEM
FE0801ACB	OEM
FE0801ACG	OEM
FE0801ACR	OEM
FE0801AD	OEM
FE0801ADB	OEM
FE0801ADG	OEM
FE0801ADR	OEM
FE0801AE	OEM
FE0801AEB	OEM
FE0801AEG	OEM
FE0801AEQ	OEM
FE0801AER	OEM
FE0801AG	OEM
FE0801AR	OEM
FE0801ATF	OEM
FE0801ATFB	OEM
FE0801ATFG	OEM
FE0801ATFQ	OEM
FE0801ATFR	OEM
FE0801ATG	OEM
FE0801ATGB	OEM
FE0801ATGG	OEM
FE0801ATGQ	OEM
FE0801ATGR	OEM
FE0801ATH	OEM
FE0801ATHB	OEM
FE0801ATHG	OEM
FE0801ATHR	OEM
FE0801ATIB	OEM
FE0801ATIG	OEM
FE0801ATIQ	OEM
FE0801ATIR	OEM
FE0801ATJ	OEM
FE0801ATJB	OEM
FE0801ATJG	OEM
FE0801ATJQ	OEM
FE0801ATJR	OEM
FE0801B	OEM
FE0801BB	OEM
FE0801BG	OEM
FE0801BQ	OEM
FE0801C	OEM
FE0801CB	OEM
FE0801CG	OEM
FE0801CR	OEM
FE0801D	OEM
FE0801DB	OEM
FE0801DG	OEM
FE0801DQ	OEM
FE0801DR	OEM
FE0801E	OEM
FE0801EB	OEM
FE0801EG	OEM
FE0801EQ	OEM
FE0801ER	OEM
FE0801TF	OEM
FE0801TFB	OEM
FE0801TFG	OEM
FE0801TFQ	OEM
FE0801TFR	OEM
FE0801TG	OEM
FE0801TGB	OEM
FE0801TGG	OEM
FE0801TGQ	OEM
FE0801TGR	OEM
FE0801TH	OEM
FE0801THB	OEM
FE0801THG	OEM
FE0801THR	OEM
FE0801TI	OEM
FE0801TIB	OEM
FE0801TIG	OEM
FE0801TIR	OEM
FE0801TJ	OEM
FE0801TJG	OEM
FE0801TJQ	OEM
FE0801TJR	OEM
FE0801TKQ	OEM
FE0802	OEM
FE0802A	OEM
FE0802AA	OEM
FE0802AAB	OEM
FE0802AAG	OEM
FE0802AAQ	OEM
FE0802AAR	OEM
FE0802AB	OEM
FE0802ABB	OEM
FE0802ABG	OEM
FE0802ABR	OEM
FE0802AC	OEM
FE0802ACB	OEM
FE0802ACG	OEM
FE0802ACR	OEM
FE0802AD	OEM
FE0802ADB	OEM
FE0802ADG	OEM
FE0802ADQ	OEM
FE0802ADR	OEM
FE0802AE	OEM
FE0802AEB	OEM
FE0802AEG	OEM
FE0802AEQ	OEM
FE0802AER	OEM
FE0802AG	OEM
FE0802ATF	OEM
FE0802ATFB	OEM
FE0802ATFG	OEM
FE0802ATFQ	OEM
FE0802ATFR	OEM
FE0802ATG	OEM
FE0802ATGB	OEM
FE0802ATGG	OEM
FE0802ATGQ	OEM
FE0802ATGR	OEM
FE0802ATH	OEM
FE0802ATHB	OEM
FE0802ATHG	OEM
FE0802ATHR	OEM
FE0802ATI	OEM
FE0802ATIB	OEM
FE0802ATIG	OEM
FE0802ATIQ	OEM
FE0802ATIR	OEM
FE0802ATJ	OEM
FE0802ATJB	OEM
FE0802ATJG	OEM
FE0802ATJQ	OEM
FE0802ATJR	OEM
FE0802B	OEM
FE0802BB	OEM
FE0802BG	OEM
FE0802C	OEM
FE0802CB	OEM
FE0802D	OEM
FE0802DB	OEM
FE0802DG	OEM
FE0802E	OEM
FE0802EB	OEM
FE0802EG	OEM
FE0802TF	OEM
FE0802TFB	OEM
FE0802TFG	OEM
FE0802TG	OEM
FE0802TGB	OEM
FE0802TGG	OEM
FE0802TH	OEM
FE0802THB	OEM
FE0802THG	OEM
FE0802THR	OEM
FE0802TI	OEM
FE0802TIB	OEM
FE0802TIR	OEM
FE0802TJ	OEM
FE0802TJB	OEM
FE0802TJR	OEM
FE0803	OEM
FE0803ATF	OEM
FE0803ATFB	OEM
FE0803ATFG	OEM
FE0803ATFR	OEM
FE0803ATG	OEM
FE0803ATGB	OEM
FE0803ATGG	OEM
FE0803ATGQ	OEM
FE0803ATGR	OEM
FE0803ATH	OEM
FE0803ATHB	OEM
FE0803ATHG	OEM
FE0803ATHR	OEM
FE0803ATI	OEM
FE0803ATIB	OEM
FE0803ATIG	OEM
FE0803ATIQ	OEM
FE0803ATIR	OEM
FE0803ATJ	OEM
FE0803ATJB	OEM
FE0803ATJG	OEM
FE0803ATJQ	OEM
FE0803ATJR	OEM
FE0803TF	OEM
FE0803TFB	OEM
FE0803TFG	OEM
FE0803TFQ	OEM
FE0803TFR	OEM
FE0803TG	OEM
FE0803TGB	OEM
FE0803TGG	OEM
FE0803TGQ	OEM
FE0803TGR	OEM
FE0803TH	OEM
FE0803THB	OEM
FE0803THG	OEM
FE0803THR	OEM
FE0803TI	OEM
FE0803TIB	OEM
FE0803TIF	OEM
FE0803TIG	OEM
FE0803TIQ	OEM
FE0803TJ	OEM
FE0803TJB	OEM
FE0803TJG	OEM
FE0803TJQ	OEM
FE0803TJR	OEM
FE0804A	OEM
FE0804B	OEM
FE0804C	OEM
FE0804D	OEM
FE0804E	OEM
FE0804F	OEM
FE0804G	OEM
FE0804H	OEM
FE0804I	OEM
FE0804J	OEM
FE0820ABQ	OEM
FE1A	0499
FE1B	0499
FE1C	0499
FE1D	0499
FE2A	0031
FE2B	0031
FE2C	0031
FE2D	0031
FE3A	0031
FE3B	0031
FE3C	0031
FE3D	0031
FE5A	1362
FE5B	1362
FE5C	1362
FE5D	1362
FE6A	OEM
FE6B	OEM
FE6C	OEM
FE6D	OEM
FE8A	1127
FE8B	1127
FE8C	1127
FE8D	1127
FE8F	1664
FE8G	1664
FE13A	OEM
FE13B	OEM
FE13C	OEM
FE13D	OEM
FE13E	OEM
FE13F	OEM
FE13G	OEM
FE14A	OEM
FE14B	OEM
FE14C	OEM
FE14D	OEM
FE14E	OEM
FE14F	OEM
FE14G	OEM
FE15A	OEM
FE15B	OEM
FE15C	OEM
FE15D	OEM
FE15E	OEM
FE15F	OEM
FE15G	OEM
FE16A	0909
FE16B	0909
FE16C	0909
FE16D	0909
FE16F	OEM
FE16G	OEM
FE30A	OEM
FE30B	OEM
FE30C	OEM
FE30D	OEM
FE30F	OEM
FE30G	OEM
FE-100	0325
FE100	0325
FE-100(IR)	0325
FE100(IR)	0325
FE-100A	0325
FE100A	0325
FE-102	0325
FE102	0325
FE-102A	0325
FE102A	0325
FE103	OEM
FE104	OEM
FE-104A	0325
FE104A	0325
FE200	OEM
FE202	OEM
FE204	OEM
FE250	OEM
FE252	OEM
FE254	OEM
FE300	OEM
FE302	OEM
FE304	OEM
FE350	OEM
FE352	OEM
FE354	OEM
FE-400	0325
FE400	0325
FE400A	OEM
FE-402	0302
FE402	0325
FE-402A	0325
FE402A	0325
FE-404A	0325
FE404A	0325
FE1001	OEM
FE1001A	OEM
FE1001B	OEM
FE1001C	OEM
FE1001D	OEM
FE1001E	OEM
FE1001F	OEM
FE1001G	OEM
FE1001H	OEM
FE1001I	OEM
FE1001J	OEM
FE1202A	OEM
FE1202B	OEM
FE1202C	OEM
FE1202D	OEM
FE1202E	OEM
FE1202F	OEM
FE1202G	OEM
FE1202H	OEM
FE1202I	OEM
FE1202J	OEM
FE1600	OEM
FE1901	OEM
FE1901A	OEM
FE1901AA	OEM
FE1901AAB	OEM
FE1901AAG	OEM
FE1901AAQ	OEM
FE1901AAR	OEM
FE1901AB	OEM
FE1901ABB	OEM
FE1901ABG	OEM
FE1901ABQ	OEM
FE1901ABR	OEM
FE1901AC	OEM
FE1901ACB	OEM
FE1901ACG	OEM
FE1901ACR	OEM
FE1901AD	OEM
FE1901ADB	OEM
FE1901ADG	OEM
FE1901ADQ	OEM
FE1901ADR	OEM
FE1901AE	OEM
FE1901AEB	OEM
FE1901AEG	OEM
FE1901AEQ	OEM
FE1901AG	OEM
FE1901AQ	OEM
FE1901AR	OEM
FE1901B	OEM
FE1901BB	OEM
FE1901BG	OEM
FE1901BQ	OEM
FE1901BR	OEM
FE1901C	OEM
FE1901CB	OEM
FE1901CG	OEM
FE1901CR	OEM
FE1901D	OEM
FE1901DB	OEM
FE1901DG	OEM
FE1901DQ	OEM
FE1901DR	OEM
FE1901E	OEM
FE1901EB	OEM
FE1901EG	OEM
FE1901EQ	OEM
FE1901ER	OEM
FE2201	OEM
FE2201ATF	OEM
FE2201ATFB	OEM
FE2201ATFG	OEM
FE2201ATFQ	OEM
FE2201ATFR	OEM
FE2201ATG	OEM
FE2201ATGB	OEM
FE2201ATGG	OEM
FE2201ATGQ	OEM
FE2201ATGR	OEM
FE2201ATHB	OEM
FE2201ATHG	OEM
FE2201ATHR	OEM
FE2201ATI	OEM
FE2201ATIB	OEM
FE2201ATIG	OEM
FE2201ATIQ	OEM
FE2201ATIR	OEM
FE2201ATJ	OEM
FE2201ATJB	OEM
FE2201ATJG	OEM
FE2201ATJQ	OEM
FE2201ATJR	OEM
FE2201TF	OEM
FE2201TFB	OEM
FE2201TFG	OEM
FE2201TFQ	OEM
FE2201TFR	OEM
FE2201TG	OEM
FE2201TGB	OEM
FE2201TGG	OEM
FE2201TGQ	OEM
FE2201TGR	OEM
FE2201TH	OEM
FE2201THB	OEM
FE2201THG	OEM
FE2201THR	OEM
FE2201TI	OEM
FE2201TIB	OEM
FE2201TIG	OEM
FE2201TIQ	OEM
FE2201TIR	OEM
FE2201TJ	OEM
FE2201TJB	OEM
FE2201TJG	OEM
FE2201TJQ	OEM
FE2201TJR	OEM
FE3000A	OEM
FE3010A	OEM
FE3020	OEM
FE3030	OEM
FE4302	2950
FE4303	3249
FE4304	3249
FE5245	0325
FE5246	3249
FE5247	3249
FE5457	3249
FE5458	3249
FE5459	3249
FE5484	3249
FE5485	3249
FE5486	3249
FE13301	OEM
FED070JA	OEM
FED070JA1	OEM
FED071W	OEM
FED080JA	OEM

DEVICE TYPE	REPL CODE
FED080JA1	OEM
FED080JA2	OEM
FED080JH	OEM
FED080JL	OEM
FED081SA	OEM
FED081SH	OEM
FED081W	OEM
FED0202C	OEM
FED0202D	OEM
FED0203C	OEM
FED0203D	OEM
FED0206C	OEM
FED0206D	OEM
FED0401C	OEM
FED0401D	OEM
FED0501C	OEM
FED0501D	OEM
FED0601C	OEM
FED0601D	OEM
FED0701C	OEM
FED0701D	OEM
FED-1D	OEM
FED-2D	OEM
FED-3D	OEM
FED-4D	OEM
FED16AT	OEM
FED16BT	OEM
FED16CT	OEM
FED16DT	OEM
FED16FT	OEM
FED16GT	OEM
FED16HT	OEM
FED16JT	OEM
FED30AM	OEM
FED30AP	OEM
FED30BM	OEM
FED30BP	OEM
FED30CM	OEM
FED30CP	OEM
FED30DM	OEM
FED30DP	OEM
FED30FM	OEM
FED30FP	OEM
FED30GM	OEM
FED30GP	OEM
FED30HM	OEM
FED30HP	OEM
FED30JP	OEM
FED110JA	OEM
FED130JA	OEM
FED131S	OEM
FED131X	OEM
FED150JA	OEM
FELA400	OEM
FEN16AT	1727
FEN16BT	1727
FEN16CT	1727
FEN16DT	1727
FEN16FT	OEM
FEN16GT	OEM
FEN16HT	OEM
FEN16JT	OEM
FEN30AM	OEM
FEN30AP	OEM
FEN30BM	OEM
FEN30BP	OEM
FEN30CM	OEM
FEN30CP	OEM
FEN30DM	OEM
FEN30DP	OEM
FEN30FM	OEM
FEN30FP	OEM
FEN30GM	OEM
FEN30GP	OEM
FEN30HM	OEM
FEN30HP	OEM
FEN30JP	OEM
FE0654A	OEM
FE0654B	OEM
FE0654C	OEM
FEP5DT	1236
FEP16AT	1236
FEP16BT	1236
FEP16CT	1236
FEP16DT	1236
FEP16FT	5796
FEP16GT	5796
FEP16HT	5796
FEP16JT	5796
FEP30AM	OEM
FEP30AP	2234
FEP30BM	OEM
FEP30BP	2234
FEP30CM	OEM
FEP30CP	2234
FEP30DM	OEM
FEP30DP	2234
FEP30FM	OEM
FEP30FP	4488
FEP30GM	OEM
FEP30GP	4488
FEP30HM	OEM
FEP30HP	4488
FEP30JM	OEM
FEP30JP	4488
FER8AT	OEM
FER8BT	OEM
FER8CT	OEM
FER8DT	OEM
FER8FT	OEM
FER8GT	OEM
FER8HT	OEM
FER8JT	OEM
FER16AT	OEM
FER16BT	OEM
FER16CT	OEM
FER16DT	OEM
FER16FT	OEM
FER16GT	OEM
FER16HT	OEM
FER16JT	OEM
FES8AT	1145
FES8BT	1145
FES8CT	1145
FES8DT	1145
FES8FT	1145
FES8GT	1145
FES8HT	1664
FES8JT	1664
FES16AT	1664
FES16BT	1664
FES16CT	1664
FES16DT	1664
FES16FT	OEM
FES16GT	OEM
FES16HT	OEM
FES16JT	OEM
FF05	OEM
FF023900	6736
FF10	OEM
FF13A	OEM
FF13B	OEM
FF13C	OEM
FF13D	OEM
FF13E	OEM
FF13F	OEM
FF13G	OEM
FF14A	OEM
FF14B	OEM
FF14C	OEM
FF14D	OEM
FF14E	OEM
FF14F	OEM
FF14G	OEM
FF15	OEM
FF15A	OEM
FF15B	OEM
FF15C	OEM
FF15D	OEM
FF15E	OEM
FF15F	OEM
FF15G	OEM
FF102	OEM
FF108	OEM
FF274	0354
FF400	0325
FF409	OEM
FF411	OEM
FF412	OEM
FF413	OEM
FF600	OEM
FF617	OEM
FF626	OEM
FF627	OEM
FFC-4030	0723
FFD-1	OEM
FFD11	OEM
FFD12	OEM
FFD21	OEM
FFD22	OEM
FFD41	OEM
FFD51	OEM
FFD71	OEM
FFD81	OEM
FFZB9	OEM
FG-01	OEM
FG-2N	0071
FG2N	5309
FG-2N/10D-4	0102
FG-2N2	0102
FG-2NA	0015
FG2NA	OEM
FG2PC	0102
FG13A	OEM
FG13B	OEM
FG13C	OEM
FG13D	OEM
FG13E	OEM
FG13F	OEM
FG13G	OEM
FG14A	OEM
FG14B	OEM
FG14C	OEM
FG14D	OEM
FG14E	OEM
FG14F	OEM
FG14G	OEM
FG15A	OEM
FG15B	OEM
FG15C	OEM
FG15DFG15E	OEM
FG15F	OEM
FG15G	OEM
FG24SC1	OEM
FG24SD1GLR	OEM
FG24SD1GRY	OEM
FG24SG1	OEM
FG28SB1	OEM
FG28SB1GR	OEM
FG28SD1	OEM
FG28SJ1	OEM
FG28SL1	OEM
FG34	OEM
FG35	OEM
FG36	OEM
FG36A2	OEM
FG37	OEM
FG38C2	OEM
FG46A5	OEM
FG46C5	OEM
FG47A2	OEM
FG48A2	OEM
FG48B2	OEM
FG48C2	OEM
FG48D6	OEM
FG48E1	OEM
FG48E1A	OEM
FG48K6	OEM
FG48P2	OEM
FG48SA1	OEM
FG48SD1BGLR	OEM
FG48SD1BGRY	OEM
FG48SD1GR	OEM
FG48W2	OEM
FG48X2	OEM
FG48Y2A	OEM
FG58E2	OEM
FG58GW	OEM
FG78A1	OEM
FG78C2	OEM
FG79B6	OEM
FG79C6	OEM
FG80SB1GLR	OEM
FG80SB1GRY	OEM
FG80SB6	OEM
FG85A1	OEM
FG85C1	OEM
FG85D2	OEM
FG85E2	OEM
FG94B1	OEM
FG94C1	OEM
FG94G	OEM
FG95A1	OEM
FG95B1	OEM
FG95F6	OEM
FG95H6	OEM
FG95J6	OEM
FG95K2	OEM
FG95M1	OEM
FG97C6	OEM
FG97D6	OEM
FG97L6	OEM
FG99A2	OEM
FG99C1	OEM
FG99D1	OEM
FG99E1	OEM
FG100SA1	OEM
FG100SB1	OEM
FG101E4S1	OEM
FG101SB	OEM
FG105E1	OEM
FG106A2	OEM
FG108A1	OEM
FG108A1A	OEM
FG108M1	OEM
FG115A6	OEM
FG115C1	OEM
FG115E7	OEM
FG115F7	OEM
FG116A6	OEM
FG116B7	OEM
FG116C6	OEM
FG116D6	OEM
FG117C6	OEM
FG117C7	OEM
FG117D6	OEM
FG118A1	OEM
FG118E7	OEM
FG119A1	OEM
FG119B7	OEM
FG120S1	OEM
FG124B2	OEM
FG125A2	OEM
FG126A2	OEM
FG134A2	OEM
FG135B5	OEM
FG135C1	OEM
FG135E7	OEM
FG135F	OEM
FG135G	OEM
FG136C6	OEM
FG137A1	OEM
FG137C6	OEM
FG137D6	OEM
FG137E7	OEM
FG137F6	OEM
FG138A1	OEM
FG138B7	OEM
FG138G7	OEM
FG139A1	OEM
FG139A3	OEM
FG139D1	OEM
FG139E1	OEM
FG139E7	OEM
FG139F7	OEM
FG139N7	OEM
FG144B2	OEM
FG146A6	OEM
FG157C1	OEM
FG157C7	OEM
FG157D7	OEM
FG159A2	OEM
FG159B7	OEM
FG165A	OEM
FG166A2	OEM
FG169B2	OEM
FG169RA2	OEM
FG179A2	OEM
FG202SA2	OEM
FG206A2	OEM
FG209A2	OEM
FG209M3	OEM
FG210B9	OEM
FG213B9	OEM
FG213C10	OEM
FG213D10	OEM
FG326A2	OEM
FG326B2	OEM
FG326B2A	OEM
FG405A2	OEM
FG410E2	OEM
FG410F1	OEM
FG412D1	OEM
FG413D1	OEM
FG415C1	OEM
FG415E6	OEM
FG425A1	OEM
FG510A1	OEM
FG510B1	OEM
FG511E6	OEM
FG512A1	OEM
FG512F1	OEM
FG514A1	OEM
FG515B6	OEM
FG515B7	OEM
FG515F1	OEM
FG516B1	OEM
FG518B1	OEM
FG520A1	OEM
FG610A1	OEM
FG612A1	OEM
FG614A1	OEM
FG619B1	OEM
FG620A1	OEM
FG620F1	OEM
FG710B1	OEM
FG710C1	OEM
FG713ES1	OEM
FG715B1	OEM
FG715BRS1	OEM
FG715D1	OEM
FG715G2	OEM
FG816A	OEM
FG912E1	OEM
FG913B1	OEM
FG913ES1	OEM
FG914RB	OEM
FG914RB1	OEM
FG915R1	OEM
FG1115RC1	OEM
FG1210RA1	OEM
FG1211RA1	OEM
FG1213A1	OEM
FG1213RB5	OEM
FG1213RE1	OEM
FG1213RF1	OEM
FG1215RB6	OEM
FG1712A2	OEM
FGG1001	OEM
FGM03-2	0190
FGM03-4	0190
FGT3055	OEM
FH100	0917
FH340M	0755
FH1100	0917
FH1109	OEM
FH1200	OEM
FH70401E	2571
FH2000571-4	4094
FHB5	OEM
FHB6	OEM
FHB7	OEM
FHB8	OEM
FHE5	OEM
FHE6	OEM
FHE7	OEM
FHE8	OEM
FHE9	OEM
FHV1B	OEM
FHV1C	OEM
FHV1D	OEM
FHV1E	OEM
FHV2C	OEM
FHV2D	OEM
FHV2E	OEM
FHV2F	OEM
FHV2G	OEM
FHV2H	OEM
FHV3G	OEM
FHV3H	OEM
FHV3M	OEM
FHV4R	OEM
FHV4V	OEM
FHV4W	OEM
FHV4X	OEM
FHV4Y	OEM
FI	OEM
FI0049	OEM
FI-06	OEM
FI100	0844
FI1023	0127
FIB5	OEM
FIB6	OEM
FIB7	OEM
FIB8	OEM
FIB9	OEM
FID151W	OEM
FID080MA	OEM
FID080ML	OEM
FID081W	OEM
FIL-2SD	OEM
FIL-3C	OEM
FIL-3V	OEM
FIL-5C	OEM
FIL-5V	OEM
FIL-20C	OEM
FIL-20V	OEM
FIL-24D	OEM
FIL-29D	OEM
FIL-100C	OEM
FIL-100V	OEM
FIL-C4D	OEM
FIL-C10D	OEM
FIP4B8	OEM
FIP4B15S	OEM
FIP4C9	OEM
FIP4C9B	OEM
FIP4E8S	OEM
FIP4F8S	OEM
FIP4ST41	OEM
FIP5B13S	OEM
FIP5B15	OEM
FIP5D8	OEM
FIP5D10	OEM
FIP5D15S	OEM
FIP5F8S	OEM
FIP5H15	OEM
FIP5LT32	OEM
FIP5LT36ZY1	OEM
FIP5LT61ZK	OEM
FIP6A8S	OEM
FIP6A13	OEM
FIP6C8	OEM
FIP6CM7	OEM
FIP6E6	OEM
FIP6F13	OEM
FIP7BM10	OEM
FIP7BM10A	OEM
FIP7BT13ZRK	OEM
FIP7BT25ZYK	OEM
FIP7CM8	OEM
FIP7D8	OEM
FIP7EM11	OEM
FIP7GM7	OEM
FIP7GM10	OEM
FIP8A5	OEM
FIP8A5R	OEM
FIP8BT30ZK	OEM
FIP8MT05ZYK	OEM
FIP9AM10	OEM
FIP9B7	OEM
FIP9B8	OEM
FIP9BM10	OEM
FIP9C5	OEM
FIP9D5	OEM
FIP9HM7	OEM
FIP9J5	OEM
FIP9J5A	OEM
FIP9MT19ZK	OEM
FIP10A20	OEM
FIP10AM10	OEM
FIP10BT16ZK	OEM
FIP10KM7	OEM
FIP11B8A	OEM
FIP11B10	OEM
FIP11BT23GK	OEM
FIP11BT33ZK	OEM
FIP11CM6	OEM
FIP11DM6	OEM
FIP11HM6	OEM
FIP11YM6	OEM
FIP12KM7	OEM
FIP13B13	OEM
FIP13C10	OEM
FIP13C10A	OEM
FIP14DM8	OEM
FIP16A5R	OEM
FIP16B6X	OEM
FIP16B11X	OEM
FIP18A14R	OEM
FIP18BT12Z	OEM
FIP20B9X	OEM
FIP20C9X	OEM
FIP24A15YS	OEM
FIP24A17YS	OEM
FIP24BW16YS	OEM
FIP40A5X	OEM
FIP60B30T	OEM
FIP100D	0540
FIP101B8A6	OEM
FIP307	OEM
FIP325	OEM
FIP574	OEM
FIPBC8S	OEM
FIP-PEL1009	OEM
FIT5609C	OEM
FJ001A	OEM
FJ001B	OEM
FJ002B	OEM
FJ04A2-28	OEM
FJ04A2-28GG	OEM
FJ0204-12AA	OEM
FJ0230-12	OEM
FJ0230-12FF	OEM
FJ0240-12GG	OEM
FJ0402-12	OEM
FJ0402-12FF	OEM
FJ0402-28FF	OEM
FJ0402E	OEM
FJ0403-28	OEM
FJ0403-28RR	OEM
FJ0406-28	OEM
FJ0406-28FF	OEM
FJ0406-28RR	OEM
FJ0408-12	OEM
FJ0408-12FF	OEM
FJ0410-28	OEM
FJ0410-28FF	OEM
FJ0410-28RR	OEM
FJ0415-12	OEM
FJ0415-12FF	OEM
FJ0425-28	OEM
FJ0425-28FF	OEM
FJ0425-28RR	OEM
FJ0430-12	OEM
FJ0430-12GG	OEM
FJ0480-28	OEM
FJ0480-28GG	OEM
FJ0801-12	OEM
FJ0801-12LL	OEM
FJ0804-12	OEM
FJ0804-12LL	OEM
FJ0807B12LL	OEM
FJ0812-12	OEM
FJ0812-12LL	OEM
FJ0820B12LL	OEM
FJ20P5E18	OEM
FJ101E	OEM
FJ101M	OEM
FJ101R	OEM
FJ201E	OEM
FJ201F	OEM
FJ201M	OEM
FJ201R	OEM
FJ202E	OEM
FJ202F	OEM
FJ203E	OEM
FJ203F	OEM
FJ401E	OEM
FJ401F	OEM
FJ401J	OEM
FJ402F	OEM
FJ402J	OEM
FJ450E	OEM
FJ451E	OEM
FJ451J	OEM
FJ451LE	OEM
FJ451X	OEM
FJ901E	OEM
FJ901F	OEM
FJ901M	OEM
FJ901R	OEM
FJ902E	OEM
FJ902F	OEM
FJ902M	OEM
FJ902R	OEM
FJ951E	OEM
FJ952E	OEM
FJ954DD	OEM
FJ954E	OEM
FJ2301B-24	OEM
FJ2301B24BB	OEM
FJ2302B-24	OEM
FJ2302B24BB	OEM
FJ2304B-24	OEM
FJ2304B24BB	OEM
FJ2306B-24	OEM
FJ2306B24BB	OEM
FJ7201BB	OEM
FJ7201CC	OEM
FJ9203BB	OEM
FJ9203CC	OEM
FJ9203DD	OEM
FJ9208BB	OEM
FJ9208CC	OEM
FJ9208DD	OEM
FJ9215BB	OEM
FJ9215CC	OEM
FJ9215DD	OEM
FJ9225BB	OEM
FJ9225CC	OEM
FJ9225DD	OEM
FJ9235BB	OEM
FJ9235CC	OEM
FJ70351E	OEM
FJA705716	OEM
FJH101	0873
FJH101-7430	0873
FJH111	0697
FJH111-7420	0697
FJH121	0512
FJH121-7410	0512
FJH131	0232
FJH131-7400	0232
FJH141	1024
FJH141-7440	1024
FJH151	0743
FJH151-7450	0743
FJH161	1168
FJH161-7451	1168
FJH171	1185
FJH171-7453	1185
FJH181	1202
FJH181-7454	1202
FJH191	1535
FJH201	1574
FJH211	0117
FJH221	0314
FJH221-7402	0314
FJH231	0268
FJH231-7401	0268
FJH241	0361
FJH251	0385
FJH261	1052
FJH261-7442	1052
FJH291	0335
FJH291-7403	0335
FJH301	0803
FJH301-7426	0803
FJH311	0268
FJH311-7401	0268
FJH311-7401S1	OEM
FJH321	0385
FJJ101	1404
FJJ111	1426
FJJ111-7472	1426
FJJ121	1172
FJJ121-7473	1172
FJJ131	1313
FJJ131-7474	1313
FJJ141	1208
FJJ141-7490	1208
FJJ152	0834
FJJ181	1432
FJJ181-7475	1432
FJJ191	1158
FJJ191-7476	1158
FJJ211	0569
FJJ211-7493	0569
FJJ251-7492	0834
FJJ261	0942
FJJ261-74107	0942
FJK101	0175
FJK101-74121	0175
FJL101	1038
FJT1100	0133
FJT1101	0920
FJT1118	0920
FJT1134	0920
FJY101	1275
FJY101-7460	1275
FK914	0144
FK918	0144
FK2369A	0144
FK2484	0144
FK2894	0150
FK3014	0144
FK3299	0144
FK3300	0144
FK3484	0079
FK3494	0079
FK3502	0861
FK3503	0861
FK3962	OEM
FK3964	OEM
FL1S1	OEM
FL1S2	OEM
FL1S5	OEM
FL1S10	OEM
FL1S15	OEM
FL1S20	OEM
FL1S25	OEM
FL2S1	OEM
FL2S2	OEM
FL2S5	OEM
FL2S10	OEM
FL2S15	OEM
FL2S20	OEM
FL2S25	OEM
FL3S1	OEM
FL3S2	OEM
FL3S5	OEM
FL3S10	OEM
FL3S15	OEM
FL3S20	OEM
FL3S25	OEM
FL101	OEM
FL274	0167
FLC02WF	OEM
FLC08	OEM
FLC08ME	OEM
FLC053WG	OEM
FLC081WF	OEM
FLC081X	OEM
FLC081XP	OEM
FLC091WF	OEM
FLC15	OEM
FLC15ME	OEM
FLC30	OEM
FLC30ME	OEM
FLC103WG	OEM
FLC151MF	OEM

DEVICE TYPE	REPL CODE
FLC151WF	OEM
FLC151X	OEM
FLC161WF	OEM
FLC181XP	OEM
FLC253MH-6	OEM
FLC253MH-8	OEM
FLC301MG-4	OEM
FLC301MG-6	OEM
FLC301MG6	OEM
FLC301MG-8	OEM
FLC301MG8	OEM
FLC311MG-4	OEM
FLD073WB	OEM
FLD080WA	OEM
FLD081WA	OEM
FLD083JA	OEM
FLD083WB	OEM
FLD130WA	OEM
FLD131WA	OEM
FLD133JA	OEM
FLD133WA	OEM
FLH101	0232
FLH111	0512
FLH121	0697
FLH131	0873
FLH141	1024
FLH151	0743
FLH161	1168
FLH171	1185
FLH181	1202
FLH191	0314
FLH201	0268
FLH211	0361
FLH221	1535
FLH231	1574
FLH241	0117
FLH271	0385
FLH281	1052
FLH291	0335
FLH291U	0803
FLH341	1368
FLH351	1441
FLH361	1060
FLH371	1072
FLH381	0467
FLH391	0492
FLH401	1842
FLH411	1856
FLH421	1829
FLH431	0374
FLH441	2579
FLH451	4399
FLH481	1206
FLH481T	1349
FLH491	1339
FLH491T	1352
FLH501	2242
FLH511	3466
FLH521	3477
FLH531	3520
FLH541	0996
FLH551	1125
FLH581	0527
FLH601	1271
FLH611	4603
FLH621	0818
FLH631	0899
FLJ101	1404
FLJ111	1426
FLJ121	1172
FLJ131	1158
FLJ141	1313
FLJ151	1432
FLJ161	1208
FLJ171	0834
FLJ181	0569
FLJ191	1486
FLJ201	1912
FLJ211	1917
FLJ221	0980
FLJ231	1702
FLJ241	1921
FLJ251	1926
FLJ261	1715
FLJ271	0942
FLJ311	1964
FLJ321	1971
FLJ331	2940
FLJ381	1950
FLJ391	1956
FLJ401	1631
FLJ411	1645
FLJ421	1013
FLJ431	1666
FLJ441	0734
FLJ451	1685
FLJ461	0231
FLJ531	1770
FLJ541	1787
FLJ561	1943
FLK012WF	OEM
FLK012XP	OEM
FLK022WG	OEM
FLK022XP	OEM
FLK022XV	OEM
FLK052WG	OEM
FLK052XV	OEM
FLK101	0175
FLK102MH-14	OEM

DEVICE TYPE	REPL CODE
FLK102XV	OEM
FLK111	1139
FLK121	1157
FLK202MH-14	OEM
FLK202XV	OEM
FLL10ME	OEM
FLL17ME	OEM
FLL35ME	OEM
FLL50MK	OEM
FLL100MK	OEM
FLL101	1377
FLL111	1082
FLL111T	0619
FLL121	1098
FLL121T	1108
FLL121U	1098
FLL121V	1108
FLM0910-2	OEM
FLM0910-4C	OEM
FLM1011-2	OEM
FLM1011-4C	OEM
FLM1112-4C	OEM
FLM1213-4C	OEM
FLM1414-2	OEM
FLM1414-4C	OEM
FLM3742-3	OEM
FLM3742-4B	OEM
FLM3742-5	OEM
FLM3742-6	OEM
FLM3742-8B	OEM
FLM4450-3	OEM
FLM4450-4B	OEM
FLM4450-5	OEM
FLM4450-6	OEM
FLM4450-8B	OEM
FLM5359-8B	OEM
FLM5964-3	OEM
FLM5964-3B	OEM
FLM5964-4C	OEM
FLM5964-5	OEM
FLM5964-5IM	OEM
FLM5964-8C	OEM
FLM6472-3B	OEM
FLM6472-4C	OEM
FLM6472-5	OEM
FLM6472-8C	OEM
FLM7177-3BN	OEM
FLM7177-4C	OEM
FLM7177-5	OEM
FLM7177-8C	OEM
FLM7785-4	OEM
FLM7785-4C	OEM
FLM7785-8C	OEM
FLQ101	5455
FLQ111	3121
FLQ131	1722
FLR014XP	OEM
FLR024XP	OEM
FLR024XV	OEM
FLR054XV	OEM
FLS02W	OEM
FLS09	OEM
FLS09ME	OEM
FLS16	OEM
FLS16ME	OEM
FLS31	OEM
FLS31ME	OEM
FLS50	OEM
FLS50ME	OEM
FLT-U2	OEM
FLV100	OEM
FLV101	OEM
FLV104A	OEM
FLV110	OEM
FLV111	OEM
FLV117	1358
FLV140	OEM
FLV150	1358
FLV-160	OEM
FLV160	OEM
FLV310	OEM
FLV311	OEM
FLV340	OEM
FLV350	OEM
FLV360	OEM
FLV410	OEM
FLV411	OEM
FLV440	OEM
FLV450	3157
FLV460	OEM
FLV510	OEM
FLV511	OEM
FLV540	OEM
FLV550	OEM
FLV551	OEM
FLV560	OEM
FLX03MB	OEM
FLX06MB	OEM
FLX12MB	OEM
FLX30MB	OEM
FLX102MH-12	OEM
FLX202MH-12	OEM
FLZ80	OEM
FM009M	0196
FM1J2	0015
FM2A	OEM
FM50	OEM
FM75	OEM
FM100	OEM

DEVICE TYPE	REPL CODE
FM150	OEM
FM200	OEM
FM201	OEM
FM211N	0015
FM250	OEM
FM708	0144
FM709	0144
FM720A	0144
FM870	0144
FM871	0144
FM910	0144
FM911	0144
FM914	0144
FM1100	OEM
FM1100A	OEM
FM1101	OEM
FM1101A	OEM
FM1102	OEM
FM1102A	OEM
FM1103	OEM
FM1103A	OEM
FM1104	OEM
FM1104A	OEM
FM1105	OEM
FM1105A	OEM
FM1106	OEM
FM1106A	OEM
FM1107	OEM
FM1107A	OEM
FM1108	OEM
FM1108A	OEM
FM1109	OEM
FM1109A	OEM
FM1110	OEM
FM1110A	OEM
FM1111	OEM
FM1111A	OEM
FM1200	OEM
FM1201	OEM
FM1202	OEM
FM1203	OEM
FM1204	OEM
FM1205	OEM
FM1206	OEM
FM1207	OEM
FM1208	OEM
FM1209	OEM
FM1210	OEM
FM1211	OEM
FM1613	0144
FM1711	0144
FM1893	0233
FM2242	0016
FM2368	0144
FM2369	0144
FM2846	0144
FM2894	0150
FM3014	0144
FM3954	OEM
FM3954A	OEM
FM3955	OEM
FM3955A	OEM
FM3956	OEM
FM3957	OEM
FM3958	3206
FMC2	OEM
FMG1	OEM
FMG2	OEM
FMG2(G2)	OEM
FMG4T98	OEM
FMGG26	OEM
FMG-G26S	1664
FMG-G36S	OEM
FML12S	5952
FMLG12	OEM
FML-G12S	1127
FMLG12S	1127
FML-G12S(F)	1127
FML-G12SF	OEM
FML-G16S	1179
FMMD109	OEM
FMMD914	OEM
FMMD3102	OEM
FMMT2222	OEM
FMMT2222A	OEM
FMMT2369	OEM
FMMT2907	OEM
FMMT3903	OEM
FMMT3904	OEM
FMMT3905	OEM
FMMT3906	OEM
FMMTA05	OEM
FMMTA06	OEM
FMMTA12	OEM
FMMTA13	OEM
FMMTA14	OEM
FMMTA20	OEM
FMMTA55	OEM
FMMTA56	OEM
FMN1	OEM
FMP-G12S	OEM
FMPS-A20	0079
FMPSA20	0079
FMS-3FU	OEM
FMU-11SLF-F	5952
FMU32S	OEM
FMU-115LF-F	OEM
FMUG165	OEM
FMW1	OEM

DEVICE TYPE	REPL CODE
FN1L4M-M31	OEM
FN-51-1A	0079
FN651	OEM
FN651(0)	OEM
FN651-B	OEM
FN651-0	OEM
FN651-R	OEM
FN651-Y	OEM
FN1024	OEM
FN1034	OEM
FN2907	0037
FN4117	OEM
FN4117A	OEM
FN4118A	OEM
FN4119	OEM
FN4119A	OEM
FN4392	OEM
FN4393	OEM
FNA12	OEM
FND100	OEM
FND350	OEM
FND357	OEM
FND358	OEM
FND360	OEM
FND367	OEM
FND368	OEM
FND500	2199
FND501	OEM
FND507	2206
FND508	OEM
FND530	OEM
FND531	OEM
FND537	OEM
FND538	OEM
FND540	OEM
FND541	OEM
FND547	OEM
FND548	OEM
FND550	OEM
FND551	OEM
FND557	OEM
FND560	OEM
FND561	OEM
FND567	OEM
FND568	OEM
FND800	OEM
FND807	2378
F05	0015
FOD-12	OEM
FOD100	OEM
FOE-7	OEM
FOE-16	OEM
FOP40A5AX	OEM
FOR1A411E	0901
FOR1B	0901
FOR2D	1137
FOR2G	OEM
FOR-2G41	OEM
FOR2G41	OEM
FOR3G	0407
FOR36	OEM
FORIB	OEM
FOS100	0111
FOS101	0111
FOS104	0079
FP50BG2	OEM
FP54ALS16L8J	OEM
FP54ALS16L8N	OEM
FP54ALS16R4J	OEM
FP54ALS16R4N	OEM
FP54ALS16R6J	OEM
FP54ALS16R6N	OEM
FP54ALS16R8J	OEM
FP54ALS16R8N	OEM
FP54AS839JT	OEM
FP54AS839NT	OEM
FP54AS840JT	OEM
FP54AS840NT	OEM
FP54LS333JT	OEM
FP54LS333NT	OEM
FP54LS335JT	OEM
FP54LS335NT	OEM
FP74ALS16L8J	OEM
FP74ALS16L8N	OEM
FP74ALS16R4J	OEM
FP74ALS16R4N	OEM
FP74ALS16R6J	OEM
FP74ALS16R6N	OEM
FP74ALS16R8J	OEM
FP74ALS16R8N	OEM
FP74AS839JT	OEM
FP74AS839NT	OEM
FP74AS840JT	OEM
FP74AS840NT	OEM
FP74LS333JT	OEM
FP74LS333NT	OEM
FP74LS335JT	OEM
FP74LS335NT	OEM
FP4339	OEM
FP4340	OEM
FP50801	0127
FP100002S	OEM
FP100110S	OEM
FP100111S	OEM
FP100112S	OEM
FP100113S	OEM
FPA100	OEM
FPA102	OEM
FPA103	OEM

DEVICE TYPE	REPL CODE
FPA104	OEM
FPA105	OEM
FPA106	OEM
FPA107	OEM
FPA108	OEM
FPA720	OEM
FPB-A	OEM
FPD080CA	OEM
FPD080MA	OEM
FPD081MA	OEM
FPD140M3	OEM
FPD150C	OEM
FPD150M	OEM
FPD150M3	OEM
FPE104	OEM
FPE700	OEM
FPLAS93459PC	OEM
FPM100	OEM
FPM200	OEM
FPN100	OEM
FPN2907	0037
FPN3563	0144
FPO100	OEM
FPQ2222	0544
FPQ2907	0282
FPQ3467	OEM
FPQ3468	OEM
FPQ3724	OEM
FPQ3725	OEM
FPR40-1001	0016
FPR40-1003	0144
FPR40-1004	0004
FPR40-1005	0004
FPR40-1006	0143
FPR50-1001	0016
FPR50-1002	0016
FPR50-1003	0144
FPR50-1004	0144
FPR50-1005	0004
FPR50-1006	0004
FPT100	OEM
FPT100A	OEM
FPT100B	OEM
FPT101	OEM
FPT110	OEM
FPT110A	OEM
FPT120	OEM
FPT120A	OEM
FPT120B	OEM
FPT130	OEM
FPT130A	OEM
FPT130B	OEM
FPT220	OEM
FPT320	OEM
FPT330	OEM
FPT500	2313
FPT510	2313
FPT510A	2313
FPT540A	OEM
FPT550	OEM
FPT700	OEM
FPT720	OEM
FPZ5V6	OEM
FPZ6V2	0636
FPZ6V8	OEM
FPZ7V5	OEM
FPZ8V2	0244
FPZ9V1	0012
FPZ10	OEM
FPZ11	OEM
FPZ12	0137
FQ3467	OEM
FQ3468	OEM
FQ3724	OEM
FQ3725	OEM
FQB01-35	0820
FQL0008A	OEM
FQL0008B	OEM
FR061	0023
FR062	0023
FR063	0023
FR064	0023
FR065	0023
FR066	0017
FR067	0017
FR0401AAQ	OEM
FR-1	0015
FR1	0015
FR1-54A	0015
FR1D	0133
FR-1H	0015
FR1H	0015
FR-1H(M)	0015
FR-1HM	0015
FR-1M	0015
FR1M	0015
FR1MB	0015
FR-1MD	0015
FR1MD	7150
FR-1N	0015
FR-1P	0015
FR1P	0023
FR-1U	0015
FR1U	0015
FR-2	0015
FR2	0023
FR2-005	0102
FR2-02	0023
FR2-02(DIO)	0015

DEVICE TYPE	REPL CODE
FR2-02(RECT)	0102
FR2-02C	0015
FR2-04	0102
FR2-06	0023
FR2-08	0102
FR2-020	0015
FR2-0Z	0015
FR2-10	0102
FR-2M	0015
FR2M	0015
FR2P	4240
FR2PC	0540
FR2U	OEM
FR7	OEM
FR-10	0102
FR10	0015
FR15	OEM
FR20	OEM
FR25	OEM
FR30	OEM
FR40	OEM
FR101	0023
FR101G	1090
FR102	0023
FR102G	1090
FR103	0023
FR103G	1090
FR104	0023
FR104G	0271
FR105	0023
FR105G	0023
FR106	0017
FR106G	0017
FR107	0017
FR107G	0017
FR150AX10	OEM
FR150DX12	OEM
FR150DX16	OEM
FR150DX20	OEM
FR150DX24	OEM
FR150DY4	OEM
FR150DY6	OEM
FR150DY8	OEM
FR150DY10	OEM
FR150DY12	OEM
FR150DY16	OEM
FR151	2079
FR151G	OEM
FR152	2079
FR152G	OEM
FR153	0023
FR153G	OEM
FR154	3176
FR154G	OEM
FR155	0031
FR155G	OEM
FR155P	0023
FR156	0102
FR156G	OEM
FR157	0102
FR157G	OEM
FR201	2079
FR201G	OEM
FR-202	0015
FR202	2079
FR202C	OEM
FR202G	OEM
FR203	2079
FR203G	OEM
FR204	2280
FR204G	OEM
FR205	0031
FR205G	OEM
FR206	0102
FR206G	OEM
FR207	0102
FR207G	OEM
FR251	0031
FR252	0031
FR253	0031
FR254	0031
FR255	0031
FR256	1928
FR257	1928
FR300AX10	OEM
FR300DX12	OEM
FR300DX16	OEM
FR300DX20	OEM
FR300DX24	OEM
FR300DY4	OEM
FR300DY6	OEM
FR300DY8	OEM
FR300DY10	OEM
FR300DY12	OEM
FR300DY16	OEM
FR301	2079
FR301G	OEM
FR302	2079
FR302G	3549
FR303	2079
FR303G	OEM
FR304	2280
FR304G	OEM
FR305	2280
FR305G	OEM
FR306	3188
FR306G	OEM
FR307	3188
FR307G	OEM

DEVICE TYPE	REPL CODE
FR401	OEM
FR500AW12	OEM
FR500AW16	OEM
FR500AW20	OEM
FR500AW24	OEM
FR500AX12	OEM
FR500AX16	OEM
FR500AX20	OEM
FR500AX24	OEM
FR500AX26	OEM
FR500AY4	OEM
FR500AY6	OEM
FR500AY8	OEM
FR500AY10	OEM
FR500AY12	OEM
FR500AY16	OEM
FR500AY20	OEM
FR500AY24	OEM
FR501	1362
FR600AW40	OEM
FR600AW50	OEM
FR600AX40	OEM
FR600AX50	OEM
FR601	OEM
FR602	OEM
FR603	OEM
FR604	OEM
FR605	OEM
FR606	OEM
FR607	OEM
FR801	1145
FR801R	OEM
FR802	1145
FR802R	OEM
FR803	1145
FR803R	OEM
FR804	1145
FR804R	OEM
FR805	1664
FR805R	OEM
FR806	OEM
FR806R	OEM
FR807	OEM
FR807R	OEM
FR1000BW50	OEM
FR1000BX50	OEM
FR1001	OEM
FR-1033	0774
FR1033	0774
FR1502E10	OEM
FR1502E11	OEM
FR1502F10	OEM
FR1502F11	OEM
FR2001	OEM
FR3001	2274
FR4001	OEM
FRA81-004	OEM
FRB12-01RK	0609
FRB24-06B	OEM
FRB80-004	OEM
FRB81-004	OEM
FRC26-13L	0102
FRH-101	0015
FRP1610	3551
FRR7	OEM
FRR10	OEM
FRR15	OEM
FRR20	OEM
FRR25	OEM
FRR30	OEM
FRR40	OEM
FRR80-3	OEM
FR-U	0015
FS	OEM
FS1-2AA	OEM
FS1-2CA	OEM
FS1-4AA	OEM
FS1-4CA	OEM
FS1-6AA	OEM
FS1-6CA	OEM
FS1-8AA	OEM
FS1-10AA	OEM
FS1-12AA	OEM
FS19	0143
FS21	OEM
FS23	OEM
FS24	OEM
FS32B	OEM
FS36	OEM
FS125	OEM
FS340M	0755
FS503PA	0657
FS503RA	0657
FS503VA	0657
FS703PA	0657
FS703RA	0657
FS703VA	0657
FS-1133	0079
FS1168E641	0079
FS1168F813	0079
FS1221	0016
FS1308	0127
FS1331	0086
FS1682	0007
FS1974	0016
FS1978	0698
FS2003-1	0103
FS2042	0595
FS2043	0079

DEVICE TYPE	REPL CODE
FS-2299	0150
FS2299	0050
FS3266	0144
FS3683	0144
FS7805	4338
FS-7812	0334
FS7812	0334
FS24226	0037
FS24954	0037
FS26382	0150
FS27233	0086
FS27604	0079
FS32669	0127
FS35529	0144
FS326690	0144
FSA1169	1080
FSA1177	0196
FSA1178	0196
FSA1202	1080
FSA1410M	OEM
FSA1411M	OEM
FSA2002M	OEM
FSA2003M	OEM
FSA2500M	OEM
FSA2501M	OEM
FSA2501P	OEM
FSA2502M	OEM
FSA2503M	OEM
FSA2503P	OEM
FSA2504M	OEM
FSA2509M	OEM
FSA2509P	OEM
FSA2510M	OEM
FSA2510P	OEM
FSA2563M	OEM
FSA2563P	OEM
FSA2564M	OEM
FSA2564P	OEM
FSA2565M	OEM
FSA2565P	OEM
FSA2566M	OEM
FSA2566P	OEM
FSA2610M	OEM
FSA2619M	OEM
FSA2619P	OEM
FSA2620P	OEM
FSA2621M	OEM
FSA2702M	OEM
FSA2703M	OEM
FSA2704M	OEM
FSA2705M	OEM
FSA2719M	OEM
FSA2719P	OEM
FSA2720M	OEM
FSA2720P	OEM
FSA2721M	OEM
FSC05FA	OEM
FSC10LF	OEM
FSC10X	OEM
FSC11LF	OEM
FSC9015D	0037
FSE1001	0144
FSE3001	0144
FSE5002	0144
FSN6-2AA	OEM
FSN6-2CA	OEM
FSN6-4AA	OEM
FSN6-4CA	OEM
FSN6-6AA	OEM
FSN6-6CA	OEM
FSN6-8AA	OEM
FSN6-8CA	OEM
FSN6-10AA	OEM
FSN6-10CA	OEM
FSP-1	0144
FSP1	0127
FSP2	OEM
FSP5	OEM
FSP22	OEM
FSP-42	0144
FSP42	OEM
FSP-42-1	0144
FSP42-1	0144
FSP55	OEM
FSP-164	0144
FSP164	0144
FSP-165	0144
FSP165	0144
FSP-166	0144
FSP166	0144
FSP-166-1	0144
FSP166-1	0144
FSP-215	0144
FSP215	OEM
FSP220	OEM
FSP-242-1	0144
FSP242-1	0144
FSP-270-1	0144
FSP270-1	OEM
FSP-288-1	0015
FSP288-1	4240
FSP-289-1	0144
FSP289-1	OEM
FSP400	OEM
FSP598	OEM
FSS226A	OEM
FSS226B	OEM
FSS226C	OEM
FSS226D	OEM

DEVICE TYPE	REPL CODE
FST2	0015
FST3	0015
FSX01WF	OEM
FSX11X	OEM
FSX51W	OEM
FSX51WF	OEM
FSX51X	OEM
FSX52W	OEM
FSX52WF	OEM
FSX52X	OEM
FSX53W	OEM
FSX53WF	OEM
FSX53X	OEM
FSY07	OEM
FT001	0086
FT002	0086
FT003	0086
FT004	0086
FT004A	0086
FT005	0016
FT006	0016
FT008	0016
FT008A	0016
FT0019H	0086
FT0019M	0037
FT012	OEM
FT020	OEM
FT023	0016
FT024	0016
FT025	0016
FT026	0016
FT027	0086
FT044	OEM
FT052	0037
FT053	0016
FT-0601	0212
FT0601	OEM
FT-1	0015
FT1	0133
FT-1M	0102
FT1M	0102
FT-1N	0015
FT1N	0015
FT-1P	0102
FT1P	0102
FT-10	0015
FT10	0071
FT14A	0015
FT19H	0037
FT19M	0037
FT30A1C	OEM
FT30A1D	OEM
FT30A1N	OEM
FT30A2C	OEM
FT30A2D	OEM
FT30A2N	OEM
FT30AAC	OEM
FT30AAD	OEM
FT30AAN	OEM
FT34C	0233
FT34D	0233
FT34G	0325
FT34Y	0325
FT40	0079
FT45	0007
FT47	0392
FT48	0168
FT49	0168
FT50	0168
FT57	OEM
FT100B14	OEM
FT100BX14	OEM
FT100BY14	OEM
FT100DM4	OEM
FT100DM6	OEM
FT100DM8	OEM
FT100DM10	OEM
FT100DM12	OEM
FT100DM16	OEM
FT100DM20	OEM
FT100DM24	OEM
FT100DM28	OEM
FT100DM32	OEM
FT100DM36	OEM
FT100DX4	OEM
FT100DX6	OEM
FT100DX8	OEM
FT100DX10	OEM
FT100DX12	OEM
FT100DX16	OEM
FT100DX20	OEM
FT100DY4	OEM
FT100DY6	OEM
FT100DY8	OEM
FT100DY10	OEM
FT100DY12	OEM
FT100DY16	OEM
FT101BX2	OEM
FT101BX4	OEM
FT101BX6	OEM
FT107A	0400
FT107B	0111
FT118	0007
FT150B2	OEM
FT150B4	OEM
FT150B8	OEM
FT150B12	OEM
FT150B16	OEM
FT150B20	OEM

DEVICE TYPE	REPL CODE
FT150B24	OEM
FT150B28	OEM
FT150B32	OEM
FT150BX2	OEM
FT150BY2	OEM
FT150D68	OEM
FT150DX4	OEM
FT150DX6	OEM
FT150DX8	OEM
FT150DX10	OEM
FT150DX12	OEM
FT150DX16	OEM
FT150DX20	OEM
FT150DY4	OEM
FT150DY6	OEM
FT150DY10	OEM
FT150DY12	OEM
FT150DY16	OEM
FT150EM16	OEM
FT150EM20	OEM
FT150EM24	OEM
FT150EM28	OEM
FT150EM32	OEM
FT150EM36	OEM
FT150EX16	OEM
FT150EX20	OEM
FT150EX24	OEM
FT150EX28	OEM
FT150EX32	OEM
FT150EX36	OEM
FT151BX2	OEM
FT151BX4	OEM
FT151BX6	OEM
FT250BX2	OEM
FT250BY2	OEM
FT250DM4	OEM
FT250DM6	OEM
FT250DM8	OEM
FT250DM10	OEM
FT250DM12	OEM
FT250DM16	OEM
FT250DM20	OEM
FT250DM24	OEM
FT250DM28	OEM
FT250DM32	OEM
FT250DM36	OEM
FT250DX4	OEM
FT250DX6	OEM
FT250DX8	OEM
FT250DX10	OEM
FT250DX12	OEM
FT250DX16	OEM
FT250DX20	OEM
FT250DY4	OEM
FT250DY6	OEM
FT250DY8	OEM
FT250DY10	OEM
FT250DY12	OEM
FT250DY16	OEM
FT251BX2	OEM
FT251BX4	OEM
FT251BX6	OEM
FT251BX8	OEM
FT251BY4	OEM
FT251BY6	OEM
FT251BY8	OEM
FT300	0142
FT300A2	OEM
FT300A4	OEM
FT300A6	OEM
FT300A8	OEM
FT300A10	OEM
FT300A12	OEM
FT300A16	OEM
FT300A20	OEM
FT300A24	OEM
FT300A28	OEM
FT300A32	OEM
FT300A36	OEM
FT300A40	OEM
FT300A50	OEM
FT300AH40	OEM
FT300AH50	OEM
FT300C2	OEM
FT300C4	OEM
FT300C6	OEM
FT300C8	OEM
FT300C10	OEM
FT300C12	OEM
FT300DL4	OEM
FT300DL6	OEM
FT300DL8	OEM
FT300DL10	OEM
FT300DL12	OEM
FT300DL16	OEM
FT300DL20	OEM
FT300DL24	OEM
FT300DM28	OEM
FT300DM32	OEM
FT300DM36	OEM
FT300DX4	OEM
FT300DX6	OEM
FT300DX8	OEM
FT300DX10	OEM
FT300DX12	OEM
FT300DX16	OEM
FT300DX20	OEM
FT300DX42	OEM
FT300DY4	OEM

DEVICE TYPE	REPL CODE
FT300DY6	OEM
FT300DY8	OEM
FT300DY10	OEM
FT300DY12	OEM
FT300DY16	OEM
FT300EM16	OEM
FT300EM20	OEM
FT300EM24	OEM
FT300EM28	OEM
FT300EM32	OEM
FT300EM36	OEM
FT300EX16	OEM
FT300EX20	OEM
FT300EX24	OEM
FT300EX28	OEM
FT300EX32	OEM
FT300EX36	OEM
FT301AX2	OEM
FT301AX4	OEM
FT301AX6	OEM
FT317	OEM
FT317A	OEM
FT317B	5542
FT340	0560
FT359	OEM
FT401	0642
FT402	0065
FT410	0103
FT411	0270
FT413	0074
FT417	OEM
FT417A	OEM
FT417B	5680
FT423	0074
FT427	1984
FT430	1991
FT431	0642
FT500-32	OEM
FT500-36	OEM
FT500-40	OEM
FT500-50	OEM
FT500-60	OEM
FT500A16	OEM
FT500A20	OEM
FT500A24	OEM
FT500A28	OEM
FT500A32	OEM
FT500A36	OEM
FT500A40	OEM
FT500A50	OEM
FT500A60	OEM
FT500A80	OEM
FT500AH60	OEM
FT500AH70	OEM
FT500AH80	OEM
FT500AL4	OEM
FT500AL6	OEM
FT500AL8	OEM
FT500AL10	OEM
FT500AL12	OEM
FT500AL16	OEM
FT500AL20	OEM
FT500AL24	OEM
FT500AL28	OEM
FT500AM16	OEM
FT500AM20	OEM
FT500AM24	OEM
FT500AM28	OEM
FT500AM32	OEM
FT500AM36	OEM
FT500AM40	OEM
FT500AM50	OEM
FT500AW12	OEM
FT500AW16	OEM
FT500AW20	OEM
FT500AW24	OEM
FT500AW26	OEM
FT500AX8	OEM
FT500AX10	OEM
FT500AX12	OEM
FT500AX16	OEM
FT500AX20	OEM
FT500AY2	OEM
FT500AY4	OEM
FT500AY6	OEM
FT500AY8	OEM
FT500AY10	OEM
FT500AY12	OEM
FT500AY14	OEM
FT500AY16	OEM
FT500BX20	OEM
FT500BX24	OEM
FT500BX26	OEM
FT500BY20	OEM
FT500BY24	OEM
FT500BY26	OEM
FT500C8	OEM
FT500C10	OEM
FT500C12	OEM
FT500C16	OEM
FT500C20	OEM
FT500C24	OEM
FT500C28	OEM
FT500DL4	OEM
FT500DL6	OEM
FT500DL8	OEM
FT500DL10	OEM
FT500DL12	OEM
FT500DL16	OEM

DEVICE TYPE	REPL CODE
FT500DL20	OEM
FT500DL24	OEM
FT500DL28	OEM
FT500DX16	OEM
FT500DX20	OEM
FT500DX24	OEM
FT500DY16	OEM
FT500DY20	OEM
FT500DY24	OEM
FT500EX20	OEM
FT500EX24	OEM
FT500EY20	OEM
FT500EY24	OEM
FT500GZ16	OEM
FT500GZ20	OEM
FT500GZ24	OEM
FT520	1085
FT527	0936
FT600A6	OEM
FT600A8	OEM
FT600A10	OEM
FT600A12	OEM
FT701	OEM
FT704	0844
FT709	0144
FT800C4	OEM
FT800C6	OEM
FT800C8	OEM
FT800C10	OEM
FT800C12	OEM
FT800C16	OEM
FT800C20	OEM
FT800C24	OEM
FT800DL4	OEM
FT800DL6	OEM
FT800DL8	OEM
FT800DL10	OEM
FT800DL12	OEM
FT800DL16	OEM
FT800DL20	OEM
FT800DL24	OEM
FT1000A16	OEM
FT1000A20	OEM
FT1000A24	OEM
FT1000A28	OEM
FT1000A32	OEM
FT1000A36	OEM
FT1000A40	OEM
FT1000A50	OEM
FT1000B80	OEM
FT1000BV60	OEM
FT1000BV70	OEM
FT1000BV80	OEM
FT1000CM16	OEM
FT1000CM20	OEM
FT1000CM24	OEM
FT1000CM28	OEM
FT1000CM32	OEM
FT1000CM36	OEM
FT1000CX16	OEM
FT1000CX20	OEM
FT1000CX24	OEM
FT1000CX28	OEM
FT1000CX32	OEM
FT1000CX36	OEM
FT1200A2	OEM
FT1200A4	OEM
FT1200A6	OEM
FT1200A8	OEM
FT1200A10	OEM
FT1200A12	OEM
FT1210	OEM
FT1315	0144
FT1324B	0144
FT1324C	0144
FT1341	0037
FT1500AL4	OEM
FT1500AL6	OEM
FT1500AL8	OEM
FT1500AL12	OEM
FT1500AL16	OEM
FT1500AL20	OEM
FT1500AL24	OEM
FT1500DL4	OEM
FT1500DL6	OEM
FT1500DL8	OEM
FT1500DL12	OEM
FT1500DL16	OEM
FT1500DL20	OEM
FT1500DL24	OEM
FT1500DV60	OEM
FT1500DV70	OEM
FT1500DV80	OEM
FT1500EX16	OEM
FT1500EX20	OEM
FT1500EX24	OEM
FT1500EY16	OEM
FT1500EY20	OEM
FT1500EY24	OEM
FT1607R	OEM
FT1608R	OEM
FT1702	0150
FT1717AR	OEM
FT1746	0037
FT1869	OEM
FT1870	OEM
FT1871	OEM
FT1872	OEM
FT1873	OEM

DEVICE TYPE	REPL CODE
FT1874	OEM
FT1881	OEM
FT1882	OEM
FT1883	OEM
FT1884	OEM
FT1885	OEM
FT2009	OEM
FT2010	OEM
FT2011	OEM
FT2012	OEM
FT2013	OEM
FT2014	OEM
FT2483	OEM
FT2484	OEM
FT2500BH16	OEM
FT2500BH20	OEM
FT2500BH24	OEM
FT2500BH28	OEM
FT2500BH32	OEM
FT2500BH36	OEM
FT2500BH40	OEM
FT2500BH50	OEM
FT2500BH56	OEM
FT2955	1369
FT2974	OEM
FT2978	OEM
FT3001C	OEM
FT3001D	OEM
FT3001N	OEM
FT3002C	OEM
FT3002D	OEM
FT3002N	OEM
FT3003D	OEM
FT3003N	OEM
FT3055	0482
FT3567	0079
FT3568	0079
FT3569	0079
FT3638	0037
FT3641	0233
FT3642	0018
FT3643	0016
FT3644	0791
FT3645	0532
FT3820	OEM
FT3838	OEM
FT3909	OEM
FT4017	OEM
FT4018	OEM
FT4019	OEM
FT4020	OEM
FT4021	OEM
FT4022	OEM
FT4023	OEM
FT4024	OEM
FT4025	OEM
FT4354	0037
FT4355	0037
FT4356	OEM
FT5040	0037
FT5041	0037
FT5315	OEM
FT5320	OEM
FT5321	OEM
FT5701	OEM
FT5701AM	OEM
FT5702	OEM
FT5702M	OEM
FT5709	OEM
FT5709M	OEM
FT5712M	0544
FT5713M	0544
FT5714M	0282
FT5722R	OEM
FT5723M	OEM
FT5726DR	OEM
FT5727BR	OEM
FT5727DR	OEM
FT5728BR	OEM
FT5742M	OEM
FT5743M	OEM
FT5748M	OEM
FT5751M	OEM
FT5753M	OEM
FT5754M	OEM
FT5764M	OEM
FT5828DR	OEM
FT6200	OEM
FT7207A	OEM
FT7207B	OEM
FTD3440	0338
FTIM	0102
FT0654A	OEM
FT0654B	OEM
FT0654C	OEM
FT0654D	OEM
FT0654E	OEM
FTR129	0224
FTR1003A	OEM
FU1K	0102
FU-1M	0102
FU1M	0102
FU-1MA	0023
FU-1N	0023
FU1N	0102
FU-1NA	0023
FU1NA	0023
FU1P	0102
FU1U	0015

DEVICE TYPE	REPL CODE
FU2	OEM
FU5D-703E	OEM
FU5F7715393	1554
FU6A7709393	1705
FU-1043	0554
FU-IM	0023
FUIU	0015
FUL914-28	1069
FUL923	OEM
FUN14LH026	0443
FV5D770339	0632
FV-11	0139
FV21	0649
FV-22	0636
FV-23	0143
FV23	0143
FV-24	0301
FV24	0133
FV123	OEM
FV214	0120
FV310G	OEM
FV914	0144
FV918	0144
FV1006	OEM
FV1006A	OEM
FV1006B	OEM
FV1006C	OEM
FV1006D	OEM
FV1008	OEM
FV1008A	OEM
FV1008B	OEM
FV1008C	OEM
FV1008D	OEM
FV1010	OEM
FV1010A	OEM
FV1010B	OEM
FV1010C	OEM
FV1010D	OEM
FV1012	OEM
FV1012A	OEM
FV1012B	OEM
FV1012C	OEM
FV1012D	OEM
FV1014	OEM
FV1014A	OEM
FV1014B	OEM
FV1014C	OEM
FV1014D	OEM
FV1016	OEM
FV1016A	OEM
FV1016B	OEM
FV1016C	OEM
FV1016D	OEM
FV-1043	0554
FV1043	0554
FV2369A	0144
FV2484	0144
FV2747C	0280
FV2894	0150
FV3014	0144
FV3299	0144
FV3300	0144
FV3502	0861
FV3503	0861
FV3962	OEM
FV3964	OEM
FW50	0277
FW100	0015
FW200	0289
FW300	0296
FW400	0015
FW500	0015
FW600	0015
FW600A	0015
FW800	0250
FW1000	0250
FWB3001	0289
FWB3002	0289
FWB3003	0296
FWB3004	0296
FWB3005	0302
FWB3006	0302
FWL50	0277
FWL100	0015
FWL200	0015
FWL300	0015
FWL400	0296
FWL500	0302
FWL600	0302
FWL700	0250
FWL800	0250
FWL1000	0250
FWLA50	0277
FWLA100	0277
FWLA200	0289
FWLA300	0296
FWLA400	0296
FWLA500	0302
FWLA600	0302
FWLA700	0250
FWLA800	0250
FWLA1000	0250
FWLC50	0289
FWLC100	0289
FWLC200	0289
FWLC300	0296
FWLC400	0296
FWLC500	0737
FWLC600	0302

If replacement code is OEM, contact original manufacturer for replacement.

DEVICE TYPE	REPL CODE
FWLC700	OEM
FWLC800	0250
FWLC1000	0250
FWLD50	0323
FWLD100	0323
FWLD200	1414
FWLD300	OEM
FWLD400	0473
FWLD500	OEM
FWLD600	0446
FWLD700	OEM
FWLD800	1422
FWLD1000	OEM
FX0003CE	OEM
FX0007CE	1166
FX0011CE	1166
FX0014CE	1291
FX41	OEM
FX45	OEM
FX47	OEM
FX274	0352
FX709	0127
FX914	0224
FX918	0224
FX2368	0079
FX2369A	0224
FX2483	0861
FX2894	0037
FX3013	0224
FX3014	0224
FX3299	0018
FX3300	0079
FX3502	0861
FX3724	0018
FX3962	0791
FX3964	0693
FX4046	0079
FX4960	0595
FZ-2	OEM
FZ3.6T5	0188
FZ3.6T10	0188
FZ-4	OEM
FZ5.1T5	0162
FZ5.1T10	0162
FZ5.6T5	0157
FZ5.6T10	0157
FZ6.2T10	0636
FZ6.5T10	0636
FZ10A	0170
FZ10B	0098
FZ12A	0137
FZ12B	0186
FZ12T5	0137
FZ12T10	0137
FZ14T5	0100
FZ14T10	0100
FZ15A	0002
FZ15T5	0002
FZ15T10	0002
FZ18A	0495
FZ22A	0565
FZ27A	0441
FZ27T5	0441
FZ27T10	0441
FZ33A	0039
FZ39A	0350
FZ47A	0999
FZ56A	0869
FZ68A	2157
FZ82A	0331
FZ101	0079
FZ309B	4235
FZ901	0162
FZ1215	0137
FZA-1D	OEM
FZA-2D	OEM
FZD6V8	2133
FZD8V2	OEM
FZD10	OEM
FZD12	OEM
FZD15	OEM
FZD18	1767
FZD22	OEM
FZD27	1952
FZD33	OEM
FZD39	OEM
FZD47	OEM
FZD56	OEM
FZD62	OEM
FZD68	1987
FZD82	OEM
FZD100	OEM
FZD120	OEM
FZD150	OEM
FZD180	OEM
FZH101	OEM
FZH101A	OEM
FZH105A	OEM
FZH111	OEM
FZH111A	OEM
FZH115B	OEM
FZH121	OEM
FZH125	OEM
FZH131	OEM
FZH135	OEM
FZH141	OEM
FZH145	OEM
FZH151	OEM
FZH155	OEM

DEVICE TYPE	REPL CODE
FZH161	OEM
FZH165	OEM
FZH165B	OEM
FZH171	OEM
FZH175	OEM
FZH181	OEM
FZH185	OEM
FZH191	OEM
FZH195	OEM
FZH201	OEM
FZH205	OEM
FZH211	OEM
FZH211S	OEM
FZH215B	OEM
FZH231	OEM
FZH235	OEM
FZH241	OEM
FZH245	OEM
FZH245B	OEM
FZH251	OEM
FZH255B	OEM
FZH261	OEM
FZH265B	OEM
FZH271	OEM
FZH275	OEM
FZH281	OEM
FZH285B	OEM
FZH291	OEM
FZH295B	OEM
FZH301	OEM
FZH305	OEM
FZJ101	OEM
FZJ105	OEM
FZJ111	OEM
FZJ115	OEM
FZJ121	OEM
FZJ125	OEM
FZJ131	OEM
FZJ135	OEM
FZJ141	OEM
FZJ141A	OEM
FZJ145A	OEM
FZJ151	OEM
FZJ151A	OEM
FZJ155A	OEM
FZJ161	OEM
FZJ165	OEM
FZK101	OEM
FZK105	OEM
FZL101	OEM
FZL105	OEM
FZL111	5028
FZL121	OEM
FZL121S	OEM
FZL125	OEM
FZL125X	OEM
FZL131	OEM
FZL131S	OEM
FZL135	OEM
FZL135S	OEM
FZL141	OEM
FZL141S	OEM
FZL145	OEM
FZL145S	OEM
FZY101	OEM
FZY105	OEM
G00001B	OEM
G00004A	0143
G00005B	OEM
G00008A	0019
G0000.5B	OEM
G0001	0136
G0003	0050
G0004	0164
G0005	0164
G0006	0211
G0007	0004
G0009	0004
G00010B	OEM
G00020B	OEM
G00050B	OEM
G00-003-A	0019
G00-004-A	0143
G00-008-A	0143
G00-009-A	0143
G00-012-A	0133
G00-012A	0133
G00-013-8	0143
G00-014-A	0133
G00-073-8	OEM
G00-502-A	0015
G00-502A	0015
G00-534-A	0015
G00-535-B	0023
G00-535A	0015
G00-536-A	0015
G00-536A	0015
G00-543-A	0015
G00-551-A	0015
G00-803-A	0133
G001A0A	OEM
G001A0B	OEM
G001A0C	OEM
G001A0D	OEM
G0011	0600
G00100B	OEM
G00200B	OEM
G00500A	OEM
G00500B	OEM

DEVICE TYPE	REPL CODE
G00500C	OEM
G00500D	OEM
G00502A	0023
G00515A	0198
G00535	0015
G00535B	4240
G01	0015
G01-012-F	0170
G01-036-G	0057
G01-036-H	0012
G01-036A	0636
G01-037-A	0137
G01-083-A	0133
G01-209-B	0133
G01-209B	0133
G01-217-A	0139
G01-401-A	0030
G01-406-A	0628
G01-407-A	0030
G01-603-A	OEM
G01-803-A	0133
G01-803A	0133
G01A	0917
G02	0015
G02A	0917
G03-007	OEM
G03-007C	0161
G03-014	OEM
G03-017-B	0006
G03-404-B	0126
G03-404-C	0126
G03-406-C	0460
G03-406-C(5)	1367
G03-407-Y	0006
G03-703C	0681
G04-041B	0127
G04-701-A	0160
G04-704-A	0222
G04-711-E	0164
G04-711-F	0164
G04-711-F,E,G,H	0164
G04-711-G	0164
G04-711-H	0164
G05-003-A	0330
G05-003-B	0151
G05-004A	0144
G05-010-A	0016
G05-011-A	0016
G05-012-G	0111
G05-015-D	0016
G05-015C	0016
G05-034-D	0016
G05-035-D	0016
G05-035-D,E	0111
G05-035-E	0111
G05-035D	0111
G05-035E	0016
G05-036-B	0079
G05-036-C	1220
G05-036-C,D,E	0016
G05-036-D	1220
G05-036-E	1220
G05-036B	0286
G05-036C	0286
G05-036D	0016
G05-036E	0016
G05-037-A	0224
G05-037-B	0151
G05-037-D	0151
G05-037B	0016
G05-050-C	0127
G05-055-C	0155
G05-055-D	0155
G05-055-E	0155
G05-059	OEM
G05-063-R	1144
G05-064-A	0079
G05-065-A	1066
G05-066A	0076
G05-406-C	0460
G05-413A	0155
G05-413B	0155
G05-413C	0155
G05-413D	0155
G05-415	OEM
G05-415-B	0018
G05-416-C	1790
G05-705	OEM
G05-706-D	0537
G05-706-E	0537
G06-711-B	0560
G06-714C	0219
G06-717-B	0161
G06-717-B(10)	0566
G06-717-B,C,D	0566
G06-717-C	0161
G06-717-D	0161
G08-005L	0325
G08-007-B	1280
G08D	0087
G09-006-A	2262
G09-006-B	2262
G09-006-C	2262
G09-007-A	1994
G09-007-B	1994
G09-008-B	1018
G09-008-C	1018
G09-008-D	1018
G09-008-E	1018

DEVICE TYPE	REPL CODE
G09-009-A	1215
G09-009-C	OEM
G09-010-A	2262
G09-011-A	4409
G09-012-A	4409
G09-012-B	4409
G09-012-C	4409
G09-013-A	2124
G09-013-A(1)	2124
G09-015-B	0579
G09-017-A	2656
G09-017-B	2656
G09-017-C	2656
G09-017-D	2656
G09-018-A	2755
G09-028-A	0520
G09-029-B	0433
G09-029-C	1994
G09-029-D	1994
G09-029B	1994
G09-034-A	0520
G09-502A	OEM
G09-505-A	OEM
G09-508-A	OEM
G0500.5B	OEM
G01036	0012
G01209	0133
G01211	0015
G01401A	0720
G01803	0133
G03001B-2R	OEM
G03007	0037
G03014	0532
G05001B	OEM
G05005B	OEM
G05010B	OEM
G05015	0198
G05015D	0198
G05015E	0198
G05020B	OEM
G05035	0016
G05036	0016
G05041A	OEM
G05050B	OEM
G05059	0016
G05415	0018
G05704	OEM
G05704B	OEM
G05704C	OEM
G05705	0042
G08005L	3046
G09007	1994
G09009	1215
G09015	0579
G030010-1R	OEM
G030010-2R	OEM
G030010-3R	OEM
G030012-1R	OEM
G030012-2R	OEM
G030012-3R	OEM
G030015-1R	OEM
G030015-2R	OEM
G030015-3R	OEM
G030018-1R	OEM
G030018-3R	OEM
G030022-1R	OEM
G030022-2R	OEM
G030022-3R	OEM
G030027-1R	OEM
G030027-2R	OEM
G030027-3R	OEM
G060007-1R	OEM
G060007-2R	OEM
G060010-1R	OEM
G060010-2R	OEM
G060010-3R	OEM
G060012-1R	OEM
G060012-2R	OEM
G060012-3R	OEM
G060015-1R	OEM
G060015-2R	OEM
G060015-3R	OEM
G060018-1R	OEM
G060018-2R	OEM
G060018-3R	OEM
G060022-1R	OEM
G060022-2R	OEM
G060022-3R	OEM
G060027-1R	OEM
G060027-2R	OEM
G060027-3R	OEM
G0Q-535-B	0015
G1	0015
G1A	3911
G1B	0023
G1D	3911
G1G	3911
G1HA	0143
G1J	0229
G1J-082	OEM
G1K	2637
G1M	2637
G1RG3K	1217
G2	1228
G2A	0015
G2B	0087
G2D	0023
G2DG1	0071
G2G	0087

DEVICE TYPE	REPL CODE
G2J	0087
G2K	0087
G2M	0087
G2SC1175-D	OEM
G3	0015
G3A	0110
G3A3	0599
G3B	0953
G3B3	0599
G3C3	0599
G3D	0953
G3D3	0599
G3E3	0599
G3F3	0599
G3G	0540
G3G3	0599
G3H3	1986
G3J	0817
G3K	0817
G3K3	1986
G3M	0922
G3M3	0657
G3N3	0657
G3.0T5	1713
G3.0T10	1713
G3.0T20	1713
G3.3T5	0292
G3.3T10	0292
G3.3T20	0292
G3.6T5	0188
G3.6T10	0188
G3.6T20	1713
G3.9T5	0456
G3.9T10	0456
G3.9T20	0456
G4	0015
G4A	0087
G4A3	0599
G4B	0953
G4B3	0599
G4C3	0599
G4D	0242
G4D3	0599
G4E3	0599
G4F3	0599
G4G	0540
G4G3	0599
G4H3	0657
G4HZ	OEM
G4J	0965
G4K	0965
G4K3	1986
G4M	0922
G4M3	0657
G4N1	0074
G4N3	0657
G4.3T5	0533
G4.3T10	0533
G4.3T20	0533
G4.7T5	0451
G4.7T10	0451
G4.7T20	0451
G5	0015
G5A	OEM
G5A3	0599
G5B	OEM
G5B3	0599
G5C	0143
G5C3	0599
G5D	OEM
G5D3	0599
G5E3	0599
G5F	0143
G5F3	0599
G5G3	0599
G5H1	OEM
G5H3	1986
G5K	0143
G5K3	1986
G5M3	0657
G5N3	0657
G5R	OEM
G5U	OEM
G5.1T5	0162
G5.1T10	0162
G5.1T20	0162
G5.6T5	0157
G5.6T10	0157
G5.6T20	0157
G6	0015
G6A	OEM
G6A3	OEM
G6B	OEM
G6B3	OEM
G6C	OEM
G6D	OEM
G6D3	OEM
G6E3	OEM
G6F	OEM
G6F3	OEM
G6G3	OEM
G6H3	OEM
G6HZ	0122
G6K3	OEM
G6M3	OEM
G6N3	OEM
G6P3	OEM
G6R3	OEM
G6S3	OEM

DEVICE TYPE	REPL CODE
G6U	OEM
G6.2T5	0636
G6.2T10	0636
G6.2T20	0636
G6.8T5	0025
G6.8T10	0025
G6.8T20	0025
G7	0917
G7A	0917
G7D	0143
G7E	0143
G7F	0143
G7G	0143
G7SE211	0491
G7.5T5	0649
G7.5T10	0649
G7.5T20	0649
G8	0071
G8HZ	OEM
G8.2T5	0244
G8.2T10	0244
G8.2T20	0244
G9.1T5	0012
G9.1T10	0012
G9.1T20	0012
G10	0071
G10HZ	OEM
G10T5	0170
G10T10	0170
G10T20	0170
G11	0004
G11T5	0317
G11T10	0317
G11T20	0317
G12	0004
G12T5	0137
G12T10	0137
G12T20	0137
G13	0050
G13T5	0365
G13T10	0365
G13T20	0365
G14	0004
G15R4	OEM
G15T5	0137
G15T10	0002
G15T20	0002
G16	0038
G16T5	0420
G16T10	0420
G16T20	0420
G17	0038
G18	0038
G18T5	0495
G18T10	0495
G18T20	0495
G19	0160
G20T5	0531
G20T10	0531
G20T20	0531
G22T5	0565
G22T10	0565
G22T20	0565
G24T5	0402
G24T10	0402
G24T20	0402
G25X01G	OEM
G25X02G	OEM
G25X15G	OEM
G27T5	0441
G27T10	0441
G27T20	0441
G30R4	OEM
G30T5	0726
G30T10	0726
G30T20	0726
G50X01D	OEM
G50X02D	OEM
G50X15D	OEM
G60R4	OEM
G65HZ	OEM
G65SC02C-3	OEM
G65SC02C-4	OEM
G65SC02CI-3	OEM
G65SC02CI-4	OEM
G65SC02CM-2	OEM
G65SC02CM-3	OEM
G65SC02CM-4	OEM
G65SC02D-3	OEM
G65SC02D-4	OEM
G65SC02DI-3	OEM
G65SC02DI-4	OEM
G65SC02DM-2	OEM
G65SC02DM-3	OEM
G65SC02DM-4	OEM
G65SC02E-3	OEM
G65SC02E-4	OEM
G65SC02EI-3	OEM
G65SC02EI-4	OEM
G65SC02EM-2	OEM
G65SC02EM-3	OEM
G65SC02EM-4	OEM
G65SC02P-3	OEM
G65SC02P-4	OEM
G65SC02PI-3	OEM
G65SC02PI-4	OEM
G65SC02PM-3	OEM
G65SC02X-3	OEM
G65SC02X-4	OEM

DEVICE TYPE	REPL CODE
G65SC02XI-4	OEM
G65SC02XM-2	OEM
G65SC02XM-3	OEM
G65SC02XM-4	OEM
G65SC03C-1	OEM
G65SC03C-2	OEM
G65SC03C-3	OEM
G65SC03C-4	OEM
G65SC03CI-2	OEM
G65SC03CI-3	OEM
G65SC03CI-4	OEM
G65SC03CM-1	OEM
G65SC03CM-2	OEM
G65SC03CM-3	OEM
G65SC03CM-4	OEM
G65SC03D-1	OEM
G65SC03D-2	OEM
G65SC03D-3	OEM
G65SC03D-4	OEM
G65SC03DI-1	OEM
G65SC03DI-2	OEM
G65SC03DI-3	OEM
G65SC03DI-4	OEM
G65SC03DM-1	OEM
G65SC03DM-2	OEM
G65SC03DM-3	OEM
G65SC03DM-4	OEM
G65SC03E-1	OEM
G65SC03E-3	OEM
G65SC03E-4	OEM
G65SC03EI-1	OEM
G65SC03EI-2	OEM
G65SC03EI-3	OEM
G65SC03EI-4	OEM
G65SC03EM-1	OEM
G65SC03EM-2	OEM
G65SC03EM-3	OEM
G65SC03EM-4	OEM
G65SC03P-1	OEM
G65SC03P-2	OEM
G65SC03P-3	OEM
G65SC03P-4	OEM
G65SC03PI-1	OEM
G65SC03PI-2	OEM
G65SC03PI-3	OEM
G65SC03PI-4	OEM
G65SC03PM-1	OEM
G65SC03PM-2	OEM
G65SC03PM-3	OEM
G65SC03PM-4	OEM
G65SC03X-1	OEM
G65SC03X-2	OEM
G65SC03X-3	OEM
G65SC03X-4	OEM
G65SC03XI-1	OEM
G65SC03XI-2	OEM
G65SC03XI-3	OEM
G65SC03XI-4	OEM
G65SC03XM-1	OEM
G65SC03XM-2	OEM
G65SC03XM-3	OEM
G65SC03XM-4	OEM
G65SC04C04	OEM
G65SC04C-3	OEM
G65SC04CI-2	OEM
G65SC04CI-3	OEM
G65SC04CI-4	OEM
G65SC04CM-2	OEM
G65SC04CM-3	OEM
G65SC04CM-4	OEM
G65SC04D-3	OEM
G65SC04D-4	OEM
G65SC04DI-2	OEM
G65SC04DI-3	OEM
G65SC04DI-4	OEM
G65SC04DM-2	OEM
G65SC04DM-3	OEM
G65SC04DM-4	OEM
G65SC04E-3	OEM
G65SC04E-4	OEM
G65SC04EI-2	OEM
G65SC04EI-3	OEM
G65SC04EI-4	OEM
G65SC04EM-2	OEM
G65SC04EM-3	OEM
G65SC04EM-4	OEM
G65SC04P-3	OEM
G65SC04P-4	OEM
G65SC04PI-2	OEM
G65SC04PI-3	OEM
G65SC04PI-4	OEM
G65SC04PM-2	OEM
G65SC04PM-3	OEM
G65SC04PM-4	OEM
G65SC04X-3	OEM
G65SC04X-4	OEM
G65SC04XI-2	OEM
G65SC04XI-3	OEM
G65SC04XI-4	OEM
G65SC04XM-2	OEM
G65SC04XM-3	OEM
G65SC04XM-4	OEM
G65SC05C-1	OEM
G65SC05C-2	OEM
G65SC05C-3	OEM
G65SC05C-4	OEM
G65SC05CI-1	OEM
G65SC05CI-2	OEM
G65SC05CI-3	OEM

DEVICE TYPE	REPL CODE
G65SC05CI-4	OEM
G65SC05CM-1	OEM
G65SC05CM-2	OEM
G65SC05CM-3	OEM
G65SC05CM-4	OEM
G65SC05D-1	OEM
G65SC05D-2	OEM
G65SC05D-3	OEM
G65SC05D-4	OEM
G65SC05DI-1	OEM
G65SC05DI-2	OEM
G65SC05DI-3	OEM
G65SC05DI-4	OEM
G65SC05DM-1	OEM
G65SC05DM-2	OEM
G65SC05DM-3	OEM
G65SC05DM-4	OEM
G65SC05E-1	OEM
G65SC05E-2	OEM
G65SC05E-3	OEM
G65SC05E-4	OEM
G65SC05EI-1	OEM
G65SC05EI-2	OEM
G65SC05EI-3	OEM
G65SC05EI-4	OEM
G65SC05EM-1	OEM
G65SC05EM-2	OEM
G65SC05EM-3	OEM
G65SC05EM-4	OEM
G65SC05P-1	OEM
G65SC05P-2	OEM
G65SC05P-3	OEM
G65SC05P-4	OEM
G65SC05PI-1	OEM
G65SC05PI-2	OEM
G65SC05PI-3	OEM
G65SC05PI-4	OEM
G65SC05PM-1	OEM
G65SC05PM-2	OEM
G65SC05PM-3	OEM
G65SC05PM-4	OEM
G65SC05X-1	OEM
G65SC05X-2	OEM
G65SC05X-4	OEM
G65SC05XI-1	OEM
G65SC05XI-2	OEM
G65SC05XI-3	OEM
G65SC05XM-1	OEM
G65SC05XM-2	OEM
G65SC05XM-3	OEM
G65SC06C-2	OEM
G65SC06C-3	OEM
G65SC06C-4	OEM
G65SC06CI-1	OEM
G65SC06CI-2	OEM
G65SC06CI-3	OEM
G65SC06CI-4	OEM
G65SC06CM-1	OEM
G65SC06CM-2	OEM
G65SC06CM-3	OEM
G65SC06CM-4	OEM
G65SC06D-1	OEM
G65SC06D-2	OEM
G65SC06D-3	OEM
G65SC06D-4	OEM
G65SC06DI-1	OEM
G65SC06DI-2	OEM
G65SC06DI-3	OEM
G65SC06DI-4	OEM
G65SC06DM-1	OEM
G65SC06DM-2	OEM
G65SC06DM-3	OEM
G65SC06DM-4	OEM
G65SC06E-1	OEM
G65SC06E-2	OEM
G65SC06E-3	OEM
G65SC06E-4	OEM
G65SC06EI-1	OEM
G65SC06EI-2	OEM
G65SC06EI-3	OEM
G65SC06EI-4	OEM
G65SC06EM-1	OEM
G65SC06EM-2	OEM
G65SC06EM-3	OEM
G65SC06EM-4	OEM
G65SC06P-2	OEM
G65SC06P-3	OEM
G65SC06P-4	OEM
G65SC06PI-1	OEM
G65SC06PI-2	OEM
G65SC06PI-3	OEM
G65SC06PI-4	OEM
G65SC06PM-1	OEM
G65SC06PM-2	OEM
G65SC06PM-3	OEM
G65SC06PM-4	OEM
G65SC06X-1	OEM
G65SC06X-2	OEM
G65SC06X-3	OEM
G65SC06X-4	OEM
G65SC06XI-1	OEM
G65SC06XI-2	OEM
G65SC06XI-3	OEM
G65SC06XI-4	OEM
G65SC06XM-1	OEM
G65SC06XM-2	OEM
G65SC06XM-3	OEM
G65SC06XM-4	OEM
G65SC07C-1	OEM
G65SC07C-2	OEM
G65SC07C-3	OEM
G65SC07CI-1	OEM
G65SC07CI-2	OEM
G65SC07CI-3	OEM
G65SC07CM-1	OEM
G65SC07CM-2	OEM
G65SC07D-1	OEM
G65SC07D-2	OEM
G65SC07D-3	OEM
G65SC07DI-1	OEM
G65SC07DI-2	OEM
G65SC07DI-3	OEM
G65SC07DM-1	OEM
G65SC07DM-2	OEM
G65SC07E-1	OEM
G65SC07E-2	OEM
G65SC07E-3	OEM
G65SC07EI-1	OEM
G65SC07EI-2	OEM
G65SC07EI-3	OEM
G65SC07EM-1	OEM
G65SC07EM-2	OEM
G65SC07P-1	OEM
G65SC07P-2	OEM
G65SC07P-3	OEM
G65SC07PI-1	OEM
G65SC07PI-2	OEM
G65SC07PI-3	OEM
G65SC07PM-1	OEM
G65SC07X-1	OEM
G65SC07X-2	OEM
G65SC07X-3	OEM
G65SC07XI-1	OEM
G65SC07XI-2	OEM
G65SC07XI-3	OEM
G65SC07XM-1	OEM
G65SC07XM-2	OEM
G65SC12C-1	OEM
G65SC12C-2	OEM
G65SC12C-3	OEM
G65SC12C-4	OEM
G65SC12CI-1	OEM
G65SC12CI-2	OEM
G65SC12CI-3	OEM
G65SC12CI-4	OEM
G65SC12CM-1	OEM
G65SC12CM-2	OEM
G65SC12CM-3	OEM
G65SC12CM-4	OEM
G65SC12D-1	OEM
G65SC12D-2	OEM
G65SC12D-3	OEM
G65SC12D-4	OEM
G65SC12DI-1	OEM
G65SC12DI-2	OEM
G65SC12DI-4	OEM
G65SC12DM-1	OEM
G65SC12DM-2	OEM
G65SC12DM-3	OEM
G65SC12DM-4	OEM
G65SC12E-1	OEM
G65SC12E-2	OEM
G65SC12E-3	OEM
G65SC12E-4	OEM
G65SC12EI-1	OEM
G65SC12EI-2	OEM
G65SC12EI-3	OEM
G65SC12EI-4	OEM
G65SC12EM-1	OEM
G65SC12EM-2	OEM
G65SC12EM-3	OEM
G65SC12EM-4	OEM
G65SC12P-1	OEM
G65SC12P-2	OEM
G65SC12P-3	OEM
G65SC12P-4	OEM
G65SC12PI-1	OEM
G65SC12PI-2	OEM
G65SC12PI-3	OEM
G65SC12PI-4	OEM
G65SC12PM-1	OEM
G65SC12PM-2	OEM
G65SC12PM-3	OEM
G65SC12PM-4	OEM
G65SC12X-1	OEM
G65SC12X-2	OEM
G65SC12X-3	OEM
G65SC12X-4	OEM
G65SC12XI-1	OEM
G65SC12XI-2	OEM
G65SC12XI-3	OEM
G65SC12XI-4	OEM
G65SC12XM-1	OEM
G65SC12XM-2	OEM
G65SC12XM-3	OEM
G65SC12XM-4	OEM
G65SC13C-1	OEM
G65SC13C-2	OEM
G65SC13C-3	OEM
G65SC13C-4	OEM
G65SC13CI-1	OEM
G65SC13CI-2	OEM
G65SC13CI-4	OEM
G65SC13CM-1	OEM
G65SC13CM-2	OEM
G65SC13CM-3	OEM
G65SC13CM-4	OEM
G65SC13D-1	OEM
G65SC13D-2	OEM
G65SC13D-3	OEM
G65SC13D-4	OEM
G65SC13DI-1	OEM
G65SC13DI-2	OEM
G65SC13DI-3	OEM
G65SC13DI-4	OEM
G65SC13DM-1	OEM
G65SC13DM-2	OEM
G65SC13DM-3	OEM
G65SC13DM-4	OEM
G65SC13E-1	OEM
G65SC13E-2	OEM
G65SC13E-3	OEM
G65SC13E-4	OEM
G65SC13EI-1	OEM
G65SC13EI-2	OEM
G65SC13EI-3	OEM
G65SC13EI-4	OEM
G65SC13EM-1	OEM
G65SC13EM-2	OEM
G65SC13EM-3	OEM
G65SC13EM-4	OEM
G65SC13P-1	OEM
G65SC13P-2	OEM
G65SC13P-3	OEM
G65SC13PI-1	OEM
G65SC13PI-2	OEM
G65SC13PI-3	OEM
G65SC13PI-4	OEM
G65SC13PM-1	OEM
G65SC13PM-2	OEM
G65SC13PM-3	OEM
G65SC13PM-4	OEM
G65SC13X-1	OEM
G65SC13X-2	OEM
G65SC13X-3	OEM
G65SC13X-4	OEM
G65SC13XI-1	OEM
G65SC13XI-2	OEM
G65SC13XI-3	OEM
G65SC13XI-4	OEM
G65SC13XM-1	OEM
G65SC13XM-2	OEM
G65SC13XM-3	OEM
G65SC13XM-4	OEM
G65SC14C-1	OEM
G65SC14C-2	OEM
G65SC14C-3	OEM
G65SC14C-4	OEM
G65SC14CI-1	OEM
G65SC14CI-2	OEM
G65SC14CI-3	OEM
G65SC14CI-4	OEM
G65SC14CM-1	OEM
G65SC14CM-2	OEM
G65SC14CM-3	OEM
G65SC14CM-4	OEM
G65SC14D-1	OEM
G65SC14D-2	OEM
G65SC14D-3	OEM
G65SC14D-4	OEM
G65SC14DI-1	OEM
G65SC14DI-2	OEM
G65SC14DI-3	OEM
G65SC14DI-4	OEM
G65SC14DM-1	OEM
G65SC14DM-2	OEM
G65SC14DM-3	OEM
G65SC14DM-4	OEM
G65SC14E-1	OEM
G65SC14E-2	OEM
G65SC14E-3	OEM
G65SC14E-4	OEM
G65SC14EI-1	OEM
G65SC14EI-2	OEM
G65SC14EI-3	OEM
G65SC14EI-4	OEM
G65SC14EM-1	OEM
G65SC14EM-2	OEM
G65SC14EM-3	OEM
G65SC14EM-4	OEM
G65SC14P-1	OEM
G65SC14P-2	OEM
G65SC14P-3	OEM
G65SC14P-4	OEM
G65SC14PI-1	OEM
G65SC14PI-2	OEM
G65SC14PI-3	OEM
G65SC14PI-4	OEM
G65SC14PM-1	OEM
G65SC14PM-2	OEM
G65SC14PM-3	OEM
G65SC14PM-4	OEM
G65SC14X-1	OEM
G65SC14X-2	OEM
G65SC14X-3	OEM
G65SC14XI-1	OEM
G65SC14XI-2	OEM
G65SC14XI-3	OEM
G65SC14XI-4	OEM
G65SC14XM-1	OEM
G65SC14XM-2	OEM
G65SC14XM-3	OEM
G65SC14XM-4	OEM
G65SC15C-1	OEM
G65SC15C-2	OEM
G65SC15C-3	OEM
G65SC15C-4	OEM
G65SC15CI-1	OEM
G65SC15CI-2	OEM
G65SC15CI-3	OEM
G65SC15CI-4	OEM
G65SC15CM-1	OEM
G65SC15CM-2	OEM
G65SC15CM-3	OEM
G65SC15CM-4	OEM
G65SC15D-1	OEM
G65SC15D-2	OEM
G65SC15D-3	OEM
G65SC15D-4	OEM
G65SC15DI-1	OEM
G65SC15DI-2	OEM
G65SC15DI-3	OEM
G65SC15DI-4	OEM
G65SC15DM-1	OEM
G65SC15DM-2	OEM
G65SC15DM-3	OEM
G65SC15DM-4	OEM
G65SC15E-1	OEM
G65SC15E-2	OEM
G65SC15E-3	OEM
G65SC15E-4	OEM
G65SC15EI-1	OEM
G65SC15EI-2	OEM
G65SC15EI-3	OEM
G65SC15EI-4	OEM
G65SC15EM-1	OEM
G65SC15EM-2	OEM
G65SC15EM-3	OEM
G65SC15EM-4	OEM
G65SC15P-1	OEM
G65SC15P-2	OEM
G65SC15P-3	OEM
G65SC15P-4	OEM
G65SC15PI-1	OEM
G65SC15PI-2	OEM
G65SC15PI-3	OEM
G65SC15PI-4	OEM
G65SC15PM-1	OEM
G65SC15PM-2	OEM
G65SC15PM-3	OEM
G65SC15PM-4	OEM
G65SC15X-1	OEM
G65SC15X-2	OEM
G65SC15X-3	OEM
G65SC15X-4	OEM
G65SC15XI-1	OEM
G65SC15XI-2	OEM
G65SC15XI-3	OEM
G65SC15XI-4	OEM
G65SC15XM-1	OEM
G65SC15XM-2	OEM
G65SC15XM-3	OEM
G65SC15XM-4	OEM
G65SC22P-2	OEM
G65SC102C-1	OEM
G65SC102C-2	OEM
G65SC102C-3	OEM
G65SC102C-4	OEM
G65SC102CI-1	OEM
G65SC102CI-2	OEM
G65SC102CI-3	OEM
G65SC102CI-4	OEM
G65SC102CM-1	OEM
G65SC102CM-2	OEM
G65SC102CM-3	OEM
G65SC102CM-4	OEM
G65SC102D-1	OEM
G65SC102D-2	OEM
G65SC102D-3	OEM
G65SC102D-4	OEM
G65SC102DI-1	OEM
G65SC102DI-2	OEM
G65SC102DI-3	OEM
G65SC102DI-4	OEM
G65SC102DM-1	OEM
G65SC102DM-2	OEM
G65SC102DM-3	OEM
G65SC102DM-4	OEM
G65SC102E-1	OEM
G65SC102E-2	OEM
G65SC102E-3	OEM
G65SC102E-4	OEM
G65SC102EI-1	OEM
G65SC102EI-2	OEM
G65SC102EI-3	OEM
G65SC102EI-4	OEM
G65SC102EM-2	OEM
G65SC102EM-3	OEM
G65SC102EM-4	OEM
G65SC102P-1	OEM
G65SC102P-2	OEM
G65SC102P-3	OEM
G65SC102P-4	OEM
G65SC102PI-1	OEM
G65SC102PI-2	OEM
G65SC102PI-3	OEM
G65SC102PI-4	OEM
G65SC102PM-1	OEM
G65SC102PM-2	OEM
G65SC102PM-3	OEM
G65SC102PM-4	OEM
G65SC102X-1	OEM
G65SC102X-2	OEM
G65SC102X-3	OEM
G65SC102X-4	OEM
G65SC102XI-1	OEM
G65SC102XI-2	OEM
G65SC102XI-3	OEM
G65SC102XI-4	OEM
G65SC102XM-1	OEM
G65SC102XM-2	OEM
G65SC102XM-3	OEM
G65SC102XM-4	OEM
G65SC103C-1	OEM
G65SC103C-2	OEM
G65SC103C-3	OEM
G65SC103C-4	OEM
G65SC103CI-1	OEM
G65SC103CI-2	OEM
G65SC103CI-3	OEM
G65SC103CI-4	OEM
G65SC103CM-1	OEM
G65SC103CM-2	OEM
G65SC103CM-3	OEM
G65SC103CM-4	OEM
G65SC103D-1	OEM
G65SC103D-2	OEM
G65SC103D-3	OEM
G65SC103D-4	OEM
G65SC103DI-1	OEM
G65SC103DI-2	OEM
G65SC103DI-3	OEM
G65SC103DI-4	OEM
G65SC103DM-1	OEM
G65SC103DM-2	OEM
G65SC103DM-3	OEM
G65SC103DM-4	OEM
G65SC103E-1	OEM
G65SC103E-2	OEM
G65SC103E-3	OEM
G65SC103E-4	OEM
G65SC103EI-1	OEM
G65SC103EI-2	OEM
G65SC103EI-3	OEM
G65SC103EI-4	OEM
G65SC103EM-1	OEM
G65SC103EM-2	OEM
G65SC103EM-3	OEM
G65SC103P-1	OEM
G65SC103P-2	OEM
G65SC103P-3	OEM
G65SC103PI-1	OEM
G65SC103PI-2	OEM
G65SC103PI-3	OEM
G65SC103PI-4	OEM
G65SC103PM-1	OEM
G65SC103PM-2	OEM
G65SC103PM-3	OEM
G65SC103PM-4	OEM
G65SC103X-1	OEM
G65SC103X-2	OEM
G65SC103X-3	OEM
G65SC103XI-1	OEM
G65SC103XI-2	OEM
G65SC103XI-3	OEM
G65SC103XI-4	OEM
G65SC103XM-1	OEM
G65SC103XM-2	OEM
G65SC103XM-3	OEM
G65SC103XM-4	OEM
G65SC104C-1	OEM
G65SC104C-3	OEM
G65SC104C-4	OEM
G65SC104CI-1	OEM
G65SC104CI-2	OEM
G65SC104CI-3	OEM
G65SC104CM-1	OEM
G65SC104CM-2	OEM
G65SC104CM-3	OEM
G65SC104CM-4	OEM
G65SC104D-1	OEM
G65SC104D-2	OEM
G65SC104D-3	OEM
G65SC104D-4	OEM
G65SC104DI-1	OEM
G65SC104DI-2	OEM
G65SC104DI-3	OEM
G65SC104DM-1	OEM
G65SC104DM-2	OEM
G65SC104DM-3	OEM
G65SC104DM-4	OEM
G65SC104E-1	OEM
G65SC104E-2	OEM
G65SC104E-3	OEM
G65SC104E-4	OEM
G65SC104EI-1	OEM
G65SC104EI-2	OEM
G65SC104EI-3	OEM
G65SC104EM-1	OEM
G65SC104EM-2	OEM
G65SC104EM-3	OEM
G65SC104EM-4	OEM
G65SC104P-1	OEM
G65SC104P-2	OEM
G65SC104P-3	OEM
G65SC104P-4	OEM
G65SC104PI-1	OEM
G65SC104PI-2	OEM
G65SC104PI-3	OEM
G65SC104PM-1	OEM
G65SC104PM-2	OEM
G65SC104PM-4	OEM
G65SC104X-1	OEM
G65SC104X-2	OEM
G65SC104X-3	OEM
G65SC104XI-1	OEM
G65SC104XI-2	OEM
G65SC104XI-3	OEM
G65SC104XM-1	OEM
G65SC104XM-2	OEM
G65SC104XM-3	OEM
G65SC104XM-4	OEM
G65SC105C-1	OEM
G65SC105C-2	OEM
G65SC105C-3	OEM
G65SC105C-4	OEM
G65SC105CI-1	OEM
G65SC105CI-2	OEM
G65SC105CI-3	OEM
G65SC105CI-4	OEM
G65SC105CM-1	OEM
G65SC105CM-2	OEM
G65SC105CM-3	OEM
G65SC105CM-4	OEM
G65SC105D-1	OEM
G65SC105D-2	OEM
G65SC105D-3	OEM
G65SC105D-4	OEM
G65SC105DI-1	OEM
G65SC105DI-2	OEM
G65SC105DI-3	OEM
G65SC105DI-4	OEM
G65SC105DM-1	OEM
G65SC105DM-2	OEM
G65SC105DM-3	OEM
G65SC105DM-4	OEM
G65SC105E-1	OEM
G65SC105E-2	OEM
G65SC105E-3	OEM
G65SC105E-4	OEM
G65SC105EI-1	OEM
G65SC105EI-2	OEM
G65SC105EI-3	OEM
G65SC105EM-1	OEM
G65SC105EM-2	OEM
G65SC105EM-4	OEM
G65SC105P-1	OEM
G65SC105P-2	OEM
G65SC105P-3	OEM
G65SC105P-4	OEM
G65SC105PI-1	OEM
G65SC105PI-2	OEM
G65SC105PI-3	OEM
G65SC105PI-4	OEM
G65SC105PM-1	OEM
G65SC105PM-2	OEM
G65SC105PM-3	OEM
G65SC105PM-4	OEM
G65SC105X-1	OEM
G65SC105X-2	OEM
G65SC105X-3	OEM
G65SC105X-4	OEM
G65SC105XI-1	OEM
G65SC105XI-2	OEM
G65SC105XI-3	OEM
G65SC105XI-4	OEM
G65SC105XM-1	OEM
G65SC105XM-2	OEM
G65SC105XM-3	OEM
G65SC105XM-4	OEM
G65SC106C-1	OEM
G65SC106C-2	OEM
G65SC106C-3	OEM
G65SC106C-4	OEM
G65SC106CI-1	OEM
G65SC106CI-2	OEM
G65SC106CI-3	OEM
G65SC106CI-4	OEM
G65SC106CM-1	OEM
G65SC106CM-2	OEM
G65SC106CM-3	OEM
G65SC106CM-4	OEM
G65SC106D-1	OEM
G65SC106D-2	OEM
G65SC106D-3	OEM
G65SC106D-4	OEM
G65SC106DI-1	OEM
G65SC106DI-2	OEM
G65SC106DI-3	OEM
G65SC106DI-4	OEM
G65SC106DM-1	OEM
G65SC106DM-2	OEM
G65SC106DM-3	OEM
G65SC106DM-4	OEM
G65SC106E-1	OEM
G65SC106E-2	OEM
G65SC106E-3	OEM
G65SC106E-4	OEM
G65SC106EI-1	OEM
G65SC106EI-2	OEM
G65SC106EI-3	OEM
G65SC106EI-4	OEM
G65SC106EM-1	OEM
G65SC106EM-2	OEM
G65SC106EM-3	OEM
G65SC106EM-4	OEM
G65SC106P-1	OEM
G65SC106P-2	OEM
G65SC106P-3	OEM
G65SC106P-4	OEM
G65SC106PI-1	OEM
G65SC106PI-2	OEM
G65SC106PI-3	OEM
G65SC106PI-4	OEM
G65SC106PM-1	OEM
G65SC106PM-2	OEM
G65SC106PM-3	OEM
G65SC106PM-4	OEM
G65SC106X-1	OEM
G65SC106X-2	OEM
G65SC106X-3	OEM
G65SC106X-4	OEM
G65SC106XI-1	OEM
G65SC106XI-2	OEM
G65SC106XI-3	OEM
G65SC106XI-4	OEM
G65SC106XM-1	OEM
G65SC106XM-2	OEM
G65SC106XM-3	OEM
G65SC106XM-4	OEM
G65SC107C-1	OEM
G65SC107C-2	OEM
G65SC107C-3	OEM
G65SC107C-4	OEM
G65SC107CI-1	OEM
G65SC107CI-2	OEM
G65SC107CI-3	OEM
G65SC107CI-4	OEM
G65SC107CM-1	OEM
G65SC107CM-2	OEM
G65SC107CM-3	OEM
G65SC107CM-4	OEM
G65SC107D-1	OEM
G65SC107D-2	OEM
G65SC107D-3	OEM
G65SC107D-4	OEM
G65SC107DI-1	OEM
G65SC107DI-2	OEM
G65SC107DI-3	OEM
G65SC107DI-4	OEM
G65SC107DM-1	OEM
G65SC107DM-2	OEM
G65SC107DM-3	OEM
G65SC107DM-4	OEM
G65SC107E-1	OEM
G65SC107E-2	OEM
G65SC107E-3	OEM
G65SC107E-4	OEM
G65SC107EI-1	OEM
G65SC107EI-2	OEM
G65SC107EI-3	OEM
G65SC107EI-4	OEM
G65SC107EM-1	OEM
G65SC107EM-2	OEM
G65SC107EM-3	OEM
G65SC107EM-4	OEM
G65SC107P-1	OEM
G65SC107P-2	OEM
G65SC107P-3	OEM
G65SC107P-4	OEM
G65SC107PI-1	OEM
G65SC107PI-2	OEM
G65SC107PI-3	OEM
G65SC107PI-4	OEM
G65SC107PM-1	OEM
G65SC107PM-2	OEM
G65SC107PM-3	OEM
G65SC107PM-4	OEM
G65SC107X-1	OEM
G65SC107X-2	OEM
G65SC107X-4	OEM
G65SC107XI-1	OEM
G65SC107XI-2	OEM
G65SC107XI-3	OEM
G65SC107XI-4	OEM
G65SC107XM-1	OEM
G65SC107XM-2	OEM
G65SC107XM-3	OEM
G65SC107XM-4	OEM
G65SC112C-1	OEM
G65SC112C-2	OEM
G65SC112C-3	OEM
G65SC112C-4	OEM
G65SC112CI-1	OEM
G65SC112CI-2	OEM
G65SC112CI-3	OEM
G65SC112CI-4	OEM
G65SC112CM-1	OEM
G65SC112CM-2	OEM
G65SC112CM-3	OEM
G65SC112CM-4	OEM
G65SC112D-1	OEM
G65SC112D-2	OEM
G65SC112D-3	OEM
G65SC112D-4	OEM
G65SC112DI-1	OEM
G65SC112DI-3	OEM
G65SC112DI-4	OEM
G65SC112DM-1	OEM
G65SC112DM-2	OEM
G65SC112DM-3	OEM
G65SC112DM-4	OEM
G65SC112E-1	OEM
G65SC112E-2	OEM
G65SC112E-3	OEM
G65SC112E-4	OEM
G65SC112EI-1	OEM
G65SC112EI-2	OEM
G65SC112EI-3	OEM
G65SC112EI-4	OEM
G65SC112EM-1	OEM
G65SC112EM-2	OEM

If replacement code is OEM, contact original manufacturer for replacement.

DEVICE TYPE	REPL CODE
G65SC112EM-3	OEM
G65SC112EM-4	OEM
G65SC112P-1	OEM
G65SC112P-2	OEM
G65SC112P-3	OEM
G65SC112P-4	OEM
G65SC112PI-1	OEM
G65SC112PI-2	OEM
G65SC112PI-3	OEM
G65SC112PM-1	OEM
G65SC112PM-2	OEM
G65SC112PM-3	OEM
G65SC112PM-4	OEM
G65SC112X-1	OEM
G65SC112X-2	OEM
G65SC112X-3	OEM
G65SC112X-4	OEM
G65SC112XI-1	OEM
G65SC112XI-2	OEM
G65SC112XI-3	OEM
G65SC112XI-4	OEM
G65SC112XM-1	OEM
G65SC112XM-2	OEM
G65SC112XM-3	OEM
G65SC112XM-4	OEM
G65SC115C-1	OEM
G65SC115C-2	OEM
G65SC115C-3	OEM
G65SC115C-4	OEM
G65SC115CI-1	OEM
G65SC115CI-2	OEM
G65SC115CI-3	OEM
G65SC115CI-4	OEM
G65SC115CM-1	OEM
G65SC115CM-2	OEM
G65SC115CM-3	OEM
G65SC115CM-4	OEM
G65SC115D-2	OEM
G65SC115D-3	OEM
G65SC115D-4	OEM
G65SC115DI-1	OEM
G65SC115DI-2	OEM
G65SC115DI-3	OEM
G65SC115DI-4	OEM
G65SC115DM-1	OEM
G65SC115DM-2	OEM
G65SC115DM-3	OEM
G65SC115DM-4	OEM
G65SC115E-1	OEM
G65SC115E-2	OEM
G65SC115E-3	OEM
G65SC115E-4	OEM
G65SC115EI-1	OEM
G65SC115EI-2	OEM
G65SC115EI-3	OEM
G65SC115EI-4	OEM
G65SC115EM-1	OEM
G65SC115EM-2	OEM
G65SC115EM-3	OEM
G65SC115EM-4	OEM
G65SC115P-1	OEM
G65SC115P-2	OEM
G65SC115P-3	OEM
G65SC115P-4	OEM
G65SC115PI-1	OEM
G65SC115PI-2	OEM
G65SC115PI-3	OEM
G65SC115PI-4	OEM
G65SC115PM-1	OEM
G65SC115PM-2	OEM
G65SC115PM-3	OEM
G65SC115PM-4	OEM
G65SC115X-1	OEM
G65SC115X-2	OEM
G65SC115X-3	OEM
G65SC115X-4	OEM
G65SC115XI-1	OEM
G65SC115XI-2	OEM
G65SC115XI-3	OEM
G65SC115XI-4	OEM
G65SC115XM-1	OEM
G65SC115XM-2	OEM
G65SC115XM-3	OEM
G65SC115XM-4	OEM
G65SC140PM-3	OEM
G65SC150EM-3	OEM
G65SC1103EM-4	OEM
G65SCP7PM-2	OEM
G65XC05X-3	OEM
G74SC137D	OEM
G74SC137P	OEM
G74SC138D	OEM
G74SC138P	OEM
G74SC139D	OEM
G74SC139P	OEM
G74SC240D	OEM
G74SC240P	OEM
G74SC241D	OEM
G74SC241P	OEM
G74SC244D	OEM
G74SC244P	OEM
G74SC245D	OEM
G74SC245P	OEM
G74SC373D	OEM
G74SC373P	OEM
G74SC374D	OEM
G74SC374P	OEM
G85HZ	OEM
G95G	0023

DEVICE TYPE	REPL CODE
G100A	0015
G100B	0015
G100D	0015
G100G	0015
G100J	0015
G100K	0071
G100M	0071
G105	OEM
G105A	OEM
G105B	OEM
G105HZ	OEM
G106A1SD	OEM
G106A1SH	OEM
G106A2SD	OEM
G106A2SG	OEM
G106A2SH	OEM
G106A12G	OEM
G106B1SD	OEM
G106B1SG	OEM
G106B1SH	OEM
G106B2SD	OEM
G106B2SG	OEM
G106B2SH	OEM
G106C1SD	OEM
G106C1SG	OEM
G106C1SH	OEM
G106C2SD	OEM
G106C2SG	OEM
G106C2SH	OEM
G106D1SD	OEM
G106D1SG	OEM
G106D1SH	OEM
G106D2SD	OEM
G106D2SG	OEM
G106D2SH	OEM
G106E1SD	OEM
G106E1SG	OEM
G106E1SH	OEM
G106E2SD	OEM
G106E2SG	OEM
G106E2SH	OEM
G106F1SD	OEM
G106F1SG	OEM
G106F1SH	OEM
G106F2SD	OEM
G106F2SG	OEM
G106F2SH	OEM
G106M1SD	OEM
G106M1SG	OEM
G106M1SH	OEM
G106M2SD	OEM
G106M2SG	OEM
G106M2SH	OEM
G110	0440
G110A	0005
G110B	OEM
G112A	OEM
G115	OEM
G115A	OEM
G115AP	OEM
G115B	OEM
G115BP	OEM
G115CDE	OEM
G115CFE	OEM
G115IDE	OEM
G115IFE	OEM
G115MDE	OEM
G115MFE	OEM
G116AL	OEM
G116AP	OEM
G116CDD	OEM
G116CFD	OEM
G116IDD	OEM
G116IFD	OEM
G116MDD	OEM
G116MFD	OEM
G117AL	OEM
G117CDD	OEM
G117CFD	OEM
G117IDD	OEM
G117IFD	OEM
G117MDD	OEM
G117MFD	OEM
G118AL	OEM
G118AP	OEM
G118CDD	OEM
G118CFD	OEM
G118IDD	OEM
G118IFD	OEM
G118MDD	OEM
G118MFD	OEM
G119AL	OEM
G119CDD	OEM
G119CFD	OEM
G119IDD	OEM
G119IFD	OEM
G119MDD	OEM
G119MFD	OEM
G122	0628
G122AL	OEM
G122B	OEM
G123AL	OEM
G123AP	OEM
G123BP	OEM
G123CDD	OEM
G123CFD	OEM
G123IDD	OEM
G123IFD	OEM
G123MDD	OEM

DEVICE TYPE	REPL CODE
G123MFD	OEM
G125IFD	OEM
G125MFD	OEM
G126IFD	OEM
G126MFD	OEM
G127IFD	OEM
G127MFD	OEM
G128IFD	OEM
G128MFD	OEM
G129	OEM
G129IFD	OEM
G129MFD	OEM
G130	0015
G130IFD	OEM
G130MFD	OEM
G131IFD	OEM
G131MFD	OEM
G132IFD	OEM
G132MFD	OEM
G133	0030
G133A	0030
G133B	OEM
G147	OEM
G147A	OEM
G147B	OEM
G154	0143
G155	0143
G156	0143
G157	0143
G158	0143
G159	0143
G168	OEM
G168A	OEM
G168B	OEM
G182AG	0917
G198	0143
G199	0143
G200	0143
G206	0554
G206A	0554
G206B	OEM
G210	0005
G210A	0005
G210B	OEM
G215	OEM
G215A	OEM
G215B	OEM
G222	0015
G222A	0628
G222B	OEM
G233	0030
G233A	0030
G233B	OEM
G247	OEM
G247A	OEM
G247B	OEM
G268	OEM
G268A	OEM
G268B	OEM
G296	0015
G297	0143
G306	0554
G306A	0554
G306B	OEM
G310	0005
G310A	0005
G310B	OEM
G315	OEM
G315A	OEM
G315B	OEM
G322	0628
G322A	0628
G322B	OEM
G333	0030
G333A	0030
G333B	OEM
G347	OEM
G347A	OEM
G347B	OEM
G406	0554
G406A	0554
G406B	OEM
G409	0143
G410	0005
G410A	0005
G410B	OEM
G415	OEM
G415A	OEM
G415B	OEM
G422	0628
G422A	0628
G422B	OEM
G433	0030
G433A	0030
G433B	OEM
G498	0143
G506	0970
G506A	0554
G506B	OEM
G510	0005
G510A	0005
G510B	OEM
G511A	OEM
G511B	OEM
G511C	OEM
G511D	OEM
G511F	OEM
G511U	OEM
G515	OEM

DEVICE TYPE	REPL CODE
G515A	OEM
G515B	OEM
G522A	0628
G522B	OEM
G522C	OEM
G580	0143
G603A	OEM
G603B	OEM
G603C	OEM
G604A	OEM
G604B	OEM
G604C	OEM
G605A	OEM
G605B	OEM
G605C	OEM
G610A	0005
G610B	OEM
G610C	OEM
G611A	OEM
G611B	OEM
G611C	OEM
G611D	OEM
G611F	OEM
G611U	OEM
G615A	OEM
G615B	OEM
G615C	OEM
G618	0604
G657	0015
G659	0015
G700	0015
G701	0015
G702	0015
G702A	OEM
G702B	OEM
G702C	OEM
G704	0143
G766	0143
G788	0143
G789	0143
G790	0143
G814	0143
G815	0143
G816	0143
G820	0143
G821	0143
G822	0143
G823	0143
G824	0143
G825	0143
G844	0143
G845	0143
G846	0143
G847	0143
G868	0143
G869	0143
G900	OEM
G1006	0989
G-1010	0143
G1010	0143
G1010A	0133
G1106	0586
G1110	0145
G1149	0015
G1179	0015
G1200	OEM
G1206	OEM
G1210	OEM
G1210-3	OEM
G1220-3	OEM
G1230-3	OEM
G1242	0015
G1288	0143
G1330IFD	OEM
G1330MFD	OEM
G1340IFD	OEM
G1340MFD	OEM
G1350IFD	OEM
G1350MFD	OEM
G1360IFD	OEM
G1360MFD	OEM
G1501	0609
G1506	OEM
G1810-3	OEM
G1820-3	OEM
G1830-3	OEM
G1850	0031
G1851	0031
G1852	0031
G1854	0031
G1856	0031
G2006	0989
G2010	0097
G3006	0197
G3010	0109
G3010-3	OEM
G3020-3	OEM
G3030-3	OEM
G4006	0197
G4010	0109
G4090	0239
G4104CI	OEM
G4104CM	OEM
G4325CI	OEM
G4325CM	OEM
G4325P	OEM
G4326CM	OEM
G4326P	OEM
G5006	0204

DEVICE TYPE	REPL CODE
G5010	0122
G5010-3	OEM
G5019	0196
G5020-3	OEM
G5030-3	OEM
G5080PI(A)	OEM
G5087DI(A)	OEM
G5087P(A)	OEM
G5089DI(A)	OEM
G5089P(A)	OEM
G5316D3	OEM
G5316P3	OEM
G5364D-3	OEM
G5364D-4	OEM
G5364P-3	OEM
G5364P-4	OEM
G5365D-3	OEM
G5365D-4	OEM
G5365P-3	OEM
G5365P-4	OEM
G5815	OEM
G5927	OEM
G6000	2717
G6002	4012
G6003	0085
G6004	0440
G6005	0160
G6006	0440
G6007	0975
G6008	0975
G6009	0604
G6010	4985
G6011	1424
G6012	3298
G6013	0160
G6014	0604
G6015	0604
G6017	0975
G6018	0604
G8006	0206
G8010	0545
G8860DI	OEM
G8860DI(A)	OEM
G8860P	OEM
G8860P(A)	OEM
G8860PI	OEM
G8860PI(A)	OEM
G8862P	OEM
G8862PI	OEM
G8863DI	OEM
G8863PI	OEM
G8865DI	OEM
G8865PI	OEM
G8912DI(A)	OEM
G9423	0224
G9600	0016
G9600(G.E.)	0016
G9623	0016
G9625	0224
G9696	0016
G10001D	OEM
G10002D	OEM
G10015D	OEM
G10119	0015
G13403	OEM
G13638	0037
G15001D	OEM
G15002D	OEM
G15015D	OEM
G16506	0600
G18337	OEM
G18341	OEM
G18403	OEM
G18410	OEM
G18415	OEM
G18419	OEM
G20001G	OEM
G20002G	OEM
G20015G	OEM
G25001D	OEM
G25002D	OEM
G25015D	OEM
G30001G	OEM
G30002G	OEM
G30015G	OEM
G40001G	OEM
G40002G	OEM
G40015G	OEM
G50001D	OEM
G50002D	OEM
G50015D	OEM
G59019	0196
G59122	0196
G59815	OEM
G101079	0600
G390503S1	1554
G612994	3861
G65010042	OEM
G65020054	OEM
GA1	OEM
GA2	OEM
GA100	OEM
GA101	OEM
GA102	OEM
GA200	OEM
GA200A	OEM
GA201	OEM
GA201A	OEM
GA300	OEM

DEVICE TYPE	REPL CODE
GA300A	OEM
GA301	OEM
GA301A	OEM
GA500	0200
GA504A	OEM
GA504B	OEM
GA504V	OEM
GA4558	OEM
GA5087PI(A)	OEM
GA5319	OEM
GA52609	OEM
GA-52829	0136
GA52829	0280
GA52830	OEM
GA52837	OEM
GA52996	OEM
GA53080	OEM
GA53104	OEM
GA-53149	0136
GA53149	0280
GA53194	OEM
GA53213	OEM
GA53233	OEM
GA-53242	0136
GA53242	0280
GA53270	0600
GAD2000	OEM
GAD2001	OEM
GAD2002	OEM
GAD2003	OEM
GAD3000	OEM
GAD3001	OEM
GAD3002	OEM
GAD3003	OEM
GAL10	OEM
GAL11	OEM
GAL12	OEM
GA02D-3	OEM
GA05D-D	OEM
GA010C-D	OEM
GA010C-H	OEM
GA010D-D	OEM
GA010D-H	OEM
GA010E-D	OEM
GA010E-H	OEM
GA010F-D	OEM
GA010F-H	OEM
GA010G-D	OEM
GA010G-H	OEM
GA010I-D	OEM
GA010I-H	OEM
GA020C-D	OEM
GA020C-H	OEM
GA020D-D	OEM
GA020D-H	OEM
GA020E-D	OEM
GA020E-H	OEM
GA020F-D	OEM
GA020F-H	OEM
GA020G-D	OEM
GA020G-H	OEM
GA020I-D	OEM
GA020I-H	OEM
GA050C-D	OEM
GA050C-H	OEM
GA050D-D	OEM
GA050D-H	OEM
GA050E-D	OEM
GA050E-H	OEM
GA050F-D	OEM
GA050F-H	OEM
GA050G-D	OEM
GA050G-H	OEM
GA050I-D	OEM
GA050I-H	OEM
GA0100C-D	OEM
GA0100C-H	OEM
GA0100D-D	OEM
GA0100D-H	OEM
GA0100E-D	OEM
GA0100E-H	OEM
GA0100F-D	OEM
GA0100F-H	OEM
GA0100G-D	OEM
GA0100G-H	OEM
GA0100I-D	OEM
GA0100I-H	OEM
GA0200C-D	OEM
GA0200C-H	OEM
GA0200D-D	OEM
GA0200D-H	OEM
GA0200E-D	OEM
GA0200E-H	OEM
GA0200F-D	OEM
GA0200F-H	OEM
GA0200G-D	OEM
GA0200G-H	OEM
GA0200I-D	OEM
GA0200I-E	OEM
GA0200I-H	OEM
GAP01AX	OEM
GAP01BX	OEM
GAP01EP	OEM
GAP01EX	OEM
GAP01FP	OEM
GAP01FX	OEM
GAP01N	OEM
GARE	0636
GAU401	OEM

DEVICE TYPE	REPL CODE
GB08D	0609
GB-1	0143
GB1	0133
GB107	OEM
GB110	OEM
GB120	OEM
GB200	OEM
GB200A	OEM
GB201	OEM
GB201A	OEM
GB207	OEM
GB210	OEM
GB220	OEM
GB300	OEM
GB300A	OEM
GB301	OEM
GB301A	OEM
GB307	OEM
GB310	OEM
GB320	OEM
GB407	OEM
GB410	OEM
GB420	OEM
GB507	OEM
GB510	OEM
GB520	OEM
GBD100A	OEM
GBD250A	OEM
GBDX25A	OEM
GBDX50A	OEM
GBG1000	5114
GBI1	OEM
GBS176D	OEM
GBS210E	OEM
GBS216	OEM
GBS266E	0411
GBS276D	OEM
GBS410E	OEM
GBS416	OEM
GBS466E	0415
GBS476D	OEM
GBS3110F	0408
GBS3115F	0154
GBS3125F	3148
GBS3210F	0411
GBS3215F	0154
GBS3225F	3150
GBS3310F	0415
GBS3315F	0147
GBS3325F	3151
GBS3410F	0325
GBS3415F	0147
GBX10422AD	OEM
GC31	0280
GC32	0280
GC33	0280
GC34	0280
GC35	0280
GC60	0280
GC61	0280
GC148	0208
GC150RE	OEM
GC181	0280
GC182	0280
GC250	0004
GC282	0050
GC283	0050
GC284	0050
GC285	0208
GC286	0208
GC343	0004
GC360	0280
GC387	0050
GC388	0050
GC389	0050
GC408	0004
GC452	0600
GC453	0600
GC454	0600
GC460	0050
GC461	0050
GC462	0050
GC463	0208
GC464	0004
GC465	0208
GC466	0004
GC467	0208
GC520	0004
GC521	0004
GC532	0280
GC551	0004
GC552	0004
GC578	0004
GC579	0004
GC580	0004
GC581	0004
GC588	0004
GC608	0208
GC609	0208
GC630	0050
GC630A	0050
GC631	0050
GC639	0004
GC640	0004
GC641	0160
GC655A	OEM
GC666	OEM
GC680	0004
GC681	0004

DEVICE TYPE	REPL CODE
GC682	0004
GC691	0160
GC692	0160
GC733B	0211
GC783	0198
GC784	0198
GC816	OEM
GC817	OEM
GC818	OEM
GC841	OEM
GC856	0004
GC864	0004
GC927	OEM
GC928	OEM
GC929	OEM
GC968	OEM
GC971	OEM
GC1000	OEM
GC1003	0050
GC1004	0050
GC1005	0050
GC1006	0050
GC1007	0050
GC1034	0600
GC1035	0600
GC1036	0600
GC1081	0136
GC1092	0050
GC1093	OEM
GC1093X3	0050
GC1097	0004
GC1101	0004
GC1134	0004
GC1136	0004
GC1137	0208
GC1142	0050
GC1143	0004
GC1144	0016
GC1145	0004
GC1146	0050
GC1148	0050
GC1149	0050
GC1150	0004
GC1155	0050
GC1159	0280
GC1163	OEM
GC1164	OEM
GC1182	0050
GC1183	0004
GC1184	0004
GC1185	0208
GC1186	0004
GC1187	0004
GC1200	OEM
GC1201	OEM
GC1202	OEM
GC1203	OEM
GC1204	OEM
GC1205	OEM
GC1206	OEM
GC1207	OEM
GC1208	OEM
GC1209	OEM
GC1210	OEM
GC1247	OEM
GC1257	0004
GC1300	OEM
GC1301	OEM
GC1302	0280
GC1303	OEM
GC1304	OEM
GC1305	OEM
GC1306	OEM
GC1307	OEM
GC1308	OEM
GC1309	OEM
GC1310	OEM
GC1422	0004
GC1423	0208
GC1500	OEM
GC1500A	OEM
GC1500B	OEM
GC1500D	OEM
GC1500E	OEM
GC1500H	OEM
GC1501	OEM
GC1501A	OEM
GC1501B	OEM
GC1501D	OEM
GC1501E	OEM
GC1501H	OEM
GC1502	OEM
GC1502A	OEM
GC1502B	OEM
GC1502D	OEM
GC1502E	OEM
GC1502H	OEM
GC1503	OEM
GC1503A	OEM
GC1503B	OEM
GC1503D	OEM
GC1503E	OEM
GC1503H	OEM
GC1504	OEM
GC1504A	OEM
GC1504B	OEM
GC1504D	OEM
GC1504H	OEM
GC1505	OEM
GC1505A	OEM
GC1505B	OEM
GC1505D	OEM
GC1505E	OEM
GC1505H	OEM
GC1506	OEM
GC1506A	OEM
GC1506B	OEM
GC1506D	OEM
GC1506E	OEM
GC1506H	OEM
GC1507	OEM
GC1507A	OEM
GC1507B	OEM
GC1507D	OEM
GC1507E	OEM
GC1507H	OEM
GC1508	OEM
GC1508A	OEM
GC1508B	OEM
GC1508D	OEM
GC1508E	OEM
GC1508H	OEM
GC1509	OEM
GC1509A	OEM
GC1509B	OEM
GC1509D	OEM
GC1509E	OEM
GC1509H	OEM
GC1510	OEM
GC1510A	OEM
GC1510B	OEM
GC1510D	OEM
GC1510E	OEM
GC1510H	OEM
GC1511	OEM
GC1511A	OEM
GC1511B	OEM
GC1511D	OEM
GC1511E	OEM
GC1511H	OEM
GC1512	OEM
GC1512A	OEM
GC1512B	OEM
GC1512D	OEM
GC1512H	OEM
GC1513	OEM
GC1513A	OEM
GC1513B	OEM
GC1513D	OEM
GC1513E	OEM
GC1513H	OEM
GC1514A	OEM
GC1514B	OEM
GC1514D	OEM
GC1514E	OEM
GC1514H	OEM
GC1573	0050
GC1600	OEM
GC1600A	OEM
GC1600B	OEM
GC1600D	OEM
GC1600E	OEM
GC1600H	OEM
GC1601	OEM
GC1601A	OEM
GC1601B	OEM
GC1601D	OEM
GC1601E	OEM
GC1601H	OEM
GC1602	OEM
GC1602A	OEM
GC1602B	OEM
GC1602D	OEM
GC1602E	OEM
GC1602H	OEM
GC1603	OEM
GC1603A	OEM
GC1603B	OEM
GC1603D	OEM
GC1603E	OEM
GC1603H	OEM
GC1604	OEM
GC1604A	OEM
GC1604B	OEM
GC1604D	OEM
GC1604E	OEM
GC1604H	OEM
GC1605	OEM
GC1605A	OEM
GC1605B	OEM
GC1605D	OEM
GC1605E	OEM
GC1605H	OEM
GC1606	OEM
GC1606A	OEM
GC1606B	OEM
GC1606D	OEM
GC1606E	OEM
GC1606H	OEM
GC1607	OEM
GC1607A	OEM
GC1607B	OEM
GC1607D	OEM
GC1607E	OEM
GC1607H	OEM
GC1608	OEM
GC1608A	OEM
GC1608B	OEM
GC1608D	OEM
GC1608E	OEM
GC1608H	OEM
GC1609	OEM
GC1609A	OEM
GC1609B	OEM
GC1609D	OEM
GC1609E	OEM
GC1609H	OEM
GC1610	OEM
GC1611	OEM
GC1612	OEM
GC1613	OEM
GC1615-1	0086
GC1700	OEM
GC1700A	OEM
GC1701	OEM
GC1702	OEM
GC1702A	OEM
GC1702B	OEM
GC1702C	OEM
GC1702D	OEM
GC1702E	OEM
GC1702H	OEM
GC1703	OEM
GC1703A	OEM
GC1703B	OEM
GC1703C	OEM
GC1703D	OEM
GC1703E	OEM
GC1703H	OEM
GC1704	OEM
GC1704A	OEM
GC1704B	OEM
GC1704C	OEM
GC1704D	OEM
GC1704E	OEM
GC1704H	OEM
GC1705	OEM
GC1705A	OEM
GC1705B	OEM
GC1705C	OEM
GC1705D	OEM
GC1705E	OEM
GC1705H	OEM
GC1706	OEM
GC1706A	OEM
GC1706B	OEM
GC1706C	OEM
GC1706D	OEM
GC1706E	OEM
GC1706H	OEM
GC1707	OEM
GC1707A	OEM
GC1707B	OEM
GC1707C	OEM
GC1707D	OEM
GC1707E	OEM
GC1707H	OEM
GC1708	OEM
GC1708A	OEM
GC1708B	OEM
GC1708C	OEM
GC1708D	OEM
GC1708E	OEM
GC1708H	OEM
GC1709	OEM
GC1709A	OEM
GC1709B	OEM
GC1709C	OEM
GC1709D	OEM
GC1709E	OEM
GC1709H	OEM
GC1710	OEM
GC1710A	OEM
GC1710B	OEM
GC1710C	OEM
GC1710D	OEM
GC1710E	OEM
GC1710H	OEM
GC1711	OEM
GC1711A	OEM
GC1711B	OEM
GC1711C	OEM
GC1711D	OEM
GC1711E	OEM
GC1711H	OEM
GC1712	OEM
GC1712A	OEM
GC1712B	OEM
GC1712C	OEM
GC1712D	OEM
GC1712E	OEM
GC1712H	OEM
GC1713	OEM
GC1713A	OEM
GC1713B	OEM
GC1713C	OEM
GC1713D	OEM
GC1713E	OEM
GC1713H	OEM
GC1714	OEM
GC1714A	OEM
GC1714B	OEM
GC1714C	OEM
GC1714D	OEM
GC1714E	OEM
GC1714H	OEM
GC1715	OEM
GC1715A	OEM
GC1715B	OEM
GC1715C	OEM
GC1715D	OEM
GC1715E	OEM
GC1715H	OEM
GC1716	OEM
GC1716A	OEM
GC1716B	OEM
GC1716C	OEM
GC1716D	OEM
GC1716E	OEM
GC1716H	OEM
GC1717	OEM
GC1717A	OEM
GC1717B	OEM
GC1717C	OEM
GC1717D	OEM
GC1717E	OEM
GC1717H	OEM
GC1718	OEM
GC1719	OEM
GC1720	OEM
GC1754	OEM
GC1754B	OEM
GC1754C	OEM
GC1755	OEM
GC1755B	OEM
GC1755C	OEM
GC1756	OEM
GC1756B	OEM
GC1756C	OEM
GC1757	OEM
GC1757B	OEM
GC1757C	OEM
GC1758	OEM
GC1758B	OEM
GC1758C	OEM
GC1759	OEM
GC1759A	OEM
GC1759B	OEM
GC1759C	OEM
GC1760	OEM
GC1760A	OEM
GC1760B	OEM
GC1760C	OEM
GC1761	0554
GC1761A	OEM
GC1761B	OEM
GC1761C	OEM
GC1762	OEM
GC1762A	OEM
GC1762B	OEM
GC1762C	OEM
GC1763	0005
GC1763A	OEM
GC1763B	OEM
GC1763C	OEM
GC1764	0720
GC1764A	OEM
GC1764B	OEM
GC1764C	OEM
GC1765	OEM
GC1765A	OEM
GC1765B	OEM
GC1765C	OEM
GC1766	OEM
GC1766A	OEM
GC1766B	OEM
GC1766C	OEM
GC1767	0628
GC1767A	OEM
GC1767B	OEM
GC1767C	OEM
GC1768	OEM
GC1768A	OEM
GC1768B	OEM
GC1768C	OEM
GC1769	OEM
GC1769A	OEM
GC1769B	OEM
GC1769C	OEM
GC1770	OEM
GC1770A	OEM
GC1770B	OEM
GC1770C	OEM
GC1800	OEM
GC1801	OEM
GC1802	OEM
GC1803	OEM
GC1804	OEM
GC1805	OEM
GC1806	OEM
GC1807	OEM
GC1808	OEM
GC1809	OEM
GC1810	OEM
GC1811	OEM
GC1812	OEM
GC1813	OEM
GC1814	OEM
GC1815	OEM
GC1816	OEM
GC2510	OEM
GC2511	OEM
GC2512	OEM
GC2513	OEM
GC2514	OEM
GC2515	OEM
GC2516	OEM
GC2520	OEM
GC2521	OEM
GC2522	OEM
GC2523	OEM
GC2524	OEM
GC2525	OEM
GC2526	OEM
GC2530	OEM
GC2531	OEM
GC2532	OEM
GC2533	OEM
GC2534	OEM
GC2535	OEM
GC2536	OEM
GC2540	OEM
GC2541	OEM
GC2542	OEM
GC2543	OEM
GC2544	OEM
GC2545	OEM
GC2546	OEM
GC3202	OEM
GC3203	OEM
GC3204	OEM
GC3205	OEM
GC3206	OEM
GC3207	OEM
GC3208	OEM
GC3209	OEM
GC3302	OEM
GC3303	OEM
GC3304	OEM
GC3304A	OEM
GC3305	OEM
GC3306	OEM
GC3307	OEM
GC3307A	OEM
GC3308	OEM
GC4020	0160
GC4021	OEM
GC4022	0280
GC4023	OEM
GC4025	OEM
GC4025X	OEM
GC4032	OEM
GC4045	0160
GC4050	OEM
GC4051	OEM
GC4052	OEM
GC4053	OEM
GC4054	OEM
GC4057	0160
GC4062	0160
GC4087	0160
GC4094	0222
GC4097	0160
GC4100	OEM
GC4101	OEM
GC4102	OEM
GC4103	OEM
GC4111	0160
GC4144	0004
GC4156	0160
GC4162	OEM
GC4182	OEM
GC4210	OEM
GC4211	OEM
GC4212	OEM
GC4213	0847
GC4214	OEM
GC4215	OEM
GC4220	OEM
GC4221	OEM
GC4222	OEM
GC4223	OEM
GC4224	OEM
GC4225	OEM
GC4251	0160
GC4267-2	0160
GC4271	OEM
GC4272	OEM
GC4273	OEM
GC4274	OEM
GC4275	OEM
GC4310	OEM
GC4311	OEM
GC4312	OEM
GC4313	OEM
GC4314	OEM
GC4315	OEM
GC4320	OEM
GC4321	OEM
GC4322	OEM
GC4323	OEM
GC4324	OEM
GC4325	OEM
GC4371	OEM
GC4372	OEM
GC4373	OEM
GC4374	OEM
GC4375	OEM
GC4410	OEM
GC4411	OEM
GC4412	OEM
GC4413	OEM
GC4430	OEM
GC4431	OEM
GC4432	OEM
GC4433	OEM
GC4490	OEM
GC4491	OEM
GC4492	OEM
GC4493	OEM
GC4510	OEM
GC4511	OEM
GC4512	OEM
GC4513	OEM
GC4530	OEM
GC4531	OEM
GC4532	OEM
GC4533	OEM
GC4590	OEM
GC4591	OEM
GC4592	OEM
GC4593	OEM
GC4710	OEM
GC4711	OEM
GC4712	OEM
GC4720	OEM
GC4721	OEM
GC4722	OEM
GC5000	0004
GC5010	0211
GC5012	0143
GC5411	OEM
GC5412	OEM
GC5413	OEM
GC5414	OEM
GC5421	OEM
GC5422	OEM
GC5423	OEM
GC5424	OEM
GC5431	OEM
GC5432	OEM
GC5441	OEM
GC5511A	OEM
GC5511B	OEM
GC5511C	OEM
GC5511D	OEM
GC5511E	OEM
GC5511F	OEM
GC5511G	OEM
GC5511H	OEM
GC5511J	OEM
GC5511K	OEM
GC5511L	OEM
GC5511M	OEM
GC5512A	OEM
GC5512B	OEM
GC5512C	OEM
GC5512D	OEM
GC5512E	OEM
GC5512F	OEM
GC5512G	OEM
GC5512H	OEM
GC5512J	OEM
GC5512K	OEM
GC5512L	OEM
GC5512M	OEM
GC5513A	OEM
GC5513B	OEM
GC5513C	OEM
GC5513D	OEM
GC5513E	OEM
GC5513F	OEM
GC5513G	OEM
GC5513H	OEM
GC5513J	OEM
GC5513K	OEM
GC5513L	OEM
GC5513M	OEM
GC5514A	OEM
GC5514B	OEM
GC5514C	OEM
GC5514D	OEM
GC5514E	OEM
GC5514F	OEM
GC5514G	OEM
GC5514J	OEM
GC5514K	OEM
GC5514L	OEM
GC5514M	OEM
GC5515A	OEM
GC5515B	OEM
GC5515C	OEM
GC5515D	OEM
GC5515E	OEM
GC5515F	OEM
GC5515G	OEM
GC5515H	OEM
GC5515J	OEM
GC5515K	OEM
GC5515L	OEM
GC5516A	OEM
GC5516B	OEM
GC5516C	OEM
GC5516D	OEM
GC5516E	OEM
GC5516F	OEM
GC5516G	OEM
GC5516H	OEM
GC5516J	OEM
GC5516K	OEM
GC5517A	OEM
GC5517B	OEM
GC5517C	OEM
GC5517D	OEM
GC5517E	OEM
GC5517F	OEM
GC5517G	OEM
GC5517H	OEM
GC5517J	OEM
GC5518A	OEM
GC5518B	OEM
GC5518C	OEM
GC5518D	OEM
GC5518E	OEM
GC5518F	OEM
GC5518G	OEM
GC5518H	OEM
GC5519A	OEM
GC5519B	OEM
GC5519C	OEM
GC5519D	OEM
GC5519E	OEM
GC5519F	OEM
GC5519G	OEM
GC5531C	OEM
GC5531D	OEM
GC5531E	OEM
GC5531F	OEM
GC5531G	OEM
GC5532C	OEM
GC5532D	OEM
GC5532E	OEM
GC5532F	OEM
GC5532G	OEM
GC5533C	OEM
GC5533D	OEM
GC5533E	OEM
GC5533F	OEM
GC5533G	OEM
GC5534C	OEM
GC5534D	OEM
GC5534E	OEM
GC5534F	OEM
GC5534G	OEM
GC5535C	OEM
GC5535D	OEM
GC5535E	OEM
GC5535F	OEM
GC5535G	OEM
GC5536C	OEM
GC5536D	OEM
GC5536E	OEM
GC5536F	OEM
GC5536G	OEM
GC5537C	OEM
GC5537D	OEM
GC5537E	OEM
GC5537F	OEM
GC5538C	OEM
GC5538D	OEM
GC5538E	OEM
GC5539C	OEM
GC5539D	OEM
GC5601A	OEM
GC5601B	OEM
GC5601C	OEM
GC5601D	OEM
GC5601E	OEM
GC5601F	OEM
GC5601G	OEM
GC5601H	OEM
GC5602A	OEM
GC5602B	OEM
GC5602C	OEM
GC5602D	OEM
GC5602E	OEM
GC5602F	OEM
GC5602G	OEM
GC5602H	OEM
GC5605A	OEM
GC5605B	OEM
GC5605C	OEM
GC5605D	OEM
GC5605E	OEM
GC5605F	OEM
GC5605G	OEM
GC5605H	OEM
GC5607A	OEM
GC5607B	OEM
GC5607C	OEM
GC5607D	OEM
GC5607E	OEM
GC5607F	OEM
GC5607G	OEM
GC5610A	OEM
GC5610B	OEM
GC5610C	OEM
GC5610D	OEM
GC5610E	OEM
GC5610F	OEM
GC5610G	OEM
GC5615A	OEM
GC5615B	OEM
GC5615C	OEM
GC5615D	OEM
GC5615E	OEM
GC5615F	OEM
GC5620A	OEM
GC5620B	OEM
GC5620C	OEM
GC5620D	OEM
GC5620E	OEM
GC5620F	OEM

If replacement code is OEM, contact original manufacturer for replacement.

DEVICE TYPE	REPL CODE
GC5625A	OEM
GC5625B	OEM
GC5625C	OEM
GC5625D	OEM
GC5625E	OEM
GC5630A	OEM
GC5630B	OEM
GC5630C	OEM
GC5630D	OEM
GC5630E	OEM
GC5640A	OEM
GC5640B	OEM
GC5640C	OEM
GC5640D	OEM
GC5640E	OEM
GC5650A	OEM
GC5650B	OEM
GC5650C	OEM
GC5650D	OEM
GC5730A	OEM
GC5730B	OEM
GC5730C	OEM
GC5730D	OEM
GC5730E	OEM
GC5740A	OEM
GC5740B	OEM
GC5740C	OEM
GC5740D	OEM
GC5740E	OEM
GC5750A	OEM
GC5750B	OEM
GC5750C	OEM
GC5750D	OEM
GC5830A	OEM
GC5830B	OEM
GC5830C	OEM
GC5830D	OEM
GC5830E	OEM
GC5840A	OEM
GC5840B	OEM
GC5840C	OEM
GC5840D	OEM
GC5840E	OEM
GC5850A	OEM
GC5850B	OEM
GC5850C	OEM
GC5850D	OEM
GC5901	OEM
GC5902	OEM
GC5903	OEM
GC5904	OEM
GC51101	OEM
GC51102	OEM
GC51103	OEM
GC51104	OEM
GC51105	OEM
GC51106	OEM
GC51107	OEM
GC51108	OEM
GC51109	OEM
GC51110	OEM
GC51111	OEM
GC51112	OEM
GC51113	OEM
GC51301	OEM
GC51302	OEM
GC51303	OEM
GC51304	OEM
GC51305	OEM
GC51306	OEM
GC51307	OEM
GC51308	OEM
GC51309	OEM
GC51310	OEM
GC51311	OEM
GC51312	OEM
GC51313	OEM
GC51401	OEM
GC51402	OEM
GC51403	OEM
GC51404	OEM
GC51405	OEM
GC51406	OEM
GC51407	OEM
GC51408	OEM
GC51409	OEM
GC51410	OEM
GC51411	OEM
GC51412	OEM
GC51413	OEM
GC51601	OEM
GC51602	OEM
GC51603	OEM
GC51604	OEM
GC51605	OEM
GC51606	OEM
GC51607	OEM
GC51608	OEM
GC51609	OEM
GC51610	OEM
GC51611	OEM
GC51612	OEM
GC55010A	OEM
GC55010B	OEM
GC55010C	OEM
GC55010D	OEM
GC55010E	OEM
GC55010F	OEM
GC55015A	OEM

DEVICE TYPE	REPL CODE
GC55015B	OEM
GC55015C	OEM
GC55015D	OEM
GC55015E	OEM
GC55015F	OEM
GC55020A	OEM
GC55020B	OEM
GC55020C	OEM
GC55020D	OEM
GC55020E	OEM
GC55020F	OEM
GC55025A	OEM
GC55025B	OEM
GC55025C	OEM
GC55025D	OEM
GC55025E	OEM
GC55025F	OEM
GC55030A	OEM
GC55030B	OEM
GC55030C	OEM
GC55030D	OEM
GC55030E	OEM
GC55030F	OEM
GC55035A	OEM
GC55035B	OEM
GC55035C	OEM
GC55035D	OEM
GCC830	OEM
GCC830A	OEM
GCC830B	OEM
GC0201	OEM
GC0205	OEM
GC0501	OEM
GC0505	OEM
GC01001	OEM
GC01005	OEM
GC02001	OEM
GC02005	OEM
GC03001	OEM
GC03005	OEM
GC04001	OEM
GC04005	OEM
GC05001	OEM
GC05005	OEM
GCS10	OEM
GCS20	OEM
GCS30	OEM
GCS40	OEM
GCT2402C	OEM
GD1E	0143
GD1P	0143
GD1Q	0143
GD2E	0143
GD3E	OEM
GD4E	0143
GD5E	0143
GD6E	0143
GD8E	0143
GD10	0143
GD11E	0143
GD12	0015
GD12E	0143
GD13E	0143
GD-25	0143
GD25	0143
GD-26	0143
GD-29	0143
GD-30	0143
GD72E/3	0143
GD72E/4	0143
GD72E/5	0143
GD72E3	0143
GD72E4	0143
GD72E5	0143
GD73E/3	0143
GD73E/4	0143
GD73E/5	0143
GD73E3	0143
GD73E4	0143
GD73E5	0143
GD74E/3	0143
GD74E/4	0143
GD74E/5	0143
GD74E3	0143
GD74E4	0143
GD74E5	0143
GD74LS00	1527
GD74LS04	1595
GD74LS06	OEM
GD74LS07	OEM
GD74LS21	1763
GD74LS32	0088
GD74LS74A	0243
GD74LS125A	0075
GD74LS138	0427
GD74LS157	1161
GD74LS158	1656
GD74LS175	1672
GD74LS240	0452
GD74LS244	0458
GD74LS245	0463
GD74LS373	0709
GD74S04	2263
GD74S08	2569
GD74S74	2504
GD107A	0019
GD308A	OEM
GD308AA	OEM
GD308B	OEM

DEVICE TYPE	REPL CODE
GD308C	OEM
GD308D	OEM
GD308E	OEM
GD311A	OEM
GD311AA	OEM
GD311B	OEM
GD311C	OEM
GD311D	OEM
GD311E	OEM
GD314A	OEM
GD314AA	OEM
GD314B	OEM
GD314C	OEM
GD314D	OEM
GD314E	OEM
GD318A	OEM
GD318AA	OEM
GD318B	OEM
GD318C	OEM
GD318D	OEM
GD400	0143
GD401	0143
GD402	0143
GD402A	OEM
GD403	0019
GD404	0019
GD405	0019
GD406	0143
GD407	0143
GD408	0143
GD408A	OEM
GD408B	OEM
GD408C	OEM
GD408D	OEM
GD408E	OEM
GD409	0143
GD410	OEM
GD411	0143
GD411A	OEM
GD411B	OEM
GD411C	OEM
GD411D	OEM
GD411E	OEM
GD507A	OEM
GD508A	OEM
GD508AA	OEM
GD508B	OEM
GD508C	OEM
GD508D	OEM
GD508E	OEM
GD-510	0133
GD510	0143
GD511A	OEM
GD511AA	OEM
GD511B	OEM
GD511C	OEM
GD511D	OEM
GD511E	OEM
GD514A	OEM
GD514AA	OEM
GD514B	OEM
GD514C	OEM
GD514D	OEM
GD514E	OEM
GD518A	OEM
GD518AA	OEM
GD518B	OEM
GD518C	OEM
GD518D	OEM
GD663	0143
GD731	OEM
GD732	OEM
GD733	OEM
GD741	OEM
GD742	OEM
GD743	OEM
GD744	OEM
GD800-1	OEM
GD800-2	OEM
GD800-3	OEM
GD800-4	OEM
GD800-5	OEM
GD800-6	OEM
GD800-7	OEM
GD800-8	OEM
GD800-9	OEM
GD800-10	OEM
GD800-12	OEM
GD800-14	OEM
GD800-16	OEM
GD800-18	OEM
GD800-20	OEM
GD800-22	OEM
GD800-24	OEM
GD800-26	OEM
GD800-28	OEM
GD800-30	OEM
GD1001	0143
GD1250-2	OEM
GD1250-4	OEM
GD1250-6	OEM
GD1250-12	OEM
GD1250-16	OEM
GD1250-20	OEM
GD1250-24	OEM
GD1400-1	OEM
GD1400-2	OEM
GD1400-3	OEM
GD1400-4	OEM

DEVICE TYPE	REPL CODE
GD1400-5	OEM
GD1400-6	OEM
GD1400-7	OEM
GD1400-8	OEM
GD1400-9	OEM
GD1400-10	OEM
GD1400-12	OEM
GD1400-14	OEM
GD1400-16	OEM
GD1400-18	OEM
GD1400-20	OEM
GD3638	0133
GD4011B	OEM
GD4052B	0024
GD4066	0101
GD4066B	0101
GD4066BD	0101
GD4066BP	0101
GD4207GD	OEM
GD7815	OEM
GD7818	OEM
GD75188	0508
GD75189A	0511
GE-1	0280
GE1C-166	2889
GE-1N506	0015
GE-1N5061	0023
GE1.14CX511	OEM
GE-2	0211
GE-3	0160
GE-4	0440
GE-5	0600
GE5ZD-3.3	0782
GE5ZD-3.6	0796
GE5ZD-3.9	0806
GE5ZD-4.3	0821
GE5ZD-4.7	0833
GE5ZD-5.0	0442
GE5ZD-5.1	0442
GE5ZD-5.6	0876
GE5ZD-5.8	6225
GE5ZD-6.0	3127
GE5ZD-6.2	0185
GE5ZD-6.6	0205
GE5ZD-6.8	0205
GE5ZD-7.5	0480
GE5ZD-8.0	0504
GE5ZD-8.1	0504
GE5ZD-8.2	0504
GE5ZD-8.7	3317
GE5ZD-9.1	0684
GE5ZD-10	0225
GE5ZD-11	0230
GE5ZD-12	0234
GE5ZD-13	0237
GE5ZD-14	1397
GE5ZD-15	0247
GE5ZD-16	0251
GE5ZD-17	1178
GE5ZD-18	0256
GE5ZD-19	2395
GE5ZD-20	0262
GE5ZD-21	0269
GE5ZD-22	0269
GE5ZD-24	0274
GE5ZD-25	2399
GE5ZD-27	0294
GE5ZD-28	1177
GE5ZD-30	0309
GE5ZD-33	0318
GE-6	0600
GE-6GC1	0196
GE-6GD1	0196
GE-6GX1	1080
GE6.2	0636
GE6.2A	0636
GE6.2B	0636
GE-7	0600
GE-8	0600
GE-9	0050
GE-9A	0050
GE-10	0079
GE-10A	0111
GE-11	0127
GE-12	0142
GE-13MP	3033
GE-14	0103
GE-15MP	1514
GE-16	0085
GE-17	0937
GE-18	0086
GE-19	0103
GE-20	0016
GE-21	0150
GE-21A	0126
GE-22	0037
GE-23	0178
GE-24MP	0178
GE-25	0975
GE-26	0925
GE-27	0285
GE-28	0560
GE-29	1267
GE-30	0222
GE-31MP	0822
GE-32	0168
GE-33	0561
GE-34	1199
GE-35	0074

DEVICE TYPE	REPL CODE
GE-36	0363
GE-37	0003
GE-38	0065
GE-39	0007
GE-40	0233
GE-41	4465
GE-42	1030
GE42-7	0015
GE-43	2763
GE-44	0222
GE-45	4246
GE-46	1411
GE-47	0018
GE-47MP	3049
GE-48	0532
GE-49	1010
GE-50	0050
GE-51	0050
GE-52	0136
GE-53	0164
GE-54	2866
GE-55	0561
GE-56	1199
GE-57	0161
GE-58	0460
GE-59	0208
GE-60	2209
GE-61	0224
GE-62	0111
GE-63	0218
GE-63A	0595
GE-64	0400
GE-65	0693
GE-66	0042
GE-66A	0042
GE-67	1242
GE-67A	0791
GE-69	1308
GE-69A	0925
GE-72	0074
GE-73	0642
GE-74	1681
GE-75	0130
GE-76	0604
GE-77	0177
GE-80	0847
GE-81	0016
GE-82	0037
GE-83	0551
GE-84	0382
GE-85	0111
GE-86	0127
GE-88	1386
GE-89	0477
GE-90	0030
GE-91	0493
GE-92	1751
GE-117	0470
GE-121	0160
GE-121MP	0265
GE-123	0198
GE-123AP	0079
GE-129	0127
GE129	0127
GE-130	0015
GE-131MP	0822
GE-163A	0642
GE-166	0277
GE-167	0289
GE-168	0296
GE-169	0302
GE-178MP	0913
GE-195A	0698
GE-205	2624
GE-208	0050
GE-210	0286
GE-211	0368
GE-212	0151
GE-213	0076
GE-214	0259
GE-215	0936
GE-216	0836
GE-217	0264
GE-218	0382
GE-219	6050
GE-220	0861
GE-221	2741
GE-222	0715
GE-223	0342
GE-224	0338
GE-225	3758
GE-226	0551
GE-227	4205
GE-228	0439
GE-229	0187
GE-230	6051
GE-231	6052
GE-232	0836
GE-233	6053
GE-234	0905
GE-235	0233
GE-236	2050
GE-237	0065
GE-238	2847
GE-239	0604
GE-240	0040
GE-241	0423
GE-242	OEM
GE-243	0086

DEVICE TYPE	REPL CODE
GE-244	0126
GE-245	0050
GE-246	0178
GE-247	0219
GE-248	1051
GE-249	0338
GE-250	0854
GE-251	0168
GE-252	0566
GE-253	1367
GE-254	0437
GE-255	3081
GE-256	3280
GE-257	1708
GE-258	3519
GE-259	0313
GE-260	1751
GE-261	0418
GE-262	0177
GE-262MP	2140
GE-263	2013
GE-264	0622
GE-265	0543
GE-265MP	2371
GE-265P	0543
GE-266	1598
GE-267	1027
GE-268	0155
GE-268MP	1224
GE-269	0006
GE-270	1591
GE-271	0324
GE-271MP	3247
GE-272	0436
GE-273	2046
GE-273MP	3376
GE-274	1430
GE-275	1173
GE-276	1946
GE-277	0493
GE-278	0475
GE-279	1978
GE-280	0259
GE-281	2946
GE-282	2312
GE-283	0008
GE-284	1974
GE-285	6050
GE-286	3576
GE-287	1233
GE-288	1974
GE-289A	0155
GE-291	OEM
GE-292	OEM
GE-293	OEM
GE-294	3421
GE-295	1233
GE-296	1974
GE-297	2041
GE-298	1974
GE-299	0418
GE-300	0133
GE-301	1708
GE-302	0236
GE-303	0681
GE-304	3377
GE-305	0293
GE-309K	1922
GE-312	0325
GE-314	1825
GE-320	0169
GE-321	OEM
GE-322	0936
GE-323	0892
GE-324	OEM
GE-325	0168
GE-326	2258
GE-327	0224
GE-328	0620
GE-329	0936
GE-330	2987
GE-331	1591
GE-332	0836
GE-333	6058
GE-334	0532
GE-335	OEM
GE-336	1591
GE-337	1984
GE-338	OEM
GE-339	6059
GE-340	OEM
GE-341	2997
GE-342	3165
GE-343	2427
GE-344	2277
GE-345	1212
GE-346	0178
GE-347	0849
GE-348	3370
GE-349	2235
GE-350	2439
GE-351	3373
GE-352	3374
GE-353	0602
GE-354	2835
GE-355	3525
GE-356	3528
GE-357	3539
GE-358	2432

DEVICE TYPE	REPL CODE
GE-359	3526
GE-360	0558
GE-361	3529
GE-362	0134
GE-363	2485
GE-367	2868
GE-368	3536
GE-375	0392
GE-380	2939
GE-381	3571
GE-389	0223
GE-414	0133
GE-504	0015
GE-504A	0015
GE-505	0071
GE-509	2637
GE-509A	0071
GE-510	0071
GE-511	0102
GE-512	0087
GE-513	1323
GE-514	0124
GE-515	1217
GE-516	1501
GE-517	1501
GE-518	0190
GE-519	0409
GE-520	0128
GE-521	1706
GE-522	1997
GE-522A	1997
GE-523	1501
GE-524	2804
GE-525	0409
GE-526	1710
GE-527	0190
GE-528	1067
GE-529	2546
GE-530	0348
GE-531	0071
GE-532	1293
GE-533	0348
GE-534	2982
GE-535	2985
GE-536	1197
GE-537	2984
GE-538	1054
GE-539	3858
GE-540	3655
GE-548	3662
GE-550	6075
GE-551	0182
GE-601	0139
GE-663	0595
GE-700	0239
GE-701	2329
GE-702	1712
GE-703	6083
GE-704	0061
GE-705	1652
GE-706	2571
GE-720	3550
GE-721	2876
GE-722	1187
GE-723	2861
GE-724	4559
GE-725	4626
GE-726	5080
GE-727	OEM
GE-730	2743
GE-750	3652
GE-761	1876
GE-792	2167
GE-793	2043
GE-794	3380
GE-812	3770
GE-818	2501
GE-822	2812
GE-961	1285
GE-962	0923
GE-972	2239
GE-973	0417
GE-984	3270
GE-1005	2568
GE-1021	2155
GE-1038	4409
GE-1110	2305
GE-1127	4420
GE-1162	4038
GE-1164	3269
GE-1169	4401
GE-1175	2786
GE-1178	3705
GE-1179	2572
GE-1183	4196
GE-1192	1540
GE-1193	3318
GE-1194	1816
GE-1196	1055
GE-1198	2253
GE-1211	2193
GE-1218	1571
GE-1219	1612
GE-1228	3190
GE-1231	1170
GE-1232	1048
GE-1239	3262
GE-1257	4641
GE-1258	4642

DEVICE TYPE	REPL CODE
GE-1259	5618
GE-1260	4644
GE-3229	0042
GE-3265	0016
GE3265	0079
GE-3638	0143
GE3638	0143
GE-4000	2024
GE-4001	0478
GE-4002	2055
GE-4007	2846
GE-4009	1999
GE-4011	0215
GE-4012	0498
GE-4013	0413
GE-4015	1014
GE-4016	1143
GE-4017	0513
GE-4019	1525
GE-4020	1661
GE-4021	1749
GE-4023	0519
GE-4024	1957
GE-4025	2072
GE-4027	1949
GE-4030	0500
GE-4040	0056
GE-4042	0121
GE-4049	0001
GE-4050	0398
GE-4051	0366
GE-4052	0024
GE-4055	3304
GE-4081	0626
GE-4300A1	2998
GE-4415A	3000
GE-4518	1043
GE-5004	1000
GE-5005	1012
GE-5008	2081
GE-5009	1073
GE-5012	0612
GE-5013	1188
GE-5016	0989
GE-5017	0990
GE-5020	0197
GE-5021	0997
GE-5024	0204
GE-5025	0515
GE-5028	0586
GE-5029	0948
GE-5032	0097
GE-5033	0765
GE-5036	0109
GE-5037	0538
GE-5040	0122
GE-5041	0816
GE-5044	0145
GE-5045	0550
GE-5048	1250
GE-5056	1586
GE-5057	3294
GE-5060	2626
GE5060	2626
GE-5061	2626
GE5061	2626
GE-5062	1852
GE5062	2626
GE-5062A	OEM
GE-5063A	1312
GE-5064	2900
GE-5065	1241
GE-5068	3751
GE-5069	2365
GE-5072	1112
GE-5073	2376
GE-5076	3011
GE-5077	0449
GE-5096	1124
GE-5097	1107
GE-5100	0805
GE-5101	0258
GE-5104	0319
GE-5105	0267
GE-5108	0044
GE-5109	0281
GE-5128	2840
GE-5129	2182
GE-5132	2850
GE-5133	2191
GE-5136	2871
GE-5137	2197
GE-5140	2473
GE-5141	2340
GE-5142	3117
GE-5151A	0405
GE-5155A	0238
GE-5190A	1153
GE-5304	0106
GE-5312	0323
GE-5314	0473
GE-5322	2363
GE-5400	1137
GE-5401	0344
GE-5402	0901
GE-5404	0058
GE-5414	3620
GE-5415	3324
GE-5442	4588
GE-5444	2270
GE-5448	3404
GE-5453	1259
GE-5455	0940
GE-5457	0095
GE-5463	0505
GE-5512	3383
GE-5642	1413
GE-5801	0953
GE6060	2626
GE6061	2626
GE6062	2626
GE6063	0133
GE6251	1991
GE6252	2626
GE6253	2626
GE-6366	0015
GE6366	0015
GE-6400A	1669
GE-6401	2136
GE-6402	0316
GE-7400	0232
GE-7402	0314
GE-7404	0361
GE-7406	1206
GE7406	OEM
GE-7408	0467
GE-7410	0512
GE-7413	1441
GE-7416	1349
GE-7417	1352
GE-7420	0697
GE-7427	0818
GE-7430	0873
GE-7432	0899
GE-7441	1038
GE-7447	1108
GE-7448	1125
GE-7451	1168
GE-7473	1172
GE-7474	1313
GE-7475	1432
GE-7476	1158
GE-7485	0374
GE-7486	1368
GE-7490	1208
GE-7492	0834
GE10000	2626
GE10001	2626
GE10002	2626
GE10003	2626
GE10006	2626
GE10007	2626
GE10020	OEM
GE10021	OEM
GE10022	4013
GE10023	2626
GE-74123	1157
GE-74145	0619
GE-74150	1492
GE-74154	1555
GE-74192	1921
GE-74193	1926
GE-74196	1950
GEA6.2(E)B2	0471
GEA15A	1282
GEA-107	OEM
GEA-109	OEM
GEA-110	OEM
GEA-114	OEM
GEA114B	0031
GEA-121	3327
GEAU-1-40A	OEM
GEAU-1-60A	OEM
GEAU-1-520A	OEM
GEAU-1-520AS	OEM
GEAU-1-40742	OEM
GEAU-1-40742HS	OEM
GEAU-1-44782	OEM
GEAU-1-44782HS	OEM
GEAU-2-40A	OEM
GEAU-2-60A	OEM
GEAU-2-520A	OEM
GEAU-2-520AS	OEM
GEAU-2-40742	OEM
GEAU-2-40742HS	OEM
GEAU-2-44782	OEM
GEAU-2-44782HS	OEM
GEAU-5-40A	OEM
GEAU-5-60A	OEM
GEAU-5-520A	OEM
GEAU-5-520AS	OEM
GEAU-5-40742	OEM
GEAU-5-40742HS	OEM
GEAU-5-44782	OEM
GEAU-5-44782HS	OEM
GEB100	0323
GEB101	0323
GEB102	1414
GEB104	0473
GEB106	0446
GEB108	1422
GEBR-206	1414
GEBR-425	2369
GEBR-600	0302
GEBR-1000	0250
GECD-1-9145-1	OEM
GECD-1-9145-2HS	OEM
GECD-2-9145-1	OEM
GECD-2-9145-2HS	OEM
GECD-5-9145-1	OEM
GECD-5-9145-2HS	OEM
GE-CR1	0774
GECR-1	0774
GE-CR2	1303
GECR-2	1303
GE-CR3	0474
GECR-3	0474
GECR-4	0201
GECR-5	0288
GECR-6	0378
GECR-7	0757
GECU-1-9145-1	OEM
GECU-1-9145-2HS	OEM
GECU-2-9145-1	OEM
GECU-2-9145-2HS	OEM
GECU-5-9145-1	OEM
GECU-5-9145-2HS	OEM
GED05B850	2232
GED05B850	0143
GE-FET-1	3046
GE-FET-2	0325
GE-FET-3	0849
GE-FET-4	0212
GE-FET-5	0212
GE-FET-6	0353
GEFR-4	2979
GE-FR8	2977
GEFR-8	2977
GE-FR9	2978
GEFR-9	2978
GEFR-10	2980
GEFR-12	5362
GEHG-1-9145-1	OEM
GEHG-1-9145-2HS	OEM
GEHG-2-9145-1	OEM
GEHG-2-9145-2HS	OEM
GEHG-5-9145-1	OEM
GEHG-5-9145-2HS	OEM
GE-IC2	0167
GEIC-2	0167
GEIC-3	2161
GEIC-4	0352
GEIC-5	0666
GE-IC6	0354
GEIC-6	0354
GE-IC7	0443
GEIC-7	0443
GE-IC8	0654
GEIC-8	0654
GEIC-9	0701
GEIC-10	0664
GE-IC11	0678
GEIC-11	0678
GEIC-12	0632
GEIC-13	2457
GEIC-14	2279
GEIC-15	1345
GEIC-16	1443
GEIC-17	1759
GEIC-18	0349
GEIC-19	0692
GEIC-20	0364
GEIC-21	0802
GEIC-22	2549
GEIC-23	0328
GEIC-24	1421
GEIC-25	2552
GEIC-26	0789
GEIC-27	2557
GEIC-28	1940
GEIC-29	0856
GEIC-30	0751
GEIC-31	0379
GEIC-32	1395
GEIC-33	2560
GEIC-34	1611
GEIC-35	0520
GEIC-36	0878
GEIC-37	0864
GEIC-38	1215
GEIC-39	0389
GEIC-40	0607
GEIC-41	0855
GEIC-42	2568
GEIC-43	2571
GEIC43	OEM
GEIC-44	2573
GEIC-45	2576
GEIC-46	2577
GEIC-47	2578
GEIC-48	1837
GEIC-49	2124
GEIC-50	2581
GEIC-51	2582
GEIC-52	2583
GEIC-53	2584
GEIC-54	2585
GEIC-55	2586
GEIC-56	2587
GEIC-57	2588
GEIC-58	2590
GEIC-59	2591
GEIC-60	2592
GEIC-61	2593
GEIC-62	2594
GEIC-63	2596
GEIC-64	2599
GEIC-65	2601
GEIC-66	2602
GEIC-67	2603
GEIC-68	2604
GEIC-69	2605
GEIC-70	0621
GEIC-71	0881
GEIC-72	0884
GEIC-73	0043
GEIC-74	2532
GEIC-75	2572
GEIC-76	0416
GEIC-77	0843
GEIC-79	0882
GEIC-80	2614
GEIC-81	2617
GEIC-82	2618
GEIC-83	2617
GEIC-84	2624
GEIC-86	0823
GEIC-87	1393
GEIC-88	2582
GEIC-89	0829
GEIC-90	2630
GEIC-91	2057
GEIC-92	0911
GEIC-93	2632
GEIC-94	1994
GEIC-95	2634
GEIC-96	0878
GEIC-97	0395
GEIC-98	0855
GEIC-99	2639
GEIC-100	OEM
GEIC-101	4582
GEIC-102	4556
GEIC-103	1018
GEIC-104	2631
GEIC-105	3915
GEIC-106	2949
GEIC-107	2231
GEIC-109	4594
GEIC-110	4595
GEIC-111	1326
GEIC-112	6233
GEIC-113	4634
GEIC-114	2284
GEIC-115	0789
GEIC-116	2882
GEIC-117	0470
GEIC-118	0528
GEIC-119	4543
GEIC-120	0638
GEIC-121	0653
GEIC-122	0673
GEIC-123	2891
GEIC-124	0930
GEIC-125	0946
GEIC-126	0964
GEIC-127	2886
GEIC-128	1478
GEIC-130	2316
GEIC-131	3785
GEIC-132	4575
GEIC-133	2891
GEIC-135	0433
GEIC-136	3352
GEIC-137	4612
GEIC-138	2872
GEIC-139	1479
GEIC-140	2261
GEIC-142	2887
GEIC-143	2582
GEIC-147	0167
GEIC-148	0167
GEIC-149	0878
GEIC-150	4627
GEIC-153	4632
GEIC-154	3651
GEIC-155	0520
GEIC-156	0607
GEIC-157	0043
GEIC-158	0389
GEIC-159	1621
GEIC-160	4570
GEIC-161	4569
GEIC-162	2262
GEIC-163	4561
GEIC-164	4563
GEIC-165	4564
GEIC-166	2889
GEIC-167	4565
GEIC-168	4566
GEIC-169	4390
GEIC-170	4093
GEIC-171	2656
GEIC-172	1696
GEIC-173	0093
GEIC-175	0852
GEIC-176	0857
GEIC-178	2931
GEIC-179	1050
GEIC-180	2689
GEIC-181	2019
GEIC-182	3366
GEIC-183	0789
GEIC-184	0403
GEIC-185	1301
GEIC-190	0624
GEIC-191	1196
GEIC191	1196
GEIC-193	1313
GEIC-194	0268
GEIC-196	3695
GEIC-197	1160
GEIC197	2914
GEIC-199	1379
GEIC-201	4578
GEIC-202	4631
GEIC-203	2669
GEIC-204	2670
GEIC-205	2624
GEIC-207	2715
GEIC-208	3722
GEIC-209	3724
GEIC-210	2529
GEIC-211	OEM
GEIC-212	3008
GEIC-213	3735
GEIC-214	1395
GEIC-215	1421
GEIC-216	2930
GEIC-217	0785
GEIC-218	1808
GEIC-219	3219
GEIC-220	0360
GEIC-221	0372
GEIC-222	0364
GEIC-223	2747
GEIC-224	0377
GEIC-225	0802
GEIC-226	2707
GEIC-227	2714
GEIC-228	2257
GEIC-229	2755
GEIC-230	0349
GEIC-231	1337
GEIC-232	3195
GEIC-233	0521
GEIC-234	1753
GEIC-235	3761
GEIC-236	2699
GEIC-237	3763
GEIC-238	3410
GEIC-239	3422
GEIC-240	3766
GEIC-241	2480
GEIC-242	3770
GEIC-243	3006
GEIC-244	0848
GEIC-245	2687
GEIC-246	2701
GEIC-247	2711
GEIC-248	2705
GEIC-249	1301
GEIC-250	1705
GEIC-251	1797
GEIC-252	1800
GEIC-253	1890
GEIC-254	1897
GEIC-255	3808
GEIC-256	2727
GEIC-257	2749
GEIC-258	2722
GEIC-259	1192
GEIC-260	0026
GEIC-261	2890
GEIC-262	1919
GEIC-263	0410
GEIC-264	1975
GEIC-265	0312
GEIC-266	3663
GEIC-267	2368
GEIC-268	2368
GEIC-269	0973
GEIC-271	3020
GEIC-272	4557
GEIC-273	4562
GEIC-274	4409
GEIC-275	2854
GEIC-276	3670
GEIC-277	4586
GEIC-278	1248
GEIC-279	4147
GEIC-280	4003
GEIC-281	4034
GEIC-282	6531
GEIC-284	3135
GEIC-285	4597
GEIC-286	4598
GEIC-287	4600
GEIC-288	2537
GEIC-289	1854
GEIC-290	2698
GEIC-291	1509
GEIC-292	1260
GEIC-293	2960
GEIC-294	OEM
GEIC-295	1524
GEIC-296	4086
GEIC-297	OEM
GEIC-298	4068
GEIC-299	4047
GEIC-300	2942
GEIC-301	3269
GEIC-302	3273
GEIC-303	4055
GEIC-304	1300
GEIC-310	OEM
GEIC-311	3958
GEIC-312	1634
GEIC-313	4593
GEIC-314	2062
GEIC-315	2773
GEIC-316	2781
GEIC-317	1714
GEIC-318	2062
GEIC-319	2305
GEIC-320	4038
GEIC-321	3269
GEIC-322	4401
GEIC-323	3705
GEIC-324	2572
GEIC-325	4196
GEIC-326	1540
GEIC-327	1816
GEIC-328	1055
GEIC-329	2193
GEIC-330	1571
GEIC-331	1612
GEIC-332	3190
GEIC-333	1170
GEIC-334	1048
GEIC-335	3262
GEIC-336	4644
GEL211AL1	0664
GEL211F1	0664
GEL2111	0664
GEL2111AL1	0664
GEL2111F1	0664
GEL2113	0678
GEL2113AL1	0678
GEL2113F1	0678
GEL2114	0666
GEL3072F1	0349
GEM1	OEM
GEM2	OEM
GEM3	OEM
GEM4	OEM
GEM8	OEM
GEM9	OEM
GE-M100	0050
GEMR-1	0097
GEMR-2	0871
GEMR-3	0722
GEMR-4	1487
GEMR-5	1396
GEMR-6	1690
GEN11760	0924
GEPS2001	0541
GER4001	0071
GER4002	0071
GER4003	0071
GER4004	0071
GER4005	0071
GER4006	0071
GER4007	0071
GER-A	0004
GER-A-D	0050
GES92	0076
GES93	0037
GES97	0079
GES98	0861
GES929	0079
GES930	0079
GES2221	0079
GES2221A	0079
GES2222	0079
GES2222A	0079
GES2483	0079
GES2646	2136
GES2647	1175
GES2904	0037
GES2905	0037
GES2905A	2260
GES2906	0037
GES2907	0037
GES2923	0079
GES2924	0079
GES2925	0079
GES2926	0079
GES3053	0079
GES3390	OEM
GES3391	OEM
GES3391A	OEM
GES3392	0079
GES3393	0079
GES3395	0079
GES3396	0079
GES3397	0079
GES3414	0079
GES3415	0079
GES3416	0079
GES3417	0079
GES3565	0079
GES3566	0079
GES3567	0861
GES3568	0861
GES3569	0861
GES4121	0037
GES4122	0037
GES4248	0037
GES5305	0400
GES5306	0400
GES5306A	0400
GES5307	0400
GES5308	0400
GES5308A	6984
GES5368	0079
GES5369	0079
GES5370	0079
GES5371	0079
GES5372	0037
GES5373	0037
GES5374	0037
GES5375	0037
GES5447	0037
GES5448	0037
GES5449	0016
GES5450	0016
GES5451	0079
GES5810	0324
GES5811	0037
GES5812	0324
GES5813	0037
GES5814	0079
GES5815	0037
GES5816	0079
GES5817	0037
GES5818	0079
GES5819	0037
GES5820	0324
GES5821	0037
GES5822	0079
GES5823	0037
GES5824	0079
GES5825	0079
GES5826	0079
GES5827	0079
GES5827A	0079
GES5828	0111
GES5828A	0111
GES6000	0079
GES6001	0037
GES6002	0079
GES6003	0037
GES6004	0079
GES6005	0037
GES6006	0079
GES6007	0037
GES6010	0086
GES6011	0037
GES6012	0086
GES6013	0037
GES6014	0086
GES6015	0037
GES6016	0086
GES6017	0037
GES6218	0715
GES6219	0715
GES6220	0715
GES6221	0861
GES6222	0079
GES6224	0079
GES6426	3799
GES6427	3799
GES6560	0079
GES6561	0079
GES6562	OEM
GES6563	0037
GES60002#	0079
GESS-2AV2	1449
GESS-3A3	1306
GESS-3AT2	2083
GESS-3DB3	2084
GESS-6DW4	2085
GET0-50P	0164
GET3	OEM
GET4	OEM
GET5	OEM
GET6	OEM
GET7	OEM
GET8	OEM
GET9	OEM
GET15	OEM
GET102	OEM
GET-103	0211
GET103	0004
GET104	OEM
GET105	OEM
GET106	OEM
GET110	OEM
GET111	OEM
GET-113	0211
GET113	0004
GET-113A	0211
GET113A	0004
GET-114	0211
GET114	0004
GET115	OEM
GET116	OEM
GET120	OEM
GET535	0164
GET536	0164
GET538	0164
GET571	OEM
GET-572	0160
GET572	0160
GET573	OEM
GET574	OEM
GET581	0160
GET582	0160
GET583	0160
GET584	0160
GET585	OEM
GET586	OEM

If replacement code is OEM, contact original manufacturer for replacement.

Original Device Types

DEVICE TYPE	REPL CODE
GET671	0050
GET-672	0050
GET672	0050
GET-672A	0050
GET672A	0050
GET-673	0050
GET673	0050
GET691	0050
GET-692	0050
GET692	0050
GET693	0050
GET706	0016
GET708	0016
GET870	OEM
GET871	0136
GET872	0136
GET873	0136
GET-873A	0050
GET873A	0136
GET874	0136
GET875	0136
GET880	0280
GET881	0280
GET882	0280
GET-883	0050
GET883	0136
GET884	0004
GET885	0136
GET887	0280
GET888	0280
GET889	0280
GET890	0280
GET891	0280
GET892	0280
GET895	0280
GET896	0280
GET897	0280
GET898	0211
GET914	0016
GET929	0086
GET930	0086
GET931	0050
GET2221	0016
GET2221A	0079
GET2222	0016
GET2222A	0079
GET2369	0016
GET2483	0111
GET2904	0150
GET2905	0150
GET2906	0532
GET2907	0532
GET3013	0016
GET3014	0016
GET3562	0111
GET3563	0224
GET3638	0037
GET3638A	0037
GET3646	0016
GET3903	0079
GET3904	0079
GET3905	0150
GET3906	0150
GET4870	2136
GET4871	1175
GET5116	0050
GET5117	0050
GET5305	0400
GET5306	0400
GET5306A	OEM
GET5307	0400
GET5308	0400
GET5308A	0400
GET5457	OEM
GET5458	OEM
GET5459	OEM
GET-FET-6	0353
GETO-50P	0164
GEVR-100	1292
GEVR-101	0624
GEVR-102	0624
GEVR-104	1285
GEVR-105	1285
GEVR-106	0083
GEVR-107	1196
GEVR-108	1196
GEVR-109	1828
GEVR-110	0334
GEVR-111	0334
GEVR-112	1836
GEVR-113	1838
GEVR-114	1838
GEVR-115	1192
GEVR-116	1192
GE-X1	0240
GE-X3	2188
GE-X4	0469
GE-X5	1259
GE-X8	0038
GEX8	0038
GE-X9	0004
GE-X10	2136
GE-X11	0244
GE-X12	0154
GE-X13	2150
GE-X16	0676
GE-X16A1938	0016
GE-X17	0316
GE-X18	0560
GE-X36	0015
GEX36	0015
GE-X66	0143
GEX66	0015
GEX541	OEM
GEX542	OEM
GEXM66	OEM
GE-Z83	0475
GEZD-3.3	0292
GEZD-3.6	0188
GEZD-3.9	0456
GEZD-4.3	0533
GEZD-4.7	0451
GEZD-5.0	0162
GEZD-5.1	0162
GEZD-5.6	0157
GEZD-5.8	3127
GEZD-6.0	0301
GEZD-6.2	0636
GEZD-6.6	0025
GEZD-6.8	0025
GEZD-7.5	0649
GEZD-8.0	0244
GEZD-8.1	0244
GEZD-8.2	0244
GEZD-8.7	1083
GEZD-9.1	0012
GEZD-10	0170
GEZD-10-4	0170
GEZD-10.4	0064
GEZD-11	0317
GEZD-11.5	0794
GEZD-12	0137
GEZD-13	0365
GEZD-14	0100
GEZD-15	0002
GEZD-16	0420
GEZD-17	1649
GEZD-18	0495
GEZD-19	0949
GEZD-20	0531
GEZD-22	0565
GEZD-24	0402
GEZD-25	1606
GEZD-27	0441
GEZD-28	1674
GEZD-30	0726
GEZD-33	0039
GEZD-35	0820
GEZD-36	0820
GEZD-39	0350
GEZD-43	0931
GEZD-47	0999
GEZD-51	0502
GEZD-55	1834
GEZD-56	0869
GEZD-60	1156
GEZD-62	0783
GEZD-68	1268
GEZD-75	1189
GEZD-82	0331
GEZD-87	3026
GEZD-91	1311
GEZD-100	0098
GEZD-110	0149
GEZD-120	0186
GEZD-130	0213
GEZD-140	0245
GEZD-150	0028
GEZD-160	0255
GEZD-170	0877
GEZD-180	0367
GEZD-190	2858
GEZD-200	0421
GEZJ252A	0015
GEZJ252B	0015
GEZN-1-9145-1	OEM
GEZN-2-9145-1	OEM
GEZN-3-9145-1	OEM
GEZN-5-9145-1	OEM
GF021A2	OEM
GF0D1A1	OEM
GF20	0211
GF21	0211
GF32	0211
GF45017	OEM
GFA03L	0017
GFA03R	0017
GFB30B	0031
GFB30C	0031
GFB30E	0031
GFB-30G	4948
GFB30G	0023
GFB30G-FD1	0031
GFB7496D	OEM
GFD10B	0023
GFD10C	0023
GFD10E	0023
GFD10G	0023
GFE10	0017
GFE10R	0182
GFE10RGET0-50P	OEM
GFH6001	OEM
GFH6002	OEM
GFH6003	OEM
GFOD1A1	OEM
GFOD1B1	OEM
GFOD1B2	OEM
GFOE1A1	OEM
GFOE1A2	OEM
GFT20	OEM
GFT20/15	0004
GFT20/30	0004
GFT20/60	OEM
GFT20R	0004
GFT21	OEM
GFT21/15	0004
GFT21/30	0004
GFT21/60	OEM
GFT21R	0004
GFT22	OEM
GFT22/15	0164
GFT22/30	0164
GFT22/60	OEM
GFT22R	0164
GFT25	OEM
GFT25/15	0004
GFT25/30	0004
GFT25/60	OEM
GFT25R	0004
GFT26	OEM
GFT30	OEM
GFT31	0004
GFT31/15	0164
GFT31/30	0004
GFT31/60	OEM
GFT32	0004
GFT32/15	OEM
GFT32/30	OEM
GFT32/60	OEM
GFT34	0004
GFT34/15	OEM
GFT34/30	OEM
GFT34/60	OEM
GFT41	OEM
GFT42A	OEM
GFT42B	OEM
GFT43	OEM
GFT43A	OEM
GFT43B	OEM
GFT44	0050
GFT44/15E	OEM
GFT44/30	OEM
GFT45	0050
GFT45/30	OEM
GFT2006	OEM
GFT2006/30	OEM
GFT2006/60	OEM
GFT2006/90	OEM
GFT3008/20	OEM
GFT3008/40	0211
GFT3008/60	OEM
GFT3008/80	OEM
GFT3408/20	OEM
GFT3408/40	OEM
GFT3408/60	OEM
GFT3408/80	OEM
GFT4012	OEM
GFT4012/30	OEM
GFT4012/60	OEM
GFT4308/40	OEM
GFT4308/60	OEM
GFT4308/80	OEM
GFT4412/30	OEM
GFT4412/60	OEM
GFT4608/40	OEM
GFT4608/60	OEM
GFT4608/80	OEM
GFT8024	OEM
GH1D	OEM
GH-1E	0102
GH-1F	0182
GH1F	0348
GH2-021	OEM
GH-3E	0182
GH-3F	0102
GH3F	0158
GH3FLF	0158
GHC801	OEM
GHC801A	OEM
GHC801B	OEM
GHV-02SSN	OEM
GHV06BSS	OEM
GHV-06SSN	OEM
GHV-2	OEM
GHV2	OEM
GHV-3	OEM
GHV3	OEM
GHV-4	OEM
GHV4	OEM
GHV-5	OEM
GHV5	OEM
GHV-6	OEM
GHV-7	OEM
GHV7	OEM
GHV-8	OEM
GHV8	OEM
GHV-9	OEM
GHV-10	OEM
GHV-10SL	OEM
GHV-10SS	0201
GHV-10SSN	OEM
GHV-11	OEM
GHV-12	OEM
GHV-12SL	OEM
GHV-12SS	0201
GHV-12SSN	OEM
GHV-13	OEM
GHV-14	OEM
GHV-14SL	OEM
GHV-14SS	0288
GHV-14SSN	OEM
GHV-15	OEM
GHV-16	OEM
GHV-16SL	OEM
GHV-16SSN	OEM
GHV-18SS	0378
GHV-18SSN	OEM
GHV-20SL	OEM
GHV-24SL	OEM
GI08B	0015
GI1	0280
GI-1N4385	0015
GI2	0211
GI3	0136
GI4	0211
GI5	0600
GI6	0600
GI7	0600
GI8	0038
GI9	0440
GI10	0016
GI82AG	OEM
GI-100-1	0015
GI103A	OEM
GI103B	OEM
GI103G	OEM
GI103V	OEM
GI212-152-04	0446
GI254	0023
GI263	OEM
GI298-03	0023
GI-300B	0015
GI-300D	0015
GI304A	OEM
GI304B	OEM
GI305A	OEM
GI305B	OEM
GI307A	OEM
GI308A	OEM
GI308B	OEM
GI308D	OEM
GI308E	OEM
GI308G	OEM
GI308I	OEM
GI308K	OEM
GI308SZ	OEM
GI308V	OEM
GI323-04	0023
GI339-04	0102
GI356-03	1236
GI401A	OEM
GI401B	OEM
GI402D	OEM
GI403A	OEM
GI411	0015
GI415B	0023
GI419-0	0015
GI419B	0015
GI420	0015
GI428B	0015
GI437B	0031
GI500	0110
GI501	0953
GI502	0242
GI504	0540
GI506	0965
GI508	0817
GI510	0922
GI750	1282
GI751	1282
GI752	1287
GI754	1287
GI756	1293
GI758	1296
GI810	0023
GI811	0023
GI812	0023
GI814	0023
GI816	0023
GI817	0017
GI818	0102
GI820	1362
GI821	1362
GI822	1362
GI824	1366
GI826	1372
GI828	1372
GI850	0031
GI851	0031
GI852	0031
GI854	0031
GI856	0031
GI910	0031
GI911	0031
GI912	0031
GI914	0031
GI916	0031
GI917	1191
GI918	OEM
GI1001	0499
GI1002	0499
GI1003	0499
GI1004	0499
GI1101	0031
GI1102	0031
GI1103	0031
GI1104	0031
GI1301	OEM
GI1302	OEM
GI1303	OEM
GI1304	OEM
GI1401	1145
GI1401R	OEM
GI1402	1145
GI1402R	OEM
GI1403	1145
GI1403R	OEM
GI1404	1145
GI1404R	OEM
GI2401	1236
GI2401R	1727
GI2402	1236
GI2402R	1727
GI2403	1236
GI2403R	1727
GI2404	1236
GI2404R	1727
GI2500	OEM
GI2501	OEM
GI2502	OEM
GI2504	OEM
GI2506	OEM
GI2508	OEM
GI2510	OEM
GI2601	OEM
GI2601R	OEM
GI2602	OEM
GI2602R	OEM
GI2603	OEM
GI2603R	OEM
GI2604	OEM
GI2604R	OEM
GI2605	OEM
GI2605R	OEM
GI2606	OEM
GI2606R	OEM
GI-2711	0079
GI2711	0016
GI-2712	0079
GI2712	0016
GI-2713	0861
GI2713	0079
GI-2714	0079
GI2714	0016
GI-2715	0007
GI2715	0016
GI-2716	0007
GI2716	0016
GI-2921	0007
GI2921	0016
GI-2922	0007
GI2922	0016
GI-2923	0007
GI2923	0016
GI-2924	0111
GI2924	0016
GI-2925	0111
GI2925	OEM
GI-2926	0008
GI2926	OEM
GI3002	0015
GI-3008	0015
GI-3391	0111
GI-3391A	0111
GI-3392	0111
GI3392	OEM
GI-3393	0111
GI-3394	0111
GI-3395	0111
GI-3396	0111
GI-3397	0111
GI-3398	0111
GI-3402	0861
GI-3403	0079
GI-3404	0861
GI-3405	0079
GI-3414	0861
GI-3415	0079
GI-3416	0079
GI-3417	0079
GI-3566	0861
GI-3605	0007
GI-3606	0007
GI-3607	0007
GI-3638	0861
GI3638	0037
GI-3638A	0861
GI3638A	0037
GI-3641	0079
GI3641	0016
GI-3642	0079
GI-3643	0079
GI3643	0016
GI-3644	0532
GI3644	0037
GI-3702	0861
GI3702	0037
GI-3703	0861
GI3703	0037
GI-3704	0079
GI3704	0016
GI-3705	0079
GI3705	0016
GI-3706	0079
GI3706	0016
GI-3707	0111
GI3707	0016
GI-3708	0111
GI3708	0016
GI-3709	0111
GI3709	0016
GI-3710	0111
GI3710	0016
GI-3711	0111
GI3711	0016
GI-3721	0111
GI-3793	0861
GI3793	0007
GI-3794	0861
GI3794	OEM
GI-3900	0111
GI-3900A	0111
GI3992-17	0015
GI5823	OEM
GI5824	OEM
GI5825	OEM
GI6506	0600
GI8301	OEM
GI12605	OEM
GI12605R	OEM
GI12606	OEM
GI12606R	OEM
GI405919-4	OEM
GIB2500	5760
GIB2501	5760
GIB2502	5761
GIB2504	5762
GIB2506	5764
GIB2508	5766
GIB2510	5768
GIB3500	5926
GIB3501	5926
GIB3502	5926
GIB3504	5927
GIB3506	5927
GIB3508	5929
GIB3510	5929
GID	0015
GIFE16A	1236
GIFE16D	OEM
GI-P100-D	0015
GIP100-D	0087
GIP100D	0015
GIRG3K	1217
GI-TVC3	0474
GJ3M	OEM
GJ4M	0015
GJ5M	OEM
GJ6M	OEM
GL1HD51	OEM
GL1HD101	OEM
GL1HY102	OEM
GL-1N4385	0015
GL1PR101	OEM
GL1PR102	OEM
GL1PR111	OEM
GL2AR1	OEM
GL2HY1	OEM
GL2PG1	OEM
GL2PR1	OEM
GL2PR6	OEM
GL2PR9	OEM
GL3AR1	OEM
GL3AR2	OEM
GL3AR3	OEM
GL3D403	OEM
GL3D405	OEM
GL3HD1	OEM
GL3HD5	OEM
GL3HY1	OEM
GL3HY2	OEM
GL3HY3	OEM
GL3HY8	OEM
GL3N403	OEM
GL3N405	OEM
GL3N407	OEM
GL3NG1	OEM
GL3NG5	OEM
GL3P01	OEM
GL3P403	OEM
GL3P405	OEM
GL3P407	OEM
GL3P503	OEM
GL3PG1	OEM
GL3PG2	OEM
GL3PG3	OEM
GL3PG5	OEM
GL3PR1	OEM
GL3PR2	OEM
GL3PR3	OEM
GL-3PR7	OEM
GL3PR8	OEM
GL4AR2	OEM
GL4HY2	OEM
GL4NG2	OEM
GL4PG2	OEM
GL4PR2	OEM
GL4R04A	OEM
GL4R06A	OEM
GL4R10A	OEM
GL5AR1	3182
GL5AR2	1962
GL-5HD5	OEM
GL5HD5	OEM
GL5HD10	OEM
GL5HD21	OEM
GL-5HD23	OEM
GL5HY1	OEM
GL5HY2	OEM
GL5HY5	OEM
GL5HY8	OEM
GL5HY41	OEM
GL-5IID23	OEM
GL5ND5	OEM
GL5NG6	5381
GL5NG10	OEM
GL5NP5	OEM
GL5PG1	OEM
GL5PG2	OEM
GL-5PG5	OEM
GL5PG5	OEM
GL5PG22	OEM
GL5PR002	OEM
GL5PR1	OEM
GL-5PR2	OEM
GL5PR2	OEM
GL5PR5	OEM
GL5PR6	1962
GL6D201	OEM
GL6N201	OEM
GL6N202	OEM
GL6N402	OEM
GL6P201	4332
GL6P202	OEM
GL6P401	OEM
GL6R201	OEM
GL6R202	OEM
GL7D212	OEM
GL7N201	OEM
GL-7N202	OEM
GL8D04	OEM
GL8D06	OEM
GL8D303D	OEM
GL8G03	OEM
GL8G04	OEM
GL8N03D	OEM
GL8N04	OEM
GL8N06	OEM
GL8P03	OEM
GL8P03D	OEM
GL8P04	OEM
GL8PR22	OEM
GL8R03	OEM
GL8R03D	OEM
GL8R06	OEM
GL8R10	OEM
GL9HD2	OEM
GL9HD4	5443
GL9HD6	OEM
GL9HD9	OEM
GL9HD10	OEM
GL9HD22	OEM
GL9HD23	OEM
GL9HD24	OEM
GL9HP2	OEM
GL9HY2	OEM
GL9HY4	5063
GL9HY6	OEM
GL9HY9	OEM
GL9HY10	OEM
GL9HY22	OEM
GL9HY23	OEM
GL9HY24	OEM
GL9N06	OEM
GL9N06A	OEM
GL9ND2	OEM
GL9ND24	OEM
GL9NG2	OEM
GL9NG4	OEM
GL9NG5	OEM
GL9NG9	OEM
GL9NG10	OEM
GL9NG22	OEM
GL9NG23	OEM
GL9NG24	OEM
GL9NP2	OEM
GL9NP24	OEM
GL9PG2	OEM
GL9PG3	OEM
GL9PG4	5444
GL9PG9	OEM
GL9PG12	OEM
GL9PG24	OEM
GL9PG34	OEM
GL9PR2	OEM
GL9PR3	OEM
GL9PR4	5443
GL9PR5	OEM
GL9PR6	OEM
GL9PR9	OEM
GL9PR10	OEM
GL9PR12	OEM
GL9PR19	OEM
GL9PR22	OEM
GL9PR23	OEM
GL-9PR24	OEM
GL9PR24	OEM
GL9PR34	OEM
GL19G102	OEM
GL27A	OEM
GL27B	OEM
GL27D	OEM
GL27G	OEM

If replacement code is OEM, contact original manufacturer for replacement.

DEVICE TYPE	REPL CODE
GL27J	OEM
GL27K	OEM
GL27M	OEM
GL31AR	3019
GL-32AR	3019
GL-40-RG	OEM
GL41A	OEM
GL41B	OEM
GL41D	OEM
GL41G	OEM
GL41J	OEM
GL41K	OEM
GL41M	OEM
GL50G	OEM
GL-52B	OEM
GL56	1983
GL68	OEM
GL105G11	OEM
GL105H5	OEM
GL105H11	OEM
GL105M5	OEM
GL105M11	OEM
GL105N5	OEM
GL105N11	OEM
GL105R5	OEM
GL105R11	OEM
GL105R53	OEM
GL106H5	OEM
GL106M5	OEM
GL106N5	OEM
GL-106NS	OEM
GL106R5	OEM
GL106R50	OEM
GL107G12	OEM
GL107H12	OEM
GL107N12	OEM
GL107R12	OEM
GL107S12	OEM
GL112G9	OEM
GL112G13	OEM
GL112H9	OEM
GL112H13	OEM
GL112M2	OEM
GL112M9	OEM
GL112M13	OEM
GL112N9	OEM
GL112N13	OEM
GL112R3	OEM
GL112R9	OEM
GL112R13	OEM
GL112S9	OEM
GL112S13	OEM
GL112T9	OEM
GL112T13	OEM
GL211	0841
GL324	OEM
GL348	OEM
GL350	OEM
GL358	0770
GL393	OEM
GL410	OEM
GL430	OEM
GL-450	2703
GL450	2703
GL-450V	2703
GL450V	OEM
GL451	2703
GL452S	OEM
GL504	OEM
GL513F	OEM
GL514	OEM
GL520	OEM
GL-521	OEM
GL-526	OEM
GL528	3334
GL552	OEM
GL-750S	OEM
GL-751RS	OEM
GL-950S	OEM
GL-951RS	OEM
GL-1001	OEM
GL-1002	OEM
GL-1003	OEM
GL1010	0496
GL1050S	OEM
GL1051RS	OEM
GL1130	0732
GL3101A	2116
GL3101A/B	OEM
GL3101B	2116
GL3120	0539
GL-3201	0167
GL3201	0167
GL3301	2122
GL3301B	2122
GL3320	0557
GL3401	OEM
GL3711	OEM
GL3812	OEM
GL3820	OEM
GL4484	OEM
GL4558	3440
GL4850	1778
GL4950	1778
GL7445	OEM
GL7805	0624
GL7805A	OEM
GL7806	3463
GL7809	OEM
GL7812	0334
GL7815	1321
GLA28	OEM
GLA28A	OEM
GLA28B	OEM
GLA31	OEM
GLA31A	OEM
GLA31B	OEM
GLA35	OEM
GLA35A	OEM
GLA35B	OEM
GLA39	0188
GLA39A	0188
GLA39B	0456
GLA43	0533
GLA43A	0533
GLA43B	0533
GLA47	0162
GLA47A	0162
GLA47B	0451
GLA51	0162
GLA51A	0162
GLA51B	0162
GLA56	0157
GLA56A	0157
GLA56B	0157
GLA62	0636
GLA62A	0636
GLA62B	0636
GLA68	0025
GLA68A	0025
GLA68B	0025
GLA75	0649
GLA75A	0649
GLA75B	0649
GLA82	0244
GLA82A	0244
GLA82B	0244
GLA91	0012
GLA91A	0012
GLA91B	0012
GLA100	0170
GLA100A	0170
GLA100B	0170
GLE102	OEM
GLE502	OEM
GLE703	OEM
GLPHY4	5063
GLT12	OEM
GLT24	OEM
GLT47	OEM
GLT120	OEM
GM0290	0050
GM0375	0050
GM0376	0050
GM0377	0050
GM0378	0296
GM0380	0144
GM0384	OEM
GM-1	0087
GM-1A	0087
GM-1B	0071
GM-1C	0071
GM1J2	0071
GM-1Z	0087
GM-3	0087
GM3	0087
GM-3A	0087
GM-3B	0087
GM3B	0564
GM-3C	0922
GM3DM	OEM
GM-3Y	0087
GM3Y	0023
GM-3Z	0087
GM3Z	0087
GM30A	3676
GM30B	3676
GM30D	3676
GM30G	3679
GM30J	3680
GM30K	OEM
GM290	0050
GM290A	0050
GM308	0007
GM320MP-5.2	1285
GM378	0050
GM378A	0050
GM-380	OEM
GM380	0136
GM384	OEM
GM428	0160
GM656A	0050
GM760	0037
GM-770	0127
GM770	0127
GM770GM0375	OEM
GM875	0050
GM876	0050
GM877	0050
GM878	0050
GM878A	0050
GM878AB	0050
GM878B	0050
GM3043	OEM
GM3742-01	OEM
GM3742-02	OEM
GMA-01	0124
GMA01	0124
GMA01-01-BT	0124
GMA01-4-BT	OEM
GMA01-8T	OEM
GMA01-AT	OEM
GMA-01-BT	0124
GMA01-BT	0124
GMA01-FY1	OEM
GMA01-L	0133
GMA01L	0133
GMA-10-BT	0124
GMB01	0124
GMB01B	0124
GMB01BT	0124
GMB01L	0124
GMD1	OEM
GMD2	OEM
GMD5	OEM
GME040-1	0037
GME0404	0150
GME0404-1	0150
GME0404-2	0150
GME404-1	0150
GME1001	0016
GME1002	0016
GME2001	0016
GME2002	0016
GME3001	0144
GME3002	0144
GME4001	0016
GME4002	0016
GME4003	0016
GME6001	0127
GME6002	0127
GME6003	0016
GME9001	0144
GME9002	0144
GME9021	0144
GME9022	0144
GM0290	0050
GM0375	0050
GM0376	0050
GM0377	0050
GM0378	0050
GMO-380	0079
GM0380	0007
GMP-5	OEM
GMP5	OEM
GMP-5A	OEM
GMP5A	OEM
GMP-5B	OEM
GMP5B	OEM
GN13G	1291
GN1036E	OEM
GN-1039E-006	OEM
GN2050	OEM
GN3002U	OEM
GN3100E	OEM
GN3210	OEM
GN3500	OEM
GN3600	OEM
G01-803A	0162
G03-007C	0560
G03-404-B	0126
G03PR5	OEM
G04-041B	0079
G04-701-A	0085
G04-703-A	0847
G04-704-A	0222
G04-711-E	0164
G04-711-F	0164
G04-711-G	0164
G04-711-H	0164
G05-003-A	0079
G05-003-B	0079
G05-004A	0127
G05-010-A	0016
G05-011-A	0016
G05-015-C	0079
G05-015-D	0086
G05-034-D	0086
G05-035-D	0111
G05-035-D,E	0111
G05-035-E	0111
G05-036-C,D,E	0079
G05-036B	0079
G05-036C	0079
G05-036D	0086
G05-036E	0079
G05-037B	0079
G05-050-C	0079
G05-055-D	0086
G05-413A	0086
G06-714C	0560
G06-717-B	0560
G08-005L	0325
G00-004-A	0143
G00-008-A	0143
G00-009-A	0143
G00-012-A	0162
G00-013-8	0143
G00-014-A	0162
G00-535-B	0540
G00-536A	0540
G00-543-A	0540
GP05A	0015
GP08B	0015
GP-08D	0087
GP08D	0087
GP08DPKG3	OEM
GP08DPKG23	0087
GP08G	0965
GP1	0015
GP1A52HR	OEM
GP1A521	OEM
GP1F31T	OEM
GP1L01	OEM
GP1L01F	OEM
GP1L02	OEM
GP1S01F	OEM
GP1S02	OEM
GP1S03	OEM
GP1S04	OEM
GP1S53	OEM
GP1U50X	OEM
GP1U52	OEM
GP1U521H	OEM
GP1U521Y	OEM
GP1U561	OEM
GP1U721Q	OEM
GP1U721Y	OEM
GP1U781Q	OEM
GP1U901X	OEM
GP-2	OEM
GP2	OEM
GP2-015	OEM
GP2-345	0133
GP2-354	0133
GP2L01	OEM
GP2L01F	OEM
GP2L02	OEM
GP2S01	OEM
GP2S01F	OEM
GP2S02	5662
GP2S04A	OEM
GP2S04B	OEM
GP2S09-B	OEM
GP2S09-C	OEM
GP3A21	OEM
GP-6-019	OEM
GP6C10	OEM
GP6C20	OEM
GP6C30	OEM
GP10	OEM
GP10A	1335
GP10B	0080
GP10C	0015
GP10D	0023
GP10G	0795
GP-10G-018	0023
GP10G5020	OEM
GP10J	0015
GP10K	0072
GP10M	0071
GP10N	0348
GP10Q	0348
GP10T	0348
GP10V	0348
GP10W	0348
GP10Y	0348
GP15	0087
GP15A	0071
GP15B	0015
GP15B(L)	0071
GP15BL	0071
GP15D	0015
GP15G	0071
GP15J	0071
GP15J-16	0071
GP15JTP	0071
GP15K	0071
GP15M	0071
GP16	OEM
GP20A	0071
GP20B	0071
GP20D	0071
GP20G	0023
GP20J	0071
GP20K	0071
GP20M	0071
GP25A	0110
GP-25B	0015
GP25B	0953
GP25D	0242
GP25G	0540
GP25J	0965
GP25K	0817
GP25M	0922
GP30A	0110
GP-30B	0953
GP30B	0953
GP-30D	0922
GP30D	0242
GP30G	0540
GP30J	0965
GP30JG	0965
GP30K	0817
GP30M	0922
GP80A	1145
GP80B	1145
GP80D	1145
GP80G	1145
GP80J	1664
GP80K	OEM
GP80M	OEM
GP106	0071
GP130	OEM
GP139	0004
GP139A	0004
GP139B	0004
GP150	OEM
GP175	OEM
GP199	OEM
GP204	4415
GP230	0015
GP250	0015
GP252	0087
GP-300	0242
GP330	0124
GP340	0124
GP350	OEM
GP360	OEM
GP411	OEM
GP411F	OEM
GP415	OEM
GP450	OEM
GP450F	OEM
GP500	OEM
GP1432	0160
GP1448	OEM
GP1462	OEM
GP1493	0160
GP1494	0160
GP1600	OEM
GP1622	0440
GP1882	0160
GP2354	0143
GP2364	0143
GP-6019	OEM
GP6019	OEM
GPA1100	OEM
GPC215	OEM
GPC215A	OEM
GPD020	OEM
GPD040	OEM
GPD050	OEM
GPD060	OEM
GPD070	OEM
GPD1	OEM
GPD5	OEM
GPD201	OEM
GPD202	OEM
GPD251	OEM
GPD252	OEM
GPD401	OEM
GPD402	OEM
GPD403	OEM
GPD404	OEM
GPD411	OEM
GPD420	OEM
GPD426	OEM
GPD430	OEM
GPD440	OEM
GPD461	OEM
GPD462	OEM
GPD463	OEM
GPD464	OEM
GPD470	OEM
GPD520	OEM
GPD540	OEM
GPD550	OEM
GPD620	OEM
GPD620C	OEM
GPD620P	OEM
GPD630C	OEM
GPD630P	OEM
GPD640C	OEM
GPD640P	OEM
GPD660C	OEM
GPD660P	OEM
GPD670C	OEM
GPD670P	OEM
GPD720P	OEM
GPD726	OEM
GPD730P	OEM
GPD740P	OEM
GPD760P	OEM
GPD770P	OEM
GPD886	OEM
GPD1001	OEM
GPD1002	OEM
GPD1003	OEM
GPD1061	OEM
GPD1062	OEM
GPD1063	OEM
GPIB11V-2	OEM
GPIU5	OEM
GPL4A	OEM
GPL1001	OEM
GPM1NA	0143
GPM1NB	0143
GPM1NC	0143
GPM2NA	0143
GPM2NB	0143
GPP1A	OEM
GPP1B	OEM
GPP1D	OEM
GPP1G	OEM
GPP1J	OEM
GPP1K	OEM
GPP1M	OEM
GPP5A	OEM
GPP5B	OEM
GPP5D	OEM
GPP5G	OEM
GPP5J	OEM
GPP5K	OEM
GPP10A	0182
GPP10B	0182
GPP10D	0182
GPP10G	0182
GPP10J	0182
GPP10K	0182
GPP10M	0182
GPP15A	0102
GPP15B	0102
GPP15D	0102
GPP15G	0102
GPP15J	0102
GPP15K	0102
GPP15M	0102
GPP30A	0031
GPP30B	0031
GPP30D	0031
GPP30G	0031
GPP30J	0031
GPP30K	1217
GPP30M	1928
GPP60A	OEM
GPP60B	OEM
GPP60D	OEM
GPP60G	OEM
GPP60J	OEM
GPP60K	OEM
GPP60M	OEM
GPR-1	OEM
GPS203	OEM
GPT	OEM
GPT-16	0160
GPT16	0085
GQ5Y	OEM
GR10-602GB	OEM
GR16	0071
GR17	OEM
GR18	OEM
GR19	OEM
GR20	OEM
GR21	OEM
GR22	0087
GR23	0087
GR24	0087
GR606	0711
GR612	0711
GR727	OEM
GR806	OEM
GR812	OEM
GRE5036	OEM
GRE5036-ZB	OEM
GRE7159	OEM
GRE7189	OEM
GREA4	OEM
GREA7	OEM
GREA7A	OEM
GREA10B	OEM
GREA14	OEM
GREA14A	OEM
GREA19B	OEM
GREA29A	OEM
GREA35	OEM
GREA38B	OEM
GREA70	OEM
GREA72A	OEM
GREA95B	OEM
GREA145A	OEM
GREA190B	OEM
GRL-LF2	OEM
GRU-2A	0102
GRU2A	0102
GS2C117	OEM
GS6.2	OEM
GS6.2A	OEM
GS13	OEM
GS23	OEM
GS43	OEM
GS63	OEM
GS73E/3	0143
GS83	OEM
GS100	OEM
GS100L	OEM
GS101A	OEM
GS102	OEM
GS103	OEM
GS120	OEM
GS135	OEM
GS150	OEM
GS160	OEM
GS161	OEM
GS162	OEM
GS163	OEM
GS164	OEM
GS165	OEM
GS170	OEM
GS172	OEM
GS300	OEM
GS301	OEM
GS302	OEM
GS370	OEM
GS372	OEM
GS400L	OEM
GS401L	OEM
GS402L	OEM
GS403L	OEM
GS404L	OEM
GS405L	OEM
GS420L	OEM
GS450L	OEM
GS463L	OEM
GS468L	OEM
GS469L	OEM
GS600L	OEM
GS601	OEM
GS611	OEM
GS614	OEM
GS660	OEM
GS1020-1	OEM
GS1020-2	OEM
GS1020-3	OEM
GS1020-4	OEM
GS1030-1	OEM
GS1030-2	OEM
GS1030-3	OEM
GS1140-1	OEM
GS1140-2	OEM
GS1140-3	OEM
GS2012	0042
GS2012(G)	0042
GS2012(H)	0042
GS2012G	0042
GS2012H	0042
GS2013	2520
GS2013(G)	0042
GS2013(H)	0042
GS2013-G	0042
GS2013-H	0042
GS2013-I	0042
GS2013-O	0042
GS2013-Y	0042
GS2013G	0461
GS2013H	0461
GS2014	0168
GS2014(G)	0168
GS2014-G	0168
GS2014-H	0168
GS2014-I	0168
GS2014G	0168
GS2014H	0168
GS2014I	0168
GS2015	2146
GS2015(D)	0392
GS2015(E)	0392
GS2015D	0392
GS2015E	0392
GS2015H	0392
GS2016(H)	0392
GS2016-H	0392
GS2016H	0392
GS2017	1085
GS2017(G)	1085
GS2017(H)	1085
GS2017-G	1085
GS2017-H	1085
GS2017G	1085
GS2017H	0261
GS2020-1	OEM
GS2020-2	OEM
GS2023-G	OEM
GS2052(G)	3752
GS2052(H)	0925
GS2052D	0925
GS2052G	0925
GS2052H	0925
GS2053-G	0925
GS2053-H	0925
GS2053-I	0718
GS2053G	3752
GS2053H	3752
GS3020-1	OEM
GS3020-2	OEM
GS3020-3	OEM
GS3020-4	OEM
GS3030-1	OEM
GS3030-2	OEM
GS3030-3	OEM
GS3040-1	OEM
GS3040-2	OEM
GS3040-3	OEM
GS3040-4	OEM
GS4021-1	OEM
GS4021-2	OEM
GS4021-3	OEM
GS4021-4	OEM
GS4123-1	OEM
GS4123-2	OEM
GS4123-3	OEM
GS4123-4	OEM
GS5040-1	OEM
GS5040-2	OEM
GS5040-3	OEM
GS5628	0373
GS7812	OEM
GS8008-09	OEM
GS8008-09A	OEM
GS8008-09B	OEM
GS8038-06A	OEM
GS8108-01A	OEM
GS8208-10B	OEM
GS8208-15C0	OEM
GS8308-16B	OEM
GS8408-19A	OEM
GS8908-01A	OEM
GS8908-02A	OEM
GS8908-02B	OEM
GS8908-02C	OEM
GS8908-02D	OEM
GS8908-04D	OEM
GS8908-041A	OEM

If replacement code is OEM, contact original manufacturer for replacement.

DEVICE TYPE	REPL CODE
GS9011(I)	0079
GS9011(J)	0079
GS9011-F	0079
GS9011-G	0079
GS9011-H	0079
GS9011-I	0079
GS9011F	0079
GS9011G	1954
GS9011H	0079
GS9011I	0079
GS9011J	0079
GS9012-G	0037
GS9012-H	0037
GS9012-I	0037
GS9012G	0037
GS9012H	0037
GS9012I	0037
GS9012J	0037
GS9012N	OEM
GS9014	0079
GS9014-G	0168
GS9014-H	0079
GS9014-I	0079
GS9014-J	0079
GS9014-K	0079
GS9014H	0286
GS9014I	0079
GS9014J	0079
GS9014K	0079
GS9015-H	0037
GS9015-I	0037
GS9015-J	0037
GS9015H	0037
GS9015I	0037
GS9015J	0037
GS9018-G	0079
GS9018-H	0079
GS9018-I	0079
GS9018F	0016
GS9018G	0079
GS9019	6199
GS9019-G	0715
GS9019-H	0286
GS9019-I	0286
GS9019-J	0286
GS9019G	2012
GS9019H	0076
GS9019I	0715
GS9022	0037
GS9022(G)	0342
GS9022(H)	0342
GS9022(I)	0342
GS9022-G	0286
GS9022-H	0286
GS9022F	0037
GS9022G	0342
GS9022H	0342
GS9022I	0342
GS9022J	0342
GS9022P	0342
GS9022Q	0342
GS9023	0079
GS9023(H)	0079
GS9023-1	0079
GS9023-G	0079
GS9023-H	0079
GS9023-I	0079
GS9023G	0079
GS9023H	0079
GS9023I	0079
GS9023J	0079
GS9023K	0016
GS9024	0286
GS9024(J)	0286
GS9024(K)	0286
GS9024-I	0286
GS9024J	0286
GS9024K	0286
GS9025(J)	0079
GS9025H	0079
GS9025I	0079
GS9025J	0079
GS9025J-1	0286
GS9027	2335
GS9027(G)	0261
GS9027(H)	0261
GS9027G	0261
GS9027H	0261
GS9032H	0342
GS9032I	0342
GS9032J	0342
GS9033	0079
GS9033(I)	0079
GS9033(J)	0079
GS9033H	0715
GS9033I	0715
GS9033J	0715
GS9038H	OEM
GS9038I	OEM
GS9038J	OEM
GS578030	OEM
GS830805C	OEM
GSA15B	0071
GSA15C	0071
GSA15E	0071
GSA15G	0071
GSA30B	0087
GSA30C	0087
GSA30E	0087
GSA30G	0087
GSA30J	0087
GSB1A	OEM
GSB1B	OEM
GSB1C	OEM
GSB2	OEM
GSB3A	OEM
GSB3B	OEM
GSB3C	OEM
GSB10B	0071
GSB10C	0071
GSB10E	0071
GSB10G	0023
GSB10J	0071
GSB10L	0071
GSB100	OEM
GSB10142	OEM
GSB10231D	OEM
GSB10231P	OEM
GSBX011	OEM
GSC2052G	0925
GSC2052H	0925
GSDB10008	OEM
GSDS50018	OEM
GSDS50020	OEM
GSDU7530	3388
GSDU7535	3388
GSDU7540	3388
GSH047B	OEM
GSH051A/AN3231K	OEM
GSH3007B	OEM
GSH3009	OEM
GSH3013	OEM
GSH3015	OEM
GSH6026B	OEM
GSH6234	OEM
GSH6235	OEM
GSH6235C	OEM
GSH7323A	OEM
GSH8701	OEM
GSH8702	OEM
GSH8703	OEM
GSH8705	OEM
GSH9025	0079
GS-I9011	0079
GSM51	1335
GSM52	1335
GSM53	0015
GSM54	0015
GSM56	OEM
GSM57	OEM
GSM482	0015
GSM483	0015
GSM914	OEM
GSR1	0071
GSR2	OEM
GSR3	OEM
GSR4	OEM
GSR5	OEM
GSR6	OEM
GSR7	OEM
GSR8	OEM
GSR9	OEM
GSR10	OEM
GSR11	OEM
GSR12	OEM
GSR13	OEM
GSR14	OEM
GSR15	OEM
GSRU15030	OEM
GSRU15035	OEM
GSRU15040	OEM
GSRU20030	OEM
GSRU20035	OEM
GSRU20040	OEM
GSTU4030	1341
GSTU4035	1341
GSTU4040	1341
GSTU6030	1341
GSTU6035	1341
GSTU6040	1341
GSTU8035	4386
GSTU8040	4386
GSTU10030	4386
GSTU10035	4386
GSTU10040	4386
GSTU15018	4386
GSTU15020	4386
GSTU30018	OEM
GSTU30020	OEM
GSV101	OEM
GSV102	OEM
GSV103	OEM
GSZ2.7	0760
GSZ2.7A	0760
GSZ3.0	0118
GSZ3.0A	0118
GSZ3.3	0299
GSZ3.3A	0299
GSZ3.6	0376
GSZ3.6A	0376
GSZ3.9	0036
GSZ3.9A	0036
GSZ4.3	0275
GSZ4.3A	0275
GSZ4.7	0140
GSZ4.7A	0140
GSZ5.1	0041
GSZ5.1A	0041
GSZ5.6	0253
GSZ5.6A	0253
GSZ6.2	0471
GSZ6.2A	0471
GSZ6.8	0062
GSZ6.8A	0062
GSZ7.5	0077
GSZ7.5A	0077
GSZ8.2	0165
GSZ8.2A	0165
GSZ9.1	0057
GSZ9.1A	0057
GSZ10	0064
GSZ10A	0064
GSZ11	0181
GSZ11A	0181
GSZ12	0052
GSZ12A	0052
GSZ13	0053
GSZ13A	0053
GSZ15	0686
GSZ15A	0686
GSZ16	0445
GSZ16A	0445
GSZ18	0375
GSZ18A	0375
GT06	0593
GT08	OEM
GT015	OEM
GT-1	OEM
GT1	0050
GT2	0050
GT3	0050
GT11	0280
GT12	0280
GT13	0280
GT-14	0280
GT14	0004
GT-14H	0280
GT14H	0004
GT15T5	0002
GT16	0772
GT18	0004
GT-20	0280
GT20	0211
GT-20H	0280
GT20H	0004
GT20R	0004
GT24H	0050
GT26	0772
GT28	OEM
GT31	0004
GT32	0004
GT33	0004
GT34	0211
GT-34HV	0136
GT34HV	0004
GT-34S	0211
GT34S	0280
GT-35	2730
GT35	0208
GT36	0772
GT38	OEM
GT40	0004
GT41	0004
GT42	0004
GT43	0004
GT44	0004
GT45	0004
GT46	0004
GT47	0004
GT48	OEM
GT56	0617
GT58	OEM
GT-66	0136
GT66	0136
GT68	OEM
GT74	0211
GT75	0211
GT81	0211
GT-81H	0280
GT81H	0280
GT-81HS	0280
GT81HS	0004
GT81R	0004
GT82	0211
GT-83	0280
GT83	0280
GT-87	0280
GT87	0280
GT-88	0280
GT88	0280
GT100	0280
GT-109	0211
GT109	0004
GT109R	0004
GT115	OEM
GT115A	OEM
GT115B	OEM
GT115D	OEM
GT115G	OEM
GT115V	OEM
GT122	0211
GT122A	OEM
GT122B	OEM
GT122G	OEM
GT122V	OEM
GT123	0211
GT124A	OEM
GT124B	OEM
GT124V	OEM
GT125A	OEM
GT125B	OEM
GT125D	OEM
GT125E	OEM
GT125G	OEM
GT125I	OEM
GT125K	OEM
GT125L	OEM
GT125SZ	OEM
GT125V	OEM
GT132	0004
GT-153	0280
GT153	0280
GT167	0600
GT200	OEM
GT-210H	0280
GT210H	0280
GT215	OEM
GT222	0211
GT229	0600
GT-269	0211
GT269	0280
GT315	OEM
GT322M4	OEM
GT322M5	OEM
GT322M6	OEM
GT322M7	OEM
GT328A	0050
GT328B	0050
GT336	0208
GT346A	0050
GT364	0038
GT365	0038
GT366	0038
GT402D	OEM
GT402E	OEM
GT402I	OEM
GT402SZ	OEM
GT404D	OEM
GT404E	OEM
GT404I	OEM
GT404SZ	OEM
GT406A	OEM
GT415	OEM
GT515	OEM
GT600	OEM
GT705A	OEM
GT705B	OEM
GT705D	OEM
GT705G	OEM
GT705V	OEM
GT751	0004
GT758	0211
GT-759	0280
GT759	0211
GT-759R	0211
GT759R	0004
GT-760	0004
GT760	0280
GT-760R	0211
GT760R	0004
GT-761	0004
GT761	0280
GT-761R	0211
GT761R	0280
GT-762	0004
GT762	0280
GT-762R	0211
GT762R	0280
GT-763	0004
GT763	0004
GT-764	0280
GT764	0280
GT766	0280
GT766A	0280
GT792	0600
GT792R	OEM
GT811	OEM
GT812	OEM
GT832	0280
GT-903	0038
GT903	0208
GT904	0600
GT-905	0038
GT905	0600
GT905R	0600
GT906AM	OEM
GT-947	0038
GT947	0600
GT948	0600
GT948R	0038
GT949	0038
GT949R	0600
GT1079	0600
GT-1200	0600
GT1200	0600
GT1201	0600
GT1202	0600
GT1223	0004
GT1604	0280
GT1605	0280
GT1606	0280
GT1607	0280
GT1608	0600
GT1609	0600
GT1624	OEM
GT1644	0037
GT1658	0208
GT1665	0004
GT2693	0211
GT2694	0280
GT2695	0211
GT2696	0211
GT2765	0600
GT2766	0600
GT2767	0600
GT2768	0208
GT2883	0211
GT2884	0600
GT2885	0211
GT2886	0600
GT2887	0211
GT2888	0600
GT2906	0600
GT3150	0600
GT5116	0136
GT5117	0136
GT5148	0136
GT5149	0136
GT5151	0136
GT5153	0280
GT8000	2626
GT8001	2626
GT8002	2626
GT8003	1991
GT8004	1991
GT8005	1991
GT8100	2432
GT8101	2432
GT8102	2432
GT8103	2432
GTA1	OEM
GTA2	OEM
GTA3	OEM
GTE1	0280
GTE-2	0211
GTE2	0280
GTJ33141	0136
GTJ33229	0136
GTJ33230	0136
GTJ33231	0004
GTJ33232	0004
GTJ33350	0917
GTL1	OEM
GTL3	OEM
GTSMPA	OEM
GTSMPB	OEM
GTSMPC	OEM
GTSMPD	OEM
GTSMPE	OEM
GTV	0280
GTX-2001	0160
GTX2001	0160
GU-1	0023
GU-1A	0023
GU1A100SB	OEM
GU-1B	0102
GU-1C	0102
GU-1E	0102
GU-1F	0182
GU-1Z	0023
GU-3	0023
GU3	0023
GU-3A	0023
GU3A	0023
GU-3B	0102
GU3B	0102
GU-3C	0023
GU3C	0023
GU-3SY	0031
GU3SY	OEM
GU-3SZ	0242
GU3SZ	0242
GU-3Z	0023
GV-3SY	0031
GV5760	0133
GV6063	0016
GW1	OEM
GW2	OEM
GW3	OEM
GW5	OEM
GW10	OEM
GW11	OEM
GXB10100	OEM
GXB10100D	OEM
GXB10100P	OEM
GXB10101	OEM
GXB10101D	OEM
GXB10101P	OEM
GXB10102	OEM
GXB10102D	OEM
GXB10102P	OEM
GXB10103	OEM
GXB10104	OEM
GXB10104D	OEM
GXB10104P	OEM
GXB10105	OEM
GXB10105D	OEM
GXB10105P	OEM
GXB10106	OEM
GXB10106D	OEM
GXB10106P	OEM
GXB10107	OEM
GXB10107D	OEM
GXB10107P	OEM
GXB10108	OEM
GXB10109	OEM
GXB10109D	OEM
GXB10109P	OEM
GXB10110	OEM
GXB10110D	OEM
GXB10110P	OEM
GXB10111	OEM
GXB10111D	OEM
GXB10111P	OEM
GXB10112	OEM
GXB10113	OEM
GXB10113D	OEM
GXB10113P	OEM
GXB10117	OEM
GXB10117D	OEM
GXB10117P	OEM
GXB10118	OEM
GXB10118D	OEM
GXB10118P	OEM
GXB10119	OEM
GXB10119D	OEM
GXB10119P	OEM
GXB10121	OEM
GXB10121D	OEM
GXB10121P	OEM
GXB10123D	OEM
GXB10123P	OEM
GXB10124	OEM
GXB10124P	OEM
GXB10125P	OEM
GXB10129P	OEM
GXB10130	OEM
GXB10130D	OEM
GXB10130P	OEM
GXB10131	OEM
GXB10131D	OEM
GXB10131P	OEM
GXB10133	OEM
GXB10135	OEM
GXB10136	OEM
GXB10136D	OEM
GXB10136P	OEM
GXB10137	OEM
GXB10139	OEM
GXB10140	OEM
GXB10140D	OEM
GXB10140P	OEM
GXB10141	OEM
GXB10141D	OEM
GXB10141P	OEM
GXB10142D	OEM
GXB10142P	OEM
GXB10144	OEM
GXB10144F	OEM
GXB10145	OEM
GXB10145E	OEM
GXB10148	OEM
GXB10148D	OEM
GXB10148P	OEM
GXB10149	OEM
GXB10149D	OEM
GXB10149F	OEM
GXB10155	OEM
GXB10155F	OEM
GXB10155N	OEM
GXB10160	OEM
GXB10160D	OEM
GXB10160P	OEM
GXB10161	OEM
GXB10161P	OEM
GXB10162	OEM
GXB10162P	OEM
GXB10165	OEM
GXB10170	OEM
GXB10171P	OEM
GXB10172	OEM
GXB10172D	OEM
GXB10175	OEM
GXB10175D	OEM
GXB10175P	OEM
GXB10176	OEM
GXB10176D	OEM
GXB10176P	OEM
GXB10179	OEM
GXB10179D	OEM
GXB10179P	OEM
GXB10180	OEM
GXB10180D	OEM
GXB10180P	OEM
GXB10181	OEM
GXB10181D	OEM
GXB10181P	OEM
GXB10191P	OEM
GXB10210	OEM
GXB10210D	OEM
GXB10210P	OEM
GXB10211	OEM
GXB10211D	OEM
GXB10211P	OEM
GXB10231	OEM
GXB10405	OEM
GXB10405D	OEM
GXB10415	OEM
GXB10415A	OEM
GXB10415AD	OEM
GXB10415AE	OEM
GXB10415ALD	OEM
GXB10415BD	OEM
GXB10415BLD	OEM
GXB10415D	OEM
GXB10415E	OEM
GXB10415LD	OEM
GXB10422AD	OEM
GXB10422AF	OEM
GXB10422BD	OEM
GXB10422D	OEM
GXB100415	OEM
GXB100422AF	OEM
GXB100473	OEM
GXI3.3G	0292
GXI3.6	0188
GXI3.9	0456
GXI4.3	0533
GXI4.7	0451
GXI5.1	0162
GXI5.6	0157
GXI6.2	0636
GXI6.8	0025
GXI7.5	0649
GXI8.2	0244
GXI9.1	0012
GXI10	0170
GXI11	0317
GXI12G	0137
GXI14	0100
GXI15	0002
GXI20G	0531
GXI22	0565
GXI24	0402
GXI26	0441
GXI27	0441
GXI29	0726
GXI33G	0039
GXI36	0820
GXI39	0350
GXI43	0931
GXI47	0999
GXI51	0502
GXI55	1834
GXI56	0869
GXI62L	0783
GXI68	1268
GXI82	0331
GXI91	1311
GXI111	0149
GXI115	0794
GXI120	0186
GXI128	0365
GXI130	0502
GXI140	0245
GXI150G	0028
GXI160	0255
GXI161	0420
GXI170	0877
GXI178	0495
GXI180	0367
GXI190	2858
GXI200	0421
GXI200G	0421
GYMTZ9.1-C	0057
GYW76	0023
GZ1-3.3G	0292
GZ1-3.6	0292
GZ1-3.9	0456
GZ1-4.3	0533
GZ1-4.7	0451
GZ1-5.1	0162
GZ1-5.6	0157
GZ1-6.2	0636
GZ1-6.8	0025
GZ1-7.5	0649
GZ1-8.2	0244
GZ1-9.1	0012
GZ1-10	0170
GZ1-11	0317
GZ1-12G	0137
GZ1-14	0100
GZ1-15	0002
GZ1-20G	0531
GZ1-22	0565
GZ1-24	0402
GZ1-26	0441
GZ1-27	0441
GZ1-29	0726
GZ1-33G	0039
GZ1-36	0820
GZ1-39	0350
GZ1-43	0931
GZ1-47	0999
GZ1-51	0502
GZ1-55	1834
GZ1-56	0869
GZ1-62P	0783
GZ1-68	1268
GZ1-91	1311
GZ1-110A	0149
GZ1-115G	0794
GZ1-120	0186
GZ1-127	0365
GZ1-130G	0213
GZ1-140	0245
GZ1-150G	0028
GZ1-160	0255
GZ1-161	0420
GZ1-170	0877
GZ1-178	0495
GZ1-180	0367
GZ1-190	2858

DEVICE TYPE	REPL CODE
GZ1-200	0421
GZ1-200G	0421
GZ5.1	2407
GZ6A	OEM
GZ6.2	0636
GZ8A	OEM
GZ9.1	1618
GZ10A	OEM
GZ10B	OEM
GZ12A	OEM
GZ12B	OEM
GZ15A	OEM
GZ15B	OEM
GZ18A	OEM
GZ18B	OEM
GZ22A	OEM
GZ27	0441
GZ27A	OEM
GZ33A	6136
GZ39A	OEM
GZ47A	OEM
GZ56A	OEM
GZ68A	OEM
GZ82A	OEM
GZA056U	0253
GZA2.0	0375
GZA2.0X	OEM
GZA2.2	OEM
GZA2.2X	OEM
GZA2.2Z	OEM
GZA2.4	1276
GZA2.4X	OEM
GZA2.4Z	1276
GZA2.7	0760
GZA2.7X	0760
GZA2.7Y	OEM
GZA2.7Z	OEM
GZA3.0	0118
GZA3.0Y	OEM
GZA3.0Z	OEM
GZA3.3	0299
GZA3.3-Y-BT	OEM
GZA3.3X	0299
GZA3.3Y	OEM
GZA3.3Z	OEM
GZA3.6	0376
GZA3.6A1	OEM
GZA3.6X	OEM
GZA3.6Z	OEM
GZA3.9	0036
GZA3.9(Y)	0036
GZA3.9-X	0036
GZA3.9X	0036
GZA3.9Y	OEM
GZA3.9Z	OEM
GZA4.3	0275
GZA4.3(Y)	0275
GZA4.3-Y	0275
GZA4.3-Y-BT	0533
GZA4.3Y	0275
GZA4.3Z	OEM
GZA4.7	0140
GZA4.7-X-BT	0140
GZA4.7X	0140
GZA4.7Y	0140
GZA4.7Z	0041
GZA5.1	0253
GZA5.1(X)	0041
GZA5.1(Y)	0041
GZA5.1(Z)	0041
GZA5.1-BT	0041
GZA5.1-X-BT	0041
GZA5.1-Y	0589
GZA5.1-Y-BT	0041
GZA5.1-Y-BT-T	0041
GZA5.1L	0041
GZA5.1U	5953
GZA5.1X	0041
GZA5.1Y	0041
GZA5.1Y-AT	OEM
GZA5.1Z	5953
GZA5.6	0253
GZA5.6-Y	0253
GZA5.6-Y-BT	0253
GZA5.6L	0883
GZA5.6U	5445
GZA5.6X	OEM
GZA5.6Y	0253
GZA5.6Z	OEM
GZA6.2	0471
GZA6.2(X)	0471
GZA6.2(XYZ)	0091
GZA6.2(Y)	0471
GZA6.2-BT	0471
GZA6.2-X	0471
GZA6.2-Y	0471
GZA6.2-Y-BT	0471
GZA6.2-Z-BT	0295
GZA6.2BT	0471
GZA6.2L	0091
GZA6.2U	0471
GZA6.2X	OEM
GZA6.2Y	0471
GZA6.2Y-BT	0471
GZA6.2Z	0295
GZA6.2Z-AT	OEM
GZA6.8	0062
GZA6.8-Y	0062
GZA6.8-Y-BT	0062
GZA6.8L	3369
GZA6.8X	OEM
GZA6.8Y	0062
GZA6.8Z	0062
GZA7.5	0077
GZA7.5(Y)	0077
GZA7.5-Y	0077
GZA7.5-Z-BT	0077
GZA7.5V	0077
GZA7.5X	6377
GZA7.5Y	0077
GZA7.5Z	0077
GZA8.2	0165
GZA8.2(Y)	0165
GZA8.2-X-BT	0165
GZA8.2-Y	0165
GZA8.2X	OEM
GZA8.2Y	4309
GZA8.2Z	OEM
GZA9.1	0057
GZA9.1-Y-BT	OEM
GZA9.1-Z	0057
GZA9.1E	0057
GZA9.1L	6564
GZA9.1X	OEM
GZA9.1Y	OEM
GZA9.1Z	0057
GZA10	0064
GZA10(Y)	0064
GZA10-(Y)	0064
GZA10-X-BT	OEM
GZA10-Y	0064
GZA10-Z	0064
GZA10L	1258
GZA10X	OEM
GZA10Y	0064
GZA10Z	1258
GZA11	0181
GZA11-Y-BT	0181
GZA11BT	0181
GZA11X	OEM
GZA11Y	0181
GZA11YBT	0181
GZA11Z	0181
GZA12	0052
GZA12-Y-BT	1005
GZA12BT	0052
GZA12X	0052
GZA12Y	1005
GZA12YBT	0052
GZA12Z	OEM
GZA13	0053
GZA13(X)	0053
GZA13-A-BT	0053
GZA13-X-BT	0053
GZA13-Y	0053
GZA13-Y-BT	0053
GZA13U	0053
GZA13X	0053
GZA13Y	0053
GZA13Z	0053
GZA15	0686
GZA15U	1105
GZA15Y	OEM
GZA15Z	OEM
GZA16	0445
GZA16-Y	0445
GZA16Y	0445
GZA16Z	OEM
GZA18	0375
GZA18-Y-BT	0375
GZA18U	4848
GZA18X	OEM
GZA18Y	0375
GZA18Y-BT	0375
GZA18Z	OEM
GZA20	0700
GZA20-X-BT	0700
GZA20X	0700
GZA20Y	OEM
GZA22	0705
GZA22(Z)	0705
GZA24	0494
GZA24-Y	0494
GZA24Z	OEM
GZA27	0455
GZA-27-0Z	0455
GZA27-Y	0455
GZA27-Y-BT	0455
GZA27BT	0455
GZA-27X	0455
GZA27X	0455
GZA27Y	0455
GZA27YBT	0455
GZA27Z	OEM
GZA30	0195
GZA30-Y	0195
GZA30-Y-BT	0195
GZA30BT	0195
GZA30Y	0195
GZA30Z	OEM
GZA33	0166
GZA33X	OEM
GZA33Y	0166
GZA36	0010
GZA36(Y)	0010
GZA36-Y	0010
GZA36Y	0010
GZA39	0032
GZA39-Y	0032
GZA43	0054
GZA43X	OEM
GZA47	0068
GZA51	0092
GZB2.0	OEM
GZB2.2	OEM
GZB2.4	OEM
GZB2.7	1312
GZB3.0	1713
GZB3.3	0292
GZB3.6	0188
GZB3.6B	0376
GZB3.9	0456
GZB4.3	0533
GZB4.7	0451
GZB5.1	0162
GZB5.6	0157
GZB6.2	0636
GZB6.8	0025
GZB7.5	0649
GZB7.5C	0649
GZB8.2	0244
GZB9.1	0012
GZB9.1B	0012
GZB9.1C	0012
GZB10	0170
GZB10C	0170
GZB11	0317
GZB11C	0317
GZB12	0137
GZB12B	0137
GZB12C	0137
GZB13	0365
GZB13C	0365
GZB15	0002
GZB15C	0002
GZB16	0420
GZB16C	0420
GZB18	0495
GZB18B	0495
GZB18C	0495
GZB20	0531
GZB20C	0531
GZB22	0565
GZB22C	0565
GZB24	0402
GZB24C	0402
GZB27	1674
GZB27C	1674
GZB30	0726
GZB30B	0726
GZB30C	0726
GZB33	0039
GZB36	0820
GZF1100D	OEM
GZF1100P	OEM
GZF1106D	OEM
GZF1106P	OEM
GZF1201P	OEM
GZF1202P	OEM
GZF1400D	OEM
GZF1400P	OEM
GZM5.6	0253
GZM6.2	0471
GZM6.2(B2)	0471
GZM6.2B2	OEM
GZM6.8	0062
GZM6.8-B2	OEM
GZM6.8B2	OEM
GZM7.5	0077
GZM7.5(B2)	0077
GZM7.5-B2	0077
GZM7.5B2	OEM
GZM7.5C	OEM
GZM8.2	0165
GZM8.2(B2)	0165
GZM9.1	0057
GZM9.1(B2)	0057
GZM9.1B2	OEM
GZM10	0064
GZM11	0181
GZM12	0052
GZM13	0053
GZM15	0686
GZM16	0445
GZM18	0375
GZM20	0700
GZM20(B2)	0700
GZM20-B2	0700
GZM20B2	OEM
GZM22	0705
GZM24	0494
GZM24(B2)	0494
GZM24-B2	0494
GZS2.0	OEM
GZS2.0X-BT	OEM
GZS2.2	OEM
GZS2.2X	OEM
GZS2.2Z	OEM
GZS2.4	1276
GZS2.7	0760
GZS3.0	0118
GZS3.3	0299
GZS3.6	0376
GZS3.9	0036
GZS4.3	0275
GZS4.3Z	0275
GZS4.7	0140
GZS5.1	0140
GZS5.1Y	6839
GZS5.6	0253
GZS5.6Y	OEM
GZS5.6Z	OEM
GZS6.2	0471
GZS6.2X	OEM
GZS6.2Y	OEM
GZS6.2Z	OEM
GZS6.8	0062
GZS6.8Y	0062
GZS7.5	0077
GZS7.5Z	OEM
GZS8.2	0165
GZS8.2Y	OEM
GZS9.1	0057
GZS9.1BT	OEM
GZS9.1X	OEM
GZS9.1Y	0057
GZS9.1Z	OEM
GZS10	0064
GZS10X	OEM
GZS10Y	OEM
GZS11	0181
GZS11Y	OEM
GZS12	0052
GZS12Y	OEM
GZS13	0053
GZS15	0686
GZS16	0445
GZS18	0375
GZS20	0700
GZS20Z	OEM
GZS22	0705
GZS24	0494
GZS24Z	0494
GZS27	0455
GZS27R	OEM
GZS27R-BT	OEM
GZS30	0195
GZS30R	OEM
GZS30Y-BT	OEM
GZS33	0166
GZS36	0010
GZS39	0032
GZS39X	OEM
GZS39Y	OEM
GZS39Z	OEM
H*22A	0705
H*51A	0092
H(P-Z0424	0441
H0004	OEM
H0005	OEM
H01G	OEM
H01J	OEM
H01L	OEM
H01M	OEM
H03CF	OEM
H03DA	OEM
H03EA	OEM
H07	OEM
H078	1303
H087	0143
H089	0137
H091	0143
H0R3D	OEM
H1V	0144
H1Z	0023
H2P15	0050
H2SE303	OEM
H3A	OEM
H3D42	OEM
H3MLR	OEM
H5A3	1023
H5B2N3	OEM
H5B3	1023
H5C3	1023
H5D3	1023
H5E3	1023
H5F3	1023
H5G3	1023
H5H3	1036
H5K3	1036
H5M3	1046
H5N3	1046
H6A3	OEM
H6B3	OEM
H6C3	OEM
H6D3	OEM
H6E3	OEM
H6F3	OEM
H6G3	OEM
H6H3	OEM
H6K3	OEM
H6M3	OEM
H6N3	OEM
H6P3	OEM
H6R3	OEM
H6S3	OEM
H8D1133	OEM
H8D1134	OEM
H8D1147	OEM
H8D1177	OEM
H8D1193	OEM
H10	0004
H11A1	0315
H11A2	0273
H11A3	0273
H11A4	0315
H11A5	0315
H11A10	OEM
H11A520	0315
H11A550	0315
H11A5100	0315
H11AA1	5425
H11AA2	5425
H11AA3	OEM
H11AA4	OEM
H11AV1	OEM
H11AV1A	OEM
H11AV2	OEM
H11AV3	OEM
H11AX	2108
H11B1	1053
H11B2	1053
H11B3	1053
H11B255	1109
H11B256	OEM
H11BX522	1053
H11C1	2236
H11C2	2236
H11C3	2236
H11C4	2236
H11C5	2236
H11C6	2236
H11D1	5424
H11D2	5424
H11D3	5424
H11D4	5424
H11F1	5421
H11F2	5421
H11F3	5421
H11G1	OEM
H11G2	OEM
H11G3	OEM
H11J1	5051
H11J2	5051
H11J3	5051
H11J4	5051
H11J5	5051
H11L1	5426
H11L2	5426
H12	0440
H12A	0440
H15B	OEM
H15D	OEM
H15M	OEM
H15N	OEM
H18CPU	OEM
H21A1	1417
H21A2	1417
H21A3	1417
H21A4	1417
H21A5	1417
H21A6	1417
H21B1	2636
H21B2	2636
H21B3	2636
H21B4	2636
H21B5	2636
H21B6	2636
H22A1	2640
H22A2	2640
H22A3	2640
H22A4	2640
H22A5	2640
H22A6	2640
H22B1	5433
H22B2	5433
H22B3	5433
H22B4	5433
H22B5	5433
H22B6	5433
H23A1	OEM
H23B1	OEM
H24A1	3367
H24A2	3367
H24B1	5142
H24B2	5142
H36A2U57	OEM
H36A2U57D	OEM
H50	0015
H62SC01-1	OEM
H62SC03-1	OEM
H62SC06-1	OEM
H62SC09-1	OEM
H62SC12-1	OEM
H62SC15-1	OEM
H62SC18-1	OEM
H68AN01-1	OEM
H68AT01-1	OEM
H68CC01-1	OEM
H68CC02-1	OEM
H68CM08-1	OEM
H68DB03-1	OEM
H68DM48-1	OEM
H68DN01-1	OEM
H68DT01-1	OEM
H68DX01-1	OEM
H68FD01-1(A)	OEM
H68FD03-1	OEM
H68FD04-1(A)	OEM
H68MV01-1	OEM
H680CC01-1	OEM
H680DM12	OEM
H680MN01	OEM
H68PM32-1	OEM
H68PR03-1	OEM
H68PW03-1	OEM
H68SB02-1	OEM
H68SB03-1	OEM
H68SM16-1	OEM
H68XM68-1	OEM
H69DB01-1	OEM
H69SB01-1	OEM
H70	OEM
H74A1	OEM
H74C1	OEM
H74C2	OEM
H74SC137C(A)	OEM
H74SC137D(A)	OEM
H74SC137P(A)	OEM
H74SC138C(A)	OEM
H74SC138D(A)	OEM
H74SC138P(A)	OEM
H74SC139C(A)	OEM
H74SC139D(A)	OEM
H74SC139P(A)	OEM
H74SC237C(A)	OEM
H74SC237D(A)	OEM
H74SC237P(A)	OEM
H74SC238C(A)	OEM
H74SC238D(A)	OEM
H74SC238P(A)	OEM
H74SC239C(A)	OEM
H74SC239D(A)	OEM
H74SC239P(A)	OEM
H74SC240C(MA)	OEM
H74SC240D(MA)	OEM
H74SC240P(MA)	OEM
H74SC241C(MA)	OEM
H74SC241D(MA)	OEM
H74SC241P(MA)	OEM
H74SC242C(MA)	OEM
H74SC242D(MA)	OEM
H74SC242P(MA)	OEM
H74SC243C(MA)	OEM
H74SC243D(MA)	OEM
H74SC243P(MA)	OEM
H74SC244C(MA)	OEM
H74SC244D(MA)	OEM
H74SC244P(MA)	OEM
H74SC245C(MA)	OEM
H74SC245D(MA)	OEM
H74SC245P(MA)	OEM
H74SC540C(MA)	OEM
H74SC540D(MA)	OEM
H74SC540P(MA)	OEM
H74SC541C(MA)	OEM
H74SC541D(MA)	OEM
H74SC541P(MA)	OEM
H89/1	0774
H100	0015
H101	0196
H102	0016
H102D1	0078
H102D2	0078
H102D6	0078
H103A	0160
H103D1	4149
H103D2	4149
H103D6	4149
H103SH	2383
H104	0079
H104D1	0618
H104D2	0618
H104D6	0618
H105D1	0800
H105D2	0800
H105D6	0800
H109	0087
H109D1	OEM
H109D6	OEM
H110D1	OEM
H110D2	OEM
H110D6	OEM
H111D1	OEM
H111D2	OEM
H111D6	OEM
H112D1	3045
H112D6	3045
H113SH	2383
H114B	OEM
H114D	OEM
H114M	OEM
H115B	OEM
H115C	OEM
H115D	OEM
H115D1	4173
H115D6	4173
H117D1	OEM
H117D6	OEM
H118D1	4178
H118D6	4178
H119D1	4171
H119D6	4171
H121	OEM
H122D1	2434
H122D2	2434
H122D6	2434
H123SH	2394
H124D1	2597
H124D2	2597
H124D6	2597
H142	0124
H156D1	OEM
H156D2	OEM
H156D6	OEM
H157D1	OEM
H157D2	OEM
H157D6	OEM
H158D1	OEM
H158D2	OEM
H158D6	OEM
H159D1	OEM
H160D1	OEM
H165D1	OEM
H166D1	OEM
H167D1	OEM
H168D1	OEM
H200	0015
H200E	OEM
H200EA	OEM
H200EB	OEM
H202B1	0078
H203B1	0078
H204B1	0618
H205B1	0800
H209B1	OEM
H210B1	OEM
H211B1	OEM
H215	2743
H218-10	OEM
H221	0802
H222B1	2434
H224B1	2597
H227B1	OEM
H254B4	0015
H254E4	0015
H256B1	OEM
H257B1	OEM
H258B1	OEM
H284D4	0023
H300	0015
H-300B	OEM
H312K4	OEM
H315F4	0015
H315M2	0015
H315Y6	OEM
H316	0143
H326B4	0023
H386-9	0757
H386C-5	OEM
H400	0015
H410	0920
H431	1323
H438	OEM
H439	OEM
H440	OEM
H441	OEM
H442	0144
H443	OEM
H445	1323
H463	OEM
H466	OEM
H467	OEM
H468	OEM
H475	0015
H-484	1323
H484	1323
H484-1	1323
H485-62	0378
H486W	0378
H500	0015
H501	0196
H505W	0288
H521	0201
H521-35	0288
H521-55	0288
H540	OEM
H541	OEM
H542	OEM
H580B1A	OEM
H580B1B	OEM
H585	0015
H590B1A	OEM
H590B1B	OEM
H598	0288
H598-10	0288
H600	0015
H614	0143
H615	0196
H616	0015
H617	0015
H618	0015
H619	0015
H620	0015
H621	0071
H622	0774
H623	0133
H624	0143
H625	0015
H626	0015
H629B1	OEM
H632B1	OEM
H635B1	OEM
H635F1	OEM
H680BS01(A)	OEM
H680FD01(A)	OEM
H680SB01	OEM
H680TR01(A)	OEM
H701	OEM
H702	OEM
H703	OEM
H770B1	OEM
H771B1	OEM

DEVICE TYPE	REPL CODE
H772B1	OEM
H773B1	OEM
H781	0015
H783	0015
H800	0071
H815	0102
H816	0378
H850	OEM
H-881	0071
H881	0015
H889	0133
H890	0920
H891	0015
H909	6371
H931	0016
H932	0233
H933	0016
H934	0016
H1000	0071
H1017A	0540
H1144-1	OEM
H1152	OEM
H1153	1323
H-1567	0079
H1567	0016
H1701-9	0201
H1701-10	0201
H1701-11	0201
H1701-12	0201
H1701-15	OEM
H1703-13	0288
H1703-14	0288
H1703-15	0288
H1703-16	0378
H1802	5905
H1809	OEM
H2535-5	OEM
H2535-6	OEM
H2621	OEM
H2622	OEM
H4439A06	OEM
H5535-5	OEM
H5535-6	OEM
H5535-G	OEM
H5535-M	OEM
H5535-X	OEM
H5540-5	OEM
H5540-6	OEM
H5540-G	OEM
H5540-M	OEM
H5540-X	OEM
H5545-5	OEM
H5545-6	OEM
H5545-G	OEM
H5545-M	OEM
H5545-X	OEM
H5560-G	OEM
H5560-M	OEM
H5560-X	OEM
H6010	OEM
H6010C	OEM
H6020	OEM
H7126-3	0015
H7618	0551
H8287	0143
H8287-4	0143
H8513A	0143
H8513D	0143
H9275-A	OEM
H9423	0016
H9623	0016
H9625	0144
H9696	0016
H10174	0015
H20052	0015
H61220-C	OEM
H70420M	OEM
H120029A	OEM
H801193	OEM
H21100015	OEM
H21100016	OEM
H21100018	OEM
H21400014	0133
H21400015	0124
H21400018	0023
H21400026	0133
H21500029	0091
H21500086	0118
H23100024	0148
H23100045	0693
H23100051	0342
H23200019	0532
H25300013	OEM
H26200009	3133
H28900001	OEM
H31100125	OEM
H31100127	1014
H31100128	0626
H31100129	0001
H31100130	2504
HA00052	0280
HA00053	0280
HA-0054	0150
HA-00102	0136
HA-00354	0050
HA-00495	0037
HA-00496	0460
HA00562	0006
HA-00564	0693

DEVICE TYPE	REPL CODE
HA-00610	0006
HA-00634	0681
HA-00636	2021
HA-00643	1242
HA-00699	0681
HA-00733	0006
HA-1	0102
HA1-2600-2	OEM
HA1-2602-2	OEM
HA1-2605-5	OEM
HA1-2620-2	OEM
HA1-2622-2	OEM
HA1-2625-5	OEM
HA1-4741-2	OEM
HA1-4741-5	2823
HA1-4741-8	OEM
HA1-4741-B	OEM
HA2-2500-2	OEM
HA2-2502-2	OEM
HA2-2505-5	OEM
HA2-2510-2	OEM
HA2-2512-2	OEM
HA2-2515-5	OEM
HA2-2520-2	OEM
HA2-2522-2	OEM
HA2-2525-5	OEM
HA2-2600-2	OEM
HA2-2602-2	OEM
HA2-2605-5	OEM
HA2-2620-2	OEM
HA2-2622-2	OEM
HA2-2625-5	OEM
HA3-4741-5	2823
HA3-4741-5DB	OEM
HA6A1	OEM
HA9-2500-2	OEM
HA9-2502-2	OEM
HA9-2505-5	OEM
HA9-2510-2	OEM
HA9-2512-2	OEM
HA9-2515-5	OEM
HA9-2520-2	OEM
HA9-2522-2	OEM
HA9-2525-5	OEM
HA9-2600-2	OEM
HA9-2602-2	OEM
HA9-2605-5	OEM
HA9-2620-2	OEM
HA9-2622-2	OEM
HA9-2625-5	OEM
HA10	OEM
HA-11A	0181
HA-11B	0181
HA-11C	0181
HA-12	0280
HA12	0050
HA-15	0136
HA15	0280
HA20	OEM
HA25	OEM
HA30	0050
HA40	OEM
HA-49	0633
HA49	0280
HA50	0015
HA-52	0136
HA52	0136
HA-53	0633
HA53	0136
HA54	0004
HA56	0004
HA60K	OEM
HA60P	OEM
HA70	0050
HA75	OEM
HA100	0015
HA-101	0136
HA102	0136
HA103	0136
HA104	0136
HA120EL	OEM
HA120KL	OEM
HA120TL	OEM
HA178M05	0624
HA178M05P	0624
HA178M06P	0923
HA178M07P	OEM
HA178M08	1196
HA178M08P	1196
HA178M059A	OEM
HA178M12P	0334
HA178M15P	1321
HA178M18P	2259
HA178M20P	OEM
HA178M24P	2239
HA180EN	OEM
HA180KN	OEM
HA180LN	OEM
HA180LO	OEM
HA180TN	OEM
HA200	0015
HA-201	0211
HA201	0136
HA202	0280
HA-234	0050
HA234	0050
HA-234B	0050
HA235	0136
HA-235A	0050

DEVICE TYPE	REPL CODE
HA235A	0050
HA-235C	0050
HA240	0050
HA240E0	OEM
HA240L0	OEM
HA250T0	OEM
HA266	0050
HA267	0050
HA-268	0050
HA268	0050
HA-269	0136
HA269	0050
HA300	0015
HA300EP	OEM
HA300LP	OEM
HA300TP	OEM
HA321J	0050
HA-330	0211
HA330	0136
HA342	0136
HA-350	0136
HA350	0050
HA-350A	0136
HA-353	0136
HA353	0050
HA-353C	0136
HA-354	0136
HA354	0050
HA-354B	0136
HA400	0015
HA471	0136
HA500	0015
HA-505	1267
HA505	0925
HA525	0050
HA600	0015
HA800	0071
HA1000	0071
HA1040	0050
HA1075R	OEM
HA1077R	OEM
HA1105R	OEM
HA1106R	OEM
HA1107R	OEM
HA1108	2582
HA1108-0M	OEM
HA1108R	OEM
HA1108.IF	OEM
HA1108.0M	OEM
HA1111SW	0443
HA1112	4395
HA1113	4616
HA1115	0443
HA1115(W)	0443
HA1115W	0443
HA1117	0607
HA1118	0043
HA1119	0389
HA1124	0167
HA1124D	0167
HA1124DS	0167
HA1124S	0167
HA1124Z	0167
HA1125	0167
HA1125D	0167
HA1125S	0167
HA1126	0878
HA1126AW	0878
HA1126D	0878
HA1126DW	0878
HA-1128	OEM
HA1128	0167
HA1128E	0167
HA1133	2743
HA1137	2755
HA1137(W)	2755
HA1137P	2755
HA1137W	2755
HA1138	4810
HA1139A	4401
HA1141	0167
HA1141G	OEM
HA11410	OEM
HA1141R	OEM
HA1141Y	OEM
HA1142	OEM
HA1142A	OEM
HA1142G	OEM
HA11420	OEM
HA1142R	OEM
HA1142Y	OEM
HA1143G	OEM
HA11430	OEM
HA1143R	OEM
HA1143Y	OEM
HA1144	4627
HA1144G	OEM
HA11440	OEM
HA1144R	OEM
HA1144Y	OEM
HA1146	0878
HA1147	OEM
HA1148	4632
HA1150	0520
HA1151	3651
HA1152	0855
HA1154	0167
HA1155	OEM
HA1156	0520

DEVICE TYPE	REPL CODE
HA1156(W)	0520
HA1156-6C	0520
HA1156-612	OEM
HA1156H	0520
HA1156W	0520
HA1156WP	0520
HA1157	0607
HA1157P	OEM
HA1158	0043
HA1158-O	0043
HA1159	0389
HA1162Z	OEM
HA1166	OEM
HA1166A	OEM
HA1166X	OEM
HA1166Y	OEM
HA1166Z	OEM
HA1167	OEM
HA1174ANT	OEM
HA1177	1055
HA1178	OEM
HA1179	OEM
HA1180	OEM
HA1181G	OEM
HA11810	OEM
HA1181R	OEM
HA1181Y	OEM
HA1182G	OEM
HA11820	OEM
HA1182R	OEM
HA1182Y	OEM
HA1183G	OEM
HA11830	OEM
HA1183R	OEM
HA1183Y	OEM
HA1184G	OEM
HA11840	OEM
HA1184R	OEM
HA1184Y	OEM
HA1196	4799
HA1197	6387
HA1197W	0651
HA1199	4089
HA1199P	6685
HA-1201	OEM
HA1201	1621
HA-1202	4570
HA1202	4570
HA1203	4571
HA1211	4569
HA1215A	OEM
HA1215B	OEM
HA1215C	OEM
HA1216A	OEM
HA1216B	OEM
HA1216C	OEM
HA1217A	OEM
HA1217B	OEM
HA1217C	OEM
HA1218A	OEM
HA1218B	OEM
HA1218C	OEM
HA1219A	OEM
HA1219B	OEM
HA1219C	OEM
HA1220A	0554
HA1220B	OEM
HA1220C	OEM
HA1222A	OEM
HA1222B	OEM
HA1222C	OEM
HA1224A	0005
HA1224B	OEM
HA1224C	OEM
HA1226A	0720
HA1226B	OEM
HA1226C	OEM
HA1228A	OEM
HA1228B	OEM
HA1228C	OEM
HA1230A	OEM
HA1230B	OEM
HA1230C	OEM
HA1234A	0628
HA1234B	OEM
HA1234C	OEM
HA1236A	OEM
HA1236B	OEM
HA1236C	OEM
HA1238A	0030
HA1238B	OEM
HA1238C	OEM
HA1301	2537
HA1304	0410
HA-1306	OEM
HA1306	2262
HA-1306P	2262
HA1306P	2262
HA1306PU	2262
HA1306W	2262
HA1306WU	2262
HA1308	4561
HA-1310	OEM
HA1310	4562
HA-1311	OEM
HA1311	4563
HA1311W	4563
HA1312	4564
HA1313	2889

DEVICE TYPE	REPL CODE
HA1316	4566
HA1318PU	4409
HA1318PU-1	4409
HA1318PU-2	4409
HA1318PU-3	4409
HA1319	4390
HA1322	4093
HA1322C	4093
HA1325	4018
HA1338	4561
HA-1339	5598
HA1339	4401
HA1339A	4401
HA1342	3262
HA1342A	3262
HA1348	OEM
HA1350	0004
HA1353	0855
HA1360	0004
HA1364	1170
HA1366	4644
HA1366R	1128
HA1366W	4644
HA1366WR	1128
HA1374	6401
HA1374A	6401
HA1377	5110
HA1377A	5110
HA1385	OEM
HA1388	5111
HA1389R	4739
HA1392	5188
HA1393	5187
HA1394	OEM
HA1396	OEM
HA1397	OEM
HA1403	OEM
HA1406	2656
HA1406-2	2656
HA1406-3	2656
HA1406-4	2656
HA1452	4560
HA1452W	4560
HA1702A	OEM
HA1702B	OEM
HA1702D	OEM
HA1702E	OEM
HA1702H	OEM
HA1703A	OEM
HA1703B	OEM
HA1703D	OEM
HA1703E	OEM
HA1703H	OEM
HA1704A	OEM
HA1704B	OEM
HA1704D	OEM
HA1704E	OEM
HA1704H	OEM
HA1705A	OEM
HA1705B	OEM
HA1705D	OEM
HA1705E	OEM
HA1705H	OEM
HA1706A	OEM
HA1706B	OEM
HA1706D	OEM
HA1706E	OEM
HA1706H	OEM
HA1707A	OEM
HA1707B	OEM
HA1707D	OEM
HA1707E	OEM
HA1707H	OEM
HA1708A	OEM
HA1708B	OEM
HA1708D	OEM
HA1708E	OEM
HA1708G	OEM
HA1708H	OEM
HA1709A	OEM
HA1709B	OEM
HA1709D	OEM
HA1709E	OEM
HA1709H	OEM
HA1710A	OEM
HA1710B	OEM
HA1710D	OEM
HA1710H	OEM
HA1711A	OEM
HA1711B	OEM
HA1711D	OEM
HA1711E	OEM
HA1711H	OEM
HA1712A	OEM
HA1712B	OEM
HA1712D	OEM
HA1712E	OEM
HA1712H	OEM
HA1713A	OEM
HA1713B	OEM
HA1713D	OEM
HA1713E	OEM
HA1713H	OEM
HA1714A	OEM
HA1714B	OEM
HA1714D	OEM
HA1714E	OEM
HA1714H	OEM
HA1715A	OEM

DEVICE TYPE	REPL CODE
HA1715B	OEM
HA1715D	OEM
HA1715E	OEM
HA1715H	OEM
HA1716A	OEM
HA1716B	OEM
HA1716D	OEM
HA1716E	OEM
HA1716H	OEM
HA1717A	OEM
HA1717B	OEM
HA1717D	OEM
HA1717E	OEM
HA1717H	OEM
HA1807	OEM
HA1812GS	OEM
HA1812PS	OEM
HA1813PS	OEM
HA1822A	OEM
HA1822B	OEM
HA1822C	OEM
HA1824A	OEM
HA1824B	OEM
HA1824C	OEM
HA1826A	OEM
HA1826B	OEM
HA1826C	OEM
HA1836A	OEM
HA1836B	OEM
HA1836C	OEM
HA1838A	OEM
HA1838B	OEM
HA1838C	OEM
HA1844A	OEM
HA1844B	OEM
HA1844C	OEM
HA1848A	OEM
HA1848B	OEM
HA1848C	OEM
HA1850A	OEM
HA1850B	OEM
HA1850C	OEM
HA1915A	OEM
HA1915B	OEM
HA1915C	OEM
HA1916A	OEM
HA1916B	OEM
HA1916C	OEM
HA1917A	OEM
HA1917B	OEM
HA1917C	OEM
HA1918A	OEM
HA1918B	OEM
HA1918C	OEM
HA1919A	OEM
HA1919B	OEM
HA1919C	OEM
HA1920A	0554
HA1920B	OEM
HA1920C	OEM
HA1922A	OEM
HA1922B	OEM
HA1922C	OEM
HA1924A	0005
HA1924B	OEM
HA1924C	OEM
HA1926A	0720
HA1926B	OEM
HA1926C	OEM
HA1928A	OEM
HA1928B	OEM
HA1928C	OEM
HA1930A	OEM
HA1930B	OEM
HA1930C	OEM
HA1934A	0628
HA1934B	OEM
HA1934C	OEM
HA1936A	OEM
HA1936B	OEM
HA1936C	OEM
HA1938A	0030
HA1938B	OEM
HA1938C	OEM
HA1940A	OEM
HA1940B	OEM
HA1940C	OEM
HA1942A	OEM
HA1942B	OEM
HA1942C	OEM
HA1944A	OEM
HA1944B	OEM
HA1944C	OEM
HA1946A	OEM
HA1946B	OEM
HA1946C	OEM
HA1948A	OEM
HA1948B	OEM
HA1948C	OEM
HA1950A	OEM
HA1950B	OEM
HA1950C	OEM
HA2000	OEM
HA2001	0325
HA2010	0325
HA2020	OEM
HA2132R	OEM
HA2133R	OEM
HA2134R	OEM

DEVICE TYPE	REPL CODE
HA2137R	OEM
HA21420	OEM
HA2142R	OEM
HA21430	OEM
HA2143R	OEM
HA21440	OEM
HA2144R	OEM
HA21470	OEM
HA2147R	OEM
HA2190	0050
HA2356	0050
HA2400	OEM
HA2404	OEM
HA2405	OEM
HA2420	OEM
HA2425	OEM
HA2500	OEM
HA2502	OEM
HA2505	OEM
HA2507	OEM
HA2510	OEM
HA2512	OEM
HA2515	OEM
HA2517	OEM
HA2520	OEM
HA2522	OEM
HA2525	OEM
HA2526	OEM
HA2527	OEM
HA2539-2	OEM
HA2539-5	OEM
HA2539-8	OEM
HA2540-2(A)	OEM
HA2540-5(A)	OEM
HA2540-B	OEM
HA2600	OEM
HA2602	OEM
HA2605	OEM
HA2607	OEM
HA2620	OEM
HA2622	OEM
HA2625	OEM
HA2627	OEM
HA2630	OEM
HA2635	OEM
HA2640	OEM
HA2645	OEM
HA2655	OEM
HA2720	OEM
HA2740-2(A)	OEM
HA2740-5(A)	OEM
HA3210	0050
HA3480	0050
HA3670	0050
HA4071	OEM
HA4400	0050
HA4602-2	OEM
HA4605-5	OEM
HA4622-2	OEM
HA4625-5	OEM
HA4741-2	OEM
HA4741-5	2823
HA4900-2	OEM
HA4900-8	OEM
HA4905-5	OEM
HA4920-2	OEM
HA4925-5	OEM
HA4950-2	OEM
HA4950-5	OEM
HA4950-8	OEM
HA5001	0600
HA5002	0600
HA5003	0600
HA5005	0600
HA5009	0600
HA5010	0208
HA5011	0600
HA5012	0600
HA5014	0600
HA5016	0600
HA5020	0600
HA5021	0600
HA5022	0600
HA5023	0600
HA5024	0600
HA5025	0600
HA5026	0600
HA5062-2	OEM
HA5062-5	3741
HA5062A5	OEM
HA5062B5	OEM
HA5064-2	OEM
HA5064-5(A)	OEM
HA5064A5(A)	OEM
HA5064B5(A)	OEM
HA5082-2	OEM
HA5082-5	OEM
HA5082A5	OEM
HA5082B5	OEM
HA5084-2	OEM
HA5084-5	3391
HA5084-8	OEM
HA5084A5	OEM
HA5084B5	OEM
HA5100-2	OEM
HA5100-5	OEM
HA5105-5	OEM
HA5110-2	OEM
HA5110-5	OEM

DEVICE TYPE	REPL CODE
HA5115-5	OEM
HA5130-2	OEM
HA5130-5	OEM
HA5130-8	OEM
HA5135-2	OEM
HA5135-5	OEM
HA5135-8	OEM
HA5160-2	OEM
HA5160-5	OEM
HA5160-8	OEM
HA5162-5	OEM
HA5170-2(A)	OEM
HA5170-5(A)	OEM
HA5170-8(A)	OEM
HA5190-2	OEM
HA5190-8	OEM
HA5195-5	OEM
HA5905-6	OEM
HA7206	0861
HA7207	0861
HA7501	0037
HA7502	0037
HA7506	0037
HA7507	0037
HA7510	0037
HA7520	0791
HA7521	0791
HA7522	0791
HA7523	0791
HA7524	0791
HA7526	0791
HA7527	0791
HA7528	0791
HA7530	0126
HA7531	0126
HA7532	0532
HA7533	0037
HA7534	0037
HA7535	0342
HA7536	0037
HA7537	0037
HA7538	0037
HA7539	0342
HA7540	0342
HA7541	0342
HA7542	0342
HA7543	0037
HA7597	0037
HA7598	0037
HA7599	0037
HA7630	0126
HA7631	0126
HA7632	0126
HA7633	0037
HA7723	0126
HA7725	0126
HA7730	0791
HA7731	0126
HA7732	0791
HA7733	0126
HA7734	0791
HA7735	0791
HA7736	0791
HA7737	0791
HA7804	0037
HA7806	0037
HA7807	OEM
HA7808	0037
HA7809	OEM
HA7810	0037
HA7815	0037
HA8832	OEM
HA9048	0861
HA9049	0861
HA9054	0861
HA9055	0861
HA9056	0861
HA9057	0861
HA9058	0861
HA9059	0861
HA9078	0861
HA9079	0861
HA9500	0126
HA9501	0126
HA9502	0126
HA9531	0861
HA9531A	0861
HA9532	0861
HA9532A	0861
HA9532B	0037
HA11103	0167
HA11107	0167
HA11112	4395
HA11113	4616
HA11115W	0443
HA11120-1	OEM
HA11123	OEM
HA11211	3716
HA11215A	4782
HA11218	OEM
HA11219	4853
HA11220	4744
HA11221	4808
HA11222	OEM
HA11223W	4798
HA11225	2252
HA11226	5306
HA11227	3057
HA11228	4627

DEVICE TYPE	REPL CODE
HA11229	2184
HA11230	OEM
HA11231	3719
HA11235	1429
HA11238	3282
HA11238N	3282
HA11244	4854
HA11247	0067
HA11251	4805
HA11251A	4782
HA11252A	OEM
HA-11401	OEM
HA11401	3351
HA11408A	OEM
HA11409	5182
HA11410	OEM
HA11412	4787
HA11412A	4787
HA11413	3816
HA11413A	OEM
HA11414	3359
HA11417	5185
HA11418	OEM
HA11423	1201
HA11431	0450
HA11431NT	0450
HA11433	OEM
HA11434	4787
HA11436	2122
HA11436A	2122
HA11440	2116
HA11440A	2116
HA11441	6620
HA11442	OEM
HA11445	OEM
HA11446	2122
HA11480	2122
HA11480A	2122
HA11508	OEM
HA11510NT	4532
HA11511CNT	OEM
HA11511CNT-01	OEM
HA11511NT	OEM
HA11513	OEM
HA11516C4	3651
HA11516E1	3651
HA11516F1	OEM
HA11516J	3651
HA11518	OEM
HA11525MP	OEM
HA11525NT	OEM
HA11528NT	OEM
HA11529	OEM
HA11529NT	OEM
HA11532MP	OEM
HA11532NT	OEM
HA11539NT	OEM
HA11541	OEM
HA11541NT	OEM
HA11559NT	OEM
HA11561	OEM
HA11566	OEM
HA11566NT	OEM
HA11569FS	OEM
HA11569W	OEM
HA11574	OEM
HA11574FS	OEM
HA11580	1055
HA11701	5180
HA11702	5770
HA11703	5655
HA11704	OEM
HA11705	OEM
HA11706	5183
HA11710	OEM
HA11711	5181
HA11712	OEM
HA11712A	OEM
HA11713	OEM
HA11714	5185
HA11715	5158
HA11716	OEM
HA11718	OEM
HA11719	OEM
HA11720	OEM
HA11721	OEM
HA11722	OEM
HA11724	OEM
HA11725	OEM
HA11726	OEM
HA11727	OEM
HA11732	OEM
HA-11741	OEM
HA11741	OEM
HA11741NT	OEM
HA-11744	OEM
HA11744	OEM
HA-11745	OEM
HA11745	OEM
HA11747	6658
HA11747-A	5793
HA11747A	5793
HA11747ANT	5793
HA11747BNT	2489
HA11749	5790
HA11752	OEM
HA11753NT	OEM
HA11757ANT	OEM
HA11757NT	OEM

DEVICE TYPE	REPL CODE
HA11758NT	OEM
HA11787	OEM
HA11788	OEM
HA11793	OEM
HA11793ANT	OEM
HA11793NT	OEM
HA11797	OEM
HA11797NT	OEM
HA11803	OEM
HA11811NT2	OEM
HA11816NT	OEM
HA11825NT	OEM
HA11837	OEM
HA11837NT	OEM
HA11839NT	OEM
HA11845NT	OEM
HA11847NT	OEM
HA11853ANT	OEM
HA11853BNT	OEM
HA11853NT	OEM
HA11856A	OEM
HA11856ANT	OEM
HA11870NT	OEM
HA11876MP	OEM
HA11876NT	OEM
HA11887	OEM
HA11887V	OEM
HA11888	OEM
HA11888V	OEM
HA12002	5199
HA12002-W	5199
HA12003	0706
HA12005	OEM
HA12006	OEM
HA12009	OEM
HA12010	OEM
HA12013	0362
HA12016	OEM
HA12017	OEM
HA12018	OEM
HA12019	OEM
HA12024	OEM
HA12025	OEM
HA12026	1260
HA12028	OEM
HA12038	OEM
HA12044	OEM
HA12045	5819
HA12046	OEM
HA12047	OEM
HA12048FP	OEM
HA12058	OEM
HA12058NT	OEM
HA12072ANT	OEM
HA12072NT	OEM
HA12084BNT	OEM
HA12085NT	OEM
HA12088NT	OEM
HA12116NT	OEM
HA12117NT	OEM
HA12118NT	OEM
HA12134	OEM
HA12136	OEM
HA12138NT	OEM
HA12150NT	OEM
HA12411	5177
HA12412	5160
HA12413	3816
HA12413-03	3816
HA12417	OEM
HA12418	OEM
HA12430	OEM
HA13006	OEM
HA13007	OEM
HA13115	OEM
HA13402	OEM
HA13403	2186
HA13403V	OEM
HA13407	OEM
HA13407A	OEM
HA13408	OEM
HA13412	OEM
HA13421	4953
HA13421A	4953
HA13430	OEM
HA13430A	OEM
HA13434	OEM
HA13434MP	OEM
HA13434NT	OEM
HA13435	OEM
HA13455	OEM
HA14066B	OEM
HA16613AP	OEM
HA16631P	2759
HA16642MP	OEM
HA17008G	OEM
HA17008P	OEM
HA17012GB	OEM
HA17012GC	OEM
HA17080GS	OEM
HA17080GSA	OEM
HA17080PS	OEM
HA17080PSA	OEM
HA17082GS	OEM
HA17082GSA	OEM
HA17082PS	OEM
HA17082PSA	OEM
HA17083G	OEM
HA17083GA	OEM

DEVICE TYPE	REPL CODE
HA17083P	OEM
HA17083PA	OEM
HA17084G	OEM
HA17084P	OEM
HA17084PA	OEM
HA17093PS	OEM
HA17094PS	0770
HA17301G	2247
HA17301P	2247
HA17339	0176
HA17358	0770
HA17393	0631
HA17408P	5312
HA17431P	OEM
HA17458GS	0360
HA17458PS	0360
HA17524G	OEM
HA17524P	6350
HA17555	0973
HA17555PS	0973
HA17558	0360
HA17715G	OEM
HA17723G	0026
HA17733G	OEM
HA17741	0312
HA17741G	0312
HA17741GS	0312
HA17741PS	0312
HA17747G	2358
HA17747P	2358
HA17805	0624
HA17805H	0624
HA17805P	0624
HA17806P	0923
HA17807P	OEM
HA17808P	1196
HA17812	6261
HA17812P	0334
HA17812W	OEM
HA17815P	1321
HA17818P	2259
HA17824P	2239
HA17901G	0176
HA17901P	6248
HA17902G	0625
HA17902P	0625
HA17903GS	2310
HA17903PS	2310
HA17904	0770
HA17904GS	OEM
HA17904PS	0770
HA18031	OEM
HA19211NT	OEM
HA19216	OEM
HA19216MP	OEM
HA19507MP	OEM
HA19508	OEM
HA19508AMP	OEM
HA19508MP	OEM
HA19510	OEM
HA44015	OEM
HA114323	1201
HA118002	OEM
HA118002FP	OEM
HA118003MP	OEM
HA118003NT	OEM
HA118014NT	OEM
HA118018NT	OEM
HA118031	OEM
HA118039NT	OEM
HA118041NT	OEM
HA118053NT	OEM
HA118059	OEM
HA118088NT	OEM
HA118104	OEM
HA118105NT	OEM
HA118108NT	OEM
HA118114NT	OEM
HA118115	OEM
HA118120	OEM
HA118121FP	OEM
HAB005	0087
HAB010	OEM
HAB020	OEM
HAB060	OEM
HAB080	OEM
HAB100	OEM
HAB1500	OEM
HAB1501	OEM
HAB1502	OEM
HAD1000A	OEM
HAF0385	OEM
HAG20011	OEM
HAG20019	OEM
HAG20021	OEM
HAG20029	OEM
HAL10H8CF	OEM
HAL10H8CJ	OEM
HAL10H8CN	OEM
HAL10H8MF	OEM
HAL10H8MJ	OEM
HAL10H8MN	OEM
HAL10L8CF	OEM
HAL10L8CJ	OEM
HAL10L8CN	OEM
HAL10L8MF	OEM
HAL10L8MJ	OEM
HAL10L8MN	OEM
HAL12H6CF	OEM

DEVICE TYPE	REPL CODE
HAL12H6CJ	OEM
HAL12H6CN	OEM
HAL12H6MF	OEM
HAL12H6MJ	OEM
HAL12H6MN	OEM
HAL12L6CF	OEM
HAL12L6CJ	OEM
HAL12L6CN	OEM
HAL12L6MF	OEM
HAL12L6MJ	OEM
HAL12L6MN	OEM
HAL12L10CF	OEM
HAL12L10CJ	OEM
HAL12L10CN	OEM
HAL12L10MF	OEM
HAL12L10MJ	OEM
HAL12L10MN	OEM
HAL14H4CF	OEM
HAL14H4CJ	OEM
HAL14H4CN	OEM
HAL14H4MF	OEM
HAL14H4MJ	OEM
HAL14H4MN	OEM
HAL14L4CF	OEM
HAL14L4CJ	OEM
HAL14L4CN	OEM
HAL14L4MF	OEM
HAL14L4MJ	OEM
HAL14L4MN	OEM
HAL14L8CF	OEM
HAL14L8CJ	OEM
HAL14L8CN	OEM
HAL14L8CNS	OEM
HAL14L8MF	OEM
HAL14L8MJ	OEM
HAL14L8MN	OEM
HAL16A4CF	OEM
HAL16A4CJ	OEM
HAL16A4CN	OEM
HAL16A4MF	OEM
HAL16A4MJ	OEM
HAL16A4MN	OEM
HAL16C1CF	OEM
HAL16C1CJ	OEM
HAL16C1CN	OEM
HAL16C1MF	OEM
HAL16C1MJ	OEM
HAL16C1MN	OEM
HAL16H2CF	OEM
HAL16H2CJ	OEM
HAL16H2CN	OEM
HAL16H2MF	OEM
HAL16H2MJ	OEM
HAL16H2MN	OEM
HAL16L2CF	OEM
HAL16L2CJ	OEM
HAL16L2CN	OEM
HAL16L2MF	OEM
HAL16L2MJ	OEM
HAL16L2MN	OEM
HAL16L6CF	OEM
HAL16L6CJ	OEM
HAL16L6CN	OEM
HAL16L6MF	OEM
HAL16L6MJ	OEM
HAL16L6MN	OEM
HAL16L8ACN	OEM
HAL16L8CF	OEM
HAL16L8CJ	OEM
HAL16L8CN	OEM
HAL16L8MF	OEM
HAL16L8MJ	OEM
HAL16L8MN	OEM
HAL16LBACN	OEM
HAL16R4CF	OEM
HAL16R4CJ	OEM
HAL16R4CN	OEM
HAL15R4MF	OEM
HAL16R4MJ	OEM
HAL16R4MN	OEM
HAL16R6CF	OEM
HAL16R6CJ	OEM
HAL16R6CN	OEM
HAL16R6MF	OEM
HAL16R6MJ	OEM
HAL16R6MN	OEM
HAL16R8CF	OEM
HAL16R8CJ	OEM
HAL16R8CN	OEM
HAL16R8MF	OEM
HAL16R8MJ	OEM
HAL16R8MN	OEM
HAL16X4CF	OEM
HAL16X4CJ	OEM
HAL16X4CN	OEM
HAL16X4MF	OEM
HAL16X4MJ	OEM
HAL16X4MN	OEM
HAL18L4CF	OEM
HAL18L4CJ	OEM
HAL18L4CN	OEM
HAL18L4MF	OEM
HAL18L4MJ	OEM
HAL18L4MN	OEM
HAL20C1CF	OEM
HAL20C1CJ	OEM
HAL20C1CN	OEM
HAL20C1MJ	OEM
HAL20L2CF	OEM

DEVICE TYPE	REPL CODE
HAL20L2CJ	OEM
HAL20L2CN	OEM
HAL20L2MF	OEM
HAL20L2MJ	OEM
HAL20L2MN	OEM
HAL20L10CF	OEM
HAL20L10CJ	OEM
HAL20L10CN	OEM
HAL20L10MF	OEM
HAL20L10MJ	OEM
HAL20L10MN	OEM
HAL20X4CF	OEM
HAL20X4CJ	OEM
HAL20X4CN	OEM
HAL20X4CNS	OEM
HAL20X4MF	OEM
HAL20X4MJ	OEM
HAL20X4MN	OEM
HAL20X8CF	OEM
HAL20X8CJ	OEM
HAL20X8CN	OEM
HAL20X8MF	OEM
HAL20X8MJ	OEM
HAL20X8MN	OEM
HAL20X10CF	OEM
HAL20X10CJ	OEM
HAL20X10CN	OEM
HAL20X10MF	OEM
HAL20X10MJ	OEM
HAL20X10MN	OEM
HAM-1	0004
HAR10	0015
HAR15	0015
HAR20	0015
HB-00054	0004
HB-00056	0004
HB-00156	0004
HB-00171	0004
HB-00172	0004
HB-00173	0004
HB-00175	0004
HB-00176	0004
HB-00178	0004
HB-00186	0004
HB-00187	0004
HB-00303	0211
HB-00324	0164
HB-00370	0211
HB-00405	0164
HB-00564	0532
HB1	OEM
HB2	0015
HB3	1335
HB-7B	0025
HB-12B	0365
HB12B	0137
HB-32	0136
HB32	0004
HB-33	0136
HB33	0004
HB50	0071
HB-54	0004
HB54	0004
HB55	0004
HB-56	0004
HB56	0004
HB-75	0004
HB75	0004
HB-75B	0211
HB75B	0004
HB75C	0004
HB-77	0004
HB77	0004
HB77B	0004
HB-77C	0211
HB77C	0004
HB-85	0004
HB100	OEM
HB-156	0211
HB156	0004
HB156C	0004
HB171	0004
HB-172	0004
HB172	0004
HB-173	0004
HB-175	0004
HB175	0004
HB176	0004
HB178	0004
HB-186	0136
HB186	0004
HB-187	0136
HB187	0004
HB200	OEM
HB-233	OEM
HB233(A)	OEM
HB263	0004
HB270	0004
HB300	OEM
HB324	0164
HB332	OEM
HB333	OEM
HB352	OEM
HB353	OEM
HB356	OEM
HB-365	0164
HB365	0164
HB367	0222
HB400	OEM

DEVICE TYPE	REPL CODE
HB-415	0211
HB415	0164
HB422	0136
HB-459	0211
HB459	0004
HB-461	0136
HB461	0847
HB-475	0164
HB475	0164
HB500	OEM
HB600	0087
HB800	0087
HB1000	0087
HB4059AD	OEM
HB-18005	OEM
HB18005	OEM
HB-18007	OEM
HBC4000AD	2024
HBC4000AF	2024
HBC4000AK	2024
HBC4001AD	0478
HBC4001AF	0478
HBC4001AK	0478
HBC4002AD	2055
HBC4002AF	2055
HBC4002AK	2055
HBC4006AD	OEM
HBC4006AF	OEM
HBC4006AK	OEM
HBC4008AD	0988
HBC4008AF	0988
HBC4008AK	0988
HBC4011AD	0215
HBC4011AF	0215
HBC4011AK	0215
HBC4012AD	0498
HBC4012AF	0498
HBC4012AK	0498
HBC4013AD	0413
HBC4013AF	0413
HBC4013AK	0413
HBC4014AD	OEM
HBC4014AF	OEM
HBC4014AK	OEM
HBC4015AD	OEM
HBC4015AF	OEM
HBC4015AK	OEM
HBC4017AD	0513
HBC4017AF	0513
HBC4017AK	0513
HBC4018AD	1391
HBC4018AF	1391
HBC4018AK	1391
HBC4020AD	1661
HBC4020AF	1661
HBC4020AK	1661
HBC4021AD	OEM
HBC4021AF	OEM
HBC4021AK	OEM
HBC4022AD	1256
HBC4022AE	1256
HBC4022AF	1256
HBC4022AK	1256
HBC4023AD	0519
HBC4023AF	0519
HBC4023AK	0519
HBC4024AD	1957
HBC4024AF	1957
HBC4024AK	1957
HBC4025AD	2072
HBC4025AF	2072
HBC4025AK	2072
HBC4026AD	OEM
HBC4026AF	OEM
HBC4026AK	OEM
HBC4027AD	1949
HBC4027AF	1949
HBC4027AK	1949
HBC4028AD	2228
HBC4028AF	2228
HBC4028AK	2228
HBC4029AD	2233
HBC4029AF	2233
HBC4029AK	2233
HBC4030AD	0500
HBC4030AF	0500
HBC4030AK	0500
HBC4031AD	OEM
HBC4031AF	OEM
HBC4031AK	OEM
HBC4032AD	OEM
HBC4032AF	OEM
HBC4032AK	OEM
HBC4033AD	OEM
HBC4033AF	OEM
HBC4033AK	OEM
HBC4034AD	OEM
HBC4034AK	OEM
HBC4035AD	OEM
HBC4035AF	OEM
HBC4035AK	OEM
HBC4036AD	OEM
HBC4036AK	OEM
HBC4037AD	OEM
HBC4037AF	OEM
HBC4037AK	OEM
HBC4038AF	OEM
HBC4038AK	OEM
HBC4039AD	OEM

If replacement code is OEM, contact original manufacturer for replacement.

DEVICE TYPE	REPL CODE
HBC4039AK	OEM
HBC4040AD	0056
HBC4040AF	0056
HBC4040AK	0056
HBC4042AD	0121
HBC4042AF	0121
HBC4042AK	0121
HBC4043AD	1553
HBC4043AF	1553
HBC4043AK	1553
HBC4044AD	2308
HBC4044AF	2308
HBC4044AK	2308
HBC4045AD	OEM
HBC4045AF	OEM
HBC4045AK	OEM
HBC4047AD	2331
HBC4047AK	2331
HBC4048AD	OEM
HBC4048AF	OEM
HBC4048AK	OEM
HBC4054AF	OEM
HBC4054AK	OEM
HBC4055AD	3304
HBC4055AF	3304
HBC4055AK	3304
HBC4056AD	OEM
HBC4056AF	OEM
HBC4056AK	OEM
HBC4057AD	OEM
HBC4059AK	OEM
HBC4060AD	0146
HBC4060AF	0146
HBC4060AK	0146
HBC4062AK	OEM
HBF4000AE	2024
HBF4000AF	2024
HBF4001	0478
HBF4001A	0478
HBF4001AE	0478
HBF4001AF	0478
HBF4002	2055
HBF4002A	2055
HBF4002AE	2055
HBF4002AF	2055
HBF4006AE	0646
HBF4006AF	0646
HBF4007	2846
HBF4008AE	0988
HBF4008AF	0988
HBF4009	1999
HBF4009A	0001
HBF4011	0215
HBF4011A	0215
HBF4011AE	0215
HBF4011AF	0215
HBF4012	0498
HBF4012A	0498
HBF4012AE	0498
HBF4012AF	0498
HBF4013	0413
HBF4013A	0413
HBF4013AE	0413
HBF4013AF	0413
HBF4014A	0860
HBF4014AE	0860
HBF4014AF	OEM
HBF4015A	1014
HBF4015AE	OEM
HBF4015AF	OEM
HBF4016	1143
HBF4016A	1143
HBF4017	0513
HBF4017A	0513
HBF4017AE	0513
HBF4017AF	0513
HBF4018AE	1391
HBF4018AF	1391
HBF4019A	1525
HBF4020	1661
HBF4020A	1661
HBF4020AE	1661
HBF4020AF	1661
HBF4021	1749
HBF4021A	1749
HBF4021AE	OEM
HBF4021AF	OEM
HBF4022AE	1256
HBF4022AF	1256
HBF4023	0519
HBF4023A	0519
HBF4023AE	0519
HBF4023AF	0519
HBF4024	1957
HBF4024A	1957
HBF4024AE	1957
HBF4024AF	1957
HBF4025	2072
HBF4025A	2072
HBF4025AE	2072
HBF4025AF	2072
HBF4026AE	OEM
HBF4026AF	OEM
HBF4027	1949
HBF4027A	1949
HBF4027AE	1949
HBF4027AF	1949
HBF4028A	2228
HBF4028AE	2228

DEVICE TYPE	REPL CODE
HBF4028AF	2228
HBF4029A	2233
HBF4029AE	2233
HBF4029AF	2233
HBF4030A	0500
HBF4030AE	0500
HBF4030AF	0500
HBF4031AE	OEM
HBF4031AF	OEM
HBF4032AE	2531
HBF4032AF	2531
HBF4033AE	2635
HBF4033AF	2635
HBF4034AD	OEM
HBF4034AE	3615
HBF4035AE	OEM
HBF4035AF	OEM
HBF4036AD	OEM
HBF4036AE	OEM
HBF4037AE	OEM
HBF4037AF	OEM
HBF4038AE	2981
HBF4038AF	2981
HBF4039AD	OEM
HBF4039AE	OEM
HBF4040A	0056
HBF4040AE	0056
HBF4040AF	0056
HBF4042A	0121
HBF4042AE	0121
HBF4042AF	0121
HBF4043AE	1553
HBF4043AF	1553
HBF4044AE	2308
HBF4044AF	2308
HBF4045AE	OEM
HBF4045AF	OEM
HBF4047AE	2331
HBF4048AE	3457
HBF4048AF	3457
HBF4049A	0001
HBF4050A	0398
HBF4051A	0366
HBF4052A	0024
HBF4054AE	OEM
HBF4054AF	OEM
HBF4055AE	3304
HBF4055AF	3304
HBF4056AE	OEM
HBF4056AF	OEM
HBF4057AD	OEM
HBF4059AD	OEM
HBF4059AE	OEM
HBF4060AE	0146
HBF4060AF	0146
HBF4062AE	OEM
HBG20011	OEM
HBG20019	OEM
HBG20021	OEM
HBG20029	OEM
HC-00184	0475
HC-00227	0016
HC-00268	0086
HC-00372	0224
HC-00373	0111
HC-00380	0127
HC-00394	0155
HC-00458	0079
HC-00460	0079
HC-00461	0151
HC-00481	0222
HC-00496	0161
HC-00509	0079
HC-00535	0224
HC-00536	0111
HC-00537	0079
HC-00644	0086
HC-00668	0144
HC-00693	0079
HC-00711	0111
HC-00730	0600
HC-00732	0111
HC-00735	0079
HC-00772	0144
HC-00784	1144
HC-00828	1220
HC-00829	0127
HC-00838	0079
HC00838	0079
HC-00839	0076
HC-00871	0079
HC-00900	1221
HC-00920	0127
HC-00921	0079
HC-00923	0552
HC-00924	0079
HC-00929	0224
HC-00930	0127
HC00930	0224
HC-00945	0076
HC-01000	0111
HC-01047	0127
HC-01060	0560
HC-01096	0390
HC-01098	0560
HC-01209	0086
HC-01226	0219
HC-01317	0086
HC-01318	0155

DEVICE TYPE	REPL CODE
HC-01327	1221
HC-01335	0111
HC-01359	0127
HC-01390	0597
HC-01417	0127
HC-01820	0079
HC-01830	0144
HC-2	0348
HC2	5351
HC5	OEM
HC8	OEM
HC10	OEM
HC12	OEM
HC14	OEM
HC15	OEM
HC-30	0015
HC30	0087
HC-39	0015
HC39	0087
HC50	0087
HC-56	0079
HC67	0015
HC67F	0071
HC-68	0015
HC68	0015
HC68F	OEM
HC69	0015
HC69F	OEM
HC70	OEM
HC70F	OEM
HC71	0015
HC71F	2645
HC72	0071
HC72F	0087
HC73	0071
HC73F	0087
HC80	0015
HC100	OEM
HC200	OEM
HC206	0144
HC300	OEM
HC300F	OEM
HC371	0079
HC372	0079
HC-373	0079
HC373	0079
HC380	0127
HC381	0127
HC394	0127
HC400	OEM
HC454	0127
HC458	0079
HC460	0127
HC461	0127
HC-495	0560
HC495	0042
HC500	0015
HC500F	OEM
HC502C	OEM
HC503C	OEM
HC515	0142
HC-521	OEM
HC521	OEM
HC-535	0079
HC535	0127
HC535A	0127
HC535B	0127
HC-537	0111
HC537	0127
HC539	0079
HC545	0127
HC-561	0079
HC561	0079
HC600	0087
HC645	0144
HC-668	0007
HC668	0144
HC670	0015
HC680	0015
HC700	0015
HC710	0015
HC720	0071
HC730	0071
HC-772	0079
HC772	0144
HC784	0127
HC800	0087
HC829	0144
HC-1000	0111
HC1000	OEM
HC1200	OEM
HC1200F	OEM
HC1240-1	OEM
HC2000H	OEM
HC2000H3	OEM
HC2000H4	OEM
HC2500	OEM
HC6809E	OEM
HC7001	OEM
HC7002	OEM
HC7004	OEM
HC7005	OEM
HC7006	OEM
HC7007	OEM
HC7008	OEM
HC7060	OEM
HC7082	OEM
HC7083	OEM
HC7084	0030

DEVICE TYPE	REPL CODE
HC7085	OEM
HC7086	OEM
HC7087	OEM
HC7088	OEM
HC7089	OEM
HC7090	OEM
HC7095	OEM
HC7096	OEM
HC7097	OEM
HC7098	OEM
HC7127	OEM
HC10125	OEM
HC-50102	0901
HC100603	2509
HC1000109	0632
HC1000111-0	0632
HC1000114-0	2617
HC1000117-0	0654
HC1000217	0410
HC1000217-0	1975
HC1000403	2568
HC1000405	OEM
HC1000417	1611
HC1000503	2509
HC1000505	2631
HC1000603	4318
HC1000703	2876
HC1001001	4390
HC1001103	OEM
HC1001203	OEM
HC1001703	OEM
HC10001090	0632
HC10001200	2534
HC10002050	0911
HC10002170	0410
HC10003060	OEM
HC10004010	0520
HC10004110	0232
HC10006170	0520
HC10008060	3815
HC10019210	2638
HC10022010	2019
HC10023010	3262
HC10034050	2019
HC10036010	3816
HC10044030	1951
HCC4000BD	2024
HCC4000BF	2024
HCC4000BK	2024
HCC4001BD	0478
HCC4001BF	0478
HCC4001BK	0478
HCC4002BD	2055
HCC4002BF	2055
HCC4002BK	2055
HCC4006BD	OEM
HCC4006BF	0646
HCC4006BK	OEM
HCC4008BD	0988
HCC4008BF	0988
HCC4008BK	0988
HCC4011BD	0215
HCC4011BF	0215
HCC4011BK	0215
HCC4012BD	0498
HCC4012BF	0498
HCC4012BK	0498
HCC4013BD	0413
HCC4013BF	0413
HCC4013BK	0413
HCC4014BD	OEM
HCC4014BF	3210
HCC4014BK	OEM
HCC4015BD	OEM
HCC4015BF	6329
HCC4015BK	OEM
HCC4016BD	OEM
HCC4016BF	1143
HCC4016BK	OEM
HCC4017BD	0513
HCC4017BF	0513
HCC4017BK	0513
HCC4018BD	1391
HCC4018BF	1391
HCC4018BK	1391
HCC4019BD	OEM
HCC4019BF	1525
HCC4019BK	OEM
HCC4020BD	1661
HCC4020BF	1661
HCC4020BK	1661
HCC4021BD	OEM
HCC4021BF	1749
HCC4021BK	OEM
HCC4022BD	1256
HCC4022BE	1256
HCC4022BF	1256
HCC4022BK	1256
HCC4023BD	0519
HCC4023BF	0519
HCC4023BK	0519
HCC4024BD	1957
HCC4024BF	1957
HCC4024BK	1957
HCC4025BD	2072
HCC4025BF	2072
HCC4025BK	2072
HCC4026BD	OEM
HCC4026BF	2152

DEVICE TYPE	REPL CODE
HCC4026BK	OEM
HCC4027BD	1949
HCC4027BF	1949
HCC4027BK	1949
HCC4028BD	2228
HCC4028BF	2228
HCC4028BK	2228
HCC4029BD	2233
HCC4029BF	2233
HCC4029BK	2233
HCC4030BD	OEM
HCC4030BF	0500
HCC4030BK	OEM
HCC4031BD	OEM
HCC4031BF	2971
HCC4031BK	OEM
HCC4032BD	OEM
HCC4032BF	2531
HCC4032BK	OEM
HCC4033BD	OEM
HCC4033BF	2635
HCC4033BK	OEM
HCC4034BD	OEM
HCC4035BD	OEM
HCC4035BF	2777
HCC4035BK	OEM
HCC4038BD	OEM
HCC4038BF	2981
HCC4038BK	OEM
HCC4040BD	0056
HCC4040BF	0056
HCC4040BK	0056
HCC4042BD	0121
HCC4042BF	0121
HCC4042BK	0121
HCC4043BD	1553
HCC4043BF	1553
HCC4043BK	1553
HCC4044BD	2308
HCC4044BF	2308
HCC4044BK	2308
HCC4045BD	OEM
HCC4045BF	3443
HCC4045BK	OEM
HCC4047BD	OEM
HCC4047BF	2331
HCC4047BK	OEM
HCC4048BD	OEM
HCC4048BF	3459
HCC4048BK	OEM
HCC4049BD	OEM
HCC4049BF	OEM
HCC4049BK	OEM
HCC4050BD	OEM
HCC4050BK	OEM
HCC4051BD	OEM
HCC4051BF	0366
HCC4051BK	OEM
HCC4052BD	OEM
HCC4052BF	0024
HCC4052BK	OEM
HCC4053BD	OEM
HCC4053BF	0035
HCC4053BK	OEM
HCC4054BD	OEM
HCC4054BF	4530
HCC4054BK	OEM
HCC4055BD	OEM
HCC4055BF	3304
HCC4055BK	OEM
HCC4056BD	3706
HCC4056BF	3706
HCC4056BK	3706
HCC4060BD	0146
HCC4060BF	0146
HCC4060BK	0146
HCC4063BD	3727
HCC4063BF	3727
HCC4063BK	3727
HCC4066BD	OEM
HCC4066BF	0101
HCC4066BK	OEM
HCC4068BD	2503
HCC4068BF	2503
HCC4068BK	2503
HCC4070BD	2516
HCC4070BF	2516
HCC4070BK	2516
HCC4071BD	0129
HCC4071BF	0129
HCC4071BK	0129
HCC4072BD	2524
HCC4072BF	2524
HCC4072BK	2524
HCC4073BD	1536
HCC4073BF	1536
HCC4073BK	1536
HCC4075BD	2540
HCC4075BF	2540
HCC4075BK	2540
HCC4076BD	OEM
HCC4076BF	3496
HCC4076BK	OEM
HCC4077BD	2558
HCC4077BF	2558
HCC4077BK	2558
HCC4078BD	0921
HCC4078BF	0921
HCC4078BK	0921

DEVICE TYPE	REPL CODE
HCC4081BD	0626
HCC4081BF	0626
HCC4081BK	0626
HCC4082BD	0300
HCC4082BF	0300
HCC4082BK	0300
HCC4085BD	0303
HCC4085BF	0303
HCC4085BK	0303
HCC4086BD	0535
HCC4086BF	0535
HCC4086BK	0535
HCC4089BD	OEM
HCC4089BF	3830
HCC4089BK	OEM
HCC4094BD	OEM
HCC4094BF	1682
HCC4094BK	OEM
HCC4095BD	OEM
HCC4095BF	3848
HCC4095BK	OEM
HCC4096BD	OEM
HCC4096BF	3851
HCC4096BK	OEM
HCC4098BD	3611
HCC4098BF	3611
HCC4098BK	3611
HCC4099BD	3330
HCC4099BF	3330
HCC4099BK	3330
HCC4508BD	OEM
HCC4508BK	OEM
HCC4510BD	1963
HCC4510BE	1963
HCC4510BF	1963
HCC4510BK	1963
HCC4511BD	1544
HCC4511BF	1544
HCC4511BK	1544
HCC4514BD	1830
HCC4515BD	3600
HCC4516BD	2347
HCC4516BF	2347
HCC4516BK	2347
HCC4518BD	1043
HCC4518BF	1043
HCC4518BK	1043
HCC4520BD	2675
HCC4520BF	2675
HCC4520BK	2675
HCC4527BD	3145
HCC4527BF	3145
HCC4527BK	3145
HCC4532BD	OEM
HCC4532BF	1016
HCC4532BK	OEM
HCC4555BD	2938
HCC4555BF	2938
HCC4555BK	2938
HCC4556BD	3431
HCC4556BF	3431
HCC4556BK	3431
HCC40100BD	3952
HCC40100BF	3952
HCC40100BK	OEM
HCC40101BD	OEM
HCC40101BF	OEM
HCC40101BK	OEM
HCC40102BD	4061
HCC40102BF	4061
HCC40102BK	OEM
HCC40103BD	4091
HCC40103BF	4091
HCC40103BK	OEM
HCC40105BD	OEM
HCC40105BF	OEM
HCC40105BK	OEM
HCC40106BD	OEM
HCC40106BF	3626
HCC40106BK	OEM
HCC40109BD	OEM
HCC40109BF	OEM
HCC40109BK	OEM
HCC40160BD	1359
HCC40160BF	1359
HCC40160BK	1359
HCC40161BD	1373
HCC40161BF	1373
HCC40161BK	1373
HCC40162BD	OEM
HCC40162BF	1388
HCC40162BK	1388
HCC40163BD	1407
HCC40163BF	1407
HCC40163BK	1407
HCC40174BD	1551
HCC40174BF	1551
HCC40174BK	1551
HCC40182BD	OEM
HCC40182BF	4660
HCC40182BK	OEM
HCCA100	OEM
HCF0088	OEM
HCF4000BE	2024
HCF4000BF	2024
HCF4001BE	0478
HCF4001BF	0478
HCF4002BE	2055
HCF4002BF	2055

DEVICE TYPE	REPL CODE
HCF4006BE	0646
HCF4006BF	0646
HCF4007UBE	4984
HCF4008BE	0988
HCF4008BF	0988
HCF4011BE	0215
HCF4011BF	0215
HCF4012BE	0498
HCF4012BF	0498
HCF4013BE	0413
HCF4013BF	0413
HCF4014BE	0860
HCF4014BF	0860
HCF4015BE	1014
HCF4015BF	1014
HCF4016BE	1143
HCF4016BF	1143
HCF4017BE	0513
HCF4017BF	0513
HCF4018BE	1391
HCF4018BF	1391
HCF4019BE	1525
HCF4019BF	1525
HCF4020BE	1661
HCF4020BF	1661
HCF4021BE	1749
HCF4021BF	1749
HCF4022BE	1256
HCF4022BF	1256
HCF4023BE	0519
HCF4023BF	0519
HCF4024BE	1957
HCF4024BF	1957
HCF4025BE	2072
HCF4025BF	2072
HCF4026BE	2152
HCF4026BF	2152
HCF4027BE	1949
HCF4027BF	1949
HCF4028BE	2228
HCF4028BF	2228
HCF4029BE	2233
HCF4029BF	2233
HCF4030BE	0500
HCF4030BF	0500
HCF4031BE	2971
HCF4031BF	2971
HCF4032BE	2531
HCF4032BF	2531
HCF4033BE	2635
HCF4033BF	2635
HCF4034BD	OEM
HCF4034BE	3615
HCF4035BE	2777
HCF4035BF	2777
HCF4038BF	2981
HCF4040BE	0056
HCF4040BF	0056
HCF4042BE	0121
HCF4042BF	0121
HCF4043BE	1553
HCF4043BF	1553
HCF4044BE	2308
HCF4044BF	2308
HCF4045BE	3443
HCF4045BF	3443
HCF4047BE	2331
HCF4047BF	2331
HCF4048BE	3459
HCF4048BF	3459
HCF4049BE	0001
HCF4049BF	0001
HCF4050BE	0398
HCF4050BF	0398
HCF4051BE	0366
HCF4051BF	0366
HCF4052BE	0024
HCF4052BF	0024
HCF4053BE	0035
HCF4053BF	0035
HCF4054BE	4530
HCF4054BF	4530
HCF4055BE	3304
HCF4055BF	3304
HCF4056BE	3706
HCF4056BF	3706
HCF4058BE	OEM
HCF4060BE	0146
HCF4060BF	0146
HCF4063BE	3727
HCF4063BF	3727
HCF4066BE	0101
HCF4066BF	0101
HCF4068BE	2503
HCF4068BF	2503
HCF4070BE	2516
HCF4070BF	2516
HCF4071BE	0129
HCF4071BF	0129
HCF4072BE	2524
HCF4072BF	2524
HCF4073BE	1536
HCF4073BF	1536
HCF4075BE	2540
HCF4075BF	2540
HCF4076BE	3496
HCF4076BF	3496
HCF4077BE	2558
HCF4077BF	2558

If replacement code is OEM, contact original manufacturer for replacement.

DEVICE TYPE	REPL CODE
HCF4078BE	0921
HCF4078BF	0921
HCF4081BE	0626
HCF4081BF	0626
HCF4082BE	0300
HCF4082BF	0300
HCF4085BE	0303
HCF4085BF	0303
HCF4086BE	0535
HCF4086BF	0535
HCF4086BK	0535
HCF4089BE	3830
HCF4089BF	3830
HCF4094BE	1682
HCF4094BF	1682
HCF4095BE	3848
HCF4095BF	3848
HCF4096BE	3851
HCF4096BF	3851
HCF4098BE	3611
HCF4098BF	3611
HCF4099BE	3330
HCF4099BF	3330
HCF4502BE	1037
HCF4508BD	OEM
HCF4510BE	1963
HCF4510BF	1963
HCF4511BE	1544
HCF4511BF	1544
HCF4514BD	1830
HCF4514BE	1830
HCF4515BD	3600
HCF4515BE	3600
HCF4516BE	2347
HCF4516BF	2347
HCF4518BE	1043
HCF4518BF	1043
HCF4520BE	2675
HCF4520BF	2675
HCF4527BE	3145
HCF4527BF	3145
HCF4532BE	1016
HCF4532BF	1016
HCF4555BE	2938
HCF4555BF	2938
HCF4556BE	3431
HCF4556BF	3431
HCF10160BE	OEM
HCF40100BE	3952
HCF40100BF	3952
HCF40101BE	OEM
HCF40101BF	4017
HCF40102BE	OEM
HCF40102BF	4061
HCF40103BE	OEM
HCF40103BF	4091
HCF40105BE	OEM
HCF40105BF	4123
HCF40106BE	3626
HCF40106BF	3626
HCF40109BE	OEM
HCF40109BF	4210
HCF40160BE	1359
HCF40160BF	1359
HCF40161BE	1373
HCF40161BF	1373
HCF40162BE	1388
HCF40162BF	1388
HCF40163BE	1407
HCF40163BF	1407
HCF40174BE	1551
HCF40174BF	1551
HCF40182BE	4660
HCF40182BF	4660
HCMP1802ACD	OEM
HCMP1802ACL	OEM
HCMP1802ACP	OEM
HCMP1802ACY	OEM
HCMP1802AD	OEM
HCMP1802AL	OEM
HCMP1802AP	OEM
HCMP1802AY	OEM
HCMP1822CD	OEM
HCMP1822CP	OEM
HCMP1823CD	OEM
HCMP1823CH	OEM
HCMP1823CP	OEM
HCMP1824CD	OEM
HCMP1824CP	OEM
HCMP1824D	OEM
HCMP1824P	OEM
HCMP1831CD	OEM
HCMP1831CP	OEM
HCMP1831D	OEM
HCMP1831P	OEM
HCMP1832CD	OEM
HCMP1832CP	OEM
HCMP1832D	OEM
HCMP1832P	OEM
HCMP1833CD	OEM
HCMP1833CP	OEM
HCMP1833D	OEM
HCMP1833P	OEM
HCMP1834CD	OEM
HCMP1834CP	OEM
HCMP1834D	OEM
HCMP1834P	OEM
HCMP1835CD	OEM
HCMP1835CP	OEM
HCMP1835D	OEM
HCMP1835P	OEM
HCMP1836CD	OEM
HCMP1836CP	OEM
HCMP1836D	OEM
HCMP1836P	OEM
HCMP1852CD	OEM
HCMP1852CP	OEM
HCMP1852CY	OEM
HCMP1852D	OEM
HCMP1852P	OEM
HCMP1852Y	OEM
HCMP1853CD	OEM
HCMP1853CP	OEM
HCMP1853D	OEM
HCMP1853P	OEM
HCMP1854CD	OEM
HCMP1854CP	OEM
HCMP1854D	OEM
HCMP1854P	OEM
HCMP1856CD	OEM
HCMP1856CP	OEM
HCMP1856D	OEM
HCMP1856P	OEM
HCMP1857CD	OEM
HCMP1857CP	OEM
HCMP1857D	OEM
HCMP1857P	OEM
HCMP1858CD	OEM
HCMP1858CP	OEM
HCMP1858D	OEM
HCMP1858P	OEM
HCMP1859CD	OEM
HCMP1859CP	OEM
HCMP1859D	OEM
HCMP1859P	OEM
HCPL-2502	OEM
HCPL-2530	OEM
HCPL2530	5428
HCPL-2531	OEM
HCPL-2601	OEM
HCPL-2602	OEM
HCPL-2630	5674
HCPL2630	5674
HCPL-2631	OEM
HCPL-2730	OEM
HCPL-2731	OEM
HCPL-3700	OEM
HCPL-4562	OEM
HCPL-4661	OEM
HCR5-100	OEM
HCR5-200	OEM
HCR5-300	OEM
HCR5-400	OEM
HCR30P	OEM
HCR50N	OEM
HCR50P	OEM
HCR100N	OEM
HCR100P	OEM
HCR150N	OEM
HCR150P	OEM
HCR200N	OEM
HCR200P	OEM
HCR300P	OEM
HCR400P	OEM
HCR1020	OEM
HCR1025	OEM
HCR1030	OEM
HCR1040	OEM
HCR1050	OEM
HCR1070	OEM
HCR1100	OEM
HCR4038BE	OEM
HCT373	6123
HCU04	OEM
HCV	0015
HD0000206	0139
HD-00072	0208
HD-00227	0016
HD-00261	0218
HD-00471	0018
HD-1	0015
HD1-74C00	2112
HD1-74C02	2114
HD1-74C08	2118
HD1-74C10	2119
HD1-74C20	2123
HD1-74C42	2143
HD1-74C48	2148
HD1-74C73	2162
HD1-74C74	2163
HD1-74C76	2164
HD1-74C85	2171
HD1-74C90	2174
HD1-74C93	2177
HD1-74C154	4014
HD1-74C161	4046
HD1-74C192	4119
HD1-74C193	4122
HD1-74C221	0353
HD1-6495-2	OEM
HD1-6495-9	OEM
HD1-6495A-2	OEM
HD1-6495A-9	OEM
HD1-15530-2	OEM
HD1-15530-9	OEM
HD1-15531-2	OEM
HD1-15531-9	OEM
HD3	OEM
HD3-4702-2	OEM
HD3-4702-9	OEM
HD4-4702-2	OEM
HD4-4702-9	OEM
HD4-6440-2	OEM
HD4-6440-9	OEM
HD4-15530-2	OEM
HD4-15530-9	OEM
HD4-15531-2	OEM
HD4-15531-9	OEM
HD9-54C20	OEM
HD9-54C93	OEM
HD9-74C00	2112
HD9-74C02	2114
HD9-74C08	2118
HD9-74C10	2119
HD9-74C20	2123
HD9-74C42	2143
HD9-74C48	2148
HD9-74C73	2162
HD9-74C74	2163
HD9-74C76	2164
HD9-74C85	2171
HD9-74C90	2174
HD9-74C93	2177
HD9-74C154	4014
HD9-74C161	4046
HD9-74C192	4119
HD9-74C193	4122
HD9-74C221	3047
HD9-15530-2	OEM
HD9-15530-9	OEM
HD10	2096
HD10-001-01	0143
HD20-003-01	0015
HD61J228P	OEM
HD63A03Y	OEM
HD63A03YF	OEM
HD63A03YP	OEM
HD63B03X	OEM
HD63B03XP	OEM
HD63B03YP	OEM
HD68A00	OEM
HD68A09	OEM
HD68A09EP	2117
HD68A21	OEM
HD68A21P	OEM
HD68A44C	OEM
HD68A44P	OEM
HD68A45SP	OEM
HD68A50	OEM
HD68A52	OEM
HD68B00	OEM
HD68B09	OEM
HD68B21	OEM
HD68B21P	OEM
HD68B40	OEM
HD68B40P	OEM
HD68B45P	OEM
HD74ALS00	OEM
HD74ALS01	OEM
HD74ALS03	OEM
HD74ALS04	OEM
HD74ALS04P	OEM
HD74ALS05	OEM
HD74ALS08	OEM
HD74ALS09	OEM
HD74ALS20	OEM
HD74ALS21	OEM
HD74ALS22	OEM
HD74ALS74	OEM
HD74ALS109	OEM
HD74ALS112	OEM
HD74ALS112P	OEM
HD74ALS113	OEM
HD74ALS113P	OEM
HD74ALS114	OEM
HD74ALS114P	OEM
HD74ALS175	OEM
HD74ALS175P	OEM
HD74HC00P	OEM
HD74HC04P	OEM
HD74HC10	OEM
HD74HC74P	1561
HD74HC374FP	OEM
HD74LS00	1527
HD74LS00P	1527
HD74LS01	1546
HD74LS01P	1546
HD74LS02	1559
HD74LS02P	1559
HD74LS03	1579
HD74LS03P	1579
HD74LS04	1595
HD74LS04P	1595
HD74LS05	1608
HD74LS05P	1608
HD74LS07P	OEM
HD74LS08	1633
HD74LS08P	1633
HD74LS09	1642
HD74LS09P	1642
HD74LS10	1662
HD74LS10P	1662
HD74LS11	1667
HD74LS11P	1667
HD74LS12	1679
HD74LS12P	1679
HD74LS13	1688
HD74LS13P	1688
HD74LS14	1698
HD74LS14P	1698
HD74LS15	1707
HD74LS15P	1707
HD74LS20	0034
HD74LS20P	0034
HD74LS21	1763
HD74LS21P	1763
HD74LS22	1775
HD74LS22P	1775
HD74LS26	1382
HD74LS26P	1382
HD74LS27	0183
HD74LS27P	0183
HD74LS30	0828
HD74LS30P	0828
HD74LS32	0088
HD74LS32P	0088
HD74LS37	1730
HD74LS37P	1730
HD74LS38	1839
HD74LS38P	1839
HD74LS40	0135
HD74LS40P	0135
HD74LS42	1841
HD74LS42P	1841
HD74LS47	1845
HD74LS47P	1845
HD74LS48	1849
HD74LS48P	1849
HD74LS49	1850
HD74LS49P	1850
HD74LS51	1033
HD74LS51P	1033
HD74LS54	1857
HD74LS54P	1857
HD74LS55	0457
HD74LS55P	0457
HD74LS73	1867
HD74LS73P	1867
HD74LS74	0243
HD74LS74A	0243
HD74LS74AP	0243
HD74LS75	1870
HD74LS75P	1870
HD74LS76	2180
HD74LS76P	2180
HD74LS77	1872
HD74LS77P	1872
HD74LS78	1873
HD74LS78P	1873
HD74LS83A	2218
HD74LS83AP	2218
HD74LS85	0431
HD74LS85P	0431
HD74LS86	0291
HD74LS86P	0291
HD74LS90	1882
HD74LS90P	1882
HD74LS91	1885
HD74LS91P	1885
HD74LS92	1887
HD74LS92P	1887
HD74LS93	1888
HD74LS93P	1888
HD74LS95B	0771
HD74LS95BP	0771
HD74LS96	OEM
HD74LS96P	OEM
HD74LS107	1602
HD74LS107P	1602
HD74LS109A	1906
HD74LS109AP	1906
HD74LS109P	4972
HD74LS112	2128
HD74LS112A	OEM
HD74LS112P	2128
HD74LS113	2256
HD74LS113P	2256
HD74LS114	2302
HD74LS114P	2302
HD74LS122	1620
HD74LS122P	1620
HD74LS123	0979
HD74LS123P	0979
HD74LS125	OEM
HD74LS125AP	0075
HD74LS132P	OEM
HD74LS136	1628
HD74LS136P	1628
HD74LS138	0427
HD74LS138P	0427
HD74LS139	0153
HD74LS139P	0153
HD74LS145	1564
HD74LS145P	1564
HD74LS148	3912
HD74LS148P	3912
HD74LS151	1646
HD74LS151P	1646
HD74LS152	OEM
HD74LS152P	OEM
HD74LS153	0959
HD74LS153P	0959
HD74LS154	5045
HD74LS154P	5045
HD74LS155	0209
HD74LS155P	0209
HD74LS156	1654
HD74LS156P	1654
HD74LS157	1161
HD74LS157P	1161
HD74LS158	1656
HD74LS158P	1656
HD74LS160	0837
HD74LS161	0858
HD74LS161P	0858
HD74LS162	0880
HD74LS162P	0880
HD74LS163	0893
HD74LS163P	0893
HD74LS164	4339
HD74LS164P	4339
HD74LS166	4371
HD74LS166P	4371
HD74LS174	0260
HD74LS174P	0260
HD74LS175	1672
HD74LS175P	1672
HD74LS181	1678
HD74LS181P	1678
HD74LS190	1686
HD74LS190P	1686
HD74LS191	1687
HD74LS191P	1687
HD74LS192	1689
HD74LS192P	1689
HD74LS193	1692
HD74LS193P	1692
HD74LS194A	1304
HD74LS194AP	1304
HD74LS195A	1315
HD74LS195AP	1315
HD74LS221	1239
HD74LS221P	1239
HD74LS240	0452
HD74LS240P	0452
HD74LS241	1726
HD74LS241P	1726
HD74LS244	0458
HD74LS244P	0458
HD74LS245WP	0463
HD74LS247	1732
HD74LS247P	1732
HD74LS248	1734
HD74LS248P	1734
HD74LS249	1735
HD74LS249P	1735
HD74LS251	1737
HD74LS251P	1737
HD74LS253	1739
HD74LS253P	1739
HD74LS257	1744
HD74LS257P	1744
HD74LS258	1746
HD74LS258P	1746
HD74LS259	3205
HD74LS259P	3205
HD74LS266	0592
HD74LS266P	0592
HD74LS273P	0894
HD74LS279	3290
HD74LS279P	3290
HD74LS280	1773
HD74LS280P	1773
HD74LS283	1779
HD74LS283P	1779
HD74LS290	4425
HD74LS290P	4425
HD74LS293	0082
HD74LS293P	0082
HD74LS295B	2227
HD74LS295BP	2227
HD74LS298	3371
HD74LS298P	3371
HD74LS299	4426
HD74LS299P	4426
HD74LS365AP	0943
HD74LS367AP	0977
HD74LS373P	0709
HD74LS374P	0713
HD74LS375	OEM
HD74LS375P	OEM
HD74LS386	1230
HD74LS386P	1230
HD74LS390	1288
HD74LS390P	1288
HD74LS393	0819
HD74LS393P	0819
HD74LS490	2213
HD74LS490P	2213
HD74LS610P	OEM
HD74LS668	OEM
HD74LS668P	OEM
HD74LS669	OEM
HD74LS669P	OEM
HD74LSS76P	2180
HD74S00	0704
HD74S00P	0704
HD74S03	2217
HD74S03P	2217
HD74S04	2263
HD74S04P	2263
HD74S05	2321
HD74S05N	OEM
HD74S05P	2321
HD74S10	2445
HD74S10P	2445
HD74S11	2447
HD74S11P	2447
HD74S11S	2447
HD74S15	2451
HD74S15P	2451
HD74S20	1017
HD74S20P	1017
HD74S22	2461
HD74S22P	2461
HD74S40	2475
HD74S40P	2475
HD74S64	2496
HD74S64P	2496
HD74S65	2497
HD74S65P	2497
HD74S74	2504
HD74S74P	2504
HD74S86	2511
HD74S86P	2511
HD74S112	1617
HD74S112P	1617
HD74S113	1623
HD74S113P	1623
HD74S114	1629
HD74S114P	1629
HD74S133	1819
HD74S133P	1819
HD74S134	1827
HD74S134P	1827
HD74S140	1886
HD74S140P	1886
HD74S151	1955
HD74S151P	1955
HD74S174	2132
HD74S174P	2132
HD74S175P	2141
HD74S181	2165
HD74S181P	2165
HD74S251	2198
HD74S251P	2198
HD74S280P	OEM
HD-187	0600
HD224E	OEM
HD237	OEM
HD268T26	OEM
HD268T26P	OEM
HD405P	OEM
HD408B05	OEM
HD468B21	OEM
HD468B21P	OEM
HD740LP	0268
HD1062	OEM
HD1075G	OEM
HD10750	OEM
HD1075R	OEM
HD1075Y	OEM
HD1076G	OEM
HD10760	OEM
HD1076R	OEM
HD1076Y	OEM
HD1077G	OEM
HD10770	OEM
HD1077R	OEM
HD1077Y	OEM
HD1078G	OEM
HD10780	OEM
HD1078R	OEM
HD1078Y	OEM
HD1105G	OEM
HD11050	OEM
HD1105R	OEM
HD1105Y	OEM
HD1106G	OEM
HD11060	OEM
HD1106R	OEM
HD1106Y	OEM
HD1107G	OEM
HD11070	OEM
HD1107R	OEM
HD1107Y	OEM
HD1108G	OEM
HD11080	OEM
HD1108R	OEM
HD1108Y	OEM
HD1111G	OEM
HD11110	OEM
HD1111R	OEM
HD1111Y	OEM
HD1112G	OEM
HD11120	OEM
HD1112R	OEM
HD1112Y	OEM
HD1113G	OEM
HD11130	OEM
HD1113R	OEM
HD1113Y	OEM
HD1115G	OEM
HD11150	OEM
HD1115R	OEM
HD1115Y	OEM
HD1131G	OEM
HD11310	OEM
HD1131R	OEM
HD1131Y	OEM
HD1132G	OEM
HD11320	OEM
HD1132R	OEM
HD1132Y	OEM
HD1133G	OEM
HD11330	OEM
HD1133R	OEM
HD1133Y	OEM
HD1134G	OEM
HD11340	OEM
HD1134R	OEM
HD1134Y	OEM
HD1310	OEM
HD1310A	OEM
HD1320	OEM
HD1320A	OEM
HD1330	OEM
HD1330A	OEM
HD1500-1	OEM
HD1500-2	OEM
HD1500-3	OEM
HD1500-4	OEM
HD1500-5	OEM
HD1500-6	OEM
HD1500-7	OEM
HD1500-8	OEM
HD1500-9	OEM
HD1500-10	OEM
HD1500-12	OEM
HD1500-14	OEM
HD1500-16	OEM
HD1500-18	OEM
HD1500-20	OEM
HD1500-22	OEM
HD1500-24	OEM
HD1500-26	OEM
HD1500-28	OEM
HD1500-30	OEM
HD1705	OEM
HD1705A	OEM
HD1706	OEM
HD1707	OEM
HD1810	OEM
HD1811	OEM
HD1812	OEM
HD1840	OEM
HD1841	OEM
HD1842	OEM
HD1870	OEM
HD1871	OEM
HD1872	OEM
HD2000	0015
HD2000-2	OEM
HD2000-4	OEM
HD2000-6	OEM
HD2000-12	OEM
HD2000-16	OEM
HD2000-20	OEM
HD2000-24	OEM
HD2000-301	0015
HD2100-1	OEM
HD2100-2	OEM
HD2100-3	OEM
HD2100-4	OEM
HD2100-5	OEM
HD2100-6	OEM
HD2100-7	OEM
HD2100-8	OEM
HD2100-9	OEM
HD2100-10	OEM
HD2100-12	OEM
HD2100-14	OEM
HD2100-16	OEM
HD2100-18	OEM
HD2100-20	OEM
HD2135A	OEM
HD2145G	OEM
HD2145R	OEM
HD2201	1041
HD2201P	1041
HD2202	2097
HD2202P	2097
HD2203	0141
HD2203P	0141
HD2204	1823
HD2204P	1823
HD2205	0081
HD2205P	0081
HD2206	1831
HD2206P	1831
HD2207	0562
HD2207P	0562
HD2208	1176
HD2208P	1176
HD2209	0033
HD2209P	0033
HD2210	0333
HD2210P	0333
HD2211	0358
HD2211P	0358
HD2213	1433
HD2213&	1433
HD2214	0466
HD2214P	0466
HD2215	0562
HD2215P	0562
HD2216	1835
HD2216P	1835
HD2500-1	OEM
HD2500-2	OEM
HD2500-3	OEM
HD2500-4	OEM
HD2500-5	OEM

If replacement code is OEM, contact original manufacturer for replacement.

DEVICE TYPE	REPL CODE	DEVICE TYPE	REPL CODE	DEVICE TYPE	REPL CODE	DEVICE TYPE	REPL CODE	DEVICE TYPE	REPL CODE	DEVICE TYPE	REPL CODE	DEVICE TYPE	REPL CODE
HD2500-6	OEM	HD2560	0803	HD7402P	0314	HD10209	OEM	HD14524B	OEM	HD46505SP-1	OEM	HD100114(A)	OEM
HD2500-7	OEM	HD2560P	0803	HD7403	0335	HD10210	OEM	HD14526B	3610	HD46508P	OEM	HD100117	OEM
HD2500-8	OEM	HD2561	1157	HD7403P	0335	HD10211	OEM	HD14527B	3145	HD46508P-1	OEM	HD100117F	OEM
HD2500-9	OEM	HD2561P	1157	HD7404	0361	HD10230	OEM	HD14530B	OEM	HD46508PA-1	OEM	HD100118	OEM
HD2500-10	OEM	HD2562	1856	HD7404P	0361	HD10231	OEM	HD14531B	OEM	HD46818P	OEM	HD100118F	OEM
HD2500-12	OEM	HD2562P	1856	HD7405	0385	HD10551	OEM	HD14532B	1016	HD46846	OEM	HD100122	OEM
HD2500-16	OEM	HD2563	4399	HD7405P	0385	HD14000UB	2024	HD14534B	OEM	HD46846P	OEM	HD100122F	OEM
HD2500-20	OEM	HD2563P	4399	HD7406	1206	HD14001B	0478	HD14536B	3704	HD48801B85	OEM	HD100123	OEM
HD2501	1024	HD2564	1539	HD7406N	1206	HD14001BP	3244	HD14537B	OEM	HD49403NT	OEM	HD100123F	OEM
HD2501P	1024	HD2564P	1539	HD7406P	1206	HD14002B	2055	HD14538B	1063	HD49409FS	OEM	HD100124(A)	OEM
HD2502	1275	HD2572	1950	HD7407	1339	HD14006B	0646	HD14538BP	1063	HD49412FS	OEM	HD100124F(A)	OEM
HD2502P	1275	HD2572P	1950	HD7407P	1339	HD14007UB	2846	HD14539B	OEM	HD49416FS	OEM	HD100125(A)	OEM
HD2503	0232	HD2573	1956	HD7410	0512	HD14008B	0988	HD14541B	5017	HD49704	OEM	HD100125F(A)	OEM
HD2503P	0232	HD2573P	1956	HD7410P	0512	HD14011B	0215	HD14543B	5020	HD49705	OEM	HD100130	OEM
HD2504	0697	HD2580	1555	HD7412	2242	HD14011BP	0215	HD14549B	OEM	HD49707	OEM	HD100130F	OEM
HD2504P	0697	HD2580P	1555	HD7412P	2242	HD14012B	0498	HD14552B	OEM	HD49714ANT	OEM	HD100131	OEM
HD2505	1168	HD2912	OEM	HD7414	2243	HD14013B	0413	HD14553B	OEM	HD49721A	OEM	HD100131F	OEM
HD2505P	1168	HD2916	OEM	HD7414P	2243	HD14013BP	6363	HD14554B	OEM	HD49721NT	OEM	HD100136	OEM
HD2506	0743	HD2923	OEM	HD7416	1349	HD14014B	0860	HD14555B	2938	HD49723ANT	OEM	HD100136F	OEM
HD2506P	0743	HD3000-1	OEM	HD7416P	1349	HD14015B	1014	HD14556B	3431	HD49723NT	OEM	HD100141	OEM
HD2507	0512	HD3000-2	OEM	HD7417	1352	HD14016B	1143	HD14557B	OEM	HD49741	OEM	HD100141F	OEM
HD2507P	0512	HD3000-3	OEM	HD7417P	1352	HD14016BP	4150	HD14558B	5049	HD49741NT	OEM	HD100142	OEM
HD2508	0873	HD3000-4	OEM	HD7420	0697	HD14017	OEM	HD14559B	OEM	HD49747N	OEM	HD100142F	OEM
HD2508P	0873	HD3000-5	OEM	HD7420P	0697	HD14017B	0513	HD14560B	6252	HD49747NT	OEM	HD100145	OEM
HD2509	0268	HD3000-6	OEM	HD7426	0803	HD14018B	1391	HD14562B	OEM	HD61905	OEM	HD100145F	OEM
HD2509P	0268	HD3863	OEM	HD7426P	0803	HD14020B	1661	HD14568B	3618	HD62990	OEM	HD100150F	OEM
HD2510	1313	HD4000	2024	HD7427	0818	HD14020BP	6799	HD14569B	3618	HD62991	OEM	HD100151	OEM
HD2510P	1313	HD4001	0478	HD7427P	0818	HD14021B	1749	HD14572UB	3480	HD68000P8	OEM	HD100151F	OEM
HD2511	0314	HD4001BP	OEM	HD7430	0873	HD14022B	1256	HD14580B	OEM	HD74107	0942	HD100155	OEM
HD2511P	0314	HD4002	2055	HD7430P	0873	HD14023B	0519	HD14581B	OEM	HD74107P	0942	HD100155F	OEM
HD2512	1185	HD4007	2846	HD7432	0899	HD14024B	1957	HD14582B	OEM	HD74121	0175	HD100156	OEM
HD2512P	1185	HD4009	1999	HD7432P	0899	HD14025B	2072	HD14584BP	3626	HD74121P	0175	HD100156F	OEM
HD2513	1574	HD4011	0215	HD7438P	0996	HD14027B	1949	HD14681BP	OEM	HD74123P	1157	HD100158	OEM
HD2513P	1574	HD4012	0498	HD7440	1024	HD14028B	2228	HD-20008	0102	HD74125	1182	HD100158F	OEM
HD2514	1202	HD4013	0413	HD7440P	1024	HD14028BP	4186	HD36990	OEM	HD74125P	1182	HD100160	OEM
HD2514P	1202	HD4015	1014	HD7442A	1052	HD14032B	2531	HD37980	OEM	HD74126	1193	HD100160F	OEM
HD2515	1172	HD4016	1143	HD7442AP	1052	HD14034B	3615	HD38480A	OEM	HD74126P	1193	HD100163	OEM
HD2515P	1172	HD4017	0513	HD7443A	1060	HD14035B	2777	HD38570A91	OEM	HD74132	1271	HD100163F	OEM
HD2516	1158	HD4019	1525	HD7443AP	1060	HD14038B	2981	HD38630A26	OEM	HD74132P	1271	HD100164	OEM
HD2516P	1158	HD4020	1661	HD7444A	1072	HD14040B	0056	HD38700A01	OEM	HD74150	1492	HD100164F	OEM
HD2517	1432	HD4021	1749	HD7444AP	1072	HD14040BP	6931	HD38701A06	OEM	HD74150P	1492	HD100165	OEM
HD2517P	1432	HD4023	0519	HD7450	0743	HD14042B	0121	HD38702A23	OEM	HD74151A	1495	HD100165F	OEM
HD2519	1208	HD4024	1957	HD7450P	0743	HD14042BP	0121	HD38702A25	OEM	HD74151AP	1495	HD100166	OEM
HD2519P	1208	HD4025	2072	HD7451	1168	HD14042P	0121	HD38750-B05	OEM	HD74153P	4444	HD100166F	OEM
HD2520	0569	HD4027	1949	HD7451P	1168	HD14043B	1553	HD38750A51	OEM	HD74155	1576	HD100170	OEM
HD2520P	0569	HD4030	0500	HD7453	1185	HD14044B	2308	HD38750A53	OEM	HD74155P	1576	HD100170F	OEM
HD2521	0834	HD4040	0056	HD7453P	1185	HD14044BP	2308	HD38750A78	OEM	HD74156	1592	HD100171	OEM
HD2521P	0834	HD4049	0001	HD7454	1202	HD14046BP	6970	HD38750A79	OEM	HD74156P	1592	HD100171F	OEM
HD2522	0361	HD4050	0398	HD7454P	1202	HD14049UB	0001	HD38750A85	OEM	HD74159P	OEM	HD100179	OEM
HD2522P	0361	HD4101	OEM	HD7460	1275	HD14050B	0398	HD38750A89	OEM	HD74160	1631	HD100179F	OEM
HD2523	0385	HD4102	OEM	HD7460P	1275	HD14050BP	0398	HD38750A91	OEM	HD74160P	1631	HD100180	OEM
HD2523P	0385	HD4285	OEM	HD7471P	1352	HD14051B	0366	HD38757A10	OEM	HD74161	1645	HD100180F	OEM
HD2524	0980	HD4518	1043	HD7472	1426	HD14051BP	0366	HD38800A50	OEM	HD74161P	1645	HD100181	OEM
HD2524P	0980	HD4702-2	OEM	HD7472P	1426	HD14052	OEM	HD38800A57	OEM	HD74162	1013	HD100181F	OEM
HD2526	1368	HD4702-9	OEM	HD7473AP	1172	HD14052B	0024	HD38800B11	OEM	HD74162P	1013	HD146818P	OEM
HD2526P	1368	HD5000	OEM	HD7473P	1172	HD14052BP	0024	HD38800B11(BUP)	OEM	HD74163	1666	HD200207	0015
HD2528	0335	HD5001	OEM	HD7474	1313	HD14053B	0035	HD38800B27	OEM	HD74163P	1666	HD200301	0015
HD2528P	0335	HD5002	OEM	HD7474P	1313	HD14053BFP	OEM	HD38800B28	OEM	HD74164	0734	HD252538	1072
HD2529	1426	HD5003	OEM	HD7475	1432	HD14053BP	0035	HD38800B28(AUP)	OEM	HD74164P	0734	HD300101-0	OEM
HD2529P	1426	HD5004	OEM	HD7475P	1432	HD14066B	0101	HD38800B32	OEM	HD74166	0231	HD404302SA02	OEM
HD2530	0942	HD6101-2	OEM	HD7485	0374	HD14066BCP	0101	HD38800R04	OEM	HD74166P	0231	HD404708H-A35	OEM
HD2530P	0942	HD6101-9	OEM	HD7485P	0374	HD14066BP	0101	HD38800R05	OEM	HD74174	1770	HD404708H-A51	OEM
HD2531	1082	HD6101C-9	OEM	HD7486	1368	HD14068B	2503	HD38805A22	OEM	HD74174P	1770	HD404709A01S	OEM
HD2531P	1082	HD6147	0015	HD7486P	1368	HD14069UB	0119	HD38805SA44	OEM	HD74175	1787	HD404709S	OEM
HD2532	1108	HD6301V1E01P	OEM	HD7490A	1208	HD14070B	2516	HD38820-L39	OEM	HD74175P	1787	HD440071	OEM
HD2532P	1108	HD6301YOPC06	OEM	HD7490AP	1208	HD14071B	0129	HD38820L26	OEM	HD74177P	1803	HD440072	OEM
HD2533	1702	HD6303YP	OEM	HD7492A	0834	HD14072B	2524	HD38820L35	OEM	HD74180	1829	HD614042-XXXX(1283M)	
HD2533P	1702	HD6305Y0A05P	OEM	HD7492AP	0834	HD14073B	1536	HD38820L38	OEM	HD74180P	1829		OEM
HD2534	1486	HD6305Y0A08P	OEM	HD7493A	0569	HD14075B	2540	HD38820L39	OEM	HD74190	1912	HD614042F	OEM
HD2534P	1486	HD6305Y0C28P	OEM	HD7493AD	0569	HD14076B	3496	HD38820L44	OEM	HD74190P	1912	HD614042FB93	OEM
HD2535	0117	HD6305Y1FC24	OEM	HD7493AP	0569	HD14077B	2558	HD38820L47	OEM	HD74191	1917	HD614042FF41	OEM
HD2535P	0117	HD6305Y2F	OEM	HD7493P	0569	HD14078B	0921	HD38825L06	OEM	HD74191P	1917	HD614042FG43	OEM
HD2536	1052	HD6305Y0A04P	OEM	HD7496	1715	HD14081B	0626	HD38825L11	OEM	HD74194	1931	HD614042FG66	OEM
HD2536P	1052	HD6305Y0A06P	OEM	HD7496P	1715	HD14081BP	0626	HD38870L46	OEM	HD74194P	1931	HD614042FG74	OEM
HD2537	1060	HD6305Y0C28P	OEM	HD7601P	OEM	HD14082B	0300	HD38941	OEM	HD74198	1964	HD614042FG86	OEM
HD2537P	1060	HD6305Y0C29P	OEM	HD10101	OEM	HD14093B	2384	HD38980A	2728	HD74198P	1964	HD614042FH48	OEM
HD2538P	1072	HD6305Y0C53P	OEM	HD10102	OEM	HD14101G	OEM	HD38980C	2728	HD74199	1971	HD614042FH55	OEM
HD2539	1404	HD6402C9	OEM	HD10104	OEM	HD14101R	OEM	HD38986	OEM	HD74199P	1971	HD614042FH60	OEM
HD2539P	1404	HD6800	OEM	HD10105	OEM	HD14160B	OEM	HD38991	2295	HD75107A	OEM	HD614042SC23	OEM
HD2540	1722	HD6802	OEM	HD10106	OEM	HD14161B	OEM	HD38991A	2728	HD75107AP	OEM	HD614042SC67	OEM
HD2540P	1722	HD6803-1	OEM	HD10107	OEM	HD14162B	OEM	HD38992	OEM	HD75108A	OEM	HD614042SD76	OEM
HD2541	1921	HD6803P-1	OEM	HD10109	OEM	HD14163B	OEM	HD38993	4378	HD75108AP	OEM	HD614042SD97	OEM
HD2541P	1921	HD6805S1P-A05	OEM	HD10110	OEM	HD14174B	1551	HD38993A	4378	HD75109	OEM	HD614042SE15	OEM
HD2542	1926	HD6805S1PA22	OEM	HD10111	OEM	HD14175B	1528	HD38998A	OEM	HD75109P	OEM	HD614042SE25	OEM
HD2542P	1926	HD6805V1B20P	OEM	HD10116	OEM	HD14194B	1769	HD42851	3815	HD75110	OEM	HD614042SE43	OEM
HD2543	0175	HD6805V1N06P	OEM	HD10117	OEM	HD14443B	OEM	HD42851A2	3815	HD75110P	OEM	HD614042SE46	OEM
HD2543P	0175	HD6805V1N13P	OEM	HD10118	OEM	HD14447B	OEM	HD42853	4639	HD75154	OEM	HD614042SE47	OEM
HD2544	0996	HD6805V1P	OEM	HD10119	OEM	HD14501UB	3809	HD42856	OEM	HD75154P	OEM	HD614042SE57	OEM
HD2544P	0996	HD6805W	OEM	HD10121	OEM	HD14502B	1037	HD43019A	OEM	HD75188	0508	HD614042SE60	OEM
HD2545	1441	HD6805W0P	OEM	HD10124	OEM	HD14503B	2053	HD44007A	OEM	HD75188P	0508	HD614042SE94	OEM
HD2545P	1441	HD6805X2P	OEM	HD10130	OEM	HD14503P	2053	HD44015	OEM	HD75189	0511	HD614042SF07	OEM
HD2546	1715	HD6809	5967	HD10131	OEM	HD14505B	OEM	HD44040	OEM	HD75189P	0511	HD614042SF11	OEM
HD2546P	1715	HD6809EP	2117	HD10132	OEM	HD14506B	3767	HD44042	OEM	HD75450A	1231	HD614042SF44	OEM
HD2547	1842	HD6821	OEM	HD10133	OEM	HD14508B	1811	HD44053	OEM	HD75451A	1244	HD614042SF45	OEM
HD2547P	1842	HD6821P	0448	HD10134	OEM	HD14510B	1963	HD44100	OEM	HD75451AP	1244	HD614042SG25	OEM
HD2548	1492	HD6843SC	OEM	HD10136	OEM	HD14511B	1544	HD44753	OEM	HD75452	1262	HD614042SG47	OEM
HD2548P	1492	HD6843SP	OEM	HD10145	OEM	HD14511BP	1670	HD44801	OEM	HD75452P	1262	HD614042SG69	OEM
HD2549	1495	HD6844C	OEM	HD10147	OEM	HD14512B	2120	HD44801A05	OEM	HD75453	1272	HD614042SH58	OEM
HD2549P	1495	HD6844P	OEM	HD10148	OEM	HD14512BP	4217	HD44801SB82	OEM	HD75453P	1272	HD614042SH75	OEM
HD2550	0467	HD6845C	OEM	HD10160	OEM	HD14514B	1830	HD44820A17	OEM	HD75454	1289	HD614042SH89	OEM
HD2550P	0467	HD6845P	OEM	HD10161	OEM	HD14515B	1830	HD44820A48	OEM	HD75454P	1289	HD614042SJ57	OEM
HD2551	0492	HD6845SP	OEM	HD10162	OEM	HD14516B	2347	HD44820B02	OEM	HD100101	OEM	HD614043SA09	OEM
HD2551P	0492	HD6850	OEM	HD10164	OEM	HD14517B	4284	HD44820B24	OEM	HD100101F	OEM	HD614043XXXX	OEM
HD2552	3520	HD6852	OEM	HD10165	OEM	HD14518B	1043	HD44820B36	OEM	HD100102	OEM	HD614045SE85	OEM
HD2552P	3520	HD6865	0015	HD10174	OEM	HD14519B	OEM	HD44860A92	OEM	HD100102F	OEM	HD614080-SA02	OEM
HD2555	0619	HD7400	0232	HD10175	OEM	HD14520B	2675	HD44860B23	OEM	HD100107	OEM	HD614080-SA05	OEM
HD2555P	0619	HD7400P	0232	HD10179	OEM	HD14520BP	5902	HD44860B39	OEM	HD100107F	OEM	HD614080-SA19	OEM
HD2558	1377	HD7401P	0268	HD10180	OEM	HD14521B	OEM	HD46505RP-2	OEM	HD100112	OEM	HD614080SA97	OEM
HD2558P	1377	HD7402	0314	HD10181	OEM	HD14522B	OEM	HD46505SP	OEM	HD100112F	OEM	HD614080SB05	OEM

If replacement code is OEM, contact original manufacturer for replacement.

DEVICE TYPE	REPL CODE
HD614080SB70	OEM
HD614085S	OEM
HD614085SA06	OEM
HD614085SA14	OEM
HD614085SA39	OEM
HD614085SA64	OEM
HD614088SB42	OEM
HD614088SC56	OEM
HD614088SC58	OEM
HD614120PA22	OEM
HD614128SA61	OEM
HD-1000101	0143
HD1000101	0143
HD1000101-0	0143
HD1000105	0123
HD1000105-0	0123
HD1000107	0143
HD1000107-0	0143
HD1000301	0143
HD1000302	0143
HD1000303	0143
HD1000303-0	0143
HD1000309	0649
HD1000320	0123
HD-2000106	0133
HD2000106	0133
HD-2000108	0102
HD2000108	0087
HD2000110	0229
HD2000110-0	0015
HD2000206	0133
HD2000301	0015
HD2000301-0	0015
HD2000305	0015
HD2000307	0015
HD-2000308	0015
HD2000308	0015
HD2000413	0015
HD2000501	0023
HD2000510	0015
HD2000610	0015
HD2000703	0015
HD2000710	0196
HD2000803	0015
HD2000903	0015
HD2001005	0133
HD2001105	0133
HD2001310	0015
HD3000101-0	0137
HD3000109-0	0317
HD3000113	0317
HD3000201-0	0726
HD3000213	0010
HD-3000301	0015
HD3000309	0649
HD3000309-0	0649
HD3000401	0012
HD-3000409	0649
HD3000409	0244
HD3001009	0317
HD3001009-0	0181
HD3001109	0002
HD3001109-0	0100
HD3001809	0317
HD3002109	0100
HD3002409	0052
HD3003109	0244
HD3003309	0041
HD4000109	0143
HD4000909	0030
HD4074709A01S	OEM
HD4074709HCF001	OEM
HD4074709S	OEM
HD10001010	0143
HD10001050	0019
HD10002020	0143
HD20001100	0015
HD20001210	0133
HD20003010	0015
HD20005100	0087
HD20007030	0015
HD20011050	0133
HD20023100	0015
HD30001090	0181
HD30001200	0170
HD30010090	0317
HD30011090	0100
HD30017090	0012
HD30019090	0649
HD30023090	0062
HD30042090	0157
HD40001060	0030
HD40009090	0030
HD300223090	OEM
HDA412	OEM
HDA420	OEM
HDA496	OEM
HDAS-8	OEM
HDAS-16	OEM
HDK8610	OEM
HDS1395	OEM
HDS9001	OEM
HDS9009	OEM
HDS9010	0396
HDSP-0760	OEM
HDSP-0761	OEM
HDSP-0762	OEM
HDSP-0763	OEM
HDSP-0770	OEM

DEVICE TYPE	REPL CODE
HDSP-0771	OEM
HDSP-0772	OEM
HDSP-0860	OEM
HDSP-0863	OEM
HDSP-0960	OEM
HDSP-0961	OEM
HDSP-0962	OEM
HDSP-0963	OEM
HDSP-2000	OEM
HDSP-2001	OEM
HDSP2002	OEM
HDSP-2010	OEM
HDSP2300	OEM
HDSP2301	OEM
HDSP2302	OEM
HDSP-2416	OEM
HDSP-2424	OEM
HDSP-2432	OEM
HDSP-2440	OEM
HDSP-2470	OEM
HDSP-2471	OEM
HDSP-2472	OEM
HDSP2490	OEM
HDSP2491	OEM
HDSP2492	OEM
HDSP-3400	OEM
HDSP-3401	OEM
HDSP-3403	OEM
HDSP-3405	OEM
HDSP-3406	OEM
HDSP-3530	5134
HDSP-3531	5127
HDSP-3533	OEM
HDSP-3536	OEM
HDSP-3730	OEM
HDSP-3731	OEM
HDSP-3733	OEM
HDSP-3736	OEM
HDSP-3900	OEM
HDSP-3901	OEM
HDSP-3903	OEM
HDSP-3905	OEM
HDSP-3906	OEM
HDSP-4030	OEM
HDSP-4031	5129
HDSP-4033	OEM
HDSP-4036	OEM
HDSP-4130	OEM
HDSP-4131	OEM
HDSP-4133	OEM
HDSP-4136	OEM
HDSP-4200	OEM
HDSP-4201	OEM
HDSP-4203	OEM
HDSP-4205	OEM
HDSP-4206	OEM
HDSP-6300	OEM
HDSP-6504	OEM
HDSP6504	OEM
HDSP-6508	OEM
HDSP6508	OEM
HDSP-8716	OEM
HDSP-8724	OEM
HDSP-8732	OEM
HDSP-8740	OEM
HDU0014	OEM
H-DX0038CE	0023
H-DX0046CE	0124
H-DX0086TA	0023
H-DX0105TA	0023
H-DX0181CE	0031
H-DX0248CE	0023
HE-0A90	0143
HE-1N34	0143
HE-1N34A	0143
HE-1N60	0019
HE-1N60P	0143
HE-1S188	0143
HE-1S426	0143
HE-1S446	0143
HE-1S555	0628
HE-56-56	0124
HE-56-58	0471
HE-56-73	0133
HE-56-93	0133
HE-56-655	6985
HE-57-27	0015
HE-57-42	0087
HE-57-64	0071
HE-57-65	0080
HE-57-614	0817
HE212	OEM
HE212E	OEM
HE215	OEM
HE215E	OEM
HE224	OEM
HE-234-264	0023
HE-234-266	0137
HE-234-267	0023
HE-234-271	0164
HE-234-272	0164
HE-234-273	0285
HE-234-274	0016
HE-234-275	0079
HE-234-276	3014
HE-234-290	0079
HE-234-299	0124
HE-234-353	0023
HE-234-521	0102

DEVICE TYPE	REPL CODE
HE-234-522	0079
HE-234-523	0079
HE-234-526	0023
HE237E	OEM
HE252	OEM
HE252E	OEM
HE381	OEM
HE381E	OEM
HE383	OEM
HE383E	OEM
HE384	OEM
HE384E	OEM
HE-417-195	0276
HE-417-282	0543
HE-417-811	0076
HE-417-821	0079
HE-417-822	0037
HE-417-823	0316
HE-417-834	0338
HE-417-864	0086
HE-417-865	0037
HE-417-874	0006
HE-417-875	0076
HE-417-923	0313
HE-417-924	0681
HE-417-926	0551
HE-417-932	0561
HE-417-937	0079
HE-433-10	OEM
HE-442-4	2707
HE-442-21	0360
HE-442-22	0312
HE-442-24	2169
HE-442-25	2222
HE-442-30	1922
HE-442-39	1300
HE-442-48	OEM
HE-442-50	OEM
HE-442-54	0624
HE-442-60	2239
HE-442-63	1321
HE-442-66	2246
HE-442-71	2247
HE-442-73	OEM
HE-442-74	0372
HE-442-75	2104
HE-442-99	1143
HE-442-602	0625
HE-442-604	3729
HE-442-613	3829
HE-442-615	0692
HE-442-622	OEM
HE-442-623	2782
HE-442-624	OEM
HE-442-627	1292
HE-442-630	1285
HE-442-635	3972
HE-442-636	4799
HE-442-640	2748
HE-442-644	1828
HE-442-647	3428
HE-442-648	2517
HE-442-650	OEM
HE-442-651	2863
HE-442-654	3381
HE-442-655	OEM
HE-442-659	3920
HE-442-663	0334
HE-442-664	1838
HE-442-665	4508
HE-442-672	4106
HE-442-673	4798
HE-442-674	0334
HE-442-675	1838
HE-442-676	OEM
HE-442-677	4727
HE-442-682	3893
HE-442-686	0379
HE-442-687	OEM
HE-442-688	3781
HE-442-691	1196
HE-442-693	OEM
HE-442-702	1916
HE-442-703	OEM
HE-442-715	2782
HE-443-3	0873
HE-443-4	1426
HE-443-5	1172
HE-443-6	1313
HE-443-7	1208
HE-443-9	OEM
HE-443-10	OEM
HE-443-12	0512
HE-443-13	1432
HE-443-15	0743
HE-443-16	1158
HE-443-17	1377
HE-443-21	OEM
HE-443-22	0175
HE-443-23	1139
HE-443-25	1495
HE-443-26	0704
HE-443-34	0834
HE-443-36	1108
HE-443-44	1441
HE-443-45	0467
HE-443-46	0314
HE-443-53	1052
HE-443-54	0335

DEVICE TYPE	REPL CODE
HE-443-66	1921
HE-443-68	0685
HE-443-70	2969
HE-443-72	1352
HE-443-73	1349
HE-443-74	1262
HE-443-77	0996
HE-443-87	0619
HE-443-89	0492
HE-443-90	1157
HE-443-460	OEM
HE-443-603	0215
HE-443-604	2846
HE-443-606	1949
HE-443-607	0413
HE-443-612	1926
HE-443-622	1451
HE-443-623	1555
HE-443-625	1271
HE-443-626	OEM
HE-443-628	1950
HE-443-629	1208
HE-443-640	0569
HE-443-642	0385
HE-443-680	1486
HE-443-683	OEM
HE-443-695	0478
HE-443-696	OEM
HE-443-703	0478
HE-443-704	2055
HE-443-706	0129
HE-443-707	3600
HE-443-708	2347
HE-443-709	2177
HE-443-711	1732
HE-443-712	2072
HE-443-713	2228
HE-443-717	1193
HE-443-718	1315
HE-443-719	0592
HE-443-720	3180
HE-443-721	OEM
HE-443-722	2267
HE-443-727	1468
HE-443-728	1527
HE-443-729	1688
HE-443-730	0243
HE-443-731	4425
HE-443-732	0828
HE-443-733	0082
HE-443-737	1043
HE-443-738	3431
HE-443-745	1579
HE-443-751	0626
HE-443-752	1672
HE-443-753	OEM
HE-443-754	0452
HE-443-755	1595
HE-443-757	0858
HE-443-762	4547
HE-443-764	2048
HE-443-769	4339
HE-443-777	3611
HE-443-778	2384
HE-443-779	1559
HE-443-780	1633
HE-443-781	1870
HE-443-782	0209
HE-443-783	1256
HE-443-784	0500
HE-443-785	3047
HE-443-791	0458
HE-443-792	1625
HE-443-794	0508
HE-443-795	0511
HE-443-797	1662
HE-443-798	0034
HE-443-799	1161
HE-443-800	0183
HE-443-801	2834
HE-443-802	1744
HE-443-804	3205
HE-443-805	0894
HE-443-807	1841
HE-443-811	0075
HE-443-813	1882
HE-443-814	OEM
HE-443-815	1692
HE-443-816	1642
HE-443-817	1689
HE-443-818	1608
HE-443-819	OEM
HE-443-822	0153
HE-443-824	1726
HE-443-827	0388
HE-443-828	1867
HE-443-829	2180
HE-443-836	1551
HE-443-837	0709
HE-443-838	1888
HE-443-839	0906
HE-443-840	4181
HE-443-841	1116
HE-443-842	OEM
HE-443-843	0448
HE-443-854	3290
HE-443-855	1779
HE-443-856	OEM
HE-443-857	0977

DEVICE TYPE	REPL CODE
HE-443-858	2243
HE-443-860	OEM
HE-443-863	0713
HE-443-864	1667
HE-443-867	0709
HE-443-871	1830
HE-443-872	1698
HE-443-874	OEM
HE-443-875	0088
HE-443-877	0427
HE-443-879	0260
HE-443-881	3481
HE-443-884	1728
HE-443-885	0463
HE-443-886	0498
HE-443-887	0519
HE-443-888	3767
HE-443-889	1564
HE-443-891	0291
HE-443-892	4371
HE-443-896	2238
HE-443-897	2263
HE-443-898	0780
HE-443-899	2497
HE-443-900	2504
HE-443-901	2134
HE-443-904	0522
HE-443-906	OEM
HE-443-909	0521
HE-443-912	3912
HE-443-913	OEM
HE-443-915	2511
HE-443-916	3611
HE-443-919	2877
HE-443-920	0431
HE-443-921	1288
HE-443-927	3308
HE-443-928	0988
HE-443-929	0513
HE-443-930	2331
HE-443-934	0893
HE-443-935	2151
HE-443-942	0979
HE-443-944	OEM
HE-443-958	0146
HE-443-967	1206
HE-443-970	2357
HE-443-973	0819
HE-443-983	2141
HE-443-985	2141
HE-443-991	0398
HE-443-998	OEM
HE-443-999	3141
HE-443-1001	1773
HE-443-1009	OEM
HE-443-1011	OEM
HE-443-1012	OEM
HE-443-1020	1339
HE-443-1021	0051
HE-443-1024	0991
HE-443-1031	OEM
HE-443-1034	1839
HE-443-1036	1654
HE-443-1037	1744
HE-443-1040	2934
HE-443-1045	OEM
HE-443-1047	OEM
HE-443-1051	OEM
HE-443-1053	2132
HE-443-1058	2547
HE-443-1066	OEM
HE-443-1070	OEM
HE-443-1072	OEM
HE-443-1073	OEM
HE-443-1078	OEM
HE-443-1080	OEM
HE-443-1082	OEM
HE-443-1084	OEM
HE-443-1097	OEM
HE-443-1098	OEM
HE-443-1112	2286
HE-443-1119	OEM
HE-443-1165	OEM
HE-443-1166	OEM
HE-443-1167	OEM
HE-443-1168	OEM
HE-443-1169	OEM
HE-443-1170	OEM
HE-443-1172	OEM
HE-443-1173	1130
HE-443-1174	OEM
HE-443-1178	OEM
HE-443-1198	OEM
HE-443-1228	OEM
HE-443-1229	OEM
HE-443-1230	OEM
HE-443-1231	OEM
HE-443-1232	OEM
HE-443-1254	OEM
HE-444-224	OEM
HE-444-225	OEM
HE-444-226	OEM
HE-444-227	OEM
HE-444-228	2224
HE-444-229-4	1638
HE-444-230-1	OEM
HE-444-230-2	OEM
HE-444-231	OEM
HE-444-232	OEM

DEVICE TYPE	REPL CODE
HE-444-234	OEM
HE-444-235	OEM
HE-444-238	OEM
HE-444-238-1	OEM
HE-444-241	OEM
HE-444-260-4	1638
HE-444-274	OEM
HE-444-287	OEM
HE-444-292	OEM
HE-444-314	OEM
HE-444-320	OEM
HE500	OEM
HE581	OEM
HE581E	OEM
HE583	OEM
HE583E	OEM
HE584	OEM
HE584E	OEM
HE-1024	0143
HE8006	OEM
HE8101	OEM
HE-10001	0143
HE-10002	0019
HE-10003	0143
HE-10024	0143
HE-10025	2232
HE-10027	0143
HE-10030	0133
HE-10044	0123
HE-15188	0143
HE-15426	0143
HE-15446	0143
HE-20011	0015
HE-20049	0649
HE98241A	OEM
HE98245	OEM
HE-CD0000	0143
HEDS-1000	OEM
HEDS-3000	OEM
HEF0175P	OEM
HEF4000	OEM
HEF4000B	2024
HEF4000BD	2024
HEF4000BP	2024
HEF4000BT	OEM
HEF4000P	2024
HEF4001	OEM
HEF4001BD	0478
HEF4001BP	0478
HEF4001BT	OEM
HEF4001P	0478
HEF4001U	OEM
HEF4001UBD	OEM
HEF4001UBP	OEM
HEF4001UBT	OEM
HEF4002	OEM
HEF4002BD	2055
HEF4002BP	2055
HEF4002BT	OEM
HEF4002P	2055
HEF4006B	0646
HEF4006BD	0646
HEF4006BP	0646
HEF4006BT	OEM
HEF4006P	0646
HEF4008	OEM
HEF4008B	0988
HEF4008BD	0988
HEF4008BP	0988
HEF4008BT	OEM
HEF4008P	0988
HEF4011	OEM
HEF4011BD	0215
HEF4011BP	0215
HEF4011BT	OEM
HEF4011P	0215
HEF4011U	OEM
HEF4011UBD	OEM
HEF4011UBP	OEM
HEF4011UBT	OEM
HEF4012	0498
HEF4012BD	0498
HEF4012BP	0498
HEF4012BT	OEM
HEF4012P	0498
HEF4013	OEM
HEF4013BD	0413
HEF4013BP	0413
HEF4013BT	OEM
HEF4013P	0413
HEF4014B	0860
HEF4014BD	0860
HEF4014BP	0860
HEF4014BT	OEM
HEF4014P	0860
HEF4015B	1014
HEF4015BD	1830
HEF4015BP	1830
HEF4015BT	OEM
HEF4015P	1014
HEF4016BD	1143
HEF4016BP	1143
HEF4016BT	OEM
HEF4017	OEM
HEF4017BD	0513
HEF4017BP	0513
HEF4017BT	OEM
HEF4017P	0513
HEF4018	OEM

DEVICE TYPE	REPL CODE
HEF4018BD	1391
HEF4018BP	1391
HEF4018BT	OEM
HEF4018P	1391
HEF4019BD	1525
HEF4019BP	1525
HEF4019BT	OEM
HEF4020	OEM
HEF4020BD	1661
HEF4020BP	1661
HEF4020BT	OEM
HEF4020P	1661
HEF4021B	1749
HEF4021BD	1749
HEF4021BP	1749
HEF4021BT	OEM
HEF4021P	1749
HEF4022	OEM
HEF4022BD	1256
HEF4022BP	1256
HEF4022BT	OEM
HEF4022P	1256
HEF4023	OEM
HEF4023BD	0519
HEF4023BP	0519
HEF4023BT	OEM
HEF4023P	0519
HEF4024	OEM
HEF4024BD	1957
HEF4024BP	1957
HEF4024BT	OEM
HEF4024P	1957
HEF4025	OEM
HEF4025BD	2072
HEF4025BP	2072
HEF4025BT	OEM
HEF4025P	2072
HEF4027	OEM
HEF4027BD	1949
HEF4027BP	1949
HEF4027BT	OEM
HEF4027P	1949
HEF4028	OEM
HEF4028B	2228
HEF4028BD	2228
HEF4028BP	2228
HEF4028BT	OEM
HEF4028P	2228
HEF4029	OEM
HEF4029B	2233
HEF4029BD	2233
HEF4029BP	2233
HEF4029BT	OEM
HEF4029P	2233
HEF4030	OEM
HEF4030BD	0500
HEF4030BP	0500
HEF4030BT	OEM
HEF4030P	0500
HEF4031B	2971
HEF4031BD	2971
HEF4031BP	2971
HEF4031BT	OEM
HEF4031P	2971
HEF4032P	OEM
HEF4035B	2777
HEF4035BD	2777
HEF4035BP	2777
HEF4035BT	OEM
HEF4035P	2777
HEF4040	OEM
HEF4040BD	0056
HEF4040BP	0056
HEF4040BT	OEM
HEF4040P	0056
HEF4041BD	3174
HEF4041BP	3174
HEF4041BT	OEM
HEF4042	OEM
HEF4042BD	0121
HEF4042BP	0121
HEF4042BT	OEM
HEF4042P	0121
HEF4043	OEM
HEF4043BD	1553
HEF4043BP	1553
HEF4043BT	OEM
HEF4043P	1553
HEF4044	OEM
HEF4044B	2308
HEF4044BD	2308
HEF4044BP	2308
HEF4044BT	OEM
HEF4044P	2308
HEF4047	OEM
HEF4047B	2331
HEF4047BD	2331
HEF4047BP	2331
HEF4047BT	OEM
HEF4049BD	0001
HEF4049BP	0001
HEF4049BT	OEM
HEF4049P	5476
HEF4050BD	0398
HEF4050BP	0398
HEF4050BT	OEM
HEF4050P	OEM
HEF4051BD	0366
HEF4051BP	0366

If replacement code is OEM, contact original manufacturer for replacement.

DEVICE TYPE	REPL CODE
HEF4051BT	OEM
HEF4052BD	0024
HEF4052BP	0024
HEF4052BT	OEM
HEF4053BD	0035
HEF4053BP	0035
HEF4053BT	OEM
HEF4059BD	OEM
HEF4059BP	OEM
HEF4060BD	OEM
HEF4060BP	OEM
HEF4060BT	OEM
HEF4066BD	0101
HEF4066BP	0101
HEF4066BT	OEM
HEF4067BD	4279
HEF4067BP	4279
HEF4068	OEM
HEF4068BD	2503
HEF4068BP	2503
HEF4068BT	OEM
HEF4068P	2503
HEF4069P	0119
HEF4069UBD	0119
HEF4069UBP	0119
HEF4069UBT	OEM
HEF4070	OEM
HEF4070BD	2516
HEF4070BP	2516
HEF4070BT	OEM
HEF4070P	2516
HEF4071	OEM
HEF4071BD	0129
HEF4071BP	0129
HEF4071BT	OEM
HEF4071P	0129
HEF4072	OEM
HEF4072BD	2524
HEF4072BP	2524
HEF4072BT	OEM
HEF4072P	2524
HEF4073	OEM
HEF4073B	1536
HEF4073BD	1536
HEF4073BP	1536
HEF4073BT	OEM
HEF4073P	1536
HEF4075	OEM
HEF4075B	2540
HEF4075BD	2540
HEF4075BP	2540
HEF4075BT	OEM
HEF4076B	3496
HEF4076BD	3496
HEF4076BP	3496
HEF4076BT	OEM
HEF4076P	3496
HEF4077	OEM
HEF4077B	2558
HEF4077BD	2558
HEF4077BP	2558
HEF4077BT	OEM
HEF4077P	2558
HEF4078	OEM
HEF4078BD	0921
HEF4078BP	0921
HEF4078BT	OEM
HEF4078P	0921
HEF4081	OEM
HEF4081BD	0626
HEF4081BP	0626
HEF4081BT	OEM
HEF4081P	0626
HEF4082	OEM
HEF4082BD	0300
HEF4082BP	0300
HEF4082BT	OEM
HEF4082P	0300
HEF4085	OEM
HEF4085BD	0303
HEF4085BP	0303
HEF4085BT	OEM
HEF4085P	0303
HEF4086	OEM
HEF4086B	0535
HEF4086BD	0535
HEF4086BP	0535
HEF4086BT	OEM
HEF4086P	0535
HEF4093BD	2384
HEF4093BP	2384
HEF4093BT	OEM
HEF4094BD	OEM
HEF4094BP	OEM
HEF4094BT	OEM
HEF4104BD	OEM
HEF4104BP	OEM
HEF4502BD	1037
HEF4502BP	1037
HEF4502BT	OEM
HEF4505BD	OEM
HEF4505BP	OEM
HEF4508	OEM
HEF4508BD	1811
HEF4508BP	1811
HEF4510	OEM
HEF4510B	1963
HEF4510BD	1963
HEF4510BP	1963

DEVICE TYPE	REPL CODE
HEF4511B	1544
HEF4511BD	1544
HEF4511BP	1544
HEF4511BT	OEM
HEF4512BD	2120
HEF4512BP	2120
HEF4512BT	OEM
HEF4514BD	3600
HEF4514BP	1830
HEF4515BD	3600
HEF4515BP	3600
HEF4516	OEM
HEF4516B	2347
HEF4516BD	2347
HEF4516BP	2347
HEF4516BT	OEM
HEF4517BD	OEM
HEF4517BP	OEM
HEF4518	OEM
HEF4518BD	1043
HEF4518BP	1043
HEF4518BT	OEM
HEF4518P	1043
HEF4519BD	OEM
HEF4519BP	OEM
HEF4519BT	OEM
HEF4520	OEM
HEF4520BD	2675
HEF4520BP	2675
HEF4520BT	OEM
HEF4520P	2675
HEF4521	OEM
HEF4521BD	OEM
HEF4521BP	OEM
HEF4521BT	OEM
HEF4522	OEM
HEF4522BD	2837
HEF4522BP	2837
HEF4522BT	OEM
HEF4526	OEM
HEF4526BD	OEM
HEF4526BP	OEM
HEF4526BT	OEM
HEF4527BD	OEM
HEF4527BP	OEM
HEF4528	OEM
HEF4528B	3198
HEF4528BD	3198
HEF4528BP	3198
HEF4528BT	OEM
HEF4531	OEM
HEF4531B	3325
HEF4531BD	3325
HEF4531BP	3325
HEF4531BT	OEM
HEF4532BD	OEM
HEF4532BP	OEM
HEF4534	OEM
HEF4534BD	OEM
HEF4534BP	OEM
HEF4538BD	OEM
HEF4538BP	1063
HEF4538BT	OEM
HEF4539BD	3656
HEF4539BP	3656
HEF4539BT	OEM
HEF4541BD	OEM
HEF4541BP	OEM
HEF4541BT	OEM
HEF4543BD	OEM
HEF4543BP	OEM
HEF4543BT	OEM
HEF4555BD	2938
HEF4555BP	2938
HEF4556BD	3431
HEF4556BP	3431
HEF4557BD	OEM
HEF4557BP	OEM
HEF4557BT	OEM
HEF4585	OEM
HEF4585BD	1375
HEF4585BP	1375
HEF4585BT	OEM
HEF4720	OEM
HEF4720B	OEM
HEF4720BD	OEM
HEF4720BP	OEM
HEF4720P	OEM
HEF4720VD	OEM
HEF4720VP	OEM
HEF4721B	OEM
HEF4721P	OEM
HEF4724	OEM
HEF4724BD	OEM
HEF4724BP	OEM
HEF4724BT	OEM
HEF4731BD	OEM
HEF4731BP	OEM
HEF4731VD	OEM
HEF4731VP	OEM
HEF4736B	OEM
HEF4736P	OEM
HEF4737	OEM
HEF4737BD	OEM
HEF4737BP	OEM
HEF4737VD	OEM
HEF4737VP	OEM
HEF4738V	OEM
HEF4738VE	OEM

DEVICE TYPE	REPL CODE
HEF4738VP	OEM
HEF4739V	OEM
HEF4752VD	OEM
HEF4752VP	OEM
HEF4753BD	OEM
HEF4753BP	OEM
HEF4754VD	OEM
HEF4754VP	OEM
HEF40097BD	4441
HEF40097BP	4441
HEF40097BT	OEM
HEF40097P	OEM
HEF40098BD	0432
HEF40098BN	0432
HEF40098BP	0432
HEF40098BT	OEM
HEF40098P	0432
HEF40106BD	OEM
HEF40106BP	OEM
HEF40106BT	OEM
HEF40160	OEM
HEF40160B	1359
HEF40160BD	1359
HEF40160BP	1359
HEF40160BT	OEM
HEF40161	OEM
HEF40161B	1373
HEF40161BD	1373
HEF40161BP	1373
HEF40161BT	OEM
HEF40162	OEM
HEF40162B	1388
HEF40162BD	1388
HEF40162BP	1388
HEF40162BT	OEM
HEF40163	OEM
HEF40163B	1407
HEF40163BD	1407
HEF40163BP	1407
HEF40163BT	OEM
HEF40174B	1551
HEF40174BD	1551
HEF40174BP	1551
HEF40174BT	OEM
HEF40174P	1551
HEF40175B	1528
HEF40175BD	1528
HEF40175BP	1528
HEF40175BT	OEM
HEF40192	OEM
HEF40192BD	1764
HEF40192BP	1764
HEF40192BT	OEM
HEF40192P	1764
HEF40193	OEM
HEF40193BD	1776
HEF40193BP	1776
HEF40193BT	OEM
HEF40193P	1776
HEF40194B	1769
HEF40194BD	1769
HEF40194BP	1769
HEF40194BT	OEM
HEF40194P	1769
HEF40195B	1784
HEF40195BD	1784
HEF40195BP	1784
HEF40195BT	OEM
HEF40195P	OEM
HEF40240B	OEM
HEF40244B	OEM
HE-M8489	0133
HEMT-3300	OEM
HEMT-6000	OEM
HEP-0700	0917
HEP1	0136
HEP1-RT	0136
HEP-2	0050
HEP2	0280
HEP2-RT	0280
HEP3	0050
HEP3-RT	0050
HEP50	0079
HEP50-RT	0079
HEP51	0126
HEP51-RT	0126
HEP52	0037
HEP52-RT	0037
HEP53	0079
HEP53-RT	0079
HEP54	0079
HEP54-RT	0079
HEP55	2550
HEP55-RT	0191
HEP56	0144
HEP56-RT	0144
HEP57	0037
HEP57-RT	0037
HEP75	0595
HEP75-RT	0595
HEP76	0791
HEP76-RT	0791
HEP101	0170
HEP102	0188
HEP103	0636
HEP104	0012
HEP105	0137
HEP134	0143
HEP135	0143

DEVICE TYPE	REPL CODE
HEP151	2002
HEP152	0871
HEP153	1250
HEP154	0015
HEP155	0080
HEP156	0609
HEP157	0015
HEP158	0015
HEP159	0071
HEP160	0071
HEP161	0110
HEP162	0242
HEP165	0196
HEP166	1080
HEP170	0071
HEP175	0277
HEP176	0289
HEP177	0296
HEP178	0302
HEP200	0085
HEP200-RT	0085
HEP201	0440
HEP230	0085
HEP230-RT	0085
HEP230X2	0142
HEP231	0440
HEP231-RT	0440
HEP232	0160
HEP232-RT	0160
HEP233	0440
HEP233-RT	0440
HEP234	0975
HEP234-RT	0975
HEP235	0975
HEP236	0604
HEP-237	0437
HEP237	0437
HEP237-RT	0437
HEP238	1424
HEP238-RT	0076
HEP239	3298
HEP239-RT	3298
HEP240	0142
HEP240-RT	0142
HEP241	0178
HEP241-RT	0178
HEP242	0126
HEP242-RT	0126
HEP243	0086
HEP243-RT	0086
HEP244	0276
HEP244-RT	0276
HEP245	0161
HEP245-RT	0161
HEP-246	0460
HEP246	0460
HEP246-RT	0681
HEP247	0543
HEP247-RT	0543
HEP-248	1681
HEP248	0491
HEP248-RT	0491
HEP250	0211
HEP250-RT	0211
HEP251	0004
HEP251-RT	0004
HEP252	0211
HEP252-RT	0211
HEP253	0164
HEP253-RT	0164
HEP254	0208
HEP254-RT	0208
HEP280	0004
HEP281	0004
HEP300	1110
HEP302	0963
HEP304	2519
HEP305-RT	OEM
HEP306	0745
HEP307-RT	OEM
HEP310	1175
HEP310-RT	1175
HEP311	3331
HEP312	0007
HEP320	1137
HEP340	0593
HEP570	OEM
HEP572	3427
HEP573	3429
HEP580	0324
HEP581	1364
HEP582	OEM
HEP583	OEM
HEP584	1069
HEP590	1077
HEP591	2624
HEP592	OEM
HEP593	OEM
HEP-594	0654
HEP594	0654
HEP-595	0443
HEP595	0443
HEP600	OEM
HEP601	2615
HEP602	0162
HEP603	0157
HEP604	0794
HEP605	0365
HEP606	0100

DEVICE TYPE	REPL CODE
HEP607	0002
HEP608	0441
HEP609	0039
HEP610	1834
HEP610-RT	0869
HEP611	0783
HEP612	0331
HEP613	0149
HEP620	1584
HEP621	1584
HEP622	3402
HEP623	0160
HEP623-RT	0160
HEP624	0085
HEP624-RT	0085
HEP-625	0604
HEP625	0975
HEP625-RT	0975
HEP-626	0604
HEP626	0085
HEP626-RT	0085
HEP627	0604
HEP627-RT	0604
HEP628	0085
HEP628-RT	0085
HEP629	0211
HEP629-RT	0211
HEP630	0164
HEP630-RT	0164
HEP631	0211
HEP631-RT	0211
HEP632	0004
HEP632-RT	0004
HEP633	0211
HEP633-RT	0211
HEP634	0004
HEP634-RT	0004
HEP635	0136
HEP635-RT	0136
HEP636	0136
HEP636-RT	0136
HEP637	0050
HEP638	0136
HEP638-RT	0136
HEP639	0136
HEP639-RT	0136
HEP640	0136
HEP640-RT	0136
HEP641	0600
HEP641-RT	0600
HEP642	0222
HEP643	0222
HEP643X2	0222
HEP644	0975
HEP644-RT	0975
HEP700	0460
HEP700-RT	0460
HEP701	0161
HEP701-RT	0161
HEP702	0905
HEP702-RT	0905
HEP703	0178
HEP703-RT	0178
HEP703X2	0178
HEP704	0103
HEP704-RT	0103
HEP704X2	1514
HEP-705	1681
HEP705	0103
HEP705-RT	0136
HEP706	0233
HEP706-RT	0233
HEP-707	0074
HEP707	0074
HEP707-RT	0074
HEP708	0037
HEP708-RT	0037
HEP709	1144
HEP709-RT	1144
HEP710	0126
HEP710-RT	0126
HEP711	0086
HEP712	0493
HEP712-RT	0493
HEP713	0233
HEP713-RT	0233
HEP-714	0168
HEP714	0264
HEP714-RT	0264
HEP715	0532
HEP715-RT	0532
HEP716	0037
HEP716-RT	0037
HEP717	0037
HEP717-RT	0037
HEP718	0144
HEP718-RT	0144
HEP719	0224
HEP719-RT	0224
HEP720	0224
HEP720-RT	0224
HEP721	0127
HEP721-RT	0127
HEP722	0144
HEP722-RT	0144
HEP723	0016
HEP723-RT	0016
HEP724	0016
HEP724-RT	0016

DEVICE TYPE	REPL CODE
HEP725	0079
HEP725-RT	0079
HEP726	0111
HEP726-RT	0111
HEP727	0144
HEP727-RT	0144
HEP728	0016
HEP728-RT	0016
HEP729	0016
HEP729-RT	0016
HEP730	0111
HEP730-RT	0111
HEP731	0127
HEP731-RT	0127
HEP732	0127
HEP732-RT	0127
HEP733	0127
HEP733-RT	0127
HEP734	0127
HEP734-RT	0127
HEP735	0079
HEP735-RT	0079
HEP736	0079
HEP736-RT	0079
HEP737	0111
HEP737-RT	0111
HEP738	0079
HEP738-RT	0079
HEP739	0037
HEP739-RT	0037
HEP-740	0363
HEP740	0642
HEP740-RT	0642
HEP801	0325
HEP801-RT	0325
HEP802	0325
HEP802-RT	0325
HEP803	0325
HEP803-RT	0325
HEP3806P	1379
HEP6005	0160
HEP6060P	0789
HEP6061P	3219
HEPC0900P	2563
HEPC0901P	2992
HEPC0902P	2434
HEPC0903P	3015
HEPC0904P	2350
HEPC0905P	2597
HEPC0906P	3032
HEPC0907P	3039
HEPC0908P	3045
HEPC0909P	0078
HEPC0910P	3054
HEPC0911P	3066
HEPC0912P	2992
HEPC74LS00P	1527
HEPC74LS02P	1559
HEPC74LS03P	1579
HEPC74LS04P	1595
HEPC74LS05P	1608
HEPC74LS08P	1633
HEPC74LS10P	1662
HEPC74LS11P	1667
HEPC74LS13P	1688
HEPC74LS14P	1698
HEPC74LS20P	0034
HEPC74LS27P	0183
HEPC74LS30P	0828
HEPC74LS32P	0088
HEPC74LS42P	1841
HEPC74LS51P	1033
HEPC74LS54P	1857
HEPC74LS55P	0457
HEPC74LS74P	0243
HEPC74LS86P	0291
HEPC74LS90P	1882
HEPC74LS93P	1888
HEPC74LS109P	1906
HEPC74LS136P	1628
HEPC74LS138P	0427
HEPC74LS151P	1646
HEPC74LS153P	0959
HEPC74LS155P	0209
HEPC74LS157P	1161
HEPC74LS158P	1656
HEPC74LS170P	2629
HEPC74LS174P	0260
HEPC74LS175P	1672
HEPC74LS181P	1678
HEPC74LS190P	1686
HEPC74LS191P	1687
HEPC74LS192P	1689
HEPC74LS193P	1692
HEPC74LS196P	2834
HEPC74LS197P	2469
HEPC74LS251P	1737
HEPC74LS253P	1739
HEPC74LS257P	1744
HEPC74LS258P	1746
HEPC74LS259P	3205
HEPC74LS279P	3290
HEPC74LS298P	3371
HEPC74LS367P	0977
HEPC74LS368P	0991
HEPC74LS670P	1130
HEPC1030P	1823
HEPC1032P	1041
HEPC1033P	2097

DEVICE TYPE	REPL CODE
HEPC1035P	1176
HEPC1036P	1831
HEPC1038P	OEM
HEPC1039P	OEM
HEPC1044P	0033
HEPC1045P	0081
HEPC1046P	0141
HEPC1052P	0333
HEPC1053P	0358
HEPC1056P	0141
HEPC1057P	1433
HEPC1058P	0466
HEPC1062P	0562
HEPC2001P	2867
HEPC2002P	OEM
HEPC2003P	OEM
HEPC2004P	3427
HEPC2005P	3429
HEPC2006G	1069
HEPC2007G	1364
HEPC2008G	OEM
HEPC2009G	OEM
HEPC2010G	1069
HEPC2012G	1069
HEPC2500P	OEM
HEPC2501P	OEM
HEPC2502P	2867
HEPC2503P	OEM
HEPC3000L	0232
HEPC3000P	0232
HEPC3001L	0268
HEPC3001P	0268
HEPC3002P	0314
HEPC3004L	0361
HEPC3004P	0361
HEPC3010L	0512
HEPC3010P	0512
HEPC3020L	0697
HEPC3020P	0697
HEPC3030L	0873
HEPC3030P	0873
HEPC3040L	1024
HEPC3040P	1024
HEPC3041L	1038
HEPC3050L	0743
HEPC3050P	0743
HEPC3073L	1172
HEPC3073P	1172
HEPC3075L	1432
HEPC3075P	1432
HEPC3401L	1038
HEPC3800L	1208
HEPC3800P	1208
HEPC3801L	0834
HEPC3801P	0834
HEPC3802P	OEM
HEPC3803P	2267
HEPC3804P	1143
HEPC3805P	OEM
HEPC3806P	1379
HEPC3807P	2286
HEPC4000P	2024
HEPC4001P	0478
HEPC4002P	2055
HEPC4003P	2072
HEPC4004P	2516
HEPC4005P	2846
HEPC4006P	3480
HEPC4007P	2846
HEPC4008P	2308
HEPC4009P	1999
HEPC4020P	1661
HEPC4021P	1749
HEPC4030P	0500
HEPC4031P	1963
HEPC4032P	2837
HEPC4033P	1043
HEPC4040P	0056
HEPC4041P	1544
HEPC4042P	0121
HEPC4050P	0398
HEPC4051P	0366
HEPC4052P	0024
HEPC4053P	3428
HEPC4054P	1830
HEPC4055P	3304
HEPC4056P	OEM
HEPC4057L	OEM
HEPC4057T	OEM
HEPC4058P	3704
HEPC4059P	0487
HEPC6001	0632
HEPC6003	0517
HEPC6003P	0517
HEPC6004	OEM
HEPC6007	0723
HEPC6009	1899
HEPC6009P	1899
HEPC6010	3779
HEPC6010P	3779
HEPC6011	1876
HEPC6013	2027
HEPC6014	1186
HEPC6015	0623
HEPC6016	1980
HEPC6016C	1980
HEPC6016P	3974
HEPC6017	1684
HEPC6019	0723

DEVICE TYPE	REPL CODE
HEPC6049G	OEM
HEPC6049R	3663
HEPC6050G	0417
HEPC6051L	OEM
HEPC6051P	OEM
HEPC6052G	0410
HEPC6052P	0312
HEPC6053G	OEM
HEPC6053L	OEM
HEPC6054G	OEM
HEPC6054R	OEM
HEP-C6055L	0692
HEPC6055L	0692
HEPC6056P	0701
HEP-C6057P	0349
HEPC6057P	0666
HEPC6058P	0354
HEP-C6059P	0785
HEPC6059P	0785
HEP-C6060P	0789
HEPC6060P	1215
HEP-C6061P	3219
HEPC6061P	3219
HEP-C6062P	0664
HEPC6062P	0664
HEP-C6063P	0167
HEPC6063P	2393
HEP-C6065P	1940
HEPC6065P	1940
HEP-C6066P	1753
HEPC6066P	1753
HEPC6067G	0753
HEPC6067P	2161
HEP-C6068P	0443
HEPC6068P	0701
HEP-C6069G	0364
HEPC6069G	0364
HEP-C6070P	0352
HEPC6070P	0352
HEP-C6071P	0354
HEPC6071P	0354
HEP-C6072P	0751
HEPC6072P	1753
HEPC6073P	0701
HEP-C6074P	0654
HEPC6074P	0654
HEP-C6075P	0856
HEPC6075P	0856
HEP-C6076P	0395
HEPC6076P	0395
HEPC6078R	OEM
HEP-C6079P	1808
HEPC6079P	1478
HEP-C6080P	2279
HEPC6080P	2279
HEP-C6082P	0664
HEPC6082P	0678
HEP-C6083P	0167
HEPC6083P	2393
HEP-C6085P	2457
HEPC6085P	2457
HEPC6087P	OEM
HEP-C6089	2161
HEPC6089	0753
HEPC6090	2557
HEPC6091	1077
HEPC6091G	1077
HEPC6092G	OEM
HEPC6093G	OEM
HEP-C6094P	0654
HEPC6094P	0654
HEP-C6095P	0443
HEPC6095P	0443
HEP-C6096P	0520
HEPC6096P	0520
HEP-C6099P	3195
HEPC6099P	3195
HEP-C6100P	0802
HEPC6100P	0802
HEP-C6101P	1345
HEPC6101P	1345
HEP-C6102P	0360
HEPC6102P	0360
HEP-C6103P	1301
HEPC6103P	1301
HEPC6104L	1192
HEPC6105P	2239
HEPC6107P	1300
HEPC6110P	0624
HEPC6111P	0923
HEPC6112P	1196
HEPC6113P	0334
HEPC6114P	1321
HEPC6115P	2259
HEPC6116P	2239
HEPC6117P	OEM
HEPC6118P	1285
HEPC6119P	OEM
HEPC6120P	2648
HEPC6121P	2791
HEPC6122P	1838
HEPC6123P	3829
HEPC6124P	3826
HEPC6125P	3575
HEPC6126P	1611
HEPC6127P	3761
HEPC6128P	3763
HEPC6129P	2823
HEPC6130P	0973
HEPC6131P	0973
HEP-C6132P	1292
HEPC6132P	1292
HEP-C6133P	1828
HEPC6133P	1828
HEPC6134P	3395
HEPC6135P	4509
HEPC6136P	1285
HEP-C6137P	1836
HEPC6137P	1836
HEPC6138P	5041
HEPC6139P	1721
HEPC6140P	OEM
HEPC6142P	1292
HEPC6143P	OEM
HEPC6144P	0083
HEPC6800P	0388
HEPC6802P	0393
HEPC6810P	2086
HEPC6821P	0448
HEPC6850P	0514
HEPC6860P	2692
HEPC7161AP	OEM
HEPC7400P	0232
HEPC7401P	0268
HEPC7402P	0314
HEPC7403P	0335
HEPC7404P	0361
HEPC7405P	0385
HEPC7406P	1206
HEPC7407P	1339
HEPC7408P	0467
HEPC7409P	0492
HEPC7410P	0512
HEPC7411P	0527
HEPC7413P	1441
HEPC7414P	2243
HEPC7416P	1349
HEPC7417P	1352
HEPC7420P	0697
HEPC7422P	4603
HEPC7423P	3466
HEPC7425P	3477
HEPC7426P	0803
HEPC7427P	0818
HEPC7430P	0873
HEPC7432P	0899
HEPC7437P	3520
HEPC7438P	0996
HEPC7440P	1024
HEPC7441AP	1038
HEPC7441P	1038
HEPC7442P	1052
HEPC7445P	1082
HEPC7446AP	1098
HEPC7447AP	1108
HEPC7447P	1108
HEPC7448P	1125
HEPC7450P	0743
HEPC7451P	1168
HEPC7453P	1185
HEPC7454P	1202
HEPC7460P	1275
HEPC7470P	1404
HEPC7472P	1426
HEPC7473P	1172
HEPC7474P	1313
HEPC7475P	1432
HEPC7476P	1158
HEPC7483P	0117
HEPC7485P	0374
HEPC7486P	1368
HEPC7490AP	1208
HEPC7490P	1208
HEPC7491AP	0980
HEPC7492AP	0834
HEPC7492P	0834
HEPC7493AP	0569
HEPC7495P	1486
HEPC7496P	1715
HEPC74107P	0942
HEPC74109P	0968
HEPC74121P	0175
HEPC74123P	1157
HEPC74125P	1182
HEPC74126P	1193
HEPC74132P	1271
HEPC74141P	1377
HEPC74145P	0619
HEPC74147P	1451
HEPC74148P	1464
HEPC74150P	1492
HEPC74151AP	1495
HEPC74153P	1539
HEPC74154P	1555
HEPC74155P	1576
HEPC74156P	1592
HEPC74157P	1605
HEPC74160AP	1631
HEPC74161AP	1645
HEPC74162AP	1013
HEPC74163AP	1666
HEPC74164P	0734
HEPC74165P	1685
HEPC74166P	0231
HEPC74170P	1722
HEPC74173P	1766
HEPC74174P	1770
HEPC74175P	1787
HEPC74176P	1795
HEPC74177P	1803
HEPC74180P	1829
HEPC74181P	1842
HEPC74182P	1856
HEPC74184P	OEM
HEPC74185AP	OEM
HEPC74190P	1912
HEPC74191P	1917
HEPC74192P	1921
HEPC74193P	1926
HEPC74194P	OEM
HEPC74195P	1943
HEPC74196P	1950
HEPC74197P	1956
HEPC74198P	1964
HEPC74199P	1971
HEPC74251P	2299
HEPC74259P	OEM
HEPC74365P	1459
HEPC74366P	1471
HEPC74367P	1488
HEPC74368P	1508
HEPD0002	4873
HEPD0012	5131
HEPD0101	4988
HEPD1002	OEM
HEPD1201	4332
HEPF0010	3712
HEP-F0015	0325
HEPF0015	0325
HEP-F0021	0325
HEPF0021	0325
HEPF1035	2987
HEPF1036	0325
HEPF2004	0212
HEP-F2005	0849
HEPF2005	0849
HEPF2007	0212
HEPF2007A	0212
HEP-G0001	0136
HEPG0001	0136
HEP-G0002	0050
HEPG0002	0050
HEP-G0003	0050
HEPG0003	0050
HEP-G0004	0164
HEPG0004	0280
HEP-G0005	0164
HEPG0005	0280
HEPG0005P/G	0280
HEP-G0006	0164
HEPG0006	0280
HEPG0006P/G	0280
HEP-G0007	0164
HEPG0007	0280
HEPG0007P/G	0280
HEP-G0008	0136
HEPG0008	0136
HEP-G0009	0136
HEPG0009	0136
HEP-G0010	0136
HEPG0010	0136
HEP-G0011	0600
HEPG0011	0600
HEPG630	0211
HEPG631	0164
HEP-G6000	0085
HEPG6000	0085
HEPG6000P/G	0085
HEPG6001	0604
HEPG6001-RT	0604
HEPG6001P/G	0604
HEPG6002	0437
HEP-G6003	0085
HEPG6003	0085
HEPG6003P/G	0085
HEP-G6004	0440
HEPG6004	0440
HEPG6005	0160
HEPG6005P/G	0085
HEP-G6006	0437
HEPG6006	0437
HEP-G6007	0975
HEPG6007	0975
HEPG6007P/G	0975
HEPG6008	0975
HEPG6008P/G	0975
HEP-G6009	0604
HEPG6009	0604
HEPG6009P/G	0604
HEP-G6010	0437
HEPG6010	0437
HEP-G6011	0847
HEPG6011	0847
HEPG6012	3298
HEPG6013	0085
HEPG6013P/G	0085
HEP-G6014	0604
HEPG6014	0604
HEPG6014P/G	0604
HEP-G6015	0604
HEPG6015	0604
HEPG6015P/G	0604
HEPG6016	0222
HEPG6017	0975
HEPG6017P/G	0975
HEP-G6018	0604
HEPG6018	0604
HEPG6018P/G	0604
HEPP0001	2313
HEPP0002	OEM
HEPP0003	0007
HEP-P0160	2002
HEPP1001	3185
HEPP2000	OEM
HEPP2001	OEM
HEPP2002	OEM
HEPP2005	1358
HEPP5000	0541
HEPP5001	1053
HEPP5002	5051
HEPR0050	0015
HEPR0051	0015
HEPR0052	0015
HEPR0053	0015
HEPR0054	0015
HEPR0055	0071
HEPR0056	0071
HEPR0070	0953
HEPR0071	0242
HEPR0072	0540
HEPR0074	0965
HEPR0076	0817
HEPR0078	0922
HEPR0080	0087
HEPR0081	0087
HEPR0082	0087
HEPR0084	0087
HEPR0086	0087
HEPR0090	0087
HEPR0091	0087
HEPR0092	0087
HEPR0094	0087
HEPR0096	0087
HEPR0097	0087
HEPR0098	0087
HEPR0100	1282
HEPR0101	1282
HEPR0102	1282
HEPR0104	1287
HEPR0106	1293
HEP-R0130	0097
HEPR0130	0097
HEPR0130-RT	2900
HEPR0130R	0534
HEP-R0131	0097
HEPR0131	0097
HEPR0131-RT	2900
HEPR0131R	0748
HEP-R0132	0097
HEPR0132	0097
HEPR0132R	0765
HEP-R0134	0109
HEPR0134	0109
HEPR0134R	0538
HEP-R0136	0122
HEPR0136	0122
HEPR0136R	0816
HEP-R0137	0145
HEPR0137	0131
HEPR0137R	0545
HEP-R0138	0145
HEPR0138	0145
HEPR0138R	0550
HEPR0160	0805
HEPR0160R	2003
HEPR0161	0805
HEPR0161R	3265
HEPR0162	0805
HEPR0162R	1577
HEPR0164	0805
HEPR0164R	3281
HEPR0166	1586
HEPR0166R	3294
HEPR0170	0071
HEPR0180	0087
HEPR0220	2615
HEPR0220R	2566
HEPR0222	2615
HEPR0222R	OEM
HEP-R0250	1124
HEPR0250	0805
HEPR0250R	0550
HEP-R0251	1124
HEPR0251	0805
HEPR0251R	1238
HEP-R0253	1124
HEPR0253	0805
HEPR0253R	1107
HEP-R0254	0805
HEPR0254	0805
HEPR0254R	1111
HEP-R0255	0805
HEPR0255	0805
HEPR0255R	0258
HEP-R0256	0319
HEPR0256	1195
HEPR0256R	1644
HEP-R0257	0319
HEPR0257	0319
HEPR0257R	0267
HEPR0500	0735
HEPR0505	OEM
HEPR0510	OEM
HEPR0515	OEM
HEPR0520	1545
HEPR0600	0102
HEPR0602	0102
HEP-R0604	0087
HEPR0604	0102
HEP-R0606	0102
HEPR0606	0102
HEP-R0700	0917
HEPR0700	0917
HEPR0801	0277
HEPR0802	0277
HEPR0803	0289
HEPR0804	0296
HEPR0805	0302
HEPR0841	0277
HEPR0842	0277
HEPR0843	0289
HEPR0845	0296
HEPR0846	0302
HEPR0851	0277
HEPR0852	0277
HEPR0853	0289
HEPR0855	0296
HEPR0856	0302
HEPR0861	2363
HEPR0862	2363
HEPR0863	2363
HEPR0865	0473
HEPR0866	2370
HEPR0875	2369
HEPR0876	2369
HEPR0877	2369
HEPR0878	2369
HEPR0879	1589
HEPR0880	2363
HEPR0881	2363
HEPR0882	2363
HEPR0883	2369
HEPR0884	2370
HEPR0890	1643
HEPR0891	1643
HEPR0892	1643
HEPR0894	1673
HEPR0896	1673
HEPR1001	1137
HEPR1002	0344
HEPR1003	0901
HEPR1004	2342
HEPR1005	0058
HEPR1101	5469
HEPR1102	5469
HEPR1103	5469
HEPR1202	1584
HEPR1205	1584
HEPR1215	1396
HEPR1216	0447
HEPR1217	0447
HEPR1218	0940
HEPR1219	3324
HEPR1220	4588
HEPR1221	2270
HEPR1222	3402
HEPR1223	3404
HEPR1241	1487
HEPR1242	1487
HEPR1243	1487
HEPR1244	1487
HEPR1245	1487
HEPR1246	2809
HEPR1247	0336
HEPR1300	0722
HEPR1301	0722
HEPR1302	0722
HEPR1304	0722
HEPR1306	0722
HEPR1307	0722
HEPR1471	0722
HEPR1472	0722
HEPR1473	0722
HEPR1475	0722
HEPR1701	0422
HEPR1702	0422
HEPR1703	4455
HEPR1711	0593
HEPR1712	0593
HEPR1713	0593
HEPR1715	0485
HEPR1721	1089
HEPR1722	1093
HEPR1723	0411
HEPR1725	0415
HEPR1750	0404
HEPR1751	0411
HEPR1752	0415
HEPR1781	3147
HEPR1782	3148
HEPR1783	3150
HEPR1785	3152
HEPR2001	4233
HEP-R2002	3331
HEPR2002	3331
HEP-R2021	3407
HEPR2021	3407
HEPR2500	0554
HEPR2501	0005
HEPR2502	0720
HEPR2503	0030
HEPR2504	OEM
HEPR2505	OEM
HEPR3001	0774
HEPR3010	0102
HEPR3012	0348
HEPR3100	1323
HEPR3201	1501
HEPR3203	0190
HEPR4012	0348
HEPR9001	0196
HEPR9002	1080
HEPR9003	0484
HEP-R9134	2232
HEPR9134	0143
HEPR9134A	0143
HEP-R9135	2232
HEPR9135	0143
HEPR9136	0143
HEPR9137	0124
HEP-RO138	0145
HEP-S0001	0861
HEPS0001	0861
HEPS0001-RT	0861
HEP-S0002	0595
HEPS0002	0079
HEPS0002-RT	0079
HEP-S0003	0595
HEPS0003	0595
HEPS0003-RT	0595
HEPS0004	0079
HEPS0004-RT	0079
HEP-S0005	0861
HEPS0005	0861
HEP-S0006	0693
HEPS0006	0693
HEPS0006-RT	0693
HEPS0007	0861
HEPS0007-RT	0861
HEPS0008	0224
HEPS0009	1497
HEPS0010	0224
HEPS0011	0079
HEPS0012	0126
HEPS0013	0150
HEPS0014	0595
HEPS0015	0079
HEP-S0016	0127
HEPS0016	0144
HEPS0017	0224
HEPS0019	0150
HEPS0020	0144
HEP-S0021	0016
HEPS0021	0144
HEPS0022	0079
HEP-S0023	0111
HEPS0023	0111
HEP-S0024	0111
HEPS0024	0111
HEPS0025	0079
HEP-S0026	0037
HEPS0026	0037
HEPS0027	0715
HEPS0028	0342
HEPS0029	0342
HEP-S0030	0016
HEPS0030	0079
HEP-S0031	0693
HEPS0031	0693
HEP-S0032	0037
HEPS0032	0037
HEP-S0033	0224
HEPS0033	0224
HEPS0034	0016
HEPS0035	0079
HEPS0036	0079
HEPS0037	0037
HEPS0038	0079
HEPS0039	0259
HEPS0040	0079
HEP-S3001	1411
HEPS-3001	4246
HEPS3001	0630
HEPS3002	0630
HEPS3003	0126
HEPS3004	0630
HEPS3005	3576
HEPS3006	1974
HEPS3007	1974
HEPS3008	0493
HEPS3009	1974
HEPS3010	0630
HEPS3011	0698
HEP-S3012	0126
HEPS3012	0126
HEPS3013	0493
HEPS3014	0493
HEP-S3019	0264
HEPS3019	0264
HEPS3020	0630
HEP-S3021	0338
HEPS3021	0338
HEP-S3022	0338
HEPS3022	0338
HEPS3022-RT	0338
HEP-S3023	0566
HEPS3023	0566
HEPS3023-RT	0566
HEP-S3024	0566
HEPS3024	0566
HEP-S3025	0566
HEPS3025	0566
HEPS3025-RT	0566
HEP-S3026	0551
HEPS3026	0551
HEPS3026-RT	0551
HEP-S3027	1367
HEPS3027	1367
HEPS3027-RT	1367
HEP-S3028	1367
HEPS3028	1367
HEP-S3029	1367
HEPS3029	1367
HEPS3029-RT	1367
HEP-S3030	0382
HEPS3030	0382
HEPS3030-RT	0382
HEP-S3031	0382
HEPS3031	0382
HEPS3031-RT	0382
HEP-S3032	0382
HEPS3032	0382
HEP-S3033	0233
HEPS3033	0233
HEP-S3034	0233
HEPS3034	0233
HEP-S3035	0233
HEPS3035	0233
HEPS3036	2312
HEPS3037	2946
HEPS3038	0418
HEPS3039	1233
HEPS3040	1974
HEPS3041	1974
HEPS3042	2700
HEPS3043	1984
HEPS3044	1591
HEPS3045	0560
HEPS3046	3194
HEPS3047	0566
HEPS3048	0338
HEPS3049	0338
HEPS3050	1367
HEPS3051	0338
HEPS3052	2835
HEPS3053	3565
HEPS3054	0854
HEPS3055	0854
HEPS3060	0423
HEPS3061	0423
HEPS3062	0936
HEP-S5000	0161
HEPS5000	0161
HEP-S5001	0561
HEPS5001	0561
HEPS5001-RT	0561
HEP-S5002	1199
HEPS5002	1199
HEPS5002-RT	1199
HEP-S5003	0161
HEPS5003	0161
HEPS5003-RT	0161
HEP-S5004	0561
HEPS5004	0561
HEP-S5005	1199
HEPS5005	1199
HEP-S5006	0460
HEPS5006	0460
HEP-S5007	0460
HEPS5007	0460
HEPS5007-RT	0460
HEP-S5008	1199
HEPS5008	1199
HEPS5008-RT	1199
HEP-S5009	1199
HEPS5009	1199
HEPS5009-RT	1199
HEP-S5010	1199
HEPS5010	1199
HEPS5010-RT	1199
HEP-S5011	0142
HEPS5011	0142
HEP-S5012	0178
HEPS5012	0178
HEP-S5013	0126
HEPS5013	0126
HEP-S5014	0086
HEPS5014	0630
HEP-S5015	0276
HEPS5015	0276
HEP-S5018	0905
HEPS5018	0905
HEP-S5019	0178
HEPS5019	0178
HEP-S5020	0074
HEPS5020	0642
HEPS5021	0065
HEPS5022	0532
HEP-S5023	0126
HEPS5023	0126
HEP-S5024	0233
HEPS5024	0233
HEP-S5025	0233
HEPS5025	0233
HEP-S5026	0233
HEPS5026	0086
HEPS5027	0161
HEPS5028	1430
HEP-S7000	0130
HEPS7000	0130
HEP-S7001	1681
HEPS7001	1681
HEP-S7002	0103

If replacement code is OEM, contact original manufacturer for replacement.

DEVICE TYPE	REPL CODE
HEPS7002	0103
HEP-S7003	1681
HEPS7003	0491
HEP-S7004	0103
HEPS7004	0130
HEPS7005	0065
HEPS7006	0313
HEPS7007	1681
HEPS7008	0130
HEPS9001	0316
HEPS9002	2136
HEPS9002UNIJ	2136
HEP-S9100	0400
HEPS9100	0400
HEPS9101	0558
HEPS9102	3377
HEPS9103	3538
HEP-S9120	3519
HEPS9120	3519
HEPS9121	2897
HEPS9122	2237
HEPS9123	3539
HEPS9140	2427
HEPS9141	2277
HEPS9142	3525
HEPS9143	3526
HEPS9144	2277
HEPS9145	2427
HEPS9146	3373
HEPS9147	2432
HEPS9148	2439
HEPS9149	2432
HEPS9150	1212
HEPS9151	OEM
HEPS9152	3528
HEPS9153	OEM
HEPTO-36	OEM
HEP-Z0206	0292
HEPZ0206	0188
HEPZ0206A	0299
HEP-Z0208	0456
HEPZ0208	0188
HEPZ0208A	0036
HEP-Z0210	0451
HEPZ0210	0451
HEPZ0210A	0140
HEP-Z0211	0162
HEPZ0211	0162
HEPZ0211A	0041
HEP-Z0212	0157
HEPZ0212	0157
HEPZ0212A	0253
HEP-Z0214	0636
HEPZ0214	0636
HEPZ0214A	0471
HEP-Z0215	0025
HEPZ0215	0025
HEPZ0215A	0062
HEP-Z0216	0649
HEPZ0216	0649
HEPZ0216A	0077
HEP-Z0217	0244
HEPZ0217	0244
HEPZ0217A	0165
HEP-Z0219	0012
HEPZ0219	0012
HEPZ0219A	0057
HEP-Z0220	0170
HEPZ0220	0170
HEPZ0220A	0064
HEP-Z0222	0137
HEPZ0222	0137
HEPZ0222A	0052
HEP-Z0225	0002
HEPZ0225	0002
HEPZ0225A	0686
HEP-Z0228	0495
HEPZ0228	0495
HEPZ0228A	0375
HEP-Z0230	0531
HEPZ0230	0531
HEPZ0230A	0700
HEP-Z0231	0565
HEPZ0231	0565
HEPZ0231A	0705
HEP-Z0234	0441
HEPZ0234	0441
HEPZ0234A	0455
HEP-Z0254	0213
HEPZ0254	0287
HEPZ0254A	0287
HEP-Z0255	0245
HEPZ0255	0028
HEPZ0255A	0757
HEP-Z0401	0292
HEPZ0401	0188
HEP-Z0402	0188
HEPZ0402	0188
HEP-Z0403	0456
HEPZ0403	0188
HEP-Z0405	0451
HEPZ0405	0451
HEP-Z0406	0162
HEPZ0406	0162
HEP-Z0407	0157
HEPZ0407	0157
HEP-Z0408	0636
HEPZ0408	0636
HEP-Z0409	0025
HEPZ0409	0025
HEP-Z0410	0649
HEPZ0410	0649
HEP-Z0411	0244
HEPZ0411	0244
HEPZ0412	0012
HEPZ0413	0170
HEPZ0414	0317
HEPZ0415	0137
HEPZ0416	0137
HEP-Z0417	0100
HEPZ0417	0002
HEP-Z0418	0002
HEPZ0418	0002
HEP-Z0419	0420
HEPZ0419	0002
HEP-Z0420	0495
HEPZ0420	0495
HEP-Z0421	0531
HEPZ0421	0531
HEP-Z0422	0565
HEPZ0422	0565
HEP-Z0423	0402
HEPZ0423	0402
HEP-Z0424	0441
HEPZ0424	0441
HEP-Z0425	1674
HEPZ0425	1674
HEP-Z0426	0039
HEPZ0426	0039
HEP-Z0427	0820
HEPZ0427	0039
HEP-Z0428	0350
HEPZ0428	0350
HEP-Z0430	0999
HEPZ0430	0999
HEP-Z0432	0869
HEPZ0432	0999
HEP-Z0433	0783
HEPZ0433	0783
HEP-Z0436	0331
HEPZ0436	0331
HEP-Z0438	0098
HEPZ0438	0098
HEP-Z0439	0149
HEPZ0439	0149
HEP-Z0440	0186
HEPZ0440	0186
HEP-Z0442	0028
HEPZ0442	0028
HEP-Z0444	0367
HEPZ0444	0367
HEP-Z0445	0421
HEPZ0445	0421
HEPZ1409	0451
HEPZ2500	0782
HEPZ2502	0806
HEPZ2504	0833
HEPZ2506	0876
HEPZ2508	0185
HEP-Z2510	0480
HEPZ2510	0480
HEP-Z2513	0684
HEPZ2513	0684
HEP-Z2514	0225
HEPZ2514	0225
HEP-Z2516	0234
HEPZ2516	0234
HEP-Z2519	0247
HEPZ2519	0251
HEP-Z2522	0256
HEPZ2522	0256
HEP-Z2525	0269
HEPZ2525	0269
HEP-Z2526	0274
HEPZ2526	0274
HEP-Z2528	0294
HEPZ2528	0294
HEP-Z2530	0309
HEPZ2530	0309
HEP-Z2531	0318
HEPZ2531	0318
HEPZ2537	0266
HEPZ2542	0444
HEPZ2545	1180
HEPZ2547	1207
HEPZ2548	1218
HEPZ2551	0028
HEPZ3500	0547
HEPZ3500R	0317
HEPZ3502	2113
HEPZ3502R	1438
HEPZ3504	1901
HEPZ3504R	1445
HEPZ3505	0696
HEPZ3507	1616
HEPZ3511	1618
HEPZ3512	0510
HEPZ3512R	0992
HEP-Z3514	0870
HEPZ3514	0870
HEPZ3514R	1263
HEP-Z3516	1274
HEPZ3516	1274
HEPZ3516R	1639
HEPZ3518	1078
HEPZ3518R	1716
HEPZ3520	1723
HEPZ3520R	0727
HEPZ3522	1761
HEPZ3522R	1782
HEPZ3524	1788
HEPZ3524R	1799
HEPZ3530	0207
HEPZ3530R	1884
HEPZ3534	0356
HEPZ3534R	1909
HEPZ3536	0438
HEPZ3536R	1163
HEPZ4026	OEM
HER101	0546
HER102	0546
HER103	0546
HER104	OEM
HER105	OEM
HER106	OEM
HER107	OEM
HER151	0546
HER152	0546
HER153	0546
HER154	1362
HER155	1362
HER156	OEM
HER157	OEM
HER301	3549
HER302	3549
HER303	3549
HER304	1362
HER305	1362
HER306	1372
HER307	1372
HER601	OEM
HER602	OEM
HER603	OEM
HER604	OEM
HER605	OEM
HER801	1127
HER802	1127
HER803	1127
HER804	1664
HER805	1664
HER1601	0909
HER1602	0909
HER1603	0909
HER1604	0972
HER1605	0972
HER3001	2274
HER3002	2274
HER3003	2274
HER3004	4488
HER3005	4488
HER-R0257	0319
HERR3201	1501
HERR3203	0190
HER-R0257	0319
HES8011B	OEM
HES8018	OEM
HES8039A	OEM
HES8042	OEM
HES8042B	OEM
HES8045	OEM
HES8045A	OEM
HES-8053	OEM
HES8053	OEM
HES8102	OEM
HES8109	OEM
HES8110	OEM
HES8120	OEM
HES8121	OEM
HES8121B	OEM
HES8121C	OEM
HES8601B	OEM
HE-SD1	0015
HESD-1	0102
H-EX0019GE	0686
H-EX0021TA	0181
H-EX0047CE	0052
H-EX0049CE	0041
H-EX0061CE	0137
H-EX0089CE	0165
H-EX0091CE	0705
H-EX0247CE	0636
HEX3000	OEM
HF-0SW05	0015
HF0W05	0015
HF-1	0102
HF1	0102
HF-1A	0023
HF1A	0102
HF-1B	0102
HF1B	0102
HF-1C	0102
HF1C	0102
HF-1S85	0720
HF-1S334	0012
HF-1S339	0012
HF-1S553	0760
HF-1Z	0102
HF1Z	0102
HF2	0079
HF3	0079
HF3H	0050
HF3M	0050
HF4	0079
HF4A	OEM
HF4B	OEM
HF4C	OEM
HF5	0079
HF6	0079
HF6H	0050
HF6M	0050
HF7	0079
HF8	0079
HF9	0086
HF10	0126
HF11	0560
HF12	1267
HF12H	0050
HF12M	0050
HF12N	0050
HF15	0560
HF15-04D03	OEM
HF16	1267
HF17	0111
HF19	0160
HF-19D	0604
HF20	0160
HF20H	0050
HF20M	0050
HF-35	0050
HF35	0136
HF40	0861
HF43	0285
HF-47	0150
HF47	0693
HF50	0079
HF50H	0050
HF50M	0050
HF51	1267
HF57	0042
HF58	0925
HF100	OEM
HF200	OEM
HF-10024	0143
HF-20002	0030
HF-20004	0012
HF-20005	0030
HF-20007	0760
HF-20008	0143
HF-20011	0012
HF-20014	0133
HF-20032	0015
HF20032	0133
HF-20033	0025
HF-20034	0133
HF-20035	0030
HF-20040	0030
HF-20041	0137
HF-20042	0015
HF-20046	0030
HF-20047	0015
HF-20048	0133
HF-20050	0015
HF-20052	0015
HF-20060	0133
HF20060	0133
HF-20061	0133
HF-20062	0554
HF-20063	0124
HF-20064	0133
HF-20065	0012
HF-20066	0120
HF20066	0015
HF-20067	0015
HF-20071	0071
HF-20072	0005
HF-20075	OEM
HF-20080	0720
HF-20083	0015
HF-20084	0015
HF-20088	0143
HF-20094	0720
HF-20095	0133
HF-20105	0139
HF-20116	0030
HF-20123	0133
HF-20124	0133
HF200191A	0325
HF200191A/	0325
HF200191A-0	0325
HF200191A0	0325
HF200191B-0	0325
HF200191B0	0325
HF200191BO	0325
HF200301B	0325
HF200301B0	0325
HF200301C-0	0325
HF200301C-O	0325
HF200301E	0212
HF200411B	1211
HF200411C0	1211
HF309301E	0111
HF2000411B	0325
HFB7	OEM
HFBR-0010	OEM
HFBR-1001	OEM
HFBR-1002	OEM
HFBR-2001	OEM
HFBR-3099	OEM
HF-DS410	0133
HFET-1001	OEM
HFET1101	OEM
HFET1102	OEM
HFET-2001	OEM
HFET2201	OEM
HFET2202	OEM
HFET-2204	OEM
HFET5001	OEM
HF-MV2	0139
HFR5	OEM
HFR10	OEM
HFR15	OEM
HFR20	OEM
HFS-04	6013
HFS23	OEM
HFSA3	0102
HF-SD1	0023
HFSD-1	0102
HFSD1	0102
HF-SD1/BB-4	0102
HF-SD-1A	0015
HFSD-1A	0102
HFSD1A	0015
HFSD-1B	0102
HFSD1B	0015
HF-SD-1C	0015
HFSD-1C	0102
HFSD1C	6452
HF-SD-1Z	0102
HFSD-1Z	0102
HFSD1Z	0102
HFSD12	0087
HFSD-14	0102
HFSD18	OEM
HFSG005	0015
HG61H20B24F	OEM
HG62B40B41F	OEM
HG1001	OEM
HG1002	OEM
HG1003	OEM
HG1004	OEM
HG1005	OEM
HG1006	OEM
HG1007	OEM
HG1008	OEM
HG1009	OEM
HG1010	OEM
HG1011	OEM
HG1012	OEM
HG1090	0143
HG5001	0143
HG5002	0143
HG5003	0143
HG5004	0143
HG5005	0143
HG5006	0143
HG5007	0143
HG5008	0143
HG5009	0143
HG5020	0143
HG5078	0143
HG5079	0143
HG5085	0143
HG5088	0143
HG5808	0143
HG100101	OEM
HGCDTE-1-40A	OEM
HGCDTE-1-40AS	OEM
HGCDTE-1-60A	OEM
HGCDTE-1-60AS	OEM
HGCDTE-1-520A	OEM
HGCDTE-1-520AS	OEM
HGCDTE-1-40742	OEM
HGCDTE-1-44782	OEM
HGCDTE-2-520A	OEM
HGCDTE-2-520AS	OEM
HGCDTE-2-40742	OEM
HGCDTE-2-44782	OEM
HGCDTE-5-40A	OEM
HGCDTE-5-40AS	OEM
HGCDTE-5-52-AS	OEM
HGCDTE-5-60A	OEM
HGCDTE-5-60AS	OEM
HGCDTE-5-520A	OEM
HGCDTE-5-40742	OEM
HGCDTE-5-44782	OEM
HGR1	0015
HGR2	0015
HGR3	0015
HGR4	0015
HGR-5	0015
HGR5	1335
HG-R10	0015
HGR-10	0015
HGR10	0080
HGR-20	0015
HGR20	0133
HGR-30	0015
HGR30	0133
HGR-40	0015
HGR40	0795
HGR-60	0015
HGR60	0015
HH1157	OEM
HH7372	OEM
HI0-524-6	OEM
HI03SC	2352
HI03SD	2352
HI03SG	2352
HI03SH	2352
HI03SS	2352
HI1-200-2	OEM
HI1-200-4	OEM
HI1-200-5	OEM
HI1-201-1	OEM
HI1-201-2	OEM
HI1-201-4	OEM
HI1-201-5	OEM
HI1-300-2	OEM
HI1-300-5	OEM
HI1-300-8	OEM
HI1-301-2	OEM
HI1-301-5	OEM
HI1-301-8	OEM
HI1-302-2	OEM
HI1-302-5	OEM
HI1-302-8	OEM
HI1-303-2	OEM
HI1-303-5	OEM
HI1-303-8	OEM
HI1-304-2	OEM
HI1-304-5	OEM
HI1-304-8	OEM
HI1-305-2	OEM
HI1-305-5	OEM
HI1-305-8	OEM
HI1-306-2	OEM
HI1-306-5	OEM
HI1-306-8	OEM
HI1-307-2	OEM
HI1-307-8	OEM
HI1-381-2(A)	OEM
HI1-381-5(A)	OEM
HI1-381-8(A)	OEM
HI1-384-2(A)	OEM
HI1-384-5(A)	OEM
HI1-384-8(A)	OEM
HI1-387-2(A)	OEM
HI1-387-5(A)	OEM
HI1-387-8(A)	OEM
HI1-390-2(A)	OEM
HI1-390-5(A)	OEM
HI1-390-8(A)	OEM
HI1-506-2	OEM
HI1-506-5	OEM
HI1-506A-2	OEM
HI1-506A-5	OEM
HI1-507-2	OEM
HI1-507-5	OEM
HI1-507A-2	OEM
HI1-507A-5	OEM
HI1-508A-2	OEM
HI1-508A-5	OEM
HI1-509A-2	OEM
HI1-509A-5	OEM
HI1-524-2	OEM
HI1-524-5	OEM
HI1-524-8	OEM
HI1-539-2	OEM
HI1-539-4	OEM
HI1-539-5	OEM
HI1-539-8	OEM
HI1-562-2	OEM
HI1-562-4	OEM
HI1-562-5	OEM
HI1-562-8	OEM
HI1-562A2	OEM
HI1-562A4	OEM
HI1-562A5	OEM
HI1-562A8	OEM
HI1-1800A2	OEM
HI1-1800A5	OEM
HI1-1818A-2	OEM
HI1-1818A-5	OEM
HI1-1828A-2	OEM
HI1-1828A-5	OEM
HI1-1840-2	OEM
HI1-5040-2	OEM
HI1-5040-5	OEM
HI1-5047-2	OEM
HI1-5047-5	OEM
HI1-5047A-2	OEM
HI1-5047A-5	OEM
HI1-5048-2	OEM
HI1-5048-5	OEM
HI1-5049-2	OEM
HI1-5049-5	OEM
HI1-5050-2	OEM
HI1-5050-5	OEM
HI1-5051-2	OEM
HI1-5051-5	OEM
HI1-5607-2(A)	OEM
HI1-5607-5(A)	OEM
HI1-5607-8(A)	OEM
HI1-5608-2(A)	OEM
HI1-5608-5(A)	OEM
HI1-5608-8(A)	OEM
HI1-5609-2(A)	OEM
HI1-5609-5(A)	OEM
HI1-5609-8(A)	OEM
HI1-5610-2	OEM
HI1-5610-5	OEM
HI1-5610-8	OEM
HI1-5612-2	OEM
HI1-5612-5	OEM
HI1-5612-8	OEM
HI2-300-2	OEM
HI2-301-2	OEM
HI2-304-2	OEM
HI2-305-2	OEM
HI2-381-2(A)	OEM
HI2-381-8(A)	OEM
HI2-387-2(A)	OEM
HI2-387-5(A)	OEM
HI2-387-8(A)	OEM
HI3-300-5	OEM
HI3-301-5	OEM
HI3-302-5	OEM
HI3-303-5	OEM
HI3-304-5	OEM
HI3-305-5	OEM
HI3-306-5	OEM
HI3-307-5	OEM
HI3-381-5(A)	OEM
HI3-384-5(A)	OEM
HI3-387-5(A)	OEM
HI3-390-5(A)	OEM
HI13SC	2352
HI13SD	2352
HI13SG	2352
HI13SH	2383
HI13SS	2352
HI23SC	2353
HI23SD	2353
HI23SG	2353
HI23SH	2394
HI23SS	2353
HI33SD	2356
HI33SG	2356
HI33SH	1413
HI33SS	2356
HI43SC	2356
HI43SD	2356
HI43SG	2356
HI43SH	1413
HI43SS	2356
HI53SC	OEM
HI53SD	OEM
HI53SG	OEM
HI53SH	2387
HI53SS	OEM
HI63SD	OEM
HI63SG	OEM
HI63SH	2387
HI63SS	OEM
HI516-5(A)	OEM
HI516-8(A)	OEM
HI518-5(A)	OEM
HI518-8(A)	OEM
HIB005	7115
HIB010	OEM
HIDIV-1	2977
HIDIV-2	2978
HIDIV-3	2980
HIFI	0015
HILA7016//	OEM
HI053SC	OEM
HI053SD	OEM
HI053SG	OEM
HI053SH	OEM
HIT5609A	OEM
HIT5609C	0018
HIT9011	OEM
HIT9014N	OEM
HIT9014N-C	0079
HIT9015N	OEM
H-IX0065CE	1429
HJ-00191	OEM
HJ15	0004
HJ15D	0050
HJ17	0004
HJ17D	0004
HJ22	0004
HJ22D	0004
HJ23	0004
HJ23D	0004
HJ32	0050
HJ34	0050
HJ34A	0050
HJ35	0085
HJ37	0050
HJ41	0280
HJ43	0004
HJ50	0004
HJ51	0004
HJ54	0004
HJ55	0050
HJ56	0050
HJ57	0050
HJ60	0004
HJ60A	0050
HJ60C	0050
HJ62	0004
HJ70	0050
HJ71	0164
HJ72	0164
HJ73	0164
HJ74	0164
HJ75	0050
HJ226	0164
HJ228	0164
HJ230	0164
HJ315	0164
HJ606	0004
HJX2	0004
HK-00049	0325
HK-00330	0325
HK103A	OEM
HK103K	OEM
HK203A	OEM
HK203K	OEM
HK403A	OEM

If replacement code is OEM, contact original manufacturer for replacement.

DEVICE TYPE	REPL CODE
HK403K	OEM
HK603A	OEM
HK603K	OEM
HK803A	OEM
HK803K	OEM
HK1000	OEM
HK1001	OEM
HK1003A	OEM
HK1003K	OEM
HK1203A	OEM
HK1203K	OEM
HKT-158	0079
HKT-161	0079
HKZ101	OEM
HKZ101S	OEM
HL-00005	OEM
HL103A	2900
HL103K	2291
HL203A	2900
HL203K	2291
HL403A	1112
HL403K	0476
HL603A	1112
HL603K	0476
HL803A	3011
HL803K	0449
HL1003A	3011
HL1003K	0449
HL1203A	OEM
HL1203K	OEM
HL1403A	OEM
HL1403K	OEM
HL7801E	OEM
HL7801G	OEM
HL18998	0232
HL18999	1313
HL19000	0361
HL19001	0512
HL19002	1172
HL19003	0697
HL19004	0314
HL19005	0335
HL19006	0569
HL19008	0175
HL19009	1052
HL19010	1158
HL19011	1024
HL19012	1432
HL19013	0873
HL19014	1368
HL19015	1208
HL53424	1823
HL53426	1176
HL53428	0033
HL53429	0141
HL55660	0527
HL-55661	0704
HL55663	2138
HL55723	2151
HL55763	0467
HL55764	1539
HL55861	1492
HL55862	0361
HL56320	2445
HL56420	0232
HL56421	0361
HL56422	0697
HL56423	0873
HL56424	0996
HL56425	1313
HL56426	0374
HL56427	1486
HL56429	1921
HL56430	1926
HL56431	5253
HL56842	0117
HL56899	0512
HLC9694-01-2210	OEM
HLCD0437	OEM
HLCD0437(A)	OEM
HLCD0438	OEM
HLCD0438A(A)	OEM
HLCD0488	OEM
HLCD0515D	OEM
HLCD0515P	OEM
HLCD0538A	OEM
HLCD0538A(A)	OEM
HLCD0539A	OEM
HLCD0539A(A)	OEM
HLCD0540	OEM
HLCD0540(A)	OEM
HLCD0541	OEM
HLCD0542	OEM
HLCD0548	OEM
HLCD0550D	OEM
HLCD0550P	OEM
HLCD0551D	OEM
HLCD0551P	OEM
HLCD0607	OEM
HLCD07211-1P	OEM
HLCD07211-2P	OEM
HLCD07211-3P	OEM
HLCD07211-4P	OEM
HLCD7211	OEM
HLMP-0300	OEM
HLMP-0301	OEM
HLMP-0400	OEM
HLMP-0401	OEM
HLMP-0500	OEM

DEVICE TYPE	REPL CODE
HLMP-0501	OEM
HLMP301	OEM
HLMP401	OEM
HLMP504	OEM
HLMP-1300	OEM
HLMP-1301	OEM
HLMP-1302	OEM
HLMP-1400	OEM
HLMP-1401	OEM
HLMP-1402	OEM
HLMP-1500	OEM
HLMP-1501	OEM
HLMP-1502	OEM
HLMP-2300	OEM
HLMP-2350	OEM
HLMP-2400	OEM
HLMP-2450	OEM
HLMP-2500	OEM
HLMP-2550	OEM
HLMP-2600	OEM
HLMP-2620	OEM
HLMP-2635	OEM
HLMP-2655	OEM
HLMP-2670	OEM
HLMP-2685	OEM
HLMP-2700	OEM
HLMP-2720	OEM
HLMP-2735	OEM
HLMP-2755	OEM
HLMP-2770	OEM
HLMP-2785	OEM
HLMP-2800	OEM
HLMP-2835	OEM
HLMP-2855	OEM
HLMP-2870	OEM
HLMP-2885	OEM
HLMP-3105	OEM
HLMP-3112	OEM
HLMP-3600	OEM
HLMP-3650	OEM
HLMP-3680	OEM
HLMP-6203	OEM
HLMP-6204	OEM
HLMP-6205	OEM
HLMP-6600	OEM
HLMP-6620	OEM
HLP0TA	OEM
HLP0WTA	OEM
HLP20RA	OEM
HLP20RGA	OEM
HLP20RLA	OEM
HLP20TA	OEM
HLP20WRA	OEM
HLP20WTA	OEM
HLP30RA	OEM
HLP30RC	OEM
HLP30RD	OEM
HLP30RGA	OEM
HLP30RGB	OEM
HLP30RGC	OEM
HLP30RGD	OEM
HLP30RLA	OEM
HLP30RLB	OEM
HLP30RLC	OEM
HLP30RLD	OEM
HLP30TA	OEM
HLP30TB	OEM
HLP30TC	OEM
HLP30TD	OEM
HLP30WRA	OEM
HLP30WRB	OEM
HLP30WRC	OEM
HLP30WRD	OEM
HLP30WTA	OEM
HLP30WTB	OEM
HLP30WTC	OEM
HLP30WTD	OEM
HLP40RA	OEM
HLP40RB	OEM
HLP40RD	OEM
HLP40RGA	OEM
HLP40RGB	OEM
HLP40RGC	OEM
HLP40RGD	OEM
HLP40RLA	OEM
HLP40RLB	OEM
HLP40RLC	OEM
HLP40RLD	OEM
HLP40TB	OEM
HLP40TC	OEM
HLP40TD	OEM
HLP40WRA	OEM
HLP40WRB	OEM
HLP40WRC	OEM
HLP40WRD	OEM
HLP40WTB	OEM
HLP40WTC	OEM
HLP40WTD	OEM
HLP50RB	OEM
HLP50RC	OEM
HLP50RD	OEM
HLP50RGB	OEM
HLP50RGC	OEM
HLP50RGD	OEM
HLP50RLB	OEM
HLP50RLC	OEM
HLP50RLD	OEM
HLP50TB	OEM

DEVICE TYPE	REPL CODE
HLP50TC	OEM
HLP50TD	OEM
HLP50WRB	OEM
HLP50WRC	OEM
HLP50WRD	OEM
HLP50WTB	OEM
HLP50WTC	OEM
HLP50WTD	OEM
HLP60RB	OEM
HLP60RC	OEM
HLP60RD	OEM
HLP60RGB	OEM
HLP60RGC	OEM
HLP60RGD	OEM
HLP60RLB	OEM
HLP60RLC	OEM
HLP60RLD	OEM
HLP60TB	OEM
HLP60TC	OEM
HLP60TD	OEM
HLP60WRC	OEM
HLP60WRD	OEM
HLP60WTB	OEM
HLP60WTC	OEM
HLP60WTD	OEM
HLP1400	OEM
HLP1500	OEM
HLP1600	OEM
HLP2400	OEM
HLP2500	OEM
HLP2600	OEM
HLP3400	OEM
HLP3500	OEM
HLP3600	OEM
HLP5400	OEM
HLP5420	OEM
HLP5500	OEM
HLP5600	OEM
HLS1-25	OEM
HLS1-50	OEM
HLS1-100	OEM
HLS1-150	OEM
HLS1-200	OEM
HLS1-300	OEM
HLS2-25	OEM
HLS2-50	OEM
HLS2-100	OEM
HLS2-150	OEM
HLS2-200	OEM
HLS2-300	OEM
HLS25	OEM
HLSS-0533D	OEM
HLSS-0533H	OEM
HM-00049	0280
HM01301	OEM
HM-08014	0004
HM1-768-5	OEM
HM1-6322-2	OEM
HM1-6501-2	OEM
HM1-6501-5	OEM
HM1-6503-5	OEM
HM1-6504-2	OEM
HM1-6504-5	OEM
HM1-6504B2	OEM
HM1-6505-2	OEM
HM1-6505-5	OEM
HM1-6508-2	OEM
HM1-6508-5	3193
HM1-6512-2	OEM
HM1-6513-5	OEM
HM1-6514-2	OEM
HM1-6514-5	OEM
HM1-6514B2	OEM
HM1-6516-2	OEM
HM1-6516-9	OEM
HM1-6518-2	OEM
HM1-6518-5	OEM
HM1-6518B2	OEM
HM1-6551-2	OEM
HM1-6551-5	OEM
HM1-6551B2	OEM
HM1-6561-2	OEM
HM1-6561-5	OEM
HM1-6561B2	OEM
HM1-6562-2	OEM
HM1-6562-5	OEM
HM1-6562B2	OEM
HM1-6611-2	OEM
HM1-6611-5	OEM
HM1-7602-2	OEM
HM1-7602-5	3168
HM1-7603-2	OEM
HM1-7603-5	2175
HM1-7608-2	OEM
HM1-7608-5	OEM
HM1-7610-2	OEM
HM1-7610-5	1918
HM1-7610A2	OEM
HM1-7610A5	OEM
HM1-7611-2	OEM
HM1-7611-5	2224
HM1-7611A2	OEM
HM1-7611A5	OEM
HM1-7616-2	OEM
HM1-7616-5	OEM
HM1-7620-2	OEM
HM1-7620-5	2388
HM1-7620A2	OEM
HM1-7620A5	OEM

DEVICE TYPE	REPL CODE
HM1-7621-2	OEM
HM1-7621-5	2389
HM1-7621A2	OEM
HM1-7621A5	OEM
HM1-7640-2	2324
HM1-7640-5	7087
HM1-7640A2	OEM
HM1-7640A5	OEM
HM1-7641-2	OEM
HM1-7641-5	2322
HM1-7641A2	OEM
HM1-7641A5	OEM
HM1-7642-2	OEM
HM1-7642-5	2390
HM1-7642A-2	OEM
HM1-7642A-5	2390
HM1-7642P-2	OEM
HM1-7642P-5	OEM
HM1-7643-2	OEM
HM1-7643-5	2392
HM1-7643A-2	OEM
HM1-7643A-5	2392
HM1-7643P-2	OEM
HM1-7643P-5	OEM
HM1-7648-2	OEM
HM1-7648-5	3288
HM1-7649-2	OEM
HM1-7649-5	2320
HM1-7680-2	OEM
HM1-7680-5	2862
HM1-7680A5	OEM
HM1-7680P2	OEM
HM1-7680P5	OEM
HM1-7680R2	OEM
HM1-7680R5	OEM
HM1-7681-2	OEM
HM1-7681-5	2909
HM1-7681A5	OEM
HM1-7681P2	OEM
HM1-7681P5	OEM
HM1-7681R2	OEM
HM1-7681R5	OEM
HM1-7684-2	OEM
HM1-7684P2	OEM
HM1-7684P5	OEM
HM1-7685-2	OEM
HM1-7685-5	3166
HM1-7685P2	OEM
HM1-7685P5	OEM
HM1-76160-2	OEM
HM1-76160-5	OEM
HM1-76161-2	OEM
HM1-76161-5	OEM
HM1-76165-2	OEM
HM1-76165-5	OEM
HM1-76165-8	OEM
HM1-76321-2	OEM
HM1-76321-5	OEM
HM1-76321-8	OEM
HM1-76641-2	OEM
HM1-76641-5	OEM
HM1-76641-8	OEM
HM-2	0139
HM3-6322-2	OEM
HM3-6501-2	OEM
HM3-6501-5	OEM
HM3-6503-5	OEM
HM3-6504-2	OEM
HM3-6504-5	OEM
HM3-6504B2	OEM
HM3-6505-2	OEM
HM3-6505-5	OEM
HM3-6508-2	OEM
HM3-6508-5	3193
HM3-6512-2	OEM
HM3-6513-5	OEM
HM3-6514-2	OEM
HM3-6514-5	OEM
HM3-6514B2	OEM
HM3-6516-2	OEM
HM3-6516-9	OEM
HM3-6518-2	OEM
HM3-6518-5	OEM
HM3-6518B2	OEM
HM3-6551-2	OEM
HM3-6551-5	OEM
HM3-6551B2	OEM
HM3-6561-2	OEM
HM3-6561-5	OEM
HM3-6561B2	OEM
HM3-6562-2	OEM
HM3-6562-5	OEM
HM3-6562B2	OEM
HM3-7602-2	OEM
HM3-7602-5	3168
HM3-7603-2	OEM
HM3-7603-5	2175
HM3-7608-2	OEM
HM3-7608-5	OEM
HM3-7610-2	OEM
HM3-7610-5	1918
HM3-7610A2	OEM
HM3-7610A5	1918
HM3-7611-2	OEM
HM3-7611-5	2224
HM3-7611A2	OEM
HM3-7611A5	2224
HM3-7620-2	OEM
HM3-7620-5	2388

DEVICE TYPE	REPL CODE
HM3-7620A2	OEM
HM3-7620A5	2388
HM3-7621-2	OEM
HM3-7621-5	2389
HM3-7621A2	OEM
HM3-7621A5	2389
HM3-7640-2	OEM
HM3-7640-5	2324
HM3-7640A2	OEM
HM3-7640A5	2324
HM3-7641-2	OEM
HM3-7641-5	2322
HM3-7641A2	OEM
HM3-7641A5	2322
HM3-7642-2	OEM
HM3-7642-5	2390
HM3-7642A-2	OEM
HM3-7642P-2	OEM
HM3-7642P-5	OEM
HM3-7643-2	OEM
HM3-7643-5	2392
HM3-7643A-2	OEM
HM3-7643A-5	2392
HM3-7643P-2	OEM
HM3-7643P-5	OEM
HM3-7643Q-5	OEM
HM3-7648-2	OEM
HM3-7648-5	3288
HM3-7649-2	OEM
HM3-7649-5	2320
HM3-7680-2	OEM
HM3-7680-5	2862
HM3-7680A5	OEM
HM3-7680P2	OEM
HM3-7680P5	OEM
HM3-7680R2	OEM
HM3-7680R5	OEM
HM3-7681-2	OEM
HM3-7681-5	2909
HM3-7681A5	OEM
HM3-7681P2	OEM
HM3-7681P5	OEM
HM3-7681R2	OEM
HM3-7681R5	OEM
HM3-7684-2	OEM
HM3-7684-5	OEM
HM3-7684P2	OEM
HM3-7684P5	OEM
HM3-7685-2	OEM
HM3-7685-5	3166
HM3-7685P2	OEM
HM3-7685P5	OEM
HM4-6503-5	OEM
HM4-6504-2	OEM
HM4-6504-5	OEM
HM4-6504B2	OEM
HM4-6505-2	OEM
HM4-6505-5	OEM
HM4-6512-2	OEM
HM4-6513-5	OEM
HM4-6514-2	OEM
HM4-6514-5	OEM
HM4-6514B2	OEM
HM4-6516-2	OEM
HM4-6516-9	OEM
HM4-6518-2	OEM
HM4-6518-5	OEM
HM4-6518B2	OEM
HM4-6561-2	OEM
HM4-6561-5	OEM
HM4-6561B2	OEM
HM5-6564-2	OEM
HM5-6564-5	OEM
HM6.8	0062
HM6.8A	0062
HM6.8B	0062
HM7.5	0077
HM7.5A	0077
HM7.5B	0077
HM8.2	0165
HM8.2A	0165
HM8.2B	0165
HM9-0168-8	OEM
HM9-6501-2	OEM
HM9-6501-5	OEM
HM9-6503-5	OEM
HM9-6504-2	OEM
HM9-6504-5	OEM
HM9-6504B2	OEM
HM9-6505-2	OEM
HM9-6505-5	OEM
HM9-6508-2	OEM
HM9-6508-5	OEM
HM9-6513-5	OEM
HM9-6514-2	OEM
HM9-6514-5	OEM
HM9-6514B2	OEM
HM9-6518-2	OEM
HM9-6518-5	OEM
HM9-6518B2	OEM
HM9-6551-2	OEM
HM9-6551-5	OEM
HM9-6551B2	OEM
HM9-6561-2	OEM
HM9-6561-5	OEM
HM9-6561B2	OEM
HM9-6562-2	OEM
HM9-6562-5	OEM
HM9-6562B2	OEM

DEVICE TYPE	REPL CODE
HM9-6611-2	OEM
HM9-6611-5	OEM
HM9-7602-2	OEM
HM9-7602-5	OEM
HM9-7603-2	OEM
HM9-7603-5	OEM
HM9-7608-2	OEM
HM9-7608-5	OEM
HM9-7610-2	OEM
HM9-7610-5	OEM
HM9-7610A2	OEM
HM9-7610A5	OEM
HM9-7611-2	OEM
HM9-7611-5	OEM
HM9-7611A2	OEM
HM9-7611A5	OEM
HM9-7616-2	OEM
HM9-7616-5	OEM
HM9-7620-2	OEM
HM9-7620-5	OEM
HM9-7620A2	OEM
HM9-7620A5	OEM
HM9-7621-2	OEM
HM9-7621-5	OEM
HM9-7621A2	OEM
HM9-7621A5	OEM
HM9-7640-2	OEM
HM9-7640-5	OEM
HM9-7640A2	OEM
HM9-7640A5	OEM
HM9-7641-2	OEM
HM9-7641-5	OEM
HM9-7641A2	OEM
HM9-7641A5	OEM
HM9-7642-2	OEM
HM9-7642-5	OEM
HM9-7642A-2	OEM
HM9-7642A-5	OEM
HM9-7642P-2	OEM
HM9-7642P-5	OEM
HM9-7643-2	OEM
HM9-7643-5	OEM
HM9-7643A-2	OEM
HM9-7643A-5	OEM
HM9-7643P-2	OEM
HM9-7643P-5	OEM
HM9-7648-2	OEM
HM9-7648-5	OEM
HM9-7649-2	OEM
HM9-7649-5	OEM
HM9-7680-2	OEM
HM9-7680-5	OEM
HM9-7680A5	OEM
HM9-7680P2	OEM
HM9-7680P5	OEM
HM9-7680R2	OEM
HM9-7680R5	OEM
HM9-7681-2	OEM
HM9-7681-5	OEM
HM9-7681A5	OEM
HM9-7681P2	OEM
HM9-7681P5	OEM
HM9-7681R2	OEM
HM9-7681R5	OEM
HM9-7684-2	OEM
HM9-7684-5	OEM
HM9-7684P2	OEM
HM9-7684P5	OEM
HM9-7685-2	OEM
HM9-7685-5	OEM
HM9-7685P2	OEM
HM9-7685P5	OEM
HM9-76160-2	OEM
HM9-76160-5	OEM
HM9-76161-2	OEM
HM9-76161-5	OEM
HM9.1	0057
HM9.1A	0057
HM9.1B	0012
HM10	0064
HM10A	0064
HM10B	0170
HM11	0181
HM11A	0181
HM11A1A	OEM
HM11B	0181
HM12	0052
HM12A	0052
HM12B	0052
HM13	0053
HM13A	0053
HM13B	0053
HM15	0686
HM15A	0686
HM15B	0686
HM16	0445
HM16A	0445
HM16B	0445
HM18	0375
HM18A	0375
HM18B	0375
HM20	0700
HM20A	0700
HM20B	0700
HM22	0705
HM22A	0565
HM22B	0705
HM24	0494
HM24A	0494

DEVICE TYPE	REPL CODE
HM24B	0494
HM27	0455
HM27A	0455
HM27B	0441
HM30	0195
HM30A	0195
HM30B	0195
HM33	0166
HM33A	0166
HM33B	0166
HM36	0010
HM36A	0010
HM36B	0010
HM39	0032
HM39A	0032
HM39B	0032
HM43	0054
HM43A	0054
HM43B	0054
HM47	0068
HM47A	0068
HM47B	0068
HM51	0092
HM51A	0502
HM51B	0092
HM56	0125
HM56A	0125
HM56B	0125
HM62	0152
HM62A	0152
HM62B	0152
HM68	0173
HM68A	0173
HM68B	0173
HM75	0094
HM75A	0094
HM75B	0094
HM82	0049
HM82A	0049
HM82B	0049
HM91	0156
HM91A	0156
HM91B	0156
HM100	0189
HM100A	0189
HM100B	0189
HM110	0099
HM110A	0099
HM110B	0099
HM120	0089
HM120A	0089
HM120B	0089
HM130	0287
HM130A	0287
HM130B	0287
HM150	0340
HM150A	0340
HM150B	0340
HM160	0370
HM160A	0370
HM160B	0370
HM180	0425
HM180A	0425
HM180B	0425
HM200	1473
HM200A	1473
HM200B	1473
HM331	OEM
HM383B01	0190
HM468A10	OEM
HM468A10P	OEM
HM2110	OEM
HM2110-1	OEM
HM2112	OEM
HM2112-1	OEM
HM2504	OEM
HM2504-1	OEM
HM2510	OEM
HM2510-1	OEM
HM2510-2	OEM
HM2511	OEM
HM2511-1	OEM
HM4315P	OEM
HM4334-3	OEM
HM4334-4	OEM
HM4334P-3	OEM
HM4334P3L	OEM
HM4334P-4	OEM
HM4334P4L	OEM
HM4716A-1	OEM
HM4716A-2	OEM
HM4716A-3	OEM
HM4716A-4	OEM
HM4716AP-1	OEM
HM4716AP-2	OEM
HM4716AP-3	OEM
HM4716AP-4	OEM
HM4816A3	OEM
HM4816A3E	OEM
HM4816A4	OEM
HM4816A7	OEM
HM4816AP3	OEM
HM4816AP3E	OEM
HM4816AP4	OEM
HM4816AP7	OEM
HM4864-2	4505
HM4864-3	2357
HM4864A12	OEM
HM4864A15	OEM

DEVICE TYPE	REPL CODE
HM4864A20	OEM
HM4864AP12	OEM
HM4864AP15	OEM
HM4864AP20	OEM
HM4864I2	OEM
HM4864I3	OEM
HM4864K2	OEM
HM4864K3	OEM
HM4864P-2	2357
HM4864P2	OEM
HM4864P-3	OEM
HM4864P3	OEM
HM6116	1898
HM6116-2	OEM
HM6116-3	1898
HM6116-4	1898
HM6116FP2	OEM
HM6116FP-3	OEM
HM6116FP3	1898
HM6116FP-4	OEM
HM6116FP4	1898
HM6116FP-200	OEM
HM6116L2	OEM
HM6116L3	OEM
HM6116L4	OEM
HM6116LFP2	OEM
HM6116LFP-3	OEM
HM6116LFP3	OEM
HM6116LFP4(A)	OEM
HM6116LI2	OEM
HM6116LI3	OEM
HM6116LI4	OEM
HM6116LP-2	OEM
HM6116LP-3	OEM
HM6116LP-4	1898
HM6116P	OEM
HM6116P-2	OEM
HM6116P-3	1898
HM6116P-4	6181
HM6116P4	1898
HM6116PI2	OEM
HM6116PI3	OEM
HM6116PI4	OEM
HM6117FP3	OEM
HM6117FP4	OEM
HM6117LFP3(A)	OEM
HM6117LFP4(A)	OEM
HM6117LP3	OEM
HM6117LP4	OEM
HM6117P3	OEM
HM6147	OEM
HM6147-3	OEM
HM6147H35	OEM
HM6147H45(A)	OEM
HM6147HLP35	OEM
HM6147HLP45	OEM
HM6147HP35	OEM
HM6147HP45	OEM
HM6147LP	OEM
HM6147LP-3	OEM
HM6147P	OEM
HM6147P-3	OEM
HM6148	OEM
HM6148-6	OEM
HM6148LP	OEM
HM6148LP-6	OEM
HM6148P	OEM
HM6148P-6	OEM
HM6167	OEM
HM6167-6	OEM
HM6167-8	OEM
HM6167LP	OEM
HM6167LP6	OEM
HM6167LP8	OEM
HM6167P	OEM
HM6167P6	OEM
HM6167P8	OEM
HM6264	OEM
HM6264-150	OEM
HM6264ALP-10	OEM
HM6264ALP-15	OEM
HM6264ALSP	OEM
HM6264ALSP-12	OEM
HM6264ALSP-15	OEM
HM6264LP-15	OEM
HM6264P	OEM
HM6264P-15	OEM
HM6264P-100NS	OEM
HM6264P-150	OEM
HM6322A2	OEM
HM6501B2	OEM
HM6504B2	OEM
HM6504B9	OEM
HM6511-2	OEM
HM6533-2	OEM
HM6543-2	OEM
HM6611A2	OEM
HM6641-5	OEM
HM6661-2	OEM
HM6661-5	OEM
HM6661A2	OEM
HM7091	OEM
HM-7101	OEM
HM7101HD	OEM
HM7603	OEM
HM9101	OEM
HM9102	OEM
HM9105	OEM
HM9306	OEM
HM10414	OEM
HM10414-1	OEM
HM10422	OEM
HM10422-6(A)	OEM
HM10470	OEM
HM10470-1	OEM
HM10474	OEM
HM10474-1	OEM
HM10480	OEM
HM16100-2	OEM
HM16100-5	OEM
HM16100A2	OEM
HM16100A5	OEM
HM16100C2	OEM
HM16100C5	OEM
HM46810	OEM
HM46810P	OEM
HM50256-15	1472
HM50256-20	1472
HM50256P-12	1472
HM50256P-15	1472
HM50256P-20	1472
HM50464P-12	OEM
HM50464P-15	OEM
HM53051P-45	OEM
HM53461	OEM
HM53461-10	OEM
HM53461P-12B	OEM
HM53461ZP-10	OEM
HM53461ZP-12	OEM
HM65256	OEM
HM65256BLP	OEM
HM65256BLP-12	OEM
HM65256BLP-120NS	OEM
HM65256BLSP	OEM
HM65256BLSP-12	OEM
HM65256BP	OEM
HM65256BSP	OEM
HM100415	OEM
HM100422	OEM
HM100422F	OEM
HM100470	OEM
HM100470-1	OEM
HM100474	OEM
HM100474-1	OEM
HM100474F	OEM
HM472114-3	4176
HM472114-4	4176
HM472114A-1	OEM
HM472114A-2	OEM
HM472114AP-1	OEM
HM472114AP-2	OEM
HM472114P-3	2048
HM472114P-4	4176
HM514400BZ7	OEM
HMB917	OEM
HMCS42	OEM
HMCS42C	OEM
HMCS43	OEM
HMCS43C	OEM
HMCS44A	OEM
HMCS44C	OEM
HMCS45A	OEM
HMCS45C	OEM
HMG251	0124
HMG252	0322
HMG625	0124
HMG625A	0124
HMG626	OEM
HMG626A	OEM
HMG627	OEM
HMG659	OEM
HMG662	OEM
HMG662A	OEM
HMG663	OEM
HMG663A	OEM
HMG789	0124
HMG790	0124
HMG791	0124
HMG792	0124
HMG793	OEM
HMG794	OEM
HMG795	OEM
HMG796	OEM
HMG810	OEM
HMG811	0124
HMG812	0124
HMG813	0124
HMG814	OEM
HMG815	OEM
HMG818	OEM
HMG837A	OEM
HMG840	OEM
HMG844	OEM
HMG891	OEM
HMG897	OEM
HMG898	OEM
HMG899	OEM
HMG900	OEM
HMG901	OEM
HMG903	0124
HMG903A	0124
HMG904	0124
HMG904A	0124
HMG905	0124
HMG905A	0124
HMG906	0124
HMG906A	0124
HMG907	0124
HMG907A	0124
HMG908	0124
HMG908A	OEM
HMG914	OEM
HMG914A	OEM
HMG914B	OEM
HMG915	OEM
HMG916	OEM
HMG916A	OEM
HMG916B	OEM
HMG925	OEM
HMG926	OEM
HMG927	OEM
HMG929	OEM
HMG930	OEM
HMG948	OEM
HMG993	0124
HMG3062	OEM
HMG3063	OEM
HMG3064	OEM
HMG3065	OEM
HMG3066	OEM
HMG3067	0124
HMG3068	0124
HMG3069	OEM
HMG3123	0124
HMG3124	0124
HMG3206	OEM
HMG3207	OEM
HMG3257	OEM
HMG3258	OEM
HMG3567	OEM
HMG3593	0124
HMG3594	OEM
HMG3596	0124
HMG3598	OEM
HMG3600	OEM
HMG3601	OEM
HMG3602	OEM
HMG3603	OEM
HMG3604	OEM
HMG3605	OEM
HMG3606	OEM
HMG3607	OEM
HMG3608	OEM
HMG3609	OEM
HMG3669	OEM
HMG3722	OEM
HMG3731	OEM
HMG3872	OEM
HMG3873	OEM
HMG3954	OEM
HMG3956	OEM
HMG4009	OEM
HMG4043	OEM
HMG4086	OEM
HMG4092	OEM
HMG4147	0124
HMG4148	OEM
HMG4149	OEM
HMG4150	OEM
HMG4151	OEM
HMG4152	OEM
HMG4153	OEM
HMG4154	OEM
HMG4305	OEM
HMG4308	OEM
HMG4313	OEM
HMG4314	OEM
HMG4317	OEM
HMG4319	OEM
HMG4322	OEM
HMG4373	OEM
HMG4375	OEM
HMG9001	OEM
HMG9002	OEM
HMG9003	OEM
HMG9004	OEM
HMG9005	OEM
HMG9006	OEM
HMG9007	OEM
HMG9008	OEM
HMG9009	OEM
HMG9010	0124
HMMP1802ACD	OEM
HMMP1802ACL	OEM
HMMP1802ACY	OEM
HMMP1802AD	OEM
HMMP1802AL	OEM
HMMP1802AY	OEM
HN-00002	0015
HN-00003	0143
HN-00005	0052
HN-00008	0015
HN-00012	0057
HN-00018	0023
HN-00024	0057
HN-00025	0705
HN-00032	0015
HN-00033	0030
HN-00045	0030
HN-00052	0030
HN-00053	0139
HN-00061	0170
HN-0047	0133
HN100	0543
HN101	OEM
HN102	OEM
HN103	OEM
HN104	OEM
HN106	OEM
HN107	OEM
HN108	OEM
HN109	OEM
HN110	OEM
HN111	OEM
HN112	OEM
HN113	OEM
HN2756G25	OEM
HN25044	2390
HN25045	4340
HN25084	OEM
HN25084S	OEM
HN25085	3166
HN25085S	3166
HN25088	2862
HN25088S	2862
HN25089	2909
HN25089S	2909
HN25168S	OEM
HN25169S	OEM
HN27256G	OEM
HN27256G-25	OEM
HN27256G-XX	OEM
HN43128	OEM
HN43128P	OEM
HN48016P	OEM
HN48364P	OEM
HN61256	OEM
HN61256P	OEM
HN61364P	OEM
HN462532	5098
HN462532G	5098
HN462532G2	5098
HN462716G	2278
HN462716G1	OEM
HN462716GI	OEM
HN462732	2697
HN462732G	2697
HN462732G2	OEM
HN482764G	0812
HN482764G(A)	OEM
HN482764G-3	0812
HN482764G3(A)	OEM
HN482764G4	OEM
HN613128P	OEM
HN4827128G-25	1638
HN4827232AG	OEM
HND1	OEM
HNVM3004	OEM
HNVM3008D	OEM
HNVM3008H	OEM
HNVM3008P	OEM
HNVM3108D	OEM
HNVM3108H	OEM
HNVM3108P	OEM
HNVM3704D	OEM
HNVM3704H	OEM
HNVM3704P	OEM
HNVM3708	OEM
HO70	0774
HO91	0143
HO99	0474
HO300	0975
HOR3D	OEM
HOR3D4-L	OEM
HOS050	OEM
HOS050A	OEM
HOS050C	OEM
HP-3FR2	OEM
HP3FR2	OEM
HP3K	OEM
HP3ML	OEM
HP3MLRII	OEM
HP-5A	0015
HP-5FR2	OEM
HP-5FR3	OEM
HP5FR3	OEM
HP-5FRS	OEM
HP6	OEM
HP8	OEM
HP10	OEM
HP74LS240P	0452
HP80	0143
HP100	OEM
HP101	OEM
HP102	OEM
HP104	OEM
HP105	OEM
HP106	OEM
HP107	OEM
HP108	OEM
HP109	OEM
HP110	OEM
HP111	OEM
HP205	0015
HP3209	OEM
HP4510	1291
HP4562	OEM
HP5082-0001	OEM
HP5082-0008	OEM
HP5082-0009	OEM
HP5082-0012	OEM
HP5082-0013	OEM
HP5082-0015	OEM
HP5082-0017	OEM
HP5082-0018	OEM
HP5082-0020	OEM
HP5082-0021	OEM
HP5082-0023	OEM
HP5082-0024	OEM
HP5082-0025	OEM
HP5082-0029	OEM
HP5082-0030	OEM
HP5082-0031	OEM
HP5082-0032	OEM
HP5082-0034	OEM
HP5082-0039	OEM
HP5082-0047	OEM
HP5082-0049	OEM
HP5082-0057	OEM
HP5082-0058	OEM
HP5082-0087	OEM
HP5082-0090	OEM
HP5082-0094	OEM
HP5082-0097	OEM
HP5082-0103	0124
HP5082-0112	OEM
HP5082-0113	OEM
HP5082-0114	OEM
HP5082-0132	0124
HP5082-0133	0124
HP5082-0134	0124
HP5082-0151	OEM
HP5082-0152	OEM
HP5082-0153	OEM
HP5082-0154	OEM
HP5082-0180	OEM
HP5082-0181	OEM
HP5082-0182	OEM
HP5082-0240	OEM
HP5082-0241	OEM
HP5082-0242	OEM
HP5082-0243	OEM
HP5082-0251	OEM
HP5082-0252	OEM
HP5082-0253	OEM
HP5082-0254	OEM
HP5082-0299	OEM
HP5082-0300	OEM
HP5082-0305	OEM
HP5082-0306	OEM
HP5082-0307	OEM
HP5082-0308	OEM
HP5082-0309	OEM
HP5082-0310	OEM
HP5082-0320	OEM
HP5082-0321	OEM
HP5082-0335	OEM
HP5082-0340	OEM
HP5082-0360	OEM
HP5082-0364	OEM
HP5082-0365	OEM
HP5082-0370	OEM
HP5082-0375	OEM
HP5082-0386	OEM
HP5082-0437	OEM
HP5082-0438	OEM
HP5082-0439	OEM
HP5082-0800	OEM
HP5082-0803	OEM
HP5082-0805	OEM
HP5082-0810	OEM
HP5082-0815	OEM
HP5082-0820	OEM
HP5082-0821	OEM
HP5082-0825	OEM
HP5082-0830	OEM
HP5082-0833	OEM
HP5082-0835	OEM
HP5082-0840	OEM
HP5082-0885	OEM
HP5082-1001	OEM
HP5082-1002	OEM
HP5082-1006	OEM
HP5082-2080	OEM
HP5082-2200	OEM
HP5082-2201	OEM
HP5082-2202	OEM
HP5082-2203	OEM
HP5082-2207	OEM
HP5082-2208	OEM
HP5082-2209	OEM
HP5082-2210	OEM
HP5082-2213	OEM
HP5082-2214	OEM
HP5082-2215	OEM
HP5082-2221	OEM
HP5082-2223	OEM
HP5082-2225	OEM
HP5082-2351	OEM
HP5082-2353	OEM
HP5082-2354	OEM
HP5082-2366	OEM
HP5082-2374	OEM
HP5082-2500	OEM
HP5082-2511	OEM
HP5082-2521	OEM
HP5082-2523	OEM
HP5082-2524	OEM
HP5082-2544	OEM
HP5082-2551	OEM
HP5082-2552	OEM
HP5082-2553	OEM
HP5082-2554	OEM
HP5082-2555	OEM
HP5082-2563	OEM
HP5082-2564	OEM
HP5082-2566	OEM
HP5082-2577	OEM
HP5082-2579	OEM
HP5082-2602	OEM
HP5082-2707	OEM
HP5082-2740	OEM
HP5082-2752	OEM
HP5082-2758	OEM
HP5082-2759	OEM
HP5082-2800	0133
HP5082-2827	OEM
HP5082-2833	OEM
HP5082-3001	0133
HP5082-3006	OEM
HP5082-3007	OEM
HP5082-3008	OEM
HP5082-3009	OEM
HP5082-3051	OEM
HP5082-3052	OEM
HP5082-3080	0133
HP5082-4104	OEM
HP5082-4107	OEM
HP5082-4120	OEM
HP5082-4201	OEM
HP5082-4303	OEM
HP5082-4309	OEM
HP5082-4310	OEM
HP5082-4320	OEM
HP5082-4351	OEM
HP5082-4400	OEM
HP5082-4405	OEM
HP5082-4417	OEM
HP33001A	OEM
HP33002A	OEM
HP33101A	OEM
HP33102A	OEM
HP33201A	OEM
HP33202A	OEM
HP35860A	0016
HP35867B	OEM
HP35867E	OEM
HPA4202	OEM
HPF4519	OEM
HPS1610	OEM
HPS1611	OEM
HPS1612	OEM
HPS1640	OEM
HPS1641	OEM
HPS1642	OEM
HPS1670	OEM
HPS1671	OEM
HPS1672	OEM
HR	OEM
HR-05A	0015
HR-1	0004
HR1	0050
HR-1S85	0030
HR-2	0004
HR2	0050
HR-2A	0004
HR2A	0211
HR-2A(PENNCREST)	0015
HR2.3	OEM
HR2.8	OEM
HR-3	0004
HR3	0211
HR-3FR	OEM
HR3N0187	OEM
HR3N0200	OEM
HR3.8	OEM
HR-4	0004
HR4	0211
HR-4A	0004
HR4A	0280
HR4.4	OEM
HR-5	0164
HR5	0280
HR-5A	0015
HR5A	0023
HR5A8E	0015
HR-5AX2	0015
HR-5B	0015
HR5B	0015
HR5.4	OEM
HR-6	0004
HR6	0211
HR6.5	OEM
HR-7	0004
HR7	0050
HR-7A	0004
HR7A	0211
HR-8	0004
HR8	0211
HR-8A	0004
HR8A	0280
HR-9	0004
HR9	0211
HR-9A	0004
HR9A	0280
HR9.0	OEM
HR10	0015
HR-11	0016
HR11	0015
HR-11A	0016
HR11A	0079
HR-11B	0016
HR11B	0079
HR-12	0086
HR-13	0016
HR13	0015
HR-13A	0016
HR13A	0079
HR-14	0111
HR14	0079
HR-14A	0016
HR14A	0079
HR-15	0111
HR15	0136
HR-15A	0016
HR15A	0079
HR-16	0016
HR16	0079
HR-16A	0016
HR16A	0079
HR-17	0016
HR17	0079
HR-17A	0016
HR17A	0079
HR-18	0016
HR18	0079
HR-18A	0016
HR18A	0079
HR-19	0016
HR19	0079
HR-19A	0016
HR19A	0079
HR-19E	0604
HR-20	0050
HR20	0050
HR-20A	0050
HR20A	0050
HR-21	0050
HR21	0050
HR-21A	0050
HR21A	0050
HR-22	0050
HR22	0050
HR-22A	0050
HR22A	0050
HR-22B	0050
HR22B	0050
HR-24	0050
HR24	0050
HR-24A	0050
HR24A	0050
HR-25	0050
HR25	0050
HR-25A	0050
HR25A	0050
HR-26	0050
HR26	0050
HR-26A	0050
HR26A	0050
HR-27	0050
HR27	0050
HR-27A	0050
HR27A	0050
HR28	0086
HR29	0086
HR-30	0136
HR30	0164
HR-32	0079
HR32	0079
HR-36	0079
HR36	0079
HR-37	0016
HR37	0079
HR-38	0079
HR38	0595
HR-39	0004
HR39	0127
HR-40	0050
HR40	0050
HR41	0050
HR42	0050
HR-43	0050
HR43	0050
HR44	0050
HR-45	0050
HR45	0050
HR46	0050
HR-47	0111
HR47	0111
HR-48	0079
HR48	0079
HR50	0050
HR-51	0050
HR51	0050
HR52	0050
HR53	0004
HR-58	0127
HR58	0144
HR-59	0127
HR59	0144
HR-60	0079
HR60	0144
HR-61	0211
HR61	0004
HR62	0079
HR63	0079
HR64	0079
HR65	0079
HR66	0079
HR-67	0595
HR67	0595
HR-68	0423
HR-69	0560

DEVICE TYPE	REPL CODE
HR69	0042
HR-70	1267
HR70	0925
HR-71	0037
HR71	0037
HR72	0791
HR-73	0566
HR-74	1367
HR76	0127
HR77	0224
HR78	0224
HR79	0127
HR80	0127
HR81	0086
HR82	0595
HR83	0595
HR-84	0150
HR84	0150
HR85	0566
HR86	1367
HR87	0224
HR101	0085
HR-101A	0160
HR101A	0160
HR102	0160
HR102C	0160
HR103	0160
HR103A	0160
HR105	0160
HR105A	0160
HR105B	0160
HR-106	0142
HR106	0142
HR107	0142
HR-107(PHILCO)	0178
HR107H	0178
HR-200	0242
HR200	1606
HR400	6265
HR600	0087
HR761	OEM
HR762	OEM
HR763	OEM
HR764	OEM
HR771	OEM
HR772	OEM
HR773	OEM
HR774	OEM
HR800	0564
HR878A	OEM
HR878B	OEM
HR879A	OEM
HR879B	OEM
HR881A	OEM
HR881B	OEM
HR932	OEM
HR952	OEM
HR952F	OEM
HR953	OEM
HR953F	OEM
HR954	OEM
HR954F	OEM
HR961F	OEM
HR982	OEM
HR982F-1	OEM
HR982F-2	OEM
HR1000	0087
HR1301	OEM
HR1301F	OEM
HR1870B	OEM
HR1871B	OEM
HR1872B	OEM
HR1874B	OEM
HR3027	OEM
HR3028	OEM
HR3029	OEM
HR3030	OEM
HR3031	OEM
HR3032	OEM
HR40836	0050
HR40837	0050
HR43835	0050
HR43838	0050
HR45838	0050
HR45910	0050
HR45913	0050
HR104961C	1267
HR448636	0050
HRD100	OEM
HRD100F	OEM
HRD101	OEM
HRD102	OEM
HRD200	OEM
HRD200F	OEM
HRE16	OEM
HRE16A	OEM
HRE18.5	OEM
HRE18.5A	OEM
HRE25.5	OEM
HRE25.5A	OEM
HRE28	OEM
HRE28A	OEM
HRE32	OEM
HRE32A	OEM
HRE34.5	OEM
HRE34.5A	OEM
HRE41.5	OEM
HRE41.5A	OEM
HRE48	OEM

DEVICE TYPE	REPL CODE
HRE48A	OEM
HRE50.5	OEM
HRE50.5A	OEM
HRE53.5	OEM
HRE53.5A	OEM
HRE56.5	OEM
HRE56.5A	OEM
HRE60	OEM
HRE60A	OEM
HRE62.5	OEM
HRE62.5A	OEM
HRE64	OEM
HRE64A	OEM
HRE66.5	OEM
HRE66.5A	OEM
HRE69	OEM
HRE69A	OEM
HRE73.5	OEM
HRE73.5A	OEM
HRE76	OEM
HRE76A	OEM
HRE80	OEM
HRE80A	OEM
HRE82.5	OEM
HRE82.5A	OEM
HRE89	OEM
HRE89A	OEM
HRE96	OEM
HRE96A	OEM
HRP22	0023
HRP32	0023
HRW34	1236
HRW36	1236
HS04	OEM
HS04S	OEM
HS07	6861
HS08	1004
HS034	OEM
HS034S	OEM
HS1	OEM
HS1/7	0774
HS1/9	0774
HS-1B	0378
HS2	OEM
HS3	OEM
HS3/1	OEM
HS5	0004
HS6	OEM
HS6/1	0201
HS7/1	0774
HS7/17M	0774
HS7/17N	0774
HS8	0774
HS8/1	0774
HS8-1	0774
HS-9/1	0774
HS9/1	0774
HS9/17M	0757
HS9/17N	0757
HS10	OEM
HS10/1	0288
HS11	OEM
HS12	0394
HS13	OEM
HS14	OEM
HS14S	OEM
HS-15	0211
HS15	0211
HS15/1	0757
HS15/1A(SONY)	0201
HS15/1B	0201
HS-15/1C	0288
HS15/1C	0288
HS-15/12	0201
HS15/16	0288
HS15-1C	0288
HS17	6861
HS17D	0004
HS18	1004
HS-20	0757
HS20	0201
HS-20/1	0201
HS20/1	0757
HS20/1A5	0288
HS20/1B	0757
HS20/1C	0757
HS20/1D	OEM
HS20/16	0288
HS20-1	0757
HS20-1A	0757
HS20-1BS	0757
HS20-1C	0201
HS20-12	0288
HS20-16S	0288
HS20M	0288
HS-22D	0211
HS22D	0136
HS23D	0211
HS24	OEM
HS24S	OEM
HS25	0378
HS25/1	0378
HS25/1B	0378
HS25/1C	0757
HS25/12S	0288
HS25-1B	0015
HS27	6861
HS28	1004
HS29D	0136

DEVICE TYPE	REPL CODE
HS30	0378
HS-30/1B	0378
HS30/1B	5359
HS30/1BS	OEM
HS30/1C	0757
HS30-1B	0378
HS30-1C	0757
HS30-16	0378
HS34	OEM
HS35S	OEM
HS37	6870
HS38	1034
HS40	OEM
HS40A	OEM
HS41A	OEM
HS42A	OEM
HS43A	OEM
HS44	OEM
HS44S	OEM
HS47	6870
HS48	1034
HS50	OEM
HS52	OEM
HS53	OEM
HS54	OEM
HS54S	OEM
HS57	6871
HS58	1148
HS60	OEM
HS64	OEM
HS64S	OEM
HS67	6871
HS68	1148
HS75	OEM
HS102	0004
HS133	0917
HS151B	0201
HS170	0211
HS290	0211
HS811	0774
HS816	0595
HS891	OEM
HS902A	0917
HS903	OEM
HS1001	0015
HS1002	0015
HS1003	0015
HS1004	0133
HS1005	0133
HS1006	0133
HS1007	0015
HS1008	0015
HS1009	0015
HS1010	0015
HS1011	0015
HS1012	0015
HS1020	0015
HS-1168	0016
HS-1225	0144
HS-1226	0144
HS-1227	0144
HS-1229	0016
HS1395	OEM
HS1516	0201
HS2013-I	OEM
HS2027	0760
HS2030	0118
HS2033	0299
HS2036	0188
HS2038	0188
HS2039	0188
HS2047	0162
HS2051	0162
HS2056	0157
HS2062	0471
HS2068	0062
HS2075	0077
HS2082	0165
HS2091	0057
HS2100	0170
HS2110	0181
HS2120	0052
HS2135	0879
HS2150	0686
HS2165	0420
HS2180	0375
HS2200	0700
HS2220	0705
HS2240	0494
HS2270	0441
HS2300	0195
HS2330	0039
HS3043	OEM
HS3103	0015
HS3104	0015
HS3105	0133
HS3106	OEM
HS3108	0071
HS3110	0071
HS5305	OEM
HS5306	OEM
HS5306A	OEM
HS5307	OEM
HS5308	OEM
HS5308A	OEM
HS5810	0595
HS5811	0791
HS5812	0595
HS5813	0791

DEVICE TYPE	REPL CODE
HS5814	0595
HS5815	0791
HS5816	0595
HS5817	0791
HS5818	0595
HS5819	0791
HS5820	0595
HS5821	0791
HS5822	0595
HS5823	0791
HS7027	0760
HS7030	0118
HS7033	0299
HS7036	0188
HS7039	0188
HS7043	0275
HS7047	0140
HS7051	0162
HS7056	0157
HS7062	0471
HS7068	0062
HS7075	0077
HS7082	0165
HS7091	0057
HS7100	0170
HS7110	0181
HS7120	0052
HS7135	0879
HS7150	0686
HS7165	0210
HS7180	0375
HS7200	0700
HS7220	0705
HS7240	0494
HS7270	0441
HS7300	0195
HS7330	0039
HS8014	OEM
HS8407	0065
HS9001	OEM
HS9009	OEM
HS9010	0124
HS9096	OEM
HS9501	OEM
HS9502	0124
HS9503	0124
HS9504	OEM
HS9505	0124
HS9506	0124
HS9508	0124
HS9509	0124
HS-40014	0079
HS-40016	0079
HS-40017	0127
HS-40019	0144
HS40019	OEM
HS-40020	0144
HS40021	0007
HS40022	0007
HS40023	0007
HS40024	0007
HS40025	0007
HS40026	0086
HS-40027	0037
HS-40030	0111
HS-40031	0037
HS40032	0126
HS-40035	0037
HS-40037	0016
HS-40039	0079
HS-40040	0037
HS-40044	0016
HS-40045	0144
HS-40046	0016
HS-40047	0144
HS-40049	0144
HS40049	0144
HS-40050	0037
HS-40053	2481
HS-40054	0079
HS-40055	0144
HS-40057	0006
HS52005	OEM
HS950707	OEM
HSA1015	0148
HSC015	OEM
HSC-1	0087
HSC1	0071
HSC2	OEM
HSC3	OEM
HSC4	OEM
HSC5	OEM
HSC6	OEM
HSC7	OEM
HSC8	OEM
HSC9	OEM
HSC10	OEM
HSC11	OEM
HSC12	OEM
HSC15	OEM
HSC18	OEM
HSC20	OEM
HSC25	OEM
HSC1815	0076
HSC3921	OEM
HSC3954	OEM
HSC4391	OEM
HSC4392	OEM
HSC4393	OEM

DEVICE TYPE	REPL CODE
HSC4416	OEM
HSC4416A	OEM
HSC5163	OEM
HSC5457	OEM
HSC5457A	OEM
HSC5458	OEM
HSC5458A	OEM
HSC5459	OEM
HSC5459A	OEM
HSC5484	OEM
HSC5485	OEM
HSC5486	OEM
HSC5638	OEM
HSC5639	OEM
HSC5640	OEM
HSCH-1001	0080
HSFD-1	0071
HSFD1	0102
HSFD-1A	0015
HSFD-1A(SONY-58	0102
HSFD-1Z	0015
HSFD-1Z(SONY-BLANKIN	0139
HSFD-1Z(SONY-STANDBY	0015
HSG5140	OEM
HSG5420	OEM
HSK23	OEM
HSK83	OEM
HSK110	OEM
HSK120	OEM
HSK121	OEM
HSK122	OEM
HSK140/14	OEM
HSK141/10	OEM
HSK142/07	OEM
HSK143/45	OEM
HSK144/37	OEM
HSK145/27	OEM
HSK146/20	OEM
HSK147/14	OEM
HSK148/10	OEM
HSK149/20	OEM
HSK150/14	OEM
HSK152/10	OEM
HSK227	OEM
HSK277	OEM
HSK816	OEM
HSK866A	OEM
HSK869B	OEM
HSK872A	OEM
HSK8008	OEM
HSKE5200	OEM
HSKE11000	OEM
HSM276S	OEM
HSM2692	OEM
HSM2693	OEM
HSM2694	OEM
HSM2835	0903
HSM2836	4917
HSM2837	0907
HSM2838	4917
HSM2838C	OEM
HSP1001	0133
HSP1007	0133
HSP1010	0133
HSP1011	0133
HSP1013	0133
HSP1020	0609
HSR276	OEM
HSR276S	OEM
HSSH3D4	OEM
HST1050	0074
HST1051	0642
HST1052	0642
HST1053	0642
HST1054	0642
HST1055	0074
HST1056	0642
HST1057	0642
HST1058	0642
HST1059	0642
HST1060	0074
HST1061	0642
HST1062	0642
HST1063	0642
HST1064	0642
HST4451	0285
HST4452	0285
HST4453	0285
HST4454	0285
HST4483	0630
HST5001	0111
HST5002	0285
HST5003	0285
HST5004	0285
HST5005	0285
HST5006	0285
HST5007	0285
HST5008	0285
HST5009	0285
HST5010	0285
HST5051	0285
HST5052	0285
HST5053	0285
HST5054	0285
HST5055	0285
HST5056	0285
HST5501	0630

DEVICE TYPE	REPL CODE
HST5502	0285
HST5503	0285
HST5504	0285
HST5505	0285
HST5506	0630
HST5507	0285
HST5508	0285
HST5509	0285
HST5510	0285
HST5511	0630
HST5551	0285
HST5552	0285
HST5553	0285
HST5554	0285
HST5555	0285
HST5556	0285
HST5906	0836
HST7401	0630
HST7402	0285
HST7403	0285
HST7411	0630
HST7412	0285
HST7413	0285
HST7414	0630
HST7415	0285
HST7416	0285
HST7417	0630
HST7418	0168
HST7419	0168
HST7901	0168
HST7902	0168
HST7903	0168
HST7904	0168
HST7905	0168
HST7907	0168
HST7908	0168
HST7909	0168
HST7910	0168
HST9001	0630
HST9002	0285
HST9003	0285
HST9004	0630
HST9005	0285
HST9006	0285
HST9007	0630
HST9008	0285
HST9009	0285
HST9010	0630
HST-9201	0103
HST-9205	0103
HST-9206	0103
HST-9210	0103
HSZ12EB3	1005
HT	OEM
HT040519C	2866
HT040519C(2SB405)	0164
HT040519C(2SD72)	0208
HT0405190C	2866
HT1-050	OEM
HT1-050A	OEM
HT1-050B	OEM
HT1-050C	OEM
HT1-050T	OEM
HT2-050	OEM
HT2-050A	OEM
HT2-050B	OEM
HT2-050C	OEM
HT2-050T	OEM
HT3-050	OEM
HT3-050A	OEM
HT3-050B	OEM
HT3-050C	OEM
HT3-050T	OEM
HT4-050	OEM
HT4-050A	OEM
HT4-050B	OEM
HT4-050C	OEM
HT4-050T	OEM
HT5-050	OEM
HT5-050A	OEM
HT5-050B	OEM
HT5-050C	OEM
HT5-050T	OEM
HT6-100	OEM
HT6-100A	OEM
HT6-100B	OEM
HT6-100C	OEM
HT6-100T	OEM
HT7-100	OEM
HT7-100A	OEM
HT7-100B	OEM
HT7-100C	OEM
HT7-100T	OEM
HT8-100	OEM
HT8-100A	OEM
HT8-100B	OEM
HT8-100C	OEM
HT8-100T	OEM
HT25	OEM
HT25A	OEM
HT33-02	OEM
HT90A	OEM
HT91A	OEM
HT92A	OEM
HT93A	OEM
HT94A	OEM
HT95A	OEM
HT96A	OEM
HT100	0037

DEVICE TYPE	REPL CODE
HT101	0037
HT102	0155
HT103	0155
HT321B-0	OEM
HT400	0016
HT401	0016
HT559	OEM
HT1067A	OEM
HT1077	OEM
HT4204	OEM
HT4205	OEM
HT4206	OEM
HT4214	OEM
HT4214-1L	OEM
HT4215	OEM
HT4215-1L	OEM
HT4216-1L	OEM
HT4226	OEM
HT4233	OEM
HT-4234	OEM
HT4234	OEM
HT4235	OEM
HT4235A	OEM
HT4235B	OEM
HT4236	OEM
HT4245	OEM
HT4245A	OEM
HT4245B	OEM
HT4245C	OEM
HT4255	OEM
HT4256A	OEM
HT4264	OEM
HT4265B	OEM
HT4265C	OEM
HT4273	OEM
HT4275	OEM
HT4285	OEM
HT4364	OEM
HT4401	OEM
HT4403	OEM
HT4404	OEM
HT4407	OEM
HT4412	OEM
HT4413	OEM
HT4414	OEM
HT4415A	OEM
HT4421A	OEM
HT4422	OEM
HT4425	OEM
HT4429	OEM
HT4437	OEM
HT4437-5A	OEM
HT4438	OEM
HT4438A	OEM
HT4440	OEM
HT4442	OEM
HT4443	OEM
HT4444	OEM
HT4444A	OEM
HT4445	OEM
HT4446	OEM
HT4448	OEM
HT4453	OEM
HT4457	OEM
HT4468	OEM
HT4495	OEM
HT4497	OEM
HT4501	OEM
HT4503A	OEM
HT4504	OEM
HT4504A	OEM
HT4504A-5B2	OEM
HT4505	OEM
HT4506	OEM
HT4506A	OEM
HT4506AS	OEM
HT4511	OEM
HT4515	OEM
HT4516	OEM
HT4532A	OEM
HT4541	OEM
HT4609	OEM
HT4611	OEM
HT4611A	OEM
HT4612A	OEM
HT4613A	OEM
HT4614A	OEM
HT4615A	OEM
HT4615B	OEM
HT4616A	OEM
HT4616B	OEM
HT4620A	OEM
HT4620C	OEM
HT4629	OEM
HT4635	OEM
HT4635A	OEM
HT4636	OEM
HT4637C	OEM
HT4648	OEM
HT4651A	OEM
HT4652	OEM
HT4653	OEM
HT4673	OEM
HT4677D	OEM
HT4705C	OEM
HT4706	OEM
HT4706A	OEM
HT4712B	OEM
HT4712C	OEM

If replacement code is OEM, contact original manufacturer for replacement.

DEVICE TYPE	REPL CODE
HT4714	OEM
HT4723	OEM
HT4724	OEM
HT4724A	OEM
HT4724B	OEM
HT4734B	OEM
HT4736	OEM
HT4753A	OEM
HT4763	OEM
HT4763-6D	OEM
HT4763-7A	OEM
HT4765	OEM
HT4765A	OEM
HT4765B	OEM
HT4765C	OEM
HT4773	OEM
HT4784A	OEM
HT4785	OEM
HT4785B	OEM
HT4785D	OEM
HT4793F	OEM
HT4795	OEM
HT4795A	OEM
HT4806B	OEM
HT4814C	OEM
HT4815C	OEM
HT4816B	OEM
HT4822	OEM
HT4832	OEM
HT4832C	OEM
HT4842C	OEM
HT4843B	OEM
HT4871	OEM
HT4885	OEM
HT4885S	OEM
HT4886	OEM
HT4895	OEM
HT4904	OEM
HT4944A	OEM
HT4946D	OEM
HT4964	OEM
HT4971D	OEM
HT4974C	OEM
HT5111A	OEM
HT5125	OEM
HT5126	OEM
HT5163	OEM
HT5164A	OEM
HT5165	OEM
HT5166	OEM
HT5181	OEM
HT5197A	OEM
HT5198	OEM
HT5201	OEM
HT5202	OEM
HT5221A	OEM
HT5239	OEM
HT5263	OEM
HT5271E	OEM
HT5280E	OEM
HT5291C	OEM
HT5292	OEM
HT5293	OEM
HT5317	OEM
HT5318	OEM
HT5404	OEM
HT5408	OEM
HT5417A	OEM
HT5422A	OEM
HT5424	OEM
HT5424C	OEM
HT5425	OEM
HT5425A	OEM
HT5426	OEM
HT5426A	OEM
HT5428	OEM
HT5434B	OEM
HT5438	OEM
HT5465A	OEM
HT5467A	OEM
HT5469B	OEM
HT5478E	OEM
HT5490B	OEM
HT5493	OEM
HT5498S	OEM
HT5505	OEM
HT5506B	OEM
HT5511	OEM
HT5513	OEM
HT5516	OEM
HT5516B	OEM
HT5520	OEM
HT5520B	OEM
HT5523	OEM
HT5523D	OEM
HT5523E	OEM
HT5523F	OEM
HT5523G	OEM
HT5525B	OEM
HT5535A	OEM
HT5535B	OEM
HT5535C	OEM
HT5539C	OEM
HT5545	OEM
HT5550	OEM
HT5563A	OEM
HT5566A	OEM
HT5580-8F	OEM
HT5588	OEM

DEVICE TYPE	REPL CODE
HT5648D-8K01	OEM
HT5654-9B24	OEM
HT5681F	OEM
HT5684B	OEM
HT5687C	OEM
HT5688E	OEM
HT5718A	OEM
HT5751B-0A27	OEM
HT5751B-9J24	OEM
HT5755A	OEM
HT6202	OEM
HT6210B	OEM
HT6213A	OEM
HT6217A	OEM
HT6502A	OEM
HT6506B	OEM
HT6506C	OEM
HT6507B	OEM
HT6508A	OEM
HT6508B	OEM
HT7109B	OEM
HT7120-8D13	OEM
HT7127B	OEM
HT7188	OEM
HT7222B	OEM
HT7222C	OEM
HT9001P	OEM
HT10234UB	0050
HT20436A	0222
HT20451A	0164
HT30491C	0086
HT30494	0103
HT30494X	0103
HT31041CO	0113
HT36441B	0111
HT38281D	0079
HT-49356	OEM
HT55706C	OEM
HT101011X	0136
HT101021A	0136
HT102341	0050
HT102341A	0050
HT102341B	0050
HT102341C	0050
HT102351	0050
HT102351A	0050
HT103501	0050
HT103501A	0136
HT103531C	0136
HT103541B	0136
HT104861	0086
HT104861A	0086
HT104861B	0086
HT104941B-0	0037
HT104941B-0	0037
HT104941C-0	0037
HT104941C0	0037
HT104941C-0	0693
HT104941CO	0150
HT104942A	0693
HT104951A-0	0037
HT104951A-0	0150
HT104951B	0037
HT104951B-0	0037
HT104951C	0037
HT104971A	0126
HT104971A-0	0126
HT104971A0	0126
HT104971A-0	0126
HT104971A0	0126
HT105611	0037
HT105611A	0037
HT105611B	0037
HT105611B0	0037
HT105611BO	0037
HT105611C	0037
HT105612B	0006
HT105621B-0	0037
HT105621B0	0037
HT105621B-0	0150
HT105621BO	0037
HT105641C	0693
HT105641D	0693
HT105641H	0203
HT105642B	0693
HT105642CO	0203
HT106731B	0037
HT106731B0	0148
HT107211T	0693
HT107331Q0	0006
HT200540	0004
HT200540A	0004
HT200541	0004
HT200541A	0004
HT200541B	0004
HT200541B-0	0004
HT200541C	0050
HT200561	0004
HT200561A	0004
HT200561B	0004
HT200561C	0004
HT200561C-0	0004
HT200751B	0004
HT200770B	0004
HT200771	0004
HT200771A	0004
HT200771B	0004
HT200771C	0004
HT201721A	0004

DEVICE TYPE	REPL CODE
HT201721D	0004
HT201725A	0004
HT201752A	OEM
HT201782A	0004
HT201861A	0004
HT201871L	0004
HT203243A	1062
HT203701	0004
HT203701A	0004
HT203701B	0004
HT204051	0164
HT204051B	0164
HT204051C	0164
HT204051D	0164
HT204051E	0164
HT204053	0164
HT204053A	0164
HT204053B	0164
HT204071D	0160
HT204736	0222
HT303620B	0016
HT303711A	0191
HT303711A-0	0144
HT303711A0	0079
HT303711A-0	0079
HT303711AO	0079
HT303711B	0191
HT303711B-0	0079
HT303711B0	0079
HT303711B-0	0079
HT303711BO	0079
HT303711C	0127
HT303720A	0144
HT303721-0	0016
HT303721D	0016
HT303730	0016
HT303730A	0016
HT303801	0127
HT303801A	0127
HT303801A0	0127
HT303801AO	0079
HT303801B	0127
HT303801B-0	0127
HT303801B0	0127
HT303801B-0	0079
HT303801BO	0079
HT303801C	0127
HT303801C0	0127
HT303801CO	0079
HT303871OO	0549
HT303941	0144
HT303941A	0144
HT303941B	0144
HT304508K	0079
HT304531	0016
HT304531A	0016
HT304531B	0016
HT304531C	0016
HT304540A0	0144
HT304540AO	0079
HT304540B0	0016
HT304540BO	0079
HT304580	0016
HT304580A	0016
HT304580B	0016
HT304580C0	0016
HT304580CO	0079
HT304580K	0016
HT304580Y0	0016
HT304580YO	0016
HT304580Z	0016
HT304581	0016
HT304581A	0016
HT304581B	0016
HT304581B-0	0016
HT304581B-0	0079
HT304581C	0016
HT304601B0	0151
HT304601BO	0079
HT304601C0	0144
HT304601CO	0079
HT304611B	0144
HT304861	0086
HT304861A	0086
HT304861B	0016
HT304911	0178
HT304911A	0178
HT304911B	0178
HT304941X	0103
HT304961C-0	0560
HT304961C-0	0560
HT304971A	0086
HT304971A0	0086
HT304971AO	0086
HT304971B	0086
HT305351B0	0127
HT305351C0	0144
HT305351CO	0079
HT305361E	0537
HT305361G	0537
HT305371E	0016
HT305642B	0693
HT306441	0079
HT306441A	0111
HT306441A0	0086
HT306441AO	0086
HT306441B	0016
HT306441B-0	0016
HT306441B0	0016

DEVICE TYPE	REPL CODE
HT306441B-0	0086
HT306441BO	0086
HT306441C	0016
HT306441C-0	0016
HT306441C-0	0086
HT306442A	0111
HT306442B	0111
HT306451	0144
HT306451A	0127
HT306451B	0144
HT306451H	0079
HT306681C	0127
HT306962A-0	0144
HT306962A-0	0079
HT307321A	0016
HT307321B-0	0016
HT307321B-0	0079
HT307322A	0111
HT307331B	0016
HT307331C	0111
HT307331C0	0079
HT307331CO	0111
HT307341	0086
HT307341A	0086
HT307341B	0079
HT307341C-0	0079
HT307341C-0	0086
HT307342B	0086
HT307342C	0076
HT307720B	0127
HT307721C	0127
HT307721D	0127
HT307902B	1496
HT308281B	1220
HT308281B0	1220
HT308281C	0079
HT308281D	1220
HT308281G	1220
HT308281H	1220
HT308281O	0079
HT308282A	0079
HT308282A-0	0079
HT308282A-0	0079
HT308282B	0111
HT308291A	0016
HT308291A-0	0127
HT308291A-0	0016
HT308291B	0016
HT308291B-0	0016
HT308291B0	0151
HT308291B-0	0079
HT308291BO	0151
HT308291C	0151
HT308291D0	0151
HT308301B0	0042
HT308301BO	0178
HT309002A0	1221
HT309291C	1066
HT309291E	0127
HT309301C	2209
HT309301D	2209
HT309301E	0111
HT309301F	0111
HT309451L0	0076
HT309451R0	0076
HT309452A0	0076
HT309680B	0086
HT309714A-0	0086
HT309714A-0	2952
HT309841B0	0086
HT309841BO	0086
HT309842A-0	0016
HT309842A-0	0086
HT310001F	0111
HT310002A	0111
HT311621B	0042
HT312131C0	0016
HT313171R	0155
HT313172A	0155
HT313181C	0086
HT313271T	1221
HT313272B	0111
HT313591C0	0151
HT313592B	0151
HT313681B0	1790
HT313831X	0018
HT313832C	0018
HT313841R	0018
HT314071Q	2287
HT316751M0	0076
HT318291D0	0016
HT319091OO	0076
HT319571OO	2710
HT340519C	0038
HT400721	0208
HT400721A	0208
HT400721B	0208
HT400721C	0208
HT400721D	0208
HT400721E	0208
HT400723	0208
HT400723A	0208
HT400723B	0208
HT400770B	0208
HT401191	0103
HT401191A	0103
HT401191B	0103
HT401193A0	0177
HT401301B0	0178

DEVICE TYPE	REPL CODE
HT401301BO	0178
HT402352B	0042
HT403131E0	0042
HT403131EO	0423
HT403132AO	0042
HT403151E	4251
HT403152A	4251
HT403152B	0178
HT403523A	0208
HT404001E	0595
HT600011F	0037
HT600011H	0037
HT70011100	3165
HT800011F	0016
HT800011G	0016
HT800011H	0016
HT800011K	0016
HT800012F	0086
HT800191H	0086
HT1001510	0004
HT2000771C	0004
HT2040710A	0160
HT2046710	0222
HT3036201	0079
HT3036210	0016
HT3037010	0144
HT3037201	0144
HT3037201A	0144
HT3037201B	0144
HT3037210	0016
HT3037210-0	0016
HT3037210-0	0079
HT3037310	0127
HT3038013-B	0127
HT3073310	0079
HT3082810	0086
HT6000210	0037
HT7000310	0905
HT7000410-0	OEM
HT7000410-0	0126
HT8000101	0233
HT8000710	0015
HT8001210	0086
HT8001310	0086
HT8001610	0007
HT8001710	0007
HT8001810	0016
HT9000310	0103
HT9000410-0	0103
HT9000410-0	0103
HT30373100	0016
HT30387100	0549
HT30733100	0111
HT30945120	0079
HT31909100	0076
HT31957100	2710
HT32362280	1227
HT70003100	OEM
HT90003100	OEM
HTK0022	OEM
HTK0054	3133
HTK0062	0325
HTS1068C	OEM
HTS4916	OEM
HTS4966	OEM
HTS4966B	OEM
HTS5637	OEM
HTS5706	OEM
HTS5706C	OEM
HTS5726	OEM
HTS5776B	OEM
HTS7115	OEM
HTS7115T	OEM
HTV80	OEM
HTV90	OEM
HTV100	OEM
HTV300	OEM
HTV350	OEM
HTV400	OEM
HTV450	OEM
HTZ11BL	0181
HTZ-33	5591
HTZ33-02	OEM
HTZ33-12	OEM
HV0000102	0004
HV0000105	0139
HV0000105-0	0015
HV0000202	0143
HV0000206	0120
HV0000212	OEM
HV0000302	0127
HV0000405	0164
HV0000405-0	0136
HV0000705	0015
HV00001050	0133
HV00001060	0139
HV-03SS	OEM
HV-05	0293
HV05	OEM
HV-06SS	OEM
HV-08	OEM
HV-08SS	OEM
HV2.5-10	OEM
HV2.5-10F	OEM
HV2.5-12	OEM
HV2.5-12F	OEM
HV2.5-15	OEM
HV2.5-15F	OEM
HV2.5-18	OEM

DEVICE TYPE	REPL CODE
HV2.5-18F	OEM
HV5	OEM
HV5-5	1791
HV5-5F	1791
HV5-7	1791
HV5-7F	1791
HV5-10	1791
HV5-10F	1791
HV5-12	OEM
HV5-12F	OEM
HV5-15	OEM
HV5-15F	OEM
HV5-20	OEM
HV5-20F	OEM
HV5-25	OEM
HV5-25F	OEM
HV5-30	OEM
HV5-30F	OEM
HV5-40	OEM
HV5-40F	OEM
HV8	3659
HV10	3659
HV10-5	1791
HV10-5F	1791
HV10-7	1791
HV10-7F	1791
HV10-10	1791
HV10-10F	1791
HV10-12	OEM
HV10-12F	OEM
HV10-15	OEM
HV10-15F	OEM
HV10-20	OEM
HV10-20F	OEM
HV10-25	OEM
HV10-25F	OEM
HV10-30	OEM
HV10-30F	OEM
HV10-40	OEM
HV10-40F	OEM
HV10E1	OEM
HV10P1	OEM
HV12	0004
HV14	3659
HV15	0143
HV-15(RECT)	0348
HV15E1	OEM
HV-15F	OEM
HV15P1	OEM
HV16	0004
HV17	0004
HV17B	0004
HV17C	OEM
HV18	OEM
HV19	0004
HV-20	0348
HV20E1	OEM
HV20E2	OEM
HV-20F	OEM
HV20G	OEM
HV20P1	OEM
HV20P2	OEM
HV-21	0348
HV-21F	OEM
HV-23	0139
HV23	0139
HV-23(ELGIN)	1925
HV23-G	0139
HV-23BL	0143
HV23F	0139
HV-23G	0139
HV23G	0139
HV-23G(BL)	0133
HV23G(WT)	1925
HV-23GB	0139
HV23GB	0139
HV-23GBL	1925
HV23GBL	1925
HV23GW	0139
HV23GWT	1925
HV23GY	0139
HV-23GYL	0139
HV-25	0133
HV25	0016
HV-25(DIODE)	0133
HV25(HITACHI)	0133
HV-25(RCA)	0133
HV25-5	1791
HV25-5F	1791
HV25-7	1791
HV25-7F	1791
HV25-10	1791
HV25-10F	1791
HV25-12	OEM
HV25-12F	OEM
HV25-15	OEM
HV25-15F	OEM
HV25-20	OEM
HV25-20F	OEM
HV25-25	OEM
HV25-25F	OEM
HV25-30	OEM
HV25-30F	OEM
HV25-40	OEM
HV25-40F	OEM
HV25E2	OEM
HV25P2	OEM
HV-26	0015
HV26	1925

DEVICE TYPE	REPL CODE
HV-26G	0015
HV26G	OEM
HV-27	0133
HV-30	OEM
HV30E2	OEM
HV30E3	OEM
HV-30F	OEM
HV30G	OEM
HV30P2	OEM
HV30P3	OEM
HV-40	OEM
HV40E3	OEM
HV40E4	OEM
HV-40F	OEM
HV40G	OEM
HV40P3	OEM
HV40P4	OEM
HV40TN	OEM
HV-46	0120
HV46	0120
HV46G	0120
HV-46GR	0015
HV460	0120
HV46R	0120
HV-50	OEM
HV50E4	OEM
HV50E6	OEM
HV-50F	OEM
HV50G	OEM
HV50ML	OEM
HV-60	OEM
HV60E6	OEM
HV-60F	OEM
HV60G	OEM
HV60TN	OEM
HV70	OEM
HV70E6	OEM
HV70G	OEM
HV70ML	OEM
HV75	OEM
HV-80	0139
HV80	0139
HV80E	OEM
HV80E6	OEM
HV80G	OEM
HV90E	OEM
HV90G	OEM
HV-100	0015
HV100	OEM
HV100E	OEM
HV100G	OEM
HV106	OEM
HV106F	OEM
HV108	OEM
HV108F	OEM
HV110	OEM
HV110F	OEM
HV112	OEM
HV112F	OEM
HV115	OEM
HV115F	OEM
HV120E	OEM
HV140E	OEM
HV160E	OEM
HV180E	OEM
HV200E	OEM
HV206	OEM
HV206F	OEM
HV208	OEM
HV208F	OEM
HV210	OEM
HV210F	OEM
HV212	OEM
HV212F	OEM
HV215	OEM
HV215F	OEM
HV236WT	OEM
HV-801	0143
HV9601	0030
HVB2	OEM
HVB4	OEM
HVB6	OEM
HVB10	OEM
HVB20	OEM
HVB30	OEM
HVB40	OEM
HVB50	OEM
HVC2	OEM
HVC4	OEM
HVC6	OEM
HVC10	OEM
HVC-15	0348
HVC-15F	OEM
HVC-20	0348
HVC-20F	OEM
HVC-25	OEM
HVC-25F	OEM
HVC-30	OEM
HVC-30F	OEM
HVC-40	OEM
HVC-40F	OEM
HVC-50	OEM
HVC-50F	OEM
HVC-60	OEM
HVC-60F	OEM
HVC-V87	OEM
HVD-1	2977
HVD-2	2978
HVD2	OEM

DEVICE TYPE	REPL CODE
HVD-3	2980
HVD-4	2979
HVD4	OEM
HVD6	OEM
HVD10	OEM
HVD-12	5362
HVD20	OEM
HVD30	OEM
HVD40	OEM
HVD50	OEM
HVE2	OEM
HVE4	OEM
HVE6	OEM
HVK89	OEM
HVK1109-866	OEM
HVK1110-872	OEM
HVM16	OEM
HVM89	OEM
HVM1600	0190
HVMRA	0774
HVP5	3658
HVP8	3658
HVP10	3658
HVP12	3662
HVP14	OEM
HVP15	OEM
HVP16	OEM
HVPR10-12	3662
HVR-1X-01A	OEM
HVR-3B	6806
HVR-3H	OEM
HVR-3H2	OEM
HVR5	OEM
HVR-5A	6806
HVT-5	OEM
HVT-15SS	0201
HVT-20SS	0288
HVT-22	0288
HVT-22DA	0201
HVT22SS	OEM
HVT-25SS	0378
HVT30	OEM
HVT-30S	0378
HVT-30SS	0378
HVT200	0233
HVT400	1067
HVT401	OEM
HVT600	OEM
HVT800	OEM
HVT900	OEM
HVT1000	OEM
HVT-DF18	OEM
HVT-DF25	OEM
HVTR60	OEM
HVU359	OEM
HVU359TRF	OEM
HW6.8	0025
HW6.8A	0025
HW6.8B	0025
HW7.5	0649
HW7.5A	0649
HW7.5B	0649
HW8.2	0244
HW8.2A	0244
HW8.2B	0244
HW9.1	0012
HW9.1A	0012
HW9.1B	0012
HW9.81B	0012
HW10	0170
HW10A	0170
HW10B	0170
HW11	0317
HW11A	0317
HW11B	0794
HW12	0137
HW12A	0137
HW12B	0137
HW13	0365
HW13A	0365
HW13B	0365
HW15	0002
HW15A	0002
HW15B	0002
HW16	0420
HW16A	0420
HW16B	0420
HW18	0495
HW18A	0495
HW18B	0495
HW20	0531
HW20A	0531
HW20B	0531
HW22	0565
HW22A	0565
HW22B	0565
HW24	0402
HW24A	0402
HW24B	0402
HW27	0441
HW27A	0441
HW27B	0441
HW30	0726
HW30A	0726
HW30B	0726
HW33	0039
HW33A	0039
HW33B	0039
HW36	0820
HW36A	0820
HW36B	0820
HW39	0350
HW39A	0350
HW39B	0350
HW43	0931
HW43A	0931
HW43A3A	OEM
HW43B	0931
HW47	0999
HW47A	0999
HW47B	0999
HW51	0502
HW51A	0502
HW51B	0502
HW56	0869
HW56A	0869
HW56B	1834
HW62	0783
HW62A	0783
HW62B	0783
HW68	1268
HW68A	1268
HW68B	2157
HW75	1189
HW75A	1189
HW75B	1189
HW82	0331
HW82A	0331
HW82B	0331
HW91	1311
HW91A	1311
HW91B	1311
HW100	0098
HW100A	0098
HW100B	0402
HW110	0149
HW110A	0149
HW110B	0149
HW120	0186
HW120A	0186
HW120B	0186
HW130	0213
HW130A	0213
HW130B	0213
HW150	0028
HW150A	0028
HW150B	0028
HW160	0255
HW160A	0255
HW160B	0255
HW180	0367
HW180A	0367
HW180B	0367
HW200	0421
HW200A	0421
HW200B	0421
HW2061	OEM
HW2063	OEM
HW2063A	OEM
HW9062	OEM
HW9362	OEM
HW9364	OEM
HX20G	OEM
HX30G	OEM
HX40G	OEM
HX50G	OEM
HX50ML	OEM
HX60G	OEM
HX70G	OEM
HX70ML	OEM
HX80E	OEM
HX80G	OEM
HX90E	OEM
HX90G	OEM
HX100E	OEM
HX100G	OEM
HX120E	OEM
HX140E	OEM
HX160E	OEM
HX180E	OEM
HX200E	OEM
HX50001	0144
HX50002	0016
HX50003	0127
HX-50063	0079
HX-50072	0079
HX-50091	0224
HX-50092	0079
HX-50094	0037
HX-50097	0079
HX-50102	0901
HX-50103	0595
HX-50104	0791
HX-50105	0037
HX-50106	2050
HX-50107	0079
HX-50108	0086
HX-50109	0436
HX-50110	0127
HX-50112	0037
HX-50113	0079
HX-50128	0037
HX-50161	0079
HX-50172	0016
HX-50176	0037
HX-50180	0400
HX-50181	0079
HY5C2	0589
HY93C46	OEM
HY-101	OEM
HY-101SHINEI	OEM
HY6589	OEM
HY6590	OEM
HY3045801C	0016
HYB4116-P2	0522
HYB4116-P3	0522
HYB4164-P2	2357
HZ2	OEM
HZ2(H)	OEM
HZ2ALL	4320
HZ2B	OEM
HZ2B1	OEM
HZ2BLL	OEM
HZ2C	1278
HZ2C1	1276
HZ2C2	1278
HZ2CLL	1276
HZ2LL	OEM
HZ2.0P	OEM
HZ2.2P	OEM
HZ2.4P	OEM
HZ2.7E	0760
HZ2.7P	1312
HZ3	0118
HZ3(H)	0118
HZ-3A	OEM
HZ3A	OEM
HZ3A07	0118
HZ3A2	5971
HZ-3A3	OEM
HZ3A3	6993
HZ3ALL	0677
HZ3B	0118
HZ3B2	0118
HZ3B3	6993
HZ3BLL	0118
HZ3C	OEM
HZ3C1	0299
HZ3C-2	0299
HZ3C2	0299
HZ3C2-01	0299
HZ3C3	0299
HZ3D	OEM
HZ3H	0299
HZ3LL	0118
HZ3.0E	0118
HZ3.0P	1713
HZ3.3E	0299
HZ3.3P	0188
HZ3.6	0376
HZ3.6CP	0376
HZ3.6CPTA	0376
HZ3.6E	0376
HZ3.6P	0188
HZ3.9E	0036
HZ3.9P	0456
HZ4	0036
HZ-4(A)	0376
HZ-4(A)1	0376
HZ4(A1)	0376
HZ-4(A1)1	0376
HZ4(A1)1	0376
HZ4(H)	0036
HZ4A	OEM
HZ4A04	0036
HZ-4A1	0376
HZ4A-1	0376
HZ4A1	0376
HZ-4A2	0376
HZ4A-2	0376
HZ4A2	0376
HZ4A3	6042
HZ-4B	0036
HZ4B	0275
HZ4B-1	0036
HZ4B1	0036
HZ4B1S3	OEM
HZ-4B2	0036
HZ4B-2	0036
HZ4B2	0036
HZ4B3	0036
HZ4BC	OEM
HZ4C	0275
HZ4C1	0275
HZ4C2	0275
HZ4C3	5235
HZ4C11	OEM
HZ4CLL	OEM
HZ4CZ	OEM
HZ4H	0456
HZ4H4	0376
HZ4HB	0036
HZ4HC	0275
HZ4LL	0036
HZ4.3E	0275
HZ4.3P	0533
HZ4.7CPTA	OEM
HZ4.7E	0140
HZ4.7P	0451
HZ-5	0162
HZ5	0140
HZ5(C)1	0589
HZ5(C)3	0589
HZ5(H)	0140
HZ5A	OEM
HZ5A1	5235
HZ5A2	OEM
HZ5ALL	OEM
HZ-5B	0140
HZ5B	0140
HZ5B-1	0140
HZ5B1	6125
HZ5B-1-05	0140
HZ5B-2	0140
HZ5B2	0140
HZ5B3	0041
HZ5BLL	OEM
HZ-5C	0589
HZ5C	0041
HZ-5C1	0041
HZ5C-1	0041
HZ5C1	0248
HZ5C1TD	0041
HZ5C-2	0041
HZ5C2	0041
HZ5C2TA	OEM
HZ-5C3	0589
HZ5C-3	0589
HZ5C3	0589
HZ5CLL	0589
HZ5LL	0140
HZ5.0C2	0589
HZ5.1E	0041
HZ5.1P	0162
HZ5.6E	0253
HZ5.6P	0157
HZ-6	OEM
HZ6	0091
HZ-6(A)	0883
HZ6(A)1	0883
HZ6(B)	0091
HZ6(B)1	0883
HZ6(B)2	0157
HZ6(B)3	0091
HZ6(B2)	0157
HZ6(C)	OEM
HZ6(C)-1	0091
HZ6(C)1	0091
HZ6(C)3	0295
HZ-6(C3)	0471
HZ6(H)	0091
HZ6(H)L	0091
HZ-6A	0883
HZ6A	0589
HZ6A-1	0883
HZ6A1	0883
HZ6A1L	0589
HZ6A1V	0041
HZ-6A2	0253
HZ6A2	0041
HZ6A2L	OEM
HZ6A-3	0883
HZ6A3	5445
HZ6A19	0091
HZ-6B	0253
HZ6B	0253
HZ-6B1	0883
HZ6B1	0253
HZ-6B2	0883
HZ6B2	0157
HZ6B2/TA	0883
HZ6B2L	OEM
HZ-6B3	0091
HZ6B3	0091
HZ6B3L	0091
HZ6B20	0091
HZ-6C	0091
HZ6C	0471
HZ-6C-1	0091
HZ-6C1	0091
HZ6C-1	0091
HZ6C1	0091
HZ6C2	0471
HZ6C2L	0471
HZ-6C-3	0471
HZ-6C3	0295
HZ6C3	0471
HZ6C21	0295
HZ6CV	0091
HZ6L	0091
HZ-6L-B2	0091
HZ6L-B2	0091
HZ6LB2	5445
HZ6V	0091
HZ-6Z	0091
HZ6.2E	0471
HZ6.2P	0636
HZ6.2PB	0636
HZ6.3	0636
HZ6.8E	0062
HZ6.8P	0025
HZ-7	0062
HZ7	0062
HZ7(A)	0062
HZ-7(B)	0062
HZ7(B)	0062
HZ7(B)1	0062
HZ7(B)2	0062
HZ7(B)3	0062
HZ7(C)	0077
HZ7(C)1	0077
HZ7(C)3	0077
HZ7(H)	0062
HZ7(H)L	0062
HZ-7A	0062
HZ7A	0062
HZ-7A(B)	0649
HZ7A-1	0062
HZ7A1	0471
HZ7A2	OEM
HZ7A2L	OEM
HZ7A-3	0062
HZ7A3	0062
HZ7A33	0025
HZ-7AGR	0025
HZ-7B	0025
HZ7B	0062
HZ7B-1	0062
HZ7B1	3369
HZ-7B2	0062
HZ7B2	0062
HZ7B2L	OEM
HZ7B2LV1	0062
HZ-7B3	0062
HZ7B3	OEM
HZ7B3L	OEM
HZ-7C	0062
HZ7C	0077
HZ-7C-1	0077
HZ-7C1	0077
HZ7C1	0077
HZ7C2	6377
HZ7C2L	OEM
HZ7C3	0077
HZ-7GR	0062
HZ7H	0649
HZ7L	0062
HZ7LB2	OEM
HZ-7NB	0062
HZ7.5E	0077
HZ7.5P	0649
HZ8.2E	0165
HZ8.2P	0244
HZ-9	0012
HZ9	0057
HZ9(A)1	0165
HZ9(A)3	0165
HZ9(C)	0057
HZ9(C)1	0057
HZ9(C)3	0057
HZ9(H)	0322
HZ9(H)L	0322
HZ9-C1	0057
HZ-9A	0322
HZ9A	0322
HZ-9A1	0165
HZ9A-1	0165
HZ9A1	0165
HZ-9A2	0165
HZ9A-2	0165
HZ9A2	0165
HZ9A3	0165
HZ9A26	OEM
HZ9AL	OEM
HZ-9B	0012
HZ9B	1083
HZ9B1	7066
HZ9B-2	0057
HZ9B2	7066
HZ9B-3	0057
HZ9B3	0322
HZ9B24	0057
HZ9BZ	0322
HZ-9C	0057
HZ9C	OEM
HZ-9C1	0057
HZ9C1	0057
HZ9C-2	0057
HZ9C2	0057
HZ-9C3	0057
HZ9C-3	0057
HZ9C3	6564
HZ-9C-GR	OEM
HZ9H	0012
HZ9L	0057
HZ9.1	0057
HZ9.1B1	0012
HZ9.1E	0057
HZ9.1P	0012
HZ9.1PB	0012
HZ10E	0064
HZ10P	0170
HZ-11	0317
HZ11	0181
HZ-11(A)	0317
HZ11(A)	0248
HZ11(A1)	0064
HZ-11(B)	0317
HZ11(B)	0248
HZ11(B)2	0181
HZ-11(C)	0317
HZ11(C)	0248
HZ-11(C)-T2	0248
HZ-11(C)T2	0181
HZ-11(C)TZ	0181
HZ11(H)	0181
HZ11(H)L	0181
HZ-11A	0248
HZ11A	0064
HZ11A/B	OEM
HZ11A/B/C	OEM
HZ-11A-1	0181
HZ-11A1	0181
HZ11A1	0064
HZ11A2	1258
HZ11A-3	0248
HZ11A3	1258
HZ-11ABC	0181
HZ-11B	0248
HZ11B	0181
HZ-11B1	0064
HZ11B1	0064
HZ11B1V	0064
HZ-11B2	0181
HZ11B-2	0181
HZ11B2	1258
HZ11B-2L	0181
HZ11B2L	0181
HZ11B-2LTA	0181
HZ-11B-3	OEM
HZ-11C	0248
HZ11C	0181
HZ11C1L	OEM
HZ11C2	0181
HZ-11C2L	0181
HZ11C2L	0181
HZ11C-3	0181
HZ11C25	0052
HZ11E	0181
HZ11H	0317
HZ11L	0181
HZ-11LC2	0181
HZ11LC2	0181
HZ11P	0317
HZ-12	0137
HZ12	0053
HZ-12(A)	0052
HZ12(A)	0052
HZ12(A)1	1005
HZ12(A)2	1005
HZ-12(B)	0052
HZ12(B)	0052
HZ-12(B)1	1005
HZ12(B)1	1005
HZ12(B)2	0053
HZ-12(C)	0052
HZ12(C)	0052
HZ-12(C1)	0052
HZ12(C1)	0052
HZ-12(C2)	1005
HZ12(C2)	1005
HZ-12(C3)	1005
HZ12(C3)	1005
HZ12(H)	0053
HZ12(H)L	0053
HZ12-3	0053
HZ12-C	0052
HZ-12A	0052
HZ12A	0052
HZ12A1	1378
HZ-12A1-3	OEM
HZ12A2	1005
HZ-12A3	0052
HZ12A-3	0052
HZ12A3	1005
HZ-12ABC	0052
HZ-12B	0053
HZ12B	0053
HZ-12B1	1005
HZ12B-1	0053
HZ12B1	1005
HZ-12B1-3	OEM
HZ-12B2	0052
HZ12B-2	0053
HZ12B2	0053
HZ12B2L	OEM
HZ12B-3	0053
HZ12B3	1399
HZ12B14	1005
HZ12B14(TP)	1005
HZ12BL2	OEM
HZ-12C	0052
HZ12C	0052
HZ12C1	1399
HZ-12C3	1005
HZ12C-3	0879
HZ12C3	1005
HZ12E	0052
HZ12H	0365
HZ12L	0052
HZ12P	0137
HZ12PB	1005
HZ13E	0053
HZ13P	0100
HZ-15	0686
HZ15	0686
HZ15(H)	0686
HZ15(H)L	0686
HZ15-1	0879
HZ15-2	1105
HZ15-3	0686
HZ15E	0686
HZ15H	0002
HZ15L	0686
HZ15P	0002
HZ15PB	0002
HZ16	0445
HZ16(H)	0445
HZ16(H)L	0445
HZ16-2	1229
HZ16-3	0210
HZ16E	0445
HZ16H	0420
HZ16L	0445
HZ16P	0420
HZ18	0375
HZ18(H)	0375
HZ18(H)L	0375
HZ18-09	0375
HZ18-1	0210
HZ18-2	0375
HZ18-3	0375
HZ18-3L	OEM
HZ18E	0375
HZ18L	0375
HZ18P	0495
HZ20	0700
HZ20(H)	0700
HZ20(H)L	0700
HZ20-04	0700
HZ20-04(AU)	0700
HZ20-1L	0700
HZ20-2	0700
HZ20-3	0700
HZ20-3L	0700
HZ20-3T2	0700
HZ20-3TS	0700
HZ20-BL	OEM
HZ20E	0700
HZ20L	0700
HZ-20L1	OEM
HZ20P	0531
HZ20TE	0700
HZ-20VC	0531
HZ22	0705
HZ22(H)	0705
HZ22(H)L	0705
HZ22-01	4942
HZ22-01T2	0705
HZ22-04	0705
HZ22-1	0705
HZ22-2	0705
HZ22-2L	0705
HZ22-3	0705
HZ22-3L	OEM
HZ22E	0705
HZ22H	0565
HZ22L	0705
HZ22L-3	0705
HZ22L3	0705
HZ22P	0565
HZ24	0494
HZ24(H)	0494
HZ24(H)L	0494
HZ24-1L	0494
HZ24-2	0494
HZ-24-2L	0494
HZ24-2L	0494
HZ24-12	0494
HZ24BP	OEM
HZ24E	0494
HZ24H	0402
HZ24L	0494
HZ24L-2	0494
HZ24L2	2467
HZ24P	0402
HZ27	0455
HZ27(H)	0455
HZ27(H)L	0455
HZ-27-1	0714
HZ27-1	2244
HZ27-2	0455
HZ-27BP	0455
HZ27BP	0455
HZ27E	0455
HZ27H	0441
HZ27L	0455
HZ27P	0441
HZ30	0195
HZ30(H)	0195
HZ30(H)L	0195
HZ30-1	0195
HZ30-1L	0195
HZ30-1T	OEM
HZ30-2	2436
HZ30-2L	0195
HZ30E	0195
HZ30H	0726
HZ30L	0195
HZ30P	0726
HZ30PB	0726
HZ33	0166
HZ33(H)	0166
HZ33(H)L	0166
HZ33-2	2684
HZ33-2L	0166
HZ33-3L	OEM
HZ33E	0166
HZ33H	0039
HZ33L	0166
HZ33P	0039
HZ33S1	0166
HZ-35	0010
HZ35	0010
HZ35-TP	0010
HZ35TP	2904
HZ-36	0010
HZ36	0010
HZ36(H)	0010
HZ36(H)L	0010
HZ36-1	0010
HZ36-2	2904
HZ36-3	2904

If replacement code is OEM, contact original manufacturer for replacement.

DEVICE TYPE	REPL CODE
HZ36E	0010
HZ36H	0820
HZ36L	0010
HZ36P	0820
HZ39E	0032
HZ43	0275
HZ47	OEM
HZ55C1	OEM
HZ56	OEM
HZ68	OEM
HZ82	OEM
HZ100	OEM
HZ120	OEM
HZ150	OEM
HZ242L	0494
HZ272	0455
HZ303B	0118
HZ304B	0036
HZ310B	0064
HZ312A	0052
HZ331L	OEM
HZ361	OEM
HZ414-B3	0100
HZ502	0041
HZ603	OEM
HZ2201TZ	0565
HZ5226	0292
HZ5227	0188
HZ5228	0456
HZ5229	0533
HZ5230	0451
HZ53461ZP-12	OEM
HZD7.5	OEM
HZEH	0292
HZK2	OEM
HZK3	OEM
HZK4	OEM
HZK5	OEM
HZK6	OEM
HZK6L	OEM
HZK7	OEM
HZK7L	OEM
HZK9	OEM
HZK9L	OEM
HZK11	OEM
HZK11L	OEM
HZK12	OEM
HZK12L	OEM
HZK15	OEM
HZK15L	OEM
HZK16	OEM
HZK16L	OEM
HZK18	OEM
HZK18L	OEM
HZK20	OEM
HZK20L	OEM
HZK22	OEM
HZK22L	OEM
HZK24	OEM
HZK24L	OEM
HZK27	OEM
HZK27L	OEM
HZK30	OEM
HZK30L	OEM
HZK33	OEM
HZK33L	OEM
HZK36	OEM
HZK36L	OEM
HZL004	OEM
HZM3A	OEM
HZM3B	OEM
HZM3C	OEM
HZM4A	OEM
HZM4B	OEM
HZM4C	OEM
HZM5A	OEM
HZM5B	OEM
HZM5C	OEM
HZM6A	OEM
HZM6B	OEM
HZM6C	OEM
HZM7A	OEM
HZM7B	OEM
HZM7C	OEM
HZM9A	OEM
HZM9B	OEM
HZM9C	OEM
HZM11A	OEM
HZM11B	OEM
HZM11C	OEM
HZM12A	OEM
HZM12B	OEM
HZM12C	OEM
HZM15	OEM
HZM16	OEM
HZM18	OEM
HZM20	OEM
HZM22	OEM
HZM24	OEM
HZM27	OEM
HZM30	OEM
HZM33	OEM
HZM36	OEM
HZS2-B	1276
HZS2ALL	OEM
HZS2B	1276
HZS2B1	OEM
HZS2B1TA	OEM
HZS2B2	OEM
HZS2BLL	OEM
HZS2BT	OEM
HZS2C2	OEM
HZS2.0N	OEM
HZS2.2N	OEM
HZS2.4N	1276
HZS2.7E	0760
HZS2.7N	0760
HZS3A2	OEM
HZS3A2TA	OEM
HZS3B1	OEM
HZS3B2	OEM
HZS3C1	OEM
HZS3C3	OEM
HZS3C3TA	OEM
HZS3.0E	0118
HZS3.0N	0118
HZS3.0NB2	6743
HZS3.3E	0299
HZS3.3N	0299
HZS3.6E	0376
HZS3.6EB1	0376
HZS3.6EB1-TD	0376
HZS3.6N	0376
HZS3.9E	0036
HZS3.9N	0036
HZS3.0NB2	OEM
HZS4A1	OEM
HZS4A2	OEM
HZS4B1	OEM
HZS4B2	OEM
HZS4BLL	OEM
HZS4C	OEM
HZS4C1-TD	OEM
HZS4R3EB1	OEM
HZS4R7EB2	OEM
HZS4.3E	0275
HZS4.3EB1	OEM
HZS4.3EB2	OEM
HZS4.3N	0275
HZS4.7E	0140
HZS4.7EB1	OEM
HZS4.7EB2	OEM
HZS4.7MB2-1TJ	OEM
HZS4.7N	0140
HZS4.7NB2-1TJ	0140
HZS4.7NBZ-ITJ	OEM
HZS5-C1	OEM
HZS5A2	OEM
HZS5A3	OEM
HZS5B1	OEM
HZS5B2	OEM
HZS5B2TA	OEM
HZS-5B3	OEM
HZS5B3	OEM
HZS-5C1	OEM
HZS5C1	OEM
HZS5C-2	0589
HZS5C2	OEM
HZS5C2T	OEM
HZS5C3	OEM
HZS5R1EB2-T	0041
HZS5R6JB2-TE	0883
HZS5.1B3	0041
HZS5.1E	0041
HZS5.1EB	0041
HZS5.1EB1	OEM
HZS5.1EB2	0041
HZS5.1JB2	0589
HZS5.1N	0041
HZS5.1NB1	OEM
HZS5.1NB2	5953
HZS5.6	OEM
HZS5.6B	OEM
HZS5.6E	0253
HZS5.6EB1	0883
HZS5.6EB2	0253
HZS5.6J	0253
HZS5.6JB2	OEM
HZS5.6N	0253
HZS-6-C2	OEM
HZS6A-1L	0589
HZS6A1L	0883
HZS6A2	OEM
HZS6AL	OEM
HZS6B1	OEM
HZS6B1L	OEM
HZS6B2	3002
HZS6B2/TA	0883
HZS6B2L	OEM
HZS6B3	OEM
HZS6B3L	OEM
HZS6B3TA	OEM
HZS6C	0295
HZS6C1	OEM
HZS6C1L	OEM
HZS6C1LTA	OEM
HZS6C-2	0295
HZS6C2	OEM
HZS6C2L	OEM
HZS6C3	0295
HZS6C3L	OEM
HZS6LB1	OEM
HZS6R2EB2-T	OEM
HZS6R2JB2-T	0295
HZS6R8JB1	OEM
HZS6R8JB2-TE	0062
HZS6.2E	0471
HZS6.2J	0471
HZS6.2JB2	0295
HZS6.2N	0471
HZS6.2NB2	OEM
HZS6.8E	0062
HZS6.8EB2	OEM
HZS6.8J	0062
HZS6.8JB1	OEM
HZS6.8JB2	OEM
HZS6.8N	0062
HZS7-A1	OEM
HZS7A-1	OEM
HZS7A1	OEM
HZS-7A2	OEM
HZS7B1	OEM
HZS7B1L	OEM
HZS7B2	OEM
HZS7B2L	OEM
HZS7B3L	OEM
HZS7C1	OEM
HZS7C-3	OEM
HZS7C3	OEM
HZS7.5E	0077
HZS7.5EB1	OEM
HZS7.5EB2	OEM
HZS7.5J	0077
HZS7.5N	0077
HZS8.2E	0165
HZS8.2J	0165
HZS8.2N	0165
HZS8.2NB2	OEM
HZS-9-C1	OEM
HZS9-C3	OEM
HZS9A	OEM
HZS9A1L	OEM
HZS-9A2	OEM
HZS9A-2	0165
HZS9A2	0165
HZS9A3	OEM
HZS9AL	OEM
HZS9B-3	0322
HZS9B3	OEM
HZS9C1	OEM
HZS9C1L	OEM
HZS9C1LTA	OEM
HZS9C2	OEM
HZS9C2TA	OEM
HZS9C3	OEM
HZS9C3L	OEM
HZS9R1EB3	OEM
HZS9R1JB2-TE	OEM
HZS9.1E	0057
HZS9.1EB2	OEM
HZS9.1EB3	OEM
HZS9.1F	OEM
HZS9.1J	0057
HZS9.1JB1	OEM
HZS9.1JB3	OEM
HZS9.1N	0057
HZS10E	0064
HZS10EB1	0248
HZS10J	0064
HZS10N	0064
HZS11A2	OEM
HZS11A2L	OEM
HZS11A2LTA	OEM
HZS11A3	OEM
HZS11B1	OEM
HZS11B2	OEM
HZS11B3	OEM
HZS11B3L	OEM
HZS11C1L	OEM
HZS11C2L	OEM
HZS11E	0181
HZS11J	0181
HZS11JB3	OEM
HZS11JB3-TE	0181
HZS11N	0181
HZS12	OEM
HZS12-C3T	0879
HZS12A1L	OEM
HZS-12A2	OEM
HZS12A2	OEM
HZS12A2L	OEM
HZS12A2LTA	OEM
HZS-12B	OEM
HZS12B	OEM
HZS12B1L	OEM
HZS12B2	2421
HZS12B2L	1005
HZS12B2TA	0053
HZS12B3	OEM
HZS12BL	OEM
HZS12C1	OEM
HZS12C1L	OEM
HZS12C1TA	OEM
HZS12C2L	OEM
HZS12C2LTA	OEM
HZS12C-3	0879
HZS12C3	2431
HZS12C3L	OEM
HZS12C3T	OEM
HZS12E	0052
HZS12EB2	1005
HZS12EB2-T	1005
HZS12EB3	1005
HZS12J	0052
HZS12JB2	1005
HZS12JB2-TE	1005
HZS12JB3	1005
HZS12JB3-T	1005
HZS12L	1005
HZS12L(B2)	1005
HZS12LC1	OEM
HZS12LC2	OEM
HZS12N	0052
HZS13E	0053
HZS13J	0053
HZS13JB2-TE	OEM
HZS13N	0053
HZS15-1L	OEM
HZS15-3	2265
HZS15-3L-TB	OEM
HZS15-3TD	0686
HZS15E	0686
HZS15J	0686
HZS15N	0686
HZS15NB2	OEM
HZS16E	0445
HZS16J	0445
HZS16N	0445
HZS18E	0375
HZS18J	0375
HZS18N	0375
HZS20-1	OEM
HZS20-1L	OEM
HZS20-1LTA	OEM
HZS20-2	0700
HZS20-2TD	0700
HZS20E	0700
HZS20J	0700
HZS20N	0700
HZS22-1L	OEM
HZS22-2L	OEM
HZS22-3L	0705
HZS22E	0705
HZS22J	0705
HZS22L1	OEM
HZS22N	0705
HZS24-1L	0494
HZS24-2L	OEM
HZS24-3	OEM
HZS24E	0494
HZS24J	0494
HZS24N	0494
HZS24NB2	OEM
HZS24NB3	OEM
HZS27	OEM
HZS27-2L	0455
HZS27-3-TD	OEM
HZS27-3L	0714
HZS27E	0455
HZS27L3	0714
HZS27N	0455
HZS30	0195
HZS30-1	0195
HZS30-2	OEM
HZS30-2L	OEM
HZS30-3L	OEM
HZS30E	0195
HZS30N	0195
HZS30T	OEM
HZS33-1	OEM
HZS33-2	OEM
HZS33-2-TD	OEM
HZS33-2L	OEM
HZS33E	0166
HZS33L2	OEM
HZS33N	0166
HZS36-1	OEM
HZS36-1L	0010
HZS36-2L	0010
HZS36-L2	0010
HZS36E	0010
HZS36JB2-TE	0010
HZS36L1	0010
HZS36L2	0010
HZS36N	0010
HZS39E	0032
HZS39N	0032
HZS331L	OEM
HZS682	OEM
HZSA32	OEM
HZT5	OEM
HZT6C2	OEM
HZT7A1	OEM
HZT7A2	OEM
HZT7A3	OEM
HZT7B1	OEM
HZT7B2	OEM
HZT7B3	OEM
HZT7C1	OEM
HZT7C2	OEM
HZT9	OEM
HZT22	OEM
HZT31	OEM
HZT-33	1329
HZT33	3452
HZT33-01	3452
HZT33-02	1329
HZT33-02-TE	1329
HZT33-02T	1329
HZT33-04	1329
HZT33-04T	1329
HZT33-05	3452
HZT33-10	1329
HZT33-12	3452
HZT33S1	1329
HZT-3301	3452
HZTS33-1TA	OEM
HsK151/10	OEM
I	OEM
I00DF01120	2870
I00DF10110	3083
I00DF11240	OEM
I00SE01250	OEM
I01B98M090	1786
I01S058360	1529
I01SD57530	3759
I02A98L050	1292
I02A978120	0334
I02DB13820	2896
I02DC13710	2623
I02DE14020	0583
I02DF148C	3884
I02DF1480C	3365
I02DF1481C	3884
I02DF12350	OEM
I02DF13970	OEM
I02DF18700	OEM
I02I90574J	1329
I02S010310	0904
I02SD13780	2042
I02SD14980	5054
I02SF12520	2269
I02SF12530	1604
I02SP12410	OEM
I02SP12770	4840
I03A98M050	0624
I03A98M060	0923
I03A98M090	1786
I03A98M120	0334
I03B98M050	0624
I03B98M060	0923
I03B98M090	1786
I03B98M120	1894
I03B97812V	OEM
I03D040660	0101
I03D063240	0625
I03D072200	3027
I03D078200	OEM
I03D079100	2666
I03D079130	4380
I03D079400	0683
I03D37323A	OEM
I03D373170	4277
I03DA7530N	3817
I03DD78000	0332
I03DE75200	0539
I03DE76520	OEM
I03DE79550	OEM
I03DG73200	OEM
I03S06358T	OEM
I03S063580	3233
I03S063930	7045
I03S064580	2913
I03S070160	1028
I03S072220	OEM
I03S073080	3488
I03S073560	OEM
I03S073580	OEM
I03S079100	2666
I03S973100	OEM
I03SD78350	2346
I03SD78380	OEM
I03SP41400	0362
I03SP44450	3335
I04D317870	OEM
I04D380140	OEM
I04TE25110	OEM
I05D076300	3772
I05D386040	OEM
I05DA7607A	0912
I05DB7680	OEM
I05DB76780	OEM
I05DC7676A	OEM
I05DC8608A	OEM
I05DC86080	OEM
I05DC86200	OEM
I05DE7777P	3236
I05DE8601B	OEM
I05DE8725N	OEM
I05DE76440	0663
I05DE76810	4986
I05DE77780	OEM
I05DE86540	OEM
I05DE86550	OEM
I05DE86770	3236
I05DE87820	OEM
I05S925080	2194
I05SP72220	4652
I05SP73680	OEM
I06DA365SP	3635
I06DE20250	OEM
I06S51954A	OEM
I07D003240	OEM
I07D069930	0631
I07D767AS0	OEM
I07DG7212S	OEM
I07S06239N	OEM
I07S010358	OEM
I07S062470	OEM
I07S077550	OEM
I07T67751L	OEM
I08D369643	OEM
I01901280L	4141
I01901280M	3521
I02190574J	1329
I02990574J	1329
I031956310	1329
I039956300	1329
I079907070	OEM
I2AU3A0010	OEM
I2B4900410	3344
I2B4901100	OEM
I2B4910410	2289
I2LSI50030	OEM
I2MH92012A	OEM
I3PS035	0226
I3PS135	0226
I3PS235	0226
I3PS335	0226
I3PS435	0226
I3PS535	0226
I3PS635	0226
I3PT030	0908
I3PT040	0908
I3PT130	0908
I3PT140	0908
I3PT230	0908
I3PT240	0908
I3PT330	0908
I3PT340	0908
I3PT430	0908
I3PT440	0908
I3PT530	0908
I3PT540	0908
I3PT630	0908
I3PT640	0908
I4DDS82500	OEM
I4GD400780	OEM
I4QDL80860	OEM
I4QDL82590	OEM
I4SDZ765A0	OEM
I5PD14053B	0035
I5PD468180	OEM
I6PDC14880	0508
I6PDC14890	0511
I6UD91C360	OEM
I8WP400390	OEM
I8WP400400	OEM
I8WP400410	OEM
I23S953320	2303
I23S953420	OEM
I23S954220	OEM
I27D09001A	OEM
I31D012250	OEM
I41D41256C	1472
I42D400420	OEM
I42D412640	OEM
I42DD82370	OEM
I42DD82530	OEM
I43D030060	OEM
I43D030080	OEM
I51D08031B	OEM
I51DT80020	OEM
I52D001130	OEM
I52D003930	0631
I52D040530	0035
I52D061430	OEM
I52DE61040	OEM
I52DT1011A	4524
I53D040010	0478
I53D0HC050	OEM
I53D3664NA	OEM
I53D890600	OEM
I53D890660	OEM
I53DHC1740	OEM
I54D50010C	OEM
I55D04052B	0024
I55D04066B	0101
I55D04538B	1063
I55D06015C	OEM
I55D040010	0478
I55D060070	OEM
I55D083820	OEM
I55D090200	OEM
I55D0H0000	OEM
I55D0HC020	1452
I55D4HC040	1455
I55D4HC140	3870
I55DHC123P	3907
I55DHC2400	4227
I55DHCU04D	4602
I56D004550	OEM
I56D086300	OEM
I56DT7002A	OEM
I61D4LS140	1698
I61DLS1250	0075
I61DLS1740	0260
I61DLS2400	0452
I61DLS2440	0458
I61DLS2450	0463
I61DLS3730	0709
I64D7406P0	1206
I83F030110	OEM
I97D49005A	OEM
I97D49008B	OEM
I431D	OEM
I538	0015
I818	0062
I964(6	0133
I2764	0812
I2764-2	0812
I6289	0576
I8085	OEM
I8085A	OEM
I8085AH-2	OEM
I8155H-2	OEM
I8251-A	OEM
I8251A	OEM
I8255	0051
I9623	0144
I9631	0086
I10008-01	OEM
I27128	1638
I98763-3	0097
I98936-2	0147
I473688-1	0058
I5450015B	OEM
IA5D400431	OEM
IA5D400441	OEM
IAM-1	OEM
IAM-3	OEM
IAM1048	OEM
IAPST523E	OEM
IAZ33	OEM
IB	0288
IB198	0460
IC00000172	OEM
IC00001267	OEM
IC00001292	OEM
IC00001450	OEM
IC00001759	OEM
IC00001760	OEM
IC00001802	OEM
IC00001814	OEM
IC00001838	OEM
IC-00028	OEM
IC2(ECS)	1854
IC-2(RCA)	0829
IC2DF13930	OEM
IC3	3724
IC3(PHILCO)	3724
IC-4	2582
IC-5	0377
IC5	0632
IC-5(ELCOM)	0632
IC-5(PHILCO)	0377
IC-6	2715
IC6	2715
IC-6(PHILCO)	2715
IC-7	0632
IC7	0632
IC-8	0789
IC8	0789
IC-8B	2624
IC8B	0789
IC-8B#	0789
IC-8T26	4516
IC-8T28	0581
IC-8T95	4647
IC-12	0664
IC12	0753
IC-12(PHILCO)	0352
IC-13	0354
IC13	3724
IC-13(PHILCO)	0354
IC-14	0751
IC14	0751
IC-14(PHILCO)	0751
IC-15	0167
IC15	0167
IC-15(PHILCO)	0167
IC-16	0364
IC-16(PHILCO)	0364
IC-17	0785
IC17	0654
IC-17(ELCOM)	0654
IC-17(PHILCO)	0785
IC-18	0443
IC18	0443
IC-18(PHILCO)	1808
IC-19	0779
IC19	0753
IC-19#	2161
IC-19(PHILCO)	0372
IC-20	0026
IC-20(PHILCO)	0026
IC21	0377
IC-21(ELCOM)	0517
IC21(ELCOM)	0517
IC-23	0379
IC23	0379
IC-23(PHILCO)	0379
IC23(PHILCO)	0379
IC-25	2161
IC-25(ELCOM)	0753
IC25(ELCOM)	0753
IC-26	0789
IC26	2393
IC-26(ELCOM)	0789
IC-27(PHILCO)	0372
IC-28	0395
IC29(ELCOM)	1854
IC30(ELCOM)	1823
IC31(ELCOM)	1831
IC-32	0372
IC32(ELCOM)	0141
IC33(ELCOM)	0358
IC-34	0785
IC34(ELCOM)	0562
IC35(ELCOM)	1554
IC41(ELCOM)	2537
IC-55(ELCOM)	2867
IC-56(ELCOM)	3418

DEVICE TYPE	REPL CODE
IC-58(ELCOM)	1257
IC-59(ELCOM)	1364
IC-61(ELCOM)	1069
IC-62(ELCOM)	1069
IC-63(ELCOM)	OEM
IC-65(ELCOM)	3429
IC-68B21	OEM
IC-68B50	OEM
IC-71(ELCOM)	0873
IC-74(ELCOM)	0268
IC-74L03	1579
IC-74LS02	1559
IC-74LS04	1595
IC-74LS12	1679
IC-74LS20	0034
IC-74LS75	1870
IC-74LS86	0291
IC-74LS138	0427
IC-74LS139	0153
IC-74LS157	1161
IC-74LS163	0893
IC-74LS165	4356
IC-74LS193	1692
IC-74LS390	1288
IC-74S138	2138
IC-75(ELCOM)	0743
IC-79(ELCOM)	1486
IC-80(ELCOM)	0232
IC-81(ELCOM)	0569
IC-82(ELCOM)	0314
IC-83(ELCOM)	0335
IC-84(ELCOM)	0361
IC-85(ELCOM)	0385
IC-86(ELCOM)	0512
IC-87(ELCOM)	0697
IC-88(ELCOM)	1024
IC-90(ELCOM)	1052
IC-91	0632
IC91	0632
IC-91(ELCOM)	1168
IC-92(ELCOM)	1185
IC-93(ELCOM)	1202
IC-94(ELCOM)	1426
IC-95(ELCOM)	1172
IC-96(ELCOM)	1432
IC-97(ELCOM)	1313
IC-98(ELCOM)	1208
IC-99(ELCOM)	1158
IC-100(ELCOM)	0834
IC101(COLUMBIA)	0400
IC-101(ELCOM)	1108
IC-101-004	3958
IC-102(ELCOM)	0467
IC-103(ELCOM)	1441
IC-104(ELCOM)	1206
IC-105(ELCOM)	1349
IC-106(ELCOM)	0175
IC-112	0855
IC112	0855
IC126	4582
IC132	2592
IC136	3785
IC142	2891
IC146	0878
IC-201(ELCOM)	0701
IC-211(ELCOM)	2537
IC-213(ELCOM)	3972
IC213(ELCOM)	3972
IC-217(ELCOM)	1684
IC217(ELOCM)	1684
IC-220(ELCOM)	2563
IC-221(ELCOM)	3054
IC-222(ELCOM)	4142
IC-223(ELCOM)	3066
IC-224(ELCOM)	0078
IC-225(ELCOM)	2992
IC229(ELCOM)	0081
IC-234(ELCOM)	0333
IC-237(ELCOM)	0962
IC-243(ELCOM)	1899
IC-244(ELCOM)	0623
IC-276(ELCOM)	0033
IC-279(ELCOM)	3399
IC-280(ELCOM)	1041
IC-281(ELCOM)	2097
IC-282(ELCOM)	OEM
IC-285	0723
IC285	0723
IC-285(ELCOM)	0723
IC285(ELCOM)	0723
IC-286(ELCOM)	1876
IC286(ELCOM)	1876
IC-288(ELCOM)	0372
IC288(ELCOM)	0372
IC-290(ELCOM)	2027
IC290(ELCOM)	2027
IC-298(ELCOM)	2100
IC298(ELCOM)	2100
IC311(ELCOM)	2743
IC-312	2830
IC320(ELCOM)	1843
IC-409(ELCOM)	2992
IC-410(ELCOM)	2434
IC-412(ELCOM)	2350
IC-413(ELCOM)	2597
IC-414(ELCOM)	3032
IC-415(ELCOM)	3039
IC-419(ELCOM)	1433
IC-420(ELCOM)	0466
IC-423(ELCOM)	OEM
IC-432(ELCOM)	OEM
IC-433(ELCOM)	1186
IC-435(ELCOM)	2027
IC-441(ELCOM)	3015
IC-442(ELCOM)	2027
IC-443(ELCOM)	1186
IC-444(ELCOM)	3931
IC-445(ELCOM)	1671
IC-446(ELCOM)	2282
IC-447(ELCOM)	3935
IC-448(ELCOM)	3935
IC-449(ELCOM)	3940
IC-450(ELCOM)	2906
IC-451(ELCOM)	1980
IC-453(ELCOM)	3944
IC-455(ELCOM)	0358
IC500	0632
IC-500(ELCOM)	1574
IC500(IR)	0632
IC501	2624
IC501(ELCOM)	2100
IC-502	2161
IC502	2161
IC-503	2161
IC503	0753
IC503(IR)	0753
IC-504	0664
IC504	0664
IC-505	0678
IC505	0678
IC505(IR)	0678
IC506	0829
IC-507	0167
IC507	0167
IC507(IR)	0167
IC-508	0666
IC508	0666
IC-508(ELCOM)	0470
IC508(IR)	0666
IC-509	0352
IC509	0352
IC509(IR)	0352
IC-510	0354
IC510	0354
IC510(IR)	0354
IC-511	0654
IC511	0654
IC512	0443
IC-512(ELCOM)	1621
IC512(IR)	0443
IC-513	0701
IC513	0701
IC-517(ELCOM)	4571
IC-520(ELCOM)	0973
IC-524(ELCOM)	0379
IC-526(ELCOM)	3810
IC-547(ELCOM)	0138
IC-548(ELCOM)	2404
IC-549(ELCOM)	0021
IC-550(ELCOM)	2505
IC-551(ELCOM)	2476
IC-555	0973
IC-603(ELCOM)	2557
IC-605(ELCOM)	0379
IC703E	OEM
IC724	2867
IC-903	0753
IC903	0753
IC904	OEM
IC908	OEM
IC908-PC	OEM
IC918	OEM
IC-1489	0511
IC-2114-L3	2048
IC-3130	2782
IC4001BP	0478
IC5026BP	OEM
IC-6205A	OEM
IC-6502	3065
IC-6850	0514
IC-7157	OEM
IC-7400	0232
IC-7402	0314
IC-7403	0335
IC-7404	0361
IC-7408	0467
IC-7408N	0467
IC-7410	0512
IC-7417	1352
IC-7420	0697
IC7420	0697
IC-7430	0873
IC-7438	0996
IC-7474	1313
IC-7475	1432
IC7475	1432
IC-7476	1158
IC-7492	0834
IC9012F	OEM
IC25681	2985
IC-74123	1157
IC-74148	1464
IC-74151	1495
IC-74165	1685
IC-74174	1770
IC-74175	1787
IC-74390	3240
IC743024	OEM
IC743038	0698
IC743039	0698
IC743040	0224
IC743041	0224
IC743042	0016
IC743043	0006
IC743044	0155
IC743045	0936
IC743046	0325
IC743047	0012
IC743048	0015
IC743049	0124
IC743050	0143
IC743051	0133
IC3002271D	0261
IC30022710	0261
IC-BASIC1	OEM
IC-BASIC2	OEM
IC-BASIC3	OEM
IC-BASIC4	OEM
ICBASIC4	OEM
ICC6401	OEM
ICC6402	OEM
ICC6403	OEM
ICC6404	OEM
ICC6405	OEM
ICC6406	OEM
ICC6407	OEM
IC-CARCEN	OEM
IC-CARGEN	OEM
ICF-1	0632
ICF-201	OEM
ICF-202	OEM
ICF-203	OEM
IC-L2114-550	2048
ICL7106CM44	OEM
ICL7106CPL	4952
ICL7106RCPL	OEM
ICL7107CDL	5392
ICL7107CJL	OEM
ICL7107CM44	OEM
ICL7107CPL	4951
ICL7107RCPL	OEM
ICL7116CM44	OEM
ICL7116CPL	5866
ICL7116RCPL	OEM
ICL7117CPL	OEM
ICL7117RCPL	OEM
ICL7126CDL	OEM
ICL7126CJL	OEM
ICL7126CPL	0506
ICL7126RCPL	5980
ICL7129CM44	OEM
ICL7129CPL	3213
ICL7129RCPL	OEM
ICL7660	OEM
ICL8038	0304
ICM7555CBA	2771
ICM7555IPA	0585
ICM7555ITV	OEM
ICM7555MTV	OEM
ICM7556IPD	OEM
ICM7556MJD	OEM
ICP50	OEM
ICP75516-063	OEM
ICP-F10	6128
ICP-F15	5356
ICP-F20	5000
ICP-F25	4629
ICP-F38	5681
ICP-F50	OEM
ICP-F75	OEM
ICP-N5	1807
ICP-N10	3914
ICP-N15	6129
ICP-N20	6130
ICP-N25	6132
ICP-N38	OEM
ICP-N50	OEM
ICP-N75	OEM
ICR1001	2079
ICR1002	2079
ICR1003	2079
ICR1004	OEM
ICR1005	0023
ICR1006	0017
ICR1007	0017
ICR1501	2079
ICR1502	2079
ICR1503	2079
ICR1504	OEM
ICR1505	0031
ICR1506	0102
ICR1507	0102
ICR3001	OEM
ICR3002	OEM
ICR3003	OEM
ICR3004	OEM
ICR3005	0031
ICR3006	1928
ICR3007	1928
ICR6001	OEM
ICR6002	OEM
ICR6003	OEM
ICR6004	OEM
ICR6005	OEM
ICR6006	OEM
ICR6007	OEM
IC-SYN600	OEM
IC-SYNMON	OEM
ICT-5	OEM
ICT-8	OEM
ICT-10	OEM
ICT-12	OEM
ICT-15	OEM
ICT-18	OEM
ICT-22	OEM
ICT-36	OEM
ICT-45	OEM
ICTE-5	OEM
ICTE5	OEM
ICTE-5C	5459
ICTE5C	OEM
ICTE5S	OEM
ICTE5SC	OEM
ICTE-8	OEM
ICTE8	OEM
ICTE-8C	OEM
ICTE8C	OEM
ICTE-10	OEM
ICTE-10C	OEM
ICTE-12	OEM
ICTE12	OEM
ICTE-12C	OEM
ICTE12C	OEM
ICTE-15	OEM
ICTE15	OEM
ICTE-15C	OEM
ICTE15C	OEM
ICTE-18	OEM
ICTE18	OEM
ICTE-18C	OEM
ICTE18C	OEM
ICTE19	OEM
ICTE19C	OEM
ICTE-22	OEM
ICTE22	OEM
ICTE-22C	OEM
ICTE22C	OEM
ICTE-36	OEM
ICTE36	OEM
ICTE-36C	OEM
ICTE36C	OEM
ICTE-45	OEM
ICTE45	OEM
ICTE-45C	OEM
ICTE45C	OEM
ID20-0308	0385
ID100	1137
ID101	0344
ID102	0901
ID103	0058
ID104	0058
ID105	0407
ID106	0407
ID200	1137
ID201	0344
ID202	OEM
ID203	0058
ID300	0407
ID301	OEM
ID600	OEM
ID601	OEM
ID602	OEM
ID603	OEM
ID604	OEM
ID605	OEM
ID606	OEM
ID2732A-3	2697
ID8088	OEM
ID8284A	OEM
ID8288	OEM
ID27128A-3	1638
IDAZ472	0036
IDAZ472A	0036
IDAZ824	0471
IDAZ824A	0471
IDAZ824B	0471
IDAZ3397	0036
IDAZ3397A	0036
IDAZ3397B	0036
IDAZ3397C	OEM
IDAZ3398	0140
IDAZ3398A	0140
IDAZ3398B	0140
IDAZ3398C	OEM
IDAZ3399	0041
IDAZ3399A	0041
IDAZ3399B	0162
IDAZ3399C	OEM
IDAZ3403	0052
IDAZ3403A	0052
IDAZ3403B	0052
IDAZ3403C	0052
IDAZ3404	0686
IDAZ3404A	0686
IDAZ3404B	0686
IDH0742	0101
IDH0802	0035
IDUPC7912H	1838
IE12Z5	0137
IE460B	0224
IE535A	OEM
IE535B	OEM
IE-850	0208
IE850	0079
IF-65	0004
IF65	0211
IF6008T	OEM
IFE-450	OEM
IFEWC108	OEM
IFF-450A	OEM
IFSANYO-002	OEM
IFSANYO002	OEM
IFSANYO002(SANYO002)	OEM
IFSANYO-003	OEM
IFSANYO003	OEM
IFSANYO-004	OEM
IFSANYO004	OEM
IFSI-3122P	4329
IG-100	0440
IG-15630	OEM
IGD-2	OEM
IGL78M05	0624
IGL78M06	0923
IGLA3150	2530
IGLA6324	0625
IGLA6458DF	0360
IGLA7030	OEM
IGLA7031	OEM
IGLA7032	OEM
IGLA7033	OEM
IGLA7042	OEM
IGLA7224	4266
IGLA7905	OEM
IGLB1214	OEM
IGLB1240	OEM
IGLB1290	3893
IGLB1620	OEM
IGLB1645	OEM
IGLB1649	5251
IGLC4011	0215
IGLC4011B	0215
IGLC4013B	0413
IGLC4069UB	0119
IGLC4081B	0332
IGLC6502B-633	OEM
IGLC7900	OEM
IGLM6402H-158	OEM
IGLM6402H-238	OEM
IGLM6402H-329	OEM
IGLM6405H-215	OEM
IGLM6405H-330	OEM
IGLM6416E-159	OEM
IGLM6416E-209	OEM
IGLM6416E-217	OEM
IGLM6416E-239	OEM
IGLM6417E-302	OEM
IGLM8854-1912	OEM
IGSPT6402L04	OEM
IGSPT640203	OEM
IGSTK5321SL	OEM
IGSTK5322	2297
IGSTK5322H	OEM
IGSTK5327SL	OEM
IGSTK5346	OEM
IGSTK5352	OEM
IGSTK5431	2374
IGSTK5431SL	2374
IGSTK5431ST	2374
IGSTK5461ST	OEM
IGSTK6962	3348
IGSTK6972	3355
IGT8D20	OEM
IGT8D21	OEM
IGT8E20	OEM
IGT8E21	OEM
IHHA1174ANT	OEM
IHHA11711	5181
IHHA11713	OEM
IHHA11714	5185
IHHA11715	5158
IHHA11722	OEM
IHHA11724	OEM
IHHA11725	OEM
IHHA11726	OEM
IHHA-11741	OEM
IHHA11741	OEM
IHHA-11744	OEM
IHHA11744	OEM
IHHA-11745	OEM
IHHA11745	OEM
IHHA11747ANT	5793
IHHA11749	5790
IHHA11847NT	OEM
IHHD38701A06	OEM
IHHD44801A05	OEM
IHTS4966	OEM
I-HY-101	OEM
IISPI1050-01	OEM
IISPR1050-01	OEM
IISPT2300L01	OEM
IISPT2370H07	OEM
IISPT3240L05	OEM
IITL431CLP	OEM
IITMS1024	OEM
IITMS1025N2LC	OEM
IJ4B41	OEM
IJNJM4558DM	0360
IKSANYO0002	OEM
IL081ACDP	OEM
IL1	0315
IL5	0273
IL12	0315
IL15	0315
IL16	0315
IL74	0315
ILA30	1053
ILA55	1109
ILA7621	0391
ILCA2-30	1053
ILD1020	0021
ILH0096	1181
ILH0111	1201
ILL0052	0362
ILM0292	2169
ILM340T12	0334
ILMC14013B	0413
ILMC14081B	0626
ILT0014	1018
IM15/1B	OEM
IM4004	0795
IMAN262	2960
IMC1358P	0167
IMDN6838	0489
IMI1	OEM
IMMN1280-M	OEM
IMS171	OEM
IMS2620P-15	OEM
IMSG171S-35	OEM
IMT1	OEM
IMX3	OEM
IMZ1	OEM
IN60-1	0143
IN962B	0317
IN967B	OEM
IN2054-68	1023
IN3161-74	1023
IN4002	OEM
IN4051	1036
IN4066BP	0101
IN4532	OEM
IN4587-96	0599
IN5936B	OEM
IN6264	OEM
IN6265	OEM
IN6266	OEM
IN8250N-B	OEM
INC61684	0920
INF	OEM
INJ33348	0196
INJ33349	0143
INJ60034	0143
INJ60284	0143
INJ61224	0123
INJ61225	0636
INJ61227	0015
INJ61433	0133
INJ61434	0015
INJ61435	0196
INJ61675	0143
INJ61677	0133
INJ61725	0133
INJ61726	0015
INJ70972	0196
INJ70973	0143
INJ70980	0143
INMOSS348A	OEM
INMPC1363C	0683
INMPC1363CA	1945
INMPC1373H	2026
INMPC1373HA	2026
INMPC1513HA	3159
INMPD552C	OEM
INMPD4011BC	0215
INMPD4013BC	0413
INMPD4066BC	0101
INMPD4081C	0332
INMPD4093BC	2384
INMPD7519G-530	OEM
INMPD14027BC	1949
INS82LS05N	6025
INS8039N-11	OEM
INS8246N	OEM
INS8250AN	OEM
INS8250AV	OEM
INS8250BN	OEM
INS8250N	OEM
INS8250N-B	OEM
INUPC1373H	2026
INUPD4011BC	0215
INUPD4013BC	0413
INUPD4066BC	0101
INUPD4081C	0332
INUPD4093BC	2384
INUPD7519G-53	OEM
INV5000SNE	OEM
IOB1680	OEM
IP	OEM
IP20-0001	0368
IP20-0002	0368
IP20-0005	1908
IP20-0006	0076
IP20-0007	0042
IP20-006	0224
IP20-0010	3155
IP20-0012	0688
IP20-0015	0123
IP20-0016	0123
IP20-0017	0143
IP20-0018	0002
IP20-0019	0012
IP20-0020	0143
IP20-0021	0133
IP20-0022	0023
IP20-0023	0139
IP20-0025	0795
IP20-0027	0025
IP20-0028	2793
IP20-0029	0368
IP20-0032	0076
IP20-0034	5416
IP20-0035	0688
IP20-0036	0228
IP20-0037	1144
IP20-0038	0155
IP20-0039	0076
IP20-0040	0191
IP20-0041	0191
IP20-0046	0006
IP20-0048	1400
IP20-0054	0023
IP20-0055	0139
IP20-0060	0019
IP20-0061	0133
IP20-0076	0208
IP20-0078	3046
IP20-0083	0042
IP20-0086	0162
IP20-0103	2710
IP20-0110	0127
IP20-0120	0080
IP20-0122	0076
IP20-0123	2050
IP20-0131	2050
IP20-0132	1591
IP20-0135	2495
IP20-0145	0133
IP20-0151	0030
IP20-0154	0836
IP20-0155	0839
IP20-0156	0161
IP20-0157	0414
IP20-0159	0006
IP20-0160	0208
IP20-0161	1050
IP20-0164	0564
IP20-0165	0155
IP20-0166	0030
IP20-0167	0015
IP20-0172	2046
IP20-0174	0632
IP20-0179	0151
IP20-0184	0133
IP20-0185	0720
IP20-0186	0025
IP20-0191	0076
IP20-0192	0148
IP20-0203	0301
IP20-0204	0005
IP20-0205	0232
IP20-0206	1313
IP20-0207	0569
IP20-0208	4596
IP20-0209	1645
IP20-0210	0011
IP20-0211	0532
IP20-0212	4251
IP20-0213	0006
IP20-0214	0155
IP20-0216	0124
IP20-0217	0693
IP20-0218	1031
IP20-0219	OEM
IP20-0220	2301
IP20-0230	0563
IP20-0231	0191
IP20-0233	0317
IP20-0234	OEM
IP20-0244	0139
IP20-0253	2301
IP20-0266	OEM
IP20-0281	4069
IP20-0282	0133
IP20-0283	0143
IP20-0284	0471
IP20-0305	1280
IP20-0308	0385
IP20-0310	0268
IP20-0315	0569
IP20-0316	1313
IP20-0317	1128
IP20-0318	OEM
IP20-0319	OEM
IP20-0323	0042
IP20-0331	2495
IP20-0429	1540
IP20-0431	OEM
IP20-0447	2295
IP20-0448	0563
IP20-0519	OEM
IP20-2010	0011
IP28-0006	OEM
IP28-0016	OEM
IP28-0017	OEM
IP100	1137
IP101	0344
IP102	0901
IP103	2342
IP104	0058
IP105	0407
IP106	0407
IP117	OEM
IP117A	OEM
IP117AHV	OEM

If replacement code is OEM, contact original manufacturer for replacement.

DEVICE TYPE	REPL CODE
IP117HV	OEM
IP120-5	OEM
IP120-12	OEM
IP120-15	OEM
IP120A-5	OEM
IP120A-12	OEM
IP120A-15	OEM
IP123-5	OEM
IP123-12	OEM
IP123-15	OEM
IP123A-5	OEM
IP123A-12	OEM
IP123A-15	OEM
IP137	OEM
IP137A	OEM
IP137AHV	OEM
IP137HV	OEM
IP138	OEM
IP138A	OEM
IP140-5	OEM
IP140-12	OEM
IP140-15	OEM
IP140A-5	OEM
IP140A-12	OEM
IP140A-15	OEM
IP150	OEM
IP150A	OEM
IP217	OEM
IP217A	OEM
IP217AHV	OEM
IP217HV	OEM
IP220-5	OEM
IP220-12	OEM
IP220-15	OEM
IP220A-5	OEM
IP220A-12	OEM
IP220A-15	OEM
IP223-5	OEM
IP223-12	OEM
IP223-15	OEM
IP223A-5	OEM
IP223A-12	OEM
IP223A-15	OEM
IP237	OEM
IP237A	OEM
IP237AHV	OEM
IP237HV	OEM
IP238	OEM
IP238A	OEM
IP240-5	OEM
IP240-12	OEM
IP240-15	OEM
IP240A-5	OEM
IP240A-12	OEM
IP240A-15	OEM
IP250	OEM
IP250A	OEM
IP293	OEM
IP293D	OEM
IP293DA	OEM
IP293E	OEM
IP317	2723
IP317A	OEM
IP317AHV	OEM
IP317HV	OEM
IP320-5	OEM
IP320-12	OEM
IP320-15	OEM
IP320A-5	OEM
IP320A-12	OEM
IP320A-15	OEM
IP323-5	OEM
IP323-12	OEM
IP323-15	OEM
IP323A-5	OEM
IP323A-12	OEM
IP323A-15	OEM
IP337	OEM
IP337A	OEM
IP337AHV	OEM
IP337HV	OEM
IP338	OEM
IP338A	OEM
IP340-5	1922
IP340-12	3963
IP340-15	OEM
IP340A-5	OEM
IP340A-12	OEM
IP340A-15	OEM
IP350	OEM
IP350A	OEM
IP494A	OEM
IP494AC	OEM
IP1524	OEM
IP1524B	OEM
IP1525A	OEM
IP1526	OEM
IP1526A	OEM
IP1527A	OEM
IP1543	OEM
IP1544	OEM
IP1842	OEM
IP1843	OEM
IP2064	OEM
IP2065	OEM
IP2066	OEM
IP2067	OEM
IP2068	OEM
IP2069	OEM
IP2070	OEM
IP2071	OEM
IP2074	OEM
IP2075	OEM
IP2076	OEM
IP2077	OEM
IP2524	OEM
IP2524B	OEM
IP2525A	OEM
IP2526	OEM
IP2526A	OEM
IP2527A	OEM
IP2543	OEM
IP2544	OEM
IP2842	OEM
IP2843	OEM
IP3524	OEM
IP3524B	OEM
IP3525A	OEM
IP3526	OEM
IP3526A	OEM
IP3527A	OEM
IP3543	OEM
IP3544	OEM
IP3842	OEM
IP3843	OEM
IP5560	OEM
IP5560C	OEM
IP5561	OEM
IP5561C	OEM
IP7805	0624
IP7805A	0624
IP7812	0334
IP7812A	OEM
IP7815	1321
IP7815A	OEM
IP7905	1285
IP7905A	OEM
IP7912	1838
IP7912A	OEM
IP7915	3829
IP7915A	OEM
IP8051	OEM
IP8085AH	OEM
IP8085AH-2	OEM
IP8088	OEM
IP8251A	OEM
IP8255A-5	0051
IP8259A	OEM
IP33063	OEM
IP34060	OEM
IP34063	OEM
IP35060	OEM
IP35063	OEM
IPL32	OEM
IPSV01DEAN	OEM
IPSV7031AN	OEM
IPSV7031BN	OEM
IPSV7031CN	OEM
IPSV7032AN	OEM
IPSV7032CN	OEM
IPSV7033AN	OEM
IPSV7034DN	OEM
IPX	OEM
IR1F	0102
IR2A	0015
IR2C07	OEM
IR2C32	OEM
IR2E	0015
IR2E01	OEM
IR2E02	OEM
IR2E16	OEM
IR2E27A	OEM
IR2E31	OEM
IR2E31A	OEM
IR2P02	OEM
IR2P02T	OEM
(IR)2SA495	0006
(IR)2SA564A	0203
(IR)2SB186	0004
(IR)2SB187	0004
(IR)2SB367A	0222
(IR)2SB405	0164
(IR)2SB474	1010
(IR)2SB481	0222
(IR)2SC281B	0536
(IR)2SC372	0076
(IR)2SC373	0076
(IR)2SC380AO	0286
(IR)2SC403A	0155
(IR)2SC454B	0076
(IR)2SC458B	0076
(IR)2SC460B	0151
(IR)2SC535B	0127
(IR)2SC536	0537
(IR)2SC537B	0191
(IR)2SC538A	0079
(IR)2SC633A	2075
(IR)2SC634A	2075
(IR)2SC644	0111
(IR)2SC682A	0047
(IR)2SC710	0368
(IR)2SC717	0127
(IR)2SC772	1144
(IR)2SC828A	1220
(IR)2SC829B	0151
(IR)2SC838	0191
(IR)2SC945	0076
IR3C02A	OEM
IR3M02A	OEM
IR3N05	6791
IR3N06	7061
IR3N34	OEM
IR3T24	OEM
IR4N09A	OEM
IR5A	0344
IR5B	0058
IR5C	0407
IR5D	OEM
IR5F	1137
IR5G	OEM
IR5GZ61	0071
IR5H	OEM
IR5JA	0143
IR5LC10P04Q	1021
IR5U	OEM
IR6A	0179
IR6B	0179
IR6C	0346
IR6D	0346
IR6F	OEM
IR6G	OEM
IR6H	OEM
IR6U	OEM
IR10E6J	0015
IR10E6X	0015
IR20	0015
IR20HB5	1600
IR20HB20	0871
IR20HB40	0853
IR30A	7124
IR30A2	0469
IR30B	0745
IR30B2	0469
IR30C	0722
IR30C2	0725
IR30D	0740
IR30D2	0725
IR30DF	0087
IR30E	0752
IR30E2	0725
IR30F	2519
IR30F2	0469
IR30M2	0725
IR30U	2519
IR30U2	0469
IR31A	7127
IR31B	7127
IR31C	7128
IR31D	7128
IR31D8	OEM
IR31E	OEM
IR31F	7127
IR31U	7127
IR32A	1650
IR32B	OEM
IR32C	1584
IR32D	OEM
IR32E	1665
IR32F	OEM
IR32U	2519
IR33A	OEM
IR33B	OEM
IR33C	OEM
IR33D	OEM
IR33E	OEM
IR33F	OEM
IR33U	OEM
IR106A1	0447
IR106A1-C	0447
IR106A2	0447
IR106A3	0447
IR106A4	0447
IR106A41	0447
IR106B1	0940
IR106B1-C	0940
IR106B2	0940
IR106B3	0940
IR106B4	0940
IR106B41	0940
IR106C1	1222
IR106C1-C	1222
IR106C2	1222
IR106C3	1222
IR106C4	1222
IR106C41	1222
IR106D1	0095
IR106D2	0095
IR106D3	0095
IR106D4	0095
IR106D41	0095
IR106F1	1259
IR106F2	1259
IR106F3	1259
IR106F4	1259
IR106F41	1259
IR106Q1	1396
IR106Q2	1396
IR106Q3	1396
IR106Q4	1396
IR106Q41	1396
IR106Y1	1396
IR106Y1-C	1396
IR106Y2	1396
IR106Y3	1396
IR106Y4	1396
IR106Y41	OEM
IR122A	2089
IR122A-C	2089
IR122B	0505
IR122B-C	0505
IR122C	0710
IR122C-C	0710
IR122D	0710
IR122D-C	0710
IR122F	2095
IR122F-C	2095
IR122M4	OEM
IR122M5	OEM
IR122M6	OEM
IR140A	0804
IR140B	0804
IR140C	0804
IR140D	0804
IR140F	0804
IR141A	0804
IR141B	0804
IR141C	0804
IR141D	0804
IR141F	0804
IR401	1966
IR402	0363
IR403	0363
IR409	0074
IR410	0270
IR411	0270
IR413	0065
IR423	0074
IR424	3058
IR425	3058
IR515	0363
IR516	0363
IR517	0363
IR518	0363
IR519	0363
IR640	3373
IR641	3373
IR642	0134
IR645	3374
IR646	3374
IR647	0073
IR660	OEM
IR663	OEM
IR665	OEM
IR682	2188
IR683	0567
IR684	3258
IR685	0762
IR685#	0469
IR686	3268
IR687	3271
IR688	0740
IR689	3291
IR690	0764
IR692	0766
IR701	OEM
IR708	OEM
IR709	OEM
IR710	OEM
IR721	OEM
IR801	0223
IR802	OEM
IR806	OEM
IR900	2277
IR901	2277
IR1000	2427
IR1001	2427
IR1010	OEM
IR1020	OEM
IR1600	1110
IR1601	2932
IR1602	0963
IR1603	2933
IR1604	2936
IR1771	1103
IR1772	2491
IR1774	0240
IR1776	2659
IR1777	0676
IR1778	2809
IR1843A	2519
IR1844A	0741
IR1846A	0745
IR1848A	2917
IR1849A	0747
IR1850A	3243
IR2160	1669
IR2432	OEM
IR2434	OEM
IR3702	OEM
IR3771	OEM
IR3772	OEM
IR3773	OEM
IR4039	2626
IR4040	2626
IR4041	2626
IR4045	2626
IR4050	2626
IR4055	2626
IR4059	2626
IR4060	2626
IR4061	2626
IR4065	2626
IR4502	OEM
IR5000	2626
IR5001	3374
IR5002	6739
IR5060	2626
IR5061	2626
IR5062	2626
IR5063	OEM
IR5064	OEM
IR5065	OEM
IR5066	OEM
IR5251	OEM
IR5252	OEM
IR5253	6454
IR6000	2626
IR6001	2626
IR6002	2626
IR6060	2626
IR6061	2626
IR6062	2626
IR6251	1991
IR6252	2626
IR6253	2626
IR6302	OEM
IR6542	OEM
IR6543	OEM
IR6544	OEM
IR6545	OEM
IR6546	OEM
IR6547	1852
IR6582	OEM
IR6583	OEM
IR9358	6322
IR9393	0631
IR9431	OEM
IR13006	OEM
IR13007	OEM
IR13066	OEM
IR94558	OEM
IRC-5	2270
IRC5	1259
IRC10	0447
IRC20	0940
IRD-54	2730
IRD54	2730
IRD54-C	2730
IR-D431	0102
IR-D818	0293
IRD3899	1530
IRD3900	1530
IRD3901	1530
IRD3902	0029
IRD3903	0029
IRD3909	1530
IRD3910	1530
IRD3911	1530
IRD3912	0029
IRD3913	0029
IRE-150	OEM
IRE-151	OEM
IRE-160	OEM
IRE-160FA	OEM
IRE-160FB	OEM
IRE161	OEM
IRE165	OEM
IRE170	OEM
IRF120	OEM
IRF121	OEM
IRF122	OEM
IRF123	OEM
IRF130	OEM
IRF131	OEM
IRF132	OEM
IRF133	OEM
IRF140	OEM
IRF141	OEM
IRF142	OEM
IRF143	OEM
IRF150	OEM
IRF151	OEM
IRF152	OEM
IRF153	OEM
IRF153R	OEM
IRF220	5656
IRF221	5656
IRF222	5656
IRF223	5656
IRF230	OEM
IRF231	OEM
IRF232	OEM
IRF233	OEM
IRF233R	OEM
IRF240	OEM
IRF241	OEM
IRF242	OEM
IRF243	OEM
IRF250	OEM
IRF251	OEM
IRF252	OEM
IRF253	OEM
IRF253R	OEM
IRF320	1668
IRF321	1668
IRF322	1668
IRF323	1668
IRF330	OEM
IRF331	OEM
IRF332	OEM
IRF333	OEM
IRF340	OEM
IRF341	OEM
IRF342	OEM
IRF343	OEM
IRF350	OEM
IRF351	OEM
IRF352	OEM
IRF352R	OEM
IRF353	OEM
IRF353R	OEM
IRF420	OEM
IRF421	OEM
IRF422	OEM
IRF423	OEM
IRF430	OEM
IRF431	OEM
IRF432	OEM
IRF433	OEM
IRF440	OEM
IRF441	OEM
IRF442	OEM
IRF443	OEM
IRF450	OEM
IRF451	OEM
IRF452	OEM
IRF452R	OEM
IRF453	OEM
IRF453R	OEM
IRF510	3088
IRF511	2341
IRF512	3088
IRF513	2341
IRF513R	OEM
IRF520	3088
IRF521	3088
IRF521R	OEM
IRF522	3088
IRF522R	OEM
IRF523	2341
IRF523R	OEM
IRF530	5060
IRF531	5060
IRF532	5060
IRF533	5060
IRF540	OEM
IRF540Y	OEM
IRF541	OEM
IRF542	OEM
IRF543	OEM
IRF610	1465
IRF611	1465
IRF612	1465
IRF613	1465
IRF614	OEM
IRF620	5723
IRF621	2176
IRF622	1465
IRF623	2176
IRF630	5723
IRF631	2176
IRF632	1465
IRF633	5726
IRF640	OEM
IRF641	OEM
IRF642	5728
IRF643	OEM
IRF643R	OEM
IRF710	1465
IRF711	1465
IRF712	1465
IRF713	1465
IRF713R	OEM
IRF720	1465
IRF721	1465
IRF722	1465
IRF723	1465
IRF730	OEM
IRF731	OEM
IRF732	1465
IRF733	1465
IRF740	OEM
IRF741	OEM
IRF742	OEM
IRF743	OEM
IRF820	5757
IRF821	5757
IRF821R	OEM
IRF822	1466
IRF823	1466
IRF823R	OEM
IRF830	OEM
IRF831	1465
IRF832	OEM
IRF833	5765
IRF840	5769
IRF840I	5769
IRF841	5769
IRF842	1653
IRF843	1653
IRF9130	OEM
IRF9131	2271
IRF9132	OEM
IRF9133	2271
IRF9140	OEM
IRF9141	OEM
IRF9142	OEM
IRF9143	OEM
IRF9230	OEM
IRF9231	OEM
IRF9232	OEM
IRF9233	OEM
IRF9241	OEM
IRF9242	OEM
IRF9243	OEM
IRF9340	OEM
IRF9510	6343
IRF9511	6343
IRF9512	6343
IRF9513	6343
IRF9520	2271
IRF9521	2271
IRF9522	2271
IRF9523	2271
IRF9530	2271
IRF9531	2271
IRF9532	2271
IRF9533	2271
IRF9540	OEM
IRF9541	OEM
IRF9542	OEM
IRF9543	OEM
IRF9610	OEM
IRF9611	6343
IRF9612	OEM
IRF9613	6343
IRF9620	OEM
IRF9621	6343
IRF9622	OEM
IRF9623	6343
IRF9630	OEM
IRF9631	OEM
IRF9632	OEM
IRF9633	6343
IRF9640	OEM
IRF9641	OEM
IRF9642	OEM
IRF9643	OEM
IRFBC40	OEM
IRFD1Z0	OEM
IRFD1Z3	OEM
IRFD110	OEM
IRFD111	OEM
IRFD112	OEM
IRFD113	OEM
IRFD120	OEM
IRFD123	OEM
IRFD210	OEM
IRFD213	OEM
IRFD220	OEM
IRFD321R	OEM
IRFD322R	OEM
IRFD323R	OEM
IRFD9110	OEM
IRFD9113	OEM
IRFD9120	OEM
IRFD9121	OEM
IRFD9122	OEM
IRFD9123	OEM
IRFD9210	OEM
IRFD9213	OEM
IRFE100	0325
IRFF110	OEM
IRFF110R	OEM
IRFF111	OEM
IRFF112	OEM
IRFF113	OEM
IRFF113R	OEM
IRFF120	OEM
IRFF121	OEM
IRFF122	OEM
IRFF123	OEM
IRFF123R	OEM
IRFF130	OEM
IRFF131	OEM
IRFF132	OEM
IRFF133	OEM
IRFF210	OEM
IRFF211	OEM
IRFF212	OEM
IRFF213	OEM
IRFF213R	OEM
IRFF220	OEM
IRFF221	OEM
IRFF222	OEM
IRFF223	OEM
IRFF223R	OEM
IRFF230	OEM
IRFF231	OEM
IRFF232	OEM
IRFF233	OEM
IRFF310	OEM
IRFF311	OEM
IRFF312	OEM
IRFF313	OEM
IRFF320	OEM
IRFF321	OEM
IRFF322	OEM
IRFF323	OEM
IRFF330	OEM
IRFF331	OEM
IRFF332	OEM
IRFF333	OEM
IRFF423R	OEM
IRFF430	OEM
IRFF431	OEM
IRFF432	OEM
IRFF433	OEM
IRFP353R	OEM
IRFP448	OEM
IRFP453R	OEM
IRFS1Z0	OEM
IRFS1Z3	OEM

DEVICE TYPE	REPL CODE
IRFU121	OEM
IRFZ20	OEM
IRFZ22	OEM
IRFZ30	OEM
IRFZ32	OEM
IRFZ40	OEM
IRFZ42	OEM
IRIF	0102
IRKD71-12	1798
IRKD101-14	2192
IRKH26-04	2248
IRKH26-06	2248
IRKH26-08	2248
IRKH26-10	2248
IRKH26-12	2248
IRKH26-14	OEM
IRKH26-16	OEM
IRKH26-18	OEM
IRKH41-04	2248
IRKH41-06	2248
IRKH41-08	2248
IRKH41-10	2248
IRKH41-12	2248
IRKH41-14	OEM
IRKH41-16	OEM
IRKH41-18	OEM
IRKH56-04	2248
IRKH56-06	2248
IRKH56-08	2248
IRKH56-10	2248
IRKH56-12	2248
IRKH56-14	OEM
IRKH56-16	OEM
IRKH56-18	OEM
IRKH57-12	4364
IRKH71-04	2241
IRKH71-06	2241
IRKH71-08	2241
IRKH71-10	2241
IRKH71-12	2241
IRKH71-14	OEM
IRKH71-16	OEM
IRKH71-18	OEM
IRKH91-04	2241
IRKH91-06	2241
IRKH91-08	2241
IRKH91-10	2241
IRKH91-12	2241
IRKH91-14	OEM
IRKH91-16	OEM
IRKH91-18	OEM
IRKH92-12	2994
IRKH132-14D20	OEM
IRKH132-16D25	OEM
IRKH162-14D20	OEM
IRKH162-16D25	OEM
IRKH180-14D20	OEM
IRKH180-16D25	OEM
IRKH210-14D20	OEM
IRKH210-16D20	OEM
IRKH250-14D20	OEM
IRKH250-16D25	OEM
IRKK26-04	OEM
IRKK26-06	OEM
IRKK26-08	OEM
IRKK26-10	OEM
IRKK26-12	OEM
IRKK26-14	OEM
IRKK26-16	OEM
IRKK26-18	OEM
IRKK41-04	OEM
IRKK41-06	OEM
IRKK41-08	OEM
IRKK41-10	OEM
IRKK41-12	OEM
IRKK41-14	OEM
IRKK41-16	OEM
IRKK41-18	OEM
IRKK56-04	OEM
IRKK56-06	OEM
IRKK56-08	OEM
IRKK56-10	OEM
IRKK56-12	OEM
IRKK56-14	OEM
IRKK56-16	OEM
IRKK56-18	OEM
IRKK71-04	OEM
IRKK71-06	OEM
IRKK71-08	OEM
IRKK71-10	OEM
IRKK71-12	OEM
IRKK71-14	OEM
IRKK71-16	OEM
IRKK71-18	OEM
IRKK91-04	OEM
IRKK91-06	OEM
IRKK91-08	OEM
IRKK91-10	OEM
IRKK91-12	OEM
IRKK91-14	OEM
IRKK91-16	OEM
IRKK91-18	OEM
IRKL57-12	4574
IRKL92-12	2878
IRKL132-14D20	OEM
IRKL132-16D25	OEM
IRKL162-14D20	OEM
IRKL162-16D25	OEM
IRKL180-14D20	OEM
IRKL180-16D25	OEM
IRKL210-14D20	OEM
IRKL210-16D20	OEM
IRKL250-14D20	OEM
IRKL250-16D25	OEM
IRKN26-04	OEM
IRKN26-06	OEM
IRKN26-08	OEM
IRKN26-10	OEM
IRKN26-12	OEM
IRKN26-14	OEM
IRKN26-16	OEM
IRKN26-18	OEM
IRKN41-04	OEM
IRKN41-06	OEM
IRKN41-08	OEM
IRKN41-10	OEM
IRKN41-12	OEM
IRKN41-14	OEM
IRKN41-16	OEM
IRKN41-18	OEM
IRKN56-04	OEM
IRKN56-06	OEM
IRKN56-08	OEM
IRKN56-10	OEM
IRKN56-12	OEM
IRKN56-14	OEM
IRKN56-16	OEM
IRKN56-18	OEM
IRKN71-04	OEM
IRKN71-06	OEM
IRKN71-08	OEM
IRKN71-10	OEM
IRKN71-12	OEM
IRKN71-14	OEM
IRKN71-16	OEM
IRKN71-18	OEM
IRKN91-04	OEM
IRKN91-06	OEM
IRKN91-08	OEM
IRKN91-10	OEM
IRKN91-12	OEM
IRKN91-14	OEM
IRKN91-16	OEM
IRKN91-18	OEM
IRKT26-04	1948
IRKT26-06	OEM
IRKT26-08	1948
IRKT26-10	1948
IRKT26-12	1948
IRKT26-14	OEM
IRKT26-16	OEM
IRKT26-18	OEM
IRKT41-04	1948
IRKT41-06	1948
IRKT41-08	1948
IRKT41-10	1948
IRKT41-12	1948
IRKT41-14	OEM
IRKT41-16	OEM
IRKT41-18	OEM
IRKT56-04	1948
IRKT56-06	1948
IRKT56-08	1948
IRKT56-10	1948
IRKT56-12	1948
IRKT56-14	OEM
IRKT56-16	OEM
IRKT56-18	OEM
IRKT57-12	0652
IRKT71-04	1939
IRKT71-06	1939
IRKT71-08	1939
IRKT71-10	1939
IRKT71-12	1939
IRKT71-14	OEM
IRKT71-16	OEM
IRKT71-18	OEM
IRKT91-04	1939
IRKT91-06	1939
IRKT91-08	1939
IRKT91-10	1939
IRKT91-12	1939
IRKT91-14	OEM
IRKT91-16	OEM
IRKT91-18	OEM
IRKT92-12	2311
IRKU26-04	OEM
IRKU26-06	OEM
IRKU26-08	OEM
IRKU26-10	OEM
IRKU26-12	OEM
IRKU26-14	OEM
IRKU26-16	OEM
IRKU26-18	OEM
IRKU41-04	OEM
IRKU41-06	OEM
IRKU41-08	OEM
IRKU41-10	OEM
IRKU41-12	OEM
IRKU41-14	OEM
IRKU41-16	OEM
IRKU41-18	OEM
IRKU56-04	OEM
IRKU56-06	OEM
IRKU56-08	OEM
IRKU56-10	OEM
IRKU56-12	OEM
IRKU56-14	OEM
IRKU56-16	OEM
IRKU56-18	OEM
IRKU71-04	OEM
IRKU71-06	OEM
IRKU71-08	OEM
IRKU71-10	OEM
IRKU71-12	OEM
IRKU71-14	OEM
IRKU71-16	OEM
IRKU71-18	OEM
IRKU91-04	OEM
IRKU91-06	OEM
IRKU91-08	OEM
IRKU91-10	OEM
IRKU91-12	OEM
IRKU91-14	OEM
IRKU91-16	OEM
IRKU91-18	OEM
IRKV26-04	OEM
IRKV26-06	OEM
IRKV26-08	OEM
IRKV26-10	OEM
IRKV26-12	OEM
IRKV26-14	OEM
IRKV26-16	OEM
IRKV26-18	OEM
IRKV41-04	OEM
IRKV41-06	OEM
IRKV41-08	OEM
IRKV41-10	OEM
IRKV41-12	OEM
IRKV41-14	OEM
IRKV41-16	OEM
IRKV41-18	OEM
IRKV56-04	OEM
IRKV56-06	OEM
IRKV56-08	OEM
IRKV56-10	OEM
IRKV56-12	OEM
IRKV56-14	OEM
IRKV56-16	OEM
IRKV56-18	OEM
IRKV71-04	OEM
IRKV71-06	OEM
IRKV71-08	OEM
IRKV71-10	OEM
IRKV71-12	OEM
IRKV71-14	OEM
IRKV71-16	OEM
IRKV71-18	OEM
IRKV91-04	OEM
IRKV91-06	OEM
IRKV91-08	OEM
IRKV91-10	OEM
IRKV91-12	OEM
IRKV91-14	OEM
IRKV91-16	OEM
IRKV91-18	OEM
IRL60	OEM
IRSANYO0002	OEM
IRSANYO-002	OEM
IRSANYO0002	OEM
IRSANYO-003	OEM
IRSANYO0003	OEM
IRSANYO-004	OEM
IRSANYO0004	OEM
IRSANYO-012	OEM
IRT26-06	1948
IRT82	0411
IRTR01	0085
IRTR02	0085
IRTR03	0440
IRTR04	0211
IRTR05	0280
IRTR06	0050
IRTR07	0050
IRTR08	0600
IRTR09	0038
IRTR10	0600
IRTR11	0280
IRTR12	0050
IRTR14	0164
IRTR16	0160
IRTR16X2	0142
IRTR17	0050
IRTR18	0050
IRTR19	0150
IRTR20	0037
IRTR21	0018
IRTR22	0079
IRTR23	0142
IRTR24	0079
IRTR25	0698
IRTR26	0103
IRTR27	0975
IRTR28	0150
IRTR30	0150
IRTR31	0006
IRTR32	0276
IRTR36MP	0130
IRTR37	0792
IRTR-50	0222
IRTR50	0222
IRTR50X2	0822
IRTR-51	0111
IRTR51	0111
IRTR-52	0791
IRTR52	0037
IR-TR53	0086
IRTR-53	0595
IRTR53	0086
IRTR-54	0150
IRTR54	0532
IRTR-55	0560
IRTR55	0560
IRTR-56	1267
IRTR56	1267
IRTR-57	0178
IRTR57	0423
IRTR-58	1027
IRTR58	0905
IRTR-59	0103
IRTR59	0103
IRTR-60	0285
IRTR60	0285
IRTR-61	0103
IRTR61	0103
IRTR61X2	1514
IRTR-62	0079
IRTR62	0016
IRTR-63	0936
IRTR63	0016
IRTR-64	0560
IRTR64	0698
IRTR-65	0698
IRTR65	0698
IRTR-66	0622
IRTR-67	0642
IRTR67	0642
IRTR-68	0003
IRTR68	0003
IRTR-69	0400
IRTR69	0400
IRTR-70	0224
IRTR70	0007
IRTR-71	0007
IRTR71	0007
IRTR-72	0551
IRTR72	0551
IRTR-73	0382
IRTR73	0382
IRTR-74	0264
IRTR74	0264
IRTR-75	0338
IRTR75	0338
IRTR-76	0042
IRTR76	0042
IRTR-77	0925
IRTR77	0925
IRTR-78	0233
IRTR78	0233
IRTR-79	0285
IRTR79	0285
IRTR-80	0144
IRTR80	0144
IRTR81	1027
IRTR82	0847
IRTR-83	0144
IRTR83	0144
IRTR84	0164
IRTR-85	0164
IRTR85	0004
IRTR-86	0861
IRTR86	0198
IRTR-87	0630
IRTR87	0086
IRTR-88	0382
IRTR88	0126
IRTR91	2763
IRTR92	0423
IRTR93	0065
IRTR94MP	1862
IRTR-95	0007
IRTR95	0144
IRTRFE100	0325
IRW106A1	0447
IRW106C1	1222
IRW106Y1	1396
IRW122A	2089
IRW122B	0505
IRW122F	2095
IRWC470-1300	OEM
IRZ1300	0547
IRZ1302	2113
IRZ1304	1901
IRZ1306	1601
IRZ1308	1622
IRZ1310	0992
IS002	OEM
IS003	OEM
IS08	1004
IS010	2521
IS015	OEM
IS020	7160
IS01209	0139
IS18	1004
IS28	4458
IS38	1034
IS48	0710
IS58	1148
IS68	1148
IS86	0030
IS110	0023
IS115	OEM
IS120	0114
IS-136	OEM
IS188AM	OEM
IS188FM	OEM
IS210	0397
IS215	OEM
IS220	0397
IS310	0611
IS315	OEM
IS320	0611
IS352M	0030
IS410	0611
IS415	OEM
IS420	0611
IS426	OEM
IS-446D	1149
IS510	0952
IS515	OEM
IS520	0952
IS553	OEM
IS554	OEM
IS610	0952
IS615	OEM
IS620	0952
IS1209	0139
IS1210	0139
IS1211	0139
IS-1212	0162
IS1212	0133
IS2095	OEM
ISB473	0222
ISB-3101-01	OEM
ISB-3101-02	OEM
ISB3600	OEM
ISB-3812-01	OEM
ISB-3812-02	OEM
ISB-3812-03	OEM
ISB-3812-04	OEM
ISBF1	0004
ISC71	OEM
ISC72	OEM
ISC82	OEM
ISC92	OEM
ISC302A	OEM
ISC302B	OEM
ISC302C	OEM
ISC302D	OEM
ISC302E	OEM
ISC302F	OEM
ISC363	OEM
ISC363A	OEM
ISC363B	OEM
ISC363C	OEM
ISC363D	OEM
ISC363E	OEM
ISC386	OEM
ISC386A	OEM
ISD-162	0038
ISD162	0038
ISO100AP(A)	OEM
ISO100BP(A)	OEM
ISO100CP(A)	OEM
ISO162	0600
ISPS035	OEM
ISPS135	OEM
ISPS235	OEM
ISPS335	OEM
ISPS435	OEM
ISPS535	OEM
ISPS635	OEM
ISPT030	3199
ISPT040	3199
ISPT130	3199
ISPT140	3199
ISPT230	3199
ISPT240	3199
ISPT330	3207
ISPT340	3207
ISPT430	3207
ISPT440	3207
ISPT530	3222
ISPT540	3222
ISPT630	3222
ISPT640	3222
ISS54	OEM
ISS106	OEM
ISS119	OEM
ISS120	0124
ISS122	0023
ISS-8053ALR	OEM
ISSRM2114C3	2048
ISSRM211403	2048
ISV87	4391
ISV97	OEM
ISV121	0133
IT06	0555
IT08	0555
IT08A	0555
IT08B	OEM
IT08HA	0555
IT08HX	0555
IT010	0555
IT010A	0555
IT010B	OEM
IT010HA	0555
IT010HX	0555
IT015	0772
IT015A	OEM
IT015B	OEM
IT015HA	OEM
IT015HX	OEM
IT4B42	1875
IT10D4K	0015
IT16	0555
IT18	0555
IT18A	0555
IT18B	OEM
IT18HA	0555
IT18HX	0555
IT22	0143
IT22A	0143
IT22AM	0143
IT22B	0143
IT23	0143
IT23A	0143
IT23G	0143
IT23S	0143
IT25	OEM
IT25-4	OEM
IT26	0555
IT28	0555
IT28HA	0555
IT28HX	0555
IT32	OEM
IT33	OEM
IT36	0555
IT38	0576
IT38B	OEM
IT38HA	0555
IT38HX	0555
IT40	0133
IT46	0555
IT48	0576
IT48A	0555
IT48B	OEM
IT48HA	0555
IT48HX	0555
IT56	0555
IT58	0555
IT58A	0555
IT58B	OEM
IT58HA	0555
IT58HX	0555
IT66	0555
IT68	0555
IT68A	0555
IT68B	OEM
IT68HA	0555
IT68HX	0555
IT100	OEM
IT101	OEM
IT108	0325
IT110	0555
IT110A	0555
IT110B	OEM
IT110HA	0555
IT115	0772
IT115A	OEM
IT115B	OEM
IT115HA	OEM
IT115HX	OEM
IT120	0079
IT120A	OEM
IT121	0079
IT122	0111
IT124	0400
IT125	OEM
IT126	OEM
IT127	OEM
IT128	OEM
IT129	OEM
IT130	OEM
IT130A	OEM
IT131	OEM
IT132	OEM
IT136	OEM
IT137	OEM
IT138	OEM
IT139	OEM
IT205A	0050
IT210	0555
IT210A	0555
IT210B	OEM
IT210HA	0555
IT210HX	0555
IT215	0772
IT215A	OEM
IT215B	OEM
IT215HA	OEM
IT215HX	OEM
IT243	0162
IT261	0143
IT262	0143
IT263	0143
IT310	0555
IT310A	0555
IT310B	OEM
IT310HA	0555
IT310HX	0555
IT315	0744
IT315A	OEM
IT315B	OEM
IT315HA	OEM
IT315HX	OEM
IT410	0555
IT410A	0555
IT410B	OEM
IT410HA	0555
IT410HX	0555
IT415	0744
IT415B	OEM
IT415HA	OEM
IT415HX	OEM
IT500	OEM
IT501	OEM
IT502	OEM
IT503	1228
IT510	0555
IT510A	0555
IT510B	OEM
IT510HA	0555
IT510HX	0555
IT515	2044
IT515A	OEM
IT515B	OEM
IT515HA	OEM
IT515HX	OEM
IT610	0555
IT610A	0555
IT610B	OEM
IT610HA	0555
IT610HX	0555
IT615	0617
IT615A	OEM
IT615B	OEM
IT615HA	OEM
IT615HX	OEM
IT918	0007
IT918A	0111
IT929	0007
IT930	0007
IT1700	OEM
IT1750	OEM
IT2218	0079
IT2219	0079
IT2221	0079
IT2222	0079
IT2483	0861
IT2484	OEM
IT2604	0861
IT2605	0861
IT2904	0861
IT2905	0861
IT2906	0861
IT2907	0861
IT5911	OEM
IT5912	OEM
ITA7176AP	0167
ITA7607AP	0912
ITA7608CP5	2054
ITA7609P-2	4134
ITA7609P2	4134
ITC103-1	1137
ITC103-2	0344
ITC103-3	0901
ITC103-4	0058
ITC103-5	0407
ITC103-6	0407
ITC918A	0111
ITE4091	OEM
ITE4092	OEM
ITE4093	OEM
ITE4391	3130
ITE4392	OEM
ITE4393	OEM
ITE4416	0325
ITH61	0071
ITR3T24	OEM
ITS5817	0110
ITS5818	0110
ITS5823	1282
ITS5824	1282
ITS5825	1282
ITT10HX	OEM
ITT33	0124
ITT44	0124
ITT73	0229
ITT73N	0124
ITT74H00N	0682
ITT74H01N	5338
ITT74H04N	1907
ITT74H05N	3251
ITT74H10N	0685
ITT74H11N	2398
ITT74H20N	3715
ITT74H21N	4856
ITT74H30N	5383
ITT74H40N	0559
ITT74H50N	1792
ITT74H51N	1944
ITT74H53N	2101
ITT74H54N	2172
ITT74H60N	5410
ITT74H72N	3313
ITT74H73N	2463
ITT74H74N	2492
ITT74H76N	5302
ITT102	0143
ITT142	0133
ITT181-1	0133
ITT200	0124
ITT210	0030
ITT-243	OEM
ITT301	0143
ITT-310	0030
ITT310	0628
ITT310G	0030
ITT-310S	0030
ITT310S	0030
ITT310Y	0030
ITT350	0015

If replacement code is OEM, contact original manufacturer for replacement.

DEVICE TYPE	REPL CODE
ITT-401	0030
ITT401	OEM
ITT-410	0720
ITT410	0030
ITT410B	OEM
ITT410S	OEM
ITT410ST	0030
ITT413	0133
ITT417	0137
ITT600	0124
ITT601	0124
ITT718	0143
ITT895	0079
ITT920	0124
ITT921	0133
ITT922	0182
ITT923	0182
ITT930-5	1823
ITT930N	1823
ITT932-5	1041
ITT932N	1041
ITT933-5	2097
ITT933N	2097
ITT935-5	1176
ITT935N	1176
ITT936-5	1831
ITT936N	1831
ITT937-5	1835
ITT937N	1835
ITT941-5	2622
ITT941N	2622
ITT944-5	0033
ITT944N	0033
ITT945-5	0081
ITT945N	0081
ITT946-5	0141
ITT946N	0141
ITT948-5	3399
ITT948N	3399
ITT949-5	1844
ITT949N	1844
ITT951-5	2078
ITT951N	2078
ITT961-5	1859
ITT961N	1859
ITT962-5	0562
ITT962N	0562
ITT963-5	0341
ITT963N	0341
ITT973	0037
ITT975	0079
ITT1330	1808
ITT1352	0395
ITT1800-5	0850
ITT1800N	0850
ITT1801N	0874
ITT1806-5	0945
ITT1806N	0945
ITT1807-5	0945
ITT1807N	0962
ITT1808-5	0982
ITT1808N	0982
ITT1809-5	0998
ITT1809N	0998
ITT1810-5	1011
ITT1810N	1011
ITT1811-5	1025
ITT1811N	1025
ITT2001	0133
ITT2002	0133
ITT2003	2527
ITT2018	OEM
ITT2147-55D	OEM
ITT2147-55J	OEM
ITT2147-55N	OEM
ITT2147-70D	OEM
ITT2147-70J	OEM
ITT2147-70LD	OEM
ITT2147-70LJ	OEM
ITT2147-70LN	OEM
ITT2147-70N	OEM
ITT2147-90D	OEM
ITT2147-90J	OEM
ITT2147-90N	OEM
ITT2147H-55D	OEM
ITT2147H-55J	OEM
ITT2147H-55N	OEM
ITT2167-45D	OEM
ITT2167-45N	OEM
ITT2167-55D	OEM
ITT2167-55N	OEM
ITT2167-70D	OEM
ITT2167-70N	OEM
ITT2168-45D	OEM
ITT2168-45N	OEM
ITT2168-55D	OEM
ITT2168-55N	OEM
ITT2168-70D	OEM
ITT2168-70N	OEM
ITT3001	0133
ITT3002	0609
ITT3003	0133
ITT3064	0802
ITT3065	0167
ITT-3105	0030
ITT3105	OEM
ITT4000G	1335
ITT4027-2	OEM
ITT4027-2D	OEM
ITT4027-2N	OEM
ITT4027-3	OEM
ITT4027-3D	OEM
ITT4027-3N	OEM
ITT4027-4	OEM
ITT4027-4D	OEM
ITT4027-4N	OEM
ITT4027-6	OEM
ITT4027-6D	OEM
ITT4027-6N	OEM
ITT4116-2D	OEM
ITT4116-2N	OEM
ITT4116-3	0522
ITT4116-3D	OEM
ITT4116-3N	OEM
ITT4116-4	0522
ITT4116-4D	OEM
ITT4116-4N	OEM
ITT4164-15	OEM
ITT4164-20	OEM
ITT5059	0080
ITT5060	0609
ITT5061	0795
ITT5062	0015
ITT5101S	OEM
ITT5116-2	OEM
ITT5401	OEM
ITT7163	OEM
ITT7164	OEM
ITT7215	0133
ITT7400N	0232
ITT7401N	0268
ITT7402N	0314
ITT7403N	0335
ITT7404N	0361
ITT7405N	0385
ITT7406N	1206
ITT7407N	1339
ITT7408N	0467
ITT7409N	0492
ITT7410N	0512
ITT7411N	0527
ITT7412N	2242
ITT7413N	1441
ITT7416N	1349
ITT7417N	1352
ITT7420N	0697
ITT7421N	1357
ITT7425N	3477
ITT7426N	0803
ITT7428N	4180
ITT7430N	0873
ITT7432N	0899
ITT7433N	4193
ITT7437N	3520
ITT7438N	0996
ITT7440N	1024
ITT7442N	1052
ITT7443N	1060
ITT7444N	1072
ITT7445N	1082
ITT7446AN	1098
ITT7447AN	1108
ITT7448N	1125
ITT7450N	0743
ITT7451N	1168
ITT7453N	1185
ITT7454N	1202
ITT7460N	1275
ITT7470N	1404
ITT7472N	1426
ITT7473N	1172
ITT7475N	1432
ITT7476N	1158
ITT7480N	1535
ITT7482N	1574
ITT7483N	0117
ITT7486N	1368
ITT7490N	1208
ITT7491AN	0980
ITT7492N	0834
ITT7493N	0569
ITT7494N	1702
ITT7495AN	1486
ITT7496N	1715
ITT9093-5	0358
ITT9093N	0358
ITT9094-5	1632
ITT9094N	1632
ITT9097-5	1481
ITT9097N	1481
ITT9099-5	0333
ITT9099N	0333
ITT9900JL	OEM
ITT9900JL-40	OEM
ITT74107N	0942
ITT74109N	0968
ITT74121N	0175
ITT74122N	1139
ITT74123N	1157
ITT74141N	1377
ITT74145N	0619
ITT74150N	1492
ITT74151N	1495
ITT74152N	1517
ITT74153N	1539
ITT74154N	1555
ITT74155N	1576
ITT74156N	1592
ITT74160N	1631
ITT74161N	1645
ITT74162N	1013
ITT74163N	1666
ITT74164N	0734
ITT74165N	1685
ITT74174N	1770
ITT74175N	1787
ITT74180N	1829
ITT74181N	1842
ITT74182N	1856
ITT74190N	1912
ITT74192N	1921
ITT74193N	1926
ITT74195N	1943
ITT330419	0795
ITT412142	0133
ITTA75902P	0829
ITTC4027B	1949
ITTC4030BP	0500
ITTC4049BP	0001
ITTC4052BP	0024
ITTC5012BP	2053
ITTC5066BP	OEM
ITTD62504P	OEM
ITTMS1025N2LC	OEM
IULM1131CN	OEM
IUPC1363C	0683
IUPD1937C	2090
IUPD1986C	4788
IVM-1	OEM
IVM1	OEM
IVM-3	OEM
IVN5000AND	OEM
IVN5000ANE	OEM
IVN5000ANF	OEM
IVN5000ANH	OEM
IVN5000BND	OEM
IVN5000BNE	OEM
IVN5000BNF	OEM
IVN5000SND	OEM
IVN5000SNE	OEM
IVN5000SNF	OEM
IVN5000SNH	OEM
IVN5000TND	OEM
IVN5000TNE	OEM
IVN5000TNF	OEM
IVN5000TNG	OEM
IVN5000TNH	OEM
IVN5001AND	OEM
IVN5001ANE	OEM
IVN5001ANF	OEM
IVN5001ANH	OEM
IVN5001BND	OEM
IVN5001BNE	OEM
IVN5001BNF	OEM
IVN5001SND	OEM
IVN5001SNE	OEM
IVN5001SNF	OEM
IVN5001SNH	OEM
IVN5001TND	OEM
IVN5001TNE	OEM
IVN5001TNF	OEM
IVN5001TNG	OEM
IVN5001TNH	OEM
IVN5200HND	OEM
IVN5200HNE	OEM
IVN5200HNF	OEM
IVN5200HNH	OEM
IVN5200KND	OEM
IVN5200KNE	OEM
IVN5200KNF	OEM
IVN5200KNH	OEM
IVN5200TND	OEM
IVN5200TNE	OEM
IVN5200TNF	OEM
IVN5200TNH	OEM
IVN5201CND	OEM
IVN5201CNE	5060
IVN5201CNF	OEM
IVN5201CNH	3088
IVN5201HND	OEM
IVN5201HNE	OEM
IVN5201HNF	OEM
IVN5201HNH	OEM
IVN5201KND	OEM
IVN5201KNE	OEM
IVN5201KNF	OEM
IVN5201KNH	OEM
IVN5201TND	OEM
IVN5201TNE	OEM
IVN5201TNF	OEM
IVN5201TNH	OEM
IVN6000CNS	1465
IVN6000CNT	1466
IVN6000CNU	1466
IVN6000KNR	OEM
IVN6000KNS	OEM
IVN6000KNT	OEM
IVN6000KNU	OEM
IVN6000TNS	OEM
IVN6000TNT	OEM
IVN6000TNU	OEM
IVN6100TNS	OEM
IVN6100TNT	OEM
IVN6100TNU	OEM
IVN6200CNE	5060
IVN6200CNF	5060
IVN6200CNH	5060
IVN6200CNM	OEM
IVN6200CNP	OEM
IVN6200CNS	1465
IVN6200CNT	OEM
IVN6200CNU	OEM
IVN6200CNW	OEM
IVN6200CNX	OEM
IVN6200KNE	OEM
IVN6200KNF	OEM
IVN6200KNH	OEM
IVN6200KNM	OEM
IVN6200KNP	OEM
IVN6200KNS	1668
IVN6200KNT	OEM
IVN6200KNU	OEM
IVN6200KNW	OEM
IVN6200KNX	OEM
IVN6300ANE	OEM
IVN6300ANF	OEM
IVN6300ANH	OEM
IVN6300ANM	OEM
IVN6300ANP	OEM
IVN6300ANS	OEM
IVN6300ANT	OEM
IVN6300ANU	OEM
IVN6300SNE	OEM
IVN6300SNF	OEM
IVN6300SNH	OEM
IVN6300SNM	OEM
IVN6300SNP	OEM
IVN6300SNS	OEM
IVN6300SNT	OEM
IVN6300SNU	OEM
IVN6660	OEM
IVN6661	OEM
IW	OEM
IW01-08J	0244
IWM5-1454L	OEM
IWM51454L	OEM
IX0027CE	OEM
IX0037CE	1329
IX0043CE	0167
IX0054CE	2872
IX0058GE	OEM
IX0069CE	0478
IX0084TA	OEM
IX0085TA	0603
IX0086TA	4410
IX0088TA	OEM
IX047GE	OEM
IX0137CE	2194
IX0137CE2SA562T(Q)	OEM
IX0144CE	OEM
IX0178CE	1590
IX0199CE	2090
IX0212CE	3685
IX0212CEZZ	3685
IX0213CE	2896
IX0213CEZZ	OEM
IX0214CE	3772
IX0225CE	3333
IX0232CE	1846
IX0233CEZZ	OEM
IX0238CE	2042
IX0238CEZZ	OEM
IX0243CE	0167
IX0248CE	OEM
IX0252CE	0069
IX0252CEZZ	0069
IX0260CE	2666
IX0275CE	4109
IX0280CE	OEM
IX0308C	OEM
IX0308CE	OEM
IX0314CE	OEM
IX0325C	5703
IX0325CE	OEM
IX0359CE	OEM
IX0359CEZZ	OEM
IX0406GE	OEM
IX0416CE	3254
IX0424CE	OEM
IX0430GE	OEM
IX0487CE	2269
IX0488CE	0416
IX0499CE	1604
IX0508GE	OEM
IX0539CE	OEM
IX0580CE	OEM
IX0581CE	OEM
IX0600CE	3236
IX0614CE	OEM
IX0633CE	4336
IX0634CE	OEM
IX0638CE	0624
IX0669CE	OEM
IX0692CE	OEM
IX0735CE	OEM
IX0800CE	OEM
IX0820CE	OEM
IX0943CE	OEM
IX0948CE	4129
IX0950CE	OEM
IX103AF	OEM
IX1030CE	OEM
IX1031AF	OEM
IX1077CE	OEM
IX1132CE	OEM
IX1190CE	3515
IX1231CE	OEM
IX1247CE	OEM
IX1525AF	OEM
IX1712CE	OEM
IX1890CE	OEM
IX1891CE	OEM
IX5004CE	OEM
IXD2201A	OEM
IXD2210	OEM
IXD2212D	OEM
IXMSL9371RS	OEM
IXMSL9372RS	OEM
IXMSL9373RS	OEM
IXMSL9378RS	5335
IXXEEA00020	OEM
IXXEEA00030	OEM
IYBA5102AALC	2105
IYBA5112LS	OEM
IYBA5115	OEM
IYBA6109U	OEM
IYBA6304A	2425
IYBA6305	OEM
IYBA6334	OEM
IZ0008CE	OEM
IZ0027CE	OEM
IZ0073CE	OEM
IZA169DR	OEM
IZZ0008	OEM
J005	2363
J01	5400
J02	5400
J04	5402
J-05	0015
J05	0102
J05G	0087
J06	5403
J-1	0015
J1	0102
J1G	0087
J-2	0015
J2	0102
J2-4570	0015
J2G	0087
J-4	0015
J4	0102
J4-1000	0232
J4-1002	0314
J4-1004	0361
J4-1010	0512
J4-1047	1108
J4-1075	1432
J4-1076	1158
J4-1090	1208
J4-1092	0834
J4-1121	0175
J4-1203	0379
J4-1215	0312
J4-1555	0973
J4-1600	0198
J4-1601	0071
J4-1602	0540
J4-1605	0289
J4-1610	0124
J4-1615	0162
J4-1619	0301
J4-1620	0012
J4-1622	0137
J4-1625	0211
J4-1626	0155
J4-1628	0016
J4-1630	0016
J4-1632	0037
J4-1634	0400
J4-1636	0330
J4-1638	0085
J4-1640	0103
J4-1642	0543
J4-1644	0111
J4-1645	0037
J4-1648	0126
J4-1649	0086
J4-1650	0037
J4-1651	0861
J4-1654	0423
J4-1655	0854
J4-1664	1361
J4-1665	1369
J4-1668	3374
J4-1669	3373
J4-1672	1681
J4-1673	0130
J4-1700	1893
J4-1710	3384
J4-1711	0325
J4-1713	0212
J4-1725	0940
J4-1730	0593
J4-1730(DIAC)	2730
J4-1730(TRIAC)	0593
J4A3	1023
J4B3	1023
J4C3	1023
J4D3	1023
J4E3	1023
J4EB	1023
J4F3	1023
J4G	0087
J4G3	1023
J4H3	1036
J4K3	1036
J4M3	1046
J4N3	1046
J5A3	1023
J5B3	1023
J5C3	1023
J5CB	1023
J5D3	1023
J5E3	1023
J5EB	1023
J5F3	1023
J5G3	1023
J5H3	1036
J5K3	1036
J5M3	1046
J5N3	1046
J5P3	OEM
J5R3	OEM
J5S3	OEM
J-6	0015
J6	0102
J6G	0087
J-8	0071
J8	0102
J8G	0087
J-10	0071
J10	0102
J10G	0087
J11J	OEM
J11L	OEM
J11X	OEM
J20	OEM
J39B	0005
J39C	0030
J100	0015
J101	OEM
J102	2271
J103	OEM
J105	0065
J105-18	OEM
J105GR	OEM
J106	OEM
J106-18	OEM
J107	0127
J107-18	OEM
J108	0144
J108-18	OEM
J109	OEM
J109-18	OEM
J110	OEM
J110-18	OEM
J111	3842
J111-18	5647
J112	0325
J112-18	OEM
J113	3843
J113-18	5968
J114	OEM
J139A	0016
J147	OEM
J174	OEM
J174-18	OEM
J175	OEM
J175-18	OEM
J176	OEM
J176-18	OEM
J177	OEM
J177-18	OEM
J187	0144
J200	OEM
J201	1758
J201-18	OEM
J202	1758
J202-18	OEM
J203	OEM
J203-18	OEM
J204	OEM
J204-18	OEM
J210	OEM
J211	OEM
J212	OEM
J230	OEM
J230-18	OEM
J231	OEM
J232	OEM
J232-18	OEM
J241	0123
J242	0143
J243	0143
J245X	OEM
J270	2455
J270-18	6559
J271	OEM
J300	OEM
J304	0325
J305	OEM
J308	0325
J309	0325
J310	OEM
J400	0142
J401	OEM
J402	OEM
J403	OEM
J404	OEM
J405	OEM
J406	OEM
J410CN	OEM
J412CN	OEM
J422CN	OEM
J460	OEM
J461	OEM
J462	OEM
J463	OEM
J464	OEM
J465	OEM
J466	OEM
J503	OEM
J504	OEM
J505	OEM
J506	OEM
J507	OEM
J508	OEM
J509	OEM
J510	OEM
J511	OEM
J527-0A	OEM
J527-0C	OEM
J527-0D	OEM
J527-1A	OEM
J527-1C	OEM
J527-1D	OEM
J527-2A	OEM
J527-2C	OEM
J527-2D	OEM
J527-3A	OEM
J527-3C	OEM
J527-3D	OEM
J527-4A	OEM
J527-4C	OEM
J527-4D	OEM
J527-5A	OEM
J527-5C	OEM
J527-5D	OEM
J527-6A	OEM
J527-6C	OEM
J527-6D	OEM
J527-7A	OEM
J527-7C	OEM
J527-7D	OEM
J527-8A	OEM
J527-8C	OEM
J527-8D	OEM
J527-9A	OEM
J527-9C	OEM
J527-9D	OEM
J527-10A	OEM
J527-10C	OEM
J527-10D	OEM
J581	0861
J582	0861
J583	0861
J584	0861
J585	0861
J586	0861
J587	0861
J588	0861
J589	0861
J594	0861
J595	0861
J596	0861
J600	OEM
J623	0007
J624	0007
J625	0861
J626	0007
J627	0007
J628	0861
J629	0007
J630	0007
J631	0861
J650-1	OEM
J650-2	OEM
J650-3	OEM
J650-4	OEM
J650-5	OEM
J650-6	OEM
J650-7	OEM
J650-8	OEM
J650-9	OEM
J650-10	OEM
J650-11	OEM
J650-12	OEM
J650-13	OEM
J650-14	OEM
J650-15	OEM
J685	0019
J961B(G.E.)	0016
J1000-7400	0232
J1000-7402	0314
J1000-7404	0361
J1000-7410	0512
J1000-7447	1108
J1000-7476	1158
J1000-7490	1208
J1000-7492	0834
J1000-74121	0175
J1000-NE555	0973
J2441	0143
J4091	OEM
J4092	OEM
J4093	OEM
J4220	OEM
J4221	OEM
J4222	OEM
J4223	OEM
J4224	OEM
J4302	OEM
J4303	OEM

If replacement code is OEM, contact original manufacturer for replacement.

DEVICE TYPE	REPL CODE	DEVICE TYPE	REPL CODE	DEVICE TYPE	REPL CODE	DEVICE TYPE	REPL CODE	DEVICE TYPE	REPL CODE	DEVICE TYPE	REPL CODE	DEVICE TYPE	REPL CODE
J4304	OEM	J24915	0144	JAN1N3289	OEM	JB8253P-5	OEM	JD4000-4	OEM	JE9018I	0144	JT-E1014	0143
J4338	OEM	J24916	0079	JAN1N3291	OEM	JB8255AP-5	0051	JD4000-5	OEM	JE9042	0715	JT-E1024D	0015
J4339	OEM	J24919	0229	JAN1N3293	OEM	JB8259AP	OEM	JD4000-6	OEM	JE9042B	0715	JT-E1031	0143
J4391	OEM	J24920	0015	JAN1N3294	OEM	JB-BB1A	0015	JD4000-7	OEM	JE9042C	0715	JT-E1064	0015
J4392	OEM	J24921	0144	JAN1N3295	OEM	JBM5W1793-02P	OEM	JD4000-8	OEM	JE9100	OEM	JT-E1095	0774
J4393	OEM	J24923	0144	JAN1N3713	OEM	JC	OEM	JD4000-9	OEM	JE9100A	0079	JU0001	OEM
J4416	OEM	J24932	0111	JAN1N3715	OEM	JC-00012	0015	JD4000-10	OEM	JE9101	OEM	JU0002	OEM
J4856	OEM	J24933	0144	JAN1N3719	OEM	JC-00014	0015	JD4000-11	OEM	JE9112	5034	JU0010	OEM
J4857	OEM	J24934	0004	JAN1N3721	OEM	JC-00017	0102	JD4000-12	OEM	JE9112A	OEM	JU0011	OEM
J4858	OEM	J24935	0015	JAN1N3766	OEM	JC-00025	0015	JD4000-13	OEM	JE9112B	0037	JU0012	OEM
J4859	OEM	J24939	0015	JAN1N3766R	OEM	JC-00028	0015	JD4000-14	OEM	JE9112C	0037	JU0013	OEM
J4860	OEM	J24940	0015	JAN1N3768	OEM	JC-00032	0015	JD4000-15	OEM	JE9112D	OEM	JU0014	OEM
J4861	OEM	J24950	0030	JAN1N3768R	OEM	JC00032	0540	JD4000-16	OEM	JE9113A	OEM	JU0015	OEM
J5062	0050	J34155	OEM	JAN1N3890	OEM	JC-00033	0015	JD4000-17	OEM	JE9113B	OEM	JU0016	OEM
J5063	0004	J41625	0211	JAN1N3890R	OEM	JC-00035	0015	JD4000-18	OEM	JE9113C	0079	JU0017	OEM
J5064	0004	J41642	0103	JAN1N3891	OEM	JC-00037	0015	JD4000-19	OEM	JE9113D	OEM	JU0018	OEM
J6014	OEM	J41645	0037	JAN1N3891R	OEM	JC-00042	0023	JD4000-20	OEM	JEM1	0440	JU0019	OEM
J6014A	OEM	J41700	1669	JAN1N3893	OEM	JC-00044	0015	JD4000-21	OEM	JEM2	0440	JU0020	OEM
J6015	OEM	J101183	0015	JAN1N3893R	OEM	JC-00045	0015	JD4000-22	OEM	JEM3	0440	JU0021	OEM
J6015A	OEM	J241015	4163	JAN1N3909	OEM	JC-00047	0023	JD4000-23	OEM	JEM4	0440	JU0023	OEM
J9400	5537	J241054	0079	JAN1N3909R	OEM	JC-00049	0023	JD4000-24	OEM	JEM5	0440	JU0025	OEM
J9680	0037	J241099	0079	JAN1N3910	OEM	JC00049	0015	JD4000-25	OEM	JF20	OEM	JU0027	OEM
J9697	0037	J241100	0015	JAN1N3910R	OEM	JC00050	0015	JD4500-1	OEM	JF20D	OEM	JU0028	OEM
J20104	OEM	J241101	OEM	JAN1N3911	OEM	JC-00051	0015	JD4500-2	OEM	JF23B	OEM	JU0029	OEM
J20104A	OEM	J241102	0015	JAN1N3911R	OEM	JC-00055	0080	JD4500-3	OEM	JF71	OEM	JU0045	OEM
J20437	0015	J241104	OEM	JAN1N3912	OEM	JC-00059	0087	JD4500-4	OEM	JF72	OEM	JU0046	OEM
J20438	0137	J241105	0030	JAN1N3912R	OEM	JC-10D1	0015	JD4500-5	OEM	JF232	OEM	JU0053	OEM
J24186	0157	J241111	2866	JAN1N3913	OEM	JC74HC151ND	OEM	JD4500-6	OEM	JF421L	OEM	JU0056	OEM
J24262	0301	J241111(2SB370)	0004	JAN1N3913R	OEM	JC100	0440	JD4500-7	OEM	JF444L	OEM	JU0063	OEM
J24366	0222	J241111(2SD170)	0208	JAN1N4148-1	OEM	JC501	0076	JD4500-8	OEM	JF447L	OEM	JU0069	OEM
J24458	0016	J241142	0015	JAN1N4150	OEM	JC501(P)	0286	JD4500-9	OEM	JF-1033	0325	JU0070	OEM
J24561	0127	J241164	0004	JAN1N4153	OEM	JC501(Q)	0286	JD4500-10	OEM	JF1033	0325	JU0077	OEM
J24562	0127	J241177	0127	JAN1N4454	OEM	JC501-0	0286	JD4500-11	OEM	JF1033B	0325	JU0080	OEM
J24563	0127	J241178	0164	JAN1N4459	OEM	JC501P	0076	JD4500-12	OEM	JF1033G	0325	JU0081	OEM
J24564	0079	J241179	0636	JAN1N4531	OEM	JC501PQ	0076	JD5000-1	OEM	JF1033S	0325	JU0083	OEM
J24565	0079	J241181	0030	JAN1N4532	OEM	JC501Q	0076	JD5000-2	OEM	JGLC651483005	OEM	JU0089	OEM
J24566	0142	J241182	0133	JAN2N489A	OEM	JC501R	0076	JD5000-3	OEM	JH2101	OEM	JU0097	OEM
J24567	0123	J241183	0102	JAN2N490A	OEM	JCC1003B	OEM	JD5000-4	OEM	JH2102	OEM	JU0104	OEM
J24569	0030	J241184	0030	JAN2N491A	OEM	JC-DS16E	0015	JD5000-5	OEM	JH2103	OEM	JU0108	OEM
J24570	0015	J241185	0208	JAN2N492A	OEM	JC-KS05	0015	JD5000-6	OEM	JH2104	OEM	JU0109	OEM
J24596	0144	J241186	0157	JAN2N493A	OEM	JCMBL8088	OEM	JD6000-1	OEM	JH2105	OEM	JU0110	OEM
J24620	0050	J241188	0127	JAN2N494A	OEM	JCMBL8288	OEM	JD6000-2	OEM	JH2106	OEM	JU0123	OEM
J24621	0050	J241189	0127	JAN2N1021	OEM	JCN1	0015	JD6000-3	OEM	JJ80	0774	JU0125	OEM
J24622	0050	J241190	0004	JAN2N1559A	0160	JCN2	0015	JD6000-4	OEM	JJ650	0774	JU0126	OEM
J24623	0050	J241212	0229	JAN2N2323	OEM	JCN3	0015	JD6000-5	OEM	JK10B	OEM	JU0128	OEM
J24624	0016	J241219	0395	JAN2N2323A	OEM	JCN4	0015	JD6000-6	OEM	JK11B	OEM	JU0132	OEM
J24625	0016	J241221	0638	JAN2N2324	OEM	JCN5	0015	JD6101C054	OEM	JK30A	OEM	JU0133	OEM
J24626	0004	J241222	0167	JAN2N2324A	OEM	JCN6	0015	JD-8049HC-029	OEM	JK60A	OEM	JU0135	OEM
J24628	0123	J241236	0061	JAN2N2326	OEM	JCN7	0071	JD-8088D	OEM	JKV8	OEM	JU0163	OEM
J24630	0015	J241239	0030	JAN2N2326A	OEM	JC-SD-1X	0015	JD8251AC	OEM	JKV8F	OEM	JU0173	OEM
J24631	0636	J241241	0042	JAN2N2328	OEM	JC-SD-1Z	0015	JD8288D	OEM	JKV10	OEM	JU0178	OEM
J24632	0012	J241245	0143	JAN2N2328A	OEM	JC-SD1Z	0015	JD-BB1A	0015	JKV10F	OEM	JU0180	OEM
J24635	0144	J241250	0161	JAN2N2329	OEM	JC-SG005	0015	JD-SD1D	0015	JKV12	OEM	JU0191	OEM
J24636	0144	J241251	0079	JANDOD001	OEM	JC-V03C	0015	JDSD1D	0015	JKV12F	OEM	JU0192	OEM
J24637	0144	J241252	0423	JANDOD002	OEM	JCV1	OEM	JD-SD1Z	0015	JL-40A	0133	JU0193	OEM
J24639	0004	J241253	0037	JANDOD003	OEM	JCV-2	0015	JE90NF	0079	JL40A	0133	JU0199	OEM
J24640	0037	J241255	0086	JANDOD004	OEM	JCV2	0015	JE108	0161	JL41A	0015	JU0200	OEM
J24641	0016	J241256	0086	JANDOD005	OEM	JCV-3	0015	JE181	0161	JL41AM	0015	JU0210	OEM
J24642	0178	J241257	OEM	JANDOD007	OEM	JCV3	0015	JE210	0854	JM40	0133	JU0211	OEM
J24643	0123	J241258	0905	JANDOD008	OEM	JCV4	OEM	JE1033B	0325	JM401	0133	JU0212	OEM
J24645	0015	J241259	0037	JANDOD009	OEM	JCV5	OEM	JE4011	OEM	JNHD46505SP-1	OEM	JU0214	OEM
J24647	0015	J241260	0015	JANDOD010	OEM	JCV6	OEM	JE7001	2309	JNJ61673	0325	JU0219	OEM
J24658	0076	J241270	0789	JANDOD011	OEM	JCV7	0071	JE8050	5034	JO2015A	OEM	JU0223	OEM
J24701	0127	J241275	3019	JANDOD012	OEM	JCV10	OEM	JE8050C	OEM	JP2-22D	OEM	JU0229	OEM
J24752	0016	J310159	0004	JANDOD013	OEM	JD-00040	0015	JE8550	1597	JP8-02B	OEM	JU0239	OEM
J24753	0016	J310224	0004	JANDOD015	OEM	JD180L	0378	JE8550B	0532	JP9-2.5	OEM	JU0240	OEM
J24754	1657	J310249	0016	JANDOD016	OEM	JD200L	0378	JE8550C	0532	JP9-2.5C	OEM	JU0241	OEM
J24755	0133	J310250	0016	JANDOD017	OEM	JD2300-1	OEM	JE8550D	OEM	JP9-2.5D	OEM	JU0266	OEM
J24756	0015	J310251	0050	JANDOD019	OEM	JD2300-2	OEM	JE9011	4463	JP9-2.5E	OEM	JU0267	OEM
J24812	0111	J310252	0004	JANDOD020	OEM	JD2300-4	OEM	JE9011B	0111	JP9-2.5F	OEM	JU0280	OEM
J24813	0144	J320020	0015	JANDOD021	OEM	JD2300-6	OEM	JE9011C	0111	JP9-2.5H	OEM	JU0285	OEM
J24814	0144	J320041	0143	JANDOD022	OEM	JD2300-8	OEM	JE9011D	0079	JP9-2.5L	OEM	JU0286	OEM
J24817	0079	JA101	0006	JANDOD023	OEM	JD2300-10	OEM	JE9011E	0079	JP9-7D	OEM	JU0303	OEM
J24820	0123	JA101(P)	0006	JANDOD024	OEM	JD2300-12	OEM	JE9011F	0079	JP9-7L	OEM	JU0309	OEM
J24832	0006	JA101(Q)	0006	JANDOD025	OEM	JD2300-14	OEM	JE9011G	0016	JP9-7T	OEM	JU0318	OEM
J24833	0004	JA101-R	OEM	JANDOD028	OEM	JD2300-16	OEM	JE9011H	0079	JP9-15B	OEM	JU0323	OEM
J24834	0004	JA101P	0006	JANDOD029	OEM	JD2300-18	OEM	JE9011I	0079	JP9-15D	OEM	JU0332	OEM
J24838	0143	JA101PQ	5442	JANDOD030	OEM	JD2300-20	OEM	JE9012D	0037	JP9-18	OEM	JU0348	OEM
J24840	0030	JA101PQR	6828	JANDOD031	OEM	JD2300-22	OEM	JE9012E	0037	JP9-22B	OEM	JUD286	OEM
J24842	0151	JA101Q	0006	JANDOD032	OEM	JD2300-24	OEM	JE9012F	0037	JP9-22L	OEM	JUD318	OEM
J24843	0151	JA101R	OEM	JANDOD033	OEM	JD2300-26	OEM	JE9012G	0037	JP9-50A	OEM	JUD164	OEM
J24844	0286	JA1050	0037	JANDOD034	OEM	JD2300-28	OEM	JE9012H	0037	JP40	0085	JZ5.6	0157
J24845	0191	JA1050G	0037	JANDOD035	OEM	JD2300-30	OEM	JE9013	0079	JP5062	0050	JZ5.6A	0157
J24846	0076	JA1050W	0037	JANDOD036	OEM	JD2300-32	OEM	JE9013D	0079	JP5063	0004	JZ5.6B	0157
J24852	0144	JA1200	0079	JANDOD037	OEM	JD2300-34	OEM	JE9013E	0079	JP5064	0004	JZ5.6C	OEM
J24855	0079	JA1350	0079	JANDOD039	OEM	JD2300-36	OEM	JE9013F	0079	JP575005	0123	JZ5.6D	OEM
J24858	0030	JA1350A	0079	JANDOD0141	OEM	JD2300-38	OEM	JE9013G	0079	JP575995	0123	JZ6.2	0636
J24863	0144	JA1350B	0111	JANTX1N4150	OEM	JD2300-40	OEM	JE9013H	0079	JR05	0136	JZ6.2A	0636
J24868	0208	JA1350W	0111	JANTX1N4153	OEM	JD2300-42	OEM	JE9014	0079	JR5	0004	JZ6.2B	0636
J24869	0004	JA1350Y	0111	JANTX1N4454	OEM	JD2300-44	OEM	JE9014A	0079	JR10	0050	JZ6.2C	0636
J24870	0004	JA-1555	0973	JANTX1N4532	OEM	JD3000-1	OEM	JE9014B	0715	JR15	0004	JZ6.2D	0636
J24871	0015	JA1600G	0079	JANTX2N489A	OEM	JD3000-2	OEM	JE9014C	4898	JR30	0050	JZ6.8	0025
J24872	0170	JA7010	0191	JANTX2N490A	OEM	JD3000-4	OEM	JE9014D	0111	JR30X	0050	JZ6.8A	0025
J24874	0079	JA7072	0161	JANTX2N491A	OEM	JD3000-6	OEM	JE9015	0037	JR40	0085	JZ6.8B	0025
J24875	0111	JA-H	0079	JANTX2N492A	OEM	JD3000-8	OEM	JE9015A	0037	JR100	0050	JZ6.8C	OEM
J24877	0023	JA-KCDP	0087	JANTX2N493A	OEM	JD3000-10	OEM	JE9015B	0532	JR200	0050	JZ6.8D	OEM
J24878	0111	JA-L	0079	JANTX2N494A	OEM	JD3000-12	OEM	JE9015C	0037	JRC386D	6607	JZ7.5	0649
J24903	0127	JAM702C	0015	JANTX2N2323	OEM	JD3000-14	OEM	JE9016	0144	JRC733	OEM	JZ7.5A	0649
J24904	0144	JAN1N1184	OEM	JANTX2N2324	OEM	JD3000-16	OEM	JE9016D	0144	JRC8686	OEM	JZ7.5B	0649
J24905	0144	JAN1N1184R	OEM	JANTX2N2326	OEM	JD3000-18	OEM	JE9016E	0144	JS2737AS	OEM	JZ7.5C	OEM
J24906	0079	JAN1N1186	OEM	JANTX2N2328	OEM	JD3000-20	OEM	JE9016F	0144	JS8804AS	OEM	JZ7.5D	OEM
J24907	0079	JAN1N1186R	OEM	JANTX2N2329	OEM	JD3000-22	OEM	JE9016G	0144	JSP6009	3165	JZ8.2	0244
J24908	0126	JAN1N1188	OEM	JANTX1184/R	0015	JD3000-24	OEM	JE9016H	0144	JSP7001	0224	JZ8.2A	0244
J24909	0079	JAN1N1188R	OEM	JB-00030	0015	JD3000-26	OEM	JE9016I	0144	JSP7001B	0079	JZ8.2B	0244
J24910	0030	JAN1N1190	OEM	JB00030	0540	JD3000-28	OEM	JE9018D	0144	JSP7005	0144	JZ8.2C	3100
J24911	0143	JAN1N1190R	OEM	JB-00036	0229	JD3000-30	OEM	JE9018E	0144	JSP7006	0144	JZ8.2D	3100
J24912	0133	JAN1N1614	OEM	JB00036	0277	JD4000-1	OEM	JE9018F	0144	JSP7010	OEM	JZ9.1	0012
J24913	0143	JAN1N1615	OEM	JB-7B3B1	0302	JD4000-2	OEM	JE9018G	0144	JT-1601-40	0016	JZ9.1A	0012
J24914	0143	JAN1N1616	OEM	JB6101C-058	OEM	JD4000-3	OEM	JE9018H	0144	JT-1601-41	0917	JZ9.1B	0012

DEVICE TYPE	REPL CODE
JZ9.1C	0012
JZ9.1D	OEM
JZ10	0170
JZ10A	0170
JZ10B	0170
JZ10C	OEM
JZ10D	OEM
JZ11	0317
JZ11A	0317
JZ11B	0317
JZ11C	OEM
JZ11D	OEM
JZ12	0137
JZ12A	0137
JZ12B	0137
JZ12C	0137
JZ12D	OEM
JZ13	0365
JZ13A	0365
JZ13B	0365
JZ13C	OEM
JZ13D	OEM
JZ14	0100
JZ14A	0100
JZ14B	0100
JZ14C	OEM
JZ14D	OEM
JZ15	0002
JZ15A	0002
JZ15B	0002
JZ15C	0002
JZ15D	OEM
JZ16	0420
JZ16A	0420
JZ16B	0420
JZ16C	OEM
JZ16D	OEM
JZ17	1649
JZ17A	1649
JZ17B	1649
JZ17C	OEM
JZ17D	OEM
JZ18	0495
JZ18A	0495
JZ18B	0495
JZ18C	OEM
JZ18D	OEM
JZ19	0949
JZ19A	0949
JZ19B	0949
JZ19C	OEM
JZ19D	OEM
JZ20	0531
JZ20A	0531
JZ20B	0531
JZ20C	OEM
JZ20D	OEM
JZ22	0565
JZ22A	0565
JZ22B	0565
JZ22C	OEM
JZ22D	OEM
JZ24	0402
JZ24A	0402
JZ24B	0402
JZ24C	OEM
JZ24D	OEM
JZ25	1606
JZ25A	1606
JZ25B	1606
JZ25C	OEM
JZ25D	OEM
JZ27	0441
JZ27A	0441
JZ27B	0441
JZ27C	OEM
JZ27D	OEM
JZ30	0726
JZ30A	0726
JZ30B	0726
JZ30C	OEM
JZ30D	OEM
JZ33	0039
JZ33A	0039
JZ33B	0039
JZ33C	OEM
JZ33D	OEM
JZ36	0820
JZ36A	0820
JZ36B	0820
JZ36C	OEM
JZ36D	OEM
JZ39	0350
JZ39A	0350
JZ39B	0350
JZ39C	OEM
JZ39D	OEM
JZ43	0931
JZ43A	0931
JZ43B	0931
JZ43C	OEM
JZ43D	OEM
JZ45	0999
JZ45A	0999
JZ45B	0999
JZ45C	OEM
JZ45D	OEM
JZ47	0999
JZ47A	0999

DEVICE TYPE	REPL CODE
JZ47B	0999
JZ47C	OEM
JZ50	0502
JZ50A	0502
JZ50B	0502
JZ50C	OEM
JZ50D	OEM
JZ51	0502
JZ51A	0502
JZ51B	0502
JZ51C	OEM
JZ51D	OEM
JZ52	0502
JZ52A	0502
JZ52B	0502
JZ52C	OEM
JZ52D	OEM
JZ56	0869
JZ56A	0869
JZ56B	0869
JZ56C	OEM
JZ56D	OEM
JZ62	0783
JZ62A	0783
JZ62B	0783
JZ62C	OEM
JZ62D	OEM
JZ68	2157
JZ68A	2157
JZ68B	2157
JZ68C	OEM
JZ68D	OEM
JZ75	1189
JZ75A	1189
JZ75B	1189
JZ75C	OEM
JZ75D	OEM
JZ82	0331
JZ82A	0331
JZ82B	0331
JZ82C	OEM
JZ82D	OEM
JZ91	1311
JZ91A	1311
JZ91B	1311
JZ91C	OEM
JZ91D	OEM
JZ100	0098
JZ100A	0098
JZ100B	0098
JZ100C	OEM
JZ100D	OEM
JZ105	0149
JZ105A	0149
JZ105B	0149
JZ105C	OEM
JZ105D	OEM
JZ110	0149
JZ110A	0149
JZ110B	0149
JZ110C	OEM
JZ110D	OEM
JZ120	0186
JZ120A	0186
JZ120B	0186
JZ120C	OEM
JZ120D	OEM
JZ130	0213
JZ130A	0213
JZ130B	0213
JZ130C	OEM
JZ130D	OEM
JZ140	0245
JZ140A	0245
JZ140B	0245
JZ140C	OEM
JZ140D	OEM
JZ150	0028
JZ150A	0028
JZ150B	0028
JZ150C	OEM
JZ150D	OEM
JZ160	0255
JZ160A	0255
JZ160B	0255
JZ160C	OEM
JZ160D	OEM
JZ175	0367
JZ175A	0367
JZ175B	0367
JZ175C	OEM
JZ175D	OEM
JZ180	0367
JZ180A	0367
JZ180B	0367
JZ180C	OEM
JZ180D	OEM
JZ200A	0421
JZ200B	0421
JZ200C	OEM
JZ200D	OEM
JZ220	OEM
JZ220A	OEM
JZ220B	OEM
JZ220C	OEM
K-005	OEM
K-01	OEM
K-02	OEM
K-04	OEM

DEVICE TYPE	REPL CODE
K-06	OEM
K0201E	1292
K04774	0085
K1	OEM
K1A5	0015
K1B5	0015
K1C5	0015
K1D5	0015
K1E5	0015
K1F5	0015
K1G5	0015
K1H5	0015
K1K5	0071
K1M5	0071
K1N5	OEM
K1N4748A518	OEM
K1V5	5645
K1V6	5646
K1V10	5649
K1V11	5650
K1V12	5651
K1.3G22/1A	0296
K1.3G22-1A	0277
K1.3G22A	0015
K2	OEM
K2A5	0015
K2B5	0015
K2C5	0015
K2CDP22/1B	0787
K2CDP221B	0277
K2D5	0015
K2DP12	0015
K2E5	0015
K2F5	0015
K2G	0015
K2G5	0015
K2H5	0071
K2K5	0071
K2M5	0071
K2N5	OEM
K3	0143
K3A	OEM
K3A5	OEM
K3AST47	0917
K3B5	0015
K3C5	0015
K3D5	0015
K3E	0917
K3E5	OEM
K3F	OEM
K3F5	0015
K3G5	0015
K3H5	0071
K3K5	0071
K3M5	0071
K3N5	OEM
K4-500	0164
K4-501	0208
K4-505	0037
K4-506	0016
K4-507	0198
K4-510	0144
K4-520	0085
K4-521	0085
K4-525	0103
K4-550	0143
K4-555	0015
K4-557	0015
K4-584	0940
K4-586(DIAC)	2730
K4-586(TRIAC)	0593
K4-590	0312
K4-598	0379
K4-632	3384
K4-634	0325
K4A5	0015
K4B5	0015
K4C5	0015
K4D5	0015
K4E5	0015
K4F5	0015
K4G5	0015
K4H5	0071
K4K5	0071
K4M5	0071
K4N5	OEM
K5	OEM
K5A5	0015
K5B5	0015
K5C5	0015
K5D5	0015
K5E5	0015
K5F5	0015
K5G5	0015
K5H5	0071
K5K5	0071
K5M5	0071
K5N5	OEM
K6	0143
K6A5	0087
K6B5	0087
K6C5	OEM
K6D5	OEM
K6E5	OEM
K6F5	OEM
K6G5	0087
K6H5	0087
K6K5	2645
K6M5	2645

DEVICE TYPE	REPL CODE
K6N5	OEM
K7A5	0087
K7B5	0087
K7C5	0087
K7D5	OEM
K7E5	OEM
K7F5	OEM
K7G5	0087
K7H5	0087
K7K5	0087
K7M5	0087
K7N5	OEM
K8A5	0087
K8B5	0087
K8D5	OEM
K8E5	OEM
K8F5	OEM
K8G5	0087
K8H5	0087
K8K5	0087
K8M5	0087
K8N5	OEM
K11	0325
K12	0325
K12-0(1-GATE)	0325
K13	0325
K14-0066-1	0164
K14-0066-4	0280
K14-0066-6	0224
K14-0066-12	0079
K14-0066-13	0086
K15	3384
K15-GR(1-GATE)	0325
K15-0(1-GATE)	0325
K16HA	0325
K17(2S)	0325
K17-0	0325
K17-0(1-GATE)	0325
K17-OR	0325
K17A	0325
K17B	0325
K17BL	0325
K17GR	0325
K170	0325
K17R	0325
K17Y	0325
K19	0688
K19-14	0688
K19-BL	0688
K19-GR	0688
K19-Y	0688
K19A	0688
K19B	0688
K19BB	0688
K19BL	0688
K19FET	0688
K19GB	0688
K19GE	0688
K19GR(2S)	0688
K19H	0688
K19K	0688
K19V	0688
K19Y(2S)	0688
K20	2458
K22Y	0325
K23	3717
K23A(2S)	3717
K23A8	3717
K23A9	3717
K23A-540	3717
K23A540	3717
K23G	3717
K24	0325
K24(1-GATE)	0325
K24C	0325
K24C(1-GATE)	0325
K24D	0325
K24D(1-GATE)	0325
K24DR	0325
K24DR(1-GATE)	0325
K24E	0325
K24E(1-GATE)	0325
K24F	0325
K24F(1-GATE)	0325
K24G	0325
K24G(1-GATE)	0325
K25	0325
K25A	OEM
K25B	OEM
K25C	0325
K25D	0325
K25E	0325
K25ET	0325
K25F	0325
K25G	0325
K26	OEM
K27	OEM
K27A	OEM
K29	OEM
K29M	OEM
K30	3046
K30-0	3046
K30-0(1-GATE)	0325
K30-Y	3046
K30A	3046
K30A-1	3046
K30A-GR	3046
K30AGR	3046
K30AGR(JAPAN)	3046

DEVICE TYPE	REPL CODE
K30A-0	3046
K30A0	3046
K30A0(JAPAN)	3046
K30A-R	3046
K30ATM-0	3046
K30A-Y	3046
K30AY	3046
K30AY(JAPAN)	3046
K30D	3046
K30GR	3046
K30M	OEM
K300	3046
K30R	3046
K30Y(2S)	3046
K31	3717
K31C(2S)	0325
K32	0325
K32A(2-GATE)	0353
K32B	0325
K32B(2-GATE)	0353
K32C(2-GATE)	0353
K32D(2-GATE)	0353
K33	1280
K33D	1280
K33E(2S)	1280
K33F(2S)	1280
K33H	1280
K33M	OEM
K34	0325
K34(2S)	3155
K34A	3155
K34B	3155
K34C(2S)	3155
K34D(2S)	3155
K34E(2S)	3155
K34M	OEM
K35	0325
K35(1-GATE)	0325
K35-0	0325
K35-0(1-GATE)	0325
K35-1	0325
K35-1(1-GATE)	0325
K35-2	0325
K35-2(1-GATE)	0325
K35A	0325
K35A(1-GATE)	0325
K35BL	0325
K35BL(1-GATE)	0325
K35C	0325
K35C(1-GATE)	0325
K35G(1-GATE)	0353
K35GN	0325
K35GN(1-GATE)	0325
K35M	OEM
K35R	0325
K35R(1-GATE)	0325
K35Y	0325
K35Y(1-GATE)	0325
K37	0325
K37H	0325
K39	0212
K39(2-GATE)	0212
K39A	0212
K39B	0212
K39P	0212
K39P(2-GATE)	0212
K39Q	0212
K39Q(2-GATE)	0353
K39R(2-GATE)	0353
K39R(2GATE)	0353
K40	3354
K40(2-GATE)	0353
K40-3	3354
K40A	3354
K40A(2-GATE)	0353
K40B	3354
K40B(2-GATE)	0353
K40C	3354
K40C(2-GATE)	0353
K40C(JAPAN)	3354
K40D	3354
K40D(2-GATE)	0353
K40M(2-GATE)	0353
K41	1211
K41C	1211
K41D	1211
K41E	1211
K41E1	1211
K41F	1211
K42	0325
K42(2S)	0325
K42-CM1	0325
K42-CMI	0325
K43	1758
K43-3	1758
K43-4	1758
K43-OR	1758
K43A	1758
K43B	1758
K43C	1758
K43D	1758
K43E	1758
K43F	1758
K43GN	1758
K43H	1758
K43J	1758
K43K	1758
K43L	1758
K43M	1758

DEVICE TYPE	REPL CODE
K430R	1758
K43R	1758
K43S	1758
K43X	1758
K43Y	1758
K44	0325
K44C	0325
K44D	0325
K45	0849
K45(2-GATE)	0212
K45B	0849
K45B(09)(2-GATE)	0212
K45B(2-GATE)	0849
K46B15	OEM
K46B30	OEM
K46B60	OEM
K46B100	OEM
K46B150	OEM
K46B200	OEM
K46HZ	OEM
K47(2S)	2950
K48	3249
K49(2S)	1280
K49E2	1280
K49F(2S)	1280
K49H(2S)	1280
K49H1	1280
K49H2	1280
K49I(2S)	1280
K49M	1280
K52	0143
K54	1280
K54B	1280
K54C	1280
K55	0325
K55C	0325
K55CD	0325
K55D	0325
K55DE	0325
K55R	0325
K57-130751	OEM
K57-569745	OEM
K60	0143
K61	1211
K61GR	1211
K61Y	1211
K66	3133
K66HZ	OEM
K68	3342
K68-L	3342
K68-M	3342
K68A	3342
K68AL	3342
K68A-M	3342
K68AM	3342
K68L	3342
K68M	3342
K68Q	3342
K72	2945
K81	OEM
K81(JAPAN)	OEM
K83	0325
K84	3133
K86HZ	OEM
K87HB	3133
K87HD	3133
K104	0325
K104H	0325
K105	1758
K105(JAPAN)	1758
K105E	1758
K105F	1758
K106	0325
K107	0325
K107-1	0325
K107-2	0325
K107-4	0325
K108	OEM
K112	0196
K-112C	0196
K112C	0196
K112D	0484
K113	1155
K115J510-1	0196
K115J510-1,2	0196
K115J510-2	0196
K115J511-2	0143
K117	3133
K117-BL	3133
K117A	3133
K117BL	3133
K117GR	3133
K117J460-1	0196
K117J460-2	0196
K117Y	3133
K118	1758
K118-0	1758
K118-R	1758
K118I	0126
K118J966-1	0484
K118J966-2	1080
K118J966-3	1080
K118J966-4	1080
K118J9663	1080
K118R	1758
K119	0325
K120	1392
K121	1758
K121-2	1758

DEVICE TYPE	REPL CODE
K-122	0196
K122	0196
K122-J176-1	0774
K122-J176-2	0774
K122-J177P1	1303
K-122C	0196
K122C	0196
K122D	0484
K122J176-1	0774
K122J176-2	0774
K122J177-P1	1303
K136S(JAPAN)	OEM
K141	1155
K141A	1155
K144IE6	OEM
K147D	OEM
K152	0325
K152-3	0325
K152LA	0325
K155AG1	OEM
K155ID3	OEM
K155IE1	OEM
K155IE2	OEM
K155IE4	OEM
K155IE7	OEM
K155IE8	OEM
K155IE9	OEM
K155IM1	OEM
K155IM2	OEM
K155IM3	OEM
K155IP3	OEM
K155IR13	OEM
K155KP5	OEM
K155KP7	OEM
K155LA1	OEM
K155LA2	OEM
K155LA3	OEM
K155LA6	OEM
K155LA7	OEM
K155LA8	OEM
K155LA11	OEM
K155LA12	OEM
K155LD1	OEM
K155LD3	OEM
K155LE1	OEM
K155LE2	OEM
K155LE3	OEM
K155LE6	OEM
K155LI1	OEM
K155LI5	OEM
K155LL1	OEM
K155LN1	OEM
K155LN2	OEM
K155LN3	OEM
K155LN4	OEM
K155LP7	OEM
K155LR1	OEM
K155LR3	OEM
K155LR4	OEM
K155PR6	OEM
K155PR7	OEM
K155RE3	OEM
K155RE21	OEM
K155RE22	OEM
K155RE23	OEM
K155RE24	OEM
K155RP1	OEM
K155RU1	OEM
K155RU2	OEM
K155RU5	OEM
K155TM2	OEM
K155TM5	OEM
K155TM7	OEM
K155TV1	OEM
K161	3622
K168D	0325
K168F	0325
K170	0325
K175LE10	OEM
K176ID1	OEM
K176IE1	OEM
K176IE2	OEM
K176IE3	OEM
K176IE4	OEM
K176IE5	OEM
K176IE8	OEM
K176IR2	OEM
K176IR3	OEM
K176IR10	OEM
K176LA7	OEM
K176LA8	OEM
K176LA9	OEM
K176LE5	OEM
K176LE6	OEM
K176LI1	OEM
K176LP1	OEM
K176LP2	OEM
K176LP4	OEM
K176LP11	OEM
K176LP12	OEM
K176LS1	OEM
K176RM1	OEM
K176RU2	OEM
K176TM2	OEM
K176TV1	OEM
K191	3249
K192	0325
K194	0325
K195H	1392

DEVICE TYPE	REPL CODE
K196R	3249
K197	0325
K197YC	0325
K197YD	0325
K197YE	0325
K200	1228
K212	1392
K212AE	1392
K212E	1392
K218	OEM
K249-1608722	OEM
K301R(JAPAN)	OEM
K304	0325
K-304SRC-BC/R	OEM
K308	OEM
K314	OEM
K314A	OEM
K315	OEM
K316	OEM
K316A	OEM
K319	1465
K335	OEM
K347A	OEM
K351	OEM
K352	OEM
K359	OEM
K370	OEM
K370W	OEM
K371	OEM
K371W	OEM
K372	OEM
K372W	OEM
K376L	OEM
K377L	OEM
K381	1758
K381-E	1758
K383	5060
K384	OEM
K385	OEM
K386	OEM
K386W	OEM
K387	OEM
K387L	OEM
K391	1334
K391A	OEM
K408	OEM
K408A	OEM
K408B	OEM
K415	OEM
K417-68	0136
K510Z	OEM
K540	OEM
K656	OEM
K678-3M180100RC	OEM
K678-4M180/5	OEM
K700-1	OEM
K728	OEM
K728T	OEM
K743A	OEM
K743B	OEM
K744	OEM
K745	OEM
K746	OEM
K765	OEM
K766	OEM
K767	OEM
K768	OEM
K769	OEM
K770	OEM
K771	OEM
K774	OEM
K805	OEM
K806	OEM
K824	OEM
K882	0143
K901	OEM
K902	OEM
K903	OEM
K904	OEM
K905	OEM
K917	OEM
K918	OEM
K920	OEM
K928	OEM
K1001	OEM
K1002	OEM
K1003	OEM
K1004	OEM
K1004(CXK1004L)	OEM
K1005	OEM
K1005(SM)	OEM
K1011	OEM
K1011(IC)	OEM
K1038	OEM
K1039	OEM
K1040	OEM
K1140	OEM
K1201	OEM
K1202	OEM
K1240	OEM
K1440	OEM
K1501	OEM
K1502	OEM
K1504	0844
K1507B	OEM
K1615	0196
K1615AC	0196
K1616	0484
K1616AC	0484
K1617	1080
K1617AC	1080
K2001	0127
K2040	OEM
K2070	0144
K2071	0144
K2072	0007
K2073	0007
K2101	0418
K2101A	0418
K2102	0418
K2102A	0418
K2103	0418
K2103A	0418
K2104	0418
K2104A	0418
K2105	0418
K2105A	0418
K2106	0418
K2106A	0418
K2107	0418
K2107A	0418
K2108	0418
K2108A	0418
K2109	0144
K2109A	0259
K2110	0144
K2110A	0259
K2111	0144
K2111A	0259
K2112	0144
K2112A	0259
K2113	0144
K2113A	0259
K2114	0144
K2114A	0259
K2115	0007
K2115A	0259
K2116	0144
K2116A	0259
K2117	0144
K2117A	0259
K2118	0144
K2118A	0259
K2119	0127
K2119A	0259
K2120	0127
K2120A	0259
K2121	0127
K2121A	0259
K2122	0127
K2122A	0259
K2123	0127
K2123A	0259
K2124	0127
K2124A	0259
K2125	0127
K2125A	0259
K2126	0127
K2126A	0259
K2127	0127
K2127A	0259
K2501	0127
K2502	0127
K2503	0127
K2507	0418
K2509	0127
K2523	0418
K2524	0418
K2525	0259
K2601	0418
K2601A	0418
K2601C	0127
K2602	0418
K2602A	0418
K2602C	0127
K2603	0418
K2603A	0418
K2603C	0127
K2604	0127
K2604A	0259
K2604C	0127
K2607	0418
K2607A	0418
K2608	0418
K2608A	0418
K2609	0418
K2609A	0418
K2610	0418
K2610A	0418
K2611	0418
K2611A	0418
K2612	0418
K2612A	0418
K2613	0418
K2613A	0418
K2614	0418
K2614A	0418
K2615	0127
K2615A	0259
K2616	0127
K2616A	0259
K2857C	0007
K2857P	0007
K3004	OEM
K3005	OEM
K3006	OEM
K3007	OEM
K3010P	5051
K3020P	5065
K3040	OEM
K3069	OEM
K3073	OEM
K3076	OEM
K3077	OEM
K3078-6975	OEM
K3079	OEM
K3080	OEM
K3081	OEM
K3091	OEM
K3094	OEM
K3097	OEM
K3098	OEM
K3111	OEM
K3118	OEM
K3217H	OEM
K3270	OEM
K3271	OEM
K3276H	OEM
K3282	OEM
K3283	OEM
K3284	OEM
K3286	OEM
K3287	OEM
K3342W92K	OEM
K3382	OEM
K3383	OEM
K3384	OEM
K3683C	0007
K3683P	0007
K3880C	0007
K3880P	0144
K4002	0127
K4040	OEM
K4540	OEM
K4560	1323
K4586	0593
K4590	OEM
K4634	0325
K5001	0144
K5002	0144
K5003	0144
K5010	OEM
K5011	OEM
K5012	0330
K5020	OEM
K5040	OEM
K5100	OEM
K5150	OEM
K5151	OEM
K5200	OEM
K5250	OEM
K5256	OEM
K5258	OEM
K5500	OEM
K5502	OEM
K5503C	0259
K5504	OEM
K5550	OEM
K5552	OEM
K5554	OEM
K5560-5	OEM
K5560-6	OEM
K6001N	0861
K6002	0861
K6002N	0861
K6003	0861
K6009	0861
K6012	0861
K6022	0861
K6040	OEM
K6100	OEM
K6150	OEM
K6200	OEM
K6202	OEM
K6203	OEM
K6250	OEM
K6251	OEM
K6252	OEM
K6253	OEM
K6500	OEM
K6502	OEM
K6550	OEM
K6552	OEM
K7100	OEM
K7150	OEM
K7200	OEM
K7250	OEM
K8010	OEM
K8013P	OEM
K8020	OEM
K8030	OEM
K8031P	OEM
K8040	OEM
K-8073	OEM
K8100	OEM
K8101	OEM
K8102	OEM
K8103	OEM
K8105	OEM
K8110	OEM
K8111	OEM
K8112	OEM
K8113	OEM
K8114	OEM
K8120	OEM
K8121	OEM
K8130	OEM
K8138E	OEM
K8139E	1234
K8140	OEM
K8141	OEM
K8150	OEM
K8151	OEM
K8152	OEM
K8153	OEM
K8154	OEM
K8155	OEM
K8160	OEM
K8161	OEM
K8162	OEM
K8163	OEM
K8164	OEM
K8165	OEM
K8170	OEM
K8171	OEM
K8180	OEM
K8181	OEM
K8190	OEM
K8191	OEM
K8324	OEM
K8700	OEM
K8701	OEM
K8702	OEM
K8703	OEM
K8710	OEM
K8711	OEM
K8712	OEM
K8713	OEM
K8713-16	OEM
K8720	OEM
K8730	OEM
K8740	OEM
K8900	OEM
K8920	OEM
K8940	OEM
K8941	OEM
K8950	OEM
K9000	OEM
K9001	OEM
K9002	OEM
K9010	OEM
K9011	OEM
K9012	OEM
K9014	OEM
K9015	OEM
K9020	OEM
K9021	OEM
K9022	OEM
K9682	0037
K18046	0946
K26044	0964
K26174	2887
K53342W98K	OEM
K75508-1	0050
K76026	4639
K1038345-22A	0196
K1038345-22C	1080
K1038345-22E	0484
K8532799	0015
K8533058-1	0015
K8533137	0196
K8534992P1	0196
K8534992P2	0196
KA1Z25	0086
KA33V	1329
KA431CD	5294
KA507A	OEM
KA507V	OEM
KA509A	OEM
KA509B	OEM
KA509V	OEM
KA510A	OEM
KA510B	OEM
KA510D	OEM
KA510E	OEM
KA510G	OEM
KA510V	OEM
KA513A	OEM
KA513B	OEM
KA562-Y	0006
KA562Y	0006
KA602A	OEM
KA602B	OEM
KA602D	OEM
KA602G	OEM
KA602V	OEM
KA651	OEM
KA701A	OEM
KA1002R	OEM
KA1002RF	OEM
KA1003RF	OEM
KA1004R	OEM
KA1004RF	OEM
KA1006R	OEM
KA1006RF	OEM
KA1008R	OEM
KA1008RF	OEM
KA1010R	OEM
KA1010RF	OEM
KA1015-0	0006
KA1330	OEM
KA1390	OEM
KA1450	OEM
KA1510	OEM
KA1570	OEM
KA1630	OEM
KA1690	OEM
KA1750	OEM
KA2101	0167
KA2107	OEM
KA2131	0976
KA2153	0663
KA2155	0557
KA2181	2026
KA2192	OEM
KA2201	6445
KA2201N	OEM
KA2221	OEM
KA2224	2595
KA2247	OEM
KA2249	OEM
KA2261	1260
KA2281	OEM
KA2284	OEM
KA2911	0912
KA2914A	4109
KA2915	OEM
KA2918	OEM
KA2919	0539
KA4559	0325
KA7809C	OEM
KA7812	0334
KA7812C	OEM
KA8301	OEM
KA22062	OEM
KAP1151	OEM
KAP1152	OEM
KAQ6502	3005
KB9CVV	0016
KB102	0133
KB-162	0139
KB162	0133
KB-162C5	0139
KB-162C5A	0139
KB-162C5B	OEM
KB162N	0139
KB162W	0139
KB-165	0139
KB165	0139
KB167	1925
KB-169	0139
KB169	0139
KB-182	0015
KB202	0289
KB209	OEM
KB-262	0120
KB262	0120
KB-262C4	OEM
KB-262M	OEM
KB-265	0139
KB265	0139
KB265A	0133
KB-265C4	OEM
KB-269	0120
KB269	0139
KB296	3230
KB-362	OEM
KB-362M	OEM
KB-365	0139
KB365	OEM
KB-369	OEM
KB369	OEM
KB-462	OEM
KB462F	3442
KB-465	OEM
KB-469	OEM
KB3600-PRO	OEM
KB4400	OEM
KB4409	0520
KB4419	OEM
KB4420B	OEM
KB4424	OEM
KB4431	OEM
KB4432	OEM
KB4433	OEM
KB4436	OEM
KB4437	OEM
KB4438	OEM
KB4441	OEM
KB4446	OEM
KB4448	OEM
KB4454	OEM
KB8339	0016
KB8416	0079
KBCTD15-005	OEM
KBCTD15-02	OEM
KBCTD15-04	OEM
KBCTD15-06	OEM
KBCTD15-08	OEM
KBCTD15-10	OEM
KBCTP15-005	OEM
KBCTP15-02	OEM
KBCTP15-04	OEM
KBCTP15-06	OEM
KBCTP15-08	OEM
KBCTP15-10	OEM
KBF005	4829
KBF-02	0729
KBF02	4829
KBF04	4833
KBF06	4833
KBF08	4837
KBF10	4837
KBH2502	2363
KBH2504	2369
KBH2506	2370
KBL005	0729
KBL01	0729
KBL-02	OEM
KBL02	3023
KBL04	0737
KBL05	OEM
KBL06	3035
KBL08	0742
KBL10	3050
KBP-005	0277
KBP005	0277
KBP01	0277
KBP-02	OEM
KBP02	0289
KBP04	0296
KBP06	0302
KBP08	0250
KBP-10	0250
KBP10	0250
KBPC005	0323
KBPC02	1414
KBPC02A	OEM
KBPC04	0473
KBPC04A	OEM
KBPC06	0446
KBPC06A	OEM
KBPC08	1422
KBPC08A	OEM
KBPC10	2443
KBPC10-005	5483
KBPC10-01	2363
KBPC10-02	2363
KBPC10-04	2369
KBPC10-06	2370
KBPC10-08	2372
KBPC10-10	2373
KBPC12-005	2363
KBPC12-02	2363
KBPC12-04	2369
KBPC12-06	2370
KBPC12-08	2372
KBPC12-10	2373
KBPC15-005	2363
KBPC15-02	2363
KBPC15-04	2369
KBPC15-06	2370
KBPC15-08	2372
KBPC15-10	2373
KBPC25-005	2363
KBPC25-02	2363
KBPC25-04	2369
KBPC25-06	2370
KBPC25-08	2372
KBPC25-10	2373
KBPC35-005	1643
KBPC35-02	1643
KBPC35-04	1673
KBPC35-06	1673
KBPC35-08	1589
KBPC35-10	1589
KBPC101	0323
KBPC102	1414
KBPC102A	OEM
KBPC104	0473
KBPC104A	OEM
KBPC106	0446
KBPC106A	OEM
KBPC108	1422
KBPC108A	OEM
KBPC110	2443
KBPC601	0323
KBPC602	1414
KBPC604	0473
KBPC606	0446
KBPC608	1422
KBPC610	2443
KBPC-801	0323
KBPC801	0323
KBPC802	1414
KBPC802T	2363
KBPC804	0473
KBPC804T	2369
KBPC806	0446
KBPC806T	2370
KBPC808	1422
KBPC808T	2372
KBPC810	2443
KBPC810T	2373
KBPC1005	0323
KBPC1201	2363
KBPC1202	2363
KBPC1204	2369
KBPC1206	2370
KBPC1208	2372
KBPC1210	2373
KBPC1501	2363
KBPC1502	2363
KBPC1504	2369
KBPC1506	2370
KBPC1508	2372
KBPC1510	2373
KBPC2501	2363
KBPC2502	2363
KBPC2502W	OEM
KBPC2504	2369
KBPC2506	2370
KBPC2508	2372
KBPC2510	2373
KBPC3501	1643
KBPC3502	1643
KBPC3504	1673
KBPC3506	1673
KBPC3508	1589
KBPC3510	1589
KBPC6005	0323
KBPC8005	0323
KBPC8005T	2363
KBPC12005	2363
KBPC15005	2363
KBPC25005	2363
KBPC35005	1643
KBPS-005	1335
KBPS-02	OEM
KBPS-04	OEM
KBPS-06	OEM
KBPS-08	OEM
KBPS-10	OEM
KBU4A	1042
KBU4B	1042
KBU4D	1042
KBU4G	1045
KBU4J	1045
KBU4K	5150
KBU4M	5150
KBU6A	1042
KBU6B	1042
KBU6D	1042
KBU6G	1045
KBU6J	1045
KBU6K	1047
KBU6M	1047
KBU8A	OEM
KBU8B	OEM
KBU8D	OEM
KBU8G	OEM
KBU8J	OEM
KBU8K	OEM
KBU8M	OEM
KC0-8CP	0015
KC0-8CP11/1+12/1	0015
KC06E11/8	0015
KC08C11/8	0015
KC08C11/10	0015
KC08C21/5	0015
KC08C22/19	0015
KC08C215	0015
KC08C221	0015
KC08C1110	0015
KC08C2219	0277
KC0911/8	0015
KC06911	0015
KC0.8C22/1	0277
KC0.8CP	0015
KC0.8CP11	0015
KC0.8CP11/1	0015
KC0.8CP11/1+121Y	0015
KC0.8CP12/1	0015
KC1.3C	0015
KC1.3C3X11/1	0015
KC1.3C22/1	0015
KC1.3G	0015
KC1.3G12/1X2	0015
KC1.3G22/1A	0015
KC2AP22/1B	0277
KC2AP221B	0277
KC2AP22116	0787
KC2BP22/1B	0015
KC2D22/1	0277
KC2D22/1A	0787
KC2D221	0015
KC2DP	0015
KC2DP11/1	0015
KC2DP12/1	0015
KC2DP12/1N	0015
KC2DP12/2	0015
KC2DP22/1	0015
KC2DP22/1B	0277
KC2DP22/1C	0787
KC2DP22/16	0787
KC2DP121N	0015
KC2DP122	0015
KC2DP221	0015
KC2DP221B	0015
KC2DP221C	0277
KC2G	0133
KC2G11/1	0015
KC2G11/1+12/1	0015
KC2G11/1&12/1	0015
KC2G12/1	0015
KC13C221	0015
KC112A	0196
KC383C	4641
KC388A	0286
KC580C	OEM
KC580C1	1055
KC581C	4642
KC582C	4637
KC583C	4641
KC850C1	0043
KC1322/1	0015
KC1815-GR	0286
KC1815-Y	0286
KC1815Y	0286
KC1959-O	0286
KCD-80P11/1+12/1	0102
KCDP12-1	0087
KC06E11/8	0540

If replacement code is OEM, contact original manufacturer for replacement.

DEVICE TYPE	REPL CODE
KC08C11/8	0540
KC08C21/5	0540
KC08C22/19	0540
KCO-8CP11/1-12/1	0015
KC0911/8	0540
KC0.8C22/1	0787
KC0.8CP11	0015
KC0.8CP11/1+12/1	0015
KCS1674R	OEM
KD11HA	OEM
KD11HF	OEM
KD27	0143
KD102A	OEM
KD102B	OEM
KD103A	0133
KD103B	0133
KD104A	OEM
KD202A	OEM
KD202B	OEM
KD202D	OEM
KD202E	OEM
KD202G	OEM
KD202I	OEM
KD202K	OEM
KD202L	OEM
KD202M	OEM
KD202N	OEM
KD202R	OEM
KD202S	OEM
KD202SZ	OEM
KD202V	OEM
KD204A	OEM
KD204B	OEM
KD204V	OEM
KD205A	OEM
KD205B	OEM
KD205D	OEM
KD205E	OEM
KD205G	OEM
KD205V	OEM
KD206A	OEM
KD206B	OEM
KD206V	OEM
KD208A	OEM
KD209A	OEM
KD209B	OEM
KD209V	OEM
KD229D	OEM
KD229E	OEM
KD300A	0133
KD407A	OEM
KD409A	OEM
KD503A	OEM
KD503B	OEM
KD514A	OEM
KD518A	OEM
KD519A	OEM
KD520A	OEM
KD521A	OEM
KD521V	OEM
KD522A	OEM
KD522B	OEM
KD1925P	OEM
KD1926P	OEM
KD2101	0004
KD2102	0016
KD2103	0015
KD2104	0015
KD2114	1854
KD2118	0086
KD2119	0144
KD2120	0126
KD2121	0847
KD2124	0038
KD2130	0212
KD2131	6422
KD2501	0170
KD2503	0636
KD2504	0012
KD2505	0137
KD2541	0418
KD4002	0418
KD4025	0418
KD5000	0144
KD5051	OEM
KD5052	OEM
KD5053	OEM
KD5054	OEM
KD5055	OEM
KD6000-1	OEM
KD6000-2	OEM
KD6000-3	OEM
KD6000-4	OEM
KD6000-5	OEM
KD6000-6	OEM
KD6311	0528
KDA4002	OEM
KDC-80P11/1912/1	0102
KDD-0013	0143
KDD0015	0057
KDD0032	0080
KDD0041	0133
KDF11-AA	OEM
KDF11-HK	OEM
KDF11-LK	OEM
KDS288Y	OEM
KDS362	0392
KDS362N	0728
KDS1553	0133
KDS1554	0133
KDS1555	0133
KDS1555-T	0133
KDS2236	OEM
KDT-16	0540
KE-262	0015
KE3586	OEM
KE3684	2950
KE3685	2950
KE3687	2950
KE3970	3130
KE3971	OEM
KE3972	OEM
KE4091	3130
KE4092	OEM
KE4093	OEM
KE4220	2950
KE4221	OEM
KE4222	OEM
KE4223	3249
KE4224	3249
KE4391	3130
KE4392	OEM
KE4393	OEM
KE4416	0325
KE4856	3130
KE4857	OEM
KE4858	OEM
KE4859	3130
KE4860	OEM
KE4861	OEM
KE5103	2950
KE5104	3249
KE5105	0325
KE5460	OEM
KEY-C00CV	OEM
KEY-C00SV	OEM
KEY-C00SV-F	OEM
KF-104	OEM
KF104	OEM
KF3509	0493
KFLD20	OEM
KFLD20A	OEM
KFLD30	OEM
KFLD30A	OEM
KFLD40	OEM
KFLD40A	OEM
KFLD50	OEM
KFLD50A	OEM
KFLD60	OEM
KFLD60A	OEM
KG100D	OEM
KG100F	OEM
KG100G	OEM
KG100H	OEM
KGB0430SX	OEM
KGB0431SX	OEM
KGB0440SX	OEM
KGB0521SX	OEM
KGB0522SX	OEM
KGB0523SX	OEM
KGB0524SX	OEM
KGB0611SX	OEM
KGB0612SX	OEM
KGB0613SX	OEM
KGB0614SX	OEM
KGB0711SX	OEM
KGB0712SX	OEM
KGB0713SX	OEM
KGB0714SX	OEM
KGD1420SX	OEM
KGD1421SX	OEM
KGD1422SX	OEM
KGD1423SX	OEM
KGD1424SX	OEM
KGD1430SX	OEM
KGD1431SX	OEM
KGD1440SX	OEM
KGD1521SX	OEM
KGD1522SX	OEM
KGD1523SX	OEM
KGD1524SX	OEM
KGD1611SX	OEM
KGD1612SX	OEM
KGD1613SX	OEM
KGD1614SX	OEM
KGD1711SX	OEM
KGD1712SX	OEM
KGD1714SX	OEM
KGE41007	0015
KGE41013	4578
KGE41054	0380
KGE41055	0151
KGE41061	2046
KGE41414	0018
KGE41959	0143
KGE46109	0133
KGE46146	0076
KGE46338	0076
KGE46441	1018
KGE46442	1540
KGE46465	0133
KGS1000	0004
KGS1001	0164
KGS1002	OEM
KGS1003	OEM
KGS1004	OEM
KGS1005	OEM
KH3207	OEM
KH5015	OEM
KH5196	OEM
KH5197	OEM
KH5198	OEM
KH5199	OEM
KHA109	OEM
KHA125	OEM
KHA136	OEM
KHP3	OEM
KHP4	OEM
KHP5	OEM
KHP6	OEM
KHP7	OEM
KHP8	OEM
KHP9	OEM
KHP10	OEM
KHP15	OEM
KHP20	OEM
KHP25	OEM
KHP30	OEM
KHP35	OEM
KHP40	OEM
KHP45	OEM
KHP50	OEM
KIA78L005AP	5105
KIA78L05AP	5105
KIA78L05BP	5105
KIA574P(ZENER)	OEM
KIA4558P	OEM
KIA6930P	3772
KIA7033	OEM
KIA7033P	OEM
KIA7042	OEM
KIA7042P	OEM
KIA7137P	OEM
KIA7205	OEM
KIA7205AP	1050
KIA7205P	1050
KIA7310P	1540
KIA7313AP	0362
KIA7313AP(B/C)	OEM
KIA7343P	OEM
KIA7358P	OEM
KIA7366P	OEM
KIA7373AP	0362
KIA7630	3772
KIA7630P	3772
KIA7640AP	3725
KIA7668BP	OEM
KIA7812P	0334
KIA65393S	OEM
KIA75558P	OEM
KIA75902P	OEM
KIA78005AP	OEM
KIA78006P	OEM
KIA78012AP	3485
KIA78012AP(KEC)	0334
KIA78012P	OEM
KIU12AM	OEM
KIU12BM	OEM
KJ6001	OEM
KJ6007N	OEM
KJA733C-G-T	OEM
KJA733C-L-T	OEM
KL5	OEM
KL10	OEM
KL15	OEM
KL20	OEM
KL25	OEM
KL30	OEM
KL35	OEM
KL40	OEM
KL45	OEM
KL50	OEM
KL101A	OEM
KL101B	OEM
KL101V	OEM
KL105A	OEM
KL105B	OEM
KL105V	OEM
KLG114	OEM
KLG124E	OEM
KLG205E	OEM
KLG206E	OEM
KLG-208E	OEM
KLH704	0079
KLH1422	0079
KLH4567	0015
KLH4577	0015
KLH4746	0037
KLH4747	3538
KLH4763	0133
KLH4781	0854
KLH4792	0127
KLH4793	0701
KLH5489	0654
KLK1422	OEM
KLN1068	OEM
KLR5	OEM
KLR10	OEM
KLR15	OEM
KLR20	OEM
KLR25	OEM
KLR114	OEM
KLR114E	1615
KLR114F	1615
KLR124E	OEM
KLR-130L	OEM
KLR162E	OEM
KLR205E	OEM
KLR206E	OEM
KLR208E	OEM
KLR226E	OEM
KLR226ERD	OEM
KLY124E	OEM
KM41C464P-10	OEM
KM155KP5	OEM
KM155KP7	OEM
KM858C	2253
KM858C/11	OEM
KM904	OEM
KM905	OEM
KM917E	0079
KM917F	0076
KM917G	0079
KM918	OEM
KM918E	0144
KM918F	0144
KM934	OEM
KM935	OEM
KM3502P	OEM
KM4164-15	2357
KM4164A-15	OEM
KM5624	4639
KM6264BL-10L	OEM
KM7245BP	OEM
KM7245P	OEM
KM41256-12	OEM
KM41256-15	OEM
KM41256AP-12	5740
KM41256AP-15	1472
KN2	0196
KN1201	OEM
KN1201A	OEM
KN1203	OEM
KN1204	OEM
KN1301	OEM
KN1401	OEM
KN2203	OEM
KN2207	OEM
KN2209	OEM
KN2222	0079
KN2907	0037
KNL8251LA	OEM
KO20-01	OEM
KO20-02	OEM
KO20-03	OEM
KO25-51	OEM
KO25-52	OEM
KO25-53	OEM
KO25-181	OEM
KO4774	0085
KOS25201-1	0143
KOS25201-2	0143
KOS25211-2	0025
KOS25221-1	0030
KOS25642-10	0136
KOS25642-20	0136
KOS25642-40	0136
KOS25643-10	0136
KOS25643-15	0136
KOS25651-20	0004
KOS25657-53	0004
KOS25661-20	0085
KOS25671-20	0216
KOS25671-21	0216
KOS25671-23	0216
KP20-01	OEM
KP20-02	OEM
KP20-03	OEM
KP202D	OEM
KP202E	OEM
KP302A	OEM
KP302B	OEM
KP302G	OEM
KP302V	OEM
KP303A	OEM
KP303B	OEM
KP303D	OEM
KP303E	OEM
KP303G	OEM
KP303I	OEM
KP303SZ	OEM
KP303V	OEM
KP307A	OEM
KP307B	OEM
KP307D	OEM
KP307E	OEM
KP307G	OEM
KP307SZ	OEM
KP307V	OEM
KP308A	OEM
KP308B	OEM
KP308D	OEM
KP308G	OEM
KP308V	OEM
KP312A	OEM
KP312B	OEM
KP313A	OEM
KP313B	OEM
KP313V	OEM
KP314A	OEM
KP350A	OEM
KP350B	OEM
KP350V	OEM
KP901A	OEM
KP901B	OEM
KP902A	OEM
KP902B	OEM
KP902V	OEM
KP903A	OEM
KP903B	OEM
KP903V	OEM
KP904A	OEM
KPB02	0289
KPB04	0296
KPB08	0250
KPB10	0250
KPC107	OEM
KPC107A	OEM
KPC107B	OEM
KPC107C	OEM
KPC107D	OEM
KPC112	OEM
KPC112A	OEM
KPC112B	OEM
KPC112C	OEM
KPC112D	OEM
KPC113	OEM
KPC113A	OEM
KPC113B	OEM
KPC113C	OEM
KPC113D	OEM
KPC114	OEM
KPC114A	OEM
KPC114B	OEM
KPC114C	OEM
KPC114D	OEM
KPC115	OEM
KPC115A	OEM
KPC115B	OEM
KPC115C	OEM
KPC115D	OEM
KPC116	OEM
KPC116A	OEM
KPC116B	OEM
KPC116C	OEM
KPC116D	OEM
KPC117	OEM
KPC117A	OEM
KPC117B	OEM
KPC117C	OEM
KPC117D	OEM
KPC122	OEM
KPC122A	OEM
KPC122B	OEM
KPC122C	OEM
KPC122D	OEM
KPC123	OEM
KPC123A	OEM
KPC123B	OEM
KPC123C	OEM
KPC123D	OEM
KPC124	OEM
KPC124A	OEM
KPC124B	OEM
KPC124C	OEM
KPC124D	OEM
KPC125	OEM
KPC125A	OEM
KPC125B	OEM
KPC125C	OEM
KPC125D	OEM
KPC126	OEM
KPC126A	OEM
KPC126B	OEM
KPC126C	OEM
KPC126D	OEM
KPC127	OEM
KPC127A	OEM
KPC127B	OEM
KPC127C	OEM
KPC127D	OEM
KPC128	OEM
KPC128A	OEM
KPC128B	OEM
KPC128C	OEM
KPC128D	OEM
KPC129	OEM
KPC129A	OEM
KPC129B	OEM
KPC129C	OEM
KPC129D	OEM
KPC130	OEM
KPC130A	OEM
KPC130B	OEM
KPC130C	OEM
KPC130D	OEM
KPC132	OEM
KPC132A	OEM
KPC132B	OEM
KPC132C	OEM
KPC132D	OEM
KPC133	OEM
KPC133A	OEM
KPC133B	OEM
KPC133C	OEM
KPC133D	OEM
KPC134	OEM
KPC134A	OEM
KPC134B	OEM
KPC134C	OEM
KPC134D	OEM
KPC135	OEM
KPC135A	OEM
KPC135B	OEM
KPC135C	OEM
KPC135D	OEM
KPC136	OEM
KPC136A	OEM
KPC136B	OEM
KPC136C	OEM
KPC136D	OEM
KPC137	OEM
KPC137A	OEM
KPC137B	OEM
KPC137C	OEM
KPC137D	OEM
KPC138	OEM
KPC138A	OEM
KPC138B	OEM
KPC138C	OEM
KPC138D	OEM
KPC139	OEM
KPC139A	OEM
KPC139B	OEM
KPC139C	OEM
KPC139D	OEM
KPC140	OEM
KPC140A	OEM
KPC140B	OEM
KPC140C	OEM
KPC140D	OEM
KPC141	OEM
KPC141A	OEM
KPC141B	OEM
KPC141C	OEM
KPC141D	OEM
KPD1404	0270
KPD2502M	OEM
KPG107	OEM
KPG107A	OEM
KPG107B	OEM
KPG110	OEM
KPG110A	OEM
KPG110B	OEM
KPG115	OEM
KPG115A	OEM
KPG115B	OEM
KPG122	OEM
KPG122A	OEM
KPG122B	OEM
KPG133	OEM
KPG133A	OEM
KPG133B	OEM
KPG139	OEM
KPG139A	OEM
KPG139B	OEM
KPG147	OEM
KPG147A	OEM
KPG147B	OEM
KPG207	OEM
KPG207A	OEM
KPG207B	OEM
KPG210	OEM
KPG210A	OEM
KPG210B	OEM
KPG215	OEM
KPG215A	OEM
KPG215B	OEM
KPG222	OEM
KPG222A	OEM
KPG222B	OEM
KPG233	OEM
KPG233A	OEM
KPG233B	OEM
KPG239	OEM
KPG239A	OEM
KPG239B	OEM
KPG247	OEM
KPG247A	OEM
KPG247B	OEM
KPG307	OEM
KPG307A	OEM
KPG307B	OEM
KPG310	OEM
KPG310A	OEM
KPG310B	OEM
KPG315	OEM
KPG315A	OEM
KPG315B	OEM
KPG322	OEM
KPG322A	OEM
KPG322B	OEM
KPG333	OEM
KPG333A	OEM
KPG333B	OEM
KPG339	OEM
KPG339A	OEM
KPG339B	OEM
KPG6682	0016
KPS202A	OEM
KPS202B	OEM
KPS202G	OEM
KPS202V	OEM
KQ20-01	OEM
KQ20-02	OEM
KQ20-03	OEM
KR50	OEM
KR51	OEM
KR52	OEM
KR53	OEM
KR54	OEM
KR55	OEM
KR56	OEM
KR57	OEM
KR58	OEM
KR59	OEM
KR60	OEM
KR123AA	OEM
KR123BA	OEM
KR123CA	OEM
KR123DA	OEM
KR123EA	OEM
KR123FA	OEM
KR123GA	OEM
KR123HA	OEM
KR123KA	OEM
KR123MA	OEM
KR123PA	OEM
KR123RA	OEM
KR123TA	OEM
KR123VA	OEM
KR-162	0124
KR162AA	OEM
KR162BA	OEM
KR162CA	OEM
KR162DA	OEM
KR162EA	OEM
KR162FA	OEM
KR162GA	OEM
KR162HA	OEM
KR162KA	OEM
KR162MA	OEM
KR162PA	OEM
KR162RA	OEM
KR162TA	OEM
KR162VA	OEM
KR252AA	OEM
KR252BA	OEM
KR252CA	OEM
KR252DA	OEM
KR252EA	OEM
KR252FA	OEM
KR252GA	OEM
KR252HA	OEM
KR252KA	OEM
KR252MA	OEM
KR252PA	OEM
KR252RA	OEM
KR252TA	OEM
KR252VA	OEM
KR302AA	OEM
KR302BA	OEM
KR302CA	OEM
KR302DA	OEM
KR302EA	OEM
KR302FA	OEM
KR302GA	OEM
KR302HA	OEM
KR302KA	OEM
KR302MA	OEM
KR302PA	OEM
KR302RA	OEM
KR302TA	OEM
KR302VA	OEM
KR502AA	OEM
KR502BA	OEM
KR502CA	OEM
KR502DA	OEM
KR502EA	OEM
KR502FA	OEM
KR502GA	OEM
KR502HA	OEM
KR502KA	OEM
KR502MA	OEM
KR502PA	OEM
KR502RA	OEM
KR502TA	OEM
KR502VA	OEM
KR602AA	OEM
KR602BA	OEM
KR602CA	OEM
KR602DA	OEM
KR602EA	OEM
KR602FA	OEM
KR602GA	OEM
KR602HA	OEM
KR602KA	OEM
KR602MA	OEM
KR602PA	OEM
KR602RA	OEM
KR602TA	OEM
KR602VA	OEM
KR866	OEM
KR867	OEM
KR2376	OEM
KR-2376-XX	OEM
KR2376-XX	OEM
KR3600	OEM
KR-3600-PRO	OEM
KR3600-PRO	OEM
KR-3600-ST	OEM
KR-3600-STD	OEM
KR-3600-XX	OEM
KR3600PRO	OEM
KR3600ST	OEM
KR3600STD	OEM
KR6003	0440
KR8417	0079
KRA106M	OEM
KRA1080	OEM
KRA1081	OEM
KRA2206	OEM
KRC105M	OEM

DEVICE TYPE	REPL CODE
KRC111M	OEM
KRC1205	OEM
KRC1211	OEM
KR-Q0001	0004
KR-Q0002	0004
KR-Q0004	0004
KR-Q0005	0143
KR-Q1010	0004
KR-Q1011	0004
KR-Q1012	0004
KR-Q1013	0016
KS-05	0015
KS05	0015
KS030A	1713
KS033A	0292
KS033B	0299
KS036A	0188
KS039A	0456
KS039B	0036
KS043A	0533
KS047A	0451
KS047B	0140
KS051A	0162
KS056A	0157
KS056B	0157
KS062A	0636
KS9-20B	OEM
KS9-20D	OEM
KS9-20G	OEM
KS30A	0188
KS30AF	0188
KS30B	0188
KS30BF	0188
KS31A	0188
KS31AF	OEM
KS32A	0188
KS32AF	0188
KS32B	0188
KS32BF	0188
KS33A	OEM
KS33AF	OEM
KS34A	0162
KS34AF	0162
KS34B	0162
KS34BF	0162
KS35A	0162
KS35A2	OEM
KS35AF	0162
KS36A	0157
KS36AF	0157
KS36B	0157
KS36BF	0157
KS37A	0471
KS37AF	0471
KS38A	OEM
KS38AF	OEM
KS38B	OEM
KS38BF	OEM
KS39A	OEM
KS39AF	OEM
KS40A	OEM
KS40AF	OEM
KS40B	OEM
KS40BF	OEM
KS41A	0057
KS41AF	0057
KS42A	0170
KS42AF	0170
KS42B	0170
KS42BF	0170
KS43A	OEM
KS43AF	OEM
KS44A	0052
KS44AF	0052
KS44B	0052
KS44BF	0052
KS45AF	OEM
KS46	0002
KS46AF	0686
KS46BF	0686
KS47AF	OEM
KS48AF	OEM
KS48BF	OEM
KS74HCTLS245	OEM
KS77	OEM
KS77B	OEM
KS78	OEM
KS79B	OEM
KS100A	0170
KS100B	0170
KS102AA	0708
KS102BA	0580
KS102CA	0580
KS102DA	0580
KS102EA	1000
KS102FA	1000
KS102GA	1000
KS102HA	1000
KS102KA	2081
KS102MA	2081
KS102PA	2088
KS102RA	2088
KS102TA	0612
KS102VA	0612
KS120A	0137
KS120B	0137
KS123AA	0084
KS123BA	0097
KS123CA	0097
KS123DA	0097
KS123EA	0109
KS123FA	0109
KS123GA	0109
KS123HA	0109
KS123KA	0122
KS123MA	0122
KS123PA	0131
KS123RA	0131
KS123TA	0145
KS123VA	0145
KS150A	0002
KS150B	0002
KS162AA	0708
KS162BA	0580
KS162CA	0580
KS162DA	0580
KS162EA	1000
KS162FA	1000
KS162GA	1000
KS162HA	1000
KS162KA	2081
KS162MA	2081
KS162PA	2088
KS162RA	2088
KS162TA	0612
KS162VA	0612
KS196A	OEM
KS196B	OEM
KS196G	OEM
KS302AA	0708
KS302BA	0580
KS302CA	0580
KS302DA	0580
KS302EA	1000
KS302FA	1000
KS302GA	1000
KS302HA	1000
KS302KA	2081
KS302MA	2081
KS302PA	2088
KS302RA	2088
KS302TA	0612
KS302VA	0612
KS433A	OEM
KS439A	OEM
KS471A	OEM
KS502AA	0970
KS502BA	0989
KS502CA	0989
KS502DA	0989
KS502EA	0197
KS502FA	0197
KS502GA	0197
KS502HA	0197
KS502KA	0204
KS502MA	0204
KS502PA	0206
KS502RA	0206
KS502TA	0586
KS502VA	0586
KS512A	0137
KS515A	0002
KS560	OEM
KS602AA	0970
KS602BA	0989
KS602CA	0989
KS602DA	0989
KS602EA	0197
KS602FA	0197
KS602GA	0197
KS602HA	0197
KS602KA	0204
KS602MA	0204
KS602PA	0206
KS602RA	0206
KS602TA	0586
KS602VA	0586
KS751AA	0708
KS751BA	0580
KS751CA	0580
KS751DA	0580
KS751EA	1000
KS751FA	1000
KS751GA	1000
KS751HA	1000
KS751KA	2081
KS751MA	2081
KS751PA	2088
KS751RA	2088
KS751TA	0612
KS751VA	0612
KS1060	OEM
KS1160	OEM
KS1260	OEM
KS1460	OEM
KS2033A	0292
KS2033B	0299
KS2036A	0188
KS2036B	0376
KS2039A	0456
KS2039B	0188
KS2043A	0533
KS2043B	0275
KS2047A	0451
KS2047B	0162
KS2051A	0162
KS2051B	0162
KS2056A	0157
KS2056B	0157
KS2060	OEM
KS2062A	0636
KS2062B	0636
KS2091A	0012
KS2091B	0012
KS2100A	0170
KS2100B	0170
KS2120A	0137
KS2120B	0137
KS2150A	0002
KS2150B	0002
KS3060	OEM
KS-3302	OEM
KS3302	OEM
KS4060	OEM
KS5060	OEM
KS5302	OEM
KS5410	OEM
KS5803B	OEM
KS6060	OEM
KS8060	OEM
KS10969-L5	1349
KS20967-L1	0232
KS20967-L2	0361
KS20967-L3	1368
KS20969-L2	0117
KS20969-L3	0834
KS20969-L4	1486
KS21282-L1	0467
KS21282-L2	3477
KS21282-L3	0899
KS51900-11	OEM
KSA495Y	0037
KSA539	0037
KSA539-0	0148
KSA539-O	0037
KSA539-Y	0037
KSA539Y	0037
KSA542-0	0006
KSA542-O	0037
KSA542Y	0006
KSA562	0006
KSA562-0	0006
KSA562TM	0037
KSA562TM-0	0006
KSA614	0681
KSA614(Y)	0681
KSA614-0	6068
KSA614-Y	0681
KSA614Y	0681
KSA634	1051
KSA634(0)	1051
KSA634(Y)	1051
KSA634-0	0681
KSA634-O	1051
KSA634-Y	1051
KSA634Y	0681
KSA639-0	OEM
KSA642	0037
KSA642-0	0006
KSA642-O	0037
KSA642-Y	0037
KSA6420	0037
KSA643	0037
KSA643(0)	0037
KSA643(Y)	0037
KSA643-0	0532
KSA643-O	0037
KSA643-OY	0791
KSA643-Y	0037
KSA6430	1242
KSA643Y	0532
KSA653	0791
KSA708	0037
KSA708(0)	0037
KSA708(Y)	0037
KSA708-0	1597
KSA708-Y	1597
KSA708D	OEM
KSA708Y	0037
KSA733	0037
KSA733(0)	0037
KSA733(Y)	0037
KSA733-0	0006
KSA733-O	0037
KSA733-Y	0037
KSA733C-G-T	OEM
KSA733C-L-T	OEM
KSA733C-Y	0807
KSA733Y	0006
KSA735-0	0037
KSA928AY	OEM
KSA931	1597
KSA931-0	OEM
KSA931Y	OEM
KSA940	1911
KSA950Y	OEM
KSA1013ACY	OEM
KSA1015	0006
KSA1015Y	0006
KSA1175	0013
KSA1266Y	OEM
KSA6643-0	1242
KSA7339	OEM
KSB546	1648
KSB546-0	1911
KSB546Y	OEM
KSB564(Y)	0532
KSB564-Y	0532
KSB564A	0532
KSB564A-Y	0436
KSB564Y	0532
KSC105G	OEM
KSC184(0)	0079
KSC539-Y	OEM
KSC581C	OEM
KSC582C	OEM
KSC815	0079
KSC815(0)	0079
KSC815(Y)	0079
KSC815-0	0076
KSC815-O	0079
KSC815-OY	0079
KSC815-Y	0079
KSC815D	0079
KSC815D-0	0079
KSC815Y	0079
KSC838	0079
KSC838(0)	0155
KSC838-0	0155
KSC838-O	0155
KSC838A-0	OEM
KSC838CY	OEM
KSC853(Y)	0079
KSC945	0079
KSC945(G)	0286
KSC945(0)	0079
KSC945(Y)	0079
KSC945-G	0079
KSC945-0	0079
KSC945-OY	0079
KSC945-Y	0079
KSC945C	OEM
KSC945C-G	OEM
KSC945C-K	OEM
KSC945C-Y	0079
KSC945G	0076
KSC945Y	0076
KSC1008	0079
KSC1008(Y)	0079
KSC1008-0	0079
KSC1008-O	5133
KSC1008-Y	0079
KSC1008C	0079
KSC1008Y	0079
KSC1096	0219
KSC1096(0)	0219
KSC1096-0	0236
KSC1096-O	0219
KSC1096-Y	0219
KSC1096Y	0236
KSC1187	0224
KSC1187(0)	0224
KSC1187(R)	0224
KSC1187-0	0224
KSC1187-O	0224
KSC1187-R	0224
KSC1187C	0224
KSC11870	0224
KSC1187R	0224
KSC1189R	0079
KSC1394	2309
KSC1395	0144
KSC1507	0168
KSC1507(0)	0168
KSC1507-0	0643
KSC-1507-0	0168
KSC1507-O	0168
KSC-1507-Y	0168
KSC1507-Y	0168
KSC1507A0	0168
KSC1507A-0	0168
KSC1507AO	0168
KSC1507A-Y	0643
KSC1507AY	0168
KSC1507K	0643
KSC1507L	0168
KSC15070	0342
KSC1507Y	0643
KSC1509-0	OEM
KSC1520	0168
KSC1520(0)	0168
KSC1520(Y)	0168
KSC1520-0	0168
KSC1520-R	0168
KSC1520Y	0643
KSC1674	0144
KSC1674(0)	0224
KSC1674(R)	0144
KSC1674-0	0144
KSC1674-O	0144
KSC16740	0224
KSC1674R	0144
KSC1675	0079
KSC1675-0	0842
KSC1675-O	0079
KSC1675-Y	0079
KSC1675R	0286
KSC1675Y	0079
KSC1684	OEM
KSC1815-Y	OEM
KSC1815Y	1954
KSC2073	1284
KSC2120Y	0155
KSC2230-0	0261
KSC2310	0261
KSC2310-0	0261
KSC2310-OY	0261
KSC2310-Y	0261
KSC2310Y	0261
KSC2328	3938
KSC2328A-Y	3938
KSC2328AY	3938
KSC2330	0261
KSC2330-0	0261
KSC2330-Y	0261
KSC23300	0261
KSC2330Y	0261
KSC2331	1624
KSC2331-0	1624
KSC2331-OY	0018
KSC2331-Y	1624
KSC23310	1624
KSC2331Y	1624
KSC2335-0	OEM
KSC2340	0261
KSC2785	2954
KSC3198-TP-Y	OEM
KSC3198Y	OEM
KSC3229	OEM
KSD73	1165
KSD73(0)	1165
KSD73-0	1165
KSD73-O	1165
KSD73-Y	1165
KSD73Y	1165
KSD227(Y)	0079
KSD227-0	0286
KSD261	0079
KSD261(0)	0595
KSD261(Y)	0595
KSD261-0	0595
KSD261-O	0079
KSD261-OY	0595
KSD261-Y	0595
KSD261Y	0595
KSD288	2997
KSD361N	0392
KSD362	1165
KSD362(R)	0392
KSD362-C	1165
KSD362-N	0392
KSD362-NR	0392
KSD362-R	0392
KSD362-Y	1165
KSD362C	1165
KSD362C(Y)	1165
KSD362C-Y	1165
KSD362N	1165
KSD362R	1165
KSD362Y	1165
KSD363	0728
KSD401	0392
KSD401(0)	0392
KSD401(Y)	0392
KSD401-7	0392
KSD401-0	0392
KSD401-Y	0392
KSD401A	0392
KSD401Y	0392
KSD415-0	0261
KSD415-O	0715
KSD415-Y	0715
KSD415D	0715
KSD4150	0715
KSD415R	0261
KSD415Y	0261
KSD471-Y	0324
KSD471Y	0018
KSD807M	1493
KSD1051	0103
KSD1052	0177
KSD1053	0177
KSD1054	0177
KSD1055	0103
KSD1056	0103
KSD1057	0177
KSD1058	0177
KSD2101	0178
KSD2102	0178
KSD2103	0178
KSD2201	0177
KSD2202	0177
KSD2203	0130
KSD3055	0130
KSD3771	0130
KSD3772	0130
KSD3773	0363
KSD4150	0715
KSD4154	0715
KSD9701	0543
KSD9701A	0543
KSD9702	0543
KSD9702A	0543
KSD9703	0543
KSD9703A	0543
KSD9704	0543
KSD9705	0543
KSD9706	0543
KSD9707	0543
KSDT7601	OEM
KSKE12C500	0071
KSKE40C200	0015
KSKE40C500	0015
KSKE125C200	0071
KSP1171	0620
KSP1172	0620
KSP1173	0620
KSP1174	0620
KSP1175	0620
KSP1176	0620
KSR1001	OEM
KSR1004	0898
KSR1204	OEM
KSR2001	OEM
KSR2004	4130
KSS240A	OEM
KSS9808	OEM
KT102AA	0933
KT102BA	0947
KT102CA	0947
KT102DA	0947
KT102EA	1000
KT102FA	1012
KT102GA	1012
KT102HA	1012
KT102KA	1073
KT102MA	1073
KT102PA	1138
KT102RA	1138
KT102TA	1188
KT102VA	1188
KT123AA	0534
KT123BA	0765
KT123CA	0765
KT123DA	0765
KT123EA	0538
KT123FA	0538
KT123GA	0538
KT123HA	0538
KT123KA	0816
KT123MA	0816
KT123PA	0545
KT123RA	0545
KT123TA	0550
KT123VA	0550
KT162AA	0933
KT162BA	0947
KT162CA	0947
KT162DA	0947
KT162EA	1000
KT162FA	1012
KT162GA	1012
KT162HA	1012
KT162KA	1073
KT162MA	1073
KT162PA	1138
KT162RA	1138
KT162TA	1188
KT162VA	1188
KT208A	OEM
KT208B	OEM
KT208D	OEM
KT208E	OEM
KT208G	OEM
KT208I	OEM
KT208K	OEM
KT208L	OEM
KT208M	OEM
KT208S	OEM
KT208V	OEM
KT218	0079
KT218F	0861
KT302A	OEM
KT302AA	0933
KT302B	OEM
KT302BA	0947
KT302CA	0947
KT302DA	0947
KT302EA	1000
KT302FA	1012
KT302G	OEM
KT302GA	1012
KT302HA	1012
KT302KA	1073
KT302MA	1073
KT302PA	1138
KT302RA	1138
KT302TA	1188
KT302V	OEM
KT302VA	1188
KT326A	OEM
KT326B	OEM
KT339B	OEM
KT339D	OEM
KT339G	OEM
KT339V	OEM
KT345A	OEM
KT345B	OEM
KT345V	OEM
KT375A	OEM
KT375B	OEM
KT502AA	0985
KT502BA	0990
KT502CA	0990
KT502DA	0990
KT502EA	0997
KT502FA	0997
KT502GA	0997
KT502HA	0997
KT502KA	0515
KT502MA	0515
KT502PA	1008
KT502RA	1008
KT502TA	0948
KT502VA	0948
KT600	0018
KT600F	0018
KT600G	0168
KT600T	0018
KT602AA	0985
KT602BA	0990
KT602CA	0990
KT602DA	0990
KT602EA	0997
KT602FA	0997
KT602GA	0997
KT602HA	0997
KT602KA	0515
KT602MA	0515
KT602PA	1008
KT602RA	1008
KT602TA	0948
KT602VA	0948
KT751AA	0933
KT751BA	0947
KT751CA	0947
KT751DA	0947
KT751EA	1000
KT751FA	1012
KT751GA	1012
KT751HA	1012
KT751KA	1073
KT751MA	1073
KT751PA	1138
KT751RA	1138
KT751TA	1188
KT751VA	1188
KT1017	0160
KT1815-GR	0286
KT8740	OEM
KTA473	0925
KTA473-0	0925
KTA473A	OEM
KTA473Y	OEM
KTA562	0006
KTA562(Y)	0006
KTA562-0	0006
KTA562-TM	0006
KTA562-Y	0006
KTA562A	0006
KTA5620	0006
KTA562TM	0006
KTA562TM-0	0006
KTA562TM-0(TA)	0006
KTA562TM-0-TA	OEM
KTA562TM-Y	0006
KTA562TMY	0006
KTA562TM-Y(TA)	0006
KTA562TM-Y-0	0006
KTA562Y	0006
KTA733	0037
KTA817	4090
KTA817-Y	OEM
KTA817A	0477
KTA817A-0	0477
KTA-940	1648
KTA940	1911
KTA949-0	1522
KTA950	5954
KTA950-0	0006
KTA950-Y	0532
KTA950-Y(TA)	0532
KTA950Y	0006
KTA965	OEM
KTA965-Y	1522
KTA965Y	1522
KTA966-Y	0532
KTA966A	OEM
KTA966AY	0532
KTA966Y	OEM
KTA968-Y	1648
KTA968A-Y	1648
KTA968Y	1648
KTA1013-0	1522
KTA1015	0148
KTA1015(0)	0006
KTA1015(Y)	0006
KTA1015-0	3108
KTA1015-G	0006
KTA1015-GR	0148
KTA1015-O	0006
KTA1015-Y	0148
KTA1015-Y(TA)	0148
KTA1015A	0006
KTA1015GR	0148
KTA10150	0006
KTA1015Y	0148
KTA1023-YTP	OEM
KTA1024-0	OEM
KTA1162(SG)	OEM
KTA1266-TP-Y	4519
KTA1266-Y	6612
KTA1266GR	4519
KTA1266Y	4519
KTA1267GR	OEM
KTA1270-TA-Y	0006
KTA1270-TP-Y	0006
KTA1270Y	0006
KTA1271-TP-Y	0477
KTA1271Y	0477
KTA1274-0	6682
KTA1275-O-AT	7118
KTA4558P	OEM

If replacement code is OEM, contact original manufacturer for replacement.

DEVICE TYPE	REPL CODE
KTA7805	OEM
KTA7812	OEM
KTB595	0718
KTB595-O	OEM
KTB595-O	0718
KTB595Y	0718
KTB596-O	3752
KTB834	6186
KTB834-O	6186
KTB834-Y	6186
KTB834Y	0854
KTB988-O	OEM
KTB1366-Y	OEM
KTB1366Y	OEM
KTC0141(TM-O)	OEM
KTC200-Y	0016
KTC200Y	0016
KTC380-O	0286
KTC380TM-O	0144
KTC382	0224
KTC388	0286
KTC388(TA)	0286
KTC388A	0842
KTC388A(TA)	0286
KTC388A-TM	0286
KTC388ATM	0286
KTC388ATMFA	0286
KTC388AY	OEM
KTC732TM-BL	0507
KTC1001	OEM
KTC1026Y	OEM
KTC1027-YTP	OEM
KTC1096-O	0219
KTC1173	0042
KTC1173-O	0042
KTC1173Y	OEM
KTC1627	1386
KTC1627-O	1386
KTC1627-Y	1386
KTC1627A	1386
KTC1627A-O	1386
KTC1627A-TP	1386
KTC1627A-Y	1386
KTC1627AY	1386
KTC1627A-Y(TA)	1386
KTC1627AY(TA)	1386
KTC1627Y	1386
KTC1815	0076
KTC1815(Y)	0286
KTC1815-BL	0286
KTC1815-D	0286
KTC1815-G	0006
KTC1815-GR	0286
KTC1815-GRFA-1	0286
KTC1815-O	0286
KTC1815-O(TA)	0286
KTC1815-O-TA	OEM
KTC1815-TP-Y	0076
KTC1815-Y	0286
KTC1815-Y(TA)	0286
KTC-1815BL	0286
KTC1815BL	0076
KTC1815D	0286
KTC1815G	OEM
KTC1815GR	0076
KTC1815O	0286
KTC-1815Y	0286
KTC1815Y	0076
KTC1923	OEM
KTC1923(O)	0224
KTC1923-O	0224
KTC1923Y	OEM
KTC1959	0383
KTC1959-O	0286
KTC1959-Y	0286
KTC1959Y	0286
KTC2068	0643
KTC2068(FA-1)	0643
KTC2068-1	0643
KTC2068-FA1	OEM
KTC2068FA-1	0643
KTC2068I	0643
KTC2073	0392
KTC2073FA-2	0392
KTC2120	0866
KTC2120(Y)	0155
KTC2120-O	0155
KTC2120-Y	0155
KTC2120-Y(TA)	0866
KTC2120-Y(TR)	0155
KTC2120Y	0866
KTC2229	0261
KTC-2229-O	0261
KTC2229-O	0261
KTC2229-Y	0261
KTC2229-Y(TA)	0261
KTC2229D	0261
KTC-2229K	0261
KTC2229O	0261
KTC2229Y	0261
KTC2230	0261
KTC2230(GR)	OEM
KTC2230-GR	0261
KTC2230-TP	OEM
KTC2230-Y	0261
KTC2230A	0261
KTC2230A(Y)	0261
KTC2230A-GR	0261
KTC2230AGR	0261
KTC2230A-Y	0261
KTC2230AY	0261
KTC2230GR	0261
KTC2230TP	0261
KTC2230Y	0261
KTC2233	0392
KTC2233FA-1	0392
KTC2235	4614
KTC2235-Y	6241
KTC2235Y	1562
KTC2236-Y	6242
KTC2236A	OEM
KTC2236AY	6193
KTC2236Y	OEM
KTC2238A	0392
KTC2238A-Y	0392
KTC2240	OEM
KTC2240BL	OEM
KTC2240GR	OEM
KTC2310-O	0261
KTC2330-O	OEM
KTC2383	4250
KTC2383-Y	1562
KTC2383Y	1562
KTC2482	0066
KTC2482(TA)	0261
KTC2482-TP	0261
KTC2482FA-1	0261
KTC2482K	0261
KTC2482TA	2335
KTC2482Y	0066
KTC2712(LG)	OEM
KTC3197	1092
KTC3197-AT	1092
KTC3197-TP	1092
KTC3197,TP	1092
KTC3197TP	1092
KTC3198-TA-Y	0286
KTC3198-TP-Y	0286
KTC3198-Y	6814
KTC3198Y	0286
KTC3199GR	OEM
KTC3202-TP-Y	0383
KTC3203-TP-Y	1386
KTC3203-Y	OEM
KTC3203Y	1386
KTC3205-Y	6850
KTC3205Y	OEM
KTC3206-Y	0261
KTC3207	0261
KTC3207K	6085
KTC3207TP	0261
KTC3227-OTP	0324
KTC3227-O	OEM
KTC3227-OTP	OEM
KTC3227-TP	0324
KTC3227-Y	0324
KTC3227Y	0324
KTC3228-O	4250
KTC3228-O-AT	4614
KTC3228-Y	OEM
KTC3229	OEM
KTC3230Y	OEM
KTC9011	OEM
KTC9011A-G	OEM
KTC9013A	0079
KTC9013A(H)	0079
KTC9013A(I)	0079
KTC9014(C)	0079
KTC9014A	0079
KTC9014A(C)	0079
KTC9014A(D)	0079
KTC9014A-B	OEM
KTC9014A-C	OEM
KTC9014AC	0079
KTC9014C	0079
KTC9015(C)	0037
KTC9015A	0037
KTC9015A(B)	0037
KTC9015A-B	OEM
KTC9016A-G	OEM
KTC9016A-H	OEM
KTC9018F	OEM
KTC9018G	OEM
KTD19TM(GR)	OEM
KTD525-O	OEM
KTD718	OEM
KTD880	0042
KTD880(GR)	0042
KTD880(Y)	0042
KTD880-O	0461
KTD880-Y	0042
KTD880GR	OEM
KTD880Y	0461
KTD1351-O	OEM
KTD1351-Y	0406
KTD1351Y	0406
KTD1406	OEM
KTD1406Y	0060
KTD1414	OEM
KTD1554	1541
KTD1555	0556
KTD1555A	5305
KTD1651	1541
KTD1879	OEM
KTD2058-Y	OEM
KTD2058Y	OEM
KTD2059-O	4007
KTD7180	OEM
KTD8800	0042
KTK19(GR)	OEM
KTK117A	OEM
KTK161YR	OEM
KTR0710C	0368
KTR0815C	0155
KTR0839C	0076
KTR0945C	6709
KTR1017	2050
KTR1096C	0390
KTR1687C	0224
KTTA1015O	0006
KU6HZ	OEM
KU902	OEM
KU904	OEM
KU906	OEM
KU908	OEM
KU910	OEM
KU912	OEM
KU914	OEM
KU1302	OEM
KU1304	OEM
KU1306	OEM
KU1308	OEM
KU1310	OEM
KU1312	OEM
KU1314	OEM
KV0714-2	OEM
KV-1	0004
KV1	0004
KV-2	0211
KV2	0004
KV-3	7125
KV3	OEM
KV-4	0004
KV4	0004
KV-4D	0211
KV-5	OEM
KV5	OEM
KV-14	0280
KV107	OEM
KV107A	OEM
KV107B	OEM
KV107C	OEM
KV107D	OEM
KV123	OEM
KV123A	OEM
KV123B	OEM
KV123C	OEM
KV123D	OEM
KV124	OEM
KV124A	OEM
KV124B	OEM
KV124C	OEM
KV124D	OEM
KV127	OEM
KV128	OEM
KV132	OEM
KV133	OEM
KV134	OEM
KV135	OEM
KV136	OEM
KV137	OEM
KV138	OEM
KV139	OEM
KV156	OEM
KV157	OEM
KV158	OEM
KV159	OEM
KV160	OEM
KV161	OEM
KV169	OEM
KV171	OEM
KV172	OEM
KV173	OEM
KV1225	OEM
KV1226	OEM
KV1230Z	OEM
KV1234Z	OEM
KV1235	OEM
KV1235Z	OEM
KV1235Z3	OEM
KV1235Z3-1	OEM
KV1235Z3-2	OEM
KV1235Z3-3	OEM
KV1235Z3-4	OEM
KV1235Z3-5	OEM
KV1235Z3-A	OEM
KV1235Z3-B	OEM
KV1235Z3-C	OEM
KV1235Z3-D	OEM
KV1235Z3-E	OEM
KV1235Z5-A	OEM
KV1235Z5-B	OEM
KV1235Z5-C	OEM
KV1235Z5-D	OEM
KV1235Z5-E	OEM
KV1236	3688
KV1236Z	OEM
KV1250	OEM
KV1250M	OEM
KV1260	OEM
KV1260M	OEM
KV1260Z	OEM
KV1310	OEM
KV1310-6	OEM
KV1310A	OEM
KV1310A-3	OEM
KV1320	OEM
KV1330A-2	OEM
KV1340A-3	OEM
KV2003	OEM
KV2160	OEM
KV2203	OEM
KV2303	OEM
KV2403	OEM
KV2503	OEM
KV2603	OEM
KV2703	OEM
KV5139	OEM
KV5140	OEM
KV5141	OEM
KV5142	OEM
KV5143	OEM
KV13120A-3	OEM
KVR10	0170
KVTKDS288Y	OEM
KVTKSA928AY	OEM
KX-1	0133
KX-1005	OEM
KY5DPF	7154
KY1011	OEM
KY4042	OEM
KY4043	OEM
KY4099	OEM
KY8607	OEM
KYA1015Y	OEM
KZ5DPF	OEM
KZ6	0025
KZ6A	0025
KZ6ZA	0025
KZ-8A	0244
KZ62A	0471
KZ91A	0057
KZ1075	0077
KZ1075-M	0077
KZ1075M	OEM
KZL083	OEM
L05-E	0071
L-0650	OEM
L-0651	OEM
L-0662	OEM
L1/1H	OEM
L1/2H	0133
L1/3H	0133
L1/4H	OEM
L1/5H	OEM
L1/6H	OEM
L1/8H	OEM
L1/10H	OEM
L1/15H	OEM
L1/20H	OEM
L1A0091	OEM
L1H1	OEM
L1V	OEM
L3	OEM
L3/1H	OEM
L3/2H	0015
L3/3H	OEM
L3/4H	OEM
L3/5H	OEM
L3/6H	OEM
L3/8H	OEM
L3/10H	OEM
L3/15H	OEM
L3/20H	OEM
L4	0111
L5	0111
L6	0724
L6/1H	OEM
L6/2H	0015
L6/3H	OEM
L6/4H	OEM
L6/5H	OEM
L6/6H	OEM
L6/8H	OEM
L6/10H	OEM
L6/15H	OEM
L6/20H	OEM
L6-SAM	0724
L6A	OEM
L6B	OEM
L6C	OEM
L6D	OEM
L6F	OEM
L6U	OEM
L6X12	OEM
L7	0111
L7A	OEM
L7B	OEM
L7C	OEM
L7D	OEM
L7F	OEM
L7U	OEM
L7X12	OEM
L8/1H	OEM
L8/2H	0015
L8/3H	OEM
L8/4H	OEM
L8/5H	OEM
L8/6H	OEM
L8/8H	OEM
L8/10H	OEM
L8/15H	OEM
L8/20H	OEM
L8H6.2	OEM
L14B	OEM
L14B1	OEM
L14B2	OEM
L14B3	OEM
L14B4	OEM
L14F1	3552
L14F2	3552
L14G1	3878
L14G2	2313
L14G3	2313
L14H1	3878
L14H2	3878
L14H3	3878
L14H4	3878
L14T1	OEM
L14T2	OEM
L14T3	OEM
L14T4	OEM
L14T5	OEM
L15C1	OEM
L15C2	OEM
L15D1	OEM
L15D2	OEM
L20	OEM
L32H	0015
L33-SAM	3482
L34	OEM
L62H	0015
L78LR05C	OEM
L78LR05D	OEM
L78LR05D-MA	OEM
L78M05	0624
L78M05-A	0624
L78M05-LU	0624
L78M05-RA	0624
L78M05ABV	0624
L78M05C-V	0624
L78M05CV	0624
L78M05N000	OEM
L78M05P	0624
L78M06	0923
L78M06SA	0923
L78M08ABV	OEM
L78M09	1786
L78M09-RA	1786
L78M09-SA	1786
L78M09ABV	OEM
L78M09SA	1786
L78M12	0334
L78M12-LU	0334
L78M12-RA	0334
L78M12-SA	1894
L78M12CV	0334
L78M12RL	0334
L78MR05	OEM
L78MR05-FA	OEM
L78MR06-FA	OEM
L78N05	1292
L78N06	5678
L78N06-SE	0923
L78N12	0334
L78N12-RA	0334
L79M05	OEM
L79M05N000	OEM
L82H	0071
L120	0201
L123CB	0026
L150LA2	OEM
L180	0378
L241	OEM
L272	6122
L272B	6122
L272M	6098
L272MB	6098
L274	2393
L292	6160
L292V	6160
L293B	5241
L295	5242
L296	5214
L296P	OEM
L296V	5214
L298	OEM
L298N	OEM
L347N	OEM
L-417-29BLK	0160
L-417-29GRN	0160
L-417-29WHT	0160
L-417-60	0160
L563G	OEM
L780S05-FA	OEM
L842	0144
L1010	OEM
L1010A	OEM
L1010B	OEM
L1014	OEM
L1014A	OEM
L1014B	OEM
L1015	OEM
L1015A	OEM
L1015B	OEM
L1016	OEM
L1016A	OEM
L1016B	OEM
L1019	OEM
L1019A	OEM
L1019B	OEM
L1020	OEM
L1020A	OEM
L1020B	OEM
L1021	OEM
L1021A	OEM
L1021B	OEM
L1022	OEM
L1022A	OEM
L1022B	OEM
L1025	OEM
L1025A	OEM
L1025B	OEM
L1026	OEM
L1026A	OEM
L1026B	OEM
L1029	OEM
L1029A	OEM
L1029B	OEM
L1030	OEM
L1030A	OEM
L1030B	OEM
L1031	OEM
L1031A	OEM
L1031B	OEM
L1032	OEM
L1032A	OEM
L1032B	OEM
L1035	OEM
L1035A	OEM
L1035B	OEM
L1036	OEM
L1036A	OEM
L1036B	OEM
L1039	OEM
L1039A	OEM
L1039B	OEM
L1040	OEM
L1040A	OEM
L1040B	OEM
L1041	OEM
L1041A	OEM
L1041B	OEM
L1042	OEM
L1042A	OEM
L1042B	OEM
L1045	OEM
L1045A	OEM
L1045B	OEM
L1046	OEM
L1046A	OEM
L1046B	OEM
L1049	OEM
L1049A	OEM
L1049B	OEM
L1060	OEM
L1060A	OEM
L1060B	OEM
L1061	OEM
L1061A	OEM
L1061B	OEM
L1062	OEM
L1062A	OEM
L1062B	OEM
L1065	OEM
L1065A	OEM
L1065B	OEM
L1066	OEM
L1066A	OEM
L1066B	OEM
L1069	OEM
L1069A	OEM
L1069B	OEM
L1100	OEM
L1100A	OEM
L1100B	OEM
L1101	OEM
L1101A	OEM
L1101B	OEM
L1102	OEM
L1102A	OEM
L1102B	OEM
L1105	OEM
L1105A	OEM
L1105B	OEM
L1106	OEM
L1106A	OEM
L1106B	OEM
L1141	OEM
L1141A	OEM
L1141B	OEM
L1145	OEM
L1145A	OEM
L1145B	OEM
L1146	OEM
L1146A	OEM
L1146B	OEM
L1147	OEM
L1147A	OEM
L1147B	OEM
L1149	OEM
L1149A	OEM
L1149B	OEM
L1150	OEM
L1150A	OEM
L1150B	OEM
L1151	OEM
L1151A	OEM
L1151B	OEM
L1152	OEM
L1152A	OEM
L1152B	OEM
L1155	OEM
L1155A	OEM
L1155B	OEM
L1156	OEM
L1156A	OEM
L1156B	OEM
L1165	OEM
L1165A	OEM
L1165B	OEM
L1166	OEM
L1166A	OEM
L1166B	OEM
L1169	OEM
L1169A	OEM
L1169B	OEM
L1220	OEM
L1220A	OEM
L1220B	OEM
L1221	OEM
L1221A	OEM
L1221B	OEM
L1222	OEM
L1222A	OEM
L1222B	OEM
L1225	OEM
L1225A	OEM
L1225B	OEM
L-1481	OEM
L-1540	OEM
L-1541	OEM
L1946	OEM
L1946A	OEM
L1946B	OEM
L1949	OEM
L1949A	OEM
L1949B	OEM
L2114-550	2048
L2732	OEM
L3010	OEM
L3010A	OEM
L3010B	OEM
L3015	OEM
L3015A	OEM
L3015B	OEM
L3016	OEM
L3016A	OEM
L3016B	OEM
L3019	OEM
L3019A	OEM
L3019B	OEM
L3020	OEM
L3020A	OEM
L3020B	OEM
L3021	OEM
L3021A	OEM
L3021B	OEM
L3022	OEM
L3022A	OEM
L3022B	OEM
L3025	OEM
L3025A	OEM
L3025B	OEM
L3026	OEM
L3026A	OEM
L3026B	OEM
L3029	OEM
L3029A	OEM
L3029B	OEM
L3030	OEM
L3030A	OEM
L3030B	OEM
L3031	OEM
L3031A	OEM
L3031B	OEM
L3032	OEM
L3032A	OEM
L3032B	OEM
L3035	OEM
L3035A	OEM
L3035B	OEM
L3036	OEM
L3036A	OEM
L3036B	OEM
L3039	OEM
L3039A	OEM
L3039B	OEM
L3040	OEM
L3040A	OEM
L3040B	OEM
L3041	OEM
L3041A	OEM
L3041B	OEM
L3042	OEM
L3042A	OEM
L3042B	OEM
L3045	OEM
L3045A	OEM
L3045B	OEM
L3046	OEM
L3046A	OEM
L3046B	OEM
L3049	OEM
L3049A	OEM
L3049B	OEM
L3059	OEM
L3059A	OEM
L3059B	OEM
L3060	OEM
L3060A	OEM
L3060B	OEM
L3061A	OEM

DEVICE TYPE	REPL CODE
L3061B	OEM
L3062	OEM
L3062A	OEM
L3062B	OEM
L3065	OEM
L3065A	OEM
L3065B	OEM
L3066	OEM
L3066A	OEM
L3066B	OEM
L3100	OEM
L3100A	OEM
L3100B	OEM
L3101	OEM
L3101A	OEM
L3101B	OEM
L3102	OEM
L3102A	OEM
L3102B	OEM
L3105	OEM
L3105A	OEM
L3105B	OEM
L3106	OEM
L3106A	OEM
L3106B	OEM
L3140	OEM
L3140A	OEM
L3140B	OEM
L3141A	OEM
L3141B	OEM
L3145	OEM
L3145A	OEM
L3145B	OEM
L3146	OEM
L3146A	OEM
L3146B	OEM
L3149	OEM
L3149A	OEM
L3149B	OEM
L3150	OEM
L3150A	OEM
L3150B	OEM
L3151	OEM
L3151A	OEM
L3152	OEM
L3152A	OEM
L3152B	OEM
L3155	OEM
L3155A	OEM
L3155B	OEM
L3156	OEM
L3156A	OEM
L3156B	OEM
L3165	OEM
L3165A	OEM
L3165B	OEM
L3166	OEM
L3166A	OEM
L3166B	OEM
L3220	OEM
L3220A	OEM
L3220B	OEM
L3221	OEM
L3221A	OEM
L3221B	OEM
L3222	OEM
L3222A	OEM
L3222B	OEM
L3225A	OEM
L3225B	OEM
L4000KS	2356
L4066B	OEM
L4120	OEM
L4122	OEM
L4130	OEM
L4131	OEM
L4132	OEM
L4134	OEM
L4135	OEM
L4136	OEM
L4138	OEM
L4139	OEM
L4140	OEM
L4141	OEM
L4142	OEM
L4143	OEM
L4144	OEM
L4145	OEM
L4146	OEM
L4147	OEM
L4150	OEM
L4153	OEM
L4154	OEM
L4154A	OEM
L4154B	OEM
L4155	OEM
L4155A	OEM
L4155B	OEM
L4156	OEM
L4156A	OEM
L4156B	OEM
L4157	OEM
L4157A	OEM
L4157B	OEM
L4160	OEM
L4164	OEM
L4164A	OEM
L4164B	OEM
L4165	OEM
L4165A	OEM
L4165B	OEM
L4166	OEM
L4166A	OEM
L4166B	OEM
L4167	OEM
L4167A	OEM
L4167B	OEM
L4170	OEM
L4171	OEM
L4172	OEM
L4173	OEM
L4174	OEM
L4175	OEM
L4178	OEM
L4181	OEM
L4181A	OEM
L4185	OEM
L4185A	OEM
L4185B	OEM
L4190	OEM
L4190A	OEM
L4190B	OEM
L4190C	OEM
L4221	OEM
L4221A-E	OEM
L4221F	OEM
L4221G	OEM
L4221H	OEM
L4221J-L	OEM
L4222	OEM
L4222A-E	OEM
L4222F	OEM
L4222G	OEM
L4222H	OEM
L4222J	OEM
L4222K	OEM
L4223	OEM
L4223A-E	OEM
L4223F	OEM
L4223G	OEM
L4223H	OEM
L4223J	OEM
L4224	OEM
L4224A-E	OEM
L4224F	OEM
L4224G	OEM
L4224H	OEM
L4231	OEM
L4231A-E	OEM
L4231F	OEM
L4231G	OEM
L4231H	OEM
L4231J-L	OEM
L4232	OEM
L4232A-E	OEM
L4232F	OEM
L4232G	OEM
L4232H	OEM
L4232J	OEM
L4232K	OEM
L4233	OEM
L4233A-E	OEM
L4233F	OEM
L4233G	OEM
L4233H	OEM
L4233J	OEM
L4234	OEM
L4234A-E	OEM
L4234F	OEM
L4234G	OEM
L4234H	OEM
L4235	OEM
L4235A-E	OEM
L4235F	OEM
L4242	OEM
L4242A-E	OEM
L4242F	OEM
L4242G	OEM
L4242H	OEM
L4242J	OEM
L4243	OEM
L4243A-E	OEM
L4243F	OEM
L4243G	OEM
L4243H	OEM
L4244	OEM
L4244A-E	OEM
L4244F	OEM
L4245	OEM
L4245A-E	OEM
L4252	OEM
L4252A-E	OEM
L4252F	OEM
L4252G	OEM
L4252H	OEM
L4253	OEM
L4253A-E	OEM
L4253F	OEM
L4254	OEM
L4254A-E	OEM
L4255	OEM
L4255A-D	OEM
L4356	OEM
L4356A	OEM
L4356B	OEM
L4357	OEM
L4357A	OEM
L4358	OEM
L4366	OEM
L4366A	OEM
L4366B	OEM
L4367	OEM
L4367A	OEM
L4368	OEM
L4377	OEM
L4377A	OEM
L4412	OEM
L4413	OEM
L4414	OEM
L4503	OEM
L4505	OEM
L4506	OEM
L4507	OEM
L4510	OEM
L4520	OEM
L4530	OEM
L4540	OEM
L4650	OEM
L4650A	OEM
L4651	OEM
L4651A	OEM
L4652	OEM
L4652A	OEM
L4653	OEM
L4653A	OEM
L4654	OEM
L4654A	OEM
L4655	OEM
L4655A	OEM
L4656	OEM
L4656A	OEM
L4660	OEM
L4660A	OEM
L4661	OEM
L4661A	OEM
L4662	OEM
L4662A	OEM
L4663	OEM
L4663A	OEM
L4664	OEM
L4664A	OEM
L4665	OEM
L4665A	OEM
L4666	OEM
L4666A	OEM
L4700	OEM
L4701	OEM
L4702	OEM
L4703	OEM
L4704	OEM
L4705	OEM
L4706	OEM
L4707	OEM
L4708	OEM
L4709	OEM
L4710	OEM
L4711	OEM
L4712	OEM
L4713	OEM
L4714	OEM
L4715	OEM
L4716	OEM
L4717	OEM
L4718	OEM
L4719	OEM
L4720	OEM
L4721	OEM
L4722	OEM
L4750	OEM
L4751	OEM
L4752	OEM
L4753	OEM
L4754	OEM
L4755	OEM
L4756	OEM
L4757	OEM
L4758	OEM
L4759	OEM
L4765	OEM
L4766	OEM
L4767	OEM
L4768	OEM
L4769	OEM
L4770	OEM
L4771	OEM
L4772	OEM
L4773	OEM
L4774	OEM
L4775	OEM
L4776	OEM
L4818B	OEM
L4821	OEM
L4821A-E	OEM
L4821F	OEM
L4821G	OEM
L4821H	OEM
L4822	OEM
L4822A-E	OEM
L4822F	OEM
L4822G	OEM
L4822H	OEM
L4822J	OEM
L4822K	OEM
L4823	OEM
L4823A-E	OEM
L4823F	OEM
L4823G	OEM
L4823H	OEM
L4823J	OEM
L4824	OEM
L4824A-E	OEM
L4824F	OEM
L4824G	OEM
L4824H	OEM
L4831	OEM
L4831A-E	OEM
L4831F	OEM
L4831G	OEM
L4831H	OEM
L4831J-L	OEM
L4832	OEM
L4832A-E	OEM
L4832F	OEM
L4832G	OEM
L4832H	OEM
L4832J	OEM
L4832K	OEM
L4833	OEM
L4833A-E	OEM
L4833F	OEM
L4833G	OEM
L4833H	OEM
L4833J	OEM
L4834	OEM
L4834A-E	OEM
L4834F	OEM
L4834G	OEM
L4834H	OEM
L4835	OEM
L4835A-E	OEM
L4835F	OEM
L4842	OEM
L4842A-E	OEM
L4842F	OEM
L4842G	OEM
L4842H	OEM
L4842J	OEM
L4843	OEM
L4843A-E	OEM
L4843F	OEM
L4843G	OEM
L4843H	OEM
L4844	OEM
L4844A-E	OEM
L4844F	OEM
L4844G	OEM
L4845	OEM
L4845A-E	OEM
L4852	OEM
L4852A-E	OEM
L4852F	OEM
L4852G	OEM
L4852H	OEM
L4853	OEM
L4853A-E	OEM
L4853F	OEM
L4853G	OEM
L4853H	OEM
L4854	OEM
L4854A-E	OEM
L4854F	OEM
L4854G	OEM
L4855	OEM
L4855A-D	OEM
L4856	OEM
L4856A	OEM
L4856B	OEM
L4856C	OEM
L4866	OEM
L4866A-C	OEM
L4867	OEM
L4867A	OEM
L4867B	OEM
L4877	OEM
L4877A	OEM
L4877B	OEM
L4921	OEM
L4921A-E	OEM
L4921F	OEM
L4921G	OEM
L4921H	OEM
L4921J-L	OEM
L4921M	OEM
L4922	OEM
L4922A-E	OEM
L4922F	OEM
L4922G	OEM
L4922H	OEM
L4922J-L	OEM
L4923	OEM
L4923A-E	OEM
L4923F	OEM
L4923G	OEM
L4923H	OEM
L4923J-L	OEM
L4924	OEM
L4924A-E	OEM
L4924F	OEM
L4924G	OEM
L4924H	OEM
L4924J	OEM
L4924K	OEM
L4931	OEM
L4931A-K	OEM
L4931F	OEM
L4931G	OEM
L4931H	OEM
L4931J-L	OEM
L4932	OEM
L4932A-E	OEM
L4932F	OEM
L4932G	OEM
L4932H	OEM
L4932J	OEM
L4932K	OEM
L4933	OEM
L4933A-E	OEM
L4933G	OEM
L4933H	OEM
L4933J	OEM
L4933K	OEM
L4934	OEM
L4934A-E	OEM
L4934F	OEM
L4934G	OEM
L4934H	OEM
L4935	OEM
L4935A-E	OEM
L4935F	OEM
L4942	OEM
L4942A-E	OEM
L4942F	OEM
L4942G	OEM
L4942H	OEM
L4942J	OEM
L4943	OEM
L4943A-E	OEM
L4943F	OEM
L4943G	OEM
L4943H	OEM
L4944	OEM
L4944A-E	OEM
L4944F	OEM
L4945	OEM
L4945A-E	OEM
L4946	OEM
L4946A-D	OEM
L4947	OEM
L4947A	OEM
L4947B	OEM
L4947C	OEM
L4952	OEM
L4952A-E	OEM
L4952F	OEM
L4952G	OEM
L4952H	OEM
L4953	OEM
L4953A-E	OEM
L4953F	OEM
L4953G	OEM
L4953H	OEM
L4954	OEM
L4954A-E	OEM
L4955	OEM
L4955A-D	OEM
L4956	OEM
L4956A-D	OEM
L4957	OEM
L4957A-B	OEM
L4966	OEM
L4966A	OEM
L4966B	OEM
L4966C	OEM
L4967	OEM
L4967A	OEM
L4967B	OEM
L4970	OEM
L4971	OEM
L4972	OEM
L4973	OEM
L4974	OEM
L4975	OEM
L4976	OEM
L4977	OEM
L4978	OEM
L4980	OEM
L4981	OEM
L4982	OEM
L4983	OEM
L4984	OEM
L4985	OEM
L4986	OEM
L4987	OEM
L4988	OEM
L5005	OEM
L5021	0004
L5022	0004
L5022A	0004
L5025	0004
L5025A	0004
L5108	0050
L5121	0050
L5122	0050
L5181	0050
L5431	OEM
L5630	1329
L5631	1329
L5631-AA	1329
L6210	OEM
L6506	OEM
L7801A	OEM
L7805ACCV	OEM
L7805ACV	OEM
L7805CV	1812
L7806	1343
L7808CY	1196
L7809	1346
L7809CV	OEM
L7812	1351
L7812CV	0334
L7812V-SA	OEM
L7815CV	1321
L7815V	OEM
L7818CT	3650
L7824CT	3881
L7912CV	6668
L7918CT	5290
L7924CT	3573
L8156	OEM
L8156A-D	OEM
L8157	OEM
L8157A-C	OEM
L8158	OEM
L8158A	OEM
L8158B	OEM
L8166	OEM
L8166A-D	OEM
L8167	OEM
L8167A	OEM
L8167B	OEM
L8168	OEM
L8168A	OEM
L8168B	OEM
L8176	OEM
L8176A	OEM
L8176B	OEM
L8176C	OEM
L8177	OEM
L8177A	OEM
L8177B	OEM
L8178	OEM
L8178A	OEM
L8187	OEM
L8187A	OEM
L8187B	OEM
L8188	OEM
L8188A	OEM
L8200	OEM
L8200A	OEM
L8200B	OEM
L8200C	OEM
L8201	OEM
L8201A	OEM
L8201B	OEM
L8201C	OEM
L8202	OEM
L8202A	OEM
L8202B	OEM
L8202C	OEM
L8202D	OEM
L8202E	OEM
L8203A	OEM
L8203B	OEM
L8203C	OEM
L8204	OEM
L8204A	OEM
L8204B	OEM
L8204C	OEM
L8205	OEM
L8205A	OEM
L8205B	OEM
L8206	OEM
L8206A	OEM
L8206B	OEM
L8206C	OEM
L8206D	OEM
L8206E	OEM
L8207	OEM
L8207A	OEM
L8207B	OEM
L8207C	OEM
L8207D	OEM
L8207E	OEM
L8208	OEM
L8208A	OEM
L8208B	OEM
L8208C	OEM
L8209	OEM
L8209A	OEM
L8209B	OEM
L8209C	OEM
L8210	OEM
L8210A	OEM
L8210B	OEM
L8210C	OEM
L8211	OEM
L8211A	OEM
L8211B	OEM
L8211C	OEM
L8211D	OEM
L8211E	OEM
L8212	OEM
L8212A	OEM
L8212B	OEM
L8212C	OEM
L8213	OEM
L8213A	OEM
L8213B	OEM
L8214	OEM
L8214A	OEM
L8214B	OEM
L8215	OEM
L8215A	OEM
L8215B	OEM
L8216	OEM
L8216A	OEM
L8216B	OEM
L8313A	OEM
L8313B	OEM
L8314A	OEM
L8314B	OEM
L8315A	OEM
L8315B	OEM
L8316A	OEM
L8316B	OEM
L8317A	OEM
L8317B	OEM
L8323A	OEM
L8323B	OEM
L8324A	OEM
L8324B	OEM
L8325B	OEM
L8326A	OEM
L8326B	OEM
L8327A	OEM
L8327B	OEM
L8333A	OEM
L8333B	OEM
L8334A	OEM
L8334B	OEM
L8335A	OEM
L8335B	OEM
L8336A	OEM
L8336B	OEM
L8337A	OEM
L8337B	OEM
L8343A	OEM
L8343B	OEM
L8344A	OEM
L8344B	OEM
L8345	OEM
L8345A	OEM
L8345B	OEM
L8346A	OEM
L8346B	OEM
L8347A	OEM
L8347B	OEM
L8350	OEM
L8351	OEM
L8352	OEM
L8353	OEM
L8354	OEM
L8355	OEM
L8356	OEM
L8370	OEM
L8380	OEM
L8413A	OEM
L8413B	OEM
L8414A	OEM
L8414B	OEM
L8415A	OEM
L8415B	OEM
L8416A	OEM
L8416B	OEM
L8417A	OEM
L8417B	OEM
L8423A	OEM
L8423B	OEM
L8424A	OEM
L8424B	OEM
L8425A	OEM
L8425B	OEM
L8426A	OEM
L8426B	OEM
L8427A	OEM
L8427B	OEM
L8433A	OEM
L8433B	OEM
L8434A	OEM
L8434B	OEM
L8435A	OEM
L8435B	OEM
L8436A	OEM
L8436B	OEM
L8437A	OEM
L8437B	OEM
L8443A	OEM
L8443B	OEM
L8444A	OEM
L8444B	OEM
L8445A	OEM
L8445B	OEM
L8446A	OEM
L8446B	OEM
L8447A	OEM
L8447B	OEM
L8503	OEM
L8504	OEM
L8505	OEM
L8512	OEM
L8513	OEM
L8514	OEM
L8551	OEM
L8552	OEM
L8553	OEM
L8554	OEM
L8555	OEM
L8563	OEM
L8564	OEM
L8565	OEM
L8566	OEM
L8567	OEM
L8621	OEM

If replacement code is OEM, contact original manufacturer for replacement.

DEVICE TYPE	REPL CODE
L8621A	OEM
L8621B-E	OEM
L8621F-H	OEM
L8621J-L	OEM
L8621M	OEM
L8622	OEM
L8622A	OEM
L8622B-E	OEM
L8622F-H	OEM
L8622J2J	OEM
L8622K	OEM
L8631	OEM
L8631A	OEM
L8631B-E	OEM
L8631F-H	OEM
L8631J-L	OEM
L8632	OEM
L8632A	OEM
L8632B-E	OEM
L8632F-H	OEM
L8632J	OEM
L8632K	OEM
L8633	OEM
L8633A	OEM
L8633B-E	OEM
L8633F-H	OEM
L8633J	OEM
L8642	OEM
L8642A	OEM
L8642B-E	OEM
L8642F-H	OEM
L8642J	OEM
L8643	OEM
L8643A	OEM
L8643B-E	OEM
L8643F-H	OEM
L8644	OEM
L8644A	OEM
L8644B-E	OEM
L8644F	OEM
L8644G	OEM
L8645	OEM
L8645A	OEM
L8645B-E	OEM
L8653	OEM
L8653A	OEM
L8653B-3	OEM
L8653F	OEM
L8653G	OEM
L8654	OEM
L8654A	OEM
L8654B-E	OEM
L8654F	OEM
L8655	OEM
L8655A	OEM
L8655B-E	OEM
L8656	OEM
L8656A	OEM
L8656B-D	OEM
L8666	OEM
L8666A	OEM
L8666B-D	OEM
L8667	OEM
L8667A	OEM
L8667B	OEM
L8667C	OEM
L8668	OEM
L8668A	OEM
L8676	OEM
L8676A	OEM
L8676B	OEM
L8676C	OEM
L8677	OEM
L8677A	OEM
L8677B	OEM
L8678	OEM
L8678A	OEM
L8687	OEM
L8687A	OEM
L8687B	OEM
L8700	OEM
L8700A	OEM
L8701	OEM
L8701A	OEM
L8702	OEM
L8702A	OEM
L8703	OEM
L8703A	OEM
L8704	OEM
L8704A	OEM
L8850	OEM
L8851	OEM
L8852	OEM
L8853	OEM
L8860	OEM
L8861	OEM
L8862	OEM
L8870	OEM
L8871	OEM
L8872	OEM
L8900	OEM
L8901	OEM
L8902	OEM
L8903	OEM
L8904	OEM
L8905	OEM
L8906	OEM
L8910	OEM
L8911	OEM

DEVICE TYPE	REPL CODE
L8912	OEM
L8913	OEM
L8914	OEM
L8915	OEM
L8916	OEM
L314141	OEM
L-612099	0467
L-612106	3477
L-612107	0899
L-612150	1539
L-612158	1349
L-612161	0734
L2091241-2	0050
L2091241-3	0050
LA60	0201
LA-71	4870
LA-72	4989
LA-73	4871
LA-74	4873
LA78LR05D-MA	OEM
LA80	0201
LA111	OEM
LA174B	OEM
LA300	0015
LA-301MB	OEM
LA-301MC	OEM
LA-301ML	OEM
LA-301MM	OEM
LA-3010B	OEM
LA-3010C	OEM
LA-3010L	OEM
LA-3010M	OEM
LA-301VB	OEM
LA-301VC	OEM
LA-301VL	OEM
LA-301VM	OEM
LA-301YB	OEM
LA-301YC	OEM
LA-301YL	OEM
LA-301YM	OEM
LA350	0416
LA361	OEM
LA600	0015
LA703	OEM
LA703E	0632
LA733(P)	OEM
LA780Z	OEM
LA800	0015
LA1014	OEM
LA1017	OEM
LA1024	OEM
LA1027	OEM
LA1034	OEM
LA1038	OEM
LA1041	0138
LA1110	0355
LA1111	0387
LA1111P	0387
LA1135	OEM
LA1140	0465
LA1140B	OEM
LA1150	OEM
LA1150N	OEM
LA1152N	1446
LA1175	OEM
LA1177	OEM
LA1180	OEM
LA1185	OEM
LA1186N	OEM
LA1200	0579
LA-1201	0579
LA1201	0579
LA1201(B)	0579
LA1201(C)	0579
LA1201-22	OEM
LA1201B	0579
LA1201C	0579
LA1201C-W	0579
LA1201L	OEM
LA-1201T	0579
LA1201T	0579
LA1201W	0579
LA1210	0603
LA1210L	0603
LA1221	OEM
LA1222	0621
LA1230	0637
LA1231	0639
LA1231N	0639
LA1235	OEM
LA1240	0651
LA1245	0662
LA1260	OEM
LA1260S	OEM
LA1261	OEM
LA1265	OEM
LA1265S	OEM
LA1265T	OEM
LA1266	OEM
LA1267	OEM
LA1320A	3246
LA1342	0829
LA1352	0852
LA1353	0855
LA1354	0857
LA1357	0865
LA1357A	OEM
LA1357AB	0865
LA1357B	0865

DEVICE TYPE	REPL CODE
LA1357N	0865
LA1357NA	OEM
LA1357NB	0865
LA1357R	3403
LA1363	0167
LA1363W	0167
LA1364	0878
LA1364N	0878
LA1365	0167
LA1365N	0167
LA1365V	OEM
LA1366	0881
LA1366N	0881
LA1367	0884
LA1368	0856
LA1368BP	4453
LA1368CW	4454
LA1369	0886
LA1373	0043
LA1374	0389
LA1375	0607
LA1376	0607
LA1385	0904
LA1390	0916
LA1390B	0916
LA1390C	0916
LA1425	0971
LA1460	1022
LA1461	1026
LA1503	OEM
LA1805	OEM
LA1810	OEM
LA1810SSK	OEM
LA1851N	OEM
LA2010	OEM
LA2015	OEM
LA2026	OEM
LA2037	OEM
LA2110	OEM
LA2400	OEM
LA2600S	3103
LA2730	OEM
LA2746	OEM
LA2770	OEM
LA2800	OEM
LA3018	1854
LA3086N	1696
LA3115	2514
LA3148	2529
LA3150	2530
LA3155	2532
LA3160	2534
LA3161	2538
LA3166N	0881
LA3190	0916
LA3201	2572
LA3210	2063
LA3210ALC	2063
LA3220	2595
LA3225T	OEM
LA3226T	OEM
LA3245	OEM
LA3246	OEM
LA-3300	2568
LA3300	2568
LA-3301	1215
LA3301	1215
LA3310	2669
LA3311	2670
LA3320	OEM
LA3350	0416
LA3350A	0416
LA3361	1260
LA3361A	1260
LA3365	OEM
LA3370	2712
LA3375	OEM
LA3376P	OEM
LA3390	OEM
LA3400	OEM
LA3400N	OEM
LA3401	OEM
LA-3410	5128
LA3410	OEM
LA-3420	5135
LA-3430	OEM
LA3430	OEM
LA3430P	OEM
LA-3440	5131
LA3600	OEM
LA3605	OEM
LA-3610	5127
LA-3620	5134
LA-3630	OEM
LA-3640	5130
LA-3810	5129
LA-3820	5137
LA-3830	OEM
LA-3840	5132
LA-3910	OEM
LA-3920	OEM
LA-3930	OEM
LA-3940	OEM
LA4000	OEM
LA4030	OEM
LA4030-7	3020
LA4030P	0843
LA4031P	0864
LA4032P	0882

DEVICE TYPE	REPL CODE
LA4100	2614
LA4100DN	OEM
LA4101	2614
LA4102	3190
LA4102KN	OEM
LA4108R	OEM
LA4112	3196
LA4125	1951
LA4125T	1951
LA4126	1951
LA4126T	1960
LA4140	0362
LA4140-1	OEM
LA4145	OEM
LA4160	OEM
LA4170	3221
LA4180	OEM
LA4182	3224
LA4183	3224
LA4190	2386
LA4190T	OEM
LA4192	2386
LA4192S	2386
LA4201	3228
LA4220	2193
LA4230	3239
LA4250	3239
LA4260	3253
LA4261	3253
LA4261T	2918
LA4265	OEM
LA4270	3097
LA4275	OEM
LA4280	OEM
LA4280-TV	OEM
LA4282	OEM
LA4400	3318
LA4400FR	3318
LA4400Y	2193
LA4420	2193
LA4440	3333
LA4440-R	3333
LA4440R	3333
LA4445	3335
LA4460N	OEM
LA4465	OEM
LA4500	OEM
LA4505	3363
LA4507	OEM
LA4512B	OEM
LA4555	OEM
LA4570	OEM
LA4597	OEM
LA5003LP	OEM
LA5005	OEM
LA5100	OEM
LA5110N	OEM
LA5112	1446
LA5112N	1446
LA5112N-B	OEM
LA5112N-G	1446
LA5112R	OEM
LA5115N	1446
LA5192P	OEM
LA5511	OEM
LA5512	OEM
LA5515	OEM
LA5522	OEM
LA5523	OEM
LA5535N	OEM
LA5536N	OEM
LA5601	OEM
LA5666	OEM
LA5667	OEM
LA6324	0625
LA6324N	OEM
LA6339	0176
LA6358	0770
LA6358S	3233
LA6358T	OEM
LA6393D	6728
LA6393S	OEM
LA6458	OEM
LA6458D	0360
LA6458DF	0360
LA6458DS	OEM
LA6458M	OEM
LA6458S	2913
LA6458SS	OEM
LA-6460	OEM
LA6462D	OEM
LA6462S	OEM
LA-6480	OEM
LA6500	OEM
LA6510	OEM
LA6510L	OEM
LA6515	OEM
LA6520	4120
LA6532M	OEM
LA-6660	OEM
LA-6680	OEM
LA-6760	2206
LA-6780	2199
LA-6860	OEM
LA-6880	OEM
LA-6960	OEM
LA-6980	OEM
LA7000	OEM
LA7000S	OEM

DEVICE TYPE	REPL CODE
LA7003	OEM
LA7003S	OEM
LA7005	OEM
LA7005S	OEM
LA7007	OEM
LA7007S	OEM
LA7009	OEM
LA7009S	OEM
LA7011	OEM
LA7013	OEM
LA7016	1028
LA7016//-1	1028
LA7017	OEM
LA7018	OEM
LA7019	OEM
LA7031	OEM
LA7032	OEM
LA7033	OEM
LA7034	OEM
LA7040	2105
LA7042	OEM
LA7051	OEM
LA7053	OEM
LA7054	OEM
LA7054Z	OEM
LA7055	OEM
LA7058	OEM
LA7058R	OEM
LA7060	OEM
LA7070	OEM
LA7071	OEM
LA7090	OEM
LA7095	OEM
LA7096	OEM
LA7110	OEM
LA7113	OEM
LA7116	OEM
LA7117	OEM
LA7122	OEM
LA7122S	OEM
LA7124	OEM
LA7210	OEM
LA7212	OEM
LA7213	OEM
LA7220	3027
LA7221	OEM
LA7222	4265
LA7224	4266
LA7225	OEM
LA7227	OEM
LA7237	OEM
LA7270M	OEM
LA7280MA	OEM
LA7293M	OEM
LA7295	OEM
LA7296-PK	OEM
LA7297	OEM
LA7300	OEM
LA7301M	OEM
LA7305	OEM
LA7305A	OEM
LA7305N	OEM
LA7306	OEM
LA7306N	OEM
LA7309	OEM
LA7309N	OEM
LA7315	OEM
LA7316A	OEM
LA7317	4277
LA7318	OEM
LA7320	OEM
LA7320M	OEM
LA7321	OEM
LA7321M	OEM
LA7322	OEM
LA7323	OEM
LA7323A	OEM
LA7325	OEM
LA7505	OEM
LA7507	4307
LA7510	OEM
LA7520	0539
LA7521	OEM
LA7522	OEM
LA7525	OEM
LA-7530	3817
LA7530	3817
LA7530N	3817
LA7531	OEM
LA7545	OEM
LA7550	OEM
LA7555	OEM
LA7575	OEM
LA7577	OEM
LA7577N	OEM
LA7600N	OEM
LA7601	1988
LA7601G	1988
LA-7601W	1988
LA7601W	1988
LA7610	OEM
LA7620	0557
LA7621	0391
LA7625	0486
LA7626	0462
LA7628	OEM
LA7629	0523
LA7650	OEM
LA7650K	4094

DEVICE TYPE	REPL CODE
LA7650K-N	OEM
LA7650KN	4094
LA7651N	OEM
LA7651P	OEM
LA7652	OEM
LA7655	4094
LA7655N	4094
LA7655N(SMC)	OEM
LA7670	4326
LA7671	OEM
LA7672A	OEM
LA7673	OEM
LA7690	4174
LA7696	OEM
LA7751	OEM
LA7760	3365
LA7761	3884
LA7765	OEM
LA7800	0332
LA7800R	0332
LA7802	1590
LA7806	4351
LA7808	OEM
LA7809	OEM
LA7810	OEM
LA7811	4353
LA7820	OEM
LA7822	OEM
LA7823	4355
LA7830	0732
LA7831	2957
LA7833	OEM
LA7835	2346
LA7835-TV	2346
LA7835K	2346
LA7836	1266
LA7836)TV	OEM
LA7836-TV	1266
LA7837	4359
LA7838	4360
LA7845	OEM
LA7900	OEM
LA7905	OEM
LA7910	2666
LA7911	OEM
LA7912	4377
LA7913	4380
LA7915	4381
LA7916	4382
LA7920	4384
LA7925	OEM
LA7926	OEM
LA7930	OEM
LA7934	OEM
LA7935	OEM
LA7938	OEM
LA7940	0683
LA7950	OEM
LA7951	OEM
LA7952	OEM
LA7953	OEM
LA7954	OEM
LA7955	OEM
LA7956	OEM
LA7957	OEM
LA7970	OEM
LA7990	OEM
LA9000	OEM
LA9000N	OEM
LA9010	OEM
LA9100	OEM
LA9200N	OEM
LA9200NM	OEM
LA13301	1215
LAA300	0015
LAA600	0015
LAA800	0015
LAB1363	0167
LAC010	OEM
LAD010	4589
LAD-011	2031
LAD011	2031
LAG570	OEM
LAG600	OEM
LAG624	OEM
LAP-011	1235
LAP011	1235
LAP-012	4437
LAS723	1192
LAS723B	1192
LAS1505	1922
LASER128ROM	OEM
LB120	OEM
LB-156	OEM
LB156	2189
LB160	0378
LB-202MB	OEM
LB-202ML	OEM
LB-2020B	OEM
LB-2020L	OEM
LB-202VL	OEM
LB-202VR	OEM
LB-202YB	OEM
LB-202YL	OEM
LB-203MB	OEM
LB-203ML	OEM
LB-2030B	OEM
LB-2030L	OEM
LB-203VB	OEM

DEVICE TYPE	REPL CODE
LB-203VL	OEM
LB230	OEM
LB-402MA	OEM
LB-402MK	OEM
LB-4020A	OEM
LB-4020K	OEM
LB-402VA	OEM
LB-402VK	OEM
LB402YL1	OEM
LB602-MA	OEM
LB-602MA	OEM
LB602MA-2	OEM
LB602MA2	OEM
LB-602MA2Z	OEM
LB-603MA	OEM
LB-603VA	OEM
LB1021	OEM
LB1200	OEM
LB1205	OEM
LB1214	OEM
LB1240	OEM
LB1259	OEM
LB1274	4963
LB1274K	4963
LB1274KF	OEM
LB1274KFT	OEM
LB1275	4962
LB1275K	OEM
LB1287	4138
LB1287K	OEM
LB1288	5284
LB1290	3893
LB1331	5159
LB1332	5159
LB1403	5162
LB1403N	OEM
LB1405	4830
LB1408	OEM
LB1409	4826
LB1412	OEM
LB1413N	OEM
LB1415	OEM
LB1416	4280
LB1419	OEM
LB1423	OEM
LB1423N	OEM
LB1433N	OEM
LB1500	OEM
LB1501	OEM
LB1515	OEM
LB1550	OEM
LB1551	OEM
LB-1620	OEM
LB1620	OEM
LB1624D	OEM
LB1624H	OEM
LB1630	OEM
LB1631MTP	OEM
LB1639	OEM
LB1641	OEM
LB1642	OEM
LB1645	OEM
LB1645N	OEM
LB1646	OEM
LB1649	5251
LB1684H	OEM
LB1731	OEM
LB-2000	0141
LB2000	0141
LB2001	0562
LB2002	1823
LB2003	1041
LB2004	0033
LB2005	2097
LB2006	1831
LB2007	1176
LB2030	0081
LB2031	0358
LB2032	0333
LB2100	1844
LB2101	0341
LB2102	1859
LB2106	1835
LB2130	3399
LB2131	1632
LB2132	1481
LB3000	0232
LB3001	0512
LB3001(FANON)	2406
LB3002	0697
LB3003	0873
LB3004	0743
LB3005	1275
LB3006	0361
LB3008	0314
LB3009	1024
LB3060	1535
LB3150	1208
LB3175	0980
LB3500	OEM
LB-6410	OEM
LB-6430	OEM
LB-6440	OEM
LB-6610	OEM
LB-6630	OEM
LB-6640	OEM
LB-6650	OEM
LB-6710	4327
LB-6730	4331

If replacement code is OEM, contact original manufacturer for replacement.

DEVICE TYPE	REPL CODE
LB-6740	4332
LB-6750	4333
LB-6810	OEM
LB-6830	OEM
LB-6840	OEM
LB-6850	OEM
LB-6910	OEM
LB-6930	OEM
LB-6940	OEM
LB-6950	OEM
LB8555S	OEM
LBA-01Z1	0277
LBA-02	0289
LBA-02Z1	0277
LBA-04	0296
LBA-04Z1	0296
LBA-06	0302
LBA-08	0250
LBA-10	0250
LBC547B	OEM
LBC547C	OEM
LC0.09M11/15	1303
LC0.09M1115	0015
LC6.5A	OEM
LC7.0A	OEM
LC7.5A	OEM
LC8.0A	OEM
LC8.5A	OEM
LC9.0A	OEM
LC10A	OEM
LC11A	OEM
LC12A	OEM
LC13A	OEM
LC14A	OEM
LC15	OEM
LC16	OEM
LC17	OEM
LC18	OEM
LC20	OEM
LC22	OEM
LC24	OEM
LC26	OEM
LC28	OEM
LC30	OEM
LC33	OEM
LC36	OEM
LC40	OEM
LC43	OEM
LC45	OEM
LC48	OEM
LC51	OEM
LC54	OEM
LC58	OEM
LC60	OEM
LC64	OEM
LC70	OEM
LC74HC02M	OEM
LC74HC05	OEM
LC74HC138	OEM
LC74HC174	OEM
LC74HCU04	OEM
LC75	OEM
LC80	OEM
LC90	OEM
LC100	OEM
LC110	OEM
LC120	OEM
LC130	OEM
LC150	OEM
LC160	OEM
LC170	OEM
LC180	OEM
LC-204MB	OEM
LC-204MD	OEM
LC-204ML	OEM
LC-204MN	OEM
LC-2040B	OEM
LC-2040D	OEM
LC-2040N	OEM
LC-204PL	OEM
LC-204RB	OEM
LC-204RD	OEM
LC-204RL	OEM
LC-204RN	OEM
LC-289(ELCOM)	1753
LC654H4069	OEM
LC945(P)	OEM
LC3019	OEM
LC3019A	OEM
LC3019B	OEM
LC3029	OEM
LC3029A	OEM
LC3029B	OEM
LC3039	OEM
LC3039A	OEM
LC3039B	OEM
LC3049	OEM
LC3049A	OEM
LC3049B	OEM
LC3059	OEM
LC3059A	OEM
LC3059B	OEM
LC3149	OEM
LC3149A	OEM
LC3149B	OEM
LC3514A	OEM
LC3514AD	OEM
LC3514D	OEM
LC3517A-15	OEM
LC3517AM-15	OEM
LC3517AS-15	OEM
LC3517BM-15	OEM
LC3517BML15T	OEM
LC3517BS	OEM
LC3517BS-15	OEM
LC3664N-10	OEM
LC3664N-12	OEM
LC4001B	0478
LC4001BP	0478
LC4011	0215
LC4011B	0215
LC4013B	0413
LC4013BM	OEM
LC4016B	4150
LC4066B	0101
LC4066BH	OEM
LC4069UB	0119
LC4071B	0129
LC4075B	6414
LC4081B	0626
LC4966	OEM
LC4969	OEM
LC6502B-633	OEM
LC6510C773	OEM
LC6512A3377	OEM
LC6512A3997	OEM
LC6512D797	OEM
LC6514B-729	OEM
LC6514B-3357	OEM
LC6521H-3257	OEM
LC6521H-3351	OEM
LC6521H-3755	OEM
LC6522C-3433	OEM
LC6523H304	OEM
LC6523H-3043	OEM
LC6526CPA	OEM
LC6527C-3570	OEM
LC6528C-3175	OEM
LC6528C-3421	OEM
LC6528C-3664	OEM
LC6538D-3753	OEM
LC6538D-3830	OEM
LC6538D-3992	OEM
LC6543H	OEM
LC6543H-3313	OEM
LC6543H-3384	OEM
LC6543H-3390	OEM
LC6543H-4016	OEM
LC6554H-3683	OEM
LC6554H3683	OEM
LC6554H-3906	OEM
LC6554H-4069	OEM
LC6554H4069	OEM
LC6554H-4082	OEM
LC6558D-3253	OEM
LC6568H-3350	OEM
LC6568H-3379	OEM
LC6658C	OEM
LC6658C-3395	OEM
LC7110	2362
LC7113	OEM
LC7120	4850
LC7130	OEM
LC7131	4791
LC7135	OEM
LC7136	OEM
LC7137	OEM
LC7150	OEM
LC7180	OEM
LC7181	OEM
LC7191	OEM
LC7217	OEM
LC7218	OEM
LC7220	OEM
LC7250	OEM
LC7265	OEM
LC7351	OEM
LC7360	OEM
LC7363	OEM
LC7385	OEM
LC7401	OEM
LC7410	OEM
LC7411-05	OEM
LC7411-08	OEM
LC7411-8011	OEM
LC7412	OEM
LC7412-8015	OEM
LC7413	OEM
LC7413C-8022	OEM
LC7414	OEM
LC7415	OEM
LC7416-8065	OEM
LC7416-8071	OEM
LC7416-8083	OEM
LC7418-8067	OEM
LC7418-8073	OEM
LC7418-8081	OEM
LC7418-8086	OEM
LC7419	OEM
LC7419-8090	OEM
LC7431	OEM
LC7431M	OEM
LC7432	OEM
LC7432M	OEM
LC7440	OEM
LC7440E	OEM
LC7450	OEM
LC7456A	OEM
LC7457A-02	OEM
LC7458A-02	OEM
LC7458A-03	OEM
LC7458B-03	OEM
LC7458B-04	OEM
LC7460	OEM
LC7461-8101	OEM
LC7461M	OEM
LC7461M-8100	OEM
LC7462M	OEM
LC7463M	OEM
LC7464M	OEM
LC7465M	OEM
LC7480	OEM
LC7512	OEM
LC7522	OEM
LC7533	OEM
LC7550	OEM
LC7566	OEM
LC7570	OEM
LC7580	OEM
LC7582	OEM
LC7582B	OEM
LC7583	OEM
LC7640	OEM
LC7800	OEM
LC7816	OEM
LC7818	OEM
LC7821	OEM
LC7860J	OEM
LC7860N	OEM
LC7863A	OEM
LC7880	OEM
LC7880M	OEM
LC7881-B	OEM
LC7881-C	OEM
LC7881B	OEM
LC7881C	OEM
LC7900	OEM
LC8640-FS9	OEM
LC8991	OEM
LC35188ML	OEM
LC66508B-3721	OEM
LC66508B-3762	OEM
LC66508B-4142	OEM
LC66508B-4197	OEM
LC74402	OEM
LC75280	OEM
LC89060	OEM
LC89060M	OEM
LC89066	OEM
LC89066M	OEM
LC89080	OEM
LC89086	OEM
LC864016B	OEM
LC864020A	OEM
LC864024-FSG	OEM
LC864024A	OEM
LC864024A-5413	OEM
LC864032A-5414	OEM
LC864116K	OEM
LC864156A-5730	OEM
LC864156A-5830	OEM
LC864156A-5838	OEM
LC8640116B	OEM
LC651483005	OEM
LCC-2	OEM
LCC-3	OEM
LCC-4	OEM
LCC-5	OEM
LCC-6	OEM
LCC-7	OEM
LCC-8	OEM
LCC-9	OEM
LCC-10	OEM
LCC-11	OEM
LCC-12	OEM
LCC-13	OEM
LCC-14	OEM
LCC-15	OEM
LCC-16	OEM
LCC-FS	OEM
LCD15M	OEM
LCD915-51	OEM
LCD9898	OEM
LCE6.5A	OEM
LCE7.0A	OEM
LCE7.5A	OEM
LCE8.0A	OEM
LCE8.5A	OEM
LCE9.0A	OEM
LCE10A	OEM
LCE11A	3172
LCE12A	OEM
LCE13A	OEM
LCE14A	OEM
LCE15A	3191
LCE16A	OEM
LCE17A	3201
LCE18A	OEM
LCE20A	OEM
LCE22A	1915
LCE24A	OEM
LCE26A	OEM
LCE28A	OEM
LCE30A	OEM
LCE33A	OEM
LCE36A	OEM
LCE40A	OEM
LCE43A	0568
LCE45A	OEM
LCE48A	OEM
LCE51A	OEM
LCE54A	OEM
LCE58A	0831
LCE60A	OEM
LCE64A	OEM
LCE70A	OEM
LCE75A	OEM
LCE80A	OEM
LCE90A	OEM
LCE100A	OEM
LCE110A	OEM
LCE120A	OEM
LCE130A	OEM
LCE150A	OEM
LCE160A	OEM
LCE170A	1405
LCO.09M11/15	1303
LCS1008P	0167
LC.09M	0015
LC.09M11/15	0015
LD	OEM
LD-001DU	OEM
LD-001MG	OEM
LD-001VR	OEM
LD-001YY	OEM
LD-002GB	OEM
LD-002RB	OEM
LD-002UB	OEM
LD-002YB	OEM
LD-003GB	OEM
LD-003RB	OEM
LD-003UB	OEM
LD-003YB	OEM
LD34R	0124
LD37A	OEM
LD-64R	OEM
LD100	OEM
LD101	OEM
LD-101DU	OEM
LD-101MG	OEM
LD-101VR	OEM
LD-101YY	OEM
LD102	OEM
LD128	0917
LD-201DU	OEM
LD-201MC	OEM
LD-201MG	OEM
LD-201VR	OEM
LD-201YY	OEM
LD271	0516
LD271E	OEM
LD271E7317	OEM
LD274	OEM
LD300	2406
LD-701DU	OEM
LD-701MG	OEM
LD-701VR	OEM
LD-701YY	OEM
LD-702DU	OEM
LD-702MG	OEM
LD-702VR	OEM
LD-702YY	OEM
LD-706DU	OEM
LD-706MG	OEM
LD-706VR	OEM
LD-706YY	OEM
LD1020	0021
LD1020A	0021
LD1041	0138
LD-1110	0355
LD1110	0355
LD3000	2404
LD3001	2406
LD3001W	2406
LD-3001X	OEM
LD3001X	2406
LD3040	2440
LD3061	2464
LD3080	2476
LD3110	2505
LD3110A	2505
LD3120	2509
LD3141	OEM
LD-3150	2155
LD3150	2155
LD-3150C	OEM
LD3150C	2155
LD315009-308025	OEM
LDA400	0079
LDA400MP	0079
LDA401	0079
LDA401MP	0079
LDA402	0079
LDA404	0086
LDA405	0086
LDA406	0086
LDA408	0086
LDA410	0111
LDA412	0418
LDA420	0400
LDA450	0126
LDA451	0532
LDA452	0150
LDA453	0532
LDA454	0693
LDA455	0693
LDD10	OEM
LDD50	OEM
LDF603	3622
LDF604	3622
LDF605	4214
LDF691	OEM
LDF692	OEM
LDF693	OEM
LDR05	OEM
LDR07	OEM
LDS200	0086
LDS201	0086
LDS202	0150
LDS203	0150
LDS206	0079
LDS207	0111
LDS208	0400
LDS210	0079
LDS257	0693
LDZ70/6A2	0471
LDZ70/9A1	0057
LE93CS56B1	OEM
LE2668E	OEM
LED55B	0591
LED55BF	0591
LED55C	0591
LED55CF	0591
LED56	0591
LED56F	0591
LED-15005	1069
LED21085	1257
LED21092	2078
LEDU-1	OEM
LEDU-4	OEM
LEDU-9	OEM
LEDU-12	OEM
LEDV3	OEM
LF155AH	OEM
LF155H	OEM
LF155J	OEM
LF156AH	OEM
LF156H	OEM
LF156J	OEM
LF157AH	OEM
LF157H	OEM
LF157J	OEM
LF255H	OEM
LF255J	OEM
LF255N	OEM
LF256H	OEM
LF256J	OEM
LF256N	OEM
LF257H	OEM
LF257J	OEM
LF257N	OEM
LF347	3391
LF347BN	OEM
LF347N	3391
LF351N	5877
LF353	3741
LF353N	3741
LF355	3672
LF355AH	6016
LF355BH	OEM
LF355BJ	OEM
LF355BN	6017
LF355H	3621
LF355J	6039
LF355N	6039
LF356	6039
LF356AH	6016
LF356BH	OEM
LF356BJ	OEM
LF356BN	6017
LF356H	3621
LF356J	6039
LF356N	6039
LF357AH	6016
LF357BH	OEM
LF357BJ	OEM
LF357BN	6017
LF357H	3621
LF357J	6039
LF357N	6039
LF11201D	OEM
LF11202D	OEM
LF11331D	OEM
LF11332D	OEM
LF11333D	OEM
LF13201D	OEM
LF13201N	OEM
LF13202D	OEM
LF13202M	OEM
LF13202N	OEM
LF13331D	OEM
LF13331M	OEM
LF13331N	3989
LF13332D	OEM
LF13332M	OEM
LF13332N	OEM
LF13333D	OEM
LF13333M	OEM
LF13333N	OEM
LFB01	OEM
LFB-01L	OEM
LFB01L	OEM
LF-F10	OEM
LFM5	OEM
LFRC-27-15	0023
LG1353G	OEM
LG1354G	OEM
LG1454G	OEM
LH0115A	OEM
LH5060	OEM
LH5060B	OEM
LH5116-15	OEM
LH5116N	OEM
LH5116N-15	OEM
LH21256-12	1472
LH534187	OEM
LHV1B	0348
LHV1C	0348
LHV2D	0348
LHV2E	OEM
LHV2F	OEM
LHV3G	OEM
LHV3H	OEM
LHV3K	OEM
LHV4M	OEM
LHV4Q	OEM
LHV4R	OEM
LHV5T	OEM
LHV5V	OEM
LI201	OEM
LI210G1	OEM
LI2108	OEM
LID6.8	OEM
LID6.8A	OEM
LID7.5	OEM
LID7.5A	OEM
LID8.2	OEM
LID8.2A	OEM
LID9.1	OEM
LID9.1A	OEM
LID10	OEM
LID10A	OEM
LID11	OEM
LID11A	OEM
LID12	OEM
LID12A	OEM
LID13	OEM
LID13A	OEM
LID15	OEM
LID15A	OEM
LID16	OEM
LID16A	OEM
LID18	OEM
LID18A	OEM
LID20	OEM
LID20A	OEM
LID22	OEM
LID22A	OEM
LID24	OEM
LID24A	OEM
LID27	OEM
LID27A	OEM
LID30	OEM
LID30A	OEM
LID33	OEM
LID33A	OEM
LID929	0111
LID930	0111
LJ-152	0037
LJ152	0037
LJ152(0)	0037
LJ152(0)	0150
LJ152G	0037
LK1352P	0395
LKA4020	OEM
LKB52P	0395
LL-2	0015
LL2	0015
LL4148	0124
LL4821J-L	OEM
LM-035MG-A	OEM
LM-035MG-B	OEM
LM-035VR-A	OEM
LM-035VR-B	OEM
LM-035YY-A	OEM
LM-035YY-B	OEM
LM7-28	OEM
LM11	OEM
LM11CH	OEM
LM11CJ	OEM
LM11CJ-8	OEM
LM11CLH	OEM
LM11CLI	OEM
LM11CLI-8	OEM
LM11CLJ	OEM
LM11CLJ-8	OEM
LM11CLN	OEM
LM11CLN-14	OEM
LM11CN	OEM
LM11CN-14	OEM
LM11H	OEM
LM11J	OEM
LM11J-8	OEM
LM31GCPHLM-U	OEM
LM55SCN	3285
LM59NV	OEM
LM78CN	OEM
LM78L05	5644
LM78L05A	1292
LM78L05ACH	1292
LM78L05ACI	1292
LM78L05ACZ	1292
LM78L08ACH	0083
LM78L08ACZ	0083
LM78L08CH	0083
LM78L12	1828
LM78L12ACZ	1828
LM78MR05-FA	OEM
LM101A	OEM
LM101ADE	OEM
LM101ADE/883B	OEM
LM101AGC	OEM
LM101AH	0093
LM101AH/883B	OEM
LM101AJ	OEM
LM101H	0093
LM107	OEM
LM108AF	OEM
LM108AH	OEM
LM108AJ	OEM
LM108AJ-8	OEM
LM108AN	OEM
LM108F	OEM
LM108H	OEM
LM108J	OEM
LM108J-8	OEM
LM108N	OEM
LM111	OEM
LM111H	0419
LM111J-8	OEM
LM112H	OEM
LM117H	OEM
LM117K	OEM
LM117LH	OEM
LM117MR	OEM
LM123AK	OEM
LM123K	0660
LM124	0661
LM124A	0661
LM124J	OEM
LM124J/883B	OEM
LM1310N	0520
LM137H	OEM
LM137K	OEM
LM137MR	OEM
LM139	OEM
LM139AJ	OEM
LM139AJ/883B	OEM
LM139J	OEM
LM139J/883B	OEM
LM140K-5.0	0660
LM140K-6.0	OEM
LM140K-8.0	OEM
LM140K-12	1927
LM140K-15	1930
LM140K-18	OEM
LM140K-24	OEM
LM148	OEM
LM148J	OEM
LM148J/883B	OEM
LM150K	OEM
LM158	OEM
LM158A	OEM
LM158AJ	OEM
LM158H	OEM
LM158J	OEM
LM193	OEM
LM193A	OEM
LM193AH	OEM
LM193H	OEM
LM201A	OEM
LM201ADE	OEM
LM201AH	0093
LM201AJ	OEM
LM201AN	1300
LM201AND	OEM
LM201H	0093
LM207H	1781
LM207N	0312
LM208AD	OEM
LM208AF	OEM
LM208AH	OEM
LM208AJ	OEM
LM208AJ-8	OEM
LM208AN	OEM
LM208F	OEM
LM208H	OEM
LM208J	OEM
LM208J-8	OEM
LM208N	OEM
LM209H	OEM
LM209K	1801
LM211	OEM
LM211H	1815
LM211J-8	OEM
LM212H	OEM
LM217H	OEM
LM217MR	OEM
LM218	OEM
LM223AK	OEM
LM223K	1916
LM224	0661
LM224A	0661
LM224AD	0625
LM224AF	0625
LM224AJ	0625
LM224AN	0625
LM224D	0625
LM224J	0625
LM224N	0625
LM237H	OEM
LM237K	OEM

If replacement code is OEM, contact original manufacturer for replacement.

DEVICE TYPE	REPL CODE
LM237MR	OEM
LM238K	OEM
LM239	0176
LM239A	0176
LM239AA	OEM
LM239AF	OEM
LM239AJ	0176
LM239AN	0176
LM239F	OEM
LM239J	0176
LM239N	0176
LM246J	OEM
LM246N	OEM
LM248	OEM
LM248J	OEM
LM248N	OEM
LM249J	OEM
LM250K	OEM
LM258	OEM
LM258A	OEM
LM258AH	1677
LM258AJ	6018
LM258AN	1677
LM258AT	1677
LM258D	OEM
LM258FE	OEM
LM258H	1677
LM258J	OEM
LM258JG	0770
LM258N	0770
LM258P	0770
LM258T	1677
LM260H	OEM
LM260N14	OEM
LM261H	OEM
LM261J	OEM
LM275D	OEM
LM292H	OEM
LM292J	OEM
LM293	OEM
LM293A	OEM
LM293AFE	OEM
LM293AH	OEM
LM293AN	OEM
LM293AT	OEM
LM293FE	OEM
LM293H	OEM
LM293JG	OEM
LM293N	OEM
LM293P	OEM
LM293T	OEM
LM299AH	OEM
LM299AH-20	OEM
LM299H	OEM
LM301A	OEM
LM301AD	2355
LM301ADE	1300
LM301AF	OEM
LM301AFE	OEM
LM301AH	0093
LM301AJ	OEM
LM301AJD	OEM
LM301AJDS	OEM
LM301AJG	OEM
LM301AJS	OEM
LM301AN	1300
LM301AN-14	OEM
LM301AND	OEM
LM301ANDS	OEM
LM301ANS	OEM
LM301AP	1300
LM301AT	1300
LM301AU	OEM
LM301AV	1300
LM301F	OEM
LM302H	2222
LM304F	OEM
LM304H	OEM
LM304J	OEM
LM304N	OEM
LM305AH	OEM
LM305AJG	OEM
LM305AP	2169
LM305H	2470
LM305JG	OEM
LM305P	OEM
LM305T	2470
LM306	OEM
LM306H	OEM
LM306J	OEM
LM306JG	OEM
LM306N	OEM
LM306P	OEM
LM306U	OEM
LM307	OEM
LM307DE	2283
LM307F	2283
LM307H	2486
LM307J	2283
LM307J14	OEM
LM307JG	OEM
LM307N	2283
LM307P	2283
LM307T	1781
LM307U	OEM
LM307V	OEM
LM308AD	OEM
LM308AF	OEM
LM308AH	0896
LM308AH-1	OEM
LM308AH-2	OEM
LM308AJ	OEM
LM308AJ-8	0896
LM308AJ8	OEM
LM308AJ-8D	OEM
LM308AJ-8DS	OEM
LM308AJ-8S	OEM
LM308AN	2246
LM308AND	OEM
LM308ANDS	OEM
LM308ANS	OEM
LM308AT	0896
LM308D	2246
LM308DE	OEM
LM308F	OEM
LM308H	0896
LM308J	OEM
LM308J-8	2246
LM308J8	OEM
LM308J-8D	OEM
LM308J-8DS	OEM
LM308J-8S	OEM
LM308N	2246
LM308ND	OEM
LM308NDS	OEM
LM308NS	OEM
LM308T	0896
LM309DA	1922
LM309DB	OEM
LM309H	OEM
LM309K	1922
LM310D	OEM
LM310H	2222
LM310J	OEM
LM310J8	OEM
LM310N	2517
LM311	1815
LM311D	2104
LM311DE	OEM
LM311F	2104
LM311H	1815
LM311J	OEM
LM311J-8	OEM
LM311J8	OEM
LM311J-8D	OEM
LM311J-8DS	OEM
LM311J-8S	OEM
LM311JG	6448
LM311N	2104
LM311N-14	OEM
LM311N14	OEM
LM311ND	OEM
LM311NDS	2104
LM311NS	OEM
LM311P	2104
LM311T	1815
LM311U	OEM
LM311V	OEM
LM312H	OEM
LM313H	OEM
LM316AH	OEM
LM316H	OEM
LM317	2723
LM317H	OEM
LM317HVH	OEM
LM317HVK	OEM
LM317K	1192
LM317KA	OEM
LM317KC	2564
LM317LH	OEM
LM317LZ	5280
LM317MP	OEM
LM317MR	OEM
LM317MT	OEM
LM317T	2564
LM318	2565
LM318DE	OEM
LM318H	2562
LM318J	OEM
LM318J-8	2567
LM318JG	2567
LM318M	2565
LM318N	2567
LM318P	2567
LM318U	OEM
LM319F	2570
LM319H	2574
LM319J	2570
LM319K	2574
LM319N	2570
LM320H5.0	OEM
LM320H12	OEM
LM320H15	OEM
LM320K5.0	2004
LM320K12	5287
LM320K15	5289
LM320KC5.0	OEM
LM320KC12	OEM
LM320KC15	OEM
LM320L-12	1836
LM320LZ5.0	OEM
LM320LZ-12	1836
LM320LZ12	1836
LM320LZ15	5041
LM320MLP5.0	OEM
LM320MLP5.0TB	OEM
LM320MLP12	OEM
LM320MLP12TB	OEM
LM320MLP15	OEM
LM320MLP15TB	OEM
LM320MP5.0	OEM
LM320MP5.0TB	OEM
LM320MP-5.2	1285
LM320MP-12	1838
LM320MP12	OEM
LM320MP12TB	OEM
LM320MP15	OEM
LM320MP15TB	OEM
LM320MP-24	3575
LM320T-5.0	1285
LM320T5.0	1285
LM320T-5.2	1285
LM320T-6.0	2648
LM320T6.0	2648
LM320T8.0	1196
LM320T-12	1838
LM320T12	1838
LM320T-15	3829
LM320T15	3829
LM320T-24	3575
LM320T24	3575
LM321H	OEM
LM322H	OEM
LM322N	OEM
LM323AK	OEM
LM323AT	OEM
LM323K	1916
LM323T	OEM
LM324	0625
LM324A	0625
LM324AD	0625
LM324AF	0625
LM324AJ	0625
LM324AN	0625
LM324D	0625
LM324F	0661
LM324J	0625
LM324JD	OEM
LM324JDS	OEM
LM324JS	OEM
LM324M	OEM
LM324N	0625
LM324ND	OEM
LM324NDS	OEM
LM324NS	OEM
LM325AN	OEM
LM325H	OEM
LM325N	OEM
LM326H	OEM
LM329AH	OEM
LM329BH	OEM
LM329BZ	OEM
LM329CH	OEM
LM329CZ	OEM
LM329DH	OEM
LM329DZ	OEM
LM329N	OEM
LM330T5.0	OEM
LM331AH	OEM
LM331AN	OEM
LM331H	OEM
LM331N	OEM
LM334H	OEM
LM334Z	OEM
LM335AH	OEM
LM335AZ	OEM
LM335H	OEM
LM335Z	OEM
LM336BH	OEM
LM336BZ	OEM
LM336H	OEM
LM336Z	OEM
LM337	OEM
LM337H	OEM
LM337HVH	OEM
LM337HVK	OEM
LM337K	2723
LM337LZ	5281
LM337MP	OEM
LM337MR	OEM
LM337MT	6726
LM337T	2736
LM338K	2737
LM339	0176
LM339A	0176
LM339AA	OEM
LM339AF	OEM
LM339AJ	0176
LM339AM	OEM
LM339AN	0176
LM339D	2742
LM339DP	0176
LM339F	OEM
LM339J	0176
LM339JD	OEM
LM339JDS	OEM
LM339JS	OEM
LM339M	OEM
LM339M(SM)	OEM
LM339N	0176
LM339ND	OEM
LM339NDS	OEM
LM339NS	OEM
LM340	1812
LM340-5	1922
LM340-5DA	1922
LM340-5KC	0624
LM340-5U	0624
LM340-6	OEM
LM340-6DA	OEM
LM340-6KC	0923
LM340-6U	0923
LM340-8	OEM
LM340-8DA	OEM
LM340-8KC	1196
LM340-8U	1196
LM340-10	OEM
LM340-10KC	OEM
LM340-12	3963
LM340-12DA	3881
LM340-12KC	0334
LM340-12U	0334
LM340-15	OEM
LM340-15DA	2000
LM340-15KC	OEM
LM340-15U	1321
LM340-18	OEM
LM340-18DA	OEM
LM340-18KC	2259
LM340-18U	2259
LM340-24	OEM
LM340-24DA	OEM
LM340-24KC	2239
LM340-24U	2239
LM340AK5.0	1801
LM340AK12	3963
LM340AK15	2000
LM340AT5.0	0624
LM340AT12	1351
LM340AT15	4341
LM340CT	OEM
LM340CT0.2	OEM
LM340K-5.0	0660
LM340K5.0	1922
LM340K-6.0	OEM
LM340K-8.0	OEM
LM340K-12	3963
LM340K12	1998
LM340K-15	2000
LM340K15	2000
LM340K-18	3650
LM340K-24	3881
LM340KC5.0	1801
LM340KC12	3963
LM340KC15	2000
LM340T	0624
LM340T-5	0624
LM340T5	0624
LM340T-5.0	0624
LM340T5.0	0624
LM340T-5.0R	0624
LM340T-6	0923
LM340T-6.0	0923
LM340T6.0	0923
LM340T-6.0R	0923
LM340T-8.0	1196
LM340T8.0	1196
LM340T-12	0334
LM340T12	0334
LM340T-12R	0334
LM340T-15	1321
LM340T15	1321
LM340T-15R	1321
LM340T-18	2259
LM340T-24	2239
LM340T24	2239
LM340T-24R	2239
LM340TS	OEM
LM340U5	0624
LM340U6	0923
LM340U8	1196
LM340U12	0334
LM340U15	1321
LM340U24	2239
LM341-5	0624
LM341-6	0923
LM341-12	0334
LM341-15	1321
LM341-24	2239
LM341P5	OEM
LM341P-5.0	0624
LM341P5.0	0624
LM341P5.0TB	OEM
LM341P-6.0	0923
LM341P-8.0	1196
LM341P8.0	1196
LM341P-12	0334
LM341P12	0334
LM341P12TB	OEM
LM341P-15	1321
LM341P15	1321
LM341P15TB	OEM
LM341P-24	2239
LM342P-5	OEM
LM342P5.0	6046
LM342P12	0334
LM342P15	1321
LM343H	OEM
LM343J	OEM
LM343N	OEM
LM344H	OEM
LM345K5.0	OEM
LM345K5.2	OEM
LM346J	OEM
LM346N	OEM
LM348	2823
LM348AJ	OEM
LM348AN	OEM
LM348D	2816
LM348J	2823
LM348N	2823
LM349	OEM
LM349J	OEM
LM349N	OEM
LM350K	2838
LM350T	OEM
LM350T12	OEM
LM358	0770
LM358A	OEM
LM358AH	1677
LM358AJ	OEM
LM358AN	0770
LM358AT	0770
LM358C	OEM
LM358D	2881
LM358FE	OEM
LM358H	1677
LM358J	0770
LM358JD	OEM
LM358JDS	OEM
LM358JG	0770
LM358JS	OEM
LM358L	1677
LM358M	2881
LM358N	0770
LM358NB	1677
LM358ND	OEM
LM358NDS	OEM
LM358NS	OEM
LM358P	0770
LM358PS	OEM
LM358PS-T1	OEM
LM358S	2892
LM358T	1677
LM359J	2893
LM359N	2893
LM360H	OEM
LM360N	OEM
LM360N14	OEM
LM361H	OEM
LM361J	OEM
LM361N	OEM
LM363AD	OEM
LM363AH10	OEM
LM363AH100	OEM
LM363AH500	OEM
LM363D	OEM
LM363H10	OEM
LM363H100	OEM
LM363H500	OEM
LM370	3810
LM373N	2975
LM375D	OEM
LM375N	2975
LM376JG	OEM
LM376N	2169
LM376P	OEM
LM377	6763
LM377-N	1491
LM377KC	2736
LM-377N	2557
LM377N	1491
LM377N10	2557
LM378N	3018
LM379S	OEM
LM380	0379
LM380N	0379
LM381A	3029
LM381AA	OEM
LM381AN	3029
LM381N	3029
LM382A	OEM
LM382N	OEM
LM383	1048
LM383AT	1048
LM383T	1048
LM384N	3056
LM386	3063
LM386(SM)	OEM
LM386A	3063
LM386F	OEM
LM386LM	3063
LM386N	3063
LM386N-1	3063
LM386N-2	OEM
LM386N-3	3063
LM386N-4	OEM
LM387	3079
LM387AN	3079
LM387N	3079
LM387V	OEM
LM388N	3086
LM388N-1	3086
LM388N-2	OEM
LM388N-3	OEM
LM388NE1	OEM
LM388NE2	OEM
LM388NE3	OEM
LM389N	OEM
LM390AH	0791
LM390AI	0791
LM390AJ	0791
LM390N	3099
LM391N	3110
LM391N-60	3110
LM391N-80	3110
LM392H	OEM
LM392J	OEM
LM392N	OEM
LM393	OEM
LM393A	OEM
LM393AFE	OEM
LM393AH	OEM
LM393AJ	OEM
LM393AN	OEM
LM393AT	OEM
LM393FE	OEM
LM393H	OEM
LM393J	OEM
LM393JG	7144
LM393M	3118
LM393N	0631
LM393NB	0631
LM393P	0631
LM393T	OEM
LM396K	OEM
LM399AH	0595
LM399AH-50	OEM
LM399AI	0595
LM399AJ	0595
LM399H	OEM
LM529CH	OEM
LM529CN	OEM
LM555	0973
LM555C	0973
LM555CH	3637
LM555CJ	0973
LM555CN	0973
LM555CN#1	0973
LM555CN#2	0973
LM555H	3637
LM555J	OEM
LM556CJ	OEM
LM556CN	3285
LM556D	2606
LM556J	OEM
LM556M	2606
LM565CN	3381
LM566CH	3825
LM566CN	3827
LM566H	3825
LM566N	3827
LM567CM	6080
LM640SG-2031	OEM
LM703	0632
LM703E	0632
LM703L	0632
LM703LH	0632
LM-703LN	0632
LM703LN	0632
LM709AH	OEM
LM709C	4239
LM709CH	1301
LM709CN	1705
LM709H	OEM
LM710CH	1797
LM710CN	1800
LM710H	OEM
LM710N	OEM
LM711CH	1890
LM711CN	1897
LM711H	4245
LM723CG	1192
LM723CH	1192
LM723CJ	OEM
LM723CN	0026
LM723H	1192
LM723J	OEM
LM723N	4269
LM725AH	OEM
LM725CH	2890
LM725CN	OEM
LM725H	OEM
LM733	OEM
LM733CH	1121
LM733CN	1118
LM733H	OEM
LM741	0312
LM741AH	0410
LM741AJ14	OEM
LM741C	0410
LM741CH	0410
LM741CJ	0312
LM741CJ-14	OEM
LM741CJ14	OEM
LM741CM	1976
LM741CN	0312
LM741EH	0410
LM741EN	0312
LM741H883	OEM
LM741J14	OEM
LM741N	0312
LM746	0666
LM746H	0753
LM746N	0349
LM747AH	2368
LM747AJ	2358
LM747CH	2368
LM747CJ	2358
LM747CN	2358
LM747EH	2368
LM747EJ	2358
LM747EN	OEM
LM747H	2368
LM747H883	OEM
LM747J	2358
LM748CH	0093
LM748CJ	OEM
LM748CN	3917
LM748CN7	6246
LM748H	2454
LM748J	OEM
LM748N	2452
LM758N	OEM
LM760CH	OEM
LM760CN	OEM
LM832(A)	OEM
LM833	OEM
LM833M	OEM
LM833M63	OEM
LM833N	OEM
LM856C	OEM
LM1011	OEM
LM1011A	OEM
LM1011AN	6963
LM1011N	4933
LM1014AN	OEM
LM1014N	OEM
LM1019N	OEM
LM1030	OEM
LM1035	OEM
LM1037	OEM
LM1038	OEM
LM1090E	0079
LM1090F	0079
LM1090G	0079
LM1110A	0144
LM1110B	0127
LM1111A	OEM
LM1111B	OEM
LM1111C	OEM
LM1111CN	OEM
LM1112B	OEM
LM1112C	OEM
LM1112CN	OEM
LM1117	0111
LM1117(A)	OEM
LM1117(B)	OEM
LM1117C	0111
LM1117D	0079
LM1120B	0127
LM1120C	0127
LM1121A	OEM
LM1121B	OEM
LM1121C	OEM
LM1123H	0144
LM-1129	0079
LM-1130	0079
LM1131	OEM
LM1131A	OEM
LM1131B	OEM
LM1131C	OEM
LM-1132	0079
LM-1133	0079
LM1133	0330
LM1138	0224
LM1138E/F	0224
LM1138G/F	0224
LM1138G/H	0144
LM1138H/I	0224
LM-1147	0079
LM-1148	0079
LM-1149	0037
LM-1150	0037
LM-1151	0037
LM-1153	0037
LM-1154	0285
LM-1155	0079
LM-1156	0338
LM-1157	0561
LM-1158	0015
LM-1159	0133
LM1159	0133
LM-1160	0015
LM1160	0015
LM1274	OEM
LM1304	0654
LM1304N	0654
LM1305	0443
LM1305M	0443
LM1305N	0443
LM1305N01	0443
LM1307	0701
LM1307A	0701
LM1307E	0701
LM-1307N	0701
LM1307N	0701
LM1307N01	0701
LM1307P	0701
LM1307PQ	0701
LM1307W	0701
LM1310	0520
LM-1310N	0520
LM1310N	0520
LM1310N01	0520
LM1329	0521
LM1351#	0789
LM1351N	0789
LM1391N	0848
LM1403	0079
LM1404	0037
LM1408J-6	OEM
LM1408J-7	OEM

DEVICE TYPE	REPL CODE
LM1408J-8	OEM
LM1408N-6	4316
LM1408N-7	4316
LM1408N-8	OEM
LM1414J	OEM
LM1414N	OEM
LM1415-6	0016
LM1415-7	0016
LM1415J	OEM
LM1458	0360
LM1458AN	OEM
LM1458C	2476
LM1458H	3137
LM1458J	OEM
LM1458M	0360
LM1458N	0360
LM1496	0417
LM1496H	0417
LM1496J	3801
LM1496N	3801
LM1501H	0086
LM1502H	0126
LM1508D-8	OEM
LM1508J-8	OEM
LM1540	0079
LM1540(B)	OEM
LM1540C	0111
LM1558AH	OEM
LM1558AJ	OEM
LM1558H	3137
LM1558J	OEM
LM1566F	0079
LM1596H	0417
LM1614D	0079
LM1614M	0079
LM1795	0037
LM1796	0261
LM1800	1395
LM1800A	1395
LM1800N	1395
LM1801	OEM
LM1812A	OEM
LM1815N	OEM
LM1818	0079
LM1818N	4552
LM1819N	4553
LM1820	1421
LM1820A	1421
LM1820N	1421
LM1821S	4568
LM1823	1425
LM1823N	1425
LM1828N	OEM
LM1829	0521
LM1830N	OEM
LM1834	0030
LM1837(A)	OEM
LM1841	1443
LM1845M	2457
LM1848N	OEM
LM1850	OEM
LM-1862	0015
LM1862	0015
LM1865(A)	OEM
LM1866(A)	OEM
LM1868N	OEM
LM1870	1482
LM1877	1491
LM1877N	1491
LM1877N-1	OEM
LM1877N-2	OEM
LM1877N-3	1491
LM1877N-4	OEM
LM1877N-5	OEM
LM1877N-6	OEM
LM1877N-7	OEM
LM1877N-8	OEM
LM1877N-9	4132
LM1877N-10	1491
LM1880F	OEM
LM1889	1516
LM1889N	1516
LM1894N	OEM
LM1895	OEM
LM1896N	4890
LM1897	OEM
LM1900J	OEM
LM1932	0102
LM1946H	0417
LM1965(A)	OEM
LM1996H	OEM
LM2000N	OEM
LM2001N	OEM
LM2002AT	OEM
LM2002T	OEM
LM2020	0150
LM-2111	0664
LM2111	0664
LM2111A	0664
LM2111M	0664
LM-2111N	0664
LM2111N	0664
LM2111N01	0664
LM2113N01	0678
LM2152	0079
LM2576	OEM
LM2576T-ADJ	OEM
LM-2589	0037
LM2682	0079
LM2701	0264
LM2808N	OEM
LM2877P	OEM
LM2895	OEM
LM2896	2304
LM2896P1	2304
LM2896P2	OEM
LM2900	2247
LM2900J	OEM
LM2900N	2247
LM2901	0176
LM2901D	2742
LM2901F	OEM
LM2901J	0176
LM2901N	0176
LM2901NDS	OEM
LM2901NS	OEM
LM2902	0625
LM2902D	3202
LM2902J	0625
LM2902N	0625
LM2903	2310
LM2903D	OEM
LM2903FE	OEM
LM2903H	OEM
LM2903JG	OEM
LM2903M	OEM
LM2903N	2310
LM2903P	OEM
LM2903T	OEM
LM2904	OEM
LM2904CT-12	OEM
LM2904D	OEM
LM2904H	OEM
LM2904J	0770
LM2904JG	0770
LM2904L	1677
LM2904M	2881
LM2904N	0770
LM2904P	0770
LM2905N	OEM
LM2907N	OEM
LM2907N-8	OEM
LM2907N8	OEM
LM2908J	OEM
LM2908N	OEM
LM2917	2318
LM2917-8	2318
LM2917J	2318
LM2917M	OEM
LM2917N	2318
LM2917N-8	4551
LM2917N8	4551
LM2924J	OEM
LM2924N	OEM
LM2931AT5.0	OEM
LM2931AZ5.0	OEM
LM2931T	OEM
LM2931T5.0	OEM
LM2931Z5.0	OEM
LM2936Z-5	OEM
LM2936Z-5.0	OEM
LM2940CT	OEM
LM2940CT-5.0	OEM
LM2940CT-12	OEM
LM2940T5	OEM
LM2940T5M	OEM
LM3028AH	0823
LM3028BH	0823
LM3045D	1696
LM3045J	1696
LM3045N	1696
LM3046N	6359
LM3053H	0823
LM3064M	0364
LM3064N	0802
LM3064N-01	0802
LM3064N01	OEM
LM3065	0167
LM3065N	0167
LM3065N01	0167
LM3067N	0328
LM3070A	0352
LM3070N	0352
LM3070N01	0352
LM3071A	0354
LM3071N	0354
LM3071N01	0354
LM3072	0666
LM3072N	0666
LM3075N01	1345
LM3075N01A	1345
LM3080AN	3836
LM3080N	6392
LM3086	OEM
LM3086N	1696
LM3086W	1696
LM3089N	2755
LM3105	0443
LM3189N	6459
LM3216	OEM
LM3231	OEM
LM3301N	2247
LM3302	0176
LM3302J	0176
LM3302N	0176
LM3401N	6696
LM3429-50	OEM
LM3524J	OEM
LM3524N	5233
LM3820N	7002
LM3862M	3063
LM3900	2247
LM3900J	OEM
LM3900N	2247
LM3905N	OEM
LM3909	3062
LM3909N	3062
LM3911N	OEM
LM3914	3068
LM3914D	OEM
LM3914N	3068
LM3915	3069
LM3916	3071
LM3999Z	OEM
LM4250	OEM
LM4250C	OEM
LM4250CH	3820
LM4250CJ	OEM
LM4250CN	3434
LM4250F	OEM
LM4250H	OEM
LM4250J	OEM
LM4500AN	OEM
LM5047	OEM
LM6402A	OEM
LM6402A-022	OEM
LM6402A022	OEM
LM6402A-095	OEM
LM6402A095	OEM
LM6402A127	OEM
LM6402A-220	OEM
LM6402A220	OEM
LM6402A-221	OEM
LM6402A221	OEM
LM6402G-582	OEM
LM6402G-1844	OEM
LM6402G-1957	OEM
LM6402G-1999	OEM
LM6402H-158	OEM
LM6402H-238	OEM
LM6402H-274	OEM
LM6402H-329	OEM
LM6402H-344	OEM
LM6402H-350	OEM
LM6402H-562	OEM
LM6402H1871	OEM
LM6405G031	OEM
LM6405H-215	OEM
LM6405H-330	OEM
LM6405H-SA069	OEM
LM6405H-SB173	OEM
LM6413-1830(TN2001A)	OEM
LM6413E	OEM
LM6413E-342	OEM
LM6413E-554	OEM
LM6413E-563	OEM
LM6413E-1807	OEM
LM6413E-1968	OEM
LM6413E-1969	OEM
LM6413E-2068	OEM
LM6413E-2069	OEM
LM6413E-2088	OEM
LM6413E-2099	OEM
LM6413E-2129	OEM
LM6413E-2130	OEM
LM6416E	OEM
LM6416E-159	OEM
LM6416E-202	OEM
LM6416E-209	OEM
LM6416E-217	OEM
LM6416E-239	OEM
LM6416E-351	OEM
LM6416E-363	OEM
LM6416E-410	OEM
LM6416E-557	OEM
LM6416E-1917	OEM
LM6416E-1942	OEM
LM6416E-2084	OEM
LM6416E-2085	OEM
LM6417E-302	OEM
LM6417E-1838	OEM
LM7000	OEM
LM7001	OEM
LM7805ACZ	1292
LM7805CT	1812
LM7805CV	1812
LM7808A-8	0083
LM7812	0334
LM7812CT	6021
LM7905CT	1285
LM7912CT	1838
LM8000	0595
LM8050I/J	0079
LM8071	OEM
LM-8360	OEM
LM8360	2295
LM8361	2295
LM8361D	2295
LM8361DH	2728
LM8362	OEM
LM8363	1247
LM8363D	1247
LM8363DH	OEM
LM8363H	1247
LM8364	OEM
LM8365	OEM
LM8368	OEM
LM8372	OEM
LM8460	OEM
LM8471	OEM
LM8521	OEM
LM8523	OEM
LM8560	3982
LM8560B	OEM
LM8561	OEM
LM8562	OEM
LM8569	OEM
LM8854-576	OEM
LM8854-1851	OEM
LM8854-1912	OEM
LM8942	OEM
LM8972	OEM
LM9016F	0127
LM9018F	1954
LM9018G/F	0079
LM11123H	OEM
LM11700AJ	OEM
LM13080N	OEM
LM13080P	OEM
LM13600	OEM
LM13600A	OEM
LM13600AD	OEM
LM13600AN	OEM
LM13600D	OEM
LM13600M	OEM
LM13600N	2168
LM13700AN	OEM
LM13700J	OEM
LM13700N	OEM
LM24102	OEM
LM33256N-15	OEM
LM39000A30	OEM
LM40101	OEM
LMA-595	OEM
LMC555CM	2771
LMC-1992	OEM
LMC1992	OEM
LMC1992N	OEM
LMH8	OEM
LMP15	0002
LMP15-20	0002
LMP15A	0002
LMQ5	OEM
LMS15	OEM
LMS20	OEM
LMS25	OEM
LMS30	OEM
LMS40	OEM
LMS50	OEM
LMS60	OEM
LMS80	OEM
LMS100	OEM
LMS120	OEM
LMS150	OEM
LMS180	OEM
LMS8560	OEM
LMZ3.3	0782
LMZ3.3-20	6232
LMZ3.3A	0782
LMZ3.6	0796
LMZ3.6-20	0796
LMZ3.6A	0796
LMZ3.9	0806
LMZ3.9-20	0806
LMZ3.9A	0806
LMZ4.3	0821
LMZ4.3-20	0821
LMZ4.3A	0821
LMZ4.7	0833
LMZ4.7-20	0833
LMZ4.7A	0833
LMZ5.1	0442
LMZ5.1-20	0442
LMZ5.1A	0442
LMZ5.6	0442
LMZ5.6-20	0876
LMZ5.6A	0876
LMZ6.2	0185
LMZ6.2-20	0185
LMZ6.2A	0185
LMZ6.8	0205
LMZ6.8A	0205
LMZ7.5	0480
LMZ7.5A	0480
LMZ8.2	0504
LMZ8.2A	0504
LMZ9.1	0684
LMZ9.1A	0684
LMZ10	0225
LMZ10A	0225
LMZ11	0230
LMZ11A	0230
LMZ12	0234
LMZ12A	0234
LMZ13	0237
LMZ13A	0237
LMZ14	1397
LMZ14A	1397
LMZ15	0247
LMZ15A	0247
LMZ16	0251
LMZ16A	0251
LMZ18	0256
LMZ18A	0256
LMZ20	0262
LMZ20A	0262
LMZ22	0269
LMZ22A	0269
LMZ24	0274
LMZ24A	0274
LMZ27	0294
LMZ27A	0294
LMZ30	0309
LMZ30A	0309
LMZ33	0318
LMZ33A	0318
LMZ36	0320
LMZ36A	0320
LMZ39	0326
LMZ39A	0326
LMZ43	0337
LMZ43A	0337
LMZ47	0347
LMZ47A	0347
LMZ51	0027
LMZ51A	0027
LMZ56	0266
LMZ56A	0266
LMZ62	0205
LMZ62A	0386
LMZ68	0405
LMZ68A	0405
LMZ75	0426
LMZ75A	0426
LMZ82	0444
LMZ82A	0444
LMZ91	0238
LMZ91A	0238
LMZ100	1180
LMZ100A	1180
LMZ110	1190
LMZ110-20	1190
LMZ110A	1190
LMZ120	1207
LMZ120-20	1207
LMZ120A	1207
LMZ130	1218
LMZ130-20	1218
LMZ130A	1218
LMZ150	0647
LMZ150-20	0647
LMZ150A	0647
LMZ160	1255
LMZ160-20	1255
LMZ160A	1255
LMZ180	1279
LMZ180-20	1279
LMZ180A	1279
LMZ200	0605
LMZ200-20	0605
LMZ200A	0605
LN0312CP3	OEM
LN0312P3	OEM
LN02202P	4443
LN02302P	2047
LN02402P	5037
LN03202P	4443
LN03302P	2047
LN03402P	5037
LN04202P	4443
LN04302P	2047
LN04402P	5037
LN05101P1	OEM
LN05101P2	OEM
LN05103P	OEM
LN05201P	OEM
LN05202P	4443
LN05203P	OEM
LN05301P	OEM
LN05302P	2047
LN05401P	OEM
LN05402P	5037
LN06202P	4443
LN06302P	2047
LN06402P	5037
LN07202P	4443
LN07302P	2047
LN07402P	5037
LN014210P	OEM
LN018315P	OEM
LN021616PH	OEM
LN033417P	OEM
LN041395P1	OEM
LN051219P2	OEM
LN061219P2	OEM
LN072163P	OEM
LN072198P	OEM
LN072198P4	OEM
LN072203P	OEM
LN078328P	OEM
LN12	OEM
LN15BP	OEM
LN15WP	OEM
LN15WP.F	OEM
LN16BP	OEM
LN16WP.F	OEM
LN21CALULFKT	OEM
LN21CP.H	OEM
LN21CP.HL	OEM
LN21GCPHLM	OEM
LN21RCPH	1962
LN21RCPHL	1962
LN21RCPHLM	OEM
LN21RCPHLM-V	OEM
LN21RCPHLMV	1962
LN21RCPHV	1962
LN21RCP.H	OEM
LN21RCP.HL	OEM
LN21RPH	OEM
LN21RPH-(C2)	OEM
LN21RPH-(CF1)	OEM
LN21RPH-(TA)	OEM
LN21RPH-C	OEM
LN21RPHL	1962
LN21RPHLM	OEM
LN21RPSLLS	OEM
LN21RP.H	OEM
LN21RP.HL	OEM
LN21WP.H	OEM
LN21WP.HL	OEM
LN23SCP	OEM
LN23SRP	OEM
LN23SWP	OEM
LN25CP	OEM
LN25RCP	OEM
LN25RP	OEM
LN25WP	OEM
LN26RP	OEM
LN28RCP	OEM
LN-28RP	3019
LN28RP	5566
LN28WP	OEM
LN28WPVT	OEM
LN29RCP	OEM
LN29RP	OEM
LN29RPP	OEM
LN31GCPH(U)	1778
LN31GCPH(U)-(C)	1778
LN31GCPHILMU	1778
LN31GCPHL	1778
LN31GCPHLM	1778
LN31GCPHLM(U)	1778
LN31GCPHLM-U	1778
LN31GCPHLMU	1778
LN31GCPHL-U	1778
LN31GCPHLU	1778
LN31GCPILMU	1778
LN31GCP-UH	1778
LN31GCP-UHL	1778
LN31GCPUHL	1778
LN31GCP.H	1778
LN31GCP.HL	1778
LN31GCUPHL-U	1778
LN31GPH	OEM
LN31GPHL	1778
LN31GP.H	1778
LN31GP.HL	1778
LN33SGP	OEM
LN33SGP-H	OEM
LN35BP	OEM
LN35GCP	OEM
LN35GP	OEM
LN36BP	OEM
LN38GCP	OEM
LN38GCPH	OEM
LN38GP	OEM
LN38GP(C)	OEM
LN38GPL	0841
LN39GCP	OEM
LN39GP	OEM
LN40GCP	OEM
LN41TPHL	OEM
LN41YCPH	3157
LN41YCPHL	OEM
LN41YCPHLM	OEM
LN41YCP.H	OEM
LN41YCP.HL	OEM
LN41YPHL	OEM
LN41YP.H	OEM
LN41YP.HL	OEM
LN43SYP	OEM
LN45YCP	OEM
LN45YP	OEM
LN46YP	OEM
LN48YCPH	OEM
LN48YP	OEM
LN49YP	OEM
LN51F	OEM
LN51L	OEM
LN52	OEM
LN55	OEM
LN57	OEM
LN58	OEM
LN59	2703
LN59NV	2703
LN59SAN	OEM
LN62S	OEM
LN65	OEM
LN66	5710
LN66-S	OEM
LN66C	5710
LN66NC	OEM
LN66S	5710
LN66SNC	OEM
LN70	OEM
LN71	OEM
LN81CPH	OEM
LN81CPHL	OEM
LN81CP.H	OEM
LN81CP.HL	OEM
LN81RCPH	OEM
LN81RCPH-(C)	OEM
LN81RCPHL	OEM
LN81RCPSL	OEM
LN81RCP.H	OEM
LN81RCP.HL	OEM
LN81RP.H	OEM
LN81RP.HL	OEM
LN85	OEM
LN85RCP	OEM
LN88RCPH	OEM
LN210RP	OEM
LN210WP	OEM
LN211RP	OEM
LN211WP	OEM
LN212RP	4859
LN213RP	OEM
LN216RP-LS	OEM
LN217RP	4443
LN217RPH	OEM
LN219RP	OEM
LN220RP	OEM
LN220RP-LS	OEM
LN221RP	OEM
LN222RP	OEM
LN222RPH	OEM
LN224	OEM
LN224RP	OEM
LN224RPH	OEM
LN226RP	OEM
LN227RP	OEM
LN228RP	OEM
LN229RP	OEM
LN229RPH	OEM
LN230RA	OEM
LN230RP	OEM
LN233RP	OEM
LN235RPH	OEM
LN238RPH	OEM
LN242RP	OEM
LN242RPH	OEM
LN248RP-LS	OEM
LN249RP	OEM
LN250RP	OEM
LN260RCPP	OEM
LN310GP	OEM
LN311GP	OEM
LN311GPHL	OEM
LN312GP	1970
LN313GP	2047
LN316GP-LS	OEM
LN317GPH	OEM
LN319GP	OEM
LN320GP	OEM
LN320GP-LS	OEM
LN320GPLS	OEM
LN321GP	OEM
LN322GP	OEM
LN322GPH	OEM
LN324GP	OEM
LN324GP2	OEM
LN324GP3	OEM
LN324GPH	OEM
LN326GP	OEM
LN327GP	OEM
LN328GP	OEM
LN329GP	OEM
LN329GPH	OEM
LN333GP	OEM
LN335GPH	OEM
LN338GPH	OEM
LN340GCP	OEM
LN342RPH	OEM
LN348GP-LS	OEM
LN410YP	OEM
LN411YP	OEM
LN412YP	4861
LN413YP	OEM
LN416YP-LS	OEM
LN417YP	5037
LN417YPH	OEM
LN419YP	OEM
LN420YP	OEM
LN420YP(U)MS	OEM
LN421YP	OEM
LN422YP	OEM
LN422YPH	OEM
LN424YP	OEM
LN424YPH	OEM
LN426YP	OEM
LN427YP	OEM
LN428YP	OEM
LN429YP	OEM
LN429YPH	OEM
LN433YP	OEM
LN435YPH	OEM
LN438YPH	OEM
LN440YCP	OEM
LN442RPH	OEM
LN442YP	OEM
LN448YP-LS	OEM
LN512GA	OEM
LN513GK	OEM
LN513GK-S	OEM
LN5130A	OEM
LN5130K	OEM
LN5130K-S	OEM
LN513RA	OEM
LN513RK	OEM
LN513RK-S	OEM
LN513YA	OEM
LN513YK	OEM
LN513YK-S	OEM

If replacement code is OEM, contact original manufacturer for replacement.

DEVICE TYPE	REPL CODE
LN514GA	OEM
LN514GK	OEM
LN5140A	OEM
LN5140K	OEM
LN514RA	5138
LN514RK	OEM
LN514YA	OEM
LN514YK	OEM
LN516GA	OEM
LN516GK	OEM
LN5160A	OEM
LN5160K	OEM
LN516RA	OEM
LN516RK	OEM
LN516YA	OEM
LN516YK	OEM
LN524GA	OEM
LN524GAS	OEM
LN524GK	OEM
LN524RA	4327
LN524RAL1	OEM
LN524RAS	OEM
LN524RK	OEM
LN524YA	OEM
LN524YK	OEM
LN526GA	OEM
LN526GADP	OEM
LN526GA-L	OEM
LN526GK	OEM
LN526MI	4327
LN526RA	4327
LN526RK	4332
LN526YA	OEM
LN526YK	OEM
LN543GA	OEM
LN543GAH	OEM
LN543GK	OEM
LN543GKH	OEM
LN543RA	OEM
LN543RAH	OEM
LN543RK	OEM
LN543RKH	OEM
LN543YA	OEM
LN543YAH	OEM
LN543YK	OEM
LN543YKH	OEM
LN655	OEM
LN831RP	OEM
LN840RCP	OEM
LN846RP-LS	OEM
LN1304N01	0654
LN1310	0520
LN1328A	OEM
LN2064B	4868
LN5240A	OEM
LN5240K	OEM
LN5260A	OEM
LN5260K	OEM
LN5430A	OEM
LN5430A.H	OEM
LN5430K	OEM
LN5430K.H	OEM
LN10204P	OEM
LN108399P	OEM
LN142154P	OEM
LN261478PH	OEM
LNA0410	OEM
LNA328	OEM
LNA328B	OEM
LNA331	OEM
LNA331A	OEM
LNA331B	OEM
LNA335	OEM
LNA335A	OEM
LNA335B	OEM
LNA339	OEM
LNA339A	0036
LNA339B	OEM
LNA343	0275
LNA343A	0275
LNA343B	OEM
LNA347	0140
LNA347A	0140
LNA347B	OEM
LNA351	OEM
LNA351A	0442
LNA351B	OEM
LNA356	OEM
LNA356A	0157
LNA356B	OEM
LNA362	0636
LNA362A	0636
LNA362B	OEM
LNA368	OEM
LNA368A	0025
LNA368B	OEM
LNA375	OEM
LNA375A	0649
LNA375B	OEM
LNA382	3100
LNA382A	0244
LNA382B	3100
LNA391	0012
LNA391A	0012
LNA391B	OEM
LNA1100	OEM
LNA3100	OEM
LNA3100A	0170
LNA3100B	OEM
LO100BV	OEM
LO100DV	OEM
LO100MV	OEM
LO3001	OEM
LOG100JP	OEM
LP/HP2100	OEM
LP1H	0015
LP1S	OEM
LP1SU	OEM
LP2H	0015
LP2S	OEM
LP2SU	OEM
LP3H	0015
LP4H	0015
LP100	0124
LP101	OEM
LP102	OEM
LP271	OEM
LP1014A	OEM
LP1014B	OEM
LP1014D	OEM
LP1014E	OEM
LP1014H	OEM
LP1016A	OEM
LP1016B	OEM
LP1016D	OEM
LP1016E	OEM
LP1016H	OEM
LP1017A	OEM
LP1017B	OEM
LP1017D	OEM
LP1017E	OEM
LP1017H	OEM
LP1018A	OEM
LP1018B	OEM
LP1018D	OEM
LP1018E	OEM
LP1018H	OEM
LP1019B	OEM
LP1019E	OEM
LP1019H	OEM
LP1020A	OEM
LP1020B	OEM
LP1020D	OEM
LP1020E	OEM
LP1020H	OEM
LP1022A	OEM
LP1022B	OEM
LP1022D	OEM
LP1022E	OEM
LP1022H	OEM
LP1024A	OEM
LP1024B	OEM
LP1024D	OEM
LP1024E	OEM
LP1024H	OEM
LP1026A	OEM
LP1026B	OEM
LP1026D	OEM
LP1026E	OEM
LP1026H	OEM
LP1028A	OEM
LP1028B	OEM
LP1028D	OEM
LP1028E	OEM
LP1028H	OEM
LP1030A	OEM
LP1030B	OEM
LP1030D	OEM
LP1030E	OEM
LP1030H	OEM
LP1032A	OEM
LP1032B	OEM
LP1032D	OEM
LP1032E	OEM
LP1032H	OEM
LP1034A	OEM
LP1034B	OEM
LP1034D	OEM
LP1034E	OEM
LP1034H	OEM
LP1036A	OEM
LP1036B	OEM
LP1036D	OEM
LP1036E	OEM
LP1036H	OEM
LP1038A	OEM
LP1038B	OEM
LP1038D	OEM
LP1038E	OEM
LP1038H	OEM
LP1057R	OEM
LP1919A	OEM
LP9510I	OEM
LP9510J	OEM
LPA1H	1228
LPA2H	OEM
LPA3H	OEM
LPCL11	OEM
LPI11	OEM
LPM3.3	0292
LPM3.3-20	0292
LPM3.3A	0292
LPM3.6	0188
LPM3.6-20	0188
LPM3.6A	0188
LPM3.9	0456
LPM3.9-20	0456
LPM3.9A	0456
LPM4.3	0533
LPM4.3-20	0533
LPM4.3A	0533
LPM4.7	0162
LPM4.7-20	0451
LPM4.7A	0162
LPM5.1	0162
LPM5.1-20	0162
LPM5.1A	0162
LPM5.6	0157
LPM5.6-20	0157
LPM5.6A	0157
LPM6.2	0636
LPM6.2-20	0636
LPM6.2A	0636
LPM6.8	0025
LPM6.8-20	0025
LPM6.8A	0025
LPM7.5	0649
LPM7.5-20	0649
LPM7.5A	0649
LPM8.2	0244
LPM8.2-20	0244
LPM8.2A	0244
LPM9.1	0012
LPM9.1-20	0012
LPM9.1A	0012
LPM10	0170
LPM10-20	0170
LPM10A	0170
LPM11	0794
LPM11-20	0317
LPM11A	0317
LPM12	0137
LPM12-20	0137
LPM12A	0137
LPM13	0365
LPM13-20	0365
LPM13A	0365
LPM15	0002
LPM15-20	0002
LPM15A	0002
LPM16	0420
LPM16-20	0420
LPM16A	0420
LPM18	0495
LPM18-20	0495
LPM18A	0495
LPM20	0531
LPM20-20	0531
LPM20A	0531
LPM22	0565
LPM22-20	0565
LPM22A	0565
LPM24	0402
LPM24-20	0402
LPM24A	0402
LPM27	0441
LPM27-20	0441
LPM27A	0441
LPM30	0726
LPM30-20	0726
LPM30A	0726
LPM33	0039
LPM33-20	0039
LPM33A	0039
LPM36	0820
LPM36-20	0820
LPM36A	0820
LPM39	0350
LPM39-20	0350
LPM39A	0350
LPM43	0931
LPM43-20	0931
LPM43A	0931
LPM47	0999
LPM47-20	0999
LPM47A	0999
LPM51	0502
LPM51-20	0502
LPM51A	0502
LPM56	1834
LPM56-20	0869
LPM56A	0869
LPM62	0783
LPM62-20	0783
LPM62A	0783
LPM68	1268
LPM68-20	2157
LPM68A	1268
LPM75	1189
LPM75-20	1189
LPM75A	1189
LPM82	0331
LPM82-20	0331
LPM82A	0331
LPM91	1311
LPM91-20	1311
LPM91A	1311
LPM100	0098
LPM100-20	OEM
LPM100A	0098
LPM110	0149
LPM110-20	0149
LPM110A	0149
LPM120	0186
LPM120-20	0186
LPM120A	0186
LPM130	0213
LPM130-20	0213
LPM130A	0213
LPM150	0028
LPM150-20	0028
LPM150A	0028
LPM160	0255
LPM160-20	0255
LPM160A	0255
LPM180	0367
LPM180-20	0367
LPM180A	0367
LPM200	0421
LPM200-20	0421
LPM200A	0421
LPM-596	OEM
LPT100	OEM
LPT100A	OEM
LPT100B	OEM
LPT110	OEM
LPT110A	OEM
LPT110B	OEM
LPZT8.2	0244
LPZT10	0170
LPZT12	0137
LPZT15	0002
LPZT18	0495
LPZT22	0565
LPZT27	0441
LPZT33	0039
LQV4M1957	OEM
LR5GZ61	0071
LR33H	0299
LR36H	OEM
LR39CH	0188
LR43CH	0275
LR47CH	0162
LR50	7115
LR51CH	0162
LR51D	0041
LR51E	0041
LR56CH	0162
LR56D	0253
LR56E	0253
LR62CH	0471
LR62D	0471
LR62E	0471
LR68CH	0062
LR68D	0062
LR68E	0062
LR75CH	0077
LR75D	0077
LR75E	0077
LR82CH	0165
LR82D	0165
LR82E	0165
LR91CH	0057
LR91D	0057
LR91E	0057
LR100CH	0064
LR100D	0064
LR100E	0064
LR110CH	0181
LR110D	0181
LR110E	0181
LR120CH	0052
LR120D	0140
LR120E	0052
LR130CH	0053
LR130D	0053
LR130E	0053
LR150CH	0686
LR150D	0686
LR150E	0686
LR160CH	0445
LR160D	0445
LR160E	0445
LR180CH	0375
LR180D	0375
LR180E	0375
LR200CH	0700
LR200D	0700
LR200E	0700
LR220D	0705
LR220E	0705
LR240D	0494
LR240E	0494
LR270D	0455
LR270E	0455
LR300D	0195
LR300E	0195
LR301R	OEM
LR330D	0166
LR330E	0166
LR360D	0010
LR360E	0010
LR390D	0032
LR390E	0032
LR505P	OEM
LR1351G	OEM
LR1352G	OEM
LR1361E	OEM
LR1362E	OEM
LR1363E	OEM
LR1364E	OEM
LR1371R	OEM
LR1372R	OEM
LR1373R	OEM
LR1374R	OEM
LR1381Y	OEM
LR1382Y	OEM
LR1383Y	OEM
LR1384Y	OEM
LR1451G	OEM
LR1455G	OEM
LR1461E	OEM
LR1464E	OEM
LR1465E	OEM
LR1471R	OEM
LR1474R	OEM
LR1475R	OEM
LR1481Y	OEM
LR1484Y	OEM
LR1485Y	OEM
LR1704R	OEM
LR1705R	OEM
LR1706R	OEM
LR1707R	OEM
LR1717R	OEM
LR1720R	OEM
LR1723R	OEM
LR1737R	OEM
LR1738R	OEM
LR1739R	OEM
LR1740R	OEM
LR1743R	OEM
LR1800	OEM
LR1801	OEM
LR1804	OEM
LR1805	OEM
LR2601A	OEM
LR2665E	OEM
LR2666E	OEM
LR2671R	OEM
LR2673P	OEM
LR2674R	OEM
LR2675R	OEM
LR2676R	OEM
LR2678R	OEM
LR2703	OEM
LR3419	OEM
LR3714	OEM
LR3714M	OEM
LR3784R	OEM
LR3785R	OEM
LR4089	OEM
LR4803	OEM
LR4806	OEM
LR4806B	OEM
LR4806R	OEM
LR4816B	OEM
LR6504R	OEM
LR6508R	OEM
LR7164R	OEM
LR8321	OEM
LR8322	OEM
LR8330	OEM
LR37631	OEM
LR37632	OEM
LR40992	2761
LR40993	6389
LR48066T	OEM
LRQ849	0079
LRR-50	0015
LRR50	0015
LRR-100	0015
LRR100	0015
LRR-200	0015
LRR200	0015
LRR-300	0015
LRR300	0015
LRR-400	0015
LRR400	0015
LRR-500	0015
LRR500	0015
LS00	1527
LS-0095-AR-213	0111
LS02	1559
LS02(SM)	OEM
LS05	1608
LS05(MIN)	OEM
LS05FP(MIN)	OEM
LS05FP(SM)	OEM
LS06	OEM
LS045T	OEM
LS-0142	1443
LS14	1698
LS14D	OEM
LS15	1707
LS16-1	OEM
LS20	0034
LS25	OEM
LS32	0088
LS32N	0088
LS40	0135
LS50	OEM
LS51N	1033
LS52	0085
LS60	OEM
LS74	0243
LS80	OEM
LS86	0291
LS100	OEM
LS101AT	OEM
LS101T	OEM
LS107T	OEM
LS120	OEM
LS125	0075
LS126AD	OEM
LS139	0153
LS141AT	OEM
LS141CB	OEM
LS141CM	OEM
LS141CT	OEM
LS141T	OEM
LS148AT	OEM
LS148CB	OEM
LS148CM	OEM
LS148CT	OEM
LS148T	OEM
LS155	0209
LS157N	1161
LS166	0231
LS174	0260
LS201AT	OEM
LS201B	OEM
LS201M	OEM
LS201T	OEM
LS204AT	OEM
LS204CB	OEM
LS204CM	OEM
LS204CT	OEM
LS204M	OEM
LS204T	OEM
LS207T	OEM
LS240	0452
LS240M	OEM
LS244	0458
LS245	0463
LS245(SM)	OEM
LS285A	OEM
LS301AB	OEM
LS301AM	OEM
LS301AT	0093
LS307B	2283
LS307M	OEM
LS307T	OEM
LS342	OEM
LS373	0709
LS374	0713
LS400	OEM
LS500	OEM
LS505P	OEM
LS600	OEM
LS602	OEM
LS610	OEM
LS656	OEM
LS656B	OEM
LS703L	0632
LS709AT	OEM
LS709CB	OEM
LS709CT	OEM
LS709T	OEM
LS776CB	OEM
LS776CM	OEM
LS776CT	OEM
LS776T	OEM
LS953	OEM
LS3705	0079
LS5484	0325
LS5485	0325
LS7030	OEM
LS7031	OEM
LS7040	OEM
LS7056	OEM
LS7060	OEM
LS7061	OEM
LS7062	OEM
LS7100	OEM
LS7110	OEM
LS7210	OEM
LS7220	OEM
LS7225	OEM
LS7231	OEM
LS7232	OEM
LS7233	OEM
LS7234	OEM
LS7235	OEM
LS7236	OEM
LS7240	OEM
LS7270	OEM
LS7709	OEM
LS7717	OEM
LS7718	OEM
LS7721	OEM
LS7722	OEM
LS7723	OEM
LS7724	OEM
LS7726	OEM
LS7727	OEM
LS7728	OEM
LS7729	OEM
LS7730	OEM
LS7738	OEM
LS7739	OEM
LS7746	OEM
LS7747	OEM
LS7748	OEM
LS7749	OEM
LS7750	OEM
LS7751	OEM
LS7754	OEM
LS8045M	OEM
LSA13	OEM
LSA17	OEM
LSC5D	OEM
LSC14	OEM
LSC30D	OEM
LSC1008	0167
LSC1008P	0167
LSC89919B	OEM
LSC89921B	OEM
LSI-11	OEM
LSM88441Y	OEM
LSM88492Y	OEM
LST153S	OEM
LST154R	OEM
LST515AR10	OEM
LST700	OEM
LST700L	OEM
LST710	OEM
LST710L	OEM
LST711	OEM
LST711L	OEM
LST712	OEM
LST712L	OEM
LST713	OEM
LST713L	OEM
LST722	OEM
LST722L	OEM
LST733	OEM
LST733L	OEM
LST1052	OEM
LST1056	OEM
LST3300S	OEM
LST4053	OEM
LST4053F	OEM
LST4054-1	OEM
LST4153	OEM
LST4153F	OEM
LST4153S	OEM
LST4154	OEM
LST4154S	OEM
LST4253	OEM
LST4253F	OEM
LST4254	OEM
LST4353	OEM
LST4354	OEM
LST4753	OEM
LST4754	OEM
LST5053	OEM
LST5053R	OEM
LST5053R5	OEM
LST5053R10	OEM
LST5054-1	OEM
LST5054-1R5	OEM
LST5054-1R10	OEM
LST5054R	OEM
LST5152	OEM
LST5152R	OEM
LST5152R5	OEM
LST5152R10	OEM
LST5152S	OEM
LST5153	OEM
LST5153R	OEM
LST5153R5	OEM
LST5153R10	OEM
LST5154	OEM
LST5154R5	OEM
LST5154S	OEM
LST5252	OEM
LST5252R	OEM
LST5252R5	OEM
LST5252R10	OEM
LST5253	1778
LST5253R	OEM
LST5253R5	OEM
LST5253R10	OEM
LST5254	OEM
LST5254R	OEM
LST5254R5	OEM
LST5352	OEM
LST5352R	OEM
LST5352R5	OEM
LST5352R10	OEM
LST5353	3157
LST5353R	OEM
LST5353R5	OEM
LST5353R10	OEM
LST5354	OEM
LST5354R	OEM
LST5354R5	OEM
LST5354R10	OEM
LST5752	OEM
LST5752R	OEM
LST5752R5	OEM
LST5752R10	OEM
LST5753	1962
LST5753R	OEM
LST5753R5	OEM
LST5753R10	OEM
LST5754	OEM
LST5754R	OEM
LST5754R5	OEM
LST5754R10	OEM
LST525410	OEM
LSX900	OEM
LT0028	OEM
LT0038	OEM
LT0047	OEM
LT-1E21A	OEM
LT-1H11A	OEM
LT3P8D	OEM
LT51	OEM
LT55	OEM
LT67T-24	OEM
LT101AR	OEM

If replacement code is OEM, contact original manufacturer for replacement.

Original Device Types

DEVICE TYPE	REPL CODE	DEVICE TYPE	REPL CODE	DEVICE TYPE	REPL CODE	DEVICE TYPE	REPL CODE	DEVICE TYPE	REPL CODE	DEVICE TYPE	REPL CODE	DEVICE TYPE	REPL CODE
LT101R	OEM	LT297EW	OEM	LT515R	OEM	LT637A-12	OEM	LT678G-12	OEM	LT755P-24	OEM	LT1221	OEM
LT106RC	OEM	LT298EC	OEM	LT516	OEM	LT637A-24	OEM	LT678G-24	OEM	LT756	OEM	LT1222	OEM
LT111AR	OEM	LT299E	OEM	LT516R	OEM	LT637AG	OEM	LT678PG-12	OEM	LT765R	OEM	LT1223	OEM
LT111R	OEM	LT299ED	OEM	LT517	OEM	LT637AG-12	OEM	LT678PG-24	OEM	LT767A	OEM	LT1224	OEM
LT116RC	OEM	LT301E	OEM	LT517R	OEM	LT637AG-24	OEM	LT687A	OEM	LT767A-12	OEM	LT1231	OEM
LT131AG	OEM	LT301G	OEM	LT518	OEM	LT637AP	OEM	LT687A-12	OEM	LT767AP	OEM	LT1232	OEM
LT131G	OEM	LT301P	OEM	LT518R	OEM	LT637AP-12	OEM	LT687A-24	OEM	LT767AP-12	OEM	LT1233	OEM
LT136GC	OEM	LT301R	OEM	LT519	OEM	LT637AP-24	OEM	LT687AG	OEM	LT781	OEM	LT1233G	OEM
LT151AY	OEM	LT301Y	OEM	LT519R	OEM	LT637D	OEM	LT687AP	OEM	LT782	OEM	LT1234	OEM
LT151Y	OEM	LT302A	OEM	LT522R	OEM	LT637D-12	OEM	LT687B	OEM	LT783	OEM	LT1251	OEM
LT156YC	OEM	LT302E	OEM	LT522RA	OEM	LT637DG	OEM	LT687B-12	OEM	LT784	OEM	LT1252	OEM
LT181Y	OEM	LT302G	OEM	LT523E	OEM	LT637DG-12	OEM	LT687B-24	OEM	LT797	OEM	LT1253	OEM
LT191AE	OEM	LT302P	OEM	LT523G	OEM	LT637DP	OEM	LT687BG	OEM	LT797-12	OEM	LT1253Y	OEM
LT191E	OEM	LT302R	OEM	LT523GA	OEM	LT637DP-12	OEM	LT687BG-12	OEM	LT797-24	OEM	LT1254	OEM
LT196EC	OEM	LT302Y	OEM	LT523R	OEM	LT647A	OEM	LT687BG-24	OEM	LT797A-12	OEM	LT1291	OEM
LT201CRD	OEM	LT303A	OEM	LT523RA	OEM	LT647A-12	OEM	LT687BP	OEM	LT797AP	OEM	LT1292	OEM
LT201R	OEM	LT303E	OEM	LT523Y	OEM	LT647AG	OEM	LT687BP-12	OEM	LT797AP-12	OEM	LT1293	OEM
LT201RD	OEM	LT303G	OEM	LT524E	OEM	LT647AG-12	OEM	LT687BP-24	OEM	LT797G	OEM	LT1293E	OEM
LT202CRT	OEM	LT303P	OEM	LT524G	OEM	LT647AP	OEM	LT687D	OEM	LT797G-12	OEM	LT1294	OEM
LT202RT	OEM	LT303R	OEM	LT524GA	OEM	LT647AP-12	OEM	LT687D-12	OEM	LT797G-24	OEM	LT1466	OEM
LT203	OEM	LT303Y	OEM	LT524R	OEM	LT647D	OEM	LT687DG	OEM	LT797P	OEM	LT1478	OEM
LT203R	OEM	LT304E	OEM	LT524RA	OEM	LT647D-12	OEM	LT687DG-12	OEM	LT797P-12	OEM	LT1488	OEM
LT203RD	OEM	LT304G	OEM	LT524Y	OEM	LT647DG	OEM	LT687DP	OEM	LT797P-24	OEM	LT1498	OEM
LT204RT	OEM	LT304P	OEM	LT527E	OEM	LT647DG-12	OEM	LT687DP-12	OEM	LT816AK	OEM	LT1588A	OEM
LT205CWD	OEM	LT304R	OEM	LT527G	OEM	LT647DP	OEM	LT696A	OEM	LT816BK	OEM	LT1604	OEM
LT205RW	OEM	LT304Y	OEM	LT527Ga	OEM	LT647DP-12	OEM	LT696AG	OEM	LT816CK	OEM	LT1720E	OEM
LT206CWC	OEM	LT305E	OEM	LT-527R	OEM	LT656-12	OEM	LT696APG	OEM	LT816DK	OEM	LT1720G	OEM
LT206RC	OEM	LT305G	OEM	LT527R	OEM	LT656-24	OEM	LT697-12	OEM	LT829	OEM	LT1720P	OEM
LT207RW	OEM	LT305HR	OEM	LT527RA	OEM	LT656G-12	OEM	LT697-24	OEM	LT831	OEM	LT1720R	OEM
LT208RC	OEM	LT305P	OEM	LT527Y	OEM	LT656G-24	OEM	LT697A	OEM	LT832	OEM	LT1720Y	OEM
LT209R	OEM	LT305R	OEM	LT528E	OEM	LT656PG-12	OEM	LT697A-12	OEM	LT834	OEM	LT1723E	OEM
LT209RD	OEM	LT305Y	OEM	LT528G	OEM	LT656PG-24	OEM	LT697A-24	OEM	LT917A	OEM	LT1723G	OEM
LT-211	OEM	LT311G	OEM	LT528GA	OEM	LT656T	OEM	LT697AG	OEM	LT917A-12	OEM	LT1723P	OEM
LT211CRD	OEM	LT311HR	OEM	LT528R	OEM	LT656T-12	OEM	LT697AP	OEM	LT917AG	OEM	LT1723R	OEM
LT211R	OEM	LT311R	OEM	LT528RA	OEM	LT656T-24	OEM	LT697B	OEM	LT917AG-12	OEM	LT1723Y	OEM
LT211RD	OEM	LT311Y	OEM	LT532G	OEM	LT656TG	OEM	LT697B-12	OEM	LT917AR	OEM	LT1740E	OEM
LT212CRT	OEM	LT312E	OEM	LT532R	OEM	LT656TG-12	OEM	LT697B-24	OEM	LT917AR-12	OEM	LT1740G	OEM
LT212RT	OEM	LT312G	OEM	LT533G	OEM	LT656TG-24	OEM	LT697BG	OEM	LT927A	OEM	LT1740P	OEM
LT213R	OEM	LT312HR	OEM	LT533R	OEM	LT656TP	OEM	LT697BG-12	OEM	LT927A-12	OEM	LT1740R	OEM
LT213RD	OEM	LT312R	OEM	LT534	OEM	LT656TPG-12	OEM	LT697BG-24	OEM	LT927AG	OEM	LT1740Y	OEM
LT214RT	OEM	LT312Y	OEM	LT542HR	OEM	LT656TPG-24	OEM	LT697BP	OEM	LT927AG-12	OEM	LT1743E	OEM
LT215CWD	OEM	LT313G	OEM	LT542R	OEM	LT667-12	OEM	LT697BP-12	OEM	LT927AR	OEM	LT1743G	OEM
LT215RW	OEM	LT313HR	OEM	LT543HR	OEM	LT667-24	OEM	LT697BP-24	OEM	LT927AR-12	OEM	LT1743P	OEM
LT216CWC	OEM	LT313R	OEM	LT543R	OEM	LT667A	OEM	LT697D	OEM	LT937A	OEM	LT1743R	OEM
LT216RC	OEM	LT313Y	OEM	LT544HR	OEM	LT667A-12	OEM	LT697D-12	OEM	LT937A-12	OEM	LT1743Y	OEM
LT217RW	OEM	LT314E	OEM	LT544R	OEM	LT667A-24	OEM	LT697DG	OEM	LT937AG	OEM	LT1784E	OEM
LT218RC	OEM	LT314G	OEM	LT545HR	OEM	LT667AG	OEM	LT697DG-12	OEM	LT937AG-12	OEM	LT1784P	OEM
LT219R	OEM	LT314R	OEM	LT545R	OEM	LT667AG-12	OEM	LT697DP	OEM	LT937AR	OEM	LT1784R	OEM
LT219RD	OEM	LT314Y	OEM	LT546E	OEM	LT667AG-24	OEM	LT697DP-12	OEM	LT937AR-12	OEM	LT1785E	OEM
LT221CRD	OEM	LT315E	OEM	LT546G	OEM	LT667AP	OEM	LT697G-12	OEM	LT955	OEM	LT1785P	OEM
LT221R	OEM	LT315G	OEM	LT546P	OEM	LT667AP-12	OEM	LT697G-24	OEM	LT955-12	OEM	LT1785R	OEM
LT221RD	OEM	LT315R	OEM	LT546R	OEM	LT667AP-24	OEM	LT697PG-12	OEM	LT955-24	OEM	LT1800E	OEM
LT222CRT	OEM	LT315Y	OEM	LT546Y	OEM	LT667D	OEM	LT697PG-24	OEM	LT955A	OEM	LT1800G	OEM
LT222RT	OEM	LT316G	OEM	LT547E	OEM	LT667D-12	OEM	LT698-12	OEM	LT955A-12	OEM	LT1800HR	OEM
LT223R	OEM	LT316R	OEM	LT547G	OEM	LT667DG	OEM	LT698-24	OEM	LT955AG	OEM	LT1800P	OEM
LT223RD	OEM	LT317G	OEM	LT547P	OEM	LT667DG-12	OEM	LT698G-12	OEM	LT955AG-12	OEM	LT1800R	OEM
LT224RT	OEM	LT317R	OEM	LT547R	OEM	LT667DP	OEM	LT698G-24	OEM	LT955AR	OEM	LT1800Y	OEM
LT229R	OEM	LT318G	OEM	LT547Y	OEM	LT667DP-12	OEM	LT698PG-12	OEM	LT955AR-12	OEM	LT1801E	OEM
LT229RD	OEM	LT318R	OEM	LT548E	OEM	LT667G-12	OEM	LT698PG-24	OEM	LT955G	OEM	LT1801G	OEM
LT231CGD	OEM	LT319G	OEM	LT548G	OEM	LT667G-24	OEM	LT699-12	OEM	LT955G-12	OEM	LT1801HR	OEM
LT231G	OEM	LT319R	OEM	LT548P	OEM	LT667N	OEM	LT699-24	OEM	LT955G-24	OEM	LT1801P	OEM
LT231GD	OEM	LT322G	OEM	LT548R	OEM	LT667N-12	OEM	LT699G-12	OEM	LT975	OEM	LT1801R	OEM
LT232CGT	OEM	LT322P	OEM	LT548Y	OEM	LT667N-24	OEM	LT699G-24	OEM	LT975-12	OEM	LT1801Y	OEM
LT232GT	OEM	LT322RA	OEM	LT549E	OEM	LT667NG	OEM	LT699PG-12	OEM	LT975-24	OEM	LT1802E	OEM
LT233G	OEM	LT323G	OEM	LT549G	OEM	LT667NP	OEM	LT699PG-24	OEM	LT975A	OEM	LT1802G	OEM
LT233GD	OEM	LT323P	OEM	LT549P	OEM	LT667PG-12	OEM	LT707A	OEM	LT975A-12	OEM	LT1802HR	OEM
LT234CWD	OEM	LT323R	OEM	LT549R	OEM	LT667PG-24	OEM	LT707AK	OEM	LT975AG	OEM	LT1802P	OEM
LT234GT	OEM	LT323RA	OEM	LT549Y	OEM	LT668-12	OEM	LT707AS	OEM	LT975AG-12	OEM	LT1802R	OEM
LT235GW	OEM	LT332R	OEM	LT581	OEM	LT668-24	OEM	LT707B	OEM	LT975AR	OEM	LT1802Y	OEM
LT236CWC	OEM	LT334	OEM	LT582	OEM	LT668G-24	OEM	LT707BK	OEM	LT975AR-12	OEM	LT1803E	OEM
LT236GC	OEM	LT373	OEM	LT583	OEM	LT668PG-12	OEM	LT707C	OEM	LT975G	OEM	LT1803G	OEM
LT237GW	OEM	LT374	OEM	LT584	OEM	LT668PG-24	OEM	LT707CK	OEM	LT975G-12	OEM	LT1803HR	OEM
LT238GC	OEM	LT381	OEM	LT612A	OEM	LT676-12	OEM	LT707CS	OEM	LT975G-24	OEM	LT1803P	OEM
LT239G	OEM	LT382	OEM	LT612AP	OEM	LT676-24	OEM	LT707D	OEM	LT997A	OEM	LT1803R	OEM
LT239GD	OEM	LT502P	OEM	LT612D	OEM	LT676G-12	OEM	LT707DK	OEM	LT997A-12	OEM	LT1803Y	OEM
LT251CYD	OEM	LT502R	OEM	LT612DG	OEM	LT676G-24	OEM	LT707DS	OEM	LT997AG	OEM	LT1804AR	OEM
LT251Y	OEM	LT503P	OEM	LT612DP	OEM	LT676PG-12	OEM	LT707GA	OEM	LT997AG-12	OEM	LT1804E	OEM
LT251YD	OEM	LT503R	OEM	LT617A	OEM	LT676PG-24	OEM	LT707GAK	OEM	LT997AR	OEM	LT1804G	OEM
LT252CYT	OEM	LT504P	OEM	LT617A-12	OEM	LT676T	OEM	LT707GAS	OEM	LT997AR-12	OEM	LT1804HR	OEM
LT252YT	OEM	LT504R	OEM	LT617AG	OEM	LT676T-12	OEM	LT707GB	OEM	LT1000	OEM	LT1804P	OEM
LT253Y	OEM	LT505P	OEM	LT617AG-12	OEM	LT676TG	OEM	LT707GBK	OEM	LT1001A	OEM	LT1804R	OEM
LT253YD	OEM	LT505R	OEM	LT617AP	OEM	LT676TP	OEM	LT707GBS	OEM	LT1014	OEM	LT1804Y	OEM
LT254YT	OEM	LT506E	OEM	LT617AP-12	OEM	LT677	OEM	LT707GC	OEM	LT1016(E)	0079	LT1805AR	OEM
LT255CWD	OEM	LT506G	OEM	LT617D	OEM	LT677-12	OEM	LT707GCK	OEM	LT1016D	0144	LT1805E	OEM
LT255YW	OEM	LT506P	OEM	LT617D-12	OEM	LT677-24	OEM	LT707GCS	OEM	LT1016E	0144	LT1805G	OEM
LT256CWC	OEM	LT506R	OEM	LT617DG	OEM	LT677A-12	OEM	LT707GD	OEM	LT1016H	0224	LT1805HR	OEM
LT256YC	OEM	LT506Y	OEM	LT617DG-12	OEM	LT677A-24	OEM	LT707GDK	OEM	LT1016I	0144	LT1805P	OEM
LT257YW	OEM	LT507E	OEM	LT617DP	OEM	LT677AG	OEM	LT707GDS	OEM	LT1016I,H	0079	LT1805R	OEM
LT258YC	OEM	LT507G	OEM	LT617DP-12	OEM	LT677AG-12	OEM	LT734	OEM	LT1016T	OEM	LT1805Y	OEM
LT259Y	OEM	LT507P	OEM	LT621AG	OEM	LT677AG-24	OEM	LT737	OEM	LT1016T,H	0079	LT1806E	OEM
LT259YD	OEM	LT507R	OEM	LT622A	OEM	LT677AP	OEM	LT737-12	OEM	LT1017	OEM	LT1806G	OEM
LT271A	OEM	LT507Y	OEM	LT622AG	OEM	LT677AP-12	OEM	LT737-24	OEM	LT1024	OEM	LT1806HR	OEM
LT271AD	OEM	LT508E	OEM	LT622AP	OEM	LT677AP-24	OEM	LT737A	OEM	LT1034	OEM	LT1806P	OEM
LT272AT	OEM	LT508G	OEM	LT622D	OEM	LT677D	OEM	LT737AP	OEM	LT1038	OEM	LT1806R	OEM
LT273A	OEM	LT508P	OEM	LT622DP	OEM	LT677D-12	OEM	LT737AP-12	OEM	LT1057E	OEM	LT1806Y	OEM
LT273AD	OEM	LT508R	OEM	LT627A	OEM	LT677DG-12	OEM	LT737G	OEM	LT1057G	OEM	LT1807G	OEM
LT274AT	OEM	LT508Y	OEM	LT627A-12	OEM	LT677DP	OEM	LT737G-12	OEM	LT1057HR	OEM	LT1807HR	OEM
LT291CED	OEM	LT509E	OEM	LT627AG	OEM	LT677DP-12	OEM	LT737G-24	OEM	LT1057P	OEM	LT1807P	OEM
LT291E	OEM	LT509G	OEM	LT627AG-12	OEM	LT677G-12	OEM	LT737P	OEM	LT1057R	OEM	LT1807R	OEM
LT291ED	OEM	LT509P	OEM	LT627AP	OEM	LT677G-24	OEM	LT737P-12	OEM	LT1057Y	OEM	LT1807Y	OEM
LT292CET	OEM	LT509R	OEM	LT627AP-12	OEM	LT677N	OEM	LT737P-24	OEM	LT1201	OEM	LT1808E	OEM
LT292ET	OEM	LT509Y	OEM	LT627D	OEM	LT677N-12	OEM	LT755	OEM	LT1202	OEM	LT1808G	OEM
LT293E	OEM	LT512	OEM	LT627D-12	OEM	LT677N-24	OEM	LT755-12	OEM	LT1203	OEM	LT1808HR	OEM
LT293ED	OEM	LT512R	OEM	LT627DG	OEM	LT677NG	OEM	LT755-24	OEM	LT1203R	OEM	LT1808P	OEM
LT294ET	OEM	LT513	OEM	LT627DG-12	OEM	LT677NP	OEM	LT755G	OEM	LT1211	OEM	LT1808R	OEM
LT295CWD	OEM	LT513R	OEM	LT627DP	OEM	LT677PG-12	OEM	LT755G-12	OEM	LT1212	OEM	LT1808Y	OEM
LT295EW	OEM	LT514	OEM	LT627DP-12	OEM	LT677PG-24	OEM	LT755G-24	OEM	LT1213	OEM	LT1809E	OEM
LT296CWD	OEM	LT514R	OEM	LT633DG	OEM	LT678-12	OEM	LT755P	OEM	LT1213R	OEM	LT1809G	OEM
LT296EC	OEM	LT515	OEM	LT637A	OEM	LT678-24	OEM	LT755P-12	OEM	LT1214	OEM	LT1809HR	OEM

If replacement code is OEM, contact original manufacturer for replacement.

DEVICE TYPE	REPL CODE
LT1809P	OEM
LT1809Y	OEM
LT1810AP	OEM
LT1810E	OEM
LT1810G	OEM
LT1810HR	OEM
LT1810P	OEM
LT1810R	OEM
LT1810Y	OEM
LT1811AP	OEM
LT1811E	OEM
LT1811G	OEM
LT1811HR	OEM
LT1811P	OEM
LT1811R	OEM
LT1811Y	OEM
LT1812E	OEM
LT1812G	OEM
LT1812HR	OEM
LT1812P	OEM
LT1812R	OEM
LT1812Y	OEM
LT1813E	OEM
LT1813G	OEM
LT1813HR	OEM
LT1813P	OEM
LT1813R	OEM
LT1813Y	OEM
LT1814E	OEM
LT1814G	OEM
LT1814HR	OEM
LT1814P	OEM
LT1814R	OEM
LT1814Y	OEM
LT1815E	OEM
LT1815G	OEM
LT1815HR	OEM
LT1815P	OEM
LT1815R	OEM
LT1815Y	OEM
LT1816E	OEM
LT1816G	OEM
LT1816HR	OEM
LT1816P	OEM
LT1816R	OEM
LT1816Y	OEM
LT1817E	OEM
LT1817G	OEM
LT1817HR	OEM
LT1817P	OEM
LT1817R	OEM
LT1817Y	OEM
LT1818E	OEM
LT1818G	OEM
LT1818HR	OEM
LT1818P	OEM
LT1818R	OEM
LT1819E	OEM
LT1819G	OEM
LT1819HR	OEM
LT1819P	OEM
LT1819R	OEM
LT1819Y	OEM
LT1820AR	OEM
LT1820E	OEM
LT1820G	OEM
LT1820HR	OEM
LT1820P	OEM
LT1820R	OEM
LT1820Y	OEM
LT1821AR	OEM
LT1821E	OEM
LT1821G	OEM
LT1821HR	OEM
LT1821P	OEM
LT1821R	OEM
LT1821Y	OEM
LT1824AP	OEM
LT1824E	OEM
LT1824G	OEM
LT1824HR	OEM
LT1824P	OEM
LT1824R	OEM
LT1824Y	OEM
LT1825AP	OEM
LT1825E	OEM
LT1825G	OEM
LT1825HR	OEM
LT1825P	OEM
LT1825R	OEM
LT1825Y	OEM
LT2001	OEM
LT2001AR	OEM
LT2001R	OEM
LT2015	OEM
LT2037	OEM
LT2203R	OEM
LT2211AR	OEM
LT2211R	OEM
LT2213R	OEM
LT2231AG	OEM
LT2231G	OEM
LT2233G	OEM
LT2251AY	OEM
LT2251Y	OEM
LT2253Y	OEM
LT2291AE	OEM
LT2291E	OEM
LT2293E	OEM
LT3005	OEM
LT3014	OEM
LT3046	OEM
LT3156	OEM
LT3156G	OEM
LT3167	OEM
LT3201R	OEM
LT3202R	OEM
LT3203R	OEM
LT3204R	OEM
LT3207R	OEM
LT3208R	OEM
LT3209R	OEM
LT-3211B	OEM
LT3211R	OEM
LT3212R	OEM
LT3213R	OEM
LT3214R	OEM
LT3215R	OEM
LT3216R	OEM
LT3217R	OEM
LT3218R	OEM
LT3219R	OEM
LT3223R	OEM
LT3226R	OEM
LT3229R	OEM
LT3231G	OEM
LT3232G	OEM
LT3233G	OEM
LT3234G	OEM
LT3235G	OEM
LT3236G	OEM
LT3237G	OEM
LT3238G	OEM
LT3239G	OEM
LT3251Y	OEM
LT3252Y	OEM
LT3253Y	OEM
LT3254Y	OEM
LT3255Y	OEM
LT3256Y	OEM
LT3257Y	OEM
LT3258Y	OEM
LT3259Y	OEM
LT3291E	OEM
LT3292E	OEM
LT3293E	OEM
LT3294E	OEM
LT3295E	OEM
LT3296E	OEM
LT3297E	OEM
LT3298E	OEM
LT3299E	OEM
LT3382	OEM
LT3400E	OEM
LT3400G	OEM
LT3400P	OEM
LT3400R	OEM
LT3400Y	OEM
LT3401AP	OEM
LT3401E	OEM
LT3401G	OEM
LT3401HAR	OEM
LT3401P	OEM
LT3401R	OEM
LT3401Y	OEM
LT3403AP	OEM
LT3403E	OEM
LT3403G	OEM
LT3403HAr	OEM
LT3403P	OEM
LT3403R	OEM
LT3403Y	OEM
LT3405E	OEM
LT3405G	OEM
LT3405P	OEM
LT3405R	OEM
LT3405Y	OEM
LT3406AP	OEM
LT3406E	OEM
LT3406G	OEM
LT3406HAR	OEM
LT3406P	OEM
LT3406R	OEM
LT3406Y	OEM
LT3700	OEM
LT3746	OEM
LT3772	OEM
LT3784E	OEM
LT3784P	OEM
LT3784R	OEM
LT3785	OEM
LT3785E	OEM
LT3785P	OEM
LT3785R	OEM
LT3867	OEM
LT3867-12	OEM
LT3867-24	OEM
LT3867G	OEM
LT3867G-12	OEM
LT3867G-24	OEM
LT3877	OEM
LT3877-12	OEM
LT3877-24	OEM
LT3877G	OEM
LT3877G-12	OEM
LT3877G-24	OEM
LT3881	OEM
LT3882	OEM
LT4201R	OEM
LT4201RD	OEM
LT4202RT	OEM
LT4203B	OEM
LT4203P	OEM
LT4203R	OEM
LT4203RD	OEM
LT4204RT	OEM
LT4205RW	OEM
LT4206RC	OEM
LT4207RW	OEM
LT4208RC	OEM
LT4211R	OEM
LT4211RD	OEM
LT4211RT	OEM
LT4213R	OEM
LT4213RD	OEM
LT4214RT	OEM
LT4215RW	OEM
LT4216RC	OEM
LT4217RW	OEM
LT4218RC	OEM
LT4221R	OEM
LT4221RD	OEM
LT4222RT	OEM
LT4223B	OEM
LT4223P	OEM
LT4223R	OEM
LT4223RD	OEM
LT4224RT	OEM
LT4231G	OEM
LT4231GD	OEM
LT4232GT	OEM
LT4233B	OEM
LT4233G	OEM
LT4233P	OEM
LT-4234G	OEM
LT4234GT	OEM
LT4235GW	OEM
LT4236GC	OEM
LT4237GW	OEM
LT4238GC	OEM
LT4251Y	OEM
LT4252YT	OEM
LT4253B	OEM
LT4253P	OEM
LT4253Y	OEM
LT4254YT	OEM
LT4255YW	OEM
LT4256YC	OEM
LT4257YW	OEM
LT4258YC	OEM
LT4271A	OEM
LT4272AT	OEM
LT4273A	OEM
LT4274AT	OEM
LT4291E	OEM
LT4292ET	OEM
LT4293E	OEM
LT4294ET	OEM
LT4295EW	OEM
LT4296EC	OEM
LT4297EW	OEM
LT4298EC	OEM
LT4400	OEM
LT4446	OEM
LT4485	OEM
LT4700	OEM
LT4746	OEM
LT4772	OEM
LT4785	OEM
LT5020	OEM
LT5022	0604
LT5025	0604
LT5028	0604
LT5031	0604
LT5034	0604
LT5038	OEM
LT5039	OEM
LT5042	OEM
LT5045	OEM
LT5048	OEM
LT5051	OEM
LT5054	0604
LT5057	0604
LT5060	0604
LT5063	0604
LT5066	0604
LT5069	0604
LT5072	0160
LT5075	0160
LT5077B	OEM
LT5077C	OEM
LT5078	0160
LT5081	OEM
LT5084	OEM
LT5087	OEM
LT5090	0160
LT5093	0160
LT5096	OEM
LT5099	0160
LT5102	0160
LT5105	OEM
LT5108	0160
LT5111	0160
LT5114	OEM
LT5117	OEM
LT5120	OEM
LT5123	OEM
LT5140	OEM
LT5157	OEM
LT5158	OEM
LT5159	OEM
LT5160	OEM
LT5161	OEM
LT5162	OEM
LT5164	OEM
LT5165	OEM
LT5177B	OEM
LT5201	OEM
LT5201RD	OEM
LT5202	OEM
LT5202RT	OEM
LT5203LR	OEM
LT5203R	OEM
LT5204LRT	OEM
LT5204RT	OEM
LT5209	OEM
LT5210	OEM
LT5211RD	OEM
LT5212RT	OEM
LT5213R	OEM
LT5214RT	OEM
LT5221RD	OEM
LT5222RT	OEM
LT5223R	OEM
LT5231GD	OEM
LT5233G	OEM
LT5234GT	OEM
LT5251YD	OEM
LT5252YT	OEM
LT5253Y	OEM
LT5254YT	OEM
LT5277B	OEM
LT5291ED	OEM
LT5292ET	OEM
LT5293E	OEM
LT5294ET	OEM
LT5377B	OEM
LT5382	OEM
LT5388	OEM
LT5415	OEM
LT5491	OEM
LT5515	OEM
LT5777B	OEM
LT5881	OEM
LT5882	OEM
LT5917	OEM
LT5918	OEM
LT5921	OEM
LT5922	OEM
LT6203R	OEM
LT6223R	OEM
LT6233G	OEM
LT6253Y	OEM
LT6293E	OEM
LT6510	OEM
LT6530	OEM
LT6540	OEM
LT6550	OEM
LT6610	OEM
LT6630	OEM
LT6640	OEM
LT6650	OEM
LT6710	OEM
LT6730	OEM
LT6740	OEM
LT-6740R	OEM
LT6750	OEM
LT6810	OEM
LT6830	OEM
LT6840	OEM
LT6850	OEM
LT7120	OEM
LT7201R	OEM
LT7211R	OEM
LT7231G	OEM
LT7231PG	OEM
LT7251Y	OEM
LT7291E	OEM
LT7382	OEM
LT7650R	OEM
LT7653R	OEM
LT7656R	OEM
LT7660Y	OEM
LT7661Y	OEM
LT7663Y	OEM
LT7666Y	OEM
LT7670G	OEM
LT7671G	OEM
LT7673G	OEM
LT7676G	OEM
LT7680E	OEM
LT7681E	OEM
LT7683E	OEM
LT7686E	OEM
LT7750R	OEM
LT7751R	OEM
LT7756R	OEM
LT7760R	OEM
LT7881	OEM
LT7882	OEM
LT8211	OEM
LT8212	OEM
LT8213	OEM
LT8214	OEM
LT8215	OEM
LT8231	OEM
LT8232	OEM
LT8233	OEM
LT8234	OEM
LT8235	OEM
LT8251	OEM
LT8252	OEM
LT8253	OEM
LT8254	OEM
LT8255	OEM
LT8291	OEM
LT8292	OEM
LT8293	OEM
LT8294	OEM
LT8295	OEM
LT8500R1	OEM
LT9213R	OEM
LT9223A	OEM
LT9230D	OEM
LT9233G	OEM
LT9233PG	OEM
LT9253Y	OEM
LT9293E	OEM
LT9320H	OEM
LT9323H	OEM
LT10203-1	OEM
LT10203-2	OEM
LT10207	OEM
LT10208	OEM
LT10223	OEM
LT10223B1	OEM
LT10223B2	OEM
LT10223C2	OEM
LT10233	OEM
LT10233B1	OEM
LT10233B2	OEM
LT10233C1	OEM
LT10233C2	OEM
LT10253	OEM
LT10253B1	OEM
LT10253B2	OEM
LT10253C1	OEM
LT10253C2	OEM
LT13202	OEM
LT13204	OEM
LT13211	OEM
LT13212	OEM
LT13213	OEM
LT13214	OEM
LT13223	OEM
LT13231	OEM
LT13232	OEM
LT13233	OEM
LT13234	OEM
LT13251	OEM
LT13252	OEM
LT13253	OEM
LT13254	OEM
LT13291	OEM
LT13293	OEM
LT13294	OEM
LT52124	OEM
LT53124	OEM
LT57124	OEM
LT57173	OEM
LTA320	OEM
LTA709	OEM
LTA709C	OEM
LTA741	OEM
LTA741C	OEM
LTA747	OEM
LTA747C	OEM
LTC3211A	OEM
LTC53173	5448
LTD482	OEM
LTE1016	0224
LTE1016(G.E.)	0144
LTG312A	OEM
LTG313A	OEM
LTG314A	OEM
LTG315A	OEM
LTG320	OEM
LTG321	OEM
LTG324	OEM
LTG325	OEM
LTG326	OEM
LTG327	OEM
LTG352	OEM
LTG353	OEM
LTG370	OEM
LTG371	OEM
LTG2110	OEM
LTG2120	OEM
LTG4115	OEM
LTG4125	OEM
LTG8101	OEM
LTH1016	0224
LTH1016(G.E.)	0144
LTK817	4924
LTL-2234GT	OEM
LTL3211A	OEM
LTL9213A	5443
LTL9233A	5444
LTL9253A	5063
LTL42135F	OEM
LTL42735F	OEM
LTL54173	5447
LTL57173	5446
LTR306	OEM
LTR307	OEM
LTR308	OEM
LTR309	OEM
LTR312	OEM
LTR313	OEM
LTR314	OEM
LTR315	OEM
LTR320	OEM
LTR321	OEM
LTR322	OEM
LTR323	OEM
LTR324	OEM
LTR325	OEM
LTR326	OEM
LTR327	OEM
LTR352	OEM
LTR353	OEM
LTR370	OEM
LTR371	OEM
LTR-546E	OEM
LTR2110	OEM
LTR2120	OEM
LTR4115	OEM
LTR4125	OEM
LTR4224	OEM
LTR8101	OEM
LTV817B	OEM
LTZ-G15	OEM
LTZ-MG15	OEM
LTZ-MR15	OEM
LTZ-MR15-T77	OEM
LTZ-MU15	OEM
LTZ-R15	OEM
LTZ-U15	OEM
LU2N544	0136
LU547P	OEM
L-UPD8155C	OEM
LV1000NA	OEM
LV1140	OEM
LV1200	OEM
LV4100	OEM
LVA43A	0275
LVA43B	OEM
LVA43C	OEM
LVA47A	0140
LVA47B	OEM
LVA47C	OEM
LVA51A	0041
LVA51B	OEM
LVA51C	OEM
LVA56A	0253
LVA56B	OEM
LVA56C	OEM
LVA62A	0471
LVA62B	OEM
LVA62C	OEM
LVA68A	0062
LVA68B	OEM
LVA68C	OEM
LVA75A	0077
LVA75B	OEM
LVA75C	OEM
LVA82A	0165
LVA82B	3100
LVA82C	OEM
LVA91A	0057
LVA91B	OEM
LVA91C	OEM
LVA100A	0350
LVA100B	OEM
LVA100C	OEM
LVA150	0041
LVA150A	OEM
LVA150B	OEM
LVA150C	OEM
LVA153	OEM
LVA153A	OEM
LVA153B	OEM
LVA153C	OEM
LVA156	0253
LVA156A	0157
LVA156B	OEM
LVA156C	OEM
LVA159	OEM
LVA159A	OEM
LVA159B	OEM
LVA159C	OEM
LVA162	0471
LVA162A	1338
LVA162B	OEM
LVA162C	OEM
LVA165	OEM
LVA165A	OEM
LVA165B	OEM
LVA165C	OEM
LVA168	0062
LVA168A	OEM
LVA168B	OEM
LVA168C	OEM
LVA171	OEM
LVA171A	OEM
LVA171B	OEM
LVA171C	OEM
LVA174	OEM
LVA174A	OEM
LVA174C	OEM
LVA177	OEM
LVA177A	OEM
LVA177B	OEM
LVA177C	OEM
LVA180	0244
LVA180A	OEM
LVA180B	OEM
LVA180C	OEM
LVA183	OEM
LVA183A	OEM
LVA183B	OEM
LVA183C	OEM
LVA186	OEM
LVA186A	OEM
LVA186B	OEM
LVA186C	OEM
LVA189	OEM
LVA189A	OEM
LVA189B	OEM
LVA189C	OEM
LVA192	OEM
LVA192A	OEM
LVA192B	OEM
LVA192C	OEM
LVA195	OEM
LVA195A	OEM
LVA195B	OEM
LVA195C	OEM
LVA198	OEM
LVA198A	OEM
LVA198B	OEM
LVA198C	OEM
LVA343A	0275
LVA343B	OEM
LVA343C	OEM
LVA347A	0140
LVA347B	OEM
LVA347C	OEM
LVA351A	0162
LVA351B	OEM
LVA351C	OEM
LVA356A	0157
LVA356B	OEM
LVA356C	OEM
LVA362A	0636
LVA362B	OEM
LVA362C	OEM
LVA368A	0025
LVA368B	OEM
LVA368C	OEM
LVA375A	0649
LVA375B	OEM
LVA375C	OEM
LVA382A	0244
LVA382B	3100
LVA382C	OEM
LVA391A	0012
LVA391B	OEM
LVA391C	OEM
LVA450	0041
LVA450A	OEM
LVA450B	OEM
LVA450C	OEM
LVA453A	OEM
LVA453B	OEM
LVA453C	OEM
LVA456A	0157
LVA456B	OEM
LVA456C	OEM
LVA459A	OEM
LVA459B	OEM
LVA459C	OEM
LVA462A	1338
LVA462B	OEM
LVA462C	OEM
LVA465A	OEM
LVA465B	OEM
LVA465C	OEM
LVA468A	3465
LVA468B	OEM
LVA468C	OEM
LVA471A	OEM
LVA471B	OEM
LVA471C	OEM
LVA474A	OEM
LVA474B	OEM
LVA474C	OEM
LVA477A	OEM
LVA477B	OEM
LVA477C	OEM
LVA480A	OEM
LVA480B	OEM
LVA480C	OEM
LVA483A	OEM
LVA483B	OEM
LVA483C	OEM
LVA486A	OEM
LVA486B	OEM
LVA486C	OEM
LVA489A	OEM
LVA489B	OEM
LVA489C	OEM
LVA492A	OEM
LVA492B	OEM
LVA492C	OEM
LVA495A	OEM
LVA495B	OEM
LVA495C	OEM
LVA498A	OEM
LVA498B	OEM
LVA498C	OEM
LVA513	OEM
LVA513A	OEM

DEVICE TYPE	REPL CODE
LVA519S2	OEM
LVA521	OEM
LVA521S	OEM
LVA522A	OEM
LVA522S	OEM
LVA522S-2	OEM
LVA522SA-2	OEM
LVA523A	OEM
LVA523S	OEM
LVA1043A	0533
LVA1043B	OEM
LVA1043C	OEM
LVA1047A	0451
LVA1047B	OEM
LVA1047C	OEM
LVA1051A	0162
LVA1051B	OEM
LVA1051C	OEM
LVA1056A	0157
LVA1056B	OEM
LVA1056C	OEM
LVA1062A	0636
LVA1062B	OEM
LVA1062C	OEM
LVA1068A	0025
LVA1068B	OEM
LVA1068C	OEM
LVA1075A	0649
LVA1075B	OEM
LVA1075C	OEM
LVA1082A	0244
LVA1082B	OEM
LVA1082C	OEM
LVA1091A	0012
LVA1091B	OEM
LVA1091C	OEM
LVA3100A	0350
LVA3100B	OEM
LVA3100C	OEM
LVA10100A	0350
LVA10100B	OEM
LVA10100C	OEM
LWK-1A	OEM
LWK-2	OEM
LWK-3	OEM
LX738	OEM
LY30	OEM
LY50	OEM
LY61	OEM
LY70	OEM
LY100	OEM
LZ4.7	0162
LZ5.6	0157
LZ6.8	OEM
LZ10	0170
LZZ1-3.3	0292
LZZ1-3.6	0188
LZZ1-3.9	0456
LZZ1-4.3	0533
LZZ1-4.7	0451
LZZ1-5.1	0162
LZZ1-5.6	0157
LZZ1-6.2	0636
LZZ1-6.8	0025
LZZ1-7.5	0649
LZZ1-8.2	0244
LZZ1-9.1	0012
LZZ1-10	0170
LZZ1-11	0317
LZZ1-12	0137
LZZ1-14	0100
LZZ1-15	0002
LZZ1-20	0531
LZZ1-22	0565
LZZ1-24	0402
LZZ1-26	0441
LZZ1-27	0441
LZZ1-29	0726
LZZ1-33G	0039
LZZ1-36	0820
LZZ1-39	0350
LZZ1-43	0931
LZZ1-47	0999
LZZ1-51	0502
LZZ1-55	1834
LZZ1-56	0869
LZZ1-62	0783
LZZ1-68	1268
LZZ1-82	0331
LZZ1-91	1311
LZZ1-110B	0149
LZZ1-115G	0794
LZZ1-120	0186
LZZ1-128	0365
LZZ1-130	0213
LZZ1-140	0245
LZZ1-150	0028
LZZ1-160	0255
LZZ1-161	0420
LZZ1-170	0877
LZZ1-180	0367
LZZ1-190	2858
LZZ1-200	0421
LZZ1-200G	0421
M005T1	OEM
M006B1	OEM
M006D1	OEM
M009T1	OEM
M-0027	0015
M0027	0015
M012	0144
M024	0144
M054B1	OEM
M055B1	OEM
M083B1	4599
M086B1	4772
M087B1	OEM
M089B1	OEM
M089D1	OEM
M089F1	OEM
M0810	OEM
M0816	OEM
M0818	OEM
M0870	OEM
M02010GA	OEM
M02010GA-J245X	OEM
M02010GA-J319X	OEM
M02011GA	OEM
M02011GA-J301X	OEM
M06075B200Z	OEM
M050430-582SP	OEM
M050430-583SP	OEM
M0C604A	OEM
M0C1000	0541
M0TMJE371	0460
M0TMJE521	0161
M0Z	0015
M1	OEM
M1-152	0071
M1-301	0133
M1A1	0071
M1A5	0071
M1A9	0071
M1B1	0071
M1B5	0071
M1B9	0071
M1BF	OEM
M1C1	0071
M1C5	0071
M1C9	0071
M1D	OEM
M1D1	0071
M1D5	0071
M1D9	0071
M1E1	0071
M1E5	0071
M1E9	0071
M1F1	0071
M1F5	0071
M1F9	0071
M1G1	0071
M1G5	0071
M1G9	0071
M1H	0015
M1H1	0071
M1H5	0071
M1H9	0071
M1K1	0071
M1K5	0071
M1K9	0071
M1M1	0071
M1M5	0071
M1M9	0071
M1MA151K-W	OEM
M1S-12795B	0103
M1X	0086
M1Z	0023
M1Z30	OEM
M1Z43	OEM
M-2	0196
M2	OEM
M2-100	OEM
M2-100R	OEM
M2-200	OEM
M2-200R	OEM
M2-400	OEM
M2-400R	OEM
M2-600	OEM
M2-600R	OEM
M2-800	OEM
M2-800R	OEM
M2-1000	OEM
M2-1000R	OEM
M2-1200	OEM
M2-1200R	OEM
M2A1	0071
M2A5	0071
M2A9	0071
M2B	0071
M2B1	0071
M2B5	0071
M2B9	0071
M2C1	0071
M2C5	0071
M2C9	0071
M2D1	0071
M2D5	0071
M2D9	0071
M2E1	0071
M2E5	0071
M2E9	0071
M2F1	0071
M2F5	0071
M2F9	0071
M2G1	0071
M2G5	0071
M2G9	0071
M2GH1	0071
M2H1	2645
M2H5	0071
M2H9	0071
M2K1	0071
M2K5	0071
M2K9	0071
M2M1	0071
M2M5	0071
M2M9	0071
M2N168A	0600
M2.5A	0087
M-3	OEM
M3	OEM
M3A1	0071
M3A5	0071
M3A9	0071
M3B1	0071
M3B5	0071
M3B9	0071
M3C1	0071
M3C5	0071
M3C9	0071
M3D1	0071
M3D5	0071
M3D9	0071
M3E1	0071
M3E5	0071
M3E9	0071
M3F1	0071
M3F5	0071
M3F9	0071
M3G	0485
M3G1	0071
M3G5	0071
M3G9	0071
M3H1	0071
M3H5	0071
M3H9	2645
M3K1	0071
M3K5	0071
M3K9	0071
M3M1	0071
M3M5	0071
M3M9	0071
M3025BM	OEM
M3Z6.8	0205
M3Z6.8A	0205
M3Z7.5	0480
M3Z7.5A	0480
M3Z8.2	0504
M3Z8.2A	0504
M3Z9.1	0684
M3Z9.1A	0684
M3Z10	0225
M3Z10A	0225
M3Z11	0237
M3Z11A	0230
M3Z12	0234
M3Z12A	0234
M3Z13	0237
M3Z13A	0237
M3Z15	0247
M3Z15A	0247
M3Z16	0251
M3Z16A	0251
M3Z18	0256
M3Z18A	0256
M3Z20	0262
M3Z20A	0262
M3Z22	0269
M3Z22A	0269
M3Z24	0274
M3Z24A	0274
M3Z27	0294
M3Z27A	0294
M3Z30	0309
M3Z30A	0309
M3Z33	0318
M3Z33A	0318
M3Z36	0320
M3Z36A	0320
M3Z39	0326
M3Z39A	0326
M3Z43	0337
M3Z43A	0337
M3Z47	0347
M3Z47A	0347
M3Z51	0027
M3Z51A	0027
M3Z56	0266
M3Z56A	0266
M3Z62	0386
M3Z62A	0386
M3Z68	0405
M3Z68A	0405
M3Z75	0426
M3Z75A	0426
M3Z82	0444
M3Z82A	0444
M3Z91	0238
M3Z91A	0238
M3Z100	1180
M3Z100A	1180
M3Z110	1190
M3Z110A	1190
M3Z120	1207
M3Z120A	1207
M3Z130	1218
M3Z130A	1218
M3Z150	0647
M3Z150A	0647
M3Z160	1255
M3Z160A	1255
M3Z180	1279
M3Z180A	1279
M3Z200	0605
M3Z200A	0605
M4	0861
M4A1	0087
M4A5	0087
M4A9	0087
M4B1	0087
M4B5	0087
M4B9	0087
M4B-31	0071
M4B31	0015
M4B31-13	0015
M4B-31-22	0323
M4B32	0323
M4B41	5499
M4B41-13	0087
M4B41-S	OEM
M4C1	0087
M4C5	0087
M4C9	0087
M4C-31	0071
M4C-31-32	0071
M4D1	0087
M4D5	0087
M4D9	0087
M4E1	0087
M4E9	0087
M4F1	0087
M4F5	0087
M4F9	0087
M4G1	0087
M4G5	0087
M4G9	0087
M4H1	0087
M4H5	0087
M4H9	0087
M4HZ	0015
M4K1	0087
M4K5	0087
M4K9	0087
M4L20-3	OEM
M4L20-8	OEM
M4L20-28	OEM
M4L20A	OEM
M4L20M3	OEM
M4L20M8	OEM
M4L20M28	OEM
M4L30-3	OEM
M4L30-8	OEM
M4L30-28	OEM
M4L30A	OEM
M4L30M3	OEM
M4L30M8	OEM
M4L30M28	OEM
M4L40-3	OEM
M4L40-8	OEM
M4L40-28	OEM
M4L40A	OEM
M4L40M3	OEM
M4L40M8	OEM
M4L40M28	OEM
M4L50-3	OEM
M4L50-8	OEM
M4L50-28	OEM
M4L50A	OEM
M4L50M3	OEM
M4L50M8	OEM
M4L50M28	OEM
M4L2052	OEM
M4L2053	OEM
M4L2054	OEM
M4M1	0087
M4M5	0087
M4M9	0087
M4Z3.3	0188
M4Z3.3-20	0299
M4Z3.3A	0188
M4Z3.6	0376
M4Z3.6A	0376
M4Z3.9	0188
M4Z3.9-20	0036
M4Z3.9A	0188
M4Z4.3	0275
M4Z4.3-20	0275
M4Z4.3A	0275
M4Z4.7	0162
M4Z4.7-20	0162
M4Z4.7A	0162
M4Z5.1	0162
M4Z5.1-20	0162
M4Z5.1A	0162
M4Z5.6	0157
M4Z5.6-20	0157
M4Z5.6A	0157
M4Z6.2	0471
M4Z6.2-20	0636
M4Z6.2A	0636
M4Z6.8	0062
M4Z6.8-20	0062
M4Z6.8A	0062
M4Z7.5	0077
M4Z7.5-20	0077
M4Z7.5A	0077
M4Z8.2	0165
M4Z8.2-20	0165
M4Z8.2A	0165
M4Z9.1	0057
M4Z9.1-20	0057
M4Z9.1A	0057
M4Z10	0170
M4Z10-20	0170
M4Z10A	0170
M4Z11	0181
M4Z11-20	0181
M4Z11A	0181
M4Z12	0052
M4Z12-20	0052
M4Z12A	0052
M4Z13	0053
M4Z13-20	0053
M4Z13A	0053
M4Z15	0686
M4Z15-20	0686
M4Z15A	0686
M4Z16	0445
M4Z16-20	0445
M4Z16A	0445
M4Z18	0375
M4Z18-20	0375
M4Z18A	0375
M4Z20	0700
M4Z20-20	0700
M4Z20A	0700
M4Z22	0705
M4Z22-20	0705
M4Z22A	0705
M4Z24	OEM
M4Z24-20	0494
M4Z24A	0494
M4Z27	0441
M4Z27-20	0441
M4Z27A	0441
M4Z30	1083
M4Z30-20	0195
M4Z30A	0195
M4Z33	0039
M4Z33-20	0166
M4Z33A	0166
M4Z36	0010
M4Z36-20	0010
M4Z36A	0010
M4Z39	0032
M4Z39-20	0032
M4Z39A	0032
M4Z43	0054
M4Z43-20	0054
M4Z43A	0054
M4Z47	0068
M4Z47-20	0068
M4Z47A	0068
M4Z51	0092
M4Z51-20	0092
M4Z51A	0092
M4Z56	0125
M4Z56-20	0125
M4Z56A	0125
M4Z62	0783
M4Z62-20	0152
M4Z62A	0783
M4Z68	0173
M4Z68-20	0173
M4Z68A	0173
M4Z75	0094
M4Z75-20	0094
M4Z75A	0094
M4Z82	0331
M4Z82-20	0049
M4Z82A	0049
M4Z91	0156
M4Z91-20	0156
M4Z91A	0156
M4Z100	0189
M4Z100-20	0189
M4Z100A	0189
M4Z110	0149
M4Z110-20	7012
M4Z110A	0099
M4Z120	0089
M4Z120-20	0089
M4Z120A	0089
M4Z130	0287
M4Z130-20	0287
M4Z130A	0287
M4Z150	0340
M4Z150-20	0340
M4Z150A	0340
M4Z160	0370
M4Z160-20	0370
M4Z160A	0370
M4Z180	0425
M4Z180-20	0425
M4Z180A	0425
M4Z200	1473
M4Z200-20	1473
M4Z200A	1473
M5	0861
M5-1454L	OEM
M5A	0363
M5A5	OEM
M5B	0363
M5B5	OEM
M5C	0363
M5C5	OEM
M5D	0363
M5D5	OEM
M5E5	OEM
M5F5	OEM
M5F78M09L	OEM
M5F78M12L	OEM
M5F7805	0624
M5F7805L	OEM
M5F7806	OEM
M5F7807	OEM
M5F7808	OEM
M5F7809	OEM
M5F7812	OEM
M5F7812L	OEM
M5F7907	OEM
M5F7909	OEM
M5F7924	OEM
M5G5	OEM
M5G1400P	1846
M5H5	OEM
M5K5	OEM
M5K4116P-2	OEM
M5K4116P-3	OEM
M5K4116S-2	OEM
M5K4164-15	OEM
M5K4164ANP	OEM
M5K4164ANP-12	2357
M5K4164ANP-15	2357
M5K4164AP	OEM
M5K4164NP12	OEM
M5K4164NP15	OEM
M5K4164NP20	OEM
M5K4164NS15	OEM
M5K4164NS20	OEM
M5K4164P12	OEM
M5K4164P15	OEM
M5K4164P20	OEM
M5K4164S-15	4315
M5K4164S-20	4315
M5K41165-3	OEM
M5L2101AP-4	OEM
M5L2102AP-4	OEM
M5L2102AS-4	OEM
M5L2111AP-4	OEM
M5L2112AP-4	OEM
M5L2114LP	OEM
M5L2114LP-2	OEM
M5L2114LP-3	2048
M5L2114LS-2	OEM
M5L2114LS-3	OEM
M5L2716K	2278
M5L2732K	2697
M5L2732K-6	2697
M5L2764K	0812
M5L2764K-2	0812
M5L2764K2	OEM
M5L2764K3	OEM
M5L2764K-FA552	0812
M5L5101LP-1	OEM
M5L8039P-11	OEM
M5L8049-124P-6	OEM
M5L8049-565P	OEM
M5L8049-570P	OEM
M5L8088S-2	OEM
M5L8155P	OEM
M5L8216P	1863
M5L8226P	OEM
M5L8243P	OEM
M5L8251AP	OEM
M5L8251AP-5	OEM
M5L8253P-5	OEM
M5L8255AP-5	0051
M5L8259AP	OEM
M5L8279P	OEM
M5L8279P-5	OEM
M5L27128K	1638
M5L27128K-2	OEM
M5L27256K	OEM
M5L27256K-2	OEM
M5L27256K-X	OEM
M5L27512K-17	OEM
M5M4C264	OEM
M5M4C264L-10	OEM
M5M4C264L-12	OEM
M5M4C500AL	OEM
M5M4C500L	OEM
M5M4C500L-10	OEM
M5M5	OEM
M5M8N10	OEM
M5M2167S55	OEM
M5M2167S70	OEM
M5M2364-169P	OEM
M5M4164ANP	OEM
M5M4256P	OEM
M5M4257P-15	OEM
M5M4416P	OEM
M5M4416P-15	OEM
M5M4464AP	OEM
M5M4464P	OEM
M5M5116P	OEM
M5M5116P15	OEM
M5M5117P	OEM
M5M5117P12	OEM
M5M5117P15	OEM
M5M5118P	OEM
M5M5118P12	OEM
M5M5118P15	OEM
M5M5165P-10	OEM
M5M5165P-15	OEM
M5M5165P-70	OEM
M5M5165P70	OEM
M5M5178P-55	OEM
M5N5	OEM
M5P5	OEM
M5R4558P	3440
M5T4044P-20	OEM
M5T4044P-30	OEM
M5T4044P-45	OEM
M5T4044S-30	OEM
M5T4044S-45	OEM
M5T40445-20	OEM
M5W1793-02P	OEM
M5Z6.8	0205
M5Z6.8A	0205
M5Z7.5	0480
M5Z7.5A	0480
M5Z8.2	0504
M5Z8.2A	0504
M5Z9.1	0684
M5Z9.1A	0684
M5Z10	0225
M5Z10A	0225
M5Z11	0230
M5Z11A	0230
M5Z12	0234
M5Z12A	0234
M5Z13	0237
M5Z13A	0237
M5Z15	0247
M5Z15A	0247
M5Z16	0251
M5Z16A	0251
M5Z18	0256
M5Z18A	0256
M5Z20	0262
M5Z20A	0262
M5Z22	0269
M5Z22A	0269
M5Z24	0274
M5Z24A	0274
M5Z27	0294
M5Z27A	0294
M5Z33	0318
M5Z33A	0318
M5Z36	0320
M5Z36A	0320
M5Z39	0326
M5Z39A	0326
M5Z43	0337
M5Z43A	0337
M5Z47	0347
M5Z47A	0347
M5Z51	0027
M5Z51A	0027
M5Z56	0266
M5Z56A	0266
M5Z62	0386
M5Z62A	0386
M5Z68	0405
M5Z68A	0405
M5Z75	0426
M5Z75A	0426
M5Z82	0444
M5Z82A	0444
M5Z91	0238
M5Z91A	0238
M5Z100	1180
M5Z100A	1180
M5Z110	1190
M5Z120	1207
M5Z120A	1207
M5Z130	1218
M5Z130A	1218
M5Z150	0647
M5Z150A	0647
M5Z160	1255
M5Z160A	1255
M5Z180	1279
M5Z180A	1279
M5Z200	0605
M5Z200A	0605
M5.6Z	0253
M6	0861
M6-100	OEM
M6-100R	OEM
M6-200	OEM
M6-200R	OEM
M6-400	OEM
M6-400R	OEM
M6-600	OEM
M6-600R	OEM
M6-800	OEM
M6-800R	OEM
M6-1000	OEM
M6-1000R	OEM
M6-1200	OEM
M6-1200R	OEM
M6-SAM	1742
M6HZ	0015
M6M80011AL	OEM
M6M80011AP	OEM
M6M80011L	OEM
M6M80011P	OEM
M6M80021L	OEM
M6M80021P	OEM
M6M80022P	OEM
M6M80041P	OEM

DEVICE TYPE	REPL CODE
M6.2Z	0471
M6.8	0030
M6.8Z	0062
M7	0693
M7A5	OEM
M7B5	OEM
M7C5	OEM
M7D5	OEM
M7E5	OEM
M7F5	OEM
M7G5	OEM
M7H5	OEM
M7K5	OEM
M7M5	OEM
M7N5	OEM
M7P5	OEM
M7.5Z	0077
M8HZ	0071
M8T28B	OEM
M8.2Z	0165
M9Z	0057
M10A	0620
M10A10KA	OEM
M10A10LA	OEM
M10A17EA	OEM
M10B	0620
M10D	0074
M10HZ	OEM
M10Z	0170
M11Z	0181
M12	0015
M12A02RA	OEM
M12H	OEM
M12Z	0137
M13Z	0053
M14	0015
M14H	OEM
M15H	OEM
M15HZ	OEM
M15K	OEM
M15Z	0686
M16-100	OEM
M16-100R	OEM
M16-200	OEM
M16-200R	OEM
M16-400	OEM
M16-400R	OEM
M16-600	OEM
M16-600R	OEM
M16-800	OEM
M16-800R	OEM
M16-1000	OEM
M16-1000R	OEM
M16-1200	OEM
M16-1200R	OEM
M16Z	0445
M18TB7	OEM
M18Z	0375
M20	OEM
M20H	OEM
M20HZ	OEM
M20K	OEM
M20Z	0700
M21C	0058
M21C/Q	0058
M21C/R	0058
M21C5	0058
M21CA	0407
M21C-K	OEM
M21CQ	0058
M21C-R	0058
M21CR	0058
M21CY	0058
M21F	OEM
M22	0015
M22P2	OEM
M22P3	OEM
M22P4	OEM
M22Z	0705
M23C	0940
M23CA	0940
M23P-X504	OEM
M23P-X509	OEM
M23P-X516	OEM
M24	0079
M24A	0079
M24B	0079
M24P-X502	OEM
M24Z	0494
M25	0079
M25-100	OEM
M25-100R	OEM
M25-103A	OEM
M25-200	OEM
M25-200R	OEM
M25-203A	OEM
M25-400	OEM
M25-400R	OEM
M25-403A	OEM
M25-600	OEM
M25-600R	OEM
M25-603A	OEM
M25-800	OEM
M25-800R	OEM
M25-803A	OEM
M25-1000	OEM
M25-1000R	OEM
M25-1200	OEM
M25-1200R	OEM
M25A	0079
M25A2	0079
M25A10PA	OEM
M25B	0079
M25B2	0079
M25Hz	OEM
M26	0133
M26P-X504	OEM
M26P-X505	OEM
M26P-X516	OEM
M26P-X517	OEM
M26P-X531	OEM
M26P-X558	OEM
M26P-X560	OEM
M26W0	0139
M26WA	0120
M26X256	OEM
M27Z	0441
M28P-X507	OEM
M28P-X508	OEM
M30	OEM
M30G	OEM
M30HZ	OEM
M30K	OEM
M30Z	0195
M-31	0102
M32P-X503	OEM
M32P-X506	OEM
M32P-X508	OEM
M32P-X509	OEM
M-33	OEM
M33Z	0166
M34A	0143
M40	OEM
M40K	OEM
M41-100	OEM
M41-101A	OEM
M41-200	OEM
M41-201A	OEM
M41-400	OEM
M41-401A	OEM
M41-600	OEM
M41-601A	OEM
M41-800	OEM
M41-801A	OEM
M41-1000	OEM
M41-1200	OEM
M42	0015
M46F	OEM
M47F	OEM
M50A	OEM
M50K	OEM
M51	0143
M54	0079
M54A	0079
M54B	0079
M54BLK	0079
M54BLU	0079
M54BRN	0079
M54C	0102
M54D	0079
M54E	0079
M54GRN	0079
M54ORN	0079
M54RED	0079
M54WHT	0079
M54YEL	0079
M60	0143
M60A	OEM
M60H	OEM
M60K	OEM
M62	0015
M63P-X503	OEM
M65-031	OEM
M65-032	OEM
M65-045	OEM
M65A	0037
M65B	0037
M65C	0037
M65D	0037
M65E	0037
M65F	0037
M67	0015
M67A	0015
M67B	0015
M67C	0015
M67P-X504	OEM
M68	0015
M68A	0015
M68B	0015
M68BSA1	OEM
M68BSA2	OEM
M68BSA4	OEM
M68C	0015
M68D1M2A	OEM
M68KVAM	OEM
M68KVM01A1	OEM
M68KVM01A2	OEM
M68KVM02-1	OEM
M68KVM02-3	OEM
M68KVM11-1	OEM
M68KVM11-2	OEM
M68KVM21(A)	OEM
M68KVM40(A)	OEM
M68KVM80-1	OEM
M68KVM80-2	OEM
M68KVM80-4	OEM
M68KVMCC1	OEM
M68KVMPM1	OEM
M68KVMPS1	OEM
M68KWW	OEM
M68MM01	OEM
M68MM01A	OEM
M68MM01A1	OEM
M68MM01A2	OEM
M68MM01B	OEM
M68MM01B1A	OEM
M68MM01D	OEM
M68MM02	OEM
M68MM03	OEM
M68MM03-1	OEM
M68MM03-2	OEM
M68MM04	OEM
M68MM04A	OEM
M68MM05A	OEM
M68MM05B	OEM
M68MM05C	OEM
M68MM06	OEM
M68MM07	OEM
M68MM09	OEM
M68MM10A	OEM
M68MM10B	OEM
M68MM10C	OEM
M68MM13A	OEM
M68MM13B	OEM
M68MM13C	OEM
M68MM13D	OEM
M68MM14	OEM
M68MM14A	OEM
M68MM15A	OEM
M68MM15A1	OEM
M68MM16-1	OEM
M68MM16-2	OEM
M68MM16-3	OEM
M68MM17	OEM
M68MM18(A)	OEM
M68MM19	OEM
M68MM19A	OEM
M68MM19SB	OEM
M68MM22	OEM
M68MMFLC1	OEM
M68RWIN1-1(A)	OEM
M68RWIN1-2(A)	OEM
M69	0015
M69A	0015
M69B	0015
M69C	0015
M70	0015
M70-101A	OEM
M70-101K	OEM
M70-201A	OEM
M70-201K	OEM
M70-401A	OEM
M70-401K	OEM
M70-601A	OEM
M70-601K	OEM
M70-801A	OEM
M70-801K	OEM
M70A	0015
M70B	0015
M70C	0015
M71	0015
M71-100	OEM
M71-100R	OEM
M71-101A	OEM
M71-200	OEM
M71-200R	OEM
M71-201A	OEM
M71-400	OEM
M71-400R	OEM
M71-401A	OEM
M71-600	OEM
M71-600R	OEM
M71-601A	OEM
M71-800	OEM
M71-800R	OEM
M71-801A	OEM
M71-1000	OEM
M71-1000R	OEM
M71-1200	OEM
M71-1200R	OEM
M71A	0015
M71B	0015
M71C	0015
M72	0071
M72A	0071
M72B	0071
M72C	0071
M72D	0087
M73	0071
M73A	0071
M73B	0071
M73C	0071
M73P1	OEM
M73P-X502	OEM
M74ALS00AP	OEM
M74ALS02P	OEM
M74ALS04P	OEM
M74ALS27P	OEM
M74ALS32P	OEM
M74ALS109P	OEM
M74ALS244P	OEM
M74ALS573P	OEM
M74H002	OEM
M74HC02P	OEM
M74HC08P	OEM
M74HC14P	OEM
M74HC74P	OEM
M74HC138P	OEM
M74HC251P	OEM
M74HC374P	OEM
M74HC4053B1	OEM
M74LS00	3774
M74LS00P	1527
M74LS02P	1559
M74LS03P	1579
M74LS04	6179
M74LS04P	1595
M74LS05	6180
M74LS05P	1608
M74LS06P	OEM
M74LS08	6182
M74LS08P	1633
M74LS09P	1642
M74LS10P	1662
M74LS11P	1667
M74LS12P	1679
M74LS13P	1688
M74LS14	OEM
M74LS14P	1698
M74LS15P	1707
M74LS20P	0034
M74LS21P	1763
M74LS22P	1775
M74LS27P	0183
M74LS30P	0828
M74LS32P	0088
M74LS37P	1730
M74LS38P	1839
M74LS40P	0135
M74LS42P	1841
M74LS47P	1845
M74LS48P	1849
M74LS51P	1033
M74LS73AP	1867
M74LS73P	1867
M74LS74	6190
M74LS74AP	0243
M74LS74P	0243
M74LS75P	1870
M74LS76AP	2180
M74LS83AP	2218
M74LS83P	2218
M74LS85P	0431
M74LS86BP	0291
M74LS86P	0291
M74LS90P	1882
M74LS91P	1885
M74LS92P	1887
M74LS93P	1888
M74LS95BP	0771
M74LS96P	OEM
M74LS107AP	1602
M74LS107P	1602
M74LS109AP	1906
M74LS109P	1906
M74LS112AP	2128
M74LS113AP	2256
M74LS114AP	2302
M74LS122P	1620
M74LS123P	0979
M74LS125AP	0075
M74LS126AP	2877
M74LS132P	1625
M74LS133P	3400
M74LS136P	1628
M74LS138BP	0427
M74LS138P	0427
M74LS139	OEM
M74LS139P	0153
M74LS145P	1564
M74LS148P	3912
M74LS151P	1646
M74LS153P	0959
M74LS155P	0209
M74LS156P	1654
M74LS157P	1161
M74LS158P	1656
M74LS160AP	0837
M74LS161AP	0858
M74LS161P	0858
M74LS162AP	0880
M74LS163AP	0893
M74LS163P	0893
M74LS164P	4339
M74LS165P	4356
M74LS166AP	4371
M74LS166P	4371
M74LS170P	2629
M74LS173AP	5217
M74LS174P	0260
M74LS175P	1672
M74LS183P	OEM
M74LS190P	1686
M74LS191P	1687
M74LS192P	1689
M74LS193P	1692
M74LS194AP	1304
M74LS195AP	1315
M74LS196P	2834
M74LS197P	2469
M74LS221P	1239
M74LS240P	0452
M74LS241P	1726
M74LS242P	1728
M74LS243P	0906
M74LS244P	0458
M74LS245P	0463
M74LS247P	1732
M74LS248P	1734
M74LS251P	1737
M74LS253P	1739
M74LS256P	OEM
M74LS257AP	1744
M74LS258AP	1746
M74LS259P	3205
M74LS266P	0592
M74LS273P	0894
M74LS279P	3290
M74LS280P	1773
M74LS283P	1779
M74LS290P	4425
M74LS293P	0082
M74LS295AP	2227
M74LS295BP	2227
M74LS298P	3371
M74LS299P	4426
M74LS323P	OEM
M74LS352P	0761
M74LS353P	0773
M74LS365AP	0943
M74LS366AP	0956
M74LS367AP	0977
M74LS368AP	0991
M74LS373P	0709
M74LS374P	0713
M74LS375P	OEM
M74LS377P	1120
M74LS386P	1230
M74LS390P	1288
M74LS393P	0819
M74LS395AP	1330
M74LS395P	1330
M74LS490P	2213
M74LS620P	OEM
M74LS640-1P	0669
M74LS641-1P	0690
M74LS642-1P	0719
M74LS643-1P	2056
M74LS644-1P	OEM
M74LS645-1P	0775
M74LS668P	OEM
M74LS669P	OEM
M74LS670P	1130
M75	2900
M76	0050
M77	0050
M78	0050
M78A	0050
M78B	0050
M78BLK	0050
M78C	0050
M78D	0050
M78E	0050
M78GRN	0050
M78L12CP	OEM
M78RED	0050
M78YEL	0050
M79H048	OEM
M80	OEM
M80C51-45	OEM
M80C51F-49	OEM
M80C51F-520	OEM
M80C154	OEM
M80C154-1	OEM
M80C154V-1	OEM
M80K	OEM
M81	OEM
M82	0071
M82P-X500	OEM
M83C154-228	OEM
M84	0085
M84B	0160
M89	OEM
M91	0127
M91A	0079
M91A01	0015
M91A02	0015
M91A03	0015
M91A06	0071
M91B	0079
M91BGRN	0079
M91C	0079
M91CM624	0079
M91D	0079
M91E	0079
M91F	0079
M91FM624	0079
M95	0143
M100	0325
M100-101A	OEM
M100-101K	OEM
M100-201A	OEM
M100-201K	OEM
M100-401A	OEM
M100-401K	OEM
M100-601A	OEM
M100-601K	OEM
M100-801A	OEM
M100-801K	OEM
M100A	1335
M100B	0080
M100D	0609
M100G	0795
M100J	0015
M100K	0072
M100M	0071
M101	3628
M101B2R	OEM
M101B4R	OEM
M101B6R	OEM
M101T2R	OEM
M101T4R	OEM
M101T6R	OEM
M102	0071
M106B1	OEM
M108	0004
M116	6774
M116A	OEM
M119	OEM
M120DB1	OEM
M120DF1	OEM
M120K	OEM
M124J779-1	0015
M128	OEM
M128J422	2965
M128J422-1	0474
M128J753	0474
M128J753-1	0474
M132	6778
M135	0649
M140-1	0127
M140-3	0079
M142A-B1	OEM
M142A-D1	OEM
M142B1	OEM
M142D1	OEM
M147B1	OEM
M150	0015
M150-1	0133
M150K	OEM
M151-100	OEM
M151-100R	OEM
M151-200	OEM
M151-200R	OEM
M151-400	OEM
M151-400R	OEM
M151-600	OEM
M151-600R	OEM
M151-800	OEM
M151-800R	OEM
M151-1000	OEM
M151-1000R	OEM
M151-1200	OEM
M151-1200R	OEM
M-152R	0071
M159A	OEM
M163	OEM
M164	OEM
M172A	0015
M190	OEM
M190B	OEM
M193	OEM
M193AB1	OEM
M193B1	OEM
M200K	OEM
M203	OEM
M204	OEM
M-204B	0015
M204B	0015
M206	OEM
M207	OEM
M250D1	OEM
M252B1AA	OEM
M252B1AD	OEM
M252D1AA	OEM
M253B1AA	OEM
M253B1AC	OEM
M253D1AA	OEM
M254B1AD	OEM
M254B1AM	OEM
M255B1AB	OEM
M258B1	OEM
M259B1	OEM
M300K	OEM
M304	OEM
M304/4050	2027
M310	OEM
M311	OEM
M351	0050
M366-1	0065
M380	0379
M404	OEM
M405	OEM
M423N	OEM
M-433	0400
M433	0400
M440	OEM
M441	OEM
M484	0079
M500	0015
M500A	0015
M500B	0015
M500C	0015
M501	0160
M502	OEM
M503A	OEM
M503DA	OEM
M504A	OEM
M505	OEM
M509	OEM
M510	OEM
M511	OEM
M511A	OEM
M513B	OEM
M517	OEM
M5190P	0856
M521	OEM
M523	OEM
M529	OEM
M530	0086
M530J	OEM
M530K	OEM
M530L	OEM
M537A	OEM
M538A	OEM
M539	OEM
M540J	OEM
M540K	OEM
M546	0007
M549	OEM
M554	OEM
M565	OEM
M566	OEM
M569	OEM
M570	OEM
M573	OEM
M574	OEM
M575	OEM
M575B	OEM
M577B	OEM
M578B	OEM
M579	OEM
M586	OEM
M595B	OEM
M596	OEM
M597	OEM
M598B	OEM
M604	0277
M604HT	0015
M605	OEM
M608	OEM
M612	0007
M613	0007
M614	0007
M644	0037
M652/PIC	0150
M652P1C	0126
M652PIC	0126
M670	0015
M670A	0015
M670B	0015
M670C	0015
M671	0079
M680	0015
M680A	0015
M680B	0015
M680C	0015
M690	0015
M690A	0015
M690B	0015
M690C	0015
M700	0015
M700A	0015
M700B	0015
M700C	0015
M701B	0015
M702	0071
M702B	0071
M702B1	OEM
M702C	0015
M702D1	OEM
M702D2	OEM
M706B1	OEM
M710	0015
M710A	0015
M710B	0015
M710C	0015
M714B1	OEM
M714D1	OEM
M720	0071
M720A	0071
M720B	0071
M720C	0071
M726	3536
M730	0071
M730A	0071
M730B	0071
M730B1	OEM
M730C	0071
M730D1	OEM
M731B1	OEM
M731D1	OEM
M738	OEM
M738B1	OEM
M740	6812
M740B1	OEM
M741	OEM
M741B1	OEM
M747	OEM
M747B1	OEM
M750B1	OEM
M750D1	OEM
M751B1	OEM
M751D1	OEM
M752B1	OEM
M752D1	OEM
M754B1	OEM
M754D1	OEM
M758	0157
M773	0079
M773RED	0079
M774	0079
M774ORN	0079

DEVICE TYPE	REPL CODE
M775	0079
M775BRN	0079
M776	0079
M776GRN	0079
M779	0016
M779BLU	0079
M780	0016
M780WHT	0079
M783	0079
M783RED	0079
M784	0079
M784ORN	0079
M785	0079
M785YEL	0079
M786	0016
M787	0016
M787BLU	0079
M791	0079
M813A	0139
M818	0079
M818WHT	0079
M819	0233
M822	0079
M822A	0079
M822A-BLU	0079
M822B	0079
M823	0079
M823B	0079
M823WHT	0079
M827	0079
M827BRN	0079
M828	0126
M828GRN	0079
M829A	0037
M829B	0037
M829C	0037
M829D	0037
M829E	0037
M829F	0037
M833	0037
M841	OEM
M844	0161
M847	0079
M847BLK	0079
M895	OEM
M912	0276
M924	0079
M954-644	OEM
M1000B	OEM
M1001B	OEM
M1020	OEM
M1020A	OEM
M1020B	OEM
M1021	OEM
M1021A	OEM
M1021B	OEM
M1022	OEM
M1022A	OEM
M1022B	OEM
M1024B5	OEM
M1025	OEM
M1025A	OEM
M1025B	OEM
M1025B5	OEM
M1025B5BA	OEM
M1025B5BB	OEM
M1025B5C5	OEM
M1025B5CA	OEM
M1026	OEM
M1026A	OEM
M1026B	OEM
M1029	OEM
M1029A	OEM
M1029B	OEM
M1030	OEM
M1030A	OEM
M1030B	OEM
M1031	OEM
M1031A	OEM
M1031B	OEM
M1032	OEM
M1032A	OEM
M1032B	OEM
M1035	OEM
M1035A	OEM
M1035B	OEM
M1036	OEM
M1036B	OEM
M1039	OEM
M1039A	OEM
M1039B	OEM
M1040	OEM
M1040A	OEM
M1040B	OEM
M1041	OEM
M1041A	OEM
M1041B	OEM
M1042	OEM
M1042A	OEM
M1042B	OEM
M1045	OEM
M1045A	OEM
M1045B	OEM
M1046	OEM
M1046A	OEM
M1046B	OEM
M1049	OEM
M1049A	OEM
M1049B	OEM
M1060	OEM
M1060A0A	OEM
M1060B	OEM
M1061	OEM
M1061A	OEM
M1061B	OEM
M1062	OEM
M1062A	OEM
M1062B	OEM
M1065	OEM
M1065A	OEM
M1065B	OEM
M1066	OEM
M1066A	OEM
M1066B	OEM
M1069	OEM
M1069A	OEM
M1069B	OEM
M1100	OEM
M1100A	OEM
M1101	OEM
M1101A	OEM
M1102	2835
M1102A	OEM
M1102B	OEM
M1105	OEM
M1105A	OEM
M1105B	OEM
M1106	OEM
M1106A	OEM
M1106B	OEM
M1109	OEM
M1109A	OEM
M1109B	OEM
M1150	OEM
M1150A	OEM
M1150B	OEM
M1151	3598
M1151A	OEM
M1151B	OEM
M1152	OEM
M1152A	3598
M1152B	OEM
M1155	OEM
M1155A	OEM
M1155B	OEM
M1156	OEM
M1156A	OEM
M1156B	OEM
M1159	OEM
M1159A	OEM
M1159B	OEM
M1200	OEM
M1220	OEM
M1220A	OEM
M1220B	OEM
M1221	OEM
M1221A	OEM
M1221B	OEM
M1222	OEM
M1222A	OEM
M1222B	OEM
M1223	OEM
M1224	OEM
M1225	OEM
M1225A	OEM
M1225B	OEM
M1226	OEM
M1226A	OEM
M1226B	OEM
M1227	OEM
M1229	OEM
M1229A	OEM
M1229B	OEM
M1230	OEM
M1233	OEM
M1243	OEM
M1300	OEM
M1302	OEM
M1304	OEM
M1305	OEM
M1306	OEM
M1308	OEM
M1310	OEM
M1311	OEM
M1312	OEM
M1313	OEM
M1315	OEM
M1316	OEM
M1320	OEM
M1321	OEM
M1323	OEM
M1325	OEM
M1326	OEM
M1327	OEM
M1328	OEM
M1330	OEM
M1330A	OEM
M1330B	OEM
M1331	OEM
M1331A	OEM
M1331B	OEM
M1332	OEM
M1332A	OEM
M1332B	OEM
M1335	OEM
M1335A	OEM
M1335B	OEM
M1336	OEM
M1336A	OEM
M1336B	OEM
M1337	OEM
M1339	OEM
M1339A	OEM
M1339B	OEM
M1341	OEM
M1347	OEM
M1348	OEM
M1350	OEM
M1353	OEM
M1354	OEM
M1355	OEM
M1358P	0395
M1359	OEM
M1369	OEM
M1373	OEM
M1379	OEM
M1400	OEM
M1400-1	0127
M1406	OEM
M1407	OEM
M1470	OEM
M1470A	OEM
M1470B	OEM
M1471	OEM
M1471A	OEM
M1471B	OEM
M1472	OEM
M1472A	OEM
M1472B	OEM
M1475	OEM
M1475A	OEM
M1475B	OEM
M1476	OEM
M1476A	OEM
M1476B	OEM
M1506	OEM
M1507	OEM
M1680	OEM
M1680A	OEM
M1680B	OEM
M1681	OEM
M1681A	OEM
M1681B	OEM
M1682	OEM
M1682A	OEM
M1682B	OEM
M1685	OEM
M1685A	OEM
M1685B	OEM
M1713DP	0740
M1713MP	0764
M1713NP	0766
M1713PP	OEM
M1751	OEM
M1800	OEM
M1800A	OEM
M1800B	OEM
M1801	OEM
M1801A	OEM
M1801B	OEM
M1802	OEM
M1802A	OEM
M1802B	OEM
M2002	0823
M2003	1696
M2009	1980
M2012ALB1-2	OEM
M2016	1696
M2032	2749
M2102AB1	OEM
M2102AB1-2	OEM
M2102AB1-4	OEM
M2102AB1-6	OEM
M2102AD1	OEM
M2102AD1-2	OEM
M2102AD1-4	OEM
M2102AD1-6	OEM
M2102AF1	OEM
M2102AF1-2	OEM
M2102AF1-4	OEM
M2102AF1-6	OEM
M2102ALB1	OEM
M2102ALB1-4	OEM
M2102ALD1	OEM
M2102ALD1-2	OEM
M2102ALD1-4	OEM
M2102ALF1	OEM
M2102ALF1-2	OEM
M2102ALF1-4	OEM
M-2115	OEM
M2128-20	1898
M2147-3B1	OEM
M2147-3F1	OEM
M2147-B1	OEM
M2147-F1	OEM
M2163	3192
M2167	3192
M2169	3192
M2207	0025
M2316E-D1	OEM
M2332B1	OEM
M2332D1	OEM
M2333/250	OEM
M2333B1	OEM
M2333D1	OEM
M2400	OEM
M2497	0015
M2513DP	0740
M2513MP	0764
M2513NP	0766
M2513PP	OEM
M2532F1	OEM
M2537-1F1	OEM
M2716	2278
M2716-1F1	OEM
M2716F1	2278
M2716M	2278
M2732-2F1	OEM
M2732AF1	2697
M2764A-20	OEM
M2797-02/L2376-01	OEM
M2816	OEM
M3013DP	0722
M3013MP	0725
M3013NP	0750
M3013PP	OEM
M3016	0015
M3020	OEM
M3020A	OEM
M3020B	OEM
M3021	OEM
M3021A	OEM
M3021B	OEM
M3022	OEM
M3022A	OEM
M3022B	OEM
M3025	OEM
M3025A	OEM
M3025B	OEM
M3026	OEM
M3026A	OEM
M3026B	OEM
M3029	OEM
M3029A	OEM
M3029B	OEM
M3030	OEM
M3030A	OEM
M3030B	OEM
M3031	OEM
M3031A	OEM
M3031B	OEM
M3032	OEM
M3032A	OEM
M3032B	OEM
M3035	OEM
M3035A	OEM
M3035B	OEM
M3036	OEM
M3036A	OEM
M3036B	OEM
M3039	OEM
M3039A	OEM
M3039B	OEM
M3040	OEM
M3040A	OEM
M3040B	OEM
M3041	OEM
M3041A	OEM
M3041B	OEM
M3042	OEM
M3042A	OEM
M3042B	OEM
M3045	OEM
M3045A	OEM
M3045B	OEM
M3046	OEM
M3046A	OEM
M3046B	OEM
M3049	OEM
M3049A	OEM
M3049B	OEM
M3059	OEM
M3059A	OEM
M3059B	OEM
M3060	OEM
M3060A	OEM
M3060B	OEM
M3061	OEM
M3061A	OEM
M3061B	OEM
M3062	OEM
M3062A	OEM
M3062B	OEM
M3065	OEM
M3065A	OEM
M3065B	OEM
M3066	OEM
M3066A	OEM
M3066B	OEM
M3100	OEM
M3100A	OEM
M3100B	OEM
M3101	OEM
M3101A	OEM
M3101B	OEM
M3102	OEM
M3102A	OEM
M3102B	OEM
M3105	OEM
M3105A	OEM
M3105B	OEM
M3106	OEM
M3106A	OEM
M3106B	OEM
M3109	OEM
M3109A	OEM
M3109B	OEM
M3150	OEM
M3150A	OEM
M3150B	OEM
M3151	OEM
M3151A	OEM
M3151B	OEM
M3152	OEM
M3152A	OEM
M3152B	OEM
M3155	OEM
M3155A	OEM
M3155B	OEM
M3156	OEM
M3156A	OEM
M3156B	OEM
M3159	OEM
M3159A	OEM
M3159B	OEM
M3220	OEM
M3220A	OEM
M3220B	OEM
M3221	OEM
M3221A	OEM
M3221B	OEM
M3222	OEM
M3222A	OEM
M3222B	OEM
M3225	OEM
M3225A	OEM
M3225B	OEM
M3226	OEM
M3226A	OEM
M3226B	OEM
M3230A	OEM
M3330	OEM
M3330A	OEM
M3330B	OEM
M3331	OEM
M3331A	OEM
M3331B	OEM
M3332	OEM
M3332A	OEM
M3332B	OEM
M3335	OEM
M3335A	OEM
M3335B	OEM
M3451M8-561SP	OEM
M3470	OEM
M3470A	OEM
M3470B	OEM
M3471	OEM
M3471A	OEM
M3471B	OEM
M3472	OEM
M3472A	OEM
M3472B	OEM
M3519	0079
M3567-2	0161
M3636	OEM
M3681	OEM
M3681A	OEM
M3681B	OEM
M3682	OEM
M3682A	OEM
M3682B	OEM
M3764-15RS	2357
M3764-20RS	2357
M3764A-15	OEM
M3800	OEM
M3800A	OEM
M3800B	OEM
M3802	OEM
M3802A	OEM
M3802B	OEM
M3864B-21	OEM
M4001	OEM
M4001BP	0478
M4002BP	2055
M4011	OEM
M4011BP	0215
M4011UBP	OEM
M4012BP	0498
M4013DP	OEM
M4013MP	OEM
M4013NP	OEM
M4013PP	OEM
M4015B	1014
M4016BP	1143
M4020BP	1661
M4023BP	0519
M4024BP	1957
M4025BP	2072
M4027P-2B1	OEM
M4027P-2D1	OEM
M4027P-2F1	OEM
M4027P-3B1	OEM
M4027P-3D1	OEM
M4027P-3F1	OEM
M4027P-4B1	OEM
M4027P-4D1	OEM
M4027P-4F1	OEM
M4030BP	OEM
M4040BP	0056
M4042BP	0121
M4043BP	1553
M4049BP	4343
M4050BP	0398
M4051BP	0366
M4052BP	0024
M4053BP	0035
M4066BP	0101
M4069UBP	0119
M4071BP	0129
M4081BP	0626
M4116P-2B1	OEM
M4116P-2D1	OEM
M4116P-2F1	OEM
M4116P-3B1	OEM
M4116P-3D1	OEM
M4116P-3F1	OEM
M4116P-4B1	OEM
M4116P-4D1	OEM
M4116P-4F1	OEM
M4313	0004
M4315	0004
M4327	0211
M4331	0160
M4363	0136
M4363BLU	0136
M4363GRN	0136
M4363ORN	0136
M4363WHT	0136
M4364	0136
M4365	0136
M4366	0136
M4367	0136
M4368	0136
M4388	0136
M4389	0280
M4398	0004
M4439	0050
M4442	0037
M4450	0164
M4454	0136
M4456	0136
M4457	0136
M4459	0975
M4462	0004
M4463	0085
M4464	0079
M4465	0079
M4466	0004
M4466ORN	0164
M4468	0004
M4468BRN	0164
M4469	0004
M4469RED	0164
M4470	0004
M4470ORN	0164
M4471	0004
M4471YEL	0164
M4472	0004
M4472GRN	0164
M4473	0164
M4474	0004
M4474YEL	0164
M4475	0004
M4475GRN	0164
M4476	0004
M4476BLU	0164
M4477	0004
M4477PUR	0164
M4478	0126
M4484	0050
M4485	0050
M4486	0050
M4501	0136
M4504	0050
M4505E	OEM
M4506	0050
M4507	0050
M4508B	1811
M4509	0136
M4510	0164
M4510BP	1963
M4512BP	2120
M4514BP	1830
M4515BP	3600
M4516BP	2347
M4518BP	1043
M4520BP	2675
M4524	0050
M4525	0037
M4526	0050
M4528BP	3198
M4537	OEM
M4539BP	3656
M4543	OEM
M4545	0136
M4545BLU	0136
M4545WHT	0136
M4552	0100
M4553	0004
M4553BLU	0004
M4553BRN	0004
M4553GRN	0004
M4553ORN	0004
M4553PUR	0440
M4553RED	0004
M4553YEL	0004
M4562	0164
M4563	0004
M4564	0004
M4565	0004
M4567	0211
M4570	0085
M4573	0004
M4582	0160
M4582BRN	0160
M4583	0160
M4583RED	0160
M4584	0160
M4584BP	OEM
M4584GRN	0160
M4586	0136
M4589	0136
M4590	0037
M4594	0079
M4595	0004
M4596	0211
M4597	0211
M4597GRN	0211
M4597RED	0211
M4603	0136
M4604	0136
M4605	0136
M4605RED	0136
M4606	0160
M4607	0211
M4608	0160
M4619	0085
M4619RED	0160
M4620	0085
M4620GRN	0160
M4621	0136
M4622	0440
M4623	0975
M4624	0079
M4627	0211
M4628A	OEM
M4630	0079
M4632	0136
M4634	OEM
M4640	0604
M4640P	0604
M4648	0233
M4649	0085
M4652	0975
M4653	0012
M4659	0100
M4663	0039
M4689	0086
M4697	0050
M4699	0331
M4700	0600
M4701	0604
M4702	0604
M4704	0331
M4705	0079
M4706	0079
M4709	0144
M4709RED/RED	0144
M4714	0079
M4715	0103
M4719	OEM
M-4721	0079
M-4722	OEM
M4722	0160
M4722BLU	0160
M4722GRN	0160
M4722ORN	0240
M4722PUR	0160
M4722RED	0160
M4722YEL	0160
M4727	0085
M4728	0149
M4730	0160
M4732	0079
M-4733	OEM
M4733	0144
M4734	0079
M4736	0015
M4737	0079
M4739	0079
M4745	0037
M4746	0127
M4756	0007
M4757	0127
M4765	0079
M4766	0160
M4767	0160
M4768	0079
M4789	0079
M4815	0037
M4815D	0037
M4816	0012
M4818	OEM
M4819	0233
M4820	0144
M4821	0079
M4825	0127
M4826	0144
M4826F	OEM
M4834	0079
M4837	0144
M4838	0233
M4839	0233
M4840	0079
M4840A	0127
M4840E	0004
M4841	0079
M4842	0079
M4842A	0079
M4842C	0079
M4843	0233
M4844	0079

DEVICE TYPE	REPL CODE
M4845	0144
M4847	OEM
M4850	0794
M4851	0794
M4852	0079
M4853	0233
M4854	0079
M4855	0144
M4857	0144
M4858	0039
M4860	0050
M4872	0142
M4882	0103
M4885	0142
M4887A	0085
M4888	0160
M4888A	0160
M4888B	0160
M4898	0079
M4900	0074
M4901	0642
M4906	0079
M4910	0150
M4917	0196
M4918	0086
M4919	0086
M4920	0142
M4926	0079
M4927	0233
M4933	0079
M4936	0178
M4937	0079
M4941	0079
M4952	0079
M4953	0079
M4967	OEM
M4970	0079
M4995	0642
M4998	0276
M5005	OEM
M5005A	OEM
M5016	OEM
M5019	OEM
M5021	OEM
M5023	OEM
M5024	OEM
M5025	OEM
M5028	OEM
M5031	OEM
M5032	OEM
M5033	OEM
M5034A	OEM
M5035	OEM
M5042S	OEM
M5048	OEM
M5051	OEM
M5052	OEM
M5053	OEM
M5054	OEM
M5055	OEM
M5057	OEM
M5059	OEM
M5060	OEM
M5061	OEM
M5062	OEM
M5063-2J70B	OEM
M5064H	OEM
M5065	OEM
M5067H	OEM
M5075	OEM
M5076	OEM
M5077	OEM
M5079A	OEM
M5080	OEM
M5081	OEM
M5083A	OEM
M5084	OEM
M5085	OEM
M5086	OEM
M5087	OEM
M5089	OEM
M5091A	OEM
M5094	OEM
M5097	OEM
M5100	OEM
M5101	4485
M5101-245	OEM
M5101P	4485
M5105	OEM
M5106	3263
M5106P	3263
M5108	4260
M5108-9170	4260
M5108P	4260
M5109	2722
M5109P	6381
M5111	OEM
M5112	3958
M5112Y	3958
M5113	2624
M5113P	OEM
M5113T	2624
M5114B	OEM
M5115	2305
M5115P	2305
M5115P-9085	2305
M5115PA	2305
M5115PR	2305
M5115PRA	3017

DEVICE TYPE	REPL CODE
M5115R	OEM
M5115RP	3017
M5117A	OEM
M5117B	OEM
M5117C	OEM
M5117D	OEM
M5117E	OEM
M5117F	OEM
M5117G	OEM
M5117H	OEM
M5117L	4669
M5118L	6188
M5120L	OEM
M5123	OEM
M5124	3958
M5125	OEM
M5125X	OEM
M5126A	OEM
M5127	OEM
M5128-20	OEM
M5128-20(SM)	OEM
M5130P	OEM
M5131P	1634
M5132P	OEM
M5133	OEM
M5134	1634
M5134-8266	1634
M5134P	1634
M5135	4643
M5135P	0878
M5136	OEM
M5137	OEM
M5138P	OEM
M5142	OEM
M-5142P	OEM
M5142P	OEM
M5143	0167
M5143P	0167
M5144BP	0167
M5144P	0167
M5146P	3948
M5149	OEM
M5151RE	OEM
M5152L	2534
M5153L	0520
M5153P	0520
M5154	OEM
M5155	OEM
M5155L	OEM
M5155P	4593
M5156	OEM
M5157A	OEM
M5157B	OEM
M5157C	OEM
M5157D	OEM
M5162	OEM
M5163	OEM
M5165LL	OEM
M5167	OEM
M5168	OEM
M5169	1808
M5169P	1808
M5170	OEM
M5172L	OEM
M5174P	OEM
M5176P	1808
M5182	OEM
M5183	0395
M5183P	0395
M5185AP	0767
M5185P	0767
M5186AP	0918
M5186AP(RED)	OEM
M5186BP	0918
M5186P	0918
M5186P/AP	OEM
M5187	OEM
M5187P	OEM
M5190P	0856
M5191	OEM
M5191P	2891
M5192P	4196
M5193	OEM
M5193P	4389
M5194AP	OEM
M5194L	OEM
M5194P	OEM
M5195	4672
M5195P	4672
M5196A	OEM
M5196B	OEM
M5196F	OEM
M5196P	OEM
M5197	OEM
M5198	OEM
M5199	OEM
M5201FP	OEM
M5201L	OEM
M5201P	OEM
M5204P	OEM
M5210P	OEM
M5212L	OEM
M5213L	OEM
M5214L	OEM
M5215L	OEM
M5218AL	OEM
M5218L	1699
M5218P	3440
M5218PF	OEM

DEVICE TYPE	REPL CODE
M5220L	OEM
M5220P	OEM
M5221P	OEM
M5222L	OEM
M5223L	OEM
M5223P	3440
M5224P	0625
M5226P	OEM
M5227P	OEM
M5228P	OEM
M5229P	OEM
M5230L	OEM
M5231	OEM
M5231L	OEM
M5231TL	OEM
M5233P	0631
M5236L	5827
M5236ML	4118
M5237L	OEM
M5238L	OEM
M5238PF	OEM
M5239L	OEM
M5248P	OEM
M5251L	OEM
M5256	OEM
M5258	OEM
M5278D05	OEM
M5278L-05	1292
M5278L05	1292
M5278L10	OEM
M5278L-56	OEM
M5278L56	OEM
M5285	0004
M5290P	OEM
M5294P	OEM
M5299A	OEM
M5304	1275
M5304P	1275
M5310	0873
M5310P	0873
M5311	OEM
M5323BP	0996
M5323P	1060
M5340	OEM
M5340P	OEM
M5352	0743
M5352P	0743
M5362	1052
M5362P	1052
M5372	1426
M5372P	1426
M5373	OEM
M5374	1313
M5374P	1313
M5375	1404
M5375P	1404
M5395	1486
M5395P	1486
M5441-0P	OEM
M5468L-D	OEM
M5508K	OEM
M5701	OEM
M5709	OEM
M5730	OEM
M5730A	OEM
M5735-5	OEM
M5735-6	OEM
M5735-G	OEM
M5735-M	OEM
M5735-X	OEM
M5740-5	OEM
M5740-6	OEM
M5740-G	OEM
M5740-M	OEM
M5740-X	OEM
M5745-5	OEM
M5745-6	OEM
M5745-G	OEM
M5745-M	OEM
M5745-X	OEM
M5930	1823
M5930P	1823
M5932	1041
M5932P	1041
M5933	2097
M5933P	2097
M5935	1176
M5935P	1176
M5936	1831
M5936P	1831
M5937	1835
M5937P	1835
M5944	0033
M5944P	0033
M5945	0081
M5945P	0081
M5946	0141
M5946P	0141
M5948	3399
M5948P	3399
M5949	1844
M5949P	1844
M5951	OEM
M5953	0358
M5953P	0358
M5955P	1481
M5956	1632
M5956P	1632
M5961	1859

DEVICE TYPE	REPL CODE
M5961P	1859
M5962	0562
M5962P	0562
M5963	0341
M5963P	0341
M6400A	OEM
M6840	0280
M6931	0429
M6964-2	OEM
M6964-3	OEM
M6984	OEM
M6990	OEM
M6990-7203V1	OEM
M7002	0233
M7003	0016
M7006	0079
M7014	0198
M7015	0016
M7032H	OEM
M7033	0016
M7366	OEM
M7611AP	0069
M7641	1172
M7644BP	0663
M7800B	OEM
M7880	OEM
M7895	OEM
M8000H	OEM
M-8008	OEM
M8014	0030
M-8016(A)	OEM
M8051-178	OEM
M8062A	0004
M8062B	0004
M8062C	0004
M8073B	0004
M8073C	0280
M8082B1	OEM
M8100	OEM
M8105	0016
M8106	OEM
M8108	OEM
M8108A	OEM
M8108B	OEM
M8116	0136
M8120	0600
M-8124	0917
M8124	0050
M8128	OEM
M8221	0079
M8222	0015
M8303	0728
M8317B	0378
M8320	1791
M-8371-10000	OEM
M8394-100	OEM
M8399	0015
M8411	OEM
M8444	OEM
M8482	0917
M8482A	OEM
M8482C	6732
M8482F	0917
M8482FA-4	0917
M8484	OEM
M8486	OEM
M8487	OEM
M8489	0124
M8489-A	0143
M-8489A	0133
M8489A	0124
M8500	OEM
M8504	OEM
M8511	0030
M8513	0139
M8513(LAFAYETTE)	0133
M8513-0	0124
M8513-O	0139
M-8513A	0133
M8513A	0139
M8513A(O)	1925
M8513A0	0139
M8513A-4	OEM
M8513A-C	OEM
M8513A-O	0139
M8513AO	1925
M8513A-R	0139
M-8513R	0019
M8513R	0139
M8515	0012
M8550	OEM
M8555	0124
M8569	0913
M8604	0004
M8604A	0004
M8628	OEM
M8640	0164
M8640A	0164
M8640E	0136
M8640E(C-M)	0133
M8640E(DIO)	0133
M8640E(TRANSISTOR)	0164
M8640E(XSTR)	0211
M-8641A	0164
M8652	OEM
M8700	OEM
M8714	OEM
M8724LM	OEM
M8830	OEM

DEVICE TYPE	REPL CODE
M9002	0164
M9010	0007
M9010A	OEM
M9032	0016
M9063	OEM
M9090	0975
M9092	0038
M9093	0038
M9095	0016
M9134	0560
M9138	0086
M9142	0160
M9148	0211
M9159	0016
M9170	0198
M9184	0086
M9197	0111
M9198	0004
M-9202-15KV	OEM
M9206	1228
M9209	0086
M9213-100	OEM
M9213-200	OEM
M9213-450	OEM
M9216	OEM
M9220	OEM
M9225	0178
M9228	0086
M9235	0015
M9237	0160
M-9242-10KV	OEM
M9244	0103
M9248	0079
M9249	0004
M9255	0160
M9256	2136
M9257A	OEM
M9257B	OEM
M9257C	OEM
M9257D	OEM
M9259	0103
M9263	0160
M9264	1669
M9266	0007
M9274	0178
M9278	0103
M9301	0178
M9302	0103
M9305	OEM
M9306	1825
M9306B1	OEM
M9308	0126
M9312	0015
M9314	0015
M9317	0015
M9319	0015
M9321	0103
M9328	0037
M9329	0111
M9330	OEM
M9338	0111
M9342	0160
M9344	1681
M9348	0925
M9359	1681
M9380	0086
M9389	0155
M9393	0178
M9400	0126
M9408	0074
M9436	0160
M9442BP	OEM
M9442DV	OEM
M9450	0007
M9481	0007
M9482	0144
M9491	0198
M9514	0037
M9519	0086
M9525	0016
M9526	0037
M9527	0037
M9531	0037
M9532	0016
M9556	0161
M9561	0086
M9570	0016
M9571	0037
M9582	0460
M9715	0130
M9731	OEM
M9732	OEM
M9736	OEM
M9737	5580
M9741	3589
M9742	OEM
M9746	OEM
M9747	OEM
M12050CA	OEM
M12051CA	OEM
M14818A	OEM
M16010C	OEM
M16911	OEM
M22100B1	OEM
M22100D1	OEM
M23163B1	OEM
M27256F1	OEM
M31001	0016
M33611	0253

DEVICE TYPE	REPL CODE
M34300-221SP	OEM
M34300-223SP	OEM
M34300-227SP	OEM
M34302M8-511	OEM
M34302M8-511SP	OEM
M34302M8-512SP	OEM
M34302M8-514SP	OEM
M34350N6-522SP	OEM
M34350N6-524SP	OEM
M34350N6-527SP	OEM
M34350N6-528SP	OEM
M34350N6-561SP	OEM
M34350N6-562SP	OEM
M34350N6-566SP	OEM
M34350N6-567SP	OEM
M34351M8-011SP	OEM
M36000-5B1	OEM
M36000-5D1	OEM
M37100	OEM
M37100M8	OEM
M37100M8-110SP	OEM
M37100M8-115SP	OEM
M37100M8-222SP	OEM
M37100M8-521SP	OEM
M37100M8-711SP	OEM
M37100M8-717SP	OEM
M37100M8-919SP	OEM
M37100M8-8155P	OEM
M37100M8-B16SP	OEM
M37100M8-C16SP	OEM
M37100M8-C17SP	OEM
M37101M4-521SP	OEM
M37101M4-522SP	OEM
M37102M8-AA5SP	OEM
M37200M6-A16SP	OEM
M37201M6-A18FP	OEM
M37204M8	OEM
M37204M8-652SP	OEM
M37204M8-657SP	OEM
M37211M2-520SP	OEM
M37250M6-561SP	OEM
M37262MB-561SP	OEM
M37263M3-507SP	OEM
M37263M3-609SP	OEM
M37263M4-571SP	OEM
M37263MB-561SP	OEM
M37265M4-SV4812	OEM
M38128A-43	OEM
M40175BP	1528
M40308B	OEM
M41032A	0071
M41223-2	0015
M41256-15	OEM
M41256A	OEM
M41464-10	OEM
M41464-12	OEM
M41464-15	OEM
M50110AP	OEM
M50110BP	OEM
M50110CP	OEM
M50110P	OEM
M50111AP	OEM
M50111BP	OEM
M50111CP	OEM
M50111P	OEM
M50112P	OEM
M50115AP	OEM
M50115BP	OEM
M50115CP	OEM
M50115P	OEM
M50116AP	OEM
M50116BP	OEM
M50116CP	OEM
M50116P	OEM
M50117AP	OEM
M50117BP	OEM
M50117CP	OEM
M50117P	OEM
M50119P	OEM
M50143-003P	OEM
M50143-007P	OEM
M50143-007SP	OEM
M50161-020SP	OEM
M50161-059SP	OEM
M50161-151SP	OEM
M50161-152SP	OEM
M50161-154SP	OEM
M50161-251	OEM
M50161-251SP	OEM
M50161-452SP	OEM
M50161-453SP	OEM
M50161-454SP	OEM
M50161-459SP	OEM
M50161-542SP	OEM
M50163-55SP	OEM
M50199P	OEM
M50422P	OEM
M50430-582SP	OEM
M50430-583SP	OEM
M50433B-501S	OEM
M50433B-502SP	OEM
M50434-122-SP	OEM
M50436-594SP	OEM
M50439-113SP	OEM
M50439-411SP	OEM
M50439-513SP	OEM
M50439-517SP	OEM
M50439-519SP	OEM
M50439-614SP	OEM

DEVICE TYPE	REPL CODE
M50439-711SP	OEM
M50439-919FP	OEM
M50441-540SP	OEM
M50441-561BSP	OEM
M50441-561SP	OEM
M50442-505SP	OEM
M50442-506SP	OEM
M50442-508SP	OEM
M50442-522SP	OEM
M50442-525SP	OEM
M50442-526SP	OEM
M50442-562SP	OEM
M50442-563SP	OEM
M50442-564SP	OEM
M50442-577SP	OEM
M50442-620SP	OEM
M50442-622SP	OEM
M50442-623SP	OEM
M50442-624SP	OEM
M50442-625SP	OEM
M50442-626SP	OEM
M50442-671SP	OEM
M50442-674SP	OEM
M50445-010SP	OEM
M50450-001P	OEM
M50450-007P	OEM
M50450-031P	OEM
M50452-003	OEM
M50452-019P	OEM
M50452A-003P	OEM
M50452A-006P	OEM
M50453-015P	OEM
M50454-054SP	OEM
M50455-001SP	OEM
M50455-003SP	OEM
M50455-007SP	OEM
M50459-001SP	OEM
M50460-044P	OEM
M50460-055P	OEM
M50461-56FB	OEM
M50467-033FP	OEM
M50541FP	OEM
M50542-003	OEM
M50554-001SP	OEM
M50554-003SP	OEM
M50560-001P	OEM
M50560-125P	OEM
M50560-145	OEM
M50560-145P	OEM
M50560-155P	OEM
M50560-198P	OEM
M50560-391GP	OEM
M50561-021P	OEM
M50572ASP	OEM
M50574BSP	OEM
M50720-408SP	OEM
M50725-116SP	OEM
M50725-151SP	OEM
M50725-648SP	OEM
M50730-601SP	OEM
M50731-608SP	OEM
M50731-611SP	OEM
M50734SP	OEM
M50734SP-10	OEM
M50742-404SP	OEM
M50742-653SP	OEM
M50743FP	OEM
M50746-145SP	OEM
M50746-412SP	OEM
M50747-408SP	OEM
M50747-458SP	OEM
M50747-A71	OEM
M50752-101SP	OEM
M50752-620SP	OEM
M50752-622SP	OEM
M50752-626SP	OEM
M50752-635SP	OEM
M50754-115SP	OEM
M50754-360SP	OEM
M50754-363SP	OEM
M50754-644SP	OEM
M50754-673SP	OEM
M50754-679SP	OEM
M50754-680SP	OEM
M50754-688SP	OEM
M50754-689SP	OEM
M50757-301SP	OEM
M50757-311SP	OEM
M50757-610SP	OEM
M50757-621SP	OEM
M50757-640SP	OEM
M50757-660SP	OEM
M50757-671SP	OEM
M50757-690SP	OEM
M50757-691SP	OEM
M50757-692SP	OEM
M50760-670P	OEM
M50761-258P	OEM
M50761-417P	OEM
M50761-418P	OEM
M50761-420P	OEM
M50761-439P	OEM
M50761-668P	OEM
M50763-610SP	OEM
M50765-441SP	OEM
M50769-330P	OEM
M50925-302SP	OEM
M50930-410FP	OEM
M50930-432FP	OEM

DEVICE TYPE	REPL CODE
M50930-921SP	OEM
M50940-123SP	OEM
M50940-440SP	OEM
M50940-441SP	OEM
M50941-404SP	OEM
M50941-413SP	OEM
M50942-261FP	OEM
M50950-621SP	OEM
M50950-623SP	OEM
M50950-626SP	OEM
M50950-661SP	OEM
M50954-325SP	OEM
M50954-644SP	OEM
M50954-662SP	OEM
M50954-666SP	OEM
M50954-668SP	OEM
M50954-683SP	OEM
M50955-233SP	OEM
M50955-237SP	OEM
M50955-669SP	OEM
M50955-701SP	OEM
M50955-941SP	OEM
M50955-943SP	OEM
M50957-191SP	OEM
M50957-192SP	OEM
M50957-197SP	OEM
M50957-657SP	OEM
M50958-302SP	OEM
M50964-212SP	OEM
M51003L	OEM
M51014L	OEM
M51017AP	OEM
M51018AP	OEM
M51029P	OEM
M51104L	OEM
M51124	3958
M51131L	OEM
M51132L	OEM
M51132L-600	OEM
M51140AP	OEM
M51146P	OEM
M51164L	OEM
M51171L	4669
M51172P	4428
M51173AP	OEM
M51177L	OEM
M51182	4757
M51182L	4757
M51201L	OEM
M51202	1910
M51202L	1910
M51203L	OEM
M51204	0872
M51204L	0872
M51205L	OEM
M51206L	OEM
M51207L	6653
M51231P	OEM
M51232L	OEM
M51232L-600	OEM
M51232P	OEM
M51233P	OEM
M51240P	OEM
M51242P	6030
M51247	3273
M51247P	3273
M51251P	OEM
M51271SP	OEM
M51301P	OEM
M51307BSP	3343
M51308SP	OEM
M51320P	OEM
M51321P	OEM
M51326P	OEM
M51329P	OEM
M51340P	OEM
M51342P	OEM
M51345AP	OEM
M51346AP	OEM
M51354AP	1442
M51354APMN1217A	OEM
M51354APO	1442
M51354P	1442
M51355P	5786
M51356	5874
M51356P	5874
M51358P	0912
M51363	OEM
M51363P	OEM
M51364P	OEM
M51365-SP	OEM
M51365P	6809
M51365SP	3635
M51366P	2557
M51366SP	2111
M51375P	4165
M51376BSP	6162
M51376BZT	OEM
M51376SP	4336
M51378L	OEM
M51380P	OEM
M51381P	OEM
M51382P	OEM
M51383P	OEM
M51386L	4102
M51389P	4549
M51392P	OEM
M51401K	OEM
M51401P	OEM
M51408SP	OEM
M51411SP	OEM
M51414BSP	OEM
M51414SP	OEM
M51454L	OEM
M51460P	OEM
M51475G-3B	OEM
M51483P	OEM
M51485L	OEM
M51489L	OEM
M51494L	OEM
M51496P	OEM
M51501L	OEM
M51502L	1910
M51503L	OEM
M51513L	4640
M51513LB	4640
M51515L	4663
M51516L	4713
M51517L	4715
M51518L	5104
M51521L	2534
M51522AL	OEM
M51522L	OEM
M51523AL	OEM
M51523L	OEM
M51542AP	OEM
M51544AL	6969
M51544L	4834
M51564P	OEM
M51565P	OEM
M51567P	OEM
M51654AP	OEM
M51654P	OEM
M51660L	OEM
M51709T	1301
M51796P	OEM
M51802L	OEM
M51802P	OEM
M51841P	0973
M51843P	OEM
M51845L	OEM
M51846L	OEM
M51847P	OEM
M51848L	3956
M51848P	0973
M51849L	OEM
M51901P	OEM
M51903L	0810
M51904L	OEM
M51910P	OEM
M51953A	OEM
M51953AL	OEM
M51953B	OEM
M51953BL	OEM
M51954AL	OEM
M51955AL	OEM
M51957AL	OEM
M51970L	OEM
M51971L	OEM
M51976FP	OEM
M52005P	OEM
M52025SP	OEM
M52029FP	OEM
M52030ASP	OEM
M52030SP	OEM
M52041SP	OEM
M52054FP	OEM
M52055FP	OEM
M52055P	OEM
M52303ASP	OEM
M52303SP	OEM
M52470AP	OEM
M52470P	OEM
M52472P	OEM
M52678P	OEM
M52679FP	OEM
M52679P	OEM
M52682FP	OEM
M52684AFP	OEM
M52684AP	OEM
M52686AP	OEM
M52686FP	OEM
M52686P	OEM
M52687L	OEM
M52690SP	OEM
M52691FP	OEM
M52692SP	OEM
M52693SP	OEM
M52800FP	OEM
M52803P	OEM
M52812FP	OEM
M53200	0232
M53200P	0232
M53201	0268
M53201P	0268
M53202	0314
M53202P	0314
M53203	0335
M53203P	0335
M53204	0361
M53204P	0361
M53205	0385
M53205P	0385
M53206	6586
M53206P	1206
M53207P	1339
M53208P	0467
M53209P	0492
M53210	0512
M53210P	0512
M53213P	1441
M53214P	2243
M53216P	1349
M53217P	1352
M53220	0697
M53220P	0697
M53225P	3477
M53227P	0818
M53230	0873
M53230P	0873
M53232	OEM
M53237P	3520
M53238P	0996
M53240	1024
M53240P	1024
M53241	1038
M53241P	1038
M53242	1052
M53242P	1052
M53243	1060
M53243P	1060
M53244	1072
M53244P	1072
M53245P	1082
M53247	1108
M53247P	1108
M53248	1125
M53248A	1125
M53248P	1125
M53250	0743
M53250P	0743
M53253	1185
M53253P	1185
M53260	1275
M53260P	1275
M53270	1404
M53270P	1404
M53272	1426
M53272P	1426
M53273	1172
M53273P	1172
M53274	1313
M53274P	1313
M53275	1432
M53275P	1432
M53276	1158
M53276P	1158
M53280	1535
M53280P	1535
M53283	0117
M53283P	0117
M53284P	OEM
M53285P	0374
M53286	1368
M53286P	1368
M53289P	5455
M53290	1208
M53290P	1208
M53291	0980
M53291P	0980
M53292	0834
M53292P	0834
M53293	0569
M53293P	0569
M53295	1486
M53295P	1486
M53296	1715
M53296P	1715
M53307	0942
M53307P	0942
M53321	0175
M53321P	0175
M53322P	1139
M53323P	1157
M53325P	1182
M53326P	1193
M53332P	1271
M53335P	OEM
M53345P	0619
M53350P	1492
M53351	1495
M53351P	1495
M53353P	1539
M53354P	1555
M53355P	1576
M53356P	1592
M53357P	OEM
M53358P	OEM
M53360P	1631
M53361P	1645
M53362P	1013
M53363P	1666
M53364P	0734
M53365P	1685
M53366P	0231
M53370P	1722
M53374P	1770
M53375P	1787
M53380	1829
M53380P	1829
M53381P	1842
M53382P	1856
M53385P	OEM
M53390P	1912
M53391P	1917
M53392	1921
M53392P	1921
M53393	1926
M53393P	1926
M53398P	1964
M53399P	1971
M53478P	OEM
M54401AP	OEM
M54402P	OEM
M54405P	OEM
M54406P	OEM
M54408P	OEM
M54410	OEM
M54410P	OEM
M54454P	OEM
M54455L	OEM
M54456L	OEM
M54459L	5239
M54462	OEM
M54470	OEM
M54471	OEM
M54472	OEM
M54477	OEM
M54477L	OEM
M54502P	OEM
M54503P	OEM
M54504P	OEM
M54512L	OEM
M54513P	OEM
M54514AP	OEM
M54514P	OEM
M54515P	OEM
M54516P	4138
M54517P	0969
M54519P	0969
M54521P	4965
M54522P	4940
M54523P	1134
M54524P	0845
M54525P	1007
M54526P	1261
M54527P	4963
M54528P	4962
M54529P	4961
M54530P	4536
M54531P	4536
M54532P	OEM
M54533P	4959
M54534P	4957
M54535P	4956
M54536P	4955
M54537P	OEM
M54538P	OEM
M54539P	OEM
M54542L	OEM
M54543AL	OEM
M54543L	2073
M54543L-B	2073
M54543LB	2073
M54544AL	OEM
M54545L	OEM
M54548L	1938
M54549AL	OEM
M54549L	3080
M54549LA	OEM
M54560P	OEM
M54561P	OEM
M54562P	OEM
M54563P	OEM
M54564P	OEM
M54565P	OEM
M54566P	OEM
M54567P	OEM
M54568P	OEM
M54569P	OEM
M54571P	OEM
M54576FP	OEM
M54577FP	OEM
M54577P	OEM
M54578P	OEM
M54580	OEM
M54580P	OEM
M54600P	OEM
M54601P	OEM
M54602P	OEM
M54603P	OEM
M54604P	OEM
M54605P	OEM
M54610P	OEM
M54641L	OEM
M54644BL	OEM
M54644L	OEM
M54645AL	OEM
M54646P	OEM
M54648AL-B	OEM
M54648L-8	OEM
M54648L-B	OEM
M54648L-C	OEM
M54648L-D	OEM
M54649L	OEM
M54654P	OEM
M54700K	OEM
M54700P	OEM
M54700S	OEM
M54730K	OEM
M54730P	OEM
M54730P-2	OEM
M54730S	OEM
M54740AP	OEM
M54740AS	OEM
M54741AP	OEM
M54741AS	OEM
M54802P	OEM
M54817P	OEM
M54818L	OEM
M54819L	OEM
M54832P	OEM
M54872L	OEM
M54873P	OEM
M54874P	OEM
M54876-141P	OEM
M54899	OEM
M54899P	OEM
M54940P	OEM
M54955P	OEM
M54955P-A	OEM
M56529AP	OEM
M56529P	OEM
M58321	OEM
M58333-XXXP	OEM
M58334-XXXP	OEM
M58371	OEM
M58412P	OEM
M58413P	OEM
M58418P	4112
M58434P	OEM
M58435P	OEM
M58436-001P	OEM
M58437-001P	OEM
M58472P	1172
M58473P	OEM
M58476-141P	OEM
M58478P	4112
M58479P	OEM
M58480P	OEM
M58481P	OEM
M58482P	OEM
M58483P	OEM
M58484P	OEM
M58485P	2879
M58487P	OEM
M58609-04P	OEM
M58609-04S	OEM
M58609-09P	OEM
M58609-09S	OEM
M58630P	OEM
M58653P	OEM
M58657P	OEM
M58658	OEM
M58658P	OEM
M58659P	OEM
M58725P	1898
M58725P-15	OEM
M58725S-15	OEM
M58731-001S	OEM
M58735-064P	OEM
M58735-065P	OEM
M58735XXXP	OEM
M58741P	OEM
M58839-606P	OEM
M58839-608P	OEM
M58839-608SP	OEM
M58839-609P	OEM
M58839-622SP	OEM
M58839-623P	OEM
M58839-642P	OEM
M58839-660P	OEM
M58839-670P	OEM
M58872P	OEM
M58981P30	OEM
M58981P45	OEM
M58981S-45	OEM
M60002A	OEM
M60002A-0107	OEM
M60002A-0107SP	OEM
M60025-1001SP	OEM
M60032-1006FP	OEM
M60201	OEM
M60306	OEM
M60307	OEM
M61048	OEM
M61203L	OEM
M62358P	OEM
M62359P	OEM
M64100BA	OEM
M64100BB086	OEM
M64100KB	OEM
M64104CA	OEM
M64200CA	OEM
M65607SP	OEM
M66310P	OEM
M66320P	OEM
M67154SL	OEM
M73003	OEM
M75205-1	0084
M75205-2	0534
M75516-2	0050
M75516-2P	0050
M75516-2R	0050
M-75517-1	0211
M75517-1	0004
M75517-2	0004
M75537-2	0847
M-75543-1	0178
M75543-1	0042
M75545-1	0007
M75547-1	0144
M75547-2	0144
M75561-7	0004
M75561-8	0126
M75561-10RK	0849
M75561-17	0126
M75561-23	0353
M75561-23RN	0353
M81461B-12RS-PSZ	OEM
M100412	0015
M100449	1030
M105064	0474
M105330	1030
M106379	0015
M109474	0196
M112826	0015
M138841-180	OEM
M138841-180J	OEM
M442256AJ-10	OEM
M449192	OEM
M450072	OEM
M467857	OEM
M468487-8545H	OEM
M501121P	OEM
M502030SP	OEM
M506868	OEM
M509040	OEM
M514940	OEM
M755162-P	0050
M755162-R	0050
M840176P	OEM
M840176P-SH	OEM
M881464-12	OEM
M8534992	0196
M8534992-2	0196
M50439517S	OEM
M50560198P	OEM
M69907209V1	OEM
MA00017	OEM
MA00087CP	OEM
MA00088CP	OEM
MA00287	OEM
MA00318CP	OEM
MA00357CP	OEM
MA00357MD	OEM
MA0401	0126
MA0402	0126
MA0404	0037
MA0404-1	0150
MA0404-2	0150
MA0411	0150
MA0412	0532
MA0413	0150
MA0414	0037
MA03080	OEM
MA1	0050
MA1Z	0015
MA2	0015
MA3.0EB	0118
MA4B300	OEM
MA4C11	OEM
MA4C100	OEM
MA4C100B	OEM
MA4C100C	OEM
MA4C102	OEM
MA4C102B	OEM
MA4C102C	OEM
MA4C103	OEM
MA4C103B	OEM
MA4C103C	OEM
MA4C104	OEM
MA4C104B	OEM
MA4C104C	OEM
MA4C105	OEM
MA4C105B	OEM
MA4C105C	OEM
MA4C106	OEM
MA4C106B	OEM
MA4C106C	OEM
MA4C107	OEM
MA4C107B	OEM
MA4C107C	OEM
MA4C108	OEM
MA4C108B	OEM
MA4C108C	OEM
MA4C109	OEM
MA4C109B	OEM
MA4C109C	OEM
MA4C110	OEM
MA4C110B	OEM
MA4C110C	OEM
MA4C111	OEM
MA4C111B	OEM
MA4C111C	OEM
MA4C112	OEM
MA4C112B	OEM
MA4C112C	OEM
MA4C113	OEM
MA4C113B	OEM
MA4C113C	OEM
MA4C114B	OEM
MA4C114C	OEM
MA4C157	OEM
MA4C157B	OEM
MA4C157C	OEM
MA4C158	OEM
MA4C158B	OEM
MA4C158C	OEM
MA4C250D	OEM
MA4C250H	OEM
MA4C251D	OEM
MA4C251H	OEM
MA4C252D	OEM
MA4C252H	OEM
MA4C253D	OEM
MA4C253H	OEM
MA4C254D	OEM
MA4C254H	OEM
MA4C255D	OEM
MA4C255H	OEM
MA4C256D	OEM
MA4C256H	OEM
MA4C257D	OEM
MA4C257H	OEM
MA4C258D	OEM
MA4C258H	OEM
MA4C260D	OEM
MA4C260H	OEM
MA4C261D	OEM
MA4C261H	OEM
MA4C262D	OEM
MA4C262H	OEM
MA4C263D	OEM
MA4C263H	OEM
MA4C264H	OEM
MA4C265D	OEM
MA4C265H	OEM
MA4C266D	OEM
MA4C266H	OEM
MA4C267D	OEM
MA4C267H	OEM
MA4C268D	OEM
MA4C268H	OEM
MA4C270D	OEM
MA4C270H	OEM
MA4C271D	OEM
MA4C271H	OEM
MA4C272D	OEM
MA4C272H	OEM
MA4C273D	OEM
MA4C273H	OEM
MA4C274D	OEM
MA4C274H	OEM
MA4C275D	OEM
MA4C275H	OEM
MA4C276D	OEM
MA4C276H	OEM
MA4C277D7D	OEM
MA4C277H	OEM
MA4C278D	OEM
MA4C278H	OEM
MA4C400	OEM
MA4C400A	OEM
MA4C400B	OEM
MA4C400C	OEM
MA4C400D	OEM
MA4C400D1	OEM
MA4C400D2	OEM
MA4C401	OEM
MA4C401A	OEM
MA4C401B	OEM
MA4C401C	OEM
MA4C401D	OEM
MA4C401D1	OEM
MA4C401D2	OEM
MA4C402	OEM
MA4C402A	OEM
MA4C402B	OEM
MA4C402C	OEM
MA4C402D	OEM
MA4C402D1	OEM
MA4C402D2	OEM
MA4C403	OEM
MA4C403A	OEM
MA4C403B	OEM
MA4C403C	OEM
MA4C403D	OEM
MA4C403D1	OEM
MA4C403D2	OEM
MA4C404	OEM
MA4C404A	OEM
MA4C404B	OEM
MA4C404C	OEM
MA4C404D	OEM
MA4C404D1	OEM
MA4C404D2	OEM
MA4C405	OEM
MA4C405A	OEM
MA4C405B	OEM
MA4C405C	OEM
MA4C405D	OEM
MA4C405D1	OEM
MA4C405D2	OEM
MA4C550	OEM
MA4C550B	OEM
MA4C550C	OEM
MA4C551	OEM
MA4C551B	OEM
MA4C551C	OEM
MA4C552	OEM
MA4C552B	OEM
MA4C552C	OEM
MA4C553	OEM
MA4C553B	OEM
MA4C553C	OEM
MA4C554	OEM
MA4C554B	OEM
MA4C554C	OEM
MA4C555	OEM
MA4C555B	OEM
MA4C555C	OEM
MA4C556	OEM

DEVICE TYPE	REPL CODE	DEVICE TYPE	REPL CODE	DEVICE TYPE	REPL CODE	DEVICE TYPE	REPL CODE	DEVICE TYPE	REPL CODE	DEVICE TYPE	REPL CODE	DEVICE TYPE	REPL CODE
MA4C556B	OEM	MA4P101	OEM	MA40	OEM	MA165	0124	MA266	OEM	MA443B	OEM	MA701A	OEM
MA4C556C	OEM	MA4P102	OEM	MA41	OEM	MA165-T2	0124	MA270	OEM	MA444	OEM	MA702	OEM
MA4C557	OEM	MA4P103-30	OEM	MA47	0143	MA165-Y	0124	MA271	OEM	MA444A	OEM	MA704	OEM
MA4C557B	OEM	MA4P202-30	OEM	MA48	OEM	MA165SAN	OEM	MA272	OEM	MA444B	OEM	MA704A	OEM
MA4C557C	OEM	MA4P203-30	OEM	MA49	OEM	MA165TA	0124	MA273	OEM	MA444C	OEM	MA709AMJ	OEM
MA4C558	OEM	MA4P303-30	OEM	MA50	0015	MA165TA5	0124	MA274	OEM	MA444D	OEM	MA709AMJG	OEM
MA4C558B	OEM	MA4P404-30	OEM	MA51A	0143	MA165TA5-VT	0124	MA277	OEM	MA445	OEM	MA709CJ	OEM
MA4C558C	OEM	MA4P504-30	OEM	MA53	2518	MA165.SAN	0124	MA277-A	OEM	MA445A	OEM	MA709CJG	OEM
MA4C559	OEM	MA4P505-30	OEM	MA56	0133	MA166	0124	MA278	OEM	MA445B	OEM	MA709CN	OEM
MA4C559B	OEM	MA4P506-30	OEM	MA56T	OEM	MA166(C)	0124	MA279	OEM	MA449B	OEM	MA709CP	OEM
MA4C559C	OEM	MA4P604-30	OEM	MA57	0124	MA166C	0124	MA281	OEM	MA449C	OEM	MA709MJ	OEM
MA4C560	OEM	MA4P606-30	OEM	MA60	2645	MA167	0124	MA285	OEM	MA449D	OEM	MA709MP	OEM
MA4C560B	OEM	MA4P607	OEM	MA61	2730	MA167A	0124	MA286	0280	MA449E	OEM	MA710	OEM
MA4C560C	OEM	MA4P608	OEM	MA62	5911	MA167TA	OEM	MA287	0280	MA449F	OEM	MA711	OEM
MA4C561	OEM	MA4P709	OEM	MA72	2518	MA170	0124	MA288	0280	MA450A	OEM	MA711CJ	OEM
MA4C561B	OEM	MA4c264D	OEM	MA73	2518	MA171	0124	MA291	OEM	MA450B	OEM	MA711CN	OEM
MA4C561C	OEM	MA5	OEM	MA74HC4053F	OEM	MA171TA5	0124	MA293	OEM	MA450C	OEM	MA711MJ	OEM
MA4C562B	OEM	MA6Y	OEM	MA77	OEM	MA175WA	2203	MA294	OEM	MA450D	OEM	MA715	1791
MA4C562C	OEM	MA8	0143	MA80	2645	MA175WK	0694	MA300CD	OEM	MA450E	OEM	MA720	0720
MA4C563B	OEM	MA-10	0400	MA90	0143	MA176WA	2203	MA301	0720	MA450F	OEM	MA721TW	OEM
MA4C563C	OEM	MA10	0400	MA100	0280	MA176WK	0694	MA302	0628	MA450G	OEM	MA723CJ	OEM
MA4C700D	OEM	MA11	OEM	MA101	0015	MA178	0124	MA303	0030	MA450H	OEM	MA723CN	OEM
MA4C700H	OEM	MA13	OEM	MA102	0015	MA179	0124	MA308CD	OEM	MA451A	OEM	MA723MJ	OEM
MA4C701D	OEM	MA15	OEM	MA102A	OEM	MA180	OEM	MA311	0030	MA451B	OEM	MA723MN	OEM
MA4C701H	OEM	MA17	OEM	MA102B	OEM	MA181	0023	MA319	6567	MA451C	OEM	MA723TA	OEM
MA4C702D	OEM	MA18	OEM	MA102C	OEM	MA182	0023	MA320	0720	MA451D	OEM	MA725	0288
MA4C702H	OEM	MA20	OEM	MA102E	OEM	MA183	0023	MA320B	0720	MA451E	OEM	MA730	0288
MA4C703D	OEM	MA21	0143	MA102F	OEM	MA184	0023	MA320B2	0720	MA451F	OEM	MA733CJ	OEM
MA4C703H	OEM	MA23	3683	MA102G	OEM	MA185	0023	MA320G1	0720	MA452	OEM	MA733CN	OEM
MA4C704D	OEM	MA23(B)	0143	MA102H	OEM	MA186	0023	MA320G1NR	0720	MA452A	OEM	MA733MN	OEM
MA4C704H	OEM	MA23A	3683	MA102I	OEM	MA188	1090	MA320GIVR	OEM	MA452B	OEM	MA741	0312
MA4C705D	OEM	MA-23B	3683	MA102J	OEM	MA190	0124	MA321	OEM	MA458B	OEM	MA741CJ	OEM
MA4C705H	OEM	MA23B	3683	MA103	0015	MA190WK	OEM	MA322	OEM	MA458C	OEM	MA741CJG	OEM
MA4C710D	OEM	MA25	7123	MA106	OEM	MA194	OEM	MA322-PM	OEM	MA458D	OEM	MA741CN	OEM
MA4C710H	OEM	MA25A	3683	MA-110	0302	MA195	0124	MA323	OEM	MA458E	OEM	MA741CP	OEM
MA4C711D	OEM	MA25B	OEM	MA110	OEM	MA195-5(TA5)	OEM	MA324	1029	MA458F	OEM	MA741MJ	OEM
MA4C711H	OEM	MA-26	0139	MA111	0023	MA199	OEM	MA325	1029	MA459B	OEM	MA741MJG	OEM
MA4C712D	OEM	MA26	0139	MA111O	OEM	MA203	0023	MA325B	1029	MA459C	OEM	MA741MP	OEM
MA4C712H	OEM	MA-26-1	0139	MA112	0211	MA203A	0017	MA325G	1029	MA459D	OEM	MA747-1CJ	OEM
MA4C713D	OEM	MA26-2	0139	MA113	0211	MA206	0004	MA326	OEM	MA459F	OEM	MA747-1CN	OEM
MA4C713H	OEM	MA26A	0139	MA114	0211	MA207CP	OEM	MA327	OEM	MA460A	OEM	MA747-1MJ	OEM
MA4C714D	OEM	MA26G	0139	MA115	0211	MA208	OEM	MA328	OEM	MA460B	OEM	MA747CJ	OEM
MA4C714H	OEM	MA26T	0120	MA116	0211	MA209	1282	MA329	OEM	MA460C	OEM	MA747CN	OEM
MA4C715D	OEM	MA26T-A	0120	MA117	0211	MA211	0015	MA330	0005	MA460D	OEM	MA747MJ	OEM
MA4C715H	OEM	MA26TA	0120	MA120	OEM	MA211S	1296	MA332	0628	MA460E	OEM	MA748CJ	OEM
MA4C720D	OEM	MA26TB	OEM	MA121	OEM	MA212	OEM	MA332A	OEM	MA460F	OEM	MA748CJG	OEM
MA4C720H	OEM	MA26TO	0120	MA150	0133	MA212A	OEM	MA332B	OEM	MA460G	OEM	MA748CL	OEM
MA4C721D	OEM	MA26TO-A	0120	MA150DD	0133	MA212B	OEM	MA332C	OEM	MA460H	OEM	MA748CN	OEM
MA4C721H	OEM	MA26TOA	0120	MA150FV	0124	MA212C	OEM	MA332CP	OEM	MA461	OEM	MA748CP	OEM
MA4C722D	OEM	MA26TOA-B	0120	MA150LF	0124	MA212D	OEM	MA333CP	OEM	MA461A	OEM	MA748MJ	OEM
MA4C722H	OEM	MA26TO-B	0120	MA150TA	0124	MA214	OEM	MA334	OEM	MA461B	OEM	MA748MJG	OEM
MA4C723D	OEM	MA26W	0120	MA150TA5	0133	MA214B	OEM	MA335	OEM	MA490	OEM	MA748MP	OEM
MA4C723H	OEM	MA26W0	0120	MA150TP	0133	MA214C	OEM	MA336CP	OEM	MA490A	OEM	MA767	0701
MA4C724D	OEM	MA26W0-B	0120	MA150TR	OEM	MA214D	OEM	MA337	OEM	MA490B	OEM	MA767PC	0701
MA4C724H	OEM	MA26W0B	0120	MA150WK	0124	MA215	0015	MA339CP	OEM	MA490C	OEM	MA790	0143
MA4C725D	OEM	MA26WA	0139	MA151A	0124	MA215A	1296	MA340	0554	MA490D	OEM	MA815	0004
MA4C725H	OEM	MA26W-B	0120	MA151A(MA)	OEM	MA217	OEM	MA341	OEM	MA490E	OEM	MA840	OEM
MA4C850	OEM	MA26WB	0120	MA151K	3749	MA217A	OEM	MA342	OEM	MA490F	OEM	MA856	0124
MA4C850A	OEM	MA26WD	0120	MA151K(MH)	3749	MA217B	OEM	MA342-B	OEM	MA490G	OEM	MA856TV	0124
MA4C850B	OEM	MA26WO	0120	MA151KPSK	OEM	MA217C	OEM	MA342-M	OEM	MA490H	OEM	MA858	0124
MA4C850C	OEM	MA26WO-A	0120	MA151K-TW	0124	MA218	OEM	MA342CP	OEM	MA491	OEM	MA859	0124
MA4C850D	OEM	MA26WO-B	0120	MA151K-W	OEM	MA219G	OEM	MA345	6569	MA491A	OEM	MA860	OEM
MA4C850D1	OEM	MA26WOB	0120	MA151WA	0903	MA221	OEM	MA345B	OEM	MA491B	OEM	MA862	OEM
MA4C850D2	OEM	MA26Y	OEM	MA151WA(MN)	0903	MA221A	OEM	MA350	0023	MA491C	OEM	MA881	0004
MA4C851	OEM	MA27	0343	MA151WK	0907	MA221B	OEM	MA351	0023	MA491D	OEM	MA882	0004
MA4C851A	OEM	MA27-A	0120	MA151WK(MT)	OEM	MA221C	OEM	MA356	6039	MA491E	OEM	MA883	0004
MA4C851B	OEM	MA27-B	0120	MA151WKPSK	5484	MA221D	OEM	MA360	OEM	MA491F	OEM	MA884	0004
MA4C851C	OEM	MA27-C	0120	MA152	OEM	MA222	OEM	MA370	OEM	MA491G	OEM	MA885	0004
MA4C851D	OEM	MA27A	6566	MA152A	OEM	MA223	OEM	MA393	0004	MA491H	OEM	MA886	0004
MA4C851D1	OEM	MA27B	0120	MA152K	3749	MA231	OEM	MA393A	0004	MA492C	OEM	MA887	0004
MA4C851D2	OEM	MA27Q-A	0120	MA152WA	0903	MA231A	OEM	MA393B	0004	MA492D	OEM	MA888	0004
MA4C852	OEM	MA27QA	OEM	MA152WA-TX	OEM	MA231B	OEM	MA393C	0004	MA492E	OEM	MA889	0004
MA4C852A	OEM	MA27Q-B	0120	MA152WK	0907	MA232	OEM	MA393E	0004	MA492F	OEM	MA890	0004
MA4C852B	OEM	MA27T	0343	MA152WK(MU)	0907	MA232A	OEM	MA393G	0004	MA492G	OEM	MA891	0004
MA4C852C	OEM	MA27T-A	0343	MA152WKTA	OEM	MA232B	OEM	MA393R	0004	MA493B	OEM	MA892	0004
MA4C852D	OEM	MA27TA	0343	MA152WK-TW	0907	MA233	OEM	MA400	OEM	MA493C	OEM	MA893	0004
MA4C852D1	OEM	MA27T-ATA	0343	MA152WKTX	0907	MA238	0280	MA401	OEM	MA494	OEM	MA894	0211
MA4C852D2	OEM	MA27T-B	0343	MA152WK-W	OEM	MA239	OEM	MA404A1	OEM	MA494A	OEM	MA895	0211
MA4C853	OEM	MA27TB	0343	MA153	OEM	MA239A	OEM	MA404B1	OEM	MA494B	OEM	MA896	0211
MA4C853A	OEM	MA27TBTA	0343	MA153(MC)	0133	MA239B	OEM	MA404C1	OEM	MA494C	OEM	MA897	0211
MA4C853B	OEM	MA27W	0120	MA153A	OEM	MA240	0004	MA404D1	OEM	MA494D	OEM	MA898	0211
MA4C853C	OEM	MA27W-A	0120	MA153TW	OEM	MA240A	OEM	MA404E1	OEM	MA500CP	OEM	MA899	0211
MA4C853D	OEM	MA27WA	0343	MA154	0694	MA240B	OEM	MA404F1	OEM	MA522	OEM	MA-900	0143
MA4C853D1	OEM	MA27WATA	0343	MA154WA	2203	MA240C	OEM	MA404G1	OEM	MA522VT	OEM	MA900	0211
MA4C853D2	OEM	MA27WATP	0120	MA154WE	OEM	MA241	OEM	MA405(H)-T2	OEM	MA550	OEM	MA901	0280
MA4E003	OEM	MA27W-B	0120	MA154WK	0694	MA242	0015	MA408A	OEM	MA551	3749	MA902	0004
MA4E004	OEM	MA27WB	0343	MA155WA	2203	MA242BC	0087	MA408B	OEM	MA553	OEM	MA903	0004
MA4E005	OEM	MA27WBTA	0343	MA155WK	3192	MA242C	0015	MA412	OEM	MA615	1791	MA904	0004
MA4E180	OEM	MA28	OEM	MA156	1091	MA242CR	OEM	MA414	OEM	MA619	0201	MA909	0004
MA4E181	OEM	MA28-A	OEM	MA157	OEM	MA242R	OEM	MA417	OEM	MA622	0288	MA910	0004
MA4E182	OEM	MA28-B	OEM	MA157-M	OEM	MA242RC	0087	MA418	OEM	MA625	0288	MA1024	1276
MA4E183	OEM	MA28T-A	OEM	MA157A	OEM	MA245	OEM	MA418A	OEM	MA630	0378	MA1027	0760
MA4E184	OEM	MA28T-B	OEM	MA158	3895	MA246	OEM	MA418B	OEM	MA649	1236	MA1028L	0677
MA4E185	OEM	MA28W-A	OEM	MA159	0124	MA246A	OEM	MA419	OEM	MA650	1236	MA1030	0118
MA4E186	OEM	MA29-A	OEM	MA159A	OEM	MA249	OEM	MA419A	OEM	MA651	2234	MA1030H	OEM
MA4E187	OEM	MA29-B	OEM	MA160	OEM	MA250	OEM	MA421A	OEM	MA653	0972	MA1030M	0118
MA4E188	OEM	MA29Q-A	OEM	MA160A	OEM	MA251	OEM	MA421B	OEM	MA654	0972	MA1033	0299
MA4E189	OEM	MA29T-A	OEM	MA160MID	OEM	MA252	OEM	MA423A	OEM	MA655	4488	MA1033L	0299
MA4E190	OEM	MA29TA	OEM	MA-161	0124	MA253	OEM	MA425	OEM	MA661	2234	MA1036	0376
MA4E191	OEM	MA29T-B	OEM	MA161	0133	MA257	OEM	MA428	OEM	MA669	1236	MA1036H	OEM
MA4E200	OEM	MA29TB	0343	MA161C	0124	MA257T	OEM	MA432	OEM	MA670	1236	MA1039	0036
MA4E201	OEM	MA29W-A	OEM	MA161TA	0124	MA258	OEM	MA433	0917	MA671	2234	MA1039L	0036
MA4E202	OEM	MA29W-B	OEM	MA161TA5	0124	MA259	OEM	MA433D2	OEM	MA673	0972	MA1043	0054
MA4E203	OEM	MA29WB	OEM	MA161TP	0124	MA260	0343	MA435	OEM	MA674	0972	MA1043H	0275
MA4E204	OEM	MA29WBA	OEM	MA162	0124	MA261	OEM	MA437	OEM	MA675	4488	MA1043M	0275
MA4E205	OEM	MA29W-BTA	OEM	MA162A	0124	MA262	OEM	MA439	OEM	MA700	0339	MA1047	0451
MA4E206	OEM	MA30	OEM	MA162LF	0124	MA263	OEM	MA441	OEM	MA700A	0339	MA1047A	0140
MA4E207	OEM	MA30S	OEM	MA162TA5	0124	MA264	OEM	MA443	OEM	MA700ATA	OEM	MA1047H	0140
MA4E208	OEM	MA32	OEM	MA162TP	0124	MA265	OEM	MA443A	OEM	MA701	OEM	MA1047L	OEM

DEVICE TYPE	REPL CODE
MA1047M	0140
MA1051	0041
MA1051H	0041
MA1051HTV	0041
MA1051L	0041
MA1051M	0589
MA1051TP	0041
MA1056	0157
MA1056H	0157
MA1056M	0157
MA1062	0471
MA1062H	0471
MA1062L	OEM
MA1062LF	0471
MA1062LM	0471
MA1062M	0471
MA1068	0062
MA1068H	0062
MA1068L	0062
MA1068T	0062
MA1075	0077
MA1075M	0077
MA1075TA	0077
MA1082	0244
MA1082L	0165
MA1082M	0165
MA1091	0057
MA1091-M	0057
MA1091M	0057
MA1100	0064
MA1100H	0064
MA1100M	0248
MA1110	0181
MA1110(DIODE)	0023
MA1110M	0181
MA1110TA	0181
MA1114	5186
MA1120	1005
MA1120-L	0052
MA1120A	0052
MA1120L	0052
MA1120M	0052
MA1125	0167
MA1128	0167
MA1130	0053
MA1130H	0053
MA1130LF	0053
MA1130M	0053
MA1130MH	0053
MA1140	OEM
MA1140-M	0879
MA1150	0686
MA1150A	0686
MA1150M	OEM
MA1150TA	0686
MA1160	0445
MA1180	0375
MA1180M	OEM
MA1190	5252
MA1200	0700
MA1220	0705
MA1240	0494
MA1270	0455
MA1270H	0455
MA1300	0053
MA1300M	0195
MA1306W	2262
MA1318	0004
MA1330	0166
MA1330M	0195
MA1360	0010
MA1360H	5663
MA1360M	OEM
MA1390H	OEM
MA1420(M)	OEM
MA1501TP	0041
MA1545	OEM
MA1700	0211
MA1701	0080
MA1701(DIODE)	0102
MA1702	0164
MA1702(DIODE)	0133
MA1703	0211
MA1703(DIODE)	0133
MA1704	0211
MA1704(DIODE)	0124
MA1705	0211
MA1706	0211
MA1707	0211
MA1708	0211
MA2043	0533
MA2051	0162
MA2056	0157
MA2062	0636
MA2062B	OEM
MA2068	0025
MA2075	0649
MA2082	0244
MA2091	0012
MA2100	0170
MA2100B	0170
MA2110	0317
MA2110BLF	OEM
MA2120	0137
MA2120B	OEM
MA2130	0365
MA2150	0002
MA2160	0420
MA2160B	0420

DEVICE TYPE	REPL CODE
MA2169	OEM
MA2180	0495
MA2180A	OEM
MA2180B	OEM
MA2200	0531
MA2220	0565
MA2240	0402
MA2270	0441
MA2300	0195
MA2330	0039
MA2360	0820
MA2390	0350
MA2430	0931
MA2470	0999
MA2510	0502
MA2560	0869
MA2801	OEM
MA2802	OEM
MA2803	OEM
MA2804	OEM
MA2805	OEM
MA2807	OEM
MA2808	OEM
MA2810	OEM
MA2811	OEM
MA2812	OEM
MA2813	OEM
MA2820	OEM
MA2822	OEM
MA2823	OEM
MA2824	OEM
MA3024	OEM
MA3027	OEM
MA3030	OEM
MA3033	OEM
MA3036	OEM
MA3039	OEM
MA3043	OEM
MA3047	OEM
MA3047-L	OEM
MA3047H	OEM
MA3047L	OEM
MA3047M	OEM
MA3051	OEM
MA3051-H	OEM
MA3051H	OEM
MA3051L	OEM
MA3056	OEM
MA3056L	OEM
MA3056M	OEM
MA3062	0471
MA3062(M)-W	OEM
MA3062-M	OEM
MA3062L	0471
MA3062M	OEM
MA3062TW	OEM
MA3065	0167
MA3068	OEM
MA3068(M)-W	OEM
MA3068H	OEM
MA3068L	OEM
MA3068M	OEM
MA3068N	OEM
MA3075	OEM
MA3075L	OEM
MA3075M	OEM
MA3082	OEM
MA3082L	OEM
MA3082M	OEM
MA3091	0057
MA3091L	0057
MA3091M	OEM
MA3100	OEM
MA3100(M)	OEM
MA3100L	OEM
MA3100M	OEM
MA3110	OEM
MA3110L	OEM
MA3120	OEM
MA3120-W	OEM
MA3120L	0052
MA3130	OEM
MA3130L	OEM
MA3150	0686
MA3150L	0686
MA3160	OEM
MA3160L	OEM
MA3180	OEM
MA3180L	OEM
MA3200	OEM
MA3200L	OEM
MA3220	OEM
MA3220L	OEM
MA3227	OEM
MA3228	OEM
MA3229	OEM
MA3230	OEM
MA3231	OEM
MA3232	OEM
MA3233	OEM
MA3234	OEM
MA3240	OEM
MA3240L	OEM
MA3240M	OEM
MA3270	OEM
MA3300	OEM
MA3330	OEM
MA3330(L)-W	OEM
MA3360	OEM

DEVICE TYPE	REPL CODE
MA4011B	OEM
MA4011C	OEM
MA4011D	OEM
MA4019B	OEM
MA4019C	OEM
MA4019D	OEM
MA4019E	OEM
MA4019F	OEM
MA4019G	OEM
MA4024	1276
MA4027	0760
MA4027M	5173
MA4027MTA	5173
MA4030	0118
MA4030(M)	OEM
MA4030(M)-Y	OEM
MA4030-M	0118
MA4030M	0118
MA4030MTA	OEM
MA4033	0299
MA4033L	OEM
MA4033MTA	OEM
MA4034	OEM
MA4035	OEM
MA4036	0376
MA4036-M	0376
MA4036L	OEM
MA4036M	0376
MA4037	OEM
MA4038	OEM
MA4039	0036
MA4039(H)	0036
MA4039(H)-T2	0036
MA4039H	OEM
MA4039HTA	OEM
MA4041A	OEM
MA4041B	OEM
MA4041C	OEM
MA4041D	OEM
MA4041E	OEM
MA4041F	OEM
MA4041G	OEM
MA4042	OEM
MA4042A	OEM
MA4042A1	OEM
MA4042A2	OEM
MA4042B	OEM
MA4042B1	OEM
MA4042C	OEM
MA4042D	OEM
MA4042D1	OEM
MA4042E	OEM
MA4042E1	OEM
MA4042F	OEM
MA4042F1	OEM
MA4042G	OEM
MA4042G1	OEM
MA4043	0275
MA4043(H)	0275
MA4043(M)	OEM
MA4043(M)-T2	OEM
MA4043(M)-Y	OEM
MA4043-H	0275
MA4043H	5235
MA4043M	0275
MA4044A	OEM
MA4044A1	OEM
MA4044A2	OEM
MA4044B	OEM
MA4044B1	OEM
MA4044B2	OEM
MA4044C	OEM
MA4044C1	OEM
MA4044D	OEM
MA4044D1	OEM
MA4044E	OEM
MA4044E1	OEM
MA4044F	OEM
MA4044F1	OEM
MA4044G	OEM
MA4044G1	OEM
MA4045A	OEM
MA4045A1	OEM
MA4045A2	OEM
MA4045B	OEM
MA4045B1	OEM
MA4045B2	OEM
MA4045C	OEM
MA4045C1	OEM
MA4045D	OEM
MA4045D1	OEM
MA4045E	OEM
MA4045E1	OEM
MA4045F	OEM
MA4045F1	OEM
MA4045G	OEM
MA4046A	OEM
MA4046A1	OEM
MA4046B	OEM
MA4046B1	OEM
MA4046C	OEM
MA4046C1	OEM
MA4046D	OEM
MA4046D1	OEM
MA4046E	OEM
MA4046E1	OEM
MA4046F	OEM
MA4046F1	OEM
MA4046G	OEM

DEVICE TYPE	REPL CODE
MA4047	0140
MA4047(H)	0140
MA4047(H)-Y	0140
MA4047(M)	0140
MA4047(M)-Y	0140
MA4047-H	0140
MA4047A	OEM
MA4047A1	OEM
MA4047B	OEM
MA4047B1	OEM
MA4047C	OEM
MA4047C1	OEM
MA4047D	OEM
MA4047D1	OEM
MA4047E	OEM
MA4047E1	OEM
MA4047F	OEM
MA4047F1	OEM
MA4047G	OEM
MA4047G1	OEM
MA4047H	0140
MA4047L	0140
MA4047M	0140
MA4047MTA	0140
MA4048A	OEM
MA4048A1	OEM
MA4048B	OEM
MA4048B1	OEM
MA4048C	OEM
MA4048C1	OEM
MA4048D	OEM
MA4048D1	OEM
MA4048E	OEM
MA4048E1	OEM
MA4048F	OEM
MA4048G	OEM
MA4051	0041
MA4051(H)	0041
MA4051(H)-T2	OEM
MA4051(H)-Y	0041
MA4051(L)	OEM
MA4051(L)-Y	OEM
MA4051(M)	0041
MA4051(M)-T2	OEM
MA4051-M	0589
MA4051A	OEM
MA4051A1	OEM
MA4051A2	OEM
MA4051AA	OEM
MA4051AA1	OEM
MA4051B	OEM
MA4051B1	OEM
MA4051C	OEM
MA4051C1	OEM
MA4051D	OEM
MA4051D1	OEM
MA4051E	OEM
MA4051E1	OEM
MA4051F	OEM
MA4051F1	OEM
MA4051G	OEM
MA4051G1	OEM
MA4051H	0041
MA4051L	OEM
MA4051LM	0041
MA4051M	0041
MA4051MTA	0041
MA4051N-M	0589
MA4052A	OEM
MA4052A1	OEM
MA4052A2	OEM
MA4052AA	OEM
MA4052AA1	OEM
MA4052B	OEM
MA4052B1	OEM
MA4052C	OEM
MA4052C1	OEM
MA4052D	OEM
MA4052D1	OEM
MA4052E	OEM
MA4052E1	OEM
MA4052F	OEM
MA4052F1	OEM
MA4052G	OEM
MA4052G1	OEM
MA4054A	OEM
MA4054A1	OEM
MA4054A2	OEM
MA4054AA	OEM
MA4054AA1	OEM
MA4054AA2	OEM
MA4054B	OEM
MA4054B1	OEM
MA4054B2	OEM
MA4054C	OEM
MA4054C1	OEM
MA4054D	OEM
MA4054D1	OEM
MA4054E	OEM
MA4054E1	OEM
MA4054F	OEM
MA4054F1	OEM
MA4054G	OEM
MA4054G1	OEM
MA4055A	OEM
MA4055A1	OEM
MA4055A2	OEM
MA4055AA	OEM
MA4055AA1	OEM

DEVICE TYPE	REPL CODE
MA4055AA2	OEM
MA4055B	OEM
MA4055B1	OEM
MA4055B2	OEM
MA4055C	OEM
MA4055C1	OEM
MA4055D	OEM
MA4055D1	OEM
MA4055E	OEM
MA4055E1	OEM
MA4055F	OEM
MA4055F1	OEM
MA4055G	OEM
MA4056	0253
MA4056(M)	0253
MA4056(M)-T2	0253
MA4056(M)-Y	0253
MA4056-M	0253
MA4056A	OEM
MA4056B	OEM
MA4056B1	OEM
MA4056C	OEM
MA4056C1	OEM
MA4056D	OEM
MA4056D1	OEM
MA4056E	OEM
MA4056E1	OEM
MA4056F	OEM
MA4056F1	OEM
MA4056G	OEM
MA4056H	0253
MA4056MTA	OEM
MA4057A	OEM
MA4057A1	OEM
MA4057B	OEM
MA4057B1	OEM
MA4057C	OEM
MA4057C1	OEM
MA4057D	OEM
MA4057D1	OEM
MA4057E	OEM
MA4057E1	OEM
MA4057F	OEM
MA4057F1	OEM
MA4057G	OEM
MA4057G1	OEM
MA4058A	OEM
MA4058A1	OEM
MA4058B	OEM
MA4058B1	OEM
MA4058C	OEM
MA4058C1	OEM
MA4058D	OEM
MA4058D1	OEM
MA4058E	OEM
MA4058E1	OEM
MA4058F	OEM
MA4058F1	OEM
MA4058G	OEM
MA4060A	OEM
MA4060AA	OEM
MA4060B	OEM
MA4060C	OEM
MA4060D	OEM
MA4061A	OEM
MA4061AA	OEM
MA4061B	OEM
MA4061C	OEM
MA4061D	OEM
MA4062	0471
MA4062(H)	0471
MA4062(H)-T2	0471
MA4062(L)	1005
MA4062(M)	1005
MA4062(M)-T2	0471
MA4062(M)-Y	1005
MA4062-L	0471
MA4062A	OEM
MA4062B	OEM
MA4062C	OEM
MA4062D	OEM
MA4062H	0471
MA4062HTA	0471
MA4062L	0091
MA4062LTA	OEM
MA4062LTV	0091
MA4062M	0471
MA4062MTA	OEM
MA4062N-H	OEM
MA4063	OEM
MA4064	OEM
MA4065	OEM
MA4068	0062
MA4068(L)	0062
MA4068(L)-T2	0062
MA4068(L)-Y	0062
MA4068(M)	0062
MA4068(M)C1-T2	0062
MA4068(M)-T2	0062
MA4068(M)-Y	0062
MA4068(N)	0062
MA4068(N)C1-T2	0062
MA4068(N)C1-Y	0062
MA4068(N)V1	0062
MA4068(N)V1-Y	0062
MA4068-M	0062
MA4068-MM	0062
MA4068H	0062
MA4068L	0062

DEVICE TYPE	REPL CODE
MA4068M	0062
MA4071B	OEM
MA4072B	OEM
MA4075	0077
MA4075(M)	0077
MA4075(M)-T2	OEM
MA4075-H	0077
MA4075-M	0077
MA4075H	0077
MA4075L	0077
MA4075LTA	0077
MA4075M	0077
MA4082	0165
MA4082(M)	OEM
MA4082(M)-T2	OEM
MA4082(M)-Y	OEM
MA4082-H	0165
MA4082-L	4309
MA4082-M	0165
MA4082L	OEM
MA4082M	0165
MA4082MTA	OEM
MA4083	OEM
MA4084	OEM
MA4085	OEM
MA4086	OEM
MA4091	0057
MA4091(M)	0057
MA4091(M)-T2	0057
MA4091(M)-Y	0057
MA4091-L	0057
MA4091-M	0057
MA4091L	0057
MA4091M	0057
MA4091N-H	OEM
MA4091N-L	0057
MA4091N-M	0057
MA4098AA	OEM
MA4099D	OEM
MA4100	0064
MA4100(M)	0064
MA4100(M)-T2	0064
MA4100(M)-Y	0064
MA4100-H	0064
MA4100-M	0064
MA4100H	0064
MA4100M	0064
MA4100MTA	OEM
MA4100N	0064
MA4101	0016
MA4102	0016
MA4103	0016
MA4104	0016
MA4110	0181
MA4110(M)	OEM
MA4110(M)-Y	OEM
MA4110-L	0181
MA4110-M	0181
MA4110H	OEM
MA4110L	OEM
MA4110M	OEM
MA4110MTA	OEM
MA4110TA	0181
MA4116	OEM
MA4120	0052
MA4120(M)	0052
MA4120(M)-T2	0052
MA4120(M)-Y	0052
MA4120-H	0052
MA4120H	0053
MA4120L	OEM
MA4120M	1005
MA4120MTA	1005
MA4123	OEM
MA4123A	OEM
MA4123B	OEM
MA4124	OEM
MA4124A	OEM
MA4125	OEM
MA4125A	OEM
MA4126	OEM
MA4126A	OEM
MA4127	OEM
MA4127A	OEM
MA4128	OEM
MA4129	OEM
MA4130	0053
MA4130(H)	1005
MA4130(H)-T2	1005
MA4130(H)-Y	1005
MA4130(L)	0053
MA4130(L)-Y	0053
MA4130A	OEM
MA4130H	0053
MA4130L	0053
MA4130M	0053
MA4131	OEM
MA4131A	OEM
MA4133	OEM
MA4135	OEM
MA4136	OEM
MA4136A	OEM
MA4137	OEM
MA4139	OEM
MA4140-H	0879
MA4142	OEM
MA4142A	OEM
MA4146	OEM
MA4147	OEM

DEVICE TYPE	REPL CODE
MA4148	OEM
MA4150	0686
MA4150(L)	0686
MA4150(L)-T2	0686
MA4150(L)-Y	0686
MA4150(M)	0686
MA4150(M)-T2	6751
MA4150(M)-Y	0686
MA4150-M	0686
MA4150H	OEM
MA4150M	0686
MA4150MTA	OEM
MA4160	0445
MA4160-M	0445
MA4160M	0445
MA4160N-M	0445
MA4171	OEM
MA4172	OEM
MA4173	OEM
MA4174	OEM
MA4180	0375
MA4180(M)-T2	OEM
MA4180-H	4848
MA4180M	0375
MA4180MTA	OEM
MA4200	0700
MA4200(M)	OEM
MA4200(M)-Y	OEM
MA4200-M	0700
MA4200H	OEM
MA4200M	OEM
MA4220	0705
MA4220-T	OEM
MA4220H	0705
MA4220M	OEM
MA4220T	OEM
MA4230	0396
MA4240	0494
MA4240-M	0548
MA4240H	OEM
MA4245	OEM
MA4251	OEM
MA4253	OEM
MA4254	OEM
MA4255	OEM
MA4256	OEM
MA4257	OEM
MA4259	OEM
MA4260	OEM
MA4261	OEM
MA4263	OEM
MA4264	OEM
MA4265	OEM
MA4266	OEM
MA4267	OEM
MA4268	OEM
MA4270	0455
MA4270(M)	0441
MA4270(M)-Y	0455
MA4270-H	0455
MA4270M	0455
MA4273B	OEM
MA4273C	OEM
MA4273D	OEM
MA4273E	OEM
MA4273F	OEM
MA4273G	OEM
MA4280	OEM
MA4281	OEM
MA4282	OEM
MA4283	OEM
MA4284	OEM
MA4285	OEM
MA4286	OEM
MA4287	OEM
MA4288	OEM
MA4289	OEM
MA4290	OEM
MA4291	OEM
MA4292	OEM
MA4296	OEM
MA4297	OEM
MA4298	OEM
MA4300	0195
MA4300H	OEM
MA4300M	0195
MA4300MTA	OEM
MA4321A	OEM
MA4321A1	OEM
MA4321AA	OEM
MA4321AA1	OEM
MA4321B	OEM
MA4321B1	OEM
MA4321C	OEM
MA4321C1	OEM
MA4321D	OEM
MA4321D1	OEM
MA4321E	OEM
MA4321E1	OEM
MA4321F	OEM
MA4321G	OEM
MA4322A	OEM
MA4322A1	OEM
MA4322AA	OEM
MA4322AA1	OEM
MA4322B	OEM
MA4322B1	OEM
MA4322C	OEM
MA4322C1	OEM

If replacement code is OEM, contact original manufacturer for replacement.

DEVICE TYPE	REPL CODE
MA4322D	OEM
MA4322D1	OEM
MA4322E	OEM
MA4322E1	OEM
MA4322F	OEM
MA4322G	OEM
MA4324A	OEM
MA4324A1	OEM
MA4324B	OEM
MA4324B1	OEM
MA4324C	OEM
MA4324C1	OEM
MA4324D	OEM
MA4324D1	OEM
MA4324E	OEM
MA4324E1	OEM
MA4324F	OEM
MA4324F1	OEM
MA4324G	OEM
MA4325A	OEM
MA4325A1	OEM
MA4325AA	OEM
MA4325AA1	OEM
MA4325B	OEM
MA4325B1	OEM
MA4325C	OEM
MA4325C1	OEM
MA4325D	OEM
MA4325D1	OEM
MA4325E	OEM
MA4325E1	OEM
MA4325F	OEM
MA4325F1	OEM
MA4325G	OEM
MA4326A	OEM
MA4326A1	OEM
MA4326AA	OEM
MA4326AA1	OEM
MA4326B	OEM
MA4326B1	OEM
MA4326C	OEM
MA4326C1	OEM
MA4326D	OEM
MA4326D1	OEM
MA4326E	OEM
MA4326E1	OEM
MA4326F	OEM
MA4326F1	OEM
MA4326G	OEM
MA4327A	OEM
MA4327A1	OEM
MA4327B	OEM
MA4327B1	OEM
MA4327C	OEM
MA4327C1	OEM
MA4327D	OEM
MA4327D1	OEM
MA4327E	OEM
MA4327E1	OEM
MA4328A	OEM
MA4328A1	OEM
MA4328B	OEM
MA4328B1	OEM
MA4328C	OEM
MA4328C1	OEM
MA4328D	OEM
MA4328D1	OEM
MA4330	0166
MA4330(L)	0166
MA4330(L)-Y	0166
MA4330(M)	0166
MA4330(M)-T2	0166
MA4330(M)-Y	0166
MA4330H	OEM
MA4330L	OEM
MA4330M	OEM
MA4331A	OEM
MA4331B	OEM
MA4331C	OEM
MA4331D	OEM
MA4331E	OEM
MA4331F	OEM
MA4331G	OEM
MA4332A	OEM
MA4332B	OEM
MA4332C	OEM
MA4332D	OEM
MA4332E	OEM
MA4332F	OEM
MA4332G	OEM
MA4333A	OEM
MA4333B	OEM
MA4333C	OEM
MA4333D	OEM
MA4333E	OEM
MA4333F	OEM
MA4333G	OEM
MA4334A	OEM
MA4334A1	OEM
MA4334A2	OEM
MA4334B	OEM
MA4334B1	OEM
MA4334B2	OEM
MA4334C	OEM
MA4334C1	OEM
MA4334C2	OEM
MA4334D	OEM
MA4334D1	OEM
MA4334E	OEM
MA4334E1	OEM
MA4334G	OEM
MA4335A	OEM
MA4335A1	OEM
MA4335A2	OEM
MA4335AA	OEM
MA4335AA1	OEM
MA4335AA2	OEM
MA4335B	OEM
MA4335B1	OEM
MA4335B2	OEM
MA4335D	OEM
MA4335D1	OEM
MA4335D2	OEM
MA4335E	OEM
MA4335E1	OEM
MA4335F	OEM
MA4335G	OEM
MA4336A	OEM
MA4336A1	OEM
MA4336A2	OEM
MA4336AA	OEM
MA4336AA1	OEM
MA4336AA2	OEM
MA4336B	OEM
MA4336B1	OEM
MA4336B2	OEM
MA4336C	OEM
MA4336C1	OEM
MA4336C2	OEM
MA4336D	OEM
MA4336D1	OEM
MA4336D2	OEM
MA4336E	OEM
MA4336E1	OEM
MA4336F	OEM
MA4336G	OEM
MA4337A	OEM
MA4337A1	OEM
MA4337B	OEM
MA4337B1	OEM
MA4337C	OEM
MA4337C1	OEM
MA4337D	OEM
MA4337D1	OEM
MA4337E	OEM
MA4337E1	OEM
MA4337F	OEM
MA4337F1	OEM
MA4337G	OEM
MA4338A	OEM
MA4338A1	OEM
MA4338B	OEM
MA4338B1	OEM
MA4338C	OEM
MA4338C1	OEM
MA4338D	OEM
MA4338D1	OEM
MA4338E	OEM
MA4338E1	OEM
MA4338F	OEM
MA4341A	OEM
MA4341A1	OEM
MA4341A2	OEM
MA4341AA	OEM
MA4341AA1	OEM
MA4341AA2	OEM
MA4341B	OEM
MA4341B1	OEM
MA4341B2	OEM
MA4341C	OEM
MA4341C1	OEM
MA4341C2	OEM
MA4341D	OEM
MA4341D1	OEM
MA4341D2	OEM
MA4341E	OEM
MA4341E1	OEM
MA4341F	OEM
MA4341F1	OEM
MA4341G	OEM
MA4341G1	OEM
MA4342A	OEM
MA4342A1	OEM
MA4342A2	OEM
MA4342AA	OEM
MA4342AA1	OEM
MA4342AA2	OEM
MA4342B	OEM
MA4342B1	OEM
MA4342B2	OEM
MA4342C	OEM
MA4342C1	OEM
MA4342C2	OEM
MA4342D	OEM
MA4342D1	OEM
MA4342D2	OEM
MA4342E	OEM
MA4342E1	OEM
MA4342F	OEM
MA4342F1	OEM
MA4342G	OEM
MA4342G1	OEM
MA4343A	OEM
MA4343A1	OEM
MA4343A2	OEM
MA4343AA	OEM
MA4343AA1	OEM
MA4343AA2	OEM
MA4343B	OEM
MA4343B1	OEM
MA4343B2	OEM
MA4343C	OEM
MA4343C1	OEM
MA4343C2	OEM
MA4343D	OEM
MA4343D1	OEM
MA4343D2	OEM
MA4343E	OEM
MA4343E1	OEM
MA4343F	OEM
MA4343F1	OEM
MA4343G	OEM
MA4344A	OEM
MA4344A1	OEM
MA4344A2	OEM
MA4344B	OEM
MA4344B1	OEM
MA4344B2	OEM
MA4344C	OEM
MA4344C1	OEM
MA4344C2	OEM
MA4344D	OEM
MA4344D1	OEM
MA4344D2	OEM
MA4344E	OEM
MA4344E1	OEM
MA4344E2	OEM
MA4344F	OEM
MA4344F1	OEM
MA4344G	OEM
MA4344G1	OEM
MA4345A	OEM
MA4345A1	OEM
MA4345A2	OEM
MA4345AA	OEM
MA4345AA1	OEM
MA4345AA2	OEM
MA4345B	OEM
MA4345B1	OEM
MA4345B2	OEM
MA4345C	OEM
MA4345C1	OEM
MA4345C2	OEM
MA4345D	OEM
MA4345D1	OEM
MA4345D2	OEM
MA4345E	OEM
MA4345E1	OEM
MA4345F	OEM
MA4345F1	OEM
MA4345G	OEM
MA4346A	OEM
MA4346A1	OEM
MA4346A2	OEM
MA4346AA	OEM
MA4346AA1	OEM
MA4346AA2	OEM
MA4346B	OEM
MA4346B1	OEM
MA4346B2	OEM
MA4346C	OEM
MA4346C1	OEM
MA4346C2	OEM
MA4346D	OEM
MA4346D1	OEM
MA4346D2	OEM
MA4346E	OEM
MA4346E1	OEM
MA4346F	OEM
MA4346F1	OEM
MA4346G	OEM
MA4347A	OEM
MA4347A1	OEM
MA4347B	OEM
MA4347B1	OEM
MA4347C	OEM
MA4347C1	OEM
MA4347D	OEM
MA4347D1	OEM
MA4347E	OEM
MA4347E1	OEM
MA4347F	OEM
MA4347G	OEM
MA4348A	OEM
MA4348A1	OEM
MA4348B	OEM
MA4348B1	OEM
MA4348C	OEM
MA4348C1	OEM
MA4348D	OEM
MA4348D1	OEM
MA4348E	OEM
MA4348E1	OEM
MA4348F	OEM
MA4351A	OEM
MA4351A1	OEM
MA4351A2	OEM
MA4351AA	OEM
MA4351AA1	OEM
MA4351AA2	OEM
MA4351B	OEM
MA4351B1	OEM
MA4351B2	OEM
MA4351C	OEM
MA4351C1	OEM
MA4351C2	OEM
MA4351D	OEM
MA4351D1	OEM
MA4351D2	OEM
MA4351E	OEM
MA4351E1	OEM
MA4351F	OEM
MA4351F1	OEM
MA4351G	OEM
MA4351G1	OEM
MA4352A	OEM
MA4352A1	OEM
MA4352A2	OEM
MA4352AA	OEM
MA4352AA1	OEM
MA4352AA2	OEM
MA4352B	OEM
MA4352B1	OEM
MA4352B2	OEM
MA4352C	OEM
MA4352C1	OEM
MA4352C2	OEM
MA4352D	OEM
MA4352D1	OEM
MA4352D2	OEM
MA4352E	OEM
MA4352E1	OEM
MA4352F	OEM
MA4352F1	OEM
MA4352G	OEM
MA4352G1	OEM
MA4353A	OEM
MA4353A1	OEM
MA4353A2	OEM
MA4353AA	OEM
MA4353AA1	OEM
MA4353AA2	OEM
MA4353B	OEM
MA4353B1	OEM
MA4353B2	OEM
MA4353C	OEM
MA4353C1	OEM
MA4353C2	OEM
MA4353D	OEM
MA4353D1	OEM
MA4353D2	OEM
MA4353E	OEM
MA4353E1	OEM
MA4353F	OEM
MA4353F1	OEM
MA4353G	OEM
MA4354A	OEM
MA4354A1	OEM
MA4354A2	OEM
MA4354B	OEM
MA4354B1	OEM
MA4354B2	OEM
MA4354C	OEM
MA4354C1	OEM
MA4354C2	OEM
MA4354D	OEM
MA4354D1	OEM
MA4354D2	OEM
MA4354E	OEM
MA4354E1	OEM
MA4354E2	OEM
MA4354F	OEM
MA4354F1	OEM
MA4354G	OEM
MA4354G1	OEM
MA4355A	OEM
MA4355A1	OEM
MA4355A2	OEM
MA4355AA	OEM
MA4355AA1	OEM
MA4355AA2	OEM
MA4355B	OEM
MA4355B1	OEM
MA4355B2	OEM
MA4355C	OEM
MA4355C1	OEM
MA4355C2	OEM
MA4355D	OEM
MA4355D1	OEM
MA4355D2	OEM
MA4355E	OEM
MA4355E1	OEM
MA4355F	OEM
MA4355F1	OEM
MA4355G	OEM
MA4356AA	OEM
MA4356AA1	OEM
MA4356AA2	OEM
MA4356B	OEM
MA4356B1	OEM
MA4356B2	OEM
MA4356C	OEM
MA4356C1	OEM
MA4356C2	OEM
MA4356D	OEM
MA4356D1	OEM
MA4356D2	OEM
MA4356E	OEM
MA4356E1	OEM
MA4356F	OEM
MA4356F1	OEM
MA4356G	OEM
MA4357A	OEM
MA4357A1	OEM
MA4357B	OEM
MA4357B1	OEM
MA4357C	OEM
MA4357C1	OEM
MA4357D	OEM
MA4357D1	OEM
MA4357E	OEM
MA4357E1	OEM
MA4357F	OEM
MA4357F1	OEM
MA4357G	OEM
MA4358A	OEM
MA4358A1	OEM
MA4358B	OEM
MA4358B1	OEM
MA4358C	OEM
MA4358C1	OEM
MA4358D	OEM
MA4358D1	OEM
MA4358E	OEM
MA4358E1	OEM
MA4358F	OEM
MA4360	0010
MA4360H	OEM
MA4360M	0010
MA4361	OEM
MA4362	OEM
MA4372C	OEM
MA4373C	OEM
MA4374C	OEM
MA4375C	OEM
MA4375E	OEM
MA4380	OEM
MA4381	OEM
MA4382	OEM
MA4383	OEM
MA4384	OEM
MA4385	OEM
MA4386	OEM
MA4387	OEM
MA4388	5676
MA4400	OEM
MA4400A	OEM
MA4404	0211
MA4404A	0211
MA4472	OEM
MA4473	OEM
MA4497A	OEM
MA4497B	OEM
MA4497C	OEM
MA4501	0041
MA4534	OEM
MA4535	OEM
MA4536	OEM
MA4537	OEM
MA4538	OEM
MA4539	OEM
MA4540	OEM
MA4541	OEM
MA4542	OEM
MA4544	OEM
MA4546	OEM
MA4547	OEM
MA4551	OEM
MA4552	OEM
MA4553	OEM
MA4554	OEM
MA4555	OEM
MA4556	OEM
MA4557	OEM
MA4562	OEM
MA4570A	OEM
MA4570A1	OEM
MA4570B	OEM
MA4570B1	OEM
MA4570C	OEM
MA4570C1	OEM
MA4570D	OEM
MA4570D1	OEM
MA4570E	OEM
MA4570E1	OEM
MA4571A	OEM
MA4571A1	OEM
MA4571A2	OEM
MA4571B	OEM
MA4571B1	OEM
MA4571B2	OEM
MA4571C	OEM
MA4571C1	OEM
MA4571C2	OEM
MA4571D	OEM
MA4571D1	OEM
MA4571D2	OEM
MA4571E	OEM
MA4571E1	OEM
MA4571E2	OEM
MA4572A	OEM
MA4572A1	OEM
MA4572B	OEM
MA4572B1	OEM
MA4573A	OEM
MA4573A1	OEM
MA4573A2	OEM
MA4573B	OEM
MA4573B1	OEM
MA4573B2	OEM
MA4583A	OEM
MA4583B	OEM
MA4583C	OEM
MA4604A	OEM
MA4604A1	OEM
MA4604B	OEM
MA4604C	OEM
MA4605A	OEM
MA4605B	OEM
MA4605C	OEM
MA4606A	OEM
MA4606A1	OEM
MA4606A2	OEM
MA4606B	OEM
MA4606C	OEM
MA4607A	OEM
MA4607A1	OEM
MA4607A2	OEM
MA4607B	OEM
MA4607C	OEM
MA4608A	OEM
MA4608B	OEM
MA4608C	OEM
MA4609A	OEM
MA4609B	OEM
MA4609C	OEM
MA4610	OEM
MA4610A	OEM
MA4613	OEM
MA4614	OEM
MA4615	OEM
MA4616	OEM
MA4617	OEM
MA4618	OEM
MA4619	OEM
MA4619A	OEM
MA4623	OEM
MA4623A	OEM
MA4624	OEM
MA4624A	OEM
MA4625	OEM
MA4625A	OEM
MA4631A	OEM
MA4631A1	OEM
MA4631A2	OEM
MA4633A	OEM
MA4633A1	OEM
MA4644A-E	OEM
MA4645B-E	OEM
MA4646C-E	OEM
MA4651A	OEM
MA4651B	OEM
MA4651C	OEM
MA4652A	OEM
MA4652B	OEM
MA4652C	OEM
MA4653A	OEM
MA4653B	OEM
MA4653C	OEM
MA4654A	OEM
MA4654A1	OEM
MA4654B	OEM
MA4654C	OEM
MA4655	OEM
MA4663A2	OEM
MA4670	0160
MA4684	OEM
MA4685	OEM
MA4686	OEM
MA4687	OEM
MA4688	OEM
MA4701B	OEM
MA4701B1	OEM
MA4701C	OEM
MA4701C1	OEM
MA4701D	OEM
MA4701D1	OEM
MA4701E	OEM
MA4701E1	OEM
MA4701F	OEM
MA4701F1	OEM
MA4701G	OEM
MA4701G1	OEM
MA4701H	OEM
MA4701H1	OEM
MA4701I	OEM
MA4701I1	OEM
MA4702A	OEM
MA4702A1	OEM
MA4702B	OEM
MA4702B1	OEM
MA4702C	OEM
MA4702C1	OEM
MA4702D	OEM
MA4702D1	OEM
MA4702E	OEM
MA4702E1	OEM
MA4702F	OEM
MA4702F1	OEM
MA4702G	OEM
MA4702G1	OEM
MA4703B	OEM
MA4703C	OEM
MA4703D	OEM
MA4703E	OEM
MA4703F	OEM
MA4703G	OEM
MA4704B	OEM
MA4704C	OEM
MA4704D	OEM
MA4704E	OEM
MA4704F	OEM
MA4704G	OEM
MA4732A	OEM
MA4732B	OEM
MA4732C	OEM
MA4733A	OEM
MA4733B	OEM
MA4733C	OEM
MA4748	OEM
MA4749	OEM
MA4750	OEM
MA4751	OEM
MA4752	OEM
MA4753	OEM
MA4754	OEM
MA4755	OEM
MA4756	OEM
MA4760A	OEM
MA4760AA	OEM
MA4760B	OEM
MA4760C	OEM
MA4760D	OEM
MA4761A	OEM
MA4761AA	OEM
MA4761B	OEM
MA4761C	OEM
MA4761D	OEM
MA4762	OEM
MA4763	OEM
MA4764	OEM
MA4765	OEM
MA4811	OEM
MA4811A	OEM
MA4811B	OEM
MA4811C	OEM
MA4811D	OEM
MA4812	OEM
MA4812A	OEM
MA4812B	OEM
MA4812C	OEM
MA4813	OEM
MA4813A	OEM
MA4813B	OEM
MA4814	OEM
MA4814A	OEM
MA4814B	OEM
MA4815	OEM
MA4815A	OEM
MA4825B1	OEM
MA4825B2	OEM
MA4825C1	OEM
MA4825C2	OEM
MA4825D1	OEM
MA4825D2	OEM
MA4825E1	OEM
MA4825E2	OEM
MA4826B1	OEM
MA4826B2	OEM
MA4826C1	OEM
MA4826C2	OEM
MA4826D1	OEM
MA4826D2	OEM
MA4850	OEM
MA4851	OEM
MA4852	OEM
MA4853	OEM
MA4854	OEM
MA4855	OEM
MA4856	OEM
MA4861C	OEM
MA4861D	OEM
MA4861E	OEM
MA4861F	OEM
MA4861G	OEM
MA4861H	OEM
MA4882	OEM
MA4883	OEM
MA4942H	OEM
MA4944H	OEM
MA4946J	OEM
MA4948J	OEM
MA4950A	OEM
MA4950B	OEM
MA4951A	OEM
MA4951B	OEM
MA4952B	OEM
MA4952C	OEM
MA4953B	OEM
MA4953C	OEM
MA4954B	OEM
MA4954C	OEM
MA4954D	OEM
MA4955C	OEM
MA4955D	OEM
MA4955E	OEM
MA4956C	OEM
MA4956D	OEM
MA4956E	OEM
MA4956F	OEM
MA4957D	OEM
MA4957E	OEM
MA4957F	OEM
MA4958E	OEM
MA4958F	OEM
MA4960A	OEM
MA4960B	OEM
MA4961A	OEM
MA4961B	OEM
MA4962B	OEM
MA4962C	OEM
MA4963B	OEM
MA4963C	OEM

DEVICE TYPE	REPL CODE
MA4964B	OEM
MA4964C	OEM
MA4964D	OEM
MA4965B	OEM
MA4965C	OEM
MA4965D	OEM
MA4965E	OEM
MA4966C	OEM
MA4966D	OEM
MA4966E	OEM
MA4966F	OEM
MA4967D	OEM
MA4967E	OEM
MA4967F	OEM
MA4967G	OEM
MA4968E	OEM
MA4968F	OEM
MA4968G	OEM
MA4969E	OEM
MA4969F	OEM
MA4969G	OEM
MA4980	OEM
MA4984	OEM
MA4985	OEM
MA4986	OEM
MA4987	OEM
MA4988	OEM
MA4989	OEM
MA4990	OEM
MA4992	OEM
MA4998	OEM
MA5047	OEM
MA5051	OEM
MA5056	OEM
MA5062	OEM
MA5068	OEM
MA5075	OEM
MA5082	OEM
MA5091	OEM
MA5100	OEM
MA5110	OEM
MA5113	2617
MA5120	OEM
MA5130	OEM
MA5150	OEM
MA5160	OEM
MA5180	OEM
MA5186AP	0918
MA5200	OEM
MA5220	OEM
MA5240	OEM
MA5292C	OEM
MA6001	0016
MA6002	0016
MA6003	0016
MA6082	OEM
MA6101	0086
MA6102	0016
MA-6301	OEM
MA6301	2031
MA6301T	2031
MA7300	4219
MA7700A-0004	OEM
MA7700A-0005	OEM
MA7700K-0001	OEM
MA7700K-0002	OEM
MA7700K-0003	OEM
MA7705	OEM
MA7707H	OEM
MA7707J	OEM
MA7707J-0003	OEM
MA7707J-0012	OEM
MA7707J-0013	OEM
MA7707N-0006	OEM
MA7707N-0008	OEM
MA7707N-0009	OEM
MA7707N-0012	OEM
MA7707N-0013	OEM
MA7707N-0014	OEM
MA7707N-0015	OEM
MA7708	OEM
MA7709A-0001	OEM
MA7709A-0003	OEM
MA7709A-0005	OEM
MA7709A-0007	OEM
MA7709A-0009	OEM
MA7709A-0011	OEM
MA7709A-0012	OEM
MA7709A-0013	OEM
MA7709A-0014	OEM
MA7709A-0015	OEM
MA7709U-0012	OEM
MA7709U-0013	OEM
MA7709U-0014	OEM
MA7709U-0015	OEM
MA7710	OEM
MA7715A-000003	OEM
MA7715A-0001	OEM
MA7715A-0002	OEM
MA7715A-0004	OEM
MA7715A-0005	OEM
MA7715A-0006	OEM
MA7715A-0007	OEM
MA7715A-0008	OEM
MA7715A-0009	OEM
MA7715A-0012	OEM
MA7715A-0013	OEM
MA7715U-0012	OEM
MA7715U-0013	OEM
MA7717J-0009	OEM
MA7717N-0001	OEM
MA7717N-0002	OEM
MA7717N-0003	OEM
MA7717N-0004	OEM
MA7717N-0005	OEM
MA7717N-0006	OEM
MA7717N-0007	OEM
MA7717N-0008	OEM
MA7717N-0012	OEM
MA7717N-0013	OEM
MA7742J-0001	OEM
MA7742N-0005	OEM
MA7742N-0007	OEM
MA7742N-0009	OEM
MA7742N-0011	OEM
MA7742N-0012	OEM
MA7744A-0001	OEM
MA7744A-0003	OEM
MA7744A-0005	OEM
MA7744A-0007	OEM
MA7744A-0009	OEM
MA7744A-0012	OEM
MA7757J-0012	OEM
MA7757J-0013	OEM
MA7805	0624
MA7805UC	OEM
MA7807	OEM
MA7809	OEM
MA7811	OEM
MA7816	OEM
MA7817	OEM
MA8001	0086
MA8002	0086
MA8003	0324
MA8047M	OEM
MA8051-M	OEM
MA8051M	OEM
MA8334	OEM
MA8342-0001	OEM
MA8342-0002	OEM
MA8342-0003	OEM
MA8342-0004	OEM
MA8342-0005	OEM
MA8342-0006	OEM
MA8342-0007	OEM
MA8342-0008	OEM
MA8342-0009	OEM
MA8342-0010	OEM
MA8342-0011	OEM
MA8343-0001	OEM
MA8343-0005	OEM
MA8343-0006	OEM
MA8343-0050	OEM
MA8343-0051	OEM
MA8343-0052	OEM
MA8343-0053	OEM
MA8343-0054	OEM
MA8343-0055	OEM
MA8345-0001	OEM
MA8345-0002	OEM
MA8345-0003	OEM
MA8345-0056	OEM
MA8345-0057	OEM
MA8345-0058	OEM
MA8345-0059	OEM
MA9001	0493
MA9002	0493
MA9003	0493
MA11238N	OEM
MA13080	OEM
MA34022	OEM
MA40000	OEM
MA40001	OEM
MA40002	OEM
MA40003	OEM
MA40006	OEM
MA40007	OEM
MA40008	OEM
MA40009	OEM
MA40012	OEM
MA40013	OEM
MA40014	OEM
MA40015	OEM
MA40024	OEM
MA40025	OEM
MA40026	OEM
MA40027	OEM
MA40028	OEM
MA40029	OEM
MA40033	OEM
MA40034	OEM
MA40035	OEM
MA40040	OEM
MA40041	OEM
MA40042	OEM
MA40043	OEM
MA40051D	OEM
MA40051E	OEM
MA40051F	OEM
MA40051G	OEM
MA40051H	OEM
MA40052	OEM
MA40053	OEM
MA40054	OEM
MA40061	OEM
MA40062	OEM
MA40071-1	OEM
MA40071D	OEM
MA40071E	OEM
MA40071F	OEM
MA40071G	OEM
MA40071H	OEM
MA40071I	OEM
MA40072	OEM
MA40073	OEM
MA40074	OEM
MA40075	OEM
MA40081	OEM
MA40082	OEM
MA40100	OEM
MA40101	OEM
MA40102	OEM
MA40103	OEM
MA40104	OEM
MA40105	OEM
MA40106	OEM
MA40107	OEM
MA40110	OEM
MA40111	OEM
MA40115	OEM
MA40116	OEM
MA40120	OEM
MA40121	OEM
MA40122	OEM
MA40123	OEM
MA40124	OEM
MA40126	OEM
MA40127	OEM
MA40128	OEM
MA40140	OEM
MA40150	OEM
MA40151	OEM
MA40152	OEM
MA40153	OEM
MA40154	OEM
MA40155	OEM
MA40156	OEM
MA40157	OEM
MA40160	OEM
MA40161	OEM
MA40162	OEM
MA40163	OEM
MA40165	OEM
MA40166	OEM
MA40167	OEM
MA40168	OEM
MA40170	OEM
MA40171	OEM
MA40172	OEM
MA40173	OEM
MA40174	OEM
MA40175	OEM
MA40176	OEM
MA40177	OEM
MA40178	OEM
MA40182	OEM
MA40183	OEM
MA40184	OEM
MA40190	OEM
MA40191	OEM
MA40192	OEM
MA40193	OEM
MA40194	OEM
MA40195	OEM
MA40201	OEM
MA40202	OEM
MA40203	OEM
MA40204	OEM
MA40205	OEM
MA40206	OEM
MA40207	OEM
MA40208	OEM
MA40215	OEM
MA40216	OEM
MA40220	OEM
MA40221	OEM
MA40222	OEM
MA40227	OEM
MA40228	OEM
MA40230	OEM
MA40231	OEM
MA40232	OEM
MA40233	OEM
MA40234	OEM
MA40235	OEM
MA40236	OEM
MA40237	OEM
MA40240	OEM
MA40241	OEM
MA40242	OEM
MA40243	OEM
MA40244	OEM
MA40245	OEM
MA40246	OEM
MA40248	OEM
MA40249	OEM
MA40251	OEM
MA40252	OEM
MA40253	OEM
MA40254	OEM
MA40255	OEM
MA40256	OEM
MA40257	OEM
MA40258	OEM
MA40265	OEM
MA40266	OEM
MA40267	OEM
MA40268	OEM
MA40270	OEM
MA40271	OEM
MA40272	OEM
MA40273	OEM
MA40274	OEM
MA40291	OEM
MA40292	OEM
MA40293	OEM
MA40294	OEM
MA40295	OEM
MA40296	OEM
MA40410	OEM
MA40412	OEM
MA40417	OEM
MA40423	OEM
MA40426	OEM
MA40427	OEM
MA40428	OEM
MA40430	OEM
MA40431	OEM
MA40432	OEM
MA40433	OEM
MA40434	OEM
MA40435	OEM
MA40436	OEM
MA40439	OEM
MA40440	OEM
MA40441	OEM
MA40442	OEM
MA40443	OEM
MA40444	OEM
MA40445	OEM
MA40446	OEM
MA40447	OEM
MA40449	OEM
MA40480	OEM
MA40490	OEM
MA40491	OEM
MA40492	OEM
MA40493-226	OEM
MA40494-227	OEM
MA40495-228	OEM
MA40496-227	OEM
MA40497-227	OEM
MA40499	OEM
MA40543	OEM
MA40995	OEM
MA40996	OEM
MA40997	OEM
MA40998	OEM
MA40999	OEM
MA41201C	OEM
MA41201D	OEM
MA41201E	OEM
MA41201F	OEM
MA41201G	OEM
MA41202C	OEM
MA41202D	OEM
MA41202E	OEM
MA41202F	OEM
MA41202G	OEM
MA41205	OEM
MA41206	OEM
MA41207	OEM
MA41220C	OEM
MA41220E	OEM
MA41220F	OEM
MA41220G	OEM
MA41221D	OEM
MA41221E	OEM
MA41221F	OEM
MA41222	OEM
MA41223	OEM
MA41224	OEM
MA41225	OEM
MA41500	OEM
MA41501	OEM
MA41502	OEM
MA41503	OEM
MA41504	OEM
MA41505	OEM
MA41506	OEM
MA41507	OEM
MA41508	OEM
MA41509	OEM
MA41510	OEM
MA41511	OEM
MA41512	OEM
MA41513	OEM
MA41514	OEM
MA41515	OEM
MA41516	OEM
MA41801	OEM
MA41802	OEM
MA41803	OEM
MA41804	OEM
MA41805	OEM
MA41806	OEM
MA41807	OEM
MA41808	OEM
MA41809	OEM
MA41810	OEM
MA41811	OEM
MA41812	OEM
MA41813	OEM
MA41814	OEM
MA41816	OEM
MA41817	OEM
MA41819	OEM
MA41820	OEM
MA42001-509	OEM
MA42002-509	OEM
MA42003-509	OEM
MA42004-509	OEM
MA42005-509	OEM
MA42006-510	OEM
MA42008-511	OEM
MA42009-509	OEM
MA42010-509	OEM
MA42010-510	OEM
MA42011-510	OEM
MA42012-510	OEM
MA42014-509	OEM
MA42015-510	OEM
MA42016-510	OEM
MA42020-509	OEM
MA42021-509	OEM
MA42022-509	OEM
MA42023-509	OEM
MA42024-509	OEM
MA42025-509	OEM
MA42026-509	OEM
MA42027-509	OEM
MA42028-509	OEM
MA42051-509	OEM
MA42051-510	OEM
MA42051-511	OEM
MA42052-509	OEM
MA42052-510	OEM
MA42052-511	OEM
MA42055-510	OEM
MA42055-511	OEM
MA42056-510	OEM
MA42056-511	OEM
MA42057-510	OEM
MA42057-511	OEM
MA42060	OEM
MA42061	OEM
MA42062	OEM
MA42063	OEM
MA42100-509	OEM
MA42100-510	OEM
MA42100-511	OEM
MA42100-512	OEM
MA42110-509	OEM
MA42110-510	OEM
MA42110-511	OEM
MA42110-512	OEM
MA42111-509	OEM
MA42111-510	OEM
MA42111-511	OEM
MA42112-509	OEM
MA42112-510	OEM
MA42112-511	OEM
MA42113-509	OEM
MA42113-510	OEM
MA42113-511	OEM
MA42120-508	OEM
MA42121-508	OEM
MA42122-509	OEM
MA42123-509	OEM
MA42141-509	OEM
MA42141-510	OEM
MA42141-511	OEM
MA42142-509	OEM
MA42142-510	OEM
MA42142-511	OEM
MA42143-509	OEM
MA42143-510	OEM
MA42143-511	OEM
MA42151-511	OEM
MA42161-511	OEM
MA42162-511	OEM
MA42181-510	OEM
MA42191-510	OEM
MA43000	OEM
MA43001	OEM
MA43002	OEM
MA43003	OEM
MA43004	OEM
MA43005	OEM
MA43007	OEM
MA43030	OEM
MA43031	OEM
MA43032	OEM
MA43033	OEM
MA43034	OEM
MA43035	OEM
MA43036	OEM
MA43037	OEM
MA43543	OEM
MA43556A	OEM
MA43556A1	OEM
MA43556A2	OEM
MA43592	OEM
MA44010	OEM
MA44020	OEM
MA44030	OEM
MA44040	OEM
MA44050	OEM
MA44051	OEM
MA44052	OEM
MA44053	OEM
MA44058	OEM
MA44060	OEM
MA44061	OEM
MA44070	OEM
MA44071	OEM
MA44100	OEM
MA44110	OEM
MA44120	OEM
MA44130	OEM
MA44140	OEM
MA44150	OEM
MA44200	OEM
MA44210	OEM
MA44220	OEM
MA44230	OEM
MA44240	OEM
MA44250	OEM
MA44300	OEM
MA44310	OEM
MA44320	OEM
MA44900A	OEM
MA44900B	OEM
MA44900C	OEM
MA44900D	OEM
MA44900E	OEM
MA44900F	OEM
MA44901A	OEM
MA44901B	OEM
MA44901C	OEM
MA44901D	OEM
MA44901E	OEM
MA44901F	OEM
MA44902A	OEM
MA44902B	OEM
MA44902C	OEM
MA44902D	OEM
MA44902E	OEM
MA44902F	OEM
MA44903A	OEM
MA44903B	OEM
MA44903C	OEM
MA44903D	OEM
MA44903E	OEM
MA44903F	OEM
MA44904A	OEM
MA44904B	OEM
MA44904C	OEM
MA44904D	OEM
MA44904E	OEM
MA44904F	OEM
MA44905A	OEM
MA44905B	OEM
MA44905C	OEM
MA44905D	OEM
MA44905E	OEM
MA44905F	OEM
MA44906A	OEM
MA44906B	OEM
MA44906C	OEM
MA44906D	OEM
MA44906E	OEM
MA44906F	OEM
MA44907A	OEM
MA44907B	OEM
MA44907C	OEM
MA44907D	OEM
MA44907E	OEM
MA44907F	OEM
MA44908A	OEM
MA44909A	OEM
MA44910A	OEM
MA44911A	OEM
MA45000	OEM
MA45001	OEM
MA45002	OEM
MA45003	OEM
MA45004	OEM
MA45005	OEM
MA45006	OEM
MA45007	OEM
MA45008	OEM
MA45009	OEM
MA45010	OEM
MA45011	OEM
MA45012	OEM
MA45013	OEM
MA45014	OEM
MA45015	OEM
MA45054	OEM
MA45055	OEM
MA45056	OEM
MA45062	OEM
MA45063	OEM
MA45064	OEM
MA45065	OEM
MA45066	OEM
MA45067	OEM
MA45068	OEM
MA45070	OEM
MA45071	0720
MA45072	OEM
MA45073	OEM
MA45074	OEM
MA45075	OEM
MA45101	OEM
MA45102	OEM
MA45103	OEM
MA45104	OEM
MA45105	OEM
MA45106	OEM
MA45107	OEM
MA45108	OEM
MA45109	OEM
MA45110	OEM
MA45111	0554
MA45112	0005
MA45113	OEM
MA45114	0628
MA45115	0030
MA45116	OEM
MA45121	OEM
MA45122	OEM
MA45123	OEM
MA45124	OEM
MA45125	OEM
MA45126	OEM
MA45127	OEM
MA45128	OEM
MA45129	OEM
MA45130	OEM
MA45131	0554
MA45132	0005
MA45133	OEM
MA45134	0628
MA45135	OEM
MA45136	OEM
MA45141	OEM
MA45142	OEM
MA45143	OEM
MA45144	OEM
MA45145	OEM
MA45146	OEM
MA45147	OEM
MA45148	OEM
MA45149	OEM
MA45150	OEM
MA45151	0720
MA45152	OEM
MA45153	OEM
MA45154	OEM
MA45156	OEM
MA45157	OEM
MA45158	0554
MA45159	OEM
MA45160	0005
MA45161	0720
MA45162	OEM
MA45164	OEM
MA45165	OEM
MA45166	OEM
MA45167	OEM
MA45168	0554
MA45169	OEM
MA45170	OEM
MA45171	0720
MA45172	OEM
MA45225	OEM
MA45226	OEM
MA45226C	OEM
MA45227	OEM
MA45227C	OEM
MA45228	OEM
MA45228C	OEM
MA45229	OEM
MA45229C	OEM
MA45230	OEM
MA45230C	OEM
MA45231	OEM
MA45231C	OEM
MA45232	OEM
MA45232C	OEM
MA45233	OEM
MA45233C	OEM
MA45234	OEM
MA45234C	OEM
MA45235	OEM
MA45235C	OEM
MA45236	OEM
MA45236C	OEM
MA45237	OEM
MA45237C	OEM
MA45238	OEM
MA45238C	OEM
MA45239C	OEM
MA45240C	OEM
MA45241C	OEM
MA45242C	OEM
MA45245	OEM
MA45246	OEM
MA45246C	OEM
MA45247	OEM
MA45247C	OEM
MA45248	OEM
MA45248C	OEM
MA45249	OEM
MA45249C	OEM
MA45250	OEM
MA45250C	OEM
MA45251	OEM
MA45251C	OEM
MA45252	OEM
MA45252C	OEM
MA45253	OEM
MA45253C	OEM
MA45254	OEM
MA45254C	OEM
MA45255	OEM
MA45255C	OEM
MA45256	OEM
MA45256C	OEM
MA45257	OEM
MA45257C	OEM

If replacement code is OEM, contact original manufacturer for replacement.

DEVICE TYPE	REPL CODE
MA45258	OEM
MA45258C	OEM
MA45259	OEM
MA45259C	OEM
MA45260	OEM
MA45260C	OEM
MA45261	OEM
MA45261C	OEM
MA45262	OEM
MA45262C	OEM
MA45263	OEM
MA45263C	OEM
MA45264	OEM
MA45264C	OEM
MA45265	OEM
MA45265C	OEM
MA45266	OEM
MA45266C	OEM
MA45267	OEM
MA45267C	OEM
MA45268	OEM
MA45268C	OEM
MA45269	OEM
MA45269C	OEM
MA45270	OEM
MA45270C	OEM
MA45271	OEM
MA45271C	OEM
MA45272	OEM
MA45272C	OEM
MA45273	OEM
MA45274	OEM
MA45275	OEM
MA45276	OEM
MA45277	OEM
MA45278	OEM
MA45279	OEM
MA45280	OEM
MA45290	OEM
MA45290C	OEM
MA45291	OEM
MA45291C	OEM
MA45292	OEM
MA45293	OEM
MA45293C	OEM
MA45294	OEM
MA45294C	OEM
MA45295	OEM
MA45295C	OEM
MA45296	OEM
MA45296C	OEM
MA45297	OEM
MA45297C	OEM
MA45298	OEM
MA45298C	OEM
MA45299	OEM
MA45299C	OEM
MA45330	OEM
MA45331	OEM
MA45332	OEM
MA45333	OEM
MA45334	OEM
MA45335	OEM
MA45336	OEM
MA45337	OEM
MA45338	OEM
MA45339	OEM
MA45340	OEM
MA45341	OEM
MA45342	OEM
MA45343	OEM
MA45345	OEM
MA45346	OEM
MA45347	OEM
MA45348	OEM
MA45349	OEM
MA45350	OEM
MA45351	OEM
MA45352	OEM
MA45353	OEM
MA45354	OEM
MA45355	OEM
MA45356	OEM
MA45357	OEM
MA45358	OEM
MA45360	OEM
MA45361	OEM
MA45362	OEM
MA45363	OEM
MA45364	OEM
MA45365	OEM
MA45366	OEM
MA45367	OEM
MA45368	OEM
MA45369	OEM
MA45370	OEM
MA45371	OEM
MA45372	OEM
MA46021	OEM
MA46022	OEM
MA46023	OEM
MA46024	OEM
MA46025	OEM
MA46026	OEM
MA46027	OEM
MA46028	OEM
MA46029	OEM
MA46030	OEM
MA46031	OEM
MA46032	OEM
MA46033	OEM
MA46034	OEM
MA46035	OEM
MA46520A	OEM
MA46520B	OEM
MA46520C	OEM
MA46520D	OEM
MA46522A	OEM
MA46522B	OEM
MA46522C	OEM
MA46522D	OEM
MA46523A	OEM
MA46523B	OEM
MA46523C	OEM
MA46523D	OEM
MA46524A	OEM
MA46524B	OEM
MA46524C	OEM
MA46524D	OEM
MA46525A	OEM
MA46525B	OEM
MA46525C	OEM
MA46525D	OEM
MA46533A	OEM
MA46533B	OEM
MA46533C	OEM
MA46533D	OEM
MA46533E	OEM
MA46533F	OEM
MA46534A	OEM
MA46534B	OEM
MA46534C	OEM
MA46534D	OEM
MA46534E	OEM
MA46534F	OEM
MA46540C	OEM
MA46540D	OEM
MA46540E	OEM
MA46540F	OEM
MA46541C	OEM
MA46541D	OEM
MA46541E	OEM
MA46541F	OEM
MA46542C	OEM
MA46542D	OEM
MA46542E	OEM
MA46543B	OEM
MA46543C	OEM
MA46543D	OEM
MA46543E	OEM
MA46544B	OEM
MA46544C	OEM
MA46544D	OEM
MA46545B	OEM
MA46545C	OEM
MA46545D	OEM
MA46546A	OEM
MA46546B	OEM
MA46546C	OEM
MA46547A	OEM
MA46547B	OEM
MA46547C	OEM
MA46550C	OEM
MA46550D	OEM
MA46550E	OEM
MA46550F	OEM
MA46551C	OEM
MA46551D	OEM
MA46551E	OEM
MA46551F	OEM
MA46552C	OEM
MA46556A	OEM
MA46556B	OEM
MA46556C	OEM
MA46557A	OEM
MA46557B	OEM
MA46557C	OEM
MA46560C	OEM
MA46560D	OEM
MA46560E	OEM
MA46560F	OEM
MA46561C	OEM
MA46561D	OEM
MA46561E	OEM
MA46561F	OEM
MA46562C	OEM
MA46562D	OEM
MA46562E	OEM
MA46563B	OEM
MA46563C	OEM
MA46563D	OEM
MA46563E	OEM
MA46564B	OEM
MA46564C	OEM
MA46564D	OEM
MA46565B	OEM
MA46565C	OEM
MA46565D	OEM
MA46566A	OEM
MA46566B	OEM
MA46566C	OEM
MA46567A	OEM
MA46567B	OEM
MA46567C	OEM
MA46570C	OEM
MA46570D	OEM
MA46570E	OEM
MA46570F	OEM
MA46571C	OEM
MA46571D	OEM
MA46571E	OEM
MA46571F	OEM
MA46572C	OEM
MA46572D	OEM
MA46572E	OEM
MA46573B	OEM
MA46573C	OEM
MA46573D	OEM
MA46573E	OEM
MA46574B	OEM
MA46574C	OEM
MA46574D	OEM
MA46575B	OEM
MA46575C	OEM
MA46575D	OEM
MA46576A	OEM
MA46576B	OEM
MA46576C	OEM
MA46577A	OEM
MA46577B	OEM
MA46577C	OEM
MA46600D	OEM
MA46600E	OEM
MA46600F	OEM
MA46600G	OEM
MA46600H	OEM
MA46600J	OEM
MA46601D	OEM
MA46601E	OEM
MA46601F	OEM
MA46601G	OEM
MA46602D	OEM
MA46602E	OEM
MA46602F	OEM
MA46603D	OEM
MA46603E	OEM
MA46604D	OEM
MA46610D	OEM
MA46610E	OEM
MA46610F	OEM
MA46610G	OEM
MA46610H	OEM
MA46611D	OEM
MA46611E	OEM
MA46611F	OEM
MA46611G	OEM
MA46612D	OEM
MA46612E	OEM
MA46612F	OEM
MA46613D	OEM
MA46613E	OEM
MA46614D	OEM
MA46620D	OEM
MA46620E	OEM
MA46620F	OEM
MA46620G	OEM
MA46621D	OEM
MA46621E	OEM
MA46621F	OEM
MA46622D	OEM
MA46622E	OEM
MA46623D	OEM
MA46623E	OEM
MA46624D	OEM
MA47000	OEM
MA47001	OEM
MA47002	OEM
MA47003	OEM
MA47004	OEM
MA47005	OEM
MA47006	OEM
MA47007	OEM
MA47008	OEM
MA47009	OEM
MA47010	OEM
MA47011	OEM
MA47012	OEM
MA47013	OEM
MA47014	OEM
MA47015	OEM
MA47016	OEM
MA47017	OEM
MA47018	OEM
MA47019	OEM
MA47020	OEM
MA47021	OEM
MA47022	OEM
MA47023	OEM
MA47024	OEM
MA47025	OEM
MA47026	OEM
MA47027	OEM
MA47028	OEM
MA47029	OEM
MA47030	OEM
MA47031	OEM
MA47032	OEM
MA47033	OEM
MA47034	OEM
MA47035	OEM
MA47036	OEM
MA47037	OEM
MA47038	OEM
MA47039	OEM
MA47040	OEM
MA47041	OEM
MA47042	OEM
MA47043	OEM
MA47044	OEM
MA47045	OEM
MA47046	OEM
MA47047	OEM
MA47051	OEM
MA47052	OEM
MA47053	OEM
MA47054	OEM
MA47057	OEM
MA47068	OEM
MA47070	OEM
MA47075	OEM
MA47076	OEM
MA47077	OEM
MA47078	OEM
MA47079	OEM
MA47080	OEM
MA47081	OEM
MA47082	OEM
MA47083	OEM
MA47084	OEM
MA47085	OEM
MA47086	OEM
MA47088	OEM
MA47089	OEM
MA47090	OEM
MA47091	OEM
MA47100	OEM
MA47101	OEM
MA47102	OEM
MA47103	OEM
MA47104	OEM
MA47105	OEM
MA47106	OEM
MA47107	OEM
MA47108	OEM
MA47109	OEM
MA47110	OEM
MA47111	OEM
MA47120	OEM
MA47121	OEM
MA47122	OEM
MA47123	OEM
MA47126	OEM
MA47127	OEM
MA47128	OEM
MA47129	OEM
MA47130	OEM
MA47160	OEM
MA47161	OEM
MA47162	OEM
MA47200	OEM
MA47201	OEM
MA47202	OEM
MA47203	OEM
MA47204	OEM
MA47205	OEM
MA47206	OEM
MA47207	OEM
MA47208	OEM
MA47220	OEM
MA47221	OEM
MA47222	OEM
MA47223	OEM
MA47266	OEM
MA47301	OEM
MA47302	OEM
MA47400	OEM
MA47401	OEM
MA47402	OEM
MA47403	OEM
MA47404	OEM
MA47405	OEM
MA47406	OEM
MA47407	OEM
MA47408	OEM
MA47409	OEM
MA47410	OEM
MA47411	OEM
MA47412	OEM
MA47413	OEM
MA47414	OEM
MA47416	OEM
MA47418	OEM
MA47419	OEM
MA47420	OEM
MA47421	OEM
MA47422	OEM
MA47423	OEM
MA47424	OEM
MA47425	OEM
MA47426	OEM
MA47427	OEM
MA47600	OEM
MA47865	OEM
MA47867	OEM
MA47868	OEM
MA47869	OEM
MA47870	OEM
MA47876	OEM
MA47878	OEM
MA47879	OEM
MA47880	OEM
MA47881	OEM
MA47883	OEM
MA47890-131	OEM
MA47890-150	OEM
MA47891-109	OEM
MA47891-131	OEM
MA47891-150	OEM
MA47892-109	OEM
MA47892-131	OEM
MA47892-150	OEM
MA47893-30	OEM
MA47893-109	OEM
MA47893-131	OEM
MA47894-30	OEM
MA47894-109	OEM
MA47894-131	OEM
MA47895-30	OEM
MA47895-54	OEM
MA47895-109	OEM
MA47895-131	OEM
MA47896-30	OEM
MA47896-54	OEM
MA47896-131	OEM
MA47897-30	OEM
MA47897-54	OEM
MA47897-131	OEM
MA47898-30	OEM
MA47898-54	OEM
MA47898-131	OEM
MA47899-30	OEM
MA47899-54	OEM
MA47899-131	OEM
MA48501A	OEM
MA48501B-F	OEM
MA48501G	OEM
MA48501H-K	OEM
MA48502A	OEM
MA48502B-F	OEM
MA48502G	OEM
MA48502H-K	OEM
MA48503A	OEM
MA48503B-F	OEM
MA48503G	OEM
MA48503H	OEM
MA48503I	OEM
MA48504A	OEM
MA48504B-F	OEM
MA48504G	OEM
MA48504H	OEM
MA48505A	OEM
MA48505B-F	OEM
MA48505G	OEM
MA48506A	OEM
MA48506B-F	OEM
MA48507A	OEM
MA48507B-F	OEM
MA48508A	OEM
MA48508B-E	OEM
MA48509A	OEM
MA48509B-E	OEM
MA48510A	OEM
MA48510B-E	OEM
MA48511A	OEM
MA48511B-E	OEM
MA48701A	OEM
MA48701B-F	OEM
MA48701G	OEM
MA48701H-K	OEM
MA48702A	OEM
MA48702B-F	OEM
MA48702G	OEM
MA48702H-K	OEM
MA48703A	OEM
MA48703B-F	OEM
MA48703G	OEM
MA48703H-K	OEM
MA48704A	OEM
MA48704B-F	OEM
MA48704G	OEM
MA48704H-J	OEM
MA48705A	OEM
MA48705B-F	OEM
MA48705G	OEM
MA48705H	OEM
MA48705I	OEM
MA48706A	OEM
MA48706AB-F	OEM
MA48707A	OEM
MA48707B-F	OEM
MA48708A	OEM
MA48708B-E	OEM
MA48709A	OEM
MA48709B-E	OEM
MA48710A	OEM
MA48710B-E	OEM
MA48711A	OEM
MA48711B-E	OEM
MA49104	OEM
MA49105	OEM
MA49107	OEM
MA49109	OEM
MA49110	OEM
MA49117-118	OEM
MA49121	OEM
MA49122	OEM
MA49123	OEM
MA49124	OEM
MA49126-118	OEM
MA49126-138	OEM
MA49128-138	OEM
MA49135	OEM
MA49136	OEM
MA49137	OEM
MA49138	OEM
MA49139	OEM
MA49140-118	OEM
MA49145	OEM
MA49146	OEM
MA49147	OEM
MA49151	OEM
MA49152	OEM
MA49153	OEM
MA49154	OEM
MA49156	OEM
MA49157	OEM
MA49158	OEM
MA49159	OEM
MA49161	OEM
MA49162	OEM
MA49163	OEM
MA49164	OEM
MA49166	OEM
MA49167	OEM
MA49168	OEM
MA49169	OEM
MA49170	OEM
MA49171	OEM
MA49172	OEM
MA49172-138	OEM
MA49173	OEM
MA49173-138	OEM
MA49177-138	OEM
MA49178-118	OEM
MA49179-118	OEM
MA49179-138	OEM
MA49180-118	OEM
MA49180-138	OEM
MA49181-138	OEM
MA49182-138	OEM
MA49183	OEM
MA49184	OEM
MA49185	OEM
MA49186	OEM
MA49187	OEM
MA49188	OEM
MA49189	OEM
MA49190	OEM
MA49191	OEM
MA49192	OEM
MA49193	OEM
MA49195	OEM
MA49260	OEM
MA49265	OEM
MA49508	OEM
MA49618	OEM
MA49628	OEM
MA51355P	OEM
MA60391CP	OEM
MA-80391CP	OEM
MA4914848	OEM
MAB8021	OEM
MAB8421P-F016	OEM
MAB8421PF016	OEM
MAB8421PF047	OEM
MAB8421P-F065	OEM
MAB8441P	OEM
MAB8441P/T158	OEM
MAB8441P-T012	OEM
MAB8441P-T018	OEM
MAB8441P-T023	OEM
MAB8441P-T042	OEM
MAB8441P-T050	OEM
MAB8441P-T065	OEM
MAB8441P-T082	OEM
MAB8441P-T105	OEM
MAB8441P-T107	OEM
MAB8441P-T120	OEM
MAB-8461P-W077	OEM
MAB8461P-W077	OEM
MAB8461P-W086	OEM
MAC2-1	0399
MAC2-2	0404
MAC2-3	0408
MAC2-4	0411
MAC2-5	0415
MAC2-6	0415
MAC2-8	0422
MAC5-1	0399
MAC5-2	0404
MAC5-3	0408
MAC5-4	0411
MAC5-5	0415
MAC5-6	0415
MAC5-7	OEM
MAC5-8	0422
MAC5-T	1056
MAC6-1	OEM
MAC6-2	OEM
MAC6-3	OEM
MAC6-4	OEM
MAC6-5	OEM
MAC6-6	OEM
MAC6-7	OEM
MAC6-8	OEM
MAC7C7.5D5	OEM
MAC7C7.5D10	OEM
MAC10-1	0593
MAC10-2	0593
MAC10-3	0954
MAC10-4	0572
MAC10-5	0957
MAC10-6	0960
MAC10-7	0961
MAC10-8	0966
MAC11-1	0593
MAC11-2	1089
MAC11-3	1093
MAC11-4	0411
MAC11-5	1095
MAC11-6	0415
MAC11-7	1099
MAC11-8	1100
MAC15-4	0772
MAC15-4FP	2202
MAC15-6	0744
MAC15-6FP	2202
MAC15-8	0617
MAC15-8FP	2202
MAC15-10	0875
MAC15-10FP	2202
MAC15A-4	0772
MAC15A4	0772
MAC15A4FP	2202
MAC15A-6	0744
MAC15A6	0744
MAC15A6FP	2202
MAC15A-8	0875
MAC15A8	0617
MAC15A8FP	2202
MAC15A-10	0875
MAC15A10	0875
MAC15A10FP	2202
MAC15A-B	OEM
MAC20-4	0908
MAC20-5	0908
MAC20-6	0908
MAC20-7	0908
MAC20-8	0908
MAC20-9	OEM
MAC20-10	OEM
MAC20A-4	5673
MAC20A4	0908
MAC20A-5	5673
MAC20A5	0908
MAC20A-6	5673
MAC20A6	0908
MAC20A-7	5673
MAC20A7	0908
MAC20A-8	5673
MAC20A8	0908
MAC20A-9	OEM
MAC20A9	OEM
MAC20A-10	OEM
MAC20A10	OEM
MAC21-1	OEM
MAC21-2	OEM
MAC21-3	OEM
MAC21-4	OEM
MAC21-5	OEM
MAC21-6	OEM
MAC21-7	OEM
MAC25-4	2348
MAC25-5	2348
MAC25-6	2348
MAC25-7	0908
MAC25-8	0908
MAC25-9	OEM
MAC25-10	OEM
MAC25A-4	5673
MAC25A4	0908
MAC25A-5	5673
MAC25A5	2348
MAC25A-6	5673
MAC25A6	0908
MAC25A-7	5673
MAC25A7	0908
MAC25A-8	5673
MAC25A8	0908
MAC25A-9	OEM
MAC25A9	OEM
MAC25A-10	OEM
MAC25A10	OEM
MAC35-1	1064
MAC35-2	1064
MAC35-3	1064
MAC35-4	1064
MAC35-5	1317
MAC35-6	1317
MAC35-7	1891
MAC35-8	1891
MAC35-9	OEM
MAC35-10	OEM
MAC36-1	3146
MAC36-2	3147
MAC36-3	3148
MAC36-4	3150
MAC36-5	3151
MAC36-6	3152
MAC36-7	3153
MAC36-8	2018
MAC36-9	OEM
MAC36-10	OEM
MAC37-1	1064
MAC37-2	3153
MAC37-3	1064
MAC37-4	1064
MAC37-5	1317
MAC37-6	1317
MAC37-7	1891
MAC37-8	OEM
MAC37-9	OEM
MAC37-10	OEM

If replacement code is OEM, contact original manufacturer for replacement.

DEVICE TYPE	REPL CODE
MAC38-1	3234
MAC38-2	2015
MAC38-3	2015
MAC38-4	2015
MAC38-5	2017
MAC38-6	2017
MAC38-7	2018
MAC38-8	3238
MAC38-9	OEM
MAC38-10	OEM
MAC50-4	0908
MAC50-5	0908
MAC50-6	0908
MAC50-7	0908
MAC50-8	0908
MAC50-9	OEM
MAC50-10	OEM
MAC50A-4	5673
MAC50A4	0908
MAC50A-5	5673
MAC50A5	0908
MAC50A-6	5673
MAC50A6	0908
MAC50A-7	5673
MAC50A7	0908
MAC50A-8	5673
MAC50A8	0908
MAC50A-9	OEM
MAC50A9	OEM
MAC50A-10	OEM
MAC50A10	OEM
MAC77-1	4412
MAC77-2	1755
MAC77-3	4414
MAC77-4	3499
MAC77-5	4121
MAC77-6	0485
MAC77-7	4416
MAC77-8	4417
MAC80-6	OEM
MAC80-8	OEM
MAC81-6	OEM
MAC81-8	OEM
MAC91-1	OEM
MAC91-2	OEM
MAC91-3	OEM
MAC91-4	OEM
MAC91-5	OEM
MAC91-6	OEM
MAC91-7	OEM
MAC91-8	OEM
MAC91A-1	OEM
MAC91A1	OEM
MAC91A-2	OEM
MAC91A2	OEM
MAC91A-3	OEM
MAC91A3	OEM
MAC91A-4	OEM
MAC91A4	OEM
MAC91A-5	OEM
MAC91A5	OEM
MAC91A-6	OEM
MAC91A6	OEM
MAC91A-7	OEM
MAC91A7	OEM
MAC91A-8	OEM
MAC91A8	OEM
MAC92-1	4455
MAC92-2	4455
MAC92-3	4455
MAC92-4	4455
MAC92-5	4456
MAC92-6	4456
MAC92-7	4482
MAC92-8	4482
MAC92A-1	1934
MAC92A1	4455
MAC92A-2	1934
MAC92A2	4455
MAC92A-3	1934
MAC92A3	4455
MAC92A-4	1934
MAC92A4	4455
MAC92A-5	1935
MAC92A5	4456
MAC92A-6	1935
MAC92A6	4456
MAC92A-7	1937
MAC92A7	4482
MAC92A-8	1937
MAC92A8	4482
MAC93-1	OEM
MAC93-2	OEM
MAC93-3	OEM
MAC93-4	OEM
MAC93-5	OEM
MAC93-6	OEM
MAC93-7	OEM
MAC93-8	OEM
MAC93A-1	OEM
MAC93A1	OEM
MAC93A-2	OEM
MAC93A2	OEM
MAC93A-3	OEM
MAC93A3	OEM
MAC93A-4	OEM
MAC93A4	OEM
MAC93A-5	OEM
MAC93A5	OEM

DEVICE TYPE	REPL CODE
MAC93A-6	OEM
MAC93A6	OEM
MAC93A-7	OEM
MAC93A7	OEM
MAC93A-8	OEM
MAC93A8	OEM
MAC94-1	4455
MAC94-2	4455
MAC94-3	4455
MAC94-4	4455
MAC94-5	4482
MAC94-6	4456
MAC94-7	4482
MAC94-8	4482
MAC94-A2	4455
MAC94A1	4455
MAC94A2	4455
MAC94A3	4455
MAC94A4	4455
MAC94A5	4456
MAC94A6	4456
MAC94A7	4482
MAC94A8	4482
MAC95-1	4455
MAC95-2	4455
MAC95-3	4455
MAC95-4	4456
MAC95-5	4456
MAC95-6	4482
MAC95-7	4482
MAC95-8	4482
MAC95A1	4455
MAC95A2	4455
MAC95A3	4455
MAC95A4	4455
MAC95A5	4456
MAC95A6	4456
MAC95A7	4482
MAC95A8	4482
MAC96-1	OEM
MAC96-2	OEM
MAC96-3	OEM
MAC96-4	OEM
MAC96-5	OEM
MAC96-6	OEM
MAC96-7	OEM
MAC96-8	OEM
MAC96A1	OEM
MAC96A2	OEM
MAC96A3	OEM
MAC96A4	OEM
MAC96A5	OEM
MAC96A6	OEM
MAC96A7	OEM
MAC96A8	OEM
MAC97-4	5013
MAC154	OEM
MAC156	OEM
MAC158	OEM
MAC216-4	OEM
MAC216-6	OEM
MAC216-7	OEM
MAC216-8	OEM
MAC216A-4	OEM
MAC216A4	OEM
MAC216A-6	OEM
MAC216A6	OEM
MAC216A-7	OEM
MAC216A7	OEM
MAC216A-8	OEM
MAC216A8	OEM
MAC218-2	OEM
MAC218-3	OEM
MAC218-4	0614
MAC218-5	OEM
MAC218-6	3235
MAC218-7	OEM
MAC218-8	2145
MAC218-8FP	3593
MAC218-9	OEM
MAC218-10	0596
MAC218-10FP	3593
MAC218A2	OEM
MAC218A3	OEM
MAC218A-4	OEM
MAC218A4	0614
MAC218A-5	OEM
MAC218A5	OEM
MAC218A-6	OEM
MAC218A6	3235
MAC218A-7	OEM
MAC218A7	OEM
MAC218A-8	OEM
MAC218A8	2145
MAC218A8FP	3593
MAC218A-9	OEM
MAC218A9	OEM
MAC218A-10	OEM
MAC218A10	0596
MAC218A10FP	3593
MAC220-2	5394
MAC220-3	5394
MAC220-5	5394
MAC220-7	0596
MAC220-9	0596
MAC221-2	5394
MAC221-3	5394
MAC221-5	5394
MAC221-7	0596
MAC221-9	0596

DEVICE TYPE	REPL CODE
MAC222-1	OEM
MAC222-2	OEM
MAC222-3	OEM
MAC222-4	OEM
MAC222-5	OEM
MAC222-6	OEM
MAC222-7	OEM
MAC222-8	OEM
MAC222-9	OEM
MAC222-10	OEM
MAC222A-1	OEM
MAC222A1	OEM
MAC222A-2	OEM
MAC222A2	OEM
MAC222A-3	OEM
MAC222A3	OEM
MAC222A-4	OEM
MAC222A4	OEM
MAC222A-5	OEM
MAC222A5	OEM
MAC222A-6	OEM
MAC222A6	OEM
MAC222A-7	OEM
MAC222A7	OEM
MAC222A-8	OEM
MAC222A8	OEM
MAC222A-9	OEM
MAC222A9	OEM
MAC222A-10	OEM
MAC222A10	OEM
MAC223-3	5323
MAC223-4	5323
MAC223-4FP	3467
MAC223-5	5326
MAC223-6	5326
MAC223-6FP	3467
MAC223-7	1659
MAC223-8	1659
MAC223-8FP	3467
MAC223-9	1967
MAC223-10	1967
MAC223-10FP	3467
MAC223A-3	OEM
MAC223A3	5323
MAC223A-4	5323
MAC223A4	5323
MAC223A4FP	3467
MAC223A-5	5326
MAC223A5	5326
MAC223A-6	5326
MAC223A6	5326
MAC223A6FP	3467
MAC223A-7	OEM
MAC223A7	1659
MAC223A-8	1967
MAC223A8	1659
MAC223A8FP	3467
MAC223A-9	1967
MAC223A9	1967
MAC223A-10	1967
MAC223A10	1967
MAC223A10FP	0793
MAC224-4	1136
MAC224-5	OEM
MAC224-6	1136
MAC224-7	OEM
MAC224-8	1136
MAC224-9	OEM
MAC224-10	OEM
MAC224-A10	6152
MAC224A-4	OEM
MAC224A4	1136
MAC224A-6	OEM
MAC224A6	1136
MAC224A-7	OEM
MAC224A7	OEM
MAC224A-8	OEM
MAC224A8	1136
MAC224A-9	OEM
MAC224A9	OEM
MAC224A-10	OEM
MAC224A10	1136
MAC225A5	OEM
MAC228-2	OEM
MAC228-3	OEM
MAC228-4	OEM
MAC228-5	OEM
MAC228-6	0572
MAC228-7	OEM
MAC228-8	0935
MAC228-9	OEM
MAC228-10	0059
MAC228A2	OEM
MAC228A3	OEM
MAC228A4	5534
MAC228A5	OEM
MAC228A6	5534
MAC228A7	OEM
MAC228A8	6138
MAC228A9	OEM
MAC228A10	0059
MAC229-4	5534
MAC229-6	6138
MAC229-8	0851
MAC229-10	0851
MAC229A4	5534
MAC229A6	6138
MAC229A8	0851
MAC229A10	0059

DEVICE TYPE	REPL CODE
MAC244A-5	OEM
MAC320-4	5323
MAC320-4FP	0793
MAC320-6	5326
MAC320-6FP	0793
MAC320-8	1659
MAC320-8FP	0793
MAC320-10	1967
MAC320-10FP	0793
MAC320A4	5323
MAC320A4FP	0793
MAC320A6	5326
MAC320A6FP	0793
MAC320A8	1659
MAC320A8FP	0793
MAC320A10	1967
MAC320A10FP	0793
MAC515-4	OEM
MAC515-5	OEM
MAC515-6	OEM
MAC515-7	OEM
MAC515-8	OEM
MAC515-9	OEM
MAC515-10	OEM
MAC515A-4	OEM
MAC515A4	OEM
MAC515A-5	OEM
MAC515A5	OEM
MAC515A-6	OEM
MAC515A6	OEM
MAC515A-7	OEM
MAC515A7	OEM
MAC515A-8	OEM
MAC515A8	OEM
MAC515A-9	OEM
MAC515A9	OEM
MAC515A-10	OEM
MAC525-4	OEM
MAC525-5	OEM
MAC525-6	OEM
MAC525-7	OEM
MAC525-8	OEM
MAC525-9	OEM
MAC525-10	OEM
MAC525A-4	OEM
MAC525A4	OEM
MAC525A-5	OEM
MAC525A5	OEM
MAC525A-6	OEM
MAC525A6	OEM
MAC525A-7	OEM
MAC525A7	OEM
MAC525A-8	OEM
MAC525A8	OEM
MAC525A-9	OEM
MAC525A9	OEM
MAC525A-10	OEM
MAC525A10	OEM
MAC800-02	OEM
MAC800-05	OEM
MAC800-10	OEM
MAC800-20	OEM
MAC800-40	OEM
MAC800-60	OEM
MAC800-80	OEM
MAC800A02	OEM
MAC800A05	OEM
MAC800A10	OEM
MAC800A20	OEM
MAC800A40	OEM
MAC800A60	OEM
MAC800A80	OEM
MAC800B02	OEM
MAC800B05	OEM
MAC800B10	OEM
MAC800B20	OEM
MAC800B40	OEM
MAC800B60	OEM
MAC800B80	OEM
MAC3010-4	5575
MAC3010-8	5394
MAC3010-15	3235
MAC3010-25	0306
MAC3010-40	5657
MAC3010-40I	OEM
MAC3010-401	4335
MAC3020-4	5575
MAC3020-8	4335
MAC3020-15	3235
MAC3020-25	0306
MAC3020-40	3245
MAC3020-40I	OEM
MAC3020-401	4335
MAC3030-4	5575
MAC3030-8	5394
MAC3030-15	3235
MAC3030-25	0306
MAC3030-40	3245
MAC3030-40I	OEM
MAC3030-401	4335
MAC3040-4	5575
MAC3040-8	5394
MAC3040-15	3235
MAC3040-25	0306
MAC3040-40	3245
MAC3040-40I	OEM
MAC3040-401	4335
MAC4688	3199
MAC4689	3207

DEVICE TYPE	REPL CODE
MAC4690	3222
MAC5441	1064
MAC5442	1317
MAC5443	1891
MAC5444	2015
MAC5445	2017
MAC5446	2018
MAC5569	0154
MAC5570	0147
MAC5573	0154
MAC5574	0147
MAC40688	3199
MAC40689	3207
MAC40690	3222
MAC40795	1891
MAC40796	0279
MAC40797	1891
MAC40798	0279
MAC40799	3199
MAC40800	3207
MAC40801	3222
MACAZ28D5	OEM
MACCZA19D5	OEM
MACZ9.1D	0057
MACZ9.1D5	0057
MACZ9.1D10	0057
MACZA6.8D	0062
MACZA6.8D5	0062
MACZA6.8D10	0062
MACZA7.5D	0077
MACZA7.5D5	0077
MACZA7.5D10	0077
MACZA8.2D	0165
MACZA8.2D5	0165
MACZA8.2D10	0165
MACZA9.1D	0057
MACZA9.1D5	0057
MACZA9.1D10	0057
MACZA10D	0064
MACZA10D1010	OEM
MACZA11D	0181
MACZA11D5	0181
MACZA11D10	0181
MACZA12D	0052
MACZA12D5	0052
MACZA12D10	0052
MACZA13D	0053
MACZA13D5	0053
MACZA13D10	0053
MACZA14D	0879
MACZA14D5	0879
MACZA14D10	0879
MACZA15D	0686
MACZA15D5	0686
MACZA15D10	0686
MACZA16D	0445
MACZA16D5	0445
MACZA16D10	0445
MACZA17D	0210
MACZA17D5	0210
MACZA17D10	0210
MACZA18D	0375
MACZA18D5	0375
MACZA18D10	0375
MACZA19D	0671
MACZA19D5	0671
MACZA19D10	0671
MACZA20D	0700
MACZA20D5	0700
MACZA20D10	0700
MACZA22D	0705
MACZA22D5	0705
MACZA22D10	0705
MACZA24D	0494
MACZA24D5	0494
MACZA24D1010	0494
MACZA25D	0714
MACZA25D5	0714
MACZA25D10	0714
MACZA27D	0455
MACZA27D5	0455
MACZA27D10	0455
MACZA28D	0257
MACZA28D10	0257
MACZA30D	0195
MACZA30D5	0195
MACZA30D10	0195
MACZA33D	0166
MACZA33D5	0166
MACZA33D10	0166
MACZA36D	0010
MACZA36D5	0010
MACZA36D10	0010
MACZA39D	0032
MACZA39D5	0032
MACZA39D10	0032
MACZA43D	0054
MACZA43D5	0054
MACZA43D10	0054
MACZA47D	0068
MACZA47D5	0068
MACZA47D10	0068
MACZA51D	0092
MACZA51D5	0092
MACZA51D10	0092
MACZA56D	0125
MACZA56D5	0125
MACZA56D10	0125
MACZA60D	2317

DEVICE TYPE	REPL CODE
MACZA60D5	2317
MACZA60D10	2317
MACZA62D	0152
MACZA62D5	0152
MACZA62D10	0152
MACZA68D	0173
MACZA68D5	0173
MACZA68D10	0173
MACZA75D	0094
MACZA75D5	0094
MACZA75D10	0094
MACZA82D	0049
MACZA82D5	0049
MACZA82D1010	0049
MACZA87D	0104
MACZA87D5	0104
MACZA87D10	0104
MACZA91D	0156
MACZA91D5	0156
MACZA91D10	0156
MACZA100D	0189
MACZA100D5	0189
MACZA100D10	0189
MACZA105D	0099
MACZA105D5	0099
MACZA105D10	0099
MACZA110D	0099
MACZA110D5	0099
MACZA110D10	0099
MACZA120D	0089
MACZA120D5	0089
MACZA120D10	0089
MACZA130D	0455
MACZA130D5	0455
MACZA130D10	0455
MACZA140D	0252
MACZA140D5	0252
MACZA140D10	0252
MACZA150D	0340
MACZA150D5	0340
MACZA150D10	0340
MACZA160D	0370
MACZA160D5	0370
MACZA160D10	0370
MACZA175D	0425
MACZA175D5	0425
MACZA175D10	0425
MACZA180D	0425
MACZA180D5	0425
MACZA180D10	0425
MACZA200D	1473
MACZA200D5	1473
MACZA200D10	1473
MACZB6.8D	0062
MACZB6.8D5	0062
MACZB6.8D10	0062
MACZB7.5D	0077
MACZB7.5D5	0077
MACZB7.5D10	0077
MACZB8.2D	0165
MACZB8.2D5	0165
MACZB8.2D10	0165
MACZB9.1D	0057
MACZB9.1D5	0057
MACZB9.1D10	0057
MACZB10D	0064
MACZB10D5	0064
MACZB10D10	0064
MACZB11D	0181
MACZB11D5	0181
MACZB11D10	0181
MACZB12D	0052
MACZB12D5	0052
MACZB12D10	0052
MACZB13D	0053
MACZB13D5	0053
MACZB13D10	0053
MACZB14D	0879
MACZB14D5	0879
MACZB14D10	0879
MACZB15D	0686
MACZB15D5	0686
MACZB15D10	0686
MACZB16D	0445
MACZB16D5	0445
MACZB16D10	0445
MACZB17D	0210
MACZB17D5	0210
MACZB17D10	0210
MACZB18D	0375
MACZB18D5	0375
MACZB18D10	0375
MACZB19D	0671
MACZB19D5	0671
MACZB19D10	0671
MACZB20D	0700
MACZB20D5	0700
MACZB20D10	0700
MACZB22D	0705
MACZB22D5	0705
MACZB22D10	0705
MACZB24D	0494
MACZB24D5	0494
MACZB24D10	0494
MACZB25D	0714
MACZB25D5	0714
MACZB25D10	0714
MACZB27D	0455
MACZB27D5	0455

DEVICE TYPE	REPL CODE
MACZB27D10	0455
MACZB30D	0195
MACZB30D5	0195
MACZB30D10	0195
MACZB33D	0166
MACZB33D5	0166
MACZB33D10	0166
MACZB36D	0010
MACZB36D5	0010
MACZB36D10	0010
MACZB39D	0032
MACZB39D5	0032
MACZB39D10	0032
MACZB43D	0054
MACZB43D5	0054
MACZB43D10	0054
MACZB47D	0068
MACZB47D5	0068
MACZB47D10	0068
MACZB51D	0092
MACZB51D5	0092
MACZB51D10	0092
MACZB56D	0125
MACZB56D5	0125
MACZB56D10	0125
MACZB62D	0152
MACZB62D5	0152
MACZB62D10	0152
MACZB68D	0173
MACZB68D5	0173
MACZB68D10	0173
MACZB75D	0094
MACZB75D5	0094
MACZB75D10	0094
MACZB82D	0049
MACZB82D5	0049
MACZB82D10	0049
MACZB91D	0156
MACZB91D5	0156
MACZB91D10	0156
MACZB100D	0189
MACZB100D5	0189
MACZB100D10	0189
MACZB105D	0099
MACZB105D5	0099
MACZB105D10	0099
MACZB110D	0099
MACZB110D5	0099
MACZB110D10	0099
MACZB120D	0089
MACZB120D5	0089
MACZB120D10	0089
MACZB130D	0287
MACZB130D5	0287
MACZB130D10	0287
MACZB140D	0252
MACZB140D5	0252
MACZB140D10	0252
MACZB150D	0340
MACZB150D5	0340
MACZB150D10	0340
MACZB160D	0370
MACZB160D5	0370
MACZB160D10	0370
MACZB175D	0425
MACZB175D5	0425
MACZB175D10	0425
MACZB180D	0425
MACZB180D5	0425
MACZB180D10	0425
MACZB200D	1473
MACZB200D5	1473
MACZB200D10	1473
MACZC6.8D	0062
MACZC6.8D5	0062
MACZC6.8D10	0062
MACZC7.5D	0077
MACZC8.2D	0165
MACZC8.2D5	0165
MACZC8.2D10	0165
MACZC9.1D	0057
MACZC9.1D5	0057
MACZC9.1D10	0057
MACZC10D	0064
MACZC10D5	0064
MACZC10D10	0064
MACZC11D	0181
MACZC11D10	0181
MACZC12D	0052
MACZC12D5	0052
MACZC12D10	0052
MACZC13D	0053
MACZC13D5	0053
MACZC13D10	0053
MACZC14D	0879
MACZC14D5	0879
MACZC14D10	0879
MACZC15D	0686
MACZC15D5	0686
MACZC15D10	0686
MACZC16D	0445
MACZC16D5	0445
MACZC16D10	0445
MACZC17D	0210
MACZC17D5	0210
MACZC17D10	0210
MACZC18D	0375
MACZC18D5	0375
MACZC18D10	0375

If replacement code is OEM, contact original manufacturer for replacement.

DEVICE TYPE	REPL CODE
MACZC19D	0671
MACZC19D5	0671
MACZC19D10	0671
MACZC20D	0700
MACZC20D5	0700
MACZC20D10	0700
MACZC22D	0705
MACZC22D10	0705
MACZC24D	0494
MACZC24D5	0494
MACZC24D10	0494
MACZC25D	0714
MACZC25D5	0714
MACZC25D10	0714
MACZC27D	0455
MACZC27D5	0455
MACZC27D10	0455
MACZC30D	0195
MACZC30D5	0195
MACZC30D10	0195
MACZC33D	0166
MACZC33D5	0166
MACZC33D10	0166
MACZC36D	0010
MACZC36D5	0010
MACZC36D10	0010
MACZC39D	0032
MACZC39D5	0032
MACZC39D10	0032
MACZC43D	0054
MACZC43D5	0054
MACZC43D10	0054
MACZC47D	0068
MACZC47D5	0068
MACZC47D10	0068
MACZC51D	0092
MACZC51D5	0092
MACZC51D10	0092
MACZC56D	0125
MACZC56D5	0125
MACZC56D10	0125
MACZC62D	0152
MACZC62D5	0152
MACZC62D10	0152
MACZC68D	0173
MACZC68D5	0173
MACZC68D10	0173
MACZC75D	0094
MACZC75D5	0094
MACZC75D10	0094
MACZC82D	0049
MACZC82D5	0049
MACZC82D10	0049
MACZC91D	0156
MACZC91D5	0156
MACZC91D10	0156
MACZC100D	0189
MACZC100D5	0189
MACZC100D10	0189
MACZC105D	0099
MACZC105D5	0099
MACZC105D10	0099
MACZC110D	0099
MACZC110D5	0099
MACZC110D10	0099
MACZC120D	0089
MACZC120D5	0089
MACZC120D10	0089
MACZC130D	0287
MACZC130D5	0287
MACZC130D10	0287
MACZC140D	0252
MACZC140D5	0252
MACZC140D10	0252
MACZC150D	0340
MACZC150D5	0340
MACZC150D10	0340
MACZC160D	0370
MACZC160D5	0370
MACZC160D10	0370
MACZC175D	0425
MACZC175D5	0425
MACZC175D10	0425
MACZC180D	0425
MACZC180D5	0425
MACZC180D10	0425
MACZC200D	1473
MACZC200D5	1473
MACZC200D10	1473
MACZC220D5	OEM
MACZD6.8D	0062
MACZD6.8D5	0062
MACZD6.8D10	0062
MACZD7.5D	0077
MACZD7.5D5	0077
MACZD7.5D10	0077
MACZD8.2D	0165
MACZD8.2D5	0165
MACZD8.2D10	0165
MACZD9.1D	0057
MACZD9.1D5	0057
MACZD9.1D10	0057
MACZD10D	0064
MACZD10D5	0064
MACZD10D10	0064
MACZD11D	0181
MACZD11D5	0181
MACZD11D10	0181
MACZD12D	0052
MACZD12D5	0052
MACZD12D10	0052
MACZD13D	0053
MACZD13D5	0053
MACZD13D10	0053
MACZD14D	0879
MACZD14D5	0879
MACZD14D10	0879
MACZD15D	0686
MACZD15D5	0686
MACZD15D10	0686
MACZD16D	0445
MACZD16D5	0445
MACZD16D10	0445
MACZD17D	0210
MACZD17D5	0210
MACZD17D10	0210
MACZD18D	0375
MACZD18D5	0375
MACZD18D10	0375
MACZD19D	0671
MACZD19D5	0671
MACZD19D10	0671
MACZD20D	0700
MACZD20D5	0700
MACZD20D10	0700
MACZD22D	0705
MACZD22D5	0705
MACZD22D10	0705
MACZD24D	0494
MACZD24D5	0494
MACZD24D10	0494
MACZD25D	0714
MACZD25D5	0714
MACZD25D10	0714
MACZD27D	0455
MACZD27D5	0455
MACZD27D10	0455
MACZD28D	0257
MACZD28D10	0257
MACZD30D	0195
MACZD30D5	0195
MACZD30D10	0195
MACZD33D	0166
MACZD33D5	0166
MACZD33D10	0166
MACZD36D	0010
MACZD36D5	0010
MACZD36D10	0010
MACZD39D	0032
MACZD39D5	0032
MACZD39D10	0032
MACZD43D	0054
MACZD43D5	0054
MACZD43D10	0054
MACZD47D	0068
MACZD47D5	0068
MACZD47D10	0068
MACZD51D	0092
MACZD51D5	0092
MACZD51D10	0092
MACZD56D	0125
MACZD56D5	0125
MACZD56D10	0125
MACZD60D	2317
MACZD60D5	2317
MACZD60D10	2317
MACZD62D	0152
MACZD62D5	0152
MACZD62D10	0152
MACZD68D	0173
MACZD68D5	0173
MACZD68D10	0173
MACZD75D	0094
MACZD75D5	0094
MACZD75D10	0094
MACZD82D	0049
MACZD82D5	0049
MACZD82D10	0049
MACZD87D	0104
MACZD87D5	0104
MACZD87D10	0104
MACZD91D	0156
MACZD91D5	0156
MACZD91D10	0156
MACZD100D	0189
MACZD100D5	0189
MACZD100D10	0189
MACZD105D	0099
MACZD105D5	0099
MACZD105D10	0099
MACZD110D	0099
MACZD110D5	0099
MACZD110D10	0099
MACZD120D	0089
MACZD120D5	0089
MACZD120D10	0089
MACZD130D	0287
MACZD130D5	0287
MACZD130D10	0287
MACZD140D	0252
MACZD140D5	0252
MACZD140D10	0252
MACZD150D	0340
MACZD150D5	0340
MACZD150D10	0340
MACZD160D	0370
MACZD160D5	0370
MACZD160D10	0370
MACZD175D	0425
MACZD175D5	0425
MACZD175D10	0425
MACZD180D	0425
MACZD180D5	0425
MACZD180D10	0425
MACZD200D	1473
MACZD200D5	1473
MACZD200D10	1473
MACZD2805	0257
MACZE6.8D	0062
MACZE6.8D5	0062
MACZE6.8D10	0062
MACZE7.5D	0077
MACZE7.5D5	0077
MACZE7.5D10	0077
MACZE8.2D	0165
MACZE8.2D5	0165
MACZE8.2D10	0165
MACZE9.1D	0057
MACZE9.1D5	0057
MACZE9.1D10	0057
MACZE10D	0064
MACZE10D5	0064
MACZE10D10	0064
MACZE11D	0181
MACZE11D5	0181
MACZE11D10	0181
MACZE12D	0052
MACZE12D5	0052
MACZE12D10	0052
MACZE13D	0053
MACZE13D5	0053
MACZE13D10	0287
MACZE14D	0879
MACZE14D5	0879
MACZE14D10	0879
MACZE15D	0686
MACZE15D5	0686
MACZE15D10	0686
MACZE16D	0445
MACZE16D5	0445
MACZE16D10	0445
MACZE17D	0210
MACZE17D5	0210
MACZE17D10	0210
MACZE18D	0375
MACZE18D5	0375
MACZE18D10	0375
MACZE19D	0671
MACZE19D5	0671
MACZE19D10	0671
MACZE20D	0700
MACZE20D5	0700
MACZE20D10	0700
MACZE22D	0705
MACZE22D5	0705
MACZE22D10	0705
MACZE24D	0494
MACZE24D5	0494
MACZE24D10	0494
MACZE25D	0714
MACZE25D5	0714
MACZE25D10	0714
MACZE27D	0455
MACZE27D5	0455
MACZE27D10	0455
MACZE30D	0195
MACZE30D5	0195
MACZE30D10	0195
MACZE33D	0166
MACZE33D5	0166
MACZE33D10	0166
MACZE36D	0010
MACZE36D5	0010
MACZE36D10	0010
MACZE39D	0032
MACZE39D5	0032
MACZE39D10	0032
MACZE43D	0054
MACZE43D5	0054
MACZE43D10	0054
MACZE47D	0068
MACZE47D5	0068
MACZE47D10	0068
MACZE51D	0092
MACZE51D5	0092
MACZE51D10	0092
MACZE56D	0125
MACZE56D5	0125
MACZE56D10	0125
MACZE62D	0152
MACZE62D5	0152
MACZE62D10	0152
MACZE68D	0173
MACZE68D5	0173
MACZE68D10	0173
MACZE75D	0094
MACZE75D5	0094
MACZE75D10	0094
MACZE82D	0049
MACZE82D5	0049
MACZE82D10	0049
MACZE91D	0156
MACZE91D5	0156
MACZE91D10	0156
MACZE100D	0189
MACZE100D5	0189
MACZE100D10	0189
MACZE105D	0099
MACZE105D5	0099
MACZE105D10	0099
MACZE110D	0099
MACZE110D5	0099
MACZE110D10	0099
MACZE120D	0089
MACZE120D5	0089
MACZE120D10	0089
MACZE130D	0287
MACZE130D5	0287
MACZE130D10	0287
MACZE140D	0252
MACZE140D5	0252
MACZE140D10	0252
MACZE150D	0340
MACZE150D5	0340
MACZE150D10	0340
MACZE160D	0370
MACZE160D5	0370
MACZE160D10	0370
MACZE175D	0425
MACZE175D5	0425
MACZE175D10	0425
MACZE180D	0425
MACZE180D5	0425
MACZE180D10	0425
MACZE200D	1473
MACZE200D5	1473
MACZE200D10	1473
MACZF1.8D	OEM
MACZF1.8D5	OEM
MACZF1.8D10	OEM
MACZF2.0D	OEM
MACZF2.0D5	OEM
MACZF2.0D10	OEM
MACZF2.2D	OEM
MACZF2.2D5	OEM
MACZF2.2D10	OEM
MACZF2.4D	1276
MACZF2.4D5	1276
MACZF2.4D10	1276
MACZF2.7D	0760
MACZF2.7D5	0760
MACZF2.7D10	0760
MACZF3.0D	0118
MACZF3.0D5	0118
MACZF3.0D10	0118
MACZF3.3D	0299
MACZF3.3D5	0299
MACZF3.3D10	0299
MACZF3.6D	0376
MACZF3.6D5	0299
MACZF3.6D10	0376
MACZF3.9D	0036
MACZF3.9D5	0036
MACZF3.9D10	0036
MACZF4.3D	0275
MACZF4.3D5	0275
MACZF4.3D10	0275
MACZF4.7D	0140
MACZF4.7D5	0140
MACZF4.7D10	0140
MACZF5.1D	0041
MACZF5.1D5	0041
MACZF5.1D10	0041
MACZF5.6D	0253
MACZF5.6D5	0253
MACZF5.6D10	0253
MACZF6.2D	0471
MACZF6.2D5	0471
MACZF6.2D10	0471
MACZF6.8D	0062
MACZF6.8D5	0062
MACZF6.8D10	0062
MACZF7.5D	0077
MACZF7.5D5	0077
MACZF7.5D10	0077
MACZF8.2D	0165
MACZF8.2D5	0165
MACZF8.2D10	0165
MACZF9.1D	0057
MACZF9.1D5	0057
MACZF9.1D10	0057
MACZF10D	0064
MACZF10D5	0064
MACZF10D10	0064
MACZF12D	0052
MACZF12D5	0052
MACZF12D10	0052
MACZG2.4D	1276
MACZG2.4D5	1276
MACZG2.4D10	1276
MACZG2.7D	0760
MACZG2.7D5	0760
MACZG2.7D10	0760
MACZG3.0D	0118
MACZG3.0D5	0118
MACZG3.0D10	0118
MACZG3.3D	0299
MACZG3.3D5	0299
MACZG3.3D10	0299
MACZG3.6D	0376
MACZG3.6D5	0376
MACZG3.6D10	0376
MACZG3.9D	0036
MACZG3.9D5	0036
MACZG3.9D10	0036
MACZG4.3D	0275
MACZG4.3D5	0275
MACZG4.3D10	0275
MACZG4.7D	0140
MACZG4.7D5	0140
MACZG4.7D10	0140
MACZG5.1D	0041
MACZG5.1D5	0041
MACZG5.1D10	0041
MACZG5.6D	0253
MACZG5.6D5	0253
MACZG5.6D10	0253
MACZG5.8D5	OEM
MACZG6.2D	0471
MACZG6.2D5	0471
MACZG6.2D10	0471
MACZG6.8D	0062
MACZG6.8D10	0062
MACZG7.5D	0077
MACZG7.5D5	0077
MACZG7.5D10	0077
MACZG8.2D	0165
MACZG8.2D5	0165
MACZG8.2D10	0165
MACZG9.1D	0057
MACZG9.1D5	0057
MACZG9.1D10	0057
MACZG10D	0064
MACZG10D5	0064
MACZG10D10	0064
MACZG12D	0052
MACZG12D5	0052
MACZG12D10	0052
MAD-835	OEM
MAG001B	OEM
MAG15	OEM
MAG17	OEM
MAG19	OEM
MAG20	OEM
MAG21A	OEM
MAG21B	OEM
MAG21C	OEM
MAG23A	OEM
MAG23C	OEM
MAG23D	OEM
MAH0801-2-2	OEM
MAI01	OEM
MAI02	OEM
MAI03	OEM
MAI04	OEM
MAI10	OEM
MAI11	OEM
MAI20	OEM
MAL100	OEM
MAN1A	4987
MAN2A	OEM
MAN10A	4987
MAN51A	5437
MAN52A	5439
MAN53A	OEM
MAN54A	5441
MAN58A	OEM
MAN71A	4870
MAN72A	4989
MAN73A	4871
MAN74	4873
MAN74A	4873
MAN78A	OEM
MAN81A	5129
MAN82A	5137
MAN83A	OEM
MAN84A	5132
MAN88A	OEM
MAN101A	4988
MAN1001A	4988
MAN2815	OEM
MAN3610A	5127
MAN3620A	5134
MAN3630A	OEM
MAN3640A	5130
MAN3680A	OEM
MAN3910A	OEM
MAN3920A	OEM
MAN3930A	OEM
MAN3940A	OEM
MAN3980A	OEM
MAN4505	OEM
MAN4510	OEM
MAN4540	OEM
MAN4580A	OEM
MAN4605	OEM
MAN4610	OEM
MAN4630	OEM
MAN4640	OEM
MAN4680A	OEM
MAN4705	OEM
MAN4710	6744
MAN4740	6750
MAN4780A	OEM
MAN4805	OEM
MAN4810	OEM
MAN4840	OEM
MAN4880A	OEM
MAN4905A	OEM
MAN4910A	OEM
MAN4940A	OEM
MAN4980A	OEM
MAN6410	OEM
MAN6610	OEM
MAN6630	OEM
MAN6640	OEM
MAN6650	OEM
MAN6660	OEM
MAN6680	OEM
MAN6710	4327
MAN6730	4331
MAN6740	4332
MAN6750	4333
MAN6760	2206
MAN6780	2199
MAN6910	OEM
MAN6930	OEM
MAN6940	OEM
MAN6950	OEM
MAN6960	OEM
MAN6980	OEM
MAN8610	2378
MAN8630	OEM
MAN8640	OEM
MAN8650	OEM
MAN8910	OEM
MAN8930	OEM
MAN8940	OEM
MAN8950	OEM
MAS20	0136
MAS21	0136
MAS22	0136
MAS23	0136
MAS32	OEM
MAS39	OEM
MAS-839	OEM
MAS-842	OEM
MAT0000100	OEM
MAT0000102	OEM
MAT0000133	OEM
MAT01AH	OEM
MAT01FH	OEM
MAT01GH	OEM
MAT01H	OEM
MAY3877S	OEM
MAZCA10D5	OEM
MAg23B	OEM
MB-01	0277
MB01	0015
MB-02	0289
MB-04	0296
MB-06	0302
MB-08	0250
MB0103M	OEM
MB1B	0071
MB1C	OEM
MB-1D	0348
MB1D	0017
MB-1DS	OEM
MB1E	OEM
MB-1F	0182
MB1F	0348
MB-1FS	0017
MB1G	OEM
MB-4-01	0277
MB-4-02	0289
MB-4-04	0296
MB-4-06	0302
MB-4-08	0250
MB6Y	OEM
MB8A	OEM
MB-10	0102
MB10	0102
MB11A02V05	0323
MB11A02V10	0323
MB11A02V20	1414
MB11A02V30	0473
MB11A02V40	0473
MB11A02V60	0446
MB11A02V80	1422
MB11A02W10	2443
MB12A10V05	2363
MB12A10V10	2363
MB12A10V20	2363
MB12A10V30	2369
MB12A10V40	2369
MB12A10V60	2370
MB12A10V80	2372
MB12A10W10	2373
MB12A25V05	2363
MB12A25V10	2363
MB12A25V20	2363
MB12A25V30	2369
MB12A25V40	2369
MB12A25V60	2370
MB12A25V80	2372
MB12A25W10	2373
MB25-100	OEM
MB25-400	OEM
MB25-600	OEM
MB25-800	OEM
MB25-1000	OEM
MB25-1200	OEM
MB60BH302A	OEM
MB61	0323
MB62	1414
MB62H118	OEM
MB64	0473
MB64H153	OEM
MB64H303	OEM
MB66	0446
MB68	1422
MB74LS00	1527
MB74LS00M	1527
MB74LS00P	OEM
MB74LS01	1546
MB74LS01M	1546
MB74LS02	1559
MB74LS02M	1559
MB74LS03	1579
MB74LS03M	1579
MB74LS04	1595
MB74LS04M	1595
MB74LS05	1608
MB74LS05M	1608
MB74LS06	OEM
MB74LS08	1633
MB74LS08M	1633
MB74LS09	1642
MB74LS09M	1642
MB74LS10	1662
MB74LS10M	1662
MB74LS11	1667
MB74LS11M	1667
MB74LS12	1679
MB74LS12M	1679
MB74LS13	1688
MB74LS13M	1688
MB74LS14	1698
MB74LS14M	1698
MB74LS15	1707
MB74LS15M	1707
MB74LS20	0034
MB74LS20M	0034
MB74LS21	1763
MB74LS21M	1763
MB74LS22	1775
MB74LS22M	1775
MB74LS26	1382
MB74LS26M	1382
MB74LS27	0183
MB74LS27M	0183
MB74LS28	0472
MB74LS28M	0472
MB74LS30	0828
MB74LS30M	0828
MB74LS32	0088
MB74LS32M	0088
MB74LS33	1832
MB74LS33M	1832
MB74LS37	1730
MB74LS37M	1730
MB74LS38	1839
MB74LS38M	1839
MB74LS38P	OEM
MB74LS40	0135
MB74LS40M	0135
MB74LS42	1841
MB74LS42M	1841
MB74LS47	1845
MB74LS47M	1845
MB74LS48	1849
MB74LS48M	1849
MB74LS49	1850
MB74LS49M	1850
MB74LS51	1033
MB74LS51M	1033
MB74LS54	1857
MB74LS54M	1857
MB74LS55	0457
MB74LS55M	0457
MB74LS73A	1867
MB74LS73AM	1867
MB74LS74A	0243
MB74LS74AM	0243
MB74LS76A	2180
MB74LS76AM	2180
MB74LS76P	2180
MB74LS78	1873
MB74LS78AM	1873
MB74LS83A	2218
MB74LS83AM	2218
MB74LS85	0431
MB74LS85M	0431
MB74LS86	0291
MB74LS86M	0291
MB74LS86P	0291
MB74LS107	1602
MB74LS107A	1602
MB74LS107AM	1602
MB74LS107M	1602
MB74LS109A	1906
MB74LS109AM	1906
MB74LS112A	2128
MB74LS112AM	2128
MB74LS113A	2256
MB74LS113AM	2256
MB74LS114A	2302
MB74LS114AM	2302
MB74LS122	1620
MB74LS122M	1620
MB74LS123	0979
MB74LS123M	0979
MB74LS125A	0075
MB74LS125AM	0075
MB74LS126A	2877
MB74LS126AM	2877
MB74LS132	1625
MB74LS132M	1625
MB74LS136	1628
MB74LS136M	1628
MB74LS138	0427
MB74LS138M	0427

DEVICE TYPE	REPL CODE
MB74LS139	0153
MB74LS139M	0153
MB74LS145	1564
MB74LS145M	1564
MB74LS151	1646
MB74LS151M	1646
MB74LS153	0959
MB74LS153M	0959
MB74LS155	0209
MB74LS155M	0209
MB74LS156	1654
MB74LS156M	1654
MB74LS157	1161
MB74LS157M	1161
MB74LS158	1656
MB74LS158M	1656
MB74LS160A	0837
MB74LS160AM	0837
MB74LS161A	0858
MB74LS161AM	0858
MB74LS162A	0880
MB74LS162AM	0880
MB74LS163A	0893
MB74LS163AM	0893
MB74LS174	0260
MB74LS174M	0260
MB74LS175	1672
MB74LS175M	1672
MB74LS181	1678
MB74LS181M	1678
MB74LS183	OEM
MB74LS183M	OEM
MB74LS190	1686
MB74LS190M	1686
MB74LS191	1687
MB74LS191M	1687
MB74LS192	1689
MB74LS192M	1689
MB74LS193	1692
MB74LS193M	1692
MB74LS221	1239
MB74LS221M	1239
MB74LS240	0452
MB74LS240M	0452
MB74LS241	1726
MB74LS242	1728
MB74LS242M	1728
MB74LS243	0906
MB74LS243M	0906
MB74LS244	0458
MB74LS244M	0458
MB74LS245	0463
MB74LS245M	0463
MB74LS247	1732
MB74LS247M	1732
MB74LS248	1734
MB74LS248M	1734
MB74LS249	1735
MB74LS249M	1735
MB74LS251	1737
MB74LS251M	1737
MB74LS253	1739
MB74LS253M	1739
MB74LS257	1744
MB74LS257M	1744
MB74LS258	1746
MB74LS258M	1746
MB74LS266	0592
MB74LS266M	0592
MB74LS273M	0894
MB74LS280	1773
MB74LS280M	1773
MB74LS283	1779
MB74LS283M	1779
MB74LS352	0761
MB74LS352M	0761
MB74LS353	0773
MB74LS353M	0773
MB74LS365A	0943
MB74LS365AM	0943
MB74LS366A	0956
MB74LS366AM	0956
MB74LS367A	0977
MB74LS367AM	0977
MB74LS368A	0991
MB74LS368AM	0991
MB74LS373	0709
MB74LS373P	0709
MB74LS374	0713
MB74LS386	1230
MB74LS386M	1230
MB74LS640	0669
MB74LS640M	0669
MB74LS641	0690
MB74LS641M	0690
MB74LS642	0719
MB74LS642M	0719
MB74LS643	2056
MB74LS643M	2056
MB74LS644	OEM
MB74LS644M	OEM
MB74LS645	0775
MB74LS645M	0775
MB74LS25858	OEM
MB74S00	0704
MB74S38	0780
MB75LS54	1857
MB75LS54M	1857
MB81	0323
MB81G	OEM
MB82	1414
MB82G	OEM
MB82HS181	OEM
MB82HS191	OEM
MB82HS321	OEM
MB82S131	OEM
MB82S321	OEM
MB84	0473
MB84G	OEM
MB84LS85M	OEM
MB86	0446
MB86G	OEM
MB88	1422
MB88G	OEM
MB101	OEM
MB101G	OEM
MB102	OEM
MB102G	OEM
MB104	OEM
MB104G	OEM
MB106	OEM
MB106G	OEM
MB108	OEM
MB108G	OEM
MB200	OEM
MB201	OEM
MB202	0133
MB204	OEM
MB206	OEM
MB207	OEM
MB208	OEM
MB209	OEM
MB211	OEM
MB211BAL2	OEM
MB213	OEM
MB244	0015
MB251	2363
MB251W	OEM
MB252	2363
MB252W	OEM
MB254	2369
MB254W	OEM
MB256	2370
MB256W	OEM
MB257	0015
MB258	2372
MB258W	OEM
MB269	0015
MB270	0015
MB335	OEM
MB336	OEM
MB337	OEM
MB338	OEM
MB339	OEM
MB340	OEM
MB351	1643
MB352	1643
MB354	1673
MB356	1673
MB358	1589
MB366	OEM
MB367	OEM
MB368	OEM
MB369	OEM
MB370	OEM
MB371	OEM
MB400	0232
MB400M	OEM
MB401	0512
MB401M	OEM
MB402	0697
MB402M	OEM
MB403	0873
MB403M	OEM
MB404	1024
MB404M	OEM
MB405	0743
MB405M	OEM
MB406	1275
MB406M	OEM
MB407	1426
MB407M	OEM
MB408	1535
MB408M	OEM
MB410	0942
MB410M	OEM
MB411	1185
MB411M	OEM
MB416	0268
MB416M	OEM
MB417	0314
MB417M	OEM
MB418	0361
MB418M	OEM
MB419	OEM
MB419M	OEM
MB420	1313
MB420M	OEM
MB421	OEM
MB421M	OEM
MB422	OEM
MB422M	OEM
MB423	OEM
MB423M	OEM
MB424	4516
MB424M	OEM
MB425	1863
MB425M	1863
MB426	OEM
MB426M	OEM
MB427	OEM
MB427M	OEM
MB428	OEM
MB428M	OEM
MB429	OEM
MB429M	OEM
MB430	OEM
MB430M	OEM
MB431	OEM
MB431M	OEM
MB432	OEM
MB432M	OEM
MB433	0996
MB433M	OEM
MB434	0996
MB434M	OEM
MB435	3520
MB435M	OEM
MB436	OEM
MB436M	OEM
MB437	OEM
MB437M	OEM
MB438	OEM
MB438M	OEM
MB439	OEM
MB439M	OEM
MB440	1157
MB440M	OEM
MB442	1052
MB442M	OEM
MB443	0619
MB443M	OEM
MB444	OEM
MB444M	OEM
MB445	1495
MB445M	OEM
MB446	OEM
MB446M	OEM
MB447	1829
MB447M	OEM
MB448	0374
MB448M	OEM
MB449	1368
MB449M	OEM
MB450	1645
MB450M	OEM
MB451	1631
MB451M	OEM
MB452	1715
MB453	1486
MB454	0980
MB455	1964
MB456	1917
MB456M	OEM
MB457	1912
MB457M	OEM
MB458	1842
MB458M	OEM
MB459	1856
MB459M	OEM
MB460	1722
MB513AR	0133
MB601	0232
MB602	0512
MB603	0697
MB604	0873
MB605	1024
MB605(BRIDGE)	0323
MB606	0743
MB607	1275
MB609	1426
MB610	2443
MB613	2101
MB614	4856
MB618	5418
MB805	0323
MB805G	OEM
MB810	2443
MB884-180J	OEM
MB1005	OEM
MB1005G	OEM
MB1010	OEM
MB1101	OEM
MB1151	OEM
MB2114A4	OEM
MB2114A5	OEM
MB2114AL1	OEM
MB2114AL2	OEM
MB2114AL3	OEM
MB2114AL4	OEM
MB2115A	OEM
MB2115A2	OEM
MB2115AL	OEM
MB2115H2	OEM
MB2115H3	OEM
MB2115H4	OEM
MB2118-10	OEM
MB2118-12	OEM
MB2118-15	OEM
MB2125A	OEM
MB2125A2	OEM
MB2125AL	OEM
MB2125AL2	OEM
MB2125H1	OEM
MB2125H2	OEM
MB2125H3	OEM
MB2125H4	OEM
MB2128-15	OEM
MB2128-20	OEM
MB2141-3	OEM
MB2141-4	OEM
MB2141-5	OEM
MB2141L3	OEM
MB2141L4	OEM
MB2141L5	OEM
MB2142-2	OEM
MB2142-3	OEM
MB2142L	OEM
MB2142L3	OEM
MB2147A(A)	OEM
MB2147A3(A)	OEM
MB2147AL(A)	OEM
MB2147AL3(A)	OEM
MB2147H	OEM
MB2147H1	OEM
MB2147H2	OEM
MB2147H3	OEM
MB2147HL	OEM
MB2147HL3	OEM
MB2148H	OEM
MB2148H2	OEM
MB2148H3	OEM
MB2148HL	OEM
MB2148HL3	OEM
MB2149H	OEM
MB2149H2	OEM
MB2149H3	OEM
MB2149HL	OEM
MB2149HL3	OEM
MB2164-15(A)	OEM
MB2164-20(A)	OEM
MB2164-25(A)	OEM
MB2167-10(A)	OEM
MB2167-70(A)	OEM
MB2167L70(A)	OEM
MB2186(A)	OEM
MB2505	2363
MB2505W	OEM
MB2510	2373
MB2510W	OEM
MB2716	OEM
MB2716-1	OEM
MB2716-2	OEM
MB2716-5	OEM
MB2716-6	OEM
MB2732A	OEM
MB2732A2	OEM
MB2732A3	OEM
MB2732A4	OEM
MB2732A20	OEM
MB2732A25	OEM
MB2732A30	OEM
MB2764-2	OEM
MB2764-3	OEM
MB2764-4	OEM
MB2764-25	OEM
MB2764-30	OEM
MB2764-45	OEM
MB2815	OEM
MB2815-3	OEM
MB2815-4	OEM
MB2816	OEM
MB2816-3	OEM
MB2816-4	OEM
MB2817(A)	OEM
MB2817-3(A)	OEM
MB2817-4(A)	OEM
MB3106M	OEM
MB3110	OEM
MB3110A	OEM
MB3110APS-G-SNY	OEM
MB3110A-SNY	OEM
MB3202	2261
MB3501K	OEM
MB3505	1643
MB3510	1589
MB3511	OEM
MB3511P-SH	OEM
MB3512	OEM
MB3512PF-EF	OEM
MB3512PFQ	OEM
MB3512PFQ-G-BND-EF	OEM
MB3602C	OEM
MB3602K	OEM
MB3602M	OEM
MB3603C	OEM
MB3603K	OEM
MB3603M	OEM
MB3604C	OEM
MB3607C	OEM
MB3607K	OEM
MB3607M	OEM
MB3608C	OEM
MB3608K	OEM
MB3608M	OEM
MB3609C	OEM
MB3609M	OEM
MB3612C	OEM
MB3612M	OEM
MB3614	0661
MB3614C	OEM
MB3614M	OEM
MB3615M	OEM
MB3628A1	OEM
MB3628A3	OEM
MB3628A4(A)	OEM
MB3632	OEM
MB3632-1	OEM
MB3636B	OEM
MB3636B1	OEM
MB3636B2	OEM
MB3705M	OEM
MB3710	1816
MB3712	4291
MB3712HM	4291
MB3713	4476
MB3713HM	4476
MB3730M	OEM
MB3752C	OEM
MB3752K	OEM
MB3752M	OEM
MB3756	4648
MB3756M	OEM
MB3759	1824
MB3759C	OEM
MB3759M	OEM
MB3759P	1824
MB3760C	OEM
MB3760M	OEM
MB3761	OEM
MB3761M	OEM
MB3763	OEM
MB3763PF	OEM
MB3773	OEM
MB3773PS	OEM
MB3775PF	OEM
MB4001K	OEM
MB4002K	OEM
MB4111	OEM
MB4112	OEM
MB4204	0176
MB4204C	0176
MB4204F	OEM
MB4204M	0176
MB6001	OEM
MB6002	OEM
MB6356	1083
MB7051C	2175
MB7051G3	OEM
MB7052P	OEM
MB7052Z	2224
MB7053P	OEM
MB7053Z	2389
MB7054C	OEM
MB7055C	OEM
MB7056C	3168
MB7057P	OEM
MB7057Z	1918
MB7058P	OEM
MB7058Z	2388
MB7059C	OEM
MB7060C	OEM
MB7071ET	OEM
MB7071HT	OEM
MB7071NT	OEM
MB7072EC	OEM
MB7114-40	OEM
MB7114E	OEM
MB7121EC	2390
MB7121EP	OEM
MB7121HC	OEM
MB7121HP	OEM
MB7122EC	2392
MB7122EP	OEM
MB7122HC	OEM
MB7122HP	OEM
MB7123EZ	3288
MB7123HZ	OEM
MB7124EZ	2320
MB7124HZ	OEM
MB7127EP	OEM
MB7127EZ	OEM
MB7127HP	OEM
MB7127HZ	OEM
MB7128EP	OEM
MB7128EZ	3166
MB7128HP	OEM
MB7128HZ	3166
MB7128YP	OEM
MB7128YZ	OEM
MB7130EC	OEM
MB7130HC	OEM
MB7131EC	2862
MB7131HC	2862
MB7132EC	2909
MB7132HC	2909
MB7134EC	OEM
MB7134HC	OEM
MB7136E	OEM
MB7136ET	OEM
MB7136H	OEM
MB7136HT	OEM
MB7137EC	OEM
MB7137HC	OEM
MB7138EC	OEM
MB7138HC	OEM
MB7141EC	OEM
MB7141HC	OEM
MB7142EC	OEM
MB7142HC	OEM
MB7142NC	OEM
MB8116E	0522
MB8116EC	0522
MB8116EP	0522
MB8116EZ	OEM
MB8116HC	OEM
MB8116HP	OEM
MB8116HZ	OEM
MB8116NC	OEM
MB8116NP	OEM
MB8116NZ	OEM
MB8117-10P	OEM
MB8117-10Z	OEM
MB8117-12P	OEM
MB8117-12Z	OEM
MB8117-15C	OEM
MB8117-15P	OEM
MB8117-15Z(A)	OEM
MB8118-10P	OEM
MB8118-10Z	OEM
MB8118-12	OEM
MB8118-12P	OEM
MB8118-12Z	OEM
MB8118-15P	OEM
MB8118-15Z	OEM
MB8128-10C	OEM
MB8128-10P0P	OEM
MB8128-10Z	OEM
MB8128-15	1898
MB8128-15C	6181
MB8128-15P	OEM
MB8128-15Z	OEM
MB8148(A)	OEM
MB8167	OEM
MB8167-55C	OEM
MB8167-70C	OEM
MB8167A45Z	OEM
MB8167A55Z	OEM
MB8168-55Z	OEM
MB8168-70	OEM
MB8168-70Z	OEM
MB8168-90Z	OEM
MB8216E	OEM
MB8216EC	OEM
MB8216N	OEM
MB8264-15	OEM
MB8264-15C	OEM
MB8264-15Z	OEM
MB8264-20	OEM
MB8264-20C	OEM
MB8264-20Z	OEM
MB8264A10P	OEM
MB8264A10Z	OEM
MB8264A12P	OEM
MB8264A12Z	OEM
MB8265-15Z	OEM
MB8265-20Z	OEM
MB8265A10P	OEM
MB8265A10P(A)	OEM
MB8265A10Z	OEM
MB8265A12P	OEM
MB8265A12P(A)	OEM
MB8265A12Z	OEM
MB8266A10Z	OEM
MB8266A12Z	OEM
MB8364A	OEM
MB8364M	OEM
MB8404EC	OEM
MB8404EP	OEM
MB8414EC	OEM
MB8414EP	OEM
MB8416-15P	OEM
MB8416-15Z	OEM
MB8416-20LPF	OEM
MB8416-20P	OEM
MB8416-20Z	OEM
MB8416-25P	OEM
MB8416-25Z	OEM
MB8416A12P	OEM
MB8416A12Z	OEM
MB8416A15P	OEM
MB8416A15Z	OEM
MB8416P	OEM
MB8416XP	OEM
MB8416XZ	OEM
MB8416Z	OEM
MB8417-15P	OEM
MB8417-15Z	OEM
MB8417-20P	OEM
MB8417-20Z	OEM
MB8417-25P	OEM
MB8417-25Z	OEM
MB8417A12P	OEM
MB8417A12Z	OEM
MB8417A15P	OEM
MB8417A15Z	OEM
MB8417P	OEM
MB8417XP	OEM
MB8417XZ	OEM
MB8417Z	OEM
MB8418	OEM
MB8418-20P	OEM
MB8418-20Z	OEM
MB8418-25P	OEM
MB8418-25Z	OEM
MB8418A12P	OEM
MB8418A12Z	OEM
MB8418A15P	OEM
MB8418A15Z	OEM
MB8418B	OEM
MB8418P	OEM
MB8418XP	OEM
MB8418XZ	OEM
MB8418Z	OEM
MB8464-15L	OEM
MB8464-15LLP	OEM
MB8464A-15LL	OEM
MB8464A-15LLP-SK-G	OEM
MB8464A-80L-SK	OEM
MB8719	OEM
MB-8726	OEM
MB8726	OEM
MB8734	OEM
MB8747	OEM
MB8752	OEM
MB8758	OEM
MB8805-403N	OEM
MB8805-417	OEM
MB8805-417N	OEM
MB8841-180	OEM
MB8841-180J	OEM
MB8841-410J	OEM
MB8843	OEM
MB8844	OEM
MB8844-1436K	OEM
MB8845/529M	OEM
MB8845-495M	OEM
MB8845-527	OEM
MB8845-527M	OEM
MB8845-528	OEM
MB8845-528M	OEM
MB8845-549	OEM
MB8845-591	OEM
MB8845-592	OEM
MB8845-595	OEM
MB8847-1257K	OEM
MB8849	OEM
MB8851	OEM
MB8851-514L	OEM
MB8851A	OEM
MB-8851A-123A	OEM
MB8853	OEM
MB8854	OEM
MB8855-139M	OEM
MB8855-140	OEM
MB8855-141	OEM
MB8855-161	OEM
MB8855-176M	OEM
MB8855-333L	OEM
MB8855-406L	OEM
MB8866	OEM
MB8866C	OEM
MB8866M	OEM
MB8868AC	OEM
MB8868AM	OEM
MB8876	OEM
MB8876C	OEM
MB8876M	OEM
MB8877	OEM
MB8877C	OEM
MB8877M	OEM
MB8881	OEM
MB8881N	OEM
MB10101	OEM
MB10101M	OEM
MB10102	OEM
MB10102M	OEM
MB10103	OEM
MB10104	OEM
MB10104M	OEM
MB10105	OEM
MB10105M	OEM
MB10106	OEM
MB10106M	OEM
MB10107	OEM
MB10107M	OEM
MB10109	OEM
MB10109M	OEM
MB10110	OEM
MB10110M	OEM
MB10111	OEM
MB10111M	OEM
MB10115	OEM
MB10115M	OEM
MB10116	OEM
MB10116M	OEM
MB10117	OEM
MB10117M	OEM
MB10118	OEM
MB10118M	OEM
MB10119	OEM
MB10119M	OEM
MB10121	OEM
MB10121M	OEM
MB10124	OEM
MB10124M	OEM
MB10125	OEM
MB10125M	OEM
MB10130	OEM
MB10130M	OEM
MB10131	OEM
MB10131M	OEM
MB10133	OEM
MB10133M	OEM
MB10135	OEM
MB10135M	OEM
MB10136	OEM
MB10136M	OEM
MB10137	OEM
MB10137M	OEM
MB10160	OEM
MB10160M	OEM
MB10161	OEM

If replacement code is OEM, contact original manufacturer for replacement.

DEVICE TYPE	REPL CODE
MB10161M	OEM
MB10162	OEM
MB10162M	OEM
MB10164	OEM
MB10164M	OEM
MB10174	OEM
MB10174M	OEM
MB10175	OEM
MB10175M	OEM
MB10179	OEM
MB10179M	OEM
MB10180	OEM
MB10180M	OEM
MB10181	OEM
MB10181M	OEM
MB27128-2(A)	1638
MB40166	OEM
MB40166PF-EF	OEM
MB40176	OEM
MB40176P	OEM
MB40176PF	OEM
MB40176PF-EF	OEM
MB40176PF-G-BND-EF	OEM
MB40176P-SH	OEM
MB40576	OEM
MB62131	OEM
MB81256-15	1472
MB81416-12	OEM
MB81416-15	OEM
MB81461	OEM
MB81461-12	OEM
MB81461-12-PSZ-G-BF2	OEM
MB81461B	OEM
MB81461B-12	OEM
MB81461B-12RS-PSZ	OEM
MB83256	OEM
MB84001B	0478
MB84001BM	0478
MB84002B	2055
MB84002BM	2055
MB84008B	0988
MB84008BM	0988
MB84011	0215
MB84011-U	0215
MB84011B	0215
MB84011BM	0215
MB84011M	0215
MB84011U	0215
MB84011V	0478
MB84012B	0498
MB84012BM	0498
MB84013B	0413
MB84013BM	0413
MB84016B	1143
MB84016BM	1143
MB84017B	0513
MB84017BM	0513
MB84019B	1525
MB84019BM	1525
MB84020B	1661
MB84020BM	1661
MB84022B	1256
MB84022BM	1256
MB84023B	0519
MB84023BM	0519
MB84025B	2072
MB84025BM	2072
MB84027B	1949
MB84027BM	OEM
MB84028B	2228
MB84028BM	2228
MB84029B	2233
MB84029BM	2233
MB84040B	0056
MB84040BM	0056
MB84049B	0001
MB84049BM	0001
MB84050B	0398
MB84050BM	0398
MB84051B	0366
MB84051BM	0366
MB84052B	0024
MB84052BM	0024
MB84053B	0035
MB84053BM	0035
MB84060B	0146
MB84060BM	0146
MB84066B	0101
MB84066BM	0101
MB84068B	0101
MB84068BM	2503
MB84069B	0119
MB84069BM	0119
MB84070B	2516
MB84070BM	2516
MB84071B	0129
MB84071BM	0129
MB84072B	2524
MB84072BM	2524
MB84073B	1536
MB84073BM	1536
MB84075B	2540
MB84075BM	2540
MB84077B	2558
MB84077BM	2558
MB84078B	0921
MB84078BM	0921
MB84081B	0626

DEVICE TYPE	REPL CODE
MB84081BM	0626
MB84082B	0300
MB84082BM	0300
MB84082BMBM	OEM
MB84518BM	OEM
MB84520B	2675
MB84520BM	OEM
MB84524B	OEM
MB84524BM	OEM
MB86140	OEM
MB86140P-SH	OEM
MB86144	OEM
MB86144B	OEM
MB86144BPF	OEM
MB86144BPF-G-BND	OEM
MB88201-128L	OEM
MB88201-196L	OEM
MB88201-638L	OEM
MB88201P-127K	OEM
MB88301A	OEM
MB88303	OEM
MB88303-P	OEM
MB88303P	OEM
MB88323-K-1	OEM
MB88341PFV	OEM
MB88421-147L	OEM
MB88421-147P	OEM
MB88501-114M	OEM
MB88501-120M	OEM
MB88501-249L	OEM
MB88503-115M	OEM
MB88503-119M	OEM
MB88505-340N	OEM
MB88505-403N	OEM
MB88505-417N	OEM
MB88511-108N	OEM
MB88514B-640L	OEM
MB88541-130M	OEM
MB88733-123	OEM
MB88733-137	OEM
MB88733-143	OEM
MB89009-107	OEM
MB89009F-102	OEM
MB89009G-107	OEM
MB89009P-102	OEM
MB89009PG-107	OEM
MB89010-101	OEM
MB89010A-112	OEM
MB89012-102	OEM
MB90062-103A	OEM
MB90062P-106	OEM
MB90063-103K	OEM
MB653839	OEM
MB654308	OEM
MB654308-8825	OEM
MB654839	OEM
MB672312U	OEM
MB673198U	OEM
MB825191	OEM
MB885505-403N	OEM
MB88201196L	OEM
MBA7767S	OEM
MBB11	OEM
MBB80	OEM
MBB88/01	OEM
MBB88/04	OEM
MBB88/06	OEM
MBB88/08	OEM
MBB88/0202	OEM
MBB100	OEM
MBC008	OEM
MBC008L	OEM
MBC008L3	OEM
MBC0083	OEM
MBC01A2	OEM
MBC010-65	OEM
MBC010-68	OEM
MBC016	OEM
MBC016D	OEM
MBC016L	OEM
MBC016L3	OEM
MBC020-65	OEM
MBC020-68	OEM
MBC-032/16	OEM
MBC-032/16C	OEM
MBC032D	OEM
MBC048D	OEM
MBC-064/16	OEM
MBC-064/16C	OEM
MBC064D	OEM
MBC-096/16	OEM
MBC-096/16C	OEM
MBC0163	OEM
MBC11	OEM
MBC-86/12(A)	OEM
MBC86/12-032/5	OEM
MBC-86/12-032/5(A)	OEM
MBC86/12-032/8	OEM
MBC-86/12-032/8(A)	OEM
MBC86/12-128/5	OEM
MBC-86/12-128/5(A)	OEM
MBC86/12-128/8	OEM
MBC-86/12-128/8(A)	OEM
MBC100	OEM
MBC-128/16	OEM
MBC-128/16C	OEM
MBC-256/16	OEM
MBC-256/16C	OEM
MBC-384/16	OEM

DEVICE TYPE	REPL CODE
MBC-384/16C	OEM
MBC-512/16	OEM
MBC-512/16C	OEM
MBCBCZC75D10	OEM
MBCTF1	OEM
MBCZA6.8D	0062
MBCZA6.8D5	0062
MBCZA6.8D10	0062
MBCZA7.5D	0077
MBCZA7.5D5	0077
MBCZA7.5D10	0077
MBCZA8.2D	0165
MBCZA8.2D5	0165
MBCZA8.2D10	0165
MBCZA9.1D	0057
MBCZA9.1D5	0057
MBCZA9.1D10	0057
MBCZA10D	0064
MBCZA10D5	0064
MBCZA10D10	0064
MBCZA11D	0181
MBCZA11D5	0181
MBCZA11D10	0181
MBCZA12D	0052
MBCZA12D5	0052
MBCZA12D10	0052
MBCZA13D	0053
MBCZA13D5	0053
MBCZA13D10	0053
MBCZA14D	0879
MBCZA14D5	0879
MBCZA14D10	0879
MBCZA15D	0686
MBCZA15D5	0686
MBCZA15D10	0686
MBCZA16D	0445
MBCZA16D5	0445
MBCZA16D10	0445
MBCZA17D	0210
MBCZA17D5	0210
MBCZA17D10	0210
MBCZA18D	0375
MBCZA18D5	0375
MBCZA18D10	0375
MBCZA19D	0671
MBCZA19D5	0671
MBCZA19D10	0671
MBCZA20D	0700
MBCZA20D5	0700
MBCZA20D10	0700
MBCZA22D	0705
MBCZA22D5	0705
MBCZA22D10	0705
MBCZA24D	0494
MBCZA24D5	0494
MBCZA24D10	0494
MBCZA25D	0714
MBCZA25D5	0714
MBCZA25D10	0714
MBCZA27D	0455
MBCZA27D5	0455
MBCZA27D10	0455
MBCZA28D	0257
MBCZA28D5	0257
MBCZA28D10	0257
MBCZA30D	0195
MBCZA30D5	0195
MBCZA30D10	0195
MBCZA33D	0166
MBCZA33D5	0166
MBCZA33D10	0166
MBCZA36D	0010
MBCZA36D5	0010
MBCZA36D10	0010
MBCZA39D	0032
MBCZA39D5	0032
MBCZA39D10	0032
MBCZA43D	0054
MBCZA43D5	0054
MBCZA43D10	0054
MBCZA47D	0068
MBCZA47D5	0068
MBCZA47D10	0068
MBCZA51D	0092
MBCZA51D5	0092
MBCZA51D10	0092
MBCZA56D	0125
MBCZA56D5	0125
MBCZA56D10	0125
MBCZA60D	2317
MBCZA60D5	2317
MBCZA60D10	2317
MBCZA62D	0152
MBCZA62D5	0152
MBCZA62D10	0152
MBCZA68D	0173
MBCZA68D5	0173
MBCZA68D10	0173
MBCZA75D	0094
MBCZA75D5	0094
MBCZA75D10	0094
MBCZA82D	0049
MBCZA82D5	0049
MBCZA82D10	0049
MBCZA87D	0104
MBCZA87D5	0104
MBCZA87D10	0104
MBCZA91D	0156
MBCZA91D5	0156

DEVICE TYPE	REPL CODE
MBCZA91D10	0156
MBCZA100D	0189
MBCZA100D5	0189
MBCZA100D10	0189
MBCZA105D	0099
MBCZA105D5	0099
MBCZA105D10	0099
MBCZA110D	0099
MBCZA110D5	0099
MBCZA110D10	0099
MBCZA120D	0089
MBCZA120D5	0089
MBCZA120D10	0089
MBCZA130D	0287
MBCZA130D5	0287
MBCZA130D10	0287
MBCZA140D	0252
MBCZA140D5	0252
MBCZA140D10	0252
MBCZA150D	0340
MBCZA150D5	0340
MBCZA150D10	0340
MBCZA160D	0370
MBCZA160D5	0370
MBCZA160D10	0370
MBCZA175D	0425
MBCZA175D5	0425
MBCZA175D10	0425
MBCZA180D	0425
MBCZA180D5	0425
MBCZA180D10	0425
MBCZA200D	1473
MBCZA200D5	1473
MBCZA200D10	1473
MBCZB6.8D	0062
MBCZB6.8D5	0062
MBCZB6.8D10	0062
MBCZB7.5D	0077
MBCZB7.5D5	0077
MBCZB7.5D10	0077
MBCZB8.2D	0165
MBCZB8.2D5	0165
MBCZB8.2D10	0165
MBCZB9.1D	0057
MBCZB9.1D5	0057
MBCZB9.1D10	0057
MBCZB10D	0064
MBCZB10D5	0064
MBCZB10D10	0064
MBCZB11D	0181
MBCZB11D5	0181
MBCZB11D10	0181
MBCZB12D	0052
MBCZB12D5	0052
MBCZB12D10	0052
MBCZB13D	0053
MBCZB13D5	0053
MBCZB13D10	0053
MBCZB14D	0879
MBCZB14D5	0879
MBCZB14D10	0879
MBCZB15D	0686
MBCZB15D5	0686
MBCZB15D10	0686
MBCZB16D	0445
MBCZB16D5	0445
MBCZB16D10	0445
MBCZB17D	0210
MBCZB17D5	0210
MBCZB17D10	0210
MBCZB18D	0375
MBCZB18D5	0375
MBCZB18D10	0375
MBCZB19D	0671
MBCZB19D5	0671
MBCZB19D10	0671
MBCZB20D	0700
MBCZB20D5	0700
MBCZB20D10	0700
MBCZB22D	0705
MBCZB22D5	0705
MBCZB22D10	0705
MBCZB24D	0494
MBCZB24D5	0494
MBCZB24D10	0494
MBCZB25D	0714
MBCZB25D5	0714
MBCZB25D10	0714
MBCZB27D	0455
MBCZB27D5	0455
MBCZB27D10	0455
MBCZB30D	0195
MBCZB30D5	0195
MBCZB30D10	0195
MBCZB33D	0166
MBCZB33D5	0166
MBCZB33D10	0166
MBCZB36D	0010
MBCZB36D5	0010
MBCZB36D10	0010
MBCZB39D	0032
MBCZB39D5	0032
MBCZB39D10	0032
MBCZB43D	0054
MBCZB43D5	0054
MBCZB43D10	0054
MBCZB47D	0068
MBCZB47D5	0068
MBCZB47D10	0068

DEVICE TYPE	REPL CODE
MBCZB51D	0092
MBCZB51D5	0092
MBCZB51D10	0092
MBCZB56D	0125
MBCZB56D5	0125
MBCZB56D10	0125
MBCZB62D	0152
MBCZB62D5	0152
MBCZB62D10	0152
MBCZB68D	0173
MBCZB68D5	0173
MBCZB68D10	0173
MBCZB75D	0094
MBCZB75D5	0165
MBCZB75D10	0094
MBCZB82D	0049
MBCZB82D5	0049
MBCZB82D10	0049
MBCZB91D	0156
MBCZB91D5	0156
MBCZB91D10	0156
MBCZB100D	0189
MBCZB100D5	0189
MBCZB100D10	0189
MBCZB105D	0099
MBCZB105D5	0099
MBCZB105D10	0099
MBCZB110D	0099
MBCZB110D5	0099
MBCZB110D10	0099
MBCZB120D	0089
MBCZB120D5	0089
MBCZB120D10	0089
MBCZB130D	0287
MBCZB130D5	0287
MBCZB130D10	0287
MBCZB140D	0152
MBCZB140D5	0152
MBCZB140D10	0152
MBCZB150D	0340
MBCZB150D5	0394
MBCZB150D10	0370
MBCZB160D	0370
MBCZB160D5	0370
MBCZB160D10	0370
MBCZB175D	0425
MBCZB175D5	0425
MBCZB175D10	0425
MBCZB180D	0425
MBCZB180D5	0425
MBCZB180D10	0425
MBCZB200D	1473
MBCZB200D5	1473
MBCZB200D10	1473
MBCZC6.8D	0062
MBCZC6.8D5	0062
MBCZC6.8D10	0062
MBCZC7.5D	0077
MBCZC7.5D5	0077
MBCZC7.5D10	0077
MBCZC8.2D	0165
MBCZC8.2D5	0165
MBCZC8.2D10	0165
MBCZC9.1D	0057
MBCZC9.1D5	0057
MBCZC9.1D10	0057
MBCZC10D	0064
MBCZC10D5	0064
MBCZC10D10	0064
MBCZC11D	0181
MBCZC11D5	0181
MBCZC11D10	0181
MBCZC12D	0052
MBCZC12D5	0052
MBCZC12D10	0052
MBCZC13D	0053
MBCZC13D5	0053
MBCZC13D10	0053
MBCZC14D	0879
MBCZC14D5	0879
MBCZC14D10	0879
MBCZC15D	0686
MBCZC15D5	0686
MBCZC15D10	0686
MBCZC16D	0445
MBCZC16D5	0445
MBCZC16D10	0445
MBCZC17D	0210
MBCZC17D5	0210
MBCZC17D10	0210
MBCZC18D	0375
MBCZC18D5	0375
MBCZC18D10	0375
MBCZC19D	0671
MBCZC19D5	0671
MBCZC19D10	0671
MBCZC20D	0700
MBCZC20D5	0700
MBCZC20D10	0700
MBCZC22D	0705
MBCZC22D5	0705
MBCZC22D10	0705
MBCZC24D	0494
MBCZC24D5	0494
MBCZC24D10	0494
MBCZC25D	0714
MBCZC25D5	0714
MBCZC25D10	0714
MBCZC27D	0455

DEVICE TYPE	REPL CODE
MBCZC27D5	0455
MBCZC27D10	0455
MBCZC30D	0195
MBCZC30D5	0195
MBCZC30D10	0195
MBCZC33D	0166
MBCZC33D5	0166
MBCZC33D10	0166
MBCZC36D	0010
MBCZC36D5	0010
MBCZC36D10	0010
MBCZC39D	0032
MBCZC39D5	0032
MBCZC39D10	0032
MBCZC43D	0054
MBCZC43D5	0054
MBCZC43D10	0054
MBCZC47D	0068
MBCZC47D5	0068
MBCZC47D10	0068
MBCZC51D	0092
MBCZC51D5	0092
MBCZC51D10	0092
MBCZC56D	0125
MBCZC56D5	0125
MBCZC56D10	0125
MBCZC62D	0152
MBCZC62D5	0152
MBCZC62D10	0152
MBCZC68D	0173
MBCZC68D5	0173
MBCZC68D10	0173
MBCZC75D	0094
MBCZC75D5	0094
MBCZC82D	0049
MBCZC82D5	0049
MBCZC82D10	0049
MBCZC91D	0156
MBCZC91D5	0156
MBCZC91D10	0156
MBCZC100D	0189
MBCZC100D5	0189
MBCZC100D10	0189
MBCZC105D	0099
MBCZC105D5	0099
MBCZC105D10	0099
MBCZC110D	0099
MBCZC110D5	0099
MBCZC110D10	0099
MBCZC120D	0089
MBCZC120D5	0089
MBCZC120D10	0089
MBCZC130D	0287
MBCZC130D5	0287
MBCZC130D10	0287
MBCZC140D	0252
MBCZC140D5	0252
MBCZC140D10	0252
MBCZC150D	0340
MBCZC150D5	0340
MBCZC150D10	0340
MBCZC160D	0370
MBCZC160D5	0370
MBCZC160D10	0370
MBCZC175D	0425
MBCZC175D5	0425
MBCZC175D10	0425
MBCZC180D	0425
MBCZC180D5	0425
MBCZC180D10	0425
MBCZC200D	1473
MBCZC200D5	1473
MBCZC200D10	1473
MBCZD6.8D	0062
MBCZD6.8D5	0062
MBCZD6.8D10	0062
MBCZD7.5D	0077
MBCZD7.5D5	0077
MBCZD7.5D10	0077
MBCZD8.2D	0165
MBCZD8.2D5	0165
MBCZD8.2D10	0165
MBCZD9.1D	0057
MBCZD9.1D5	0057
MBCZD9.1D10	0057
MBCZD10D	0064
MBCZD10D5	0064
MBCZD10D10	0064
MBCZD11D5	0181
MBCZD11D10	0181
MBCZD12D	0052
MBCZD12D5	0052
MBCZD12D10	0052
MBCZD13D	0053
MBCZD13D5	0053
MBCZD13D10	0053
MBCZD14D	0879
MBCZD14D5	0879
MBCZD14D10	0879
MBCZD15D	0686
MBCZD15D5	0686
MBCZD15D10	0686
MBCZD16D	0445
MBCZD16D5	0445
MBCZD16D10	0445
MBCZD17D	0210
MBCZD17D5	0210
MBCZD17D10	0210
MBCZD18D	0375

DEVICE TYPE	REPL CODE
MBCZD18D5	0375
MBCZD18D10	0375
MBCZD19D	0671
MBCZD19D5	0671
MBCZD19D10	0671
MBCZD20D	0700
MBCZD20D5	0700
MBCZD20D10	0700
MBCZD22D	0705
MBCZD22D5	0705
MBCZD22D10	0705
MBCZD24D	0494
MBCZD24D5	0494
MBCZD24D10	0494
MBCZD25D	0714
MBCZD25D5	0714
MBCZD25D10	0714
MBCZD27D	0455
MBCZD27D5	0455
MBCZD27D10	0455
MBCZD28D	0257
MBCZD28D5	0257
MBCZD28D10	0257
MBCZD30D	0195
MBCZD30D5	0195
MBCZD30D10	0195
MBCZD33D	0166
MBCZD33D5	0166
MBCZD33D10	0166
MBCZD36D	0010
MBCZD36D5	0010
MBCZD36D10	0010
MBCZD39D	0032
MBCZD39D5	0032
MBCZD39D10	0032
MBCZD43D	0054
MBCZD43D5	0054
MBCZD43D10	0054
MBCZD47D	0068
MBCZD47D5	0068
MBCZD47D10	0068
MBCZD51D	0092
MBCZD51D5	0092
MBCZD51D10	0092
MBCZD56D	0125
MBCZD56D5	0125
MBCZD56D10	0125
MBCZD60D	2317
MBCZD60D5	2317
MBCZD60D10	2317
MBCZD62D	0152
MBCZD62D5	0152
MBCZD62D10	0152
MBCZD68D	0173
MBCZD68D5	0173
MBCZD68D10	0173
MBCZD75D	0094
MBCZD75D5	0094
MBCZD75D10	0094
MBCZD82D	0049
MBCZD82D5	0049
MBCZD82D10	0049
MBCZD87D	0104
MBCZD87D5	0104
MBCZD87D10	0104
MBCZD91D	0156
MBCZD91D5	0156
MBCZD91D10	0156
MBCZD100D	0189
MBCZD100D5	0189
MBCZD100D10	0189
MBCZD105D	0099
MBCZD105D5	0099
MBCZD105D10	0099
MBCZD110D	0099
MBCZD110D5	0099
MBCZD110D10	0099
MBCZD120D	0089
MBCZD120D5	0089
MBCZD120D10	0089
MBCZD130D	0287
MBCZD130D5	0287
MBCZD130D10	0287
MBCZD140D	0252
MBCZD140D5	0252
MBCZD140D10	0252
MBCZD150D	0340
MBCZD150D5	0340
MBCZD150D10	0340
MBCZD160D	0370
MBCZD160D5	0370
MBCZD160D10	0370
MBCZD175D	0425
MBCZD175D5	0425
MBCZD175D10	0425
MBCZD180D	0425
MBCZD180D5	0425
MBCZD180D10	0425
MBCZD200D	1473
MBCZD200D5	1473
MBCZD200D10	1473
MBCZE6.8D	0062
MBCZE6.8D5	0062
MBCZE6.8D10	0062
MBCZE7.5D	0077
MBCZE7.5D5	0077
MBCZE7.5D10	0077
MBCZE8.2D	0165
MBCZE8.2D5	0165

DEVICE TYPE	REPL CODE
MBCZE8.2D10	0165
MBCZE9.1D	0057
MBCZE9.1D5	0057
MBCZE9.1D10	0057
MBCZE10D	0064
MBCZE10D5	0064
MBCZE10D10	0064
MBCZE11D	0181
MBCZE11D5	0181
MBCZE11D10	0181
MBCZE12D	0052
MBCZE12D5	0052
MBCZE12D10	0052
MBCZE13D	0053
MBCZE13D5	0053
MBCZE13D10	0053
MBCZE14D	0879
MBCZE14D5	0879
MBCZE14D10	0879
MBCZE15D	0686
MBCZE15D5	0686
MBCZE15D10	0686
MBCZE16D	0445
MBCZE16D5	0445
MBCZE16D10	0445
MBCZE17D	0210
MBCZE17D5	0210
MBCZE17D10	0210
MBCZE18D	0375
MBCZE18D5	0375
MBCZE18D10	0375
MBCZE19D	0671
MBCZE19D5	0671
MBCZE19D10	0671
MBCZE20D	0700
MBCZE20D5	0700
MBCZE20D10	0700
MBCZE22D	0705
MBCZE22D5	0705
MBCZE22D10	0705
MBCZE24D	0494
MBCZE24D5	0494
MBCZE24D10	0494
MBCZE25D	0714
MBCZE25D5	0714
MBCZE25D10	0714
MBCZE27D	0455
MBCZE27D5	0455
MBCZE27D10	0455
MBCZE30D	0195
MBCZE30D5	0195
MBCZE30D10	0195
MBCZE33D	0195
MBCZE33D5	0195
MBCZE33D10	0195
MBCZE36D	0010
MBCZE36D5	0010
MBCZE36D10	0010
MBCZE39D	0032
MBCZE39D5	0032
MBCZE39D10	0032
MBCZE43D	0054
MBCZE43D5	0054
MBCZE43D10	0054
MBCZE47D	0068
MBCZE47D5	0068
MBCZE47D10	0068
MBCZE51D	0092
MBCZE51D5	0092
MBCZE51D10	0092
MBCZE56D	0125
MBCZE56D5	0125
MBCZE56D10	0125
MBCZE62D	0152
MBCZE62D5	0152
MBCZE62D10	0152
MBCZE68D	0173
MBCZE68D5	0173
MBCZE68D10	0173
MBCZE75D	0094
MBCZE75D5	0094
MBCZE75D10	0094
MBCZE82D	0049
MBCZE82D5	0049
MBCZE82D10	0049
MBCZE91D	0156
MBCZE91D5	0156
MBCZE91D10	0156
MBCZE100D	0189
MBCZE100D5	0189
MBCZE100D10	0189
MBCZE105D	0099
MBCZE105D5	0099
MBCZE105D10	0099
MBCZE110D	0099
MBCZE110D5	0099
MBCZE110D10	0099
MBCZE120D	0089
MBCZE120D5	0089
MBCZE120D10	0089
MBCZE130D	0287
MBCZE130D5	0287
MBCZE130D10	0287
MBCZE140D	0252
MBCZE140D5	0252
MBCZE140D10	0252
MBCZE150D	0340
MBCZE150D5	0340
MBCZE150D10	0340

DEVICE TYPE	REPL CODE
MBCZE160D	0370
MBCZE160D5	0370
MBCZE160D10	0370
MBCZE175D	0425
MBCZE175D5	0425
MBCZE175D10	0425
MBCZE180D	0425
MBCZE180D5	0425
MBCZE180D10	0425
MBCZE200D	1473
MBCZE200D5	1473
MBCZE200D10	1473
MBCZF1.8D	OEM
MBCZF1.8D5	OEM
MBCZF1.8D10	OEM
MBCZF2.0D	OEM
MBCZF2.0D5	OEM
MBCZF2.0D10	OEM
MBCZF2.2D	OEM
MBCZF2.2D5	OEM
MBCZF2.2D10	OEM
MBCZF2.4D	1276
MBCZF2.4D5	1276
MBCZF2.4D10	1276
MBCZF2.7D	0760
MBCZF2.7D5	0760
MBCZF2.7D10	0760
MBCZF3.0D	0118
MBCZF3.0D5	0118
MBCZF3.0D10	0118
MBCZF3.3D	0299
MBCZF3.3D5	0299
MBCZF3.3D10	0299
MBCZF3.6D	0376
MBCZF3.6D5	0376
MBCZF3.6D10	0376
MBCZF3.9D	0036
MBCZF3.9D5	0036
MBCZF3.9D10	0036
MBCZF4.3D	0275
MBCZF4.3D5	0275
MBCZF4.3D10	0275
MBCZF4.7D	0140
MBCZF4.7D5	0140
MBCZF4.7D10	0140
MBCZF5.1D	0041
MBCZF5.1D5	0041
MBCZF5.1D10	0041
MBCZF5.6D	0253
MBCZF5.6D5	0253
MBCZF5.6D10	0253
MBCZF6.2D	0471
MBCZF6.2D5	0471
MBCZF6.2D10	0471
MBCZF6.8D	0062
MBCZF6.8D5	0062
MBCZF6.8D10	0062
MBCZF7.5D	0077
MBCZF7.5D5	0165
MBCZF7.5D10	0077
MBCZF8.2D	0165
MBCZF8.2D5	0165
MBCZF8.2D10	0165
MBCZF9.1D	0057
MBCZF9.1D5	0057
MBCZF9.1D10	0057
MBCZF10D	0064
MBCZF10D5	0064
MBCZF10D10	0064
MBCZF12D	0052
MBCZF12D5	0052
MBCZF12D10	0052
MBCZG2.4D	1276
MBCZG2.4D5	1276
MBCZG2.4D10	1276
MBCZG2.7D	0760
MBCZG2.7D5	0760
MBCZG2.7D10	0760
MBCZG3.0D	0118
MBCZG3.0D5	0118
MBCZG3.0D10	0118
MBCZG3.3D	0299
MBCZG3.3D5	0299
MBCZG3.3D10	0299
MBCZG3.6D	0376
MBCZG3.6D5	0376
MBCZG3.6D10	0376
MBCZG3.9D	0036
MBCZG3.9D5	0036
MBCZG3.9D10	0036
MBCZG4.3D	0275
MBCZG4.3D5	0275
MBCZG4.3D10	0275
MBCZG4.7D	0140
MBCZG4.7D5	0140
MBCZG4.7D10	0140
MBCZG5.1D	0041
MBCZG5.1D5	0041
MBCZG5.1D10	0041
MBCZG5.6D	0253
MBCZG5.6D5	0253
MBCZG5.6D10	0253
MBCZG6.2D	0471
MBCZG6.2D5	0471
MBCZG6.2D10	0471
MBCZG6.8D	0062
MBCZG6.8D5	0062
MBCZG6.8D10	0062
MBCZG7.5D	0077

DEVICE TYPE	REPL CODE
MBCZG7.5D5	OEM
MBCZG7.5D10	0077
MBCZG8.2D	0165
MBCZG8.2D5	0165
MBCZG8.2D10	0165
MBCZG9.1D	0057
MBCZG9.1D5	0057
MBCZG9.1D10	0057
MBCZG10D	0064
MBCZG10D5	0064
MBCZG10D10	0064
MBCZG12D	0052
MBCZG12D5	0052
MBCZG12D10	0052
MBCZZD11D	OEM
MBD101	0917
MBD102	0917
MBD201	OEM
MBD301	OEM
MBD501	OEM
MBD502	OEM
MBD701	OEM
MBD702	OEM
MBD5300	OEM
MBD5400	OEM
MBD5500	OEM
MBD5500A	OEM
MBD5550	OEM
MBD5550A	OEM
MBI-49-LPC	OEM
MBI-101	0917
MBI101	OEM
MBL6821N	0448
MBL8042H	OEM
MBL8049H	OEM
MBL8051AH	OEM
MBL8086-2	OEM
MBL8086-2P	OEM
MBL8088	OEM
MBL8088-1	OEM
MBL8259A-2	OEM
MBL8259A-2P	OEM
MBL8288	OEM
MBL82284-8	OEM
MBM27C64-20	OEM
MBM27C64-25CV	OEM
MBM27C64-25Z	OEM
MBM27C64-30CV	OEM
MBM27C64-30Z	OEM
MBM27C512-20	OEM
MBM2114A10LP	OEM
MBM2114A15LP	OEM
MBM2114A20LP	OEM
MBM2114A20P	OEM
MBM2147E	1693
MBM2147F35Z	OEM
MBM2147F45Z	OEM
MBM2147H35C	OEM
MBM2147H35F	OEM
MBM2147H35Z	OEM
MBM2147H45C	OEM
MBM2147H45F	OEM
MBM2147H45Z	OEM
MBM2147H55Z	1693
MBM2147H70Z	1693
MBM2148-55	OEM
MBM2148-55LZ	OEM
MBM2148-55Z(A)	OEM
MBM2148-70	OEM
MBM2148-70LZ	OEM
MBM2148-70Z(A)	OEM
MBM2149-45Z	OEM
MBM2149-55LZ	OEM
MBM2149-55Z	OEM
MBM2149-70LZ	OEM
MBM2149-70Z	OEM
MBM2716C	OEM
MBM2716HC	OEM
MBM2716HZ	OEM
MBM2716XZ	OEM
MBM2716Z	OEM
MBM2732-35Z	2697
MBM2732-45Z	2697
MBM2732A-20-QGA0-1	OEM
MBM2732A-20-QGA1	2697
MBM2732A20Z	2697
MBM2732A25	5967
MBM2732A25Z	2697
MBM2732A30	2697
MBM2732A30XZ	OEM
MBM2732A30Z	2697
MBM2732A35XZ	OEM
MBM2732A35Z	2697
MBM2764-20	0812
MBM2764-20Z	OEM
MBM2764-25	0812
MBM2764-25Z	OEM
MBM2764-30	0812
MBM2764-30XZ	OEM
MBM2764-30Z	OEM
MBM10415AC	OEM
MBM10415AHC	OEM
MBM10422A7C	OEM
MBM10422C	OEM
MBM10470-20	OEM
MBM10470A20C	OEM
MBM10474A15C	OEM
MBM10474C	OEM
MBM10480	OEM

DEVICE TYPE	REPL CODE
MBM27128-25	1638
MBM27128-25TED-K	1638
MBM27256	OEM
MBM27256-20	OEM
MBM27256-XX	OEM
MBM93419C	OEM
MBM100422A7C	OEM
MBM100422C	OEM
MBM100422TF	OEM
MBM100470C	OEM
MBM100470CF	OEM
MBM100474-15C	OEM
MBM100474-20C	OEM
MBN101	OEM
MBR115P	1335
MBR120P	0735
MBR130P	0735
MBR140P	0735
MBR160	3888
MBR170	3888
MBR180	3888
MBR190	3888
MBR320M	OEM
MBR320P	OEM
MBR330M	OEM
MBR330P	OEM
MBR335M	OEM
MBR340M	OEM
MBR340P	OEM
MBR370	3604
MBR380	3604
MBR390	3604
MBR735	OEM
MBR745	OEM
MBR920	3372
MBR1020	OEM
MBR1035	OEM
MBR1045	OEM
MBR1060	OEM
MBR1100	OEM
MBR1520	0615
MBR1530	0615
MBR1535	0615
MBR1535CT	4313
MBR1540	0615
MBR1545CT	4313
MBR1635	OEM
MBR1645	OEM
MBR2035CT	4313
MBR2045CT	4313
MBR2060CT	OEM
MBR2520	OEM
MBR2530	OEM
MBR2535	OEM
MBR2535CT	2515
MBR2540	OEM
MBR2545CT	2515
MBR3020CT	OEM
MBR3035CT	OEM
MBR3035PT	6070
MBR3045CT	OEM
MBR3045HI	OEM
MBR3045PT	1942
MBR3100	OEM
MBR3520	5608
MBR3530	OEM
MBR3535	0615
MBR3545	0615
MBR3545H	OEM
MBR3545H1	OEM
MBR4020	1545
MBR4030	1545
MBR4030PF	OEM
MBR4035	1545
MBR4040	1545
MBR5825H	OEM
MBR5825H1	OEM
MBR5831H	OEM
MBR5831H1	OEM
MBR6020	1545
MBR6020B	OEM
MBR6020B0B	OEM
MBR6035	1545
MBR6035B	OEM
MBR6035CT	OEM
MBR6035PF	OEM
MBR6045	1545
MBR6045B	OEM
MBR6045CT	OEM
MBR6045H	OEM
MBR6045H1	OEM
MBR6045PF	OEM
MBR6535	OEM
MBR6545	OEM
MBR7535	OEM
MBR7545	OEM
MBR7545H	OEM
MBR7545H1	OEM
MBR8035	OEM
MBR8045	OEM
MBR12035CT	OEM
MBR12045CT	OEM
MBR20035CT	OEM
MBR20045CT	OEM
MBR30035CT	OEM
MBR30045CT	OEM
MBR583131H	OEM
MBRL030	OEM
MBRL040	OEM

DEVICE TYPE	REPL CODE
MBS4C	OEM
MBS100	OEM
MBS4991	4233
MBS4992	4233
MC	OEM
MC001	OEM
MC001A	OEM
MC002	OEM
MC005	OEM
MC0	OEM
MC010	0015
MC015	0015
MC020	0609
MC020A	0609
MC021	1335
MC021A	1335
MC022	0080
MC022A	0080
MC023	0609
MC023A	0609
MC025	0015
MC030	0795
MC030A	0609
MC030B	0015
MC031	OEM
MC031A	OEM
MC032	OEM
MC032A	OEM
MC035	0015
MC040	0795
MC040A	0795
MC041	OEM
MC041A	OEM
MC042	OEM
MC042A	OEM
MC050	0015
MC050A	0015
MC051	OEM
MC051A	OEM
MC052	OEM
MC052A	OEM
MC060	0015
MC060A	0015
MC061	OEM
MC061A	OEM
MC062	OEM
MC062A	OEM
MC070	0071
MC070A	0071
MC080	0071
MC080A	0071
MC081	OEM
MC081A	OEM
MC082	OEM
MC082A	OEM
MC090	0071
MC090A	0071
MC091	OEM
MC092	OEM
MC092A	OEM
MC093	OEM
MC093A	OEM
MC094	OEM
MC094A	OEM
MC095	OEM
MC095A	OEM
MC096	OEM
MC096A	OEM
MC097A	OEM
MC098	OEM
MC098A	OEM
MC0101PD	OEM
MC02323	OEM
MC09797	OEM
MC1	OEM
MC1B	OEM
MC1C	OEM
MC1D	OEM
MC1E	OEM
MC1F	OEM
MC1G	OEM
MC2	0133
MC3	OEM
MC4B1	OEM
MC4B2	OEM
MC4C1	OEM
MC4C2	OEM
MC5	OEM
MC5B1	OEM
MC5B2	OEM
MC5C1	OEM
MC5C2	OEM
MC6A1	OEM
MC6A2	OEM
MC6B1	OEM
MC6B2	OEM
MC6C1	OEM
MC6C2	OEM
MC7	OEM
MC7A	OEM
MC7A1	OEM
MC7A2	OEM
MC7B	OEM
MC7B1	OEM
MC7B2	OEM
MC7C	OEM
MC7C1	OEM
MC7C2	OEM
MC7D	OEM

DEVICE TYPE	REPL CODE
MC8A1	OEM
MC8A2	OEM
MC8B1	OEM
MC8B2	OEM
MC8T2LD	OEM
MC8T2LDS	OEM
MC8T13L	OEM
MC8T13LD	OEM
MC8T13LDS	OEM
MC8T13LS	OEM
MC8T13P	OEM
MC8T13PD	OEM
MC8T13PDS	OEM
MC8T13PS	OEM
MC8T14L	OEM
MC8T14LD	OEM
MC8T14LDS	OEM
MC8T14LS	OEM
MC8T14P	OEM
MC8T14PD	OEM
MC8T14PDS	OEM
MC8T14PS	OEM
MC8T23L	OEM
MC8T23LD	OEM
MC8T23LDS	OEM
MC8T23LS	OEM
MC8T23P	OEM
MC8T23PD	OEM
MC8T23PDS	OEM
MC8T23PS	OEM
MC8T24L	OEM
MC8T24LS	OEM
MC8T24P	OEM
MC8T24PD	OEM
MC8T24PDS	OEM
MC8T24PS	OEM
MC8T26AL	4516
MC8T26ALD	OEM
MC8T26ALDS	OEM
MC8T26ALS	OEM
MC8T26AP	4516
MC8T26APD	OEM
MC8T26APDS	OEM
MC8T26APS	OEM
MC8T28L	0581
MC8T28LD	OEM
MC8T28LDS	OEM
MC8T28LS	OEM
MC8T28P	0581
MC8T28PD	OEM
MC8T28PDS	OEM
MC8T28PS	OEM
MC8T95L	4647
MC8T95LD	OEM
MC8T95LDS	OEM
MC8T95LS	OEM
MC8T95P	4647
MC8T95PD	OEM
MC8T95PDS	OEM
MC8T95PS	OEM
MC8T96L	4651
MC8T96LD	OEM
MC8T96LDS	OEM
MC8T96LS	OEM
MC8T96P	4651
MC8T96PD	OEM
MC8T96PDS	OEM
MC8T96PS	OEM
MC8T97L	4658
MC8T97LD	OEM
MC8T97LDS	OEM
MC8T97LS	OEM
MC8T97P	4658
MC8T97PD	OEM
MC8T97PDS	OEM
MC8T97PS	OEM
MC8T98L	4667
MC8T98LD	OEM
MC8T98LDS	OEM
MC8T98LS	OEM
MC8T98P	4667
MC8T98PD	OEM
MC8T98PDS	OEM
MC8T98PS	OEM
MC9	OEM
MC9A1	OEM
MC9A2	OEM
MC9B1	OEM
MC9B2	OEM
MC10	0005
MC10A1	OEM
MC10A2	OEM
MC10B1	OEM
MC10B2	OEM
MC10F	OEM
MC10H016L	OEM
MC10H016P	OEM
MC10H100L	OEM
MC10H100P	OEM
MC10H101L	OEM
MC10H101P	OEM
MC10H102L	OEM
MC10H102P	OEM
MC10H103L	OEM
MC10H103P	OEM
MC10H104L	OEM
MC10H104P	OEM
MC10H105L	OEM
MC10H105P	OEM

DEVICE TYPE	REPL CODE
MC10H106L	OEM
MC10H106P	OEM
MC10H107L	OEM
MC10H107P	OEM
MC10H109L	OEM
MC10H109P	OEM
MC10H115L	OEM
MC10H115P	OEM
MC10H116L	OEM
MC10H116P	OEM
MC10H117L	OEM
MC10H117P	OEM
MC10H118L	OEM
MC10H118P	OEM
MC10H119L	OEM
MC10H119P	OEM
MC10H121L	OEM
MC10H121P	OEM
MC10H124L	OEM
MC10H124P	OEM
MC10H125L	OEM
MC10H125P	OEM
MC10H130L	OEM
MC10H130P	OEM
MC10H131L	OEM
MC10H131P	OEM
MC10H136L	OEM
MC10H136P	OEM
MC10H141L	OEM
MC10H141P	OEM
MC10H145L	OEM
MC10H145P	OEM
MC10H155L	OEM
MC10H155P	OEM
MC10H158L	OEM
MC10H158P	OEM
MC10H159L	OEM
MC10H159P	OEM
MC10H160L	OEM
MC10H160P	OEM
MC10H161L	OEM
MC10H161P	OEM
MC10H162L	OEM
MC10H162P	OEM
MC10H164L	OEM
MC10H164P	OEM
MC10H165L	OEM
MC10H165P	OEM
MC10H166L	OEM
MC10H166P	OEM
MC10H173L	OEM
MC10H173P	OEM
MC10H174L	OEM
MC10H174P	OEM
MC10H175L	OEM
MC10H175P	OEM
MC10H176L	OEM
MC10H176P	OEM
MC10H179L	OEM
MC10H179P	OEM
MC10H180L	OEM
MC10H180P	OEM
MC10H181L	OEM
MC10H181P	OEM
MC10H186AL	OEM
MC10H186AP	OEM
MC10H188L	OEM
MC10H188P	OEM
MC10H189L	OEM
MC10H189P	OEM
MC10H209L	OEM
MC10H209P	OEM
MC10H210L	OEM
MC10H210P	OEM
MC10H211L	OEM
MC10H211P	OEM
MC10H330L	OEM
MC10H330P	OEM
MC10H332L	OEM
MC10H332P	OEM
MC10H334L	OEM
MC10H334P	OEM
MC10H350L	OEM
MC10H350P	OEM
MC10H424L	OEM
MC10H424P	OEM
MC11A1	OEM
MC11A2	OEM
MC11B1	OEM
MC11B2	OEM
MC12A1	OEM
MC12A2	OEM
MC12B1	OEM
MC12B2	OEM
MC13A1	OEM
MC13A2	OEM
MC13B1	OEM
MC13B2	OEM
MC14	OEM
MC14A1	OEM
MC14A2	OEM
MC14B1	OEM
MC14B2	OEM
MC14HC247ND	OEM
MC15	OEM
MC15A1	OEM
MC15A2	OEM
MC15B1	OEM
MC15B2	OEM

If replacement code is OEM, contact original manufacturer for replacement.

DEVICE TYPE	REPL CODE
MC16	OEM
MC17	OEM
MC18	OEM
MC19	0609
MC20	OEM
MC20F	OEM
MC21	OEM
MC22	7137
MC22CL	OEM
MC22CP	OEM
MC22CS	OEM
MC22L	OEM
MC22P	OEM
MC22S	OEM
MC25	OEM
MC25F	OEM
MC26S10L	OEM
MC26S10P	OEM
MC30	0071
MC30F	OEM
MC34F19	OEM
MC34F19A	OEM
MC42	OEM
MC43	OEM
MC44	OEM
MC45	OEM
MC46	OEM
MC47	OEM
MC48	OEM
MC49	OEM
MC50	OEM
MC50F	OEM
MC51	0795
MC54F00J	OEM
MC54F00J(A)	OEM
MC54F00N	OEM
MC54F00N(A)	OEM
MC54F02J	OEM
MC54F02J(A)	OEM
MC54F02N	OEM
MC54F02N(A)	OEM
MC54F04J	OEM
MC54F04J(A)	OEM
MC54F04N	OEM
MC54F08J	OEM
MC54F08N	OEM
MC54F8N	OEM
MC54F10J	OEM
MC54F10J(A)	OEM
MC54F10N	OEM
MC54F10N(A)	OEM
MC54F11J	OEM
MC54F11J(A)	OEM
MC54F11N	OEM
MC54F11N(A)	OEM
MC54F20J	OEM
MC54F20J(A)	OEM
MC54F20N	OEM
MC54F20N(A)	OEM
MC54F32J	OEM
MC54F32J(A)	OEM
MC54F32N	OEM
MC54F32N(A)	OEM
MC54F64J	OEM
MC54F64N	OEM
MC54F86J	OEM
MC54F86N	OEM
MC54F109J	OEM
MC54F109N	OEM
MC54F112	OEM
MC54F113	OEM
MC54F138J(A)	OEM
MC54F138N(A)	OEM
MC54F139J	OEM
MC54F139J(A)	OEM
MC54F139N	OEM
MC54F139N(A)	OEM
MC54F151	OEM
MC54F153J	OEM
MC54F153N	OEM
MC54F174	OEM
MC54F190J	OEM
MC54F190N	OEM
MC54F191J	OEM
MC54F191N	OEM
MC54F192	OEM
MC54F193	OEM
MC54F253J	OEM
MC54F253N	OEM
MC54F258J	OEM
MC54F258N	OEM
MC54F350J	OEM
MC54F350N	OEM
MC54F352J	OEM
MC54F352N	OEM
MC54F374J	OEM
MC54F374N	OEM
MC54F378	OEM
MC54F379	OEM
MC54F521J	OEM
MC54F521N	OEM
MC54F533J	OEM
MC54F533N	OEM
MC54HC00J	OEM
MC54HC02J	OEM
MC54HC03J	OEM
MC54HC03J(A)	OEM
MC54HC08BCAN(A)	OEM
MC54HC08J	OEM
MC54HC10J	OEM
MC54HC20J	OEM
MC54HC20J(A)	OEM
MC54HC27J	OEM
MC54HC30J	OEM
MC54HC42J	OEM
MC54HC42J(A)	OEM
MC54HC51J	OEM
MC54HC51J(A)	OEM
MC54HC58J	OEM
MC54HC58J(A)	OEM
MC54HC73J	OEM
MC54HC74J	OEM
MC54HC74J(A)	OEM
MC54HC75J	OEM
MC54HC76J	OEM
MC54HC85J	OEM
MC54HC86J	OEM
MC54HC107J	OEM
MC54HC109J	OEM
MC54HC109J(A)(A)	OEM
MC54HC112J	OEM
MC54HC113J	OEM
MC54HC132J	OEM
MC54HC133J	OEM
MC54HC137J	OEM
MC54HC138J	OEM
MC54HC138J(A)	OEM
MC54HC139J	OEM
MC54HC139J(A)	OEM
MC54HC147J	OEM
MC54HC151J	OEM
MC54HC157J	OEM
MC54HC158J	OEM
MC54HC160J	OEM
MC54HC161J	OEM
MC54HC162J	OEM
MC54HC163J	OEM
MC54HC164J	OEM
MC54HC164J(A)	OEM
MC54HC165J	OEM
MC54HC165J(A)	OEM
MC54HC173J	OEM
MC54HC174J	OEM
MC54HC174J(A)	OEM
MC54HC175J(A)	OEM
MC54HC194J	OEM
MC54HC195J	OEM
MC54HC237J	OEM
MC54HC240J	OEM
MC54HC241J	OEM
MC54HC242J	OEM
MC54HC242J(A)	OEM
MC54HC243J	OEM
MC54HC243J(A)	OEM
MC54HC244J	OEM
MC54HC251J	OEM
MC54HC257J	OEM
MC54HC266J	OEM
MC54HC273J	OEM
MC54HC273J(A)	OEM
MC54HC280J	OEM
MC54HC280J(A)	OEM
MC54HC365J	OEM
MC54HC366J	OEM
MC54HC367J	OEM
MC54HC368J	OEM
MC54HC373J	OEM
MC54HC374J	OEM
MC54HC374J(A)	OEM
MC54HC390J	OEM
MC54HC393J	OEM
MC54HC533J	OEM
MC54HC534J	OEM
MC54HC534J(A)	OEM
MC54HC595J	OEM
MC54HC4002J	OEM
MC54HC4017J	OEM
MC54HC4020J	OEM
MC54HC4020J(A)	OEM
MC54HC4024J	OEM
MC54HC4024J(A)	OEM
MC54HC4040J	OEM
MC54HC4040J(A)	OEM
MC54HC4049J	OEM
MC54HC4050J	OEM
MC54HC4075J	OEM
MC54HC4078J	OEM
MC54HC4511J	OEM
MC54HC4514J	OEM
MC54HC4514J(A)	OEM
MC54HC4538J	OEM
MC54HC4543J	OEM
MC54HC4543J(A)	OEM
MC54HCU04J	OEM
MC54HCU04J(A)	OEM
MC60	OEM
MC61	OEM
MC68A00CL	OEM
MC68A00CP	OEM
MC68A00Cs	OEM
MC68A00L	OEM
MC68A00P	OEM
MC68A00S	OEM
MC68A01L1	OEM
MC68A01P1	OEM
MC68A02CL	OEM
MC68A02CP	OEM
MC68A02L	OEM
MC68A02P	OEM
MC68A03G	OEM
MC68A03L	OEM
MC68A03P	OEM
MC68A05P6L	OEM
MC68A05P6P	OEM
MC68A05P6S	OEM
MC68A08L	OEM
MC68A08P	OEM
MC68A09CL	OEM
MC68A09CP	OEM
MC68A09CS	OEM
MC68A09ECL	OEM
MC68A09ECP	OEM
MC68A09ECS	OEM
MC68A09EL	OEM
MC68A09EP	OEM
MC68A09ES	OEM
MC68A09L	OEM
MC68A09P	OEM
MC68A09S	OEM
MC68A21CL	OEM
MC68A21CP	OEM
MC68A21CS	OEM
MC68A21L	OEM
MC68A21P	OEM
MC68A21S	OEM
MC68A22L	OEM
MC68A22P	OEM
MC68A29CL	OEM
MC68A29CP	OEM
MC68A29CS	OEM
MC68A29L	OEM
MC68A29P	OEM
MC68A29S	OEM
MC68A35CL	OEM
MC68A35CP	OEM
MC68A35CS	OEM
MC68A35L	OEM
MC68A35P	OEM
MC68A35S	OEM
MC68A39L	OEM
MC68A39P	OEM
MC68A40CL	OEM
MC68A40CP	OEM
MC68A40CS	OEM
MC68A40L	OEM
MC68A40P	OEM
MC68A40S	OEM
MC68A44CL	OEM
MC68A44CP	OEM
MC68A44CS	OEM
MC68A44L	OEM
MC68A44P	OEM
MC68A44S	OEM
MC68A45CL	OEM
MC68A45CP	OEM
MC68A45CS	OEM
MC68A45L	OEM
MC68A45P	OEM
MC68A45R1CL	OEM
MC68A45R1CP	OEM
MC68A45R1CS	OEM
MC68A45R1L	OEM
MC68A45R1P	OEM
MC68A45R1S	OEM
MC68A45S	OEM
MC68A46L	OEM
MC68A46P	OEM
MC68A46S	OEM
MC68A50CL	OEM
MC68A50CP	OEM
MC68A50CS	OEM
MC68A50L	OEM
MC68A50P	OEM
MC68A50S	OEM
MC68A52CL	OEM
MC68A52CP	OEM
MC68A52CS	OEM
MC68A52L	OEM
MC68A52P	OEM
MC68A52S	OEM
MC68A54CL	OEM
MC68A54CP	OEM
MC68A54CS	OEM
MC68A54L	OEM
MC68A54P	OEM
MC68A54S	OEM
MC68A488L	OEM
MC68A488P	OEM
MC68A701L	OEM
MC68B00L	OEM
MC68B00P	OEM
MC68B00S	OEM
MC68B01L1	OEM
MC68B01P1	OEM
MC68B02L	OEM
MC68B02P	OEM
MC68B03G	OEM
MC68B03L	OEM
MC68B03P	OEM
MC68B05P6L	OEM
MC68B05P6P	OEM
MC68B05P6S	OEM
MC68B08L	OEM
MC68B08P	OEM
MC68B09CL	OEM
MC68B09CP	OEM
MC68B09CS	OEM
MC68B09ECL	OEM
MC68B09ECP	OEM
MC68B09ECS	OEM
MC68B09EL	OEM
MC68B09EP	OEM
MC68B09ES	OEM
MC68B09L	OEM
MC68B09P	OEM
MC68B09S	OEM
MC68B21L	OEM
MC68B21P	OEM
MC68B21S	OEM
MC68B22L	OEM
MC68B22P	OEM
MC68B29L	OEM
MC68B29P	OEM
MC68B29S	OEM
MC68B35CL	OEM
MC68B35CP	OEM
MC68B35CS	OEM
MC68B35L	OEM
MC68B35P	OEM
MC68B35S	OEM
MC68B39L	OEM
MC68B39P	OEM
MC68B40CL	OEM
MC68B40CP	OEM
MC68B40CS	OEM
MC68B40L	OEM
MC68B40P	OEM
MC68B40S	OEM
MC68B44L	OEM
MC68B44P	OEM
MC68B44S	OEM
MC68B45L	OEM
MC68B45P	OEM
MC68B45R1L	OEM
MC68B45R1P	OEM
MC68B45R1S	OEM
MC68B45S	OEM
MC68B50C	OEM
MC68B50L	OEM
MC68B50P	OEM
MC68B50S	OEM
MC68B52L	OEM
MC68B52P	OEM
MC68B52S	OEM
MC68B54L	OEM
MC68B54P	OEM
MC68B54S	OEM
MC68B488L	OEM
MC68B488P	OEM
MC68B701L	OEM
MC68C68A1	OEM
MC68HC04P2P	OEM
MC68HC04P2Z	OEM
MC68HC05C4-SC403717	OEM
MC68HC05C4P	OEM
MC68HC05C4Z	OEM
MC68HC05N4-SC406656	OEM
MC68HC05T7	OEM
MC68HC05T7-LSC89919B	OEM
MC68HC09EL	OEM
MC68HC09EP	OEM
MC68HC51P	OEM
MC68HC53P	OEM
MC68HC68R1	OEM
MC68HC68R2	OEM
MC68HC68T1P	OEM
MC70	OEM
MC71	OEM
MC72	OEM
MC73	0052
MC73HC158JD	OEM
MC74D2970	OEM
MC74F00J	OEM
MC74F00J(A)	OEM
MC74F00N	OEM
MC74F00N(A)	OEM
MC74F02J	OEM
MC74F02J(A)	OEM
MC74F02N	OEM
MC74F02N(A)	OEM
MC74F04J	OEM
MC74F04N	OEM
MC74F08J	OEM
MC74F08J(A)	OEM
MC74F08N	OEM
MC74F08N(A)	OEM
MC74F0404N	OEM
MC74F10J	OEM
MC74F10J(A)	OEM
MC74F10N	OEM
MC74F10N(A)	OEM
MC74F11J	OEM
MC74F11J(A)	OEM
MC74F11N	OEM
MC74F11N(A)	OEM
MC74F20J	OEM
MC74F20J(A)	OEM
MC74F20N	OEM
MC74F20N(A)	OEM
MC74F32J	OEM
MC74F32J(A)	OEM
MC74F32N	OEM
MC74F32N(A)	OEM
MC74F64J	OEM
MC74F64N	OEM
MC74F74J	OEM
MC74F74N	OEM
MC74F86J	OEM
MC74F86N	OEM
MC74F109J	OEM
MC74F109N	OEM
MC74F112	OEM
MC74F112J	OEM
MC74F112N	OEM
MC74F113	OEM
MC74F113J	OEM
MC74F113N	OEM
MC74F114J	OEM
MC74F114N	OEM
MC74F138J	OEM
MC74F138J(A)	OEM
MC74F138N	OEM
MC74F138N(A)	OEM
MC74F139J	OEM
MC74F139J(A)	OEM
MC74F139N	OEM
MC74F139N(A)	OEM
MC74F151	OEM
MC74F151J	OEM
MC74F151N	OEM
MC74F153J	OEM
MC74F153N	OEM
MC74F157J	OEM
MC74F157N	OEM
MC74F158J	OEM
MC74F158N	OEM
MC74F160J	OEM
MC74F160N	OEM
MC74F161J	OEM
MC74F161N	OEM
MC74F162J	OEM
MC74F162N	OEM
MC74F163J	OEM
MC74F163N	OEM
MC74F168J	OEM
MC74F168N	OEM
MC74F169J	OEM
MC74F169N	OEM
MC74F174	OEM
MC74F174J	OEM
MC74F174N	OEM
MC74F175J	OEM
MC74F175N	OEM
MC74F181J	OEM
MC74F181N	OEM
MC74F182J	OEM
MC74F182N	OEM
MC74F189J	OEM
MC74F189N	OEM
MC74F190J	OEM
MC74F190N	OEM
MC74F191J	OEM
MC74F191N	OEM
MC74F192	OEM
MC74F192J	OEM
MC74F192N	OEM
MC74F193	OEM
MC74F193J	OEM
MC74F193N	OEM
MC74F194J	OEM
MC74F194N	OEM
MC74F195J	OEM
MC74F195N	OEM
MC74F240J	OEM
MC74F240N	OEM
MC74F241J	OEM
MC74F241N	OEM
MC74F242J	OEM
MC74F242N	OEM
MC74F243J	OEM
MC74F243N	OEM
MC74F244J	OEM
MC74F244N	OEM
MC74F245J	OEM
MC74F245N	OEM
MC74F251J	OEM
MC74F251N	OEM
MC74F253J	OEM
MC74F253N	OEM
MC74F257J	OEM
MC74F257N	OEM
MC74F258J	OEM
MC74F258N	OEM
MC74F280J	OEM
MC74F280N	OEM
MC74F283J	OEM
MC74F283N	OEM
MC74F289J	OEM
MC74F289N	OEM
MC74F299J	OEM
MC74F299N	OEM
MC74F323J	OEM
MC74F323N	OEM
MC74F350J	OEM
MC74F350N	OEM
MC74F352J	OEM
MC74F352N	OEM
MC74F353J	OEM
MC74F353N	OEM
MC74F373J	OEM
MC74F373N	OEM
MC74F374J	OEM
MC74F374N	OEM
MC74F378	OEM
MC74F378J	OEM
MC74F378N	OEM
MC74F379	OEM
MC74F379J	OEM
MC74F379N	OEM
MC74F381J	OEM
MC74F381N	OEM
MC74F382J	OEM
MC74F382N	OEM
MC74F521J	OEM
MC74F521N	OEM
MC74F533J	OEM
MC74F533N	OEM
MC74F534J	OEM
MC74F534N	OEM
MC74F537J	OEM
MC74F537N	OEM
MC74F538J	OEM
MC74F538N	OEM
MC74F539J	OEM
MC74F539N	OEM
MC74F620J	OEM
MC74F620N	OEM
MC74F623J	OEM
MC74F623N	OEM
MC74F640J	OEM
MC74F640N	OEM
MC74F643J	OEM
MC74F643N	OEM
MC74F2960	OEM
MC74F2960J	OEM
MC74F2960N	OEM
MC74F2961J	OEM
MC74F2961N	OEM
MC74F2962J	OEM
MC74F2962N	OEM
MC74F2968	OEM
MC74F2968J	OEM
MC74F2968N	OEM
MC74F2969	OEM
MC74F2969J	OEM
MC74F2969N	OEM
MC74F2970J	OEM
MC74F2970N	OEM
MC74H58JS	OEM
MC74H74AF	2492
MC74H74AL	2492
MC74H74P	2492
MC74H87F	2579
MC74H87L	2579
MC74H87P	2579
MC74H237JDS	OEM
MC74H237JS	OEM
MC74HC00F	OEM
MC74HC00J	OEM
MC74HC00JD	OEM
MC74HC00JDS	OEM
MC74HC00JS	OEM
MC74HC00N	1448
MC74HC00ND	OEM
MC74HC00NDS	OEM
MC74HC00NS	OEM
MC74HC02	OEM
MC74HC02J	OEM
MC74HC02JD	OEM
MC74HC02JDS	OEM
MC74HC02JS	OEM
MC74HC02N	1452
MC74HC02ND	OEM
MC74HC02NDS	OEM
MC74HC02NS	OEM
MC74HC03J	OEM
MC74HC03J(A)	OEM
MC74HC03JD	OEM
MC74HC03JDS	OEM
MC74HC03JS	OEM
MC74HC03N	1453
MC74HC03N(A)	OEM
MC74HC03ND	OEM
MC74HC03NDS	OEM
MC74HC03NS	OEM
MC74HC04F	1455
MC74HC08	4289
MC74HC08BCAJ(A)	OEM
MC74HC08BCAN(A)	OEM
MC74HC08J	OEM
MC74HC08JD	OEM
MC74HC08JDS	OEM
MC74HC08JS	OEM
MC74HC08N	2408
MC74HC08ND	OEM
MC74HC08NDS	OEM
MC74HC08NS	OEM
MC74HC10F	OEM
MC74HC10J	OEM
MC74HC10JD	OEM
MC74HC10JDS	OEM
MC74HC10JS	OEM
MC74HC10N	1462
MC74HC10ND	OEM
MC74HC10NDS	OEM
MC74HC10NS	OEM
MC74HC20J	OEM
MC74HC20J(A)	OEM
MC74HC20JD	OEM
MC74HC20JDS	OEM
MC74HC20N	1475
MC74HC20N(A)	OEM
MC74HC20ND	OEM
MC74HC20NDS	OEM
MC74HC20NS	OEM
MC74HC27J	OEM
MC74HC27JD	OEM
MC74HC27JS	OEM
MC74HC27N	1485
MC74HC27ND	OEM
MC74HC27NDS	OEM
MC74HC27NS	OEM
MC74HC30J	OEM
MC74HC30N	1489
MC74HC42J	OEM
MC74HC42J(A)	OEM
MC74HC42JD	OEM
MC74HC42JDS	OEM
MC74HC42JS	OEM
MC74HC42N	1515
MC74HC42N(A)	OEM
MC74HC42ND	OEM
MC74HC42NDS	OEM
MC74HC42NS	OEM
MC74HC51J	OEM
MC74HC51J(A)	OEM
MC74HC51JD	OEM
MC74HC51JDS	OEM
MC74HC51JS	OEM
MC74HC51N	OEM
MC74HC51N(A)	OEM
MC74HC51ND	OEM
MC74HC51NDS	OEM
MC74HC51NS	OEM
MC74HC58J	OEM
MC74HC58J(A)	OEM
MC74HC58JD	OEM
MC74HC58JDS	OEM
MC74HC58N	OEM
MC74HC58N(A)	OEM
MC74HC58ND	OEM
MC74HC58NDS	OEM
MC74HC58NS	OEM
MC74HC73J	OEM
MC74HC73JD	OEM
MC74HC73JDS	OEM
MC74HC73JS	OEM
MC74HC73N	1558
MC74HC73ND	OEM
MC74HC73NDS	OEM
MC74HC73NS	OEM
MC74HC74J	OEM
MC74HC74J(A)	OEM
MC74HC74JD	OEM
MC74HC74JDS	OEM
MC74HC74JS	OEM
MC74HC74N	1561
MC74HC74N(A)	OEM
MC74HC74ND	OEM
MC74HC74NDS	OEM
MC74HC74NS	OEM
MC74HC75J	OEM
MC74HC75N	1565
MC74HC76J	OEM
MC74HC76JD	OEM
MC74HC76JDS	OEM
MC74HC76N	OEM
MC74HC76ND	OEM
MC74HC76NDS	OEM
MC74HC76NS	OEM
MC74HC85J	OEM
MC74HC85JD	OEM
MC74HC85JDS	OEM
MC74HC85JS	OEM
MC74HC85N	1583
MC74HC85ND	OEM
MC74HC85NDS	OEM
MC74HC85NS	OEM
MC74HC86J	OEM
MC74HC86JD	OEM
MC74HC86JDS	OEM
MC74HC86JS	OEM
MC74HC86N	1588
MC74HC86ND	OEM
MC74HC86NDS	OEM
MC74HC86NS	OEM
MC74HC107J	OEM
MC74HC107JD	OEM
MC74HC107JDS	OEM
MC74HC107JS	OEM
MC74HC107N	4692
MC74HC107ND	OEM
MC74HC107NDS	OEM
MC74HC107NS	OEM
MC74HC109J	OEM
MC74HC109J(A)	OEM
MC74HC109JD	OEM
MC74HC109JDS	OEM
MC74HC109JS	OEM
MC74HC109N	0115
MC74HC109N(A)	OEM
MC74HC109ND	OEM
MC74HC109NDS	OEM
MC74HC109NS	OEM
MC74HC112J	OEM
MC74HC112JD	OEM
MC74HC112JDS	OEM
MC74HC112JS	OEM
MC74HC112N	0214
MC74HC112ND	OEM

DEVICE TYPE	REPL CODE
MC74HC112NDS	OEM
MC74HC112NS	OEM
MC74HC113J	OEM
MC74HC113JD	OEM
MC74HC113JDS	OEM
MC74HC113JS	OEM
MC74HC113N	OEM
MC74HC113ND	OEM
MC74HC113NDS	OEM
MC74HC113NS	OEM
MC74HC132J	OEM
MC74HC132JDS	OEM
MC74HC132JS	OEM
MC74HC132N	0680
MC74HC132ND	OEM
MC74HC132NDS	OEM
MC74HC132NS	OEM
MC74HC133J	OEM
MC74HC133N	OEM
MC74HC137J	OEM
MC74HC137JD	OEM
MC74HC137JDS	OEM
MC74HC137JS	OEM
MC74HC137N	0784
MC74HC137ND	OEM
MC74HC137NDS	OEM
MC74HC137NS	OEM
MC74HC138J	OEM
MC74HC138J(A)	OEM
MC74HC138JD	OEM
MC74HC138JDS	OEM
MC74HC138JS	OEM
MC74HC138N	0797
MC74HC138N(A)	OEM
MC74HC138ND	OEM
MC74HC138NDS	OEM
MC74HC138NS	OEM
MC74HC139J	OEM
MC74HC139JD	OEM
MC74HC139JDS	OEM
MC74HC139JS	OEM
MC74HC139N	0809
MC74HC139N(A)	OEM
MC74HC139ND	OEM
MC74HC139NDS	OEM
MC74HC139NS	OEM
MC74HC147J	OEM
MC74HC147N	0938
MC74HC151	OEM
MC74HC151J	OEM
MC74HC151JD	OEM
MC74HC151JDS	OEM
MC74HC151JS	OEM
MC74HC151N	1003
MC74HC151NDS	OEM
MC74HC151NS	OEM
MC74HC157AN	OEM
MC74HC157J	OEM
MC74HC157JD	OEM
MC74HC157JDS	OEM
MC74HC157JS	OEM
MC74HC157N	1075
MC74HC157ND	OEM
MC74HC157NDS	OEM
MC74HC157NS	OEM
MC74HC158J	OEM
MC74HC158JDS	OEM
MC74HC158JS	OEM
MC74HC158N	1088
MC74HC158ND	OEM
MC74HC158NDS	OEM
MC74HC158NS	OEM
MC74HC160J	OEM
MC74HC160JD	OEM
MC74HC160JDS	OEM
MC74HC160JS	OEM
MC74HC160N	1117
MC74HC160ND	OEM
MC74HC160NDS	OEM
MC74HC160NS	OEM
MC74HC161	5413
MC74HC161J	OEM
MC74HC161JD	OEM
MC74HC161JDS	OEM
MC74HC161JS	OEM
MC74HC161N	1129
MC74HC161ND	OEM
MC74HC161NDS	OEM
MC74HC161NS	OEM
MC74HC162J	OEM
MC74HC162JD	OEM
MC74HC162JDS	OEM
MC74HC162JS	OEM
MC74HC162N	1142
MC74HC162ND	OEM
MC74HC162NDS	OEM
MC74HC162NS	OEM
MC74HC163J	OEM
MC74HC163JD	OEM
MC74HC163JDS	OEM
MC74HC163JS	OEM
MC74HC163N	1159
MC74HC163ND	OEM
MC74HC163NDS	OEM
MC74HC163NS	OEM
MC74HC164J	OEM
MC74HC164J(A)	OEM
MC74HC164JD	OEM
MC74HC164JDS	OEM
MC74HC164JS	OEM
MC74HC164N	1171
MC74HC164N(A)	OEM
MC74HC164ND	OEM
MC74HC164NDS	OEM
MC74HC164NS	OEM
MC74HC165J	OEM
MC74HC165J(A)	OEM
MC74HC165JD	OEM
MC74HC165JDS	OEM
MC74HC165JS	OEM
MC74HC165N	1183
MC74HC165N(A)	OEM
MC74HC165ND	OEM
MC74HC165NDS	OEM
MC74HC165NS	OEM
MC74HC173J	OEM
MC74HC173JD	OEM
MC74HC173JDS	OEM
MC74HC173JS	OEM
MC74HC173N	1281
MC74HC173ND	OEM
MC74HC173NDS	OEM
MC74HC173NS	OEM
MC74HC174J	OEM
MC74HC174J(A)	OEM
MC74HC174JD	OEM
MC74HC174JDS	OEM
MC74HC174JS	OEM
MC74HC174N	1298
MC74HC174N(A)	OEM
MC74HC174ND	OEM
MC74HC174NDS	OEM
MC74HC174NS	OEM
MC74HC175J	OEM
MC74HC175JD	OEM
MC74HC175JDS	OEM
MC74HC175JS	OEM
MC74HC175N	1305
MC74HC175N(A(	OEM
MC74HC175ND	OEM
MC74HC175NDS	OEM
MC74HC175NS	OEM
MC74HC194J	OEM
MC74HC194JD	OEM
MC74HC194JDS	OEM
MC74HC194JS	OEM
MC74HC194N	1557
MC74HC194ND	OEM
MC74HC194NDS	OEM
MC74HC194NS	OEM
MC74HC195J	OEM
MC74HC195JD	OEM
MC74HC195JDS	OEM
MC74HC195JS	OEM
MC74HC195N	1578
MC74HC195ND	OEM
MC74HC195NDS	OEM
MC74HC195NS	OEM
MC74HC237J	OEM
MC74HC237JD	OEM
MC74HC237N	1958
MC74HC237ND	OEM
MC74HC237NDS	OEM
MC74HC237NS	OEM
MC74HC240J	OEM
MC74HC240JD	OEM
MC74HC240JDS	OEM
MC74HC240JS	OEM
MC74HC240N	1979
MC74HC240ND	OEM
MC74HC240NDS	OEM
MC74HC240NS	OEM
MC74HC241J	OEM
MC74HC241JD	OEM
MC74HC241JDS	OEM
MC74HC241JS	OEM
MC74HC241N	1990
MC74HC241NDS	OEM
MC74HC241NS	OEM
MC74HC242J	OEM
MC74HC242J(A)	OEM
MC74HC242JDS	OEM
MC74HC242JS	OEM
MC74HC242N	1996
MC74HC242N(A)	OEM
MC74HC242ND	OEM
MC74HC242NDS	OEM
MC74HC242NS	OEM
MC74HC243J	OEM
MC74HC243J(A)	OEM
MC74HC243JD	OEM
MC74HC243JDS	OEM
MC74HC243JS	OEM
MC74HC243N	2001
MC74HC243N(A)	OEM
MC74HC243NDS	OEM
MC74HC243NS	OEM
MC74HC244J	OEM
MC74HC244JD	OEM
MC74HC244JDS	OEM
MC74HC244JS	OEM
MC74HC244N	2005
MC74HC244ND	OEM
MC74HC244NDS	OEM
MC74HC244NS	OEM
MC74HC245N	2014
MC74HC251J	OEM
MC74HC251JD	OEM
MC74HC251JDS	OEM
MC74HC251JS	OEM
MC74HC251N	2065
MC74HC251ND	OEM
MC74HC251NDS	OEM
MC74HC251NS	OEM
MC74HC257J	OEM
MC74HC257JD	OEM
MC74HC257JDS	OEM
MC74HC257JS	OEM
MC74HC257N	2115
MC74HC257NDS	OEM
MC74HC257NS	OEM
MC74HC266J	OEM
MC74HC266JD	OEM
MC74HC266JDS	OEM
MC74HC266JS	OEM
MC74HC266N	OEM
MC74HC266ND	OEM
MC74HC266NDS	OEM
MC74HC266NS	OEM
MC74HC273J	OEM
MC74HC273J(A)	OEM
MC74HC273JD	OEM
MC74HC273JDS	OEM
MC74HC273JS	OEM
MC74HC273N	2210
MC74HC273N(A)	OEM
MC74HC273ND	OEM
MC74HC273NDS	OEM
MC74HC273NS	OEM
MC74HC280J	OEM
MC74HC280J(A)	OEM
MC74HC280JD	OEM
MC74HC280JDS	OEM
MC74HC280JS	OEM
MC74HC280N	2249
MC74HC280N(A)	OEM
MC74HC280ND	OEM
MC74HC280NDS	OEM
MC74HC280NS	OEM
MC74HC365J	OEM
MC74HC365N	2880
MC74HC366J	OEM
MC74HC366N	2885
MC74HC367J	OEM
MC74HC367N	2894
MC74HC368J	OEM
MC74HC368N	2898
MC74HC373AN	OEM
MC74HC373J	OEM
MC74HC373JD	OEM
MC74HC373JDS	OEM
MC74HC373JS	OEM
MC74HC373N	2920
MC74HC373ND	OEM
MC74HC373NDS	OEM
MC74HC373NS	OEM
MC74HC374J	OEM
MC74HC374J(A)	OEM
MC74HC374JD	OEM
MC74HC374JDS	OEM
MC74HC374JS	OEM
MC74HC374N	2927
MC74HC374N(A)	OEM
MC74HC374ND	OEM
MC74HC374NDS	OEM
MC74HC374NS	OEM
MC74HC374P	OEM
MC74HC390J	OEM
MC74HC390N	3043
MC74HC393J	OEM
MC74HC393N	3064
MC74HC533J	OEM
MC74HC533JD	OEM
MC74HC533JDS	OEM
MC74HC533JS	OEM
MC74HC533N	3677
MC74HC533ND	OEM
MC74HC533NDS	OEM
MC74HC533NS	OEM
MC74HC534J	OEM
MC74HC534J(A)	OEM
MC74HC534JD	OEM
MC74HC534JDS	OEM
MC74HC534JS	OEM
MC74HC534N	3681
MC74HC534ND	OEM
MC74HC534NDS	OEM
MC74HC534NS	OEM
MC74HC589J	OEM
MC74HC589JD	OEM
MC74HC589JDS	OEM
MC74HC589JS	OEM
MC74HC589N	OEM
MC74HC589ND	OEM
MC74HC589NDS	OEM
MC74HC589NS	OEM
MC74HC595J	OEM
MC74HC595JD	OEM
MC74HC595JDS	OEM
MC74HC595JS	OEM
MC74HC595N	OEM
MC74HC595ND	OEM
MC74HC595NDS	OEM
MC74HC595NS	OEM
MC74HC597J	OEM
MC74HC597JD	OEM
MC74HC597JDS	OEM
MC74HC597JS	OEM
MC74HC4002J	OEM
MC74HC4002JD	OEM
MC74HC4002JDS	OEM
MC74HC4002N	5488
MC74HC4002ND	OEM
MC74HC4002NDS	OEM
MC74HC4017J	OEM
MC74HC4017JD	OEM
MC74HC4017JDS	OEM
MC74HC4017JS	OEM
MC74HC4017N	5511
MC74HC4017ND	OEM
MC74HC4017NDS	OEM
MC74HC4017NS	OEM
MC74HC4020J	OEM
MC74HC4020J(A)	OEM
MC74HC4020JD	OEM
MC74HC4020JDS	OEM
MC74HC4020JS	OEM
MC74HC4020N	6347
MC74HC4020N(A)	OEM
MC74HC4020ND	OEM
MC74HC4020NDS	OEM
MC74HC4020NS	OEM
MC74HC4024J	OEM
MC74HC4024J(A)	OEM
MC74HC4024JD	OEM
MC74HC4024JDS	OEM
MC74HC4024JS	OEM
MC74HC4024N	5515
MC74HC4024N(A)	OEM
MC74HC4024ND	OEM
MC74HC4024NDS	OEM
MC74HC4024NS	OEM
MC74HC4040J	OEM
MC74HC4040J(A)	OEM
MC74HC4040JD	OEM
MC74HC4040JDS	OEM
MC74HC4040JS	OEM
MC74HC4040N	6357
MC74HC4040N(A)	OEM
MC74HC4040ND	OEM
MC74HC4040NDS	OEM
MC74HC4040NS	OEM
MC74HC4049J	OEM
MC74HC4049N	5152
MC74HC4050J	OEM
MC74HC4050N	5540
MC74HC4053F	OEM
MC74HC4053N	3806
MC74HC4060J	OEM
MC74HC4060JD	OEM
MC74HC4060JDS	OEM
MC74HC4060JS	OEM
MC74HC4060N	6372
MC74HC4060ND	OEM
MC74HC4060NDS	OEM
MC74HC4060NS	OEM
MC74HC4066N	5571
MC74HC4075J	OEM
MC74HC4075JD	OEM
MC74HC4075JDS	OEM
MC74HC4075JS	OEM
MC74HC4075N	5576
MC74HC4075ND	OEM
MC74HC4075NDS	OEM
MC74HC4075NS	OEM
MC74HC4078J	OEM
MC74HC4078N	OEM
MC74HC4511J	OEM
MC74HC4511JD	OEM
MC74HC4511JDS	OEM
MC74HC4511JS	OEM
MC74HC4511N	6088
MC74HC4511ND	OEM
MC74HC4511NDS	OEM
MC74HC4511NS	OEM
MC74HC4514J	OEM
MC74HC4514J(A)	OEM
MC74HC4514JD	OEM
MC74HC4514JDS	OEM
MC74HC4514JS	OEM
MC74HC4514N	6093
MC74HC4514N(A)	OEM
MC74HC4514ND	OEM
MC74HC4514NDS	OEM
MC74HC4514NS	OEM
MC74HC4538J	OEM
MC74HC4538JD	OEM
MC74HC4538JDS	OEM
MC74HC4538JS	OEM
MC74HC4538N	6126
MC74HC4538ND	OEM
MC74HC4538NDS	OEM
MC74HC4538NS	OEM
MC74HC4543J	OEM
MC74HC4543J(A)	OEM
MC74HC4543JD	OEM
MC74HC4543JDS	OEM
MC74HC4543JS	OEM
MC74HC4543N	6137
MC74HC4543N(A)	OEM
MC74HC4543ND	OEM
MC74HC4543NDS	OEM
MC74HC4543NS	OEM
MC74HCU04J	OEM
MC74HCU04JD	OEM
MC74HCU04JDS	OEM
MC74HCU04JS	OEM
MC74HCU04N	4602
MC74HCU04N(A)	OEM
MC74HCU04ND	OEM
MC74HCU04NDS	OEM
MC74HCU04NS	OEM
MC74HD27JDS	OEM
MC75	3555
MC75F	OEM
MC75HC597N	OEM
MC75HC597ND	OEM
MC75HC597NDS	OEM
MC75HC597NS	OEM
MC75S110L	OEM
MC75S110P	OEM
MC78L02ACG	OEM
MC78L02ACP	OEM
MC78L05	1292
MC78L05,CP	1292
MC78L05ACG	1292
MC78L05ACP	1292
MC78L05ACPRP	1292
MC78L05C	1292
MC78L05CG	1292
MC78L05CP	1292
MC78L06AV	2301
MC78L08ACG	0083
MC78L08ACP	0083
MC78L08AGC	OEM
MC78L08CG	0083
MC78L08CP	0083
MC78L12ACG	1828
MC78L12ACP	1828
MC78L12CG	1828
MC78L12CP	1828
MC78L15ACG	3395
MC78L15ACP	3395
MC78L15CG	3395
MC78L15CP	3395
MC78L15CT	OEM
MC78L18ACG	OEM
MC78L18ACP	4506
MC78L18CG	OEM
MC78L18CP	4506
MC78L24ACG	OEM
MC78L24ACP	4509
MC78L24CG	OEM
MC78L24CP	4509
MC78L25ACP	OEM
MC78M05	OEM
MC78M05BT	OEM
MC78M05CDT	OEM
MC78M05CG	OEM
MC78M05CT	0624
MC78M05CTD	OEM
MC78M06BT	OEM
MC78M06CG	OEM
MC78M06CT	0923
MC78M06CTD	OEM
MC78M08BT	OEM
MC78M08CG	OEM
MC78M08CT	1196
MC78M08CTD	OEM
MC78M12	0334
MC78M12BT	OEM
MC78M12CG	OEM
MC78M12CGCG	OEM
MC78M12CT	0334
MC78M12CTD	OEM
MC78M15BT	OEM
MC78M15CG	OEM
MC78M15CT	1321
MC78M15CTD	OEM
MC78M18BT	OEM
MC78M18CG	OEM
MC78M18CT	2259
MC78M18CTD	OEM
MC78M20BT	OEM
MC78M20CG	OEM
MC78M20CT	OEM
MC78M24BT	OEM
MC78M24CG	OEM
MC78M24CP	2239
MC78M24CT	2239
MC78M24CTD	OEM
MC78T05ACK	OEM
MC78T05ACT	OEM
MC78T05AK	OEM
MC78T05CK	1738
MC78T05CT	OEM
MC78T05K	1916
MC78T06CK	OEM
MC78T06CT	OEM
MC78T06K	OEM
MC78T08CK	OEM
MC78T08CT	OEM
MC78T08K	OEM
MC78T12ACK	OEM
MC78T12ACKCT	OEM
MC78T12ACT	OEM
MC78T12AK	OEM
MC78T12CK	2254
MC78T12CT	0334
MC78T12K	3393
MC78T15ACK	OEM
MC78T15ACT	OEM
MC78T15AK	OEM
MC78T15CK	2462
MC78T15CT	OEM
MC78T15K	OEM
MC78T18CK	OEM
MC78T18CT	OEM
MC78T18K	OEM
MC78T24CK	OEM
MC78T24CT	OEM
MC78T24K	OEM
MC79L00AC	OEM
MC79L03ACG	OEM
MC79L03ACP	OEM
MC79L03CG	OEM
MC79L03CP	OEM
MC79L05AC	4508
MC79L05ACG	4508
MC79L05ACP	4508
MC79L05ACPCP	OEM
MC79L05CG	4508
MC79L05CP	1285
MC79L06	OEM
MC79L12AC	1836
MC79L12ACG	1836
MC79L12ACP	1836
MC79L12CG	1836
MC79L12CP	1836
MC79L15AC	6546
MC79L15ACG	5041
MC79L15ACP	5041
MC79L15CG	5041
MC79L15CP	5041
MC79L18ACG	OEM
MC79L18ACP	4886
MC79L18CG	OEM
MC79L18CP	4886
MC79L24ACG	OEM
MC79L24ACP	1721
MC79L24CG	OEM
MC79L24CP	1721
MC82HS181	OEM
MC82HS191	OEM
MC82HS321	OEM
MC82S181	OEM
MC82S191	OEM
MC82S321	OEM
MC85-001	OEM
MC85-002	OEM
MC85-011	OEM
MC85-012	OEM
MC100	0071
MC100A	0087
MC100F	OEM
MC101	0050
MC103	0050
MC103A	0163
MC104	OEM
MC105	OEM
MC106	0554
MC110	0373
MC110A	0373
MC110AS	0373
MC110S	0373
MC120	0102
MC120A	0102
MC120AS	0102
MC120S	0102
MC125F	OEM
MC130	OEM
MC130A	OEM
MC130AS	OEM
MC130S	OEM
MC1330P	1808
MC140A	OEM
MC140AS	OEM
MC140S	OEM
MC145BL	OEM
MC150A	OEM
MC150AS	OEM
MC150F	OEM
MC150S	OEM
MC153	OEM
MC154	OEM
MC155	OEM
MC155DG	1077
MC156	OEM
MC157	OEM
MC158	OEM
MC159	OEM
MC160BCP	OEM
MC162	OEM
MC170	0133
MC200	OEM
MC201	0030
MC202	OEM
MC203	OEM
MC204	OEM
MC205	OEM
MC206	OEM
MC207	OEM
MC208	OEM
MC209	OEM
MC210	OEM
MC211	OEM
MC212	OEM
MC213	OEM
MC214	OEM
MC215	OEM
MC216	OEM
MC217	OEM
MC218	OEM
MC219	OEM
MC220	OEM
MC221	OEM
MC222	OEM
MC223	OEM
MC224	OEM
MC225	OEM
MC226	OEM
MC227	OEM
MC228	OEM
MC229	OEM
MC230	OEM
MC231	OEM
MC300	OEM
MC-301	0133
MC301	0163
MC302	0163
MC303	OEM
MC303P	2463
MC305	OEM
MC308	0143
MC328	OEM
MC338	OEM
MC350	1367
MC400	OEM
MC401	OEM
MC402	OEM
MC402L	OEM
MC403	OEM
MC405	OEM
MC407	0554
MC410	0005
MC415	OEM
MC422	OEM
MC433	OEM
MC456	1335
MC456A	0133
MC457	0080
MC457A	0133
MC458	0609
MC458A	0133
MC459	0133
MC459A	0133
MC461	0133
MC461A	0133
MC462	0133
MC462A	0080
MC463	0133
MC463A	0609
MC464	0133
MC464A	0609
MC482	0133
MC482A	0133
MC482B	0133
MC483	0133
MC483A	0133
MC483B	0124
MC484	0133
MC484A	0133
MC484B	0609
MC485	0133
MC485A	0133
MC485B	0609
MC486	0795
MC486A	0795
MC486B	0795
MC487	0795
MC487A	0795
MC487B	0795
MC488	0795
MC488A	0795
MC488B	0795
MC500A	OEM
MC500B	OEM
MC501A	OEM
MC501B	OEM
MC502A	OEM
MC502B	OEM
MC503A	OEM
MC503B	OEM
MC504A	OEM
MC504B	OEM
MC505A	OEM
MC505B	OEM
MC506A	OEM
MC506B	OEM
MC507A	OEM
MC507B	OEM
MC508A	OEM
MC508B	OEM
MC509A	OEM
MC509B	OEM
MC510A	OEM
MC510B	OEM
MC511A	OEM
MC511B	OEM
MC512A	OEM
MC512B	OEM
MC513A	OEM
MC513B	OEM
MC514A	OEM
MC514B	OEM
MC515A	OEM
MC515B	OEM
MC516A	OEM
MC516B	OEM
MC517A	OEM
MC517B	OEM
MC518A	OEM
MC518B	OEM

DEVICE TYPE	REPL CODE
MC519A	OEM
MC519B	OEM
MC567	OEM
MC600	OEM
MC600L	OEM
MC600P	OEM
MC601	OEM
MC602	OEM
MC603	OEM
MC605	OEM
MC607	0554
MC610	0005
MC615	OEM
MC622	OEM
MC643	OEM
MC643A	OEM
MC656P	OEM
MC658	OEM
MC658A	OEM
MC659	OEM
MC659A	OEM
MC660	OEM
MC660A	OEM
MC660L	OEM
MC660P	0618
MC660TL	OEM
MC661	OEM
MC661A	OEM
MC661L	OEM
MC661P	2597
MC661TL	OEM
MC662	OEM
MC662A	OEM
MC662L	OEM
MC662P	2563
MC662TL	OEM
MC663	OEM
MC663A	OEM
MC663L	OEM
MC663P	2992
MC663TL	OEM
MC664	OEM
MC664L	OEM
MC664L4L	OEM
MC664P	3032
MC664TL	OEM
MC665L	OEM
MC665P	3054
MC665TL	OEM
MC666L	OEM
MC666P	4142
MC666TL	OEM
MC667	OEM
MC667L	OEM
MC667P	3066
MC667P7P	OEM
MC667TL	OEM
MC668	OEM
MC668L	OEM
MC668P	2434
MC668TL	OEM
MC669	OEM
MC669L	OEM
MC669P	3015
MC670	OEM
MC670L	OEM
MC670P	2350
MC670TL	OEM
MC671	OEM
MC671L	OEM
MC671P	4149
MC671TL	OEM
MC672	0078
MC672L	OEM
MC672P	0078
MC672TL	OEM
MC673	OEM
MC673L	OEM
MC673P	0800
MC673TL	OEM
MC674	OEM
MC674L	OEM
MC674P	4161
MC674RL	OEM
MC674TL	OEM
MC675	OEM
MC675L	OEM
MC675P	4164
MC675TL	OEM
MC676	OEM
MC676L	OEM
MC676P	4170
MC676TL	OEM
MC677	OEM
MC677L	OEM
MC677P	4171
MC677TL	OEM
MC678	0609
MC678L	OEM
MC678P	4173
MC678TL	OEM
MC679	OEM
MC679B	OEM
MC679BL	OEM
MC679BP	OEM
MC679L	OEM
MC679P	4175
MC679TL	OEM
MC680	OEM

DEVICE TYPE	REPL CODE
MC680L	OEM
MC680P	4178
MC680TL	OEM
MC681L	OEM
MC681P	3045
MC681TL	OEM
MC682	OEM
MC682L	OEM
MC682P	3039
MC682TL	OEM
MC683	OEM
MC683L	OEM
MC683P	4185
MC683TL	OEM
MC684	OEM
MC684L	OEM
MC684P	4187
MC684TL	OEM
MC685	OEM
MC685L	OEM
MC685P	4191
MC685TL	OEM
MC686	OEM
MC686L	OEM
MC686P	4195
MC686TL	OEM
MC688	OEM
MC688L	OEM
MC688P	4199
MC688TL	OEM
MC689	OEM
MC689L	OEM
MC689P	4201
MC689TL	OEM
MC690	OEM
MC690L	OEM
MC690P	4204
MC690TL	OEM
MC691	OEM
MC691L	OEM
MC691P	4207
MC691TL	OEM
MC693L	OEM
MC693P	OEM
MC693TL	OEM
MC696	OEM
MC696L	OEM
MC696P	4221
MC696TL	OEM
MC697L	OEM
MC697P	OEM
MC697TL	OEM
MC699L	OEM
MC700G	3418
MC703G	3861
MC704G	3833
MC705G	3714
MC706G	4234
MC707G	4236
MC708G	3845
MC709G	3850
MC710G	1257
MC711G	1364
MC712G	3897
MC713G	3901
MC714G	1069
MC715G	3899
MC717P	OEM
MC720	OEM
MC721G	3337
MC722	OEM
MC723G	OEM
MC723P	OEM
MC723PQ	OEM
MC724	OEM
MC724P	2867
MC726	OEM
MC726G	3924
MC727G	4218
MC728	OEM
MC730	OEM
MC732	OEM
MC774G	1810
MC776P	3427
MC778	OEM
MC779	OEM
MC789	3429
MC789P	3429
MC790P	OEM
MC791P	OEM
MC794	OEM
MC795	OEM
MC796	OEM
MC797	OEM
MC798	OEM
MC799	OEM
MC799P	OEM
MC800	0264
MC800G	3418
MC801	OEM
MC802	OEM
MC803	OEM
MC803G	3861
MC804	0133
MC804G	3833
MC805G	3714
MC806	OEM
MC806G	4234
MC807	OEM

DEVICE TYPE	REPL CODE
MC807G	4236
MC808	OEM
MC808G	3845
MC809	OEM
MC809G	3850
MC810	OEM
MC810G	1257
MC811G	1364
MC812G	3897
MC813G	3901
MC814G	1069
MC815G	3899
MC816P	OEM
MC817P	OEM
MC821G	3337
MC823G	OEM
MC824AP	OEM
MC824P	2867
MC826G	3924
MC827G	4218
MC830F	OEM
MC830G	OEM
MC830L	1823
MC830P	1823
MC831L	3061
MC831P	3061
MC832F	OEM
MC832G	OEM
MC832L	1041
MC832N	1823
MC832P	1041
MC833L	2097
MC833P	2097
MC834L	OEM
MC834P	OEM
MC835F	OEM
MC835L	1176
MC835P	1176
MC836F	OEM
MC836L	1831
MC836P	1831
MC837	OEM
MC837A	OEM
MC837F	OEM
MC837L	1835
MC837P	1835
MC838	OEM
MC838F	OEM
MC838L	OEM
MC838P	OEM
MC839	OEM
MC839F	OEM
MC839L	OEM
MC839P	OEM
MC840	OEM
MC840L	1176
MC840P	1176
MC841	OEM
MC841L	OEM
MC841P	2622
MC842	OEM
MC843	OEM
MC844	OEM
MC844F	OEM
MC844G	OEM
MC844L	0033
MC844P	0033
MC845	OEM
MC845F	OEM
MC845G	OEM
MC845L	0081
MC845P	0081
MC846F	OEM
MC846G	OEM
MC846L	0141
MC846P	0141
MC848F	OEM
MC848G	OEM
MC848L	3399
MC848P	3399
MC849F	OEM
MC849G	OEM
MC849L	1844
MC849P	1844
MC850L	2619
MC850P	2619
MC851F	OEM
MC851G	OEM
MC851L	2078
MC851P	2078
MC852F	OEM
MC852L	0333
MC852P	0333
MC853F	OEM
MC853L	0358
MC853P	0358
MC855F	OEM
MC855L	1481
MC855P	1481
MC856F	OEM
MC856L	1632
MC856P	1632
MC857F	OEM
MC857L	1433
MC857P	1433
MC-858	0466
MC858	0466
MC858F	OEM
MC858L	0466

DEVICE TYPE	REPL CODE
MC858P	0466
MC861F	OEM
MC861G	OEM
MC861L	1859
MC861P	1859
MC862F	OEM
MC862G	OEM
MC862L	0562
MC862P	0562
MC863F	OEM
MC863G	OEM
MC863L	0341
MC863P	0341
MC870P	OEM
MC874G	1810
MC876P	3427
MC886P	OEM
MC887P	OEM
MC889P	3429
MC890	0133
MC890P	OEM
MC891	OEM
MC892	OEM
MC893	OEM
MC899G	OEM
MC899P	OEM
MC900G	OEM
MC903	OEM
MC903A	OEM
MC903G	OEM
MC904	OEM
MC904A	OEM
MC904G	OEM
MC905	OEM
MC905A	OEM
MC905G	OEM
MC906	OEM
MC906A	OEM
MC906G	OEM
MC907	OEM
MC907A	OEM
MC907G	OEM
MC908	OEM
MC908A	OEM
MC908G	OEM
MC909G	OEM
MC910G	OEM
MC911	2203
MC911G	1364
MC912G	OEM
MC913G	OEM
MC914	OEM
MC914A	OEM
MC914G	OEM
MC915G	OEM
MC916	OEM
MC916A	OEM
MC916G	OEM
MC921	3501
MC926G	OEM
MC927G	OEM
MC930F	OEM
MC930G	OEM
MC930L	OEM
MC931	1091
MC932F	OEM
MC932G	OEM
MC932L	OEM
MC933L	OEM
MC934L	OEM
MC935F	OEM
MC935L	OEM
MC936F	OEM
MC936L	OEM
MC937F	OEM
MC937L	OEM
MC938F	OEM
MC938L	OEM
MC939F	OEM
MC939L	OEM
MC940L	OEM
MC941L	OEM
MC944F	OEM
MC944G	OEM
MC944L	0033
MC945F	OEM
MC945G	OEM
MC945L	OEM
MC946F	OEM
MC946G	OEM
MC946L	OEM
MC948F	OEM
MC948G	OEM
MC948L	OEM
MC949F	OEM
MC949G	OEM
MC949L	OEM
MC950L	OEM
MC951F	OEM
MC951G	OEM
MC951L	OEM
MC952F	OEM
MC952L	OEM
MC953F	OEM
MC953L	OEM
MC955F	OEM
MC955L	OEM
MC956F	OEM
MC956L	OEM

DEVICE TYPE	REPL CODE
MC957F	OEM
MC957L	OEM
MC958F	OEM
MC958L	OEM
MC961F	OEM
MC961G	OEM
MC961L	OEM
MC962F	OEM
MC962G	OEM
MC962L	OEM
MC963F	OEM
MC963G	OEM
MC963L	OEM
MC981G	OEM
MC982G	OEM
MC1017P	OEM
MC1026	OEM
MC1031	OEM
MC1031A	OEM
MC1032	OEM
MC1032A	OEM
MC1041	OEM
MC1041A	OEM
MC1042	OEM
MC1042A	OEM
MC1044P	0033
MC1051	OEM
MC1051A	OEM
MC1052	OEM
MC1052A	OEM
MC1055A	OEM
MC1055B	OEM
MC1055C	OEM
MC1055D	OEM
MC1055E	OEM
MC1055F	OEM
MC1055G	OEM
MC1055H	OEM
MC1055I	OEM
MC1055J	OEM
MC1055K	OEM
MC1055L	OEM
MC1055M	OEM
MC1055N	OEM
MC1061	OEM
MC1061A	OEM
MC1062	OEM
MC1062A	OEM
MC1081	OEM
MC1081A	OEM
MC1082	OEM
MC1082A	OEM
MC1091	OEM
MC1091A	OEM
MC1092	OEM
MC1092A	OEM
MC1093	OEM
MC1093A	OEM
MC1094	OEM
MC1094A	OEM
MC1095	OEM
MC1095A	OEM
MC1096	OEM
MC1096A	OEM
MC1097	OEM
MC1097A	OEM
MC1098	OEM
MC1098A	OEM
MC1125G	OEM
MC1158P	OEM
MC1226	OEM
MC1302P	OEM
MC1303	0692
MC1303L	0692
MC1303P	0692
MC1304	0654
MC1304P	0654
MC1304PQ	0654
MC1305	0443
MC1305P	0443
MC1305P-C	0443
MC1305PC	0443
MC1305PQ	0443
MC1306P	2930
MC1307	0701
MC-1307-P	0701
MC1307P	0701
MC1307PQ	0701
MC1309	0706
MC1309P	0706
MC1310	0520
MC1310A	0520
MC-1310P	0520
MC1310P	0520
MC1311	0632
MC1312P	1611
MC1313P	OEM
MC1314G	2624
MC1314P	2986
MC1315P	2995
MC1316P	3006
MC1317P	0701
MC1324	0751
MC1324P	0751
MC-1326	0751
MC1326	0753
MC1326P	0751
MC1326PQ	0751
MC1327AP	OEM

DEVICE TYPE	REPL CODE
MC1327P	OEM
MC1328	0666
MC-1328G	0753
MC1328G	0753
MC1328GT	OEM
MC1328P	0349
MC1328PQ	0349
MC1329P	0666
MC1330	0443
MC1330A	1808
MC1330A1P	1808
MC1330A2P	6986
MC1330P	1808
MC1331	OEM
MC1331P	OEM
MC1335P	OEM
MC1339P	2279
MC1344	0372
MC1344P	0372
MC1345	0779
MC1345P	0372
MC1345PQ	0372
MC1346P	OEM
MC1349P	3195
MC1350	0785
MC1350P	0785
MC1351	0789
MC1351N	0789
MC1351P	0789
MC1352	0395
MC1352P	0395
MC1353P	0855
MC1354	OEM
MC1355P	3219
MC-1356P	0664
MC1356P	1443
MC1357	0664
MC1357A	0664
MC1357P	0664
MC1357PQ	0664
MC1358	0167
MC1358P	0167
MC1358PQ	0167
MC1362P	OEM
MC-1364	0802
MC1364	0802
MC1364G	0364
MC1364P	0802
MC1364PQ	0802
MC1370	0352
MC1370P	0352
MC1371	0354
MC1371P	0354
MC1372	0814
MC1372P	0814
MC1373P	3297
MC1374P	OEM
MC-1375P	1345
MC1375P	1345
MC1375PQ	1345
MC1376	OEM
MC1376P	OEM
MC1377DW	OEM
MC1377P	3312
MC1385P	OEM
MC1391P	0848
MC1393A	OEM
MC1393AP	3380
MC1393P	3380
MC1394	0848
MC1394P	2493
MC1398	0856
MC1398P	0856
MC1399P	3397
MC1399PW	OEM
MC1400AG2	OEM
MC1400AG5	OEM
MC1400AG6	OEM
MC1400AG10	OEM
MC1400AU2	OEM
MC1400AU5	OEM
MC1400AU6	OEM
MC1400AU10	OEM
MC1400G2	OEM
MC1400G5	OEM
MC1400G6	OEM
MC1400G10	OEM
MC1400U2	OEM
MC1400U5	OEM
MC1400U6	OEM
MC1400U10	OEM
MC1401AU5	OEM
MC1403AU	OEM
MC1403D	OEM
MC1403U	OEM
MC1403UD	OEM
MC1403UDS	OEM
MC1403US	OEM
MC1404AU5	OEM
MC1404AU6	OEM
MC1404AU10	OEM
MC1404L	OEM
MC1404U5	OEM
MC1404U6	OEM
MC1404U10	OEM
MC1405L	OEM
MC1406	OEM
MC1406L	OEM
MC1406LD	OEM

DEVICE TYPE	REPL CODE
MC1406LDS	OEM
MC1406LS	OEM
MC1408-7F	OEM
MC1408-7N	4316
MC1408-8D	OEM
MC1408-8F	OEM
MC1408-8N	OEM
MC1408D6	OEM
MC1408D7	OEM
MC1408D8	OEM
MC1408L-6	OEM
MC1408L6	OEM
MC1408L6D	OEM
MC1408L6DS	OEM
MC1408L6S	OEM
MC1408L-7	OEM
MC1408L7	OEM
MC1408L7D	OEM
MC1408L7DS	OEM
MC1408L7S	OEM
MC1408L-8	OEM
MC1408L8	OEM
MC1408L8D	OEM
MC1408L8DS	OEM
MC1408L8S	OEM
MC1408P6	OEM
MC1408P7	OEM
MC1408P8	OEM
MC1411L	OEM
MC1411P	0845
MC1411P1P	OEM
MC1412L	OEM
MC1412P	1007
MC1413L	OEM
MC1413P	1134
MC1413PD	OEM
MC1413PDS	OEM
MC1413PS	OEM
MC1413TL	OEM
MC1413TP	OEM
MC1414L	OEM
MC1414LD	OEM
MC1414LDS	OEM
MC1414LS	OEM
MC1414P	OEM
MC1416L	OEM
MC1416P	1261
MC1422P1	1543
MC1435G	OEM
MC1435L	OEM
MC1436CD	OEM
MC1436CG	OEM
MC1436CU	OEM
MC1436G	OEM
MC1436U	OEM
MC1437L	OEM
MC1437P	OEM
MC1438R	OEM
MC1439G	OEM
MC1439L	OEM
MC1439P1	OEM
MC1439P2	OEM
MC1441P	OEM
MC1444	OEM
MC1444L	OEM
MC1445	OEM
MC1445F	OEM
MC1445G	OEM
MC1445J	OEM
MC1445L	OEM
MC1445N	OEM
MC1445W	OEM
MC1446L	OEM
MC1454G	OEM
MC1455D	3634
MC1455G	3637
MC1455P	OEM
MC1455P1	0973
MC1455P1D	OEM
MC1455P1DS	OEM
MC1455P1S	OEM
MC1455U	OEM
MC1455UD	OEM
MC1455UDS	OEM
MC1455US	OEM
MC1456CG	OEM
MC1456CL	OEM
MC1456CP1	OEM
MC1456CU	OEM
MC1456CUCU	OEM
MC1456F	OEM
MC1456FE	OEM
MC1456G	OEM
MC1456L	OEM
MC1456N	OEM
MC1456P1	OEM
MC1456T	OEM
MC1456U	OEM
MC1456V	OEM
MC1458	0360
MC1458A	OEM
MC1458CD	OEM
MC1458CG	2552
MC1458CL	OEM
MC1458CP	4035
MC1458CP1	0360
MC1458CP1D	OEM
MC1458CP1DS	OEM
MC1458CP1S	OEM

DEVICE TYPE	REPL CODE
MC1458CP2	OEM
MC1458CU	OEM
MC1458CUD	OEM
MC1458CUDS	OEM
MC1458CUS	OEM
MC1458D	OEM
MC1458F	OEM
MC1458FE	OEM
MC1458G	0360
MC1458H	OEM
MC1458JG	OEM
MC1458L	3137
MC1458N	3440
MC1458N-14	OEM
MC1458NG	OEM
MC1458NL	OEM
MC1458NP1	OEM
MC1458NP2	OEM
MC1458NU	OEM
MC1458P	0360
MC1458P1	0360
MC1458P1D	OEM
MC1458P1DS	OEM
MC1458P1S	OEM
MC1458P2	OEM
MC1458SG	OEM
MC1458SL	OEM
MC1458SP1	OEM
MC1458SP2	OEM
MC1458SU	OEM
MC1458T	3137
MC1458U	OEM
MC1458UD	OEM
MC1458UDS	OEM
MC1458US	OEM
MC1458V	0360
MC1461R	3663
MC1462	OEM
MC1463G	OEM
MC1463R	OEM
MC1465	OEM
MC1466L	OEM
MC1468BAL	OEM
MC1468G	OEM
MC1468L	3686
MC1468LD	OEM
MC1468LDS	OEM
MC1468LS	OEM
MC1468R	OEM
MC1469G	OEM
MC1469PS	OEM
MC1469R	3663
MC1472P1	OEM
MC1472U	OEM
MC1476G	OEM
MC1488	0508
MC1488F	0508
MC1488L	0508
MC1488LD	OEM
MC1488LDS	OEM
MC1488LS	OEM
MC1488N	0508
MC1488P	0508
MC1488PD	0508
MC1488PDS	OEM
MC1488PS	OEM
MC1489	0511
MC1489/SN75189N	0511
MC1489A	0511
MC1489AF	OEM
MC1489AL	OEM
MC1489ALD	OEM
MC1489ALDS	OEM
MC1489ALS	OEM
MC1489AN	OEM
MC1489AP	0511
MC1489APD	OEM
MC1489APDS	OEM
MC1489APS	OEM
MC1489F	0511
MC1489L	0511
MC1489LD	OEM
MC1489LDS	OEM
MC1489LS	OEM
MC1489N	0511
MC1489P	0511
MC1489PD	OEM
MC1489PDS	OEM
MC1489PS	OEM
MC1493BCLD	OEM
MC1494L	OEM
MC1495L	OEM
MC1496	0417
MC1496A	3801
MC1496D	OEM
MC1496G	0417
MC1496H	OEM
MC1496K	0417
MC1496L	3801
MC1496N	3801
MC1496P	3801
MC1500AG2	OEM
MC1500AG5	OEM
MC1500AG6	OEM
MC1500AU2	OEM
MC1500AU5	OEM
MC1500AU6	OEM
MC1500AU10	OEM
MC1500G2	OEM

DEVICE TYPE	REPL CODE
MC1500G5	OEM
MC1500G6	OEM
MC1500G10	OEM
MC1500U2	OEM
MC1500U5	OEM
MC1500U6	OEM
MC1500U10	OEM
MC1503AU	OEM
MC1503U	OEM
MC1504AU5	OEM
MC1504AU6	OEM
MC1504AU10	OEM
MC1504U5	OEM
MC1504U6	OEM
MC1504U10	OEM
MC1505L	OEM
MC1506L	OEM
MC1508-8F	OEM
MC1508-8N	OEM
MC1508L-8	OEM
MC1508L8	OEM
MC1511	OEM
MC1512	OEM
MC1513	OEM
MC1514	OEM
MC1514L	OEM
MC1515	OEM
MC1519G	OEM
MC1520	0080
MC1521	0609
MC1522	0609
MC1523	0795
MC1524	0795
MC1525	0015
MC1526	0015
MC1527	0071
MC1528	0071
MC1529	0071
MC1536G	OEM
MC1536U	OEM
MC1537L	OEM
MC1538R	OEM
MC1539G	OEM
MC1539L	OEM
MC1545	OEM
MC1545F	OEM
MC1545G	OEM
MC1545L	OEM
MC1546	OEM
MC1546L	OEM
MC1547	OEM
MC1548	OEM
MC1549	OEM
MC1550	1077
MC1550F	OEM
MC1550G	1077
MC-1550P	1913
MC1550P	1913
MC1554G	OEM
MC1555G	3637
MC1555U	3780
MC1556F	OEM
MC1556FE	OEM
MC1556G	OEM
MC1556L	OEM
MC1556N	OEM
MC1556T	OEM
MC1556U	OEM
MC1558	OEM
MC1558F	OEM
MC1558FE	OEM
MC1558G	0360
MC1558H	OEM
MC1558JG	OEM
MC1558L	OEM
MC1558N	OEM
MC1558N-14	OEM
MC1558NG	OEM
MC1558NL	OEM
MC1558NU	OEM
MC1558P	2393
MC1558SG	OEM
MC1558SL	OEM
MC1558SU	OEM
MC1558T	3137
MC1558U	OEM
MC1563G	OEM
MC1563R	OEM
MC1566L	OEM
MC1568G	OEM
MC1568L	OEM
MC1568R	OEM
MC1569G	OEM
MC1569R	OEM
MC1580L	OEM
MC1581L	OEM
MC1582L	OEM
MC1583L	OEM
MC1584L	OEM
MC1590G	OEM
MC1594L	OEM
MC1595L	OEM
MC1596F	OEM
MC1596G	OEM
MC1596H	OEM
MC1596L	4131
MC1600	OEM
MC1648F	OEM
MC1648L	OEM

DEVICE TYPE	REPL CODE
MC1648LD	OEM
MC1648LDS	OEM
MC1648LS	OEM
MC1648ML	OEM
MC1648MLD	OEM
MC1648MLDS	OEM
MC1648MP	OEM
MC1648P	OEM
MC1648PD	OEM
MC1648PDS	OEM
MC1648PS	OEM
MC1650F	OEM
MC1650L	OEM
MC1650LD	OEM
MC1650LDS	OEM
MC1650LS	OEM
MC1651F	OEM
MC1651L	OEM
MC1651LD	OEM
MC1651LDS	OEM
MC1651LS	OEM
MC1654L	OEM
MC1654LD	OEM
MC1654LDS	OEM
MC1654LS	OEM
MC1658F	OEM
MC1658L	OEM
MC1658LD	OEM
MC1658LDS	OEM
MC1658LS	OEM
MC1658P	OEM
MC1658PD	OEM
MC1658PDS	OEM
MC1658PS	OEM
MC1660F	OEM
MC1660L	OEM
MC1660LD	OEM
MC1660LDS	OEM
MC1660LS	OEM
MC1662F	OEM
MC1662L	OEM
MC1662LD	OEM
MC1662LDS	OEM
MC1662LS	OEM
MC1664F	OEM
MC1664L	OEM
MC1664LD	OEM
MC1664LDS	OEM
MC1664LS	OEM
MC1666F	OEM
MC1666L	OEM
MC1666LD	OEM
MC1666LDS	OEM
MC1666LS	OEM
MC1668F	OEM
MC1668L	OEM
MC1668LD	OEM
MC1668LDS	OEM
MC1668LS	OEM
MC1670F	OEM
MC1670L	OEM
MC1670LD	OEM
MC1670LDS	OEM
MC1670LS	OEM
MC1672F	OEM
MC1672L	OEM
MC1672LD	OEM
MC1672LDS	OEM
MC1672LS	OEM
MC1674F	OEM
MC1674L	OEM
MC1674LD	OEM
MC1674LDS	OEM
MC1674LS	OEM
MC1678L	OEM
MC1678LD	OEM
MC1678LDS	OEM
MC1678LS	OEM
MC1679L	OEM
MC1679LD	OEM
MC1679LDS	OEM
MC1679LS	OEM
MC1684MLS	OEM
MC1688F	OEM
MC1688L	OEM
MC1688LD	OEM
MC1688LDS	OEM
MC1688LS	OEM
MC1690F	OEM
MC1690L	OEM
MC1690LD	OEM
MC1690LDS	OEM
MC1690LS	OEM
MC1692F	OEM
MC1692L	OEM
MC1692LD	OEM
MC1692LDS	OEM
MC1692LS	OEM
MC1694L	OEM
MC1694LD	OEM
MC1694LDS	OEM
MC1694LS	OEM
MC1697L	OEM
MC1697LD	OEM
MC1697LDS	OEM
MC1697LS	OEM
MC1697P	OEM
MC1697PD	OEM
MC1697PDS	OEM

DEVICE TYPE	REPL CODE
MC1697PS	OEM
MC1699F	OEM
MC1699L	OEM
MC1699LD	OEM
MC1699LDS	OEM
MC1699LS	OEM
MC1709AF	OEM
MC1709AG	OEM
MC1709AL	OEM
MC1709C	OEM
MC1709CF	OEM
MC1709CG	1301
MC1709CL	1705
MC1709CP1	1301
MC1709CP2	1705
MC1709F	OEM
MC1709G	1301
MC1709L	OEM
MC1710CF	OEM
MC1710CG	1797
MC1710CL	1800
MC1710CP	1800
MC1710F	OEM
MC1710G	OEM
MC1710L	OEM
MC1710P	1800
MC1711CF	OEM
MC1711CG	1890
MC1711CL	1897
MC1711CP	1897
MC1711F	OEM
MC1711G	OEM
MC1711L	1897
MC1723CD	OEM
MC1723CG	1192
MC1723CL	0026
MC1723CLD	OEM
MC1723CLDS	OEM
MC1723CLS	OEM
MC1723CP	0026
MC1723CPD	OEM
MC1723CPDS	OEM
MC1723CPS	OEM
MC1723G	1192
MC1723L	OEM
MC1731G	0410
MC1733CG	1121
MC1733CL	1118
MC1733CP	1118
MC1733G	1121
MC1733L	OEM
MC1741	0410
MC1741-5C	0410
MC1741C2P	1982
MC1741CD	1976
MC1741CF	OEM
MC1741CG	0410
MC1741CL	1975
MC1741CP	0312
MC1741CP1	0312
MC1741CP1D	OEM
MC1741CP1DS	OEM
MC1741CP1S	OEM
MC1741CP2	1975
MC1741CPI	OEM
MC1741CU	OEM
MC1741CUCU	OEM
MC1741CUD	OEM
MC1741CUDS	OEM
MC1741CUS	OEM
MC1741F	OEM
MC1741G	0410
MC1741L	OEM
MC1741NCF	OEM
MC1741NCG	0410
MC1741NCL	OEM
MC1741NCP1	0312
MC1741NCP2	OEM
MC1741NCU	OEM
MC1741NF	OEM
MC1741NG	OEM
MC1741NL	OEM
MC1741NU	OEM
MC1741P1	0312
MC1741SCG	0410
MC1741SCP1	3803
MC1741SG	OEM
MC1741U	OEM
MC1747CG	2368
MC1747CL	2358
MC1747CLD	OEM
MC1747CLDS	OEM
MC1747CLS	OEM
MC1747CP2	2358
MC1747CP2D	OEM
MC1747CP2DS	OEM
MC1747CP2S	OEM
MC1747G	2368
MC1747L	2358
MC1748CG	4369
MC1748CP1	3917
MC1748CU	3917
MC1748G	2454
MC1748U	3917
MC1776CG	OEM
MC1776CP1	3942
MC1776CU	OEM
MC1776G	OEM
MC1776U	OEM

DEVICE TYPE	REPL CODE
MC1800F	OEM
MC1800L	OEM
MC1800P	0850
MC1801F	OEM
MC1801L	OEM
MC1801P	0874
MC1802L	OEM
MC1802P	0888
MC1803L	OEM
MC1803P	0900
MC1804L	OEM
MC1804P	0919
MC1805	OEM
MC1805L	OEM
MC1805P	0929
MC1806L	OEM
MC1806P	0945
MC1807L	OEM
MC1807P	0962
MC1808L	OEM
MC1808P	0982
MC1809L	OEM
MC1809P	0998
MC1810L	OEM
MC1810P	1011
MC1811L	OEM
MC1811P	1025
MC1812L	OEM
MC1812P	1039
MC1813L	OEM
MC1813P	4538
MC1814L	OEM
MC1814P	4541
MC1816L	2959
MC1816P	2959
MC1818L	OEM
MC1818P	OEM
MC1820L	OEM
MC1820P	1929
MC1880P	OEM
MC1900F	OEM
MC1900L	OEM
MC1901F	OEM
MC1901L	OEM
MC1902F	OEM
MC1902L	OEM
MC1903F	OEM
MC1903L	OEM
MC1904F	OEM
MC1904L	OEM
MC1905F	OEM
MC1905L	OEM
MC1906F	OEM
MC1906L	OEM
MC1907F	OEM
MC1907L	OEM
MC1908F	OEM
MC1908L	OEM
MC1909F	OEM
MC1909L	OEM
MC1910F	OEM
MC1910L	OEM
MC1911F	OEM
MC1911L	OEM
MC1912F	OEM
MC1912L	OEM
MC1914F	OEM
MC1914L	OEM
MC1918F	OEM
MC1918L	OEM
MC1920L	OEM
MC1946A	0417
MC2001	OEM
MC2114A4	OEM
MC2114A5	OEM
MC2114AL1	OEM
MC2114AL2	OEM
MC2114AL3	OEM
MC2114AL4	OEM
MC2115A	OEM
MC2115A2	OEM
MC2115AL	OEM
MC2115AL2	OEM
MC2115H2	OEM
MC2115H3	OEM
MC2115H4	OEM
MC2118-10	OEM
MC2118-12	OEM
MC2118-15	OEM
MC2125A	OEM
MC2125A2	OEM
MC2125AL	OEM
MC2125AL2	OEM
MC2125H1	OEM
MC2125H2	OEM
MC2125H3	OEM
MC2125H4	OEM
MC2128-15	OEM
MC2128-20	OEM
MC2141-2	OEM
MC2141-3	OEM
MC2141-4	OEM
MC2141-5	OEM
MC2141L3	OEM
MC2141L4	OEM
MC2141L5	OEM
MC2142	OEM
MC2142-2	OEM
MC2142-3	OEM

DEVICE TYPE	REPL CODE
MC2142L	OEM
MC2142L3	OEM
MC2147A	OEM
MC2147A3	OEM
MC2147AL	OEM
MC2147AL3	OEM
MC2147H	OEM
MC2147H1	OEM
MC2147H2	OEM
MC2147H3	OEM
MC2147HL	OEM
MC2147HL3	OEM
MC2148H	OEM
MC2148H2	OEM
MC2148H3	OEM
MC2148HL	OEM
MC2148HL3	OEM
MC2149H	OEM
MC2149H2	OEM
MC2149H3	OEM
MC2149HL	OEM
MC2149HL3	OEM
MC2164-15	OEM
MC2164-20	OEM
MC2164-25	OEM
MC2167-10	OEM
MC2167-55	OEM
MC2167-70	OEM
MC2167L70	OEM
MC2186	OEM
MC2187	OEM
MC2315-3	2491
MC2321	0124
MC2325	0133
MC2326	0133
MC2335	0133
MC2336	0133
MC2338	OEM
MC2355	0124
MC2356	0124
MC2501L	OEM
MC2502L	OEM
MC2503L	OEM
MC2525	0133
MC2526	0133
MC2535	0133
MC2536	0133
MC2555	0124
MC2556	0124
MC2625	0133
MC2626	0133
MC2635	0124
MC2636	0133
MC2652L2(A)	OEM
MC2655	OEM
MC2656	OEM
MC2661A	OEM
MC2661B	OEM
MC2661C	OEM
MC2670L	OEM
MC2670P	OEM
MC2670S	OEM
MC2671L	OEM
MC2671P	OEM
MC2671S	OEM
MC2672L	OEM
MC2672P	OEM
MC2672S	OEM
MC2673L	OEM
MC2673P	OEM
MC2674L	OEM
MC2674P	OEM
MC2674S	OEM
MC2675L	OEM
MC2675P	OEM
MC2675S	OEM
MC2716	OEM
MC2716-1	OEM
MC2716-2	OEM
MC2716-5	OEM
MC2716-6	OEM
MC2732A	OEM
MC2732A2	OEM
MC2732A3	OEM
MC2732A4	OEM
MC2732A20	OEM
MC2732A25	OEM
MC2732A30	OEM
MC2764	OEM
MC2764-2	OEM
MC2764-3	OEM
MC2764-4	OEM
MC2764-25	OEM
MC2764-30	OEM
MC2764-45	OEM
MC2801P	OEM
MC2815	OEM
MC2815-3	OEM
MC2815-4	OEM
MC2816	OEM
MC2816-3	OEM
MC2816-4	OEM
MC2817	OEM
MC2817-3	OEM
MC2817-4	OEM
MC2833	OEM
MC2833P	OEM
MC2836	OEM
MC2838	OEM

DEVICE TYPE	REPL CODE
MC2840-T12-1	OEM
MC2870P	OEM
MC2871	OEM
MC2871A	OEM
MC3000L	0682
MC3000P	0682
MC3001	1540
MC3001L	5355
MC3001P	6332
MC3002	OEM
MC3002L	OEM
MC3002P	OEM
MC3004L	5338
MC3004P	5338
MC3005L	0685
MC3005P	0685
MC3006L	2398
MC3006P	2398
MC3008L	1907
MC3008P	1907
MC3009L	3251
MC3009P	3251
MC3010L	3715
MC3010P	3715
MC3011L	4856
MC3011P	4856
MC3012L	4596
MC3012P	4596
MC3015L	OEM
MC3015P	OEM
MC3016L	5383
MC3016P	5383
MC3018L	2731
MC3018P	2731
MC3019L	2663
MC3019P	2663
MC3020L	1792
MC3020P	1792
MC3021L	5791
MC3021P	5791
MC3022L	OEM
MC3022P	OEM
MC3023L	1944
MC3023P	1944
MC3024L	0559
MC3024P	0559
MC3025L	OEM
MC3025P	OEM
MC3026L	OEM
MC3026P	OEM
MC3028L	OEM
MC3028P	OEM
MC3029L	OEM
MC3029P	OEM
MC3030L	5410
MC3030P	5410
MC3031L	2020
MC3031P	2020
MC3032L	2101
MC3032P	2101
MC3033L	2172
MC3033P	2172
MC3034L	3158
MC3034P	3158
MC3051L	OEM
MC3051P	OEM
MC3052L	OEM
MC3052P	OEM
MC3054L	3264
MC3054P	3264
MC3055L	3313
MC3055P	3313
MC3060L	OEM
MC3060P	OEM
MC3061L	OEM
MC3061P	1629
MC3062L	OEM
MC3062P	1623
MC3063L	2463
MC3101F	OEM
MC3101L	OEM
MC3102F	OEM
MC3102L	OEM
MC3103F	OEM
MC3103L	OEM
MC3107F	OEM
MC3107L	OEM
MC3115F	OEM
MC3115L	OEM
MC3121F	OEM
MC3121L	OEM
MC3122F	OEM
MC3122L	OEM
MC3125F	OEM
MC3125L	OEM
MC3126F	OEM
MC3126L	OEM
MC3128F	OEM
MC3128L	OEM
MC3129F	OEM
MC3129L	OEM
MC3151F	OEM
MC3151L	OEM
MC3152F	OEM
MC3152L	OEM
MC3160F	OEM
MC3160L	OEM
MC3161F	OEM
MC3161L	OEM

If replacement code is OEM, contact original manufacturer for replacement.

DEVICE TYPE	REPL CODE
MC3162F	OEM
MC3162L	OEM
MC3232AL	OEM
MC3232AP	OEM
MC3242AL	OEM
MC3242ALD	OEM
MC3242AP	OEM
MC3242APD	OEM
MC3301P	2247
MC3302	0176
MC3302A	OEM
MC3302F	OEM
MC3302L	OEM
MC3302N	OEM
MC3302P	0176
MC3302PD	OEM
MC3302PDS	OEM
MC3302PS	OEM
MC3303	OEM
MC3303F	OEM
MC3303J	0829
MC3303L	0829
MC3303N	0829
MC3303P	0829
MC3310P	3972
MC3320P	3974
MC3321P	OEM
MC3324AL	OEM
MC3324AP	OEM
MC3324L	OEM
MC3324P	OEM
MC3325P	OEM
MC3333P	OEM
MC3334P	OEM
MC3340	1899
MC-3340P	1899
MC3340P	1899
MC3340P1	0623
MC3340PA	1899
MC3344L	OEM
MC3344P	OEM
MC3346P	1696
MC3350P	OEM
MC3356P	OEM
MC3357	2673
MC3357P	3791
MC3358P1	OEM
MC3359P	3795
MC3360P	0517
MC3361D	OEM
MC3361P	OEM
MC3370P	3944
MC3373	2686
MC3373P	OEM
MC3386P	1696
MC3393P	OEM
MC3393PN	OEM
MC3396P	OEM
MC3401L	2247
MC3401P	2247
MC3403	2704
MC3403D	OEM
MC3403F	OEM
MC3403J	0829
MC3403L	0829
MC3403LD	OEM
MC3403LDS	OEM
MC3403LS	OEM
MC3403N	0829
MC3403P	6698
MC3403PD	OEM
MC3403PDS	OEM
MC3403PS	OEM
MC3405L	OEM
MC3405P	OEM
MC3408L	OEM
MC3410CF	OEM
MC3410CL	OEM
MC3410CN	OEM
MC3410F	OEM
MC3410L	OEM
MC3410N	OEM
MC3412L	OEM
MC3412P	OEM
MC3416L	OEM
MC3416P	OEM
MC3417L	OEM
MC3418L	OEM
MC3419AL	OEM
MC3419CL	OEM
MC3419L	OEM
MC3420L	OEM
MC3420P	OEM
MC3423	2718
MC3423JG	OEM
MC3423P	OEM
MC3423P1	2718
MC3423P1D	OEM
MC3423P1DS	OEM
MC3423P1S	OEM
MC3423U	2718
MC3423UDS	OEM
MC3423US	OEM
MC3424AL	OEM
MC3424AP	OEM
MC3424L	OEM
MC3424P	OEM
MC3425AP1	OEM
MC3425AU	OEM
MC3425P1	OEM
MC3425U	OEM
MC3430L	OEM
MC3430P	OEM
MC3431L	OEM
MC3431P	OEM
MC3432L	OEM
MC3432P	OEM
MC3433L	OEM
MC3433P	OEM
MC3437L	OEM
MC3437P	OEM
MC3438L	OEM
MC3438P	OEM
MC3440AP	OEM
MC3441AP	OEM
MC3443AP	OEM
MC3446	OEM
MC3446AP	OEM
MC3446J	OEM
MC3446N	OEM
MC3447L	OEM
MC3447P	OEM
MC3447P3	OEM
MC3448AL	OEM
MC3448AP	OEM
MC3450L	OEM
MC3450LD	OEM
MC3450LDS	OEM
MC3450LS	OEM
MC3450P	OEM
MC3450PD	OEM
MC3450PDS	OEM
MC3450PS	OEM
MC3452L	OEM
MC3452P	OEM
MC3453L	OEM
MC3453LD	OEM
MC3453P	OEM
MC3453PD	OEM
MC3456L	OEM
MC3456P	3285
MC3456PC	OEM
MC3456PD	OEM
MC3458G	4005
MC3458P1	2330
MC3458U	OEM
MC3459L	OEM
MC3459P	OEM
MC3460L	OEM
MC3460P	OEM
MC3467L	OEM
MC3467P	OEM
MC3468L	OEM
MC3468P	OEM
MC3469P	OEM
MC3470	2759
MC3470AP	2759
MC3470N	2759
MC3470P	2759
MC3476G	3820
MC3476P1	3434
MC3476U	OEM
MC3476U6U	OEM
MC3479P	5823
MC3480L	OEM
MC3480P	OEM
MC3481L	OEM
MC3481P	OEM
MC3482AL	OEM
MC3482BL	OEM
MC3484V2	OEM
MC3484V4	OEM
MC3485L	OEM
MC3485P	OEM
MC3486	OEM
MC3486J	OEM
MC3486L	OEM
MC3486N	OEM
MC3486P	OEM
MC3487	OEM
MC3487J	OEM
MC3487L	OEM
MC3487LD	OEM
MC3487N	OEM
MC3487P	OEM
MC3487PD	OEM
MC3488AP1	OEM
MC3488AU	OEM
MC3488BU	OEM
MC3490P	4947
MC3491P	1877
MC3492P	1877
MC3492PD	OEM
MC3494P	4949
MC3503	OEM
MC3503F	OEM
MC3503J	OEM
MC3503L	OEM
MC3505L	OEM
MC3510F	OEM
MC3510L	OEM
MC3510N	OEM
MC3512L	OEM
MC3517L	OEM
MC3518L	OEM
MC3520L	OEM
MC3523JG	OEM
MC3523U	OEM
MC3524AL	OEM
MC3524L	OEM
MC3525AU	OEM
MC3525U	OEM
MC3556L	OEM
MC3558G	OEM
MC3558U	OEM
MC3628A1	OEM
MC3628A3	OEM
MC3628A4(A)	OEM
MC3632	OEM
MC3632-1	OEM
MC3636B	OEM
MC3636B1	OEM
MC3636B2	OEM
MC3870L	OEM
MC3870P	OEM
MC3870S	OEM
MC4000F	OEM
MC4000L	OEM
MC4000P	OEM
MC4001L	OEM
MC4001P	OEM
MC4002F	OEM
MC4002L	OEM
MC4002P	OEM
MC4004	3121
MC4004F	OEM
MC4004L	OEM
MC4004P	7148
MC4005L	OEM
MC4005P	7148
MC4006L	OEM
MC4006P	OEM
MC4007L	OEM
MC4007P	OEM
MC4008L	OEM
MC4008P	OEM
MC4010L	OEM
MC4010P	OEM
MC4012L	OEM
MC4012P	0498
MC4015L	OEM
MC4015P	OEM
MC4016L	OEM
MC4016P	1143
MC4017L	OEM
MC4017P	OEM
MC4018L	1391
MC4018P	1391
MC4019L	OEM
MC4019P	OEM
MC4021L	OEM
MC4021P	OEM
MC4022L	OEM
MC4022P	OEM
MC4023L	OEM
MC4023P	OEM
MC4024	OEM
MC4024L	OEM
MC4024P	OEM
MC4026L	OEM
MC4026P	OEM
MC4028F	OEM
MC4028L	OEM
MC4028P	OEM
MC4030L	OEM
MC4030P	OEM
MC4032F	OEM
MC4032L	OEM
MC4032P	OEM
MC4035L	OEM
MC4037L	OEM
MC4037P	OEM
MC4038L	OEM
MC4038P	OEM
MC4039L	OEM
MC4039P	OEM
MC4040L	OEM
MC4040P	0011
MC4041P	OEM
MC4042L	OEM
MC4042P	0121
MC4044	0011
MC4044CP	0011
MC4044D	0011
MC4044F	OEM
MC4044L	0011
MC4044P	0011
MC4047	2331
MC4048L	OEM
MC4048P	OEM
MC4048P8P	OEM
MC4050	2027
MC4050L	OEM
MC4050P	OEM
MC4051B	0366
MC4051L	OEM
MC4051P	OEM
MC4056L	OEM
MC4056P	OEM
MC4060L	OEM
MC4060P	OEM
MC4062P	OEM
MC-4080	OEM
MC4300F	OEM
MC4300L	OEM
MC4301L	OEM
MC4302F	OEM
MC4302L	OEM
MC4304F	OEM
MC4304L	OEM
MC4305F	OEM
MC4305L	OEM
MC4306F	OEM
MC4306L	OEM
MC4307F	OEM
MC4307L	OEM
MC4308F	OEM
MC4308L	OEM
MC4310F	OEM
MC4310L	OEM
MC4312F	OEM
MC4312L	OEM
MC4315F	OEM
MC4315L	OEM
MC4316F	OEM
MC4316L	OEM
MC4317F	OEM
MC4317L	OEM
MC4318F	OEM
MC4318L	OEM
MC4319F	OEM
MC4319L	OEM
MC4321F	OEM
MC4321L	OEM
MC4322F	OEM
MC4322L	OEM
MC4323F	OEM
MC4323L	OEM
MC4324F	OEM
MC4324L	OEM
MC4326F	OEM
MC4326L	OEM
MC4328F	OEM
MC4328L	OEM
MC4330F	OEM
MC4330L	OEM
MC4332F	OEM
MC4332L	OEM
MC4335F	OEM
MC4335L	OEM
MC4337F	OEM
MC4337L	OEM
MC4342F	OEM
MC4342L	OEM
MC4344F	OEM
MC4344L	OEM
MC4348L	OEM
MC4350F	OEM
MC4350L	OEM
MC4356F	OEM
MC4356L	OEM
MC4360F	OEM
MC4360L	OEM
MC4558ACP1	0360
MC4558C	4035
MC4558CD	5531
MC4558CG	OEM
MC4558CP	0360
MC4558CP1	0360
MC4558CU	0360
MC4558G	OEM
MC4558NCG	OEM
MC4558NCP1	OEM
MC4558NCU	OEM
MC4558NG	OEM
MC4558NU	OEM
MC4558S	OEM
MC4558U	OEM
MC4741CD	OEM
MC4741CL	2823
MC4741CLD	OEM
MC4741CP	2823
MC4741CPD	OEM
MC4741L	2823
MC4741LD	OEM
MC5201	3609
MC5223	OEM
MC5321	0124
MC5404F	OEM
MC5480L	OEM
MC5600	OEM
MC5601	OEM
MC5602	OEM
MC5603	OEM
MC5604	OEM
MC5605	OEM
MC5606	OEM
MC5607	OEM
MC5610	OEM
MC5611	OEM
MC5612	OEM
MC5613	OEM
MC5614	OEM
MC5615	OEM
MC5616	OEM
MC5617	OEM
MC5618	OEM
MC5619	OEM
MC6007	0157
MC6007A	0157
MC6008	0471
MC6008A	0471
MC6009	OEM
MC6009A	OEM
MC6010	0062
MC6010A	0062
MC6011	0077
MC6011A	0077
MC6012	0165
MC6012A	0165
MC6013	0057
MC6013A	0057
MC6014	0170
MC6014A	0170
MC6015	0181
MC6015A	0181
MC6016	0052
MC6016A	0052
MC6017	0053
MC6017A	0053
MC6018	0686
MC6018A	0686
MC6019	0445
MC6019A	0445
MC6020	0375
MC6020A	0375
MC6021	0700
MC6021A	0700
MC6022	0705
MC6022A	0705
MC6023	0494
MC6023A	0494
MC6024	0441
MC6024A	0441
MC6025	0195
MC6025A	0195
MC6026	0039
MC6026A	0039
MC6027	0010
MC6027A	0010
MC6028	0032
MC6028A	0032
MC6029	0054
MC6029A	0054
MC6030	0068
MC6030A	0068
MC6107	0157
MC6107A	0157
MC6108	0636
MC6108A	0636
MC6108L	OEM
MC6109	OEM
MC6109A	OEM
MC6110	0025
MC6110A	0025
MC6111	0649
MC6111A	0649
MC6112	0244
MC6112A	0244
MC6113	0012
MC6113A	0012
MC6114	0170
MC6114A	0170
MC6115	0317
MC6115A	0317
MC6116	0137
MC6116A	0137
MC6117	0365
MC6117A	0365
MC6118	0002
MC6118A	0002
MC6119	0420
MC6119A	0420
MC6120	0495
MC6120A	0495
MC6121	0531
MC6121A	0531
MC6122	0565
MC6122A	0565
MC6123	0565
MC6123A	0402
MC6124	0441
MC6124A	0441
MC6125	0726
MC6125A	0726
MC6126	0039
MC6126A	0039
MC6127	0820
MC6127A	0820
MC6128	0350
MC6128A	0350
MC6129	0931
MC6129A	0931
MC6130	0999
MC6130A	0999
MC6172L	OEM
MC6172P	OEM
MC6173L	OEM
MC6173P	OEM
MC6187P	OEM
MC6190L	OEM
MC6190P	OEM
MC6191L	OEM
MC6191P	OEM
MC6192L	OEM
MC6192P	OEM
MC6193L	OEM
MC6193P	OEM
MC6194L	OEM
MC6194P	OEM
MC6195L	OEM
MC6195P	OEM
MC6196L	OEM
MC6196P	OEM
MC6308	0185
MC6308A	0185
MC6309	OEM
MC6309A	OEM
MC6310	0205
MC6310A	0205
MC6311	0480
MC6311A	0480
MC6312	0504
MC6312A	0504
MC6313	0684
MC6313A	0684
MC6314	0225
MC6314A	0225
MC6315	0230
MC6315A	0230
MC6316	0234
MC6316A	0234
MC6317	0237
MC6317A	0237
MC6318	0247
MC6318A	0247
MC6319	0251
MC6319A	0251
MC6320	0256
MC6320A	0256
MC6321	0262
MC6321A	0262
MC6322	0269
MC6322A	0269
MC6323	0274
MC6323A	0274
MC6324	0294
MC6324A	0294
MC6325	0309
MC6325A	0309
MC6326	0318
MC6326A	0318
MC6327	0320
MC6327A	0320
MC6328	0326
MC6328A	0326
MC6329	0337
MC6329A	0337
MC6330	0347
MC6330A	0347
MC6400	OEM
MC6400A	OEM
MC6400B	OEM
MC6400C	OEM
MC6401A	OEM
MC6401B	OEM
MC6401C	OEM
MC6402A	OEM
MC6402B	OEM
MC6402C	OEM
MC6403A	OEM
MC6403B	OEM
MC6403C	OEM
MC6404A	OEM
MC6404B	OEM
MC6404C	OEM
MC6405A	OEM
MC6405B	OEM
MC6405C	OEM
MC6406A	OEM
MC6406B	OEM
MC6406C	OEM
MC6407A	OEM
MC6407B	OEM
MC6407C	OEM
MC6408A	OEM
MC6408B	OEM
MC6408C	OEM
MC6409A	OEM
MC6409B	OEM
MC6409C	OEM
MC6410A	OEM
MC6410B	OEM
MC6410C	OEM
MC6411A	OEM
MC6411B	OEM
MC6411C	OEM
MC6412A	OEM
MC6412B	OEM
MC6412C	OEM
MC6413A	OEM
MC6413B	OEM
MC6413C	OEM
MC6414A	OEM
MC6414B	OEM
MC6414C	OEM
MC6415B	OEM
MC6415C	OEM
MC6416	OEM
MC6416A	OEM
MC6416B	OEM
MC6416C	OEM
MC6417	OEM
MC6417A	OEM
MC6417B	OEM
MC6417C	OEM
MC6418	OEM
MC6418A	OEM
MC6418B	OEM
MC6418C	OEM
MC6419	OEM
MC6419A	OEM
MC6419B	OEM
MC6419C	OEM
MC6420	OEM
MC6420A	OEM
MC6420B	OEM
MC6420C	OEM
MC6421	OEM
MC6421A	OEM
MC6421B	OEM
MC6421C	OEM
MC6422	OEM
MC6422A	OEM
MC6422B	OEM
MC6422C	OEM
MC6423	OEM
MC6423A	OEM
MC6423B	OEM
MC6423C	OEM
MC6424	OEM
MC6424A	OEM
MC6424B	OEM
MC6424C	OEM
MC6425	OEM
MC6425A	OEM
MC6425B	OEM
MC6425C	OEM
MC6426	OEM
MC6426A	OEM
MC6426B	OEM
MC6426C	OEM
MC6427	OEM
MC6427A	OEM
MC6427B	OEM
MC6427C	OEM
MC6428	OEM
MC6428A	OEM
MC6428B	OEM
MC6428C	OEM
MC6429	OEM
MC6429A	OEM
MC6429B	OEM
MC6429C	OEM
MC6525P	OEM
MC6800BQCS	OEM
MC6800CL	0388
MC6800CP	0388
MC6800CQCS	OEM
MC6800CS	OEM
MC6800L	OEM
MC6800P	0388
MC6800S	OEM
MC6801-1G	OEM
MC6801-1L	OEM
MC6801C-1G	OEM
MC6801C-1L	OEM
MC6801CG	OEM
MC6801CL	OEM
MC6801CL1	OEM
MC6801CL1-1	OEM
MC6801CP1	OEM
MC6801G	OEM
MC6801G1	OEM
MC6801L	OEM
MC6801L-1	OEM
MC6801L1	OEM
MC6801L1-1	OEM
MC6801P1	OEM
MC6801U4CL1	OEM
MC6801U4CP1	OEM
MC6801U4L1	OEM
MC6801U4L1-1	OEM
MC6801U4P1	OEM
MC6801U4P1-1	OEM
MC6801VL1	OEM
MC6801VP1	OEM
MC6802CL	OEM
MC6802CP	OEM
MC6802L	OEM
MC6802NSL	OEM
MC6802NSP	OEM
MC6802P	0393
MC6802P2P	OEM
MC6803-1L	OEM
MC6803C-1G	OEM
MC6803C-1L	OEM
MC6803CG	OEM
MC6803CL	OEM
MC6803CL-1	OEM
MC6803CP	OEM
MC6803CP-1	OEM
MC6803CP1-1	OEM
MC6803EG	OEM
MC6803EG-1	OEM
MC6803EL	OEM
MC6803EL-1	OEM
MC6803EP	OEM
MC6803EP-1	OEM
MC6803G	OEM
MC6803L	OEM
MC6803L-1	OEM
MC6803L1	OEM
MC6803P	OEM
MC6803P1	OEM
MC6803P1-1	OEM
MC6803U4CL1	OEM
MC6803U4CP1	OEM
MC6803U4L	OEM
MC6803U4L-1	OEM
MC6803U4P	OEM
MC6803U4P-1	OEM
MC6803VL	OEM

DEVICE TYPE	REPL CODE
MC6803VP	OEM
MC6804	OEM
MC6805P2	OEM
MC6805P2L	OEM
MC6805P2L1	OEM
MC6805P2P	OEM
MC6805P2P1	OEM
MC6805P2S	OEM
MC6805P4L	OEM
MC6805P4P	OEM
MC6805P4S	OEM
MC6805P6L	OEM
MC6805P6P	OEM
MC6805P6S	OEM
MC6805R2L	OEM
MC6805R2L1	OEM
MC6805R2P	OEM
MC6805R2P1	OEM
MC6805R2S	OEM
MC6805R3L	OEM
MC6805R3P	OEM
MC6805R3S	OEM
MC6805T2L	OEM
MC6805T2P	OEM
MC6805U2G	OEM
MC6805U2L	OEM
MC6805U2L2	OEM
MC6805U2P	OEM
MC6805U2P2	OEM
MC6805U2S	OEM
MC6805U3L	OEM
MC6805U3P	OEM
MC6805U3S	OEM
MC6808L	OEM
MC6808P	OEM
MC6809CL	OEM
MC6809CP	OEM
MC6809CS	OEM
MC6809E	2117
MC6809ECL	OEM
MC6809ECP	OEM
MC6809ECS	OEM
MC6809EL	2117
MC6809EP	2117
MC6809ES	4354
MC6809L	5967
MC6809P	5967
MC6809S	6992
MC6820	0448
MC6821	0448
MC6821BQCS	OEM
MC6821CL	OEM
MC6821CP	0448
MC6821CQCS	OEM
MC6821CS	OEM
MC6821L	OEM
MC6821P	0448
MC6821S	OEM
MC6822L	OEM
MC6822P	OEM
MC6822S	OEM
MC6828L	OEM
MC6828P	OEM
MC6829CL	OEM
MC6829CP	OEM
MC6829CS	OEM
MC6829L	OEM
MC6829P	OEM
MC6829S	OEM
MC6835CL	OEM
MC6835CP	OEM
MC6835CS	OEM
MC6835L	OEM
MC6835P	OEM
MC6835S	OEM
MC6840CL	OEM
MC6840CP	OEM
MC6840CS	OEM
MC6840L	OEM
MC6840P	OEM
MC6840S	OEM
MC6843	OEM
MC6843P	OEM
MC6844CL	OEM
MC6844CP	OEM
MC6844CS	OEM
MC6844L	OEM
MC6844P	OEM
MC6844S	OEM
MC6845	OEM
MC6845CL	OEM
MC6845CP	OEM
MC6845CS	OEM
MC6845L	OEM
MC6845P	OEM
MC6845R1CL	OEM
MC6845R1CP	OEM
MC6845R1CS	OEM
MC6845R1L	OEM
MC6845R1P	OEM
MC6845R1S	OEM
MC6845S	OEM
MC6846CL	OEM
MC6846CP	OEM
MC6846CS	OEM
MC6846L	OEM
MC6846L1	OEM
MC6846P	OEM
MC6846P1	OEM
MC6846P3	OEM
MC6846S	OEM
MC6847	OEM
MC6847L	OEM
MC6847P	OEM
MC6847S	OEM
MC6847YL	OEM
MC6847YP	OEM
MC6847YS	OEM
MC6850BJCS	OEM
MC6850CJCS	OEM
MC6850CL	OEM
MC6850CP	0514
MC6850CS	OEM
MC6850L	OEM
MC6850P	0514
MC6850S	OEM
MC6852CL	OEM
MC6852CP	OEM
MC6852CS	OEM
MC6852L	OEM
MC6852P	OEM
MC6852S	OEM
MC6854CL	OEM
MC6854CP	OEM
MC6854CS	OEM
MC6854L	OEM
MC6854P	OEM
MC6854S	OEM
MC6859L	OEM
MC6859L(A)	OEM
MC6859S(A)	OEM
MC6860CL	OEM
MC6860L	2692
MC6860P	2692
MC6860S	OEM
MC6862CL	OEM
MC6862CP	OEM
MC6862CS	OEM
MC6862L	OEM
MC6862P	OEM
MC6862S	OEM
MC6863L	OEM
MC6863P	OEM
MC6875	4181
MC6875AL	OEM
MC6875L	4181
MC6880AL	4516
MC6880AP	4516
MC6882AL	OEM
MC6882BL	OEM
MC6883	OEM
MC6883J(A)	OEM
MC6883L	OEM
MC6883N(A)	OEM
MC6883P	OEM
MC6885L	4647
MC6885P	4647
MC6886L	4651
MC6886P	4651
MC6887L	4658
MC6887P	4658
MC6888L	4667
MC6888P	4667
MC6889L	0581
MC6889P	0581
MC6890AL	OEM
MC6890AL(A)	OEM
MC6890L	OEM
MC6890L(A)	OEM
MC-7000	OEM
MC7000	OEM
MC7001	OEM
MC7002	OEM
MC7003	OEM
MC7004	OEM
MC7005	OEM
MC7006	OEM
MC7007	OEM
MC7008	OEM
MC7009	OEM
MC7400	0232
MC7400F	0232
MC7400L	0232
MC7400N	0232
MC7400P	0232
MC7401F	0268
MC7401L	0268
MC7401P	0268
MC-7402	0314
MC7402	4283
MC7402F	0314
MC7402L	0314
MC7402P	0314
MC7403L	0335
MC7403P	0335
MC7404F	OEM
MC7404L	0361
MC7404P	0361
MC7405L	0385
MC7405P	0385
MC7406L	1206
MC7406P	1206
MC7407L	1339
MC7407P	1339
MC7408L	0467
MC7408P	0467
MC7409L	0492
MC7409P	0492
MC7410F	0512
MC7410L	0512
MC7410P	0512
MC7416L	1349
MC7416P	1349
MC7417L	1352
MC7417P	1352
MC7420F	0697
MC7420L	0697
MC7420P	0697
MC7426L	0803
MC7426P	0803
MC7430F	0873
MC7430L	0873
MC7430P	0873
MC7437F	3520
MC7437L	3520
MC7437P	3520
MC7438F	0996
MC7438L	0996
MC7438P	0996
MC7440F	1024
MC7440L	1024
MC7440P	1024
MC7441AL	1038
MC7441AP	1038
MC7442L	1052
MC7442P	1052
MC7443L	1060
MC7443P	1060
MC7444L	1072
MC7444P	1072
MC7445L	1082
MC7445P	1082
MC7446L	1098
MC7446P	1098
MC7447L	1108
MC7447P	1108
MC7448L	1125
MC7448P	1125
MC7450F	0743
MC7450L	0743
MC7450P	0743
MC7451F	1168
MC7451L	1168
MC7451P	1168
MC7453F	1185
MC7453L	1185
MC7453P	1185
MC7454F	1202
MC7454L	1202
MC7454P	1202
MC7460F	1275
MC7460L	1275
MC7460P	1275
MC7470L	1404
MC7470P	1404
MC7472F	1426
MC7472L	1426
MC7472P	1426
MC7473F	1172
MC7473L	1172
MC7473P	1172
MC7474P	1313
MC7475L	1432
MC7475P	1432
MC7476L	1158
MC7476P	1158
MC7480L	1535
MC7480P	1535
MC7483L	0117
MC7483P	0117
MC7486F	1368
MC7486L	1368
MC7486P	1368
MC7490AP	1208
MC7490F	1208
MC7490L	1208
MC7490P	1208
MC7491AL	0980
MC7491AP	0980
MC7492F	0834
MC7492L	0834
MC7492P	0834
MC7493F	0569
MC7493L	0569
MC7493P	0569
MC7494L	1702
MC7494P	1702
MC7495L	1486
MC7495P	1486
MC7496L	1715
MC7496P	1715
MC7805	0624
MC7805ACK	1916
MC7805ACT	0624
MC7805AK	1916
MC7805BK	1916
MC7805BT	0624
MC7805C	0624
MC7805CK	1916
MC7805CP	0624
MC7805CT	0624
MC7805CTD	OEM
MC7805CTDS	0624
MC7805K	1916
MC7805UC	0624
MC7806ACK	OEM
MC7806ACT	0923
MC7806AK	OEM
MC7806BK	OEM
MC7806BT	0923
MC7806CK	OEM
MC7806CT	0923
MC7806CTD	OEM
MC7806CTS	OEM
MC7806K	OEM
MC7808ACK	OEM
MC7808ACT	6701
MC7808AK	OEM
MC7808BK	OEM
MC7808BT	1196
MC7808C	1196
MC7808CK	OEM
MC7808CP	6071
MC7808CT	1196
MC7808CTD	OEM
MC7808CTDS	OEM
MC7808CTS	OEM
MC7808K	OEM
MC7808UC	OEM
MC7809C	1346
MC7809CT	1346
MC7812	0334
MC7812AC	0334
MC7812ACK	3393
MC7812ACT	0334
MC7812AK	3393
MC7812BK	3393
MC7812BT	0334
MC7812C	0334
MC7812CK	3393
MC7812CT	0334
MC7812CTD	OEM
MC7812CTDS	OEM
MC7812CTS	OEM
MC7812K	3393
MC7815	1321
MC7815ACK	1309
MC7815ACT	1321
MC7815AK	1309
MC7815BK	1309
MC7815BT	1321
MC7815C	4341
MC7815CK	2000
MC7815CP	1321
MC7815CT	1321
MC7815CTD	OEM
MC7815CTDS	OEM
MC7815CTS	OEM
MC7815K	1309
MC7815P	1321
MC7818A	4342
MC7818AC	4342
MC7818ACK	OEM
MC7818ACT	2259
MC7818AK	OEM
MC7818B	OEM
MC7818BK	OEM
MC7818BT	2259
MC7818C	4342
MC7818CK	3650
MC7818CT	2259
MC7818CTD	OEM
MC7818CTDS	OEM
MC7818CTS	OEM
MC7818K	OEM
MC7819CT	3826
MC7824	2239
MC7824ACK	OEM
MC7824ACT	2239
MC7824AK	OEM
MC7824BK	OEM
MC7824BT	2239
MC7824CK	3881
MC7824CT	2239
MC7824CTD	OEM
MC7824CTDS	OEM
MC7824CTS	OEM
MC7824K	OEM
MC7902CK	OEM
MC7902CT	OEM
MC7902CTD	OEM
MC7905	1285
MC7905ACK	OEM
MC7905ACT	1285
MC7905C	1285
MC7905CK	2004
MC7905CT	1285
MC7905CTD	OEM
MC7905CTDS	OEM
MC7905CTS	OEM
MC7905.2ACT	OEM
MC7905.2CK	OEM
MC7905.2CT	1285
MC7905.2CTD	OEM
MC7906ACT	2211
MC7906CK	OEM
MC7906CT	2648
MC7906CTD	OEM
MC7906CTDS	OEM
MC7908ACT	3424
MC7908CK	OEM
MC7908CT	2791
MC7908CTD	OEM
MC7908CTDS	OEM
MC7908CTS	OEM
MC7912ACK	OEM
MC7912ACT	1838
MC7912CK	5287
MC7912CT	1838
MC7912CTD	OEM
MC7912CTDS	OEM
MC7912CTS	OEM
MC7915ACK	OEM
MC7915ACT	3829
MC7915CK	5289
MC7915CP	3829
MC7915CT	3829
MC7915CTD	OEM
MC7915CTDS	OEM
MC7915CTS	OEM
MC7918	3346
MC7918ACT	3346
MC7918C	3346
MC7918CK	3829
MC7918CT	3826
MC7918CTD	OEM
MC7918CTDS	OEM
MC7918CTS	OEM
MC7924ACT	3599
MC7924CK	3573
MC7924CT	3575
MC7924CTD	OEM
MC7924CTDS	OEM
MC7924CTS	OEM
MC8035L	OEM
MC8048	OEM
MC8085A	OEM
MC8086	OEM
MC8088B	OEM
MC8148	OEM
MC8155	OEM
MC8180A	OEM
MC8212	OEM
MC8214	OEM
MC8216	OEM
MC8224	OEM
MC8226	OEM
MC8228	OEM
MC8251	OEM
MC8253	OEM
MC8255A	OEM
MC8257	OEM
MC8259A	OEM
MC8284	OEM
MC8288	OEM
MC8500	OEM
MC8500L	OEM
MC8500P	OEM
MC8501	OEM
MC8501L	OEM
MC8501P	OEM
MC8502	OEM
MC8502L	OEM
MC8502P	OEM
MC8503	OEM
MC8503L	OEM
MC8503P	OEM
MC8504L	OEM
MC8504P	OEM
MC8506	OEM
MC8506L	OEM
MC8506P	OEM
MC8507L	OEM
MC8507P	OEM
MC8520L	OEM
MC8601P	2267
MC8748	OEM
MC9346	OEM
MC9601L	2267
MC9930	1823
MC10100F	OEM
MC10100L	OEM
MC10100LD	OEM
MC10100LDS	OEM
MC10100LS	OEM
MC10100P	OEM
MC10100PD	OEM
MC10100PDS	OEM
MC10100PS	OEM
MC10101F	OEM
MC10101L	OEM
MC10101LD	OEM
MC10101LDS	OEM
MC10101LS	OEM
MC10101P	OEM
MC10101PDS	OEM
MC10101PS	OEM
MC10102	OEM
MC10102F	OEM
MC10102L	OEM
MC10102LD	OEM
MC10102LDS	OEM
MC10102LS	OEM
MC10102P	OEM
MC10102PD	OEM
MC10102PDS	OEM
MC10102PS	OEM
MC10103F	OEM
MC10103L	OEM
MC10103LD	OEM
MC10103LDS	OEM
MC10103LS	OEM
MC10103P	OEM
MC10103PD	OEM
MC10103PDS	OEM
MC10103PS	OEM
MC10104F	OEM
MC10104L	OEM
MC10104LD	OEM
MC10104LDS	OEM
MC10104LS	OEM
MC10104P	OEM
MC10104PD	OEM
MC10104PDS	OEM
MC10104PS	OEM
MC10105F	OEM
MC10105L	OEM
MC10105LD	OEM
MC10105LDS	OEM
MC10105LS	OEM
MC10105P	OEM
MC10105PD	OEM
MC10105PDS	OEM
MC10105PS	OEM
MC10106F	OEM
MC10106L	OEM
MC10106LD	OEM
MC10106LDS	OEM
MC10106LS	OEM
MC10106P	OEM
MC10106PD	OEM
MC10106PDS	OEM
MC10106PS	OEM
MC10107F	OEM
MC10107L	OEM
MC10107LD	OEM
MC10107LDS	OEM
MC10107LS	OEM
MC10107P	OEM
MC10107PD	OEM
MC10107PDS	OEM
MC10107PS	OEM
MC10109F	OEM
MC10109L	OEM
MC10109LD	OEM
MC10109LDS	OEM
MC10109LS	OEM
MC10109P	OEM
MC10109PD	OEM
MC10109PDS	OEM
MC10109PS	OEM
MC10110L	OEM
MC10110LD	OEM
MC10110LDS	OEM
MC10110LS	OEM
MC10110P	OEM
MC10110PD	OEM
MC10110PDS	OEM
MC10110PS	OEM
MC10111L	OEM
MC10111LD	OEM
MC10111LDS	OEM
MC10111LS	OEM
MC10111P	OEM
MC10111PD	OEM
MC10111PDS	OEM
MC10111PS	OEM
MC10113F	OEM
MC10113L	OEM
MC10113LD	OEM
MC10113LDS	OEM
MC10113LS	OEM
MC10113P	OEM
MC10113PD	OEM
MC10113PDS	OEM
MC10113PS	OEM
MC10114F	OEM
MC10114L	OEM
MC10114LD	OEM
MC10114LDS	OEM
MC10114LS	OEM
MC10114P	OEM
MC10114PD	OEM
MC10114PDS	OEM
MC10114PS	OEM
MC10115F	OEM
MC10115L	OEM
MC10115LD	OEM
MC10115LDS	OEM
MC10115LS	OEM
MC10115P	OEM
MC10115PD	OEM
MC10115PDS	OEM
MC10115PS	OEM
MC10116F	OEM
MC10116L	OEM
MC10116LD	OEM
MC10116LDS	OEM
MC10116LS	OEM
MC10116P	OEM
MC10116PD	OEM
MC10116PDS	OEM
MC10116PS	OEM
MC10117F	OEM
MC10117L	OEM
MC10117LD	OEM
MC10117LDS	OEM
MC10117LS	OEM
MC10117P	OEM
MC10117PD	OEM
MC10117PDS	OEM
MC10117PS	OEM
MC10118F	OEM
MC10118L	OEM
MC10118LD	OEM
MC10118LDS	OEM
MC10118LS	OEM
MC10118P	OEM
MC10118PD	OEM
MC10118PDS	OEM
MC10118PS	OEM
MC10119F	OEM
MC10119L	OEM
MC10119LD	OEM
MC10119LDS	OEM
MC10119LS	OEM
MC10119P	OEM
MC10119PD	OEM
MC10119PDS	OEM
MC10119PS	OEM
MC10121F	OEM
MC10121L	OEM
MC10121LD	OEM
MC10121LDS	OEM
MC10121LS	OEM
MC10121P	OEM
MC10121PD	OEM
MC10121PDS	OEM
MC10121PS	OEM
MC10123L	OEM
MC10123LD	OEM
MC10123LDS	OEM
MC10123LS	OEM
MC10123P	OEM
MC10123PD	OEM
MC10123PDS	OEM
MC10123PS	OEM
MC10124F	OEM
MC10124L	OEM
MC10124LD	OEM
MC10124LDS	OEM
MC10124LS	OEM
MC10124P	OEM
MC10124PD	OEM
MC10124PDS	OEM
MC10124PS	OEM
MC10125F	OEM
MC10125L	OEM
MC10125LD	OEM
MC10125LDS	OEM
MC10125LS	OEM
MC10125P	OEM
MC10125PD	OEM
MC10125PDS	OEM
MC10125PS	OEM
MC10128L	OEM
MC10128LD	OEM
MC10128LDS	OEM
MC10128LS	OEM
MC10129L	OEM
MC10129LD	OEM
MC10129LDS	OEM
MC10129LS	OEM
MC10130F	OEM
MC10130L	OEM
MC10130LD	OEM
MC10130LDS	OEM
MC10130LS	OEM
MC10130P	OEM
MC10130PD	OEM
MC10130PDS	OEM
MC10130PS	OEM
MC10131F	OEM
MC10131L	OEM
MC10131LD	OEM
MC10131LDS	OEM
MC10131LS	OEM
MC10131P	OEM
MC10131PD	OEM
MC10131PDS	OEM
MC10131PS	OEM
MC10132L	OEM
MC10132LD	OEM
MC10132LDS	OEM
MC10132LS	OEM
MC10132P	OEM
MC10132PD	OEM
MC10132PDS	OEM
MC10132PS	OEM
MC10133F	OEM
MC10133L	OEM
MC10133LD	OEM
MC10133LDS	OEM
MC10133LS	OEM
MC10133P	OEM
MC10133PD	OEM
MC10133PDS	OEM
MC10133PS	OEM
MC10134L	OEM
MC10134LD	OEM
MC10134LDS	OEM
MC10134LS	OEM
MC10134P	OEM
MC10134PD	OEM
MC10134PDS	OEM
MC10134PS	OEM
MC10135F	OEM
MC10135L	OEM
MC10135LD	OEM
MC10135LDS	OEM
MC10135LS	OEM
MC10135P	OEM
MC10135PD	OEM

If replacement code is OEM, contact original manufacturer for replacement.

DEVICE TYPE	REPL CODE
MC10135PDS	OEM
MC10135PS	OEM
MC10136L	OEM
MC10136LD	OEM
MC10136LDS	OEM
MC10136LS	OEM
MC10136P	OEM
MC10136PD	OEM
MC10136PDS	OEM
MC10136PS	OEM
MC10137L	OEM
MC10137LD	OEM
MC10137LDS	OEM
MC10137LS	OEM
MC10137P	OEM
MC10137PD	OEM
MC10137PDS	OEM
MC10137PS	OEM
MC10138F	OEM
MC10138L	OEM
MC10138LD	OEM
MC10138LDS	OEM
MC10138LS	OEM
MC10138P	OEM
MC10138PD	OEM
MC10138PDS	OEM
MC10138PS	OEM
MC10141F	OEM
MC10141L	OEM
MC10141LD	OEM
MC10141LDS	OEM
MC10141LS	OEM
MC10141P	OEM
MC10141PD	OEM
MC10141PDS	OEM
MC10141PS	OEM
MC10153F	OEM
MC10153L	OEM
MC10153LD	OEM
MC10153LDS	OEM
MC10153LS	OEM
MC10153P	OEM
MC10153PD	OEM
MC10153PDS	OEM
MC10153PS	OEM
MC10154L	OEM
MC10154LD	OEM
MC10154LDS	OEM
MC10154LS	OEM
MC10154P	OEM
MC10154PD	OEM
MC10154PDS	OEM
MC10154PS	OEM
MC10158F	OEM
MC10158L	OEM
MC10158LD	OEM
MC10158LDS	OEM
MC10158LS	OEM
MC10158P	OEM
MC10158PD	OEM
MC10158PDS	OEM
MC10158PS	OEM
MC10159L	OEM
MC10159LD	OEM
MC10159LDS	OEM
MC10159LS	OEM
MC10159P	OEM
MC10159PD	OEM
MC10159PDS	OEM
MC10159PS	OEM
MC10160F	OEM
MC10160L	OEM
MC10160LD	OEM
MC10160LDS	OEM
MC10160LS	OEM
MC10160P	OEM
MC10160PD	OEM
MC10160PDS	OEM
MC10160PS	OEM
MC10161F	OEM
MC10161L	OEM
MC10161LD	OEM
MC10161LDS	OEM
MC10161LS	OEM
MC10161P	OEM
MC10161PD	OEM
MC10161PDS	OEM
MC10161PS	OEM
MC10162F	OEM
MC10162L	OEM
MC10162LD	OEM
MC10162LDS	OEM
MC10162LS	OEM
MC10162P	OEM
MC10162PD	OEM
MC10162PDS	OEM
MC10162PS	OEM
MC10163F	OEM
MC10163L	OEM
MC10163LD	OEM
MC10163LDS	OEM
MC10163LS	OEM
MC10163P	OEM
MC10163PD	OEM
MC10163PDS	OEM
MC10163PS	OEM
MC10164F	OEM
MC10164L	OEM
MC10164LD	OEM
MC10164LDS	OEM
MC10164LS	OEM
MC10164P	OEM
MC10164PD	OEM
MC10164PDS	OEM
MC10164PS	OEM
MC10165F	OEM
MC10165L	OEM
MC10165LD	OEM
MC10165LDS	OEM
MC10165LS	OEM
MC10165P	OEM
MC10165PD	OEM
MC10165PDS	OEM
MC10165PS	OEM
MC10166L	OEM
MC10166LD	OEM
MC10166LDS	OEM
MC10166LS	OEM
MC10166P	OEM
MC10166PD	OEM
MC10166PDS	OEM
MC10166PS	OEM
MC10168F	OEM
MC10168L	OEM
MC10168LD	OEM
MC10168LDS	OEM
MC10168LS	OEM
MC10168P	OEM
MC10168PD	OEM
MC10168PDS	OEM
MC10168PS	OEM
MC10169F	OEM
MC10170L	OEM
MC10170LD	OEM
MC10170LDS	OEM
MC10170LS	OEM
MC10170P	OEM
MC10170PD	OEM
MC10170PDS	OEM
MC10170PS	OEM
MC10171F	OEM
MC10171L	OEM
MC10171LD	OEM
MC10171LDS	OEM
MC10171LS	OEM
MC10171P	OEM
MC10171PD	OEM
MC10171PDS	OEM
MC10171PS	OEM
MC10172F	OEM
MC10172L	OEM
MC10172LD	OEM
MC10172LDS	OEM
MC10172LS	OEM
MC10172P	OEM
MC10172PD	OEM
MC10172PDS	OEM
MC10172PS	OEM
MC10173L	OEM
MC10173LD	OEM
MC10173LDS	OEM
MC10173LS	OEM
MC10173P	OEM
MC10173PD	OEM
MC10173PDS	OEM
MC10173PS	OEM
MC10174F	OEM
MC10174L	OEM
MC10174LD	OEM
MC10174LDS	OEM
MC10174LS	OEM
MC10174P	OEM
MC10174PD	OEM
MC10174PDS	OEM
MC10174PS	OEM
MC10175F	OEM
MC10175L	OEM
MC10175LD	OEM
MC10175LDS	OEM
MC10175LS	OEM
MC10175P	OEM
MC10175PD	OEM
MC10175PDS	OEM
MC10175PS	OEM
MC10176F	OEM
MC10176L	OEM
MC10176LD	OEM
MC10176LDS	OEM
MC10176LS	OEM
MC10176P	OEM
MC10176PD	OEM
MC10176PDS	OEM
MC10176PS	OEM
MC10177L	OEM
MC10177LD	OEM
MC10177LDS	OEM
MC10177LS	OEM
MC10178F	OEM
MC10178L	OEM
MC10178LD	OEM
MC10178LDS	OEM
MC10178LS	OEM
MC10178P	OEM
MC10178PD	OEM
MC10178PDS	OEM
MC10178PS	OEM
MC10179F	OEM
MC10179L	OEM
MC10179LD	OEM
MC10179LDS	OEM
MC10179LS	OEM
MC10179P	OEM
MC10179PD	OEM
MC10179PDS	OEM
MC10179PS	OEM
MC10180F	OEM
MC10180L	OEM
MC10180LD	OEM
MC10180LDS	OEM
MC10180LS	OEM
MC10180P	OEM
MC10180PD	OEM
MC10180PDS	OEM
MC10180PS	OEM
MC10181F	OEM
MC10181L	OEM
MC10181LD	OEM
MC10181LDS	OEM
MC10181LS	OEM
MC10181P	OEM
MC10181PD	OEM
MC10181PDS	OEM
MC10181PS	OEM
MC10182F	OEM
MC10182L	OEM
MC10182LD	OEM
MC10182LDS	OEM
MC10182LS	OEM
MC10182P	OEM
MC10182PD	OEM
MC10182PDS	OEM
MC10182PS	OEM
MC10183L	OEM
MC10183LD	OEM
MC10183LDS	OEM
MC10183LS	OEM
MC10186F	OEM
MC10186L	OEM
MC10186LD	OEM
MC10186LDS	OEM
MC10186LS	OEM
MC10186P	OEM
MC10186PD	OEM
MC10186PDS	OEM
MC10186PS	OEM
MC10188L	OEM
MC10188LD	OEM
MC10188LDS	OEM
MC10188LS	OEM
MC10188P	OEM
MC10188PD	OEM
MC10188PDS	OEM
MC10188PS	OEM
MC10189L	OEM
MC10189LD	OEM
MC10189LDS	OEM
MC10189LS	OEM
MC10189P	OEM
MC10189PD	OEM
MC10189PDS	OEM
MC10189PS	OEM
MC10190F	OEM
MC10190L	OEM
MC10190LD	OEM
MC10190LDS	OEM
MC10190LS	OEM
MC10190P	OEM
MC10190PD	OEM
MC10190PDS	OEM
MC10190PS	OEM
MC10191F	OEM
MC10191L	OEM
MC10191LD	OEM
MC10191LDS	OEM
MC10191LS	OEM
MC10191P	OEM
MC10191PD	OEM
MC10191PDS	OEM
MC10191PS	OEM
MC10192L	OEM
MC10192LD	OEM
MC10192LDS	OEM
MC10192LS	OEM
MC10192P	OEM
MC10192PD	OEM
MC10192PDS	OEM
MC10192PS	OEM
MC10193F	OEM
MC10193L	OEM
MC10193LD	OEM
MC10193LDS	OEM
MC10193LS	OEM
MC10193P	OEM
MC10193PD	OEM
MC10193PDS	OEM
MC10193PS	OEM
MC10194F	OEM
MC10194L	OEM
MC10194LD	OEM
MC10194LDS	OEM
MC10194LS	OEM
MC10194P	OEM
MC10194PD	OEM
MC10194PDS	OEM
MC10194PS	OEM
MC10195F	OEM
MC10195L	OEM
MC10195LD	OEM
MC10195LDS	OEM
MC10195LS	OEM
MC10195P	OEM
MC10195PD	OEM
MC10195PDS	OEM
MC10195PS	OEM
MC10197F	OEM
MC10197L	OEM
MC10197LD	OEM
MC10197LDS	OEM
MC10197LS	OEM
MC10197P	OEM
MC10197PD	OEM
MC10197PDS	OEM
MC10197PS	OEM
MC10198L	OEM
MC10198LD	OEM
MC10198LDS	OEM
MC10198LS	OEM
MC10198P	OEM
MC10198PD	OEM
MC10198PDS	OEM
MC10198PS	OEM
MC10210F	OEM
MC10210L	OEM
MC10210LD	OEM
MC10210LDS	OEM
MC10210LS	OEM
MC10210P	OEM
MC10210PD	OEM
MC10210PDS	OEM
MC10210PS	OEM
MC10211F	OEM
MC10211L	OEM
MC10211LD	OEM
MC10211LDS	OEM
MC10211LS	OEM
MC10211P	OEM
MC10211PD	OEM
MC10211PDS	OEM
MC10211PS	OEM
MC10212F	OEM
MC10212L	OEM
MC10212LD	OEM
MC10212LDS	OEM
MC10212LS	OEM
MC10212P	OEM
MC10212PD	OEM
MC10212PDS	OEM
MC10212PS	OEM
MC10216F	OEM
MC10216L	OEM
MC10216LD	OEM
MC10216LDS	OEM
MC10216LS	OEM
MC10216P	OEM
MC10216PD	OEM
MC10216PDS	OEM
MC10216PS	OEM
MC10231F	OEM
MC10231L	OEM
MC10231LD	OEM
MC10231LDS	OEM
MC10231LS	OEM
MC10231P	OEM
MC10231PD	OEM
MC10231PDS	OEM
MC10231PS	OEM
MC10287F	OEM
MC10287L	OEM
MC10287LD	OEM
MC10287LDS	OEM
MC10287LS	OEM
MC10287P	OEM
MC10287PD	OEM
MC10287PDS	OEM
MC10287PS	OEM
MC10315L	OEM
MC10315L(A)	OEM
MC10317L	OEM
MC10317L(A)	OEM
MC10318CL6	OEM
MC10318CL7	OEM
MC10318L	OEM
MC10318L9	OEM
MC10425UBALD	OEM
MC10434BAL	OEM
MC10434BCL	OEM
MC10438BCP	OEM
MC10442BAL	OEM
MC10493BALD	OEM
MC10500F	OEM
MC10500L	OEM
MC10501F	OEM
MC10501L	OEM
MC10502F	OEM
MC10502L	OEM
MC10503F	OEM
MC10503L	OEM
MC10504F	OEM
MC10504L	OEM
MC10505F	OEM
MC10505L	OEM
MC10506F	OEM
MC10506L	OEM
MC10507F	OEM
MC10507L	OEM
MC10509F	OEM
MC10509L	OEM
MC10513F	OEM
MC10513L	OEM
MC10514F	OEM
MC10514L	OEM
MC10515F	OEM
MC10515L	OEM
MC10516F	OEM
MC10516L	OEM
MC10517F	OEM
MC10517L	OEM
MC10518F	OEM
MC10518L	OEM
MC10519F	OEM
MC10519L	OEM
MC10521F	OEM
MC10521L	OEM
MC10524F	OEM
MC10524L	OEM
MC10525F	OEM
MC10525L	OEM
MC10530F	OEM
MC10530L	OEM
MC10531F	OEM
MC10531L	OEM
MC10533F	OEM
MC10533L	OEM
MC10535F	OEM
MC10535L	OEM
MC10538F	OEM
MC10538L	OEM
MC10541F	OEM
MC10541L	OEM
MC10550	OEM
MC10553F	OEM
MC10553L	OEM
MC10558F	OEM
MC10558L	OEM
MC10560F	OEM
MC10560L	OEM
MC10561F	OEM
MC10562F	OEM
MC10563F	OEM
MC10563L	OEM
MC10564F	OEM
MC10564L	OEM
MC10565F	OEM
MC10565L	OEM
MC10568F	OEM
MC10568L	OEM
MC10570F	OEM
MC10570L	OEM
MC10571F	OEM
MC10571L	OEM
MC10572F	OEM
MC10572L	OEM
MC10574F	OEM
MC10574L	OEM
MC10575F	OEM
MC10575L	OEM
MC10578L	OEM
MC10579F	OEM
MC10579L	OEM
MC10580F	OEM
MC10580L	OEM
MC10581F	OEM
MC10581L	OEM
MC10582F	OEM
MC10582L	OEM
MC10590F	OEM
MC10590L	OEM
MC10591F	OEM
MC10591L	OEM
MC10593F	OEM
MC10593L	OEM
MC10594F	OEM
MC10594L	OEM
MC10595F	OEM
MC10595L	OEM
MC10597L	OEM
MC10616F	OEM
MC10616L	OEM
MC10631F	OEM
MC10631L	OEM
MC10800	OEM
MC10803L	OEM
MC10804L	OEM
MC10805L	OEM
MC10806	OEM
MC10807L	OEM
MC10808L	OEM
MC10900Z	OEM
MC10901Z	OEM
MC10902	OEM
MC10904	OEM
MC10905	OEM
MC10905(A)	OEM
MC12000L	OEM
MC12002L	OEM
MC12002LD	OEM
MC12002LDS	OEM
MC12002LS	OEM
MC12002P	OEM
MC12002PD	OEM
MC12002PDS	OEM
MC12002PS	OEM
MC12009L	OEM
MC12009LD	OEM
MC12009LDS	OEM
MC12009LS	OEM
MC12009P	OEM
MC12009PD	OEM
MC12009PDS	OEM
MC12009PS	OEM
MC12011L	OEM
MC12011LD	OEM
MC12011LDS	OEM
MC12011LS	OEM
MC12011P	OEM
MC12011PD	OEM
MC12011PDS	OEM
MC12011PS	OEM
MC12012L	OEM
MC12013L	OEM
MC12013LD	OEM
MC12013LDS	OEM
MC12013LS	OEM
MC12013P	OEM
MC12013PD	OEM
MC12013PDS	OEM
MC12013PS	OEM
MC12014L	OEM
MC12014LD	OEM
MC12014LDS	OEM
MC12014LS	OEM
MC12014P	OEM
MC12014PD	OEM
MC12014PDS	OEM
MC12015P	OEM
MC12015P(A)	OEM
MC12016P	OEM
MC12016P(A)	OEM
MC12017P	OEM
MC12017P(A)	OEM
MC12018P	OEM
MC12019P	OEM
MC12020L	OEM
MC12020P	OEM
MC12021L	OEM
MC12021P	OEM
MC12022P	OEM
MC12023P	OEM
MC12040L	OEM
MC12040LD	OEM
MC12040LDS	OEM
MC12040LS	OEM
MC12040P	OEM
MC12040PD	OEM
MC12040PDS	OEM
MC12040PS	OEM
MC12060L	OEM
MC12060LS	OEM
MC12060P	OEM
MC12061L	OEM
MC12061LD	OEM
MC12061LDS	OEM
MC12061LS	OEM
MC12061P	OEM
MC12061PD	OEM
MC12061PDS	OEM
MC12061PS	OEM
MC12071P	OEM
MC12071PD	OEM
MC12071PDS	OEM
MC12071PS	OEM
MC12073P	OEM
MC12074P	OEM
MC12075P	OEM
MC12090L	OEM
MC12090LD	OEM
MC12090LDS	OEM
MC12090LS	OEM
MC12104PS	OEM
MC12509L	OEM
MC12509P	OEM
MC12511L	OEM
MC12513L	OEM
MC12513P	OEM
MC12520L	OEM
MC12521L	OEM
MC12560L	OEM
MC12561L	OEM
MC12663	OEM
MC13001P	OEM
MC13001XP	OEM
MC13002P	OEM
MC13002XP	OEM
MC13010P	OEM
MC13010P(A)	OEM
MC13020P	OEM
MC13030P	OEM
MC13074P	OEM
MC13301P	3772
MC13503T4	OEM
MC13513T4	OEM
MC13516	OEM
MC13516T2	OEM
MC13528P	OEM
MC13580SP	OEM
MC14000	2024
MC14000ABCBS	OEM
MC14000ADCBS	OEM
MC14000BCL	OEM
MC14000UBAL	4295
MC14000UBALD	OEM
MC14000UBALDS	OEM
MC14000UBALS	OEM
MC14000UBCL	2024
MC14000UBCLD	OEM
MC14000UBCLDS	OEM
MC14000UBCLS	OEM
MC14000UBCP	2024
MC14000UBCPD	OEM
MC14000UBCPDS	OEM
MC14000UBCPS	OEM
MC14001	0478
MC14001ABCBS	OEM
MC14001ADCBS	OEM
MC14001B	0478
MC14001BAL	3244
MC14001BALD	OEM
MC14001BALDS	OEM
MC14001BALS	OEM
MC14001BBCBS	OEM
MC14001BCL	0478
MC14001BCP	0478
MC14001BCPD	OEM
MC14001BCPDS	OEM
MC14001BCPS	OEM
MC14001BDCBS	OEM
MC14001BP	3244
MC14001CP	0478
MC14001UBAL	OEM
MC14001UBALD	OEM
MC14001UBALDS	OEM
MC14001UBALS	OEM
MC14001UBCL	OEM
MC14001UBCLD	OEM
MC14001UBCLDS	OEM
MC14001UBCLS	OEM
MC14001UBCP	0478
MC14001UBCPD	OEM
MC14001UBCPDS	OEM
MC14001UBCPS	OEM
MC14002	2055
MC14002ABCBS	OEM
MC14002ADCBS	OEM
MC14002BAL	2055
MC14002BALD	OEM
MC14002BALDS	OEM
MC14002BALS	OEM
MC14002BBCBS	OEM
MC14002BCL	2055
MC14002BCLD	OEM
MC14002BCLDS	OEM
MC14002BCLS	OEM
MC14002BCP	2055
MC14002BCPD	OEM
MC14002BCPDS	OEM
MC14002BCPS	OEM
MC14002BDCBS	OEM
MC14002CP	2055
MC14002P	2055
MC14002UBAL	OEM
MC14002UBALD	OEM
MC14002UBALDS	OEM
MC14002UBALS	OEM
MC14002UBCL	OEM
MC14002UBCLD	OEM
MC14002UBCLDS	OEM
MC14002UBCLS	OEM
MC14002UBCP	OEM
MC14002UBCPD	OEM
MC14002UBCPDS	OEM
MC14002UBCPS	OEM
MC14006BAL	0646
MC14006BALD	OEM
MC14006BALDS	OEM
MC14006BALS	OEM
MC14006BBCBS	OEM
MC14006BCL	0646
MC14006BCLD	OEM
MC14006BCLDS	OEM
MC14006BCLS	OEM
MC14006BCP	0646
MC14006BCPD	OEM
MC14006BCPDS	OEM
MC14006BCPS	OEM
MC14006BDCBS	OEM
MC14007	2846
MC14007ABCBS	OEM
MC14007ADCBS	OEM
MC14007UBAL	2846
MC14007UBALD	OEM
MC14007UBALDS	OEM
MC14007UBALS	OEM
MC14007UBCL	4984
MC14007UBCLD	OEM
MC14007UBCLDS	OEM
MC14007UBCLS	OEM
MC14007UBCP	2846
MC14007UBCPD	OEM
MC14007UBCPDS	OEM
MC14007UBCPS	OEM
MC14007UBL	OEM
MC14007UBP	OEM
MC14008BAL	0988
MC14008BALD	OEM
MC14008BALDS	OEM
MC14008BCL	0988
MC14008BCLD	OEM
MC14008BCLDS	OEM
MC14008BCLS	OEM
MC14008BCP	0988
MC14008BCPD	OEM
MC14008BCPDS	OEM
MC14008BCPS	OEM

If replacement code is OEM, contact original manufacturer for replacement.

DEVICE TYPE	REPL CODE
MC14008BDEBS	OEM
MC14008BEBS	OEM
MC14008CL	OEM
MC14011	0215
MC14011ABCBS	OEM
MC14011ADCBS	OEM
MC14011B	0215
MC14011BAL	0215
MC14011BALD	OEM
MC14011BALDS	OEM
MC14011BALS	OEM
MC14011BCL	0215
MC14011BCP	0215
MC14011BCPD	OEM
MC14011BDCBS	OEM
MC14011BF	OEM
MC-14011CP	0215
MC14011CP	0215
MC14011UB	OEM
MC14011UBAL	OEM
MC14011UBALD	OEM
MC14011UBALDS	OEM
MC14011UBALS	OEM
MC14011UBCL	OEM
MC14011UBCLD	OEM
MC14011UBCLDS	OEM
MC14011UBCLS	OEM
MC14011UBCP	OEM
MC14011UBCPD	OEM
MC14011UBCPDS	OEM
MC14011UBCPS	OEM
MC14012	0498
MC14012ABCBS	OEM
MC14012ADCBS	OEM
MC14012BAL	0498
MC14012BALD	OEM
MC14012BALDS	OEM
MC14012BALS	OEM
MC14012BBCBS	OEM
MC14012BCL	0498
MC14012BCLD	OEM
MC14012BCLDS	OEM
MC14012BCLS	OEM
MC14012BCP	0498
MC14012BCPD	OEM
MC14012BCPDS	OEM
MC14012BCPS	OEM
MC14012BDCBS	OEM
MC14012CP	0498
MC14012UBAL	OEM
MC14012UBALD	OEM
MC14012UBALDS	OEM
MC14012UBALS	OEM
MC14012UBCL	OEM
MC14012UBCLD	OEM
MC14012UBCLDS	OEM
MC14012UBCLS	OEM
MC14012UBCP	OEM
MC14012UBCPD	OEM
MC14012UBCPDS	OEM
MC14012UBCPS	OEM
MC14013	0413
MC14013A1	0413
MC14013B	0413
MC14013BAL	0413
MC14013BALD	OEM
MC14013BALDS	OEM
MC14013BALS	OEM
MC14013BBCBS	OEM
MC14013BCL	0413
MC14013BCLD	OEM
MC14013BCLDS	OEM
MC14013BCLS	OEM
MC14013BCP	0413
MC14013BCPD	OEM
MC14013BCPDS	OEM
MC14013BCPS	OEM
MC14013BDCBS	OEM
MC14013BL	OEM
MC14013BP	OEM
MC14013CP	0413
MC14014BAL	0860
MC14014BALD	OEM
MC14014BALDS	OEM
MC14014BALS	OEM
MC14014BBEBS	OEM
MC14014BCL	0860
MC14014BCLD	OEM
MC14014BCLDS	OEM
MC14014BCLS	OEM
MC14014BCP	0860
MC14014BCPD	OEM
MC14014BCPS	OEM
MC14014BDEBS	OEM
MC14014BL	OEM
MC14014BP	OEM
MC14014CP	0860
MC14015	1014
MC14015BAL	1014
MC14015BALD	OEM
MC14015BALDS	OEM
MC14015BALS	OEM
MC14015BBEBS	OEM
MC14015BCL	1014
MC14015BCLD	OEM
MC14015BCLDS	OEM
MC14015BCLS	OEM
MC14015BCP	1014
MC14015BCPD	OEM
MC14015BCPDS	OEM
MC14015BCPS	OEM
MC14015BDEBS	OEM
MC14015BL	OEM
MC14015BP	OEM
MC14015CP	1014
MC14016	0101
MC14016B	4150
MC14016BAL	1143
MC14016BALD	OEM
MC14016BALDS	OEM
MC14016BALS	OEM
MC14016BBCBS	OEM
MC14016BCL	1143
MC14016BCLD	OEM
MC14016BCLDS	OEM
MC14016BCLS	OEM
MC14016BCP	1143
MC14016BCPDS	OEM
MC14016BCPS	OEM
MC14016BDCBS	OEM
MC14016CP	1143
MC14017	0513
MC14017BAL	0513
MC14017BALD	OEM
MC14017BALDS	OEM
MC14017BALS	OEM
MC14017BBEBS	OEM
MC14017BCL	0513
MC14017BCLD	OEM
MC14017BCLDS	OEM
MC14017BCLS	OEM
MC14017BCP	0513
MC14017BCPD	OEM
MC14017BCPDS	OEM
MC14017BCPS	OEM
MC14017BDEBS	OEM
MC14017CP	0513
MC14018BAL	1391
MC14018BALD	OEM
MC14018BALDS	OEM
MC14018BALS	OEM
MC14018BBEBS	OEM
MC14018BCL	1391
MC14018BCLS	OEM
MC14018BCP	1391
MC14018BCPD	OEM
MC14018BCPDS	OEM
MC14018BCPS	OEM
MC14018BDEBS	OEM
MC14018BL	3923
MC14018BP	OEM
MC14018CL	OEM
MC14018CLD	OEM
MC14018CLDS	OEM
MC14020	1661
MC14020BAL	1661
MC14020BALD	OEM
MC14020BALDS	OEM
MC14020BALS	OEM
MC14020BBEBS	OEM
MC14020BCL	1661
MC14020BCLD	OEM
MC14020BCLDS	OEM
MC14020BCLS	OEM
MC14020BCP	1661
MC14020BCPD	OEM
MC14020BCPDS	OEM
MC14020BCPS	OEM
MC14020BDEBS	OEM
MC14020CP	1661
MC14021	1749
MC14021BAL	1749
MC14021BALD	OEM
MC14021BALDS	OEM
MC14021BALS	OEM
MC14021BBEBS	OEM
MC14021BCL	1749
MC14021BCLD	OEM
MC14021BCLDS	OEM
MC14021BCLS	OEM
MC14021BCP	1749
MC14021BCPD	OEM
MC14021BCPDS	OEM
MC14021BCPS	OEM
MC14021BDEBS	OEM
MC14021BL	OEM
MC14021BP	OEM
MC14021CP	1749
MC14022BAL	1256
MC14022BALD	OEM
MC14022BALDS	OEM
MC14022BALS	OEM
MC14022BBEBS	OEM
MC14022BCL	1256
MC14022BCLD	OEM
MC14022BCLDS	OEM
MC14022BCLS	OEM
MC14022BCP	1256
MC14022BCPD	OEM
MC14022BCPDS	OEM
MC14022BCPS	OEM
MC14022BDEBS	OEM
MC14023	0519
MC14023ABCBS	OEM
MC14023ADCBS	OEM
MC14023B	0519
MC14023BAL	0519
MC14023BALD	OEM
MC14023BALDS	OEM
MC14023BALS	OEM
MC14023BBCBS	OEM
MC14023BCL	0519
MC14023BCLD	OEM
MC14023BCLDS	OEM
MC14023BCLS	OEM
MC14023BCP	0519
MC14023BCPD	OEM
MC14023BCPDS	OEM
MC14023BCPS	OEM
MC14023BDCBS	OEM
MC14023CP	0519
MC14023UBAL	OEM
MC14023UBALD	OEM
MC14023UBALDS	OEM
MC14023UBALS	OEM
MC14023UBCL	OEM
MC14023UBCLD	OEM
MC14023UBCLDS	OEM
MC14023UBCLS	OEM
MC14023UBCP	OEM
MC14023UBCPD	OEM
MC14023UBCPDS	OEM
MC14023UBCPS	OEM
MC14024	1957
MC14024B	1957
MC14024BAL	1957
MC14024BALD	OEM
MC14024BALDS	OEM
MC14024BALS	OEM
MC14024BBCBS	OEM
MC14024BCL	1957
MC14024BCLD	OEM
MC14024BCLDS	OEM
MC14024BCLS	OEM
MC14024BCP	1957
MC14024BCPD	OEM
MC14024BCPDS	OEM
MC14024BCPS	OEM
MC14024BDCBS	OEM
MC14024CP	1957
MC14025	2072
MC14025ABCBS	OEM
MC14025ADCBS	OEM
MC14025AL	2072
MC14025B	2072
MC14025BAL	2072
MC14025BALD	OEM
MC14025BALDS	OEM
MC14025BALS	OEM
MC14025BBCBS	OEM
MC14025BCL	2072
MC14025BCLD	OEM
MC14025BCLDS	OEM
MC14025BCLS	OEM
MC14025BCP	2072
MC14025BCPD	OEM
MC14025BCPDS	OEM
MC14025BCPS	OEM
MC14025BDCBS	OEM
MC14025CP	2072
MC14025UBAL	OEM
MC14025UBALDS	OEM
MC14025UBALS	OEM
MC14025UBCL	OEM
MC14025UBCLD	OEM
MC14025UBCLDS	OEM
MC14025UBCLS	OEM
MC14025UBCP	OEM
MC14025UBCPD	OEM
MC14025UBCPDS	OEM
MC14025UBCPS	OEM
MC14027	1949
MC14027BAL	1949
MC14027BALD	OEM
MC14027BALDS	OEM
MC14027BALS	OEM
MC14027BBEBS	OEM
MC14027BCL	1949
MC14027BCLD	OEM
MC14027BCLDS	OEM
MC14027BCLS	OEM
MC14027BCP	1949
MC14027BCPD	OEM
MC14027BCPDS	OEM
MC14027BCPS	OEM
MC14027BDEBS	OEM
MC14027CP	1949
MC14028	2228
MC14028B	4186
MC14028BAL	2228
MC14028BALD	OEM
MC14028BALDS	OEM
MC14028BALS	OEM
MC14028BBEBS	OEM
MC14028BCL	2228
MC14028BCLD	OEM
MC14028BCLDS	OEM
MC14028BCLS	OEM
MC14028BCP	2228
MC14028BCPD	OEM
MC14028BCPDS	OEM
MC14028BCPS	OEM
MC14028BDEBS	OEM
MC14028CP	2228
MC14029BAL	2233
MC14029BALD	OEM
MC14029BALDS	OEM
MC14029BALS	OEM
MC14029BBEBS	OEM
MC14029BCL	2233
MC14029BCLD	OEM
MC14029BCLDS	OEM
MC14029BCLS	OEM
MC14029BCP	2233
MC14029BCPD	OEM
MC14029BCPDS	OEM
MC14029BCPS	OEM
MC14029BDEBS	OEM
MC14029BL	OEM
MC14029BP	OEM
MC14029CP	2233
MC14032BAL	2531
MC14032BALD	OEM
MC14032BALS	OEM
MC14032BALDS	OEM
MC14032BBEBS	OEM
MC14032BCL	2531
MC14032BCLD	OEM
MC14032BCLDS	OEM
MC14032BCLS	OEM
MC14032BCP	2531
MC14032BCPDS	OEM
MC14032BCPS	OEM
MC14032BDEBS	OEM
MC14034BAL	3615
MC14034BALD	OEM
MC14034BALDS	OEM
MC14034BALS	OEM
MC14034BBJBS	OEM
MC14034BCL	5581
MC14034BCLD	OEM
MC14034BCLDS	OEM
MC14034BCLS	OEM
MC14034BCP	3615
MC14034BCPD	OEM
MC14034BCPDS	OEM
MC14034BDJBS	OEM
MC14035BAL	2777
MC14035BALD	OEM
MC14035BALDS	OEM
MC14035BALS	OEM
MC14035BBEBS	OEM
MC14035BCL	2777
MC14035BCLD	OEM
MC14035BCLDS	OEM
MC14035BCLS	OEM
MC14035BCP	2777
MC14035BCPD	OEM
MC14035BCPDS	OEM
MC14035BCPS	OEM
MC14035BDEBS	OEM
MC14038BAL	2981
MC14038BALD	OEM
MC14038BALDS	OEM
MC14038BALS	OEM
MC14038BBEBS	OEM
MC14038BCL	2981
MC14038BCLD	OEM
MC14038BCLDS	OEM
MC14038BCLS	OEM
MC14038BCP	2981
MC14038BCPD	OEM
MC14038BCPS	OEM
MC14038BDEBS	OEM
MC14040	0056
MC14040B	0056
MC14040BAL	0056
MC14040BALD	OEM
MC14040BALDS	OEM
MC14040BALS	OEM
MC14040BBEBS	OEM
MC14040BCL	0056
MC14040BCLD	OEM
MC14040BCLDS	OEM
MC14040BCLS	OEM
MC14040BCP	0056
MC14040BCPD	OEM
MC14040BCPDS	OEM
MC14040BCPS	OEM
MC14040BDEBS	OEM
MC14040BL	OEM
MC14040BP	OEM
MC14040CP	0056
MC14042	0121
MC14042BAL	0121
MC14042BALD	OEM
MC14042BALDS	OEM
MC14042BALS	OEM
MC14042BB3EBS	OEM
MC14042BCL	0121
MC14042BCLD	OEM
MC14042BCLDS	OEM
MC14042BCLS	OEM
MC14042BCP	0121
MC14042BCPD	OEM
MC14042BCPDS	OEM
MC14042BCPS	OEM
MC14042BDEBS	OEM
MC14042CP	0121
MC14043BAL	1553
MC14043BALD	OEM
MC14043BALDS	OEM
MC14043BALS	OEM
MC14043BBEBS	OEM
MC14043BCL	1553
MC14043BCLD	OEM
MC14043BCLDS	OEM
MC14043BCLS	OEM
MC14043BCP	1553
MC14043BCPD	OEM
MC14043BCPDS	OEM
MC14043BCPS	OEM
MC14043BDEBS	OEM
MC14043BL	OEM
MC14043BP	OEM
MC14044BAL	2308
MC14044BALD	OEM
MC14044BALDS	OEM
MC14044BALS	OEM
MC14044BBEBS	OEM
MC14044BCL	2308
MC14044BCLD	OEM
MC14044BCLDS	OEM
MC14044BCLS	OEM
MC14044BCP	2308
MC14044BCPD	OEM
MC14044BCPDS	OEM
MC14044BCPS	OEM
MC14044BDEBS	OEM
MC14044BL	OEM
MC14044BP	OEM
MC14046BAL	3831
MC14046BALD	OEM
MC14046BALDS	OEM
MC14046BALS	OEM
MC14046BBEBS	OEM
MC14046BCL	6797
MC14046BCLD	OEM
MC14046BCLDS	OEM
MC14046BCLS	OEM
MC14046BCP	3428
MC14046BCPD	OEM
MC14046BCPDS	OEM
MC14046BCPS	OEM
MC14046BDEBS	OEM
MC14046BL	OEM
MC14046BP	OEM
MC14046CP	0056
MC14049	0001
MC14049ABEBS	OEM
MC14049ADEBS	OEM
MC14049B	0001
MC14049CP	0001
MC14049UB	0001
MC14049UBAL	0001
MC14049UBALD	OEM
MC14049UBALDS	OEM
MC14049UBALS	OEM
MC14049UBCL	0001
MC14049UBCLD	OEM
MC14049UBCLDS	OEM
MC14049UBCLS	OEM
MC14049UBCP	0001
MC14049UBCPD	OEM
MC14049UBCPDS	OEM
MC14049UBCPS	OEM
MC14049UBL	OEM
MC14049UBP	OEM
MC14050	0398
MC14050B	0398
MC14050BAL	0398
MC14050BALD	OEM
MC14050BALDS	OEM
MC14050BALS	OEM
MC14050BBEBS	OEM
MC14050BCL	0398
MC14050BCLD	OEM
MC14050BCLDS	OEM
MC14050BCLS	OEM
MC14050BCP	0398
MC14050BCPD	OEM
MC14050BCPDS	OEM
MC14050BCPS	OEM
MC14050BDEBS	OEM
MC14050BL	OEM
MC14050BP	OEM
MC14050CP	0398
MC14051	0366
MC14051BAL	0366
MC14051BALD	OEM
MC14051BALDS	OEM
MC14051BALS	OEM
MC14051BBEBS	OEM
MC14051BCL	0366
MC14051BCLD	OEM
MC14051BCLDS	OEM
MC14051BCLS	OEM
MC14051BCP	0366
MC14051BCPD	OEM
MC14051BCPDS	OEM
MC14051BCPS	OEM
MC14051BDEBS	OEM
MC14051BL	OEM
MC14051BP	OEM
MC14052	0024
MC14052B	0024
MC14052BAL	0024
MC14052BALD	OEM
MC14052BALDS	OEM
MC14052BALS	OEM
MC14052BBEBS	OEM
MC14052BCL	0024
MC14052BCLD	OEM
MC14052BCLDS	OEM
MC14052BCLS	OEM
MC14052BCP	0024
MC14052BCPD	OEM
MC14052BCPDS	OEM
MC14052BCPS	OEM
MC14052BDEBS	OEM
MC14052BL	OEM
MC14052BP	OEM
MC14053	0035
MC14053B	0035
MC14053BAL	0035
MC14053BALD	OEM
MC14053BALDS	OEM
MC14053BALS	OEM
MC14053BBEBS	OEM
MC14053BCL	0035
MC14053BCLD	OEM
MC14053BCLDS	OEM
MC14053BCLS	OEM
MC14053BCP	0035
MC14053BCPD	OEM
MC14053BCPDS	OEM
MC14053BCPS	OEM
MC14053BDEBS	OEM
MC14053BF-T2	OEM
MC14053BL	OEM
MC14053BP	OEM
MC14060B	0146
MC14060BAL	0146
MC14060BALD	OEM
MC14060BALDS	OEM
MC14060BALS	OEM
MC14060BCL	4911
MC14060BCLD	OEM
MC14060BCLDS	OEM
MC14060BCLS	OEM
MC14060BCP	0146
MC14060BCPD	OEM
MC14060BCPDS	OEM
MC14060BCPS	OEM
MC14066	0101
MC14066B	0101
MC14066BAL	0101
MC14066BALD	OEM
MC14066BALDS	OEM
MC14066BALS	OEM
MC14066BBCBS	OEM
MC14066BCL	0101
MC14066BCLD	OEM
MC14066BCLDS	OEM
MC14066BCLS	OEM
MC14066BCP	0101
MC14066BCPD	OEM
MC14066BCPDS	OEM
MC14066BCPS	OEM
MC14066BDCBS	OEM
MC14066BF-T1	OEM
MC14066BL	OEM
MC14066BP	OEM
MC14067BAL	3742
MC14067BALD	OEM
MC14067BALDS	OEM
MC14067BALS	OEM
MC14067BCL	OEM
MC14067BCLD	OEM
MC14067BCLDS	OEM
MC14067BCLS	OEM
MC14067BCP	4279
MC14067BCPD	OEM
MC14067BCPDS	OEM
MC14067BCPS	OEM
MC14067BL	OEM
MC14067BP	OEM
MC14068B	2503
MC14068BAL	2503
MC14068BALD	OEM
MC14068BALDS	OEM
MC14068BALS	OEM
MC14068BBCBS	OEM
MC14068BCL	2503
MC14068BCLD	OEM
MC14068BCLDS	OEM
MC14068BCLS	OEM
MC14068BCP	2503
MC14068BCPD	OEM
MC14068BCPDS	OEM
MC14068BCPS	OEM
MC14068BDCBS	OEM
MC14069ABCBS	OEM
MC14069ADCBS	OEM
MC14069B	3311
MC14069U	0119
MC14069UB	3311
MC14069UBAL	0119
MC14069UBALD	OEM
MC14069UBALDS	OEM
MC14069UBALS	OEM
MC14069UBAP	OEM
MC14069UBCL	0119
MC14069UBCLD	OEM
MC14069UBCLDS	OEM
MC14069UBCLS	OEM
MC14069UBCP	0119
MC14069UBCPD	OEM
MC14069UBCPDS	OEM
MC14069UBCPS	OEM
MC14069UBF	OEM
MC14070BAL	2516
MC14070BALD	OEM
MC14070BALDS	OEM
MC14070BALS	OEM
MC14070BBCBS	OEM
MC14070BCL	2516
MC14070BCLD	OEM
MC14070BCLDS	OEM
MC14070BCLS	OEM
MC14070BCP	2516
MC14070BCPD	OEM
MC14070BCPDS	OEM
MC14070BCPS	OEM
MC14070BDCBS	OEM
MC14070BL	5245
MC14070BP	OEM
MC14071	0129
MC14071BAL	0129
MC14071BALD	OEM
MC14071BALDS	OEM
MC14071BALS	OEM
MC14071BBCBS	OEM
MC14071BCL	0129
MC14071BCLD	OEM
MC14071BCLDS	OEM
MC14071BCLS	OEM
MC14071BCP	0129
MC14071BCPD	OEM
MC14071BCPDS	OEM
MC14071BCPS	OEM
MC14071BDCBS	OEM
MC14071CP	0129
MC14072BAL	2524
MC14072BALD	OEM
MC14072BALDS	OEM
MC14072BALS	OEM
MC14072BBCBS	OEM
MC14072BCL	6410
MC14072BCLD	OEM
MC14072BCLDS	OEM
MC14072BCLS	OEM
MC14072BCP	6410
MC14072BCPD	OEM
MC14072BCPDS	OEM
MC14072BCPS	OEM
MC14072BDCBS	OEM
MC14073B	1536
MC14073BAL	1536
MC14073BALD	OEM
MC14073BALDS	OEM
MC14073BALS	OEM
MC14073BBCBS	OEM
MC14073BCL	1536
MC14073BCLD	OEM
MC14073BCLDS	OEM
MC14073BCLS	OEM
MC14073BCP	1536
MC14073BCPD	OEM
MC14073BCPDS	OEM
MC14073BCPS	OEM
MC14073BDCBS	OEM
MC14073BF	OEM
MC14075BAL	2540
MC14075BALD	OEM
MC14075BALDS	OEM
MC14075BALS	OEM
MC14075BBCBS	OEM
MC14075BCL	2540
MC14075BCLD	OEM
MC14075BCLDS	OEM
MC14075BCLS	OEM
MC14075BCP	2540
MC14075BCPD	OEM
MC14075BCPDS	OEM
MC14075BCPS	OEM
MC14075BDCBS	OEM
MC14076BAL	3496
MC14076BALD	OEM
MC14076BALDS	OEM
MC14076BALS	OEM
MC14076BBEBS	OEM
MC14076BCL	3496
MC14076BCLD	OEM
MC14076BCLDS	OEM
MC14076BCLS	OEM
MC14076BCP	3496
MC14076BCPD	OEM
MC14076BCPDS	OEM
MC14076BCPS	OEM
MC14076BDEBS	OEM
MC14077	OEM
MC14077BAL	2558
MC14077BALD	OEM
MC14077BALDS	OEM
MC14077BALS	OEM
MC14077BBCBS	OEM
MC14077BCL	2558
MC14077BCLD	OEM
MC14077BCLDS	OEM
MC14077BCLS	OEM
MC14077BCP	2558
MC14077BCPD	OEM
MC14077BCPDS	OEM
MC14077BCPS	OEM
MC14077BDCBS	OEM
MC14077BL	OEM
MC14077BP	OEM
MC14078B	4281
MC14078BAL	0921
MC14078BALD	OEM
MC14078BALDS	OEM
MC14078BALS	OEM

If replacement code is OEM, contact original manufacturer for replacement.

DEVICE TYPE	REPL CODE
MC14078BBCBS	OEM
MC14078BCL	0921
MC14078BCLD	OEM
MC14078BCLDS	OEM
MC14078BCLS	OEM
MC14078BCP	0921
MC14078BCPD	OEM
MC14078BCPDS	OEM
MC14078BCPS	OEM
MC14078BDCBS	OEM
MC14081	0626
MC14081ABCBS	OEM
MC14081ADCBS	OEM
MC14081B	0626
MC14081BAL	0626
MC14081BALD	OEM
MC14081BALDS	OEM
MC14081BALS	OEM
MC14081BBCBS	OEM
MC14081BCL	0626
MC14081BCLD	OEM
MC14081BCLDS	OEM
MC14081BCLS	OEM
MC14081BCP	0332
MC14081BCPD	OEM
MC14081BCPDS	OEM
MC14081BCPS	OEM
MC14081BDCBS	OEM
MC14081CP	0626
MC14082B	0300
MC14082BAL	0300
MC14082BALD	OEM
MC14082BALDS	OEM
MC14082BALS	OEM
MC14082BBCBS	OEM
MC14082BCL	0300
MC14082BCLD	OEM
MC14082BCLDS	OEM
MC14082BCLS	OEM
MC14082BCP	0300
MC14082BCPD	OEM
MC14082BCPDS	OEM
MC14082BCPS	OEM
MC14082BDCBS	OEM
MC14088B	OEM
MC14093	2384
MC14093B	2384
MC14093BAL	2384
MC14093BALDS	OEM
MC14093BALS	OEM
MC14093BBCBS	OEM
MC14093BCL	2384
MC14093BCLDS	OEM
MC14093BCLS	OEM
MC14093BCP	2384
MC14093BCPD	OEM
MC14093BCPDS	OEM
MC14093BCPS	OEM
MC14093BDCBS	OEM
MC14093BL	OEM
MC14093BP	OEM
MC14094BAL	1682
MC14094BALD	OEM
MC14094BALDS	OEM
MC14094BALS	OEM
MC14094BBEBS	OEM
MC14094BCL	1682
MC14094BCLD	OEM
MC14094BCLDS	OEM
MC14094BCLS	OEM
MC14094BCP	1682
MC14094BCPD	OEM
MC14094BCPDS	OEM
MC14094BCPS	OEM
MC14094BDEBS	OEM
MC14097BAL	3855
MC14097BALD	OEM
MC14097BALDS	OEM
MC14097BALS	OEM
MC14097BCL	OEM
MC14097BCLD	OEM
MC14097BCLDS	OEM
MC14097BCLS	OEM
MC14097BCP	5999
MC14097BCPD	OEM
MC14097BCPDS	OEM
MC14097BCPS	OEM
MC14097BL	OEM
MC14097BP	OEM
MC14099B	4312
MC14099BAL	3330
MC14099BALD	OEM
MC14099BALDS	OEM
MC14099BALS	OEM
MC14099BBEBS	OEM
MC14099BCL	3330
MC14099BCLD	OEM
MC14099BCLDS	OEM
MC14099BCLS	OEM
MC14099BCP	3330
MC14099BCPD	OEM
MC14099BCPDS	OEM
MC14099BCPS	OEM
MC14099BDEBS	OEM
MC14160BAL	1359
MC14160BALD	OEM
MC14160BALDS	OEM
MC14160BALS	OEM
MC14160BBEBS	OEM

DEVICE TYPE	REPL CODE
MC14160BCL	1359
MC14160BCLD	OEM
MC14160BCLDS	OEM
MC14160BCLS	OEM
MC14160BCP	1359
MC14160BCPD	OEM
MC14160BCPDS	OEM
MC14160BCPS	OEM
MC14160BDEBS	OEM
MC14160BL	OEM
MC14160BP	OEM
MC14161BAL	1373
MC14161BALD	OEM
MC14161BALDS	OEM
MC14161BALS	OEM
MC14161BBEBS	OEM
MC14161BCL	1373
MC14161BCLD	OEM
MC14161BCLDS	OEM
MC14161BCLS	OEM
MC14161BCP	1373
MC14161BCPD	OEM
MC14161BCPDS	OEM
MC14161BCPS	OEM
MC14161BDEBS	OEM
MC14161BL	OEM
MC14161BP	OEM
MC14162BAL	1388
MC14162BALD	OEM
MC14162BALDS	OEM
MC14162BALS	OEM
MC14162BBEBS	OEM
MC14162BCL	1388
MC14162BCLD	OEM
MC14162BCLDS	OEM
MC14162BCLS	OEM
MC14162BCP	1388
MC14162BCPD	OEM
MC14162BCPDS	OEM
MC14162BCPS	OEM
MC14162BDEBS	OEM
MC14162BL	OEM
MC14162BP	OEM
MC14163BAL	1407
MC14163BALD	OEM
MC14163BALDS	OEM
MC14163BALS	OEM
MC14163BBEBS	OEM
MC14163BCL	1407
MC14163BCLD	OEM
MC14163BCLDS	OEM
MC14163BCLS	OEM
MC14163BCP	1407
MC14163BCPD	OEM
MC14163BCPDS	OEM
MC14163BCPS	OEM
MC14163BDEBS	OEM
MC14163BL	OEM
MC14163BP	OEM
MC14174B	1551
MC14174BAL	1551
MC14174BALD	OEM
MC14174BALDS	OEM
MC14174BALS	OEM
MC14174BBEBS	OEM
MC14174BCL	1551
MC14174BCLD	OEM
MC14174BCLDS	OEM
MC14174BCLS	OEM
MC14174BCP	1551
MC14174BCPD	OEM
MC14174BCPDS	OEM
MC14174BCPS	OEM
MC14174BDEBS	OEM
MC14174BL	OEM
MC14174BP	OEM
MC14175BAL	OEM
MC14175BALD	OEM
MC14175BALDS	OEM
MC14175BALS	OEM
MC14175BBEBS	OEM
MC14175BCL	1528
MC14175BCLD	OEM
MC14175BCLDS	OEM
MC14175BCLS	OEM
MC14175BCP	1528
MC14175BCPD	OEM
MC14175BCPDS	OEM
MC14175BCPS	OEM
MC14175BDEBS	OEM
MC14175BL	OEM
MC14175BP	OEM
MC14194BAL	1769
MC14194BALD	OEM
MC14194BALDS	OEM
MC14194BALS	OEM
MC14194BBEBS	OEM
MC14194BCL	1769
MC14194BCLD	OEM
MC14194BCLDS	OEM
MC14194BCLS	OEM
MC14194BCP	1769
MC14194BCPD	OEM
MC14194BCPDS	OEM
MC14194BCPS	OEM
MC14194BDEBS	OEM
MC14194BL	OEM
MC14194BP	OEM
MC14400L	OEM

DEVICE TYPE	REPL CODE
MC14401L	OEM
MC14402L	OEM
MC14402Z	OEM
MC14403L	OEM
MC14403L1(A)	OEM
MC14403L2(A)	OEM
MC14404L	OEM
MC14404P	OEM
MC14404Z	OEM
MC14405L	OEM
MC14406L	OEM
MC14406P	OEM
MC14407L	OEM
MC14407P	OEM
MC14407Z	OEM
MC14408L	OEM
MC14408LD	OEM
MC14408P	OEM
MC14408PD	OEM
MC14409L	OEM
MC14409LD	OEM
MC14409P	OEM
MC14409PD	OEM
MC14410ABEBS	OEM
MC14410ADEBS	OEM
MC14410L	OEM
MC14410LD	OEM
MC14410P	OEM
MC14410PD	OEM
MC14411BBJBS	OEM
MC14411BDJBS	OEM
MC14411L	OEM
MC14411LD	OEM
MC14411P	OEM
MC14411PD	OEM
MC14412FL	5311
MC14412FLD	OEM
MC14412FP	5311
MC14412FPD	OEM
MC14412L	5097
MC14412P	5097
MC14412VL	5311
MC14412VLD	OEM
MC14412VP	5311
MC14412VPD	OEM
MC14413-1L	OEM
MC14413-1P	OEM
MC14413-1S	OEM
MC14413-2L	OEM
MC14413-2P	OEM
MC14413-2S	OEM
MC14413L1	OEM
MC14413L2	OEM
MC14413P1	OEM
MC14413P2	OEM
MC14413Z1	OEM
MC14413Z2	OEM
MC14414-1L	OEM
MC14414-1P	OEM
MC14414-1S	OEM
MC14414-2L	OEM
MC14414-2P	OEM
MC14414-2S	OEM
MC14414L1	OEM
MC14414L2	OEM
MC14414P1	OEM
MC14414P2	OEM
MC14414Z1	OEM
MC14414Z2	OEM
MC14415EFL	OEM
MC14415EFLD	OEM
MC14415EVL	OEM
MC14415EVLD	OEM
MC14415FL	OEM
MC14415FLD	OEM
MC14415FP	OEM
MC14415MC14415FPD	OEM
MC14415VL	OEM
MC14415VLD	OEM
MC14415VP	OEM
MC14415VPD	OEM
MC14416L	OEM
MC14416P	OEM
MC14417L	OEM
MC14417P	OEM
MC14418L	OEM
MC14418P	OEM
MC14419L	OEM
MC14419LD	OEM
MC14419P	OEM
MC14419PD	OEM
MC14424L	OEM
MC14424P	OEM
MC14433L	0487
MC14433LD	OEM
MC14433P	0487
MC14433PD	OEM
MC14435EFL	OEM
MC14435EVL	OEM
MC14435FL	OEM
MC14435FP	OEM
MC14435VL	OEM
MC14435VP	OEM
MC14440L	OEM
MC14442L	OEM
MC14442P	OEM
MC14443L	OEM
MC14443LD	OEM
MC14443P	OEM

DEVICE TYPE	REPL CODE
MC14443PD	OEM
MC14444L	OEM
MC14444L(A)	OEM
MC14444P	OEM
MC14444P(A)	OEM
MC14447CP	OEM
MC14447L	OEM
MC14447LD	OEM
MC14447P	OEM
MC14447PD	OEM
MC14457L	OEM
MC14457LD	OEM
MC14457LDS	OEM
MC14457LS	OEM
MC14457P	OEM
MC14457PD	OEM
MC14457PDS	OEM
MC14457PS	OEM
MC14458L	OEM
MC14458P	OEM
MC14460L	OEM
MC14460LD	OEM
MC14460LDS	OEM
MC14460LS	OEM
MC14460P	OEM
MC14460PD	OEM
MC14460PDS	OEM
MC14460PS	OEM
MC14465-1P	OEM
MC14466P	OEM
MC14467L	OEM
MC14467P	OEM
MC14469L	OEM
MC14469LD	OEM
MC14469P	OEM
MC14469PD	OEM
MC14490ABEBS	OEM
MC14490ADEBS	OEM
MC14490EFL	OEM
MC14490EFLD	OEM
MC14490EFLDS	OEM
MC14490EFLS	OEM
MC14490EVL	OEM
MC14490EVLD	OEM
MC14490EVLDS	OEM
MC14490EVLS	OEM
MC14490FL	OEM
MC14490FLD	OEM
MC14490FLDS	OEM
MC14490FLS	OEM
MC14490FP	OEM
MC14490FPD	OEM
MC14490FPDS	OEM
MC14490FPS	OEM
MC14490L	OEM
MC14490LD	OEM
MC14490LDS	OEM
MC14490LS	OEM
MC14490P	OEM
MC14490PD	OEM
MC14490PDS	OEM
MC14490PS	OEM
MC14490VL	OEM
MC14490VLD	OEM
MC14490VLDS	OEM
MC14490VLS	OEM
MC14490VP	OEM
MC14490VPD	OEM
MC14490VPDS	OEM
MC14490VPS	OEM
MC14495-1L	OEM
MC14495L1	OEM
MC14495P1	OEM
MC14500BAL	OEM
MC14500BALD	OEM
MC14500BCL	OEM
MC14500BCLD	OEM
MC14500BCP	OEM
MC14500BCPD	OEM
MC14500BL	OEM
MC14500BP	OEM
MC14501ABEBS	OEM
MC14501ADEBS	OEM
MC14501UBAL	OEM
MC14501UBALD	OEM
MC14501UBALDS	OEM
MC14501UBALS	OEM
MC14501UBCL	3809
MC14501UBCLD	OEM
MC14501UBCLDS	OEM
MC14501UBCLS	OEM
MC14501UBCP	3809
MC14501UBCPD	OEM
MC14501UBCPDS	OEM
MC14501UBCPS	OEM
MC14501UBD	OEM
MC14502BAL	1037
MC14502BALD	OEM
MC14502BALDS	OEM
MC14502BALS	OEM
MC14502BBEBS	OEM
MC14502BCL	1037
MC14502BCLD	OEM
MC14502BCLDS	OEM
MC14502BCLS	OEM
MC14502BCP	1037
MC14502BCPD	OEM
MC14502BCPDS	OEM
MC14502BCPS	OEM

DEVICE TYPE	REPL CODE
MC14502BDEBS	OEM
MC14503BAL	2053
MC14503BALD	OEM
MC14503BALDS	OEM
MC14503BALS	OEM
MC14503BBEBS	OEM
MC14503BCL	2098
MC14503BCLD	OEM
MC14503BCLDS	OEM
MC14503BCLS	OEM
MC14503BCP	2053
MC14503BCPD	OEM
MC14503BCPDS	OEM
MC14503BCPS	OEM
MC14503BDEBS	OEM
MC14503BL	OEM
MC14503BP	6677
MC14504BAL	OEM
MC14504BALD	OEM
MC14504BALDS	OEM
MC14504BALS	OEM
MC14504BBEBS	OEM
MC14504BCL	OEM
MC14504BCLD	OEM
MC14504BCLDS	OEM
MC14504BCLS	OEM
MC14504BCP	OEM
MC14504BCPD	OEM
MC14504BCPDS	OEM
MC14504BCPS	OEM
MC14504BDEBS	OEM
MC14504BL	OEM
MC14504BP	OEM
MC14506	OEM
MC14506BAL	OEM
MC14506BALD	OEM
MC14506BALDS	OEM
MC14506BALS	OEM
MC14506BBEBS	OEM
MC14506BCL	3767
MC14506BCLD	OEM
MC14506BCLDS	OEM
MC14506BCLS	OEM
MC14506BCP	3767
MC14506BCPD	OEM
MC14506BCPDS	OEM
MC14506BCPS	OEM
MC14506BDEBS	OEM
MC14506UBCL	OEM
MC14506UBCP	3767
MC14506UBD	OEM
MC14508BAL	1811
MC14508BALD	OEM
MC14508BALDS	OEM
MC14508BALS	OEM
MC14508BBJBS	OEM
MC14508BCL	1811
MC14508BCP	1811
MC14508BCPD	OEM
MC14508BCPDS	OEM
MC14508BCPS	OEM
MC14508BDJBS	OEM
MC14508CL	OEM
MC14508CLD	OEM
MC14508CLDS	OEM
MC14508CLS	OEM
MC14510BAL	1963
MC14510BALD	OEM
MC14510BALDS	OEM
MC14510BALS	OEM
MC14510BBEBS	OEM
MC14510BCL	4879
MC14510BCLD	OEM
MC14510BCLDS	OEM
MC14510BCLS	OEM
MC14510BCP	1963
MC14510BCPD	OEM
MC14510BCPDS	OEM
MC14510BCPS	OEM
MC14510BDEBS	OEM
MC14510CL	OEM
MC14511	1544
MC14511B	1544
MC14511B1	1544
MC14511BAL	1544
MC14511BALD	OEM
MC14511BALDS	OEM
MC14511BALS	OEM
MC14511BBEBS	OEM
MC14511BCL	1544
MC14511BCLD	OEM
MC14511BCLS	OEM
MC14511BCP	1544
MC14511BCPD	OEM
MC14511BCPDS	OEM
MC14511BCPS	OEM
MC14511BDEBS	OEM
MC14511BP	1544
MC14511L	1544
MC14512B	4217
MC14512BAL	2120
MC14512BALD	OEM
MC14512BALDS	OEM
MC14512BALS	OEM
MC14512BBEBS	OEM
MC14512BCL	2120
MC14512BCLD	OEM
MC14512BCLDS	OEM
MC14512BCLS	OEM

DEVICE TYPE	REPL CODE
MC14512BCP	2120
MC14512BCPD	OEM
MC14512BCPDS	OEM
MC14512BCPS	OEM
MC14512BDEBS	OEM
MC14512BL	OEM
MC14512BP	OEM
MC14512CP	2120
MC14513BAL	OEM
MC14513BALD	OEM
MC14513BALDS	OEM
MC14513BALS	OEM
MC14513BCL	4976
MC14513BCLD	OEM
MC14513BCLDS	OEM
MC14513BCLS	OEM
MC14513BCP	4976
MC14513BCPD	OEM
MC14513BCPDS	OEM
MC14513BCPS	OEM
MC14514	1830
MC14514B	1830
MC14514BAL	1830
MC14514BALD	OEM
MC14514BALDS	OEM
MC14514BALS	OEM
MC14514BBJBS	OEM
MC14514BCL	1830
MC14514BCLD	OEM
MC14514BCLDS	OEM
MC14514BCLS	OEM
MC14514BCP	1830
MC14514BCPD	OEM
MC14514BCPDS	OEM
MC14514BCPS	OEM
MC14514BDJBS	OEM
MC14515	3600
MC14515BAL	3600
MC14515BALD	OEM
MC14515BALDS	OEM
MC14515BALS	OEM
MC14515BBJBS	OEM
MC14515BCL	3600
MC14515BCLD	OEM
MC14515BCLDS	OEM
MC14515BCLS	OEM
MC14515BCP	3600
MC14515BCPD	OEM
MC14515BCPDS	OEM
MC14515BCPS	OEM
MC14515BDJBS	OEM
MC14516BAL	2347
MC14516BALD	OEM
MC14516BALDS	OEM
MC14516BALS	OEM
MC14516BBEBS	OEM
MC14516BCL	2347
MC14516BCLD	OEM
MC14516BCLDS	OEM
MC14516BCLS	OEM
MC14516BCP	2347
MC14516BCPD	OEM
MC14516BCPDS	OEM
MC14516BCPS	OEM
MC14516BDEBS	OEM
MC14516CP	2347
MC14517BAL	4284
MC14517BALD	OEM
MC14517BALDS	OEM
MC14517BALS	OEM
MC14517BBEBS	OEM
MC14517BCL	4284
MC14517BCLD	OEM
MC14517BCLDS	OEM
MC14517BCLS	OEM
MC14517BCP	4284
MC14517BCPD	OEM
MC14517BCPDS	OEM
MC14517BCPS	OEM
MC14517BDEBS	OEM
MC14518	1043
MC14518B	1043
MC14518BAL	1043
MC14518BALD	OEM
MC14518BALDS	OEM
MC14518BALS	OEM
MC14518BBEBS	OEM
MC14518BCL	1043
MC14518BCLD	OEM
MC14518BCLDS	OEM
MC14518BCLS	OEM
MC14518BCP	1043
MC14518BCPD	OEM
MC14518BCPDS	OEM
MC14518BCPS	OEM
MC14518BDEBS	OEM
MC14518BL	OEM
MC14518BP	OEM
MC14518CP	1043
MC14519BAL	1525
MC14519BALD	OEM
MC14519BALDS	OEM
MC14519BALS	OEM
MC14519BBEBS	OEM
MC14519BCL	OEM
MC14519BCLD	OEM
MC14519BCLDS	OEM
MC14519BCLS	OEM
MC14519BCP	1525

DEVICE TYPE	REPL CODE
MC14519BCPD	OEM
MC14519BCPDS	OEM
MC14519BCPS	OEM
MC14519BDEBS	OEM
MC14520BAL	2675
MC14520BALD	OEM
MC14520BALDS	OEM
MC14520BALS	OEM
MC14520BBEBS	OEM
MC14520BCL	2675
MC14520BCLD	OEM
MC14520BCLDS	OEM
MC14520BCLS	OEM
MC14520BCP	2675
MC14520BCPD	OEM
MC14520BCPDS	OEM
MC14520BDEBS	OEM
MC14520BL	OEM
MC14520BP	OEM
MC14520CP	5902
MC14521BAL	OEM
MC14521BALD	OEM
MC14521BALDS	OEM
MC14521BALS	OEM
MC14521BBEBS	OEM
MC14521BCL	4990
MC14521BCLD	OEM
MC14521BCLDS	OEM
MC14521BCLS	OEM
MC14521BCP	4990
MC14521BCPD	OEM
MC14521BCPDS	OEM
MC14521BCPS	OEM
MC14521BDEBS	OEM
MC14522BAL	OEM
MC14522BALD	OEM
MC14522BALDS	OEM
MC14522BALS	OEM
MC14522BBEBS	OEM
MC14522BCL	2837
MC14522BCLD	OEM
MC14522BCLDS	OEM
MC14522BCLS	OEM
MC14522BCP	2837
MC14522BCPD	OEM
MC14522BCPDS	OEM
MC14522BCPS	OEM
MC14522BDEBS	OEM
MC14522BL	OEM
MC14522BP	OEM
MC14526	3610
MC14526BAL	OEM
MC14526BALD	OEM
MC14526BALDS	OEM
MC14526BALS	OEM
MC14526BBEBS	OEM
MC14526BCL	3610
MC14526BCLD	OEM
MC14526BCLDS	OEM
MC14526BCLS	OEM
MC14526BCP	3610
MC14526BCPD	OEM
MC14526BCPDS	OEM
MC14526BCPS	OEM
MC14526BDEBS	OEM
MC14526BL	OEM
MC14526BP	OEM
MC14527BAL	3145
MC14527BALD	OEM
MC14527BALDS	OEM
MC14527BALS	OEM
MC14527BBEBS	OEM
MC14527BCL	3145
MC14527BCLD	OEM
MC14527BCLDS	OEM
MC14527BCLS	OEM
MC14527BCP	3145
MC14527BCPD	OEM
MC14527BCPS	OEM
MC14527BDEBS	OEM
MC14527BL	OEM
MC14527BP	OEM
MC14527CP	3145
MC14528	3611
MC14528BAL	3198
MC14528BALD	OEM
MC14528BALDS	OEM
MC14528BALS	OEM
MC14528BCL	3198
MC14528BCLD	OEM
MC14528BCLDS	OEM
MC14528BCLS	OEM
MC14528BCP	3198
MC14528BCPD	OEM
MC14528BCPDS	OEM
MC14528BCPS	OEM
MC14528BF	OEM
MC14528BL	OEM
MC14528BP	OEM
MC14529B	4531
MC14529BAL	OEM
MC14529BALD	OEM
MC14529BALDS	OEM
MC14529BALS	OEM
MC14529BBEBS	OEM
MC14529BCL	4531
MC14529BCLD	OEM
MC14529BCLDS	OEM
MC14529BCLS	OEM

DEVICE TYPE	REPL CODE
MC14529BCP	4531
MC14529BCPD	OEM
MC14529BCPDS	OEM
MC14529BDEBS	OEM
MC14529PCPS	OEM
MC14530BAL	OEM
MC14530BALD	OEM
MC14530BALDS	OEM
MC14530BALS	OEM
MC14530BBEBS	OEM
MC14530BCL	OEM
MC14530BCLD	OEM
MC14530BCLDS	OEM
MC14530BCLS	OEM
MC14530BCP	OEM
MC14530BCPD	OEM
MC14530BCPDS	OEM
MC14530BCPS	OEM
MC14530BDEBS	OEM
MC14531BAL	OEM
MC14531BALD	OEM
MC14531BALDS	OEM
MC14531BALS	OEM
MC14531BBEBS	OEM
MC14531BCL	3325
MC14531BCLD	OEM
MC14531BCLDS	OEM
MC14531BCP	3325
MC14531BCPD	OEM
MC14531BCPDS	OEM
MC14531BCPS	OEM
MC14531BD	OEM
MC14531BDEBS	OEM
MC14532BAL	1016
MC14532BALD	OEM
MC14532BALDS	OEM
MC14532BALS	OEM
MC14532BBEBS	OEM
MC14532BCL	1016
MC14532BCLD	OEM
MC14532BCLDS	OEM
MC14532BCLS	OEM
MC14532BCP	1016
MC14532BCPD	OEM
MC14532BCPDS	OEM
MC14532BCPS	OEM
MC14532BDEBS	OEM
MC14534BAL	OEM
MC14534BALD	OEM
MC14534BALDS	OEM
MC14534BALS	OEM
MC14534BBJBS	OEM
MC14534BCL	OEM
MC14534BCLD	OEM
MC14534BCLDS	OEM
MC14534BCLS	OEM
MC14534BCP	OEM
MC14534BCPD	OEM
MC14534BCPDS	OEM
MC14534BCPS	OEM
MC14534BDJBS	OEM
MC14536BAL	3704
MC14536BALD	OEM
MC14536BALDS	OEM
MC14536BALS	OEM
MC14536BBEBS	OEM
MC14536BCL	3760
MC14536BCLD	OEM
MC14536BCLDS	OEM
MC14536BCLS	OEM
MC14536BCP	3704
MC14536BCPD	OEM
MC14536BCPS	OEM
MC14536BDEBS	OEM
MC14536BL	OEM
MC14536BP	OEM
MC14538	3611
MC14538B	1063
MC14538BAL	1063
MC14538BALD	OEM
MC14538BALDS	OEM
MC14538BALS	OEM
MC14538BBEBS	OEM
MC14538BCL	4302
MC14538BCLD	OEM
MC14538BCLDS	OEM
MC14538BCLS	OEM
MC14538BCP	1063
MC14538BCPD	OEM
MC14538BCPDS	OEM
MC14538BCPS	OEM
MC14538BDEBS	OEM
MC14538BF	OEM
MC14538BL	OEM
MC14538BP	1063
MC14539BAL	OEM
MC14539BALD	OEM
MC14539BALDS	OEM
MC14539BALS	OEM
MC14539BBEBS	OEM
MC14539BCL	3656
MC14539BCLD	OEM
MC14539BCLDS	OEM
MC14539BCLS	OEM
MC14539BCP	3656
MC14539BCPD	OEM
MC14539BCPDS	OEM
MC14539BCPS	OEM
MC14539BD	OEM

DEVICE TYPE	REPL CODE
MC14539BDEBS	OEM
MC14539CP	3656
MC14541BAL	5017
MC14541BALD	OEM
MC14541BALDS	OEM
MC14541BALS	OEM
MC14541BBCBS	OEM
MC14541BCL	5017
MC14541BCLD	OEM
MC14541BCLDS	OEM
MC14541BCLS	OEM
MC14541BCP	5017
MC14541BCPD	OEM
MC14541BCPDS	OEM
MC14541BCPS	OEM
MC14541BDCBS	OEM
MC14543BAL	5020
MC14543BALD	OEM
MC14543BALDS	OEM
MC14543BALS	OEM
MC14543BBEBS	OEM
MC14543BCL	5020
MC14543BCLD	OEM
MC14543BCLDS	OEM
MC14543BCLS	OEM
MC14543BCP	5020
MC14543BCPD	OEM
MC14543BCPDS	OEM
MC14543BCPS	OEM
MC14543BDEBS	OEM
MC14544BAL	OEM
MC14544BALD	OEM
MC14544BALDS	OEM
MC14544BALS	OEM
MC14544BCL	OEM
MC14544BCLD	OEM
MC14544BCLDS	OEM
MC14544BCLS	OEM
MC14544BCP	OEM
MC14544BCPD	OEM
MC14544BCPDS	OEM
MC14544BCPS	OEM
MC14547BAL	OEM
MC14547BALD	OEM
MC14547BALDS	OEM
MC14547BCL	5032
MC14547BCLD	OEM
MC14547BCP	5032
MC14547BCPD	OEM
MC14547BDW	OEM
MC14549BAL	OEM
MC14549BALD	OEM
MC14549BALDS	OEM
MC14549BALS	OEM
MC14549BBEBS	OEM
MC14549BCL	OEM
MC14549BCLD	OEM
MC14549BCLDS	OEM
MC14549BCLS	OEM
MC14549BCP	OEM
MC14549BCPD	OEM
MC14549BCPDS	OEM
MC14549BCPS	OEM
MC14549BDEBS	OEM
MC14551BAL	OEM
MC14551BALD	OEM
MC14551BALDS	OEM
MC14551BALS	OEM
MC14551BBEBS	OEM
MC14551BCL	5039
MC14551BCLD	OEM
MC14551BCLDS	OEM
MC14551BCLS	OEM
MC14551BCP	5039
MC14551BCPD	OEM
MC14551BCPDS	OEM
MC14551BCPS	OEM
MC14551BD	OEM
MC14551BDEBS	OEM
MC14553BAL	OEM
MC14553BALD	OEM
MC14553BALDS	OEM
MC14553BALS	OEM
MC14553BBEBS	OEM
MC14553BCL	5040
MC14553BCLD	OEM
MC14553BCLDS	OEM
MC14553BCLS	OEM
MC14553BCP	5040
MC14553BCPD	OEM
MC14553BCPDS	OEM
MC14553BCPS	OEM
MC14553BDEBS	OEM
MC14553BDW	OEM
MC14554BAL	OEM
MC14554BALD	OEM
MC14554BALDS	OEM
MC14554BALS	OEM
MC14554BBEBS	OEM
MC14554BCL	OEM
MC14554BCLD	OEM
MC14554BCLDS	OEM
MC14554BCLS	OEM
MC14554BCP	OEM
MC14554BCPD	OEM
MC14554BCPDS	OEM
MC14554BCPS	OEM
MC14554BDEBS	OEM
MC14555BAL	2938

DEVICE TYPE	REPL CODE
MC14555BALD	OEM
MC14555BALDS	OEM
MC14555BALS	OEM
MC14555BBEBS	OEM
MC14555BCL	2938
MC14555BCLD	OEM
MC14555BCLDS	OEM
MC14555BCLS	OEM
MC14555BCP	2938
MC14555BCPD	OEM
MC14555BCPDS	OEM
MC14555BCPS	OEM
MC14555BDEBS	OEM
MC14555CP	2938
MC14556	3431
MC14556BAL	3431
MC14556BALD	OEM
MC14556BALDS	OEM
MC14556BALS	OEM
MC14556BBEBS	OEM
MC14556BCL	3431
MC14556BCLD	OEM
MC14556BCLDS	OEM
MC14556BCLS	OEM
MC14556BCP	3431
MC14556BCPD	OEM
MC14556BCPDS	OEM
MC14556BCPS	OEM
MC14556BDEBS	OEM
MC14556CP	3431
MC14557BAL	OEM
MC14557BALD	OEM
MC14557BALDS	OEM
MC14557BALS	OEM
MC14557BBEBS	OEM
MC14557BCL	OEM
MC14557BCLD	OEM
MC14557BCLDS	OEM
MC14557BCP	OEM
MC14557BCPD	OEM
MC14557BCPDS	OEM
MC14557BCPS	OEM
MC14557BDEBS	OEM
MC14557BDLS	OEM
MC14558BAL	OEM
MC14558BALD	OEM
MC14558BALDS	OEM
MC14558BALS	OEM
MC14558BBEBS	OEM
MC14558BCL	5049
MC14558BCLD	OEM
MC14558BCLDS	OEM
MC14558BCLS	OEM
MC14558BCP	5049
MC14558BCPD	OEM
MC14558BCPDS	OEM
MC14558BCPS	OEM
MC14558BD	OEM
MC14558BDEBS	OEM
MC14559BAL	OEM
MC14559BALD	OEM
MC14559BALDS	OEM
MC14559BALS	OEM
MC14559BBEBS	OEM
MC14559BCL	OEM
MC14559BCLD	OEM
MC14559BCLDS	OEM
MC14559BCLS	OEM
MC14559BCP	OEM
MC14559BCPD	OEM
MC14559BCPDS	OEM
MC14559BCPS	OEM
MC14559BDEBS	OEM
MC14560BAL	OEM
MC14560BALD	OEM
MC14560BALDS	OEM
MC14560BALS	OEM
MC14560BBEBS	OEM
MC14560BCL	6252
MC14560BCLDS	OEM
MC14560BCP	6252
MC14560BCPD	OEM
MC14560BCPDS	OEM
MC14560BCPS	OEM
MC14560BDEBS	OEM
MC14561BAL	OEM
MC14561BALD	OEM
MC14561BALDS	OEM
MC14561BALS	OEM
MC14561BBCBS	OEM
MC14561BCL	OEM
MC14561BCLD	OEM
MC14561BCLDS	OEM
MC14561BCLS	OEM
MC14561BCP	OEM
MC14561BCPD	OEM
MC14561BCPDS	OEM
MC14561BCPS	OEM
MC14561BDCBS	OEM
MC14562BAL	OEM
MC14562BALD	OEM
MC14562BALDS	OEM
MC14562BALS	OEM
MC14562BBCBS	OEM
MC14562BCL	1035
MC14562BCLD	OEM
MC14562BCLDS	OEM
MC14562BCLS	OEM
MC14562BCP	1035

DEVICE TYPE	REPL CODE
MC14562BCPD	OEM
MC14562BCPDS	OEM
MC14562BCPS	OEM
MC14562BD	OEM
MC14562BDCBS	OEM
MC14566BAL	OEM
MC14566BALD	OEM
MC14566BALDS	OEM
MC14566BALS	OEM
MC14566BBEBS	OEM
MC14566BCL	3504
MC14566BCLD	OEM
MC14566BCLDS	OEM
MC14566BCLS	OEM
MC14566BCP	3504
MC14566BCPD	OEM
MC14566BCPDS	OEM
MC14566BCPS	OEM
MC14566BDEBS	OEM
MC14568	3618
MC14568B	3618
MC14568BALD	OEM
MC14568BALDS	OEM
MC14568BALS	OEM
MC14568BBEBS	OEM
MC14568BCL	3618
MC14568BCLD	OEM
MC14568BCLDS	OEM
MC14568BCLS	OEM
MC14568BCP	3618
MC14568BCPD	OEM
MC14568BCPDS	OEM
MC14568BCPS	OEM
MC14568BD	OEM
MC14568BDEBS	OEM
MC14568BL	OEM
MC14568BP	OEM
MC14568CP	3618
MC14569BAL	OEM
MC14569BALD	OEM
MC14569BALDS	OEM
MC14569BALS	OEM
MC14569BBEBS	OEM
MC14569BCL	4487
MC14569BCLD	OEM
MC14569BCLDS	OEM
MC14569BCLS	OEM
MC14569BCP	4487
MC14569BCPD	OEM
MC14569BCPDS	OEM
MC14569BCPS	OEM
MC14569BDEBS	OEM
MC14569BL	OEM
MC14569BP	OEM
MC14572ABEBS	OEM
MC14572ADEBS	OEM
MC14572UBAL	OEM
MC14572UBALD	OEM
MC14572UBALDS	OEM
MC14572UBALS	OEM
MC14572UBCL	3480
MC14572UBCLD	OEM
MC14572UBCLDS	OEM
MC14572UBCLS	OEM
MC14572UBCP	3480
MC14572UBCPD	OEM
MC14572UBCPDS	OEM
MC14572UBCPS	OEM
MC14573CL	OEM
MC14573CLD	OEM
MC14573CP	OEM
MC14573CPD	OEM
MC14574CL	OEM
MC14574CLD	OEM
MC14574CP	OEM
MC14574CPD	OEM
MC14575CL	OEM
MC14575CLD	OEM
MC14575CP	OEM
MC14575CPD	OEM
MC14580BAL	OEM
MC14580BALD	OEM
MC14580BALDS	OEM
MC14580BALS	OEM
MC14580BBJBS	OEM
MC14580BCL	OEM
MC14580BCLD	OEM
MC14580BCLDS	OEM
MC14580BCLS	OEM
MC14580BCP	OEM
MC14580BCPD	OEM
MC14580BCPDS	OEM
MC14580BDJBS	OEM
MC14581BAL	3470
MC14581BALD	OEM
MC14581BALDS	OEM
MC14581BALS	OEM
MC14581BBJBS	OEM
MC14581BCL	5638
MC14581BCLD	OEM
MC14581BCLDS	OEM
MC14581BCLS	OEM
MC14581BCP	5638
MC14581BCPD	OEM
MC14581BCPDS	OEM
MC14581BCPSW	OEM
MC14581BDJBS	OEM
MC14582BAL	4660
MC14582BALD	OEM

DEVICE TYPE	REPL CODE
MC14582BALDS	OEM
MC14582BALS	OEM
MC14582BBEBS	OEM
MC14582BCL	4660
MC14582BCLD	OEM
MC14582BCLDS	OEM
MC14582BCLS	OEM
MC14582BCP	4660
MC14582BCPD	OEM
MC14582BCPDS	OEM
MC14582BCPS	OEM
MC14582BDEBS	OEM
MC14583BAL	OEM
MC14583BALD	OEM
MC14583BALDS	OEM
MC14583BALS	OEM
MC14583BBEBS	OEM
MC14583BCL	1297
MC14583BCLD	OEM
MC14583BCLDS	OEM
MC14583BCLS	OEM
MC14583BCP	1297
MC14583BCPD	OEM
MC14583BCPDS	OEM
MC14583BCPS	OEM
MC14583BD	OEM
MC14583BDEBS	OEM
MC14584	3626
MC14584B	3626
MC14584BAL	3626
MC14584BALD	OEM
MC14584BALDS	OEM
MC14584BALS	OEM
MC14584BBCBS	OEM
MC14584BCL	3626
MC14584BCLD	OEM
MC14584BCLDS	OEM
MC14584BCLS	OEM
MC14584BCP	6597
MC14584BCPD	OEM
MC14584BCPDS	OEM
MC14584BCPS	OEM
MC14584BD	OEM
MC14584BDCBS	OEM
MC14584BL	OEM
MC14584BP	OEM
MC14585BAL	1375
MC14585BALD	OEM
MC14585BALDS	OEM
MC14585BALS	OEM
MC14585BBEBS	OEM
MC14585BCL	1375
MC14585BCLD	OEM
MC14585BCLDS	OEM
MC14585BCLS	OEM
MC14585BCP	3727
MC14585BCPD	OEM
MC14585BCPDS	OEM
MC14585BCPS	OEM
MC14585BDEBS	OEM
MC14597BAL	OEM
MC14597BALD	OEM
MC14597BALDS	OEM
MC14597BALS	OEM
MC14597BCL	4110
MC14597BCLD	OEM
MC14597BCLDS	OEM
MC14597BCLS	OEM
MC14597BCP	4110
MC14597BCPD	OEM
MC14597BCPDS	OEM
MC14597BCPS	OEM
MC14597BDW	OEM
MC14597BL	OEM
MC14597BP	OEM
MC14598BAL	OEM
MC14598BALD	OEM
MC14598BALDS	OEM
MC14598BALS	OEM
MC14598BCL	4301
MC14598BCLD	OEM
MC14598BCLDS	OEM
MC14598BCLS	OEM
MC14598BCP	4301
MC14598BCPD	OEM
MC14598BCPDS	OEM
MC14598BCPS	OEM
MC14598BDW	OEM
MC14598BL	OEM
MC14598BP	OEM
MC14599BAL	OEM
MC14599BALD	OEM
MC14599BALDS	OEM
MC14599BALS	OEM
MC14599BCL	OEM
MC14599BCLD	OEM
MC14599BCLDS	OEM
MC14599BCLS	OEM
MC14599BCP	OEM
MC14599BCPD	OEM
MC14599BCPDS	OEM
MC14599BCPS	OEM
MC14669BAL	OEM
MC14669BCP	OEM
MC14669BDW	OEM
MC14681BP	OEM
MC27128-2	1638
MC33063P1	OEM
MC33063U	OEM

DEVICE TYPE	REPL CODE
MC33172ML	OEM
MC33174M	OEM
MC33423UD	OEM
MC34001AG	OEM
MC34001AP	OEM
MC34001AU	OEM
MC34001BG	OEM
MC34001BP	OEM
MC34001BU	OEM
MC34001D	OEM
MC34001G	OEM
MC34001P	6017
MC34001U	OEM
MC34002AG	OEM
MC34002AP	OEM
MC34002AU	OEM
MC34002BG	OEM
MC34002BP	OEM
MC34002BU	OEM
MC34002D	OEM
MC34002G	OEM
MC34002P	4107
MC34002U	OEM
MC34004BL	OEM
MC34004BP	OEM
MC34004D	OEM
MC34004L	OEM
MC34004P	5788
MC34012-1	OEM
MC34012-2	OEM
MC34012-3	OEM
MC34014P	OEM
MC34017	OEM
MC34018	OEM
MC34018P	OEM
MC34060L	5243
MC34060P	5243
MC34061AP	OEM
MC34061AP1	OEM
MC34061AU	OEM
MC34061P	OEM
MC34061P1	OEM
MC34061U	OEM
MC34062P1	OEM
MC34062U	OEM
MC34063AM	OEM
MC34063P1	OEM
MC34063U	OEM
MC34065P	5805
MC34065P-H	OEM
MC34072P	OEM
MC34084	OEM
MC34084P	OEM
MC35001AG	OEM
MC35001AU	OEM
MC35001BG	OEM
MC35001BU	OEM
MC35002AG	OEM
MC35002AU	OEM
MC35002BG	OEM
MC35002BU	OEM
MC35004BL	OEM
MC35060L	OEM
MC35061AU	OEM
MC35061U	OEM
MC35062U	OEM
MC35063U	OEM
MC38050N-25	OEM
MC41014BCPDS	OEM
MC41016BCPD	OEM
MC53200	0232
MC54802P	OEM
MC55107L	OEM
MC55108L	OEM
MC55325L	OEM
MC68000G6	OEM
MC68000G8	OEM
MC68000G10	OEM
MC68000L	OEM
MC68000L4	OEM
MC68000L6	OEM
MC68000L8	OEM
MC68000L10	OEM
MC68000L12	OEM
MC68000P8	OEM
MC68000Z	OEM
MC68020	OEM
MC68120-1L	OEM
MC68120L	OEM
MC68121-1	OEM
MC68121-1L	OEM
MC68121L	OEM
MC68153L	OEM
MC68153P	OEM
MC68230L8L	OEM
MC68230L8P	OEM
MC68230L10L	OEM
MC68230L10P	OEM
MC68451L4	OEM
MC68451L6	OEM
MC68451L8	OEM
MC68452L	OEM
MC68452P	OEM
MC68486	OEM
MC68488L	OEM
MC68488P	OEM
MC68652L(A)	OEM
MC68652L2(A)	OEM
MC68652P(A)	OEM

DEVICE TYPE	REPL CODE
MC68652P2(A)	OEM
MC68652S(A)	OEM
MC68661A	OEM
MC68661B	OEM
MC68661C	OEM
MC68681	OEM
MC68701-1L	OEM
MC68701CL	OEM
MC68701CL-1	OEM
MC68701L	OEM
MC68701L-1	OEM
MC68701L1	OEM
MC68701U4L	OEM
MC68701U4L-1	OEM
MC68705P3L	OEM
MC68705P3S	OEM
MC68705PP	OEM
MC68705R3L	OEM
MC68705U3L	OEM
MC74107P	0942
MC74121P	0175
MC74145P	0619
MC74150P	1492
MC74152P	1517
MC74153P	1539
MC74155P	1576
MC74156P	1592
MC74164AP	0734
MC74165P	1685
MC74176P	1795
MC74177P	1803
MC74180P	1829
MC74181P	1842
MC74182P	1856
MC74192P	1921
MC74193P	1926
MC74195P	1943
MC74415P	1495
MC75107	OEM
MC75107L	OEM
MC75107P	OEM
MC75108L	OEM
MC75108P	OEM
MC75110L	OEM
MC75110P	OEM
MC75125L	OEM
MC75125P	OEM
MC75127L	OEM
MC75127P	OEM
MC75128L	OEM
MC75128P	OEM
MC75129L	OEM
MC75129P	OEM
MC75325L	OEM
MC75452P	1262
MC75452P1	1262
MC75462P	OEM
MC75463P	OEM
MC75491P	6379
MC75492P	6382
MC140007BCL	OEM
MC140008BALS	OEM
MC141000L	OEM
MC141000P	OEM
MC141000S	OEM
MC141099L	OEM
MC141200L	OEM
MC141200P	OEM
MC141200S	OEM
MC142100AL	OEM
MC142100ALD	OEM
MC142100CL	OEM
MC142100CLD	OEM
MC142100CP	OEM
MC142100CPD	OEM
MC142100L	OEM
MC142100P	OEM
MC143403L	OEM
MC143403P	OEM
MC143404L	OEM
MC143404P	OEM
MC144111P	OEM
MC144115P	OEM
MC144143P1	OEM
MC144143P2	OEM
MC145000	OEM
MC145000BCL	OEM
MC145000BCP	OEM
MC145000L	OEM
MC145000P	OEM
MC145001	OEM
MC145001BCL	OEM
MC145001BCP	OEM
MC145001L	OEM
MC145001P	OEM
MC145026L	OEM
MC145026P	OEM
MC145026PP	OEM
MC145027L	OEM
MC145027P	OEM
MC145028	OEM
MC145028L	OEM
MC145028P	OEM
MC145035CL	OEM
MC145100AL	OEM
MC145100ALD	OEM
MC145100CL	OEM
MC145100CLD	OEM
MC145100CP	OEM

DEVICE TYPE	REPL CODE
MC145100CPD	OEM
MC145100L	OEM
MC145100P	OEM
MC145104	2616
MC145104L	OEM
MC145104P	2616
MC145104PD	OEM
MC145106	OEM
MC145106L	OEM
MC145106P	OEM
MC145106PD	OEM
MC145107L	OEM
MC145107P	OEM
MC145107PD	OEM
MC145109	1714
MC145109L	OEM
MC145109P	1714
MC145109PD	OEM
MC145112L	OEM
MC145112P	OEM
MC145112PD	OEM
MC145143L	OEM
MC145143P	OEM
MC145144L	OEM
MC145144P	OEM
MC145145L	OEM
MC145145P	OEM
MC145146L	OEM
MC145146P	OEM
MC145151L	OEM
MC145151P	OEM
MC145152L	OEM
MC145152P	OEM
MC145155L	OEM
MC145155P	OEM
MC145156L	OEM
MC145156P	OEM
MC145157L	OEM
MC145157P	OEM
MC145158L	OEM
MC145158P	OEM
MC145411P	OEM
MC145414L	OEM
MC145414P	OEM
MC145415L	OEM
MC145415P	OEM
MC145422L	OEM
MC145423L	OEM
MC145426L	OEM
MC145428L	OEM
MC145428P	OEM
MC145429L	OEM
MC145429P	OEM
MC145431L	OEM
MC145431P	OEM
MC145432L	OEM
MC145432P	OEM
MC145433L	OEM
MC145433P	OEM
MC145434L	OEM
MC145434P	OEM
MC145436BCPDS	OEM
MC145440L	OEM
MC145440P	OEM
MC145441L	OEM
MC145441P	OEM
MC145445L	OEM
MC145445P	OEM
MC145450L	OEM
MC145450P	OEM
MC145668BCP	OEM
MC146805E2CL	OEM
MC146805E2CP	OEM
MC146805E2CS	OEM
MC146805E2CZ	OEM
MC146805E2L	OEM
MC146805E2P	OEM
MC146805E2S	OEM
MC146805E2Z	OEM
MC146805F2L	OEM
MC146805F2P	OEM
MC146805F2S	OEM
MC146805F2Z	OEM
MC146805G2L	OEM
MC146805G2P	OEM
MC146805G2S	OEM
MC146805G2Z	OEM
MC146805H2L	OEM
MC146805H2P	OEM
MC146805H2S	OEM
MC146805H2Z	OEM
MC146818AL	OEM
MC146818AP	OEM
MC146818APN	OEM
MC146818AS	OEM
MC146818L	OEM
MC146818P	OEM
MC146818S	OEM
MC146818Z	OEM
MC146823L	OEM
MC146823P	OEM
MC146823S	OEM
MC146823Z	OEM
MC147805	OEM
MC6800048	OEM
MCA	OEM
MCA7	OEM
MCA8	2636
MCA22CL	OEM

DEVICE TYPE	REPL CODE
MCA22CP	OEM
MCA22CS	OEM
MCA22L	OEM
MCA22P	OEM
MCA22S	OEM
MCA81	2636
MCA230	1053
MCA231	0315
MCA255	1109
MCA2500ECL	OEM
MCB	OEM
MCB22CL	OEM
MCB22CP	OEM
MCB22CS	OEM
MCB22L	OEM
MCB22P	OEM
MCB22S	OEM
MCB-302	OEM
MCB-304	OEM
MCB-308	OEM
MCB-316	OEM
MCB-332	OEM
MCB-512-01	OEM
MCB-512-03	OEM
MCB-512-04	OEM
MCC	OEM
MCC-06I01	0652
MCC12	OEM
MCC15	OEM
MCC16	OEM
MCC17	OEM
MCC19	OEM
MCC20	OEM
MCC21	OEM
MCC555A	0973
MCC555B	0973
MCC670	OEM
MCC671	OEM
MCCZA6.8D	0062
MCCZA6.8D5	0062
MCCZA6.8D10	0062
MCCZA7.5D	0077
MCCZA7.5D5	0077
MCCZA7.5D10	0077
MCCZA8.2D	0165
MCCZA8.2D5	0165
MCCZA8.2D10	0165
MCCZA9.1D	0057
MCCZA9.1D5	0057
MCCZA9.1D10	0057
MCCZA10D	0064
MCCZA10D5	0064
MCCZA10D10	0064
MCCZA11D	0181
MCCZA11D5	0181
MCCZA11D10	0181
MCCZA12D	0052
MCCZA12D5	0052
MCCZA12D10	0052
MCCZA13D	0053
MCCZA13D5	0053
MCCZA13D10	0053
MCCZA14D	0879
MCCZA14D5	0879
MCCZA14D10	0879
MCCZA15D	0686
MCCZA15D5	0686
MCCZA15D10	0686
MCCZA16D	0445
MCCZA16D5	0445
MCCZA16D10	0445
MCCZA17D	0210
MCCZA17D5	0210
MCCZA17D10	0210
MCCZA18D	0375
MCCZA18D5	0375
MCCZA18D10	0375
MCCZA19D	0671
MCCZA19D10	0671
MCCZA20D	0700
MCCZA20D5	0700
MCCZA20D10	0700
MCCZA22D	0705
MCCZA22D5	0705
MCCZA22D10	0705
MCCZA24D	0494
MCCZA24D5	0494
MCCZA24D10	0494
MCCZA25D	0714
MCCZA25D5	0714
MCCZA25D10	0714
MCCZA27D	0455
MCCZA27D5	0455
MCCZA27D10	0455
MCCZA28D	0257
MCCZA28D5	0257
MCCZA28D10	0257
MCCZA30D	0195
MCCZA30D5	0195
MCCZA30D10	0195
MCCZA33D	0166
MCCZA33D5	0166
MCCZA33D10	0166
MCCZA36D	0010
MCCZA36D5	0010
MCCZA36D10	0010
MCCZA39D	0032
MCCZA39D5	0032
MCCZA39D10	0032

DEVICE TYPE	REPL CODE
MCCZA43D	0054
MCCZA43D5	0054
MCCZA43D10	0054
MCCZA47D	0068
MCCZA47D5	0068
MCCZA47D10	0068
MCCZA51D	0092
MCCZA51D5	0092
MCCZA51D10	0092
MCCZA56D	0125
MCCZA56D5	0125
MCCZA56D10	0125
MCCZA60D	2317
MCCZA60D5	2317
MCCZA60D10	2317
MCCZA62D	0152
MCCZA62D5	0152
MCCZA62D10	0152
MCCZA68D	0173
MCCZA68D5	0173
MCCZA68D10	0173
MCCZA75D	0094
MCCZA75D5	0094
MCCZA75D10	0094
MCCZA82D	0049
MCCZA82D5	0049
MCCZA82D10	0049
MCCZA87D	0104
MCCZA87D5	0104
MCCZA87D10	0104
MCCZA91D	0156
MCCZA91D5	0156
MCCZA91D10	0156
MCCZA100D	0189
MCCZA100D5	0189
MCCZA100D10	0189
MCCZA105D	0099
MCCZA105D5	0099
MCCZA105D10	0099
MCCZA110D	0099
MCCZA110D5	0099
MCCZA110D10	0099
MCCZA120D	0089
MCCZA120D5	0089
MCCZA120D10	0089
MCCZA130D	0287
MCCZA130D5	0287
MCCZA130D10	0287
MCCZA140D	0252
MCCZA140D5	0252
MCCZA140D10	0252
MCCZA150D	0340
MCCZA150D5	0340
MCCZA150D10	0340
MCCZA160D	0370
MCCZA160D5	0370
MCCZA160D10	0370
MCCZA175D	0425
MCCZA175D5	0425
MCCZA175D10	0425
MCCZA180D	0425
MCCZA180D5	0425
MCCZA180D10	0425
MCCZA200D	1473
MCCZA200D5	1473
MCCZA200D10	1473
MCCZB6.8D	0062
MCCZB6.8D5	0062
MCCZB6.8D10	0062
MCCZB7.5D	0077
MCCZB7.5D5	0077
MCCZB7.5D10	0077
MCCZB8.2D	0165
MCCZB8.2D5	0165
MCCZB8.2D10	0165
MCCZB9.1D	0057
MCCZB9.1D5	0057
MCCZB9.1D10	0057
MCCZB10D	0064
MCCZB10D5	0064
MCCZB10D10	0064
MCCZB11D	0181
MCCZB11D5	0181
MCCZB11D10	0181
MCCZB12D	0052
MCCZB12D5	0052
MCCZB12D10	0052
MCCZB13D	0053
MCCZB13D5	0053
MCCZB13D10	0053
MCCZB14D	0879
MCCZB14D5	0879
MCCZB14D10	0879
MCCZB15D	0686
MCCZB15D5	0686
MCCZB15D10	0686
MCCZB16D	0445
MCCZB16D5	0445
MCCZB16D10	0445
MCCZB17D	0210
MCCZB17D5	0210
MCCZB17D10	0210
MCCZB18D	0375
MCCZB18D5	0375
MCCZB18D10	0375
MCCZB19D	0671
MCCZB19D5	0671
MCCZB19D10	0671
MCCZB20D	0700

DEVICE TYPE	REPL CODE
MCCZB20D5	0700
MCCZB20D10	0700
MCCZB22D5	0705
MCCZB22D10	0705
MCCZB24D	0494
MCCZB24D5	0494
MCCZB24D10	0494
MCCZB25D	0714
MCCZB25D5	0714
MCCZB25D10	0714
MCCZB27D	0455
MCCZB27D5	0455
MCCZB27D10	0455
MCCZB30D	0195
MCCZB30D5	0195
MCCZB30D10	0195
MCCZB33D	0166
MCCZB33D5	0166
MCCZB33D10	0166
MCCZB36D	0010
MCCZB36D5	0010
MCCZB36D10	0010
MCCZB39D	0032
MCCZB39D5	0032
MCCZB39D10	0032
MCCZB43D	0054
MCCZB43D5	0054
MCCZB43D10	0054
MCCZB47D	0068
MCCZB47D5	0068
MCCZB47D10	0068
MCCZB51D	0092
MCCZB51D5	0092
MCCZB51D10	0092
MCCZB56D	0125
MCCZB56D5	0125
MCCZB56D10	0125
MCCZB62D	0152
MCCZB62D5	0152
MCCZB62D10	0152
MCCZB68D	0173
MCCZB68D5	0173
MCCZB68D10	0173
MCCZB75D	0094
MCCZB75D5	0094
MCCZB75D10	0094
MCCZB82D	0049
MCCZB82D5	0049
MCCZB82D10	0049
MCCZB91D	0156
MCCZB91D5	0156
MCCZB91D10	0156
MCCZB100D	0189
MCCZB100D5	0189
MCCZB100D10	0189
MCCZB105D	0099
MCCZB105D5	0099
MCCZB105D10	0099
MCCZB110D	0099
MCCZB110D5	0099
MCCZB110D10	0099
MCCZB120D	0089
MCCZB120D5	0089
MCCZB120D10	0089
MCCZB130D	0287
MCCZB130D5	0287
MCCZB130D10	0287
MCCZB140D	0252
MCCZB140D5	0252
MCCZB140D10	0252
MCCZB150D	0340
MCCZB150D5	0340
MCCZB150D10	0340
MCCZB160D	0370
MCCZB160D5	0370
MCCZB160D10	0370
MCCZB175D	0425
MCCZB175D5	0425
MCCZB175D10	0425
MCCZB180D	0425
MCCZB180D5	0425
MCCZB180D10	0425
MCCZB200D	1473
MCCZB200D5	1473
MCCZB200D10	1473
MCCZC6.8D	0062
MCCZC6.8D5	0062
MCCZC6.8D10	0062
MCCZC7.5D	0077
MCCZC7.5D5	0077
MCCZC7.5D10	0077
MCCZC8.2D	0165
MCCZC8.2D5	0165
MCCZC8.2D10	0165
MCCZC9.1D	0057
MCCZC9.1D5	0057
MCCZC9.1D10	0057
MCCZC10D	0064
MCCZC10D5	0064
MCCZC10D10	0064
MCCZC11D	0181
MCCZC11D5	0181
MCCZC11D10	0181
MCCZC12D	0052
MCCZC12D5	0052
MCCZC12D10	0052
MCCZC13D	0125
MCCZC13D5	0053
MCCZC13D10	0053

DEVICE TYPE	REPL CODE
MCCZC14D	0879
MCCZC14D5	0879
MCCZC14D10	0879
MCCZC15D	0686
MCCZC15D5	0686
MCCZC15D10	0686
MCCZC16D	0445
MCCZC16D5	0445
MCCZC16D10	0445
MCCZC17D	0210
MCCZC17D5	0210
MCCZC17D10	0210
MCCZC18D	0375
MCCZC18D5	0375
MCCZC18D10	0375
MCCZC19D	0671
MCCZC19D5	0671
MCCZC19D10	0671
MCCZC20D	0700
MCCZC20D5	0700
MCCZC20D10	0700
MCCZC22D	0705
MCCZC22D5	0705
MCCZC22D10	0705
MCCZC24D	0494
MCCZC24D5	0494
MCCZC24D10	0494
MCCZC25D	0714
MCCZC25D5	0714
MCCZC25D10	0714
MCCZC27D	0455
MCCZC27D5	0455
MCCZC27D10	0455
MCCZC30D	0195
MCCZC30D5	0195
MCCZC30D10	0195
MCCZC33D	0166
MCCZC33D5	0166
MCCZC33D10	0166
MCCZC36D	0010
MCCZC36D5	0010
MCCZC36D10	0010
MCCZC39D	0032
MCCZC39D5	0032
MCCZC39D10	0032
MCCZC43D	0054
MCCZC43D5	0054
MCCZC43D10	0054
MCCZC47D	0068
MCCZC47D5	0068
MCCZC47D10	0068
MCCZC51D	0092
MCCZC51D5	0092
MCCZC51D10	0092
MCCZC56D	0125
MCCZC56D5	0125
MCCZC56D10	0125
MCCZC62D	0152
MCCZC62D5	0152
MCCZC62D10	0152
MCCZC68D	0173
MCCZC68D5	0173
MCCZC68D10	0173
MCCZC75D	0094
MCCZC75D5	0094
MCCZC75D10	0094
MCCZC82D	0049
MCCZC82D5	0049
MCCZC82D10	0049
MCCZC91D	0156
MCCZC91D10	0156
MCCZC92D5	0156
MCCZC100D	0189
MCCZC100D5	0189
MCCZC100D10	0064
MCCZC105D	0099
MCCZC105D5	0099
MCCZC105D10	0099
MCCZC110D	0099
MCCZC110D5	0099
MCCZC110D10	0099
MCCZC120D	0089
MCCZC120D5	0089
MCCZC120D10	0089
MCCZC130D	0287
MCCZC130D5	0287
MCCZC130D10	0287
MCCZC140D	0152
MCCZC140D5	0152
MCCZC140D10	0152
MCCZC150D	0340
MCCZC150D5	0340
MCCZC150D10	0340
MCCZC160D	0370
MCCZC160D5	0370
MCCZC160D10	0370
MCCZC175D	0425
MCCZC175D5	0425
MCCZC175D10	0425
MCCZC180D	0671
MCCZC180D5	0671
MCCZC180D10	0671
MCCZC200D	1473
MCCZC200D5	1473
MCCZC200D10	1473
MCCZD6.8D	0062
MCCZD6.8D5	0062
MCCZD6.8D10	0062
MCCZD7.5D	0077

DEVICE TYPE	REPL CODE
MCCZD7.5D5	0077
MCCZD7.5D10	0077
MCCZD8.2D	0165
MCCZD8.2D5	0165
MCCZD8.2D10	0165
MCCZD9.1D	0057
MCCZD9.1D5	0057
MCCZD9.1D10	0057
MCCZD10D	0064
MCCZD10D5	0064
MCCZD10D10	0064
MCCZD11D	0181
MCCZD11D5	0181
MCCZD11D10	0181
MCCZD12D	0052
MCCZD12D5	0052
MCCZD12D10	0052
MCCZD13D	0053
MCCZD13D5	0053
MCCZD13D10	0053
MCCZD14D	0879
MCCZD14D5	0879
MCCZD14D10	0879
MCCZD15D	0686
MCCZD15D5	0686
MCCZD15D10	0686
MCCZD16D	0445
MCCZD16D5	0445
MCCZD16D10	0445
MCCZD17D	0210
MCCZD17D5	0210
MCCZD17D10	0210
MCCZD18D5	0375
MCCZD18D10	0375
MCCZD19D	0671
MCCZD19D5	0671
MCCZD19D10	0671
MCCZD20D	0700
MCCZD20D5	0700
MCCZD20D10	0700
MCCZD22D	0705
MCCZD22D5	0705
MCCZD22D10	0705
MCCZD24D	0494
MCCZD24D5	0494
MCCZD24D10	0494
MCCZD25D	0714
MCCZD25D5	0714
MCCZD25D10	0714
MCCZD27D	0455
MCCZD27D5	0455
MCCZD27D10	0455
MCCZD28D	0257
MCCZD28D5	0257
MCCZD28D10	0257
MCCZD30D	0195
MCCZD30D5	0195
MCCZD30D10	0195
MCCZD33D	0166
MCCZD33D5	0166
MCCZD33D10	0166
MCCZD36D	0010
MCCZD36D5	0010
MCCZD36D10	0010
MCCZD39D	0032
MCCZD39D5	0032
MCCZD39D10	0032
MCCZD43D	0054
MCCZD43D5	0054
MCCZD43D10	0054
MCCZD47D	0068
MCCZD47D5	0068
MCCZD47D10	0068
MCCZD51D	0092
MCCZD51D5	0092
MCCZD51D10	0092
MCCZD56D	0125
MCCZD56D5	0125
MCCZD56D10	0125
MCCZD60D	2317
MCCZD60D5	2317
MCCZD60D10	2317
MCCZD62D	0152
MCCZD62D5	0152
MCCZD62D10	0152
MCCZD68D	0173
MCCZD68D5	0173
MCCZD68D10	0173
MCCZD75D	0094
MCCZD75D5	0094
MCCZD75D10	0094
MCCZD82D	0049
MCCZD82D5	0049
MCCZD82D10	0049
MCCZD87D	0104
MCCZD87D5	0104
MCCZD87D10	0104
MCCZD91D	0156
MCCZD91D5	0156
MCCZD91D10	0156
MCCZD100D	0189
MCCZD100D5	0189
MCCZD100D10	0189
MCCZD105D	0099
MCCZD105D5	0099
MCCZD105D10	0099
MCCZD110D	0099
MCCZD110D5	0099
MCCZD110D10	0099

DEVICE TYPE	REPL CODE
MCCZD120D	0089
MCCZD120D5	0089
MCCZD120D10	0089
MCCZD130D	0455
MCCZD130D5	0455
MCCZD130D10	0455
MCCZD140D	0252
MCCZD140D5	0252
MCCZD140D10	0252
MCCZD150D	0340
MCCZD150D5	0340
MCCZD150D10	0340
MCCZD160D	0370
MCCZD160D5	0370
MCCZD160D10	0370
MCCZD175D	0425
MCCZD175D5	0425
MCCZD175D10	0425
MCCZD180D	0425
MCCZD180D5	0425
MCCZD180D10	0425
MCCZD200D	1473
MCCZD200D5	1473
MCCZD200D10	1473
MCCZE6.8D	0062
MCCZE6.8D5	0062
MCCZE6.8D10	0062
MCCZE7.5D	0077
MCCZE7.5D5	0077
MCCZE7.5D10	0077
MCCZE8.2D	0165
MCCZE8.2D5	0165
MCCZE8.2D10	0165
MCCZE9.1D1D	0057
MCCZE9.1D5	0057
MCCZE9.1D10	0057
MCCZE10D	0064
MCCZE10D5	0064
MCCZE10D10	0064
MCCZE11D	0181
MCCZE11D5	0181
MCCZE11D10	0181
MCCZE12D	0052
MCCZE12D5	0052
MCCZE12D10	0052
MCCZE13D	0053
MCCZE13D5	0053
MCCZE13D10	0053
MCCZE14D	0879
MCCZE14D5	0879
MCCZE14D10	0879
MCCZE15D	0686
MCCZE15D5	0686
MCCZE15D10	0686
MCCZE16D	0445
MCCZE16D5	0445
MCCZE16D10	0445
MCCZE17D	0210
MCCZE17D5	0210
MCCZE17D10	0210
MCCZE18D	0375
MCCZE18D5	0375
MCCZE18D10	0375
MCCZE19D	0671
MCCZE19D5	0671
MCCZE19D10	0671
MCCZE20D	0700
MCCZE20D5	0700
MCCZE20D10	0700
MCCZE22D	0705
MCCZE22D5	0705
MCCZE22D10	0705
MCCZE24D	0494
MCCZE24D5	0494
MCCZE24D10	0494
MCCZE25D	0714
MCCZE25D5	0714
MCCZE25D10	0714
MCCZE27D	0455
MCCZE27D5	0455
MCCZE27D10	0455
MCCZE30D	0195
MCCZE30D5	0195
MCCZE30D10	0195
MCCZE33D	0166
MCCZE33D5	0166
MCCZE33D10	0166
MCCZE36D	0010
MCCZE36D5	0010
MCCZE36D10	0010
MCCZE39D	0032
MCCZE39D5	0032
MCCZE39D10	0032
MCCZE43D	0054
MCCZE43D5	0054
MCCZE43D10	0054
MCCZE47D	0068
MCCZE47D5	0068
MCCZE47D10	0068
MCCZE51D	0092
MCCZE51D5	0092
MCCZE51D10	0092
MCCZE56D	0125
MCCZE56D5	0125
MCCZE56D10	0125
MCCZE62D	0152
MCCZE62D5	0152
MCCZE62D10	0152
MCCZE68D	0173

DEVICE TYPE	REPL CODE
MCCZE68D5	0173
MCCZE68D10	0173
MCCZE75D	0094
MCCZE75D5	0094
MCCZE75D10	0094
MCCZE82D	0049
MCCZE82D5	0049
MCCZE82D10	0049
MCCZE91D	0156
MCCZE91D5	0156
MCCZE91D10	0156
MCCZE100D	0189
MCCZE100D5	0189
MCCZE100D10	0189
MCCZE105D	0099
MCCZE105D5	0099
MCCZE105D10	0099
MCCZE110D	0099
MCCZE110D5	0099
MCCZE110D10	0099
MCCZE120D	0089
MCCZE120D5	0089
MCCZE120D10	0089
MCCZE130D	0287
MCCZE130D5	0287
MCCZE130D10	0287
MCCZE140D	0252
MCCZE140D5	0252
MCCZE140D10	0252
MCCZE150D	0340
MCCZE150D5	0340
MCCZE150D10	0340
MCCZE160D	0370
MCCZE160D5	0370
MCCZE160D10	0370
MCCZE175D	0425
MCCZE175D5	0425
MCCZE175D10	0425
MCCZE180D	0425
MCCZE180D5	0425
MCCZE180D10	0425
MCCZE200D	1473
MCCZE200D5	1473
MCCZE200D10	1473
MCCZF1.8D	OEM
MCCZF2.0D	OEM
MCCZF2.0D5	OEM
MCCZF2.0D10	OEM
MCCZF2.2D	OEM
MCCZF2.2D5	OEM
MCCZF2.2D10	OEM
MCCZF2.4D	OEM
MCCZF2.4D5	OEM
MCCZF2.4D10	OEM
MCCZF2.7D	0760
MCCZF2.7D5	0760
MCCZF2.7D10	0760
MCCZF2.8D5	0677
MCCZF2.8D10	0677
MCCZF3.0D	0118
MCCZF3.0D5	0118
MCCZF3.0D10	0118
MCCZF3.3D	0299
MCCZF3.3D5	0299
MCCZF3.3D10	0299
MCCZF3.6D	0376
MCCZF3.6D5	0376
MCCZF3.6D10	0376
MCCZF3.9D	0036
MCCZF3.9D5	0036
MCCZF3.9D10	0036
MCCZF4.3D	0275
MCCZF4.3D5	OEM
MCCZF4.3D10	0275
MCCZF4.7D	0140
MCCZF4.7D5	0140
MCCZF4.7D10	0140
MCCZF5.1D	0041
MCCZF5.1D5	0041
MCCZF5.1D10	0041
MCCZF5.6D	0253
MCCZF5.6D5	0253
MCCZF5.6D10	0253
MCCZF6.2D	0471
MCCZF6.2D5	0471
MCCZF6.2D10	0471
MCCZF6.8D	0062
MCCZF6.8D5	0062
MCCZF6.8D10	0062
MCCZF7.5D	0077
MCCZF7.5D5	0077
MCCZF7.5D10	0077
MCCZF8.2D	0165
MCCZF8.2D5	0165
MCCZF8.2D10	0165
MCCZF9.1D	0057
MCCZF9.1D5	0057
MCCZF9.1D10	0057
MCCZF10D	0064
MCCZF10D5	0064
MCCZF12D	0052
MCCZF12D5	0052
MCCZF12D10	0052
MCCZG2.4D	1276
MCCZG2.4D5	1276
MCCZG2.4D10	1276
MCCZG2.7D	0760
MCCZG2.7D5	0760
MCCZG2.7D10	0760

DEVICE TYPE	REPL CODE
MCCZG3.0D	0118
MCCZG3.0D5	0118
MCCZG3.0D10	0118
MCCZG3.3D	0299
MCCZG3.3D5	0299
MCCZG3.3D10	0299
MCCZG3.6D	0376
MCCZG3.6D5	0376
MCCZG3.6D10	0376
MCCZG3.9D	0036
MCCZG3.9D5	0036
MCCZG3.9D10	0036
MCCZG4.3D	0275
MCCZG4.3D5	0275
MCCZG4.3D10	0275
MCCZG4.7D	0140
MCCZG4.7D5	0140
MCCZG4.7D10	0140
MCCZG5.1D	0041
MCCZG5.1D5	0041
MCCZG5.1D10	0041
MCCZG5.6D	0253
MCCZG5.6D5	0253
MCCZG5.6D10	0253
MCCZG6.2D	0471
MCCZG6.2D5	0471
MCCZG6.2D10	0471
MCCZG6.8D	0062
MCCZG6.8D5	0062
MCCZG6.8D10	0062
MCCZG7.5D	0077
MCCZG7.5D5	0077
MCCZG7.5D10	0077
MCCZG8.2D	0165
MCCZG8.2D5	0165
MCCZG8.2D10	0165
MCCZG9.1D	0057
MCCZG9.1D5	0057
MCCZG9.1D10	0057
MCCZG10D	0064
MCCZG10D5	0064
MCCZG10D10	0064
MCCZG12D	0052
MCCZG12D5	0052
MCCZG12D10	0052
MCD	OEM
MCD521H	OEM
MCD521L	OEM
MCD522H	OEM
MCD522L	OEM
MCD526	OEM
MCD527	OEM
MCD537	OEM
MCD725	OEM
MCD735	OEM
MCF1145	OEM
MCF1165	OEM
MCF1166	OEM
MCF1327	OEM
MCF1329	OEM
MCF1330	OEM
MCF1331	OEM
MCF1492	OEM
MCF1493	OEM
MCF1510	OEM
MCF6021	1876
MCH-01	OEM
MCH2005F	2258
MCH5862	OEM
MCH5867	OEM
MCH5871	OEM
MCH5872	OEM
MCH5875	OEM
MCH5880	OEM
MCH5883	OEM
MCI00	OEM
MCI01	OEM
MCI12	OEM
MCL45	OEM
MCL601	OEM
MCL611	OEM
MCL701N	OEM
MCL702A	OEM
MCL702L	OEM
MCL703C	OEM
MCL703L	OEM
MCL716A	OEM
MCL723C	OEM
MCL1300	OEM
MCL1301	OEM
MCL1302	OEM
MCL1303	OEM
MCL1304	OEM
MCLTC6010	OEM
MCLTC6025	OEM
MCLTC6050	OEM
MCLTC6100	OEM
MCM10	OEM
MCM11	OEM
MCM20	OEM
MCM21	OEM
MCM21L14C20	OEM
MCM21L14C25	OEM
MCM21L14C30	OEM
MCM21L14C45	OEM
MCM21L14P20	OEM
MCM21L14P25	OEM
MCM21L14P30	OEM
MCM21L14P45	OEM

DEVICE TYPE	REPL CODE
MCM21L15AC-45	OEM
MCM21L15AC-70	OEM
MCM21L25AC-45	OEM
MCM21L25AC-70	OEM
MCM27A08C	OEM
MCM27A08L	OEM
MCM27L16-25L	OEM
MCM27S19DC	OEM
MCM27S19PC	OEM
MCM27S25DC	OEM
MCM27S25PC	OEM
MCM27S27DC	OEM
MCM27S27PC	OEM
MCM27S29DC	OEM
MCM27S29PC	OEM
MCM27S31DC	OEM
MCM27S31PC	OEM
MCM27S35DC	OEM
MCM27S35PC	OEM
MCM27S37DC	OEM
MCM27S37PC	OEM
MCM27S45DC	OEM
MCM27S45PC	OEM
MCM27S47DC	OEM
MCM27S47PC	OEM
MCM27S181DC	OEM
MCM27S181PC	OEM
MCM27S191DC	OEM
MCM27S191PC	OEM
MCM27S281DC	OEM
MCM27S281PC	OEM
MCM27S291DC	OEM
MCM27S291PC	OEM
MCM40	OEM
MCM41	OEM
MCM51L01C45	OEM
MCM51L01C65	OEM
MCM51L01P45	OEM
MCM51L01P65	OEM
MCM66L41-20P	OEM
MCM66L41-25C	OEM
MCM66L41-25P	OEM
MCM66L41-30C	OEM
MCM66L41-30P	OEM
MCM66L41-45C	OEM
MCM66L41-45P	OEM
MCM68A	OEM
MCM68A10CL	OEM
MCM68A10CP	OEM
MCM68A10CS	OEM
MCM68A10L	OEM
MCM68A10P	OEM
MCM68A10S	OEM
MCM68A30AC	OEM
MCM68A30AP	OEM
MCM68A308C	OEM
MCM68A308P	OEM
MCM68A316E	OEM
MCM68A364	OEM
MCM68A708C	OEM
MCM68A708L	OEM
MCM68B10L	OEM
MCM68B10P	OEM
MCM68B10S	OEM
MCM68HC34L	OEM
MCM68HC34P	OEM
MCM80	OEM
MCM81	OEM
MCM93L422AD	OEM
MCM93L422AP	OEM
MCM93L422D	OEM
MCM93L422DC	OEM
MCM93L422P	OEM
MCM93L422PC	OEM
MCM101C80	OEM
MCM332	OEM
MCM364	OEM
MCM2016H	OEM
MCM2114	2048
MCM2114-20L	OEM
MCM2114-20P	OEM
MCM2114-25L	OEM
MCM2114-25P	OEM
MCM2114-30	2048
MCM2114-30L	OEM
MCM2114-30P	OEM
MCM2114-45	2048
MCM2114-45L	OEM
MCM2114-45P	OEM
MCM2114-P20	OEM
MCM2114C20	OEM
MCM2114C25	OEM
MCM2114C30	OEM
MCM2114C45	OEM
MCM2114P	5300
MCM2114P20	2048
MCM2114P25	OEM
MCM2114P-30	2048
MCM2114P30	OEM
MCM2114P-45	2048
MCM2114P45	2048
MCM2115AC-45	OEM
MCM2115AC-55	OEM
MCM2115AC-70	OEM
MCM2125AC-55	OEM
MCM2125AC-70	OEM
MCM2125AC-4545	OEM
MCM2147C55	1693
MCM2147C70	1693

DEVICE TYPE	REPL CODE
MCM2147C85	1693
MCM2147C100	1693
MCM2167H-35L	OEM
MCM2167H-35P	OEM
MCM2167H-35Z	OEM
MCM2167H-45L	OEM
MCM2167H-45P	OEM
MCM2167H-45Z	OEM
MCM2167H-55L	OEM
MCM2167H-55P	OEM
MCM2167H-55Z	OEM
MCM2532C	5098
MCM2708C	4424
MCM2708L	4424
MCM2716C	2278
MCM2716L	6403
MCM2801	OEM
MCM2832-15L	OEM
MCM2832-15P	OEM
MCM2832-20L	OEM
MCM2832-20P	OEM
MCM2832P	OEM
MCM2833P	OEM
MCM4027AC2	OEM
MCM4027AC3	4974
MCM4027AC4	4974
MCM4116AC15	OEM
MCM4116AC20	0522
MCM4116AC25	0522
MCM4116AC30	0522
MCM4116BC15	OEM
MCM4116BC-20	OEM
MCM4116BC20	0522
MCM4116BC-25	OEM
MCM4116BC25	0522
MCM4116BC-30	OEM
MCM4116BC30	0522
MCM4116BC35	OEM
MCM4116BP-15	OEM
MCM4116BP15	OEM
MCM4116BP20	0522
MCM4116BP25	0522
MCM4116BP30	0522
MCM4116BP35	0522
MCM4517-12P	OEM
MCM4517-15P	OEM
MCM4517-20	OEM
MCM4517P	OEM
MCM4517P10	OEM
MCM4517P12	OEM
MCM4517P-15	OEM
MCM4517P15	OEM
MCM4517P15B1	OEM
MCM4517P20	OEM
MCM5101C65	OEM
MCM5101P65	OEM
MCM5101P80	OEM
MCM6116P15	OEM
MCM6116P20	OEM
MCM6256-10L	OEM
MCM6256-12L	OEM
MCM6256-15L	OEM
MCM6257L	OEM
MCM6257P	OEM
MCM6604L	OEM
MCM6604L2	OEM
MCM6604L4	OEM
MCM6604P	OEM
MCM6604P2	OEM
MCM6604P4	OEM
MCM6632AL15	OEM
MCM6632AL15(A)	OEM
MCM6632AL20	OEM
MCM6632AL20(A)	OEM
MCM6632AP12(A)	OEM
MCM6633AL	OEM
MCM6633AL15	OEM
MCM6633AL20	OEM
MCM6633AP	OEM
MCM6633AP12	OEM
MCM6641-20P	OEM
MCM6641-25P	OEM
MCM6641-30P	OEM
MCM6641-45P	OEM
MCM6664-20L	OEM
MCM6664A-12L	OEM
MCM6664A-12P	OEM
MCM6664A-15L	OEM
MCM6664A-15P	OEM
MCM6664A-20L	OEM
MCM6664A-20P	OEM
MCM6664AL	OEM
MCM6664AL15	OEM
MCM6664AL20	4315
MCM6664AL25	OEM
MCM6664AP	4315
MCM6664AP15	4315
MCM6664AP20	4315
MCM6664AP25	OEM
MCM6665	2357
MCM6665-20L	OEM
MCM6665A-12L	OEM
MCM6665A-12P	OEM
MCM6665A-15	OEM
MCM6665A-15L	OEM
MCM6665A-15P	OEM
MCM6665A-20	OEM
MCM6665A-20L	OEM
MCM6665A-20P	OEM

DEVICE TYPE	REPL CODE
MCM6665AL15	2357
MCM6665AL20	2357
MCM6665AL25	OEM
MCM6665AP	4505
MCM6665AP15	2357
MCM6665AP20	2357
MCM6665AP25	OEM
MCM6665BP-20	OEM
MCM6665L25	2357
MCM6670L	OEM
MCM6670P	OEM
MCM6674P	OEM
MCM6810	4249
MCM6810BJCS	OEM
MCM6810CJCS	OEM
MCM6810CL	2086
MCM6810CP	2086
MCM6810CS	OEM
MCM6810L	OEM
MCM6810P	2086
MCM6810S	OEM
MCM7621AD	OEM
MCM7621ADC	OEM
MCM7621AP	OEM
MCM7621APC	OEM
MCM7621D	5086
MCM7621DC	OEM
MCM7621P	OEM
MCM7621PC	OEM
MCM7641AD	OEM
MCM7641ADC	OEM
MCM7641AP	OEM
MCM7641APC	OEM
MCM7641D	5109
MCM7641DC	OEM
MCM7641P	OEM
MCM7641PC	OEM
MCM7643AD	OEM
MCM7643ADC	OEM
MCM7643AP	OEM
MCM7643APC	OEM
MCM7643D	4340
MCM7643DC	OEM
MCM7643P	OEM
MCM7643PC	OEM
MCM7649ADC	OEM
MCM7649APC	OEM
MCM7649DC	OEM
MCM7649PC	OEM
MCM7681	2909
MCM7681A	OEM
MCM7681AD	OEM
MCM7681ADC	OEM
MCM7681AP	OEM
MCM7681APC	OEM
MCM7681D	3783
MCM7681DC	OEM
MCM7681P	2909
MCM7681PC	OEM
MCM7685AD	OEM
MCM7685ADC	OEM
MCM7685AP	OEM
MCM7685APC	OEM
MCM7685D	3166
MCM7685DC	OEM
MCM7685P	OEM
MCM7685PC	OEM
MCM10139L	OEM
MCM10143L	OEM
MCM10144L	OEM
MCM10145AL	OEM
MCM10145L	OEM
MCM10146L	OEM
MCM10147L	OEM
MCM10148L	OEM
MCM10149L	OEM
MCM10152L	OEM
MCM10415AF	OEM
MCM10415AL	OEM
MCM10415F	OEM
MCM10415L	OEM
MCM10422A	OEM
MCM10422F	OEM
MCM10422L	OEM
MCM10470AF	OEM
MCM10470AL	OEM
MCM10470F	OEM
MCM10470L	OEM
MCM10474AL	OEM
MCM10474L	OEM
MCM12090D	OEM
MCM12090P	OEM
MCM14505AL	OEM
MCM14505ALD	OEM
MCM14505CL	OEM
MCM14505CLD	OEM
MCM14505CP	OEM
MCM14505CPD	OEM
MCM14524AL	OEM
MCM14524CL	OEM
MCM14524CP	OEM
MCM14537AL	OEM
MCM14537ALD	OEM
MCM14537CL	OEM
MCM14537CLD	OEM
MCM14537CP	OEM
MCM14552AL	OEM
MCM14552ALD	OEM
MCM14552CL	OEM

DEVICE TYPE	REPL CODE
MCM14552CLD	OEM
MCM14552CP	OEM
MCM14552CPD	OEM
MCM63128P	OEM
MCM63256P	OEM
MCM65116-12C(A)	OEM
MCM65116-12P(A)	OEM
MCM65116-15C(A)	OEM
MCM65116-15P(A)	OEM
MCM65116-20C(A)	OEM
MCM65116-20P(A)	OEM
MCM65116C	6181
MCM65116P	6181
MCM65116P15	1898
MCM65116P20	1898
MCM65147-55C(A)	OEM
MCM65147-55P(A)	OEM
MCM65147-70C	OEM
MCM65147-70P(A)	OEM
MCM65147C	OEM
MCM65147P	OEM
MCM65147P55	OEM
MCM65147P70	OEM
MCM65256P	OEM
MCM65256P(A)	OEM
MCM65516-43L	OEM
MCM65516-43P	OEM
MCM65516-55L	OEM
MCM65516-55P	OEM
MCM65516L	OEM
MCM65516L43M	OEM
MCM65516P	OEM
MCM65516P43M	OEM
MCM66320AL15	OEM
MCM66320AL15(A)	OEM
MCM66320AL20	OEM
MCM66320AL20(A)	OEM
MCM66321AL15	OEM
MCM66321AL15(A)	OEM
MCM66321AL20	OEM
MCM66321AL20(A)	OEM
MCM66330AL15	OEM
MCM66330AL20	OEM
MCM66331AL15	OEM
MCM66331AL20	OEM
MCM66700C	OEM
MCM66700P	OEM
MCM66710C	OEM
MCM66710P	OEM
MCM66714C	OEM
MCM66714P	OEM
MCM66720C	OEM
MCM66720P	OEM
MCM66730C	OEM
MCM66730P	OEM
MCM66734C	OEM
MCM66734P	OEM
MCM66740C	OEM
MCM66740P	OEM
MCM66750C	OEM
MCM66750P	OEM
MCM66760C	OEM
MCM66760P	OEM
MCM66770C	OEM
MCM66770P	OEM
MCM66780C	OEM
MCM66780P	OEM
MCM66790C	OEM
MCM66790P	OEM
MCM68364-20C	OEM
MCM68364-20L	OEM
MCM68364-20P	OEM
MCM68364-25C	OEM
MCM68364-25L	OEM
MCM68364-25P	OEM
MCM68364-30C	OEM
MCM68364-30L	OEM
MCM68364-30P	OEM
MCM68364C	OEM
MCM68364P	OEM
MCM68365-25C	OEM
MCM68365-25L	OEM
MCM68365-25P	OEM
MCM68365-35C	OEM
MCM68365-35L	OEM
MCM68365-35P	OEM
MCM68365P	OEM
MCM68366-25C	OEM
MCM68366-25L	OEM
MCM68366-25P	OEM
MCM68366-35C	OEM
MCM68366-35L	OEM
MCM68366-35P	OEM
MCM68366P	OEM
MCM68367P	OEM
MCM68368P	OEM
MCM68369P	OEM
MCM68370P	OEM
MCM68708C	OEM
MCM68708L	OEM
MCM68764C	OEM
MCM68764L	OEM
MCM68766-35C	OEM
MCM68766C	OEM
MCM68766C35	OEM
MCM68766L	OEM
MCM76161AD	OEM
MCM76161ADC	OEM
MCM76161AP	OEM

DEVICE TYPE	REPL CODE
MCM76161APC	OEM
MCM76161D	OEM
MCM76161DC	OEM
MCM76161P	OEM
MCM76161PC	OEM
MCM76165ADC	OEM
MCM76165APC	OEM
MCM76165DC	OEM
MCM76165PC	OEM
MCM93415DC	OEM
MCM93415DM	OEM
MCM93415PC	OEM
MCM93422AD	OEM
MCM93422AP	OEM
MCM93422D	OEM
MCM93422DC	OEM
MCM93422P	OEM
MCM93422PC	OEM
MCM93425DC	OEM
MCM93425DM	OEM
MCM93425PC	OEM
MCN701A	OEM
MCN721A	OEM
MCP-893	OEM
MCPU-800	OEM
MCPU800-02	OEM
MCPU800-03	OEM
MCR051	OEM
MCR052	OEM
MCR053	OEM
MCR054	OEM
MCR16L4L6	OEM
MCR22-2	OEM
MCR22-3	OEM
MCR22-4	OEM
MCR22-5	OEM
MCR22-6	OEM
MCR22-7	OEM
MCR22-8	OEM
MCR23-2	OEM
MCR23-3	OEM
MCR23-4	OEM
MCR23-5	OEM
MCR23-6	OEM
MCR23-7	OEM
MCR23-8	OEM
MCR32-05	OEM
MCR32-20	OEM
MCR32-30	OEM
MCR32-40	OEM
MCR32-50	OEM
MCR32-60	OEM
MCR39-05	2412
MCR39-20	2412
MCR39-30	6390
MCR39-40	6390
MCR39-50	OEM
MCR39-60	OEM
MCR45-10	0217
MCR45-20	0217
MCR45-30	0217
MCR45-40	0217
MCR45-50	0217
MCR45-60	0217
MCR45-70	0658
MCR45-80	0658
MCR45-90	0658
MCR45-100	0658
MCR45-110	0658
MCR45-120	0658
MCR46-10	OEM
MCR46-20	OEM
MCR46-30	OEM
MCR46-40	OEM
MCR46-50	OEM
MCR46-60	OEM
MCR46-70	OEM
MCR46-80	OEM
MCR46-90	OEM
MCR46-100	OEM
MCR46-110	OEM
MCR46-120	OEM
MCR50-10	OEM
MCR50-20	OEM
MCR50-30	OEM
MCR50-40	OEM
MCR50-50	OEM
MCR50-60	OEM
MCR50-70	OEM
MCR50-80	OEM
MCR50-90	OEM
MCR50C-10	OEM
MCR50C-20	OEM
MCR50C-30	OEM
MCR50C-40	OEM
MCR50C-50	OEM
MCR50C-60	OEM
MCR50C-70	OEM
MCR50C-80	OEM
MCR50C-90	OEM
MCR50C-100	OEM
MCR50D-10	OEM
MCR50D-20	OEM
MCR50D-30	OEM
MCR50D-40	OEM
MCR50D-50	OEM
MCR50D-60	OEM
MCR50D-70	OEM
MCR50D-80	OEM

If replacement code is OEM, contact original manufacturer for replacement.

DEVICE TYPE	REPL CODE
MCR50D-90	OEM
MCR50D-100	OEM
MCR50D-110	OEM
MCR50D-120	OEM
MCR52-10	OEM
MCR52-20	OEM
MCR52-30	OEM
MCR52-40	OEM
MCR52-50	OEM
MCR52-60	OEM
MCR52-70	OEM
MCR52-80	OEM
MCR52-90	OEM
MCR60-10	OEM
MCR60-20	OEM
MCR60-30	OEM
MCR60-40	OEM
MCR60-50	OEM
MCR62-10	OEM
MCR62-20	OEM
MCR62-30	OEM
MCR62-40	OEM
MCR62-50	OEM
MCR63-1	OEM
MCR63-2	OEM
MCR63-3	OEM
MCR63-4	OEM
MCR63-5	OEM
MCR63-6	OEM
MCR63-7	OEM
MCR63-8	OEM
MCR63-9	OEM
MCR63-10	OEM
MCR64-1	OEM
MCR64-2	6846
MCR64-3	6846
MCR64-4	6846
MCR64-5	OEM
MCR64-6	6847
MCR64-7	OEM
MCR64-8	6847
MCR64-9	OEM
MCR64-10	6082
MCR65-1	OEM
MCR65-2	OEM
MCR65-3	OEM
MCR65-4	OEM
MCR65-5	OEM
MCR65-6	OEM
MCR65-7	OEM
MCR65-8	OEM
MCR65-9	OEM
MCR65-10	OEM
MCR67-1	OEM
MCR67-2	OEM
MCR67-3	OEM
MCR68-1	6875
MCR68-2	6875
MCR68-3	6875
MCR69-1	OEM
MCR69-2	OEM
MCR69-3	OEM
MCR69-6	OEM
MCR70-1	OEM
MCR70-2	OEM
MCR70-3	OEM
MCR70-6	OEM
MCR71-1	OEM
MCR71-2	OEM
MCR71-3	OEM
MCR72-1	1396
MCR72-2	1259
MCR72-3	0447
MCR72-4	0940
MCR72-5	1222
MCR72-6	0717
MCR72-7	0308
MCR72-8	0308
MCR80-0.5	OEM
MCR80-5	OEM
MCR80-10	OEM
MCR80-20	OEM
MCR80-30	OEM
MCR80-40	OEM
MCR80-50	OEM
MCR80-60	OEM
MCR80-70	OEM
MCR80-80	OEM
MCR81-0.5	OEM
MCR81-5	OEM
MCR81-10	OEM
MCR81-20	OEM
MCR81-30	OEM
MCR81-40	OEM
MCR81-50	OEM
MCR81-60	OEM
MCR81-70	OEM
MCR81-80	OEM
MCR82-0.5	OEM
MCR82-5	OEM
MCR82-10	OEM
MCR82-20	OEM
MCR82-30	OEM
MCR82-40	OEM
MCR82-50	OEM
MCR82-60	OEM
MCR82-70	OEM
MCR82-80	OEM
MCR83-5	OEM
MCR83-10	OEM
MCR83-20	OEM
MCR83-30	OEM
MCR83-40	OEM
MCR83-50	OEM
MCR83-60	OEM
MCR83-70	OEM
MCR83-80	OEM
MCR100-3	0901
MCR100-4	0058
MCR100-5	0407
MCR100-6	0407
MCR100-7	1683
MCR100-8	1683
MCR100-B	1683
MCR101	1137
MCR102	1137
MCR103	0344
MCR104	0901
MCR106-1	4458
MCR106-2	0174
MCR106-3	3854
MCR106-4	3620
MCR106-5	3324
MCR106-6	3324
MCR106-7	1502
MCR106-8	0308
MCR107-1	4458
MCR107-2	0174
MCR107-3	3854
MCR107-4	3620
MCR107-5	0717
MCR107-6	0710
MCR107-7	6720
MCR107-8	0308
MCR115	2342
MCR120	0058
MCR130	OEM
MCR150-10	OEM
MCR150-20	OEM
MCR150-30	OEM
MCR150-40	OEM
MCR150-50	0468
MCR150-60	0468
MCR150-70	0468
MCR150-80	0468
MCR150-90	0468
MCR150-100	0468
MCR150-110	0468
MCR150-120	0468
MCR150-130	OEM
MCR150-140	OEM
MCR150-150	OEM
MCR152-10	OEM
MCR152-20	OEM
MCR152-30	OEM
MCR152-40	OEM
MCR152-50	OEM
MCR152-60	OEM
MCR152-70	OEM
MCR152-80	OEM
MCR152-90	OEM
MCR152-100	OEM
MCR152-110	OEM
MCR152-120	OEM
MCR152-130	OEM
MCR152-140	OEM
MCR152-150	OEM
MCR154-10	0526
MCR154-20	0526
MCR154-30	0526
MCR154-40	0526
MCR154-50	0526
MCR154-60	0526
MCR155-10	0526
MCR155-20	0526
MCR155-30	0526
MCR155-40	0526
MCR155-50	0526
MCR155-60	0526
MCR156-10	OEM
MCR156-20	OEM
MCR156-30	OEM
MCR156-40	OEM
MCR156-50	OEM
MCR156-60	OEM
MCR157-10	OEM
MCR157-20	OEM
MCR157-30	OEM
MCR157-40	OEM
MCR157-50	OEM
MCR157-60	OEM
MCR158-50	OEM
MCR158-60	OEM
MCR158-70	OEM
MCR158-80	OEM
MCR158-90	OEM
MCR158-100	OEM
MCR158-110	OEM
MCR158-120	OEM
MCR159-50	OEM
MCR159-60	OEM
MCR159-70	OEM
MCR159-80	OEM
MCR159-90	OEM
MCR159-100	OEM
MCR159-110	OEM
MCR159-120	OEM
MCR201	OEM
MCR202	4015
MCR203	4016
MCR204	0573
MCR205	OEM
MCR206	4020
MCR218-2	2095
MCR218-3	2089
MCR218-4	0505
MCR218-5	OEM
MCR218-6	0863
MCR218-7	OEM
MCR218-8	0327
MCR218-9	OEM
MCR218-10	0327
MCR220-5	0611
MCR220-7	0952
MCR220-9	4010
MCR221-5	4053
MCR221-7	4054
MCR221-9	4010
MCR225-2FP	2611
MCR225-4FP	2611
MCR225-5	4053
MCR225-6FP	OEM
MCR225-7	4054
MCR225-8FP	4576
MCR225-9	4010
MCR225-10FP	4576
MCR225-12	OEM
MCR235-10	OEM
MCR235-20	OEM
MCR235-30	OEM
MCR235-40	OEM
MCR235-50	OEM
MCR235-60	OEM
MCR235-70	OEM
MCR235-80	OEM
MCR235-90	OEM
MCR235-100	OEM
MCR235-110	OEM
MCR235-120	OEM
MCR235-130	OEM
MCR235-140	OEM
MCR235-150	OEM
MCR235A-10	OEM
MCR235A-20	OEM
MCR235A-30	OEM
MCR235A-40	OEM
MCR235A-50	OEM
MCR235A-60	OEM
MCR235B-10	OEM
MCR235B-20	OEM
MCR235B-30	OEM
MCR235B-40	OEM
MCR235B-50	OEM
MCR235B-60	OEM
MCR235B-70	OEM
MCR235B-80	OEM
MCR235C-10	OEM
MCR235C-20	OEM
MCR235C-30	OEM
MCR235C-40	OEM
MCR235C-50	OEM
MCR235C-60	OEM
MCR235C-70	OEM
MCR235C-80	OEM
MCR235C-90	OEM
MCR235C-100	OEM
MCR251B-10	OEM
MCR251B-20	OEM
MCR251B-30	OEM
MCR251B-40	OEM
MCR251B-50	OEM
MCR251B-60	OEM
MCR251B-70	OEM
MCR251B-80	OEM
MCR251B-90	OEM
MCR251B-100	OEM
MCR251B-110	OEM
MCR251B-120	OEM
MCR251B-130	OEM
MCR251B-140	OEM
MCR251B-150	OEM
MCR251L-10	OEM
MCR251L-20	OEM
MCR251L-30	OEM
MCR251L-40	OEM
MCR251L-50	OEM
MCR251L-60	OEM
MCR251L-70	OEM
MCR251L-80	OEM
MCR251L-90	OEM
MCR251L-100	OEM
MCR251L-110	OEM
MCR251L-120	OEM
MCR251L-130	OEM
MCR251L-140	OEM
MCR251L-150	OEM
MCR264-2	5077
MCR264-3	5077
MCR264-4	5077
MCR264-6	5077
MCR264-8	5079
MCR264-10	5079
MCR264-12	OEM
MCR265-2	OEM
MCR265-4	OEM
MCR265-6	OEM
MCR265-8	OEM
MCR265-10	OEM
MCR320-1	OEM
MCR320-2	OEM
MCR320-3	OEM
MCR320-4	OEM
MCR320-5	OEM
MCR320-6	OEM
MCR320-7	OEM
MCR320-8	OEM
MCR380-10	OEM
MCR380-20	OEM
MCR380-30	OEM
MCR380-40	OEM
MCR380-50	OEM
MCR380-60	OEM
MCR380-70	OEM
MCR380-80	OEM
MCR380-90	OEM
MCR380-100	OEM
MCR380-110	OEM
MCR380-120	OEM
MCR380-130	OEM
MCR380-140	OEM
MCR380-150	OEM
MCR380B-10	OEM
MCR380B-20	OEM
MCR380B-30	OEM
MCR380B-40	OEM
MCR380B-50	OEM
MCR380B-60	OEM
MCR380B-70	OEM
MCR380B-80	OEM
MCR380C-10	OEM
MCR380C-20	OEM
MCR380C-30	OEM
MCR380C-40	OEM
MCR380C-50	OEM
MCR380C-60	OEM
MCR380C-70	OEM
MCR380C-80	OEM
MCR380C-90	OEM
MCR380C-100	OEM
MCR380D-10	OEM
MCR380D-20	OEM
MCR380D-30	OEM
MCR380D-40	OEM
MCR380D-50	OEM
MCR380D-60	OEM
MCR380D-70	OEM
MCR380D-80	OEM
MCR380D-90	OEM
MCR380D-100	OEM
MCR380D-110	OEM
MCR380D-120	OEM
MCR406-1	6169
MCR406-2	3284
MCR406-3	1923
MCR406-4	3620
MCR407-1	3283
MCR407-2	6171
MCR407-3	6172
MCR407-4	3620
MCR470-10	OEM
MCR470-20	OEM
MCR470-30	OEM
MCR470-40	OEM
MCR470-50	OEM
MCR470-60	OEM
MCR470-70	OEM
MCR470-80	OEM
MCR470-90	OEM
MCR470-100	OEM
MCR470-110	OEM
MCR470-120	OEM
MCR470-130	OEM
MCR470-140	OEM
MCR470-150	OEM
MCR470C-10	OEM
MCR470C-20	OEM
MCR470C-30	OEM
MCR470C-40	OEM
MCR470C-50	OEM
MCR470C-60	OEM
MCR470C-70	OEM
MCR470C-80	OEM
MCR470D-10	OEM
MCR470D-20	OEM
MCR470D-30	OEM
MCR470D-40	OEM
MCR470D-50	OEM
MCR470D-60	OEM
MCR470D-70	OEM
MCR470D-80	OEM
MCR470D-90	OEM
MCR470D-100	OEM
MCR470E-10	OEM
MCR470E-20	OEM
MCR470E-30	OEM
MCR470E-40	OEM
MCR470E-50	OEM
MCR470E-60	OEM
MCR470E-70	OEM
MCR470E-80	OEM
MCR470E-90	OEM
MCR470E-100	OEM
MCR470E-110	OEM
MCR470E-120	OEM
MCR525-4	OEM
MCR525-5	OEM
MCR525-6	OEM
MCR525-7	OEM
MCR525-8	OEM
MCR525-9	OEM
MCR525-10	OEM
MCR568-1	OEM
MCR568-2	OEM
MCR568-3	OEM
MCR568-6	OEM
MCR569-1	OEM
MCR569-2	OEM
MCR569-3	OEM
MCR569-6	OEM
MCR606-2	OEM
MCR606-3	OEM
MCR606-4	OEM
MCR606-5	OEM
MCR606-6	OEM
MCR606-7	OEM
MCR606-8	OEM
MCR649-1	2519
MCR649-2	OEM
MCR649-3	1650
MCR649-4	1651
MCR649-5	1584
MCR649-6	OEM
MCR649-7	1665
MCR649-8	OEM
MCR649-9	OEM
MCR649-10	OEM
MCR649A1	OEM
MCR649A2	OEM
MCR649A3	OEM
MCR649A4	OEM
MCR649A5	OEM
MCR649A6	OEM
MCR649A7	OEM
MCR649A8	OEM
MCR649A9	OEM
MCR649A10	OEM
MCR649AP1	OEM
MCR649AP2	OEM
MCR649AP3	OEM
MCR649AP4	OEM
MCR649AP5	OEM
MCR649AP6	OEM
MCR649AP7	OEM
MCR649AP8	OEM
MCR649AP9	OEM
MCR649AP10	OEM
MCR649P1	OEM
MCR649P2	OEM
MCR649P3	OEM
MCR649P4	OEM
MCR649P5	OEM
MCR649P6	OEM
MCR649P7	OEM
MCR649P8	OEM
MCR649P9	OEM
MCR649P10	OEM
MCR729-5	OEM
MCR729-6	OEM
MCR729-7	OEM
MCR729-8	OEM
MCR729-9	OEM
MCR729-10	OEM
MCR800-10	OEM
MCR800-20	OEM
MCR800-30	OEM
MCR800-40	OEM
MCR800-50	OEM
MCR800-60	OEM
MCR800-70	OEM
MCR800-80	OEM
MCR800-90	OEM
MCR800-100	OEM
MCR800-110	OEM
MCR800-120	OEM
MCR800-130	OEM
MCR800-140	OEM
MCR800-150	OEM
MCR808-1	OEM
MCR808-2	OEM
MCR808-3	OEM
MCR808-4	OEM
MCR808-5	OEM
MCR808-6	OEM
MCR846-1	1103
MCR846-2	1103
MCR846-3	0240
MCR846-4	0240
MCR914-1	OEM
MCR914-2	OEM
MCR914-3	OEM
MCR914-5	OEM
MCR914-6	OEM
MCR1000-4	OEM
MCR1000-6	OEM
MCR1000-8	OEM
MCR1304-1	OEM
MCR1304-2	OEM
MCR1304-3	OEM
MCR1304-4	OEM
MCR1304-5	OEM
MCR1305-6	OEM
MCR1305R1	OEM
MCR1305R2	OEM
MCR1305R3	OEM
MCR1305R4	OEM
MCR1305R5	OEM
MCR1305R6	OEM
MCR1308-1	0469
MCR1308-2	0469
MCR1308-3	0469
MCR1308-4	0469
MCR1308-5	0722
MCR1308-6	0722
MCR1336-5	OEM
MCR1336-6	OEM
MCR1336-7	OEM
MCR1336-8	OEM
MCR1336-9	OEM
MCR1336-10	OEM
MCR1604R1	OEM
MCR1604R2	OEM
MCR1604R3	OEM
MCR1604R4	OEM
MCR1604R5	OEM
MCR1604R6	OEM
MCR1605R1	OEM
MCR1605R2	OEM
MCR1605R3	OEM
MCR1605R4	OEM
MCR1605R5	OEM
MCR1605R6	OEM
MCR1718-5	0804
MCR1718-6	0804
MCR1718-7	0804
MCR1718-8	0804
MCR1819-8	OEM
MCR1906-1	0179
MCR1906-2	0179
MCR1906-3	0179
MCR1906-4	2412
MCR1906-5	0346
MCR1906-6	0346
MCR1907-1	0722
MCR1907-2	0722
MCR1907-3	0722
MCR1907-4	0722
MCR1907-5	0722
MCR1907-6	0722
MCR2064-6	1487
MCR2080-4	OEM
MCR2080-5	OEM
MCR2080-6	OEM
MCR2080-7	OEM
MCR2080-8	OEM
MCR2080-9	OEM
MCR2080-10	OEM
MCR2080A-4	OEM
MCR2080A4	OEM
MCR2080A-5	OEM
MCR2080A5	OEM
MCR2080A-6	OEM
MCR2080A6	OEM
MCR2080A-7	OEM
MCR2080A7	OEM
MCR2080A-8	OEM
MCR2080A8	OEM
MCR2080A-9	OEM
MCR2080A9	OEM
MCR2080A-10	OEM
MCR2080A10	OEM
MCR2150-4	OEM
MCR2150-5	OEM
MCR2150-6	OEM
MCR2150-7	OEM
MCR2150-8	OEM
MCR2150-9	OEM
MCR2150-10	OEM
MCR2150A-4	OEM
MCR2150A4	OEM
MCR2150A-5	OEM
MCR2150A5	OEM
MCR2150A-6	OEM
MCR2150A6	OEM
MCR2150A-7	OEM
MCR2150A7	OEM
MCR2150A-8	OEM
MCR2150A8	OEM
MCR2150A-9	OEM
MCR2150A9	OEM
MCR2150A-10	OEM
MCR2150A10	OEM
MCR2305-1	0469
MCR2305-2	0240
MCR2305-3	0469
MCR2305-4	0240
MCR2305-5	0722
MCR2305-6	0722
MCR2305-7	OEM
MCR2305-8	OEM
MCR2305-9	OEM
MCR2305-10	OEM
MCR2315-1	3419
MCR2315-2	1103
MCR2315-3	0240
MCR2315-4	0240
MCR2315-5	2659
MCR2315-6	0676
MCR2604-1	1487
MCR2604-2	1487
MCR2604-3	1487
MCR2604-4	1584
MCR2604-5	1584
MCR2604-6	1584
MCR2604-7	OEM
MCR2604-8	OEM
MCR2605-1	1487
MCR2605-2	1487
MCR2605-3	1487
MCR2605-4	1487
MCR2605-5	1487
MCR2605-6	1487
MCR2605-7	OEM
MCR2605-8	OEM
MCR2614L1	OEM
MCR2614L2	OEM
MCR2614L3	OEM
MCR2614L-4	1584
MCR2614L4	OEM
MCR2614L-5	1584
MCR2614L5	OEM
MCR2614L-6	1584
MCR2818-1	OEM
MCR2818-2	OEM
MCR2818-3	OEM
MCR2818-4	OEM
MCR2818-5	OEM
MCR2818-6	OEM
MCR2818-7	OEM
MCR2818-8	OEM
MCR2819-1	OEM
MCR2819-2	OEM
MCR2819-3	OEM
MCR2819-4	OEM
MCR2819-5	OEM
MCR2819-6	OEM
MCR2819-7	OEM
MCR2835-1	1650
MCR2835-2	1650
MCR2835-3	1650
MCR2835-4	1650
MCR2835-5	2647
MCR2835-6	2647
MCR2835-7	2649
MCR2835-8	2649
MCR2918-1	0722
MCR2918-2	OEM
MCR2918-3	0722
MCR2918-4	OEM
MCR2918-5	0722
MCR2918-6	OEM
MCR2918-7	OEM
MCR2918-8	OEM
MCR2935-1	0469
MCR2935-2	0469
MCR2935-3	0469
MCR2935-4	0469
MCR2935-5	0722
MCR2935-6	0722
MCR2935-7	0725
MCR2935-8	0725
MCR3000-1	4588
MCR3000-2	2270
MCR3000-3	2270
MCR3000-4	2270
MCR3000-5	3402
MCR3000-6	3402
MCR3000-7	3404
MCR3000-8	3404
MCR3000-9	0327
MCR3000-10	OEM
MCR3818-1	1651
MCR3818-2	1651
MCR3818-3	1651
MCR3818-4	1651
MCR3818-5	1584
MCR3818-6	1584
MCR3818-7	1665
MCR3818-8	1665
MCR3818-9	OEM
MCR3818-10	OEM
MCR3835-1	1650
MCR3835-2	1650
MCR3835-3	1650
MCR3835-4	1650
MCR3835-5	2647
MCR3835-6	2647
MCR3835-7	2649
MCR3835-8	2649
MCR3835-9	OEM
MCR3835-10	OEM
MCR3918-1	3307
MCR3918-2	2188
MCR3918-3	0567
MCR3918-4	0762
MCR3918-5	3271
MCR3918-6	0740
MCR3918-7	3291
MCR3918-8	0764
MCR3918-9	OEM
MCR3918-10	OEM
MCR3935-1	3277
MCR3935-2	0731
MCR3935-3	0712
MCR3935-4	0469
MCR3935-5	0721
MCR3935-6	0722
MCR3935-7	0778
MCR3935-8	0725
MCR3935-9	0750
MCR3935-10	0750
MCR3936-10	OEM
MCR4018-1	OEM
MCR4018-2	OEM
MCR4018-3	OEM

DEVICE TYPE	REPL CODE
MCR4018-4	OEM
MCR4018-5	OEM
MCR4018-6	OEM
MCR4018-7	OEM
MCR4018-8	OEM
MCR4018-9	OEM
MCR4018-10	OEM
MCR4035-1	OEM
MCR4035-2	OEM
MCR4035-3	OEM
MCR4035-4	OEM
MCR4035-5	OEM
MCR4035-6	OEM
MCR4035-7	OEM
MCR4035-8	OEM
MCR4035-9	OEM
MCR4035-10	OEM
MCS00	OEM
MCS2	0235
MCS31	OEM
MCS35	OEM
MCS101	0133
MCS105	0133
MCS2027	OEM
MCS2114-30	2048
MCS2114-35	2048
MCS2114-45	OEM
MCS2114L-30	OEM
MCS2114L-35	OEM
MCS2114L-45	2048
MCS2135	0861
MCS2137	0861
MCS2316	OEM
MCS2332	OEM
MCS2400	2236
MCS6200	OEM
MCS6201	OEM
MCS8001	OEM
MCS8004	OEM
MCS8007	OEM
MCT00	OEM
MCT02	OEM
MCT2	0273
MCT-2E	0273
MCT2E	0273
MCT4	OEM
MCT4-R	OEM
MCT6	0297
MCT8	0305
MCT10	OEM
MCT11	OEM
MCT12	OEM
MCT15	OEM
MCT18	OEM
MCT19	OEM
MCT20	OEM
MCT21	OEM
MCT24	OEM
MCT26	0315
MCT27	OEM
MCT28	OEM
MCT29	OEM
MCT36	OEM
MCT38	OEM
MCT40	OEM
MCT41	OEM
MCT43	OEM
MCT44	OEM
MCT45	OEM
MCT46	OEM
MCT66	1403
MCT78T08K	OEM
MCT81	1417
MCT210	4419
MCT271	OEM
MCT272	OEM
MCT273	4431
MCT274	OEM
MCT275	4431
MCT276	OEM
MCT277	OEM
MCT492P	0834
MCT2114-45	OEM
MCT2114L-45	OEM
MCT2201	4529
MCT3737	OEM
MCT7809BT	OEM
MCT7809CT	3332
MCV	0015
MCV45	OEM
MCV1055E	OEM
MCV1055F5F	OEM
MCV1055G	OEM
MCV1125	OEM
MCV1170	OEM
MCV1300	OEM
MCV1352	OEM
MCV1353	OEM
MCV1420	OEM
MCV1421	OEM
MCX6809P	OEM
MCX14032BCPD	OEM
MD04	1335
MD06	OEM
MD08	OEM
MD0232	OEM
MD0233	OEM
MD0234	OEM
MD0235	OEM

DEVICE TYPE	REPL CODE
MD1	OEM
MD1D914	OEM
MD1D914A	OEM
MD1D914B	OEM
MD1F3066	OEM
MD1F3067	OEM
MD1F3068	OEM
MD1F3069	OEM
MD1F3070	OEM
MD1F3071	OEM
MD1F3458	OEM
MD1F3459	OEM
MD1F3460	OEM
MD1F3823	OEM
MD1F4391	OEM
MD1F4392	OEM
MD1F4393	OEM
MD1F4416	OEM
MD1T918	OEM
MD1T1893	OEM
MD1T2222	OEM
MD1T2369	OEM
MD1T2484	OEM
MD1T2605	OEM
MD1T2907	OEM
MD1T3251	OEM
MD1T3704	OEM
MD2	OEM
MD2D914	OEM
MD2D914A	OEM
MD2D914B	OEM
MD2F	OEM
MD3	OEM
MD3F1	OEM
MD3H1	OEM
MD3N1	OEM
MD4	OEM
MD4F	OEM
MD-4H50	OEM
MD-4N50	OEM
MD5	OEM
MD6F	OEM
MD6F1	OEM
MD6H1	OEM
MD6N1	OEM
MD8F	OEM
MD-8H10	OEM
MD-8N10	OEM
MD10	OEM
MD10F	OEM
MD10F1	OEM
MD-10F10	OEM
MD10H1	OEM
MD10N1	OEM
MD-10N10	OEM
MD-12H10	OEM
MD12H10	OEM
MD-12N10	OEM
MD12N10	OEM
MD15F1	OEM
MD15H1	OEM
MD15N1	OEM
MD20	OEM
MD20F1	OEM
MD20H1	OEM
MD20N1	OEM
MD20SH05K	OEM
MD21SC14AE10	OEM
MD21SC14AE15	OEM
MD21SC14AE25	OEM
MD-22N01	OEM
MD22N01	OEM
MD23SC16AE	OEM
MD24	OEM
MD25	OEM
MD25-0	OEM
MD25-1	OEM
MD25-2	OEM
MD25-3	OEM
MD25-4	OEM
MD-25X1	OEM
MD25X1	OEM
MD-25X1-4-1-1	OEM
MD-25X1-5-1-1	OEM
MD-25X1-6-1-1	OEM
MD31	OEM
MD32	OEM
MD-34	0143
MD34	0143
MD34A	0143
MD34B	OEM
MD35	0143
MD-36N4	OEM
MD36SH05K	OEM
MD38	0143
MD-45H01	OEM
MD45H01	OEM
MD46	0143
MD50	0133
MD50SH05K	OEM
MD54	0143
MD54SC137AF	OEM
MD54SC138AF	OEM
MD54SC139AF	OEM
MD54SC237AF	OEM
MD54SC238AF	OEM
MD54SC239AF	OEM
MD54SC240AF	OEM
MD54SC241AF	OEM

DEVICE TYPE	REPL CODE
MD54SC244AF	OEM
MD54SC245AF	OEM
MD54SC540AF	OEM
MD54SC541AF	OEM
MD54SC545AF	OEM
MD56	0143
MD58	OEM
MD60	0143
MD-60A	0123
MD60A	0123
MD-60E01	OEM
MD68SC02AC	OEM
MD68SC02AF	OEM
MD74HCT374	OEM
MD74LS37M	1730
MD74LS183M	OEM
MD74SC137AC	OEM
MD74SC138AC	OEM
MD74SC139AC	OEM
MD74SC237AC	OEM
MD74SC238AC	OEM
MD74SC239AC	OEM
MD74SC240AC	OEM
MD74SC241AC	OEM
MD74SC244AC	OEM
MD74SC245AC	OEM
MD74SC373AC	OEM
MD74SC373AE	OEM
MD74SC374AC	OEM
MD74SC374AE	OEM
MD74SC533AC	OEM
MD74SC533AE	OEM
MD74SC534AC	OEM
MD74SC534AE	OEM
MD74SC540AC	OEM
MD74SC541AC	OEM
MD74SC545AC	OEM
MD74SC563AC	OEM
MD74SC563AE	OEM
MD74SC564AC	OEM
MD74SC564AE	OEM
MD74SC573AE	OEM
MD74SC574AC	OEM
MD74SC574AE	OEM
MD82HS181	OEM
MD82HS191	OEM
MD82HS321	OEM
MD82S181	OEM
MD82S191	OEM
MD82S321	OEM
MD90	OEM
MD100	OEM
MD100-0	OEM
MD100-1	OEM
MD100-2	OEM
MD100-3	OEM
MD100-4	OEM
MD-100X08C	OEM
MD101	OEM
MD102	OEM
MD106	OEM
MD121	OEM
MD-125N08C	OEM
MD125N08C	OEM
MD134	0124
MD135	0124
MD136	0124
MD137	0023
MD138	0023
MD139	0023
MD189	OEM
MD189F	OEM
MD190	OEM
MD190F	OEM
MD191	OEM
MD191F	OEM
MD210CP	OEM
MD219	OEM
MD219F	OEM
MD220	OEM
MD220F	OEM
MD221	OEM
MD221F	OEM
MD234	0124
MD235	0124
MD236	0124
MD237	0023
MD238	0023
MD358	OEM
MD402	OEM
MD420	0050
MD-430	OEM
MD-430DP	OEM
MD-430PM	OEM
MD-440	OEM
MD-440DP	OEM
MD-440PM	OEM
MD-450	OEM
MD-450DP	OEM
MD-450PM	OEM
MD501	0004
MD501B	0004
MD630	OEM
MD-630DP	OEM
MD-630PM	OEM
MD631	OEM
MD632	OEM
MD633	OEM
MD634	OEM

DEVICE TYPE	REPL CODE
MD635	OEM
MD636	OEM
MD640	OEM
MD-640DP	OEM
MD-640EW	OEM
MD-640LR	OEM
MD-640NS	OEM
MD-640PM	OEM
MD-640XY	OEM
MD650	OEM
MD-650DP	OEM
MD-650EW	OEM
MD-650LRLR	OEM
MD-650NS	OEM
MD-650PM	OEM
MD-650XY	OEM
MD695	6023
MD695A	6023
MD696	6024
MD696A	6024
MD697	6023
MD697A	6023
MD698	6024
MD698A	6024
MD700	OEM
MD700Y	OEM
MD701	OEM
MD701Y	OEM
MD702	OEM
MD702Y	OEM
MD703	OEM
MD703B	OEM
MD703Y	OEM
MD704	OEM
MD704Y	OEM
MD705	OEM
MD705Y	OEM
MD706	OEM
MD706Y	OEM
MD707	OEM
MD707Y	OEM
MD708	5377
MD708A	5377
MD708AF	5377
MD708B	5377
MD708BF	5377
MD708F	5377
MD708Y	OEM
MD709	OEM
MD709Y	OEM
MD710	OEM
MD710Y	OEM
MD711	OEM
MD711Y	OEM
MD712	OEM
MD712Y	OEM
MD713	OEM
MD713Y	OEM
MD714	OEM
MD714Y	OEM
MD715	OEM
MD715Y	OEM
MD752	0157
MD752A	0157
MD753	0636
MD753A	0636
MD757	0012
MD757A	0012
MD759	0137
MD759A	0137
MD918	2045
MD918A	2045
MD918AF	2045
MD918B	2045
MD918BF	2045
MD918F	2045
MD920A8	2400
MD981	OEM
MD981F	OEM
MD982	0037
MD982F	0037
MD984	2468
MD984F	OEM
MD985	5085
MD985F	OEM
MD986	OEM
MD986F	OEM
MD990	OEM
MD1120	OEM
MD1120F	OEM
MD1121	OEM
MD1121F	OEM
MD1122	2045
MD1122F	OEM
MD1123	5228
MD1123F	OEM
MD1124	OEM
MD1124F	OEM
MD1125	OEM
MD1125F	OEM
MD1126	2045
MD1127	2045
MD1128	OEM
MD1129	2045
MD1129F	OEM
MD1130	2468
MD1130F	OEM
MD1131	2045
MD1131F	OEM

DEVICE TYPE	REPL CODE
MD1132	2045
MD1132F	OEM
MD1133	OEM
MD1133F	OEM
MD1134	OEM
MD1668	OEM
MD1670	OEM
MD1672	OEM
MD1674	OEM
MD1676	OEM
MD1678	OEM
MD1680	OEM
MD1682	OEM
MD1684	OEM
MD2060F	OEM
MD2114A4	OEM
MD2114A5	OEM
MD2114AL1	OEM
MD2114AL2	OEM
MD2114AL3	OEM
MD2114AL4	OEM
MD2114L	OEM
MD2114L2	OEM
MD2114L3	OEM
MD2114L4	OEM
MD2114L5	OEM
MD2115A	OEM
MD2115A2	OEM
MD2115AL	OEM
MD2115AL2	OEM
MD2115H2	OEM
MD2115H3	OEM
MD2115H4	OEM
MD2118-4	OEM
MD2118-7	OEM
MD2118-10	OEM
MD2118-12	OEM
MD2118-15	OEM
MD2125A	OEM
MD2125A2	OEM
MD2125AL	OEM
MD2125AL2	OEM
MD2125H1	OEM
MD2125H2	OEM
MD2125H3	OEM
MD2125H4	OEM
MD2128-15	OEM
MD2128-20	OEM
MD2141-2	OEM
MD2141-3	OEM
MD2141-4	OEM
MD2141-5	OEM
MD2141L3	OEM
MD2141L4	OEM
MD2141L5	OEM
MD2142	OEM
MD2142-2	OEM
MD2142-3	OEM
MD2142L	OEM
MD2142L3	OEM
MD2147	OEM
MD2147-3	OEM
MD2147A	OEM
MD2147A3	OEM
MD2147AL	OEM
MD2147AL3	OEM
MD2147H	OEM
MD2147H1	OEM
MD2147H2	OEM
MD2147H3	OEM
MD2147HL	OEM
MD2147HL3	OEM
MD2148	OEM
MD2148-3	OEM
MD2148H	OEM
MD2148H2	OEM
MD2148H3	OEM
MD2148HL	OEM
MD2148HL3	OEM
MD2149H	OEM
MD2149H2	OEM
MD2149H3	OEM
MD2149HL	OEM
MD2149HL3	OEM
MD2164-15	OEM
MD2164-20	OEM
MD2164-25	OEM
MD2167-10	OEM
MD2167-55	OEM
MD2167-70	OEM
MD2167L70	OEM
MD2186	OEM
MD2187	OEM
MD2218	2045
MD2218A	2045
MD2218AF	OEM
MD2218F	OEM
MD2219	2045
MD2219A	2045
MD2219AF	OEM
MD2219F	OEM
MD2363B	2045
MD2369	OEM
MD2369A	2045
MD2369B	6662
MD2716	OEM
MD2716-1	OEM
MD2716-2	OEM
MD2716-5	OEM

DEVICE TYPE	REPL CODE
MD2716-6	OEM
MD2732A	OEM
MD2732A2	OEM
MD2732A3	OEM
MD2732A4	OEM
MD2732A20	OEM
MD2732A25	OEM
MD2732A30	OEM
MD2764	OEM
MD2764-2	OEM
MD2764-3	OEM
MD2764-4	OEM
MD2764-25	OEM
MD2764-30	OEM
MD2764-45	OEM
MD2815	OEM
MD2815-3	OEM
MD2815-4	OEM
MD2816	OEM
MD2816-3	OEM
MD2816-4	OEM
MD2817	OEM
MD2817-3	OEM
MD2817-4	OEM
MD2904	2468
MD2904A	2468
MD2904AF	OEM
MD2904F	OEM
MD2905	2468
MD2905A	2468
MD2905AF	OEM
MD2905F	OEM
MD2974	OEM
MD2975	OEM
MD2978	OEM
MD2979	OEM
MD3133	2468
MD3133F	0150
MD3134	2468
MD3134F	0150
MD3250	2468
MD3250A	2468
MD3250AF	OEM
MD3250F	OEM
MD3251	2468
MD3251A	2468
MD3251AF	OEM
MD3251F	OEM
MD3409	2045
MD3410	2045
MD3467	OEM
MD3467F	OEM
MD3628A1	OEM
MD3628A3	OEM
MD3628A4	OEM
MD3632	OEM
MD3632-1	OEM
MD3636B	OEM
MD3636B1	OEM
MD3636B2	OEM
MD3725	OEM
MD3725F	OEM
MD3762	OEM
MD3762F	OEM
MD4260	OEM
MD4261	OEM
MD4330BC	OEM
MD4330BE	OEM
MD4332BD	OEM
MD4332BE	4934
MD4957	OEM
MD5000	OEM
MD5000A	OEM
MD5000B	OEM
MD6001	OEM
MD6001F	OEM
MD6002	OEM
MD6002F	OEM
MD6003	OEM
MD6003F	OEM
MD6100	OEM
MD6100F	OEM
MD6900	OEM
MD7000	2045
MD7001	2468
MD7001F	OEM
MD7002	2045
MD7002A	5085
MD7002B	5085
MD7003	2468
MD7003A	OEM
MD7003AF	OEM
MD7003B	OEM
MD7003F	OEM
MD7004	OEM
MD7004F	OEM
MD7005	OEM
MD7005F	OEM
MD7007	OEM
MD7007A	OEM
MD7007B	OEM
MD7007BF	OEM
MD7007F	OEM
MD7021	OEM
MD7021F	OEM
MD8001	2045
MD8002	2045
MD8003	2045
MD8035L	OEM

DEVICE TYPE	REPL CODE
MD8048	OEM
MD8085A	OEM
MD8086	OEM
MD8148	OEM
MD8155	OEM
MD8212	OEM
MD8214	OEM
MD8216	OEM
MD8224	OEM
MD8226	OEM
MD8228	OEM
MD8251	OEM
MD8253	OEM
MD8255A	OEM
MD8257	OEM
MD8259A	OEM
MD8284	OEM
MD8288	OEM
MD8748	OEM
MD27128-2	1638
MD50459	OEM
MDA3U3	OEM
MDA10Z-BIN	OEM
MDA65SN1K	OEM
MDA100	0277
MDA100A	0277
MDA101	0277
MDA101A	0277
MDA102	0289
MDA102A	0289
MDA104	0296
MDA104A	0296
MDA106	0302
MDA106A	0302
MDA108	0250
MDA108A	0250
MDA110	0250
MDA110A	0302
MDA200	0277
MDA201	0277
MDA202	0289
MDA204	0296
MDA206	0302
MDA208	0250
MDA210	0250
MDA220	OEM
MDA431	OEM
MDA431SMA	OEM
MDA435	OEM
MDA435SMA	OEM
MDA438	OEM
MDA438SMA	OEM
MDA800	2363
MDA801	0323
MDA802	1414
MDA804	0473
MDA806	0446
MDA-836-1	OEM
MDA-836-2	OEM
MDA920-1	0277
MDA920-2	0015
MDA920-4	0289
MDA920-5	0296
MDA920-6	0296
MDA920-7	0302
MDA920A1	OEM
MDA920A2	4192
MDA920A3	OEM
MDA920A4	OEM
MDA920A5	OEM
MDA920A6	0106
MDA920A7	2010
MDA920A8	2400
MDA920A9	3546
MDA942-1	0289
MDA942-2	0277
MDA942-3	0289
MDA942-4	0296
MDA942-5	0296
MDA942-6	0302
MDA942A-1	0277
MDA942A1	0277
MDA942A-2	0277
MDA942A2	0289
MDA942A-3	0289
MDA942A3	0289
MDA942A-4	0296
MDA942A4	OEM
MDA942A-5	0296
MDA942A5	0296
MDA942A-6	0302
MDA942A6	0302
MDA942A7	0250
MDA952FR-1	0323
MDA952FR-2	0323
MDA952FR-3	0289
MDA952FR-4	0473
MDA952FR-5	0473
MDA960-1	0323
MDA960-2	0323
MDA960-3	1414
MDA960-4	OEM
MDA960-5	0473
MDA970-1	OEM
MDA970-2	OEM
MDA970-3	OEM
MDA970-4	OEM
MDA970-5	OEM
MDA970A1	1042

DEVICE TYPE	REPL CODE
MDA970A2	1042
MDA970A3	1042
MDA970A5	1045
MDA970A6	1045
MDA980-1	1643
MDA980-2	1643
MDA980-3	1643
MDA980-4	2369
MDA980-5	2369
MDA980-6	1589
MDA990-1	2363
MDA990-2	2363
MDA990-3	2363
MDA990-4	2369
MDA990-5	2369
MDA990-6	2370
MDA1200	2363
MDA1201	2363
MDA1202	2363
MDA1204	2369
MDA1206	2370
MDA1330H	OEM
MDA1331H	OEM
MDA2061	OEM
MDA2500	5483
MDA2501	2363
MDA2502	2369
MDA2504	2369
MDA2506	2370
MDA2550	2363
MDA2551	2363
MDA3500	1643
MDA3501	1643
MDA3502	1643
MDA3504	1673
MDA3506	1673
MDA3508	4159
MDA3510	1589
MDA3550	1643
MDA3551	1501
MDA3552	1501
MDA3661	0190
MDA3662	0190
MDA7705	OEM
MDA7705SMA	OEM
MDA7708	OEM
MDA7708SMA	OEM
MDA7709SMA	OEM
MDA7710	OEM
MDA9901	2363
MDA9902	2363
MDA9903	2363
MDA9905	2369
MDA9906	2370
MDAS-8D	OEM
MDAS-16	OEM
MDB-42XX	OEM
MDB-46-206	OEM
MDB510	OEM
MDB510-500	OEM
MDB510-1000	OEM
MDB511	OEM
MDB-CR11	OEM
MDB-DA11-B0I	OEM
MDB-DA11-BJ	OEM
MDB-DA528	OEM
MDB-DT1711	OEM
MDB-DUP11	OEM
MDB-DZ11-A	OEM
MDB-DZ11-A/422	OEM
MDB-DZ11-B	OEM
MDB-DZ11-E	OEM
MDB-KW11-P	OEM
MDB-LP11/LLO	OEM
MDB-MDU11	OEM
MDB-PC11	OEM
MDB-TA528	OEM
MDB-XY11	OEM
MDC5FA2	1236
MDC11	OEM
MDC12FA2	2234
MDC20FA2	2234
MDCZ8.2D	0244
MDCZ8.2D10	0244
MDCZ1511OUT	3415
MDCZ15109UL69-536	OEM
MDCZ15110UL	3415
MDCZA6.8D	0062
MDCZA6.8D5	0062
MDCZA6.8D10	0062
MDCZA7.5D	0077
MDCZA7.5D5	0077
MDCZA7.5D10	0077
MDCZA8.2D	0165
MDCZA8.2D5	0244
MDCZA8.2D10	0244
MDCZA9.1D	0057
MDCZA9.1D5	0057
MDCZA9.1D10	0057
MDCZA10D	0064
MDCZA10D5	0064
MDCZA10D10	0064
MDCZA11D	0181
MDCZA11D5	0181
MDCZA11D10	0181
MDCZA12D	0052
MDCZA12D5	0052
MDCZA12D10	0052
MDCZA13D	0053
MDCZA13D5	0053
MDCZA13D10	0053
MDCZA14D	0879
MDCZA14D5	0879
MDCZA14D10	0879
MDCZA15D	0686
MDCZA15D5	0686
MDCZA15D10	0686
MDCZA16D	0445
MDCZA16D5	0445
MDCZA16D10	0445
MDCZA17D	0210
MDCZA17D5	0210
MDCZA17D10	0210
MDCZA18D	0375
MDCZA18D5	0375
MDCZA18D10	0375
MDCZA19D	0671
MDCZA19D5	0671
MDCZA19D10	0671
MDCZA20D	0700
MDCZA20D5	0700
MDCZA20D10	0700
MDCZA22D	0705
MDCZA22D5	0705
MDCZA22D10	0705
MDCZA24D	OEM
MDCZA24D5	OEM
MDCZA24D10	OEM
MDCZA25D	0714
MDCZA25D5	0714
MDCZA25D10	0714
MDCZA27D	0455
MDCZA27D5	0455
MDCZA27D10	0455
MDCZA28D	0257
MDCZA28D5	0257
MDCZA28D10	0257
MDCZA30D	0195
MDCZA30D5	0195
MDCZA30D10	0195
MDCZA33D	0166
MDCZA33D5	0166
MDCZA33D10	0166
MDCZA36D	0010
MDCZA36D5	0010
MDCZA36D10	0010
MDCZA39D	0032
MDCZA39D5	0032
MDCZA39D10	0032
MDCZA43D	0054
MDCZA43D5	0054
MDCZA43D10	0054
MDCZA47D	0068
MDCZA47D5	0068
MDCZA47D10	0068
MDCZA51D	0092
MDCZA51D5	0092
MDCZA51D10	0092
MDCZA56D	0125
MDCZA56D5	0125
MDCZA56D10	0125
MDCZA60D	OEM
MDCZA60D5	2317
MDCZA60D10	2317
MDCZA62D	0152
MDCZA62D5	0152
MDCZA62D10	0152
MDCZA68D	0173
MDCZA68D5	0173
MDCZA68D10	0173
MDCZA75D	0094
MDCZA75D5	0094
MDCZA75D10	0094
MDCZA82D	0049
MDCZA82D5	0049
MDCZA82D10	0049
MDCZA87D	0104
MDCZA87D5	0104
MDCZA87D10	0104
MDCZA91D	0156
MDCZA91D5	0156
MDCZA91D10	0156
MDCZA100D	0189
MDCZA100D5	0189
MDCZA100D10	0189
MDCZA105D	OEM
MDCZA105D5	OEM
MDCZA105D10	OEM
MDCZA110D	0099
MDCZA110D5	0099
MDCZA110D10	0099
MDCZA120D	0089
MDCZA120D5	0089
MDCZA120D10	0089
MDCZA130D	0287
MDCZA130D5	0287
MDCZA130D10	0287
MDCZA140D	0252
MDCZA140D5	0252
MDCZA140D10	0252
MDCZA150D	0340
MDCZA150D5	0340
MDCZA150D10	0340
MDCZA160D	0370
MDCZA160D5	0370
MDCZA160D10	0370
MDCZA175D	OEM
MDCZA175D5	OEM
MDCZA175D10	OEM
MDCZA180D	0425
MDCZA180D5	0425
MDCZA180D10	0425
MDCZA200D	1473
MDCZA200D5	1473
MDCZA200D10	1473
MDCZB6.8D	0062
MDCZB6.8D5	0062
MDCZB6.8D10	0062
MDCZB7.5D	0077
MDCZB7.5D5	0077
MDCZB7.5D10	0077
MDCZB8.2D	0244
MDCZB8.2D5	0244
MDCZB8.2D10	0165
MDCZB9.1D	0057
MDCZB9.1D5	0057
MDCZB9.1D10	0057
MDCZB10D	0064
MDCZB10D5	0064
MDCZB10D10	0064
MDCZB11D	0052
MDCZB11D5	0052
MDCZB11D10	0052
MDCZB12D	0052
MDCZB12D5	0052
MDCZB12D10	0052
MDCZB13D	0053
MDCZB13D5	0053
MDCZB13D10	0053
MDCZB14D	0053
MDCZB14D5	0053
MDCZB14D10	0053
MDCZB15D	0686
MDCZB15D5	0686
MDCZB15D10	0686
MDCZB16D	0445
MDCZB16D5	0445
MDCZB16D10	0370
MDCZB17D	0210
MDCZB17D5	0210
MDCZB17D10	0210
MDCZB18D	0375
MDCZB18D5	0375
MDCZB18D10	0375
MDCZB19D	0671
MDCZB19D5	0671
MDCZB19D10	0671
MDCZB20D	0700
MDCZB20D5	0700
MDCZB20D10	0700
MDCZB22D	0705
MDCZB22D5	0705
MDCZB22D10	0705
MDCZB25D	0714
MDCZB25D5	0714
MDCZB25D10	0714
MDCZB27D	0455
MDCZB27D5	0455
MDCZB27D10	0455
MDCZB30D	0195
MDCZB30D5	0195
MDCZB30D10	0195
MDCZB33D	0166
MDCZB33D5	0166
MDCZB33D10	0166
MDCZB36D	0010
MDCZB36D5	0010
MDCZB36D10	0010
MDCZB39D	0032
MDCZB39D5	0032
MDCZB39D10	0032
MDCZB43D	0054
MDCZB43D5	0054
MDCZB43D10	0054
MDCZB47D	0068
MDCZB47D5	0068
MDCZB47D10	0068
MDCZB51D	0092
MDCZB51D5	0092
MDCZB51D10	0092
MDCZB56D	0125
MDCZB56D5	0125
MDCZB56D10	0125
MDCZB62D	0152
MDCZB62D5	0152
MDCZB62D10	0152
MDCZB68D	0173
MDCZB68D5	0173
MDCZB68D10	0173
MDCZB75D	0094
MDCZB75D5	0094
MDCZB75D10	0094
MDCZB82D	0049
MDCZB82D5	0049
MDCZB82D10	0049
MDCZB91D	0156
MDCZB91D5	OEM
MDCZB91D10	0156
MDCZB100D	0189
MDCZB100D5	0189
MDCZB100D10	0189
MDCZB105D	OEM
MDCZB105D5	OEM
MDCZB105D10	OEM
MDCZB110D	0099
MDCZB110D5	0099
MDCZB110D10	0099
MDCZB120D	0089
MDCZB120D5	0089
MDCZB120D10	0089
MDCZB130D	0287
MDCZB130D5	0287
MDCZB130D10	0287
MDCZB140D	0252
MDCZB140D5	0252
MDCZB140D10	0252
MDCZB150D	0340
MDCZB150D5	0340
MDCZB150D10	0340
MDCZB160D	0370
MDCZB160D5	0370
MDCZB160D10	OEM
MDCZB175D	OEM
MDCZB175D5	OEM
MDCZB175D10	OEM
MDCZB180D	0425
MDCZB180D5	0425
MDCZB180D10	0425
MDCZB200D	1473
MDCZB200D5	1473
MDCZB200D10	1473
MDCZC6.8D	0062
MDCZC6.8D5	0062
MDCZC6.8D10	0062
MDCZC7.5D	0077
MDCZC7.5D5	0077
MDCZC7.5D10	0077
MDCZC8.2D	0165
MDCZC8.2D5	0244
MDCZC8.2D10	0244
MDCZC9.1D	0057
MDCZC9.1D5	0057
MDCZC9.1D10	0057
MDCZC10D	0064
MDCZC10D5	0064
MDCZC10D10	0064
MDCZC11D	0181
MDCZC11D5	0181
MDCZC11D10	0181
MDCZC12D	0052
MDCZC12D5	0052
MDCZC12D10	0052
MDCZC13D	0053
MDCZC13D5	0053
MDCZC13D10	0053
MDCZC14D	0879
MDCZC14D5	0879
MDCZC14D10	0879
MDCZC15D	0686
MDCZC15D5	0686
MDCZC15D10	0686
MDCZC16D	0445
MDCZC16D5	0445
MDCZC16D10	0445
MDCZC17D	0210
MDCZC17D5	0210
MDCZC17D10	0210
MDCZC18D	0375
MDCZC18D5	0375
MDCZC18D10	0375
MDCZC19D	0671
MDCZC19D5	0671
MDCZC19D10	0671
MDCZC20D	0700
MDCZC20D5	0700
MDCZC20D10	0700
MDCZC22D	0705
MDCZC22D5	0705
MDCZC22D10	0705
MDCZC24D	0494
MDCZC24D5	0494
MDCZC24D10	0494
MDCZC25D	0714
MDCZC25D5	0714
MDCZC25D10	0714
MDCZC27D	0455
MDCZC27D5	0455
MDCZC27D10	0455
MDCZC30D	0195
MDCZC30D5	0195
MDCZC30D10	0195
MDCZC33D	0166
MDCZC33D5	0166
MDCZC33D10	0166
MDCZC36D	0010
MDCZC36D5	0010
MDCZC36D10	0010
MDCZC39D	0032
MDCZC39D5	0032
MDCZC39D10	0032
MDCZC43D	0054
MDCZC43D5	0054
MDCZC43D10	0054
MDCZC47D	0068
MDCZC47D5	0068
MDCZC47D10	0068
MDCZC51D	0092
MDCZC51D5	0092
MDCZC51D10	0092
MDCZC56D	0125
MDCZC56D5	0125
MDCZC56D10	0125
MDCZC62D	0152
MDCZC62D5	0152
MDCZC62D10	0152
MDCZC68D	0173
MDCZC68D5	0173
MDCZC68D10	0173
MDCZC75D	0094
MDCZC75D5	0094
MDCZC75D10	0094
MDCZC82D	0049
MDCZC82D5	0049
MDCZC82D10	0049
MDCZC91D	0156
MDCZC91D5	0156
MDCZC91D10	0156
MDCZC100D	0189
MDCZC100D5	0189
MDCZC100D10	0189
MDCZC105D	OEM
MDCZC105D5	OEM
MDCZC105D10	OEM
MDCZC110D	0099
MDCZC110D5	OEM
MDCZC110D10	0099
MDCZC120D	0089
MDCZC120D5	0089
MDCZC120D10	0089
MDCZC130D	0287
MDCZC130D5	0287
MDCZC130D10	0287
MDCZC140D	0252
MDCZC140D5	0252
MDCZC140D10	0252
MDCZC150D	0340
MDCZC150D5	0340
MDCZC150D10	0340
MDCZC160D	0370
MDCZC160D5	0370
MDCZC160D10	0370
MDCZC175D	OEM
MDCZC175D5	OEM
MDCZC175D10	OEM
MDCZC180D	0425
MDCZC180D5	0425
MDCZC180D10	0425
MDCZC200D	1473
MDCZC200D5	1473
MDCZC200D10	1473
MDCZD6.8D	0062
MDCZD6.8D5	0062
MDCZD6.8D10	0062
MDCZD7.5D	0077
MDCZD7.5D5	0077
MDCZD7.5D10	0077
MDCZD8.2D	0165
MDCZD8.2D5	0244
MDCZD8.2D10	0244
MDCZD9.1D	0057
MDCZD9.1D5	0057
MDCZD9.1D10	0057
MDCZD10D	0064
MDCZD10D5	0064
MDCZD10D10	0064
MDCZD11D	0181
MDCZD11D5	0181
MDCZD11D10	0181
MDCZD12D	0052
MDCZD12D5	0052
MDCZD12D10	0052
MDCZD13D	0053
MDCZD13D5	0053
MDCZD13D10	0053
MDCZD14D	0879
MDCZD14D5	0879
MDCZD14D10	0879
MDCZD15D	0686
MDCZD15D5	0686
MDCZD15D10	0686
MDCZD16D	0445
MDCZD16D5	0445
MDCZD16D10	0445
MDCZD17D	0210
MDCZD17D5	0210
MDCZD17D10	0210
MDCZD18D	0375
MDCZD18D5	0375
MDCZD18D10	0375
MDCZD19D	0671
MDCZD19D5	0671
MDCZD19D10	0671
MDCZD20D	0700
MDCZD20D5	0700
MDCZD20D10	0700
MDCZD22D	0705
MDCZD22D5	0705
MDCZD22D10	OEM
MDCZD24D	0494
MDCZD24D5	0494
MDCZD24D10	0494
MDCZD25D	0714
MDCZD25D5	0714
MDCZD25D10	0714
MDCZD27D	0455
MDCZD27D5	0455
MDCZD27D10	0455
MDCZD28D	0257
MDCZD28D5	0257
MDCZD28D10	0257
MDCZD30D	0195
MDCZD30D5	0195
MDCZD30D10	0195
MDCZD33D	0166
MDCZD33D5	0166
MDCZD33D10	0166
MDCZD36D	0010
MDCZD36D5	0010
MDCZD36D10	0010
MDCZD39D	0032
MDCZD39D5	0032
MDCZD39D10	0032
MDCZD43D	0054
MDCZD43D5	0054
MDCZD43D10	0054
MDCZD47D	0068
MDCZD47D5	0068
MDCZD47D10	0068
MDCZD51D	0092
MDCZD51D5	0092
MDCZD51D10	0092
MDCZD56D	0125
MDCZD56D5	0125
MDCZD56D10	0125
MDCZD60D	2317
MDCZD60D5	2317
MDCZD60D10	2317
MDCZD62D	0152
MDCZD62D5	0152
MDCZD62D10	0152
MDCZD68D	0173
MDCZD68D5	0173
MDCZD68D10	0173
MDCZD75D	0094
MDCZD75D5	0094
MDCZD75D10	0094
MDCZD82D	0049
MDCZD82D5	0049
MDCZD82D10	0049
MDCZD87D	0104
MDCZD87D5	0104
MDCZD87D10	0104
MDCZD91D	0156
MDCZD91D5	0156
MDCZD91D10	0156
MDCZD100D	0189
MDCZD100D5	0189
MDCZD100D10	0189
MDCZD105D	OEM
MDCZD105D5	OEM
MDCZD105D10	OEM
MDCZD110D	0099
MDCZD110D5	0099
MDCZD110D10	0099
MDCZD120D	0089
MDCZD120D5	0089
MDCZD120D10	0089
MDCZD130D	0287
MDCZD130D5	0287
MDCZD130D10	0287
MDCZD140D	0252
MDCZD140D5	0252
MDCZD140D10	0252
MDCZD150D	0340
MDCZD150D5	0340
MDCZD150D10	0340
MDCZD160D	0370
MDCZD160D5	0370
MDCZD160D10	0370
MDCZD175D	OEM
MDCZD175D5	OEM
MDCZD175D10	OEM
MDCZD180D	0425
MDCZD180D5	0425
MDCZD180D10	0425
MDCZD200D	1473
MDCZD200D5	1473
MDCZD200D10	1473
MDCZE6.8D	0062
MDCZE6.8D5	0062
MDCZE6.8D10	0062
MDCZE7.5D	0077
MDCZE7.5D5	0077
MDCZE7.5D10	0077
MDCZE8.2D	0165
MDCZE8.2D5	0244
MDCZE8.2D10	0244
MDCZE9.1D	0057
MDCZE9.1D5	0057
MDCZE9.1D10	0057
MDCZE10D	0064
MDCZE10D5	0064
MDCZE10D10	0064
MDCZE11D	0181
MDCZE11D5	0181
MDCZE11D10	0181
MDCZE12D	0052
MDCZE12D5	0052
MDCZE12D10	0052
MDCZE13D	0053
MDCZE13D5	0053
MDCZE13D10	0053
MDCZE14D	0879
MDCZE14D5	0879
MDCZE14D10	OEM
MDCZE15D	0686
MDCZE15D5	0686
MDCZE15D10	0686
MDCZE16D	0445
MDCZE16D5	0445
MDCZE16D10	0445
MDCZE17D	0210
MDCZE17D5	0210
MDCZE17D10	0210
MDCZE18D	0375
MDCZE18D5	0375
MDCZE18D10	0375
MDCZE19D	0671
MDCZE19D5	0671
MDCZE19D10	0671
MDCZE20D	0700
MDCZE20D5	0700
MDCZE20D10	0700
MDCZE22D	0705
MDCZE22D5	0705
MDCZE22D10	0705
MDCZE24D	0494
MDCZE24D5	0494
MDCZE24D10	0494
MDCZE25D	0714
MDCZE25D5	0714
MDCZE25D10	0714
MDCZE27D	0455
MDCZE27D5	0455
MDCZE27D10	0455
MDCZE30D	0195
MDCZE30D5	0195
MDCZE30D10	0195
MDCZE33D	0166
MDCZE33D5	0166
MDCZE33D10	0166
MDCZE36D	0010
MDCZE36D5	0010
MDCZE36D10	0010
MDCZE39D	0032
MDCZE39D5	0032
MDCZE39D10	0032
MDCZE43D	0054
MDCZE43D5	0054
MDCZE43D10	0054
MDCZE47D	0068
MDCZE47D5	0068
MDCZE47D10	0068
MDCZE51D	0092
MDCZE51D5	0092
MDCZE51D10	0092
MDCZE56D	0125
MDCZE56D5	0125
MDCZE56D10	0125
MDCZE62D	0152
MDCZE62D5	0152
MDCZE62D10	0152
MDCZE68D	0173
MDCZE68D5	0173
MDCZE68D10	0173
MDCZE75D	0094
MDCZE75D5	0094
MDCZE75D10	0094
MDCZE82D	0049
MDCZE82D5	0049
MDCZE82D10	0049
MDCZE91D	0156
MDCZE91D5	0156
MDCZE91D10	0156
MDCZE100D	0189
MDCZE100D5	0189
MDCZE100D10	0189
MDCZE105D	OEM
MDCZE105D5	OEM
MDCZE105D10	OEM
MDCZE110D	0099
MDCZE110D5	0099
MDCZE110D10	0099
MDCZE120D	0089
MDCZE120D5	0089
MDCZE120D10	0089
MDCZE130D	0287
MDCZE130D5	0287
MDCZE130D10	0287
MDCZE140D	0252
MDCZE140D5	0252
MDCZE140D10	0252
MDCZE150D	0340
MDCZE150D5	0340
MDCZE150D10	0340
MDCZE160D	0370
MDCZE160D5	0370
MDCZE160D10	0370
MDCZE175D	OEM
MDCZE175D5	OEM
MDCZE175D10	OEM
MDCZE180D	0425
MDCZE180D5	0425
MDCZE180D10	0425
MDCZE200D	1473
MDCZE200D10	1473
MDCZE200E5	1473
MDCZF1.8D	OEM
MDCZF1.8D5	OEM
MDCZF1.8D10	OEM
MDCZF2.0D	OEM
MDCZF2.0D5	OEM
MDCZF2.0D10	OEM
MDCZF2.2D	OEM
MDCZF2.2D5	OEM
MDCZF2.2D10	OEM
MDCZF2.4D	1276
MDCZF2.4D5	1276
MDCZF2.4D10	1276
MDCZF2.7D	0760
MDCZF2.7D5	0760
MDCZF2.7D10	0760
MDCZF3.0D	0118

DEVICE TYPE	REPL CODE
MDCZF3.0D5	0118
MDCZF3.0D10	0118
MDCZF3.3D	0299
MDCZF3.3D5	0299
MDCZF3.3D10	0299
MDCZF3.6D	0376
MDCZF3.6D5	0376
MDCZF3.6D10	0299
MDCZF3.9D	0036
MDCZF3.9D5	0036
MDCZF3.9D10	0036
MDCZF4.3D	0275
MDCZF4.3D5	0275
MDCZF4.3D10	0275
MDCZF4.7D	0140
MDCZF4.7D5	0140
MDCZF4.7D10	0140
MDCZF5.1D	0041
MDCZF5.1D5	0041
MDCZF5.1D10	0041
MDCZF5.6D	0253
MDCZF5.6D5	0253
MDCZF5.6D10	0253
MDCZF6.2D	0471
MDCZF6.2D5	0471
MDCZF6.2D10	0471
MDCZF6.8D	0062
MDCZF6.8D5	0062
MDCZF6.8D10	0062
MDCZF7.5D	0077
MDCZF7.5D5	0077
MDCZF7.5D10	0077
MDCZF8.2D	0244
MDCZF8.2D5	0244
MDCZF8.2D10	0244
MDCZF9.1D	0057
MDCZF9.1D5	0057
MDCZF9.1D10	0057
MDCZF10D	0064
MDCZF10D5	0064
MDCZF10D10	0064
MDCZF12D	0052
MDCZF12D5	0052
MDCZF12D10	0052
MDCZG2.4D	1276
MDCZG2.4D5	1276
MDCZG2.7D	0760
MDCZG2.7D5	0760
MDCZG2.7D10	0760
MDCZG3.0D	0118
MDCZG3.0D5	0118
MDCZG3.0D10	0118
MDCZG3.3D	0299
MDCZG3.3D5	0299
MDCZG3.3D10	0299
MDCZG3.6D	0376
MDCZG3.6D5	0376
MDCZG3.6D10	0376
MDCZG3.9D	0036
MDCZG3.9D5	0036
MDCZG3.9D10	0036
MDCZG4.3D	0275
MDCZG4.3D5	0275
MDCZG4.3D10	0275
MDCZG4.7D	0140
MDCZG4.7D5	0140
MDCZG4.7D10	0140
MDCZG5.1D	0041
MDCZG5.1D5	0041
MDCZG5.1D10	0041
MDCZG5.6D	0253
MDCZG5.6D5	0253
MDCZG5.6D10	0253
MDCZG6.2D	0471
MDCZG6.2D5	0471
MDCZG6.2D10	0471
MDCZG6.8D	0062
MDCZG6.8D5	0062
MDCZG6.8D10	0062
MDCZG7.5D	0077
MDCZG7.5D5	0077
MDCZG7.5D10	0077
MDCZG8.2D	0244
MDCZG8.2D5	0244
MDCZG8.2D10	0244
MDCZG9.1D	0057
MDCZG9.1D5	0057
MDCZG9.1D10	0057
MDCZG10D	0064
MDCZG10D5	0064
MDCZG10D10	0064
MDCZG12D	0052
MDCZG12D5	0052
MDCZG12D10	0052
MDD25-04N1	1798
MDD25-06N1	1798
MDD25-08N1	1798
MDD25-12N1	1798
MDD25-14N1	OEM
MDD25-16N1	OEM
MDD41-12N1	OEM
MDD42-04N1	1798
MDD42-06N1	1798
MDD42-08N1	1798
MDD42-14N1	OEM
MDD42-16N1	OEM
MDD-43	OEM
MDD-44	OEM
MDD-45	OEM
MDD-63	OEM
MDD-64	OEM
MDD-65	OEM
MDD-450	OEM
MDD-640	OEM
MDD-650	OEM
MDG-01	OEM
MDL-11	OEM
MDL-11W	OEM
MDL221	OEM
MDL235	OEM
MDL235-TV	OEM
MDL236-TV	OEM
MDL238	OEM
MDL255-TV	OEM
MDL258	OEM
MDL259	OEM
MDL420	OEM
MDL421	OEM
MDL422	OEM
MDL2255-TV	OEM
MDL2265-TV	OEM
MDL2788SKP	OEM
MDL4755SKP	OEM
MDL4777SKP	OEM
MDP173	0133
MDS-01	OEM
MDS20	0285
MDS21	0285
MDS26	0560
MDS27	0560
MDS31	0050
MDS32	0050
MDS33	0050
MDS33A	0050
MDS33C	0050
MDS33D	0050
MDS34	0050
MDS35	0136
MDS36	0050
MDS37	0050
MDS38	0050
MDS39	0050
MDS40	0050
MDS60	2200
MDS76	1267
MDS77	1267
MDS88	OEM
MDS503	OEM
MDS1678	0936
MDS6518	0791
MDS9400	0161
MDS9401	0161
MDS9450	0460
MDS9451	0460
MDX580C	OEM
MDX580D	OEM
MDX580M	OEM
MDX580N	OEM
MDX581C	OEM
MDX581D	OEM
MDX581M	OEM
MDX581N	OEM
MDX582C	OEM
MDX582D	OEM
MDX582M	OEM
MDX582N	OEM
MDX583C	OEM
MDX583D	OEM
MDX583M	OEM
MDX583N3N	OEM
MDX584C	OEM
MDX584D	OEM
MDX584M	OEM
MDX584N	OEM
MDX585C	OEM
MDX585D	OEM
MDX585M	OEM
MDX585N	OEM
MDX586C	OEM
MDX586D	OEM
MDX586M	OEM
MDX586N	OEM
MDX587C	OEM
MDX587D	OEM
MDX587M	OEM
MDX587N	OEM
MDX588C	OEM
MDX588D	OEM
MDX588M	OEM
MDX588N	OEM
MDX589C	OEM
MDX589D	OEM
MDX589M	OEM
MDX589N	OEM
MDX590C	OEM
MDX590D	OEM
MDX590M	OEM
MDX590N	OEM
MDX591C	OEM
MDX591D	OEM
MDX591M	OEM
MDX591N	OEM
MDX592C	OEM
MDX592D	OEM
MDX592M	OEM
MDX592N	OEM
MDX593C	OEM
MDX593D	OEM
MDX593M	OEM
MDX593N	OEM
MDX594C4C	OEM
MDX594D	OEM
MDX594M	OEM
MDX594N	OEM
MDX595C	OEM
MDX595D	OEM
MDX595M	OEM
MDX595N	OEM
MDX596C	OEM
MDX596D	OEM
MDX596M	OEM
MDX596N	OEM
MDX597C	OEM
MDX597D	OEM
MDX597M	OEM
MDX597N	OEM
MDX598C	OEM
MDX598D	OEM
MDX598M	OEM
MDX598N	OEM
MDX599C	OEM
MDX599D	OEM
MDX599M	OEM
MDX599N	OEM
MDX623	OEM
MDX624	OEM
MDX625	OEM
MDX640	OEM
MDX641	OEM
MDX642	OEM
MDX643	OEM
MDX644	OEM
MDX645	OEM
MDX646	OEM
MDX647	OEM
MDX653	OEM
MDX654	OEM
MDX655	OEM
MDXP-32	OEM
MDXP-32-1	OEM
MDX-SV1-1	OEM
MDZ-01	OEM
MDZ-02	OEM
MDZ-03	OEM
MDZ-04	OEM
MDZ-05	OEM
MDZ-06	OEM
MDZ-07	OEM
MDZ-08	OEM
MDZ-09	OEM
MDZ-10	OEM
MDZ-11	OEM
MDZ-12	OEM
MDZ-13	OEM
MDZ-14	OEM
MDZ-15	OEM
MDZ-16	OEM
MDZ-17	OEM
MDZ-18	OEM
MDZ-19	OEM
MDZ-20	OEM
MDZ-21	OEM
MDZ-22	OEM
MDZ-23	OEM
MDZ-24	OEM
MDZ-25	OEM
MDZ-26	OEM
MDZ-27	OEM
MDZ-28	OEM
MDZ-29	OEM
MDZ-30	OEM
MDZ-31	OEM
MDZ-32	OEM
MDZ-33	OEM
MDZG2.4D10	OEM
ME0401	0150
ME0402	0150
ME0404	0037
ME0404-1	0037
ME0404-2	0037
ME0404A	0006
ME0411	0861
ME0412	0861
ME0413	0861
ME0414	0861
ME0461	0037
ME0462	0037
ME0463	0150
ME0475	0861
ME-1	0016
ME1D821	0471
ME1D821A	0471
ME1D823	0471
ME1D823A	0471
ME1D825	0471
ME1D825A	0471
ME1D826	OEM
ME1D827	0471
ME1D827A	0471
ME1D828	OEM
ME1D829	0471
ME1D935	OEM
ME1D935A	OEM
ME1D935B	OEM
ME1D936	OEM
ME1D936A	OEM
ME1D936B	OEM
ME1D937	OEM
ME1D937A	OEM
ME1D937B	OEM
ME1D938	OEM
ME1D938A	OEM
ME1D938B	OEM
ME1D939	OEM
ME1D939A	OEM
ME1D939B	OEM
ME1D940	OEM
ME1D940A	OEM
ME1D940B	OEM
ME1D941	OEM
ME1D941A	OEM
ME1D941B	OEM
ME1D942	OEM
ME1D942A	OEM
ME1D942B	OEM
ME1D943	OEM
ME1D943A	OEM
ME1D943B	OEM
ME1D944	OEM
ME1D944A	OEM
ME1D944B	OEM
ME1D945	OEM
ME1D945A	OEM
ME1D945B	OEM
ME1D946	OEM
ME1D946A	OEM
ME1D946B	OEM
ME1D4565	OEM
ME1D4565A	OEM
ME1D4566	OEM
ME1D4566A	OEM
ME1D4567	OEM
ME1D4567A	OEM
ME1D4568	OEM
ME1D4568A	OEM
ME1D4569	OEM
ME1D4569A	OEM
ME1D4570	OEM
ME1D4570A	OEM
ME1D4571	OEM
ME1D4571A	OEM
ME1D4572	OEM
ME1D4572A	OEM
ME1D4573	OEM
ME1D4573A	OEM
ME1D4574	OEM
ME1D4574A	OEM
ME1D4575	OEM
ME1D4575A	OEM
ME1D4576	OEM
ME1D4576A	OEM
ME1D4577	OEM
ME1D4577A	OEM
ME1D4578	OEM
ME1D4578A	OEM
ME1D4579	OEM
ME1D4579A	OEM
ME1D4580	OEM
ME1D4580A	OEM
ME1D4581	OEM
ME1D4581A	OEM
ME1D4582	OEM
ME1D4582A	OEM
ME1D4583	OEM
ME1D4583A	OEM
ME1D4584	OEM
ME1D4584A	OEM
ME1Z935	OEM
ME1Z935A	OEM
ME1Z935B	OEM
ME1Z936	OEM
ME1Z936A	OEM
ME1Z936B	OEM
ME1Z937	OEM
ME1Z937A	OEM
ME1Z937B	OEM
ME1Z938	OEM
ME1Z938A	OEM
ME1Z938B	OEM
ME1Z939	OEM
ME1Z939A	OEM
ME1Z939B	OEM
ME1Z940	OEM
ME1Z940A	OEM
ME1Z940B	OEM
ME1Z941	OEM
ME1Z941A	OEM
ME1Z941B	OEM
ME1Z942	OEM
ME1Z942A	OEM
ME1Z942B	OEM
ME1Z943	OEM
ME1Z943A	OEM
ME1Z943B	OEM
ME1Z944	OEM
ME1Z944A	OEM
ME1Z944B	OEM
ME1Z945	OEM
ME1Z945A	OEM
ME1Z945B	OEM
ME1Z946	OEM
ME1Z946A	OEM
ME1Z946B	OEM
ME1.2	OEM
ME1.5	OEM
ME-2	0079
ME2	OEM
ME-3	0079
ME-4	0139
ME-5	0566
ME30	OEM
ME60	5113
ME61	OEM
ME100	OEM
ME100D	OEM
ME-116	OEM
ME116R	OEM
ME120	OEM
ME200	OEM
ME209	OEM
ME213	0016
ME213A	0016
ME214	OEM
ME216	0016
ME217	0016
ME408-02C	0244
ME409-02B	0012
ME495	OEM
ME501	0037
ME502	0037
ME503	0150
ME504	OEM
ME509	OEM
ME510	OEM
ME511	0150
ME512	0150
ME513	0791
ME900	0079
ME900A	0016
ME901	0016
ME901A	0016
ME1001	0016
ME1002	0016
ME1013	OEM
ME1075	0086
ME1100	0861
ME1110	0233
ME1120	0233
ME1138	0224
ME1303	OEM
ME2001	0016
ME2002	0016
ME3001	0259
ME3002	0127
ME3011	0127
ME3440	0168
ME4001	0016
ME4002	0016
ME4003	0016
ME4003C	0016
ME4101	0016
ME4102	0016
ME4103	0016
ME4104	0016
ME5001	0144
ME6001	0016
ME6002	0016
ME6003	0016
ME6021	OEM
ME6022	OEM
ME6101	0037
ME7021	4183
ME7024	OEM
ME7121	6874
ME7124	OEM
ME7161	5114
ME8001	0086
ME8002	0086
ME8003	0086
ME8101	0144
ME8201	0144
ME9001	0079
ME9002	0127
ME9003	0127
ME9021	0144
ME9022	0127
ME32865-00001-B	2699
MEE-888	OEM
MEF68	OEM
MEF69	OEM
MEF70	OEM
MEF101	OEM
MEF102	OEM
MEGA-4/64	OEM
MEGA-4/128	OEM
MEGA-4/256	OEM
MEGA-4/512	OEM
MEK6800D2	OEM
MEK6800D2AT	OEM
MEK6800D2C	OEM
MEK6800D2CAT	OEM
MEK6800D2D	OEM
MEK6800D2DAT	OEM
MEL11	OEM
MEL12	OEM
MEL31	OEM
MEL32	OEM
MEL100	OEM
MEM8N08	OEM
MEM11	OEM
MEM-88	OEM
MEM100	OEM
MEM101	OEM
MEM102	OEM
MEM200	OEM
MEM201	OEM
MEM202	OEM
MEM300	OEM
MEM301	OEM
MEM302	OEM
MEM400	OEM
MEM401	OEM
MEM402	OEM
MEM515	OEM
MEM519	OEM
MEM554	0212
MEM564C	0212
MEM620	OEM
MEM621	OEM
MEM622	OEM
MEM630	0212
MEM640	OEM
MEM641	OEM
MEM642	OEM
MEM656A	OEM
MEM680	0212
MEM680Y	0212
MEM803	OEM
MEM804	OEM
MEM805	OEM
MEM808	OEM
MEM809	OEM
MEM4001	0478
MEM4007	2846
MEM4011	0215
MEM4013	0413
MEM4016	1143
MEM4049	0001
MEM4050	0398
MEM4051	0366
MEMORY-32	OEM
MEP04	OEM
MEU21	OEM
MEU22	OEM
MEX27T10	0441
MEX68KDM	OEM
MEX68KECB	OEM
MEX848-22	OEM
MEX6801	OEM
MEX6801EVM	OEM
MEX6802-46	OEM
MEX6805	OEM
MEX6816-22D	OEM
MEX6816-22S	OEM
MEX6820	OEM
MEX6821-2	OEM
MEX6832-22	OEM
MEX6850	OEM
MEX6864-1HR	OEM
MEX6864-22	OEM
MEZ1-3.3	0292
MEZ1-3.6	0188
MEZ1-3.9	0456
MEZ1-4.3	0533
MEZ1-4.7	0451
MEZ1-5.1	0162
MEZ1-5.6	0157
MEZ1-6.2	0636
MEZ1-6.8	0025
MEZ1-7.5	0649
MEZ1-8.2	0244
MEZ1-9.1	0012
MEZ1-10	0170
MEZ1-11	0317
MEZ1-12G	0137
MEZ1-14	0100
MEZ1-15	0002
MEZ1-20	0531
MEZ1-22	0565
MEZ1-24	0402
MEZ1-26	0441
MEZ1-27	0441
MEZ1-29	0726
MEZ1-33G	0039
MEZ1-36	0820
MEZ1-39	0931
MEZ1-43	0931
MEZ1-47	0999
MEZ1-51	0502
MEZ1-55	1834
MEZ1-56	0869
MEZ1-62	0783
MEZ1-68	1268
MEZ1-82	0331
MEZ1-91	1311
MEZ1-110F	0149
MEZ1-115	0794
MEZ1-120	0186
MEZ1-128	0365
MEZ1-130	0213
MEZ1-140	0245
MEZ1-150	0028
MEZ1-160	0255
MEZ1-162	0420
MEZ1-170	0877
MEZ1-178	0495
MEZ1-180	0367
MEZ1-190	2858
MEZ1-200	0421
MEZ1-200G	0421
MEZ5.6T5	0157
MEZ5.6T10	0157
MEZ12T5	0137
MEZ12T10	0137
MEZ15T5	0002
MEZ-15T5A	0002
MEZ15T10	0002
MEZ15T10C	0002
MEZ27T5	0441
MEZ27T5A	0441
MEZ27T10	0441
MEZ27T10A	0441
MF03104F	OEM
MF0C600	OEM
MF0D100	OEM
MF0D102F	OEM
MF0D104F	OEM
MF0D200	OEM
MF0D202F	OEM
MF0D300	OEM
MF0D302F	OEM
MF0D402F	OEM
MF0D404F	OEM
MF0D405F	OEM
MF0D624F	OEM
MF0D2202	OEM
MF0D2302	OEM
MF0D2404	OEM
MF0D2405	OEM
MF0E100	OEM
MF0E102F	OEM
MF0E103F	OEM
MF0E106F	OEM
MF0E107F	OEM
MF0E108F	OEM
MF0E200	OEM
MF0E1200	OEM
MF0L02R	OEM
MF0L02T	OEM
MF2/1B	OEM
MF2A40F	OEM
MF2A100F	OEM
MF2A103F	OEM
MF2A200F	OEM
MF2A203F	OEM
MF2A403F	OEM
MF2A600F	OEM
MF2A603F	OEM
MF2A800	OEM
MF2B100F	OEM
MF2B103F	OEM
MF2B200F	OEM
MF2B203F	OEM
MF2B400F	OEM
MF2B403F	OEM
MF2B600F	OEM
MF2B603F	OEM
MF2B800	OEM
MF2B800F	OEM
MF3CN1	OEM
MF3CN3	OEM
MF3M1	OEM
MF3M3	OEM
MF3N6	OEM
MF3N12	OEM
MF5CN1	OEM
MF5CN3	OEM
MF5M1	OEM
MF5M3	OEM
MF5N6	OEM
MF5N12	OEM
MF7N6	OEM
MF7N12	OEM
MF9N6	OEM
MF9N12	OEM
MF10	OEM
MF10-100	OEM
MF10-200	OEM
MF10A103A	OEM
MF10A203A	OEM
MF10A403A	OEM
MF10A603A	OEM
MF10B103A	OEM
MF10B203A	OEM
MF10B403A	OEM
MF10B603A	OEM
MF10B803A	OEM
MF12/1B	0201
MF12/16	0201
MF-15/1B	0201
MF15/1B	0288
MF20/1B	0288
MF20/16	0288
MF25/1B	0757
MF25/16	0378
MF-30/1B	0378
MF30/1B	0378
MF-30/16	0378
MF35-600	OEM
MF35-800	OEM
MF35-1000	OEM
MF35-1200	OEM
MF-55-62	0160
MF71-100	OEM
MF71-200	OEM
MF71-600	OEM
MF71-800	OEM
MF71-1000	OEM
MF85	OEM
MF85MATH11A	OEM
MF85MATH12	OEM
MF100	OEM
MF101	OEM

If replacement code is OEM, contact original manufacturer for replacement.

DEVICE TYPE	REPL CODE
MF201-02	OEM
MF201-03	OEM
MF201-04	OEM
MF201-05	OEM
MF201-06	OEM
MF201-07	OEM
MF201-08	OEM
MF201-10	OEM
MF201-11	OEM
MF201-12	OEM
MF433	OEM
MF434	OEM
MF435	OEM
MF521	0325
MF1161	0127
MF1162	0127
MF1163	0127
MF1164	0127
MF1237SB	OEM
MF2147	OEM
MF2147-3	OEM
MF2148	OEM
MF2148-3	OEM
MF3304	0037
MF4501B-F	OEM
MFC01	OEM
MFC04	OEM
MFC05D240M	OEM
MFC05D300M	OEM
MFC05D390M	OEM
MFC05D510M	OEM
MFC05D680M	OEM
MFC08D121M	OEM
MFC08D150M	OEM
MFC08D200M	OEM
MFC08D201K	OEM
MFC08D240M	OEM
MFC08D241K	OEM
MFC08D281K	OEM
MFC08D300M	OEM
MFC08D331K	OEM
MFC08D390M	OEM
MFC08D391K	OEM
MFC08D431K	OEM
MFC08D510M	OEM
MFC08D511K	OEM
MFC08D611K	OEM
MFC08D680M	OEM
MFC08D820M	OEM
MFC10D201K	OEM
MFC10D241K	OEM
MFC10D281K	OEM
MFC10D331K	OEM
MFC10D391K	OEM
MFC10D431K	OEM
MFC10D511K	OEM
MFC10D611K	OEM
MFC14D201K	OEM
MFC14D241K	OEM
MFC14D281K	OEM
MFC14D331K	OEM
MFC14D391K	OEM
MFC14D431K	OEM
MFC14D511K	OEM
MFC14D611K	OEM
MFC4000	0517
MFC4000A	0517
MFC4000B	0517
MFC4010	3779
MFC4010A	3779
MFC4030	0723
MFC4040	0723
MFC4050	2027
MFC4052	2027
MFC4060	1186
MFC4060A	1186
MFC4062	1186
MFC4062A	1186
MFC4063	3931
MFC4063A	3931
MFC4064	1671
MFC4064A	1671
MFC6010	2100
MFC6020	1876
MFC6030	OEM
MFC6030A	OEM
MFC6032	2282
MFC6032A	2282
MFC6033	3935
MFC6033A	3935
MFC6034	3935
MFC6034A	3935
MFC6040	1899
MFC6050	0623
MFC6060	3940
MFC6080	2906
MFC8020	1980
MFC8021	1980
MFC8021A	1980
MFC8030	1684
MFC8070	3944
MFCH04D200	OEM
MFCH04D300	OEM
MFCH04D510	OEM
MFCH06D200	OEM
MFCH06D300	OEM
MFCH06D510	OEM
MFCH125A200	OEM
MFE16-50	OEM
MFE16-100	OEM
MFE16-150	OEM
MFE41-50	OEM
MFE41-100	OEM
MFE41-150	OEM
MFE71-50	OEM
MFE71-100	OEM
MFE71-150	OEM
MFE120	OEM
MFE121	0212
MFE122	OEM
MFE130	0212
MFE130-712	0212
MFE131	0212
MFE132	0212
MFE140	OEM
MFE521	OEM
MFE590	0212
MFE591	0212
MFE823	OEM
MFE825	OEM
MFE910	OEM
MFE930	OEM
MFE960	OEM
MFE990	OEM
MFE2000	0325
MFE2001	0325
MFE2004	1155
MFE2005	1155
MFE2006	1155
MFE2007	1155
MFE2008	OEM
MFE2009	OEM
MFE2010	OEM
MFE2011	OEM
MFE2012	OEM
MFE2093	0325
MFE2094	0325
MFE2095	0325
MFE2097	OEM
MFE2098	1155
MFE2133	OEM
MFE3001	4214
MFE3002	0849
MFE3003	OEM
MFE3004	0849
MFE3005	0353
MFE3006	0212
MFE3007	0212
MFE-3008	0212
MFE3008	0212
MFE3020	OEM
MFE3021	OEM
MFE3954	OEM
MFE3954A	OEM
MFE3955	OEM
MFE3955A	OEM
MFE3956	OEM
MFE3957	OEM
MFE3958	OEM
MFE4007	3384
MFE4008	0325
MFE4009	0325
MFE4010	0325
MFE4011	0325
MFE4012	0325
MFE4150	OEM
MFE6659	OEM
MFE6660	OEM
MFE6661	OEM
MFE9200	OEM
MFEC2010	OEM
MFEC3002	OEM
MFEC3003	OEM
MFJ1033S	0325
MFOD100	OEM
MFOD102F	OEM
MFOD200	OEM
MFOD202F	OEM
MFOD300	OEM
MFOD302F	OEM
MFOD402F	OEM
MFOD404F	OEM
MFOD405F	OEM
MFOE71&MFOD72	OEM
MFOE100	OEM
MFOE102F	OEM
MFOE103F	OEM
MFOE106F	OEM
MFOE200	OEM
MFP121	0212
MFPT100	OEM
MFPT100A	OEM
MFPT100B	OEM
MFQ930C	OEM
MFQ960C	OEM
MFQ990C	OEM
MFR221	OEM
MFR222	OEM
MFR224	OEM
MG15G1AL2	OEM
MG21D	OEM
MG30G1BL2	OEM
MG30G2CL2	OEM
MG31C	OEM
MG31D	OEM
MG31T	OEM
MG31W	OEM
MG32D	OEM
MG50G1BL2	OEM
MG50G2CL2	OEM
MG51C	OEM
MG51D	OEM
MG51T	OEM
MG51W	OEM
MG52C	OEM
MG52D	OEM
MG52T	OEM
MG52W	OEM
MG60T1	OEM
MG61D	OEM
MG62D	OEM
MG71D	OEM
MG81D	OEM
MG91T2	OEM
MG100G1AL2	OEM
MG131A	OEM
MG131AF	OEM
MG131C	OEM
MG131CF	OEM
MG132A	OEM
MG132C	OEM
MG133A	OEM
MG133C	OEM
MG134A	OEM
MG134C	OEM
MG141A	OEM
MG141C	OEM
MG143A	OEM
MG143C	OEM
MG151A	OEM
MG151C	OEM
MG200-1	OEM
MG200-2	OEM
MG241A	OEM
MG241C	OEM
MG251A	OEM
MG251C	OEM
MG5200	OEM
MG5201	OEM
MG5213	OEM
MG5222	OEM
MG5223	OEM
MG5239	OEM
MG5240	OEM
MG7712	OEM
MGA100	OEM
MGA300	OEM
MGA500	OEM
MGA600	OEM
MGA700	OEM
MGB51C	OEM
MGB51D	OEM
MGB51T	OEM
MGB51W	OEM
MGB52C	OEM
MGB52D	OEM
MGB52T	OEM
MGB52W	OEM
MGB61D	OEM
MGB71D	OEM
MGB81D	OEM
MGC31D	OEM
MGCS821-1	OEM
MGCS821-2	OEM
MGCS821-3	OEM
MGCS821-4	OEM
MGCS821-5	OEM
MGCS821-6	OEM
MGCS924-1	OEM
MGCS924-2	OEM
MGCS924-3	OEM
MGCS924-4	OEM
MGCS924-5	OEM
MGCS924-6	OEM
MGCS925-2	OEM
MGCS925-3	OEM
MGCS925-4	OEM
MGCS925-5	OEM
MGCS925-6	OEM
MGD25	OEM
MGD47	OEM
MGD50	OEM
MGD72	0124
MGD73	OEM
MGD91	OEM
MGF1202	OEM
MGF1400	OEM
MGF1402	OEM
MGF1403	OEM
MGF1404	OEM
MGF1412	OEM
MGF1801	OEM
MGF1802	OEM
MGF2116	OEM
MGF2117	OEM
MGF2124	OEM
MGF2124F	OEM
MGF2124G	OEM
MGF2148	OEM
MGF2148F	OEM
MGF2148G	OEM
MGF2172	OEM
MGF2205	OEM
MGF-X34M	OEM
MGKD72	OEM
MGLA8.2	0244
MGLA8.2A	0244
MGLA8.2B	0244
MGLA28	OEM
MGLA28A	OEM
MGLA28B	OEM
MGLA31	OEM
MGLA31A	OEM
MGLA31B	OEM
MGLA35	OEM
MGLA35A	OEM
MGLA35B	OEM
MGLA39	0188
MGLA39A	0456
MGLA39B	0188
MGLA43	0275
MGLA43A	0275
MGLA43B	0533
MGLA47	0162
MGLA47A	0451
MGLA47B	0162
MGLA51	0162
MGLA51A	0162
MGLA51B	0162
MGLA56	0157
MGLA56A	0157
MGLA56B	0157
MGLA62	0636
MGLA62A	0636
MGLA62B	0636
MGLA68	OEM
MGLA68A	OEM
MGLA68B	0025
MGLA75	OEM
MGLA75A	OEM
MGLA75B	OEM
MGLA82	0244
MGLA82A	0244
MGLA82B	0244
MGLA91	0012
MGLA91A	0012
MGLA91B	0012
MGLA100	0170
MGLA100A	0170
MGLA100B	0170
MGM5N45	OEM
MGM5N50	OEM
MGM20N45	OEM
MGM20N50	OEM
MGP00	OEM
MGP5N45	OEM
MGP5N50	OEM
MGP10N10	OEM
MGP20N45	OEM
MGP20N50	OEM
MGT108A	OEM
MGT108B	OEM
MGT108D	OEM
MGT108G	OEM
MGT108V	OEM
MH0007CH	OEM
MH0007H	OEM
MH0009CG	OEM
MH0009G	OEM
MH0012CG	OEM
MH0012G	OEM
MH0013CG	OEM
MH0013G	OEM
MH0810	OEM
MH0816	OEM
MH0818	OEM
MH0870	OEM
MH2F	OEM
MH4F	OEM
MH-5-015-D	0229
MH6F	OEM
MH8F	OEM
MH10F	OEM
MH12F	OEM
MH14F	OEM
MH16F	OEM
MH18F	OEM
MH20F	OEM
MH-26	3684
MH60	OEM
MH63	OEM
MH65	OEM
MH67	0015
MH68	0015
MH69	2645
MH70	0015
MH71	0015
MH72	0071
MH73	0087
MH100	OEM
MH104A	0161
MH110	OEM
MH111	OEM
MH-163	OEM
MH-166	OEM
MH-167	OEM
MH214	OEM
MH320	OEM
MH321F	OEM
MH321N	OEM
MH323	OEM
MH353	1997
MH353A	1997
MH353A01	1997
MH354	0190
MH383-B01	0190
MH383B01	0190
MH401	0348
MH500	0015
MH670	0015
MH680	0015
MH700	0015
MH710	0015
MH720	0071
MH730	0071
MH745	1285
MH746	1735
MH913	1501
MH913-5	OEM
MH913A01	1501
MH914	0190
MH914A01	0190
MH915	1067
MH915A01	0409
MH915A09	0409
MH915A12	2804
MH915A13	2804
MH915C	2804
MH915C01	0409
MH915C02	2804
MH915C04	2804
MH915G01	0409
MH915G02	2804
MH919	1501
MH919A01	1501
MH919D01	1501
MH920	0190
MH920A01	0190
MH920A06	0190
MH920A07	0190
MH920A10	3858
MH920G01	1501
MH931	1501
MH931A01	1501
MH931A04	1501
MH931A07	1501
MH932	0190
MH932A01	0190
MH943	2982
MH943A01	2982
MH955A01	1501
MH955B01	1501
MH966A01	0190
MH970A02	1710
MH970C01	1197
MH970C02	1710
MH970G01	1197
MH970G02	1710
MH-983	1501
MH983A01	1501
MH983A02	1501
MH983A03	1501
MH983A04	1501
MH983D	1501
MH983D-01	1501
MH983D-02	1501
MH983D-03	1501
MH983D-04	1501
MH983G01	1501
MH984	0409
MH984A01	0409
MH984C02	2804
MH984D	0409
MH984D-01	0409
MH-985	2982
MH985	2982
MH985A	2982
MH985A01	2982
MH985D-01	2982
MH985D-02	2982
MH985G01	2982
MH-987	0190
MH987	0190
MH987A01	0190
MH987A02	0190
MH987A03	0190
MH987A04	3858
MH987D-01	0190
MH987D-02	0190
MH987D-03	0190
MH987D-04	0190
MH987G01	0190
MH987G02	OEM
MH988A03	1997
MH988B03	1997
MH988G02	1997
MH1001	2984
MH1002	1501
MH1002A01	1501
MH1011	2984
MH1011B01	1054
MH1014	2984
MH1030	2984
MH1030A01	2984
MH1030A02	1054
MH1030A03	1054
MH1030B01	2984
MH1030D-01	2984
MH1030D-02	2984
MH1030D-03	2984
MH1030G01	2984
MH1030G02	1054
MH1031C01	1054
MH1200	OEM
MH1200G01	3655
MH1201	2546
MH1201A01	2546
MH1201A03	2546
MH1201C01	2546
MH1203	1067
MH1203A01	1067
MH1203A02	1067
MH1203B01	1067
MH1203B02	1067
MH1203B03	1067
MH1203B10	1067
MH1203B13	1067
MH1204	1067
MH1204A01	1067
MH1205	1067
MH1205A01	1067
MH1206	3655
MH1206A01	3655
MH1207	3655
MH1207A01	3655
MH1209	3655
MH1209A01	3655
MH1209H02	1054
MH-1220	3655
MH1220	3655
MH1220A01	3655
MH1220G01	3655
MH-1221	1067
MH1221	1067
MH1221A	1706
MH1221A01	1067
MH1221A02	1706
MH1221D-01	1067
MH1221D-02	1067
MH1221E02	1706
MH1221G01	OEM
MH1221G02	OEM
MH-1222	2546
MH1222	2546
MH1222A01	2546
MH1222A02	2546
MH1222A03	2985
MH1222D-01	2546
MH1222D-02	2546
MH1222D-03	2546
MH1222D-04	2546
MH1222D-05	2546
MH1222D-06	2546
MH1222G01	2546
MH1222G03	2985
MH1501	0595
MH1502	0791
MH7301	OEM
MH7302	OEM
MH7303	OEM
MH8100	OEM
MH8106	OEM
MH8108	OEM
MH8111	0161
MH8112	0161
MH8113	0161
MH8121	0161
MH8122	0161
MH8123	0161
MH8131	0161
MH8132	0161
MH8133	0161
MH8134	0161
MH8136	0161
MH8500	OEM
MH8700	OEM
MH9830A01	1501
MH987001	0190
MHD611	OEM
MHD612	OEM
MHD613	OEM
MHD614	OEM
MHD615	OEM
MHD616	OEM
MHD617	OEM
MHD618	OEM
MHD619	0124
MHF15-05D03	OEM
MHM1001	0144
MHM1101	0144
MHM1201	OEM
MHM2001	OEM
MHM2011	OEM
MHM2012	OEM
MHM2013	OEM
MHM2014	OEM
MHM2015	OEM
MHM2016	OEM
MHM2017	OEM
MHM2021	OEM
MHM2101	OEM
MHM2111	OEM
MHM2112	OEM
MHM2113	OEM
MHM2114	OEM
MHM2115	OEM
MHM2116	OEM
MHM2117	OEM
MHM2201	OEM
MHM2211	OEM
MHM2212	OEM
MHM2213	OEM
MHM2214	OEM
MHM2215	OEM
MHM2216	OEM
MHM2217	OEM
MHO15-02D03	OEM
MHO15-04D03	OEM
MHO15-05D03	OEM
MHO15-06D03	OEM
MHO811	0460
MHO812	0460
MHO813	0460
MHO821	0460
MHO822	0460
MHO823	0460
MHO831	0460
MHO832	0460
MHO833	0460
MHO834	0460
MH-OKOA	1997
MHQ918	OEM
MHQ2221	OEM
MHQ2222	0544
MHQ2369	OEM
MHQ2483	OEM
MHQ2484	OEM
MHQ2906	OEM
MHQ2907	OEM
MHQ3467	OEM
MHQ3546	OEM
MHQ3798	OEM
MHQ3799	OEM
MHQ4001A	OEM
MHQ4002A	OEM
MHQ4013	OEM
MHQ4014	3210
MHQ6001	OEM
MHQ6002	OEM
MHQ6100	OEM
MHQ6842	OEM
MHR	OEM
MHT180	0440
MHT181	0440
MHT230	0440
MHT-1802	0440
MHT1802	0440
MHT-1803	0440
MHT1803	0440
MHT-1804	0440
MHT1804	0440
MHT1807	0440
MHT1808	0440
MHT1809	0440
MHT2002	0600
MHT2003	0600
MHT2004	0600
MHT2008	0600
MHT2009	0600
MHT2010	0600
MHT2305	0440
MHT2414	0086
MHT2418	0086
MHT4401	0086
MHT4402	0233
MHT4411	0086
MHT4412	0086
MHT4413	0086
MHT4414	0168
MHT4415	0168
MHT4416	0168
MHT4417	0168
MHT4418	0168
MHT4419	0168
MHT4451	0086
MHT4452	0264
MHT4453	0285
MHT4454	0285
MHT4483	0086
MHT4511	0086
MHT4512	0086
MHT4513	0086
MHT5001	0285
MHT5002	0285
MHT5003	0285
MHT5004	0285
MHT5005	0285
MHT5501	0630
MHT5506	0630
MHT5507	0161
MHT5508	0161
MHT5511	0630
MHT5901	0178
MHT5906	0042
MHT5911	0178
MHT6901	OEM
MHT6902	OEM
MHT6903	OEM
MHT6904	OEM
MHT6905	OEM
MHT6906	OEM
MHT6907	OEM
MHT6908	OEM
MHT7206	0074
MHT7207	0074
MHT7208	0074
MHT7209	0642
MHT7401	0086
MHT7411	0086
MHT7412	0086
MHT7413	0285
MHT7414	0086
MHT7415	0285

If replacement code is OEM, contact original manufacturer for replacement.

DEVICE TYPE	REPL CODE
MHT7416	0285
MHT7417	0086
MHT7601	0103
MHT7602	0103
MHT7603	0103
MHT7607	0103
MHT7608	0103
MHT7609	0103
MHT8920	OEM
MHT8921	OEM
MHT8922	OEM
MHT8923	OEM
MHT9001	0086
MHT9002	0086
MHT9003	0264
MHT9004	0086
MHT9005	0086
MHT9006	0264
MHT9007	0630
MHT9008	0285
MHT9009	0285
MHT9010	0630
MHT18010	0440
MHV1.5	OEM
MHV2	OEM
MHV2.5	OEM
MHV3	OEM
MHV3.5	OEM
MHV4	OEM
MHV5	OEM
MHV6	OEM
MHV7	OEM
MHV8	OEM
MHV9	OEM
MHV10	OEM
MHVG10	OEM
MHVG20	OEM
MHVG30	OEM
MHVG40	OEM
MHVG50	OEM
MHVG60	OEM
MHVS10	OEM
MHVS20	OEM
MHVS30	OEM
MHVS40	OEM
MHVS50	OEM
MHVS60	OEM
MHW252	OEM
MHW401-1	OEM
MHW401-2	OEM
MHW401-3	OEM
MHW590	OEM
MHW591	OEM
MHW592	OEM
MHW593	OEM
MHW612	OEM
MHW612A	OEM
MHW613A	OEM
MHW709-1	OEM
MHW709-2	OEM
MHW709-3	OEM
MHW710-1	OEM
MHW710-2	OEM
MHW710-3	OEM
MHW720-1	OEM
MHW720-2	OEM
MHW1121	OEM
MHW1122	OEM
MHW1134	OEM
MHW1171	OEM
MHW1171R	OEM
MHW1172	OEM
MHW1172R	OEM
MHW1182	OEM
MHW1184	OEM
MHW1221	OEM
MHW1222	OEM
MHW1224	OEM
MHW1244	OEM
MHW1341	OEM
MHW1342	OEM
MHW1343	OEM
MHW1344	OEM
MHW2172	OEM
MHW4171	OEM
MHW4172	OEM
MHW5122	OEM
MHW5171	OEM
MHW5172	OEM
MHW5222	OEM
MHW5342	OEM
MHW6171	OEM
MHW6172	OEM
MHZ016	OEM
MHZ018	OEM
MHZ018Y	OEM
MI12	OEM
MI13	OEM
MI14	OEM
MI-15	1802
MI15	OEM
MI-15MR	1097
MI-15MS	1802
MI-15R	1097
MI15RC	0015
MI-15S	1802
MI15S	0015
MI15SC	0015
MI16	OEM

DEVICE TYPE	REPL CODE
MI22	OEM
MI23	OEM
MI24	OEM
MI25	OEM
MI31T	OEM
MI51	OEM
MI51T	OEM
MI52	OEM
MI53	OEM
MI54	OEM
MI89	OEM
MI90	OEM
MI95	OEM
MI-101	0196
MI101	OEM
MI-101R	1097
MI101R	OEM
MI-102	1802
MI102	OEM
MI-102R	1097
MI102R	OEM
MI-104	0199
MI104	OEM
MI-104R	1015
MI104R	OEM
MI105	OEM
MI-106	0199
MI106	OEM
MI-106R	1015
MI106R	OEM
MI120	OEM
MI137	OEM
MI-151	0199
MI151	0196
MI-151R	1015
MI151R	1080
MI-152	0199
MI152	0199
MI-152A	0199
MI152A	0071
MI-152DR	3501
MI-152DS	0199
MI-152R	1015
MI152R	1097
MI-152RA	1015
MI152RA	0071
MI-154	0199
MI-154R	1015
MI154R	OEM
MI-156	0199
MI-156R	1015
MI156R	OEM
MI158-1	OEM
MI202	OEM
MI204	OEM
MI206	OEM
MI268	OEM
MI-301	0133
MI301	0133
MI303	OEM
MI308	OEM
MI402	0023
MI407	OEM
MI501M	OEM
MI505	OEM
MI507	OEM
MI508	OEM
MI509A	OEM
MI509B	OEM
MI509D	OEM
MI509E	OEM
MI509G	OEM
MI509V	OEM
MI521A2	OEM
MI521B2	OEM
MI521D2	OEM
MI521G2	OEM
MI521V2	OEM
MI522	OEM
MI588AM	OEM
MI589A	OEM
MI589B	OEM
MI589V	OEM
MI1546	0144
MI7001	OEM
MI7002	OEM
MI7022	OEM
MIC33T	OEM
MIC33TSA	OEM
MIC33TSB	OEM
MIC33TSC	OEM
MIC33TSD	OEM
MIC33TSE	OEM
MIC723-1	1192
MIC741-5C	0410
MIC930-5D	1823
MIC930-5P	1823
MIC931-5D	3061
MIC931-5P	3061
MIC932-5D	1041
MIC932-5P	1041
MIC933-5D	2097
MIC933-5P	2097
MIC935-5D	1176
MIC935-5P	1176
MIC936-5D	1831
MIC936-5P	1831
MIC937-5D	1835
MIC937-5P	1835

DEVICE TYPE	REPL CODE
MIC941-5D	2622
MIC941-5P	2622
MIC944-5D	0033
MIC944-5P	0033
MIC945-5D	0081
MIC945-5P	0081
MIC946-5D	0141
MIC946-5P	0141
MIC948-5D	3399
MIC948-5P	3399
MIC949-5D	1844
MIC949-5P	1844
MIC950-5D	2619
MIC950-5P	2619
MIC951-5D	2078
MIC951-5P	2078
MIC961-5D	1859
MIC961-5P	1859
MIC962-5D	0562
MIC962-5P	0562
MIC963-5D	0341
MIC963-5P	0341
MIC7400J	0232
MIC7400N	0232
MIC7401J	0268
MIC7401N	0268
MIC7402J	0314
MIC7402N	0314
MIC7403J	0335
MIC7403N	0335
MIC7404J	0361
MIC7404N	0361
MIC7405J	0385
MIC7405N	0385
MIC7410J	0512
MIC7410N	0512
MIC7413J	1441
MIC7413N	1441
MIC7420J	0697
MIC7420N	0697
MIC7426J	0803
MIC7426N	0803
MIC7428J	4180
MIC7428N	4180
MIC7430J	0873
MIC7430N	0873
MIC7440J	1024
MIC7440N	1024
MIC7441AJ	1038
MIC7441AN	1038
MIC7442J	1052
MIC7442N	1052
MIC7443J	1060
MIC7443N	1060
MIC7444J	1072
MIC7444N	1072
MIC7445J	1082
MIC7445N	1082
MIC7450J	0743
MIC7450N	0743
MIC7451J	1168
MIC7451N	1168
MIC7453J	1185
MIC7453N	1185
MIC7454J	1202
MIC7454N	1202
MIC7460J	1275
MIC7460N	1275
MIC7470J	1404
MIC7470N	1404
MIC7472J	1426
MIC7472N	1426
MIC7473J	1172
MIC7473N	1172
MIC7474J	1313
MIC7474N	1313
MIC7475J	1432
MIC7475N	1432
MIC7476J	1158
MIC7476N	1158
MIC7481J	3121
MIC7481N	3121
MIC7482J	1574
MIC7482N	1574
MIC7483J	0117
MIC7483N	0117
MIC7486J	1368
MIC7486N	1368
MIC7490J	1208
MIC7490N	1208
MIC7491AJ	0980
MIC7491AN	0980
MIC7492J	0834
MIC7492N	0834
MIC7493J	0569
MIC7493N	0569
MIC7494J	1702
MIC7494N	1702
MIC7495J	1486
MIC7495N	1486
MIC7496J	1715
MIC9093-5D	0358
MIC9093-5P	0358
MIC9094-5D	1632
MIC9094-5P	1632
MIC9097-5D	1481
MIC9097-5P	1481
MIC9099-5D	0333
MIC9099-5P	0333

DEVICE TYPE	REPL CODE
MIC74107J	0942
MIC74107N	0942
MIC74121J	0175
MIC74121N	0175
MIC74145J	0619
MIC74145N	0619
MIC74150J	1492
MIC74150N	1492
MIC74151J	1495
MIC74151N	1495
MIC74154J	1555
MIC74154N	1555
MIC74155J	1576
MIC74155N	1576
MIC74156J	1592
MIC74156N	1592
MIC74180J	1829
MIC74180N	1829
MIC77413J	1441
MID241	OEM
MID400	OEM
MII66003-001	OEM
MII66003-002	OEM
MII66003-003	OEM
MII66004-001	OEM
MII66004-002	OEM
MII66005-001	OEM
MII66005-002	OEM
MIKUL500	OEM
MIKUL509	OEM
MIKUL514M	OEM
MIKUL519	OEM
MIKUL576	OEM
MIKUL580	OEM
MIKUL598-4	OEM
MIKUL598-6	OEM
MIKUL598-8	OEM
MIKUL900-4	OEM
MIKUL900-6	OEM
MIKUL902	OEM
MIKUL991	OEM
MIKUL993	OEM
MIKUL995	OEM
MIKUL996	OEM
MIKUL997	OEM
MIKUL998	OEM
MIKUL6001	OEM
MIKUL6004	OEM
MIKUL6009-5	OEM
MIKUL6009-6	OEM
MIKUL6009-10	OEM
MIKUL6009-12	OEM
MIKUL6017	OEM
MIKUL6019	OEM
MIKUL6032	OEM
MIKUL6045	OEM
MIKUL6047	OEM
MIKUL6077	OEM
MIKUL6809-2	OEM
MIKUL6809-3	OEM
MIKUL6882	OEM
MIKUL60640	OEM
MIKUL60641	OEM
MINT-01-01	OEM
MINT-01-02	OEM
MIOV42092-056	OEM
MIOV42092-066	OEM
MIOV42092-076	OEM
MIOV42092-086	OEM
MIOV42092-096	OEM
MIOV42092-106	OEM
MIOV42092-126	OEM
MIOV42092-146	OEM
MIOV42092-166	OEM
MIOV42092-186	OEM
MIOV42092-206	OEM
MIOV42092-510	OEM
MIOV42092-515	OEM
MIOV42092-610	OEM
MIOV42092-615	OEM
MIOV42092-710	OEM
MIOV42092-715	OEM
MIOV42092-810	OEM
MIOV42092-815	OEM
MIOV42092-910	OEM
MIOV42092-915	OEM
MIOV42092-1010	OEM
MIOV42092-1015	OEM
MIOV42092-1210	OEM
MIOV42092-1215	OEM
MIOV42092-1410	OEM
MIOV42092-1415	OEM
MIOV42092-1610	OEM
MIOV42092-1615	OEM
MIOV42092-1810	OEM
MIOV42092-1815	OEM
MIOV42092-2010	OEM
MIOV42092-2015	OEM
MIOV42093-0520	OEM
MIOV42093-0525	OEM
MIOV42093-0530	OEM
MIOV42093-0535	OEM
MIOV42093-0620	OEM
MIOV42093-0625	OEM
MIOV42093-0630	OEM
MIOV42093-0635	OEM
MIOV42093-0720	OEM
MIOV42093-0725	OEM
MIOV42093-0730	OEM

DEVICE TYPE	REPL CODE
MIOV42093-0735	OEM
MIOV42093-0820	OEM
MIOV42093-0825	OEM
MIOV42093-0830	OEM
MIOV42093-0835	OEM
MIOV42093-0920	OEM
MIOV42093-0925	OEM
MIOV42093-0930	OEM
MIOV42093-0935	OEM
MIOV42093-1020	OEM
MIOV42093-1025	OEM
MIOV42093-1030	OEM
MIOV42093-1035	OEM
MIOV42093-1220	OEM
MIOV42093-1225	OEM
MIOV42093-1230	OEM
MIOV42093-1235	OEM
MIOV42093-1420	OEM
MIOV42093-1425	OEM
MIOV42093-1430	OEM
MIOV42093-1435	OEM
MIOV42093-1620	OEM
MIOV42093-1625	OEM
MIOV42093-1630	OEM
MIOV42093-1635	OEM
MIOV42093-1820	OEM
MIOV42093-1825	OEM
MIOV42093-1830	OEM
MIOV42093-1835	OEM
MIOV42093-2020	OEM
MIOV42093-2025	OEM
MIOV42093-2030	OEM
MIOV42093-2035	OEM
MIS-181C1-4	1041
MIS-18101-3	1823
MIS-18101-5	0081
MIV-843	OEM
MIV42050-284	OEM
MIVR42050-055	OEM
MIVR42050-109	OEM
MIVR42050-128	OEM
MIVR42050-148	OEM
MIVR42050-158	OEM
MIVR42050-168	OEM
MIVR42050-188	OEM
MIVR42050-208	OEM
MIVR42050-224	OEM
MIVR42050-244	OEM
MIVR42050-264	OEM
MIVR42050-304	OEM
MIVR42050-324	OEM
MIVR42050-344	OEM
MIVR42050-510	OEM
MIVR42050-610	OEM
MIVR42050-710	OEM
MIVR42050-810	OEM
MIVR42050-910	OEM
MIVR42051-055	OEM
MIVR42051-065	OEM
MIVR42051-075	OEM
MIVR42051-085	OEM
MIVR42051-095	OEM
MIVR42051-105	OEM
MIVR42051-124	OEM
MIVR42051-144	OEM
MIVR42051-154	OEM
MIVR42051-164	OEM
MIVR42051-184	OEM
MIVR42051-204	OEM
MIVR42051-223	OEM
MIVR42051-243	OEM
MIVR42051-263	OEM
MIVR42051-283	OEM
MIVR42051-303	OEM
MIVR42051-323	OEM
MIVR42051-343	OEM
MIVR42052-055	OEM
MIVR42052-109	OEM
MIVR42052-128	OEM
MIVR42052-148	OEM
MIVR42052-158	OEM
MIVR42052-168	OEM
MIVR42052-188	OEM
MIVR42052-208	OEM
MIVR42052-224	OEM
MIVR42052-244	OEM
MIVR42052-264	OEM
MIVR42052-284	OEM
MIVR42052-304	OEM
MIVR42052-324	OEM
MIVR42052-344	OEM
MIVR42052-510	OEM
MIVR42052-610	OEM
MIVR42052-710	OEM
MIVR42052-810	OEM
MIVR42052-910	OEM
MIVR42055-0520	OEM
MIVR42055-0620	OEM
MIVR42055-0720	OEM
MIVR42055-0820	OEM
MIVR42055-0920	OEM
MIVR42055-1020	OEM
MIVR42055-1216	OEM
MIVR42055-1416	OEM
MIVR42055-1516	OEM
MIVR42055-1612	OEM
MIVR42055-1812	OEM
MIVR42055-2010	OEM
MIVR42055-2210	OEM

DEVICE TYPE	REPL CODE
MIVR42055-2410	OEM
MIVR42055-2608	OEM
MIVR42055-2808	OEM
MIVR42094-005	OEM
MIVR42094-012	OEM
MIVR42094-015	OEM
MIVR42094-018	OEM
MIVR42094-024	OEM
MIVR42094-030	OEM
MIVR42095-005	OEM
MIVR42095-012	OEM
MIVR42095-015	OEM
MIVR42095-018	OEM
MIVR42095-024	OEM
MIVR42095-030	OEM
MJ105	0065
MJ139A	0155
MJ205	OEM
MJ320	OEM
MJ335	OEM
MJ340	0276
MJ400	0142
MJ410	0270
MJ411	0270
MJ413	0363
MJ420	0233
MJ420S	0233
MJ-421	0187
MJ421	0187
MJ421S	0187
MJ423	1966
MJ424	0642
MJ425	0642
MJ431	0363
MJ432	0363
MJ440	0630
MJ450	1681
MJ480	0103
MJ481	0103
MJ490	2013
MJ491	2013
MJ802	0130
MJ804	OEM
MJ900	2277
MJ901	2277
MJ920	2277
MJ921	2277
MJ1000	2427
MJ1001	2427
MJ1200	2427
MJ1201	2427
MJ1800	0270
MJ2249	0429
MJ2250	0429
MJ-2251	0086
MJ2251	0142
MJ2252	0142
MJ2253	0905
MJ2254	0905
MJ2267	1598
MJ2268	1598
MJ2300	OEM
MJ2305	OEM
MJ2500	3374
MJ2500M	OEM
MJ2501	3374
MJ2800	0103
MJ2801	0103
MJ2802	0103
MJ2840	0543
MJ2841	0543
MJ2901	0491
MJ2940	1681
MJ2941	1681
MJ2955	0491
MJ2955A	4466
MJ3000	3373
MJ3001	3373
MJ-3010	0074
MJ3010	0313
MJ3011	0313
MJ3026	0074
MJ3027	0074
MJ3028	0270
MJ3029	1991
MJ3030	0363
MJ3040	2626
MJ3042	2626
MJ3101	0429
MJ3201	0142
MJ3202	0142
MJ3237	OEM
MJ3238	OEM
MJ3247	OEM
MJ3248	OEM
MJ3260	0363
MJ3430	1966
MJ3480	0065
MJ3520	3373
MJ3521	3373
MJ3583	1027
MJ3584	1027
MJ3701	0905
MJ3738	OEM
MJ3739	OEM
MJ3760	0363
MJ3761	0363
MJ3771	0130
MJ3772	0130

DEVICE TYPE	REPL CODE
MJ3773	0543
MJ4000	2427
MJ4001	2427
MJ4010	2277
MJ4011	2277
MJ4030	2448
MJ4031	2448
MJ4032	2448
MJ4033	1394
MJ4034	1394
MJ4035	1394
MJ4101	0178
MJ4102	0178
MJ4237	5639
MJ4238	5639
MJ4247	5640
MJ4248	5640
MJ4360	OEM
MJ4361	OEM
MJ4380	OEM
MJ4381	OEM
MJ4400	OEM
MJ4401	OEM
MJ4502	1681
MJ4645	0126
MJ4646	5711
MJ4648	5711
MJ5202	0178
MJ5203	0178
MJ5204	0178
MJ5257	0130
MJ5415	0439
MJ5416	0126
MJ5602	0543
MJ5603	0543
MJ6002	0543
MJ6257	0130
MJ6302	0177
MJ6502	3388
MJ6503	3388
MJ7160	0074
MJ7161	0074
MJ8100	1267
MJ8101	0622
MJ8400	0065
MJ8500	OEM
MJ8501	3486
MJ8502	3486
MJ8503	3486
MJ8504	0642
MJ8505	OEM
MJ9000	0363
MJ10000	2626
MJ10001	2626
MJ10002	1991
MJ10003	1991
MJ10004	2626
MJ10005	2626
MJ10006	1991
MJ10007	1991
MJ10008	2626
MJ10009	2626
MJ10010	OEM
MJ10011	5043
MJ10012	2626
MJ10013	OEM
MJ10014	OEM
MJ10015	4013
MJ10016	OEM
MJ10020	OEM
MJ10021	OEM
MJ10022	4013
MJ10023	4013
MJ10024	OEM
MJ10025	OEM
MJ10041	OEM
MJ10042	OEM
MJ10044	OEM
MJ10045	OEM
MJ10047	OEM
MJ10048	OEM
MJ10050	OEM
MJ10051	OEM
MJ10052	OEM
MJ10100	OEM
MJ10101	OEM
MJ10102	OEM
MJ10200	OEM
MJ10201	OEM
MJ10202	OEM
MJ11011	2420
MJ11012	2428
MJ11013	2420
MJ11014	2428
MJ11015	2420
MJ11016	2428
MJ11017	OEM
MJ11018	OEM
MJ11019	OEM
MJ11020	OEM
MJ11021	OEM
MJ11022	OEM
MJ11028	2428
MJ11029	2420
MJ11030	2428
MJ11031	2420
MJ11032	2428
MJ11033	2420
MJ12002	0313

DEVICE TYPE	REPL CODE
MJ12003	0223
MJ12004	0313
MJ12005	0223
MJ12005D	OEM
MJ12010	3387
MJ12020	5414
MJ12021	5414
MJ12022	5414
MJ13014	1966
MJ13015	3388
MJ13070	4591
MJ13071	4591
MJ13080	3388
MJ13081	3388
MJ13090	3388
MJ13091	3388
MJ13100	OEM
MJ13101	OEM
MJ13330	4386
MJ13331	4386
MJ13332	1852
MJ13333	1852
MJ13334	1852
MJ13335	1852
MJ14000	OEM
MJ14001	OEM
MJ14002	OEM
MJ14003	OEM
MJ15001	0177
MJ15002	2013
MJ15003	3701
MJ15004	4144
MJ15011	0867
MJ15012	3606
MJ15015	2414
MJ15016	3606
MJ15022	0543
MJ15023	1598
MJ15024	2414
MJ15025	3606
MJ15026	OEM
MJ15027	OEM
MJ16002	5414
MJ16004	5414
MJ16006	3486
MJ16008	3486
MJ16010	5414
MJ16012	5414
MJ16014	OEM
MJ16016	OEM
MJ212007	OEM
MJ336941	OEM
MJ338978	OEM
MJ338986	OEM
MJ338987	OEM
MJ338995	OEM
MJ373034	OEM
MJA22219	OEM
MJC574J	1329
MJC10007	OEM
MJC10009	OEM
MJC10011	OEM
MJC10012	OEM
MJE29	0423
MJE29A	0423
MJE29B	0423
MJE29C	0423
MJE30	0854
MJE30A	0854
MJE30B	0854
MJE30C	1199
MJE31	0423
MJE31A	0423
MJE31B	0423
MJE31C	0423
MJE32	0854
MJE32A	0854
MJE-32B	0854
MJE32B	0854
MJE32C	1199
MJE33	0423
MJE33A	0423
MJE33B	0423
MJE33C	0423
MJE34	0925
MJE34A	1199
MJE34B	1199
MJE34C	1199
MJE41	0423
MJE41A	0423
MJE41B	0423
MJE41C	0423
MJE-42	0854
MJE42	0854
MJE42A	0854
MJE42B	0854
MJE42C	1199
MJE-47	0392
MJE47	OEM
MJE-48	0392
MJE48	OEM
MJE-49	0168
MJE49	OEM
MJE-50	0168
MJE51T	2819
MJE52T	2819
MJE53T	2819
MJE101	0925
MJE102	0925
MJE103	0925
MJE104	1199
MJE105	1199
MJE105K	3165
MJE170	0854
MJE171	0460
MJE172	0681
MJE180	0161
MJE181	0561
MJE182	0561
MJE-200	0219
MJE200	0161
MJE-200E	0219
MJE200E	0161
MJE201	0042
MJE202	0042
MJE203	0561
MJE204	0561
MJE205	0561
MJE205K	2997
MJE210	0854
MJE-220	0161
MJE220	0161
MJE221	0161
MJE222	0161
MJE223	0161
MJE224	0161
MJE225	0161
MJE230	0460
MJE231	0460
MJE232	0460
MJE233	0460
MJE234	0460
MJE235	0460
MJE240	0060
MJE241	0060
MJE242	0060
MJE250	1308
MJE251	1308
MJE252	1308
MJE270	2551
MJE271	2555
MJE340	0276
MJE340K	0276
MJE341	0276
MJE341K	0276
MJE344	0276
MJE344D	0168
MJE344K	0276
MJE345	0285
MJE350	0525
MJE370	0460
MJE370K	3165
MJE371	0460
MJE371K	3165
MJE423	0074
MJE488	0161
MJE520	0161
MJE520K	0042
MJE-521	0161
MJE521	0161
MJE521K	2997
MJE700	2897
MJE700T	0602
MJE701	2897
MJE701T	0602
MJE702	2897
MJE702T	0602
MJE703	2897
MJE703T	0602
MJE710	0460
MJE711	0460
MJE712	1199
MJE720	0161
MJE721	0161
MJE722	0561
MJE800	0558
MJE800T	1212
MJE801	0558
MJE801T	1212
MJE802	0558
MJE802T	1212
MJE803	0558
MJE803T	1212
MJE1033G	OEM
MJE1090	3528
MJE1091	3528
MJE1092	3528
MJE1093	3528
MJE1100	3377
MJE1101	3377
MJE1102	3377
MJE1103	3377
MJE1290	1199
MJE1291	1199
MJE1305	OEM
MJE1660	0561
MJE1661	0561
MJE2010	0854
MJE-2011	0561
MJE2011	0561
MJE-2020	1199
MJE2020	1199
MJE2021	0423
MJE2050	0161
MJE2090	2237
MJE2091	0602
MJE2092	0602
MJE2093	0602
MJE2100	2235
MJE2101	2235
MJE2102	2235
MJE2103	2235
MJE2150	0460
MJE2360	0168
MJE2360T	0168
MJE2361	0168
MJE2361T	0168
MJE2370	3165
MJE2371	3165
MJE2380	0423
MJE2381	0423
MJE2382	0423
MJE2383	0423
MJE2480	2997
MJE2481	2997
MJE2482	2997
MJE2483	2997
MJE2490	3165
MJE2491	3165
MJE2520	2997
MJE2521	2997
MJE2522	2997
MJE2523	2997
MJE2801	0561
MJE2801K	0561
MJE2801T	0482
MJE2901	1199
MJE2901K	1199
MJE2901T	1369
MJE2940	0103
MJE2955	1199
MJE2955K	1199
MJE2955T	1369
MJE2995T	OEM
MJE3054	2997
MJE3055	0561
MJE3055K	0561
MJE3055T	0482
MJE3300	6591
MJE3301	6592
MJE3302	6592
MJE3310	6593
MJE3311	6593
MJE3312	6593
MJE3370	0460
MJE3371	0460
MJE3439	0847
MJE3440	0276
MJE3520	0042
MJE3521	0161
MJE3730	0392
MJE3738	0276
MJE3739	0276
MJE3740	3165
MJE3741	3165
MJE4340	3435
MJE4341	3435
MJE4342	3435
MJE4343	3435
MJE4350	3673
MJE4351	3673
MJE4352	3673
MJE4353	3673
MJE4918	1051
MJE4919	1051
MJE4920	0460
MJE4921	0161
MJE4922	0161
MJE4923	0161
MJE5170	OEM
MJE5171	OEM
MJE5172	OEM
MJE5180	OEM
MJE5181	OEM
MJE5182	OEM
MJE5190	0161
MJE5191	0161
MJE5192	0161
MJE5192J	OEM
MJE5193	0460
MJE5194	0460
MJE5195	0460
MJE5655	0276
MJE5656	0276
MJE5657	0168
MJE5740	OEM
MJE5741	OEM
MJE5742	OEM
MJE5850	5637
MJE5851	5637
MJE5852	5637
MJE5974	1199
MJE5975	1199
MJE5976	1199
MJE5977	0561
MJE5978	0561
MJE5979	0561
MJE5980	1199
MJE5981	1199
MJE5982	1199
MJE5983	0561
MJE5984	0561
MJE5985	0561
MJE6040	3530
MJE6041	3530
MJE6042	3530
MJE6043	3529
MJE6044	3529
MJE6045	3529
MJE8500	6988
MJE8501	OEM
MJE8502	OEM
MJE8503	OEM
MJE9400	0161
MJE9400-2	0161
MJE9411T	0144
MJE9730	0392
MJE9742	0276
MJE10000	OEM
MJE10001	OEM
MJE12004	OEM
MJE12007	2766
MJE13002	OEM
MJE13003	OEM
MJE13004	3014
MJE13005	0728
MJE13006	0728
MJE13007	0728
MJE13007A	2908
MJE13008	6742
MJE13009	0728
MJE13070	6748
MJE13071	6146
MJE15028	1165
MJE15029	6876
MJE15030	0392
MJE15031	0681
MJE16002	6146
MJE16002P	OEM
MJE16004	6146
MJE16004P	OEM
MJE16080	6146
MJE16081	4590
MJEC244	OEM
MJEC254	OEM
MJEC3302	OEM
MJEC3312	OEM
MJEC13009	OEM
MJEC15030	OEM
MJEC15031	OEM
MJF1033G	0325
MJF3370	0925
MJF10335	0325
MJG194	1199
MJM2229	OEM
MJM4558D	OEM
MK-10	0325
MK10	0325
MK-10-2	0325
MK10-2	0325
MK-10-E	0325
MK10-E	0325
MK38C70/20J	OEM
MK38C70/20N	OEM
MK38C70/20P	OEM
MK38P70/02R-00	OEM
MK38P70/02R-05	OEM
MK38P70/02R-10	OEM
MK38P70/02R-15	OEM
MK38P73/02R-10	OEM
MK38P73/02R-15	OEM
MK102	0325
MK104	OEM
MK-111	OEM
MK279-14	0062
MK279-27	0495
MK433	OEM
MK585	0325
MK1002L	OEM
MK1002P	OEM
MK3805N	OEM
MK3807	OEM
MK3850N3F8	OEM
MK3850N13F8	OEM
MK3850P13F8	OEM
MK3850P23F8	OEM
MK3851N	OEM
MK3851P	OEM
MK3852N	OEM
MK3852N10	OEM
MK3852P	OEM
MK3852P10	OEM
MK3853N	OEM
MK3853N10	OEM
MK3853N20	OEM
MK3853P	OEM
MK3853P10	OEM
MK3853P20	OEM
MK3854N	OEM
MK3854N10	OEM
MK3854P	OEM
MK3854P10	OEM
MK3861	OEM
MK3870/10J-05	OEM
MK3870/10J-10	OEM
MK3870/10J-15	OEM
MK3870/10N-05	OEM
MK3870/10N-10	OEM
MK3870/10N-15	OEM
MK3870/10P-05	OEM
MK3870/10P-10	OEM
MK3870/10P-15	OEM
MK3870/12J-05	OEM
MK3870/12J-10	OEM
MK3870/12J-15	OEM
MK3870/12N-05	OEM
MK3870/12N-10	OEM
MK3870/12N-15	OEM
MK3870/12P-05	OEM
MK3870/12P-10	OEM
MK3870/12P-15	OEM
MK3870/20	OEM
MK3870/20J-05	OEM
MK3870/20J-10	OEM
MK3870/20J-15	OEM
MK3870/20N-05	OEM
MK3870/20N-10	OEM
MK3870/20N-15	OEM
MK3870/20P-05	OEM
MK3870/20P-10	OEM
MK3870/20P-15	OEM
MK3870/22J-05	OEM
MK3870/22J-10	OEM
MK3870/22J-15	OEM
MK3870/22N-05	OEM
MK3870/22N-10	OEM
MK3870/22N-15	OEM
MK3870/22P-05	OEM
MK3870/22P-10	OEM
MK3870/22P-15	OEM
MK3870/30J-05	OEM
MK3870/30J-10	OEM
MK3870/30J-15	OEM
MK3870/30N-05	OEM
MK3870/30N-10	OEM
MK3870/30N-15	OEM
MK3870/30P-05	OEM
MK3870/30P-10	OEM
MK3870/30P-15	OEM
MK3870/32J-05	OEM
MK3870/32J-10	OEM
MK3870/32J-15	OEM
MK3870/32N-05	OEM
MK3870/32N-10	OEM
MK3870/32N-15	OEM
MK3870/32P-05	OEM
MK3870/32P-10	OEM
MK3870/32P-15	OEM
MK3870/40J-05	OEM
MK3870/40J-10	OEM
MK3870/40J-15	OEM
MK3870/40N-05	OEM
MK3870/40N-15	OEM
MK3870/40N-1010	OEM
MK3870/40P-05	OEM
MK3870/40P-10	OEM
MK3870/40P-15	OEM
MK3870/42J-05	OEM
MK3870/42J-10	OEM
MK3870/42J-15	OEM
MK3870/42N-00	OEM
MK3870/42N-05	OEM
MK3870/42N-10	OEM
MK3870/42N-15	OEM
MK3870/42P-05	OEM
MK3870/42P-10	OEM
MK3870N	OEM
MK3870P	OEM
MK3870SCU1J-00	OEM
MK3870SCU1J-05	OEM
MK3870SCU1J-15	OEM
MK3870SCU1N-00	OEM
MK3870SCU1N-05	OEM
MK3870SCU1N-10	OEM
MK3870SCU1N-15	OEM
MK3870SCU1P-00	OEM
MK3870SCU1P-05	OEM
MK3870SCU1P-10	OEM
MK3870SCU1P-15	OEM
MK3871N	OEM
MK3871P	OEM
MK3873/10J-05	OEM
MK3873/10J-10	OEM
MK3873/10J-15	OEM
MK3873/10P-00	OEM
MK3873/10P-05	OEM
MK3873/10P-10	OEM
MK3873/10P-15	OEM
MK3873/12J-00	OEM
MK3873/12J-0505	OEM
MK3873/12J-10	OEM
MK3873/12J-15	OEM
MK3873/12N-00	OEM
MK3873/12N-05	OEM
MK3873/12N-15	OEM
MK3873/12P-05	OEM
MK3873/12P-10	OEM
MK3873/12P-15	OEM
MK3873/20J-00	OEM
MK3873/20J-05	OEM
MK3873/20J-10	OEM
MK3873/20J-15	OEM
MK3873/20P-00	OEM
MK3873/20P-05	OEM
MK3873/20P-10	OEM
MK3873/20P-15	OEM
MK3873/22P-00	OEM
MK3873/22P-05	OEM
MK3873/22P-15	OEM
MK3873EPC1J-00	OEM
MK3873EPC1N-00	OEM
MK3873EPC1P-00	OEM
MK3873SCU20N	OEM
MK3875/22J-00	OEM
MK3875/22J-05	OEM
MK3875/22J-10	OEM
MK3875/22J-15	OEM
MK3875/22N-00	OEM
MK3875/22N-05	OEM
MK3875/22N-10	OEM
MK3875/22N-15	OEM
MK3875/22P-00	OEM
MK3875/22P-05	OEM
MK3875/22P-10	OEM
MK3875/22P-15	OEM
MK3875/42J-00	OEM
MK3875/42J-05	OEM
MK3875/42J-10	OEM
MK3875/42J-15	OEM
MK3875/42N-00	OEM
MK3875/42N-05	OEM
MK3875/42N-10	OEM
MK3875/42N-15	OEM
MK3875/42P-00	OEM
MK3875/42P-05	OEM
MK3875/42P-10	OEM
MK3875/42P-15	OEM
MK3880N-4	6229
MK3880N4Z80	OEM
MK3880NZ80	OEM
MK3880P4Z80	OEM
MK3880P10Z80	OEM
MK3880P20Z80	OEM
MK3880PZ80	OEM
MK3881N	1904
MK3881N4	1904
MK3881P	6198
MK3881P4	OEM
MK3881P10	OEM
MK3881P20	OEM
MK3882N	5339
MK3882N4	OEM
MK3882N10	OEM
MK3882P	5339
MK3882P4	OEM
MK3882P10	OEM
MK3883N	OEM
MK3883N4	OEM
MK3883P	OEM
MK3883P4	OEM
MK3884N	OEM
MK3884N4	OEM
MK3884N10	OEM
MK3884P	OEM
MK3884P4	OEM
MK3884P10	OEM
MK3885N	OEM
MK3885N10	OEM
MK3885P	OEM
MK3885P10	OEM
MK3887N	OEM
MK3887N4	OEM
MK3887N10	OEM
MK3887P	OEM
MK3887P4	OEM
MK3887P10	OEM
MK4027J-1	OEM
MK4027J-2	OEM
MK4027J-3	4974
MK4027J-4	4974
MK4027N-1	OEM
MK4027N-2	OEM
MK4027N-3	4974
MK4027N-4	4974
MK4027P-1	OEM
MK4027P-2	OEM
MK4027P-3	4974
MK4104E-3	OEM
MK4104E-4	OEM
MK4104E-5	OEM
MK4104E-6	OEM
MK4104E-34	OEM
MK4104E-35	OEM
MK4104J-3	OEM
MK4104J-4	OEM
MK4104J-5	OEM
MK4104J-6	OEM
MK4104J-33	OEM
MK4104J-34	OEM
MK4104J-35	OEM
MK4104N-3	OEM
MK4104N-4	OEM
MK4104N-5	OEM
MK4104N-6	OEM
MK4104N-33	OEM
MK4104N-34	OEM
MK4104N-35	OEM
MK4116	0522
MK4116-4	0522
MK4116E-2	0522
MK4116E-3	0522
MK4116E-4	0522
MK4116J-2	0522
MK4116J-3	0522
MK4116J-4	0522
MK4116J44GP	OEM
MK4116J-53GP	0522
MK4116N-2	0522
MK4116N-3	0522
MK4116N-3GP	0522
MK4116N-4	0522
MK4116N-44GP	0522
MK4116P-2	0522
MK4116P-3	0522
MK4116P-4	0522
MK4118	OEM
MK4118AJ-1	OEM
MK4118AJ-2	OEM
MK4118AJ-3	OEM
MK4118AJ-4	OEM
MK4118AN-1	OEM
MK4118AN-2	OEM
MK4118AN-3	OEM
MK4118AN-4	OEM
MK4118AP-1	OEM
MK4118AP-2	OEM
MK4118AP-3	OEM
MK4118AP-4	OEM
MK4167P55	OEM
MK4332D-3	4934
MK4516E-15	OEM
MK4516J-12	OEM
MK4516N-12	OEM
MK4516N-15	OEM
MK4528D-15	OEM
MK4528D-20	OEM
MK4528D-25	OEM
MK4564J15	OEM
MK4564J20	OEM
MK4564J25	OEM
MK4564N-15	2357
MK4564N-20	OEM
MK4801AJ-55	OEM
MK4801AJ-70	OEM
MK4801AJ-90	OEM
MK4801AN-55	OEM
MK4801AN-70	OEM
MK4801AN-90	OEM
MK4801AP-55	OEM
MK4801AP-70	OEM
MK4801AP-90	OEM
MK4802J-1	OEM
MK4802J-3	OEM
MK4802J-70	OEM
MK4802J-90	OEM
MK4802N-1	OEM
MK4802N-3	OEM
MK4802N-70	OEM
MK4802N-90	OEM
MK4802P-1	OEM
MK4802P-3	1898
MK4802P-70	OEM
MK4802P-90	OEM
MK4808-5	OEM
MK4816J-3	OEM
MK4816J-4	OEM
MK4816J-5	OEM
MK4816N-4	OEM
MK4816N-5	OEM
MK5002N	OEM
MK5002P	OEM
MK5005N	OEM
MK5005P	OEM
MK5007N	OEM
MK5007P	OEM
MK5009N	OEM
MK5009P	OEM
MK5037N	OEM
MK5087J	OEM
MK5087N	OEM
MK5087P	OEM
MK5089J	0570
MK5089K	0570
MK5089N	OEM
MK5089P	0570
MK5091N	OEM
MK5092N	OEM
MK5094N	OEM
MK5099N	OEM
MK5102N-5	OEM
MK5103N-5	OEM
MK5116P	OEM
MK5151J	OEM
MK5151P	OEM
MK5156J	OEM
MK5156P	OEM
MK5168N-1	OEM
MK5173N(A)	OEM
MK5175N	OEM
MK5380J	OEM
MK5380N	OEM
MK5380P	OEM
MK5389J(A)	OEM
MK5389N(A)	OEM
MK5389P(A)	OEM
MK5912J3	OEM
MK9150-1	OEM
MK9150-2	OEM
MK9350-1	OEM
MK9350-2	OEM
MK12716J77	OEM
MK12716J78	OEM
MK14116J72	OEM
MK14116J73	OEM
MK14116J74	OEM
MK14680-00	OEM
MK14802J71	OEM
MK14802J73	OEM
MK14802J790	OEM
MK14802P71	OEM
MK14802P73	OEM
MK14802P790	OEM

If replacement code is OEM, contact original manufacturer for replacement.

DEVICE TYPE	REPL CODE
MK18106-00	OEM
MK30884	OEM
MK34000	OEM
MK34000J-3	OEM
MK34000N-3	OEM
MK34000P-3	OEM
MK34073N-3	OEM
MK34073P-3	OEM
MK36000	OEM
MK36000J3	OEM
MK36000J4	OEM
MK36000J5	OEM
MK36000N3	OEM
MK36000N-4	OEM
MK36000N-5	OEM
MK36000P3	OEM
MK36000P-4	OEM
MK36000P-5	OEM
MK36906N-4	OEM
MK37000-4	OEM
MK37000J-4	OEM
MK37000J-5	OEM
MK37000N-4	OEM
MK37000N-5	OEM
MK37000P-5	OEM
MK38000N25	OEM
MK38000P25	OEM
MK38036N-25	OEM
MK38097N-21	OEM
MK41164-3GP	0522
MK50116J	OEM
MK50240N	4599
MK50240P	4599
MK50241N	4599
MK50241P	4599
MK50242N	OEM
MK50242P	OEM
MK50366N	OEM
MK50372	OEM
MK50372B	OEM
MK50372N	OEM
MK50375N	OEM
MK50395N	OEM
MK50396N	OEM
MK50397N	OEM
MK50398N	OEM
MK50399N	OEM
MK50808N	OEM
MK50808N-1	OEM
MK50808P	OEM
MK50808P-1	OEM
MK50816N	OEM
MK50816N-1	OEM
MK50816P	OEM
MK50816P-1	OEM
MK50981N	2683
MK50982N	OEM
MK50991N	OEM
MK50992N	2761
MK77650-4	OEM
MK77651-0	OEM
MK77651-4	OEM
MK77652-0	OEM
MK77653	OEM
MK77654	OEM
MK77655-0	OEM
MK77665-0	OEM
MK77666-0	OEM
MK77669-0	OEM
MK77750-0	OEM
MK77751	OEM
MK77752-0	OEM
MK77752-4	OEM
MK77753-0	OEM
MK77753-4	OEM
MK77754-0	OEM
MK77754-4	OEM
MK77755	OEM
MK77756	OEM
MK77757	OEM
MK77758	OEM
MK77759	OEM
MK77760	OEM
MK77850-0	OEM
MK77850-4	OEM
MK77851-0	OEM
MK77851-4	OEM
MK77852-0	OEM
MK77853-0	OEM
MK77853-4	OEM
MK77950-0	OEM
MK77950-4	OEM
MK77958	OEM
MK77963-0	OEM
MK77963-4	OEM
MK77967	OEM
MK77969	OEM
MK77973	OEM
MK78033	OEM
MK78035	OEM
MK78106	OEM
MK78109	OEM
MK78110	OEM
MK78122	OEM
MK78124	OEM
MK78146	OEM
MK78172-42	OEM
MK78172-56	OEM
MK78175-40	OEM

DEVICE TYPE	REPL CODE
MK78177-26	OEM
MK78182-1	OEM
MK78182-2	OEM
MK78192	OEM
MK78194	OEM
MK78198	OEM
MK78199	OEM
MK78207	OEM
MK776534	OEM
MKB1-2H49	OEM
MKB1-7H26P	OEM
MKB1-7H28P	OEM
MKB2-12H49	OEM
MKB4H38	OEM
MKB5H38	OEM
MKB5H69	OEM
MKB7H38	OEM
MKB7H69	OEM
MKB2716E88	OEM
MKB2716E90	OEM
MKB2716J-87	OEM
MKB2716J-88	OEM
MKB2716J-90	OEM
MKB4027F-84	OEM
MKB4027J-83	OEM
MKB4027J-84	OEM
MKB4104E-84(M)	OEM
MKB4104E-85	OEM
MKB4104E-86	OEM
MKB4104J-84(M)	OEM
MKB4104J-85	OEM
MKB4104J-86	OEM
MKB4104P-84(M)	OEM
MKB4104P-85	OEM
MKB4104P-86	OEM
MKB4116E-83	OEM
MKB4116E-84	OEM
MKB4116F-84	OEM
MKB4116J-82	OEM
MKB4116J-83	OEM
MKB4116J-84	OEM
MKB4116P-82	OEM
MKB4116P-83	OEM
MKB4116P-84	OEM
MKB4118AJ-82(MD)	OEM
MKB4118AJ-83(MD)	OEM
MKB4118AJ-84(MD)	OEM
MKB4118AP-82(MD)	OEM
MKB4118AP-83(MD)	OEM
MKB4118AP-84(MD)	OEM
MKB12040W	OEM
MKB12100W	OEM
MKB36000J-80(M)	OEM
MKB36000J-83(M)	OEM
MKB36000J-84(M)	OEM
MKB36000P-80	OEM
MKB36000P-83(M)	OEM
MKB36000P-84	OEM
MKB37000E-84(M)	OEM
MKB37000E-85(M)	OEM
MKB37000J-84(M)	OEM
MKB37000J-85(M)	OEM
MKB37000P-84(M)	OEM
MKB37000P-85(M)	OEM
MKB38000E-84(M)	OEM
MKB38000E-85(M)	OEM
MKB38000J-84(M)	OEM
MKB38000J-85(M)	OEM
MKB38000P-84(M)	OEM
MKB38000P-85(M)	OEM
MKK12	OEM
MKK16	OEM
MKK22	OEM
MKK25	OEM
MKK26	OEM
MKK26A	OEM
MKK29	OEM
MKK37	OEM
MKP9V120	OEM
MKP9V130	OEM
MKP9V240	OEM
MKP9V260	OEM
MKP9V270	OEM
MKT25	OEM
MKT27	OEM
MKT32	OEM
MKT33	OEM
MKT34	OEM
MKT35	OEM
MKT36	OEM
MKT37	OEM
MKT42	OEM
MKY1-4H37	OEM
MKY1-4H48	OEM
MKY1-5C38E	OEM
MKY1-5H26	OEM
MKY1-5H37	OEM
MKY1-5H38	OEM
MKY1-5H48	OEM
MKY1-5H49	OEM
MKY1-7C38E	OEM
MKY1-7H26	OEM
MKY1-7H37	OEM
MKY1-7H38	OEM
MKY1-7H39	OEM
MKY1-7H48	OEM
MKY1-7H49	OEM
ML	OEM
ML10A	OEM

DEVICE TYPE	REPL CODE
ML20A	OEM
ML25	OEM
ML25P	OEM
ML25PS	OEM
ML25P-SF3	OEM
ML25P-TS	OEM
ML30A	OEM
ML33	OEM
ML37	OEM
ML39A	OEM
ML51A	OEM
ML60A	OEM
ML78L05	5644
ML78L05A	5513
ML78L12	5518
ML78L12A	5518
ML78L15	5522
ML78L15A	5522
ML78L18A	4506
ML78L24A	4509
ML78P05	OEM
ML78P05A	OEM
ML78P12	OEM
ML78P12A	OEM
ML78P15	OEM
ML78P15A	OEM
ML78P18	OEM
ML78P18A	OEM
ML79L24A	1721
ML82A	OEM
ML100A	OEM
ML101A	OEM
ML101B	OEM
ML102A	OEM
ML102B	OEM
ML103GM	OEM
ML103HM	OEM
ML110A	OEM
ML111B	OEM
ML120	OEM
ML131B	OEM
ML132A	OEM
ML132B	OEM
ML153A	OEM
ML153B	OEM
ML154A	OEM
ML154B	OEM
ML157A	OEM
ML157B	OEM
ML201AT	0093
ML231B-DG	OEM
ML231B-DP	OEM
ML232B-DG	OEM
ML232B-DP	OEM
ML236B-DP	OEM
ML237B-DP	OEM
ML238B-DP	OEM
ML239B-DP	OEM
ML301T	0093
ML307S	2283
ML530B	OEM
ML709CT	1301
ML723CM	0026
ML723CP	0026
ML723CT	1192
ML723T	1192
ML741CS	0312
ML741CT	0410
ML747CP	2358
ML747CT	2368
ML748CS	1300
ML920DP	OEM
ML922DG	OEM
ML923	OEM
ML924	3484
ML925	OEM
ML926	OEM
ML927	OEM
ML928DP	OEM
ML929DP	OEM
ML1307P	0701
ML1458S	0360
ML2201	OEM
ML2205	OEM
ML3001	OEM
ML3101	OEM
ML3401	OEM
ML4101	OEM
ML4202	OEM
ML4204	OEM
ML4206	OEM
ML4310	OEM
ML4311	OEM
ML4312	OEM
ML4313	OEM
ML4314	OEM
ML4315	OEM
ML4316	OEM
ML4317	OEM
ML4318	OEM
ML4319	OEM
ML4331	OEM
ML4332	OEM
ML4333	OEM
ML4334	OEM
ML4335	OEM
ML4336	OEM
ML4337	OEM
ML4338	OEM

DEVICE TYPE	REPL CODE
ML4339	OEM
ML4340	OEM
ML4341	OEM
ML4342	OEM
ML4343	OEM
ML4351	OEM
ML4352	OEM
ML4353	OEM
ML4354	OEM
ML4355	OEM
ML4356	OEM
ML4357	OEM
ML4358	OEM
ML4359	OEM
ML4360	OEM
ML4361	OEM
ML4362	OEM
ML4363	OEM
ML4364	OEM
ML4365	OEM
ML4372	OEM
ML4373	OEM
ML4374	OEM
ML4375	OEM
ML4376	OEM
ML4377	OEM
ML4378	OEM
ML4379	OEM
ML4380	OEM
ML4381	OEM
ML4382	OEM
ML4383	OEM
ML4384	OEM
ML4385	OEM
ML4401	OEM
ML4402	OEM
ML4404	OEM
ML4405	OEM
ML4406	OEM
ML4407	OEM
ML4408	OEM
ML4409	OEM
ML4602	OEM
ML4603	OEM
ML4604	OEM
ML4605	OEM
ML4610	OEM
ML4611	OEM
ML4612	OEM
ML4613	OEM
ML4614	OEM
ML4615	OEM
ML4616	OEM
ML4617	OEM
ML4618	OEM
ML4619	OEM
ML4620	OEM
ML4621	OEM
ML4622	OEM
ML4623	OEM
ML4624	OEM
ML4625	OEM
ML4626	OEM
ML4627	OEM
ML4628	OEM
ML4629	OEM
ML4630	OEM
ML4631	OEM
ML4640	OEM
ML4641	OEM
ML4642	OEM
ML4643	OEM
ML4644	OEM
ML4645	OEM
ML4646	OEM
ML4647	OEM
ML4648	OEM
ML4649	OEM
ML4650	OEM
ML4651	OEM
ML4652	OEM
ML4653	OEM
ML4654	OEM
ML4655	OEM
ML4660-114	OEM
ML4660-115	OEM
ML4660-144	OEM
ML4661-114	OEM
ML4661-115	OEM
ML4661-144	OEM
ML4662-114	OEM
ML4662-115	OEM
ML4662-144	OEM
ML4663-114	OEM
ML4663-115	OEM
ML4663-144	OEM
ML4664-114	OEM
ML4664-115	OEM
ML4664-144	OEM
ML4665-114	OEM
ML4665-115	OEM
ML4665-144	OEM
ML4666-114	OEM
ML4666-115	OEM
ML4666-144	OEM
ML4667-114	OEM
ML4667-115	OEM
ML4667-144	OEM
ML4668-114	OEM

DEVICE TYPE	REPL CODE
ML4668-115	OEM
ML4668-144	OEM
ML4669-114	OEM
ML4669-115	OEM
ML4669-144	OEM
ML4670-114	OEM
ML4670-115	OEM
ML4670-144	OEM
ML4671-114	OEM
ML4671-115	OEM
ML4671-144	OEM
ML4672-114	OEM
ML4672-115	OEM
ML4672-144	OEM
ML4673-114	OEM
ML4673-115	OEM
ML4673-144	OEM
ML4674-114	OEM
ML4674-115	OEM
ML4674-144	OEM
ML4703	OEM
ML4703S	OEM
ML4704	OEM
ML4704S	OEM
ML4705	OEM
ML4705S	OEM
ML4706	OEM
ML4707	OEM
ML4708	OEM
ML4709	OEM
ML4750	OEM
ML4803	OEM
ML4804	OEM
ML4805	OEM
ML5620C	OEM
ML5620CJ	OEM
ML5620J	OEM
ML5620PL	OEM
ML5620PS	OEM
ML5620S	OEM
ML5620X	OEM
ML5621C	OEM
ML5621CJ	OEM
ML5621J	OEM
ML5621PL	OEM
ML5621PS	OEM
ML5621S	OEM
ML5621X	OEM
ML5624C	OEM
ML5624J	OEM
ML5624PL	OEM
ML5624PS	OEM
ML5624S	OEM
ML5624SJ	OEM
ML5624X	OEM
ML5625C	OEM
ML5625J	OEM
ML5625PL	OEM
ML5625PS	OEM
ML5625S	OEM
ML5625SJ	OEM
ML5625X	OEM
ML5640C	OEM
ML5640J	OEM
ML5640PL	OEM
ML5640PS	OEM
ML5640S	OEM
ML5640X	OEM
ML5640XJ	OEM
ML5641C	OEM
ML5641J	OEM
ML5641PL	OEM
ML5641PS	OEM
ML5641S	OEM
ML5641X	OEM
ML5641XJ	OEM
ML5642J	OEM
ML5642X	OEM
ML5643J	OEM
ML5643X	OEM
ML5644C	OEM
ML5644J	OEM
ML5644PL	OEM
ML5644PS	OEM
ML5644S	OEM
ML5644X	OEM
ML5644XJ	OEM
ML5645C	OEM
ML5645J	OEM
ML5645PL	OEM
ML5645PS	OEM
ML5645S	OEM
ML5645X	OEM
ML5645XJ	OEM
ML5646J	OEM
ML5646X	OEM
ML5647J	OEM
ML5647X	OEM
ML5710C	OEM
ML5710J	OEM
ML5710LJ	OEM
ML5710PL	OEM
ML5710PX	OEM
ML5710S	OEM
ML5710X	OEM
ML5720C	OEM
ML5720J	OEM
ML5720LJ	OEM
ML5720PL	OEM

DEVICE TYPE	REPL CODE
ML5720PX	OEM
ML5720S	OEM
ML5720X	OEM
ML5730C	OEM
ML5730J	OEM
ML5730PL	OEM
ML5730S	OEM
ML5730X	OEM
ML5740C	OEM
ML5740PL	OEM
ML5740S	OEM
ML5740X	OEM
ML7805	4338
ML7805A	4338
ML7812	1351
ML7812A	1351
ML7815	4341
ML7815A	4341
ML7815P	1321
ML7818	4342
ML7818A	4342
ML8204AE	6621
ML8205AE	OEM
ML17310M2	OEM
ML17310M3	OEM
ML17310M4	OEM
ML17320M2	OEM
ML17320M3	OEM
ML17320M4	OEM
ML17330M2	OEM
ML17330M3	OEM
ML17330M4	OEM
ML17340M2	OEM
ML17340M3	OEM
ML17340M4	OEM
ML17350M2	OEM
ML17350M3	OEM
ML17370M1	OEM
ML17370M2	OEM
ML17380M1	OEM
ML17380M2	OEM
MLA12	OEM
MLB.2A	OEM
MLC74HC00A	6625
MLC74HC00AM	OEM
MLC74HC02A	6634
MLC74HC02AM	OEM
MLC74HC03A	OEM
MLC74HC03AM	OEM
MLC74HC04A	6637
MLC74HC04AM	OEM
MLC74HC08A	4289
MLC74HC08AM	OEM
MLC74HC10	OEM
MLC74HC10M	OEM
MLC74HC11	5530
MLC74HC11M	OEM
MLC74HC14A	6650
MLC74HC14AM	OEM
MLC74HC20	OEM
MLC74HC20M	OEM
MLC74HC27	OEM
MLC74HC27M	OEM
MLC74HC30	OEM
MLC74HC30M	OEM
MLC74HC32A	6672
MLC74HC32AM	OEM
MLC74HC42	OEM
MLC74HC42M	OEM
MLC74HC51	OEM
MLC74HC51M	OEM
MLC74HC74	OEM
MLC74HC74AM	OEM
MLC74HC76	OEM
MLC74HC76M	OEM
MLC74HC86	5535
MLC74HC86M	OEM
MLC74HC107	OEM
MLC74HC107M	OEM
MLC74HC109	6642
MLC74HC109M	OEM
MLC74HCU04	OEM
MLC74HCU04M	OEM
MLC4066B	OEM
MLC41005	OEM
MLE-7541	OEM
MLED15	OEM
MLED45	OEM
MLED50	OEM
MLED55	OEM
MLED60	OEM
MLED71	3232
MLED90	OEM
MLED92	3232
MLED93	OEM
MLED94	OEM
MLED95	OEM
MLED500	OEM
MLED600	OEM
MLED610	OEM
MLED630	OEM
MLED640	OEM
MLED650	1358
MLED655	2215
MLED660	OEM
MLED750	OEM
MLED900	OEM
MLED910	OEM
MLED930	0591

DEVICE TYPE	REPL CODE
MLH5G	OEM
MLH5S	OEM
MLH5Y	OEM
MLH10G	OEM
MLH10S	OEM
MLH10Y	OEM
MLI8G	OEM
MLI8S	OEM
MLI8Y	OEM
MLI80G	OEM
MLI80S	OEM
MLI80Y	OEM
MLM101AG	1300
MLM111AG	1815
MLM139	0176
MLM139AL	0176
MLM139L	0176
MLM201AG	0093
MLM211AG	1815
MLM224L	0625
MLM224P	0625
MLM239AL	0176
MLM239L	0176
MLM239P	OEM
MLM301AG	0093
MLM301AU	1300
MLM307P1	2283
MLM307U	2283
MLM309K	1922
MLM311P1	2104
MLM324	0625
MLM324L	0625
MLM324P	0625
MLM324P1	0625
MLM339(P)	0176
MLM339AL	0176
MLM339L	0176
MLM339P	2780
MLM2901P	6248
MLNA328	3010
MLNA328A	3010
MLNA328B	OEM
MLNA331	1713
MLNA331A	OEM
MLNA331B	OEM
MLNA335	0188
MLNA335A	OEM
MLNA335B	OEM
MLNA339	0456
MLNA339A	0456
MLNA339B	0456
MLNA343	0533
MLNA343A	0533
MLNA343B	OEM
MLNA347	0451
MLNA347A	0451
MLNA347B	OEM
MLNA351	0162
MLNA351A	0162
MLNA351B	OEM
MLNA356	0157
MLNA356A	0157
MLNA356B	OEM
MLNA362	0636
MLNA362A	0636
MLNA362B	OEM
MLNA368	0025
MLNA368A	0025
MLNA368B	OEM
MLNA375	0649
MLNA375A	0649
MLNA375B	OEM
MLNA382	0244
MLNA382A	0244
MLNA382B	0244
MLNA391	0012
MLNA391A	0012
MLNA391B	OEM
MLNA3100	0170
MLNA3100A	0170
MLNA3100B	OEM
MLS101	OEM
MLS102	OEM
MLS103	OEM
MLS104	OEM
MLS105	OEM
MLS201	OEM
MLS202	OEM
MLS203	OEM
MLS204	OEM
MLS205	OEM
MLSI-11B	OEM
MLSI-512	OEM
MLSI-1710	OEM
MLSI2480	OEM
MLSI-BPA84	OEM
MLSI-CR11	OEM
MLSI-DLV11	OEM
MLSI-DLV11E	OEM
MLSI-DR11-B	OEM
MLSI-DRV11C	OEM
MLSI-DRV11P	OEM
MLSI-DT1761	OEM
MLSI-DUPV11	OEM
MLSI-DUV11	OEM
MLSI-DZ11-A	OEM
MLSI-DZ11-B	OEM
MLSI-IBV11	OEM
MLSI-LP11	OEM

DEVICE TYPE	REPL CODE
MLSI-LP11-A	OEM
MLSI-MRV000	OEM
MLSI-MRV002	OEM
MLSI-MRV003	OEM
MLSI-MRV-004	OEM
MLSI-MRV-005	OEM
MLSI-PC11	OEM
MLSI-SMU	OEM
MLSI-TEV	OEM
MLSI-VRV001	OEM
MLSI-XYV11	OEM
MLT29	OEM
MLT33	OEM
MLT40	OEM
MLT47	OEM
MLT49	OEM
MLT438	OEM
MLT439	OEM
MLT478	OEM
MLV746A	0292
MLV747A	0376
MLV748A	0456
MLV749A	0533
MLV750A	0451
MLV751A	0162
MLV752A	0157
MLV753A	0636
MLV754A	0025
MLV755A	0649
MLV756A	0244
MLV757A	0012
MLV758A	0170
MLV759A	0137
MLV4370A	3003
MLV4371A	1312
MLV4372A	1713
MLZ-90	OEM
MLZ-90A-MF16C-1	OEM
MLZ-90A-MF16S-1	OEM
MLZ-90A-MF16S-2	OEM
MLZ-91A-32P-MFGDR	OEM
MLZ-91A-64P-MFGDR	OEM
MLZ-92A-16P-MFGDR	OEM
MLZ-92A-32P-MFGDR	OEM
MLZ-92A-64P-MFGDR	OEM
MLZ-93A	OEM
MLZ312	OEM
MLZ-DAQ	OEM
MM0	0015
MM1	OEM
MM1-A1	OEM
MM1-ACPU	OEM
MM1-AOS-4	OEM
MM1-AOS-8	OEM
MM1-CPU	OEM
MM1-DIO	OEM
MM1-ENC-RM	OEM
MM1-MSC	OEM
MM1-MVRAM-1	OEM
MM1-NVRAM-2	OEM
MM1-OPT	OEM
MM1-PBA	OEM
MM1-PROM	OEM
MM1-RAM	OEM
MM1-ZCPU	OEM
MM2	0015
MM3	0015
MM4	0015
MM5	0015
MM6	0015
MM7	0071
MM8	0071
MM9	0071
MM10	0071
MM54C00	OEM
MM54C00D	OEM
MM54C02	OEM
MM54C02D	OEM
MM54C04	OEM
MM54C04J	OEM
MM54C04W	OEM
MM54C08	OEM
MM54C08D	OEM
MM54C10	OEM
MM54C10D	OEM
MM54C14J	OEM
MM54C14W	OEM
MM54C20	OEM
MM54C20D	OEM
MM54C30D	OEM
MM54C32D	OEM
MM54C42	OEM
MM54C42D	OEM
MM54C42J	OEM
MM54C42W	OEM
MM54C48D	OEM
MM54C48N	OEM
MM54C73D	OEM
MM54C74	OEM
MM54C74D	OEM
MM54C76D	OEM
MM54C83D	OEM
MM54C85	OEM
MM54C85D	OEM
MM54C86	OEM
MM54C86D	OEM
MM54C89J	OEM
MM54C90	OEM
MM54C90D	OEM

DEVICE TYPE	REPL CODE
MM54C93	OEM
MM54C93D	OEM
MM54C107D	OEM
MM54C150J	OEM
MM54C151J	OEM
MM54C151W	OEM
MM54C154J	OEM
MM54C157J	OEM
MM54C157W	OEM
MM54C160D	OEM
MM54C161D	OEM
MM54C162D	OEM
MM54C163D	OEM
MM54C173D	OEM
MM54C174D	OEM
MM54C175D	OEM
MM54C192D	OEM
MM54C193D	OEM
MM54C200J	OEM
MM54C221D	OEM
MM54C901J	OEM
MM54C901W	OEM
MM54HC14	OEM
MM54HCT00	OEM
MM74C00	2112
MM74C00N	2112
MM74C02	6210
MM74C02N	2114
MM74C04	6211
MM74C04J	2958
MM74C04N	2958
MM74C08	6212
MM74C08N	2118
MM74C10	6214
MM74C10N	2119
MM74C14J	3031
MM74C14N	3031
MM74C20	6217
MM74C20N	2123
MM74C30N	3128
MM74C32N	3134
MM74C42	6218
MM74C42J	2143
MM74C42N	2143
MM74C48N	2148
MM74C73N	2162
MM74C74	6219
MM74C74N	2163
MM74C76N	2164
MM74C83N	OEM
MM74C85	6221
MM74C85N	2171
MM74C86	6222
MM74C86N	2999
MM74C89J	OEM
MM74C89N	OEM
MM74C90	2174
MM74C90N	2174
MM74C93	2177
MM74C93N	2177
MM74C95N	3401
MM74C107N	3834
MM74C150J	OEM
MM74C150N	OEM
MM74C151J	4001
MM74C151N	4001
MM74C154J	4014
MM74C154N	4014
MM74C157J	4033
MM74C157N	4033
MM74C160N	4045
MM74C161N	4046
MM74C162N	1142
MM74C163N	4057
MM74C164N	4062
MM74C173N	4088
MM74C174N	4092
MM74C175N	4096
MM74C192N	4119
MM74C193N	4122
MM74C200J	OEM
MM74C200N	OEM
MM74C221	3047
MM74C221N	3047
MM74C240N	2956
MM74C244N	4230
MM74C373N	4462
MM74C374N	4467
MM74C901N	4607
MM74C902N	4608
MM74C903N	4609
MM74C904N	4610
MM74C906	OEM
MM74C906N	OEM
MM74C922N	5597
MM74C923N	5599
MM74C925N	4621
MM74HC14	OEM
MM74HC240N	1979
MM74HC243N	2001
MM74HC245N	2014
MM74HC595	OEM
MM74HC595N	OEM
MM74HCT00	OEM
MM306	0086
MM380	0050
MM404	0164
MM404A	0136
MM420	0233

DEVICE TYPE	REPL CODE
MM421	OEM
MM430	OEM
MM439	3607
MM440	OEM
MM486	0086
MM487	0086
MM488	0086
MM511	0086
MM512	0086
MM513	0086
MM709	0144
MM719	OEM
MM741	0312
MM799	OEM
MM800	OEM
MM801	OEM
MM869B	0079
MM999	0037
MM1002	OEM
MM1008	OEM
MM1028BF	OEM
MM1031	OEM
MM1031XS	OEM
MM1111XS	OEM
MM1114XEF	OEM
MM1114XFF	OEM
MM1139	0050
MM1151	0004
MM1152	0004
MM1153	0004
MM1154	0004
MM1161	OEM
MM1162	OEM
MM1163	OEM
MM1164	OEM
MM1199	0050
MM1367/2SC684	0144
MM1382	0144
MM1387	OEM
MM1461	OEM
MM1462	OEM
MM1500	0418
MM1501	0418
MM1501A	0418
MM1505	0079
MM1510	OEM
MM1511	OEM
MM1557	3586
MM1558	3587
MM1559	2506
MM1601	3576
MM1602	2170
MM1603	1977
MM1606	2170
MM1612	0698
MM1613	0079
MM1614	0037
MM1619	0042
MM1620	3568
MM1702AQ	OEM
MM1711	0079
MM1712	OEM
MM1736	OEM
MM1737	OEM
MM1738	OEM
MM1739	OEM
MM1742	0004
MM1748	0330
MM1755	0016
MM1756	0016
MM1757	0016
MM1758	0016
MM1803	0493
MM1809	0086
MM1809A	0086
MM1810	0698
MM1810A	0086
MM1812	0285
MM1874A	OEM
MM1893	0630
MM1941	0493
MM1943	0086
MM1945	0144
MM2005-2	0016
MM2102	OEM
MM2103	0844
MM2112-2	OEM
MM2114N2L	OEM
MM2114N-3	2048
MM2193A	0595
MM2258	0168
MM2259	0168
MM2260	0233
MM2261	0595
MM2262	0630
MM2263	0168
MM2264	0630
MM2266	0086
MM2270	0595
MM2316ED	OEM
MM2316EN	OEM
MM2484	0016
MM2503	0050
MM2550	0050
MM2552	0050
MM2554	0050
MM2613	0079
MM2614	0037
MM2708Q	4424

DEVICE TYPE	REPL CODE
MM2708Q-1	OEM
MM2711	0079
MM2712	0037
MM2716Q	2278
MM2716Q-1	OEM
MM2716Q-2	OEM
MM2716QE	OEM
MM2716QM	OEM
MM2758Q-A	OEM
MM2758Q-B	OEM
MM2894	0050
MM3000	0233
MM3001	0168
MM3002	0233
MM-3003	0233
MM3003	0285
MM3004	0560
MM3005	4225
MM3006	0016
MM3007	0622
MM3008	0264
MM-3009	0264
MM3009	0233
MM3014	0264
MM3025	OEM
MM3053	0630
MM3100	0233
MM3101	0233
MM3200-2C	OEM
MM3200-3C	OEM
MM3200-C	OEM
MM3724	0936
MM3725	0936
MM3726	0037
MM3734	0086
MM3735	0086
MM3736	0086
MM3737	0086
MM3903	0016
MM3904	0016
MM3905	0037
MM3906	0037
MM4000	0439
MM4000(IC)	2024
MM4001	0439
MM4002	0439
MM4003	0439
MM4005	0126
MM4006	0126
MM4007	2846
MM4008	0126
MM4009	1999
MM4010	0126
MM4011	0215
MM4012	0498
MM4013	0413
MM4015	1014
MM4016A	1143
MM4017	0513
MM4018	0126
MM4019	0126
MM4019(IC)	1525
MM4020	3580
MM4020(IC)	1661
MM4021	3582
MM4022	3583
MM4023	3584
MM4024	1957
MM4025	2072
MM4027	1949
MM4030	0500
MM4031	0126
MM4032	0126
MM4033	0126
MM4036	0126
MM4037	0126
MM4040	0056
MM4042	0121
MM4048	0037
MM4049	0001
MM4050	0398
MM4051	0366
MM4052	0037
MM4052(IC)	0024
MM4257	3607
MM4258	3607
MM4429	0560
MM4430	0560
MM4511	1544
MM4518	1043
MM5000	0050
MM5001	0050
MM5002	0050
MM5005	0126
MM5006	0892
MM5007	0892
MM5107N	OEM
MM5189	0086
MM5262	0086
MM5290N-4	1286
MM5311N	OEM
MM5314N	OEM
MM5316	2295
MM5316D	OEM
MM5316N	2728
MM5322N	OEM
MM5369	OEM
MM5369AA/N	5863
MM5387	2295

DEVICE TYPE	REPL CODE
MM5387/HD38991A	OEM
MM5387AA	2295
MM5387AA/N	2295
MM5387AA-N	2295
MM5387AAN	4378
MM5387AB/N	2295
MM5387ABN	OEM
MM5387N	2295
MM5402	2295
MM-5402N	OEM
MM5402N	2295
MM5407N	OEM
MM5415	5861
MM5416	0439
MM5430	OEM
MM5430N	OEM
MM5431N	OEM
MM5439N	OEM
MM5450N	OEM
MM5455N	OEM
MM5456N	OEM
MM5457N	OEM
MM5462N	OEM
MM5504D	OEM
MM5504F	OEM
MM5554N	OEM
MM5555N	OEM
MM5556N	OEM
MM5601AN	5539
MM5602AN	5559
MM5611AN	0215
MM5611BN	OEM
MM5612AN	5692
MM5613AN	OEM
MM5613BN	OEM
MM5617AN	5789
MM5619AN	4370
MM5620AN	5844
MM5622AN	5882
MM5623AN	5892
MM5624AN	5897
MM5625AN	5908
MM5627AN	1949
MM5628BN	OEM
MM5630AN	5972
MM5650BN	OEM
MM5660BN	6209
MM5673BN	OEM
MM5725N	OEM
MM5734N	OEM
MM5736N	OEM
MM5737N	OEM
MM5738N	OEM
MM5739N	OEM
MM5740AAC	OEM
MM5758N	OEM
MM5760N	OEM
MM5763N	OEM
MM5764N	OEM
MM5765N	OEM
MM5766N	OEM
MM5767N	OEM
MM5777N	OEM
MM5780N	OEM
MM5791N	OEM
MM5794N	OEM
MM5795N	OEM
MM5799N	OEM
MM5823N	OEM
MM5824N	OEM
MM5829	OEM
MM5832N	OEM
MM5833N	OEM
MM5839	OEM
MM5840N	OEM
MM5841N	OEM
MM5860	OEM
MM5863N	OEM
MM5871N	OEM
MM5879	OEM
MM5880	OEM
MM5885	OEM
MM5886	OEM
MM5889	OEM
MM5889AB	OEM
MM5890	OEM
MM5899	OEM
MM6400A-C	OEM
MM6400C	OEM
MM6427	5979
MM6800	OEM
MM6800/16	OEM
MM6800D	OEM
MM7087	0233
MM7088	0233
MM7218BN	OEM
MM7317BN	OEM
MM8000	0418
MM8001	0418
MM8002	0418
MM8003	2070
MM8004	0698
MM8006	0007
MM8007	0007
MM8008	0418
MM8010	0418
MM8011	0418
MM8012	2070
MM8080/16	OEM

DEVICE TYPE	REPL CODE
MM8080AL	OEM
MM8080B	OEM
MM8086	OEM
MM8086/16	OEM
MM8086D	OEM
MM14511BCN	1544
MM14511CN	1544
MM14538BCN	OEM
MM52116D	OEM
MM52116FDWD	OEM
MM52116FDWN	OEM
MM52116FDXD	OEM
MM52116FDXN	OEM
MM52116N	OEM
MM52132D	OEM
MM52132N	OEM
MM52164D	OEM
MM52164N	OEM
MM53100	OEM
MM53104	OEM
MM53105	OEM
MM53108N	OEM
MM53110AAN	OEM
MM53110ABN	OEM
MM53113N	2728
MM53118AA	OEM
MM53124N	OEM
MM53125N	OEM
MM53130N	OEM
MM53143J	OEM
MM53144J	OEM
MM53190N	OEM
MM53214NGQ/N	OEM
MM53224N	OEM
MM54240	OEM
MM55104	2616
MM55104N	2616
MM55106	OEM
MM55106N	OEM
MM55107N	OEM
MM55108	6766
MM55110	OEM
MM55114	OEM
MM55116	OEM
MM55116N	OEM
MM55121N	OEM
MM55122N	OEM
MM55123N	OEM
MM55124N	OEM
MM55126N	OEM
MM57103N	OEM
MM57104N	OEM
MM57123N	OEM
MM57135N	OEM
MM57136N	OEM
MM57140	OEM
MM57150	OEM
MM57150FHJ	OEM
MM57150N	OEM
MM57150PHJ	OEM
MM57409D	OEM
MM57409J	OEM
MM57409N	OEM
MM57455N	OEM
MM57459N	OEM
MM58104	OEM
MM58106N	OEM
MM58115	OEM
MM58117	OEM
MM58118	OEM
MM58119	OEM
MM58120	OEM
MM58127	OEM
MM58128	OEM
MM58129	OEM
MM58130	OEM
MM58142N	OEM
MM58143N	OEM
MM58144N	OEM
MM58146N	OEM
MM58167AN	OEM
MM58174AJM	OEM
MM58174AN	OEM
MM58183N	OEM
MM58184N	OEM
MM58201N	OEM
MM58274N	OEM
MM58313N	OEM
MM58601	OEM
MM58801	OEM
MM587100N	OEM
MM681031	OEM
MMBA811C5	3886
MMBA811C6	3886
MMBA811C7	3886
MMBA811C8	OEM
MMBA812M3	3886
MMBA812M4	3886
MMBA812M5	3886
MMBA812M6	3886
MMBA812M7	3886
MMBA813S2	OEM
MMBA813S3	OEM
MMBA813S4	OEM
MMBA956H3	OEM
MMBA956H4	OEM
MMBA956H5	OEM
MMBA956H6	OEM
MMBC1009F1	5385

DEVICE TYPE	REPL CODE
MMBC1009F2	5385
MMBC1009F3	5385
MMBC1009F4	5385
MMBC1009F5	5385
MMBC1321Q2	OEM
MMBC1321Q3	OEM
MMBC1321Q4	OEM
MMBC1321Q5	OEM
MMBC1621B2	OEM
MMBC1621B3	OEM
MMBC1621B4	OEM
MMBC1622D6	5385
MMBC1622D7	5385
MMBC1622D8	OEM
MMBC1623L3	5385
MMBC1623L4	5385
MMBC1623L5	5385
MMBC1623L6	5385
MMBC1623L7	5385
MMBC1653N2	2332
MMBC1653N3	2332
MMBC1653N4	2332
MMBC1654N5	2332
MMBC1654N6	2332
MMBC1654N7	2332
MMBC2107G3	OEM
MMBC2107G4	OEM
MMBC2107G5	OEM
MMBC2107G6	OEM
MMBD101	OEM
MMBD501	OEM
MMBD914	5493
MMBD2835	0903
MMBD2836	4917
MMBD2837	0907
MMBD2838	0907
MMBD6050	3749
MMBD6100	0907
MMBD7000	OEM
MMBF170	OEM
MMBF4393	OEM
MMBF4416	OEM
MMBF4860	OEM
MMBF5457	OEM
MMBF5459	OEM
MMBF5460	OEM
MMBF5484	OEM
MMBF5486	OEM
MMBFU310	OEM
MMBR901	3372
MMBR920	OEM
MMBR930	3372
MMBR931	3372
MMBR2060	OEM
MMBR2857	OEM
MMBR4957	OEM
MMBR5031	3372
MMBR5179	OEM
MMBS5060	OEM
MMBS5061	OEM
MMBS5062	OEM
MMBT290Z	OEM
MMBT404	1435
MMBT404A	1435
MMBT918	OEM
MMBT918R	OEM
MMBT2222	1733
MMBT2222A	1733
MMBT2222AR	OEM
MMBT2222R	OEM
MMBT2369	1435
MMBT2369R	OEM
MMBT2907A	1499
MMBT2907AR	OEM
MMBT3014	OEM
MMBT3640	1499
MMBT3903	1435
MMBT3904	1435
MMBT3906	1499
MMBT3960	OEM
MMBT3960A	OEM
MMBT4260	OEM
MMBT4261	OEM
MMBT4403	1499
MMBT5086	1499
MMBT5087	1499
MMBT5088	1435
MMBT5089	1435
MMBT5401	1503
MMBT5550	2332
MMBT6427	OEM
MMBT6428	1435
MMBT6429	1435
MMBTA05	2332
MMBTA06	2332
MMBTA13	2338
MMBTA14	2338
MMBTA20	1435
MMBTA42	OEM
MMBTA42R	OEM
MMBTA43	OEM
MMBTA43R	OEM
MMBTA55	1503
MMBTA56	1503
MMBTA63	OEM
MMBTA64	OEM
MMBTA70	1499
MMBTA92	OEM
MMBTA93	OEM

DEVICE TYPE	REPL CODE
MMBTH-10	OEM
MMBTH10	OEM
MMBTH-24	OEM
MMBTH24	OEM
MMBTH81	OEM
MMBV105G	OEM
MMBV109	OEM
MMBV2097	OEM
MMBV2098	OEM
MMBV2101	OEM
MMBV2103	OEM
MMBV2108	OEM
MMBV2109	OEM
MMBV3102	OEM
MMBV3401	OEM
MMBZ5226	OEM
MMBZ5227	OEM
MMBZ5228	OEM
MMBZ5229	OEM
MMBZ5230	OEM
MMBZ5231	OEM
MMBZ5232	OEM
MMBZ5233	OEM
MMBZ5234	0471
MMBZ5235	OEM
MMBZ5236	OEM
MMBZ5237	OEM
MMBZ5238	OEM
MMBZ5239	0057
MMBZ5240	OEM
MMBZ5241	OEM
MMBZ5242	0052
MMBZ5243	OEM
MMBZ5244	OEM
MMBZ5245	OEM
MMBZ5246	OEM
MMBZ5247	OEM
MMBZ5248	OEM
MMBZ5249	OEM
MMBZ5250	OEM
MMBZ5251	OEM
MMBZ5252	OEM
MMBZ5253	OEM
MMBZ5254	OEM
MMBZ5255	OEM
MMBZ5256	OEM
MMBZ5257	OEM
MMBZ5263	OEM
MMBZ5265	OEM
MMBZ5281	OEM
MMCM918	0144
MMCM930	0111
MMCM2221	OEM
MMCM2222	2446
MMCM2484	0861
MMCM2857	OEM
MMCM2906	OEM
MMCM2907	0150
MMCM3798	OEM
MMCM3799	OEM
MMCM3903	OEM
MMCM3904	OEM
MMCM3905	OEM
MMCM3906	OEM
MMCM3960A	OEM
MMD70	OEM
MMD6050	OEM
MMD6100	OEM
MMD6150	OEM
MMD7000	OEM
MMH0026CG	OEM
MMH0026CL	OEM
MMH0026CPI	OEM
MMH0026CPID	OEM
MMH0026CPIDS	OEM
MMH0026CPIS	OEM
MMH0026G	OEM
MMH0026L	OEM
MM-S100	OEM
MM-SBC-80F	OEM
MM-SBC-80H	OEM
MMSD914	OEM
MMST918	OEM
MMST2222	OEM
MMST2222A	1435
MMST2907	1499
MMST2907A	OEM
MMST3903	OEM
MMST3904	OEM
MMST3905	OEM
MMST3906	OEM
MMST4123	OEM
MMST4124	OEM
MMST4125	OEM
MMST4126	OEM
MMST4400	OEM
MMST4401	OEM
MMST4402	OEM
MMST4403	OEM
MMST5086	OEM
MMST5087	OEM
MMST5088	OEM
MMST5089	OEM
MMST6428	OEM
MMST6429	OEM
MMSTA05	OEM
MMSTA06	OEM
MMSTA12	OEM
MMSTA13	OEM
MMSTA14	OEM
MMSTA20	OEM
MMSTA55	OEM
MMSTA56	OEM
MMSTA63	OEM
MMSTA64	OEM
MMSTA70	OEM
MMSTH24	OEM
MMT70	0111
MMT71	0861
MMT72	0079
MMT73	0079
MMT74	0418
MMT75	0861
MMT76	0861
MMT918	0224
MMT2060	OEM
MMT2222	0079
MMT2222A	OEM
MMT2857A	OEM
MMT2907	0532
MMT2907A	OEM
MMT3014	0079
MMT3546	4433
MMT3798	0791
MMT3799	0006
MMT3823	6614
MMT3903	0079
MMT3904	0079
MMT3905	0150
MMT3906	0532
MMT4123	OEM
MMT4124	OEM
MMT4125	OEM
MMT4126	OEM
MMT4261	OEM
MMT4400	OEM
MMT4401	OEM
MMT4402	OEM
MMT4403	OEM
MMT4416	OEM
MMT5031	OEM
MMT5031A	OEM
MMT5086	OEM
MMT5087	OEM
MMT5088	OEM
MMT5089	OEM
MMT6428	OEM
MMT6429	OEM
MMT8006	OEM
MMT8007	OEM
MMT8008	OEM
MMT8015	0127
MMTA05	OEM
MMTA06	OEM
MMTA12	OEM
MMTA13	OEM
MMTA14	OEM
MMTA20	OEM
MMTA55	OEM
MMTA56	OEM
MMTA63	OEM
MMTA64	OEM
MMTA70	OEM
MMTH24	OEM
MMZ3.3	0292
MMZ3.3-20	0292
MMZ3.3A	0292
MMZ3.6	0188
MMZ3.6-20	0188
MMZ3.6A	0188
MMZ3.9	0456
MMZ3.9-20	0456
MMZ3.9A	0456
MMZ4.3	0533
MMZ4.3-20	0533
MMZ4.3A	0533
MMZ4.7	0451
MMZ4.7-20	0451
MMZ4.7A	0451
MMZ5.1	0162
MMZ5.1-20	0162
MMZ5.1A	0162
MMZ5.6	0157
MMZ5.6-20	0157
MMZ5.6A	0157
MMZ6.2	0636
MMZ6.2-20	0636
MMZ6.2A	0636
MMZ6.8	0025
MMZ6.8-20	0025
MMZ6.8A	0025
MMZ7.5	0649
MMZ7.5-20	0649
MMZ7.5A	0649
MMZ8.2	0244
MMZ8.2-20	0244
MMZ8.2A	0244
MMZ9.1	0012
MMZ9.1-20	0012
MMZ9.1A	0012
MMZ10	0170
MMZ10-20	0170
MMZ10A	0170
MMZ11	0317
MMZ11-20	0317
MMZ11A	0317
MMZ12	0137
MMZ12(06)	0365
MMZ12-20	0137
MMZ12A	0137
MMZ13	0365
MMZ13-20	0365
MMZ13A	0365
MMZ15	0002
MMZ15-20	0002
MMZ15A	0002
MMZ15A-20	0002
MMZ16	0420
MMZ16-20	0420
MMZ16A	0420
MMZ18	0495
MMZ18-20	0495
MMZ18A	0495
MMZ20	0531
MMZ20-20	0531
MMZ20A	0531
MMZ22	0565
MMZ22-20	0565
MMZ22A	0565
MMZ24	0402
MMZ24-20	0402
MMZ24A	0402
MMZ27	0441
MMZ27-20	0441
MMZ27A	0441
MMZ30	0195
MMZ30-20	0195
MMZ30A	0195
MMZ33	0039
MMZ33-20	0039
MMZ33A	0039
MMZ36	0820
MMZ36-20	0820
MMZ36A	0820
MMZ39	0350
MMZ39-20	0350
MMZ39A	0350
MMZ43	0931
MMZ43-20	0931
MMZ43A	0931
MMZ47	0999
MMZ47-20	0999
MMZ47A	0999
MMZ51	0502
MMZ51-20	0502
MMZ51A	0502
MMZ56	0869
MMZ56-20	0869
MMZ56A	0869
MMZ62	0783
MMZ62-20	0783
MMZ62A	0783
MMZ68	2157
MMZ68-20	2157
MMZ68A	2157
MMZ75	1189
MMZ75-20	1189
MMZ75A	1189
MMZ82	0331
MMZ82-20	0331
MMZ82A	0331
MMZ91	1311
MMZ91-20	1311
MMZ91A	1311
MMZ100	0098
MMZ100-20	0098
MMZ100A	0098
MMZ110	OEM
MMZ110-20	0149
MMZ110A	0149
MMZ120	0186
MMZ120-20	0186
MMZ120A	0186
MMZ130	0213
MMZ130-20	0213
MMZ130A	0213
MMZ150	0028
MMZ150-20	0028
MMZ150A	0028
MMZ160	0255
MMZ160-20	0255
MMZ160A	0255
MMZ180	0367
MMZ180-20	0367
MMZ180A	0367
MMZ200	0421
MMZ200-20	0421
MMZ200A	0421
MN12C25	OEM
MN12C25D	OEM
MN12C201D	OEM
MN12C261D	OEM
MN13A	OEM
MN13B	OEM
MN13C	OEM
MN19	OEM
MN21	0604
MN22	0160
MN23	0160
MN24	0160
MN25	0160
MN26	0160
MN28	OEM
MN29	0160
MN29BLK	0160
MN29GRN	0160
MN29PUR	0160
MN29WHT	0160
MN32	0160
MN34A	0143
MN46	0160
MN48	0160
MN49	0160
MN51	0143
MN52	0004
MN-53	0164
MN53	0004
MN53BLU	0004
MN53GRN	0004
MN53RED	0004
MN54	0079
MN60	0004
MN60(DIODE)	0143
MN61	0975
MN61A	0604
MN62A	0604
MN63	0975
MN63A	0604
MN64	0975
MN73	0160
MN73BLK	0160
MN73WHT	0160
MN74HC00	OEM
MN74HC04	OEM
MN74HC08	OEM
MN74HC14	OEM
MN74HC32	OEM
MN74HC138	OEM
MN74HC373	OEM
MN74HC4066S	OEM
MN76	0085
MN93C-46N	OEM
MN115P	OEM
MN194	0160
MN380	0379
MN380H	0379
MN512K	OEM
MN520	OEM
MN864A	OEM
MN1114XFF	OEM
MN1118XFF	OEM
MN1201A	OEM
MN1202	OEM
MN1203	OEM
MN1204B	OEM
MN1204E	OEM
MN1205A	OEM
MN1205D	OEM
MN1205E	OEM
MN1205F	OEM
MN1205H	OEM
MN1205P	OEM
MN1206A	5888
MN1207	OEM
MN1208	OEM
MN1215P	OEM
MN1217A	OEM
MN1218	OEM
MN1218A	OEM
MN1219	OEM
MN1224	OEM
MN1225	OEM
MN1226	OEM
MN1227(A)	OEM
MN1227A	OEM
MN1228	OEM
MN1228Q	OEM
MN1228T	OEM
MN1230	OEM
MN1257CQ	OEM
MN1280	3521
MN1280-K	3521
MN1280-L	4141
MN1280-M	OEM
MN1280-P	OEM
MN1280-Q	OEM
MN1280-R	3521
MN1280-S	OEM
MN1280L	4141
MN1280M	OEM
MN1280P	OEM
MN1280PQR	3521
MN1280R	3521
MN1280S	OEM
MN1297HNA	OEM
MN1301	4278
MN1380-M	OEM
MN1380-R	OEM
MN1381-Q	OEM
MN1400	OEM
MN1400ES	OEM
MN1400SE	OEM
MN1400VFA	OEM
MN1400VL	OEM
MN1400VN	OEM
MN1400VP	OEM
MN1402	OEM
MN1402ST0	OEM
MN1402STO	OEM
MN1403	OEM
MN1403-TN	OEM
MN1403SB	OEM
MN1403TN	OEM
MN1404	OEM
MN1405	OEM
MN1405GB	OEM
MN1405GH	OEM
MN1405MF	OEM
MN1405SF	OEM
MN1405SH	OEM
MN1405SHB	OEM
MN1405V1	OEM
MN1405VDG	OEM
MN1405VI	OEM
MN1405VK	OEM
MN1405VKL	OEM
MN1405VM	OEM
MN1405VS	OEM
MN1405VT	OEM
MN1405VXAO	OEM
MN1414HE	OEM
MN1414HF	OEM
MN1416SAA	OEM
MN1418BQR	OEM
MN1418BQS	OEM
MN1421FPC	OEM
MN1425VKI	OEM
MN1425VKL	OEM
MN1425VKM	OEM
MN1430	OEM
MN1432	OEM
MN1435	OEM
MN1450	OEM
MN1453	OEM
MN1454	OEM
MN1455	OEM
MN1455BVC	OEM
MN1483AH	OEM
MN1498	OEM
MN1499	OEM
MN1499A	OEM
MN1513VTF	1065
MN1527KMA	OEM
MN1534VRD	OEM
MN1542	OEM
MN1544	OEM
MN1547VSG	OEM
MN1550PDT	OEM
MN1550PEB	OEM
MN1551ASSA-1	OEM
MN1551KVA	OEM
MN1551SSA	OEM
MN1562	OEM
MN1564	OEM
MN1599	OEM
MN1610	OEM
MN1611	OEM
MN2020	OEM
MN2114	5300
MN2114-2	OEM
MN2200	OEM
MN2200H	OEM
MN2332	OEM
MN2716	6491
MN3001	5780
MN3002	OEM
MN3003	OEM
MN3004	OEM
MN3005	OEM
MN3006	OEM
MN3007	5201
MN3008	OEM
MN3009	OEM
MN3010	OEM
MN3011	5780
MN3012	OEM
MN3013	OEM
MN3101	4278
MN3102	OEM
MN3107CS	OEM
MN3204	OEM
MN3206	OEM
MN3207	OEM
MN3209	OEM
MN3210	OEM
MN3349	OEM
MN3349H	OEM
MN3412	OEM
MN3412H	OEM
MN3735F	OEM
MN3735FC	OEM
MN3735FCW	OEM
MN3735FS	OEM
MN3739FC	OEM
MN3810K	OEM
MN3810S	OEM
MN3817S	OEM
MN3818S	OEM
MN3830S	OEM
MN4001B	0478
MN4011B	0215
MN4011BS	OEM
MN4013B	0413
MN4013BS	OEM
MN4027B	1949
MN4030L	OEM
MN4040BS	OEM
MN4049B	5476
MN4052B	0024
MN4052BS	OEM
MN4053B	0035
MN4053BS	OEM
MN4053S	OEM
MN4066B	0101
MN4066BP	0101
MN4069UB	0119
MN4069UBS	0119
MN4071B	0129
MN4071BS	OEM
MN4081B	0626
MN4116	0522
MN4164P-12A	OEM
MN4164P-15A	2357
MN4264-15	OEM
MN4264P-15	OEM
MN4364-15	OEM
MN4503B	2053
MN4503BS	OEM
MN4528B	3198
MN4700	OEM
MN5101	4604
MN5104	OEM
MN5104S	OEM
MN5128	OEM
MN5150	OEM
MN5150H	OEM
MN5245	OEM
MN5260	OEM
MN5613AN	6654
MN6010	OEM
MN6010A	OEM
MN6013AS	OEM
MN6013BS	OEM
MN6013C	OEM
MN6013G	OEM
MN6013N	OEM
MN6014JS	OEM
MN6014W	OEM
MN6016A	4746
MN6016AK	OEM
MN6016AS	OEM
MN6021	OEM
MN6024	OEM
MN6025D	OEM
MN6027	OEM
MN6027B	OEM
MN6028	OEM
MN6028R	OEM
MN6030C	OEM
MN6030CA	OEM
MN6030HA	OEM
MN6030Z	OEM
MN6040	2616
MN6040A	2616
MN6044	OEM
MN6044A	OEM
MN6049	5275
MN6053	OEM
MN6054	OEM
MN6056	OEM
MN6057	OEM
MN6058	OEM
MN6059	OEM
MN6061A	4746
MN6062	OEM
MN6064	OEM
MN6064R	OEM
MN6069	OEM
MN6076	4750
MN6076L	OEM
MN6091	OEM
MN6092	OEM
MN6092A	OEM
MN6093	OEM
MN6163	5263
MN6163A	5263
MN6165VAA	OEM
MN6165VCA	OEM
MN6168IB	OEM
MN6168V1B	OEM
MN6168VIB	OEM
MN6168VIH	OEM
MN6175VCAK	OEM
MN6178CC	OEM
MN6178VAD	OEM
MN6178VAF	OEM
MN6179VBH	OEM
MN6221	OEM
MN6250	OEM
MN6251	OEM
MN6366	6328
MN6550B	OEM
MN6617	OEM
MN6617S	OEM
MN6618A	OEM
MN6623	OEM
MN6625	OEM
MN6632A	OEM
MN6632AK	OEM
MN6633	OEM
MN6634	OEM
MN6636	OEM
MN6636S	OEM
MN6712VLQ	OEM
MN6746VDAK	OEM
MN6746VDBK	OEM
MN6748FVAN	OEM
MN6790S	OEM
MN6790S-E2	OEM
MN6790S-T2	OEM
MN8027	OEM
MN8036	OEM
MN8303	0194
MN8303(IC)	OEM
MN12872A	OEM
MN12872FVAB	OEM
MN12973NAB	OEM
MN12973NAC	OEM
MN14821VWZ	OEM
MN14823QS	OEM
MN14823QSA	OEM
MN14823QU	OEM
MN14831JTD	OEM
MN14831JTZ	OEM
MN14833JTY	OEM
MN14834H	OEM
MN14834HH	OEM
MN14843JTX	OEM
MN14844HK	OEM
MN15151KWDC	OEM
MN15151Q14	OEM
MN15151Q14A	OEM
MN15151Q14N	OEM
MN15151Q14R	OEM
MN15151SAMA-2	OEM
MN15212HX	OEM
MN15221HP	OEM
MN15241QT	OEM
MN15241QTA	OEM
MN15241QTB2	OEM
MN15243VSN	OEM
MN15247Q10	OEM
MN15247Q10A	OEM
MN15247Q12	OEM
MN15261D01	OEM
MN15261D05	OEM
MN15261D01	OEM
MN15261PDU	OEM
MN15261SVA	OEM
MN15263VAC	OEM
MN15266HM	OEM
MN15266HN	OEM
MN15266JMC	OEM
MN15266JMC-1	OEM
MN15281VAE	OEM
MN15283PER	OEM
MN15283PJG-2	OEM
MN15283PJL	OEM
MN15284HQ	OEM
MN15284HS	OEM
MN15286Q9	OEM
MN15286Q11	OEM
MN15287-SNE3	OEM
MN15287HM	OEM
MN15287HW	OEM
MN15287HY	OEM
MN15287JMM	OEM
MN15287SNE-4	OEM
MN15341-VKR	OEM
MN15522VLDS	OEM
MN15522VLMS	OEM
MN15831VSE	OEM
MN15832VRB	OEM
MN15832VSR	OEM
MN15841VKP	OEM
MN15841VKQ	OEM
MN15842-VXC	OEM
MN15842VUD	OEM
MN15842VXC	OEM
MN15843VRA	OEM
MN15844VSA	OEM
MN15844VSA-4	OEM
MN15844VSB	OEM
MN15844VSC	OEM
MN15845PEQ	OEM
MN15846VRC	OEM
MN15846VRG	OEM
MN15846VRG3	OEM
MN15847VSG	OEM
MN15847VSJ	OEM
MN15847VSL	OEM
MN15847VSL3	OEM
MN15847VSP	OEM
MN18781D22	OEM
MN18781D25	OEM
MN18781D26	OEM
MN18781D36	OEM
MN18788VRN2	OEM
MN18788VZL3	OEM
MN40098B	0432
MN41256-15	1472
MN41464-15	OEM
MN47464L-12	OEM
MN51003NJB	OEM
MN51003SVB	OEM
MN52040KVF	OEM
MN53015	OEM
MN53015XBM	OEM
MN53015XBR	OEM
MN60111AK	OEM
MN61753VCBK	OEM
MN61753VCEK	OEM
MN61754VCCK	OEM
MN67461VDFK	OEM
MN67461VDHK	OEM
MN67461VDLK	OEM
MN67461VDMF	OEM
MN67461VDNF	OEM
MN67512VLQ	OEM
MN67520IFVCS	OEM
MN67520VLS	OEM
MN67520VLT	OEM
MN67520VLW	OEM

If replacement code is OEM, contact original manufacturer for replacement.

DEVICE TYPE	REPL CODE
MN67520VLX4	OEM
MN67603NS	OEM
MN73203SFB	OEM
MN152121JGM2	OEM
MN152121JM	OEM
MN152121JMT-3	OEM
MN152121JMT3	OEM
MN152121JMT-4	OEM
MN152121JMT4	OEM
MN152121JMT-5	OEM
MN152121JMT5	OEM
MN152121MTB-2	OEM
MN152121MTC	OEM
MN152611D07	OEM
MN152611PEN	OEM
MN152810Q16N	OEM
MN152811HYE	OEM
MN158241VAK	OEM
MN158461VAA2	OEM
MN158461VAL	OEM
MN158471VAB	OEM
MN187124FVCW	OEM
MN187129VLP	OEM
MN187164VRQ	OEM
MN188166VLC3	OEM
MN617531VCEK	OEM
MN617541VCDK	OEM
MN675201VZE	OEM
MN1872012AHHB-5	OEM
MN1872012HHB-1	OEM
MN1872012HHB-3	OEM
MN1872013JGU4	OEM
MN1872419Q17	OEM
MN1872432QAB	OEM
MN1872432QAD	OEM
MN1873234HYC	OEM
MN1873234QAM	OEM
MN1873234QAP	OEM
MN1873234QAS	OEM
MN1873237JH6	OEM
MN1876466	OEM
MN1876466JKN1	OEM
MN18720113JGU4	OEM
MN18872013JGU4	OEM
MNADC85C	OEM
MNADC87	OEM
MNADC87B	OEM
MNADC87H	OEM
MNADC87H/B	OEM
MO066A	OEM
MO101AL	OEM
MO101AM	OEM
MO101AS	OEM
MO101BL-FL	OEM
MO101BM-FM	OEM
MO101BS-FS	OEM
MO101GL	OEM
MO101GM	OEM
MO101GP	OEM
MO101GS	OEM
MO101HL	OEM
MO101HM	OEM
MO101HP	OEM
MO101HS	OEM
MO101JL-ML	OEM
MO101JM-MM	OEM
MO101JP-MP	OEM
MO101JS-MS	OEM
MO102AL	OEM
MO102AM	OEM
MO102AP	OEM
MO102AS	OEM
MO102BL-FL	OEM
MO102BM-FM	OEM
MO102BP	OEM
MO102BP-FP	OEM
MO102BS-FS	OEM
MO102GL	OEM
MO102GM	OEM
MO102GP	OEM
MO102GS	OEM
MO102HL	OEM
MO102HM	OEM
MO102HP	OEM
MO102HS	OEM
MO102JL-LL	OEM
MO102JM-LM	OEM
MO102JP-LP	OEM
MO102JS-LS	OEM
MO103AL	OEM
MO103AM	OEM
MO103AP	OEM
MO103AS	OEM
MO103BL-FL	OEM
MO103BM-EM	OEM
MO103BP-FP	OEM
MO103BS-FS	OEM
MO103FM	OEM
MO103GL	OEM
MO103GP	OEM
MO103GS	OEM
MO103HL	OEM
MO103HP	OEM
MO103HS	OEM
MO111AJ	OEM
MO111AL	OEM
MO111AM	OEM
MO111BJ	OEM
MO111BL	OEM
MO111BM	OEM
MO112AJ	OEM
MO112AL	OEM
MO112AM	OEM
MO113AJ	OEM
MO113AL	OEM
MO113AM	OEM
MO114AJ	OEM
MO114AL	OEM
MO120AJ	OEM
MO120AL	OEM
MO120AM	OEM
MO120BJ	OEM
MO120BL	OEM
MO120BM	OEM
MO121AJ	OEM
MO121AL	OEM
MO121AM	OEM
MO121BJ	OEM
MO121BL	OEM
MO121BM	OEM
MO122AJ	OEM
MO122AL	OEM
MO122AM	OEM
MO122BJ	OEM
MO122BL	OEM
MO122BM	OEM
MO123AJ	OEM
MO123AL	OEM
MO123BJ	OEM
MO123BL	OEM
MO124AJ	OEM
MO124AL	OEM
MO124BJ	OEM
MO124BL	OEM
MO130C	OEM
MO130L	OEM
MO131C	OEM
MO131L	OEM
MO132C	OEM
MO132L	OEM
MO133C	OEM
MO133L	OEM
MO140C	OEM
MO140L	OEM
MO141C	OEM
MO141L	OEM
MO142C	OEM
MO142L	OEM
MO143C	OEM
MO143L	OEM
MO150C	OEM
MO150L	OEM
MO151C	OEM
MO151L	OEM
MO152C	OEM
MO152L	OEM
MO153C	OEM
MO153L	OEM
MO201AL	OEM
MO201AM	OEM
MO201AP	OEM
MO201AS	OEM
MO201BL-EL	OEM
MO201BM-EM	OEM
MO201BP-EP	OEM
MO201BS-ES	OEM
MO201FL	OEM
MO201FM	OEM
MO201FP	OEM
MO201FS	OEM
MO201GL	OEM
MO201GM-HM	OEM
MO201GP	OEM
MO201GS	OEM
MO201HL	OEM
MO201HP	OEM
MO201HS	OEM
MO201JL	OEM
MO201JM	OEM
MO201JP	OEM
MO201JS	OEM
MO201KL-NL	OEM
MO201KM-NM	OEM
MO201KP-NP	OEM
MO201KS-NS	OEM
MO202AL	OEM
MO202AM	OEM
MO202AP	OEM
MO202AS	OEM
MO202BL-FL	OEM
MO202BM-FM	OEM
MO202BP-FP	OEM
MO202BS-FS	OEM
MO202GL	OEM
MO202GM	OEM
MO202GP	OEM
MO202GS	OEM
MO202HL	OEM
MO202HM	OEM
MO202HP	OEM
MO202HS	OEM
MO202JL-LL	OEM
MO202JM-LM	OEM
MO202JP-LP	OEM
MO202JS-LS	OEM
MO203AL	OEM
MO203AM	OEM
MO203AP	OEM
MO203AS	OEM
MO203BL-EL	OEM
MO203BM-EM	OEM
MO203BP-EP	OEM
MO203BS-ES	OEM
MO203FL-HL	OEM
MO203FM-HM	OEM
MO203FP-HP	OEM
MO203FS-HS	OEM
MO261C	OEM
MO261J	OEM
MO261L	OEM
MO262C	OEM
MO262J	OEM
MO262L	OEM
MO263C	OEM
MO263J	OEM
MO263L	OEM
MO264C	OEM
MO264J	OEM
MO264L	OEM
MO265C	OEM
MO265J	OEM
MO265L	OEM
MO266C	OEM
MO266J	OEM
MO266L	OEM
MO267C	OEM
MO267J	OEM
MO267L	OEM
MO268C	OEM
MO268J	OEM
MO268L	OEM
MO269C	OEM
MO269J	OEM
MO269L	OEM
MO270C	OEM
MO270J	OEM
MO270L	OEM
MO271C	OEM
MO271J	OEM
MO271L	OEM
MO272C	OEM
MO272J	OEM
MO272L	OEM
MO273C	OEM
MO273J	OEM
MO273L	OEM
MO274C	OEM
MO274J	OEM
MO274L	OEM
MO275C	OEM
MO275J	OEM
MO275L	OEM
MO276C	OEM
MO276J	OEM
MO276L	OEM
MO277C	OEM
MO277J	OEM
MO277L	OEM
MO280C	OEM
MO280J	OEM
MO280L	OEM
MO281C	OEM
MO281J	OEM
MO281L	OEM
MO282C	OEM
MO282J	OEM
MO282L	OEM
MO283C	OEM
MO283J	OEM
MO283L	OEM
MO284C	OEM
MO284J	OEM
MO284L	OEM
MO285C	OEM
MO285J	OEM
MO285L	OEM
MO286C	OEM
MO286J	OEM
MO286L	OEM
MO287C	OEM
MO287J	OEM
MO287L	OEM
MO301AD	OEM
MO301AL	OEM
MO301AM	OEM
MO301AP	OEM
MO301AS	OEM
MO301BD-DD	OEM
MO301BL-DL	OEM
MO301BM-DM	OEM
MO301BP-DP	OEM
MO301BS-DS	OEM
MO302AD	OEM
MO302AL	OEM
MO302AM	OEM
MO302AP	OEM
MO302AS	OEM
MO302BD-DD	OEM
MO302BL-DL	OEM
MO302BM-DM	OEM
MO302BP-DP	OEM
MO302BS-DS	OEM
MO303AD	OEM
MO303AL	OEM
MO303AM	OEM
MO303AP	OEM
MO303AS	OEM
MO303BD	OEM
MO303BL	OEM
MO303BM	OEM
MO303BP	OEM
MO303BS	OEM
MO304AD	OEM
MO304AL	OEM
MO304AM	OEM
MO304AP	OEM
MO304AS	OEM
MO304BD	OEM
MO304BL	OEM
MO304BM	OEM
MO304BP	OEM
MO304BS	OEM
MO305AD	OEM
MO305AL	OEM
MO305AM	OEM
MO305AP	OEM
MO305AS	OEM
MO305BD	OEM
MO305BL	OEM
MO305BM	OEM
MO305BP	OEM
MO305BS	OEM
MO311AD	OEM
MO311AL	OEM
MO311AM	OEM
MO311AP	OEM
MO311AS	OEM
MO311BD-DD	OEM
MO311BL-DL	OEM
MO311BM-DM	OEM
MO311BP-DP	OEM
MO311BS-DS	OEM
MO312AD	OEM
MO312AL	OEM
MO312AM	OEM
MO312AP	OEM
MO312AS	OEM
MO312BD-DD	OEM
MO312BL-DL	OEM
MO312BM-DM	OEM
MO312BP-DP	OEM
MO312BS-DS	OEM
MO313AD	OEM
MO313AL	OEM
MO313AM	OEM
MO313AP	OEM
MO313AS	OEM
MO313BD	OEM
MO313BL	OEM
MO313BM	OEM
MO313BP	OEM
MO313BS	OEM
MO314AD	OEM
MO314AL	OEM
MO314AM	OEM
MO314AP	OEM
MO314AS	OEM
MO314BD	OEM
MO314BL	OEM
MO314BM	OEM
MO314BP	OEM
MO314BS	OEM
MO315AD	OEM
MO315AL	OEM
MO315AM	OEM
MO315AP	OEM
MO315AS	OEM
MO315BD	OEM
MO315BL	OEM
MO315BM	OEM
MO315BP	OEM
MO315BS	OEM
MO321AD	OEM
MO321AL	OEM
MO321AM	OEM
MO321AP	OEM
MO321AS	OEM
MO321BD	OEM
MO321BL	OEM
MO321BM	OEM
MO321BP	OEM
MO321BS	OEM
MO321CD	OEM
MO321CL	OEM
MO321CM	OEM
MO321CP	OEM
MO321CS	OEM
MO322AD	OEM
MO322AL	OEM
MO322AM	OEM
MO322AP	OEM
MO322AS	OEM
MO322BD	OEM
MO322BL	OEM
MO322BM	OEM
MO322BP	OEM
MO322BS	OEM
MO322CD	OEM
MO322CL	OEM
MO322CM	OEM
MO322CP	OEM
MO322CS	OEM
MO323AD	OEM
MO323AL	OEM
MO323AM	OEM
MO323AP	OEM
MO323AS	OEM
MO323BD	OEM
MO323BL	OEM
MO323BM	OEM
MO323BS	OEM
MO323CD	OEM
MO323CL	OEM
MO323CM	OEM
MO323CP	OEM
MO323CS	OEM
MO324AD	OEM
MO324AL	OEM
MO324AM	OEM
MO324AP	OEM
MO324AS	OEM
MO324BD	OEM
MO324BL	OEM
MO324BM	OEM
MO324BP	OEM
MO324BS	OEM
MO325AD	OEM
MO325AL	OEM
MO325AM	OEM
MO325AP	OEM
MO325AS	OEM
MO325BD	OEM
MO325BL	OEM
MO325BM	OEM
MO325BP	OEM
MO325BS	OEM
MO326AD	OEM
MO326AL	OEM
MO326AM	OEM
MO326AP	OEM
MO326AS	OEM
MO326BD	OEM
MO326BL	OEM
MO326BM	OEM
MO326BP	OEM
MO326BS	OEM
MO327AD	OEM
MO327AL	OEM
MO327AM	OEM
MO327AP	OEM
MO327AS	OEM
MO327BD	OEM
MO327BL	OEM
MO327BM	OEM
MO327BP	OEM
MO327BS	OEM
MO328AD	OEM
MO328AL	OEM
MO328AM	OEM
MO328AP	OEM
MO328AS	OEM
MO328BD	OEM
MO328BL	OEM
MO328BM	OEM
MO328BP	OEM
MO328BS	OEM
MO331AD	OEM
MO331AL	OEM
MO331AM	OEM
MO331AP	OEM
MO331AS	OEM
MO331BD	OEM
MO331BL	OEM
MO331BM	OEM
MO331BP	OEM
MO331BS	OEM
MO331CD	OEM
MO331CL	OEM
MO331CM	OEM
MO331CP	OEM
MO331CS	OEM
MO332AD	OEM
MO332AL	OEM
MO332AM	OEM
MO332AP	OEM
MO332AS	OEM
MO332BD	OEM
MO332BL	OEM
MO332BM	OEM
MO332BP	OEM
MO332BS	OEM
MO332CD	OEM
MO332CL	OEM
MO332CM	OEM
MO332CP	OEM
MO332CS	OEM
MO333AD	OEM
MO333AL	OEM
MO333AM	OEM
MO333AP	OEM
MO333AS	OEM
MO333BD	OEM
MO333BL	OEM
MO333BM	OEM
MO333BP	OEM
MO333BS	OEM
MO333CD	OEM
MO333CL	OEM
MO333CM	OEM
MO333CP	OEM
MO333CS	OEM
MO334AD	OEM
MO334AL	OEM
MO334AM	OEM
MO334AP	OEM
MO334AS	OEM
MO334BD	OEM
MO334BL	OEM
MO334BM	OEM
MO334BP	OEM
MO334BS	OEM
MO335AD	OEM
MO335AL	OEM
MO335AM	OEM
MO335AP	OEM
MO335AS	OEM
MO335BD	OEM
MO335BL	OEM
MO335BM	OEM
MO335BP	OEM
MO335BS	OEM
MO336AD	OEM
MO336AL	OEM
MO336AM	OEM
MO336AP	OEM
MO336AS	OEM
MO336BD	OEM
MO336BL	OEM
MO336BM	OEM
MO336BP	OEM
MO336BS	OEM
MO337AD	OEM
MO337AL	OEM
MO337AM	OEM
MO337AP	OEM
MO337AS	OEM
MO337BD	OEM
MO337BL	OEM
MO337BM	OEM
MO337BP	OEM
MO337BS	OEM
MO338AD	OEM
MO338AL	OEM
MO338AP	OEM
MO338AS	OEM
MO342AD	OEM
MO342AL	OEM
MO342AM	OEM
MO342AP	OEM
MO342AS	OEM
MO342BD	OEM
MO342BL	OEM
MO342BM	OEM
MO342BP	OEM
MO342BS	OEM
MO342CD	OEM
MO342CL	OEM
MO342CM	OEM
MO342CP	OEM
MO342CS	OEM
MO343AD	OEM
MO343AL	OEM
MO343AM	OEM
MO343AP	OEM
MO343AS	OEM
MO343BD	OEM
MO343BL	OEM
MO343BM	OEM
MO343BP	OEM
MO343BS	OEM
MO344AD	OEM
MO344AL	OEM
MO344AM	OEM
MO344AP	OEM
MO344AS	OEM
MO344BD	OEM
MO344BL	OEM
MO344BM	OEM
MO344BP	OEM
MO344BS	OEM
MO345AD	OEM
MO345AL	OEM
MO345AM	OEM
MO345AP	OEM
MO345AS	OEM
MO345BD	OEM
MO345BL	OEM
MO345BM	OEM
MO345BP	OEM
MO345BS	OEM
MO346AD	OEM
MO346AL	OEM
MO346AM	OEM
MO346AP	OEM
MO346AS	OEM
MO346BD	OEM
MO346BL	OEM
MO346BM	OEM
MO346BP	OEM
MO346BS	OEM
MO347AD	OEM
MO347AL	OEM
MO347AM	OEM
MO347AP	OEM
MO347AS	OEM
MO347BD	OEM
MO347BL	OEM
MO347BM	OEM
MO347BP	OEM
MO347BS	OEM
MO348AD	OEM
MO348AL	OEM
MO348AP	OEM
MO348AS	OEM
MO352AD	OEM
MO352AL	OEM
MO352AM	OEM
MO352AP	OEM
MO352AS	OEM
MO352BD	OEM
MO352BL	OEM
MO352BM	OEM
MO352BP	OEM
MO352BS	OEM
MO353AD	OEM
MO353AL	OEM
MO353AM	OEM
MO353AP	OEM
MO353AS	OEM
MO353BD	OEM
MO353BL	OEM
MO353BM	OEM
MO353BP	OEM
MO353BS	OEM
MO354AD	OEM
MO354AL	OEM
MO354AM	OEM
MO354AP	OEM
MO354AS	OEM
MO354BD	OEM
MO354BL	OEM
MO354BM	OEM
MO354BP	OEM
MO354BS	OEM
MO355AD	OEM
MO355AL	OEM
MO355AM	OEM
MO355AP	OEM
MO355AS	OEM
MO355BD	OEM
MO355BM	OEM
MO355BO	OEM
MO355BP	OEM
MO355BS	OEM
MO356AD	OEM
MO356AL	OEM
MO356AM	OEM
MO356AP	OEM
MO356AS	OEM
MO356BD	OEM
MO356BL	OEM
MO356BM	OEM
MO356BP	OEM
MO356BS	OEM
MO357AD	OEM
MO357AL	OEM
MO357AM	OEM
MO357AP	OEM
MO357AS	OEM
MO357BD	OEM
MO357BL	OEM
MO357BM	OEM
MO357BP	OEM
MO357BS	OEM
MO358AD	OEM
MO358AL	OEM
MO358AP	OEM
MO358AS	OEM
MO358BD	OEM
MO358BL	OEM
MO358BM	OEM
MO358BP	OEM
MO358BS	OEM
MO362AD	OEM
MO362AL	OEM
MO362AM	OEM
MO362AP	OEM
MO362AS	OEM
MO362BD	OEM
MO362BL	OEM
MO362BM	OEM
MO362BP	OEM
MO362BS	OEM
MO363AD	OEM
MO363AL	OEM
MO363AM	OEM
MO363AP	OEM
MO363AS	OEM
MO363BD	OEM
MO363BL	OEM
MO363BM	OEM
MO363BP	OEM
MO363BS	OEM
MO364AD	OEM
MO364AL	OEM
MO364AM	OEM
MO364AP	OEM
MO364AS	OEM
MO364BD	OEM
MO364BL	OEM
MO364BM	OEM
MO364BP	OEM
MO364BS	OEM
MO365AD	OEM
MO365AL	OEM
MO365AM	OEM
MO365AP	OEM
MO365AS	OEM
MO365BD	OEM
MO365BL	OEM
MO365BM	OEM
MO365BP	OEM
MO365BS	OEM

DEVICE TYPE	REPL CODE
MO366AD	OEM
MO366AL	OEM
MO366AM	OEM
MO366AP	OEM
MO366AS	OEM
MO366BD	OEM
MO366BL	OEM
MO366BM	OEM
MO366BP	OEM
MO366BS	OEM
MO367AD	OEM
MO367AL	OEM
MO367AM	OEM
MO367AP	OEM
MO367AS	OEM
MO368AD	OEM
MO368AL	OEM
MO368AP	OEM
MO368AS	OEM
MO1124	OEM
MO1125	OEM
MO1126	OEM
MO1201	OEM
MO1202	OEM
MO1203	OEM
MO1204	OEM
MO1205	OEM
MO1206	OEM
MO1207	OEM
MO1208	OEM
MO1209	OEM
MO1210	OEM
MO1902	OEM
MO1902A	OEM
MO1902B	OEM
MO1903	OEM
MO1903A	OEM
MO1903B	OEM
MO1903C	OEM
MO-2001	OEM
MO2002B	OEM
MO2002C	OEM
MO2002D	OEM
MO2002E	OEM
MO2002F	OEM
MO2002G	OEM
MO2003B	OEM
MO2003C	OEM
MO2003D	OEM
MO2003E	OEM
MO2003F	OEM
MO2003G	OEM
MO2004	OEM
MO2004A	OEM
MO-2005	OEM
MO-2005A	OEM
MO2014	OEM
MO2014A	OEM
MO2014B	OEM
MO2014C	OEM
MO2014D	OEM
MO2014E	OEM
MO2014F	OEM
MO2111	OEM
MO2120B	OEM
MO2120C	OEM
MO2120D	OEM
MO2120E	OEM
MO2120F	OEM
MO2120G	OEM
MO2121B	OEM
MO2121C	OEM
MO2121D	OEM
MO2121F	OEM
MO2121G	OEM
MO2122	OEM
MO2123	OEM
MO2130B	OEM
MO2130C	OEM
MO2130D	OEM
MO2130E	OEM
MO2130F	OEM
MO2130G	OEM
MO2131B	OEM
MO2131C	OEM
MO2131D	OEM
MO2131F	OEM
MO2131G	OEM
MO2132	OEM
MO2132A	OEM
MO2132B	OEM
MO2140	OEM
MO2140A	OEM
MO2141	OEM
MO2141A	OEM
MO2150	OEM
MO2151	OEM
MO2160	OEM
MO2160A	OEM
MO2170	OEM
MO2171	OEM
MO-2200	OEM
MO-2200A	OEM
MO-2201	OEM
MO-2201A	OEM
MO-2210A	OEM
MO-2211	OEM
MO-2211A	OEM
MO-2211B	OEM
MO-2212	OEM
MO-2212A	OEM
MO-2213	OEM
MO-2213A	OEM
MO-2220	OEM
MO-2220A	OEM
MO-2221	OEM
MO-2221A	OEM
MO-2230	OEM
MO-2230A	OEM
MO-2231	OEM
MO-2231A	OEM
MO-2231B	OEM
MO-2232	OEM
MO-2232A	OEM
MO-2233	OEM
MO-2233A	OEM
MO-2240	OEM
MO-2240A	OEM
MO-2241	OEM
MO-2241A	OEM
MO-2250	OEM
MO-2250A	OEM
MO-2250B	OEM
MO-2251	OEM
MO-2251A	OEM
MO-2251B	OEM
MO-2252	OEM
MO-2252A	OEM
MO-2253	OEM
MO-2253A	OEM
MO-2260	OEM
MO-2260A	OEM
MO-2261	OEM
MO-2261A	OEM
MO-2270	OEM
MO-2270A	OEM
MO-2270B	OEM
MO-2271	OEM
MO-2271A	OEM
MO-2271B	OEM
MO-2272	OEM
MO-2272A	OEM
MO-2273	OEM
MO-2273A	OEM
MO-2290	OEM
MO2301	OEM
MO2301A	OEM
MO2301B	OEM
MO2301C	OEM
MO2302	OEM
MO2302A	OEM
MO2302B	OEM
MO2302C	OEM
MO2303	OEM
MO2304	OEM
MO2305	OEM
MO2305A	OEM
MO-2306	OEM
MO-2306A	OEM
MO-2307	OEM
MO-2307A	OEM
MO2308	OEM
MO-2309	OEM
MO-2310	OEM
MO2360	OEM
MO2370	OEM
MO-2420	OEM
MO-2420A	OEM
MO-2460	OEM
MO-2470	OEM
MO2490	OEM
MO2491	OEM
MO2500	OEM
MO2500A	OEM
MO2500B	OEM
MO2500C	OEM
MO2501	OEM
MO2501A	OEM
MO2501B	OEM
MO2501C	OEM
MO2510	OEM
MO2510A	OEM
MO2510B	OEM
MO2510C	OEM
MO2511	OEM
MO2511A	OEM
MO2511B	OEM
MO2511C	OEM
MO-2620	OEM
MO-2620A	OEM
MO-2621	OEM
MO-2621A	OEM
MO-2630	OEM
MO-2630A	OEM
MO-2630B	OEM
MO-2631	OEM
MO-2631A	OEM
MO-2631B	OEM
MO2691	OEM
MO2701	OEM
MO2701A	OEM
MO2701B	OEM
MO2701C	OEM
MO2701D	OEM
MO2701E	OEM
MO2701F	OEM
MO2703	OEM
MO2703A	OEM
MO2703B	OEM
MO2703C	OEM
MO2703D	OEM
MO2703E	OEM
MO2703F	OEM
MO2707	OEM
MO2707B	OEM
MO2707C	OEM
MO2707D	OEM
MO2707E	OEM
MO2707F	OEM
MO2707G	OEM
MO2708B	OEM
MO2708C	OEM
MO2708D	OEM
MO2708E	OEM
MO2708F	OEM
MO2708G	OEM
MO2800	OEM
MO2800A	OEM
MO2800B	OEM
MO2800C	OEM
MO2800D	OEM
MO2801	OEM
MO2801A	OEM
MO2801B	OEM
MO2801C	OEM
MO2801D	OEM
MO2821	OEM
MO2822	OEM
MO2823	OEM
MO2824	OEM
MO2825	OEM
MOC31D	OEM
MOC31DR	OEM
MOC64A	OEM
MOC119	OEM
MOC601A	OEM
MOC602A	OEM
MOC603A	OEM
MOC604A	OEM
MOC622A	OEM
MOC623A	OEM
MOC624A	OEM
MOC625A	OEM
MOC626A	OEM
MOC627A	OEM
MOC628A	OEM
MOC629A	OEM
MOC633A	OEM
MOC635A	OEM
MOC640A	OEM
MOC1000	0541
MOC1001	0541
MOC1002	0541
MOC1003	0541
MOC1005	0315
MOC1006	0315
MOC1200	1053
MOC3002	0235
MOC3003	0235
MOC3007	0235
MOC3009	5051
MOC3010	5051
MOC3010MTE	5051
MOC3011	5051
MOC3020	5065
MOC3021	5065
MOC3022	5065
MOC3023	5065
MOC3030	5126
MOC3031	5126
MOC3032	5126
MOC3033	OEM
MOC3040	6538
MOC3041	5430
MOC3042	OEM
MOC3043	OEM
MOC5003	OEM
MOC5004	OEM
MOC5007	5426
MOC5008	5426
MOC5009	5426
MOC5010	OEM
MOC7811	1417
MOC7812	1417
MOC7813	1417
MOC7821	2640
MOC7822	2640
MOC7823	2640
MOC8020	OEM
MOC8021	OEM
MOC8030	4932
MOC8050	2108
MOC8100	5429
MOC8111	0541
MOC8112	0315
MOC8113	0315
MOC8204	OEM
MOC8205	OEM
MOC8206	OEM
MOD634A	OEM
MOS3635	0353
MOS6502	3065
MOS6502A	3065
MOS6510	OEM
MOS6510CBM	OEM
MOS6522	OEM
MOS6526	OEM
MOS6526R4	OEM
MOS6529B	OEM
MOS6551CBM	OEM
MOS6560-101	OEM
MOS6567R8	OEM
MOS6567R56A	OEM
MOS6581	OEM
MOS7360R7	OEM
MOS7501	OEM
MOS7712	1633
MOS8250P9	OEM
MOS8360R1	OEM
MOS8501R1	OEM
MOS8701	OEM
MOS8712	1633
MOS8713	1595
MOS65245	OEM
MOS251641-02	OEM
MOS318006-01	OEM
MOS901225-01	OEM
MOS901226-01	OEM
MOS901227-02	OEM
MOS901460-03	OEM
MOS901486-01	OEM
MOS901486-06	OEM
MOS906114-01	OEM
MOZ	0133
MP/100	OEM
MP/100-4K/8K	OEM
MP/100-8K	OEM
MP/100-16K/32K	OEM
MP/200	OEM
MP/200-8K	OEM
MP/200-16K/32K	OEM
MP/200MFC	OEM
MP-01	0015
MP01	0015
MP-02	0289
MP-04	0296
MP-06	0302
MP-08	0250
MP-055	OEM
MP066	OEM
MP067	OEM
MP067A	OEM
MP068	OEM
MP068A	OEM
MP0574J	1329
MP2	0124
MP-10	0250
MP20	OEM
MP40	OEM
MP50A	OEM
MP60	OEM
MP80	OEM
MP81	0323
MP82	1414
MP82HS181	OEM
MP82HS191	OEM
MP82HS321	OEM
MP82S181	OEM
MP82S191	OEM
MP82S321	OEM
MP84	0473
MP86	0446
MP88	1422
MP-100	OEM
MP100	0015
MP101	OEM
MP102	OEM
MP106	OEM
MP108	OEM
MP110	0160
MP110B	0604
MP110B-BLU	0604
MP110B-GRN	0604
MP110B-RED	0604
MP125	OEM
MP200DIAA	OEM
MP200DIAP	OEM
MP200DIBA	OEM
MP200DIBP	OEM
MP200DICJ	OEM
MP201DIAP	OEM
MP201DIBP	OEM
MP201DICJ	OEM
MP225	0015
MP249A	OEM
MP259	OEM
MP260	OEM
MP261	OEM
MP277	OEM
MP278	OEM
MP279	OEM
MP280	OEM
MP281	OEM
MP282	OEM
MP300	0015
MP301	OEM
MP302	OEM
MP303	OEM
MP310	OEM
MP311	OEM
MP312	OEM
MP313	OEM
MP318	OEM
MP-324	OEM
MP350	OEM
MP351	OEM
MP352	OEM
MP358	OEM
MP360	OEM
MP361	OEM
MP362	OEM
MP377B18	OEM
MP377C18	OEM
MP400	0015
MP500	0437
MP500(RECT.)	0015
MP500A	0440
MP501	0440
MP501A	0440
MP502	0440
MP502A	0440
MP503	0440
MP503A	0440
MP504	0440
MP504A	0440
MP505	0440
MP505A	0440
MP506	0440
MP506A	0440
MP507	0440
MP507A	0440
MP525	0085
MP525-1	0604
MP525-2	0604
MP525-3	0604
MP525-4	0604
MP525-5	0604
MP525-6	0604
MP549	0242
MP561JD	OEM
MP561JN	OEM
MP561KD	OEM
MP561KN	OEM
MP561SD	OEM
MP561TD	OEM
MP562ADBCD	OEM
MP562ADBIN	OEM
MP562KDBIN	OEM
MP562SDBCD	OEM
MP562SDBIN	OEM
MP600	0604
MP600A	OEM
MP601	0604
MP601A	OEM
MP602	0604
MP602A	OEM
MP603	0604
MP603A	OEM
MP651	0015
MP701	OEM
MP702	OEM
MP710	OEM
MP802	OEM
MP805	0323
MP810	2443
MP825	OEM
MP830	OEM
MP831	OEM
MP832	OEM
MP833	OEM
MP835	OEM
MP840	OEM
MP841	OEM
MP842	OEM
MP1003-1	0015
MP1003-2	0015
MP1003-4	0015
MP1005	OEM
MP1010	OEM
MP1014	0160
MP1014-1	0004
MP1014-2	0016
MP1014-4	0004
MP1014-5	0004
MP1014-6	0004
MP1077	OEM
MP1104	OEM
MP1111	OEM
MP1111B	OEM
MP1216	OEM
MP1304P	0654
MP1304PQ	0654
MP1509-1	0160
MP1509-2	0160
MP1509-3	0160
MP1529	0604
MP1529A	0604
MP1530	0604
MP1530A	0604
MP1531	0604
MP1531A	0604
MP1532	0160
MP1534	0604
MP1534A	0604
MP1535	0604
MP1535A	0604
MP1536	0604
MP1536A	0604
MP1537	0160
MP1537SL	OEM
MP1538SL	OEM
MP1539	OEM
MP1539A	0604
MP1539SL	OEM
MP1540	OEM
MP1540A	0604
MP1541	OEM
MP1541A	0604
MP1542	OEM
MP1542A	0160
MP1544	OEM
MP1544A	0604
MP1545	OEM
MP1545A	0604
MP1546	OEM
MP1546A	0604
MP1547	OEM
MP1547A	0160
MP1549	0604
MP1550	0604
MP1551	0604
MP1553	0604
MP1553A	0604
MP1554	0604
MP1554A	0604
MP1555	0604
MP1555A	0604
MP1557	0604
MP1557A	0604
MP1558	0604
MP1558A	0604
MP1559	0604
MP1559A	0604
MP1612	0975
MP1612A	0975
MP1612B	0975
MP1613	0975
MP2000A	0604
MP2000B	OEM
MP2060	0160
MP2060-1	0160
MP2060-2	OEM
MP2060-3	OEM
MP2060-4	OEM
MP2060-5	OEM
MP2060-6	OEM
MP2060-7	OEM
MP2061	0160
MP2061-1	OEM
MP2061-2	OEM
MP2061-3	OEM
MP2061-4	OEM
MP2061-5	OEM
MP2061-6	OEM
MP2061-7	OEM
MP2062	0160
MP2062-1	OEM
MP2062-3	OEM
MP2062-4	OEM
MP2062-5	OEM
MP2062-6	OEM
MP2062-7	OEM
MP2063	0604
MP2063-1	OEM
MP2063-2	OEM
MP2063-3	OEM
MP2063-4	OEM
MP2063-5	OEM
MP2063-6	OEM
MP2063-7	OEM
MP2100A	0604
MP2137	OEM
MP2137A	0160
MP2138	OEM
MP2138A	0160
MP2139	0160
MP2139A	0160
MP2140	0160
MP2140A	0604
MP2141	0160
MP2141-2	OEM
MP2141-3	OEM
MP2141-4	OEM
MP2141-5	OEM
MP2141A	0604
MP2141L3	OEM
MP2141L4	OEM
MP2141L5	OEM
MP2142	0085
MP2142-2	OEM
MP2142-3	OEM
MP2142A	0160
MP2142L	OEM
MP2142L3	OEM
MP2143	0160
MP2143A	0160
MP2144	0160
MP2144A	0160
MP2145	0160
MP2145A	0160
MP2146	0160
MP2146A	0160
MP2200A	0604
MP2200B	OEM
MP2216	OEM
MP2216-AO	OEM
MP2300A	0604
MP2526	0604
MP2810SL	OEM
MP2832	0604
MP3495A-N1LL	OEM
MP3495A-NILL	OEM
MP3611	0160
MP3612	0160
MP3613	0160
MP3614	0160
MP3615	0160
MP3616	0160
MP3617	0160
MP3618	0160
MP3730	0975
MP3730A	4297
MP3730B	4297
MP3731	0975
MP4001	OEM
MP4004	OEM
MP4301	OEM
MP5106P	3263
MP5113	0015
MP-5115	0015
MP5186AP	7152
MP5190P	0856
MP6359L	OEM
MP6566	0079
MP7572	OEM
MP8111	0042
MP8112	0042
MP8211	0936
MP8212	0936
MP8213	0936
MP8214	OEM
MP8216	OEM
MP8221	0936
MP8222	0936
MP8223	0936
MP8224	OEM
MP8226	OEM
MP8228	OEM
MP8231	0936
MP8232	0936
MP8251	OEM
MP8253	OEM
MP8255	OEM
MP8255A	OEM
MP8257	OEM
MP8259A	OEM
MP8284	OEM
MP8288	OEM
MP8304	OEM
MP8408	OEM
MP8416	OEM
MP8417	OEM
MP8418	OEM
MP8430	OEM
MP8608	OEM
MP8616	OEM
MP8632	OEM
MP8748	OEM
MP9377-16-4	OEM
MP9377-16-5	OEM
MP9377-16-6	OEM
MP13421	OEM
MP27128-2	1638
MP88303-P	OEM
MP88303P	OEM
MPB2-4H48	OEM
MPB2-5H49	OEM
MPB2-5H59	OEM
MPB2-7C39	OEM
MPB2-7C49	OEM
MPB2-7H39	OEM
MPB2-7H49	OEM
MPB2-12H49	OEM
MPB2-4859	OEM
MPB-100	OEM
MPB121D	2434
MPB123D	2597
MPB124D	4173
MPB125D	2563
MPC-4	OEM
MPC4D	OEM
MPC8D	OEM
MPC8S	OEM
MPC16S	OEM
MPC20C	0470
MPC23C	0528
MPC29C	0638
MPC29C2	0638
MPC30C	0653
MPC-31C	0673
MPC31C	0673
MPC33TS	OEM
MPC33TSAMPC800KG	OEM
MPC46C	0930
MPC47C	0946
MPC48C	0964
MPC100	OEM
MPC558C	3785
MPC561C	2592
MPC562C	2891
MPC566C	OEM
MPC566HB	1994
MPC566HC	1994
MPC566HD	1994
MPC570C	2887
MPC-574J	1329
MPC574J	1329
MPC575C	3840
MPC575C2	2872
MPC577H	2261
MPC595C	0852
MPC596C	0857
MPC596C2	0857

If replacement code is OEM, contact original manufacturer for replacement.

DEVICE TYPE	REPL CODE
MPC596C2B	0857
MPC800SG	OEM
MPC801KG	OEM
MPC801SG	OEM
MPC1001	OEM
MPC1001H2	4420
MPC1003C2	OEM
MPC1051	OEM
MPC1052	OEM
MPC1053	OEM
MPC1054	OEM
MPC1094C	OEM
MPC1228H	OEM
MPC1352C	0490
MPC1355C	0395
MPC1356C	0865
MPC1356C2	0865
MPC1363	0683
MPC1363C	0683
MPC1367C	2916
MPC1368CA	OEM
MPC1372C	2623
MPC1373H	2026
MPC1373HA	2026
MPC1378	OEM
MPC1382C	2896
MPC1458C	0360
MPC1715FU	OEM
MPC1725MR	OEM
MPC1870CA-001	OEM
MPC1891ACY	OEM
MPC3500	0133
MPC-5888	OEM
MPC7812H	0334
MPC7847	OEM
MPC14305	0624
MPD200	0133
MPD201	0133
MPD202	0133
MPD203	0133
MPD300	0133
MPD301	0133
MPD302	0133
MPD400	0133
MPD401	0133
MPD402	0133
MPD552C	OEM
MPD554C-118	OEM
MPD650C109	OEM
MPD858	2253
MPD858C	2253
MPD861C	3815
MPD861CE	3815
MPD1937	OEM
MPD1937C	2090
MPD1986C	OEM
MPD2801	OEM
MPD2801C	OEM
MPD4011B	OEM
MPD4011BC	0215
MPD4013BC	OEM
MPD4013C	OEM
MPD4049C	OEM
MPD4066BC	0101
MPD4066C	OEM
MPD4081C	OEM
MPD4093BC	2384
MPD4555	5690
MPD6104C	OEM
MPD6117C	OEM
MPD6145	OEM
MPD7801	OEM
MPD7810H	OEM
MPD14027BC	1949
MPD42274LE10	OEM
MPD42274LE12	OEM
MPDA	OEM
MPF101	0325
MPF102	1392
MPF103	0325
MPF104	0325
MPF105	0325
MPF-106	0325
MPF106	1392
MPF107	0325
MPF108	0325
MPF109	3133
MPF109B	OEM
MPF109G	OEM
MPF109N	OEM
MPF109R	OEM
MPF109V	OEM
MPF109W	OEM
MPF109Y	OEM
MPF110	OEM
MPF111	2950
MPF112	0325
MPF120	0212
MPF-121	0212
MPF121	0212
MPF122	OEM
MPF130	0212
MPF131	0212
MPF132	0212
MPF161	2987
MPF201	OEM
MPF202	OEM
MPF203	OEM
MPF211	OEM
MPF212	OEM
MPF213	OEM
MPF230	OEM
MPF231	OEM
MPF232	OEM
MPF256	0325
MPF480	OEM
MPF481	OEM
MPF521	OEM
MPF820	0325
MPF910	OEM
MPF930	OEM
MPF960	OEM
MPF970	OEM
MPF971	OEM
MPF990	OEM
MPF1000	0212
MPF2608	OEM
MPF2609	OEM
MPF3330	OEM
MPF3821	OEM
MPF3822	OEM
MPF3823	OEM
MPF3824	OEM
MPF3970	OEM
MPF3971	OEM
MPF3972	OEM
MPF3993	OEM
MPF3994	OEM
MPF4091	OEM
MPF4092	OEM
MPF4093	OEM
MPF4117	OEM
MPF4117A	OEM
MPF4118	OEM
MPF4118A	OEM
MPF4119	OEM
MPF4119A	OEM
MPF4150	OEM
MPF4220	OEM
MPF4220A	OEM
MPF4221	OEM
MPF4221A	OEM
MPF4222	OEM
MPF4222A	OEM
MPF4223	OEM
MPF4224	OEM
MPF4391	OEM
MPF4392	OEM
MPF4393	OEM
MPF4416	OEM
MPF4416A	OEM
MPF4856	OEM
MPF4856A	OEM
MPF4857	OEM
MPF4857A	OEM
MPF4858	OEM
MPF4858A	OEM
MPF4859	OEM
MPF4859A	OEM
MPF4860	OEM
MPF4860A	OEM
MPF4861	OEM
MPF4861A	OEM
MPF6659	OEM
MPF6660	OEM
MPF6661	OEM
MPF9200	OEM
MPG06D	0023
MPG3877S	OEM
MPI3401	OEM
MPL1000	0037
MPM5006	0224
MPN3201	0133
MPN3202	OEM
MPN3208	OEM
MPN3209	OEM
MPN-3401	0133
MPN3401	2518
MPN3402	2518
MPN3404	OEM
MPN3411	OEM
MPN3412	0124
MPN3500	OEM
MPN3503	OEM
MPN3504	OEM
MPQ918	0544
MPQ1500	0290
MPQ2221	0544
MPQ2221A	OEM
MPQ2222	0544
MPQ2222A	OEM
MPQ2369	0544
MPQ2483	2905
MPQ2483-1	OEM
MPQ2483-2	OEM
MPQ2484	OEM
MPQ2484-1	OEM
MPQ2484-2	OEM
MPQ2906	3673
MPQ2906A	OEM
MPQ2907	3673
MPQ2907A	OEM
MPQ3303	OEM
MPQ3467	0290
MPQ3546	OEM
MPQ3724	OEM
MPQ3725	0544
MPQ3725A	OEM
MPQ3762	OEM
MPQ3798	0282
MPQ3798-1	OEM
MPQ3798-2	OEM
MPQ3799	0282
MPQ3799-1	OEM
MPQ3799-2	OEM
MPQ3904	0544
MPQ3906	0282
MPQ6001	OEM
MPQ6002	OEM
MPQ6100	OEM
MPQ6100A	OEM
MPQ6426	OEM
MPQ6427	OEM
MPQ6501	3004
MPQ6502	3005
MPQ6600	OEM
MPQ6600-1	OEM
MPQ6600-2	OEM
MPQ6600A	OEM
MPQ6600A1	OEM
MPQ6600A2	OEM
MPQ6700	3004
MPQ6842	OEM
MPQ7041	5348
MPQ7042	2219
MPQ7043	5348
MPQ7051	OEM
MPQ7052	OEM
MPQ7053	OEM
MPQ7091	OEM
MPQ7092	OEM
MPQ7093	OEM
MPR10	OEM
MPR12	OEM
MPR15	OEM
MPR20	OEM
MPR40	OEM
MPR60	OEM
MPR80	OEM
MPR100	OEM
MPR125	OEM
MPR3877S	OEM
MPS-22	OEM
MPS25	0160
MPS-42	OEM
MPS-44	OEM
MPS-82	OEM
MPS-82-CL	OEM
MPS393	0079
MPS404	0037
MPS404A	0037
MPS407	OEM
MPS551M	0715
MPS654	0127
MPS-706	0079
MPS706	0079
MPS706A	0079
MPS706AK	OEM
MPS706AL	OEM
MPS706AM	OEM
MPS706K	OEM
MPS706L	OEM
MPS706M	OEM
MPS708	0144
MPS751	OEM
MPS751A	OEM
MPS805	0144
MPS834	0079
MPS835	0079
MPS918	0144
MPS929	0037
MPS929A	0037
MPS930	0037
MPS930A	0037
MPS960T	0079
MPS1097	0050
MPS1572	0037
MPS2114-30	2048
MPS2114-35	2048
MPS2114-45	2048
MPS2114-50	2048
MPS2114L-30	OEM
MPS2114L-35	OEM
MPS2114L-45	OEM
MPS2221	0079
MPS2221A	0079
MPS2222	0079
MPS2222A	0079
MPS2222AK	OEM
MPS2222AL	OEM
MPS2222AM	OEM
MPS2222L	OEM
MPS2222M	OEM
MPS2223K	OEM
MPS2316	OEM
MPS2332	OEM
MPS2364	OEM
MPS2369	0079
MPS2369A	0739
MPS2369AK	OEM
MPS2369AL	OEM
MPS2369AM	OEM
MPS2369K	OEM
MPS2369L	OEM
MPS2369M	OEM
MPS2484	OEM
MPS2484K	OEM
MPS2484L	OEM
MPS2484M	OEM
MPS2709	0165
MPS2711	0079
MPS2711K	OEM
MPS2711L	OEM
MPS2711M	OEM
MPS2712	0079
MPS2712K	OEM
MPS2712L	OEM
MPS2712M	OEM
MPS2713	0079
MPS2713K	OEM
MPS2713L	OEM
MPS2713M	OEM
MPS2714	0079
MPS2714K	OEM
MPS2714L	OEM
MPS2714M	OEM
MPS2715	0079
MPS-2716	0079
MPS2716	0079
MPS2823	0127
MPS2894	0127
MPS2906	0079
MPS2907	0037
MPS2907A	0037
MPS2907AK	OEM
MPS2907AL	OEM
MPS2907AM	OEM
MPS2907K	OEM
MPS2907L	OEM
MPS2907M	OEM
MPS2923	0079
MPS2923K	OEM
MPS2923L	OEM
MPS2923M	OEM
MPS2924	0079
MPS2924K	OEM
MPS2924L	OEM
MPS2924M	OEM
MPS2925	0079
MPS2925K	OEM
MPS2925L	OEM
MPS2925M	OEM
MPS2926	0079
MPS2926-B	0086
MPS2926-BRN	0079
MPS2926-G	0079
MPS2926-GRN	0079
MPS2926-O	0086
MPS2926-ORG	0079
MPS2926-R	0086
MPS2926-RED	0079
MPS2926-Y	0086
MPS2926-YEL	0079
MPS2926BRN	0079
MPS2926GRN	0079
MPS2926ORN	0079
MPS2926RED	0079
MPS2926YEL	0079
MPS2990	OEM
MPS3013	0037
MPS3390	0079
MPS3391	0079
MPS3391A	0079
MPS3392	0079
MPS3392K	OEM
MPS3392L	OEM
MPS3392M	OEM
MPS3393	0079
MPS3393K	OEM
MPS3393L	OEM
MPS3393M	OEM
MPS3394	0079
MPS3394K	OEM
MPS3394L	OEM
MPS3394M	OEM
MPS3395	0079
MPS3395K	OEM
MPS3395L	OEM
MPS3395M	OEM
MPS3396	0079
MPS3397	0111
MPS3398	0079
MPS3402	0595
MPS3403	0595
MPS3404	0595
MPS3405	0595
MPS3414	OEM
MPS3414K	OEM
MPS3414L	OEM
MPS3414M	OEM
MPS3415	OEM
MPS3415K	OEM
MPS3415L	OEM
MPS3415M	OEM
MPS3416	OEM
MPS3416K	OEM
MPS3416L	OEM
MPS3416M	OEM
MPS3417	OEM
MPS3417K	OEM
MPS3417L	OEM
MPS3417M	OEM
MPS3534	0150
MPS3536	0144
MPS-3563	0079
MPS3563	0007
MPS3564	OEM
MPS3565	0144
MPS3566	0079
MPS3567	0861
MPS3568	0861
MPS3569	0861
MPS3634	0126
MPS3638	0037
MPS-3638A	0037
MPS3638A	0037
MPS3638AK	OEM
MPS3638AL	OEM
MPS3638AM	OEM
MPS3638K	OEM
MPS3638L	3108
MPS3638M	OEM
MPS3639	0693
MPS-3640	0150
MPS3640	0037
MPS3641	OEM
MPS3641K	OEM
MPS3641L	OEM
MPS3641M	OEM
MPS3642	0079
MPS3642K	OEM
MPS3642L	OEM
MPS3642M	OEM
MPS3643	0016
MPS3644	0006
MPS3644K	OEM
MPS3644L	OEM
MPS3644M	OEM
MPS3645	0037
MPS3645K	OEM
MPS3645L	OEM
MPS3645M	OEM
MPS3646	0079
MPS3693	0144
MPS3693K	OEM
MPS3693L	OEM
MPS3693M	OEM
MPS3694	0076
MPS-3702	0037
MPS3702	0006
MPS3702K	OEM
MPS3702L	OEM
MPS3702M	OEM
MPS3703	0133
MPS3703K	OEM
MPS3703L	OEM
MPS3703M	OEM
MPS3704	0079
MPS3704K	OEM
MPS3704L	OEM
MPS3704M	OEM
MPS-3705	0079
MPS3705	0079
MPS3705K	OEM
MPS3705L	OEM
MPS3705M	OEM
MPS3706	0079
MPS3706K	OEM
MPS3706L	OEM
MPS3706M	OEM
MPS3707	1221
MPS3707K	OEM
MPS3707L	OEM
MPS3707M	OEM
MPS3708	0079
MPS3708K	OEM
MPS3708L	OEM
MPS3708M	OEM
MPS3709	0076
MPS3709K	OEM
MPS3709L	OEM
MPS3709M	OEM
MPS3710	0079
MPS3710K	OEM
MPS3710L	OEM
MPS3710M	OEM
MPS3711	0079
MPS3711K	OEM
MPS3711L	OEM
MPS3711M	OEM
MPS3721	0079
MPS3731	0975
MPS3826	0079
MPS3827	0079
MPS3866	OEM
MPS3866A	OEM
MPS3900	0079
MPS3900A	0079
MPS3901	0079
MPS3903	0079
MPS3904	4376
MPS3905	0037
MPS3906	0037
MPS3992	0079
MPS4145	0016
MPS4248	0037
MPS4249	0037
MPS4250	0037
MPS4250A	0037
MPS4257	OEM
MPS4258	5634
MPS4274	0079
MPS4275	0079
MPS4354	0037
MPS4355	0037
MPS4356	0006
MPS4888	OEM
MPS4889	OEM
MPS5015	OEM
MPS5041	OEM
MPS5086	0037
MPS5127	0079
MPS5128	0144
MPS5129	0144
MPS5131	0144
MPS5132	0144
MPS5133	0079
MPS5134	0079
MPS5135	0079
MPS5136	0079
MPS5137	0079
MPS5138	0037
MPS5138K	OEM
MPS5139	OEM
MPS5142	0037
MPS5143	0037
MPS-5172	0111
MPS5172	0079
MPS5172K	OEM
MPS5172L	OEM
MPS5172M	OEM
MPS5179	OEM
MPS5192	OEM
MPS5306	0400
MPS5308	0400
MPS5551M	0715
MPS5668	0325
MPS5855	0037
MPS5856	OEM
MPS5857	0037
MPS5858	OEM
MPS6076	0037
MPS6134	0037
MPS6172	0037
MPS6351	0016
MPS6413	0016
MPS6434	0150
MPS6502	3065
MPS6502A	3065
MPS6507	0144
MPS6511	0144
MPS6511-S	0144
MPS6512	0076
MPS6513	0079
MPS6514	0076
MPS6515	0076
MPS6516	0006
MPS6517	0037
MPS6518	0006
MPS6519	0006
MPS6520	0079
MPS6521	0079
MPS6522	0037
MPS6522(IC)	OEM
MPS6523	0693
MPS6528	0127
MPS6529	0127
MPS6530	0076
MPS6530K	OEM
MPS6530L	OEM
MPS6530M	OEM
MPS6531	0079
MPS6531K	OEM
MPS6531L	OEM
MPS6531M	OEM
MPS6532	0076
MPS6532K	OEM
MPS6532L	OEM
MPS6532M	OEM
MPS6533	0037
MPS6533K	OEM
MPS6533L	OEM
MPS6533M	0037
MPS6534	0037
MPS6534K	OEM
MPS6534L	OEM
MPS6534M	0037
MPS6535	0006
MPS6535K	OEM
MPS6535L	OEM
MPS6535M	0037
MPS6539	0224
MPS6540	0016
MPS6540-012	OEM
MPS6541	0224
MPS6542	0224
MPS6543	0224
MPS6544	0076
MPS6545	0076
MPS6546	0224
MPS6547	0144
MPS6548	0224
MPS6552	0079
MPS6553	0037
MPS6554	0079
MPS6555	0079
MPS6556	0079
MPS6560	0324
MPS6561	0324
MPS6562	0006
MPS6563	0006
MPS6564	0079
MPS6565	0076
MPS6565K	OEM
MPS6565L	OEM
MPS6565M	OEM
MPS6566	0079
MPS6566K	OEM
MPS6566L	OEM
MPS6566M	OEM
MPS6567	0079
MPS6568	0224
MPS6568A	0224
MPS6569	0224
MPS6569A	0330
MPS6570	0224
MPS6570A	0330
MPS-6571	0079
MPS6571	1221
MPS6572	0111
MPS6573	0079
MPS6574	0079
MPS6574-B	0086
MPS6574-BLUE	0086
MPS6574-G	0086
MPS6574-GREEN	0079
MPS6574-S	0086
MPS6574-SIL	0086
MPS6574-Y	0086
MPS6574-YEL	0086
MPS6575	0079
MPS6575GREEN	0155
MPS6576	0079
MPS6576-B	0079
MPS6576-BLUE	0079
MPS6576-G	0079
MPS6576-GREEN	0079
MPS6576-S	0079
MPS6576-SIL	0079
MPS6576-Y	0079
MPS6576-YEL	0155
MPS6576GREEN	0155
MPS6576SIL	0016
MPS6579	0037
MPS6580	0037
MPS6590	0086
MPS6591	0037
MPS6595	OEM
MPS6601	5034
MPS6602	5034
MPS6651	1597
MPS6652	1597
MPS6714	5034
MPS6715	5034
MPS6716	5034
MPS6717	5034
MPS6724	3799
MPS6725	3799
MPS6726	1597
MPS6727	1597
MPS6728	1597
MPS6729	1597
MPS6733	3764
MPS6734	3764
MPS6735	3764
MPS7156	OEM
MPS7418T	OEM
MPS7427B	OEM
MPS7751(A)	OEM
MPS7751B	3519
MPS8000	4463
MPS8001	0079
MPS8092	OEM
MPS8093	2260
MPS8097	0079
MPS8098	0079
MPS8099	0076
MPS8598	0037
MPS8599	0037
MPS9185	0016
MPS9246A.B	0079
MPS9406(C)	OEM
MPS9410	0086
MPS9410A	0086
MPS9410AJ	0086
MPS9410AK	0086
MPS9410H	0086
MPS9411	2050
MPS9411(AI)	0715
MPS9411A	2050
MPS9411A1	0715
MPS9411AI	0715
MPS9411T	2050
MPS9412	OEM
MPS9412(J)	OEM
MPS9416	0324
MPS9416A	0324
MPS9416AS	0324
MPS9416AT	0324
MPS9416S	0324
MPS9417	5701
MPS9417A	0324
MPS9417AS	0324
MPS9417A-T	0079
MPS9417AT	0324
MPS9418	0086
MPS9418AS	0324
MPS9418AT	0324
MPS9418S	0324
MPS9418T	0086
MPS9423	0079
MPS9423G	0144
MPS9423H	0144

If replacement code is OEM, contact original manufacturer for replacement.

DEVICE TYPE	REPL CODE
MPS9423I	0144
MPS9426	0224
MPS9426(B)	OEM
MPS9426-B	0224
MPS9426A.B	0079
MPS9426B	0224
MPS9426B,C	0224
MPS9426BC	0224
MPS9426C	0224
MPS9427	0224
MPS9427B	0224
MPS9427B.C	0079
MPS9427C	0224
MPS9433	0079
MPS9433I	0079
MPS9433J	0111
MPS9433K	0079
MPS9433S	0079
MPS9433T	0079
MPS9434	0079
MPS9434J	0079
MPS9434K	0079
MPS9440	OEM
MPS9444	0133
MPS9460	0126
MPS9460A	0126
MPS9460H	0126
MPS9461	OEM
MPS9461(AI)	OEM
MPS9461A1	0342
MPS9461AI	0342
MPS9462	OEM
MPS9466	2033
MPS9466A	0018
MPS9466AT	0436
MPS9467	2033
MPS9467A	0436
MPS9467AS	0436
MPS9467A-T	0037
MPS9467AT	0436
MPS9467T	0436
MPS9468	0436
MPS9468A	0037
MPS9468AS	0436
MPS9468AT	0436
MPS9468S	0436
MPS9468T	0037
MPS9483T	0006
MPS9484	0037
MPS9600	0079
MPS9600(G)	0079
MPS9600-5	0079
MPS9600F	0079
MPS9600G	0079
MPS9600G/H	0144
MPS9601	0144
MPS9602	0015
MPS9604	0224
MPS9604(E)	0079
MPS9604D	0079
MPS9604E	0079
MPS9604F	0079
MPS9604FG	0079
MPS9604I	0079
MPS9604R	0079
MPS9606(H)	0124
MPS9606(I)	0124
MPS9606I	0133
MPS9611-5	0079
MPS9616	0079
MPS9616A	0018
MPS9616J	0079
MPS9618	0079
MPS9618(J)	0079
MPS9618H	0079
MPS9618I	0079
MPS9618J	0079
MPS9623	0079
MPS9623C	0079
MPS9623C(F)	0079
MPS9623E	0079
MPS9623E.G	0079
MPS9623F	0079
MPS-9623G	0079
MPS9623G	0079
MPS9623G/H	0079
MPS9623H	0079
MPS9623H/I	0079
MPS9623I	0079
MPS9623I/J	0079
MPS9623J	0079
MPS9624	OEM
MPS9625	0224
MPS9625D	0224
MPS9625E	0079
MPS9625F	0224
MPS9625G	0224
MPS9625H	0224
MPS9626	0079
MPS9626(E)	0079
MPS9626(F)	OEM
MPS9626G	0079
MPS9626H	0079
MPS9626I	0079
MPS9630	0079
MPS9630H	0079
MPS9630H.I	0079
MPS-9630I	0079
MPS9630I	0079

DEVICE TYPE	REPL CODE
MPS9630K	1954
MPS9630T	0079
MPS9631	0079
MPS9631(I)	0016
MPS9631(S)	0079
MPS9631(T)	0079
MPS9631(U)	OEM
MPS9631I	0079
MPS9631J	0079
MPS9631K	0079
MPS9631S	0079
MPS9631T	0079
MPS9632	0079
MPS9632(I)	0079
MPS9632(K)	0079
MPS9632H	0079
MPS9632I	0079
MPS9632J	0079
MPS9632K	0079
MPS9632T	0079
MPS9633	0076
MPS9633B	0079
MPS9633C	0079
MPS9633D	0111
MPS9633G	0111
MPS9633X	0079
MPS9634	0079
MPS9634-O	1010
MPS9634B	0286
MPS9634C	0079
MPS9634D	0079
MPS9644	0133
MPS9646	0133
MPS9646G	0133
MPS9646H	0133
MPS9646I	0133
MPS9646J	0133
MPS9646V	OEM
MPS9666	0037
MPS9680	0037
MPS9680(T)	0037
MPS9680H	0037
MPS9680H/E	0693
MPS9680H/I	0037
MPS9680I	0037
MPS9680I/J	0037
MPS9680J	0037
MPS9680T	0037
MPS9681	0037
MPS9681(T)	0037
MPS9681J	0037
MPS9681K	0037
MPS9681T	0037
MPS9682(I)	0037
MPS9682I	0037
MPS9682J	0037
MPS9682K	0037
MPS9682T	0037
MPS9696	0079
MPS9696F	0079
MPS9696G	0155
MPS9696H	0079
MPS9700D	0079
MPS9700E	0079
MPS9700F	0079
MPS9700G	0079
MPS9700T	0079
MPS9700U	0079
MPS9750D	0037
MPS9750E	0037
MPS9750F	0037
MPS9750G	0037
MPS97500F	0037
MPS9750T	0037
MPS96800	0037
MPS775113	OEM
MPS-A05	0111
MPSA05	0086
MPSA05K	OEM
MPSA05M	OEM
MPS-A06	0715
MPSA06	0079
MPSA06K	OEM
MPSA06L	OEM
MPSA06M	OEM
MPS-A09	0111
MPSA09	0079
MPSA2	OEM
MPS-A10	0016
MPSA10	0079
MPSA10-BLU	0079
MPSA10-GRN	0079
MPSA10-RED	0079
MPSA10-WHT	0079
MPSA10-YEL	0079
MPS-A12	3799
MPSA12	0016
MPSA12K	OEM
MPS-A13	2797
MPSA-13	2797
MPSA13	0400
MPSA13(11)	0400
MPSA13K	OEM
MPS-A14	3799
MPSA14	0400
MPSA14K	OEM
MPSA16	0111
MPSA17	0076
MPS-A18	2860

DEVICE TYPE	REPL CODE
MPSA18	0111
MPS-A20	2446
MPSA20	0198
MPSA20-BLU	0111
MPSA20-GRN	0111
MPSA20-RED	0111
MPSA20-WHT	0111
MPSA20-YEL	0111
MPSA20K	OEM
MPSA20L	OEM
MPSA20M	OEM
MPS-A25	3025
MPSA25	3025
MPS-A26	3025
MPSA26	3025
MPS-A27	3025
MPSA27	3025
MPS-A28	2797
MPSA28	5979
MPS-A29	2797
MPSA29	2797
MPS-A42	0715
MPSA42	0715
MPS-A43	0715
MPSA43	0715
MPS-A44	OEM
MPSA44	OEM
MPS-A45	OEM
MPSA45	OEM
MPS-A55	0037
MPSA55	0037
MPSA55K	OEM
MPS-A56	0037
MPSA56	0037
MPSA56K	OEM
MPS-A62	3519
MPSA62	3519
MPS-A63	3519
MPSA-63	3519
MPSA63	3519
MPS-A64	3519
MPSA64	3519
MPS-A65	3519
MPSA65	3519
MPS-A66	3519
MPSA66	3519
MPS-A70	0037
MPSA70	0006
MPSA70-BLU	0861
MPSA70-GRN	0861
MPSA70-RED	0861
MPSA70-WHT	0861
MPSA70-YEL	0693
MPS-A75	3519
MPSA75	3519
MPS-A76	3519
MPSA76	3519
MPS-A77	3519
MPSA77	3519
MPS-A92	0342
MPSA92	0342
MPSA92K	OEM
MPS-A93	0342
MPSA93	0342
MPSA93K	OEM
MPSA9416A	OEM
MPSAY2	0715
MPS-D01	0715
MPSD01	0715
MPSD02	0715
MPSD03	0861
MPS-D04	0400
MPSD04	0400
MPSD05	0079
MPSD06	0079
MPS-D51	0342
MPSD51	0342
MPSD52	0342
MPSD53	0342
MPS-D54	3519
MPSD54	3519
MPSD55	0037
MPSD56	0037
MPSEL239	0079
MPSH02	0224
MPS-H04	0861
MPSH04	0861
MPSH05	0076
MPSH07	0144
MPSH08	0144
MPSH09	0127
MPSH10	0224
MPS-H11	0144
MPSH11	0224
MPS-H17	0224
MPSH17	0224
MPS-H19	0144
MPSH19	0224
MPS-H20	0144
MPSH20	0224
MPS-H24	0144
MPSH24	0144
MPSH30	0224
MPSH31	0224
MPS-H32	0079
MPSH32	0937
MPSH33	0493
MPS-H34	0224
MPSH34	0016

DEVICE TYPE	REPL CODE
MPS-H37	0224
MPSH37	0224
MPSH54	0037
MPSH55	0037
MPSH69	2033
MPSH81	0037
MPS-H83	0150
MPSH85	0037
MPSHC05	OEM
MPSHC08	OEM
MPSHC17	OEM
MPSHC24	OEM
MPSHC55	OEM
MPSHC81	OEM
MPS-K20	0079
MPSK20	0079
MPS-K21	0079
MPSK21	0079
MPS-K22	0079
MPSK22	0079
MPSK70	0037
MPSK71	0037
MPSK72	0037
MPSL01	0076
MPS-L51	5489
MPSL51	0006
MPSM-10	OEM
MPSN-24	OEM
MPS-U01	0551
MPSU01	0551
MPS-U01A	4477
MPSU01A	0551
MPS-U02	5588
MPSU02	0551
MPS-U03	0264
MPSU03	0264
MPS-U04	0264
MPSU04	0264
MPS-U05	5588
MPSU05	0018
MPS-U06	5588
MPSU06	0551
MPS-U07	0551
MPSU07	0551
MPS-U10	0338
MPSU-10	0338
MPSU10	0338
MPS-U10F	0338
MPSU11	0338
MPS-U-31	OEM
MPS-U31	1173
MPSU31	1984
MPS-U45	3538
MPSU45	3538
MPS-U51	0382
MPSU51	0382
MPS-U51A	4584
MPSU51A	0382
MPS-U52	5592
MPSU52	0382
MPS-U55	0382
MPSU55	0148
MPS-U56	0382
MPSU56	0382
MPS-U57	4205
MPSU57	0382
MPS-U60	5593
MPSU60	3758
MPS-U95	5600
MPSU95	3539
MPSW01	5034
MPSW01A	5034
MPSW05	5034
MPSW06	5034
MPSW10	3764
MPSW13	3799
MPSW14	3799
MPSW42	3764
MPSW43	3764
MPSW45	3799
MPSW51	1597
MPSW51A	1597
MPSW55	1597
MPSW56	1597
MPSW60	OEM
MPSW63	OEM
MPSW64	OEM
MPSW92	OEM
MPSW93	OEM
MPT12N05	OEM
MPT12N10	OEM
MPT20	3648
MPT24	OEM
MPT28	3331
MPT32	2730
MPTE-5	OEM
MPTE-5C	5459
MPTE-8	OEM
MPTE-8C	OEM
MPTE-10	OEM
MPTE-10C	OEM
MPTE-12	OEM
MPTE-12C	OEM
MPTE-15	OEM
MPTE-15C	OEM
MPTE-18	OEM
MPTE-18C	OEM
MPTE-22	OEM
MPTE-22C	OEM

DEVICE TYPE	REPL CODE
MPTE-36	OEM
MPTE-36C	OEM
MPTE-45	OEM
MPTE-45C	OEM
MPU-131	0316
MPU131	0316
MPU132	0316
MPU133	0316
MPU138	OEM
MPU231	OEM
MPU232	OEM
MPU233	OEM
MPU6027	0316
MPU6028	0316
MPV-131	OEM
MPV131	OEM
MPX-25	0015
MPX25	0015
MPX215	0015
MPX9623	0079
MPX9623H	0079
MPX9623H/I	0079
MPX9623I	0079
MPZ5-16A	OEM
MPZ5-16B	OEM
MPZ5-32A	OEM
MPZ5-32B	OEM
MPZ5-32C	OEM
MPZ5-180A	OEM
MPZ5-180B	OEM
MPZ5-180C	OEM
MQ1	0079
MQ1/1	OEM
MQ1/2	OEM
MQ1/3	OEM
MQ1/4	OEM
MQ1/5	OEM
MQ2	0079
MQ3/1	OEM
MQ3/2	0015
MQ3/3	OEM
MQ3/4	OEM
MQ3/5	OEM
MQ6/1	OEM
MQ6/2	0015
MQ6/3	OEM
MQ6/4	OEM
MQ6/5	OEM
MQ8/1	OEM
MQ8/2	0015
MQ8/3	OEM
MQ8/4	OEM
MQ32	0015
MQ62	0015
MQ82	0015
MQ918	OEM
MQ930	OEM
MQ982	OEM
MQ1120	OEM
MQ1129	OEM
MQ2218	OEM
MQ2219A	OEM
MQ2369	OEM
MQ2484	OEM
MQ2904	OEM
MQ2905A	OEM
MQ3251	OEM
MQ3467	OEM
MQ3725	OEM
MQ3798	OEM
MQ3799	OEM
MQ3799A	3095
MQ6001	OEM
MQ6002	OEM
MQ7001	OEM
MQ7003	OEM
MQ7004	OEM
MQ7005	OEM
MQ7007	OEM
MQ7021	OEM
MR05	OEM
MR-1	0015
MR1	1760
MR1-1000	1059
MR1-1200	0182
MR1-1400	0102
MR1-1600	0348
MR1A	OEM
MR-1C	0071
MR1C	0071
MR-1M	0071
MR1M	0071
MR2A	OEM
MR3A	OEM
MR4A	OEM
MR5	OEM
MR5-45B	OEM
MR5A	OEM
MR6A	OEM
MR7A	OEM
MR10	OEM
MR10A	OEM
MR11	OEM
MR11A	OEM
MR20	OEM
MR21	OEM
MR21CR	0058
MR30	OEM
MR31	OEM

DEVICE TYPE	REPL CODE
MR33C-H	0299
MR35	OEM
MR36C-H	OEM
MR36E-H	OEM
MR36H	0188
MR39C-H	0188
MR40	0133
MR41	OEM
MR43C-H	OEM
MR43E-H	OEM
MR47C-H	0162
MR50	1637
MR51C-H	0162
MR51E-H	0162
MR54Z	OEM
MR56C-H	0157
MR60	0015
MR62C-H	0471
MR62E-H	0636
MR62H	0636
MR64Z	OEM
MR68C-H	OEM
MR70	0373
MR75C-H	OEM
MR75E-H	OEM
MR80	0373
MR82C-H	OEM
MR90	0373
MR91C-H	0057
MR91E-H	0057
MR100	0373
MR100C-H	0170
MR100H	0170
MR110C-H	OEM
MR1130R	0550
MR120C-H	0052
MR130C-H	OEM
MR132B	0087
MR-150-01	0015
MR150C-H	0686
MR160C-H	OEM
MR180C-H	OEM
MR200C-H	OEM
MR250-1	0102
MR322	5026
MR322R	2559
MR323	2615
MR323R	2566
MR324	2615
MR324R	2566
MR325	5026
MR325R	2559
MR326	5026
MR326R	2566
MR327	2615
MR327R	2615
MR328	2615
MR328R	2566
MR330	2615
MR330R	2566
MR331	OEM
MR331R	OEM
MR400	0424
MR-461	OEM
MR500	0110
MR501	0242
MR502	0242
MR504	0540
MR506	0965
MR508	0817
MR509	0922
MR510	0922
MR600	0424
MR750	1282
MR751	1282
MR752	1287
MR754	1287
MR756	1293
MR758	1296
MR760	1296
MR800	4257
MR801	0102
MR802	OEM
MR804	OEM
MR806	OEM
MR810	0102
MR811	0102
MR812	0102
MR813	0102
MR814	0102
MR816	0102
MR817	0102
MR818	0102
MR820	1362
MR821	1362
MR822	1362
MR824	1362
MR826	1372
MR830	1191
MR831	1191
MR832	1191
MR834	1191
MR836	1191
MR850	1145
MR851	1145
MR-852	0102
MR852	0023
MR854	1145
MR856	1191

DEVICE TYPE	REPL CODE
MR860	1530
MR860R	4515
MR861	1530
MR861R	1520
MR862	1530
MR862R	1520
MR864	0029
MR864R	1847
MR866	0601
MR866R	1851
MR870	OEM
MR870R	OEM
MR871	OEM
MR871R	OEM
MR872	OEM
MR872R	OEM
MR874	OEM
MR874R	OEM
MR876	OEM
MR876R	OEM
MR910	0087
MR911	0031
MR912	0031
MR914	0031
MR916	0031
MR917	OEM
MR918	OEM
MR990	0071
MR990A	OEM
MR991	OEM
MR991A	OEM
MR992	OEM
MR992A	OEM
MR993	OEM
MR993A	OEM
MR994	OEM
MR994A	OEM
MR995A	OEM
MR996A	OEM
MR1030A	0087
MR1030AR	0087
MR1030B	0087
MR1030BR	0087
MR1031	0071
MR1031A	0087
MR1031AR	0087
MR1031B	0087
MR1031BR	0087
MR1032A	0087
MR1032AR	0087
MR1032B	0087
MR1032BR	0087
MR1033A	0087
MR1033AR	0087
MR1033B	0087
MR1033BR	0087
MR1034A	0087
MR1034AR	0087
MR1034B	0087
MR1034BR	0087
MR1035A	0087
MR1035AR	0087
MR1035B	0087
MR1035BR	0087
MR1036A	0087
MR1036AR	0087
MR1036B	0087
MR1036BR	0087
MR1038A	0087
MR1038AR	0087
MR1038B	0087
MR1038BR	0087
MR1040A	0087
MR1040AR	0087
MR1040B	0087
MR1040BR	0087
MR1120	0084
MR1120R	0534
MR1121	0090
MR1121R	0748
MR1122	0097
MR1122R	0765
MR1123	0105
MR1123R	0777
MR1124	0109
MR1124R	0538
MR1125	0116
MR1125R	0801
MR1126	0122
MR1126R	0816
MR1128	0131
MR1128R	0545
MR1130	0145
MR1130R	0550
MR1150A	OEM
MR1183	0319
MR1183R	0267
MR1184	0319
MR1184R	0267
MR1185	0319
MR1185R	0267
MR1186	0319
MR1186R	0267
MR1187	0319
MR1187R	0267
MR1188	0319
MR1188R	0267
MR1189	0319
MR1189R	0267

If replacement code is OEM, contact original manufacturer for replacement.

DEVICE TYPE	REPL CODE
MR1190	0319
MR1190R	0267
MR1191	0319
MR1191R	0267
MR1192	0319
MR1192R	0267
MR1193	0319
MR1193R	0267
MR1194	0319
MR1194R	0267
MR1195	0319
MR1195A	0319
MR1195R	0267
MR1195RA	0267
MR1196	0319
MR1196A	0319
MR1196R	0267
MR1196RA	0267
MR1197	0319
MR1197A	0319
MR1197R	0267
MR1197RA	0267
MR1198	0319
MR1198A	0319
MR1198R	0267
MR1198RA	0267
MR1200	OEM
MR1200FL	OEM
MR1201	OEM
MR1201FL	OEM
MR1202	OEM
MR1202FL	OEM
MR1203	OEM
MR1203FL	OEM
MR1204	OEM
MR1205	OEM
MR1206	OEM
MR1207	OEM
MR1207FL	OEM
MR1208	OEM
MR1208FL	OEM
MR1209	OEM
MR1210SB	OEM
MR1210SL	OEM
MR1211SB	OEM
MR1211SL	OEM
MR1212SB	OEM
MR1212SL	OEM
MR1213SB	OEM
MR1213SL	OEM
MR1214SB	OEM
MR1214SL	OEM
MR1215SB	OEM
MR1216SB	OEM
MR1216SL	OEM
MR1217SB	OEM
MR1217SL	OEM
MR1218SB	OEM
MR1218SL	OEM
MR1219SB	OEM
MR1220R	1023
MR1220SB	1023
MR1220SL	OEM
MR1221FB	OEM
MR1221FL	OEM
MR1221R	0501
MR1221SB	1023
MR1221SL	OEM
MR1222FB	OEM
MR1222FL	OEM
MR1222R	0501
MR1222SB	1023
MR1222SL	OEM
MR1223FB	OEM
MR1223FL	OEM
MR1223R	0501
MR1223SB	1023
MR1223SL	OEM
MR1224FB	OEM
MR1224FL	OEM
MR1224SB	OEM
MR1224SL	OEM
MR1225FB	OEM
MR1225FL	OEM
MR1225R	0501
MR1225SB	1023
MR1225SL	OEM
MR1226FB	OEM
MR1226FL	OEM
MR1226SB	OEM
MR1226SL	OEM
MR1227FB	OEM
MR1227FL	OEM
MR1227R	1777
MR1227SB	1036
MR1227SL	OEM
MR1228FB	OEM
MR1228FL	OEM
MR1228R	1777
MR1228SB	1036
MR1228SL	OEM
MR1229FB	OEM
MR1229FL	OEM
MR1229R	1777
MR1229SB	1036
MR1229SL	OEM
MR1230FB	OEM
MR1230FL	OEM
MR1230R	0501

DEVICE TYPE	REPL CODE
MR1230SB	1023
MR1230SL	OEM
MR1231FB	OEM
MR1231FL	OEM
MR1231R	0501
MR1231SB	1023
MR1231SL	OEM
MR1232FB	OEM
MR1232FL	OEM
MR1232R	0501
MR1232SB	1023
MR1232SL	OEM
MR1233FB	OEM
MR1233FL	OEM
MR1233R	0501
MR1233SB	1023
MR1233SL	OEM
MR1234FB	OEM
MR1234FL	OEM
MR1234R	0501
MR1234SB	1023
MR1234SL	OEM
MR1235FB	OEM
MR1235FL	OEM
MR1235R	0501
MR1235SB	1023
MR1235SL	OEM
MR1236FB	OEM
MR1236FL	OEM
MR1236R	1777
MR1236SB	1036
MR1236SL	OEM
MR1237FB	0229
MR1237FL	0229
MR1237R	1777
MR1237SB	1023
MR1237SL	0229
MR1238FB	OEM
MR1238FL	OEM
MR1238R	1777
MR1238SB	1036
MR1238SL	OEM
MR1239FB	OEM
MR1239FL	OEM
MR1239R	1777
MR1239SB	1036
MR1239SL	OEM
MR1240FB	OEM
MR1240FL	OEM
MR1240SB	OEM
MR1240SL	OEM
MR1241FB	OEM
MR1241FL	OEM
MR1241SB	OEM
MR1241SL	OEM
MR1242FB	OEM
MR1242FL	OEM
MR1242SB	OEM
MR1242SL	OEM
MR1243FB	OEM
MR1243FL	OEM
MR1243SB	OEM
MR1243SL	OEM
MR1244FB	OEM
MR1244FL	OEM
MR1244SB	OEM
MR1244SL	OEM
MR1245FB	OEM
MR1245SB	OEM
MR1246FB	OEM
MR1246FL	OEM
MR1246SB	OEM
MR1246SL	OEM
MR1247FB	0229
MR1247FL	0229
MR1247SB	0229
MR1247SL	0229
MR1248FB	OEM
MR1248FL	OEM
MR1248SB	OEM
MR1248SL	OEM
MR1249FB	OEM
MR1249SB	OEM
MR1260	OEM
MR1260FL	OEM
MR1261	OEM
MR1261FL	OEM
MR1262	OEM
MR1262FL	OEM
MR1263	OEM
MR1263FL	OEM
MR1264	OEM
MR1265	OEM
MR1266	OEM
MR1267	0229
MR1267FL	OEM
MR1268	OEM
MR1268FL	OEM
MR1269	OEM
MR1290	OEM
MR1291	OEM
MR1292	OEM
MR1293	OEM
MR1294	OEM
MR1295	OEM
MR1296	OEM
MR1297	OEM
MR1299	OEM
MR1337-1	0023

DEVICE TYPE	REPL CODE
MR1337-2	0023
MR1337-3	0023
MR1337-4	0023
MR1337-5	0023
MR1366	0711
MR1366R	2967
MR1376	0711
MR1376R	2967
MR1386	0601
MR1386R	1851
MR1396	0601
MR1396R	1851
MR1810R	1347
MR1810SB	0599
MR1810SL	OEM
MR1811R	1347
MR1811SB	0599
MR1811SL	OEM
MR1812R	1347
MR1812SB	0599
MR1812SL	OEM
MR1813R	1347
MR1813SB	0599
MR1813SL	OEM
MR1814R	1347
MR1814SB	0599
MR1814SL	OEM
MR1815R	1347
MR1815SB	0599
MR1816R	1905
MR1816SB	1986
MR1816SL	OEM
MR1817R	1905
MR1817SB	1986
MR1817SL	OEM
MR1818R	1905
MR1818SB	1986
MR1818SL	OEM
MR1819R	1905
MR1819SB	1986
MR2000	1600
MR2000R	5005
MR2000S	1600
MR2000SR	1238
MR2001	OEM
MR2001R	OEM
MR2001S	0871
MR2001SR	1240
MR2002	0871
MR2002R	OEM
MR2002S	0871
MR2002SR	1635
MR2004	0853
MR2004R	1251
MR2004S	0853
MR2004SR	1251
MR2006	1609
MR2006R	OEM
MR2006S	1609
MR2006SR	1205
MR2008	1610
MR2008R	2137
MR2008S	1610
MR2008SR	2137
MR2010	1614
MR2010R	2251
MR2010S	1614
MR2010SR	2251
MR2064	0015
MR2065	0015
MR2066	OEM
MR2067	OEM
MR2068	OEM
MR2069	0087
MR2070	0087
MR2071	OEM
MR2072	OEM
MR2073	OEM
MR2074	OEM
MR2075	0087
MR2080HA	OEM
MR2081HA	OEM
MR2082HA	OEM
MR2084HA	OEM
MR2100HA	OEM
MR2101HA	OEM
MR2102HA	OEM
MR2103HA	OEM
MR2104HA	OEM
MR2261	0015
MR2262	0229
MR2263	OEM
MR2264	OEM
MR2265	OEM
MR2266	0182
MR2271	0015
MR2272	1217
MR2273	0182
MR2360	0133
MR2361	OEM
MR2369	0087
MR2370	OEM
MR2371	OEM
MR2372	OEM
MR2373	OEM
MR2374	OEM
MR2375	0087
MR2400	OEM
MR2400F	OEM

DEVICE TYPE	REPL CODE
MR2401	OEM
MR2401F	OEM
MR2402	2884
MR2402F	OEM
MR2404	OEM
MR2404F	OEM
MR2406	OEM
MR2406F	OEM
MR2500	OEM
MR2501	OEM
MR2502	OEM
MR2504	OEM
MR2506	OEM
MR2508	OEM
MR2510	OEM
MR2525	2049
MR2525L	OEM
MR2525R	OEM
MR3932	0079
MR3933	0086
MR3934	0126
MR4001	OEM
MR4002	OEM
MR4003	OEM
MR4004	OEM
MR4005	OEM
MR4006	OEM
MR4007	OEM
MR5005	3517
MR5005R	5012
MR5010	3517
MR5010R	5012
MR5020	3517
MR5020R	5012
MR5030	3517
MR5030R	5012
MR5040	3517
MR5040R	5012
MR5059	OEM
MR5060	3541
MR5061	OEM
MR5102	3517
MR9600	0015
MR9601	0023
MR9602	0015
MR9803	0015
MRB-20C	0015
MRD14B	3185
MRD100	1497
MRD148	OEM
MRD150	0007
MRD160	OEM
MRD200	OEM
MRD210	OEM
MRD250	OEM
MRD300	3878
MRD310	2313
MRD360	3552
MRD370	3552
MRD450	0007
MRD500	OEM
MRD510	OEM
MRD601	OEM
MRD602	OEM
MRD603	OEM
MRD604	OEM
MRD611	OEM
MRD612	OEM
MRD613	OEM
MRD614	OEM
MRD701	OEM
MRD711	6473
MRD721	OEM
MRD821	0471
MRD823	OEM
MRD825	0471
MRD827	0471
MRD829	0471
MRD920	OEM
MRD921	0471
MRD923	0471
MRD925	0471
MRD927	0471
MRD929	0471
MRD933	OEM
MRD3010	OEM
MRD3011	OEM
MRD3050	2313
MRD3051	2313
MRD3052	2313
MRD3053	2313
MRD3054	2313
MRD3055	2313
MRD3056	2313
MRF100	OEM
MRF134	OEM
MRF140	OEM
MRF148	OEM
MRF150	OEM
MRF171	OEM
MRF172	OEM
MRF174	OEM
MRF200	OEM
MRF207	2041
MRF208	1198
MRF209	1233
MRF212	1233
MRF216	3119
MRF221	5303

DEVICE TYPE	REPL CODE
MRF222	OEM
MRF223	OEM
MRF224	2884
MRF225	0418
MRF227	2039
MRF229	3572
MRF230	3326
MRF231	3587
MRF232	1198
MRF233	1233
MRF234	1974
MRF237	3572
MRF238	3041
MRF239	6647
MRF240	OEM
MRF243	OEM
MRF244	3194
MRF245	3194
MRF247	3636
MRF250	OEM
MRF250A	OEM
MRF260	2719
MRF261	OEM
MRF262	2720
MRF264	OEM
MRF309	OEM
MRF313	OEM
MRF313A	OEM
MRF314	OEM
MRF314A	OEM
MRF315	OEM
MRF315A	OEM
MRF316	OEM
MRF317	OEM
MRF321	OEM
MRF323	OEM
MRF325	OEM
MRF326	OEM
MRF327	OEM
MRF328	OEM
MRF329	OEM
MRF331	OEM
MRF338	OEM
MRF340	OEM
MRF342	OEM
MRF344	OEM
MRF400	OEM
MRF402	3326
MRF404	2700
MRF406	5202
MRF412	2946
MRF412A	3563
MRF420	3565
MRF421	3631
MRF426	OEM
MRF426A	OEM
MRF427	OEM
MRF427A	OEM
MRF428	OEM
MRF428A	OEM
MRF429	OEM
MRF432	OEM
MRF433	5202
MRF435	OEM
MRF448	OEM
MRF449	2312
MRF449A	1198
MRF450	3565
MRF450A	1198
MRF452	2835
MRF452A	OEM
MRF453	2835
MRF453A	3563
MRF454	3565
MRF454A	3567
MRF455	2835
MRF455A	3563
MRF458	3565
MRF458A	3567
MRF460	2312
MRF463	2544
MRF464	2545
MRF464A	OEM
MRF466	OEM
MRF472	1591
MRF475	0936
MRF476	0936
MRF477	OEM
MRF479	OEM
MRF485	2495
MRF486	OEM
MRF492	OEM
MRF492A	OEM
MRF497	OEM
MRF501	0007
MRF502	0007
MRF511	2070
MRF515	0689
MRF517	OEM
MRF519	OEM
MRF525	OEM
MRF526	OEM
MRF531	0086
MRF534	OEM
MRF536	OEM
MRF539	OEM
MRF559	OEM
MRF571	OEM
MRF572	OEM

DEVICE TYPE	REPL CODE
MRF573	OEM
MRF580	OEM
MRF581	OEM
MRF587	OEM
MRF597	OEM
MRF600	OEM
MRF603	1198
MRF604	2700
MRF605	OEM
MRF606	2700
MRF607	3421
MRF618	2093
MRF619	3589
MRF620	3589
MRF621	2868
MRF626	2700
MRF627	OEM
MRF629	2039
MRF630	OEM
MRF641	2093
MRF644	3589
MRF646	2868
MRF648	3591
MRF652	OEM
MRF660	OEM
MRF750	OEM
MRF752	OEM
MRF754	OEM
MRF800	OEM
MRF817	3560
MRF835	OEM
MRF838	OEM
MRF838A	OEM
MRF840	OEM
MRF841	OEM
MRF842	OEM
MRF844	3644
MRF846	OEM
MRF870	OEM
MRF870A	OEM
MRF890	OEM
MRF892	OEM
MRF894	OEM
MRF901	3007
MRF904	OEM
MRF905	OEM
MRF911	2844
MRF914	OEM
MRF931	OEM
MRF961	OEM
MRF962	OEM
MRF965	OEM
MRF966	OEM
MRF967	OEM
MRF1000MA	OEM
MRF1000MB	OEM
MRF1002MA	OEM
MRF1002MB	OEM
MRF1004MA	OEM
MRF1004MB	OEM
MRF1008MA	OEM
MRF1008MB	OEM
MRF1015MA	OEM
MRF1015MB	OEM
MRF1035MA	OEM
MRF1035MB	OEM
MRF1090MA	OEM
MRF1090MB	OEM
MRF1150MA	OEM
MRF1150MB	OEM
MRF1250M	OEM
MRF1325M	OEM
MRF2001	OEM
MRF2001B	OEM
MRF2001M	OEM
MRF2003	OEM
MRF2003B	OEM
MRF2003M	OEM
MRF2005	OEM
MRF2005B	OEM
MRF2005M	OEM
MRF2010	OEM
MRF2010B	OEM
MRF2010M	OEM
MRF2016M	OEM
MRF4070	3194
MRF5177A	OEM
MRF8003	2700
MRF8004	0698
MRS6548	0127
MRS-H10	OEM
MRV-20C	0015
MS-1	0080
MS1H	0015
MS-2	0015
MS2H	0015
MS3H	0015
MS-4	0795
MS4H	0015
MS5	0084
MS5H	0015
MS9/1X3	OEM
MS10	OEM
MS11H	0015
MS12H	0015
MS13H	0015
MS14H	0015
MS15	OEM
MS15/1C	0288

DEVICE TYPE	REPL CODE
MS20	OEM
MS20K	OEM
MS21	OEM
MS22B	0016
MS30	OEM
MS30K	OEM
MS35H	0015
MS36H	0015
MS37H	0795
MS38H	0015
MS40	OEM
MS40K	OEM
MS41	OEM
MS50	0015
MS50K	OEM
MS51	OEM
MS60	1637
MS60K	OEM
MS80	OEM
MS80K	OEM
MS81-4	OEM
MS100	OEM
MS100A	OEM
MS115P	2305
MS120K	OEM
MS166	OEM
MS222	OEM
MS223	OEM
MS224	OEM
MS225	OEM
MS232	OEM
MS233	OEM
MS235	OEM
MS242	OEM
MS300	OEM
MS301	0396
MS302	0396
MS302M	OEM
MS303	OEM
MS303M	OEM
MS401	OEM
MS401M	OEM
MS500	OEM
MS501	OEM
MS510	0595
MS701T	0144
MS801X	OEM
MS802A	OEM
MS802B	OEM
MS802X	OEM
MS803X	OEM
MS804X	OEM
MS805X	OEM
MS806X	OEM
MS807X	OEM
MS808X	OEM
MS809X	OEM
MS810A	OEM
MS810B	OEM
MS812B	OEM
MS813A	OEM
MS813B	OEM
MS831A	OEM
MS831B	OEM
MS832A	OEM
MS832B	OEM
MS838A	OEM
MS838B	OEM
MS851X	OEM
MS852A	OEM
MS852B	OEM
MS852X	OEM
MS853X	OEM
MS854X	OEM
MS855X	OEM
MS856X	OEM
MS857A	OEM
MS857B	OEM
MS859B	OEM
MS874A	OEM
MS874B	OEM
MS893B	OEM
MS918F	OEM
MS1005	OEM
MS1006	OEM
MS1007	OEM
MS1008	OEM
MS1010	0595
MS1015	OEM
MS1016	OEM
MS1020	OEM
MS1021	OEM
MS1022	OEM
MS1022A	OEM
MS1023	OEM
MS1023A	OEM
MS1035	OEM
MS1036	OEM
MS1037	OEM
MS1040	OEM
MS1041	OEM
MS1042	OEM
MS1043	OEM
MS1049	OEM
MS1050	OEM
MS1051	OEM
MS1052	OEM
MS1053	OEM
MS1054	OEM

DEVICE TYPE	REPL CODE
MS1055	OEM
MS1056	OEM
MS1057	OEM
MS1058	OEM
MS1059	OEM
MS1080	OEM
MS1081	OEM
MS1081A	OEM
MS1082	OEM
MS1083	OEM
MS1083A	OEM
MS1084	OEM
MS1085	OEM
MS1085A	OEM
MS1086	OEM
MS1087	OEM
MS1087A	OEM
MS1088	OEM
MS1089	OEM
MS1090	OEM
MS1100	OEM
MS1101	OEM
MS1102	OEM
MS1103	OEM
MS1103A	OEM
MS1151	OEM
MS1153	OEM
MS1155	OEM
MS1156	OEM
MS1157	OEM
MS1158	OEM
MS1162	OEM
MS1164	OEM
MS1166	OEM
MS1168	OEM
MS1169	OEM
MS1170	OEM
MS1202	OEM
MS1203	OEM
MS1204	OEM
MS1205	OEM
MS1206	OEM
MS1207	OEM
MS1208	OEM
MS1209	OEM
MS1220	OEM
MS1221	OEM
MS1222	OEM
MS1223	OEM
MS1504	OEM
MS1505	OEM
MS1506	OEM
MS1507	OEM
MS1508	OEM
MS1509	OEM
MS1510	OEM
MS1511	OEM
MS1513A	OEM
MS1526A	OEM
MS1530	OEM
MS1530A	OEM
MS1531	OEM
MS1531A	OEM
MS1532	OEM
MS1532-16A	OEM
MS1532A	OEM
MS1533	OEM
MS1533A	OEM
MS1534	OEM
MS1534A	OEM
MS1535	OEM
MS1535A	OEM
MS1536A	OEM
MS1540	OEM
MS1541	OEM
MS1542	OEM
MS1550	OEM
MS1550A	OEM
MS1551	OEM
MS1551A	OEM
MS1552	OEM
MS1552A	OEM
MS1553	OEM
MS1553A	OEM
MS1554	OEM
MS1554A	OEM
MS1555	OEM
MS1555A	OEM
MS1560	OEM
MS1561	OEM
MS1562	OEM
MS1563	OEM
MS1570	OEM
MS1571	OEM
MS1572	OEM
MS1573	OEM
MS1580	OEM
MS1620	OEM
MS1621	OEM
MS1625	OEM
MS1626	OEM
MS1700	OEM
MS1701	OEM
MS1702	OEM
MS1703	OEM
MS1704	OEM
MS1705	OEM
MS2022	OEM
MS2023	OEM

DEVICE TYPE	REPL CODE
MS2024	OEM
MS2025	OEM
MS2501	OEM
MS2502	OEM
MS2503	OEM
MS2504	OEM
MS2505	OEM
MS2542	OEM
MS2543	OEM
MS2544	OEM
MS2545	OEM
MS2562	OEM
MS2562A	OEM
MS2563	OEM
MS2563A	OEM
MS2564	OEM
MS2564A	OEM
MS2565	OEM
MS2565A	OEM
MS2602	OEM
MS2603	OEM
MS2604	OEM
MS2605	OEM
MS2606	OEM
MS2620	OEM
MS2621	OEM
MS2623	OEM
MS2630	OEM
MS2631	OEM
MS2632	OEM
MS2991	0086
MS3001	OEM
MS3002	OEM
MS3003	OEM
MS3004	OEM
MS3005	OEM
MS3006	OEM
MS3012	OEM
MS3013	OEM
MS3014	OEM
MS3015	OEM
MS3016	OEM
MS3017	OEM
MS3062	OEM
MS3063	OEM
MS3064	OEM
MS3065	OEM
MS3130	OEM
MS3131	OEM
MS3132	OEM
MS3133	OEM
MS3134	OEM
MS3135	OEM
MS3136	OEM
MS3140	OEM
MS3141	OEM
MS3142	OEM
MS3143	OEM
MS3145	OEM
MS3146	OEM
MS3147	OEM
MS3150	OEM
MS3151-5	OEM
MS3160	OEM
MS3161-5	OEM
MS3166	OEM
MS3170	OEM
MS3171-5	OEM
MS3180	OEM
MS3181	OEM
MS3182	OEM
MS3183	OEM
MS3190	OEM
MS3191	OEM
MS3192	OEM
MS3200	OEM
MS3201-5	OEM
MS3210	OEM
MS3210-4	OEM
MS3220	OEM
MS3221	OEM
MS3222	OEM
MS3223	OEM
MS3230	OEM
MS3231	OEM
MS3250	OEM
MS3251-5	OEM
MS3256	OEM
MS3260	OEM
MS3261-5	OEM
MS3266	OEM
MS3270	OEM
MS3271-4	OEM
MS3280	OEM
MS3281-2	OEM
MS3694	0079
MS4044	OEM
MS4108	OEM
MS4109	OEM
MS4110	OEM
MS4111	OEM
MS4112	OEM
MS4170	OEM
MS4171-4	OEM
MS4180	OEM
MS4181	OEM
MS4182	OEM
MS4220	OEM
MS4221-3	OEM

DEVICE TYPE	REPL CODE
MS4230	OEM
MS4231	OEM
MS4232	OEM
MS4270	OEM
MS4271	OEM
MS4272	OEM
MS4273	OEM
MS4280	OEM
MS4281	OEM
MS4282	OEM
MS4320	OEM
MS4321	OEM
MS4322	OEM
MS4330	OEM
MS4331	OEM
MS4408	OEM
MS4409	OEM
MS4410	OEM
MS4411	OEM
MS4412	OEM
MS4500A	OEM
MS4500B-F	OEM
MS4500G	OEM
MS4501A	OEM
MS4501G	OEM
MS4502A	OEM
MS4502B-E	OEM
MS4503A	OEM
MS4503B	OEM
MS4503C	OEM
MS4510A	OEM
MS4510B-F	OEM
MS4510G	OEM
MS4511A	OEM
MS4511B-F	OEM
MS4511G	OEM
MS4512A	OEM
MS4512B-E	OEM
MS4513A	OEM
MS4513B	OEM
MS4513C	OEM
MS4520A	OEM
MS4520B-E	OEM
MS4520F	OEM
MS4520G	OEM
MS4521A	OEM
MS4521B-F	OEM
MS4522A	OEM
MS4522B-D	OEM
MS4523A	OEM
MS4523B	OEM
MS4523C	OEM
MS4530A	OEM
MS4530B-D	OEM
MS4531A	OEM
MS4531B-D	OEM
MS4532A	OEM
MS4532B	OEM
MS4532C	OEM
MS4533A	OEM
MS4533B	OEM
MS4540A	OEM
MS4540B	OEM
MS4540C	OEM
MS4541A	OEM
MS4541B	OEM
MS4541C	OEM
MS4542A	OEM
MS4542B	OEM
MS4542C	OEM
MS4543A	OEM
MS4543B	OEM
MS4543C	OEM
MS5000-8	OEM
MS5000-8A	OEM
MS5000-12	OEM
MS5000-16	OEM
MS5000-16A	OEM
MS5000A	OEM
MS5000B-F	OEM
MS5000G	OEM
MS5001A	OEM
MS5001B-F	OEM
MS5001G	OEM
MS5002A	OEM
MS5002B-F	OEM
MS5002G	OEM
MS5003A	OEM
MS5003B-F	OEM
MS5003G	OEM
MS5010A	OEM
MS5010B-F	OEM
MS5010G	OEM
MS5011A	OEM
MS5011B-F	OEM
MS5011G	OEM
MS5012A	OEM
MS5012B-F	OEM
MS5012G	OEM
MS5013A	OEM
MS5013B-F	OEM
MS5013G	OEM
MS5015	OEM
MS5020A	OEM
MS5020B-F	OEM
MS5021A	OEM
MS5021B-F	OEM
MS5022A	OEM
MS5022B-F	OEM

DEVICE TYPE	REPL CODE
MS5023A	OEM
MS5023B-F	OEM
MS5030A	OEM
MS5030B-D	OEM
MS5031A	OEM
MS5031B-D	OEM
MS5032A	OEM
MS5032B-D	OEM
MS5033A	OEM
MS5033B-D	OEM
MS5040A	OEM
MS5040B	OEM
MS5040C	OEM
MS5041A	OEM
MS5041B	OEM
MS5041C	OEM
MS5042A	OEM
MS5042B	OEM
MS5042C	OEM
MS5043A	OEM
MS5043B	OEM
MS5043C	OEM
MS5132-4A	OEM
MS5132-8	OEM
MS5133-4A	OEM
MS5133-8	OEM
MS5133-16A	OEM
MS5134-4A	OEM
MS5134-8	OEM
MS5134-16A	OEM
MS5142-8	OEM
MS5142-16A	OEM
MS5200-4A	OEM
MS5200-8	OEM
MS5200-8A	OEM
MS5240-4A	OEM
MS5240-8	OEM
MS5240-16A	OEM
MS5241-4A	OEM
MS5242-4A	OEM
MS5242-8	OEM
MS5242-16A	OEM
MS5243-4A	OEM
MS5243-8	OEM
MS5243-16A	OEM
MS5251-4A	OEM
MS5251-8	OEM
MS5251-16A	OEM
MS5252-4A	OEM
MS5252-8	OEM
MS5252-16A	OEM
MS5253-4A	OEM
MS5253-8	OEM
MS5253-16A	OEM
MS5254-4A	OEM
MS5254-8	OEM
MS5254-16A	OEM
MS6000	OEM
MS6001	OEM
MS6002	OEM
MS6003	OEM
MS6004	OEM
MS6005	OEM
MS6006	OEM
MS6007	OEM
MS6008	OEM
MS6009	OEM
MS6010	OEM
MS6011	OEM
MS6012	OEM
MS6013	OEM
MS6014	OEM
MS6015	OEM
MS6016	OEM
MS6017	OEM
MS6018	OEM
MS6019	OEM
MS6030-1	OEM
MS6030-8	OEM
MS6031-1	OEM
MS6031-8	OEM
MS6032-1	OEM
MS6032-8	OEM
MS6033-1	OEM
MS6033-8	OEM
MS6034-1	OEM
MS6034-8	OEM
MS6035-1	OEM
MS6035-8	OEM
MS6040-1	OEM
MS6040-8	OEM
MS6041-1	OEM
MS6041-8	OEM
MS6042-1	OEM
MS6042-8	OEM
MS6043-1	OEM
MS6043-8	OEM
MS6044-1	OEM
MS6044-8	OEM
MS6045-1	OEM
MS6045-8	OEM
MS6101-8	OEM
MS6101-8A	OEM
MS6101-22	OEM
MS6101-23	OEM
MS6101-24	OEM
MS6103-8	OEM
MS6103-8A	OEM
MS6103-22	OEM

DEVICE TYPE	REPL CODE
MS6103-23	OEM
MS6103-24	OEM
MS6105-8	OEM
MS6105-8A	OEM
MS6105-22	OEM
MS6105-23	OEM
MS6105-24	OEM
MS6107-8	OEM
MS6107-8A	OEM
MS6107-22	OEM
MS6107-23	OEM
MS6107-24	OEM
MS7000	OEM
MS7001	OEM
MS7101	OEM
MS7102	OEM
MS7111	OEM
MS7112	OEM
MS7500	0030
MS7501S	0144
MS7501T	0144
MS7502R	0016
MS7502S	0144
MS7502T	0144
MS7503R	0016
MS7504	0400
MS7505	0037
MS7506G	0126
MS7506H	0086
MS7506J	0086
MS8101	OEM
MS8102	OEM
MS8108	OEM
MS8112	OEM
MS8201	OEM
MS8202	OEM
MS8204	OEM
MS8208	OEM
MS8212	OEM
MSA137MS	OEM
MSA970A3	OEM
MSA7505	0042
MSA8503	0400
MSA8505	0042
MSA8506	0560
MSA8507	0400
MSA8508	0560
MSB709R	OEM
MSC6458-20SS	OEM
MSC6458-32SS	OEM
MSC11371RS	OEM
MSC645832SS	OEM
MSD601(R)	OEM
MSD601(S)	OEM
MSD601R	OEM
MSD601RT1	OEM
MSD6100	0196
MSD6101	1652
MSD6102	0196
MSD6150	1080
MSD7000	1652
MSG7506	0086
MSJ7505	0126
MSK110	OEM
MSK5405	0086
MSL915RS	OEM
MSL2301	OEM
MSL2514	OEM
MSL9371RS	OEM
MSL9372RS	OEM
MSL9373RS	OEM
MSL9378	OEM
MSL9378RS	5335
MSM54C864-80ZS	OEM
MSM63	0037
MSM74HC4094RS	OEM
MSM81C55RS	OEM
MSM311EL	2104
MSM561	OEM
MSM631	0037
MSM2114	2048
MSM2114L-3RS	2048
MSM2114L3RS	2048
MSM2764RS	0812
MSM3980	OEM
MSM3982	OEM
MSM4001	0478
MSM4002	2055
MSM4008	0988
MSM4011	0215
MSM4011RS	0215
MSM4012	0498
MSM4013	0413
MSM4013RS	0413
MSM4016	1143
MSM4016B	OEM
MSM4016RS	1143
MSM4017	0513
MSM4017AN	0513
MSM4019	1525
MSM4020	1661
MSM4020RS	OEM
MSM4023	0519
MSM4025	2072
MSM4027	1949
MSM4028	2228
MSM4030	0500
MSM4040	0056
MSM4042	0121

DEVICE TYPE	REPL CODE
MSM4043	1553
MSM4044	2308
MSM4050	0398
MSM4050RS	0398
MSM4064	0119
MSM4066RS	0101
MSM4068	2503
MSM4069	0119
MSM4069RS	0119
MSM4069UBRU	0119
MSM4071	0129
MSM4071BN	0129
MSM4072	2524
MSM4073	1536
MSM4075	2540
MSM4078	0921
MSM4081	0626
MSM4081RS	0626
MSM4082	0300
MSM4085	0303
MSM4086	0535
MSM4514RS	1830
MSM4518	1043
MSM4520	2675
MSM5258RS	OEM
MSM5550RSK	OEM
MSM5562	OEM
MSM5565-01RS	OEM
MSM5565RS	OEM
MSM5807	OEM
MSM5816RS	OEM
MSM5830RS	OEM
MSM5907	OEM
MSM5951	OEM
MSM6205GS	OEM
MSM6219AS	OEM
MSM6219GS-VK	OEM
MSM6234RS	OEM
MSM6404A-115RS	OEM
MSM6404A-120RS	OEM
MSM6404A-180RS	OEM
MSM6408-26SS	OEM
MSM6408-32SS	OEM
MSM6964	OEM
MSM6964-3RS	OEM
MSM6964MS	OEM
MSM6984	OEM
MSM6990-7203V1	OEM
MSM7400R	OEM
MSM7400RS	OEM
MSM16911RS	OEM
MSM40192	1764
MSM40193	1776
MSM55271RS	OEM
MSM58371RS	OEM
MSM84444Y	OEM
MSM87182	OEM
MSM512815RS	OEM
MSP3	OEM
MSP6	OEM
MSP10	0264
MSP10A	3280
MSP15	0264
MSP15A	3280
MSP20	0168
MSP20A	3280
MSP25	0168
MSP25A	3280
MSP30	0168
MSP31	OEM
MSP32	OEM
MSP33	OEM
MSP34	OEM
MSP1161	0437
MSP999058-1	0086
MSR2S28	OEM
MSR5	OEM
MSR10	OEM
MSR20	OEM
MSR40	OEM
MSR60	OEM
MSR80	OEM
MSR100	OEM
MSR-500	0015
MSR2502G	OEM
MSR2509A	OEM
MSR2520	OEM
MSR2528	OEM
MSR2530	OEM
MSR7502	0144
MSR7503	0016
MSR-V5	0015
MSS100-4A	OEM
MSS-1000	0015
MSS1000	0015
MSS1001	OEM
MSS1002	OEM
MSS7501	0144
MSS7502	0144
MST001RS-KSS	OEM
MST-10	0086
MST10	0264
MST10S	0233
MST15	0264
MST16	OEM
MST20	0168
MST20B	0168
MST20S	0233
MST22	OEM

DEVICE TYPE	REPL CODE
MST25	0168
MST30	0168
MST30B	0168
MST30S	0233
MST35	0168
MST40	0168
MST40B	0168
MST40S	0187
MST45	0168
MST48	OEM
MST50	0168
MST50B	0168
MST50S	0168
MST55	0168
MST105	0168
MST7501	0144
MSZ9.1	0057
MSZ10	OEM
MT01	OEM
MT020A	OEM
MT021	0015
MT021A	0015
MT022	0015
MT022A	0015
MT0404	0037
MT0404-1	0037
MT0404-2	0037
MT0411	0037
MT0412	0037
MT0413	0037
MT0414	0861
MT0461	0037
MT0462	0037
MT0463	0150
MT0493	0224
MT1	5122
MT2	5516
MT5	5514
MT10	OEM
MT14	0015
MT20	OEM
MT24	0015
MT30	OEM
MT30LA	OEM
MT32	OEM
MT44	0015
MT50	OEM
MT64	0015
MT84	0071
MT100	0144
MT101	0144
MT101B	OEM
MT102	0144
MT102B	OEM
MT104	0016
MT106	0144
MT107	0144
MT131B	OEM
MT132A	OEM
MT132B	OEM
MT170	0609
MT181	OEM
MT-301	0126
MT-302	0086
MT456	1335
MT456A	0133
MT457	0080
MT457A	0080
MT458	0609
MT458A	0133
MT459	0609
MT459A	0609
MT461	0133
MT461A	0133
MT462	0133
MT462A	0080
MT463	0609
MT463A	0609
MT464	0609
MT464A	0609
MT482	1335
MT482A	0133
MT482B	0133
MT483	0080
MT483A	0080
MT483B	0080
MT484	0609
MT484A	0133
MT484B	0133
MT485	0609
MT485A	0609
MT485B	0609
MT486	0795
MT486A	0795
MT486B	0795
MT696	0016
MT697	0016
MT698	0233
MT699	0233
MT706	0016
MT706A	0016
MT706B	0016
MT707	0016
MT708	0016
MT726	0037
MT743	0144
MT744	0144
MT753	0144
MT869	0037

If replacement code is OEM, contact original manufacturer for replacement.

DEVICE TYPE	REPL CODE
MT870	0233
MT871	0233
MT890	0080
MT910	0233
MT911	0233
MT912	0233
MT914	OEM
MT995	OEM
MT1038	0144
MT1038A	0144
MT1039	0144
MT1060	0259
MT1060A	0144
MT1061	0007
MT1061A	0144
MT1062	0418
MT1070	0560
MT1075	0861
MT1100	0861
MT1131	0037
MT1131A	0037
MT1132	0037
MT1132A	0037
MT1132B	0037
MT1254	0037
MT1255	0037
MT1256	0037
MT1257	0037
MT1258	0037
MT1259	0037
MT1259-12	OEM
MT1420	0037
MT1613	0086
MT1711	0086
MT1889	0143
MT1893	0233
MT1991	0037
MT2303	0037
MT2411	0037
MT2412	0037
MT3001	0007
MT3002	0007
MT3011	0144
MT3202	0142
MT4101	0079
MT4102	0079
MT4102A	0079
MT4103	0079
MT4104	0007
MT4264	OEM
MT4264-15	2357
MT4264-20	2357
MT5010	0080
MT5011	0080
MT5050	0015
MT5051	0015
MT5100	0609
MT5101	0609
MT5102	0080
MT5103	0080
MT6001	0079
MT6002	0079
MT6003	0079
MT6007	OEM
MT6011	OEM
MT6021	OEM
MT9001	0079
MT9002	0079
MT9003	0144
MT102351A	0050
MTA001	OEM
MTA001M	OEM
MTC1370PQ	0354
MTD1N40	OEM
MTD1N45	OEM
MTD2N50	OEM
MTD4N20	OEM
MTD4P05	OEM
MTD4P06	OEM
MTD5N05	OEM
MTD5N06	OEM
MTD6N08	OEM
MTD6N10	OEM
MTD6N15	OEM
MTD10N05E	OEM
MTD2972	OEM
MTD2973	OEM
MTD2974	OEM
MTD2975	OEM
MTD2978	OEM
MTD2979	OEM
MTD3055E	OEM
MTE13B	OEM
MTF101	OEM
MTF102	OEM
MTH5N95	OEM
MTH5N100	OEM
MTH6N55	OEM
MTH6N60	OEM
MTH6N85	OEM
MTH6N90	OEM
MTH7N45	OEM
MTH7N50	OEM
MTH8N35	OEM
MTH8N40	OEM
MTH8N45	OEM
MTH8N55	OEM
MTH8N60	OEM
MTH8P18	OEM
MTH8P20	OEM
MTH13N45	OEM
MTH13N50	OEM
MTH15N20	OEM
MTH15N35	OEM
MTH15N40	OEM
MTH20N15	OEM
MTH20P08	OEM
MTH20P10	OEM
MTH25N08	OEM
MTH25N10	OEM
MTH25P05	OEM
MTH25P06	OEM
MTH30N20	OEM
MTH35N05	OEM
MTH35N06	OEM
MTH35N15	OEM
MTH40N05	OEM
MTH40N06	OEM
MTH40N08	OEM
MTH40N10	OEM
MTH50N05E	OEM
MTM1N95	OEM
MTM1N100	OEM
MTM2N45	OEM
MTM2N50	OEM
MTM2N55	OEM
MTM2N60	OEM
MTM2N85	OEM
MTM2N90	OEM
MTM2P45	5360
MTM2P50	OEM
MTM3N35	OEM
MTM3N40	OEM
MTM3N55	OEM
MTM3N60	OEM
MTM3N75	OEM
MTM3N80	OEM
MTM3N95	OEM
MTM3N100	OEM
MTM3P25	OEM
MTM4N45	OEM
MTM4N50	OEM
MTM4N85	OEM
MTM4N90	OEM
MTM5N18	OEM
MTM5N20	OEM
MTM5N35	1668
MTM5N40	1668
MTM5N45	OEM
MTM5N95	OEM
MTM5N100	OEM
MTM5P18	OEM
MTM5P20	OEM
MTM5P25	OEM
MTM6N55	OEM
MTM6N60	1668
MTM6N85	OEM
MTM6N90	OEM
MTM7N12	OEM
MTM7N15	OEM
MTM7N18	OEM
MTM7N20	OEM
MTM7N45	OEM
MTM7N50	OEM
MTM8N08	OEM
MTM8N10	OEM
MTM8N12	OEM
MTM8N15	OEM
MTM8N18	OEM
MTM8N20	OEM
MTM8N35	OEM
MTM8N40	OEM
MTM8N60	OEM
MTM8P08	OEM
MTM8P10	OEM
MTM8P18	OEM
MTM8P20	OEM
MTM8P25	OEM
MTM8P80	OEM
MTM8P08	OEM
MTM10N05	OEM
MTM10N06	OEM
MTM10N08	OEM
MTM10N5	OEM
MTM10N10	OEM
MTM10N12	OEM
MTM10N12L	OEM
MTM10N15	OEM
MTM10N15L	OEM
MTM10N25	OEM
MTM12N05	OEM
MTM12N06	OEM
MTM12N08	OEM
MTM12N10	OEM
MTM12N12	OEM
MTM12N15	OEM
MTM12N18	OEM
MTM12N20	OEM
MTM12P05	OEM
MTM12P06	OEM
MTM12P08	OEM
MTM12P10	OEM
MTM15N05	OEM
MTM15N05L	OEM
MTM15N06	OEM
MTM15N06E	OEM
MTM15N06L	OEM
MTM15N08	OEM
MTM15N10	OEM
MTM15N12	OEM
MTM15N15	OEM
MTM15N18	OEM
MTM15N20	OEM
MTM15N35	OEM
MTM15N40	OEM
MTM15N45	OEM
MTM15N50	OEM
MTM20N08	OEM
MTM20N10	OEM
MTM20N12	OEM
MTM20N15	OEM
MTM20P05	OEM
MTM20P08	OEM
MTM20P10	OEM
MTM25N05	OEM
MTM25N05L	OEM
MTM25N06	OEM
MTM25N06L	OEM
MTM25N08	OEM
MTM25N10	OEM
MTM25N10E	OEM
MTM25P05	OEM
MTM25P06	OEM
MTM35N05	OEM
MTM35N06	OEM
MTM35N06E	OEM
MTM40N20	OEM
MTM45N05E	OEM
MTM45N15	OEM
MTM50N05E	OEM
MTM55N08	OEM
MTM55N10	OEM
MTM60N05	OEM
MTM60N06	OEM
MTM360	OEM
MTM474	OEM
MTM475	OEM
MTM564	1668
MTM565	1668
MTM814	2271
MTM815	2271
MTM1034	OEM
MTM1035	OEM
MTM1224	OEM
MTM1225	OEM
MTMP5N35	OEM
MTN5N20	OEM
MTN6N55	OEM
MTN712	OEM
MTN1130-ASR	OEM
MTN1130-CSR	OEM
MTN2130-AG	OEM
MTN2130-CG	OEM
MTN3000-ASR	OEM
MTN3100-CSR	4190
MTN3200-AG	OEM
MTN3300-CG	OEM
MTN3600-AHR	OEM
MTN3600-AO	OEM
MTN3700-CHR	OEM
MTN3700-CO	OEM
MTN4130-AHR	OEM
MTN4130-AO	OEM
MTN4130-CHR	OEM
MTN4130-CO	OEM
MTP1N45	5757
MTP1N50	5757
MTP1N55	OEM
MTP1N60	OEM
MTP1N95	OEM
MTP1N100	OEM
MTP2N18	OEM
MTP2N20	OEM
MTP2N25	OEM
MTP2N35	OEM
MTP2N40	OEM
MTP2N45	1466
MTP2N50	1466
MTP2N55	OEM
MTP2N60	OEM
MTP2N85	3718
MTP2N90	OEM
MTP2P45	5360
MTP2P50	5360
MTP3N08L	OEM
MTP3N10L	OEM
MTP3N12	OEM
MTP3N15	OEM
MTP3N18	OEM
MTP3N20	1465
MTP3N35	1465
MTP3N40	1465
MTP3N45	OEM
MTP3N50	OEM
MTP3N55	OEM
MTP3N60	3718
MTP3N75	3718
MTP3N80	3718
MTP3N95	OEM
MTP3N100	OEM
MTP3P25	OEM
MTP4N05L	OEM
MTP4N06L	OEM
MTP4N08	OEM
MTP4N10	OEM
MTP4N12	OEM
MTP4N15	1465
MTP4N25M	OEM
MTP4N45	7111
MTP4N50	1653
MTP4N50M	OEM
MTP4N85	3718
MTP4N90	OEM
MTP5N05	OEM
MTP5N06	OEM
MTP5N08	OEM
MTP5N10	OEM
MTP5N12	OEM
MTP5N18	OEM
MTP5N20	1465
MTP5N35	1465
MTP5N40	1465
MTP5N50	OEM
MTP5P18	OEM
MTP5P20	OEM
MTP5P25	OEM
MTP6N05	OEM
MTP6N06	OEM
MTP6N10	OEM
MTP6N55	1653
MTP6N60	OEM
MTP7N05	OEM
MTP7N06	2341
MTP7N12	OEM
MTP7N15	2176
MTP7N18	OEM
MTP7N20	OEM
MTP7N45	OEM
MTP7N50	OEM
MTP7P05	OEM
MTP7P06	OEM
MTP8N08	3088
MTP8N10	3088
MTP8N10E	OEM
MTP8N12	3092
MTP8N15	3092
MTP8N18	OEM
MTP8N20	OEM
MTP8N35	OEM
MTP8N40	OEM
MTP8N45	1653
MTP8N50	1653
MTP8N50M	OEM
MTP8P08	2271
MTP8P10	2271
MTP8P25	OEM
MTP10N05	5948
MTP10N06	5948
MTP10N06E	OEM
MTP10N08	OEM
MTP10N10	OEM
MTP10N10E	OEM
MTP10N10M	OEM
MTP10N12	3092
MTP10N12L	5723
MTP10N15	3092
MTP10N15L	5723
MTP10N25	OEM
MTP10N25M	OEM
MTP10N35	OEM
MTP10N40	OEM
MTP12N05	5361
MTP12N05E	OEM
MTP12N06	5361
MTP12N08	5060
MTP12N08L	OEM
MTP12N10	5060
MTP12N10L	OEM
MTP12N18	OEM
MTP12N20	5723
MTP12P05	OEM
MTP12P06	OEM
MTP12P08	OEM
MTP12P10	OEM
MTP14N05A	OEM
MTP15N05	5948
MTP15N05E	OEM
MTP15N05L	OEM
MTP15N06	5948
MTP15N06E	OEM
MTP15N06L	OEM
MTP15N12	OEM
MTP15N15	OEM
MTP16N05A	OEM
MTP20N08	OEM
MTP20N10	OEM
MTP20N10E	OEM
MTP20P06	OEM
MTP25N05	OEM
MTP25N05E	OEM
MTP25N05L	OEM
MTP25N06	OEM
MTP25N06E	OEM
MTP25N06L	OEM
MTP25N10	OEM
MTP25N10E	OEM
MTP25N10M	OEM
MTP30N05E	OEM
MTP35N06E	OEM
MTP40N06M	OEM
MTP45N05E	OEM
MTP50N05E	OEM
MTP50N05M	OEM
MTP474	OEM
MTP475	OEM
MTP564	1465
MTP565	1465
MTP814	2271
MTP815	2271
MTP1034	OEM
MTP1035	OEM
MTP1224	5060
MTP1225	5060
MTP3055E	OEM
MTW8N50E	5033
MTZ2.3B	OEM
MTZ2.4B	OEM
MTZ3.6A	OEM
MTZ3.6B	2160
MTZ4.7	0140
MTZ4.7A	OEM
MTZ4.7B	0140
MTZ4.7JA	OEM
MTZ5.1	0041
MTZ5.1(B)	0041
MTZ5.1(C)	0041
MTZ5.1-B	0589
MTZ5.1A	OEM
MTZ5.1B	0041
MTZ5.1B(TA)	0041
MTZ5.1BT-77	0041
MTZ5.1BTA	0589
MTZ5.1BTP	OEM
MTZ5.1C	OEM
MTZ5.1TP	4111
MTZ5.6	0253
MTZ5.6B	0253
MTZ5.6B-T77	0253
MTZ5.6BT-77	0253
MTZ5.6C	OEM
MTZ5.6V	0253
MTZ6.2	0471
MTZ6.2(C)	0471
MTZ6.2A	0471
MTZ6.2A-T77	OEM
MTZ6.2B	0471
MTZ6.2B(TA)	OEM
MTZ6.2BT	0471
MTZ6.2BT-77	0471
MTZ6.2C	OEM
MTZ6.2CT	OEM
MTZ6.8	0062
MTZ6.8(A)-T2	OEM
MTZ6.8-C	OEM
MTZ6.8B	0062
MTZ6.8BT-77	0471
MTZ6.8C	0062
MTZ6.8JA	OEM
MTZ6.8JB	OEM
MTZ7.5	0077
MTZ7.5(C)	0077
MTZ7.5-C	0077
MTZ7.5A	OEM
MTZ7.5B	0077
MTZ7.5C	OEM
MTZ7.5S-T2	OEM
MTZ8.2	0244
MTZ8.2(C)	0165
MTZ8.2-C	0165
MTZ8.2B	OEM
MTZ8.2C	4309
MTZ8.2T-77	OEM
MTZ9.1	0057
MTZ9.1(C)	0057
MTZ9.1A	OEM
MTZ9.1B	0057
MTZ9.1BTP	5223
MTZ9.1C	OEM
MTZ9.1JA	OEM
MTZ9.1JC	OEM
MTZ9.1TP	5030
MTZ10	0537
MTZ10(C)	0064
MTZ10A	OEM
MTZ10B	OEM
MTZ10C	OEM
MTZ11	0181
MTZ11B	0181
MTZ11BT-77	0181
MTZ11BTP	0181
MTZ11TP	0181
MTZ12	0052
MTZ12B	1005
MTZ12BT-77	1005
MTZ12C	OEM
MTZ13	0053
MTZ13A	0053
MTZ13B	OEM
MTZ13BT	OEM
MTZ13C	OEM
MTZ15	0686
MTZ15A	OEM
MTZ15B	OEM
MTZ15C	OEM
MTZ16	0445
MTZ16C	0445
MTZ18	0375
MTZ18A	OEM
MTZ20	0700
MTZ20-D	0700
MTZ20C	OEM
MTZ20D	OEM
MTZ22	0705
MTZ24	0494
MTZ24(C)	0494
MTZ24-C	0494
MTZ24A	OEM
MTZ24B	OEM
MTZ24C	2467
MTZ27B	OEM
MTZ27C	OEM
MTZ27D	OEM
MTZ30A	0195
MTZ30B	0195
MTZ30B-T77	OEM
MTZ30BTP	0195
MTZ30D	OEM
MTZ30TP	OEM
MTZ33B	OEM
MTZ33B-T77	OEM
MTZ33C	OEM
MTZ39A	OEM
MTZ607	0157
MTZ607A	0157
MTZ608	0471
MTZ608A	0471
MTZ610	0062
MTZ610A	0025
MTZ611	0077
MTZ611A	0649
MTZ612	0165
MTZ612A	0244
MTZ613	0057
MTZ613A	0057
MTZ614	0170
MTZ614A	0170
MTZ615	0181
MTZ615A	0181
MTZ616	0052
MTZ616A	0052
MTZ617	0053
MTZ617A	0053
MTZ618	0686
MTZ618A	0686
MTZ619	0445
MTZ619A	0445
MTZ620	0375
MTZ620A	0375
MTZ621	0700
MTZ621A	0700
MTZ622A	0705
MTZ623	0494
MTZ623A	0494
MTZ624	0441
MTZ624A	0441
MTZ625	0195
MTZ625A	0726
MTZ626	0166
MTZ626A	0039
MTZ627	0010
MTZ627A	0820
MTZ628	0032
MTZ628A	0350
MTZ629	0054
MTZ629A	0931
MTZ630	0068
MTZ630A	0999
MTZ6722	OEM
MTZJ2.7(B)-T2	OEM
MTZJ3.3A	OEM
MTZ-J3.6B	OEM
MTZJ3.9(B)-T2	OEM
MTZJ4.3(A)-T2	OEM
MTZJ4.3B	OEM
MTZJ4.7(A)-T2	6906
MTZJ4.7C	OEM
MTZ-J5.1C	OEM
MTZJ5.6(A)-T2	OEM
MTZJ5.6B	OEM
MTZJ5.6C	OEM
MTZ-J6.2A	OEM
MTZJ6.2B	2661
MTZJ6.2C	OEM
MTZJ6.8(C)-T2	OEM
MTZJ6.8C	OEM
MTZJ7.5	OEM
MTZJ7.5(B)-T2	OEM
MTZ-J7.5A	OEM
MTZJ7.5C	OEM
MTZJ7.5S-T2	OEM
MTZJ7.5S-TS	OEM
MTZJ8.2A	OEM
MTZJ8.2B	OEM
MTZJ9.1(B)-T2	OEM
MTZJ9.1A	OEM
MTZJ9.1B	OEM
MTZ-J9.1C	OEM
MTZJ10(C)-T2	OEM
MTZJ10A	0057
MTZJ11(A)-T2	OEM
MTZ-J11A	OEM
MTZJ11A	OEM
MTZ-J11C	OEM
MTZJ12C	OEM
MTZJ13A	OEM
MTZJ13C	OEM
MTZJ15(A)-T2	OEM
MTZJ15(C)-T2	OEM
MTZJ15A-T2	OEM
MTZJ16A	OEM
MTZJ20B	OEM
MTZ-J22B	OEM
MTZJ27C	OEM
MTZJ33(A)-T2	OEM
MTZ-J33A	OEM
MTZJ36(A)-T2	OEM
MTZJ-T-77-10	OEM
MU10	1893
MU20	1175
MU851	OEM
MU852	OEM
MU853	OEM
MU2646	2136
MU2646M	2136
MU3060PT	5992
MU4851	OEM
MU4852	OEM
MU4853	OEM
MU4891	1893
MU4892	2136
MU4893	1893
MU4894	1893
MU9610T	0551
MU9611(G)	OEM
MU9611T	0551
MU9660T	0382
MU9661T	0382
MUR105	1090
MUR110	1090
MUR115	1090
MUR120	1090
MUR405	1366
MUR410	1366
MUR415	1366
MUR420	1366
MUR430	1366
MUR440	1366
MUR450	3887
MUR460	3887
MUR470	3887
MUR480	3887
MUR490	3887
MUR605CT	0909
MUR610CT	0909
MUR615CT	0909
MUR620CT	0909
MUR805	1127
MUR810	1127
MUR815	1127
MUR820	1127
MUR860	1664
MUR1505	3551
MUR1510	3551
MUR1515	3551
MUR1520	3551
MUR1605CT	1236
MUR1610CT	1236
MUR1615CT	1236
MUR1620CT	1236
MUR1630CT	0972
MUR1640CT	0972
MUR1650CT	0972
MUR1660CT	0972
MUR3005PT	2234
MUR3010PT	2234
MUR3015PT	2234
MUR3020PT	2234
MUR3030PT	5992
MUR3040PT	5992
MUR3050PT	5992
MUR3060PT	5992
MUR4100	3887
MUS4987	3407
MUS4988	3407
MUTE-X7B	OEM
MV0009	0030
MV009	0030
MV-1	0139
MV-1Y	0139
MV1Y	0139
MV-2	0139
MV-3	0139
MV3	0139
MV-4	1925
MV4	0133
MV-5	0015
MV5	0139
MV-5T	OEM
MV5T	OEM
MV-5W	OEM
MV5W	OEM
MV-5WY	OEM
MV-11	0139
MV11	0139
MV11T	5632
MV-12	0720
MV12A	0120
MV-12H	0139
MV12H	0139
MV-12NV	0120
MV-13	0120
MV13	0343
MV-13(BIAS)	0015
MV-13(DIO)	0133
MV-13(HV-46)	0540
MV13-YH	0120
MV-13H	0120
MV13Y	0120
MV-13YH	0120
MV13YH	0120
MV20C	0348
MV50	1981
MV52	1983

DEVICE TYPE	REPL CODE
MV53	1985
MV-54	1989
MV54	1989
MV102	OEM
MV-103	OEM
MV103	OEM
MV104	OEM
MV104G	OEM
MV104GM2	OEM
MV104GM3	OEM
MV104GM4	OEM
MV104M2	OEM
MV104M3	OEM
MV104M4	OEM
MV-104V	OEM
MV109	OEM
MV121LF	0139
MV201	0005
MV202	OEM
MV-203	OEM
MV203	OEM
MV209	0030
MV209M2	OEM
MV209M3	OEM
MV209M4	OEM
MV-303	OEM
MV306	OEM
MV830	OEM
MV831	OEM
MV832	0628
MV833	OEM
MV834	0030
MV835	OEM
MV836	OEM
MV837	OEM
MV838	OEM
MV839	OEM
MV840	OEM
MV1002	OEM
MV1211F	OEM
MV1212F	0139
MV1401	OEM
MV1403H	OEM
MV1404	OEM
MV1404H	OEM
MV1405	1029
MV1405H	OEM
MV1405VT	1029
MV1410	0628
MV1410A	0628
MV1620	0554
MV1620A	OEM
MV1622	OEM
MV1622A	OEM
MV1624	0005
MV1624A	OEM
MV1626	0720
MV1626A	OEM
MV1628	OEM
MV1628A	OEM
MV1630	OEM
MV1630A	OEM
MV1632	OEM
MV1634	0628
MV1634A	OEM
MV1636	OEM
MV1636A	OEM
MV1638	0030
MV1638A	OEM
MV1640	OEM
MV1640A	OEM
MV1642	OEM
MV1642A	OEM
MV1644	OEM
MV1644A	OEM
MV1646	OEM
MV1646A	OEM
MV1648	OEM
MV1648A	OEM
MV1650	OEM
MV1650A	OEM
MV1652	OEM
MV1654	OEM
MV1656	OEM
MV1658	OEM
MV1660	OEM
MV1662	OEM
MV1664	OEM
MV1666	OEM
MV1720A	0554
MV1724	0005
MV1724A	0005
MV1726	0720
MV1726A	0720
MV1734	0628
MV1734A	0628
MV1738	0030
MV1738A	0030
MV1805C	OEM
MV1805J	OEM
MV1806C	OEM
MV1807C1	OEM
MV1807J	OEM
MV1808A	OEM
MV1808B	OEM
MV1808B1	OEM
MV1808C	OEM
MV1808C1	OEM
MV1809C	OEM
MV1809C1	OEM
MV1810A	OEM
MV1810B	OEM
MV1810B1	OEM
MV1812A	OEM
MV1812B	OEM
MV1816A1	OEM
MV1816B	OEM
MV1816B1	OEM
MV1817A	OEM
MV1817A1	OEM
MV1817B	OEM
MV1817B1	OEM
MV1866	OEM
MV1868	OEM
MV1870	OEM
MV1871	OEM
MV1872	OEM
MV1874	OEM
MV1876	OEM
MV1877	OEM
MV1878	OEM
MV2101	0554
MV2102	OEM
MV2103	0005
MV2104	0720
MV2105	0720
MV2106	OEM
MV2107	0628
MV2108	OEM
MV2109	0030
MV2110	OEM
MV2111	0030
MV2112	OEM
MV2113	OEM
MV2114	OEM
MV-2115	OEM
MV2115	OEM
MV2201	0554
MV2203	0005
MV2205	0030
MV2205(8)	0030
MV2207	OEM
MV-2209	0554
MV2209	0030
MV2209WC	0030
MV2211	OEM
MV2213	OEM
MV2215	OEM
MV2301	OEM
MV2302	OEM
MV2303	OEM
MV2304	OEM
MV2305	OEM
MV2306	OEM
MV2307	OEM
MV2308	OEM
MV2401	OEM
MV2402	OEM
MV2403	OEM
MV3007	0554
MV3007E	0554
MV3010	0005
MV3010E	0005
MV3012	0720
MV3012E	0720
MV3015	0720
MV3015E	0720
MV3020	0628
MV3020E	0628
MV3027	0628
MV3027E	0628
MV3033	0030
MV3033E	0030
MV3039	0030
MV3039E	0030
MV3047	OEM
MV3047E	OEM
MV3056	OEM
MV3056E	OEM
MV3068	OEM
MV3082	OEM
MV3100	OEM
MV3102	OEM
MV3103	OEM
MV3140	OEM
MV3141	OEM
MV3142	OEM
MV5023	1615
MV5024	5119
MV5053	1358
MV5054-2	3182
MV5074B	3019
MV5153	3276
MV5174B	5115
MV5253	1778
MV5274B	0841
MV5353	3157
MV5374B	3124
MV5491	2195
MV5753	1962
MV5774B	3096
MV6113	0628
MV6114	OEM
MV6202	OEM
MV8066	OEM
MV8092A	OEM
MV8094	OEM
MV8096	OEM
MV8103B	OEM
MV8103C	OEM
MV8103D	OEM
MV8107B	OEM
MV8107C	OEM
MV8122	OEM
MV8122A	OEM
MV8124	OEM
MV8152	OEM
MV8152A	OEM
MV8153B	OEM
MV8153C	OEM
MV8153D	OEM
MV8154	OEM
MV8157B	OEM
MV8157C	OEM
MV8202	OEM
MV8203A	OEM
MV8203B	OEM
MV8203C	OEM
MV8203D	OEM
MV8207B	OEM
MV8207C	OEM
MV8253A	OEM
MV8253B	OEM
MV8253C	OEM
MV8253D	OEM
MV8303A	OEM
MV8303B	OEM
MV8303C	OEM
MV8303D	OEM
MV9051	0005
MV9052	0554
MV9053	0554
MV9054	0554
MV9111S	OEM
MV9112S	OEM
MV9151	0720
MV9152	0720
MV9153	0720
MV9154	0720
MV9251	0628
MV9252	0628
MV9253	0628
MV9254	0030
MV9351	0030
MV9352	0030
MV9353	0030
MV9354	0030
MV9600	0030
MV9601	0030
MV9602	0030
MV12097	OEM
MV12098	OEM
MV12099	OEM
MV12100	OEM
MV12101	OEM
MV12102	OEM
MV12103	OEM
MV12104	OEM
MV12105	OEM
MV12106	OEM
MV12107	OEM
MV12108	OEM
MV12109	OEM
MV53173	5448
MV54173	6626
MV57164	5153
MV57173	5446
MVA-05A	0015
MVA-05A(DIO)	0015
MVA-05A(RECT)	0015
MVAM1	OEM
MVAM109	OEM
MVB6113	0628
MVB6114	OEM
MVB6116	0628
MVB6117	OEM
MVB6118	OEM
MVB6124	0720
MVB6125	0720
MVB6126	0720
MVE6003	OEM
MVE6005	OEM
MVE6115	6802
MVE6117	OEM
MVE6118	OEM
MVS240	OEM
MVS460	5880
MVT210	4529
MWA110	4613
MWA110H	OEM
MWA120	4618
MWA120H	OEM
MWA130	OEM
MWA130H	OEM
MWA210	OEM
MWA210H	OEM
MWA220	4666
MWA220H	OEM
MWA230	OEM
MWA230H	OEM
MWA310	OEM
MWA310H	OEM
MWA320	OEM
MWA320H	OEM
MWA330	OEM
MWA330H	OEM
MWS5101AE3	OEM
MWS5101D	4604
MWS5101E	4604
MWS5101EL3	6196
MWS5105EL2	OEM
MX06092	OEM
MX-201	0136
MX-2269	2666
MX-2273	2026
MX-2279	OEM
MX-2986	OEM
MX-3013	0579
MX3057	2362
MX-3059	1910
MX3059	1910
MX-3108	0079
MX-3122	0224
MX-3123	0224
MX3138	OEM
MX3139	1645
MX-3198	1292
MX-3235	3815
MX-3240	4669
MX-3243	0520
MX-3256	1816
MX3256	1540
MX-3260	1540
MX-3267	0079
MX-3276	0324
MX3336	0785
MX-3364	1248
MX-3369	3801
MX-3370	4209
MX-3372	1816
MX-3379	4484
MX-3389	2261
MX-3393	2872
MX-3399	2253
MX-3452	1196
MX-3481	0324
MX-3540	4639
MX-3545	1540
MX-3587	1816
MX3587	1816
MX3619	2828
MX-3631	2828
MX-3632	3804
MX-3634	0215
MX3634	0215
MX-3733	0224
MX3733	0224
MX-3734	0324
MX-3735	0436
MX3735	0436
MX-3743	0079
MX-3808	0243
MX-3820	OEM
MX-3937	0079
MX-3944	0006
MX-3948	2534
MX-3976	1260
MX-3977	4728
MX-3978	0079
MX-3979	3057
MX3979	3057
MX-4158	1010
MX-4160	0079
MX-4234	OEM
MX-4264	OEM
MX-4294	OEM
MX-4295	0894
MX-4297	0803
MX-4313	1349
MX-4314	0153
MX-4339	2806
MX4340	3236
MX-4342	OEM
MX-4376	0024
MX-4404	3772
MX-4465	4280
MX-4476	2926
MX-4478	1329
MX-4479	4635
MX-4491	0079
MX-4492	0037
MX-4493	0324
MX-4504	0155
MX-4539	0224
MX-4540	0224
MX-4541	0079
MX-4542	3816
MX-4712	5172
MX-4821	1134
MX-4873	0155
MX-4875	4831
MX-4969	4930
MX-4972	0079
MX-4973	0037
MX-4974	0706
MX-5200	5193
MX5200	OEM
MX-5201	3609
MX5201	3609
MX-5202	2019
MX5202	2019
MX-5279	0490
MX-5280	0912
MX-5281	0332
MX-5352	2654
MX-5367	6072
MX-5368	0155
MX-5369	0006
MX-5370	0155
MX-5371	0155
MX-5372	0079
MX-5375	2623
MX-5389	3982
MX-5421	3333
MX-5429	3816
MX-5430	1260
MX-5435	OEM
MX-5436	2063
MX-5451	5188
MX-5452	OEM
MX5452	OEM
MX-5473	OEM
MX-5474	4406
MX-5475	2386
MX-5476	4834
MX5560	2117
MX-5594	1260
MX-5595	2595
MX-5596	0362
MX-5679	0770
MX5741	0973
MX-5925	2759
MX-5926	0832
MX-5927	OEM
MX-6092	2194
MX-6093	3685
MX-6159	1028
MX-6160	2042
MX6198	OEM
MX6199	OEM
MX6200	OEM
MX6201	OEM
MX6202	OEM
MX6203	OEM
MX6211	OEM
MX-6387	2194
MX-6406	2447
MX-6452	1429
MX-6536	3437
MX-6829	4109
MX-6973	OEM
MX-6974	OEM
MXC-1312	1611
MXC1312A	1611
MXF-5375	2623
MXR3866	OEM
MXR5160	OEM
MXR5583	OEM
MXR5943	OEM
MXT3904	5510
MXT3906	5406
MXTA14	OEM
MXTA27	OEM
MXTA42	4512
MXTA43	OEM
MXTA44	OEM
MXTA64	OEM
MXTA77	OEM
MXTA92	4514
MXTA93	4514
MY-1	0015
MZ-00	0133
MZ-08	0244
MZ090	0012
MZ3.9T5	0456
MZ3.9T10	0456
MZ3.9T20	0456
MZ4A	0140
MZ-4HV	OEM
MZ4.3T5	0533
MZ4.3T10	0533
MZ4.3T20	0533
MZ4.7T5	0451
MZ4.7T10	0451
MZ4.7T20	0451
MZ-5	0157
MZ5A	0157
MZ5.1T5	0162
MZ5.1T10	0162
MZ5.1T20	0162
MZ5.6T5	0157
MZ5.6T10	0157
MZ5.6T20	0157
MZ-6	0025
MZ6	0025
MZ-6HV	OEM
MZ6.2	0636
MZ6.2A	0471
MZ6.2B	0636
MZ6.2T5	0636
MZ6.2T10	0636
MZ6.2T20	0636
MZ6.8T5	0025
MZ6.8T10	0025
MZ6.8T20	0025
MZ-7	0025
MZ7.5T5	0649
MZ7.5T10	0649
MZ7.5T20	0649
MZ-8	0244
MZ8	0244
MZ-8HTE1	OEM
MZ-8HV	OEM
MZ8.2T5	0244
MZ8.2T10	0244
MZ8.2T20	0244
MZ8.6	0322
MZ-9	0012
MZ9	0222
MZ9.1B	0012
MZ9.1T5	0012
MZ9.1T10	0012
MZ9.1T20	0012
MZ-10	0170
MZ10	0170
MZ10A	0170
MZ-10HV	OEM
MZ10T5	0170
MZ10T20	0170
MZ-11	0137
MZ11	0317
MZ11B	OEM
MZ11T5	0317
MZ11T10	0317
MZ11T20	0317
MZ-12	0137
MZ12	0137
MZ12A	0137
MZ-12B	0137
MZ12B	0137
MZ12T5	0137
MZ12T10	0137
MZ12T20	0137
MZ13T5	0365
MZ13T10	0365
MZ13T20	0365
MZ14	0100
MZ15A	0002
MZ15T5	0002
MZ15T10	0002
MZ15T20	0002
MZ16T5	0420
MZ16T10	0420
MZ16T20	0420
MZ18T5	0495
MZ18T10	0495
MZ18T20	0495
MZ20T5	0531
MZ20T10	0531
MZ20T20	0531
MZ22T5	0565
MZ22T10	0565
MZ22T20	0565
MZ24T5	0402
MZ24T10	0402
MZ24T20	0402
MZ27	0441
MZ27A	0441
MZ27T5	0441
MZ27T10	0441
MZ27T20	0441
MZ30T5	0726
MZ30T10	0726
MZ30T20	0726
MZ33A	0039
MZ70-2.4	1276
MZ70-2.4A	1276
MZ70-2.4B	1276
MZ70-2.5	2874
MZ70-2.5A	2874
MZ70-2.5B	2874
MZ70-2.7	0760
MZ70-2.7A	0760
MZ70-2.7B	0760
MZ70-2.8	0677
MZ70-2.8A	0677
MZ70-2.8B	0677
MZ70-3.0	0118
MZ70-3.0A	0118
MZ70-3.0B	0118
MZ70-3.3	0299
MZ70-3.3A	0299
MZ70-3.3B	0299
MZ70-3.6	0376
MZ70-3.6A	0376
MZ70-3.6B	0376
MZ70-3.9	0036
MZ70-3.9A	0036
MZ70-3.9B	0036
MZ70-4.3	0275
MZ70-4.3A	0275
MZ70-4.3B	0275
MZ70-4.7	0140
MZ70-4.7A	0140
MZ70-5.1	0041
MZ70-5.1A	0041
MZ70-5.1B	0041
MZ70-5.6	0253
MZ70-5.6A	0253
MZ70-5.6B	0253
MZ70-6.0	0091
MZ70-6.0A	0091
MZ70-6.0B	0091
MZ70-6.2	0471
MZ70-6.2A	0471
MZ70-6.2B	0471
MZ70-6.8	0062
MZ70-6.8A	0062
MZ70-6.8B	0062
MZ70-7.5	0077
MZ70-7.5A	0077
MZ70-7.5B	0077
MZ70-8.2	0165
MZ70-8.2A	0165
MZ70-8.2B	0165
MZ70-8.7	0322
MZ70-8.7A	0322
MZ70-8.7B	0322
MZ70-9.1	0057
MZ70-9.1A	0057
MZ70-9.1B	0057
MZ70-10	0064
MZ70-10A	0064
MZ70-10B	0064
MZ70-11	0181
MZ70-11A	0181
MZ70-11B	0181
MZ70-12	0052
MZ70-12A	0052
MZ70-12B	0052
MZ70-13	0053
MZ70-13A	0053
MZ70-13B	0053
MZ70-14	0879
MZ70-14A	0879
MZ70-14B	0879
MZ70-15	0686
MZ70-15A	0686
MZ70-15B	0686
MZ70-16	0445
MZ70-16A	0445
MZ70-16B	0445
MZ70-17	0210
MZ70-17A	0210
MZ70-17B	0210
MZ70-18	0375
MZ70-18A	0375
MZ70-18B	0375
MZ70-19	0671
MZ70-19A	0671
MZ70-19B	0671
MZ70-20	0700
MZ70-20A	0700
MZ70-20B	0700
MZ70-22	0705
MZ70-22A	0705
MZ70-22B	0705
MZ70-24	0494
MZ70-24A	0494
MZ70-24B	0494
MZ70-25	0714
MZ70-25A	0714
MZ70-25B	0714
MZ70-27	OEM
MZ70-27A	0455
MZ70-27B	0455
MZ70-28	0257
MZ70-28A	0257
MZ70-28B	0257
MZ70-30	0195
MZ70-30A	0195
MZ70-30B	0195
MZ70-33	0166
MZ70-33A	0166
MZ70-33B	0166
MZ70-36	0010
MZ70-36A	0010
MZ70-36B	0010
MZ70-39	0032
MZ70-39A	0032
MZ70-39B	0032
MZ70-43	0054
MZ70-43A	0054
MZ70-43B	0054
MZ70-47	0068
MZ70-47A	0068
MZ70-47B	0068
MZ70-51	0092
MZ70-51A	0092
MZ70-51B	0092
MZ70-56	0125
MZ70-56A	0125
MZ70-56B	0125
MZ70-60	2317
MZ70-60A	2317
MZ70-60B	2317
MZ70-62	0152
MZ70-62A	OEM
MZ70-62B	0152
MZ70-68	0173
MZ70-68A	0173
MZ70-68B	0173
MZ70-75	0094
MZ70-75A	0094
MZ70-75B	0094
MZ70-82	0049
MZ70-82A	0049
MZ70-82B	0049
MZ70-87	0104
MZ70-87A	0104
MZ70-87B	0104
MZ70-91	0156
MZ70-91A	0156
MZ70-91B	0156
MZ70-100	0189
MZ70-100A	0189
MZ70-100B	0189
MZ70-110	0099
MZ70-110A	0099
MZ70-110B	0099
MZ70-120	0089
MZ70-120A	0089
MZ70-120B	0089

DEVICE TYPE	REPL CODE
MZ70-130	0287
MZ70-130A	0287
MZ70-130B	0287
MZ70-140	0252
MZ70-140A	0252
MZ70-140B	0252
MZ70-150	0340
MZ70-150A	0340
MZ70-150B	0340
MZ70-160	0370
MZ70-160A	0370
MZ70-160B	0370
MZ70-170	0394
MZ70-170A	OEM
MZ70-170B	0394
MZ70-180	0425
MZ70-180A	0425
MZ70-180B	0425
MZ70-190	0453
MZ70-190A	0453
MZ70-190B	0453
MZ70-200	1473
MZ70-200A	1473
MZ70-200B	1473
MZ70MA	0301
MZ82A	0331
MZ92-2.4	1276
MZ92-2.4A	1276
MZ92-2.4B	1276
MZ92-2.5	2874
MZ92-2.5A	2874
MZ92-2.5B	2874
MZ92-2.7	0760
MZ92-2.7A	0760
MZ92-2.7B	0760
MZ92-2.8	0677
MZ92-2.8A	0677
MZ92-2.8B	0677
MZ92-3.0	0118
MZ92-3.0A	0118
MZ92-3.0B	0118
MZ92-3.3	0299
MZ92-3.3A	0299
MZ92-3.3B	0299
MZ92-3.6	0376
MZ92-3.6A	0376
MZ92-3.6B	0376
MZ92-3.9	0036
MZ92-3.9A	0188
MZ92-3.9B	0036
MZ92-4.3	0275
MZ92-4.3A	0275
MZ92-4.3B	0275
MZ92-4.7	0140
MZ92-4.7A	0451
MZ92-4.7B	0140
MZ92-5.1	0041
MZ92-5.1A	0162
MZ92-5.1B	0041
MZ92-5.6	0253
MZ92-5.6A	0157
MZ92-6,0B	0636
MZ92-6.0	0091
MZ92-6.0A	0091
MZ92-6.0B	0301
MZ92-6.2	0471
MZ92-6.2A	0471
MZ92-6.2B	0471
MZ92-6.8	0062
MZ92-6.8A	0062
MZ92-6.8B	0062
MZ92-7.5	0077
MZ92-7.5A	0077
MZ92-7.5B	0077
MZ92-8.2	0165
MZ92-8.2A	0165
MZ92-8.2B	0244
MZ92-8.7	0322
MZ92-8.7A	0322
MZ92-8.7B	0322
MZ92-9.1	0057
MZ92-9.1A	0057
MZ-92-9.1B	0012
MZ92-9.1B	0057
MZ92-10	0064
MZ92-10A	0170
MZ92-10B	0064
MZ92-11	0181
MZ92-11A	0181
MZ92-11B	0181
MZ92-12	0052
MZ92-12A	0052
MZ92-12B	0137
MZ92-13	0053
MZ92-13A	0053
MZ92-13B	0053
MZ92-14	0879
MZ92-14A	0879
MZ92-14B	0879
MZ92-15	0686
MZ92-15A	0686
MZ92-15B	0686
MZ92-16	0445
MZ92-16A	0445
MZ92-16B	0420
MZ92-17	0210
MZ92-17A	0210
MZ92-17B	0210
MZ92-18	0375

DEVICE TYPE	REPL CODE
MZ92-18A	0375
MZ92-18B	0375
MZ92-19	0671
MZ92-19A	0671
MZ92-19B	0671
MZ92-20	0700
MZ92-20A	0700
MZ92-20B	0700
MZ92-22	0705
MZ92-22A	0705
MZ92-22B	0705
MZ92-24	0494
MZ92-24A	0494
MZ92-24B	0565
MZ92-25	0445
MZ92-25A	0445
MZ92-25B	0445
MZ92-27	0455
MZ92-27A	0441
MZ92-27B	0455
MZ92-28	0257
MZ92-28A	0257
MZ92-28B	0257
MZ92-30	0195
MZ92-30A	0195
MZ92-30B	0195
MZ92-33	0166
MZ92-33A	0166
MZ92-33B	0166
MZ92-36	0010
MZ92-36A	0010
MZ92-36B	0820
MZ92-39	0032
MZ92-39A	0032
MZ92-39B	0032
MZ92-43	0054
MZ92-43A	0054
MZ92-43B	0054
MZ92-47	0068
MZ92-47A	0068
MZ92-47B	0068
MZ92-51	0092
MZ92-51A	0092
MZ92-51B	0092
MZ92-56	0125
MZ92-56A	0125
MZ92-56B	0125
MZ92-60	2317
MZ92-60A	2317
MZ92-60B	2317
MZ92-62	0152
MZ92-62A	0152
MZ92-62B	0152
MZ92-68	0173
MZ92-68A	0173
MZ92-68B	0173
MZ92-75	0094
MZ92-75A	0094
MZ92-75B	0094
MZ92-82	0049
MZ92-82A	0049
MZ92-82B	0049
MZ92-87	0104
MZ92-87A	0104
MZ92-87B	0104
MZ92-91	0156
MZ92-91A	0156
MZ92-91B	0156
MZ92-100	0189
MZ92-100A	0189
MZ92-100B	OEM
MZ92-110	0099
MZ92-110A	0099
MZ92-110B	0099
MZ92-120	0089
MZ92-120A	0089
MZ92-120B	0089
MZ92-130	0287
MZ92-130A	0287
MZ92-140	0252
MZ92-140A	0252
MZ92-140B	0252
MZ92-150	0340
MZ92-150A	0340
MZ92-150B	0340
MZ92-160	0370
MZ92-160A	0370
MZ92-160B	0370
MZ92-170	0394
MZ92-170A	0394
MZ92-170B	0394
MZ92-180	0425
MZ92-180A	0425
MZ92-180B	0425
MZ92-190	0453
MZ92-190A	OEM
MZ92-190B	0453
MZ92-200	1473
MZ92-200A	1473
MZ92-200B	1473
MZ110	1180
MZ111	1190
MZ112	1207
MZ113	1218
MZ114	1881
MZ115	0647
MZ116	1255
MZ117	2102
MZ118	1279

DEVICE TYPE	REPL CODE
MZ120	0605
MZ130	OEM
MZ203	0118
MZ204	0036
MZ-204B	0036
MZ204B	0456
MZ204S	0456
MZ205	0041
MZ-206	0157
MZ206	0091
MZ207	0062
MZ207-02A	0025
MZ207A	0025
MZ207B	0025
MZ207C	0025
MZ208	0165
MZ-209	0012
MZ209	0057
MZ209A	0012
MZ209B	0012
MZ209C	0012
MZ210	0064
MZ-210B	0170
MZ-212	0137
MZ212	0052
MZ214	0879
MZ215	OEM
MZ216	0445
MZ218	0375
MZ220	0700
MZ222	0705
MZ224	0494
MZ224A	0494
MZ250	0057
MZ303	0118
MZ303A	0118
MZ303B	0118
MZ303C	0299
MZ304	0036
MZ304A	0376
MZ304B	0036
MZ304C	0376
MZ305	0041
MZ305A	0589
MZ-305B	0041
MZ305B	0041
MZ305C	0041
MZ306	0091
MZ306-A1	0091
MZ306-A2	0091
MZ306-B2	0091
MZ306A	0091
MZ306B	0091
MZ306B2	OEM
MZ306C	0301
MZ306C3	0091
MZ307	0062
MZ307-A1	0062
MZ307A	0062
MZ307A1	0062
MZ307B	0062
MZ307C	0062
MZ308	0165
MZ308-B2	OEM
MZ308A	0165
MZ308B	0165
MZ308C	0165
MZ309	0057
MZ309-B2	OEM
MZ309A	0057
MZ309B	0012
MZ309B2	0057
MZ309C	0057
MZ310	0064
MZ310-B	0064
MZ310A	0064
MZ310B	0064
MZ310C	0064
MZ310C1	0064
MZ311	1190
MZ312	0052
MZ312A	0052
MZ312B	0052
MZ312C	0052
MZ313	1218
MZ314	0879
MZ314A	0879
MZ314B	0879
MZ314C	0879
MZ315	0647
MZ316	0445
MZ316A	0445
MZ316B	0445
MZ316C	0445
MZ317	2102
MZ318	0375
MZ318A	0375
MZ318B	0375
MZ319	2225
MZ320	0700
MZ320A	0700
MZ320B	0700
MZ322	0705
MZ322A	0705
MZ322B	0705
MZ324	0494
MZ-324A	0494
MZ324A	0494
MZ324B	0494

DEVICE TYPE	REPL CODE
MZ327	0455
MZ330	0195
MZ-330-23-B	0726
MZ330-23-B	0195
MZ330A	0195
MZ330B	0195
MZ333	0166
MZ333A	0166
MZ333B	0166
MZ336	0010
MZ339	0032
MZ343	0054
MZ406	0301
MZ407	0025
MZ408	0244
MZ408-02C	0165
MZ408A	0244
MZ408B	0244
MZ408C	0244
MZ409	0012
MZ409-02B	0012
MZ409A	0012
MZ409B	0012
MZ409C	0012
MZ410	0170
MZ410A	0170
MZ410B	0170
MZ410C	0170
MZ412	0137
MZ412A	0137
MZ412B	0137
MZ412C	0137
MZ414	0100
MZ414-B	0100
MZ414-B3	0100
MZ414A	0100
MZ414B	0100
MZ414C	0002
MZ416	0420
MZ416A	0420
MZ416B	0420
MZ416C	0420
MZ418	0495
MZ418A	0495
MZ418B	0495
MZ420	0531
MZ420A	0531
MZ420B	0531
MZ422	0565
MZ422A	0565
MZ422B	0565
MZ424	0402
MZ424A	0402
MZ424B	0402
MZ500(5.1V)	OEM
MZ500(10V)	OEM
MZ500-1	1276
MZ500-2	0760
MZ500-3	0118
MZ500-4	0299
MZ500-5	0188
MZ500-6	0036
MZ500-7	0275
MZ500-8	0140
MZ500-9	0041
MZ500-10	0253
MZ500-11	0471
MZ500-12	0062
MZ500-13	0077
MZ500-14	0165
MZ500-15	0057
MZ500-16	0064
MZ500-17	0181
MZ500-18	0052
MZ500-19	0053
MZ500-20	0686
MZ500-21	0445
MZ500-22	0375
MZ500-23	0700
MZ500-24	0705
MZ500-25	0494
MZ500-26	0455
MZ500-27	0195
MZ500-28	0166
MZ500-29	0010
MZ500-30	0032
MZ500-31	0054
MZ500-32	0068
MZ500-33	0092
MZ500-34	0125
MZ500-35	0152
MZ500-36	0173
MZ500-37	0094
MZ500-38	0049
MZ500-39	0156
MZ500-40	0189
MZ500.5	0188
MZ500.9	0162
MZ500.10	0157
MZ500.15	0012
MZ500.16	0170
MZ500.18	0137
MZ500.20	0531
MZ500.26	0441
MZ605	0471
MZ610	0471
MZ620	0471
MZ640	0471
MZ706	0205

DEVICE TYPE	REPL CODE
MZ707	0480
MZ708	0504
MZ709	0684
MZ710	0225
MZ711	0230
MZ712	0234
MZ713	0237
MZ714	1397
MZ715	0247
MZ716	0251
MZ718	0256
MZ720	0262
MZ722	0269
MZ724	0274
MZ727	0294
MZ730	0309
MZ733	0318
MZ736	0320
MZ739	0326
MZ743	0337
MZ747	0347
MZ751	0027
MZ756	0266
MZ760	2856
MZ762	0386
MZ768	0405
MZ775	0426
MZ782	0444
MZ791	0238
MZ829	0471
MZ829A	0471
MZ920-130B	OEM
MZ939	0057
MZ939A	0057
MZ939B	0057
MZ943	0317
MZ945	0181
MZ945A	0181
MZ945B	0181
MZ946	0064
MZ946A	0173
MZ946B	0181
MZ1000-1	0292
MZ1000-2	0188
MZ1000-3	0456
MZ1000-4	0533
MZ1000-5	0451
MZ1000-6	0162
MZ1000-7	0157
MZ1000-8	0636
MZ1000-9	0025
MZ1000-10	0649
MZ1000-11	0244
MZ1000-12	0012
MZ1000-13	0170
MZ1000-14	0317
MZ-1000-15	0137
MZ1000-15	0137
MZ1000-16	0365
MZ1000-17	0002
MZ1000-18	0420
MZ1000-19	0495
MZ1000-20	0531
MZ1000-21	0565
MZ1000-22	0402
MZ1000-23	0441
MZ1000-24	0726
MZ1000-25	0039
MZ1000-26	0320
MZ1000-27	0350
MZ1000-28	0931
MZ1000-29	0999
MZ1000-30	0502
MZ1000-31	0869
MZ1000-32	0783
MZ1000-33	2157
MZ1000-34	1189
MZ1000-35	3026
MZ1000-36	1311
MZ1000-37	0098
MZ1002	OEM
MZ1003	0118
MZ1004	0036
MZ1005	0162
MZ1006	0301
MZ1007	0062
MZ1008	0165
MZ1009	0057
MZ1010	0064
MZ1012	0052
MZ1014	0879
MZ1016	0445
MZ1018	0375
MZ1020	0700
MZ1022	0705
MZ2360	0133
MZ2361	0133
MZ2362	1276
MZ3154	0165
MZ3154A	0165
MZ3155	0165
MZ3155A	0165
MZ3156	0165
MZ3156A	0165
MZ3157	0165
MZ3157A	0165
MZ4617	1276
MZ4618	2874
MZ4619	0118

DEVICE TYPE	REPL CODE
MZ4620	0188
MZ4621	0376
MZ4622	0188
MZ4623	0275
MZ4624	0162
MZ4625	0162
MZ4626	0157
MZ4627	0636
MZ4682	0760
MZ5110	1180
MZ5111	1190
MZ5112	1207
MZ5113	1218
MZ5114	1881
MZ5115	0647
MZ5116	1255
MZ5117	2102
MZ5119	2225
MZ5120	0605
MZ5706	0205
MZ5707	0480
MZ5708	0504
MZ5709	0684
MZ5710	0225
MZ5712	0234
MZ5713	0237
MZ5714	1397
MZ5715	0247
MZ5716	0251
MZ5718	0256
MZ5720	0262
MZ5722	0269
MZ5724	0274
MZ5727	0294
MZ5730	0309
MZ5733	0318
MZ5736	0320
MZ5760	2856
MZ5766	OEM
MZ5775	0426
MZ5915	0247
MZ5918	0256
MZ5922	0269
MZ5927	0294
MZ5933	0318
MZ5956	0266
MZ7002.4B	OEM
MZ10510	OEM
MZC25B10	1606
MZC28B10	1674
MZC87B10	3026
MZG746	OEM
MZG746A	OEM
MZG747	OEM
MZG747A	OEM
MZG748	OEM
MZG748A	OEM
MZG749	OEM
MZG749A	OEM
MZG750	OEM
MZG750A	OEM
MZG751	OEM
MZG751A	OEM
MZG752	OEM
MZG752A	OEM
MZG753	OEM
MZG753A	OEM
MZG754	OEM
MZG754A	OEM
MZG755	OEM
MZG755A	OEM
MZG756	OEM
MZG756A	OEM
MZG757	OEM
MZG757A	OEM
MZG758	OEM
MZG758A	OEM
MZG759	OEM
MZG759A	OEM
MZG957	OEM
MZG957A	OEM
MZG957B	OEM
MZG958	OEM
MZG958A	OEM
MZG958B	OEM
MZG959	OEM
MZG959A	OEM
MZG959B	OEM
MZG960	OEM
MZG960A	OEM
MZG960B	OEM
MZG961	OEM
MZG961A	OEM
MZG961B	OEM
MZG962	OEM
MZG962A	OEM
MZG962B	OEM
MZG963	OEM
MZG963A	OEM
MZG963B	OEM
MZG964	OEM
MZG964A	OEM
MZG964B	OEM
MZG965	OEM
MZG965A	OEM
MZG965B	OEM
MZG966	OEM
MZG966A	OEM
MZG966B	OEM

DEVICE TYPE	REPL CODE
MZG967	OEM
MZG967A	OEM
MZG967B	OEM
MZG968	OEM
MZG968A	OEM
MZG968B	OEM
MZG969	OEM
MZG969A	OEM
MZG969B	OEM
MZG970	OEM
MZG970A	OEM
MZG970B	OEM
MZG971	OEM
MZG971A	OEM
MZG971B	OEM
MZG972	OEM
MZG972A	OEM
MZG972B	OEM
MZG973	OEM
MZG973A	OEM
MZG973B	OEM
MZG974	OEM
MZG974A	OEM
MZG974B	OEM
MZG975	OEM
MZG975A	OEM
MZG975B	OEM
MZG976	OEM
MZG976A	OEM
MZG976B	OEM
MZG977	OEM
MZG977A	OEM
MZG977B	OEM
MZG978	OEM
MZG978A	OEM
MZG978B	OEM
MZG979	OEM
MZG979A	OEM
MZG979B	OEM
MZG980	OEM
MZG980A	OEM
MZG980B	OEM
MZG981	OEM
MZG981A	OEM
MZG981B	OEM
MZG982	OEM
MZG982A	OEM
MZG982B	OEM
MZG983	OEM
MZG983A	OEM
MZG983B	OEM
MZG984	OEM
MZG984A	OEM
MZG984B	OEM
MZG4370	OEM
MZG4370A	OEM
MZG4371	OEM
MZG4371A	OEM
MZG4372	OEM
MZG4372A	OEM
MZL1.8B10	OEM
MZL2.0B10	OEM
MZL2.2B10	OEM
MZL2.4A10	1276
MZL2.4B10	1276
MZL2.5A10	2874
MZL2.7A10	0760
MZL2.7B10	0760
MZL2.8A10	0677
MZL3.0A10	0118
MZL3.0B10	0118
MZL3.3A10	0299
MZL3.3B10	0299
MZL3.6A10	0376
MZL3.6B10	0376
MZL3.9A10	0036
MZL3.9B10	0036
MZL4.3A10	0275
MZL4.3B10	0275
MZL4.7A10	0140
MZL4.7B10	0140
MZL5.1A10	0041
MZL5.1B10	0041
MZL5.6A10	0253
MZL5.6B10	0253
MZL6.0A10	0091
MZL6.2A10	0471
MZL6.2B10	0471
MZL6.8A10	0062
MZL6.8B10	0062
MZL7.5A10	0077
MZL7.5B10	0077
MZL8.2A10	0165
MZL8.2B10	0165
MZL8.7A10	0322
MZL8.7B10	0322
MZL9.1A10	0057
MZL9.1B10	0057
MZL10A10	0064
MZL10B10	0064
MZL11A10	0181
MZL11B10	0181
MZL12A10	0052
MZL12B10	0052
MZL13A10	0053
MZL13B10	0053
MZL14A10	0879
MZL14B10	0879
MZL15A10	0686

DEVICE TYPE	REPL CODE
MZL15B10	0686
MZL16A10	0445
MZL16B10	0445
MZL17A10	0210
MZL17B10	0210
MZL18A10	0375
MZL18B10	0375
MZL19A10	0671
MZL19B10	0671
MZL20A10	0700
MZL20B10	0700
MZL22A10	0705
MZL22B10	0705
MZL24A10	0494
MZL24B10	0494
MZL25A10	0714
MZL25B10	0714
MZL27A10	0455
MZL27B10	0455
MZL28A10	0257
MZL28B10	0257
MZL30A10	0195
MZL30B10	0195
MZL33A10	0166
MZL33B10	0166
MZL36A10	0010
MZL36B10	0010
MZL39A10	0032
MZL39B10	0032
MZL43A10	0054
MZL43B10	0054
MZL47A10	0068
MZL47B10	0068
MZL51A10	0092
MZL51B10	0092
MZL56A10	0125
MZL56B10	0125
MZL60A10	2317
MZL60B10	2317
MZL62A10	0152
MZL62B10	0152
MZL68A10	0173
MZL68B10	0173
MZL75A10	0094
MZL75B10	0094
MZL82A10	0049
MZL82B10	0049
MZL87A10	0104
MZL87B10	0104
MZL91A10	0156
MZL91B10	0156
MZL100A10	0189
MZL100B10	0189
MZL110A10	0099
MZL110B10	0099
MZL120A10	0089
MZL120B10	0089
MZL130A10	0287
MZL130B10	0287
MZL140A10	0252
MZL140B10	0252
MZL150A10	0340
MZL150B10	0340
MZL160A10	0370
MZL160B10	0370
MZL170A10	0394
MZL170B10	0394
MZL180A10	0425
MZL180B10	0425
MZL190A10	0453
MZL190B10	0453
MZL200A10	1473
MZL200B10	1473
MZL306	0091
MZL307	0062
MZL308	0165
MZL309	0057
MZL310	0064
MZL312	0052
MZL314	0879
MZL316	0686
MZL318	0375
MZL320	0700
MZL322	0705
MZL324	0494
MZL327	0455
MZL330	0195
MZL333	0166
MZL336	0010
MZL339	0032
MZL343	0054
MZ-00	0124
MZS1282L	OEM
MZT33-01	3452
MZT39B	OEM
MZT39C	OEM
MZU1.8B10	OEM
MZU2.0B10	OEM
MZU2.2B10	OEM
MZU2.4A10	1276
MZU2.4B10	1276
MZU2.5A10	2874
MZU2.7A10	0760
MZU2.7B10	0760
MZU2.8A10	0677
MZU3.0A10	0118
MZU3.0B10	0118
MZU3.3A10	0299
MZU3.3B10	0299

DEVICE TYPE	REPL CODE
MZU3.6A10	0376
MZU3.6B10	0376
MZU3.9A10	0036
MZU3.9B10	0036
MZU4.3A10	0275
MZU4.3B10	0275
MZU4.7A10	0140
MZU4.7B10	0140
MZU5.1A10	0041
MZU5.1B10	0041
MZU5.6A10	0253
MZU5.6B10	0253
MZU6.0A10	0091
MZU6.2A10	0471
MZU6.2B10	0471
MZU6.8A10	0062
MZU6.8B10	0062
MZU7.5A10	0077
MZU7.5B10	0077
MZU8.2A10	0165
MZU8.2B10	0165
MZU8.7A10	0322
MZU8.7B10	0322
MZU9.1A10	0057
MZU9.1B10	0057
MZU10A10	0064
MZU10B10	0064
MZU11A10	0181
MZU11B10	0181
MZU12A10	0052
MZU12B10	0052
MZU13A10	0053
MZU13B10	0053
MZU14A10	0879
MZU14B10	0879
MZU15A10	0686
MZU15B10	0686
MZU16A10	0445
MZU16B10	0445
MZU17A10	0210
MZU17B10	0210
MZU18A10	0375
MZU18B10	0375
MZU19A10	0671
MZU19B10	0671
MZU20A10	0700
MZU20B10	0700
MZU22A10	0705
MZU22B10	0705
MZU24A10	0494
MZU24B10	0494
MZU25A10	0714
MZU25B10	0714
MZU27A10	0455
MZU27B10	0455
MZU28A10	0257
MZU28B10	0257
MZU30B10	0195
MZU33A10	0166
MZU33B10	0166
MZU36A10	0010
MZU36B10	0010
MZU39A10	0032
MZU39B10	0032
MZU43A10	0054
MZU43B10	0054
MZU47A10	0068
MZU47B10	0068
MZU51A10	0092
MZU51B10	0092
MZU56A10	0125
MZU56B10	0125
MZU60A10	2317
MZU60B10	2317
MZU62A10	0152
MZU62B10	0152
MZU68A10	0173
MZU68B10	0173
MZU75A10	0094
MZU75B10	0094
MZU82A10	0322
MZU82B10	0049
MZU87A10	0104
MZU87B10	0104
MZU91A10	0156
MZU91B10	0156
MZU100A10	0189
MZU100B10	0189
MZU110A10	0099
MZU110B10	0099
MZU120A10	0089
MZU120B10	0089
MZU130A10	0287
MZU130B10	0287
MZU140A10	0252
MZU140B10	0252
MZU150A10	0340
MZU150B10	0340
MZU160A10	0370
MZU160B10	0370
MZU170A10	0394
MZU170B10	0394
MZU180A10	0425
MZU180B10	0425
MZU190A10	0453
MZU190B10	0453
MZU200A10	1473
MZU200B10	1473
MZX9.1	0012

DEVICE TYPE	REPL CODE
MZZ3.6-20	OEM
Ma408	OEM
Mc1496F	OEM
Mc10561L	OEM
Mc10562L	OEM
Mc10578F	OEM
Mc10597F	OEM
Mk3870/42P-15	OEM
Mk3873/12N-10	OEM
N	OEM
N-020	0050
N020	0136
N1S7261A	OEM
N1X	0086
N2A	0143
N2A(DIODE)	0143
N2A(TRIPLER)	2985
N2A-1	2546
N2A-2	2546
N2XA	0233
N5	OEM
N5C2U	OEM
N7N	OEM
N8H80A	0682
N8H80J	0682
N8T13	OEM
N8T14	OEM
N8T22A	2267
N8T22N	1139
N8T23	OEM
N8T24	OEM
N8T26	4084
N8T26A	4084
N8T26AB	OEM
N8T26AF	4516
N8T26AN	4516
N8T28F	0581
N8T28N	0581
N8T95B	4646
N8T95F	4647
N8T95N	4647
N8T96B	4646
N8T96F	4651
N8T96N	4651
N8T97	4658
N8T97B	4655
N8T97F	4658
N8T97N	4658
N8T98F	4667
N8T98N	4667
N9HHB	0143
N9NDO	0182
N9NUS	0133
N9T22F	OEM
N10	OEM
N10R	OEM
N13T	0316
N13T1	0316
N13T2	0316
N19C	OEM
N19E	OEM
N19G	OEM
N19J	OEM
N20	OEM
N20C	OEM
N20E	OEM
N20G	OEM
N20J	OEM
N24	OEM
N25	OEM
N35A	OEM
N-41	0015
N48	0143
N57B2-3	0004
N57B2-6	0004
N57B2-7	0004
N57B2-8	0208
N57B2-11	0050
N57B2-13	0050
N57B2-14	0050
N57B2-15	0004
N57B2-17	0136
N57B2-18	0136
N57B2-19	0136
N57B2-22	0050
N57B2-23	0136
N57B2-25	0004
N57B4-2	0160
N57B4-4	0160
N60R006B	OEM
N60R006B16	OEM
N74H00A	0682
N74H00F	0682
N74H00N	0682
N74H01A	5338
N74H01F	5338
N74H01N	5338
N74H04A	1907
N74H04F	1907
N74H04N	1907
N74H05A	3251
N74H05F	3251
N74H05N	3251
N74H08A	5355
N74H08F	5355
N74H08N	5355
N74H10A	0685
N74H10F	0685
N74H10N	0685

DEVICE TYPE	REPL CODE
N74H11A	2398
N74H11F	2398
N74H11N	2398
N74H20A	3715
N74H20F	3715
N74H20N	3715
N74H21A	4856
N74H21F	4856
N74H21N	4856
N74H22A	4596
N74H22F	4596
N74H22N	4596
N74H30A	5383
N74H30F	5383
N74H30N	5383
N74H30W	5383
N74H40A	0559
N74H40F	0559
N74H40N	0559
N74H50A	1792
N74H50F	1792
N74H50N	1792
N74H51A	1944
N74H51F	1944
N74H51N	1944
N74H52A	2020
N74H52F	2020
N74H52N	2020
N74H53A	2101
N74H53F	2101
N74H53N	2101
N74H54A	2172
N74H54F	2172
N74H54N	2172
N74H55A	3158
N74H55F	3158
N74H55N	3158
N74H60A	5410
N74H60F	5410
N74H60N	5410
N74H61A	2663
N74H61F	2663
N74H61N	2663
N74H62A	2731
N74H62F	2731
N74H62N	2731
N74H71A	3264
N74H71F	3264
N74H71N	3264
N74H72A	3313
N74H72F	3313
N74H72N	3313
N74H73A	2463
N74H73F	2463
N74H73N	2463
N74H74A	2492
N74H74F	2492
N74H74N	2492
N74H76B	5302
N74H76F	5302
N74H76N	5302
N74H101A	5521
N74H101F	5521
N74H101N	5521
N74H102A	5523
N74H102F	5523
N74H102N	5523
N74H103A	2969
N74H103F	2969
N74H103N	2969
N74H106B	5253
N74H106F	5253
N74H106N	5253
N74H108A	0180
N74H108F	0180
N74H108N	0180
N74LS00	1527
N74LS00F	1527
N74LS00N	1527
N74LS01F	1546
N74LS01N	1546
N74LS02F	1559
N74LS02N	1559
N74LS03F	1579
N74LS03N	1579
N74LS04F	1595
N74LS04N	1595
N74LS05F	1608
N74LS05N	1608
N74LS08F	1633
N74LS08N	1633
N74LS09F	1642
N74LS09N	1642
N74LS10F	1662
N74LS10N	1662
N74LS11F	1667
N74LS11N	1667
N74LS12F	1679
N74LS12N	1679
N74LS13F	1688
N74LS13N	1688
N74LS14F	1698
N74LS14N	1698
N74LS15F	1707
N74LS15N	1707
N74LS20F	0034
N74LS20N	0034
N74LS21F	1763
N74LS21N	1763

DEVICE TYPE	REPL CODE
N74LS22F	1775
N74LS22N	1775
N74LS26F	1382
N74LS26N	1382
N74LS27F	0183
N74LS27N	0183
N74LS28F	0472
N74LS28N	0472
N74LS30F	0828
N74LS30N	0828
N74LS32F	0088
N74LS32N	0088
N74LS33F	1832
N74LS33N	1832
N74LS37F	1730
N74LS37N	1730
N74LS38F	1839
N74LS38N	1839
N74LS40F	0135
N74LS40N	0135
N74LS42F	1841
N74LS42N	1841
N74LS51F	1033
N74LS51N	1033
N74LS54F	1857
N74LS54N	1857
N74LS55F	0457
N74LS55N	0457
N74LS73F	1867
N74LS73N	1867
N74LS74AF	0243
N74LS74AN	0243
N74LS74F	0243
N74LS74N	0243
N74LS75F	1870
N74LS75N	1870
N74LS76F	2180
N74LS76N	2180
N74LS78F	1873
N74LS78N	1873
N74LS83AF	2218
N74LS83AN	2218
N74LS83F	2218
N74LS85F	0431
N74LS85N	0431
N74LS86F	0291
N74LS86N	0291
N74LS89F	OEM
N74LS89N	OEM
N74LS90F	1882
N74LS90N	1882
N74LS92F	1887
N74LS92N	1887
N74LS93F	1888
N74LS93N	1888
N74LS95BF	0771
N74LS95BN	0771
N74LS96F	OEM
N74LS96N	OEM
N74LS107F	1602
N74LS107N	1602
N74LS109AF	1906
N74LS109AN	1906
N74LS109F	1906
N74LS109N	1906
N74LS112F	2128
N74LS112N	2128
N74LS113F	2256
N74LS113N	2256
N74LS114F	2302
N74LS114N	2302
N74LS123AF	0979
N74LS123AN	0979
N74LS125F	0075
N74LS125N	0075
N74LS126F	2877
N74LS126N	2877
N74LS132F	1625
N74LS132N	1625
N74LS136F	1628
N74LS136N	1628
N74LS138F	0427
N74LS138N	0427
N74LS139F	0153
N74LS139N	0153
N74LS145F	1564
N74LS145N	1564
N74LS151F	1646
N74LS151N	1646
N74LS153F	0959
N74LS153N	0959
N74LS154F	5045
N74LS154N	5045
N74LS155F	0209
N74LS155N	0209
N74LS156F	1654
N74LS156N	1654
N74LS157D	OEM
N74LS157F	1161
N74LS157N	1161
N74LS158F	1656
N74LS158N	1656
N74LS160AF	0837
N74LS160AN	0837
N74LS161AF	0858
N74LS161AN	0858
N74LS162AF	0880
N74LS162AN	0880
N74LS163AF	0893

DEVICE TYPE	REPL CODE
N74LS163AN	0893
N74LS164F	4339
N74LS164N	4339
N74LS168AF	0967
N74LS168AN	0967
N74LS169AF	0986
N74LS169AN	0986
N74LS170B	2629
N74LS170F	2629
N74LS170N	2629
N74LS173D	OEM
N74LS173N	6842
N74LS174B	0260
N74LS174F	0260
N74LS174N	0260
N74LS175B	1672
N74LS175F	1672
N74LS175N	1672
N74LS181F	1678
N74LS181N	1678
N74LS190F	1686
N74LS190N	1686
N74LS191F	1687
N74LS191N	1687
N74LS192F	1689
N74LS192N	1689
N74LS193F	1692
N74LS193N	1692
N74LS194AF	1304
N74LS194AN	1304
N74LS195AF	1315
N74LS195AN	1315
N74LS196F	2834
N74LS196N	2834
N74LS197F	2469
N74LS197N	2469
N74LS221F	1239
N74LS221N	1239
N74LS240F	0452
N74LS240N	0452
N74LS241F	1726
N74LS241N	1726
N74LS242F	1728
N74LS242N	1728
N74LS243F	0906
N74LS243N	0906
N74LS244F	0458
N74LS244N	0458
N74LS245F	0463
N74LS245N	0463
N74LS251AF	1737
N74LS251AN	1737
N74LS253F	1739
N74LS253N	1739
N74LS256F	OEM
N74LS256N	OEM
N74LS257AF	1744
N74LS257AN	1744
N74LS258AF	1746
N74LS258AN	1746
N74LS259F	3205
N74LS259N	3205
N74LS260F	5995
N74LS260N	5995
N74LS261F	OEM
N74LS261N	OEM
N74LS266F	0592
N74LS266N	0592
N74LS273F	0894
N74LS273N	0894
N74LS279F	3290
N74LS279N	3290
N74LS283F	1779
N74LS283N	1779
N74LS289N	OEM
N74LS290F	4425
N74LS290N	4425
N74LS293F	0082
N74LS293N	0082
N74LS295BF	2227
N74LS295BN	2227
N74LS298F	3371
N74LS298N	3371
N74LS299F	4426
N74LS299N	4426
N74LS301F	OEM
N74LS301N	OEM
N74LS323F	OEM
N74LS323N	OEM
N74LS352N	0761
N74LS353N	0773
N74LS363F	6005
N74LS363N	6005
N74LS364F	6006
N74LS364N	6006
N74LS365AD	OEM
N74LS365AF	0943
N74LS365AN	0943
N74LS366AN	7112
N74LS367AD	OEM
N74LS367AF	0977
N74LS367AN	0977
N74LS368AD	OEM
N74LS368AF	0991
N74LS368AN	0991
N74LS373F	0709
N74LS373N	0709
N74LS374F	0713
N74LS374N	0713

DEVICE TYPE	REPL CODE
N74LS375F	OEM
N74LS375N	OEM
N74LS377F	1120
N74LS377N	1120
N74LS378F	1133
N74LS378N	1133
N74LS379F	1151
N74LS379N	1151
N74LS386F	1230
N74LS386N	1230
N74LS390F	1288
N74LS390N	1288
N74LS393F	0819
N74LS393N	0819
N74LS395AF	1330
N74LS395AN	1330
N74LS398F	1383
N74LS398N	1383
N74LS399F	1398
N74LS399N	1398
N74LS445F	OEM
N74LS445N	OEM
N74LS490F	2213
N74LS490N	2213
N74LS568F	OEM
N74LS568N	OEM
N74LS569F	OEM
N74LS569N	OEM
N74LS640-1F	0669
N74LS640-1N	0669
N74LS640F	0669
N74LS640N	0669
N74LS641-1F	0690
N74LS641-1N	0690
N74LS641F	0690
N74LS641N	0690
N74LS642-1N	0719
N74LS642F	0719
N74LS642N	0719
N74LS645-1F	0775
N74LS645F	0775
N74LS645N	0775
N74LS670B	1130
N74LS670F	1130
N74LS670N	1130
N74S00A	0704
N74S00F	0704
N74S00N	0704
N74S02F	2238
N74S02N	2238
N74S03A	2217
N74S03F	2217
N74S03N	2217
N74S04A	2263
N74S04F	2263
N74S04N	2263
N74S05A	2321
N74S05F	2321
N74S05N	2321
N74S08F	2569
N74S08N	2569
N74S09F	2667
N74S09N	2667
N74S10A	2445
N74S10F	2445
N74S10N	2445
N74S11A	2447
N74S11F	2447
N74S11N	2447
N74S15A	2451
N74S15F	2451
N74S15N	2451
N74S20A	1017
N74S20F	1017
N74S20N	1017
N74S22A	2461
N74S22F	2461
N74S22N	2461
N74S32F	OEM
N74S32N	OEM
N74S37F	5756
N74S37N	OEM
N74S38F	0780
N74S38N	OEM
N74S40A	2475
N74S40F	2475
N74S40N	2475
N74S51F	4305
N74S51N	4305
N74S64A	2496
N74S64F	2496
N74S64N	2496
N74S65A	2497
N74S65F	2497
N74S65N	2497
N74S74A	2504
N74S74F	2504
N74S74N	2504
N74S85F	5773
N74S85N	5773
N74S86A	2511
N74S86F	2511
N74S86N	2511
N74S86W	2511
N74S89F	OEM
N74S89N	OEM
N74S112B	1617
N74S112F	1617
N74S112N	1617

If replacement code is OEM, contact original manufacturer for replacement.

DEVICE TYPE	REPL CODE	DEVICE TYPE	REPL CODE	DEVICE TYPE	REPL CODE	DEVICE TYPE	REPL CODE	DEVICE TYPE	REPL CODE	DEVICE TYPE	REPL CODE	DEVICE TYPE	REPL CODE
N74S113A	1623	N82S137BN	2392	N7420A	0697	N7485A	0374	N74141B	1377	N74221D	OEM	NA22	0015
N74S113F	1623	N82S153N	OEM	N7420F	0697	N7485B	0374	N74145	0619	N74221F	2142	NA25	0015
N74S113N	1623	N104B	OEM	N7420N	0697	N7485F	0374	N74145A	0619	N74221N	2142	NA30	0208
N74S114A	1629	N110-2	OEM	N7421A	1357	N7485N	0374	N74145B	0619	N74279F	OEM	NA-32	0015
N74S114F	1629	N110-3	OEM	N7421F	1357	N7486A	1368	N74145F	0619	N74279N	OEM	NA32	0015
N74S114N	1629	N203A	0901	N7421N	1357	N7486F	1368	N74145N	0619	N74298F	OEM	NA33	0015
N74S133B	1819	N203B	0058	N7423F	3466	N7486N	1368	N74147F	1451	N74298N	OEM	NA-35	0015
N74S133F	1819	N203C	0407	N7423N	3466	N7488B	OEM	N74147N	1451	N74365AF	1459	NA35	0015
N74S133N	1819	N203D	1683	N7425F	3477	N7488W	OEM	N74148	1464	N74365AN	1459	NA36	0015
N74S133W	1819	N203YY	0344	N7425N	3477	N7489B	5455	N74148F	1464	N74366AF	1471	NA-42	0015
N74S134B	1827	N413	OEM	N7426A	0803	N7490A	1208	N74148N	1464	N74366AN	1471	NA42	0015
N74S134F	1827	N413L	3331	N7426F	0803	N7490F	1208	N74150A	1492	N74367AF	1471	NA-45	0015
N74S134N	1827	N450CH06	OEM	N7426N	0803	N7490N	1208	N74150B	1492	N74367AN	1488	NA45	0015
N74S134W	1827	N-756A	0322	N7427A	0818	N7491A	0980	N74150F	1492	N74368AF	1508	NA46	0015
N74S135F	OEM	N874S40F	OEM	N7427F	0818	N7491AF	0980	N74150N	1492	N74368AN	1508	NA-62	0015
N74S135N	OEM	N1112	1032	N7427N	0818	N7491AN	0980	N74151A	1495	N74478AN	OEM	NA62	0229
N74S138B	2138	N1212	0887	N7428F	4180	N7491F	0980	N74151B	1495	N80286-8	OEM	NA-63	0015
N74S138F	2138	N1213	0898	N7428N	4180	N7492A	0834	N74151F	1495	N1716580001	OEM	NA63	0229
N74S138N	2138	N3563	0016	N7430A	0873	N7492F	0834	N74151N	1495	N5301571130	0053	NA-65	0015
N74S139F	OEM	N4000	2024	N7430F	0873	N7492N	0834	N74152F	1517	N5301571220	0705	NA65	0229
N74S139N	OEM	N4001	0478	N7430N	0873	N7493A	0569	N74152N	1517	N5301571330	0166	NA-66	0015
N74S140A	1886	N4002	2055	N7432A	0899	N7493F	0569	N74153B	1539	N5301571399	0036	NA66	0229
N74S140F	1886	N4007	2846	N7432F	0899	N7493N	0569	N74153F	1539	N5301571439	0275	NA-74	0071
N74S140N	1886	N4009	1999	N7432N	0899	N7494B	1702	N74153N	1539	N5301571569	0253	NA74	0071
N74S151B	1955	N4011	0215	N7433F	4193	N7494F	1702	N74154A	1555	N5301810001	0124	NA-75	0071
N74S151F	1955	N4012	0498	N7433N	4193	N7494N	1702	N74154F	1555	N5301811001	0124	NA75	0071
N74S151N	2165	N4013	0413	N7437A	3520	N7495A	1486	N74154N	1555	N5302300003	OEM	NA-76	0071
N74S153F	2151	N4015	1014	N7437F	3520	N7495AF	1486	N74155B	1576	N5302350001	2628	NA76	0071
N74S153J	2151	N4016	1143	N7437N	3520	N7495AN	1486	N74155F	1576	N5302470001	0917	NA-84	0071
N74S153N	2151	N4019	1525	N7438A	0996	N7495F	1486	N74155N	1576	N5302471001	0917	NA84	0071
N74S157B	1695	N4021	1749	N7438F	0996	N7495N	1486	N74156B	1592	N5302491160	0445	NA-85	0071
N74S157F	1695	N4023	0519	N7438N	0996	N7496B	1715	N74156F	1592	N5302491519	0589	NA85	0071
N74S157N	1695	N4025	2072	N7439A	5836	N7496F	1715	N74156N	1592	N5302491759	0077	NA-86	0071
N74S158B	2154	N4027	1949	N7439F	5836	N7496N	1715	N74157F	1605	N5302541039	0471	NA86	0071
N74S158F	2154	N4030	0500	N7439N	5836	N8233F	2074	N74157N	1605	N5302600001	0031	NA-104	0071
N74S158N	2154	N4049	0001	N7440A	1024	N8233N	2074	N74158F	4446	N5302600002	0031	NA104	0071
N74S158d	OEM	N4050	0398	N7440F	1024	N8234F	2080	N74158N	4446	N5302620001	0071	NA-105	0071
N74S172F	OEM	N4081	0626	N7440N	1024	N8234N	2080	N74160B	1631	N5302660001	0124	NA105	0071
N74S172N	OEM	N4111	4172	N7441B	1038	N8235F	2087	N74160F	1631	N5302740001	0516	NA603	0970
N74S174B	2132	N4114	OEM	N7441F	1038	N8235N	2087	N74160N	1631	N5302980001	OEM	NA615	0989
N74S174F	2132	N4212	0887	N7442A	1052	N8250F	2187	N74161B	1645	N5302990001	1090	NA1005	1250
N74S174N	2132	N4213	0898	N7442B	OEM	N8250N	2187	N74161F	1645	N5302990002	1090	NA1022-1001	0050
N74S175B	2141	N4214	OEM	N7442BA	1052	N8252F	2196	N74161N	1645	N5303010002	OEM	NA1022-1007	0004
N74S175F	2141	N4770F	OEM	N7442BF	1052	N8252N	2196	N74162B	1013	N5303020003	0052	NA-1114-1001	0050
N74S175N	2141	N5065A	0167	N7442F	1052	N8259A-2	OEM	N74162F	1013	N5303040001	OEM	NA-1114-1002	0050
N74S178A	OEM	N5070B	0352	N7442N	1052	N8266F	2275	N74162N	1013	N5303051003	0031	NA-1114-1004	0004
N74S178F	OEM	N5070N	0352	N7443A	1060	N8266N	2275	N74163B	1666	N5303100001	0023	NA-1114-1005	0004
N74S179B	OEM	N5071A	0354	N7443F	1060	N8268F	1535	N74163F	1666	N5303100003	0031	NA-1114-1006	0004
N74S179F	OEM	N5072A	0666	N7443N	1060	N8270F	1813	N74163N	1666	N5303101003	0031	NA-1114-1007	0004
N74S181F	2165	N5111	0664	N7444B	1072	N8270N	1813	N74164A	0734	N5303150001	OEM	NA-1114-1008	0004
N74S181J	2165	N5111A	0664	N7444F	1072	N8271N	1820	N74164F	0734	N5303150002	OEM	NA-1114-1009	0004
N74S181N	2165	N5406	0133	N7444N	1072	N8280F	1795	N74164N	0734	N5303150003	OEM	NA-1114-1010	0004
N74S181W	2165	N5596K	0417	N7445B	1082	N8280N	1795	N74165B	1685	N5303260003	0546	NA-1114-1011	0004
N74S182F	2166	N5723T	1192	N7445F	1082	N8281F	1803	N74165F	1685	N6101580003	0126	NA1302	2699
N74S182N	2166	N5741V	0312	N7445N	1082	N8281N	1803	N74165N	1685	N6102230001	0037	NA1505	1250
N74S189F	OEM	N-7400A	0232	N7446AF	1098	N8290F	1950	N74166B	0231	N6102240001	0076	NA2002	1048
N74S189N	OEM	N7400A	0232	N7446AN	1098	N8290N	1950	N74166F	0231	N6102320002	0198	NA2005	1250
N74S189W	OEM	N7400F	0232	N7446B	1098	N8291F	1956	N74166N	0231	N6102500003	0285	NA5015-1012	0136
N74S194B	1931	N7400N	0232	N7447A	1108	N8293F	2469	N74170B	1722	N6103600001	0342	NA5018-1001	0050
N74S194F	1931	N7401	0268	N7447AF	1108	N8293N	2469	N74170F	1722	N6103620001	OEM	NA5018-1002	0136
N74S194J	1931	N7401A	0232	N7447B	1108	N8455F	1024	N74170N	1722	N6103690001	0037	NA5018-1003	0136
N74S194N	1931	N7401F	0232	N7447F	1108	N8470F	0512	N74172F	OEM	N6104330002	5345	NA5018-1004	0136
N74S194W	1931	N7401N	0268	N7448A	1125	N8526AN	OEM	N74172N	OEM	N6104340001	0037	NA5018-1005	0136
N74S195B	OEM	N7402A	0314	N7448B	1125	N8598F	OEM	N74173N	1766	N6104350001	0079	NA5018-1006	0136
N74S195F	OEM	N7402F	0314	N7448F	1125	N8815N	3477	N74174	1770	N6104350004	0079	NA5018-1007	0136
N74S195J	OEM	N7402N	0314	N7448N	1125	N8828F	1313	N74174F	1770	N6104420001	0008	NA5018-1008	0136
N74S195N	OEM	N7403A	0335	N7450A	0743	N8840F	0743	N74174N	1770	N6104430001	0016	NA5018-1009	0136
N74S195W	OEM	N7403F	0335	N7450F	0743	N8848F	2492	N74175B	1787	N6104980001	0037	NA5018-1010	0136
N74S196F	OEM	N7403N	0335	N7450N	0743	N8875F	0268	N74175F	1787	N6105000001	0079	NA5018-1011	0136
N74S197F	OEM	N-7404A	0361	N7451A	1168	N8881F	0268	N74175N	1787	N6105000004	0079	NA5018-1012	0136
N74S200F	OEM	N7404A	0361	N7451F	1168	N8881N	0268	N74178A	1813	N6105150002	OEM	NA5018-1013	0004
N74S200I	OEM	N7404F	0361	N7451N	1168	N9308F	1070	N74179B	1820	N6105210001	OEM	NA5018-1014	0004
N74S200N	OEM	N7404N	0361	N7453A	1185	N9308N	1070	N74179F	1820	N6105310001	0168	NA5018-1015	0004
N74S206B	OEM	N7405A	0385	N7453F	1185	N9309F	5528	N74180A	1829	N6105320003	1511	NA5018-1016	0004
N74S206I	OEM	N7405F	0385	N7453N	1185	N9309N	5528	N74180F	1829	N6105360001	OEM	NA5018-1219	0136
N74S251B	2198	N7405N	0385	N7454A	1202	N9314F	2488	N74180N	1829	N6105510001	3014	NA5018-1220	0136
N74S251F	2198	N7406A	1206	N7454F	1202	N9314N	2488	N74181F	1842	N6110180001	0863	NAM383	1048
N74S251N	2198	N7406F	1206	N7454N	1202	N9334N	5565	N74181N	1842	N6110190001	0617	NAP-TZ-8	0160
N74S253F	OEM	N7406N	1206	N7460A	1275	N9602B	2286	N74182B	1856	N6121860001	1143	NAP-T-Z-10	0004
N74S253N	OEM	N7407A	1339	N7460F	1275	N9602N	2286	N74182F	1856	N6123270001	4601	NAP-TZ-10	0004
N74S257F	OEM	N7407F	1339	N7460N	1275	N74107A	0942	N74182N	1856	N6123700001	0351	NAS29	0168
N74S257N	OEM	N7407N	1339	N7470A	1404	N74107F	0942	N74190B	1912	N6124120002	3161	NAS29A	0168
N74S258B	2205	N-7408A	0467	N7470F	1404	N74107N	0942	N74190F	1912	N6124400001	OEM	NAS29B	0168
N74S258F	2205	N7408A	0467	N7470N	1404	N74109B	0968	N74190N	1912	N6124440001	2178	NAS29C	0168
N74S258J	2205	N7408F	0467	N7472A	1426	N74109F	0968	N74190W	OEM	N6124670001	3495	NAS815R	0745
N74S258N	2205	N7408N	0467	N7472F	1426	N74109N	0968	N74191B	1917	N6124720001	OEM	NB011	0532
N74S260F	OEM	N7409A	0492	N7472N	1426	N74109W	OEM	N74191F	1917	N6124790001	0624	NB011EN	0079
N74S260N	OEM	N7409F	0492	N7473	1172	N74116F	4439	N74191N	1917	N6124850002	OEM	NB013	0111
N74S273F	OEM	N7409N	0492	N-7473A	1172	N74116N	6545	N74191W	OEM	N6124930001	0035	NB021	0532
N74S273N	OEM	N7410A	0512	N7473A	1172	N74121A	0175	N74192A	1921	N6124990002	OEM	NB021EV	0037
N74S280F	2220	N7410F	0512	N7473F	1172	N74121F	0175	N74192B	1921	N6125070001	3044	NB021EV(N.S.)	0037
N74S280N	2220	N7410N	0512	N7473N	1172	N74121N	0175	N74192F	1921	N6125080001	5391	NB-07014	0167
N74S301	OEM	N7411A	0527	N7474A	1313	N74121W	OEM	N74192N	1921	N6125450001	3386	NB121	0037
N74S301N	OEM	N7411F	0527	N7474F	1313	N74122A	1139	N74192P	1921	N6125560002	OEM	NB211	0324
N74S301R	OEM	N7411N	0527	N7474N	1313	N74122F	1139	N74193A	1926	N6125590001	OEM	NB211E1	0324
N74S350F	OEM	N7412F	2242	N7475A	1432	N74122N	1139	N74193B	1926	N6125630001	3365	NB211EI	0324
N74S350N	OEM	N7412N	2242	N7475B	1432	N74123A	1157	N74193F	1926	N6125640001	3884	NB211ET	OEM
N74S373F	2264	N7413A	1441	N7475F	1432	N74123B	1157	N74193N	1926	N6125660001	OEM	NC29	0143
N74S373N	2264	N7413F	1441	N7475N	1432	N74123F	1157	N74194B	OEM	N6125790001	OEM	NC30	0004
N74S374F	2266	N7413N	1441	N7476A	1158	N74123N	1157	N74194F	OEM	N6125880001	OEM	NC32	0164
N74S374N	2266	N7414B	2243	N7476B	1158	N74125F	1182	N74194N	OEM	N6125990001	OEM	NC33	0208
N74S534F	OEM	N7414F	2243	N7476F	1158	N74125N	1182	N74195B	1943	NA0305	2002	NC34	0847
N74S534N	OEM	N7414N	2243	N7476N	1158	N74126F	1193	N74195F	1943	NA0505	2002	NC-629	OEM
N74S551F	4305	N7416A	1349	N7480A	1535	N74126N	1193	N74195N	1943	NA1	2002	NCB14	1591
N74SL289F	OEM	N7416F	1349	N7480F	1535	N74128F	1219	N74196A	1950	NA11	0580	NCB35	0936
N75E	OEM	N7416N	1349	N7480N	1535	N74129N	OEM	N74198F	1964	NA-13	0015	NCBJ35	0839
N82S123F	OEM	N7417A	1352	N7483B	0117	N74132B	1271	N74198N	1964	NA13	0015	NCBV14	2710
N82S123N	2175	N7417F	1352	N7483F	0117	N74132F	1271	N74199F	1971	NA20	0600	NCBW35	0936
N82S129N	2224	N7417N	1352	N7483N	0117	N74132N	1271	N74199N	1971	NA21	0580	NCF150E	OEM

DEVICE TYPE	REPL CODE
NCF200E	OEM
NCF250E	OEM
NCF300E	OEM
NCF400E	OEM
NCF600E	OEM
NCM5A	OEM
NCM6A	OEM
NCM7A	OEM
NCM8A	OEM
NCM9A	OEM
NCP704	OEM
NCR022D	OEM
NCR025D	OEM
NCR050D	OEM
NCR052B	OEM
NCR052D	OEM
NCR100D	OEM
NCR102B	OEM
NCR102D	OEM
NCR150D	OEM
NCR200D	OEM
NCR202D	OEM
NCR250D	OEM
NCR300D	OEM
NCR302B	OEM
NCR302D	OEM
NCR400D	OEM
NCR402B	OEM
NCR402D	OEM
NCR609-0380855	OEM
NCR2364-30	OEM
NCR23128-30	OEM
NCRF811521A	OEM
NCRF815103G	OEM
NCS9018D	0127
NCT200	0315
NCT260	0315
ND-07001	0232
ND10	0373
ND10A	0373
ND16AT	OEM
ND16BT	OEM
ND16DT	OEM
ND16GT	OEM
ND16JT	OEM
ND16KT	OEM
ND16MT	OEM
ND30AM	OEM
ND30BM	OEM
ND30DM	OEM
ND30GM	OEM
ND30JM	OEM
ND30KM	OEM
ND30MM	OEM
ND74LS01F	1546
ND487C1-3P	OEM
ND487C1-3R	OEM
ND487C2-3P	OEM
ND487C2-3R	OEM
ND487R1-3P	OEM
ND487R1-3R	OEM
ND487R2-3P	OEM
ND487R2-3R	OEM
ND587R-3P	OEM
ND587R-3R	OEM
ND587T-3P	OEM
ND587T-3R	OEM
ND5558-00	OEM
ND9862	OEM
NDAN601	OEM
NDC40013	OEM
NE02112	OEM
NE02132	OEM
NE02133	OEM
NE02135	OEM
NE02137	OEM
NE106J	OEM
NE112J	OEM
NE116J	OEM
NE124J	OEM
NE156J	OEM
NE161J	OEM
NE170J	OEM
NE180J	OEM
NE527H	3261
NE532AN	0770
NE532AT	1677
NE532N	0770
NE532T	1677
NE536T	1919
NE542N	3721
NE545B	3729
NE550A	0026
NE550F	0026
NE550N	0026
NE555	0973
NE555C	3780
NE555JG	0973
NE555N	0973
NE555P	0973
NE555V	0973
NE556	3285
NE556A	3285
NE556N	3285
NE558	3798
NE558CP	3798
NE558N	3798
NE565A	3381
NE566T	3825

DEVICE TYPE	REPL CODE
NE566V	3827
NE567N	3781
NE571	OEM
NE571N	OEM
NE575D	OEM
NE575N	OEM
NE591A	OEM
NE592	1118
NE592A	1118
NE592D	1118
NE592D14	4273
NE592F	1118
NE592FH	OEM
NE592H	1121
NE592K	1121
NE592N	1118
NE592N14	1118
NE645(B)	4106
NE645BN	4106
NE646B	OEM
NE646N	4106
NE657N	OEM
NE4304	0325
NE5532AP	OEM
NE5534	OEM
NE5534A	OEM
NE5534N	OEM
NE5560D	OEM
NE5560F	5907
NE5560N	5212
NE5568N	OEM
NE8131	OEM
NE21935	OEM
NE21937	OEM
NE22120	2070
NE41137	OEM
NE41612	OEM
NE41632	OEM
NE41635	OEM
NE46734	OEM
NE49433	OEM
NE59312	OEM
NE64535	OEM
NE64614B	OEM
NE66912	OEM
NE68033	OEM
NE68035	OEM
NE68037	OEM
NE68132	OEM
NE68133	OEM
NE68135	OEM
NE68137	OEM
NE68333	OEM
NE68337	OEM
NE71111	OEM
NE73412	OEM
NE73432	OEM
NE73433	OEM
NE73435	OEM
NE73437	OEM
NE74014	0418
NE85632	OEM
NE85633	OEM
NE85634	OEM
NE85635	OEM
NE85637	OEM
NE87112	OEM
NE88912	OEM
NE88933	OEM
NE88935	OEM
NE99532	OEM
N-EA15X130	0127
N-EA15X131	0144
N-EA15X132	0144
N-EA15X133	0136
N-EA15X134	0127
N-EA15X135	0127
N-EA15X136	0016
N-EA15X137	0016
N-EA15X138	0016
N-EA15X139	0822
N-EA16X27	0123
N-EA16X29	0137
N-EA16X30	0015
N-EA2136	0164
NEC303-1	OEM
NEZ0910-2A	OEM
NEZ0910-4A	OEM
NEZ0910-6A	OEM
NEZ1011-2A	OEM
NEZ1011-4A	OEM
NEZ1011-6A	OEM
NEZ1414-2A	OEM
NEZ1414-4A	OEM
NF520	0325
NF522	0325
NF523	0325
NF531	0325
NF533	0325
NF550	0015
NF580	OEM
NF581	OEM
NF582	OEM
NF583	OEM
NF584	OEM
NF3819	OEM
NF4302	2950
NF4303	2950
NF4304	5652

DEVICE TYPE	REPL CODE
NF5163	OEM
NF5457	OEM
NF5458	OEM
NF5459	OEM
NF5485	0325
NF5486	0325
NF5638	OEM
NF5639	OEM
NF5640	OEM
NF5653	OEM
NF5654	OEM
NGP3002	0143
NGP3003	0242
NGP5002	0636
NGP5007	0012
NGP5010	0137
NHM78M05A	OEM
NIM4558D	0360
NIS7261A	OEM
NIS7264	OEM
NIS7264B	OEM
NJ100A	0127
NJ100B	0016
NJ-101B	OEM
NJ101B	0037
NJ102C	0016
NJ107	0086
NJ181B	0004
NJ-202B	OEM
NJ202B	0127
NJ703N	0632
NJD6505S	OEM
NJD6506S	OEM
NJL1102L	OEM
NJL5121DC	0315
NJL5141EA	5662
NJL5141E-AB	OEM
NJL5141F-AB	OEM
NJL5161K	OEM
NJL6130A	OEM
NJL6145L	OEM
NJM072	OEM
NJM072-FP	OEM
NJM072D	4107
NJM082	OEM
NJM082M	4105
NJM58L05	3024
NJM78L02AA	OEM
NJM78L05	1292
NJM78L05A	1292
NJM78L05AV	1292
NJM78L05UA	OEM
NJM78L06A	2301
NJM78L08A	6204
NJM78L09	1786
NJM78L09A	1786
NJM78L09K	4480
NJM78L12A	6215
NJM78M05	0624
NJM78M05A	6046
NJM78M05FA	0624
NJM78M09A	OEM
NJM78M09E	OEM
NJM78M09FA	3332
NJM78M12	0334
NJM78M12A	6261
NJM78M13A	OEM
NJM78M15	OEM
NJM78M18A	2259
NJM78M93D	OEM
NJM78M93FD	OEM
NJM79L05A	4884
NJM79L06A	OEM
NJM79L12A	7063
NJM79M05A	7121
NJM79M05FA	OEM
NJM79M06A	6878
NJM79M06FA	OEM
NJM79M09A	OEM
NJM79M12A	6668
NJM386	OEM
NJM386BS	OEM
NJM386D	6607
NJM387A	OEM
NJM387DA	OEM
NJM555D	0973
NJM703	0632
NJM-703N	4101
NJM703N	0632
NJM2041DD	OEM
NJM2043S	OEM
NJM2043SE	OEM
NJM2058D	OEM
NJM2061D	OEM
NJM2068	OEM
NJM2068DD	OEM
NJM2068S	OEM
NJM2068SD	OEM
NJM2068SDN	OEM
NJM2074	OEM
NJM2074AD	OEM
NJM2074D	OEM
NJM2100MAT1	OEM
NJM2201	2261
NJM2207	OEM
NJM2207S	OEM
NJM2217L	OEM
NJM2220S	OEM
NJM2225M	OEM

DEVICE TYPE	REPL CODE
NJM2225S	OEM
NJM2227AL	OEM
NJM2227L	OEM
NJM2229S	OEM
NJM-2233BS	OEM
NJM2233BS	OEM
NJM2234	OEM
NJM2234D	OEM
NJM2234L	OEM
NJM2234S	OEM
NJM2235M	OEM
NJM2245S	OEM
NJM2248S	OEM
NJM2249L	OEM
NJM2355	OEM
NJM2406FTE1	OEM
NJM2901	0631
NJM2901D	0631
NJM2901M	4254
NJM2901MB	OEM
NJM2901N	0176
NJM2902M	4255
NJM2902N	1074
NJM2903	OEM
NJM2903D	0631
NJM2903M	OEM
NJM2903S	4258
NJM2904	0770
NJM2904D	0770
NJM2904M	4259
NJM2904MTE2	OEM
NJM3404AD	OEM
NJM3404S	OEM
NJM3414M	OEM
NJM3414MT1	OEM
NJM3415D	OEM
NJM4556D	OEM
NJM4556M	OEM
NJM4558C	0360
NJM4558D	0360
NJM4558D-1	0360
NJM4558DD	0360
NJM4558D-K	0360
NJM4558DM	0360
NJM4558DMA	0360
NJM4558DV	0360
NJM4558DX	0360
NJM4558K	0360
NJM4558M	5531
NJM4558MDA	0360
NJM-4558S	2913
NJM4558S	2913
NJM4558SD	2913
NJM4559D	0360
NJM4559S	OEM
NJM4560D	OEM
NJM4560DD	OEM
NJM4560F	OEM
NJM4560M	OEM
NJM4560S	OEM
NJM4562DDR	OEM
NJM4562S	OEM
NJM4562SD	OEM
NJM4565SB	OEM
NJM5532	OEM
NJM5532DD	OEM
NJM7805A	3458
NJM7805FA	0624
NJM7812A	3485
NJM7812B	OEM
NJM7812FA	6352
NJM7818FA	2259
NJM7905FA	OEM
NJM7906A	6878
NJM7906FA	OEM
NJM7912	OEM
NJU74HC04M	OEM
NJU74HCU04D	OEM
NJU4051BD	OEM
NJU4053B	OEM
NJU4053BM	OEM
NK1302	0050
NK1404	0050
NK5252	OEM
NKT4	0004
NKT5	0004
NKT11	0004
NKT12	0004
NKT24	0004
NKT25	0004
NKT32	0004
NKT33	0004
NKT42	0280
NKT43	0280
NKT52	0004
NKT53	0004
NKT54	0004
NKT62	0280
NKT63	0280
NKT64	0280
NKT72	0280
NKT73	0280
NKT74	0280
NKT101	OEM
NKT102	0164
NKT103	0136
NKT104	0164
NKT105	0164
NKT106	0164

DEVICE TYPE	REPL CODE
NKT107	0164
NKT108	0164
NKT109	0164
NKT121	0050
NKT122	0050
NKT123	0004
NKT124	0050
NKT125	0050
NKT126	0004
NKT127	0136
NKT128	0280
NKT129	0280
NKT131	0136
NKT132	0136
NKT133	0211
NKT141	0280
NKT142	0280
NKT143	0280
NKT144	0280
NKT151	0136
NKT152	0136
NKT153/25	0164
NKT154/25	0164
NKT162	0280
NKT163	0280
NKT163/25	0164
NKT164	0280
NKT164/25	0164
NKT201	OEM
NKT202	0280
NKT203	0280
NKT204	0280
NKT205	0280
NKT206	0280
NKT207	0280
NKT208	0004
NKT211	0164
NKT212	0164
NKT213	0164
NKT214	0164
NKT215	0164
NKT216	0164
NKT217	0164
NKT218	0164
NKT219	0164
NKT221	0211
NKT222	0211
NKT222S1	0280
NKT222S2	0280
NKT223	0211
NKT223A	0004
NKT224	0211
NKT225	0211
NKT226	0211
NKT227	0211
NKT228	0211
NKT229	0164
NKT231	0211
NKT232	0211
NKT240	0847
NKT241	0847
NKT242	0136
NKT243	0280
NKT244	0004
NKT245	0211
NKT246	0004
NKT247	0211
NKT249	0136
NKT251	0050
NKT252	0136
NKT253	0136
NKT254	0136
NKT255	0136
NKT261	0280
NKT262	0280
NKT263	0280
NKT264	0280
NKT265	0136
NKT270	0136
NKT271	0164
NKT272	0164
NKT273	0004
NKT274	0164
NKT275	0164
NKT275A	0164
NKT275E	0164
NKT275J	0164
NKT278	0004
NKT281	0164
NKT301A	OEM
NKT302A	OEM
NKT303	0004
NKT351	0164
NKT352	0164
NKT361	OEM
NKT362	OEM
NKT-401	0160
NKT401	0160
NKT-402	0160
NKT402	0160
NKT-403	0160
NKT403	0160
NKT-404	0160
NKT404	0160
NKT-405	0160
NKT405	0160
NKT406	0160
NKT-415	0160
NKT415	0160

DEVICE TYPE	REPL CODE
NKT-416	0160
NKT416	0160
NKT450	0160
NKT450X2	OEM
NKT-451	0160
NKT451	0222
NKT-452	0160
NKT452	0222
NKT452-S1	0222
NKT452S1	OEM
NKT-453	0160
NKT453	0222
NKT-454	0160
NKT454	0160
NKT-501	0160
NKT501	0160
NKT502	OEM
NKT-503	0160
NKT503	0160
NKT-504	0160
NKT504	0160
NKT618	0136
NKT674F	0136
NKT675	0136
NKT676	0136
NKT677	0136
NKT677F	0136
NKT701	0208
NKT703	0208
NKT713	0208
NKT717	0208
NKT732	0038
NKT734	0600
NKT735	OEM
NKT736	0600
NKT738	0600
NKT751	0208
NKT752	0208
NKT753	0600
NKT773	0208
NKT774	0600
NKT781	0208
NKT4054	0178
NKT4055	0491
NKT10339	0016
NKT10419	0016
NKT10439	0016
NKT10519	0016
NKT12329	0016
NKT12429	0016
NKT13329	0016
NKT13429	0016
NKT15325	0136
NKT15425	0136
NKT16229	0007
NKT16325	0280
NKT16425	0280
NKT20329	0037
NKT20339	0037
NKT35219	0007
NKT80111	2950
NKT80112	2950
NKT80113	2950
NKT80211	0325
NKT80212	0325
NKT80213	0325
NKT80214	0325
NKT80215	2950
NKT80216	2950
NKT800112	0325
NKT800113	0325
NL-2N683	0151
NL-2N684	3258
NL-2N685	0762
NL-2N686	3268
NL-2N687	3271
NL-2N688	0740
NL-2N689	3291
NL-2N690	0764
NL-2N691	2875
NL-2N692	0766
NL2N2023	OEM
NL2N2024	OEM
NL2N2025	OEM
NL2N2026	OEM
NL2N2027	OEM
NL2N2028	OEM
NL2N2029	OEM
NL2N2030	OEM
NL5	0015
NL10	0015
NL15	0015
NL20	0015
NL25	0015
NL30	0015
NL40	0015
NL50	0015
NL60	0015
NL100B	0127
NL-511-3	0679
NL-511-4	0679
NL-511-6	0679
NL516M	OEM
NL516S	OEM
NL548A	OEM
NL548B	OEM
NL548C	0655
NL548D	OEM
NL548E	OEM

DEVICE TYPE	REPL CODE
NL548M	OEM
NL548N	OEM
NL548S	OEM
NL549A	OEM
NL549B	OEM
NL549C	OEM
NL549D	OEM
NL549E	OEM
NL549M	OEM
NL549N	OEM
NL549P	OEM
NL549S	OEM
NL549T	OEM
NL551A	OEM
NL551B	OEM
NL551C	OEM
NL551D	OEM
NL551E	OEM
NL551M	OEM
NL551N	OEM
NL551S	OEM
NL552A	OEM
NL552B	OEM
NL552C	OEM
NL552D	OEM
NL552E	OEM
NL552M	OEM
NL552N	OEM
NL552P	OEM
NL552S	OEM
NL552T	OEM
NL555A	0468
NL555B	0468
NL555C	0468
NL555D	0468
NL555E	0468
NL555F	OEM
NL555G	OEM
NL555H	OEM
NL555M	0468
NL555N	0468
NL555P	0468
NL555PA	0468
NL555PB	0468
NL555PC	OEM
NL555R	OEM
NL555S	0658
NL556A	0658
NL556B	0217
NL556C	0217
NL556D	0217
NL556E	0217
NL556F	OEM
NL556G	OEM
NL556H	OEM
NL556M	0217
NL556N	0468
NL556P	OEM
NL556PA	OEM
NL556PB	OEM
NL556PC	OEM
NL556S	0468
NL556T	OEM
NL570A	OEM
NL570B	OEM
NL570C	OEM
NL570D	0740
NL570E	0764
NL570M	0764
NL571A	OEM
NL571B	OEM
NL571C	OEM
NL571D	OEM
NL571E	OEM
NL571F	OEM
NL571M	OEM
NL576B	1704
NL576C	4031
NL576D	4031
NL576E	4031
NL576M	4031
NL576N	0679
NL576P	0679
NL576PA	0679
NL576PB	0679
NL576S	0679
NL576T	0679
NL577B	1704
NL577C	4031
NL577D	4031
NL577E	4031
NL577M	4031
NL577N	0679
NL577P	0679
NL577P7P	OEM
NL577PA	0679
NL577PB	0679
NL577S	0679
NL577T	0679
NL-578	OEM
NL578B	1704
NL578C	4031
NL578D	4031
NL578E	4031
NL578M	4031
NL578N	0679
NL578P	0679
NL578PA	0679
NL578PB	0679

DEVICE TYPE	REPL CODE
NL578S	0679
NL578T	0679
NL-579	OEM
NL579B	1704
NL579C	4031
NL579D	4031
NL579E	4031
NL579M	4031
NL579N	0679
NL579P	0679
NL579PA	0679
NL579PB	0679
NL579S	0679
NL579T	0679
NL-1580	OEM
NL1580A	2843
NL1580B	2843
NL1580D	1765
NL1580E	1765
NL1580M	1765
NL1580N	1814
NL1580P	1814
NL1580PA	1814
NL1580PB	1814
NL1580S	1814
NL1580T	1814
NL-C35A	0712
NLC35A	0712
NL-C35B	0469
NLC35B	0469
NL-C35C	0778
NLC35C	0721
NL-C35D	0722
NLC35D	0722
NL-C35E	0725
NLC35E	0725
NL-C35F	0731
NL-C35G	0469
NLC35G	0731
NL-C35H	0721
NLC35H	0722
NL-C35M	0725
NLC35M	0725
NLC35N	0750
NLC35S	0750
NL-C35U	3277
NL-C36A	0722
NLC36A	0741
NL-C36A/2N1844	0741
NL-C36B	0722
NLC36B	0745
NL-C36B/2N1846	0745
NL-C36C	0722
NLC36C	0747
NL-C36C/2N1848	2917
NL-C36D	0722
NLC36D	0747
NL-C36D/2N1849	0747
NLC36E	0752
NL-C36E/2N1850	3243
NL-C36F	0722
NLC36F	0469
NL-C36G	0722
NLC36G	0745
NL-C36G2N1845	3105
NLC36H	0747
NLC36M	0764
NLC36N	0766
NLC36S	0766
NL-C36U	0722
NLC36U	0469
NL-C37A	0567
NLC37A	0567
NL-C37B	0762
NLC37B	0762
NL-C37C	3271
NLC37C	0740
NL-C37D	0740
NLC37D	0740
NL-C37E	3291
NLC37E	0764
NL-C37F	0722
NLC37F	0469
NL-C37G	0722
NL-C37M	0764
NLC37M	0764
NL-C37N	0766
NLC37N	0679
NL-C37S	2875
NLC37S	0679
NL-C37U	0722
NLC37U	0469
NL-C38A	0731
NLC38A	0731
NL-C38B	0469
NLC38B	0469
NL-C38C	0778
NLC38C	0722
NL-C38D	0722
NLC38D	0722
NL-C38E	0725
NLC38E	0778
NLC38F	OEM
NL-C38G	0469
NLC38G	0469
NL-C38H	0721
NLC38H	0722
NL-C38M	0725
NL-C38U	3277

DEVICE TYPE	REPL CODE
NLC38U	0731
NLC40E	0804
NL-C45	OEM
NLC45A	0608
NLC45B	0608
NLC45C	0610
NLC45D	0610
NLC45E	0610
NL-C45F	OEM
NLC45G	0608
NLC45H	0610
NLC45M	0610
NLC45N	0468
NLC45P	0468
NLC45PA	0468
NLC45PB	0468
NLC45S	0468
NLC45T	0468
NL-C45U	OEM
NL-C46	OEM
NLC46A	0641
NLC46B	0641
NLC46C	0217
NLC46D	0217
NLC46E	0217
NL-C46F	OEM
NLC46G	0641
NLC46H	0217
NLC46M	0217
NLC46N	0658
NLC46P	0658
NLC46PA	0658
NLC46PB	0658
NLC46S	0658
NLC46T	0658
NL-C46U	OEM
NL-C50	OEM
NLC50A	0217
NLC50B	0217
NLC50C	0217
NLC50D	0217
NLC50E	0217
NLC50G	0217
NLC50H	0217
NLC50M	0217
NLC50N	0468
NLC50S	0468
NL-C52	OEM
NLC52A	0217
NLC52B	0217
NLC52C	0217
NLC52D	0217
NLC52E	0217
NLC52G	0217
NLC52H	0217
NLC52M	0217
NLC52N	0658
NLC52S	0658
NL-C52U	OEM
NL-C55	OEM
NL-C55F	OEM
NL-C55U	OEM
NL-C56	OEM
NL-C56F	OEM
NL-C56U	OEM
NL-C60	OEM
NL-C62	OEM
NL-C135A2N3754	0712
NL-C135B2N3755	0469
NL-C135C2N3756	0721
NL-C135D2N3757	0722
NL-C135E2N3758	0778
NL-C135F2N3759	0731
NL-C135M2N3760	0725
NLC135N	OEM
NL-C137E	0778
NL-C137M	0725
NL-C150	OEM
NLC150E	0468
NLC150M	0468
NLC150N	0468
NLC150P	0468
NLC150PA	0468
NLC150PB	0468
NLC150S	0468
NLC150T	0468
NL-C151	OEM
NL-C151PA	OEM
NL-C151PB	OEM
NL-C151PC	OEM
NL-C152	OEM
NL-C153	OEM
NL-C153PA	OEM
NL-C153PB	OEM
NL-C153PC	OEM
NL-C154	OEM
NL-C154/C155B	OEM
NL-C155	OEM
NL-C156	OEM
NL-C157	OEM
NL-C178	OEM
NLC178A	0738
NLC178B	0738
NLC178C	0738
NLC178D	0738
NLC178E	0738
NLC178M	0738
NLC178N	0738
NLC178P	0738

DEVICE TYPE	REPL CODE
NLC178PA	0738
NLC178PAA	OEM
NLC178PB	0738
NLC178S	0738
NLC178T	0738
NL-C180	OEM
NLC180A	0738
NLC180B	0738
NLC180C	0738
NLC180D	0738
NLC180E	0738
NLC180M	0738
NLC180M0M	OEM
NLC180N	0738
NLC180P	0738
NLC180P0P	OEM
NLC180PA	0738
NLC180PB	0738
NLC180PC	OEM
NLC180S	0738
NLC180T	0738
NL-C181	OEM
NLC181A	1084
NLC181B	1084
NLC181C	1086
NLC181D	1086
NLC181E	1086
NLC181M	1086
NLC181N	1102
NLC181P	1102
NLC181S	1102
NLC181T	1102
NL-C185	OEM
NLC290A	1900
NLC290B	1900
NLC290C	1900
NLC290D	0738
NLC290E	0738
NLC290G	1900
NLC290H	1900
NLC290M	0738
NLC290N	0738
NLC290P	0738
NLC290PA	0738
NLC290PB	0738
NLC290S	0738
NLC290T	0738
NL-C350	OEM
NLC350A	0192
NLC350B	0192
NLC350C	0096
NLC350D	0096
NLC350E	0096
NLC350M	0096
NLC350N	0096
NLC350P	0096
NLC350PA	0096
NLC350PB	0096
NLC350S	0096
NLC350T	0096
NL-C354	OEM
NL-C354/C355M	OEM
NL-C355	OEM
NL-C380	OEM
NLC380A	0192
NLC380B	0192
NLC380C	0096
NLC380D	0096
NLC380E	0096
NLC380M	0096
NLC380N	2843
NLC380P	2843
NLC380PA	2843
NLC380PB	2843
NLC380S	2843
NLC380T	2843
NL-C385	OEM
NL-C501	OEM
NLC501A	2843
NLC501B	2843
NLC501C	1765
NLC501D	1765
NLC501E	1765
NLC501M	1765
NLC501N	1814
NLC501P	1814
NLC501PA	1814
NLC501PB	1814
NLC501S	1814
NLC501T	1814
NL-C601	OEM
NLC601N	1725
NLC601P	1725
NLC601T	1725
NLE45B	OEM
NLE45C	OEM
NLE45D	OEM
NLE45E	OEM
NLE45M	OEM
NLE45N	OEM
NLE45S	OEM
NLE45T	OEM
NLE46B	OEM
NLE46C	OEM
NLE46D	OEM
NLE46M	OEM
NLE46N	OEM
NLE46S	OEM
NLE46T	OEM

DEVICE TYPE	REPL CODE
NLE150B	OEM
NLE150C	OEM
NLE150D	OEM
NLE150E	OEM
NLE150M	OEM
NLE150N	OEM
NLE150P	OEM
NLE150S	OEM
NLE150T	OEM
NLE151E	OEM
NLE151M	OEM
NLE151N	OEM
NLE151P	OEM
NLE151S	OEM
NLE151T	OEM
NLE152B	OEM
NLE152C	OEM
NLE152D	OEM
NLE152E	OEM
NLE152M	OEM
NLE152N	OEM
NLE152P	OEM
NLE152S	OEM
NLE152T	OEM
NLE153E	OEM
NLE153M	OEM
NLE153N	OEM
NLE153P	OEM
NLE153S	OEM
NLE153T	OEM
NLE155A	OEM
NLE155B	OEM
NLE155C	OEM
NLE155D	OEM
NLE155E	OEM
NLE157A	OEM
NLE157B	OEM
NLE157C	OEM
NLE157D	OEM
NLE157E	OEM
NLF45B	OEM
NLF45C	OEM
NLF45D	OEM
NLF45E	OEM
NLF45M	OEM
NLF45N	OEM
NLF45S	OEM
NLF45T	OEM
NLF46B	OEM
NLF46C	OEM
NLF46D	OEM
NLF46E	OEM
NLF46M	OEM
NLF46N	OEM
NLF46S	OEM
NLF46T	OEM
NL-F150	OEM
NLF150B	0608
NLF150C	0610
NLF150D	0610
NLF150E	0610
NLF150M	0610
NLF150N	0468
NLF150P	0468
NLF150S	0468
NLF150T	0468
NL-F151	OEM
NLF151B	0526
NLF151C	0526
NLF151D	0526
NLF151E	0526
NLF151M	0526
NLF151N	0108
NLF151P	0108
NLF151PA	OEM
NLF151S	0108
NLF151T	0108
NL-F152	OEM
NLF152B	0641
NLF152C	0217
NLF152D	0217
NLF152E	0217
NLF152M	0217
NLF152N	0658
NLF152P	0658
NLF152S	0658
NLF152T	0658
NL-F153	OEM
NLF153B	0655
NLF153C	0655
NLF153D	0655
NLF153E	0655
NLF153M	0655
NLF153P	0655
NLF153PA	OEM
NLF153S	0655
NLF153T	0655
NL-F154	OEM
NL-F154/F155D	OEM
NL-F155	OEM
NL-F156	OEM
NL-F157	OEM
NL-F158	OEM
NLF158A	0526
NLF158B	0526
NLF158C	0526
NLF158D	0526
NLF158E	0526
NLF158M	0526

DEVICE TYPE	REPL CODE
NL-F159	OEM
NLF159A	0655
NLF159B	0655
NLF159C	0655
NLF159D	0655
NLF159E	0655
NLF159M	0655
NLF159N	0020
NLF159P	0020
NLF159S	0020
NLF159T	0020
NL-F180	OEM
NLF180A	1084
NLF180B	1084
NLF180C	1086
NLF180D	1086
NLF180E	1086
NLF180M	1086
NLF180N	1102
NLF180P	1102
NLF180PA	OEM
NLF180PB	OEM
NLF180PC	OEM
NLF180S	1102
NLF180T	1102
NL-F185	OEM
NLF290A	1900
NLF290B	1900
NLF290C	1900
NLF290D	1900
NLF290E	1900
NLF290M	1900
NLF290N	0738
NLF290P	0738
NLF290PA	OEM
NLF290PB	OEM
NLF290S	0738
NLF290T	0738
NLF291PA	OEM
NLF291PB	OEM
NLF350A	0192
NLF350B	0192
NLF350C	0159
NLF350D	0159
NLF350E	0159
NLF350M	0159
NLF350N	0096
NLF350P	0096
NLF350S	0096
NLF350T	0096
NL-F358	OEM
NLF358A	0459
NLF358B	0459
NLF358C	0459
NLF358D	0459
NLF358E	0459
NLF358M	0459
NLF358N	0587
NLF358P	0587
NLF358S	0587
NLF358T	0587
NL-F380	OEM
NLF380A	0192
NLF380B	0192
NLF380C	0096
NLF380D	0096
NLF380E	0096
NLF380M	0096
NLF380N	2843
NLF380P	2843
NLF380S	2843
NL-F385	OEM
NL-F390	OEM
NLF390E	1765
NLF390M	1765
NLF390N	1814
NLF390P	1814
NLF390PA	1814
NLF390PB	1814
NLF390S	1814
NLF390T	1814
NL-F394	OEM
NL-F395	OEM
NL-F397	OEM
NL-F398	OEM
NL-F701	OEM
NLF701E	4148
NLF701M	4148
NLF701N	1725
NLF701P	1725
NLF701PA	1725
NLF701PB	1725
NLF701S	1725
NLF701T	1725
NLH150B	OEM
NLH150C	OEM
NLH150D	OEM
NLH150E	OEM
NLH150M	OEM
NLH150N	OEM
NLH150P	OEM
NLH150S	OEM
NLH150T	OEM
NLH152B	OEM
NLH152C	OEM
NLH152D	OEM
NLH152E	OEM
NLH152M	OEM
NLH152N	OEM

DEVICE TYPE	REPL CODE
NLH152P	OEM
NLH152S	OEM
NLH152T	OEM
NM93C46	OEM
NMC93C-46N	OEM
NMC2147H-3	1693
NMC2148HJ-3	OEM
NMC2148HN-3	OEM
NMC2148J-3	OEM
NMC6508J-5	3193
NMC6508J-9	3193
NMC6508N-5	3193
NMC6508N-9	3193
NMC9306M	OEM
NMC9306N	OEM
NMC9313BN	OEM
NMC9346	OEM
NMC9346N	OEM
NMJ2245S	OEM
NML16L8LNC	OEM
NML16L8LNJ	OEM
NN16AT	1727
NN16BT	1727
NN16DT	1727
NN16GT	1727
NN16JT	OEM
NN16KT	OEM
NN16MT	OEM
NN30AM	OEM
NN30BM	OEM
NN30DM	OEM
NN30GM	OEM
NN30JM	OEM
NN30KM	OEM
NN30MM	OEM
NN50	0015
NN650	0037
NN1199	0224
NN5762N	OEM
NN7000	0595
NN7001	0218
NN7002	0595
NN7003	0595
NN7004	0595
NN7005	0595
NN7500	0791
NN7501	0791
NN7502	0791
NN7503	0791
NN7504	0791
NN7505	0791
NN7511	1242
NN9017	0079
NP04S49	0715
NP9ZOKLR133L	OEM
NP16AT	1236
NP16BT	1236
NP16DT	1236
NP16GT	1236
NP16JT	OEM
NP16KT	OEM
NP16MT	OEM
NP30AM	2234
NP30AP	3676
NP30BM	2234
NP30BP	3676
NP30DM	2234
NP30DP	3676
NP30GM	2234
NP30GP	3679
NP30JM	6599
NP30KM	OEM
NP30KP	OEM
NP30MM	OEM
NP30MP	OEM
NP50A	0015
NP60A	0015
NP389	0015
NP1000	OEM
NP1001	OEM
NP2000	OEM
NP2001	OEM
NP2222AP	0079
NP3000	OEM
NP4510	2616
NP5114	OEM
NP5116	OEM
NPC00050	OEM
NPC0010	0015
NPC0050	0015
NPC069	0079
NPC069-98	0079
NPC079	0037
NPC079-98	0861
NPC0100	0015
NPC0200	OEM
NPC0400	OEM
NPC0600	OEM
NPC0800	OEM
NPC06998	0007
NPC12-1A	0178
NPC12-1B	0178
NPC12-2	0142
NPC108	1392
NPC108A	1392
NPC115	0086
NPC151	OEM
NPC167	0007
NPC173	0144

DEVICE TYPE	REPL CODE
NPC187	0086
NPC188	0144
NPC189	0086
NPC200C	OEM
NPC211N	2950
NPC212N	2950
NPC213N	2950
NPC214N	2950
NPC215N	2950
NPC216N	2950
NPC290	0086
NPC312N	0325
NPC544	4003
NPC634	0460
NPC636	0460
NPC737	0016
NPC750	0693
NPC1000	OEM
NPC1075	0330
NPC1096	0161
NPC1098	0161
NPC1500	OEM
NPC5104	OEM
NPC5107	OEM
NPC6069S	OEM
NPC7629	1714
NPC7837	OEM
NPC8021	OEM
NPC76315	4224
NPC76365	2616
NPC-C1006A	OEM
NPD0531	0133
NPQ199	OEM
NPQ300	OEM
NPQ6261	OEM
NPQ8317	OEM
NPS404	0037
NPS404A	0037
NPS6512	0079
NPS6513	0079
NPS6514	0079
NPS6516	0150
NPS6517	0150
NPS6518	0925
NPS6519	0079
NPS6520	0079
NPS6522	0079
NPS6523	0079
NPSA20	0111
NPT800	OEM
NQS10KTA1267	OEM
NQS10KTC3199	0286
NQS40KTC3198	0286
NQSOOKTA1275	5835
NQSOOKTC3228	4250
NR05	0600
NR041	0079
NR041E	0079
NR-071AU	0016
NR091ET	0079
NR5	0600
NR8AT	OEM
NR8BT	OEM
NR8DT	OEM
NR8GT	OEM
NR8JT	OEM
NR8KT	OEM
NR8MT	OEM
NR10	0600
NR20	0038
NR30	0600
NR201AY	0016
NR-261AS	0079
NR261AS	0016
NR271AY	0016
NR401	0016
NR401EG	0016
NR401EH	0016
NR421	0127
NR421DG	0127
NR-431AS	0016
NR441	0016
NR441EG	0016
NR441EN	0016
NR461	0079
NR461AA	0007
NR461AF	0127
NR-461AS	0016
NR461EG	0079
NR461EH	0079
NR601BT	0037
NR621AT	0037
NR621EU	0037
NR631AY	0037
NR700	0600
NR7916	0325
NR49333	OEM
NS8AT	1145
NS8BT	1145
NS8DT	1145
NS8GT	1145
NS8JT	1969
NS8KT	OEM
NS8MT	OEM
NS20-42	0233
NS32	0211
NS100	OEM
NS101	OEM
NS121	0211

DEVICE TYPE	REPL CODE
NS142	0167
NS200	0016
NS316	0325
NS345	0007
NS381	0144
NS382	0144
NS383	0016
NS384	0016
NS404	0037
NS430	1480
NS431	1480
NS432	1480
NS433	1480
NS434	1480
NS435	1480
NS436	1480
NS437	1480
NS438	1480
NS475	0016
NS476	0016
NS477	0016
NS478	0016
NS479	0016
NS480	0016
NS500	OEM
NS501	OEM
NS502	OEM
NS503	OEM
NS504	OEM
NS505	OEM
NS506	OEM
NS507	OEM
NS508	OEM
NS661	0037
NS662	0037
NS663	0037
NS664	0037
NS665	0037
NS666	0037
NS667	0037
NS668	0037
NS731	0016
NS731A	0016
NS732	0037
NS732A	0037
NS733	0016
NS733A	0016
NS734	0016
NS734A	0016
NS746	0666
NS792	0936
NS793	0936
NS949	0016
NS950	0595
NS1000	1335
NS1001	0037
NS1002	OEM
NS1110	0142
NS1111	OEM
NS1112	OEM
NS1116	OEM
NS1234	OEM
NS1355	0595
NS1356	0144
NS1500	0016
NS1510	0127
NS1672	0037
NS1673	0037
NS1674	0037
NS1675	0037
NS1861	0037
NS1862	0037
NS1863	0037
NS1864	0037
NS1900	0595
NS1960	0595
NS1972	0016
NS1973	0016
NS1974	0016
NS1975	0016
NS2000	0071
NS2006	0087
NS2007	0087
NS2008	0087
NS2100	0086
NS2101	0086
NS2310	OEM
NS2311	OEM
NS2525	OEM
NS3000	0087
NS3001	0087
NS3006	0087
NS3007	0087
NS3008	0087
NS3039	0127
NS3040	0127
NS3041	0127
NS3050	OEM
NS3051	OEM
NS3052	OEM
NS3053	OEM
NS3108	OEM
NS3109	OEM
NS3110	OEM
NS3300	0144
NS3762	0892
NS3903	0016
NS3904	0016
NS3905	0037
NS3906	0037
NS5387	OEM
NS6000	1567
NS6001	0037
NS6002	1567
NS6003	2153
NS6004	2153
NS6062	0037
NS6063	0037
NS6064	0037
NS6065	0037
NS6112	0144
NS6113	0144
NS6114	0016
NS6115	0016
NS6201	OEM
NS6203	OEM
NS6205	OEM
NS6207	0016
NS6208	OEM
NS6209	OEM
NS6210	0016
NS6211	0037
NS6212	0233
NS6213	OEM
NS6214	OEM
NS6241	0037
NS7000	OEM
NS7001	OEM
NS7070	OEM
NS7100	OEM
NS7200	OEM
NS7201	OEM
NS7261	0144
NS7262	0016
NS7267	0144
NS7300	OEM
NS7301	OEM
NS7302	OEM
NS7303	OEM
NS7304	OEM
NS7305	OEM
NS7630	OEM
NS8000	OEM
NS8003	OEM
NS8250AN	OEM
NS8250AV	OEM
NS8250N-B	OEM
NS9210	OEM
NS9211	OEM
NS9400	0086
NS9420	0086
NS9500	0086
NS9540	0086
NS9608	OEM
NS9609	OEM
NS9609A	OEM
NS9710	0144
NS9713	OEM
NS9726	OEM
NS9728	0086
NS9729	0086
NS9730	0086
NS9731	0086
NS12000	2153
NS12001	2153
NS12002	2153
NS12006	0711
NS16450N	OEM
NS16550V	OEM
NS45006	0008
NS48004	0233
NSD102	0824
NSD103	0824
NSD104	0824
NSD105	0824
NSD106	0824
NSD131	0285
NSD132	0285
NSD133	0285
NSD134	0285
NSD151	3535
NSD152	3535
NSD202	0164
NSD203	0164
NSD204	3327
NSD205	3327
NSDU01	0560
NSDU01A	0560
NSDU05	0824
NSDU07	0824
NSDU51	1367
NSDU51A	1367
NSDU52	3327
NSDU55	3327
NSDU57	3327
NSE170	1267
NSE-181	0219
NSL5023	1615
NSL5024	5119
NSL5027	3182
NSL5053	3182
NSL5057	1962
NSL5076A	3019
NSL5253	1778
NSL5253A	1778
NSL5274	0841
NSL5353	0841
NSL5353A	3157
NSL5374	3124
NSL5753	3182
NSL5774	3096
NS060	OEM
NS061	OEM
NS063	OEM
NS064	OEM
NS066	OEM
NS067	OEM
NS069	OEM
NS070	OEM
NS072	OEM
NS073	OEM
NS075	OEM
NS078	OEM
NSQ4110L1	0415
NSS1021	0071
NT0.6C0	OEM
NT0.6C12	0137
NT0.6C13	OEM
NT0.6C15	0002
NT0.6C16	OEM
NT0.6C18	OEM
NT0.6C20	OEM
NT0.6C22	OEM
NT0.6C24	OEM
NT0.6C27	OEM
NT0.6C30	OEM
NT0.6C33	OEM
NT0.6C36	OEM
NT0.6C39	OEM
NT0.6C43	OEM
NT0.6C47	OEM
NT0.6C51	OEM
NT0.6C56	OEM
NT3C10	0992
NT3C11	0995
NT3C12	1263
NT3C13	1249
NT3C15	1703
NT3C16	1703
NT3C18	1716
NT3C20	1731
NT3C22	0727
NT3C24	1756
NT3C27	1782
NT3C30	1772
NT3C33	1799
NT3C36	1809
NT3C39	1817
NT3C43	1826
NT3C47	1853
NT3C51	1866
NT3C56	1884
NT3C62	1895
NT3C68	1902
NT3C75	0736
NT3C82	1909
NT3C91	1914
NT3C100	0438
NT50C4V7	0451
NT55C3V3	0292
NT55C3V6	0188
NT55C3V9	0456
NT55C4V3	0533
NT55C4V7	0451
NT101	0344
NT102	0316
NT161	OEM
NT162	OEM
NT163	OEM
NT164	OEM
NT165	OEM
NT251	OEM
NT252	OEM
NT253	OEM
NT254	OEM
NT255	OEM
NT481EY	0715
NT826AB	4084
NT1056	0016
NT1503	0111
NT2222A	0079
NT2907	2260
NT3000	2309
NT5007	0037
NT5508	0292
NT5509	0188
NT5510	0456
NT5511	0533
NT5512	0451
NT6034	0111
NTC-4	0224
NTC-5	0127
NTC-6	0037
NTC-7	0076
NTC-8	0066
NTC-9	0955
NTC-10	0006
NTC-11	0076
NTC-12	0065
NTC-13	0019
NTC-14	0143
NTC-15	0124
NTC-16	0288
NTC-17	0023
NTC-18	0102
NTC-19	0015
NTC-20	0317
NTC-21	0167
NTE1V010	1691
NTE1V014	2008
NTE1V015	1314
NTE1V017	2325
NTE1V020	2790
NTE1V025	4367
NTE1V030	5744
NTE1V035	2935
NTE1V040	4263
NTE1V050	4472
NTE1V075	4921
NTE1V095	5157
NTE1V115	4135
NTE1V130	1374
NTE1V150	0246
NTE1V250	1332
NTE1V275	1658
NTE1V300	4471
NTE2V010	1694
NTE2V014	2009
NTE2V015	1319
NTE2V017	2326
NTE2V020	2796
NTE2V025	3073
NTE2V030	3075
NTE2V035	2941
NTE2V040	3078
NTE2V050	3085
NTE2V075	3104
NTE2V095	3116
NTE2V115	3123
NTE2V130	0838
NTE2V150	1993
NTE2V250	1334
NTE2V275	1660
NTE2V300	2034
NTE2V420	3257
NTE2V480	1101
NTE10	2851
NTE11	1500
NTE12	4011
NTE13	0014
NTE14	5061
NTE15	2223
NTE16	0582
NTE17	2484
NTE18	2471
NTE19	0359
NTE20	0009
NTE21	4104
NTE22	3608
NTE23	1806
NTE24	3938
NTE25	4548
NTE26	1270
NTE27	4099
NTE28	4012
NTE29	4345
NTE30	4344
NTE31	1562
NTE32	1522
NTE33	0046
NTE34	3630
NTE35	5056
NTE36	0194
NTE36MP	5393
NTE37	3111
NTE37MCP	4475
NTE38	0792
NTE39	4479
NTE40	2924
NTE41	5057
NTE42	2411
NTE43	5058
NTE44	2947
NTE45	5059
NTE46	2797
NTE47	2860
NTE48	3799
NTE49	0824
NTE50	3327
NTE51	3014
NTE52	2912
NTE53	3038
NTE54	1165
NTE54MP	5603
NTE55	0718
NTE55MCP	3789
NTE56	2069
NTE57	0629
NTE58	3226
NTE59	3646
NTE60	3701
NTE60MP	5693
NTE61	4144
NTE61MCP	2011
NTE61MP	5704
NTE62	2847
NTE63	2844
NTE64	3007
NTE65	1342
NTE66	5060
NTE67	1465
NTE68	3606
NTE68MCP	4864
NTE69	2309
NTE70	3489
NTE71	3512
NTE72	0939
NTE73	2662
NTE74	2873
NTE74C00	2112
NTE74C02	2114
NTE74C04	2958
NTE74C08	2118
NTE74C10	2119
NTE74C14	3031
NTE74C20	2123
NTE74C30	3128
NTE74C32	3134
NTE74C42	2143
NTE74C48	2148
NTE74C73	2162
NTE74C74	2163
NTE74C76	2164
NTE74C85	2171
NTE74C90	2174
NTE74C93	2177
NTE74C95	3401
NTE74C107	3834
NTE74C151	4001
NTE74C154	4014
NTE74C157	4033
NTE74C160	4045
NTE74C161	4046
NTE74C164	4062
NTE74C173	4088
NTE74C174	4092
NTE74C175	4096
NTE74C192	4119
NTE74C193	4122
NTE74C221	3047
NTE74C240	2956
NTE74C244	4230
NTE74C373	4462
NTE74C374	4467
NTE74C901	4607
NTE74C902	4608
NTE74C903	4609
NTE74C904	4610
NTE74C922	5597
NTE74C923	5599
NTE74C925	4621
NTE74H00	0682
NTE74H01	5338
NTE74H04	1907
NTE74H05	3251
NTE74H08	5355
NTE74H10	0685
NTE74H11	2398
NTE74H20	3715
NTE74H21	4856
NTE74H22	4596
NTE74H30	5383
NTE74H40	0559
NTE74H50	1792
NTE74H51	1944
NTE74H52	2020
NTE74H53	2101
NTE74H54	2172
NTE74H55	3158
NTE74H60	5410
NTE74H61	2663
NTE74H62	2731
NTE74H71	3264
NTE74H72	3313
NTE74H73	2463
NTE74H74	2492
NTE74H76	5302
NTE74H78	5418
NTE74H87	2579
NTE74H101	5521
NTE74H102	5523
NTE74H103	2969
NTE74H106	5253
NTE74H108	0180
NTE74H183	4399
NTE74L93	0656
NTE74LS00	1527
NTE74LS01	1546
NTE74LS02	1559
NTE74LS03	1579
NTE74LS04	1595
NTE74LS05	1608
NTE74LS08	1633
NTE74LS09	1642
NTE74LS10	1662
NTE74LS11	1667
NTE74LS12	1679
NTE74LS13	1688
NTE74LS14	1698
NTE74LS15	1707
NTE74LS20	0034
NTE74LS21	1763
NTE74LS22	1775
NTE74LS26	1382
NTE74LS27	0183
NTE74LS28	0472
NTE74LS30	0828
NTE74LS32	0088
NTE74LS33	1832
NTE74LS37	1730
NTE74LS38	1839
NTE74LS40	0135
NTE74LS42	1841
NTE74LS47	1845
NTE74LS48	1849
NTE74LS49	1850
NTE74LS51	1033
NTE74LS54	1857
NTE74LS55	0457
NTE74LS63	1864
NTE74LS73	1867
NTE74LS74A	0243
NTE74LS75	1870
NTE74LS76A	2180
NTE74LS77	1872
NTE74LS78	1873
NTE74LS83A	2218
NTE74LS85	0431
NTE74LS86	0291
NTE74LS90	1882
NTE74LS91	1885
NTE74LS92	1887
NTE74LS93	1888
NTE74LS95B	0771
NTE74LS107	1602
NTE74LS109A	1906
NTE74LS112A	2128
NTE74LS113	2256
NTE74LS114	2302
NTE74LS122	1620
NTE74LS123	0979
NTE74LS124	5988
NTE74LS125A	0075
NTE74LS126A	2877
NTE74LS132	1625
NTE74LS133	3400
NTE74LS136	1628
NTE74LS138	0427
NTE74LS139	0153
NTE74LS145	1564
NTE74LS147	4400
NTE74LS148	3912
NTE74LS151	1646
NTE74LS153	0959
NTE74LS155	0209
NTE74LS156	1654
NTE74LS157	1161
NTE74LS158	1656
NTE74LS160A	0837
NTE74LS161A	0858
NTE74LS162A	0880
NTE74LS163A	0893
NTE74LS164	4339
NTE74LS164A	OEM
NTE74LS165	4356
NTE74LS166	4371
NTE74LS168A	0967
NTE74LS169A	0986
NTE74LS170	2629
NTE74LS173A	5217
NTE74LS174	0260
NTE74LS175	1672
NTE74LS181	1678
NTE74LS190	1686
NTE74LS191	1687
NTE74LS192	1689
NTE74LS193	1692
NTE74LS194	1304
NTE74LS194A	1304
NTE74LS195A	1315
NTE74LS196	2834
NTE74LS197	2469
NTE74LS221	1239
NTE74LS240	0452
NTE74LS241	1726
NTE74LS242	1728
NTE74LS243	0906
NTE74LS244	0458
NTE74LS245	0463
NTE74LS246A	OEM
NTE74LS247	1732
NTE74LS248	1734
NTE74LS249	1735
NTE74LS251	1737
NTE74LS253	1739
NTE74LS257	1744
NTE74LS258	1746
NTE74LS259	3205
NTE74LS260	5995
NTE74LS266	0592
NTE74LS273	0894
NTE74LS279	3290
NTE74LS280	1773
NTE74LS283	5997
NTE74LS290	4425
NTE74LS293	0082
NTE74LS295	5998
NTE74LS295A	2227
NTE74LS298	3371
NTE74LS299	4426
NTE74LS324	6000
NTE74LS327	6002
NTE74LS348	0665
NTE74LS352	0761
NTE74LS353	0773
NTE74LS363	6005
NTE74LS364	6006
NTE74LS365A	0943
NTE74LS366A	0956
NTE74LS367	0977
NTE74LS368	0991
NTE74LS373	0709
NTE74LS374	0713
NTE74LS377	1120
NTE74LS378	1133
NTE74LS379	1151
NTE74LS386	1230
NTE74LS390	1288
NTE74LS393	0819
NTE74LS395A	1330
NTE74LS396	6008
NTE74LS398	1383
NTE74LS445	6011
NTE74LS490	2213
NTE74LS540	2541
NTE74LS541	2547
NTE74LS624	3141
NTE74LS625	3149
NTE74LS626	3154
NTE74LS627	3162
NTE74LS629	3175
NTE74LS640	0669
NTE74LS641	0690
NTE74LS642	0719
NTE74LS643	2056
NTE74LS645	0775
NTE74LS670	1130
NTE74S00	0704
NTE74S02	2238
NTE74S03	2217
NTE74S04	2263
NTE74S05	2321
NTE74S08	2569
NTE74S09	2667
NTE74S10	2445
NTE74S11	2447
NTE74S15	2451
NTE74S20	1017
NTE74S22	2461
NTE74S30	3726
NTE74S37	5756
NTE74S38	0780
NTE74S40	2475
NTE74S51	4305
NTE74S64	2496
NTE74S65	2497
NTE74S74	2504
NTE74S85	5773
NTE74S86	2511
NTE74S112	1617
NTE74S113	1623
NTE74S114	1629
NTE74S124	2126
NTE74S132	2134
NTE74S133	1819
NTE74S134	1827
NTE74S138	2138
NTE74S140	1886
NTE74S151	1955
NTE74S153	2151
NTE74S157	1695
NTE74S158	2154
NTE74S163	2156
NTE74S174	2132
NTE74S175	2141
NTE74S181	2165
NTE74S182	2166
NTE74S194	1931
NTE74S251	2198
NTE74S257	2204
NTE74S258	2205
NTE74S260	2207
NTE74S280	2220
NTE74S287	2224
NTE74S288	2175
NTE74S373	2264
NTE74S374	2266
NTE74S387	1918
NTE74S472	2320
NTE74S474	2322
NTE74S475	2324
NTE74S570	2388
NTE74S571	2389
NTE74S572	2390
NTE74S573	2392
NTE75	2964
NTE76	2070
NTE76MP	5956
NTE77	2208
NTE78	5062
NTE79	3797
NTE80	2598
NTE80C95	3177
NTE80C96	3179
NTE80C97	3180
NTE81	2045
NTE82	2468
NTE83	5018
NTE84	5064
NTE85	0286
NTE86	2548
NTE87	0867
NTE87MP	6034
NTE88	3500
NTE88MCP	4500
NTE88MP	6041
NTE89	0055
NTE90	0530
NTE91	0648
NTE92	2276
NTE93	3581
NTE93L08	2127
NTE93L16	2685

DEVICE TYPE	REPL CODE
NTE93MCP	3303
NTE94	0270
NTE95	4252
NTE96	4029
NTE96L02	1468
NTE96LS02	2934
NTE96S02	4292
NTE97	1991
NTE98	2626
NTE99	4013
NTE100	0280
NTE101	0600
NTE102	0211
NTE102A	0004
NTE103	0038
NTE103A	0208
NTE104	0085
NTE104MP	3033
NTE105	0440
NTE106	0150
NTE107	0127
NTE108	0144
NTE108-1	3777
NTE109	0143
NTE110MP	0123
NTE112	0917
NTE113A	0196
NTE114	0484
NTE115	1080
NTE116	0015
NTE116(4)	OEM
NTE117	4630
NTE118	0774
NTE119	OEM
NTE120	0474
NTE121	0160
NTE121MP	0265
NTE123	0198
NTE123A	0016
NTE123AP	0076
NTE124	0142
NTE125	0071
NTE125(2)	OEM
NTE126	0050
NTE126A	0633
NTE127	0975
NTE128	0086
NTE128P	5034
NTE129	0126
NTE129MCP	4163
NTE129P	1597
NTE130	0103
NTE130MP	1514
NTE131	0222
NTE131MP	0822
NTE132	3712
NTE133	3483
NTE134A	0188
NTE135A	0162
NTE136A	0157
NTE137A	0636
NTE138A	0649
NTE139A	0012
NTE140A	0170
NTE141A	0794
NTE142A	0137
NTE143A	0365
NTE144A	0100
NTE145A	0002
NTE146A	0441
NTE147A	0039
NTE148A	1834
NTE149A	0783
NTE150A	0331
NTE151A	0149
NTE152	0236
NTE152MP	5088
NTE153	0925
NTE153MCP	4220
NTE154	0233
NTE155	2763
NTE156	0087
NTE156A	5190
NTE157	0276
NTE158	0164
NTE159	0037
NTE159MCP	1381
NTE160	0050
NTE161	0007
NTE162	0074
NTE163A	0642
NTE164	0003
NTE165	0065
NTE166	0277
NTE167	0289
NTE168	0296
NTE169	0302
NTE170	0250
NTE171	0285
NTE172A	0400
NTE173BP	0293
NTE174	0924
NTE175	0178
NTE176	0847
NTE177	0133
NTE178MP	0913
NTE179	0604
NTE179MP	5536
NTE180	1681
NTE180MCP	3122
NTE181	0130
NTE181MP	5474
NTE182	0561
NTE183	1199
NTE184	0161
NTE184MP	5595
NTE185	0460
NTE185MCP	4605
NTE186	0560
NTE186A	0219
NTE187	1267
NTE187A	1051
NTE188	0551
NTE189	0382
NTE190	0264
NTE191	0338
NTE192	0595
NTE192A	0218
NTE193	0791
NTE193A	1242
NTE194	0861
NTE195A	0698
NTE196	0423
NTE197	0854
NTE198	0168
NTE199	0111
NTE209	3509
NTE210	0566
NTE211	1367
NTE213	0437
NTE214	3746
NTE215	3748
NTE216	1526
NTE217	3750
NTE218	0905
NTE219	0491
NTE219MCP	5498
NTE220	0849
NTE221	0353
NTE222	0212
NTE224	0630
NTE225	3280
NTE226	1010
NTE226MP	1862
NTE227	3754
NTE228A	1708
NTE229	0224
NTE230	0239
NTE231	0061
NTE232	3519
NTE233	0330
NTE234	0693
NTE235	0936
NTE236	0836
NTE237	1411
NTE238	0313
NTE239	2336
NTE240	3758
NTE241	2997
NTE242	3165
NTE243	2427
NTE244	2277
NTE245	3373
NTE246	3374
NTE247	2439
NTE248	2432
NTE249	1394
NTE250	2448
NTE251	3525
NTE252	3526
NTE253	0558
NTE253MCP	6764
NTE254	2897
NTE255	3764
NTE256	3765
NTE257	3377
NTE258	3528
NTE259	3529
NTE260	3530
NTE261	1212
NTE262	0602
NTE263	2235
NTE264	2237
NTE265	2258
NTE266	3533
NTE267	3534
NTE268	3535
NTE269	3536
NTE270	0134
NTE271	0073
NTE272	3538
NTE273	3539
NTE274	3370
NTE275	3540
NTE276	1652
NTE277	2214
NTE278	0418
NTE280	0177
NTE280MP	2140
NTE281	2013
NTE281MCP	5778
NTE282	0622
NTE283	0363
NTE284	0543
NTE284MP	2371
NTE285	1598
NTE285MCP	6141
NTE286	1027
NTE287	0715
NTE288	0342
NTE289	0155
NTE289A	0155
NTE289AMP	1224
NTE289MP	2767
NTE290	0006
NTE290A	0006
NTE290AMCP	3689
NTE291	0236
NTE292	0681
NTE292MCP	3255
NTE293	0018
NTE293MP	3049
NTE294	0532
NTE295	1591
NTE296	2200
NTE297	0324
NTE297MP	3247
NTE298	0436
NTE299	2050
NTE300	2046
NTE300MP	3376
NTE302	1173
NTE306	1946
NTE307	1430
NTE308	1712
NTE308P	4615
NTE309K	1922
NTE310	2329
NTE310P	4606
NTE311	0493
NTE312	0325
NTE313	0475
NTE314	1825
NTE315	1978
NTE316	0259
NTE317	2946
NTE318	2312
NTE319	0008
NTE319P	0008
NTE320	1974
NTE320F	2884
NTE321	1751
NTE322	1984
NTE323	0892
NTE324	1480
NTE325	1198
NTE326	2987
NTE327	2485
NTE328	0620
NTE329	2700
NTE330	3561
NTE331	0482
NTE331MP	4184
NTE332	1369
NTE332MCP	3949
NTE333	2835
NTE334	3563
NTE335	3565
NTE336	3567
NTE337	3568
NTE338	3569
NTE338F	5202
NTE339	3570
NTE340	2725
NTE341	3572
NTE342	2719
NTE343	2720
NTE344	2526
NTE345	3041
NTE346	2041
NTE347	3576
NTE348	3578
NTE349	2170
NTE350	1233
NTE350F	5303
NTE351	1977
NTE352	3194
NTE353	3580
NTE354	3582
NTE355	3583
NTE356	3584
NTE357	3586
NTE358	1030
NTE358A	4465
NTE358B	5036
NTE358C	5503
NTE359	3587
NTE360	2506
NTE361	2039
NTE362	3560
NTE363	2091
NTE364	1420
NTE365	2093
NTE366	3589
NTE367	2868
NTE368	3591
NTE369	2096
NTE373	0563
NTE374	0525
NTE375	0392
NTE376	1085
NTE377	0060
NTE378	1308
NTE379	0728
NTE380	2939
NTE380MP	2735
NTE381	3571
NTE382	1386
NTE383	0477
NTE384	0429
NTE385	1966
NTE386	1852
NTE387	2433
NTE387MP	3187
NTE388	2414
NTE389	0223
NTE390	3081
NTE391	0859
NTE392	3602
NTE393	3603
NTE394	2613
NTE395	3607
NTE396	0187
NTE397	0439
NTE398	1648
NTE399	0261
NTE451	1392
NTE452	2888
NTE453	3072
NTE454	2458
NTE455	0371
NTE456	3622
NTE457	2950
NTE458	1758
NTE459	3133
NTE460	1141
NTE461	2945
NTE462	3628
NTE464	0844
NTE465	0983
NTE466	1155
NTE467	3130
NTE470	3631
NTE471	2545
NTE472	3421
NTE473	3326
NTE474	3632
NTE475	3321
NTE476	3322
NTE477	3119
NTE478	3636
NTE480	3638
NTE481	3640
NTE482	3641
NTE483	3642
NTE484	3643
NTE485	3644
NTE486	0689
NTE488	2702
NTE500A	0190
NTE502	0201
NTE503	0288
NTE504	0378
NTE505	0757
NTE506	0102
NTE507	0920
NTE508	1306
NTE509	2083
NTE510	2084
NTE511	1449
NTE512	2085
NTE513	1323
NTE514	3649
NTE515	1217
NTE518	1791
NTE519	0124
NTE521	3858
NTE522	1501
NTE523	1067
NTE524V13	0830
NTE524V15	3415
NTE524V25	1336
NTE524V27	4867
NTE524V30	4872
NTE524V42	3859
NTE524V48	0781
NTE525	0348
NTE526A	1706
NTE527A	5731
NTE528	2109
NTE529	2546
NTE530	3655
NTE531	2804
NTE532	0409
NTE533	1710
NTE534	2982
NTE535	2985
NTE536A	1997
NTE537	1197
NTE538	2984
NTE539	1054
NTE541	3658
NTE542	3659
NTE544	3660
NTE546	3661
NTE548	3662
NTE549	6588
NTE551	0182
NTE552	3537
NTE553	0163
NTE555	2518
NTE556	3666
NTE557	3667
NTE558	0017
NTE559	3871
NTE565	3669
NTE568	2671
NTE568A	5816
NTE578	3888
NTE579	3604
NTE580	0031
NTE581	1145
NTE582	3889
NTE583	3891
NTE584	0339
NTE585	0735
NTE586	2542
NTE587	1090
NTE588	0546
NTE590	3192
NTE591	3894
NTE592	3895
NTE593	3749
NTE594	3896
NTE595	0907
NTE596	0903
NTE597	1127
NTE598	1664
NTE599	3551
NTE600	1925
NTE601	0139
NTE604	3683
NTE605	0120
NTE605A	0343
NTE606	3684
NTE607	3442
NTE610	0554
NTE611	0005
NTE612	0720
NTE613	0628
NTE614	0030
NTE615	3452
NTE615A	OEM
NTE615P	1329
NTE616	1029
NTE617	3687
NTE618	3688
NTE684	0304
NTE700	3719
NTE701	3720
NTE702	2801
NTE703	0632
NTE703A	5986
NTE704	2624
NTE705A	2161
NTE706	2571
NTE707	0753
NTE708	0664
NTE709	0678
NTE710	0829
NTE711	2715
NTE712	0167
NTE713	0666
NTE714	0352
NTE715	0354
NTE716	3722
NTE718	0654
NTE720	0443
NTE721	2279
NTE722	0701
NTE723	1345
NTE724	0823
NTE725	0692
NTE726	2617
NTE727	2529
NTE728	2549
NTE729	0328
NTE730	2743
NTE731	2457
NTE734	6605
NTE736	1759
NTE737	1443
NTE738	0856
NTE739	0751
NTE740A	0379
NTE742	3735
NTE743	1395
NTE744	1421
NTE745	2930
NTE746	0785
NTE747	1808
NTE748	0789
NTE749	0395
NTE750	3219
NTE754	0723
NTE760	2100
NTE778A	0360
NTE778S	1699
NTE778SM	5531
NTE780	0364
NTE781	2747
NTE783	0802
NTE784	2699
NTE786	2714
NTE787	2257
NTE788	2755
NTE789	1843
NTE790	0349
NTE791	1337
NTE793	2043
NTE795	3195
NTE797	0521
NTE798	1753
NTE799	1611
NTE801	0520
NTE802	3761
NTE803	3763
NTE804	2557
NTE806	3422
NTE807	3766
NTE809	2688
NTE810A	2480
NTE815	0848
NTE818	2501
NTE819	1878
NTE820	2817
NTE821	2831
NTE822	2812
NTE823	3063
NTE824	3079
NTE825	3099
NTE826	3962
NTE828	3086
NTE829	1899
NTE832	3781
NTE832SM	6080
NTE834	0176
NTE834SM	2742
NTE835	3397
NTE836	2493
NTE841	3784
NTE842	3976
NTE843	2502
NTE844	3977
NTE845	3978
NTE846	1516
NTE848	3981
NTE849	3788
NTE850	3983
NTE851	3790
NTE852	2022
NTE853	3791
NTE854	3986
NTE855	0814
NTE856	3793
NTE857M	3672
NTE857SM	6365
NTE858M	3741
NTE858SM	4105
NTE859	3391
NTE859SM	4136
NTE860	3795
NTE861	3989
NTE862	3056
NTE863	3990
NTE864	0304
NTE867	3995
NTE868	3997
NTE869	2893
NTE871	3394
NTE873	2107
NTE874	4002
NTE875	1104
NTE876	3062
NTE877	4005
NTE878	1425
NTE887M	3181
NTE888M	3942
NTE889M	3320
NTE900	2687
NTE901	2689
NTE902	2748
NTE903	2537
NTE904	1854
NTE905	2698
NTE906	2701
NTE907	2711
NTE908	2705
NTE909	1301
NTE909D	1705
NTE910	1797
NTE910D	1800
NTE911	1890
NTE911D	1897
NTE912	1696
NTE913	3808
NTE914	2727
NTE915	1554
NTE916	2749
NTE917	2722
NTE918	2562
NTE918M	2567
NTE918SM	2565
NTE919	2574
NTE919D	2570
NTE921	3686
NTE922	1815
NTE922M	2104
NTE922SM	4169
NTE923	1192
NTE923D	0026
NTE924	2222
NTE924M	2517
NTE925	2890
NTE926	3798
NTE927	1121
NTE927D	1118
NTE927SM	4273
NTE928	1677
NTE928M	0770
NTE928S	3233
NTE928SM	2881
NTE929	2751
NTE930	2782
NTE931	1916
NTE932	2863
NTE933	3393
NTE934	3813
NTE935	2737
NTE936	3814
NTE937	3621
NTE937M	6039
NTE938	0896
NTE938M	2246
NTE939	3818
NTE940	1919
NTE941	0410
NTE941D	1975
NTE941M	0312
NTE941S	6185
NTE941SM	1976
NTE942	3029
NTE943	3819
NTE943M	0631
NTE943SM	3118
NTE944	3820
NTE944M	3434
NTE946	3663
NTE947	2368
NTE947D	2358
NTE948	2823
NTE948SM	2816
NTE949	2552
NTE950	1828
NTE951	3395
NTE952	3821
NTE953	3823
NTE954	3824
NTE955M	0973
NTE955MC	0585
NTE955S	3956
NTE955SM	2771
NTE956	2564
NTE957	2736
NTE958	2259
NTE959	3826
NTE960	0624
NTE961	1285
NTE962	0923
NTE963	2648
NTE964	1196
NTE965	2791
NTE966	0334
NTE967	1838
NTE968	1321
NTE969	3829
NTE970	2838
NTE971	3575
NTE972	2239
NTE973	0417
NTE973D	3801
NTE974	0011
NTE975	1300
NTE975SM	2355
NTE976	2283
NTE977	1292
NTE978	3285
NTE978C	2869
NTE978SM	2606
NTE979	OEM
NTE980	3831
NTE981	0083
NTE982	2830
NTE983	3425
NTE984	3270
NTE985	3832
NTE986	3089
NTE987	0625
NTE987SM	3202
NTE988	2301
NTE989	3381
NTE990	1491
NTE991	3594
NTE992	2247
NTE993	3835
NTE994M	3827
NTE995	2318
NTE995M	4551
NTE996	3836
NTE997	3024
NTE998	4058
NTE1000	1235
NTE1002	0387
NTE1003	0579
NTE1004	0878
NTE1005	2568
NTE1006	1215
NTE1008	OEM
NTE1009	0843
NTE1010	0864
NTE1011	0882
NTE1012	0021
NTE1013	0138
NTE1014	0355
NTE1015	2404
NTE1016	2406
NTE1017	2440
NTE1018	2476
NTE1019	2509
NTE1022	4556
NTE1023	4557
NTE1024	3550
NTE1025	1187

If replacement code is OEM, contact original manufacturer for replacement.

DEVICE TYPE	REPL CODE
NTE1027	2876
NTE1028	4559
NTE1029	2262
NTE1030	4561
NTE1031	4562
NTE1032	4563
NTE1033	4564
NTE1034	2889
NTE1035	4565
NTE1036	4566
NTE1037	4093
NTE1038	4409
NTE1039	1621
NTE1040	4569
NTE1041	4570
NTE1042	4571
NTE1043	4390
NTE1045	2393
NTE1046	0528
NTE1047	2882
NTE1048	0673
NTE1049	0653
NTE1050	2891
NTE1051	3785
NTE1052	0433
NTE1053	2573
NTE1054	2576
NTE1055	2578
NTE1056	1837
NTE1057	2577
NTE1058	2124
NTE1060	2581
NTE1061	2582
NTE1062	0403
NTE1064	2585
NTE1065	2586
NTE1066	2587
NTE1067	2588
NTE1068	2590
NTE1069	2599
NTE1070	4076
NTE1071	2603
NTE1072	2591
NTE1073	2596
NTE1074	2592
NTE1075A	0470
NTE1078	3352
NTE1079	4575
NTE1080	0855
NTE1081A	1160
NTE1082	2261
NTE1083	2854
NTE1084	2886
NTE1085	2631
NTE1086	2887
NTE1087	1018
NTE1089	0638
NTE1090	2861
NTE1091	0964
NTE1092	2316
NTE1093	4578
NTE1094	0043
NTE1095	OEM
NTE1096	1634
NTE1097	3263
NTE1098	3958
NTE1099	4260
NTE1100	0911
NTE1101	2634
NTE1102	2632
NTE1103	1994
NTE1104	2057
NTE1105	4582
NTE1106	2630
NTE1107	3670
NTE1108	1393
NTE1109	2639
NTE1110	2305
NTE1112	2482
NTE1114	4586
NTE1115	1248
NTE1115A	3922
NTE1116	4147
NTE1117	4003
NTE1119	4589
NTE1120	2031
NTE1121	0389
NTE1122	0607
NTE1123	4034
NTE1124	4543
NTE1126	4593
NTE1127	4420
NTE1128	3915
NTE1130	1326
NTE1131	4594
NTE1132	4595
NTE1133	2231
NTE1134	2949
NTE1135	2062
NTE1137	2594
NTE1139	3936
NTE1140	2872
NTE1141	3840
NTE1142	1478
NTE1148	2864
NTE1149	4597
NTE1150	4598
NTE1151	5080
NTE1152	2505

DEVICE TYPE	REPL CODE
NTE1153	3366
NTE1154	4600
NTE1155	1050
NTE1156	3017
NTE1158	0881
NTE1159	0884
NTE1160	3695
NTE1161	2942
NTE1162	4038
NTE1162A	OEM
NTE1163	2604
NTE1164	5082
NTE1165	2773
NTE1166	2781
NTE1167	1714
NTE1168	2602
NTE1169	4401
NTE1170	2534
NTE1171	0093
NTE1172	1379
NTE1173	3273
NTE1174	2049
NTE1175	2786
NTE1176	2789
NTE1177	2798
NTE1178	3705
NTE1179	2572
NTE1180	2614
NTE1181	2593
NTE1183	4196
NTE1184	4612
NTE1185	1479
NTE1186	0852
NTE1187	0857
NTE1188	2618
NTE1189	4395
NTE1191	4087
NTE1192	1540
NTE1193	3318
NTE1194	1816
NTE1195	4209
NTE1196	1055
NTE1197	4620
NTE1198	2253
NTE1199	3805
NTE1200	4238
NTE1201	4622
NTE1202	2380
NTE1203	2306
NTE1204	4484
NTE1205	2931
NTE1206	4623
NTE1209	4625
NTE1210	2633
NTE1211	2193
NTE1212	4626
NTE1213	4627
NTE1214	0651
NTE1215	2532
NTE1216	0916
NTE1217	0416
NTE1218	1571
NTE1219	1612
NTE1221	4631
NTE1223	1524
NTE1224	4086
NTE1226	0706
NTE1227	0621
NTE1228	3190
NTE1229	4632
NTE1230	2669
NTE1231	1170
NTE1231A	1057
NTE1232	1048
NTE1233	4224
NTE1234	2019
NTE1235	4633
NTE1236	2284
NTE1237	3651
NTE1239	3262
NTE1240	4069
NTE1241	2638
NTE1242	1509
NTE1243	4634
NTE1245	0904
NTE1246	4635
NTE1247	4636
NTE1248	1260
NTE1249	4042
NTE1250	4072
NTE1251	4637
NTE1252	4638
NTE1254	3815
NTE1255	2616
NTE1256	4640
NTE1257	4641
NTE1258	4642
NTE1260	4644
NTE1261	1128
NTE1262	4645
NTE1263	2960
NTE1264	4064
NTE1265	4065
NTE1266	4067
NTE1267	4070
NTE1268	4071
NTE1269	4079
NTE1271	4648
NTE1272	3559

DEVICE TYPE	REPL CODE
NTE1273	4649
NTE1278	4652
NTE1281	4653
NTE1282	4654
NTE1283	4656
NTE1284	4657
NTE1285	0754
NTE1286	0756
NTE1287	4659
NTE1288	4573
NTE1289	4661
NTE1290	4089
NTE1291	4663
NTE1292	3451
NTE1293	4665
NTE1294	4006
NTE1295	5094
NTE1296	0067
NTE1297	4389
NTE1298	4672
NTE1299	5095
NTE1300	3933
NTE1301	4669
NTE1302	3992
NTE1303	4670
NTE1304	3996
NTE1305	3998
NTE1306	4000
NTE1307	4004
NTE1308	3948
NTE1309	4671
NTE1310	3988
NTE1311	3932
NTE1312	3984
NTE1313	3985
NTE1319	OEM
NTE1320	4673
NTE1321	4674
NTE1322	4675
NTE1323	4676
NTE1324	4677
NTE1325	4678
NTE1326	4680
NTE1327	4681
NTE1328	4682
NTE1329	4683
NTE1330	4684
NTE1331	4685
NTE1332	4686
NTE1333	4687
NTE1334	4688
NTE1335	4689
NTE1336	4690
NTE1337	4691
NTE1338	1437
NTE1339	4693
NTE1340	4694
NTE1341	4695
NTE1342	4696
NTE1343	4698
NTE1344	4699
NTE1345	1903
NTE1346	4701
NTE1347	4702
NTE1348	4703
NTE1349	4704
NTE1350	4705
NTE1351	4706
NTE1352	4707
NTE1354	4708
NTE1355	4710
NTE1356	5099
NTE1357	5100
NTE1358	4151
NTE1359	5101
NTE1361	4713
NTE1362	5104
NTE1363	4714
NTE1364	4715
NTE1365	4716
NTE1366	4717
NTE1367	4718
NTE1368	4719
NTE1369	4720
NTE1370	4722
NTE1371	4723
NTE1372	4724
NTE1373	3916
NTE1374	3710
NTE1375	2794
NTE1376	3711
NTE1377	3333
NTE1378	4725
NTE1379	4726
NTE1380	4727
NTE1381	4728
NTE1382	1951
NTE1383	4729
NTE1384	4730
NTE1385	4732
NTE1386	3239
NTE1387	4372
NTE1388	4734
NTE1389	4735
NTE1390	5107
NTE1391	4736
NTE1392	4737
NTE1393	4739
NTE1394	4740

DEVICE TYPE	REPL CODE
NTE1395	4097
NTE1396	0496
NTE1397	5108
NTE1398	5110
NTE1399	5111
NTE1400	1582
NTE1401	3351
NTE1402	3862
NTE1403	3359
NTE1404	4743
NTE1405	4744
NTE1406	4060
NTE1407	4745
NTE1408	4746
NTE1409	0683
NTE1409N	1945
NTE1410	2623
NTE1411	0767
NTE1412	4747
NTE1413	0069
NTE1414	4750
NTE1415	4495
NTE1416	0490
NTE1417	2625
NTE1418	4165
NTE1419	4751
NTE1420	0553
NTE1421	4752
NTE1422	1585
NTE1423	4291
NTE1424	4476
NTE1425	4758
NTE1426	0675
NTE1427	4759
NTE1428	1580
NTE1429	4760
NTE1430	0918
NTE1431	4761
NTE1432	5118
NTE1433	4762
NTE1434	0872
NTE1435	4763
NTE1436	4428
NTE1437	4549
NTE1438	4764
NTE1439	3103
NTE1440	3738
NTE1441	4765
NTE1442	4766
NTE1443	5121
NTE1444	4068
NTE1445	1022
NTE1446	4063
NTE1447	4768
NTE1448	4051
NTE1449	4770
NTE1450	4047
NTE1451	4771
NTE1452	4773
NTE1453	2514
NTE1454	3246
NTE1455	4776
NTE1457	4777
NTE1458	4778
NTE1459	2530
NTE1460	4779
NTE1461	2712
NTE1462	4780
NTE1463	3196
NTE1464	4781
NTE1465	0362
NTE1466	2063
NTE1467	4757
NTE1468	3980
NTE1469	4782
NTE1470	3228
NTE1471	1201
NTE1472	4785
NTE1473	4787
NTE1474	4789
NTE1475	4791
NTE1476	4793
NTE1477	4794
NTE1478	4795
NTE1479	1446
NTE1480	4560
NTE1481	4797
NTE1482	4798
NTE1483	0865
NTE1484	4799
NTE1485	3609
NTE1486	3716
NTE1487	4044
NTE1488	2252
NTE1489	4803
NTE1490	4805
NTE1491	4807
NTE1492	4808
NTE1493	4809
NTE1494	4810
NTE1495	4811
NTE1496	4134
NTE1497	4812
NTE1498	4813
NTE1499	4814
NTE1500	4815
NTE1501	4817
NTE1502	4819
NTE1503	4820

DEVICE TYPE	REPL CODE
NTE1504	4821
NTE1505	4823
NTE1507	4824
NTE1508	3068
NTE1509	3069
NTE1510	0810
NTE1511	4825
NTE1512	4280
NTE1513	4826
NTE1514	4827
NTE1515	4830
NTE1516	4831
NTE1517	4834
NTE1518	4835
NTE1519	4836
NTE1520	1627
NTE1521	3945
NTE1522	2533
NTE1523	4838
NTE1525	2694
NTE1526	4841
NTE1527	4842
NTE1528	2328
NTE1529	2913
NTE1530	4844
NTE1531	4846
NTE1532	2054
NTE1533	4847
NTE1534	4849
NTE1535	5156
NTE1536	4850
NTE1537	4851
NTE1538	0332
NTE1539	1590
NTE1540	4351
NTE1541	2693
NTE1542	4307
NTE1543	4853
NTE1544	4854
NTE1545	0912
NTE1546	3503
NTE1547	0663
NTE1548	3502
NTE1549	3071
NTE1550	1429
NTE1551	3744
NTE1552	2916
NTE1553	3518
NTE1554	5158
NTE1556	5159
NTE1557	3816
NTE1558	2105
NTE1559	5160
NTE1560	1318
NTE1561	5162
NTE1562	5163
NTE1563	2926
NTE1564	OEM
NTE1565	3057
NTE1566	4601
NTE1567	3713
NTE1568	5169
NTE1569	0107
NTE1570	4986
NTE1571	5172
NTE1572	4109
NTE1573	1676
NTE1575	2184
NTE1576	3772
NTE1577	5175
NTE1578	3423
NTE1579	5177
NTE1580	0351
NTE1581	4112
NTE1582	3973
NTE1583	OEM
NTE1584	OEM
NTE1585	5180
NTE1586	OEM
NTE1588	OEM
NTE1589	0450
NTE1590	OEM
NTE1591	5181
NTE1593	5182
NTE1594	OEM
NTE1595	OEM
NTE1596	3964
NTE1597	5183
NTE1598	OEM
NTE1599	OEM
NTE1600	5185
NTE1601	OEM
NTE1602	5187
NTE1603	OEM
NTE1604	0465
NTE1605	OEM
NTE1606	5188
NTE1607	5189
NTE1608	0662
NTE1609	2535
NTE1610	3084
NTE1611	5191
NTE1612	2787
NTE1613	5192
NTE1615	4857
NTE1616	2896
NTE1617	OEM
NTE1618	3074
NTE1619	2799

DEVICE TYPE	REPL CODE
NTE1620	5193
NTE1621	5194
NTE1624	5195
NTE1625	3753
NTE1626	4840
NTE1627	2808
NTE1628	2073
NTE1629	3759
NTE1630	0759
NTE1631	5197
NTE1632	5198
NTE1633	4274
NTE1634	2595
NTE1635	5199
NTE1636	3012
NTE1637	5200
NTE1639	4278
NTE1641	5201
NTE1648	5203
NTE1649	5204
NTE1650	2122
NTE1651	4520
NTE1654	5205
NTE1655	2842
NTE1656	1442
NTE1657	4410
NTE1658	2666
NTE1659	5206
NTE1660	2656
NTE1661	4350
NTE1662	5207
NTE1663	5208
NTE1664	5209
NTE1666	5210
NTE1667	2386
NTE1668	3776
NTE1669	3282
NTE1670	4553
NTE1671	3685
NTE1672	5212
NTE1673	1988
NTE1674	0976
NTE1675	3363
NTE1676	2042
NTE1677	5011
NTE1678	5048
NTE1679	2425
NTE1680	3126
NTE1681	5214
NTE1682	5215
NTE1683	3853
NTE1684	4927
NTE1685	3253
NTE1686	OEM
NTE1687	2304
NTE1688	5218
NTE1689	0699
NTE1690	5219
NTE1691	5220
NTE1692	5221
NTE1693	5222
NTE1700	5225
NTE1701	5226
NTE1702	5227
NTE1703	3731
NTE1704	5229
NTE1705	0489
NTE1706	OEM
NTE1707	3335
NTE1708	4384
NTE1709	5230
NTE1710	2066
NTE1711	2553
NTE1712	2067
NTE1713	2068
NTE1714M	2686
NTE1714S	2026
NTE1715	2025
NTE1716	3828
NTE1717	5232
NTE1718	4258
NTE1719	2103
NTE1720	5233
NTE1721	2829
NTE1722	2841
NTE1723	2839
NTE1724	4796
NTE1725	2116
NTE1726	5234
NTE1727	3747
NTE1728	0539
NTE1729	1824
NTE1730	5236
NTE1731	5237
NTE1732	1572
NTE1733	2374
NTE1734	5238
NTE1735	2385
NTE1736	3361
NTE1737	1596
NTE1738	2879
NTE1739	1389
NTE1740	4146
NTE1741	4650
NTE1742	1641
NTE1743	3445
NTE1744	3437
NTE1745	5239
NTE1747	4346

DEVICE TYPE	REPL CODE
NTE1748	3348
NTE1749	5241
NTE1750	5242
NTE1751	2194
NTE1752	3523
NTE1753	5243
NTE1754	2178
NTE1755	1094
NTE1757	5244
NTE1758	4788
NTE1759	2090
NTE1761	5246
NTE1762	5247
NTE1764	5250
NTE1771	0391
NTE1772	4353
NTE1773	0732
NTE1774	5251
NTE1775	4137
NTE1776	3955
NTE1777	2364
NTE1778	3959
NTE1779	3902
NTE1780	1529
NTE1781	1028
NTE1782	4129
NTE1783	4194
NTE1785	5254
NTE1786	3386
NTE1787	3159
NTE1788	4385
NTE1789	4422
NTE1790	3236
NTE1791	5256
NTE1792	3254
NTE1794	2269
NTE1795	1604
NTE1796	5257
NTE1797	2957
NTE1798	3097
NTE1799	3027
NTE1800	3365
NTE1801	3884
NTE1803	3161
NTE1804	1254
NTE1805	4379
NTE1808	5258
NTE1809	5259
NTE1810	3925
NTE1811	5260
NTE1812	5261
NTE1813	5262
NTE1814	5263
NTE1815	2512
NTE1816	5265
NTE1817	5267
NTE1818	5268
NTE1819	5269
NTE1820	5270
NTE1822	2405
NTE1823	2418
NTE1824	3340
NTE1826	4383
NTE1827	3817
NTE1828	4925
NTE1829	5271
NTE1830	5272
NTE1831	5273
NTE1834	4117
NTE1835	4310
NTE1836	4381
NTE1838	0462
NTE1839	3953
NTE1840	3574
NTE1842	3725
NTE1843	4041
NTE1844	3250
NTE1845	0557
NTE1847	2111
NTE1848	2921
NTE1849	5275
NTE1852	3495
NTE1855	2346
NTE1856	0486
NTE1859	3083
NTE1860	2870
NTE1861	3927
NTE1863	4094
NTE1871	5276
NTE1872	2419
NTE1874	5277
NTE1875	5278
NTE1876	5279
NTE1877	1483
NTE1879	1439
NTE1900	5280
NTE1901	5281
NTE1902	1786
NTE1903	1836
NTE1905	5041
NTE1906	4506
NTE1907	4886
NTE1908	4509
NTE1909	1721
NTE1911	2723
NTE1912	2254
NTE1913	2004
NTE1914	3963
NTE1915	5287

If replacement code is OEM, contact original manufacturer for replacement.

DEVICE TYPE	REPL CODE
NTE1916	2000
NTE1917	4508
NTE1918	2462
NTE1919	5289
NTE1920	3650
NTE1923	5290
NTE1924	3881
NTE1925	3573
NTE1926	5291
NTE1927	5292
NTE1928	2470
NTE1930	2169
NTE1932	2806
NTE1934	5293
NTE1934X	2539
NTE1936	4329
NTE1938	5296
NTE1940	5297
NTE2000	5306
NTE2001	3729
NTE2003	4933
NTE2004	4106
NTE2011	0845
NTE2012	1007
NTE2013	1134
NTE2014	1261
NTE2015	4935
NTE2016	4936
NTE2017	4937
NTE2018	4939
NTE2019	4940
NTE2020	4941
NTE2021	3893
NTE2022	3920
NTE2023	2750
NTE2024	2827
NTE2025	4943
NTE2026	4081
NTE2027	1701
NTE2028	1793
NTE2029	4947
NTE2030	1877
NTE2031	4949
NTE2032	4950
NTE2047	5311
NTE2050	4951
NTE2051	4952
NTE2053	3792
NTE2054	4954
NTE2055	0487
NTE2056	5312
NTE2060	2295
NTE2061	2728
NTE2062	3982
NTE2070	4955
NTE2071	4956
NTE2072	4957
NTE2073	4959
NTE2074	4536
NTE2075	4961
NTE2076	4962
NTE2077	4963
NTE2078	4965
NTE2079	0969
NTE2080	4966
NTE2081	4967
NTE2082	4968
NTE2083	4969
NTE2084	4138
NTE2085	4868
NTE2086	5314
NTE2087	4970
NTE2088	5315
NTE2090	OEM
NTE2102	4973
NTE2104	4974
NTE2107	4902
NTE2114	2048
NTE2117	0522
NTE2128	1898
NTE2147	1693
NTE2164	2357
NTE2300	3360
NTE2301	3476
NTE2302	1541
NTE2303	2766
NTE2304	2367
NTE2305	3435
NTE2306	3673
NTE2307	3001
NTE2308	2185
NTE2309	2784
NTE2310	4450
NTE2311	1506
NTE2312	2908
NTE2313	3547
NTE2314	5341
NTE2315	1511
NTE2316	5342
NTE2317	4334
NTE2318	5345
NTE2319	3388
NTE2320	3005
NTE2321	0544
NTE2322	0282
NTE2323	5348
NTE2324	3494
NTE2325	3357
NTE2326	OEM

DEVICE TYPE	REPL CODE
NTE2327	3217
NTE2328	3436
NTE2329	5349
NTE2330	3516
NTE2331	2129
NTE2332	0278
NTE2334	2366
NTE2336	3723
NTE2338	1061
NTE2340	5350
NTE2349	2428
NTE2350	2420
NTE2351	5352
NTE2352	5353
NTE2355	0832
NTE2356	4172
NTE2357	0887
NTE2358	3143
NTE2359	0898
NTE2360	4130
NTE2361	2954
NTE2362	0013
NTE2380	1466
NTE2381	5360
NTE2382	3088
NTE2383	2271
NTE2386	1668
NTE2390	5361
NTE2392	5363
NTE2401	4489
NTE2402	4498
NTE2403	4498
NTE2404	5364
NTE2405	5365
NTE2406	1733
NTE2407	1752
NTE2408	0724
NTE2409	1742
NTE2410	OEM
NTE2411	5366
NTE2412	3379
NTE2413	5368
NTE2414	3482
NTE2415	1892
NTE2416	0981
NTE2417	0703
NTE2418	3478
NTE2419	3272
NTE2426	5369
NTE2427	5370
NTE2428	0667
NTE2429	3645
NTE2430	3471
NTE2431	5371
NTE2532	5098
NTE2708	4424
NTE2716	2278
NTE2732	2697
NTE2732A	6437
NTE2764	0812
NTE2800	1846
NTE2909	OEM
NTE3000	1981
NTE3001	1989
NTE3002	1985
NTE3003	1983
NTE3004	5112
NTE3005	5113
NTE3006	5114
NTE3007	3019
NTE3008	3096
NTE3009	5115
NTE3010	0841
NTE3011	3124
NTE3012	5116
NTE3012A	OEM
NTE3013	5117
NTE3014	3455
NTE3015	3456
NTE3016	3739
NTE3018	1615
NTE3019	5119
NTE3020	1358
NTE3021	3157
NTE3022	1962
NTE3023	3276
NTE3024	1778
NTE3025	3182
NTE3026	2195
NTE3027	4183
NTE3028	0591
NTE3029	3232
NTE3029A	3232
NTE3030	4023
NTE3031	5122
NTE3032	2313
NTE3033	2628
NTE3034	3878
NTE3034A	3878
NTE3035	3185
NTE3035A	3185
NTE3036	3552
NTE3037	5123
NTE3038	5124
NTE3039	5125
NTE3040	0541
NTE3041	0315
NTE3042	0273
NTE3043	4431

DEVICE TYPE	REPL CODE
NTE3044	4932
NTE3045	2108
NTE3046	2236
NTE3047	5051
NTE3048	5065
NTE3049	5126
NTE3050	4987
NTE3051	4988
NTE3052	4870
NTE3053	5127
NTE3054	5128
NTE3055	5129
NTE3056	4873
NTE3057	4190
NTE3058	5130
NTE3059	5131
NTE3060	5132
NTE3061	4989
NTE3062	5134
NTE3063	5135
NTE3064	5137
NTE3065	4871
NTE3068	5138
NTE3069	5139
NTE3070	5140
NTE3071	5141
NTE3074	4327
NTE3075	4332
NTE3076	4331
NTE3077	4333
NTE3078	2206
NTE3079	2199
NTE3080	2378
NTE3081	3367
NTE3082	5142
NTE3083	1109
NTE3084	1053
NTE3085	5421
NTE3086	1403
NTE3087	5423
NTE3088	5424
NTE3089	5425
NTE3090	5426
NTE3091	0235
NTE3092	1291
NTE3093	5427
NTE3095	5428
NTE3096	5429
NTE3097	5430
NTE3098	0112
NTE3099	2703
NTE3100	1417
NTE3101	2636
NTE3102	2640
NTE3103	5433
NTE3111	5434
NTE3112	5435
NTE3113	5436
NTE3114	5151
NTE3115	5153
NTE3116	5154
NTE3117	5155
NTE3120	2094
NTE3121	3987
NTE3130	5438
NTE3131	5016
NTE3150	5165
NTE3151	5167
NTE3152	5168
NTE3153	5170
NTE3154	5211
NTE3155	5171
NTE3160	5174
NTE3161	1076
NTE3162	5176
NTE3163	4767
NTE3164	3491
NTE3165	5178
NTE3166	4052
NTE3167	1131
NTE3168	5179
NTE3169	5443
NTE3170	5444
NTE3171	5063
NTE3180	5446
NTE3181	5447
NTE3182	5448
NTE3470	2759
NTE3880	3481
NTE3881	1904
NTE3882	5339
NTE4000	2024
NTE4001B	0478
NTE4002B	2055
NTE4006B	0646
NTE4007	2846
NTE4008B	0988
NTE4011B	0215
NTE4012B	0498
NTE4013B	0413
NTE4014B	0860
NTE4015B	1014
NTE4016B	1143
NTE4017B	0513
NTE4018B	1391
NTE4019B	1525
NTE4020B	1661
NTE4021B	1749
NTE4022B	1256

DEVICE TYPE	REPL CODE
NTE4023B	0519
NTE4024B	1957
NTE4025B	2072
NTE4026B	2152
NTE4027B	1949
NTE4028B	2228
NTE4029B	2233
NTE4030B	0500
NTE4031B	2971
NTE4032B	2531
NTE4033B	2635
NTE4034B	3615
NTE4035B	2777
NTE4038B	2981
NTE4040B	0056
NTE4041	3174
NTE4042B	0121
NTE4043B	1553
NTE4044B	2308
NTE4045B	3443
NTE4046B	3428
NTE4047B	2331
NTE4048B	3459
NTE4049	0001
NTE4050B	0398
NTE4051B	0366
NTE4052B	0024
NTE4053B	0035
NTE4055B	3304
NTE4056B	3706
NTE4060B	0146
NTE4063B	3727
NTE4066B	0101
NTE4067B	3742
NTE4068B	2503
NTE4069	0119
NTE4070B	2516
NTE4071B	0129
NTE4072B	2524
NTE4073B	1536
NTE4075B	2540
NTE4076B	3496
NTE4077B	2558
NTE4078B	0921
NTE4081B	0626
NTE4082B	0300
NTE4085B	0303
NTE4086B	0535
NTE4089B	3830
NTE4093B	2384
NTE4094B	1682
NTE4095B	3848
NTE4096B	3851
NTE4097B	3855
NTE4098B	3611
NTE4099B	3330
NTE4164	5573
NTE4501	3809
NTE4502B	1037
NTE4503B	2053
NTE4506B	3767
NTE4508B	1811
NTE4510B	1963
NTE4511B	1544
NTE4512B	2120
NTE4513B	4976
NTE4514B	1830
NTE4515B	3600
NTE4516B	2347
NTE4517B	4284
NTE4518B	1043
NTE4520B	2675
NTE4521B	4990
NTE4522B	2837
NTE4526B	3610
NTE4527B	3145
NTE4528B	3198
NTE4529B	4531
NTE4531B	3325
NTE4532B	1016
NTE4536B	3704
NTE4538B	1063
NTE4539B	3656
NTE4541B	5017
NTE4543B	5020
NTE4547B	5032
NTE4551B	5039
NTE4553B	5040
NTE4555B	2938
NTE4556B	3431
NTE4558B	5049
NTE4566B	3504
NTE4569B	4487
NTE4583B	1297
NTE4584B	5091
NTE4585B	1375
NTE4597B	4110
NTE4598B	4301
NTE4900	2396
NTE4902	3114
NTE4903	2133
NTE4918	3172
NTE4919	1169
NTE4926	3191
NTE4927	1767
NTE4928	3201
NTE4929	1932
NTE4934	1915
NTE4935	1952

DEVICE TYPE	REPL CODE
NTE4950	0568
NTE4951	1972
NTE4958	0831
NTE4959	1987
NTE4988	1405
NTE4989	1408
NTE5000A	1276
NTE5001A	2874
NTE5002A	0760
NTE5003A	0677
NTE5004A	0118
NTE5005A	0299
NTE5006A	0376
NTE5007A	0036
NTE5008A	0275
NTE5009A	0140
NTE5010A	0041
NTE5010T1	0589
NTE5011A	0253
NTE5011T1	0883
NTE5012A	0091
NTE5013A	0471
NTE5013T1	0295
NTE5014A	0062
NTE5015A	0077
NTE5016A	0165
NTE5017A	0322
NTE5018A	0057
NTE5019A	0064
NTE5019T1	0248
NTE5020A	0181
NTE5021A	0052
NTE5021T1	1005
NTE5022A	0053
NTE5023A	0879
NTE5024A	0686
NTE5025A	0445
NTE5026A	0210
NTE5027A	0375
NTE5028A	0671
NTE5029A	0700
NTE5030A	0705
NTE5031A	0494
NTE5032A	0714
NTE5033A	0455
NTE5034A	0257
NTE5034C	OEM
NTE5035A	0195
NTE5036A	0166
NTE5037A	0010
NTE5038A	0032
NTE5039A	0054
NTE5040A	0068
NTE5041A	0092
NTE5042A	0125
NTE5043A	2317
NTE5044A	0152
NTE5045A	0173
NTE5046A	0094
NTE5047A	0049
NTE5048A	0104
NTE5049A	0156
NTE5050A	0189
NTE5051A	0099
NTE5052A	0089
NTE5053A	0287
NTE5054A	0252
NTE5055A	0340
NTE5056A	0370
NTE5057A	0394
NTE5058A	0425
NTE5059A	0453
NTE5060A	1473
NTE5061A	3003
NTE5062A	5872
NTE5063A	1312
NTE5064A	3010
NTE5065A	1713
NTE5066A	0292
NTE5067A	0456
NTE5068A	0533
NTE5069A	0451
NTE5070A	0301
NTE5071A	0025
NTE5072A	0244
NTE5073A	1083
NTE5074A	0317
NTE5075A	0420
NTE5076A	1649
NTE5077A	0495
NTE5078A	0949
NTE5079A	0531
NTE5080A	0565
NTE5081A	0402
NTE5082A	1606
NTE5083A	1674
NTE5084A	0726
NTE5085A	0820
NTE5086A	0350
NTE5087A	0931
NTE5088A	0999
NTE5089A	0502
NTE5090A	0869
NTE5091A	1156
NTE5092A	1268
NTE5093A	1189
NTE5093T	OEM
NTE5094A	3026
NTE5095A	1311

DEVICE TYPE	REPL CODE
NTE5096A	0098
NTE5097A	0186
NTE5098A	0213
NTE5099A	0245
NTE5100A	0028
NTE5101A	0255
NTE5102A	0877
NTE5103A	0367
NTE5104A	2858
NTE5105A	0421
NTE5111A	0782
NTE5112A	0796
NTE5113A	0806
NTE5114A	0821
NTE5115A	0833
NTE5116A	0442
NTE5117A	0876
NTE5118A	3127
NTE5119A	0185
NTE5120A	0205
NTE5121A	0480
NTE5122A	0504
NTE5123A	3317
NTE5124A	0684
NTE5125A	0225
NTE5126A	0230
NTE5127A	0234
NTE5128A	0237
NTE5129A	1397
NTE5130A	0247
NTE5131A	0251
NTE5132A	1178
NTE5133A	0256
NTE5134A	2395
NTE5135A	0262
NTE5136A	0269
NTE5137A	0274
NTE5138A	2399
NTE5139A	0294
NTE5140A	1177
NTE5141A	0309
NTE5142A	0318
NTE5143A	0320
NTE5144A	0326
NTE5145A	0337
NTE5146A	0347
NTE5147A	0027
NTE5148A	0266
NTE5149A	2856
NTE5150A	0386
NTE5151A	0405
NTE5152A	0426
NTE5153A	0444
NTE5154A	3028
NTE5155A	0238
NTE5156A	1180
NTE5157A	1190
NTE5158A	1207
NTE5159A	1218
NTE5160A	1881
NTE5161A	0647
NTE5162A	1255
NTE5163A	2102
NTE5164A	1279
NTE5165A	2225
NTE5166A	0605
NTE5172A	4231
NTE5172AK	1006
NTE5173A	2397
NTE5173AK	1380
NTE5174A	2035
NTE5174AK	0547
NTE5175A	2401
NTE5175AK	2403
NTE5176A	1438
NTE5176AK	2113
NTE5177A	2407
NTE5177AK	2410
NTE5178A	1445
NTE5178AK	1901
NTE5179A	2415
NTE5179AK	2416
NTE5180A	2221
NTE5180AK	0696
NTE5181A	1458
NTE5181AK	1601
NTE5182A	0221
NTE5182AK	1616
NTE5183A	1490
NTE5183AK	1622
NTE5184A	2422
NTE5184AK	2424
NTE5185A	1618
NTE5185AK	0627
NTE5186A	0510
NTE5186AK	0992
NTE5187A	0691
NTE5187AK	0995
NTE5188A	0870
NTE5188AK	1263
NTE5189A	1020
NTE5189AK	1249
NTE5190A	1153
NTE5190AK	1636
NTE5191A	1274
NTE5191AK	1639
NTE5192A	1402
NTE5192AK	1703
NTE5193A	1532

DEVICE TYPE	REPL CODE
NTE5193AK	1640
NTE5194A	1078
NTE5194AK	1716
NTE5195A	1711
NTE5195AK	1720
NTE5196A	1717
NTE5196AK	1731
NTE5197A	1723
NTE5197AK	0727
NTE5198A	1736
NTE5198AK	1756
NTE5199A	1748
NTE5199AK	1768
NTE5200A	1761
NTE5200AK	1782
NTE5201A	2450
NTE5201AK	2453
NTE5202A	1772
NTE5202AK	1794
NTE5203A	1788
NTE5203AK	1799
NTE5204A	1796
NTE5204AK	1809
NTE5205A	1804
NTE5205AK	1817
NTE5206A	1194
NTE5206AK	1826
NTE5207A	1821
NTE5207AK	1840
NTE5208A	0022
NTE5208AK	1853
NTE5209A	0070
NTE5209AK	1861
NTE5210A	0132
NTE5210AK	1866
NTE5211A	0172
NTE5211AK	1874
NTE5212A	0207
NTE5212AK	1884
NTE5213A	0227
NTE5213AK	2474
NTE5214A	0263
NTE5214AK	1895
NTE5215A	0310
NTE5215AK	1902
NTE5216A	0329
NTE5216AK	0736
NTE5217A	0356
NTE5217AK	1909
NTE5218A	0381
NTE5218AK	2479
NTE5219A	0412
NTE5219AK	1914
NTE5220A	0438
NTE5220AK	1163
NTE5221A	0464
NTE5221AK	1924
NTE5222A	0488
NTE5222AK	1933
NTE5223A	0509
NTE5223AK	1941
NTE5224A	0524
NTE5224AK	1947
NTE5225A	0542
NTE5225AK	1953
NTE5226A	0063
NTE5226AK	1961
NTE5227A	0401
NTE5227AK	0357
NTE5228A	0598
NTE5228AK	2500
NTE5229A	0616
NTE5229AK	0670
NTE5230A	0634
NTE5230AK	0776
NTE5231A	0650
NTE5231AK	2507
NTE5232A	0668
NTE5232AK	1071
NTE5240A	0815
NTE5240AK	2721
NTE5241A	0827
NTE5241AK	2724
NTE5242A	0846
NTE5242AK	2726
NTE5243A	0868
NTE5243AK	2729
NTE5244A	0885
NTE5244AK	2732
NTE5245A	0897
NTE5245AK	2734
NTE5246A	0914
NTE5246AK	2738
NTE5247A	0926
NTE5247AK	2740
NTE5248A	0944
NTE5248AK	2744
NTE5249A	0958
NTE5249AK	1225
NTE5250A	0978
NTE5250AK	2746
NTE5251A	0994
NTE5251AK	1237
NTE5252A	1009
NTE5252AK	1252
NTE5253A	1019
NTE5253AK	1269
NTE5254A	0889
NTE5254AK	1277

If replacement code is OEM, contact original manufacturer for replacement.

DEVICE TYPE	REPL CODE
NTE5255A	1049
NTE5255AK	1294
NTE5256A	1058
NTE5256AK	2752
NTE5257A	0932
NTE5257AK	1302
NTE5258A	1079
NTE5258AK	1310
NTE5259A	1096
NTE5259AK	2756
NTE5260A	1106
NTE5260AK	1324
NTE5261A	1123
NTE5261AK	2758
NTE5262A	1135
NTE5262AK	1333
NTE5263A	1152
NTE5263AK	1344
NTE5264A	1164
NTE5264AK	1356
NTE5265A	1174
NTE5265AK	2760
NTE5266A	1184
NTE5266AK	1371
NTE5267A	1200
NTE5267AK	2762
NTE5268A	1210
NTE5268AK	1387
NTE5269A	1223
NTE5269AK	1406
NTE5270A	1232
NTE5270AK	1415
NTE5271A	1246
NTE5271AK	1428
NTE5272A	1265
NTE5272AK	1440
NTE5273A	1273
NTE5273AK	2772
NTE5274A	1290
NTE5274AK	1447
NTE5275A	1299
NTE5275AK	2776
NTE5276A	1307
NTE5276AK	1461
NTE5277A	1322
NTE5277AK	2779
NTE5278A	1331
NTE5278AK	1474
NTE5279A	1340
NTE5279AK	2783
NTE5280A	1353
NTE5280AK	0613
NTE5281A	1365
NTE5281AK	1510
NTE5282A	1384
NTE5282AK	1523
NTE5283A	1401
NTE5283AK	1537
NTE5284A	1412
NTE5284AK	1550
NTE5285A	1423
NTE5285AK	1575
NTE5286A	1436
NTE5286AK	2802
NTE5287A	1444
NTE5287AK	2803
NTE5288A	1457
NTE5288AK	1507
NTE5289A	1470
NTE5289AK	2805
NTE5290A	1484
NTE5290AK	2807
NTE5291A	1505
NTE5291AK	2810
NTE5292A	1521
NTE5292AK	2815
NTE5293A	1531
NTE5293AK	2818
NTE5294A	1548
NTE5294AK	2820
NTE5295A	1568
NTE5295AK	2822
NTE5296A	1587
NTE5296AK	2824
NTE5304	0106
NTE5305	2010
NTE5306	2400
NTE5307	0787
NTE5309	0729
NTE5310	0737
NTE5311	0742
NTE5312	0323
NTE5313	1414
NTE5314	0473
NTE5315	0446
NTE5316	1422
NTE5317	2443
NTE5318	3023
NTE5319	3035
NTE5320	3050
NTE5322	2363
NTE5322W	OEM
NTE5324	2369
NTE5324W	OEM
NTE5326	2370
NTE5326W	OEM
NTE5327	2372
NTE5327W	OEM
NTE5328	2373
NTE5328W	OEM
NTE5329	1042
NTE5330	1045
NTE5332	1875
NTE5334	3546
NTE5335	5734
NTE5338	5736
NTE5340	1643
NTE5342	1673
NTE5344	1589
NTE5346	5738
NTE5348	5739
NTE5351	4337
NTE5360	0804
NTE5368	0655
NTE5369	0020
NTE5371	0526
NTE5372	0108
NTE5374	0428
NTE5375	1871
NTE5377	1700
NTE5378	0345
NTE5380	0459
NTE5381	0587
NTE5386	0769
NTE5387	0483
NTE5400	1137
NTE5401	0344
NTE5402	0901
NTE5403	2342
NTE5404	0058
NTE5405	0407
NTE5406	1683
NTE5408	0179
NTE5409	0346
NTE5410	3349
NTE5411	4458
NTE5412	0174
NTE5413	3854
NTE5414	3620
NTE5415	3324
NTE5416	1502
NTE5417	1004
NTE5418	1034
NTE5419	1148
NTE5424	1992
NTE5426	1044
NTE5427	5750
NTE5428	5751
NTE5429	5469
NTE5437	0717
NTE5438	0308
NTE5442	4588
NTE5444	2270
NTE5446	3402
NTE5448	3404
NTE5452	1396
NTE5453	1259
NTE5454	0447
NTE5455	0940
NTE5456	1222
NTE5457	0095
NTE5458	1243
NTE5461	2095
NTE5462	2089
NTE5463	0505
NTE5465	0710
NTE5466	0863
NTE5468	0327
NTE5470	1110
NTE5471	2932
NTE5472	0963
NTE5473	2933
NTE5474	2936
NTE5475	3671
NTE5476	3439
NTE5480	3419
NTE5481	1103
NTE5482	2491
NTE5483	0240
NTE5484	2659
NTE5485	0676
NTE5486	2809
NTE5487	0336
NTE5491	2449
NTE5492	0435
NTE5494	1487
NTE5496	0687
NTE5500	3102
NTE5501	2519
NTE5502	0741
NTE5503	3105
NTE5504	0745
NTE5505	3109
NTE5506	2917
NTE5507	0747
NTE5508	3243
NTE5509	0752
NTE5511	3696
NTE5512	3383
NTE5513	3624
NTE5514	1651
NTE5515	1584
NTE5516	1665
NTE5517	1650
NTE5518	2647
NTE5519	2649
NTE5520	3307
NTE5521	2188
NTE5522	0567
NTE5523	3258
NTE5524	0762
NTE5525	3268
NTE5526	3271
NTE5527	0740
NTE5528	3291
NTE5529	0764
NTE5530	2875
NTE5531	0766
NTE5534	0226
NTE5540	3277
NTE5541	0731
NTE5542	0712
NTE5543	0469
NTE5544	0721
NTE5545	0722
NTE5546	0778
NTE5547	0725
NTE5548	0750
NTE5550	2521
NTE5552	0397
NTE5554	0611
NTE5556	0952
NTE5558	1512
NTE5562	1848
NTE5564	1855
NTE5566	3215
NTE5567	1704
NTE5568	4031
NTE5569	0679
NTE5570	0608
NTE5572	0610
NTE5574	0468
NTE5575	0641
NTE5577	0217
NTE5579	0658
NTE5580	1084
NTE5582	1086
NTE5584	1102
NTE5587	1900
NTE5589	0738
NTE5590	0192
NTE5591	0159
NTE5592	0096
NTE5594	2843
NTE5595	1765
NTE5596	1814
NTE5598	4148
NTE5599	1725
NTE5600	4412
NTE5601	1755
NTE5602	4414
NTE5603	3499
NTE5604	4121
NTE5605	0485
NTE5606	4416
NTE5607	4417
NTE5608	0572
NTE5609	0935
NTE5610	0059
NTE5611	5549
NTE5612	5550
NTE5613	0954
NTE5614	4440
NTE5615	0957
NTE5616	0960
NTE5617	0961
NTE5618	0966
NTE5621	5551
NTE5622	1089
NTE5623	1093
NTE5624	0411
NTE5625	1095
NTE5626	0415
NTE5627	1099
NTE5628	1100
NTE5629	2300
NTE5631	0578
NTE5632	0571
NTE5633	0593
NTE5634	0574
NTE5635	0576
NTE5636	0577
NTE5637	0594
NTE5638	5322
NTE5640	2383
NTE5641	2394
NTE5642	1413
NTE5643	2387
NTE5645	0555
NTE5646	1456
NTE5649	5553
NTE5650	2352
NTE5651	2353
NTE5652	2356
NTE5653	5555
NTE5655	4455
NTE5656	4456
NTE5657	4482
NTE5661	0404
NTE5665	5798
NTE5673	0154
NTE5675	0147
NTE5676	4545
NTE5677	0279
NTE5679	0908
NTE5680	3146
NTE5681	3147
NTE5682	3148
NTE5683	3150
NTE5684	3151
NTE5685	3152
NTE5686	3153
NTE5687	4100
NTE5688	1064
NTE5689	1317
NTE5690	1891
NTE5693	2015
NTE5695	2017
NTE5697	2018
NTE5700	1434
NTE5701	1680
NTE5702	2294
NTE5703	2298
NTE5704	5812
NTE5705	5813
NTE5710	0652
NTE5711	4364
NTE5712	4574
NTE5720	2311
NTE5721	2994
NTE5722	2878
NTE5800	0110
NTE5801	0953
NTE5802	0242
NTE5803	1747
NTE5804	0540
NTE5805	1771
NTE5806	0965
NTE5808	0817
NTE5809	0922
NTE5812	1282
NTE5814	1287
NTE5815	1293
NTE5817	1296
NTE5818	1567
NTE5819	1547
NTE5820	2153
NTE5821	3139
NTE5822	0711
NTE5823	2967
NTE5826	3517
NTE5827	5012
NTE5828	3541
NTE5829	5577
NTE5830	0708
NTE5831	0933
NTE5832	4140
NTE5833	5517
NTE5834	0580
NTE5835	0947
NTE5836	2060
NTE5837	4523
NTE5838	1000
NTE5839	1012
NTE5840	2076
NTE5841	5567
NTE5842	2081
NTE5843	1073
NTE5846	2088
NTE5847	1138
NTE5848	0612
NTE5849	1188
NTE5850	0970
NTE5851	0985
NTE5852	3733
NTE5853	0910
NTE5854	0989
NTE5855	0990
NTE5856	3743
NTE5857	0993
NTE5858	0197
NTE5859	0997
NTE5860	0200
NTE5861	1001
NTE5862	0204
NTE5863	0515
NTE5866	0206
NTE5867	1008
NTE5868	0586
NTE5869	0948
NTE5870	0084
NTE5871	0534
NTE5872	0090
NTE5873	0748
NTE5874	0097
NTE5875	0765
NTE5876	0105
NTE5877	0777
NTE5878	0109
NTE5879	0538
NTE5880	0116
NTE5881	0801
NTE5882	0122
NTE5883	0816
NTE5886	0131
NTE5887	0545
NTE5890	0145
NTE5891	0550
NTE5892	5514
NTE5893	1238
NTE5894	5582
NTE5895	1240
NTE5896	2900
NTE5897	1241
NTE5898	5583
NTE5899	1245
NTE5900	3751
NTE5901	2365
NTE5902	3757
NTE5903	1253
NTE5904	1112
NTE5905	2376
NTE5908	3768
NTE5909	1264
NTE5910	3011
NTE5911	0449
NTE5912	1600
NTE5913	5005
NTE5914	5282
NTE5915	5491
NTE5916	0871
NTE5917	1635
NTE5918	5283
NTE5919	5495
NTE5920	0853
NTE5921	1251
NTE5922	5286
NTE5923	5496
NTE5924	1609
NTE5925	1205
NTE5928	1610
NTE5929	2137
NTE5932	1614
NTE5933	2251
NTE5940	2002
NTE5941	2003
NTE5942	0590
NTE5943	3265
NTE5944	1250
NTE5945	1577
NTE5946	5587
NTE5947	3275
NTE5948	1581
NTE5949	3281
NTE5950	5589
NTE5951	3287
NTE5952	1586
NTE5953	3294
NTE5962	5026
NTE5963	2559
NTE5966	2615
NTE5967	2566
NTE5980	3189
NTE5981	1630
NTE5982	2901
NTE5983	0254
NTE5986	1124
NTE5987	1107
NTE5988	1126
NTE5989	1111
NTE5990	0805
NTE5991	0258
NTE5992	1195
NTE5993	1644
NTE5994	0319
NTE5995	0267
NTE5998	1132
NTE5999	1119
NTE6002	0044
NTE6003	0281
NTE6006	1530
NTE6007	1520
NTE6008	0029
NTE6009	1847
NTE6010	0601
NTE6011	1851
NTE6020	3762
NTE6021	2665
NTE6022	2653
NTE6023	2695
NTE6026	2657
NTE6027	2768
NTE6030	2664
NTE6031	2855
NTE6034	2006
NTE6035	2907
NTE6038	2677
NTE6039	2974
NTE6040	2682
NTE6041	3051
NTE6042	3900
NTE6043	3117
NTE6044	2655
NTE6045	3208
NTE6048	4306
NTE6049	3771
NTE6050	1560
NTE6051	2179
NTE6054	2840
NTE6055	2182
NTE6058	4308
NTE6059	3601
NTE6060	2850
NTE6061	2191
NTE6064	2871
NTE6065	2197
NTE6068	2833
NTE6069	2216
NTE6072	2473
NTE6073	2340
NTE6074	2778
NTE6075	1745
NTE6076	2813
NTE6077	1783
NTE6078	2643
NTE6079	1818
NTE6084	0615
NTE6085	1021
NTE6087	2515
NTE6090	1942
NTE6094	1545
NTE6102	5046
NTE6103	2315
NTE6104	0749
NTE6105	2792
NTE6106	4286
NTE6107	4285
NTE6110	1226
NTE6112	1204
NTE6113	2159
NTE6115	0479
NTE6116	4922
NTE6118	2139
NTE6154	0599
NTE6155	1347
NTE6156	1986
NTE6157	1905
NTE6158	0657
NTE6159	0193
NTE6162	3309
NTE6163	0202
NTE6200	3676
NTE6202	3679
NTE6204	3680
NTE6206	2836
NTE6208	4188
NTE6220	1798
NTE6230	2192
NTE6240	1236
NTE6244	1727
NTE6246	2234
NTE6354	1023
NTE6355	0501
NTE6356	1036
NTE6357	1777
NTE6358	1046
NTE6359	1789
NTE6362	2832
NTE6363	3167
NTE6400	5909
NTE6400A	1669
NTE6401	2136
NTE6402	0316
NTE6403	4233
NTE6404	3407
NTE6405	4405
NTE6406	3648
NTE6407	3331
NTE6408	5911
NTE6409	1175
NTE6410	1893
NTE6411	5389
NTE6412	5643
NTE6415	5645
NTE6416	5646
NTE6417	5649
NTE6418	5650
NTE6419	5651
NTE6502	3065
NTE6507	3070
NTE6508	3193
NTE6532	1973
NTE6800	0388
NTE6802	0393
NTE6809	5967
NTE6809E	2117
NTE6810	2086
NTE6821	0448
NTE6850	0514
NTE6860	2692
NTE6875	4181
NTE6880	4516
NTE6885	4647
NTE6886	4651
NTE6887	4658
NTE6888	4667
NTE6889	0581
NTE7010	3515
NTE7214	5994
NTE7400	0232
NTE7401	0268
NTE7402	0314
NTE7403	0335
NTE7404	0361
NTE7405	0385
NTE7406	1206
NTE7407	1339
NTE7408	0467
NTE7409	0492
NTE7410	0512
NTE7411	0527
NTE7412	2242
NTE7413	1441
NTE7414	2243
NTE7416	1349
NTE7417	1352
NTE7420	0697
NTE7421	1357
NTE7422	4603
NTE7423	3466
NTE7425	3477
NTE7426	0803
NTE7427	0818
NTE7428	4180
NTE7430	0873
NTE7432	0899
NTE7433	4193
NTE7437	3520
NTE7438	0996
NTE7439	5836
NTE7440	1024
NTE7441	1038
NTE7442	1052
NTE7443	1060
NTE7444	1072
NTE7445	1082
NTE7446	1098
NTE7447	1108
NTE7448	1125
NTE7450	0743
NTE7451	1168
NTE7453	1185
NTE7454	1202
NTE7460	1275
NTE7470	1404
NTE7472	1426
NTE7473	1172
NTE7474	1313
NTE7475	1432
NTE7476	1158
NTE7480	1535
NTE7481	3121
NTE7482	1574
NTE7483	0117
NTE7485	0374
NTE7486	1368
NTE7488A	4904
NTE7489	5455
NTE7490	1208
NTE7491	0980
NTE7492	0834
NTE7493A	0569
NTE7494	1702
NTE7495	1486
NTE7496	1715
NTE7497	2940
NTE7785M	OEM
NTE8070	2943
NTE8076	2944
NTE8080A	4547
NTE8081	2948
NTE8085	2951
NTE8090	2955
NTE8096	2961
NTE8098	2379
NTE8103	2963
NTE8108	2968
NTE8115	2972
NTE8118	2973
NTE8123	5939
NTE8125	2976
NTE8139	2990
NTE8149	2998
NTE8167	3034
NTE8181	3036
NTE8212	1860
NTE8213	3000
NTE8214	6064
NTE8216	1863
NTE8219	6067
NTE8223	5951
NTE8224	1709
NTE8226	3013
NTE8228	1869
NTE8234	2080
NTE8235	2087
NTE8242	3021
NTE8255	0051
NTE8266	2275
NTE8301	2196
NTE8308	1070
NTE8309	5528
NTE8314	2488
NTE8316	5957
NTE8318	4145
NTE8321	5958
NTE8328	5959
NTE8345M	OEM
NTE8368	5963
NTE8370	5964
NTE8374	5965
NTE8520	5975
NTE8542	5976
NTE8546	5977
NTE8556	5978
NTE8613	6091
NTE9093	0358
NTE9094	1632
NTE9097	1481
NTE9099	0333
NTE9135	1619
NTE9157	1433
NTE9158	0466
NTE9200	4158
NTE9221	OEM
NTE9285	OEM
NTE9370	1295
NTE9402	4162
NTE9403	1599
NTE9415	OEM
NTE9601	2267
NTE9602	2286
NTE9660	0618

If replacement code is OEM, contact original manufacturer for replacement.

DEVICE TYPE	REPL CODE
NTE9661	2597
NTE9663	2992
NTE9664	3032
NTE9666	4142
NTE9668	2434
NTE9669	3015
NTE9670	2350
NTE9671	4149
NTE9672	0078
NTE9675	4164
NTE9676	4170
NTE9678	4173
NTE9679	4175
NTE9680	4178
NTE9681	3045
NTE9682	3039
NTE9689	4201
NTE9691	4207
NTE9800	0850
NTE9801	0874
NTE9802	0888
NTE9803	0900
NTE9804	0919
NTE9805	0929
NTE9806	0945
NTE9807	0962
NTE9808	0982
NTE9809	0998
NTE9810	1011
NTE9811	1025
NTE9812	1039
NTE9813	4538
NTE9814	4541
NTE9910	1257
NTE9924	2867
NTE9926	3924
NTE9930	1823
NTE9932	1041
NTE9933	2097
NTE9935	1176
NTE9936	1831
NTE9937	1835
NTE9944	0033
NTE9945	0081
NTE9946	0141
NTE9948	3399
NTE9949	1844
NTE9950	2619
NTE9951	2078
NTE9961	1859
NTE9962	0562
NTE9963	0341
NTE9989	3429
NTE15005	3159
NTE15019E	1807
NTE15020E	3914
NTE15027	4383
NTE15030	OEM
NTE15038	4277
NTE15039	4929
NTE15040	3953
NTE15041	3957
NTE15042	4930
NTE15043	4931
NTE15044	3521
NTE15045	4496
NTE15046	3927
NTE16007	5103
NTE17145	OEM
NTE21128	1638
NTE30001	2703
NTE40085B	3450
NTE40097B	4441
NTE40098B	0432
NTE40100B	3952
NTE40106B	3626
NTE40160B	1359
NTE40161B	1373
NTE40162B	1388
NTE40163B	1407
NTE40174B	1551
NTE40175B	1528
NTE40182B	4660
NTE40192B	1764
NTE40193B	1776
NTE40194B	1769
NTE40195B	1784
NTE56004	0772
NTE56006	0744
NTE56008	0617
NTE56010	0875
NTE56014	6340
NTE56015	5323
NTE56016	5326
NTE56017	1659
NTE56018	1967
NTE56019	6139
NTE56020	5612
NTE56022	3199
NTE56024	3207
NTE56026	3222
NTE65101	4604
NTE74100	4554
NTE74101	OEM
NTE74105	4555
NTE74107	0942
NTE74109	0968
NTE74110	0987
NTE74111	1002
NTE74116	4439
NTE74120	1116
NTE74121	0175
NTE74122	1139
NTE74123	1157
NTE74125	1182
NTE74126	1193
NTE74128	1219
NTE74132	1271
NTE74134	OEM
NTE74136	1316
NTE74138	OEM
NTE74141	1377
NTE74142	1390
NTE74143	1409
NTE74144	1418
NTE74145	0619
NTE74147	1451
NTE74148	1464
NTE74150	1492
NTE74151	1495
NTE74152	1517
NTE74153	1539
NTE74154	1555
NTE74155	1576
NTE74156	1592
NTE74157	1605
NTE74158	4446
NTE74160	1631
NTE74161	1645
NTE74162	1013
NTE74163	1666
NTE74164	0734
NTE74165	1685
NTE74166	0231
NTE74170	1722
NTE74173	1766
NTE74174	1770
NTE74175	1787
NTE74176	1795
NTE74177	1803
NTE74178	1813
NTE74179	1820
NTE74180	1829
NTE74181	1842
NTE74182	1856
NTE74185	1868
NTE74190	1912
NTE74191	1917
NTE74192	1921
NTE74193	1926
NTE74195	1943
NTE74196	1950
NTE74197	1956
NTE74198	1964
NTE74199	1971
NTE74221	2142
NTE74249	2290
NTE74251	2299
NTE74265	4611
NTE74278	4617
NTE74279	4619
NTE74290	2612
NTE74293	2644
NTE74298	4628
NTE74365	1459
NTE74366	1471
NTE74367	1488
NTE74368	1508
NTE74374	4662
NTE74376	4664
NTE74390	3240
NTE74393	3256
NTE74426	4697
NTE74490	3664
NTE75188	0508
NTE75189	0511
NTE75322	5001
NTE75450B	1231
NTE75451B	1244
NTE75452B	1262
NTE75453B	1272
NTE75454B	1289
NTE75491B	1729
NTE75492B	1740
NTE75493	5021
NTE75494	5022
NTE75498	5024
NTM3906	OEM
NTP5V6	OEM
NTP6V2	0471
NTP6V8	OEM
NTP7V5	OEM
NTP8V2	3100
NTP9V1	0057
NTP10	OEM
NTP11	OEM
NTP12	0052
NTP13	OEM
NTP15	0686
NTT275	0004
NU34	0143
NV004	0030
NV009	0030
NV-8B	OEM
NV10A	0373
NV-11-B	OEM
NV20	OEM
NV20A	OEM
NV20F	OEM
NV30	OEM
NV30A	OEM
NV30F	OEM
NV40A	OEM
NV200DIO	OEM
NVM3060	OEM
NX25A	OEM
NX50	OEM
NZ309	OEM
02Z5.6A	0157
02Z6.2A	0471
02Z-10A	0170
02Z12A	0137
02Z12GR	0052
03P2M	0058
04-8054-3	0015
04-8054-4	0015
04-8054-7	0015
05Z6.2Y	6877
05Z8.2U	0244
0-43	OEM
043	OEM
0-44	OEM
0-45	OEM
055C	0324
0-63	OEM
063	OEM
0-64	OEM
064	OEM
0-65	OEM
065	OEM
0101	0015
0200	1250
0201	1250
0210	1250
0211	1250
0220	1250
0221	1250
0573510	0007
0575005	0143
OA2	0920
OA6	0143
OA7	0143
OA9	0143
OA9D	0143
OA10	0110
OA47	0143
OA50	0143
OA-70	0019
OA70	0143
OA71	0143
OA71C	0143
OA72	0143
OA73	0143
OA73C	0143
OA74	0143
OA74A	0143
OA79	0143
OA81	0143
OA81C	0143
OA85	0143
OA85C	0143
OA86	1936
OA-90	0143
OA90	0143
OA-90(G)	0143
OA90-FM	0143
OA90-G	0143
OA90-M	0143
OA90-R	0143
OA90A-G	0019
OA90AG	0019
OA90A-M	0143
OA90A-R	0143
OA90FM	0123
OA90FV	0143
OA90G	0019
OA90GA	0143
OA90LF	0143
OA90M	0143
OA90MLF	0143
OA90S-G	0143
OA90S-R	0143
OA90Z	0143
OA91	0019
OA91A	0019
OA92	0143
OA-95	0133
OA95	0143
OA95A	0143
OA99	0019
OA99A	0143
OA126/5	0451
OA126/10	0181
OA126/12	0137
OA126/14	0100
OA127	0015
OA128	0015
OA129	0015
OA130	0015
OA131	0015
OA132	0015
OA134Q	0143
OA150	0143
OA159	0143
OA160	0143
OA161	0143
OA172	0143
OA174	0143
OA180	0015
OA182	0143
OA200	0133
OA201	OEM
OA202	0133
OA205	0133
OA210	0015
OA211	0015
OA214	0015
OA541	0143
OA626-50	OEM
OA626-300	OEM
OA626-500	OEM
OA626-600	OEM
OA636-600	OEM
OA645-100	OEM
OA645-200	OEM
OA645-300	OEM
OA645-400	OEM
OA645-500	OEM
OA909	0143
OA2002	0133
OA2005	0133
OA7812UC	OEM
OA12610	0170
OA12612	0137
OAZ200	0451
OAZ201	0162
OAZ202	0157
OAZ203	0636
OAZ204	0025
OAZ206	0244
OAZ207	0012
OAZ209	0162
OAZ210	0636
OAZ212	0012
OAZ213	0137
OAZ222	1901
OAZ223	0696
OAZ224	1458
OAZ225	1616
OAZ227	1618
OAZ228	0510
OAZ230	0870
OAZ232	1274
OAZ234	1078
OAZ236	1723
OAZ240	0140
OAZ241	0041
OAZ242	0253
OAZ243	0471
OAZ244	0062
OAZ245	0077
OAZ246	0165
OAZ247	0057
OAZ268	0275
OAZ269	0041
OAZ270	0471
OAZ271	0077
OAZ272	0057
OAZ273	0052
OAZ290	0696
OAZ291	1616
OAZ292	1618
OB2764K-2	0812
OC3H	0004
OC3K	OEM
OC3L	OEM
OC3LP	OEM
OC3LR	OEM
OC3N	OEM
OC4-0	OEM
OC4H	0004
OC4K	0004
OC4L	OEM
OC4LP	OEM
OC4LR	OEM
OC4N	OEM
OC5-0	OEM
OC5K	OEM
OC5L	OEM
OC5LP	OEM
OC5LR	OEM
OC5N	OEM
OC-16	0160
OC16	0085
OC19	0160
OC20	0160
OC-22	0160
OC22	0085
OC-23	0160
OC23	0085
OC-24	0160
OC24	0085
OC-25	0160
OC25	0085
OC-26	0160
OC26	0085
OC27	0085
OC-28	0160
OC28	0085
OC-29	0160
OC29	0085
OC30	0222
OC-30A	0222
OC30A	0222
OC30B	0222
OC32	0211
OC33	0004
OC34	0004
OC-35	0160
OC35	0085
OC-36	0160
OC36	0085
OC38	0004
OC40	0136
OC41	0211
OC41N	0004
OC42	0136
OC42N	0004
OC43	0136
OC43N	0136
OC44	0136
OC44N	0136
OC45	0004
OC45N	0136
OC46	0004
OC46N	0136
OC47	0004
OC47N	0136
OC50	0136
OC53	0050
OC54	0050
OC55	0050
OC56	0004
OC57	0004
OC58	0004
OC59	0004
OC60	0004
OC65	0004
OC66	0004
OC70	0004
OC70N	0004
OC71	0004
OC71A	0004
OC71N	0004
OC72	0004
OC73	0004
OC74	0004
OC74N	0004
OC75	0004
OC75N	0004
OC76	0004
OC77	0004
OC77M	0004
OC78	0004
OC79	0004
OC80	0004
OC81	0004
OC81D	0004
OC81DD	0004
OC81DN	0004
OC81N	0164
OC83	0164
OC83N	0164
OC84	0164
OC84N	0164
OC110	0004
OC120	0004
OC122	0004
OC123	0004
OC130	0136
OC139	0208
OC140	0208
OC141	0600
OC169	0050
OC169R	0050
OC170	0050
OC170N	0050
OC170R	0050
OC170V	0050
OC171	0050
OC171N	0050
OC171R	0050
OC171V	0050
OC174	0050
OC200	0037
OC201	0037
OC202	0037
OC203	0037
OC204	0037
OC205	0037
OC206	0037
OC207	0037
OC302	0004
OC303	0004
OC304	0004
OC304/1	0004
OC304/2	OEM
OC304/3	0004
OC304-1	0004
OC304-2	0004
OC304-3	0004
OC304N	0004
OC305	0004
OC305/1	0164
OC305/2	OEM
OC305-1	0004
OC305-2	0004
OC306	0004
OC306/1	0004
OC306/2	0004
OC306/3	0004
OC306-1	0004
OC306-2	0004
OC306-3	0004
OC307	0004
OC307-1	0004
OC307-2	0004
OC307-3	0004
OC308	0004
OC309	0004
OC309-1	0004
OC309-2	0004
OC309-3	0004
OC318	0004
OC320	0050
OC330	0004
OC331	0136
OC340	0004
OC341	0136
OC342	0136
OC343	0136
OC350	0004
OC351	0136
OC360	0004
OC361	0136
OC362	0136
OC363	0136
OC364	0004
OC390	0136
OC400	0050
OC410	0050
OC430	0037
OC430K	0037
OC440	0037
OC440K	0037
OC443	0037
OC443K	0037
OC445	0037
OC445K	0037
OC449	0037
OC449K	0037
OC450	0037
OC450K	OEM
OC460	0037
OC460K	0037
OC463	0037
OC463K	0037
OC465	0037
OC465K	0037
OC466	0037
OC466K	0037
OC467	0037
OC467K	0037
OC468	0037
OC468K	0037
OC469	0037
OC469K	0037
OC470	0037
OC470K	0037
OC480	OEM
OC480K	OEM
OC601	0004
OC602	0004
OC602-SPEZ	0164
OC602SP	0004
OC602SQ	0004
OC603	0004
OC604	0004
OC604-SPEZ	0164
OC604SP	0004
OC612	0136
OC613	0136
OC614	0136
OC615	0136
OC615N	0050
OC622	OEM
OC623	OEM
OC624	OEM
OC700	0037
OC700A	0037
OC700B	0037
OC701	OEM
OC702	0037
OC702A	0037
OC702B	0037
OC703	OEM
OC703A	OEM
OC704	0037
OC711	0004
OC740	0037
OC740G	0037
OC740M	0037
OC7400	0037
OC740Q	OEM
OC742	0037
OC742G	0037
OC742M	0037
OC7420	0037
OC800	OEM
OC810	0004
OC975	0633
OC3403Z1	OEM
OC7400	0037
OC7420	0037
OCD81DN	0164
OCP71	OEM
OD650	OEM
OD650B	OEM
OD651	OEM
OD651A	OEM
OD652	OEM
ODD247109AA	OEM
ODG-204A	OEM
OEC0010C	OEM
OEC0015B	OEM
OEC1011	4524
OEC2012A	OEM
OEC3006	OEM
OEC3008	OEM
OEC3011	OEM
OEC5003	OEM
OEC6007A	OEM
OEC6010B	OEM
OEC6015C	OEM
OEC6015C911E	OEM
OEC6015C911EAI	OEM
OEC6015C9111EAI	OEM
OEC6015C9115EA1	OEM
OEC7002	OEM
OEC8002	OEM
OEC8031B	OEM
OEC9001	OEM
OEC9005	OEM
OEC9008B	OEM
OECA001	OEM
OF66	0102
OF-129	0004
OF129	0004
OF156	0133
OF160	0015
OF162	0133
OF164	0015
OF173	0143
OF305	0019
OF449	0124
OF612	OEM
OF643	OEM
OFWM1953	OEM
OG336938	OEM
OG372884	OEM
OG851380	OEM
OG851794	6207
OGA00300	OEM
O-I5551	0861
OIKE780050B	OEM
OIS67	OEM
OLD-415	OEM
OMK7003/TK140	OEM
OMK7004/6940/TK76	OEM
ON67A	0143
ON74LS03N	1579
ON74LS05N	OEM
ON74LS08N	1633
ON74LS125AN	OEM
ON74LS145N	1564
ON74LS163AN	0893
ON74LS174N	0260
ON74LS175N	1672
ON74LS240N	0452
ON74LS244N	0458
ON74LS245N	0463
ON74LS373N	0709
ON74LS374N	0713
ON74LS670N	1130
ON74S00N	0704
ON74S10N	2445
ON74S20N	1017
ON74S138N	0427
ON74S240N	OEM
ON74S241N	OEM
ON150NV	OEM
ON174	0050
ON271	0079
ON274	0016
ON285A	OEM
ON769	2860
ON1102	OEM
ON1105	OEM
ON1301	OEM
ON1350	OEM
ON2160	OEM
ON2170-R	OEM
ON2170-RLF	OEM
ON2170-RS	OEM
ON2170-S	OEM
ON2170LH	OEM
ON2170L-R	OEM
ON2170R	OEM
ON3111	0112
ON3111Q	OEM
ON3111R	4924
ON3131	OEM
ON3131-R	OEM
ON3131S	OEM
ON3161	0112
ON3161(Q)	0112
ON3161(R)	0112
ON3161-Q	OEM
ON3161-QR	0112
ON3161-R	0112
ON3161QR	OEM
ON3161R	0321
ON4047	5345
ON4409(508D)	OEM
ON7406N	OEM
ON7407N	1339
ON7438N	0996
OP-05ADE	OEM
OP-05ADE/883B	OEM
OP-05AT	OEM
OP-05AT/883B	OEM
OP-05CDE	2246
OP-05CNB	2246

If replacement code is OEM, contact original manufacturer for replacement.

DEVICE TYPE	REPL CODE
OP-05CT	0896
OP-05DE	OEM
OP-05DE/883B	OEM
OP-05EDE	2246
OP-05ENB	2246
OP-05ET	0896
OP-05T	OEM
OP-05T/883B	OEM
OP-07/883B	OEM
OP-07ADE	OEM
OP-07ADE/883B	OEM
OP-07AT	OEM
OP-07AT/883B	OEM
OP-07C	OEM
OP-07CDE	2246
OP-07CNB	2246
OP-07CT	0896
OP-07D	OEM
OP-07DDE	2246
OP-07DE	OEM
OP-07DNB	2246
OP-07DT	0896
OP-07E	OEM
OP-07EDE	2246
OP-07ENB	2246
OP-07ET	0896
OP-07T	OEM
OP-07T/883B	OEM
OP-27ADE	OEM
OP-27ADE/883B	OEM
OP-27AT	OEM
OP-27AT/883B	OEM
OP-27BDE	OEM
OP-27BDE/883B	OEM
OP-27BT	OEM
OP-27BT/883B	OEM
OP-27CDE	OEM
OP-27CDE/883B	OEM
OP-27CT	OEM
OP-27CT/883B	OEM
OP-27EDE	2246
OP-27ENB	2246
OP-27ET	0896
OP-27FDE	2246
OP-27FNB	2246
OP-27FT	0896
OP-27GDE	2246
OP-27GNB	2246
OP-27GT	0896
OP-37ADE	OEM
OP-37ADE/883B	OEM
OP-37AT	OEM
OP-37AT/883B	OEM
OP-37BDE	OEM
OP-37BDE/883B	OEM
OP-37BT	OEM
OP-37BT/883B	OEM
OP-37CDE	OEM
OP-37CDE/883B	OEM
OP-37CT	OEM
OP-37CT/883B	OEM
OP-37EDE	2246
OP-37ENB	2246
OP-37ET	0896
OP-37FDE	2246
OP-37FNB	2246
OP-37FT	0896
OP-37GDE	2246
OP-37GNB	2246
OP-37GT	0896
OP-47BT	OEM
OP-47BT/883B	OEM
OP-47FT	0896
OP-47GNB	2246
OP-47GT	0896
OP130	0591
OP130W	0591
OP131	0591
OP131W	0591
OP132	0591
OP132W	0591
OP133	0591
OP133W	0591
OP135	0591
OP135W	0591
OP136	0591
OP136W	0591
OP137	0591
OP137W	0591
OP215BZ/883	OEM
OP800	2313
OP801	2313
OP811	2313
OP812	3878
OP813	2313
OP8242A	OEM
OP13152	1053
OP13153	2108
OP13251	2108
OP13252	2108
OP13253	1053
OPB814	1417
OPF54	OEM
OPF56	OEM
OPI1264A	OEM
OPI2100	0315
OPI2150	0315
OPI2151	0315
OPI2152	0541
OPI2153	0315
OPI2154	0541
OPI2155	0541
OPI2250	0315
OPI2251	0315
OPI2252	0541
OPI2253	0315
OPI2254	0541
OPI2255	0315
OPI2500	5425
OPI3009	6441
OPI3010	5051
OPI3011	6441
OPI3150	0315
OPI3151	0315
OPI3152	1109
OPI3153	1053
OPI3250	1053
OPI3251	1053
OPI3252	1109
OPI3253	1053
OPI5000	0315
OPI6100	6760
OPI7002	3367
OPI7010	3367
OPL-211	OEM
OPP28	OEM
OR06NXZ31	0201
OR06YXZ31	0378
OR6JXZ33A	OEM
OR6LXZ33A	OEM
OR6QXZ33A	OEM
OR6UXZ33A	OEM
ORP30	OEM
ORP50	OEM
ORP63	OEM
ORP94	OEM
OS-1D	0015
OS-4D	0015
OS-6D	0015
OS-8D	0071
OS13	OEM
OS15	OEM
OS16	OEM
OS17	OEM
OS70	0143
OS-100A	OEM
OS492	0847
OS536G	0079
OS643	OEM
OS16308	0015
OSD	OEM
OSD-0033	0649
OSD0033	0649
OSF02	OEM
OSF03	OEM
OSF04	OEM
OSF05	OEM
OSI5JHGA	OEM
OSL3-1	OEM
OSL3-2	OEM
OSL3-3	OEM
OSL3L1	OEM
OSL3L2	OEM
OSL3L3	OEM
OSL6-1	OEM
OSL6L1	OEM
OSL6L2	OEM
OSL6L3	OEM
OSL16-1	OEM
OSL16-2	OEM
OSL16-3	OEM
OSL16L1	OEM
OSL16L2	OEM
OSL16L3	OEM
OSL26-1	OEM
OSL26L1	OEM
OSL50	OEM
OSM9510-12	OEM
OSS-16308	0015
OSS16308	0143
OSS-16685	0143
OSS16685	0143
OSS36503	0015
OSS36685	0015
OSS-36885	0015
OSS36885	0015
OV02	0102
OV-1	OEM
OV-2	OEM
OV-2-1	OEM
OV-2-2	OEM
OV-2-3	OEM
OV2764KFA552	0812
OVSSAC431-1	OEM
OY101	0015
OY1011	OEM
OY1021	OEM
OY5061	0015
OY5062	0015
OY5063	0015
OY5064	0015
OY5065	0015
OY5066	0015
OY5067	0071
OZ3.6T5	0188
OZ3.6T10	0188
OZ5.1T5	0162
OZ5.1T10	0162
OZ5.6T5	0157
OZ5.6T10	0157
OZ6.2T5	0636
OZ6.2T10	0636
OZ10T5	0170
OZ10T10	0170
OZ12T5	0137
OZ12T10	0137
OZ15T5	0002
OZ15T10	0002
OZ27T5	0441
OZ27T10	0441
P	OEM
P006001D	OEM
P006001E-H	OEM
P006004D	OEM
P006004E-H	OEM
P006010D	OEM
P006010E	OEM
P006010F	OEM
P006010G	OEM
P027PH02F20	OEM
P027QH05FR0	OEM
P055I	0812
P056I	0812
P093	0170
P095PH04FG0	OEM
P0100AD	0901
P0100BA	0058
P0100BB	0058
P0100BD	0058
P0100CD	0407
P0100DA	0407
P0100DB	0407
P0100DD	0407
P0100MA	1683
P0100MB	1683
P0100NA	OEM
P0100NB	OEM
P0102AA	0901
P0102AB	0901
P0102BA	0058
P0102BB	0058
P0102CA	0407
P0102CB	0407
P0102DA	0407
P0102DB	0407
P0103AD	0901
P0103BA	0058
P0103BB	0058
P0103BD	0058
P0103CD	0407
P0103DA	0407
P0103DB	0407
P0103DD	0407
P0103MA	1683
P0103MB	1683
P0103NA	OEM
P0103NB	OEM
P0108EXP03A	OEM
P0117RD01B	OEM
P0165STD01E	2697
P015010D	OEM
P015010E	OEM
P015010F	OEM
P015010G	OEM
P015020C	OEM
P015020D	OEM
P015020E	OEM
P024020C	OEM
P024020D	OEM
P024020E	OEM
P030010C	OEM
P030010D	OEM
P030010E	OEM
P030010F	OEM
P045010B	OEM
P045010C-F	OEM
P045020B	OEM
P045020C	OEM
P045020D	OEM
P060003-1R	OEM
P060003-2R	OEM
P060003-3R	OEM
P060005-1R	OEM
P060005-2R	OEM
P060005-3R	OEM
P060007-1R	OEM
P060007-2R	OEM
P060010A	OEM
P060010B-E	OEM
P090020A	OEM
P090020B	OEM
P090020C	OEM
P1/1H	OEM
P1/2H	OEM
P1/3H	OEM
P1/4H	OEM
P1/5H	OEM
P1A	0160
P1A5	0015
P1B	0037
P1B5	0015
P1C	0037
P1C5	0015
P1CG	0037
P1D	0037
P1D5	0015
P1E	0491
P1E-1	0160
P1E-1BLK	0491
P1E-1BLU	0491
P1E-1GRN	0491
P1E-1RED	0491
P1E-1V10	1681
P1E-1VIO	0491
P1E-2BLK	0491
P1E-2BLU	0491
P1E-2GRN	0491
P1E-2RED	0491
P1E-2V10	1681
P1E-2VIO	0491
P1E-3BLK	0491
P1E-3BLU	0491
P1E-3GRN	0491
P1E-3RED	0491
P1E-3V10	1681
P1E-3VIO	0491
P1F	0975
P1G	0160
P1H	0037
P1J	0037
P1K	0160
P1KBLK	0160
P1KBLU	0160
P1KBRN	0160
P1KGRN	0160
P1KORN	0160
P1KRED	0160
P1KYEL	0160
P1L	0004
P1L4956	0004
P1M	0126
P1N	0037
P1N-1	0037
P1N-2	0037
P1N-3	0037
P1P	0037
P1P-1	0037
P1R	0085
P1RF100	OEM
P1RP15	OEM
P1RP24	OEM
P1RP36	OEM
P1RP48	OEM
P1RP60	OEM
P1RP72	OEM
P1RP80	OEM
P1RP100	OEM
P1T	0160
P1V	0460
P1V-2	0460
P1V-3	0460
P1V-4	0925
P1W	0037
P1Y	0975
P2A	0037
P2A5	0015
P2B	0925
P2B5	0015
P2C	0085
P2C5	0015
P2D	0160
P2D5	0015
P2DBLU	0160
P2DBRN	0160
P2DGRN	0160
P2DORN	0160
P2DRED	0160
P2DYEL	0160
P2E	0037
P2F	0847
P2G	0037
P2GE	0037
P2H	0037
P2J	1681
P2K	0925
P2L	0037
P2M	0058
P2M-1	0037
P2M-2	0037
P2M-3	0037
P2M3Z	0058
P2M49	OEM
P2P	0037
P2R	0085
P2S	0037
P2T	0460
P2T-1	0460
P2T-2	0460
P2T-3	0460
P2T-4	0460
P2U	0382
P2U-1	0382
P2U-2	1367
P2V	0382
P2W	0037
P2Y	0037
P2Z	0604
P3/1H	OEM
P3/2H	0015
P3/3H	OEM
P3/4H	OEM
P3/5H	OEM
P3A	0854
P3A5	0015
P3B	0164
P3B5	0015
P3C	0037
P3C5	0015
P3D	0004
P3D5	0015
P3E	0222
P3E5	OEM
P3EBLK	0085
P3EBLU	0085
P3EGRN	0085
P3ERED	0085
P3H	0975
P3J	0975
P3K	0382
P3M	0460
P3N-1	0460
P3N-2	0460
P3N-3	0460
P3N-4	0460
P3N-5	1199
P3N45	OEM
P3P	0460
P3P-1	0460
P3P-2	0460
P3P-3	0460
P3P-4	0460
P3P-5	0460
P3R	0222
P3R-1	0222
P3R-2	0222
P3R-3	0222
P3R-4	0222
P3S	0460
P3T	0222
P3T-1	0222
P3T-2	0222
P3U	0854
P3V	3165
P3W	0491
P3Y	0382
P3Z	0037
P4A5	0015
P4B	0791
P4B5	0015
P4C	0037
P4C5	0015
P4D	0085
P4D5	0015
P4E	0460
P4E-1	1199
P4E-2	1199
P4E-3	3165
P4E-4	3165
P4E5	OEM
P4F	0975
P4G	0037
P4G2X	OEM
P4H	0975
P4J	3165
P4J/48	0460
P4J148	0042
P4K	0037
P4L	0085
P4M	0160
P4N	0085
P4P	0037
P4R	0037
P4S	0460
P4T	1199
P4U	0460
P4V	0460
P4V-1	0460
P4V-2	1199
P4W	0925
P4W-1	0460
P4W-2	0460
P4Y	0037
P4Z	0086
P5A	OEM
P5A5	0015
P5B	0037
P5B5	0015
P5C	0037
P5C5	0015
P5D	0037
P5D5	0015
P5E5	OEM
P5F	3165
P5H	0925
P5L	0460
P5M	1267
P5R	0460
P5S	3165
P5U	0925
P6/1H	OEM
P6/2H	0015
P6/3H	OEM
P6/4H	OEM
P6/5H	OEM
P6A5	0015
P6B5	0015
P6C5	0015
P6D5	0015
P6E5	OEM
P6KE6.8	3114
P6KE6.8A	3114
P6KE6.8C	2133
P6KE6.8CA	2133
P6KE7.5	OEM
P6KE7.5A	OEM
P6KE7.5C	OEM
P6KE7.5CA	OEM
P6KE8.2	OEM
P6KE8.2A	OEM
P6KE8.2C	OEM
P6KE8.2CA	OEM
P6KE9.1	OEM
P6KE9.1A	OEM
P6KE9.1C	OEM
P6KE9.1CA	OEM
P6KE10	OEM
P6KE10A	OEM
P6KE10C	OEM
P6KE10CA	OEM
P6KE11	OEM
P6KE11A	OEM
P6KE11C	OEM
P6KE11CA	OEM
P6KE12	OEM
P6KE12A	OEM
P6KE12C	OEM
P6KE12CA	OEM
P6KE13	3172
P6KE13A	3172
P6KE13C	1169
P6KE13CA	1169
P6KE15	OEM
P6KE15A	OEM
P6KE15C	OEM
P6KE15CA	OEM
P6KE16	OEM
P6KE16A	OEM
P6KE16C	OEM
P6KE16CA	OEM
P6KE18	3191
P6KE18A	3191
P6KE18C	1767
P6KE18CA	1767
P6KE20	3201
P6KE20A	3201
P6KE20C	1932
P6KE20CA	1932
P6KE22	OEM
P6KE22A	OEM
P6KE22C	OEM
P6KE22CA	OEM
P6KE24	OEM
P6KE24A	OEM
P6KE24C	OEM
P6KE24CA	OEM
P6KE27	1915
P6KE27A	1915
P6KE27C	1952
P6KE27CA	1952
P6KE30	OEM
P6KE30A	OEM
P6KE30C	OEM
P6KE30CA	OEM
P6KE33	OEM
P6KE33A	OEM
P6KE33C	OEM
P6KE33CA	OEM
P6KE36	OEM
P6KE36A	OEM
P6KE36C	OEM
P6KE36CA	OEM
P6KE39	OEM
P6KE39A	OEM
P6KE39C	OEM
P6KE39CA	OEM
P6KE43	OEM
P6KE43A	OEM
P6KE43C	OEM
P6KE43CA	OEM
P6KE47	OEM
P6KE47A	OEM
P6KE47C	OEM
P6KE47CA	OEM
P6KE51	0568
P6KE51A	0568
P6KE51C	1972
P6KE51CA	1972
P6KE56	OEM
P6KE56A	OEM
P6KE56C	OEM
P6KE56CA	OEM
P6KE62	OEM
P6KE62A	OEM
P6KE62C	OEM
P6KE62CA	OEM
P6KE68	0831
P6KE68A	0831
P6KE68C	1987
P6KE68CA	1987
P6KE75	OEM
P6KE75A	OEM
P6KE75C	OEM
P6KE75CA	OEM
P6KE82	OEM
P6KE82A	OEM
P6KE82C	OEM
P6KE82CA	OEM
P6KE91	OEM
P6KE91A	OEM
P6KE91C	OEM
P6KE91CA	OEM
P6KE100	OEM
P6KE100A	OEM
P6KE100C	OEM
P6KE100CA	OEM
P6KE110	OEM
P6KE110A	OEM
P6KE110C	OEM
P6KE110CA	OEM
P6KE120	OEM
P6KE120A	OEM
P6KE120C	OEM
P6KE120CA	OEM
P6KE130	OEM
P6KE130A	OEM
P6KE130C	OEM
P6KE130CA	OEM
P6KE150	OEM
P6KE150A	OEM
P6KE150C	OEM
P6KE150CA	OEM
P6KE160	OEM
P6KE160A	OEM
P6KE160C	OEM
P6KE160CA	OEM
P6KE170	OEM
P6KE170A	OEM
P6KE170C	OEM
P6KE170CA	OEM
P6KE180	OEM
P6KE180A	OEM
P6KE180C	OEM
P6KE180CA	OEM
P6KE200	1405
P6KE200A	1405
P6KE200C	1408
P6KE200CA	1408
P6KE220A	OEM
P6KE250A	OEM
P6KE300A	OEM
P6KE350A	OEM
P6KE400A	OEM
P6RP8	0071
P6RP10	0071
P6RP12	OEM
P6SE6.8	OEM
P6SE6.8A	6770
P6SE7.5	OEM
P6SE7.5A	OEM
P6SE8.2	OEM
P6SE8.2A	OEM
P6SE9.1	OEM
P6SE9.1A	OEM
P6SE10	OEM
P6SE10A	OEM
P6SE11	OEM
P6SE11A	OEM
P6SE12	OEM
P6SE12A	OEM
P6SE13	OEM
P6SE13A	6629
P6SE15	OEM
P6SE15A	OEM
P6SE16	OEM
P6SE16A	OEM
P6SE18	OEM
P6SE18A	6641
P6SE20	OEM
P6SE20A	6644
P6SE22	OEM
P6SE22A	OEM
P6SE24	OEM
P6SE24A	OEM
P6SE27	OEM
P6SE27A	6652
P6SE30	OEM
P6SE30A	OEM
P6SE33	OEM
P6SE33A	OEM
P6SE36	OEM
P6SE36A	OEM
P6SE39	OEM
P6SE39A	OEM
P6SE43	OEM
P6SE43A	OEM
P6SE47	OEM
P6SE47A	OEM
P6SE51	OEM
P6SE51A	6676
P6SE56	OEM
P6SE56A	OEM
P6SE62	OEM
P6SE62A	OEM
P6SE68	OEM
P6SE68A	6694
P6SE75	OEM
P6SE75A	OEM
P6SE82	OEM
P6SE82A	OEM
P6SE91	OEM
P6SE91A	OEM
P6SE100	OEM
P6SE100A	OEM
P6SE110	OEM
P6SE110A	OEM
P6SE120	OEM
P6SE120A	OEM
P6SE130	OEM
P6SE130A	OEM
P6SE150	OEM
P6SE150A	OEM
P6SE160	OEM
P6SE160A	OEM

If replacement code is OEM, contact original manufacturer for replacement.

DEVICE TYPE	REPL CODE
P6SE170	OEM
P6SE170A	OEM
P6SE180	OEM
P6SE180A	OEM
P6SE200	OEM
P6SE200A	6395
P7A5	0015
P7B5	0015
P7C5	0015
P7D5	0015
P8/1H	OEM
P8/2H	0015
P8/3H	OEM
P8/4H	OEM
P8/5H	OEM
P8B	0854
P8C	0681
P8H	0925
P13T1	0316
P13T2	0316
P15	0196
P16	0484
P17	1080
P19C	1107
P19E	0258
P19G	0267
P19J	1119
P20	0015
P20C	1124
P20E	0805
P20G	0319
P20J	1132
P24	OEM
P32H	0015
P44	OEM
P51A	1520
P51B	1520
P51C	1520
P52A	1530
P52B	1530
P52C	1530
P61AM6116H	1898
P61CP7811N	OEM
P61LS00**H	1527
P61LS06**H	OEM
P61LS14**H	1698
P61LS32**H	0088
P61LS74**H	0243
P61LS139*H	0153
P61LS373*H	0709
P61LS374*H	0713
P61XX0002	0973
P61XX0026	OEM
P61XX0027	0176
P62D1S1588	0133
P62H	0015
P62T2B920	0718
P62X005	OEM
P62X010	OEM
P62X011	0023
P62X018	OEM
P62X020	0148
P62X045	0323
P62X046	0323
P62X048	0441
P62X049	0299
P62X050	4172
P67	0037
P82A05	OEM
P82A203	OEM
P82A204	OEM
P82A205	OEM
P82C201	OEM
P82C202	OEM
P82H	0015
P100	0015
P100A	0015
P100B	0015
P100D	0015
P100G	0015
P100J	0015
P101	1434
P101K	OEM
P101KW	OEM
P101W	1434
P102	1434
P102K	OEM
P102KW	OEM
P102W	1434
P103	1434
P103K	OEM
P103KW	OEM
P103W	1434
P104	1434
P104K	OEM
P104KW	OEM
P104W	1434
P105	1434
P105K	OEM
P105KW	OEM
P105W	1434
P111	1680
P111K	OEM
P111KW	OEM
P111W	1680
P112	1680
P112K	OEM
P112KW	OEM
P112W	1680
P113	1680
P113K	OEM
P113KW	OEM
P113W	1680
P114	1680
P114K	OEM
P114KW	OEM
P114W	1680
P115	1680
P115K	OEM
P115KW	OEM
P115W	1680
P121	OEM
P121K	OEM
P121KW	OEM
P121W	OEM
P122	OEM
P122K	OEM
P122KW	OEM
P122W	OEM
P123	OEM
P123K	OEM
P123KW	OEM
P123W	OEM
P124	OEM
P124K	OEM
P124KW	OEM
P124W	OEM
P125	OEM
P125K	OEM
P125KW	OEM
P125W	OEM
P131	6977
P131K	OEM
P132	6977
P132K	OEM
P133	6977
P133K	OEM
P134	6977
P134K	OEM
P135	6977
P135K	2294
P141	6978
P141K	OEM
P142	6978
P142K	OEM
P143	6978
P143K	OEM
P144	6978
P144K	OEM
P145	6978
P145K	2298
P150A	0015
P150B	0015
P150D	0015
P150G	0015
P150J	0015
P161	6980
P162	6980
P163	6980
P164	6980
P165	5812
P171	6982
P172	6982
P173	6982
P174	6982
P175	5813
P200	0015
P200A	0015
P201D-5R	OEM
P218-1	0551
P218-2	0382
P300A	0110
P300B	0953
P300D	0242
P300DL	OEM
P300G	0540
P300GG1	0015
P300J	0965
P300K	0817
P300M	0922
P346	0144
P346A	0144
P400	0015
P400A	0015
P401	OEM
P401K	OEM
P401KW	OEM
P401W	OEM
P402	OEM
P402K	OEM
P402KW	OEM
P402W	OEM
P403	OEM
P403K	OEM
P403KW	OEM
P403W	OEM
P404	OEM
P404K	OEM
P404KW	OEM
P404W	OEM
P405	OEM
P405K	OEM
P405KW	OEM
P405W	OEM
P411	OEM
P411K	OEM
P411KW	OEM
P411W	OEM
P412	OEM
P412K	OEM
P412KW	OEM
P412W	OEM
P413	OEM
P413K	OEM
P413KW	OEM
P413W	OEM
P414	OEM
P414K	OEM
P414KW	OEM
P414W	OEM
P415	OEM
P415K	OEM
P415KW	OEM
P415W	OEM
P421	OEM
P421K	OEM
P421KW	OEM
P421W	OEM
P422	OEM
P422K	OEM
P422KW	OEM
P422W	OEM
P423	OEM
P423K	OEM
P423KW	OEM
P423W	OEM
P424	OEM
P424K	OEM
P424KW	OEM
P424W	OEM
P425	OEM
P425K	OEM
P425KW	OEM
P425W	OEM
P431	OEM
P431K	OEM
P432	OEM
P432K	OEM
P433	OEM
P433K	OEM
P434	OEM
P434K	OEM
P435	OEM
P435K	OEM
P441	OEM
P441K	OEM
P442	OEM
P442K	OEM
P443	OEM
P443K	OEM
P444	OEM
P444K	OEM
P445	OEM
P445K	OEM
P461	OEM
P462	OEM
P463	OEM
P464	OEM
P465	OEM
P471	OEM
P472	OEM
P473	OEM
P474	OEM
P475	OEM
P504	2002
P506	2002
P521	3367
P580	0071
P600	0015
P600A	1282
P600B	1282
P600D	1287
P600G	1287
P600J	1293
P600K	1296
P600M	1296
P621	0112
P621B	4924
P621GB	1166
P800	0071
P800A	0071
P850	OEM
P1000	0071
P1000A	0037
P1003	OEM
P1004	0590
P1005	OEM
P1006	0590
P1027	OEM
P1028	OEM
P1029	OEM
P1055	0143
P1062-02	0315
P1087	0325
P1087E	0325
P1103	3559
P1172	0133
P1172-1	0133
P1201-01	OEM
P1603-02	0315
P1760	OEM
P2004	0590
P2006	0590
P2010	3918
P2114	2048
P2114-3	2048
P2163	OEM
P2221	0079
P2221A	0079
P2222	0079
P2222A	0079
P2271	0103
P2440	0087
P2906	0037
P2906A	0037
P2907	0037
P2907A	0037
P3139	0103
P3172	0178
P3309	0531
P4069	0086
P4326	0133
P5034H080S	OEM
P5058H	OEM
P5058H-546	OEM
P5100	0071
P5148	0178
P5149	0103
P5152	0016
P5153	0016
P-6006	0133
P6009	3165
P6502B	3065
P6520A	OEM
P7109	0087
P7394	0133
P7776	0229
P8035	OEM
P8048AH	OEM
P8049AH	OEM
P8051	OEM
P8051AH	OEM
P8051AH-0178	OEM
P8051AH-0552	OEM
P8088	OEM
P8088-2	OEM
P8212	1860
P8216	1863
P8224	1709
P8226	3013
P8237A-4	OEM
P8237A-5	OEM
P8237A5	OEM
P8253-5	OEM
P8254	OEM
P8254-2	OEM
P8255A-5	0051
P8257-5	OEM
P8259A	OEM
P8272A	OEM
P8314	OEM
P8393	0016
P8394	0016
P8412A	OEM
P8452	OEM
P8612	OEM
P8748	OEM
P8870	0160
P8890	0160
P8890A	0160
P8890L	0160
P9459	0015
P9797	2312
P10115	0015
P10115A	0071
P10155	0143
P10156	0015
P10156A	0071
P21309	0917
P21316	0015
P21317	0015
P21344	0012
P21443	0015
P25011LB	OEM
P-31898	0160
P31898	0160
P38103/507-10	0002
P75534	0160
P75534-1	0160
P75534-2	0085
P75534-3	0085
P75534-4	0160
P75534-5	0160
P633024C(SYLVANIA)	0086
P633024G	0086
P86421505	OEM
PA0011	OEM
PA0016	OEM
PA0017	OEM
PA0021	OEM
PA0021A	OEM
PA0022	OEM
PA0030	OEM
PA0034	OEM
PA0034A	OEM
PA069	0015
PA070	0015
PA071	0015
PA2	OEM
PA2C	OEM
PA3	0015
PA4	OEM
PA4C	OEM
PA5	OEM
PA5C	OEM
PA7-703E	OEM
PA9D522/1	0277
PA40N200	OEM
PA40N300	OEM
PA53C	4138
PA54H	OEM
PA56C	OEM
PA75N85	OEM
PA75N120	OEM
PA75N150	OEM
PA125N40	OEM
PA125N60	OEM
PA200	0015
PA234	3724
PA239	2279
PA277	2557
PA300	0015
PA305	0015
PA305A	0015
PA306	2557
PA310	0015
PA310A	0015
PA315	0015
PA315A	0015
PA-320	0015
PA320	0015
PA320A	0015
PA320B	0015
PA325	0015
PA325A	0015
PA325B	0015
PA330	0015
PA330A	0015
PA330B	0015
PA340	0015
PA340A	0015
PA340B	0015
PA350	0015
PA350A	0015
PA360	0015
PA360A	0015
PA380	0071
PA400	0015
PA401	2861
PA401X	2861
PA-500	2861
PA500	2861
PA501	2864
PA501X	2861
PA600	0015
PA1000	0037
PA1001	0037
PA2002	OEM
PA2017	OEM
PA4009-A	OEM
PA4010	OEM
PA5010	OEM
PA5012	OEM
PA6015C	OEM
PA7001/518	0873
PA7001/519	0697
PA7001/520	0512
PA7001/521	0232
PA7001/522	1024
PA7001/523	1168
PA7001/524	1185
PA7001/525	0314
PA7001/526	0268
PA7001/527	0361
PA7001/528	0385
PA7001/529	1313
PA7001/530	1404
PA7001/531	1172
PA7001/532	1426
PA7001/533	1275
PA7001/539	1202
PA7001/593	0619
PA7615	0015
PA7703	0632
PA7703E	0632
PA8260	0111
PA8261	0002
PA8543	0111
PA8645	0015
PA8900	0560
PA9004	0111
PA9005	0111
PA9006	0016
PA9154	0050
PA9155	0050
PA9156	0004
PA9157	0004
PA9158	0164
PA9160	0015
PA9267	0636
PA9483	0595
PA10556	0015
PA10880	0050
PA10887	0015
PA-10889-1	0160
PA10889-1	0164
PA-10889-2	0160
PA10889-2	0164
PA-10890	0160
PA10890	0085
PA-10890-1	0160
PA10890-1	0160
PADT20	0050
PADT21	0050
PADT22	0050
PADT23	0050
PADT24	0050
PADT25	0050
PADT26	0050
PADT27	0050
PADT28	0050
PADT29	OEM
PADT30	0050
PADT31	0050
PADT35	0050
PADT40	0050
PADT50	0085
PADT51	0050
PAL10L8CN	OEM
PAL12L6NC	OEM
PAL12L10CNS	OEM
PAL14H4CN	OEM
PAL14L8CNS	OEM
PAL16L8	OEM
PAL16L8ACN	OEM
PAL16L8CN	OEM
PAL16R4CN	OEM
PAL20L10	OEM
PAL20X8CNS	OEM
PAR-12	0160
PAR12	0160
PAVIMB625167	OEM
PAVITD62083A	OEM
PAWI1W1510MA	OEM
PAWI2W1500C	OEM
PB	0208
PB2	OEM
PB4	OEM
PB5	OEM
PB40N280	OEM
PB40N400	OEM
PB61	0323
PB61G	OEM
PB62	1414
PB62G	OEM
PB64	0473
PB64G	OEM
PB66	0446
PB66G	OEM
PB68	1422
PB68G	OEM
PB75N140	OEM
PB75N180	OEM
PB-99S005	0136
PB110	0160
PB125N60	OEM
PB125N80	OEM
PB401	0323
PB553(C)	OEM
PB605	0323
PB605G	OEM
PB610	2443
PB610G	OEM
PB6013	0037
PB6013A	0037
PB6013B	0037
PB6014	0037
PB6014A	0037
PB6015C	OEM
PBC107	0111
PBC107A	0111
PBC107B	0111
PBC108	0111
PBC108A	0111
PBC108B	0111
PBC108C	0111
PBC109	0111
PBC109B	0111
PBC109C	0111
PBC182	0144
PBC183	0155
PBC184	0595
PBE3014-1	0164
PBE3014-2	0164
PBE3020-1	0164
PBE3020-2	0208
PBE3162	0164
PBE3162-1	0004
PBE3162-2	0004
PBE3322	0143
PBL3717	OEM
PBL3717A	OEM
PBM74LS30P	0828
PBM74LS74AP	0243
PBM74LS86P	0291
PBM74LS244P	0458
PBM74LS375P	OEM
PBM74LS393P	0819
PBM152	OEM
PBM6004	0037
PBT5504	0652
PBU05	OEM
PBU10	OEM
PBU20	OEM
PBU40	OEM
PBU60	OEM
PBU80	OEM
PBU100	OEM
PBX103	0211
PBX113	0211
PC02P1/2	0015
PC0211/2	0015
PC0600A	0554
PC0601A	0005
PC0603A	0628
PC0604A	0030
PC0610A	0554
PC0611A	0005
PC0620A	0554
PC0.2P11/2	0015
PC2	OEM
PC4	OEM
PC5	OEM
PC40N500	OEM
PC40N800	OEM
PC50B187-27	OEM
PC74HC157P	OEM
PC74HC4046AP	OEM
PC74HC4053P	OEM
PC74HCT164P	OEM
PC74HCT244P	OEM
PC74HCT245P	OEM
PC75N250	OEM
PC75N400	OEM
PC78L05	1292
PC78L05A	OEM
PC78M08H	OEM
PC78M12H	OEM
PC100	0423
PC110	0423
PC111	4529
PC125N130	OEM
PC125N180	OEM
PC127	0030
PC127A	0030
PC127B	OEM
PC127C	OEM
PC127D	OEM
PC128	0030
PC128A	0030
PC128B	OEM
PC128C	OEM
PC128D	OEM
PC133	0628
PC133A	0628
PC133B	OEM
PC133C	OEM
PC133D	OEM
PC135	0005
PC135A	0005
PC135B	OEM
PC135C	OEM
PC135D	OEM
PC136	0628
PC136A	0628
PC136B	OEM
PC136C	OEM
PC136D	OEM
PC138	0554
PC138A	0554
PC138B	OEM
PC138C	OEM
PC138D	OEM
PC139	0554
PC139A	0554
PC139B	OEM
PC139C	OEM
PC139D	OEM
PC140	0554
PC140A	0554
PC140B	OEM
PC140C	OEM
PC140D	OEM
PC141	0554
PC141A	0554
PC141B	OEM
PC141C	OEM
PC141D	OEM
PC200	0854
PC210	0854
PC339C	0176
PC358C	OEM
PC393C	0631
PC563H2	1160
PC574J	1329
PC702V	OEM
PC713	0315
PC-713U	0315
PC713U	0315
PC713V	0746
PC714	0315
PC714U	OEM
PC814	OEM
PC817	1166
PC817-B	1166
PC817-C	1166
PC817A	1166
PC817B	1166
PC-817BC	1166
PC817BC	1166
PC817C	1166
PC900	OEM
PC1007	0037
PC1007A	0037
PC1008	0037
PC1008B	0037
PC1026C	0706
PC1066T	0164
PC1067T	0164
PC1068T	0164
PC1167C2	2252
PC1228H	OEM

DEVICE TYPE	REPL CODE
PC1363C	0683
PC1873CT	OEM
PC1879-004	0170
PC-2007D	OEM
PC3002	0164
PC3003	0164
PC3004	0160
PC3005	0164
PC3006	0164
PC3007	0164
PC3009	0164
PC3010	0222
PC3345	OEM
PC3346	OEM
PC-3350	OEM
PC4004	0015
PC4558C	0360
PC7327C	1069
PC7805H	0624
PC7805HF	0624
PC7812H	OEM
PC7812HF	0334
PC17805H	0624
PC20003	0579
PC-20004	2155
PC20004	2155
PC-20005	2876
PC-20006	3550
PC-20007	1215
PC-20008	0443
PC-20009	OEM
PC-20010	OEM
PC-20011	OEM
PC-20012	4561
PC-20015	3770
PC-20018	0443
PC20018	1215
PC-20022	OEM
PC-20023	0864
PC-20024	0701
PC-20030	0701
PC-20045	0520
PC-20046	3079
PC-20051	2261
PC-20066	0144
PC-20069	2581
PC-20071	2062
PC-20082	0416
PC-20083	1612
PC-20084	OEM
PC-20085	2019
PC-20086	6188
PC-20087	2295
PC-20110	0362
PC-20143	2295
PC-20151	0706
PCA8510T-012-T	OEM
PCB17A	OEM
PCD5582	OEM
PCD8582	OEM
PCF2N05	OEM
PCF2N08	OEM
PCF2N12	OEM
PCF2N18	OEM
PCF3N45	OEM
PCF5P12	OEM
PCF6P08	OEM
PCF8N18	OEM
PCF8P08	OEM
PCF10N12	OEM
PCF10N45	OEM
PCF12N08	OEM
PCF12N18	OEM
PCF12P08	OEM
PCF15N05	OEM
PCF15N12	OEM
PCF18N08	OEM
PCF25N18	OEM
PCF30N12	OEM
PCF35N08	OEM
PCF45N05	OEM
PCF2111P	OEM
PCF8571P	OEM
PCF8573P	OEM
PCF8583P	OEM
PCM53JP-V-2	OEM
PCM54HP	OEM
PCM54HP-V	OEM
PCM55HP-S	OEM
PCM56P	OEM
PCM56P(RED)	OEM
PCM56P-S	OEM
PCMB74LS00M	1527
PCMB74LS02M	1559
PCMB74LS04M	1595
PCMB74LS05M	1608
PCMB74LS10M	1662
PCMB74LS32M	0088
PCV50	OEM
PCV50A	OEM
PCV56	OEM
PCV58	OEM
PCV60	OEM
PCV61	OEM
PCV62	OEM
PCV63	OEM
PCV67	OEM
PCV68	OEM
PCV69	OEM

DEVICE TYPE	REPL CODE
PCV70	OEM
PCV71	OEM
PCV73	OEM
PCV76	OEM
PCV84	OEM
PCV85	OEM
PCV88	OEM
PCV97	OEM
PCV104	OEM
PCX-010	OEM
PCX010	OEM
PD0011A	OEM
PD0012A	OEM
PD0026A	OEM
PD0031	OEM
PD0034	OEM
PD05	0277
PD2	OEM
PD3L	OEM
PD4	OEM
PD5	OEM
PD6	OEM
PD10	0277
PD20	0289
PD31	OEM
PD32	OEM
PD40	0296
PD49P1	OEM
PD49PI	OEM
PD60	0302
PD80	0250
PD100	0250
PD101	4240
PD102	4240
PD103	0015
PD104	0015
PD105	0015
PD106	0015
PD107	0015
PD107A	0015
PD108	0015
PD110	0015
PD111	0015
PD114	0071
PD115	0071
PD116	0071
PD122	0015
PD125	0015
PD129	0015
PD130	0015
PD131	0015
PD132	0015
PD133	0015
PD134	0015
PD135	0015
PD151	OEM
PD154	0015
PD155	0015
PD301	OEM
PD302	OEM
PD303	OEM
PD304	OEM
PD305	OEM
PD306	OEM
PD307	OEM
PD308	OEM
PD309	OEM
PD310	OEM
PD311	OEM
PD323BHT	OEM
PD401	OEM
PD546C037	OEM
PD553C042	OEM
PD651G-517	OEM
PD651G-518	OEM
PD910	0015
PD913	0071
PD914	0071
PD915	0071
PD916	0071
PD1001	0079
PD1002	0079
PD1011	0277
PD1020	0302
PD-1567	0302
PD1703(C)	OEM
PD1705C	OEM
PD1913C	OEM
PD3068-B	OEM
PD3068B	OEM
PD3089A	OEM
PD3093	OEM
PD3093A	OEM
PD3095	OEM
PD3113	OEM
PD3113A	OEM
PD4035(C)	OEM
PD4051	OEM
PD4066BC	0101
PD4073A	OEM
PD4073B	OEM
PD4113A	OEM
PD4132	OEM
PD4150	OEM
PD4150A	OEM
PD4151A	OEM
PD4151B	OEM
PD4175B	OEM
PD4184	OEM

DEVICE TYPE	REPL CODE
PD4185	OEM
PD4185A	OEM
PD4192	OEM
PD4193	OEM
PD4199	OEM
PD5030	OEM
PD5039	OEM
PD5039-A	OEM
PD5039A	OEM
PD5068-A	OEM
PD5068A	OEM
PD5115	OEM
PD6000	0760
PD6000A	0760
PD6001	0118
PD6001A	0118
PD6002	0299
PD6002A	0188
PD6003	0188
PD6003A	0188
PD6004	0188
PD6004A	0456
PD6005	0275
PD6005A	0533
PD6006	0162
PD6006A	0451
PD6007	0162
PD6007A	0162
PD6008	0157
PD6008A	0157
PD6009	0471
PD6009A	0471
PD6010	0062
PD6010A	0062
PD6011	0077
PD6011A	0077
PD6012	0165
PD6012A	0165
PD6013	0057
PD6013A	0057
PD6014	0170
PD6014A	0170
PD6015	0181
PD6015A	0181
PD6016	0052
PD6016A	0052
PD6017	0053
PD6017A	0053
PD6018	0686
PD6018A	0686
PD6019	0445
PD6019A	0445
PD6020	0375
PD6020A	0375
PD6037	OEM
PD6041	0760
PD6042	0118
PD6043	0299
PD6044	0188
PD6045	0456
PD6046	0275
PD6047	0451
PD6048	0162
PD6049	0157
PD6049C	0157
PD6050	0636
PD6051	0062
PD6052	0077
PD6053	0165
PD6054	0057
PD6055	0170
PD6056	0181
PD6057	0052
PD6058	0053
PD6059	0686
PD6060	0445
PD6061	0375
PD6064A	OEM
PD6105	OEM
PD6105C	OEM
PD6201	0275
PD6202	0140
PD6209	0057
PD6336C	OEM
PD7507G-514	OEM
PD9093-59	0358
PD9094-59	1632
PD9097-59	1481
PD9099-59	0333
PD9930-59	1823
PD9932-59	1041
PD9933-59	2097
PD9935-59	1176
PD9936-59	1831
PD9937-59	1835
PD9944-59	0033
PD9945-59	0081
PD9946-59	0141
PD9948-59	3399
PD9949-59	1844
PD9950-59	2619
PD9951-59	2078
PD9961-59	1859
PD9962-59	0562
PD9963-59	0341
PDB004	OEM
PDE001	OEM
PDE003	OEM
PDE009	OEM

DEVICE TYPE	REPL CODE
PDE018	OEM
PDE018B	OEM
PDE024	OEM
PDG002B	OEM
PDG005C	OEM
PDG013-A	OEM
PDG017-A	OEM
PDG017A	OEM
PDG023-A	OEM
PDG023A	OEM
PDG035	OEM
PDG037	OEM
PDG041	OEM
PDJ002	OEM
PDR25	1335
PDR50	1335
PE40	0473
PE55ST3	OEM
PE210	0224
PE210C	0224
PE254A	0224
PE254B	0224
PE254C	0224
PE401	0015
PE401N	0015
PE402	0015
PE403	0015
PE404	0015
PE405	0015
PE406	0015
PE407	0071
PE408	0071
PE409	0071
PE410	0071
PE501	0015
PE502	0015
PE503	0015
PE504	0015
PE505	0015
PE506	0015
PE507	0071
PE508	0071
PE509	0071
PE510	0071
PE3001	0079
PE3002	0007
PE3015	0079
PE3100	0418
PE5010	0079
PE5013	0079
PE5015	0079
PE5025	0330
PE5029	0418
PE5030	0155
PE5030A	0155
PE5030B	0155
PE5031	0127
PE8050B	0018
PE8550B	0532
PE9093-59	0358
PE9094-59	1632
PE9097-59	1481
PE9099-59	0333
PE9930-59	1823
PE9932-59	1041
PE9933-59	2097
PE9935-59	1176
PE9936-59	1831
PE9937-59	1835
PE9944-59	0033
PE9945-59	0081
PE9946-59	0141
PE9948-59	3399
PE9949-59	1844
PE9950-59	2619
PE9951-59	2078
PE9961-59	1859
PE9962-59	0562
PE9963-59	0341
PEB79	OEM
PEL1009	OEM
PEL1018	OEM
PEL1021	OEM
PEL1022	OEM
PEP2	0016
PEP5	0016
PEP6	0016
PEP7	0016
PEP8	0016
PEP9	0016
PEP95	0400
PEP1001	0144
PEP2001	0086
PET0404	0861
PET0404-1	OEM
PET0404-2	0861
PET-101-1	0144
PET1001	0595
PET1002	0016
PET1075	0144
PET1075A	0233
PET2001	0016
PET2002	0016
PET3001	0144
PET3002	0007
PET3702	0037
PET3703	0037
PET3704	0016
PET3705	0079

DEVICE TYPE	REPL CODE
PET3706	0016
PET3903	0079
PET3904	0079
PET3905	0037
PET3906	0037
PET4001	0016
PET4002	0016
PET4003	0111
PET4058	0037
PET4059	0037
PET4060	0037
PET4061	0037
PET4062	0037
PET4123	OEM
PET4124	0079
PET4125	0037
PET4126	0037
PET6001	0016
PET6002	0016
PET6003	1480
PET8000	0016
PET8001	0016
PET8002	0016
PET8003	0016
PET8004	0016
PET8005	0016
PET8005A	0016
PET8006	0016
PET8006A	0016
PET8007	0016
PET8007A	0016
PET8101	0007
PET8200	0086
PET8201	0144
PET8202	0086
PET8203	0086
PET8204	0155
PET8250	0144
PET8251	0144
PET8300	0144
PET8301	0037
PET8302	OEM
PET8303	0037
PET8304	0037
PET8350	OEM
PET8351	OEM
PET8352	OEM
PET8353	OEM
PET9001	0079
PET9001A	OEM
PET9002	0016
PET9002A	OEM
PET9003	OEM
PET9004	OEM
PET9021	0086
PET9022	0086
PF110R	OEM
PF110S	OEM
PF125R	OEM
PF125S	OEM
PF140R	OEM
PF140S	OEM
PF310R	OEM
PF310S	OEM
PF325R	OEM
PF325S	OEM
PF340R	OEM
PF340S	OEM
PF510	OEM
PF810R	OEM
PF820R	OEM
PF825R	OEM
PFB1400	OEM
PFIGD	OEM
PFN3066	OEM
PFN3069	OEM
PFN3458	OEM
PFR851	3549
PFR856	0071
PFWB2	OEM
PFZ6V8	3114
PFZ6V8A	3114
PFZ13	3172
PFZ13A	3172
PFZ18	3191
PFZ18A	3191
PFZ20	3201
PFZ20A	3201
PFZ27	1915
PFZ27A	1915
PFZ51	0568
PFZ51A	0568
PFZ68	0831
PFZ68A	0831
PFZ200	1405
PFZ200A	1405
PG010	OEM
PG107A	0554
PG152R	OEM
PG207A	0554
PG307A	0554
PG310A	0005
PG322A	0628
PG333A	0030
PG2232SY-B1	OEM
PG3433SX	OEM
PG4001	1335
PG4002	0080
PG4003	0609

DEVICE TYPE	REPL CODE
PG4004	0795
PG4005	0015
PG4006	0072
PG4007	0071
PG5531K	OEM
PG5533KY-GR	OEM
PG5553K-RE	OEM
PG5553KY	OEM
PG5628SY	OEM
PG5638SY	OEM
PG5724SY	OEM
PG6100A	OEM
PG6100B	OEM
PG6100C	OEM
PG6100D	OEM
PG6101A	OEM
PG6101B	OEM
PG6101C	OEM
PG6101D	OEM
PG6150A	OEM
PG6150B	OEM
PG6150C	OEM
PG6150D	OEM
PG6151A	OEM
PG6151B	OEM
PG6151C	OEM
PG6151D	OEM
PG6220A	OEM
PG6220B	OEM
PG6220C	OEM
PG6220D	OEM
PG6221A	OEM
PG6221B	OEM
PG6221C	OEM
PG6221D	OEM
PG6330A	OEM
PG6330B	OEM
PG6330C	OEM
PG6330D	OEM
PG6331A	OEM
PG6331B	OEM
PG6331C	OEM
PG6331D	OEM
PG6470A	OEM
PG6470B	OEM
PG6470C	OEM
PG6470D	OEM
PG6680A	OEM
PG6680B	OEM
PG6680C	OEM
PG6680D	OEM
PG22225X	OEM
PGM1	OEM
PGM2	OEM
PGMI	OEM
PGR-24	0242
PH-1	OEM
PH9-221	0277
PH9D5	0071
PH9D522/1	0302
PH9D522/11	0296
PH9D522M	0277
PH9D5221	0277
PH9DS22	0015
PH9DS221	0277
PH11C22/1	0540
PH25C22	0015
PH25C22/1	0015
PH25C22/21	0015
PH25C221	0277
PH-70	OEM
PH101	OEM
PH101T	OEM
PH103	OEM
PH108	0102
PH109	0071
PH110	OEM
PH204	0015
PH208	0015
PH244N	OEM
PH-302	2628
PH302	2628
PH302B	OEM
PH302B(1)	OEM
PH309	OEM
PH310	OEM
PH310A-C	OEM
PH404	0015
PH421-6	1369
PH1021	0015
PH2222A	0079
PH2907A	0037
PH6659	OEM
PH6660	OEM
PH6661	OEM
PIC	0037
PIC500	OEM
PIC501	OEM
PIC502	OEM
PIC1640	OEM
PIC1650	OEM
PIK	0160
PIL/4956	0050
PIL74956	0050
PIP8.4	OEM
PIP24	OEM
PIP30	OEM
PIP60	OEM
PIP120	OEM

DEVICE TYPE	REPL CODE
PIP208	OEM
PIP250	OEM
PIP440	OEM
PIP500	OEM
PJVD05AZ12	OEM
PJVD05AZ24	OEM
PJVIBA12003	OEM
PJVIHD63B3XP	OEM
PJVIL78LR05C	OEM
PJVIM70H036	OEM
PJVIM5165LL	OEM
PJVIM7032H	OEM
PJVIM54532P	OEM
PJVIMSM5299A	OEM
PJVISED1330F	OEM
PJWI1W1000M	OEM
PJWI2XR430M	OEM
PJWI2XR440M	OEM
PJWIR440M	OEM
PJWIXR340M	OEM
PJWIXR350M	OEM
PJWIXR430M	OEM
PL3V6Z	OEM
PL4V3Z	0533
PL6V2Z	0636
PL9V1Z	0012
PL12Z	0137
PL15Z	0002
PL51Z	0502
PL56Z	0869
PL62Z	0783
PL68Z	1268
PL75Z	1189
PL82Z	0331
PL91Z	1311
PL100Z	0098
PL102	1714
PL110Z	0149
PL130Z	0213
PL-150-001-9-00	0143
PL-150-001-9-005	0019
PL-150-006-9-001	0123
PL-151-032-9-004	0120
PL-151-035-9-001	0133
PL-151-040-9-001	0133
PL-151-040-9-002	0133
PL-151-040-9-003	0015
PL-151-045-9-001	0015
PL-151-045-9-002	0133
PL-151-045-9-003	0143
PL-151-045-9-004	0015
PL-152-044-9-001	0012
PL-152-051-9-001	0012
PL-152-052-9-002	0091
PL-152-054-9-001	0244
PL-154-001-9-001	0030
PL160Z	0255
PL-172-010-9-001	0042
PL-172-014-9-001	0042
PL-172-014-9-002	0622
PL-172-024-9-004	0390
PL-176-025-9-001	0076
PL-176-029-9-001	0016
PL-176-029-9-002	0698
PL-176-042-9-001	0155
PL-176-042-9-002	0076
PL-176-042-9-003	0076
PL-176-042-9-004	0111
PL-176-042-9-005	0219
PL-176-042-9-006	0191
PL-176-047-9-002	0368
PL-176-049-9-002	0076
PL-177-006-9-002	0006
PL180Z	0367
PL-182-014-9-002	3046
PL-182-014-9-002(FET	0325
PL-307-047-9-001	0417
PL-307-047-9-002	0823
PL1021	0007
PL1022	0007
PL1023	0007
PL1024	0127
PL1025	0127
PL1026	0127
PL1031	0037
PL1032	0861
PL1033	0037
PL1033-1	OEM
PL1034	0037
PL1036-1	OEM
PL1051	0144
PL1052	0016
PL1053	0144
PL1054	0016
PL1055	0144
PL1061	0144
PL1062	0144
PL1063	0144
PL1064	0144
PL1065	0144
PL1066	0007
PL1067	0007
PL1068	OEM
PL1081	0144
PL1082	0144
PL1083	0086
PL1084	0086
PL1085	0285
PL1091	0325

If replacement code is OEM, contact original manufacturer for replacement.

DEVICE TYPE	REPL CODE
PL1092	0325
PL1093	0325
PL1094	0325
PL1101	0037
PL1102	0037
PL1103	0037
PL1104	0037
PL1111	0007
PL1112	0007
PL1113	0127
PL4001	OEM
PL4001L	1335
PL4002	OEM
PL4002L	OEM
PL4003	OEM
PL4003L	0609
PL4004	OEM
PL4004L	0795
PL4005	OEM
PL4005L	OEM
PL4006	OEM
PL4006L	0072
PL4007	OEM
PL4007L	0373
PL4021	0079
PL4023	0418
PL4031	0150
PL4032	0150
PL4033	0150
PL4034	0150
PL4051	0079
PL4052	0079
PL4053	0079
PL4054	0079
PL4055	0079
PL4061	0111
PL4062	0111
PL4112	0418
PL20015	3770
PLA552	OEM
PLE-48	0042
PLE202	1591
PLE534N	OEM
PLL01A	OEM
PLL02	1714
PLL02A	1714
PLL02A-C	1714
PLL02A-F	1714
PLL02A-G	4224
PLL02AG	4224
PLL03	3559
PLL03A	3559
PLQ0.8	0080
PLR501	OEM
PLR1001	OEM
PLR2001	OEM
PLR3001	OEM
PLR4001	OEM
PLS153	OEM
PLS153N	OEM
PLS173	OEM
PLS173N	OEM
PM0001	OEM
PM-3	OEM
PM3	OEM
PM-4	OEM
PM8	OEM
PM-9	OEM
PM9	OEM
PM101-1	4721
PM101-2	4721
PM101-4	4721
PM101-6	4721
PM102-1	4783
PM102-2	4783
PM102-4	4783
PM102-6	4783
PM194	0127
PM195	0127
PM308AP	2246
PM355AJ	3621
PM355J	3621
PM356AJ	3621
PM356J	3621
PM357AJ	3621
PM357J	3621
PM725CJ	2890
PM741CJ	0410
PM741CY	1975
PM1210B	OEM
PM1220B	OEM
PM4550C	OEM
PM5006	0079
PMBT2222	1435
PMBT2222A	1435
PMBT2369	1435
PMBT2907	OEM
PMBT2907A	1499
PMBT3640	OEM
PMBT3903	1435
PMBT3904	1435
PMBT3906	1499
PMBT4401	1435
PMBT4403	1499
PMBT5086	OEM
PMBT5087	OEM
PMBT5088	5385
PMBT5089	OEM
PMBT5401	1503
PMBT5550	2332
PMBT5551	2332
PMBT6427	OEM
PMBT6428	OEM
PMBT6429	OEM
PMBTA05	1435
PMBTA06	1435
PMBTA13	OEM
PMBTA14	OEM
PMBTA20	OEM
PMBTA42	OEM
PMBTA43	OEM
PMBTA55	1499
PMBTA56	1503
PMBTA63	OEM
PMBTA64	OEM
PMBTA70	OEM
PMBTA92	OEM
PMBTA93	OEM
PMC1367P	2916
PMD10K40	2439
PMD10K60	2439
PMD10K80	2439
PMD10K100	2439
PMD11K40	2432
PMD11K60	2432
PMD11K80	2432
PMD11K100	2432
PMD12K40	2427
PMD12K60	2427
PMD12K80	2427
PMD12K100	2439
PMD13K40	2277
PMD13K60	2277
PMD13K80	2277
PMD13K100	2432
PMD16K60	3525
PMD16K80	3525
PMD16K100	3525
PMD17K40	3526
PMD17K60	3526
PMD17K80	3526
PMD17K100	3526
PMK16D40	OEM
PMM8713	OEM
PMP-G12S	OEM
PMT011	0086
PMT012	0086
PMT013	0079
PMT014	0079
PMT015	OEM
PMT016	0086
PMT018	0086
PMT019	0086
PMT020	0086
PMT021	OEM
PMT022	OEM
PMT023	0007
PMT024	0086
PMT025	OEM
PMT111	0079
PMT112	0086
PMT113	0079
PMT114	0086
PMT115	OEM
PMT116	0086
PMT117	OEM
PMT118	0086
PMT119	0086
PMT120	0086
PMT121	OEM
PMT122	OEM
PMT123	OEM
PMT124	OEM
PMT125	OEM
PMT211	OEM
PMT212	OEM
PMT213	OEM
PMT214	OEM
PMT215	OEM
PMT216	OEM
PMT217	OEM
PMT218	OEM
PMT219	OEM
PMT220	OEM
PMT221	OEM
PMT222	OEM
PMT223	OEM
PMT224	OEM
PMT225	OEM
PMT1767	0144
PMT1767M	OEM
PMT1767P	OEM
PMT1767T	OEM
PMT1787M	OEM
PMT1787P	OEM
PMT1787T	OEM
PMTT18501	OEM
PMV107	OEM
PMV107A	OEM
PMV112	OEM
PMV112A	OEM
PMV113	OEM
PMV113A	OEM
PMV114	OEM
PMV114A	OEM
PMV115	OEM
PMV115A	OEM
PMV116	OEM
PMV116A	OEM
PMV117	OEM
PMV117A	OEM
PMV122	OEM
PMV122A	OEM
PMV123	OEM
PMV123A	OEM
PMV124	OEM
PMV124A	OEM
PMV125	OEM
PMV125A	OEM
PMV126	OEM
PMV126A	OEM
PMV127	OEM
PMV127A	OEM
PMV128	OEM
PMV128A	OEM
PMV129	OEM
PMV129A	OEM
PMV130	OEM
PMV130A	OEM
PMV132	OEM
PMV132A	OEM
PMV133	OEM
PMV133A	OEM
PMV134	OEM
PMV134A	OEM
PMV135	OEM
PMV135A	OEM
PMV136	OEM
PMV136A	OEM
PMV137	OEM
PMV137A	OEM
PMV138	OEM
PMV138A	OEM
PMV139	OEM
PMV139A	OEM
PMV140	OEM
PMV140A	OEM
PMV141	OEM
PMV141A	OEM
PN2	OEM
PN26	0178
PN66	0042
PN70	0037
PN71	0037
PN72	0037
PN107	0016
PN108	0079
PN109	0079
PN111W	OEM
PN150	2094
PN150-TV	OEM
PN150NV	2094
PN150SAN	OEM
PN150TV	2094
PN158	OEM
PN158NV	OEM
PN204	0015
PN205	3987
PN268R-NC	OEM
PN302	OEM
PN303	OEM
PN313	2628
PN323	OEM
PN323B-HT	OEM
PN323BHT	OEM
PN350	0103
PN917	0144
PN918	0144
PN929	0111
PN930	0079
PN930(NSC)	0079
PN930(NTLB)	0079
PN1185	OEM
PN1190	OEM
PN1613	0086
PN1711	0086
PN2025	OEM
PN2218A	0739
PN2219A	0739
PN2221	0018
PN2221A	0079
PN2222	0079
PN2222A	0079
PN2369	0079
PN2369A	0079
PN2484	0127
PN2904	0037
PN2904A	0037
PN2905	0150
PN2905A	0037
PN2906	0037
PN2906A	0037
PN2907	0037
PN2907A	0037
PN2908	0037
PN3054	0561
PN3238-HT	OEM
PN3250	0037
PN3250A	0037
PN3251	0037
PN3251A	0037
PN3548	OEM
PN3563	0144
PN3564	0144
PN3565	0079
PN3566	0079
PN3567	0861
PN3568	0861
PN3569	0861
PN3638	0037
PN3638A	0037
PN3639	6594
PN3640	6594
PN3641	0079
PN3642	0079
PN3643	0079
PN3644	0037
PN3645	0037
PN3646	0079
PN3662	0144
PN3663	0144
PN3684	OEM
PN3685	OEM
PN3686	OEM
PN3687	OEM
PN3691	0079
PN3692	0079
PN3693	OEM
PN3694	0079
PN3946	OEM
PN3947	OEM
PN3962	OEM
PN4013	0079
PN4014	OEM
PN4054	0681
PN4121	0037
PN4122	0037
PN4140	0079
PN4141	0079
PN4142	0037
PN4143	0037
PN4248	0037
PN4249	0037
PN4258	0037
PN4275	0079
PN4302	OEM
PN4302-18	OEM
PN4303	OEM
PN4303-18	OEM
PN4304	OEM
PN4304-18	OEM
PN4342	OEM
PN4354	0037
PN4355	0037
PN4356	0037
PN4391	6655
PN4392	OEM
PN4393	OEM
PN4416	0325
PN4423	0037
PN4916	0037
PN4917	OEM
PN5127	0079
PN5128	0079
PN5129	0079
PN5130	0144
PN5131	0079
PN5132	0079
PN5133	0079
PN5134	0079
PN5135	0079
PN5136	0079
PN5137	0079
PN5138	0037
PN5139	0144
PN5142	0037
PN5143	0037
PN5179	0224
PO74LS04	1595
PO74LS368A	0991
PO93	0143
PO117RD01B	OEM
POSN74ALS08N	OEM
POSN74LS14N	1698
POSN74LS123N	0979
POSN74LS138N	0427
POSN74LS157N	1161
POSN74LS161AN	0858
POSN74LS166N	4371
POSN74LS221N	1239
POSN74LS245N	0463
POSN74LS368AN	0991
POSN74LS373N	0709
POSN74LS628N	OEM
POSN74S153N	2151
POSN7407N	1339
POSN7416N	1349
POSN7417N	1352
POSN7445N	1082
POSN74145N	0619
POSN74159N	OEM
POWER-12	0160
POWER12	0160
POWER-25	0160
POWER25	0160
POWER-40	0440
POWER40	0440
POWER-60	0440
POWER60	0440
POWER-80	0440
POWER80	0440
POWER-99	0160
POWER99	0160
POWER-299	0160
POWER299	0160
POWER500	0440
PP74F153	OEM
PP601-1	OEM
PP3000	0103
PP3001	0130
PP3002	0130
PP3003	0103
PP3004	0130
PP3006	0103
PP3007	0130
PP3008	0177
PP3083	0561
PP3084	0561
PP3085	0561
PP3086	0561
PP3087	0561
PP3088	0561
PP3250	0042
PP3310	0042
PP3312	0042
PP3647	0130
PPC1	OEM
PPR1006	0560
PPR1008	0560
PPT720	0016
PQ05R04	OEM
PQ05R041	OEM
PQ09RA1	OEM
PQ09RF11	OEM
PQ09RF21	OEM
PQ27	0050
PQ28	0164
PQ29	0164
PQ30	0050
PQ31	0085
PQ09RF21	OEM
PR3-3	0222
PR515	0162
PR605	0162
PR617	0137
PR620	0002
PR804	0162
PR9010R	0550
PR1005	7153
PR3005	2280
PR3423S	OEM
PR3433S	OEM
PR4524K	OEM
PR5400	0087
PR5401	0087
PR5402	0087
PR5403	0087
PR5404	0087
PR5405	0087
PR5406	0087
PR5407	0087
PR5408	0087
PR5505S	OEM
PR5532B	OEM
PR5533K-RE	OEM
PR-5534S	5666
PR5534S	5666
PR5553K	OEM
PR5608S	OEM
PR5628S	OEM
PR5638S	OEM
PR5704S	OEM
PR5705S-BS	OEM
PR5724S	OEM
PR9000	2088
PR9000R	1138
PR9001	2088
PR9001R	1138
PR9002	0612
PR9002R	1188
PR9003	0612
PR9003R	1188
PR9004	0206
PR9004R	1008
PR9005	0206
PR9005R	1008
PR9006	0586
PR9006R	0948
PR9007	0586
PR9007R	0948
PR9008	0131
PR9008R	0545
PR9009	0131
PR9009R	0545
PR9010R	0550
PR9011	0145
PR9011R	0550
PR9012	1132
PR9012R	1119
PR9013	1132
PR9013R	1119
PR9014	0044
PR9014R	0281
PR9015	0044
PR9015R	0281
PR9023	1132
PR9024	1132
PR9025	0044
PR9026	0044
PR9034	3762
PR9034R	2665
PR9035	2657
PR9035R	2768
PR9036	2657
PR9036R	2768
PR9037	2006
PR9037R	2907
PR9038	2006
PR9038R	2907
PR9039	2682
PR9039R	3051
PR9040	2682
PR9040R	3051
PR9041	3900
PR9041R	3117
PR9042	3900
PR9042R	3117
PR9043	2655
PR9043R	3208
PR9044	2655
PR9044R	3208
PRB738	OEM
PRB8128	OEM
PRC2A	0240
PRC5A	0240
PRC10A	0240
PRC15A	0240
PRC20A	0240
PRF851	OEM
PRO-050	OEM
PRO5	OEM
PRS3017	0162
PRT-38PT3F(M)	OEM
PRT-101	0016
PRT101	0233
PRT-104	0016
PRT-104-1	0016
PRT-104-2	0016
PRT-104-3	0016
PRT-104-4	0086
PRV94	OEM
PRV100	OEM
PS	OEM
PS005	0015
PS08	1651
PS010	0015
PS015	0015
PS020	0015
PS025	0015
PS030	0015
PS035	0015
PS040	0015
PS050	0015
PS060	0015
PS0125	0015
PS-1	0160
PS1	0160
PS1A	0023
PS-1.5-2	0071
PS12	OEM
PS18	1651
PS24	OEM
PS28	1651
PS28(SCR)	1651
PS38	1584
PS48	1584
PS58	1665
PS68	1665
PS105	0015
PS110	0015
PS120	0015
PS125	0015
PS130	0015
PS135	0015
PS140	0015
PS150	0015
PS152R	0102
PS152RL	0071
PS160	0015
PS220	1651
PS320	1584
PS330	2647
PS405	0015
PS410	0015
PS415	0015
PS420	0015
PS425	0015
PS430	0015
PS435	0015
PS440	0015
PS450	0015
PS460	0015
PS520	1665
PS530	2649
PS603	0015
PS604	0015
PS605	0015
PS609	0015
PS610	0015
PS611	0015
PS615	0015
PS616	0015
PS617	0015
PS620	1665
PS621	0015
PS622	0015
PS623	0015
PS627	0015
PS628	0015
PS629	0015
PS630	2649
PS632	0015
PS633	0015
PS636	0015
PS637	0015
PS1140	0071
PS1325	0471
PS1511	0170
PS1512	0170
PS1513	0170
PS1514	0170
PS1515	0170
PS1516	0170
PS1517	0170
PS2001B	0273
PS2006B	1291
PS2021	0315
PS2207	0080
PS2208	0080
PS2209	0080
PS2247	0015
PS2249	0015
PS2346	0087
PS2347	0087
PS2401-2	OEM
PS2411	0015
PS2412	0015
PS2413	0015
PS2414	OEM
PS2415	0015
PS2416	0071
PS2417	0071
PS2501	OEM
PS2501-1	OEM
PS2501-1LB	OEM
PS3534	2874
PS3534A	2874
PS4005	OEM
PS4559	0080
PS4560	0795
PS4725	0080
PS5175	0330
PS5300	0795
PS5301	0795
PS5302	0795
PS5303	0015
PS5304	0015
PS6325	0331
PS6326	OEM
PS6327	OEM
PS6465	0760
PS6466	OEM
PS6467	OEM
PS6468	0162
PS6469	3127
PS6470	OEM
PS7267	OEM
PS7268	OEM
PS7269	OEM
PS7270	OEM
PS8900	1713
PS8901	0299
PS8902	0376
PS8903	0036
PS8904	0275
PS8905	0451
PS8906	0162
PS8907	0157
PS8908	0471
PS8909	0062
PS8910	0649
PS8911	0165
PS8912	0057
PS8913	0170
PS8914	0317
PS8915	0052
PS8916	0365
PS8917	0686
PS8918	0420
PS8919	0375
PS10016B	0025
PS10017B	0649
PS10018B	0244
PS10019B	0012
PS10020B	0170
PS10021B	0317
PS10022B	0137
PS10023B	0365
PS10024B	0002
PS10025B	0365
PS10026B	0495
PS10060	0025
PS10061	0649
PS10062	0649
PS10063	0012
PS10064	0170
PS10065	0317
PS10066	0137
PS10067	0365
PS10068	0002
PS10070	0495
PS25011LB	OEM
PSA06	0861
PSA20	0079
PSA56	0037
PSA70	0037
PSIB110	OEM
PSIBD125	0599
PSIBD150	0599
PSIC160	OEM
PSIC235	OEM
PSICD160	1023
PSICD250	1023

DEVICE TYPE	REPL CODE
PSIF180	OEM
PSIF220	OEM
PSIF300	OEM
PSIF400	OEM
PSIF500	OEM
PSIF600	OEM
PSIFD600	OEM
PSIFD750	OEM
PSIFD900	OEM
PSIFT1000	OEM
PSIG300	OEM
PSIG400	OEM
PSIG500	OEM
PSIG650	OEM
PSIG850	OEM
PSIG950	OEM
PSIGD800	OEM
PSIGD1400	OEM
PSIGD1500	OEM
PSIH800	OEM
PSIH1000	OEM
PSIH1200	OEM
PSIH1400	OEM
PSIH1600	OEM
PSIH1600-16	OEM
PSIH1600-18	OEM
PSIH1800	OEM
PSIH1800-8	OEM
PSIH1800-10	OEM
PSIH1800-12	OEM
PSIH1800-14	OEM
PSIH1800-16	OEM
PSIH1800-18	OEM
PSIH2000	OEM
PSIH2000-1	OEM
PSIH2000-2	OEM
PSIH2000-3	OEM
PSIH2000-4	OEM
PSIH2000-5	OEM
PSIH2000-6	OEM
PSIH2300-1	OEM
PSIH2300-2	OEM
PSIH2300-3	OEM
PSIH2300-4	OEM
PSIHD1500	OEM
PSIHD2000	OEM
PSIHD2500	OEM
PSIHD3000	OEM
PSIHH1400-1	OEM
PSIHH1400-2	OEM
PSIHH1400-3	OEM
PSIHH1400-4	OEM
PSIHH1400-5	OEM
PSIHH1400-6	OEM
PSIHH1400-7	OEM
PSIHH1400-8	OEM
PSIHH1400-10	OEM
PSIHH1400-12	OEM
PSIHH1600-1	OEM
PSIHH1600-2	OEM
PSIHH1600-3	OEM
PSIHH1600-4	OEM
PSIHH1600-5	OEM
PSIHH1600-6	OEM
PSIHH1600-7	OEM
PSIHH1600-8	OEM
PSIHH1600-10	OEM
PSIHH1600-12	OEM
PSIHH1800-1	OEM
PSIHH1800-2	OEM
PSIHH1800-3	OEM
PSIHH1800-4	OEM
PSIHH1800-5	OEM
PSIHH1800-6	OEM
PSIHH1800-7	OEM
PSIHH1800-8	OEM
PSIHH1800-10	OEM
PSIHH1800-12	OEM
PSIHH2000-1	OEM
PSIHH2000-2	OEM
PSIHH2000-3	OEM
PSIHH2000-4	OEM
PSIHH2000-5	OEM
PSIHH2000-6	OEM
PSIHH2000-7	OEM
PSIHH2000-8	OEM
PSIHH2000-10	OEM
PSIHH2000-12	OEM
PSIHH2200-1	OEM
PSIHH2200-2	OEM
PSIHH2200-3	OEM
PSIHH2200-4	OEM
PSIHH2200-5	OEM
PSIHH2200-6	OEM
PSIHH2200-7	OEM
PSIHH2200-8	OEM
PSIHH2200-10	OEM
PSIHH2200-12	OEM
PSIJ3900-1	OEM
PSIJ3900-2	OEM
PSIJ3900-3	OEM
PSIJ3900-4	OEM
PSIJ3900-5	OEM
PSIJ3900-6	OEM
PSIJ4000-1	OEM
PSIJ4000-2	OEM
PSIJ4000-3	OEM
PSIJ4000-4	OEM
PSIJ4000-5	OEM

DEVICE TYPE	REPL CODE
PSIJ4000-6	OEM
PSIJ4200-1	OEM
PSIJ4200-2	OEM
PSIJ4200-3	OEM
PSIJ4200-4	OEM
PSIJ4200-5	OEM
PSIJ4200-6	OEM
PSIJ4900-1	OEM
PSIJ4900-2	OEM
PSIJ4900-3	OEM
PSIJ4900-4	OEM
PSIK2500-2	OEM
PSIK2500-4	OEM
PSIK2500-6	OEM
PSIK3500-2	OEM
PSIK3500-4	OEM
PSIK3500-6	OEM
PSL11A	OEM
PSL12A	OEM
PST518	OEM
PST519A	OEM
PST520	OEM
PST520E	OEM
PST520G	OEM
PST523C	OEM
PST523D	OEM
PST523E	OEM
PST523E-2	OEM
PST523G	OEM
PST523H-2	OEM
PST524D	OEM
PST524D2	OEM
PST529C	OEM
PST529C2	OEM
PST529D	OEM
PST529D-2	OEM
PST529E	OEM
PST529H2	OEM
PST532	OEM
PST532A	OEM
PST572D-2	OEM
PSU06	0264
PSV111-08	0005
PSV111-14	0005
PSV111-16	0005
PSV114-06	0628
PSV114-14	0628
PSV123-08	0005
PSV123-14	0005
PSV123-16	0005
PSV126-06	0628
PSV126-14	0628
PT06	0160
PT08	1064
PT010	1064
PT015	1064
PT025	1064
PT-029	0042
PT030	1064
PT040	1064
PT0139	0164
PT2A	0050
PT2S	0050
PT-3	0015
PT3	0015
PT-3A	0160
PT3A	0160
PT4	2763
PT5	0015
PT-5B	0015
PT5B	0015
PT-6	0160
PT6	0160
PT6D22/1	0277
PT6D22-1	0277
PT9C22/1	0071
PT-12	0160
PT12	0160
PT13	OEM
PT16	1064
PT18	1064
PT23	OEM
PT23F	OEM
PT23F/WITHHOLDER	OEM
PT-23F-F	5695
PT-23F-HLD	3185
PT-25	0160
PT25	0160
PT26	1064
PT28	1064
P-T-30	0160
PT30	0160
PT32	0222
PT36	1317
PT38	1317
PT-40	0160
PT40	0160
PT46	1317
PT48	1317
PT-50	0160
PT50	0160
PT52	OEM
PT53	OEM
PT56	1891
PT58	1891
PT60	OEM
PT66	1891
PT68	1891
PT110	1064

DEVICE TYPE	REPL CODE
PT115	1064
PT125	1064
PT130	1064
PT140	1064
PT-150	0160
PT150	0160
PT-155	0160
PT155	0160
PT-176	0160
PT176	0160
PT200	OEM
PT201	0440
PT210	1064
PT215	1064
PT225	1064
PT230	1064
PT-234	0160
PT234	0160
PT-235	0160
PT235	0160
PT-235A	0160
PT235A	0085
PT-236	0160
PT236	0085
PT-236A	0160
PT236A	0085
PT-236B	0160
PT236B	0160
PT236C	0160
PT240	1064
PT-242	0160
PT242	0160
PT250	0440
PT-255	0160
PT255	0160
PT-256	0160
PT256	0160
PT-285	0160
PT285	0160
PT-285A	0160
PT285A	0160
PT-301	0160
PT301	0160
PT-301A	0160
PT301A	0160
PT-307	0160
PT307	0160
PT-307A	0160
PT307A	0160
PT310	1317
PT315	1317
PT320	OEM
PT325	1317
PT330	1317
PT336B	0085
PT340	1317
PT-350C	OEM
PT-350D	OEM
PT360FS	OEM
PT366B	0160
PT410	1317
PT415	1317
PT421FH	OEM
PT421FI-F	OEM
PT425	1317
PT430	1317
PT431	OEM
PT440	1317
PT481F	OEM
PT483F1	OEM
PT483FI	OEM
PT501	0160
PT-505	0015
PT505	0015
PT-510	0015
PT510	0015
PT-515	0015
PT515	0440
PT515(SEMITRON)	0437
PT516	OEM
PT517	OEM
PT518	OEM
PT519	OEM
PT-520	0015
PT520	0015
PT522	OEM
PT523	OEM
PT-525	0015
PT525	0015
PT-530	0015
PT530	0015
PT530-1	OEM
PT-530A	0280
PT530A	0004
PT531	0630
PT-540	0015
PT540	0015
PT-550	0015
PT550	0015
PT-554	0160
PT554	0085
PT-555	0160
PT555	0160
PT-560	0015
PT560	0015
PT580	0071
PT600	0560
PT601	0560
PT610	1891

DEVICE TYPE	REPL CODE
PT612	0086
PT615	1891
PT622/1	0302
PT625	1891
PT627	0016
PT630	1891
PT640	1891
PT657	OEM
PT665	0042
PT703	0016
PT706	OEM
PT706-1	OEM
PT706A	OEM
PT706A-1	OEM
PT709	OEM
PT720	0016
PT801	OEM
PT802	OEM
PT822	OEM
PT850	0086
PT850A	0086
PT851	0016
PT852	OEM
PT853	OEM
PT855	0050
PT856	0050
PT886	0016
PT887	0016
PT888	0086
PT896	0086
PT897	0016
PT898	0016
PT900	OEM
PT900-1	OEM
PT901	OEM
PT901-1	OEM
PT902	OEM
PT902-1	OEM
PT903	OEM
PT903-1	OEM
PT1515	0168
PT1544	0086
PT1545	0086
PT1558	0016
PT1559	0016
PT1610	0016
PT1835	0016
PT1836	0016
PT1837	0016
PT1941	0103
PT2040A	7064
PT2523	0285
PT2524	0285
PT2525	0285
PT2525A	0168
PT2540	0595
PT2575	0016
PT2600	OEM
PT2620	0560
PT2630	0168
PT2634	OEM
PT2635	0042
PT2640	0560
PT2660	0560
PT2670	0168
PT-2677C	2170
PT2677C	0698
PT2760	0016
PT3141	0016
PT3141A	0016
PT3141B	0016
PT3151A	0079
PT3151B	0016
PT3151C	0016
PT3473	OEM
PT3500	0016
PT3501	3326
PT3502	0086
PT3503	0560
PT3539	OEM
PT3540	OEM
PT3760	OEM
PT4144	OEM
PT4579	0418
PT4690	0560
PT4800	0016
PT4816	0127
PT4830	0127
PT5690	OEM
PT5692	OEM
PT5693	0042
PT5694	OEM
PT6001	OEM
PT6007	OEM
PT6011	OEM
PT6021	OEM
PT6022/1	0302
PT6618	0560
PT6669	0560
PT6696	0560
PT6905	OEM
PT6905A	OEM
PT6905B	OEM
PT6905C	OEM
PT6941	0620
PT6942	0620
PT6994	0620
PT6995	0620
PT6996	2485

DEVICE TYPE	REPL CODE
PT7015	OEM
PT7016	OEM
PT7903	0620
PT7904	0620
PT7905	0620
PT7906	0620
PT7907	0620
PT7908	0620
PT7909	2485
PT7910	2485
PT7930	0130
PT7931	0130
PT8809	3560
PT8810	2091
PT8811	1420
PT8838	1974
PT8874F	2884
PT9784	2312
PT9795	2312
PT9796	2312
PT9796A	1974
PT9799	2312
PT9803	2312
PT-9816	1974
PT31961	0698
PT72130	0015
PT-72130-1	0540
PT1011601-A	OEM
PTC101	0086
PTC102	0280
PTC103	0037
PTC104	0142
PTC105	0160
PTC105A	0160
PTC106	0440
PTC107	0050
PTC108	0600
PTC109	0211
PTC110	0161
PTC111	0460
PTC112	0178
PTC113	0905
PTC114	0085
PTC115	0079
PTC116	0130
PTC117	0233
PTC118	0074
PTC119	0103
PTC120	0222
PTC121	0079
PTC122	0975
PTC123	1480
PTC124	1708
PTC125	0861
PTC126	0007
PTC127	0861
PTC128	0622
PTC128-RT	OEM
PTC129	0642
PTC129A	0065
PTC130	0003
PTC131	0037
PTC132	0007
PTC133	2525
PTC134	0208
PTC135	0164
PTC136	0016
PTC137	0561
PTC137-RT	OEM
PTC138	0604
PTC139	0079
PTC140	0103
PTC141	0382
PTC142	1267
PTC143	0560
PTC144	0630
PTC145	0136
PTC146	0065
PTC147	0604
PTC148	0178
PTC149	0491
PTC150	0338
PTC151	3712
PTC152	0325
PTC153	0400
PTC154	0423
PTC155	0164
PTC156	0164
PTC157	0925
PTC158	0698
PTC160	0050
PTC161	0325
PTC162	0925
PTC163	0042
PTC164	1199
PTC165	0561
PTC166	0854
PTC167	0423
PTC168	1199
PTC169	0423
PTC170	0715
PTC171	0342
PTC172	1681
PTC173	0130
PTC174	1681
PTC175	0130
PTC176	0630
PTC177	0791
PTC178	0595

DEVICE TYPE	REPL CODE
PTC179	0142
PTC180	1591
PTC181	0212
PTC182	0212
PTC183	0849
PTC184	0212
PTC186	0836
PTC191	0698
PTC192	2046
PTC193	2046
PTC194	0222
PTC195	0236
PTC196	0681
PTC197	0551
PTC198	0382
PTC201	0540
PTC202	0015
PTC203	0071
PTC204	0071
PTC205	0071
PTC206	0143
PTC206M	0123
PTC207	0143
PTC207M	0143
PTC208	0774
PTC209	1303
PTC210	0201
PTC211	0288
PTC212	0757
PTC213	0757
PTC214	0162
PTC214M	0133
PTC215	0162
PTC216	0102
PTC217	0917
PTC218	0293
PTC219	OEM
PTC301	0139
PTC302	0120
PTC311	0005
PTC401	0296
PTC402	0302
PTC403	0015
PTC404	0474
PTC405	2966
PTC406	0913
PTC407	0196
PTC501	0188
PTC502	0157
PTC503	0471
PTC504	0649
PTC505	0057
PTC506	0170
PTC507	0052
PTC508	0365
PTC509	0686
PTC510	0495
PTC511	0100
PTC512	0039
PTC513	0869
PTC514	0783
PTC515	0331
PTC516	0149
PTC517	0350
PTC518	0999
PTC601	1997
PTC602	0128
PTC603	1501
PTC604	1030
PTC605	1030
PTC606	4465
PTC607	OEM
PTC608	1710
PTC651	0901
PTC658	OEM
PTC660	OEM
PTC665	OEM
PTC666	OEM
PTC701	0664
PTC703	0678
PTC705	0666
PTC707	0753
PTC708	2161
PTC709	0654
PTC711	1940
PTC713	0443
PTC715	0352
PTC717	2279
PTC719	0354
PTC721	0701
PTC723	1345
PTC726	0167
PTC728	0349
PTC729	1395
PTC730	0802
PTC731	2457
PTC732	2557
PTC733	1337
PTC734	1421
PTC735	0520
PTC736	0360
PTC737	1443
PTC738	2557
PTC739	0751
PTC740	5320
PTC741	0856
PTC742	2557
PTC743	0632
PTC744	0823

DEVICE TYPE	REPL CODE
PTC745	0789
PTC746	0395
PTC747	5321
PTC754	0878
PTC756	OEM
PTC757	2124
PTC758	OEM
PTC759	OEM
PTC762	OEM
PTC765	0692
PTC767	2755
PTC769	0521
PTC776	0328
PTC780	1050
PTC781	0911
PTC787	2755
PTC801	OEM
PTC802	OEM
PTC803	OEM
PTC804	OEM
PTC805	OEM
PTC806	OEM
PTC809	OEM
PTC810	OEM
PTC904	0566
PTC905	1367
PTC906	0236
PTC1101	0785
PTC1123	OEM
PTC1423	2427
PTC1623	0785
PTC1645	OEM
PTC1667	OEM
PTC1689	OEM
PTC1701	OEM
PTC1723	OEM
PTC1745	0212
PTC1778	OEM
PTC1956	2747
PTC1978	0382
PTC6000	OEM
PTC6001	OEM
PTC6002	OEM
PTC6003	OEM
PTC6060	2626
PTC6061	2626
PTC6062	OEM
PTC6063	OEM
PTC7000	4013
PTC7001	4013
PTC7002	4013
PTC7003	4013
PTC8080	OEM
PTC8081	OEM
PTC8082	OEM
PTC8083	OEM
PTC9045	OEM
PTCL11	OEM
PT-H3	OEM
PTH631D01	OEM
PTIN141S	OEM
PT-L3	OEM
PT-M3	OEM
PTO-6	0160
PT0139	0004
PTS520G	OEM
PTSN7417N	1352
PU2	OEM
PU6C22	0015
PU42C26	OEM
PU42C26NC	OEM
PU131	0316
PU4120	OEM
PU-4120A	OEM
PU4120A	OEM
PU4120B	OEM
PU4124	OEM
PU4320	OEM
PU4411A	OEM
PU4411B	OEM
PU4417	OEM
PU4520	OEM
PU22106A	OEM
PU22198-2A	OEM
PUA3228	OEM
PV-8	0015
PV8	0015
PV509	OEM
PV1043	0124
PV1500	OEM
PV1503	OEM
PV1506	OEM
PV1511	OEM
PVGA-1A	OEM
PW300	OEM
PW400	OEM
PX0016CE	OEM
PX0020GE	OEM
PX0032GE	OEM
PX0207CE	OEM
PX0219CE	OEM
PX0226CE	OEM
PX0238CE	OEM
PX0242CE	OEM
PX5822	OEM
PXB-103	0004
PXB103	0211
PXB-113	0004
PXB113	0211

Original Device Types

DEVICE TYPE	REPL CODE
PXC-101	0004
PXC101	0211
PXC101A	0211
PXC-101AB	0004
PXC101AB	0211
PXT2222A	OEM
PXT2907A	OEM
PXT3904	OEM
PXT3906	OEM
PXT4401	OEM
PXT4403	OEM
PXTA14	OEM
PXTA42	OEM
PXTA64	OEM
PXTA92	OEM
PY-5	0015
PY5608S	OEM
PZ10A	0510
PZ18A	1078
PZ22A	1723
PZ140B	OEM
PZ140D	OEM
PZ140F	OEM
PZ140H	OEM
PZ140K	OEM
PZT8.2	OEM
PZT10	OEM
PZT12	OEM
PZT15	OEM
PZT18	1078
PZT22	1723
PZT27	1761
PZT33	1788
PZT2222A	OEM
PZT2907A	OEM
PZT3904	OEM
PZT3906	OEM
PZTA13	OEM
PZTA14	OEM
PZTA42	OEM
PZTA43	OEM
PZTA63	OEM
PZTA64	OEM
PZTA92	OEM
PZTA93	OEM
Q001	2185
Q-00169C	1280
Q00169C	0325
Q-00184R	1280
Q00184R	0325
Q-00269C	0076
Q-00284R	0127
Q-00369C	0191
Q-00384R	0151
Q-00469C	0076
Q-00484R	0076
Q-00569C	0076
Q-00584R	0151
Q-00669C	1419
Q-00769C	1147
Q-00784R	0076
Q-00869C	0016
Q-00969C	0698
Q-00984R	0037
Q0-419	0037
Q01E	0853
Q02E	0853
Q05B	OEM
Q05C	OEM
Q05D	OEM
Q06B	OEM
Q06C	OEM
Q06D	OEM
Q-0115C	0079
Q-01069C	1411
Q-01084R	0222
Q-01115C	0368
Q-01169C	0086
Q-01184R	0016
Q-01269C	0178
Q-01284R	0622
Q-01369	2057
Q-01384R	1411
Q-02115C	0368
Q-03115C	0368
Q-04115C	0368
Q-05115C	0368
Q-06115C	0368
Q-07115C	0368
Q-08115C	0380
Q-09115C	0380
Q-013696	2057
Q0V60526	0004
Q0V60528	0004
Q0V60538	0016
Q-1	0211
Q1	0004
Q1/1	OEM
Q1/2	OEM
Q1/3	OEM
Q1/4	OEM
Q1/5	OEM
Q1-7C	0004
Q-1A	0280
Q1B	0015
Q1H	0015
Q-1N914	0124
Q-2	0600
Q2	0007

DEVICE TYPE	REPL CODE
Q2-7C	0004
Q2N406	0004
Q2N1526	0050
Q2N2428	0004
Q2N2613	0004
Q2N4105	0038
Q2N4106	0164
Q2N4107	2866
Q-2N5225	0076
Q-2N5226	0037
Q2SC2621D***	0276
Q2T2222	0544
Q2T2905	0282
Q2T3725	0544
Q2Z10A	0064
Q-3	0600
Q3	0007
Q3/1	OEM
Q3/2	0015
Q3/3	OEM
Q3/4	OEM
Q3/5	OEM
Q-4	0211
Q4	0004
Q4B	0015
Q4S4	OEM
Q-5	0600
Q5	0007
Q-6	0004
Q6	0164
Q6/1	OEM
Q6/2	0015
Q6/3	OEM
Q6/4	OEM
Q6/5	OEM
Q-7	0004
Q7	0164
Q7-C0710XBE	0368
Q7-C0735XBT	0191
Q7-C1318XDN	0086
Q7-K0033XBE	1280
Q-8	0004
Q8	0164
Q8/1	OEM
Q8/2	0015
Q8/3	OEM
Q8/4	OEM
Q8/5	OEM
Q-9	0600
Q10	OEM
Q-16	0004
Q16	0164
Q19C	4992
Q19D	7084
Q19E	4994
Q19G	4995
Q19J	4996
Q19L	4997
Q20C	OEM
Q20D	7085
Q20E	7086
Q20G	5002
Q20J	5003
Q20L	5004
Q32	0015
Q-35	0016
Q-36	0037
Q49	0143
Q50	0143
Q51	0143
Q52	1335
Q53	0015
Q54	0015
Q55	0015
Q56	0015
Q57	0015
Q58	0015
Q59	0015
Q59A	OEM
Q60	0015
Q61	0015
Q62	0015
Q70-500	OEM
Q70-500A	OEM
Q80-500	OEM
Q80-500A	OEM
Q80-750	OEM
Q80-950	OEM
Q82	0015
Q90-500	OEM
Q100-500	OEM
Q100-500A	OEM
Q100-750	OEM
Q100-950	OEM
Q110-500	OEM
Q205	0701
Q301	0127
Q402	OEM
Q506R4	OEM
Q510ZQ	0150
Q2001LT	5395
Q2001MS2	1413
Q2001P	2394
Q2001PT	2394
Q2003L3	0555
Q2003L4	5322
Q2003L5	OEM
Q2003LT	5395
Q2004A	0908

DEVICE TYPE	REPL CODE
Q2004B	3199
Q2004F31	2300
Q2004F41	2300
Q2004FT1	OEM
Q2004L3	5322
Q2004L4	0555
Q2004LT	1456
Q2004R4	OEM
Q2004RT	OEM
Q2006	OEM
Q2006A	0908
Q2006AT	0593
Q2006B	3199
Q2006BT	0593
Q2006F41	OEM
Q2006FT1	OEM
Q2006G	1064
Q2006H	0154
Q2006L4	0555
Q2006LT	1456
Q2006N	3199
Q2006R4	5255
Q2006RT	OEM
Q2006T	0593
Q2008	OEM
Q2008A	0908
Q2008AT	OEM
Q2008B	3199
Q2008F41	OEM
Q2008FT1	OEM
Q2008G	1064
Q2008H	0154
Q2008L4	0555
Q2008L4A	0555
Q2008LT	1456
Q2008LTA	OEM
Q2008N	3199
Q2008R4	5255
Q2008RT	OEM
Q2008T	OEM
Q2010	OEM
Q2010A	0908
Q2010AT	OEM
Q2010B	3199
Q2010F41	OEM
Q2010F51	OEM
Q2010FT1	OEM
Q2010G	1064
Q2010H	0154
Q2010L4	0555
Q2010L5	0555
Q2010LT	5395
Q2010N	3199
Q2010R4	5322
Q2010R5	OEM
Q2010RT	OEM
Q2010T	OEM
Q2012L5	0758
Q2012R5	0772
Q2015	OEM
Q2015A	0908
Q2015B	3199
Q2015G	1064
Q2015H	0154
Q2015L5	0772
Q2015LT	OEM
Q2015N	3199
Q2015R5	0772
Q2015RT	OEM
Q2015U	OEM
Q2025	OEM
Q2025C	0908
Q2025D	3199
Q2025G	1064
Q2025H	2015
Q2025J6	5610
Q2025K6	5610
Q2025L6	6139
Q2025N	3199
Q2025P	0908
Q2025R6	0298
Q2025U	OEM
Q2040	OEM
Q2040C	0908
Q2040D	3199
Q2040J7	5610
Q2040K7	5610
Q2040P	1064
Q2040U	OEM
Q2040W7	5610
Q4001L3	5322
Q4001L4	0485
Q4001LT	1456
Q4001M	1413
Q4001P	0485
Q4003L3	0555
Q4003L4	5322
Q4003LT	OEM
Q4003P3	0485
Q4004A	0908
Q4004B	3207
Q4004F41	2300
Q4004FT1	OEM
Q4004L3	5322
Q4004L4	0555
Q4004LT	1456
Q4004R4	OEM
Q4004RT	OEM
Q4006	OEM

DEVICE TYPE	REPL CODE
Q4006A	0908
Q4006AT	OEM
Q4006B	3207
Q4006F41	2300
Q4006FT1	OEM
Q4006G	1317
Q4006H	0147
Q4006L4	0555
Q4006L5	OEM
Q4006LT	5395
Q4006N	3207
Q4006R4	5394
Q4006RT	OEM
Q4006T	OEM
Q4008	OEM
Q4008A	0908
Q4008AT	OEM
Q4008B	3207
Q4008F41	OEM
Q4008FT1	OEM
Q4008G	1317
Q4008H	0908
Q4008L4	0555
Q4008L4A	0555
Q4008LT	1456
Q4008N	3207
Q4008R4	5394
Q4008RT	OEM
Q4008T	OEM
Q4010	OEM
Q4010A	0908
Q4010AT	OEM
Q4010B	3207
Q4010F41	OEM
Q4010F51	OEM
Q4010FT1	OEM
Q4010G	1317
Q4010H	0908
Q4010L4	0555
Q4010L5	0555
Q4010LT	1456
Q4010N	3207
Q4010R4	OEM
Q4010R5	OEM
Q4010RT	OEM
Q4010T	OEM
Q4012L5	5460
Q4012R5	0744
Q4015	OEM
Q4015A	0908
Q4015B	3207
Q4015G	1317
Q4015H	0147
Q4015L5	0744
Q4015LT	OEM
Q4015N	3207
Q4015R5	0744
Q4015RT	OEM
Q4015U	OEM
Q4025	OEM
Q4025C	0908
Q4025D	3207
Q4025G	1317
Q4025H	2017
Q4025J6	5610
Q4025K6	5610
Q4025L6	5612
Q4025N	3207
Q4025P	0908
Q4025R6	0306
Q4025U	OEM
Q4040	OEM
Q4040C	0908
Q4040D	3207
Q4040J7	5610
Q4040K7	5878
Q4040P	0908
Q4040U	OEM
Q4040W7	3755
Q4994F31	OEM
Q5004A	0908
Q5004B	3222
Q5004F41	OEM
Q5004FT1	OEM
Q5004L4	0555
Q5004LT	1456
Q5004R4	OEM
Q5004RT	OEM
Q5006	OEM
Q5006A	0908
Q5006AT	OEM
Q5006B	3222
Q5006FT1	OEM
Q5006G	1891
Q5006H	0279
Q5006L4	0555
Q5006LT	1456
Q5006N	3222
Q5006R4	0594
Q5006RT	OEM
Q5006T	OEM
Q5008	OEM
Q5008A	0908
Q5008AT	OEM
Q5008B	3222
Q5008FT1	OEM
Q5008G	1891
Q5008H	0279
Q5008L4	0555

DEVICE TYPE	REPL CODE
Q5008LT	1456
Q5008N	3222
Q5008R4	0594
Q5008RT	OEM
Q5008T	OEM
Q5010	OEM
Q5010A	0908
Q5010AT	OEM
Q5010B	3222
Q5010F51	OEM
Q5010FT1	OEM
Q5010G	1891
Q5010H	0279
Q5010L4	0710
Q5010L5	0555
Q5010LT	1456
Q5010N	3222
Q5010R4	OEM
Q5010R5	2037
Q5010RT	OEM
Q5010T	OEM
Q5012L5	2145
Q5012R5	0617
Q5015	OEM
Q5015A	0908
Q5015B	3222
Q5015G	1891
Q5015H	0279
Q5015L5	0617
Q5015LT	OEM
Q5015N	3222
Q5015R5	0617
Q5015RT	OEM
Q5015U	OEM
Q5025	OEM
Q5025C	0908
Q5025D	3222
Q5025G	1891
Q5025H	2018
Q5025J6	3040
Q5025K6	3040
Q5025L6	1659
Q5025N	3222
Q5025P	0908
Q5025R6	0311
Q5025U	OEM
Q5040	OEM
Q5040C	0908
Q5040D	3222
Q5040J7	3040
Q5040K7	3040
Q5040P	0908
Q5040U	OEM
Q5040W7	3755
Q5044	0050
Q5053	0537
Q5053A	0537
Q5053C	0079
Q5053D	0537
Q5053E	0537
Q5053F	0537
Q5053G	0537
Q5055	OEM
Q5073D	0079
Q5073E	0079
Q5073F	0079
Q5075C	0142
Q5075CLY	0142
Q-5075CXY	0142
Q5075D	0142
Q5075DLY	0142
Q5075DXY	0142
Q5075E	0142
Q5075ELY	0142
Q5075EXY	0142
Q5075F	0142
Q5075FXY	0142
Q5075K	0142
Q5075X	0142
Q5075Y	0142
Q5075ZMY	0142
Q5077	0006
Q5077A	0006
Q5077E	0006
Q5077EA	0006
Q5078Z	0076
Q5083B	0103
Q5083C	0074
Q5094Z	0003
Q5095ZEE	0003
Q5099E	0086
Q5099F	0086
Q5100A	0126
Q5101D	0178
Q5102	0203
Q5102A	0203
Q5102P	0203
Q5102Q	0203
Q5102QA	0203
Q5102R	0203
Q5104K	0142
Q5104Z	0142
Q5104ZXF	0142
Q5110Z	0103
Q5111ZK	0313
Q5113ZLM	0142
Q5113ZMM	0142
Q5116C	0148
Q5116CA	0148

DEVICE TYPE	REPL CODE
Q5119	0220
Q5119D	0220
Q5120P	0042
Q5120Q	0042
Q5120R	0074
Q51210	0111
Q5121Q	0076
Q5121R	0111
Q5123E	0076
Q5123F	0076
Q5124	0155
Q5132Z	0891
Q5135	0006
Q5135-OY	0006
Q5137BA	0392
Q5137CA	0392
Q5138	0955
Q5138K	0955
Q5138L	0955
Q5138M	0955
Q5140XP	2051
Q5140Z	0065
Q5140ZX-O	0065
Q5140ZXP	2051
Q5140ZXQ	2051
Q5140ZXR	0065
Q5141ZM	0142
Q5149	0042
Q5160	0324
Q5160-0	0168
Q5160-O	0643
Q5160R	0643
Q5160Y	0643
Q5161Z	0313
Q5163	0392
Q5163D	0392
Q5169Z	OEM
Q5175	0643
Q5175TM	0643
Q5178	0955
Q5180	0552
Q5182	0016
Q5183	0076
Q5183C	0076
Q5183P	0076
Q5196	0532
Q5199	0220
Q5202	0392
Q5205	0006
Q5206	0270
Q5207	0313
Q5207Z	0313
Q5209	0532
Q5210	0066
Q5217	0261
Q6004A	0908
Q6004B	3222
Q6004F41	OEM
Q6004FT1	OEM
Q6004L4	0710
Q6004LT	1456
Q6004R4	OEM
Q6004RT	OEM
Q6006	OEM
Q6006A	0908
Q6006AT	OEM
Q6006B	3222
Q6006FT1	OEM
Q6006G	1891
Q6006H	0279
Q6006L4	0710
Q6006LT	1456
Q6006N	3222
Q6006R4	OEM
Q6006RT	OEM
Q6006T	OEM
Q6008	OEM
Q6008A	0908
Q6008B	3222
Q6008F51	OEM
Q6008FT1	OEM
Q6008G	1891
Q6008H	0279
Q6008L4	0710
Q6008LT	1456
Q6008N	3222
Q6008R4	0555
Q6008RT	OEM
Q6010	OEM
Q6010A	0908
Q6010AT	OEM
Q6010B	3222
Q6010F51	OEM
Q6010FT1	OEM
Q6010G	1891
Q6010H	0279
Q6010L4	0710
Q6010L5	0555
Q6010LT	1456
Q6010N	3222
Q6010R4	OEM
Q6010R5	2037
Q6010RT	OEM
Q6010T	OEM
Q6012L5	2145
Q6012R5	0617
Q6015	OEM
Q6015A	0908
Q6015B	3222

DEVICE TYPE	REPL CODE
Q6015G	1891
Q6015H	0279
Q6015L5	0617
Q6015LT	OEM
Q6015N	3222
Q6015R5	0617
Q6015RT	OEM
Q6015U	OEM
Q6025	OEM
Q6025C	0908
Q6025D	3222
Q6025G	1891
Q6025H	2018
Q6025J6	3040
Q6025K6	3040
Q6025L6	1659
Q6025N	3222
Q6025P	0908
Q6025R6	0311
Q6025U	OEM
Q6040	OEM
Q6040C	0908
Q6040D	3222
Q6040J7	3040
Q6040K7	3708
Q6040P	0908
Q6040U	OEM
Q6040W7	3755
Q6521	0342
Q6522	0342
Q6526Z	0261
Q6531	1327
Q6531ZM	1327
Q6536	0261
Q6536P	0261
Q6536Q	0261
Q6536R	0261
Q6541	0066
Q6541ZM	0066
Q6543	0261
Q6543A	0261
Q6562	0148
Q6562G	0148
Q6565	0076
Q6582Z	2851
Q6597	0055
Q6597Z	0055
Q6609	0643
Q6609Z	0643
Q6627	1392
Q6640Z	2912
Q6649	0563
Q6649Z	0563
Q6650	0525
Q6650Z	0525
Q6651Z	3880
Q6697	3046
Q6730D	1562
Q6730R	1562
Q6780	2902
Q6780Z	2902
Q6783	0276
Q6783Z	0276
Q6784	0276
Q6784K	0276
Q6784L	0276
Q6784M	0276
Q6785	0055
Q6793	2910
Q6803	0261
Q6805	0276
Q6805R	0276
Q6829	2185
Q6831	3267
Q7010L5	5412
Q7010R5	0596
Q7012L5	5412
Q7012R5	0875
Q7015L5	0875
Q7015R5	0875
Q7025J6	OEM
Q7025K6	OEM
Q7025L6	3467
Q7025P	OEM
Q7025R6	5607
Q7040J7	OEM
Q7040K7	OEM
Q7040P	OEM
Q7040W7	OEM
Q7219	0360
Q7220	OEM
Q7227	0332
Q7227R	0332
Q7256	2666
Q7293	3502
Q7296	3502
Q7327A	0035
Q7350	0291
Q7368	5216
Q7372A	0035
Q7393	0413
Q7402	0291
Q7406	0024
Q7424	4377
Q7448	0035
Q7478A	0101
Q7652	0101
Q7653A	0101
Q7668	0416

If replacement code is OEM, contact original manufacturer for replacement.

DEVICE TYPE	REPL CODE
Q7672	OEM
Q7682	3772
Q7748	0621
Q8004	OEM
Q8004A	OEM
Q8004AT	OEM
Q8004B	OEM
Q8004BT	OEM
Q8004L5	OEM
Q8004LT	OEM
Q8004R5	OEM
Q8004RT	OEM
Q8004T	OEM
Q8006	OEM
Q8006A	OEM
Q8006AT	OEM
Q8006B	OEM
Q8006BT	OEM
Q8006G	OEM
Q8006H	OEM
Q8006L5	OEM
Q8006LT	OEM
Q8006N	OEM
Q8006R5	OEM
Q8006RT	OEM
Q8006T	OEM
Q8008	OEM
Q8008A	OEM
Q8008AT	OEM
Q8008B	OEM
Q8008BT	OEM
Q8008G	OEM
Q8008H	OEM
Q8008L5	6435
Q8008LT	OEM
Q8008N	OEM
Q8008R5	OEM
Q8008RT	OEM
Q8008T	OEM
Q8010	OEM
Q8010A	OEM
Q8010AT	OEM
Q8010B	OEM
Q8010BT	OEM
Q8010G	OEM
Q8010H	OEM
Q8010L5	5412
Q8010LT	OEM
Q8010N	OEM
Q8010R5	0875
Q8010RT	OEM
Q8010T	OEM
Q8012L5	5412
Q8012R5	0875
Q8015	OEM
Q8015A	OEM
Q8015AT	OEM
Q8015B	OEM
Q8015BT	OEM
Q8015G	OEM
Q8015H	OEM
Q8015L5	0875
Q8015LT	OEM
Q8015N	OEM
Q8015R5	0875
Q8015RT	OEM
Q8015U	OEM
Q8025	OEM
Q8025C	OEM
Q8025D	OEM
Q8025G	OEM
Q8025H	OEM
Q8025J6	OEM
Q8025K6	OEM
Q8025L6	3467
Q8025N	OEM
Q8025P	OEM
Q8025R6	5607
Q8025U	OEM
Q8040	OEM
Q8040C	OEM
Q8040D	OEM
Q8040J7	OEM
Q8040K7	OEM
Q8040P	OEM
Q8040U	OEM
Q8040W7	OEM
Q8266R	0155
Q8266S	0155
Q8266T	0155
Q-10115C	0286
Q-11115C	0042
Q-12115C	0042
Q-13115C	0380
Q-14115C	0368
Q-15115C	0368
Q-16115C	0076
Q-17115C	2050
Q-18115C	2296
Q-20115C	0143
Q-21115C	0133
Q-22115C	0019
Q-23115C	0133
Q-24115C	0133
Q-25115C	0012
Q-26115C	0015
Q34450	0142
Q35218	0004
Q35242	0079
Q35259	0007
Q40263	0004
Q40359	0050
Q50782	OEM
Q50787	0330
Q50787Z	0330
Q51112K	OEM
Q51210	0111
Q500614	0555
QA01-06SB	0301
QA01-07R	0002
QA01-07RE	0025
QA01-08R	0244
QA01-11SE	0317
QA01-11.5E	0317
QA01-12S	0137
QA01-14RD	0100
QA01-25A	0402
QA01-25R	0402
QA01-25R-A	0402
QA01-25RA	0402
QA01-25RB	OEM
QA0313AB	OEM
QA0313B	4964
QA-1	0050
QA1-11M	0317
QA-8	0086
QA8	0086
QA-9	0126
QA-10	0086
QA-11	0126
QA11	0142
QA-12	0016
QA-13	0016
QA-14	0016
QA-15	0016
QA-16	0016
QA-17	0126
QA-18	0325
QA-19	0016
QA-20	0325
QA-21	0037
QA105R	0041
QA105RA	0589
QA105S	0589
QA106S	0295
QA106SB	0301
QA106SBV	0471
QA107R3	OEM
QA107RA	0062
QA107RE	0025
QA107REV	0025
QA107SP	0062
QA108SH	0165
QA111M	0317
QA111SE	0317
QA111SEV	0317
QA112R1	0137
QA112RI	0137
QA112RN-1	0052
QA112RN-2	0052
QA116R	0445
QA121R2	0705
QA135R2	0032
QA205AB	0140
QA205ABV	0140
QA205C	0041
QA205D	0041
QA205E	0041
QA205F	0041
QA205G	0041
QA205GV3	0041
QA206A	0091
QA206AB	0253
QA206B	0091
QA206BV3	0091
QA206CD	0091
QA206G	0295
QA206M	0091
QA206MV	0091
QA207D	0077
QA207M	0062
QA207M3	0062
QA208	0165
QA208C	0165
QA208G	0165
QA208GV3	0165
QA208M	OEM
QA209B	0057
QA209C	0057
QA209M	0057
QA209M9	0057
QA210D	0248
QA211B	0181
QA211CD	1005
QA211N1	0181
QA212B	0053
QA213A	0053
QA213B	0053
QA213M2	0053
QA220AB	0700
QA225	0714
QA225A	0714
QA225ABC	0714
QA225B	0714
QA225C	0714
QA228B	0257
QA230B	0195
QA230C	0195
QA232M2	OEM
QA232RC	OEM
QA233C	0166
QA233CV	0166
QA233G	0010
QA235A	0714
QA235AC	0714
QA235C	0010
QA235G	0010
QA703E	0632
QA2086M	OEM
QB01-05M	0162
QB01-11ZB	0317
QB01-15ZB	0002
QB01-18	0495
QB105	OEM
QB105M	0162
QB105N	0162
QB106P	0301
QB106PV	0301
QB107D	0025
QB107S	0025
QB107SV	0025
QB109SA	6724
QB110D	0317
QB110DV	0170
QB111	0317
QB111Z	0317
QB112	0137
QB115ZB	0002
QB115ZBV	0002
QB115ZD	0002
QB118	0495
QB400428	1069
QC	OEM
QC0008	OEM
QC0009	OEM
QC0011	1028
QC0014	OEM
QC0031A	1442
QC0032	4869
QC0042	0334
QC0048	OEM
QC0063	OEM
QC0065	4650
QC0069	0770
QC0073	4353
QC0074	4129
QC0080	3253
QC0081	0024
QC0085	0539
QC0093	1988
QC0234	0129
QC0235	0129
QC0237	OEM
QC0239	0129
QC0253	OEM
QC0275	0557
QC0291	5011
QC0300	0732
QC0309	OEM
QC0343	0413
QC0351	OEM
QC0352	3097
QC0353	OEM
QC0353(LA7018)	OEM
QC0380	3437
QC0543	3913
QC0554	0391
QC0554(LA7621)	OEM
QC0573	2957
QC0587	OEM
QC0679	OEM
QC0680	3083
QC0713	2111
QC0840	OEM
QC0846	1529
QC0852	0462
QC0908	OEM
QC-1	1030
QC1074	OEM
QC1238	OEM
QC1258	OEM
QC1280	OEM
QC1400	OEM
QC1872D1709ACT755	OEM
QCPL-3209	OEM
QD-2M205XE	0041
QD100-71	OEM
QD100-78	OEM
QD101-71	OEM
QD101-78	OEM
QD102-71	OEM
QD102-78	OEM
QD103-71	OEM
QD103-78	OEM
QD104-78	OEM
QD150-71	OEM
QD150-78	OEM
QD200-71	OEM
QD200-78	OEM
QD201-71	OEM
QD201-78	OEM
QD203-71	OEM
QD203-78	OEM
QD204-71	OEM
QD204-78	OEM
QD400-71	OEM
QD400-78	OEM
QD401-71	OEM
QD401-78	OEM
QD402-71	OEM
QD402-78	OEM
QD403-71	OEM
QD403-78	OEM
QD404-71	OEM
QD404-78	OEM
QD-CS2688D1	OEM
QD-CS2688DJ	0030
QD-CTT310XQ	0628
QDCTT410XQ	0030
QD-G1N60PXT	0123
QD-G1N60XXT	0019
QD-G1S32XXT	0143
QDNZOERA1502	0023
QDPZEA1504G	0023
QDQZD2SBA40	0296
QDQZOERB4404	0023
QD-S1S953XA	0133
QD-SMA150XN	0133
QDSOERA2204	0795
QD-SS1555XT	0133
QD-SS1885XT	0023
QD-SS15555XT	OEM
QD-SSR1KX4P	0023
QDSSR3AMBE	0031
QD-SSS53XXA	0133
QD-SV06CXXB	0023
QDTBOOUZ33BS	OEM
QDTCOUZ9R1BS	OEM
QDTZ01N4148M	0124
QDY31	OEM
QD-ZBZ102XJ	0420
QD-ZBZ162XJ	0420
QD-ZMZ205XE	0041
QDZMZ205XE	0041
QD-ZMZ306CE	0301
QD-ZMZ408CE	0165
QD-ZMZ409BE	0012
QD-ZRD9EXAA	0057
QD-ZRD56EAA	0091
QD-ZRD56FAA	0253
QG-0074	0160
QG0074	0160
QG-0076	0004
QG0076	0004
QG0254	0079
QGE-1	OEM
QH3091	OEM
QKH1199	OEM
QKH1214	OEM
QKH1302	OEM
QKH1325	OEM
QKH1374	OEM
QKH1381	OEM
QKH1448	OEM
QKH1495	OEM
QKH1512	OEM
QKH1516	OEM
QKH1527	OEM
QKH1528	OEM
QKH1535	OEM
QKH1553	OEM
QKH1560	OEM
QKH1569	OEM
QKH1578	OEM
QKH1579	OEM
QKH1589	OEM
QKH1592	OEM
QKH1599	OEM
QKH1663	OEM
QKH1665	OEM
QKH1666	OEM
QKH1667	OEM
QKH1693	OEM
QKH1708	OEM
QKH1709	OEM
QKH1744	OEM
QKH1757	OEM
QKH1763	OEM
QKH1777	OEM
QKH1793	OEM
QKH1811	OEM
QKH1817	OEM
QKH1820	OEM
QKH1832	OEM
QKH1833	OEM
QKH1850	OEM
QKH1851	OEM
QKH1865	OEM
QKH1866	OEM
QKH1977	OEM
QKH1978	OEM
QKH1982	OEM
QKH1990	OEM
QKH2039	OEM
QKH2040	OEM
QKK404	OEM
QKK412	OEM
QKK531	OEM
QKK532	OEM
QKK549	OEM
QKK623	OEM
QKK752	OEM
QKK753	OEM
QKK754A	OEM
QKK755A	OEM
QKK756A	OEM
QKK758A	OEM
QKK759A	OEM
QKK822	OEM
QKK826	OEM
QKK833A	OEM
QKK860	OEM
QKK869	OEM
QKK873	OEM
QKK877	OEM
QKK878	OEM
QKK879	OEM
QKK910	OEM
QKK965	OEM
QKK967	OEM
QKK968	OEM
QKK978	OEM
QKT-0033XBE	1280
QKT0033XBE	0325
QLMP-D268	OEM
QMC144143P1-N	OEM
QN2613	0211
QNJ937	OEM
QO1C	0871
QO1D	0853
QO2C	0871
QO2D	0853
QO7E	0853
QO8E	0853
QOV60526	0164
QOV60527	0208
QOV60528	0004
QOV60529	0079
QOV60530	0016
QOV60537	0208
QOV60538	0164
QOV60539	0004
QP-1	0604
QP1	0604
QP-1A	0604
QP1A	0604
QP-2	0604
QP2	0604
QP-3	0604
QP-3L	0925
QP-4	0604
QP-5	0604
QP-6	0604
QP6	0604
QP-7	0604
QP7	0604
QP-8	0103
QP8	0103
QP8-6623N	0440
QP-8-P	0103
QP-10	0604
QP-11	0103
QP-12	0103
QP-13	0460
QP-14	0161
QP-31	0854
QPCL-3209	OEM
QQ-0PLL01A0	OEM
QQ-0PLL01A0	OEM
QQ0PLL02AN	1714
QQ-0PPL02A0	1714
QQ-5311NA1	OEM
QQ-5311NAL	OEM
QQC61210	0004
QQ-M07205AT	1050
QQM07205AT	1050
QQ-MAN612AN	4042
QQMAN612AN	4042
QQ-MBA511BX	2773
QQ-MBA521AX	2781
QQ-MC3001AN	1540
QQ-MC3001AT	1540
QQ-MC3001DT	1540
QQ-M07205AT	1050
QQ-0PLL02A0	1714
QQ-0PLL02AN	1714
QQ-0PLL02A0	1714
QQ-0PLL020A	4224
QQ-0PLL02AN	1714
QQ0PLL02AN	4224
QQ-0PPL02A0	1714
QQ-PLL02A0	OEM
QQV60528	0004
QQV60529	0086
QQV60539	0004
QR2378	0004
QRF-2	0086
QRF3	0325
QRG-3	0325
QRT-100	0198
QRT-100A	6731
QRT-101	0006
QRT-101A	0436
QRT-102	0016
QRT-102A	0016
QRT-103	0037
QRT-103A	0037
QRT-104	0111
QRT-105	0111
QRT-106	0006
QRT-107	0079
QRT-107A	0324
QRT-108	0007
QRT-108A	0079
QRT-109	0079
QRT-110	0233
QRT-111	OEM
QRT-112	0111
QRT-113	0330
QRT-114	0622
QRT-115	OEM
QRT-116	0178
QRT-117	0004
QRT-118	0150
QRT-119	0038
QRT-120	0150
QRT-121	1242
QRT-122	0164
QRT-123	0164
QRT-124	0085
QRT-125	0440
QRT-126	0037
QRT-126A	0037
QRT-127	0085
QRT-127A	OEM
QRT-128	0178
QRT-129	0208
QRT-130	0975
QRT-131	0103
QRT-132	OEM
QRT-133	OEM
QRT-134	0127
QRT-135	0127
QRT-136	0164
QRT-137	2046
QRT-139	0313
QRT-140	0313
QRT-141	0324
QRT-143	2046
QRT-144	0400
QRT-145	0178
QRT-146	0042
QRT-147	0604
QRT-148	1681
QRT-149	0130
QRT-150	0561
QRT-151	1199
QRT-152	0042
QRT-153	0925
QRT-154	0236
QRT-155	0681
QRT-156	0423
QRT-157	0854
QRT-158	0423
QRT-159	0338
QRT-159A	0338
QRT-160	0936
QRT-162	OEM
QRT-163	OEM
QRT-164	1361
QRT-164A	OEM
QRT-165	1369
QRT-165A	OEM
QRT-166	1361
QRT-167	1369
QRT-172	0016
QRT-174	1908
QRT-175	0325
QRT-176	0325
QRT-177	2987
QRT-180	0353
QRT-181	0212
QRT-182	4133
QRT-183	0136
QRT-185	0892
QRT-186	0338
QRT-187	0155
QRT-188	0136
QRT-189	0905
QRT-190	0178
QRT-191	0936
QRT-192	0316
QRT-193	2136
QRT-194	1361
QRT-195	1369
QRT-196	1199
QRT-197	0561
QRT-200	0143
QRT-201	0922
QRT-202	0917
QRT-203	0071
QRT-204	0378
QRT-206	OEM
QRT-210	0071
QRT-212	0015
QRT-213	0015
QRT-214	0015
QRT-215	0015
QRT-216	0293
QRT-216D	0293
QRT-218	0133
QRT-220	0474
QRT-221	0196
QRT-222	0484
QRT-223	1080
QRT-224	1303
QRT-225	0774
QRT-226	0201
QRT-227	0239
QRT-228	0752
QRT-229	0288
QRT-230	0289
QRT-231	0302
QRT-232	0250
QRT-234	0188
QRT-235	0162
QRT-236	0157
QRT-237	0636
QRT-238	0025
QRT-239	0649
QRT-240	0012
QRT-241	0170
QRT-242	0317
QRT-243	0137
QRT-244	0365
QRT-245	0002
QRT-246	0420
QRT-247	0531
QRT-248	0565
QRT-249	0441
QRT-250	0039
QRT-251	0350
QRT-252	0999
QRT-253	0869
QRT-254	0783
QRT-255	0331
QRT-256	0149
QRT-257	0244
QRT-258	0495
QRT-259	0402
QRT-260	0186
QRT-261	0028
QRT-262	0030
QRT-263	0143
QRT-264	1791
QRT-265	1323
QRT-266	0421
QRT-267	0016
QRT-268	0016
QRT-269	0037
QRT-270	0037
QRT-271	0006
QRT-272	1890
QRT-300	0079
QRT-301	0595
QRT-302	0079
QRT-303	0037
QRT-304	0324
QRT-305	0436
QRT-306	5893
QRT-306F	0630
QRT-307	0630
QRT-308	0079
QRT-309	0079
QRT-310	1984
QRT-311	0791
QRT-312	0016
QRT-313	0144
QRT-314	0342
QRT-315	0715
QRT-600	1306
QRT-601	2083
QRT-602	2084
QRT-603	1449
QRT-604	2085
QRT-605	3649
QRT-700	OEM
QRT-701	OEM
QRT-702	OEM
QRT-703	OEM
QRT-704	OEM
QRT-705	OEM
QRT-708	OEM
QRT-709	OEM
QRT-710	OEM
QRT-711	OEM
QRT-712	OEM
QRT-713	OEM
QRT-714	OEM
QRT-715	OEM
QRT-716	OEM
QRT-717	OEM
QRT-718	OEM
QRT-719	OEM
QRT-720	OEM
QS054	0079
QS-0254	0155
QS0254	0079
QS23	OEM
QS24	OEM
QS25	0288
QS33	OEM
QS34	OEM
QS102AA	0015
QS102BA	0015
QS102CA	0015
QS102DA	0015
QS102EA	0015
QS102FA	0015
QS102GA	0015
QS102HA	0015
QS102KA	0015
QS102MA	0015
QS102PA	0071
QS102RA	0071
QS102TA	0071
QS102VA	0071
QS132AA	0071
QS132BA	0071
QS132CA	0071
QS132DA	0071
QS132EA	0071
QS132FA	0071

If replacement code is OEM, contact original manufacturer for replacement.

DEVICE TYPE	REPL CODE
QS132GA	0071
QS132HA	0071
QS132KA	0071
QS132MA	0071
QS132PA	0071
QS132RA	0071
QS132TA	0071
QS132VA	0071
QS316	0037
QS317	OEM
QS318	OEM
QS319	OEM
QS320	OEM
QS751AA	0015
QS751BA	0015
QS751CA	0015
QS751DA	0015
QS751EA	0015
QS751FA	0015
QS751GA	0015
QS751HA	0015
QS751KA	0015
QS751MA	0015
QS751PA	0071
QS751RA	0071
QS751TA	0071
QS751VA	0071
QS1018	2050
QS1306	0936
QSA1318TAANP	0037
QSBLAOSSN016	OEM
QSBLAOZSQ019	3913
QSBLAOZSY003	OEM
QSBLAOZSY018	OEM
QSC372	0532
QSC380	0079
QSC509	0324
QSC784	0079
QSE1001	0079
QSE3001	0144
QSE5020	0007
QSI-5022	3954
QSMQBOSMB018	OEM
QT0013Z	1541
QT0036	1541
QT0036Z	1541
QT0072Z	3516
QT0074Z	1541
QT0085	0556
QT0085Z	0556
QT0095	0148
QT0100	0151
QT0100R	0151
QT0144Z	1541
QT0185R	OEM
QT0190Z	1506
QT0191	0563
QT0192	0525
QT0226R	1386
QT0226S	1386
QT0267Q	0042
QT0267R	0042
QT0267S	0042
QT0268Z	1506
QT-A0719AXN	0006
QT-A0719XAN	0006
QT-A0719XCN	0006
QT-A0719XHN	0006
QT-A0733XAA	0006
QT-A0733XDN	0006
QT-A0733XON	0006
QTA562	0037
QT-A0719XAN	0006
QT-C0372XAT	0076
QT-C0460CBB	0151
QT-C0710XAE	0368
QT-C0710XBE	0368
QT-C0710XEE	0368
QT-C0735XBT	0191
QT-C0828XAN	1220
QT-C0828XDN	1220
QT-C0829XAN	0151
QT-C0829XBN	0151
QT-C0829XON	0151
QT-C0839XDA	0076
QT-C0900XBA	1221
QT-C0900XBD	1221
QT-C0900XCA	0111
QT-C0945ACA	0076
QT-C0945AGA	0076
QT-C131BXDN	0155
QT-C1047SAN	0113
QT-C1047XAN	0113
QT-C1047XBN	0113
QT-C1306	0839
QT-C1306XZA	0839
QT-C1307XZA	0836
QT-C1307ZXA	0836
QT-C1318XAN	0286
QT-C1318XDN	0155
QT-C1318XON	0324
QT-C1359XAN	0151
QT-C1687XAN	0224
QT-C1760XAS	1946
QT-C1760XCS	1946
QT-C1846XAN	0786
QT-CBC546AA	0111
QT-C0372XAT	0144
QT-C0710XBE	0079
QT-C0828XDN	0144
QT-C0900XCA	0111
QT-C0945ACA	0111
QT-CQ460CBB	0151
QT-D0313XAC	0042
QT-D0325XAC	0042
QT-D0525XDT	0423
QTD0704XAE	0060
QT-K0023AAS	0325
QT-K0033XAE	0325
QT-K0033XBE	1280
QT-R0033XBE	OEM
QTVCM-5	0678
QTVCM-16	1345
QTVCM-44	0632
QTVCM-45	0692
QTVCM-48	0823
QTVCM-59	0785
QTVCM-73	0520
QTVCM-75	0579
QTVCM-77	2057
QTVCM-78	0911
QTVCM-79	1994
QTVCM-81	1050
QTVCM-500	0232
QTVCM-502	1313
QTVCM-505	1208
QU01-20	OEM
QU01-24	OEM
QU01-27	OEM
QV3508	OEM
QV3512	OEM
QV3518	OEM
QV3522	OEM
QV3530	OEM
QV3540	OEM
QV4508	OEM
QV4512	OEM
QV4518	OEM
QV4522	OEM
QV4530	OEM
QV4540	OEM
QV6008	OEM
QV6012	OEM
QV6018	OEM
QV6022	OEM
QV6030	OEM
QV6040	OEM
QV9008	OEM
QV9012	OEM
QV9018	OEM
QV9025	OEM
QV9035	OEM
QV9045	OEM
QV-BD11246A	0139
QVD1KF114	0143
QVI1330	OEM
QVIM51181L	OEM
QVM800B	0143
QVM8513ART	0139
QZ3.6T5	0188
QZ3.6T10	0188
QZ5.1T5	0162
QZ5.1T10	0162
QZ5.6T5	0157
QZ5.6T10	0157
QZ6.2T5	0636
QZ6.2T10	0636
QZ10T5	0170
QZ10T10	0170
QZ12T5	0137
QZ12T10	0137
QZ14T5	0100
QZ14T10	0100
QZ15T5	0002
QZ15T10	0002
QZ27T5	0441
QZ27T10	0441
QZ1575	0002
R002Z	0071
R0050	0071
R0051	0071
R0052	0071
R0053	3911
R0054	0229
R0055	0072
R0056	0373
R0070	0953
R0071	0242
R0072	0540
R0074	0817
R0076	0817
R0078	0922
R0080	0087
R0081	0087
R0082	0087
R0084	OEM
R0086	0087
R0090	0087
R0091	0087
R0092	0222
R0094	0087
R0096	0087
R0097	0087
R0098	0087
R0-2AV	0071
R02	0965
R02A	0564
R02AU	0071
R02AV	0071
R02B	0087
R02C	0071
R02K	0071
R02Z	OEM
R06-1001	0144
R06-1002	0144
R06-1003	0144
R06-1004	0144
R06-1005	0144
R06-1006	0144
R06-1007	0004
R06-1008	0004
R06-1009	0004
R06-1010	0004
R07A	0025
R080	0071
R0100	OEM
R0100(HEP)	1282
R0101	OEM
R0101(HEP)	1282
R0102	0023
R0102(HEP)	1282
R0103AA	4455
R0103BA	4455
R0103CA	4456
R0103DA	4456
R0104	OEM
R0104(HEP)	1287
R0106	OEM
R0106(HEP)	1293
R0106AA	4455
R0106BA	4455
R0106CA	4456
R0106DA	4456
R0107AA	4455
R0107BA	4455
R0107CA	4456
R0107DA	4456
R0130	0122
R0130(HEP)	1600
R0131	0122
R0132	0122
R0132(HEP)	0871
R0134	0122
R0134(HEP)	0853
R0136	0122
R0137	OEM
R0138	OEM
R0160	0319
R0161	0319
R0162	0319
R0164	0319
R0166	6192
R0186	OEM
R0250	0319
R0251	0319
R0253	0319
R0254	0319
R0255	0319
R0256	0319
R0257	0319
R0801	6529
R0802	6529
R0803	6529
R0804	0106
R0805	0787
R0851	0277
R0852	0277
R0853	0289
R0855	0296
R0856	0302
R0861	OEM
R0862	OEM
R0863	OEM
R0865	OEM
R0866	OEM
R0875	2363
R0876	2363
R0877	2363
R0878	OEM
R0879	OEM
R0880	OEM
R0881	OEM
R0882	OEM
R0883	OEM
R0884	OEM
R0890	OEM
R0891	OEM
R0892	OEM
R0894	OEM
R0896	OEM
R07105	0794
R-1	0015
R1	0015
R1/1H	OEM
R1/2H	OEM
R1/3H	OEM
R1/4H	OEM
R-1A	0015
R1A	0015
R-1B	0242
R1B	0015
R1C	0609
R-1S188	0143
R-1S333Y	0244
R1Z	0102
R2	1303
R-2A	1250
R-2AV2	1449
R-2B	1250
R2D	0137
R2E	0002
R2H	0609
R2KN	OEM
R2M	5505
R-2SA222	0136
R-2SB186	0004
R-2SB187	0004
R-2SB303	0004
R-2SB405	0164
R-2SB474	0111
R-2SB492	0847
R2SB492	0847
R-2SC535	0127
R-2SC537	0191
R-2SC545	0127
R-2SC668	0127
R-2SC772	1144
R-2SC858	0111
R-2SD187	0208
R2SZ	0015
R3	0474
R3/1H	OEM
R3/2H	0015
R3/3H	OEM
R3/4H	OEM
R-3A	1250
R-3A3	1306
R-3AT2	2083
R-3DB3	2084
R4	OEM
R-4A	0540
R4A	0015
R4HZ	0805
R5	0071
R-5B	0242
R5B	0071
R-5C	0242
R5C	0071
R5.1JSB2X	OEM
R6	0071
R6/1H	OEM
R6/2H	0015
R6/3H	OEM
R6/4H	OEM
R-6A	1250
R-6B	1250
R6HZ	0319
R8	0071
R8/1H	OEM
R8/2H	0071
R8/3H	OEM
R8/4H	OEM
R9-56	2195
R9A	0015
R9H02015	OEM
R9H02024	OEM
R9H02215	OEM
R9H02415	OEM
R9H02615	OEM
R9H02815	OEM
R10D1	0015
R10DC	0015
R10G	OEM
R10J	0023
R12	0600
R14	0600
R15	0142
R16	0004
R18S12A	OEM
R18S12A2	OEM
R18S12A3	OEM
R18S12B	3309
R18S12B2	OEM
R18S12B3	OEM
R18S14A	OEM
R18S14A2	OEM
R18S14A3	OEM
R18S14B	3309
R18S14B2	OEM
R18S14B3	OEM
R18S16A	OEM
R18S16A2	OEM
R18S16A3	OEM
R18S18A	OEM
R18S18A2	OEM
R18S18A3	OEM
R18SG4A	OEM
R18SG4A2	OEM
R18SG4A3	OEM
R18SG4B	OEM
R18SG4B2	OEM
R18SG4B3	OEM
R18SG6A	OEM
R18SG6A2	OEM
R18SG6A3	OEM
R18SG6B	OEM
R18SG6B2	OEM
R18SG6B3	OEM
R18SG8A	OEM
R18SG8A2	OEM
R18SG8A3	OEM
R18SG8B	OEM
R18SG8B2	OEM
R18SG8B3	OEM
R18SG10A	OEM
R18SG10A2	OEM
R18SG10A3	OEM
R18SG10B	OEM
R18SG10B2	OEM
R18SG10B3	OEM
R18SGR4A	OEM
R18SGR4B	OEM
R18SGR6A	OEM
R18SGR6B	OEM
R18SGR8A	OEM
R18SGR8B	OEM
R18SGR10A	OEM
R18SGR10B	OEM
R18SR12A	OEM
R18SR12B	0202
R18SR14A	OEM
R18SR14B	0202
R18SR16A	OEM
R18SR18A	OEM
R23-1003	0004
R23-1004	0004
R23A4A	OEM
R23A4B	1226
R23A6A	OEM
R23A6B	1226
R23A8A	OEM
R23A8B	OEM
R23A10A	OEM
R23A10B	OEM
R23A12A	OEM
R23A12B	1204
R23A14A	OEM
R23A14B	OEM
R23A16A	OEM
R23A16B	OEM
R23A18A	OEM
R23A18B	OEM
R23A20A	OEM
R23A20B	OEM
R23A22A	OEM
R23A22B	OEM
R23A24A	OEM
R23A26A	OEM
R23D16A	OEM
R23DR16A	5397
R24-1001	0004
R24-1002	0004
R24-1003	0004
R24-1004	0004
R30D20B	OEM
R30DR20B	6156
R30E4A	OEM
R30E4B	OEM
R30E6A	OEM
R30E6B	OEM
R30E8A	OEM
R30E8B	OEM
R30E10A	OEM
R30E10B	OEM
R30E12A	OEM
R30E12B	OEM
R30E14A	OEM
R30E14B	OEM
R30E16A	OEM
R30E16B	OEM
R30E18A	OEM
R30E18B	OEM
R30E20A	OEM
R30E20B	OEM
R30E22A	OEM
R30E22B	OEM
R30E24A	OEM
R30E24B	OEM
R30E26A	OEM
R30E26B	OEM
R30E28A	OEM
R30E30A	OEM
R33	0208
R34	0208
R34B4A	OEM
R34B4B	2159
R34B6A	OEM
R34B6B	2159
R34B8A	OEM
R34B8B	OEM
R34B10A	OEM
R34B10B	OEM
R34B12A	OEM
R34B12B	0479
R34B14A	OEM
R34B14B	OEM
R34B16A	OEM
R34B16B	OEM
R34B18A	OEM
R34B18B	OEM
R34B20A	OEM
R34B20B	OEM
R34B22A	OEM
R34B22B	OEM
R34B24A	OEM
R34B24B	OEM
R34B26A	OEM
R34B26B	OEM
R34B28A	OEM
R34B30A	OEM
R35	0004
R38B4A	OEM
R38B4B	OEM
R38B6A	OEM
R38B6B	OEM
R38B8A	OEM
R38B8B	OEM
R38B10A	OEM
R38B10B	OEM
R38B12A	OEM
R38B12B	OEM
R38B14A	OEM
R38B14B	OEM
R38B16A	OEM
R38B16B	OEM
R38B18A	OEM
R38B18B	OEM
R38B20A	OEM
R38B20B	OEM
R38B22A	OEM
R38B22B	OEM
R38B24A	OEM
R38B24B	OEM
R38B26A	OEM
R38B26B	OEM
R38B28A	OEM
R38B28B	OEM
R38B30A	OEM
R38B30B	OEM
R38B32A	OEM
R38B34A	OEM
R38B36B	OEM
R38B38B	OEM
R38B40A	OEM
R38B42A	OEM
R38B44B	OEM
R38B46B	OEM
R38B48A	OEM
R38B50A	OEM
R41	0600
R43HZ	0805
R46	0211
R52	0004
R52K4A	OEM
R52K4B	4922
R52K6A	OEM
R52K6B	4922
R52K8A	OEM
R52K8B	2139
R52K10A	OEM
R52K10B	2139
R52K12A	2139
R52K12B	OEM
R52K14A	OEM
R52K14B	OEM
R52K16A	OEM
R52K16B	OEM
R52K18A	OEM
R52K18B	OEM
R52K20A	OEM
R52K20B	OEM
R52K22A	OEM
R52K22B	OEM
R52K24A	OEM
R52K24B	OEM
R52K26A	OEM
R52K26B	OEM
R52K28A	OEM
R52K28B	OEM
R52K30A	OEM
R52K30B	OEM
R52K32A	OEM
R52K34A	OEM
R52K36B	OEM
R52K38B	OEM
R52K40A	OEM
R52K42A	OEM
R52K44B	OEM
R52K46B	OEM
R52K48A	OEM
R52K50A	OEM
R56	0004
R60-1001	0136
R60-1002	0050
R60-1003	0050
R60-1004	0004
R60-1005	0004
R60-1006	0004
R60-1007	0143
R61	0208
R62	0208
R63	0208
R-63HZ	0319
R64	0004
R65	0004
R66	0004
R66-8504	0015
R67	0004
R79	0208
R80	0208
R83	0004
R87	0004
R98	0004
R100-1	0004
R100-8	0004
R100-9	0004
R101-2	0004
R101-3	0004
R101-4	0004
R104-5	0136
R104-6	0136
R104-7	0136
R104-8	0136
R117	0600
R118	0127
R119	0280
R120	0004
R122C	0015
R123	0086
R125	0600
R135	0600
R136	0600
R137	0600
R142	0133
R145B	0302
R152	0004
R154B	0277
R163	0280
R164	0004
R177	0208
R186	0280
R200	OEM
R202	0600
R203	0600
R204B	0277
R205	0071
R210	0087
R212	OEM
R220	OEM
R227	0280
R242	0004
R244	0280
R245	0004
R250	0087
R250-H	OEM
R250F	0087
R255	0004
R258	0280
R265A	0160
R289	0004
R290	0004
R291	0004
R324	0004
R336	0136
R337	0136
R338	0050
R339	0050
R340	0016
R341	0050
R350	0087
R350F	0087
R364	0004
R400	OEM
R-424	0280
R424	0050
R424-1	0050
R425	0050
R428	0004
R440M	OEM
R488	0280
R497	0050
R499-01-C	0079
R506	0208
R515	0050
R515A	0050
R516	0160
R516(T.I.)	0050
R516A	0050
R530	0004
R537	0004
R539	0050
R558	0050
R558(T.I.)	0004
R563	0050
R563(T.I.)	0004
R564	0050
R565	0050
R579	0050
R579(T.I.)	0004
R581	0136
R582	0016
R592	0600
R593	0050
R593A	0050
R600**20	1023
R600**25	1023
R608	0004
R608A	0004
R610**20	1023
R610**25	1023
R621-1	0042
R623-1	0423
R684	0136
R702	OEM
R704	OEM
R706	OEM
R710XPT	2234
R-711	0120
R711XPT	2234
R712XPT	2234
R713XPT	OEM
R714	0136
R714XPT	2234
R715	0136
R855-2	0015
R868	0004
R895	0079
R1000	0087
R1001	1137
R1002	0344
R1003	0901
R1004	2342
R1005	0058

If replacement code is OEM, contact original manufacturer for replacement.

DEVICE TYPE	REPL CODE
R1019	0037
R1035	0015
R1106	0143
R1107	0143
R1109	0143
R1117-1	0142
R1202	1584
R1205	1584
R1215	OEM
R1215(HEP)	1396
R1216	OEM
R1217	OEM
R1217(HEP)	0447
R1218	3620
R1219	OEM
R1220	4588
R1220(HEP)	2095
R1221	2270
R1221(HEP)	0505
R1222	3402
R1223	3404
R1241	0764
R1242	0764
R1243	0764
R1243(MOTOROLA)	0240
R1244	0764
R1245	0764
R1246	0764
R1247	0764
R1273	0004
R1274	0004
R1300	0764
R1301	0764
R1302	0764
R1303	0764
R1304	0745
R1305	1630
R1306	0764
R1307	0764
R1329	0015
R1331	OEM
R1332	OEM
R1333	OEM
R1334	OEM
R1335	OEM
R1336	OEM
R1337	OEM
R-1348	0137
R1471	0725
R1472	0712
R1473	0725
R1475	0725
R1530	0208
R1531	0208
R1532	0208
R1533	0600
R1534	0208
R1537	0208
R1538	0208
R1539	0050
R1540	0004
R1541	0004
R1542	0004
R1543	0004
R1544	0004
R1545	0208
R1546	0004
R1547	0208
R1548	0004
R1549	0208
R1550	0050
R1553	0208
R1554	0050
R1555	0004
R1667	0143
R1701	OEM
R1702	OEM
R1703	OEM
R1711	1755
R1712	4414
R1713	3499
R1715	0485
R1721	1089
R1722	1093
R1723	0593
R1725	0415
R1750	0147
R1751	0147
R1752	0147
R1781	3152
R1782	3152
R1783	3152
R1785	3152
R1889	0143
R1901C-6L31	OEM
R1903	OEM
R1903-7609	OEM
R2001	0975
R2003	0975
R2005	5005
R2010	1635
R2015	1635
R2020	1635
R2025	1251
R2030	1251
R2035	1251
R2040	1251
R2045	1205
R2050	1205

DEVICE TYPE	REPL CODE
R2060	1205
R2070	2137
R2080	2137
R2090	2251
R2096	0178
R2105	5005
R2110	1635
R2115	1635
R2120	1635
R2125	1251
R2130	2137
R2135	1251
R2140	1251
R2145	1205
R2150	1205
R2159	0015
R2160	1205
R2164	0143
R2170	2137
R2180	2137
R2190	2251
R2252	0015
R2270-7693	0126
R2350	0004
R2351	0004
R2352	0004
R2353	0004
R2355	0004
R2356	0208
R2359	0208
R2360	0208
R2364	0208
R2365	0208
R2366	0004
R2367	0004
R2373	0004
R2374	0208
R2375	0208
R2432-1	0829
R2434-1	2571
R-2436	0313
R2442	0015
R2444	0142
R2445-1	2715
R2460-1	0229
R-2460-3	0975
R2460-3	0975
R2460-4	0229
R2460-9(RCA)	0975
R2473	0144
R2474-2	0233
R2476	0144
R2477	0144
R2482-1	0004
R2516	1345
R2516-1	0167
R2675	0004
R2677	0004
R2683	0050
R2684	0050
R2685	0050
R2686	0050
R2687	0050
R2688	0050
R2689	0004
R2694	0050
R2695	0050
R2696	0050
R2697	0050
R2749	0004
R2749M	0004
R2964	0975
R2982	0103
R3012	0102
R3012(HEP)	0102
R3057	0196
R3105	0267
R3110	1107
R3115	1107
R3120	1107
R3125	0258
R3130	0258
R3135	0258
R3140	0258
R3145	0267
R3150	0267
R3160	0267
R3170	1119
R3180	1119
R3190	0281
R3205	1630
R3210	1107
R3215	1107
R3220	1107
R3225	0258
R3230	0258
R3235	0258
R3240	0258
R3245	0267
R3250	0267
R3260	0267
R3270	1119
R3275	0004
R3276	0004
R3277	0050
R3278	0050
R3279	0050
R3279-13	OEM
R3280	1119

DEVICE TYPE	REPL CODE
R3280(RCA)	0004
R3282	0004
R3283	0016
R3284	0004
R3285	0015
R3286	0004
R3287	0050
R3288	0050
R3290	0281
R3293	0208
R3293(GE)	0016
R3299	0004
R3301	0004
R3309	0050
R3314	0196
R3405	2665
R3410	2768
R3415	2768
R3420	2768
R3425	2907
R3430	2907
R3435	2907
R3440	2907
R3445	3051
R3450	3051
R3460	3051
R3470	3117
R3480	3117
R3490	3208
R3503	0142
R3505	1783
R3508	0086
R3510	1783
R3514-1	0975
R3515	0085
R3515(RCA)	0160
R3515(RECT)	1783
R3520	0142
R3520(RECT)	1783
R3520-1	0142
R3525	1783
R3530	1783
R3533	0196
R3535	1783
R3540	1783
R3545	1783
R3550	1783
R3560	1783
R3573-1	0208
R3578-1	0004
R3583-7	0239
R3583-8	0239
R3585-5	0239
R3585-6	0061
R3585-7	0239
R3585-8	0061
R3598-2	0004
R3605	2665
R3608	0086
R3608-1	0086
R3608-2	0086
R3610	2182
R3613-3(RCA)	0338
R3615	2182
R3619-3	2394
R3620	2182
R3625	2191
R3630	2191
R3635	2191
R3640	2191
R3645	2197
R3650	2197
R3651-1	0212
R3660	2197
R3670	2216
R3680	2216
R3690	2340
R3705	1783
R3710	1783
R3715	1783
R3720	1783
R3725	1783
R3730	1783
R3740	1783
R3750	1783
R3760	1783
R3780	1818
R4057	0016
R4105	1347
R4110	1347
R4115	1347
R4120	1347
R4125	1347
R4130	1347
R4135	1905
R4140	1905
R4145	1905
R4150	1905
R4160	1905
R4170	0193
R4180	0193
R4190	0193
R4192	0142
R4193	0142
R4194	0142
R4195	0142
R4196	0142
R4210	1347
R4215	1347

DEVICE TYPE	REPL CODE
R4220	1347
R4225	1347
R4230	1347
R4235	1905
R4240	1905
R4245	1905
R4250	1905
R4260	1905
R4270	0193
R4280	0193
R4290	0193
R4295-1	1186
R4295-1(RCA)	1186
R4295-2	1186
R4305	1347
R4310	1347
R4315	1347
R4320	1347
R4325	1347
R4330	1347
R4335	1905
R4340	1905
R4345	1905
R4348	0004
R4349	0004
R4350	1905
R4360	1905
R4369	0103
R4370	0193
R4380	0193
R4390	0193
R4409	0196
R4439-1	1843
R4439-1(RCA)	1843
R4439-2	1843
R4439-2(RCA)	1843
R4666	0196
R5048	0086
R5050	0208
R5051	0004
R5052	0004
R5053	0004
R5054	0208
R5055	0004
R5056	0208
R5096	0143
R5097	0004
R5098	0004
R5099	0004
R5100	0004
R5101	0164
R5102	0050
R5103	0050
R5105	0501
R5110	0501
R5115	0501
R5120	0501
R5125	0501
R5130	0501
R5135	1777
R5140	1777
R5150	1777
R5160	1777
R5170	1789
R5179	0208
R5180	0208
R5181	0004
R5182	0164
R5190	1789
R5221AD	OEM
R5305	0501
R5310	0501
R5315	0501
R5320	0501
R5325	0501
R5330	0501
R5335	1777
R5340	1777
R5345	1777
R5350	1777
R5360	1777
R5370	1789
R5380	1789
R5390	1789
R5405	0501
R5410	0501
R5415	0501
R5420	0501
R5425	0501
R5430	0501
R5435	1777
R5440	1777
R5450	1777
R5460	1777
R5470	1789
R5480	1789
R5490	1789
R5522	0143
R5523	0004
R5524	0004
R5525	0004
R5708	0004
R5740-1	0280
R5862	0123
R5970	0015
R5971	0015
R5983	0030
R6048	0015
R6109	0123

DEVICE TYPE	REPL CODE
R6110	0015
R6171-11	1217
R6171-12	1217
R6171-19	1217
R6171-20	1217
R6422	0015
R6502-40	3065
R6502C	3065
R6502E	OEM
R6502P	3065
R6507C	3070
R6507P	3070
R6520-13	OEM
R6520-26	OEM
R6520AP	OEM
R6532-12	OEM
R6532P	1973
R6551-11	OEM
R6551P	OEM
R6553	0004
R6765-5P	OEM
R6765P	OEM
R6789	0030
R6922	0004
R-7026	0133
R-7027	0133
R7028	0143
R-7029	0143
R7029	0143
R-7030	0649
R7030	0649
R7048	0004
R-7051	0143
R-7056	OEM
R-7092	0139
R-7093	0471
R-7094	0077
R-7096	0023
R-7097	0064
R-7103	0137
R-7110	0139
R7124	0004
R7127	0004
R7162	0015
R7163	0016
R7164	0004
R7165	0016
R7166	0004
R7167	0085
R7248	0015
R7249	0016
R7253	0160
R7271	0015
R7343	0016
R7359	0016
R7360	0016
R7361	0016
R7362	0208
R7363	0004
R7489	0004
R7490	0004
R7491	0004
R7582	0016
R7612	0004
R7613	0086
R7615	0087
R7620	0085
R7682	0015
R7743	0143
R7885	0050
R7886	0050
R7887	0016
R7888	0004
R7889	0004
R7890	0440
R7891	0050
R7892	0143
R7893	0143
R7894	0137
R7953	0016
R7954	0015
R7962	0050
R8022	0133
R8023	0133
R8024	0015
R8060	0143
R8061	0143
R8066	0016
R8067	0016
R8068	0016
R8069	0016
R8070	0016
R8115	0016
R8116	0016
R8117	0016
R8118	0016
R8119	0016
R8120	0016
R8121	0004
R8158	0142
R8219	0143
R8223	0016
R8224	0016
R8225	0016
R8240	0050
R8241	0050
R8242	0050
R8243	0016
R8244	0016

DEVICE TYPE	REPL CODE
R8257	0143
R8259	0016
R8260	0016
R8261	0016
R8305	0016
R8310	0004
R8311	0004
R8312	0016
R8313	0085
R8314	0133
R8364	0137
R8470	0015
R8471	0774
R8472	1303
R8473	0015
R8474	0196
R8475	0143
R8477	0474
R8528	0016
R8529	0127
R8530	0016
R8543	0016
R8551	0016
R8552	0016
R8553	0016
R8554	0016
R8555	0016
R8556	0016
R8557	0016
R8559	0050
R8560	0474
R8620	0016
R8645	0087
R8646	0016
R8647	0016
R8648	0016
R8649	0086
R8658	0016
R8659	0160
R8685	0050
R8686	0050
R8687	0004
R8688	0004
R8692	0050
R8693	0050
R8694	0050
R8695	0004
R8697	0004
R8703	0050
R8704	0050
R8705	0050
R8706	0004
R8707	0004
R8721	0229
R8881	0050
R8882	0050
R8883	0004
R8884	0004
R8885	0004
R8886	0004
R8887	0143
R8889	0016
R8900	0016
R8914	0016
R8915	0086
R8916	0016
R8963	0016
R8964	0016
R8965	0016
R8966	0016
R8967	0037
R8968	0016
R8969	0037
R8970	0143
R8971	0004
R8989	0142
R9004	0016
R9005	0016
R9006	0016
R9025	0016
R9071	0016
R9381	0531
R9382	0285
R9383	0285
R9384	0016
R9385	0016
R9470	0015
R9483	0016
R9531	0050
R9532	0050
R9533	0164
R9534	0164
R9590	0143
R9597	0015
R9600	0224
R9601	0050
R9602	0050
R9603	0164
R9604	0164
R10254P936	1831
R10254P945	0081
R10254P945B	0081
R10254P946	0141
R10255P946B	0141
R10256P962	0562
R10256P962B	0562
R10564	0474
R20100	2251
R21100	2251

DEVICE TYPE	REPL CODE
R24451	2715
R31100	0281
R32100	0281
R34100	3208
R36100	2340
R37100	1818
R41100	0193
R42100	0193
R43100	0193
R50305	0501
R50310	0501
R50315	1777
R50320	0501
R50325	0501
R50330	0501
R50335	1777
R50340	1777
R50345	1777
R50350	1777
R50360	1777
R50370	1789
R50380	1789
R50390	1789
R50405	0501
R50410	0501
R50415	0501
R50420	0501
R50425	0501
R50430	0501
R50435	1777
R50440	1777
R50445	1036
R50450	1777
R50460	1777
R50470	1789
R50480	1789
R50490	1789
R51100	1789
R53100	1789
R54100	1789
R80286-8	OEM
R105064	0774
R105330	1030
R106379	0015
R109180	0196
R109328	0196
R109474	0196
R112524	0143
R113321	0015
R113391	1303
R113392	0015
R113397	0774
R503100	1789
R504100	1789
R600428R	0501
R1500715	0087
R2400730	0087
R3020506	0711
R3020512	0711
R3020606	0711
R3020612	0711
R3030506	2967
R3030512	2967
R3030606	2967
R3030612	2967
R4020520	0601
R4020530	0601
R4020620	0601
R4020630	0601
R4030520	1851
R4030530	1851
R4030620	1851
R4030630	1851
R4040060	2813
R4040070	2813
R4040160	2813
R4040170	2813
R4040260	2813
R4040270	2813
R4040360	2813
R4040370	2813
R4040460	2813
R4040470	2813
R4040560	2813
R4040570	2813
R4040660	2813
R4040670	2813
R4040760	2845
R4040770	2845
R4040860	2845
R4040870	2845
R4040960	2845
R4040970	2845
R4041060	2845
R4041070	2845
R4050060	1783
R4050070	1783
R4050160	1783
R4050170	1783
R4050260	1783
R4050270	1783
R4050360	1783
R4050370	1783
R4050460	1783
R4050470	1783
R4050560	1783
R4050570	1783
R4050660	1783
R4050670	1783

If replacement code is OEM, contact original manufacturer for replacement.

DEVICE TYPE	REPL CODE
R4050760	1818
R4050770	1818
R4050860	1818
R4050870	1818
R4050960	1818
R4050970	1818
R4051060	1818
R4051070	1818
R4100022	0319
R4100025	3189
R4100040	2813
R4100070	2840
R4100122	0319
R4100140	2813
R4100170	2840
R4100222	0319
R4100225	1124
R4100240	2813
R4100322	0319
R4100325	0805
R4100340	2813
R4100422	0319
R4100425	0805
R4100440	2813
R4100522	0319
R4100540	2813
R4100622	0319
R4100640	2813
R4100822	1132
R4100840	1132
R4100860	3900
R4101022	0044
R4101040	0044
R4101060	2655
R4101070	2473
R4101822	1119
R4108070	2833
R4110022	0267
R4110040	1783
R4110070	2182
R4110122	0267
R4110140	1783
R4110170	2182
R4110222	0267
R4110240	1783
R4110322	0267
R4110340	1783
R4110422	0267
R4110440	1783
R4110522	0267
R4110540	1783
R4110622	0267
R4110640	1783
R4110822	0281
R4110840	1119
R4110860	3117
R4111022	0281
R4111040	0281
R4111060	3208
R4111070	2340
R4118070	2216
R4140060	3762
R4140070	3771
R4140160	2653
R4140170	1560
R4140260	2657
R4140270	2840
R4140360	OEM
R4140370	OEM
R4140460	2006
R4140470	2850
R4140560	2677
R4140570	2871
R4140660	2682
R4140670	2871
R4140760	3900
R4140770	2833
R4140860	3900
R4140870	2833
R4140960	2655
R4140970	2473
R4141060	2655
R4141070	2473
R5000110	0599
R5000110XXWA	0599
R5000110XXWC	0599
R5000115	0599
R5000115XXWA	0599
R5000115XXWC	0599
R5000210	0599
R5000210XXWA	0599
R5000210XXWC	0599
R5000215	0599
R5000215XXWA	0599
R5000215XXWC	0599
R5000310	0599
R5000310XXWA	0599
R5000315	0599
R5000315XXWA	0599
R5000315XXWC	0599
R5000410	1986
R5000410XXWA	1986
R5000410XXWC	1986
R5000415	1986
R5000415XXWA	1986
R5000415XXWC	1986
R5000510	1986
R5000510XXWA	1986
R5000510XXWC	1986

DEVICE TYPE	REPL CODE
R5000515	1986
R5000515XXWA	1986
R5000515XXWC	1986
R5000610	1986
R5000610XXWA	1986
R5000610XXWC	1986
R5000615	1986
R5000615WA	1986
R5000615XXWA	1986
R5000615XXWC	1986
R5000710	0657
R5000710XXWA	0657
R5000710XXWC	0657
R5000715	0657
R5000715XXWA	0657
R5000715XXWC	0657
R5000810	0657
R5000810XXWA	0657
R5000810XXWC	0657
R5000815	0657
R5000815XXWA	0657
R5000815XXWC	0657
R5000910	0657
R5000910XXWA	0657
R5000910XXWC	0657
R5000915	0657
R5000915XXWA	0657
R5000915XXWC	0657
R5001010	0657
R5001010XXWA	0657
R5001010XXWC	0657
R5001015	0657
R5001015XXWA	0657
R5001015XXWC	0657
R5010110	1347
R5010110XXWA	1347
R5010110XXWC	1347
R5010115	1347
R5010115XXWA	1347
R5010115XXWC	1347
R5010210	1347
R5010210XXWA	1347
R5010210XXWC	1347
R5010215	1347
R5010215XXWA	1347
R5010215XXWC	1347
R5010310	1347
R5010310XXWA	1347
R5010310XXWC	1347
R5010315	1347
R5010315XXWA	1347
R5010315XXWC	1347
R5010410	1905
R5010410XXWA	1905
R5010410XXWC	1905
R5010415	1905
R5010415XXWA	1905
R5010415XXWC	1905
R5010510	1905
R5010510XXWA	1905
R5010510XXWC	1905
R5010515	1905
R5010515XXWA	1905
R5010515XXWC	1905
R5010610	1905
R5010610XXWA	1905
R5010610XXWC	1905
R5010615	1905
R5010615XXWA	1905
R5010615XXWC	1905
R5010710	0193
R5010710XXWA	0193
R5010710XXWC	0193
R5010715	0193
R5010715XXWA	0193
R5010715XXWC	0193
R5010810	0193
R5010810XXWA	0193
R5010810XXWC	0193
R5010815	0193
R5010815XXWA	0193
R5010815XXWC	0193
R5010910	0193
R5010910XXWA	0193
R5010910XXWC	0193
R5010915	0193
R5010915XXWA	0193
R5010915XXWC	0193
R5011010	0193
R5011010XXWA	0193
R5011010XXWC	0193
R5011015	0193
R5011015XXWA	0193
R5011015XXWC	0193
R5020108CJ	OEM
R5020108CJWA	OEM
R5020108EJ	OEM
R5020108EJWA	OEM
R5020108FJ	OEM
R5020108FJWA	OEM
R5020110CJ	OEM
R5020110EJ	OEM
R5020110FJ	OEM
R5020208CJ	OEM
R5020208EJ	OEM
R5020208FJ	OEM
R5020210CJ	OEM
R5020210EJ	OEM
R5020210FJ	OEM

DEVICE TYPE	REPL CODE
R5020308CJ	OEM
R5020308EJ	OEM
R5020308FJ	OEM
R5020310CJ	OEM
R5020310EJ	OEM
R5020310FJ	OEM
R5020408CJ	OEM
R5020408EJ	OEM
R5020408FJ	OEM
R5020410CJ	OEM
R5020410EJ	OEM
R5020410FJ	OEM
R5020508CJ	OEM
R5020508EJ	OEM
R5020508FJ	OEM
R5020510CJ	OEM
R5020510EJ	OEM
R5020510FJ	OEM
R5020608CJ	OEM
R5020608EJ	OEM
R5020608FJ	OEM
R5020610CJ	OEM
R5020610EJ	OEM
R5020610FJ	OEM
R5020708CJ	OEM
R5020708EJ	OEM
R5020708FJ	OEM
R5020710CJ	OEM
R5020710EJ	OEM
R5020710FJ	OEM
R5020808CJ	OEM
R5020808EJ	OEM
R5020808FJ	OEM
R5020810CJ	OEM
R5020810EJ	OEM
R5020810FJ	OEM
R5020908CJ	OEM
R5020908EJ	OEM
R5020908FJ	OEM
R5020910CJ	OEM
R5020910EJ	OEM
R5020910FJ	OEM
R5021008CJ	OEM
R5021008EJ	OEM
R5021008FJ	OEM
R5021010CJ	OEM
R5021010EJ	OEM
R5021010FJ	OEM
R5021108CJ	OEM
R5021108EJ	OEM
R5021108FJ	OEM
R5021110CJ	OEM
R5021110EJ	OEM
R5021110FJ	OEM
R5021208CJ	OEM
R5021208EJ	OEM
R5021208FJ	OEM
R5021210CJ	OEM
R5021210EJ	OEM
R5021210FJ	OEM
R5021408CJ	OEM
R5021408EJ	OEM
R5021408FJ	OEM
R5021410CJ	OEM
R5021410EJ	OEM
R5021410FJ	OEM
R5021608CJ	OEM
R5021608EJ	OEM
R5021608FJ	OEM
R5021610CJ	OEM
R5021610EJ	OEM
R5021610FJ	OEM
R5030108CJ	OEM
R5030108EJ	OEM
R5030108FJ	OEM
R5030110CJ	OEM
R5030110EJ	OEM
R5030110FJ	OEM
R5030208CJ	OEM
R5030208EJ	OEM
R5030208FJ	OEM
R5030210CJ	OEM
R5030210EJ	OEM
R5030210FJ	OEM
R5030308CJ	OEM
R5030308EJ	OEM
R5030308FJ	OEM
R5030310CJ	OEM
R5030310EJ	OEM
R5030310FJ	OEM
R5030408CJ	OEM
R5030408EJ	OEM
R5030408FJ	OEM
R5030410CJ	OEM
R5030410EJ	OEM
R5030410FJ	OEM
R5030508CJ	OEM
R5030508EJ	OEM
R5030508FJ	OEM
R5030510CJ	OEM
R5030510EJ	OEM
R5030510FJ	OEM
R5030608CJ	OEM
R5030608EJ	OEM
R5030608FJ	OEM
R5030610CJ	OEM
R5030610EJ	OEM
R5030610FJ	OEM
R5030708CJ	OEM

DEVICE TYPE	REPL CODE
R5030708EJ	OEM
R5030708FJ	OEM
R5030710CJ	OEM
R5030710EJ	OEM
R5030710FJ	OEM
R5030808CJ	OEM
R5030808EJ	OEM
R5030808FJ	OEM
R5030810CJ	OEM
R5030810EJ	OEM
R5030810FJ	OEM
R5030908CJ	OEM
R5030908EJ	OEM
R5030908FJ	OEM
R5030910CJ	OEM
R5030910EJ	OEM
R5030910FJ	OEM
R5031008CJ	OEM
R5031008EJ	OEM
R5031008FJ	OEM
R5031010CJ	OEM
R5031010EJ	OEM
R5031010FJ	OEM
R5031108CJ	OEM
R5031108EJ	OEM
R5031108FJ	OEM
R5031110CJ	OEM
R5031110EJ	OEM
R5031110FJ	OEM
R5031208CJ	OEM
R5031208EJ	OEM
R5031208FJ	OEM
R5031210CJ	OEM
R5031210EJ	OEM
R5031210FJ	OEM
R5031408CJ	OEM
R5031408EJ	OEM
R5031408FJ	OEM
R5031410CJ	OEM
R5031410EJ	OEM
R5031410FJ	OEM
R5031608CJ	OEM
R5031608EJ	OEM
R5031608FJ	OEM
R5031610CJ	OEM
R5031610EJ	OEM
R5031610FJ	OEM
R5100110	0599
R5100110XXWA	0599
R5100110XXWC	0599
R5100115	0599
R5100115XXWA	0599
R5100115XXWC	0599
R5100210	0599
R5100210XXWA	0599
R5100210XXWC	0599
R5100215	0599
R5100215WA	0599
R5100215XXWA	0599
R5100215XXWC	0599
R5100310	0599
R5100310XXWA	0599
R5100310XXWC	0599
R5100315	0599
R5100315XXWA	0599
R5100315XXWC	0599
R5100410	1986
R5100410XXWA	1986
R5100410XXWC	1986
R5100415	1986
R5100415XXWA	1986
R5100415XXWC	1986
R5100510	1986
R5100510XXWA	1986
R5100510XXWC	1986
R5100515	1986
R5100515XXWA	1986
R5100515XXWC	1986
R5100610	1986
R5100610XXWA	1986
R5100610XXWC	1986
R5100615	1986
R5100615XXWA	1986
R5100615XXWC	1986
R5100710	0657
R5100710XXWA	0657
R5100710XXWC	0657
R5100715	0657
R5100715XXWA	0657
R5100715XXWC	0657
R5100810	0657
R5100810XXWA	0657
R5100810XXWC	0657
R5100815	0657
R5100815XXWA	0657
R5100815XXWC	0657
R5100910	0657
R5100910XXWA	0657
R5100910XXWC	0657
R5100915	0657
R5100915XXWA	0657
R5100915XXWC	0657
R5101010	0657
R5101010XXWA	0657
R5101010XXWC	0657
R5101015	0657
R5101015XXWA	0657
R5101015XXWC	0657
R5110110	1347

DEVICE TYPE	REPL CODE
R5110110XXWA	1347
R5110110XXWC	1347
R5110115	1347
R5110115XXWA	1347
R5110115XXWC	1347
R5110210	1347
R5110210XXWA	1347
R5110210XXWC	1347
R5110215	1347
R5110215XXWA	1347
R5110215XXWC	1347
R5110310	1347
R5110310XXWA	1347
R5110310XXWC	1347
R5110315	1347
R5110315XXWA	1347
R5110315XXWC	1347
R5110410	1905
R5110410XXWA	1905
R5110410XXWC	1905
R5110415	1905
R5110415XXWA	1905
R5110415XXWC	1905
R5110510	1905
R5110510XXWA	1905
R5110510XXWC	1905
R5110515	1905
R5110515XXWA	1905
R5110515XXWC	1905
R5110610	1905
R5110610XXWA	1905
R5110610XXWC	1905
R5110615	1905
R5110615XXWA	1905
R5110615XXWC	1905
R5110710	0193
R5110710XXWA	0193
R5110710XXWC	0193
R5110715	0193
R5110715XXWA	0193
R5110715XXWC	0193
R5110810	0193
R5110810XXWA	0193
R5110810XXWC	0193
R5110815	0193
R5110815XXWA	0193
R5110815XXWC	0193
R5110910	0193
R5110910XXWA	0193
R5110910XXWC	0193
R5110915	0193
R5110915XXWA	0193
R5110915XXWC	0193
R5111010	0193
R5111010XXWA	0193
R5111015	0193
R5111015XXWA	0193
R5111015XXWC	0193
R6000124	1023
R6000124R	0501
R6000128	1023
R6000128R	0501
R6000224	1023
R6000224R	0501
R6000228	1023
R6000228R	0501
R6000324	1023
R6000324R	0501
R6000328	1023
R6000328R	0501
R6000424	1023
R6000424R	0501
R6000428	1023
R6000524	1036
R6000524R	1777
R6000528	1036
R6000528R	1777
R6000624	1036
R6000624R	1777
R6000628	1036
R6000628R	1777
R6000724	1046
R6000724R	1789
R6000728	1046
R6000728R	1789
R6000824	1046
R6000824R	1789
R6000828	1046
R6000828R	1789
R6000924	1046
R6000924R	1789
R6000928	1046
R6000928R	1789
R6001024	1046
R6001024R	1789
R6001028	1046
R6001028R	1789
R6001220	3928
R6001220YA	2832
R6001225	3928
R6001225YA	2832
R6001230	3928
R6001230YA	2832
R6001420	2832
R6001420YA	2832
R6001425	2832
R6001430YA	2832
R6001620	OEM
R6001625	OEM

DEVICE TYPE	REPL CODE
R6001820	OEM
R6001825	OEM
R6002020	OEM
R6002025	OEM
R6002220	OEM
R6002420	OEM
R6002620	OEM
R6002820	OEM
R6002830YA	OEM
R6003020	OEM
R6003030YA	OEM
R6011220	3531
R6011225	3531
R6011230	3531
R6011420	3167
R6011425	3167
R6011620	OEM
R6011625	OEM
R6011820	OEM
R6011825	OEM
R6012020	OEM
R6012025	OEM
R6012220	OEM
R6012420	OEM
R6012620	OEM
R6012820	OEM
R6013020	OEM
R6020108CJ	OEM
R6020108EJ	OEM
R6020108FJ	OEM
R6020110CJ	OEM
R6020110EJ	OEM
R6020110FJ	OEM
R6020208CJ	OEM
R6020208EJ	OEM
R6020208FJ	OEM
R6020210CJ	OEM
R6020210EJ	OEM
R6020210FJ	OEM
R6020225CJ	OEM
R6020225EJ	OEM
R6020225FJ	OEM
R6020308CJ	OEM
R6020308EJ	OEM
R6020308FJ	OEM
R6020310CJ	OEM
R6020310EJ	OEM
R6020310FJ	OEM
R6020408CJ	OEM
R6020408EJ	OEM
R6020408FJ	OEM
R6020410CJ	OEM
R6020410EJ	OEM
R6020410FJ	OEM
R6020425CJ	OEM
R6020425EJ	OEM
R6020425FJ	OEM
R6020508CJ	OEM
R6020508EJ	OEM
R6020508FJ	OEM
R6020510CJ	OEM
R6020510EJ	OEM
R6020510FJ	OEM
R6020608CJ	OEM
R6020608EJ	OEM
R6020608FJ	OEM
R6020610CJ	OEM
R6020610EJ	OEM
R6020610FJ	OEM
R6020625CJ	OEM
R6020625EJ	OEM
R6020625FJ	OEM
R6020708CJ	OEM
R6020708EJ	OEM
R6020708FJ	OEM
R6020710CJ	OEM
R6020710EJ	OEM
R6020710FJ	OEM
R6020808CJ	OEM
R6020808EJ	OEM
R6020808FJ	OEM
R6020810CJ	OEM
R6020810EJ	OEM
R6020810FJ	OEM
R6020825CJ	OEM
R6020825EJ	OEM
R6020825FJ	OEM
R6020908CJ	OEM
R6020908EJ	OEM
R6020908FJ	OEM
R6020910CJ	OEM
R6020910EJ	OEM
R6020910FJ	OEM
R6021008CJ	OEM
R6021008EJ	OEM
R6021008FJ	OEM
R6021010CJ	OEM
R6021010EJ	OEM
R6021010FJ	OEM
R6021025CJ	OEM
R6021025EJ	OEM
R6021025FJ	OEM
R6021108CJ	OEM
R6021108EJ	OEM
R6021108FJ	OEM
R6021110CJ	OEM
R6021110EJ	OEM
R6021110FJ	OEM
R6021208CJ	OEM

DEVICE TYPE	REPL CODE
R6021208EJ	OEM
R6021208FJ	OEM
R6021210CJ	OEM
R6021210EJ	OEM
R6021210FJ	OEM
R6021225CJ	OEM
R6021225EJ	OEM
R6021225FJ	OEM
R6021408CJ	OEM
R6021408EJ	OEM
R6021408FJ	OEM
R6021410CJ	OEM
R6021410EJ	OEM
R6021410FJ	OEM
R6021420FJYA	OEM
R6021425FJYA	OEM
R6021608CJ	OEM
R6021608EJ	OEM
R6021608FJ	OEM
R6021610CJ	OEM
R6021610EJ	OEM
R6021610FJ	OEM
R6021620FJYA	OEM
R6021625FJYA	OEM
R6030108CJ	OEM
R6030108EJ	OEM
R6030108FJ	OEM
R6030110CJ	OEM
R6030110EJ	OEM
R6030110FJ	OEM
R6030208CJ	OEM
R6030208EJ	OEM
R6030208FJ	OEM
R6030210CJ	OEM
R6030210EJ	OEM
R6030210FJ	OEM
R6030308CJ	OEM
R6030308EJ	OEM
R6030308FJ	OEM
R6030310CJ	OEM
R6030310EJ	OEM
R6030310FJ	OEM
R6030408CJ	OEM
R6030408EJ	OEM
R6030408FJ	OEM
R6030410CJ	OEM
R6030410EJ	OEM
R6030410FJ	OEM
R6030508CJ	OEM
R6030508EJ	OEM
R6030508FJ	OEM
R6030510CJ	OEM
R6030510EJ	OEM
R6030510FJ	OEM
R6030608CJ	OEM
R6030608EJ	OEM
R6030608FJ	OEM
R6030610CJ	OEM
R6030610EJ	OEM
R6030610FJ	OEM
R6030708CJ	OEM
R6030708EJ	OEM
R6030708FJ	OEM
R6030710CJ	OEM
R6030710EJ	OEM
R6030710FJ	OEM
R6030808CJ	OEM
R6030808EJ	OEM
R6030808FJ	OEM
R6030810CJ	OEM
R6030810EJ	OEM
R6030810FJ	OEM
R6030908CJ	OEM
R6030908EJ	OEM
R6030908FJ	OEM
R6030910CJ	OEM
R6030910EJ	OEM
R6030910FJ	OEM
R6031008CJ	OEM
R6031008EJ	OEM
R6031008FJ	OEM
R6031010CJ	OEM
R6031010EJ	OEM
R6031010FJ	OEM
R6031108CJ	OEM
R6031108EJ	OEM
R6031108FJ	OEM
R6031110CJ	OEM
R6031110EJ	OEM
R6031110FJ	OEM
R6031208CJ	OEM
R6031208EJ	OEM
R6031208FJ	OEM
R6031210CJ	OEM
R6031210EJ	OEM
R6031210FJ	OEM
R6031408CJ	OEM
R6031408EJ	OEM
R6031408FJ	OEM
R6031410CJ	OEM
R6031410EJ	OEM
R6031410FJ	OEM
R6031608CJ	OEM
R6031608EJ	OEM
R6031608FJ	OEM
R6031610CJ	OEM
R6031610EJ	OEM
R6031610FJ	OEM
R6100120	1023

DEVICE TYPE	REPL CODE
R6100120XXYA	1023
R6100120YA	1023
R6100125	1023
R6100125XXYA	1023
R6100125YA	1023
R6100130	1023
R6100130XXYA	1023
R6100130YA	1023
R6100220	1023
R6100220XXYA	1023
R6100220YA	1023
R6100225	1023
R6100225XXYA	1023
R6100225YA	1023
R6100230	1023
R6100230XXYA	1023
R6100230YA	1023
R6100320	1023
R6100325	1023
R6100330	1023
R6100420	1023
R6100420XXYA	1036
R6100420YA	1023
R6100425	1036
R6100425XXYA	1036
R6100425YA	1023
R6100430	1023
R6100430XXYA	1036
R6100430YA	1023
R6100520	1036
R6100525	1036
R6100530	1036
R6100620	1036
R6100620XXYA	1036
R6100620YA	1036
R6100625	1036
R6100625XXYA	1036
R6100625YA	1036
R6100630	1036
R6100630XXYA	1036
R6100630YA	1036
R6100720	1046
R6100725	1046
R6100730	1046
R6100820	1046
R6100820XXYA	1046
R6100820YA	1046
R6100825	1046
R6100825XXYA	1046
R6100825YA	1046
R6100830	1046
R6100830XXYA	1046
R6100830YA	1046
R6100920	1046
R6100925	1046
R6100930	1046
R6101020	1046
R6101020XXYA	1046
R6101020YA	1046
R6101025	1046
R6101025XXYA	1046
R6101025YA	1046
R6101030	1046
R6101030XXYA	1046
R6101030YA	1046
R6110120	0501
R6110120XXYA	0501
R6110125	0501
R6110125XXYA	0501
R6110130	0501
R6110130XXYA	0501
R6110220	0501
R6110220XXYA	0501
R6110225	0501
R6110225XXYA	0501
R6110230	0501
R6110230XXYA	0501
R6110320	0501
R6110325	0501
R6110330	0501
R6110420	1777
R6110420XXYA	1777
R6110425	0501
R6110425XXYA	1777
R6110430	0501
R6110430XXYA	1777
R6110520	1777
R6110525	1777
R6110530	1777
R6110620	1777
R6110620XXYA	1777
R6110625	1777
R6110625XXYA	1777
R6110630	1777
R6110630XXYA	1777
R6110720	1789
R6110725	1789
R6110730	1789
R6110820	1789
R6110820XXYA	1789
R6110825	1789
R6110825XXYA	1789
R6110830	1789
R6110830XXYA	1789
R6110920	1789
R6110925	1789
R6110930	1789
R6111020	1789
R6111020XXYA	1789
R6111025	1789
R6111025XXYA	1789
R6111030	1789
R6111030XXYA	1789
R6200130	1226
R6200140	1226
R6200150	1226
R6200230	1226
R6200240	1226
R6200250	1226
R6200330	1226
R6200340	1226
R6200350	1226
R6200430	1226
R6200440	1226
R6200450	1226
R6200530	1226
R6200540	1226
R6200550	1226
R6200630	1226
R6200640	1226
R6200650	1226
R6200830	1204
R6200840	1204
R6200850	1204
R6201030	1204
R6201040	1204
R6201050	1204
R6201230	1204
R6201240	1204
R6201250	1204
R6201430	OEM
R6201440	OEM
R6201630	OEM
R6201640	OEM
R6201830	OEM
R6201840	OEM
R6202030	OEM
R6202040	OEM
R6202230	OEM
R6202430	OEM
R6202630	OEM
R6202830	OEM
R6203030	OEM
R6220135CJ	OEM
R6220135EJ	OEM
R6220135FJ	OEM
R6220140CJ	OEM
R6220140EJ	OEM
R6220140FJ	OEM
R6220235CJ	OEM
R6220235EJ	OEM
R6220235FJ	OEM
R6220240CJ	OEM
R6220240EJ	OEM
R6220240FJ	OEM
R6220335CJ	OEM
R6220335EJ	OEM
R6220335FJ	OEM
R6220340CJ	OEM
R6220340EJ	OEM
R6220340FJ	OEM
R6220435CJ	OEM
R6220435EJ	OEM
R6220435FJ	OEM
R6220440CJ	OEM
R6220440EJ	OEM
R6220440FJ	OEM
R6220535CJ	OEM
R6220535EJ	OEM
R6220535FJ	OEM
R6220540CJ	OEM
R6220540EJ	OEM
R6220540FJ	OEM
R6220635CJ	OEM
R6220635EJ	OEM
R6220635FJ	OEM
R6220640CJ	OEM
R6220640EJ	OEM
R6220640FJ	OEM
R6220735CJ	OEM
R6220735EJ	OEM
R6220735FJ	OEM
R6220740CJ	OEM
R6220740EJ	OEM
R6220740FJ	OEM
R6220835CJ	OEM
R6220835EJ	OEM
R6220835FJ	OEM
R6220840CJ	OEM
R6220840EJ	OEM
R6220840FJ	OEM
R6220935CJ	OEM
R6220935EJ	OEM
R6220935FJ	OEM
R6220940CJ	OEM
R6220940EJ	OEM
R6220940FJ	OEM
R6221035CJ	OEM
R6221035EJ	OEM
R6221035FJ	OEM
R6221040CJ	OEM
R6221040EJ	OEM
R6221040FJ	OEM
R6221135CJ	OEM
R6221135EJ	OEM
R6221135FJ	OEM
R6221140CJ	OEM
R6221140EJ	OEM
R6221140FJ	OEM
R6221235CJ	OEM
R6221235EJ	OEM
R6221235FJ	OEM
R6221240CJ	OEM
R6221240EJ	OEM
R6221240FJ	OEM
R6221435CJ	OEM
R6221435EJ	OEM
R6221435FJ00	OEM
R6221440CJ	OEM
R6221440EJ	OEM
R6221440FJ00	OEM
R6221635CJ	OEM
R6221635EJ	OEM
R6221635FJ00	OEM
R6221640CJ	OEM
R6221640EJ	OEM
R6221640FJ00	OEM
R7000103	4286
R7000103UA	5046
R7000104	4286
R7000104UA	5046
R7000105	5046
R7000203	4286
R7000203UA	5046
R7000204	4286
R7000204UA	5046
R7000205	5046
R7000205UA	5046
R7000303	4286
R7000304	4286
R7000305	5046
R7000403	4286
R7000403UA	5046
R7000404	4286
R7000404UA	5046
R7000405	5046
R7000405UA	5046
R7000503	4286
R7000503UA	OEM
R7000504	4286
R7000504UA	OEM
R7000505	5046
R7000505UA	OEM
R7000603	4286
R7000603UA	5046
R7000604	4286
R7000604UA	5046
R7000605	5046
R7000605UA	5046
R7000703	4286
R7000704	4286
R7000705	0749
R7000803	4286
R7000804	4286
R7000805	0749
R7000903	4286
R7000904	4286
R7000905	0749
R7001003	4286
R7001003UA	0749
R7001004	4286
R7001004UA	0749
R7001005	0749
R7001005UA	0749
R7001203	4286
R7001203UA	0749
R7001204	4286
R7001204UA	0749
R7001205	0749
R7001205UA	0749
R7001403	4286
R7001403UA	4286
R7001404	4286
R7001404UA	4286
R7001603	4286
R7001603UA	4286
R7001604	4286
R7001604UA	4286
R7001803	OEM
R7001804	OEM
R7002003	OEM
R7002004	OEM
R7002203	OEM
R7002403	OEM
R7002603	OEM
R7002803	OEM
R7003003	OEM
R7003203	OEM
R7003403	OEM
R7003603	OEM
R7003803	OEM
R7004003	OEM
R7010103	4285
R7010104	4285
R7010105	2315
R7010203	4285
R7010204	4285
R7010205	2315
R7010303	4285
R7010304	4285
R7010305	2315
R7010403	4285
R7010404	4285
R7010405	2315
R7010503	4285
R7010504	4285
R7010505	2315
R7010603	4285
R7010604	4285
R7010605	2315
R7010703	4285
R7010704	4285
R7010705	2792
R7010803	4285
R7010804	4285
R7010805	2792
R7010903	4285
R7010904	4285
R7010905	2792
R7011003	4285
R7011004	4285
R7011005	2792
R7011203	4285
R7011204	4285
R7011205	2792
R7011403	4285
R7011404	4285
R7011603	4285
R7011604	4285
R7011803	OEM
R7011804	OEM
R7012003	OEM
R7012004	OEM
R7012203	OEM
R7012403	OEM
R7012603	OEM
R7012803	OEM
R7013003	OEM
R7013203	OEM
R7013403	OEM
R7013603	OEM
R7013803	OEM
R7014003	OEM
R7200106	2159
R7200108	OEM
R7200109	2159
R7200110	OEM
R7200112	2159
R7200205	OEM
R7200206	2159
R7200208	OEM
R7200209	2159
R7200210	OEM
R7200212	2159
R7200305	OEM
R7200306	2159
R7200308	OEM
R7200309	2159
R7200310	OEM
R7200312	2159
R7200405	OEM
R7200406	2159
R7200408	OEM
R7200409	2159
R7200410	OEM
R7200412	2159
R7200505	OEM
R7200506	2159
R7200508	OEM
R7200509	2159
R7200510	OEM
R7200512	2159
R7200605	OEM
R7200606	2159
R7200608	OEM
R7200609	2159
R7200610	OEM
R7200612	2159
R7200705	OEM
R7200708	OEM
R7200710	OEM
R7200805	OEM
R7200806	0479
R7200808	OEM
R7200809	0479
R7200810	OEM
R7200812	0479
R7200905	OEM
R7200908	OEM
R7200910	OEM
R7201005	OEM
R7201006	0479
R7201008	OEM
R7201009	0479
R7201010	OEM
R7201012	0479
R7201205	OEM
R7201206	0479
R7201208	OEM
R7201209	0479
R7201210	OEM
R7201212	0479
R7201405	OEM
R7201406	OEM
R7201408	OEM
R7201409	OEM
R7201605	OEM
R7201606	OEM
R7201608	OEM
R7201609	OEM
R7201805	OEM
R7201806	OEM
R7201808	OEM
R7201809	OEM
R7202005	OEM
R7202006	OEM
R7202008	OEM
R7202009	OEM
R7202205	OEM
R7202206	OEM
R7202405	OEM
R7202406	OEM
R7202605	OEM
R7202606	OEM
R7202805	OEM
R7202806	OEM
R7203005	OEM
R7203006	OEM
R7203015	OEM
R7203205	OEM
R7203405	OEM
R7203605	OEM
R7203805	OEM
R7204005	OEM
R7220105AJ	OEM
R7220105CJ	OEM
R7220105EJ	OEM
R7220106CJ	OEM
R7220106EJ	OEM
R7220108CJ	OEM
R7220108EJ	OEM
R7220205AJ	OEM
R7220205CJ	OEM
R7220205EJ	OEM
R7220206CJ	OEM
R7220206EJ	OEM
R7220208CJ	OEM
R7220208EJ	OEM
R7220305AJ	OEM
R7220305CJ	OEM
R7220305EJ	OEM
R7220306CJ	OEM
R7220306EJ	OEM
R7220308CJ	OEM
R7220308EJ	OEM
R7220405AJ	OEM
R7220405CJ	OEM
R7220405EJ	OEM
R7220406CJ	OEM
R7220406EJ	OEM
R7220408CJ	OEM
R7220408EJ	OEM
R7220505AJ	OEM
R7220505CJ	OEM
R7220505EJ	OEM
R7220506CJ	OEM
R7220506EJ	OEM
R7220508CJ	OEM
R7220508EJ	OEM
R7220605AJ	OEM
R7220605CJ	OEM
R7220605EJ	OEM
R7220606CJ	OEM
R7220606EJ	OEM
R7220608CJ	OEM
R7220608EJ	OEM
R7220705AJ	OEM
R7220705CJ	OEM
R7220705EJ	OEM
R7220706CJ	OEM
R7220706EJ	OEM
R7220708CJ	OEM
R7220708EJ	OEM
R7220805AJ	OEM
R7220805CJ	OEM
R7220805EJ	OEM
R7220806CJ	OEM
R7220806EJ	OEM
R7220808CJ	OEM
R7220808EJ	OEM
R7220905AJ	OEM
R7220905CJ	OEM
R7220905EJ	OEM
R7220906CJ	OEM
R7220906EJ	OEM
R7220908CJ	OEM
R7220908EJ	OEM
R7221005AJ	OEM
R7221005CJ	OEM
R7221005EJ	OEM
R7221006CJ	OEM
R7221006EJ	OEM
R7221008CJ	OEM
R7221008EJ	OEM
R7221105AJ	OEM
R7221105CJ	OEM
R7221105EJ	OEM
R7221106CJ	OEM
R7221106EJ	OEM
R7221108CJ	OEM
R7221108EJ	OEM
R7221205AJ	OEM
R7221205CJ	OEM
R7221205EJ	OEM
R7221206CJ	OEM
R7221206EJ	OEM
R7221208CJ	OEM
R7221208EJ	OEM
R7221305EJ	OEM
R7221306EJ	OEM
R7221308EJ	OEM
R7221405AJ	OEM
R7221405CJ	OEM
R7221405EJ	OEM
R7221406CJ	OEM
R7221406EJ	OEM
R7221408CJ	OEM
R7221408EJ	OEM
R7221505EJ	OEM
R7221506EJ	OEM
R7221508EJ	OEM
R7221605AJ	OEM
R7221605CJ	OEM
R7221605EJ	OEM
R7221606CJ	OEM
R7221606EJ	OEM
R7221608CJ	OEM
R7221608EJ	OEM
R7221705CJ	OEM
R7221705EJ	OEM
R7221706AJ	OEM
R7221706CJ	OEM
R7221706EJ	OEM
R7221708AJ	OEM
R7221708CJ	OEM
R7221708EJ	OEM
R7221805AJ	OEM
R7221805CJ	OEM
R7221805EJ	OEM
R7221806AJ	OEM
R7221806CJ	OEM
R7221806EJ	OEM
R7221808AJ	OEM
R7221808CJ	OEM
R7221808EJ	OEM
R7221905CJ	OEM
R7221905EJ	OEM
R7221906AJ	OEM
R7221906CJ	OEM
R7221906EJ	OEM
R7221908AJ	OEM
R7221908CJ	OEM
R7221908EJ	OEM
R7222005AJ	OEM
R7222005CJ	OEM
R7222005EJ	OEM
R7222006AJ	OEM
R7222006CJ	OEM
R7222006EJ	OEM
R7222008AJ	OEM
R7222008CJ	OEM
R7222008EJ	OEM
R7222105CJ	OEM
R7222105EJ	OEM
R7222106AJ	OEM
R7222106CJ	OEM
R7222106EJ	OEM
R7222108AJ	OEM
R7222108CJ	OEM
R7222108EJ	OEM
R7222205AJ	OEM
R7222205CJ	OEM
R7222205EJ	OEM
R7222206AJ	OEM
R7222206CJ	OEM
R7222206EJ	OEM
R7222208AJ	OEM
R7222208CJ	OEM
R7222208EJ	OEM
R7222305CJ	OEM
R7222305EJ	OEM
R7222306AJ	OEM
R7222306CJ	OEM
R7222306EJ	OEM
R7222308AJ	OEM
R7222308CJ	OEM
R7222308EJ	OEM
R7222405AJ	OEM
R7222405CJ	OEM
R7222405EJ	OEM
R7222406AJ	OEM
R7222406CJ	OEM
R7222406EJ	OEM
R7222408AJ	OEM
R7222408CJ	OEM
R7222408EJ	OEM
R7222505AJ	OEM
R7222505CJ	OEM
R7222505EJ	OEM
R7222506AJ	OEM
R7222506CJ	OEM
R7222506EJ	OEM
R7222508AJ	OEM
R7222508CJ	OEM
R7222508EJ	OEM
R9200116	OEM
R9200120	OEM
R9200216	OEM
R9200220	OEM
R9200316	OEM
R9200320	OEM
R9200416	OEM
R9200420	OEM
R9200516	OEM
R9200520	OEM
R9200616	OEM
R9200620	OEM
R9200816	OEM
R9200820	OEM
R9201016	OEM
R9201020	OEM
R9201216	OEM
R9201220	OEM
R9201416	OEM
R9201420	OEM
R9201616	OEM
R9201620	OEM
R9201816	OEM
R9201820	OEM
R9202016	OEM
R9202020	OEM
R40501060	1818
R40501070	1818
R620013000	1226
R620014000	1226
R620015000	1226
R620023000	1226
R620024000	1226
R620025000	1226
R620043000	1226
R620044000	1226
R620045000	1226
R620063000	1226
R620064000	1226
R620065000	1226
R620083000	1204
R620085000	1204
R620103000	1204
R620104000	1204
R620105000	1204
R620124000	1204
R620125000	1204
R620145000	OEM
R620165000	OEM
R620185000	OEM
R620205000	OEM
R620224000	OEM
R620225000	OEM
R620244000	OEM
R620245000	OEM
R620264000	OEM
R620265000	OEM
R620284000	OEM
R620285000	OEM
R620304000	OEM
R620305000	OEM
R720010600	2159
R720010900	2159
R720011200	2159
R720020600	2159
R720021200	2159
R720040600	2159
R720040900	2159
R720041200	2159
R720060600	2159
R720060900	2159
R720061200	2159
R720080600	0479
R720080900	0479
R720081200	0479
R720100600	0479
R720100900	0479
R720101200	0479
R720120900	0479
R720121200	0479
R720141200	OEM
R720161200	OEM
R720181200	OEM
R720201200	OEM
R720220900	OEM
R720221200	OEM
R720240900	OEM
R720241200	OEM
R720260900	OEM
R720261200	OEM
R720280900	OEM
R720281200	OEM
R720300900	OEM
R720301200	OEM
R920211200	OEM
R920211600	OEM
R920221600	OEM
R920222000	OEM
R920231600	OEM
R920232000	OEM
R920241600	OEM
R920242000	OEM
R920251600	OEM
R920252000	OEM
R920261600	OEM
R920262000	OEM
R920271600	OEM
R920281600	OEM
R920282000	OEM
R920291600	OEM
R920292000	OEM
R920301600	OEM
R920302000	OEM
RA-1	0015
RA1	0102
RA-1A	0015
RA1A	0133
RA-1B	0072
RA1B	0015
RA-1C	0071
RA1C	0071
RA-1D	0102
RA-1E	0102
RA-1F	0348
RA-1Y	0080
RA-1Z	0023
RA1Z	0023

If replacement code is OEM, contact original manufacturer for replacement.

DEVICE TYPE	REPL CODE
RA-1ZC	0015
RA1ZC	0015
RA-2	0071
RA2	0502
RA2A	0071
RA2B	OEM
RA-2C	0071
RA2C	0071
RA3	OEM
RA-3-9600	OEM
RA-3-9600A	OEM
RA3A	OEM
RA3B	OEM
RA4	OEM
RA5E	0710
RA21	0502
RA25	0502
RA-26	1834
RA26	0502
RA26(50V)	0531
RA102AA	0015
RA102BA	0015
RA102CA	0015
RA102DA	0015
RA102EA	0015
RA102FA	0015
RA102GA	0015
RA102HA	0015
RA102KA	0015
RA102MA	0015
RA102PA	0071
RA102RA	0071
RA102TA	0071
RA102VA	0071
RA132AA	0071
RA132BA	0071
RA132CA	0071
RA132DA	0071
RA132EA	0071
RA132FA	0071
RA132GA	0071
RA132HA	0071
RA132JA	OEM
RA132KA	0071
RA132MA	0071
RA132PA	0071
RA132RA	0071
RA132TA	0071
RA132VA	0071
RA162AA	0071
RA162BA	0071
RA162CA	0071
RA162DA	0071
RA162EA	0071
RA162FA	0071
RA162GA	0071
RA162HA	0071
RA162KA	0071
RA162MA	0071
RA162PA	0071
RA162RA	0071
RA162TA	0071
RA162VA	0071
RA251	OEM
RA252	OEM
RA254	OEM
RA256	OEM
RA258	OEM
RA751AA	0015
RA751BA	0015
RA751CA	0015
RA751DA	0015
RA751EA	0015
RA751FA	0015
RA751GA	0015
RA751HA	0015
RA751KA	0015
RA751MA	0015
RA751PA	0071
RA751RA	0071
RA751TA	0071
RA751VA	0071
RA2505	OEM
RA2510	OEM
RA4153DC	OEM
RA39600	OEM
RANDY	6235
RA-Z	0102
RB7-M331	OEM
RB7M-17	OEM
RB7M-37	OEM
RB40	0473
RB-40C	0473
RB40C	0473
RB40N	0473
RB60	OEM
RB-100A	OEM
RB100A	OEM
RB-150	0106
RB150	0106
RB150Z	OEM
RB-151	0106
RB151	0106
RB151Z	OEM
RB-152	0106
RB152	0106
RB-152LF-F	OEM
RB152LF-F	OEM
RB152Z	OEM
RB153	0106
RB-154	0196
RB154	0106
RB154Z	OEM
RB155	2010
RB-156	2400
RB156	2400
RB-156-LFB	2400
RB156LB	2400
RB-156LFB	2400
RB156LFB	2010
RB156Z	OEM
RB157	0787
RB158	OEM
RB158Z	OEM
RB401	0323
RB402	1414
RB-402U	1414
RB404	0473
RB-404N	0473
RB406	0446
RB-406N	0446
RB406N	0446
RB406NH	0473
RB-601	0323
RB601	OEM
RB-602	1414
RB602	5694
RB602-A	1414
RB602A	OEM
RB-602U	1414
RB-604	0473
RB-606	0446
RBA-401	OEM
RBA402	OEM
RBA406B	OEM
RBS520T	OEM
RBS530T	OEM
RBS535T	OEM
RBS540T	OEM
RBS545T	OEM
RBS820T	OEM
RBS830T	OEM
RBS835T	OEM
RBS840T	OEM
RBS845T	OEM
RBS850T	OEM
RBS860T	OEM
RBS1020T	OEM
RBS1030T	OEM
RBS1035T	OEM
RBS1040T	OEM
RBS1045T	OEM
RBV-40C	0742
RBV-401	0729
RBV-402	2089
RBV402	OEM
RBV-402L	OEM
RBV402L	OEM
RBV-406	1045
RBV406	1045
RBV-406H	1045
RBV406H	5461
RBV-406H-01	1045
RBV-406M	1045
RBV406M	1045
RBV406M-LAF	1774
RBV-406M-LFA	1045
RBV-406MLFA	1774
RBV406M-LFA	1045
RBV-408	0742
RBV-602	1040
RBV602	OEM
RBV602-01	OEM
RBV-604	1774
RBV-606	1045
RBV606	1045
RBV606LF-A	OEM
RBV-608	OEM
RC	OEM
RC080	0071
RC-2	0348
RC2	0348
RC-2V	0348
RC2V	0348
RC78L05A	1292
RC78L09A	OEM
RC78L12A	1828
RC78M05A	OEM
RC78M05E	OEM
RC78M05FA	0624
RC78M09A	OEM
RC78M09FA	OEM
RC78M93FD	OEM
RC185-6	3537
RC201	OEM
RC202	OEM
RC203	OEM
RC204	OEM
RC205	OEM
RC206	OEM
RC207	OEM
RC529	OEM
RC529A	OEM
RC529B	OEM
RC529C	OEM
RC555	0973
RC555DE	0973
RC555NB	0973
RC555T	6772
RC556DB	3285
RC556DC	3285
RC709D	1705
RC709DC	1705
RC709T	1301
RC710DC	1800
RC710T	1797
RC714CDE	2246
RC714CH	0896
RC714EDE	2246
RC714EH	0896
RC714LDE	2246
RC714LH	0896
RC723D	0026
RC723T	1192
RC741D	1982
RC741DC	1982
RC741DE	0312
RC741DN	0312
RC741H	0410
RC741N	0312
RC741NB	0312
RC741T	0410
RC747DB	2358
RC747DC	2358
RC747T	2368
RC1458DE	0360
RC1458DN	0360
RC1458H	3137
RC1458NB	0360
RC1458T	3137
RC1700	0103
RC2041M	0360
RC2041NB	OEM
RC2043M	OEM
RC2043NB	0360
RC2403M	OEM
RC2403NB	0631
RC3078DE	OEM
RC3078NB	OEM
RC3078T	OEM
RC3302DB	0176
RC3403ADB	OEM
RC3403ADC	OEM
RC3414M	OEM
RC4136	3024
RC4136DB	3024
RC4136DC	3024
RC4151DE	OEM
RC4151NB	OEM
RC4151T	OEM
RC4152DE	OEM
RC4152NB	OEM
RC4152T	OEM
RC4153ADC	OEM
RC4153DC	OEM
RC4156DB	2823
RC4156DC	2823
RC4191DE	OEM
RC4191NB	OEM
RC4192DE	OEM
RC4192NB	OEM
RC4193DE	OEM
RC4193NB	OEM
RC4194DB	OEM
RC4194DC	OEM
RC4194TK	OEM
RC4195NB	5837
RC4195T	OEM
RC4195TK	OEM
RC4200ADE	OEM
RC4200ANB	OEM
RC4200DE	OEM
RC4200NB	OEM
RC4260DB	OEM
RC4260DC	OEM
RC4391DE	OEM
RC4391NB	OEM
RC4444PU	OEM
RC4444R	OEM
RC4556M	OEM
RC4556NB	OEM
RC4558	0360
RC4558DE	0360
RC4558DQ	0360
RC4558JG	0360
RC4558M	OEM
RC4558NB	0360
RC4558P	0360
RC4558PS	OEM
RC4558PS-E20	OEM
RC4558PS-T1	OEM
RC4558T	3137
RC4559DE	0360
RC4559M	OEM
RC4559N	OEM
RC4559NB	0360
RC4559T	3137
RC4560M	OEM
RC4560NB	OEM
RC4562M	OEM
RC4562NB	OEM
RC4739DB	0692
RC4739DP	0692
RC4805DE	OEM
RC4805EDE	OEM
RC4805ET	OEM
RC4805T	OEM
RC5258	OEM
RC5532ADE	OEM
RC5532ANB	OEM
RC5532AT	OEM
RC5532DE	OEM
RC5532NB	OEM
RC5532T	OEM
RC5534ADE	OEM
RC5534ANB	OEM
RC5534AT	OEM
RC5534DE	OEM
RC5534NB	OEM
RC5534T	OEM
RC7809FA	OEM
RC7812FA	0334
RCA1A01	0086
RCA1A02	0126
RCA1A03	3329
RCA1A04	4587
RCA1A05	0126
RCA1A06	0086
RCA1A07	0086
RCA1A08	0126
RCA1A09	0187
RCA1A10	0439
RCA1A11	0187
RCA1A15	0187
RCA1A16	0439
RCA1A17	0086
RCA1A18	0086
RCA1A19	0126
RCA1B01	0543
RCA1B04	0363
RCA1B05	0363
RCA1B06	0177
RCA1B07	3373
RCA1B08	3374
RCA1B09	0363
RCA1C03	0236
RCA1C04	0681
RCA1C05	0423
RCA1C06	0854
RCA1C07	0423
RCA1C08	1199
RCA1C09	0423
RCA1C10	0423
RCA1C11	0854
RCA1C12	0236
RCA1C13	0681
RCA1C14	0423
RCA1C15	2235
RCA1C16	2237
RCA1E02	0178
RCA1E03	0792
RCA29	0042
RCA29A	0042
RCA29ASDH	0423
RCA29B	0236
RCA29BSDH	0236
RCA29C	0236
RCA29CSDH	0236
RCA29SDH	0423
RCA30	0925
RCA30A	0925
RCA30B	0681
RCA30C	0681
RCA31	0423
RCA31A	0423
RCA31ASDH	0423
RCA31B	0423
RCA31BSDH	0423
RCA31C	0423
RCA31CSDH	0423
RCA31SDH	0423
RCA32	0854
RCA32A	0854
RCA32C	1199
RCA41	0561
RCA41A	0423
RCA41B	0423
RCA41C	0423
RCA41DSH	0561
RCA41SDH	5735
RCA42	1199
RCA42A	1199
RCA42B	1199
RCA42C	1199
RCA105	1308
RCA120	1212
RCA121	1212
RCA125	0602
RCA126	0602
RCA205	1308
RCA370	1308
RCA371	1308
RCA410	0103
RCA411	0270
RCA413	0074
RCA423	0074
RCA431	0074
RCA508	4786
RCA520	0060
RCA521	0060
RCA900	2277
RCA901	2277
RCA-1000	3373
RCA1000	2427
RCA-1001	3373
RCA1001	2427
RCA3054	0423
RCA3055	0423
RCA3517	0004
RCA3773	6450
RCA3858	0004
RCA6263	3113
RCA6340	6725
RCA6341	6725
RCA8203	0602
RCA8203A	2237
RCA8203B	2237
RCA8350	3374
RCA8350A	3374
RCA8350B	3374
RCA8638C	3701
RCA8638D	3701
RCA8638E	3701
RCA8766	5419
RCA8766A	5419
RCA8766B	5419
RCA8766C	5419
RCA8766D	5419
RCA8766E	5419
RCA9116C	4144
RCA9116D	4144
RCA9116E	4144
RCA9213A	1465
RCA9213B	1465
RCA9213C	OEM
RCA34098	0136
RCA34099	0136
RCA34100	0136
RCA34101	0004
RCA34106	0004
RCA35953	0004
RCA35954	0004
RCA40231	0208
RCA40245	0007
RCA40246	0007
RCA40250	0178
RCA40395	0004
RCA40396N	0208
RCA40396P	0004
RCA44098	0136
RCA45190	0561
RCA45191	0561
RCA45192	0561
RCA45193	1199
RCA45194	1199
RCA45195	1199
RCA1466860-2	1197
RCC7022	0474
RCL10580	OEM
RCP111A	0285
RCP111B	0285
RCP111C	0285
RCP111D	0285
RCP113A	0285
RCP113B	0285
RCP113C	0285
RCP113D	0285
RCP115	0285
RCP115B	0285
RCP117	0285
RCP117B	0285
RCP700A	3327
RCP700B	3327
RCP700C	0164
RCP700D	0164
RCP701A	0824
RCP701B	0824
RCP701C	0824
RCP701D	0824
RCP702A	0164
RCP702B	3327
RCP702C	3327
RCP702D	3327
RCP703A	0824
RCP703B	0824
RCP703C	0824
RCP703D	0824
RCP704	3327
RCP704B	3327
RCP705	0824
RCP705B	0824
RCP706	3327
RCP706B	3327
RCP707	0824
RCP707B	0824
RCR43U05JM	OEM
RCR70AX12	OEM
RCR70AX16	OEM
RCR70AX20	OEM
RCR70AY4	OEM
RCR70AY6	OEM
RCR70AY8	OEM
RCR70AY10	OEM
RCR70AY12	OEM
RCR150AX12	OEM
RCR150AX16	OEM
RCR150AX20	OEM
RCR150AX24	OEM
RCR150AY4	OEM
RCR150AY6	OEM
RCR150AY10	OEM
RCR150AY12	OEM
RCR300AX12	OEM
RCR300AX16	OEM
RCR300AX20	OEM
RCR300AX24	OEM
RCR300AY4	OEM
RCR300AY6	OEM
RCR300AY8	OEM
RCR300AY10	OEM
RCR300AY12	OEM
RCS29	0934
RCS29A	0178
RCS29B	0178
RCS29C	0178
RCS30	0905
RCS30A	0905
RCS30B	0905
RCS30C	1199
RCS31	0178
RCS31A	0178
RCS31B	0178
RCS31C	0178
RCS32	1199
RCS32A	1199
RCS32B	1199
RCS32C	1199
RCS242	0103
RCS258	0130
RCS559	0792
RCS560	0792
RCS564	0074
RCS579	0642
RCS617	OEM
RCS618	1681
RCS880	0439
RCS881	0439
RCS882	0728
RCV2	0102
RD,11EBTA21R	0181
RD027	0441
RD082	0244
RD10E-N1	0064
RD2R4E	2874
RD2R4ESB1	OEM
RD2R7EB1	0760
RD2R7EB2	0760
RD2.0E	OEM
RD2.0EB	OEM
RD2.0EC	OEM
RD2.0EK	OEM
RD2.0ES	OEM
RD2.0ES-B1	OEM
RD2.0ESB1	OEM
RD2.0F	OEM
RD2.0FB	OEM
RD2.0FC	OEM
RD2.0M	OEM
RD2.0P	OEM
RD-2.2E	0244
RD2.2E	OEM
RD2.2EB	OEM
RD2.2EC	OEM
RD2.2EK	OEM
RD2.2ES	OEM
RD2.2ES-B2	OEM
RD2.2F	OEM
RD2.2FB	OEM
RD2.2FC	OEM
RD2.2M	OEM
RD2.2P	OEM
RD2.4E	1276
RD2.4EB	OEM
RD2.4EB2	OEM
RD2.4EC	OEM
RD2.4EK	OEM
RD2.4ES	1276
RD2.4ESB1	OEM
RD2.4F	1278
RD2.4FB	1278
RD2.4FC	OEM
RD2.4M	OEM
RD2.4M-B	OEM
RD2.4MB	OEM
RD2.4P	OEM
RD2.7E	0760
RD2.7E-B	0760
RD2.7EB	0760
RD2.7EB1	0760
RD2.7E-B2	0760
RD2.7EB2	0760
RD2.7EC	OEM
RD2.7EL	OEM
RD2.7EL1	0760
RD2.7EL2	OEM
RD2.7EN	OEM
RD2.7EN-1	0760
RD2.7EN1	0760
RD2.7EN2	OEM
RD2.7ES	0760
RD2.7ESAB1	OEM
RD2.7ESB	OEM
RD2.7F	1312
RD2.7FB	1312
RD2.7FB1	OEM
RD2.7FB2	OEM
RD2.7FC	OEM
RD2.7M	OEM
RD2.7MB	OEM
RD2.7MB1	OEM
RD2.7MB2	OEM
RD2.7P	OEM
RD-3	0015
RD3A/-1B4	OEM
RD3A-1B4	0015
RD3R0EB	0118
RD3R3EB	0299
RD3R6EB	0376
RD3R9EC	0036
RD3.0BB	0118
RD3.0E	0118
RD3.0E-B	0292
RD3.0EB	0118
RD3.0EB1	0118
RD3.0EB2	0118
RD3.0EC	OEM
RD3.0EL	OEM
RD3.0EL-1	0118
RD3.0EL1	0118
RD3.0EL2	OEM
RD3.0EN	OEM
RD3.0EN2	0118
RD3.0ES	0118
RD3.0ESB	OEM
RD3.0ES-B1	0118
RD3.0F	1713
RD3.0FB	6044
RD3.0FB1	OEM
RD3.0FB2	OEM
RD3.0FC	OEM
RD3.0M	OEM
RD3.0MB	OEM
RD3.0MB1	OEM
RD3.0MB2	OEM
RD3.0P	OEM
RD3.3E	0299
RD3.3E-B	0299
RD3.3EB	0299
RD3.3EB1	0299
RD3.3EB2	OEM
RD-3.3EC	0376
RD3.3EC	OEM
RD3.3EL	OEM
RD3.3EL1	OEM
RD3.3EL2	OEM
RD3.3EN	OEM
RD3.3EN1	OEM
RD3.3EN2	OEM
RD3.3ES	0299
RD3.3ESAB1	OEM
RD3.3ES-B2	OEM
RD3.3ESB2	OEM
RD3.3F	0292
RD3.3FB	0292
RD3.3FB1	OEM
RD3.3FB2	OEM
RD3.3M	OEM
RD3.3MB	OEM
RD3.3MB1	OEM
RD3.3MB2	OEM
RD3.3P	OEM
RD3.6E	0376
RD3.6E-B	0188
RD3.6EB	0376
RD3.6EB1	0376
RD3.6EB2	OEM
RD3.6EB2HF	OEM
RD3.6EC	0376
RD3.6EL	0376
RD3.6E-L1	0376
RD3.6EL1	0376
RD3.6EL1T	0376
RD3.6EL1Z	0376
RD3.6E-L2	0376
RD3.6EL2	OEM
RD3.6EN	0376
RD3.6E-N1	0376
RD3.6EN1	0376
RD3.6E-N2	0376
RD3.6EN2	0376
RD3.6ES	0376
RD3.6ES-B1	0376
RD3.6ESB1	0376
RD3.6ES-B2	0376
RD3.6ES-L1	OEM
RD3.6ESL1	OEM
RD3.6ES-T1B1	0376
RD3.6EST2L	0376
RD3.6F	0188
RD3.6FB	0376
RD3.6FB1	0376
RD3.6FB2	OEM
RD3.6FBI	0376
RD3.6FC	OEM
RD3.6M	OEM
RD3.6MB	OEM
RD3.6M-B1	OEM
RD3.6MB1	OEM
RD3.6MB2	OEM
RD3.6P	OEM
RD3.8E-B	OEM
RD3.9E	0036
RD3.9E-B	0036
RD3.9EB	0036
RD3.9E-B1	0036
RD3.9EB1	OEM
RD3.9EB2	0036
RD3.9EC	0036
RD3.9EL	0036
RD3.9EL1	0036
RD3.9EL2	0036
RD3.9EN	0036
RD3.9EN1	OEM

If replacement code is OEM, contact original manufacturer for replacement.

DEVICE TYPE	REPL CODE
RD3.9E-N2	0036
RD3.9EN2	0036
RD3.9E-N3	0036
RD3.9EN3	OEM
RD3.9ES	0036
RD3.9ESB	OEM
RD3.9ES-B1	OEM
RD3.9ESB1	OEM
RD3.9ES-B2	0036
RD3.9ESB2	OEM
RD3.9ES-T1B	OEM
RD3.9F	0456
RD3.9FB	0036
RD3.9FB1	OEM
RD3.9FB2	OEM
RD3.9FC	OEM
RD3.9M	OEM
RD3.9MB	OEM
RD3.9MB1	OEM
RD3.9MB2	OEM
RD3.9P	OEM
RD-4A	0188
RD4A	0036
RD-4AM	0188
RD4AM	0036
RD4B	0456
RD4R5ESB2	OEM
RD4R7	0140
RD4R7EB	0140
RD4R7EB2	0140
RD4R7JB2	0140
RD4R7JB3	0140
RD4R7JSB1	OEM
RD4R7JSB2	0140
RD4R7JSB3	0140
RD4Z	OEM
RD4.3B	0275
RD4.3E	0275
RD4.3E-B	0275
RD4.3EB	0533
RD4.3EB1	0275
RD4.3EB2	0275
RD4.3EB3	OEM
RD4.3EC	0451
RD4.3EL	0275
RD4.3EL2	OEM
RD4.3EL3	OEM
RD4.3EN	OEM
RD4.3EN1	OEM
RD4.3EN2	OEM
RD4.3EN3	OEM
RD4.3ES	0275
RD4.3ESAB1	OEM
RD4.3ESB	0275
RD4.3ES-B1	0275
RD4.3ESB1	0275
RD4.3ESB2	OEM
RD4.3ES-T1B1	0275
RD4.3E.1	OEM
RD4.3F	0533
RD4.3FB	6907
RD4.3FB1	OEM
RD4.3FB2	OEM
RD4.3FB3	OEM
RD4.3FC	OEM
RD4.3M	OEM
RD4.3MB	OEM
RD4.3MB1	OEM
RD4.3MB2	OEM
RD4.3MB3	OEM
RD4.5ESB2	OEM
RD4.7BE	0451
RD4.7E	0140
RD-4.7EB	0140
RD4.7EB	0140
RD4.7EB1	0140
RD4.7E-B2	0140
RD4.7EB2	0140
RD4.7EB3	0140
RD4.7EB3T	0140
RD4.7EB23	0140
RD4.7EC	0140
RD4.7EL	OEM
RD4.7EL1	OEM
RD4.7EL2	OEM
RD4.7EL3	OEM
RD4.7EN	OEM
RD4.7EN-1	0140
RD4.7EN1	0140
RD4.7EN2	OEM
RD4.7EN3	OEM
RD4.7ES	0140
RD4.7ESAB2	0140
RD4.7ESB	7151
RD4.7ES-B1	0140
RD4.7ESB1	0140
RD4.7ES-B2	0140
RD4.7ESB2	0140
RD4.7F	0451
RD4.7FB	0451
RD4.7FB1	OEM
RD4.7FB2	OEM
RD4.7FB3	OEM
RD4.7FC	OEM
RD4.7J	0140
RD4.7JB	OEM
RD4.7J-B1	0140
RD4.7JB1	0140
RD4.7J-B2	0140
RD4.7JB2	0140
RD4.7JB3	0140
RD4.7JS	0140
RD4.7JSB1	OEM
RD4.7JSB2	0140
RD4.7JSB3	0140
RD4.7M	0140
RD4.7MB	OEM
RD4.7MB1	OEM
RD4.7MB2	OEM
RD4.7MB3	OEM
RD4.7P	0451
RD5-6ESB3	OEM
RD-5A	0162
RD5A	0041
RD5AK	6019
RD5AL	0140
RD5AM	0140
RD5AN	0041
RD5B	0162
RD5BK	OEM
RD5BL	OEM
RD5BM	OEM
RD5BN	OEM
RD5C	2410
RD5CK	OEM
RD5CL	2403
RD5CM	2403
RD5CN	2113
RD5D	2410
RD5DK	OEM
RD5DL	2403
RD5DLM	OEM
RD5DM	2113
RD5DMN	OEM
RD5DN	2113
RD5E-C	0041
RD5H32A	OEM
RD5H32C	OEM
RD5R1EB	0041
RD5R1EB2	0041
RD5R1EBB1	0041
RD5R1JB	0041
RD5R1JB1	0041
RD5R1JB3	0041
RD5R6EB	0253
RD5R6EB2	0253
RD5R6JB	0253
RD5R6JB2	0253
RD5R6JB3	0253
RD5.1(E)B3	0041
RD5.1E	0041
RD-5.1EB	0041
RD5.1E-B	0041
RD5.1EB	0041
RD5.1EB(2)	0041
RD5.1EB(2)-T4	0041
RD5.1EB(3)	0041
RD5.1EB1	0041
RD5.1E-B2	0041
RD5.1EB-2	0041
RD5.1EB2	0041
RD5.1E-B3	0041
RD5.1EB3	0041
RD-5.1EC	0162
RD5.1E-C	0041
RD5.1EC	0041
RD5.1EL	OEM
RD5.1EL1	0041
RD5.1EL2	OEM
RD5.1EL2Z	0589
RD5.1EN	OEM
RD5.1E-N1	0041
RD5.1EN1	0041
RD5.1E-N2	0041
RD5.1EN2	0041
RD5.1E-N3	0041
RD5.1EN3	0041
RD5.1ES	0041
RD5.1ES(B3)	0589
RD5.1ES(B3)-Y	0589
RD5.1ESB	0041
RD5.1ES-B1	0041
RD5.1ESB1	0041
RD5.1ES-B2	0041
RD5.1ESB2	0041
RD5.1ES-B3	0589
RD5.1ESB3	0589
RD5.1ES-L2	OEM
RD5.1ES-T1B	0589
RD5.1ES-T1B1	0589
RD5.1ES-T2B2	0041
RD5.1ES-TIB	0589
RD5.1EV	0041
RD5.1E.3	OEM
RD5.1F	0162
RD5.1FB	0041
RD5.1FB1	0162
RD5.1FB2	0041
RD5.1F-B3	0162
RD5.1FB3	OEM
RD5.1FC	OEM
RD5.1J	0041
RD5.1JB	0041
RD5.1JB1	0041
RD5.1JB2	0041
RD5.1JB3	OEM
RD5.1JS	OEM
RD5.1JS2	OEM
RD5.1JSAB2	OEM
RD5.1JSAB2X	0041
RD5.1JSB2	5953
RD5.1JSB2X	0041
RD5.1M	OEM
RD5.1MB	OEM
RD5.1MB1	OEM
RD5.1M-B2	OEM
RD5.1MB2	OEM
RD5.1MB2-3	OEM
RD5.1M-B3	OEM
RD5.1P	OEM
RD5.2EC	OEM
RD5.2MB3	OEM
RD5.6-B2Z	0253
RD5.6B	0157
RD5.6E	0253
RD-5.6EB	0253
RD5.6E-B	0253
RD5.6EB	0253
RD5.6EB(1)	0253
RD5.6EB(2)	0253
RD5.6EB(2)-T4	0253
RD5.6EB1	0041
RD5.6E-B2	0253
RD5.6EB2	0253
RD5.6E-B2Z	0253
RD5.6EB2Z	0253
RD5.6E-B3	0253
RD5.6EB3	0253
RD5.6EB7ZS	0253
RD5.6EB23	0253
RD5.6EB-T4	0253
RD5.6E-BZ	0253
RD5.6EBZ	OEM
RD5.6E-BZ7	0253
RD5.6E-BZ7S	0253
RD5.6EB-Z7S	0253
RD5.6EBZ7S	0253
RD5.6EBZ7S1	0253
RD5.6EBZ7STN	0253
RD5.6E-C	0253
RD5.6EC	0091
RD5.6ED	0253
RD5.6EK	0253
RD5.6EL	OEM
RD5.6EL1	OEM
RD5.6E-L2	0253
RD5.6EL2	0253
RD5.6EL3	OEM
RD5.6EN	OEM
RD5.6E-N1	0253
RD5.6EN1	0253
RD5.6E-N1TN	0253
RD5.6E-N2	0253
RD5.6EN2	0253
RD5.6E-N2TN	0253
RD5.6E-N3	0253
RD5.6EN3	0253
RD5.6ER2TA21R	0253
RD5.6ES	0253
RD5.6ESAB1	OEM
RD5.6ESAB2	0253
RD5.6ESB	OEM
RD5.6ES-B1	0253
RD5.6ESB1	0253
RD5.6ES-B2	0253
RD5.6ESB2	0253
RD5.6ES-B3	0253
RD5.6ESB3	0253
RD5.6ES-T1B1	0253
RD5.6ES-T1B2	0253
RD5.6ES-T1B3	0253
RD5.6ES-T2B	0253
RD5.6ES-T2B2	0253
RD5.6ES-T2B3	0253
RD5.6ES-TIB3	OEM
RD5.6EZ7S	0253
RD5.6F	0157
RD5.6FA	0157
RD5.6FB	0157
RD5.6FB1	0157
RD5.6FB2	OEM
RD5.6FB3	OEM
RD5.6F-BZ7STN	0253
RD5.6J	0253
RD5.6JB	0253
RD5.6JB1	0253
RD5.6JB2	0253
RD5.6JB3	OEM
RD5.6JB-3B	OEM
RD5.6JB-B3	OEM
RD5.6JS	0253
RD5.6JSB2	0253
RD5.6M	0253
RD5.6MB	OEM
RD5.6MB1	OEM
RD5.6M-B2	OEM
RD5.6MB2	OEM
RD5.6MB3	OEM
RD5.6P	OEM
RD5.6SB	OEM
RD5.7FC	OEM
RD5.M-B2	OEM
RD6	0157
RD-6A	0157
RD6A	0091
RD-6A(M)	0157
RD6A(M)	0157
RD6AK	OEM
RD6AL	0253
RD-6AM	0157
RD6AM	0157
RD6AN	0301
RD6B	0301
RD6BK	OEM
RD6BL	OEM
RD6BM	OEM
RD6BN	OEM
RD6C	2416
RD6CK	OEM
RD6CL	1901
RD6CM	1901
RD6CN	2416
RD6D	2416
RD6DK	OEM
RD6DL	1901
RD6DLM	OEM
RD6DM	1901
RD6DN	2416
RD6R2B2	0471
RD6R2EB	0471
RD6R2EB1	0471
RD6R2JSB1	OEM
RD6R2JSB2	OEM
RD6R8EB	0062
RD6R8EE	0025
RD6R8JSB1	OEM
RD6R8JSB3	OEM
RD6.2-EM	0471
RD6.2B2	0091
RD6.2E	0471
RD6.2E(4)	0471
RD6.2E(B1-B3)	0471
RD6.2E(B2)	0471
RD6.2E(B3)	0471
RD6.2E(FA-1)	0471
RD-6.2EB	0471
RD6.2E-B	0471
RD6.2EB	0471
RD6.2EB(1)	0471
RD6.2EB(3)	0471
RD6.2EB(3)-T4	0471
RD6.2E-B1	0471
RD6.2EB1	0091
RD6.2EB1Z	0091
RD6.2E-B2	0471
RD6.2EB2	0471
RD6.2EB2-0.4W	0471
RD6.2EB-2-TA11R	0471
RD6.2E-B3	0471
RD6.2EB-3	0471
RD6.2EB3	0471
RD6.2EB23	0471
RD6.2EC	0471
RD6.2EFA-1	0471
RD6.2EFA1	0471
RD6.2EL	OEM
RD6.2EL1	OEM
RD6.2EL2	0471
RD6.2EL3	OEM
RD6.2EN	OEM
RD6.2E-N1	0471
RD6.2EN1	OEM
RD6.2E-N1TN	0471
RD6.2E-N2	0471
RD6.2EN2	0471
RD6.2EN3	OEM
RD6.2ES	0471
RD6.2ESA2	OEM
RD6.2ESAB3	0295
RD6.2ESB	0471
RD6.2ES-B1	0471
RD6.2ESB1	0091
RD6.2ES-B2	0471
RD6.2ESB2	0471
RD6.2ES-B3	0471
RD6.2ESB3	OEM
RD6.2F	0636
RD6.2FA-1	0471
RD6.2FB	0636
RD6.2FB1	0636
RD6.2FB2	0471
RD6.2FB-3	0636
RD6.2FB3	OEM
RD6.2FC	OEM
RD6.2J	0471
RD6.2JB	0471
RD6.2JB1	OEM
RD6.2JB2	0471
RD6.2JB3	OEM
RD6.2JS	0471
RD6.2JSB1	OEM
RD6.2JSB2	OEM
RD6.2L	0471
RD6.2M	0471
RD6.2MB	0471
RD6.2M-B1	OEM
RD6.2MB1	OEM
RD6.2M-B2	OEM
RD6.2MB2	OEM
RD6.2MB3	OEM
RD6.2P	OEM
RD6.2SB	OEM
RD6.3EB1	0471
RD6.5EB	OEM
RD6.8E	0062
RD6.8E8	0062
RD-6.8EB	0025
RD6.8E-B	0062
RD6.8EB	0062
RD6.8EB(3)	0062
RD6.8EB(3)-T4	0062
RD6.8E-B1	0025
RD6.8EB1	0471
RD6.8EB2	0062
RD6.8EB3	0062
RD6.8EBZ	OEM
RD6.8E-C	0062
RD6.8EC	0062
RD6.8EL	OEM
RD6.8EL1	OEM
RD6.8EL2	OEM
RD6.8EL3	OEM
RD6.8EN	OEM
RD6.8E-N1	0062
RD6.8EN1	0471
RD6.8EN2	0062
RD6.8EN3	OEM
RD6.8ENZ	0062
RD6.8ES	0062
RD6.8ESAB2	0062
RD6.8ESB	OEM
RD6.8ES-B1	0295
RD6.8ESB1	0295
RD6.8ES-B2	0062
RD6.8ESB2	0062
RD6.8ES-B3	0062
RD6.8ESB3	OEM
RD6.8ES-T1B1	0295
RD6.8ES-T1B2	0062
RD6.8ES-TIB2	OEM
RD6.8E-V3-Y	0062
RD6.8F	0025
RD6.8FA	0025
RD6.8FB	0025
RD6.8FB1	OEM
RD6.8FB2	OEM
RD6.8FB3	OEM
RD6.8FC	OEM
RD6.8J	0062
RD6.8JB	OEM
RD6.8JB1	OEM
RD6.8JB2	0062
RD6.8JB3	0062
RD6.8J-N3	0062
RD6.8JS	0062
RD6.8JSB1	OEM
RD6.8JSB2	0062
RD6.8JSB3	OEM
RD6.8M	OEM
RD6.8MB	OEM
RD6.8M-B1	OEM
RD6.8MB1	OEM
RD6.8MB2	0062
RD6.8MB3	OEM
RD6.8N2	OEM
RD6.8P	OEM
RD6.ESB2	OEM
RD-7A	0649
RD7A	0062
RD7AK	OEM
RD-7AM	0025
RD7AM	0062
RD7AN	0062
RD7B	0025
RD7BK	OEM
RD7BL	OEM
RD7BM	OEM
RD7BN	OEM
RD7C	1616
RD7CK	OEM
RD7CL	0696
RD7CM	1601
RD7CN	1616
RD7D	1616
RD7DK	OEM
RD7DL	0696
RD7DLM	OEM
RD7DM	1601
RD7DMN	OEM
RD7DN	1616
RD-7E	0025
RD7H	0062
RD7R5EB	0649
RD7R5EB2	0077
RD7R5EB3	0077
RD7R5FB	0077
RD7R5JB2	0077
RD7R5JSB2	0077
RD7.0EB	0062
RD7.2EB	0077
RD7.5B2	0649
RD7.5E	0077
RD7.5E(B3)	0077
RD-7.5EB	0077
RD7.5E-B	0077
RD7.5EB	0077
RD7.5E-B1	0077
RD7.5EB-1	0077
RD7.5EB1	0077
RD7.5EB2	0077
RD7.5E-B3	0077
RD7.5EB3	0077
RD7.5E-B3Z	0077
RD7.5EBM	0077
RD7.5EBMV	0077
RD7.5EB-Y	0077
RD7.5EC	0077
RD7.5ED	0077
RD7.5EL	OEM
RD7.5EL1	OEM
RD7.5E-L2	0077
RD7.5EL2	0077
RD7.5EL3	OEM
RD7.5EN	OEM
RD7.5EN1	OEM
RD7.5EN2	OEM
RD7.5EN3	0077
RD7.5ES	0077
RD7.5ES(B2)-T2	0077
RD7.5ES(B3)	0077
RD7.5ES(B3)-T2	0077
RD7.5ES(B3)-Y	0077
RD7.5ES-B2	OEM
RD7.5F	0649
RD7.5FA	0649
RD7.5FB	0649
RD7.5FB1	OEM
RD7.5FB2	0649
RD7.5FB3	OEM
RD7.5FC	OEM
RD7.5J	0077
RD7.5JB	0077
RD7.5JB1	OEM
RD7.5JB2	0077
RD7.5JB3	OEM
RD7.5JS	0077
RD7.5JSB2	0077
RD7.5M	0649
RD7.5MB	OEM
RD7.5MB1	OEM
RD7.5MB2	OEM
RD7.5MB3	OEM
RD7.5P	OEM
RD7.6E-B	0077
RD8	0244
RD8H	0165
RD8R2EV	0165
RD8R2FB3	0244
RD8R2JB1	0165
RD8.2	0165
RD-8.2#E	0244
RD-8.2-E	0165
RD-8.2A	0165
RD8.2A	0244
RD-8.2E	0165
RD8.2E	0165
RD-8.2E(B)	0165
RD8.2E(B1)	0165
RD8.2E(B3)	0165
RD-8.2E(C)	0165
RD-8.2EB	0165
RD8.2E-B	0165
RD8.2EB	0165
RD8.2EB1	0077
RD8.2EB2	0165
RD8.2EB2Z	0165
RD8.2EB3	0165
RD8.2EB3T	0165
RD8.2EBS2	OEM
RD-8.2EC	0165
RD8.2EC	0165
RD8.2EK	0165
RD8.2EL	OEM
RD8.2EL1	OEM
RD8.2EL2	OEM
RD8.2EL3	OEM
RD8.2EN	OEM
RD8.2EN1	OEM
RD8.2EN2	OEM
RD8.2EN3	OEM
RD8.2ES	0165
RD8.2ES(B1)	0165
RD8.2ES(B1)-T2	0165
RD8.2ES(B1)-Y	0165
RD8.2ESB	OEM
RD8.2ES-B1	OEM
RD8.2ESB1	OEM
RD8.2ESB2	OEM
RD8.2ESB3	0165
RD8.2EV	0165
RD8.2EW	0165
RD8.2F	1083
RD-8.2FB	0165
RD8.2FB	0165
RD8.2FB1	OEM
RD8.2FB2	0244
RD8.2FB3	0244
RD8.2FC	0244
RD8.2J	0165
RD8.2JB	0165
RD8.2JB1	0165
RD8.2JB2	OEM
RD8.2JB3	OEM
RD8.2JS	0165
RD8.2JSB1	0165
RD8.2M	0165
RD8.2MB	OEM
RD8.2MB1	OEM
RD8.2M-B2	OEM
RD8.2MB3	OEM
RD8.2M-T2BB3	OEM
RD8.2P	OEM
RD8.2V	0165
RD8.3EB	0165
RD9	0012
RD-9A	0012
RD9A	0057
RD9A(10)	0012
RD9AK	OEM
RD-9AL	0012
RD9AL	1083
RD9AM	0057
RD9A-N	0012
RD9AN	0057
RD9B	1083
RD9BK	OEM
RD9BL	OEM
RD9BM	OEM
RD9BN	OEM
RD9C	1622
RD9CK	OEM
RD9CL	1622
RD9CM	2424
RD9CN	0627
RD9D	1622
RD9DK	OEM
RD9DL	1622
RD9DLM	OEM
RD9DM	2424
RD9DN	0627
RD-9E	0170
RD9R1EB	0057
RD9R1EW	0057
RD9R1JSB1	OEM
RD-9.1	0012
RD-9.1E	0012
RD9.1E	0057
RD9.1EB	0057
RD-9.1EB(1)	0012
RD9.1EB1	0057
RD9.1EB2	0057
RD9.1E-B3	0057
RD9.1EB3	0057
RD-9.1EBD	0057
RD9.1EBD	0057
RD9.1EBT	0057
RD9.1EC	0057
RD9.1ED	0012
RD9.1EK	0057
RD9.1EL	OEM
RD9.1EL1	OEM
RD9.1EL2	OEM
RD9.1EL3	0057
RD9.1EN	OEM
RD9.1EN1	OEM
RD9.1EN2	OEM
RD9.1E-N3	0057
RD9.1EN3	0057
RD9.1ES	0057
RD9.1ES(B2)-T2	OEM
RD9.1ES(BS)-T2	OEM
RD9.1ESAB1	OEM
RD9.1ESAB2	OEM
RD9.1ESAB3	5030
RD9.1ESB	OEM
RD9.1ES-B1	OEM
RD9.1ESB1	OEM
RD9.1ES-B2	0057
RD9.1ESB2	5030
RD9.1ES-B3	OEM
RD9.1ESB3	OEM
RD9.1EW	0057
RD9.1F	0012
RD9.1FA	0012
RD-9.1FB	0012
RD9.1FB	0012
RD9.1FB1	OEM
RD9.1FB2	OEM
RD9.1F-B3	0057
RD9.1FB3	0012
RD9.1FC	0057
RD9.1J	0057
RD9.1JB	0057
RD9.1JB1	OEM
RD9.1JB2	0057
RD9.1JB3	0057
RD9.1JS	0057
RD9.1JSAB1	OEM
RD9.1JSAB3	OEM
RD9.1JSB	OEM
RD9.1JSB1	OEM
RD9.1JSB2	OEM
RD9.1JSB3	OEM
RD9.1M	0057
RD9.1MB	0057
RD9.1M-B1	OEM
RD9.1MB1	OEM
RD9.1MB2	OEM
RD9.1MB3	OEM
RD9.1P	OEM
RD10(E)B3	0064
RD10E	0064
RD-10E(B)	0064
RD10EA	0064
RD-10EB	0064
RD10E-B	0064
RD10EB	0064
RD10EB(2)	0064
RD10EB-0.4W	5653
RD10E-B1	0064
RD10EB1	0057
RD10EB-2	0170
RD10EB2	0064

DEVICE TYPE	REPL CODE
RD10E-B3	0064
RD10EB3	0064
RD10EB23	0064
RD10EB23T	0064
RD10EC	OEM
RD10EL	OEM
RD10EL1	OEM
RD10EL2	OEM
RD10EL3	0248
RD10EL4	OEM
RD10EN	OEM
RD10E-N1	0064
RD10EN1	0064
RD10EN1T	0057
RD10E-N2	0064
RD10EN2	0064
RD10E-N3	0064
RD10EN3	0064
RD10E-N3TN	0064
RD10ENI	OEM
RD10ES	0064
RD10ES-AB	OEM
RD10ESB	3441
RD10ES-B1	0064
RD10ESB1	0057
RD10ES-B2	0064
RD10ESB2	0064
RD10ES-B3	0064
RD10ESB3	0064
RD10ES-T1B1	OEM
RD10ES-T1B2	0064
RD10ES-T2B2	0064
RD10ES-TIB2	0064
RD10E-T1B1	0064
RD10E-TIBI	OEM
RD10F	0170
RD10FA	0170
RD10FB	0170
RD10FB1	OEM
RD10FB2	0170
RD10FB3	OEM
RD10FBD	0170
RD10FC	0170
RD10H	0064
RD10J	0064
RD10JB	OEM
RD10JB(2)	0064
RD10JB(3)	0064
RD10JB1	OEM
RD10JB2	0064
RD10JB3	0064
RD10JS	0064
RD10JSAB3	0064
RD10JSB	0064
RD10JSB(2)	0064
RD10JSB(3)	0064
RD10JSB2	0064
RD10JSB2T	0064
RD10JSB3	0064
RD10M	0170
RD10MB	OEM
RD10MB1	OEM
RD10MB2	OEM
RD10MB3	OEM
RD10P	OEM
RD10S-B	OEM
RD10.0E	0170
RD11A	0181
RD11AK	OEM
RD11AL	0064
RD11AM	0064
RD11AN	0181
RD-11B	0181
RD11B	0317
RD11BK	OEM
RD11BL	OEM
RD11BM	OEM
RD11BN	OEM
RD11C	0995
RD11CK	OEM
RD11CL	0992
RD11CM	0992
RD11CN	0995
RD11D	0995
RD11DK	OEM
RD11DL	0992
RD11DLM	OEM
RD11DM	0992
RD11DMN	OEM
RD11DN	0995
RD-11E	0317
RD11E	0181
RD11E(B)	0181
RD11E(B3)	0181
RD-11EB	0181
RD11EB	0181
RD11E-B1	0181
RD11EB1	0064
RD11E-B2	0181
RD11EB2	0181
RD11EB2-TA21R	0181
RD11EB2-TR21R	0181
RD11EB2TA21R	0181
RD-11EB3	0181
RD11EB3	0181
RD11EB-Y	0181
RD11EC	0052
RD11EE	0023
RD11EL	OEM
RD11EL1	OEM
RD11EL2	0181
RD11EL3	OEM
RD11EM	0317
RD11EN	OEM
RD11EN1	OEM
RD11EN2	0317
RD11EN3	OEM
RD11ES	0181
RD11ESAB1	OEM
RD11ESB2	0181
RD11F	0317
RD11F(B)	0317
RD11FB	0317
RD11FB02-TA21-R	0317
RD11FB02TA	OEM
RD11FB1	OEM
RD11FB2	0317
RD11FB3	OEM
RD11FBD2	0317
RD11FBD2TA	0317
RD11FBD3	OEM
RD11FC	OEM
RD11GB02	0023
RD11J	0181
RD11JB	0317
RD11JB1	OEM
RD11JB2	OEM
RD11JB3	OEM
RD11JD	0181
RD11JS	0181
RD11M	0181
RD11MB	OEM
RD11MB1	OEM
RD11MB2	OEM
RD11M-B3	OEM
RD11MB3	OEM
RD11P	OEM
RD12	0137
RD12-OF	0137
RD12B	0052
RD12E	0052
RD12E(B1)	0052
RD12E(B2)-T2	OEM
RD12E(B3)-T2	OEM
RD12EA	0052
R-D12EB	0052
RD-12EB	0052
RD12E-B	0052
RD12EB	1005
RD12EB(1)	0052
RD-12EB(3)	0052
RD12EB(3)	0052
RD12EB(3)-T4	0052
RD12EB(AU)	0052
RD12EB0(2)	0052
RD12E-B1	0052
RD12EB-1	0052
RD12EB1	0052
RD12E-B1Z	0052
RD12EB1Z	0052
RD12E-B2	0052
RD12EB-2	0052
RD12EB2	1005
RD12EB2M	0052
RD12E-B2Z	0052
RD12E-B3	0052
RD12EB3	0052
RD12EBD	0052
RD12EBH	0137
RD12EBM	0052
RD12EB-T4	0052
RD12EC	0137
RD12EL	OEM
RD12EL1	OEM
RD12EL2	OEM
RD12EL3	OEM
RD12EN	OEM
RD12EN1	OEM
RD12EN2	1005
RD12EN3	OEM
RD12ES	0052
RD12ESAB1	1005
RD12ES-AB3-T1	OEM
RD12ES-B2	0052
RD12ESB2	3328
RD12ES-B3	0052
RD12ESB3	0052
RD12ES-T2B2	0052
RD12E-T2	0052
RD12F	0137
RD12F(B1)	0137
RD12FA	0137
RD12FB	0137
RD12FB1	0137
RD12FB2	0137
RD12FB3	0137
RD12FBD	0137
RD12FBT	0137
RD12FC	OEM
RD12J	0052
RD12JB	0052
RD12JB1	OEM
RD12JB2	0052
RD12JB3	OEM
RD12JS	OEM
RD12JSB1	OEM
RD12JSB2	1005
RD12M	0052
RD12MB	0052
RD12M-B1	OEM
RD12MB1	OEM
RD12MB2	OEM
RD12MB3	OEM
RD12P	OEM
RD13	0365
RD13(E)B2	0053
RD-13A	0365
RD13A	0053
RD-13A(M)	0365
RD-13A(N)	OEM
RD-13AD	0365
RD13AK	0365
RD-13AL	0244
RD13AL	0052
RD-13AM	0365
RD13AM	0053
RD-13AN	0137
RD13AN	0365
RD13ANP	0365
RD13B	0365
RD13BK	OEM
RD13BL	0052
RD13BM	OEM
RD13BN	OEM
RD13C	0237
RD13CK	OEM
RD13CL	0052
RD13CM	1249
RD13CN	1249
RD13D	0237
RD13DK	OEM
RD13DL	1263
RD13DLM	OEM
RD13DM	1249
RD13DMN	OEM
RD13DN	1249
RD-13E	0053
RD13E	0053
RD13E-B	0053
RD13EB	0053
RD13EB(2)	0053
RD13EB-1	0053
RD13EB1	0053
RD13E-B1Z8	0053
RD-13EB2	0053
RD13E-B2	0053
RD13EB2	0053
RD13E-B2TN	0053
RD13E-B2Z	0053
RD13EB2Z	0053
RD13EB3	0053
RD13EC	0053
RD13ED1	0053
RD13EL	OEM
RD13EL1	0053
RD13EL2	0053
RD13EL3	OEM
RD13EN	0053
RD13E-N1	0053
RD13EN1	0053
RD13E-N2	0053
RD13EN2	0053
RD13EN3	0053
RD13ES	0053
RD13ES-B	0053
RD13ESB	0053
RD13ES-B1	OEM
RD13ESB1	OEM
RD13ESB1M4	OEM
RD13ES-B2	0053
RD13ESB2	OEM
RD13ESB3	OEM
RD13ES-T1B2	0053
RD13ES-TIB2	OEM
RD13F	0365
RD13FB	0365
RD13FB1	OEM
RD13FB2	OEM
RD13FB3	OEM
RD13FC	OEM
RD13H	OEM
RD13J	0053
RD13JB	OEM
RD13JB1	OEM
RD13JB2	OEM
RD13JB3	OEM
RD13JS	0053
RD13JSB3	OEM
RD13JS-Y	0053
RD13K	0137
RD13M	0365
RD13MB	OEM
RD13MB1	OEM
RD13MB2	OEM
RD13MB3	OEM
RD13P	OEM
RD15E	0002
RD15EA	0686
RD15EB	0879
RD15EB1	OEM
RD15EB2	4526
RD15E-B3	0686
RD15EB3	0686
RD15EB3-0.4W	0686
RD15EB3Z	0686
RD15EB23	0686
RD15EC	0686
RD15EL	OEM
RD15EN	OEM
RD15E-N1	0686
RD15EN1	OEM
RD15EN2	OEM
RD15E-N3	0686
RD15EN3	0686
RD15ES	0686
RD15ESB	OEM
RD15ESB2	0686
RD15F	0002
RD15F1	0002
RD15FA	0002
RD15FB	0002
RD15FB1	OEM
RD15FB3	0002
RD15FC	OEM
RD15FM	0002
RD15FMZ	0002
RD15J	0686
RD15JB1	OEM
RD15JB2	OEM
RD15JB3	0686
RD15JS	0686
RD15JS(B2)-T	0686
RD15JSB2	OEM
RD15JSB3	OEM
RD15L	OEM
RD15M	0686
RD15MB	0686
RD15M-B1	OEM
RD15MB1	OEM
RD15MB2	OEM
RD15MB3	OEM
RD15P	OEM
RD15SB	OEM
RD16A	0445
RD16AK	OEM
RD16AL	0879
RD16AM	0686
RD16AN	0210
RD16AT	OEM
RD16B	0420
RD16BK	OEM
RD16BL	OEM
RD16BM	OEM
RD16BN	OEM
RD16BT	OEM
RD16C	1703
RD16CK	OEM
RD16CL	1636
RD16CM	1639
RD16CN	1640
RD16C-Y	0420
RD16D	1703
RD16DK	OEM
RD16DL	1636
RD16DLM	OEM
RD16DM	1639
RD16DMN	OEM
RD16DN	1640
RD16DT	OEM
RD16E	0445
RD16E(B3)	0445
RD16EB	0445
RD16EB1	0445
RD16EB2	0445
RD16E-B3	0445
RD16EB3	0445
RD16EBS	OEM
RD16EC	1649
RD16EL	OEM
RD16EL1	OEM
RD16EL2	OEM
RD16EL3	0445
RD16E-M	0420
RD16EN	0445
RD16EN1	5089
RD16E-N2	0420
RD16EN2	OEM
RD16E-N3	0420
RD16EN3	OEM
RD16ES	0445
RD16ESB	OEM
RD16F	0420
RD16FB	0445
RD16FB1	0420
RD16FB2	OEM
RD16FB3	OEM
RD16FC	OEM
RD16GT	OEM
RD-16H	0002
RD16H	0002
RD16J	0445
RD16JB	OEM
RD16JB1	0445
RD16JB2	0445
RD16JB3	OEM
RD16JS	0445
RD16JSAB2	0077
RD16JT	OEM
RD16KT	OEM
RD16M	0445
RD16MB	OEM
RD16MB1	OEM
RD16MB2	OEM
RD16MB3	OEM
RD16P	OEM
RD18E	0375
RD18EB	0375
RD18E-B1	0375
RD18EB1	0375
RD18EB2	OEM
RD18EB3	0375
RD18EC	0375
RD18EL	OEM
RD18EL1	OEM
RD18EL2	OEM
RD18EL3	OEM
RD18EN	OEM
RD18EN1	OEM
RD18EN2	OEM
RD18EN3	OEM
RD18ES	0375
RD18ESB	OEM
RD18ES-B1	0210
RD18ESB1	0210
RD18ES-B2	0375
RD18ES-T1B1	OEM
RD18EST1B1	0210
RD18ES-TIBI	OEM
RD18F	0495
RD18FA	0495
RD18FB	0495
RD18FB1	OEM
RD18FB2	0495
RD18FB3	0495
RD18FC	OEM
RD18J	0375
RD18JB	OEM
RD18JB1	OEM
RD18JB2	OEM
RD18JB3	OEM
RD18JB-B3	OEM
RD18JS	0375
RD18M	OEM
RD18MB	OEM
RD18M-B1	OEM
RD18MB1	OEM
RD18M-B3	OEM
RD18MB3	OEM
RD18P	OEM
RD19	0949
RD19A	0671
RD19AK	OEM
RD19AL	0671
RD19AM	0671
RD19B	0949
RD19BK	OEM
RD19BL	OEM
RD19BM	OEM
RD19C	1720
RD19CK	OEM
RD19CL	1716
RD19CM	1731
RD19D	1720
RD19DK	OEM
RD19DL	1716
RD19DLM	OEM
RD19DM	1731
RD-20E	0531
RD20E	0700
RD20E(B1)	0700
RD20EB	0700
RD20EB1	OEM
RD20EB2	0671
RD20EB3	0210
RD20EC	OEM
RD20EL	OEM
RD20EL1	OEM
RD20EL2	OEM
RD20EL3	OEM
RD20EL4	OEM
RD20EN	OEM
RD20EN1	OEM
RD20EN2	OEM
RD20EN3	OEM
RD20EN4	OEM
RD20ES	0700
RD20ESAB2	OEM
RD20ES-B1	0671
RD20ESB1	0671
RD20ES-B2	0671
RD20ESB2	0671
RD20ES-T1B2	0671
RD20ES-TIB2	OEM
RD20EV2	0700
RD20EV2-Y	0700
RD20F	0531
RD20FB	0531
RD20FB1	OEM
RD20FB2	0531
RD20FB3	OEM
RD20FC	OEM
RD20J	0700
RD20JB	OEM
RD20JB1	OEM
RD20JB2	OEM
RD20JB3	OEM
RD20JS	0700
RD20JSB2	6544
RD20M	OEM
RD20MB	OEM
RD20MB1	OEM
RD20MB2	OEM
RD20MB3	OEM
RD20P	OEM
RD22E	0705
RD22E-B	0705
RD22EB	0705
RD22EB1	0705
RD22EB2	OEM
RD22EB3	OEM
RD22EB4	OEM
RD22EC	0205
RD22EL	OEM
RD22EL1	OEM
RD22EL2	OEM
RD22EL3	OEM
RD22EL4	OEM
RD22ES	0705
RD22F	0565
RD22FA	0565
RD22FB	0565
RD22FB1	OEM
RD22FB2	5993
RD22FC	0565
RD22J	0700
RD22JB	OEM
RD22JB1	OEM
RD22JB2	OEM
RD22JB3	OEM
RD22JS	0700
RD22M	OEM
RD22MB	OEM
RD22MB1	OEM
RD22MB2	OEM
RD22MB3	OEM
RD22P	OEM
RD23D12A	OEM
RD23D12A3	OEM
RD23D12AF	OEM
RD23D12B	OEM
RD23D12B3	OEM
RD23D12BF	OEM
RD23D14A	OEM
RD23D14A3	OEM
RD23D14AF	OEM
RD23D14B	OEM
RD23D14B3	OEM
RD23D14BF	OEM
RD23D16A	OEM
RD23D16A3	OEM
RD23D16AF	OEM
RD23D16B	OEM
RD23D16B3	OEM
RD23D16BF	OEM
RD23D18A	OEM
RD23D18A3	OEM
RD23D18AF	OEM
RD23D18B	OEM
RD23D18B3	OEM
RD23D18BF	OEM
RD23D20A	OEM
RD23D20A3	OEM
RD23D20AF	OEM
RD23D20B	OEM
RD23D20B3	OEM
RD23D20BF	OEM
RD23D22A	OEM
RD23D22A3	OEM
RD23D22AF	OEM
RD23D22B	OEM
RD23D22B3	OEM
RD23D22BF	OEM
RD23D24A	OEM
RD23D24A3	OEM
RD23D24AF	OEM
RD23D26A	OEM
RD23D26A3	OEM
RD23D26AF	OEM
RD23DG4A	OEM
RD23DG4A3	OEM
RD23DG4AF	OEM
RD23DG4B	OEM
RD23DG4B3	OEM
RD23DG4BF	OEM
RD23DG6A	OEM
RD23DG6AF	OEM
RD23DG6B	OEM
RD23DG6B3	OEM
RD23DG6BF	OEM
RD23DG8A	OEM
RD23DG8A3	OEM
RD23DG8AF	OEM
RD23DG8B	OEM
RD23DG8B3	OEM
RD23DG8BF	OEM
RD23DG10A	OEM
RD23DG10A3	OEM
RD23DG10AF	OEM
RD23DG10B	OEM
RD23DG10B3	OEM
RD23DG10BF	OEM
RD23DR12A	OEM
RD23DR12B	OEM
RD23DR14A	OEM
RD23DR14B	OEM
RD23DR16A	OEM
RD23DR16B	OEM
RD23DR18A	OEM
RD23DR18B	OEM
RD23DR20A	OEM
RD23DR20B	OEM
RD23DR22A	OEM
RD23DR22B	OEM
RD23DR24A	OEM
RD23DR26A	OEM
RD23DRG4A	OEM
RD23DRG4B	OEM
RD23DRG6A	OEM
RD23DRG6B	OEM
RD23DRG8A	OEM
RD23DRG8B	OEM
RD23DRG10A	OEM
RD23DRG10B	OEM
RD-24A	0402
RD24A	0494
RD24AK	OEM
RD24AL	0494
RD24AM	0494
RD24AN	0494
RD24B	0402
RD24BK	OEM
RD24BL	OEM
RD24BM	OEM
RD24BN	OEM
RD24C	1756
RD24CK	OEM
RD24CL	0727
RD24CM	0727
RD24CN	1768
RD24D	1756
RD24DK	OEM
RD24DL	0727
RD24DLM	OEM
RD24DM	0727
RD24DMN	OEM
RD24DN	1768
RD24E	0494
RD24E(B2)	0494
RD24E(B3)	0494
RD24E-B	0494
RD24EB	0494
RD24EB1	0494
RD24EB2	0494
RD24EB2VF	OEM
RD24EB3	0494
RD24EB4	OEM
RD24EB23	0494
RD24E-B27	0494
RD24E-BZ7	0494
RD24EB-Z7	0494
RD24EBZ7	0494
RD24E-BZ7TN	0494
RD24EC	1606
RD24EE	OEM
RD24EL	OEM
RD24EL1	OEM
RD24EL2	OEM
RD24EL3	OEM
RD24EL4	OEM
RD24ES	0494
RD24ES-AB1	OEM
RD24ESB	OEM
RD24ES-B1	0494
RD24F	0402
RD24FB	0274
RD24FB1	OEM
RD24FB2	OEM
RD24FB3	OEM
RD24F-BS	0402
RD24FC	1606
RD24J	0494
RD24JB	OEM
RD24JB1	OEM
RD24JB2	OEM
RD24JB3	OEM
RD24JS	0494
RD24M	0494
RD24MB	OEM
RD24MB1	OEM
RD24MB2	OEM
RD24MB3	OEM
RD24P	OEM
RD27	0441
RD27B1	OEM
RD27E	0455
RD27E(B4)	0441
RD27EB	0441
RD27EB1	0455
RD27EB2	0455
RD27E-B2Z	0455
RD27EB3	0455
RD27E-B4	0441
RD27EB4	0455
RD27EB4TN	0455
RD27EC	OEM
RD27ED	0441
RD27EL	OEM
RD27EL1	OEM
RD27EL2	OEM
RD27EL3	OEM
RD27EL4	OEM
RD27ES	0455
RD27ESAB3	OEM
RD27ES-B1	0714
RD27ESB1	0714
RD27F	0441
RD27FA	0441
RD27FB	0441
RD27FB1	1606
RD27FB2	0441
RD27FB3	OEM

DEVICE TYPE	REPL CODE
RD27FC	0441
RD27FCD	0441
RD27J	0257
RD27JB	OEM
RD27JB1	OEM
RD27JB2	OEM
RD27JB3	OEM
RD27JS	0455
RD27JS-B3	0455
RD27JST	0441
RD27JS-T1	0455
RD27JST1	0455
RD27M	0455
RD27P	OEM
RD29A	0257
RD29AK	OEM
RD29AL	0455
RD29AM	0257
RD29AN	0195
RD29B	1674
RD29BK	OEM
RD29BL	OEM
RD29BM	OEM
RD29BN	OEM
RD29C	2453
RD29CK	OEM
RD29CL	1782
RD29CM	2453
RD29CN	1794
RD29D	2453
RD29DK	OEM
RD29DL	1782
RD29DLM	OEM
RD29DM	2453
RD29DMN	OEM
RD29DN	1799
RD30AM	OEM
RD30B3	OEM
RD30BM	OEM
RD30D4A	OEM
RD30D4A3	OEM
RD30D4AF	OEM
RD30D4B	OEM
RD30D4B3	OEM
RD30D4BF	OEM
RD30D6A	OEM
RD30D6A3	OEM
RD30D6AF	OEM
RD30D6B	OEM
RD30D6B3	OEM
RD30D6BF	OEM
RD30D8A	OEM
RD30D8A3	OEM
RD30D8AF	OEM
RD30D8B	OEM
RD30D8B3	OEM
RD30D8BF	OEM
RD30D10A	OEM
RD30D10A3	OEM
RD30D10AF	OEM
RD30D10B	OEM
RD30D10B3	OEM
RD30D10BF	OEM
RD30D12A	OEM
RD30D12A3	OEM
RD30D12AF	OEM
RD30D12B	OEM
RD30D12B3	OEM
RD30D12BF	OEM
RD30D14A	OEM
RD30D14A3	OEM
RD30D14AF	OEM
RD30D14B	OEM
RD30D14B3	OEM
RD30D14BF	OEM
RD30D16A	OEM
RD30D16A3	OEM
RD30D16AF	OEM
RD30D16B	OEM
RD30D16B3	OEM
RD30D16BF	OEM
RD30D18A	OEM
RD30D18A3	OEM
RD30D18AF	OEM
RD30D18B	OEM
RD30D18B3	OEM
RD30D18BF	OEM
RD30D20A	OEM
RD30D20A3	OEM
RD30D20AF	OEM
RD30D20B	6155
RD30D20B3	OEM
RD30D20BF	OEM
RD30D22A	OEM
RD30D22A3	OEM
RD30D22AF	OEM
RD30D22B	OEM
RD30D22B3	OEM
RD30D22BF	OEM
RD30D24A	OEM
RD30D24A3	OEM
RD30D24AF	OEM
RD30D24B	OEM
RD30D24B3	OEM
RD30D24BF	OEM
RD30D26A	OEM
RD30D26A3	OEM
RD30D26AF	OEM
RD30D26B	OEM
RD30D26B3	OEM
RD30D26BF	OEM
RD30D28A	OEM
RD30D28A3	OEM
RD30D28AF	OEM
RD30D30A	OEM
RD30D30A3	OEM
RD30D30AF	OEM
RD30DM	OEM
RD30DR4A	OEM
RD30DR4B	OEM
RD30DR6A	OEM
RD30DR6B	OEM
RD30DR8A	OEM
RD30DR8B	OEM
RD30DR10A	OEM
RD30DR10B	OEM
RD30DR12A	OEM
RD30DR12B	OEM
RD30DR14A	OEM
RD30DR14B	OEM
RD30DR16A	OEM
RD30DR16B	OEM
RD30DR18A	OEM
RD30DR18B	OEM
RD30DR20A	OEM
RD30DR20B	OEM
RD30DR22A	OEM
RD30DR22B	OEM
RD30DR24A	OEM
RD30DR24B	OEM
RD30DR26A	OEM
RD30DR26B	OEM
RD30DR28A	OEM
RD30DR30A	OEM
RD30E	0726
RD-30EB	0166
RD30EB	0195
RD30EB1	OEM
RD30E-B2	0195
RD30EB2	0195
RD30E-B2Z	0195
RD30EB3	0195
RD30EB4	0195
RD30EC	OEM
RD30EL	OEM
RD30EL1	OEM
RD30EL2	OEM
RD30EL3	OEM
RD30EL4	OEM
RD30ES	0195
RD30ESB1	0195
RD30ES-B2	0195
RD30ESB2	OEM
RD30ESB4	0195
RD30F	0726
RD30FB	0726
RD30FB(1)	1674
RD30FB(2)	0726
RD30FB(3)	0726
RD30FB-1	1674
RD30FB1	1674
RD30FB2	0726
RD30FB3	0726
RD30FB3-1	OEM
RD30FB3-2	OEM
RD30FB31	OEM
RD30FB32	OEM
RD30FC	OEM
RD30GM	OEM
RD30J	0195
RD30JB	OEM
RD30JB1	OEM
RD30JB2	OEM
RD30JB3	OEM
RD30JM	OEM
RD30JS	0195
RD30KM	OEM
RD30M	0195
RD30MM	OEM
RD30P	OEM
RD33	0166
RD33B-E1	0166
RD33E	0166
RD33E(B1)-T2	OEM
RD33E(B1)-Y	OEM
RD33EB	0166
RD33EB*5*	0166
RD33EB(5)	0166
RD33E-B1	0166
RD33EB1	0195
RD33E-B2	0166
RD33EB2	0166
RD33EB2T	0166
RD33E-B2TN	0166
RD33EB3	0166
RD33EB-4	0166
RD33EB4	0166
RD33EB5	0166
RD33EC	0820
RD33EL	OEM
RD33EL1	OEM
RD33EL2	OEM
RD33EL3	OEM
RD33EL4	OEM
RD33ES	0166
RD33ESB1	OEM
RD33ES-B2	0166
RD33ESB2	0166
RD33ESB3	0166
RD33ES-L3	OEM
RD33ES-T1B2	0195
RD33ES-TIB2	OEM
RD33F	0039
RD33FA	0039
RD33FB	0039
RD33FB1	0166
RD33FB2	0039
RD33FB3	OEM
RD33FC	OEM
RD33J	0166
RD33JB	OEM
RD33JB1	OEM
RD33JB2	OEM
RD33JB3	OEM
RD33JS	0166
RD33JSB1	0166
RD33JSB2	0166
RD33M	0166
RD33P	OEM
RD-35	0820
RD-35A	0039
RD35A	0010
RD35AE	0032
RD35AK	OEM
RD35AL	0166
RD35AM	0010
RD35AN	OEM
RD35B	0820
RD35BK	OEM
RD35BL	OEM
RD35BM	OEM
RD35BN	OEM
RD35C	1809
RD35CK	OEM
RD35CL	1799
RD35CLM	OEM
RD35CM	1809
RD35CN	1817
RD35D	1809
RD35DK	OEM
RD35DL	1799
RD35DM	1809
RD35DMN	OEM
RD35DN	1817
RD36E	0166
RD-36EB	OEM
RD36E-B	0010
RD36EB	0010
RD36E-B1	0010
RD36EB1	0010
RD36EB2	0166
RD36EB3	0010
RD36EB3V	0010
RD36EB4	OEM
RD36EB-T1	0010
RD36EB-TA11R	0010
RD36EC	OEM
RD36EL	OEM
RD36EL1	OEM
RD36EL2	OEM
RD36EL3	OEM
RD36EL4	OEM
RD36ES	0010
RD36ESB	0010
RD36ES-B-T1	0010
RD36ES-T1	0010
RD36EV3	0166
RD36EV8	0010
RD36F	0820
RD36FB	0820
RD36FB1	OEM
RD36FB2	OEM
RD36FB3	OEM
RD36FC	OEM
RD36J	0010
RD36JB	OEM
RD36JB1	OEM
RD36JB2	OEM
RD36JB3	OEM
RD36JS	0010
RD36M	0010
RD36P	OEM
RD39E	0032
RD39EB	0032
RD39E-B1	0032
RD39EB1	0032
RD39EB2	0032
RD39EB3	OEM
RD39EB4	OEM
RD39EC	OEM
RD39EC1	OEM
RD39E-C2	0032
RD39EC2	OEM
RD39EC3	OEM
RD39EL	OEM
RD39EL1	OEM
RD39EL2	OEM
RD39EL3	OEM
RD39EL4	OEM
RD39EL5	OEM
RD39EL6	OEM
RD39EL7	OEM
RD39ES	0032
RD39ESB	OEM
RD39ES-B2	OEM
RD39ES-B4	OEM
RD39ESB4	OEM
RD39F	0350
RD39FA	0350
RD39FB	0350
RD39FB1	OEM
RD39FB2	OEM
RD39FB3	OEM
RD39FC	OEM
RD39J	0032
RD39JB	OEM
RD39JB1	OEM
RD39JB2	OEM
RD39JB3	OEM
RD39JS	0032
RD39M	0032
RD39P	OEM
RD-41PI	OEM
RD42EB	0275
RD43E	0054
RD43EB	3413
RD43F	0931
RD43FB	3420
RD43M	OEM
RD47D	OEM
RD47E	0068
RD47EB	3592
RD47EB2	0999
RD47EL1	OEM
RD47F	0999
RD47FB	3595
RD47FBD	0999
RD47F-TB	0068
RD47M	OEM
RD47P	OEM
RD51E	0092
RD51EB	0162
RD51E-B2	OEM
RD51F	0502
RD51FB	3756
RD51F-T7B	OEM
RD51P	OEM
RD56B	0869
RD56E	0125
RD56EB	0125
RD56E-B2Z	0253
RD56E-BZ7S	0253
RD56F	0869
RD56FB	0869
RD56P	OEM
RD62E	0152
RD62EB	0152
RD62F	0783
RD62FB	4128
RD62P	OEM
RD68E	0173
RD68EB	0173
RD68F	1268
RD68FB	4253
RD68P	OEM
RD75E	0094
RD75E(B)	2895
RD75E(B)-T5	2895
RD75EB	4361
RD75F	1189
RD75FB	4363
RD75P	OEM
RD82E	0049
RD82EB	4510
RD82EC	OEM
RD82F	0331
RD82FB	4513
RD-91	0012
RD-91E	0012
RD91E	1311
RD91EB	0156
RD-91F	OEM
RD91P	OEM
RD92EB	OEM
RD100E	0189
RD100EB	7062
RD100P	OEM
RD110E	0099
RD110EB	7126
RD110P	OEM
RD120E	0089
RD120EB	0702
RD120P	OEM
RD130E	0287
RD140E	0252
RD150E	0340
RD151	0106
RD152	0106
RD153	0106
RD154	0106
RD155	2010
RD156	2400
RD157	0787
RD160E	0340
RD165S	OEM
RD170E	0394
RD180E	0425
RD190E	0453
RD200E	1473
RD250	0015
RD316	OEM
RD362EB2	0166
RD515	0133
RD900	OEM
RD1015	OEM
RD1212EB-T4	0052
RD2015	OEM
RD2512	OEM
RD3015	OEM
RD3472	0015
RD4015	OEM
RD4258	OEM
RD4502	OEM
RD6551-11	OEM
RD9037	0015
RD26235-1	0015
RD29799P	0015
RD31903P	0015
RD-EX0096CEZZ	0705
RDF02M	OEM
RDG.2EBZ	OEM
RDI2E-B	0052
RDR.1JB3	OEM
RDS5.6EB3	OEM
R-DW4	2085
RE-1	0280
RE1	0280
RE-2	0600
RE2	0015
RE-3	0211
RE3	0015
RE-4	0136
RE4	0004
RE-5	0038
RE5	0038
RE-6	0208
RE6	0208
RE7	0085
RE-7MP	3033
RE7MP	3033
RE-8	0440
RE8	0440
RE-9	0127
RE9	0127
RE-10	0144
RE10	0144
RE11	0160
RE-11MP	0160
RE11MP	0160
RE-12	0198
RE12	0198
RE-13	0016
RE13	0016
RE-14	0142
RE14	0142
RE15	0136
RE-16	0975
RE16	0975
RE-17	0086
RE17	0086
RE-18	0126
RE18	0126
RE-19	0103
RE19	0103
RE-20	0222
RE20	0222
RE-20MP	0822
RE20MP	0222
RE-21	0423
RE21	0423
RE-22	0854
RE22	0854
RE-23	0233
RE23	0233
RE-24	0187
RE24	0276
RE-25	0164
RE25	0164
RE-26	0037
RE26	0037
RE-27	0050
RE27	0050
RE-28	0007
RE28	0007
RE-29	0074
RE29	0074
RE-30	0642
RE30	0642
RE-31	0289
RE31	0003
RE-32	0065
RE32	0065
RE-33	0400
RE33	0400
RE-34	0178
RE34	0178
RE-35	0847
RE35	0847
RE-36	0604
RE36	0604
RE-37	0130
RE37	0130
RE38	0561
RE39	1199
RE-40	0161
RE40	0161
RE-41	0460
RE41	0460
RE-42	0560
RE42	0560
RE-43	1267
RE43	1267
RE44	0264
RE45	0325
RE46	0325
RE-47	0015
RE47	0143
RE-48	0917
RE48	0917
RE-49	0015
RE49	0015
RE-50	0229
RE50	0229
RE-51	0071
RE51	0071
RE-52	0015
RE52	0133
RE-53	0150
RE53	0150
RE-54	2763
RE54	2763
RE-55	0102
RE55	0102
RE56	0792
RE57	0892
RE58	0892
RE59	1480
RE60	0715
RE-61	2013
RE61	2013
RE62	0693
RE63	0006
RE64	0111
RE65	0353
RE66	0079
RE-67	0111
RE67	0111
RE-68	0222
RE68	0905
RE69	0050
RE-70	0595
RE70	0086
RE71	0622
RE72	0037
RE-73	0222
RE73	0285
RE-74	1681
RE74	1681
RE-75	0551
RE75	0551
RE-76	0382
RE76	0382
RE-77	0338
RE77	0338
RE78	0861
RE-79	0150
RE79	0698
RE79A	0698
RE-80	2050
RE80	0566
RE81	1367
RE-82	0491
RE82	0491
RE-83	1010
RE83	1010
RE83MP	1010
RE85	1862
RE86	0123
RE-87	0196
RE87	0196
RE-88	0484
RE88	0484
RE-89	1080
RE89	1080
RE-90	2637
RE90	0071
RE-91	0474
RE91	0474
RE-92	0922
RE92	0087
RE-93	0293
RE93	0293
RE93P	0293
RE-94	0124
RE94	0124
RE95	3003
RE96	OEM
RE97	1312
RE98	3010
RE99	1713
RE-100	0292
RE100	0292
RE-101	0188
RE101	0188
RE-102	0456
RE102	0456
RE-103	0533
RE103	0533
RE-104	0451
RE104	0451
RE-105	0162
RE105	0162
RE106	0157
RE-107	0157
RE107	0157
RE-108	0301
RE108	0301
RE-109	0636
RE109	0636
RE-110	0025
RE110	0025
RE-111	0649
RE111	0649
RE-112	0244
RE112	0244
RE113	1083
RE-114	0012
RE114	0012
RE-115	0170
RE115	0170
RE-116	0317
RE116	0317
RE-117	0794
RE117	0794
RE-118	0137
RE118	0137
RE-119	0365
RE119	0365
RE-120	0100
RE120	0100
RE-121	0002
RE121	0002
RE-122	0420
RE122	0420
RE-123	1649
RE123	1649
RE-124	0495
RE124	0495
RE-125	0949
RE125	0949
RE-126	0531
RE126	0531
RE-127	0565
RE127	0565
RE-128	0402
RE128	0402
RE-129	1606
RE129	1606
RE-130	0039
RE130	0039
RE-131	0350
RE131	0350
RE-132	0783
RE132	0783
RE-133	0098
RE133	0098
RE134	4231
RE135	2397
RE136	2035
RE137	2401
RE138	1438
RE140	2407
RE141	1445
RE142	2415
RE143	2221
RE144	1458
RE145	0221
RE146	1490
RE147	2422
RE148	1618
RE149	0510
RE150	0691
RE152	0870
RE153	1020
RE154	1153
RE155	1274
RE156	1402
RE157	1532
RE158	1078
RE159	1711
RE160	1717
RE161	1723
RE162	1736
RE163	1748
RE164	1772
RE165	1804
RE166	0070
RE-167	0774
RE167	0774
RE-168	0447
RE168	0447
RE169	2491
RE170	0712
RE-171	0940
RE171	0940
RE172	0240
RE-173	0469
RE173	0469
RE174	0095
RE-175	0676
RE175	0676
RE-176	0722
RE176	0722
RE177	0336
RE178	0725
RE179	2383
RE180	0408
RE181	3148
RE182	2394
RE183	0411
RE184	3150
RE185	1413
RE186	0415
RE187	3152
RE188	2387
RE189	0422
RE-190	2730
RE190	5911
RE-191	0168
RE191	0168
RE-192	0111
RE192	0111
RE-193	0693

If replacement code is OEM, contact original manufacturer for replacement.

DEVICE TYPE	REPL CODE
RE193	0693
RE-194	0936
RE194	0936
RE-195	0554
RE195	0628
RE196	0595
RE197	0791
RE198	1411
RE199	0212
RE200	0288
RE201	0836
RE202	0630
RE203	0936
RE204	1173
RE205	0423
RE206	0622
RE207	0715
RE208	0342
RE209	1591
RE210	0324
RE211	0436
RE212	2050
RE213	2046
RE214	0190
RE215	1067
RE216	0201
RE217	0378
RE218	0058
RE219	0532
RE220	0018
RE221	0854
RE222	0236
RE223	0681
RE224	1946
RE225	1323
RE226	2136
RE227	0316
RE228	0400
RE229	0219
RE230	0849
RE231	0239
RE232	0061
RE233	0313
RE234	1652
RE235	0006
RE236	1712
RE237	2329
RE238	1978
RE239	1984
RE240	6031
RE241	6032
RE242	1198
RE243	3858
RE244	1925
RE245	OEM
RE246	1137
RE247	0901
RE248	0901
RE249	3404
RE250	0139
RE251	0441
RE252	1834
RE253	0331
RE254	0149
RE255	0277
RE256	0905
RE257	0353
RE258	0103
RE259	3280
RE260	0224
RE261	0330
RE262	0620
RE263	0155
RE264	0493
RE265	1030
RE266	0720
RE-300-IC	0632
RE300-IC	0632
RE301-IC	0678
RE302-IC	2161
RE-303-IC	0802
RE303-IC	0802
RE-304-IC	0395
RE304-IC	0395
RE-305-IC	0167
RE305-IC	0167
RE-306-IC	0666
RE306-IC	0666
RE-307-IC	0352
RE307-IC	0352
RE308-IC	0354
RE309-IC	0701
RE-310-IC	1808
RE310-IC	1808
RE-311-IC	0664
RE311-IC	0664
RE-312-IC	0751
RE312-IC	0751
RE-313-IC	0856
RE313-IC	0856
RE-314-IC	3735
RE314-IC	3735
RE-315-IC	1443
RE315-IC	1443
RE-316-IC	1759
RE316-IC	1759
RE-317-IC	1395
RE317-IC	1395
RE-318-IC	2279
RE318-IC	2279
RE-319-IC	0654
RE319-IC	0654
RE-320-IC	0443
RE320-IC	0443
RE-321-IC	0379
RE321-IC	0379
RE-322-IC	2557
RE322-IC	2557
RE-323-IC	0349
RE323-IC	0349
RE324-IC	1337
RE325-IC	0878
RE326-IC	1215
RE327-IC	0433
RE328-IC	0855
RE329-IC	1994
RE330-IC	1478
RE331-IC	0789
RE332-IC	2755
RE333-IC	0521
RE334-IC	0520
RE335-IC	0579
RE336-IC	2568
RE337-IC	0528
RE338-IC	2124
RE339-IC	1160
RE340-M	1018
RE340M	1018
RE341-M	2261
RE342-M	2316
RE343-IC	1634
RE344-M	0911
RE345-IC	1345
RE346-IC	0328
RE347-IC	6702
RE347M	3550
RE348-IC	1187
RE348-M	1187
RE348M	1187
RE349-IC	2876
RE349-M	2876
RE350M	OEM
RE351-IC	4093
RE352-IC	2596
RE353-M	2861
RE354-IC	1326
RE355-IC	2231
RE356-IC	2949
RE357-IC	1050
RE358-IC	2773
RE359-IC	2781
RE360-IC	4224
RE361-IC	2457
RE362-IC	1421
RE363-IC	OEM
RE364-IC	0364
RE367-IC	2891
RE368-IC	1837
RE369-IC	OEM
RE370-IC	2632
RE371-IC	OEM
RE372-IC	3915
RE373-IC	4594
RE374-IC	4595
RE375-IC	2062
RE376-IC	3366
RE377-IC	2942
RE378-IC	0167
RE379-IC	OEM
RE380-IC	3269
RE382-IC	0232
RE383-IC	1172
RE384-IC	1313
RE385-IC	1172
RE386-IC	1645
RE387-IC	1292
RE388-IC	1157
RE389-IC	2049
RE390-IC	1540
RE504	0229
RE664	OEM
RE670	OEM
RE896	OEM
RE1001	0144
RE1002	0144
RE2001	0007
RE2002	0007
RE3001	0127
RE3002	0127
RE4001	0111
RE4002	0111
RE4010	0111
RE5001	0007
RE5002	0007
RE136137	OEM
REC86345	OEM
RECT-SI-154	0023
RECT-SI-154(AU)	0023
RECT-SI-1001	0795
RECT-SI-1001(AU)	0795
RECT-SI-1002	0023
RECT-SI-1002(AU)	0023
RECT-SI-1005	0023
RECT-SI-1019	0348
RECT-SI-1020	0023
RECT-SI-1021	0023
RECT-SI-1021(AU)	0023
RECT-SI-1026(AU)	0023
RECT-SI-1030	0015
RECT-SI-1039	0795
RECT-SI-1039(AU)	0795
RECT-SI-1044	0015
RECT-SI-1044(AU)	0015
RECT-U-007	OEM
RECT-U-1001	OEM
RECT-U-1002	OEM
RECT-U-1003	OEM
RECT-U-1004	OEM
RECT-U-1012	2010
RECT-UG-1002	2010
RECT-UG-1003	0277
RECT-UG-1004	0106
REF-01ADE	OEM
REF-01ADE/883B	OEM
REF-01AT	OEM
REF-01AT/883B	OEM
REF-01CDE	OEM
REF-01CT	OEM
REF-01DDE	OEM
REF-01DE	OEM
REF-01DE/883B	OEM
REF-01DT	OEM
REF-01EDE	OEM
REF-01ET	OEM
REF-01HDE	OEM
REF-01HT	OEM
REF-01T	OEM
REF-01T/883B	OEM
REF-02ADE	OEM
REF-02ADE/883B	OEM
REF-02AT	OEM
REF-02AT/883B	OEM
REF-02CDE	OEM
REF-02CT	OEM
REF-02DDE	OEM
REF-02DE	OEM
REF-02DE/883B	OEM
REF-02DT	OEM
REF-02EDE	OEM
REF-02ET	OEM
REF-02HDE	OEM
REF-02HT	OEM
REF-02T	OEM
REF-02T/883B	OEM
REF-03CDE	OEM
REF-03CNB	OEM
REF-03CT	OEM
REF-03DDE	OEM
REF-03DE/883B	OEM
REF-03DNB	OEM
REF-03DT	OEM
REF-03T	OEM
REF-03T/883B	OEM
REJ70148	0196
REJ70432	OEM
REJ70643	0015
REJ70931	0015
REJ71253	0133
REN56	2096
REN57	0892
REN58	0892
REN59	1480
REN60	0715
REN61	2013
REN62	0693
REN63	0006
REN64	0111
REN65	0353
REN66	0861
REN67	0111
REN68	0905
REN69	0050
REN70	0086
REN71	0622
REN72	0126
REN85	2987
REN90	0071
REN94	0124
REN100	0280
REN101	0600
REN102	0211
REN102A	0004
REN103	0038
REN103A	0208
REN104	0085
REN104MP	3033
REN105	0440
REN106	0150
REN107	0127
REN108	0144
REN109	0143
REN110	1114
REN112	0917
REN113	0196
REN114	0484
REN115	1080
REN116	0015
REN117	0229
REN118	0774
REN120	0474
REN121	0160
REN121MP	0265
REN123	0198
REN123A	0079
REN124	0142
REN125	0071
REN126	0136
REN127	0975
REN128	0086
REN129	0126
REN130	0103
REN131	0222
REN131MP	0822
REN132	0325
REN133	3483
REN134	0188
REN135	0162
REN136	0157
REN137	0636
REN138	0649
REN139	0012
REN140	0170
REN141	0794
REN142	0137
REN143	0365
REN144	0100
REN145	0002
REN146	0441
REN147	0039
REN148	1834
REN149	0783
REN150	0331
REN151	0149
REN152	0042
REN153	0925
REN154	0233
REN155	2763
REN156	0087
REN157	0276
REN158	0164
REN159	0037
REN160	0050
REN161	0007
REN162	0074
REN163	0642
REN164	0003
REN165	0065
REN166	0277
REN171	0285
REN172	0400
REN172A	0400
REN173	0293
REN173A	0293
REN173P	0293
REN175	0178
REN176	0847
REN177	0133
REN179	0604
REN180	1681
REN181	0130
REN182	0561
REN183	1199
REN184	0161
REN185	0460
REN186	0560
REN186A	0219
REN187	1267
REN188	0551
REN189	0382
REN190	0264
REN191	0338
REN192	0595
REN193	0791
REN194	0861
REN195A	0698
REN196	0423
REN197	0854
REN198	0168
REN199	0111
REN203	OEM
REN209	3509
REN210	0566
REN211	1367
REN218	0905
REN219	0491
REN220	0849
REN221	0353
REN222	0212
REN223	0103
REN224	0630
REN225	3280
REN226	1010
REN226MP	1862
REN229	0224
REN230	0239
REN231	0061
REN233	0330
REN234	0693
REN235	0936
REN236	0836
REN237	1411
REN238	0313
REN276	1652
REN280	0177
REN282	0622
REN287	0715
REN288	0342
REN289	0155
REN290	0006
REN291	0236
REN292	0681
REN293	0018
REN294	0532
REN295	1591
REN297	0324
REN298	0436
REN299	2050
REN300	2046
REN302	1173
REN311	0493
REN315	1978
REN322	1984
REN323	0892
REN324	1480
REN325	1198
REN358	1030
REN500A	0190
REN502	0201
REN503	0288
REN504	0378
REN506	0102
REN513	1323
REN521	3858
REN522	1501
REN523	1067
REN526A	1706
REN600	1925
REN601	0139
REN604	3683
REN612	0720
REN614	0030
REN703A	5986
REN705A	2161
REN708	0664
REN709	0678
REN712	0167
REN713	0666
REN714	0352
REN715	0354
REN718	0654
REN720	0443
REN721	2279
REN722	0701
REN723	1345
REN729	0328
REN731	2457
REN736	1759
REN737	1443
REN738	0856
REN739	0751
REN740	0379
REN742	3735
REN743	1395
REN744	1421
REN747	1808
REN748	0789
REN749	0395
REN755	2027
REN780	0364
REN783	0802
REN788	2755
REN790	0349
REN791	1337
REN797	0521
REN801	0520
REN804	2557
REN977	1292
REN1003	0579
REN1004	0878
REN1005	2568
REN1006	1215
REN1024	3550
REN1025	1187
REN1027	2876
REN1028	4559
REN1037	4093
REN1046	0528
REN1050	2891
REN1052	0433
REN1056	1837
REN1058	2124
REN1073	2596
REN1080	0855
REN1081A	1160
REN1082	2261
REN1087	1018
REN1090	2861
REN1092	2316
REN1096	1634
REN1099	4260
REN1100	0911
REN1102	2632
REN1103	1994
REN1127	4420
REN1128	3915
REN1130	1326
REN1131	4594
REN1132	4595
REN1133	2231
REN1134	2949
REN1135	2062
REN1142	1478
REN1153	3366
REN1155	1050
REN1161	2942
REN1162	4038
REN1163	2604
REN1164	3269
REN1165	2773
REN1166	2781
REN1167	1714
REN1174	2049
REN1192	1540
REN5061	3003
REN5062	OEM
REN5063	1312
REN5064	3010
REN5065	1713
REN5066	0292
REN5067	0456
REN5068	0533
REN5069	0451
REN5070	0301
REN5071	0025
REN5072	0244
REN5073	1083
REN5074	0317
REN5075	0420
REN5076	1649
REN5077	0495
REN5078	0949
REN5079	0531
REN5080	0565
REN5081	0402
REN5082	1606
REN5086	0350
REN5096	0098
REN5174	2035
REN5175	2401
REN5176	1438
REN5177	2407
REN5178	1445
REN5179	2415
REN5180	2221
REN5181	1458
REN5182	0221
REN5183	1490
REN5184	2422
REN5185	1618
REN5186	0510
REN5187	0691
REN5188	0870
REN5189	1020
REN5190	1153
REN5191	1274
REN5192	1402
REN5193	1532
REN5194	1078
REN5195	1711
REN5196	1717
REN5197	1723
REN5198	1736
REN5199	1748
REN5202	1772
REN5205	1804
REN5209	0070
REN5400	1137
REN5401	0344
REN5402	0901
REN5403	2342
REN5404	0058
REN5448	3404
REN5454	0447
REN5455	0940
REN5457	0095
REN5482	2491
REN5483	0240
REN5485	0676
REN5487	0336
REN5542	0712
REN5543	0469
REN5545	0722
REN5547	0725
REN5640	2383
REN5641	2394
REN5642	1413
REN5643	2387
REN5662	0408
REN5663	0411
REN5665	0415
REN5667	0422
REN5682	3148
REN5683	3150
REN5685	3152
REN6401	2136
REN6402	0316
REN6406	OEM
REN6408	5911
REN7400	0232
REN7473	1172
REN7474	1313
REN7493A	0569
REN74123	1157
REN74161	1645
RER-023	0913
RER023	0196
RET20	0015
RET-UG-1004	OEM
RF01F	0017
RF03C06	0288
RF-1	0023
RF1	0023
RF-1A	0023
RF1A	0023
RF1AV	0102
RF1B	0023
RF-1Z	0023
RF1Z	0023
RF10K35	0102
RF100	OEM
RF101	1335
RF200	0007
RF300	0609
RF414	OEM
RF1811	0143
RF2065	3572
RF2073	OEM
RF2084	OEM
RF3160	0015
RF3472	0015
RF5464-1P	0196
RF5465	0196
RF5465-1&	0196
RF5465-1P	0196
RF5794	0484
RF-6235-1	0015
RF7313	0142
RF26231-1	0015
RF26234-1	0015
RF26235-1	0015
RF26235-2	0015
RF26235-5	0015
RF29799P	0015
RF31903P	0015
RF32101-8	0015
RF32101-9	0015
RF32101R	0015
RF32102R	1303
RF32103-1	0774
RF32103R	0774
RF32412-3	0015
RF32426-7	0913
RF32645	0015
RF33426-7	0196
RF33550-1	0143
RF33976	0015
RF34383	0015
RF34661	0133
RF34720	0015
RF35123	0143
RF60034	0143
RFA70597	0015
RFA70600	0015
RFC61197	0015
RFC86345	OEM
RFH10N45	OEM
RFH12N35	OEM
RFH25N18	OEM
RFH25N20	OEM
RFH25P08	OEM
RFH25P10	OEM
RFH30N12	OEM
RFH30N15	OEM
RFH45N05	OEM
RFJ6134	0015
RFJ30704	0015
RFJ31218	0015
RFJ31362	0015
RFJ31363	0015
RFJ33292	0015
RFJ60033	0917
RFJ60172	0143
RFJ60173	0015
RFJ60174	0015
RFJ60286	0015
RFJ60313	0196
RFJ60366	0015
RFJ60614	0143
RFJ60869	0015
RFJ70147	0540
RFJ70148	0196
RFJ70149	0102
RFJ70161	OEM
RFJ70431	0196
RFJ70432	0015
RFJ70487	0015
RFJ70643	1080
RFJ70703	0015
RFJ70931	0015
RFJ70970	0015
RFJ70971	0102
RFJ70974	0015
RFJ70976	0102
RFJ70977	0015
RFJ71122	0015
RFJ71123	0087
RFJ71480	0012
RFJ72360	0015
RFJ72787	0015
RFK10N45	OEM
RFK10N50	OEM
RFK12N35	OEM
RFK12N40	OEM
RFK25N18	OEM
RFK25N20	OEM
RFK25P08	OEM
RFK25P10	OEM
RFK30N12	OEM
RFK30N15	OEM
RFK35N08	OEM
RFK35N10	OEM
RFK45N05	OEM
RFK45N06	OEM
RFL1N08	OEM
RFL1N08L	OEM
RFL1N10	OEM
RFL1N10L	OEM
RFL1N12	OEM
RFL1N12L	OEM
RFL1N15	OEM
RFL1N15L	OEM
RFL1N18	OEM
RFL1N18L	OEM

DEVICE TYPE	REPL CODE
RFL1N20	OEM
RFL1N20L	OEM
RFL1P08	5913
RFL1P10	5913
RFL2N05	OEM
RFL2N06	OEM
RFL2N06L	OEM
RFL4N12	OEM
RFL4N15	OEM
RFL30596	0015
RFM3N45	OEM
RFM3N50	OEM
RFM4N35	OEM
RFM4N40	OEM
RFM5P12	OEM
RFM5P15	OEM
RFM6N45	OEM
RFM6N50	OEM
RFM6P08	OEM
RFM6P10	OEM
RFM8N18	OEM
RFM8N18L	OEM
RFM8N20	OEM
RFM8N20L	OEM
RFM8P08	OEM
RFM8P10	OEM
RFM10N12	OEM
RFM10N12L	OEM
RFM10N15	OEM
RFM10N15L	OEM
RFM10P12	OEM
RFM10P15	OEM
RFM12N08	OEM
RFM12N08L	OEM
RFM12N10	OEM
RFM12N10L	OEM
RFM12N18	2176
RFM12N20	2176
RFM12P08	OEM
RFM12P10	OEM
RFM15N05	OEM
RFM15N05L	OEM
RFM15N06	OEM
RFM15N06L	OEM
RFM15N12	OEM
RFM15N15	OEM
RFM18N08	4572
RFM18N10	4572
RFM25N05	5363
RFM25N06	5363
RFM33160	0015
RFP1N35	OEM
RFP1N40	OEM
RFP2N05L	OEM
RFP2N08	OEM
RFP2N08L	OEM
RFP2N10	OEM
RFP2N10L	OEM
RFP2N12	OEM
RFP2N12L	OEM
RFP2N15	6843
RFP2N15L	OEM
RFP2N18	OEM
RFP2N18L	OEM
RFP2N20	OEM
RFP2N20L	OEM
RFP2P08	2271
RFP2P10	3088
RFP3N45	5757
RFP3N50	5757
RFP4N05	OEM
RFP4N05L	OEM
RFP4N06	OEM
RFP4N06L	OEM
RFP4N35	OEM
RFP4N40	OEM
RFP5P12	6343
RFP5P15	6343
RFP6N45	7111
RFP6N50	1653
RFP6P08	7059
RFP6P10	3088
RFP8N18	2176
RFP8N18L	OEM
RFP8N20	2176
RFP8N20L	OEM
RFP8P08	2271
RFP8P10	2271
RFP10N12	2271
RFP10N12L	OEM
RFP10N15	2176
RFP10N15L	OEM
RFP10P12	OEM
RFP10P15	OEM
RFP12N08	5060
RFP12N08L	OEM
RFP12N10	5060
RFP12N10L	OEM
RFP12N18	2176
RFP12N20	2176
RFP12P08	5524
RFP12P10	5524
RFP15N05	5948
RFP15N05L	OEM
RFP15N06	5948
RFP15N06L	OEM
RFP15N12	OEM
RFP15N15	OEM
RFP18N08	OEM

DEVICE TYPE	REPL CODE
RFP18N10	OEM
RFP25N05	OEM
RFP25N06	OEM
RFP33118	0015
RFS61436	0229
RFV60500	0015
RG010G	0023
RG1	0015
RG1A	0023
RG1B	0023
RG1D	0023
RG1G	0023
RG1J	0023
RG1K	0017
RG1M	0017
RG-2	0031
RG2	0031
RG2A	0031
RG2B	0031
RG2D	0031
RG2G	0031
RG2J	0031
RG2K	0102
RG2M	0102
RG-2V	0031
RG2V	0031
RG-2Y	0031
RG2Y	0031
RG2Z	0031
RG3	0031
RG3A	0031
RG3B	0031
RG3D	0031
RG3G	0031
RG3G-5007L	0031
RG3J	0031
RG3K	3188
RG3M	3188
RG4	0031
RG4A	0031
RG4B	0031
RG4C	3887
RG4D	0031
RG4G	0031
RG4J	0031
RG4K	3188
RG4LF-M1	OEM
RG4LFM1	0031
RG4M	3188
RG4Y	0031
RG4Z	0031
RG5LFM1	0031
RG100B	0015
RG100D	0015
RG100G	0015
RG100J	0015
RG600	OEM
RG601	OEM
RG602	3766
RG602T	3766
RG604	2688
RG604T	2688
RG606	0110
RG606T	0110
RG608	3900
RG608T	3900
RG610	2655
RG610T	2655
RG1004	0015
RG1127	0015
RGL27A	OEM
RGL27B	OEM
RGL27D	OEM
RGL27G	OEM
RGL27J	OEM
RGL27K	OEM
RGL27M	OEM
RGL41A	OEM
RGL41B	OEM
RGL41D	OEM
RGL41G	OEM
RGL41J	OEM
RGL41K	OEM
RGL41M	OEM
RGP01-08	0023
RGP01-10	0017
RGP01-12	0017
RGP01-14	0017
RGP01-16	0348
RGP01-17	0348
RGP01-17PKG23	0017
RGP01-18	0348
RGP01-20	0348
RGP02-17	0348
RGP02-17EL	OEM
RGP02-20EL-6394	OEM
RGP10	0023
RGP10-DG1	0080
RGP10A	3077
RGP-10B	0023
RGP10B	0023
RGP-10D	0023
RGP10D	3077
RGP10DG1	0609
RGP-10DGI	OEM
RGP-10G	0023
RGP10G	0023
RGP10GL	OEM
RGP10GL-6391	OEM

DEVICE TYPE	REPL CODE
RGP10GPKG3	OEM
RGP10GPKG23	0023
RGP-10J	0023
RGP10J	0023
RGP10J(C)-T3	OEM
RGP10J(C1)-T3	0023
RGP10J,G23	0023
RGP10JG23	0023
RGP10J-T3	0023
RGP10J-Z	0023
RGP10K	0017
RGP10M	0017
RGP15-10	0017
RGP15-12	0017
RGP15-14	0017
RGP15-15	0017
RGP15-16	0348
RGP15-18	0348
RGP15-20	0348
RGP15A	0031
RGP15B	0031
RGP15D	0023
RGP-15G	2527
RGP15G	0031
RGP15G-004	0023
RGP15GPKG23	0031
RGP15J	0031
RGP15JTP	0031
RGP15K	0102
RGP15K-6179	OEM
RGP15M	0102
RGP18G	OEM
RGP20A	0031
RGP20B	0031
RGP20D	0031
RGP20G	0031
RGP20GL	0071
RGP20J	0031
RGP20K	0102
RGP20M	0102
RGP25A	0031
RGP25B	0031
RGP25D	0031
RGP25G	0031
RGP25J	0031
RGP25K	0071
RGP25M	0071
RGP30A	0546
RGP30B	0546
RGP30D	0546
RGP30G	0017
RGP30G-015	0031
RGP30G-5001L	0017
RGP30GJ	OEM
RGP30J	0031
RGP30K	3188
RGP30M	3188
RGP-100	0023
RGP100	0023
RGP150	OEM
RGP5040	0023
RGP5080	0071
RGP01-17PKG23	0017
RGPP1A	OEM
RGPP1B	OEM
RGPP1D	OEM
RGPP1G	OEM
RGPP1J	OEM
RGPP1K	OEM
RGPP5A	OEM
RGPP5B	OEM
RGPP5D	OEM
RGPP5G	OEM
RGPP5J	OEM
RGPP5K	OEM
RGPP10A	2079
RGPP10B	2079
RGPP10D	2079
RGPP10G	OEM
RGPP10J	0023
RGPP10K	0017
RGPP10M	0017
RGPP15A	2079
RGPP15B	2079
RGPP15D	2079
RGPP15G	OEM
RGPP15J	0031
RGPP15K	0102
RGPP15M	0102
RGPP30A	OEM
RGPP30B	OEM
RGPP30D	OEM
RGPP30G	OEM
RGPP30J	0031
RGPP30K	1928
RGPP30M	1928
RGPP60A	OEM
RGPP60B	OEM
RGPP60D	OEM
RGPP60G	OEM
RGPP60J	OEM
RGPP60K	OEM
RGPP60M	OEM
RH-1	0023
RH1	0023
RH-1A	0023
RH1A	0023
RH-1AV1	0023
RH-1B	0102

DEVICE TYPE	REPL CODE
RH1B	0102
RH-1BV	0023
RH-1B-Z	0102
RH-1C	0102
RH1C	0102
RH-1K0027CEZZ	OEM
RH-1M	0023
RH1M	0023
RH-1MV	0023
RH-1S	0023
RH1S	0023
RH1S-FAL	0023
RH1S-LFA1	0023
RH1S-T3	0023
RH1S-TS	0023
RH1SV	0023
RH-1S-Z	0023
RH1S-Z	0023
RH-1V	0023
RH-1X0001TAZZ	0789
RH-1X0004CEZZ	0395
RH-1X0005PAZZ	1158
RH-1X0015TAZZ	0904
RH-1X0018TAZZ	2284
RH-1X0020CEZZ	0878
RH-1X0021CEZZ	0855
RH-1X0022CEZZ	3785
RH-1X0032CEZZ	0638
RH-1X0038CEZZ	0167
RH-1X0043CEZZ	0167
RH-1X0047CEZZ	4395
RH-1X0092CEZZ	0912
RH-1X1020AFZZ	1248
RH-1X1039AFZZ	4620
RH-1X1311AF	OEM
RH-1X1465AF	OEM
RH-1Z	0499
RH1Z	0023
RH-1ZM	0023
RH-1ZV	0023
RH1ZV	OEM
RH-2	0124
RH-2D	0546
RH2D	0546
RH-2F	0017
RH2F	0017
RH2FM	0348
RH-2FS	0017
RH2FV	0017
RH10FV1	OEM
RH-12	OEM
RH-13	0424
RH120	0016
RH3761	0418
RH-DX0004TAZZ	5313
RH-DX0008CEZZ	0015
RH-DX0014-CEZZ	0023
RH-DX0017CEZZ	0102
RH-DX0025CEZZ	0023
RH-DX0026AGZZ	0015
RH-DX0028CEZZ	0182
RH-DX0029CEZZ	0182
RH-DX0033TAZZ	0133
RH-DX0038CEZZ	0023
RH-DX0039TAZZ	0102
RH-DX0041CEZZ	0196
RH-DX0042CEZZ	0015
RH-DX0043TAZZ	0023
RH-DX0045CEZZ	0124
RH-DX0045GEZZ	0124
RH-DX0046CEZZ	0133
RH-DX0047GEZZ	0735
RH-DX0048CEZZ	0133
RH-DX0048GEZZ	0133
RH-DX0049GEZZ	0499
RH-DX0051CEZZ	0102
RH-DX0052GEZZ	0023
RH-DX0053GEZZ	0124
RH-DX0054CEZZ	0133
RH-DX0055TAZZ	0015
RH-DX0056CEZZ	0015
RH-DX0056TAZZ	0102
RH-DX0059TAZZ	0015
RH-DX0062CEZZ	0102
RH-DX0063CEZZ	0102
RH-DX0064CEZZ	0023
RH-DX0065CEZZ	0102
RH-DX0065CZZ	0015
RH-DX0066TAZZ	0015
RH-DX0067TAZZ	0023
RH-DX0068TAZZ	0015
RH-DX0069TAZZ	0795
RH-DX0072CEZZ	0015
RH-DX0072TAZZ	OEM
RH-DX0073CEZZ	0023
RH-DX0073TAZZ	1414
RH-DX0077CEZZ	0348
RH-DX0079TAZZ	0015
RH-DX0081CEZZ	0015
RH-DX0081TAZZ	0015
RH-DX0083CEZZ	0015
RH-DX0085TAZZ	0253
RH-DX0086TAZZ	0023
RHDX0086TAZZ	0023
RH-DX0090CEZZ	0348
RH-DX0091CEZZ	0087
RH-DX0092CEZZ	0071
RH-DX0094CEZZ	OEM
RH-DX0096CEZZ	0348

DEVICE TYPE	REPL CODE
RH-DX0100CEZZ	0348
RH-DX0101-CEZZ	0023
RH-DX0101CEZZ	0023
RHDX0101CEZZ	0023
RH-DX0103CEZZ	OEM
RH-DX0104CEZZ	0182
RH-DX0105TAZZ	0023
RH-DX0105TEZZ	OEM
RH-DX0106CEZZ	0023
RH-DX0110CEZZ	0795
RH-DX0114CEZZ	0023
RH-DX0115CEZZ	0023
RH-DX0115TAZZ	0071
RH-DX0117TAZZ	0023
RH-DX0123CEZZ	0023
RHDX0123CEZZ	0023
RH-DX0126CEZZ	0023
RH-DX0130CEZZ	0023
RH-DX0131CEZZ	0023
RH-DX0132CEZZ	6354
RH-DX0136TAZZ	0071
RH-DX0142CEZZ	0133
RH-DX0152CEZZ	0473
RH-DX0154CEZZ	0023
RH-DX0155CEZZ	0023
RH-DX0179CEZZ	0124
RH-DX0181CEZZ	0031
RH-DX0190CEZZ	0017
RH-DX0200CEZZ	1875
RH-DX0202CEZZ	0023
RH-DX0203CEZZ	0031
RH-DX0204CEZZ	1091
RH-DX0208CEZZ	1875
RH-DX0214CEZZ	3549
RH-DX0220CEZZ	0609
RH-DX0224CEZZ	OEM
RH-DX0226CEZZ	0023
RH-DX0229CEZZ	0102
RH-DX0234CEZZ	0031
RH-DX0235CEZZ	0023
RH-DX0236CEZZ	0133
RH-DX0237CEZZ	0124
RH-DX0240CEZZ	OEM
RH-DX0247CEZZ	0283
RH-DX0248CEZZ	0023
RH-DX0255CEZZ	0017
RH-DX0259CEZZ	OEM
RH-DX0284CEZZ	OEM
RH-DX0302CEZZ	0017
RH-DX0382CEZZ	3301
RH-DX0403GEZZ	OEM
RH-DX0417CEZZ	OEM
RH-DX0441CEZZ	OEM
RH-DX0444CEZZ	OEM
RH-DX0446CEZZ	OEM
RH-DXOL05TAZZ	1217
RH-DX1005AFZZ	0015
RH-DXO045GEZZ	OEM
RH-DY0003SEZZ	0015
RH-DZA008WREO	OEM
RH-EX0002AEZZ	6038
RH-EX0003CEZZ	OEM
RH-EX002YTAZZ	0162
RH-EX0010GEZZ	0253
RH-EX0011CEZZ	0365
RH-EX0011GEZZ	0686
RH-EX0012CEZZ	0402
RH-EX0013CEZZ	0949
RH-EX0015CEZZ	0137
RH-EX0017CEZZ	0137
RH-EX0019CEZZ	0025
RH-EX0019GEZZ	0686
RH-EX0019TAZZ	0053
RH-EX0020CEZZ	OEM
RH-EX0020GEZZ	0041
RH-EX0021TAZZ	0181
RH-EX0022CEZZ	0949
RH-EX0022TAZZ	0181
RH-EX0023GEZZ	0170
RH-EX0024CEZZ	0471
RH-EX0024GEZZ	2522
RH-EX0024TAZZ	5904
RH-EX0033CEZZ	0039
RH-EX0034CEZZ	0402
RH-EX0037CEZZ	0531
RH-EX0038CEZZ	0137
RH-EX0041TAZZ	0062
RH-EX0042TAZZ	0140
RH-EX0043CEZZ	0195
RH-EX0043TAZZ	0041
RH-EX0044CEZZ	0064
RH-EX0045GEZZ	OEM
RH-EX0045TAZZ	OEM
RH-EX0046TAZZ	0077
RH-EX0047CEZZ	1005
RH-EX0047GEZZ	0140
RH-EX0047TAZZ	0064
RH-EX0048CEZZ	0471
RH-EX0049CEZZ	0041
RH-EX0049TAZZ	0883
RH-EX0053CEZZ	0010
RH-EX0053GEZZ	0253
RH-EX0053TAZZ	0471
RH-EX0054CEZZ	0186
RH-EX0057CEZZ	1834
RH-EX0057GEZZ	0471
RH-EX0061CEZZ	0137
RH-EX0062CEZZ	0102
RH-EX0063GEZZ	OEM

DEVICE TYPE	REPL CODE
RH-EX0065CEZZ	0445
RH-EX0072CEZZ	0531
RH-EX0074CEZZ	0647
RH-EX0084CEZZ	1189
RH-EX0084GEZZ	OEM
RH-EX0085CEZZ	0062
RH-EX0088CEZZ	0062
RH-EX0089CEZZ	0165
RH-EX0091CEZZ	0705
RH-EX0092CEZZ	0376
RH-EX0094CEZZ	4577
RH-EX0095GEZZ	OEM
RH-EX0098CEZZ	0705
RH-EX013CEZZ	0949
RH-EX022CEZZ	0949
RH-EX0101CEZZ	0760
RH-EX0102CEZZ	6813
RH-EX0103CEZZ	0253
RH-EX0106CEZZ	0049
RH-EX0112CEZZ	0375
RH-EX0114CEZZ	OEM
RH-EX0116CEZZ	6175
RH-EX0116GEZZ	0062
RH-EX0130CEZZ	0700
RH-EX0130GEZZ	0140
RH-EX0131CEZZ	0041
RH-EX0136CEZZ	0820
RH-EX0137CEZZ	0041
RH-EX0137GEZZ	OEM
RH-EX0137TAZZ	0140
RH-EX0138CEZZ	0062
RH-EX0139CEZZ	0162
RH-EX0140CEZZ	0062
RH-EX0145CEZZ	OEM
RH-EX0152CEZZ	OEM
RH-EX0152GEZZ	OEM
RH-EX0153CEZZ	OEM
RH-EX0154CEZZ	4528
RH-EX0155CEZZ	0062
RH-EX0156CEZZ	OEM
RH-EX0161CEZZ	0077
RH-EX0163CEZZ	0156
RH-EX0163GEZZ	1005
RH-EX0166CEZZ	0686
RH-EX0188GEZZ	OEM
RH-EX0192CEZZ	0636
RH-EX0193CEZZ	0181
RH-EX0198GEZZ	2106
RH-EX0200CEZZ	0052
RH-EX0204CEZZ	OEM
RH-EX0206CEZZ	OEM
RH-EX0207CEZZ	0195
RH-EX0217CEZZ	0879
RH-EX0219CEZZ	0166
RH-EX0223CEZZ	OEM
RH-EX0232CEZZ	1276
RH-EX0238CEZZ	4469
RH-EX0247CEZZ	0636
RH-EX0255CEZZ	5089
RH-EX0280CEZZ	1713
RH-EX0284CEZZ	OEM
RH-EX0293CEZZ	OEM
RH-EX0294CEZZ	6960
RH-EX0296CEZZ	5700
RH-EX0297CEZZ	OEM
RH-EX0298CEZZ	OEM
RH-EX0300CEZZ	OEM
RH-EX0300CEZZ0	OEM
RH-EX0302CEZZ	OEM
RH-EX0309CEZZ	OEM
RH-EX0311CEZZ	OEM
RH-EX0312CEZZ	5030
RH-EX0313CEZZ	4407
RH-EX0314CEZZ	OEM
RH-EX0316CEZZ	OEM
RH-EX0318CEZZ	OEM
RH-EX0322CEZZ	OEM
RH-EX0323CEZZ	OEM
RH-EX0324CEZZ	OEM
RH-EX0328CEZ	OEM
RH-EX0334CEZZ	OEM
RH-EX0374GEZZ	0883
RH-EX0375GEZZ	0883
RH-EX0397CEZZ	4347
RH-EX0398CEZZ	OEM
RH-EX0413GEZZ	OEM
RH-EX0417GEZZ	OEM
RH-EX0443CEZZ	0036
RH-EX0444CEZZ	0376
RH-EX0586CEZZ	4235
RH-EX0592CEZZ	4232
RH-EX0598CEZZ	OEM
RH-EX0625GEZZ	OEM
RH-EX0647CEZZ	OEM
RH-EX0655CEZZ	OEM
RH-EX0664GEZZ	0195
RH-EX0701GEZZ	OEM
RH-FX0003CEZZ	0315
RH-FX0007CEZZ	1166
RH-FX0008CEZZ	1291
RH-FX0011CEZZ	1166
RH-FX0014CEZZ	1291
RH-FX0018CEZZ	0112
RH-I2157CEN1	OEM
RH-I2157CEZZ	OEM
RH-IX00004CEZZ	0395
RH-IX0001PAZZ	0857
RH-IX0001TAZZ	0789
RH-IX0004CE	0395

DEVICE TYPE	REPL CODE
RH-IX0004CEZZ	0395
RH-IX0005PAZZ	1158
RH-IX0007TAZZ	OEM
RH-IX0012PAZZ	0361
RH-IX0013AEZZ	OEM
RH-IX0014PAZZ	0624
RH-IX0015TAZZ	4620
RH-IX0017AAZZ	OEM
RH-IX0017TAZZ	0167
RH-IX0018TAZZ	2284
RH-IX0020CE	0878
RH-IX0020CEZZ	0878
RH-IX0021CEZZ	0855
RH-IX0021PAZZ	OEM
RH-IX0022CEZZ	3785
RH-IX0022PAZZ	OEM
RH-IX0023CEZZ	2592
RH-IX0023PAZZ	OEM
RH-IX0024CEZZ	4986
RH-IX0024PAZZ	3921
RH-IX0025CEZZ	2891
RH-IX0025PAZZ	2144
RH-IX0027CEZZ	OEM
RH-IX0027PAZZ	OEM
RH-IX0028CEZZ	OEM
RH-IX0028PAZZ	2135
RH-IX0032CEZZ	0638
RH-IX0034PAZZ	OEM
RH-IX0035GAZZ	0904
RH-IX0035PAZZ	OEM
RH-IX0035TAZZ	0904
RH-IX0037CEZZ	1329
RH-IX0038CEZZ	OEM
RH-IX0038PAZZ	1206
RH-IX0039PAZZ	0942
RH-IX0040PAZZ	0175
RH-IX0041PAZZ	1157
RH-IX0042PAZZ	1271
RH-IX0043CE	0167
RH-IX0043CEZZ	0167
RH-IX0045CEZZ	OEM
RH-IX0047CE	4395
RH-IX0047CEZZ	4395
RH-IX0047PAZZ	OEM
RH-IX0048CEZZ	4616
RH-IX0050PAZZ	OEM
RH-IX0054CEZZ	2872
RH-IX0056CEZZ	OEM
RH-IX0058GEZZ	OEM
RH-IX0061CEZZ	OEM
RH-IX0065CEZZ	1429
RH-IX0069CEZZ	0478
RH-IX0072CEZZ	0215
RH-IX0072PAZZ	1579
RH-IX0092CEZZ	0069
RH-IX0092CFZZ	OEM
RH-IX0093CEZZ	2054
RH-IX0094CEZZ	4134
RH-IX0108CEZZ	OEM
RH-IX0111CEZZ	1292
RH-IX0111CFZZ	OEM
RH-IX0113CEZZ	OEM
RH-IX0114PAZZ	1830
RH-IX0116CEZZ	1339
RH-IX0130CEZZ	OEM
RH-IX0137CEZZ	2194
RH-IX0144CEZZ	OEM
RH-IX0199CEZZ	2090
RH-IX0200CEZZ	OEM
RH-IX0203GEZZ	2666
RH-IX0205GEZZ	OEM
RH-IX0212CEZZ	3685
RH-IX0212GEZZ	OEM
RH-IX0213CEZZ	2896
RH-IX0214CEZZ	3772
RH-IX0223CEZZ	4109
RH-IX0224CEZZ	4986
RH-IX0225CEZZ	3333
RH-IX0228CEZZ	OEM
RH-IX0234GEZZ	OEM
RH-IX0238CEZZ	2042
RH-IX0252CEZZ	0069
RH-IX0258GEZZ	OEM
RH-IX0260CEZZ	2666
RH-IX0269CEZZ	OEM
RH-IX0270CEZZ	OEM
RH-IX0275CEZZ	4109
RH-IX0280CEZZ	OEM
RH-IX0308CEZZ	OEM
RH-IX0314CEZZ	OEM
RH-IX0321CEZZ	OEM
RH-IX0325CEZZ	6187
RH-IX0359CEZZ	OEM
RH-IX0360PAZZ	OEM
RH-IX0406CEZZ	OEM
RH-IX0416CEZZ	3254
RH-IX0424CEZZ	OEM
RH-IX0430GEZZ	OEM
RH-IX0450CEZZ	4310
RH-IX0453CEZZ	OEM
RH-IX0454CEZZ	OEM
RH-IX0468GEZZ	OEM
RH-IX0470GEZZ	OEM
RH-IX0474GEZZ	OEM
RH-IX0478PAZZ	OEM
RH-IX0487CEZZ	2269
RH-IX0488CEZZ	0416
RH-IX0499CEZZ	1604
RH-IX0508GEZZ	OEM

If replacement code is OEM, contact original manufacturer for replacement.

DEVICE TYPE	REPL CODE
RH-IX0528GEZZ	OEM
RH-IX0539CEZZ	OEM
RH-IX0550GEZZ	OEM
RHIX0550GEZZ	OEM
RH-IX0564CEZZ	OEM
RH-IX0579CEZZ	OEM
RH-IX0580CEZZ	OEM
RH-IX0581CEZZ	OEM
RH-IX0600CEZZ	3236
RH-IX0614CEZZ	OEM
RH-IX0633CEZZ	4336
RH-IX0634CEZZ	OEM
RH-IX0669CEZZ	OEM
RH-IX0692CEZZ	OEM
RH-IX0715CEZZ	OEM
RH-IX0717CEZZ	OEM
RH-IX0733PAZZ	OEM
RH-IX0735CEZZ	OEM
RH-IX0740CEZZ	OEM
RH-IX0740PAZZ	OEM
RH-iX0758CEZZ	OEM
RH-IX0800CEZZ	OEM
RH-IX0820CEZZ	OEM
RH-IX0854CEZZ	OEM
RH-IX0920PAZZ	OEM
RH-IX0930CEZZ	OEM
RH-iX0943CEZZ	OEM
RH-IX0948CEZZ	4129
RH-IX0950CEZZ	OEM
RH-IX0996CEZZ	OEM
RH-IX212CEZZ	OEM
RH-iX1011CEZZ	OEM
RH-IX1016AFZZ	OEM
RH-IX1018AFZZ	4793
RH-IX1020AFZZ	1248
RH-IX1023CEZZ	OEM
RH-IX1030AFZZ	3651
RH-IX1030CEZZ	OEM
RH-IX1031AFZZ	OEM
RH-IX1036AFZZ	OEM
RH-IX1037AFZZ	OEM
RH-IX1038AFZZ	3293
RH-IX1039AFZZ	4624
RH-IX1067AFZZ	OEM
RH-IX1068AFZZ	1540
RH-IX1077CEZZ	OEM
RH-IX1078CEZZ	OEM
RH-IX1112CEZZ	6810
RH-IX1113CEZZ	OEM
RH-IX1132CEZZ	OEM
RH-IX1169CEZZ	OEM
RH-IX1190CEZZ	3515
RH-iX1224CEZZ	OEM
RH-IX1231CEZZ	OEM
RH-IX1247CEZZ	OEM
RH-IX1311AF	OEM
RHIX1311AF	OEM
RHIX1311AFZZ	OEM
RH-IX1332AFZZ	OEM
RH-IX1505CEZZ	OEM
RH-iX1535CEZZ	OEM
RH-IX1553CEZZ	OEM
RH-IX1590CEZZ	OEM
RH-IX1675CEZZ	OEM
RH-IX1676CEZZ	OEM
RH-IX1705CEZZ	OEM
RH-IX1709CEZZ	OEM
RH-IX1712CEZZ	3515
RH-IX1750CEZZ	OEM
RH-IX2092CEZZ	OEM
RH-iX2157CEN1	OEM
RH-iX2157CEN2	OEM
RH-iX2157CEN3	OEM
RH-iX2157CEZZ	OEM
RH-iX2163CEZZ	OEM
RH-IX2195CEZZ	OEM
RH-IX2281CEZZ	OEM
RH-iX2428CEN1	OEM
RH-iX2447CEZZ	OEM
RH-IX2582CEN2	OEM
RH-iX2599CEZZ	OEM
RH-IX0137CEZZ	OEM
RH-IZ0008CEZZ	OEM
RH-IZ008CEZZ	OEM
RH-IZ0027CEZZ	OEM
RH-IZ0073CEZZ	OEM
RH-PH0167CEZZ	OEM
RH-PX0003GEZZ	OEM
RH-PX0012CEZZ	OEM
RH-PX0016CEZZ	OEM
RH-PX0033GEZZ	OEM
RH-PX0034GEZZ	OEM
RH-PX0040GEZZ	OEM
RH-PX0053GEZZ	OEM
RH-PX0068PAZZ	OEM
RH-PX0069CEZZ	OEM
RH-PX0085CEZZ	OEM
RH-PX0092CEZZ	OEM
RH-PX0095CEZZ	OEM
RH-PX0097PAZZ	OEM
RH-PX0102CEZZ	OEM
RH-PX0109PAZZ	OEM
RH-PX0119CEZZ	OEM
RH-PX0127CEZZ	OEM
RH-PX0133CEZZ	OEM
RH-PX0135GEZZ	OEM
RH-PX0140CEZZ	OEM
RH-PX0142PAZZ	OEM
RH-PX0143CEZZ	OEM
RH-PX0153CEZZ	OEM
RH-PX0159GEZZ	OEM
RH-PX0161CEZZ	OEM
RH-PX0167CEZZ	OEM
RH-PX0168CEZZ	OEM
RH-PX0183CEZZ	OEM
RH-PX0195CEZZ	OEM
RH-PX0207CEZZ	OEM
RH-PX0218CEZZ	OEM
RH-PX0219CEZZ	OEM
RH-PX0222CEZZ	OEM
RH-PX0223CEZZ	OEM
RH-PX0226CEZZ	OEM
RH-PX0238CEZZ	OEM
RH-PX0242CEZZ	OEM
RH-PX0265CEZZ	OEM
RH-PX0304CEZZ	OEM
RH-PX0378CEZZ	OEM
RH-VX0004TAZZ	0120
RH-VX0006CEZZ	OEM
RH-VX0009TAZZ	0139
RH-VX0014TAZZ	0139
RI1A	0023
RI1B	OEM
RI1C	OEM
RI2	OEM
RI2B	OEM
RI4010658	0124
RJF60313	0196
RK-13	2542
RK13	2542
RK14	2542
RK34	2542
RK39	3604
RK43	0031
RK44	0031
RK46	3604
RK46LF-L1	3604
RKBP005	OEM
RKBP01	OEM
RKBP02	OEM
RKBP04	OEM
RKBP06	OEM
RKBP08	OEM
RKBP10	OEM
RKBPC101	OEM
RKBPC102	OEM
RKBPC104	OEM
RKBPC106	OEM
RKBPC108	OEM
RKBPC110	OEM
RKBPC601	OEM
RKBPC602	OEM
RKBPC604	OEM
RKBPC606	OEM
RKBPC608	OEM
RKBPC610	OEM
RKBPC801	OEM
RKBPC802	OEM
RKBPC804	OEM
RKBPC806	OEM
RKBPC808	OEM
RKBPC810	OEM
RKBPC1005	OEM
RKBPC6005	OEM
RKBPC8005	OEM
RL005	0071
RL010	0071
RL020	0071
RL040	0071
RL060	0071
RL1N4002	0080
RL-2Z	0546
RL2Z	0546
RL22LF-B2	OEM
RL31	0143
RL32	0143
RL32G	0143
RL34	0143
RL34G	0143
RL41	0143
RL41G	0143
RL42	0143
RL43	0143
RL43G	0143
RL44	OEM
RL44G	0143
RL52	0143
RL55-5	1989
RL101F	0023
RL101G	0015
RL102F	0023
RL102G	0015
RL103F	0023
RL103G	0015
RL104F	0023
RL104G	0015
RL105F	0023
RL105G	0015
RL106F	0017
RL106G	0087
RL107F	0017
RL107G	0087
RL151	0102
RL151G	OEM
RL-152	0953
RL152	0102
RL152G	OEM
RL153	0102
RL153G	OEM
RL154	0102
RL154G	OEM
RL155	0102
RL155G	OEM
RL156	0102
RL156G	OEM
RL157	0102
RL157G	OEM
RL201	0102
RL201G	0087
RL202	0102
RL202-M11	OEM
RL202G	0087
RL203	0102
RL203G	0087
RL204	0102
RL204G	0087
RL205	0102
RL205G	0087
RL206	0102
RL206G	0087
RL207	0102
RL207G	0087
RL209-1	3019
RL232	OEM
RL232B	OEM
RL232G	0143
RL246	0143
RL247	0143
RL247G	0143
RL251	0071
RL252	0071
RL253	0071
RL254	0071
RL255	0071
RL256	0071
RL257	0071
RL709T	1301
RLC312	OEM
RLF1G	0015
RLH111	0512
RLM2902M	OEM
RLM2904M	OEM
RLR4001	OEM
RLR4002	OEM
RLR4003	OEM
RLR4004	OEM
RLS-71	OEM
RLS-72	OEM
RLS72	OEM
RLS-73	OEM
RLS73	OEM
RLS73TE11	OEM
RLS-92	OEM
RLS-93	OEM
RLS-94	OEM
RLS-135	OEM
RLS135	OEM
RLS-139	OEM
RLS-140	OEM
RLS-141	OEM
RLS-142	OEM
RLS-143	OEM
RLS-144	OEM
RLS914	OEM
RLS914A	OEM
RLS914B	OEM
RLS916	OEM
RLS916A	OEM
RLS916B	OEM
RLS3064	OEM
RLS3600	OEM
RLS3604	OEM
RLS3605	OEM
RLS3606	OEM
RLS-4148	OEM
RLS4148	OEM
RLS-4149	OEM
RLS4149	OEM
RLS-4150	OEM
RLS4150	OEM
RLS-4151	OEM
RLS4151	OEM
RLS-4152	OEM
RLS4152	OEM
RLS-4153	OEM
RLS4153	OEM
RLS-4154	OEM
RLS4154	OEM
RLS-4446	OEM
RLS4446	OEM
RLS-4447	OEM
RLS4447	OEM
RLS-4448	OEM
RLS4448	OEM
RLS-4449	OEM
RLS4449	OEM
RLS-4450	OEM
RLS4450	OEM
RLS-4454	OEM
RLS4454	OEM
RLS-4606	OEM
RLS4606	OEM
RLV11	OEM
RLZ-4.7	OEM
RLZ-5.1	OEM
RLZ5.1B	OEM
RLZ5.1B-TE-11	OEM
RLZ-5.6	OEM
RLZ5.6B	OEM
RLZ5.6B-TE-11	OEM
RLZ-6.2	OEM
RLZ-6.8	OEM
RLZ6.8B	OEM
RLZ6.8B-TE-11	OEM
RLZ-7.5	OEM
RLZ-8.2	OEM
RLZ8.2	OEM
RLZ8.2-TE-11	OEM
RLZ8.2A	OEM
RLZ8.2A-TE-11	OEM
RLZ8.2ATE-11	OEM
RLZ-9.1	OEM
RLZ9.1A	OEM
RLZ-10	OEM
RLZ-11	OEM
RLZ-12	OEM
RLZ-13	OEM
RLZ-15	OEM
RLZ-16	OEM
RLZ-18	OEM
RLZ-20	OEM
RLZ-22	OEM
RLZ-24	OEM
RLZ746	OEM
RLZ746A	OEM
RLZ747	OEM
RLZ747A	OEM
RLZ748	OEM
RLZ748A	OEM
RLZ749	OEM
RLZ749A	OEM
RLZ750	OEM
RLZ750A	OEM
RLZ751	OEM
RLZ751A	OEM
RLZ752	OEM
RLZ752A	OEM
RLZ753	OEM
RLZ753A	OEM
RLZ754	OEM
RLZ754A	OEM
RLZ755	OEM
RLZ755A	OEM
RLZ756	OEM
RLZ756A	OEM
RLZ757	OEM
RLZ757A	OEM
RLZ758	OEM
RLZ758A	OEM
RLZ759	OEM
RLZ759A	OEM
RLZ957	OEM
RLZ957A	OEM
RLZ957B	OEM
RLZ958	OEM
RLZ958A	OEM
RLZ958B	OEM
RLZ959	OEM
RLZ959A	OEM
RLZ959B	OEM
RLZ960	OEM
RLZ960A	OEM
RLZ960B	OEM
RLZ961	OEM
RLZ961A	OEM
RLZ961B	OEM
RLZ962	OEM
RLZ962A	OEM
RLZ962B	OEM
RLZ963	OEM
RLZ963A	OEM
RLZ963B	OEM
RLZ964	OEM
RLZ964A	OEM
RLZ964B	OEM
RLZ965	OEM
RLZ965A	OEM
RLZ965B	OEM
RLZ966	OEM
RLZ966A	OEM
RLZ966B	OEM
RLZ967	OEM
RLZ967A	OEM
RLZ967B	OEM
RLZ968	OEM
RLZ968A	OEM
RLZ968B	OEM
RLZ969	OEM
RLZ969A	OEM
RLZ969B	OEM
RLZ970	OEM
RLZ970A	OEM
RLZ970B	OEM
RLZ971	OEM
RLZ971A	OEM
RLZ971B	OEM
RLZ972	OEM
RLZ972A	OEM
RLZ972B	OEM
RLZ973	OEM
RLZ973A	OEM
RLZ973B	OEM
RLZ5226	OEM
RLZ5226A	OEM
RLZ5226B	OEM
RLZ5226C	OEM
RLZ5226D	OEM
RLZ5227	OEM
RLZ5227A	OEM
RLZ5227B	OEM
RLZ5227C	OEM
RLZ5227D	OEM
RLZ5228	OEM
RLZ5228A	OEM
RLZ5228B	OEM
RLZ5228C	OEM
RLZ5228D	OEM
RLZ5229	OEM
RLZ5229A	OEM
RLZ5229B	OEM
RLZ5229C	OEM
RLZ5229D	OEM
RLZ5230	OEM
RLZ5230A	OEM
RLZ5230B	OEM
RLZ5230C	OEM
RLZ5230D	OEM
RLZ5231	OEM
RLZ5231A	OEM
RLZ5231B	OEM
RLZ5231C	OEM
RLZ5231D	OEM
RLZ5232	OEM
RLZ5232A	OEM
RLZ5232B	OEM
RLZ5232C	OEM
RLZ5232D	OEM
RLZ5233	OEM
RLZ5233A	OEM
RLZ5233B	OEM
RLZ5233C	OEM
RLZ5233D	OEM
RLZ5234	OEM
RLZ5234A	OEM
RLZ5234B	OEM
RLZ5234C	OEM
RLZ5234D	OEM
RLZ5235	OEM
RLZ5235A	OEM
RLZ5235B	OEM
RLZ5235C	OEM
RLZ5235D	OEM
RLZ5236	OEM
RLZ5236A	OEM
RLZ5236B	OEM
RLZ5236C	OEM
RLZ5236D	OEM
RLZ5237	OEM
RLZ5237A	OEM
RLZ5237B	OEM
RLZ5237C	OEM
RLZ5237D	OEM
RLZ5238	OEM
RLZ5238A	OEM
RLZ5238B	OEM
RLZ5238C	OEM
RLZ5238D	OEM
RLZ5239	OEM
RLZ5239A	OEM
RLZ5239B	OEM
RLZ5239C	OEM
RLZ5239D	OEM
RLZ5240	OEM
RLZ5240A	OEM
RLZ5240B	OEM
RLZ5240C	OEM
RLZ5240D	OEM
RLZ5241	OEM
RLZ5241A	OEM
RLZ5241B	OEM
RLZ5241C	OEM
RLZ5241D	OEM
RLZ5242	OEM
RLZ5242A	OEM
RLZ5242B	OEM
RLZ5242C	OEM
RLZ5242D	OEM
RLZ5243	OEM
RLZ5243A	OEM
RLZ5243B	OEM
RLZ5243C	OEM
RLZ5243D	OEM
RLZ5244	OEM
RLZ5244A	OEM
RLZ5244B	OEM
RLZ5244C	OEM
RLZ5244D	OEM
RLZ5245	OEM
RLZ5245A	OEM
RLZ5245B	OEM
RLZ5245C	OEM
RLZ5245D	OEM
RLZ5246	OEM
RLZ5246A	OEM
RLZ5246B	OEM
RLZ5246C	OEM
RLZ5246D	OEM
RLZ5247	OEM
RLZ5247A	OEM
RLZ5247B	OEM
RLZ5247C	OEM
RLZ5247D	OEM
RLZ5248	OEM
RLZ5248A	OEM
RLZ5248B	OEM
RLZ5248C	OEM
RLZ5248D	OEM
RLZ5249	OEM
RLZ5249A	OEM
RLZ5249B	OEM
RLZ5249C	OEM
RLZ5249D	OEM
RLZ5250	OEM
RLZ5250A	OEM
RLZ5250B	OEM
RLZ5250C	OEM
RLZ5250D	OEM
RLZ5251	OEM
RLZ5251A	OEM
RLZ5251B	OEM
RLZ5251C	OEM
RLZ5251D	OEM
RLZ5252	OEM
RLZ5252A	OEM
RLZ5252B	OEM
RLZ5252C	OEM
RLZ5252D	OEM
RLZ5253	OEM
RLZ5253A	OEM
RLZ5253B	OEM
RLZ5253C	OEM
RLZ5253D	OEM
RLZ5254	OEM
RLZ5254A	OEM
RLZ5254B	OEM
RLZ5254C	OEM
RLZ5254D	OEM
RLZ5255	OEM
RLZ5255A	OEM
RLZ5255B	OEM
RLZ5255C	OEM
RLZ5255D	OEM
RLZ5256	OEM
RLZ5256A	OEM
RLZ5256B	OEM
RLZ5256C	OEM
RLZ5256D	OEM
RLZ5257	OEM
RLZ5257A	OEM
RLZ5257B	OEM
RLZ5257C	OEM
RLZ5257D	OEM
RLZ5258	OEM
RLZ5258A	OEM
RLZ5258B	OEM
RLZ5258C	OEM
RLZ5258D	OEM
RLZ5259	OEM
RLZ5259A	OEM
RLZ5259B	OEM
RLZ5259C	OEM
RLZ5259D	OEM
RLZ5260	OEM
RLZ5260A	OEM
RLZ5260B	OEM
RLZ5260C	OEM
RLZ5260D	OEM
RLZ5261	OEM
RLZ5261A	OEM
RLZ5261B	OEM
RLZ5261C	OEM
RLZ5261D	OEM
RLZ5262	OEM
RLZ5262A	OEM
RLZ5262B	OEM
RLZ5262C	OEM
RLZ5262D	OEM
RLZ5263	OEM
RLZ5263A	OEM
RLZ5263B	OEM
RLZ5263C	OEM
RLZ5263D	OEM
RLZ5264	OEM
RLZ5264A	OEM
RLZ5264B	OEM
RLZ5264C	OEM
RLZ5264D	OEM
RLZ5265	OEM
RLZ5265A	OEM
RLZ5265B	OEM
RLZ5265C	OEM
RLZ5265D	OEM
RLZ5266	OEM
RLZ5266A	OEM
RLZ5266B	OEM
RLZ5266C	OEM
RLZ5266D	OEM
RM01	OEM
RM-1	0023
RM1	0023
RM-1A	0023
RM1A	0023
RM-1AV	0015
RM-1B	0072
RM1B	0072
RM1C	0071
RM1D	0102
RM1E	0102
RM1F	0017
RM1S	0023
RM-1V	0015
RM1V	0023
RM-1Z	0015
RM1Z	0023
RM1ZM	0015
RM-1ZMV	0023
RM1ZMV	0023
RM-1ZV	0023
RM1ZV	0015
RM-2	0071
RM2	0071
RM2-C-Z	0071
RM-2A	0564
RM2A	0564
RM-2AV	0564
RM2AV	0071
RM-2AVTA	0564
RM2B	0071
RM-2C	2637
RM2C	0023
RM2C-LAF1	OEM
RM2C-LFA1	0071
RM2CS	0071
RM2CV	0071
RM2C-Z	0071
RM-2SC	0071
RM-2V	0071
RM2Z	0015
RM2ZLFB1	OEM
RM-4	OEM
RM4	0071
RM-4A	0965
RM4A	4114
RM4B	0071
RM4C	0071
RM4Y	0071
RM4Z	0071
RM4Z-LFK5	OEM
RM10	0023
RM10A	0071
RM10B	0071
RM10F	OEM
RM10M	0071
RM10MV	0071
RM10N	OEM
RM10TA-H	OEM
RM10TA-M	OEM
RM10Z	0071
RM11	0564
RM11A	0564
RM11AV	OEM
RM11AVM	OEM
RM-11B	0071
RM11B	0071
RM11BV	0564
RM11C	0071
RM12M	OEM
RM15	0023
RM15TA-H	OEM
RM15TA-M	OEM
RM23T	OEM
RM-25	0502
RM25	0502
RM25V	0502
RM-26	2157
RM26	1268
RM26(50V)	0502
RM-26(ZENER)	2157
RM26(ZENER)	0502
RM26V	1268
RM78M93FD	OEM
RM-103	0071
RM103	0133
RM257	1834
RM260EMC	OEM
RM555DE	3780
RM555DE/883B	OEM
RM555T	3637
RM555T/883B	OEM
RM556DC	OEM
RM556DC/883B	OEM
RM714DE	OEM
RM714DE/883B	OEM
RM714H	OEM
RM714H/883B	OEM
RM723T	1192
RM725DE	OEM
RM741DC	OEM
RM741DC/883B	OEM
RM741DE	OEM
RM741DE/883B	OEM
RM741T	4262
RM741T/883B	OEM
RM741TE	0410
RM747DC	OEM
RM747DC/883B	OEM
RM747T	4272
RM747T/883B	OEM
RM1558DE	OEM
RM1558DE/883B	OEM
RM1558T	3137
RM1558T/883B	OEM
RM3001	OEM
RM3002	OEM
RM3005	0042
RM3010	OEM
RM3022	2258
RM3036	OEM
RM3078ADE	OEM
RM3078AT	OEM
RM3078AT/883B	OEM
RM3503ADC	OEM
RM3503ADE/883	OEM

DEVICE TYPE	REPL CODE
RM4136	OEM
RM4136DC	OEM
RM4136DC/883B	OEM
RM4151DE	OEM
RM4151DE/883B	OEM
RM4151T	OEM
RM4151T/883B	OEM
RM4152DE	OEM
RM4152DE/883B	OEM
RM4152T	OEM
RM4152T/883B	OEM
RM4153DC	OEM
RM4153DE/883B	OEM
RM4156DC	OEM
RM4156DC/883B	OEM
RM4191DE	OEM
RM4191DE/883B	OEM
RM4192DE	OEM
RM4192DE/883B	OEM
RM4193DE	OEM
RM4193DE/883B	OEM
RM4194DC	OEM
RM4194DC/883B	OEM
RM4194TK	OEM
RM4195T	OEM
RM4195TK	OEM
RM4200ADE	OEM
RM4200DE	OEM
RM4260DC	OEM
RM4260DC/883B	OEM
RM4391DE/883B	OEM
RM4558	0360
RM4558DE	OEM
RM4558DE/883B	OEM
RM4558T	OEM
RM4558T/883B	OEM
RM4559DE	OEM
RM4559DE/883B	OEM
RM4559T	OEM
RM4559T/883B	OEM
RM4805ADE	OEM
RM4805ADE/883B	OEM
RM4805AT	OEM
RM4805AT/883B	OEM
RM4805DE	OEM
RM4805DE/883B	OEM
RM4805T	OEM
RM4805T/883B	OEM
RM5008D	OEM
RM5532ADE	OEM
RM5532ADE/883B	OEM
RM5532AT	OEM
RM5532AT/883B	OEM
RM5532DE	OEM
RM5532DE/883B	OEM
RM5532T	OEM
RM5532T/883B	OEM
RM5534ADE	OEM
RM5534ADE/883B	OEM
RM5534AT	OEM
RM5534AT/883B	OEM
RM5534DE	OEM
RM5534DE/883B	OEM
RM5534T	OEM
RM5534T/883B	OEM
RM8007D	OEM
RM9391DE	OEM
RMIV	0015
RMP5020	0015
RMPG06D	1090
RMPG06G	0023
RMPTC0028CEZZ	3852
RMPTC0059CEZZ	6443
RMPTC0128CEZZ	OEM
RMPTC0135CEZZ	OEM
RMPTJ0027CEZZ	OEM
RMPTJ0028CEZZ	3852
RMPTJ0029CEZZ	5903
RMPTJ0037CEZZ	3841
RMPTJ0155CEZZ	OEM
RMPTJ0156CEZZ	OEM
RMPTJ0157CEZZ	OEM
RMVTC00921-3	0484
RN0008	OEM
RN0018	OEM
RN0020	OEM
RN0029	OEM
RN16AT	1727
RN16BT	1727
RN16DT	1727
RN16GT	OEM
RN16JT	OEM
RN16KT	OEM
RN30AM	OEM
RN30BM	OEM
RN30DM	OEM
RN30GM	OEM
RN30JM	OEM
RN30KM	OEM
RN30MM	OEM
RN515	2002
RN835	OEM
RN1001	OEM
RN1002	0832
RN1003	0887
RN1004	0898
RN1015	1250
RN1020	OEM
RN1030	OEM
RN1030A	OEM
RN1115	OEM
RN1120	1124
RN1135	OEM
RN1201	5161
RN1202	0832
RN1203	0887
RN1204	0898
RN1205	OEM
RN1206	OEM
RN1402	OEM
RN1403	5249
RN1404	3478
RN2003	1032
RN2005	OEM
RN2006	OEM
RN2015	1250
RN2201	OEM
RN2202	4172
RN2202TPE4	4172
RN2203	1032
RN2204	4130
RN2206	OEM
RN2403	0703
RN3015	1581
RN3020	OEM
RN3020R	OEM
RN3030	OEM
RN3030R	OEM
RN4015	1581
RN5015	1586
RN5016	1586
RN6015	1586
RN8015	OEM
RNUR8558571	0854
RO2	0071
RO2A	0071
RO2AV	0071
RO2Z	0071
RO-3-9502-011	OEM
RO-3-9502-025-8223	OEM
RO-3-9503	OEM
RO-3-9503-003	OEM
RO-3-9504-026-8210	OEM
RO-3-9504-026-8211	OEM
RO-3-9506	OEM
RO-3-9506-010	OEM
RO102	0015
RO130	0122
RO131	0122
RO132	0122
RO134	0122
RO136	0122
ROA-12015	OEM
RP1A	0102
RP1H	OEM
RP4Z	0071
RP16AT	1236
RP16BT	1236
RP16DT	1236
RP16GT	0909
RP16JT	OEM
RP16KT	OEM
RP30AM	2234
RP30AP	OEM
RP30BM	2234
RP30BP	OEM
RP30DM	2234
RP30DP	OEM
RP30GM	2234
RP30GP	OEM
RP30JM	2709
RP30JP	OEM
RP30KM	OEM
RP30KP	OEM
RP30MM	OEM
RP30MP	OEM
RP104	OEM
RP300G	0031
RP520	OEM
RP1020	2901
RP1040	2901
RP1120	0044
RP1140	0044
RP2020	1124
RP2040	1124
RP3020	OEM
RP4040	0805
RP5020	OEM
RP6020	0319
RP6040	0319
RP8020	1132
RP8040	1132
RP23256E	OEM
RP23256E-2110	OEM
RPG10J	0023
RPG10J(C1)-T3	OEM
RPG10J-T3	0023
RPG15G	OEM
RPR540	1530
RPR1040	1530
RPR2040	1530
RPR3040	0029
RPR4040	0029
RPX50	OEM
RPX100	OEM
RPX200	OEM
RPX400	OEM
RPX600	OEM
RPX800	OEM
RPY13	OEM
RPY14	OEM
RPY17	OEM
RPY23	OEM
RPY41	OEM
RQ-409S	0015
RQ-444S	0151
RR0102DM	1282
RR0104DM	1293
RR-1	0031
RR-1Z	0031
RR8AT	OEM
RR8BT	OEM
RR8DT	OEM
RR8GT	OEM
RR8JT	OEM
RR8KT	OEM
RR7504	0016
RR8068	0016
RR8070	0144
RR8116	0144
RR8118	0144
RR8119	0144
RR8914	0016
RR8989	0144
RR8999	0144
RR9036R	2768
RRB24-06	0015
RRET-0004CEZZ	OEM
RRET-0006CEZZ	OEM
RRET-0011CEZZ	OEM
RRET-0021CEZZ	OEM
RRET-0025CEZZ	OEM
RRET-0033CEZZ	OEM
RS-05	0110
RS05	0110
RS1	0023
RS1A	0017
RS1B	0017
RS1C	0182
RS1FM-12	0015
RS1Z	0023
RS3FS	0023
RS4F	0348
RS4FS	1570
RS6	OEM
RS8AT	1145
RS8BT	1145
RS8DT	1145
RS8GT	1145
RS8JT	1969
RS8KT	OEM
RS10	0015
RS20	OEM
RS30	OEM
RS-35	0039
RS40	OEM
RS50	OEM
RS52	OEM
RS52B	OEM
RS53B	OEM
RS60	OEM
RS80	OEM
RS100	OEM
RS-101	0050
RS101	0277
RS-102	0164
RS102	0277
RS-103	0050
RS103	0289
RS-104	0600
RS104	0296
RS-105	0085
RS105	0302
RS-106	0440
RS106	0250
RS-107	0016
RS107	0250
RS-108	0016
RS-109	0144
RS-110	0037
RS128	0016
RS132	0086
RS136	0016
RS201	0277
RS202	0277
RS203	0289
RS204	0296
RS205	0302
RS206	0250
RS207	0250
RS220AF	0015
RS230AF	0015
RS-253C	0133
RS-253S	0133
RS-267S	0133
RS-272US	0133
RS-279US	0111
RS-280S	0133
RS322	0004
RS372	OEM
RS380	0379
RS401	0729
RS401L	0729
RS402	0729
RS402L	0729
RS403	0729
RS403L	3023
RS404	0737
RS404L	0737
RS405	0737
RS405L	3035
RS406	0742
RS406L	0742
RS407	0742
RS407L	3050
RS501	OEM
RS502	OEM
RS503	OEM
RS504	OEM
RS505	OEM
RS506	OEM
RS507	OEM
RS555	0973
RS593	0050
RS601	1040
RS602	1040
RS603	1042
RS604	1045
RS605	1045
RS606	1047
RS607	1047
RS684	0050
RS685	0050
RS686	0050
RS687	0050
RS723	1192
RS741	0410
RS-805US	1220
RS-862S	2592
RS1049	0016
RS1059	0016
RS1192	0004
RS1234	0229
R-S1264	0015
RS1264	0015
RS1290	1338
RS1296	0015
RS-1347	0015
RS1347	0015
RS1348	0137
RS1428	0133
RS1458	0360
RS1513	0600
RS1524	0208
RS1530	0600
RS1531	0600
RS1532	0600
RS1533	0208
RS1534	0600
RS1536	0600
RS1537	0600
RS1538	0600
RS1539	0050
RS1540	0004
RS1541	0004
RS1542	0004
RS1543	0004
RS1544	0004
RS1545	0004
RS1546	0004
RS1547	0600
RS1548	0004
RS1549	0208
RS1550	0050
RS1553	0600
RS1554	0050
RS1555	0004
R-S-1720	0015
RS1720	0015
RS1726	0144
RS1749	0015
R-S1805	0540
RS1805	0015
RS1811	0143
RS1823	0229
RS1832	0015
RS-2001	0600
RS2001	0600
RS-2002	0050
RS-2003	0136
RS2003	0050
RS-2004	0211
RS-2005	0211
RS-2006	0085
RS-2007	0164
RS-2008	0233
RS2008	0715
RS-2009	0079
RS2009	0016
RS-2010	0079
RS2010	0079
RS-2011	0007
RS-2012	0715
RS2012	0715
RS-2013	0079
RS2013	0079
RS-2014	0086
RS-2015	0144
RS-2016	0079
RS2016	0079
RS-2017	0161
RS2017	0042
RS-2018	0566
RS2018	0042
RS-2019	0561
RS2019	1361
RS-2020	0161
RS-2021	0037
RS-2022	0037
RS2022	0037
RS-2023	0037
RS-2024	0037
RS-2025	0460
RS2025	0925
RS-2026	1367
RS2026	1367
RS-2027	1199
RS2027	1369
RS-2028	0325
RS-2029	1175
RS-2030	0016
RS-2031	0330
RS-2032	0037
RS-2033	0016
RS-2034	0037
RS2038	0493
RS2039	0103
RS2040	0491
RS2041	0103
RS2042	1394
RS2043	0491
RS2350	0004
RS2351	0004
RS2352	0004
RS2353	0004
RS2354	0004
RS2355	0004
RS2356	0600
RS2359	0600
RS2360	0600
RS2364	0600
RS2365	0600
RS2366	0600
RS2367	0004
RS2373	0004
RS2374	0004
RS2375	0600
RS2675	0004
RS2677	0004
RS2679	0050
RS2680	0050
RS2683	0050
RS2684	0050
RS2685	0050
RS2686	0050
RS2687	0050
RS2688	0050
RS2689	0004
RS2690	0280
RS2691	0280
RS2692	0280
RS2694	0050
RS2695	0050
RS2696	0280
RS2697	0004
RS2801	0143
RS2867	0004
RS2914	0016
RS3211	0004
RS3275	0004
RS3276	0211
RS3277	0050
RS3278	0050
RS3279	0050
RS3280	0004
RS3281	0280
RS3282	0004
RS3283	0004
RS3284	0004
RS3285	0004
RS3286	0004
RS3287	0280
RS3288	0050
RS3289	0004
RS3293	0004
RS3299	0004
RS3301	0004
RS3306	0600
RS3308	0004
RS3309	0050
RS3310	0004
RS3316	0004
RS3316-1	0004
RS3316-2	0004
RS3318	0004
RS3322	0050
RS3323	0050
RS3324	0050
RS3358-1	0160
RS3359-1	0160
RS3570	0015
RS3668	0050
RS3717	0004
RS3726	0004
RS3727	0015
RS3857	0004
RS3858	0160
RS-3858-1	0160
RS3858-1	0160
RS3862	0050
RS3863	0050
RS3864	0050
RS3866	0004
RS3867	0164
RS3868	0050
RS3880	0004
RS3892	0280
RS3897	0004
RS3898	0050
RS3900	0050
RS-3901	0050
RS3901	0050
RS3902	0050
RS3903	0050
RS3904	0004
RS3905	0050
RS3906	0050
RS3907	0050
RS3911	0050
RS3912	0050
RS3913	0004
RS-3914	0280
RS3914	0280
RS-3915	0280
RS3915	0280
RS3925	0004
RS-3926	0050
RS3926	0004
RS-3929	0280
RS3929	0050
RS3931	0208
RS3959	0160
RS3959-1	0160
RS3986	0050
RS3995	0050
RS4001	0478
RS4011	0215
RS4013	0413
RS4017	0513
RS4020	1661
RS4027	1949
RS4049	0001
RS4050	0398
RS4518	1043
RS5008	0004
RS5101	0050
RS5102	0004
RS5103	0004
RS5104	0280
RS5105	0280
RS5106	0050
RS5107	0004
RS5108	0050
RS5109	0050
RS5201	0050
RS5202	0004
RS5203	0004
RS5204	0050
RS5205	0050
RS5206	0050
RS5207	0050
RS5208	0050
RS5209	0050
RS5243-2	0004
RS5301	0050
RS5302	0280
RS5303	0280
RS5305	0050
RS5306	0050
RS5311	0050
RS5312	0050
RS5313	0050
RS5314	0050
RS5317	0050
RS5401	0004
RS5402	0280
RS5403	0280
RS5406	0004
RS5502	0004
RS5503	0004
RS5504	0280
RS5505	0004
RS5506	0004
RS5507	0004
RS5511	0280
RS5530	0004
RS5531	0004
RS5532	0004
RS5533	0004
RS5534	0004
RS5535	0004
RS5536	0004
RS5540	0280
RS5541	0004
RS5542	0004
RS5543	0004
RS5544	0004
RS5545	0004
RS5551	0004
RS5552	0004
RS5553	0004
RS5554	0004
RS5555	0004
RS5556	0004
RS5557	0004
RS5558	0004
RS5563	0004
RS5564	0004
RS5565	0004
RS5566	0004
RS5567	0004
RS5568	0004
RS5602	0004
RS5603	0004
RS5605	0004
RS5607	0004
RS5608	0004
RS5610	0004
RS5612	0160
RS-5613	0160
RS5613	0160
RS5614	0160
RS5616	0160
RS5704	0004
RS5704-2	0004
RS5708	0004
RS5708-2	0004
RS5709	0004
RS5711	0004
RS5717	0004
RS5717-1	0004
RS5717-3	0004
RS5717-6	0004
RS5720	0004
RS5731	0004
RS5732	0004
RS5733	0004
RS5734	0004
RS5735	0004
RS5736	0004
RS5737	0004
RS5738	0004
RS5740	0004
RS5740-1	0004
RS5742	0004
RS5743	0004
RS-5743-1	0004
RS5743-1	0004
RS-5743-2	0004
RS5743-2	0004
RS-5743-3	0004
RS5743-3	0004
RS5743.3	0280
RS-5744	0004
RS5744	0004
RS-5744-3	0004
RS5744-3	0004
RS5745	0004
RS5746	0004
RS5747	0004
RS5748	0004
RS5749	0004
RS5750	0004
RS5751	0004
RS5752	0004
RS5753	0136
RS-5753-2	0211
RS5753-2	0136
RS5754	0136
RS-5755	0136
RS5755	0136
RS5756	0136
RS5757	0136
RS5758	0136
RS5759	0136
RS5760	0136
RS5761	0136
RS5762	0136
RS5765	0004
RS5766	0004
RS5767	0004
RS5768	0004
RS5788	0847
RS5802	0136
RS5818	0050
RS5825	0211
RS-5835	0160
RS5835	0160
RS5851	0016
RS-5852	0004
RS5852	0004
RS5853	0016
RS5854	0004
RS-5855	0160
RS5855	0265
RS5856	0079
RS5857	0079
RS6344	0015
RS6461	0015
RS6471	0015
RS6523	0144
RS6705	0015
RS6821	0136
RS6822	0136
RS6824	0280
RS6840	0004
RS6843	0211
RS6846	0211
RS7101	0144
RS7102	0144
RS7103	0016
RS7104	0144
RS7105	0016
RS7106	0144
RS7107	0144
RS-7108	0144
RS7108	0016
RS7109	0144
RS7110	0144
RS7111	0016
RS7112	0144
RS7113	0144
RS7114	0224

If replacement code is OEM, contact original manufacturer for replacement.

DEVICE TYPE	REPL CODE
RS7115	0144
RS7116	0144
RS7117	0144
RS7118	0144
RS7119	0144
RS7120	0144
RS7121	0016
RS7122	0144
RS7123	0144
RS-7124	0079
RS7124	0144
RS7125	0144
RS7126	0144
RS-7127	0079
RS7127	0079
RS7128	0144
RS-7129	0079
RS7129	0079
RS7132	0016
RS7133	0016
RS7135	0144
RS7136	0016
RS7138	0144
RS7139	0144
RS7140	0144
RS7141	0144
RS7142	0144
RS7143	0127
RS7144	0144
RS7145	0144
RS7160	0016
RS7161	0144
RS7162	0144
RS7163	0144
RS7164	0144
RS7165	0144
RS7166	0144
RS7167	0144
RS7168	0144
RS7169	0144
RS7170	0144
RS7173	0224
RS7174	0144
RS7175	0144
RS7176	0144
RS7177	0144
RS-7201	0127
RS7201	0144
RS-7202	0144
RS7202	0144
RS7209	0144
RS7210	0144
RS7211	0144
RS-7212	0127
RS7212	0144
RS7214	0144
RS7215	0144
RS7216	0144
RS7217	0144
RS7218	0144
RS7219	0144
RS7220	0144
RS7221	0144
RS7222	0127
RS7223	0016
RS7224	0016
RS7225	0144
RS7226	0016
RS7227	0144
RS7228	0144
RS7229	0144
RS7230	0144
RS7231	0144
RS7232	0016
RS7233	0224
RS7234	0079
RS7235	0079
RS7236	0016
RS7237	0144
RS7238	0079
RS7241	0079
RS7242	0079
RS7310	0142
RS7311	0142
RS7312	0142
RS7313	0142
RS7315	0142
RS7316	0142
RS7317	0142
RS7318	0142
RS7320	0142
RS7321	0142
RS7327	0142
RS7328	0142
RS7329	0142
RS7330	0142
RS7333	0144
RS7334	0144
RS7365	0142
RS7366	0142
RS7367	0142
RS7368	0142
RS7400	0232
RS7402	0314
RS7404	0361
RS7405	0016
RS7406	0016
RS7406(IC)	1206
RS7407	0016
RS7408	0016
RS7408(IC)	0467
RS7409	0016
RS7410	0016
RS7410(IC)	0512
RS7411	0016
RS7412	0016
RS7413	0016
RS7413(IC)	1441
RS7415	0016
RS7420	0697
RS7421	0016
RS7427	0818
RS7432	0899
RS7441	1038
RS7447	1108
RS7448	1125
RS7451	1168
RS7473	1172
RS7474	1313
RS7475	1432
RS7476	1158
RS7485	0374
RS7486	1368
RS7490	1208
RS7492	0834
RS7504	0016
RS7510	0016
RS7511	0144
RS7512	0144
RS7513	0079
RS7513-15	0079
RS7514	0079
RS7515	0079
RS7516	0079
RS7517	0079
RS7517-19	0079
RS7518	0079
RS7519	0079
RS7520	0144
RS7521	0079
RS7522	0144
RS7523	0127
RS7524	0144
RS7525	0079
RS7526	0079
RS7527	0079
RS7528	0079
RS7529	0079
RS7530	0079
RS7532	0144
RS7533	0144
RS7542	0079
RS7543	0079
RS7544	0079
RS7555	0079
RS7568	0004
RS7606	0016
RS7607	0016
RS7609	0016
RS7610	0016
RS7611	0016
RS7612	0016
RS7613	0016
RS7614	0016
RS7620	0016
RS7621	0016
RS7622	0016
RS7623	0016
RS7624	0079
RS7625	0079
RS7626	0079
RS7627	0079
RS7628	0079
RS7634	0079
RS7635	0079
RS7636	0079
RS7637	0079
RS7638	0079
RS7639	0079
RS7640	0079
RS7641	0079
RS7642	0079
RS7643	0079
RS7665	0037
RS7678	0086
RS7814	0079
RS7916	0325
RS8100	0126
RS8101	0086
RS8102	0126
RS8103	0086
RS8104	0126
RS8105	0086
RS8106	0126
RS8107	0086
RS8108	0126
RS8109	0086
RS8110	0126
RS8111	0086
RS8112	0126
RS8113	0086
RS8406	0164
RS8407	0208
RS8420	0208
RS8421	0004
RS8424	0164
RS8430	0015
RS8441	0208
RS8442	0079
RS8443	0208
RS8444	0004
RS8446	0164
RS8470	OEM
RS8471	OEM
RS8476	OEM
RS8478	OEM
RS8487	OEM
RS8493	OEM
RS8503	0079
RS8609	OEM
RS9510	0127
RS9511	0127
RS9512	0127
RS57042	0004
RS57062	0004
RS57433	0004
RS74123	1157
RS74145	0619
RS74150	1492
RS74154	1555
RS74192	1921
RS74193	1926
RS74196	1950
RSL0003F	OEM
RSL0004-F	OEM
RSL0007F	OEM
RSL0014-F	OEM
RSL0014F	OEM
RSLNA0004CEZZ	0474
RS-LNB0001CEZZ	0196
RSLNB0001CEZZ	0196
RSLND0003CEZZ	0102
RSWY/16L7Z	0143
RT05	OEM
RT1N241S	2323
RT1N441C	4049
RT1N441C-T12-A1	OEM
RT1N441S	OEM
RT1P241S	1032
RT1P441S	OEM
RT6	OEM
RT10	OEM
RT20	OEM
RT30	OEM
RT40	OEM
RT60	OEM
RT80	OEM
RT-100	0016
RT100	0016
RT-100A	OEM
RT-101	0016
RT-101A	0436
RT-102	0079
RT102	0324
RT-102A	0324
RT-103	0037
RT103	0016
RT-103A	0436
RT-104	0111
RT104	0079
RT-105	0111
RT105	0016
RT-106	0861
RT106	0037
RT-107	0111
RT107	0144
RT-107A	0324
RT-108	0111
RT108	0144
RT-108A	0015
RT-109	0861
RT109	0861
RT-110	0233
RT110	0233
RT-111	0285
RT-112	0111
RT112	0127
RT-113	0007
RT113	0016
RT-114	0086
RT114	0016
RT-115	0150
RT115	0150
RT-116	0178
RT116	1690
RT-117	0004
RT-118	0861
RT-119	0038
RT119	0038
RT-120	0861
RT-121	0791
RT121	0164
RT-122	0038
RT-123	0164
RT-124	0085
RT124	0160
RT-125	0440
RT-126	0150
RT126	0016
RT-126A	0037
RT-127	0085
RT127	0160
RT-128	0142
RT128	0142
RT-129	0208
RT-130	0975
RT-131	0103
RT131	0103
RT-133	0042
RT133	0561
RT-134	0127
RT-135	0168
RT135	1708
RT-136	0164
RT-137	2046
RT-139	0003
RT-140	0065
RT-141	0086
RT141	0086
RT-143	2046
RT-144	0400
RT-145	0178
RT-146	0936
RT-147	0604
RT-148	1681
RT-149	0130
RT-150	0042
RT150	0042
RT150A	OEM
RT150B	OEM
RT-151	0925
RT151	0925
RT-152	0560
RT152	0560
RT-153	0854
RT-154	0103
RT154	0086
RT-155	0925
RT155	0925
RT-156	0168
RT156	0161
RT-157	1367
RT157	0460
RT-158	0423
RT-159	0285
RT159	0285
RT-159A	0338
RT-160	0936
RT-164	1361
RT-165	1369
RT-166	1361
RT-167	1369
RT-172	0015
RT-174	1908
RT-175	0325
RT175	0325
RT-176	3712
RT176	0325
RT-177	3384
RT-180	0212
RT180	0212
RT-181	0212
RT-182	0698
RT-183	0136
RT-185	0004
RT185	0004
RT-186	0338
RT-187	0155
RT-188	0086
RT188	0086
RT-189	0905
RT-190	0178
RT-191	0936
RT-192	0316
RT-193	2136
RT-194	0103
RT-195	1369
RT-196	1369
RT-197	1361
RT-200	0143
RT-201	0922
RT-202	0917
RT-203	0071
RT-204	0378
RT-206	OEM
RT-210	0071
RT210	0071
RT-212	0015
RT-213	0015
RT213	0015
RT-214	0015
RT214	0015
RT-215	0015
RT215	0015
RT-216	0293
RT-216D	0293
RT-218	0133
RT218	0133
RT-220	0474
RT-221	0196
RT-222	0484
RT-223	1080
RT-224	1303
RT-225	0774
RT-226	0201
RT-227	0239
RT-228	0752
RT-229	0288
RT-230	0289
RT-231	0302
RT-232	0250
RT-234	0188
RT-235	0162
RT235	0162
RT-236	0157
RT-237	0636
RT-238	0025
RT-239	0649
RT-240	0012
RT-241	0170
RT-242	0317
RT-243	0137
RT-244	0365
RT-245	0002
RT-246	0420
RT-247	0531
RT-248	0565
RT248	0565
RT-249	0441
RT-250	0039
RT-251	0350
RT-252	0999
RT-253	0869
RT-254	0783
RT-255	0331
RT-256	0149
RT-257	0244
RT-258	0495
RT-259	0402
RT-260	0186
RT-261	0028
RT-262	0030
RT-263	0143
RT-264	1791
RT-265	1323
RT-266	0421
RT-267	0016
RT-268	0016
RT-269	0037
RT-270	0037
RT-271	1890
RT-272	1890
RT-300	0324
RT-301	0595
RT-302	0155
RT-303	0006
RT-304	0324
RT304	0079
RT-305	0436
RT-306	2700
RT-306F	0630
RT-307	0630
RT-308	0111
RT-309	2075
RT-310	1984
RT-311	0791
RT-312	0016
RT-313	0144
RT-314	0342
RT-315	0715
RT409E	1480
RT476	0079
RT482	0086
RT483	0086
RT484	0086
RT497M	OEM
RT498M	OEM
RT-600	1306
RT-601	2083
RT601	2083
RT-602	2084
RT602	2084
RT-603	1449
RT603	1449
RT-604	2085
RT604	2085
RT-605	3649
RT656M	OEM
RT657M	OEM
RT679M	OEM
RT696AM	1480
RT696M	OEM
RT697AM	OEM
RT697M	0016
RT698	OEM
RT698M	OEM
RT699AM	OEM
RT699M	0086
RT-700	OEM
RT-701	OEM
RT-702	OEM
RT-703	OEM
RT-704	4296
RT-705	OEM
RT-709	OEM
RT-710	OEM
RT-711	OEM
RT-712	OEM
RT-713	OEM
RT-714	OEM
RT-715	OEM
RT-716	OEM
RT-717	OEM
RT717M	OEM
RT-718	OEM
RT718AM	OEM
RT718M	OEM
RT-719	OEM
RT719M	OEM
RT-720	OEM
RT720M	OEM
RT730M	1480
RT731M	1480
RT910M	1480
RT-929-H	0016
RT929H	0016
RT930H	0144
RT1008	0143
RT1106	0143
RT1108	0143
RT1110	OEM
RT1111	OEM
RT1115	0861
RT1116	0285
RT1184	0143
RT1210	0233
RT1252M	1480
RT1253M	1480
RT1306	0157
RT1306(G.E.)	0157
RT1395	0016
RT1409M	1480
RT1410M	1480
RT1420M	OEM
RT1595	0015
RT1613M	OEM
RT1667	0123
RT1669	0133
RT1686	0102
RT-1689	0143
RT1689	0133
RT1840	0015
RT1893	0285
RT1899	OEM
RT2016	0016
RT2061(G.E.)	0133
RT2221	OEM
RT2230	0004
RT2309	0552
RT2329	0004
RT2330	0004
RT2331	0004
RT2332	0016
RT2334	0143
RT2451	0123
RT2452	0143
RT2459	OEM
RT2460	OEM
RT2461	OEM
RT2462	OEM
RT2463	OEM
RT2669	0015
RT2694	0143
RT2709	0004
RT2914	0079
RT2915	0144
RT3061	OEM
RT3062	OEM
RT3063	0016
RT3064	0016
RT3065	0037
RT3069	0144
RT3070	0144
RT3071	0037
RT3072	0143
RT3095	0144
RT3096	0208
RT3097	0004
RT3098	0004
RT3099	0143
RT3101	OEM
RT3103	OEM
RT3111	OEM
RT3170	OEM
RT3171	OEM
RT3225	0144
RT3226	0144
RT3227	0144
RT3228	0016
RT3229	0004
RT3230	0004
RT3231	0004
RT3232	0144
RT3233	0143
RT3336	0143
RT3361	0136
RT3362	0136
RT3363	0004
RT3364	0004
RT3365	0004
RT3443	0015
RT3449	0004
RT3466	0050
RT3467	0004
RT3468	0004
RT3469	0143
RT3500	OEM
RT3501	OEM
RT3558	0015
RT3564	0004
RT3565	0016
RT3566	0004
RT3567	0079
RT3568	0004
RT3574	0649
RT3585	0015
RT3671	0002
RT3858	0015
RT3981	0015
RT4050	0015
RT4069	0015
RT4230	OEM
RT4232	0015
RT4293	0143
RT4525	0050
RT4584	OEM
RT4624	0004
RT4625	0164
RT4644	0143
RT4760	0111
RT4761	0191
RT4762	0222
RT4762MHF25	0085
RT-4764	0087
RT4764	0015
RT4767	0015
RT4880	0123
RT5001	0630
RT5002	0630
RT5003	0630
RT5004	0630
RT5061	0144
RT5063	0136
RT5070	0015
RT5108	0630
RT5151	0086
RT5152	0086
RT5200	0144
RT5201	0144
RT5202	0079
RT5203	0086
RT5204	0286
RT5205	0144
RT5206	0079
RT5207	0006
RT5208	0111
RT5211	0030
RT5212	0123
RT5213	0019
RT5214	0143
RT5215	0025
RT5216	0229
RT5217	0133
RT5230	0126
RT5321AD	OEM
RT5379	0123
RT5385	0015
RT5401	0086
RT5402	0086
RT5403	0086
RT5404	0086
RT5411	OEM
RT5412	OEM
RT5413	OEM
RT5418	OEM
RT5435	0111
RT5464	0127
RT5465	0127
RT5466	0136
RT5467	0136
RT5468	0004
RT5470	0123
RT5471	0025
RT5472	0015
RT5473	0030
RT5520	0136
RT5521	0004
RT5522	0004
RT5551	0079
RT5554	0133
RT5637	0004
RT5738	0123
RT5765	OEM
RT5793	0636
RT5804	1480
RT5900	0076
RT5901	0016
RT5902	0144
RT5903	0144
RT5904	0144
RT5905	0076
RT5906	0086
RT5907	0086
RT5908	0143
RT5909	0133
RT5911	0015
RT5912	0123
RT5939	0143
RT6105	0244
RT6119	0143
RT6157	0144
RT6158	0144
RT6159	0144
RT6160	0144
RT6178	0030
RT6179	0143
RT6180	0143
RT6181	0143
RT6182	0143
RT6183	0143
RT6184	0143
RT6189	0143
RT6201	0224
RT6202	0224
RT6203	0144
RT-6204	0191
RT6204	0151
RT6205	0004
RT6322	0015
RT6332	0015
RT6600	0127
RT6600MHF25	0079
RT6601	1144

DEVICE TYPE	REPL CODE
RT6602	1144
RT6603	0030
RT6604	0004
RT6605	0071
RT6619	0143
RT6728	0123
RT6729	0015
RT6731	0123
RT6732	0151
RT6733	0151
RT6734	0004
RT6735	0103
RT6736	0004
RT6737	1220
RT6787	2209
RT6789	0030
RT6790	0579
RT6791	0015
RT6921	0191
RT6921MHF25	0079
RT-6922	0162
RT6922	0041
RT6923	0244
RT6988	0136
RT6989	0079
RT6990	0004
RT6991	0144
RT7007E	OEM
RT7320	0144
RT7321	0144
RT7322	0079
RT7323	0144
RT7324	0144
RT7325	0079
RT7326	0537
RT7327	0079
RT7329	0030
RT7330	0143
RT-7399	0579
RT7399	0579
RT7400	0208
RT7401	0004
RT7402	0277
RT7511	0016
RT7514	0079
RT7515	0016
RT7517	0016
RT7518	0016
RT7528	0016
RT7538	0123
RT7539	0636
RT7557	0076
RT7558	0164
RT7559	0111
RT7634	0015
RT7636	0143
RT7638	0016
RT7689	0143
RT7703	2209
RT7704	0127
RT7705	OEM
RT7845	0016
RT7846	1062
RT7848	0110
RT7849	0015
RT7850	0015
RT7851	0143
RT7943	0191
RT7944	0208
RT7945	0086
RT7946	0133
RT8047	0111
RT8101-2	OEM
RT8193	0016
RT8195	0016
RT8197	0191
RT8198	0076
RT8199	0229
RT8200	0053
RT8201	0079
RT8231	0015
RT8330	0079
RT8331	0325
RT8332	0016
RT8333	2209
RT8334	OEM
RT8335	0042
RT8336	0925
RT8337	0111
RT8338	0925
RT8339	0057
RT8340	0015
RT8442	0004
RT8527	0113
RT8528	OEM
RT8601-2	OEM
RT8602	0004
RT8664	OEM
RT8665	0071
RT8666	0111
RT-8667	0325
RT8667	0688
RT8668	1144
RT8669	0113
RT8670	0006
RT8671	0143
RT8701-2	OEM
RT8779	0030
RT8801-2	OEM
RT8838	OEM
RT8839	0015
RT8840	0015
RT8841	0015
RT8842	0164
RT8863	0111
RT8895	0037
RT8901-2	OEM
RT61012	0143
RT-61014	0136
RT61014	0136
RT-61015	0164
RT61015	0004
RT-61016	0164
RT61016	0004
RTB0103	1137
RTB0106	0344
RTB0110	0901
RTB0120	0058
RTB0130	0407
RTB0140	OEM
RTC05	OEM
RTC0103	1137
RTC0106	0344
RTC0110	0901
RTC0120	0058
RTC0130	0407
RTC0140	OEM
RTC10	OEM
RTC20	OEM
RTC40	OEM
RTC60	OEM
RTC80	OEM
RTC101L	OEM
RTC102L	OEM
RTC103L	OEM
RTC106L	OEM
RTC107L	OEM
RTC114L	OEM
RTC115L	OEM
RTC1001	OEM
RTC1002	OEM
RTC1003	OEM
RTC1005	OEM
RTC1100	OEM
RTC3001	OEM
RTC3003	OEM
RTC3005	OEM
RTF0101	OEM
RTF0103	OEM
RTF0106	OEM
RTF0110	OEM
RTF0115	OEM
RTF0120	OEM
RTF0125	OEM
RTF0130	OEM
RTF0135	OEM
RTF0140	OEM
RTF0201	OEM
RTF0203	OEM
RTF0206	OEM
RTF0210	OEM
RTF0215	OEM
RTF0220	OEM
RTF0225	OEM
RTF0230	OEM
RTF0235	OEM
RTF0240	OEM
RTF0301	OEM
RTF0303	OEM
RTF0306	OEM
RTF0310	OEM
RTF0315	OEM
RTF0320	OEM
RTF0325	OEM
RTF0330	OEM
RTF0335	OEM
RTF0340	OEM
RTF0401	OEM
RTF0403	OEM
RTF0406	OEM
RTF0410	OEM
RTF0415	OEM
RTF0420	OEM
RTF0425	OEM
RTF0430	OEM
RTF0435	OEM
RTF0440	OEM
RTF0501	OEM
RTF0503	OEM
RTF0506	OEM
RTF0510	OEM
RTF0515	OEM
RTF0520	OEM
RTF0525	OEM
RTF0530	OEM
RTF0535	OEM
RTF0540	OEM
RTF0601	OEM
RTF0603	OEM
RTF0606	OEM
RTF0610	OEM
RTF0615	OEM
RTF0620	OEM
RTGC0101	OEM
RTGC0103	OEM
RTGC0106	OEM
RTGC0110	OEM
RTGC0115	OEM
RTGC0120	OEM
RTGD0101	OEM
RTGD0103	OEM
RTGD0106	OEM
RTGD0110	OEM
RTGD0115	OEM
RTGD0120	OEM
RTGD0125	OEM
RTGD0130	OEM
RTH0101	OEM
RTH0103	OEM
RTH0106	OEM
RTH0110	OEM
RTH0115	OEM
RTH0120	OEM
RTH0125	OEM
RTH0130	OEM
RTH0201	OEM
RTH0203	OEM
RTH0206	OEM
RTH0210	OEM
RTH0215	OEM
RTH0220	OEM
RTH0225	OEM
RTH0230	OEM
RTH0301	OEM
RTH0303	OEM
RTH0306	OEM
RTH0310	OEM
RTH0315	OEM
RTH0320	OEM
RTH0325	OEM
RTH0330	OEM
RTH0401	OEM
RTH0403	OEM
RTH0406	OEM
RTH0410	OEM
RTH0415	OEM
RTH0420	OEM
RTH0425	OEM
RTH0430	OEM
RTH0501	OEM
RTH0503	OEM
RTH0506	OEM
RTH0510	OEM
RTH0515	OEM
RTH0520	OEM
RTH0525	OEM
RTH0530	OEM
RTH0601	OEM
RTH0603	OEM
RTH0606	OEM
RTH0610	OEM
RTH0615	OEM
RTH0620	OEM
RTIN241S	2323
RTIP241S	1032
RTJ0103	1137
RTJ0106	0344
RTJ0110	0901
RTJ0120	0901
RTJ0130	0407
RTJ0140	0407
RTJ0201	1137
RTJ0203	1137
RTJ0206	0344
RTJ0210	0901
RTJ0215	0407
RTJ0220	0058
RTJ0225	0407
RTJ0230	0407
RTL0605	2095
RTL0610	2089
RTL0620	0505
RTL0640	0710
RTL0660	OEM
RTL0805	2095
RTL0810	2089
RTL0820	0505
RTL0840	0089
RTL0860	OEM
RTL1005	OEM
RTL1010	OEM
RTL1020	OEM
RTL1040	OEM
RTL1060	OEM
RTL1510	OEM
RTL1520	OEM
RTL1540	OEM
RTL1560	OEM
RTL1610	OEM
RTL1620	OEM
RTL1640	OEM
RTL1660	OEM
RTN0102	1103
RTN0105	1103
RTN0110	2491
RTN0120	0240
RTN0130	0676
RTN0140	0676
RTN0150	0336
RTN0160	0336
RTN0202	2449
RTN0205	2449
RTN0210	2449
RTN0220	0435
RTN0230	1487
RTN0240	1487
RTN0250	0687
RTN0260	0687
RTN0302	2449
RTN0305	2449
RTN0310	2449
RTN0320	0435
RTN0330	1487
RTN0340	1487
RTN0350	0687
RTN0360	0687
RTN0802	1103
RTN0805	1103
RTN0810	2491
RTN0820	0240
RTN0840	0676
RTN0860	0336
RTPC0103	OEM
RTPC0103-1	OEM
RTPC0106	OEM
RTPC0106-1	OEM
RTPC0110	OEM
RTPC0110-1	OEM
RTPC0115	OEM
RTPC0115-1	OEM
RTPC0120	OEM
RTPC0120-1	OEM
RTR0202	1004
RTR0205	2095
RTR0210	2089
RTR0220	0505
RTR0230	0710
RTR0240	0710
RTR0250	0863
RTR0260	0863
RTR0302	1004
RTR0305	2095
RTR0310	2089
RTR0320	0505
RTR0330	0710
RTR0340	0710
RTR0350	0863
RTR0360	0863
RTR0620	0505
RTR0660	OEM
RTR1005	OEM
RTR1010	OEM
RTR1020	OEM
RTR1040	OEM
RTR1060	OEM
RTS0202	1651
RTS0205	1651
RTS0210	1651
RTS0220	1651
RTS0230	1584
RTS0240	1584
RTS0250	1665
RTS0260	1665
RTS0702	1650
RTS0705	1650
RTS0710	1650
RTS0720	1650
RTS0730	2647
RTS0740	2647
RTS0750	2649
RTS0760	2649
RTS1002	1651
RTS1005	1651
RTS1010	1651
RTS1020	1651
RTS1040	1584
RTS1060	1665
RTS2505	1650
RTS2510	1650
RTS2520	1650
RTS2540	2647
RTS2560	2649
RTT0102	0469
RTT0105	0469
RTT0110	0469
RTT0120	0469
RTT0130	0722
RTT0140	0722
RTT0150	0725
RTT0160	0725
RTT0202	0469
RTT0205	0469
RTT0210	0469
RTT0220	0469
RTT0230	0722
RTT0240	0722
RTT0250	0725
RTT0260	0725
RTT0302	0469
RTT0305	0469
RTT0310	0469
RTT0320	0469
RTT0330	0722
RTT0340	0722
RTT0350	0725
RTT0360	0725
RTT0402	0469
RTT0405	0469
RTT0410	0469
RTT0420	0469
RTT0430	0722
RTT0440	0722
RTT0450	0725
RTT0460	0725
RTT0502	0804
RTT0505	0804
RTT0510	0804
RTT0520	0804
RTT0530	0804
RTT0540	0804
RTT0550	0804
RTT0560	0804
RTT0602	0469
RTT0605	0469
RTT0610	0469
RTT0620	0469
RTT0630	0722
RTT0640	0722
RTT0650	0725
RTT0660	0725
RTT2505	0731
RTT2510	0712
RTT2520	0469
RTT2540	0722
RTT2560	0725
RTU023#	0747
RTU0102	2519
RTU0105	2519
RTU0110	6949
RTU0120	0745
RTU0130	0747
RTU0140	0740
RTU0150	0752
RTU0160	0764
RTU0202	3307
RTU0205	2188
RTU0210	0567
RTU0220	0762
RTU0230	0740
RTU0240	0740
RTU0250	0764
RTU0260	0764
RTU0302	0469
RTU0305	0469
RTU0310	0469
RTU0320	0469
RTU0330	0722
RTU0340	0722
RTU0350	0725
RTU0360	0725
RTU0402	0469
RTU0405	0469
RTU0410	0469
RTU0420	0469
RTU0430	0722
RTU0440	0722
RTU0450	0725
RTU0460	0725
RTU0502	0469
RTU0505	0469
RTU0510	0469
RTU0520	0469
RTU0530	0722
RTU0540	0722
RTU0550	0725
RTU0560	0725
RTU0705	0731
RTU0710	0712
RTU0720	0469
RTU0730	0722
RTU0740	0722
RTU0750	0725
RTU0760	0725
RTU102	2519
RTU1005	2449
RTU1010	2449
RTU1020	0435
RTU1040	1487
RTU1060	0687
RTU2505	0731
RTU2510	0712
RTU2520	0469
RTU2540	0722
RTU2560	0725
RTV0102	OEM
RTV0105	OEM
RTV0110	OEM
RTV0120	OEM
RTV0130	OEM
RTV0140	OEM
RTV0150	OEM
RTV0160	OEM
RTV0202	OEM
RTV0205	4402
RTV0210	OEM
RTV0220	OEM
RTV0230	OEM
RTV0240	OEM
RTV0250	OEM
RTV0260	OEM
RTV0302	OEM
RTV0305	OEM
RTV0310	OEM
RTV0320	OEM
RTV0330	OEM
RTV0340	OEM
RTV0350	OEM
RTV0360	OEM
RTV0402	OEM
RTV0405	OEM
RTV0410	OEM
RTV0420	OEM
RTV0430	OEM
RTV0440	OEM
RTV0450	OEM
RTV0460	OEM
RTV0502	OEM
RTV0505	OEM
RTV0510	OEM
RTV0520	OEM
RTV0530	OEM
RTV0540	OEM
RTV0550	OEM
RTV0560	OEM
RTV0602	OEM
RTV0605	OEM
RTV0610	OEM
RTV0620	OEM
RTV0630	OEM
RTV0640	OEM
RTV0650	OEM
RTV0660	OEM
RU-1	0023
RU1	0023
RU-1A	0023
RU1A	0023
RU-1AV	0023
RU1B	0017
RU1C	0017
RU-1P	0023
RU1P	0023
RU-1V	0023
RU1V	0023
RU1Z	0023
RU-2	0023
RU2	0023
RU2-LFA1	0023
RU2-T3	OEM
RU2AM	0023
RU2AMV	0023
RU2AN	0023
RU-2AV	0023
RU2B	0023
RU2C	0023
RU-2M	0023
RU2M	0023
RU-2M(V)	0023
RU-2M(V)TA	0023
RU-2MV	0023
RU2MV	0023
RU-2MVTA	0023
RU2N	0023
RU-2V	0023
RU2V	0102
RU2YX	0546
RU2Z	0023
RU-3	0283
RU3	0283
RU3-AM	0283
RU3A	0283
RU3AH	0283
RU-3AM	0283
RU3AM	0283
RU3AM-LFB1	0283
RU3AM-LFC4	0283
RU3AMV	0283
RU3AN	0023
RU-3B	0640
RU3B	0640
RU-3M	0283
RU3M	0283
RU3N	0031
RU3V	0283
RU3YX	0546
RU3YX-M	OEM
RU3YX-MLF-C2	OEM
RU3YX-MLF-C4	OEM
RU3YX-MV1	OEM
RU4	0102
RU-4A	0102
RU4A	0102
RU4A-J1	0102
RU-4AM	0031
RU4AM	0031
RU4AM(LF-L1)	0031
RU4AM-C1	OEM
RU4AM-K1	0031
RU4AMK1	0031
RU4AM-LF-J3	0031
RU4AMLF-J3	0031
RU4AM-LFK2	OEM
RU4AMLF-K2	0031
RU4AM-T3	0031
RU4B	0102
RU4C	0102
RU-4D	0102
RU4D	0102
RU-4DS	0031
RU4DS	0102
RU4DS-C1	OEM
RU4DS-LFK2	0102
RU4H	OEM
RU4M	0031
RU4M-LF-J3	0031
RU4MLF-K2	0031
RU4Y	0102
RU4YX	0546
RU4YX-C1	OEM
RU4YX-LF-J3	0546
RU4YXLF-J3	0546
RU4YX-LFK2	OEM
RU4YXLF-K2	0546
RU-4Z	0031
RU4Z	0031
RU4ZLF-L1	0031
RU30-C1	OEM
RU30A	OEM
RU30LF-S1	OEM
RU800	OEM
RU801	OEM
RU802	OEM
RU803	OEM
RU804	OEM
RU805	OEM
RU806	OEM
RU808	OEM
RU810	OEM
RU-A1	OEM
RUD2DP22/1B	0015
RUNTK0152CEZZ	OEM
RUTFK505	OEM
RUZYX	OEM
RV06	0015
RV1	0023
RV1BA1330	0416
RV1LA3301	1215
RV1LA4126	1951
RV1S1998	2295
RV1SM5104	2616
RV2	0023
RV6.2	0471
RV15	OEM
RV101	OEM
RV555NB	0973
RV556DB	3285
RV741NB	0312
RV1017	0133
RV1059	0006
RV1068	0127
RV1180	0004
RV1181	0157
RV1189	0023
RV1226	0133
RV1395	0191
RV1424	0015
RV1458NB	3137
RV1467	0144
RV1468	0144
RV1469	0144
RV1470	0144
RV1471	0079
RV1472	0126
RV1473	0086
RV1474	0079
RV1475	0004
RV1476	0540
RV1477	0030
RV1478	0015
RV1479	0143
RV2068	0018
RV2069	0126
RV2070	0111
RV2071	0133
RV2072	0015
RV2213	0057
RV2248	0111
RV2249	0079
RV2250	0015
RV2260	0006
RV-2289	0015
RV2289	0015
RV2351	0037
RV2353	0203
RV2354	1220
RV2355	0006
RV2356	1267
RV3403ADB	OEM
RV4136DB	3024
RV4136DC	3024
RV4143NB	OEM
RV4144NB	OEM
RV4151DE	OEM
RV4151NB	OEM
RV4152DE	OEM
RV4153DC	OEM
RV4156DC	2823
RV4191DE	OEM
RV4191NB	OEM
RV4192DE	OEM
RV4192NB	OEM
RV4193DE	OEM
RV4193NB	OEM
RV4200ADE	OEM
RV4200ANB	OEM
RV4200DE	OEM
RV4200NB	OEM
RV4391DE	OEM
RV4391NB	OEM
RV4558DE	0360
RV4558NB	0360
RV4559DE	0360
RV4559NB	0360
RV2347042	OEM
RV2347043	OEM
RVB406	OEM
RVB-406MLFA	OEM
RVCFV-212A	0120
RVD0.8C22/1A	0015
RVD1E1LF	0015
RVD1K110	0143
RVD1N34A	0143
RVD1N4733A	0442
RVD1N4738	0244

If replacement code is OEM, contact original manufacturer for replacement.

DEVICE TYPE	REPL CODE
RVD1N4739	0012
RVD1N4740	0170
RVD1SS54	0133
RVD1SV50S	0030
RVD1SV505	0030
RVD2-1K110	0143
RVD2DP	0015
RVD2DP22/1B	0015
RVD2DP22/1C	0277
RVD2DP22/18	0015
RVD2DP221B	0277
RVD2DP221P	0015
RVD2P22/1B	0015
RVD4B265J2	0015
RVD8MB4	0277
RVD8MBA	0015
RVD10D1	0015
RVD10DC1	0015
RVD10DC1R	0015
RVD10E1	0015
RVD10E1LF	0015
RVD10E11F	0015
RVD12B-1	0277
RVD1051	OEM
RVD14740	0170
RVDC08P1	0015
RVDC08P1R	0196
RVDCD0033	0301
RVDD124B	0277
RVDD1245	0277
RVDDS-410	0015
RVDEQA0106S	0301
RVDEQA0107S	0649
RVDEQA0108S	0649
RVDFV211	0139
RVDFV-212	0120
RVDFV212	0120
RVDFV214	0120
RVDGP05A	0015
RVDHZ3C1	0299
RVDK	0120
RVDKB162A	0133
RVDKB162C5	0133
RVDKB167	1925
RVDKB265J2	0139
RVDKB265J3	OEM
RVDMZ206	0091
RVDMZ208	0165
RVDMZ209	0057
RVD0.8C2211A	0540
RVDR5R6EB	0301
RVDR154B	0277
RVDRD5A1E	0041
RVDRD5R1EB	0041
RVDRD5R1EBB1	0041
RVDRD5R6EB	0253
RVDRD7AN	0062
RVDRD7R5E	0077
RVDRD7R5EB	0649
RVDRD7R5FB	0077
RVDRD9R1E	0057
RVDRD11AN	0181
RVDRD11EB	0064
RVDRD11EB,1	0181
RVDRD11EB1	0064
RVDS5277B	0015
RVDSC-15	0628
RVDSC20	3658
RVDSC-L5	0030
RVDSD-1	0102
RVDSD-1U	0015
RVDSD-1Y	0015
RVDSD113	0030
RVDSD113TA	0030
RVDSG-5N	0087
RVDSG-5P	0087
RVDSL55VR3FT	OEM
RVDSMIA02TPA	0015
RVDSR3AM2N	0071
RVDVD1150L	0139
RVDVD1150M	0139
RVDVD1210L	0120
RVDVD1210M	0120
RVDVD1211L	0120
RVDVD1212L	0120
RVDVD1213	0120
RVDVD1250M	0133
RVDVD1251L	0139
RVDVD1253	3683
RVDVD1260L	0139
RVDVD1262MF	0139
RVDVD12102	OEM
RVIBA328MR	2534
RVIBA338	2654
RVIBA532S	2794
RVIBA1330	0416
RVIBA6133	OEM
RVIHD36990	OEM
RVIHD38980A	2728
RVILM703	0632
RVILM8364DH	OEM
RVIM5102L	1910
RVIM51182	4757
RVIM51202	1910
RVIM51502L	1910
RVIMC4080	2854
RVIMK50372N	OEM
RVIS1998	2295
RVITK3200	OEM
RVITMS1943	2295
RVITMS1943N2	2295
RVITMS3451NC	OEM
RVITMS3451NL	OEM
RVITMS3454NR	OEM
RVIUPC20C2	0470
RVIUPC22C	3352
RVIUPC78L05	1292
RVIUPC575	2872
RVIUPC1018CE	2926
RVIUPD861	3815
RVS2SC645	0144
RVTCS1381	0016
RVTCS1382	0037
RVTCS1383	0016
RVTCS1384	0144
RVTCS1473	0111
RVTM51202	1910
RVTMC00921-3	0484
RVTMK10-2	0325
RVTMK10-E	0325
RVTS22410	0016
RVTS22411	0037
RW005M	OEM
RW01M	OEM
RW02M	OEM
RW04M	OEM
RW06M	OEM
RW08M	OEM
RW10M	OEM
RW15	OEM
RX-0034	OEM
RX-0043	OEM
RX090	0012
RX15	OEM
RXC192	OEM
RX-DX0041CEZZ	0196
RX-DX0131CEZZ	0023
RX-EX0019CEZZ	0025
RX-IX0035GAZZ	0904
RY10U	OEM
RY15	OEM
RY23	0609
RY24	0795
RZ08	OEM
RZ-3	0087
RZ3	OEM
RZ3.3	0292
RZ3.6	0188
RZ3.9	0456
RZ4.7	0451
RZ5.1	0041
RZ5.6	0157
RZ6.2	0636
RZ6.8	0025
RZ8.2	0244
RZ9.1	0012
RZ10	0170
RZ12	0137
RZ12A	0052
RZ14	0879
RZ15	0002
RZ15A	0686
RZ16	0445
RZ18	0495
RZ20	0700
RZ22	0565
RZ22A	0705
RZ25	0714
RZ27	0441
RZ27A	0441
RZ30	0195
RZ33A	0166
RZ39A	0032
RZ40	OEM
RZ47A	0068
RZ50	0092
RZ56A	0125
RZ68A	0173
RZ82A	0049
RZZ3.6	0188
RZZ3.9	0456
RZZ4.7	0451
RZZ5.1	0041
RZZ5.6	0157
RZZ6.2	0636
RZZ6.8	0025
RZZ8.2	0244
RZZ9.1	0012
RZZ10	0170
RZZ12	0137
RZZ15	0879
RZZ18	0375
RZZ22	0705
RZZ27	0455
S	0012
S0012	0126
S0013	0037
S0019	3095
S0026	0037
S0028	0342
S0029	0342
S0031	3107
S0032	3108
S0037	OEM
S00601DF	1567
S00602DF	1567
S00603DF	2153
S00604DF	2153
S00605DF	0711
S00606DF	0711
S001683	0086
S0-632	OEM
S0-4295-10	0103
S01	0050
S01E	0853
S02	0133
S02C	0871
S02D	0853
S02E	0853
S03	0050
S04	0133
S-05	0015
S05	OEM
S-05/01	0071
S-05-005	0015
S-05-01	0015
S05-01	OEM
S-05-02	0015
S05-02	OEM
S-05-04	0015
S05-04	OEM
S-05-06	0015
S05-06	OEM
S-05-08	0071
S05-08	OEM
S05-1	OEM
S-05-10	0071
S05-10	OEM
S-05-12	0102
S05K60	4222
S05K75	4921
S05K95	5157
S05K130	1374
S05K140	OEM
S05K150	0246
S05K175	OEM
S05K230	OEM
S05K250	1332
S05K275	1658
S05K300	4471
S07K60	4222
S07K75	4921
S07K95	5157
S07K130	1374
S07K140	OEM
S07K150	0246
S07K175	OEM
S07K230	OEM
S07K250	1332
S07K275	1658
S07K300	4471
S010G	0071
S025	0004
S028	0600
S031A	0016
S035	OEM
S035AADF	1530
S037	0016
S042F	OEM
S042P	0325
S056F3	0469
S058F3	0469
S065	0164
S065A	0004
S074-007-001	0133
S082A	0917
S088	0004
S0300KS2	1137
S0301K	OEM
S0301M	2412
S0301MS2	1137
S0301MS3	OEM
S0303MS2	1396
S0303MS3	1396
S0303RS1	OEM
S0303RS2	1396
S0303RS3	1396
S0304F1	1396
S0306B	OEM
S0306G	1651
S0306H	2449
S0306L	1004
S0306LS2	1044
S0306RS2	OEM
S0306RS3	OEM
S0308B	OEM
S0308G	1651
S0308H	2449
S0308L	1004
S0308LS2	1044
S0308LS3	1044
S0308RS2	OEM
S0308RS3	OEM
S0310B	OEM
S0310G	1651
S0310H	2449
S0310L	1004
S0310LS2	1044
S0310LS3	1044
S0310RS2	OEM
S0310RS3	OEM
S0315G	1651
S0315H	2519
S0315L	2521
S0316B	OEM
S0316G	2519
S0316H	2519
S0316L	OEM
S0320L	2521
S0325B	OEM
S0325C	0226
S0325D	1848
S0325G	2519
S0325H	2188
S0325L	2611
S0325R	2621
S0325V	OEM
S0325Y	OEM
S0325Z	OEM
S0335C	0226
S0335D	1848
S0335G	1650
S0335H	0731
S0335J	2690
S0335K	2690
S0335W	2696
S0402BH	0717
S0402DH	0717
S0402MH	0308
S0402NH	0327
S0405BH	0717
S0405DH	0717
S0405MH	0308
S0405NH	0327
S0407BH	0717
S0407DH	0717
S0407MH	0308
S0407NH	0327
S0410BH	0717
S0410DH	0717
S0410MH	0308
S0410NH	0327
S0500KS2	0344
S-0501	0015
S0501	0015
S0501K	OEM
S0501L	0015
S0501LS1	1044
S0501LS2	1044
S0501LS3	5106
S0501M	0179
S0501MS2	1137
S0501MS3	OEM
S0503L	1004
S0503LS1	1044
S0503LS2	1044
S0503LS3	4630
S0503M	0179
S0503MS2	1259
S0503MS3	1259
S0503RS1	OEM
S0503RS2	1259
S0503RS3	0174
S0504F1	1259
S0506B	OEM
S0506G	1651
S0506H	2449
S0506L	2095
S0506LS2	1044
S0506LS3	1044
S0506RS2	OEM
S0506RS3	OEM
S0508B	OEM
S0508G	1651
S0508H	2449
S0508L	2095
S0508RS2	OEM
S0508RS3	OEM
S0510B	OEM
S0510F3	0469
S0510G	1651
S0510H	2449
S0510L	1004
S0510LS2	1044
S0510LS3	1044
S0510RS2	OEM
S0510RS3	OEM
S0515G	3620
S0515H	0344
S0515L	2521
S0516B	OEM
S0516G	2521
S0516H	2519
S0516L	OEM
S0520F4	0469
S0520L	2521
S0525B	OEM
S0525C	0226
S0525D	1848
S0525G	1650
S0525H	2188
S0525L	2611
S0525R	2621
S0525V	OEM
S0525Y	OEM
S0525Z	OEM
S0535	OEM
S0535C	0226
S0535D	1848
S0535G	1650
S0535H	0731
S0535J	2690
S0535K	2690
S0535W	2696
S0555M	2696
S0555W	2696
S0565J	OEM
S0565K	OEM
S0570W	OEM
S0602BH	0717
S0602DH	0717
S0602MH	0308
S0602NH	0327
S0605BH	0717
S0605DH	0717
S0605MH	0308
S0605NH	0327
S0607BH	0717
S0607DH	0717
S0607MH	0308
S0607NH	0327
S0610BH	0717
S0610DH	0717
S0610MH	0308
S0610NH	0327
S0702	0012
S0704	0016
S0802BH	0717
S0802DH	0717
S0802MH	0308
S0802NH	0327
S0805BH	0717
S0805DH	0717
S0805MH	0308
S0805NH	0327
S0807BH	0717
S0807DH	0717
S0807MH	0308
S0807NH	0327
S0810BH	0717
S0810DH	0717
S0810MH	0308
S0810NH	0327
S02501DF	6616
S02502DF	6616
S02503DF	6638
S02504DF	6638
S02505DF	6663
S02506DF	6663
S020803	OEM
S1	0023
S1-1	0015
S1-3090CA	OEM
S1-RECT-35	0133
S1-RECT-102	0015
S1A	0015
S1A05F	2079
S1A06	0015
S1A060	0015
S1A1F	2079
S1A2F	2079
S1A3	1338
S1A3F	0023
S1A4F	0023
S1A5F	0023
S1A6F	0023
S1A8F	0017
S1A10	OEM
S1A10F	0017
S1A12F	0017
S1A20	OEM
S1A40	OEM
S1A60	0015
S1A80	0071
S1A100	0071
S1A150	0102
S1AN6	0989
S1AN12	0097
S1AN15	0871
S1AN15R	1635
S1AN20	0871
S1AN20R	1635
S1AN31	1124
S1AN31R	1107
S1AN40	1124
S1AN40R	1107
S1AN41	OEM
S1AN55	2657
S1AN55R	2768
S1AN70	OEM
S1AN71	OEM
S1AR1	0015
S1AR2	0015
S1B	0015
S1B01	0015
S1B01-01	0015
S1B-01-02	0015
S1B01-02	0015
S1B01-04	0965
S1B01-0226	0015
S1B01-1	0015
S1B02	0015
S1B02-0	0015
S1B02-03C	0087
S1B02-06CE	0015
S1B02-06CRE	0015
S1B02-C	0015
S1B02-CR	0015
S1B02-0	0015
S1B02C	0164
S1B0101CR	0015
S1B0102	0015
S1B0201B	0277
S1B0201CR	0015
S1B-0306	0071
S1B1	0015
S1BD1-02	0015
S1BN15	0871
S1BN15R	1635
S1BN20	0871
S1BN20R	1635
S1BN31	1124
S1BN31R	1107
S1BN40	1124
S1BN40R	1107
S1BN55	2657
S1BN55R	2768
S1BR2	0989
S1BR5	1250
S1BR30	OEM
S1BR70	OEM
S1BR512	OEM
S1C	0071
S1CN1	0015
S1CR5	1250
S1CR8	OEM
S1CR25	OEM
S1CR30	OEM
S1D	0071
S1D2A-1	0913
S1D2A-3	0913
S1D23-13	0102
S1D23-15	0102
S1D26	0102
S1D30-13	0102
S1D50B851-A	0133
S1D51C052-19	0015
S1D51C169-1	0015
S1D153	0042
S1DR5	1250
S1DR8	OEM
S1DR25	OEM
S1DR30	OEM
S1ER5	1250
S1ER8	OEM
S1ER25	OEM
S1ER30	OEM
S1F	0017
S1FN270	OEM
S1FN290	OEM
S1G1	OEM
S1G2	OEM
S1G4	OEM
S1G6Z	OEM
S1G8Z	OEM
S1G10	0023
S1G20	0015
S1G40	0015
S1G40Z	0015
S1G60Z	0015
S1G80Z	0072
S1GN270	OEM
S1GR2	0953
S1K20	0023
S1K20H	0023
S1K40	0023
S1L200	0015
S1M	0110
S1M1	0110
S1M2	0110
S1N1189	OEM
S1N1204A	OEM
S1N1616	OEM
S1N3911	OEM
S1NB20	OEM
S10AN6	0145
S1P	0102
S1P20	0102
S1P20Z	0102
S1P40Z	0102
S1P60	0017
S1P80	0017
S1P100	0017
S1P150	0017
S1PB15	0289
S1Q20Z	0289
S1QB10	0277
S1QB20	0289
S1QB20Z	0289
S1QB40	0296
S1QB40Z	0296
S1QB60	0302
S1R12B	0137
S1R13B	0137
S1R20	0133
S1R20Z	0102
S1R40Z	0102
S1R50	OEM
S1R60	0023
S1R80	2528
S1R100	0017
S1R150	0017
S1RB	0277
S1RB-10	0277
S1RB10	0277
S1RB20	0289
S1RB20Z	0289
S1RB40	0296
S1RB40Z	0296
S1RB60	0302
S1RBA	0277
S1RBA-10	6665
S1RBA10	0106
S1RBA10-F10	OEM
S1RBA20	0106
S1RBA20Z	0106
S1RBA40	0106
S1RBA40Z	0106
S1RBA60	2010
S1RBA80	2400
S1RC20	0015
S1RC20R	0015
S1S	OEM
S1S2	0133
S1S3H	0023
S1S4M	0023
S1S11	0143
S1S20	0143
S1SD-1	0015
S1SD-1HF	0015
S1SD-1X	0015
S1SM-150-01	0015
S1SM-150-02	0015
S1SS	OEM
S1T20H	6114
S1T40H	OEM
S1T60H	6027
S1U-1N	OEM
S1U-1P	OEM
S1U-12	OEM
S1V1D	OEM
S1V20	OEM
S1VB	0289
S1VB-10	0277
S1VB10	0277
S1VB10-S	0277
S1VB10S	0277
S1VB20	0289
S1VB20Z	0289
S1VB40	0296
S1VB40Z	0296
S1VB60	0302
S1VB60Z	0302
S1VN0104	OEM
S1WB	0241
S1WB(A)10	0241
S1WB(A)10D	0241
S1WB10	0241
S1WB20	0241
S1WB40	4581
S1WB60	1875
S1WBA10	OEM
S1WBA20	OEM
S1WBA40	1875
S1WBA40A	1875
S1WBA60	1875
S1WBS10	0241
S1WBS40	OEM
S1WBS4095	OEM
S1Y	1137
S1Y1D	OEM
S1YB10	OEM
S1YB20	OEM
S1YB40	OEM
S1YB60	OEM
S1Z	1303
S1.5	0015
S-1.5-0	0015
S1.5-01	0015
S-1.5-01FR	0102
S-1.5-02	0015
S1.5-02	OEM
S-1.5-02FR	0102
S-1.5-04	0015
S1.5-04	OEM
S-1.5-04FR	0102
S-1.5-06	0015
S1.5-06	OEM
S-1.5-06FR	0102
S-1.5-08	0071
S1.5-08	OEM
S-1.5-08FR	0102
S-1.5-10	0071
S1.5-10	OEM
S-1.5-10FR	0102
S-2	0015
S2	0150
S2A06	0015
S2A2F	OEM
S2A10	0015
S2A20	0242
S2A30	0540
S2A40	0540
S2A42	OEM
S2A50	0965
S2A60	0965
S2A80	0817
S2A100	0922
S2AN6	0097
S2AN12	0097
S2AN31	OEM
S2AN40	OEM
S2AN41	OEM
S2AN55	OEM
S2AN70	OEM
S2AN71	OEM
S2A06	0110
S2AR1	0015
S2AR2	0015
S2BR30	OEM
S2BR70	OEM
S2BR512	OEM
S2C-10	OEM

DEVICE TYPE	REPL CODE
S2C30	0015
S2C40	0015
S2C40A	0015
S2CN1	0242
S2CR8	OEM
S2CR25	OEM
S2CR30	OEM
S2DR8	OEM
S2DR25	OEM
S2DR30	OEM
S2E20	0015
S2E60	0015
S2E60-1	0015
S2E100	0071
S2ER8	OEM
S2ER25	OEM
S2ER30	OEM
S2F10	0015
S2F20	0015
S2FN270	OEM
S2FN290	OEM
S2GN270	OEM
S2GN290	OEM
S2GR2	0242
S2HB20Z	OEM
S2K-20	OEM
S2K20	0640
S2K20H	0546
S2K40	0609
S2K45	0031
S2K49	0087
S2L20UF	OEM
S2LA20	0546
S2LA20F	OEM
S2M1	0953
S2M2	0110
S2PB	0277
S2PB10	0277
S2PB20	0289
S2PB40	0296
S2Q20	0015
S2Q40	0015
S2Q60	0015
S2Q80	0071
S2Q100	0071
S2RB40	0296
S2TB	0296
S2TB10	0277
S2TB20	0289
S2TB40	0296
S2TB80	0250
S2V	0071
S2V2D	OEM
S2V3D	OEM
S2V10	0071
S2V20	0564
S2V40	0071
S2V60	0071
S2VB	1414
S2VB10	0323
S2VB-10-1	0323
S2VB20	1414
S2VB40	0473
S2VB60	0446
S2VC	0015
S2VC10	0071
S2VC10R	0071
S2Y1D	OEM
S2Z20	0242
S2Z40	0540
S2Z60	0965
S2Z80	0817
S2Z100	0922
S3	0150
S3-3AT2	2083
S3A05	0110
S3A05F	0546
S3A06	0015
S3A025	0110
S3A1	0953
S3A1F	0546
S3A2	0242
S3A2F	0546
S-3A3	1306
S3A3	0540
S3A3F	1362
S3A4	0540
S3A4F	1362
S3A5	0965
S3A5F	0031
S3A6	0965
S3A6F	0031
S3A7	0817
S3A8	0817
S3A8F	1372
S3A9	0922
S3A10	0922
S3A10F	1372
S3A12	OEM
S3A12F	OEM
S3A15	OEM
S3AN6	0109
S3AN12	0109
S3AN15	OEM
S3AN20	OEM
S3AN31	OEM
S3AN40	OEM
S3AN41	OEM
S3AN55	OEM
S3AN70	OEM
S3AN71	OEM
S3AR1	0015
S3AR2	OEM
S-3AT2	2083
S3B20Z	OEM
S3B40-02	OEM
S3B40-04	OEM
S3B40-06	OEM
S3B40-08	OEM
S3B40-10	OEM
S3B40-12	OEM
S3B40Z	OEM
S3BA05	0110
S3BN15	OEM
S3BN20	OEM
S3BN31	OEM
S3BN40	OEM
S3BN55	OEM
S3BN71	OEM
S3BR30	OEM
S3BR70	OEM
S3BR502	OEM
S3BR511	OEM
S3BR512	OEM
S3C321	OEM
S3C471	OEM
S3C661	OEM
S3CN1	0540
S3CN3	OEM
S3CR8	OEM
S3CR25	OEM
S3CR30	OEM
S-3DB3	2084
S3DR8	OEM
S3DR25	OEM
S3DR30	OEM
S3DR601	OEM
S3EN310	OEM
S3ER8	OEM
S3ER25	OEM
S3ER30	OEM
S3FN270	OEM
S3FN290	OEM
S3FN310	OEM
S3FN350	OEM
S3G1	OEM
S3G2	OEM
S3G4	0071
S3G4Z	OEM
S3G6Z	OEM
S3G8	OEM
S3G8Z	OEM
S3G10	0953
S3G20	0242
S3G40	0540
S3G40Z	0540
S3G41	0710
S3G60Z	0965
S3G80Z	0817
S3GN55	OEM
S3GN71	OEM
S3GN270	OEM
S3GN290	OEM
S3GN310	OEM
S3GN350	OEM
S3GR2	0540
S3GR311	OEM
S3H-02	OEM
S3HN310	OEM
S3HR311	OEM
S3K40	0031
S3M1	0242
S3M2	OEM
S3M3	OEM
S3MX	0015
S3S3M	2542
S3S4M	0564
S3S6M	3604
S3V1D	OEM
S3V-10	0087
S3V10	0087
S3V10SS	OEM
S3V20	0564
S3V40	0087
S3V60	0087
S3V60Z	0031
S3VC10	0199
S3VC10R	1015
S3VC20	0199
S3VC20R	1015
S3VC40	0199
S3VC40R	1015
S3VC60	OEM
S3WB	0446
S3WB10	0729
S3WB20	0729
S3WB40	0737
S3WB60	0737
S3WB60Z	0737
S4	0150
S4A06	0015
S4AN6	0109
S4AN12	0109
S4AN31	OEM
S4AN40	OEM
S4AN41	OEM
S4AN55	OEM
S4AN70	OEM
S4AN71	OEM
S4A06	0015
S4AR1	0015
S4AR2	0015
S4AR30	0015
S4B01	OEM
S4BR30	OEM
S4BR70	OEM
S4BR502	OEM
S4BR512	OEM
S-4C	0017
S4C	0015
S4CN1	0242
S4CR8	OEM
S4CR25	OEM
S4CR30	OEM
S4DR8	OEM
S4DR25	OEM
S4DR30	OEM
S4ER8	OEM
S4ER25	OEM
S4ER30	OEM
S4FN270	OEM
S4FN290	OEM
S4FN300	0015
S4GN270	OEM
S4GN290	OEM
S4GR2	0242
S4LS132N	1625
S4LS221N	1239
S4LS374N	0713
S4M1	0540
S4M2	OEM
S4VB	1414
S4VB-10	0323
S4VB10	0323
S4VB-10-1	0323
S4VB20	1414
S4VB20A	OEM
S4VB20F	1414
S4VB40	0473
S4VB40F1	0473
S4VB60	0446
S4VB60-L15	OEM
S-5A05	1282
S5A05	1282
S5A05F	1362
S5A025	1282
S-5A1	1282
S5A1	1282
S5A1F	1362
S-5A2	1282
S5A2	1282
S5A2F	1362
S-5A3	1287
S5A3	1287
S5A3F	1362
S-5A4	1287
S5A4	1287
S5A4F	1362
S-5A5	1293
S5A5	1293
S5A5F	1372
S-5A6	1293
S5A6	1293
S5A6F	1372
S-5A8	1296
S5A8	1296
S5A8F	1372
S-5A10	1296
S5A10	1296
S5A10F	1372
S5A12	OEM
S5A12F	OEM
S5A15	OEM
S5AN6	0122
S5AN12	0122
S5AN15	OEM
S5AN20	OEM
S5AN31	OEM
S5AN40	OEM
S5AN41	OEM
S5AN55	OEM
S5AN70	OEM
S5AN71	OEM
S5AR1	0015
S5AR2	0015
S5B-01	OEM
S5B10	OEM
S5B20	OEM
S5B30	OEM
S5B40	OEM
S5B50	OEM
S5B60	OEM
S5B80	OEM
S5B100	OEM
S5BN15	OEM
S5BN20	OEM
S5BN31	OEM
S5BN40	OEM
S5BN55	OEM
S5BN71	OEM
S5BR30	OEM
S5BR70	OEM
S5BR502	OEM
S5BR511	OEM
S5BR512	OEM
S5C321	OEM
S5C471	OEM
S5C661	OEM
S5CN1	0540
S5CN3	OEM
S5CR8	OEM
S5CR25	OEM
S5CR30	OEM
S5DR8	OEM
S5DR25	OEM
S5DR30	OEM
S5DR601	OEM
S5EN310	OEM
S5ER8	OEM
S5ER25	OEM
S5ER30	OEM
S5FN270	OEM
S5FN290	OEM
S5FN310	OEM
S5FN350	OEM
S5GN55	OEM
S5GN71	OEM
S5GN270	OEM
S5GN290	OEM
S5GN310	OEM
S5GN350	OEM
S5GR2	0965
S5GR311	OEM
S5HN310	OEM
S5HN311	OEM
S5HR311	OEM
S5KC20	1236
S5KC20H	1236
S5KC20R	1727
S5KC20RH	1727
S5KC40	OEM
S5KC40R	OEM
S5KD20	OEM
S5KD20H	OEM
S5KD40	OEM
S5KP5.0	OEM
S5KP5.0A	OEM
S5KP6.0	OEM
S5KP6.0A	OEM
S5KP6.5	OEM
S5KP6.5A	OEM
S5KP7.0	OEM
S5KP7.0A	OEM
S5KP7.5	OEM
S5KP7.5A	OEM
S5KP8.0	OEM
S5KP8.0A	OEM
S5KP8.5	OEM
S5KP8.5A	OEM
S5KP9.0	OEM
S5KP9.0A	OEM
S5KP10	OEM
S5KP10A	OEM
S5KP11	OEM
S5KP11A	OEM
S5KP12	OEM
S5KP12A	OEM
S5KP13	OEM
S5KP13A	OEM
S5KP14	OEM
S5KP14A	OEM
S5KP15	OEM
S5KP15A	OEM
S5KP16	OEM
S5KP16A	OEM
S5KP17	OEM
S5KP17A	OEM
S5KP18	OEM
S5KP18A	OEM
S5KP20	OEM
S5KP20A	OEM
S5KP22	OEM
S5KP22A	OEM
S5KP24	OEM
S5KP24A	OEM
S5KP26	OEM
S5KP26A	OEM
S5KP28	OEM
S5KP28A	OEM
S5KP30	OEM
S5KP30A	OEM
S5KP33	OEM
S5KP33A	OEM
S5KP36	OEM
S5KP36A	OEM
S5KP40	OEM
S5KP40A	OEM
S5KP43	OEM
S5KP43A	OEM
S5KP45	OEM
S5KP45A	OEM
S5KP48	OEM
S5KP48A	OEM
S5KP51	OEM
S5KP51A	OEM
S5KP54	OEM
S5KP54A	OEM
S5KP58	OEM
S5KP58A	OEM
S5KP60	OEM
S5KP60A	OEM
S5KP64	OEM
S5KP64A	OEM
S5KP70	OEM
S5KP70A	OEM
S5KP75	OEM
S5KP75A	OEM
S5KP78	OEM
S5KP78A	OEM
S5KP85	OEM
S5KP85A	OEM
S5KP90	OEM
S5KP90A	OEM
S5KP100	OEM
S5KP100A	OEM
S5KP110	OEM
S5KP110A	OEM
S5M1	0540
S5M2	OEM
S5M3	OEM
S5S	0015
S5S3	2077
S5S3M	OEM
S5S4	2077
S5S4M	OEM
S5SR	0015
S5VB	1414
S5VB-10	1414
S5VB10	0323
S5VB10F	OEM
S5VB20	1414
S5VB40	0473
S5VB40-F1	0473
S5VB60	0446
S6-3	0277
S6-10	0170
S6-103A	2900
S6-103K	2291
S6-203A	2900
S6-203K	2291
S6-403A	1112
S6-403K	0476
S6-603A	1112
S6-603K	0476
S6-803A	3011
S6-803K	0449
S6-1003A	3011
S6-1003K	0449
S6-1203K	OEM
S6-Z603A	1112
S6-Z603K	0476
S6-Z803A	3011
S6-Z803K	0449
S6-Z1003A	3011
S6-Z1003K	0449
S6A05	1282
S6A1	1282
S6A2	1287
S6A3	1287
S6A4	1287
S6A5	1293
S6A6	1293
S6A8	1296
S6A10	1296
S6AN6	0122
S6AN12	0122
S6AN31	OEM
S6AN40	OEM
S6AN41	OEM
S6AN55	OEM
S6AN70	OEM
S6AN71	OEM
S6AR1	0015
S6AR2	0015
S6BR30	OEM
S6BR70	OEM
S6BR512	OEM
S6CR8	OEM
S6CR25	OEM
S6CR30	OEM
S6DR8	OEM
S6DR25	OEM
S6DR30	OEM
S6ER8	OEM
S6ER25	OEM
S6ER30	OEM
S6FN270	OEM
S6GN270	OEM
S6GR2	0965
S6K20	5915
S6K20H	1127
S6K20R	OEM
S6K40	5915
S6K40R	OEM
S6M1	0965
S6M2	OEM
S6Z803A	OEM
S6Z1003A	OEM
S6Z1203A	OEM
S7	OEM
S7-8	0133
S7AN6	0131
S7AN12	0131
S7AN15	OEM
S7AN20	OEM
S7AN31	OEM
S7AN40	OEM
S7AN55	OEM
S7AN70	OEM
S7AN71	OEM
S7AR1	0071
S7BN15	OEM
S7BN20	OEM
S7BN31	OEM
S7BN40	OEM
S7BN55	OEM
S7BN200	OEM
S7BR30	OEM
S7BR70	OEM
S7BR125	OEM
S7BR502	OEM
S7BR511	OEM
S7BR512	OEM
S7C321	OEM
S7C452	OEM
S7C471	OEM
S7C661	OEM
S7CN3	OEM
S7CR8	OEM
S7CR25	OEM
S7CR30	OEM
S7DR8	OEM
S7DR25	OEM
S7DR30	OEM
S7DR601	OEM
S7EN310	OEM
S7ER8	OEM
S7ER25	OEM
S7ER30	OEM
S7FN270	OEM
S7FN290	OEM
S7FN310	OEM
S7FN350	OEM
S7GN55	OEM
S7GN71	OEM
S7GN270	OEM
S7GN290	OEM
S7GN310	OEM
S7GN350	OEM
S7GR311	OEM
S7HN310	OEM
S7HR311	OEM
S7KN125	OEM
S7LN125	OEM
S7PN125	OEM
S-8-01	0871
S8-01	1250
S8-01F	OEM
S-8-01FR	0871
S-8-02	0871
S8-02	OEM
S8-02F	OEM
S-8-02FR	0871
S-8-03	0853
S8-03	OEM
S8-03F	OEM
S-8-04	0853
S8-04	OEM
S8-04F	OEM
S-8-04FR	0853
S-8-06	4322
S8-06	0087
S8-06F	OEM
S-8-06FR	OEM
S-8-08	4323
S8-08	OEM
S8-08F	OEM
S-8-08FR	OEM
S8-10	0087
S8-10F	OEM
S-8-10FR	OEM
S-8-12	OEM
S8AN6	0131
S8AN12	0131
S8AN31	OEM
S8AN40	OEM
S8AN55	OEM
S8AN70	OEM
S8AN71	OEM
S8AR1	0071
S8AR2	0071
S8B10	OEM
S8B20	OEM
S8B30	OEM
S8B40	OEM
S8B50	OEM
S8B60	OEM
S8B80	OEM
S8B100	OEM
S8BN200	OEM
S8BR30	OEM
S8BR70	OEM
S8BR503	OEM
S8BR512	OEM
S8CN1	0817
S8CR8	OEM
S8CR25	OEM
S8CR30	OEM
S8DR8	OEM
S8DR25	OEM
S8DR30	OEM
S8ER8	OEM
S8ER25	OEM
S8ER30	OEM
S8GN270	OEM
S8GR2	0817
S8H80J	0682
S8M1	0965
S8M2	OEM
S-8R-01	5943
S-8R-01FR	OEM
S-8R-02	5943
S-8R-02FR	OEM
S-8R-03	OEM
S-8R-04	5944
S-8R-04FR	OEM
S-8R-06	5945
S-8R-06FR	OEM
S-8R-08	5946
S-8R-08FR	OEM
S-8R-10	5947
S-8R-10FR	OEM
S-8R-12	OEM
S8T28F	0581
S8T95F	OEM
S8T97F	OEM
S8T98F	OEM
S9AH12	0145
S9AN6	0145
S9AN12	0145
S9AN15	OEM
S9AN20	OEM
S9AN31	OEM
S9AN40	OEM
S9AN55	OEM
S9AN70	OEM
S9AN71	OEM
S9AR1	0071
S9BN15	OEM
S9BN20	OEM
S9BN31	OEM
S9BN40	OEM
S9BN55	OEM
S9BN71	OEM
S9BN200	OEM
S9BR30	OEM
S9BR70	OEM
S9BR125	OEM
S9BR502	OEM
S9BR503	OEM
S9BR511	OEM
S9BR512	OEM
S9C321	OEM
S9C452	OEM
S9C471	OEM
S9C661	OEM
S9CN3	OEM
S9CR8	OEM
S9CR25	OEM
S9CR30	OEM
S9DR8	OEM
S9DR25	OEM
S9DR30	OEM
S9DR601	OEM
S9EN310	OEM
S9ER8	OEM
S9ER25	OEM
S9ER30	OEM
S9FN270	OEM
S9FN290	OEM
S9FN310	OEM
S9FN350	OEM
S9GN55	OEM
S9GN71	OEM
S9GN270	OEM
S9GN290	OEM
S9GN310	OEM
S9GN350	OEM
S9GR311	OEM
S9HN310	OEM
S9HR311	OEM
S9K1003K3	OEM
S9K1003K4	OEM
S9K1003K5	OEM
S9K1003K35	OEM
S9K1003K45	OEM
S9K1103K3	OEM
S9K1103K4	OEM
S9K1103K5	OEM
S9K1103K35	OEM
S9K1103K45	OEM
S9K1203K3	OEM
S9K1203K4	OEM
S9K1203K5	OEM
S9K1203K35	OEM
S9K1203K45	OEM
S9K1303K3	OEM
S9K1303K4	OEM
S9K1303K5	OEM
S9K1303K35	OEM
S9K1303K45	OEM
S9K1403K3	OEM
S9K1403K4	OEM
S9K1403K5	OEM
S9K1403K35	OEM
S9K1403K45	OEM
S9K1503K3	OEM
S9K1503K4	OEM
S9K1503K5	OEM
S9K1503K35	OEM
S9K1503K45	OEM
S9K1603K3	OEM
S9K1603K4	OEM
S9K1603K5	OEM
S9K1603K35	OEM
S9K1603K45	OEM
S9K1703K3	OEM
S9K1703K4	OEM
S9K1703K5	OEM
S9K1703K35	OEM
S9K1703K45	OEM
S9K1803K3	OEM
S9K1803K4	OEM

DEVICE TYPE	REPL CODE
S9K1803K5	OEM
S9K1803K35	OEM
S9K1803K45	OEM
S9K2003K4	OEM
S9K2003K5	OEM
S9K2003K35	OEM
S9K2003K45	OEM
S9K2103K4	OEM
S9K2103K5	OEM
S9K2103K45	OEM
S9K2203K4	OEM
S9K2203K5	OEM
S9K2203K45	OEM
S9K2303K4	OEM
S9K2303K5	OEM
S9K2303K45	OEM
S9K2403A4	OEM
S9K2403K5	OEM
S9K2403K45	OEM
S9K2503K5	OEM
S9K2503K45	OEM
S9K2603K5	OEM
S9K2703K5	OEM
S9K2803K5	OEM
S9K2903K5	OEM
S9K3003K5	OEM
S9KN125	OEM
S9LN125	OEM
S9PN125	OEM
S10	0015
S10-12	2312
S10A	0015
S10AN6	0145
S10AN12	0145
S10AN31	OEM
S10AN40	OEM
S10AN55	OEM
S10AN70	OEM
S10AN71	OEM
S10AR1	0071
S10AR2	0071
S10BN200	OEM
S10BR30	OEM
S10BR70	OEM
S10BR503	OEM
S10BR512	OEM
S10BV1069	0323
S10C	0871
S10CN1	0817
S10CR8	OEM
S10CR25	OEM
S10CR30	OEM
S10DR8	OEM
S10DR25	OEM
S10DR30	OEM
S10E	0853
S10ER8	OEM
S10ER25	OEM
S10ER30	OEM
S10FFD01	OEM
S10FFD02	OEM
S10FFD04	OEM
S10FFD06	OEM
S10FFD08	OEM
S10FFD10	OEM
S10FFD12	OEM
S10FHD01	OEM
S10FHD02	OEM
S10FHD04	OEM
S10FHD06	OEM
S10FHD08	OEM
S10FN270	OEM
S10FND02	OEM
S10FND04	OEM
S10FND06	OEM
S10FND08	OEM
S10FND10	OEM
S10FND12	OEM
S10FUD01	OEM
S10FUD02	OEM
S10FUD04	OEM
S10GN270	OEM
S10GN302	OEM
S10GR2	0922
S10GR302	OEM
S10HN302	OEM
S10HR302	OEM
S-10K	OEM
S10K14	2009
S10K17	2326
S10K20	2796
S10K25	3073
S10K30	3075
S10K35	2941
S10K40	3078
S10K50	3085
S10K60	3090
S10K75	3104
S10K95	3116
S10K130	0838
S10K140	OEM
S10K150	1993
S10K175	OEM
S10K230	OEM
S10K250	1334
S10K275	1660
S10K300	2034
S10K385	OEM
S10K420	3257

DEVICE TYPE	REPL CODE
S10K460	OEM
S10K510	OEM
S10K550	OEM
S10K625	OEM
S10K680	OEM
S10M1	0817
S10M2	OEM
S10N500E	OEM
S10N500E3	OEM
S10N500E4	OEM
S10N500E5	OEM
S10N600E	OEM
S10N600E3	OEM
S10N600E4	OEM
S10N600E5	OEM
S10N700E	OEM
S10N700E3	OEM
S10N700E4	OEM
S10N700E5	OEM
S10N800E	OEM
S10N800E3	OEM
S10N800E4	OEM
S10N800E5	OEM
S10N900E	OEM
S10N900E3	OEM
S10N900E4	OEM
S10N900E5	OEM
S10N1000E	OEM
S10N1000E3	OEM
S10N1000E4	OEM
S10N1000E5	OEM
S10N1100E	OEM
S10N1100E3	OEM
S10N1100E4	OEM
S10N1100E5	OEM
S10N1200E	OEM
S10N1200E3	OEM
S10N1200E4	OEM
S10N1200E5	OEM
S10N1300E	OEM
S10N1300E3	OEM
S10N1300E4	OEM
S10N1300E5	OEM
S10N1400E	OEM
S10N1400E3	OEM
S10N1400E4	OEM
S10N1400E5	OEM
S10N1500E	OEM
S10N1500E3	OEM
S10N1500E4	OEM
S10N1500E5	OEM
S10N1600E	OEM
S10N1600E3	OEM
S10N1600E4	OEM
S10N1600E5	OEM
S10N1700E	OEM
S10N1800E	OEM
S10SC3	OEM
S10SC3M	1236
S10SC3MR	1727
S10SC4	OEM
S10SC4M	1236
S10SC4MR	1727
S10VB	0323
S10VB10	0323
S10VB20	1414
S10VB40	0473
S10VB60	0446
S10V-S05K150	0246
S11	0133
S11AN15	OEM
S11AN20	OEM
S11AN31	OEM
S11AN40	OEM
S11AN55	OEM
S11AN70	OEM
S11AN71	OEM
S11AN125	OEM
S11B	OEM
S11BN15	OEM
S11BN20	OEM
S11BN31	OEM
S11BN40	OEM
S11BN55	OEM
S11BN71	OEM
S11BN200	OEM
S11BR125	OEM
S11BR502	OEM
S11BR503	OEM
S11BR511	OEM
S11BR512	OEM
S11C	OEM
S11C321	OEM
S11C452	OEM
S11C471	OEM
S11C661	OEM
S11D	OEM
S11DR601	OEM
S11EN310	OEM
S11FN270	OEM
S11FN290	OEM
S11FN310	OEM
S11FN350	OEM
S11GN55	OEM
S11GN71	OEM
S11GN270	OEM
S11GN290	OEM
S11GN302	OEM
S11GN310	OEM

DEVICE TYPE	REPL CODE
S11GN350	OEM
S11GR302	OEM
S11GR311	OEM
S11HN302	OEM
S11HN310	OEM
S11HR302	OEM
S11HR311	OEM
S11KN125	OEM
S11LN125	OEM
S11P500E	OEM
S11P600E	OEM
S11P700E	OEM
S11P800E	OEM
S11P900E	OEM
S11P1000E	OEM
S11P1100E	OEM
S11P1200E	OEM
S11P1300E	OEM
S11P1400E	OEM
S11P1500E	OEM
S11P1600E	OEM
S11P1700E	OEM
S11P1800E	OEM
S11PN125	OEM
S-12	0242
S-12A	0242
S12AN6	OEM
S12AN31	OEM
S12AN40	OEM
S12AN55	OEM
S12AN70	OEM
S12AN71	OEM
S12AR2	OEM
S12B	OEM
S12BN200	OEM
S12BR30	OEM
S12BR70	OEM
S12BR503	OEM
S12C	OEM
S12C20	OEM
S12C40	OEM
S12C60	OEM
S12C80	OEM
S12C100	OEM
S12CN1	0817
S12CR8	OEM
S12CR25	OEM
S12CR30	OEM
S12D	OEM
S12DR8	OEM
S12DR25	OEM
S12DR30	OEM
S12ER8	OEM
S12ER25	OEM
S12ER30	OEM
S12FN270	OEM
S12GN270	OEM
S12GN302	OEM
S12GR2	OEM
S12GR302	OEM
S12HN302	OEM
S12HR302	OEM
S12KC20	2234
S12KC20H	2234
S12KC40	4488
S12M1	0922
S12M2	OEM
S12P500E	OEM
S12P600E	OEM
S12P700E	OEM
S12P800E	OEM
S12P900E	OEM
S12P1000E	OEM
S12P1100E	OEM
S12P1200E	OEM
S12P1300E	OEM
S12P1400E	OEM
S12P1500E	OEM
S12P1600E	OEM
S12P1700E	OEM
S12P1800E	OEM
S13	0015
S13A	0540
S13AN15	OEM
S13AN20	OEM
S13AN31	OEM
S13AN40	OEM
S13AN70	OEM
S13AN125	OEM
S13BN15	OEM
S13BN20	OEM
S13BN31	OEM
S13BN40	OEM
S13BN55	OEM
S13BN71	OEM
S13BN200	OEM
S13BR125	OEM
S13BR503	OEM
S13BR512	OEM
S13C322	OEM
S13C453	OEM
S13C472	OEM
S13C661	OEM
S13DR601	OEM
S13EN310	OEM
S13FN270	OEM
S13FN290	OEM
S13FN310	OEM
S13FN350	OEM

DEVICE TYPE	REPL CODE
S13GN55	OEM
S13GN71	OEM
S13GN270	OEM
S13GN290	OEM
S13GN302	OEM
S13GN310	OEM
S13GN350	OEM
S13GR302	OEM
S13GR312	OEM
S13GR313	OEM
S13HN302	OEM
S13HN310	OEM
S13HR302	OEM
S13HR312	OEM
S13HR313	OEM
S13KN125	OEM
S13KV	0201
S13LN125	OEM
S13N500E	OEM
S13N600E	OEM
S13N700E	OEM
S13N800E	OEM
S13N900E	OEM
S13N1000E	OEM
S13N1100E	OEM
S13N1200E	OEM
S13N1300E	OEM
S13N1400E	OEM
S13N1500E	OEM
S13N1600E	OEM
S13N1700E	OEM
S13N1800E	OEM
S13PN125	OEM
S14	0015
S14A	0540
S14AN31	OEM
S14AN40	OEM
S14AN70	OEM
S14AN125	OEM
S14BN200	OEM
S14BR125	OEM
S14C322	OEM
S14FN270	OEM
S14GN270	OEM
S14GN302	OEM
S14GR302	OEM
S14GR312	OEM
S14HN302	OEM
S14HR302	OEM
S14HR312	OEM
S14K14	2009
S14K17	2326
S14K20	2796
S14K25	3073
S14K30	3075
S14K35	2941
S14K40	3078
S14K50	3085
S14K60	3090
S14K75	3104
S14K95	3116
S14K130	0838
S14K140	OEM
S14K150	1993
S14K175	OEM
S14K230	OEM
S14K250	1334
S14K275	1660
S14K300	2034
S14K385	OEM
S14K420	3257
S14K460	OEM
S14K510	OEM
S14K550	OEM
S14K625	OEM
S14K680	OEM
S14K1000	OEM
S14KN125	OEM
S14LN125	OEM
S14PN125	OEM
S-15	0015
S15	0102
S-15-10	0015
S15A	0015
S15AN6	OEM
S15AN31	OEM
S15AN40	OEM
S15AN70	OEM
S15AR2	OEM
S15AR504	OEM
S15B	OEM
S15BN200	OEM
S15BR30	OEM
S15BR512	OEM
S-15C	0017
S15C	1059
S15C322	OEM
S15C323	OEM
S15C453	OEM
S15C472	OEM
S15CN1	OEM
S15CN3	OEM
S15CR8	OEM
S15CR25	OEM
S15CR30	OEM
S15D	OEM
S15DR8	OEM
S15DR25	OEM
S15DR30	OEM

DEVICE TYPE	REPL CODE
S15ER8	OEM
S15ER25	OEM
S15ER30	OEM
S15FN270	OEM
S15GN270	OEM
S15GN302	OEM
S15GR2	OEM
S15GR302	OEM
S15GR312	OEM
S15GR313	OEM
S-15H	0102
S15H	0133
S15HN302	OEM
S15HR302	OEM
S15HR312	OEM
S15HR313	OEM
S15M1	OEM
S15M3	OEM
S15S3	0615
S15S4	0615
S15S6	OEM
S15SC3M	1942
S15SC4M	1942
S15VB10	2363
S15VB20	2363
S15VB40	2369
S15VB60	2370
S16	0015
S16A	0015
S16AR504	OEM
S16B	0015
S16B10	OEM
S16B20	OEM
S16B30	OEM
S16B40	OEM
S16B50	OEM
S16B60	OEM
S16B80	OEM
S16B100	OEM
S16C	OEM
S16C322	OEM
S16D	OEM
S16EN270	OEM
S16EN290	OEM
S16EN310	OEM
S16F3	0469
S16FN270	OEM
S16FN290	OEM
S16FN310	OEM
S16GN270	OEM
S16GN290	OEM
S16GN302	OEM
S16GN310	OEM
S16GR302	OEM
S16GR312	OEM
S16HN270	OEM
S16HN290	OEM
S16HN302	OEM
S16HN310	OEM
S16HN350	OEM
S16HR302	OEM
S16HR312	OEM
S17	0015
S17A	0015
S17AR504	OEM
S17BR503	OEM
S17BR512	OEM
S17C322	OEM
S17C323	OEM
S17C453	OEM
S17C472	OEM
S17C662	OEM
S17DR602	OEM
S17GN302	OEM
S17GR302	OEM
S17GR312	OEM
S17GR313	OEM
S17HN302	OEM
S17HR302	OEM
S17HR312	OEM
S17HR313	OEM
S18	0015
S18A	0015
S18AR2	OEM
S18B	0015
S18C322	OEM
S18F3	0469
S18GN302	OEM
S18GR2	OEM
S18GR302	OEM
S18GR312	OEM
S18HN302	OEM
S18HR302	OEM
S18HR312	OEM
S18KV	0378
S19	0015
S19A	0015
S19AR505	OEM
S19BR512	OEM
S19C	1107
S19C322	OEM
S19C323	OEM
S19C454	OEM
S19C472	OEM
S19C662	OEM
S19D	1111
S19DR602	OEM
S19E	0258
S19G	0267

DEVICE TYPE	REPL CODE
S19GN302	OEM
S19GR302	OEM
S19GR312	OEM
S19GR313	OEM
S19HN302	OEM
S19HR302	OEM
S19HR312	OEM
S19HR313	OEM
S19J	1119
S19L	0281
S20	0071
S-20-01	0871
S20-01	OEM
S20-01F	OEM
S-20-01FR	0871
S-20-02	0871
S20-02	OEM
S20-02F	OEM
S-20-02FR	0871
S-20-04	0853
S20-04	OEM
S20-04F	OEM
S-20-04FR	0853
S-20-06	5002
S20-06	OEM
S20-06F	OEM
S-20-06FR	OEM
S-20-08	5003
S20-08	OEM
S20-08F	OEM
S-20-08FR	OEM
S-20-10	5004
S20-10	OEM
S-20-10-FR	OEM
S20-10F	OEM
S20AR505	OEM
S20C	1124
S20C322	OEM
S20D	1126
S20E	0805
S20FFA01	OEM
S20FFA02	OEM
S20FFA04	OEM
S20FFA06	OEM
S20FFA08	OEM
S20FFA10	OEM
S20FFA12	OEM
S20FFD01	OEM
S20FFD02	OEM
S20FFD04	OEM
S20FFD06	OEM
S20FFD08	OEM
S20FFD10	OEM
S20FFD12	OEM
S20FFK01	OEM
S20FFK02	OEM
S20FFK04	OEM
S20FFK06	OEM
S20FFK08	OEM
S20FFK10	OEM
S20FFK12	OEM
S20FHA01	OEM
S20FHA02	OEM
S20FHA04	OEM
S20FHA06	OEM
S20FHA08	OEM
S20FHD01	OEM
S20FHD02	OEM
S20FHD04	OEM
S20FHD06	OEM
S20FHD08	OEM
S20FHK01	OEM
S20FHK02	OEM
S20FHK04	OEM
S20FHK06	OEM
S20FHK08	OEM
S20FNA02	OEM
S20FNA04	OEM
S20FNA06	OEM
S20FNA08	OEM
S20FNA10	OEM
S20FNA12	OEM
S20FND02	OEM
S20FND04	OEM
S20FND06	OEM
S20FND08	OEM
S20FND10	OEM
S20FND12	OEM
S20FNK02	OEM
S20FNK04	OEM
S20FNK06	OEM
S20FNK08	OEM
S20FNK10	OEM
S20FNK12	OEM
S20FUA01	OEM
S20FUA02	OEM
S20FUA04	OEM
S20FUD01	OEM
S20FUD02	OEM
S20FUD04	OEM
S20FUK01	OEM
S20FUK02	OEM
S20FUK04	OEM
S20G	0319
S20GN302	OEM
S20GR302	OEM
S20GR312	OEM
S20HN302	OEM
S20HR302	OEM

DEVICE TYPE	REPL CODE
S20HR312	OEM
S20J	1132
S20K60	OEM
S20K75	OEM
S20K95	OEM
S20K130	0830
S20K140	OEM
S20K150	3415
S20K175	OEM
S20K230	OEM
S20K250	1336
S20K275	4867
S20K300	4872
S20K385	OEM
S20K420	3859
S20K460	OEM
S20K510	OEM
S20K550	OEM
S20K625	OEM
S20K680	OEM
S20K1000	OEM
S20KV	0378
S20L	0044
S20N400D3	OEM
S20N400D4	OEM
S20N400D5	OEM
S20N400G3	OEM
S20N400G4	OEM
S20N400G5	OEM
S20N400H	OEM
S20N500J3	OEM
S20N500J4	OEM
S20N500J5	OEM
S20N600J3	OEM
S20N600J4	OEM
S20N600J5	OEM
S20N700J3	OEM
S20N700J4	OEM
S20N700J5	OEM
S20N750D3	OEM
S20N750D4	OEM
S20N750D5	OEM
S20N750G3	OEM
S20N750G4	OEM
S20N750G5	OEM
S20N800J3	OEM
S20N800J4	OEM
S20N800J5	OEM
S20N900J3	OEM
S20N900J4	OEM
S20N900J5	OEM
S20N1000D3	OEM
S20N1000D4	OEM
S20N1000D5	OEM
S20N1000G3	OEM
S20N1000G4	OEM
S20N1000G5	OEM
S20N1000J3	OEM
S20N1000J4	OEM
S20N1000J5	OEM
S20N1100J3	OEM
S20N1100J4	OEM
S20N1100J5	OEM
S20N1200J3	OEM
S20N1200J4	OEM
S20N1200J5	OEM
S20N1250D3	OEM
S20N1250D4	OEM
S20N1250D5	OEM
S20N1250G3	OEM
S20N1250G4	OEM
S20N1250G5	OEM
S20N1300J3	OEM
S20N1300J4	OEM
S20N1300J5	OEM
S20N1400J3	OEM
S20N1400J4	OEM
S20N1400J5	OEM
S20N1500D3	OEM
S20N1500D4	OEM
S20N1500D5	OEM
S20N1500G3	OEM
S20N1500G4	OEM
S20N1500G5	OEM
S20N1500J3	OEM
S20N1500J4	OEM
S20N1500J5	OEM
S20N1600J3	OEM
S20N1600J4	OEM
S20N1600J5	OEM
S20ND400	0015
S20NH400	0015
S-20R-01	4991
S-20R-01FR	OEM
S-20R-02	4992
S-20R-02FR	OEM
S-20R-04	4994
S-20R-04FR	OEM
S-20R-06	4995
S-20R-06FR	OEM
S-20R-08	4996
S-20R-08FR	OEM
S-20R-10	4997
S-20R-10FR	OEM
S21	0015
S21AR505	OEM
S21BR512	OEM
S21C322	OEM
S21C323	OEM

DEVICE TYPE	REPL CODE
S21C454	OEM
S21C472	OEM
S21C662	OEM
S21DR602	OEM
S21GN302	OEM
S21GR302	OEM
S21GR312	OEM
S21GR313	OEM
S21H4302	OEM
S21HN302	OEM
S21HR313	OEM
S21N100H	OEM
S21N200H	OEM
S21N300H	OEM
S21N400H	OEM
S21N500H	OEM
S21N600H	OEM
S21N700H	OEM
S21N800H	OEM
S22	0015
S22A	0015
S22C322	OEM
S22GN302	OEM
S22GR302	OEM
S22GR312	OEM
S22HN302	OEM
S22HR302	OEM
S22HR312	OEM
S22MD1	0235
S23	0015
S23A	0015
S23C322	OEM
S23C455	OEM
S23C662	OEM
S23DR602	OEM
S23GR312	OEM
S23HR312	OEM
S24	0071
S24AR506	OEM
S25A05	3189
S25A1	2901
S25A2	1124
S25A3	1126
S25A4	0805
S25A6	0319
S25A7	OEM
S25A8	1132
S25A10	0044
S25AR506	OEM
S25C424	OEM
S25C455	OEM
S25GR314	OEM
S25HR314	OEM
S25VB10	2363
S25VB20	2363
S25VB40	2369
S25VB60	2370
S25X01D	OEM
S25X02D	OEM
S25X2	OEM
S25X15D	OEM
S26	0015
S26F3	0469
S27AR506	OEM
S27C424	OEM
S27C456	OEM
S27C663	OEM
S27CN506	OEM
S27CR506	OEM
S27GR314	OEM
S27HR314	OEM
S28	0071
S28CN506	OEM
S28CR506	OEM
S28F3	0469
S29AR506	OEM
S29C424	OEM
S29C456	OEM
S29C663	OEM
S29CN506	OEM
S29CR506	OEM
S29GR314	OEM
S29HR314	OEM
S30	0015
S30-12	2312
S30CN506	OEM
S30CR506	OEM
S30N500JA3	OEM
S30N600JA3	OEM
S30N700JA3	OEM
S30N800JA3	OEM
S30N900JA3	OEM
S30N1000JA3	OEM
S30N1100JA3	OEM
S30N1200JA3	OEM
S30N1300JA3	OEM
S30N1400JA3	OEM
S30N1500JA3	OEM
S30N1600JA3	OEM
S30S3	4314
S30S3A	1545
S30S4	OEM
S30S4A	1545
S30S6	OEM
S30SC3	OEM
S30SC3F	OEM
S30SC3M	1942
S30SC4	OEM
S30SC4F	OEM
S30SC4M	1942
S31	0015
S31AR506	OEM
S31C456	OEM
S31C663	OEM
S31CN506	OEM
S31CR506	OEM
S32	0015
S32CN506	OEM
S32CR506	OEM
S33	0015
S33AR506	OEM
S33C456	OEM
S33C663	OEM
S33CN506	OEM
S33CR506	OEM
S-34	0102
S34	0023
S34AR507	OEM
S34CN506	OEM
S34CR506	OEM
S35	0015
S35AR506	OEM
S35AR507	OEM
S35C456	OEM
S35CN506	OEM
S35CR506	OEM
S36	0015
S36CN506	OEM
S36CR506	OEM
S36F3	0722
S37	0084
S37C457	OEM
S38	0084
S38DC014PG02	OEM
S38F3	0722
S39	0097
S39AR507	OEM
S39C457	OEM
S-39T	0160
S40	0015
S40A	0015
S40A05	3189
S40A1	2901
S40A2	1124
S40A3	1126
S40A4	0805
S40A5	0319
S40A6	0319
S40A8	1132
S40A10	0044
S40FFA01	OEM
S40FFA02	OEM
S40FFA04	OEM
S40FFA06	OEM
S40FFA08	OEM
S40FFA10	OEM
S40FFA12	OEM
S40FFK01	OEM
S40FFK02	OEM
S40FFK04	OEM
S40FFK06	OEM
S40FFK08	OEM
S40FFK10	OEM
S40FFK12	OEM
S40FHA01	OEM
S40FHA02	OEM
S40FHA04	OEM
S40FHA06	OEM
S40FHA08	OEM
S40FHA12	OEM
S40FHK04	OEM
S40FHK06	OEM
S40FHK08	OEM
S40FNA02	OEM
S40FNA04	OEM
S40FNA06	OEM
S40FNA08	OEM
S40FNA10	OEM
S40FNA12	OEM
S40FNK02	OEM
S40FNK04	OEM
S40FNK06	OEM
S40FNK08	OEM
S40FNK10	OEM
S40FNK12	OEM
S40FUA01	OEM
S40FUA02	OEM
S40FUA04	OEM
S40FUK01	OEM
S40FUK02	OEM
S40FUK04	OEM
S-40T	0160
S-40TB	0160
S41AR507	OEM
S41C457	OEM
S-41T	0160
S-42T	0160
S43	0015
S-43T	0160
S44	0015
S45AR508	OEM
S45C458	OEM
S46	0015
S-46T	0160
S47	0015
S47AR508	OEM
S47C458	OEM
S48	0015
S48F3	0722
S-48T	0160
S49	0015
S49AR508	OEM
S49C458	OEM
S-49T	0160
S50	0131
S50-12	3565
S50-28	2946
S51	0131
S51A	1520
S51AR508	OEM
S51B	1520
S51C	1520
S51C458	OEM
S52	0097
S52A	1530
S52B	1530
S52C	1530
S53	0109
S53A	1547
S53B	1547
S54	OEM
S54A	1567
S54B	1567
S54B-10	1414
S54LS20F	OEM
S54LS20W	OEM
S54S20F	OEM
S54S20W	OEM
S55	0122
S-55TB	0136
S56	0122
S56F3	0725
S57	0131
S58	0131
S58F3	0725
S-58TB	0160
S59	0145
S-60	OEM
S60	OEM
S60FFN01	OEM
S60FFN02	OEM
S60FFN04	OEM
S60FFN06	OEM
S60FFN08	OEM
S60FFN10	OEM
S60FFN12	OEM
S60FFR01	OEM
S60FFR02	OEM
S60FFR04	OEM
S60FFR06	OEM
S60FFR08	OEM
S60FFR10	OEM
S60FFR12	OEM
S60FHN01	OEM
S60FHN02	OEM
S60FHN04	OEM
S60FHN06	OEM
S60FHN08	OEM
S60FHR01	OEM
S60FHR02	OEM
S60FHR04	OEM
S60FHR06	OEM
S60FHR08	OEM
S60FUN01	OEM
S60FUN02	OEM
S60FUN04	OEM
S60FUR01	OEM
S60FUR02	OEM
S60FUR04	OEM
S60S3	1545
S60S4	1545
S60S6	OEM
S61	0071
S62	0071
S63	0071
S66F3	0725
S68F3	0725
S70-12	2946
S70FNN-02	OEM
S70FNN-04	OEM
S70FNN-06	OEM
S70FNN-08	OEM
S70FNN-10	OEM
S70FNN-12	OEM
S70FNR-02	OEM
S70FNR-04	OEM
S70FNR-06	OEM
S70FNR-08	OEM
S70FNR-10	OEM
S70FNR-12	OEM
S-70T	0136
S70T	0050
S72	0015
S73	0015
S75	0015
S77	0015
S78-06H02	OEM
S79	0015
S80-12	3565
S-80T	0136
S81	0015
S82	0015
S83	0015
S84	0015
S-85	0161
S85	0015
S-86	0161
S86	0015
S-87TB	0136
S87TB	0050
S-88TB	0136
S88TB	0050
S90SC4M	OEM
S91	0015
S91-A	0015
S91-H	0015
S91A	0110
S91B	0953
S91H	0015
S92	0015
S92-A	0015
S92-H	0015
S92A	0110
S92H	0015
S93	0015
S93A	0015
S93H	0015
S93SE133	0103
S93SE140	1514
S93SE165	0103
S93.20.709	0289
S93.20.714	0289
S94	0015
S95	0015
S-99	0023
S99	0133
S-99(DIODE)	0133
S100	0071
S-100-01	6635
S-100-01FR	OEM
S-100-02	6635
S-100-02FR	OEM
S-100-03	6635
S-100-04	6635
S-100-04FR	OEM
S-100-06	6076
S-100-08	6636
S-100-10	6636
S100-12	3631
S100-28	2545
S101	0015
S102	0015
S102P	OEM
S-102V	0015
S103	0015
S104	0015
S104P	OEM
S105	0023
S105A	0110
S106	0015
S106-05	1259
S106-1	0447
S106-2	0940
S106-4	0095
S106A	0447
S106A1	0447
S106A2	0447
S106B1	0940
S106B2	0940
S106C	0095
S106C1	0095
S106C2	0095
S106D	0095
S106D1	0095
S106D2	0095
S106E	0308
S106E1	0308
S106E2	OEM
S106F	1259
S106F1	1259
S106F2	1259
S106M	0308
S106M1	0308
S106M2	OEM
S106N1	OEM
S106N2	OEM
S106P1	OEM
S106P2	OEM
S106Q	1396
S106Y	1396
S106Y1	1396
S106Y2	1396
S107	0015
S107-05	1259
S107-1	0447
S107-2	0940
S107-4	0095
S107A1	0447
S107A2	0447
S107B1	5240
S107B2	0940
S107C1	0095
S107C2	0095
S107D1	0095
S107D2	0095
S107E1	0308
S107E2	OEM
S107F1	1259
S107F2	1259
S107M1	0308
S107M2	OEM
S107N1	OEM
S107N2	OEM
S107P1	OEM
S107P2	OEM
S107Y1	1396
S107Y2	1396
S108	0015
S109	OEM
S110	0927
S110A	0922
S110F3	0469
S115	0015
S118	0124
S120	OEM
S120-1/4	OEM
S120F4	0469
S125	OEM
S129	0015
S130	OEM
S130-138	0007
S130-251	0007
S133-1	0016
S135	OEM
S154	0102
S160	0133
S169N	0016
S182SDU03	OEM
S184SHU09	OEM
S191G	0015
S200	0015
S-200-02	OEM
S200-02F	OEM
S-200-02FR	OEM
S-200-03	3138
S200-03F	OEM
S-200-04	3138
S200-04F	OEM
S-200-04FR	OEM
S-200-06	3140
S200-06F	OEM
S-200-08	3142
S200-08F	OEM
S-200-10	3142
S200-10F	OEM
S201	0015
S201MN2B-02	OEM
S201MN2B-04	OEM
S201MN2B-06	OEM
S201MN2B-08	OEM
S201MN2B-10	OEM
S202	0015
S203	0015
S204	0015
S205	0015
S206	1064
S206S	0154
S208	0071
S210	1064
S210F3	0469
S210S	0154
S215	1064
S215S	0154
S217	0015
S218	0015
S219	0015
S220	0015
S220F4	0469
S221	0015
S222	0015
S223	0015
S224	0015
S225	1064
S225S	3150
S227	1091
S227(JAPAN)	1091
S229	0242
S230	0015
S231	OEM
S232	0015
S233	0015
S234	0015
S235	0015
S236	0242
S238	0015
S239	0015
S240	0015
S241	0015
S243	0015
S250	0015
S251	0015
S252	0015
S253	0015
S254	0015
S255	0015
S256	0015
S257	0071
S258	0071
S260	0071
S262	0015
S-277B	OEM
S298	0133
S300	OEM
S305	0103
S305A	0103
S306A	0178
S310F3	0722
S320F4	0722
S321	0133
S322	0133
S324	0133
S325	0133
S334	0133
S337	0133
S353	0103
S354	0178
S364S	2871
S380	0379
S398	0133
S400	0133
S401	0133
S402	0133
S403	0133
S406	1317
S406S	0147
S409F	0086
S410	1317
S410F3	0722
S410S	0147
S411	0097
S413	0109
S415	1317
S415S	0147
S417	0131
S420	0097
S420F4	0722
S421	0097
S423	0109
S425	1317
S425S	3152
S427	0131
S428	0015
S431	0015
S435	OEM
S481	1090
S-500	0071
S500	0133
S500-2	OEM
S500-3	OEM
S500-4	OEM
S500-5	OEM
S500-6	OEM
S500-7	OEM
S500-8	OEM
S500-10	OEM
S-500B	0015
S500B	0540
S-500C	0015
S500C	0540
S501	0133
S502	0133
S502-23X1	0190
S502-23XL	0190
S502J	OEM
S504	0133
S505	0133
S506	0133
S507	0133
S508	0133
S509	0133
S510F3	0725
S518T	0065
S520	0037
S520F4	0725
S525	OEM
S535	OEM
S541	OEM
S542	OEM
S550	OEM
S551	OEM
S552	OEM
S575A	OEM
S600	0122
S601T	OEM
S608L	OEM
S610	OEM
S610F3	0725
S611F	OEM
S612F	OEM
S613F	OEM
S614F	OEM
S620	OEM
S620F4	0725
S622F	OEM
S625	OEM
S630T	0065
S635	OEM
S637T	3345
S641	1626
S642	OEM
S651	OEM
S652	OEM
S653	OEM
S671T	0276
S673	OEM
S673T	4479
S684	0050
S685	0004
S686	0004
S687	0004
S708	1303
S715	0560
S750	0071
S750C	0071
S-789	1303
S-798	0015
S800	0131
S801	0560
S-810	0015
S812SDU01	OEM
S812SDU02	OEM
S812SDU04	OEM
S812SDU05	OEM
S812SDU06	OEM
S812SDU07	OEM
S812SDU08	OEM
S812SDU09	OEM
S812SDU10	OEM
S812SDU11	OEM
S812SDU12	OEM
S812SDU13	OEM
S812SDU14	OEM
S812SDU15	OEM
S812SDU16	OEM
S812SDU17	OEM
S812SDU18	OEM
S812SDU19	OEM
S812SDU20	OEM
S812SDU21	OEM
S812SDU22	OEM
S812SDU23	OEM
S812SHU01	OEM
S812SHU02	OEM
S812SHU03	OEM
S812SHU04	OEM
S812SHU05	OEM
S812SHU06	OEM
S812SHU07	OEM
S812SHU08	OEM
S812SHU09	OEM
S812SHU10	OEM
S812SHU11	OEM
S812SHU12	OEM
S812SHU13	OEM
S812SHU14	OEM
S812SHU15	OEM
S812SHU16	OEM
S812SHU17	OEM
S812SHU18	OEM
S812SHU19	OEM
S812SHU20	OEM
S812SHU21	OEM
S812SHU22	OEM
S812SHU23	OEM
S813SDU01	OEM
S813SDU02	OEM
S813SDU03	OEM
S813SDU04	OEM
S813SDU05	OEM
S813SDU06	OEM
S813SDU07	OEM
S813SDU08	OEM
S813SDU09	OEM
S813SDU10	OEM
S813SDU11	OEM
S813SDU12	OEM
S813SDU14	OEM
S813SDU15	OEM
S813SDU16	OEM
S813SDU17	OEM
S813SDU18	OEM
S813SDU19	OEM
S813SDU20	OEM
S813SDU21	OEM
S813SDU22	OEM
S813SHU01	OEM
S813SHU02	OEM
S813SHU03	OEM
S813SHU04	OEM
S813SHU05	OEM
S813SHU06	OEM
S813SHU07	OEM
S813SHU08	OEM
S813SHU09	OEM
S813SHU10	OEM
S813SHU11	OEM
S813SHU12	OEM
S813SHU13	OEM
S813SHU14	OEM
S813SHU15	OEM
S813SHU16	OEM
S813SHU17	OEM
S813SHU18	OEM
S813SHU19	OEM
S813SHU20	OEM
S813SHU21	OEM
S813SHU22	OEM
S814SDU01	OEM
S814SDU02	OEM
S814SDU03	OEM
S814SDU04	OEM
S814SDU05	OEM
S814SDU06	OEM
S814SDU07	OEM
S814SDU08	OEM
S814SDU09	OEM
S814SDU10	OEM
S814SDU11	OEM
S814SDU12	OEM
S814SDU13	OEM
S814SDU14	OEM
S814SDU15	OEM
S814SDU16	OEM
S814SDU17	OEM
S814SDU18	OEM
S814SDU19	OEM
S814SDU20	OEM
S814SDU21	OEM
S814SDU22	OEM
S814SDU23	OEM
S814SDU24	OEM
S814SDU25	OEM
S814SDU26	OEM
S814SDU27	OEM

DEVICE TYPE	REPL CODE	DEVICE TYPE	REPL CODE	DEVICE TYPE	REPL CODE	DEVICE TYPE	REPL CODE	DEVICE TYPE	REPL CODE	DEVICE TYPE	REPL CODE	DEVICE TYPE	REPL CODE
S814SDU28	OEM	S904SBU16	OEM	S953SAU05	OEM	S954SBU02	OEM	S1000	0612	S1070W	OEM	S1104SFU11	OEM
S814SDU29	OEM	S904SBU17	OEM	S953SAU06	OEM	S954SBU03	OEM	S1000-1	0071	S1074(R)	0016	S1104SFU12	OEM
S814SDU30	OEM	S904SBU18	OEM	S953SAU07	OEM	S954SBU04	OEM	S1000KS2	0901	S1074(X)	0940	S1104SFU13	OEM
S814SHU01	OEM	S904SBU19	OEM	S953SAU08	OEM	S954SBU05	OEM	S1001K	OEM	S1076	0144	S1104SFU14	OEM
S814SHU02	OEM	S904SBU20	OEM	S953SAU09	OEM	S954SBU06	OEM	S1001L	1004	S1078	0144	S1104SFU15	OEM
S814SHU03	OEM	S904SBU21	OEM	S953SAU10	OEM	S954SBU07	OEM	S1001LS2	1044	S1079	0144	S1104SFU16	OEM
S814SHU04	OEM	S904SBU22	OEM	S953SAU11	OEM	S954SBU08	OEM	S1001LS3	5025	S1080	0817	S1104SFU17	OEM
S814SHU05	OEM	S904SBU23	OEM	S953SAU12	OEM	S954SBU09	OEM	S1001M	0179	S1090	0922	S1104SFU18	OEM
S814SHU06	OEM	S904SBU24	OEM	S953SAU13	OEM	S954SBU10	OEM	S1001MS2	0344	S1101SCU11	OEM	S1104SFU19	OEM
S814SHU07	OEM	S904SBU25	OEM	S953SAU14	OEM	S954SBU11	OEM	S1001MS3	0179	S1101SFU01	OEM	S1104SFU20	OEM
S814SHU08	OEM	S904SBU26	OEM	S953SAU15	OEM	S954SBU12	OEM	S1003L	1004	S1101SFU02	OEM	S1104SFU21	OEM
S814SHU10	OEM	S904SBU27	OEM	S953SAU16	OEM	S954SBU13	OEM	S1003LS1	1044	S1101SFU03	OEM	S1104SFU22	OEM
S814SHU11	OEM	S904SBU28	OEM	S953SAU17	OEM	S954SBU14	OEM	S1003LS2	1044	S1101SFU04	OEM	S1104SFU23	OEM
S814SHU12	OEM	S904SBU29	OEM	S953SAU18	OEM	S954SBU15	OEM	S1003LS3	1044	S1101SFU05	OEM	S1104SFU24	OEM
S814SHU13	OEM	S904SCU10	OEM	S953SAU19	OEM	S954SBU16	OEM	S1003M	0179	S1101SFU06	OEM	S1104SFU25	OEM
S814SHU14	OEM	S904SCU11	OEM	S953SAU20	OEM	S954SBU17	OEM	S1003MS2	0447	S1101SFU07	OEM	S1104SFU27	OEM
S814SHU15	OEM	S904SCU12	OEM	S953SAU21	OEM	S954SBU18	OEM	S1003MS3	0447	S1101SFU08	OEM	S1104SFU28	OEM
S814SHU16	OEM	S904SCU13	OEM	S953SAU22	OEM	S954SBU19	OEM	S1003RS1	OEM	S1101SFU09	OEM	S1104SFU29	OEM
S814SHU17	OEM	S904SCU14	OEM	S953SAU23	OEM	S954SBU20	OEM	S1003RS2	0447	S1101SFU10	OEM	S1107SAU01	OEM
S814SHU18	OEM	S904SCU15	OEM	S953SBU01	OEM	S954SBU21	OEM	S1003RS3	0447	S1101SFU11	OEM	S1107SAU02	OEM
S814SHU19	OEM	S904SCU16	OEM	S953SBU02	OEM	S954SBU22	OEM	S1004F1	0447	S1101SFU12	OEM	S1107SAU03	OEM
S814SHU20	OEM	S904SCU17	OEM	S953SBU03	OEM	S954SBU23	OEM	S1005BH	0505	S1104SAU01	OEM	S1107SAU04	OEM
S814SHU21	OEM	S904SCU18	OEM	S953SBU04	OEM	S954SBU24	OEM	S1005DH	0710	S1104SAU02	OEM	S1107SAU05	OEM
S814SHU22	OEM	S904SCU19	OEM	S953SBU05	OEM	S954SBU25	OEM	S1005MH	0863	S1104SAU03	OEM	S1107SAU06	OEM
S814SHU23	OEM	S904SCU20	OEM	S953SBU06	OEM	S954SBU26	OEM	S1005NH	0327	S1104SAU04	OEM	S1107SAU07	OEM
S814SHU24	OEM	S904SCU21	OEM	S953SBU07	OEM	S954SBU27	OEM	S1006B	3499	S1104SAU05	OEM	S1107SAU08	OEM
S814SHU25	OEM	S904SCU22	OEM	S953SBU08	OEM	S954SBU28	OEM	S1006G	1651	S1104SAU06	OEM	S1107SAU09	OEM
S814SHU26	OEM	S904SCU23	OEM	S953SBU09	OEM	S954SBU29	OEM	S1006H	2449	S1104SAU07	OEM	S1107SAU10	OEM
S814SHU27	OEM	S904SCU24	OEM	S953SBU10	OEM	S954SCU01	OEM	S1006L	2089	S1104SAU08	OEM	S1107SAU11	OEM
S814SHU28	OEM	S904SCU25	OEM	S953SBU11	OEM	S954SCU02	OEM	S1006LS2	1044	S1104SAU09	OEM	S1107SAU12	OEM
S814SHU29	OEM	S904SCU26	OEM	S953SBU12	OEM	S954SCU03	OEM	S1006LS3	1044	S1104SAU10	OEM	S1107SAU13	OEM
S814SHU30	OEM	S904SCU27	OEM	S953SBU13	OEM	S954SCU04	OEM	S1006RS2	0447	S1104SAU11	OEM	S1107SAU14	OEM
S842G	OEM	S904SCU28	OEM	S953SBU14	OEM	S954SCU05	OEM	S1006RS3	3854	S1104SAU12	OEM	S1107SAU15	OEM
S855	0474	S904SCU29	OEM	S953SBU15	OEM	S954SCU06	OEM	S1007BH	1004	S1104SAU13	OEM	S1107SAU16	OEM
S-870R	OEM	S904SFU10	OEM	S953SBU16	OEM	S954SCU07	OEM	S1007DH	1034	S1104SAU14	OEM	S1107SAU17	OEM
S870R	OEM	S904SFU11	OEM	S953SBU17	OEM	S954SCU08	OEM	S1007MH	1148	S1104SAU15	OEM	S1107SAU18	OEM
S-870S	OEM	S904SFU12	OEM	S953SBU18	OEM	S954SCU09	OEM	S1007NH	0327	S1104SAU16	OEM	S1107SAU19	OEM
S870S	OEM	S904SFU13	OEM	S953SBU19	OEM	S954SCU10	OEM	S1008B	3499	S1104SAU17	OEM	S1107SAU20	OEM
S872A	OEM	S904SFU14	OEM	S953SBU20	OEM	S954SCU11	OEM	S1008G	1651	S1104SAU18	OEM	S1107SAU21	OEM
S-873TB	0050	S904SFU15	OEM	S953SBU21	OEM	S954SCU12	OEM	S1008H	2449	S1104SAU19	OEM	S1107SAU22	OEM
S873TB	0178	S904SFU16	OEM	S953SBU22	OEM	S954SCU13	OEM	S1008L	2089	S1104SAU20	OEM	S1107SAU23	OEM
S-874TB	0050	S904SFU17	OEM	S953SBU23	OEM	S954SCU14	OEM	S1008LS2	1044	S1104SAU21	OEM	S1107SAU24	OEM
S-880	0774	S904SFU18	OEM	S953SCU01	OEM	S954SCU15	OEM	S1008LS3	1044	S1104SAU22	OEM	S1107SAU25	OEM
S903SAU10	OEM	S904SFU19	OEM	S953SCU02	OEM	S954SCU16	OEM	S1008RS2	OEM	S1104SAU23	OEM	S1107SAU26	OEM
S903SAU11	OEM	S904SFU20	OEM	S953SCU03	OEM	S954SCU17	OEM	S1008RS3	OEM	S1104SAU24	OEM	S1107SAU27	OEM
S903SAU12	OEM	S904SFU21	OEM	S953SCU04	OEM	S954SCU18	OEM	S1009	0144	S1104SAU25	OEM	S1107SAU28	OEM
S903SAU13	OEM	S904SFU22	OEM	S953SCU05	OEM	S954SCU19	OEM	S1010	0953	S1104SAU26	OEM	S1107SAU29	OEM
S903SAU14	OEM	S904SFU23	OEM	S953SCU06	OEM	S954SCU20	OEM	S1010B	3499	S1104SAU27	OEM	S1107SAU30	OEM
S903SAU15	OEM	S904SFU24	OEM	S953SCU07	OEM	S954SCU21	OEM	S1010BH	0505	S1104SAU28	OEM	S1107SAU31	OEM
S903SAU16	OEM	S904SFU25	OEM	S953SCU08	OEM	S954SCU22	OEM	S1010DH	0710	S1104SAU29	OEM	S1107SAU32	OEM
S903SAU17	OEM	S904SFU26	OEM	S953SCU09	OEM	S954SCU23	OEM	S1010G	1651	S1104SBU01	OEM	S1107SAU33	OEM
S903SAU18	OEM	S904SFU27	OEM	S953SCU10	OEM	S954SCU24	OEM	S1010H	2449	S1104SBU02	OEM	S1107SAU34	OEM
S903SAU19	OEM	S904SFU28	OEM	S953SCU11	OEM	S954SCU25	OEM	S1010L	1004	S1104SBU03	OEM	S1107SBU01	OEM
S903SAU20	OEM	S904SFU29	OEM	S953SCU12	OEM	S954SCU26	OEM	S1010LS2	1044	S1104SBU04	OEM	S1107SBU02	OEM
S903SAU21	OEM	S913	0774	S953SCU13	OEM	S954SCU27	OEM	S1010LS3	1044	S1104SBU05	OEM	S1107SBU03	OEM
S903SAU22	OEM	S926	0774	S953SCU14	OEM	S954SCU28	OEM	S1010MH	0863	S1104SBU06	OEM	S1107SBU04	OEM
S903SAU23	OEM	S939	OEM	S953SCU15	OEM	S954SCU29	OEM	S1010NH	0327	S1104SBU07	OEM	S1107SBU05	OEM
S903SBU10	OEM	S951SAU01	OEM	S953SCU16	OEM	S954SFU10	OEM	S1010RS2	OEM	S1104SBU08	OEM	S1107SBU06	OEM
S903SBU11	OEM	S951SAU02	OEM	S953SCU17	OEM	S954SFU11	OEM	S1010RS3	OEM	S1104SBU09	OEM	S1107SBU07	OEM
S903SBU12	OEM	S951SAU03	OEM	S953SCU18	OEM	S954SFU12	OEM	S1016	0016	S1104SBU10	OEM	S1107SBU08	OEM
S903SBU13	OEM	S951SAU04	OEM	S953SCU19	OEM	S954SFU13	OEM	S1016B	1848	S1104SBU11	OEM	S1107SBU09	OEM
S903SBU14	OEM	S951SAU05	OEM	S953SCU20	OEM	S954SFU14	OEM	S1016G	1650	S1104SBU12	OEM	S1107SBU10	OEM
S903SBU15	OEM	S951SAU06	OEM	S953SCU21	OEM	S954SFU15	OEM	S1016H	5014	S1104SBU13	OEM	S1107SBU11	OEM
S903SBU16	OEM	S951SAU07	OEM	S953SCU22	OEM	S954SFU16	OEM	S1016L	OEM	S1104SBU14	OEM	S1107SBU12	OEM
S903SBU17	OEM	S951SAU08	OEM	S953SCU23	OEM	S954SFU17	OEM	S-1019	0127	S1104SBU15	OEM	S1107SBU13	OEM
S903SBU18	OEM	S951SAU09	OEM	S953SFU10	OEM	S954SFU18	OEM	S1019	0144	S1104SBU16	OEM	S1107SBU14	OEM
S903SBU19	OEM	S951SAU10	OEM	S953SFU11	OEM	S954SFU19	OEM	S-1019(UHF)	0127	S1104SBU17	OEM	S1107SBU15	OEM
S903SBU20	OEM	S951SAU11	OEM	S953SFU12	OEM	S954SFU20	OEM	S1020	0242	S1104SBU18	OEM	S1107SBU16	OEM
S903SBU21	OEM	S951SAU12	OEM	S953SFU13	OEM	S954SFU21	OEM	S1020L	0397	S1104SBU19	OEM	S1107SBU17	OEM
S903SBU22	OEM	S951SBU01	OEM	S953SFU14	OEM	S954SFU22	OEM	S1025B	OEM	S1104SBU20	OEM	S1107SBU18	OEM
S903SBU23	OEM	S951SBU02	OEM	S953SFU15	OEM	S954SFU23	OEM	S1025C	0226	S1104SBU21	OEM	S1107SBU19	OEM
S903SCU10	OEM	S951SBU03	OEM	S953SFU16	OEM	S954SFU24	OEM	S1025D	1848	S1104SBU22	OEM	S1107SBU20	OEM
S903SCU11	OEM	S951SBU04	OEM	S953SFU17	OEM	S954SFU25	OEM	S1025G	1650	S1104SBU23	OEM	S1107SBU21	OEM
S903SCU12	OEM	S951SBU05	OEM	S953SFU18	OEM	S954SFU26	OEM	S1025H	5029	S1104SBU24	OEM	S1107SBU22	OEM
S903SCU13	OEM	S951SBU06	OEM	S953SFU19	OEM	S954SFU27	OEM	S1025L	2611	S1104SBU25	OEM	S1107SBU23	OEM
S903SCU14	OEM	S951SBU07	OEM	S953SFU20	OEM	S954SFU28	OEM	S1025R	5031	S1104SBU26	OEM	S1107SBU24	OEM
S903SCU15	OEM	S951SBU08	OEM	S953SFU21	OEM	S954SFU29	OEM	S1025V	OEM	S1104SBU27	OEM	S1107SBU25	OEM
S903SCU17	OEM	S951SBU09	OEM	S953SFU22	OEM	S962SJU25	OEM	S1025Y	OEM	S1104SBU28	OEM	S1107SBU26	OEM
S903SCU18	OEM	S951SBU10	OEM	S953SFU23	OEM	S962SJU26	OEM	S1025Z	OEM	S1104SBU29	OEM	S1107SBU27	OEM
S903SCU19	OEM	S951SBU11	OEM	S954SAU01	OEM	S962SJU27	OEM	S1030	1747	S1104SCU01	OEM	S1107SBU28	OEM
S903SCU20	OEM	S951SBU12	OEM	S954SAU02	OEM	S962SJU28	OEM	S1035C	0226	S1104SCU02	OEM	S1107SBU29	OEM
S903SCU21	OEM	S951SCU01	OEM	S954SAU03	OEM	S962SJU29	OEM	S1035D	1848	S1104SCU03	OEM	S1107SBU30	OEM
S903SCU22	OEM	S951SCU02	OEM	S954SAU04	OEM	S962SJU30	OEM	S1035G	1650	S1104SCU04	OEM	S1107SBU31	OEM
S903SCU23	OEM	S951SCU03	OEM	S954SAU05	OEM	S962SJU31	OEM	S1035H	0712	S1104SCU05	OEM	S1107SBU32	OEM
S904SAU10	OEM	S951SCU04	OEM	S954SAU06	OEM	S962SJU32	OEM	S1035J	OEM	S1104SCU06	OEM	S1107SBU33	OEM
S904SAU11	OEM	S951SCU05	OEM	S954SAU07	OEM	S962SJU33	OEM	S1035K	2690	S1104SCU07	OEM	S1107SBU34	OEM
S904SAU12	OEM	S951SCU06	OEM	S954SAU08	OEM	S962SJU34	OEM	S1035W	2696	S1104SCU08	OEM	S1107SCU01	OEM
S904SAU13	OEM	S951SCU07	OEM	S954SAU09	OEM	S962SJU35	OEM	S1037	0144	S1104SCU09	OEM	S1107SCU02	OEM
S904SAU14	OEM	S951SCU08	OEM	S954SAU10	OEM	S962SJU36	OEM	S1040	0540	S1104SCU10	OEM	S1107SCU03	OEM
S904SAU15	OEM	S951SCU09	OEM	S954SAU11	OEM	S962SJU37	OEM	S1041	0127	S1104SCU11	OEM	S1107SCU04	OEM
S904SAU16	OEM	S951SCU10	OEM	S954SAU12	OEM	S962SJU38	OEM	S1041-16GN	0127	S1104SCU12	OEM	S1107SCU05	OEM
S904SAU17	OEM	S951SCU11	OEM	S954SAU13	OEM	S962SJU39	OEM	S1044	0144	S1104SCU13	OEM	S1107SCU06	OEM
S904SAU18	OEM	S951SCU12	OEM	S954SAU14	OEM	S962SJU40	OEM	S1050	1771	S1104SCU14	OEM	S1107SCU07	OEM
S904SAU19	OEM	S951SFU01	OEM	S954SAU15	OEM	S962SJU41	OEM	S1055M	2696	S1104SCU15	OEM	S1107SCU08	OEM
S904SAU20	OEM	S951SFU02	OEM	S954SAU16	OEM	S962SJU42	OEM	S1055W	2696	S1104SCU16	OEM	S1107SCU09	OEM
S904SAU21	OEM	S951SFU03	OEM	S954SAU17	OEM	S962SJU43	OEM	S1058	0144	S1104SCU17	OEM	S1107SCU10	OEM
S904SAU22	OEM	S951SFU04	OEM	S954SAU18	OEM	S962SJU44	OEM	S1059	0144	S1104SCU18	OEM	S1107SCU11	OEM
S904SAU23	OEM	S951SFU05	OEM	S954SAU19	OEM	S962SJU45	OEM	S1060	0965	S1104SCU19	OEM	S1107SCU12	OEM
S904SAU25	OEM	S951SFU06	OEM	S954SAU20	OEM	S962SJU46	OEM	S1061	0016	S1104SCU20	OEM	S1107SCU13	OEM
S904SAU26	OEM	S951SFU07	OEM	S954SAU21	OEM	S962SJU47	OEM	S1062	0144	S1104SCU21	OEM	S1107SCU14	OEM
S904SAU27	OEM	S951SFU08	OEM	S954SAU22	OEM	S962SJU48	OEM	S1063	OEM	S1104SCU22	OEM	S1107SCU15	OEM
S904SAU28	OEM	S951SFU09	OEM	S954SAU23	OEM	S962SJU49	OEM	S1064	OEM	S1104SCU23	OEM	S1107SCU16	OEM
S904SAU29	OEM	S951SFU10	OEM	S954SAU24	OEM	S962SJU50	OEM	S1065	0016	S1104SCU24	OEM	S1107SCU17	OEM
S904SBU10	OEM	S951SFU11	OEM	S954SAU25	OEM	S962SJU51	OEM	S1065J	OEM	S1104SCU25	OEM	S1107SCU18	OEM
S904SBU11	OEM	S951SFU12	OEM	S954SAU26	OEM	S962SJU52	OEM	S1065K	OEM	S1104SCU26	OEM	S1107SCU19	OEM
S904SBU12	OEM	S953SAU01	OEM	S954SAU27	OEM	S962SJU53	OEM	S1066	0016	S1104SCU27	OEM	S1107SCU20	OEM
S904SBU13	OEM	S953SAU02	OEM	S954SAU28	OEM	S962SJU54	OEM	S1068	0016	S1104SCU28	OEM	S1107SCU21	OEM
S904SBU14	OEM	S953SAU03	OEM	S954SAU29	OEM	S962SJU55	OEM	S1069	0016	S1104SCU29	OEM	S1107SCU22	OEM
S904SBU15	OEM	S953SAU04	OEM	S954SBU01	OEM	S962SJU56	OEM	S1070	0817	S1104SFU10	OEM	S1107SCU23	OEM

If replacement code is OEM, contact original manufacturer for replacement.

DEVICE TYPE	REPL CODE
S1107SCU24	OEM
S1107SCU25	OEM
S1107SCU26	OEM
S1107SCU27	OEM
S1107SCU28	OEM
S1107SCU29	OEM
S1107SCU30	OEM
S1107SCU31	OEM
S1107SCU32	OEM
S1107SCU33	OEM
S1107SCU34	OEM
S1107SJU24	OEM
S1107SJU25	OEM
S1107SJU26	OEM
S1107SJU27	OEM
S1107SJU28	OEM
S1107SJU29	OEM
S1107SJU30	OEM
S1107SJU31	OEM
S1107SJU32	OEM
S1107SJU33	OEM
S1107SJU34	OEM
S1107SJU35	OEM
S1107SJU36	OEM
S1109SJU30	OEM
S1109SJU31	OEM
S1109SJU32	OEM
S1109SJU33	OEM
S1109SJU34	OEM
S1109SJU35	OEM
S1109SJU36	OEM
S1109SJU37	OEM
S1109SJU38	OEM
S1109SJU40	OEM
S1109SJU41	OEM
S1109SJU42	OEM
S1109SJU43	OEM
S1109SJU44	OEM
S1122	0127
S1126	0127
S1128	0016
S1142	0144
S1143	0016
S1153	0144
S1158	OEM
S1200-1	1303
S1205BH	0397
S1205DH	0611
S1205MH	0952
S1205NH	1512
S1207BH	0397
S1207DH	0611
S1207MH	0952
S1207NH	1512
S-1210	0139
S1210	0953
S1210-3	OEM
S1210BH	0397
S1210DH	0611
S1210MH	0952
S1210NH	1512
S1211N	0325
S1212N	0325
S1213N	0325
S1214N	0325
S1215-3	OEM
S1215N	0325
S1216N	0325
S1220	0242
S1220-3	OEM
S1221	0016
S1221A	0016
S1221N	0325
S1222N	0325
S1223N	0325
S1224N	0325
S1225N	0325
S1226	0016
S1226N	0325
S1227	0144
S1230	1747
S1231N	0325
S1232N	0325
S1233N	0325
S1234N	0325
S1235N	0325
S1236N	0325
S1240	0540
S1241	0016
S1242	0016
S1243	0016
S1245	0016
S1250	1771
S1260	0965
S1270	0817
S1272	0016
S-1276	0144
S1276	0144
S1280	0817
S1286	0007
S1290	0922
S-1296	0144
S1296	0144
S1298	0285
S1307	0016
S1308	0127
S1309	0016
S-1313	0127
S1313	0144

DEVICE TYPE	REPL CODE
S-1316	0127
S1316	0144
S1317	0144
S-1318	0127
S1318	0144
S1331	0016
S1331N	0016
S1331W	0016
S1332	0136
S1341P	0861
S1342P	0861
S1343P	0861
S1348	0004
S1349	0004
S1350	0037
S1351	OEM
S1351P	0861
S1352P	0861
S1353P	0861
S1360	0144
S1361	0144
S1362	0144
S1363	0016
S1364	0016
S1366	0233
S1367	0037
S1368	0086
S1369	0016
S1373	0016
S1374	0016
S1382	OEM
S1384	0123
S1400-1	0102
S1403	0016
S1405	0016
S1407	0233
S1408	0144
S1409	0144
S1419	0016
S1420	0016
S1421	0086
S1428	0133
S1429-3	0016
S1430	0126
S1431	0126
S1432	0016
S1443	0016
S-1453	0038
S1453	0016
S1475	0079
S1476	0016
S1477	0037
S1487	0016
S1502	0016
S1510	0016
S1512	0016
S1514	0086
S1516	0086
S1517	0086
S1520	0126
S1523	0086
S1525	0086
S1526	0016
S1527	0016
S1529	0016
S1530	0016
S1533	0016
S1556-2	0160
S1559	0016
S1568	0016
S1570	0016
S1600	0015
S1610BH	0397
S1610DH	0611
S1610MH	0952
S1610NH	1512
S1612BH	0397
S1612DH	0611
S1612MH	0952
S1612NH	1512
S1619	0016
S1620	0016
S1629	0016
S1636	0144
S1639	0004
S1640	0050
S1642	0086
S1644	0086
S1671	0086
S1672	0164
S1674	0144
S1674A	0144
S1682	0144
S1683	0086
S1685	0103
S1689	0086
S1691	0103
S1692	0103
S1697	0016
S1698	0126
S1715	OEM
S1761	0016
S1761A	0016
S1761B	0016
S1761C	0016
S1762	0086
S1764	0016
S1765	0016
S1766	0016

DEVICE TYPE	REPL CODE
S1768	0016
S1769	0233
S1770	0016
S1772	0016
S1773	0086
S1777	0086
S1784	0016
S1785	0016
S1788	0079
S1801-02	0015
S1810-3	OEM
S1815-3	OEM
S1835	0016
S1854	OEM
S1854(FA2)	OEM
S1854-2	OEM
S1854FA	OEM
S1854FA-1	OEM
S1854FA-2	OEM
S1854FA2	OEM
S1854LBM-4	OEM
S1854M4	OEM
S1863	0126
S1864	0086
S1865	0103
S1871	0016
S1874	0086
S1878	0079
S1880	OEM
S1889	0037
S1891	0016
S1891A	0016
S1891B	0016
S1897	0127
S1905	0103
S1905A	0103
S1907	0103
S1955	0016
S1970	OEM
S1983	0126
S1993	0016
S1998	2295
S1998A	2295
S2000A3	OEM
S2000AF	2490
S2000KS2	0058
S2001K	OEM
S2001L	1004
S2001LS1	1044
S2001LS2	1044
S2001LS3	5025
S2001M	0179
S2001MS1	OEM
S2001MS2	0058
S2001MS3	0179
S2002	0007
S2002SPU23	OEM
S2003-1	0103
S2003L	1004
S2003LS1	1044
S2003LS2	1044
S2003LS3	5025
S2003MS2	0940
S2003MS3	0940
S2003RS1	OEM
S2003RS2	0940
S2003RS3	3620
S2004F1	0940
S2004SPU10	OEM
S2004SPU11	OEM
S2004SPU12	OEM
S2004SPU13	OEM
S2004SPU14	OEM
S2004SPU15	OEM
S2004SPU16	OEM
S2004SPU17	OEM
S2004SPU18	OEM
S2004SPU19	OEM
S2004SPU20	OEM
S2004SPU21	OEM
S2004SPU22	OEM
S2004SPU24	OEM
S2005	1600
S2006B	1848
S2006G	1651
S2006H	0435
S2006L	0505
S2006LS2	1044
S2006LS3	1044
S2006RS2	OEM
S2006RS3	OEM
S2008B	1848
S2008G	1651
S2008H	0435
S2008L	0505
S2008LS2	1044
S2008LS3	1044
S2008RS2	OEM
S2008RS3	OEM
S2010	0871
S2010B	1848
S2010G	1651
S2010H	0435
S2010L	1004
S2010LS2	1044
S2010LS3	1044
S2010RS2	OEM
S2010RS3	OEM
S2012BH	0397

DEVICE TYPE	REPL CODE
S2012DH	0611
S2012MH	0952
S2012NH	1512
S2014BH	0397
S2014DH	0611
S2014MH	0952
S2014NH	1512
S2015	0871
S2015G	1651
S2015H	0745
S2015L	0397
S2016B	OEM
S2016G	1651
S2016H	0745
S2016L	OEM
S2020	0871
S2020L	4576
S2025	0853
S2025B	OEM
S2025C	0226
S2025D	1848
S2025G	1004
S2025H	0762
S2025L	OEM
S2025R	OEM
S2025V	OEM
S2025Y	OEM
S2025Z	OEM
S2030	0853
S2030G	1650
S2030H	0469
S2034	0016
S2035	0853
S2035C	0226
S2035D	1848
S2035G	1650
S2035H	0469
S2035J	2690
S2035K	2690
S2035W	2696
S2038	0111
S2040	0853
S2041	0925
S-2041-8661	0033
S2042	0042
S2043	0016
S2044	0016
S2045	1609
S2050	1609
S2055M	2696
S2055W	2696
S2059	0142
S2060	1609
S2060A	0447
S2060B	0940
S2060C	1222
S2060D	0095
S2060E	0308
S2060F	2095
S2060M	0308
S2060Q	2095
S2060Y	2095
S2061A	0447
S2061B	0940
S2061C	1222
S2061D	0095
S2061E	0308
S2061F	2095
S2061M	0308
S2061Q	2095
S2061Y	2095
S2062A	0447
S2062B	0940
S2062C	0095
S2062D	0095
S2062E	0308
S2062F	1259
S2062M	0308
S2062Q	0717
S2062Y	1396
S2065J	OEM
S2065K	OEM
S2070	1610
S2070W	OEM
S2080	1610
S2085	0143
S2090	1614
S2091	0037
S2104	0595
S2105	1600
S2110	0871
S2114-2CE	OEM
S2114UCB	OEM
S2115	0871
S2117	0126
S2118	0086
S2120	0871
S2121	0016
S2122	0016
S2123	0016
S2124	0016
S2125	0853
S2128	0037
S2129	0037
S2130	0853
S2131	0127
S2132	0127
S2133	0127
S2134	0127

DEVICE TYPE	REPL CODE
S2135	0853
S2139B	0030
S2140	0853
S2140(RECTIFIER)	0853
S2145	1609
S2150	1609
S2150A	OEM
S2152	OEM
S2159	0144
S2160	1609
S2170	1610
S2171	0016
S2172	0016
S2180	1610
S2190	1614
S2200	OEM
S-2200-1135	1907
S2200A	OEM
S2209	0086
S2210	OEM
S2224	0144
S2225	0016
S2241	0103
S2274	0126
S2321	0178
S2350	OEM
S2368	0126
S2369	0086
S2370	0126
S2371	0086
S2392	0103
S2397	0016
S2398C	0126
S2400	0086
S2400A	0086
S2400B	0086
S2400D	OEM
S2400M	OEM
S2401	0086
S2401A	0086
S2401B	0086
S2401C	0086
S2402	0086
S2402A	0086
S2402B	0086
S2402C	0086
S2403B	0103
S2403C	0103
S2427	0086
S2438	0224
S2471	0103
S2486	0178
S2487	0086
S2501DF	1530
S2502DF	1530
S2503DF	0029
S2504DF	0029
S2505DF	0601
S2506DF	0601
S2512BK	1534
S2512DK	1534
S2512MK	1534
S2512NK	OEM
S2514BK	OEM
S2514DK	1534
S2514MK	1534
S2514NK	OEM
S2525	0037
S2526	0086
S2527	0178
S2536A	OEM
S2537A	OEM
S2581	0016
S2582	0016
S2590	0016
S2593	0016
S2600B	0505
S2600D	0710
S2600M	0863
S2610D	0747
S2610M	0764
S-2617	0127
S2617	3390
S2617(UHF)	0127
S2620B	0505
S2620D	0710
S2620M	0863
S2635	0016
S2636	0016
S2645	0037
S2648	0086
S2710B	3624
S2710D	3624
S2710M	3624
S2716	0224
S2718	0224
S2719	0144
S2741	0103
S2771	0126
S2775	0023
S2794	0086
S2800A	2089
S2800B	0505
S2800C	0710
S2800D	0710
S2800E	6950
S2800F	2095
S2800M	6950
S2800S	0327

DEVICE TYPE	REPL CODE
S2935	0016
S-2940I	OEM
S2944	0016
S2984	0016
S2985	0016
S2986	0233
S2988	0037
S2989	0016
S2991	0126
S2992	0086
S2993	0126
S2994	0126
S2995	0126
S2996	0016
S2997	0016
S2998	0016
S2999	0016
S3001	0698
S3001(HEP)	0698
S3002	0233
S3002(HEP)	0233
S3003	3698
S3004	0037
S3004(HEP)	0037
S3004-1715	0636
S3004-1716	0133
S3004-1718	0636
S3005	4198
S3005(HEP)	3576
S3006	1974
S3006(HEP)	1974
S3006C	OEM
S3006D	OEM
S3006E	OEM
S3006S	OEM
S3007	1977
S3007(HEP)	1974
S3008	0493
S3008(HEP)	0418
S3009	OEM
S3009(HEP)	1974
S3010	3329
S3010-3	OEM
S3011	OEM
S3011(HEP)	1224
S3012	0126
S3013	0493
S3014	0493
S3015-3	OEM
S3015A	OEM
S3015B	OEM
S3016	0139
S3016-R	0139
S3016R	0139
S3019	0144
S3019(DIODE)	OEM
S3019A	OEM
S3019B	OEM
S3020	0144
S3020-3	OEM
S3020A	OEM
S3021	0338
S3023	4202
S3024	4203
S3027C	OEM
S3028	4205
S3030G	0143
S3032	0382
S3033	0233
S3034	0233
S3035	0233
S3036	OEM
S3037	OEM
S3038	OEM
S3039	OEM
S3039(HEP)	1198
S3040	1974
S3040(HEP)	1974
S3041	OEM
S3041(HEP)	1974
S3042	0086
S3042(HEP)	0418
S3043	1984
S3043(HEP)	1984
S3044	OEM
S3044(HEP)	1591
S3045	0936
S3045(HEP)	0936
S3046	OEM
S3047	OEM
S3047(HEP)	0566
S3048(HEP)	0338
S3049	1591
S3049(HEP)	0338
S3050	OEM
S3050(HEP)	1367
S3051	OEM
S3051(HEP)	0338
S3052	1198
S3053	OEM
S3054	OEM
S3054(HEP)	0854
S3055	OEM
S3055(HEP)	0854
S3060	OEM
S3060(HEP)	0423
S3060D	OEM
S3060E	OEM
S3060S	OEM

DEVICE TYPE	REPL CODE
S3061	OEM
S3061(HEP)	0423
S3062	OEM
S3062(HEP)	0936
S3066	OEM
S3072	0139
S3072C	0133
S3073	0030
S3087B	2329
S3105	3189
S3106	OEM
S3110	2901
S3115	1124
S3120	1124
S3125	1126
S3130	1126
S3135	0805
S3140	0805
S3145	1195
S3150	1195
S3160	0319
S3170	1132
S3180	1132
S3190	0044
S3205	3189
S3210	2901
S3215	1124
S3220	1124
S3225	1126
S3230	1126
S3235	0805
S3240	0805
S3245	1195
S3250	0319
S3260	1195
S3270	1132
S3280	1132
S3290	0044
S3295G-TPA2	OEM
S3295GTPA2	0023
S3372G	0019
S3386	0126
S3405	3762
S3410	2653
S3415	2657
S3420	2657
S3425	2664
S3430	2664
S3435	2006
S3440	2006
S3445	2682
S3450	2677
S3460	2682
S3470	3900
S3480	3900
S3490	2655
S3505	2813
S3510	2813
S3515	2813
S3515G	OEM
S3520	2813
S3525	2813
S3530	2813
S3535	2813
S3540	2813
S3545	2813
S3550	2813
S3560	2813
S3603G	0143
S3605	4306
S3610	1560
S3615	2840
S3620	2840
S3625	4308
S3630	4308
S3635	2850
S3639	0037
S3640	2850
S3645	2871
S3650	2871
S3655	0037
S3660	2871
S3670	2833
S3680	2833
S3690	2473
S3700B	0239
S3700D	0239
S3700M	0239
S3701M	OEM
S3702S	0061
S3702SF	0061
S3703SF	0239
S3704A	0061
S3704B	0061
S3704D	0061
S3704M	0061
S3704S	0061
S3704SD	0239
S3705	2813
S3705M	0239
S3706E	0061
S3706M	OEM
S3708G	OEM
S3710	2813
S3714A	OEM
S3714B	OEM
S3714D	OEM
S3714M	OEM
S3714S	OEM

If replacement code is OEM, contact original manufacturer for replacement.

DEVICE TYPE	REPL CODE
S3715	2813
S3720	2813
S3725	2813
S3730	2813
S3735	2813
S3740	2813
S3745	2813
S3750	2813
S3760	2813
S3770	2845
S3771	0103
S3780	2845
S3790	2845
S3800D	OEM
S3800E	OEM
S3800EF	OEM
S3800M	OEM
S3800MF	OEM
S3800S	1712
S3800SF	2329
S3838G	OEM
S3838GA	0143
S3900E	4606
S3900MF	4606
S3900S	4606
S3900SF	4606
S3901M	4615
S3901MF	4615
S3901S	4615
S3902DF	OEM
S3903MF	OEM
S3953G	OEM
S3961G	0143
S3967G	OEM
S3969G	OEM
S3972G	OEM
S4000KS2	0407
S4001	0015
S4001K	OEM
S4001L	1034
S4001LS1	1044
S4001LS2	1044
S4001LS3	1044
S4001M	0346
S4001MS1	OEM
S4001MS2	OEM
S4001MS3	0346
S4002	0007
S4003	0015
S4003L	1034
S4003LS1	1044
S4003LS2	1044
S4003LS3	1044
S4003M	0346
S4003MS2	0095
S4003MS3	0095
S4003RS1	OEM
S4003RS2	0095
S4003RS3	0095
S4004F1	0095
S4006B	1855
S4006G	1584
S4006H	1487
S4006J	OEM
S4006L	0710
S4006LS2	1044
S4006LS3	1044
S4006RS2	OEM
S4006RS3	OEM
S4008B	1855
S4008F1	OEM
S4008G	1584
S4008H	1487
S4008L	0710
S4008LS2	1044
S4008LS3	1044
S4008RS2	OEM
S4008RS3	OEM
S4010	OEM
S4010-3	OEM
S4010B	1855
S4010G	1584
S4010H	1487
S4010L	1034
S4010RS2	OEM
S4010RS3	OEM
S4012BK	OEM
S4012DK	1534
S4012MK	1534
S4012NK	OEM
S4014BK	1534
S4014DK	1534
S4014MK	1534
S4014NK	OEM
S4015-3	OEM
S4015L	0611
S4016B	1855
S4016G	OEM
S4016H	0740
S4016L	OEM
S4020-3	OEM
S4025B	1855
S4025C	0226
S4025D	1855
S4025G	2647
S4025H	0740
S4025L	2611
S4025R	4053
S4025V	OEM
S4025Y	OEM
S4025Z	OEM
S4035C	0226
S4035D	1855
S4035G	2647
S4035H	0722
S4035J	2690
S4035K	2690
S4035W	2696
S4055M	5078
S4055W	2696
S4065J	OEM
S4065K	OEM
S4070W	OEM
S4105	0599
S4110	0599
S4115	0599
S4120	0599
S4125	0599
S4130	0599
S4135	1986
S4140	1986
S4145	1986
S4150	1986
S4160	1986
S4170	0657
S4180	0657
S4190	0657
S4205	0599
S4210	0599
S4215	0599
S4220	0599
S4225	0599
S4230	0599
S4235	1986
S4240	1986
S4245	1986
S4248	0004
S4249	0037
S4250	1986
S4260	1986
S4270	0657
S4280	0657
S4290	0657
S4305	0599
S4310	0599
S4315	0599
S4320	0599
S4325	0599
S4330	0599
S4335	1986
S4340	1986
S4345	1986
S4350	1986
S4360	1986
S4370	0657
S4380	0657
S4390	0657
S5000	4580
S5004	0561
S5005	4583
S5006	2580
S5010B	OEM
S5010D	OEM
S5010M	OEM
S5012	3553
S5013	4587
S5014	4202
S5014(HEP)	0086
S5015	0276
S5018	0905
S5019	0178
S5020	0007
S5021	0007
S5022	OEM
S5023	3698
S5024	0187
S5025	0730
S5026	3426
S5027(HEP)	0161
S5028	OEM
S5028(HEP)	1051
S5041T951	OEM
S5089-A	0015
S5092-4B	OEM
S5092-6A	OEM
S5102	OEM
S5103	OEM
S5104	OEM
S5105	1023
S5105R	0501
S5106	OEM
S5110	1023
S5110R	0501
S5115	1023
S5115R	0501
S5120	1023
S5120R	0501
S5125	1023
S5125R	0501
S5130	1023
S5135	1023
S5135R	0501
S5140	1023
S5140R	0501
S5145	1036
S5145R	1777
S5150	1036
S5150R	1777
S5151	1282
S5151R	1282
S5160	1036
S5160R	1777
S5162	OEM
S5166	OEM
S5170	1046
S5170R	1789
S5177B	0015
S5180	1046
S5180R	1789
S5188	2363
S5190	1046
S5190R	1789
S5226	0299
S5226A	0299
S5226B	0299
S5227	0376
S5227A	0376
S5227B	0376
S5228	0036
S5228A	0036
S5228B	0036
S5229	0275
S5229A	0275
S5229B	0275
S5230	0140
S5230A	0140
S5230B	0140
S5241	OEM
S-5277B	0015
S5277B	0023
S5277D	0023
S5277G	0795
S5277J	0023
S5277L	0072
S5277N	0071
S5295	0133
S5295-G	0023
S5295B	0023
S5295D	0023
S5295G	0023
S5295GTPA2	0023
S5295I	0023
S5295J	0023
S5295JTPA2	0023
S5305	1023
S5305R	0501
S5310	1023
S5310R	0501
S5315	1023
S5315R	0501
S5320	1023
S5320R	0501
S5325	1023
S5325R	0501
S5327E	0144
S5328E	0144
S5330	1023
S5330R	0501
S5335	1023
S5335R	0501
S5340	1023
S5340R	0501
S5343	OEM
S5344	OEM
S5345	1036
S5345R	1777
S5347	OEM
S5350	1036
S5350R	1777
S5360	1036
S5360R	1777
S5367	OEM
S5370	1046
S5370R	1789
S5373	OEM
S5380	1046
S5380R	1789
S5390	1046
S5390R	1789
S5405	1023
S5410	1023
S5415	1023
S5420	1023
S5420F	OEM
S5420W	OEM
S5425	1023
S5430	1023
S5435	1036
S5440	1036
S5440-5	3878
S5445	1036
S5450	1036
S5456J	0564
S5460	1036
S5470	1046
S5480	1046
S5490	1046
S5500	0015
S5500-D	0015
S5500B	0080
S5500D	0015
S5500DTF	OEM
S5500DTP	0080
S5500G	0795
S5566	OEM
S5566B	0015
S5566G	0795
S5566J	0015
S5566N	0071
S5670E	0144
S5688B	0015
S5688G-TPA3	OEM
S5723T	1192
S5800B	OEM
S5800C	OEM
S5800D	OEM
S5800E	OEM
S5800M	OEM
S5801B	OEM
S5801C	OEM
S5801D	OEM
S5801E	OEM
S5801M	OEM
S5802B	OEM
S5802C	OEM
S5802D	OEM
S5802E	OEM
S5802M	OEM
S5907	OEM
S6000	0774
S6000C	0611
S6000E	0952
S6000KS2	OEM
S6000KS3	OEM
S6000S	OEM
S6001K	OEM
S6001L	1148
S6001M	3349
S6001MS2	0308
S6001MS3	0308
S6003L	1148
S6003M	3349
S6003MS2	0308
S6003MS3	0308
S6003RS1	OEM
S6003RS2	0308
S6003RS3	0308
S6004F1	0095
S6005	0015
S6006B	3215
S6006G	1665
S6006H	0687
S6006L	0863
S6006RS2	OEM
S6006RS3	OEM
S6008B	3215
S6008G	1665
S6008H	0687
S6008L	1148
S6008RS2	OEM
S6008RS3	OEM
S6010	OEM
S6010B	3215
S6010G	1665
S6010H	0687
S6010L	1148
S6010RS2	OEM
S6010RS3	OEM
S6015	OEM
S6015G	1665
S6015H	0764
S6015L	0952
S6016B	OEM
S6016G	1665
S6016H	0764
S6016L	OEM
S6020	OEM
S6022	OEM
S6022B	OEM
S6025B	OEM
S6025C	0226
S6025D	1148
S6025G	2649
S6025H	0764
S6025L	2611
S6025R	4054
S6026H	0725
S6030	4983
S6035	OEM
S6035C	0226
S6035D	3215
S6035G	2649
S6035H	0725
S6035J	2690
S6035K	2690
S6035W	6325
S6040	OEM
S6045	OEM
S6050	OEM
S6055M	6325
S6055W	6325
S6060	OEM
S6065J	OEM
S6065K	OEM
S6070	OEM
S6070W	OEM
S6080	OEM
S6080A	0061
S6080B	0061
S6080C	OEM
S6087A	1712
S6087B	2329
S6089	0239
S6090	OEM
S6100C	7161
S6100E	7162
S6100S	OEM
S6121-1	0289
S6121-3	0473
S6121-5	0446
S6132A	OEM
S6132B	OEM
S6142	0095
S6142C	0095
S6142G	0095
S6200A	1650
S6200B	1651
S6200D	1584
S6200M	1665
S6210A	0741
S6210B	0745
S6210D	0747
S6210M	0764
S6220A	1848
S6220B	OEM
S6220D	1855
S6220M	OEM
S6230-AH	0289
S6230-AH1	0473
S6230-AH2	0446
S6230A	OEM
S6230B	OEM
S6230D	OEM
S6230M	OEM
S6240A	0226
S6240B	0226
S6240D	0226
S6240M	0226
S6250A	OEM
S6250B	OEM
S6250D	OEM
S6250M	OEM
S6321F	1992
S6321FLC6	6145
S6340G	0095
S6344G	2739
S6394GL6	0755
S6394GLC6	0755
S6410A	0469
S6410B	0469
S6410D	0722
S6410M	0725
S6413M	3186
S6420A	1848
S6420B	1848
S6420D	3212
S6420M	3215
S6430A	OEM
S6430B	OEM
S6430D	OEM
S6430M	OEM
S6430N	OEM
S6431M	0725
S6440A	0226
S6440B	0226
S6440D	0226
S6440M	0226
S6440N	OEM
S6450A	OEM
S6450B	OEM
S6450D	OEM
S6450M	OEM
S6450N	OEM
S6493M	OEM
S6500	0774
S6508	3193
S6508-1	3193
S6508A	3193
S6508E	3193
S6508P	3193
S6565(LB)	OEM
S6565(OEC)	OEM
S6565G	1992
S6565G-(LB)	1992
S6565G(LB)	OEM
S6565G-(OEC)	1992
S6565G(OEC)	OEM
S-6700A	OEM
S6700A	OEM
S6785G	4352
S6800	0388
S6801	0016
S6802	0393
S6804	0086
S6810P	2086
S6821P	0448
S6850	0514
S6850P	0514
S6856DL	OEM
S7000	0130
S7001	OEM
S7002	5147
S7003	OEM
S7004	0103
S7005	4960
S7006	OEM
S7006(HEP)	0925
S7007	OEM
S7007(HEP)	1681
S7008	OEM
S7008(HEP)	0130
S7230B	OEM
S7310B	OEM
S7310C	OEM
S7310D	OEM
S7310E	OEM
S7310M	OEM
S7410M	0804
S7412M	0804
S7430M	0804
S7502	0407
S7503	0407
S8001L	OEM
S8001M	6244
S8003L	OEM
S8003M	OEM
S8006B	OEM
S8006G	OEM
S8006H	0766
S8006L	6245
S8008B	OEM
S8008G	OEM
S8008H	0766
S8008L	OEM
S8010B	OEM
S8010G	OEM
S8010H	0766
S8010L	1512
S8015L	1512
S8016B	OEM
S8016G	OEM
S8016H	0766
S8016L	OEM
S8020L	1512
S8025	OEM
S8025C	OEM
S8025D	OEM
S8025G	OEM
S8025H	0766
S8025L	2611
S8025R	4010
S8033HA	OEM
S8035	OEM
S8035C	OEM
S8035D	OEM
S8035G	OEM
S8035H	0750
S8035J	2690
S8035K	2690
S8035W	OEM
S8040R	5079
S8049H4	OEM
S8050K	5974
S8052ALR	OEM
S8052ALY	OEM
S8053ALB	OEM
S-8053ALR	OEM
S8053ALR	OEM
S8054ALB	OEM
S8054ALB-LM-S	OEM
S-8054ALR	OEM
S8054ALR	OEM
S8054ALRS-LN	OEM
S8055M	3277
S8055W	6325
S8056H	OEM
S8056H1	OEM
S8065J	OEM
S8065K	OEM
S8070W	OEM
S8201	OEM
S8201A	OEM
S8202	OEM
S8202A	OEM
S8205	OEM
S8206	OEM
S8250	OEM
S8250A	OEM
S8250B	OEM
S8251	OEM
S8252	OEM
S8660	0161
S8851	0190
S8856	0917
S9100	0400
S9101	0558
S9101(HEP)	0558
S9102	3377
S9102(HEP)	3377
S9103	OEM
S9120	3519
S9121	2897
S9122	3528
S9123	OEM
S9140	2427
S9140(HEP)	2427
S9141	2277
S9141(HEP)	2277
S9142	OEM
S9143	OEM
S9144	OEM
S9144(HEP)	2277
S9145	OEM
S9145(HEP)	2427
S9146	OEM
S9146(HEP)	3373
S9147	OEM
S9147(HEP)	3374
S9148	OEM
S9148(HEP)	2439
S9149	OEM
S9149(HEP)	2432
S9150	OEM
S9151	OEM
S9152	OEM
S9153	OEM
S9308F	1070
S9314F	2488
S9516	0050
S9524	0004
S9602F	2286
S9631	0079
S9923	0004
S10001D	OEM
S10002D	OEM
S10015D	OEM
S10100	0922
S10110	0102
S10129	OEM
S10130	OEM
S10131	OEM
S10149	0015
S10153	0161
S-10386	3426
S11042FU26	OEM
S12100	0922
S12110	0102
S12120	0102
S12130	0102
S12140	0102
S12150	OEM
S15001D	OEM
S15002D	OEM
S15015D	OEM
S15649	0016
S15650	0007
S15657	0007
S15658	0007
S15659	0007
S15660	0086
S16901	OEM
S17074	0196
S17862	0233
S17900	0086
S18000	0086
S18100	0037
S18200	0086
S19386	0086
S20005	5514
S20100	3011
S20405	0084
S20410	0097
S20415	0097
S20420	0097
S20425	0109
S20430	0109
S20435	0109
S20445	1400
S-20446	0698
S20446	1411
S20450	0122
S20460	0122
S21100	1614
S21271	0143
S21520	0178
S21549	0086
S21639	0012
S21648	0127
S22543	0016
S23130	0037
S23579	0079
S24226	0037
S24591	0016
S24592	0111
S24594	0126
S24596	0016
S24597	0126
S24598	0086
S24612	0126
S24614	0086
S24615	0126
S24616	0086
S25261	0007
S25805	0144
S25941	0127
S26822	0079
S27233	0086
S27604	0007
S27893	0224
S29956	0111
S30001D	OEM
S30002D	OEM
S30015D	OEM
S30405	3189
S30410	1124
S30415	1124
S30420	1124
S30425	1126
S30430	1126
S30435	0805
S30440	0805
S30445	1195
S30450	1195
S30460	0319
S31100	0044
S31160	OEM
S31551	0037
S31866	0079
S32417	0144
S32550	0086
S32669	0144
S32903	0161
S33291	0142

DEVICE TYPE	REPL CODE
S33529	0925
S33530	0228
S33755	0079
S33886	0126
S33990	0144
S34100	2655
S34540	0016
S35232	0007
S35233	0144
S35487	0130
S36100	2473
S36951	0155
S37100	2845
S37110	2845
S37120	2643
S37162	0423
S37165	0925
S37166	0042
S37182	0079
S37214	0079
S37423	0079
S38280	OEM
S38763	0079
S38787	0079
S38789	0016
S38854	0079
S39094	0037
S39261	0925
S39262	0042
S39509	0382
S39560	0233
S39868	0264
S40204	0007
S40205	0233
S41100	0657
S42100	0657
S42110	3309
S42120	3309
S42130	3309
S42140	3309
S43100	0657
S43110	3309
S43120	3309
S43130	3309
S43140	3309
S50305	1023
S50310	1023
S50315	1023
S50320	1023
S50325	1023
S50330	1023
S50335	1036
S50340	1036
S50345	1036
S50350	1036
S50360	1036
S50370	1046
S50380	1046
S50390	1046
S50410	1023
S50415	1023
S50420	1023
S50425	1023
S50430	1023
S50435	1036
S50440	1036
S50445	1036
S50450	1036
S50460	1036
S50470	1046
S50480	1046
S50490	1046
S51100	1046
S51100R	1789
S53100	1046
S53100R	1789
S53110	1046
S53120	3928
S53130	2832
S53140	2832
S54100	1046
S60100	OEM
S60252	OEM
S63512D	OEM
S67794	0142
S67809	0160
S-81230AG	OEM
S81230AG	OEM
S81250HG	OEM
S81350AG	OEM
S-88509	2983
S88509	0190
S88568	OEM
S-88569	0190
S-88569S	OEM
S-88570	0190
S88570	0190
S-88571	0190
S88571	0190
S-88576	0917
S88576	0917
S95101	0050
S95102	0050
S95103	0050
S95104	0050
S95106	0050
S95125	0144
S95125A	0144
S95126	0144
S95126A	0144
S95201	0004
S95202	0016
S95203	0004
S95204	0004
S95206	0004
S95207	0004
S95214	0004
S95218	0004
S95252	0276
S-95253	0160
S95253	0160
S-95253-1	0160
S95253-1	0160
S97520	OEM
S97539	OEM
S99101	0050
S99102	0050
S99103	0050
S99104	0050
S99201	0004
S99203	0004
S99218	0004
S99252	0276
S326690	0144
S413796	0211
S503100	1046
S503110	5520
S503120	5520
S504100	1046
S504110	5520
S504120	5520
S1977634	0103
S2041635	0004
S2042634	0208
SA0419	0126
SA1AN12	2900
SA1CN3	OEM
SA1M1	0953
SA-2	0087
SA2	5570
SA-2A	0087
SA2AN12	2900
SA-2B	0087
SA2B	0015
SA-2C	0087
SA2CN3	OEM
SA2H	0071
SA2M1	0242
SA-2Z	0087
SA2Z	0071
SA3B	0015
SA3M1	0540
SA4AN12	1112
SA5AN12	OEM
SA5CN3	OEM
SA5M1	0965
SA5M3	OEM
SA5.0A(DIO)	2396
SA6AN12	1112
SA6CN3	OEM
SA6M1	0965
SA6.0A(DIO)	3114
SA6.5A(DIO)	OEM
SA7	0600
SA7CN3	OEM
SA7M3	OEM
SA7.0A(DIO)	OEM
SA7.5A(DIO)	OEM
SA8AN12	3011
SA8CN3	OEM
SA8M1	0817
SA8.0A(DIO)	OEM
SA8.5A(DIO)	OEM
SA9CN3	OEM
SA9.0A(DIO)	OEM
SA10A(DIO)	OEM
SA10AN12	3011
SA10CN3	OEM
SA10M1	0922
SA11A(DIO)	3172
SA12A	0280
SA12A(DIO)	OEM
SA12B	0280
SA12C	0280
SA12D	0280
SA12E	0280
SA12F	0280
SA12G	0280
SA12L	0280
SA12M	0280
SA12M1	OEM
SA12R	0280
SA12T	0280
SA12X	0280
SA12Y	0280
SA13A	0280
SA13A(DIO)	OEM
SA13B	0280
SA13C	0280
SA13D	0280
SA13E	0280
SA13F	0280
SA13G	0280
SA13L	0280
SA13M	0280
SA13R	0280
SA13T	0280
SA13X	0280
SA13Y	0280
SA14A	0280
SA14A(DIO)	OEM
SA14B	0280
SA14C	0280
SA14D	0280
SA14E	0280
SA14F	0280
SA14G	0280
SA14L	0280
SA14M	0280
SA14R	0280
SA14T	0280
SA14X	0280
SA14Y	0280
SA15A	0050
SA15A(DIO)	3191
SA15B	0050
SA15C	0050
SA15D	0050
SA15E	0050
SA15F	0050
SA15G	0050
SA15L	0050
SA15M	0050
SA15N	0050
SA15R	0050
SA15T	0050
SA15X	0050
SA15Y	0050
SA16A	0280
SA16A(DIO)	OEM
SA16B	0280
SA16C	0280
SA16D	0280
SA16E	0280
SA16F	0280
SA16G	0280
SA16L	0280
SA16M	0280
SA16R	0280
SA16T	0280
SA16X	0280
SA16Y	0280
SA17A	0280
SA17A(DIO)	3201
SA17B	0280
SA17C	0280
SA17D	0280
SA17E	0280
SA17F	0280
SA17G	0280
SA17L	0280
SA17M	0280
SA17R	0280
SA17T	0280
SA17X	0280
SA17Y	0280
SA18A	0136
SA18A(DIO)	OEM
SA18B	0136
SA18C	0136
SA18D	0136
SA18E	0136
SA18F	0136
SA18G	0136
SA18L	0136
SA18M	0136
SA18R	0136
SA18T	0136
SA18X	0136
SA18Y	0136
SA19A	0136
SA19B	0136
SA19C	0136
SA19D	0136
SA19E	0136
SA19F	0136
SA19G	0136
SA19L	0136
SA19M	0136
SA19R	0136
SA19T	0136
SA19X	0136
SA19Y	0136
SA20	OEM
SA20A	0136
SA20A(DIO)	OEM
SA20B	0136
SA20C	0136
SA20D	0136
SA20E	0136
SA20F	0136
SA20G	0136
SA20L	0136
SA20M	0136
SA20R	0136
SA20T	0136
SA20X	0136
SA20Y	0136
SA21	OEM
SA21A	0136
SA21B	0136
SA21C	0136
SA21D	0136
SA21E	0136
SA21F	0136
SA21G	0136
SA21L	0136
SA21M	0136
SA21R	0136
SA21T	0136
SA21X	0136
SA21Y	0136
SA22A(DIO)	1915
SA24A(DIO)	OEM
SA26A	0280
SA26A(DIO)	OEM
SA26B	0280
SA26C	0280
SA26D	0280
SA26E	0280
SA26F	0280
SA26G	0280
SA26L	0280
SA26M	0280
SA26R	0280
SA26T	0280
SA26X	0280
SA26Y	0280
SA-28	OEM
SA28A	0280
SA28A(DIO)	OEM
SA28B	0280
SA28C	0280
SA28D	0280
SA28E	0280
SA28F	0280
SA28G	0280
SA28L	0280
SA28M	0280
SA28R	0280
SA28T	0280
SA28X	0280
SA28Y	0280
SA29	0164
SA29A	0136
SA29B	0136
SA29C	0136
SA29D	0136
SA29E	0136
SA29F	0136
SA29G	0136
SA29L	0136
SA29M	0136
SA29R	0136
SA29T	0136
SA29X	0136
SA29Y	0136
SA30A	0280
SA30A(DIO)	OEM
SA30B	0280
SA30C	0280
SA30D	0280
SA30E	0280
SA30F	0280
SA30G	0280
SA30L	0280
SA30M	0280
SA30R	0280
SA30T	0280
SA30X	0280
SA30Y	0280
SA31A	0280
SA31B	0280
SA31C	0280
SA31D	0280
SA31E	0280
SA31F	0280
SA31G	0280
SA31L	0280
SA31M	0280
SA31R	0280
SA31T	0280
SA31X	0280
SA31Y	0280
SA32A	0280
SA32B	0280
SA32C	0280
SA32D	0280
SA32E	0280
SA32F	0280
SA32G	0280
SA32L	0280
SA32M	0280
SA32R	0280
SA32T	0280
SA32X	0280
SA32Y	0280
SA33	0004
SA33A(DIO)	OEM
SA33BRN	0004
SA33RED	0004
SA35A	0280
SA35B	0280
SA35C	0280
SA35E	0280
SA35F	0280
SA35G	0280
SA35L	0280
SA35M	0280
SA35R	0280
SA35T	0280
SA35X	0280
SA35Y	0280
SA36A	0280
SA36A(DIO)	OEM
SA36B	0280
SA36C	0280
SA36D	0280
SA36E	0280
SA36F	0280
SA36G	0280
SA36L	0280
SA36M	0280
SA36R	0280
SA36T	0280
SA36X	0280
SA36Y	0280
SA37A	0280
SA37B	0280
SA37C	0280
SA37D	0280
SA37E	0280
SA37F	0280
SA37G	0280
SA37L	0280
SA37M	0280
SA37R	0280
SA37T	0280
SA37X	0280
SA37Y	0280
SA38A	0280
SA38B	0280
SA38C	0280
SA38D	0280
SA38E	0280
SA38F	0280
SA38G	0280
SA38L	0280
SA38M	0280
SA38R	0280
SA38T	0280
SA38X	0280
SA38Y	0280
SA39A	0280
SA39B	0280
SA39C	0280
SA39D	0280
SA39E	0280
SA39F	0280
SA39G	0280
SA39L	0280
SA39M	0280
SA39R	0280
SA39T	0280
SA39X	0280
SA39Y	0280
SA40	OEM
SA40A	0280
SA40A(DIO)	OEM
SA40B	0280
SA40C	0280
SA40D	0280
SA40E	0280
SA40F	0280
SA40G	0280
SA40L	0280
SA40M	0280
SA40R	0280
SA40T	0280
SA40X	0280
SA40Y	0280
SA41A	0280
SA41B	0280
SA41C	0280
SA41D	0280
SA41E	0280
SA41F	0280
SA41G	0280
SA41L	0280
SA41M	0280
SA41R	0280
SA41T	0280
SA41X	0280
SA41Y	0280
SA42A	0847
SA42B	0847
SA42C	0847
SA42D	0847
SA42E	0847
SA42F	0847
SA42G	0847
SA42L	0847
SA42M	0847
SA42R	0847
SA42T	0847
SA42X	0847
SA42Y	0847
SA43A	0050
SA43A(DIO)	0568
SA43B	0050
SA43C	0050
SA43D	0050
SA43E	0050
SA43F	0050
SA43G	0050
SA43L	0050
SA43M	0050
SA43R	0050
SA43T	0050
SA43X	0050
SA43Y	0050
SA44A	0280
SA44B	0280
SA44C	0280
SA44D	0280
SA44E	0280
SA44F	0280
SA44G	0280
SA44L	0280
SA44M	0280
SA44R	0280
SA44T	0280
SA44X	0280
SA44Y	0280
SA45A	0050
SA45A(DIO)	OEM
SA45B	0050
SA45C	0050
SA45D	0050
SA45E	0050
SA45F	0050
SA45G	0050
SA45L	0050
SA45M	0050
SA45R	0050
SA45T	0050
SA45X	0050
SA45Y	0050
SA48A(DIO)	OEM
SA49A	0280
SA49B	0280
SA49C	0280
SA49D	0280
SA49E	0280
SA49F	0280
SA49G	0280
SA49L	0280
SA49M	0280
SA49R	0280
SA49T	0280
SA49X	0280
SA49Y	0280
SA50	0037
SA51	0037
SA51A	0037
SA51A(DIO)	OEM
SA52	0037
SA52A	0037
SA52B	0037
SA53	0037
SA54	0037
SA54A	0037
SA54A(DIO)	OEM
SA55	0037
SA56	0037
SA58A(DIO)	0831
SA60A(DIO)	OEM
SA64A(DIO)	OEM
SA70	0037
SA70A(DIO)	OEM
SA71A	0050
SA71B	0050
SA71C	0050
SA71D	0050
SA71E	0050
SA71F	0050
SA71G	0050
SA71L	0050
SA71M	0050
SA71R	0050
SA71T	0050
SA71X	0050
SA71Y	0050
SA72A	0050
SA72B	0050
SA72C	0050
SA72D	0050
SA72E	0050
SA72F	0050
SA72G	0050
SA72L	0050
SA72M	0050
SA72R	0050
SA72T	0050
SA72X	0050
SA72Y	0050
SA73A	0050
SA73B	0050
SA73C	0050
SA73D	0050
SA73E	0050
SA73F	0050
SA73G	0050
SA73L	0050
SA73M	0050
SA73R	0050
SA73T	0050
SA73X	0050
SA73Y	0050
SA74A	0050
SA74B	0050
SA74C	0050
SA74D	0050
SA74E	0050
SA74F	0050
SA74G	0050
SA74L	0050
SA74M	0050
SA74R	0050
SA74T	0050
SA74X	0050
SA74Y	0050
SA75A	0280
SA75A(DIO)	OEM
SA75B	0280
SA75C	0280
SA75D	0280
SA75E	0280
SA75F	0280
SA75G	0280
SA75L	0280
SA75M	0280
SA75R	0280
SA75T	0280
SA75X	0280
SA75Y	0280
SA76A	0050
SA76B	0050
SA76C	0050
SA76D	0050
SA76E	0050
SA76F	0050
SA76G	0050
SA76L	0050
SA76M	0050
SA76R	0050
SA76T	0050
SA76X	0050
SA76Y	0050
SA77A	0050
SA77B	0050
SA77C	0050
SA77D	0050
SA77E	0050
SA77F	0050
SA77G	0050
SA77L	0050
SA77M	0050
SA77R	0050
SA77T	0050
SA77X	0050
SA77Y	0050
SA78A	0050
SA78A(DIO)	OEM
SA78B	0050
SA78C	0050
SA78D	0050
SA78E	0050
SA78F	0050
SA78G	0050
SA78HV05CDA	1916
SA78HV05CU	0624
SA78HV12CDA	3393
SA78HV12CU	0334
SA78L	0050
SA78M	0050
SA78R	0050
SA78T	0050
SA78X	0050
SA78Y	0050
SA79A	0847
SA79B	0847
SA79C	0847
SA79D	0847
SA79E	0847
SA79F	0847
SA79G	0847
SA79L	0847
SA79M	0847
SA79R	0847
SA79T	0847
SA79X	0847
SA79Y	0847
SA80A	0050
SA80B	0050
SA80C	0050
SA80D	0050
SA80E	0050
SA80F	0050
SA80G	0050
SA80L	0050
SA80M	0050
SA80R	0050
SA80T	0050
SA80X	0050
SA80Y	0050
SA81A	0136
SA81B	0136
SA81C	0136
SA81D	0136
SA81E	0136
SA81F	0136
SA81G	0136
SA81L	0136
SA81M	0136
SA81R	0136
SA81T	0136
SA81X	0136
SA81Y	0136
SA82A	0136
SA82B	0136
SA82C	0136
SA82D	0136
SA82E	0136
SA82F	0136
SA82G	0136
SA82L	0136
SA82M	0136
SA82R	0136
SA82T	0136
SA83A	0050
SA83B	0050
SA83C	0050

If replacement code is OEM, contact original manufacturer for replacement.

DEVICE TYPE	REPL CODE	DEVICE TYPE	REPL CODE	DEVICE TYPE	REPL CODE	DEVICE TYPE	REPL CODE	DEVICE TYPE	REPL CODE	DEVICE TYPE	REPL CODE	DEVICE TYPE	REPL CODE
SA83D	0050	SA94G	0136	SA106M	0050	SA114Y	0050	SA125D	0050	SA136F	0050	SA146T	0280
SA83E	0050	SA94L	0136	SA106R	0050	SA115A	0050	SA125E	0050	SA136G	0050	SA146X	0280
SA83F	0050	SA94M	0136	SA106T	0050	SA115B	0050	SA125F	0050	SA136L	0050	SA146Y	0280
SA83G	0050	SA94R	0136	SA106X	0050	SA115C	0050	SA125G	0050	SA136M	0050	SA147A	0280
SA83L	0050	SA94T	0136	SA106Y	0050	SA115D	0050	SA125L	0050	SA136R	0050	SA147B	0280
SA83M	0050	SA94X	0136	SA107	OEM	SA115E	0050	SA125M	0050	SA136T	0050	SA147C	0280
SA83R	0050	SA94Y	0136	SA107B	0050	SA115F	0050	SA125R	0050	SA136X	0050	SA147D	0280
SA83T	0050	SA95A	0037	SA107C	0050	SA115G	0050	SA125T	0050	SA136Y	0050	SA147E	0280
SA83X	0050	SA95B	0037	SA107D	0050	SA115L	0050	SA125X	0050	SA137A	0280	SA147F	0280
SA83Y	0050	SA95C	0037	SA107E	0050	SA115M	0050	SA125Y	0050	SA137B	0280	SA147G	0280
SA84A	0050	SA95D	0037	SA107F	0050	SA115R	0050	SA128	0004	SA137C	0280	SA147L	0280
SA84B	0050	SA95E	0037	SA107G	0050	SA115T	0050	SA128-1	0004	SA137D	0280	SA147M	0280
SA84C	0050	SA95F	0037	SA107L	0050	SA115X	0050	SA129A	0136	SA137E	0280	SA147R	0280
SA84D	0050	SA95G	0037	SA107M	0050	SA115Y	0050	SA129B	0136	SA137F	0280	SA147T	0280
SA84E	0050	SA95L	0037	SA107R	0050	SA116A	0050	SA129C	0136	SA137G	0280	SA147X	0280
SA84F	0050	SA95M	0037	SA107T	0050	SA116B	0050	SA129D	0136	SA137L	0280	SA147Y	0280
SA84G	0050	SA95R	0037	SA107X	0050	SA116C	0050	SA129E	0136	SA137M	0280	SA148A	0280
SA84L	0050	SA95T	0037	SA107Y	0050	SA116D	0050	SA129F	0136	SA137R	0280	SA148B	0280
SA84M	0050	SA95X	0037	SA108A	0050	SA116E	0050	SA129G	0136	SA137T	0280	SA148C	0280
SA84R	0050	SA95Y	0037	SA108B	0050	SA116F	0050	SA129L	0136	SA137X	0280	SA148D	0280
SA84T	0050	SA100	OEM	SA108C	0050	SA116G	0050	SA129M	0136	SA137Y	0280	SA148E	0280
SA84X	0050	SA100A	0280	SA108D	0050	SA116L	0050	SA129R	0136	SA139A	0050	SA148F	0280
SA84Y	0050	SA100A(DIO)	OEM	SA108E	0050	SA116M	0050	SA129T	0136	SA139B	0050	SA148G	0280
SA85A	0136	SA100B	0280	SA108F	0050	SA116R	0050	SA129X	0136	SA139C	0050	SA148L	0280
SA85A(DIO)	OEM	SA100C	0280	SA108G	0050	SA116T	0050	SA129Y	0136	SA139D	0050	SA148M	0280
SA85B	0136	SA100D	0280	SA108L	0050	SA116X	0050	SA130A	0050	SA139E	0050	SA148R	0280
SA85C	0136	SA100E	0280	SA108M	0050	SA116Y	0050	SA130A(DIO)	OEM	SA139F	0050	SA148T	0280
SA85D	0136	SA100F	0280	SA108R	0050	SA117A	0050	SA130B	0050	SA139G	0050	SA148X	0280
SA85E	0136	SA100G	0280	SA108T	0050	SA117B	0050	SA130C	0050	SA139L	0050	SA148Y	0280
SA85F	0136	SA100L	0280	SA108X	0050	SA117C	0050	SA130D	0050	SA139M	0050	SA149A	0280
SA85G	0136	SA100M	0280	SA108Y	0050	SA117D	0050	SA130E	0050	SA139R	0050	SA149B	0280
SA85L	0136	SA100R	0280	SA109A	0050	SA117E	0050	SA130F	0050	SA139T	0050	SA149C	0280
SA85M	0136	SA100T	0280	SA109B	0050	SA117F	0050	SA130G	0050	SA139X	0050	SA149D	0280
SA85R	0136	SA100X	0280	SA109C	0050	SA117G	0050	SA130L	0050	SA139Y	0050	SA149E	0280
SA85T	0136	SA100Y	0280	SA109D	0050	SA117L	0050	SA130M	0050	SA141A	0050	SA149F	0280
SA85X	0136	SA101A	0050	SA109E	0050	SA117M	0050	SA130R	0050	SA141B	0050	SA149G	0280
SA85Y	0136	SA101B	0050	SA109F	0050	SA117R	0050	SA130T	0050	SA141C	0050	SA149L	0280
SA86A	0847	SA101C	0050	SA109G	0050	SA117T	0050	SA130X	0050	SA141D	0050	SA149M	0280
SA86B	0847	SA101D	0050	SA109L	0050	SA117X	0050	SA130Y	0050	SA141E	0050	SA149R	0280
SA86C	0847	SA101E	0050	SA109M	0050	SA117Y	0050	SA131A	0280	SA141F	0050	SA149T	0280
SA86D	0847	SA101F	0050	SA109R	0050	SA118A	0050	SA131B	0280	SA141G	0050	SA149X	0280
SA86E	0847	SA101G	0050	SA109T	0050	SA118B	0050	SA131C	0280	SA141L	0050	SA149Y	0280
SA86F	0847	SA101L	0050	SA109X	0050	SA118C	0050	SA131D	0280	SA141M	0050	SA150A	0136
SA86G	0847	SA101M	0050	SA109Y	0050	SA118D	0050	SA131E	0280	SA141R	0050	SA150A(DIO)	OEM
SA86L	0847	SA101R	0050	SA110A	0050	SA118E	0050	SA131F	0280	SA141T	0050	SA150B	0136
SA86M	0847	SA101T	0050	SA110A(DIO)	OEM	SA118F	0050	SA131G	0280	SA141X	0050	SA150C	0136
SA86R	0847	SA101X	0050	SA110B	0050	SA118G	0050	SA131L	0280	SA141Y	0050	SA150D	0136
SA86T	0847	SA101Y	0050	SA110C	0050	SA118L	0050	SA131M	0280	SA142A	0050	SA150E	0136
SA86X	0847	SA102	0136	SA110D	0050	SA118M	0050	SA131R	0280	SA142B	0050	SA150F	0136
SA86Y	0847	SA102A	0050	SA110E	0050	SA118R	0050	SA131T	0280	SA142C	0050	SA150G	0136
SA87A	0050	SA102B	0050	SA110F	0050	SA118T	0050	SA131X	0280	SA142D	0050	SA150L	0136
SA87B	0050	SA102C	0050	SA110G	0050	SA118X	0050	SA131Y	0280	SA142E	0050	SA150M	0136
SA87C	0050	SA102D	0050	SA110L	0050	SA118Y	0050	SA132A	0280	SA142F	0050	SA150R	0136
SA87D	0050	SA102E	0050	SA110M	0050	SA120A(DIO)	OEM	SA132B	0280	SA142G	0050	SA150T	0136
SA87E	0050	SA102F	0050	SA110R	0050	SA121A	0050	SA132C	0280	SA142L	0050	SA150X	0136
SA87F	0050	SA102G	0050	SA110T	0050	SA121B	0050	SA132D	0280	SA142M	0050	SA150Y	0136
SA87G	0050	SA102L	0050	SA110X	0050	SA121C	0050	SA132E	0280	SA142R	0050	SA151A	0280
SA87L	0050	SA102M	0050	SA110Y	0050	SA121D	0050	SA132F	0280	SA142T	0050	SA151B	0280
SA87M	0050	SA102R	0050	SA111A	0050	SA121E	0050	SA132G	0280	SA142X	0050	SA151C	0280
SA87R	0050	SA102T	0050	SA111B	0050	SA121F	0050	SA132L	0280	SA142Y	0050	SA151D	0280
SA87T	0050	SA102X	0050	SA111C	0050	SA121G	0050	SA132M	0280	SA143A	0050	SA151E	0280
SA87X	0050	SA102Y	0050	SA111D	0050	SA121L	0050	SA132R	0280	SA143B	0050	SA151F	0280
SA87Y	0050	SA103A	0050	SA111E	0050	SA121M	0050	SA132T	0280	SA143C	0050	SA151G	0280
SA90A	0050	SA103B	0050	SA111F	0050	SA121R	0050	SA132X	0280	SA143D	0050	SA151L	0280
SA90A(DIO)	OEM	SA103C	0050	SA111G	0050	SA121T	0050	SA132Y	0280	SA143E	0050	SA151M	0280
SA90B	0050	SA103D	0050	SA111L	0050	SA121X	0050	SA133A	0050	SA143F	0050	SA151R	0280
SA90C	0050	SA103E	0050	SA111M	0050	SA121Y	0050	SA133B	0050	SA143G	0050	SA151T	0280
SA90D	0050	SA103F	0050	SA111R	0050	SA122A	0050	SA133C	0050	SA143L	0050	SA151X	0280
SA90E	0050	SA103G	0050	SA111T	0050	SA122B	0050	SA133D	0050	SA143M	0050	SA151Y	0280
SA90F	0050	SA103L	0050	SA111X	0050	SA122C	0050	SA133E	0050	SA143R	0050	SA152A	0280
SA90G	0050	SA103M	0050	SA111Y	0050	SA122D	0050	SA133F	0050	SA143T	0050	SA152B	0280
SA90L	0050	SA103R	0050	SA112A	0050	SA122E	0050	SA133G	0050	SA143X	0050	SA152C	0280
SA90M	0050	SA103T	0050	SA112B	0050	SA122F	0050	SA133L	0050	SA143Y	0050	SA152D	0280
SA90R	0050	SA103X	0050	SA112C	0050	SA122G	0050	SA133M	0050	SA144A	0280	SA152E	0280
SA90T	0050	SA103Y	0050	SA112D	0050	SA122L	0050	SA133R	0050	SA144B	0280	SA152F	0280
SA90X	0050	SA104A	0050	SA112E	0050	SA122M	0050	SA133T	0050	SA144C	0280	SA152G	0280
SA90Y	0050	SA104B	0050	SA112F	0050	SA122R	0050	SA133X	0050	SA144D	0280	SA152L	0280
SA92A	0050	SA104C	0050	SA112G	0050	SA122T	0050	SA133Y	0050	SA144E	0280	SA152M	0280
SA92B	0050	SA104D	0050	SA112L	0050	SA122X	0050	SA134A	0050	SA144F	0280	SA152R	0280
SA92C	0050	SA104E	0050	SA112M	0050	SA122Y	0050	SA134B	0050	SA144G	0280	SA152T	0280
SA92D	0050	SA104F	0050	SA112R	0050	SA123A	0050	SA134C	0050	SA144L	0280	SA152X	0280
SA92E	0050	SA104G	0050	SA112T	0050	SA123B	0050	SA134D	0050	SA144M	0280	SA152Y	0280
SA92F	0050	SA104L	0050	SA112X	0050	SA123C	0050	SA134E	0050	SA144R	0280	SA153A	0050
SA92G	0050	SA104M	0050	SA112Y	0050	SA123D	0050	SA134F	0050	SA144T	0280	SA153B	0050
SA92L	0050	SA104R	0050	SA113A	0050	SA123E	0050	SA134G	0050	SA144X	0280	SA153C	0050
SA92M	0050	SA104T	0050	SA113B	0050	SA123F	0050	SA134L	0050	SA144Y	0280	SA153D	0050
SA92R	0050	SA104X	0050	SA113C	0050	SA123G	0050	SA134M	0050	SA145A	0280	SA153E	0050
SA92T	0050	SA104Y	0050	SA113D	0050	SA123L	0050	SA134R	0050	SA145B	0280	SA153F	0050
SA92X	0050	SA105A	0136	SA113E	0050	SA123M	0050	SA134T	0050	SA145C	0280	SA153G	0050
SA92Y	0050	SA105B	0136	SA113F	0050	SA123R	0050	SA134X	0050	SA145D	0280	SA153L	0050
SA93A	0050	SA105C	0136	SA113G	0050	SA123T	0050	SA134Y	0050	SA145E	0280	SA153M	0050
SA93B	0050	SA105D	0136	SA113L	0050	SA123X	0050	SA135A	0280	SA145F	0280	SA153R	0050
SA93C	0050	SA105E	0136	SA113M	0050	SA123Y	0050	SA135B	0280	SA145G	0280	SA153T	0050
SA93D	0050	SA105F	0136	SA113R	0050	SA124A	0050	SA135C	0280	SA145L	0280	SA153X	0050
SA93E	0050	SA105G	0136	SA113T	0050	SA124B	0050	SA135D	0280	SA145M	0280	SA153Y	0050
SA93F	0050	SA105L	0136	SA113X	0050	SA124C	0050	SA135E	0280	SA145R	0280	SA154A	0050
SA93G	0050	SA105M	0136	SA113Y	0050	SA124D	0050	SA135F	0280	SA145T	0280	SA154B	0050
SA93L	0050	SA105R	0136	SA114A	0050	SA124E	0050	SA135G	0280	SA145X	0280	SA154C	0050
SA93M	0050	SA105T	0136	SA114B	0050	SA124F	0050	SA135L	0280	SA145Y	0280	SA154D	0050
SA93R	0050	SA105X	0136	SA114C	0050	SA124G	0050	SA135M	0280	SA146A	0280	SA154E	0050
SA93T	0050	SA105Y	0136	SA114D	0050	SA124L	0050	SA135R	0280	SA146B	0280	SA154F	0050
SA93X	0050	SA106A	0050	SA114E	0050	SA124M	0050	SA135T	0280	SA146C	0280	SA154G	0050
SA93Y	0050	SA106B	0050	SA114F	0050	SA124R	0050	SA135X	0280	SA146D	0280	SA154L	0050
SA94A	0136	SA106C	0050	SA114G	0050	SA124T	0050	SA135Y	0280	SA146E	0280	SA154M	0050
SA94B	0136	SA106D	0050	SA114L	0050	SA124X	0050	SA136A	0050	SA146F	0280	SA154R	0050
SA94C	0136	SA106E	0050	SA114M	0050	SA124Y	0050	SA136B	0050	SA146G	0280	SA154T	0050
SA94D	0136	SA106F	0050	SA114R	0050	SA125A	0050	SA136C	0050	SA146L	0280	SA154X	0050
SA94E	0136	SA106G	0050	SA114T	0050	SA125B	0050	SA136D	0050	SA146M	0280	SA154Y	0050
SA94F	0136	SA106L	0050	SA114X	0050	SA125C	0050	SA136E	0050	SA146R	0280	SA155A	0050

If replacement code is OEM, contact original manufacturer for replacement.

DEVICE TYPE	REPL CODE
SA155B	0050
SA155C	0050
SA155D	0050
SA155E	0050
SA155F	0050
SA155G	0050
SA155L	0050
SA155M	0050
SA155R	0050
SA155T	0050
SA155X	0050
SA155Y	0050
SA156A	0050
SA156B	0050
SA156C	0050
SA156D	0050
SA156E	0050
SA156F	0050
SA156G	0050
SA156L	0050
SA156M	0050
SA156R	0050
SA156T	0050
SA156X	0050
SA156Y	0050
SA157A	0050
SA157B	0050
SA157C	0050
SA157D	0050
SA157E	0050
SA157F	0050
SA157G	0050
SA157L	0050
SA157M	0050
SA157R	0050
SA157T	0050
SA157X	0050
SA157Y	0050
SA159A	0050
SA159B	0050
SA159C	0050
SA159D	0050
SA159E	0050
SA159F	0050
SA159G	0050
SA159L	0050
SA159M	0050
SA159R	0050
SA159T	0050
SA159X	0050
SA159Y	0050
SA160	0280
SA160A	0280
SA160A(DIO)	OEM
SA160B	0280
SA160C	0280
SA160D	0280
SA160E	0280
SA160F	0280
SA160G	0280
SA160L	0280
SA160M	0280
SA160R	0280
SA160T	0280
SA160X	0280
SA160Y	0280
SA161A	0050
SA161B	0050
SA161C	0050
SA161D	0050
SA161E	0050
SA161F	0050
SA161G	0050
SA161L	0050
SA161M	0050
SA161R	0050
SA161T	0050
SA161X	0050
SA161Y	0050
SA162A	0050
SA162B	0050
SA162C	0050
SA162D	0050
SA162E	0050
SA162F	0050
SA162G	0050
SA162L	0050
SA162M	0050
SA162R	0050
SA162T	0050
SA162X	0050
SA162Y	0050
SA163A	0050
SA163B	0050
SA163C	0050
SA163D	0050
SA163E	0050
SA163F	0050
SA163G	0050
SA163L	0050
SA163M	0050
SA163R	0050
SA163T	0050
SA163X	0050
SA163Y	0050
SA164A	0050
SA164B	0050
SA164C	0050
SA164D	0050
SA164E	0050
SA164F	0050
SA164G	0050
SA164L	0050
SA164M	0050
SA164R	0050
SA164T	0050
SA164X	0050
SA164Y	0050
SA165A	0050
SA165B	0050
SA165C	0050
SA165D	0050
SA165E	0050
SA165F	0050
SA165G	0050
SA165L	0050
SA165M	0050
SA165R	0050
SA165T	0050
SA165X	0050
SA165Y	0050
SA166A	0050
SA166B	0050
SA166C	0050
SA166D	0050
SA166E	0050
SA166F	0050
SA166G	0050
SA166L	0050
SA166M	0050
SA166R	0050
SA166T	0050
SA166X	0050
SA166Y	0050
SA167A	0280
SA167B	0280
SA167C	0280
SA167D	0280
SA167E	0280
SA167F	0280
SA167G	0280
SA167L	0280
SA167M	0280
SA167R	0280
SA167T	0280
SA167X	0280
SA167Y	0280
SA168A	0280
SA168B	0280
SA168C	0280
SA168D	0280
SA168E	0280
SA168F	0280
SA168G	0280
SA168L	0280
SA168M	0280
SA168R	0280
SA168T	0280
SA168X	0280
SA168Y	0280
SA169A	0280
SA169B	0280
SA169C	0280
SA169D	0280
SA169E	0280
SA169F	0280
SA169G	0280
SA169L	0280
SA169M	0280
SA169R	0280
SA169T	0280
SA169X	0280
SA169Y	0280
SA170A	0280
SA170A(DIO)	1405
SA170B	0280
SA170C	0280
SA170D	0280
SA170E	0280
SA170F	0280
SA170G	0280
SA170L	0280
SA170M	0280
SA170R	0280
SA170T	0280
SA170X	0280
SA170Y	0280
SA171A	0280
SA171B	0280
SA171C	0280
SA171D	0280
SA171E	0280
SA171F	0280
SA171G	0280
SA171L	0280
SA171M	0280
SA171R	0280
SA171T	0280
SA171X	0280
SA171Y	0280
SA172A	0280
SA172B	0280
SA172C	0280
SA172D	0280
SA172E	0280
SA172F	0280
SA172G	0280
SA172L	0280
SA172M	0280
SA172R	0280
SA172T	0280
SA172X	0280
SA172Y	0280
SA173A	0280
SA173B	0280
SA173C	0280
SA173D	0280
SA173E	0280
SA173F	0280
SA173G	0280
SA173L	0280
SA173M	0280
SA173R	0280
SA173T	0280
SA173X	0280
SA173Y	0280
SA174A	0280
SA174B	0280
SA174C	0280
SA174D	0280
SA174E	0280
SA174F	0280
SA174G	0280
SA174L	0280
SA174M	0280
SA174R	0280
SA174T	0280
SA174X	0280
SA174Y	0280
SA175A	0050
SA175B	0050
SA175C	0050
SA175D	0050
SA175E	0050
SA175F	0050
SA175G	0050
SA175L	0050
SA175M	0050
SA175R	0050
SA175T	0050
SA175X	0050
SA175Y	0050
SA176A	0136
SA176B	0136
SA176C	0136
SA176D	0136
SA176E	0136
SA176F	0136
SA176G	0136
SA176L	0136
SA176M	0136
SA176R	0136
SA176T	0136
SA176X	0136
SA176Y	0136
SA180A	0280
SA180B	0280
SA180C	0280
SA180D	0280
SA180E	0280
SA180F	0280
SA180G	0280
SA180L	0280
SA180M	0280
SA180R	0280
SA180T	0280
SA180X	0280
SA180Y	0280
SA181A	0280
SA181B	0280
SA181C	0280
SA181D	0280
SA181E	0280
SA181F	0280
SA181G	0280
SA181L	0280
SA181M	0280
SA181R	0280
SA181T	0280
SA181X	0280
SA181Y	0280
SA182A	0280
SA182B	0280
SA182C	0280
SA182D	0280
SA182E	0280
SA182F	0280
SA182G	0280
SA182L	0280
SA182M	0280
SA182R	0280
SA182T	0280
SA182X	0280
SA182Y	0280
SA183A	0280
SA183B	0280
SA183C	0280
SA183D	0280
SA183E	0280
SA183F	0280
SA183G	0280
SA183L	0280
SA183M	0280
SA183R	0280
SA183T	0280
SA183X	0280
SA183Y	0280
SA188A	0280
SA188B	0280
SA188C	0280
SA188D	0280
SA188E	0280
SA188F	0280
SA188G	0280
SA188L	0280
SA188M	0280
SA188R	0280
SA188T	0280
SA188X	0280
SA188Y	0280
SA189A	0280
SA189B	0280
SA189C	0280
SA189D	0280
SA189E	0280
SA189F	0280
SA189G	0280
SA189L	0280
SA189M	0280
SA189R	0280
SA189T	0280
SA189X	0280
SA189Y	0280
SA197	0004
SA197-1	0004
SA197-2	0004
SA197-3	0004
SA198A	0280
SA198B	0280
SA198C	0280
SA198D	0280
SA198E	0280
SA198F	0280
SA198G	0280
SA198L	0280
SA198M	0280
SA198R	0280
SA198T	0280
SA198X	0280
SA198Y	0280
SA201A	0280
SA201B	0280
SA201C	0280
SA201D	0280
SA201E	0280
SA201F	0280
SA201G	0280
SA201L	0280
SA201M	0280
SA201R	0280
SA201T	0280
SA201X	0280
SA201Y	0280
SA202A	0136
SA202B	0136
SA202C	0136
SA202D	0136
SA202E	0136
SA202F	0136
SA202G	0136
SA202L	0136
SA202M	0136
SA202R	0136
SA202T	0136
SA202X	0136
SA202Y	0136
SA203A	0280
SA203B	0280
SA203C	0280
SA203D	0280
SA203E	0280
SA203F	0280
SA203G	0280
SA203L	0280
SA203M	0280
SA203R	0280
SA203T	0280
SA203X	0280
SA203Y	0280
SA204	0004
SA205BLU	0004
SA205BRN	0004
SA205GRN	0004
SA205ORN	0004
SA205RED	0004
SA205VIO	0004
SA205WHT	0004
SA205YEL	0004
SA206A	0280
SA206B	0280
SA206C	0280
SA206D	0280
SA206E	0280
SA206F	0280
SA206G	0280
SA206L	0280
SA206M	0280
SA206R	0280
SA206T	0280
SA206X	0280
SA206Y	0280
SA207A	0280
SA207B	0280
SA207C	0280
SA207D	0280
SA207E	0280
SA207F	0280
SA207G	0280
SA207L	0280
SA207M	0280
SA207R	0280
SA207T	0280
SA207X	0280
SA207Y	0280
SA208A	0280
SA208B	0280
SA208C	0280
SA208D	0280
SA208E	0280
SA208F	0280
SA208G	0280
SA208L	0280
SA208M	0280
SA208R	0280
SA208T	0280
SA208X	0280
SA208Y	0280
SA209A	0280
SA209B	0280
SA209C	0280
SA209D	0280
SA209E	0280
SA209F	0280
SA209G	0280
SA209L	0280
SA209M	0280
SA209R	0280
SA209T	0280
SA209X	0280
SA209Y	0280
SA210A	0280
SA210B	0280
SA210C	0280
SA210D	0280
SA210E	0280
SA210F	0280
SA210G	0280
SA210L	0280
SA210M	0280
SA210R	0280
SA210T	0280
SA210X	0280
SA210Y	0280
SA211A	0280
SA211B	0280
SA211C	0280
SA211D	0280
SA211E	0280
SA211F	0280
SA211G	0280
SA211L	0280
SA211M	0280
SA211R	0280
SA211T	0280
SA211X	0280
SA211Y	0280
SA212A	0280
SA212B	0280
SA212C	0280
SA212D	0280
SA212E	0280
SA212F	0280
SA212G	0280
SA212L	0280
SA212M	0280
SA212R	0280
SA212T	0280
SA212X	0280
SA212Y	0280
SA213A	0050
SA213B	0050
SA213C	0050
SA213D	0050
SA213E	0050
SA213F	0050
SA213G	0050
SA213L	0050
SA213M	0050
SA213R	0050
SA213T	0050
SA213X	0050
SA213Y	0050
SA214A	0050
SA214B	0050
SA214C	0050
SA214D	0050
SA214E	0050
SA214F	0050
SA214G	0050
SA214L	0050
SA214M	0050
SA214R	0050
SA214T	0050
SA214X	0050
SA214Y	0050
SA215A	0050
SA215B	0050
SA215C	0050
SA215D	0050
SA215E	0050
SA215F	0050
SA215G	0050
SA215L	0050
SA215M	0050
SA215R	0050
SA215T	0050
SA215X	0050
SA215Y	0050
SA216A	0050
SA216B	0050
SA216C	0050
SA216D	0050
SA216E	0050
SA216F	0050
SA216G	0050
SA216L	0050
SA216M	0050
SA216R	0050
SA216T	0050
SA216X	0050
SA216Y	0050
SA217A	0050
SA217B	0050
SA217C	0050
SA217D	0050
SA217E	0050
SA217F	0050
SA217G	0050
SA217L	0050
SA217M	0050
SA217R	0050
SA217T	0050
SA217X	0050
SA217Y	0050
SA218A	0136
SA218B	0136
SA218C	0136
SA218D	0136
SA218E	0136
SA218F	0136
SA218G	0136
SA218L	0136
SA218M	0136
SA218R	0136
SA218T	0136
SA218X	0136
SA218Y	0136
SA219A	0050
SA219B	0050
SA219C	0050
SA219D	0050
SA219E	0050
SA219F	0050
SA219G	0050
SA219L	0050
SA219M	0050
SA219R	0050
SA219T	0050
SA219X	0050
SA219Y	0050
SA220A	0136
SA220B	0136
SA220C	0136
SA220D	0136
SA220E	0136
SA220F	0136
SA220G	0136
SA220L	0136
SA220M	0136
SA220R	0136
SA220T	0136
SA220X	0136
SA220Y	0136
SA221A	0136
SA221B	0136
SA221C	0136
SA221D	0136
SA221E	0136
SA221F	0136
SA221G	0136
SA221L	0136
SA221M	0136
SA221R	0136
SA221T	0136
SA221X	0136
SA221Y	0136
SA222A	0136
SA222B	0136
SA222C	0136
SA222D	0136
SA222E	0136
SA222F	0136
SA222G	0136
SA222L	0136
SA222M	0136
SA222R	0136
SA222T	0136
SA222X	0136
SA222Y	0136
SA223A	0050
SA223B	0050
SA223C	0050
SA223D	0050
SA223E	0050
SA223F	0050
SA223G	0050
SA223L	0050
SA223M	0050
SA223R	0050
SA223T	0050
SA223X	0050
SA223Y	0050
SA224A	0136
SA224B	0136
SA224C	0136
SA224D	0136
SA224E	0136
SA224F	0136
SA224G	0136
SA224L	0136
SA224M	0136
SA224X	0136
SA224Y	0136
SA225A	0050
SA225B	0050
SA225C	0050
SA225D	0050
SA225E	0050
SA225F	0050
SA225G	0050
SA225M	0050
SA225R	0050
SA225T	0050
SA225X	0050
SA225Y	0050
SA227A	0050
SA227B	0050
SA227C	0050
SA227D	0050
SA227E	0050
SA227F	0050
SA227G	0050
SA227L	0050
SA227M	0050
SA227R	0050
SA227T	0050
SA227X	0050
SA227Y	0050
SA229A	0050
SA229B	0050
SA229C	0050
SA229D	0050
SA229E	0050
SA229F	0050
SA229G	0050
SA229L	0050
SA229M	0050
SA229R	0050
SA229T	0050
SA229X	0050
SA229Y	0050
SA230A	0050
SA230B	0050
SA230C	0050
SA230D	0050
SA230E	0050
SA230F	0050
SA230G	0050
SA230L	0050
SA230M	0050
SA230R	0050
SA230T	0050
SA230X	0050
SA230Y	0050
SA231A	0847
SA231B	0847
SA231C	0847
SA231D	0847
SA231E	0847
SA231F	0847
SA231G	0847
SA231L	0847
SA231M	0847
SA231R	0847
SA231T	0847
SA231X	0847
SA231Y	0847
SA233A	0050
SA233B	0050
SA233C	0050
SA233D	0050
SA233E	0050
SA233F	0050
SA233G	0050
SA233L	0050
SA233M	0050
SA233R	0050
SA233T	0050
SA233X	0050
SA233Y	0050
SA234A	0136
SA234B	0136
SA234C	0136
SA234D	0136
SA234E	0136
SA234F	0136
SA234G	0136
SA234L	0136
SA234M	0136
SA234R	0136
SA234T	0136
SA234X	0136
SA234Y	0136
SA235A	0050
SA235B	0050
SA235C	0050
SA235D	0050
SA235E	0050
SA235F	0050
SA235G	0050
SA235L	0050
SA235M	0050
SA235R	0050

If replacement code is OEM, contact original manufacturer for replacement.

DEVICE TYPE	REPL CODE	DEVICE TYPE	REPL CODE	DEVICE TYPE	REPL CODE	DEVICE TYPE	REPL CODE	DEVICE TYPE	REPL CODE	DEVICE TYPE	REPL CODE	DEVICE TYPE	REPL CODE
SA235T	0050	SA249B	0136	SA257G	0050	SA265X	0050	SA274D	0136	SA283G	0280	SA291X	0050
SA235X	0050	SA249C	0136	SA257L	0050	SA265Y	0050	SA274E	0136	SA283L	0280	SA291Y	0050
SA235Y	0050	SA249D	0136	SA257M	0050	SA266A	0136	SA274F	0136	SA283M	0280	SA292A	0050
SA236A	0050	SA249E	0136	SA257R	0050	SA266B	0136	SA274G	0136	SA283R	0280	SA292B	0050
SA236B	0050	SA249F	0136	SA257T	0050	SA266C	0136	SA274L	0136	SA283T	0280	SA292C	0050
SA236C	0050	SA249G	0136	SA257X	0050	SA266D	0136	SA274M	0136	SA283X	0280	SA292D	0050
SA236D	0050	SA249L	0136	SA257Y	0050	SA266E	0136	SA274R	0136	SA283Y	0280	SA292E	0050
SA236E	0050	SA249M	0136	SA258A	0050	SA266F	0136	SA274T	0136	SA284A	0280	SA292F	0050
SA236F	0050	SA249R	0136	SA258B	0050	SA266G	0136	SA274X	0136	SA284B	0280	SA292G	0050
SA236G	0050	SA249T	0136	SA258C	0050	SA266L	0136	SA274Y	0136	SA284C	0280	SA292L	0050
SA236L	0050	SA249X	0136	SA258D	0050	SA266M	0136	SA275A	0136	SA284D	0280	SA292M	0050
SA236M	0050	SA249Y	0136	SA258E	0050	SA266R	0136	SA275B	0136	SA284E	0280	SA292R	0050
SA236R	0050	SA250A	0136	SA258F	0050	SA266T	0136	SA275C	0136	SA284F	0280	SA292T	0050
SA236T	0050	SA250B	0136	SA258G	0050	SA266X	0136	SA275D	0136	SA284G	0280	SA292X	0050
SA236X	0050	SA250C	0136	SA258L	0050	SA266Y	0136	SA275E	0136	SA284L	0280	SA292Y	0050
SA236Y	0050	SA250D	0136	SA258M	0050	SA267A	0050	SA275F	0136	SA284M	0280	SA293A	0050
SA237A	0136	SA250E	0136	SA258R	0050	SA267B	0050	SA275G	0136	SA284R	0280	SA293B	0050
SA237B	0136	SA250F	0136	SA258T	0050	SA267C	0050	SA275L	0136	SA284T	0280	SA293C	0050
SA237C	0136	SA250G	0136	SA258X	0050	SA267D	0050	SA275M	0136	SA284X	0280	SA293D	0050
SA237D	0136	SA250L	0136	SA258Y	0050	SA267E	0050	SA275R	0136	SA284Y	0280	SA293E	0050
SA237E	0136	SA250M	0136	SA259A	0136	SA267F	0050	SA275T	0136	SA285A	0136	SA293F	0050
SA237F	0136	SA250R	0136	SA259B	0136	SA267G	0050	SA275X	0136	SA285B	0136	SA293G	0050
SA237G	0136	SA250T	0136	SA259C	0136	SA267L	0050	SA275Y	0136	SA285C	0136	SA293L	0050
SA237L	0136	SA250X	0136	SA259D	0136	SA267M	0050	SA277A	0280	SA285D	0136	SA293M	0050
SA237M	0136	SA250Y	0136	SA259E	0136	SA267R	0050	SA277B	0280	SA285E	0136	SA293R	0050
SA237R	0136	SA251A	0050	SA259F	0136	SA267T	0050	SA277C	0280	SA285F	0136	SA293T	0050
SA237T	0136	SA251B	0050	SA259G	0136	SA267X	0050	SA277D	0280	SA285G	0136	SA293X	0050
SA237X	0136	SA251C	0050	SA259L	0136	SA267Y	0050	SA277E	0280	SA285L	0136	SA293Y	0050
SA237Y	0136	SA251D	0050	SA259M	0136	SA268A	0050	SA277F	0280	SA285M	0136	SA294A	0050
SA238A	0050	SA251E	0050	SA259R	0136	SA268B	0050	SA277G	0280	SA285R	0136	SA294B	0050
SA238B	0050	SA251F	0050	SA259T	0136	SA268C	0050	SA277L	0280	SA285T	0136	SA294C	0050
SA238C	0050	SA251G	0050	SA259X	0136	SA268D	0050	SA277M	0280	SA285X	0136	SA294D	0050
SA238D	0050	SA251L	0050	SA259Y	0136	SA268E	0050	SA277R	0280	SA285Y	0136	SA294E	0050
SA238E	0050	SA251M	0050	SA260A	0050	SA268F	0050	SA277T	0280	SA286A	0136	SA294F	0050
SA238F	0050	SA251R	0050	SA260B	0050	SA268G	0050	SA277X	0280	SA286B	0136	SA294G	0050
SA238G	0050	SA251T	0050	SA260C	0050	SA268L	0050	SA277Y	0280	SA286C	0136	SA294L	0050
SA238L	0050	SA251X	0050	SA260D	0050	SA268M	0050	SA278A	0280	SA286D	0136	SA294M	0050
SA238M	0050	SA251Y	0050	SA260E	0050	SA268R	0050	SA278B	0280	SA286E	0136	SA294R	0050
SA238R	0050	SA252A	0050	SA260F	0050	SA268T	0050	SA278C	0280	SA286F	0136	SA294T	0050
SA238T	0050	SA252B	0050	SA260G	0050	SA268X	0050	SA278D	0280	SA286G	0136	SA294X	0050
SA238X	0050	SA252C	0050	SA260L	0050	SA268Y	0050	SA278E	0280	SA286L	0136	SA294Y	0050
SA238Y	0050	SA252D	0050	SA260M	0050	SA269A	0136	SA278F	0280	SA286M	0136	SA295A	0050
SA239A	0050	SA252E	0050	SA260R	0050	SA269B	0136	SA278G	0280	SA286R	0136	SA295B	0050
SA239B	0050	SA252F	0050	SA260T	0050	SA269C	0136	SA278L	0280	SA286T	0136	SA295C	0050
SA239C	0050	SA252G	0050	SA260X	0050	SA269D	0136	SA278M	0280	SA286X	0136	SA295D	0050
SA239D	0050	SA252L	0050	SA260Y	0050	SA269E	0136	SA278R	0280	SA286Y	0136	SA295E	0050
SA239E	0050	SA252M	0050	SA261A	0050	SA269F	0136	SA278T	0280	SA287A	0136	SA295F	0050
SA239F	0050	SA252R	0050	SA261B	0050	SA269G	0136	SA278X	0280	SA287B	0136	SA295G	0050
SA239G	0050	SA252T	0050	SA261C	0050	SA269L	0136	SA278Y	0280	SA287C	0136	SA295L	0050
SA239L	0050	SA252X	0050	SA261D	0050	SA269M	0136	SA279A	0280	SA287D	0136	SA295M	0050
SA239M	0050	SA252Y	0050	SA261E	0050	SA269R	0136	SA279B	0280	SA287E	0136	SA295R	0050
SA239R	0050	SA253A	0050	SA261F	0050	SA269T	0136	SA279C	0280	SA287F	0136	SA295T	0050
SA239T	0050	SA253B	0050	SA261G	0050	SA269X	0136	SA279D	0280	SA287G	0136	SA295X	0050
SA239X	0050	SA253C	0050	SA261L	0050	SA269Y	0136	SA279E	0280	SA287L	0136	SA295Y	0050
SA239Y	0050	SA253D	0050	SA261M	0050	SA270A	0050	SA279F	0280	SA287M	0136	SA296A	0280
SA240	0004	SA253E	0050	SA261R	0050	SA270B	0050	SA279G	0280	SA287R	0136	SA296B	0280
SA241A	0050	SA253F	0050	SA261T	0050	SA270C	0050	SA279L	0280	SA287T	0136	SA296C	0280
SA241B	0050	SA253G	0050	SA261X	0050	SA270D	0050	SA279M	0280	SA287X	0136	SA296D	0280
SA241C	0050	SA253L	0050	SA261Y	0050	SA270E	0050	SA279R	0280	SA287Y	0136	SA296E	0280
SA241D	0050	SA253M	0050	SA262A	0050	SA270F	0050	SA279T	0280	SA288A	0136	SA296F	0280
SA241E	0050	SA253R	0050	SA262B	0050	SA270G	0050	SA279X	0280	SA288B	0136	SA296G	0280
SA241F	0050	SA253T	0050	SA262C	0050	SA270L	0050	SA279Y	0280	SA288C	0136	SA296L	0280
SA241G	0050	SA253X	0050	SA262D	0050	SA270M	0050	SA280A	0050	SA288D	0136	SA296M	0280
SA241L	0050	SA253Y	0050	SA262E	0050	SA270R	0050	SA280B	0050	SA288E	0136	SA296R	0280
SA241M	0050	SA254A	0280	SA262F	0050	SA270T	0050	SA280C	0050	SA288F	0136	SA296T	0280
SA241R	0050	SA254B	0280	SA262G	0050	SA270X	0050	SA280D	0050	SA288G	0136	SA296X	0280
SA241T	0050	SA254C	0280	SA262L	0050	SA270Y	0050	SA280E	0050	SA288L	0136	SA296Y	0280
SA241X	0050	SA254D	0280	SA262M	0050	SA271A	0136	SA280F	0050	SA288M	0136	SA297A	0280
SA241Y	0050	SA254E	0280	SA262R	0050	SA271B	0136	SA280G	0050	SA288R	0136	SA297B	0280
SA246A	0050	SA254F	0280	SA262T	0050	SA271C	0136	SA280L	0050	SA288T	0136	SA297C	0280
SA246B	0050	SA254G	0280	SA262X	0050	SA271D	0136	SA280M	0050	SA288X	0136	SA297E	0280
SA246C	0050	SA254L	0280	SA262Y	0050	SA271E	0136	SA280R	0050	SA288Y	0136	SA297F	0280
SA246D	0050	SA254M	0280	SA263A	0050	SA271F	0136	SA280T	0050	SA289A	0050	SA297G	0280
SA246E	0050	SA254R	0280	SA263B	0050	SA271G	0136	SA280X	0050	SA289B	0050	SA297L	0280
SA246F	0050	SA254T	0280	SA263C	0050	SA271L	0136	SA280Y	0050	SA289C	0050	SA297M	0280
SA246G	0050	SA254X	0280	SA263D	0050	SA271M	0136	SA281A	0050	SA289D	0050	SA297R	0280
SA246L	0050	SA254Y	0280	SA263E	0050	SA271R	0136	SA281B	0050	SA289E	0050	SA297T	0280
SA246M	0050	SA255A	0280	SA263F	0050	SA271T	0136	SA281C	0050	SA289F	0050	SA297X	0280
SA246R	0050	SA255B	0280	SA263G	0050	SA271X	0136	SA281D	0050	SA289G	0050	SA297Y	0280
SA246T	0050	SA255C	0280	SA263L	0050	SA271Y	0136	SA281E	0050	SA289L	0050	SA298A	0136
SA246X	0050	SA255D	0280	SA263M	0050	SA272A	0136	SA281F	0050	SA289M	0050	SA298B	0136
SA246Y	0050	SA255E	0280	SA263R	0050	SA272B	0136	SA281G	0050	SA289R	0050	SA298C	0136
SA247A	0050	SA255F	0280	SA263T	0050	SA272C	0136	SA281L	0050	SA289T	0050	SA298D	0136
SA247B	0050	SA255G	0280	SA263X	0050	SA272D	0136	SA281M	0050	SA289X	0050	SA298E	0136
SA247C	0050	SA255L	0280	SA263Y	0050	SA272E	0136	SA281R	0050	SA289Y	0050	SA298L	0136
SA247D	0050	SA255M	0280	SA264A	0050	SA272F	0136	SA281T	0050	SA290A	0050	SA298M	0136
SA247E	0050	SA255R	0280	SA264B	0050	SA272G	0136	SA281X	0050	SA290B	0050	SA298R	0136
SA247F	0050	SA255T	0280	SA264C	0050	SA272L	0136	SA281Y	0050	SA290C	0050	SA298T	0136
SA247G	0050	SA255X	0280	SA264D	0050	SA272M	0136	SA282	0133	SA290D	0050	SA298X	0136
SA247L	0050	SA255Y	0280	SA264E	0050	SA272R	0136	SA282A	0280	SA290E	0050	SA298Y	0136
SA247M	0050	SA256A	0136	SA264F	0050	SA272T	0136	SA282B	0280	SA290F	0050	SA301	OEM
SA247R	0050	SA256B	0136	SA264G	0050	SA272X	0136	SA282C	0280	SA290G	0050	SA301A	0050
SA247T	0050	SA256C	0136	SA264L	0050	SA272Y	0136	SA282D	0280	SA290L	0050	SA301B	0050
SA247X	0050	SA256D	0136	SA264M	0050	SA273A	0136	SA282E	0280	SA290M	0050	SA301C	0050
SA247Y	0050	SA256E	0136	SA264R	0050	SA273B	0136	SA282F	0280	SA290R	0050	SA301D	0050
SA248A	0280	SA256F	0136	SA264T	0050	SA273C	0136	SA282G	0280	SA290T	0050	SA301E	0050
SA248B	0280	SA256G	0136	SA264X	0050	SA273D	0136	SA282L	0280	SA290X	0050	SA301F	0050
SA248C	0280	SA256L	0136	SA264Y	0050	SA273E	0136	SA282M	0280	SA290Y	0050	SA301G	0050
SA248D	0280	SA256M	0136	SA265A	0050	SA273F	0136	SA282R	0280	SA291A	0050	SA301L	0050
SA248E	0280	SA256R	0136	SA265B	0050	SA273G	0136	SA282T	0280	SA291B	0050	SA301M	0050
SA248F	0280	SA256T	0136	SA265C	0050	SA273L	0136	SA282X	0280	SA291C	0050	SA301R	0050
SA248G	0280	SA256X	0136	SA265D	0050	SA273M	0136	SA282Y	0280	SA291D	0050	SA301T	0050
SA248L	0280	SA256Y	0136	SA265E	0050	SA273R	0136	SA283	0133	SA291E	0050	SA301X	0050
SA248M	0280	SA257A	0050	SA265F	0050	SA273T	0136	SA283A	0280	SA291F	0050	SA301Y	0050
SA248R	0280	SA257B	0050	SA265G	0050	SA273X	0136	SA283B	0280	SA291G	0050	SA303AY	OEM
SA248T	0280	SA257C	0050	SA265L	0050	SA273Y	0136	SA283C	0280	SA291L	0050	SA304A	0847
SA248X	0280	SA257D	0050	SA265M	0050	SA274A	0136	SA283D	0280	SA291M	0050	SA304B	0847
SA248Y	0280	SA257E	0050	SA265R	0050	SA274B	0136	SA283E	0280	SA291R	0050	SA304C	0847
SA249A	0136	SA257F	0050	SA265T	0050	SA274C	0136	SA283F	0280	SA291T	0050	SA304D	0847

DEVICE TYPE	REPL CODE
SA304E	0847
SA304F	0847
SA304G	0847
SA304L	0847
SA304M	0847
SA304R	0847
SA304T	0847
SA304X	0847
SA304Y	0847
SA305A	0280
SA305B	0280
SA305C	0280
SA305D	0280
SA305E	0280
SA305F	0280
SA305G	0280
SA305L	0280
SA305M	0280
SA305R	0280
SA305T	0280
SA305X	0280
SA305Y	0280
SA307A	0136
SA307B	0136
SA307C	0136
SA307D	0136
SA307E	0136
SA307F	0136
SA307G	0136
SA307L	0136
SA307M	0136
SA307R	0136
SA307T	0136
SA307X	0136
SA307Y	0136
SA308A	0050
SA308B	0050
SA308C	0050
SA308D	0050
SA308E	0050
SA308F	0050
SA308G	0050
SA308R	0050
SA308T	0050
SA308X	0050
SA308Y	0050
SA309A	0050
SA309B	0050
SA309C	0050
SA309D	0050
SA309E	0050
SA309F	0050
SA309G	0050
SA309L	0050
SA309M	0050
SA309R	0050
SA309T	0050
SA309X	0050
SA309Y	0050
SA310	0037
SA311	0037
SA312	0037
SA313	0037
SA313A	0037
SA314	0037
SA315	0037
SA316	0037
SA318-2	0164
SA318-3	0164
SA321A	0050
SA321B	0050
SA321C	0050
SA321D	0050
SA321E	0050
SA321F	0050
SA321G	0050
SA321L	0050
SA321M	0050
SA321R	0050
SA321T	0050
SA321X	0050
SA321Y	0050
SA322A	0136
SA322B	0136
SA322C	0136
SA322D	0136
SA322E	0136
SA322F	0136
SA322G	0136
SA322L	0136
SA322M	0136
SA322R	0136
SA322T	0136
SA322X	0136
SA322Y	0136
SA323A	0136
SA323B	0136
SA323C	0136
SA323D	0136
SA323E	0136
SA323F	0136
SA323G	0136
SA323L	0136
SA323M	0136
SA323R	0136
SA323T	0136
SA323X	0136
SA323Y	0136
SA324A	0050
SA324B	0050
SA324C	0050
SA324D	0050
SA324E	0050
SA324F	0050
SA324G	0050
SA324L	0050
SA324M	0050
SA324R	0050
SA324T	0050
SA324X	0050
SA324Y	0050
SA325A	0050
SA325B	0050
SA325C	0050
SA325D	0050
SA325E	0050
SA325F	0050
SA325G	0050
SA325L	0050
SA325M	0050
SA325R	0050
SA325T	0050
SA325X	0050
SA325Y	0050
SA326A	0050
SA326B	0050
SA326C	0050
SA326D	0050
SA326E	0050
SA326F	0050
SA326G	0050
SA326L	0050
SA326M	0050
SA326R	0050
SA326T	0050
SA326X	0050
SA326Y	0050
SA329A	0050
SA329B	0050
SA329C	0050
SA329D	0050
SA329E	0050
SA329F	0050
SA329G	0050
SA329L	0050
SA329M	0050
SA329R	0050
SA329T	0050
SA329X	0050
SA329Y	0050
SA330A	0280
SA330B	0280
SA330C	0280
SA330D	0280
SA330E	0280
SA330F	0280
SA330G	0280
SA330L	0280
SA330M	0280
SA330R	0280
SA330T	0280
SA330X	0280
SA330Y	0280
SA331A	0050
SA331B	0050
SA331C	0050
SA331D	0050
SA331E	0050
SA331F	0050
SA331G	0050
SA331L	0050
SA331M	0050
SA331R	0050
SA331T	0050
SA331X	0050
SA331Y	0050
SA332A	0136
SA332B	0136
SA332C	0136
SA332D	0136
SA332E	0136
SA332F	0136
SA332G	0136
SA332L	0136
SA332M	0136
SA332R	0136
SA332T	0136
SA332X	0136
SA332Y	0136
SA334A	0050
SA334B	0050
SA334C	0050
SA334D	0050
SA334E	0050
SA334F	0050
SA334G	0050
SA334L	0050
SA334M	0050
SA334R	0050
SA334T	0050
SA334X	0050
SA334Y	0050
SA335A	0050
SA335B	0050
SA335C	0050
SA335D	0050
SA335E	0050
SA335F	0050
SA335G	0050
SA335L	0050
SA335M	0050
SA335R	0050
SA335T	0050
SA335X	0050
SA335Y	0050
SA337A	0050
SA337B	0050
SA337C	0050
SA337D	0050
SA337E	0050
SA337F	0050
SA337G	0050
SA337L	0050
SA337M	0050
SA337R	0050
SA337T	0050
SA337X	0050
SA337Y	0050
SA338A	0136
SA338B	0136
SA338C	0136
SA338D	0136
SA338E	0136
SA338F	0136
SA338G	0136
SA338L	0136
SA338M	0136
SA338R	0136
SA338T	0136
SA338X	0136
SA338Y	0136
SA339A	0136
SA339B	0136
SA339C	0136
SA339D	0136
SA339E	0136
SA339F	0136
SA339G	0136
SA339L	0136
SA339M	0136
SA339R	0136
SA339T	0136
SA339X	0136
SA339Y	0136
SA341A	0136
SA341B	0136
SA341C	0136
SA341D	0136
SA341E	0136
SA341F	0136
SA341G	0136
SA341L	0136
SA341M	0136
SA341R	0136
SA341T	0136
SA341X	0136
SA341Y	0136
SA342A	0136
SA342B	0136
SA342C	0136
SA342D	0136
SA342E	0136
SA342F	0136
SA342G	0136
SA342L	0136
SA342M	0136
SA342R	0136
SA342T	0136
SA342X	0136
SA342Y	0136
SA343A	0050
SA343B	0050
SA343C	0050
SA343D	0050
SA343E	0050
SA343F	0050
SA343G	0050
SA343L	0050
SA343M	0050
SA343R	0050
SA343T	0050
SA343X	0050
SA343Y	0050
SA344A	0050
SA344B	0050
SA344C	0050
SA344D	0050
SA344E	0050
SA344F	0050
SA344G	0050
SA344L	0050
SA344M	0050
SA344R	0050
SA344T	0050
SA344X	0050
SA344Y	0050
SA348A	0050
SA348B	0050
SA348C	0050
SA348D	0050
SA348E	0050
SA348F	0050
SA348G	0050
SA348L	0050
SA348M	0050
SA348R	0050
SA348T	0050
SA348X	0050
SA348Y	0050
SA350A	0050
SA350B	0050
SA350C	0050
SA350D	0050
SA350E	0050
SA350F	0050
SA350G	0050
SA350L	0050
SA350M	0050
SA350R	0050
SA350T	0050
SA350X	0050
SA350Y	0050
SA351A	0050
SA351B	0050
SA351C	0050
SA351D	0050
SA351E	0050
SA351F	0050
SA351G	0050
SA351L	0050
SA351M	0050
SA351R	0050
SA351T	0050
SA351X	0050
SA351Y	0050
SA352A	0050
SA352B	0050
SA352C	0050
SA352D	0050
SA352E	0050
SA352F	0050
SA352G	0050
SA352L	0050
SA352M	0050
SA352R	0050
SA352T	0050
SA352X	0050
SA352Y	0050
SA353A	0050
SA353B	0050
SA353C	0050
SA353D	0050
SA353E	0050
SA353F	0050
SA353G	0050
SA353L	0050
SA353M	0050
SA353R	0050
SA353T	0050
SA353X	0050
SA353Y	0050
SA354B	0038
SA355A	0050
SA355B	0050
SA355C	0050
SA355D	0050
SA355E	0050
SA355F	0050
SA355G	0050
SA355L	0050
SA355M	0050
SA355R	0050
SA355T	0050
SA355X	0050
SA355Y	0050
SA356A	0136
SA356B	0136
SA356C	0136
SA356D	0136
SA356E	0136
SA356F	0136
SA356G	0136
SA356L	0136
SA356M	0136
SA356R	0136
SA356T	0136
SA356X	0136
SA356Y	0136
SA357A	0136
SA357B	0136
SA357C	0136
SA357D	0136
SA357E	0136
SA357F	0136
SA357G	0136
SA357L	0136
SA357M	0136
SA357R	0136
SA357T	0136
SA357X	0136
SA357Y	0136
SA358A	0050
SA358B	0050
SA358C	0050
SA358D	0050
SA358E	0050
SA358F	0050
SA358G	0050
SA358L	0050
SA358M	0050
SA358R	0050
SA358T	0050
SA358X	0050
SA358Y	0050
SA359A	0050
SA359B	0050
SA359C	0050
SA359D	0050
SA359E	0050
SA359F	0050
SA359G	0050
SA359L	0050
SA359M	0050
SA359R	0050
SA359T	0050
SA359X	0050
SA359Y	0050
SA360A	0050
SA360B	0050
SA360C	0050
SA360D	0050
SA360E	0050
SA360F	0050
SA360G	0050
SA360L	0050
SA360M	0050
SA360R	0050
SA360T	0050
SA360X	0050
SA360Y	0050
SA361A	0050
SA361B	0050
SA361C	0050
SA361D	0050
SA361E	0050
SA361F	0050
SA361G	0050
SA361L	0050
SA361M	0050
SA361R	0050
SA361T	0050
SA361X	0050
SA361Y	0050
SA364A	0050
SA364B	0050
SA364C	0050
SA364D	0050
SA364E	0050
SA364F	0050
SA364G	0050
SA364L	0050
SA364M	0050
SA364R	0050
SA364T	0050
SA364X	0050
SA364Y	0050
SA365A	0050
SA365B	0050
SA365C	0050
SA365D	0050
SA365E	0050
SA365F	0050
SA365G	0050
SA365L	0050
SA365M	0050
SA365R	0050
SA365T	0050
SA365X	0050
SA365Y	0050
SA366A	0050
SA366B	0050
SA366C	0050
SA366D	0050
SA366E	0050
SA366F	0050
SA366G	0050
SA366L	0050
SA366M	0050
SA366R	0050
SA366T	0050
SA366X	0050
SA366Y	0050
SA367A	0050
SA367B	0050
SA367C	0050
SA367D	0050
SA367E	0050
SA367F	0050
SA367G	0050
SA367L	0050
SA367M	0050
SA367R	0050
SA367T	0050
SA367X	0050
SA367Y	0050
SA368A	0050
SA368B	0050
SA368C	0050
SA368D	0050
SA368E	0050
SA368F	0050
SA368G	0050
SA368L	0050
SA368M	0050
SA368R	0050
SA368T	0050
SA368X	0050
SA368Y	0050
SA369A	0136
SA369B	0136
SA369C	0136
SA369D	0136
SA369E	0136
SA369F	0136
SA369G	0136
SA369L	0136
SA369M	0136
SA369R	0136
SA369T	0136
SA369X	0136
SA369Y	0136
SA375A	0847
SA375B	0847
SA375C	0847
SA375D	0847
SA375E	0847
SA375F	0847
SA375G	0847
SA375L	0847
SA375M	0847
SA375R	0847
SA375T	0847
SA375X	0847
SA375Y	0847
SA376A	0050
SA376B	0050
SA376C	0050
SA376D	0050
SA376E	0050
SA376F	0050
SA376G	0050
SA376L	0050
SA376M	0050
SA376R	0050
SA376T	0050
SA376X	0050
SA376Y	0050
SA377A	0050
SA377B	0050
SA377C	0050
SA377D	0050
SA377E	0050
SA377F	0050
SA377G	0050
SA377L	0050
SA377M	0050
SA377R	0050
SA377T	0050
SA377X	0050
SA377Y	0050
SA378A	0050
SA378B	0050
SA378C	0050
SA378D	0050
SA378E	0050
SA378F	0050
SA378G	0050
SA378L	0050
SA378M	0050
SA378R	0050
SA378T	0050
SA378X	0050
SA378Y	0050
SA379A	0050
SA379B	0050
SA379C	0050
SA379D	0050
SA379E	0050
SA379F	0050
SA379G	0050
SA379L	0050
SA379M	0050
SA379R	0050
SA379T	0050
SA379X	0050
SA379Y	0050
SA380A	0050
SA380B	0050
SA380C	0050
SA380D	0050
SA380E	0050
SA380F	0050
SA380G	0050
SA380L	0050
SA380M	0050
SA380R	0050
SA380T	0050
SA380X	0050
SA380Y	0050
SA381A	0136
SA381B	0136
SA381C	0136
SA381D	0136
SA381E	0136
SA381F	0136
SA381G	0136
SA381L	0136
SA381M	0136
SA381R	0136
SA381T	0136
SA381X	0136
SA381Y	0136
SA382A	0136
SA382B	0136
SA382C	0136
SA382D	0136
SA382E	0136
SA382F	0136
SA382G	0136
SA382L	0136
SA382M	0136
SA382R	0136
SA382T	0136
SA382X	0136
SA382Y	0136
SA384A	0136
SA384B	0136
SA384C	0136
SA384D	0136
SA384E	0136
SA384F	0136
SA384G	0136
SA384L	0136
SA384M	0136
SA384R	0136
SA384T	0136
SA384X	0136
SA384Y	0136
SA385A	0050
SA385B	0050
SA385C	0050
SA385D	0050
SA385E	0050
SA385F	0050
SA385G	0050
SA385L	0050
SA385M	0050
SA385R	0050
SA385T	0050
SA385X	0050
SA385Y	0050
SA391A	0050
SA391B	0050
SA391C	0050
SA391D	0050
SA391E	0050
SA391F	0050
SA391G	0050
SA391L	0050
SA391M	0050
SA391R	0050
SA391T	0050
SA391X	0050
SA391Y	0050
SA392A	0050
SA392B	0050
SA392C	0050
SA392D	0050
SA392E	0050
SA392F	0050
SA392G	0050
SA392L	0050
SA392M	0050
SA392R	0050
SA392T	0050
SA392X	0050
SA392Y	0050
SA393A	0050
SA393B	0050
SA393C	0050
SA393D	0050
SA393E	0050
SA393F	0050
SA393G	0050
SA393L	0050
SA393M	0050
SA393R	0050
SA393T	0050
SA393X	0050
SA393Y	0050
SA394A	0050
SA394B	0050
SA394C	0050
SA394D	0050
SA394E	0050
SA394F	0050
SA394G	0050
SA394L	0050
SA394M	0050
SA394R	0050
SA394T	0050
SA394X	0050
SA394Y	0050
SA395A	0050
SA395B	0050
SA395C	0050
SA395D	0050
SA395E	0050
SA395F	0050
SA395G	0050
SA395L	0050
SA395M	0050
SA395R	0050
SA395T	0050
SA395X	0050
SA395Y	0050
SA398A	0050
SA398B	0050
SA398C	0050
SA398D	0050
SA398E	0050
SA398F	0050
SA398G	0050
SA398L	0050
SA398M	0050
SA398R	0050
SA398T	0050
SA398X	0050
SA398Y	0050
SA399A	0050
SA399B	0050
SA399C	0050
SA399D	0050

If replacement code is OEM, contact original manufacturer for replacement.

DEVICE TYPE	REPL CODE	DEVICE TYPE	REPL CODE	DEVICE TYPE	REPL CODE	DEVICE TYPE	REPL CODE	DEVICE TYPE	REPL CODE	DEVICE TYPE	REPL CODE	DEVICE TYPE	REPL CODE
SA399E	0050	SA420C	0050	SA435L	0136	SA454Y	0050	SA470M	0050	SA479A	0050	SA493F	0037
SA399F	0050	SA420D	0050	SA435M	0136	SA455A	0050	SA470R	0050	SA479B	0050	SA493G	0037
SA399G	0050	SA420E	0050	SA435R	0136	SA455B	0050	SA470T	0050	SA479C	0050	SA493L	0037
SA399L	0050	SA420F	0050	SA435T	0136	SA455C	0050	SA470X	0050	SA479D	0050	SA493M	0037
SA399M	0050	SA420G	0050	SA435X	0136	SA455D	0050	SA470Y	0050	SA479E	0050	SA493R	0037
SA399R	0050	SA420L	0050	SA435Y	0136	SA455E	0050	SA471A	0050	SA479F	0050	SA493T	0037
SA399T	0050	SA420M	0050	SA436A	0050	SA455F	0050	SA471B	0050	SA479G	0050	SA493X	0037
SA399X	0050	SA420R	0050	SA436B	0050	SA455G	0050	SA471C	0050	SA479L	0050	SA493Y	0037
SA399Y	0050	SA420T	0050	SA436C	0050	SA455L	0050	SA471D	0050	SA479M	0050	SA494A	0037
SA400A	0136	SA420X	0050	SA436D	0050	SA455M	0050	SA471E	0050	SA479R	0050	SA494B	0037
SA400B	0136	SA420Y	0050	SA436E	0050	SA455R	0050	SA471F	0050	SA479T	0050	SA494C	0037
SA400C	0136	SA426A	0050	SA436F	0050	SA455T	0050	SA471G	0050	SA479X	0050	SA494D	0037
SA400D	0136	SA426B	0050	SA436G	0050	SA455X	0050	SA471L	0050	SA479Y	0050	SA494E	0037
SA400E	0136	SA426C	0050	SA436L	0050	SA455Y	0050	SA471M	0050	SA480A	0037	SA494F	0037
SA400F	0136	SA426D	0050	SA436M	0050	SA456A	0050	SA471R	0050	SA480B	0037	SA494G	0037
SA400G	0136	SA426E	0050	SA436R	0050	SA456B	0050	SA471T	0050	SA480C	0037	SA494L	0037
SA400L	0136	SA426F	0050	SA436T	0050	SA456C	0050	SA471X	0050	SA480D	0037	SA494M	0037
SA400M	0136	SA426G	0050	SA436X	0050	SA456D	0050	SA471Y	0050	SA480E	0037	SA494R	0037
SA400R	0136	SA426L	0050	SA436Y	0050	SA456E	0050	SA472A	0050	SA480F	0037	SA494T	0037
SA400T	0136	SA426M	0050	SA437A	0050	SA456F	0050	SA472B	0050	SA480G	0037	SA494X	0037
SA400X	0136	SA426R	0050	SA437B	0050	SA456G	0050	SA472C	0050	SA480L	0037	SA494Y	0037
SA400Y	0136	SA426T	0050	SA437C	0050	SA456L	0050	SA472D	0050	SA480M	0037	SA495	0150
SA403A	0050	SA426X	0050	SA437D	0050	SA456M	0050	SA472E	0050	SA480R	0037	SA495A	0037
SA403B	0050	SA426Y	0050	SA437E	0050	SA456R	0050	SA472F	0050	SA480T	0037	SA495B	0037
SA403C	0050	SA427A	0050	SA437F	0050	SA456T	0050	SA472G	0050	SA480X	0037	SA495C	0037
SA403D	0050	SA427B	0050	SA437G	0050	SA456X	0050	SA472L	0050	SA480Y	0037	SA495D	0037
SA403E	0050	SA427C	0050	SA437L	0050	SA456Y	0050	SA472M	0050	SA482A	0037	SA495E	0037
SA403F	0050	SA427D	0050	SA437M	0050	SA457A	0050	SA472R	0050	SA482B	0037	SA495F	0037
SA403G	0050	SA427E	0050	SA437R	0050	SA457B	0050	SA472T	0050	SA482C	0037	SA495G	0037
SA403L	0050	SA427F	0050	SA437T	0050	SA457C	0050	SA472X	0050	SA482D	0037	SA495L	0037
SA403M	0050	SA427G	0050	SA437X	0050	SA457D	0050	SA472Y	0050	SA482E	0037	SA495M	0037
SA403R	0050	SA427L	0050	SA437Y	0050	SA457E	0050	SA473A	0854	SA482F	0037	SA495R	0037
SA403T	0050	SA427M	0050	SA438A	0050	SA457F	0050	SA473B	0854	SA482G	0037	SA495T	0037
SA403X	0050	SA427R	0050	SA438B	0050	SA457G	0050	SA473C	0854	SA482L	0037	SA495X	0037
SA403Y	0050	SA427T	0050	SA438C	0050	SA457L	0050	SA473D	0854	SA482M	0037	SA495Y	0037
SA404A	0050	SA427X	0050	SA438D	0050	SA457M	0050	SA473E	0854	SA482R	0037	SA496	0150
SA404B	0050	SA427Y	0050	SA438E	0050	SA457R	0050	SA473F	0854	SA482T	0037	SA496A	0150
SA404C	0050	SA428A	0050	SA438F	0050	SA457T	0050	SA473G	0854	SA482X	0037	SA496B	0150
SA404D	0050	SA428B	0050	SA438G	0050	SA457X	0050	SA473L	0854	SA482Y	0037	SA496C	0854
SA404E	0050	SA428C	0050	SA438L	0050	SA457Y	0050	SA473M	0854	SA483A	0854	SA496D	0854
SA404F	0050	SA428D	0050	SA438M	0050	SA465A	0037	SA473R	0854	SA483B	0854	SA496E	0854
SA404G	0050	SA428E	0050	SA438R	0050	SA465B	0037	SA473T	0854	SA483C	0854	SA496F	0854
SA404L	0050	SA428F	0050	SA438T	0050	SA465C	0037	SA473X	0854	SA483D	0854	SA496G	0854
SA404M	0050	SA428G	0050	SA438X	0050	SA465D	0037	SA473Y	0854	SA483E	0854	SA496L	0854
SA404R	0050	SA428L	0050	SA438Y	0050	SA465E	0037	SA474A	0050	SA483F	0854	SA496M	0854
SA404T	0050	SA428M	0050	SA440A	0050	SA465F	0037	SA474B	0050	SA483G	0854	SA496R	0854
SA404X	0050	SA428R	0050	SA440B	0050	SA465G	0037	SA474C	0050	SA483L	0854	SA496T	0854
SA404Y	0050	SA428T	0050	SA440C	0050	SA465L	0037	SA474D	0050	SA483M	0854	SA496X	0854
SA406A	0280	SA428X	0050	SA440D	0050	SA465M	0037	SA474E	0050	SA483R	0854	SA496Y	0854
SA406B	0280	SA428Y	0050	SA440E	0050	SA465R	0037	SA474F	0050	SA483T	0854	SA497A	0126
SA406C	0280	SA429A	0037	SA440F	0050	SA465T	0037	SA474G	0050	SA483X	0854	SA497B	0126
SA406D	0280	SA429B	0037	SA440G	0050	SA465X	0037	SA474L	0050	SA483Y	0854	SA497C	0126
SA406E	0280	SA429C	0037	SA440L	0050	SA465Y	0037	SA474M	0050	SA484A	0126	SA497D	0126
SA406F	0280	SA429D	0037	SA440M	0050	SA466A	0050	SA474R	0050	SA484B	0126	SA497E	0126
SA406G	0280	SA429E	0037	SA440R	0050	SA466B	0050	SA474T	0050	SA484C	0126	SA497F	0126
SA406L	0280	SA429F	0037	SA440T	0050	SA466C	0050	SA474X	0050	SA484D	0126	SA497G	0126
SA406M	0280	SA429G	0037	SA440X	0050	SA466D	0050	SA474Y	0050	SA484E	0126	SA497L	0126
SA406R	0280	SA429L	0037	SA440Y	0050	SA466E	0050	SA475A	0050	SA484F	0126	SA497M	0126
SA406T	0280	SA429M	0037	SA446A	0050	SA466F	0050	SA475B	0050	SA484G	0126	SA497R	0126
SA406X	0280	SA429R	0037	SA446B	0050	SA466G	0050	SA475C	0050	SA484L	0126	SA497T	0126
SA406Y	0280	SA429T	0037	SA446C	0050	SA466L	0050	SA475D	0050	SA484M	0126	SA497X	0126
SA407A	0280	SA429X	0037	SA446D	0050	SA466M	0050	SA475E	0050	SA484R	0126	SA497Y	0126
SA407B	0280	SA429Y	0037	SA446E	0050	SA466R	0050	SA475F	0050	SA484T	0126	SA498A	0126
SA407C	0280	SA432A	0050	SA446F	0050	SA466T	0050	SA475G	0050	SA484X	0126	SA498B	0126
SA407D	0280	SA432B	0050	SA446G	0050	SA466X	0050	SA475L	0050	SA484Y	0126	SA498C	0126
SA407E	0280	SA432C	0050	SA446L	0050	SA466Y	0050	SA475M	0050	SA485A	0126	SA498D	0126
SA407F	0280	SA432D	0050	SA446M	0050	SA467A	0037	SA475R	0050	SA485B	0126	SA498E	0126
SA407G	0280	SA432E	0050	SA446R	0050	SA467B	0037	SA475T	0050	SA485C	0126	SA498F	0126
SA407L	0280	SA432F	0050	SA446T	0050	SA467C	0037	SA475X	0050	SA485D	0126	SA498G	0126
SA407M	0280	SA432G	0050	SA446X	0050	SA467D	0037	SA475Y	0050	SA485E	0126	SA498L	0126
SA407R	0280	SA432L	0050	SA446Y	0050	SA467E	0037	SA476A	0136	SA485F	0126	SA498M	0126
SA407T	0280	SA432M	0050	SA447A	0050	SA467F	0037	SA476B	0136	SA485G	0126	SA498R	0126
SA407X	0280	SA432R	0050	SA447B	0050	SA467L	0037	SA476C	0136	SA485L	0126	SA498T	0126
SA407Y	0280	SA432T	0050	SA447C	0050	SA467M	0037	SA476D	0136	SA485M	0126	SA498X	0126
SA408A	0847	SA432X	0050	SA447D	0050	SA467R	0037	SA476E	0136	SA485R	0126	SA498Y	0126
SA408B	0847	SA432Y	0050	SA447E	0050	SA468A	0050	SA476F	0136	SA485T	0126	SA499A	0037
SA408C	0847	SA433A	0050	SA447F	0050	SA468B	0050	SA476G	0136	SA485X	0126	SA499B	0037
SA408D	0847	SA433B	0050	SA447G	0050	SA468C	0050	SA476L	0136	SA485Y	0126	SA499C	0037
SA408E	0847	SA433C	0050	SA447L	0050	SA468D	0050	SA476M	0136	SA489A	0854	SA499D	0037
SA408F	0847	SA433D	0050	SA447M	0050	SA468E	0050	SA476R	0136	SA489B	0854	SA499E	0037
SA408G	0847	SA433E	0050	SA447R	0050	SA468F	0050	SA476T	0136	SA489C	0854	SA499F	0037
SA408L	0847	SA433F	0050	SA447T	0050	SA468G	0050	SA476X	0136	SA489D	0854	SA499G	0037
SA408M	0847	SA433G	0050	SA447X	0050	SA468L	0050	SA476Y	0136	SA489E	0854	SA499L	0037
SA408R	0847	SA433L	0050	SA447Y	0050	SA468M	0050	SA477A	0050	SA489F	0854	SA499M	0037
SA408T	0847	SA433M	0050	SA453A	0050	SA468R	0050	SA477B	0050	SA489G	0854	SA499R	0037
SA408X	0847	SA433R	0050	SA453B	0050	SA468T	0050	SA477C	0050	SA489L	0854	SA499T	0037
SA408Y	0847	SA433T	0050	SA453C	0050	SA468X	0050	SA477D	0050	SA489M	0854	SA499X	0037
SA409A	0847	SA433X	0050	SA453D	0050	SA468Y	0050	SA477E	0050	SA489R	0854	SA499Y	0037
SA409B	0847	SA433Y	0050	SA453E	0050	SA469A	0050	SA477F	0050	SA489T	0854	SA500A	0037
SA409C	0847	SA434A	0050	SA453F	0050	SA469B	0050	SA477G	0050	SA489X	0854	SA500B	0037
SA409D	0847	SA434B	0050	SA453G	0050	SA469C	0050	SA477L	0050	SA489Y	0854	SA500C	0037
SA409E	0847	SA434C	0050	SA453L	0050	SA469D	0050	SA477M	0050	SA490A	0854	SA500D	0037
SA409F	0847	SA434D	0050	SA453M	0050	SA469E	0050	SA477R	0050	SA490B	0854	SA500E	0037
SA409G	0847	SA434E	0050	SA453R	0050	SA469F	0050	SA477T	0050	SA490C	0854	SA500F	0037
SA409L	0847	SA434F	0050	SA453T	0050	SA469G	0050	SA477X	0050	SA490D	0854	SA500G	0037
SA409M	0847	SA434G	0050	SA453X	0050	SA469L	0050	SA477Y	0050	SA490E	0854	SA500L	0037
SA409R	0847	SA434L	0050	SA453Y	0050	SA469M	0050	SA478A	0050	SA490F	0854	SA500M	0037
SA409T	0847	SA434M	0050	SA454A	0050	SA469R	0050	SA478B	0050	SA490G	0854	SA500R	0037
SA409X	0847	SA434R	0050	SA454B	0050	SA469T	0050	SA478C	0050	SA490L	0854	SA500T	0037
SA409Y	0847	SA434T	0050	SA454C	0050	SA469X	0050	SA478D	0050	SA490M	0854	SA500X	0037
SA410	0037	SA434X	0050	SA454D	0050	SA469Y	0050	SA478E	0050	SA490R	0854	SA500Y	0037
SA411	0037	SA434Y	0050	SA454E	0050	SA470A	0050	SA478F	0050	SA490T	0854	SA501A	0126
SA412	0037	SA435A	0136	SA454F	0050	SA470B	0050	SA478G	0050	SA490X	0854	SA501B	0126
SA413	0037	SA435B	0136	SA454G	0050	SA470C	0050	SA478L	0050	SA490Y	0854	SA501C	0126
SA414	0037	SA435C	0136	SA454L	0050	SA470D	0050	SA478M	0050	SA493A	0037	SA501D	0126
SA415	0037	SA435D	0136	SA454M	0050	SA470E	0050	SA478R	0050	SA493B	0037	SA501E	0126
SA416	0037	SA435E	0136	SA454R	0050	SA470F	0050	SA478T	0050	SA493C	0037	SA501F	0126
SA420A	0050	SA435F	0136	SA454T	0050	SA470G	0050	SA478X	0050	SA493D	0037	SA501G	0126
SA420B	0050	SA435G	0136	SA454X	0050	SA470L	0050	SA478Y	0050	SA493E	0037	SA501L	0126

DEVICE TYPE	REPL CODE
SA501M	0126
SA501R	0126
SA501T	0126
SA501X	0126
SA501Y	0126
SA502A	0126
SA502B	0126
SA502C	0126
SA502D	0126
SA502E	0126
SA502F	0126
SA502G	0126
SA502L	0126
SA502M	0126
SA502R	0126
SA502T	0126
SA502X	0126
SA502Y	0126
SA503A	0126
SA503B	0126
SA503C	0126
SA503D	0126
SA503E	0126
SA503F	0126
SA503G	0126
SA503L	0126
SA503M	0126
SA503R	0126
SA503T	0126
SA503X	0126
SA503Y	0126
SA504A	0126
SA504B	0126
SA504C	0126
SA504D	0126
SA504E	0126
SA504F	0126
SA504G	0126
SA504L	0126
SA504M	0126
SA504R	0126
SA504T	0126
SA504X	0126
SA504Y	0126
SA505A	0460
SA505B	0460
SA505C	0460
SA505D	0460
SA505E	0460
SA505F	0460
SA505G	0460
SA505L	0460
SA505M	0460
SA505R	0460
SA505T	0460
SA505X	0460
SA505Y	0460
SA507A	0050
SA507B	0050
SA507C	0050
SA507D	0050
SA507E	0050
SA507F	0050
SA507G	0050
SA507L	0050
SA507M	0050
SA507R	0050
SA507T	0050
SA507X	0050
SA507Y	0050
SA509A	0037
SA509B	0037
SA509C	0037
SA509D	0037
SA509E	0037
SA509F	0037
SA509G	0037
SA509L	0037
SA509M	0037
SA509R	0037
SA509T	0037
SA509X	0037
SA509Y	0037
SA510A	0126
SA510B	0126
SA510C	0126
SA510D	0126
SA510E	0126
SA510F	0126
SA510G	0126
SA510L	0126
SA510M	0126
SA510R	0126
SA510T	0126
SA510X	0126
SA510Y	0126
SA511A	0037
SA511B	0037
SA511C	0037
SA511D	0037
SA511E	0037
SA511F	0037
SA511G	0037
SA511L	0037
SA511M	0037
SA511R	0037
SA511T	0037
SA511X	0037
SA511Y	0037

DEVICE TYPE	REPL CODE
SA512A	0126
SA512B	0126
SA512C	0126
SA512D	0126
SA512E	0126
SA512F	0126
SA512G	0126
SA512L	0126
SA512M	0126
SA512R	0126
SA512T	0126
SA512X	0126
SA512Y	0126
SA513A	0037
SA513B	0037
SA513C	0037
SA513D	0037
SA513E	0037
SA513F	0037
SA513G	0037
SA513L	0037
SA513M	0037
SA513R	0037
SA513T	0037
SA513X	0037
SA513Y	0037
SA516A	0126
SA516B	0126
SA516C	0126
SA516D	0126
SA516E	0126
SA516F	0126
SA516G	0126
SA516L	0126
SA516M	0126
SA516R	0126
SA516T	0126
SA516X	0126
SA516Y	0126
SA517A	0136
SA517B	0136
SA517C	0136
SA517D	0136
SA517E	0136
SA517F	0136
SA517G	0136
SA517L	0136
SA517M	0136
SA517R	0136
SA517T	0136
SA517X	0136
SA517Y	0136
SA518A	0050
SA518B	0050
SA518C	0050
SA518D	0050
SA518E	0050
SA518F	0050
SA518G	0050
SA518L	0050
SA518M	0050
SA518R	0050
SA518T	0050
SA518X	0050
SA518Y	0050
SA522A	0037
SA522B	0037
SA522C	0037
SA522D	0037
SA522E	0037
SA522F	0037
SA522G	0037
SA522L	0037
SA522M	0037
SA522R	0037
SA522T	0037
SA522X	0037
SA522Y	0037
SA525A	0037
SA525B	0037
SA525C	0037
SA525D	0037
SA525E	0037
SA525F	0037
SA525G	0037
SA525L	0037
SA525M	0037
SA525R	0037
SA525T	0037
SA525X	0037
SA525Y	0037
SA527A	0126
SA527B	0126
SA527C	0126
SA527D	0126
SA527E	0126
SA527F	0126
SA527G	0126
SA527L	0126
SA527M	0126
SA527R	0126
SA527T	0126
SA527X	0126
SA527Y	0126
SA528A	0126
SA528B	0126
SA528C	0126
SA528D	0126
SA528E	0126

DEVICE TYPE	REPL CODE
SA528F	0126
SA528G	0126
SA528L	0126
SA528M	0126
SA528R	0126
SA528T	0126
SA528X	0126
SA528Y	0126
SA529	0004
SA530A	0037
SA530B	0037
SA530C	0037
SA530D	0037
SA530E	0037
SA530F	0037
SA530G	0037
SA530L	0037
SA530M	0037
SA530R	0037
SA530T	0037
SA530X	0037
SA530Y	0037
SA532A	0126
SA532B	0126
SA532C	0126
SA532D	0126
SA532E	0126
SA532F	0126
SA532G	0126
SA532L	0126
SA532M	0126
SA532R	0126
SA532T	0126
SA532X	0126
SA532Y	0126
SA537	0150
SA537A	0126
SA537B	0126
SA537C	0126
SA537D	0126
SA537E	0126
SA537F	0126
SA537G	0126
SA537L	0126
SA537M	0126
SA537R	0126
SA537T	0126
SA537X	0126
SA537Y	0126
SA538	0150
SA539	0150
SA539A	0037
SA539B	0037
SA539C	0037
SA539D	0037
SA539E	0037
SA539F	0037
SA539G	0037
SA539L	0037
SA539M	0037
SA539R	0037
SA539T	0037
SA539X	0037
SA539Y	0037
SA540	0150
SA542A	0037
SA542B	0037
SA542C	0037
SA542D	0037
SA542E	0037
SA542F	0037
SA542G	0037
SA542L	0037
SA542M	0037
SA542R	0037
SA542T	0037
SA542X	0037
SA542Y	0037
SA543A	0126
SA543B	0126
SA543C	0126
SA543D	0126
SA543E	0126
SA543F	0126
SA543G	0126
SA543L	0126
SA543M	0126
SA543R	0126
SA543T	0126
SA543X	0126
SA543Y	0126
SA544A	0037
SA544B	0037
SA544C	0037
SA544D	0037
SA544E	0037
SA544F	0037
SA544G	0037
SA544L	0037
SA544M	0037
SA544R	0037
SA544T	0037
SA544X	0037
SA544Y	0037
SA545A	0037
SA545B	0037
SA545C	0037
SA545D	0037
SA545E	0037

DEVICE TYPE	REPL CODE
SA545F	0037
SA545G	0037
SA545L	0037
SA545M	0037
SA545R	0037
SA545T	0037
SA545X	0037
SA545Y	0037
SA546A	0126
SA546B	0126
SA546C	0126
SA546D	0126
SA546E	0126
SA546F	0126
SA546G	0126
SA546L	0126
SA546M	0126
SA546R	0126
SA546T	0126
SA546X	0126
SA546Y	0126
SA548A	0150
SA548B	0150
SA548C	0150
SA548D	0150
SA548E	0150
SA548F	0150
SA548G	0150
SA548L	0150
SA548M	0150
SA548R	0150
SA548T	0150
SA548X	0150
SA548Y	0150
SA550A	0126
SA550B	0126
SA550C	0126
SA550D	0126
SA550E	0126
SA550F	0126
SA550G	0126
SA550L	0126
SA550M	0126
SA550R	0126
SA550T	0126
SA550X	0126
SA550Y	0126
SA551A	0037
SA551B	0037
SA551C	0037
SA551D	0037
SA551E	0037
SA551F	0037
SA551G	0037
SA551L	0037
SA551M	0037
SA551R	0037
SA551T	0037
SA551X	0037
SA551Y	0037
SA552A	0126
SA552B	0126
SA552C	0126
SA552D	0126
SA552E	0126
SA552F	0126
SA552G	0126
SA552L	0126
SA552M	0126
SA552R	0126
SA552T	0126
SA552X	0126
SA552Y	0126
SA555	OEM
SA556	OEM
SA560A	0126
SA560B	0126
SA560C	0126
SA560D	0126
SA560E	0126
SA560F	0126
SA560G	0126
SA560L	0126
SA560M	0126
SA560R	0126
SA560T	0126
SA560X	0126
SA560Y	0126
SA561A	0037
SA561B	0037
SA561C	0037
SA561D	0037
SA561E	0037
SA561F	0037
SA561G	0037
SA561L	0037
SA561M	0037
SA561R	0037
SA561T	0037
SA561X	0037
SA561Y	0037
SA562A	0037
SA562B	0037
SA562C	0037
SA562D	0037
SA562E	0037
SA562F	0037
SA562G	0037
SA562L	0037

DEVICE TYPE	REPL CODE
SA562M	0037
SA562R	0037
SA562T	0037
SA562X	0037
SA562Y	0037
SA564A	0037
SA564B	0037
SA564C	0037
SA564D	0037
SA564E	0037
SA564F	0037
SA564G	0037
SA564L	0037
SA564M	0037
SA564R	0037
SA564T	0037
SA564X	0037
SA564Y	0037
SA565	0004
SA565A	0037
SA565B	0037
SA565C	0037
SA565D	0037
SA565E	0037
SA565F	0037
SA565G	0037
SA565L	0037
SA565M	0037
SA565R	0037
SA565T	0037
SA565X	0037
SA565Y	0037
SA566A	0854
SA566B	0854
SA566C	0854
SA566D	0854
SA566E	0854
SA566F	0854
SA566G	0854
SA566L	0854
SA566M	0854
SA566R	0854
SA566T	0854
SA566X	0854
SA566Y	0854
SA567A	0037
SA567B	0037
SA567C	0037
SA567D	0037
SA567E	0037
SA567F	0037
SA567G	0037
SA567L	0037
SA567M	0037
SA567R	0037
SA567T	0037
SA567X	0037
SA567Y	0037
SA568A	0150
SA568B	0150
SA568C	0150
SA568D	0150
SA568E	0150
SA568F	0150
SA568G	0150
SA568L	0150
SA568M	0150
SA568R	0150
SA568T	0150
SA568X	0150
SA568Y	0150
SA569A	0037
SA569B	0037
SA569C	0037
SA569D	0037
SA569E	0037
SA569F	0037
SA569G	0037
SA569L	0037
SA569M	0037
SA569R	0037
SA569T	0037
SA569X	0037
SA569Y	0037
SA570A	0037
SA570B	0037
SA570C	0037
SA570D	0037
SA570E	0037
SA570F	0037
SA570G	0037
SA570L	0037
SA570M	0037
SA570R	0037
SA570T	0037
SA570X	0037
SA570Y	0037
SA571A	0126
SA571B	0126
SA571C	0126
SA571D	0126
SA571E	0126
SA571F	0126
SA571G	0126
SA571L	0126
SA571M	0126
SA571R	0126
SA571T	0126
SA571X	0126

DEVICE TYPE	REPL CODE
SA571Y	0126
SA592A	0037
SA592B	0037
SA592C	0037
SA592D	0037
SA592E	0037
SA592F	0037
SA592G	0037
SA592L	0037
SA592M	0037
SA592R	0037
SA592T	0037
SA592X	0037
SA592Y	0037
SA594A	0126
SA594B	0126
SA594C	0126
SA594D	0126
SA594E	0126
SA594F	0126
SA594G	0126
SA594L	0126
SA594M	0126
SA594R	0126
SA594T	0126
SA594X	0126
SA594Y	0126
SA597A	0126
SA597B	0126
SA597C	0126
SA597D	0126
SA597E	0126
SA597F	0126
SA597G	0126
SA597L	0126
SA597M	0126
SA597R	0126
SA597T	0126
SA597X	0126
SA597Y	0126
SA603A	0150
SA603B	0150
SA603C	0150
SA603D	0150
SA603E	0150
SA603F	0150
SA603G	0150
SA603L	0150
SA603M	0150
SA603R	0150
SA603T	0150
SA603X	0150
SA603Y	0150
SA604A	0126
SA604B	0126
SA604C	0126
SA604D	0126
SA604E	0126
SA604F	0126
SA604G	0126
SA604L	0126
SA604M	0126
SA604R	0126
SA604T	0126
SA604X	0126
SA604Y	0126
SA606A	0126
SA606B	0126
SA606C	0126
SA606D	0126
SA606E	0126
SA606F	0126
SA606G	0126
SA606L	0126
SA606M	0126
SA606R	0126
SA606T	0126
SA606X	0126
SA606Y	0126
SA607A	0126
SA607B	0126
SA607C	0126
SA607D	0126
SA607E	0126
SA607F	0126
SA607G	0126
SA607L	0126
SA607M	0126
SA607R	0126
SA607T	0126
SA607X	0126
SA607Y	0126
SA608A	0037
SA608B	0037
SA608C	0037
SA608D	0037
SA608E	0037
SA608F	0037
SA608G	0037
SA608L	0037
SA608M	0037
SA608R	0037
SA608T	0037
SA608X	0037
SA608Y	0037
SA609A	0037
SA609B	0037
SA609C	0037
SA609D	0037

DEVICE TYPE	REPL CODE
SA609E	0037
SA609F	0037
SA609G	0037
SA609L	0037
SA609M	0037
SA609R	0037
SA609T	0037
SA609X	0037
SA609Y	0037
SA610A	0126
SA610B	0126
SA610C	0126
SA610E	0126
SA610F	0126
SA610G	0126
SA610L	0126
SA610M	0126
SA610R	0126
SA610T	0126
SA610X	0126
SA610Y	0126
SA611A	0126
SA611B	0126
SA611C	0126
SA611D	0126
SA611E	0126
SA611F	0126
SA611G	0126
SA611L	0126
SA611M	0126
SA611R	0126
SA611T	0126
SA611X	0126
SA611Y	0126
SA612A	0126
SA612B	0126
SA612C	0126
SA612D	0126
SA612E	0126
SA612F	0126
SA612G	0126
SA612L	0126
SA612M	0126
SA612R	0126
SA612T	0126
SA612X	0126
SA612Y	0126
SA616A	0854
SA616B	0854
SA616C	0854
SA616D	0854
SA616E	0854
SA616F	0854
SA616G	0854
SA616L	0854
SA616M	0854
SA616R	0854
SA616T	0854
SA616X	0854
SA616Y	0854
SA623A	0854
SA623B	0854
SA623C	0854
SA623D	0854
SA623E	0854
SA623F	0854
SA623G	0854
SA623L	0854
SA623M	0854
SA623R	0854
SA623T	0854
SA623X	0854
SA623Y	0854
SA624A	0854
SA624B	0854
SA624C	0854
SA624D	0854
SA624E	0854
SA624F	0854
SA624G	0854
SA624L	0854
SA624M	0854
SA624R	0854
SA624T	0854
SA624X	0854
SA624Y	0854
SA628A	0037
SA628B	0037
SA628C	0037
SA628D	0037
SA628E	0037
SA628F	0037
SA628G	0037
SA628L	0037
SA628M	0037
SA628R	0037
SA628T	0037
SA628X	0037
SA628Y	0037
SA629A	0037
SA629B	0037
SA629C	0037
SA629D	0037
SA629E	0037
SA629F	0037
SA629G	0037
SA629L	0037
SA629M	0037
SA629R	0037

If replacement code is OEM, contact original manufacturer for replacement.

DEVICE TYPE	REPL CODE
SA629T	0037
SA634A	0854
SA634B	0854
SA634C	0854
SA634D	0854
SA634E	0854
SA634F	0854
SA634G	0854
SA634L	0854
SA634M	0854
SA634R	0854
SA634T	0854
SA634X	0854
SA634Y	0854
SA636A	0854
SA636B	0854
SA636C	0854
SA636D	0854
SA636E	0854
SA636F	0854
SA636G	0854
SA636L	0854
SA636M	0854
SA636R	0854
SA636T	0854
SA636X	0854
SA636Y	0854
SA639A	0037
SA639B	0037
SA639C	0037
SA639D	0037
SA639E	0037
SA639F	0037
SA639G	0037
SA639L	0037
SA639M	0037
SA639R	0037
SA639T	0037
SA640A	0150
SA640B	0150
SA640C	0150
SA640D	0150
SA640E	0150
SA640F	0150
SA640G	0150
SA640L	0150
SA640M	0150
SA640R	0150
SA640T	0150
SA640X	0150
SA640Y	0150
SA641A	0037
SA641B	0037
SA641C	0037
SA641D	0037
SA641E	0037
SA641F	0037
SA641G	0037
SA641L	0037
SA641M	0037
SA641R	0037
SA641T	0037
SA641X	0037
SA641Y	0037
SA642A	0854
SA642B	0854
SA642C	0854
SA642D	0854
SA642E	0854
SA642F	0854
SA642G	0854
SA642L	0854
SA642M	0854
SA642R	0854
SA642T	0854
SA642X	0854
SA642Y	0854
SA643A	0071
SA643B	0071
SA643C	0071
SA643D	0071
SA643E	0071
SA643F	0071
SA643G	0071
SA643L	0071
SA643M	0071
SA643R	0071
SA643T	0071
SA643X	0071
SA643Y	0071
SA645A	0566
SA645B	0566
SA645C	0566
SA645D	0566
SA645E	0566
SA645F	0566
SA645G	0566
SA645L	0566
SA645M	0566
SA645R	0566
SA645T	0566
SA645X	0566
SA645Y	0566
SA646	0004
SA646A	1051
SA646B	1051
SA646C	1051
SA646D	1051
SA646E	1051
SA646F	1051
SA646G	1051
SA646L	1051
SA646M	1051
SA646R	1051
SA646T	1051
SA646X	1051
SA646Y	1051
SA649	2013
SA657A	0491
SA657B	0491
SA657C	0491
SA657D	0491
SA657E	0491
SA657F	0491
SA657G	0491
SA657L	0491
SA657M	0491
SA657R	0491
SA657T	0491
SA657X	0491
SA657Y	0491
SA658A	0491
SA658B	0491
SA658C	0491
SA658D	0491
SA658E	0491
SA658F	0491
SA658G	0491
SA658L	0491
SA658M	0491
SA658R	0491
SA658T	0491
SA658X	0491
SA658Y	0491
SA659A	0037
SA659B	0037
SA659C	0037
SA659D	0037
SA659E	0037
SA659F	0037
SA659G	0037
SA659L	0037
SA659M	0037
SA659R	0037
SA659T	0037
SA659X	0037
SA659Y	0037
SA661A	0037
SA661B	0037
SA661C	0037
SA661D	0037
SA661E	0037
SA661F	0037
SA661G	0037
SA661L	0037
SA661M	0037
SA661R	0037
SA661T	0037
SA661X	0037
SA661Y	0037
SA663A	0491
SA663B	0491
SA663C	0491
SA663D	0491
SA663E	0491
SA663F	0491
SA663G	0491
SA663L	0491
SA663M	0491
SA663R	0491
SA663T	0491
SA663X	0491
SA663Y	0491
SA666A	0037
SA666B	0037
SA666C	0037
SA666D	0037
SA666E	0037
SA666F	0037
SA666G	0037
SA666L	0037
SA666M	0037
SA666R	0037
SA666T	0037
SA666X	0037
SA666Y	0037
SA670A	0854
SA670B	0854
SA670C	0854
SA670D	0854
SA670E	0854
SA670F	0854
SA670G	0854
SA670L	0854
SA670M	0854
SA670R	0854
SA670T	0854
SA670X	0854
SA670Y	0854
SA671A	0854
SA671B	0854
SA671C	0854
SA671D	0854
SA671E	0854
SA671F	0854
SA671G	0854
SA671L	0854
SA671M	0854
SA671R	0854
SA671T	0854
SA671X	0854
SA671Y	0854
SA672A	0037
SA672B	0037
SA672C	0037
SA672D	0037
SA672E	0037
SA672F	0037
SA672G	0037
SA672L	0037
SA672M	0037
SA672R	0037
SA672T	0037
SA672X	0037
SA672Y	0037
SA673A	0037
SA673B	0037
SA673C	0037
SA673D	0037
SA673E	0037
SA673F	0037
SA673G	0037
SA673L	0037
SA673R	0037
SA673T	0037
SA673X	0037
SA673Y	0037
SA677A	0150
SA677B	0150
SA677C	0150
SA677D	0150
SA677E	0150
SA677F	0150
SA677G	0150
SA677L	0150
SA677M	0150
SA677R	0150
SA677T	0150
SA677X	0150
SA677Y	0150
SA678A	0150
SA678B	0150
SA678C	0150
SA678D	0150
SA678E	0150
SA678F	0150
SA678G	0150
SA678L	0150
SA678M	0150
SA678R	0150
SA678T	0150
SA678X	0150
SA678Y	0150
SA680A	0491
SA680B	0491
SA680C	0491
SA680D	0491
SA680E	0491
SA680F	0491
SA680G	0491
SA680L	0491
SA680M	0491
SA680R	0491
SA680T	0491
SA680X	0491
SA680Y	0491
SA681	0164
SA682A	0126
SA682B	0126
SA682C	0126
SA682D	0126
SA682E	0126
SA682F	0126
SA682G	0126
SA682L	0126
SA682M	0126
SA682R	0126
SA682T	0126
SA682X	0126
SA682Y	0126
SA683A	0126
SA683B	0126
SA683C	0126
SA683D	0126
SA683E	0126
SA683F	0126
SA683G	0126
SA683L	0126
SA683M	0126
SA683R	0126
SA683T	0126
SA683X	0126
SA683Y	0126
SA684A	0126
SA684B	0126
SA684C	0126
SA684D	0126
SA684E	0126
SA684F	0126
SA684G	0126
SA684L	0126
SA684M	0126
SA684R	0126
SA684T	0126
SA684X	0126
SA684Y	0126
SA685A	0037
SA685B	0037
SA685C	0037
SA685D	0037
SA685E	0037
SA685F	0037
SA685G	0037
SA685L	0037
SA685M	0037
SA685R	0037
SA685T	0037
SA685X	0037
SA685Y	0037
SA695A	0126
SA695B	0126
SA695C	0126
SA695D	0126
SA695E	0126
SA695F	0126
SA695G	0126
SA695L	0126
SA695M	0126
SA695R	0126
SA695T	0126
SA695X	0126
SA695Y	0126
SA696A	0150
SA696B	0150
SA696C	0150
SA696D	0150
SA696E	0150
SA696F	0150
SA696G	0150
SA696L	0150
SA696M	0150
SA696R	0150
SA696T	0150
SA696X	0150
SA696Y	0150
SA697A	0037
SA697B	0037
SA697C	0037
SA697D	0037
SA697E	0037
SA697F	0037
SA697G	0037
SA697L	0037
SA697M	0037
SA697R	0037
SA697T	0037
SA697X	0037
SA697Y	0037
SA699A	0854
SA699B	0854
SA699C	0854
SA699D	0854
SA699E	0854
SA699F	0854
SA699G	0854
SA699L	0854
SA699M	0854
SA699R	0854
SA699T	0854
SA699X	0854
SA699Y	0854
SA700A	0854
SA700B	0854
SA700C	0854
SA700D	0854
SA700E	0854
SA700F	0854
SA700G	0854
SA700L	0854
SA700M	0854
SA700R	0854
SA700T	0854
SA700X	0854
SA700Y	0854
SA701A	0150
SA701B	0150
SA701C	0150
SA701D	0150
SA701E	0150
SA701F	0150
SA701G	0150
SA701L	0150
SA701M	0150
SA701R	0150
SA701T	0150
SA701X	0150
SA701Y	0150
SA704A	0150
SA704B	0150
SA704C	0150
SA704D	0150
SA704E	0150
SA704F	0150
SA704G	0150
SA704L	0150
SA704M	0150
SA704R	0150
SA704T	0150
SA704X	0150
SA704Y	0150
SA705A	0150
SA705B	0150
SA705C	0150
SA705D	0150
SA705E	0150
SA705F	0150
SA705G	0150
SA705L	0150
SA705M	0150
SA705R	0150
SA705T	0150
SA705X	0150
SA705Y	0150
SA707A	0037
SA707B	0037
SA707C	0037
SA707D	0037
SA707E	0037
SA707F	0037
SA707G	0037
SA707L	0037
SA707M	0037
SA707R	0037
SA707T	0037
SA707X	0037
SA707Y	0037
SA708A	1367
SA708B	1367
SA708C	1367
SA708D	1367
SA708E	1367
SA708F	1367
SA708G	1367
SA708L	1367
SA708M	1367
SA708R	1367
SA708T	1367
SA708X	1367
SA708Y	1367
SA715A	0460
SA715B	0460
SA715C	0460
SA715D	0460
SA715E	0460
SA715F	0460
SA715G	0460
SA715L	0460
SA715M	0460
SA715R	0460
SA715T	0460
SA715X	0460
SA715Y	0460
SA717A	1367
SA717B	1367
SA717C	1367
SA717D	1367
SA717E	1367
SA717F	1367
SA717G	1367
SA717L	1367
SA717M	1367
SA717R	1367
SA717T	1367
SA717X	1367
SA717Y	1367
SA719A	0126
SA719B	0126
SA719C	0126
SA719D	0126
SA719E	0126
SA719F	0126
SA719G	0126
SA719L	0126
SA719M	0126
SA719R	0126
SA719T	0126
SA719X	0126
SA719Y	0126
SA720A	0037
SA720B	0037
SA720C	0037
SA720D	0037
SA720E	0037
SA720F	0037
SA720G	0037
SA720L	0037
SA720M	0037
SA720R	0037
SA720T	0037
SA720X	0037
SA720Y	0037
SA721A	0037
SA721B	0037
SA721C	0037
SA721D	0037
SA721E	0037
SA721F	0037
SA721G	0037
SA721L	0037
SA721M	0037
SA721R	0037
SA721T	0037
SA721X	0037
SA721Y	0037
SA723CN	0026
SA733A	0126
SA733B	0126
SA733C	0126
SA733D	0126
SA733E	0126
SA733F	0126
SA733G	0126
SA733L	0126
SA733M	0126
SA733R	0126
SA733T	0126
SA733X	0126
SA733Y	0126
SA736A	0150
SA736B	0150
SA736C	0150
SA736D	0150
SA736E	0150
SA736F	0150
SA736G	0150
SA736L	0150
SA736M	0150
SA736R	0150
SA736T	0150
SA736X	0150
SA736Y	0150
SA738A	0854
SA738B	0854
SA738C	0854
SA738D	0854
SA738E	0854
SA738F	0854
SA738G	0854
SA738L	0854
SA738M	0854
SA738R	0854
SA738T	0854
SA738X	0854
SA738Y	0854
SA741A	0037
SA741B	0037
SA741C	0037
SA741D	0037
SA741E	0037
SA741F	0037
SA741G	0037
SA741L	0037
SA741M	0037
SA741R	0037
SA741T	0037
SA741X	0037
SA741Y	0037
SA742A	0126
SA742B	0126
SA742C	0126
SA742D	0126
SA742E	0126
SA742F	0126
SA742G	0126
SA742L	0126
SA742M	0126
SA742R	0126
SA742T	0126
SA742X	0126
SA742Y	0126
SA743A	0126
SA743B	0126
SA743C	0126
SA743D	0126
SA743E	0126
SA743F	0126
SA743G	0126
SA743L	0126
SA743M	0126
SA743R	0126
SA743T	0126
SA743X	0126
SA743Y	0126
SA751A	0126
SA751B	0126
SA751C	0126
SA751D	0126
SA751E	0126
SA751F	0126
SA751G	0126
SA751L	0126
SA751M	0126
SA751R	0126
SA751T	0126
SA751X	0126
SA751Y	0126
SA754A	0854
SA754B	0854
SA754C	0854
SA754D	0854
SA754E	0854
SA754F	0854
SA754G	0854
SA754L	0854
SA754M	0854
SA754R	0854
SA754T	0854
SA754X	0854
SA754Y	0854
SA755A	0854
SA755B	0854
SA755C	0854
SA755D	0854
SA755E	0854
SA755F	0854
SA755G	0854
SA755L	0854
SA755M	0854
SA755T	0854
SA755X	0854
SA755Y	0854
SA756A	0491
SA756B	0491
SA756C	0491
SA756D	0491
SA756E	0491
SA756F	0491
SA756G	0491
SA756L	0491
SA756M	0491
SA756R	0491
SA756T	0491
SA756X	0491
SA756Y	0491
SA766A	0854
SA766B	0854
SA766C	0854
SA766D	0854
SA766E	0854
SA766F	0854
SA766G	0854
SA766L	0854
SA766M	0854
SA766R	0854
SA766T	0854
SA766X	0854
SA766Y	0854
SA773A	0037
SA773B	0037
SA773C	0037
SA773D	0037
SA773E	0037
SA773F	0037
SA773G	0037
SA773L	0037
SA773M	0037
SA773R	0037
SA773T	0037
SA773X	0037
SA773Y	0037
SA775A	0854
SA775B	0854
SA775C	0854
SA775D	0854
SA775E	0854
SA775F	0854
SA775G	0854
SA775L	0854
SA775M	0854
SA775R	0854
SA775T	0854
SA775X	0854
SA775Y	0854
SA778A	0037
SA778B	0037
SA778C	0037
SA778D	0037
SA778E	0037
SA778F	0037
SA778G	0037
SA778L	0037
SA778M	0037
SA778R	0037
SA778T	0037
SA778Y	0037
SA779A	0854
SA779B	0854
SA779C	0854
SA779D	0854
SA779E	0854
SA779F	0854
SA779G	0854
SA779L	0854
SA779M	0854
SA779R	0854
SA779T	0854
SA779X	0854
SA779Y	0854
SA780A	0854
SA780B	0854
SA780C	0854
SA780D	0854
SA780E	0854
SA780F	0854
SA780G	0854
SA780L	0854
SA780M	0854
SA780R	0854
SA780T	0854
SA780X	0854
SA780Y	0854
SA816A	0854
SA816B	0854
SA816C	0854
SA816D	0854
SA816E	0854
SA816F	0854
SA816G	0854
SA816L	0854
SA816M	0854
SA816R	0854
SA816T	0854
SA816X	0854
SA816Y	0854
SA821	0636
SA821A	0636
SA823	0636
SA823A	0636
SA825	0636
SA825A	0636
SA827	0636
SA827A	0636
SA829	0636

DEVICE TYPE	REPL CODE
SA829A	0636
SA836A	0037
SA836B	0037
SA836C	0037
SA836D	0037
SA836E	0037
SA836F	0037
SA836G	0037
SA836L	0037
SA836M	0037
SA836R	0037
SA836T	0037
SA836X	0037
SA836Y	0037
SA841A	0037
SA841B	0037
SA841C	0037
SA841D	0037
SA841E	0037
SA841F	0037
SA841G	0037
SA841L	0037
SA841M	0037
SA841R	0037
SA841T	0037
SA841X	0037
SA841Y	0037
SA842A	0037
SA842B	0037
SA842C	0037
SA842D	0037
SA842E	0037
SA842F	0037
SA842G	0037
SA842L	0037
SA842M	0037
SA842R	0037
SA842T	0037
SA842X	0037
SA842Y	0037
SA909A	0037
SA909B	0037
SA909C	0037
SA909D	0037
SA909E	0037
SA909F	0037
SA909G	0037
SA909L	0037
SA909M	0037
SA909R	0037
SA909T	0037
SA909X	0037
SA909Y	0037
SA1054P	OEM
SA1270	OEM
SA2002R	OEM
SA2004R	OEM
SA2006R	OEM
SA2008R	OEM
SA2913	OEM
SA2914	OEM
SA2915	OEM
SA2916	OEM
SA2917	OEM
SA2918	OEM
SA2919	OEM
SA2920	OEM
SA7805CDA	1916
SA7805CU	0624
SA7812CDA	3393
SA7812CU	0334
SA334490	OEM
SA334498	OEM
SA338979	OEM
SA338981	OEM
SA764059	OEM
SA878163	OEM
SA879849	OEM
SAA07347	OEM
SAA2-036	OEM
SAA4-001	OEM
SAA1000	OEM
SAA1004-N	OEM
SAA1005	OEM
SAA1005-P	OEM
SAA1008	OEM
SAA1010	OEM
SAA1020	OEM
SAA1021	OEM
SAA1022	OEM
SAA1024	3921
SAA1025	2144
SAA1030	OEM
SAA1042	OEM
SAA1043P	OEM
SAA1050	OEM
SAA1051	OEM
SAA1061	OEM
SAA1071	OEM
SAA1072	OEM
SAA1073	OEM
SAA1074	OEM
SAA1075	OEM
SAA1076	OEM
SAA1080	OEM
SAA-1124	OEM
SAA1124	3921
SAA1130	OEM
SAA1350	OEM

DEVICE TYPE	REPL CODE
SAA1351	OEM
SAA2008	OEM
SAA3000	OEM
SAA3010P	OEM
SAA3027P	OEM
SAA3037	OEM
SAA3100	OEM
SAA6000	OEM
SAA6001	OEM
SAA6002	OEM
SAA7000	OEM
SAA7010	OEM
SAA7020	OEM
SAA7030	OEM
SAA7210P	OEM
SAA7220	5821
SAA7220N	5821
SAA7220PA	OEM
SAA9010P	OEM
SAA60425	OEM
SAB1018P	OEM
SAB1044	0007
SAB3034	OEM
SAB3035	OEM
SAB3036	OEM
SAB3037	3386
SAB3037N	3386
SAB3469	0007
SAB8086-2-P	OEM
SAB8237A5P	OEM
SAB8259AP	OEM
SAB80513	OEM
SAC40	0150
SAC40A	0150
SAC40B	0150
SAC42	0150
SAC42A	0150
SAC42B	0150
SAC44	0150
SAC431-1	OEM
SAC-1843	0127
SADBG497ZSK	OEM
SAF0300	OEM
SAF45MJ	OEM
SAF45MS	OEM
SAF1032PN	OEM
SAF1039P	0950
SAF1039PN	0950
SAF1055	OEM
SAH190	OEM
SAH215	OEM
SAJ110	4358
SAJ270E	OEM
SAJ300R	OEM
SAJ300S	OEM
SAJ300T	OEM
SAJ2160	2887
SAJ72155	0946
SAJ72156	0964
SAJ72157	0528
SAJ72158	2882
SAJ72159	0930
SAJ72160	2887
SAJ72161	0638
SAJ72162	0673
SAJ72184	OEM
SAK115	OEM
SAK215	OEM
SANY0002	OEM
SANYO-002	OEM
SANY0002	OEM
SANYO-003	OEM
SANY0003	OEM
SANYO-004	OEM
SANY0004	OEM
SANYO-012	OEM
SANYO-051	OEM
SANYO-091	OEM
SA0403	OEM
SA0403A	OEM
SA0419	OEM
SAW-1S1941	0023
SAW-1S1944	0023
SAW-2SB56	0004
SAW-2SC372GR	0079
SAW-2SC372Y	0076
SAW-2SC945R	0076
SAY115X	OEM
SAY115Y	OEM
SB007-03SPA	OEM
SB0015-03A	OEM
SB-01	0277
SB01	0015
SB01-02	0133
SB01-05CP	OEM
SB-02	0289
SB-03	0015
SB-04	0296
SB05-05CP	OEM
SB-06	0302
SB07-03	OEM
SB07-03N	OEM
SB-08	0250
SB0319	0042
SB0419	0905
SB-1	0102
SB1-01-04	0015
SB-1AHF	0795
SB-1B	0102

DEVICE TYPE	REPL CODE
SB-1Z	0015
SB1Z	0110
SB-2	0102
SB2	OEM
SB-2(CENTERING-58	0920
SB-2B	0102
SB-2C	0348
SB2C	1059
SB-2C-GL	OEM
SB-2CGL	0348
SB-2CH	0102
SB2CH	OEM
SB2G	0102
SB-2T	0102
SB-2Y	0348
SB-3	0015
SB-3-02	0015
SB-3F	0087
SB-3F01	0015
SB-3N	0015
SB-10	0250
SB-15	0030
SB100	0136
SB101	0050
SB102	0050
SB103	0050
SB120	1335
SB130	1335
SB140	1335
SB150	1335
SB160	0080
SB168	0004
SB169	0004
SB170	0080
SB180	0080
SB190	3888
SB200	0136
SB302	0015
SB-309A	0015
SB309A	0277
SB-309C	0015
SB309C	0277
SB315	0015
SB320	0087
SB330	0087
SB332	0015
SB333	0015
SB340	0087
SB350	0087
SB360	0087
SB370	0087
SB380	0087
SB390	3604
SB393	0015
SB520	OEM
SB530	OEM
SB540	OEM
SB550	OEM
SB560	OEM
SB561	0037
SB570	OEM
SB580	OEM
SB821	0636
SB821A	0636
SB823	0636
SB823A	0636
SB825	0636
SB825A	0636
SB827	0636
SB827A	0636
SB829	0636
SB829A	0636
SB840	OEM
SB1000	0015
SB5021	OEM
SB5122	0136
SBA10L	OEM
SBD1620T	OEM
SBD1630T	OEM
SBD1635T	OEM
SBD1640T	OEM
SBD1645T	OEM
SBD1650T	OEM
SBD1660T	OEM
SBD3020M	OEM
SBD3020P	OEM
SBD3030M	OEM
SBD3030P	OEM
SBD3035M	OEM
SBD3035P	OEM
SBD3040M	OEM
SBD3040P	OEM
SBD3045M	OEM
SBD3045P	OEM
SBD3050M	OEM
SBD3050P	OEM
SBD3060M	OEM
SBD3060P	OEM
SBEB05	0277
SBEB6	0302
SBL-121	OEM
SBL-122	OEM
SBL-221	OEM
SBL-801	OEM
SBL-801(A)	OEM
SBL-801(Q)	OEM
SBL-801(S)	OEM
SBL-802	OEM
SBL-802(A)	OEM

DEVICE TYPE	REPL CODE
SBL-802(Q)	OEM
SBL-802(S)	OEM
SBL-803	OEM
SBL-803(A)	OEM
SBL-803(Q)	OEM
SBL-803(S)	OEM
SBL-804	OEM
SBL-804(A)	OEM
SBL-804(Q)	OEM
SBL-804(S)	OEM
SBL1030CT	1021
SBL1040CT	1021
SBN1020T	OEM
SBN1030T	OEM
SBN1035T	OEM
SBN1040T	OEM
SBN1045T	OEM
SBN1620T	OEM
SBN1630T	OEM
SBN1635T	OEM
SBN1640T	OEM
SBN1645T	OEM
SBN1650T	OEM
SBN1660T	OEM
SBN3020M	OEM
SBN3020P	OEM
SBN3030M	OEM
SBN3030P	OEM
SBN3035M	OEM
SBN3035P	OEM
SBN3040M	OEM
SBN3040P	OEM
SBN3045M	OEM
SBN3045P	OEM
SBN3050M	OEM
SBN3050P	OEM
SBN3060M	OEM
SBN3060P	OEM
SBP1020T	1021
SBP1030T	1021
SBP1035T	1021
SBP1040T	1021
SBP1045T	2515
SBP1620T	2515
SBP1630T	2515
SBP1635T	2515
SBP1640T	2515
SBP1645T	2515
SBP1650T	OEM
SBP1660T	OEM
SBP3020M	OEM
SBP3020P	1942
SBP3030M	OEM
SBP3030P	1942
SBP3035M	OEM
SBP3035P	1942
SBP3040M	OEM
SBP3040P	1942
SBP3045M	OEM
SBP3045P	1942
SBP3050M	OEM
SBP3050P	OEM
SBP3060M	OEM
SBP3060P	OEM
SBP9900A	OEM
SBP9900ANJ-1	OEM
SBP9989	OEM
SBP9989NJ	OEM
SBR05	0277
SBR6A05	2363
SBR6A1	2363
SBR6A2	2363
SBR6A3	2369
SBR6A4	2369
SBR6A5	2370
SBR6A6	2370
SBR6A8	2372
SBR6A10	2373
SBR10A05	2363
SBR10A1	2363
SBR10A2	2363
SBR10A3	2369
SBR10A4	2369
SBR10A5	2370
SBR10A6	2370
SBR10A8	2372
SBR10A10	2373
SBR15	OEM
SBR-260	0015
SBR520T	OEM
SBR530T	OEM
SBR535T	OEM
SBR540T	OEM
SBR545T	OEM
SBR820T	OEM
SBR830T	OEM
SBR835T	OEM
SBR840T	OEM
SBR845T	OEM
SBR850T	OEM
SBR860T	OEM
SBR1020T	OEM
SBR1030T	OEM
SBR1035T	OEM
SBR1040T	OEM
SBR1045T	OEM
SBR1620T	OEM
SBR1630T	OEM
SBR1635T	OEM

DEVICE TYPE	REPL CODE
SBR1640T	OEM
SBR1645T	OEM
SBS520T	OEM
SBS530T	OEM
SBS535T	OEM
SBS540T	OEM
SBS545T	OEM
SBS820T	OEM
SBS830T	OEM
SBS835T	OEM
SBS840T	OEM
SBS845T	OEM
SBS850T	OEM
SBS860T	OEM
SBS1020T	OEM
SBS1030T	OEM
SBS1035T	OEM
SBS1040T	OEM
SBS1045T	OEM
SBS1620T	OEM
SBS1630T	OEM
SBS1635T	OEM
SBS1640T	OEM
SBS1645T	OEM
SBT-101A+	2813
SBT-101K+	1783
SBT-201A+	2813
SBT-201K+	1783
SBT-401A+	2813
SBT-401K+	1783
SBT-601A+	2813
SBT-601K+	1783
SBT-801A+	2845
SBT-801K+	1818
SBT-1001A+	2845
SBT-1001K+	1818
SBX1435	OEM
SBX1435-01	OEM
SBX1475	OEM
SBX1483-51	OEM
SBX1483-59	OEM
SBX1492-51	OEM
SBX1505-01	OEM
SBX1535-02	OEM
SBX1535-04	OEM
SBX1535-51	OEM
SBX1535-52	OEM
SBX1535-54	OEM
SBX1535-54V	OEM
SBX1568-5	OEM
SBX1568-51	OEM
SBX1618-11	OEM
SBX1618-51	OEM
SBX1637-11	OEM
SBX1637A	OEM
SBX1637A-01	OEM
SBX1653-01	OEM
SBX1681-01	OEM
SC	0030
SC05	0015
SC05E	0015
SC-05M	OEM
SC015-02	OEM
SC-080	OEM
SC-080U	OEM
SC0321	0130
SC0328	0130
SC0421	1681
SC0428	1681
SC0515	1600
SC1	0015
SC2	0015
SC2C	0015
SC3	OEM
SC-4	0015
SC4	0015
SC5	OEM
SC-6	0143
SC6	0229
SC-7	0095
SC7	OEM
SC-7(PHILCO)	0095
SC7C	OEM
SC7D	OEM
SC7S08F	OEM
SC8	0071
SC8A	0071
SC9	OEM
SC10	0071
SC10A	0071
SC11	OEM
SC12	0004
SC12(DIODE)	0133
SC-15	0628
SC15	0030
SC-16	0015
SC16	0015
SC16C5	OEM
SC16N	OEM
SC-20	0628
SC20	3658
SC20F	OEM
SC25	OEM
SC-29	0196
SC29	0913
SC30	3658
SC30F	OEM
SC35A	0154
SC35B	0154

DEVICE TYPE	REPL CODE
SC35D	0147
SC35F	0154
SC36B	1064
SC36D	1317
SC40	3658
SC40B	0154
SC40B-1	0147
SC40B23	OEM
SC40B33	OEM
SC40D	0147
SC40E	0279
SC40F	6769
SC41B	1064
SC41D	1317
SC41E	1891
SC43	0280
SC44	0280
SC45	0004
SC45A	0147
SC45B	0154
SC45B13	0154
SC45D	0147
SC45D13	0147
SC45E	0279
SC45E13	0279
SC45F	0147
SC46	0280
SC46B	1064
SC46D	1317
SC46E	1891
SC47	OEM
SC50	3658
SC50A	0147
SC50B	0154
SC50B13	0154
SC50D	0147
SC50D13	0147
SC50E	0279
SC50E13	0279
SC50F	6780
SC51	OEM
SC51D	1317
SC51E	1891
SC54	0143
SC-56	0208
SC56	0208
SC60	3658
SC60A	3148
SC60A13	3148
SC60B	3150
SC60B13	3150
SC60C	3151
SC60C13	3151
SC60D	3152
SC60D13	3152
SC60E	3152
SC60E13	3152
SC60F	6787
SC60F13	3147
SC61	OEM
SC63	0004
SC65	0016
SC-66	0004
SC66	0004
SC68	0004
SC69	0004
SC-70	0160
SC70	3658
SC70F	OEM
SC71	0050
SC72	0050
SC73	0004
SC74	0050
SC75	OEM
SC78	0050
SC79	0050
SC80	3658
SC80F	OEM
SC82	OEM
SC91	0012
SC92A	5013
SC92B	5013
SC92D	6807
SC92F	5013
SC100	OEM
SC107	0079
SC107A	0079
SC107B	0079
SC108	0079
SC108A	0079
SC108B	0079
SC108C	0079
SC109	0079
SC109A	0079
SC109B	0079
SC109C	OEM
SC110	0015
SC115	0871
SC116D	0593
SC120	OEM
SC126A	OEM
SC126B	OEM
SC126C	OEM
SC126D	OEM
SC126E	OEM
SC126F	OEM
SC126M	OEM
SC129B	0298
SC129D	0306

DEVICE TYPE	REPL CODE
SC129E	0311
SC129M	0311
SC136A	OEM
SC136B	0481
SC136C	OEM
SC136D	0485
SC136E	OEM
SC136M	OEM
SC140B	0555
SC140B3	OEM
SC140B5	OEM
SC140B6	OEM
SC140D	0555
SC140D3	OEM
SC140D5	OEM
SC140D6	OEM
SC140E	0555
SC140E3	OEM
SC140E5	OEM
SC140E6	OEM
SC140M	0555
SC140M3	OEM
SC140M5	OEM
SC140M6	OEM
SC141A	0571
SC141A1	0571
SC141A2	0571
SC141A3	0571
SC141A4	0571
SC141A5	0571
SC141A6	0571
SC141B	0572
SC141B1	0593
SC141B2	0593
SC141B3	0593
SC141B4	0593
SC141B5	0593
SC141B6	0593
SC141C	0574
SC141C1	0574
SC141C2	0574
SC141C3	0574
SC141C4	0574
SC141C5	0574
SC141C6	0574
SC141D	0576
SC141D1	0576
SC141D2	0576
SC141D3	0576
SC141D4	0576
SC141D5	0576
SC141D6	0576
SC141E	0577
SC141E1	0577
SC141E2	0577
SC141E3	0577
SC141E4	0577
SC141E5	0577
SC141E6	0577
SC141F	0578
SC141F1	0578
SC141F2	0578
SC141F3	0578
SC141F4	0578
SC141F5	0578
SC141F6	0578
SC141M	0594
SC141N	0596
SC141S	OEM
SC142B	0593
SC142B2	OEM
SC142B5	OEM
SC142B6	OEM
SC142D	0593
SC142D3	OEM
SC142D5	OEM
SC142D6	OEM
SC142E	0555
SC142E3	OEM
SC142E5	OEM
SC142E6	OEM
SC142M	0555
SC142M3	OEM
SC142M5	OEM
SC142M6	OEM
SC143B	0614
SC143D	0593
SC143E	0617
SC143M	0594
SC146A	0571
SC146A1	0571
SC146A2	0571
SC146A3	0571
SC146A4	0571
SC146A5	0571
SC146A6	0571
SC146B	0593
SC146B1	0593
SC146B2	0593
SC146B3	0593
SC146B4	0593
SC146B5	0593
SC146B6	0593
SC146C	0574
SC146C1	0574
SC146C2	0574
SC146C3	0574
SC146C4	0574
SC146C5	0574

If replacement code is OEM, contact original manufacturer for replacement.

DEVICE TYPE	REPL CODE
SC146C6	OEM
SC146D	0576
SC146D1	0576
SC146D2	0576
SC146D3	0576
SC146D4	0576
SC146D5	0576
SC146D6	0576
SC146E	0577
SC146E1	0577
SC146E3	0577
SC146E4	0577
SC146E5	0577
SC146E6	0577
SC146F	0578
SC146F1	0578
SC146F2	0578
SC146F3	0578
SC146F4	0578
SC146F5	0578
SC146F6	0578
SC146M	0594
SC146N	OEM
SC146S	OEM
SC147	0079
SC147A	0111
SC147B	0111
SC147B3	OEM
SC147B5	OEM
SC147B6	OEM
SC147D	0555
SC147D3	OEM
SC147D5	OEM
SC147D6	OEM
SC147E	0555
SC147M	0555
SC147M3	OEM
SC147M5	OEM
SC147M6	OEM
SC148	0111
SC148A	0111
SC148B	0111
SC148B3	OEM
SC148B5	OEM
SC148B6	OEM
SC148C	0111
SC148D	OEM
SC148D3	OEM
SC148D5	OEM
SC148D6	OEM
SC148E	OEM
SC148E5	OEM
SC148E6	OEM
SC148M	OEM
SC148M3	OEM
SC148M5	OEM
SC148M6	OEM
SC149	0111
SC149B	0111
SC149B(TRIAC)	0772
SC149C	0079
SC149D	0744
SC149E	0617
SC149M	0617
SC150	OEM
SC150B	0758
SC150B3	OEM
SC150B5	OEM
SC150B6	OEM
SC150C20	OEM
SC150C30	OEM
SC150C40	OEM
SC150C40J	OEM
SC150C50	OEM
SC150C60	OEM
SC150C80	OEM
SC150C100	OEM
SC150C120	OEM
SC150D	0763
SC150D3	OEM
SC150D5	OEM
SC150D6	OEM
SC150E	OEM
SC150E3	OEM
SC150E5	OEM
SC150E6	OEM
SC150M	OEM
SC150M3	OEM
SC150M5	OEM
SC150M6	OEM
SC151B	0772
SC151D	0744
SC151E	0617
SC151M	0617
SC157	0861
SC157A	0861
SC157V1	0037
SC157VI	0861
SC158	0861
SC158A	0861
SC158B	0693
SC158V1	0037
SC158VI	0861
SC159	0861
SC159A	0861
SC159B	0693
SC160B	0908
SC160D	0908
SC160E	0915
SC160M	OEM
SC174	0079
SC174A	0079
SC174B	0079
SC180	OEM
SC200	OEM
SC215	0871
SC240B	0154
SC240B2	3199
SC240B3	3199
SC240B12	OEM
SC240B13	0154
SC240B14	OEM
SC240B22	OEM
SC240B22/32	OEM
SC240B23	OEM
SC240B24	OEM
SC240B32	OEM
SC240B33	OEM
SC240D	0147
SC240D2	3207
SC240D3	3207
SC240D12	OEM
SC240D13	0147
SC240D14	OEM
SC240D22	OEM
SC240D23	OEM
SC240D24	OEM
SC240D32	OEM
SC240D33	OEM
SC240E	0279
SC240E2	3222
SC240E3	3222
SC240E12	OEM
SC240E13	0279
SC240E14	OEM
SC240E22	OEM
SC240E23	OEM
SC240E24	OEM
SC240E32	OEM
SC240E33	OEM
SC240M	0279
SC240M2	3222
SC240M3	3222
SC240M12	OEM
SC240M13	OEM
SC240M22	OEM
SC240M23	OEM
SC240M32	OEM
SC240M33	OEM
SC241B	1064
SC241B12	OEM
SC241B13	OEM
SC241B14	OEM
SC241D	1317
SC241D12	OEM
SC241D13	OEM
SC241D14	OEM
SC241E	1891
SC241E12	OEM
SC241E13	OEM
SC241E14	OEM
SC241M	1891
SC241M12	OEM
SC241M13	OEM
SC245B	0154
SC245B2	3199
SC245B3	3199
SC245B12	OEM
SC245B13	0154
SC245B14	OEM
SC245B22	OEM
SC245B23	OEM
SC245B24	OEM
SC245B32	OEM
SC245B33	OEM
SC245D	0147
SC245D2	3207
SC245D3	3207
SC245D12	OEM
SC245D13	0279
SC245D14	OEM
SC245D22	OEM
SC245D23	OEM
SC245D24	OEM
SC245D32	OEM
SC245D33	OEM
SC245E	0279
SC245E2	3222
SC245E3	3222
SC245E12	OEM
SC245E13	0279
SC245E22	OEM
SC245E23	OEM
SC245E24	OEM
SC245E32	OEM
SC245E33	OEM
SC245M	0279
SC245M2	3222
SC245M3	3222
SC245M12	OEM
SC245M13	OEM
SC245M22	OEM
SC245M23	OEM
SC245M32	OEM
SC245M33	OEM
SC246B	1064
SC246B12	OEM
SC246B13	OEM
SC246B14	OEM
SC246D	1317
SC246D12	OEM
SC246D13	OEM
SC246D14	OEM
SC246E	1891
SC246E12	OEM
SC246E13	OEM
SC246E14	OEM
SC246F13	OEM
SC246M	1891
SC246M12	OEM
SC246M13	1891
SC250	OEM
SC250B	0154
SC250B2	3199
SC250B3	3199
SC250B12	OEM
SC250B13	0154
SC250B14	OEM
SC250B22	OEM
SC250B23	OEM
SC250B24	OEM
SC250B32	OEM
SC250D	0147
SC250D2	3207
SC250D3	3207
SC250D12	OEM
SC250D13	0147
SC250D14	OEM
SC250D22	OEM
SC250D23	OEM
SC250D24	OEM
SC250D32	OEM
SC250D33	OEM
SC250E	0279
SC250E2	3222
SC250E3	3222
SC250E12	OEM
SC250E13	0279
SC250E14	OEM
SC250E22	OEM
SC250E23	OEM
SC250E24	OEM
SC250E32	OEM
SC250E33	OEM
SC250M	0279
SC250M2	3222
SC250M3	3222
SC250M12	OEM
SC250M13	OEM
SC250M22	OEM
SC250M23	OEM
SC250M23/33	OEM
SC250M32	OEM
SC250M33	OEM
SC251B	1064
SC251B12	OEM
SC251B13	OEM
SC251B14	OEM
SC251D	1317
SC251D12	OEM
SC251D13	OEM
SC251D14	OEM
SC251E	1891
SC251E12	OEM
SC251E13	OEM
SC251E14	OEM
SC251M	1891
SC251M12	OEM
SC251M13	OEM
SC256	0861
SC256A	0861
SC256B	0791
SC257	0861
SC257A	0861
SC257V1	0037
SC257VI	0861
SC258	0861
SC258A	0861
SC258B	0693
SC258V1	0037
SC258VI	0861
SC259	0861
SC259A	0861
SC259B	0693
SC260B	2015
SC260B2	3199
SC260B3	3199
SC260B12	OEM
SC260B13	OEM
SC260B22	OEM
SC260B23	OEM
SC260B32	OEM
SC260B33	OEM
SC260D	2017
SC260D2	3207
SC260D3	3207
SC260D12	OEM
SC260D13	OEM
SC260D22	OEM
SC260D23	OEM
SC260D32	OEM
SC260D33	OEM
SC260E	2018
SC260E2	3222
SC260E3	3222
SC260E12	OEM
SC260E13	OEM
SC260E22	OEM
SC260E23	OEM
SC260E32	OEM
SC260E33	OEM
SC260M	2018
SC260M2	3222
SC260M3	3222
SC260M12	OEM
SC260M13	OEM
SC260M22	OEM
SC260M23	OEM
SC260M32	OEM
SC260M33	OEM
SC261B	1064
SC261B12	OEM
SC261B13	OEM
SC261D	1317
SC261D12	OEM
SC261D13	OEM
SC261E	1891
SC261E12	OEM
SC261E13	OEM
SC261M	1891
SC261M12	OEM
SC261M13	OEM
SC265B	2015
SC265B2	3199
SC265B3	3199
SC265D	2017
SC265D2	3207
SC265D3	3207
SC265E	2018
SC265E2	3222
SC265E3	3222
SC265M	2018
SC265M2	3222
SC265M3	3222
SC266B	1064
SC266D	1317
SC266E	1891
SC266M	1891
SC305	0015
SC315	OEM
SC350	0016
SC365	0126
SC400Y	0142
SC415	OEM
SC441	0168
SC4960B	OEM
SC515	OEM
SC570A	OEM
SC727	0142
SC777	0142
SC785	0198
SC786	0007
SC821	0636
SC821A	0636
SC823	0636
SC823A	0636
SC825	0636
SC825A	0636
SC827	0636
SC827A	0636
SC829	0636
SC829A	0636
SC832	0016
SC842	0016
SC843	0126
SC843(T.I.)	0233
SC902	OEM
SC902A	OEM
SC969	OEM
SC1001	0016
SC1007	0050
SC1010	0016
SC1148-4	0348
SC1168G	0016
SC1168H	0016
SC1227F	0144
SC1227G	0144
SC1229E	0086
SC1229G	0016
SC1294H	0126
SC1600	OEM
SC1601	OEM
SC1611	0844
SC1612	0844
SC1613	0844
SC1614	0844
SC1625	OEM
SC3162	OEM
SC4001BH	0478
SC4001UBC	0478
SC4004	0142
SC4010	0016
SC4044	0016
SC4073	0208
SC4116	0015
SC4131	0086
SC4131-1	0086
SC4132	0142
SC4133	0161
SC4135	0086
SC4162	OEM
SC4163	OEM
SC4164	OEM
SC4167	0086
SC4195	OEM
SC4244	0086
SC4274	0222
SC4277	OEM
SC4303	0042
SC4303-1	0042
SC4303-2	0042
SC4308	0042
SC5087	OEM
SC5088	OEM
SC5117	0654
SC5118P	0443
SC5150P	3008
SC5172P	1940
SC5172PC	1940
SC5172PQ	1940
SC5175G	0410
SC5177P	0654
SC5199P	0654
SC5204P	1753
SC5219P	0354
SC5220P	0167
SC5221P	5092
SC5245P	0789
SC5282P	1913
SC5454P	OEM
SC5858	3720
SC5896	2501
SC5898	1170
SC6394GLC6	OEM
SC8705P	0701
SC8709P	0666
SC8723P	0349
SC9314P	0654
SC9341P	OEM
SC9430P	1808
SC9431P	0395
SC9436P	0167
SC9438P	0167
SC9961P	OEM
SC9963P	2867
SC9964P	3427
SC9965P	3429
SC11007	OEM
SC11014	OEM
SC41342P	OEM
SC41843P	OEM
SC42502P	3594
SC42530P	OEM
SC42682P	OEM
SC51750	0410
SC67331	OEM
SC67331P	OEM
SC68681C5140(A)	OEM
SC68681CSN40(A)	OEM
SC74186	OEM
SC77526	OEM
SC77527	OEM
SC77527P	OEM
SC77562P	OEM
SC78130U	OEM
SC80240P	OEM
SC80444P	OEM
SC80756P	OEM
SC82165FN	OEM
SC84251P	OEM
SC87898P	OEM
SC88639B	OEM
SC88658B	OEM
SC93870P	OEM
SC94815P	OEM
SC99616P	OEM
SC99630P	OEM
SC99642P	OEM
SC99643P	OEM
SC99659P	OEM
SC99666B	OEM
SC99669P	OEM
SC403602P	OEM
SC403603P	OEM
SC403604P	OEM
SC403737P	OEM
SC409006B	OEM
SC411733B	OEM
SC420111P	OEM
SC4036037P	OEM
SCA05	0015
SCA1	0015
SCA2	0015
SCA3	0015
SCA4	0015
SCA5	0015
SCA6	0015
SCA8	0071
SCA10	0071
SCA42	0715
SCA43	0715
SCA44	0715
SCA45	0715
SCA-45A	1323
SCA45A	6203
SCA92	0342
SCA95	0477
SCA1103	0015
SCA3021	0259
SCA3022	0007
SCA3023	0418
SCA3235	0007
SCA3236	0007
SCA3237	0007
SCA3238	0007
SCA3239	0007
SCA3240	0007
SCA3242	0050
SCA3243	0007
SCA3244	0007
SCA3246	0007
SCAJ2	1414
SCAJ4	0473
SCAJ6	0446
SCBA2	2363
SCBA6	2370
SCBR05F	0071
SCBR1F	0071
SCBR6	0087
SCBR6F	0071
SCBR15	OEM
SCBR35F	0071
SCC321	0130
SCC421	1681
SCC8530	OEM
SCD11B1	1053
SCD11B2	1053
SCD11B3	1053
SCD255	1109
SCD321	0130
SCD421	1681
SCD1181	OEM
SCD1182	OEM
SCD1183	OEM
SCD-1062013	0466
SCE1	0015
SCE2	0015
SCE3	0015
SCE4	0015
SCE6	0015
SCE8	0071
SCE10	0071
SCE15	OEM
SCE20	OEM
SCE25	OEM
SCE25A	OEM
SCE30	OEM
SCE30A	OEM
SCE40	OEM
SCE40A	OEM
SCE50	OEM
SCE50A	OEM
SCE75	OEM
SCE100	OEM
SCE321	0130
SCE421	1681
SCF40501	2027
SCH120	OEM
SCH130	OEM
SCH140	OEM
SCH150	OEM
SCH160	OEM
SCH320	OEM
SCH330	OEM
SCH340	OEM
SCH350	OEM
SCH360	OEM
SCH520	OEM
SCH530	OEM
SCH540	OEM
SCH550	OEM
SCH560	OEM
SCH1020	OEM
SCH1030	OEM
SCH1040	OEM
SCH1050	OEM
SCH1060	OEM
SCI148-4	0102
SCITT310YXXXX	0030
SCITT410STXX	0030
SCITT410STXXX	0030
SCL003URC3FX	OEM
SCL003URC5F	OEM
SCL5.0A	2396
SCL6.0A	3114
SCL6.5A	OEM
SCL7.0A	OEM
SCL7.5A	OEM
SCL8.0A	OEM
SCL8.5A	OEM
SCL9.0A	OEM
SCL10A	OEM
SCL11A	3172
SCL12A	OEM
SCL13A	OEM
SCL14A	OEM
SCL15A	3191
SCL16A	OEM
SCL17A	3201
SCL18A	OEM
SCL20A	OEM
SCL22A	1915
SCL24A	OEM
SCL26A	OEM
SCL28A	OEM
SCL30A	OEM
SCL33A	OEM
SCL36A	OEM
SCL40A	OEM
SCL43A	0568
SCL45A	OEM
SCL48A	OEM
SCL51A	OEM
SCL54A	OEM
SCL58A	0831
SCL60A	OEM
SCL64A	OEM
SCL70A	OEM
SCL75A	OEM
SCL78A	OEM
SCL85A	OEM
SCL90A	OEM
SCL100A	OEM
SCL110A	OEM
SCL120A	OEM
SCL130A	OEM
SCL150A	OEM
SCL160A	OEM
SCL170A	1405
SCL4000	2024
SCL4000B	2024
SCL4000BC	2024
SCL4000BD	2024
SCL4000BE	2024
SCL4000BF	2024
SCL4000BH	2024
SCL4001	0478
SCL4001B	0478
SCL4001BC	0478
SCL4001BD	0478
SCL4001BE	0478
SCL4001BF	0478
SCL4001UB	OEM
SCL4001UBD	0478
SCL4001UBE	0478
SCL4001UBF	0478
SCL4001UBH	0478
SCL4002	2055
SCL4002B	2055
SCL4002BC	2055
SCL4002BD	2055
SCL4002BE	2055
SCL4002BF	2055
SCL4002BH	2055
SCL4006BC	0646
SCL4006BD	0646
SCL4006BE	0646
SCL4006BF	0646
SCL4007	2846
SCL4007UBC	2846
SCL4007UBD	2846
SCL4007UBE	2846
SCL4007UBHN	OEM
SCL4008B	0988
SCL4008BC	0988
SCL4008BD	0988
SCL4008BE	0988
SCL4008BF	0988
SCL4008BH	0988
SCL4008BP	0988
SCL4009	1999
SCL4009UBC	1999
SCL4009UBD	1999
SCL4009UBE	0001
SCL4009UBHN	OEM
SCL4010BC	3252
SCL4010BD	3252
SCL4010BE	0398
SCL4010BHN	OEM
SCL4011	0215
SCL4011B	0215
SCL4011BC	0215
SCL4011BD	0215
SCL4011BE	0215
SCL4011BF	0215
SCL4011BH	0215
SCL4011UB	OEM
SCL4011UBC	0215
SCL4011UBD	0215
SCL4011UBE	0215
SCL4011UBF	0215
SCL4011UBH	0215
SCL4012	0498
SCL4012B	0498
SCL4012BC	0498
SCL4012BD	0498
SCL4012BE	0498
SCL4012BF	0498
SCL4012BH	0498
SCL4013	0413
SCL4013AC	0413
SCL4013AD	0413
SCL4013AE	0413
SCL4013AF	0413
SCL4013AH	0413
SCL4013BC	0413
SCL4013BD	0413
SCL4013BE	0413
SCL4013BF	0413
SCL4013BH	0413
SCL4013BP	0413
SCL4014BC	0860
SCL4014BD	0860
SCL4014BE	0860
SCL4014BF	OEM
SCL4014BHN	OEM
SCL4015	1014
SCL4015BC	1014
SCL4015BD	1014
SCL4015BE	1014
SCL4015BF	OEM
SCL4015BHN	OEM
SCL4016	1143

If replacement code is OEM, contact original manufacturer for replacement.

DEVICE TYPE	REPL CODE
SCL4016BH	OEM
SCL4016BHN	OEM
SCL4017	0513
SCL4017ABC	0513
SCL4017ABD	0513
SCL4017ABE	0513
SCL4017ABF	0513
SCL4017ABH	0513
SCL4017AC	0513
SCL4017AD	0513
SCL4017AE	0513
SCL4017AF	0513
SCL4017AH	0513
SCL4017B	0513
SCL4017BC	0513
SCL4017BD	0513
SCL4017BE	0513
SCL4017BF	0513
SCL4017BH	0513
SCL4018AC	1391
SCL4018AD	1391
SCL4018AE	1391
SCL4018AF	1391
SCL4018AH	1391
SCL4018B	1391
SCL4018BD	1391
SCL4018BE	1391
SCL4018BF	1391
SCL4018BH	1391
SCL4019	1525
SCL4020	1661
SCL4020ABC	1661
SCL4020ABD	1661
SCL4020ABE	1661
SCL4020ABF	1661
SCL4020ABH	1661
SCL4020AC	1661
SCL4020AD	1661
SCL4020AE	1661
SCL4020AF	1661
SCL4020AH	1661
SCL4020B	1661
SCL4021	1749
SCL4021BC	1749
SCL4021BD	1749
SCL4021BE	1749
SCL4021BF	OEM
SCL4021BHN	OEM
SCL4022ABC	1256
SCL4022ABD	1256
SCL4022ABE	1256
SCL4022ABF	1256
SCL4022ABH	1256
SCL4022AC	1256
SCL4022AD	1256
SCL4022AE	1256
SCL4022AF	1256
SCL4022AH	1256
SCL4022B	1256
SCL4022BC	1256
SCL4022BD	1256
SCL4022BE	1256
SCL4022BF	1256
SCL4022BH	1256
SCL4023	0519
SCL4023AC	0519
SCL4023AD	0519
SCL4023AE	0519
SCL4023AF	0519
SCL4023AH	0519
SCL4023B	0519
SCL4023BC	0519
SCL4023BD	0519
SCL4023BE	0519
SCL4023BF	0519
SCL4023BH	0519
SCL4024	1957
SCL4024AC	1957
SCL4024AD	1957
SCL4024AE	1957
SCL4024AF	1957
SCL4024AH	1957
SCL4024AT	1957
SCL4024B	1957
SCL4024BC	1957
SCL4024BD	1957
SCL4024BE	1957
SCL4024BF	1957
SCL4024BH	1957
SCL4025	2072
SCL4025AC	2072
SCL4025AD	2072
SCL4025AE	2072
SCL4025AF	2072
SCL4025AH	2072
SCL4025B	2072
SCL4025BC	2072
SCL4025BD	2072
SCL4025BE	2072
SCL4025BF	2072
SCL4025BH	2072
SCL4026AB	2152
SCL4026ABC	2152
SCL4026ABD	2152
SCL4026ABE	2152
SCL4027	1949
SCL4027B	1949
SCL4027BC	1949
SCL4027BD	1949

DEVICE TYPE	REPL CODE
SCL4027BE	1949
SCL4027BF	1949
SCL4027BH	1949
SCL4028AC	2228
SCL4028AD	2228
SCL4028AE	2228
SCL4028AF	2228
SCL4028AH	2228
SCL4028B	2228
SCL4028BC	2228
SCL4028BD	2228
SCL4028BE	2228
SCL4028BF	2228
SCL4028BH	2228
SCL4029AC	2233
SCL4029AD	2233
SCL4029AE	2233
SCL4029AF	2233
SCL4029AH	2233
SCL4029B	2233
SCL4029BC	2233
SCL4029BD	2233
SCL4029BE	2233
SCL4029BF	2233
SCL4029BH	2233
SCL4030	0500
SCL4030AC	0500
SCL4030AD	0500
SCL4030AE	0500
SCL4030AF	0500
SCL4030AH	0500
SCL4030B	0500
SCL4030BC	0500
SCL4030BD	0500
SCL4030BE	0500
SCL4030BF	0500
SCL4030BH	0500
SCL4033AB	2635
SCL4033ABC	2635
SCL4033ABD	2635
SCL4033ABE	2635
SCL4034BC	3615
SCL4034BD	3615
SCL4034BE	3615
SCL4034BH	OEM
SCL4035BC	2777
SCL4035BD	2777
SCL4035BE	2777
SCL4035BF	OEM
SCL4035BHN	OEM
SCL4040	0056
SCL4040ABC	0056
SCL4040ABD	0056
SCL4040AC	0056
SCL4040AD	0056
SCL4040AE	0056
SCL4040AF	0056
SCL4040AH	0056
SCL4040B	0056
SCL4041UBC	3174
SCL4041UBD	3174
SCL4041UBE	3174
SCL4041UBHN	OEM
SCL4042	0121
SCL4042B	0121
SCL4042BC	0121
SCL4042BD	0121
SCL4042BE	0121
SCL4042BF	0121
SCL4042BH	0121
SCL4042BHN	OEM
SCL4043ABC	1553
SCL4043ABD	1553
SCL4043ABE	1553
SCL4043ABF	1553
SCL4043ABH	1553
SCL4043B	1553
SCL4043BC	1553
SCL4043BD	1553
SCL4043BE	1553
SCL4043BF	1553
SCL4043BH	1553
SCL4044ABC	2308
SCL4044ABE	2308
SCL4044ABF	2308
SCL4044ABH	2308
SCL4044B	2308
SCL4044BC	2308
SCL4044BD	2308
SCL4044BE	2308
SCL4044BF	2308
SCL4044BH	2308
SCL4047B	2331
SCL4047BC	2331
SCL4047BD	2331
SCL4047BE	2331
SCL4047BF	2331
SCL4047BH	2331
SCL4049	0001
SCL4049UBH	OEM
SCL4049UBHN	OEM
SCL4050	0398
SCL4050BH	OEM
SCL4050BHN	OEM
SCL4051	0366
SCL4051BH	OEM
SCL4051BHN	OEM
SCL4052	0024
SCL4053BH	OEM

DEVICE TYPE	REPL CODE
SCL4053BHN	OEM
SCL4060ABC	0146
SCL4060ABD	0146
SCL4060ABE	0146
SCL4060ABF	0146
SCL4060ABH	0146
SCL4060AC	0146
SCL4060AD	0146
SCL4060AE	0146
SCL4060AF	0146
SCL4060AH	0146
SCL4060B	0146
SCL4066BH	OEM
SCL4066BHN	OEM
SCL4068B	2503
SCL4068BC	2503
SCL4068BD	2503
SCL4068BE	2503
SCL4068BF	2503
SCL4068BH	2503
SCL4070B	2516
SCL4070BC	2516
SCL4070BD	2516
SCL4070BE	2516
SCL4070BF	2516
SCL4070BH	2516
SCL4071AC	0129
SCL4071AD	0129
SCL4071AE	0129
SCL4071AF	0129
SCL4071AH	0129
SCL4071B	0129
SCL4071BC	0129
SCL4071BD	0129
SCL4071BE	0129
SCL4071BF	0129
SCL4071BH	0129
SCL4072AC	2524
SCL4072AD	2524
SCL4072AE	2524
SCL4072AF	2524
SCL4072AH	2524
SCL4072B	2524
SCL4072BC	2524
SCL4072BD	2524
SCL4072BE	2524
SCL4072BF	2524
SCL4072BH	2524
SCL4073AC	1536
SCL4073AD	1536
SCL4073AE	1536
SCL4073AF	1536
SCL4073AH	1536
SCL4073B	1536
SCL4073BC	1536
SCL4073BD	1536
SCL4073BE	1536
SCL4073BF	1536
SCL4073BH	1536
SCL4075AC	2540
SCL4075AD	2540
SCL4075AE	2540
SCL4075AF	2540
SCL4075AH	2540
SCL4075B	2540
SCL4075BC	2540
SCL4075BD	2540
SCL4075BE	2540
SCL4075BF	2540
SCL4075BH	2540
SCL4076BC	3496
SCL4076BD	3496
SCL4076BE	3496
SCL4076BF	OEM
SCL4076BHN	OEM
SCL4077B	2558
SCL4077BC	2558
SCL4077BD	2558
SCL4077BE	2558
SCL4077BF	2558
SCL4077BH	2558
SCL4078B	0921
SCL4078BC	0921
SCL4078BD	0921
SCL4078BE	0921
SCL4078BF	0921
SCL4078BH	0921
SCL4081	0626
SCL4081AC	0626
SCL4081AD	0626
SCL4081AE	0626
SCL4081AF	0626
SCL4081AH	0626
SCL4081B	0626
SCL4081BC	0626
SCL4081BD	0626
SCL4081BE	0626
SCL4081BF	0626
SCL4081BH	0626
SCL4082B	0300
SCL4082BC	0300
SCL4082BD	0300
SCL4082BE	0300
SCL4082BF	0300
SCL4082BH	0300
SCL4085B	0303
SCL4085BC	0300
SCL4085BD	0303
SCL4085BE	0303

DEVICE TYPE	REPL CODE
SCL4085BF	0303
SCL4085BH	0303
SCL4086B	0535
SCL4086BC	0535
SCL4086BD	0535
SCL4086BE	0535
SCL4086BF	0535
SCL4086BH	0535
SCL4094BD	1682
SCL4094BE	1682
SCL4160B	1359
SCL4161B	1373
SCL4162B	1388
SCL4163B	1407
SCL4174B	1551
SCL4416	OEM
SCL4416ABE	OEM
SCL4449UBE	0001
SCL4502BE	1037
SCL4510AC	1963
SCL4510AD	1963
SCL4510AE	1963
SCL4510AF	1963
SCL4510AH	1963
SCL4510B	1963
SCL4510BC	1963
SCL4510BD	1963
SCL4510BE	1963
SCL4510BF	1963
SCL4510BH	1963
SCL4511AC	1544
SCL4511AD	1544
SCL4511AE	1544
SCL4511AF	1544
SCL4511AH	1544
SCL4511BC	1544
SCL4511BD	1544
SCL4511BE	1544
SCL4511BF	1544
SCL4511BH	1544
SCL4512BC	2120
SCL4512BD	2120
SCL4512BE	2120
SCL4512BHN	OEM
SCL4514B	1830
SCL4514BD	1830
SCL4514BE	1830
SCL4514BH	1830
SCL4515B	3600
SCL4515BD	3600
SCL4515BE	3600
SCL4515BH	3600
SCL4516AC	2347
SCL4516AD	2347
SCL4516AE	2347
SCL4516AF	2347
SCL4516AH	2347
SCL4516B	2347
SCL4516BC	2347
SCL4516BD	2347
SCL4516BE	2347
SCL4516BF	2347
SCL4516BH	2347
SCL4518	1043
SCL4518AC	1043
SCL4518AD	1043
SCL4518AE	1043
SCL4518AF	1043
SCL4518AH	1043
SCL4518B	1043
SCL4518BC	1043
SCL4518BD	1043
SCL4518BE	1043
SCL4518BF	1043
SCL4518BH	1043
SCL4520AC	2675
SCL4520AD	2675
SCL4520AE	2675
SCL4520AF	2675
SCL4520AH	2675
SCL4520B	2675
SCL4520BC	2675
SCL4520BD	2675
SCL4520BE	2675
SCL4520BF	2675
SCL4520BH	2675
SCL4522B	OEM
SCL4522BC	2837
SCL4522BD	2837
SCL4522BE	2837
SCL4522BF	2837
SCL4522BH	2837
SCL4526B	3610
SCL4527B	3145
SCL4527BC	3145
SCL4527BD	3145
SCL4527BE	3145
SCL4527BF	3145
SCL4527BH	3145
SCL4528B	3611
SCL4531B	OEM
SCL4543BC	5017
SCL4543BD	OEM
SCL4543BE	5020
SCL4555B	2938
SCL4555BC	2938
SCL4555BD	2938
SCL4555BE	2938
SCL4555BF	2938

DEVICE TYPE	REPL CODE
SCL4555BH	2938
SCL4556BC	3431
SCL4556BD	3431
SCL4556BE	3431
SCL4556BF	3431
SCL4556BH	3431
SCM5.0A	2396
SCM6.0A	3114
SCM6.5A	OEM
SCM7.0A	OEM
SCM7.5A	OEM
SCM8.0A	OEM
SCM8.5A	OEM
SCM9.0A	OEM
SCM10	OEM
SCM10A	OEM
SCM11A	3172
SCM12A	OEM
SCM13A	OEM
SCM14A	OEM
SCM15	OEM
SCM15A	3191
SCM16A	OEM
SCM17A	3201
SCM18A	OEM
SCM20	OEM
SCM20A	OEM
SCM22A	1915
SCM24A	OEM
SCM25	OEM
SCM26A	OEM
SCM28A	OEM
SCM30A	OEM
SCM33A	OEM
SCM36A	OEM
SCM40A	OEM
SCM43A	0568
SCM45A	OEM
SCM48A	OEM
SCM51A	OEM
SCM54A	OEM
SCM58A	0831
SCM60A	OEM
SCM64A	OEM
SCM70A	OEM
SCM75A	OEM
SCM78A	OEM
SCM85A	OEM
SCM90A	OEM
SCM100A	OEM
SCM110A	OEM
SCM120A	OEM
SCM130A	OEM
SCM150A	OEM
SCM160A	OEM
SCM170A	1405
SCM90072C	0522
SCM90072P	0522
SCM90522P	OEM
SCM90590P	OEM
SCM91248C	OEM
SCM91619P	OEM
SCM91665P	OEM
SCM91692P	OEM
SCM92135P	OEM
SCM93007P	OEM
SCM93008P	OEM
SCM95987P	OEM
SCN8048AC6N40-A	OEM
SCN68000C8N64	OEM
SC0321	OEM
SC0328	OEM
SC0421	OEM
SC0428	OEM
SCP5E	0015
SCPH3	OEM
SCPH5	OEM
SCPH7	OEM
SCPH15	OEM
SCPH25	OEM
SCPH35	OEM
SCPH45	OEM
SCQ38068PB01	OEM
SCQ38068PB02	OEM
SCR-01	1103
SCR-01-B	1103
SCR01C	1103
SCR-02	0240
SCR-02-C	0240
SCR02C	1073
SCR-03	2519
SCR03	0722
SCR-03-C	2188
SCR-04	0745
SCR04	0745
SCR-04-C	0762
SCR04C	0745
SCR5P-05M	0505
SCR10A	OEM
SCR11A	OEM
SCR12A	OEM
SCR13A	OEM
SCR14A	OEM
SCR15	0344
SCR15A	OEM
SCR16A	OEM
SCR17A	OEM
SCR18A	OEM
SCR30A	OEM

DEVICE TYPE	REPL CODE
SCR31A	OEM
SCR32A	OEM
SCR33A	OEM
SCR34A	OEM
SCR35A	OEM
SCR36A	OEM
SCR37A	OEM
SCR38A	OEM
SCR50A	OEM
SCR51	OEM
SCR51A	OEM
SCR52	OEM
SCR52(ELCOM)	2188
SCR52A	OEM
SCR53A	OEM
SCR54	OEM
SCR54A	OEM
SCR55	1103
SCR55A	OEM
SCR56A	OEM
SCR57A	OEM
SCR58A	OEM
SCR71	OEM
SCR72	OEM
SCR74	OEM
SCR76	OEM
SCR92	OEM
SCR102	0741
SCR102(ELCOM)	0567
SCR104	2270
SCR104(ELCOM)	1923
SCR108	0901
SCR108(ELCOM)	0901
SCR111	OEM
SCR112	OEM
SCR118	0745
SCR118(ELCOM)	2491
SCR122(ELCOM)	1651
SCR150	OEM
SCR160	OEM
SCR160A	OEM
SCR161A	OEM
SCR162A	OEM
SCR163A	OEM
SCR164A	OEM
SCR165A	OEM
SCR166A	OEM
SCR167A	OEM
SCR168A	OEM
SCR200	0058
SCR202	0745
SCR202(ELCOM)	0762
SCR204	2270
SCR207	0745
SCR207(ELCOM)	0435
SCR208	0745
SCR208(ELCOM)	0240
SCR218	0058
SCR218(ELCOM)	0058
SCR222(ELCOM)	1651
SCR255	0731
SCR303	OEM
SCR315	OEM
SCR325	OEM
SCR335	OEM
SCR345	OEM
SCR350	OEM
SCR355	OEM
SCR360	OEM
SCR370	OEM
SCR380	OEM
SCR402	0747
SCR402(ELCOM)	0740
SCR404(ELCOM)	0095
SCR408	0747
SCR408(ELCOM)	0676
SCR422(ELCOM)	1584
SCR502	2519
SCR505	2809
SCR505(ELCOM)	2809
SCR508	2519
SCR508(ELCOM)	1103
SCR520	0240
SCR535	OEM
SCR535A	OEM
SCR538	2270
SCR545	OEM
SCR545A	OEM
SCR550A	OEM
SCR555	OEM
SCR555A	OEM
SCR560A	OEM
SCR570	OEM
SCR570A	OEM
SCR580	OEM
SCR580A	OEM
SCR604(ELCOM)	3404
SCR605(ELCOM)	3624
SCR608(ELCOM)	0336
SCR613	2270
SCR635(ELCOM)	0725
SCR805(ELCOM)	0061
SCR816(ELCOM)	0766
SCR835	OEM
SCR845	OEM
SCR855	OEM
SCR860	OEM
SCR865	OEM
SCR870	OEM

DEVICE TYPE	REPL CODE
SCR875	OEM
SCR880	OEM
SCR1635	OEM
SCR1635A	OEM
SCR1645	OEM
SCR1645A	OEM
SCR1655	OEM
SCR1655A	OEM
SCR1660A	OEM
SCR1670A	OEM
SCR1680A	OEM
SCR2520	0469
SCRF10A	OEM
SCRF11A	OEM
SCRF12A	OEM
SCRF13A	OEM
SCRF14A	OEM
SCRF15A	OEM
SCRF16A	OEM
SCRF17A	OEM
SCRF18A	OEM
SCRF30A	OEM
SCRF31A	OEM
SCRF32A	OEM
SCRF33A	OEM
SCRF34A	OEM
SCRF35A	OEM
SCRF36A	OEM
SCRF37A	OEM
SCRF38A	OEM
SCRF50A	OEM
SCRF51A	OEM
SCRF52A	OEM
SCRF53A	OEM
SCRF54A	OEM
SCRF55A	OEM
SCRF56A	OEM
SCRF57A	OEM
SCRF58A	OEM
SCRF160A	OEM
SCRF161A	OEM
SCRF162A	OEM
SCRF163A	OEM
SCRF164A	OEM
SCRF165A	OEM
SCRF166A	OEM
SCRF167A	OEM
SCRF168A	OEM
SCRF503	OEM
SCRF505	OEM
SCRF510	OEM
SCRF515	OEM
SCRF520	OEM
SCRF525	OEM
SCRF530	OEM
SCRF535	OEM
SCRF540	OEM
SCRF545	OEM
SCRF550	OEM
SCRF555	OEM
SCRF560	OEM
SCRF570	OEM
SCRF580	OEM
SCRF1603	OEM
SCRF1605	OEM
SCRF1610	OEM
SCRF1615	OEM
SCRF1620	OEM
SCRF1625	OEM
SCRF1630	OEM
SCRF1635	OEM
SCRF1640	OEM
SCRF1645	OEM
SCRF1650	OEM
SCRF1655	OEM
SCRF1665	OEM
SCRF1670	OEM
SCRF1680	OEM
SCRFL03	OEM
SCRFL05	OEM
SCRFL10	OEM
SCRFL15	OEM
SCRFL20	OEM
SCRFL25	OEM
SCRFL30	OEM
SCRFL35	OEM
SCRFL40	OEM
SCRFL45	OEM
SCRFL50	OEM
SCRFL55	OEM
SCRFL60	OEM
SCRFL70	OEM
SCRFL80	OEM
SCRJ03	OEM
SCRJ05	OEM
SCRJ10	OEM
SCRJ15	OEM
SCRJ20	OEM
SCRJ25	OEM
SCRJ30	OEM
SCRJ35	OEM
SCRJ40	OEM
SCRJ45	OEM
SCRJ50	OEM
SCRJ55	OEM
SCRJ60	OEM
SCRJ70	OEM
SCRJ80	OEM
SCRJL01	OEM

If replacement code is OEM, contact original manufacturer for replacement.

DEVICE TYPE	REPL CODE
SCRJL03	OEM
SCRJL05	OEM
SCRJL15	OEM
SCRJL20	OEM
SCRJL25	OEM
SCRJL30	OEM
SCRJL35	OEM
SCRJL40	OEM
SCRJL45	OEM
SCRJL50	OEM
SCRJL55	OEM
SCRJL60	OEM
SCRJL70	OEM
SCRJL80	OEM
SCRL35	OEM
SCRL40	OEM
SCRL45	OEM
SCRL50	OEM
SCRL55	OEM
SCRL60	OEM
SCRL70	OEM
SCRL80	OEM
SCT1	0071
SCT2	0071
SCT3	0071
SCT4	0071
SCT5	0071
SCT-24	2525
SD-0033	0025
SD05	0015
SD-07	0015
SD07	0015
SD014	OEM
SD020	0123
SD040	0015
SD0345	0060
SD0445	1308
SD-1	0102
SD1	0023
SD1-1	0015
SD1-11	0102
SD-1-30DA	0102
SD1-30DA	0102
SD-1-211B	0015
SD1-211B	0015
SD-1-211C	0015
SD1-X	0015
SD1-Z	0015
SD-1A	0015
SD1A	0102
SD1A(UF)	OEM
SD-1AHF	0920
SD1AHF	1303
SD-1AUF	0133
SD-1B	0015
SD1B	0102
SD-1B(HF)	0102
SD-1BHF	0102
SD1BHF	0102
SD-1C	0102
SD1C	0071
SD-1C-4F	0015
SD-1C-UF	0015
SD-1CUF	0015
SD1CUF	0015
SD1DM-4	0015
SD-1HF	0015
SD1HF	0015
SD-1L	0015
SD1L	OEM
SD-1LA	0015
SD1LA	0015
SD-1N34A	0123
SD-1N60	0019
SD-1N60P	0123
SD-1N60S	0019
SD-1N4148	0124
SD1P60XXXXFXT	0019
SD-1S295	1909
SD-1S1555	0133
SD-1S1925	0917
SD-1S2076	0133
SD-1S2207	0030
SD-1S2339	0554
SD-1S2692	2518
SD-1SS16	0917
SD-1SV58	1029
SD-1SV59	1029
SD-1SV70	1029
SD-1U	0015
SD-1UF	0015
SD-1VHF	0102
SD1W	OEM
SD-1X	0015
SD1X	0015
SD-1Y	0015
SD1Y	0023
SD-1Z	0015
SD1Z	0023
SD-1ZHF	0015
SD1ZHF	0015
SD-2	0015
SD2	0071
SD2A	0015
SD2B	0015
SD2C	0071
SD2L	OEM
SD2LE	OEM
SD2LEE	OEM
SD-2N5062	0901
SD2WE	OEM
SD4	0015
SD4(DUAL)	0196
SD-5	0133
SD5	0015
SD5(DUAL)	0484
SD-5(PHILCO)	0133
SD-6	0162
SD6	0015
SD6(DUAL)	0242
SD-7	4113
SD7	5574
SD-7(PHILCO)	0133
SD-8	0012
SD8	0071
SD-8(PHILCO)	0012
SD-8(RECTIFIER)	0817
SD10	0015
SD-11	0495
SD11F	0123
SD-12	0143
SD12	0143
SD12B	0143
SD12E	0143
SD12M	0143
SD12V	0143
SD-13	0015
SD13	0143
SD-14	0143
SD14	0143
SD-15	0012
SD15	0143
SD-16	0143
SD16	0143
SD-16A	0015
SD-16D	0015
SD17	0143
SD-18	0015
SD18	0143
SD-19	0277
SD19	0030
SD20	0015
SD21	0143
SD21A	0143
SD-22	0102
SD22	0102
SD23	0015
SD24	0030
SD-27	0012
SD30	OEM
SD-31	0531
SD31	0565
SD-32	0157
SD32	0253
SD-33	0365
SD33	0137
SD-34	0133
SD34	0143
SD38	0143
SD-39	0030
SD39	0030
SD41	OEM
SD42	0025
SD-43	0133
SD43	0133
SD45	0015
SD-46	0143
SD46	0143
SD46(4)	0123
SD-46-2	0143
SD46-2	0143
SD46R	0123
SD-49	0025
SD49	0025
SD50	1335
SD-51	0917
SD51	0917
SD53	0137
SD54	0143
SD-56	0143
SD56	0143
SD-60	0143
SD60	0143
SD-60P	0015
SD60P	1335
SD-61P	0015
SD61P	0795
SD-80	0015
SD80	0015
SD82	0917
SD82A	0917
SD82AG	0917
SD-91	0015
SD91	0110
SD-91A	0015
SD91A	0242
SD-91S	0015
SD91S	0242
SD-92	0015
SD92	0242
SD-92A	0015
SD92A	0242
SD-92S	0015
SD92S	0242
SD-93	0015
SD93	0540
SD-93A	0015
SD93A	0540
SD93S	0015
SD-94	0015
SD94	0540
SD-94A	0015
SD94A	0540
SD-94AB	0015
SD94AB	0015
SD94B	0015
SD94S	0015
SD-95	0015
SD95	0015
SD-95A	0015
SD95A	0015
SD96	0015
SD96A	0015
SD96S	0015
SD98	0071
SD98A	0071
SD98S	0071
SD100	0133
SD-101	0015
SD101	0124
SD102	0124
SD103	0124
SD103B	0917
SD103C	OEM
SD104	0023
SD105	0023
SD-109	0079
SD109	0079
SD-110	0133
SD111	0720
SD112	OEM
SD113	0030
SD113TA	0030
SD115	0760
SD-116	0005
SD116	0005
SD120	OEM
SD122	OEM
SD141	0178
SD-150	0143
SD160	0133
SD165	0133
SD184-1	OEM
SD190	3888
SD200	OEM
SD-201	0015
SD201	0015
SD202	0015
SD218	0620
SD241	OEM
SD241P	0584
SD282A	0917
SD300C02C	OEM
SD300C04C	OEM
SD300C06C	OEM
SD300C08C	OEM
SD300C10C	OEM
SD300C12C	OEM
SD300C14C	OEM
SD300C16C	OEM
SD300C18C	OEM
SD300C20C	OEM
SD300C22C	OEM
SD300C24C	OEM
SD300C26C	OEM
SD300C28C	OEM
SD300C30C	OEM
SD345	0042
SD390	3604
SD400C02C	OEM
SD400C04C	OEM
SD400C06C	OEM
SD400C08C	OEM
SD400C10C	OEM
SD400C12C	OEM
SD400C14C	OEM
SD400C16C	OEM
SD400C18C	OEM
SD400C20C	OEM
SD400C22C	OEM
SD400C24C	OEM
SD400N02MC	OEM
SD400N02PC	OEM
SD400N04MC	OEM
SD400N04PC	OEM
SD400N06MC	OEM
SD400N06PC	OEM
SD400N08MC	OEM
SD400N08PC	OEM
SD400N10MC	OEM
SD400N10PC	OEM
SD400N12MC	OEM
SD400N12PC	OEM
SD400N14MC	OEM
SD400N14PC	OEM
SD400N16MC	OEM
SD400N16PC	OEM
SD400N18MC	OEM
SD400N18PC	OEM
SD400N20MC	OEM
SD400N20PC	OEM
SD400N22MC	OEM
SD400N22PC	OEM
SD400N24MC	OEM
SD400N24PC	OEM
SD400R02PC	OEM
SD400R04PC	OEM
SD400R06PC	OEM
SD400R08PC	OEM
SD400R10PC	OEM
SD400R12PC	OEM
SD400R14PC	OEM
SD400R16PC	OEM
SD400R18PC	OEM
SD400R20PC	OEM
SD400R22PC	OEM
SD400R24PC	OEM
SD-404	0917
SD404	0917
SD445	0925
SD461	0143
SD467	0286
SD-470	0015
SD470	0015
SD471CG	OEM
SD500	0133
SD500C	0015
SD500KD	1668
SD501	0015
SD503K	OEM
SD600	0133
SD600C	0015
SD600N02P	5046
SD600N04P	5046
SD600N06P	5046
SD600N06PC	5046
SD600N08P	0749
SD600N10P	0749
SD600N12P	0749
SD600N12PC	OEM
SD600N14P	OEM
SD600N16P	OEM
SD600N18PC	OEM
SD600N20PC	OEM
SD600N22PC	OEM
SD600N24PC	OEM
SD600R02P	2315
SD600R04P	2315
SD600R06P	2315
SD600R06PC	2315
SD600R08P	2792
SD600R10P	2792
SD600R12P	2792
SD600R12PC	2792
SD600R14P	OEM
SD600R16P	OEM
SD600R18PC	OEM
SD600R20PC	OEM
SD600R22PC	OEM
SD600R24PC	OEM
SD603K	OEM
SD-630	0133
SD630	0133
SD-632	0012
SD632	0012
SD632(10)	0012
SD701-02	0133
SD703K	OEM
SD800	0071
SD803K	OEM
SD838	0917
SD903K	OEM
SD910	0071
SD910A	0071
SD910S	0071
SD950	0015
SD974	0133
SD1002KD	OEM
SD1003K	OEM
SD1005	2070
SD1005KD	OEM
SD1006	2208
SD1011KD	1668
SD1012	0286
SD1012KD	1668
SD1013	1198
SD1014	1233
SD1014-1	5303
SD1014KD	OEM
SD1015KD	OEM
SD1018	1974
SD1020	0493
SD1021KD	1668
SD1023	0560
SD1069	1233
SD1074	2312
SD1076	3565
SD1077	2700
SD1078	OEM
SD1080	2041
SD1080-2	2041
SD1080-4	2041
SD1080-7	2041
SD1088	3589
SD1089	2868
SD1094	OEM
SD1095	OEM
SD1096	OEM
SD1098	3643
SD1100C02C	OEM
SD1100C04C	OEM
SD1100C06C	OEM
SD1100C08C	OEM
SD1100C10C	OEM
SD1100C12C	OEM
SD1100C14C	OEM
SD1100C16C	OEM
SD1100C18C	OEM
SD1100C20C	OEM
SD1100C22C	OEM
SD1100C24C	OEM
SD1101	0015
SD1102	0015
SD1103	0015
SD1103K	OEM
SD1115-4	0418
SD1127	3572
SD1133	1233
SD1134	3560
SD1134-1	3421
SD1135	2091
SD1136	1420
SD1143	1233
SD1144	3560
SD1144-1	0418
SD1145	2091
SD1146	1420
SD1169	3570
SD1203K	OEM
SD1212-7	3632
SD1214	1198
SD1214-6	3797
SD1216	1977
SD1229	1977
SD1270	3421
SD1272	3041
SD1278	OEM
SD1285	5202
SD1288	3567
SD1290	3570
SD1300	0259
SD1301	0259
SD1303K	OEM
SD1335	0060
SD1345	0042
SD1347-7	6184
SD1403K	OEM
SD1405	3565
SD1407	2545
SD1410	3640
SD1412	3642
SD1416	3194
SD1418	3641
SD1428	3119
SD1433	1420
SD1434	2868
SD1444	2039
SD1445	0925
SD1451	2312
SD1479	2170
SD1484	2041
SD1503K	OEM
SD1556-3	0604
SD1603K	OEM
SD1703K	OEM
SD1803K	OEM
SD3621A	OEM
SD5010	OEM
SD5011	OEM
SD5012	OEM
SD5013	OEM
SD5014	OEM
SD5015	OEM
SD5050	OEM
SD5051	OEM
SD5171	OEM
SD5410-1	3552
SD5410-2	3552
SD5410-3	3552
SD5440-1	2313
SD5440-2	2313
SD5440-3	2313
SD5440-4	3878
SD5443-1	2313
SD5443-2	2313
SD11441	0007
SD58655	OEM
SDA345	0042
SDA445	0925
SDA980-1	2363
SDA980-1F	2363
SDA980-2	2363
SDA980-2F	2363
SDA980-3	2363
SDA980-3F	2363
SDA980-4	2369
SDA980-4F	2369
SDA980-5	2369
SDA980-5F	2369
SDA980-6	2370
SDA980-6F	2370
SDA980-8	2372
SDA980-10	2373
SDA990-1	2363
SDA990-1F	1643
SDA990-2	1643
SDA990-2F	1643
SDA990-3	1643
SDA990-3F	1643
SDA990-4	1673
SDA990-4F	1673
SDA990-5	1673
SDA990-5F	1673
SDA990-6	1673
SDA990-6F	1673
SDA990-8	1589
SDA990-10	1589
SDA3202	OEM
SDA9086-3	OEM
SDA9187X	OEM
SDA9188X	OEM
SDB345	0042
SDB445	0925
SDB501B-GD	OEM
SD-BA244	0163
SD-BA244A	0133
SD-BB409	OEM
SD-BZX9V1	2478
SDD4	0196
SDD5	0484
SDD6	1080
SDD320	OEM
SDD420	0086
SDD421	0016
SDD820	0127
SDD821	0016
SDD1220	0086
SDD3000	0016
SDD-C05	OEM
SDD-C10	0196
SDDS442XXXXXX	0124
SDDSF10XCXXXXT	0023
SD-ERB24-04D	0023
SD-ERB26-20	0348
SD-ERC26-13	0102
SDF10	OEM
SDF20	OEM
SDF30	OEM
SDF40	OEM
SDF50	OEM
SDF60	OEM
SDF9401F	OEM
SDG600	0604
SDG601	0604
SDG602	0604
SDG603	0604
SDGMA01XXXXXT	0124
SDGZA033XXXXX	0299
SDGZA062YXXXX	0471
SDH-2	0143
SDH-2HC	0917
SD-HV23GBL	1925
SDI345	0042
SDI445	0925
SD-IN4148	0124
SDJ345	0042
SDJ445	0925
SDK345	0042
SDK445	0925
SDL345	0042
SDL445	0925
SDM345	0042
SDM402	OEM
SDM445	0925
SDM4001	1394
SDM4002	1394
SDM4003	1394
SDM4004	1394
SDM4005	1394
SDM4006	1394
SDM4010	2439
SDM5001	3525
SDM5002	3525
SDM5003	3525
SDM5004	3525
SDM5005	3525
SDM5006	3525
SDM5010	2626
SDM5011	2626
SDM5012	2626
SDM5013	2626
SDN201	1212
SDN345	0042
SDN401	3370
SDN445	0925
SDN501	2235
SDN601	3374
SDN6000	2626
SDN6060	2626
SDN6061	2626
SDN6062	2626
SDN6251	1991
SDN6252	2626
SDN6253	2626
SDN22311	1991
SDN22312	1991
SDN22313	1991
SDO345	0236
SDO445	0681
SDP251	0602
SDP345	0423
SDP445	1199
SDP451	3540
SDP502	OEM
SDR3A	0087
SDR-25	0015
SDR25	0015
SDR2712	OEM
SDR2713	OEM
SDR2720	OEM
SDR2721	OEM
SDR2722	OEM
SDR2723	OEM
SDR3001	0087
SDR3002	0087
SDS113	0015
SDS240	0007
SDT234A	OEM
SDT401	0074
SDT402	0223
SDT-410	0074
SDT410	0270
SDT-411	1966
SDT411	0270
SDT-413	0363
SDT413	0074
SDT-423	0642
SDT423	1966
SDT424	0223
SDT425	0313
SDT430	0363
SDT431	1966
SDT445	0925
SDT520	OEM
SDT521	OEM
SDT522	0622
SDT525	OEM
SDT526	OEM
SDT527	0363
SDT530	OEM
SDT531	OEM
SDT532	0363
SDT535	OEM
SDT536	OEM
SDT537	0363
SDT540	OEM
SDT541	OEM
SDT542	0363
SDT545	OEM
SDT547	0363
SDT550	OEM
SDT551	OEM
SDT552	OEM
SDT555	OEM
SDT556	OEM
SDT557	OEM
SDT560	OEM
SDT561	OEM
SDT562	OEM
SDT565	OEM
SDT566	OEM
SDT567	OEM
SDT570	OEM
SDT571	OEM
SDT572	OEM
SDT1000	OEM
SDT1001	0363
SDT1002	0363
SDT1003	0363
SDT1004	0363
SDT1005	0363
SDT1006	0313
SDT1007	0313
SDT1011	0363
SDT1012	0363
SDT1013	0363
SDT1014	0363
SDT1015	0363
SDT1016	0313
SDT1017	0313
SDT1051	0363
SDT1055	0363
SDT1056	0363
SDT1060	0363
SDT1061	0363
SDT1611	1966
SDT1612	1966
SDT1613	1966
SDT1614	1966
SDT1615	1966
SDT1616	1966
SDT1617	1966
SDT1618	1966
SDT1621	0130
SDT1622	0130
SDT1623	0130
SDT1631	0130
SDT1632	0130
SDT1633	0130
SDT3321	0126
SDT3322	0126
SDT3325	1267
SDT3326	0560
SDT3421	6396
SDT3422	0622
SDT3423	0622
SDT3424	0622
SDT3425	6396
SDT3426	6396
SDT3427	0622
SDT3428	0622
SDT3429	0622
SDT3501	0126
SDT3502	0126
SDT3503	0126
SDT3504	6404
SDT3505	0382
SDT3506	0382
SDT3507	4587
SDT3508	4587
SDT3509	0925
SDT3510	0925
SDT3513	0925

DEVICE TYPE	REPL CODE
SDT3514	0925
SDT3550	0382
SDT3552	0382
SDT3553	0382
SDT3575	0905
SDT3576	0905
SDT3577	0905
SDT3578	0905
SDT3579	0905
SDT3701	0925
SDT3702	0925
SDT3703	0925
SDT3704	0925
SDT3706	0925
SDT3707	0925
SDT3708	0925
SDT3709	0925
SDT3710	0925
SDT3711	0925
SDT3712	0925
SDT3713	0925
SDT3715	0925
SDT3716	0925
SDT3717	0925
SDT3718	1199
SDT3719	1199
SDT3720	0925
SDT3721	0925
SDT3722	0925
SDT3723	1199
SDT3724	1199
SDT3725	0925
SDT3726	0925
SDT3727	0925
SDT3728	1199
SDT3729	0925
SDT3730	0925
SDT3731	1199
SDT3732	1199
SDT3733	0925
SDT3750	1598
SDT3751	1598
SDT3752	1598
SDT3753	1598
SDT3754	1681
SDT3755	1681
SDT3756	1598
SDT3757	1598
SDT3758	1598
SDT3759	1681
SDT3760	1681
SDT3761	1598
SDT3762	1598
SDT3763	1598
SDT3764	1681
SDT3765	1681
SDT3766	1681
SDT3775	1267
SDT3776	1267
SDT3778	1267
SDT3801	1199
SDT3802	1199
SDT3803	1199
SDT3804	1199
SDT3805	1199
SDT3806	1199
SDT3807	1199
SDT3825	1598
SDT3826	1681
SDT3827	1681
SDT3850	1199
SDT3851	1199
SDT3852	1199
SDT3875	1681
SDT3876	1681
SDT3877	1681
SDT4301	6396
SDT4302	6396
SDT4304	6396
SDT4305	6396
SDT4306	0622
SDT4307	6396
SDT4308	6396
SDT4309	0622
SDT4310	6396
SDT4311	6396
SDT4312	0622
SDT4455	0560
SDT4483	0560
SDT4551	0560
SDT4553	0560
SDT4583	0560
SDT4611	0560
SDT4612	0560
SDT4614	0560
SDT4615	0560
SDT4813	OEM
SDT4814	OEM
SDT5001	6396
SDT5002	3329
SDT5003	3329
SDT5006	6396
SDT5007	3329
SDT5008	3329
SDT5011	0560
SDT5012	3329
SDT5013	3329
SDT5102	0042
SDT5112	0925

DEVICE TYPE	REPL CODE
SDT5501	6396
SDT5502	3329
SDT5503	3329
SDT5504	0622
SDT5506	6396
SDT5507	3329
SDT5508	3329
SDT5509	0622
SDT5511	0560
SDT5512	3329
SDT5513	3329
SDT5635	OEM
SDT5636	OEM
SDT5901	0560
SDT5902	0560
SDT5903	2548
SDT5906	0560
SDT5907	0042
SDT5908	2548
SDT5911	2548
SDT5912	2548
SDT5913	2548
SDT6001	0042
SDT6011	0042
SDT6013	0042
SDT6031	0042
SDT6101	0560
SDT6102	0560
SDT6103	0042
SDT6104	0560
SDT6105	0560
SDT6106	0560
SDT6383	OEM
SDT6417	OEM
SDT6418	OEM
SDT6419	OEM
SDT6420	OEM
SDT6901	0178
SDT6905	0178
SDT7011	2662
SDT7012	2662
SDT7013	2662
SDT7014	2662
SDT7015	2662
SDT7016	2662
SDT7140	2662
SDT7141	2662
SDT7150	2662
SDT7151	2662
SDT7152	2662
SDT7154	2662
SDT7155	2662
SDT7206	2273
SDT7401	6396
SDT7402	0622
SDT7403	0622
SDT7411	6396
SDT7412	0622
SDT7413	0622
SDT7414	0622
SDT7415	0622
SDT7416	0622
SDT7417	0622
SDT7418	0622
SDT7419	0622
SDT7471	OEM
SDT7472	OEM
SDT7473	OEM
SDT7474	OEM
SDT7475	OEM
SDT7476	OEM
SDT7477	OEM
SDT7478	OEM
SDT7511	0042
SDT7512	0042
SDT7514	0042
SDT7515	0042
SDT7530	OEM
SDT7531	OEM
SDT7601	0103
SDT7602	0103
SDT7603	0103
SDT7604	2433
SDT7607	0103
SDT7608	0103
SDT7609	0103
SDT7610	6536
SDT7611	6536
SDT7612	6536
SDT7618	OEM
SDT7619	OEM
SDT7731	0103
SDT7732	0103
SDT7733	0103
SDT7734	4786
SDT7735	4786
SDT8002	3512
SDT8003	3512
SDT8012	3512
SDT8013	3512
SDT8015	3512
SDT8016	3512
SDT8045	3512
SDT8070	3512
SDT8071	3512
SDT8301	3489
SDT8302	3489
SDT8303	3489
SDT8304	3489

DEVICE TYPE	REPL CODE
SDT9001	0560
SDT9002	0560
SDT9003	0560
SDT9004	1267
SDT9005	0560
SDT9006	0560
SDT9007	0560
SDT9008	0560
SDT9009	0042
SDT9113	OEM
SDT9201	0103
SDT9202	0543
SDT9203	6885
SDT9204	0543
SDT9205	0103
SDT9206	0103
SDT9207	0543
SDT9208	0543
SDT9209	0543
SDT9210	0103
SDT9261	0103
SDT9301	0103
SDT9302	0103
SDT9303	0103
SDT9304	0103
SDT9305	0363
SDT9306	0103
SDT9307	0103
SDT9308	0103
SDT9309	0103
SDT9701	0130
SDT9702	2485
SDT9703	4786
SDT9704	0130
SDT9705	2485
SDT9706	4786
SDT9707	0130
SDT9804	4786
SDT9901	0543
SDT9902	0543
SDT9903	0543
SDT9904	0543
SDT12201	OEM
SDT12202	OEM
SDT12203	OEM
SDT12301	1852
SDT12302	1852
SDT12303	1852
SDT13201	OEM
SDT13202	OEM
SDT13203	OEM
SDT13204	OEM
SDT13205	OEM
SDT13301	1852
SDT13302	1852
SDT13303	1852
SDT13304	OEM
SDT13305	OEM
SDT14304	OEM
SDT14305	OEM
SD-W005M	0106
SD-W04	0289
SD-Y	0015
SE0001	0076
SE002(1)	0127
SE-05	0015
SE05	0015
SE-05-01	0015
SE-05-02	0015
SE-05-2	0229
SE-05-A	0015
SE-05-B	0015
SE-05-C	0071
SE-05-D	0071
SE-05A	0015
SE05A	0540
SE05B	0015
SE-05C	0087
SE05C	0071
SE05D	0015
SE05S	0015
SE05SS	0015
SE-05X	0015
SE0555	OEM
SE-0.5A	0540
SE-0.5B	0015
SE-1.5A	0087
SE1.5A	OEM
SE1.5B	OEM
SE1.5C	OEM
SE1.5D	OEM
SE1.5S	OEM
SE1.5SS	OEM
SE-2	0015
SE-5	0015
SE-5-0399	0975
SE-5-0819	0211
SE5-0930	2830
SE5-0933	0167
SE5-0996	0325
SE5U4GE	OEM
SE6(DUAL)	1080
SE6X4	OEM
SE19	OEM
SE21	OEM
SE30B26A	0015
SE46	0015
SE106J	OEM
SE111J	OEM

DEVICE TYPE	REPL CODE
SE112J	OEM
SE116J	OEM
SE120	OEM
SE120N	OEM
SE125J	OEM
SE125N	OEM
SE130N	OEM
SE130NH	OEM
SE135	OEM
SE-135N	OEM
SE135N	OEM
SE-135NS	OEM
SE136	OEM
SE136N	OEM
SE140	OEM
SE140N	OEM
SE140NL	OEM
SE155J	OEM
SE156J	OEM
SE161J	OEM
SE170J	OEM
SE180J	OEM
SE303	OEM
SE303A	0516
SE303AC	OEM
SE303ACK	0516
SE303ANC	OEM
SE303AX	OEM
SE303AX2	OEM
SE303AY	OEM
SE304	OEM
SE393A	OEM
SE416J	OEM
SE424J	OEM
SE455J	OEM
SE480J	OEM
SE500G	0144
SE504	0144
SE521	0144
SE555	3780
SE555V	0973
SE556	6897
SE592F	OEM
SE592FH	OEM
SE592K	OEM
SE-705G	0187
SE724J	OEM
SE1000C	OEM
SE1000P	OEM
SE1001	0076
SE1001-1	0224
SE1001-2	0224
SE-1002	0016
SE1002	0076
SE1002-1	0007
SE1002-2	0007
SE1003	OEM
SE1004C	OEM
SE-1010	0127
SE1010	0007
SE1012	0595
SE1019	0144
SE1044	0007
SE1100C	OEM
SE1331	0016
SE1400P	OEM
SE1419	0144
SE1730	0071
SE1731	OEM
SE1732	OEM
SE1733	OEM
SE1734	OEM
SE2001	0007
SE2002	0007
SE2020	0127
SE2100P	OEM
SE2130P	OEM
SE2200P	OEM
SE2382	OEM
SE2383	OEM
SE2384	OEM
SE2385	OEM
SE2397	0127
SE2400	0050
SE2400P	OEM
SE2401	0016
SE2402	0016
SE-3001	0144
SE3001	0007
SE3001R	0144
SE3002	0007
SE3003	0144
SE3005	0007
SE3019	0007
SE3030	0074
SE3031	0074
SE3032	0177
SE3033	0103
SE3034	0630
SE3035	0103
SE3036	0103
SE3040	0429
SE3041	0429
SE3100	0330
SE3450-1	0591
SE3450-2	0591
SE3450-3	0591
SE3451-1	0591
SE3451-2	0591

DEVICE TYPE	REPL CODE
SE3451-3	0591
SE3453-1	0591
SE3453-2	0591
SE3453-3	0591
SE3453-4	0591
SE3455-1	0591
SE3455-2	0591
SE3455-3	OEM
SE3455-4	OEM
SE3646	0007
SE3819	0325
SE-4001	0079
SE4001	0016
SE4002	1221
SE-4010	0016
SE4010	1221
SE-4020	0086
SE4020	1221
SE4020/6-04	0007
SE4021	1221
SE4022	0007
SE4172	0016
SE4329	OEM
SE-5001	0144
SE5001	0224
SE5002	0076
SE5003	0007
SE5004	0007
SE-5006	0016
SE5006	0144
SE5010	0144
SE5015	0144
SE5020	0007
SE5021	0007
SE5022	0007
SE-5023	0007
SE5023	0007
SE-5024	0007
SE5024	0007
SE-5025	0007
SE5025	0144
SE5029	0144
SE5030	0144
SE5030A	0076
SE5030B	0016
SE5031	0144
SE5032	0007
SE5035	0007
SE5036	0144
SE5040	0144
SE5050	0007
SE5051	0007
SE5052	0007
SE5055	0224
SE5056	0144
SE5151	0016
SE5315	0212
SE5450-1	0591
SE5450-2	0591
SE5450-3	0591
SE5451-1	0591
SE5451-2	0591
SE5451-3	0591
SE5453-1	0591
SE5453-2	0591
SE5453-3	0591
SE5453-4	0591
SE5455-1	0591
SE5455-2	0591
SE5455-3	0591
SE5455-4	0591
SE5534	OEM
SE5560F	OEM
SE5560N	OEM
SE-6001	0016
SE6001	0076
SE-6002	0016
SE6002	0076
SE6006	0086
SE6010	0016
SE6020	0016
SE6020A	0086
SE6021	0086
SE6021A	0086
SE6022	0086
SE6023	0086
SE6452-3	OEM
SE6550	OEM
SE7001	0076
SE7002	0233
SE7005	0086
SE7006	0142
SE7010	0233
SE7015	0086
SE7016	0233
SE7017	0233
SE7020	0142
SE7030	0142
SE7050	0233
SE7055	0233
SE7056	0233
SE8001	0086
SE8002	0076
SE8010	0086
SE8012	0086
SE8040	0016
SE8041	0086
SE8042	0086
SE8510	0086

DEVICE TYPE	REPL CODE
SE8520	0086
SE8521	3426
SE8540	0126
SE8541	0126
SE8542	0126
SE9002	0103
SE9020	0074
SE9060	0178
SE9061	0178
SE9062	0178
SE9063	0178
SE9070	2548
SE9071	2548
SE9080	0103
SE9081	2548
SE9200	OEM
SE9201	OEM
SE9202	OEM
SE9203	OEM
SE9205	OEM
SE9206	OEM
SE9207	OEM
SE9208	OEM
SE9220	OEM
SE9221	OEM
SE9222	OEM
SE9223	OEM
SE9225	OEM
SE9226	OEM
SE9227	OEM
SE9228	OEM
SE9230	OEM
SE9231	OEM
SE9232	OEM
SE9233	OEM
SE9235	OEM
SE9236	OEM
SE9237	OEM
SE9238	OEM
SE9300	2235
SE9301	2235
SE9302	2235
SE9303	3373
SE9304	3373
SE9305	1991
SE9331	1027
SE9400	2237
SE9401	2237
SE9402	2237
SE9403	3374
SE9404	3373
SE9405	2432
SE9560	0905
SE9561	0905
SE9562	0905
SE9563	0905
SE9570	0460
SE9571	0460
SE9572	0460
SE9573	0460
SE-40022	0160
SE40022	0160
SE50399	0975
SEB1000C	OEM
SEB1000P	OEM
SEB1004C	OEM
SEB1100C	OEM
SEB1100P	OEM
SEB1400P	OEM
SEB2100P	OEM
SEB2130P	OEM
SEB2400P	OEM
SEC1077	OEM
SEC1078	0595
SEC1079	0595
SEC1080	OEM
SEC1477	0595
SEC1478	OEM
SEC1479	0595
SEC1480	OEM
SED1330F	OEM
SED9420C	OEM
SEL-101C	OEM
SEL101R	OEM
SEL102R	OEM
SEL-103R	OEM
SEL103RF	OEM
SEL112NP-N	OEM
SEL162B	OEM
SEL550S	OEM
SEL550ST	OEM
SEL650ST	OEM
SEL1112R	OEM
SEL-1120R	4052
SEL1120R	4052
SEL1210R	OEM
SEL1210S	OEM
SEL1210SC	OEM
SEL1210W	OEM
SEL1222R-C	OEM
SEL1222R-D	OEM
SEL1320G	1131
SEL1410E	OEM
SEL1410G	OEM
SEL-1413(GRN)	OEM
SEL1413E	OEM
SEL1910D	OEM
SEL1910W	OEM
SEL2110R	OEM

DEVICE TYPE	REPL CODE
SEL2110S	OEM
SEL2201WC	OEM
SEL2210R	OEM
SEL2210S	OEM
SEL2210S-C	OEM
SEL2213C	OEM
SEL2213E	OEM
SEL2310E	OEM
SEL2413E	OEM
SEL2510C-C	OEM
SEL2810A-C	OEM
SEL2910A	OEM
SEL2910A-C	OEM
SEL2910AX	OEM
SEL2910AY	OEM
SEL2910D	OEM
SEL2910GW-C	OEM
SEL2910W-C	OEM
SEL4214R	OEM
SEL4214S	OEM
SEL4214W	OEM
SEL4410E	OEM
SEL4414E	OEM
SEL4910A-C	OEM
SEL4914A-X	OEM
SEL4914AX	OEM
SEL4914AY	OEM
SEL5500ST	OEM
SELEN-26	0196
SELEN-30	0201
SELEN30	0201
SELEN-36	OEM
SELEN-38	0196
SELEN-40	0757
SELEN-42	0474
SELEN-44	0277
SELEN-48	1303
SELEN48	1303
SELEN-52	0288
SELEN-58	0201
SELEN-64	0087
SELEN-70	0015
SELEN-92	0102
SELEN-701	0015
SEN2A1	0277
SEN2A2	0289
SEN2A4	0296
SEN2A6	0302
SEN2A8	0250
SEN2A10	0250
SEN105	1335
SEN105FR	OEM
SEN110	0080
SEN110FR	OEM
SEN120	0609
SEN120FR	OEM
SEN130	0795
SEN130FR	OEM
SEN140	0795
SEN140FR	OEM
SEN150	0015
SEN150FR	OEM
SEN160	0015
SEN160FR	OEM
SEN170FR	OEM
SEN180	0072
SEN180FR	OEM
SEN205	0087
SEN205FR	OEM
SEN210	0087
SEN210FR	OEM
SEN220	0087
SEN220FR	OEM
SEN230	0087
SEN230FR	OEM
SEN240	0087
SEN240FR	OEM
SEN250	0087
SEN250FR	OEM
SEN260	0087
SEN260FR	OEM
SEN270FR	OEM
SEN280	0087
SEN305	0087
SEN310	0087
SEN320	0087
SEN330	0087
SEN340	0087
SEN350	0087
SEN360	0087
SEN380	0087
SEN480	OEM
SEN480FR	OEM
SEN1100	0071
SEN2100	0087
SEN3100	0087
SES632	0103
SES881	0103
SES3819	0325
SES5001	0071
SF0R1A41	0901
SF0R1B42	0901
SF0R1B42IG1	0901
SF0R3G42	0407
SF0R3G42(G5H1)	0407
SF0R3G42-1G5	0407
SF0R3G42-IG5	0407
SF-1	0348
SF1	0102

If replacement code is OEM, contact original manufacturer for replacement.

DEVICE TYPE	REPL CODE
SF1A11	0174
SF1A11A	0174
SF1AN6	OEM
SF1AN12	OEM
SF1AR6	OEM
SF1AR12	OEM
SF1B11	3854
SF1B11A	3854
SF1C11	OEM
SF1C11A	OEM
SF1CN1	0071
SF1D11	3620
SF1D11A	3620
SF1E11	OEM
SF1E11A	OEM
SF1F11	OEM
SF1F11A	OEM
SF1G11	OEM
SF1G11A	OEM
SF1GH315	OEM
SF1GR316	OEM
SF1HN315	OEM
SF1HR316	OEM
SF1R-3B41	2270
SF1R3B41	0447
SF1R3D41	0940
SF1Z11	OEM
SF1Z11A	OEM
SF2AN6	OEM
SF2AN12	OEM
SF2AR6	OEM
SF2AR12	OEM
SF2B41	0447
SF2D41	0940
SF2G41	0095
SF2GR316	OEM
SF2HN315	OEM
SF2K212XXX3XX	1392
SF3	OEM
SF3AN6	OEM
SF3AN12	OEM
SF3AN126	OEM
SF3AR6	OEM
SF3AR12	OEM
SF3B12	2491
SF3B41	2089
SF3BR64	OEM
SF3C325	OEM
SF3CN1	0071
SF3D12	0240
SF3D41	0940
SF3G4L	0710
SF3G12	2809
SF3G14	3383
SF3G41	0710
SF3GN315	OEM
SF3GR64	OEM
SF3GR315	OEM
SF3GR316	OEM
SF3HN315	OEM
SF3HR315	OEM
SF3HR316	OEM
SF3J41	0863
SF4	0071
SF4AN6	OEM
SF4AN12	OEM
SF4AR6	OEM
SF4AR12	OEM
SF4CN1	0071
SF4GN315	OEM
SF4GR316	OEM
SF4H4316	OEM
SF4HN315	OEM
SF5	0071
SF5A11	1103
SF5AN12	OEM
SF5AN126	OEM
SF5AR6	OEM
SF5AR12	OEM
SF5B11	2491
SF5B12	2491
SF5B13	2491
SF5B41	2089
SF5BR64	OEM
SF5C11	OEM
SF5C325	OEM
SF5D11	0240
SF5D12	0240
SF5D13	0240
SF5D41	0505
SF5E11	OEM
SF5F11	2659
SF5F12	0676
SF5G11	0676
SF5G12	0676
SF5G13	0676
SF5G41	0710
SF5GN315	OEM
SF5GR315	OEM
SF5GR316	OEM
SF5H11	OEM
SF5HN315	OEM
SF5HR315	OEM
SF5HR316	OEM
SF5J11	0336
SF5J12	0336
SF5J13	0336
SF5J41	0863
SF5Z11	1103
SF6	0071
SF6AN6	OEM
SF6AN12	OEM
SF6AR6	OEM
SF6AR12	OEM
SF6GN315	OEM
SF6GR64	OEM
SF6GR316	OEM
SF6HN315	OEM
SF6HR316	OEM
SF7AN126	OEM
SF7BR64	OEM
SF7C325	OEM
SF7GN315	OEM
SF7GR64	OEM
SF7GR315	OEM
SF7GR316	OEM
SF7HN315	OEM
SF7HR315	OEM
SF7HR316	OEM
SF8	OEM
SF8B41	2089
SF8D41	0505
SF8G41	0710
SF8GN315	OEM
SF8GR316	OEM
SF8HN315	OEM
SF8HR316	OEM
SF8J41	0863
SF9AN126	OEM
SF9BR64	OEM
SF9C325	OEM
SF9GN315	OEM
SF9GR64	OEM
SF9GR315	OEM
SF9GR316	OEM
SF9HN315	OEM
SF9HR315	OEM
SF9HR316	OEM
SF10	OEM
SF10A11	2519
SF10B11	0741
SF10B12	0567
SF10B13	0741
SF10C11	0745
SF10D11	0745
SF10D12	0762
SF10D13	0745
SF10E11	0747
SF10F11	2917
SF10G11	0747
SF10G12	0740
SF10GN315	OEM
SF10GZ47	3908
SF10H11	0752
SF10HN315	OEM
SF10J11	0764
SF10J12	0764
SF10JZ47	3908
SF10L11	0766
SF10L12	0766
SF10N11	OEM
SF10N12	OEM
SF10Z11	2519
SF11	1090
SF11AN126	OEM
SF11BR64	OEM
SF11C325	OEM
SF11GN315	OEM
SF11GR64	OEM
SF11GR315	OEM
SF11GR316	OEM
SF11HN315	OEM
SF11HR315	OEM
SF11HR316	OEM
SF12	1090
SF12GN315	OEM
SF12HN315	OEM
SF13	1090
SF13GR316	OEM
SF13HR316	OEM
SF14	1090
SF15GR316	OEM
SF15HR316	OEM
SF16A11	2188
SF16B11	0567
SF16B12	0567
SF16B13	0567
SF16B14	0567
SF16C11	0762
SF16D11	0762
SF16D12	0447
SF16D13	0762
SF16D14	0397
SF16E11	0762
SF16F11	3271
SF16F12	0740
SF16G11	0740
SF16G12	0740
SF16G13	0740
SF16G14	0611
SF16J11	0764
SF16J12	0740
SF16J13	0764
SF16L11	0766
SF16L12	0766
SF16L13	0766
SF16N11	OEM
SF16N12	OEM
SF16Q12	OEM
SF16R11	OEM
SF16Z11	2188
SF17GR316	OEM
SF17HR316	OEM
SF22	0546
SF23	0546
SF24	0546
SF30	OEM
SF30D11	OEM
SF30F11	OEM
SF30G11	OEM
SF30J11	OEM
SF30L11	OEM
SF30N11	OEM
SF31	0546
SF32	0546
SF33	0546
SF34	0546
SF35	1362
SF36	1362
SF47C	OEM
SF50	OEM
SF50D11	OEM
SF50G11	OEM
SF50J11	OEM
SF50L11	OEM
SF50N11	OEM
SF50R11	OEM
SF51	1366
SF52	1366
SF53	1366
SF54	1366
SF55	1366
SF56	1366
SF60	OEM
SF61	OEM
SF62	OEM
SF63	OEM
SF80	OEM
SF80D15	0608
SF80G15	0610
SF80L13	0468
SF80N13	0468
SF80Q13	0468
SF81	1127
SF81R	OEM
SF82	1127
SF82R	OEM
SF83	1127
SF83R	OEM
SF84	1127
SF84R	OEM
SF85	4167
SF85R	OEM
SF86	4167
SF86R	OEM
SF91	OEM
SF93	OEM
SF100	OEM
SF100G21	0159
SF100J21	0159
SF100Q21	0096
SF101	OEM
SF102	OEM
SF103	OEM
SF110	OEM
SF111	OEM
SF112	OEM
SF113	OEM
SF115	0224
SF115A	0224
SF115B	0224
SF115C	0224
SF115D	0224
SF115E	0224
SF120	OEM
SF121	OEM
SF122	OEM
SF123	OEM
SF130	OEM
SF131	OEM
SF132	OEM
SF133	OEM
SF150G15	1086
SF150J15	1086
SF150L13	1102
SF150N13	1102
SF150Q13	1102
SF161A	0902
SF161C	0909
SF161D	OEM
SF162A	0902
SF162C	OEM
SF162D	OEM
SF163A	0902
SF163C	0909
SF163D	OEM
SF164A	0902
SF164C	0909
SF164D	OEM
SF165A	OEM
SF165C	0972
SF165D	OEM
SF166A	OEM
SF166C	0972
SF166D	OEM
SF167	0007
SF173	0007
SF194	0007
SF194B	0007
SF195	0007
SF195C	0007
SF195D	0007
SF196	0007
SF197	0007
SF200	OEM
SF201	OEM
SF202	OEM
SF203	OEM
SF210	OEM
SF211	OEM
SF212	OEM
SF213	OEM
SF251	0414
SF280	OEM
SF281	OEM
SF282	OEM
SF283	OEM
SF294	0007
SF294B	0007
SF295	0007
SF295C	0007
SF295D	0007
SF300G15	1900
SF300J15	1900
SF300L13	0738
SF300N13	0738
SF300Q13	0738
SF301A	OEM
SF301C	2274
SF301D	OEM
SF302A	OEM
SF302C	2274
SF302D	OEM
SF303A	OEM
SF303C	2274
SF303D	OEM
SF304A	2274
SF304C	2274
SF304D	OEM
SF310	0007
SF313	OEM
SF314	0007
SF334	0007
SF334B	0007
SF335	0007
SF335C	0007
SF335D	0007
SF500EX22	OEM
SF500U22	OEM
SF500Y22	OEM
SF703SF	0239
SF800J22	OEM
SF800J25	4148
SF800L21	OEM
SF800L22	OEM
SF800L25	1725
SF800N21	OEM
SF800N22	OEM
SF800N25	1725
SF800R21	OEM
SF800R22	OEM
SF1001	0016
SF1220	OEM
SF1713	0016
SF1714	0016
SF1726	0016
SF1730	0016
SF12412	OEM
SF12413	OEM
SFB8970	0212
SFC400E	0232
SFC401E	0268
SFC402E	0314
SFC408E	0467
SFC409E	0492
SFC420E	0697
SFC426E	0803
SFC430E	0873
SFC437E	3520
SFC438E	0996
SFC440E	1024
SFC442E	1052
SFC450E	0743
SFC451E	1168
SFC453E	1185
SFC454E	1202
SFC460E	1275
SFC472E	1426
SFC473E	1172
SFC474E	1313
SFC475E	1432
SFC476E	1158
SFC485E	0374
SFC486E	1368
SFC492E	0834
SFC899S	0212
SFC-1616	0212
SFC-1617	0212
SFC2109RM	1916
SFC2209R	1916
SFC2301A	0093
SFC2309R	1916
SFC2458DC	0360
SFC2709C	1301
SFC2710C	1797
SFC2711C	1890
SFC2741C	0410
SFC2748C	0093
SFC2748M	0093
SFC2805EC	0624
SFC2805RC	1922
SFC2806EC	0923
SFC2808EC	1196
SFC2812EC	0334
SFC2812RC	3393
SFC2815EC	1321
SFC2824EC	2239
SFC3012	0007
SFC3020	0007
SFC4050	2027
SFC4052	2027
SFC4107E	0942
SFC4121E	0175
SFC4122E	1139
SFC4123E	1157
SFC4141E	1377
SFC4155E	1576
SFC4156E	1592
SFC4180E	1829
SFC4181E	1842
SFC4182E	1856
SFC4192E	1921
SFC4193E	1926
SFC6050	0623
SFC8544	OEM
SFC8999	0212
SFD037A	0019
SFD12	0143
SFD43	0133
SFD46	0133
SFD48	0133
SFD79	0133
SFD83	0133
SFD102	OEM
SFD103	0143
SFD104	0019
SFD107	0143
SFD111	0143
SFD112	0143
SFD161	OEM
SFD162	OEM
SFD163	OEM
SFD164	0133
SFD165	OEM
SFD166	OEM
SFD180	0133
SFD181	0133
SFD182	0124
SFD400B	OEM
SFD400E	OEM
SFD400F	OEM
SFD400G	OEM
SFD400H	OEM
SFD401E	OEM
SFD401F	OEM
SFD401G	OEM
SFD401H	OEM
SFD401K	OEM
SFD402E	OEM
SFD402F	OEM
SFD402G	OEM
SFD402H	OEM
SFD410E	OEM
SFD410F	OEM
SFD410G	OEM
SFD410H	OEM
SFD411B	OEM
SFD411E	OEM
SFD411F	OEM
SFD411G	OEM
SFD411H	OEM
SFD412E	OEM
SFD412F	OEM
SFD412G	OEM
SFD412H	OEM
SFD420C	OEM
SFD420D	OEM
SFD420E	OEM
SFD420F	OEM
SFD421C	OEM
SFD421D	OEM
SFD421E	OEM
SFD421F	OEM
SFD421G	OEM
SFD422C	OEM
SFD422D	OEM
SFD422E	OEM
SFD422F	OEM
SFD422G	OEM
SFD423C	OEM
SFD423D	OEM
SFD423E	OEM
SFD423F	OEM
SFD430A	OEM
SFD430B	OEM
SFD430C	OEM
SFD430D	OEM
SFD431A	OEM
SFD431B	OEM
SFD431C	OEM
SFD431D	OEM
SFD432A	OEM
SFD432B	OEM
SFD432C	OEM
SFD432D	OEM
SFD433A	OEM
SFD433B	OEM
SFD433C	OEM
SFD433D	OEM
SFD440A	OEM
SFD441A	OEM
SFD450A	OEM
SFD460	OEM
SFD2285	0212
SFD4160	OEM
SFD4161	OEM
SFD4162	OEM
SFD4163	OEM
SFD4164	OEM
SFD4165	OEM
SFD4166	OEM
SFE145	0325
SFE253	0212
SFE303	0353
SFE425	1450
SFE427	0212
SFE303424	0212
SFF103	OEM
SFF121	OEM
SFH205	OEM
SFIR3B41	0447
SFIR3D41	0940
SFM10	OEM
SFM15	OEM
SFM20	OEM
SFM25	OEM
SFOR1A41	0901
SFOR1B41	0901
SFOR1B42	4877
SFOR1B42(IG1)	0901
SFOR1B421G1	OEM
SFOR2B41	0901
SFOR2D41	0058
SFOR2G2	OEM
SFOR2G41	0407
SFOR3B42	OEM
SFOR3B42(C1)-T	OEM
SFOR3G42	7082
SFOR3G42-IG5	OEM
SFOR3G42IG5	0407
SFOR3J42	OEM
SFORIB42IG1	OEM
SFPB-54VL	OEM
SFPB54VL	OEM
SFPM-52V	OEM
SFPM52V	OEM
SFR135	0015
SFR135/2	OEM
SFR151	0015
SFR152	0015
SFR153	0015
SFR154	0015
SFR155	0015
SFR156	0015
SFR164	0015
SFR251	0015
SFR252	0015
SFR253	0015
SFR254	0015
SFR255	0015
SFR256	0015
SFR258	0071
SFR264	0015
SFR266	0015
SFR268	0071
SFT104	0143
SFT106	OEM
SFT107	OEM
SFT108	0143
SFT113	OEM
SFT114	OEM
SFT115	OEM
SFT120	0050
SFT121	0164
SFT122	0164
SFT123	0164
SFT124	0164
SFT125	0164
SFT125P	0164
SFT126	OEM
SFT127	OEM
SFT128	OEM
SFT130	0164
SFT131	0164
SFT131P	0164
SFT135	OEM
SFT136	OEM
SFT141	OEM
SFT142	OEM
SFT143	0164
SFT144	0164
SFT145	0164
SFT146	0164
SFT150	OEM
SFT151	0004
SFT152	0004
SFT153	OEM
SFT155	OEM
SFT162	0050
SFT163	0050
SFT171	0050
SFT172	0050
SFT173	0050
SFT174	0050
SFT184	0600
SFT185	OEM
SFT186	0233
SFT187	0233
SFT190	0160
SFT191	0160
SFT192	0160
SFT211	OEM
SFT212	0160
SFT213	0160
SFT214	0160
SFT220	OEM
SFT221	0211
SFT221A	0004
SFT222	0211
SFT223	0280
SFT226	0280
SFT227	0280
SFT228	0280
SFT229	0280
SFT232	0004
SFT233	OEM
SFT234	OEM
SFT235	OEM
SFT237	0280
SFT238	0160
SFT239	0160
SFT240	0160
SFT241	0164
SFT242	0164
SFT243	0164
SFT244	OEM
SFT245	OEM
SFT250	0160
SFT251	0280
SFT252	0280
SFT253	0280
SFT259	0600
SFT260	0600
SFT261	0600
SFT264	0440
SFT265	0440
SFT266	0440
SFT267	0440
SFT268	0050
SFT288	0280
SFT298	0600
SFT306	0004
SFT307	0136
SFT308	0136
SFT315	0050
SFT316	0050
SFT317	0136
SFT318	0136
SFT319	0136
SFT320	0136
SFT321	0004
SFT322	0004
SFT323	0004
SFT325	0164
SFT327	0004
SFT337	0004
SFT337B	0004
SFT337V	0004
SFT351	0004
SFT352	0004
SFT353	0004
SFT354	0136
SFT357	0136
SFT357P	0136
SFT358	0050
SFT367	0164
SFT377	0208
SFT440	0086
SFT443	0086
SFT443A	0086
SFT445	0086
SFT523	0164
SFT526	0004
SFT601	OEM
SFT602	OEM
SFT603	OEM
SFT604	OEM
SFT713	0016
SFT714	0016
SFT714A	OEM
SFT715	OEM
SFT715A	OEM
SFT918	OEM
SFT918B	OEM
SFTAN6	OEM
SFZ708	0157
SFZ716	0052
SFZ720	OEM
SFZ722	OEM
SF.T124	0164
SF.T125	0004
SF.T125P	0164
SF.T130	0164
SF.T131	0004
SF.T131P	0164
SF.T163	0050
SF.T171	0280
SF.T172	0280
SF.T173	0280
SF.T174	0280
SF.T184	0600
SF.T186	0233
SF.T187	0285

DEVICE TYPE	REPL CODE
SF.T191	0085
SF.T212	0975
SF.T213	0975
SF.T214	0975
SF.T221	0164
SF.T222	0211
SF.T223	0164
SF.T227	0211
SF.T228	0280
SF.T237	0004
SF.T238	0975
SF.T239	0975
SF.T240	0975
SF.T250	0975
SF.T251	0004
SF.T252	0004
SF.T253	0004
SF.T264	0440
SF.T265	0440
SF.T266	0440
SF.T306	0004
SF.T307	0136
SF.T315	0050
SF.T316	0050
SF.T317	0050
SF.T318	0211
SF.T319	0050
SF.T320	0050
SF.T321	0164
SF.T322	0164
SF.T323	0004
SF.T337	0004
SF.T351	0164
SF.T352	0164
SF.T353	0847
SF.T354	0050
SF.T357	0050
SF.T358	0050
SF.T377	0208
SF.T440	0595
SF.T443	0595
SF.T443A	0595
SF.T445	0595
SF.T714	0595
SG	OEM
SG-005	0015
SG005	0540
SG011	OEM
SG-2BC	OEM
SG-5(-)	OEM
SG-5(+)	OEM
SG-5AR	OEM
SG-5AS	OEM
SG-5BR	OEM
SG-5BS	OEM
SG-5F	OEM
SG5F	OEM
SG-5FR	OEM
SG-5FS	OEM
SG-5L(-)	OEM
SG-5L(+)	OEM
SG-5N	0087
SG-5P	0087
SG-5T	0015
SG-5T(-)	OEM
SG-5T(+)	OEM
SG-6(-)	OEM
SG-6(+)	OEM
SG-6AR	OEM
SG-6AS	OEM
SG-6BR	OEM
SG-6BS	OEM
SG-6LR	OEM
SG-6LS	OEM
SG16H11	OEM
SG-21R	OEM
SG-21S	OEM
SG-22R	OEM
SG-22S	OEM
SG-23R	OEM
SG-23S	OEM
SG40	OEM
SG41	OEM
SG42	OEM
SG43	OEM
SG50	OEM
SG51	OEM
SG52	OEM
SG53	OEM
SG-53R	OEM
SG-53S	OEM
SG-56R	OEM
SG-56S	OEM
SG60	OEM
SG61	OEM
SG62	OEM
SG63	OEM
SG-63R	OEM
SG-66R	OEM
SG-66S	OEM
SG70	OEM
SG71	OEM
SG72	OEM
SG73	OEM
SG90	OEM
SG91	OEM
SG92	OEM
SG93	OEM
SG100	OEM
SG101	OEM
SG102	OEM
SG103	OEM
SG-105	0015
SG105	0023
SG110	OEM
SG111	OEM
SG112	OEM
SG113	OEM
SG120	OEM
SG121	OEM
SG122	OEM
SG123	0660
SG130	OEM
SG131	OEM
SG132	OEM
SG133	OEM
SG140	OEM
SG141	OEM
SG142	OEM
SG143	OEM
SG150	OEM
SG151	OEM
SG152	OEM
SG153	OEM
SG160	OEM
SG161	OEM
SG162	OEM
SG163	OEM
SG170	OEM
SG171	OEM
SG172	OEM
SG173	OEM
SG180	OEM
SG181	OEM
SG182	OEM
SG183	OEM
SG190	OEM
SG191	OEM
SG192	OEM
SG193	OEM
SG200	OEM
SG201	OEM
SG201AT	0093
SG202	OEM
SG203	OEM
SG-205	0015
SG205	0242
SG210	OEM
SG211	OEM
SG212	OEM
SG213	OEM
SG213T	OEM
SG220	OEM
SG221	OEM
SG222	OEM
SG223	0660
SG224J	0625
SG224N	0625
SG230	OEM
SG231	OEM
SG232	OEM
SG233	OEM
SG240	OEM
SG240D	OEM
SG241	OEM
SG242	OEM
SG243	OEM
SG250	OEM
SG251	OEM
SG252	OEM
SG253	OEM
SG260	OEM
SG261	OEM
SG262	OEM
SG263	OEM
SG-264A	1652
SG264A	1652
SG270	OEM
SG271	OEM
SG272	OEM
SG273	OEM
SG290	OEM
SG291	OEM
SG292	OEM
SG293	OEM
SG300	OEM
SG301	OEM
SG301AM	1300
SG301AT	0093
SG302	OEM
SG303	OEM
SG-305	0015
SG305	0540
SG308AM	2246
SG308AT	0896
SG308AY	2246
SG308M	2246
SG308T	0896
SG308Y	2246
SG309K	1922
SG310	OEM
SG310M	2517
SG310T	2222
SG311	OEM
SG312	OEM
SG313	OEM
SG320	OEM
SG321	OEM
SG322	OEM
SG323	0015
SG323K	1916
SG324N	0625
SG330	OEM
SG331	OEM
SG332	OEM
SG333	OEM
SG340	OEM
SG340-05K	1916
SG340-12K	3393
SG341	OEM
SG342	OEM
SG343	OEM
SG351	OEM
SG353	OEM
SG370	OEM
SG371	OEM
SG372	OEM
SG373	OEM
SG380	OEM
SG381	OEM
SG382	OEM
SG383	OEM
SG505	0015
SG555CM	0973
SG555M	0973
SG-608	OEM
SG609	2337
SG609-3	2337
SG613	1652
SG613-A	OEM
SG710CN	1800
SG710CT	1797
SG711CN	1897
SG711CT	1890
SG723CN	0026
SG723CT	1192
SG723T	1192
SG733CJ	1118
SG733CT	1118
SG741CM	0312
SG741CN	1975
SG741CT	0410
SG741SCT	0410
SG747CT	2368
SG-805	0015
SG805	0817
SG-870	OEM
SG1009	0591
SG1009/F	0591
SG1009A	0591
SG1009A/F	0591
SG1010	0591
SG1010/F	0591
SG1010A	0591
SG1010A/F	0591
SG-1198	0015
SG1198	0015
SG1458M	0360
SG1488J	0508
SG1489J	0511
SG1495N	OEM
SG1496	0417
SG1496T	0417
SG1691	OEM
SG1910	OEM
SG1911	OEM
SG1912	OEM
SG1920	OEM
SG1921	OEM
SG1922	OEM
SG1930	OEM
SG1931	OEM
SG1932	OEM
SG1940	OEM
SG1941	OEM
SG1942	OEM
SG2182	2866
SG2183	2866
SG3081N	2749
SG3400	0015
SG3524	5233
SG3524N	OEM
SG3525A	2829
SG3526N	2841
SG3527A	2839
SG5000	OEM
SG5009	0133
SG5013	0086
SG5018	OEM
SG5100	OEM
SG5200	OEM
SG5250	OEM
SG5260	OEM
SG5270	OEM
SG5300	OEM
SG5400	0133
SG5717	0133
SG5777	OEM
SG5800	OEM
SG5811	OEM
SG6476	OEM
SG7400N	0232
SG7402N	0314
SG7410N	0512
SG7420N	0697
SG7430N	0873
SG7440N	1024
SG7450N	0743
SG7451N	1168
SG7453N	1185
SG7454N	1202
SG7460N	1275
SG7812	0334
SG7812CK	3393
SG7812T	1828
SG7815T	3395
SG7818ACP	2259
SG7818CK	3650
SG7818CP	2259
SG7824CK	3881
SG7908ACP	2791
SG7908CP	2791
SG7918ACP	3826
SG7918CK	5290
SG7918CP	0023
SG9012I	0037
SG9072	OEM
SG-9150	0133
SG9150	0133
SG9600	OEM
SGA-2	OEM
SGA-2CN	OEM
SGA-2CP	OEM
SGA-4	OEM
SGA-4CN	OEM
SGA-4CP	OEM
SGA-5	OEM
SGA-5CN	OEM
SGA-5CP	OEM
SGB-2	OEM
SGB-2CN	OEM
SGB-2CP	OEM
SGB-4	OEM
SGB-4CN	OEM
SGB-4CP	OEM
SGB-5	OEM
SGB-5CN	OEM
SGB-5CP	OEM
SGB-8CN	OEM
SGB-8CP	OEM
SGB-9742	0144
SGC-7202	0079
SGC7202	0086
SGF-2	OEM
SGF-4	OEM
SGF-5	OEM
SGI5001	0031
SGI5002	0031
SGI5003	0031
SGI5004	0031
SGI5301	OEM
SGI5302	OEM
SGI5303	OEM
SGI5304	OEM
SGI5401	1145
SGI5401C	1236
SGI5401CR	1727
SGI5401R	OEM
SGI5402	1145
SGI5402C	1236
SGI5402CR	1727
SGI5402R	OEM
SGI5403	1145
SGI5403C	1236
SGI5403CR	1727
SGI5403R	OEM
SGI5404	1145
SGI5404C	1236
SGI5404CR	1727
SGI5404R	OEM
SGI5601C	OEM
SGI5601CR	OEM
SGI5602C	OEM
SGI5602CR	OEM
SGI5603C	OEM
SGI5603CR	OEM
SGI5604C	OEM
SGI5604CR	OEM
SGM-12R	OEM
SGM-12S	OEM
SGM-13	OEM
SGM-22	OEM
SGM-52	OEM
SGM-53	OEM
SGM-54	OEM
SGM182BL	OEM
SGR100	0015
SGS1	OEM
SGS17139XRH	2158
SGS87231	0007
SGSIF461	3510
SGT03U13	OEM
SGT06U13	OEM
SGT10S10	OEM
SGT23B13	OEM
SGT23U13	OEM
SH-1	0023
SH1	0015
SH-1A	0023
SH1A	0015
SH-1B	0017
SH1B	0071
SH1B42	OEM
SH-1C	0017
SH1C	0071
SH-1D	0017
SH-1DE	0015
SH1DE	0015
SH-1E	0017
SH-1S	0015
SH1S	0242
SH-1Z	0023
SH4D05	0015
SH4D1	0015
SH4D2	0015
SH4D3	0015
SH4D4	0015
SH4D5	OEM
SH4D6	0015
SH4D8	0071
SH5A	OEM
SH5B11	0408
SH5B12	OEM
SH5D11	OEM
SH5D12	OEM
SH5F11	0422
SH5F12	OEM
SH5G11	0422
SH5G12	OEM
SH5H11	0422
SH5H12	OEM
SH5J11	0422
SH5J12	OEM
SH-6K	0201
SH6K	OEM
SH-8K	0201
SH8K	OEM
SH-10K	0201
SH10K	OEM
SH-12K	0201
SH12K	OEM
SH15	0015
SH16B11	3148
SH16B12	OEM
SH16D11	3150
SH16D12	OEM
SH16F11	3151
SH16F12	OEM
SH16G11	3152
SH16G12	OEM
SH16H11	3153
SH16H12	OEM
SH16J11	2018
SH16J12	OEM
SH80U12A	OEM
SH200	1335
SH200A	1335
SH202	0133
SH202A	0133
SH212	0133
SH212A	0133
SH222	0133
SH222A	0133
SH242	0133
SH242A	0133
SH252	0133
SH252A	0133
SH300	0133
SH300A	0133
SH301	0133
SH301A	0133
SH302	0133
SH302A	0133
SH323SC	1916
SH400L21	OEM
SH400N21	OEM
SH400R21	OEM
SH400U23	OEM
SH1064	0016
SH5000	OEM
SH7500	OEM
SH10000	OEM
SH12500	OEM
SH15000	OEM
SH20000	OEM
SH25000	OEM
SH764059	OEM
SH779567	OEM
SHA7520	0791
SHA7521	0791
SHA7522	0791
SHA7523	0791
SHA7524	0791
SHA7526	0791
SHA7527	0791
SHA7528	0791
SHA7530	0037
SHA7531	0037
SHA7532	0037
SHA7533	0037
SHA7534	0037
SHA7535	0342
SHA7536	0037
SHA7537	0037
SHA7538	0037
SHA7539	0342
SHA7597	0791
SHA7598	0791
SHA7599	0791
SHAD-1	0015
SHF5000	OEM
SHF7500	OEM
SHF10000	OEM
SHF12500	OEM
SHF15000	OEM
SHF20000	OEM
SHF25000	OEM
SHOR3D42	OEM
SHV-03	OEM
SHV-06	OEM
SHV-06UN	OEM
SHV-08	OEM
SHV-08UN	OEM
SHV-10	OEM
SHV-12	OEM
SHV-12U	OEM
SHV-16	OEM
SHV-16U	OEM
SHV-20	OEM
SHV-24	OEM
SHVF2500	OEM
SHVF5000	OEM
SHVF7500	OEM
SHVF10000	OEM
SHVF12500	OEM
SHVF15000	OEM
SHVF20000	OEM
SHVF25000	OEM
SI03E	OEM
SI03K	OEM
SI03L	OEM
SI03N	OEM
SI05	0110
SI-05A	0071
SI05A	0110
SI0V-S05K150	0246
SI1	0953
SI-1A	0071
SI1A	0242
SI1SS54	0133
SI2	0242
SI-2A	0071
SI2A	OEM
SI3	0540
SI-3A	0071
SI3A	OEM
SI4	0540
SI-4A	0071
SI4A	OEM
SI5	0965
SI-5A	0071
SI5A	OEM
SI6	0965
SI-6A	0071
SI6A	OEM
SI7	0817
SI-7A	0071
SI7A	OEM
SI8	0817
SI-8A	0071
SI8A	0087
SI9	0922
SI10	0922
SI-10A	0071
SI10A	0087
SI11E	OEM
SI11K	OEM
SI11L	OEM
SI11N	OEM
SI12	OEM
SI12A	OEM
SI15	OEM
SI21E	OEM
SI21K	OEM
SI21L	OEM
SI21N	OEM
SI50E	0015
SI61E	OEM
SI61K	OEM
SI61L	OEM
SI61N	OEM
SI91E	OEM
SI91G	0015
SI91K	OEM
SI91L	OEM
SI91N	OEM
SI100E	0015
SI-154	0023
SI211N	OEM
SI212N	OEM
SI213N	OEM
SI214N	OEM
SI215N	OEM
SI216N	OEM
SI221N	OEM
SI222N	OEM
SI223N	OEM
SI224N	OEM
SI225N	OEM
SI226N	OEM
SI231N	OEM
SI232N	OEM
SI233N	OEM
SI234N	OEM
SI235N	OEM
SI236N	OEM
SI244N	OEM
SI245N	OEM
SI246N	OEM
SI250P12K38	OEM
SI250P14K38	OEM
SI250P16K38	OEM
SI250P20K38	OEM
SI300P10K35	OEM
SI300P12K35	OEM
SI300P14K35	OEM
SI300P16K35	OEM
SI300P16K44	OEM
SI300P18K35	OEM
SI300P18K44	OEM
SI300P20K35	OEM
SI300P20K44	OEM
SI300P22K44	OEM
SI300P26K44	OEM
SI300P28K44	OEM
SI300P30K44	OEM
SI341P	0150
SI342P	0150
SI343P	0150
SI345P	OEM
SI346P	OEM
SI351P	0150
SI352P	0150
SI353P	0150
SI-600	0302
SI1000E	0071
SI-1002	0023
SI-1006	0023
SI-1014	0023
SI-1019	0348
SI-1020	0023
SI-1021	0023
SI-1030	0015
SI-3052P	6553
SI-3052V	2539
SI3052V	2539
SI-3090CA	OEM
SI-3120C	OEM
SI-3120CA	OEM
SI-3122P	4329
SI-3122V	4329
SI-3152P	6640
SI-3152V	OEM
SI-3242P	6730
SI4101	OEM
SI-4102	OEM
SI4102	3030
SI-6102	OEM
SI-6901	OEM
SI-6901A	OEM
SI-7115B	2921
SI7115B	2921
SI-7200E	OEM
SI-7200M	OEM
SI-7230E	OEM
SI-7230M	OEM
SI-7300A	OEM
SI-7301A	OEM
SI-7330A	OEM
SI-7500A	OEM
SIB01	0015
SIB01-01	0023
SIB01-01V	4403
SIB01-01W	0023
SIB-01-02	0015
SIB01-02	0080
SIB01-02TP1	0609
SIB01-02U	OEM
SIB01-02VI	0609
SIB01-04	0023
SIB01-04VI	0965
SIB01-06	0023
SIB01-06B	OEM
SIB-01-022	0015
SIB01-022	0015
SIB01-2	0023
SIB02-03C	0015
SIB02-03CR	0015
SIB02-CR	0023
SIB02-CR1	0913
SIB03-08	0071
SIB03-10	0071
SIB03-12	0102
SIB020-1B	0277
SIB0101	4403
SIB0102	0015
SIB-0306	0071
SIB0306	0071
SIB0L-02	0015
SIB1	0015
SIB50B794-1	0277
SIB1014D	OEM
SIB01	OEM
SI-CA3064E	0802
SI-CA3065E	0167
SICGJ-1	0348
SICT-5	6899
SICT-8	OEM
SICT-10	OEM
SICT-12	OEM
SICT-15	6641
SICT-18	OEM
SICT-22	6652
SICT-36	OEM
SICT-45	OEM
SICTE-5	OEM
SICTE-8	OEM
SICTE-10	OEM
SICTE-12	OEM
SICTE-15	OEM
SICTE-18	OEM
SICTE-22	OEM
SICTE-36	OEM

DEVICE TYPE	REPL CODE
SICTE-45	OEM
SID01-01	0534
SID01-02	0015
SID01-03	0538
SID01-06	0545
SID01-09	0550
SID01-12	OEM
SID01E	0015
SID01K	0071
SID01L	0015
SID01N	OEM
SID02E	0015
SID02K	0071
SID02L	0015
SID02N	OEM
SID03E	OEM
SID03K	OEM
SID03L	OEM
SID03N	OEM
SID2A-1	0196
SID2A1	OEM
SID2A2	OEM
SID2A-3	0196
SID2A3	OEM
SID2A4	OEM
SID2A6	OEM
SID2A11	OEM
SID2A12	OEM
SID2A13	OEM
SID2A14	OEM
SID2A16	OEM
SID2B1	OEM
SID2B2	OEM
SID2B3	OEM
SID2B11	OEM
SID2B12	OEM
SID2B13	OEM
SID3A1	OEM
SID3A2	OEM
SID3A3	OEM
SID3A4	OEM
SID3A6	OEM
SID3B1	OEM
SID3B2	OEM
SID3B-3	1080
SID3B3	OEM
SID20E	OEM
SID20K	OEM
SID20L	OEM
SID20N	OEM
SID23-13	0102
SID23-15	0102
SID26-12	0102
SID26-15	0102
SID30-1	0102
SID30-13	0182
SID30-15	5196
SID50B851	0133
SID51C052-19	0087
SID51C169	0015
SID300P16K44	OEM
SID300P18K44	OEM
SID300P20K44	OEM
SID300P22K44	OEM
SID300P24K44	OEM
SID300P26K44	OEM
SID300P28K44	OEM
SID300P30K44	OEM
SID300P60K62	OEM
SID500P40K62	OEM
SID500P42K62	OEM
SID500P44K62	OEM
SID500P46K62	OEM
SID500P48K62	OEM
SID500P50K62	OEM
SID500P52K62	OEM
SID500P54K62	OEM
SID500P56K62	OEM
SID500P58K62	OEM
SID50894	0133
SID51169-2	OEM
SIDA02K	OEM
SIDA02N	OEM
SIDA02P	OEM
SIDA0.5K	0965
SIDA0.5N	0817
SIDA0.5P	0922
SIE01-01	0871
SIE01-03	0853
SIE01-06	0267
SIE01-09	0281
SIE01-12	OEM
SIE03-30	OEM
SIF4	0023
SIF14A-P	0015
SIF114E	0102
SIFN270	OEM
SIG01-01	0254
SIG01-03	0258
SIG01-06	0267
SIG01-09	0281
SIG01-12	OEM
SIG03-30	OEM
SIG1/100	0242
SIG1/200	0242
SIG1/400	0015
SIG1/600	0015
SIG1/800	0817
SIG300P10K35	OEM

DEVICE TYPE	REPL CODE
SIG300P12K35	OEM
SIG300P14K35	OEM
SIG300P16K35	OEM
SIG300P18K35	OEM
SIH31-01	1347
SIH31-01R	0599
SIH31-03	1347
SIH31-03R	0599
SIH31-06	1905
SIH31-06R	1986
SIH31-10	0193
SIH31-10R	0657
SIH31-12	0202
SIH31-12R	3309
SI-HA11580	1055
SIIR3N05XXXXX	6791
SIIR3N06XXXXX	OEM
SIL	0071
SIL1	OEM
SIL31-01	0501
SIL31-01R	1023
SIL31-03	0501
SIL31-03R	1023
SIL31-06	1777
SIL31-06R	1036
SIL31-10	1789
SIL31-10R	1046
SIL31-12	3167
SIL31-12R	6493
SIL200	0015
SILC4001BXXXX	0478
SILC4069UBXXX	0119
SI-LM3065N	0167
SILM8560BXXXX	6072
SIM2	OEM
SIM3	OEM
SIM5	OEM
SIM6	OEM
SIM8	OEM
SIM9	OEM
SIM-20ST	OEM
SI-MC1352P	0395
SI-MC1358P	0167
SI-MC1362P	OEM
SI-MC1364P	0802
SIN01-12	4285
SIN01-12R	4286
SIN03-30	OEM
SINB20	OEM
SIOV-S05K150	OEM
SIPI8.4	OEM
SIPI-24	OEM
SIPI30	OEM
SIPI60	OEM
SIPI120	OEM
SIPI208	OEM
SIPI250	OEM
SIPI440	OEM
SIPI500	OEM
SIPS08	1848
SIPS010	1848
SIPS020	1848
SIPS025	1848
SIPS030	OEM
SIPS035	1848
SIPS18	1848
SIPS28	1848
SIPS38	1855
SIPS48	1855
SIPS58	3215
SIPS68	3215
SIPS110	1848
SIPS120	1848
SIPS125	1848
SIPS130	OEM
SIPS135	1848
SIPS210	1848
SIPS220	1848
SIPS225	1848
SIPS230	OEM
SIPS235	1848
SIPS310	1855
SIPS320	1855
SIPS325	1855
SIPS330	OEM
SIPS335	1855
SIPS410	1855
SIPS420	1855
SIPS425	1855
SIPS430	OEM
SIPS435	1855
SIPS510	3215
SIPS520	3215
SIPS525	3215
SIPS530	OEM
SIPS535	3215
SIPS610	3215
SIPS620	3215
SIPS625	3215
SIPS630	OEM
SIPS635	3215
SIPT06	3199
SIPT08	3199
SIPT010	3199
SIPT015	3199
SIPT025	3199
SIPT030	3199
SIPT040	3199
SIPT16	3199

DEVICE TYPE	REPL CODE
SIPT18	3199
SIPT26	3199
SIPT28	3199
SIPT36	3207
SIPT38	3207
SIPT46	3207
SIPT48	3207
SIPT56	3222
SIPT58	3222
SIPT66	3222
SIPT68	3222
SIPT110	3199
SIPT115	3199
SIPT125	3199
SIPT130	3199
SIPT140	3199
SIPT210	3199
SIPT215	3199
SIPT225	3199
SIPT230	3199
SIPT240	3199
SIPT310	3207
SIPT315	3207
SIPT325	3207
SIPT330	3207
SIPT340	3207
SIPT410	3207
SIPT415	3207
SIPT425	3207
SIPT430	3207
SIPT440	3207
SIPT510	3222
SIPT515	3222
SIPT525	3222
SIPT530	3222
SIPT540	3222
SIPT610	3222
SIPT615	3222
SIPT625	3222
SIPT630	3222
SIPT640	3222
SIR20	0015
SIR-34ST3	OEM
SIR-56SB3F	OEM
SIR-56ST3	OEM
SIR60	0023
SIR-80	0015
SIR80	0023
SIR150	0102
SIR-562ST	OEM
SIRB-10	0277
SIRB10	0087
SIRBA00	0106
SIRBA10	0106
SI-REC-73	0015
SI-RECT-044	0015
SI-RECT-2	0015
SIRECT-2	0015
SI-RECT-20	0015
SI-RECT-23	0030
SI-RECT-25	0015
SI-RECT-27	0015
SI-RECT-33	0015
SI-RECT-34	0015
SI-RECT-35	0133
SI-RECT-36	0102
SIRECT-36	0015
SI-RECT-37	0015
SI-RECT-39	0015
SI-RECT-44	0102
SI-RECT-48	0015
SIRECT-48	0015
SI-RECT-49	0015
SI-RECT-50	0087
SI-RECT-53	0015
SI-RECT-59	0015
SI-RECT-69	0015
SI-RECT-73	0015
SI-RECT-75	0015
SI-RECT-77	0015
SI-RECT-84	0015
SI-RECT-92	0015
SIRECT-92	0015
SI-RECT-94	0015
SI-RECT-98	0924
SI-RECT-100	0015
SI-RECT-100-102	0015
SI-RECT-102	0015
SI-RECT102	0015
SIRECT-102	0015
SI-RECT-110	0087
SI-RECT-110/SB-3F	0015
SI-RECT-112	0015
SI-RECT-112/SB-3	0087
SI-RECT-114	0102
SI-RECT-122	0015
SI-RECT-126	0015
SI-RECT-136	0102
SI-RECT-140	0091
SI-RECT-140/TR-6	0636
SI-RECT-140/TR6S	0301
SI-RECT-144	0015
SI-RECT-152	0133
SI-RECT-154	0015
SI-RECT-155	0015
SI-RECT-156	0015
SI-RECT-158	0102
SI-RECT-162	0102
SI-RECT-168	0041

DEVICE TYPE	REPL CODE
SI-RECT-169	OEM
SI-RECT-170	0102
SI-RECT-174	0196
SI-RECT-178	1080
SI-RECT-178B	1080
SI-RECT-178BF	0196
SI-RECT-182	0102
SI-RECT-204	0071
SI-RECT-206	0102
SI-RECT-208	0087
SI-RECT-218	0015
SI-RECT-220	0015
SI-RECT-222	0015
SI-RECT-224	0102
SI-RECT-226	0015
SI-RECT-228	0137
SI-RECT-230	0137
SIRK514	OEM
SIRM-2C	0023
SIR-RECT-44	0015
SIS2	0143
SIS4M	0023
SIS11	0143
SISD-1	0102
SISD-1HF	0071
SISD-1X	0015
SISD-K	0015
SISM-150-01	0015
SI-SN76650N	0395
SI-SN76730N	OEM
SISW-05-02	0015
SISW-0502	0015
SIT4-50	OEM
SIT4-100	OEM
SIT4-200	OEM
SIT14-50	OEM
SIT14-100	OEM
SIT14-200	OEM
SIT14-400	OEM
SIT18-50	OEM
SIT18-100	OEM
SIT18-200	OEM
SIT18-400	OEM
SIT18-500	OEM
SIT18-600	OEM
SIT20-50	OEM
SIT20-100	OEM
SIT20-200	OEM
SIT20-300	OEM
SIT30-100	OEM
SIT30-200	OEM
SIT30-400	OEM
SIT30-500	OEM
SIT30-600	OEM
SIT50-50	OEM
SIT50-100	OEM
SIT50-200	OEM
SIT50-400	OEM
SIT50-500	OEM
SIT50-600	OEM
SIT80-50	OEM
SIT80-100	OEM
SIT80-200	OEM
SIT80-300	OEM
SIT95-50	OEM
SIT95-100	OEM
SIT95-400	OEM
SIT95-500	OEM
SIT95-600	OEM
SIT180-50	OEM
SIT180-100	OEM
SIT180-200	OEM
SIT180-500	OEM
SIT180-600	OEM
SI-TA7074P	0395
SI-TA7075P	0855
SITA7640APIXX	3725
SITO.8-50	OEM
SITO.8-100	OEM
SITO.8-200	OEM
SITO.8-300	OEM
SITR-300	OEM
SITVR-06G	0015
SITVR-066	0015
SI-UPC574J	1329
SI-UPC580C	1055
SI-UPC1031H	0904
SIV20	OEM
SIVN0104	OEM
SJ051A	2002
SJ051E	0015
SJ051F	0015
SJ052A	2002
SJ052E	0015
SJ052F	0015
SJ053	OEM
SJ053E	0970
SJ053EK	0084
SJ053F	0110
SJ053K	0970
SJ054E	0084
SJ054EK	0084
SJ054F	OEM
SJ054K	0084
SJ60F	0015
SJ101A	0580
SJ101F	0015
SJ102A	0580
SJ102F	0015

DEVICE TYPE	REPL CODE
SJ103E	0989
SJ103EK	0090
SJ103F	0953
SJ103H	OEM
SJ103K	0989
SJ104E	0084
SJ104EK	0090
SJ104F	OEM
SJ104K	0084
SJ130	0142
SJ201A	0580
SJ201F	0015
SJ202A	0580
SJ202F	0015
SJ203E	0989
SJ203EK	0097
SJ203F	0242
SJ203H	OEM
SJ203K	0989
SJ204E	0084
SJ204EK	0097
SJ204F	OEM
SJ204K	0084
SJ301F	0015
SJ302F	0015
SJ303E	OEM
SJ303F	OEM
SJ303K	OEM
SJ304E	OEM
SJ304EK	0109
SJ304F	OEM
SJ304K	OEM
SJ401F	0015
SJ402F	0015
SJ403E	OEM
SJ403F	0540
SJ403H	OEM
SJ403K	OEM
SJ404E	OEM
SJ404EK	0109
SJ404F	OEM
SJ404K	OEM
SJ501F	0015
SJ570	0016
SJ601F	0015
SJ603	0122
SJ603E	0122
SJ603EK	0122
SJ603F	0965
SJ603H	OEM
SJ603K	0122
SJ604	0122
SJ604E	0122
SJ604EK	0122
SJ604F	OEM
SJ604K	0122
SJ619	0103
SJ619-1	0103
SJ652	0905
SJ803E	OEM
SJ803F	0817
SJ803H	OEM
SJ803K	OEM
SJ805	0142
SJ806	0142
SJ807	OEM
SJ811	0178
SJ820	0103
SJ821	0491
SJ822	0905
SJ1000F	0922
SJ1003E	OEM
SJ1003F	0087
SJ1003H	OEM
SJ1003K	OEM
SJ1106	0103
SJ1152	0925
SJ1165	0142
SJ1171	0925
SJ1172	0178
SJ1201	0142
SJ1203E	OEM
SJ1203F	OEM
SJ1203H	OEM
SJ1203K	OEM
SJ1284	0925
SJ1286	0142
SJ1403H	OEM
SJ1470	0103
SJ1790	2439
SJ1903	2432
SJ1925	0325
SJ2000	0103
SJ2001	0491
SJ2008	0103
SJ2009	0905
SJ2031	0126
SJ2032	0086
SJ2095	0429
SJ3408	0178
SJ3423	0074
SJ3447	0178
SJ3477	0074
SJ3478	0074
SJ3604	0103
SJ3629	0079
SJ3648	0178
SJ3678	0103
SJ4076	OEM

DEVICE TYPE	REPL CODE
SJ5525	0003
SJ5526	0065
SJ6351	0065
SJ8701	0103
SJ9110	0103
SJ10003EK	0145
SJE42	0042
SJE100	0161
SJE103	0264
SJE106	0161
SJE108	0460
SJE111	0460
SJE112	0460
SJE113	0161
SJE114	0460
SJE133	0161
SJE202	0460
SJE203	0161
SJE205	0276
SJE210	0460
SJE211	0161
SJE218	0276
SJE220	0561
SJE221	0460
SJE222	0161
SJE227	0460
SJE228	0161
SJE229	0161
SJE231	0460
SJE232	0276
SJE237	0161
SJE241	0460
SJE242	0161
SJE243	0460
SJE244	0161
SJE245	0460
SJE246	0161
SJE248	0161
SJE253	0161
SJE254	0161
SJE255	0161
SJE256	0460
SJE257	0460
SJE261	0161
SJE262	0161
SJE265	0460
SJE267	0460
SJE271	0161
SJE272	0161
SJE273	0460
SJE274	0161
SJE275	0460
SJE276	0460
SJE277	0460
SJE278	0161
SJE279	0460
SJE280	0161
SJE283	0460
SJE284	0161
SJE288	0460
SJE289	0161
SJE305	0161
SJE320	0161
SJE340	0161
SJE400	0276
SJE401	0161
SJE402	0161
SJE403	0460
SJE404	0161
SJE405	0161
SJE407	0161
SJE408	0460
SJE513	0042
SJE514	0925
SJE515	0042
SJE527	0161
SJE583	0161
SJE584	0460
SJE633	0460
SJE634	0161
SJE-649	0161
SJE649	0161
SJE669	0161
SJE677	0925
SJE678	0042
SJE687	0382
SJE694	0042
SJE695	0925
SJE721	0161
SJE723	0460
SJE724	0161
SJE736	0460
SJE737	0161
SJE743	0460
SJE764	1681
SJE768	0460
SJE769	0161
SJE781	0161
SJE783	0042
SJE784	0161
SJE785	0161
SJE797	0460
SJE799	0460
SJE1032	0264
SJE1518	0460
SJE1519	0161
SJE1520	0161
SJE2112	OEM
SJE2123	OEM

DEVICE TYPE	REPL CODE
SJE2789	6338
SJE3478	0074
SJE5018	0561
SJE5019	0561
SJE-5038	0560
SJE5402	0161
SJE8621	0925
SJE8635	0042
SJN2M100/08	OEM
SK01	4257
SK04	0071
SK06M6	OEM
SK08	0071
SK1-K2	0609
SK1K-2	0015
SK-1W50	0502
SK-1W-55	1834
SK2A4	1276
SK2A5	2874
SK2A7	0760
SK2A8	0677
SK2V4	3003
SK2V7	1312
SK2V8	3010
SK3A0	0118
SK3A3	0299
SK3A6	0376
SK3A9	0036
SK3F01	OEM
SK3F02	OEM
SK3F04	OEM
SK3F06	OEM
SK3F08	OEM
SK3F10	OEM
SK3V0	1713
SK3V3	0292
SK3V6	0188
SK3V9	0456
SK3X3	0782
SK3X6	0796
SK3X9	0806
SK4A3	0275
SK4A7	0140
SK4V3	0533
SK4V7	0451
SK4X3	0821
SK4X7	0833
SK5A1	0041
SK5A6	0253
SK5V1	0162
SK5V6	0157
SK5X1	0442
SK5X6	0876
SK6A0	0091
SK6A2	0471
SK6A8	0062
SK6AB	OEM
SK6V0	0301
SK6V2	0636
SK6V8	0025
SK6VB	OEM
SK6X0	3127
SK6X2	0185
SK6X8	0205
SK7	0600
SK7A5	0077
SK7C00	1448
SK7C02	1452
SK7C03	1453
SK7C04	1455
SK7C08	1460
SK7C10	1462
SK7C11	1463
SK7C14	1467
SK7C20	1475
SK7C21	1477
SK7C27	1485
SK7C30	1489
SK7C32	1494
SK7C42	1515
SK7C73	1558
SK7C74	1561
SK7C75	1565
SK7C85	1583
SK7C86	1588
SK7C93	1594
SK7C107	4692
SK7C109	0115
SK7C112	0214
SK7C123	4709
SK7C125	4711
SK7C126	4712
SK7C132	0680
SK7C137	0784
SK7C138	0797
SK7C139	0809
SK7C147	0938
SK7C151	1003
SK7C153	4731
SK7C154	4733
SK7C157	1075
SK7C158	1088
SK7C160	1117
SK7C161	4738
SK7C162	1142
SK7C163	1159
SK7C164	1171
SK7C165	1183
SK7C166	4741

Original Device Types

DEVICE TYPE	REPL CODE
SK7C173	1281
SK7C174	1298
SK7C175	1305
SK7C181	4748
SK7C182	4749
SK7C190	4753
SK7C191	4754
SK7C192	4755
SK7C193	4756
SK7C194	1557
SK7C195	1578
SK7C221	4774
SK7C237	1958
SK7C238	4784
SK7C240	1979
SK7C241	1990
SK7C242	1996
SK7C243	2001
SK7C244	4792
SK7C245	2014
SK7C251	2065
SK7C253	4800
SK7C257	4802
SK7C258	4804
SK7C259	4806
SK7C273	4822
SK7C280	2249
SK7C283	4828
SK7C297	4843
SK7C299	4845
SK7C354	4862
SK7C356	4863
SK7C365	2880
SK7C366	2885
SK7C367	2894
SK7C368	2898
SK7C373	2920
SK7C374	2927
SK7C377	4865
SK7C390	3043
SK7C393	3064
SK7C404	OEM
SK7C423	4876
SK7C533	3677
SK7C534	3681
SK7C540	4888
SK7C541	4889
SK7C563	4892
SK7C564	4893
SK7C573	4896
SK7C574	4897
SK7C583	4899
SK7C597	4901
SK7C640	4912
SK7C643	4914
SK7C646	4915
SK7C648	4916
SK7C670	4919
SK7C688	4923
SK7C4002	5488
SK7C4015	5506
SK7C4016	5507
SK7C4017	5511
SK7C4020	6347
SK7C4024	5515
SK7C4040	6357
SK7C4046	5152
SK7C4046A	5152
SK7C4049	5538
SK7C4050	5540
SK7C4051	5543
SK7C4052	5545
SK7C4053	3806
SK7C4059	5558
SK7C4060	6372
SK7C4066	5571
SK7C4067	6762
SK7C4075	5576
SK7C4094	5596
SK7C4316	5955
SK7C4351	5981
SK7C4352	5982
SK7C4353	5984
SK7C4510	6087
SK7C4511	6088
SK7C4514	6093
SK7C4515	6094
SK7C4516	6097
SK7C4518	6102
SK7C4520	6106
SK7C4538	6126
SK7C4543	6137
SK7C7030	5519
SK7C7046	5148
SK7C7046A	5148
SK7C7266	5889
SK7C22106	6461
SK7C40102	5072
SK7C40103	5073
SK7C40104	5074
SK7C40105	5075
SK7CT00	5916
SK7CT02	4910
SK7CT03	5917
SK7CT04	3296
SK7CT08	3299
SK7CT10	5918
SK7CT11	5919
SK7CT14	3300

DEVICE TYPE	REPL CODE
SK7CT20	5921
SK7CT21	5922
SK7CT27	5923
SK7CT30	5925
SK7CT32	3306
SK7CT42	5928
SK7CT73	5933
SK7CT74	5934
SK7CT75	5935
SK7CT85	5936
SK7CT86	5938
SK7CT93	5940
SK7CT107	6610
SK7CT109	6611
SK7CT112	6613
SK7CT123	6617
SK7CT125	6618
SK7CT126	6619
SK7CT132	6623
SK7CT137	4398
SK7CT138	6990
SK7CT139	6628
SK7CT147	6645
SK7CT151	6649
SK7CT153	6651
SK7CT154	5045
SK7CT157	6656
SK7CT158	6095
SK7CT160	6661
SK7CT161	7071
SK7CT162	7072
SK7CT163	7073
SK7CT164	6666
SK7CT165	6667
SK7CT166	6669
SK7CT173	6673
SK7CT174	7074
SK7CT175	6675
SK7CT181	6679
SK7CT182	6120
SK7CT190	6686
SK7CT191	6687
SK7CT192	6690
SK7CT193	6691
SK7CT194	6693
SK7CT195	6100
SK7CT221	6722
SK7CT237	6733
SK7CT238	6734
SK7CT240	7080
SK7CT241	6737
SK7CT242	6738
SK7CT243	6741
SK7CT244	6108
SK7CT245	4075
SK7CT251	6749
SK7CT253	6752
SK7CT257	6754
SK7CT258	6755
SK7CT259	6757
SK7CT273	7081
SK7CT280	6779
SK7CT283	6782
SK7CT297	6793
SK7CT299	6794
SK7CT354	0790
SK7CT356	0813
SK7CT365	6821
SK7CT366	6824
SK7CT367	6825
SK7CT368	6826
SK7CT373	6123
SK7CT374	7088
SK7CT377	6831
SK7CT390	6837
SK7CT393	6840
SK7CT423	4460
SK7CT533	5719
SK7CT534	5737
SK7CT540	6954
SK7CT541	6955
SK7CT563	6965
SK7CT564	6966
SK7CT573	7097
SK7CT574	7098
SK7CT583	6981
SK7CT597	6987
SK7CT640	7029
SK7CT643	7036
SK7CT646	7039
SK7CT648	4507
SK7CT670	7069
SK7CT688	3412
SK7CT4002	3364
SK7CT4015	3405
SK7CT4016	3409
SK7CT4017	3417
SK7CT4020	3430
SK7CT4024	3448
SK7CT4040	3506
SK7CT4046A	5146
SK7CT4051	3556
SK7CT4052	3558
SK7CT4053	3564
SK7CT4059	3605
SK7CT4060	3614
SK7CT4066	3629
SK7CT4067	3633
SK7CT4075	3665

DEVICE TYPE	REPL CODE
SK7CT4094	3737
SK7CT4316	4324
SK7CT4351	4392
SK7CT4352	4394
SK7CT4353	4396
SK7CT4510	4891
SK7CT4511	4894
SK7CT4514	4903
SK7CT4515	4908
SK7CT4516	4909
SK7CT4518	4913
SK7CT4520	4918
SK7CT4538	4958
SK7CT4543	4979
SK7CT7030	3464
SK7CT7046	3527
SK7CT7046A	3527
SK7CT7266	4248
SK7CT22106	6458
SK7CT40102	5067
SK7CT40103	5068
SK7CT40104	5069
SK7CT40105	5070
SK7CU04	4602
SK7V5	0649
SK7X5	0480
SK8A2	0165
SK8A7	0322
SK8V2	0244
SK8V7	1083
SK8X2	0504
SK8X7	3317
SK9A1	0057
SK9V1	0012
SK9X1	0684
SK10A	0064
SK10V	0170
SK10X	0225
SK11A	0181
SK11V	0317
SK11V5	0794
SK11X	0230
SK12	OEM
SK12A	0052
SK12V	0137
SK12V5	0365
SK12X	0234
SK13A	0053
SK13V	0365
SK13X	0237
SK14A	0879
SK14V	0100
SK14X	1397
SK15A	0686
SK15V	0002
SK15X	0247
SK16	2008
SK16A	0445
SK16V	0420
SK16V2	6432
SK16X	0251
SK17A	0210
SK17V	1649
SK17V7	0495
SK17X	1178
SK18A	0375
SK18V	0495
SK18X	0256
SK19	0325
SK19A	0671
SK19V	0949
SK19X	2395
SK20A	0700
SK20V	0531
SK20X	0262
SK22	2790
SK22A	0705
SK22V	0565
SK22X	0269
SK24	OEM
SK24A	0494
SK24V	0402
SK24X	0274
SK25A	0714
SK25V	1606
SK25X	2399
SK26	3075
SK26V	2575
SK27A	1276
SK27V	0441
SK27X	0294
SK28A	0257
SK28V	1674
SK28X	1177
SK29V	2609
SK30	4263
SK30A	0195
SK30V	0726
SK30X	0309
SK31	3078
SK33A	0166
SK33V	0039
SK33X	0318
SK36A	0010
SK36V	0820
SK36X	0320
SK39A	0032
SK39V	0350
SK39X	0326

DEVICE TYPE	REPL CODE
SK43A	0054
SK43V	0931
SK43X	0337
SK47A	0068
SK47V	0999
SK47X	0347
SK50	0840
SK51	1374
SK51A	0092
SK51V	0502
SK51X	0027
SK52	0838
SK53	0830
SK54	4268
SK55V	1834
SK56A	0125
SK56V	0869
SK56X	0266
SK60	0246
SK60A	2317
SK60V	1156
SK60X	2856
SK62A	0152
SK62V	0783
SK62X	0386
SK63	3415
SK68A	0173
SK68V	1268
SK68X	0405
SK73	1334
SK74C02	2114
SK74C04	2958
SK74C10	2119
SK74C20	2123
SK74C30	3128
SK74C32	3134
SK74C48	2148
SK74C73	2162
SK74C76	2164
SK74C93	2177
SK74C95	3401
SK74C107	3834
SK74C151	4001
SK74C154	4014
SK74C157	4033
SK74C161	4046
SK74C164	4062
SK74C173	4088
SK74C174	4092
SK74C175	4096
SK74C192	4119
SK74C193	4122
SK74C221	3047
SK74C240	2956
SK74C244	4230
SK74C245	6477
SK74C373	4462
SK74C374	4467
SK74C901	4607
SK74C902	4608
SK74C903	4609
SK74C904	4610
SK74C922	5597
SK74C923	5599
SK74C925	4621
SK74H00	0682
SK74H04	1907
SK74LS00	1527
SK74LS01	1546
SK74LS02	1559
SK74LS03	1579
SK74LS04	1595
SK74LS05	1608
SK74LS08	1633
SK74LS10	1662
SK74LS11	1667
SK74LS12	1679
SK74LS13	1688
SK74LS14	1698
SK74LS15	1707
SK74LS20	0034
SK74LS22	1775
SK74LS26	1382
SK74LS27	0183
SK74LS28	0472
SK74LS30	0828
SK74LS32	0088
SK74LS33	1832
SK74LS37	1730
SK74LS38	1839
SK74LS40	0135
SK74LS42	1841
SK74LS47	1845
SK74LS48	1849
SK74LS49	1850
SK74LS51	1033
SK74LS54	1857
SK74LS55	0457
SK74LS73	1867
SK74LS73A	1867
SK74LS74	0243
SK74LS74A	0243
SK74LS75	1870
SK74LS76A	2180
SK74LS77	1872
SK74LS78	1873
SK74LS83	2218
SK74LS83A	2218
SK74LS85	0431

DEVICE TYPE	REPL CODE
SK74LS86	0291
SK74LS90	1882
SK74LS92	1887
SK74LS93	1888
SK74LS107	1602
SK74LS109	1906
SK74LS113	2256
SK74LS114	2302
SK74LS122	1620
SK74LS123	0979
SK74LS123P	0979
SK74LS125A	0075
SK74LS133	3400
SK74LS136	1628
SK74LS138	0427
SK74LS139	0153
SK74LS145	1564
SK74LS148	3912
SK74LS151	1646
SK74LS153	0959
SK74LS155	0209
SK74LS156	1654
SK74LS157	1161
SK74LS161	0858
SK74LS163	0893
SK74LS164	4339
SK74LS168A	0967
SK74LS169A	0986
SK74LS174	0260
SK74LS175	1672
SK74LS191	1687
SK74LS193	1692
SK74LS194	1304
SK74LS194A	1304
SK74LS195	1315
SK74LS196	2834
SK74LS221	1239
SK74LS240	0452
SK74LS241	1726
SK74LS242	1728
SK74LS243	0906
SK74LS244	0458
SK74LS245	0463
SK74LS247	1732
SK74LS248	1734
SK74LS249	1735
SK74LS253	1739
SK74LS257	1744
SK74LS258	1746
SK74LS259	3205
SK74LS260	5995
SK74LS266	0592
SK74LS273	0894
SK74LS279	3290
SK74LS280	1773
SK74LS290	4425
SK74LS293	0082
SK74LS298	3371
SK74LS322	1805
SK74LS348	0665
SK74LS352	0761
SK74LS353	0773
SK74LS364	6006
SK74LS367	0977
SK74LS373	0709
SK74LS377	1120
SK74LS378	1133
SK74LS379	1151
SK74LS386	1230
SK74LS393	0819
SK74LS398	1383
SK74LS399	1398
SK74LS490	2213
SK74LS624	3141
SK74LS625	3149
SK74LS626	3154
SK74LS627	3162
SK74LS629	3175
SK74LS641	0690
SK74LS642	0719
SK74LS645	0775
SK74S00	0704
SK74S04	2263
SK74S11	2447
SK74S20	1017
SK74S22	2461
SK74S32	3847
SK74S74	2504
SK74S86	2511
SK74S112	1617
SK74S174	2132
SK75A	0094
SK75V	1189
SK75X	0426
SK82A	0049
SK82V	0331
SK82X	0444
SK87A	0104
SK87V	3026
SK87X	3028
SK91A	0156
SK91V	1311
SK91X	0238
SK100A	0189
SK100V	0098
SK100X	1180
SK103K	OEM
SK107-4	0325
SK110A	0099

DEVICE TYPE	REPL CODE
SK110V	0149
SK110V151A	OEM
SK110X	1190
SK112	2397
SK113	1380
SK114	2035
SK115	0547
SK116	2401
SK117	2403
SK118	1438
SK119	2113
SK120	2407
SK120A	0089
SK120V	0186
SK120X	1207
SK121	2410
SK122	1445
SK123	1901
SK124	2415
SK125	2416
SK126	2221
SK127	0696
SK128	1458
SK129	1601
SK130	0221
SK130A	0287
SK130V	0213
SK130X	1218
SK131	1616
SK132	1490
SK133	1622
SK134	2422
SK135	2424
SK136	1618
SK137	0627
SK138	0510
SK139	0992
SK140	0691
SK140A	0252
SK140V	0245
SK140X	1881
SK141	0995
SK142	0870
SK143	1263
SK144	1020
SK145	1249
SK146	1153
SK147	1636
SK148	1274
SK149	1639
SK150	1402
SK150A	0340
SK150V	0028
SK150X	0647
SK151	1703
SK152	1532
SK153	1640
SK154	1078
SK155	1716
SK156	1711
SK157	1720
SK158	1717
SK159	1731
SK160	1723
SK160A	0370
SK160V	0255
SK160X	1255
SK161	0727
SK162	1736
SK163	1756
SK164	1748
SK165	1768
SK166	1761
SK167	1782
SK168	2450
SK169	2453
SK170	1772
SK170A	0394
SK170V	0877
SK170X	2102
SK171	1794
SK172	1788
SK173	1799
SK174	1796
SK175	1809
SK176	1804
SK177	1817
SK178	1194
SK179	1826
SK180	1821
SK180A	0425
SK180V	0367
SK180X	1279
SK181	1840
SK182	0022
SK183	1853
SK184	0070
SK185	1861
SK186	0132
SK187	1866
SK188	0172
SK189	1874
SK190	0207
SK190A	0453
SK190V	2858
SK190X	2225
SK191	1884
SK192	0227
SK193	2474

DEVICE TYPE	REPL CODE
SK194	0263
SK195	1895
SK196	0310
SK197	1902
SK198	0329
SK199	0736
SK200	0356
SK200A	1473
SK200V	0421
SK200X	0605
SK201	1909
SK202	0381
SK203	2479
SK203K	OEM
SK204	0412
SK205	1914
SK206	0438
SK207	1163
SK208	0464
SK209	1924
SK210	0488
SK211	1933
SK212	0509
SK213	1941
SK214	0524
SK215	1947
SK216	0542
SK217	1953
SK-218	0015
SK218	0063
SK219	1961
SK220	0401
SK221	0357
SK222	0598
SK223	2500
SK224	0616
SK225	0670
SK226	0634
SK227	0776
SK228	0650
SK229	2507
SK230	0668
SK231	1071
SK301	2543
SK342	1854
SK350	2600
SK372	2627
SK379	2360
SK400	2651
SK401	2651
SK401B	2651
SK403	2652
SK403B	2652
SK403K	OEM
SK404EK	0109
SK510	0815
SK511	2721
SK512	0827
SK513	2724
SK514	0846
SK515	2726
SK516	0868
SK517	2729
SK518	0885
SK519	2732
SK520	0897
SK521	2734
SK522	0914
SK523	2738
SK524	0926
SK525	2740
SK526	0944
SK527	2744
SK528	0958
SK529	1225
SK530	0978
SK531	2746
SK532	0994
SK533	1237
SK534	1009
SK535	1252
SK536	1019
SK537	1269
SK538	0889
SK539	1277
SK540	1049
SK541	1294
SK542	1058
SK543	2752
SK544	0932
SK545	1302
SK546	1079
SK547	1310
SK548	1096
SK549	2756
SK550	1106
SK551	1324
SK552	1123
SK553	2758
SK554	1135
SK555	1333
SK556	1152
SK557	1344
SK558	1164
SK559	1356
SK560	1174
SK561	2760
SK562	1184
SK563	1371

If replacement code is OEM, contact original manufacturer for replacement.

DEVICE TYPE	REPL CODE
SK564	1200
SK565	2762
SK566	1210
SK567	1387
SK568	1223
SK569	1406
SK570	1232
SK571	1415
SK572	1246
SK573	1428
SK574	1265
SK575	1440
SK576	1273
SK577	2772
SK578	1290
SK579	1447
SK580	1299
SK581	2776
SK582	1307
SK583	1461
SK584	1322
SK585	2779
SK586	1331
SK587	1474
SK588	1340
SK589	2783
SK590	1353
SK591	0613
SK592	1365
SK593	1510
SK594	1384
SK595	1523
SK596	1401
SK597	1537
SK598	1412
SK599	1550
SK600	1423
SK601	1575
SK602	1436
SK603	2802
SK603K	OEM
SK604	1444
SK605	2803
SK606	1457
SK607	1507
SK608	1470
SK609	2805
SK610	1484
SK611	2807
SK612	1505
SK613	2810
SK614	1521
SK615	2815
SK616	1531
SK617	2818
SK618	1548
SK619	2820
SK620	1568
SK621	2822
SK622	1587
SK623	2824
SK803K	OEM
SK814	2937
SK821	2943
SK827	2944
SK832	2948
SK836	2951
SK841	2955
SK847	2961
SK849	2379
SK854	2963
SK859	2968
SK866	2972
SK869	2973
SK876	2976
SK890	2990
SK900	2998
SK901	3000
SK914	3013
SK930	3021
SK948	3034
SK952	3036
SK1002K	OEM
SK1203K	OEM
SK1320	0144
SK1403K	OEM
SK1639	0037
SK1640	0037
SK1641	0016
SK1802	3223
SK1805	3356
SK1806	6399
SK1822	4043
SK1823	4083
SK1824	4103
SK1826	6402
SK1851	4668
SK1852	4742
SK1853	4816
SK1854	4435
SK1855	4880
SK1856	4357
SK1857	OEM
SK1858	OEM
SK1859	OEM
SK1861	5009
SK1862	5019
SK1863	OEM
SK1864	5053
SK1866	5087
SK1867	OEM
SK1868	OEM
SK1869	5145
SK1870	OEM
SK1871	5231
SK1872	5266
SK1873	5288
SK1874	5308
SK1875	5317
SK1876	5331
SK1877	5343
SK1878	6405
SK1879	6406
SK1881	6407
SK1882	6408
SK1883	6409
SK1900	0448
SK1901	0388
SK1911	3792
SK1912	2697
SK1914	1695
SK1915	2151
SK1916	OEM
SK1917	2238
SK1918	0713
SK1919	1737
SK1920	1763
SK1921	2469
SK1922	5217
SK1923	1656
SK1924	1642
SK2000	1981
SK2001	1989
SK2002	1985
SK2003	1983
SK2004	5112
SK2005	5113
SK2006	5114
SK2006A	5114
SK2007	3019
SK2008	3096
SK2009	5115
SK2009A	5115
SK2010	0841
SK2010A	0841
SK2011	3124
SK2018	1615
SK2019	5119
SK2020	1358
SK2021	3157
SK2022	1962
SK2023	3276
SK2024	1778
SK2025	3182
SK2026	2195
SK2027	4183
SK2027A	4183
SK2031	5122
SK2032	2313
SK2035	3185
SK2035A	3185
SK2040	0541
SK2041	0315
SK2042	0273
SK2043	4431
SK2046	2236
SK2048	5065
SK2049	5126
SK2052	4870
SK2053	5127
SK2054	5128
SK2055	5129
SK2056	4873
SK2058	5130
SK2059	5131
SK2060	5132
SK2061	4989
SK2062	5134
SK2063	5135
SK2064	5137
SK2065	4871
SK2068	5138
SK2069	5139
SK2070	5140
SK2071	5141
SK2074	4327
SK2075	4332
SK2076	4331
SK2077	4333
SK2078	2206
SK2079	2199
SK2083	1109
SK2084	1053
SK2086	1403
SK2087	5423
SK2111	5434
SK2112	5435
SK2113	5436
SK2114	5151
SK2115	5153
SK2116	5154
SK2117	5155
SK2131	6422
SK2132	6424
SK2150	5165
SK2151	5167
SK2152	5168
SK2153	5170
SK2154	5211
SK2160	5174
SK2161	1076
SK2162	5176
SK2163	4767
SK2164	3491
SK2165	5178
SK2166	4052
SK2167	1131
SK2168	5179
SK2169	4023
SK2170	5016
SK2171	5438
SK2214	2048
SK2222	OEM
SK2604	0037
SK2716	2278
SK2880	3481
SK2881	1904
SK2882	5339
SK3003	0004
SK3003A	0633
SK3004	0004
SK3005	5213
SK3006	0050
SK3007	0050
SK3007A	1062
SK3008	6511
SK3009	2713
SK3010	0208
SK3011	0600
SK3011A	4436
SK3012	0440
SK3013	0265
SK3014	6512
SK3015	6513
SK3016	0015
SK3016A	0965
SK3017	0229
SK3017A	0015
SK3017B	1355
SK3018	2525
SK3019	2525
SK3020	0198
SK3021	0142
SK3022	2618
SK3023	2624
SK3024	0086
SK3025	0126
SK3026	6515
SK3027	0103
SK3028	6516
SK3029	1514
SK3030	0015
SK3031	0015
SK3031A	3911
SK3032	0071
SK3032A	0072
SK3033	0071
SK3033A	2637
SK3034	0975
SK3035	1675
SK3036	2201
SK3037	6518
SK3038	6519
SK3039	0259
SK3040	0730
SK3041	0461
SK3042	0239
SK3043	1217
SK3043A	4503
SK3043B	4503
SK3044	0233
SK3045	3280
SK3046	6520
SK3047	0086
SK3047A	0086
SK3048	2700
SK3049	0630
SK3050	0353
SK3051	0087
SK3052	6521
SK3053	5711
SK3054	0423
SK3055	0188
SK3056	0162
SK3057	0157
SK3058	0636
SK3059	0649
SK3060	0012
SK3061	0170
SK3062	0137
SK3063	0002
SK3064	0441
SK3065	0212
SK3066	0774
SK3067	0201
SK3068	0288
SK3069	0128
SK3070	2715
SK3071	2529
SK3072	0167
SK3073	2549
SK3074	0328
SK3075	0352
SK3076	0354
SK3077	0666
SK3078	1843
SK3079	6525
SK3080	0373
SK3081	0071
SK3082	1010
SK3083	0854
SK3084	0854
SK3085	6526
SK3086	1862
SK3087	2232
SK3088	0019
SK3089	0917
SK3090	0143
SK3091	0143
SK3092	0794
SK3093	0365
SK3094	0100
SK3095	0039
SK3096	1834
SK3097	0783
SK3098	0331
SK3099	0149
SK3100	0124
SK3101	2571
SK3102	0829
SK3103	0187
SK3103A	0187
SK3104	2650
SK3104A	2650
SK3105	6529
SK3106	0106
SK3107	0787
SK3108	0757
SK3109	1303
SK3110	0474
SK3111	5648
SK3112	3384
SK3113	OEM
SK3114	0006
SK3114A	0006
SK3115	0065
SK3116	3249
SK3117	0047
SK3118	4433
SK3119	0196
SK3120	0484
SK3121	1080
SK3122	0191
SK3123	6530
SK3124	0155
SK3124A	0155
SK3125	0182
SK3125A	0182
SK3126	0628
SK3127	1217
SK3128	1217
SK3129	2617
SK3130	2811
SK3131	2096
SK3131A	2096
SK3132	6532
SK3133	0003
SK3134	2161
SK3135	0664
SK3136	0244
SK3137	0595
SK3138	1242
SK3139	0317
SK3140	2714
SK3141	0364
SK3142	6432
SK3143	2743
SK3144	1345
SK3145	0495
SK3146	2257
SK3147	2755
SK3148	6533
SK3149	1337
SK3150	6534
SK3151	0402
SK3152	3550
SK3153	2876
SK3154	4676
SK3155	1187
SK3156	0400
SK3157	0632
SK3158	0521
SK3159	0654
SK3160	0520
SK3161	0701
SK3162	0692
SK3163	1759
SK3164	1192
SK3165	0026
SK3166	2552
SK3167	0856
SK3168	0395
SK3169	2747
SK3170	2457
SK3171	1421
SK3172	1395
SK3173	0491
SK3174	0015
SK3175	0102
SK3175A	2528
SK3176	1233
SK3177	1977
SK3178B	0824
SK3179	4975
SK3179B	4975
SK3180	2235
SK3181	2237
SK3181A	2237
SK3182	2427
SK3183	3374
SK3183A	2277
SK3184	1248
SK3185	3270
SK3186	2049
SK3187	4511
SK3188	0561
SK3188A	0561
SK3189	1199
SK3189A	1199
SK3190	0161
SK3191	0460
SK3192	0560
SK3193	1267
SK3194	1027
SK3195	0493
SK3196	2329
SK3197	0936
SK3198	0222
SK3199	0551
SK3200	0382
SK3201	0285
SK3202	0566
SK3203	1367
SK3204	2457
SK3205	2830
SK3206	2493
SK3207	2501
SK3208	2502
SK3209	2560
SK3210	2789
SK3211	0521
SK3212	2786
SK3213	2798
SK3214	3832
SK3215	0802
SK3216	1753
SK3217	6536
SK3218	0418
SK3219	0955
SK3220	0168
SK3221	2237
SK3222	2235
SK3223	0911
SK3224	2632
SK3225	2057
SK3226	2603
SK3227	2599
SK3228	2582
SK3229	2305
SK3230	4600
SK3231	1050
SK3232	0338
SK3233	0417
SK3234	0785
SK3235	0751
SK3236	0789
SK3237	3195
SK3238	1611
SK3239	0836
SK3240	0372
SK3241	3519
SK3242	1077
SK3243	3695
SK3244	6199
SK3245	0111
SK3246	0224
SK3246A	0224
SK3247	0693
SK3248	0390
SK3249	0433
SK3250	1978
SK3251	1946
SK3252	1173
SK3253	1591
SK3254	2707
SK3255	0848
SK3256	0461
SK3257	6526
SK3258	4365
SK3259	4365
SK3260	4786
SK3261	0178
SK3262	0126
SK3263	6537
SK3264	6537
SK3265	2229
SK3266	3329
SK3267	6539
SK3268	6540
SK3269	6536
SK3270	6541
SK3271	0142
SK3272	6542
SK3273	6543
SK3274	0925
SK3275	0861
SK3276	2930
SK3277	3761
SK3278	3763
SK3279	1808
SK3280	3219
SK3281	1994
SK3282	3366
SK3283	2634
SK3284	0855
SK3285	4582
SK3286	4594
SK3287	4595
SK3288	0579
SK3289	0881
SK3290	0884
SK3291	2861
SK3292	2864
SK3293	0127
SK3294	6233
SK3295	2591
SK3296	1652
SK3297	0177
SK3298	2050
SK3299	1411
SK3300	2804
SK3301	0409
SK3302	1710
SK3303	1501
SK3304	0190
SK3305	2982
SK3306	1067
SK3307	2546
SK3308	3655
SK3309	1054
SK3310	2984
SK3311	0609
SK3312	0795
SK3313	0015
SK3314	1928
SK3315	0920
SK3315A	2523
SK3316	3077
SK3316A	3077
SK3317	0640
SK3317A	0640
SK3318	2527
SK3318A	2527
SK3319	5633
SK3320	1029
SK3322	6549
SK3323	0554
SK3324	0005
SK3325	0720
SK3326	0628
SK3327	0030
SK3328	0379
SK3329	3652
SK3330	0292
SK3331	0456
SK3332	0533
SK3333	0451
SK3334	0025
SK3335	0531
SK3336	0565
SK3337	0820
SK3338	0350
SK3339	0931
SK3340	0999
SK3341	0502
SK3342	0869
SK3343	1268
SK3344	1189
SK3345	1311
SK3346	0098
SK3347	0186
SK3348	0213
SK3349	0245
SK3350	0028
SK3351	0255
SK3352	0877
SK3353	0367
SK3354	2858
SK3355	0421
SK3356	6552
SK3357	0219
SK3358	1215
SK3359	2013
SK3360	2140
SK3361	4570
SK3362	4569
SK3363	4390
SK3364	2656
SK3365	0878
SK3366	1621
SK3367	4571
SK3368	2262
SK3369	4565
SK3370	4566
SK3371	4093
SK3372	4561
SK3373	4018
SK3374	6556
SK3375	1443
SK3376	0864
SK3377	0782
SK3378	0796
SK3379	0806
SK3380	0821
SK3381	0833
SK3382	0442
SK3383	0876
SK3384	3127
SK3385	0185
SK3386	0205
SK3387	0480
SK3388	0504
SK3389	3317
SK3390	0684
SK3391	0225
SK3392	0230
SK3393	0234
SK3394	0237
SK3395	1397
SK3396	0247
SK3397	0251
SK3398	1178
SK3399	0256
SK3400	2395
SK3401	0262
SK3402	0269
SK3403	0274
SK3404	2399
SK3405	0294
SK3406	1177
SK3407	0309
SK3408	0318
SK3409	0320
SK3410	0326
SK3411	0337
SK3412	0347
SK3413	0027
SK3414	0266
SK3415	2856
SK3416	0386
SK3417	0405
SK3418	0426
SK3419	0444
SK3420	3028
SK3421	0238
SK3422	1180
SK3423	1190
SK3424	1207
SK3425	1218
SK3426	1881
SK3427	0647
SK3428	1255
SK3429	2102
SK3430	1279
SK3431	2225
SK3432	0605
SK3433	0715
SK3434	0342
SK3435	6558
SK3436	4559
SK3437	1681
SK3438	5709
SK3439	0642
SK3439A	0642
SK3440	0236
SK3441	0681
SK3442	4615
SK3443	1323
SK3444	0016
SK3445	1540
SK3446	3263
SK3447	6188
SK3448	3717
SK3449	0324
SK3450	0436
SK3451	3766
SK3452	0144
SK3453	3735
SK3454	0349
SK3455	2557
SK3456	3734
SK3457	2576
SK3458	1837
SK3459	2124
SK3460	2581
SK3461	2261
SK3462	1292
SK3463	0139
SK3464	2046
SK3465	0360
SK3466	0037
SK3467	0363
SK3468	1479
SK3469	4209
SK3470	0653
SK3471	0528
SK3472	2316
SK3473	2872
SK3474	1160
SK3475	2891
SK3476	2631
SK3477	1018
SK3478	1326
SK3479	2272
SK3480	3705
SK3481	6561
SK3482	2931
SK3483	1509
SK3484	1816
SK3485	1478
SK3486	4612
SK3487	2019
SK3488	3915
SK3489	2949
SK3490	2231
SK3491	0678
SK3492	0882
SK3493	1524
SK3494	2578
SK3495	2592
SK3496	2596
SK3497	1260
SK3498	3352

DEVICE TYPE	REPL CODE
SK3499	0843
SK3500	0122
SK3501	0319
SK3502	3624
SK3503	3349
SK3504	3243
SK3505	0725
SK3506	1413
SK3507	6565
SK3508	0147
SK3509	2017
SK3510	0103
SK3511	0130
SK3512	3329
SK3513	6568
SK3514	0410
SK3515	2528
SK3516	2088
SK3517	0816
SK3518	0267
SK3519	2387
SK3520	0279
SK3520A	0279
SK3521	4100
SK3522	2018
SK3523	3331
SK3524	2699
SK3525	0823
SK3526	2368
SK3527	4040
SK3528	0439
SK3529	4246
SK3530	0630
SK3531	OEM
SK3532	0415
SK3533	0576
SK3534	1361
SK3535	0130
SK3536	OEM
SK3537	OEM
SK3538	6570
SK3539	2705
SK3540	2537
SK3541	2727
SK3542	1854
SK3543	1696
SK3544	2722
SK3545	2711
SK3546	2698
SK3547	2687
SK3548	2701
SK3549	2689
SK3550	2749
SK3551	3137
SK3552	0312
SK3553	0410
SK3555	3137
SK3556	2358
SK3557	6571
SK3558	0710
SK3559	0074
SK3560	4160
SK3561	3554
SK3562	6573
SK3563	6574
SK3564	0973
SK3565	0093
SK3566	1896
SK3567	1815
SK3567A	1815
SK3568	2782
SK3569	0176
SK3570	6576
SK3571	5732
SK3572	0505
SK3573	0863
SK3574	4490
SK3575	6577
SK3576	6578
SK3577	5750
SK3578	5751
SK3579	0745
SK3580	0747
SK3581	0469
SK3582	0722
SK3583	2394
SK3584	0204
SK3585	1586
SK3586	1567
SK3586A	1567
SK3587	0711
SK3587A	0711
SK3588	4212
SK3588A	4212
SK3589	5622
SK3589A	5622
SK3590	1705
SK3591	0624
SK3592	0334
SK3593	1321
SK3594	2704
SK3595	3381
SK3596	2283
SK3597	0940
SK3598	0095
SK3599	0197
SK3600	0989
SK3601	0970
SK3602	0109

DEVICE TYPE	REPL CODE
SK3603	0097
SK3604	0084
SK3605	1581
SK3606	1250
SK3607	2002
SK3608	0805
SK3609	1124
SK3610	3189
SK3611	4244
SK3612	4244
SK3613	1651
SK3614	1848
SK3615	1650
SK3616	6582
SK3617	3022
SK3618	OEM
SK3619	3554
SK3620	1361
SK3621	6584
SK3622	6585
SK3623	0792
SK3624	0792
SK3625	0905
SK3626	0934
SK3627	0058
SK3628	0316
SK3629	1922
SK3630	1196
SK3631	1093
SK3632	0411
SK3633	0415
SK3634	2270
SK3635	3402
SK3636	3404
SK3637	4633
SK3638	0901
SK3639	1282
SK3640	1293
SK3641	1300
SK3642	0604
SK3643	0625
SK3644	2452
SK3645	2454
SK3646	3394
SK3647	0289
SK3648	0296
SK3649	0250
SK3650	4152
SK3651	4159
SK3652	2015
SK3653	2647
SK3654	2649
SK3655	1848
SK3656	1855
SK3657	3215
SK3658	0772
SK3659	0744
SK3660	0617
SK3661	3199
SK3662	3207
SK3663	3222
SK3664	4414
SK3665	3499
SK3666	0485
SK3667	2827
SK3668	2104
SK3669	0923
SK3670	2239
SK3671	1285
SK3672	2648
SK3673	1838
SK3674	3829
SK3675	3575
SK3676	2010
SK3677	2400
SK3678	0302
SK3679	5483
SK3680	2363
SK3681	2369
SK3682	2370
SK3683	3696
SK3684	3383
SK3685	2095
SK3686	2089
SK3687	0710
SK3688	2247
SK3689	3285
SK3690	1781
SK3691	1677
SK3692	0770
SK3693	3637
SK3694	2750
SK3695	2751
SK3696	6590
SK3697	2499
SK3698	1107
SK3699	2259
SK3700	0416
SK3701	3318
SK3702	2514
SK3703	1634
SK3704	4627
SK3705	4395
SK3706	2593
SK3707	3651
SK3708	4401
SK3709	0123
SK3710	0313
SK3711	2639

DEVICE TYPE	REPL CODE
SK3712	4624
SK3713	4700
SK3714	4238
SK3715	5367
SK3716	0007
SK3716A	0007
SK3717	0160
SK3718	0265
SK3719	0085
SK3720	3033
SK3721	0280
SK3722	0211
SK3723	2568
SK3724	0083
SK3725	1055
SK3726	4652
SK3727	3269
SK3728	2602
SK3729	3273
SK3730	5698
SK3730(TRANS)	0050
SK3731	4634
SK3732	1714
SK3733	4620
SK3734	3422
SK3735	0916
SK3736	0651
SK3737	2572
SK3738	2532
SK3739	2193
SK3740	1571
SK3741	1612
SK3742	0857
SK3743	4034
SK3744	4644
SK3745	6596
SK3746	2987
SK3747	0276
SK3748	0622
SK3749	1083
SK3750	0365
SK3751	0420
SK3752	0495
SK3753	1606
SK3754	1674
SK3755	0726
SK3756	1306
SK3757	2083
SK3758	2084
SK3759	1449
SK3760	2085
SK3761	3649
SK3762	0621
SK3763	0706
SK3764	0975
SK3765	0698
SK3766	1276
SK3767	2874
SK3768	0760
SK3769	0677
SK3770	0118
SK3771	0299
SK3772	0376
SK3773	0036
SK3774	0275
SK3775	0140
SK3776	0041
SK3777	0253
SK3778	0091
SK3779	0471
SK3780	0062
SK3781	0077
SK3782	0165
SK3783	0322
SK3784	0057
SK3785	0064
SK3786	0181
SK3787	0052
SK3788	0053
SK3789	0879
SK3790	0686
SK3791	0445
SK3792	0210
SK3793	0375
SK3794	0671
SK3795	0700
SK3796	0705
SK3797	0494
SK3798	0714
SK3799	0455
SK3800	0257
SK3801	0195
SK3802	0166
SK3803	0010
SK3804	0032
SK3805	0054
SK3806	0068
SK3807	0092
SK3808	0125
SK3809	2317
SK3810	0152
SK3811	0173
SK3812	0094
SK3813	0049
SK3814	0104
SK3815	0156
SK3816	0189
SK3817	0099
SK3818	0089

DEVICE TYPE	REPL CODE
SK3819	0287
SK3820	0252
SK3821	0340
SK3822	0370
SK3823	0394
SK3824	0425
SK3825	0453
SK3826	1473
SK3827	2781
SK3828	2638
SK3829	2755
SK3830	2828
SK3831	4080
SK3832	1170
SK3833	3948
SK3834	3712
SK3835	0208
SK3836	0543
SK3837	1312
SK3838	1713
SK3839	2763
SK3840	0822
SK3841	0532
SK3842	6602
SK3843	6603
SK3844	1751
SK3845	0847
SK3846	1598
SK3847	6059
SK3848	0965
SK3849	0018
SK3850	5750
SK3851	5469
SK3852	1048
SK3853	3710
SK3854	0079
SK3855	1712
SK3856	2329
SK3857	0061
SK3858	3525
SK3859	3526
SK3860	2258
SK3861	0600
SK3862	0038
SK3863	1925
SK3864	0120
SK3865	6009
SK3865A	6009
SK3866	1562
SK3866A	1562
SK3867	1522
SK3867A	1522
SK3868	2977
SK3869	2978
SK3870	2980
SK3871	2979
SK3872	1128
SK3873	4640
SK3874	2534
SK3875	4663
SK3876	2062
SK3877	0470
SK3878	0904
SK3879	4635
SK3880	3815
SK3881	2328
SK3882	2831
SK3883	2362
SK3884	4850
SK3885	4549
SK3887	3425
SK3888	2614
SK3889	3190
SK3890	4598
SK3891	1899
SK3892	3801
SK3893	0042
SK3894	0429
SK3895	0620
SK3896	1212
SK3897	0602
SK3898	1825
SK3899	2465
SK3900	1997
SK3901	3867
SK3902	2985
SK3903	1197
SK3904	3667
SK3905	3666
SK3906	2109
SK3909	2671
SK3910	6606
SK3911	0672
SK3912	0825
SK3913	0629
SK3914	0362
SK3915	5618
SK3916	4649
SK3918	3089
SK3919	2812
SK3920	2960
SK3921	4071
SK3922	0754
SK3923	0756
SK3924	3029
SK3925	0348
SK3926	3444
SK3927	2817
SK3928	1878

DEVICE TYPE	REPL CODE
SK3929	2911
SK3930	5467
SK3931	0530
SK3932	0648
SK3933	2939
SK3934	3571
SK3935	0134
SK3936	0073
SK3937	0571
SK3938	0593
SK3939	0577
SK3940	1103
SK3941	2491
SK3942	0240
SK3943	0676
SK3944	0336
SK3945	2485
SK3946	1966
SK3947	2414
SK3948	2439
SK3949	2432
SK3950	1137
SK3951	0407
SK3952	1683
SK3953	0174
SK3954	3620
SK3955	3324
SK3956	1502
SK3957	2847
SK3958	3081
SK3959	0859
SK-3960	OEM
SK3960	3602
SK3961	3603
SK3962	1786
SK3963	4042
SK3964	3103
SK3965	0011
SK3966	2295
SK3967	5121
SK3968	2942
SK3969	4147
SK3970	3804
SK3971	3261
SK3973	2301
SK3974	3380
SK3975	0845
SK3976	OEM
SK3977	3622
SK3978	3529
SK3979	3530
SK3980	3003
SK3982	3010
SK3983	2613
SK3984	0150
SK3985	0323
SK3986	1414
SK3987	0473
SK3988	0446
SK3989	1422
SK3990	0849
SK3991	2458
SK3992	6340
SK3993	6340
SK3994	0908
SK3995	2928
SK3996	0558
SK3997	2897
SK3998	0017
SK3999	0293
SK4000	6109
SK4000B	6109
SK4000UB	2024
SK4001	0478
SK4001B	0478
SK4002	2055
SK4002B	2055
SK4006B	0646
SK4007	2846
SK4007UB	2846
SK4008B	0988
SK4009	1999
SK4009UB	1999
SK4010	3252
SK4010B	3252
SK4011	0215
SK4011B	0215
SK4012	0498
SK4012B	0498
SK4013	0413
SK4013B	0413
SK4014B	0860
SK4015	1014
SK4015B	1014
SK4016	1143
SK4016B	1143
SK4017	0513
SK4017B	0513
SK4018B	1391
SK4019	1525
SK4019B	1525
SK4020	1661
SK4020B	1661
SK4021	1749
SK4021B	1749
SK4022B	1256
SK4023	0519
SK4023B	0519
SK4024	1957

DEVICE TYPE	REPL CODE
SK4024B	1957
SK4025	2072
SK4025B	2072
SK4026B	2152
SK4027	1949
SK4027B	1949
SK4028B	2228
SK4029	2233
SK4029B	2233
SK4030	0500
SK4030B	0500
SK4031B	2971
SK4032B	2531
SK4033B	2635
SK4034B	3615
SK4035	2777
SK4035B	2777
SK4037A	OEM
SK4037B	OEM
SK4038B	2981
SK4040	0056
SK4040B	0056
SK4041UB	3174
SK4042	0121
SK4042B	0121
SK4043B	1553
SK4044B	2308
SK4045B	3443
SK4046	3831
SK4046B	3831
SK4047	2331
SK4047B	2331
SK4048B	3459
SK4049	0001
SK4049UB	0001
SK4050	0398
SK4050B	0398
SK4051	0366
SK4051B	0366
SK4052	0024
SK4052B	0024
SK4053B	0035
SK4054	4530
SK4054B	4530
SK4055B	3304
SK4056	3706
SK4056B	3706
SK4059A	3308
SK4059B	OEM
SK4060	0146
SK4060B	0146
SK4063B	3727
SK4066	0101
SK4066B	0101
SK4067B	3742
SK4068	2503
SK4068B	2503
SK4069	0119
SK4069UB	0119
SK4070	2516
SK4070B	2516
SK4071	0129
SK4071B	0129
SK4072	2524
SK4072B	2524
SK4073	1536
SK4073B	1536
SK4075	2540
SK4075B	2540
SK4076	3496
SK4076B	3496
SK4077	2558
SK4077B	2558
SK4078	0921
SK4078B	0921
SK4081	0626
SK4081B	0626
SK4082	0300
SK4082B	0300
SK4085B	0303
SK4086	0535
SK4086B	0535
SK4089B	3830
SK4093	2384
SK4093B	2384
SK4094	1682
SK4094B	1682
SK4095B	3848
SK4096B	3851
SK4097B	3855
SK4098	3611
SK4098B	3611
SK4099	3330
SK4099B	3330
SK4311	6627
SK4320	6631
SK4325	6632
SK4326	6633
SK4332	4934
SK4368	5963
SK4501UB	3809
SK4502B	1037
SK4503	2053
SK4503B	2053
SK4506UB	3767
SK4508B	1811
SK4510	1963
SK4510B	1963
SK4511	1544

DEVICE TYPE	REPL CODE
SK4511B	1544
SK4512	2120
SK4512B	2120
SK4513B	4976
SK4514	1830
SK4514B	1830
SK4515	3600
SK4515B	3600
SK4516B	2347
SK4517B	4284
SK4518	1043
SK4518B	1043
SK4520	2675
SK4520B	2675
SK4521B	4990
SK4522B	2837
SK4526B	3610
SK4527B	3145
SK4529B	4531
SK4531B	3325
SK4532B	1016
SK4536B	3704
SK4538B	1063
SK4539B	3656
SK4541B	5017
SK4543B	5020
SK4547B	5032
SK4551B	5039
SK4553B	5040
SK4555B	2938
SK4556B	3431
SK4558B	5049
SK4560B	6252
SK4562B	1035
SK4566B	3504
SK4568B	3618
SK4569B	4487
SK4572UB	3480
SK4583B	1297
SK4584B	6597
SK4585B	1375
SK4597B	4110
SK4598B	4301
SK4724B	3470
SK4801	0416
SK4802	4613
SK4803	4666
SK4804	4618
SK4812	5198
SK4814	1425
SK4815	1122
SK4816	4350
SK4821	4722
SK4822	3916
SK4825	4109
SK4826	3391
SK4828	5188
SK4831	2386
SK4832	2378
SK4833	5209
SK4834	2073
SK4835	0465
SK4837	4351
SK4838	4791
SK4839	4757
SK4843	3989
SK4900	4099
SK4903	3673
SK4904	3435
SK4906	3799
SK4907	3377
SK4908	2185
SK4909	1511
SK4910	4138
SK4911	1942
SK4914	2336
SK4915	0371
SK4916	5433
SK4918	2300
SK4919	4450
SK4920	5651
SK4921	5650
SK4922	5389
SK4928	3987
SK4929	0235
SK4930	1417
SK4931	2636
SK4932	2640
SK4933	5426
SK4934	5142
SK4935	4932
SK4936	3552
SK4937	0591
SK4939	0475
SK5000	OEM
SK5001	1145
SK5002	0499
SK5003	6451
SK5006	6704
SK5010	4630
SK5010A	4630
SK5012	3241
SK5014	2523
SK5016	2280
SK5020	0707
SK5028	0729
SK5030	0737
SK5031	0742
SK5032	6706

If replacement code is OEM, contact original manufacturer for replacement.

DEVICE TYPE	REPL CODE
SK5036	0031
SK5040	1928
SK5042	6707
SK5044	6708
SK5046	6710
SK5048	6711
SK5050	6712
SK5051	0015
SK5052	6713
SK5053	6714
SK5054	6715
SK5060	1236
SK5101	6196
SK5114	6719
SK5131	0578
SK5137	0594
SK5173	0154
SK5176	4545
SK5184A	0142
SK5188	0508
SK5189	0511
SK5191	OEM
SK5236	6727
SK5237	0717
SK5238	0308
SK5538	5322
SK5545	0555
SK5550	2352
SK5551	2353
SK5552	2356
SK5555	4455
SK5556	4456
SK5557	4482
SK5579	0908
SK5588	1064
SK5589	1317
SK5590	1891
SK5704	6775
SK5797	0037
SK5798	0037
SK5801	0016
SK5814A	0142
SK5915	0016
SK6006	OEM
SK6117	6804
SK6118	6805
SK6345	0037
SK6346	0037
SK6347	0037
SK6347A	0037
SK6402	5626
SK6477	OEM
SK6502	0741
SK6554	1023
SK6555	0501
SK6556	1036
SK6557	1777
SK6558	1046
SK6559	1789
SK6560	3928
SK6561	3531
SK6601	1755
SK6606	4416
SK6607	4417
SK6615	1584
SK6616	1665
SK6621	2188
SK6622	0567
SK6624	0762
SK6627	0740
SK6629	0764
SK6630	1089
SK6631	0766
SK6632	4535
SK6634	1099
SK6635	1100
SK6641	0731
SK6642	0712
SK6648	0750
SK6649	4855
SK6650	2521
SK6651	6827
SK6652	0397
SK6654	0611
SK6656	0952
SK6658	1512
SK6681	3147
SK6682	3148
SK6683	3150
SK6685	3152
SK6686	3153
SK6703	1084
SK6707	6833
SK6708	0572
SK6709	0935
SK6717	1704
SK6718	4031
SK6719	0679
SK6720	0608
SK6722	0610
SK6724	0468
SK6725	0641
SK6727	0217
SK6729	0658
SK6730	1084
SK6732	1086
SK6734	1102
SK6736	1889
SK6740	2383

DEVICE TYPE	REPL CODE
SK6744	2843
SK6745	1765
SK6746	1814
SK6747	5319
SK6748	4148
SK6749	1725
SK6752	1396
SK6753	1259
SK6754	0447
SK6760	0192
SK6761	0159
SK6762	0096
SK6768	0655
SK6769	0020
SK6771	0526
SK6772	0108
SK6774	0428
SK6775	1871
SK6777	1900
SK6778	0738
SK6780	0459
SK6781	0587
SK6786	0769
SK6787	0483
SK6788	6834
SK6791	2449
SK6792	0435
SK6794	1487
SK6796	0687
SK6810	2086
SK6848	OEM
SK7003	OEM
SK7038	1000
SK7039	4523
SK7042	2081
SK7043	1073
SK7046	2088
SK7047	1138
SK7048	0612
SK7049	1188
SK7051	0985
SK7055	0990
SK7059	0997
SK7063	0515
SK7066	0206
SK7067	1008
SK7068	0586
SK7069	0948
SK7074	2778
SK7075	1745
SK7076	2813
SK7077	1783
SK7078	2845
SK7079	1818
SK7086	0131
SK7087	0545
SK7090	0145
SK7091	0550
SK7092	5514
SK7093	1238
SK7096	2900
SK7097	1241
SK7100	3751
SK7101	2365
SK7104	1112
SK7105	2376
SK7108	3768
SK7109	1264
SK7110	3011
SK7111	0449
SK7145	1577
SK7153	3294
SK7171	0534
SK7175	0765
SK7179	0538
SK7181	1630
SK7182	2901
SK7183	0254
SK7191	0258
SK7198	1132
SK7199	1119
SK7202	0044
SK7203	0281
SK7206	1530
SK7207	1520
SK7208	0029
SK7209	1847
SK7210	0601
SK7211	1851
SK7220	3762
SK7221	2665
SK7222	2653
SK7223	2695
SK7226	2657
SK7227	2768
SK7234	2006
SK7235	2907
SK7240	2682
SK7241	3051
SK7242	3900
SK7243	3117
SK7244	2655
SK7245	3208
SK7248	4306
SK7249	3771
SK7250	1560
SK7251	2179
SK7254	2840
SK7255	2182

DEVICE TYPE	REPL CODE
SK7260	2850
SK7261	2191
SK7264	2871
SK7265	2197
SK7268	2833
SK7269	2216
SK7272	2473
SK7273	2340
SK7301	6862
SK7303	5739
SK7305	6863
SK7307	6864
SK7313	5736
SK7315	5749
SK7317	6869
SK7330	5905
SK7331	4971
SK7332	3889
SK7333	5731
SK7352	5738
SK7353	5734
SK7354	0599
SK7355	1347
SK7356	1986
SK7357	1905
SK7358	0657
SK7359	0193
SK7362	5026
SK7363	2559
SK7366	2615
SK7367	2566
SK7376	3517
SK7377	5012
SK7378	3541
SK7379	5577
SK7400	0232
SK7401	0268
SK7402	0314
SK7403	0335
SK7404	0361
SK7405	0385
SK7406	1206
SK7407	1339
SK7408	0467
SK7409	0492
SK7410	0512
SK7411	0527
SK7412	2242
SK7413	1441
SK7414	2243
SK7416	1349
SK7417	1352
SK7420	0697
SK7422	4603
SK7423	3466
SK7425	3477
SK7426	0803
SK7427	0818
SK7428	4180
SK7430	0873
SK7432	0899
SK7433	4193
SK7437	3520
SK7438	0996
SK7439	5836
SK7440	1024
SK7441	1038
SK7442	1052
SK7443	1060
SK7444	1072
SK7445	1082
SK7446	1098
SK7447	1108
SK7448	1125
SK7450	0743
SK7451	1168
SK7453	1185
SK7454	1202
SK7460	1275
SK7470	1404
SK7472	1426
SK7473	1172
SK7474	1313
SK7475	1432
SK7476	1158
SK7480	1535
SK7481	3121
SK7483	0117
SK7485	0374
SK7486	1368
SK7489	5455
SK7490	1208
SK7492	0834
SK7493	0569
SK7494	1702
SK7495	1486
SK7496	1715
SK7497	2940
SK7600	4291
SK7601	4476
SK7602	5181
SK7603	5108
SK7604	2694
SK7605	3816
SK7606	2122
SK7607	3716
SK7608	2794
SK7609	2654
SK7610	5162

DEVICE TYPE	REPL CODE
SK7611	5195
SK7612	2728
SK7613	6888
SK7614	3974
SK7615	3686
SK7616	0814
SK7617	1118
SK7618	3747
SK7619	4346
SK7620	1318
SK7621	6854
SK7622	4728
SK7623	4290
SK7626	3853
SK7627	1582
SK7628	6784
SK7629	5544
SK7630	4714
SK7632	5786
SK7633	5874
SK7634	0872
SK7635	2826
SK7636	4552
SK7637	3069
SK7638	3068
SK7641	3741
SK7642	2470
SK7643	2169
SK7644	5280
SK7645	3791
SK7646	3502
SK7647	3503
SK7648	2926
SK7649	4665
SK7650	4796
SK7651	5023
SK7652	4840
SK7653	2042
SK7654	3239
SK7655	1247
SK7656	4830
SK7657	5159
SK7658	1951
SK7659	4680
SK7660	4681
SK7661	4653
SK7662	4688
SK7663	4726
SK7664	6891
SK7665	4372
SK7666	0799
SK7667	4813
SK7668	4818
SK7669	4406
SK7670	5205
SK7671	6892
SK7672	3772
SK7673	4410
SK7674	3725
SK7675	6110
SK7676	0663
SK7677	5172
SK7678	1676
SK7679	4520
SK7680	6893
SK7681	4986
SK7682	1896
SK7683	1879
SK7686	3976
SK7687	2842
SK7690	5055
SK7691	5052
SK7692	3997
SK7693	6135
SK7694	6143
SK7695	6321
SK7696	6319
SK7697	6142
SK7698	1883
SK7699	6894
SK7700	6386
SK7701	6823
SK7702	6895
SK7703	1572
SK7704	4834
SK7705	2116
SK7706	4725
SK7707	3711
SK7708	3793
SK7709	6896
SK7710	2718
SK7711	2304
SK7712	2082
SK7713	5215
SK7714	5564
SK7715	2040
SK7717	4002
SK7718	2107
SK7719	2103
SK7720	3518
SK7721	3798
SK7722	4642
SK7723	4637
SK7724	4641
SK7725	1104
SK7726	4601
SK7727	2688
SK7728	0403
SK7729	4274

DEVICE TYPE	REPL CODE
SK7730	2480
SK7731	3795
SK7732	2759
SK7733	3967
SK7734	1442
SK7735	2896
SK7736	2666
SK7737	3986
SK7738	6898
SK7739	5293
SK7740	4329
SK7741	5296
SK7742	5297
SK7743	0351
SK7744	5616
SK7747	2279
SK7751	1543
SK7752	3821
SK7753	1235
SK7754	2404
SK7755	2406
SK7756	2509
SK7757	4557
SK7758	4563
SK7759	4564
SK7760	2882
SK7761	2573
SK7762	2577
SK7763	2584
SK7764	2587
SK7765	2588
SK7766	4575
SK7767	0638
SK7768	0964
SK7769	0043
SK7770	3958
SK7771	2630
SK7772	3670
SK7773	0389
SK7774	4593
SK7775	4420
SK7776	2594
SK7778	2604
SK7779	4055
SK7781	4631
SK7782	4086
SK7783	4632
SK7784	2284
SK7785	4069
SK7786	4638
SK7787	4065
SK7789	4648
SK7790	5706
SK7791	3998
SK7792	6901
SK7793	4761
SK7794	4763
SK7795	4765
SK7796	4773
SK7797	2530
SK7798	4779
SK7799	3228
SK7800	4785
SK7801	4560
SK7802	4820
SK7803	4836
SK7804	2913
SK7805	3713
SK7806	3823
SK7807	3824
SK7808	3333
SK7809	2374
SK7810	5238
SK7811	2385
SK7812	0701
SK7813	3962
SK7814	4736
SK7815	5107
SK7816	5890
SK7817	4146
SK7827	0759
SK7841	3685
SK7900	3407
SK7901	5911
SK7951	6913
SK7952	0343
SK8223	OEM
SK8251	0016
SK8261	0233
SK8604	OEM
SK8937	0224
SK9000	0023
SK9001	1802
SK9002	1097
SK9003	0110
SK9004	0953
SK9005	0242
SK9006	1747
SK9007	0540
SK9008	1771
SK9009	0817
SK9010	0922
SK9011	2310
SK9012	1491
SK9013	3079
SK9014	0443
SK9015	2916
SK9016	2623
SK9017	1121

DEVICE TYPE	REPL CODE
SK9018	4744
SK9019	3920
SK9020	3893
SK9021	0301
SK9022	1649
SK9023	0949
SK9024	3026
SK9025	2801
SK9026	3720
SK9027	3788
SK9028	3719
SK9029	2842
SK9030	1328
SK9031	6994
SK9032	6995
SK9033	3701
SK9034	4144
SK9035	3860
SK9036	2333
SK9037	2443
SK9038	2041
SK9039	1852
SK9040	2433
SK9041	0563
SK9042	0525
SK9043	3862
SK9044	5071
SK9045	4747
SK9046	4751
SK9047	3787
SK9048	1580
SK9049	4758
SK9050	4759
SK9051	4752
SK9052	OEM
SK9053	6621
SK9054	4079
SK9055	5562
SK9056	3738
SK9057	6997
SK9058	4495
SK9059	4060
SK9060	6998
SK9061	3609
SK9062	4162
SK9063	1599
SK9064	4746
SK9065	6999
SK9067	1916
SK9068	7001
SK9069	1482
SK9070	1890
SK9071	1897
SK9072	2888
SK9073	1910
SK9074	2633
SK9075	0277
SK9076	1051
SK9077	4936
SK9078	4937
SK9079	4939
SK9080	4940
SK9081	4941
SK9082	3648
SK9082A	4585
SK9083A	4592
SK9084	2730
SK9084A	2730
SK9085	0728
SK9086	1791
SK9087	2372
SK9088	2373
SK9089	3781
SK9090	1554
SK9091	0133
SK9092	1007
SK9093	1134
SK9094	1261
SK9095	4935
SK9096	1287
SK9097	1296
SK9098	1217
SK9099	5569
SK9100	5569
SK9101	1643
SK9102	5604
SK9103	1673
SK9104	7003
SK9105	1589
SK9106	5362
SK9107	3370
SK9108	3540
SK9109	1991
SK9110	4013
SK9111	2626
SK9112	0060
SK9113	1308
SK9114	2045
SK9115	2468
SK9116	7004
SK9117	7005
SK9118	0392
SK9119	0055
SK9120	4233
SK9121	1893
SK9122	1669
SK9123	2136
SK9124	1175
SK9125	4405

DEVICE TYPE	REPL CODE
SK9126	3099
SK9127	3086
SK9128	2222
SK9129	2517
SK9130	3423
SK9131	1506
SK9132	0148
SK9133	3058
SK9134	0130
SK9136	1681
SK9137	1386
SK9138	0477
SK9139	2851
SK9140	1341
SK9141	0270
SK9142	0595
SK9143	0791
SK9144	2562
SK9145	2567
SK9146	3621
SK9147	6039
SK9148	2945
SK9149	3133
SK9150	2518
SK9150A	2518
SK9152	7008
SK9153	6948
SK9154	7009
SK9155	6376
SK9156	6374
SK9157	0325
SK9158	0983
SK9159	0844
SK9160	3628
SK9161	2950
SK9162	3130
SK9163	1155
SK9164	1392
SK9165	5060
SK9166	0896
SK9167	2246
SK9168	2791
SK9169	1828
SK9170	3395
SK9171	1919
SK9172	3024
SK9173	2823
SK9174	2890
SK9175	1797
SK9176	1800
SK9177	1301
SK9178	1516
SK9179	1329
SK9180	0553
SK9181	3451
SK9182	4661
SK9183	4089
SK9184	3246
SK9185	0904
SK9186	4734
SK9187	1022
SK9188	1446
SK9189	4685
SK9190	2158
SK9191	5110
SK9192	3057
SK9193	4854
SK9194	1201
SK9195	2184
SK9196	4808
SK9197	2533
SK9198	0683
SK9199	0675
SK9200	2748
SK9201	3836
SK9202	3784
SK9203	4950
SK9204	4954
SK9205	3129
SK9206	3790
SK9207	3990
SK9208	3196
SK9209	2318
SK9210	3063
SK9211	0865
SK9212	2063
SK9213	6030
SK9214	4165
SK9215	2564
SK9216	2736
SK9217	2574
SK9218	2570
SK9219	4508
SK9221	1836
SK9222	5041
SK9224	3827
SK9225	0496
SK9226	4003
SK9227	5809
SK9228	4659
SK9229	0286
SK9230	0241
SK9231	4581
SK9232	1875
SK9233	3546
SK9234	0482
SK9236	1369
SK9237	0867
SK9239	3500

DEVICE TYPE	REPL CODE
SK9241	2865
SK9242	2775
SK9243	2788
SK9244	2105
SK9245	3012
SK9246	0767
SK9247	0918
SK9248	2022
SK9249	1429
SK9250	4826
SK9251	4727
SK9252	4573
SK9253	3533
SK9254	3534
SK9255	3535
SK9256	3536
SK9257	4005
SK9258	2330
SK9259	4029
SK9260	4479
SK9261	3060
SK9262	4540
SK9263	7014
SK9264	5419
SK9265	4389
SK9266	4038
SK9267	4656
SK9268	4657
SK9269	5111
SK9270	3489
SK9271	3512
SK9272	0939
SK9273	2662
SK9274	2873
SK9275	2964
SK9276	3270
SK9277	6589
SK9278	7016
SK9279	7017
SK9280	7018
SK9281	3397
SK9282	7019
SK9283	3826
SK9284	2153
SK9285	3139
SK9286	1547
SK9287	2967
SK9288	7020
SK9289	7021
SK9290	2095
SK9291	2089
SK9292	0505
SK9293	0710
SK9294	0863
SK9295	4252
SK9296	0484
SK9297	2548
SK9298	3744
SK9299	5094
SK9300	3661
SK9301	7022
SK9302	7023
SK9303	3865
SK9304	3866
SK9305	3662
SK9306	3658
SK9307	3659
SK9308	3864
SK9311	0067
SK9312	4805
SK9313	4097
SK9314	5095
SK9315	2625
SK9316	7025
SK9317	3110
SK9318	7027
SK9319	7028
SK9320	2054
SK9321	4134
SK9322	3980
SK9323	7030
SK9324	5208
SK9325	0976
SK9326	4750
SK9327	OEM
SK9328	1030
SK9329	4465
SK9330	5036
SK9331	3963
SK9332	2000
SK9333	2723
SK9334	2004
SK9335	5287
SK9336	5289
SK9337	2254
SK9338	2462
SK9339	2838
SK9340	2863
SK9341	3393
SK9342	3813
SK9343	6043
SK9344	2737
SK9345	7031
SK9346	7032
SK9347	7033
SK9348	7034
SK9349	OEM
SK9350	7035
SK9351	3758

DEVICE TYPE	REPL CODE
SK9352	0261
SK9353	3359
SK9354	3351
SK9355	4787
SK9356	4782
SK9357	0969
SK9358	6653
SK9359	2879
SK9360	4112
SK9361	4713
SK9362	1085
SK9362A	1085
SK9363	1648
SK9364	2069
SK9365	3606
SK9366	1165
SK9367	0718
SK9368	1480
SK9369	0892
SK9370	3703
SK9371	2844
SK9372	3007
SK9373	1342
SK9374	2993
SK9375	6142
SK9376	5055
SK9377	4478
SK9378	4473
SK9379	0912
SK9380	0069
SK9381	3977
SK9382	3978
SK9383	3983
SK9384	1057
SK9385	OEM
SK9386	3995
SK9387	0330
SK9388	7040
SK9389	3279
SK9390	5477
SK9391	3115
SK9392	0219
SK9393	7042
SK9394	4006
SK9395	4715
SK9396	5104
SK9397	3956
SK9398	4672
SK9399	4429
SK9400	OEM
SK9401	4890
SK9402	1988
SK9403	0332
SK9404	1590
SK9405	0603
SK9406	4684
SK9407	7043
SK9408	3350
SK9409	3498
SK9410	6777
SK9411	0223
SK9412	3764
SK9413	0194
SK9415	3111
SK9417	3225
SK9418	3406
SK9419	5018
SK9420	5064
SK9421	3627
SK9422	1541
SK9423	2924
SK9424	5057
SK9425	2411
SK9426	5058
SK9427	2947
SK9428	5059
SK9429	0046
SK9430	3630
SK9431	5307
SK9432	0008
SK9433	3203
SK9434	3617
SK9435	0264
SK9436	3538
SK9437	3539
SK9438	3373
SK9439	3374
SK9440	1394
SK9441	2448
SK9442	3025
SK9443	3519
SK9444	2382
SK9445	7047
SK9446	2276
SK9447	3581
SK9448	3226
SK9449	3646
SK9450	2997
SK9451	3165
SK9452	3014
SK9453	7048
SK9454	7049
SK9455	0400
SK9455A	0400
SK9456	0635
SK9456A	0635
SK9457	5450
SK9458	5464
SK9459	2860

DEVICE TYPE	REPL CODE
SK9460	1758
SK9461	2797
SK9462	1708
SK9463	7051
SK9464	0811
SK9465	7052
SK9466	7053
SK9467	5479
SK9468	5034
SK9469	1597
SK9470	1526
SK9471	3750
SK9472	3754
SK9473	7054
SK9474	7055
SK9475	2366
SK9476	3360
SK9477	1270
SK9478	3516
SK9479	5350
SK9480	1061
SK9481	3723
SK9482	5341
SK9483	3001
SK9484	3038
SK9485	2766
SK9486	7056
SK9487	5481
SK9488	2784
SK9489	5349
SK9495	0327
SK9500	6843
SK9501	5726
SK9502	7058
SK9503	5948
SK9504	5913
SK9505	6343
SK9506	7059
SK9507	2271
SK9508	5524
SK9509	1465
SK9600	2070
SK9601	5956
SK9602	2208
SK9603	2946
SK9604	2312
SK9605	1974
SK9606	2884
SK9607	1198
SK9608	7067
SK9609	2835
SK9610	3563
SK9611	3565
SK9612	3567
SK9613	3568
SK9614	3569
SK9615	5202
SK9616	3570
SK9617	3572
SK9618	2719
SK9619	2720
SK9620	2526
SK9621	3041
SK9622	7070
SK9623	3578
SK9624	2170
SK9625	5303
SK9626	3194
SK9627	3580
SK9628	3582
SK9629	3583
SK9630	3584
SK9631	3586
SK9632	3587
SK9633	2506
SK9634	2039
SK9635	3560
SK9636	2091
SK9637	1420
SK9638	2093
SK9639	3589
SK9640	2868
SK9641	3591
SK9642	3631
SK9643	2545
SK9644	3421
SK9645	3326
SK9646	3632
SK9647	3321
SK9648	3322
SK9649	3119
SK9650	3636
SK9651	1354
SK9652	3638
SK9653	3640
SK9654	3641
SK9655	3642
SK9656	3643
SK9657	3644
SK9658	0689
SK9659	2702
SK9660	3576
SK9661	0014
SK9662	5061
SK9663	2223
SK9664	0582
SK9665	2484
SK9666	2471
SK9667	0359

DEVICE TYPE	REPL CODE
SK9668	0009
SK9669	4104
SK9670	3608
SK9671	1806
SK9672	7076
SK9700	4658
SK9701	4516
SK9702	0514
SK9703	2909
SK9704	1623
SK9705	5125
SK9706	5199
SK9707	0662
SK9708	5051
SK9709	0615
SK9710	5325
SK9711	1456
SK9712	5819
SK9713	4102
SK9715	2026
SK9716	0583
SK9717	5256
SK9718	1880
SK9719	5335
SK9720	0645
SK9721	5304
SK9722	5258
SK9723	5328
SK9724	5259
SK9725	5218
SK9726	3902
SK9727	3925
SK9728	4310
SK9729	4927
SK9730	4129
SK9731	1529
SK9732	3437
SK9733	5260
SK9734	5261
SK9735	5262
SK9736	4925
SK9737	5271
SK9738	4141
SK9739	5263
SK9740	4117
SK9741	1032
SK9742	0887
SK9743	3955
SK9744	3959
SK9745	3097
SK9746	1028
SK9747	4266
SK9748	0539
SK9749	0391
SK9750	0462
SK9751	4174
SK9752	0732
SK9753	2957
SK9754	4377
SK9755	4380
SK9756	4381
SK9757	1437
SK9758	6418
SK9759	6416
SK9760	1533
SK9761	2745
SK9762	3348
SK9763	1166
SK9764	4496
SK9765	4383
SK9766	3159
SK9767	4137
SK9768	6295
SK9769	5011
SK9770	1291
SK9771	3680
SK9772	3679
SK9773	3676
SK9774	1545
SK9775	2515
SK9776	2234
SK9777	1727
SK9778	4197
SK9779	4188
SK9780	2836
SK9781	2174
SK9782	2171
SK9783	2163
SK9784	2143
SK9785	2118
SK9786	1729
SK9787	4441
SK9788	1481
SK9789	1632
SK9790	1863
SK9791	4667
SK9792	4651
SK9793	4647
SK9794	4181
SK9795	2692
SK9796	3967
SK9797	0393
SK9798	1021
SK9799	6365
SK9800	3672
SK9805	3031
SK9806	2112
SK9807	1244
SK9808	5612

DEVICE TYPE	REPL CODE
SK9809	0581
SK9810	2643
SK9811	1434
SK9812	2396
SK9813	4970
SK9814	1793
SK9815	3363
SK9816	2808
SK9817	2787
SK9818	2252
SK9822	3056
SK9823	3528
SK9825	5022
SK9827	1272
SK9828	1231
SK9829	4604
SK9830	1659
SK9831	4145
SK9832	0522
SK9833	1316
SK9834	1262
SK9835	3256
SK9837	1357
SK9838	4379
SK9839	5324
SK9840	3938
SK9841	4548
SK9842	1141
SK9843	5649
SK9844	5646
SK9846	5645
SK9847	3684
SK9848	3442
SK9849	6420
SK9850	3236
SK9851	0304
SK9852	1410
SK9853	0325
SK9854	3521
SK9855	3357
SK9856	5352
SK9857	5353
SK9858	1127
SK9859	1664
SK9860	3551
SK9861	1243
SK9862	3181
SK9863	3942
SK9864	3320
SK9866	4353
SK9867	5251
SK9868	4137
SK9870	2364
SK9871	3913
SK9873	4194
SK9874	3386
SK9875	4385
SK9876	4422
SK9877	3254
SK9878	2269
SK9879	1604
SK9880	5257
SK9881	3027
SK9882	3365
SK9883	3884
SK9884	3161
SK9885	3817
SK9886	6336
SK9888	5815
SK9889	5270
SK9890	5267
SK9891	2512
SK9892	5269
SK9893	5268
SK9894	1254
SK9895	5273
SK9896	5272
SK9897	5265
SK9898	3355
SK9899	3340
SK9902	2405
SK9903	4968
SK9906	4969
SK9907	5314
SK9908	5315
SK9911	0450
SK9912	1824
SK9913	4868
SK9914	4345
SK9915	4344
SK9916	3746
SK9918	1500
SK9919	4011
SK9920	5393
SK9921	4475
SK9922	5603
SK9923	3789
SK9924	5693
SK9925	2011
SK9926	5704
SK9927	4864
SK9928	6034
SK9929	4500
SK9930	6041
SK9931	3303
SK9932	4163
SK9933	4220
SK9934	0735
SK9935	2542

DEVICE TYPE	REPL CODE
SK9936	5088
SK9937	1090
SK9938	0546
SK9939	3192
SK9940	2203
SK9941	3895
SK9942	3749
SK9943	3896
SK9944	0907
SK9945	2870
SK9946	3083
SK9947	5048
SK9948	3253
SK9949	0699
SK9950	0489
SK9951	2066
SK9952	2553
SK9953	2067
SK9954	2068
SK9955	5236
SK9957	0832
SK9958	4172
SK9959	0898
SK9960	4130
SK9961	3482
SK9962	1892
SK9963	0667
SK9964	3645
SK9965	3367
SK9967	0589
SK9968	0883
SK9969	0295
SK9970	0248
SK9971	1005
SK9972	1045
SK9975	0163
SK9976	1329
SK9977	4857
SK9978	3776
SK9980	5225
SK9981	5229
SK9982	5793
SK9983	5233
SK9984	5474
SK9985	2367
SK9986	5345
SK9987	4151
SK9989	2712
SK9990	5191
SK9991	3436
SK9992	2168
SK9993	0631
SK9994	4650
SK9995	1641
SK9996	3445
SK9997	4693
SK9998	1903
SK10000	2908
SK10001	5241
SK10002	2178
SK10003	2090
SK10004	3547
SK10005	3949
SK10006	4184
SK10007	4675
SK10008	5226
SK10010	5192
SK10012	3828
SK10013	5227
SK10015	3388
SK10016	3731
SK10017	4384
SK10018	3335
SK10019	2686
SK10021	2025
SK10022	5232
SK10023	4258
SK10024	2829
SK10025	2841
SK10026	2839
SK10028	5234
SK10029	3361
SK10030	1596
SK10031	1389
SK10032	OEM
SK10033	5239
SK10035	5242
SK10036	3523
SK10037	5243
SK10038	1094
SK10041	4788
SK10042	0950
SK10044	5247
SK10045	5804
SK10047	5281
SK10048	4506
SK10049	4886
SK10050	4509
SK10051	1721
SK10052	2723
SK10053	3650
SK10054	5290
SK10055	3881
SK10056	3573
SK10058	5837
SK10059	3859
SK10060	4867
SK10061	5531
SK10062	6080

DEVICE TYPE	REPL CODE
SK10063	2742
SK10064	4105
SK10065	4136
SK10066	2893
SK10067	3062
SK10069	2565
SK10070	4169
SK10071	4273
SK10072	3233
SK10073	2881
SK10074	1976
SK10075	3118
SK10076	3820
SK10077	3434
SK10079	2816
SK10080	5230
SK10081	0107
SK10082	0557
SK10083	3343
SK10084	2111
SK10085	2346
SK10086	4869
SK10087	2539
SK10088	2129
SK10089	3892
SK10090	3890
SK10091	1959
SK10092	1968
SK10093	2954
SK10094	0013
SK10095	2919
SK10096	4498
SK10097	1733
SK10098	1752
SK10099	0724
SK10100	1742
SK10101	6116
SK10102	5366
SK10103	4489
SK10104	5364
SK10105	5365
SK10106	3468
SK10107	3379
SK10108	0981
SK10109	0703
SK10110	3478
SK10111	3272
SK10112	5369
SK10113	5370
SK10114	3471
SK10115	5371
SK10116	5342
SK10117	4334
SK10118	3005
SK10119	0544
SK10120	4077
SK10121	4078
SK10122	2428
SK10123	2420
SK10124	2323
SK10125	3143
SK10126	1466
SK10127	5360
SK10128	3088
SK10129	5899
SK10130	1668
SK10131	2176
SK10132	5361
SK10133	5363
SK10134	0282
SK10136	3494
SK10137	3136
SK10138	3217
SK10139	1699
SK10140	5276
SK10141	5277
SK10142	5818
SK10143	3953
SK10144	3574
SK10145	3305
SK10146	4041
SK10147	3250
SK10148	2921
SK10149	5275
SK10150	3156
SK10151	3495
SK10152	5821
SK10153	6122
SK10154	6098
SK10155	0486
SK10156	5823
SK10158	3927
SK10159	3773
SK10161	3488
SK10162	3888
SK10163	3604
SK10164	2419
SK10165	3023
SK10166	5278
SK10168	1483
SK10169	1439
SK10170	2303
SK10171	2413
SK10172	4143
SK10173	2334
SK10174	2417
SK10175	2765
SK10176	3837
SK10177	2703

If replacement code is OEM, contact original manufacturer for replacement.

DEVICE TYPE	REPL CODE
SK10178	0112
SK10179	5662
SK10180	6128
SK10181	5356
SK10182	5000
SK10183	4629
SK10184	5681
SK10185	1807
SK10186	3914
SK10187	6129
SK10188	6130
SK10189	6132
SK10190	5865
SK10191	1472
SK10192	6133
SK10193	3114
SK10194	2133
SK10195	3172
SK10196	1169
SK10197	3191
SK10198	1767
SK10199	3201
SK10200	1932
SK10201	1915
SK10202	1952
SK10203	0568
SK10204	1972
SK10205	0831
SK10206	1987
SK10207	1405
SK10208	1408
SK10209	5046
SK10210	2315
SK10211	0749
SK10212	2792
SK10213	1226
SK10214	1204
SK10215	2159
SK10216	0479
SK10217	4922
SK10219	3309
SK10220	0202
SK10221	5185
SK10223	5187
SK10226	2535
SK10230	5194
SK10232	3753
SK10233	5197
SK10234	2595
SK10238	5780
SK10239	5201
SK10241	4553
SK10242	5212
SK10243	3050
SK10244	2425
SK10246	5214
SK10247	5790
SK10252	5785
SK10253	2139
SK10254	1042
SK10255	1004
SK10256	1034
SK10257	1110
SK10258	1148
SK10259	2932
SK10260	0963
SK10261	2936
SK10262	3439
SK10263	3053
SK10264	3055
SK10265	0875
SK10266	5323
SK10267	5326
SK10268	1967
SK10269	6139
SK10271	3035
SK10272	5144
SK10273	1381
SK10274	4307
SK10275	5157
SK10276	4081
SK10277	1701
SK10278	0506
SK10279	4951
SK10280	4952
SK10281	5864
SK10282	0487
SK10283	5866
SK10284	5867
SK10285	5868
SK10286	4955
SK10287	4956
SK10288	4957
SK10289	4959
SK10290	4536
SK10291	4961
SK10292	4962
SK10293	4963
SK10294	4965
SK10295	3982
SK10296	1693
SK10297	2357
SK10298	5888
SK10299	4590
SK10300	3462
SK10301	3460
SK10302	0278
SK10303	5769
SK10304	5096
SK10305	2989
SK10309	2371
SK10310	6141
SK10311	1224
SK10312	3689
SK10313	3255
SK10314	3049
SK10315	3247
SK10316	3376
SK10317	3163
SK10318	3169
SK10320	3739
SK10321	0516
SK10322	3232
SK10324	2628
SK10325	5123
SK10326	5124
SK10327	2108
SK10328	4988
SK10330	5421
SK10331	5424
SK10332	5425
SK10334	5427
SK10335	5428
SK10336	5430
SK10337	5429
SK10400	1047
SK10401	3908
SK10402	4576
SK10403	4987
SK10404	5674
SK10405	5079
SK10406	5974
SK10407	5078
SK10408	3277
SK10409	5878
SK10410	3708
SK10411	6151
SK10412	6152
SK10413	0059
SK10414	1680
SK10415	5813
SK10416	0652
SK10417	2311
SK10418	4364
SK10419	2994
SK10420	4574
SK10421	2878
SK10422	1600
SK10423	5005
SK10424	0871
SK10425	1635
SK10426	0853
SK10427	1251
SK10428	1609
SK10429	1205
SK10430	1610
SK10431	2137
SK10432	1614
SK10433	2251
SK10434	1798
SK10435	2192
SK10436	2294
SK10437	2298
SK10438	5812
SK10439	5444
SK10440	5063
SK10441	5447
SK10442	5448
SK10443	6154
SK10444	5397
SK10445	5443
SK10446	5446
SK10447	4158
SK10448	4168
SK10449	0585
SK10450	2771
SK10451	4551
SK10452	3202
SK10453	2606
SK10454	2869
SK10455	2355
SK10456	2117
SK10457	6155
SK10458	6156
SK10459	0422
SK10460	6157
SK10460-5668A	OEM
SK10461	3461
SK10462	3842
SK10463	3843
SK10464	3846
SK10465	2455
SK10466	3593
SK10467	2202
SK10468	0793
SK10469	6159
SK10470	5745
SK10471	4189
SK10472	4241
SK10473	6028
SK10474	4953
SK10475	6160
SK10476	4532
SK10477	0523
SK10478	3690
SK10479	3515
SK10480	6161
SK10481	3378
SK10482	4550
SK10483	6162
SK10484	3635
SK10485	6163
SK10486	6164
SK10488	4930
SK10489	6165
SK10493	4355
SK10494	5806
SK10496	5826
SK10499	4118
SK10500	5827
SK10501	5831
SK10502	3080
SK10505	3951
SK10507	5833
SK10508	5834
SK10509	5825
SK10510	1366
SK10511	3887
SK10512	5796
SK10513	5992
SK10514	3312
SK10515	5298
SK10516	5294
SK10518	2094
SK10519	1898
SK10520	5391
SK22100	3042
SK22101	3048
SK22102	3052
SK22859	5310
SK-31024-3	0144
SK31024-3	0144
SK40061	1822
SK40098B	0432
SK40100B	3952
SK40101B	4017
SK40102B	4061
SK40103B	4091
SK40104B	4108
SK40105B	4123
SK40106B	3626
SK40107B	4153
SK40108B	4182
SK40109B	4210
SK40110B	4229
SK40114B	4298
SK40116	3106
SK40117B	6514
SK40147B	5102
SK40160B	1359
SK40161	1373
SK40161B	1373
SK40162B	1388
SK40163B	1407
SK40174B	1551
SK40175B	1528
SK40181B	5638
SK40182B	4660
SK40192B	1764
SK40193B	1776
SK40194B	1769
SK40195B	1784
SK40208B	6015
SK40251B	OEM
SK40257B	6415
SK74107	0942
SK74109	0968
SK74110	0987
SK74111	1002
SK74121	0175
SK74122	1139
SK74123	1157
SK74125	1182
SK74126	1193
SK74128	1219
SK74132	1271
SK74141	1377
SK74142	1390
SK74143	1409
SK74144	1418
SK74145	0619
SK74146	1795
SK74147	1451
SK74150	1492
SK74151	1495
SK74153	1539
SK74154	1555
SK74155	1576
SK74156	1592
SK74157	1605
SK74158	4446
SK74160	1631
SK74161	1645
SK74162	1013
SK74163	1666
SK74164	0734
SK74165	1685
SK74166	0231
SK74170	1722
SK74173	1766
SK74174	1770
SK74175	1787
SK74176	1795
SK74177	1803
SK74178	1813
SK74179	1820
SK74180	1829
SK74181	1842
SK74182	1856
SK74190	1912
SK74191	1917
SK74192	1921
SK74193	1926
SK74195	1943
SK74196	1950
SK74197	1956
SK74198	1964
SK74199	1971
SK74221	2142
SK74251	2299
SK74290	2612
SK74293	2644
SK74365	1459
SK74366	1471
SK74367	1488
SK74368	1508
SK74390	3240
SK74500	OEM
SK74511	OEM
SK74574	OEM
SK75454B	1289
SK75493	5021
SK92189	OEM
SK707380	0593
SKA0030	0079
SKA13	OEM
SKA17	OEM
SKA1079	0126
SKA1080	0016
SKA1117	0016
SKA1279	0037
SKA1395	0016
SKA1416	0127
SKA-4061	0037
SKA-4074	0127
SKA4074	0144
SKA-4075	0127
SKA4075	0144
SKA-4076	0224
SKA4076	0144
SKA4129	0037
SKA4141	0016
SKA4410	0595
SKA4525	0144
SKA-4590	0224
SKA4616	0086
SKA4621	0126
SKA4768	0007
SKA-4802	0111
SKA-5248	0127
SKA5248	0144
SKA-5541	0007
SKA5571	0285
SKA-5658	0264
SKA-5886	0144
SKA5886	0144
SKA6250	0037
SKA-6256	0079
SKA-6437	0079
SKA6556	0016
SKA-8105	0079
SKA9013	0144
SKA9096	0144
SKB8339	0016
SKH866A	OEM
SKLS30	OEM
SKMV10B	1691
SKMV10C	1694
SKMV14B	2008
SKMV14C	2009
SKMV15B	1314
SKMV15C	1319
SKMV17B	2325
SKMV17C	2326
SKMV18A	2600
SKMV20B	2790
SKMV20C	2796
SKMV25A	4367
SKMV25B	3073
SKMV30A	5744
SKMV30C	3075
SKMV35B	2935
SKMV35C	2941
SKMV40B	4263
SKMV40C	3078
SKMV50B	4472
SKMV50C	3085
SKMV60A	4222
SKMV60B	3090
SKMV75A	4921
SKMV75B	3104
SKMV95A	5157
SKMV95B	3116
SKMV115B	4135
SKMV115C	3123
SKMV130G	0840
SKMV130H	1374
SKMV130J	0838
SKMV130K	0830
SKMV130L	4268
SKMV130P	2543
SKMV132A	2627
SKMV150H	0246
SKMV150J	1993
SKMV150K	3415
SKMV250H	1332
SKMV250J	1334
SKMV250K	1336
SKMV275A	1658
SKMV275B	1660
SKMV300	4471
SKMV320B	2034
SKMV373A	2360
SKMV420	3257
SKMV480G	1101
SKMV480H	0781
SKN4F1-01	OEM
SKN4F1-02	OEM
SKN4F1-03	OEM
SKN4F1-04	OEM
SKN4F1-06	OEM
SKN4F1-08	OEM
SKN4F1-11	OEM
196433	1783
SKT1-01A	OEM
SKT1-01G	OEM
SKT1-02A	OEM
SKT1-02G	OEM
SKT1-04A	OEM
SKT1-04B	OEM
SKT1-04G	OEM
SKT1-06A	OEM
SKT1-06B	OEM
SKT1-06G	OEM
SKT1-07A	OEM
SKT1-07G	OEM
SKT1-08B	OEM
SKT1-10B	OEM
SKT1-12B	OEM
SKT3-01A	OEM
SKT3-01G	OEM
SKT3-02A	OEM
SKT3-02G	OEM
SKT3-04A	OEM
SKT3-04B	OEM
SKT3-04G	OEM
SKT3-06A	OEM
SKT3-06B	OEM
SKT3-06G	OEM
SKT3-07A	OEM
SKT3-07G	OEM
SKT3-08B	OEM
SKT3-10B	OEM
SKT3-12B	OEM
SKT12-01CR	OEM
SKT12-02C	OEM
SKT12-02CQ	OEM
SKT12-02CR	OEM
SKT12-02CS	OEM
SKT12-04C	OEM
SKT12-04CQ	OEM
SKT12-04CR	OEM
SKT12-04CS	OEM
SKT12-06C	OEM
SKT12-06CQ	OEM
SKT12-06CR	OEM
SKT12-06CS	OEM
SKT12-08C	OEM
SKT12-08CR	OEM
SKT12-08CS	OEM
SKT12-10C	OEM
SKT12-10CR	OEM
SKT12-10CS	OEM
SKT12-12C	OEM
SKT12-12CS	OEM
SKT16-01CS	OEM
SKT16-02C	OEM
SKT16-02CR	OEM
SKT16-02CS	OEM
SKT16-04C	OEM
SKT16-04CR	OEM
SKT16-04CS	OEM
SKT16-06CR	OEM
SKT16-06CS	OEM
SKT16-08C	OEM
SKT16-08CR	OEM
SKT16-08CS	OEM
SKT16-10CR	OEM
SKT16-10CS	OEM
SKT16-10E	OEM
SKT16-12CS	OEM
SKT24-02C	OEM
SKT24-04C	OEM
SKT24-06C	OEM
SKT24-08C	OEM
SKT24-08C#	0679
SKT24-10E	OEM
SKT36-02C	OEM
SKT36-04C	OEM
SKT36-06C	OEM
SKT36-08C	OEM
SKT36-10C	OEM
SKT36-10E	OEM
SKT36-12C	OEM
SKT36-12E	OEM
SKT45-02C	OEM
SKT45-04C	OEM
SKT45-06C	OEM
SKT45-08C	OEM
SKT45-10C	OEM
SKT45-10E	OEM
SKT45-12C	OEM
SKT45-12E	OEM
SKT45-14C	OEM
SKT45-14E	OEM
SKT45-16C	OEM
SKT45-16E	OEM
SKT46-09	OEM
SKT55-02C	OEM
SKT55-04C	OEM
SKT55-06C	OEM
SKT55-08C	OEM
SKT55-10C	OEM
SKT55-10E	OEM
SKT55-12C	OEM
SKT71-09	OEM
SKT80-02C	OEM
SKT80-04C	OEM
SKT80-06C	OEM
SKT80-08C	OEM
SKT80-10C	OEM
SKT80-10E	OEM
SKT90-0F	OEM
SKT100-10E	OEM
SKT111-09	OEM
SKT130-02C	OEM
SKT130-04C	OEM
SKT130-06C	OEM
SKT130-08C	OEM
SKT130-10C	OEM
SKT130-10E	OEM
SKT130-12C	OEM
SKT140-02CV	OEM
SKT140-04CV	OEM
SKT140-06CV	OEM
SKT140-08CV	OEM
SKT140-10CV	OEM
SKT140-12CV	OEM
SKT141-09	OEM
SKT160-02C	1084
SKT160-04C	1086
SKT160-06C	OEM
SKT160-08C	OEM
SKT160-10C	OEM
SKT160-10E	OEM
SKT160-12C	OEM
SKT171-09	OEM
SKT180-02CT	OEM
SKT180-02CU	OEM
SKT180-02CV	OEM
SKT180-04CT	OEM
SKT180-04CU	OEM
SKT180-04CV	OEM
SKT180-06CT	OEM
SKT180-06CU	OEM
SKT180-06CV	OEM
SKT180-08CT	OEM
SKT180-08CU	OEM
SKT180-08CV	OEM
SKT180-10CU	OEM
SKT180-10CV	OEM
SKT180-12CU	OEM
SKT180-12CV	OEM
SKT215-02C	OEM
SKT215-04C	OEM
SKT215-06C	OEM
SKT215-08C	OEM
SKT215-10C	OEM
SKT215-10E	OEM
SKT215-12C	OEM
SKT216-02C	OEM
SKT216-04C	OEM
SKT216-06C	OEM
SKT216-08C	OEM
SKT216-10C	OEM
SKT216-10E	OEM
SKT216-12C	OEM
SKT216-12E	OEM
SKT216-14C	OEM
SKT216-14E	OEM
SKT216-16C	OEM
SKT216-16E	OEM
SKT220-02F	OEM
SKT220-04F	OEM
SKT220-06F	OEM
SKT220-08F	OEM
SKT220-10	OEM
SKT220-12	OEM
SKT220-14	OEM
SKT220-16	OEM
SKT221-09	OEM
SKT230-02C	OEM
SKT230-04C	OEM
SKT230-06C	OEM
SKT230-08C	OEM
SKT230-10C	OEM
SKT230-10E	OEM
SKT230-12C	OEM
SKT240-02CT	OEM
SKT240-02CU	OEM
SKT240-02CV	OEM
SKT240-04CT	OEM
SKT240-04CU	OEM
SKT240-04CV	OEM
SKT240-06CT	OEM
SKT240-06CU	OEM
SKT240-06CV	OEM
SKT240-08CT	OEM
SKT240-08CU	OEM
SKT240-08CV	OEM
SKT240-10CU	OEM
SKT240-10CV	OEM
SKT240-12CU	OEM
SKT240-12CV	OEM
SKT250-02C	OEM
SKT250-06C	OEM
SKT250-08C	OEM
SKT250-10C	OEM
SKT250-10E	OEM
SKT250-12C	OEM
SKT251-02C	OEM
SKT251-04C	OEM
SKT251-06C	OEM
SKT251-08C	OEM
SKT251-10C	OEM
SKT251-10E	OEM
SKT251-12C	OEM
SKT251-12E	OEM
SKT251-14C	OEM
SKT251-14E	OEM
SKT251-16C	OEM
SKT251-16E	OEM
SKT300-02C	OEM
SKT300-04C	OEM
SKT300-06C	OEM
SKT300-08C	OEM
SKT300-10C	OEM
SKT300-12C	OEM
SKT301-02C	OEM
SKT301-04C	OEM
SKT301-06C	OEM
SKT301-08C	OEM
SKT301-10C	OEM
SKT301-10E	OEM
SKT301-12C	OEM
SKT301-12E	OEM
SKT301-14C	OEM
SKT301-14E	OEM
SKT301-16C	OEM
SKT301-16E	OEM
SKT330-10E	OEM
SKT450-02C	OEM
SKT450-06C	OEM
SKT450-10C	OEM
SKT450-12C	OEM
SKT300102C+	OEM
SKT300104C+	0738
SKT300106C+	0738
SKT300108C+	0738
SKT300110C+	0738
SKT300112C+	0738
SKX75000	OEM
SKX100000	OEM
SKX125000	OEM
SKX150000	OEM
SL-030	0015
SL030S	0015
SL-030T	0015
SL030T	0015
SL051A	2002
SL0305,T	0015
SL0307	OEM
SL0315	OEM
SL0325	OEM
SL0335	OEM
SL0507	OEM
SL0515	OEM
SL0525	OEM
SL0535	OEM
SL01640	1069
SL02518	3418
SL02519	1069
SL02734	1069
SL02779	1069
SL02781	OEM
SL03019	1069
SL03021	OEM
SL03667	3901
SL03911	1844
SL03912	1859
SL03913	0341
SL03914	1041
SL03915	0033
SL03916	1835
SL03917	2078
SL04194	1257
SL04217	3418
SL04218	1069
SL04563	1823
SL04567	1831
SL04568	1041
SL04570	2078
SL04732	1069
SL-2	0015
SL2	0015
SL-3	0015
SL3	0109
SL-3R	OEM
SL-5	0242
SL5	0015
SL-5R	OEM
SL10	0145
SL-25R	OEM
SL50	0084
SL53URT	OEM
SL55VR3FT	OEM
SL91	0015
SL92	0015
SL93	0015
SL100	0144
SL-103	0242
SL103	0015
SL103K	OEM

DEVICE TYPE	REPL CODE
SL119	0037
SL136-6	OEM
SL200	0150
SL201	0150
SL203A	OEM
SL203K	OEM
SL203M	OEM
SL300	0016
SL301A	0418
SL301AE	0418
SL301B	0418
SL301BE	0418
SL301C	0086
SL301CE	0086
SL301E	0418
SL301EE	0418
SL302A	OEM
SL303A	OEM
SL303AE	0418
SL303AT	0418
SL303B	OEM
SL303BE	0418
SL303BT	0418
SL354BE	0418
SL354BF	0418
SL360	OEM
SL400	0109
SL403	0155
SL403A	OEM
SL403K	OEM
SL404	OEM
SL-425G	OEM
SL-425R	OEM
SL500	0122
SL600	0122
SL603A	OEM
SL603K	OEM
SL603M	OEM
SL608	0071
SL610	0071
SL708	0071
SL710	0071
SL800	0131
SL800X	0131
SL803A	OEM
SL803K	OEM
SL803M	OEM
SL-833	0015
SL833	0015
SL-833A	0015
SL833A	0015
SL918	0144
SL1000	0145
SL1000X	0145
SL1003A	OEM
SL1003K	OEM
SL1003M	OEM
SL1041	OEM
SL1049A	OEM
SL1199	OEM
SL-1203	OEM
SL1203	OEM
SL-1203-20	OEM
SL1203-20	OEM
SL1203A	OEM
SL-1203H	OEM
SL1203K	OEM
SL1203M	OEM
SL1233A	OEM
SL1252	OEM
SL1255	OEM
SL1256	OEM
SL1259A	OEM
SL1263	OEM
SL-1263H	OEM
SL1403A	OEM
SL1424-11B	OEM
SL1424-11U	OEM
SL1462-11C	OEM
SL1489A	OEM
SL1498	OEM
SL1498IIT	OEM
SL1611C	OEM
SL1611C/DP	OEM
SL1621C	OEM
SL1622C	1645
SL1626C	1645
SL1626K	OEM
SL1640C	OEM
SL1641C	OEM
SL2010	OEM
SL-2243T	OEM
SL2243T	OEM
SL2272-05K	OEM
SL2283	OEM
SL2444-04K	OEM
SL2444-04KUL	OEM
SL2475-04K	OEM
SL2713-01K	OEM
SL3101	0126
SL3111	0126
SL3693	0144
SL7059	0632
SL7283	0632
SL7308	0632
SL7531	0632
SL7593	0632
SL8020	0632
SL8110	OEM
SL11877	1823
SL11878	1831
SL11879	0141
SL11880	3399
SL11881	0562
SL14959	0685
SL14960	2398
SL14971	0467
SL14972	0996
SL16121	OEM
SL16122	1257
SL16201	1823
SL16203	2078
SL16204	0141
SL16206	1041
SL16208	1835
SL16209	0333
SL16210	1831
SL16211	0081
SL16212	0562
SL16215	0033
SL16216	1844
SL16218	0341
SL16516	1823
SL16517	2097
SL16518	1831
SL16519	0141
SL16520	0562
SL16521	0081
SL16522	0333
SL16584	1823
SL16585	3061
SL16586	1041
SL16587	1831
SL16588	0081
SL16589	0141
SL16590	3399
SL16591	0562
SL16793	0232
SL16794	0268
SL16795	0314
SL16796	0361
SL16797	0385
SL16798	0467
SL16799	0527
SL16800	0697
SL16801	0512
SL16802	0873
SL16803	1024
SL16804	0743
SL16805	1185
SL16806	1172
SL16807	1313
SL16808	1158
SL16809	0834
SL16810	1715
SL16811	0033
SL17242	1172
SL17284	1041
SL17289	1831
SL17869	0467
SL17887	0527
SL18386	0117
SL18387	0942
SL18699	1364
SL20721	2161
SL20755	0666
SL21017	0354
SL21122	0352
SL21384	1554
SL21441	0753
SL21577	1554
SL21654	0167
SL21864	0701
SL24618	4586
SL53424	1823
SL53425	1041
SL53426	1176
SL53427	1831
SL53428	0033
SL53429	0141
SL53431	0268
SL146211C	OEM
SLA007-1	OEM
SLA009	OEM
SLA-01	3889
SLA01	0071
SLA-02	3889
SLA02	OEM
SLA-02-HC	0348
SLA-02-HF	3889
SLA02HC	OEM
SLA02HF	OEM
SLA-03	3889
SLA03	OEM
SLA-03-HC	OEM
SLA-03-HF	3889
SLA03HC	OEM
SLA03HF	OEM
SLA-04	3889
SLA04	OEM
SLA-04-HC	OEM
SLA-04-HF	3889
SLA04HC	OEM
SLA04HF	OEM
SLA-05	3889
SLA05	OEM
SLA-05-HC	OEM
SLA-05-HF	3889
SLA05HC	OEM
SLA05HF	OEM
SLA-06	3889
SLA06	OEM
SLA-08	OEM
SLA08	OEM
SLA-010	OEM
SLA010	OEM
SLA-012	OEM
SLA012	OEM
SLA-014	OEM
SLA014	OEM
SLA-016	OEM
SLA016	OEM
SLA-018	OEM
SLA018	OEM
SLA-020	OEM
SLA020	OEM
SLA-11	1335
SLA11AB	0015
SLA11C	0015
SLA-12	0080
SLA12	OEM
SLA12AB	0015
SLA12C	0015
SLA-13	0609
SLA13AB	0015
SLA13C	0015
SLA-14	0795
SLA14	OEM
SLA14AB	0015
SLA14C	0015
SLA-15	0795
SLA15AB	0015
SLA15C	0015
SLA-16	0015
SLA16	OEM
SLA16AB	0015
SLA16C	0015
SLA-17	0015
SLA17AB	0015
SLA17C	0015
SLA-18	0072
SLA18	OEM
SLA18AB	0071
SLA18C	0071
SLA-19	0071
SLA19AB	0071
SLA19C	0071
SLA-20	0182
SLA20AB	OEM
SLA20C	OEM
SLA-21	0071
SLA21A	0110
SLA21B	0110
SLA21C	0110
SLA-22	0071
SLA22A	0953
SLA22B	0953
SLA22C	0953
SLA-23	0071
SLA23A	OEM
SLA23B	0242
SLA23C	0953
SLA-24	0071
SLA24A	0540
SLA24B	0540
SLA24C	0540
SLA-25	0071
SLA25A	0540
SLA25B	0540
SLA25C	0540
SLA-26	0071
SLA26A	0965
SLA26B	0965
SLA26C	0965
SLA-27	0071
SLA27A	0965
SLA27B	0965
SLA27C	0965
SLA-28	0071
SLA28A	0817
SLA28B	0817
SLA28C	0817
SLA-29	0071
SLA29A	0922
SLA29B	0922
SLA29C	0922
SLA-30	0102
SLA30A	OEM
SLA30B	OEM
SLA30C	OEM
SLA-31	0102
SLA31A	OEM
SLA31B	OEM
SLA31C	OEM
SLA-32	0348
SLA32A	OEM
SLA32B	OEM
SLA32C	OEM
SLA-33	0348
SLA33A	OEM
SLA33B	OEM
SLA33C	OEM
SLA-34	0348
SLA34A	OEM
SLA34B	OEM
SLA34C	OEM
SLA-50-LCD	0277
SLA-100-LCD	0277
SLA-200	1414
SLA-200-LCD	0289
SLA-300	0473
SLA-400	0473
SLA-400-LCD	0296
SLA440	0015
SLA440B	0015
SLA441	0015
SLA441B	0015
SLA442	0015
SLA442B	0015
SLA443	0015
SLA443B	0015
SLA444	0015
SLA444A	0015
SLA444B	0015
SLA445	0015
SLA445B	0015
SLA-500	0446
SLA536	0015
SLA537	0015
SLA538	0015
SLA539	0015
SLA540	0015
SLA547	0015
SLA560	0071
SLA561	0071
SLA591LT3	4304
SLA599	0015
SLA599A	0015
SLA-600	0446
SLA600	0015
SLA-600-LCD	0302
SLA600A	0015
SLA601	0015
SLA601A	0015
SLA602	0015
SLA602A	0015
SLA603	0015
SLA603A	0015
SLA604	0015
SLA604A	0015
SLA605	0015
SLA605A	0015
SLA606	0015
SLA606A	0015
SLA-700	1422
SLA-800	1422
SLA-800-LCD	0250
SLA-900	2443
SLA-1000	2443
SLA-1000-LCD	0250
SLA1095	0015
SLA1096	0015
SLA1100	0015
SLA1101	0015
SLA1102	0015
SLA1103	0540
SLA1104	0015
SLA1105	0015
SLA-1400	0348
SLA1400A	OEM
SLA1400B	OEM
SLA1400C	OEM
SLA1487	0015
SLA1488	0015
SLA1489	0015
SLA1490	0015
SLA1491	0015
SLA1492	0015
SLA-1600	0348
SLA1600A	OEM
SLA1600B	OEM
SLA1600C	OEM
SLA1692	0015
SLA1693	0015
SLA1694	0015
SLA1695	0015
SLA1696	0015
SLA1697	0015
SLA-1800	0348
SLA1800A	OEM
SLA1800B	OEM
SLA1800C	OEM
SLA-2000	0348
SLA2000A	OEM
SLA2000B	OEM
SLA2000C	OEM
SLA2610	0015
SLA2611	0015
SLA2612	0015
SLA2613	0015
SLA2614	0015
SLA2615	0015
SLA2616	0071
SLA2617	0071
SLA3193	0015
SLA3194	0015
SLA3195	0015
SLA3196	0071
SLA4030C0B	OEM
SLA4061	OEM
SLA5197	0110
SLA5197HF	OEM
SLA5198	0953
SLA5198HF	OEM
SLA5199	0242
SLA5199HF	OEM
SLA5200	0540
SLA5200HF	OEM
SLA5201	0965
SLA5201HF	OEM
SLA5202	0817
SLA5202HF	OEM
SLA5203	0922
SLA5203HF	OEM
SLA5204	OEM
SLA5204HF	OEM
SLA5205	OEM
SLA5205HF	OEM
SLA6140J1P	OEM
SLA6330J0L	OEM
SLA6330J0M	OEM
SLA6430J1S	OEM
SLA6430J2A	OEM
SLA7020	OEM
SLA7020M	OEM
SLA9030M	OEM
SLA01	0071
SLB-15DU3	OEM
SLB15DU3	OEM
SLB-22DU	OEM
SLB-22GG	OEM
SLB-22UR	OEM
SLB-22YY	OEM
SLB-25DU	OEM
SLB25DU	OEM
SLB-25DU3	OEM
SLB25DU3	OEM
SLB-25MG	OEM
SLB-25MG3	OEM
SLB25MG3	OEM
SLB-25VR	OEM
SLB25VR	OEM
SLB-25VR3	OEM
SLB25VR3	OEM
SLB-26GG	OEM
SLB-26UR	OEM
SLB26UR5	OEM
SLB-26UU	OEM
SLB-26YY	OEM
SLB-55DU3	OEM
SLB55DU3	OEM
SLB55MG3	OEM
SLB55VR3	OEM
SLB-72GG5HL	OEM
SLB72UR5HL	OEM
SLB74UR3GH	OEM
SLC6.5	OEM
SLC6.5A	OEM
SLC7.0	OEM
SLC7.0A	OEM
SLC7.5	OEM
SLC7.5A	OEM
SLC8.0	OEM
SLC8.0A	OEM
SLC8.5	OEM
SLC8.5A	OEM
SLC9.0	OEM
SLC9.0A	OEM
SLC10	OEM
SLC10A	OEM
SLC11	OEM
SLC11A	OEM
SLC12	OEM
SLC12A	OEM
SLC13	OEM
SLC13A	OEM
SLC14	OEM
SLC14A	OEM
SLC15	OEM
SLC15A	OEM
SLC16	OEM
SLC16A	OEM
SLC17	OEM
SLC17A	OEM
SLC18	OEM
SLC18A	OEM
SLC20	OEM
SLC20A	OEM
SLC22	OEM
SLC22A	OEM
SLC-22DU	OEM
SLC-22GG	OEM
SLC-22UR	OEM
SLC-22YY	OEM
SLC24	OEM
SLC24A	OEM
SLC-26	OEM
SLC26	OEM
SLC26A	OEM
SLC-26GC14	OEM
SLC-26GG	OEM
SLC-26UR	OEM
SLC-26UR14	OEM
SLC26UR14	OEM
SLC-26UU	OEM
SLC-26VR5F	OEM
SLC26VR5F	OEM
SLC-26YY	OEM
SLC28	OEM
SLC28A	OEM
SLC30	OEM
SLC30A	OEM
SLC33	OEM
SLC33A	OEM
SLC36	OEM
SLC36A	OEM
SLC40	OEM
SLC40A	OEM
SLC43	OEM
SLC43A	OEM
SLC45	OEM
SLC45A	OEM
SLC48	OEM
SLC48A	OEM
SLC51	OEM
SLC51A	OEM
SLC54	OEM
SLC54A	OEM
SLC58	OEM
SLC58A	OEM
SLC60	OEM
SLC60A	OEM
SLC64	OEM
SLC64A	OEM
SLC70	OEM
SLC70A	OEM
SLC75	OEM
SLC75A	OEM
SLC90	OEM
SLC90A	OEM
SLC100	OEM
SLC100A	OEM
SLC110	OEM
SLC110A	OEM
SLC120	OEM
SLC120A	OEM
SLC130	OEM
SLC130A	OEM
SLC150	OEM
SLC150A	OEM
SLC160	OEM
SLC160A	OEM
SLC170	OEM
SLC170A	OEM
SLC50155RSN	OEM
SLCE6.5	OEM
SLCE6.5A	OEM
SLCE7.0	OEM
SLCE7.0A	OEM
SLCE7.5	OEM
SLCE7.5A	OEM
SLCE8.0	OEM
SLCE8.0A	OEM
SLCE8.5	OEM
SLCE8.5A	OEM
SLCE9.0	OEM
SLCE9.0A	OEM
SLCE10	OEM
SLCE10A	OEM
SLCE11	OEM
SLCE11A	OEM
SLCE12	OEM
SLCE12A	OEM
SLCE13	OEM
SLCE13A	OEM
SLCE14	OEM
SLCE14A	OEM
SLCE15	OEM
SLCE15A	OEM
SLCE16	OEM
SLCE16A	OEM
SLCE17	OEM
SLCE17A	OEM
SLCE18	OEM
SLCE18A	OEM
SLCE20	OEM
SLCE20A	OEM
SLCE22	OEM
SLCE22A	OEM
SLCE24	OEM
SLCE24A	OEM
SLCE26	OEM
SLCE26A	OEM
SLCE28	OEM
SLCE28A	OEM
SLCE30	OEM
SLCE30A	OEM
SLCE33	OEM
SLCE33A	OEM
SLCE36	OEM
SLCE36A	OEM
SLCE40	OEM
SLCE40A	OEM
SLCE43	OEM
SLCE43A	OEM
SLCE45	OEM
SLCE45A	OEM
SLCE48	OEM
SLCE48A	OEM
SLCE51	OEM
SLCE51A	OEM
SLCE54	OEM
SLCE54A	OEM
SLCE58	OEM
SLCE58A	OEM
SLCE60	OEM
SLCE60A	OEM
SLCE64	OEM
SLCE64A	OEM
SLCE70	OEM
SLCE70A	OEM
SLCE75	OEM
SLCE75A	OEM
SLCE90	OEM
SLCE90A	OEM
SLCE100	OEM
SLCE100A	OEM
SLCE110	OEM
SLCE110A	OEM
SLCE120	OEM
SLCE120A	OEM
SLCE130	OEM
SLCE130A	OEM
SLCE150	OEM
SLCE150A	OEM
SLCE160	OEM
SLCE160A	OEM
SLCE170	OEM
SLCE170A	OEM
SLD30AR	OEM
SLD30UR	OEM
SLEA05	2363
SLEA1	2363
SLEA2	2363
SLEA3	2369
SLEA4	2369
SLEA6	2370
SLEA8	2372
SLEA10	2373
SLED-G	OEM
SLED-R	OEM
SLG5553KB++GR	OEM
SLG5553KB++RE	OEM
SLH-34DC3	OEM
SLH-34DT3	OEM
SLH-34DU3	OEM
SLH-34MC3	OEM
SLH-34MC3F	OEM
SLH-34MG3	OEM
SLH-34MT3	OEM
SLH34MT3	OEM
SLH-34PC3	OEM
SLH-34PT3	OEM
SLH-34UR3	OEM
SLH-34URC3	OEM
SLH-34UT3	OEM
SLH-34VC3	OEM
SLH-34VR3	OEM
SLH-34VT3	OEM
SLH-34YC3	OEM
SLH-34YC3F	OEM
SLH-34YT3	OEM
SLH-34YY3	OEM
SLH54MT4	OEM
SLH-56DC3	OEM
SLH-56DT3	OEM
SLH-56DU3	OEM
SLH-56MC3	OEM
SLH56MC4F	OEM
SLH-56MC77F	OEM
SLH-56MG3	OEM
SLH-56MT3	OEM
SLH-56PC3	OEM
SLH-56PT3	OEM
SLH-56UR3	OEM
SLH-56URC3	OEM
SLH-56UT3	OEM
SLH-56VC3	OEM
SLH56VC3	OEM
SLH56VC4F	OEM
SLH56VC5	OEM
SLH-56VC77F	OEM
SLH-56VR3	OEM
SLH-56VT3	OEM
SLH-56YC3	OEM
SLH56YC3	OEM
SLH-56YT3	OEM
SLH-56YY3	OEM
SLJ-265MGHL	OEM
SLJ-265URHL	OEM
SLK2442-01	OEM
SLN2020P	OEM
SLN2030P	OEM
SLN2035P	OEM
SLN2040P	OEM
SLN2045P	OEM
SLN3020P	OEM
SLN3030P	OEM
SLN3035P	OEM
SLN3040P	OEM
SLN3045P	OEM
SLP24B	OEM
SLP-34UR5	OEM
SLP114	OEM
SLP114A	OEM
SLP114B	OEM
SLP-115R	OEM
SLP115R	OEM
SLP-123B	OEM
SLP123B	OEM
SLP131	OEM
SLP131B	OEM
SLP133B	OEM
SLP133P	OEM
SLP135A50	OEM
SLP135B	OEM
SLP135B-55	OEM
SLP135BA	OEM
SLP135BB	OEM
SLP135BC	OEM
SLP136B	OEM
SLP-136B-50	OEM
SLP136B-50	OEM

DEVICE TYPE	REPL CODE
SLP136BA	OEM
SLP-136B-B	OEM
SLP136BB	OEM
SLP136BC	OEM
SLP137B	OEM
SLP-138C-50	OEM
SLP138C-50	OEM
SLP-138C-51	OEM
SLP138C-51	OEM
SLP138C-51B	OEM
SLP-138CB-50	OEM
SLP138CB-50	OEM
SLP-141B	OEM
SLP141B	OEM
SLP144B	OEM
SLP-144B-40	OEM
SLP144B-40	OEM
SLP-144B-51	OEM
SLP144B-51	OEM
SLP-145B	OEM
SLP145B	OEM
SLP146B	OEM
SLP-146B-50	OEM
SLP146B-50	OEM
SLP-151B	5174
SLP151B	5174
SLP-151B-50	5132
SLP151B-50	OEM
SLP152B	7075
SLP152B-02	OEM
SLP-153B	OEM
SLP153B	OEM
SLP-153B-01	OEM
SLP153B-40	OEM
SLP154B	7078
SLP154B-60	OEM
SLP-155B	4052
SLP155B	7079
SLP-155B-50	4052
SLP155B-50	OEM
SLP155BRE	OEM
SLP156B	OEM
SLP-159B	OEM
SLP159B	OEM
SLP159B-50	OEM
SLP159BA	OEM
SLP159BB	OEM
SLP159BC	OEM
SLP-160C	OEM
SLP-161B	OEM
SLP161B	OEM
SLP-162B	OEM
SLP162B	OEM
SLP164BA	OEM
SLP165B	OEM
SLP166B	OEM
SLP166B-03	OEM
SLP167B	OEM
SLP167B-02	OEM
SLP167B-03	OEM
SLP167B-70	OEM
SLP169B	OEM
SLP171D	OEM
SLP171E	OEM
SLP-173	OEM
SLP173	OEM
SLP-173B-01	OEM
SLP173B-01	OEM
SLP-173B-50	4052
SLP173B-50	OEM
SLP174B	OEM
SLP175B	OEM
SLP-175B-50	OEM
SLP175B-50	OEM
SLP-177B-60	OEM
SLP177B-60	OEM
SLP-178B	OEM
SLP178B	OEM
SLP-178C-A	OEM
SLP178CA	OEM
SLP-180C-AB	OEM
SLP-181B	OEM
SLP181B	OEM
SLP181B-50	OEM
SLP181B-50A	OEM
SLP181B-50B	OEM
SLP183B-09	OEM
SLP224B	OEM
SLP-235B	0841
SLP235B	OEM
SLP-235B(GR)	0841
SLP-236B	0862
SLP-236B(G)	4679
SLP-236B(GR)	4679
SLP-236B(GREEN)	4679
SLP-236B(Y)	0862
SLP-236B(YEL)	0862
SLP-236B(YELLOW)	0862
SLP236B-50	OEM
SLP-236B-A	OEM
SLP-236P(GR)	4679
SLP237B	OEM
SLP239B	OEM
SLP244B	OEM
SLP-251B	1076
SLP251B	7133
SLP251B-50	OEM
SLP252B	7134
SLP252B-03	OEM
SLP253B	OEM
SLP-255B	1131
SLP255B	7136
SLP258B	OEM
SLP-261B	OEM
SLP261B	OEM
SLP-262B	OEM
SLP262B	OEM
SLP265B	OEM
SLP267B	OEM
SLP267B-03	OEM
SLP271D	OEM
SLP-273	OEM
SLP-273B-01	OEM
SLP273B-01	OEM
SLP273B-50U	OEM
SLP274B	OEM
SLP275B	OEM
SLP-277B-60	OEM
SLP277B-60	OEM
SLP-278C	OEM
SLP278C	OEM
SLP280F-50U	OEM
SLP280F-51	OEM
SLP280F-51P	OEM
SLP280F-51U	OEM
SLP282F-50P	OEM
SLP355B-51A	OEM
SLP355B-51B	OEM
SLP-373B-50	OEM
SLP373B-50	OEM
SLP382F-50	OEM
SLP382F-51	OEM
SLP451B	1995
SLP453B	OEM
SLP455B	2023
SLP466B	OEM
SLP475B	OEM
SLP531D	OEM
SLP651B-50A	OEM
SLP651B-50B	OEM
SLP-655B	OEM
SLP655B	OEM
SLP-662B-A	OEM
SLP662B-A	OEM
SLP-662B-B	OEM
SLP662B-B	OEM
SLP-673B-51-A	OEM
SLP-673B-51-B	OEM
SLP673B-51A	OEM
SLP673B-51B	OEM
SLP674B-50A	OEM
SLP674B-50B	OEM
SLP-678B-A	OEM
SLP678B-A	OEM
SLP678B-B	OEM
SLP-873B-51-A	OEM
SLP-888A	4368
SLP955B-50	OEM
SLP973B	OEM
SLP978C	OEM
SLP980C-51	OEM
SLP980F-51	OEM
SLP981C-50	OEM
SLP-982A-50	OEM
SLP982A-50	OEM
SLP982A-51	OEM
SLP982C	OEM
SLP982C-50	OEM
SLP2020P	1942
SLP2030P	0584
SLP2035P	1942
SLP2040P	1942
SLP2045P	1942
SLP3020P	1942
SLP3030P	1942
SLP3035P	1942
SLP3040P	1942
SLP3045P	1942
SLP3070K	OEM
SLP4070	OEM
SLP9070	OEM
SLR30UR	OEM
SLR33MGF03L/M/N	OEM
SLR-34DC3	OEM
SLR-34DU3	OEM
SLR-34GG3	OEM
SLR34MC	OEM
SLR-34MC3	OEM
SLR34MC3F	OEM
SLR34MC5	OEM
SLR-34MG3	OEM
SLR-34MT3	OEM
SLR-34PT3	OEM
SLR-34UR3	OEM
SLR-34UR3F	OEM
SLR34UR3F	OEM
SLR-34UR5	OEM
SLR34UR5	OEM
SLR-34UR5F	OEM
SLR-34URC3	OEM
SLR34URC3	OEM
SLR34URC5	OEM
SLR-34URS	OEM
SLR34UW5	OEM
SLR-34VC3	OEM
SLR34VC3F	OEM
SLR34VC3N	OEM
SLR-34VR3	OEM
SLR-34VT3	OEM
SLR-34YC3	OEM
SLR-34YY3	OEM
SLR-40MC3F	OEM
SLR40MC3F	OEM
SLR-40UR3F	OEM
SLR40UR3F	OEM
SLR-40UR5F	OEM
SLR40UR5F	OEM
SLR53GC	OEM
SLR-53GG	OEM
SLR53GG	OEM
SLR53URC	OEM
SLR-53URT	OEM
SLR53URT-1	OEM
SLR-54MC	OEM
SLR-54MT4	OEM
SLR54MT4	OEM
SLR54NPQ	OEM
SLR-54UR	OEM
SLR54UR	OEM
SLR-54UR4	OEM
SLR54UR4	OEM
SLR54URC	OEM
SLR54URC-4	OEM
SLR-54UR-H	OEM
SLR54UR-H	OEM
SLR-54URT	OEM
SLR-54UT4	OEM
SLR54VC	OEM
SLR54VC3F	OEM
SLR54VR4	OEM
SLR55GC5	OEM
SLR-55MC	OEM
SLR55MC	OEM
SLR55MC3F	OEM
SLR-55MC-H	OEM
SLR-56DC3	OEM
SLR-56DU3	OEM
SLR-56GG3	OEM
SLR-56MC3	OEM
SLR-56MG3	OEM
SLR-56MT3	OEM
SLR-56PT3	OEM
SLR-56UR3	OEM
SLR-56URC3	OEM
SLR-56UT3	OEM
SLR-56UW3	OEM
SLR-56VC3	OEM
SLR-56VR3	OEM
SLR-56VT3	OEM
SLR-56YC3	OEM
SLR-56YY3	OEM
SLR-320PG3K	OEM
SLR-320VR3K	OEM
SLR331MC70F070	OEM
SLR-342DU-T	OEM
SLR-342VR-T	OEM
SLR534RT	OEM
SLR-932A	OEM
SLR932A	OEM
SLR-933A	OEM
SLR933A	OEM
SLR-938C	OEM
SLR938C	OEM
SLR980C1	OEM
SLT-35GG	OEM
SLT-35UR	OEM
SLT-35UU	OEM
SLT-35YY	OEM
SLV56DC3	OEM
SLV56URC	OEM
SLZ-136C-07-T1	OEM
SLZ136C-14-AB-T1	OEM
SLZ146B	OEM
SLZ151B	OEM
SLZ167B	OEM
SLZ177BA	OEM
SLZ177BB	OEM
SLZ177BC	OEM
SLZ183B	OEM
SLZ-603A+	1112
SLZ-603K+	0476
SLZ-803A+	3011
SLZ-803K+	0449
SLZ-1003A+	3011
SLZ-1003K+	0449
SM-05-16FR	0348
SM-05-20FRZ	0348
SM-05A-16FR	0348
SM0843	0004
SM07275	0079
SM07286	0079
SM-1	0015
SM1	OEM
SM-1-005	0015
SM1-01C	OEM
SM-1-02	0023
SM1-02	0023
SM-1-02FR	0023
SM1-02FR	OEM
SM-1-02FRA	0023
SM-1-04	0023
SM-1-04FR	0023
SM1-04FR	OEM
SM-1-04FRA	0023
SM-1-06FRA	0023
SM-1-08	0071
SM-1-08FR	0071
SM-1-08FRA	0071
SM-1-10FR	0071
SM-1-10FRA	0071
SM-1-12	0102
SM1-12	OEM
SM-1-47	0015
SM-1A	0023
SM-1A-01	0015
SM-1A-01Z	0080
SM-1A-02	0023
SM1A-02	OEM
SM-1A-02FR	0023
SM-1A-02FRA	0023
SM-1A-02LFD	OEM
SM1A-02LFD	OEM
SM1A-02LFE	OEM
SM-1A-02Z	0609
SM-1A-04	0023
SM-1A-04FR	0023
SM-1A-04FRA	0023
SM-1A-04Z	0795
SM-1A-06	0023
SM-1A-06FR	0023
SM-1A-06FRA	0023
SM-1A-06Z	0015
SM-1A-08	0017
SM-1A-12	0017
SM-1A-15	0017
SM-1A-200	0015
SM-1K	0015
SM1K	0540
SM-1XF08	OEM
SM-1XH02	0023
SM-1XH04	0023
SM-1XH06	0023
SM-1XM02	0023
SM-1XM04	0023
SM-1XM06	0023
SM-1XM08	0017
SM-1XM10	0017
SM-1XN02	0023
SM-1XN04	0023
SM-1XN06	0023
SM-1XN08	0017
SM-1XN12	0017
SM-1XN15	0017
SM-1XP2TP	OEM
SM-1.5-02	0031
SM-1.5-02FR	0031
SM1.5-02FRT	OEM
SM-1.5-04	0031
SM-1.5-04FR	0031
SM-1.5-08	0071
SM-1.5-08FR	0071
SM-1.5-10	0071
SM-1.5-10FR	0071
SM-1.5-10Z1	0015
SM2B41	0954
SM2B42	OEM
SM2D41	4440
SM2D42	OEM
SM-3-02	0031
SM-3-02FR	0031
SM-3-04	0031
SM-3-04FR	0031
SM-3-06	0031
SM-3-06FR	0031
SM-3-08	0071
SM-3-08FR	0071
SM3G41	0485
SM3G41-LC2	0485
SM4	0015
SM5	0015
SM5B12	0408
SM5D12	0411
SM10	0015
SM10B11	0408
SM10B12	0154
SM10D11	0411
SM10D12	0154
SM10G12	0279
SM11	0015
SM12	OEM
SM12D41	0772
SM12G41	0744
SM12J41	0617
SM13	OEM
SM20	0015
SM21	OEM
SM22	OEM
SM23	OEM
SM30	0015
SM30D11	2015
SM30G11	2017
SM31	0015
SM33	OEM
SM40	0015
SM41	OEM
SM42	OEM
SM43	OEM
SM45	OEM
SM50	0015
SM51	0015
SM60	0015
SM61	OEM
SM63	1432
SM70	0071
SM71	0071
SM72	OEM
SM73	0071
SM73(I.C.)	1432
SM75	OEM
SM80	0071
SM80B11	OEM
SM80D11	OEM
SM80G11	OEM
SM81	0071
SM82	OEM
SM83	0071
SM90	OEM
SM91	OEM
SM92	OEM
SM93	OEM
SM100	0071
SM101	0071
SM103	0071
SM105	0015
SM110	0015
SM112	0015
SM113-02	OEM
SM113-03	OEM
SM118	OEM
SM118F	OEM
SM120	0015
SM121	OEM
SM121-02	OEM
SM121-03	OEM
SM123	OEM
SM123-02	OEM
SM123-03	OEM
SM130	0015
SM131-02	OEM
SM131-03	OEM
SM133-02	OEM
SM133-03	OEM
SM140	0015
SM143-02	OEM
SM150	0071
SM-150-005	0071
SM-150-01	0071
SM-150-02	0071
SM150-02	0071
SM-150-02(CENTERING)	0920
SM-150-04	0015
SM150-04	OEM
SM-150-06	0015
SM150-06	OEM
SM-150-08	0071
SM150-08	OEM
SM150-6	0071
SM-150-10	0071
SM150-10	OEM
SM150-11	0071
SM-150-12	0102
SM-150-A	0015
SM-150-B	0015
SM-150-C	0071
SM-150-D	0071
SM150A	0071
SM150B	0071
SM150C	0071
SM150D	0071
SM150S	0071
SM150SS	0071
SM153-02	OEM
SM153-03	OEM
SM153A	OEM
SM153AH	OEM
SM153B	OEM
SM153BM	OEM
SM153C	OEM
SM153CH	OEM
SM160	0015
SM163-02	OEM
SM163-03	OEM
SM170	0071
SM173-02	OEM
SM173-03	OEM
SM180	0071
SM182-02	OEM
SM182-03	OEM
SM192-02	OEM
SM192-03	OEM
SM200	0071
SM205	0015
SM210	0015
SM211-02	OEM
SM211-03	OEM
SM213-02	OEM
SM213-03	OEM
SM215A	0609
SM217	0050
SM220	0015
SM221-02	OEM
SM221-03	OEM
SM223-02	OEM
SM223-03	OEM
SM230	0015
SM240	0015
SM249I	0050
SM250	0015
SM260	0015
SM270	0071
SM280	0071
SM300	0071
SM483	0015
SM486	0015
SM487	0015
SM488	0015
SM505	0015
SM510	0015
SM512	0015
SM513	0015
SM514	0015
SM515	0015
SM516	0015
SM517	0071
SM518	0071
SM520	0071
SM576-1	0016
SM576-2	0016
SM645	0015
SM646	0015
SM705	0015
SM710	0015
SM-716	0016
SM716	0016
SM720	0015
SM730	0015
SM740	0015
SM750	0015
SM760	0015
SM770	0071
SM780	0071
SM800	0071
SM843	0004
SM862	0136
SM906	OEM
SM1297	0050
SM1507	0037
SM1600	0050
SM2491	0050
SM2492	0050
SM2700	0016
SM2701	0016
SM3014	0050
SM3104	0016
SM3117A	0016
SM3505	0016
SM3978	0086
SM3986	0016
SM3987	0126
SM-4304-S	0007
SM4304-S	0224
SM4508	OEM
SM4508-B	0016
SM4547	0037
SM4719	0037
SM5104	2616
SM5104F	2616
SM5104G	2616
SM5104P	2616
SM5107	OEM
SM5109	1714
SM5118	OEM
SM5118G	OEM
SM5205A	OEM
SM5379	0016
SM5502A	OEM
SM5564	0079
SM5601	OEM
SM5602	OEM
SM5602C	OEM
SM5643	0079
SM5796	0144
SM5807EP	OEM
SM5807ESET	OEM
SM5807FP	OEM
SM6251	0086
SM6430B	OEM
SM6442A	OEM
SM6442B	OEM
SM6727	0233
SM6728	0126
SM6762	0079
SM6773	0079
SM6814	0168
SM7545	0079
SM7815	0079
SM7836	0079
SM-7991	0698
SM7991	0086
SM8112	0079
SM8113	0079
SM8341	0004
SM8471	OEM
SM8472	OEM
SM8473	OEM
SM8475	OEM
SM8476	OEM
SM8477	OEM
SM8978	0079
SM9008	0079
SM9135	0079
SM9253	0079
SM47145	OEM
SM55450BJ	OEM
SM62186	0150
SM64442B	OEM
SM74177	OEM
SM76115N	OEM
SMA7023M	OEM
SMB-01	0277
SMB-02	0289
SMB-04	0296
SMB-08	0250
SMB-10	0250
SMB454549	0211
SMB454760	0050
SMC7400N	0232
SMC7402N	0314
SMC7408N	0467
SMC7410N	0512
SMC7420N	0697
SMC7430N	0873
SMC7440N	1024
SMC7451N	1168
SMC7473N	1172
SMC7474N	1313
SMC7475N	1432
SMC7476N	1158
SMC7490N	1208
SMC7493N	0569
SMF-7-2	OEM
SMF7-2	OEM
SMF-13-2	OEM
SMF13-2	OEM
SMH-08	OEM
SMH08	OEM
SMH-10	OEM
SMH10	OEM
SMH-16	OEM
SMH16	OEM
SMH-20	OEM
SMH-25ED	OEM
SMH204	OEM
SMHD-08	OEM
SMHD08	OEM
SMHD2-08	OEM
SMHD2-10	OEM
SMHD2-16	OEM
SMHD2-20	OEM
SMHD2-25ED	OEM
SMHD-10	OEM
SMHD10	OEM
SMHD-16	OEM
SMHD16	OEM
SMHD-20	OEM
SMHR-110	OEM
SMHR110	OEM
SMHR-135	OEM
SMHR135	OEM
SMHR-165	OEM
SMHR165	OEM
SMPAL16L8NC	OEM
SMPT-5	OEM
SMPT-8	OEM
SMPT-8C	OEM
SMPT-10	OEM
SMPT-10C	OEM
SMPT-12	OEM
SMPT-12C	OEM
SMPT-15	OEM
SMPT-15C	OEM
SMPT-18	OEM
SMPT-18C	OEM
SMPT-22	OEM
SMPT-22C	OEM
SMPT-36	OEM
SMPT-36C	OEM
SMPT-45	OEM
SMPT-45C	OEM
SMPTE-5	OEM
SMPTE-8	OEM
SMPTE-8C	OEM
SMPTE-10	OEM
SMPTE-10C	OEM
SMPTE-12	OEM
SMPTE-12C	OEM
SMPTE-15	OEM
SMPTE-15C	OEM
SMPTE-18	OEM
SMPTE-18C	OEM
SMPTE-22	OEM
SMPTE-22C	OEM
SMPTE-36	OEM
SMPTE-36C	OEM
SMPTE-45	OEM
SMT100	OEM
SMT101	OEM
SMT102	OEM
SMT103	OEM
SMT104	OEM
SMT105	OEM
SMU1275	0628
SMU1398	OEM
SMV589	OEM
SMV807	OEM
SMV1172	0554
SMZ206A	OEM
SN0303	0015
SN-1	0015
SN1	0015
SN-1Z	0015
SN7ALS12ND	OEM
SN7ALS28NDS	OEM
SN7ALS642N	OEM
SN7LS02J	OEM
SN7LS02ND	OEM
SN7S32F	OEM
SN10	0373
SN20	OEM
SN30	OEM
SN40	OEM
SN54ALS00A	OEM

If replacement code is OEM, contact original manufacturer for replacement.

DEVICE TYPE	REPL CODE
SN54ALS01	OEM
SN54ALS02	OEM
SN54ALS03A	OEM
SN54ALS04	OEM
SN54ALS05	OEM
SN54ALS08	OEM
SN54ALS09	OEM
SN54ALS10	OEM
SN54ALS11	OEM
SN54ALS12	OEM
SN54ALS15	OEM
SN54ALS20A	OEM
SN54ALS21	OEM
SN54ALS22A	OEM
SN54ALS27	OEM
SN54ALS28	OEM
SN54ALS30	OEM
SN54ALS32	OEM
SN54ALS33	OEM
SN54ALS37	OEM
SN54ALS38	OEM
SN54ALS40	OEM
SN54ALS74	OEM
SN54ALS109	OEM
SN54ALS112	OEM
SN54ALS113	OEM
SN54ALS114	OEM
SN54ALS133	OEM
SN54ALS137	OEM
SN54ALS138	OEM
SN54ALS139	OEM
SN54ALS151	OEM
SN54ALS153	OEM
SN54ALS157	OEM
SN54ALS158	OEM
SN54ALS160	OEM
SN54ALS161	OEM
SN54ALS162	OEM
SN54ALS163	OEM
SN54ALS168	OEM
SN54ALS169	OEM
SN54ALS174	OEM
SN54ALS175	OEM
SN54ALS190	OEM
SN54ALS191	OEM
SN54ALS192	OEM
SN54ALS192FH	OEM
SN54ALS192J	OEM
SN54ALS193	OEM
SN54ALS193FH	OEM
SN54ALS193J	OEM
SN54ALS240	OEM
SN54ALS241	OEM
SN54ALS242	OEM
SN54ALS243	OEM
SN54ALS244	OEM
SN54ALS245	OEM
SN54ALS251	OEM
SN54ALS253	OEM
SN54ALS257	OEM
SN54ALS258	OEM
SN54ALS259	OEM
SN54ALS273	OEM
SN54ALS299	OEM
SN54ALS323	OEM
SN54ALS352	OEM
SN54ALS353	OEM
SN54ALS373	OEM
SN54ALS374	OEM
SN54ALS465	OEM
SN54ALS466	OEM
SN54ALS467	OEM
SN54ALS468	OEM
SN54ALS521	OEM
SN54ALS538	OEM
SN54ALS540	OEM
SN54ALS541	OEM
SN54ALS560	OEM
SN54ALS561	OEM
SN54ALS563	OEM
SN54ALS564	OEM
SN54ALS568	OEM
SN54ALS569	OEM
SN54ALS573	OEM
SN54ALS574	OEM
SN54ALS576	OEM
SN54ALS580	OEM
SN54ALS620	OEM
SN54ALS621	OEM
SN54ALS622	OEM
SN54ALS623	OEM
SN54ALS636	OEM
SN54ALS637	OEM
SN54ALS638	OEM
SN54ALS639	OEM
SN54ALS640	OEM
SN54ALS641	OEM
SN54ALS642	OEM
SN54ALS643	OEM
SN54ALS644	OEM
SN54ALS645	OEM
SN54ALS688	OEM
SN54ALS689	OEM
SN54ALS804	OEM
SN54ALS1000	OEM
SN54ALS1002	OEM
SN54ALS1003	OEM
SN54ALS1004	OEM
SN54ALS1005	OEM
SN54ALS1008	OEM
SN54ALS1010	OEM
SN54ALS1011	OEM
SN54ALS1020	OEM
SN54ALS1032	OEM
SN54ALS1034	OEM
SN54ALS1035	OEM
SN54ALS1240	OEM
SN54ALS1241	OEM
SN54ALS1242	OEM
SN54ALS1243	OEM
SN54ALS1244	OEM
SN54ALS1245	OEM
SN54ALS1616	OEM
SN54ALS1620	OEM
SN54ALS1621	OEM
SN54ALS1622	OEM
SN54ALS1623	OEM
SN54ALS1638	OEM
SN54ALS1639	OEM
SN54ALS1640	OEM
SN54ALS1641	OEM
SN54ALS1642	OEM
SN54ALS1643	OEM
SN54ALS1644	OEM
SN54ALS1645	OEM
SN54AS194FH	OEM
SN54AS194J	OEM
SN54AS805	OEM
SN54AS805A	OEM
SN54AS808	OEM
SN54AS808A	OEM
SN54AS830	OEM
SN54AS832	OEM
SN54AS832A	OEM
SN54AS836	OEM
SN54AS850	OEM
SN54AS857	OEM
SN54AS859	OEM
SN54AS866	OEM
SN54AS867	OEM
SN54AS869	OEM
SN54AS870	OEM
SN54AS871	OEM
SN54AS872	OEM
SN54AS873	OEM
SN54AS874	OEM
SN54AS875	OEM
SN54AS876	OEM
SN54AS877	OEM
SN54AS880	OEM
SN54AS881A	OEM
SN54AS882	OEM
SN54AS883	OEM
SN54AS884	OEM
SN54AS885	OEM
SN54AS886	OEM
SN54AS888	OEM
SN54AS889	OEM
SN54AS890	OEM
SN54AS891	OEM
SN54AS894	OEM
SN54H00	OEM
SN54H01	OEM
SN54H04	OEM
SN54H05	OEM
SN54H10	OEM
SN54H11	OEM
SN54H15	OEM
SN54H20	OEM
SN54H21	OEM
SN54H21J	OEM
SN54H21W	OEM
SN54H22	OEM
SN54H30	OEM
SN54H40	OEM
SN54H50	OEM
SN54H51	OEM
SN54H52	OEM
SN54H53	OEM
SN54H54	OEM
SN54H55	OEM
SN54H60	OEM
SN54H61	OEM
SN54H62	OEM
SN54H71	OEM
SN54H72	OEM
SN54H73	OEM
SN54H74	OEM
SN54H76	OEM
SN54H78	OEM
SN54H87	OEM
SN54H101	OEM
SN54H102	OEM
SN54H103	OEM
SN54H106	OEM
SN54H108	OEM
SN54H183	OEM
SN54L02J	OEM
SN54L90	OEM
SN54L91	OEM
SN54L93	OEM
SN54L96	OEM
SN54L98	OEM
SN54L99	OEM
SN54L153J	OEM
SN54L164	OEM
SN54L192	OEM
SN54L192J	OEM
SN54L193	OEM
SN54L193J	OEM
SN54LS00	OEM
SN54LS01	OEM
SN54LS02FK	OEM
SN54LS02J	OEM
SN54LS02W	OEM
SN54LS03	OEM
SN54LS04	OEM
SN54LS05	OEM
SN54LS08	OEM
SN54LS09	OEM
SN54LS09J	OEM
SN54LS09N	OEM
SN54LS10	OEM
SN54LS11	OEM
SN54LS12	OEM
SN54LS13	OEM
SN54LS14	OEM
SN54LS15	OEM
SN54LS18	OEM
SN54LS19	OEM
SN54LS20	OEM
SN54LS21	OEM
SN54LS21FK	OEM
SN54LS21J	OEM
SN54LS21W	OEM
SN54LS22	OEM
SN54LS24	OEM
SN54LS26	OEM
SN54LS27	OEM
SN54LS28	OEM
SN54LS30	OEM
SN54LS32	OEM
SN54LS33	OEM
SN54LS37	OEM
SN54LS38	OEM
SN54LS38J	OEM
SN54LS40	OEM
SN54LS42	OEM
SN54LS47	OEM
SN54LS48	OEM
SN54LS49	OEM
SN54LS51	OEM
SN54LS54	OEM
SN54LS55	OEM
SN54LS63	OEM
SN54LS73A	OEM
SN54LS74A	OEM
SN54LS75	OEM
SN54LS76A	OEM
SN54LS77	OEM
SN54LS78A	OEM
SN54LS83A	OEM
SN54LS86	OEM
SN54LS90	OEM
SN54LS91	OEM
SN54LS92	OEM
SN54LS93	OEM
SN54LS96	OEM
SN54LS107A	OEM
SN54LS109A	OEM
SN54LS112A	OEM
SN54LS113A	OEM
SN54LS113AFK	OEM
SN54LS113AJ	OEM
SN54LS113AW	OEM
SN54LS114A	OEM
SN54LS122	OEM
SN54LS123	OEM
SN54LS125A	OEM
SN54LS126A	OEM
SN54LS132	OEM
SN54LS136	OEM
SN54LS136FK	OEM
SN54LS136J	OEM
SN54LS136W	OEM
SN54LS137	OEM
SN54LS138	OEM
SN54LS139	OEM
SN54LS145	OEM
SN54LS147	OEM
SN54LS148	OEM
SN54LS151	OEM
SN54LS152	OEM
SN54LS153	OEM
SN54LS153FK	OEM
SN54LS153J	OEM
SN54LS153W	OEM
SN54LS155	OEM
SN54LS156	OEM
SN54LS157	OEM
SN54LS157FK	OEM
SN54LS157J	OEM
SN54LS157W	OEM
SN54LS158	OEM
SN54LS158FK	OEM
SN54LS158J	OEM
SN54LS158N	OEM
SN54LS158W	OEM
SN54LS160A	OEM
SN54LS161A	OEM
SN54LS162A	OEM
SN54LS163A	OEM
SN54LS164	OEM
SN54LS165A	OEM
SN54LS166A	OEM
SN54LS169B	OEM
SN54LS170	OEM
SN54LS173A	OEM
SN54LS173AFK	OEM
SN54LS173AJ	OEM
SN54LS173AW	OEM
SN54LS174	OEM
SN54LS175	OEM
SN54LS181	OEM
SN54LS183	OEM
SN54LS190	OEM
SN54LS191	OEM
SN54LS192	OEM
SN54LS192FH	OEM
SN54LS192J	OEM
SN54LS193	OEM
SN54LS193FH	OEM
SN54LS193J	OEM
SN54LS194A	OEM
SN54LS194AFH	OEM
SN54LS194AJ	OEM
SN54LS195A	OEM
SN54LS196	OEM
SN54LS196J	OEM
SN54LS196W	OEM
SN54LS197	OEM
SN54LS197J	OEM
SN54LS197W	OEM
SN54LS221	OEM
SN54LS222	OEM
SN54LS224	OEM
SN54LS227	OEM
SN54LS228	OEM
SN54LS240	OEM
SN54LS241	OEM
SN54LS242	OEM
SN54LS243	OEM
SN54LS244	OEM
SN54LS245	OEM
SN54LS247	OEM
SN54LS248	OEM
SN54LS249	OEM
SN54LS251	OEM
SN54LS253	OEM
SN54LS257	OEM
SN54LS258	OEM
SN54LS259	OEM
SN54LS261	OEM
SN54LS266	OEM
SN54LS273	OEM
SN54LS275	OEM
SN54LS279	OEM
SN54LS280	OEM
SN54LS283	OEM
SN54LS290	OEM
SN54LS292	OEM
SN54LS293	OEM
SN54LS294	OEM
SN54LS295B	OEM
SN54LS297	OEM
SN54LS298	OEM
SN54LS299	OEM
SN54LS320	OEM
SN54LS321	OEM
SN54LS322A	OEM
SN54LS323	OEM
SN54LS347	OEM
SN54LS348	OEM
SN54LS352	OEM
SN54LS353	OEM
SN54LS354	OEM
SN54LS355	OEM
SN54LS356	OEM
SN54LS357	OEM
SN54LS365A	OEM
SN54LS366A	OEM
SN54LS367A	OEM
SN54LS368A	OEM
SN54LS373	OEM
SN54LS373FK	OEM
SN54LS373J	OEM
SN54LS374	OEM
SN54LS374FK	OEM
SN54LS374J	OEM
SN54LS375	OEM
SN54LS377	OEM
SN54LS378	OEM
SN54LS379	OEM
SN54LS381A	OEM
SN54LS382	OEM
SN54LS384	OEM
SN54LS385	OEM
SN54LS386	OEM
SN54LS390	OEM
SN54LS390J	OEM
SN54LS390W	OEM
SN54LS393	OEM
SN54LS393J	OEM
SN54LS393W	OEM
SN54LS395A	OEM
SN54LS396	OEM
SN54LS398	OEM
SN54LS399	OEM
SN54LS422	OEM
SN54LS423	OEM
SN54LS424	OEM
SN54LS440	OEM
SN54LS441	OEM
SN54LS442	OEM
SN54LS443	OEM
SN54LS444	OEM
SN54LS445	OEM
SN54LS446	OEM
SN54LS447	OEM
SN54LS448	OEM
SN54LS449	OEM
SN54LS465	OEM
SN54LS466	OEM
SN54LS467	OEM
SN54LS468	OEM
SN54LS490	OEM
SN54LS540	OEM
SN54LS541	OEM
SN54LS590	OEM
SN54LS591	OEM
SN54LS592	OEM
SN54LS593	OEM
SN54LS595	OEM
SN54LS596	OEM
SN54LS597	OEM
SN54LS598	OEM
SN54LS600	OEM
SN54LS601	OEM
SN54LS602	OEM
SN54LS603	OEM
SN54LS604	OEM
SN54LS605	OEM
SN54LS606	OEM
SN54LS607	OEM
SN54LS608	OEM
SN54LS610	OEM
SN54LS611	OEM
SN54LS612	OEM
SN54LS613	OEM
SN54LS620	OEM
SN54LS621	OEM
SN54LS622	OEM
SN54LS623	OEM
SN54LS624	OEM
SN54LS625	OEM
SN54LS626	OEM
SN54LS627	OEM
SN54LS628	OEM
SN54LS629	OEM
SN54LS630	OEM
SN54LS631	OEM
SN54LS636	OEM
SN54LS637	OEM
SN54LS638	OEM
SN54LS639	OEM
SN54LS640	OEM
SN54LS641	OEM
SN54LS642	OEM
SN54LS643	OEM
SN54LS644	OEM
SN54LS645	OEM
SN54LS646	OEM
SN54LS647	OEM
SN54LS648	OEM
SN54LS649	OEM
SN54LS651	OEM
SN54LS652	OEM
SN54LS668	OEM
SN54LS669	OEM
SN54LS670	OEM
SN54LS671	OEM
SN54LS672	OEM
SN54LS673	OEM
SN54LS674	OEM
SN54LS681	OEM
SN54LS682	OEM
SN54LS683	OEM
SN54LS684	OEM
SN54LS685	OEM
SN54LS686	OEM
SN54LS687	OEM
SN54LS688	OEM
SN54LS689	OEM
SN54LS690	OEM
SN54LS691	OEM
SN54LS692	OEM
SN54LS693	OEM
SN54LS696	OEM
SN54LS697	OEM
SN54LS698	OEM
SN54LS699	OEM
SN54LS1245	OEM
SN54PL16L8	OEM
SN54PL16R4	OEM
SN54PL16R6	OEM
SN54PL16R8	OEM
SN54PL333	OEM
SN54PL335	OEM
SN54PL839	OEM
SN54PL840	OEM
SN54S00	OEM
SN54S02	OEM
SN54S02FK	OEM
SN54S02J	OEM
SN54S02W	OEM
SN54S03	OEM
SN54S04	OEM
SN54S05	OEM
SN54S08	OEM
SN54S09	OEM
SN54S10	OEM
SN54S11	OEM
SN54S15	OEM
SN54S20	OEM
SN54S22	OEM
SN54S30	OEM
SN54S32	OEM
SN54S37	OEM
SN54S38	OEM
SN54S40	OEM
SN54S51	OEM
SN54S64	OEM
SN54S65	OEM
SN54S74	OEM
SN54S85	OEM
SN54S86	OEM
SN54S112	OEM
SN54S113	OEM
SN54S113FK	OEM
SN54S113J	OEM
SN54S113W	OEM
SN54S114	OEM
SN54S124	OEM
SN54S132	OEM
SN54S133	OEM
SN54S134	OEM
SN54S135	OEM
SN54S138	OEM
SN54S139	OEM
SN54S140	OEM
SN54S151	OEM
SN54S153	OEM
SN54S153FK	OEM
SN54S153J	OEM
SN54S153W	OEM
SN54S157	OEM
SN54S157FK	OEM
SN54S157J	OEM
SN54S157W	OEM
SN54S158	OEM
SN54S158FK	OEM
SN54S158J	OEM
SN54S158W	OEM
SN54S162	OEM
SN54S163	OEM
SN54S168	OEM
SN54S169	OEM
SN54S174	OEM
SN54S175	OEM
SN54S181	OEM
SN54S182	OEM
SN54S194	OEM
SN54S194FH	OEM
SN54S194J	OEM
SN54S195	OEM
SN54S196	OEM
SN54S196J	OEM
SN54S196W	OEM
SN54S197	OEM
SN54S197J	OEM
SN54S197W	OEM
SN54S226	OEM
SN54S240	OEM
SN54S241	OEM
SN54S244	OEM
SN54S251	OEM
SN54S257	OEM
SN54S260	OEM
SN54S274	OEM
SN54S275	OEM
SN54S280	OEM
SN54S281	OEM
SN54S283	OEM
SN54S299	OEM
SN54S340	OEM
SN54S341	OEM
SN54S344	OEM
SN54S373	OEM
SN54S373FK	OEM
SN54S373J	OEM
SN54S374	OEM
SN54S374FK	OEM
SN54S374J	OEM
SN54S381	OEM
SN54S412	OEM
SN54S428	OEM
SN54S436	OEM
SN54S437	OEM
SN54S438	OEM
SN54S481	OEM
SN54S482	OEM
SN54S484	OEM
SN54S485	OEM
SN60	0600
SN74ALS00A	OEM
SN74ALS00AFN	OEM
SN74ALS00AJ	OEM
SN74ALS00AN	3774
SN74ALS00J	OEM
SN74ALS00N	OEM
SN74ALS00ND	OEM
SN74ALS00NDS	OEM
SN74ALS00NS	OEM
SN74ALS01	OEM
SN74ALS01FN	OEM
SN74ALS01J	OEM
SN74ALS01N	OEM
SN74ALS01ND	OEM
SN74ALS01NDS	OEM
SN74ALS01NS	OEM
SN74ALS02	OEM
SN74ALS02FN	OEM
SN74ALS02J	OEM
SN74ALS02N	OEM
SN74ALS03A	OEM
SN74ALS03AFN	OEM
SN74ALS03AJ	OEM
SN74ALS03AN	5658
SN74ALS03J	OEM
SN74ALS03N	OEM
SN74ALS03ND	OEM
SN74ALS03NDS	OEM
SN74ALS03NS	OEM
SN74ALS04	OEM
SN74ALS04AFN	OEM
SN74ALS04AJ	OEM
SN74ALS04AN	OEM
SN74ALS04J	OEM
SN74ALS04N	OEM
SN74ALS04ND	OEM
SN74ALS04NS	OEM
SN74ALS05	OEM
SN74ALS05AFN	OEM
SN74ALS05AJ	OEM
SN74ALS05AN	OEM
SN74ALS05J	OEM
SN74ALS05N	OEM
SN74ALS05ND	OEM
SN74ALS05NDS	OEM
SN74ALS05NS	OEM
SN74ALS08	OEM
SN74ALS08FN	OEM
SN74ALS08J	OEM
SN74ALS08N	OEM
SN74ALS08ND	OEM
SN74ALS08NDS	OEM
SN74ALS08NS	OEM
SN74ALS09	OEM
SN74ALS09FN	OEM
SN74ALS09J	OEM
SN74ALS09N	OEM
SN74ALS0NDS	OEM
SN74ALS10	OEM
SN74ALS10AN	OEM
SN74ALS10FN	OEM
SN74ALS10J	OEM
SN74ALS10N	OEM
SN74ALS10ND	OEM
SN74ALS10NDS	OEM
SN74ALS10NS	OEM
SN74ALS11	OEM
SN74ALS11FN	OEM
SN74ALS11J	OEM
SN74ALS11N	OEM
SN74ALS11NDS	OEM
SN74ALS11NS	OEM
SN74ALS12	OEM
SN74ALS12FN	OEM
SN74ALS12J	OEM
SN74ALS12N	OEM
SN74ALS12NDS	OEM
SN74ALS12NS	OEM
SN74ALS13J	OEM
SN74ALS13N	OEM
SN74ALS14J	OEM
SN74ALS14N	OEM
SN74ALS15	OEM
SN74ALS15FN	OEM
SN74ALS15J	OEM
SN74ALS15N	OEM
SN74ALS15ND	OEM
SN74ALS15NDS	OEM
SN74ALS15NS	OEM
SN74ALS20A	OEM
SN74ALS20AFN	OEM
SN74ALS20AJ	OEM
SN74ALS20AN	OEM
SN74ALS20J	OEM
SN74ALS20N	OEM
SN74ALS20ND	OEM
SN74ALS20NDS	OEM
SN74ALS20NS	OEM
SN74ALS21	OEM
SN74ALS21FN	OEM
SN74ALS21J	OEM
SN74ALS21N	OEM
SN74ALS21ND	OEM
SN74ALS21NDS	OEM
SN74ALS21NS	OEM
SN74ALS22A	OEM
SN74ALS22AFN	OEM
SN74ALS22AJ	OEM
SN74ALS22AN	OEM
SN74ALS22J	OEM
SN74ALS22N	OEM
SN74ALS22ND	OEM
SN74ALS22NDS	OEM
SN74ALS22NS	OEM
SN74ALS27	OEM
SN74ALS27FN	OEM
SN74ALS27J	OEM
SN74ALS27N	OEM
SN74ALS27ND	OEM
SN74ALS27NDS	OEM
SN74ALS27NS	OEM
SN74ALS28	OEM
SN74ALS28AFN	OEM
SN74ALS28AJ	OEM
SN74ALS28AN	OEM
SN74ALS28J	OEM
SN74ALS28N	OEM
SN74ALS28ND	OEM
SN74ALS28NS	OEM

If replacement code is OEM, contact original manufacturer for replacement.

DEVICE TYPE	REPL CODE
SN74ALS30	OEM
SN74ALS30AN	OEM
SN74ALS30FN	OEM
SN74ALS30J	OEM
SN74ALS30N	OEM
SN74ALS32	OEM
SN74ALS32FN	OEM
SN74ALS32J	OEM
SN74ALS32N	3775
SN74ALS32ND	OEM
SN74ALS32NDS	OEM
SN74ALS32NS	OEM
SN74ALS33	OEM
SN74ALS33AFN	OEM
SN74ALS33AJ	OEM
SN74ALS33AN	OEM
SN74ALS33J	OEM
SN74ALS33N	OEM
SN74ALS34FN	OEM
SN74ALS34J	OEM
SN74ALS34N	OEM
SN74ALS35FN	OEM
SN74ALS35J	OEM
SN74ALS35N	OEM
SN74ALS37	OEM
SN74ALS37AFN	OEM
SN74ALS37AJ	OEM
SN74ALS37AN	OEM
SN74ALS37J	OEM
SN74ALS37N	OEM
SN74ALS38	OEM
SN74ALS38AFN	OEM
SN74ALS38AJ	OEM
SN74ALS38AN	OEM
SN74ALS38J	OEM
SN74ALS38N	3800
SN74ALS40	OEM
SN74ALS40AFN	OEM
SN74ALS40AJ	OEM
SN74ALS40AN	OEM
SN74ALS40J	OEM
SN74ALS40N	OEM
SN74ALS51J	OEM
SN74ALS51N	OEM
SN74ALS55IN	OEM
SN74ALS55IND	OEM
SN74ALS55INDS	OEM
SN74ALS55INS	OEM
SN74ALS55J	OEM
SN74ALS55N	OEM
SN74ALS55ND	OEM
SN74ALS55NDS	OEM
SN74ALS55NS	OEM
SN74ALS74	OEM
SN74ALS74AN	OEM
SN74ALS74FN	OEM
SN74ALS74J	OEM
SN74ALS74N	OEM
SN74ALS86	OEM
SN74ALS86FN	OEM
SN74ALS86J	OEM
SN74ALS86N	OEM
SN74ALS91J	OEM
SN74ALS91N	OEM
SN74ALS109	OEM
SN74ALS109AN	OEM
SN74ALS109FN	OEM
SN74ALS109J	OEM
SN74ALS109N	OEM
SN74ALS112	OEM
SN74ALS112AFN	OEM
SN74ALS112AJ	OEM
SN74ALS112AN	OEM
SN74ALS112N	OEM
SN74ALS113	OEM
SN74ALS113AFN	OEM
SN74ALS113AJ	OEM
SN74ALS113AN	OEM
SN74ALS114	OEM
SN74ALS114AFN	OEM
SN74ALS114AJ	OEM
SN74ALS114AN	OEM
SN74ALS131FN	OEM
SN74ALS131J	OEM
SN74ALS131N	OEM
SN74ALS132J	OEM
SN74ALS132N	OEM
SN74ALS133	OEM
SN74ALS133FN	OEM
SN74ALS133J	OEM
SN74ALS133N	OEM
SN74ALS137	OEM
SN74ALS137FN	OEM
SN74ALS137J	OEM
SN74ALS137N	OEM
SN74ALS138	OEM
SN74ALS138FN	OEM
SN74ALS138J	OEM
SN74ALS138N	5763
SN74ALS139	OEM
SN74ALS139FN	OEM
SN74ALS139J	OEM
SN74ALS139N	OEM
SN74ALS151	OEM
SN74ALS151FN	OEM
SN74ALS151J	OEM
SN74ALS151N	OEM
SN74ALS153	OEM
SN74ALS153FN	OEM

DEVICE TYPE	REPL CODE
SN74ALS153J	OEM
SN74ALS153N	5794
SN74ALS157	OEM
SN74ALS157FN	OEM
SN74ALS157J	OEM
SN74ALS157N	5801
SN74ALS157ND	OEM
SN74ALS157NDS	OEM
SN74ALS157NS	OEM
SN74ALS158	OEM
SN74ALS158FN	OEM
SN74ALS158J	OEM
SN74ALS158N	OEM
SN74ALS158ND	OEM
SN74ALS158NDS	OEM
SN74ALS158NS	OEM
SN74ALS160	OEM
SN74ALS160AFN	OEM
SN74ALS160AJ	OEM
SN74ALS160AN	OEM
SN74ALS160J	OEM
SN74ALS160N	OEM
SN74ALS160ND	OEM
SN74ALS160NDS	OEM
SN74ALS160NS	OEM
SN74ALS161	OEM
SN74ALS161AFN	OEM
SN74ALS161AJ	OEM
SN74ALS161AN	OEM
SN74ALS161BN	OEM
SN74ALS161J	OEM
SN74ALS161N	OEM
SN74ALS161ND	OEM
SN74ALS161NDS	OEM
SN74ALS161NS	OEM
SN74ALS162	OEM
SN74ALS162AFN	OEM
SN74ALS162AJ	OEM
SN74ALS162AN	OEM
SN74ALS162J	OEM
SN74ALS162N	OEM
SN74ALS162ND	OEM
SN74ALS162NDS	OEM
SN74ALS162NS	OEM
SN74ALS163	OEM
SN74ALS163AFN	OEM
SN74ALS163AJ	OEM
SN74ALS163AN	OEM
SN74ALS163J	OEM
SN74ALS163N	OEM
SN74ALS163ND	OEM
SN74ALS163NDS	OEM
SN74ALS163NS	OEM
SN74ALS164FN	OEM
SN74ALS164J	OEM
SN74ALS164N	OEM
SN74ALS165FN	OEM
SN74ALS165J	OEM
SN74ALS165N	OEM
SN74ALS166FN	OEM
SN74ALS166J	OEM
SN74ALS166N	OEM
SN74ALS168	OEM
SN74ALS168AFN	OEM
SN74ALS168AJ	OEM
SN74ALS168AN	OEM
SN74ALS168J	OEM
SN74ALS168N	OEM
SN74ALS169	OEM
SN74ALS169AFN	OEM
SN74ALS169AJ	OEM
SN74ALS169AN	OEM
SN74ALS169J	OEM
SN74ALS169N	OEM
SN74ALS174	OEM
SN74ALS174FN	OEM
SN74ALS174J	OEM
SN74ALS174N	OEM
SN74ALS175	OEM
SN74ALS175FN	OEM
SN74ALS175J	OEM
SN74ALS175N	OEM
SN74ALS190	OEM
SN74ALS190FN	OEM
SN74ALS190J	OEM
SN74ALS190N	OEM
SN74ALS190ND	OEM
SN74ALS190NDS	OEM
SN74ALS190S	OEM
SN74ALS191	OEM
SN74ALS191FN	OEM
SN74ALS191J	OEM
SN74ALS191N	OEM
SN74ALS191ND	OEM
SN74ALS191NDS	OEM
SN74ALS191NS	OEM
SN74ALS192	OEM
SN74ALS192FN	OEM
SN74ALS192J	OEM
SN74ALS192N	OEM
SN74ALS192ND	OEM
SN74ALS192NDS	OEM
SN74ALS192S	OEM
SN74ALS193	OEM
SN74ALS193FN	OEM
SN74ALS193J	OEM
SN74ALS193N	5876
SN74ALS193ND	OEM
SN74ALS193NDS	OEM

DEVICE TYPE	REPL CODE
SN74ALS193NS	OEM
SN74ALS217	OEM
SN74ALS218	OEM
SN74ALS238J	OEM
SN74ALS238N	OEM
SN74ALS239J	OEM
SN74ALS239N	OEM
SN74ALS240	OEM
SN74ALS240AFN	OEM
SN74ALS240AJ	OEM
SN74ALS240AN	OEM
SN74ALS240J	OEM
SN74ALS240N	5930
SN74ALS241	OEM
SN74ALS241AFN	OEM
SN74ALS241AJ	OEM
SN74ALS241AN	OEM
SN74ALS241J	OEM
SN74ALS241N	OEM
SN74ALS242	OEM
SN74ALS242AFN	OEM
SN74ALS242AJ	OEM
SN74ALS242AN	OEM
SN74ALS242J	OEM
SN74ALS242N	OEM
SN74ALS243	OEM
SN74ALS243AFN	OEM
SN74ALS243AJ	OEM
SN74ALS243AN	OEM
SN74ALS243J	OEM
SN74ALS243N	OEM
SN74ALS244	OEM
SN74ALS244AFN	OEM
SN74ALS244AJ	OEM
SN74ALS244AN	OEM
SN74ALS244J	OEM
SN74ALS244N	5941
SN74ALS245	OEM
SN74ALS245AFN	OEM
SN74ALS245AJ	OEM
SN74ALS245AN	OEM
SN74ALS245J	OEM
SN74ALS245N	OEM
SN74ALS251	OEM
SN74ALS251FN	OEM
SN74ALS251J	OEM
SN74ALS251N	OEM
SN74ALS253	OEM
SN74ALS253FN	OEM
SN74ALS253J	OEM
SN74ALS253N	OEM
SN74ALS257	OEM
SN74ALS257FN	OEM
SN74ALS257J	OEM
SN74ALS257N	OEM
SN74ALS258	OEM
SN74ALS258FN	OEM
SN74ALS258J	OEM
SN74ALS258N	OEM
SN74ALS259	OEM
SN74ALS259FN	OEM
SN74ALS259J	OEM
SN74ALS259N	OEM
SN74ALS273	OEM
SN74ALS273FN	OEM
SN74ALS273J	OEM
SN74ALS273N	OEM
SN74ALS299	OEM
SN74ALS299FN	OEM
SN74ALS299J	OEM
SN74ALS299N	OEM
SN74ALS317	OEM
SN74ALS318	OEM
SN74ALS323	OEM
SN74ALS323FN	OEM
SN74ALS323J	OEM
SN74ALS323N	OEM
SN74ALS352	OEM
SN74ALS352FN	OEM
SN74ALS352J	OEM
SN74ALS352N	OEM
SN74ALS353	OEM
SN74ALS353FN	OEM
SN74ALS353J	OEM
SN74ALS353N	OEM
SN74ALS365FN	OEM
SN74ALS365J	OEM
SN74ALS365N	OEM
SN74ALS366FN	OEM
SN74ALS366J	OEM
SN74ALS366N	OEM
SN74ALS367FN	OEM
SN74ALS367J	OEM
SN74ALS367N	6033
SN74ALS368FN	OEM
SN74ALS368J	OEM
SN74ALS368N	OEM
SN74ALS373	OEM
SN74ALS373FN	OEM
SN74ALS373J	OEM
SN74ALS373N	OEM
SN74ALS374	OEM
SN74ALS374FN	OEM
SN74ALS374J	OEM
SN74ALS374N	OEM
SN74ALS377J	OEM
SN74ALS377N	OEM
SN74ALS465	OEM
SN74ALS465AFN	OEM

DEVICE TYPE	REPL CODE
SN74ALS465AJ	OEM
SN74ALS465AN	OEM
SN74ALS466	OEM
SN74ALS466AFN	OEM
SN74ALS466AJ	OEM
SN74ALS466AN	OEM
SN74ALS467	OEM
SN74ALS467AFN	OEM
SN74ALS467AJ	OEM
SN74ALS467AN	OEM
SN74ALS468	OEM
SN74ALS468AFN	OEM
SN74ALS468AJ	OEM
SN74ALS468AN	OEM
SN74ALS518FN	OEM
SN74ALS518J	OEM
SN74ALS518N	OEM
SN74ALS519FN	OEM
SN74ALS519J	OEM
SN74ALS519N	OEM
SN74ALS520FN	OEM
SN74ALS520J	OEM
SN74ALS520N	OEM
SN74ALS521	OEM
SN74ALS521FN	OEM
SN74ALS521J	OEM
SN74ALS521N	OEM
SN74ALS522FN	OEM
SN74ALS522J	OEM
SN74ALS522N	OEM
SN74ALS526FN	OEM
SN74ALS526J	OEM
SN74ALS526N	OEM
SN74ALS527FN	OEM
SN74ALS527J	OEM
SN74ALS527N	OEM
SN74ALS528FN	OEM
SN74ALS528J	OEM
SN74ALS528N	OEM
SN74ALS533FN	OEM
SN74ALS533J	OEM
SN74ALS533N	OEM
SN74ALS534FN	OEM
SN74ALS534J	OEM
SN74ALS534N	OEM
SN74ALS537J	OEM
SN74ALS537N	OEM
SN74ALS538	OEM
SN74ALS538FN	OEM
SN74ALS538J	OEM
SN74ALS538N	OEM
SN74ALS539FN	OEM
SN74ALS539J	OEM
SN74ALS539N	OEM
SN74ALS540	OEM
SN74ALS540FN	OEM
SN74ALS540J	OEM
SN74ALS540N	OEM
SN74ALS541	OEM
SN74ALS541FN	OEM
SN74ALS541J	OEM
SN74ALS541N	OEM
SN74ALS560	OEM
SN74ALS560AFN	OEM
SN74ALS560AJ	OEM
SN74ALS560AN	OEM
SN74ALS560J	OEM
SN74ALS560N	OEM
SN74ALS561	OEM
SN74ALS561AFN	OEM
SN74ALS561AJ	OEM
SN74ALS561AN	OEM
SN74ALS561J	OEM
SN74ALS561N	OEM
SN74ALS563	OEM
SN74ALS563FN	OEM
SN74ALS563J	OEM
SN74ALS563N	OEM
SN74ALS564	OEM
SN74ALS564FN	OEM
SN74ALS564J	OEM
SN74ALS564N	OEM
SN74ALS568	OEM
SN74ALS568AFN	OEM
SN74ALS568AJ	OEM
SN74ALS568AN	OEM
SN74ALS568J	OEM
SN74ALS568N	OEM
SN74ALS569	OEM
SN74ALS569AFN	OEM
SN74ALS569AJ	OEM
SN74ALS569AN	OEM
SN74ALS569J	OEM
SN74ALS569N	OEM
SN74ALS573	OEM
SN74ALS573BN	OEM
SN74ALS573FN	OEM
SN74ALS573J	OEM
SN74ALS573N	OEM
SN74ALS574	OEM
SN74ALS574FN	OEM
SN74ALS574J	OEM
SN74ALS574N	OEM
SN74ALS575FN	OEM
SN74ALS575J	OEM
SN74ALS575N	OEM
SN74ALS576	OEM
SN74ALS576FN	OEM
SN74ALS576J	OEM

DEVICE TYPE	REPL CODE
SN74ALS576N	OEM
SN74ALS577FN	OEM
SN74ALS577J	OEM
SN74ALS577N	OEM
SN74ALS580	OEM
SN74ALS580FN	OEM
SN74ALS580J	OEM
SN74ALS580N	OEM
SN74ALS612N	OEM
SN74ALS620	OEM
SN74ALS620AFN	OEM
SN74ALS620AJ	OEM
SN74ALS620AN	OEM
SN74ALS620J	OEM
SN74ALS620N	OEM
SN74ALS621	OEM
SN74ALS621AFN	OEM
SN74ALS621AJ	OEM
SN74ALS621AN	OEM
SN74ALS621J	OEM
SN74ALS621N	OEM
SN74ALS622	OEM
SN74ALS622AFN	OEM
SN74ALS622AJ	OEM
SN74ALS622AN	OEM
SN74ALS622J	OEM
SN74ALS622N	OEM
SN74ALS623	OEM
SN74ALS623AFN	OEM
SN74ALS623AJ	OEM
SN74ALS623AN	OEM
SN74ALS623J	OEM
SN74ALS623N	OEM
SN74ALS632JD	OEM
SN74ALS633JD	OEM
SN74ALS634JD	OEM
SN74ALS635JD	OEM
SN74ALS636	OEM
SN74ALS637	OEM
SN74ALS638	OEM
SN74ALS638AFN	OEM
SN74ALS638AJ	OEM
SN74ALS638AN	OEM
SN74ALS638J	OEM
SN74ALS638N	OEM
SN74ALS639	OEM
SN74ALS639AFN	OEM
SN74ALS639AJ	OEM
SN74ALS639AN	OEM
SN74ALS639J	OEM
SN74ALS639N	OEM
SN74ALS640	OEM
SN74ALS640AFN	OEM
SN74ALS640AJ	OEM
SN74ALS640AN	OEM
SN74ALS640J	OEM
SN74ALS640N	OEM
SN74ALS641	OEM
SN74ALS641AFN	OEM
SN74ALS641AJ	OEM
SN74ALS641AN	OEM
SN74ALS641J	OEM
SN74ALS641N	OEM
SN74ALS642	OEM
SN74ALS642AFN	OEM
SN74ALS642AJ	OEM
SN74ALS642AN	OEM
SN74ALS642J	OEM
SN74ALS642N	OEM
SN74ALS643	OEM
SN74ALS643AFN	OEM
SN74ALS643AJ	OEM
SN74ALS643AN	OEM
SN74ALS643J	OEM
SN74ALS643N	OEM
SN74ALS644	OEM
SN74ALS644AFN	OEM
SN74ALS644AJ	OEM
SN74ALS644AN	OEM
SN74ALS644J	OEM
SN74ALS644N	OEM
SN74ALS645	OEM
SN74ALS645AFN	OEM
SN74ALS645AJ	OEM
SN74ALS645AN	OEM
SN74ALS646FN	OEM
SN74ALS646J	OEM
SN74ALS646JT	OEM
SN74ALS646N	OEM
SN74ALS646NT	OEM
SN74ALS647FN	OEM
SN74ALS647J	OEM
SN74ALS647JT	OEM
SN74ALS647N	OEM
SN74ALS647NT	OEM
SN74ALS648FN	OEM
SN74ALS648J	OEM
SN74ALS648JT	OEM
SN74ALS648N	OEM
SN74ALS648NT	OEM
SN74ALS649FN	OEM
SN74ALS649J	OEM
SN74ALS649JT	OEM
SN74ALS649N	OEM
SN74ALS649NT	OEM
SN74ALS651FN	OEM
SN74ALS651J	OEM
SN74ALS651JT	OEM
SN74ALS651N	OEM

DEVICE TYPE	REPL CODE
SN74ALS651NT	OEM
SN74ALS652FN	OEM
SN74ALS652J	OEM
SN74ALS652JT	OEM
SN74ALS652N	OEM
SN74ALS652NT	OEM
SN74ALS653FN	OEM
SN74ALS653J	OEM
SN74ALS653JT	OEM
SN74ALS653N	OEM
SN74ALS653NT	OEM
SN74ALS654FN	OEM
SN74ALS654J	OEM
SN74ALS654JT	OEM
SN74ALS654N	OEM
SN74ALS654NT	OEM
SN74ALS671J	OEM
SN74ALS671N	OEM
SN74ALS672J	OEM
SN74ALS672N	OEM
SN74ALS677FN	OEM
SN74ALS677JT	OEM
SN74ALS677NT	OEM
SN74ALS678FN	OEM
SN74ALS678JT	OEM
SN74ALS678NT	OEM
SN74ALS679FN	OEM
SN74ALS679J	OEM
SN74ALS679N	OEM
SN74ALS680FN	OEM
SN74ALS680J	OEM
SN74ALS680N	OEM
SN74ALS688	OEM
SN74ALS688FN	OEM
SN74ALS688J	OEM
SN74ALS688N	OEM
SN74ALS689	OEM
SN74ALS689FN	OEM
SN74ALS689J	OEM
SN74ALS689N	OEM
SN74ALS690J	OEM
SN74ALS690N	OEM
SN74ALS691J	OEM
SN74ALS691N	OEM
SN74ALS692J	OEM
SN74ALS692N	OEM
SN74ALS693J	OEM
SN74ALS693N	OEM
SN74ALS694J	OEM
SN74ALS694N	OEM
SN74ALS695J	OEM
SN74ALS695N	OEM
SN74ALS696J	OEM
SN74ALS696N	OEM
SN74ALS697J	OEM
SN74ALS697N	OEM
SN74ALS698J	OEM
SN74ALS698N	OEM
SN74ALS699J	OEM
SN74ALS699N	OEM
SN74ALS790J	OEM
SN74ALS790N	OEM
SN74ALS804	OEM
SN74ALS804FN	OEM
SN74ALS804J	OEM
SN74ALS804N	OEM
SN74ALS805	OEM
SN74ALS805FN	OEM
SN74ALS805J	OEM
SN74ALS805N	OEM
SN74ALS808	OEM
SN74ALS808FN	OEM
SN74ALS808J	OEM
SN74ALS808N	OEM
SN74ALS832	OEM
SN74ALS832FN	OEM
SN74ALS832J	OEM
SN74ALS832N	OEM
SN74ALS841FN	OEM
SN74ALS841JT	OEM
SN74ALS841NT	OEM
SN74ALS842FN	OEM
SN74ALS842JT	OEM
SN74ALS842NT	OEM
SN74ALS843FN	OEM
SN74ALS843JT	OEM
SN74ALS843NT	OEM
SN74ALS844FN	OEM
SN74ALS844JT	OEM
SN74ALS844NT	OEM
SN74ALS845FN	OEM
SN74ALS845JT	OEM
SN74ALS845NT	OEM
SN74ALS846FN	OEM
SN74ALS846JT	OEM
SN74ALS846NT	OEM
SN74ALS857	OEM
SN74ALS857FN	OEM
SN74ALS857JT	OEM
SN74ALS857NT	OEM
SN74ALS873	OEM
SN74ALS873FN	OEM
SN74ALS873J	OEM
SN74ALS873JT	OEM
SN74ALS873N	OEM
SN74ALS873NT	OEM
SN74ALS874	OEM
SN74ALS874FN	OEM
SN74ALS874J	OEM

DEVICE TYPE	REPL CODE
SN74ALS874JT	OEM
SN74ALS874N	OEM
SN74ALS874NT	OEM
SN74ALS876	OEM
SN74ALS876FN	OEM
SN74ALS876J	OEM
SN74ALS876JT	OEM
SN74ALS876N	OEM
SN74ALS876NT	OEM
SN74ALS878FN	OEM
SN74ALS878J	OEM
SN74ALS878JT	OEM
SN74ALS878N	OEM
SN74ALS878NT	OEM
SN74ALS879FN	OEM
SN74ALS879J	OEM
SN74ALS879JT	OEM
SN74ALS879N	OEM
SN74ALS879NT	OEM
SN74ALS880	OEM
SN74ALS880FN	OEM
SN74ALS880J	OEM
SN74ALS880JT	OEM
SN74ALS880N	OEM
SN74ALS880NT	OEM
SN74ALS1000	OEM
SN74ALS1000AFN	OEM
SN74ALS1000AJ	OEM
SN74ALS1000AN	OEM
SN74ALS1002	OEM
SN74ALS1002AFN	OEM
SN74ALS1002AJ	OEM
SN74ALS1002AN	OEM
SN74ALS1003	OEM
SN74ALS1003AFN	OEM
SN74ALS1003AJ	OEM
SN74ALS1003AN	OEM
SN74ALS1004	OEM
SN74ALS1004FN	OEM
SN74ALS1004J	OEM
SN74ALS1004N	OEM
SN74ALS1005	OEM
SN74ALS1005FN	OEM
SN74ALS1005J	OEM
SN74ALS1005N	OEM
SN74ALS1008	OEM
SN74ALS1008AFN	OEM
SN74ALS1008AJ	OEM
SN74ALS1008AN	OEM
SN74ALS1010	OEM
SN74ALS1010AFN	OEM
SN74ALS1010AJ	OEM
SN74ALS1010AN	OEM
SN74ALS1011	OEM
SN74ALS1011AFN	OEM
SN74ALS1011AJ	OEM
SN74ALS1011AN	OEM
SN74ALS1020	OEM
SN74ALS1020AFN	OEM
SN74ALS1020AJ	OEM
SN74ALS1020AN	OEM
SN74ALS1032	OEM
SN74ALS1032AFN	OEM
SN74ALS1032AJ	OEM
SN74ALS1032AN	OEM
SN74ALS1034	OEM
SN74ALS1034FN	OEM
SN74ALS1034J	OEM
SN74ALS1034N	OEM
SN74ALS1035	OEM
SN74ALS1035FN	OEM
SN74ALS1035J	OEM
SN74ALS1035N	OEM
SN74ALS1240	OEM
SN74ALS1240FN	OEM
SN74ALS1240J	OEM
SN74ALS1240N	OEM
SN74ALS1241	OEM
SN74ALS1241FN	OEM
SN74ALS1241J	OEM
SN74ALS1241N	OEM
SN74ALS1242	OEM
SN74ALS1242FN	OEM
SN74ALS1242J	OEM
SN74ALS1242N	OEM
SN74ALS1243	OEM
SN74ALS1243FN	OEM
SN74ALS1243J	OEM
SN74ALS1243N	OEM
SN74ALS1244	OEM
SN74ALS1244AFN	OEM
SN74ALS1244AJ	OEM
SN74ALS1244AN	OEM
SN74ALS1245	OEM
SN74ALS1245FN	OEM
SN74ALS1245J	OEM
SN74ALS1245N	OEM
SN74ALS1616	OEM
SN74ALS1620	OEM
SN74ALS1620FN	OEM
SN74ALS1620J	OEM
SN74ALS1620N	OEM
SN74ALS1621	OEM
SN74ALS1621FN	OEM
SN74ALS1621J	OEM
SN74ALS1621N	OEM
SN74ALS1622	OEM
SN74ALS1622FN	OEM
SN74ALS1622J	OEM

If replacement code is OEM, contact original manufacturer for replacement.

Original Device Types

DEVICE TYPE	REPL CODE
SN74ALS1622N	OEM
SN74ALS1623	OEM
SN74ALS1623FN	OEM
SN74ALS1623J	OEM
SN74ALS1623N	OEM
SN74ALS1638	OEM
SN74ALS1638FN	OEM
SN74ALS1638J	OEM
SN74ALS1638N	OEM
SN74ALS1639	OEM
SN74ALS1639FN	OEM
SN74ALS1639J	OEM
SN74ALS1639N	OEM
SN74ALS1640	OEM
SN74ALS1640AFN	OEM
SN74ALS1640AJ	OEM
SN74ALS1640AN	OEM
SN74ALS1641	OEM
SN74ALS1641FN	OEM
SN74ALS1641J	OEM
SN74ALS1641N	OEM
SN74ALS1642	OEM
SN74ALS1642FN	OEM
SN74ALS1642J	OEM
SN74ALS1642N	OEM
SN74ALS1643	OEM
SN74ALS1643FN	OEM
SN74ALS1643J	OEM
SN74ALS1643N	OEM
SN74ALS1644	OEM
SN74ALS1644FN	OEM
SN74ALS1644J	OEM
SN74ALS1644N	OEM
SN74ALS1645	OEM
SN74ALS1645AFN	OEM
SN74ALS1645AJ	OEM
SN74ALS1645AN	OEM
SN74ALS1670N	OEM
SN74ALS8003FN	OEM
SN74ALS8003JG	OEM
SN74ALSL191N	OEM
SN74AS00FN	OEM
SN74AS00J	OEM
SN74AS00N	OEM
SN74AS02FN	OEM
SN74AS02J	OEM
SN74AS02N	OEM
SN74AS04FN	OEM
SN74AS04J	OEM
SN74AS04N	OEM
SN74AS08FN	OEM
SN74AS08J	OEM
SN74AS08N	OEM
SN74AS10FN	OEM
SN74AS10J	OEM
SN74AS10N	OEM
SN74AS11FN	OEM
SN74AS11J	OEM
SN74AS11N	OEM
SN74AS20FN	OEM
SN74AS20J	OEM
SN74AS20N	OEM
SN74AS21FN	OEM
SN74AS21J	OEM
SN74AS21N	OEM
SN74AS27FN	OEM
SN74AS27J	OEM
SN74AS27N	OEM
SN74AS30FN	OEM
SN74AS30J	OEM
SN74AS30N	OEM
SN74AS32FN	OEM
SN74AS32J	OEM
SN74AS32N	OEM
SN74AS34FN	OEM
SN74AS34J	OEM
SN74AS34N	OEM
SN74AS74	OEM
SN74AS74FN	OEM
SN74AS74J	OEM
SN74AS74N	OEM
SN74AS94FN	OEM
SN74AS94J	OEM
SN74AS94N	OEM
SN74AS109	OEM
SN74AS109FN	OEM
SN74AS109J	OEM
SN74AS109N	OEM
SN74AS112	OEM
SN74AS112FN	OEM
SN74AS112J	OEM
SN74AS112N	OEM
SN74AS113	OEM
SN74AS113FN	OEM
SN74AS113J	OEM
SN74AS113N	OEM
SN74AS114	OEM
SN74AS114FN	OEM
SN74AS114J	OEM
SN74AS114N	OEM
SN74AS131FN	OEM
SN74AS131J	OEM
SN74AS131N	OEM
SN74AS137FN	OEM
SN74AS137J	OEM
SN74AS137N	OEM
SN74AS138FN	OEM
SN74AS138J	OEM
SN74AS138N	OEM

DEVICE TYPE	REPL CODE
SN74AS139FN	OEM
SN74AS139J	OEM
SN74AS139N	OEM
SN74AS151	OEM
SN74AS151FN	OEM
SN74AS151J	OEM
SN74AS151N	OEM
SN74AS153	OEM
SN74AS153FN	OEM
SN74AS153J	OEM
SN74AS153N	OEM
SN74AS157FN	OEM
SN74AS157J	OEM
SN74AS157N	OEM
SN74AS158FN	OEM
SN74AS158J	OEM
SN74AS158N	OEM
SN74AS160	OEM
SN74AS160FN	OEM
SN74AS160J	OEM
SN74AS160N	OEM
SN74AS161	OEM
SN74AS161FN	OEM
SN74AS161J	OEM
SN74AS161N	OEM
SN74AS162	OEM
SN74AS162FN	OEM
SN74AS162J	OEM
SN74AS162N	OEM
SN74AS163	OEM
SN74AS163FN	OEM
SN74AS163J	OEM
SN74AS163N	OEM
SN74AS168	OEM
SN74AS168FN	OEM
SN74AS168J	OEM
SN74AS168N	OEM
SN74AS169	OEM
SN74AS169FN	OEM
SN74AS169J	OEM
SN74AS169N	OEM
SN74AS174	OEM
SN74AS174FN	OEM
SN74AS174J	OEM
SN74AS174N	OEM
SN74AS175	OEM
SN74AS175FN	OEM
SN74AS175J	OEM
SN74AS175N	OEM
SN74AS181A	OEM
SN74AS181AFN	OEM
SN74AS181AJ	OEM
SN74AS181AN	OEM
SN74AS181ANT	OEM
SN74AS182FN	OEM
SN74AS182J	OEM
SN74AS182N	OEM
SN74AS194FN	OEM
SN74AS194J	OEM
SN74AS194N	OEM
SN74AS195FN	OEM
SN74AS195J	OEM
SN74AS195N	OEM
SN74AS230	OEM
SN74AS230FN	OEM
SN74AS230J	OEM
SN74AS230N	OEM
SN74AS231	OEM
SN74AS231FN	OEM
SN74AS231J	OEM
SN74AS231N	OEM
SN74AS240	OEM
SN74AS240FN	OEM
SN74AS240J	OEM
SN74AS240N	OEM
SN74AS241	OEM
SN74AS241FN	OEM
SN74AS241J	OEM
SN74AS241N	OEM
SN74AS242	OEM
SN74AS242FN	OEM
SN74AS242J	OEM
SN74AS242N	OEM
SN74AS243	OEM
SN74AS243FN	OEM
SN74AS243J	OEM
SN74AS243N	OEM
SN74AS244	OEM
SN74AS244FN	OEM
SN74AS244J	OEM
SN74AS244N	OEM
SN74AS245FN	OEM
SN74AS245J	OEM
SN74AS245N	OEM
SN74AS250FN	OEM
SN74AS250J	OEM
SN74AS250N	OEM
SN74AS251	OEM
SN74AS251FN	OEM
SN74AS251J	OEM
SN74AS251N	OEM
SN74AS253	OEM
SN74AS253FN	OEM
SN74AS253J	OEM
SN74AS253N	OEM
SN74AS257FN	OEM
SN74AS257J	OEM
SN74AS257N	OEM
SN74AS258FN	OEM

DEVICE TYPE	REPL CODE
SN74AS258J	OEM
SN74AS258N	OEM
SN74AS264FN	OEM
SN74AS264J	OEM
SN74AS264N	OEM
SN74AS280	OEM
SN74AS280FN	OEM
SN74AS280J	OEM
SN74AS280N	OEM
SN74AS282FN	OEM
SN74AS282J	OEM
SN74AS282N	OEM
SN74AS286FN	OEM
SN74AS286J	OEM
SN74AS286N	OEM
SN74AS298FN	OEM
SN74AS298J	OEM
SN74AS298N	OEM
SN74AS299FN	OEM
SN74AS299J	OEM
SN74AS299N	OEM
SN74AS323FN	OEM
SN74AS323J	OEM
SN74AS323N	OEM
SN74AS352	OEM
SN74AS352FN	OEM
SN74AS352J	OEM
SN74AS352N	OEM
SN74AS353	OEM
SN74AS353FN	OEM
SN74AS353J	OEM
SN74AS353N	OEM
SN74AS373	OEM
SN74AS373FN	OEM
SN74AS373J	OEM
SN74AS373N	OEM
SN74AS374	OEM
SN74AS374FN	OEM
SN74AS374J	OEM
SN74AS374N	OEM
SN74AS395FN	OEM
SN74AS395J	OEM
SN74AS395N	OEM
SN74AS533	OEM
SN74AS533FN	OEM
SN74AS533J	OEM
SN74AS533N	OEM
SN74AS534	OEM
SN74AS534FN	OEM
SN74AS534J	OEM
SN74AS534N	OEM
SN74AS573	OEM
SN74AS573FN	OEM
SN74AS573J	OEM
SN74AS573N	OEM
SN74AS574	OEM
SN74AS574FN	OEM
SN74AS574J	OEM
SN74AS574N	OEM
SN74AS575FN	OEM
SN74AS575J	OEM
SN74AS575N	OEM
SN74AS576	OEM
SN74AS576FN	OEM
SN74AS576J	OEM
SN74AS576N	OEM
SN74AS577FN	OEM
SN74AS577J	OEM
SN74AS577N	OEM
SN74AS580	OEM
SN74AS580FN	OEM
SN74AS580J	OEM
SN74AS580N	OEM
SN74AS598	OEM
SN74AS620FN	OEM
SN74AS620J	OEM
SN74AS620N	OEM
SN74AS621FN	OEM
SN74AS621J	OEM
SN74AS621N	OEM
SN74AS622FN	OEM
SN74AS622J	OEM
SN74AS622N	OEM
SN74AS623FN	OEM
SN74AS623J	OEM
SN74AS623N	OEM
SN74AS638FN	OEM
SN74AS638J	OEM
SN74AS638N	OEM
SN74AS639FN	OEM
SN74AS639J	OEM
SN74AS639N	OEM
SN74AS640	OEM
SN74AS640FN	OEM
SN74AS640J	OEM
SN74AS640N	OEM
SN74AS641	OEM
SN74AS641FN	OEM
SN74AS641J	OEM
SN74AS641N	OEM
SN74AS642	OEM
SN74AS642FN	OEM
SN74AS642J	OEM
SN74AS642N	OEM
SN74AS643	OEM
SN74AS643FN	OEM
SN74AS643J	OEM
SN74AS643N	OEM
SN74AS644	OEM

DEVICE TYPE	REPL CODE
SN74AS644FN	OEM
SN74AS644J	OEM
SN74AS644N	OEM
SN74AS645	OEM
SN74AS645FN	OEM
SN74AS645J	OEM
SN74AS645N	OEM
SN74AS646FN	OEM
SN74AS646JT	OEM
SN74AS646NT	OEM
SN74AS648FN	OEM
SN74AS648JT	OEM
SN74AS648NT	OEM
SN74AS651FN	OEM
SN74AS651JT	OEM
SN74AS651NT	OEM
SN74AS652FN	OEM
SN74AS652JT	OEM
SN74AS652NT	OEM
SN74AS756FN	OEM
SN74AS756J	OEM
SN74AS756N	OEM
SN74AS757FN	OEM
SN74AS757J	OEM
SN74AS757N	OEM
SN74AS758FN	OEM
SN74AS758J	OEM
SN74AS758N	OEM
SN74AS759FN	OEM
SN74AS759J	OEM
SN74AS759N	OEM
SN74AS760FN	OEM
SN74AS760J	OEM
SN74AS760N	OEM
SN74AS762FN	OEM
SN74AS762J	OEM
SN74AS762N	OEM
SN74AS763FN	OEM
SN74AS763J	OEM
SN74AS763N	OEM
SN74AS800	OEM
SN74AS800FN	OEM
SN74AS800J	OEM
SN74AS800N	OEM
SN74AS801	OEM
SN74AS802	OEM
SN74AS802FN	OEM
SN74AS802J	OEM
SN74AS802N	OEM
SN74AS804	OEM
SN74AS804A	OEM
SN74AS804AFN	OEM
SN74AS804AJ	OEM
SN74AS804AN	OEM
SN74AS805	OEM
SN74AS805A	OEM
SN74AS805AFN	OEM
SN74AS805AJ	OEM
SN74AS805AN	OEM
SN74AS808	OEM
SN74AS808A	OEM
SN74AS808AFN	OEM
SN74AS808AJ	OEM
SN74AS808AN	OEM
SN74AS821FN	OEM
SN74AS821JT	OEM
SN74AS821NT	OEM
SN74AS822FN	OEM
SN74AS822JT	OEM
SN74AS822NT	OEM
SN74AS823FN	OEM
SN74AS823JT	OEM
SN74AS823NT	OEM
SN74AS824FN	OEM
SN74AS824JT	OEM
SN74AS824NT	OEM
SN74AS825FN	OEM
SN74AS825JT	OEM
SN74AS825NT	OEM
SN74AS826FN	OEM
SN74AS826JT	OEM
SN74AS826NT	OEM
SN74AS830	OEM
SN74AS832	OEM
SN74AS832A	OEM
SN74AS832AFN	OEM
SN74AS832AJ	OEM
SN74AS832AN	OEM
SN74AS836	OEM
SN74AS841FN	OEM
SN74AS841JT	OEM
SN74AS841NT	OEM
SN74AS842FN	OEM
SN74AS842J	OEM
SN74AS842N	OEM
SN74AS843FN	OEM
SN74AS843JT	OEM
SN74AS843NT	OEM
SN74AS844FN	OEM
SN74AS844JT	OEM
SN74AS844NT	OEM
SN74AS845FN	OEM
SN74AS845JT	OEM
SN74AS845NT	OEM
SN74AS846FN	OEM
SN74AS846JT	OEM
SN74AS846NT	OEM
SN74AS850	OEM
SN74AS850FN	OEM

DEVICE TYPE	REPL CODE
SN74AS850JD	OEM
SN74AS850N	OEM
SN74AS851FN	OEM
SN74AS851JD	OEM
SN74AS851N	OEM
SN74AS852FN	OEM
SN74AS852JT	OEM
SN74AS852NT	OEM
SN74AS856FN	OEM
SN74AS856JT	OEM
SN74AS856NT	OEM
SN74AS857	OEM
SN74AS857FN	OEM
SN74AS857J	OEM
SN74AS857N	OEM
SN74AS859	OEM
SN74AS866	OEM
SN74AS866FN	OEM
SN74AS866JD	OEM
SN74AS866N	OEM
SN74AS867	OEM
SN74AS867FN	OEM
SN74AS867JT	OEM
SN74AS867NT	OEM
SN74AS869	OEM
SN74AS869FN	OEM
SN74AS869JT	OEM
SN74AS869NT	OEM
SN74AS870	OEM
SN74AS870FN	OEM
SN74AS870JT	OEM
SN74AS870NT	OEM
SN74AS871	OEM
SN74AS871FN	OEM
SN74AS871JD	OEM
SN74AS871N	OEM
SN74AS872	OEM
SN74AS873	OEM
SN74AS873FN	OEM
SN74AS873JT	OEM
SN74AS873NT	OEM
SN74AS874	OEM
SN74AS874FN	OEM
SN74AS874JT	OEM
SN74AS874NT	OEM
SN74AS875	OEM
SN74AS876	OEM
SN74AS876FN	OEM
SN74AS876JT	OEM
SN74AS876NT	OEM
SN74AS877	OEM
SN74AS877FN	OEM
SN74AS877JT	OEM
SN74AS877NT	OEM
SN74AS878FN	OEM
SN74AS878JT	OEM
SN74AS878NT	OEM
SN74AS879FN	OEM
SN74AS879JT	OEM
SN74AS879NT	OEM
SN74AS880	OEM
SN74AS880FN	OEM
SN74AS880JT	OEM
SN74AS880NT	OEM
SN74AS881A	OEM
SN74AS881AFN	OEM
SN74AS881AJT	OEM
SN74AS881ANT	OEM
SN74AS882	OEM
SN74AS882FN	OEM
SN74AS882JT	OEM
SN74AS882NT	OEM
SN74AS883	OEM
SN74AS884	OEM
SN74AS885	OEM
SN74AS885FN	OEM
SN74AS885JT	OEM
SN74AS885NT	OEM
SN74AS886	OEM
SN74AS888	OEM
SN74AS889	OEM
SN74AS890	OEM
SN74AS891	OEM
SN74AS894	OEM
SN74AS1000FN	OEM
SN74AS1000J	OEM
SN74AS1000N	OEM
SN74AS1004	OEM
SN74AS1004FN	OEM
SN74AS1004J	OEM
SN74AS1004N	OEM
SN74AS1008FN	OEM
SN74AS1008J	OEM
SN74AS1008N	OEM
SN74AS1032FN	OEM
SN74AS1032J	OEM
SN74AS1032N	OEM
SN74AS1034FN	OEM
SN74AS1034J	OEM
SN74AS1034N	OEM
SN74AS1036FN	OEM
SN74AS1036J	OEM
SN74AS1036N	OEM
SN74AS2620FN	OEM
SN74AS2620J	OEM
SN74AS2620N	OEM
SN74AS2623FN	OEM
SN74AS2623J	OEM
SN74AS2623N	OEM

DEVICE TYPE	REPL CODE
SN74AS2640FN	OEM
SN74AS2640J	OEM
SN74AS2640N	OEM
SN74AS2645FN	OEM
SN74AS2645J	OEM
SN74AS2645N	OEM
SN74H00	0682
SN74H00J	0682
SN74H00N	0682
SN74H01	5338
SN74H01J	5338
SN74H01N	5338
SN74H04	1907
SN74H04J	1907
SN74H04N	1907
SN74H05	3251
SN74H05J	3251
SN74H05N	3251
SN74H10	0685
SN74H10J	0685
SN74H10N	0685
SN74H11	2398
SN74H11J	2398
SN74H11N	2398
SN74H15	OEM
SN74H15J	OEM
SN74H15N	OEM
SN74H20	3715
SN74H20J	3715
SN74H20N	3715
SN74H21	4856
SN74H21J	4856
SN74H21N	4856
SN74H22	4596
SN74H22J	4596
SN74H22N	4596
SN74H30	5383
SN74H30J	5383
SN74H30N	5383
SN74H40	0559
SN74H40J	0559
SN74H40N	0559
SN74H50	1792
SN74H50J	1792
SN74H50N	1792
SN74H51	1944
SN74H51J	1944
SN74H51N	1944
SN74H52	2020
SN74H52J	2020
SN74H52N	2020
SN74H53	2101
SN74H53J	2101
SN74H53N	2101
SN74H54	2172
SN74H54DC	2172
SN74H54J	2172
SN74H54N	2172
SN74H55	3158
SN74H55J	3158
SN74H55N	3158
SN74H60	5410
SN74H60J	5410
SN74H60N	5410
SN74H61	2663
SN74H61J	2663
SN74H61N	2663
SN74H62	2731
SN74H62J	2731
SN74H62N	2731
SN74H71	3264
SN74H71J	3264
SN74H71N	3264
SN74H72	3313
SN74H72J	3313
SN74H72N	3313
SN74H73	2463
SN74H73J	2463
SN74H73N	2463
SN74H74	2492
SN74H74J	2492
SN74H74N	2492
SN74H76	5302
SN74H76J	5302
SN74H76N	5302
SN74H78	5418
SN74H78J	5418
SN74H78N	5418
SN74H87	2579
SN74H87J	2579
SN74H87N	2579
SN74H101	5521
SN74H101J	5521
SN74H101N	5521
SN74H102	5523
SN74H102J	5523
SN74H102N	5523
SN74H103	2969
SN74H103J	2969
SN74H103N	2969
SN74H106	5253
SN74H106J	5253
SN74H106N	5253
SN74H108	0180
SN74H108J	0180
SN74H108N	0180
SN74H183	4399
SN74H183J	4399
SN74H183N	4399

DEVICE TYPE	REPL CODE
SN74HC00NS	OEM
SN74HC02N	5384
SN74HC03N	1453
SN74HC04ANS	1455
SN74HC05ANS	OEM
SN74HC86N	1588
SN74HC240N	1979
SN74L00N	OEM
SN74L03	1579
SN74L04J	OEM
SN74L04N	OEM
SN74L46J	OEM
SN74L46N	OEM
SN74L47J	OEM
SN74L47N	OEM
SN74L77W	OEM
SN74L91J	OEM
SN74L91N	OEM
SN74L93J	0656
SN74L93N	0656
SN74L95J	OEM
SN74L95N	OEM
SN74L96J	OEM
SN74L96N	OEM
SN74L98J	OEM
SN74L98N	OEM
SN74L99J	OEM
SN74L99N	OEM
SN74L153J	OEM
SN74L153N	OEM
SN74L154J	OEM
SN74L154N	OEM
SN74L157J	OEM
SN74L157N	OEM
SN74L164J	OEM
SN74L164N	OEM
SN74L164T	OEM
SN74LD247D	OEM
SN74LD247FN	OEM
SN74LS00	1527
SN74LS00J	1527
SN74LS00JD	1527
SN74LS00JDS	1527
SN74LS00JS	1527
SN74LS00N	1527
SN74LS00ND	1527
SN74LS00NDS	1527
SN74LS00NS	1527
SN74LS00W	1527
SN74LS01	1546
SN74LS01J	1546
SN74LS01JDS	1546
SN74LS01JS	1546
SN74LS01N	1546
SN74LS01ND	1546
SN74LS01NDS	1546
SN74LS01NS	1546
SN74LS02	1559
SN74LS02D	OEM
SN74LS02FN	OEM
SN74LS02J	1559
SN74LS02JD	1559
SN74LS02JDS	1559
SN74LS02JS	1559
SN74LS02N	1559
SN74LS02NDS	1559
SN74LS02NS	1559
SN74LS02W	1559
SN74LS03	4216
SN74LS03J	1579
SN74LS03JD	1579
SN74LS03JDS	1579
SN74LS03JS	1579
SN74LS03N	1579
SN74LS03ND	1579
SN74LS03NDS	1579
SN74LS03NS	1579
SN74LS03W	1579
SN74LS04	1595
SN74LS04J	1595
SN74LS04JD	1595
SN74LS04JDS	1595
SN74LS04JS	1595
SN74LS04N	1595
SN74LS04ND	1595
SN74LS04NDS	1595
SN74LS04NS	1595
SN74LS05	1608
SN74LS05J	1608
SN74LS05JD	1608
SN74LS05JDS	1608
SN74LS05JS	1608
SN74LS05N	1608
SN74LS05ND	1608
SN74LS05NDS	1608
SN74LS05NS	1608
SN74LS06N	OEM
SN74LS07N	OEM
SN74LS08	1633
SN74LS08J	1633
SN74LS08JD	1633
SN74LS08JDS	1633
SN74LS08JS	1633
SN74LS08N	1633
SN74LS08ND	1633
SN74LS08NDS	1633
SN74LS08NS	1633
SN74LS08W	1633

If replacement code is OEM, contact original manufacturer for replacement.

DEVICE TYPE	REPL CODE
SN74LS09	1642
SN74LS09J	1642
SN74LS09JD	1642
SN74LS09JDS	1642
SN74LS09JS	1642
SN74LS09N	1642
SN74LS09ND	1642
SN74LS09NDS	1642
SN74LS09NS	1642
SN74LS10	1662
SN74LS10J	1662
SN74LS10JD	1662
SN74LS10JDS	1662
SN74LS10JS	1662
SN74LS10N	1662
SN74LS10ND	1662
SN74LS10NDS	1662
SN74LS10NS	1662
SN74LS10W	1662
SN74LS11	1667
SN74LS11J	1667
SN74LS11JD	1667
SN74LS11JDS	1667
SN74LS11JS	1667
SN74LS11N	1667
SN74LS11ND	1667
SN74LS11NDS	1667
SN74LS11NS	1667
SN74LS11W	1667
SN74LS12	1679
SN74LS12J	1679
SN74LS12JD	1679
SN74LS12JDS	1679
SN74LS12JS	1679
SN74LS12N	1679
SN74LS12ND	1679
SN74LS12NDS	1679
SN74LS12NS	1679
SN74LS13	1688
SN74LS13J	1688
SN74LS13JD	1688
SN74LS13JDS	1688
SN74LS13JS	1688
SN74LS13N	1688
SN74LS13ND	1688
SN74LS13NDS	1688
SN74LS13NS	1688
SN74LS14	1698
SN74LS14J	1698
SN74LS14JD	1698
SN74LS14JDS	1698
SN74LS14JS	1698
SN74LS14N	1698
SN74LS14ND	1698
SN74LS14NDS	1698
SN74LS14NS	1698
SN74LS15	1707
SN74LS15J	1707
SN74LS15JD	1707
SN74LS15JDS	1707
SN74LS15JS	1707
SN74LS15N	1707
SN74LS15ND	1707
SN74LS15NDS	1707
SN74LS15NS	1707
SN74LS18	OEM
SN74LS18J	OEM
SN74LS18N	OEM
SN74LS19	OEM
SN74LS19AN	OEM
SN74LS19J	OEM
SN74LS19N	OEM
SN74LS20	0034
SN74LS20J	0034
SN74LS20JD	0034
SN74LS20JDS	0034
SN74LS20JS	0034
SN74LS20N	0034
SN74LS20ND	0034
SN74LS20NDS	0034
SN74LS20NS	0034
SN74LS20W	0034
SN74LS21	1763
SN74LS21D	OEM
SN74LS21FN	OEM
SN74LS21J	1763
SN74LS21JD	1763
SN74LS21JDS	1763
SN74LS21JS	1763
SN74LS21N	1763
SN74LS21ND	1763
SN74LS21NDS	1763
SN74LS21NS	1763
SN74LS22	1775
SN74LS22J	1775
SN74LS22JD	1775
SN74LS22JDS	1775
SN74LS22JS	1775
SN74LS22N	1775
SN74LS22ND	1775
SN74LS22NDS	1775
SN74LS22NS	1775
SN74LS24	OEM
SN74LS24J	OEM
SN74LS24N	OEM
SN74LS26	1382
SN74LS26J	1382
SN74LS26JD	1382
SN74LS26JDS	1382
SN74LS26JS	1382
SN74LS26N	1382
SN74LS26ND	1382
SN74LS26NS	1382
SN74LS27J	0183
SN74LS27JD	0183
SN74LS27JDS	0183
SN74LS27N	0183
SN74LS27ND	0183
SN74LS27NDS	0183
SN74LS27NS	0183
SN74LS27W	0183
SN74LS28	0472
SN74LS28J	0472
SN74LS28JD	0472
SN74LS28JDS	0472
SN74LS28JS	0472
SN74LS28N	0472
SN74LS28ND	0472
SN74LS28NDS	0472
SN74LS28NS	0472
SN74LS30	0828
SN74LS30J	0828
SN74LS30JD	0828
SN74LS30JDS	0828
SN74LS30JS	0828
SN74LS30N	0828
SN74LS30ND	0828
SN74LS30NDS	0828
SN74LS30NS	0828
SN74LS30W	0828
SN74LS31N	OEM
SN74LS32	0088
SN74LS32J	0088
SN74LS32JD	0088
SN74LS32JDS	0088
SN74LS32JS	0088
SN74LS32N	0088
SN74LS32ND	0088
SN74LS32NDS	0088
SN74LS32NS	0088
SN74LS32W	0088
SN74LS33	1832
SN74LS33J	1832
SN74LS33JD	OEM
SN74LS33JDS	OEM
SN74LS33JS	OEM
SN74LS33N	1832
SN74LS33ND	OEM
SN74LS33NDS	OEM
SN74LS33NS	1320
SN74LS37	1730
SN74LS37J	1730
SN74LS37JD	1730
SN74LS37JDS	1730
SN74LS37JS	1730
SN74LS37N	1730
SN74LS37ND	1730
SN74LS37NDS	1730
SN74LS37NS	1730
SN74LS38	1839
SN74LS38J	1839
SN74LS38JD	1839
SN74LS38JDS	1839
SN74LS38JS	1839
SN74LS38N	1839
SN74LS38ND	1839
SN74LS38NDS	1839
SN74LS38NS	1839
SN74LS40	0135
SN74LS40J	0135
SN74LS40JD	0135
SN74LS40JDS	0135
SN74LS40JS	0135
SN74LS40N	0135
SN74LS40ND	0135
SN74LS40NDS	0135
SN74LS40NS	0135
SN74LS42	1841
SN74LS42J	1841
SN74LS42JD	1841
SN74LS42JDS	1841
SN74LS42JS	1841
SN74LS42N	1841
SN74LS42ND	1841
SN74LS42NDS	1841
SN74LS42NS	1841
SN74LS42W	1841
SN74LS47	1845
SN74LS47J	1845
SN74LS47JD	1845
SN74LS47JDS	1845
SN74LS47JS	1845
SN74LS47N	1845
SN74LS47ND	1845
SN74LS47NDS	1845
SN74LS47NS	1845
SN74LS48J	1849
SN74LS48JD	1849
SN74LS48JDS	1849
SN74LS48JS	1849
SN74LS48N	1849
SN74LS48ND	1849
SN74LS48NDS	1849
SN74LS48NS	1849
SN74LS49	1850
SN74LS49J	1850
SN74LS49JD	1850
SN74LS49JDS	1850
SN74LS49JS	1850
SN74LS49N	1850
SN74LS49ND	1850
SN74LS49NDS	1850
SN74LS49NS	1850
SN74LS51	1033
SN74LS51J	1033
SN74LS51JD	1033
SN74LS51JDS	1033
SN74LS51JS	1033
SN74LS51N	1033
SN74LS51ND	1033
SN74LS51NDS	1033
SN74LS51NS	1033
SN74LS51W	1033
SN74LS54	1857
SN74LS54J	1857
SN74LS54JD	1857
SN74LS54JDS	1857
SN74LS54JS	1857
SN74LS54N	1857
SN74LS54ND	1857
SN74LS54NDS	1857
SN74LS54NS	1857
SN74LS55	0457
SN74LS55J	0457
SN74LS55JD	0457
SN74LS55JDS	0457
SN74LS55JS	0457
SN74LS55N	0457
SN74LS55ND	0457
SN74LS55NDS	0457
SN74LS55NS	0457
SN74LS63	OEM
SN74LS63J	OEM
SN74LS63N	OEM
SN74LS68JDS	OEM
SN74LS73A	1867
SN74LS73AJ	1867
SN74LS73AJD	1867
SN74LS73AJDS	1867
SN74LS73AJS	1867
SN74LS73AN	1867
SN74LS73AND	1867
SN74LS73ANDS	1867
SN74LS73ANS	1867
SN74LS73AW	1867
SN74LS73J	1867
SN74LS73N	1867
SN74LS74	0243
SN74LS74A	0243
SN74LS74AJ	0243
SN74LS74AJD	0243
SN74LS74AJDS	0243
SN74LS74AJS	0243
SN74LS74AN	0243
SN74LS74AND	0243
SN74LS74ANDS	0243
SN74LS74ANS	0243
SN74LS74AW	0243
SN74LS74N	0243
SN74LS75	1870
SN74LS75J	1870
SN74LS75JD	1870
SN74LS75JDS	1870
SN74LS75JS	1870
SN74LS75N	1870
SN74LS75ND	1870
SN74LS75NDS	1870
SN74LS75NS	1870
SN74LS75W	1870
SN74LS76A	2180
SN74LS76AJ	2180
SN74LS76AJD	2180
SN74LS76AJDS	2180
SN74LS76AJS	2180
SN74LS76AN	2180
SN74LS76AND	2180
SN74LS76ANDS	2180
SN74LS76ANS	2180
SN74LS77J	1872
SN74LS77JD	1872
SN74LS77JDS	1872
SN74LS77JS	1872
SN74LS77N	1872
SN74LS77ND	1872
SN74LS77NDS	1872
SN74LS77NS	3993
SN74LS78A	1873
SN74LS78AJ	1873
SN74LS78AJD	1873
SN74LS78AJDS	1873
SN74LS78AJS	1873
SN74LS78AN	1873
SN74LS78AND	1873
SN74LS78ANDS	1873
SN74LS78ANS	1873
SN74LS83A	2218
SN74LS83AJ	2218
SN74LS83AJDS	2218
SN74LS83AJS	2218
SN74LS83AN	2218
SN74LS83AND	2218
SN74LS83ANDS	2218
SN74LS83ANS	2218
SN74LS83AW	2218
SN74LS83N	2218
SN74LS85J	0431
SN74LS85JD	0431
SN74LS85JDS	0431
SN74LS85JS	0431
SN74LS85N	0431
SN74LS85ND	0431
SN74LS85NDS	0431
SN74LS85NS	0431
SN74LS85W	0431
SN74LS86	0291
SN74LS86AN	0291
SN74LS86J	0291
SN74LS86JD	0291
SN74LS86JDS	0291
SN74LS86JS	0291
SN74LS86N	0291
SN74LS86ND	0291
SN74LS86NDS	0291
SN74LS86NS	0291
SN74LS86W	0291
SN74LS90	1882
SN74LS90J	1882
SN74LS90JD	1882
SN74LS90JDS	1882
SN74LS90JS	1882
SN74LS90N	1882
SN74LS90ND	1882
SN74LS90NDS	1882
SN74LS90NS	1882
SN74LS91	1885
SN74LS91J	1885
SN74LS91JD	1885
SN74LS91JDS	1885
SN74LS91JS	1885
SN74LS91N	1885
SN74LS91ND	1885
SN74LS91NDS	1885
SN74LS91NS	1885
SN74LS92	1887
SN74LS92J	1887
SN74LS92JD	1887
SN74LS92JDS	1887
SN74LS92JS	1887
SN74LS92N	1887
SN74LS92ND	1887
SN74LS92NDS	1887
SN74LS92NS	1887
SN74LS93	1888
SN74LS93J	1888
SN74LS93JD	1888
SN74LS93JDS	1888
SN74LS93N	1888
SN74LS93ND	1888
SN74LS93NDS	1888
SN74LS93NS	1888
SN74LS93W	1888
SN74LS95AJ	OEM
SN74LS95AN	OEM
SN74LS95AW	OEM
SN74LS95BJ	0771
SN74LS95BJD	0771
SN74LS95BJDS	0771
SN74LS95BJS	0771
SN74LS95BN	0771
SN74LS95BND	0771
SN74LS95BNDS	0771
SN74LS95BNS	0771
SN74LS95N	OEM
SN74LS96	OEM
SN74LS96J	OEM
SN74LS96N	OEM
SN74LS107A	1602
SN74LS107AJ	1602
SN74LS107AJD	1602
SN74LS107AJDS	1602
SN74LS107AJS	1602
SN74LS107AN	1602
SN74LS107AND	1602
SN74LS107ANDS	1602
SN74LS107ANS	1602
SN74LS107AW	1602
SN74LS107N	1602
SN74LS109A	1906
SN74LS109AJ	1906
SN74LS109AJD	1906
SN74LS109AJDS	1906
SN74LS109AJS	1906
SN74LS109AN	1906
SN74LS109AND	1906
SN74LS109ANDS	1906
SN74LS109ANS	1906
SN74LS109AW	1906
SN74LS109N	1906
SN74LS112A	2128
SN74LS112AJ	2128
SN74LS112AJD	2128
SN74LS112AJDS	2128
SN74LS112AJS	2128
SN74LS112AN	2128
SN74LS112AND	2128
SN74LS112ANDS	2128
SN74LS112ANS	2128
SN74LS112N	1617
SN74LS113A	2256
SN74LS113AD	OEM
SN74LS113AFN	OEM
SN74LS113AJ	2256
SN74LS113AJD	2256
SN74LS113AJDS	2256
SN74LS113AJS	2256
SN74LS113AN	2256
SN74LS113AND	2256
SN74LS113ANDS	2256
SN74LS113ANS	2256
SN74LS114A	2302
SN74LS114AJ	2302
SN74LS114AJD	2302
SN74LS114AJDS	2302
SN74LS114AJS	2302
SN74LS114AN	2302
SN74LS114AND	2302
SN74LS114ANDS	2302
SN74LS114ANS	2302
SN74LS122	1620
SN74LS122J	1620
SN74LS122JD	1620
SN74LS122JDS	1620
SN74LS122JS	1620
SN74LS122N	1620
SN74LS122ND	1620
SN74LS122NDS	1620
SN74LS122NS	1620
SN74LS123	0979
SN74LS123J	0979
SN74LS123JD	0979
SN74LS123JDS	0979
SN74LS123JS	0979
SN74LS123N	0979
SN74LS123ND	0979
SN74LS123NDS	0979
SN74LS123NS	0979
SN74LS123W	0979
SN74LS124J	3175
SN74LS124N	3175
SN74LS125A	0075
SN74LS125AJ	0075
SN74LS125AJD	0075
SN74LS125AJDS	0075
SN74LS125AJS	0075
SN74LS125AN	0075
SN74LS125AND	0075
SN74LS125ANDS	0075
SN74LS125ANS	0075
SN74LS126A	2877
SN74LS126AJ	2877
SN74LS126AJD	2877
SN74LS126AJDS	2877
SN74LS126AJS	2877
SN74LS126AN	2877
SN74LS126AND	2877
SN74LS126ANDS	2877
SN74LS126ANS	2877
SN74LS126J	2877
SN74LS126N	2877
SN74LS132	1625
SN74LS132J	1625
SN74LS132JD	1625
SN74LS132JDS	1625
SN74LS132JS	1625
SN74LS132N	1625
SN74LS132ND	1625
SN74LS132NDS	1625
SN74LS132NS	1625
SN74LS133J	3400
SN74LS133JD	3400
SN74LS133JDS	3400
SN74LS133JS	3400
SN74LS133N	3400
SN74LS133ND	3400
SN74LS133NDS	3400
SN74LS133NS	3400
SN74LS136	1628
SN74LS136D	OEM
SN74LS136J	1628
SN74LS136JD	1628
SN74LS136JDS	1628
SN74LS136JS	1628
SN74LS136N	1628
SN74LS136ND	1628
SN74LS136NDS	1628
SN74LS136NS	1628
SN74LS136W	1628
SN74LS137	4398
SN74LS137J	4398
SN74LS137JD	4398
SN74LS137JDS	4398
SN74LS137JS	4398
SN74LS137N	4398
SN74LS137ND	4398
SN74LS137NDS	4398
SN74LS137NS	4398
SN74LS138	0427
SN74LS138J	0427
SN74LS138JD	0427
SN74LS138JDS	0427
SN74LS138JS	0427
SN74LS138N	0427
SN74LS138ND	0427
SN74LS138NDS	0427
SN74LS138NS	0427
SN74LS138X	0427
SN74LS139	0153
SN74LS139AN	0153
SN74LS139J	0153
SN74LS139JD	0153
SN74LS139JDS	0153
SN74LS139JS	0153
SN74LS139N	0153
SN74LS139ND	0153
SN74LS139NDS	0153
SN74LS139NS	0153
SN74LS139X	0153
SN74LS145	1564
SN74LS145J	1564
SN74LS145JD	1564
SN74LS145JDS	1564
SN74LS145JS	1564
SN74LS145N	1564
SN74LS145ND	1564
SN74LS145NDS	1564
SN74LS145NS	1564
SN74LS147	4400
SN74LS147J	4400
SN74LS147JD	4400
SN74LS147JDS	4400
SN74LS147JS	4400
SN74LS147N	4400
SN74LS147ND	4400
SN74LS147NDS	4400
SN74LS147NS	4400
SN74LS148	3912
SN74LS148AJ	3912
SN74LS148AJD	3912
SN74LS148AJDS	3912
SN74LS148AJS	3912
SN74LS148AN	3912
SN74LS148AND	3912
SN74LS148ANDS	3912
SN74LS148ANS	3912
SN74LS148J	3912
SN74LS148JD	3912
SN74LS148JDS	3912
SN74LS148JS	3912
SN74LS148N	3912
SN74LS148ND	3912
SN74LS148NDS	3912
SN74LS148NS	3912
SN74LS151	1646
SN74LS151J	1646
SN74LS151JD	1646
SN74LS151JDS	1646
SN74LS151JS	1646
SN74LS151N	1646
SN74LS151ND	1646
SN74LS151NDS	1646
SN74LS151NS	1646
SN74LS151X	1646
SN74LS153	0959
SN74LS153D	OEM
SN74LS153FN	OEM
SN74LS153J	0959
SN74LS153JD	0959
SN74LS153JDS	0959
SN74LS153JS	0959
SN74LS153N	0959
SN74LS153ND	0959
SN74LS153NDS	0959
SN74LS153NS	0959
SN74LS154N	OEM
SN74LS155	0209
SN74LS155AN	0209
SN74LS155J	0209
SN74LS155JD	0209
SN74LS155JDS	0209
SN74LS155JS	0209
SN74LS155N	0209
SN74LS155ND	0209
SN74LS155NDS	0209
SN74LS155NS	0209
SN74LS156	1654
SN74LS156J	1654
SN74LS156JD	1654
SN74LS156JDS	1654
SN74LS156JS	1654
SN74LS156N	1654
SN74LS156ND	1654
SN74LS156NDS	1654
SN74LS156NS	1654
SN74LS157	1161
SN74LS157D	OEM
SN74LS157FN	OEM
SN74LS157J	1161
SN74LS157JD	1161
SN74LS157JDS	1161
SN74LS157JS	1161
SN74LS157N	1161
SN74LS157ND	1161
SN74LS157NDS	1161
SN74LS157NS	1161
SN74LS158	1656
SN74LS158D	OEM
SN74LS158FN	OEM
SN74LS158J	1656
SN74LS158JD	1656
SN74LS158JDS	1656
SN74LS158JS	1656
SN74LS158N	1656
SN74LS158ND	1656
SN74LS158NDS	1656
SN74LS158NS	1656
SN74LS160A	0837
SN74LS160AJ	0837
SN74LS160AJD	0837
SN74LS160AJDS	0837
SN74LS160AJS	0837
SN74LS160AN	0837
SN74LS160AND	0837
SN74LS160ANDS	0837
SN74LS160ANS	0837
SN74LS160J	0837
SN74LS160N	0837
SN74LS161A	0858
SN74LS161AJ	0858
SN74LS161AJD	0858
SN74LS161AJDS	0858
SN74LS161AJS	0858
SN74LS161AN	0858
SN74LS161AND	0858
SN74LS161ANDS	0858
SN74LS161ANS	0858
SN74LS161AW	0858
SN74LS161J	0858
SN74LS161N	0858
SN74LS162A	0880
SN74LS162AJ	0880
SN74LS162AJD	0880
SN74LS162AJDS	0880
SN74LS162AJS	0880
SN74LS162AN	0880
SN74LS162AND	0880
SN74LS162ANDS	0880
SN74LS162ANS	0880
SN74LS162J	0880
SN74LS162N	0880
SN74LS163A	0893
SN74LS163AJ	0893
SN74LS163AJD	0893
SN74LS163AJDS	0893
SN74LS163AJS	0893
SN74LS163AN	0893
SN74LS163AND	0893
SN74LS163ANDS	0893
SN74LS163ANS	0893
SN74LS163AW	0893
SN74LS163J	0893
SN74LS163N	0893
SN74LS164	4339
SN74LS164AN	OEM
SN74LS164J	4339
SN74LS164JD	4339
SN74LS164JDS	4339
SN74LS164JS	4339
SN74LS164N	4339
SN74LS164ND	4339
SN74LS164NDS	4339
SN74LS164NS	4339
SN74LS165A	4356
SN74LS165J	4356
SN74LS165JD	4356
SN74LS165JDS	4356
SN74LS165JS	4356
SN74LS165N	4356
SN74LS165ND	4356
SN74LS165NDS	4356
SN74LS165NS	4356
SN74LS166	4371
SN74LS166A	4371
SN74LS166AN	4371
SN74LS166J	4371
SN74LS166JD	4371
SN74LS166JDS	4371
SN74LS166JS	4371
SN74LS166N	4371
SN74LS166ND	4371
SN74LS166NDS	4371
SN74LS166NS	4371
SN74LS168N	0967
SN74LS169AJ	0986
SN74LS169AN	0986
SN74LS169B	0986
SN74LS169BN	OEM
SN74LS169J	0986
SN74LS169N	0986
SN74LS170	2629
SN74LS170J	2629
SN74LS170JD	2629
SN74LS170JDS	2629
SN74LS170JS	2629
SN74LS170N	2629
SN74LS170ND	2629
SN74LS170NDS	2629
SN74LS170NS	2629
SN74LS173A	5217
SN74LS173AD	OEM
SN74LS173AFN	OEM
SN74LS173AJ	5217
SN74LS173AJD	5217
SN74LS173AJDS	5217
SN74LS173AJS	5217
SN74LS173AN	5217
SN74LS173AND	5217
SN74LS173ANDS	5217
SN74LS173ANS	5217
SN74LS174	0260
SN74LS174J	0260
SN74LS174JD	0260
SN74LS174JDS	0260
SN74LS174JS	0260
SN74LS174N	0260
SN74LS174ND	0260
SN74LS174NDS	0260
SN74LS174NS	0260
SN74LS175	1672
SN74LS175J	1672
SN74LS175JD	1672
SN74LS175JDS	1672
SN74LS175JS	1672

If replacement code is OEM, contact original manufacturer for replacement.

DEVICE TYPE	REPL CODE
SN74LS175N	1672
SN74LS175ND	1672
SN74LS175NDS	1672
SN74LS175NS	1672
SN74LS178N	OEM
SN74LS181	1678
SN74LS181J	1678
SN74LS181JD	1678
SN74LS181JDS	1678
SN74LS181JS	1678
SN74LS181N	1678
SN74LS181ND	1678
SN74LS181NDS	1678
SN74LS181NS	1678
SN74LS182J	6120
SN74LS182JD	6120
SN74LS182JDS	6120
SN74LS182JS	6120
SN74LS182N	6120
SN74LS182ND	6120
SN74LS182NDS	6120
SN74LS182NS	6120
SN74LS183	OEM
SN74LS183J	OEM
SN74LS183JD	OEM
SN74LS183JDS	OEM
SN74LS183JS	OEM
SN74LS183N	OEM
SN74LS183ND	OEM
SN74LS183NDS	OEM
SN74LS183NS	OEM
SN74LS189A	OEM
SN74LS189N	OEM
SN74LS190	1686
SN74LS190J	1686
SN74LS190JD	1686
SN74LS190JDS	1686
SN74LS190JS	1686
SN74LS190N	1686
SN74LS190ND	1686
SN74LS190NDS	1686
SN74LS190NS	1686
SN74LS191	1687
SN74LS191J	1687
SN74LS191JD	1687
SN74LS191JDS	1687
SN74LS191JS	1687
SN74LS191N	1687
SN74LS191ND	1687
SN74LS191NDS	1687
SN74LS191NS	1687
SN74LS191W	1687
SN74LS192	1689
SN74LS192FN	OEM
SN74LS192J	1689
SN74LS192JD	1689
SN74LS192JDS	1689
SN74LS192JS	1689
SN74LS192N	1689
SN74LS192ND	1689
SN74LS192NDS	1689
SN74LS192NS	1689
SN74LS193	1692
SN74LS193FN	OEM
SN74LS193J	1692
SN74LS193JD	1692
SN74LS193JDS	1692
SN74LS193JS	1692
SN74LS193N	1692
SN74LS193ND	1692
SN74LS193NDS	1692
SN74LS193NS	1692
SN74LS193W	1692
SN74LS194A	1304
SN74LS194AFN	OEM
SN74LS194AJ	1304
SN74LS194AJD	1304
SN74LS194AJDS	1304
SN74LS194AJS	1304
SN74LS194AN	1304
SN74LS194AND	1304
SN74LS194ANDS	1304
SN74LS194ANS	1304
SN74LS194FN	OEM
SN74LS195	1315
SN74LS195A	1315
SN74LS195AJ	1315
SN74LS195AJD	1315
SN74LS195AJDS	1315
SN74LS195AJS	1315
SN74LS195AN	1315
SN74LS195AND	1315
SN74LS195ANDS	1315
SN74LS195ANS	1315
SN74LS195N	1315
SN74LS196	2834
SN74LS196D	OEM
SN74LS196FK	OEM
SN74LS196FN	OEM
SN74LS196J	2834
SN74LS196JD	2834
SN74LS196JDS	2834
SN74LS196JS	2834
SN74LS196N	2834
SN74LS196ND	2834
SN74LS196NDS	2834
SN74LS196NS	2834
SN74LS197	2469
SN74LS197AN	OEM
SN74LS197D	OEM
SN74LS197FK	OEM
SN74LS197FN	OEM
SN74LS197J	2469
SN74LS197JD	2469
SN74LS197JDS	2469
SN74LS197JS	2469
SN74LS197N	2469
SN74LS197ND	2469
SN74LS197NDS	2469
SN74LS197NS	2469
SN74LS202J	OEM
SN74LS202N	OEM
SN74LS210J	OEM
SN74LS210N	OEM
SN74LS219A	OEM
SN74LS219J	OEM
SN74LS219N	OEM
SN74LS221	1239
SN74LS221J	1239
SN74LS221JD	1239
SN74LS221JDS	1239
SN74LS221JS	1239
SN74LS221N	1239
SN74LS221ND	1239
SN74LS221NDS	1239
SN74LS221NS	1239
SN74LS221W	1239
SN74LS222	OEM
SN74LS224	OEM
SN74LS227	OEM
SN74LS228	OEM
SN74LS240	0452
SN74LS240J	0452
SN74LS240JD	0452
SN74LS240JDS	0452
SN74LS240JS	0452
SN74LS240N	0452
SN74LS240ND	0452
SN74LS240NDS	0452
SN74LS240NS	0452
SN74LS241	1726
SN74LS241J	1726
SN74LS241JD	1726
SN74LS241JDS	1726
SN74LS241JS	1726
SN74LS241N	1726
SN74LS241ND	1726
SN74LS241NDS	1726
SN74LS241NS	1726
SN74LS242	1728
SN74LS242J	1728
SN74LS242JD	1728
SN74LS242JDS	1728
SN74LS242JS	1728
SN74LS242N	1728
SN74LS242ND	1728
SN74LS242NDS	1728
SN74LS242NS	1728
SN74LS243	0906
SN74LS243J	0906
SN74LS243JD	0906
SN74LS243JDS	0906
SN74LS243JS	0906
SN74LS243N	0906
SN74LS243ND	0906
SN74LS243NDS	0906
SN74LS243NS	0906
SN74LS244	0458
SN74LS244J	0458
SN74LS244JD	0458
SN74LS244JDS	0458
SN74LS244JS	0458
SN74LS244N	0458
SN74LS244ND	0458
SN74LS244NDS	0458
SN74LS244NS	0458
SN74LS245	0463
SN74LS245J	0463
SN74LS245JD	0463
SN74LS245JDS	0463
SN74LS245JS	0463
SN74LS245N	0463
SN74LS245ND	0463
SN74LS245NDS	0463
SN74LS245NS	0463
SN74LS247	1732
SN74LS247J	1732
SN74LS247JD	1732
SN74LS247JDS	1732
SN74LS247JS	1732
SN74LS247N	1732
SN74LS247ND	1732
SN74LS247NDS	1732
SN74LS247NS	1732
SN74LS248	1734
SN74LS248D	OEM
SN74LS248FN	OEM
SN74LS248J	1734
SN74LS248JD	1734
SN74LS248JDS	1734
SN74LS248JS	1734
SN74LS248N	1734
SN74LS248ND	1734
SN74LS248NDS	1734
SN74LS248NS	1734
SN74LS249	1735
SN74LS249D	OEM
SN74LS249FN	OEM
SN74LS249J	1735
SN74LS249JD	1735
SN74LS249JDS	1735
SN74LS249JS	1735
SN74LS249N	1735
SN74LS249ND	1735
SN74LS249NDS	1735
SN74LS249NS	1735
SN74LS251	1737
SN74LS251D	OEM
SN74LS251FN	OEM
SN74LS251J	1737
SN74LS251JD	1737
SN74LS251JDS	1737
SN74LS251JS	1737
SN74LS251N	1737
SN74LS251ND	1737
SN74LS251NDS	1737
SN74LS251NS	1737
SN74LS251X	1737
SN74LS253	1739
SN74LS253J	1739
SN74LS253JD	1739
SN74LS253JDS	1739
SN74LS253JS	1739
SN74LS253N	1739
SN74LS253ND	1739
SN74LS253NDS	1739
SN74LS253NS	1739
SN74LS256J	OEM
SN74LS256JD	OEM
SN74LS256JDS	OEM
SN74LS256JS	OEM
SN74LS256N	OEM
SN74LS256ND	OEM
SN74LS256NDS	OEM
SN74LS256NS	OEM
SN74LS257	1744
SN74LS257AJ	1744
SN74LS257AJD	1744
SN74LS257AJDS	1744
SN74LS257AJS	1744
SN74LS257AN	1744
SN74LS257AND	1744
SN74LS257ANDS	1744
SN74LS257ANS	1744
SN74LS257BN	1744
SN74LS257J	1744
SN74LS257N	1744
SN74LS257X	1744
SN74LS258	1746
SN74LS258AJ	1746
SN74LS258AJD	1746
SN74LS258AJDS	1746
SN74LS258AJS	1746
SN74LS258AN	1746
SN74LS258AND	1746
SN74LS258ANDS	1746
SN74LS258ANS	1746
SN74LS258B	OEM
SN74LS258BN	OEM
SN74LS258J	1746
SN74LS258N	1746
SN74LS258X	1746
SN74LS259	3205
SN74LS259J	3205
SN74LS259JD	3205
SN74LS259JDS	3205
SN74LS259JS	3205
SN74LS259N	3205
SN74LS259ND	3205
SN74LS259NDS	3205
SN74LS259NS	3205
SN74LS259W	3205
SN74LS260J	5995
SN74LS260JD	5995
SN74LS260JDS	5995
SN74LS260JS	5995
SN74LS260N	5995
SN74LS260ND	5995
SN74LS260NDS	5995
SN74LS260NS	5995
SN74LS261	OEM
SN74LS261J	OEM
SN74LS261N	OEM
SN74LS266	0592
SN74LS266J	0592
SN74LS266JD	0592
SN74LS266JDS	0592
SN74LS266JS	0592
SN74LS266N	0592
SN74LS266ND	0592
SN74LS266NDS	0592
SN74LS266NS	0592
SN74LS266W	0592
SN74LS267NDS	OEM
SN74LS273	0894
SN74LS273J	0894
SN74LS273JD	0894
SN74LS273JDS	0894
SN74LS273JS	0894
SN74LS273N	0894
SN74LS273ND	0894
SN74LS273NDS	0894
SN74LS273NS	0894
SN74LS275	OEM
SN74LS275J	OEM
SN74LS275N	OEM
SN74LS279	3290
SN74LS279J	3290
SN74LS279JD	3290
SN74LS279JDS	3290
SN74LS279JS	3290
SN74LS279N	3290
SN74LS279ND	3290
SN74LS279NDS	3290
SN74LS279NS	3290
SN74LS280	1773
SN74LS280J	1773
SN74LS280JD	1773
SN74LS280JDS	1773
SN74LS280JS	1773
SN74LS280N	1773
SN74LS280ND	1773
SN74LS280NDS	1773
SN74LS280NS	1773
SN74LS280W	1773
SN74LS283	1779
SN74LS283J	1779
SN74LS283JD	1779
SN74LS283JDS	1779
SN74LS283JS	1779
SN74LS283N	5997
SN74LS283ND	1779
SN74LS283NDS	1779
SN74LS283NS	1779
SN74LS289A	OEM
SN74LS289J	OEM
SN74LS289N	OEM
SN74LS290	4425
SN74LS290D	OEM
SN74LS290FN	OEM
SN74LS290J	4425
SN74LS290JD	4425
SN74LS290JDS	4425
SN74LS290JS	4425
SN74LS290N	4425
SN74LS290ND	4425
SN74LS290NDS	4425
SN74LS290NS	4425
SN74LS292	OEM
SN74LS292J	OEM
SN74LS292N	OEM
SN74LS293	0082
SN74LS293D	OEM
SN74LS293FN	OEM
SN74LS293J	0082
SN74LS293JD	0082
SN74LS293JDS	0082
SN74LS293JS	0082
SN74LS293N	0082
SN74LS293ND	0082
SN74LS293NDS	0082
SN74LS293NS	0082
SN74LS294	OEM
SN74LS294J	OEM
SN74LS294N	OEM
SN74LS295AJ	2227
SN74LS295AJD	2227
SN74LS295AJDS	2227
SN74LS295AJS	2227
SN74LS295AN	2227
SN74LS295AND	2227
SN74LS295ANDS	2227
SN74LS295ANS	2227
SN74LS295B	2227
SN74LS295BJ	2227
SN74LS295BN	2227
SN74LS297	OEM
SN74LS298	3371
SN74LS298J	3371
SN74LS298JD	3371
SN74LS298JDS	3371
SN74LS298JS	3371
SN74LS298N	3371
SN74LS298ND	3371
SN74LS298NDS	3371
SN74LS298NS	3371
SN74LS299	4426
SN74LS299J	4426
SN74LS299JD	4426
SN74LS299JDS	4426
SN74LS299JS	4426
SN74LS299N	4426
SN74LS299ND	4426
SN74LS299NDS	4426
SN74LS299NS	4426
SN74LS302J	OEM
SN74LS302N	OEM
SN74LS310J	OEM
SN74LS310N	OEM
SN74LS319A	OEM
SN74LS319J	OEM
SN74LS319N	OEM
SN74LS320	OEM
SN74LS320J	OEM
SN74LS320N	OEM
SN74LS321	OEM
SN74LS321J	OEM
SN74LS321N	OEM
SN74LS322A	1805
SN74LS322AJ	1805
SN74LS322AJD	1805
SN74LS322AJDS	1805
SN74LS322AJS	1805
SN74LS322AN	1805
SN74LS322AND	1805
SN74LS322ANDS	1805
SN74LS322ANS	1805
SN74LS322J	1805
SN74LS322N	1805
SN74LS322P	OEM
SN74LS323	OEM
SN74LS323J	OEM
SN74LS323JD	OEM
SN74LS323JDS	OEM
SN74LS323JS	OEM
SN74LS323N	OEM
SN74LS323ND	OEM
SN74LS323NDS	OEM
SN74LS323NS	OEM
SN74LS340J	OEM
SN74LS340N	OEM
SN74LS341J	OEM
SN74LS341N	OEM
SN74LS344J	OEM
SN74LS344N	OEM
SN74LS347	OEM
SN74LS347J	OEM
SN74LS347N	OEM
SN74LS348	0665
SN74LS348AJ	0665
SN74LS348AJD	0665
SN74LS348AJDS	0665
SN74LS348AJS	0665
SN74LS348AN	0665
SN74LS348AND	0665
SN74LS348ANDS	0665
SN74LS348ANS	0665
SN74LS348J	0665
SN74LS348JD	0665
SN74LS348JDS	0665
SN74LS348JS	0665
SN74LS348N	0665
SN74LS348ND	0665
SN74LS348NDS	0665
SN74LS348NS	0665
SN74LS352	0761
SN74LS352J	0761
SN74LS352JD	0761
SN74LS352JDS	0761
SN74LS352JS	0761
SN74LS352N	0761
SN74LS352ND	0761
SN74LS352NDS	0761
SN74LS352NS	0761
SN74LS353	0773
SN74LS353J	0773
SN74LS353JD	0773
SN74LS353JDS	0773
SN74LS353JS	0773
SN74LS353N	0773
SN74LS353ND	0773
SN74LS353NDS	0773
SN74LS353NS	0773
SN74LS354	0790
SN74LS354J	0790
SN74LS354N	0790
SN74LS355	OEM
SN74LS355J	OEM
SN74LS355N	OEM
SN74LS356	0813
SN74LS356J	0813
SN74LS356N	0813
SN74LS357	OEM
SN74LS357J	OEM
SN74LS357N	OEM
SN74LS361N	OEM
SN74LS365A	0943
SN74LS365AJ	0943
SN74LS365AJD	0943
SN74LS365AJDS	0943
SN74LS365AJS	0943
SN74LS365AN	0943
SN74LS365AND	0943
SN74LS365ANDS	0943
SN74LS365ANS	0943
SN74LS365J	0943
SN74LS365N	0943
SN74LS366A	0956
SN74LS366AJ	0956
SN74LS366AJD	0956
SN74LS366AJDS	0956
SN74LS366AJS	0956
SN74LS366AN	0956
SN74LS366AND	0956
SN74LS366ANDS	0956
SN74LS366ANS	0956
SN74LS366J	0956
SN74LS366N	0956
SN74LS367	0977
SN74LS367A	0977
SN74LS367AJ	0977
SN74LS367AJD	0977
SN74LS367AJDS	0977
SN74LS367AJS	0977
SN74LS367AN	0977
SN74LS367AND	0977
SN74LS367ANDS	0977
SN74LS367ANS	0977
SN74LS367AN-X	0977
SN74LS367J	0977
SN74LS367N	0977
SN74LS368A	0991
SN74LS368AJ	0991
SN74LS368AJD	0991
SN74LS368AJDS	0991
SN74LS368AJS	0991
SN74LS368AN	0991
SN74LS368AND	0991
SN74LS368ANDS	0991
SN74LS368ANS	0991
SN74LS368J	0991
SN74LS368N	0991
SN74LS373	0709
SN74LS373DW	OEM
SN74LS373FN	OEM
SN74LS373J	0709
SN74LS373JD	0709
SN74LS373JDS	0709
SN74LS373JS	0709
SN74LS373N	0709
SN74LS373ND	0709
SN74LS373NDS	0709
SN74LS373NS	0709
SN74LS374	0713
SN74LS374DW	OEM
SN74LS374FN	OEM
SN74LS374J	0713
SN74LS374JD	0713
SN74LS374JDS	0713
SN74LS374JS	0713
SN74LS374N	0713
SN74LS374ND	0713
SN74LS374NDS	0713
SN74LS374NS	0713
SN74LS375	OEM
SN74LS375J	OEM
SN74LS375JD	OEM
SN74LS375JDS	OEM
SN74LS375JS	OEM
SN74LS375N	OEM
SN74LS375ND	OEM
SN74LS375NDS	OEM
SN74LS375NS	OEM
SN74LS377	1120
SN74LS377J	1120
SN74LS377JD	1120
SN74LS377JDS	1120
SN74LS377JS	1120
SN74LS377N	1120
SN74LS377ND	1120
SN74LS377NDS	1120
SN74LS377NS	1120
SN74LS378	1133
SN74LS378J	1133
SN74LS378JD	1133
SN74LS378JDS	1133
SN74LS378JS	1133
SN74LS378N	1133
SN74LS378ND	1133
SN74LS378NDS	1133
SN74LS378NS	1133
SN74LS379	1151
SN74LS379J	1151
SN74LS379JD	1151
SN74LS379JDS	1151
SN74LS379JS	1151
SN74LS379N	1151
SN74LS379ND	1151
SN74LS379NDS	1151
SN74LS379NS	1151
SN74LS381A	OEM
SN74LS382	OEM
SN74LS384	OEM
SN74LS385	OEM
SN74LS385J	OEM
SN74LS385JD	OEM
SN74LS385JDS	OEM
SN74LS385JS	OEM
SN74LS385N	OEM
SN74LS385ND	OEM
SN74LS385NDS	OEM
SN74LS385NS	OEM
SN74LS386	1230
SN74LS386J	1230
SN74LS386JD	1230
SN74LS386JDS	1230
SN74LS386JS	1230
SN74LS386N	1230
SN74LS386ND	1230
SN74LS386NDS	1230
SN74LS386NS	1230
SN74LS390	1288
SN74LS390D	OEM
SN74LS390FK	OEM
SN74LS390FN	OEM
SN74LS390J	1288
SN74LS390JD	1288
SN74LS390JDS	1288
SN74LS390JS	1288
SN74LS390N	1288
SN74LS390ND	1288
SN74LS390NDS	1288
SN74LS390NS	1288
SN74LS393	0819
SN74LS393D	OEM
SN74LS393FK	OEM
SN74LS393FN	OEM
SN74LS393J	0819
SN74LS393JD	0819
SN74LS393JDS	0819
SN74LS393JS	0819
SN74LS393N	0819
SN74LS393ND	0819
SN74LS393NDS	0819
SN74LS393NS	0819
SN74LS395A	1330
SN74LS395AN	1330
SN74LS395J	1330
SN74LS395JD	1330
SN74LS395JDS	1330
SN74LS395JS	1330
SN74LS395N	1330
SN74LS395ND	1330
SN74LS395NDS	1330
SN74LS395NS	1330
SN74LS396	OEM
SN74LS398J	1383
SN74LS398JD	1383
SN74LS398JDS	1383
SN74LS398JS	1383
SN74LS398N	1383
SN74LS398ND	1383
SN74LS398NDS	1383
SN74LS398NS	1383
SN74LS399	1398
SN74LS399J	1398
SN74LS399JD	1398
SN74LS399JDS	1398
SN74LS399JS	1398
SN74LS399N	1398
SN74LS399ND	1398
SN74LS399NDS	1398
SN74LS399NS	1398
SN74LS422	1709
SN74LS423	4460
SN74LS424	4461
SN74LS424J	1709
SN74LS424N	1709
SN74LS440	OEM
SN74LS441	OEM
SN74LS442	OEM
SN74LS443	OEM
SN74LS444	OEM
SN74LS445	OEM
SN74LS446	OEM
SN74LS447	OEM
SN74LS448	OEM
SN74LS465	OEM
SN74LS466	OEM
SN74LS467	OEM
SN74LS468	OEM
SN74LS478	OEM
SN74LS479J	OEM
SN74LS481	OEM
SN74LS490	2213
SN74LS490J	2213
SN74LS490JD	2213
SN74LS490JDS	2213
SN74LS490JS	2213
SN74LS490N	2213
SN74LS490ND	2213
SN74LS490NDS	2213
SN74LS490NS	2213
SN74LS540	2541
SN74LS540J	2541
SN74LS540JD	2541
SN74LS540JDS	2541
SN74LS540JS	2541
SN74LS540N	2541
SN74LS540ND	2541
SN74LS540NDS	2541
SN74LS540NS	2541
SN74LS541	2547
SN74LS541J	2547
SN74LS541JD	2547
SN74LS541JDS	2547
SN74LS541JS	2547
SN74LS541N	2547
SN74LS541ND	2547
SN74LS541NDS	2547
SN74LS541NS	2547
SN74LS568J	OEM
SN74LS568JD	OEM
SN74LS568JDS	OEM
SN74LS568JS	OEM
SN74LS568N	OEM
SN74LS568ND	OEM
SN74LS568NDS	OEM
SN74LS568NS	OEM
SN74LS569J	OEM
SN74LS569JD	OEM
SN74LS569JDS	OEM
SN74LS569JS	OEM
SN74LS569N	OEM
SN74LS569ND	OEM
SN74LS569NDS	OEM
SN74LS569NS	OEM
SN74LS590	OEM
SN74LS590N	OEM
SN74LS591	OEM
SN74LS592	OEM
SN74LS593	OEM
SN74LS595	OEM
SN74LS596	OEM
SN74LS597	OEM
SN74LS598	OEM
SN74LS600	OEM
SN74LS601	OEM
SN74LS602	OEM
SN74LS603	OEM
SN74LS604	OEM
SN74LS604N	OEM
SN74LS604ND	OEM

If replacement code is OEM, contact original manufacturer for replacement.

DEVICE TYPE	REPL CODE	DEVICE TYPE	REPL CODE	DEVICE TYPE	REPL CODE	DEVICE TYPE	REPL CODE	DEVICE TYPE	REPL CODE	DEVICE TYPE	REPL CODE	DEVICE TYPE	REPL CODE
SN74LS604NS	OEM	SN74LS643N	2056	SN74LS689JS	OEM	SN74S08N	2569	SN74S158J	2154	SN74S344	OEM	SN5438	OEM
SN74LS605	OEM	SN74LS643ND	2056	SN74LS689N	OEM	SN74S09	2667	SN74S158N	2154	SN74S373	2264	SN5440	OEM
SN74LS605N	OEM	SN74LS643NDS	2056	SN74LS689ND	OEM	SN74S09J	2667	SN74S160J	OEM	SN74S373DW	OEM	SN5442A	OEM
SN74LS605ND	OEM	SN74LS643NS	2056	SN74LS689NDS	OEM	SN74S09N	2667	SN74S160N	OEM	SN74S373FN	OEM	SN5443A	OEM
SN74LS605NDS	OEM	SN74LS644	OEM	SN74LS689NS	OEM	SN74S10	2445	SN74S161J	OEM	SN74S373J	OEM	SN5444A	OEM
SN74LS605NS	OEM	SN74LS644J	OEM	SN74LS690	OEM	SN74S10J	2445	SN74S161N	OEM	SN74S373N	OEM	SN5445	OEM
SN74LS606	OEM	SN74LS644JD	OEM	SN74LS691	OEM	SN74S10N	2445	SN74S162	OEM	SN74S374	2266	SN5446A	OEM
SN74LS606N	OEM	SN74LS644JDS	OEM	SN74LS692	OEM	SN74S11	2447	SN74S162J	OEM	SN74S374DW	OEM	SN5447A	OEM
SN74LS606ND	OEM	SN74LS644JS	OEM	SN74LS693	OEM	SN74S11J	2447	SN74S162N	OEM	SN74S374FN	OEM	SN5448	OEM
SN74LS606NDS	OEM	SN74LS644N	OEM	SN74LS696	OEM	SN74S11N	2447	SN74S163	0893	SN74S374J	OEM	SN5449	OEM
SN74LS606NS	OEM	SN74LS645	0775	SN74LS697	OEM	SN74S15	2451	SN74S163J	0893	SN74S374N	OEM	SN5450	OEM
SN74LS607	OEM	SN74LS645J	0775	SN74LS698	OEM	SN74S15J	2451	SN74S163N	OEM	SN74S381	OEM	SN5451	OEM
SN74LS607N	OEM	SN74LS645JD	0775	SN74LS699	OEM	SN74S15N	2451	SN74S168	OEM	SN74S387	1918	SN5453	OEM
SN74LS607ND	OEM	SN74LS645JDS	0775	SN74LS716J	OEM	SN74S20	1017	SN74S168J	OEM	SN74S412	4325	SN5454	OEM
SN74LS607NDS	OEM	SN74LS645JS	0775	SN74LS716JD	OEM	SN74S20J	1017	SN74S168N	OEM	SN74S412J	1860	SN5460	OEM
SN74LS607NS	OEM	SN74LS645N	0775	SN74LS716JDS	OEM	SN74S20N	1017	SN74S169	OEM	SN74S412N	1860	SN5470	OEM
SN74LS608	OEM	SN74LS645ND	0775	SN74LS716JS	OEM	SN74S22	2461	SN74S169J	OEM	SN74S428	4328	SN5472	OEM
SN74LS610	OEM	SN74LS645NDS	0775	SN74LS716N	OEM	SN74S22J	2461	SN74S169N	OEM	SN74S428N	1869	SN5473	OEM
SN74LS611	OEM	SN74LS645NS	0775	SN74LS716ND	OEM	SN74S22N	2461	SN74S174	2132	SN74S436	OEM	SN5474	OEM
SN74LS612	OEM	SN74LS646	OEM	SN74LS716NDS	OEM	SN74S30	3726	SN74S174J	2132	SN74S437	OEM	SN5475	OEM
SN74LS612N	OEM	SN74LS646NT	OEM	SN74LS716NS	OEM	SN74S30J	3726	SN74S174N	2132	SN74S438	OEM	SN5476	OEM
SN74LS613	OEM	SN74LS647	OEM	SN74LS718J	OEM	SN74S30N	3726	SN74S175	2141	SN74S452	OEM	SN5477	OEM
SN74LS620	OEM	SN74LS648	4507	SN74LS718JD	OEM	SN74S32	3847	SN74S175J	2141	SN74S453	OEM	SN5480	OEM
SN74LS620J	OEM	SN74LS649	OEM	SN74LS718JDS	OEM	SN74S32J	OEM	SN74S175N	2141	SN74S454	3166	SN5482	OEM
SN74LS620JD	OEM	SN74LS651	OEM	SN74LS718JS	OEM	SN74S32N	3847	SN74S181	2165	SN74S455	OEM	SN5483A	OEM
SN74LS620JDS	OEM	SN74LS652	OEM	SN74LS718N	OEM	SN74S37	OEM	SN74S181J	2165	SN74S470	OEM	SN5485	OEM
SN74LS620JS	OEM	SN74LS668	OEM	SN74LS718ND	OEM	SN74S37J	OEM	SN74S181N	2165	SN74S472	2320	SN5486	OEM
SN74LS620N	OEM	SN74LS668N	OEM	SN74LS718NDS	OEM	SN74S37N	OEM	SN74S182	2166	SN74S473	3288	SN5490A	OEM
SN74LS620ND	OEM	SN74LS669	OEM	SN74LS718NS	OEM	SN74S38	OEM	SN74S182D	OEM	SN74S474	2322	SN5491A	OEM
SN74LS620NDS	OEM	SN74LS669N	OEM	SN74LS724J	OEM	SN74S38J	OEM	SN74S182FN	OEM	SN74S475	2324	SN5492A	OEM
SN74LS620NS	OEM	SN74LS670	1130	SN74LS724N	OEM	SN74S38N	OEM	SN74S182J	2166	SN74S476	4340	SN5493A	OEM
SN74LS621	OEM	SN74LS670J	1130	SN74LS748J	OEM	SN74S40	2475	SN74S182J16	2166	SN74S477	2390	SN5493N	OEM
SN74LS621J	OEM	SN74LS670JD	1130	SN74LS748JD	OEM	SN74S40J	2475	SN74S182N	OEM	SN74S478	2909	SN5494	OEM
SN74LS621JD	OEM	SN74LS670JDS	1130	SN74LS748JDS	OEM	SN74S40N	2475	SN74S188	3168	SN74S479	2862	SN5496	OEM
SN74LS621JDS	OEM	SN74LS670JS	1130	SN74LS748JS	OEM	SN74S51	4305	SN74S189AJ	OEM	SN74S481	OEM	SN5497	OEM
SN74LS621JS	OEM	SN74LS670N	1130	SN74LS748N	OEM	SN74S51J	4305	SN74S189AW	OEM	SN74S482	OEM	SN6285	3525
SN74LS621N	OEM	SN74LS670ND	1130	SN74LS748ND	OEM	SN74S51N	4305	SN74S189B	OEM	SN74S484	OEM	SN7248L	2454
SN74LS621ND	OEM	SN74LS670NDS	1130	SN74LS748NDS	OEM	SN74S64	2496	SN74S189J	OEM	SN74S485	OEM	SN7400	0232
SN74LS621NDS	OEM	SN74LS670NS	1130	SN74LS748NS	OEM	SN74S64J	2496	SN74S189N	OEM	SN74SLA11ND	OEM	SN7400A	0232
SN74LS621NS	OEM	SN74LS671	OEM	SN74LS783N	OEM	SN74S64N	2496	SN74S194	1931	SN80	0600	SN7400J	0232
SN74LS622	OEM	SN74LS672	OEM	SN74LS795J	OEM	SN74S65	2497	SN74S194FN	OEM	SN101	OEM	SN7400N	0232
SN74LS622N	OEM	SN74LS673	OEM	SN74LS795JD	OEM	SN74S65J	2497	SN74S194J	1931	SN102	OEM	SN7400N-10	0232
SN74LS622ND	OEM	SN74LS673N	OEM	SN74LS795JDS	OEM	SN74S65N	2497	SN74S194N	1931	SN109	OEM	SN7401	0268
SN74LS622NDS	OEM	SN74LS674	OEM	SN74LS795JS	OEM	SN74S74	2504	SN74S195	OEM	SN110	OEM	SN7401J	0268
SN74LS622NS	OEM	SN74LS674N	OEM	SN74LS795N	OEM	SN74S74J	2504	SN74S195J	OEM	SN118	OEM	SN7401N	0268
SN74LS623	OEM	SN74LS681	OEM	SN74LS795ND	OEM	SN74S74N	2504	SN74S195N	OEM	SN166	0086	SN7402	0314
SN74LS623J	OEM	SN74LS682	OEM	SN74LS795NDS	OEM	SN74S85	OEM	SN74S196	OEM	SN167	0086	SN7402J	0314
SN74LS623JD	OEM	SN74LS682J	OEM	SN74LS795NS	OEM	SN74S85J	OEM	SN74S196D	OEM	SN171	OEM	SN7402N	0314
SN74LS623JDS	OEM	SN74LS682JD	OEM	SN74LS796J	OEM	SN74S85N	OEM	SN74S196FK	OEM	SN172	OEM	SN7402N-10	0314
SN74LS623JS	OEM	SN74LS682JDS	OEM	SN74LS796JD	OEM	SN74S86	2511	SN74S196FN	OEM	SN173	OEM	SN7403	0335
SN74LS623N	OEM	SN74LS682JS	OEM	SN74LS796JDS	OEM	SN74S86J	2511	SN74S196J	OEM	SN200	OEM	SN7403J	0335
SN74LS623ND	OEM	SN74LS682N	OEM	SN74LS796JS	OEM	SN74S86N	2511	SN74S196N	OEM	SN201	OEM	SN7403N	0335
SN74LS623NDS	OEM	SN74LS682ND	OEM	SN74LS796N	OEM	SN74S112	1617	SN74S197	OEM	SN202	OEM	SN7404	0361
SN74LS623NS	OEM	SN74LS682NDS	OEM	SN74LS796ND	OEM	SN74S112AN	OEM	SN74S197D	OEM	SN204	OEM	SN7404J	0361
SN74LS624	3141	SN74LS682NS	OEM	SN74LS796NDS	OEM	SN74S112J	1617	SN74S197FK	OEM	SN230	OEM	SN7404N	0361
SN74LS624J	3141	SN74LS683	OEM	SN74LS796NS	OEM	SN74S112N	1617	SN74S197FN	OEM	SN231	OEM	SN7404N-10	0361
SN74LS624N	3141	SN74LS683J	OEM	SN74LS797J	OEM	SN74S113	1623	SN74S197J	OEM	SN232	OEM	SN7405	0385
SN74LS625	3149	SN74LS683JD	OEM	SN74LS797JD	OEM	SN74S113D	OEM	SN74S197N	OEM	SN234	OEM	SN7405A	0385
SN74LS625J	3149	SN74LS683JDS	OEM	SN74LS797JDS	OEM	SN74S113FN	OEM	SN74S201	OEM	SN270	OEM	SN7405J	0385
SN74LS625N	3154	SN74LS683JS	OEM	SN74LS797JS	OEM	SN74S113J	1623	SN74S201J	OEM	SN271	OEM	SN7405N	0385
SN74LS626	3154	SN74LS683N	OEM	SN74LS797N	OEM	SN74S113N	1623	SN74S201N	OEM	SN272	OEM	SN7406	1206
SN74LS626J	3154	SN74LS683ND	OEM	SN74LS797ND	OEM	SN74S114	1629	SN74S209J	OEM	SN274	OEM	SN7406J	1206
SN74LS626N	3154	SN74LS683NDS	OEM	SN74LS797NDS	OEM	SN74S114J	1629	SN74S209N	OEM	SN423	OEM	SN7406N	1206
SN74LS627	3162	SN74LS683NS	OEM	SN74LS797NS	OEM	SN74S114N	1629	SN74S210J	OEM	SN500	OEM	SN7406N-10	1206
SN74LS627J	3162	SN74LS684	OEM	SN74LS798J	OEM	SN74S124	2126	SN74S210N	OEM	SN545LS481	OEM	SN7407	1339
SN74LS627N	3162	SN74LS684J	OEM	SN74LS798JD	OEM	SN74S124J	2126	SN74S225	OEM	SN674LS604NDS	OEM	SN7407J	1339
SN74LS628	OEM	SN74LS684JD	OEM	SN74LS798JDS	OEM	SN74S124N	2126	SN74S226	OEM	SN746AN	OEM	SN7407N	1339
SN74LS628N	OEM	SN74LS684JS	OEM	SN74LS798JS	OEM	SN74S132	2134	SN74S226J	OEM	SN784LS367AN-X	OEM	SN7408	0467
SN74LS629	3175	SN74LS684N	OEM	SN74LS798N	OEM	SN74S132J	2134	SN74S226N	OEM	SN962-2M	0562	SN7408J	0467
SN74LS629J	3175	SN74LS684ND	OEM	SN74LS798ND	OEM	SN74S132N	2134	SN74S240	OEM	SN2222	OEM	SN7408N	0467
SN74LS629N	3175	SN74LS684NDS	OEM	SN74LS798NDS	OEM	SN74S133	1819	SN74S240J	OEM	SN2222A	OEM	SN7408N-10	0467
SN74LS630	OEM	SN74LS684NS	OEM	SN74LS798NS	OEM	SN74S133J	1819	SN74S240N	OEM	SN2222AP	0079	SN7409	0492
SN74LS631	OEM	SN74LS685	OEM	SN74LS848J	OEM	SN74S133N	1819	SN74S241	OEM	SN2243	OEM	SN7409J	0492
SN74LS636	OEM	SN74LS685J	OEM	SN74LS848JD	OEM	SN74S134	1827	SN74S241J	OEM	SN2243A	OEM	SN7409N	0492
SN74LS637	OEM	SN74LS685JD	OEM	SN74LS848JDS	OEM	SN74S134J	1827	SN74S241N	OEM	SN2270	OEM	SN7410	0512
SN74LS638	OEM	SN74LS685JDS	OEM	SN74LS848JS	OEM	SN74S134N	1827	SN74S244	OEM	SN2432	OEM	SN7410J	0512
SN74LS639	OEM	SN74LS685JS	OEM	SN74LS848N	OEM	SN74S135	OEM	SN74S244J	OEM	SN2432A	OEM	SN7410N	0512
SN74LS640	0669	SN74LS685N	OEM	SN74LS848ND	OEM	SN74S135J	OEM	SN74S244N	OEM	SN2484	OEM	SN7410N-10	0512
SN74LS640J	0669	SN74LS685ND	OEM	SN74LS848NDS	OEM	SN74S135N	OEM	SN74S251	2198	SN4311N	OEM	SN7412	2242
SN74LS640JD	0669	SN74LS685NDS	OEM	SN74LS848NS	OEM	SN74S137N	OEM	SN74S251D	OEM	SN4448	0124	SN7412J	2242
SN74LS640JDS	0669	SN74LS685NS	OEM	SN74LS1245	OEM	SN74S138	2138	SN74S251FN	OEM	SN5211P	2104	SN7412N	2242
SN74LS640JS	0669	SN74LS686	OEM	SN74PL16L8	OEM	SN74S138J	2138	SN74S251J	2198	SN5400	OEM	SN7412W	2242
SN74LS640N	0669	SN74LS686J	OEM	SN74PL16R4	OEM	SN74S138N	2138	SN74S251N	2198	SN5401	OEM	SN7413	1441
SN74LS640ND	0669	SN74LS686JD	OEM	SN74PL16R6	OEM	SN74S138X	2138	SN74S251X	OEM	SN5402	OEM	SN7413J	1441
SN74LS640NDS	0669	SN74LS686JDS	OEM	SN74PL16R8	OEM	SN74S139	OEM	SN74S253J	OEM	SN5402J	OEM	SN7413N	1441
SN74LS640NS	0669	SN74LS686JS	OEM	SN74PL333	OEM	SN74S139J	OEM	SN74S253N	OEM	SN5402W	OEM	SN7413N-10	1441
SN74LS641	0690	SN74LS686N	OEM	SN74PL335	OEM	SN74S139N	OEM	SN74S253X	OEM	SN5403	OEM	SN7414	2243
SN74LS641J	0690	SN74LS686ND	OEM	SN74PL839	OEM	SN74S139X	OEM	SN74S257	OEM	SN5404	OEM	SN7414J	2243
SN74LS641JD	0690	SN74LS686NDS	OEM	SN74PL840	OEM	SN74S140	1886	SN74S257J	OEM	SN5405	OEM	SN7414N	2243
SN74LS641JDS	0690	SN74LS686NS	OEM	SN74S00	0704	SN74S140J	1886	SN74S257N	2204	SN5406	OEM	SN7415AN	OEM
SN74LS641JS	0690	SN74LS687	OEM	SN74S00J	0704	SN74S140N	1886	SN74S257X	OEM	SN5407	OEM	SN7415N	OEM
SN74LS641N	0690	SN74LS687J	OEM	SN74S00N	0704	SN74S151	1955	SN74S258J	2205	SN5408	OEM	SN7416	1349
SN74LS641ND	0690	SN74LS687JD	OEM	SN74S02	2238	SN74S151J	1955	SN74S258N	2205	SN5409	OEM	SN7416J	1349
SN74LS641NDS	0690	SN74LS687JDS	OEM	SN74S02D	OEM	SN74S151N	1955	SN74S260	OEM	SN5410	OEM	SN7416N	1349
SN74LS641NS	0690	SN74LS687JS	OEM	SN74S02FN	OEM	SN74S151X	1955	SN74S270B	OEM	SN5412	OEM	SN7417	1352
SN74LS642	0719	SN74LS687N	OEM	SN74S02J	2238	SN74S153	2151	SN74S274	OEM	SN5413	OEM	SN7417(SM)	OEM
SN74LS642J	0719	SN74LS688	3412	SN74S02N	2238	SN74S153D	OEM	SN74S275	OEM	SN5414	OEM	SN7417J	1352
SN74LS642JD	0719	SN74LS688J	3412	SN74S03	2217	SN74S153FN	OEM	SN74S280	2220	SN5416	OEM	SN7417N	1352
SN74LS642JDS	0719	SN74LS688JD	3412	SN74S03J	2217	SN74S153J	2151	SN74S280N	2220	SN5417	OEM	SN7417NS(SM)	OEM
SN74LS642JS	0719	SN74LS688JDS	3412	SN74S03N	2217	SN74S153N	2151	SN74S281	OEM	SN5420	OEM	SN7420	0697
SN74LS642N	0719	SN74LS688JS	3412	SN74S04	2263	SN74S153X	2151	SN74S283	OEM	SN5422	OEM	SN7420J	0697
SN74LS642ND	0719	SN74LS688N	3412	SN74S04AN	OEM	SN74S157	1695	SN74S287	2224	SN5425	OEM	SN7420N	0697
SN74LS642NDS	0719	SN74LS688ND	3412	SN74S04J	2263	SN74S157D	OEM	SN74S288	2175	SN5426	OEM	SN7420N-10	0697
SN74LS642NS	0719	SN74LS688NDS	3412	SN74S04N	2263	SN74S157FN	OEM	SN74S289B	OEM	SN5427	OEM	SN7422	4603
SN74LS643	2056	SN74LS688NS	3412	SN74S05	2321	SN74S157J	1695	SN74S299	OEM	SN5428	OEM	SN7422J	4603
SN74LS643J	2056	SN74LS689	OEM	SN74S05J	2321	SN74S157N	1695	SN74S299N	OEM	SN5430	OEM	SN7422N	4603
SN74LS643JD	2056	SN74LS689J	OEM	SN74S05N	2321	SN74S158	2154	SN74S301	OEM	SN5432	OEM	SN7423	3466
SN74LS643JDS	2056	SN74LS689JD	OEM	SN74S08	2569	SN74S158D	OEM	SN74S340	OEM	SN5433	OEM	SN7423J	3466
SN74LS643JS	2056	SN74LS689JDS	OEM	SN74S08J	2569	SN74S158FN	OEM	SN74S341	OEM	SN5437	OEM	SN7423N	3466

DEVICE TYPE	REPL CODE
SN7425	3477
SN7425J	3477
SN7425N	3477
SN-7426	0803
SN7426	0803
SN7426J	0803
SN7426N	0803
SN7427	0818
SN7427J	0818
SN7427N	0818
SN7427N-10	0818
SN7428	4180
SN7428J	4180
SN7428N	4180
SN7430	0873
SN7430J	0873
SN7430N	0873
SN7432	0899
SN7432J	0899
SN7432N	0899
SN7432N-10	0899
SN7433	4193
SN7433J	4193
SN7433N	4193
SN7437	3520
SN7437J	3520
SN7437N	3520
SN7438	0996
SN7438J	0996
SN7438N	0996
SN7439N	5836
SN7440	1024
SN7440J	1024
SN7440N	1024
SN7441N-10	1038
SN7442A	1052
SN7442AJ	1052
SN7442AN	1052
SN7442N	1052
SN7443A	1060
SN7443AJ	1060
SN7443AN	1060
SN7443N	1060
SN7444A	1072
SN7444AJ	1072
SN7444AN	1072
SN7444N	1072
SN7445	1082
SN7445J	1082
SN7445N	1082
SN7446A	1098
SN7446AJ	1098
SN7446AN	1098
SN7447A	1108
SN7447AJ	1108
SN7447AM	OEM
SN7447AN	1108
SN7447J	1108
SN7447N	1108
SN7447N-10	1108
SN7448	1125
SN7448J	1125
SN7448N	1125
SN7448N-10	1125
SN7449	OEM
SN7450	0743
SN7450J	0743
SN7450N	0743
SN7451	1168
SN7451J	1168
SN7451N	1168
SN7451N-10	1168
SN7453	1185
SN7453J	1185
SN7453N	1185
SN7454	1202
SN7454J	1202
SN7454N	1202
SN7460	1275
SN7460J	1275
SN7460N	1275
SN7470	1404
SN7470J	1404
SN7470N	1404
SN7472	1426
SN7472J	1426
SN7472N	1426
SN7473	1172
SN7473J	1172
SN7473N	1172
SN7473N-10	1172
SN-7474	1313
SN7474	1313
SN7474J	1313
SN7474M	1313
SN7474N	1313
SN7474N-10	1313
SN7475	1432
SN7475J	1432
SN7475N	1432
SN7475N-10	1432
SN7476	1158
SN7476B	1158
SN7476J	1158
SN7476N	1158
SN7476N-10	1158
SN7477	OEM
SN7480	1535
SN7480J	1535
SN7480N	1535
SN7481A	3121
SN7481AJ	OEM
SN7481AN	3121
SN7482	1574
SN7482J	1574
SN7482N	1574
SN7483A	0117
SN7483AJ	3121
SN7483AN	0117
SN7484A	OEM
SN7484AJ	OEM
SN7484AN	OEM
SN7485	0374
SN7485J	0374
SN7485N	0374
SN7485N-10	0374
SN7486	1368
SN7486J	1368
SN7486N	1368
SN7486N-10	1368
SN7488AN	4904
SN7489	5455
SN7489N	5455
SN7490A	1208
SN7490AJ	1208
SN7490AN	1208
SN7490N	1208
SN7490N-10	1208
SN7491A	0980
SN7491AJ	0980
SN7491AN	0980
SN7492A	0834
SN7492AJ	0834
SN7492AN	0834
SN7492N	0834
SN7492N-10	0834
SN7493	0569
SN7493A	0569
SN7493AJ	0569
SN7493AN	0569
SN7493N	0569
SN7494	1702
SN7494J	1702
SN7494N	1702
SN7495AJ	1486
SN7495AN	1486
SN7496J	1715
SN7496N	1715
SN7497	2940
SN7497J	2940
SN7497N	2940
SN7506NE	4868
SN7604	OEM
SN7630N	OEM
SN7666	0167
SN7666N	0167
SN7668IN	OEM
SN15830J	1823
SN15830N	1823
SN15831J	3061
SN15831N	3061
SN15832J	1041
SN15832N	1041
SN15833J	2097
SN15833N	2097
SN15835J	1176
SN15835N	1176
SN15836J	1831
SN15836N	1831
SN15837J	1835
SN15837N	1835
SN15838J	1619
SN15838N	1619
SN15844J	0033
SN15844N	0033
SN15845J	0081
SN15845N	0081
SN15846J	0141
SN15846N	0141
SN15848J	3399
SN15848N	3399
SN15849J	1844
SN15849N	1844
SN15850J	2619
SN15850N	2619
SN15851J	2078
SN15851N	2078
SN15857J	1433
SN15857N	1433
SN15858J	0466
SN15858N	0466
SN15861F	OEM
SN15861J	1859
SN15861N	1859
SN15861R	OEM
SN15862J	0562
SN15862N	0562
SN15863F	OEM
SN15863J	0341
SN15863N	0341
SN15863R	OEM
SN16704C	1329
SN16706N	0167
SN16726N	0354
SN16727N	0349
SN16827	OEM
SN16840C	1337
SN16859	1337
SN16869A	1337
SN16869A#	2480
SN16887A	OEM
SN16889	OEM
SN16889P	OEM
SN16890	0623
SN16901	OEM
SN29764N	OEM
SN47191	OEM
SN47248	OEM
SN52101AL	0093
SN52101L	0093
SN52301	1300
SN52723L	1192
SN52723N	0026
SN52741	0410
SN52741L	0410
SN52741P	0410
SN52747L	2368
SN52748J	2452
SN52748L	2454
SN52748N	2452
SN52748P	2452
SN54100	OEM
SN54107	OEM
SN54109	OEM
SN54110	OEM
SN54111	OEM
SN54116	OEM
SN54116J	OEM
SN54116W	OEM
SN54120	OEM
SN54120J	OEM
SN54120W	OEM
SN54121	OEM
SN54121J	OEM
SN54121W	OEM
SN54122	OEM
SN54122J	OEM
SN54122W	OEM
SN54123	OEM
SN54123J	OEM
SN54123W	OEM
SN54125	OEM
SN54125J	OEM
SN54125W	OEM
SN54126	OEM
SN54126J	OEM
SN54126W	OEM
SN54128	OEM
SN54128J	OEM
SN54128W	OEM
SN54132	OEM
SN54132J	OEM
SN54132W	OEM
SN54136	OEM
SN54136J	OEM
SN54136W	OEM
SN54141	OEM
SN54143	OEM
SN54143J	OEM
SN54143W	OEM
SN54144	OEM
SN54144J	OEM
SN54144W	OEM
SN54145	OEM
SN54145J	OEM
SN54145W	OEM
SN54147	OEM
SN54147J	OEM
SN54147W	OEM
SN54148	OEM
SN54148J	OEM
SN54148W	OEM
SN54150	OEM
SN54150J	OEM
SN54150W	OEM
SN54151A	OEM
SN54151AJ	OEM
SN54151AW	OEM
SN54152A	OEM
SN54152AW	OEM
SN54153	OEM
SN54153J	OEM
SN54153W	OEM
SN54154	OEM
SN54154J	OEM
SN54154W	OEM
SN54155	OEM
SN54155J	OEM
SN54155W	OEM
SN54156	OEM
SN54156J	OEM
SN54156W	OEM
SN54157	OEM
SN54157J	OEM
SN54157W	OEM
SN54159	OEM
SN54159J	OEM
SN54159W	OEM
SN54160	OEM
SN54160J	OEM
SN54160W	OEM
SN54161	OEM
SN54161J	OEM
SN54161W	OEM
SN54162	OEM
SN54162J	OEM
SN54162W	OEM
SN54163	OEM
SN54163J	OEM
SN54163W	OEM
SN54164	OEM
SN54164J	OEM
SN54164W	OEM
SN54165	OEM
SN54165J	OEM
SN54165W	OEM
SN54166	OEM
SN54166J	OEM
SN54166W	OEM
SN54167	OEM
SN54167J	OEM
SN54167W	OEM
SN54170	OEM
SN54170J	OEM
SN54170W	OEM
SN54173	OEM
SN54173J	OEM
SN54173W	OEM
SN54174	OEM
SN54174J	OEM
SN54174W	OEM
SN54175	OEM
SN54175J	OEM
SN54175W	OEM
SN54176	OEM
SN54176J	OEM
SN54176W	OEM
SN54177	OEM
SN54177J	OEM
SN54177W	OEM
SN54178	OEM
SN54178J	OEM
SN54179	OEM
SN54179J	OEM
SN54179W	OEM
SN54180	OEM
SN54180J	OEM
SN54180W	OEM
SN54181	OEM
SN54181J	OEM
SN54181W	OEM
SN54182	OEM
SN54182J	OEM
SN54182W	OEM
SN54184	OEM
SN54184J	OEM
SN54184W	OEM
SN54185A	OEM
SN54185AJ	OEM
SN54190	OEM
SN54191	OEM
SN54192	OEM
SN54192FH	OEM
SN54192J	OEM
SN54193	OEM
SN54193FH	OEM
SN54193J	OEM
SN54194	OEM
SN54194FH	OEM
SN54194J	OEM
SN54195	OEM
SN54196	OEM
SN54196J	OEM
SN54196W	OEM
SN54197	OEM
SN54197J	OEM
SN54197W	OEM
SN54198	OEM
SN54199	OEM
SN54221	OEM
SN54246	OEM
SN54247	OEM
SN54248	OEM
SN54249	OEM
SN54251	OEM
SN54259	OEM
SN54265	OEM
SN54273	OEM
SN54276	OEM
SN54278	OEM
SN54279	OEM
SN54283	OEM
SN54284	OEM
SN54285	OEM
SN54290	OEM
SN54293	OEM
SN54298	OEM
SN54365A	OEM
SN54366A	OEM
SN54367A	OEM
SN54376	OEM
SN54390	OEM
SN54390J	OEM
SN54390W	OEM
SN54393	OEM
SN54393J	OEM
SN54393W	OEM
SN54425	OEM
SN54490	OEM
SN55107A	OEM
SN55107B	OEM
SN55108A	OEM
SN55108B	OEM
SN55109A	OEM
SN55110A	OEM
SN55113	OEM
SN55114	OEM
SN55115	OEM
SN55116	OEM
SN55117	OEM
SN55118	OEM
SN55119	OEM
SN55121	OEM
SN55122	OEM
SN55138	OEM
SN55140	OEM
SN55141	OEM
SN55142A	OEM
SN55143A	OEM
SN55152	OEM
SN55154	OEM
SN55158	OEM
SN55160A	OEM
SN55161A	OEM
SN55182	OEM
SN55183	OEM
SN55189	OEM
SN55189A	OEM
SN55325	OEM
SN55326	OEM
SN55327	OEM
SN55426B	OEM
SN55427B	OEM
SN55446	OEM
SN55450B	OEM
SN55450BFK	OEM
SN55450J	OEM
SN55451B	OEM
SN55451BFK	OEM
SN55451BJG	OEM
SN55452B	OEM
SN55452BFK	OEM
SN55452BJG	OEM
SN55453B	OEM
SN55453BFK	OEM
SN55453BJG	OEM
SN55454B	OEM
SN55454BFK	OEM
SN55454BJG	OEM
SN55461	OEM
SN55462	OEM
SN55463	OEM
SN55464	OEM
SN55471	OEM
SN55472	OEM
SN55473	OEM
SN55474	OEM
SN57207B	OEM
SN72301	1300
SN72301AL	0093
SN72301AN	0093
SN72301AP	1300
SN72307	2283
SN72307JA	2283
SN72307JP	2283
SN72307N	2283
SN72307P	2283
SN72311P	2104
SN72555JP	0973
SN72555P	0973
SN72558L	1890
SN72709	1705
SN72709J	1705
SN72709T	1301
SN72723J	0026
SN72723L	1192
SN72723N	0026
SN72741J	1975
SN72741P	0312
SN72747	2368
SN72747J	2358
SN72747JA	2358
SN72747L	2368
SN72747N	2358
SN72748J	2452
SN72748JG	1300
SN72748L	2454
SN72748N	2452
SN72748P	2452
SN72905	1285
SN72906	2648
SN72912	1838
SN72915	3829
SN72924	2239
SN74100	OEM
SN74107	0942
SN74107N	0942
SN74109	0968
SN74109J	0968
SN74109N	0968
SN74110J	0987
SN74110N	0987
SN74111	1002
SN74111J	1002
SN74111N	1002
SN74116	1070
SN74116J	OEM
SN74116N	1070
SN74120	1116
SN74120J	1116
SN74120N	1116
SN74121	0175
SN74121J	0175
SN74121N	0175
SN74122	1139
SN74122J	1139
SN74122N	1139
SN74123	1157
SN74123J	1157
SN74123N	1157
SN74123N-10	1157
SN74125	1182
SN74125J	1182
SN74125N	1182
SN74126	1193
SN74126J	1193
SN74126N	1193
SN74128	1219
SN74128J	1219
SN74128N	1219
SN74132	1271
SN74132J	1271
SN74132N	1271
SN74136	1316
SN74136J	1316
SN74136N	1316
SN74141	1377
SN74141J	1377
SN74141N	1377
SN74142	1390
SN74142J	1390
SN74142N	1390
SN74143	1409
SN74143J	1409
SN74143N	1409
SN74144	1418
SN74144J	1418
SN74144N	1418
SN74145	0619
SN74145J	0619
SN74145N	0619
SN74145N-10	0619
SN74147	1451
SN74147J	1451
SN74147N	1451
SN74148	1464
SN74148J	1464
SN74148N	1464
SN74150	1492
SN74150J	1492
SN74150N	1492
SN74150N-10	1492
SN74151A	1495
SN74151AJ	1495
SN74151AN	1495
SN74151N	1495
SN74152A	1517
SN74152N	1517
SN74152S	1517
SN74153	1539
SN74153J	1539
SN74153N	1539
SN74154	1555
SN74154J	1555
SN74154N	1555
SN74154N-10	1555
SN74155	1576
SN74155J	1576
SN74155N	1576
SN74156	1592
SN74156J	1592
SN74156N	1592
SN74157	1605
SN74157J	1605
SN74157N	1605
SN74159	OEM
SN74159J	OEM
SN74159N	OEM
SN74160	1631
SN74160J	1631
SN74160N	1631
SN74161	1645
SN74161AN	OEM
SN74161J	1645
SN74161N	1645
SN74162	1013
SN74162J	1013
SN74162N	1013
SN74163	1666
SN74163J	1666
SN74163N	1666
SN74164	0734
SN74164J	0734
SN74164N	0734
SN74165	1685
SN74165J	1685
SN74165N	1685
SN74166	0231
SN74166J	0231
SN74166N	0231
SN74167	OEM
SN74167J	OEM
SN74167N	OEM
SN74170	1722
SN74170J	1722
SN74170N	1722
SN74172	OEM
SN74172J	OEM
SN74172N	OEM
SN74173	1766
SN74173J	1762
SN74173N	1766
SN74174	1770
SN74174J	1770
SN74174N	1770
SN74175	1787
SN74175AN	OEM
SN74175J	1787
SN74175N	1787
SN74176	1795
SN74176J	1795
SN74176N	1795
SN74177J	1803
SN74177N	1803
SN74178	1813
SN74178J	1813
SN74178N	1813
SN74179	1820
SN74179J	1820
SN74179N	1820
SN74180	1829
SN74180J	1829
SN74180N	1829
SN74181	1842
SN74181J	1842
SN74181N	1842
SN74182	1856
SN74182J	1856
SN74182N	1856
SN74184	OEM
SN74184J	OEM
SN74184N	OEM
SN74185A	1868
SN74185AJ	1868
SN74185AN	1868
SN74190	1912
SN74190J	1912
SN74190N	1912
SN74191J	1917
SN74191N	1917
SN74192	1921
SN74192J	1921
SN74192N	1921
SN74192N-10	1921
SN74193	1926
SN74193J	1926
SN74193N	1926
SN74193N-10	1926
SN74194	OEM
SN74194FN	OEM
SN74194J	OEM
SN74194N	OEM
SN74195	1943
SN74195J	1943
SN74195N	1943
SN74195W	OEM
SN74196	1950
SN74196J	1950
SN74196N	1950
SN74196N-10	1950
SN74197	1956
SN74197J	1956
SN74197N	1956
SN74198	1964
SN74198J	1964
SN74198N	1964
SN74199	1971
SN74199J	1971
SN74199N	1971
SN74221	2142
SN74221J	2142
SN74221N	2142
SN74246	OEM
SN74246J	OEM
SN74246N	OEM
SN74247	OEM
SN74247J	OEM
SN74247N	OEM
SN74248J	OEM
SN74248N	OEM
SN74249	2290
SN74249J	2290
SN74249N	2293
SN74251	2299
SN74251J	2299
SN74251N	2299
SN74259	OEM
SN74265	4611
SN74273	OEM
SN74276	OEM
SN74278	4617
SN74279	4619
SN74279N	OEM
SN74283	OEM
SN74283N	OEM
SN74284	OEM
SN74285	OEM
SN74290	2612
SN74290J	2612
SN74290N	2612
SN74293	2644
SN74293J	2644
SN74293N	2644
SN74298	4628
SN74351	OEM
SN74365A	1459
SN74365AJ	1459
SN74365AN	1459
SN74365J	1459
SN74365N	1459
SN74366A	1471
SN74366AJ	1471
SN74366AN	1471

If replacement code is OEM, contact original manufacturer for replacement.

DEVICE TYPE	REPL CODE
SN74366J	1471
SN74366N	1471
SN74367A	1488
SN74367AJ	1488
SN74367AN	1488
SN74367J	1488
SN74367N	1488
SN74368A	1508
SN74368AJ	1508
SN74368AN	1508
SN74368J	1508
SN74368N	1508
SN74376	OEM
SN74390	3240
SN74390J	3240
SN74390N	3240
SN74393	3256
SN74393J	3256
SN74393N	3256
SN74425	OEM
SN74426	4697
SN74490	3664
SN74490J	3664
SN74490N	3664
SN74500N	0704
SN74741L	0410
SN74915N	OEM
SN74921-10	0834
SN75064	OEM
SN75065	OEM
SN75066	OEM
SN75067	OEM
SN75067NE	5314
SN75068	OEM
SN75068NE	4970
SN75069	OEM
SN75069NE	4970
SN75074	OEM
SN75075	OEM
SN75107A	OEM
SN75107B	OEM
SN75108A	OEM
SN75108B	OEM
SN75109A	OEM
SN75110A	OEM
SN75112	OEM
SN75113	OEM
SN75114	OEM
SN75115	OEM
SN75115J	5316
SN75115N	5316
SN75116	OEM
SN75117	OEM
SN75118	OEM
SN75119	OEM
SN75121	OEM
SN75122	OEM
SN75123	OEM
SN75124	OEM
SN75125	OEM
SN75126	OEM
SN75127	OEM
SN75128	OEM
SN75129	OEM
SN75130	OEM
SN75136	OEM
SN75138	OEM
SN75140	OEM
SN75141	OEM
SN75142A	OEM
SN75143A	OEM
SN75150	OEM
SN75150P	OEM
SN75151	OEM
SN75152	OEM
SN75153	OEM
SN75154	OEM
SN75154D(SM)	OEM
SN75154N	OEM
SN75157	OEM
SN75158	OEM
SN75159	OEM
SN75160A	OEM
SN75161A	OEM
SN75162A	OEM
SN75163A	OEM
SN75172	OEM
SN75173	OEM
SN75173J	OEM
SN75173N	OEM
SN75174	OEM
SN75175	OEM
SN75175J	OEM
SN75175N	OEM
SN75176	OEM
SN75177	OEM
SN75178	OEM
SN75182	OEM
SN75183	OEM
SN75188	0508
SN75188J	0508
SN75188N	0508
SN75189	0511
SN75189A	0511
SN75189AN	0511
SN75189N	0511
SN75207	OEM
SN75208	OEM
SN75208B	OEM
SN75322	OEM
SN75325	OEM
SN75326	OEM
SN75327	OEM
SN75328	OEM
SN75330	OEM
SN75350	OEM
SN75357	OEM
SN75361A	OEM
SN75363	OEM
SN75365	OEM
SN75366	OEM
SN75367	OEM
SN75368	OEM
SN75369	OEM
SN75370	OEM
SN75375	OEM
SN75401	OEM
SN75402	OEM
SN75403	OEM
SN75404	OEM
SN75411	OEM
SN75412	OEM
SN75413	OEM
SN75414	OEM
SN75416	OEM
SN75417	OEM
SN75417B	OEM
SN75418	OEM
SN75419	OEM
SN75426B	OEM
SN75430	OEM
SN75431	OEM
SN75432	OEM
SN75433	OEM
SN75434	OEM
SN75436	OEM
SN75437A	OEM
SN75438	OEM
SN75441	OEM
SN75446	OEM
SN75447	OEM
SN75448	OEM
SN75449	OEM
SN75450B	1231
SN75450BN	1231
SN75451B	1244
SN75451BD	OEM
SN75451BP	1244
SN75452B	1262
SN75452BD	OEM
SN75452BP	1262
SN75452P	1262
SN75453B	1272
SN75453BD	OEM
SN75453BP	1272
SN75454BD	OEM
SN75454BN	1289
SN75454BP	OEM
SN75460	OEM
SN75461	OEM
SN75462	OEM
SN75463	OEM
SN75463P	OEM
SN75464	OEM
SN75466	OEM
SN75467	OEM
SN75468	OEM
SN75469	OEM
SN75470	OEM
SN75471	OEM
SN75472	OEM
SN75473	OEM
SN75474	OEM
SN75476	OEM
SN75477	OEM
SN75477P	OEM
SN75478	OEM
SN75479	OEM
SN75480N	1793
SN75491AN	1729
SN75491N	1729
SN75492N	1740
SN75493N	5021
SN75494N	5022
SN75497N	5857
SN75498N	5024
SN75500A	OEM
SN75501A	OEM
SN75502A	OEM
SN75503A	OEM
SN76001	2158
SN76001A	2158
SN76001AN	2158
SN-76001N	2158
SN76001N	2158
SN76007	3770
SN76007N	3770
SN76008	0785
SN76104N	0654
SN76105N	0443
SN76107	0701
SN76110	0701
SN76110N	0701
SN76110N-07	0701
SN76111	1940
SN76111N	1940
SN76113N	0701
SN76115	0520
SN76115N	0520
SN76116N	1395
SN76130	2279
SN76130N	2279
SN76131M	0692
SN76131N	0692
SN76177ND	2557
SN76242N	0352
SN76243AN	0354
SN76243N	0354
SN76246A	6523
SN76246N	0666
SN76298	0856
SN76298N	0856
SN76333	OEM
SN76334	OEM
SN76477	OEM
SN76487	OEM
SN76488	OEM
SN76489	OEM
SN76489A	OEM
SN76493	OEM
SN76494	OEM
SN76495	OEM
SN76496	OEM
SN76496N	OEM
SN76514	3999
SN76514N	OEM
SN76524N	OEM
SN76550	OEM
SN76564	0802
SN76565N	0802
SN76566	OEM
SN76591P	0848
SN76594	2493
SN76600	0785
SN76600P	0785
SN76602	OEM
SN76604	OEM
SN76635N	1421
SN76642	0678
SN76642N	0678
SN76642P	0678
SN76643	0664
SN76643A	0664
SN76643N	0664
SN76645	OEM
SN76650	0395
SN76650N	0395
SN76651N	0789
SN76653N	0664
SN76664N	0167
SN76665	0167
SN76665N	0167
SN76666	0167
SN76666N	0167
SN76666N-S	0167
SN76666NS	0167
SN76669N	1443
SN76670N	6150
SN76675	0395
SN76676HC	2747
SN76678	1759
SN76678P	1759
SN76681N	OEM
SN76689	2755
SN76689N	2755
SN76730	OEM
SN76730N	OEM
SN76760N	OEM
SN76832A	OEM
SN76862NJ	OEM
SN76862NJ-09	OEM
SN76873N	5755
SN76873NT	OEM
SN76881	OEM
SN76882	OEM
SN76889	OEM
SN76891	OEM
SN76940	OEM
SN94017P	1759
SN94018N	1345
SN94041N	OEM
SN94041NL	OEM
SN94042N	OEM
SN94042NL	OEM
SN94043N	OEM
SN94066	OEM
SN94105AN	OEM
SN94174	1337
SN94281	OEM
SN94302N	OEM
SN94310N	OEM
SN94311N	OEM
SN94312N	OEM
SN94313N	OEM
SN94314N	OEM
SN94315N	OEM
SN94316N	OEM
SN94510CN	OEM
SN94624N	OEM
SN102977AN	OEM
SN103348APG	OEM
SN103848PG	OEM
SN151800J	0850
SN151800N	0850
SN151801J	0874
SN151801N	0874
SN151802J	0888
SN151802N	0888
SN151803J	0900
SN151803N	0900
SN151804J	0919
SN151804N	0919
SN151805J	0929
SN151805N	0929
SN151806J	0945
SN151806N	0945
SN151807J	0962
SN151807N	0962
SN151808J	0982
SN151808N	0982
SN151809J	0998
SN151809N	0998
SN151810J	1011
SN151810N	1011
SN151811J	1025
SN151811N	1025
SN151812J	1039
SN151812N	1039
SN151910N	OEM
SN151911N	OEM
SN158093J	0358
SN158093N	0358
SN158093R	OEM
SN158094F	OEM
SN158094J	1632
SN158094N	1632
SN158094R	OEM
SN158097F	OEM
SN158097J	1481
SN158097N	1481
SN158097R	OEM
SN158099F	OEM
SN158099J	0333
SN158099N	0333
SN297654N	OEM
SN400091N	OEM
SN490300N	OEM
SN742471	OEM
SN745064NE	4868
SN745065NE	4868
SN745066NE	5314
SN745067NE	5314
SN745068NE	4970
SN745069NE	4970
SN764246N	0349
SN765241N	OEM
SN766008	0785
SNC5476J	OEM
SNJ54LS125AJ	OEM
SNM74S251J	2198
SNN74LS189J	OEM
SNR-7A130K	1374
SNR-7A150K	0246
SNR-7A250K	1332
SNR-7D18L	1691
SNR-7D22L	2008
SNR-7D24L	1314
SNR-7D27L	2325
SNR-7D33K	2790
SNR-7D56K	2935
SNR-7D82K	4472
SNR-14A115K	4135
SNR-14A130K	0838
SNR14A130K	0838
SNR-14A150K	1993
SNR-14A480K	1101
SNR-14D18L	1694
SNR-14D22L	2009
SNR-14D24L	1319
SNR-14D33K	2796
SNR-14D56K	2941
SNR-14D82K	3085
SNR20A130K	0830
SNR-20A150K	0830
SNR-20A250K	1336
SNR-20A480K	0781
SNR-A130L2	1374
SNR-A130L10	0838
SNR-A130L20	1374
SNR-A150L2	0246
SNR-A150L10	1993
SNR-A150L20	3415
SNR-A250L4	1332
SNR-A250L20	1334
SNR-A250L40	1336
SNR-A480L40	1101
SNR-A480L80	0781
SNR-D18M1	1691
SNR-D18M3	1694
SNR-D22L1	2008
SNR-D22L3	2009
SNR-D24K1	1314
SNR-D24K4	1319
SNR-D27L1	2325
SNR-D27L4	2326
SNR-D33K1	2790
SNR-D33K5	2796
SNR-D56K2	2935
SNR-D56K8	2941
SNR-D82K2	4472
SNR-D82K12	3085
SNR-D180K1	4135
SNR-D180K10	3123
SNT204	0037
SNW-Q-1	0211
SNW-Q-2	0038
SNW-Q-3	0038
SNW-Q-4	0211
SNW-Q-5	0038
SNW-Q-6	0211
SO-1	0050
SO1	0050
SO1C	0871
SO1D	0853
SO-2	0050
SO2	0050
SO2C	0871
SO-3	0050
SO3	0050
SO10G	0071
SO25	0004
SO42P	OEM
SO46	0123
SO65	0004
SO65A	0136
SO88	0004
SO501	0015
SO-632	0012
SOA	0071
SOC532A	OEM
SOD20AS	0708
SOD20BS	2002
SOD30AL	0110
SOD30AS	0708
SOD30BL	0110
SOD30BS	0708
SOD30CS	0708
SOD30DL	0110
SOD30DS	0708
SOD50AL	0110
SOD50AS	0708
SOD50BL	0110
SOD50BS	0708
SOD50CL	0110
SOD50DL	0110
SOD50DS	0708
SOD100AL	0242
SOD100AS	0580
SOD100BL	0242
SOD100BS	0580
SOD100CL	0242
SOD100CS	0580
SOD100DS	0242
SOD200AL	0242
SOD200BL	0242
SOD200BS	0580
SOD200CL	0242
SOD200CS	0580
SOD-200D	0087
SOD200D	0015
SOD200DL	0242
SOD200DS	0580
SOS0121	0126
SP-01	0323
SP-02	1414
SP-03	0473
SP-04	0473
SP-06	0446
SP0256	OEM
SP0256-AL2	OEM
SP-1	0015
SP1	0015
SP1K	0229
SP1K-1	0015
SP1K-2	0023
SP-1.5A	0110
SP2	0280
SP2-01	0277
SP2-02	0289
SP2-04	0296
SP2-06	0302
SP2-08	0250
SP2-10	0250
SP3	0160
SP4	0440
SP5-1	OEM
SP5-2	OEM
SP5-3	OEM
SP6	0600
SP6GC1	0196
SP6GD1	0484
SP6GX1	1080
SP7	0600
SP9	0050
SP10	0079
SP11	0127
SP12	0142
SP13MP	0265
SP14	0103
SP15MP	1514
SP16	0160
SP17	0016
SP18	0086
SP26	0085
SP34AS	0143
SP39	0975
SP44UD	0085
SP44US	0085
SP47	0085
SP60	0143
SP62	0975
SP-70	0168
SP70	0037
SP74ALS8003P	OEM
SP82	0917
SP82A	0917
SP-89RW	0160
SP89UG	0038
SP90	0037
SP100	OEM
SP101	0133
SP102	OEM
SP103	OEM
SP104	OEM
SP105	OEM
SP-148-3	0160
SP148-3	0160
SP176	0085
SP180CI	1854
SP230	0085
SP295	0143
SP334	0160
SP404	0160
SP-404T	0160
SP404T	0160
SP441	0160
SP441D	0085
SP441G	0160
SP441S	0085
SP-485	0160
SP485	0160
SP485B	0160
SP485BLK	0160
SP485BLU	0160
SP485BRN	0160
SP485W	0160
SP485WHT	0160
SP-486	0160
SP486	0160
SP486W	0160
SP486WHT	0160
SP-634	0160
SP634	0160
SP-649	0160
SP649	0160
SP-649-1	0160
SP649-1	0160
SP706F	OEM
SP708F	OEM
SP744	0160
SP819R	0160
SP-834	0160
SP834	0160
SP838	0604
SP838-1	0604
SP875	0160
SP880	0085
SP880-1	0160
SP880-3	0160
SP891	0085
SP891-B	0085
SP891B	0160
SP891BLU	0160
SP891G	0160
SP891GRN	0160
SP891R	0160
SP891W	0160
SP891WHT	0160
SP1013B	0085
SP-1108	0160
SP1108	0085
SP1118	0160
SP1137	0085
SP1271	0160
SP1273	0160
SP-1278	0160
SP1278	0847
SP1323	0085
SP1403	0085
SP1481	0160
SP1481-1	0160
SP1481-2	0160
SP1481-3	0160
SP1481-4	0160
SP1481-5	0160
SP1482	0160
SP1482-2	0160
SP1482-3	0160
SP1482-4	0160
SP-1482-5	0160
SP1482-5	0160
SP1482-6	0160
SP1482-7	0160
SP1483	0160
SP1483-1	0160
SP1483-2	0160
SP1483-3	0160
SP-1484	0160
SP1484	0160
SP1484-1	0604
SP1484-2	0604
SP1484-3	0604
SP1484-4	0604
SP1484-5	0604
SP1550-3	0160
SP1556	0160
SP1556-1	0160
SP1556-2	0160
SP1556-3	0160
SP1556-4	0160
SP1563-2	0160
SP1595BLK	0160
SP1595BLU	0160
SP1595GRN	0160
SP1595RED	0160
SP1596BLK	0160
SP1596BLU	0160
SP1596GRN	0160
SP1596RED	0160
SP1600	0085
SP-1603	0160
SP1603	0160
SP-1603-1	0160
SP1603-1	0160
SP-1603-2	0160
SP1603-2	0160
SP1603-3	0160
SP1619	0085
SP1650	0604
SP1651	0085
SP1657	0160
SP1742	0975
SP1801	0160
SP1817	0604
SP1844	0160
SP1927	0085
SP1938	0160
SP1950	0160
SP2034	OEM
SP2045	0160
SP2046	0160
SP2048	0160
SP2072	0160
SP2076	0160
SP2077	0604
SP2094	0160
SP2155	0160
SP2158	0208
SP2188	0208
SP2200	OEM
SP2234	0085
SP2247	0085
SP2341	0160
SP2358	0085
SP2361	0160
SP2361BLU	0160
SP2361BRN	0160
SP2361GRN	0160
SP2361ORN	0160
SP2361RED	0160
SP2361YEL	0160
SP2395	0160
SP2411	0975
SP2431	0085
SP2431P	0440
SP2493	0160
SP2512	0975
SP2541	0160
SP2551	0222
SP2610	0604
SP2946F	OEM
SP3003	0004
SP3004	0004
SP3005	0280
SP3006	0050
SP3007	0050
SP3008	0136
SP3009	0160
SP3010	0208
SP3011	0600
SP3012	0440
SP3013	0265
SP3014	0160
SP3015	0265
SP3016	0015
SP3017A	0229
SP3018	0144
SP3019	0144
SP3020	0198
SP3020F	OEM
SP3021	0142
SP3024	0086
SP3025	0126
SP3026	0178
SP3027	0103
SP3028	0178
SP3029	1514
SP3030	0015
SP3031	3911
SP3032	0071
SP3033	0071
SP3035	0975
SP3036	2013
SP3037	5474
SP3038	0079
SP3039	0259
SP3040	0233
SP3041	0042
SP3042	0239
SP3044	0233
SP3045	3280
SP3046	0079
SP3047	0086
SP3048	2700
SP3049	0630
SP3050	0353
SP3051	0087
SP3052	0222
SP3053	0338
SP3054	0423
SP3055	0188
SP3056	0162
SP3057	0157

DEVICE TYPE	REPL CODE
SP3058	0636
SP3059	0649
SP3060	0012
SP3061	0170
SP3062	0137
SP3063	0002
SP3064	0441
SP3065	0212
SP3066	0774
SP3067	0201
SP3068	0288
SP3079	0074
SP3080	0071
SP3081	0071
SP3082	1010
SP3083	0854
SP3084	0854
SP3085	0854
SP3086	1862
SP3087	0143
SP3088	0143
SP3089	0917
SP3090	0143
SP3091	0143
SP3092	0794
SP3093	0365
SP3094	0100
SP3095	0039
SP3096	3561
SP3096(ZENER)	0002
SP3097	0783
SP3098	0331
SP3099	0149
SP3100	0288
SP3103	3561
SP3105	0277
SP3106	0106
SP3109	1303
SP3112	0325
SP3114	0006
SP3115F	OEM
SP3116	0325
SP3116F	OEM
SP3117	0144
SP3118	0150
SP3119	0196
SP3120	0484
SP3121	1080
SP3122	0079
SP3133F	OEM
SP3134F	OEM
SP3135F	OEM
SP3136F	OEM
SP3253QD	OEM
SP3500	0122
SP3501	0319
SP3502	3696
SP3504	3243
SP3505	0725
SP3507	0415
SP3508	0147
SP3510	0103
SP3511	0130
SP3512	0086
SP3513	0126
SP3515	1771
SP3516	2088
SP4436	0007
SP4552	OEM
SP4632	OEM
SP4634	OEM
SP4653	OEM
SP4654	OEM
SP7056	0285
SP8300	OEM
SP8302	OEM
SP8303	OEM
SP8304	OEM
SP8307	OEM
SP8309	OEM
SP8310	OEM
SP8311	OEM
SP8400	0233
SP8401	0233
SP8402	0595
SP8411	OEM
SP8411A	OEM
SP8412	OEM
SP8412A	OEM
SP8413	OEM
SP8414	OEM
SP8414A	OEM
SP8416	0042
SP8588	OEM
SP8588A	OEM
SP8629	OEM
SP-8660	0161
SP8660	0042
SP8918	0161
SP10800	OEM
SP10801	OEM
SP10810	OEM
SP10811	OEM
SP22001	OEM
SPB-26MUW	OEM
SPC40	0016
SPC42	0016
SPC50	0016
SPC51	0016

DEVICE TYPE	REPL CODE
SPC52	0016
SPC401	0270
SPC402	0270
SPC410	0313
SPC411	0270
SPC413	0313
SPC423	0313
SPC423M	OEM
SPC424	0313
SPC425	0313
SPC430	0270
SPC431	0363
SPC40411	0130
SPD110	OEM
SPD-121	OEM
SPD-121(A)	OEM
SPD-121(Q)	OEM
SPD-121(S)	OEM
SPD-121(T)	OEM
SPD-221	OEM
SPD-221(A)	OEM
SPD-221(Q)	OEM
SPD-221(S)	OEM
SPD-221(T)	OEM
SPD510	OEM
SPD521	OEM
SPD550	OEM
SPD560	OEM
SPD605	1567
SPD610	1567
SPD620	1567
SPD630	1567
SPD1205	1567
SPD1210	1276
SPD1220	1567
SPD1230	2153
SPF024	0325
SPF168	3046
SPF215	0212
SPF274	0212
SPF512	0212
SPF530	0224
SPF609	0212
SPI-06FC	OEM
SPI-06PF	2628
SPI201-20	OEM
SPI1050-01	OEM
SPN-01	0015
SPN01	0015
SPN-02	0289
SPN-03	0473
SPN-04	0473
SPN-06	0446
SPR-54MUW	OEM
SPR-54MVW3F	OEM
SPR1050-01	OEM
SPR4652	OEM
SPR4652REC	OEM
SPS08	2449
SPS010	0469
SPS020	2188
SPS025	0469
SPS030	0731
SPS035	0731
SPS0121	0126
SPS0122	0086
SPS4	0144
SPS12	0037
SPS18	2449
SPS20	0144
SPS22	0037
SPS28	0435
SPS38	0127
SPS40	0144
SPS41	0016
SPS42	0037
SPS43-1	0127
SPS47	0037
SPS48	1487
SPS49	0086
SPS58	0687
SPS68	0687
SPS91	0160
SPS110	0469
SPS120	5029
SPS125	0469
SPS130	0712
SPS135	0712
SPS167	OEM
SPS210	0469
SPS220	0144
SPS220#	0469
SPS225	0469
SPS230	0469
SPS235	0469
SPS310	0722
SPS320	0740
SPS325	0722
SPS330	0722
SPS335	0469
SPS401K	0037
SPS410	0722
SPS420	0740
SPS425	0740
SPS426	OEM
SPS428	0144
SPS429	0127
SPS430	0722
SPS435	0722

DEVICE TYPE	REPL CODE
SPS477	0079
SPS-501C-1	OEM
SPS-503C-3	OEM
SPS510	0725
SPS514	0037
SPS520	0764
SPS525	0725
SPS530	0725
SPS535	0725
SPS610	0725
SPS620	0764
SPS625	0725
SPS627	0037
SPS630	0725
SPS635	0725
SPS668	0037
SPS699	0079
SPS816	OEM
SPS817	0079
SPS817N	0016
SPS-856	0127
SPS856	0127
SPS-860	0127
SPS860	0127
SPS868	0016
SPS871	0136
SPS906	0007
SPS907	0079
SPS915	0144
SPS-917	0224
SPS917	0144
SPS918	0079
SPS919	0007
SPS920	0224
SPS-934	0079
SPS934	0016
SPS-952	0079
SPS952	0079
SPS-952-2	0079
SPS952-2	0008
SPS1045	0016
SPS1082	0079
SPS1097	0037
SPS1107	0551
SPS-1351	0127
SPS1351	0144
SPS-1352	0127
SPS1352	0144
SPS-1353	0127
SPS1353	0144
SPS1359	0037
SPS1436	0551
SPS1437	0532
SPS-1473	0127
SPS1473	0144
SPS-1473RT	0224
SPS-1475(YT)	0079
SPS-1475YT	0079
SPS-1476	0111
SPS1523	0037
SPS-1539	0037
SPS-1539(WT)	0111
SPS-1539WT	0037
SPS1593WT	0037
SPS1802	0079
SPS1817	0079
SPS1846	0224
SPS1977	0079
SPS1978	0037
SPS2003	1526
SPS2104	0079
SPS2110	0144
SPS2111	0144
SPS2129	0079
SPS2130	0086
SPS2131	0126
SPS2134	0861
SPS2135	0224
SPS2142	0079
SPS2164	0079
SPS2167	0127
SPS2194	0079
SPS2195	OEM
SPS2208-1	0008
SPS2216	0111
SPS2217	0111
SPS2224	0144
SPS2225	0079
SPS2226	0126
SPS-2265	0127
SPS2265	0144
SPS2265-2	0144
SPS-2266	0127
SPS2266	0144
SPS2269	0037
SPS2270	1221
SPS2271	0111
SPS2272	0006
SPS2274	0037
SPS2279	0037
SPS2320	0144
SPS2323	3538
SPS2403	OEM
SPS2415	0079
SPS2425	0127
SPS2526	0150
SPS3003	0144
SPS3015	0079
SPS3329	0037

DEVICE TYPE	REPL CODE
SPS3370	0144
SPS3438	0037
SPS3724	0037
SPS3735	0079
SPS3786	0037
SPS3787	0127
SPS3900	0079
SPS3907	0079
SPS3908	0079
SPS3909	0079
SPS3910	OEM
SPS3912	0086
SPS3914	0086
SPS3915	0079
SPS3922	OEM
SPS3923	0079
SPS3924	0037
SPS3925	0079
SPS3926	0079
SPS3927	0037
SPS3929	0144
SPS3930	0079
SPS3931	0037
SPS3936	0079
SPS3937	0144
SPS3938	0079
SPS3940	0079
SPS3948	0144
SPS3951	0079
SPS3952	0144
SPS3957C	0079
SPS3967	0079
SPS3968	0144
SPS3971	0144
SPS3972	0079
SPS3973	0079
SPS3987	0037
SPS3988	0037
SPS3990	0037
SPS3999	0079
SPS4000	0037
SPS4002	0144
SPS4003	0079
SPS4004	0079
SPS4005	0144
SPS4006	0079
SPS4007	0037
SPS4008	0144
SPS4009	0079
SPS4010	0126
SPS4013	0037
SPS4014	0037
SPS4016	0144
SPS4017	0079
SPS4018	0037
SPS4019	0037
SPS4020	0079
SPS4025	0037
SPS4026	0037
SPS4027	0037
SPS4028	0037
SPS4029	0079
SPS4030	0144
SPS4031	0037
SPS4032	0079
SPS4034	0079
SPS4037	0079
SPS4038	0086
SPS4039	0079
SPS4040	0079
SPS4041	0079
SPS4042	0079
SPS4043	0144
SPS4044	0079
SPS4045	0079
SPS4049	0079
SPS4050	0144
SPS4051	0144
SPS4052	0079
SPS4053	0079
SPS4054	0037
SPS4055	0079
SPS4056	0037
SPS4059	0079
SPS4060	0079
SPS4061	0079
SPS4062	0079
SPS4063	0079
SPS4064	0037
SPS4066	0079
SPS4067	0079
SPS4068	0144
SPS4069	0079
SPS4072	0037
SPS4073	0037
SPS4074	0079
SPS4075	0079
SPS4076	0037
SPS4076A	0037
SPS4077	0079
SPS4078	0037
SPS4079	0144
SPS4080	0144
SPS4081	0079
SPS4082	0037
SPS4083	0079
SPS4084	0079
SPS4085	0079
SPS4086	0037

DEVICE TYPE	REPL CODE
SPS4087	0037
SPS4088	0079
SPS4089	0079
SPS4090	0037
SPS4091	0144
SPS4092	1367
SPS4095	0079
SPS4099	0382
SPS4143	0127
SPS-4145	0127
SPS4145	0144
SPS4149	OEM
SPS4167	0224
SPS4168	0144
SPS4169	0144
SPS4199	0079
SPS4201	OEM
SPS4213	OEM
SPS4236	0079
SPS4300	0086
SPS4301	0126
SPS4302	0037
SPS4303	0079
SPS4309	0086
SPS4310	0126
SPS4311	0086
SPS4312	0126
SPS4313	0079
SPS4314	0037
SPS4330	0037
SPS4331	0224
SPS4338	0037
SPS4343	0007
SPS4344	0079
SPS4345	0079
SPS4347	0079
SPS4348	0037
SPS4354	0037
SPS4355	0037
SPS4356	0079
SPS4359	0079
SPS4360	0079
SPS4361	0086
SPS4363	0079
SPS4365	0037
SPS4367	0079
SPS4368	0079
SPS4382	0079
SPS4390	0079
SPS4391	0086
SPS4392	0079
SPS-4396	0079
SPS4397	0037
SPS4399	0144
SPS4401	0150
SPS-4423	0127
SPS4423	0144
SPS4436	0224
SPS4439	OEM
SPS4443	0079
SPS4446	0079
SPS4450	0079
SPS4451	0079
SPS4452	0037
SPS4453	0079
SPS4455	0079
SPS4456	0079
SPS4457	0079
SPS4458	0037
SPS4459	0079
SPS4460	0037
SPS4461	0086
SPS4462	0126
SPS4472	0079
SPS4473	0037
SPS4476	0079
SPS4477	0126
SPS4478	0079
SPS4480	0037
SPS4489	0037
SPS4490	0086
SPS4491	0079
SPS4492	0126
SPS4493	0079
SPS4494	0079
SPS4495	0086
SPS4497	0126
SPS4498	0079
SPS4610	0144
SPS4813	0037
SPS4814	0111
SPS4815	0037
SPS4920	0079
SPS5000	0079
SPS5006	0079
SPS5006-1	0079
SPS5006-2	0079
SPS5007	0037
SPS5007-1	0037
SPS5007-2	0037
SPS5008	0037
SPS5328	0050
SPS5349	OEM
SPS5450	0861
SPS5450S	0861
SPS5451	0681
SPS5569	0127
SPS5809	0086
SPS6109	0037

DEVICE TYPE	REPL CODE
SPS6111	0079
SPS6112	0079
SPS6113	0079
SPS6124	0086
SPS6125	0126
SPS6155	0224
SPS6185	0264
SPS6571	0079
SPS6682	0127
SPS6953	0037
SPS7303	OEM
SPS7359	0079
SPS7652	0079
SPS8870	0144
SPS8871	0037
SPSB1001-01	OEM
SPT06	0154
SPT08	0154
SPT010	0154
SPT015	0154
SPT025	3147
SPT040	2015
SPT16	0154
SPT18	0154
SPT26	0154
SPT28	0154
SPT36	0147
SPT38	0147
SPT46	0147
SPT48	0147
SPT56	0279
SPT58	0279
SPT66	0279
SPT68	0279
SPT110	0154
SPT115	0154
SPT125	3148
SPT130	2015
SPT140	2015
SPT201	OEM
SPT210	0154
SPT215	0154
SPT225	3150
SPT230	2015
SPT240	2015
SPT260	OEM
SPT310	0147
SPT315	0147
SPT325	3151
SPT330	2017
SPT340	2017
SPT360	OEM
SPT410	0147
SPT415	0147
SPT425	3152
SPT430	2017
SPT440	2017
SPT460	OEM
SPT510	0279
SPT515	0279
SPT525	2018
SPT530	2018
SPT540	2018
SPT560	OEM
SPT610	0279
SPT615	0279
SPT625	2018
SPT630	2018
SPT640	2018
SPT2300L01	OEM
SPT2370H07	OEM
SPT2370H08	OEM
SPT3240L05	OEM
SPT3440	0168
SPT3713	0103
SPT5303	0130
SPT6402L04	OEM
SPT9020	OEM
SPT640203	OEM
SPX2	0315
SPX2E	0315
SPX4	0315
SPX5	0315
SPX6	0315
SPX26	0315
SPX28	0315
SPX33	0315
SPX35	0315
SPX36	0315
SPX37	0315
SPX53	0315
SPX103	0315
SPX7110	0541
SPX7130	0541
SPX7150	0315
SPX7270	0541
SPX7271	0541
SPX7272	0315
SPX7273	0315
SPX7530	0541
SPX7550	0541
SPX7590	0315
SQ7	0600
SQ46	0143
SQ1220A	0554
SQ1220B	OEM
SQ1220C	OEM
SQ1224A	0005
SQ1224B	OEM

DEVICE TYPE	REPL CODE
SQ1224C	OEM
SQ1226A	0720
SQ1226B	OEM
SQ1226C	OEM
SQ1234A	0628
SQ1234B	OEM
SQ1234C	OEM
SQ1238A	0030
SQ1238B	OEM
SQ1238C	OEM
SQ1270A	0554
SQ1270B	OEM
SQ1270C	OEM
SQ1274A	0005
SQ1274B	OEM
SQ1274C	OEM
SQ1276A	0720
SQ1276B	OEM
SQ1276C	OEM
SQ1284A	0628
SQ1284B	OEM
SQ1284C	OEM
SQ1288A	0030
SQ1288B	OEM
SQ1288C	OEM
SQ1716	OEM
SQ1716A	OEM
SQ1717	OEM
SQ1717A	OEM
SQ1718	OEM
SQ1718A	OEM
SQ1719	OEM
SQ1719A	OEM
SQ1720	0554
SQ1720A	OEM
SQ1720B	OEM
SQ1720C	OEM
SQ1722	OEM
SQ1722A	OEM
SQ1724	0005
SQ1724A	0005
SQ1724B	OEM
SQ1724C	OEM
SQ1726	0720
SQ1726A	0720
SQ1726B	OEM
SQ1726C	OEM
SQ1728	OEM
SQ1728A	OEM
SQ1730	OEM
SQ1730A	OEM
SQ1732	OEM
SQ1732A	OEM
SQ1734	0628
SQ1734A	OEM
SQ1734B	OEM
SQ1734C	OEM
SQ1736	OEM
SQ1736A	OEM
SQ1738	0030
SQ1738A	OEM
SQ1738B	OEM
SQ1738C	OEM
SQ1740	OEM
SQ1740A	OEM
SQ1742	OEM
SQ1742A	OEM
SQ1744	OEM
SQ1744A	OEM
SQ1746	OEM
SQ1746A	OEM
SQ1748	OEM
SQ1748A	OEM
SQ1750	OEM
SQ1750A	OEM
SQ2502	OEM
SQ2503	OEM
SQ2508	OEM
SQ2519	OEM
SQ2520	OEM
SQ2521	OEM
SQ2526	OEM
SQ2527	OEM
SQ2529	OEM
SQ2534	OEM
SQ2535	OEM
SQ2536	OEM
SQ2543	OEM
SQ2544	OEM
SQ2544V1	OEM
SQ2545	OEM
SQ2545V1	OEM
SQ2546	OEM
SQ2554	OEM
SQ2555	OEM
SQ2556	OEM
SQ2557	OEM
SQ2558	OEM
SQ5441A	OEM
SQ5441B	OEM
SQ5441C	OEM
SQ5442A	OEM
SQ5442B	OEM
SQ5442C	OEM
SQ5443A	OEM
SQ5443B	OEM
SQ5443C	OEM
SQ5444A	OEM
SQ5444B	OEM

DEVICE TYPE	REPL CODE
SQ5444C	OEM
SQ5445A	OEM
SQ5445B	OEM
SQ5445C	OEM
SQ5446A	OEM
SQ5446B	OEM
SQ5446C	OEM
SQ5447A	OEM
SQ5447B	OEM
SQ5447C	OEM
SQ5448A	OEM
SQ5448B	OEM
SQ5448C	OEM
SQ5449A	OEM
SQ5449B	OEM
SQ5449C	OEM
SQ5450A	OEM
SQ5450B	OEM
SQ5450C	OEM
SQ5451A	OEM
SQ5451B	OEM
SQ5451C	OEM
SQ5452A	OEM
SQ5452B	OEM
SQ5452C	OEM
SQ5453A	OEM
SQ5453B	OEM
SQ5453C	OEM
SQ5454A	OEM
SQ5454B	OEM
SQ5454C	OEM
SQ5455A	OEM
SQ5455B	OEM
SQ5455C	OEM
SQ5456A	OEM
SQ5456B	OEM
SQ5456C	OEM
SQ5461A	0554
SQ5463A	0005
SQ5464A	0720
SQ5468A	0628
SQ5470A	0030
SQB501B-GD	OEM
SR-0004	0196
SR-0005	0015
SR-0006	0196
SR-0007	0015
SR-0008	0015
SR-05K-2	0015
SR05K-2	0023
SR05K-4	0023
SR05K-8	0023
SR05K-12	0023
SR05K-16	0017
SR05K-20	0017
SR-1	0080
SR1	0004
SR1-2	0015
SR1A-1	0015
SR1A1	0015
SR1A-2	0015
SR1A2	0015
SR1A-4	0015
SR1A4	0015
SR1A-8	0015
SR1A8	0015
SR1A-12	0229
SR1A12	0015
SR1B	0277
SR1C1	OEM
SR1C-2	0609
SR1C2	OEM
SR1C-4	0795
SR1C4	OEM
SR1C-6	0795
SR1C6	OEM
SR1C-8	0015
SR1C8	OEM
SR1C-12	0015
SR1C12	OEM
SR1C-16	0071
SR1C16	OEM
SR1C-20	0071
SR1C20	OEM
SR1C-24	0102
SR1C24	OEM
SR1D-1	0015
SR1D1M	0015
SR1D-2	0015
SR1D-4	0015
SR1D-6	0015
SR1D-8	0015
SR1DM	0015
SR1DM-1	0015
SR1DM1	0015
SR1DM-2	0110
SR1DM-4	0015
SR1DM-6	0015
SR1DM-8	0015
SR1DM-10	0015
SR1DMX	0015
SR1E	0015
SR1EM	0015
SR1EM-1	0015
SR1EM1	0015
SR1EM-2	0015
SR1EM2	0015
SR1EM-4	0087
SR1EM4	0087
SR1EM-6	0015
SR1EM6	OEM
SR1EM-8	0015
SR1EM8	OEM
SR1EM-10	0015
SR1EM10	OEM
SR1EM-12	0015
SR1EM12	OEM
SR1EM-16	0071
SR1EM-20	0071
SR1EM20	OEM
SR1FM	0015
SR1FM-0	OEM
SR1FM-1	0015
SR1FM1	0015
SR1FM-2	0102
SR1FM2	0015
SR1FM-4	0015
SR1FM4	0015
SR1FM6	0015
SR1FM-8	0015
SR1FM8	0540
SR1FM-12	0071
SR1FM12	0015
SR1FM-16	0071
SR1FM16	0071
SR1FM-20	0071
SR1FM20	0071
SR1G-4	0102
SR1G4	OEM
SR1G-8	0102
SR1G8	OEM
SR1G-12	0102
SR1G12	OEM
SR1G-16	0102
SR1G16	OEM
SR1G20	OEM
SR1G24	OEM
SR-1HM-2	0015
SR1HM-2	0102
SR1HM-4	0102
SR1HM-8	0023
SR1HM-12	0102
SR1HM-16	0102
SR1K	0015
SR1K-0	0015
SR1K-1	0023
SR1K1	0015
SR1K-1K	0015
SR1K-2	0015
SR1K2	0015
SR-1K-2A	0071
SR1K-4	0023
SR1K4	0015
SR1K-8	0023
SR1K8	0015
SR1K-12	0015
SR1K-16	0071
SR1K-20	0071
SR1K-24	0348
SR1K-4100-521	OEM
SR-1K-0	OEM
SR1K-Z	0015
SR1KZ	0015
SR1S-1	1335
SR1S-2	OEM
SR1S-4	OEM
SR1S-8	OEM
SR1T	0015
SR-1Z	0015
SR1Z	0015
SR-2	0080
SR2A-1	0015
SR2A1	0015
SR2A-2	0015
SR2A2	0015
SR2A-4	0015
SR2A4	0015
SR2A-8	0071
SR2A8	0015
SR2A-12	0071
SR2A12	0015
SR2B1	OEM
SR2B2	OEM
SR2B4	OEM
SR2B6	OEM
SR2B8	OEM
SR2B-12	2081
SR2B12	0087
SR2B-16	2088
SR2B16	0087
SR2B-20	0612
SR2B20	0087
SR2B-24	OEM
SR2B24	OEM
SR2C-1	0708
SR2C1	OEM
SR2C-2	0580
SR2C2	OEM
SR2C-4	0580
SR2C4	OEM
SR2C-6	1000
SR2C6	OEM
SR2C-8	1000
SR2C8	OEM
SR2C-12	2081
SR2C12	OEM
SR2C-16	2088
SR2C16	OEM
SR2C-20	0612
SR2C20	OEM
SR2C-24	OEM
SR2C24	OEM
SR2GH315	OEM
SR2HR316	OEM
SR2K	0023
SR2KL	OEM
SR2KN	OEM
SR-2M	3883
SR2M	3883
SR-2S	OEM
SR2S	OEM
SR-2T	0080
SR-3	0015
SR3	0015
SR3AM	0133
SR3AM-1	0031
SR3AM1	0110
SR3AM-2	0031
SR3AM2	0071
SR3AM-3	0953
SR3AM3	0402
SR3AM-4	0031
SR3AM4	0071
SR3AM-6	0031
SR3AM6	0540
SR3AM-8	0031
SR3AM8	0540
SR3AM-10	0031
SR3AM10	0965
SR3AM-12	0031
SR3AM12	0965
SR3BM-6	0015
SR3U-2	0031
SR3U-4	0031
SR-4	0015
SR4	0015
SR-5	0015
SR5	0015
SR6	0196
SR-9	0015
SR10	0484
SR10-4S	0871
SR10A-2R	0765
SR10A-2S	0097
SR10A-4R	0765
SR10A-4S	0097
SR10A-6R	0538
SR10A-6S	0109
SR10A-8R	0538
SR10A-8S	0109
SR10A-10R	0816
SR10A-10S	0122
SR10A-12R	0816
SR10A-12S	0122
SR10A-16R	0545
SR10A-16S	0131
SR10A-20R	0550
SR10A-20S	0145
SR10A-24R	OEM
SR10A-24S	OEM
SR10C-2R	OEM
SR10C-2S	OEM
SR10C-4R	OEM
SR10C-4S	OEM
SR10C-8R	OEM
SR10C-8S	OEM
SR10D1	0084
SR10D2	0097
SR10D4	0590
SR10D6	OEM
SR10D8	OEM
SR10D-10	OEM
SR10D10	OEM
SR10D-12	OEM
SR10D12	OEM
SR10D-14	OEM
SR10D14	OEM
SR10D-16	OEM
SR10D16	OEM
SR10D-18	OEM
SR10D18	OEM
SR10D-20	OEM
SR10D20	OEM
SR10D-24	OEM
SR10D24	OEM
SR10F-1R	2559
SR10F-1S	5026
SR10F1S	OEM
SR10F-2R	2559
SR10F-2S	5026
SR10F2S	OEM
SR10F-4R	2559
SR10F-4S	5026
SR10F4S	OEM
SR10F-6R	2559
SR10F-6S	5026
SR10F6S	OEM
SR10F-8R	2559
SR10F-8S	5026
SR10F8S	OEM
SR10F10S	OEM
SR10H-1	1600
SR10H-2	0871
SR10H-4	0871
SR10H-6	0853
SR10H-8	0853
SR10J-1	1600
SR10J1S	OEM
SR10J-2	0871
SR10J2S	OEM
SR10J-4	0871
SR10J4S	OEM
SR10J-6	0853
SR10J6S	OEM
SR10J-8	0853
SR10J8S	OEM
SR10K2S	OEM
SR10K6S	OEM
SR10K10S	OEM
SR10L-1S	1600
SR10L-2R	1577
SR10L-2S	1250
SR10L-4S	0871
SR10L-6S	0853
SR10L-8S	0853
SR10N-2R	OEM
SR10N-2S	OEM
SR10N-4R	OEM
SR10N-4S	OEM
SR10N-6R	OEM
SR10N-6S	OEM
SR10N-8R	OEM
SR10N-8S	OEM
SR10N-10R	OEM
SR10N-10S	OEM
SR-13	2903
SR13	0196
SR13H	0015
SR-14	0015
SR14	0484
SR-15	0102
SR15	0484
SR16	0229
SR-16(PHILCO)	1217
SR16DM-2R	OEM
SR16DM-2S	OEM
SR16DM-4R	OEM
SR16DM-4S	OEM
SR16DM-6R	OEM
SR16DM-6S	OEM
SR17	0015
SR-17(PHILCO)	1217
SR-18	0015
SR18	0817
SR-20	0913
SR20	0196
SR20F1S	OEM
SR20F-2R	OEM
SR20F-2S	OEM
SR20F2S	OEM
SR20F-4R	OEM
SR20F-4S	OEM
SR20F4S	OEM
SR20F-6R	OEM
SR20F-6S	OEM
SR20F6S	OEM
SR20F-8R	OEM
SR20F-8S	OEM
SR20F8S	OEM
SR20F10S	OEM
SR20N-2R	OEM
SR20N-2S	OEM
SR20N-4R	OEM
SR20N-4S	OEM
SR20N-6R	OEM
SR20N-6S	OEM
SR20N-8R	OEM
SR20N-8S	OEM
SR20N-10R	OEM
SR20N-10S	OEM
SR20N12S	OEM
SR21	0196
SR22	0015
SR23	0015
SR24	0015
SR25	0102
SR25A-1R	OEM
SR25A-1S	OEM
SR25A1S	OEM
SR25A-2R	OEM
SR25A-2S	OEM
SR25A2S	OEM
SR25A-4R	OEM
SR25A-4S	OEM
SR25A4S	OEM
SR25A-6R	OEM
SR25A-6S	OEM
SR25A6S	OEM
SR25A-8R	OEM
SR25A-8S	OEM
SR25A8S	OEM
SR25A-10R	OEM
SR25A-10S	OEM
SR25A10S	OEM
SR25B-2R	0254
SR25B-2S	2901
SR25B-4R	1107
SR25B-4S	1124
SR25B-6R	1111
SR25B-6S	1126
SR25B-8R	0258
SR25B-8S	0805
SR25B-10R	1644
SR25B-10S	1195
SR25B12S	OEM
SR-27	0015
SR27	0015
SR-28	0015
SR28	0015
SR-29	0196
SR29	0196
SR-30	0474
SR30	0015
SR30A-2R	0254
SR30A-2S	2901
SR30A2S	OEM
SR30A-4R	1107
SR30A-4S	1124
SR30A4S	OEM
SR30A6S	OEM
SR30A-8R	0258
SR30A-8S	0805
SR30A8S	0805
SR30A10S	OEM
SR30A-12R	0267
SR30A-12S	0319
SR30A12S	OEM
SR30A-16R	1119
SR30A-16S	1132
SR30A16S	1132
SR30A-20R	0281
SR30A-20S	0044
SR30A20S	0281
SR30A-24R	OEM
SR30A-24S	OEM
SR30A24S	OEM
SR30B-1	OEM
SR30B1	OEM
SR30B-2	OEM
SR30B2	OEM
SR30B-4	OEM
SR30B4	OEM
SR30B-6	OEM
SR30B6	OEM
SR30B-8	OEM
SR30B8	OEM
SR30B-10	OEM
SR30B10	OEM
SR30B-12	OEM
SR30B12	OEM
SR30B-14	OEM
SR30B14	OEM
SR30B-16	OEM
SR30B16	OEM
SR30B-18	OEM
SR30B18	OEM
SR30B-20	OEM
SR30B20	OEM
SR30B-24	OEM
SR30B24	OEM
SR30C-2R	1520
SR30C-2S	1530
SR30C-4R	1520
SR30C-4S	1530
SR30C-6R	1847
SR30C-6S	0029
SR30C-8R	1847
SR30C-8S	0029
SR30C-10R	1851
SR30C-10S	0601
SR30C-12R	1851
SR30C-12S	0601
SR30C-16R	OEM
SR30C-16S	OEM
SR30C-20R	OEM
SR30C-20S	OEM
SR30D-8R	OEM
SR30D-8S	OEM
SR30D-16R	OEM
SR30D-16S	OEM
SR30D-24R	OEM
SR30D-24S	OEM
SR-31	1303
SR31	1303
SR-32	0774
SR32	0774
SR35	0015
SR-37	0474
SR37	0474
SR40	0015
SR40K-2R	OEM
SR40K-2S	OEM
SR40K-4R	OEM
SR40K-4S	OEM
SR40K-6R	OEM
SR40K-6S	OEM
SR40K-8R	OEM
SR40K-8S	OEM
SR40K-10R	OEM
SR40K-10S	OEM
SR40K-12R	OEM
SR40K-12S	OEM
SR50	0015
SR50A8S	2850
SR50A10S	2871
SR50A12S	2871
SR50A16S	2833
SR50A20S	2473
SR50A24S	OEM
SR60	0015
SR70C-10	2197
SR70C10	2197
SR70C-12	2179
SR70C12	2197
SR70C-16	2216
SR70C16	2216
SR70C-20	2340
SR70C20	2340
SR70C-24	OEM
SR70C24	OEM
SR70C-28	OEM
SR70C28	OEM
SR70C30	OEM
SR70C-32	OEM
SR70C32	OEM
SR70C-36	OEM
SR70C36	OEM
SR70C-40	OEM
SR70C40	OEM
SR70E2S	OEM
SR70E4S	OEM
SR70E6S	OEM
SR70E8S	OEM
SR70F12R	OEM
SR70F16R	OEM
SR70F20R	OEM
SR70K-2R	OEM
SR70K-2S	OEM
SR70K-4R	OEM
SR70K-4S	OEM
SR70K-6R	OEM
SR70K-6S	OEM
SR70K-8R	OEM
SR70K-8S	OEM
SR70K-10R	OEM
SR70K-10S	OEM
SR70K-12R	OEM
SR70K-12S	OEM
SR-76	0015
SR76	0015
SR-100	0242
SR100	0015
SR100AM-10	OEM
SR100AM-12	OEM
SR100AM-16	OEM
SR100AM-20	OEM
SR100AM-24	OEM
SR100AM-32	OEM
SR100AM-36	OEM
SR100AM-40	OEM
SR100K-2R	OEM
SR100K-2S	OEM
SR100K-4R	OEM
SR100K-4S	OEM
SR100K-6R	OEM
SR100K-6S	OEM
SR100K-8R	OEM
SR100K-8S	OEM
SR100K-10R	OEM
SR100K-10S	OEM
SR100K-12R	OEM
SR100K-12S	OEM
SR100L2S	OEM
SR100L4S	OEM
SR100L6S	OEM
SR100L8S	OEM
SR101-1	0015
SR101-2	0015
SR102	0735
SR103	0735
SR103D	3182
SR103W	1358
SR104	0735
SR105	0015
SR106	3888
SR106C	1981
SR106C4	OEM
SR106D	1981
SR108	3888
SR108D	OEM
SR112	0015
SR-114	0242
SR114	0015
SR-120	0015
SR120	0015
SR-130-1	0015
SR130-1	0144
SR131-1	0015
SR-132-1	0015
SR132-1	0110
SR135-1	0015
SR-136	0015
SR136	0110
SR144	0015
SR145	0015
SR150	0071
SR-150-1	0015
SR151	0015
SR152	0015
SR153	2002
SR200	1355
SR200B	1355
SR200DL-12R	OEM
SR200DL-12S	OEM
SR200DL-16R	OEM
SR200DL-16S	OEM
SR200DL-20R	OEM
SR200DL-20S	OEM
SR200DL-24R	OEM
SR200DL-24S	OEM
SR200DM-28R	OEM
SR200DM-28S	OEM
SR200DM-32R	OEM
SR200DM-32S	OEM
SR200DM-36R	OEM
SR200DM-36S	OEM
SR200DM-40R	OEM
SR200DM-40S	OEM
SR200H10	OEM
SR200H-12	OEM
SR200H12	OEM
SR200H-14	OEM
SR200H14	OEM
SR200H-16	OEM
SR200H16	OEM
SR200H-18	OEM
SR200H18	OEM
SR200H-20	OEM
SR200H20	OEM
SR200H-24	OEM
SR200H24	OEM
SR200P-10	OEM
SR200P-12	OEM
SR200P12	OEM
SR200P-16	OEM
SR200P16	OEM
SR200P-20	OEM
SR200P20	OEM
SR200P-24	OEM
SR200P24	OEM
SR200P-30	OEM
SR200P30	OEM
SR200P-36	OEM
SR200P36	OEM
SR200P-40	OEM
SR200P40	OEM
SR200P50	OEM
SR200P60	OEM
SR200PH-28R	4285
SR200PH-28S	4286
SR200PH-32R	4285
SR200PH-32S	4286
SR200PH-36R	OEM
SR200PH-36S	OEM
SR200PH-40R	OEM
SR200PH-40S	OEM
SR200PL-12R	1777
SR200PL-12S	1036
SR200PL-16R	1789
SR200PL-16S	1046
SR200PL-20R	1789
SR200PL-20S	1046
SR200PL-24R	4285
SR200PL-24S	4286
SR200T-12R	OEM
SR200T-12S	OEM
SR200T12S	OEM
SR200T-16R	OEM
SR200T-16S	OEM
SR200T16S	OEM
SR200T-20R	OEM
SR200T-20S	OEM
SR200T20S	OEM
SR205	0015
SR210	0242
SR220	0242
SR300A1	OEM
SR300A2	OEM
SR300A3	OEM
SR300A4	OEM
SR300A6	OEM
SR300A8	OEM
SR300A10	OEM
SR300A-12	OEM
SR300A12	OEM
SR300A-14	OEM
SR300A14	OEM
SR300A-16	OEM
SR300A16	OEM
SR300A-18	OEM
SR300A18	OEM
SR300A-20	OEM
SR300A20	OEM
SR300A-24	OEM
SR300A24	OEM
SR302	2542
SR303	2542
SR304	2542
SR305	3604
SR306	4252
SR308	OEM
SR-390	0015
SR390	0015
SR390-2	0229
SR400A30	OEM
SR400A-40	OEM
SR400A40	OEM
SR400A-50	OEM
SR400A50	OEM
SR400A-60	OEM
SR400A60	OEM
SR400A-70	OEM
SR400A70	OEM
SR400A-80	OEM
SR400A80	OEM
SR400C-10	OEM
SR400C10	OEM
SR400C-12	OEM
SR400C12	OEM
SR400C-16	OEM
SR400C16	OEM
SR400C-20	OEM
SR400C20	OEM

DEVICE TYPE	REPL CODE
SR400C-24	OEM
SR400C24	OEM
SR400C28	OEM
SR400C-30	OEM
SR400C30	OEM
SR400C32	OEM
SR400C36	OEM
SR400C-40	OEM
SR400C40	OEM
SR400C-50	OEM
SR400C50	OEM
SR400C-60	OEM
SR400C60	OEM
SR400C70	OEM
SR400C80	OEM
SR400D10	OEM
SR400D12	OEM
SR400D16	OEM
SR400D20	OEM
SR400D24	OEM
SR400D30	OEM
SR400D40	OEM
SR400D50	OEM
SR400D60	OEM
SR400D70	OEM
SR400DH-28	4285
SR400DH-32	4285
SR400DH-36	OEM
SR400DH-40	OEM
SR400DH-50	OEM
SR400DH-60	OEM
SR400DL-10	4285
SR400DL-12	4285
SR400DL-16	4285
SR400DL-20	4285
SR400DL-24	4285
SR400DV-70	OEM
SR400DV-80	OEM
SR400EL-10	OEM
SR400EL-12	OEM
SR400EL-16	OEM
SR400EL-20	OEM
SR400EL-24	OEM
SR400FH-28	OEM
SR400FH-32	OEM
SR400FH-36	OEM
SR400FH-40	OEM
SR400FH-50	OEM
SR400FH-60	OEM
SR400FV-70	OEM
SR400FV-80	OEM
SR401	0015
SR405	0015
SR475	0110
SR499	0087
SR500	0015
SR500B	0015
SR502	OEM
SR503	OEM
SR503D	OEM
SR504	OEM
SR505	OEM
SR506	OEM
SR506D	OEM
SR507	0242
SR508	OEM
SR513D	OEM
SR605	0015
SR605D	3019
SR606C	OEM
SR632D	OEM
SR632D-R	OEM
SR710	3676
SR710A	OEM
SR710CT	2234
SR710CTE	OEM
SR710CTR	2234
SR710E	OEM
SR710F	2836
SR710FA	OEM
SR710FE	OEM
SR710SD	OEM
SR711	3676
SR711A	OEM
SR711CT	2234
SR711CTE	OEM
SR711CTR	2234
SR711E	OEM
SR711F	2836
SR711FA	OEM
SR711FE	OEM
SR711SD	OEM
SR712	3676
SR712A	OEM
SR712CT	2234
SR712CTE	OEM
SR712CTR	2234
SR712E	OEM
SR712F	2836
SR712FA	OEM
SR712FE	OEM
SR712SD	OEM
SR713	3679
SR713A	OEM
SR713E	OEM
SR713F	4188
SR713FA	OEM
SR713FE	OEM
SR714	3679
SR714A	OEM
SR714CT	OEM
SR714CTE	OEM
SR714CTR	4488
SR714E	OEM
SR714F	4188
SR714FA	OEM
SR714FE	OEM
SR714SD	OEM
SR716	3680
SR716A	OEM
SR716E	OEM
SR716F	4197
SR716FA	OEM
SR716FD	OEM
SR716FE	OEM
SR718CT	OEM
SR718CTE	OEM
SR718CTR	OEM
SR718SD	OEM
SR720CT	OEM
SR720CTE	OEM
SR720CTR	OEM
SR720SD	OEM
SR802	1145
SR802R	OEM
SR803	1145
SR803R	OEM
SR804	1145
SR804R	OEM
SR805	1145
SR805R	OEM
SR806	1145
SR806-126	0015
SR806R	OEM
SR808	1145
SR808R	OEM
SR846-2	0015
SR851	0015
SR851-121	0015
SR889	0015
SR1001	2002
SR1002	OEM
SR1002R	OEM
SR1003	OEM
SR1003R	OEM
SR1004	OEM
SR1004R	OEM
SR1005	OEM
SR1005R	OEM
SR1006	OEM
SR1006R	OEM
SR1024	0015
SR1104	0015
SR1266	0015
SR1378-2	0540
SR1422	0110
SR1493	0015
SR1549	0015
SR1598	0110
SR1602A	1727
SR1602C	1236
SR1602D	OEM
SR1603A	1727
SR1603C	1236
SR1603D	OEM
SR1604A	1727
SR1604C	1236
SR1604D	OEM
SR1605A	1727
SR1605C	1236
SR1605D	OEM
SR1606A	1727
SR1606C	1236
SR1606D	OEM
SR1643A	0110
SR1650	0817
SR-1668	0015
SR1668	0015
SR1692	0015
SR1693	0015
SR1694	0015
SR1695	0015
SR1731-1	0015
SR1731-2	0015
SR1731-3	0015
SR1731-4	0015
SR1731-5	0015
SR1742	0015
SR1762	0242
SR1766	0015
SR-1849-1	0015
SR1984	0015
SR2121	0015
SR2301	0015
SR2301A	0015
SR3002A	OEM
SR3002C	1942
SR3003A	OEM
SR3003C	1942
SR3004A	OEM
SR3004C	1942
SR3005A	OEM
SR3005C	2234
SR3006A	OEM
SR3006C	2234
SR3010	0015
SR3582	0015
SR3943	0015
SR4500D80	OEM
SR5005	3433
SR5010	3433
SR5020	3433
SR5030	3433
SR5040	3433
SR5700	OEM
SR5701	OEM
SR5702	OEM
SR5703	OEM
SR5704	OEM
SR5705	OEM
SR6134	0015
SR6324	0015
SR6325	0015
SR6385	0015
SR6415	0015
SR6560	0015
SR6567	0015
SR6617	0015
SR-9000	0774
SR9000	0774
SR-9001	1303
SR9001	1303
SR-9002	0196
SR9002	0196
SR9003	0015
SR-9005	0474
SR9005	0015
SR20226	0224
SR20234	0079
SR50253-2	0289
SR50411-1	0015
SR50517	0229
SR63197	0037
SR75844	0016
SRA989A-1	OEM
SRAM2064C15	OEM
SRB-1	OEM
SRB-2	OEM
SRB-3	OEM
SRB-4	OEM
SRC220	OEM
SRC221	OEM
SRC230	OEM
SRC260	OEM
SRC290	OEM
SRC600	OEM
SRC740	OEM
SRC9231	OEM
SRC9261	OEM
SRE4-05	OEM
SRE7-05	OEM
SRE9A05	OEM
SRE12B05	OEM
SRE17A05	OEM
SRE23B05	OEM
SRE33-05	OEM
SRE86A05	OEM
SRE115B05	OEM
SREA4-05	OEM
SREA7-05	OEM
SREA9A05	OEM
SREA12B05	OEM
SREA17A05	OEM
SREA23B05	OEM
SREA33-05	OEM
SREA86A05	OEM
SREA115B05	OEM
S-RECT-174	0196
SRF-1	OEM
SRF-2	OEM
SRF-3	OEM
SRF-4	OEM
SRF2017	2312
SRH25C	OEM
SRI-155740B	0725
SRIDM	0015
SRIDM-1	0015
SRIDM-4	0015
SRIEM-1	0015
SRIEM-2	0015
SRIEM-4	0015
SRIFM-2	0102
SRIFM-4	0102
SRIFM-12	0102
SRIHM-8	0102
SR-IK-2	0015
SRIK-2	0015
SRIK-8	0015
SRK1	0015
SRK-2	0015
SRK-11ZC	OEM
SRLP1004	0892
SRM-2064C	OEM
SRM2064C	OEM
SRM2064C15	OEM
SRM2114C3	2048
SRM2264LC12	OEM
SRM2264M12	OEM
SRN02	OEM
SRN02-100	1015
SRN02-200	1015
SRN02-400	1015
SRP02	OEM
SRP02-100	0199
SRP02-200	0199
SRP02-400	0199
SRP100A	0023
SRP100B	0023
SRP100D	0023
SRP100G	0023
SRP100J	0023
SRP100K	0023
SRP300A	0031
SRP300B	0031
SRP300D	0031
SRP300G	0031
SRP300J	0031
SRP300K	1372
SRP600A	OEM
SRP600B	OEM
SRP600D	OEM
SRP600G	OEM
SRP600J	OEM
SRP600K	OEM
SRQ02-100	6457
SRQ02-200	6457
SRQ02-400	6457
SR-R13	0196
SRR13	0196
SRS105	1335
SRS1004	0264
SRS1100	OEM
SRS2004	0168
SRS2504	0168
SRS2804S	0168
SRS3014	0168
SRS3100	OEM
SRS3204S	0168
SRS3504	0168
SRS3604S	0168
SRS4014	0168
SRS4014S	0168
SRS4404S	0168
SRS4504	0168
SRS5014	0168
SRS5504	0168
SRUF15	OEM
SRUN15	OEM
SRUN38T	OEM
SRUN50T	OEM
SS0001	0004
SS0002	0004
SS0003	0164
SS0004	0004
SS0005	0004
SS0007	0143
SS0008	0143
SS0009	0015
SS00010	0015
SS000I	0004
SS-1	0015
SS1.5J4	2542
SS2J4	OEM
SS-2N5062	0901
SS-3	OEM
SS3J4	OEM
SS-3R	OEM
SS-5	OEM
SS-5R	OEM
SS-7	OEM
SS-7R	OEM
SS29A4	0693
SS29A5	0693
SS101A	OEM
SS101B	OEM
SS101C	OEM
SS151	OEM
SS151R	OEM
SS155	0050
SS201	OEM
SS277G	0795
SS301	OEM
SS321	0609
SS322	0609
SS324	0609
SS325	1335
SS334	0609
SS337	0609
SS455	0015
SS524	0233
SS1001	4691
SS1002	OEM
SS1003	OEM
SS1004	OEM
SS1005	OEM
SS1010	OEM
SS1011	OEM
SS1012	OEM
SS1015	OEM
SS1020	OEM
SS1030	OEM
SS1040	OEM
SS1123	0338
SS1202	OEM
SS1601	OEM
SS1602	OEM
SS1603	OEM
SS1606	0037
SS1621	OEM
SS1623	OEM
SS1906	3108
SS1912	0233
SS2002	OEM
SS2005	OEM
SS2010	OEM
SS2015	OEM
SS2020	OEM
SS2030	OEM
SS2040	OEM
SS2308	0016
SS2503	0037
SS2504	0016
SS3534-4	0325
SS3586	0325
SS3638	0037
SS3638A	0037
SS3672	0325
SS3694	0016
SS-3704	0325
SS3704	0325
SS3735	0325
SS3935	0698
SS4001AE	0478
SS4002AE	2055
SS4011AE	0215
SS4012AE	0498
SS4013AE	0413
SS4023AE	0519
SS4025AE	2072
SS4027AE	1949
SS4028AE	2228
SS4030AE	0500
SS4042	0007
SS6111	0233
SS6724	0037
SS9101	OEM
SS9327	0264
SS9328	0079
SSA5.0	OEM
SSA5.0A	OEM
SSA6.0	OEM
SSA6.0A	OEM
SSA6.5	OEM
SSA6.5A	OEM
SSA7.0	OEM
SSA7.0A	OEM
SSA7.5	OEM
SSA7.5A	OEM
SSA8.0	OEM
SSA8.0A	OEM
SSA8.5	OEM
SSA8.5A	OEM
SSA9.0	OEM
SSA9.0A	OEM
SSA10	OEM
SSA10A	OEM
SSA11	OEM
SSA11A	OEM
SSA12	OEM
SSA12A	OEM
SSA13	OEM
SSA13A	OEM
SSA14	OEM
SSA14A	OEM
SSA15	OEM
SSA15A	OEM
SSA16	OEM
SSA16A	OEM
SSA17	OEM
SSA17A	OEM
SSA18	OEM
SSA18A	OEM
SSA20	OEM
SSA20A	OEM
SSA22	OEM
SSA22A	OEM
SSA24	OEM
SSA24A	OEM
SSA26	OEM
SSA26A	OEM
SSA28	OEM
SSA28A	OEM
SSA30	OEM
SSA30A	OEM
SSA33	OEM
SSA33A	OEM
SSA36	OEM
SSA36A	OEM
SSA40	OEM
SSA40A	OEM
SSA43	0037
SSA43A	0037
SSA45	OEM
SSA45A	OEM
SSA46	0037
SSA46A	0037
SSA48	0037
SSA48A	0037
SSA51	OEM
SSA51A	OEM
SSA54	OEM
SSA54A	OEM
SSA58	OEM
SSA58A	OEM
SSA60	OEM
SSA60A	OEM
SSA64	OEM
SSA64A	OEM
SSA70	OEM
SSA70A	OEM
SSA75	OEM
SSA75A	OEM
SSA78	OEM
SSA78A	OEM
SSA85	OEM
SSA85A	OEM
SSA90	OEM
SSA90A	OEM
SSA100	OEM
SSA100A	OEM
SSA110	OEM
SSA110A	OEM
SSA120	OEM
SSA120A	OEM
SSA130	OEM
SSA130A	OEM
SSA150	OEM
SSA150A	OEM
SSA160	OEM
SSA160A	OEM
SSA170	OEM
SSA170A	OEM
SSB3020CT	1942
SSB3020SD	OEM
SSB3030CT	1942
SSB3030SD	OEM
SSB3040CT	1942
SSB3040SD	OEM
SSB3050CT	OEM
SSB3050SD	OEM
SSB3060CT	OEM
SSB3060SD	OEM
SSB3080CT	OEM
SSB3080SD	OEM
SSC600	OEM
SSD708	0133
SSD974	0015
SSD2404	OEM
SSF15	OEM
SSF20	OEM
SSF25	OEM
SSF30	OEM
SSF40A	OEM
SSF50A	OEM
SSF60A	OEM
SSF200	OEM
SSF201	OEM
SSF202	OEM
SSF203	OEM
SSH1A020	1335
SSH1A040	1335
SSH1A060	0080
SSH1A080	0080
SSH3A020	0087
SSH3A030	0087
SSH3A040	0087
SSH3A050	0087
SSH3A060	0087
SSH5A020	OEM
SSH5A030	OEM
SSH5A040	OEM
SSH5A050	OEM
SSH5A060	OEM
SSH30A03D	0584
SSH30A04D	0584
SSH30A020D	0584
SSH30A020S	0615
SSH30A020T	OEM
SSH30A030D	OEM
SSH30A030S	0615
SSH30A030T	OEM
SSH30A040D	OEM
SSH30A040S	0615
SSH30A040T	OEM
SSH30A050D	OEM
SSH30A050S	OEM
SSH30A050T	OEM
SSH30A060D	OEM
SSH30A060S	OEM
SSH30A060T	OEM
SSH75A020	OEM
SSH75A020V	OEM
SSH75A030	OEM
SSH75A030V	OEM
SSH75A040	OEM
SSH75A040V	OEM
SSH75A050	OEM
SSH75A050V	OEM
SSH75A060	OEM
SSH75A060V	OEM
SSH2500	OEM
SSH5000	OEM
SSH7500	OEM
SSH10000	OEM
SSH12500	OEM
SSH15000	OEM
SSH20000	OEM
SSHF2500	OEM
SSHF5000	OEM
SSHF7500	OEM
SSHF10000	OEM
SSHF12500	OEM
SSHF15000	OEM
SSHF20000	OEM
SSI202	OEM
SSI1120	0540
SSIB0140	0087
SSIB0180	0087
SSIB0640	0087
SSIB0680	0087
SSIB0710	OEM
SSIB0720	OEM
SSIB0740	OEM
SSIB0780	OEM
SSIC0810	0242
SSIC0820	0540
SSIC0840	0087
SSIC0860	0922
SSIC0880	0087
SSIC1110	0242
SSIC1140	0965
SSIC1160	0087
SSIC1180	0922
SSIC1210	0242
SSIC1220	0540
SSIC1240	0087
SSIC1260	0922
SSIC1280	0087
SSIC1660A	0922
SSIC1740	0087
SSIC1780	0087
SSIC1960A	0087
SSIC15100A	OEM
SSIC16100A	OEM
SSIC19100A	OEM
SSIC20100A	OEM
SSIE0810	OEM
SSIE0820	OEM
SSIE0840	OEM
SSIE0860	OEM
SSIE2060A	OEM
SSIE2080	OEM
SSIE2702	OEM
SSIE2705	OEM
SSIE2710	OEM
SSIE2715	OEM
SSIE2802	OEM
SSIE2805	OEM
SSIE2810	OEM
SSIE2815	OEM
SSIF0320	OEM
SSIF0340	OEM
SSIF0360	OEM
SSIH0220	OEM
SSIH0240	OEM
SSIH0260	OEM
SSIK0210	OEM
SSIK0220	OEM
SSIK0240	OEM
SSIK0260	OEM
SSIK0270	OEM
SSIK0420	OEM
SSIK0440	OEM
SSIK0460	OEM
SSIK0480	OEM
SSIK2020	OEM
SSIK2040	OEM
SSIK2060	OEM
SSIK2060A	OEM
SSIK2080	OEM
SSIK2080A	OEM
SSIL0520	OEM
SSIL0540	OEM
SSIL0560	OEM
SSIL0570	OEM
SSIL1080	OEM
SSIL1080A	OEM
SSIL1280	OEM
SSIL10100	OEM
SSIL10100A	OEM
SSIL12100	OEM
SSIL12120	OEM
SSIN2060	OEM
SSIN2080	OEM
SSIN2080A	OEM
SSIN2360	OEM
SSIN2380	OEM
SSIN2460	OEM
SSIN2483	OEM
SSIN20100	OEM
SSIN20100A	OEM
SSIN20120	OEM
SSIN23100	OEM
SSIN23120	OEM
SSIN24100	OEM
SSIN24120	OEM
SSIN98100	OEM
SSIN98110	OEM
SSIN98120	OEM
SSIN98135	OEM
SSIN98145	OEM
SSIN98155	OEM
SSIP0160	OEM
SSIP0180	OEM
SSIP01100	OEM
SSIP30120	OEM
SSIP30135	OEM
SSIP30150	OEM
SSIR-5	OEM
SSIR-5D	OEM
SSIR-SC	OEM
SSM20	OEM
SSM30	OEM
SSM40	OEM
SSM50A	OEM
SSM60A	OEM
SSM410	OEM
SSM510	OEM
SSM920SSM1020	OEM
SSN74LS243	0906
SSP6N60A	OEM
SSP310	0615
SSP320	0615

If replacement code is OEM, contact original manufacturer for replacement.

DEVICE TYPE	REPL CODE
SSP330	0615
SSP340	0615
SSP840	0615
SSP1530	0615
SSP1540	0615
SSP2020	0615
SSP2030	0615
SSP2040	0615
SSP2848	0622
SSP2849	0622
SSP2850	0622
SSP2851	0622
SSP2852	0622
SSP3005	1545
SSP3010	1545
SSP3030	1545
SSP3040	1545
SSP3051	OEM
SSRC9-001	OEM
SST610	OEM
ST/217/Q	0016
ST01	0016
ST02	0016
ST03	0016
ST04	0016
ST05	0016
ST06	0016
ST07279	0050
ST07430	0111
ST1	OEM
ST-2	3331
ST2	2150
ST2(GE)	3331
ST2-10	0097
ST2-20	0097
ST2-30	0109
ST2-40	0109
ST2B	4423
ST-2N6558	0643
ST-2SA562Y	0006
ST-2SA673C	0148
ST-2SA715C	0525
ST-2SA733Q	0006
ST-2SA844C	0693
ST-2SA1015Y	0148
ST-2SB568C	1911
ST-2SC287A	0475
ST-2SC288A	1342
ST-2SC383W	0076
ST-2SC454C	0076
ST-2SC458C	0076
ST-2SC460	0151
ST-2SC535	0127
ST-2SC535B	0127
ST-2SC605	0475
ST-2SC606	0475
ST-2SC681AYL	2437
ST-2SC1070	0007
ST-2SC1106	0220
ST-2SC1106A	0220
ST-2SC1213C	0191
ST-2SC1213D	0191
ST-2SC1213D-24	0191
ST-2SC1317Q	0155
ST-2SC1514	1085
ST-2SC1514-05	1085
ST-2SC1730	0127
ST-2SC1815GR	5686
ST-2SC1855	0224
ST-2SC1906	0127
ST-2SC1923-O	2691
ST-2SC2371L	0276
ST-2SC2371M	0276
ST-2SC2735	2919
ST-2SC2756	OEM
ST-2SC2757	OEM
ST-2SD235Y	0228
ST-2SD467B	0951
ST-2SD467C	0951
ST-2SD478C	0392
ST-2SD748A	3227
ST-2SD748A-01	3227
ST-2SD748A-01A	3227
ST-2SD781	0563
ST3	OEM
ST3-10	0805
ST3-20	0805
ST3-30	0805
ST3-40	0805
ST3A10N	0267
ST3A10P	0319
ST3A20N	0267
ST3A20P	0319
ST3A30N	0267
ST3A30P	0319
ST3A40N	0267
ST3A40P	0319
ST3A50N	0267
ST3A50P	0319
ST3A60N	0267
ST3A60P	0319
ST3B10N	0267
ST3B10P	0319
ST3B20N	0267
ST3B20P	0319
ST3B30N	0267
ST3B30P	0319
ST3B40N	0267
ST3B40P	0319
ST3B50N	0267
ST3B50P	0319
ST3B60N	0267
ST3B60P	0319
ST-3SK53	0212
ST-3SK60	2458
ST-3SK80	0371
ST-3SK87	0371
ST-4	4405
ST4	4405
ST4-10	1124
ST4-20	1124
ST4-30	0805
ST4-40	0805
ST5	OEM
ST5-	OEM
ST5A10N	1783
ST5A10P	2813
ST5A20N	1783
ST5A20P	2813
ST5A30N	1783
ST5A30P	2813
ST5A40N	1783
ST5A40P	2813
ST5A50N	1783
ST5A50P	2813
ST5A60N	1818
ST5A60P	2845
ST6	OEM
ST9	0144
ST10	0144
ST11	0127
ST12	0144
ST13	0144
ST14	0144
ST15	0144
ST16	0015
ST18	0071
ST20	OEM
ST22-05	OEM
ST22-10	OEM
ST22-20	OEM
ST22-30	OEM
ST22-40	OEM
ST22-60	OEM
ST22-80	OEM
ST22-100	OEM
ST22-120	OEM
ST24C01C1	OEM
ST24C01CB1	OEM
ST24C01CB6	OEM
ST24C02AB1	OEM
ST24C02CB1	OEM
ST24C02FM6TR	OEM
ST24C026	OEM
ST25A	0016
ST25B	0016
ST25C	0016
ST28A	0004
ST28B	0004
ST28C	0004
ST29	0144
ST30	0144
ST31	0144
ST32	0144
ST33	0144
ST34	0144
ST35	0144
ST37C	0004
ST37D	0004
ST37E	0004
ST370	0847
ST39	0036
ST40	0144
ST41	0144
ST42	0144
ST43	0144
ST44	0144
ST45	0144
ST47	0917
ST50	0016
ST51	0016
ST53	0016
ST54	0016
ST55	0016
ST56	0016
ST57	0016
ST58	0016
ST59	0016
ST60	0144
ST61	0144
ST62	0144
ST63	0016
ST64	0016
ST66	OEM
ST70	0144
ST71	0144
ST72	0144
ST80	0144
ST82	0144
ST101	0130
ST106	0440
ST107	0440
ST108	0440
ST109	0440
ST110	0440
ST111	0440
ST112	0440
ST122	0004
ST123	0004
ST150	0016
ST151	0016
ST152	0016
ST153	0016
ST154	0016
ST155	0016
ST156	0016
ST157	0016
ST160	0016
ST161	0016
ST162	0016
ST163	0016
ST172	0600
ST175	0016
ST176	0016
ST177	0016
ST178	0016
ST180	0016
ST181	0016
ST182	0016
ST185	OEM
ST186	OEM
ST187	OEM
ST210N	2291
ST210P	2900
ST220N	0476
ST220P	1112
ST230N	0476
ST230P	1112
ST235	0160
ST240N	0476
ST240P	1112
ST250	0016
ST250N	0476
ST250P	1112
ST251	0016
ST260N	0449
ST260P	3011
ST280N	0449
ST280P	3011
ST301	0004
ST302	0004
ST303	0004
ST304	0004
ST310N	0267
ST310P	0319
ST320N	0267
ST320P	0319
ST330N	0267
ST330P	0319
ST332	0004
ST340N	0267
ST340P	0319
ST350N	0267
ST350P	0319
ST360N	0281
ST360P	0044
ST382	0004
ST402	0595
ST403	0016
ST410	OEM
ST410N	1783
ST410P	2813
ST411	OEM
ST414	OEM
ST415	0144
ST420N	1783
ST420P	2813
ST430N	1783
ST430P	2813
ST440N	1783
ST440P	2813
ST450N	1783
ST450P	2813
ST460N	1818
ST460P	2845
ST480N	1818
ST480P	2845
ST501	0016
ST502	0016
ST503	0016
ST504	0016
ST510	OEM
ST510N	1347
ST510P	0599
ST520N	1905
ST520P	1986
ST530N	1905
ST530P	1986
ST540N	1905
ST540P	1986
ST550N	1905
ST550P	1986
ST560N	1986
ST560P	0193
ST610N	0657
ST610P	1347
ST615	OEM
ST620N	1347
ST620P	1347
ST630N	1347
ST630P	1905
ST640N	1986
ST640P	1986
ST650N	1905
ST650P	1986
ST660N	0193
ST660P	0193
ST710N	0501
ST710P	1023
ST720N	1777
ST720P	1036
ST721	OEM
ST722	0086
ST723	0086
ST730N	1777
ST730P	1036
ST740N	1777
ST740P	1036
ST750N	1777
ST750P	1036
ST760N	1789
ST760P	1046
ST903	0144
ST904	0144
ST904A	0144
ST905	0144
ST910	0144
ST910N	0501
ST910P	1023
ST920N	1777
ST920P	1036
ST930N	1777
ST930P	1036
ST940N	1777
ST940P	1036
ST950N	1777
ST950P	1036
ST960N	1789
ST960P	1046
ST980N	1789
ST980P	1046
ST1010N	1347
ST1010P	0599
ST1020N	1905
ST1020P	1986
ST1026	0144
ST1030N	1905
ST1030P	1986
ST1040N	1905
ST1040P	1986
ST1042	0079
ST1042C	0079
ST1050	0144
ST1050N	1905
ST1050P	1986
ST1051	0144
ST-1060	OEM
ST1060N	0193
ST1060P	0657
ST1110N	1783
ST1110P	2813
ST1120N	1783
ST1120P	1783
ST1130N	1783
ST1130P	2813
ST1140N	1783
ST1140P	2813
ST1150N	1783
ST1150P	2813
ST1160N	1818
ST1160P	2845
ST1180N	1818
ST1180P	2845
ST1242	0016
ST1243	0016
ST1244	0016
ST1290	0016
ST1336	0144
ST1402C	0016
ST1402D	0076
ST1402E	0079
ST1502B	0127
ST1502C	3807
ST1502CD	3807
ST1502D	0127
ST-1502L	0079
ST1502L	0079
ST1504	0595
ST1505	0595
ST1506	0016
ST1523	OEM
ST1524	OEM
ST1525	OEM
ST1527	OEM
ST1528	OEM
ST1543	0079
ST1602D	0037
ST1607	0016
ST1610N	1347
ST1610P	0599
ST1620N	1905
ST1620P	1986
ST1630N	1905
ST1630P	1986
ST1633	OEM
ST1640N	1905
ST1640P	1986
ST1650N	1905
ST1650P	1986
ST1660N	0193
ST1660P	0657
ST1680N	0193
ST1680P	0657
ST1694	0016
ST1700	0079
ST1702L	0079
ST1702M	0079
ST1702N	0286
ST1802M	0532
ST1802N	0532
ST1902D	OEM
ST2040P	0015
ST2110	0144
ST2120	0007
ST2130	0007
ST3030	0198
ST3031	0198
ST3042	2626
ST3043	OEM
ST4044	OEM
ST4045	OEM
ST4080	OEM
ST4081	OEM
ST4150	0086
ST4201	0086
ST4202	0086
ST4203	0086
ST4204	0086
ST4341	0086
ST4402	OEM
ST5060	0016
ST5061	0595
ST5641	0007
ST6008	OEM
ST6010	OEM
ST6110	0016
ST6120	0144
ST6125	0144
ST6130	0079
ST6398B1	OEM
ST6398B1/A	OEM
ST6398B1/B	OEM
ST6497B1	OEM
ST6497BB1/BHS	OEM
ST6510	0127
ST6511	0016
ST6512	0016
ST6573	0086
ST6574	0086
ST6593	0016
ST6594	0016
ST6600	0016
ST6601	0086
ST7100	0111
ST8014	0037
ST8033	0037
ST8034	0037
ST8035	0037
ST8036	0037
ST8065	0037
ST8181	0037
ST8182	0037
ST8183	0126
ST8184	0126
ST8190	0037
ST8191	0126
ST8500	0037
ST8509	0037
ST8700	0037
ST8704	0037
ST8705	0037
ST8709	0037
ST9001	OEM
ST9008	OEM
ST9045	OEM
ST9046	OEM
ST9047	OEM
ST10000	2626
ST10001	2626
ST10007	1991
ST10008	2626
ST10009	2626
ST27020	0925
ST29045	1681
ST29046	1681
ST29047	1681
ST47025	0144
ST48119	OEM
ST61000	0791
ST62180	0037
ST72039	0126
ST72040	0126
ST74000	0168
ST78020	OEM
ST83255	0037
ST84000	0168
ST84027	0168
ST84028	0168
ST84029	0168
ST984539-006	3399
STA301A	OEM
STA302A	OEM
STA303	OEM
STA311A	OEM
STA321A	OEM
STA341M	OEM
STA401A	OEM
STA403	OEM
STA403A	3967
STA404A	OEM
STA405A	OEM
STA434A	OEM
STA437A	OEM
STA451C	OEM
STA455C	OEM
STA608XFXXXXI	0006
STA901M	OEM
STA7604	2485
STA8012	OEM
STA8341	0363
STA9364	0074
STAR-IFPH	OEM
STB01-02	0133
STB0101	4403
STB3	OEM
STB4	0760
STB567	OEM
STB568	0133
STB569	0133
STB576	0133
STB598XEXXXXX	3812
ST-BF422	0261
ST-BF459	0276
STBF459	0276
ST-BF960	0371
ST-BU208	0065
STC389	OEM
STC536XFXXUB	0537
STC536XFXXXUB	5336
STC930XEXXXRI	2209
STC1001	OEM
STC1035	0103
STC1035A	0103
STC1036	0103
STC1036A	0103
STC1080	0103
STC1081	0103
STC1082	0103
STC1083	0103
STC1084	0103
STC-1085	0103
STC1085	0074
STC1094	0130
STC1101	OEM
STC1102	OEM
STC1103	OEM
STC1104	OEM
STC1105	OEM
STC1105A	OEM
STC1106	OEM
STC1106A	OEM
STC1300	0042
STC1311	OEM
STC1312	OEM
STC1313	OEM
STC1314	OEM
STC1331	OEM
STC1332	OEM
STC1336	0595
STC1674LXXXXX	1066
STC1725	OEM
STC1727	OEM
STC1729	OEM
STC1730	OEM
STC1732	OEM
STC1734	OEM
STC1735	OEM
STC1737	OEM
STC1739	OEM
STC1750	OEM
STC1751	OEM
STC1777	OEM
STC1800	0560
STC1850	0042
STC1860	0042
STC1862	0560
STC2100	OEM
STC2101	OEM
STC2220	0130
STC2221	0130
STC2224	0130
STC2225	0130
STC2228	0130
STC2229	0130
STC4252	0103
STC4253	0103
STC4254	0103
STC4255	0103
STC-4401	0178
STC4401	0178
STC5080	OEM
STC5081	OEM
STC5082	OEM
STC5083	OEM
STC5084	OEM
STC5085	OEM
STC5109/1	OEM
STC5110/1	OEM
STC5111/1	OEM
STC5112/1	OEM
STC5113/1	OEM
STC5114/1	OEM
STC5202	0925
STC5203	0925
STC5205	0925
STC5206	0925
STC5303	0925
STC5519/1	OEM
STC5520/1	OEM
STC5521/1	OEM
STC5522/1	OEM
STC5523/1	OEM
STC5524/1	OEM
STC5550	OEM
STC5551	OEM
STC5552	OEM
STC5553	OEM
STC5554	OEM
STC5555	OEM
STC5606	OEM
STC5608	OEM
STC5610	0126
STC5611	0126
STC5612	0126
STC5648	OEM
STC5649	OEM
STC5650	OEM
STC5651	OEM
STC5652	OEM
STC5802	0925
STC5803	0925
STC5805	0925
STC5806	0925
STD28	OEM
STD30	OEM
STD33	OEM
STD36	OEM
STD39	OEM
STD43	OEM
STD47	OEM
STD51	OEM
STD56	OEM
STD62	OEM
STD68	OEM
STD75	OEM
STD82	OEM
STD91	OEM
STD100	OEM
STD110	OEM
STD330XEXXXXX	1607
STD545XEXXXXX	0018
STD580	OEM
STD734FXFXXXX	1513
STD9007	0560
STD13301	1852
STD13302	1852
STD13303	1852
STD40304	1966
STE400	0144
STE401	2446
STFCN10	0015
ST-HIT9012H	1369
STHIT9012H	0006
ST-HIT9013H	0079
STHIT9013H	0079
ST-HIT9014C	0079
STHIT9014C	0079
ST-HIT9016G	0224
STI-10	0493
STI10	0168
STI-20	0233
STI20	0168
STI-30	0233
STI30	0168
STI40	0168
STI50	0168
STI-105	0142
STI-205	0142
STI-305	0142
STI401	0223
STI410	0223
STI411	0223
STI413	0223
STI431	0223
STI701	0003
STI1506	0142
STI2006	0142
STI2506	0142
STI3006	0142
STI3007	0142
STIP-10	0892
STIP10	0892
STIP20	0439
ST-JE9014C	4898
STK-0029	OEM
STK0050	6908
STK0050-II	OEM
STK0080	4675
STK0080-II	OEM
STK00502	OEM
STK00802	OEM
STK-011	3550
STK011	3550
STK013	4626
STK-015	2876
STK015	2876
STK016	4676
STK-018	6688
STK020	1187
STK020F	1187
STK031	4686
STK-036	2864
STK040	4677
STK041	4678
STK043	4678
STK050	4690
STK058	4694
STK070	4690
STK-075	2864
STK075	4695
STK077	4680
STK078	4680
STK080	4681

DEVICE TYPE	REPL CODE
STK-082	4681
STK082	4681
STK083	4682
STK-084	4682
STK084	4682
STK086	4691
STK0100	OEM
STK01002	OEM
STK-154	4559
STK401	5080
STK413	1571
STK415	1571
STK430	4698
STK-433	1571
STK433	1571
STK435	1571
STK-436	OEM
STK436	OEM
STK436TV	OEM
STK-437	1612
STK437	1612
STK439	1612
STK441	4683
STK443	4693
STK457	4684
STK459	4684
STK460	4702
STK461	4685
STK463	4685
STK465	1903
STK465A	1903
STK561	OEM
STK561E	OEM
STK561EA	OEM
STK-561F	1533
STK561F	2554
STK561FA	OEM
STK563A	1572
STK-563F	1572
STK563F	1572
STK583F	2745
STK-583FST	OEM
STK583FST	1572
STK-711	OEM
STK711	OEM
STK720A	OEM
STK720B	OEM
STK720C	OEM
STK730-010	OEM
STK730-030	OEM
STK740B	OEM
STK740C	OEM
STK743B	OEM
STK752	OEM
STK756	OEM
STK-756A	OEM
STK760	OEM
STK770	OEM
STK772B	OEM
STK-780	OEM
STK780	OEM
STK795	OEM
STK1050	4151
STK1050A	4151
STK2230	5278
STK2250	OEM
STK3042	1437
STK3042II	OEM
STK3042III	1437
STK3042MK2	OEM
STK3062	OEM
STK3062III	6420
STK3082	4693
STK3082II	4693
STK3102II	1410
STK3102III	1410
STK3122III	6418
STK3152III	6416
STK3633	OEM
STK4021	1122
STK4021M	1122
STK4024II	2512
STK4024V	OEM
STK4026II	OEM
STK4026V	OEM
STK4028II	OEM
STK4028V	OEM
STK4030	OEM
STK4030-2	OEM
STK4030II	OEM
STK4030V	OEM
STK4032II	OEM
STK4032V	OEM
STK4034II	OEM
STK4034V	OEM
STK4036-II	OEM
STK4036II	OEM
STK4036V	OEM
STK4038II	OEM
STK4038V	OEM
STK4040-V	OEM
STK4040II	OEM
STK4040V	OEM
STK4042II	OEM
STK4042V	OEM
STK4044-V	OEM
STK4044II	3616
STK4044V	4143
STK4046II	OEM
STK4046V	OEM
STK4046XI	OEM
STK4048-V	OEM
STK4048II	OEM
STK4048V	OEM
STK4050V	OEM
STK4112II	OEM
STK4121-II	OEM
STK4121II	5277
STK4131-II	OEM
STK4132	OEM
STK4132-II	OEM
STK4141-II	OEM
STK4141II	5752
STK4142-II	OEM
STK4142II	5752
STK4151	OEM
STK4151-II	OEM
STK4152	OEM
STK4152-2M	OEM
STK4152II	5269
STK4171-2S	OEM
STK4171II	6079
STK4181	OEM
STK4182II	3030
STK4192-II	OEM
STK4192II	5276
STK4211-5P	OEM
STK4231-5	OEM
STK4231II	OEM
STK4231M-1	OEM
STK4231MK2	OEM
STK4273	1439
STK4311	1483
STK4332	OEM
STK4362	1571
STK4362TV	OEM
STK4372	OEM
STK4392	1612
STK5314	OEM
STK5321SL	OEM
STK5322	2297
STK5322H	OEM
STK5327SL	OEM
STK5332	2303
STK5333	OEM
STK5342	OEM
STK5346	OEM
STK5352	OEM
STK5364	OEM
STK5372	2334
STK5422	OEM
STK5431	2374
STK5431SL	2374
STK5431ST	2374
STK5436	OEM
STK5443	OEM
STK5451	2385
STK5461ST	OEM
STK5464	OEM
STK5464ST	OEM
STK5466	OEM
STK5466ST	5817
STK5468	OEM
STK5468S	OEM
STK5471	2405
STK5476	2413
STK5478	OEM
STK5479	2417
STK5481	2418
STK5482	2419
STK5483	OEM
STK5487	OEM
STK5490	2426
STK5633F	OEM
STK5883F	OEM
STK-6011	OEM
STK-6021	OEM
STK6722H	2765
STK6722HZ	OEM
STK6922	OEM
STK6940	3340
STK6941	3341
STK6942	3341
STK6950	OEM
STK6951	OEM
STK6952	OEM
STK6960	3347
STK6960H	OEM
STK6961	3347
STK6961H	OEM
STK6962	3348
STK6962H	3348
STK6965	OEM
STK6965H	OEM
STK6966	OEM
STK6966H	OEM
STK6967	OEM
STK6967H	OEM
STK6968	OEM
STK6970	3353
STK6970H	OEM
STK6971	3353
STK6971H	OEM
STK6972	3355
STK6972H	OEM
STK6981B	3361
STK6982	3361
STK6982B	3361
STK7216	OEM
STK7216S	OEM
STK7217	OEM
STK7231	OEM
STK7308	3488
STK7309	3488
STK7310	OEM
STK7348	OEM
STK7356	OEM
STK7358	OEM
STK7360	OEM
STK7402	OEM
STK7402-105	OEM
STK7404	OEM
STK7406	OEM
STK7408	OEM
STK7410	OEM
STK7451	OEM
STK7452	OEM
STK7453	OEM
STK7457	OEM
STK7458	OEM
STK7459	OEM
STK7462	OEM
STK7468	OEM
STK7554	OEM
STK7563F	1596
STK7563FE	1596
STK7563GII	OEM
STK40362	OEM
STK40405	OEM
STK40442M	OEM
STK40445	OEM
STK41212	OEM
STK41312	OEM
STK66083P	OEM
STK73012	OEM
STK73405II	OEM
STK73408II	OEM
STK73409II	OEM
STK73410II	OEM
STK73605I	OEM
STK73606I	OEM
STL6961	3347
ST-LM1274	OEM
ST-LM2152	0079
ST-LM2682	0079
STM101	0190
STM-204	OEM
STM204	OEM
STM-204-1	OEM
STM204-1	OEM
STM204-8088	OEM
STM206Z86C27	OEM
STM208	OEM
STM401	2804
STM2047C432N-8097	OEM
STM20247C432-8086	OEM
STM20247C432-8097	OEM
STM-20247C432AN-8098	OEM
STM20247C432AN-8098	OEM
STM20247C432N-8086	OEM
STM20247C432N8086	OEM
STM20247C432N-8097	OEM
STM20247C432N8097	OEM
ST-MJE9742	0276
ST-MPS9410AJ	0086
ST-MPS9410AK	0086
ST-MPS9418AS	0324
ST-MPS9418AT	0324
ST-MPS9433	0079
ST-MPS9433S	0079
ST-MPS9433T	0079
ST-MPS9632J	0079
ST-MPS9633C	0079
ST-MPS9682J	0037
ST-MPS9700D	0079
ST-MPS9700E	0079
ST-MPS9700F	0079
ST-MPS9700G	0079
ST-MPS9750D	0037
ST-MPS9750F	0037
ST-MPS9750G	0037
ST-MPSH10	0224
ST-MPSH24	0144
STPT260	3552
STR101(ELCOM)	2383
STR103(ELCOM)	2383
STR106(ELCOM)	0154
STR117	3148
STR117(ELCOM)	0154
STR125	3148
STR201(ELCOM)	2394
STR202(ELCOM)	2394
STR203(ELCOM)	2394
STR206(ELCOM)	0154
STR217	3150
STR217(ELCOM)	0154
STR225	3150
STR225(ELCOM)	0154
STR230(ELCOM)	0154
STR-300	OEM
STR370	OEM
STR371	OEM
STR-380	3502
STR380	3502
STR-381	3503
STR381	3503
STR381A	3503
STR382	3518
STR383	2103
STR-384A	OEM
STR384A	OEM
STR385	3518
STR403(ELCOM)	1413
STR406(ELCOM)	0147
STR411(ELCOM)	0960
STR417	3152
STR417(ELCOM)	0147
STR418(ELCOM)	0576
STR425	3152
STR425(ELCOM)	3152
STR430(ELCOM)	0147
STR729	5784
STR2005	OEM
STR2012	1167
STR2012A	1167
STR2013	4882
STR2015	OEM
STR2024	OEM
STR-2124	OEM
STR2124	OEM
STR3030	OEM
STR3035	3445
STR3105	OEM
STR3110	OEM
STR3115	4146
STR3118	OEM
STR3123	4650
STR-3125	4650
STR3125	4650
STR3127	OEM
STR3130	1641
STR3135	3445
STR-3220	OEM
STR3220	OEM
STR3225	3310
STR-3230	OEM
STR3230	1641
STR3335	OEM
STR4142	OEM
STR5100	4869
STR-5241	OEM
STR5314	OEM
STR6431L959	OEM
STR9005	OEM
STR9012	OEM
STR9015	OEM
STR10130	OEM
STR11006	OEM
STR12006	OEM
STR13006	OEM
STR20005	OEM
STR20012	OEM
STR30110	OEM
STR30112	OEM
STR30113	OEM
STR30115	3951
STR-30120	2103
STR30120	3953
STR-30123	3955
STR30123	3955
STR30125	3957
STR30130	2364
STR30130-A	2364
STR30130A	2364
STR30134	3959
STR30135	3913
STR30135F	3913
STR50041	3344
STR50041A	3344
STR50092	OEM
STR-50092K	OEM
STR50092K	OEM
STR51041	2289
STR52100	OEM
STR52100A	OEM
STR53041	3574
STR53043	3305
STR53043A	3305
STR54041	5257
STR57051	OEM
STR60001	OEM
STR60001F51	OEM
STR90050	OEM
STR90090	OEM
STR90120	OEM
STR90150	OEM
STR-D1005	OEM
STR-D1005T	OEM
STRD1005T	OEM
STR-D1206	OEM
STR-D3030	OEM
STRD3030	OEM
STR-D3035	OEM
STRD3035	OEM
STRD6602	OEM
STR-M6511	OEM
STRM6511	OEM
STRM6513	OEM
STR-M6546	OEM
STR-S5041	OEM
STRS5041	OEM
STR-S5041G	OEM
STR-S5041T	OEM
STRS5041T	OEM
STRS5041T951	OEM
STR-S5141G	OEM
STRS5141G	OEM
STR-S5241	OEM
STRS5241	OEM
STR-S5703	OEM
STRS5703	OEM
STR-S6301	5601
STRS6301	5601
STRS6301-LF953	6317
STR-S6301A	OEM
STRS-6301A	OEM
STRS6301A	OEM
STR-S6302	OEM
STR-S6309	OEM
STRS6309S	OEM
STRS6421	OEM
STRS6431L959	OEM
STR-Z2151	OEM
STR-Z2751	OEM
STS104	0543
STS105	0543
STS106	0543
STS107	0543
STS401	0363
STS402	0363
STS403	0363
STS409	0363
STS410	0270
STS-411	0074
STS411	0270
STS413	0363
STS423	0223
STS424	0223
STS425	0223
STS430	1966
STS431	1966
STS1121	0074
STS1131	0074
STS1132	0074
STT2300	0130
STT2405	0836
STT2406	0836
STT3500	0130
STT4451	0560
STT4483	0836
STT9001	0836
STT9002	0836
STT9004	0836
STT9005	0836
STV-2H	OEM
STV-3	0015
STV-3H	0120
STV3H	OEM
STV-4H	OEM
STX0010	0178
STX0011	0126
STX0013	0161
STX0014	0103
STX0015	0142
STX0016	0178
STX0020	0460
STX0026	0161
STX0027	0103
STX0028	0233
STX0029	0925
STX0030	0925
STX0032	0103
STX0033	0136
STX0034	0136
STX0036	0050
STX0085	0050
STX0087	0050
STX0089	0050
STX0090	0050
STX0096	0004
STX0099	0004
STX0104	0004
STX0105	0004
STX0110	0004
STX0114	0004
STX0121	0164
STX0123	0004
STX0224	0164
STX0260	0004
STX0263	0004
STX0264	0004
STX0265	0004
STX0268	0004
STX0269	0004
STX5/3010	OEM
STX5/3025	OEM
STX5/5010	OEM
STX5/5025	OEM
STX5/6010	OEM
STX5/6025	OEM
STX5/7010	OEM
STX5/7025	OEM
SU012	OEM
SU26HZ	OEM
SU44	1175
SU101A	OEM
SU101K	OEM
SU110	2136
SU112	1893
SU201A	OEM
SU201K	OEM
SU212	OEM
SU300	OEM
SU305	OEM
SU306	OEM
SU312	OEM
SU314	OEM
SU315	OEM
SU316	OEM
SU320	OEM
SU331	OEM
SU332	OEM
SU401A	OEM
SU401K	OEM
SU412	OEM
SU512	OEM
SU601A	OEM
SU601K	OEM
SU801A	OEM
SU801K	OEM
SU1001A	OEM
SU1001K	OEM
SU2000	OEM
SU2002	OEM
SU2004	OEM
SU2006	OEM
SU2008	OEM
SU2010	OEM
SU2020	OEM
SU2021	OEM
SU2022	OEM
SU2023	OEM
SU2024	OEM
SU2025	OEM
SU2026	OEM
SU2027	OEM
SU2028	OEM
SU2029	OEM
SU2032	OEM
SU2034	OEM
SU2037	OEM
SU2074	OEM
SU2075	OEM
SU2076	0325
SU2077	0325
SU2080	0325
SU2081	0325
SUC650	0636
SUES501	OEM
SUES502	OEM
SUES503	OEM
SUES504	OEM
SUES505	OEM
SUES601	OEM
SUES602	OEM
SUES603	OEM
SUES604	OEM
SUES605	OEM
SUES606	OEM
SUES607	OEM
SUES608	OEM
SUES609	OEM
SUES610	OEM
SUES611	OEM
SUES701	OEM
SUES702	OEM
SUES703	OEM
SUES704	OEM
SUES705	OEM
SUES706	OEM
SUES707	OEM
SUES708	OEM
SUES709	OEM
SUES710	OEM
SUES711	OEM
SUES801	OEM
SUES802	OEM
SUES803	OEM
SUES804	OEM
SUES805	OEM
SUES806	OEM
SUES807	OEM
SUES808	OEM
SUES809	OEM
SUES810	OEM
SUES811	OEM
SUES1101	0546
SUES1102	0546
SUES1103	0546
SUES1104	1090
SUES1105	1362
SUES1106	1362
SUES1107	OEM
SUES1108	OEM
SUES1109	OEM
SUES1110	OEM
SUES1111	OEM
SUES1301	OEM
SUES1302	OEM
SUES1303	OEM
SUES1304	0546
SUES1305	1362
SUES1306	1362
SUES1307	OEM
SUES1308	OEM
SUES1309	OEM
SUES1310	OEM
SUES1311	OEM
SUES2601	OEM
SUES2602	OEM
SUES2603	OEM
SUES2604	OEM
SUES2605	OEM
SUES2606	OEM
SUES2607	OEM
SUES2608	OEM
SUES2609	OEM
SUES2610	OEM
SUES2611	OEM
SUM6010	0636
SUM6011	0636
SUM6020	0636
SUM6021	0636
SUR3005CT	2234
SUR3005CTE	OEM
SUR3005CTR	OEM
SUR3005SD	OEM
SUR3010CT	2234
SUR3010CTE	OEM
SUR3010CTR	OEM
SUR3010SD	OEM
SUR3015CT	2234
SUR3015CTE	OEM
SUR3015CTR	OEM
SUR3015SD	OEM
SUR3020CT	2234
SUR3020CTE	OEM
SUR3020CTR	OEM
SUR3020SD	OEM
SUR3030CT	OEM
SUR3030CTE	OEM
SUR3030CTR	OEM
SUR3040CT	4488
SUR3040CTE	OEM
SUR3040CTR	OEM
SUR3040SD	OEM
SUR3050CT	OEM
SUR3050CTE	OEM
SUR3050CTR	OEM
SUR3060CT	4488
SUR3060CTE	OEM
SUR3060CTR	OEM
SUR3060SD	OEM
SUR3070CT	OEM
SUR3070CTE	OEM
SUR3070CTR	OEM
SUR3080CT	OEM
SUR3080SD	OEM
SUR3100CT	OEM
SUR3100SD	OEM
SUT-101A	0319
SUT-101K	0267
SUT-201A	0319
SUT-201K	0267
SUT-401A	0319
SUT-401K	0267
SUT-601A	0319
SUT-601K	0267
SUT-801A	0044
SUT-801K	0281
SUT-1001A	0044
SUT-1001K	0281
SV-01A	0015
SV01A	0015
SV-01B	0015
SV-02	0120
SV02	0015
SV02A	0015
SV-02Y	0120
SV02YS	OEM
SV-03	0120
SV03	0120
SV-03S	0120
SV-03Y	3442
SV03YS	3684
SV-04	0120
SV-04S	3442
SV-04Y	3442
SV04YS	3442
SV-05	0015
SV05	0015
SV-05Y	OEM
SV05YS	OEM
SV-06	OEM
SV-06Y	OEM
SV06YS	OEM
SV-07	OEM
SV-08	OEM
SV-1	0080
SV1	OEM
SV-1R	0015
SV-2	0609
SV2	OEM
SV2SS	OEM
SV-3	0795
SV3	OEM
SV-3A	0120
SV3SS	OEM
SV-4R	0015
SV4SS	OEM
SV-6R	0015
SV-7	0649
SV-8R	0071
SV-9	0012
SV14B	0124
SV14C	0124
SV24A	OEM
SV24B	OEM
SV30	0143
SV-31	0143
SV31	0164

If replacement code is OEM, contact original manufacturer for replacement.

DEVICE TYPE	REPL CODE
SV-31(DIODE)	0143
SV36C	OEM
SV55A	OEM
SV55B	OEM
SV55C	OEM
SV55C-6	OEM
SV55D	OEM
SV56A	OEM
SV56B	OEM
SV56C	OEM
SV56D	OEM
SV66B	OEM
SV66B-8	OEM
SV66C	OEM
SV66D	OEM
SV67C	OEM
SV67D	OEM
SV70-14	OEM
SV77D	OEM
SV77E	OEM
SV77F	OEM
SV77G	OEM
SV77H	OEM
SV87A	OEM
SV87A(1)	OEM
SV87B	OEM
SV87B(1)	OEM
SV87C	OEM
SV87D	OEM
SV88A	OEM
SV88B	OEM
SV88C	OEM
SV88D	OEM
SV89D	OEM
SV89E	OEM
SV110	OEM
SV111	OEM
SV112	OEM
SV113	OEM
SV114	OEM
SV121	0162
SV122	0162
SV123	0157
SV135	0137
SV138	0002
SV146E	OEM
SV220	OEM
SV230	OEM
SV240	OEM
SV513C	OEM
SV513D	OEM
SV513E	OEM
SV514B	OEM
SV514C	OEM
SV514D	OEM
SV514E	OEM
SV533C	OEM
SV533D	OEM
SV533E	OEM
SV534E	OEM
SV543C	OEM
SV543D	OEM
SV543E	OEM
SV544C	OEM
SV544D	OEM
SV544E	OEM
SV566D-2	OEM
SV566D-4	OEM
SV794AN	OEM
SV800	3472
SV826AN	OEM
SV910	0995
SV1000	OEM
SV1004	OEM
SV1005	OEM
SV1006	OEM
SV1007	OEM
SV1008	OEM
SV1009	OEM
SV1010	OEM
SV1011	OEM
SV1012	OEM
SV1013	OEM
SV1014	OEM
SV1015	OEM
SV1016	OEM
SV1017	0137
SV1018	OEM
SV1019	0100
SV1020	0002
SV1021	OEM
SV1022	OEM
SV1023	OEM
SV1024	OEM
SV1025	OEM
SV1033	OEM
SV1034	OEM
SV1035	OEM
SV2002	1023
SV2004	1023
SV2006	1036
SV2008	1046
SV2010	1046
SV2052	0070
SV2058	1636
SV3140	OEM
SV3141	OEM
SV3142	OEM
SV3143	OEM
SV3144	0760
SV3145	OEM
SV3170	OEM
SV3171	OEM
SV3173	OEM
SV3174	OEM
SV3175	OEM
SV3176	OEM
SV3206	OEM
SV3207	OEM
SV4010	OEM
SV4010A	OEM
SV4011	OEM
SV4011A	OEM
SV4012	0137
SV4012A	0137
SV4013	OEM
SV4013A	OEM
SV4014	0100
SV4014A	0100
SV4015	0002
SV4015A	0002
SV4016	OEM
SV4016A	OEM
SV4017	OEM
SV4017A	OEM
SV4018	0375
SV4018A	0375
SV4019	OEM
SV4019A	OEM
SV4020	OEM
SV4020A	OEM
SV4021	OEM
SV4021A	OEM
SV4022	OEM
SV4022A	OEM
SV4023	OEM
SV4023A	OEM
SV4024	OEM
SV4024A	OEM
SV4025	OEM
SV4025A	OEM
SV4026	OEM
SV4026A	OEM
SV4027	0441
SV4027A	0441
SV4028	OEM
SV4028A	OEM
SV4029	OEM
SV4029A	OEM
SV4030	OEM
SV4030A	OEM
SV4031	OEM
SV4031A	OEM
SV4032	OEM
SV4032A	OEM
SV4033	0039
SV4033A	0039
SV4034	OEM
SV4034A	OEM
SV4035	OEM
SV4035A	OEM
SV4036	OEM
SV4036A	OEM
SV4037	OEM
SV4037A	OEM
SV4038	OEM
SV4038A	OEM
SV4039	OEM
SV4039A	OEM
SV4040	OEM
SV4040A	OEM
SV4041	OEM
SV4041A	OEM
SV4042	OEM
SV4042A	OEM
SV4043	OEM
SV4043A	OEM
SV4044	OEM
SV4044A	OEM
SV4045	OEM
SV4045A	OEM
SV4046	OEM
SV4046A	OEM
SV4047	OEM
SV4047A	OEM
SV4048	OEM
SV4048A	OEM
SV4049	OEM
SV4049A	OEM
SV4050	OEM
SV4050A	OEM
SV4051	OEM
SV4051A	OEM
SV4052	OEM
SV4052A	OEM
SV4053	OEM
SV4053A	OEM
SV4054A	OEM
SV4055	1834
SV4055A	1834
SV4056	OEM
SV4056A	OEM
SV4057	OEM
SV4057A	OEM
SV4058	OEM
SV4058A	OEM
SV4059	OEM
SV4059A	OEM
SV4060	OEM
SV4060A	OEM
SV4061	OEM
SV4061A	OEM
SV4062	0783
SV4062A	0783
SV4063	OEM
SV4063A	OEM
SV4064	OEM
SV4064A	OEM
SV4065	OEM
SV4065A	OEM
SV4066	OEM
SV4066A	OEM
SV4067	OEM
SV4067A	OEM
SV4068	OEM
SV4068A	OEM
SV4069	OEM
SV4069A	OEM
SV4070	OEM
SV4070A	OEM
SV4071	OEM
SV4071A	OEM
SV4072	OEM
SV4072A	OEM
SV4073	OEM
SV4073A	OEM
SV4074	OEM
SV4074A	OEM
SV4075	OEM
SV4075A	OEM
SV4076	OEM
SV4076A	OEM
SV4077	OEM
SV4077A	OEM
SV4078	OEM
SV4078A	OEM
SV4079	OEM
SV4079A	OEM
SV4080	OEM
SV4080A	OEM
SV4081	OEM
SV4081A	OEM
SV4082	0331
SV4082A	0331
SV4083	OEM
SV4083A	OEM
SV4084	OEM
SV4084A	OEM
SV4085	OEM
SV4085A	OEM
SV4086	OEM
SV4086A	OEM
SV4087	OEM
SV4087A	OEM
SV4088	OEM
SV4088A	OEM
SV4089	OEM
SV4089A	OEM
SV4090	OEM
SV4090A	OEM
SV4091	OEM
SV4091A	OEM
SV4092	OEM
SV4092A	OEM
SV4093	OEM
SV4093A	OEM
SV4094	OEM
SV4094A	OEM
SV4095	OEM
SV4095A	OEM
SV4096	OEM
SV4096A	OEM
SV4097	OEM
SV4097A	OEM
SV4098	OEM
SV4098A	OEM
SV4099	OEM
SV4099A	OEM
SV4100	OEM
SV4100A	OEM
SV5054	OEM
SV7031AN	OEM
SV7031BN	OEM
SV7031CN	OEM
SV7032AN	OEM
SV7032CN	OEM
SV7033AN	OEM
SV7034CN	OEM
SV7302AN	OEM
SV7307AN	OEM
SV9785	OEM
SV9786	OEM
SV10052	OEM
SV12388	0015
SV12388E	0015
SVB10-100	0277
SVB10-200	0289
SVB15-100	0277
SVB15-200	0289
SV-BZX79C10V	0064
SVC0053	0030
SVC101	0030
SVC112	1029
SVC151	OEM
SVC-201	0030
SVC201	0030
SVC201SP	OEM
SVC201SP-BJ	OEM
SVC201Y	0628
SVC201Y-AG	OEM
SVC201Y-BW	OEM
SVC202	OEM
SVC202AF	OEM
SVC203	OEM
SVC203CP	OEM
SVC211	1029
SVC211SP	OEM
SVC211SP-B	OEM
SVC211SP-B2	OEM
SVC211SP-C	OEM
SVC212	OEM
SVC-241	OEM
SVC251	0030
SVC251SP	OEM
SVC251Y	OEM
SVC303Y	3688
SVC311	3688
SVC321	3688
SVC321-C2	3688
SVC321C	OEM
SVC321C2	OEM
SVC321D2	OEM
SVC321SP	OEM
SVC321SP-C2	OEM
SVC321Y	3688
SVC333	OEM
SVC561D-14A	OEM
SVC625	0636
SVD02Z8	0244
SVD02Z8.2	0244
SVD02Z8.2A	0244
SVD02Z9.5A	0012
SVD0278.2A	0244
SVD0279.5A	0012
SVD0A70	0143
SVD0A79	0143
SVD0A90	0019
SVD1S1717	1083
SVD1S1850	0023
SVD-1S1850R	0015
SVD1S1850R	0023
SVD1S2076AT	OEM
SVD1SR35200A	OEM
SVD10D-1	0015
SVD12B2B1P-M	0277
SVD20A70	0143
SVD20A79	0143
SVDAS2HB10	0087
SVDGP1S53	OEM
SVDMA26	0139
SVDMA26-1	0139
SVDMA26-2	0139
SVD0A70	0143
SVD0A90	0143
SVDS1RB10	0277
SVDS-1S1850	0015
SVDS3V20	OEM
SVDS3V40	OEM
SVDS5688GT3	OEM
SVDSC20	3658
SVDV1121	OEM
SVDVD1121	0139
SVDVD1223	0120
SVI3102B	OEM
SVI3105A	OEM
SVI3105B	OEM
SVI3203	OEM
SVI3204	OEM
SVI3205	OEM
SVI3206	OEM
SVIBA4558F	OEM
SVIBA4560F	OEM
SVIBA6218	OEM
SVICXK5816M	OEM
SVIH8DN2041	OEM
SVIH8DN2175	OEM
SVIK4152-2M	OEM
SVILC6526CPA	OEM
SVI-LD3120	2509
SVILM833M	OEM
SVINJM4560M	OEM
SVIPCM55HP	OEM
SVI-STK018	3936
SVISTK020F	1187
SVIUPC1161C3	OEM
SV-KB262	0120
SVM61	0636
SVM601	0636
SVM602	0636
SVM605	0636
SVM6010	0471
SVM7962CBR	OEM
SVR12A	OEM
SVR12B	OEM
SVR12C	OEM
SVR19A	OEM
SVR19B	OEM
SVR34B	OEM
SVT400-3	0074
SVT450-5	OEM
SVT6000	2626
SVT6001	2626
SVT6002	2626
SVT6060	2626
SVT6061	2626
SVT6062	2626
SVT6251	1991
SVT6252	2626
SVT6253	2626
SVT6546	OEM
SVT6547	OEM
SVT7574	1852
SVVD1220XXXXX	0120
SVVD1221LXXXX	0120
SW01	0137
SW-05	0110
SW05	0015
SW-05-005	0015
SW05-005	0015
SW-05-01	0015
SW05-01	0015
SW-05-02	0015
SW05-02	0015
SW-05-A	0015
SW-05-B	0015
SW-05-C	0071
SW-05-D	0071
SW05A	0015
SW05B	0015
SW05C	0071
SW05D	0071
SW05S	0015
SW05SS	1355
SW05V	0015
SW0501	0015
SW1	0087
SW-1-01	0087
SW-1-02	0087
SW-1-03	0015
SW-1-04	0015
SW-1-06	0015
SW-1-08	0071
SW-1-10	0071
SW-1A	0015
SW1A	OEM
SW1B	OEM
SW1C	0071
SW1D	0071
SW1S	OEM
SW1SS	OEM
SW-5-005	0015
SW10	OEM
SW11	OEM
SW30	OEM
SW74LS247	1732
SW705-2M	0358
SW705-2P	0358
SW706-2M	0333
SW706-2P	0333
SW708-2M	1632
SW708-2P	1632
SW709-2M	1481
SW709-2P	1481
SW930-2M	1823
SW930-2P	1823
SW932-2M	1041
SW932-2P	1041
SW933-2M	2097
SW933-2P	2097
SW935-2M	1176
SW935-2P	1176
SW936-2M	1831
SW936-2P	1831
SW937-2M	1835
SW937-2P	1835
SW941-2M	2622
SW941-2P	2622
SW944-2M	0033
SW944-2P	0033
SW945-2M	0081
SW945-2P	0081
SW946-2M	0141
SW946-2P	0141
SW948-2M	3399
SW948-2P	3399
SW949-2M	1844
SW949-2P	1844
SW950-2M	2619
SW950-2P	2619
SW951-2M	2078
SW951-2P	2078
SW957-2M	1433
SW957-2P	1433
SW958-2M	0466
SW958-2P	0466
SW961-2M	1859
SW961-2P	1859
SW962-2P	0562
SW963-2M	0341
SW963-2P	0341
SW4001	0478
SW4002	2055
SW4011	0215
SW4012	0498
SW4013	0413
SW4015	1014
SW4016	1143
SW4017	0513
SW4019	1525
SW4020	1661
SW4021	1749
SW4023	0519
SW4024	1957
SW4025	2072
SW4027	1949
SW4030	0500
SW4049	0001
SW4050	0398
SW-10548	1835
SW-10549	0466
SWC	0071
SWD	0071
SWH-08	OEM
SWH08	OEM
SWH2-16	OEM
SWH2-20	OEM
SWH2-25ED	OEM
SWH-10	OEM
SWHD-08	OEM
SWHD2-16	OEM
SWHD2-20	OEM
SWHD2-25	OEM
SWHD3-16	OEM
SWHD3-20	OEM
SWHD-10	OEM
SWHD-16	OEM
SWHD-20	OEM
SW0.5A	0015
SWT622	2519
SWT1032	0037
SWT1042	0079
SWT1728	0136
SWT2188	0050
SWT3588	0050
SX-1.5-01	0953
SX-1.5-02	0242
SX-1.5-04	0540
SX-1.5-06	0965
SX-1.5-08	0817
SX-1.5-10	0922
SX6.8	0205
SX9.1	0684
SX10	0225
SX11	0230
SX12	0234
SX13	0237
SX15	0247
SX16	0251
SX18	0256
SX20	0262
SX22	0269
SX24	0274
SX27	0294
SX30	0309
SX33	0318
SX36	0320
SX39	0326
SX40	OEM
SX43	0337
SX47	0347
SX50	OEM
SX51	0027
SX55	0079
SX56	0266
SX-58Y	0325
SX60	OEM
SX60M	0233
SX61	0126
SX61M	0037
SX62	0386
SX64-KERN	OEM
SX68	0405
SX75	0426
SX80	OEM
SX82	0444
SX91	0238
SX100	1180
SX110	1190
SX120	1207
SX623	0015
SX631	0015
SX633	0015
SX641	0609
SX642	1355
SX643	0609
SX644	3911
SX645	3911
SX751	0090
SX752	0097
SX753	0109
SX754	0109
SX780	0124
SX781	OEM
SX782	OEM
SX3001	0144
SX3702	0037
SX3709	0079
SX3711	0079
SX3819	0325
SX3825	0144
SX3826	0224
SX3827	0144
SY20	1717
SY22	1723
SY-23	0015
SY23	0795
SY-24	0015
SY24	0795
SY27	1761
SY30	1772
SY101	0600
SY432D	OEM
SY2114AL-3	OEM
SY2128	1898
SY4320	OEM
SY6502	3065
SY6530	OEM
SY6991	OEM
SY6992	OEM
SY23128	OEM
SY68045	OEM
SYC68B20	OEM
SYC68B21	OEM
SYC1791-02	OEM
SYC1793-02	OEM
SYC2101-1	OEM
SYC2101A	OEM
SYC2101A-2	OEM
SYC2101A-4	OEM
SYC2114	2048
SYC2114-3	2048
SYC2147	1693
SYC2147-3	1693
SYC2147-6	1693
SYC2147H	OEM
SYC2147H-1	OEM
SYC2147H-2	OEM
SYC2147H-3	OEM
SYC2147HL	OEM
SYC2147HL-3	OEM
SYC2147L	1693
SYC2148H	OEM
SYC2148H-2	OEM
SYC2148H-3	OEM
SYC2148HL	OEM
SYC2148HL-3	OEM
SYC2149H	OEM
SYC2149H-2	OEM
SYC2149H-3	OEM
SYC2149HL	OEM
SYC2149HL-3	OEM
SYC2332-3	OEM
SYC2333-3	OEM
SYC2661-1	OEM
SYC2661-2	OEM
SYC2661-3	OEM
SYC3308	OEM
SYC3316	OEM
SYC3316A	OEM
SYC6502	3065
SYC6502A	OEM
SYC6502B	OEM
SYC6502C	OEM
SYC6503	OEM
SYC6503A	OEM
SYC6503B	OEM
SYC6503C	OEM
SYC6504	OEM
SYC6504A	OEM
SYC6504B	OEM
SYC6504C	OEM
SYC6505	OEM
SYC6505A	OEM
SYC6505B	OEM
SYC6505C	OEM
SYC6506	OEM
SYC6506A	OEM
SYC6506B	OEM
SYC6506C	OEM
SYC6507	3070
SYC6507A	OEM
SYC6507B	OEM
SYC6507C	OEM
SYC6512	OEM
SYC6512A	OEM
SYC6512B	OEM
SYC6512C	OEM
SYC6513	OEM
SYC6513A	OEM
SYC6513B	OEM
SYC6513C	OEM
SYC6514	OEM
SYC6514A	OEM
SYC6514B	OEM
SYC6514C	OEM
SYC6515	OEM
SYC6515A	OEM
SYC6515B	OEM
SYC6515C	OEM
SYC6520	OEM
SYC6520A	OEM
SYC6521	OEM
SYC6521A	OEM
SYC6522	OEM
SYC6522A	OEM
SYC6532	1973
SYC6532A	OEM
SYC6545	OEM
SYC6545-1	OEM
SYC6545A	OEM
SYC6545A-1	OEM
SYC6551	OEM
SYC6551A	OEM
SYC6591	OEM
SYC6591A	OEM
SYC6820	OEM
SYC6821	OEM
SYD68B20	OEM
SYD68B21	OEM
SYD1791-02	OEM
SYD1793-02	OEM
SYD2111-1	OEM
SYD2111A	OEM

DEVICE TYPE	REPL CODE
SYD2112-1	OEM
SYD2112A	OEM
SYD2112A-2	OEM
SYD2112A-4	OEM
SYD2114	2048
SYD2114-1	OEM
SYD2114-2	2048
SYD2114-3	2048
SYD2114L	2048
SYD2114L-1	OEM
SYD2114L-2	2048
SYD2114L-3	2048
SYD2114LV	OEM
SYD2114LV-2	OEM
SYD2114LV-3	OEM
SYD2142	OEM
SYD2142-2	OEM
SYD2142-3	OEM
SYD2142L	OEM
SYD2142L-2	OEM
SYD2142L-3	OEM
SYD2147	1693
SYD2147-3	1693
SYD2147-6	1693
SYD2147H	OEM
SYD2147H-1	OEM
SYD2147H-2	OEM
SYD2147H-3	OEM
SYD2147HL	OEM
SYD2147HL-3	OEM
SYD2147L	1693
SYD2148H	OEM
SYD2148H-2	OEM
SYD2148H-3	OEM
SYD2148HL	OEM
SYD2148HL-3	OEM
SYD2149H	OEM
SYD2149H-2	OEM
SYD2149H-3	OEM
SYD2149HL	OEM
SYD2149HL-3	OEM
SYD2164	OEM
SYD2164-2	OEM
SYD2164-3	OEM
SYD2164A	OEM
SYD2164A-2	OEM
SYD2164A-3	OEM
SYD2316A	OEM
SYD2316B	OEM
SYD2316B-2	OEM
SYD2316B-3	OEM
SYD2332	OEM
SYD2333	OEM
SYD2365	OEM
SYD2365-2	OEM
SYD2365-3	OEM
SYD2365A	OEM
SYD2365A-2	OEM
SYD2365A-3	OEM
SYD2661-1	OEM
SYD2661-2	OEM
SYD2661-3	OEM
SYD3308	OEM
SYD3316	OEM
SYD3316A	OEM
SYD6502	3065
SYD6502A	OEM
SYD6502B	OEM
SYD6502C	OEM
SYD6503	OEM
SYD6503A	OEM
SYD6503B	OEM
SYD6503C	OEM
SYD6504	OEM
SYD6504A	OEM
SYD6504B	OEM
SYD6504C	OEM
SYD6505	OEM
SYD6505A	OEM
SYD6505B	OEM
SYD6505C	OEM
SYD6506	OEM
SYD6506A	OEM
SYD6506B	OEM
SYD6506C	OEM
SYD6507	3070
SYD6507A	OEM
SYD6507B	OEM
SYD6507C	OEM
SYD6512	OEM
SYD6512A	OEM
SYD6512B	OEM
SYD6512C	OEM
SYD6513	OEM
SYD6513A	OEM
SYD6513B	OEM
SYD6513C	OEM
SYD6514	OEM
SYD6514A	OEM
SYD6514B	OEM
SYD6514C	OEM
SYD6515	OEM
SYD6515A	OEM
SYD6515B	OEM
SYD6515C	OEM
SYD6520	OEM
SYD6520A	OEM
SYD6521	OEM
SYD6521A	OEM
SYD6532	1973
SYD6532A	OEM
SYD6545	OEM
SYD6545-1	OEM
SYD6545A	OEM
SYD6545A-1	OEM
SYD6551	OEM
SYD6551A	OEM
SYD6591	OEM
SYD6591A	OEM
SYD6820	OEM
SYD6821	OEM
SYEC68B20	OEM
SYEC68B21	OEM
SYEC6502	OEM
SYEC6502A	OEM
SYEC6503	OEM
SYEC6503A	OEM
SYEC6504	OEM
SYEC6504A	OEM
SYEC6505	OEM
SYEC6505A	OEM
SYEC6506	OEM
SYEC6506A	OEM
SYEC6507	OEM
SYEC6507A	OEM
SYEC6512	OEM
SYEC6512A	OEM
SYEC6513	OEM
SYEC6513A	OEM
SYEC6514	OEM
SYEC6514A	OEM
SYEC6515	OEM
SYEC6515A	OEM
SYEC6520	OEM
SYEC6520A	OEM
SYEC6521	OEM
SYEC6521A	OEM
SYEC6522	OEM
SYEC6532	OEM
SYEC6532A	OEM
SYEC6551	OEM
SYEC6551A	OEM
SYEC6820	OEM
SYEC6821	OEM
SYED68B20	OEM
SYED68B21	OEM
SYED6502	OEM
SYED6502A	OEM
SYED6503	OEM
SYED6503A	OEM
SYED6504	OEM
SYED6504A	OEM
SYED6505	OEM
SYED6505A	OEM
SYED6506	OEM
SYED6506A	OEM
SYED6507	OEM
SYED6507A	OEM
SYED6512	OEM
SYED6512A	OEM
SYED6513	OEM
SYED6513A	OEM
SYED6514	OEM
SYED6514A	OEM
SYED6515	OEM
SYED6515A	OEM
SYED6520	OEM
SYED6520A	OEM
SYED6521	OEM
SYED6521A	OEM
SYED6522	OEM
SYED6532	OEM
SYED6532A	OEM
SYED6551	OEM
SYED6551A	OEM
SYED6820	OEM
SYED6821	OEM
SYEP68B20	OEM
SYEP68B21	OEM
SYEP6502	OEM
SYEP6502A	OEM
SYEP6503	OEM
SYEP6503A	OEM
SYEP6504	OEM
SYEP6504A	OEM
SYEP6505	OEM
SYEP6505A	OEM
SYEP6506	OEM
SYEP6506A	OEM
SYEP6507	OEM
SYEP6507A	OEM
SYEP6512	OEM
SYEP6512A	OEM
SYEP6513	OEM
SYEP6513A	OEM
SYEP6514	OEM
SYEP6514A	OEM
SYEP6515	OEM
SYEP6515A	OEM
SYEP6520	OEM
SYEP6520A	OEM
SYEP6521	OEM
SYEP6521A	OEM
SYEP6522	OEM
SYEP6532	OEM
SYEP6532A	OEM
SYEP6551	OEM
SYEP6551A	OEM
SYEP6820	OEM
SYEP6821	OEM
SYK4332	5501
SYL101	0600
SYL102	0600
SYL103	0038
SYL104	0038
SYL105	0280
SYL106	0280
SYL107	0211
SYL108	0211
SYL109	0160
SYL128	0123
SYL152	0198
SYL160	0280
SYL792	0600
SYL1182	0016
SYL1279	0600
SYL1297	0038
SYL1310	0600
SYL1311	0600
SYL1312	0600
SYL1313	0600
SYL1326	0600
SYL1327	0600
SYL1329	0038
SYL1380	0600
SYL1396	0038
SYL1408	0600
SYL1430	0211
SYL1454	0600
SYL1468	0208
SYL1524	0038
SYL1536	0038
SYL1537	0600
SYL1538	0038
SYL1539	0038
SYL1547	0038
SYL1583	0211
SYL1588	0280
SYL1591	0600
SYL1592	OEM
SYL1608	0280
SYL1617	0600
SYL1655	0211
SYL1665	0211
SYL1668	0211
SYL1684	OEM
SYL1690	0280
SYL1697	0280
SYL1717	0280
SYL1750	0600
SYL1986	OEM
SYL1987	0600
SYL2120	0280
SYL2130	0600
SYL2131	0600
SYL2132	0600
SYL2134	0038
SYL2135	0038
SYL2136	0038
SYL2189	0050
SYL2245	0600
SYL2246	0600
SYL2247	0280
SYL2248	0211
SYL2249	0211
SYL2250	0280
SYL2300	0211
SYL2301	OEM
SYL2494	OEM
SYL2650	0038
SYL3013	OEM
SYL3613	0211
SYL4131	0144
SYL4224	OEM
SYL4225	OEM
SYL4275	0037
SYL4280	0086
SYL4312	OEM
SYL4315	0038
SYL-4339	0038
SYL4339	0600
SYMC2147	OEM
SYMC2147-6	OEM
SYMC2148	OEM
SYMC2148-6	OEM
SYMC2149H	OEM
SYMC2149H-3	OEM
SYMC3316	OEM
SYMC3316A	OEM
SYMD2147	OEM
SYMD2147-6	OEM
SYMD2148	OEM
SYMD2148-6	OEM
SYMD2149H	OEM
SYMD2149H-3	OEM
SYMD3316	OEM
SYMD3316A	OEM
SYMF2147	OEM
SYMF2147-6	OEM
SYMF2148	OEM
SYMF2148-6	OEM
SYN600	OEM
SYP68B20	OEM
SYP68B21	OEM
SYP1791-02	OEM
SYP1793-02	OEM
SYP2101-1	OEM
SYP2101A	OEM
SYP2101A-2	OEM
SYP2101A-4	OEM
SYP2111-1	OEM
SYP2111A	OEM
SYP2111A-2	OEM
SYP2111A-4	OEM
SYP2112-1	OEM
SYP2112A	OEM
SYP2112A-2	OEM
SYP2112A-4	OEM
SYP2114	2048
SYP2114-1	OEM
SYP2114-2	2048
SYP2114-3	2048
SYP2114L	2048
SYP2114L-1	OEM
SYP2114L-2	2048
SYP2114L-3	2048
SYP2114LV	OEM
SYP2114LV-2	OEM
SYP2114LV-3	OEM
SYP2142	OEM
SYP2142-2	OEM
SYP2142-3	OEM
SYP2142L	OEM
SYP2142L-2	OEM
SYP2142L-3	OEM
SYP2142LV	OEM
SYP2142LV-2	OEM
SYP2142LV-3	OEM
SYP2164	OEM
SYP2164-2	OEM
SYP2164-3	OEM
SYP2164A	OEM
SYP2164A-2	OEM
SYP2164A-3	OEM
SYP2316A	OEM
SYP2316B	OEM
SYP2316B-2	OEM
SYP2316B-3	OEM
SYP2332	OEM
SYP2332-3	OEM
SYP2333	OEM
SYP2333-3	OEM
SYP2365	OEM
SYP2365-2	OEM
SYP2365-3	OEM
SYP2365A	OEM
SYP2365A-2	OEM
SYP2365A-3	OEM
SYP2661-1	OEM
SYP2661-2	OEM
SYP2661-3	OEM
SYP6502	3065
SYP6502A	OEM
SYP6502B	OEM
SYP6502C	OEM
SYP6503	OEM
SYP6503A	OEM
SYP6503B	OEM
SYP6503C	OEM
SYP6504	OEM
SYP6504A	OEM
SYP6504B	OEM
SYP6504C	OEM
SYP6505	OEM
SYP6505A	OEM
SYP6505B	OEM
SYP6505C	OEM
SYP6506	OEM
SYP6506A	OEM
SYP6506B	OEM
SYP6506C	OEM
SYP6507	3070
SYP6507A	OEM
SYP6507B	OEM
SYP6507C	OEM
SYP6512	OEM
SYP6512A	OEM
SYP6512B	OEM
SYP6512C	OEM
SYP6513	OEM
SYP6513A	OEM
SYP6513B	OEM
SYP6513C	OEM
SYP6514	OEM
SYP6514A	OEM
SYP6514B	OEM
SYP6514C	OEM
SYP6515	OEM
SYP6515A	OEM
SYP6515B	OEM
SYP6515C	OEM
SYP6520	OEM
SYP6520A	OEM
SYP6521	OEM
SYP6521A	OEM
SYP6522	OEM
SYP6522A	OEM
SYP6532	1973
SYP6532A	OEM
SYP6545	OEM
SYP6545-1	OEM
SYP6545A	OEM
SYP6545A-1	OEM
SYP6551	OEM
SYP6551A	OEM
SYP6591	OEM
SYP6591A	OEM
SYP6820	OEM
SYP6821	OEM
SZ090	0016
SZ1.0	OEM
SZ2.4	3003
SZ2.7	1312
SZ3.0	1713
SZ3.3	0292
SZ3.6	0188
SZ3.9	0456
SZ4.3	0533
SZ4.7	0162
SZ5	0162
SZ5.1	0162
SZ5.6	0157
SZ6	0301
SZ6.2	0636
SZ6.2A	0636
SZ6.8	0025
SZ-7	0062
SZ7	0062
SZ7.5	0649
SZ8	0244
SZ8.2	0244
SZ-9	0057
SZ9	0012
SZ9.1	0012
SZ10	0170
SZ10C	0225
SZ-11	0181
SZ11	0317
SZ11C	0230
SZ11.0	0794
SZ12	0052
SZ12C	0234
SZ12.0	0137
SZ-13	0053
SZ13	0365
SZ13C	0237
SZ13.0	0365
SZ14	0100
SZ15	0686
SZ15B	6107
SZ15C	0247
SZ15.0	0002
SZ-16	0445
SZ16	0244
SZ16C	0251
SZ16.0	0420
SZ17	0210
SZ18	0375
SZ18B	0495
SZ18C	0256
SZ19	0671
SZ20	0700
SZ22B	0565
SZ22C	0269
SZ24C	0274
SZ-25	0714
SZ27B	0441
SZ27C	0294
SZ30C	0309
SZ31	OEM
SZ33B	0039
SZ33C	0318
SZ39B	OEM
SZ47B	OEM
SZ56A	0876
SZ56B	OEM
SZ62A	0185
SZ68A	0205
SZ68B	OEM
SZ75A	0480
SZ82A	0504
SZ91A	0684
SZ119	0402
SZ200	0162
SZ200-4	0036
SZ200-5	0162
SZ200-6	0091
SZ200-7	0062
SZ-200-8	0012
SZ200-8	0012
SZ200-8.2	0244
SZ-200-9	0012
SZ200-9	0057
SZ-200-9V	0012
SZ-200-10	0170
SZ-200-11	0317
SZ200-11	0181
SZ-200-12	0137
SZ-200-13	0365
SZ200-13	0137
SZ-200-14	0100
SZ-200-15	0002
SZ-200-16	0244
SZ200-16	0445
SZ-200-17	0210
SZ200-17	1649
SZ-200-18	0375
SZ-200-19	0671
SZ-200-20	0700
SZ671-B	0137
SZ671-G	0137
SZ671-O	0317
SZ671-W	0317
SZ720C	OEM
SZ961-B	0002
SZ961-R	0100
SZ961-V	0365
SZ961-Y	0100
SZ1200	1834
SZ11763	0149
SZ14138	OEM
SZ51398	OEM
SZA-6	0301
SZA-13	0365
SZ-AW01-11	0317
SZ-BEX9V1	OEM
SZ-BZ-110	0317
SZ-BZ-120	0137
SZ-BZX9V1	OEM
SZC9	0012
SZ-EQB01-12A	0137
SZ-HZ4B	0275
SZ-HZ4B3	0036
SZ-HZ9B3	0322
SZ-HZ43	0275
SZL9	0170
SZL10	0510
SZ-LA4102	3190
SZM-206	OEM
SZM206	OEM
SZM206A	OEM
SZM-214	OEM
SZM-214D	OEM
SZM-226EC	OEM
SZM226EC	OEM
SZP-9	0012
SZP-105	OEM
SZ-RD3.9EB2	0036
SZ-RD6.2E	0471
SZ-RD6.2EB	0471
SZ-RD6.2EB3	0471
SZ-RD9.1FB	0012
SZ-RD11FB	0317
SZ-RD12EB	1005
SZ-RD12FB	0137
SZT8	0137
SZT9	0012
SZ-US-4.0C	0275
SZ-UZ-4.0C	0275
SZ-UZ-6.2B	0471
SZ-UZ6.2B	0471
SZ-UZ-12B	0052
SZ-UZP-11B	0317
SZ-UZP-12B	0137
SZ-WZ-040	0036
SZ-WZ-063	0471
SZ-WZ-120	0052
SZ-YZ-063	0471
SZ-YZ063	0471
SZ-ZD-4A2	0275
SZ-ZD-4A3	0275
T0003	0004
T0004	0004
T0005	0004
T00014	0004
T0012	0004
T0014	0004
T0015	0004
T0031	0004
T0033	0004
T0038	0004
T0039	0004
T0040	0004
T0041	0004
T0051	0004
T0-101	0280
T0-102	0280
T01-044	0325
T01-101	0016
T01-104	0016
T01-105	0016
T08N5A00	OEM
T033Z3	0280
T053A2	OEM
T054B7T	OEM
T056B7T	OEM
T058B7T	OEM
T-065	0015
T065	0015
T072F	OEM
T-075	0015
T075	0015
T082	OEM
T0101	0280
T0102	0280
T-0150	0015
T0150	0015
T0505BH	0572
T0505DH	0572
T0505MH	0935
T0505NH	0059
T0509BH	0572
T0509DH	0572
T0509MH	0935
T0509NH	0059
T0510B7T	OEM
T0510BH	0572
T0510DH	0572
T0510MH	0935
T0510NH	0059
T0512BH	0572
T0512DH	0572
T0512MH	0935
T0512NH	0059
T0515B7T	OEM
T0530G8	OEM
T0530H8	OEM
T0530J8	OEM
T0540G8	OEM
T0540H8	OEM
T0540J8	OEM
T0605BH	0572
T0605DH	0572
T0605MH	0935
T0605NH	0059
T0609BH	0572
T0609DH	0572
T0609MH	0935
T0609NH	0059
T0610BH	0572
T0610BJ	0572
T0610DH	0572
T0610DJ	0572
T0610MH	0935
T0610MJ	0935
T0610NH	0059
T0610NJ	0059
T0612BJ	0572
T0612DJ	0572
T0612MJ	0935
T0612NJ	0059
T0805BH	0572
T0805DH	0572
T0805MH	0935
T0805NH	0059
T0809BH	0572
T0809DH	0572
T0809MH	0935
T0809NH	0059
T0810BH	0572
T0810BJ	0572
T0810DH	0572
T0810DJ	0572
T0810MH	0935
T0810MJ	0935
T0810NH	0059
T0810NJ	0059
T0812BH	0572
T0812BJ	0572
T0812DH	0572
T0812DJ	0572
T0812MH	0935
T0812MJ	0935
T0812NH	0059
T0812NJ	0059
T0813BJ	0572
T0813DJ	0572
T0813MJ	0935
T0813NJ	0059
T0B	0632
T0.8/50	OEM
T0.8/100	OEM
T0.8/200	OEM
T0.8/300	OEM
T0.8/400	OEM
T0.8N0.6A00	0179
T0.8N1A00	0179
T0.8N2A00	0179
T0.8N3A00	0346
T0.8N4A00	0346
T0.8N5A00	3349
T0.8N50	OEM
T0.8N100	OEM
T0.8N200	OEM
T0.8N300	OEM
T0.8N400	OEM
T0.8N500	OEM
T1/1	OEM
T1/2	OEM
T1/3	OEM
T1/4	OEM
T1/5	OEM
T1-1A6	0016
T1-55	OEM
T1A	2624
T1B	0632
T1E	0701
T1F	0789
T1G	0143
T1H	0829
T1J	0654
T1J6G	0015
T1M	3724
T1N	0654
T1N300	0346
T1N400	0346
T1P29	0042
T1P29X	0042
T1P31	0042
T1P31A	0236
T1P33	3081
T1P33A	3081
T1S14	0325
T1S-18	0224
T1S18	0224
T1S86	0127
T1S88	0325
T1S92	0079
T1S93	0037
T1S94	0144
T1S94(PENNCREST)	0030
T1S95	0079
T1S97	0079

If replacement code is OEM, contact original manufacturer for replacement.

DEVICE TYPE	REPL CODE
T1S98	0144
T1SDS1K	0015
T1T	2027
T1V-3	0628
T1XM05	0050
T1X-M14	0007
T1X-M15	0007
T1XM15	0144
T1X-M16	0007
T1XM17	0144
T1Z	0701
T2A	0701
T2B	OEM
T2C	0701
T2D	1345
T2E	3779
T2G	1345
T2J	0664
T2K	OEM
T2L	1980
T2M	2747
T2N	1843
T2N-1	1843
T2N1	1843
T2R	0360
T2SA495-O	0006
T2SA562-O	0006
T2SA940	1911
T2SA945-O	0164
T2SC388A	0842
T2SC1195S	0103
T2SC1829S	2620
T2SC1893S	2668
T2SC1894S	0313
T2SC2068	0643
T2SC2068-1	0643
T2SC2073	1284
T2SC2120Y	0866
T2SC2229-O	0066
T2SC2229Y	0066
T2SC2230AY	0261
T2SD657S	0074
T2T	2755
T2T-1	2755
T2T-2	2755
T2V	1759
T3/1	OEM
T3/2	0015
T3/3	OEM
T3/4	OEM
T3/5	OEM
T3A	1395
T3A-1	1395
T3A-2	1395
T3B	0312
T3D	OEM
T3D-1	OEM
T3E	OEM
T3G	0143
T3H-2	OEM
T3N0.6C00	OEM
T3N1C00	OEM
T3N2C00	OEM
T3N4C00	OEM
T3N5C00	OEM
T3N6C00	OEM
T3N50	1110
T3N100	2932
T3N200	0963
T3N300	2936
T3N400	2936
T3N500	3439
T3N600	3439
T4G	OEM
T4.5/100	2491
T4.5/200	0240
T4.5/300	0676
T4.5/400	0676
T5G	OEM
T6/1	OEM
T6/2	OEM
T6/3	OEM
T6/4	OEM
T6/5	OEM
T6K40	OEM
T6N100BOC	0469
T6N200BOC	0469
T6N300BOC	0725
T6N400BOC	0725
T6N500BOC	0725
T6N600BOC	0725
T6N700BOC	0679
T6N800BOC	0679
T6.5N100	OEM
T6.5N200	OEM
T6.5N300	OEM
T6.5N400	OEM
T6.5N500	OEM
T6.5N600	OEM
T6.5N700	OEM
T6.5N800	OEM
T6.5N900	OEM
T7	0143
T8/1	OEM
T8/2	0015
T8/3	OEM
T8/4	OEM
T8/5	OEM
T8F100CCC	OEM
T8F100HCC	OEM
T8F700CCC	OEM
T8F700HCC	OEM
T8G	0143
T8N100BOC	0469
T8N200BOC	0469
T8N300BOC	0725
T8N400BOC	0725
T8N500BOC	0725
T8N600BOC	0725
T8N700BOC	0679
T8N800BOC	0679
T9	0143
T9G	0143
T9G0060803DH	4148
T9G0061003DH	4148
T9G0061203DH	4148
T9G0080803DH	1725
T9G0081003DH	1725
T9G0100803DH	1725
T9G0101003DH	1725
T9G0101203DH	1725
T9G0120803DH	1725
T9G0121003DH	1725
T9G0121203DH	1725
T9H0200603	OEM
T9H0220603	OEM
T9H0240603	OEM
T9H0260603	OEM
T9H0280603	OEM
T9H0300603	OEM
T10N100BOC+	0469
T10N200BOC+	0469
T10N300BOC+	0722
T10N400BOC+	0725
T10N500BOC+	0725
T10N600BOC+	0725
T10N700BOC+	2036
T10N800BOC+	2036
T11	0019
T12	0143
T12A6F	OEM
T12G	0143
T12N100BOC+	0469
T12N100C0B	0712
T12N200BOC+	0469
T12N400BOC+	0725
T12N500BOC+	0725
T12N600BOC+	0725
T12N700BOC+	2036
T12N800BOC+	2036
T12N900BOC+	2036
T13	0143
T13A2	OEM
T13G	0143
T14	0143
T14/50	OEM
T14/100	OEM
T14/200	OEM
T14/400	OEM
T14B7T	OEM
T14F100C	OEM
T14F100E	OEM
T14F200C	OEM
T14F200E	OEM
T14F300C	OEM
T14F300E	OEM
T14F400C	OEM
T14F400E	OEM
T14F500C	OEM
T14F500E	OEM
T14F600C	OEM
T14F600E	OEM
T14F700C	OEM
T14F700E	OEM
T14F800C	OEM
T14F800E	OEM
T14F900C	OEM
T14F900E	OEM
T14G	0143
T14N100	OEM
T14N200	OEM
T14N400	OEM
T14N500	OEM
T14N600	OEM
T14N700	OEM
T15	OEM
T15G	OEM
T15N400C	OEM
T15N500C	OEM
T15N500E	OEM
T15N600C	OEM
T15N600E	OEM
T15N700C	OEM
T15N700E	OEM
T15N800C	OEM
T15N800E	OEM
T15N900C	OEM
T15N900E	OEM
T15N1000C	OEM
T15N1000E	OEM
T15N1100C	OEM
T15N1100E	OEM
T15N1200C	OEM
T15N1200E	OEM
T15N1300C	OEM
T15N1300E	OEM
T15.LN100BOC+	0469
T15.LN200BOC+	0469
T15.LN400BOC+	0725
T15.LN500BOC+	0725
T15.LN600BOC+	0725
T15.LN700BOC+	2036
T15.LN800BOC+	2036
T15.LN900BOC+	2036
T16	0133
T16B7T	OEM
T16X19	0143
T17	0143
T17-A	0649
T17A	0649
T18	0143
T18/50	OEM
T18/100	OEM
T18/200	OEM
T18/400	OEM
T18/500	OEM
T18/600	OEM
T18/700	OEM
T18B7T	OEM
T18F100	OEM
T18F100C	OEM
T18F100E	OEM
T18F200	OEM
T18F200C	OEM
T18F200E	OEM
T18F300C	OEM
T18F300E	OEM
T18F400	OEM
T18F400C	OEM
T18F400E	OEM
T18F500	OEM
T18F500C	OEM
T18F500E	OEM
T18F600	OEM
T18F600C	OEM
T18F600E	OEM
T18F700	OEM
T18F700C	OEM
T18F700E	OEM
T18F800	OEM
T18F800C	OEM
T18F800E	OEM
T18F900	OEM
T18F900C	OEM
T18F900E	OEM
T19	0143
T19G	0143
T20	0143
T20A6	OEM
T20A6F	OEM
T20F50	OEM
T20F100	OEM
T20F150	OEM
T20F200	OEM
T20F300	OEM
T20F400	OEM
T20F500	OEM
T20F600	OEM
T20F700	OEM
T20F800	OEM
T20G	0143
T20N50R3	OEM
T20N100R3	OEM
T20N150R3	OEM
T20N200R3	OEM
T20N250R3	OEM
T20N300R3	OEM
T21	0143
T21F5	OEM
T21F10	OEM
T21F15	OEM
T21F20	OEM
T21F25	OEM
T21F30	OEM
T21F35	OEM
T21F40	OEM
T21F50	OEM
T21F60	OEM
T21G	0143
T22	0143
T22G	0143
T22N400	OEM
T22N400C	OEM
T22N400E	OEM
T22N500	OEM
T22N500C	OEM
T22N500E	OEM
T22N600	OEM
T22N600C	OEM
T22N600E	OEM
T22N700	OEM
T22N700C	OEM
T22N700E	OEM
T22N800	OEM
T22N800C	OEM
T22N800E	OEM
T22N900	OEM
T22N900C	OEM
T22N900E	OEM
T22N1000	OEM
T22N1000C	OEM
T22N1000E	OEM
T22N1100	OEM
T22N1100C	OEM
T22N1100E	OEM
T22N1200	OEM
T22N1200C	OEM
T22N1200E	OEM
T22N1300	OEM
T22N1300C	OEM
T22N1300E	OEM
T-23	0004
T23	0143
T23A2	OEM
T23G	0143
T24	OEM
T24B7T	OEM
T24G	0143
T25	OEM
T25T002080	OEM
T25T117GR5	3133
T26	OEM
T26B7T	OEM
T26G	0143
T27G	0143
T28G	0050
T30/50	OEM
T30/100	OEM
T30/200	OEM
T30/400	OEM
T30/500	OEM
T30/600	OEM
T30/700	OEM
T30A6F	OEM
T30F100	OEM
T30F100C	OEM
T30F100E	OEM
T30F200	OEM
T30F200C	OEM
T30F200E	OEM
T30F300E	OEM
T30F400	OEM
T30F400C	OEM
T30F400E	OEM
T30F500	OEM
T30F500C	OEM
T30F500E	OEM
T30F600	OEM
T30F600C	OEM
T30F600E	OEM
T30F700	OEM
T30F700C	OEM
T30F800	OEM
T30F800C	OEM
T30F800E	OEM
T30F900	OEM
T30F900C	OEM
T30F900E	OEM
T31	OEM
T31F5	OEM
T31F10	OEM
T31F15	OEM
T31F20	OEM
T31F25	OEM
T31F30	OEM
T31F35	OEM
T31F40	OEM
T31F50	OEM
T31F60	OEM
T32	OEM
T33	OEM
T33A2	OEM
T-33Z1	0004
T-33Z3	0004
T34	OEM
T34B7T	OEM
T35	OEM
T35N400B0B+	0468
T35N600B0B+	0468
T35N800B0B+	0468
T35N1000B0B+	0468
T35N1100B0B+	0468
T35N1200B0B+	0468
T36	OEM
T36B7T	OEM
T38B7T	OEM
T39	0004
T40	0150
T40R062202	OEM
T40R062203	OEM
T40R062204	OEM
T40R082202	OEM
T40R082203	OEM
T40R082204	OEM
T40R102202	OEM
T40R102203	OEM
T40R102204	OEM
T41P621	OEM
T43A2	OEM
T44B7T	OEM
T45	0004
T45N400B0B+	0468
T45N600B0B+	0468
T45N800B0B+	0468
T45N1000B0B+	0468
T45N1100B0B+	0468
T45N1200B0B+	0468
T46	0004
T46B7T	OEM
T47	0004
T47C231AN4943	OEM
T48	0004
T48B7T	OEM
T-50	0015
T50	0004
T50/50	OEM
T50/100	OEM
T50/200	OEM
T50/400	OEM
T50/500	OEM
T50/600	OEM
T50/700	OEM
T50F100	OEM
T50F200	OEM
T50F300	OEM
T50F400	OEM
T50F500	OEM
T50F600	OEM
T50F700	OEM
T50F800	OEM
T50F900	OEM
T50N400	OEM
T50N500	OEM
T50N600	OEM
T50N700	OEM
T50N800	OEM
T50N900	OEM
T50N1000	OEM
T50N1100	OEM
T50N1200	OEM
T50N1300	OEM
T51A	1520
T51B	1520
T51C	1520
T51N400B0B+	0738
T51N600B0B+	0738
T51N800B0B+	0738
T51N1000B0B+	0738
T51N1100B0B+	0738
T51N1200B0B+	0738
T52A	1530
T52B	1530
T52C	1530
T53A	1547
T53A1	OEM
T53A2	OEM
T53A3	OEM
T53B	1547
T53T	1547
T54A	1567
T54B	1567
T54B7T	OEM
T54T	1567
T56B7T	OEM
T58B7T	OEM
T59	0004
T60	0004
T61	0004
T62J011022DN	OEM
T62J011032DN	OEM
T62J011033DN	OEM
T62J011042DN	OEM
T62J011043DN	OEM
T62J011052DN	OEM
T62J011053DN	OEM
T62J011062DN	OEM
T62J011063DN	OEM
T62J011072DN	OEM
T62J011073DN	OEM
T62J011082DN	OEM
T62J011083	OEM
T62J011083DN	OEM
T62J011522DN	OEM
T62J011532DN	OEM
T62J011533DN	OEM
T62J011542DN	OEM
T62J011543DN	OEM
T62J011552DN	OEM
T62J011553DN	OEM
T62J011562DN	OEM
T62J011563DN	OEM
T62J011572DN	OEM
T62J011573	OEM
T62J011573DN	OEM
T62J011582DN	OEM
T62J011583DN	OEM
T62J012022DN	OEM
T62J012032DN	OEM
T62J012033DN	OEM
T62J012042DN	OEM
T62J012043DN	OEM
T62J012052DN	OEM
T62J012053DN	OEM
T62J012062DN	OEM
T62J012063	OEM
T62J012063DN	OEM
T62J012072DN	OEM
T62J012073DN	OEM
T62J012082DN	OEM
T62J012083DN	OEM
T62J021022DN	OEM
T62J021032DN	OEM
T62J021033DN	OEM
T62J021042DN	OEM
T62J021043DN	OEM
T62J021052DN	OEM
T62J021053DN	OEM
T62J021062DN	OEM
T62J021063DN	OEM
T62J021072DN	OEM
T62J021073DN	OEM
T62J021082DN	OEM
T62J021083	OEM
T62J021083DN	OEM
T62J021522DN	OEM
T62J021532DN	OEM
T62J021533DN	OEM
T62J021542DN	OEM
T62J021543DN	OEM
T62J021552DN	OEM
T62J021553DN	OEM
T62J021562DN	OEM
T62J021563DN	OEM
T62J021572DN	OEM
T62J021573	OEM
T62J021573DN	OEM
T62J021582DN	OEM
T62J021583DN	OEM
T62J022022DN	OEM
T62J022032DN	OEM
T62J022033DN	OEM
T62J022042DN	OEM
T62J022043DN	OEM
T62J022052DN	OEM
T62J022053DN	OEM
T62J022062DN	OEM
T62J022063	OEM
T62J022063DN	OEM
T62J022072DN	OEM
T62J022073DN	OEM
T62J022082DN	OEM
T62J022083DN	OEM
T62J031022DN	OEM
T62J031032DN	OEM
T62J031033DN	OEM
T62J031042DN	OEM
T62J031043DN	OEM
T62J031052DN	OEM
T62J031053DN	OEM
T62J031062DN	OEM
T62J031063DN	OEM
T62J031072DN	OEM
T62J031073DN	OEM
T62J031082DN	OEM
T62J031083	OEM
T62J031083DN	OEM
T62J031522DN	OEM
T62J031532DN	OEM
T62J031533DN	OEM
T62J031542DN	OEM
T62J031543DN	OEM
T62J031552DN	OEM
T62J031553DN	OEM
T62J031562DN	OEM
T62J031563DN	OEM
T62J031572DN	OEM
T62J031573	OEM
T62J031573DN	OEM
T62J031582DN	OEM
T62J031583DN	OEM
T62J032022DN	OEM
T62J032032DN	OEM
T62J032033DN	OEM
T62J032042DN	OEM
T62J032043DN	OEM
T62J032052DN	OEM
T62J032053DN	OEM
T62J032062DN	OEM
T62J032063	OEM
T62J032063DN	OEM
T62J032072DN	OEM
T62J032073DN	OEM
T62J032082DN	OEM
T62J032083DN	OEM
T62J041022DN	OEM
T62J041032DN	OEM
T62J041033DN	OEM
T62J041042DN	OEM
T62J041043DN	OEM
T62J041052DN	OEM
T62J041053DN	OEM
T62J041062DN	OEM
T62J041063DN	OEM
T62J041072DN	OEM
T62J041073DN	OEM
T62J041082DN	OEM
T62J041083	OEM
T62J041083DN	OEM
T62J041522DN	OEM
T62J041532DN	OEM
T62J041533DN	OEM
T62J041542DN	OEM
T62J041543DN	OEM
T62J041552DN	OEM
T62J041553DN	OEM
T62J041562DN	OEM
T62J041563DN	OEM
T62J041572DN	OEM
T62J041573	OEM
T62J041573DN	OEM
T62J041582DN	OEM
T62J041583DN	OEM
T62J042022DN	OEM
T62J042032DN	OEM
T62J042033DN	OEM
T62J042042DN	OEM
T62J042043DN	OEM
T62J042052DN	OEM
T62J042053DN	OEM
T62J042062DN	OEM
T62J042063	OEM
T62J042063DN	OEM
T62J042072DN	OEM
T62J042073DN	OEM
T62J042082DN	OEM
T62J042083DN	OEM
T62J051022DN	OEM
T62J051032DN	OEM
T62J051033DN	OEM
T62J051042DN	OEM
T62J051043DN	OEM
T62J051052DN	OEM
T62J051053DN	OEM
T62J051062DN	OEM
T62J051063DN	OEM
T62J051072DN	OEM
T62J051073DN	OEM
T62J051082DN	OEM
T62J051083	OEM
T62J051083DN	OEM
T62J051522DN	OEM
T62J051532DN	OEM
T62J051533DN	OEM
T62J051542DN	OEM
T62J051543DN	OEM
T62J051552DN	OEM
T62J051553DN	OEM
T62J051562DN	OEM
T62J051563DN	OEM
T62J051572DN	OEM
T62J051573	OEM
T62J051573DN	OEM
T62J051582DN	OEM
T62J051583DN	OEM
T62J052022DN	OEM
T62J052032DN	OEM
T62J052033DN	OEM
T62J052042DN	OEM
T62J052043DN	OEM
T62J052052DN	OEM
T62J052053DN	OEM
T62J052062DN	OEM
T62J052063	OEM
T62J052063DN	OEM
T62J052072DN	OEM
T62J052073DN	OEM
T62J052082DN	OEM
T62J052083DN	OEM
T62J061022DN	OEM
T62J061032DN	OEM
T62J061033DN	OEM
T62J061042DN	OEM
T62J061043DN	OEM
T62J061052DN	OEM
T62J061053DN	OEM
T62J061062DN	OEM
T62J061063DN	OEM
T62J061072DN	OEM
T62J061073DN	OEM
T62J061082DN	OEM
T62J061083	OEM
T62J061083DN	OEM
T62J061522DN	OEM
T62J061532DN	OEM
T62J061533DN	OEM
T62J061542DN	OEM
T62J061543DN	OEM
T62J061552DN	OEM
T62J061553DN	OEM
T62J061562DN	OEM
T62J061563DN	OEM
T62J061572DN	OEM
T62J061573	OEM
T62J061573DN	OEM
T62J061582DN	OEM
T62J061583DNDN	OEM
T62J062022DN	OEM
T62J062032DN	OEM
T62J062033DN	OEM
T62J062042DN	OEM
T62J062043DN	OEM
T62J062052DN	OEM
T62J062053DN	OEM
T62J062062DN	OEM
T62J062063	OEM
T62J062063DN	OEM
T62J062072DN	OEM
T62J062073DN	OEM
T62J062082DN	OEM
T62J062083DN	OEM
T62J071022DN	OEM
T62J071032DN	OEM
T62J071033DN	OEM
T62J071042DN	OEM
T62J071043DN	OEM
T62J071052DN	OEM
T62J071053DN	OEM
T62J071062DN	OEM
T62J071063DN	OEM
T62J071072DN	OEM
T62J071073DN	OEM
T62J071083	OEM
T62J071083DN	OEM
T62J071522DN	OEM
T62J071532DN	OEM
T62J071533DN	OEM
T62J071542DN	OEM
T62J071543DN	OEM
T62J071552DN	OEM
T62J071553DN	OEM
T62J071562DN	OEM
T62J071563DN	OEM
T62J071572DN	OEM

DEVICE TYPE	REPL CODE
T62J071573	OEM
T62J071573DN	OEM
T62J071583DN	OEM
T62J072022DN	OEM
T62J072032DN	OEM
T62J072033DN	OEM
T62J072042DN	OEM
T62J072043DN	OEM
T62J072052DN	OEM
T62J072053DN	OEM
T62J072062DN	OEM
T62J072063	OEM
T62J072063DN	OEM
T62J072072DN	OEM
T62J072073DN	OEM
T62J072083DN	OEM
T62J081022DN	OEM
T62J081032DN	OEM
T62J081033DN	OEM
T62J081042DN	OEM
T62J081043DN	OEM
T62J081052DN	OEM
T62J081053DN	OEM
T62J081062DN	OEM
T62J081063DN	OEM
T62J081072DN	OEM
T62J081073DN	OEM
T62J081082DN	OEM
T62J081083	OEM
T62J081083DN	OEM
T62J081522DN	OEM
T62J081532DN	OEM
T62J081533DN	OEM
T62J081542DN	OEM
T62J081543DN	OEM
T62J081552DN	OEM
T62J081553DN	OEM
T62J081562DN	OEM
T62J081563DN	OEM
T62J081572DN	OEM
T62J081573	OEM
T62J081573DN	OEM
T62J081582DN	OEM
T62J081583DN	OEM
T62J082022DN	OEM
T62J082032DN	OEM
T62J082033DN	OEM
T62J082042DN	OEM
T62J082043DN	OEM
T62J082052DN	OEM
T62J082053DN	OEM
T62J082062DN	OEM
T62J082063	OEM
T62J082063DN	OEM
T62J082072DN	OEM
T62J082073DN	OEM
T62J082082DN	OEM
T62J082083DN	OEM
T62J091022DN	OEM
T62J091032DN	OEM
T62J091033DN	OEM
T62J091042DN	OEM
T62J091043DN	OEM
T62J091052DN	OEM
T62J091053DN	OEM
T62J091062DN	OEM
T62J091063DN	OEM
T62J091072DN	OEM
T62J091073DN	OEM
T62J091083	OEM
T62J091083DN	OEM
T62J091522DN	OEM
T62J091532DN	OEM
T62J091533DN	OEM
T62J091542DN	OEM
T62J091543DN	OEM
T62J091552DN	OEM
T62J091553DN	OEM
T62J091562DN	OEM
T62J091563DN	OEM
T62J091572DN	OEM
T62J091573	OEM
T62J091573DN	OEM
T62J091583DN	OEM
T62J092022DN	OEM
T62J092032DN	OEM
T62J092033DN	OEM
T62J092042DN	OEM
T62J092043DN	OEM
T62J092052DN	OEM
T62J092053DN	OEM
T62J092062DN	OEM
T62J092063	OEM
T62J092063DN	OEM
T62J092072DN	OEM
T62J092073DN	OEM
T62J092083DN	OEM
T62J101022DN	OEM
T62J101032DN	OEM
T62J101033DN	OEM
T62J101042DN	OEM
T62J101043DN	OEM
T62J101052DN	OEM
T62J101053DN	OEM
T62J101062DN	OEM
T62J101063DN	OEM
T62J101072DN	OEM
T62J101073DN	OEM
T62J101083	OEM

DEVICE TYPE	REPL CODE
T62J101083DN	OEM
T62J101522DN	OEM
T62J101532DN	OEM
T62J101533DN	OEM
T62J101542DN	OEM
T62J101543DN	OEM
T62J101552DN	OEM
T62J101553DN	OEM
T62J101562DN	OEM
T62J101563DN	OEM
T62J101572DN	OEM
T62J101573	OEM
T62J101573DN	OEM
T62J101583DN	OEM
T62J102022DN	OEM
T62J102032DN	OEM
T62J102033DN	OEM
T62J102042DN	OEM
T62J102043DN	OEM
T62J102052DN	OEM
T62J102053DN	OEM
T62J102062DN	OEM
T62J102063	OEM
T62J102063DN	OEM
T62J102072DN	OEM
T62J102073DN	OEM
T62J102083DN	OEM
T62J111022DN	OEM
T62J111032DN	OEM
T62J111033DN	OEM
T62J111042DN	OEM
T62J111043DN	OEM
T62J111052DN	OEM
T62J111053DN	OEM
T62J111062DN	OEM
T62J111063	OEM
T62J111063DN	OEM
T62J111073DN	OEM
T62J111083DN	OEM
T62J111522DN	OEM
T62J111532DN	OEM
T62J111533DN	OEM
T62J111542DN	OEM
T62J111543DN	OEM
T62J111552DN	OEM
T62J111553DN	OEM
T62J111562DN	OEM
T62J111563	OEM
T62J111563DN	OEM
T62J111573DN	OEM
T62J111583DN	OEM
T62J112022DN	OEM
T62J112032DN	OEM
T62J112033DN	OEM
T62J112042DN	OEM
T62J112043DN	OEM
T62J112052DN	OEM
T62J112053DN	OEM
T62J112062DN	OEM
T62J112063	OEM
T62J112063DN	OEM
T62J112073DN	OEM
T62J112083DN	OEM
T62J121022DN	OEM
T62J121032DN	OEM
T62J121033DN	OEM
T62J121042DN	OEM
T62J121043DN	OEM
T62J121052DN	OEM
T62J121053DN	OEM
T62J121062DN	OEM
T62J121063	OEM
T62J121063DN	OEM
T62J121073DN	OEM
T62J121083DN	OEM
T62J121522DN	OEM
T62J121532DN	OEM
T62J121533DN	OEM
T62J121542DN	OEM
T62J121543DN	OEM
T62J121552DN	OEM
T62J121553DN	OEM
T62J121562DN	OEM
T62J121563	OEM
T62J121563DN	OEM
T62J121573DN	OEM
T62J121583DN	OEM
T62J122022DN	OEM
T62J122032DN	OEM
T62J122033DN	OEM
T62J122042DN	OEM
T62J122043DN	OEM
T62J122052DN	OEM
T62J122053DN	OEM
T62J122062DN	OEM
T62J122063	OEM
T62J122063DN	OEM
T62J122073DN	OEM
T62J122083DN	OEM
T63A011790	1752
T63A1	OEM
T63A2	OEM
T63A3	OEM
T64B7T	OEM
T66B7T	OEM
T67A1037K0	1742
T68B7T	OEM
T70F200EEB	OEM
T70F200EEC	OEM

DEVICE TYPE	REPL CODE
T70F200EFB	OEM
T70F200EFC	OEM
T70F200EGB	OEM
T70F200EGC	OEM
T70F400EEB	OEM
T70F400EEC	OEM
T70F400EFB	OEM
T70F400EFC	OEM
T70F400EGB	OEM
T70F400EGC	OEM
T70F500EEB	OEM
T70F500EEC	OEM
T70F500EFB	OEM
T70F500EFC	OEM
T70F500EGB	OEM
T70F500EGC	OEM
T70F600EEB	OEM
T70F600EEC	OEM
T70F600EFB	OEM
T70F600EFC	OEM
T70F600EGB	OEM
T70F600EGC	OEM
T70F700EEB	OEM
T70F700EEC	OEM
T70F700EFB	OEM
T70F700EFC	OEM
T70F700EGB	OEM
T70F700EGC	OEM
T70F800EEB	OEM
T70F800EEC	OEM
T70F800EFB	OEM
T70F800EFC	OEM
T70F800EGB	OEM
T70F800EGC	OEM
T70F900EFB	OEM
T70F900EFC	OEM
T70F900EGB	OEM
T70F900EGC	OEM
T70F1000EFB	OEM
T70F1000EFC	OEM
T70F1000EGB	OEM
T70F1000EGC	OEM
T70F1100EFB	OEM
T70F1100EFC	OEM
T70F1100EGB	OEM
T70F1100EGC	OEM
T70F1200EFB	OEM
T70F1200EFC	OEM
T70F1200EGB	OEM
T70F1200EGC	OEM
T70F1300EFB	OEM
T70F1300EFC	OEM
T70F1300EGB	OEM
T70F1300EGC	OEM
T70N400E0B	OEM
T70N400E0C	OEM
T70N400E0F	OEM
T70N600E0B	OEM
T70N600E0C	OEM
T70N600E0F	OEM
T70N800E0B	OEM
T70N800E0C	OEM
T70N800E0F	OEM
T70N1000E0B	OEM
T70N1000E0C	OEM
T70N1000E0F	OEM
T70N1100E0B	OEM
T70N1100E0C	OEM
T70N1100E0F	OEM
T70N1200E0B	OEM
T70N1200E0C	OEM
T70N1200E0F	OEM
T70N1300E0B	OEM
T70N1300E0C	OEM
T70N1300E0F	OEM
T70N1400E0B	OEM
T70N1400E0C	OEM
T70N1400E0F	OEM
T70N1500E0B	OEM
T70N1500E0C	OEM
T70N1500E0F	OEM
T70N1600E0B	OEM
T70N1600E0C	OEM
T70N1600E0F	OEM
T71N400B0B+	0738
T71N600B0B+	0738
T71N800B0B+	0738
T71N1000B0B+	0738
T71N1100B0B+	0738
T71N1200B0B+	0738
T72	0004
T72H012524DN	OEM
T72H012534DN	OEM
T72H012544DN	OEM
T72H012554DN	OEM
T72H012564	OEM
T72H012564DN	OEM
T72H012574DN	OEM
T72H012584	OEM
T72H012584DN	OEM
T72H013524DN	OEM
T72H013534DN	OEM
T72H013544DN	OEM
T72H013554DN	OEM
T72H013564DN	OEM
T72H013574	OEM
T72H013574DN	OEM
T72H013584DN	OEM
T72H014524DN	OEM

DEVICE TYPE	REPL CODE
T72H014534DN	OEM
T72H014544DN	OEM
T72H014554	OEM
T72H014554DN	OEM
T72H014564DN	OEM
T72H014574DN	OEM
T72H014584DN	OEM
T72H022524DN	OEM
T72H022534DN	OEM
T72H022544DN	OEM
T72H022554DN	OEM
T72H022564DN	OEM
T72H022574DN	OEM
T72H022584	OEM
T72H022584DN	OEM
T72H023524DN	OEM
T72H023534DN	OEM
T72H023544DN	OEM
T72H023554DN	OEM
T72H023564DN	OEM
T72H023574	OEM
T72H023574DN	OEM
T72H023584DN	OEM
T72H024524DN	OEM
T72H024534DN	OEM
T72H024544DN	OEM
T72H024554	OEM
T72H024554DN	OEM
T72H024564DN	OEM
T72H024574DN	OEM
T72H024584DN	OEM
T72H032524DN	OEM
T72H032534DN	OEM
T72H032544DN	OEM
T72H032554DN	OEM
T72H032564DN	OEM
T72H032574DN	OEM
T72H032584	OEM
T72H032584DN	OEM
T72H033524DN	OEM
T72H033534DN	OEM
T72H033544DN	OEM
T72H033554DN	OEM
T72H033564DN	OEM
T72H033574	OEM
T72H033574DN	OEM
T72H033584DN	OEM
T72H034524DN	OEM
T72H034534DN	OEM
T72H034544DN	OEM
T72H034554	OEM
T72H034554DN	OEM
T72H034564DN	OEM
T72H034574DN	OEM
T72H034584DN	OEM
T72H042524DN	OEM
T72H042534DN	OEM
T72H042544DN	OEM
T72H042554DN	OEM
T72H042564DN	OEM
T72H042574DN	OEM
T72H042584	OEM
T72H042584DN	OEM
T72H043524DN	OEM
T72H043534DN	OEM
T72H043544DN	OEM
T72H043554DN	OEM
T72H043564DN	OEM
T72H043574	OEM
T72H043574DN	OEM
T72H043584DN	OEM
T72H044524DN	OEM
T72H044534DN	OEM
T72H044544DN	OEM
T72H044554	OEM
T72H044554DN	OEM
T72H044564DN	OEM
T72H044574DN	OEM
T72H044584DN	OEM
T72H052524DN	OEM
T72H052534DN	OEM
T72H052544DN	OEM
T72H052554DN	OEM
T72H052564DN	OEM
T72H052574DN	OEM
T72H052584	OEM
T72H052584DN	OEM
T72H053524DN	OEM
T72H053534DN	OEM
T72H053544DN	OEM
T72H053554DN	OEM
T72H053564DN	OEM
T72H053574	OEM
T72H053574DN	OEM
T72H053584DN	OEM
T72H054524DN	OEM
T72H054534DN	OEM
T72H054544DN	OEM
T72H054554	OEM
T72H054554DN	OEM
T72H054564DN	OEM
T72H054574DN	OEM
T72H054584DN	OEM
T72H062524DN	OEM
T72H062534DN	OEM
T72H062544DN	OEM
T72H062554DN	OEM
T72H062564DN	OEM
T72H062574DN	OEM

DEVICE TYPE	REPL CODE
T72H062584	OEM
T72H062584DN	OEM
T72H063524DN	OEM
T72H063534DN	OEM
T72H063544DN	OEM
T72H063554DN	OEM
T72H063564DN	OEM
T72H063574	OEM
T72H063574DN	OEM
T72H063584DN	OEM
T72H064524DN	OEM
T72H064534DN	OEM
T72H064544DN	OEM
T72H064554	OEM
T72H064554DN	OEM
T72H064564DN	OEM
T72H064574DN	OEM
T72H064584DN	OEM
T72H072524DN	OEM
T72H072534DN	OEM
T72H072544DN	OEM
T72H072554DN	OEM
T72H072564DN	OEM
T72H072574	OEM
T72H072574DN	OEM
T72H072584DN	OEM
T72H073524DN	OEM
T72H073534DN	OEM
T72H073544DN	OEM
T72H073554DN	OEM
T72H073564DN	OEM
T72H073574	OEM
T72H073574DN	OEM
T72H073584DN	OEM
T72H074524DN	OEM
T72H074534DN	OEM
T72H074544DN	OEM
T72H074554	OEM
T72H074554DN	OEM
T72H074564DN	OEM
T72H074574DN	OEM
T72H074584DN	OEM
T72H082524DN	OEM
T72H082534DN	OEM
T72H082544DN	OEM
T72H082554DN	OEM
T72H082564DN	OEM
T72H082574	OEM
T72H082574DN	OEM
T72H082584DN	OEM
T72H083524DN	OEM
T72H083534DN	OEM
T72H083544DN	OEM
T72H083554DN	OEM
T72H083564DN	OEM
T72H083574	OEM
T72H083574DN	OEM
T72H083584DN	OEM
T72H084524DN	OEM
T72H084534DN	OEM
T72H084544DN	OEM
T72H084554	OEM
T72H084554DN	OEM
T72H084564DN	OEM
T72H084574DN	OEM
T72H084584DN	OEM
T72H092524DN	OEM
T72H092534DN	OEM
T72H092544DN	OEM
T72H092554DN	OEM
T72H092564	OEM
T72H092564DN	OEM
T72H092574DN	OEM
T72H092584DN	OEM
T72H093524DN	OEM
T72H093534DN	OEM
T72H093544DN	OEM
T72H093554DN	OEM
T72H093564	OEM
T72H093564DN	OEM
T72H093574DN	OEM
T72H093584DN	OEM
T72H094524DN	OEM
T72H094534DN	OEM
T72H094544DN	OEM
T72H094554	OEM
T72H094554DN	OEM
T72H094564DN	OEM
T72H094574DN	OEM
T72H094584DN	OEM
T72H102524DN	OEM
T72H102534DN	OEM
T72H102544DN	OEM
T72H102554DN	OEM
T72H102564DN	OEM
T72H102574DN	OEM
T72H102584DN	OEM
T72H103524DN	OEM
T72H103534DN	OEM
T72H103544DN	OEM
T72H103554DN	OEM
T72H103564	OEM
T72H103564DN	OEM
T72H103574DN	OEM
T72H103584DN	OEM
T72H104524DN	OEM
T72H104534DN	OEM
T72H104544DN	OEM
T72H104554	OEM

DEVICE TYPE	REPL CODE
T72H104554DN	OEM
T72H104564DN	OEM
T72H104574DN	OEM
T72H104584DN	OEM
T72H112524DN	OEM
T72H112534DN	OEM
T72H112544DN	OEM
T72H112554DN	OEM
T72H112564	OEM
T72H112564DN	OEM
T72H112574DN	OEM
T72H112584DN	OEM
T72H113524DN	OEM
T72H113534DN	OEM
T72H113544DN	OEM
T72H113554DN	OEM
T72H113564DN	OEM
T72H113574DN	OEM
T72H113584DN	OEM
T72H114524DN	OEM
T72H114534DN	OEM
T72H114544DN	OEM
T72H114554	OEM
T72H114554DN	OEM
T72H114564DN	OEM
T72H114574DN	OEM
T72H114584DN	OEM
T72H122524DN	OEM
T72H122534DN	OEM
T72H122544DN	OEM
T72H122554DN	OEM
T72H122564	OEM
T72H122564DN	OEM
T72H122574DN	OEM
T72H122584DN	OEM
T72H123524DN	OEM
T72H123534DN	OEM
T72H123544DN	OEM
T72H123554DN	OEM
T72H123564DN	OEM
T72H123574DN	OEM
T72H123584DN	OEM
T72H124524DN	OEM
T72H124534DN	OEM
T72H124544DN	OEM
T72H124554	OEM
T72H124554DN	OEM
T74	0004
T74H00B1	0682
T74H00D1	0682
T74H00D2	0682
T74H10B1	0685
T74H10D1	0685
T74H10D2	0685
T74H11B1	2398
T74H11D1	2398
T74H11D2	2398
T74H20B1	3715
T74H20D1	3715
T74H20D2	3715
T74H21B1	4856
T74H21D1	4856
T74H21D2	4856
T74H40B1	0559
T74H40D1	0559
T74H40D2	0559
T74H50B1	1792
T74H50D1	1792
T74H50D2	1792
T74H51B1	1944
T74H51D1	1944
T74H51D2	1944
T74H52B1	2020
T74H52D1	2020
T74H52D2	2020
T74H53B1	2101
T74H53D1	2101
T74H53D2	2101
T74H54B1	2172
T74H54D1	2172
T74H54D2	2172
T74H61B1	2663
T74H61D1	2663
T74H61D2	2663
T74H62B1	2731
T74H62D1	2731
T74H62D2	2731
T74H71B1	3264
T74H71D2	3264
T74H72B1	3313
T74H72D1	3313
T74H72D2	3313
T74LS00B1	1527
T74LS00D1	1527
T74LS01B1N	1546
T74LS02B1	1559
T74LS02D1	1559
T74LS03B1	1579
T74LS03D1	1579
T74LS04B1	1595
T74LS08B1	1633
T74LS08D1	1633
T74LS09B1	1642
T74LS09D1	1642
T74LS10B1	1662
T74LS10D1	1662
T74LS11B1	1667
T74LS11D1	1667
T74LS13B1	1688

DEVICE TYPE	REPL CODE
T74LS14B1	1698
T74LS14BI	1698
T74LS15B1	1707
T74LS15D1	1707
T74LS20B1	0034
T74LS20D1	0034
T74LS21B1	1763
T74LS21D1	1763
T74LS22B1	1775
T74LS22D1	1775
T74LS26B1	1382
T74LS26D1	1382
T74LS27B1	0183
T74LS27BI	0183
T74LS27D1	0183
T74LS28B1	0472
T74LS30B1	0828
T74LS30D1	0828
T74LS32B1	0088
T74LS32BI	0088
T74LS32D1	0088
T74LS40B1	0135
T74LS40C1	OEM
T74LS40D1	2733
T74LS40M1	OEM
T74LS51B1	1033
T74LS51D1	1033
T74LS54B1	1857
T74LS54D1	1857
T74LS55B1	0457
T74LS55B1N	0457
T74LS55C1	OEM
T74LS55D1	0457
T74LS55M1	OEM
T74LS74B1	0243
T74LS74D1	0243
T74LS83B1	2218
T74LS83D1	2218
T74LS86B1	0291
T74LS86D1	0291
T74LS109B1	1906
T74LS109D1	1906
T74LS112B1	2128
T74LS112D1	2128
T74LS113B1	2256
T74LS113D1	2256
T74LS114B1	2302
T74LS114D1	2302
T74LS125AB1	OEM
T74LS132P	OEM
T74LS136B1	1628
T74LS136D1	1628
T74LS138B1	0427
T74LS151B1	5136
T74LS155B1	0209
T74LS157B1	1161
T74LS161B1	0858
T74LS164B1	6902
T74LS174B1	0260
T74LS175B1	1672
T74LS193B1	5876
T74LS244B1	0458
T74LS257B1	1744
T74LS273B1	OEM
T74LS279B1	3290
T74LS283D1	OEM
T74LS352B1	0761
T74LS352C1	OEM
T74LS352D1	OEM
T74LS352M1	OEM
T74LS353B1	0773
T74LS353C1	OEM
T74LS353D1	OEM
T74LS353M1	OEM
T74LS373B1	0709
T74LS374B1	0713
T74LS386AB1	OEM
T74LS393B1	0819
T74LS398B1	1383
T74LS399B1	1398
T74LS399C1	OEM
T74LS399D1	OEM
T74LS399M1	OEM
T74LS670B1	1130
T76	0016
T77	0004
T78	0004
T81	0208
T-82	0280
T82	0004
T83	0004
T83A028120	0724
T83A028127	0724
T83A028140	0724
T84	0004
T84B7T	OEM
T86B7T	OEM
T87	0004
T87A	0609
T87A02412K	0724
T88B7T	OEM
T95	0004
T95/50	OEM
T95/100	OEM
T95/200	OEM
T95/400	OEM
T95/500	OEM
T95/600	OEM
T95/700	OEM

If replacement code is OEM, contact original manufacturer for replacement.

DEVICE TYPE	REPL CODE
T95F100	OEM
T95F200	OEM
T95F300	OEM
T95F400	OEM
T95F500	OEM
T95F600	OEM
T95F700	OEM
T95F800	OEM
T95F900	OEM
T99	0004
T99N400BOB+	0738
T99N600BOB+	0738
T99N800BOB+	0738
T99N1000BOB+	0738
T99N1100BOB+	0738
T99N1200BOB+	0738
T-100	0015
T100	0004
T100F200EEB	OEM
T100F200EEC	OEM
T100F200EFB	OEM
T100F200EFC	OEM
T100F200EGB	OEM
T100F200EGC	OEM
T100F400EEB	OEM
T100F400EEC	OEM
T100F400EFB	OEM
T100F400EFC	OEM
T100F400EGB	OEM
T100F400EGC	OEM
T100F500EEB	OEM
T100F500EEC	OEM
T100F500EFB	OEM
T100F500EFC	OEM
T100F500EGB	OEM
T100F500EGC	OEM
T100F600EEB	OEM
T100F600EEC	OEM
T100F600EFB	OEM
T100F600EFC	OEM
T100F600EGB	OEM
T100F600EGC	OEM
T100F700EEB	OEM
T100F700EEC	OEM
T100F700EFB	OEM
T100F700EFC	OEM
T100F700EGB	OEM
T100F700EGC	OEM
T100F800EEB	OEM
T100F800EEC	OEM
T100F800EFB	OEM
T100F800EFC	OEM
T100F800EGB	OEM
T100F800EGC	OEM
T100F900EFB	OEM
T100F900EFC	OEM
T100F900EGB	OEM
T100F900EGC	OEM
T100F1000EFB	OEM
T100F1000EFC	OEM
T100F1000EGB	OEM
T100F1000EGC	OEM
T100F1100EFB	OEM
T100F1100EFC	OEM
T100F1100EGB	OEM
T100F1100EGC	OEM
T100F1200EFB	OEM
T100F1200EFC	OEM
T100F1200EGB	OEM
T100F1200EGC	OEM
T100F1300EFB	OEM
T100F1300EFC	OEM
T100F1300EGB	OEM
T100F1300EGC	OEM
T100N400BOC	OEM
T100N400EOC	OEM
T100N600BOC	OEM
T100N600EOC	OEM
T100N800BOC	OEM
T100N800EOC	OEM
T100N1000BOC	OEM
T100N1000EOC	OEM
T100N1200BOC	OEM
T100N1200EOC	OEM
T100N1400BOC	OEM
T100N1400EOC	OEM
T100N1600BOC	OEM
T100N1600EOC	OEM
T100N1800BOC	OEM
T100N1800EOC	OEM
T-101	0160
T101	0004
T102A	0004
T104B7T	OEM
T106A1	0447
T106A1SC	2300
T106A1SD	2300
T106A1SG	2300
T106A1SH	2300
T106A1SS	2300
T106A2SC	2300
T106A2SD	2300
T106A2SG	2300
T106A2SH	2300
T106A2SS	2300
T106B1	0940
T106B1SC	2300
T106B1SD	2300
T106B1SG	2300
T106B1SH	2300
T106B1SS	2300
T106B2SC	2300
T106B2SD	2300
T106B2SG	2300
T106B2SH	2300
T106B2SS	2300
T106B7T	OEM
T106C1	0095
T106C1SC	2300
T106C1SD	2300
T106C1SG	2300
T106C1SH	2300
T106C1SS	2300
T106C2SC	2300
T106C2SD	2300
T106C2SG	2300
T106C2SH	2300
T106C2SS	2300
T106D1	0095
T106D1SC	2300
T106D1SD	2300
T106D1SG	2300
T106D1SH	2300
T106D1SS	2300
T106D2SC	2300
T106D2SD	2300
T106D2SG	2300
T106D2SH	2300
T106D2SS	2300
T106F1	1259
T106F1SC	2300
T106F1SD	2300
T106F1SG	2300
T106F1SH	2300
T106F1SS	2300
T106F2SC	2300
T106F2SD	2300
T106F2SG	2300
T106F2SH	2300
T106F2SS	2300
T106Q1	1396
T106Y1	1396
T107A1	0447
T107B1	0940
T107C1	0095
T107D1	0095
T107F1	1259
T107Q1	1396
T107Y1	1396
T108	0004
T108B7T	OEM
T109	0004
T110F200TEC	OEM
T110F800TEC	OEM
T110F900TFC	OEM
T110F1300TFC	OEM
T-112	0037
T116	0004
T125N400	OEM
T125N500	OEM
T125N600	OEM
T125N700	OEM
T125N800	OEM
T125N900	OEM
T125N1000	OEM
T125N1100	OEM
T125N1200	OEM
T125N1300	OEM
T126	0004
T-127	0160
T127	0004
T129	0004
T129N400BOB+	0738
T129N600BOB+	0738
T129N800BOB+	0738
T129N1000BOB+	0738
T129N1100BOB+	0738
T129N1200BOB+	0738
T130	0004
T130G8	OEM
T130H8	OEM
T130J8	OEM
T130N400EOC	OEM
T130N600EOC	OEM
T130N800EOC	OEM
T130N1000EOC	OEM
T130N1200EOC	OEM
T130N1400EOC	OEM
T130N1600EOC	OEM
T-131	0164
T131	0164
T139	0085
T140G8	OEM
T140H8	OEM
T140J8	OEM
T142	0160
T143	0016
T150F200TEC	OEM
T150F800TEC	OEM
T150F900TFC	OEM
T150F1300TFC	OEM
T151	0133
T152	0133
T153	0133
T154	0133
T155	0133
T155N1600	OEM
T155N1800	OEM
T155N2000	OEM
T155N2200	OEM
T156	0609
T157	0016
T158	0079
T159	0015
T159N400BOB+	0738
T159N600BOB+	0738
T159N800BOB+	0738
T159N1000BOB+	0738
T159N1100BOB+	0738
T160	0004
T160N400	OEM
T160N500	OEM
T160N600	OEM
T160N700	OEM
T160N800	OEM
T160N900	OEM
T160N1000	OEM
T160N1100	OEM
T160N1200	OEM
T160N1300	OEM
T160N1400	OEM
T160N1500	OEM
T161N400EOC	OEM
T161N600EOC	OEM
T161N800EOC	OEM
T161N1000EOC	OEM
T161N1200EOC	OEM
T161N1400EOC	OEM
T161N1600EOC	OEM
T-163	0050
T163	0050
T-164	OEM
T165N400TOC	OEM
T165N400TOF	OEM
T165N1800TOC	OEM
T165N1800TOF	OEM
T170	0079
T170F100	OEM
T170F200	OEM
T170F300	OEM
T170F400	OEM
T170F500	OEM
T170F600	OEM
T170F700	OEM
T170F800	OEM
T170F900	OEM
T170N400	OEM
T170N500	OEM
T170N600	OEM
T170N700	OEM
T170N800	OEM
T170N900	OEM
T170N1000	OEM
T170N1100	OEM
T170N1200	OEM
T170N1300	OEM
T171	0079
T180/50	OEM
T180/100	OEM
T180/200	OEM
T180/400	OEM
T180/500	OEM
T180/600	OEM
T180/700	OEM
T185	0079
T200	0015
T200N400TOC	OEM
T200N400TOF	OEM
T200N1800TOC	OEM
T200N1800TOF	OEM
T-203	0127
T210B7T	OEM
T215B7T	OEM
T220N400	OEM
T220N500	OEM
T220N600	OEM
T220N700	OEM
T220N800	OEM
T220N900	OEM
T220N1000	OEM
T220N1100	OEM
T220N1200	OEM
T220N1300	OEM
T220N1400	OEM
T220N1500	OEM
T222	0074
T225	0074
T226	OEM
T230G8	OEM
T230H8	OEM
T230J8	OEM
T-235	0160
T-235A	OEM
T235N1600	OEM
T235N1800	OEM
T235N2000	OEM
T237	0016
T240G8	OEM
T240H8	OEM
T240J8	OEM
T-246	0037
T246	0126
T247	0086
T-251	0037
T253	0050
T253(SEARS)	0050
T-255	0079
T255	0079
T256	0079
T257	0142
T260F100	OEM
T260F200	OEM
T260F300	OEM
T260F400	OEM
T260F500	OEM
T260F600	OEM
T260F700	OEM
T260F800	OEM
T260F900	OEM
T261	0142
T271	0178
T276	0037
T277	0079
T-278	0050
T278	0050
T-279	0050
T279	0050
T280	0050
T280(SEARS)	0136
T281	0050
T281(SEARS)	0136
T282	0050
T282(SEARS)	0004
T287	0276
T-291	0086
T291	0079
T300	0015
T300N400	OEM
T300N500	OEM
T300N600	OEM
T300N700	OEM
T300N800	OEM
T300N900	OEM
T300N1000	OEM
T300N1100	OEM
T300N1200	OEM
T300N1300	OEM
T301N400	OEM
T301N500	OEM
T301N600	OEM
T301N700	OEM
T301N800	OEM
T301N900	OEM
T301N1000	OEM
T301N1100	OEM
T301N1200	OEM
T301N1300	OEM
T301N1400	OEM
T302N400EOC	OEM
T302N600EOC	OEM
T302N800EOC	OEM
T302N1000EOC	OEM
T302N1200EOC	OEM
T302N1400EOC	OEM
T302N1600EOC	OEM
T308	0144
T310B7T	OEM
T313	OEM
T315B7T	OEM
T327	0079
T327-2	0079
T328	0079
T330F200TEC	OEM
T330F200TFC	OEM
T330F1200TEC	OEM
T330F1200TFC	OEM
T330G8	OEM
T330H8	OEM
T330J8	OEM
T334-2	0126
T336-2	0086
T338	0018
T-339	0086
T339	0079
T-340	0126
T340	0037
T340G8	OEM
T340H8	OEM
T340J8	OEM
T-342	0161
T342	0079
T-344	0161
T344	0161
T-345	0460
T345	0460
T346	0280
T-348	0136
T348	0050
T350N400TOC	OEM
T350N400TOF	OEM
T350N1600TOC	OEM
T350N1600TOF	OEM
T367	0050
T-368	0050
T368	0050
T370	0604
T373	0050
T374	0050
T381	0007
T381(SEARS)	0007
T386	0016
T386(SEARS)	0127
T-396	0460
T396	0460
T-399	0079
T399	0079
T400	0015
T410B7T	OEM
T411A	0408
T411F	0399
T415B7T	OEM
T416-16(SEARS)	0079
T417	0079
T422	0551
T423	0382
T430G8	OEM
T430H8	OEM
T430J8	OEM
T440G8	OEM
T440H8	OEM
T440J8	OEM
T449	0050
T449(SEARS)	0136
T450	0015
T450N400TOC	OEM
T450N400TOF	OEM
T450N1600TOC	OEM
T450N1600TOF	OEM
T452	0086
T457-16	0079
T457-16(SEARS)	0079
T458-16	0079
T459	0126
T459(SEARS)	0079
T460	0079
T460(SEARS)	0037
T461-16	0079
T461-16(SEARS)	0016
T462	0079
T462(SEARS)	0079
T464	0212
T472	0079
T472(SEARS)	0079
T475	0126
T475(SEARS)	0037
T480F100TDC	OEM
T480F500TDC	OEM
T480F600TDC	OEM
T-481	0233
T481(SEARS)	0233
T482(SEARS)	0037
T483(SEARS)	0079
T484(SEARS)	0079
T485(SEARS)	0079
T-486	0144
T486(SEARS)	0079
T500	0015
T508N1000TOB	OEM
T508N1000TOC	OEM
T508N1000TOF	OEM
T508N1100TOB	OEM
T508N1100TOC	OEM
T508N1100TOF	OEM
T508N1200TOB	OEM
T508N1200TOC	OEM
T508N1200TOF	OEM
T508N1400TOB	OEM
T508N1400TOC	OEM
T508N1400TOF	OEM
T508N1600TOB	OEM
T508N1600TOC	OEM
T508N1600TOF	OEM
T510B7T	OEM
T511F	1434
T512F	1434
T513	0212
T513F	OEM
T514F	OEM
T515B7T	OEM
T518B	OEM
T520C	OEM
T520D	OEM
T520E	OEM
T520G	OEM
T520N1100TOC	OEM
T521N800TOF	OEM
T522F	1680
T523C	OEM
T524D	OEM
T529C	OEM
T529H	OEM
T530G8	OEM
T530H8	OEM
T530J8	OEM
T540G8	OEM
T540H8	OEM
T540J8	OEM
T541	2294
T542	2294
T542-6	OEM
T550	0015
T551	2298
T552	2298
T576-1	0007
T597-1	0037
T600	0015
T600*1804	OEM
T600F600TEC	OEM
T600F800TEC	OEM
T600F1000TEC	OEM
T600F1100TEC	OEM
T600F1200TEC	OEM
T600F1300TEC	OEM
T610B7T	OEM
T611-1	0042
T611-1(RCA)	0161
T611F	OEM
T612-1	0042
T612-1(RCA)	0161
T612F	OEM
T613F	OEM
T615B7T	OEM
T621-1	0161
T621M	OEM
T625N3600TOF	OEM
T625N3800TOF	OEM
T625N4000TOF	OEM
T625N4400TOF	OEM
T630G8	OEM
T630H8	OEM
T630J8	OEM
T640G8	OEM
T640H8	OEM
T640J8	OEM
T650	0015
T651	OEM
T652	OEM
T653	OEM
T670F1600TMC	OEM
T670F1800TMC	OEM
T670F2000TMC	OEM
T671N3000TOF	OEM
T671N3200TOF	OEM
T671N3400TOF	OEM
T671N3600TOF	OEM
T698F200TCC	OEM
T698F400TCC	OEM
T698F500TCC	OEM
T699	OEM
T-700-709	0143
T700-709	0143
T700N1800TOB	OEM
T700N1800TOC	OEM
T700N1800TOF	OEM
T700N2000TOB	OEM
T700N2000TOC	OEM
T700N2000TOF	OEM
T700N2200TOB	OEM
T700N2200TOC	OEM
T700N2200TOF	OEM
T700N2400TOB	OEM
T700N2400TOC	OEM
T700N2400TOF	OEM
T700N2600TOB	OEM
T700N2600TOC	OEM
T700N2600TOF	OEM
T-750-713	0143
T-750-714	0057
T760F2000TUC	OEM
T760F2200TUC	OEM
T760F2400TUC	OEM
T800	0071
T800X	0071
T810B7T	OEM
T811	0050
T814	0164
T815	0164
T815B7T	OEM
T819	OEM
T841	0130
T842	0130
T843	0130
T844	0130
T900B1	5050
T900B1-K	OEM
T900BI-K	OEM
T911B1-T	5195
T911BIT	5195
T916BI	OEM
T951HL	OEM
T975	OEM
T-1000	0071
T1000	0164
T-1000X	0071
T1000X	0071
T1001	0004
T1002	0004
T1002A	0004
T1003	0004
T1003-521	0144
T1004	0004
T1005	0004
T1006	0004
T1007	0004
T1007(ZENITH)	0111
T1008	0004
T1008-834	0079
T1009	0004
T1010	0004
T1010B7T	OEM
T1010BH	0593
T1010BJ	0593
T1010DH	0576
T1010DJ	0576
T1010MH	0594
T1010MJ	0594
T1010NH	0875
T1010NJ	0875
T1011	0050
T1012	0050
T1012BH	0593
T1012BJ	0593
T1012DH	0576
T1012DJ	0576
T1012MH	0594
T1012MJ	0935
T1012NH	0875
T1012NJ	0059
T1013	0004
T1013BH	0593
T1013BJ	0593
T1013DH	0576
T1013DJ	0576
T1013MH	0594
T1013MJ	0594
T1013NH	0875
T1013NJ	0875
T1015B7T	OEM
T1023	0004
T1025	OEM
T1028	0050
T1032	0050
T1033	0050
T-1034	0050
T1034	0050
T1036	0004
T1037	0004
T1038	0050
T1039	OEM
T1040	0085
T1041	0085
T1042	0004
T1043	0004
T1046	0004
T1047	0004
T1050	OEM
T1076	0004
T1085	0015
T1106	0007
T1159	OEM
T1166	0050
T1167	0160
T1168	0160
T1200N1200TOF	OEM
T1200N1400TOF	OEM
T1200N1600TOF	OEM
T1202	0004
T1202(GE)	0007
T1203	0004
T1208	0325
T1210BH	0772
T1210DH	0744
T1210MH	0617
T1210NH	0875
T1212BH	0772
T1212BJ	0772
T1212DH	0744
T1212DJ	0744
T1212MH	0617
T1212MJ	0617
T1212NH	0875
T1212NJ	0875
T1213BH	0772
T1213BJ	0772
T1213DH	0744
T1213DJ	0744
T1213MH	0617
T1213MJ	0617
T1213NH	0875
T1213NJ	0875
T1224	0050
T1225	0050
T1232	0050
T1233	0050
T1250	0050
T1251	0050
T1275	0126
T1276	0126
T1282	OEM
T1289	0280
T1291	0280
T1298	0050
T1299	0050
T1300	0004
T1305	0050
T1306	0050
T1310	0004
T1312	0280
T1314	0050
T1317	OEM
T1322	0280
T1326	0280
T1327	0004
T1328	0004
T1334	0004
T1340A3H	0086
T1340A3I	0016
T1340A3J	0016
T1340A3K	0016
T1340A31	0016
T1341A3K	0111
T1342	0004
T1343	OEM
T1344	OEM
T1346	0004
T1352	0004
T1363	0004
T1364	0004
T1366	0085
T1366A	0085
T1367	0085
T1367A	0085
T1368	0085

DEVICE TYPE	REPL CODE
T1368A	0085
T1369	0085
T1369A	0085
T1370	0085
T1370A	0085
T1381	OEM
T1382	OEM
T1383	OEM
T1387	0050
T1388	0136
T1389	0050
T1390	0050
T1391	0050
T1396	0847
T1397	0847
T1400	0050
T1401	0050
T1402	0111
T1403	0050
T1408	0224
T1410	0007
T1413	0016
T1414	0016
T1415	0079
T1416	0079
T1417	0016
T1447	OEM
T1450	0015
T1454	0050
T1459	0050
T1460	0136
T1461	0050
T1474	0280
T1480E	0102
T1486	0042
T1487	0042
T1495	0016
T1501	OEM
T1510	0280
T1512BJ	0772
T1512DJ	0744
T1512MJ	0617
T1512NJ	0875
T1513BJ	0772
T1513DJ	0744
T1513MJ	0617
T1513NJ	0875
T1524BRN	0136
T1524BRN/RED	0136
T1546	0004
T1548	0050
T1558	OEM
T1559	0004
T1573	0004
T1574	0004
T1577	0004
T1581	OEM
T1583	0004
T1591	OEM
T1593	0004
T1594	0004
T1595	0004
T1596	0004
T1597	0004
T1598	0004
T1599	0004
T1601	0160
T1602	0037
T1612BH	5323
T1612DH	5326
T1612MH	1659
T1612NH	1967
T1613BH	5323
T1613DH	5326
T1613MH	1659
T1613NH	1967
T1618	0050
T1623	OEM
T1642B	0016
T1654	0136
T1654BLU	0136
T1655	OEM
T1656	OEM
T1657	0050
T1670	OEM
T1684	OEM
T1685	OEM
T1686	OEM
T1687	OEM
T1688	OEM
T1689	OEM
T1690	0050
T1691	0050
T1692	0050
T1706	0086
T1706A	0086
T1706B	0086
T1706C	0086
T1711A	0136
T1731	0212
T1737	0050
T1738	0050
T1740	0004
T1746	0016
T1746A	0016
T1746B	0016
T1746C	0016
T1748	0016
T1748A	0016
T1748B	0016
T1748C	0016
T1756	OEM
T1788	0050
T1789	OEM
T1796	OEM
T1802	0016
T1802A	0016
T1802B	0016
T1804	0016
T1805	0016
T1806	OEM
T1807	OEM
T1808	0126
T1808A	0126
T1808B	0126
T1808C	0126
T1808D	0126
T1808E	0126
T1810	0016
T1810B	0016
T1811	0086
T1811E	0086
T1811G	0086
T1814	0050
T1822	OEM
T1826	OEM
T1831	0050
T1832	OEM
T1833	OEM
T1850	OEM
T1851	OEM
T1851M	OEM
T1858	OEM
T1859	OEM
T1866	OEM
T1871	OEM
T-1877	0280
T1877	0280
T1885	OEM
T1886	OEM
T1887	OEM
T1888	OEM
T1889	OEM
T1890	OEM
T1891	OEM
T1892	OEM
T1893	OEM
T1895	OEM
T1902	0280
T1903	0004
T1904	0004
T1905	0050
T1909	0016
T1930	0050
T1937	OEM
T1947	OEM
T1961	0004
T1967	OEM
T1973	OEM
T1992	OEM
T2015	0050
T2016	0050
T2017	0050
T2018	OEM
T2019	0050
T2020	0050
T2021	0050
T2022	0050
T2023	OEM
T2024	0050
T2025	0050
T2026	0050
T-2028	0050
T2028	0050
T-2029	0050
T2029	0050
T-2030	0050
T2030	0050
T-2038	0280
T2038	0280
T-2039	0280
T2039	0280
T-2040	0280
T2040	0280
T2050	OEM
T2057	OEM
T2058	OEM
T2059	OEM
T2060	OEM
T2061	OEM
T-2062	0102
T2062	0802
T2071	OEM
T2088	OEM
T2089	OEM
T-2091	0280
T2091	0280
T2110	OEM
T2119	OEM
T-2122	0211
T2122	0280
T2144	OEM
T2145	OEM
T2159	0004
T2172	0280
T2173	0280
T2186	OEM
T2187	OEM
T2191	0050
T2198	OEM
T2211	OEM
T2256	0280
T2257	0280
T2258	0280
T2259	0280
T2260	0280
T2261	0280
T2299	OEM
T2300A	2352
T2300B	2353
T2300C	2356
T2300D	2356
T2300E	OEM
T2300F	2352
T2300PA	2352
T2300PB	2353
T2300PC	2356
T2300PD	2356
T2300PF	2352
T2301A	2352
T2301B	2353
T2301C	OEM
T2301D	2356
T2301E	OEM
T2301F	2352
T2301G	OEM
T2301PA	2352
T2301PA5	OEM
T2301PB	2353
T2301PB5	2353
T2301PC	OEM
T2301PC5	OEM
T2301PD	2356
T2301PD5	2356
T2301PE	OEM
T2301PE5	OEM
T2301PF	OEM
T2301PM	OEM
T2301PM5	OEM
T2302A	2352
T2302B	2353
T2302C	2356
T2302D	2356
T2302E	OEM
T2302F	2352
T2302PA	2352
T2302PA5	2352
T2302PB	2353
T2302PB5	2353
T2302PC	OEM
T2302PC5	OEM
T2302PD	2356
T2302PD5	2356
T2302PE	OEM
T2302PE5	OEM
T2302PF	OEM
T2302PF5	OEM
T2302PM	OEM
T2302PM5	OEM
T2303C	OEM
T2303E	OEM
T2303F	2383
T2304B	2353
T2304D	2387
T2305B	2394
T2305D	1413
T2306A	2383
T2306B	2394
T2306D	1413
T2310A	2383
T2310B	2353
T2310D	1413
T2312A	2352
T2312B	2394
T2312D	1413
T2312F	2383
T2313A	2383
T2313B	2394
T2313D	1413
T2313F	OEM
T2313M	2387
T2320A	2300
T2320B	2300
T2320C	2300
T2320D	2300
T2320E	OEM
T2320F	2300
T2322	0050
T2322A	2300
T2322B	2300
T2322C	2300
T2322D	2300
T2322E	OEM
T2322F	2300
T2323	0050
T2323A	2300
T2323B	2300
T2323C	2300
T2323D	2300
T2323E	OEM
T2323F	2300
T2323M	OEM
T2324	0050
T2327	OEM
T2327A	2300
T2327B	2300
T2327C	2300
T2327D	2300
T2327E	OEM
T2327F	2300
T2329	OEM
T2330	OEM
T2331	OEM
T2340	OEM
T2351	OEM
T2352	OEM
T-2357	0150
T2357	0150
T2363	OEM
T2364	0050
T2370	2501
T2379	0050
T2384	0050
T2392	OEM
T2393	OEM
T2401PF5	OEM
T2415	OEM
T2417	OEM
T2418	OEM
T2439	0004
T2440	0004
T2441	0004
T2446	0016
T2448	OEM
T2452	OEM
T2453	OEM
T2454	OEM
T2455	OEM
T2469	OEM
T2470	OEM
T2471	OEM
T2472	OEM
T2478	OEM
T2479	OEM
T2490	OEM
T2491	OEM
T2492	OEM
T2500B	0555
T2500D	0555
T2500E	OEM
T2500M	OEM
T2500N	OEM
T2500S	OEM
T2506B	OEM
T2506D	OEM
T-2508	2194
T2508	2194
T2512BH	5323
T2512BK	OEM
T2512DH	5326
T2512DK	0915
T2512MH	1659
T2512MK	0915
T2512NH	1967
T2512NK	0915
T2513BH	5323
T2513BK	0915
T2513DH	5326
T2513DK	0915
T2513MH	1659
T2513MK	0915
T2513NH	1967
T2513NK	OEM
T2515	0004
T2517	0004
T2560	OEM
T2578	OEM
T2579	OEM
T2580	OEM
T2588	OEM
T2589	OEM
T2606B	OEM
T2606D	OEM
T2606DF	OEM
T2610	OEM
T2611	OEM
T2616DF	OEM
T2634	0144
T2679	OEM
T2691	OEM
T2700B	0411
T2700D	0576
T2706B	4223
T2706D	4223
T2710B	0411
T2710D	0415
T2716B	4223
T2716D	4223
T2788	0050
T2800A	OEM
T2800B	0572
T2800C	0555
T2800D	0555
T2800E	0555
T2800F	OEM
T2800M	0555
T2801A	OEM
T2801B	0593
T2801C	0574
T2801D	0576
T2801DF	0577
T2801E	0577
T2801F	OEM
T2801M	0594
T2802A	OEM
T2802B	0593
T2802C	0574
T2802D	0576
T2802E	0577
T2802F	OEM
T2802M	0594
T2806B	0593
T2806C	OEM
T2806D	0576
T2806M	OEM
T2850A	0555
T2850B	0555
T2850D	0555
T2850E	OEM
T2850F	OEM
T2850M	OEM
T2851B	0555
T2851C	0555
T2851D	0555
T2851DF	OEM
T2851E	0555
T2856B	OEM
T2856C	OEM
T2856D	OEM
T2856E	OEM
T2856M	OEM
T2857	0144
T2878	0050
T2896	0050
T2945	0050
T2946	0050
T3000	OEM
T3002	OEM
T3003	OEM
T3004	OEM
T3005	0004
T3321	0004
T3322	0004
T3530	0144
T3536	0144
T3539	0144
T3565	0079
T-3568	0007
T3568	0127
T3568(RCA)	0144
T3570	0037
T3601	0079
T3601(RCA)	0127
T3889	OEM
T3890	OEM
T3891	OEM
T3892	OEM
T3893	OEM
T3899	OEM
T3900	OEM
T3901	OEM
T3902	OEM
T3902B	OEM
T3903	OEM
T3909	OEM
T3910	OEM
T3911	OEM
T3912	OEM
T3913	OEM
T4012BK	0915
T4012DK	0915
T4012MK	OEM
T4012NK	OEM
T4013BK	0915
T4013DK	0915
T4013MK	0915
T4013NK	OEM
T4100E	1891
T4100F	1891
T4100M	1891
T4101E	1891
T4101F	1064
T4101M	1891
T4103B	OEM
T4103D	OEM
T4104B	OEM
T4104D	OEM
T4105B	OEM
T4105D	OEM
T4106B	OEM
T4106D	OEM
T4106M	OEM
T4107B	OEM
T4107D	OEM
T4110E	OEM
T4110F	OEM
T4110M	0279
T4111E	OEM
T4111F	OEM
T4111M	0279
T4113B	0154
T4113D	0147
T4113E	OEM
T4113M	OEM
T4114B	0154
T4114D	0147
T4114E	OEM
T4114M	OEM
T4115B	0154
T4115D	0147
T4115E	OEM
T4115M	OEM
T4116B	0154
T4116D	0147
T4116M	OEM
T4117B	0154
T4117D	0147
T4117M	OEM
T4120B	3199
T4120D	3207
T4120E	OEM
T4120F	OEM
T4120M	3222
T4120N	OEM
T4120S	OEM
T4121B	OEM
T4121D	OEM
T4121E	OEM
T4121F	OEM
T4121M	0279
T4121N	OEM
T4121S	OEM
T4126B	OEM
T4126D	OEM
T4126M	OEM
T4127B	OEM
T4127D	OEM
T4127M	OEM
T4130B	6373
T4130D	OEM
T4130M	OEM
T4131B	OEM
T4131D	OEM
T4131M	OEM
T4140B	0908
T4140D	0908
T4140M	0908
T4141B	6373
T4141D	6373
T4141M	5673
T4150B	0908
T4150D	0908
T4150M	5673
T4151B	6373
T4151D	6373
T4151M	OEM
T4446	OEM
T4590	0123
T4700B	4244
T4700D	4223
T4700E	OEM
T4700F	OEM
T4706B	0614
T4706D	0147
T5135-5	OEM
T5135-6	OEM
T5135-G	OEM
T5135-M	OEM
T5135-X	OEM
T5140-5	OEM
T5140-6	OEM
T5140-G	OEM
T5140-M	OEM
T5140-X	OEM
T5145-5	OEM
T5145-6	OEM
T5145-G	OEM
T5145-M	OEM
T5145-X	OEM
T5627	OEM
T5901	OEM
T5902	OEM
T5903	OEM
T5909	OEM
T6000B	0593
T6000C	0576
T6000D	4980
T6000E	4981
T6000F	4982
T6000M	0594
T6001B	OEM
T6001C	OEM
T6001D	OEM
T6001E	OEM
T6001F	OEM
T6001M	OEM
T6006B	0593
T6006C	0576
T6006D	0576
T6006E	0594
T6006M	OEM
T6028	0136
T6029	0136
T6030	0136
T6031	0136
T6032	0136
T6058	0050
T6260B	5440
T6260C	5440
T6260D	5440
T6260E	0915
T6260M	0908
T6261B	0908
T6261C	0908
T6261D	0908
T6261E	0908
T6261M	0908
T6400B	1064
T6400D	1317
T6400M	1891
T6400N	OEM
T6400S	OEM
T6401B	1064
T6401D	1891
T6401E	OEM
T6401F	OEM
T6401M	1891
T6401N	OEM
T6401S	OEM
T6404B	OEM
T6404D	OEM
T6404E	OEM
T6405B	OEM
T6405D	OEM
T6405E	OEM
T6406B	OEM
T6406D	OEM
T6406E	OEM
T6406M	OEM
T6407B	OEM
T6407D	OEM
T6407E	OEM
T6407M	OEM
T6410B	5625
T6410D	3245
T6410M	5627
T6410N	5628
T6411B	2017
T6411D	2017
T6411E	5630
T6411F	5631
T6411M	2018
T6411N	OEM
T6411S	OEM
T6414B	2015
T6414D	2017
T6415B	3150
T6415D	2018
T6416B	2015
T6416D	2017
T6416M	2018
T6417B	2015
T6417D	2017
T6417M	2018
T6420	2018
T6420B	3199
T6420D	3207
T6420E	3222
T6420F	3199
T6420M	3222
T6420N	OEM
T6421B	3199
T6421D	3207
T6421E	3222
T6421F	3199
T6421M	3222
T6421N	OEM
T6421S	OEM
T6426B	OEM
T6426D	OEM
T6426M	OEM
T6427B	OEM
T6427D	OEM
T6427M	OEM
T6430B	OEM
T6430D	OEM
T6430M	OEM
T6431B	OEM
T6431D	OEM
T6431M	OEM
T6440B	5673
T6440D	5673
T6440M	5673
T6441B	0908
T6441D	0908
T6441M	5673
T6450B	5673
T6450D	5673
T6450M	OEM
T6451B	5673
T6451D	0908
T6451M	5673
T6506B	OEM
T6676AS	OEM
T7400B1	0232
T7400D1	0232
T7400D2	0232
T7401B1	0268
T7401D1	0268
T7401D2	0268
T7402B1	0314
T7402D1	0314
T7402D2	0314
T7403B1	0335
T7403D1	0335
T7403D2	0335
T7404B1	0361
T7405B1	0385
T7406B1	1206
T7407B1	1339
T7408B1	0467
T7408D1	0467
T7408D2	0467
T7409B1	0492
T7409D1	0492
T7409D2	0492
T7410B1	0512
T7410D1	0512
T7410D2	0512
T7416B1	1349
T7420B1	0697
T7420D1	0697
T7420D2	0697

If replacement code is OEM, contact original manufacturer for replacement.

DEVICE TYPE	REPL CODE
T7426B1	0803
T7426D1	0803
T7426D2	0803
T7428B1	4180
T7428D1	4180
T7428D2	4180
T7430B1	0873
T7430D1	0873
T7430D2	0873
T7433B1	4193
T7433D1	4193
T7433D2	4193
T7440B1	1024
T7440D1	1024
T7440D2	1024
T7441AB1	1038
T7441AD1	1038
T7441AD2	1038
T7442B1	1052
T7442D1	1052
T7442D2	1052
T7443B1	1060
T7443D1	1060
T7443D2	1060
T7444B1	1072
T7444D1	1072
T7444D2	1072
T7450B1	0743
T7450D1	0743
T7450D2	0743
T7451B1	1168
T7451D1	1168
T7451D2	1168
T7453B1	1185
T7453D1	1185
T7453D2	1185
T7454B1	1202
T7454D1	1202
T7454D2	1202
T7460B1	1275
T7460D1	1275
T7460D2	1275
T7472B1	1426
T7472D1	1426
T7472D2	1426
T7473B1	1172
T7473D1	1172
T7473D2	1172
T7474B1	1313
T7474D1	1313
T7474D2	1313
T7475B1	1432
T7476B1	1158
T7476D1	1158
T7476D2	1158
T7481B1	3121
T7483B1	0117
T7484B1	OEM
T7486B1	1368
T7486D1	1368
T7486D2	1368
T7490B1	1208
T7490D1	1208
T7490D2	1208
T7492B1	0834
T7492D1	0834
T7492D2	0834
T7493B1	0569
T7493D1	0569
T7493D2	0569
T7809CT	OEM
T7815A-0102	OEM
T8150	OEM
T8401B	OEM
T8401D	OEM
T8401M	OEM
T8430B	OEM
T8430D	OEM
T8430M	OEM
T8889A	OEM
T9011A1C	0079
T9011A1G	0079
T9011AZ	0079
T9011EF	0144
T9011G	0144
T9011G(CD)	0079
T9011G(EF)	0079
T9011GEF	0144
T9011H	0144
T9011H(EF)	0079
T9011I(EF)	0079
T9011J(GH)	0079
T9016F	0144
T9016H	0144
T9418	0018
T9423	0224
T9468	0532
T9631	0079
T9681	0037
T10010	0004
T10085	0004
T10144	0015
T10175	0015
T10185	0015
T10195	0277
T10453	0015
T11087T	OEM
T11618	0004
T13000	0004

DEVICE TYPE	REPL CODE
T13015	0086
T13029	0160
T14881	0023
T18218	OEM
T18231	OEM
T21237	0143
T21238	0143
T21271	0143
T21312	0015
T21313	0143
T21333	0015
T21334	0012
T21507	0015
T21600	0133
T21601	0030
T21602	0015
T21638	0229
T21639	0012
T21649	0015
T21679	0015
T30155	0229
T30155-001	0015
T30155-1	0015
T42692-001	0015
T42692-1R	0015
T50339A	0004
T50631	0004
T50818	0136
T50931B	0211
T50933B	0164
T50944	0280
T51573A	0211
T52054	0136
T52147	0280
T52147Z	0280
T52148Z	0280
T-52149	0280
T52149	0280
T52149Z	0280
T52150	0004
T52150Z	0004
T52151	0004
T52151Z	0004
T52159	0004
T59235A	0016
T59247	0004
T59249	0004
T59276	0600
T59277	0600
T69901	OEM
T74107B1	0942
T74180B1	1829
T74193B1	1926
T78010AP	2806
T78012AP	OEM
T81901	OEM
T152148	0280
T164213	0086
T250117GR5	3133
T1003521	0144
T1004671	0016
T1008834	0016
T4080016	0469
T4080116	0469
T4080216	0469
T4080316	0725
T4080416	0725
T4080516	0725
T5070140B4AQ	0526
T5070170B4AQ	0526
T5070240B4AQ	5894
T5070270B4AQ	0526
T5070280B4AQ	0526
T5070340B4AQ	0526
T5070370B4AQ	0526
T5070380B4AQ	0526
T5070440B4AQ	0526
T5070470B4AQ	0526
T5070480B4AQ	0526
T5070540B4AQ	0526
T5070570B4AQ	0526
T5070580B4AQ	0526
T5070640B4AQ	0526
T5070670B4AQ	0526
T5070680B4AQ	0526
T5070740B4AQ	0108
T5070770B4AQ	0108
T5070780B4AQ	0108
T5070840B4AQ	0108
T5070870B4AQ	0108
T5070880B4AQ	0108
T5070940B4AQ	0108
T5070970B4AQ	0108
T5070980B4AQ	0108
T5071040B4AQ	0108
T5071070B4AQ	0108
T5071080B4AQ	0108
T5071140B4AQ	0108
T5071170B4AQ	0108
T5071180B4AQ	0108
T5071240B4AQ	0108
T5071270B4AQ	0108
T5071280B4AQ	0108
T6000713	0738
T6000715	0738
T6000813	0738
T6000815	0738
T6000913	0738
T6000915	0738

DEVICE TYPE	REPL CODE
T6001013	0738
T6001015	0738
T6001113	0738
T6001115	0738
T6001213	0738
T6001215	0738
T6070113B4BT	0428
T6070115B4BT	0428
T6070118B4BT	0428
T6070213B4BT	0428
T6070215B4BT	0428
T6070218B4BT	0428
T6070313B4BT	0428
T6070315B4BT	0428
T6070413B4BT	0428
T6070415B4BT	0428
T6070418B4BT	0428
T6070513B4BT	0428
T6070515B4BT	0428
T6070613B4BT	0428
T6070615B4BT	0428
T6070618B4BT	0428
T6071013B4BT	1871
T6071015B4BT	1871
T6071113B4BT	1871
T6071115B4BT	1871
T6071118B4BT	1871
T6071213B4BT	1871
T6071215B4BT	1871
T6071218B4BT	1871
T6100113	0738
T6100115	0738
T6100213	0738
T6100215	0738
T6100313	0738
T6100315	0738
T6100413	0738
T6100415	0738
T6100513	0738
T6100515	0738
T6100613	0738
T6100615	0738
T7000125	1900
T7000130	1900
T7000225	1900
T7000230	1900
T7000425	0738
T7000430	0738
T7000625	0738
T7000630	0738
T7000825	0738
T7000830	0738
T7000835	0738
T7001025	0738
T7001030	0738
T7001225	0738
T7001230	0738
T7070128B4BY	1700
T7070228B4BY	1700
T7070328B4BY	1700
T7070428B4BY	1700
T7070528B4BY	1700
T7070625B4BY	1700
T7070628B4BY	1700
T7070630B4BY	1700
T7070725B4BY	0345
T7070728B4BY	0345
T7070730B4BY	0345
T7070825B4BY	0345
T7070828B4BY	0345
T7070830B4BY	0345
T7070925B4BY	0345
T7070930B4BY	0345
T7071025B4BY	0345
T7071030B4BY	0345
T7071125B4BY	0345
T7071130B4BY	0345
T7071225B4BY	0345
T7071230B4BY	0345
T7270135B4DN	0769
T7270235B4DN	0769
T7270335B4DN	0769
T7270435B4DN	0769
T7270535B4DN	0769
T7270635B4DN	0769
T7270735B4DN	0483
T7270835B4DN	0483
T7270935B4DN	0483
T7271035B4DN	6228
T7271135B4DN	0483
T7271235B4DN	0483
T22000104E	0325
T107108064AQ	OEM
T107108074AQ	OEM
T107114034AQ	OEM
T220001040	0325
T400001006	0469
T400001008	2519
T400001606	2188
T400001608	2188
T400002206	0731
T400002208	0731
T400011006	0741
T400011008	0741
T400011606	0567
T400011608	0567
T400012206	1584
T400012208	0712
T400021006	0469

DEVICE TYPE	REPL CODE
T400021008	0745
T400021606	0762
T400021608	0762
T400022206	5380
T400022208	0469
T400031006	0722
T400031008	0747
T400031606	0740
T400031608	0740
T400032206	1563
T400032208	0722
T400041006	0722
T400041008	0747
T400041606	0740
T400041608	0740
T400042206	1563
T400042208	0722
T400051006	0725
T400051008	0752
T400051606	0764
T400051608	0764
T400052206	3186
T400052208	0725
T400061006	0764
T400061008	0752
T400061606	0764
T400061608	0764
T400062206	3186
T400062208	0725
T400071006	0679
T400071008	0679
T400071606	0679
T400071608	0679
T400072206	1542
T400072208	0750
T400081006	0766
T400081008	0679
T400081606	0766
T400081608	0679
T400082206	1542
T400082208	0750
T400091006	0679
T400091008	0679
T400091606	0679
T400091608	0679
T400092208	OEM
T400101006	0679
T400101008	0679
T400101600	OEM
T400101606	0679
T400101608	0679
T400102208	OEM
T400111006	0679
T400111008	0679
T400111606	0679
T400111608	0679
T400112208	OEM
T400121006	0679
T400121008	0679
T400121606	0679
T400121608	0679
T400122208	OEM
T407185152	0037
T408001006	OEM
T408002206	OEM
T408011006	OEM
T408012206	OEM
T408021006	OEM
T408022206	OEM
T408031006	OEM
T408032206	OEM
T408041006	OEM
T408042206	OEM
T408051006	OEM
T408052206	OEM
T408061006	OEM
T408062206	OEM
T408071008	OEM
T408072206	OEM
T408081008	OEM
T408082206	OEM
T500014005(T083)	0655
T500014005(T094)	0526
T500014005AA	OEM
T500014005AQ	OEM
T500018005(T083)	0641
T500018005(T094)	0608
T500018005AA	OEM
T500018005AQ	OEM
T500024005(T083)	0655
T500024005(T094)	0526
T500024005AQ	OEM
T500028005(T083)	0641
T500028005(T094)	0608
T500028005AA	OEM
T500028005AQ	OEM
T500034005(T083)	0655
T500034005(T094)	0526
T500034005AA	OEM
T500034005AQ	OEM
T500038005(T083)	0217
T500038005(T094)	0610
T500038005AA	OEM
T500038005AQ	OEM
T500044005(T083)	0655
T500044005(T094)	0526
T500044005AA	OEM
T500044005AQ	OEM
T500048005(T083)	0217

DEVICE TYPE	REPL CODE
T500048005(T094)	0610
T500048005AA	OEM
T500048005AQ	OEM
T500054005(T083)	0655
T500054005(T094)	0526
T500054005AA	OEM
T500054005AQ	OEM
T500058005(T083)	0217
T500058005(T094)	0610
T500058005AA	OEM
T500058005AQ	OEM
T500064005(T083)	0655
T500064005(T094)	0526
T500064005AA	OEM
T500064005AQ	OEM
T500068005(T083)	0217
T500068005(T094)	0610
T500068005AA	OEM
T500068005AQ	OEM
T500074004AQ	0468
T500074005(T083)	0020
T500074005(T094)	0108
T500074005AB	0658
T500074005AQ	0468
T500078004AQ	0468
T500078005(T083)	0658
T500078005(T094)	0468
T500078005AB	0658
T500078005AQ	0468
T500084004AQ	0468
T500084005(T083)	0020
T500084005(T094)	0108
T500084005AB	0658
T500084005AQ	0468
T500088004AQ	0468
T500088005(T083)	0658
T500088005(T094)	0468
T500088005AB	0658
T500088005AQ	0468
T500094004AQ	0468
T500094005(T083)	0020
T500094005(T094)	0108
T500094005AB	0658
T500094005AQ	0468
T500098004AQ	0468
T500098005(T083)	0658
T500098005(T094)	0468
T500098005AB	0658
T500098005AQ	0468
T500104004AQ	0468
T500104005(T083)	0020
T500104005(T094)	0108
T500104005AB	0658
T500104005AQ	0468
T500108004AQ	0468
T500108005(T083)	0658
T500108005(T094)	0468
T500108005AB	0658
T500108005AQ	0468
T500114004AQ	0468
T500114005AB	0658
T500114005AQ	0468
T500118004AQ	0468
T500118005AB	0658
T500118005AQ	0468
T500124004AQ	0468
T500124005(T083)	OEM
T500124005(T094)	0108
T500124005AB	0658
T500124005AQ	0468
T500128004AQ	0468
T500128005(T083)	0658
T500128005(T094)	0468
T500128005AA	OEM
T500128005AB	0658
T500128005AQ	0468
T500144005(T083)	OEM
T500144005(T094)	OEM
T500148005(T083)	OEM
T500148005(T094)	OEM
T500164005(T083)	OEM
T500164005(T094)	OEM
T500168005(T083)	OEM
T500168005(T094)	OEM
T504017043AA	OEM
T504017043AQ	OEM
T504017073AA	OEM
T504017073AQ	OEM
T504027043AA	OEM
T504027043AQ	OEM
T504027073AA	OEM
T504027073AQ	OEM
T504037043AA	OEM
T504037043AQ	OEM
T504037073AA	OEM
T504037073AQ	OEM
T504047043AA	OEM
T504047043AQ	OEM
T504047073AA	OEM
T504047073AQ	OEM
T504057043AA	OEM
T504057043AQ	OEM
T504057073AA	OEM
T504057073AQ	OEM
T504067043AQ	OEM
T504067073AA	OEM
T504067073AQ	OEM
T504077043AA	OEM
T504077043AQ	OEM

DEVICE TYPE	REPL CODE
T504077073AA	OEM
T504077073AQ	OEM
T504087043AA	OEM
T504087043AQ	OEM
T504087073AA	OEM
T504087073AQ	OEM
T504097043AA	OEM
T504097043AQ	OEM
T504107043AA	OEM
T504107043AQ	OEM
T505014005AA	OEM
T507014024AA	0655
T507014024AQ	0526
T507014034AA	0655
T507014034AQ	0526
T507014044AA	0655
T507014044AQ	0526
T507014054AA	0655
T507014054AQ	0526
T507014064AA	0655
T507014064AQ	0526
T507014074AA	0655
T507014074AQ	0526
T507014084AA	0655
T507014084AQ	0526
T507017024AA	0655
T507017024AQ	0526
T507017034AA	0655
T507017034AQ	0526
T507017044AA	0655
T507017044AQ	0526
T507017054AA	0655
T507017054AQ	0526
T507017064AA	0655
T507017064AQ	0526
T507017074AA	0655
T507017074AQ	0526
T507017084AA	0655
T507017084AQ	0526
T507018024AA	0655
T507018024AQ	0526
T507018034AA	0655
T507018034AQ	0526
T507018044AA	0655
T507018044AQ	0526
T507018054AA	0655
T507018054AQ	0526
T507018064AA	0655
T507018064AQ	0526
T507018074AA	0655
T507018074AQ	0526
T507018084AA	0655
T507018084AQ	0526
T507024024AA	0655
T507024024AQ	0526
T507024034AA	0655
T507024034AQ	0526
T507024044AA	0655
T507024044AQ	0526
T507024054AA	0655
T507024054AQ	0526
T507024064AA	0655
T507024064AQ	0526
T507024074AA	0655
T507024074AQ	0526
T507024084AA	0655
T507024084AQ	0526
T507027024AA	0655
T507027024AQ	0526
T507027034AA	0655
T507027034AQ	0526
T507027044AA	0655
T507027044AQ	0526
T507027054AA	0655
T507027054AQ	0526
T507027064AA	0655
T507027064AQ	0526
T507027074AA	0655
T507027074AQ	0526
T507027084AA	0655
T507027084AQ	0526
T507028024AA	0655
T507028024AQ	0526
T507028034AA	0655
T507028034AQ	0526
T507028044AA	0655
T507028044AQ	0526
T507028054AA	0655
T507028054AQ	0526
T507028064AA	0655
T507028064AQ	0526
T507028074AA	0655
T507028074AQ	0526
T507028084AA	0655
T507028084AQ	0526
T507034024AA	0655
T507034024AQ	0526
T507034034AA	0655
T507034034AQ	0526
T507034044AA	0655
T507034044AQ	0526
T507034054AA	0655
T507034054AQ	0526
T507034064AA	0655
T507034064AQ	0526
T507034074AA	0655
T507034074AQ	0526
T507034084AA	0655
T507034084AQ	0526

DEVICE TYPE	REPL CODE
T507037024AA	0655
T507037024AQ	0526
T507037034AA	0655
T507037034AQ	0526
T507037044AA	0655
T507037044AQ	0526
T507037054AA	0655
T507037054AQ	0526
T507037064AA	0655
T507037064AQ	0526
T507037074AA	0655
T507037074AQ	0526
T507037084AA	0655
T507037084AQ	0526
T507038024AA	0655
T507038024AQ	0526
T507038034AA	0655
T507038034AQ	0526
T507038044AA	0655
T507038044AQ	0526
T507038054AA	0655
T507038054AQ	0526
T507038064AA	0655
T507038064AQ	0526
T507038074AA	0655
T507038074AQ	0526
T507038084AA	0655
T507038084AQ	0526
T507044024AA	0655
T507044024AQ	0526
T507044034AA	0655
T507044034AQ	0526
T507044044AA	0655
T507044044AQ	0526
T507044054AA	0655
T507044054AQ	0526
T507044064AA	0655
T507044064AQ	0526
T507044074AA	0655
T507044074AQ	0526
T507044084AA	0655
T507044084AQ	0526
T507047024AA	0655
T507047024AQ	0526
T507047034AA	0655
T507047034AQ	0526
T507047044AA	0655
T507047044AQ	0526
T507047054AA	0655
T507047054AQ	0526
T507047064AA	0655
T507047064AQ	0526
T507047074AA	0655
T507047074AQ	0526
T507047084AA	0655
T507047084AQ	0526
T507048024AA	0655
T507048024AQ	0526
T507048034AA	0655
T507048034AQ	0526
T507048044AA	0655
T507048044AQ	0526
T507048054AA	0655
T507048054AQ	0526
T507048064AA	0655
T507048064AQ	0526
T507048074AA	0655
T507048074AQ	0526
T507048084AA	0655
T507048084AQ	0526
T507054024AA	0655
T507054024AQ	0526
T507054034AA	0655
T507054034AQ	0526
T507054044AA	0655
T507054044AQ	0526
T507054054AA	0655
T507054054AQ	0526
T507054064AA	0655
T507054064AQ	0526
T507054074AA	0655
T507054074AQ	0526
T507054084AA	0655
T507054084AQ	0526
T507057024AA	0655
T507057024AQ	0526
T507057034AA	0655
T507057034AQ	0526
T507057044AA	0655
T507057044AQ	0526
T507057054AA	0655
T507057054AQ	0526
T507057064AA	0655
T507057064AQ	0526
T507057074AA	0655
T507057074AQ	0526
T507057084AA	0655
T507057084AQ	0526
T507058024AA	0655
T507058024AQ	0526
T507058034AA	0655
T507058034AQ	0526
T507058044AA	0655
T507058044AQ	0526
T507058054AA	0655
T507058054AQ	0526
T507058064AA	0655
T507058064AQ	0526
T507058074AA	0655

DEVICE TYPE	REPL CODE
T507058074AQ	0526
T507058084AA	0655
T507058084AQ	0526
T507064024AA	0655
T507064024AQ	0526
T507064034AA	0655
T507064034AQ	0526
T507064044AA	0655
T507064044AQ	0526
T507064054AA	0655
T507064054AQ	0526
T507064064AA	0655
T507064064AQ	0526
T507064074AA	0655
T507064074AQ	0526
T507064084AA	0655
T507064084AQ	0526
T507067024AA	0655
T507067024AQ	0526
T507067034AA	0655
T507067034AQ	0526
T507067044AA	0655
T507067044AQ	0526
T507067054AA	0655
T507067054AQ	0526
T507067064AA	0655
T507067064AQ	0526
T507067074AA	0655
T507067074AQ	0526
T507067084AA	0655
T507067084AQ	0526
T507068024AA	0655
T507068024AQ	0526
T507068034AA	0655
T507068034AQ	0526
T507068044AA	0655
T507068044AQ	0526
T507068054AA	0655
T507068054AQ	0526
T507068064AA	0655
T507068064AQ	0526
T507068074AA	0655
T507068074AQ	0526
T507068084AA	0655
T507068084AQ	0526
T507074024AA	0020
T507074024AQ	0108
T507074034AA	0020
T507074034AQ	0108
T507074044AA	0020
T507074044AQ	0108
T507074054AA	0020
T507074054AQ	0108
T507074064AA	0020
T507074064AQ	0108
T507074074AA	0020
T507074074AQ	0108
T507074084AQ	0108
T507077024AA	0020
T507077024AQ	0108
T507077034AA	0020
T507077034AQ	0108
T507077044AA	0020
T507077044AQ	0108
T507077054AA	0020
T507077054AQ	0108
T507077064AA	0020
T507077064AQ	0108
T507077074AA	0020
T507077074AQ	0108
T507077084AQ	0108
T507078024AA	0020
T507078024AQ	0108
T507078034AA	0020
T507078034AQ	0108
T507078044AA	0020
T507078044AQ	0108
T507078054AA	0020
T507078054AQ	0108
T507078064AA	0020
T507078064AQ	0108
T507078074AA	0020
T507078074AQ	0108
T507078084AQ	0108
T507084024AA	0020
T507084024AQ	0108
T507084034AA	0020
T507084034AQ	0108
T507084044AA	0020
T507084044AQ	0108
T507084054AA	0020
T507084054AQ	0108
T507084064AA	0020
T507084064AQ	0108
T507084074AA	0020
T507084074AQ	0108
T507087024AA	0020
T507087024AQ	0108
T507087034AA	0020
T507087034AQ	0108
T507087044AA	0020
T507087044AQ	0108
T507087054AA	0020
T507087054AQ	0108
T507087064AA	0020
T507087064AQ	0108
T507087074AA	0020
T507087074AQ	0108
T507087084AQ	0108
T507088024AA	0020
T507088024AQ	0108
T507088034AA	0020
T507088034AQ	0108
T507088044AA	0020
T507088044AQ	0108
T507088054AA	0020
T507088054AQ	0108
T507088064AA	0020
T507088064AQ	0108
T507088074AA	0020
T507088074AQ	0108
T507088084AQ	0108
T507094024AA	0020
T507094024AQ	0108
T507094034AA	0020
T507094034AQ	0108
T507094044AA	0020
T507094044AQ	0108
T507094054AA	0020
T507094054AQ	0108
T507094064AA	0020
T507094064AQ	0108
T507094074AA	0020
T507094074AQ	0108
T507094084AQ	0108
T507097024AA	0020
T507097024AQ	0108
T507097034AA	0020
T507097034AQ	0108
T507097044AA	0020
T507097044AQ	0108
T507097054AA	0020
T507097054AQ	0108
T507097064AA	0020
T507097064AQ	0108
T507097074AA	0020
T507097074AQ	0108
T507097084AQ	0108
T507098024AA	0020
T507098024AQ	0108
T507098034AA	0020
T507098034AQ	0108
T507098044AA	0020
T507098044AQ	0108
T507098054AA	0020
T507098054AQ	0108
T507098064AA	0020
T507098064AQ	0108
T507098074AA	0020
T507098074AQ	0108
T507104024AA	0020
T507104024AQ	0108
T507104034AA	0020
T507104034AQ	0108
T507104044AA	0020
T507104044AQ	0108
T507104054AA	0020
T507104054AQ	0108
T507104064AA	0020
T507104064AQ	0108
T507104074AA	0020
T507104074AQ	0108
T507107024AA	0020
T507107024AQ	0108
T507107034AA	0020
T507107034AQ	0108
T507107044AA	0020
T507107044AQ	0108
T507107054AA	0020
T507107054AQ	0108
T507107064AA	0020
T507107064AQ	0108
T507107074AA	0020
T507107074AQ	0108
T507108024AA	0020
T507108024AQ	0108
T507108034AA	0020
T507108034AQ	0108
T507108044AA	0020
T507108044AQ	0108
T507108054AA	0020
T507108054AQ	0108
T507108064AA	0020
T507108064AQ	0108
T507108074AA	0020
T507108074AQ	0108
T507114024AA	0020
T507114024AQ	0108
T507114034AA	0020
T507114034AQ	0108
T507114044AA	0020
T507114044AQ	0108
T507114054AA	0020
T507114054AQ	0108
T507114064AA	0020
T507114064AQ	0108
T507114074AQ	0108
T507117024AA	0020
T507117024AQ	0108
T507117034AA	0020
T507117034AQ	0108
T507117044AA	0020
T507117044AQ	0108
T507117054AA	0020
T507117054AQ	0108
T507117064AA	0020
T507117064AQ	0108
T507117074AQ	0108
T507118024AA	0020
T507118024AQ	0108
T507118034AA	0020
T507118034AQ	0108
T507118044AA	0020
T507118044AQ	0108
T507118054AA	0020
T507118054AQ	0108
T507118064AA	0020
T507118064AQ	0108
T507118074AQ	0108
T507124024AA	0020
T507124024AQ	0108
T507124034AA	0020
T507124034AQ	0108
T507124044AA	0020
T507124044AQ	0108
T507124054AA	0020
T507124054AQ	0108
T507124064AA	0020
T507124064AQ	0108
T507124074AQ	0108
T507127024AA	0020
T507127024AQ	0108
T507127034AA	0020
T507127034AQ	0108
T507127044AA	0020
T507127044AQ	0108
T507127054AA	0020
T507127054AQ	0108
T507127064AA	0020
T507127064AQ	0108
T507127074AQ	0108
T507128024AA	0020
T507128024AQ	0108
T507128034AA	0020
T507128034AQ	0108
T507128044AA	0020
T507128044AQ	0108
T507128054AA	0020
T507128054AQ	0108
T507128064AA	0020
T507128064AQ	0108
T507128074AQ	0108
T507138044AB	OEM
T507138044AQ	OEM
T507144024AA	OEM
T507144024AQ	OEM
T507144034AA	OEM
T507144034AB	OEM
T507144034AQ	OEM
T507144044AA	OEM
T507144044AQ	OEM
T507144054AA	OEM
T507144054AB	OEM
T507144054AQ	OEM
T507144064AA	OEM
T507144064AQ	OEM
T507147024AA	OEM
T507147024AQ	OEM
T507147034AA	OEM
T507147034AB	OEM
T507147034AQ	OEM
T507147044AA	OEM
T507147044AB	OEM
T507147044AQ	OEM
T507147054AA	OEM
T507147054AB	OEM
T507147054AQ	OEM
T507147064AA	OEM
T507147064AQ	OEM
T507148024AA	OEM
T507148024AQ	OEM
T507148034AA	OEM
T507148034AQ	OEM
T507148044AA	OEM
T507148044AQ	OEM
T507148054AA	OEM
T507148054AQ	OEM
T507148064AA	OEM
T507148064AQ	OEM
T507154024AA	OEM
T507154024AQ	OEM
T507154034AA	OEM
T507154034AQ	OEM
T507154044AA	OEM
T507154044AQ	OEM
T507154054AA	OEM
T507154054AQ	OEM
T507157024AA	OEM
T507157024AQ	OEM
T507157034AA	OEM
T507157034AQ	OEM
T507157044AA	OEM
T507157044AQ	OEM
T507157054AA	OEM
T507157054AQ	OEM
T507158024AA	OEM
T507158024AQ	OEM
T507158034AA	OEM
T507158034AQ	OEM
T507158044AA	OEM
T507158044AQ	OEM
T507158054AA	OEM
T507158054AQ	OEM
T508097007AA	OEM
T508107007AA	OEM
T508257007AA	OEM
T510005004AB	0217
T510005004AQ	0217
T510005005AB	0217
T510005005AQ	0217
T510005007(TO83)	0655
T510005007(TO94)	0526
T510005007AB	0217
T510008004AB	0217
T510008004AQ	0641
T510008005AB	0217
T510008005AQ	0217
T510008007(TO83)	0655
T510008007(TO94)	0526
T510008007AB	0217
T510008007AQ	0217
T510015004AB	0217
T510015004AQ	0217
T510015005AB	0217
T510015005AQ	0217
T510015007(TO83)	0655
T510015007(TO94)	0526
T510015007AB	0217
T510015007AQ	0217
T510018004AB	0217
T510018004AQ	0217
T510018005AB	0217
T510018005AQ	0217
T510018007(TO83)	0655
T510018007(TO94)	0526
T510018007AB	0217
T510018007AQ	0217
T510025004AB	0217
T510025004AQ	0217
T510025005AB	0217
T510025005AQ	0217
T510025007(TO83)	0655
T510025007(TO94)	0526
T510025007AB	0217
T510025007AQ	0217
T510028004AB	0217
T510028004AQ	0217
T510028005AB	0217
T510028005AQ	0217
T510028007(TO83)	0655
T510028007(TO94)	0526
T510028007AB	0217
T510028007AQ	0217
T510035004AB	0217
T510035004AQ	0217
T510035005AB	0217
T510035005AQ	0217
T510035007(TO83)	0655
T510035007(TO94)	0526
T510035007AB	0217
T510035007AQ	0217
T510038004AB	0217
T510038004AQ	0217
T510038005AB	0217
T510038005AQ	0217
T510038007(TO83)	0655
T510038007(TO94)	0526
T510038007AB	0217
T510038007AQ	0217
T510045004AB	0217
T510045004AQ	0217
T510045005AB	0217
T510045005AQ	0217
T510045007(TO83)	0655
T510045007(TO94)	0526
T510045007AB	0217
T510045007AQ	0217
T510048004AB	0217
T510048004AQ	0217
T510048005AB	0217
T510048005AQ	0217
T510048007(TO83)	0655
T510048007(TO94)	0526
T510048007AB	0217
T510048007AQ	0217
T510055004AB	0217
T510055004AQ	0217
T510055005AB	0217
T510055005AQ	0217
T510055007(TO83)	0655
T510055007(TO94)	0526
T510055007AB	0217
T510055007AQ	0217
T510058004AB	0217
T510058004AQ	0217
T510058005AB	0217
T510058005AQ	0217
T510058007(TO83)	0655
T510058007(TO94)	0526
T510058007AB	0217
T510058007AQ	0217
T510065004AB	0217
T510065004AQ	0217
T510065005AB	0217
T510065005AQ	0217
T510065007(TO83)	0655
T510065007(TO94)	0526
T510065007AB	0217
T510065007AQ	0217
T510068004AB	0217
T510068004AQ	0217
T510068005AB	0217
T510068005AQ	0217
T510068007(TO83)	0655
T510068007(TO94)	0526
T510068007AB	0217
T510068007AQ	0217
T520011305	0192
T520021305	0192
T520031305	0159
T520041305	0159
T520051305	0159
T520061305	0159
T520071305	0096
T520081305	0096
T520091305	0096
T520101305	0096
T520121305	0096
T520141305	OEM
T520161305	OEM
T527011324DN	0459
T527011334DN	0459
T527011344DN	0459
T527011354DN	0459
T527011364DN	0459
T527011374DN	0459
T527011384DN	0459
T527021324DN	0459
T527021334DN	0459
T527021354DN	0459
T527021364DN	0459
T527021374DN	0459
T527021384DN	0459
T527031324DN	0459
T527031334DN	0459
T527031344DN	0459
T527031354DN	0459
T527031364DN	0459
T527031374DN	0459
T527031384DN	0459
T527041324DN	0459
T527041334DN	0459
T527041344DN	0459
T527041354DN	0459
T527041364DN	0459
T527041374DN	0459
T527041384DN	0459
T527051324DN	0459
T527051334DN	0459
T527051344DN	0459
T527051354DN	0459
T527051364DN	0459
T527051374DN	0459
T527051384DN	0459
T527061324DN	0459
T527061334DN	0459
T527061344DN	0459
T527061354DN	0459
T527061364DN	0459
T527061374DN	0459
T527061384DN	0459
T527071324DN	0587
T527071334DN	0587
T527071344DN	0587
T527071354DN	0587
T527071364DN	0587
T527071374DN	0587
T527081324DN	0587
T527081334DN	0587
T527081344DN	0587
T527081354DN	0587
T527081364DN	0587
T527081374DN	0587
T527091324DN	0587
T527091334DN	0587
T527091344DN	0587
T527091354DN	0587
T527091364DN	0587
T527091374DN	0587
T527101324DN	0587
T527101334DN	0587
T527101344DN	0587
T527101354DN	0587
T527101374DN	0587
T527111324DN	0587
T527111334DN	0587
T527111344DN	0587
T527111354DN	0587
T527111364DN	0587
T527121284DN	0587
T527121324DN	0587
T527121334DN	0587
T527121344DN	0587
T527121354DN	0587
T527121364DN	0587
T527141324DN	OEM
T527141334DN	OEM
T527141344DN	OEM
T527141354DN	OEM
T527141364DN	OEM
T527151324DN	OEM
T527151334DN	OEM
T527151354DN	OEM
T600011304	1084
T600011504	1084
T600011804	1084
T600021304	1084
T600021504	1084
T600021804	1084
T600031304	1086
T600031504	1086
T600031804	1086
T600041304	1086
T600041504	1086
T600041804	1086
T600051304	1086
T600051504	1086
T600051804	1086
T600061304	1086
T600061504	1086
T600061804	1086
T600071304	1102
T600071304BT	0738
T600071504	1102
T600071504BT	0738
T600071804	1102
T600081304	1102
T600081304BT	0738
T600081504	1102
T600081504BT	0738
T600081804	1102
T600081804BT	1102
T600091304	1102
T600091304BT	0738
T600091504	1102
T600091804	1102
T600091804BT	1102
T600101304	1102
T600101304BT	0738
T600101504	1102
T600101504BT	0738
T600101804	1102
T600101804BT	1102
T600111304BT	0738
T600111504BT	0738
T600111804BT	1102
T600121304	1102
T600121304BT	0738
T600121504	1102
T600121504BT	0738
T600121804	1102
T600121804BT	1102
T600141304	OEM
T600141504	OEM
T600141804	OEM
T600161304	OEM
T600161504	OEM
T600161804	OEM
T604011842BT	OEM
T604011852BT	OEM
T604021842BT	OEM
T604021852BT	OEM
T604031842BT	OEM
T604031852BT	OEM
T604041842BT	OEM
T604041852BT	OEM
T604051842BT	OEM
T604051852BT	OEM
T604061842BT	OEM
T604061852BT	OEM
T604071842BT	OEM
T604071852BT	OEM
T604081842BT	OEM
T604081852BT	OEM
T604091842BT	OEM
T604101842BT	OEM
T607011324BT	0428
T607011334BT	0428
T607011344BT	0428
T607011354BT	0428
T607011364BT	0428
T607011374BT	0428
T607011384BT	0428
T607011524BT	0428
T607011534BT	0428
T607011544BT	0428
T607011554BT	0428
T607011564BT	0428
T607011574BT	0428
T607011584BT	0428
T607011824BT	0428
T607011834BT	0428
T607011844BT	0428
T607011854BT	0428
T607011864BT	0428
T607011874BT	0428
T607011884BT	0428
T607017084BT	0428
T607021324BT	0428
T607021334BT	0428
T607021344BT	0428
T607021354BT	0428
T607021364BT	0428
T607021374BT	0428
T607021384BT	0428
T607021524BT	0428
T607021534BT	0428
T607021544BT	0428
T607021554BT	0428
T607021564BT	0428
T607021574BT	0428
T607021584BT	0428
T607021824BT	0428
T607021834BT	0428
T607021844BT	0428
T607021854BT	0428
T607021864BT	0428
T607021874BT	0428
T607021884BT	0428
T607031324BT	0428
T607031334BT	0428
T607031344BT	0428
T607031354BT	0428
T607031364BT	0428
T607031374BT	0428
T607031384BT	0428
T607031524BT	0428
T607031534BT	0428
T607031544BT	0428
T607031554BT	0428
T607031564BT	0428
T607031574BT	0428
T607031584BT	0428
T607031824BT	0428
T607031834BT	0428
T607031844BT	0428
T607031854BT	0428
T607031864BT	0428
T607031874BT	0428
T607031884BT	0428
T607041324BT	0428
T607041334BT	0428
T607041344BT	0428
T607041354BT	0428
T607041364BT	0428
T607041374BT	0428
T607041384BT	0428
T607041524BT	0428
T607041534BT	0428
T607041544BT	0428
T607041554BT	0428
T607041564BT	0428
T607041574BT	0428
T607041584BT	0428
T607041824BT	0428
T607041834BT	0428
T607041844BT	0428
T607041854BT	0428
T607041864BT	0428
T607041874BT	0428
T607041884BT	0428
T607051324BT	0428
T607051334BT	0428
T607051344BT	0428
T607051354BT	0428
T607051364BT	0428
T607051374BT	0428
T607051384BT	0428
T607051524BT	0428
T607051534BT	0428
T607051544BT	0428
T607051554BT	0428
T607051564BT	0428
T607051574BT	0428
T607051584BT	0428
T607051824BT	0428
T607051834BT	0428
T607051844BT	0428
T607051854BT	0428
T607051864BT	0428
T607051874BT	0428
T607051884BT	0428
T607061324BT	0428
T607061334BT	0428
T607061344BT	0428
T607061354BT	0428
T607061364BT	0428
T607061374BT	0428
T607061384BT	0428
T607061524BT	0428
T607061534BT	0428
T607061544BT	0428
T607061554BT	0428
T607061564BT	0428
T607061574BT	0428
T607061584BT	0428
T607061824BT	0428
T607061834BT	0428
T607061844BT	0428
T607061854BT	0428
T607061864BT	0428
T607061874BT	0428
T607061884BT	0428
T607071324BT	1871
T607071334BT	1871
T607071344BT	1871
T607071354BT	1871
T607071364BT	1871
T607071374BT	1871
T607071384BT	1871
T607071524BT	1871
T607071534BT	1871
T607071544BT	1871
T607071554BT	1871
T607071564BT	1871
T607071574BT	1871
T607071584BT	1871
T607071824BT	1871
T607071834BT	1871
T607071844BT	1871
T607071854BT	1871
T607071864BT	1871
T607071874BT	1871
T607071884BT	1871
T607081324BT	1871
T607081334BT	1871
T607081344BT	1871
T607081354BT	1871
T607081364BT	1871
T607081374BT	1871

If replacement code is OEM, contact original manufacturer for replacement.

DEVICE TYPE	REPL CODE
T607081384BT	1871
T607081524BT	1871
T607081534BT	1871
T607081544BT	1871
T607081554BT	1871
T607081564BT	1871
T607081574BT	1871
T607081584BT	1871
T607081824BT	1871
T607081834BT	1871
T607081844BT	1871
T607081854BT	1871
T607081864BT	1871
T607081874BT	1871
T607081884BT	1871
T607091324BT	1871
T607091334BT	1871
T607091344BT	1871
T607091354BT	1871
T607091364BT	1871
T607091374BT	1871
T607091384BT	1871
T607091524BT	1871
T607091534BT	1871
T607091544BT	1871
T607091554BT	1871
T607091564BT	1871
T607091574BT	1871
T607091584BT	1871
T607091824BT	1871
T607091834BT	1871
T607091844BT	1871
T607091854BT	1871
T607091864BT	1871
T607091874BT	1871
T607091884BT	1871
T607101324BT	1871
T607101334BT	1871
T607101344BT	1871
T607101354BT	1871
T607101364BT	1871
T607101374BT	1871
T607101524BT	1871
T607101534BT	1871
T607101544BT	1871
T607101554BT	1871
T607101564BT	1871
T607101574BT	1871
T607101824BT	1871
T607101834BT	1871
T607101844BT	1871
T607101854BT	1871
T607101864BT	1871
T607101874BT	1871
T607111324BT	1871
T607111334BT	1871
T607111344BT	1871
T607111354BT	1871
T607111364BT	1871
T607111524BT	1871
T607111534BT	1871
T607111544BT	1871
T607111554BT	1871
T607111564BT	1871
T607111824BT	1871
T607111834BT	1871
T607111844BT	1871
T607111854BT	1871
T607111864BT	1871
T607121324BT	1871
T607121334BT	1871
T607121344BT	1871
T607121354BT	1871
T607121364BT	1871
T607121524BT	1871
T607121534BT	1871
T607121544BT	1871
T607121554BT	1871
T607121564BT	1871
T607121824BT	1871
T607121834BT	1871
T607121844BT	1871
T607121854BT	1871
T607121864BT	1871
T607141324BT	OEM
T607141334BT	OEM
T607141344BT	OEM
T607141354BT	OEM
T607141364BT	OEM
T607141524BT	OEM
T607141534BT	OEM
T607141544BT	OEM
T607141554BT	OEM
T607141564BT	OEM
T607141824BT	OEM
T607141834BT	OEM
T607141844BT	OEM
T607141854BT	OEM
T607141864BT	OEM
T607151324BT	OEM
T607151334BT	OEM
T607151344BT	OEM
T607151354BT	OEM
T607151524BT	OEM
T607151534BT	OEM
T607151544BT	OEM
T607151554BT	OEM
T607151824BT	OEM
T607151834BT	OEM
T607151844BT	OEM
T607151854BT	OEM
T610011304BT	1084
T610011304DT	0738
T610011504BT	1084
T610011504DT	0738
T610011804BT	1084
T610021304BT	1084
T610021304DT	0738
T610021504BT	1084
T610021504DT	0738
T610021804BT	1084
T610031304BT	1086
T610031304DT	0738
T610031504BT	1086
T610031504DT	0738
T610031804BT	1086
T610041304BT	1086
T610041304DT	0738
T610041504BT	1086
T610041504DT	0738
T610041804BT	1086
T610051304BT	0738
T610051504BT	0738
T610051804BT	1086
T610061304BT	0738
T610061504BT	0738
T610061804BT	1086
T620011304	0192
T620012004	0192
T620013004	0192
T620021304	0192
T620022004	0192
T620023004	0192
T620031304	0159
T620032004	0159
T620033004	0159
T620041304	0159
T620042004	0159
T620043004	0159
T620051304	0159
T620052004	0159
T620053004	0159
T620061304	0159
T620062004	0159
T620063004	0159
T620071304	0096
T620071504DN	0096
T620072004	0096
T620072004DN	0096
T620073004	0096
T620073004DN	0096
T620081304	0096
T620082004	0096
T620082004DN	0096
T620083004	0096
T620083004DN	0096
T620091304	0096
T620091504DN	0096
T620092004	0096
T620092004DN	0096
T620093004	0096
T620093004DN	0096
T620101304	0096
T620101504DN	0096
T620102004	0096
T620102004DN	0096
T620103004	0096
T620103004DN	0096
T620111504DN	0096
T620112004DN	0096
T620113004DN	0096
T620121304	0096
T620121504DN	0096
T620122004	0096
T620122004DN	0096
T620123004	0096
T620123004DN	0096
T620141304	OEM
T620142004	OEM
T620143004	OEM
T620161304	OEM
T620162004	OEM
T620163004	OEM
T627011524DN	0459
T627011534DN	0459
T627011544DN	0459
T627011554DN	0459
T627011564DN	0459
T627011574DN	0459
T627011584DN	0459
T627012024DN	0459
T627012034DN	0459
T627012044DN	0459
T627012054DN	0459
T627012064DN	0459
T627012074DN	0459
T627012084DN	0459
T627012524DN	0459
T627012534DN	0459
T627012544DN	0459
T627012554DN	0459
T627012564DN	0459
T627012574DN	0459
T627012584DN	0459
T627021524DN	0459
T627021534DN	0459
T627021544DN	0459
T627021554DN	0459
T627021564DN	0459
T627021574DN	0459
T627021584DN	0459
T627022024DN	0459
T627022034DN	0459
T627022044DN	0459
T627022054DN	0459
T627022064DN	0459
T627022074DN	0459
T627022084DN	0459
T627022524DN	0459
T627022534DN	0459
T627022544DN	0459
T627022554DN	0459
T627022564DN	0459
T627022574DN	0459
T627022584DN	0459
T627031524DN	0459
T627031534DN	0459
T627031544DN	0459
T627031554DN	0459
T627031564DN	0459
T627031574DN	0459
T627031584DN	0459
T627032024DN	0459
T627032034DN	0459
T627032044DN	0459
T627032054DN	0459
T627032064DN	0459
T627032074DN	0459
T627032084DN	0459
T627032524DN	0459
T627032534DN	0459
T627032544DN	0459
T627032554DN	0459
T627032564DN	0459
T627032574DN	0459
T627032584DN	0459
T627041524DN	OEM
T627041534DN	0459
T627041544DN	0459
T627041554DN	0459
T627041564DN	0459
T627041574DN	0459
T627041584DN	0459
T627042024DN	0459
T627042034DN	0459
T627042044DN	0459
T627042054DN	0459
T627042064DN	0459
T627042074DN	0459
T627042084DN	0459
T627042524DN	0459
T627042534DN	0459
T627042544DN	0459
T627042554DN	0459
T627042564DN	0459
T627042574DN	0459
T627042584DN	0459
T627051524DN	0459
T627051534DN	0459
T627051544DN	0459
T627051554DN	0459
T627051564DN	0459
T627051574DN	0459
T627051584DN	0459
T627052024DN	0459
T627052034DN	0459
T627052044DN	0459
T627052054DN	0459
T627052064DN	0459
T627052074DN	0459
T627052084DN	0459
T627052524DN	0459
T627052534DN	0459
T627052544DN	0459
T627052554DN	0459
T627052564DN	0459
T627052574DN	0459
T627052584DN	0459
T627061524DN	0459
T627061534DN	0459
T627061544DN	0459
T627061554DN	0459
T627061564DN	0459
T627061574DN	0459
T627061584DN	0459
T627062024DN	0459
T627062034DN	0459
T627062044DN	0459
T627062054DN	0459
T627062064DN	0459
T627062074DN	0459
T627062084DN	0459
T627062524DN	0459
T627062534DN	0459
T627062544DN	0459
T627062554DN	0459
T627062564DN	0459
T627062574DN	0459
T627062584DN	0459
T627071524DN	0587
T627071534DN	0587
T627071544DN	0587
T627071554DN	0587
T627071564DN	0587
T627071574DN	0587
T627071584DN	0587
T627072024DN	0587
T627072034DN	0587
T627072044DN	0587
T627072054DN	0587
T627072064DN	0587
T627072074DN	0587
T627072084DN	0587
T627072524DN	0587
T627072534DN	0587
T627072544DN	0587
T627072554DN	0587
T627072564DN	0587
T627072574DN	0587
T627072584DN	0587
T627081524DN	0587
T627081534DN	0587
T627081544DN	0587
T627081554DN	0587
T627081564DN	0587
T627081574DN	0587
T627081584DN	0587
T627082024DN	0587
T627082034DN	0587
T627082044DN	0587
T627082054DN	0587
T627082064DN	0587
T627082074DN	0587
T627082084DN	0587
T627082524DN	0587
T627082534DN	0587
T627082544DN	0587
T627082554DN	0587
T627082564DN	0587
T627082574DN	0587
T627082584DN	0587
T627091524DN	0587
T627091534DN	0587
T627091544DN	0587
T627091554DN	0587
T627091564DN	0587
T627091574DN	0587
T627091584DN	0587
T627092024DN	0587
T627092034DN	0587
T627092044DN	0587
T627092054DN	0587
T627092064DN	0587
T627092074DN	0587
T627092084DN	0587
T627092524DN	0587
T627092534DN	0587
T627092544DN	0587
T627092554DN	0587
T627092564DN	0587
T627092574DN	0587
T627092584DN	0587
T627101524DN	0587
T627101534DN	0587
T627101544DN	0587
T627101554DN	0587
T627101564DN	0587
T627101574DN	0587
T627101584DN	0587
T627102024DN	0587
T627102034DN	0587
T627102044DN	0587
T627102054DN	0587
T627102064DN	0587
T627102074DN	0587
T627102084DN	0587
T627102524DN	0587
T627102534DN	0587
T627102544DN	0587
T627102554DN	0587
T627102564DN	0587
T627102574DN	0587
T627102584DN	0587
T627111524DN	0587
T627111534DN	0587
T627111544DN	0587
T627111554DN	0587
T627111564DN	0587
T627111574DN	0587
T627111584DN	0587
T627112024DN	0587
T627112034DN	0587
T627112044DN	0587
T627112054DN	0587
T627112064DN	0587
T627112074DN	0587
T627112084DN	0587
T627112524DN	0587
T627112534DN	0587
T627112544DN	0587
T627112554DN	0587
T627112564DN	0587
T627112574DN	0587
T627112584DN	0587
T627121524DN	0587
T627121534DN	0587
T627121544DN	0587
T627121554DN	0587
T627121564DN	0587
T627121574DN	0587
T627121584DN	0587
T627122024DN	0587
T627122034DN	0587
T627122044DN	0587
T627122054DN	0587
T627122064DN	0587
T627122074DN	0587
T627122084DN	0587
T627122524DN	0587
T627122534DN	0587
T627122544DN	0587
T627122554DN	0587
T627122564DN	0587
T627122574DN	0587
T627122584DN	0587
T627141524DN	OEM
T627141534DN	OEM
T627141544DN	OEM
T627141554DN	OEM
T627141564DN	OEM
T627142024DN	OEM
T627142034DN	OEM
T627142044DN	OEM
T627142054DN	OEM
T627142064DN	OEM
T627142524DN	OEM
T627142534DN	OEM
T627142544DN	OEM
T627142554DN	OEM
T627142564DN	OEM
T627151524DN	OEM
T627151534DN	OEM
T627151544DN	OEM
T627151554DN	OEM
T627152024DN	OEM
T627152034DN	OEM
T627152044DN	OEM
T627152054DN	OEM
T627152524DN	OEM
T627152534DN	OEM
T627152544DN	OEM
T627152554DN	OEM
T630011504DN	0159
T630012004DN	0159
T630013004DN	0192
T630021504DN	0192
T630022004DN	0159
T630023004DN	0192
T630031504DN	0159
T630032004DN	0159
T630033004DN	0159
T630041504DN	0159
T630042004DN	0159
T630043004DN	0159
T630051504DN	0159
T630052004DN	0159
T630053004DN	0159
T630061504DN	0159
T630062004DN	0159
T630063004DN	0159
T660013004BY	OEM
T660023004BY	OEM
T660033004BY	OEM
T660043004BY	OEM
T660053004BY	OEM
T660063004BY	OEM
T660073004BY	OEM
T660083004BY	OEM
T660093004BY	OEM
T660103004BY	OEM
T660113004BY	OEM
T660123004BY	OEM
T660133004BY	OEM
T660143004BY	OEM
T700012504	1900
T700012504BY	1900
T700013004	1900
T700013004BY	1900
T700013504	1900
T700022504	1900
T700022504BY	1900
T700023004	1900
T700023004BY	1900
T700023504	1900
T700032504	1900
T700033004	1900
T700033504	1900
T700042504	1900
T700042504BY	0738
T700043004	1900
T700043004BY	0738
T700043504	1900
T700052504	1900
T700053004	1900
T700053504	1900
T700062504	1900
T700062504BY	0738
T700063004	1900
T700063004BY	0738
T700063504	1900
T700072504	0738
T700073004	0738
T700073504	0738
T700082504	0738
T700082504BY	0738
T700083004	0738
T700083004BY	0738
T700083504	0738
T700092504	0738
T700093004	0738
T700093504	0738
T700102504	0738
T700102504BY	0738
T700103004	0738
T700103004BY	0738
T700103504	0738
T700122504	0738
T700122504BY	0738
T700123004	0738
T700123004BY	0738
T700123504	0738
T700123504BY	0738
T700142504	OEM
T700143004	OEM
T700143504	OEM
T700162504	OEM
T700163004	OEM
T700163504	OEM
T700182504	OEM
T700183004	OEM
T700202504	OEM
T700203004	OEM
T700222504	OEM
T700222504BY	OEM
T700223004BY	OEM
T700242504	OEM
T707011324BT	1700
T707011334BT	1700
T707011344BT	1700
T707011354BT	1700
T707011364BT	1700
T707011374BT	OEM
T707011384BT	OEM
T707011524BT	1700
T707011534BT	1700
T707011544BT	1700
T707011554BT	1700
T707011564BT	1700
T707011574BT	OEM
T707011824BT	1700
T707011834BT	1700
T707011844BT	1700
T707011854BT	1700
T707011864BT	1700
T707011874BT	OEM
T707011884BT	OEM
T707012524BY	1700
T707012534BY	1700
T707012544BY	1700
T707012834BY	1700
T707012844BY	1700
T707012854BY	1700
T707013024BY	1700
T707013034BY	1700
T707013044BY	1700
T707013054BY	1700
T707017084BT	OEM
T707021324BT	1700
T707021334BT	1700
T707021344BT	1700
T707021354BT	1700
T707021364BT	1700
T707021374BT	OEM
T707021384BT	OEM
T707021524BT	1700
T707021534BT	1700
T707021544BT	1700
T707021554BT	1700
T707021564BT	1700
T707021574BT	OEM
T707021584BT	OEM
T707021824BT	1700
T707021834BT	1700
T707021844BT	1700
T707021854BT	1700
T707021864BT	1700
T707021874BT	OEM
T707021884BT	OEM
T707022524BY	1700
T707022534BY	1700
T707022544BY	1700
T707022554BY	1700
T707022834BY	1700
T707022844BY	1700
T707022854BY	1700
T707023024BY	1700
T707023034BY	1700
T707023044BY	1700
T707023054BY	1700
T707031324BT	1700
T707031334BT	1700
T707031344BT	1700
T707031354BT	1700
T707031364BT	1700
T707031374BT	OEM
T707031384BT	OEM
T707031524BT	1700
T707031534BT	1700
T707031544BT	1700
T707031554BT	1700
T707031564BT	1700
T707031574BT	OEM
T707031584BT	OEM
T707031824BT	1700
T707031834BT	1700
T707031844BT	1700
T707031854BT	1700
T707031864BT	1700
T707031874BT	OEM
T707031884BT	OEM
T707032524BY	1700
T707032534BY	1700
T707032544BY	1700
T707032554BY	1700
T707032834BY	6699
T707032844BY	OEM
T707032854BY	1700
T707033024BY	1700
T707033034BY	1700
T707033044BY	1700
T707033054BY	1700
T707041324BT	1700
T707041334BT	1700
T707041344BT	1700
T707041354BT	1700
T707041364BT	1700
T707041374BT	OEM
T707041384BT	OEM
T707041524BT	1700
T707041534BT	1700
T707041544BT	1700
T707041554BT	1700
T707041564BT	1700
T707041574BT	OEM
T707041584BT	OEM
T707041824BT	1700
T707041834BT	1700
T707041844BT	1700
T707041854BT	1700
T707041864BT	1700
T707041874BT	OEM
T707041884BT	OEM
T707042524BY	1700
T707042534BY	1700
T707042544BY	1700
T707042554BY	1700
T707042834BY	1700
T707042844BY	1700
T707042854BY	1700
T707043024BY	1700
T707043034BY	1700
T707043044BY	1700
T707043054BY	1700
T707051324BT	1700
T707051334BT	1700
T707051344BT	1700
T707051354BT	1700
T707051364BT	1700
T707051374BT	OEM
T707051384BT	OEM
T707051524BT	1700
T707051534BT	1700
T707051544BT	1700
T707051554BT	1700
T707051564BT	1700
T707051574BT	OEM
T707051584BT	OEM
T707051824BT	1700
T707051834BT	1700
T707051844BT	1700
T707051864BT	1700
T707051874BT	OEM
T707051884BT	OEM
T707052524BY	1700
T707052534BY	1700
T707052544BY	1700
T707052834BY	1700
T707052844BY	1700
T707052854BY	1700
T707053024BY	1700
T707053034BY	1700
T707053044BY	1700
T707053054BY	1700
T707061324BT	1700
T707061334BT	1700
T707061344BT	1700
T707061354BT	1700
T707061364BT	1700
T707061374BT	OEM
T707061384BT	OEM
T707061524BT	1700
T707061534BT	1700
T707061544BT	1700
T707061554BT	1700
T707061564BT	1700
T707061574BT	OEM
T707061584BT	OEM
T707061824BT	1700
T707061834BT	1700
T707061844BT	1700
T707061864BT	1700
T707061874BT	OEM
T707061884BT	OEM
T707062524BY	1700
T707062534BY	1700
T707062544BY	1700
T707062554BY	1700
T707062834BY	1700
T707062844BY	1700
T707062854BY	1700
T707063024BY	1700
T707063034BY	1700
T707063044BY	1700
T707063054BY	1700
T707071324BT	0345
T707071334BT	0345
T707071344BT	0769
T707071354BT	0345
T707071364BT	0345
T707071374BT	OEM
T707071524BT	0345
T707071534BT	0345

If replacement code is OEM, contact original manufacturer for replacement.

DEVICE TYPE	REPL CODE
T707071544BT	0345
T707071554BT	0345
T707071564BT	0345
T707071574BT	OEM
T707071824BT	0345
T707071834BT	0345
T707071844BT	0345
T707071854BT	0345
T707071864BT	0345
T707071874BT	OEM
T707072524BY	0345
T707072534BY	0345
T707072544BY	0345
T707072554BY	0345
T707072834BY	0345
T707072844BY	0345
T707072854BY	0345
T707073024BY	0345
T707073034BY	0345
T707073044BY	0345
T707073054BY	0345
T707081324BT	0345
T707081334BT	0345
T707081344BT	0345
T707081354BT	0345
T707081364BT	0345
T707081374BT	OEM
T707081524BT	0345
T707081534BT	0345
T707081544BT	0345
T707081554BT	0345
T707081564BT	0345
T707081574BT	OEM
T707081824BT	0345
T707081834BT	0345
T707081844BT	0345
T707081854BT	0345
T707081864BT	0345
T707081874BT	OEM
T707082524BY	0345
T707082534BY	0345
T707082544BY	0345
T707082554BY	0345
T707082834BY	0345
T707082844BY	0345
T707082854BY	0345
T707083024BY	0345
T707083034BY	0345
T707083044BY	0345
T707083054BY	0345
T707091324BT	0345
T707091334BT	0345
T707091344BT	0345
T707091354BT	0345
T707091364BT	0345
T707091374BT	OEM
T707091524BT	0345
T707091534BT	0345
T707091544BT	0345
T707091554BT	0345
T707091564BT	0345
T707091574BT	OEM
T707091824BT	0345
T707091834BT	0345
T707091844BT	0345
T707091854BT	0345
T707091864BT	0345
T707091874BT	OEM
T707092524BY	0345
T707092534BY	0345
T707092544BY	0345
T707092554BY	0345
T707093024BY	0345
T707093034BY	0345
T707093044BY	0345
T707093054BY	0345
T707101324BT	0345
T707101334BT	0345
T707101344BT	0345
T707101354BT	0345
T707101364BT	0345
T707101374BT	OEM
T707101524BT	0345
T707101534BT	0345
T707101544BT	0345
T707101554BT	0345
T707101564BT	0345
T707101574BT	OEM
T707101824BT	0345
T707101834BT	0345
T707101844BT	0345
T707101854BT	0345
T707101864BT	0345
T707101874BT	OEM
T707102524BY	0345
T707102534BY	0345
T707102544BY	0345
T707102554BY	0345
T707103024BY	0345
T707103034BY	0345
T707103044BY	0345
T707103054BY	0345
T707111324BT	0345
T707111334BT	0345
T707111344BT	0345
T707111354BT	0345
T707111364BT	0345
T707111524BT	0345
T707111534BT	0345
T707111544BT	0345
T707111554BT	0345
T707111564BT	0345
T707111824BT	0345
T707111834BT	0345
T707111844BT	0345
T707111854BT	0345
T707111864BT	0345
T707112524BY	0345
T707112534BY	0345
T707112544BY	0345
T707112554BY	0345
T707113024BY	0345
T707113034BY	0345
T707113044BY	0345
T707113054BY	0345
T707121324BT	0345
T707121334BT	0345
T707121344BT	0345
T707121354BT	OEM
T707121364BT	0345
T707121524BT	0345
T707121534BT	0345
T707121544BT	0345
T707121554BT	0345
T707121564BT	0345
T707121824BT	0345
T707121834BT	0345
T707121844BT	0345
T707121854BT	0345
T707121864BT	0345
T707122524BY	0345
T707122534BY	0345
T707122544BY	0345
T707122554BY	0345
T707123024BY	0345
T707123034BY	0345
T707123044BY	0345
T707123054BY	0345
T707141324BT	OEM
T707141334BT	OEM
T707141344BT	OEM
T707141354BT	OEM
T707141364BT	OEM
T707141524BT	OEM
T707141534BT	OEM
T707141544BT	OEM
T707141554BT	OEM
T707141564BT	OEM
T707141824BT	OEM
T707141834BT	OEM
T707141844BT	OEM
T707141854BT	OEM
T707141864BT	OEM
T720013504	2843
T720014504	2843
T720015504	2843
T720023504	2843
T720024504	2843
T720025504	2843
T720033504	1765
T720034504	1765
T720035504	1765
T720043504	1765
T720044504	1765
T720045504	1765
T720053504	1765
T720054504	1765
T720055504	1765
T720063504	1765
T720064504	1765
T720065504	1765
T720073504	1814
T720074504	1814
T720075504	1814
T720083504	1814
T720083504DN	1814
T720084504	1814
T720084504DN	1814
T720085504	1814
T720085504DN	1814
T720093504	1814
T720094504	1814
T720095504	1814
T720103504	1814
T720103504DN	1814
T720104504	1814
T720104504DN	1814
T720105504	1814
T720105504DN	1814
T720123504	1814
T720124504	1814
T720125504	1814
T720143504	OEM
T720144504	OEM
T720145504	OEM
T720163504	OEM
T720164504	OEM
T720165504	OEM
T720183504	OEM
T720184504	OEM
T720203504	OEM
T720204504	OEM
T720223504	OEM
T720243504	OEM
T727012524DN	0769
T727012534DN	0769
T727012544DN	0769
T727012554DN	0769
T727012564DN	0769
T727012574DN	OEM
T727012584DN	OEM
T727013524DN	0769
T727013534DN	0769
T727013544DN	0769
T727013554DN	0769
T727013564DN	0769
T727013574DN	OEM
T727013584DN	OEM
T727014524DN	0769
T727014534DN	0769
T727014544DN	0769
T727014554DN	0769
T727014564DN	0483
T727014574DN	OEM
T727014584DN	OEM
T727022524DN	0769
T727022534DN	0769
T727022544DN	0769
T727022554DN	0769
T727022564DN	0769
T727022574DN	OEM
T727022584DN	OEM
T727023524DN	0769
T727023534DN	0769
T727023544DN	0769
T727023554DN	0769
T727023564DN	0769
T727023574DN	OEM
T727023584DN	OEM
T727024524DN	0769
T727024534DN	0769
T727024544DN	0769
T727024554DN	0769
T727024564DN	0769
T727024574DN	OEM
T727024584DN	OEM
T727032524DN	0769
T727032534DN	0769
T727032544DN	0769
T727032554DN	0769
T727032564DN	0769
T727032574DN	OEM
T727032584DN	OEM
T727033524DN	0769
T727033534DN	0769
T727033544DN	0769
T727033554DN	0769
T727033564DN	0769
T727033574DN	OEM
T727033584DN	OEM
T727034524DN	0769
T727034534DN	0769
T727034544DN	0769
T727034554DN	0769
T727034564DN	0769
T727034574DN	OEM
T727034584DN	OEM
T727042524DN	0769
T727042534DN	0769
T727042544DN	0769
T727042554DN	0769
T727042564DN	0769
T727042574DN	OEM
T727042584DN	OEM
T727043524DN	0769
T727043534DN	0769
T727043544DN	0769
T727043554DN	0769
T727043564DN	0769
T727043574DN	OEM
T727043584DN	OEM
T727044524DN	0769
T727044534DN	0769
T727044544DN	0769
T727044554DN	0769
T727044564DN	0769
T727044574DN	OEM
T727044584DN	OEM
T727052524DN	0769
T727052534DN	0769
T727052544DN	0769
T727052554DN	0769
T727052564DN	0769
T727052574DN	OEM
T727052584DN	OEM
T727053524DN	0769
T727053534DN	0769
T727053544DN	0769
T727053554DN	0769
T727053564DN	0769
T727053574DN	OEM
T727053584DN	OEM
T727054524DN	0769
T727054534DN	0769
T727054544DN	0769
T727054554DN	0769
T727054564DN	0769
T727054574DN	OEM
T727054584DN	OEM
T727062524DN	0769
T727062534DN	0769
T727062544DN	0769
T727062554DN	0769
T727062564DN	0769
T727062574DN	OEM
T727062584DN	OEM
T727063524DN	0769
T727063534DN	0769
T727063544DN	0769
T727063554DN	0769
T727063564DN	0769
T727063574DN	OEM
T727063584DN	OEM
T727064524DN	0769
T727064534DN	0769
T727064544DN	0769
T727064554DN	0769
T727064564DN	0769
T727064574DN	OEM
T727064584DN	OEM
T727072524DN	0483
T727072534DN	0483
T727072544DN	0483
T727072554DN	0483
T727072564DN	0483
T727072574DN	OEM
T727073524DN	0483
T727073534DN	0483
T727073544DN	0483
T727073554DN	0483
T727073564DN	0483
T727073574DN	OEM
T727074524DN	0483
T727074534DN	0483
T727074544DN	0483
T727074554DN	0483
T727074564DN	0483
T727074574DN	OEM
T727082524DN	0483
T727082534DN	0483
T727082544DN	0483
T727082554DN	0483
T727082564DN	0483
T727082574DN	OEM
T727083524DN	0483
T727083534DN	0483
T727083544DN	0483
T727083554DN	0483
T727083564DN	0483
T727083574DN	OEM
T727084524DN	0483
T727084534DN	0483
T727084544DN	0483
T727084554DN	0483
T727084564DN	0483
T727084574DN	OEM
T727092524DN	0483
T727092534DN	0483
T727092544DN	0483
T727092554DN	0483
T727092564DN	0483
T727093524DN	0483
T727093534DN	0483
T727093544DN	0483
T727093554DN	0483
T727093564DN	0483
T727094524DN	0483
T727094534DN	0483
T727094544DN	0483
T727094554DN	0483
T727094564DN	0483
T727102524DN	0483
T727102534DN	0483
T727102544DN	0483
T727102554DN	0483
T727102564DN	0483
T727103524DN	0483
T727103534DN	0483
T727103544DN	0483
T727103554DN	0483
T727103564DN	0483
T727104524DN	0483
T727104534DN	0483
T727104544DN	0483
T727104554DN	0483
T727104564DN	0483
T727112524DN	0483
T727112534DN	0483
T727112544DN	0483
T727112554DN	0483
T727113524DN	0483
T727113534DN	0483
T727113544DN	0483
T727113554DN	0483
T727114524DN	0483
T727114534DN	0483
T727114544DN	0483
T727114554DN	0483
T727122524DN	0483
T727122534DN	0483
T727122544DN	0483
T727122554DN	0483
T727123524DN	0483
T727123534DN	0483
T727123544DN	0483
T727123554DN	0483
T727124524DN	0483
T727124534DN	0483
T727124544DN	0483
T727124554DN	0483
T727142524DN	OEM
T727142534DN	OEM
T727142544DN	OEM
T727143524DN	OEM
T727143534DN	OEM
T727143544DN	OEM
T727144524DN	OEM
T727144534DN	OEM
T727144544DN	OEM
T730013504DN	2843
T730023504DN	2843
T730024504DN	2843
T730025504DN	2843
T730043504DN	2843
T730044504DN	1765
T730045504DN	1765
T730063504DN	1765
T730064504DN	1765
T730065504DN	1765
T920010603	OEM
T920010703	OEM
T920010803	OEM
T920010903	OEM
T920011003	OEM
T920020603	OEM
T920020703	OEM
T920020803	OEM
T920020903	OEM
T920021003	OEM
T920030603	OEM
T920030703	OEM
T920030803	OEM
T920030903	OEM
T920031003	OEM
T920040603	OEM
T920040703	OEM
T920040803	OEM
T920040903	OEM
T920041003	OEM
T920050603	OEM
T920050703	OEM
T920050803	OEM
T920050903	OEM
T920051003	OEM
T920060603	OEM
T920060703	OEM
T920060803	OEM
T920060903	OEM
T920061003	OEM
T920070603	OEM
T920070703	OEM
T920070803	OEM
T920070903	OEM
T920071003	OEM
T920080603	OEM
T920080703	OEM
T920080803	OEM
T920080903	OEM
T920081003	OEM
T920090603	OEM
T920090703	OEM
T920090803	OEM
T920090903	OEM
T920091003	OEM
T920100603	OEM
T920100703	OEM
T920100803	OEM
T920100903	OEM
T920101003	OEM
T920120603	OEM
T920120703	OEM
T920120803	OEM
T920120903	OEM
T920121003	OEM
T920140603	OEM
T920140703	OEM
T920140803	OEM
T920140903	OEM
T920141003	OEM
T920160603	OEM
T920160703	OEM
T920160803	OEM
T920160903	OEM
T920161003	OEM
T920180603	OEM
T920180703	OEM
T920180803	OEM
T920180903	OEM
T920200603	OEM
T920200703	OEM
T920200803	OEM
T920220603	OEM
T920220703	OEM
T920220803	OEM
T920240603	OEM
T920260603	OEM
T920280603	OEM
T920300603	OEM
T40000100600	0469
T40000100800	0469
T40000101600	OEM
T40000101800	OEM
T40000160600	0469
T40000160800	0469
T40000161600	OEM
T40000161800	OEM
T40000220600	OEM
T40000220800	OEM
T40000221600	OEM
T40000221800	OEM
T40001100600	0469
T40001100800	0469
T40001101600	OEM
T40001101800	OEM
T40001160600	0469
T40001160800	0469
T40001161600	OEM
T40001161800	OEM
T40001220600	OEM
T40001220800	OEM
T40001221600	OEM
T40001221800	OEM
T40002100600	0469
T40002100800	0469
T40002101600	OEM
T40002101800	OEM
T40002160600	0469
T40002160800	0469
T40002161600	OEM
T40002161800	OEM
T40002220600	OEM
T40002220800	OEM
T40002221600	OEM
T40002221800	OEM
T40003100600	0722
T40003100800	0722
T40003101600	OEM
T40003101800	OEM
T40003160600	0722
T40003160800	0722
T40003161600	OEM
T40003161800	OEM
T40003220600	OEM
T40003220800	OEM
T40003221600	OEM
T40003221800	OEM
T40004100600	0722
T40004100800	0722
T40004101600	OEM
T40004101800	OEM
T40004160600	0722
T40004160800	0722
T40004161600	OEM
T40004161800	OEM
T40004220600	OEM
T40004220800	OEM
T40004221600	OEM
T40004221800	OEM
T40005100600	0725
T40005100800	0725
T40005101600	OEM
T40005101800	OEM
T40005160600	0725
T40005160800	0725
T40005161600	OEM
T40005161800	OEM
T40005220600	OEM
T40005220800	OEM
T40005221600	OEM
T40005221800	OEM
T40006100600	0725
T40006100800	0725
T40006101600	OEM
T40006101800	OEM
T40006160600	0725
T40006160800	0725
T40006161600	OEM
T40006161800	OEM
T40006220600	OEM
T40006220800	OEM
T40006221600	OEM
T40006221800	OEM
T40007100600	0679
T40007100800	0679
T40007101600	OEM
T40007101800	OEM
T40007160600	0679
T40007160800	0679
T40007161600	OEM
T40007161800	OEM
T40007220600	OEM
T40007220800	OEM
T40007221600	OEM
T40007221800	OEM
T40008100600	0679
T40008100800	0679
T40008101600	OEM
T40008101800	OEM
T40008160600	0679
T40008160800	OEM
T40008161600	OEM
T40008161800	OEM
T40008220600	OEM
T40008220800	OEM
T40008221600	OEM
T40008221800	OEM
T40009100600	0679
T40009100800	0679
T40009101600	OEM
T40009101800	OEM
T40009160600	0679
T40009160800	0679
T40009161600	OEM
T40009161800	OEM
T40009220600	OEM
T40009220800	OEM
T40009221600	OEM
T40009221800	OEM
T40010100600	0679
T40010100800	0679
T40010101600	OEM
T40010101800	OEM
T40010160600	0679
T40010160800	0679
T40010161600	OEM
T40010161800	OEM
T40010220600	OEM
T40010220800	OEM
T40010221600	OEM
T40010221800	OEM
T40011100600	0679
T40011100800	0679
T40011101600	OEM
T40011101800	OEM
T40011160600	0679
T40011160800	0679
T40011161600	OEM
T40011161800	OEM
T40011220600	OEM
T40011220800	OEM
T40011221600	OEM
T40011221800	OEM
T40012100600	0679
T40012100800	0679
T40012101600	OEM
T40012101800	OEM
T40012160600	0679
T40012160800	0679
TA1Z	0086
TA-2	0160
TA2A	0079
TA2SA817	0436
TA2T00733Q	0006
TA2T00952L	0006
TA2T007330	0006
TA-3	0440
TA3	0071
TA3T006080	0006
TA3T0608K0	0006
TA3T0608KF	0006
TA3T0984K0	0006
TA3T016240	0342
TA3T017060	OEM
TA3T0116240	0342
TA3T1371A0	OEM
TA-4	0004
TA-5	0136
TA5	0071
TA5Q1306Y0	OEM
TA5T010154	0148
TA5T013200	0342
TA5T013210	0342
TA-6	0016
TA6	0016
TA-7	0144
TA7	0016
TA7T009330	0148
TA10E	0576
TA21R	0181
TA50	0015
TA54	OEM
TA57	OEM
TA58	OEM
TA60	4270
TA61	OEM
TA70L9&	2634
TA76	OEM
TA78	OEM
TA78L	1292
TA78L005	1292
TA78L-005AP	OEM
TA78L005AP	1292
TA78L005AP-Y	1292
TA78L005P	1292
TA78L006AP	4080
TA78L007P	OEM
TA78L008P	0083
TA78L009AP	1786
TA78L009P	1786
TA78L05S	1292
TA78L09S	1786
TA78L010AP	OEM
TA78L010P	OEM
TA78L012AP	1828
TA78L018AP	4506
TA78L024P	4509
TA78L12S	OEM
TA78M	OEM
TA79L024P	1721
TA-98	OEM
TA100	0015
TA115	0924
TA200	0015
TA300	0015
TA304A-1L	OEM
TA310P	1540
TA320	OEM
TA400	0015
TA500	0015
TA600	0015
TA611CS	OEM
TA-705M	2715
TA721KP	0087
TA766P	4137
TA800	0071
TA991D	OEM
TA1000	0373
TA1008	OEM
TA1050G	OEM
TA1056M	OEM
TA1062	0015
TA1063	0015
TA1064	0015

If replacement code is OEM, contact original manufacturer for replacement.

DEVICE TYPE	REPL CODE
TA1201AN	OEM
TA1201N	OEM
TA1218N	OEM
TA1222	3696
TA1223N	OEM
TA1225	3383
TA1575	0004
TA1575B	0004
TA-1614	0160
TA1614	0160
TA1620A	0038
TA1620B	0038
TA1628	0050
TA1650A	0050
TA1655B	0004
TA1658	0050
TA1659	0050
TA1660	0050
TA1662	0050
TA-1682	0160
TA1682	0160
TA-1682A	0160
TA1682A	0160
TA1697	0004
TA1703B	OEM
TA1704	0280
TA-1705	0160
TA1705	0160
TA1706	0004
TA1730	0004
TA1731	0050
TA1755	0050
TA1756	0050
TA1757	0050
TA1759	0600
TA1763	0280
TA1763A	0280
TA1763B	OEM
TA-1765	0160
TA1765	0160
TA-1766	0160
TA1766	0160
TA1767	0600
TA1771	0600
TA1772	0600
TA-1773	0160
TA1773	0160
TA1778	0280
TA1782	0280
TA1783	0280
TA-1794	0160
TA1794	0160
TA1796	0050
TA1797	0050
TA1798	0050
TA1828	0050
TA1829	0050
TA1830	0050
TA1846	0050
TA1847	0050
TA1860	0050
TA1861	0050
TA-1881	0160
TA1881	0160
TA1882	OEM
TA-1890	0160
TA1890	0160
TA-1891	0160
TA1891	0160
TA1920	OEM
TA1920A	OEM
TA1920B	OEM
TA1928A	0975
TA1938	OEM
TA1939	OEM
TA1990	0050
TA-2003	OEM
TA2083	0975
TA2084	OEM
TA2090A	OEM
TA2124P	3915
TA2188	0975
TA2235A	3426
TA2275	OEM
TA2276	OEM
TA2277	1518
TA2278	OEM
TA2279	OEM
TA2280	OEM
TA2301	0160
TA2307	OEM
TA2322	0050
TA2333	3613
TA2359A	OEM
TA2388	OEM
TA2401	0144
TA2402A	0178
TA2403A	1833
TA2404	OEM
TA2442	0745
TA2444	0745
TA2458	6442
TA2462	OEM
TA2468A	2273
TA2469A	OEM
TA2470	6051
TA2492	OEM
TA2493	OEM
TA2494	3512
TA2495	3512
TA2501	3426
TA2503	0144
TA2510	3844
TA2511	3844
TA2512	3844
TA2513	OEM
TA2514	OEM
TA2529	OEM
TA2551	OEM
TA2554	0007
TA2555	0007
TA2577A	0103
TA2606	6449
TA2616	OEM
TA2626	OEM
TA2644	0212
TA2652	OEM
TA2653	3960
TA2654	3960
TA2655	3960
TA2672	0160
TA2700	OEM
TA2701	OEM
TA2735	OEM
TA2736	OEM
TA2750	OEM
TA2758	OEM
TA2761	2208
TA2773	3624
TA2786	OEM
TA2787	OEM
TA2800	0418
TA2840	0849
TA2871	3844
TA2888	0940
TA2889	0940
TA2892	1413
TA2911	0042
TA3002R	OEM
TA3002RF	OEM
TA3003RF	OEM
TA3004R	OEM
TA3004RF	OEM
TA3006RF	OEM
TA3008RF	OEM
TA3010RF	OEM
TA3137P-ST	OEM
TA4001	OEM
TA4004	OEM
TA4005	OEM
TA4006S	OEM
TA4017	OEM
TA4123	OEM
TA4124	OEM
TA4131	OEM
TA4163	OEM
TA4164A	OEM
TA4166	OEM
TA4168-1L	OEM
TA4301-1L1	OEM
TA4302	OEM
TA4304A-1L	OEM
TA4305A-1L	OEM
TA4306A-1K5	OEM
TA4307A-1L5	OEM
TA4308A-1L	OEM
TA4309-1L	OEM
TA4309A-1L	OEM
TA4310	OEM
TA4310-1L	OEM
TA4312A	OEM
TA4313	OEM
TA4313A	OEM
TA4313A-1L	OEM
TA4314	OEM
TA4314-1L	OEM
TA4315A	OEM
TA4316A	OEM
TA4316A-1L1	OEM
TA4317B-1L	OEM
TA4328	OEM
TA4335	OEM
TA4336	OEM
TA4337A	OEM
TA4338	OEM
TA4340A	OEM
TA4341	OEM
TA4342A	OEM
TA4343A	OEM
TA4344	OEM
TA4345	OEM
TA4351	OEM
TA4352	OEM
TA4372	OEM
TA4373	OEM
TA4374A	OEM
TA4375A-2M	OEM
TA4377A	OEM
TA4381E	OEM
TA4384	OEM
TA4385	OEM
TA4385-3D	OEM
TA4386A	OEM
TA4386B	OEM
TA4387A	OEM
TA4387B	OEM
TA4390A	OEM
TA4396	OEM
TA4398	OEM
TA4398A	OEM
TA4398A-5A	OEM
TA-4846	0050
TA5628	2755
TA5649	0352
TA5649A	0352
TA5702	0354
TA-5702B	0354
TA5702B	0354
TA5814	0167
TA5912	0666
TA6200	0086
TA6462	OEM
TA6472	3270
TA6480	2786
TA6699B	2502
TA6896	2501
TA7021P	OEM
TA-7027M	2617
TA7027M	2617
TA7028M	2618
TA7038M#	2624
TA-7045	0823
TA7045	0823
TA-7045M	0823
TA7045M	0823
TA-7046	1393
TA7046P	1393
TA7047	1854
TA7047M	1854
TA-7050M	2715
TA7050M	2715
TA7051/01	0102
TA7051P	0829
TA-7053M	0753
TA7053M	0753
TA7054	2630
TA7054P	2630
TA-7055P	2931
TA7055P	2931
TA7055P-D	2931
TA-7055P-E	2931
TA7055P-E	2931
TA7055P-F	2931
TA7057M	2698
TA-7060	2057
TA7060	2057
TA-7060P	2057
TA7060P	2057
TA-7061AP	0911
TA7061AP	0911
TA-7061B	0911
TA7061BP	OEM
TA-7061P	0911
TA7062P	2632
TA-7063	1994
TA7063	1994
TA7063P	1994
TA7063P-C	1994
TA7064P	4633
TA7064P-JA	4633
TA7066P	3980
TA7069	OEM
TA-7069P	2634
TA7069P	OEM
TA7070A	0878
TA7070B	0878
TA7070FA-1	0878
TA7070P	0878
TA7070PFA-1	0878
TA7070PFA1	0878
TA7070PGL	0878
TA7071GL	0167
TA7071P	0167
TA7072P	0789
TA7073AP	0789
TA7073P	0789
TA7074GL	0395
TA7074P	0395
TA7074PGL	0395
TA7075B	0855
TA7075P	0855
TA7075PL	0855
TA7076	2639
TA7076AP	2639
TA-7076P	2639
TA7076P	2639
TA7089P	OEM
TA-7092-P	3670
TA7092AP-B	4557
TA7092C	3670
TA7092P	3670
TA7092P-C	3670
TA-7102P	4582
TA7102P	4582
TA7103	3705
TA7103P	0403
TA7104P	OEM
TA7108P	5746
TA7109P	OEM
TA7115P	2764
TA7117P	4556
TA7119A	OEM
TA7119P	OEM
TA-7120P	1018
TA7120P	1018
TA7120P-A	OEM
TA7120P-B	1018
TA7120PB	1018
TA7120P-C	1018
TA7120PC	1018
TA7120P-D	1018
TA7120PD	OEM
TA7120P-E	1018
TA7122	2631
TA7122AP	2631
TA7122AR	2631
TA7124P	3915
TA7124P(FA-2)	OEM
TA7124PFA-3	3915
TA7124PFA-4	3915
TA7125P	OEM
TA7129AP	OEM
TA7129P	6223
TA7130	6851
TA7130A	OEM
TA7130A3	OEM
TA7130P	2019
TA7130P-8	OEM
TA7130P-B	2019
TA7130PB	2019
TA7130P-C	2019
TA7130PC	2019
TA7133P	OEM
TA7137	0042
TA7137(IC)	2633
TA7137P	2063
TA7137P-ST	2063
TA7140P	3980
TA7140PC	3980
TA7142P	OEM
TA7145P	2949
TA7146P	2231
TA7147P	OEM
TA7148P	4594
TA7149	0353
TA7149P	4595
TA7150	0353
TA7150P	1326
TA7151	0353
TA7152P	6233
TA7153P	OEM
TA7154P	OEM
TA7155P	0416
TA7156	0042
TA7157AP	OEM
TA7157P	4623
TA7158P	OEM
TA7159P	4634
TA7167	4238
TA7167P	4238
TA7174P	6240
TA7176AP	0167
TA7176AP(FA-1)	2284
TA7176AP(FA1)	OEM
TA7176APFA	0167
TA7176APFA-1	2284
TA7176P	2284
TA7176P(FA1)	OEM
TA7176PFA-1	0167
TA7177P	OEM
TA7178P	OEM
TA7184P	OEM
TA7187AP	OEM
TA7189	0212
TA7192P	1055
TA7199	0103
TA7200	0103
TA7200P	6249
TA7201	0103
TA7202	0103
TA7202P	5618
TA7203P	4600
TA7204P	3366
TA7205	1050
TA7205A	1050
TA7205AP	1050
TA7205AT	1050
TA-7205P	1050
TA7205P	1050
TA7207P	4793
TA7209P	5698
TA7215P	4726
TA7220P	1018
TA7222	4652
TA7222AP	4652
TA7222P	4652
TA7223P	5337
TA7230	OEM
TA7230P	4372
TA7232P	0799
TA7233P	OEM
TA7236AP	OEM
TA7236BP	OEM
TA7244P	OEM
TA7245BP	OEM
TA7245P	OEM
TA7250P	OEM
TA7252P	OEM
TA7256P	OEM
TA7257P	OEM
TA7259P	OEM
TA7262	0042
TA7262(RCA)	0353
TA7267P	OEM
TA7269P	OEM
TA7270P	5272
TA7273P	OEM
TA7274	0212
TA7274P	5818
TA7280P	5833
TA7281P	5834
TA7282AP	OEM
TA7283AP	OEM
TA7288	OEM
TA7288P	OEM
TA7291P	OEM
TA7291S	OEM
TA7299	OEM
TA7299P	OEM
TA7302P	4789
TA7303P	2694
TA7310	1540
TA7310P	1540
TA7310P-B	OEM
TA7310P-U	1540
TA7310P-Y	1540
TA7311	0561
TA7312	0561
TA7313	0561
TA7313(IC)	0362
TA7313AP	0362
TA7313F	0362
TA7313P	0362
TA7313P(IC)	0362
TA7314	0561
TA7314P	3945
TA7315	0561
TA7315AP	OEM
TA7315BP	OEM
TA7315P	OEM
TA7316	0561
TA7317P	OEM
TA7318	0561
TA7318P	4813
TA7319P	0362
TA7320	3804
TA7320P	3804
TA7320PD1626	OEM
TA7321P	OEM
TA7323P	0799
TA7324	OEM
TA7324P	OEM
TA7325P	5236
TA7328	4818
TA7328AP	4818
TA7328P	4818
TA7335P	4406
TA7337	OEM
TA7337P	OEM
TA7339P	OEM
TA7340P	OEM
TA7341P	6398
TA7342P	OEM
TA7343AP	4410
TA7343P	4410
TA7343PS	4410
TA7347	4496
TA7347AP	OEM
TA7347P	4496
TA7348AP	4383
TA7348P	4383
TA7350P	OEM
TA7354P	OEM
TA7357AP	OEM
TA7358AP	4041
TA7358F	OEM
TA-7358P	4041
TA7358P	4041
TA7360P	OEM
TA7361	3159
TA7361AP	3159
TA7361P	3159
TA7363	0042
TA7365P	OEM
TA7366P	OEM
TA7368P	OEM
TA7370P	OEM
TA7374	0212
TA7375P	OEM
TA7378P	OEM
TA7399	0212
TA7404	0505
TA7405	0710
TA7405P	OEM
TA7417AP	OEM
TA7420	0363
TA7461	2018
TA7462	2018
TA7500	2383
TA7501	2394
TA7502	1413
TA7503	2387
TA7504	0312
TA7504M	0410
TA7504P	0312
TA7504S	0312
TA7513	0363
TA7520	0460
TA7530	OEM
TA7554	0042
TA7555	0042
TA7556	0925
TA7557	0925
TA7557S	OEM
TA7579	2387
TA7580	2387
TA7581	2387
TA7582	2387
TA7584	2018
TA-7603	OEM
TA7604	OEM
TA7604AP	OEM
TA7604CP	OEM
TA7604P	OEM
TA7606AP	6463
TA7607	0912
TA7607A	0912
TA7607AP	0912
TA7607BP	0912
TA7607P	0912
TA7608AP	2054
TA7608BP	2054
TA7608BR	2054
TA7608CP	2054
TA7608CP(FA-5)	2054
TA7608CPFA-5	2054
TA7608CPFA-6	2054
TA7608CP-GS-4	2054
TA7608CP-GS-5	2054
TA7608P	2054
TA7608P(FA-3)	OEM
TA7608P(FA-5)	OEM
TA7608P(FA-6)	OEM
TA7608PFA-5	2054
TA7608PFA-6	2054
TA7608PGS-4	2054
TA7608PGS-5	2054
TA7609	4134
TA7609AP	4134
TA7609BP	OEM
TA7609GS-2	4134
TA7609P	4134
TA7609P(FA-2)	4134
TA7609P-FA2	OEM
TA7609PFA-2	4134
TA7609PFA2	4134
TA7609PFA-5	OEM
TA7609P-GS-2	4134
TA7609PGS-2	4134
TA7611AP	0069
TA7611AP-S	0069
TA7614	OEM
TA7614AP	5205
TA7614AP-Y	5205
TA7614P	5205
TA7619AP	OEM
TA7619P	OEM
TA7623AP	4818
TA7623P	OEM
TA7628AP	OEM
TA7628HP	OEM
TA7629P	OEM
TA7629P-DD	OEM
TA7630	3772
TA76300	OEM
TA7630P	3772
TA7631P	OEM
TA7632	OEM
TA7632FA-1	6110
TA7632P	6110
TA7632PFA-1	6110
TA7634AP	OEM
TA7634P	OEM
TA7636P(FA-1)	OEM
TA7636P(FA-2)	OEM
TA7637P	OEM
TA7639P	OEM
TA7640	OEM
TA-7640AP	3725
TA7640AP	3725
TA7640AP1	6063
TA7640AP2	OEM
TA7640P	3725
TA7643P	5775
TA7644AP	0663
TA7644BP	0663
TA7644BPFA-1	0663
TA7644P	6689
TA7658P	5172
TA7660	4137
TA7660P	4137
TA7660P(FA)	OEM
TA7660P(FA-1)	4137
TA7660PFA	4137
TA7660PFA-1	4137
TA7661P	OEM
TA7662P	1055
TA7664P	1676
TA7664P(FA-1)	1676
TA7664P(FA1)	OEM
TA7664PFA-1	1676
TA7664PFA1	1676
TA7665P	OEM
TA7666P	OEM
TA7668AP	4520
TA7668BP	6717
TA7668P	6717
TA7669	0212
TA7670P	3685
TA7671P	2284
TA7671P(FA-1)	2284
TA7671PFA-1	2284
TA7672P	OEM
TA7673	5191
TA7673P	5191
TA7676AP	OEM
TA7678AP	OEM
TA7678P	OEM
TA7680AP	4109
TA7680P	4109
TA7681AP	4986
TA7681P	4986
TA7683P	OEM
TA7684	0212
TA7685P	OEM
TA7688F	OEM
TA7688P	OEM
TA7691P	OEM
TA7693P	OEM
TA7705P	OEM
TA7717AP	OEM
TA7717P	OEM
TA7725P	OEM
TA7739	1708
TA7739F	OEM
TA7740	1708
TA7741	1267
TA7750P	3378
TA7757F	OEM
TA7757P	OEM
TA7758P	OEM
TA7763P	OEM
TA7764P	OEM
TA7769P	OEM
TA7772P	OEM
TA7774P	4953
TA7777N	3236
TA7777P	3236
TA7777P(FA-1)	3236
TA7777P(FA-3)	3236
TA7777PFA-1	3236
TA7777PFA-3	3236
TA7778AP	OEM
TA7778P	OEM
TA7784P	OEM
TA7805S	6815
TA7809S	6817
TA7840AP1	OEM
TA8102P	OEM
TA8110AP	OEM
TA8110P	OEM
TA8127NX	OEM
TA8135P	OEM
TA8158	0239
TA8159	2327
TA8160	1217
TA8161	1217
TA8162	1217
TA8164P	OEM
TA8184P	OEM
TA8200	OEM
TA8200AH	6430
TA8200H	OEM
TA8207K	OEM
TA8210AH	OEM
TA8210H	OEM
TA8210P	OEM
TA8211AH	OEM
TA8211H	OEM
TA8213K	OEM
TA8214K	OEM
TA8215H	OEM
TA8216H	OEM
TA8218H	OEM
TA8218HA	OEM
TA8242	0212
TA8246H	OEM
TA8400P	OEM
TA8403K	OEM
TA8406P	OEM
TA8410K	OEM
TA8427K	OEM
TA8432K	OEM
TA8445K	OEM
TA8515F	OEM
TA8601AN	OEM
TA8601BN	OEM
TA8601BN-FA-1	OEM
TA8601BNV	OEM
TA8601CN	OEM
TA8601CN(FA-1)	OEM
TA8603	OEM
TA8603P	OEM
TA8604	OEM
TA8604F	OEM
TA8604N	OEM
TA8606F	OEM
TA8607P	OEM
TA8609P	OEM
TA8611AN	OEM
TA8618S	OEM
TA8619P	OEM
TA8620P	OEM
TA8622N	OEM
TA8624F	OEM
TA8628N	OEM
TA8644N	OEM
TA8647S	OEM
TA8654AN	OEM
TA8654N	OEM
TA8655AN	OEM
TA-8677N	OEM
TA8677N	3236

If replacement code is OEM, contact original manufacturer for replacement.

DEVICE TYPE	REPL CODE
TA8680AN	3515
TA8680BN	3515
TA8680BNFA-1	3515
TA8680N	3515
TA8703S	3623
TA8720AN	OEM
TA8723N	OEM
TA8725AN	OEM
TA8725N	OEM
TA8725N-J	OEM
TA8728P	OEM
TA8745N	OEM
TA8747N	OEM
TA8776N	OEM
TA8782N	OEM
TA8785N	OEM
TA8801AN	OEM
TA8801N(FA)	OEM
TA8819N	OEM
TA8825AN	OEM
TA8845AN	OEM
TA8845BN	OEM
TA8851AN	OEM
TA8859AP	OEM
TA8870N	OEM
TA8879N	OEM
TA8991	0065
TA-9014P	OEM
TA9106	OEM
TA9112A	3088
TA9112B	2176
TA9192B	OEM
TA9193	OEM
TA9195B	OEM
TA9232	OEM
TA9437A	OEM
TA9437B	OEM
TA9438A	OEM
TA9438B	OEM
TA10072	OEM
TA10086B	OEM
TA10086C	OEM
TA10535	OEM
TA10641-1	OEM
TA10642	OEM
TA11868A	OEM
TA13280A	OEM
TA17044	0280
TA31002P	OEM
TA31032P	OEM
TA57745	OEM
TA70639	0433
TA75324P	OEM
TA75339	0176
TA75339P	0176
TA75358P	0770
TA75393	0631
TA75393P	0631
TA75393S	OEM
TA75458P	0360
TA75458S	OEM
TA75557P	0360
TA75557S	2913
TA75558P	0360
TA75558S	0770
TA75558S-1	0770
TA75559P	0360
TA75559P(FA-1)	0360
TA75559PFA-1	0360
TA75902	0661
TA75902P	0829
TA78005AP	0624
TA78005P	0624
TA78009AP	OEM
TA78010AP	2806
TA78012A	0334
TA78012AP	0334
TA78012M	3963
TA78055	OEM
TA78607AP	OEM
TA78805	OEM
TA20007330	0006
TA30016060	0718
TA50013040	1648
TAA320	0325
TAA500-B	OEM
TAA521	1301
TAA521A	1705
TAA522	1301
TAA550	4769
TAA550-B	1329
TAA550B	4769
TAA611A12	2158
TAA611B	2158
TAA611B1Z	OEM
TAA611B12	2158
TAA611C11	1920
TAA621	2472
TAA621A11	2472
TAA621A12	2482
TAA780	OEM
TAA790	OEM
TAA890	OEM
TAB101	OEM
TAC010	OEM
TAC047	0016
TA-D93	OEM
TAD93	OEM
TAG0630	OEM
TAG0660	OEM
TAG0690	OEM
TAG1-200	0179
TAG1-400	0346
TAG1-500	3349
TAG2-100	0179
TAG2-200	0179
TAG2-300	0346
TAG2-400	0346
TAG2N400	0179
TAG2.5-100	0179
TAG2.5-200	0179
TAG2.5-300	0346
TAG2.5-400	0346
TAG3-50	OEM
TAG3-300	OEM
TAG7-50R	2449
TAG7-100R	2449
TAG7-200R	0435
TAG7-300R	1487
TAG7-400R	1487
TAG7-500R	1487
TAG7-600R	0687
TAG7-700R	OEM
TAG7-800R	OEM
TAG7S50R	OEM
TAG7S100R	OEM
TAG7S200R	OEM
TAG7S300R	OEM
TAG7S400R	OEM
TAG7S500R	OEM
TAG7S600R	OEM
TAG7S700R	OEM
TAG7S800R	OEM
TAG10-50R	2449
TAG10-100R	2449
TAG10-200R	0435
TAG10-300R	1487
TAG10-400R	1487
TAG10-500R	0687
TAG10-600R	0687
TAG10-700R	OEM
TAG10-800R	OEM
TAG10S50R	OEM
TAG10S100R	OEM
TAG10S200R	OEM
TAG10S300R	OEM
TAG10S400R	OEM
TAG10S500R	OEM
TAG10S600R	OEM
TAG10S700R	OEM
TAG10S800R	OEM
TAG19F800DU	0750
TAG19F4000DU	0722
TAG20-50R	0731
TAG20-100R	0712
TAG20-200R	0469
TAG20-300R	0722
TAG20-400R	0722
TAG20-500R	0725
TAG20-600R	0725
TAG20-700R	OEM
TAG20-800R	OEM
TAG20S50R	OEM
TAG20S100R	OEM
TAG20S200R	OEM
TAG20S300R	OEM
TAG20S400R	OEM
TAG20S500R	OEM
TAG20S600R	OEM
TAG20S700R	OEM
TAG20S800R	OEM
TAG92-1	4455
TAG92-2	4455
TAG92-3	4455
TAG92-4	4455
TAG92-5	4456
TAG92-6	4456
TAG200-100	2383
TAG200-200	2394
TAG200-400	1413
TAG200-600	2387
TAG201-100	2352
TAG201-200	2353
TAG201-400	2356
TAG201-600	5942
TAG220-200	0555
TAG220-400	0555
TAG220-600	0555
TAG222-400	0572
TAG222-600	0935
TAG222-800	0059
TAG224-200	0772
TAG224-600	0617
TAG225-200	0555
TAG225-600	0555
TAG230-200	0555
TAG230-400	0555
TAG230-600	0555
TAG240-200	0555
TAG240-400	0555
TAG240-600	0555
TAG241-200	0555
TAG241-400	0555
TAG241-600	0555
TAG245-200	0555
TAG245-400	0555
TAG245-600	0555
TAG246-200	0555
TAG246-400	0555
TAG246-600	0555
TAG250-200	0772
TAG250-400	0744
TAG250-600	0617
TAG251-200	0772
TAG251-400	0744
TAG251-600	0617
TAG252-200	0555
TAG252-400	0555
TAG252-600	0555
TAG255-200	0772
TAG255-400	0744
TAG255-600	0617
TAG256-200	0772
TAG256-400	0744
TAG256-600	0617
TAG257-200	0555
TAG257-400	0555
TAG257-600	0555
TAG280-200	0772
TAG280-400	0744
TAG280-600	0617
TAG420-200	0555
TAG420-400	0555
TAG420-600	0555
TAG421-200	0555
TAG421-400	0555
TAG421-600	0555
TAG425-200	0555
TAG425-400	0555
TAG425-600	0555
TAG426-200	0555
TAG426-400	0555
TAG611-100	0179
TAG611-200	0179
TAG611-400	0346
TAG611-600	3349
TAG612-100	0179
TAG612-200	0179
TAG612-400	0346
TAG612-600	3349
TAG620-100	2089
TAG620-200	0505
TAG620-400	0710
TAG620-600	0863
TAG622-200	0717
TAG622-400	0717
TAG622-600	0308
TAG623-200	0717
TAG623-400	0717
TAG623-600	0308
TAG625-100	2089
TAG625-200	0505
TAG625-400	0710
TAG625-600	0863
TAG625-800	0327
TAG627-100	0717
TAG627-200	0717
TAG627-400	0717
TAG627-600	0308
TAG628-200	0717
TAG628-400	0717
TAG628-600	0308
TAG630-200	0505
TAG630-400	0710
TAG630-600	0863
TAG630-800	0327
TAG632-200	0717
TAG632-400	0717
TAG632-600	0308
TAG633-200	0717
TAG633-400	0717
TAG633-600	0717
TAG655S200	0505
TAG655S300	0710
TAG655S400	0710
TAG655S500	0863
TAG655S600	0863
TAG670-100	2089
TAG670-200	0505
TAG670-400	0710
TAG670-600	0863
TAG670-800	0327
TAG675-100	0397
TAG675-200	0397
TAG675-400	0611
TAG675-600	0952
TAG675-800	1512
TAG6151	0772
TAG6152	0744
TAG6153	0617
TAG6154	0772
TAG6155	0744
TAG6156	0617
TAIT00101P	0006
TAL004	OEM
TALT00952L	0006
TALT007330	0006
TALT009520	0006
TA-M93	OEM
TAM93	OEM
TASCA3137	2789
TAT004	OEM
TAT008	OEM
TAT010	OEM
TAT020	OEM
TAT328AP	OEM
TAT1371A0	OEM
TAU051A/AN3231K	OEM
TAU1765	OEM
TAV	4630
TB-1	0277
TB-2	0289
TB3G0632KE	0460
TB3G01134R	0925
TB3T006980	0006
TB3T009260	4011
TB-4	0296
TB4B	1875
TB4H	1875
TB4.1	0609
TB5	0110
TB5-1	1875
TB6.B	0241
TB-15H	0348
TB100	0242
TB200	0242
TB410FA	OEM
TB410FB	OEM
TB410V	OEM
TB412	OEM
TB413	OEM
TB414	OEM
TB415	OEM
TB416	OEM
TB417	OEM
TB418	OEM
TB419	OEM
TB420FA	OEM
TB420FB	OEM
TB420V	OEM
TB430FA	OEM
TB430FB	OEM
TB440FA	OEM
TB440FB	OEM
TB440V	OEM
TB450FA	OEM
TB450FB	OEM
TB460FA	OEM
TB460FB	OEM
TB460V	OEM
TB470FA	OEM
TB470FB	OEM
TB480FA	OEM
TB480FB	OEM
TB480V	OEM
TB490FA	OEM
TB490FB	OEM
TB820	OEM
TB1203BP	OEM
TB1203P	OEM
TB1207N-BA1	OEM
TB4100FA	OEM
TB4100FB	OEM
TB4100V	OEM
TB4110	OEM
TB4120V	OEM
TB4130	OEM
TB4140V	OEM
TB4150	OEM
TB4160V	OEM
TB200546AL	1648
TB3000632E	0460
TB3001134R	0925
TB30006320	0460
TBA10S-H	1248
TBA120	6177
TBA120AS	3451
TBA120S	3451
TBA120U	0351
TBA180S	OEM
TBA221	0410
TBA222	0410
TBA281	1192
TBA396	OEM
TBA641B11	4586
TBA-800	4147
TBA800	4147
TBA800C	OEM
TBA806	OEM
TBA810	1248
TBA810AS	1248
TBA810AT	OEM
TBA810DS	1248
TBA810P	3922
TBA-810S	1248
TBA810S	1248
TBA810S-H	1248
TBA-820	4003
TBA820	4003
TBA820L	4003
TBA820M	4006
TBA820MT	2623
TBA840	OEM
TBA890	OEM
TBA920MT	OEM
TBA940	OEM
TBA950	OEM
TBA950F	OEM
TBA1204B	OEM
TBA1440G	OEM
TBA1441GN	OEM
TBA2214	1982
TBA6003	OEM
TBAB10S	OEM
TBB0747	2368
TBB0747A	2358
TBB0748	0093
TBB0748B	1300
TBB1458B	0360
TBC337	0111
TBC338	0111
TBC414C	0079
TBC547	0111
TBC548	0111
TBC549	0111
TBC550	0111
TBG625	OEM
TBG625A	OEM
TBG630	OEM
TBG820	OEM
TBG820A	OEM
TBG825	OEM
TBG825A	OEM
TBG830	OEM
TBG920	OEM
TBG920A	OEM
TBG925	OEM
TBG925A	OEM
TBG930	OEM
TBG1020	OEM
TBG1020A	OEM
TBG1025	OEM
TBG1025A	OEM
TBG1030	OEM
TBI010	OEM
TBI110	OEM
TBI208	OEM
TBI210	OEM
TBI308	OEM
TBI310	OEM
TBI408	OEM
TBI410	OEM
TBI608	OEM
TBI610	OEM
TBI708	OEM
TBI808	OEM
TBI810	OEM
TBI1008	OEM
TBL005	OEM
TBL005A	OEM
TBL105	OEM
TBL105A	OEM
TBL205	OEM
TBL205A	OEM
TBL305	OEM
TBL305A	OEM
TBL405	OEM
TBL405A	OEM
TBL605	OEM
TBL605A	OEM
TBL805	OEM
TBL805A	OEM
TBL905	OEM
TBL905A	OEM
TBM54578P	OEM
TBP14S10	2224
TBP14S10J	2224
TBP14S10MJ	OEM
TBP14S10N	2224
TBP14SA10	1918
TBP14SA10J	1918
TBP14SA10MJ	OEM
TBP14SA10N	1918
TBP18S030	2175
TBP18S030J	2175
TBP18S030MJ	OEM
TBP18S030N	2175
TBP18S22	OEM
TBP18S22J	OEM
TBP18S22MJ	OEM
TBP18S22N	OEM
TBP18S42	2320
TBP18S42J	2320
TBP18S42MJ	OEM
TBP18S42N	2320
TBP18S46	2322
TBP18S46J	2322
TBP18S46MJ	OEM
TBP18S46N	2322
TBP18SA030	3168
TBP18SA030J	3168
TBP18SA030MJ	OEM
TBP18SA030N	3168
TBP18SA22	OEM
TBP18SA22J	OEM
TBP18SA22MJ	OEM
TBP18SA22N	OEM
TBP18SA42	3288
TBP18SA42J	3288
TBP18SA42MJ	OEM
TBP18SA42N	3288
TBP18SA46	2324
TBP18SA46J	2324
TBP18SA46MJ	OEM
TBP18SA46N	2324
TBP24S10	OEM
TBP24S10J	OEM
TBP24S10MJ	OEM
TBP24S10N	6716
TBP24S41	2392
TBP24S41J	2392
TBP24S41MJ	OEM
TBP24S41N	2392
TBP24S81	3166
TBP24S81-55	OEM
TBP24S81J	3166
TBP24S81N	3166
TBP24SA10	OEM
TBP24SA10J	OEM
TBP24SA10MJ	OEM
TBP24SA10N	OEM
TBP24SA41	2390
TBP24SA41J	2390
TBP24SA41N	2390
TBP24SA81	OEM
TBP24SA81-55	OEM
TBP24SA81J	OEM
TBP24SA81MJ	OEM
TBP24SA81N	OEM
TBP28L22	OEM
TBP28L22J	OEM
TBP28L22MJ	OEM
TBP28L22N	OEM
TBP28L42	OEM
TBP28L42J	OEM
TBP28L42MJ	OEM
TBP28L42N	OEM
TBP28L45J	OEM
TBP28L45MJ	OEM
TBP28L45N	OEM
TBP28L46	OEM
TBP28L85J	OEM
TBP28L85N	OEM
TBP28L86A	OEM
TBP28L86J	OEM
TBP28L86MJ	OEM
TBP28L86N	OEM
TBP28L166	OEM
TBP28L166J	OEM
TBP28L166N	OEM
TBP28LA22	OEM
TBP28P42J	OEM
TBP28P42N	OEM
TBP28P45J	OEM
TBP28P45N	OEM
TBP28P85J	OEM
TBP28P85N	OEM
TBP28P166J	OEM
TBP28P166N	OEM
TBP28R45J	OEM
TBP28R45N	OEM
TBP28R85J	OEM
TBP28R85N	OEM
TBP28R166	OEM
TBP28R166J	OEM
TBP28R166N	OEM
TBP28S42	OEM
TBP28S42J	OEM
TBP28S42MJ	OEM
TBP28S42N	OEM
TBP28S45J	OEM
TBP28S45N	OEM
TBP28S46	OEM
TBP28S46J	OEM
TBP28S46MJ	OEM
TBP28S46N	OEM
TBP28S85J	OEM
TBP28S85MJ	OEM
TBP28S85N	OEM
TBP28S86A	OEM
TBP28S86J	3783
TBP28S86MJ	OEM
TBP28S86N	3783
TBP28S166	OEM
TBP28S166-55	OEM
TBP28S166J	OEM
TBP28S166MJ	OEM
TBP28S166N	OEM
TBP28S2708	OEM
TBP28S2708J	OEM
TBP28S2708MJ	OEM
TBP28S2708N	OEM
TBP28SA86J	2862
TBP28SA86MJ	OEM
TBP28SA86N	2862
TBRC147B	0079
TBS20A	OEM
TBS20AS	OEM
TBS20B	OEM
TBS20BS	OEM
TBS30A	OEM
TBS30B	OEM
TBS30C	OEM
TBS40AS	OEM
TBS40BS	OEM
TBS50CS	OEM
TC-0-2P11/1	0102
TC02P12	0015
TC02P112	0015
TC-0363	OEM
TC0914	0144
TC0918	0144
TC-0.09M21/3	0196
TC0.09M21/3	0484
TC-0.09M21/5	0484
TC0.09M22/1	0277
TC-0.2	0102
TC-0.2P11/1	0102
TC0.2P11/2	0015
TC0.2P11/12	0920
TC1T013170	0155
TC2T009450	0076
TC3F03789E	0276
TC3F017560	0643
TC3F035030	0276
TC3F042170	0276
TC3Q026210	0276
TC3Q040750	1085
TC3R022710	0261
TC3T005360	0537
TC3T0536K0	0537
TC3T0536KF	0537
TC3T0536S0	0537
TC3T022710	0261
TC3T022740	0155
TC3T023620	1227
TC3T030000	0286
TC3T034680	0261
TC3T2274K0	0155
TC3T2362K0	1227
TC4S69	OEM
TC5F03620S	0276
TC5Q3298Y0	OEM
TC5T0388A0	0842
TC5T018154	0076
TC5T021204	0866
TC5T022294	0066
TC7SU04F	OEM
TC7SU04FEL	OEM
TC8A5A	OEM
TC8A50A	OEM
TC10A5A	0170
TC10A50A	OEM
TC10LB2	0940
TC12.4A5A	OEM
TC14.6A5A	OEM
TC15G008AP-0023	OEM
TC15G022AP-0013	OEM
TC15G022AP-0014	OEM
TC16.8A5A	OEM
TC17G014AP-0012	OEM
TC18.5A5A	OEM
TC21A5A	OEM
TC21SC160AT	OEM
TC23A5A	OEM
TC27A5A	0441
TC30A5A	0726
TC33A5A	0039
TC37A5A	OEM
TC40H000P	OEM
TC40H004P	OEM
TC40H032P	OEM
TC40H074P	OEM
TC40H244P	OEM
TC40H373P	OEM
TC40HC04	OEM
TC40HC032P	OEM
TC43A5A	0931
TC47A5A	0999
TC50	0110
TC51A5B	0502
TC56A5B	0869
TC62A5B	0783
TC68A5B	2157
TC74HC00	OEM
TC74HC00F	OEM
TC74HC02	1452
TC74HC02F	OEM
TC74HC02P	1452
TC74HC03P	1453
TC74HC04	1455
TC74HC04F	OEM
TC74HC04P	1455
TC74HC040P	OEM
TC74HC14	OEM
TC74HC14P	3870
TC74HC30P	1489
TC74HC123A	3903
TC74HC123F	3905
TC74HC123F-TP1	3905
TC74HC123P	3907
TC74HC139P	3950
TC74HC157AP	OEM
TC74HC157P	OEM
TC74HC240P	4227
TC74HC4094AP	OEM
TC74HCT244P	7122
TC74HCU04P	4602
TC75A5B	1189
TC82A5B	0331
TC87A5B	3026
TC91A5B	1311
TC100	0242
TC100A5B	OEM
TC101	OEM
TC102	OEM
TC105A5B	OEM
TC106A1	0447
TC106A2	0447
TC106A3	0447
TC106A4	0447
TC106B	0940
TC106B1	0940
TC106B2	0940
TC106B3	0940
TC106B4	0940
TC106C1	1222
TC106C2	1222
TC106C3	1222
TC106C4	1222
TC106D1	0095
TC106D2	0095
TC106D3	0095

If replacement code is OEM, contact original manufacturer for replacement.

DEVICE TYPE	REPL CODE
TC106D4	0095
TC106F1	1259
TC106F2	1259
TC106F3	1259
TC106F4	1259
TC106Q1	1396
TC106Q2	1396
TC106Q3	1396
TC106Q4	1396
TC106Y1	1396
TC106Y2	1396
TC106Y3	1396
TC106Y4	1396
TC110A5B	0149
TC120A5B	0186
TC130A5C	0213
TC136	0015
TC150A5C	0028
TC175A5C	OEM
TC200	0242
TC200A5C	OEM
TC261	OEM
TC301	OEM
TC302	OEM
TC303	OEM
TC304	OEM
TC306	OEM
TC2126	OEM
TC2369A	0144
TC2483	0144
TC2484	0144
TC3001	1540
TC3114	0144
TC4000BP	2024
TC4001B	OEM
TC4001BP	0478
TC4001P	0478
TC4001UBP	0478
TC4002BP	2055
TC4002P	2055
TC4007BP	0654
TC4008BP	0988
TC4009UBP	0001
TC4010BP	0398
TC4011	0215
TC4011B	0215
TC4011BP	0215
TC4011P	0215
TC4011UBP	0215
TC4012BP	0498
TC4013	OEM
TC4013B	0413
TC4013BAP	OEM
TC4013BCP	0413
TC4013BP	0413
TC4013P	0413
TC4015BP	1014
TC4015P	1014
TC4016BB	1143
TC4016BP	0101
TC4017BP	0513
TC4017P	0513
TC4018BP	1391
TC4019BP	1525
TC4019P	1525
TC4020BP	1661
TC4021P	1749
TC4022BP	1256
TC4023BP	0519
TC4023P	0519
TC4024BP	1957
TC4025BP	2072
TC4025P	2072
TC4027	1949
TC4027B	1949
TC4027BP	1949
TC4027P	1949
TC4028B	2228
TC4028BP	2228
TC4028P	2228
TC4029BP	2233
TC4030BP	0500
TC4030P	0500
TC4032BP	2531
TC4038BP	2981
TC4040B	0056
TC4040BP	0056
TC4040P	0056
TC4042BP	0121
TC4042P	0121
TC4043BP	1553
TC4044BP	2308
TC4047BP	2331
TC4049	0001
TC4049BF	OEM
TC4049BP	0001
TC4049P	0001
TC4050BP	0398
TC4050CP	0398
TC4050P	0398
TC4051BP	0366
TC4052	0024
TC4052BP	0024
TC4052BP-1	0024
TC4053BF	OEM
TC4053BP	0035
TC4053BP-1	OEM
TC4053BPHB	OEM
TC4054BP	4530

DEVICE TYPE	REPL CODE
TC4055BP	3304
TC4056BP	3706
TC4063BP	3727
TC4066	0101
TC4066B	0101
TC4066BC	0101
TC4066BP	0101
TC4066P	0101
TC4068BP	2503
TC4069B	3311
TC4069BP	0119
TC4069P	0119
TC4069UB	3311
TC4069UBP	0119
TC4069UPB	0119
TC4069UPN	0119
TC4071BP	0129
TC4071P.JA	0129
TC4072BP	2524
TC4073BP	1536
TC4075BP	2540
TC4077BP	OEM
TC4078BP	0921
TC4081	0626
TC4081BP	0332
TC4081P	0626
TC4082BP	0300
TC4082P	0300
TC4085BP	0303
TC4086BP	0535
TC4093BP	2384
TC4094BP	1682
TC4099BP	3330
TC4502BP	OEM
TC4508BP	1811
TC4510BP	1963
TC4511BP	1544
TC4512BP	2120
TC4512P	2120
TC4514BP	1830
TC4514P	1830
TC4515BP	2285
TC4516BP	2347
TC4518BP	1043
TC4519BP	OEM
TC4520BP	2675
TC4521BP	OEM
TC4522BP	2837
TC4526BP	OEM
TC4527BP	3145
TC4528BFHB	OEM
TC4528BP	3198
TC4531BP	3325
TC4532BP	3368
TC4538BP	1063
TC4539BP	3656
TC4539P	3656
TC4555BP	2938
TC4556BP	3431
TC4584	3626
TC4584BP	3626
TC5003P	OEM
TC5012BP	2053
TC5018P	OEM
TC5030BP	OEM
TC5036P	OEM
TC5064BP	OEM
TC5066BP	OEM
TC5067BP	OEM
TC5069BP	OEM
TC5080	4620
TC5080P	4700
TC5081	4624
TC5081AP	OEM
TC5081P	4624
TC5082	4620
TC5082P	4620
TC5082P-L	4620
TC5508P	3193
TC5508P-1	3193
TC5514AP	OEM
TC5514AP3	OEM
TC5516AFL-2T	OEM
TC5516AP	OEM
TC5518BF-25	OEM
TC5565	1898
TC5565-LL	OEM
TC5565AFL-15L	OEM
TC5565APL-15L	OEM
TC5565PL-15	OEM
TC7400	0232
TC7400BP	2112
TC7476BP	2164
TC8601F	OEM
TC9002AP	OEM
TC9002BP	OEM
TC9002DP	OEM
TC9002P	OEM
TC9003AP	OEM
TC9012	OEM
TC9012F	OEM
TC9012F-001	OEM
TC9012F-011	OEM
TC9020P	OEM
TC9020P-003	OEM
TC9029P-002	OEM
TC9029P-SPEC	OEM
TC9100C	1714
TC9100P	1714

DEVICE TYPE	REPL CODE
TC9102P	OEM
TC9103P	OEM
TC9106BP	OEM
TC9106D	2828
TC9106P	2828
TC9107P	OEM
TC9109P	OEM
TC9112	OEM
TC9112P	OEM
TC9116P	OEM
TC9125BP	OEM
TC9125P	OEM
TC9126N	OEM
TC9132P	OEM
TC9137BP	OEM
TC9137P	OEM
TC9138AP	OEM
TC9143P	OEM
TC9145P	6295
TC9147BP	OEM
TC9148P	OEM
TC9149P	OEM
TC9150P	OEM
TC9152P	OEM
TC9153AP	OEM
TC9153P	OEM
TC9154AP	OEM
TC9154P	OEM
TC9157AP	OEM
TC9157P	OEM
TC9158P	OEM
TC9159P	OEM
TC9160F	OEM
TC9160P	OEM
TC9162N	OEM
TC9163N	OEM
TC9164N	OEM
TC9171P	OEM
TC9172P	OEM
TC9174P	OEM
TC9176P	OEM
TC9177P	OEM
TC9183F	OEM
TC9184P	OEM
TC9188N	OEM
TC9212P	OEM
TC9214P	OEM
TC9302AF-005	OEM
TC9302AF-008	OEM
TC9302AF-012	OEM
TC9302AF-035	OEM
TC9303AN001	OEM
TC9304F-013	OEM
TC9304F-024	OEM
TC9306F-024	OEM
TC9310N-014	OEM
TC9310N-023	OEM
TC40160BP	1359
TC40161BP	1373
TC40162BP	1388
TC40163BP	1407
TC40174BP	1551
TC40175BP	1528
TC40192BP	3885
TC40193BP	1776
TC51832SPL-12	OEM
TC53257P-6887	OEM
TC53257P-6888	OEM
TC89101P	OEM
TC500388A0	0842
TC3002271D	0261
TC3002344E	1284
TC3002621E	0276
TC10013170	0155
TC20009450	0076
TC20019410	1327
TC30005360	0537
TC30022710	0261
TC30024560	0276
TC30031560	3385
TC30034670	0261
TC30039020	0563
TC30041590	1165
TC50018154	0076
TC50032960	0392
TC311200600	0143
TC312307222	0079
TC3112006000	0143
TC3112319300	0015
TC3123036722	0016
TC3123036900	0016
TC3123037111	0016
TC3123037222	0016
TC3123037412	0016
TC3123041557	0004
TCA0372DP2	3778
TCA250	OEM
TCA350	OEM
TCA350X	OEM
TCA350Y	OEM
TCA350Z	OEM
TCA380	0379
TCA420A	OEM
TCA430-N	OEM
TCA440	OEM
TCA440N	OEM
TCA700	OEM
TCA700Y	OEM
TCA710A	OEM

DEVICE TYPE	REPL CODE
TCA720	OEM
TCA830	3922
TCA830A	1248
TCA830S	3922
TCA860	OEM
TCB0331	0142
TCE1A66102-04E	OEM
TCG1V010	1691
TCG1V014	2008
TCG1V015	1314
TCG1V017	2325
TCG1V020	2790
TCG1V035	2935
TCG1V050	4472
TCG1V115	4135
TCG1V130	1374
TCG1V150	0246
TCG1V250	1332
TCG2V010	1694
TCG2V014	2009
TCG2V015	1319
TCG2V017	2326
TCG2V020	2796
TCG2V035	2941
TCG2V050	3085
TCG2V115	3123
TCG2V130	0838
TCG2V150	1993
TCG2V250	1334
TCG2V480	1101
TCG33	0046
TCG34	3630
TCG35	5056
TCG36	0194
TCG36MP	5393
TCG37	3111
TCG37MCP	4475
TCG38	0792
TCG39	4479
TCG40	2924
TCG40-45B	3443
TCG41	5057
TCG42	2411
TCG43	5058
TCG44	2947
TCG45	5059
TCG46	2797
TCG47	2860
TCG48	3799
TCG49	0824
TCG50	3327
TCG51	3014
TCG52	4591
TCG53	3038
TCG54	1165
TCG54MP	5603
TCG55	0718
TCG55MCP	3789
TCG56	2069
TCG57	0629
TCG58	3226
TCG59	3646
TCG60	3701
TCG60MP	5693
TCG61	4144
TCG61MCP	2011
TCG61MP	5704
TCG62	2847
TCG63	2844
TCG64	3007
TCG65	1342
TCG66	5060
TCG67	1465
TCG68	3606
TCG68MCP	4864
TCG70	3489
TCG71	3512
TCG72	0939
TCG73	2662
TCG74	2873
TCG74C00	2112
TCG74C02	2114
TCG74C04	2958
TCG74C08	2118
TCG74C10	2119
TCG74C14	3031
TCG74C20	2123
TCG74C30	3128
TCG74C32	3134
TCG74C42	2143
TCG74C48	2148
TCG74C74	2163
TCG74C76	2164
TCG74C85	2164
TCG74C90	2174
TCG74C95	3401
TCG74C107	3834
TCG74C154	4014
TCG74C157	4033
TCG74C160	4045
TCG74C161	4046
TCG74C164	4062
TCG74C173	4088
TCG74C174	4092
TCG74C175	4096
TCG74C192	4119
TCG74C193	4122
TCG74C221	3047
TCG74C240	2956

DEVICE TYPE	REPL CODE
TCG74C244	4230
TCG74C901	4607
TCG74C902	4608
TCG74C903	4609
TCG74C904	4610
TCG74C925	4621
TCG74H00	0682
TCG74H01	5338
TCG74H04	1907
TCG74H05	3251
TCG74H08	5355
TCG74H10	0685
TCG74H11	2398
TCG74H20	3715
TCG74H21	4856
TCG74H22	4596
TCG74H30	5383
TCG74H40	0559
TCG74H50	1792
TCG74H51	1944
TCG74H52	2020
TCG74H53	2101
TCG74H54	2172
TCG74H55	3158
TCG74H60	5410
TCG74H61	2663
TCG74H62	2731
TCG74H71	3264
TCG74H72	3313
TCG74H73	2463
TCG74H74	2492
TCG74H76	5302
TCG74H78	5418
TCG74H87	2579
TCG74H101	5521
TCG74H102	5523
TCG74H103	2969
TCG74H106	5253
TCG74H108	0180
TCG74H183	4399
TCG74L93	0656
TCG74LS00	1527
TCG74LS01	1546
TCG74LS02	1559
TCG74LS03	1579
TCG74LS04	1595
TCG74LS05	1608
TCG74LS08	1633
TCG74LS09	1642
TCG74LS10	1662
TCG74LS11	1667
TCG74LS12	1679
TCG74LS13	1688
TCG74LS14	1698
TCG74LS15	1707
TCG74LS20	0034
TCG74LS21	1763
TCG74LS22	1775
TCG74LS26	1382
TCG74LS27	0183
TCG74LS30	0828
TCG74LS32	0088
TCG74LS37	1730
TCG74LS38	1839
TCG74LS40	0135
TCG74LS42	1841
TCG74LS47	1845
TCG74LS48	1849
TCG74LS49	1850
TCG74LS51	1033
TCG74LS54	1857
TCG74LS55	0457
TCG74LS73	1867
TCG74LS74A	0243
TCG74LS75	1870
TCG74LS76A	2180
TCG74LS78	1873
TCG74LS83A	2218
TCG74LS85	0431
TCG74LS86	0291
TCG74LS90	1882
TCG74LS92	1887
TCG74LS93	1888
TCG74LS107	1602
TCG74LS109A	1906
TCG74LS112A	2128
TCG74LS113	2256
TCG74LS114	2302
TCG74LS123	0979
TCG74LS124	5988
TCG74LS125A	0075
TCG74LS126A	2877
TCG74LS132	1625
TCG74LS136	1628
TCG74LS138	0427
TCG74LS139	0153
TCG74LS145	1564
TCG74LS151	1646
TCG74LS153	0959
TCG74LS155	0209
TCG74LS156	1654
TCG74LS157	1161
TCG74LS158	1656
TCG74LS160A	0837
TCG74LS161A	0858
TCG74LS162A	0880
TCG74LS163A	0893
TCG74LS164	4339
TCG74LS170	2629

DEVICE TYPE	REPL CODE
TCG74LS173A	5217
TCG74LS174	0260
TCG74LS175	1672
TCG74LS190	1686
TCG74LS191	1687
TCG74LS192	1689
TCG74LS193	1692
TCG74LS194	1304
TCG74LS195A	1315
TCG74LS196	2834
TCG74LS197	2469
TCG74LS221	1239
TCG74LS240	0452
TCG74LS241	1726
TCG74LS242	1728
TCG74LS243	0906
TCG74LS244	0458
TCG74LS245	0463
TCG74LS247	1732
TCG74LS248	1734
TCG74LS249	1735
TCG74LS251	1737
TCG74LS253	1739
TCG74LS257	1744
TCG74LS258	1746
TCG74LS259	3205
TCG74LS266	0592
TCG74LS273	0894
TCG74LS279	3290
TCG74LS280	1773
TCG74LS283	1779
TCG74LS290	4425
TCG74LS293	0082
TCG74LS296	OEM
TCG74LS298	3371
TCG74LS324	6000
TCG74LS327	6002
TCG74LS348	0665
TCG74LS352	0761
TCG74LS353	0773
TCG74LS363	6005
TCG74LS364	6006
TCG74LS365A	0943
TCG74LS366A	0956
TCG74LS367	0977
TCG74LS368	0991
TCG74LS373	0709
TCG74LS374	0713
TCG74LS377	1120
TCG74LS378	1133
TCG74LS379	1151
TCG74LS393	0819
TCG74LS395A	1330
TCG74LS490	2213
TCG74LS540	2541
TCG74LS541	2547
TCG74LS624	3141
TCG74LS625	3149
TCG74LS626	3154
TCG74LS627	3162
TCG74LS629	3175
TCG74LS640	0669
TCG74LS641	0690
TCG74LS642	0719
TCG74LS643	2056
TCG74LS645	0775
TCG74LS670	1130
TCG74S00	0704
TCG74S02	2238
TCG74S03	2217
TCG74S04	2263
TCG74S05	2321
TCG74S08	2569
TCG74S09	2667
TCG74S10	2445
TCG74S11	2447
TCG74S15	2451
TCG74S20	1017
TCG74S22	2461
TCG74S30	3726
TCG74S37	5756
TCG74S38	0780
TCG74S40	2475
TCG74S51	4305
TCG74S64	2496
TCG74S65	2497
TCG74S74	2504
TCG74S85	5773
TCG74S86	2511
TCG74S112	1617
TCG74S113	1623
TCG74S114	1629
TCG74S124	2126
TCG74S132	2134
TCG74S133	1819
TCG74S134	1827
TCG74S138	2138
TCG74S140	1886
TCG74S151	1955
TCG74S153	2151
TCG74S157	1695
TCG74S158	2154
TCG74S163	0893
TCG74S174	2132
TCG74S175	2141
TCG74S181	2165
TCG74S182	2166
TCG74S194	1931
TCG74S251	2198

DEVICE TYPE	REPL CODE
TCG74S258	2205
TCG74S280	1773
TCG74S373	2264
TCG74S374	2266
TCG75	2964
TCG76	2070
TCG76MP	5956
TCG77	2208
TCG78	5062
TCG79	3797
TCG80	2598
TCG80C95	3177
TCG80C96	3179
TCG80C97	3180
TCG81	2045
TCG82	2468
TCG83	5018
TCG84	5064
TCG85	0286
TCG86	2548
TCG87	0867
TCG87MP	6034
TCG88	3500
TCG88MCP	4500
TCG88MP	6041
TCG89	0055
TCG90	0867
TCG91	0648
TCG92	2276
TCG93	3581
TCG93L08	2127
TCG93L16	2685
TCG93MCP	3303
TCG94	0270
TCG95	4252
TCG96	4029
TCG96L02	1468
TCG96LS02	2934
TCG96S02	4292
TCG97	1991
TCG98	2626
TCG99	4013
TCG100	0280
TCG101	0600
TCG102	0211
TCG102A	0004
TCG103	0038
TCG103A	0208
TCG104	0085
TCG104MP	3033
TCG105	0440
TCG106	0150
TCG107	0127
TCG108	0144
TCG108-1	3777
TCG109	0143
TCG110	1114
TCG110MP	0123
TCG112	0917
TCG113A	0196
TCG114	0484
TCG116	0015
TCG117	0229
TCG117A	4630
TCG118	0774
TCG120	0474
TCG121	0160
TCG121MP	0265
TCG122	0240
TCG123	0198
TCG123A	0016
TCG123AP	0079
TCG124	0142
TCG125	0071
TCG126	0136
TCG126A	0633
TCG127	0975
TCG128	0086
TCG129	0126
TCG129MCP	4163
TCG130	0103
TCG130MP	1514
TCG131	0222
TCG131MP	0822
TCG132	0325
TCG133	3483
TCG134A	0188
TCG135A	0162
TCG136A	0157
TCG137A	0636
TCG138A	0649
TCG139A	0012
TCG140A	0170
TCG141A	0794
TCG142A	0137
TCG143A	0365
TCG144A	0100
TCG145A	0002
TCG146A	0441
TCG147A	0039
TCG148A	1834
TCG149A	0783
TCG150A	0331
TCG151A	0149
TCG152	0042
TCG152MP	5088
TCG153	0925
TCG153MCP	4220
TCG154	0233

Original Device Types

DEVICE TYPE	REPL CODE
TCG155	2763
TCG156	0087
TCG157	0276
TCG158	0164
TCG159	0037
TCG159MCP	1381
TCG160	0050
TCG161	0007
TCG162	0074
TCG163A	0642
TCG164	0003
TCG165	0065
TCG166	0277
TCG167	0289
TCG168	0296
TCG169	0302
TCG170	0250
TCG171	0285
TCG172A	0400
TCG173BP	0293
TCG174	0924
TCG175	0178
TCG176	0847
TCG177	0133
TCG178MP	0913
TCG179	0604
TCG179MP	5536
TCG180	1681
TCG180MCP	3122
TCG181	0130
TCG181MP	5474
TCG182	0561
TCG183	1199
TCG184	0161
TCG184MP	5595
TCG185	0460
TCG185MCP	4605
TCG186	0560
TCG186A	0219
TCG187	1267
TCG187A	1051
TCG188	0551
TCG189	0382
TCG190	0264
TCG191	0338
TCG192	0595
TCG192A	0218
TCG193	0791
TCG193A	1242
TCG194	0861
TCG195A	0698
TCG196	0423
TCG197	0854
TCG198	0168
TCG199	0111
TCG209	3509
TCG210	0566
TCG211	1367
TCG213	0437
TCG218	0905
TCG219	0491
TCG219MCP	5498
TCG220	0849
TCG221	0353
TCG222	0212
TCG224	0630
TCG225	3280
TCG226	1010
TCG226MP	1862
TCG228A	1708
TCG229	0224
TCG230	0239
TCG231	0061
TCG232	3519
TCG233	0330
TCG234	0693
TCG235	0936
TCG236	0836
TCG237	1411
TCG238	0313
TCG239	2336
TCG241	2997
TCG242	3165
TCG243	2427
TCG244	2277
TCG245	3373
TCG246	3374
TCG247	2439
TCG248	2432
TCG249	1394
TCG250	2448
TCG251	3525
TCG252	3526
TCG253	0558
TCG254	2897
TCG257	3377
TCG258	3528
TCG259	3529
TCG260	3530
TCG261	1212
TCG262	0602
TCG263	2235
TCG264	2237
TCG265	2258
TCG266	3533
TCG267	3534
TCG268	3535
TCG269	3536
TCG270	0134
TCG271	0073
TCG272	3538
TCG273	3539
TCG274	3370
TCG275	3540
TCG276	1652
TCG277	2214
TCG278	0418
TCG280	0177
TCG280MP	2140
TCG281	2013
TCG281MCP	5778
TCG282	0622
TCG283	0363
TCG284	0543
TCG284MP	2371
TCG285	1598
TCG285MCP	6141
TCG286	1027
TCG287	0715
TCG288	0342
TCG289	0155
TCG289A	0155
TCG289AMP	1224
TCG289MP	2767
TCG290	0006
TCG290A	0006
TCG290AMCP	3689
TCG290AMP	OEM
TCG290Q	OEM
TCG291	0236
TCG292	0681
TCG292MCP	3255
TCG293	0018
TCG293MP	3049
TCG294	0532
TCG295	1591
TCG297	0324
TCG297MP	3247
TCG298	0436
TCG299	2050
TCG300	2046
TCG300MP	3376
TCG302	1173
TCG304Q	OEM
TCG306	1946
TCG307	1430
TCG308	1712
TCG308P	4615
TCG309K	1922
TCG310	2329
TCG310P	4606
TCG311	0493
TCG312	0325
TCG313	0475
TCG314	1825
TCG315	1978
TCG316	0259
TCG317	2946
TCG318	2312
TCG319	0008
TCG319P	0008
TCG320	1974
TCG320F	2884
TCG321	1751
TCG322	1984
TCG323	0892
TCG324	1480
TCG325	1198
TCG326	2987
TCG327	2485
TCG328	0620
TCG329	2700
TCG330	3561
TCG331	0482
TCG331MP	4184
TCG332	1369
TCG332MPC	3949
TCG333	2835
TCG334	3563
TCG335	3565
TCG336	3567
TCG337	3568
TCG338	3569
TCG338F	5202
TCG339	3570
TCG340	2725
TCG341	3572
TCG342	2719
TCG343	2720
TCG344	2526
TCG345	3041
TCG346	2041
TCG347	3576
TCG348	3578
TCG349	2170
TCG350	1233
TCG350F	5303
TCG351	1977
TCG352	3194
TCG353	3580
TCG354	3582
TCG355	3583
TCG356	3584
TCG357	3586
TCG358	1030
TCG358A	4465
TCG358B	5036
TCG359	3587
TCG360	2506
TCG361	2039
TCG362	3560
TCG363	2091
TCG364	1420
TCG365	2093
TCG366	3589
TCG367	2868
TCG368	3591
TCG369	2096
TCG373	0563
TCG374	0525
TCG375	0392
TCG376	1085
TCG377	0060
TCG378	1308
TCG379	0728
TCG380	2939
TCG380MP	2735
TCG381	3571
TCG382	1386
TCG383	0477
TCG384	0429
TCG385	1966
TCG386	1852
TCG387	2433
TCG387MP	3187
TCG388	2414
TCG389	0223
TCG390	3081
TCG391	0859
TCG392	3602
TCG393	3603
TCG394	2613
TCG395	3607
TCG396	0187
TCG397	0439
TCG398	1648
TCG399	0261
TCG451	1392
TCG452	2888
TCG453	3072
TCG454	2458
TCG455	0371
TCG456	3622
TCG457	2950
TCG458	1758
TCG459	3133
TCG460	1141
TCG461	2945
TCG462	3628
TCG464	0844
TCG465	0983
TCG466	1155
TCG467	3130
TCG470	3631
TCG471	2545
TCG472	3421
TCG473	3326
TCG474	3632
TCG475	3321
TCG476	3322
TCG477	3119
TCG478	3636
TCG480	3638
TCG481	3640
TCG482	3641
TCG483	3642
TCG484	3643
TCG485	3644
TCG486	0689
TCG500A	0190
TCG502	0201
TCG503	0288
TCG504	0378
TCG505	0757
TCG506	0102
TCG507	0920
TCG508	1306
TCG509	2083
TCG510	2084
TCG511	1449
TCG512	2085
TCG513	1323
TCG514	3649
TCG516	OEM
TCG519	0124
TCG521	2983
TCG522	1501
TCG523	1067
TCG524	3652
TCG524V13	0830
TCG524V15	3415
TCG524V25	1336
TCG524V48	0781
TCG525	0348
TCG526A	1706
TCG529	2546
TCG530	3655
TCG531	2804
TCG532	0409
TCG533	1710
TCG534	2982
TCG535	2985
TCG536A	1997
TCG537	1197
TCG538	2984
TCG539	1054
TCG541	3658
TCG542	3659
TCG544	3660
TCG546	3661
TCG548	3662
TCG551	0182
TCG552	0023
TCG555	2518
TCG556	3666
TCG557	3667
TCG558	0017
TCG565	3669
TCG600	1925
TCG601	0139
TCG604	3683
TCG605	0120
TCG606	3684
TCG607	3442
TCG610	0554
TCG611	0005
TCG612	0720
TCG613	0628
TCG614	0030
TCG617	3687
TCG618	3688
TCG700	3719
TCG701	3720
TCG702	2801
TCG703A	0632
TCG704	2624
TCG705A	2161
TCG706	2571
TCG707	0753
TCG708	0664
TCG709	0678
TCG710	0829
TCG711	2715
TCG712	0167
TCG714	0352
TCG715	0354
TCG716	3722
TCG717	3724
TCG718	0654
TCG720	0443
TCG721	2279
TCG722	0701
TCG723	1345
TCG724	0823
TCG725	0692
TCG726	2617
TCG727	2529
TCG728	2549
TCG729	0328
TCG730	2743
TCG731	2457
TCG736	1759
TCG737	1443
TCG738	0856
TCG740A	0379
TCG741	3734
TCG742	3735
TCG743	1395
TCG744	1421
TCG745	2930
TCG746	0785
TCG747	1808
TCG748	0789
TCG749	0395
TCG750	3219
TCG754	0723
TCG762	OEM
TCG768	1753
TCG778A	0360
TCG780	0364
TCG781	2747
TCG783	0802
TCG784	2699
TCG785	2707
TCG786	2714
TCG787	2257
TCG788	2755
TCG789	1843
TCG790	0349
TCG791	1337
TCG797	0521
TCG798	1753
TCG799	1611
TCG801	0520
TCG802	3761
TCG803	3763
TCG804	2557
TCG805	3410
TCG807	3766
TCG812	3770
TCG815	0848
TCG818	2501
TCG819	1878
TCG820	2817
TCG821	2831
TCG822	2812
TCG823	3063
TCG824	3079
TCG825	3099
TCG828	3086
TCG829	1899
TCG832	3781
TCG834	0176
TCG836	2493
TCG841	3784
TCG843	2502
TCG846	1516
TCG849	3788
TCG851	3790
TCG852	2022
TCG853	3791
TCG855	0814
TCG856	3793
TCG857M	3672
TCG859	3391
TCG860	3795
TCG900	2687
TCG901	2689
TCG902	2748
TCG903	2537
TCG904	1854
TCG905	2698
TCG906	2701
TCG907	2711
TCG908	2705
TCG909	1301
TCG909D	1705
TCG910	1797
TCG910D	1797
TCG911	1890
TCG911D	1897
TCG912	1696
TCG913	3808
TCG914	2727
TCG915	1554
TCG916	2749
TCG917	2722
TCG918	2562
TCG918M	2567
TCG919	2574
TCG919D	2570
TCG922	1815
TCG922M	2104
TCG923	1192
TCG923D	0026
TCG924	2222
TCG924M	2517
TCG925	2890
TCG926	3798
TCG927	1121
TCG927D	1118
TCG928	1677
TCG928M	0770
TCG930	2782
TCG931	1916
TCG932	2863
TCG933	3393
TCG934	3813
TCG935	2737
TCG936	3814
TCG937	3621
TCG937M	6039
TCG938	0896
TCG938M	0896
TCG939	3818
TCG940	1919
TCG941	0410
TCG941D	1982
TCG941M	0312
TCG942	3029
TCG943	3819
TCG943M	0631
TCG944	3820
TCG944M	3434
TCG946	3663
TCG947	2368
TCG947D	2358
TCG948	2823
TCG949	2552
TCG950	1828
TCG951	3395
TCG952	3821
TCG953	3823
TCG954	3824
TCG955M	0973
TCG956	2564
TCG957	2736
TCG958	2259
TCG959	3826
TCG960	0624
TCG961	1285
TCG962	0923
TCG963	2648
TCG964	1196
TCG965	2791
TCG966	0334
TCG967	1838
TCG968	1321
TCG969	3829
TCG970	2838
TCG971	3575
TCG972	2239
TCG973	0417
TCG973D	3801
TCG974	0011
TCG975	1300
TCG976	2283
TCG977	1292
TCG978	3285
TCG979	OEM
TCG980	3831
TCG981	0083
TCG984	3270
TCG985	3832
TCG986	3089
TCG987	0625
TCG988	2301
TCG989	3381
TCG990	1491
TCG991	3594
TCG992	2247
TCG993	3835
TCG994M	3827
TCG995	2318
TCG996	3836
TCG997	3024
TCG1000	1235
TCG1002	0387
TCG1003	0579
TCG1004	0878
TCG1005	2568
TCG1006	1215
TCG1009	0843
TCG1010	0864
TCG1011	0882
TCG1012	0021
TCG1013	0138
TCG1014	0355
TCG1015	2404
TCG1016	2406
TCG1017	2440
TCG1018	2476
TCG1019	2509
TCG1020	OEM
TCG1021	2155
TCG1022	4556
TCG1023	4557
TCG1024	3550
TCG1025	1187
TCG1027	2876
TCG1028	4559
TCG1029	2262
TCG1030	4561
TCG1031	4562
TCG1032	4563
TCG1033	4564
TCG1034	2889
TCG1035	4565
TCG1036	4566
TCG1037	4093
TCG1038	4409
TCG1039	1621
TCG1040	4569
TCG1041	4570
TCG1042	4571
TCG1043	4390
TCG1045	2393
TCG1046	0528
TCG1047	2882
TCG1048	0673
TCG1049	0653
TCG1050	2891
TCG1051	3785
TCG1052	0433
TCG1053	2573
TCG1054	2576
TCG1055	2578
TCG1056	1837
TCG1057	2577
TCG1058	2124
TCG1060	2581
TCG1061	2582
TCG1062	0403
TCG1064	2585
TCG1065	2586
TCG1066	2587
TCG1067	2588
TCG1068	2590
TCG1069	2599
TCG1070	4076
TCG1071	2603
TCG1072	2591
TCG1073	2596
TCG1074	2592
TCG1075A	0470
TCG1077	0930
TCG1078	3352
TCG1079	4575
TCG1080	0855
TCG1081A	1160
TCG1082	2261
TCG1083	2854
TCG1084	2886
TCG1085	2631
TCG1086	2887
TCG1087	1018
TCG1089	0638
TCG1090	2861
TCG1091	0964
TCG1092	2316
TCG1093	4578
TCG1094	0043
TCG1096	1634
TCG1097	3263
TCG1098	3958
TCG1099	4260
TCG1100	0911
TCG1101	2634
TCG1102	2632
TCG1103	1994
TCG1104	2057
TCG1105	4582
TCG1106	2630
TCG1107	3670
TCG1108	1393
TCG1109	2639
TCG1110	2305
TCG1110R	6247
TCG1114	4586
TCG1115	1248
TCG1115A	3922
TCG1116	4147
TCG1117	4003
TCG1119	4589
TCG1120	2031
TCG1121	0389
TCG1122	0607
TCG1123	4034
TCG1124	4543
TCG1126	4593
TCG1127	4420
TCG1128	3915
TCG1130	1326
TCG1131	4594
TCG1132	4595
TCG1133	2231
TCG1134	2949
TCG1135	2062
TCG1137	2594
TCG1139	3936
TCG1140	2872
TCG1141	3840
TCG1142	1478
TCG1148	2864
TCG1149	4597
TCG1150	4598
TCG1152	2505
TCG1153	3366
TCG1154	4600
TCG1155	1050
TCG1156	3017
TCG1158	0881
TCG1159	0884
TCG1160	3695
TCG1161	2942
TCG1162	4038
TCG1163	2604
TCG1164	3269
TCG1165	2773
TCG1166	2781
TCG1167	1714
TCG1168	2602
TCG1169	4401
TCG1170	2534
TCG1171	0093
TCG1172	1379
TCG1173	3273
TCG1174	2049
TCG1175	2786
TCG1176	2789
TCG1178	3705
TCG1179	2572
TCG1180	2614
TCG1181	2593
TCG1183	4196
TCG1184	4612
TCG1185	1479
TCG1186	0852
TCG1187	0857
TCG1188	2618
TCG1189	4395
TCG1190	4616
TCG1191	4087
TCG1192	1540
TCG1193	3318
TCG1194	1816
TCG1195	4209
TCG1196	1055
TCG1197	4620
TCG1198	2253
TCG1199	3805
TCG1200	4238
TCG1201	4622
TCG1202	2380
TCG1203	2306
TCG1204	4484
TCG1205	2931
TCG1206	4623
TCG1208	4624
TCG1209	4625
TCG1210	2633
TCG1211	2193
TCG1212	4626
TCG1213	4627
TCG1214	0651
TCG1215	2532
TCG1216	0916
TCG1217	0416
TCG1218	1571
TCG1219	1612
TCG1221	4631
TCG1223	1524
TCG1224	4086
TCG1226	0706
TCG1227	0621
TCG1228	3190
TCG1229	4632
TCG1230	2669
TCG1231	1170
TCG1231A	1057
TCG1232	1048
TCG1233	4224
TCG1234	2019

If replacement code is OEM, contact original manufacturer for replacement.

DEVICE TYPE	REPL CODE	DEVICE TYPE	REPL CODE	DEVICE TYPE	REPL CODE	DEVICE TYPE	REPL CODE	DEVICE TYPE	REPL CODE	DEVICE TYPE	REPL CODE	DEVICE TYPE	REPL CODE
TCG1235	4633	TCG1373	3916	TCG1493	4809	TCG3006	5114	TCG4021B	1749	TCG5007A	0036	TCG5117A	0876
TCG1236	2284	TCG1374	3710	TCG1494	4810	TCG3007	3019	TCG4022B	1256	TCG5008A	0275	TCG5118A	3127
TCG1237	3651	TCG1375	2794	TCG1495	4811	TCG3008	3096	TCG4023B	0519	TCG5009A	0140	TCG5119A	0185
TCG1238	4018	TCG1376	3711	TCG1496	4134	TCG3009	5115	TCG4024B	1957	TCG5010A	0041	TCG5120A	0205
TCG1239	3262	TCG1377	3333	TCG1497	4812	TCG3010	0841	TCG4025B	2072	TCG5010T1	0589	TCG5121A	0480
TCG1240	4069	TCG1378	4725	TCG1498	4813	TCG3011	3124	TCG4026B	2152	TCG5011A	0253	TCG5122A	0504
TCG1241	2638	TCG1379	4726	TCG1499	4814	TCG3012	5116	TCG4027B	1949	TCG5011T1	0883	TCG5123A	3317
TCG1242	1509	TCG1380	4727	TCG1500	4815	TCG3013	5117	TCG4028B	2228	TCG5012A	0091	TCG5124A	0684
TCG1243	4634	TCG1381	4728	TCG1501	4817	TCG3014	3455	TCG4029B	2233	TCG5013A	0471	TCG5125A	0225
TCG1245	0904	TCG1382	1951	TCG1502	4819	TCG3015	3456	TCG4030B	0500	TCG5013T1	0295	TCG5126A	0230
TCG1246	4635	TCG1383	4729	TCG1503	4820	TCG3016	3739	TCG4031B	2971	TCG5014A	0062	TCG5127A	0234
TCG1247	4636	TCG1384	4730	TCG1504	4821	TCG3018	1615	TCG4032B	2531	TCG5015A	0077	TCG5128A	0237
TCG1248	1260	TCG1385	4732	TCG1505	4823	TCG3019	5119	TCG4033B	2635	TCG5016A	0165	TCG5129A	1397
TCG1249	4042	TCG1386	3239	TCG1507	4824	TCG3020	1358	TCG4034B	3615	TCG5017A	0322	TCG5130A	0247
TCG1250	4072	TCG1387	4372	TCG1508	3068	TCG3021	3157	TCG4035B	2777	TCG5018A	0057	TCG5131A	0251
TCG1251	4637	TCG1388	4734	TCG1509	3069	TCG3022	1962	TCG4038B	2981	TCG5019A	0064	TCG5132A	1178
TCG1252	4638	TCG1389	4735	TCG1510	0810	TCG3023	3276	TCG4040B	0056	TCG5019T1	0248	TCG5133A	0256
TCG1253	4639	TCG1391	4736	TCG1511	4825	TCG3024	1778	TCG4041	3174	TCG5020A	0181	TCG5134A	2395
TCG1254	3815	TCG1392	4737	TCG1512	4280	TCG3025	3182	TCG4042B	0121	TCG5021A	0052	TCG5135A	0262
TCG1256	4640	TCG1393	4739	TCG1513	4826	TCG3026	2195	TCG4043B	1553	TCG5021T1	1005	TCG5136A	0269
TCG1257	4641	TCG1394	4740	TCG1514	4827	TCG3027	4183	TCG4044B	2308	TCG5022A	0053	TCG5137A	0274
TCG1258	4642	TCG1395	4097	TCG1515	4830	TCG3028	0591	TCG4047B	2331	TCG5023A	0879	TCG5138A	2399
TCG1260	4644	TCG1396	0496	TCG1516	4831	TCG3029	3232	TCG4048B	3457	TCG5024A	0686	TCG5139A	0294
TCG1261	1128	TCG1400	1582	TCG1517	4834	TCG3031	5122	TCG4049	0001	TCG5025A	0445	TCG5140A	1177
TCG1262	4645	TCG1401	3351	TCG1518	4835	TCG3032	2313	TCG4049B	0001	TCG5026A	0210	TCG5141A	0309
TCG1263	2960	TCG1403	3359	TCG1519	4836	TCG3034	3878	TCG4050B	0398	TCG5027A	0375	TCG5142A	0318
TCG1264	4064	TCG1404	4743	TCG1520	1627	TCG3035	3185	TCG4051B	0366	TCG5028A	0671	TCG5143A	0320
TCG1265	4065	TCG1405	4744	TCG1521	3945	TCG3037	5123	TCG4052B	0024	TCG5029A	0700	TCG5144A	0326
TCG1266	4067	TCG1406	4060	TCG1522	2533	TCG3038	5124	TCG4053B	0035	TCG5030A	0705	TCG5145A	0337
TCG1267	4070	TCG1407	4745	TCG1523	4838	TCG3039	5125	TCG4055B	3304	TCG5031A	0494	TCG5146A	0347
TCG1268	4071	TCG1408	4746	TCG1525	2694	TCG3040	0541	TCG4056B	3706	TCG5032A	0714	TCG5147A	0027
TCG1269	4079	TCG1409	0683	TCG1526	4841	TCG3041	0315	TCG4060B	0146	TCG5033A	0455	TCG5148A	0266
TCG1271	4648	TCG1410	2623	TCG1527	4842	TCG3042	0273	TCG4063B	3727	TCG5034A	0257	TCG5149A	2856
TCG1272	3559	TCG1411	0767	TCG1528	2328	TCG3043	4431	TCG4066B	0101	TCG5035A	0195	TCG5150A	0386
TCG1273	4649	TCG1412	4747	TCG1529	2913	TCG3044	4932	TCG4067B	3742	TCG5036A	0166	TCG5151A	0405
TCG1278	4652	TCG1413	0069	TCG1530	4844	TCG3045	2108	TCG4068B	2503	TCG5037A	0010	TCG5152A	0426
TCG1281	4653	TCG1414	4750	TCG1531	4846	TCG3046	2236	TCG4069	0119	TCG5038A	0032	TCG5153A	0444
TCG1282	4654	TCG1415	4495	TCG1532	2054	TCG3047	5051	TCG4069B	0119	TCG5039A	0054	TCG5154A	3028
TCG1283	4656	TCG1416	0490	TCG1533	4847	TCG3048	5065	TCG4070B	2516	TCG5040A	0068	TCG5155A	0238
TCG1284	4657	TCG1417	2625	TCG1534	4849	TCG3049	5126	TCG4071B	0129	TCG5041A	0092	TCG5156A	1180
TCG1285	0754	TCG1419	4751	TCG1536	4850	TCG3050	4987	TCG4072B	2524	TCG5042A	0125	TCG5157A	1190
TCG1286	0756	TCG1420	0553	TCG1537	4851	TCG3051	4988	TCG4073B	1536	TCG5043A	2317	TCG5158A	1207
TCG1287	4659	TCG1421	4752	TCG1538	0332	TCG3052	4870	TCG4075B	2540	TCG5044A	0152	TCG5159A	1218
TCG1288	4573	TCG1422	1585	TCG1539	1590	TCG3053	5127	TCG4076B	3496	TCG5045A	0173	TCG5160A	1881
TCG1289	4661	TCG1423	4291	TCG1540	4351	TCG3054	5128	TCG4077B	2558	TCG5046A	0094	TCG5161A	0647
TCG1290	4089	TCG1424	4476	TCG1541	2693	TCG3055	5129	TCG4078B	0921	TCG5047A	0049	TCG5162A	1255
TCG1291	4663	TCG1425	4758	TCG1543	4853	TCG3056	4873	TCG4081	0626	TCG5048A	0104	TCG5163A	2102
TCG1292	3451	TCG1426	0675	TCG1544	4854	TCG3057	4190	TCG4081B	0626	TCG5049A	0156	TCG5164A	1279
TCG1293	4665	TCG1427	4759	TCG1545	0912	TCG3058	5130	TCG4082B	0300	TCG5050A	0189	TCG5165A	2225
TCG1300	3933	TCG1428	1580	TCG1935	OEM	TCG3059	5131	TCG4085B	0303	TCG5051A	0099	TCG5166A	0605
TCG1301	4669	TCG1429	4760	TCG2003	4933	TCG3060	5132	TCG4086B	0535	TCG5052A	0089	TCG5174A	2035
TCG1302	3992	TCG1430	0918	TCG2004	4106	TCG3061	4989	TCG4089B	3830	TCG5053A	0287	TCG5174AK	0547
TCG1303	4670	TCG1431	4761	TCG2011	0845	TCG3062	5134	TCG4093B	2384	TCG5054A	0252	TCG5175A	2401
TCG1304	3996	TCG1433	4762	TCG2012	1007	TCG3063	5135	TCG4094B	1682	TCG5055A	0340	TCG5175AK	2403
TCG1305	3998	TCG1434	0872	TCG2013	1134	TCG3064	5137	TCG4095B	3848	TCG5056A	0370	TCG5176A	1438
TCG1306	4000	TCG1435	4763	TCG2014	1261	TCG3065	4871	TCG4096B	3851	TCG5057A	0394	TCG5176AK	2113
TCG1307	4004	TCG1436	4428	TCG2015	4935	TCG3068	5138	TCG4097B	3855	TCG5058A	0425	TCG5177A	2407
TCG1308	3948	TCG1437	4549	TCG2016	4936	TCG3069	5139	TCG4098B	3611	TCG5059A	0453	TCG5177AK	2410
TCG1309	4671	TCG1438	4764	TCG2017	4937	TCG3070	5140	TCG4099B	3330	TCG5060A	1473	TCG5178A	1445
TCG1310	3988	TCG1439	3103	TCG2018	4939	TCG3071	5141	TCG4501	3809	TCG5061A	3003	TCG5178AK	1901
TCG1311	3932	TCG1440	3738	TCG2019	4940	TCG3074	4327	TCG4502B	1037	TCG5063A	1312	TCG5179A	2415
TCG1312	3984	TCG1441	4765	TCG2020	4941	TCG3075	4332	TCG4503B	2053	TCG5064A	3010	TCG5179AK	2416
TCG1313	3985	TCG1442	4766	TCG2021	3893	TCG3076	4331	TCG4506B	3767	TCG5065A	1713	TCG5180A	2221
TCG1320	4673	TCG1445	1022	TCG2022	3920	TCG3077	4333	TCG4508B	1811	TCG5066A	0292	TCG5180AK	0696
TCG1321	4674	TCG1446	4063	TCG2023	2750	TCG3078	2206	TCG4510B	1963	TCG5067A	0456	TCG5181A	1458
TCG1322	4675	TCG1447	4768	TCG2025	4943	TCG3079	2199	TCG4511B	1544	TCG5068A	0533	TCG5181AK	1601
TCG1323	4676	TCG1448	4051	TCG2026	4081	TCG3080	2378	TCG4512B	2120	TCG5069A	0451	TCG5182A	0221
TCG1324	4677	TCG1449	4770	TCG2027	1701	TCG3081	3367	TCG4514B	1830	TCG5070A	0301	TCG5182AK	1616
TCG1325	4678	TCG1450	4047	TCG2028	1793	TCG3082	5142	TCG4515B	3600	TCG5071A	0025	TCG5183A	1490
TCG1326	4680	TCG1451	4771	TCG2029	4947	TCG3083	1109	TCG4516B	2347	TCG5072A	0244	TCG5183AK	1622
TCG1327	4681	TCG1452	4773	TCG2030	1877	TCG3084	1053	TCG4517B	4284	TCG5073A	1083	TCG5184A	2422
TCG1328	4682	TCG1453	2514	TCG2031	4949	TCG3114	5151	TCG4518B	1043	TCG5074A	0317	TCG5184AK	2424
TCG1329	4683	TCG1454	3246	TCG2032	4950	TCG3115	5153	TCG4520B	2675	TCG5075A	0420	TCG5185A	1618
TCG1330	4684	TCG1455	4776	TCG2050	4951	TCG3116	5154	TCG4521B	4990	TCG5076A	1649	TCG5185AK	0627
TCG1331	4685	TCG1457	4777	TCG2051	4952	TCG3117	5155	TCG4522B	2837	TCG5077A	0495	TCG5186	0510
TCG1332	4686	TCG1458	4778	TCG2053	3792	TCG3150	5165	TCG4526B	3610	TCG5078A	0949	TCG5186A	0510
TCG1333	4687	TCG1459	2530	TCG2054	4954	TCG3151	5167	TCG4527B	3145	TCG5079A	0531	TCG5186AK	0992
TCG1334	4688	TCG1460	4779	TCG2060	2295	TCG3152	5168	TCG4528B	3198	TCG5080A	0565	TCG5187A	0691
TCG1335	4689	TCG1461	2712	TCG2061	2728	TCG3153	5170	TCG4529B	4531	TCG5081A	0402	TCG5187AK	0995
TCG1336	4690	TCG1462	4780	TCG2070	4955	TCG3155	5171	TCG4531B	3325	TCG5082A	1606	TCG5188A	0870
TCG1337	4691	TCG1463	3196	TCG2071	4956	TCG3160	5174	TCG4532B	1016	TCG5083A	1674	TCG5188AK	1263
TCG1338	1437	TCG1464	4781	TCG2072	4957	TCG3161	1076	TCG4536B	4791	TCG5084A	0726	TCG5189A	1020
TCG1339	4693	TCG1465	0362	TCG2073	4959	TCG3162	5176	TCG4539B	3656	TCG5085A	0820	TCG5189AK	1249
TCG1340	4694	TCG1466	2063	TCG2074	4536	TCG3163	4767	TCG4541B	5017	TCG5086A	0350	TCG5190A	1153
TCG1341	4695	TCG1467	4757	TCG2075	4961	TCG3164	3491	TCG4543B	5020	TCG5087A	0931	TCG5190AK	1636
TCG1342	4696	TCG1468	3980	TCG2076	4962	TCG3165	5178	TCG4547B	5032	TCG5088A	0999	TCG5191A	1274
TCG1343	4698	TCG1469	4782	TCG2077	4963	TCG3166	4052	TCG4551B	5039	TCG5089A	0502	TCG5191AK	1639
TCG1344	4699	TCG1470	3228	TCG2078	4965	TCG3167	1131	TCG4555B	2938	TCG5090A	0869	TCG5192A	1402
TCG1345	1903	TCG1471	1201	TCG2079	0969	TCG3168	5179	TCG4556B	3431	TCG5091A	1156	TCG5192AK	1703
TCG1346	4701	TCG1472	4785	TCG2080	4966	TCG3880	3481	TCG4558B	5049	TCG5092A	2157	TCG5193A	1532
TCG1347	4702	TCG1473	4787	TCG2081	4967	TCG3881	1904	TCG4566B	3504	TCG5093A	1189	TCG5193AK	1640
TCG1348	4703	TCG1474	4789	TCG2082	4968	TCG3882	5339	TCG4583B	1297	TCG5094A	3026	TCG5194A	1078
TCG1349	4704	TCG1475	4791	TCG2083	4969	TCG4000	2024	TCG4585B	1375	TCG5095A	1311	TCG5194AK	1716
TCG1350	4705	TCG1476	4793	TCG2084	4138	TCG4001B	0478	TCG4900	2396	TCG5096A	0098	TCG5195A	1711
TCG1351	4706	TCG1477	4794	TCG2085	4868	TCG4002B	2055	TCG4902	3114	TCG5097A	0186	TCG5195AK	1720
TCG1352	4707	TCG1478	4795	TCG2087	4970	TCG4006B	0646	TCG4918	3172	TCG5098A	0213	TCG5196A	1717
TCG1354	4708	TCG1479	1446	TCG2102	4973	TCG4007	2846	TCG4926	3191	TCG5099A	0245	TCG5196AK	1731
TCG1355	4710	TCG1480	4560	TCG2104	4974	TCG4007B	2846	TCG4928	3201	TCG5100A	0028	TCG5197A	1723
TCG1361	4713	TCG1481	4797	TCG2114	2048	TCG4008B	0988	TCG4934	1915	TCG5101A	0255	TCG5197AK	0727
TCG1363	4714	TCG1482	4798	TCG2117	0522	TCG4011B	0215	TCG4950	0568	TCG5102A	0877	TCG5198A	1736
TCG1364	4715	TCG1483	0865	TCG2147	1693	TCG4012B	0498	TCG4958	0831	TCG5103A	0367	TCG5198AK	1756
TCG1365	4716	TCG1484	4799	TCG2708	4424	TCG4013B	0413	TCG4988	1405	TCG5104A	2858	TCG5199A	1748
TCG1366	4717	TCG1486	3716	TCG2716	2278	TCG4014B	0860	TCG5000A	1276	TCG5105A	0421	TCG5199AK	1768
TCG1367	4718	TCG1487	4044	TCG3000	1981	TCG4015B	1014	TCG5001A	2874	TCG5111A	0782	TCG5200A	1761
TCG1368	4719	TCG1488	2252	TCG3001	1989	TCG4016B	1143	TCG5002A	0760	TCG5112A	0796	TCG5200AK	1782
TCG1369	4720	TCG1489	4803	TCG3002	1985	TCG4017B	0513	TCG5003A	0677	TCG5113A	0806	TCG5201A	2450
TCG1370	4722	TCG1490	4805	TCG3003	1983	TCG4018B	1391	TCG5004A	0118	TCG5114A	0821	TCG5201AK	2453
TCG1371	4723	TCG1491	4807	TCG3004	5112	TCG4019B	1525	TCG5005A	0299	TCG5115A	0833	TCG5202A	1772
TCG1372	4724	TCG1492	4808	TCG3005	5113	TCG4020B	1661	TCG5006A	0376	TCG5116A	0442	TCG5202AK	1794

If replacement code is OEM, contact original manufacturer for replacement.

DEVICE TYPE	REPL CODE
TCG5203A	1788
TCG5203AK	1799
TCG5204A	1796
TCG5204AK	1809
TCG5205A	1804
TCG5205AK	1817
TCG5206A	1194
TCG5206AK	1826
TCG5207A	1821
TCG5207AK	1840
TCG5208A	0022
TCG5208AK	1853
TCG5209A	0070
TCG5209AK	1861
TCG5210A	0132
TCG5210AK	1866
TCG5211A	0172
TCG5211AK	1874
TCG5212A	0207
TCG5212AK	1884
TCG5213A	0227
TCG5213AK	2474
TCG5214A	0263
TCG5214AK	1895
TCG5215A	0310
TCG5215AK	1902
TCG5216A	0329
TCG5216AK	0736
TCG5217A	0356
TCG5217AK	1909
TCG5218A	0381
TCG5218AK	2479
TCG5219A	0412
TCG5219AK	1914
TCG5220A	0438
TCG5220AK	1163
TCG5221A	0464
TCG5221AK	1924
TCG5222A	0488
TCG5222AK	1933
TCG5223A	0509
TCG5223AK	1941
TCG5224A	0524
TCG5224AK	1947
TCG5225A	0542
TCG5225AK	1953
TCG5226A	0063
TCG5226AK	1961
TCG5227A	0401
TCG5227AK	0357
TCG5228A	0598
TCG5228AK	2500
TCG5229A	0616
TCG5229AK	0670
TCG5230A	0634
TCG5230AK	0776
TCG5231A	0650
TCG5231AK	2507
TCG5232A	0668
TCG5232AK	1071
TCG5234A	OEM
TCG5240A	0815
TCG5240AK	2721
TCG5241A	0827
TCG5241AK	2724
TCG5242A	0846
TCG5242AK	2726
TCG5243A	0868
TCG5243AK	2729
TCG5244A	0885
TCG5244AK	2732
TCG5245A	0897
TCG5245AK	2734
TCG5246A	0914
TCG5246AK	2738
TCG5247A	0926
TCG5247AK	2740
TCG5248A	0944
TCG5248AK	2744
TCG5249A	0958
TCG5249AK	1225
TCG5250A	0978
TCG5250AK	2746
TCG5251A	0994
TCG5251AK	1237
TCG5252A	1009
TCG5252AK	1252
TCG5253A	1019
TCG5253AK	1269
TCG5254A	0889
TCG5254AK	1277
TCG5255A	1049
TCG5255AK	1294
TCG5256A	1058
TCG5256AK	1302
TCG5257A	0932
TCG5257AK	1302
TCG5258A	1079
TCG5258AK	1310
TCG5259A	1096
TCG5259AK	2756
TCG5260A	1106
TCG5260AK	1324
TCG5261A	1123
TCG5261AK	2758
TCG5262A	1135
TCG5262AK	1333
TCG5263A	1152
TCG5263AK	1344
TCG5264A	1164
TCG5264AK	1356
TCG5265A	1174
TCG5265AK	2760
TCG5266A	1184
TCG5266AK	1371
TCG5267A	1200
TCG5267AK	2762
TCG5268A	1210
TCG5268AK	1387
TCG5269A	1223
TCG5269AK	1406
TCG5270A	1232
TCG5270AK	1415
TCG5271A	1246
TCG5271AK	1428
TCG5272A	1265
TCG5272AK	1440
TCG5273A	1273
TCG5273AK	2772
TCG5274A	1290
TCG5274AK	1447
TCG5275A	1299
TCG5275AK	2776
TCG5276A	1307
TCG5276AK	1461
TCG5277A	1322
TCG5277AK	1474
TCG5278A	1331
TCG5278AK	1474
TCG5279A	1340
TCG5279AK	2783
TCG5280A	1353
TCG5280AK	0613
TCG5281A	1365
TCG5281AK	1510
TCG5282A	1384
TCG5282AK	1523
TCG5283A	1401
TCG5283AK	1537
TCG5284A	1412
TCG5284AK	1550
TCG5285A	1423
TCG5285AK	1575
TCG5286A	1436
TCG5286AK	2802
TCG5287A	1444
TCG5287AK	2803
TCG5288A	1457
TCG5288AK	1507
TCG5289A	1470
TCG5289AK	2805
TCG5290A	1484
TCG5290AK	2807
TCG5291A	1505
TCG5291AK	2810
TCG5292A	1521
TCG5292AK	2815
TCG5293A	1531
TCG5293AK	2818
TCG5294A	1548
TCG5294AK	2820
TCG5295A	1568
TCG5295AK	2822
TCG5296A	1587
TCG5296AK	2824
TCG5304	0106
TCG5305	2010
TCG5306	2400
TCG5307	0787
TCG5312	0323
TCG5313	1414
TCG5314	0473
TCG5315	0446
TCG5316	1422
TCG5317	2443
TCG5322	2363
TCG5324	2369
TCG5326	2370
TCG5327	2372
TCG5328	2373
TCG5332	1875
TCG5334	3546
TCG5340	1643
TCG5342	1673
TCG5344	1589
TCG5351	4337
TCG5360	0804
TCG5368	0655
TCG5369	0020
TCG5371	0526
TCG5372	0108
TCG5374	0428
TCG5375	1871
TCG5377	1700
TCG5378	0345
TCG5380	0459
TCG5381	0587
TCG5386	0769
TCG5387	0483
TCG5400	1137
TCG5401	0344
TCG5402	0901
TCG5403	2342
TCG5404	0058
TCG5405	0407
TCG5406	1683
TCG5407	OEM
TCG5408	0179
TCG5409	0346
TCG5410	3349
TCG5411	4458
TCG5412	0174
TCG5413	3854
TCG5414	3620
TCG5415	3324
TCG5416	1502
TCG5417	1004
TCG5418	1034
TCG5419	1148
TCG5425	5527
TCG5426	1044
TCG5437	0717
TCG5438	0308
TCG5442	4588
TCG5444	2270
TCG5446	3402
TCG5448	3404
TCG5452	1396
TCG5453	1259
TCG5454	0447
TCG5455	0940
TCG5456	1222
TCG5457	0095
TCG5461	2095
TCG5462	2089
TCG5463	2095
TCG5465	0710
TCG5466	0863
TCG5468	0327
TCG5470	1110
TCG5471	2932
TCG5472	0963
TCG5473	2933
TCG5474	2936
TCG5475	3671
TCG5476	3439
TCG5480	3419
TCG5481	1103
TCG5482	2491
TCG5483	0240
TCG5484	2659
TCG5485	0676
TCG5486	2809
TCG5487	0336
TCG5491	2449
TCG5492	0435
TCG5494	1487
TCG5496	0687
TCG5500	3102
TCG5501	2519
TCG5502	0741
TCG5503	3105
TCG5504	0745
TCG5505	3109
TCG5506	2917
TCG5507	0747
TCG5508	3243
TCG5509	0752
TCG5511	3696
TCG5512	3383
TCG5513	3624
TCG5514	1651
TCG5515	1584
TCG5516	1665
TCG5517	1650
TCG5518	2647
TCG5520	3307
TCG5521	2188
TCG5522	0567
TCG5523	3258
TCG5524	0762
TCG5525	3268
TCG5526	3271
TCG5527	0740
TCG5528	3291
TCG5529	0764
TCG5530	2875
TCG5531	0766
TCG5534	0226
TCG5540	3277
TCG5541	0731
TCG5542	0712
TCG5543	0469
TCG5544	0721
TCG5545	0722
TCG5546	0778
TCG5547	0725
TCG5548	0750
TCG5550	2521
TCG5552	0397
TCG5554	0611
TCG5556	0952
TCG5558	1512
TCG5562	1848
TCG5564	1855
TCG5566	3215
TCG5567	1704
TCG5568	4031
TCG5569	0679
TCG5570	0608
TCG5572	0610
TCG5574	0468
TCG5575	0641
TCG5577	0217
TCG5579	0658
TCG5580	1084
TCG5582	1086
TCG5584	1102
TCG5600	4412
TCG5601	1755
TCG5602	4414
TCG5603	3499
TCG5604	4121
TCG5605	0485
TCG5606	4416
TCG5607	4417
TCG5608	0572
TCG5609	0935
TCG5610	0059
TCG5611	5549
TCG5612	5550
TCG5614	4440
TCG5615	0957
TCG5616	0960
TCG5617	0961
TCG5618	0966
TCG5621	5551
TCG5622	1089
TCG5623	0954
TCG5624	0411
TCG5625	1095
TCG5626	0415
TCG5627	0485
TCG5628	1100
TCG5631	0578
TCG5632	0571
TCG5633	0593
TCG5634	0574
TCG5635	0576
TCG5636	0577
TCG5637	0594
TCG5638	5322
TCG5640	2383
TCG5641	2394
TCG5642	1413
TCG5643	2387
TCG5645	0555
TCG5646	1456
TCG5649	5553
TCG5650	2352
TCG5651	2353
TCG5652	2356
TCG5653	5555
TCG5655	4455
TCG5656	4456
TCG5657	4482
TCG5660	0399
TCG5661	0404
TCG5662	0408
TCG5665	0422
TCG5666	0422
TCG5667	0422
TCG5672	5560
TCG5673	0154
TCG5675	0147
TCG5676	4545
TCG5677	0279
TCG5679	0908
TCG5680	3146
TCG5681	3147
TCG5682	3148
TCG5683	3150
TCG5684	3151
TCG5685	3152
TCG5686	3153
TCG5687	4100
TCG5688	1064
TCG5689	1317
TCG5690	1891
TCG5693	2015
TCG5695	2017
TCG5697	2018
TCG5800	0110
TCG5801	0953
TCG5802	0242
TCG5803	1747
TCG5804	0540
TCG5805	1771
TCG5806	0965
TCG5808	0817
TCG5809	0922
TCG5812	1282
TCG5814	1287
TCG5815	1293
TCG5817	1296
TCG5818	1567
TCG5819	1547
TCG5820	2153
TCG5821	3139
TCG5822	0711
TCG5823	2967
TCG5826	3517
TCG5827	5012
TCG5828	3541
TCG5829	5577
TCG5830	0708
TCG5831	0933
TCG5832	4140
TCG5833	5517
TCG5834	0580
TCG5835	0947
TCG5836	2060
TCG5837	4523
TCG5838	1000
TCG5839	1012
TCG5840	2076
TCG5841	5567
TCG5842	2081
TCG5843	1073
TCG5846	2088
TCG5847	1138
TCG5848	0612
TCG5849	1188
TCG5850	0970
TCG5851	0985
TCG5852	3733
TCG5853	0910
TCG5854	0989
TCG5855	0990
TCG5856	3743
TCG5857	0993
TCG5858	0197
TCG5859	0997
TCG5860	0200
TCG5861	1001
TCG5862	0204
TCG5863	0515
TCG5866	0206
TCG5867	1008
TCG5868	0586
TCG5869	0948
TCG5870	0084
TCG5871	0534
TCG5872	0090
TCG5873	0748
TCG5874	0097
TCG5875	0765
TCG5876	0105
TCG5877	0545
TCG5878	0109
TCG5879	0538
TCG5880	0116
TCG5881	0801
TCG5882	0122
TCG5883	0816
TCG5886	0131
TCG5887	0545
TCG5890	0145
TCG5891	0550
TCG5892	5514
TCG5893	1238
TCG5894	5582
TCG5895	1240
TCG5896	2900
TCG5897	1241
TCG5898	5583
TCG5899	1245
TCG5900	3751
TCG5901	2365
TCG5902	3757
TCG5903	1253
TCG5904	1112
TCG5905	2376
TCG5908	3768
TCG5909	1264
TCG5910	3011
TCG5911	0449
TCG5912	1600
TCG5913	5005
TCG5914	5282
TCG5915	5491
TCG5916	0871
TCG5917	1635
TCG5918	5283
TCG5919	5495
TCG5920	0853
TCG5921	1251
TCG5922	5286
TCG5923	5496
TCG5924	1609
TCG5925	1205
TCG5928	1610
TCG5929	2137
TCG5932	1614
TCG5933	2251
TCG5940	2002
TCG5941	2003
TCG5942	0590
TCG5943	3265
TCG5944	1250
TCG5945	1577
TCG5946	5587
TCG5947	3275
TCG5948	1581
TCG5949	3281
TCG5950	5589
TCG5951	3287
TCG5952	1586
TCG5953	3294
TCG5960	OEM
TCG5962	5026
TCG5963	2559
TCG5966	2615
TCG5967	2566
TCG5980	3189
TCG5981	1630
TCG5982	2901
TCG5983	0254
TCG5986	1124
TCG5987	1107
TCG5988	1126
TCG5989	1111
TCG5990	0805
TCG5991	0258
TCG5992	1195
TCG5993	1644
TCG5994	0319
TCG5995	0267
TCG5998	1132
TCG5999	1119
TCG6002	0044
TCG6003	0281
TCG6006	1530
TCG6008	0029
TCG6009	1847
TCG6010	0601
TCG6011	1851
TCG6020	3762
TCG6021	2665
TCG6022	2653
TCG6023	2695
TCG6026	2657
TCG6027	2768
TCG6030	2664
TCG6031	2855
TCG6034	2006
TCG6035	2907
TCG6038	2677
TCG6039	2974
TCG6040	2682
TCG6041	3051
TCG6042	3900
TCG6043	3117
TCG6044	2655
TCG6045	3208
TCG6048	4306
TCG6049	3771
TCG6050	1560
TCG6051	2179
TCG6054	2840
TCG6055	2182
TCG6058	4308
TCG6059	3601
TCG6060	2850
TCG6061	2191
TCG6064	2871
TCG6065	2197
TCG6068	2833
TCG6069	2216
TCG6072	2473
TCG6073	2340
TCG6074	2778
TCG6075	1745
TCG6076	2813
TCG6077	1783
TCG6078	2643
TCG6079	1818
TCG6084	0615
TCG6094	1545
TCG6154	0599
TCG6155	1347
TCG6156	1986
TCG6157	1905
TCG6158	0657
TCG6159	0193
TCG6354	1023
TCG6355	0501
TCG6356	1036
TCG6357	1777
TCG6358	1046
TCG6359	1789
TCG6400	1669
TCG6400A	1669
TCG6401	2136
TCG6402	0316
TCG6403	4233
TCG6404	3407
TCG6405	4405
TCG6406	3648
TCG6407	3331
TCG6408	2730
TCG6409	1175
TCG6410	1893
TCG6411	5389
TCG6412	5643
TCG6415	5645
TCG6416	5646
TCG6417	5649
TCG6418	5650
TCG6419	5651
TCG6502	3065
TCG6507	3070
TCG6508	3193
TCG6532	1973
TCG6800	0388
TCG6802	0393
TCG6810	2086
TCG6821	0448
TCG6850	0514
TCG6875	4181
TCG6880	4516
TCG6885	4647
TCG6886	4651
TCG6887	4658
TCG6888	4667
TCG6889	0581
TCG7400	0232
TCG7401	0268
TCG7402	0314
TCG7403	0335
TCG7404	0361
TCG7405	0385
TCG7406	1206
TCG7407	1339
TCG7408	0467
TCG7409	0492
TCG7410	0512
TCG7411	0527
TCG7412	2242
TCG7413	1441
TCG7414	2243
TCG7416	1349
TCG7417	1352
TCG7420	0697
TCG7421	1357
TCG7422	4603
TCG7423	3466
TCG7425	3477
TCG7426	0803
TCG7427	0818
TCG7428	4180
TCG7430	0873
TCG7432	0899
TCG7433	4193
TCG7437	3520
TCG7438	0996
TCG7439	5836
TCG7440	1024
TCG7441	1038
TCG7442	1052
TCG7443	1060
TCG7444	1072
TCG7445	1082
TCG7446	1098
TCG7447	1108
TCG7448	1125
TCG7450	0743
TCG7451	1168
TCG7453	1185
TCG7454	1202
TCG7460	1275
TCG7470	1404
TCG7472	1426
TCG7473	1172
TCG7474	1313
TCG7475	1432
TCG7476	1158
TCG7480	1535
TCG7481	3121
TCG7482	1574
TCG7483	0117
TCG7485	0374
TCG7486	1368
TCG7486A	1368
TCG7488A	4904
TCG7489	5455
TCG7490	1208
TCG7491	0980
TCG7492	0834
TCG7493A	0569
TCG7494	1702
TCG7495	1486
TCG7496	1715
TCG7497	2940
TCG8070	2943
TCG8076	2944
TCG8080A	4547
TCG8081	2948
TCG8085	2951
TCG8090	2955
TCG8096	2961
TCG8098	2379
TCG8108	2968
TCG8115	2972
TCG8118	2973
TCG8123	5939
TCG8125	2976
TCG8139	2990
TCG8149	2998
TCG8167	3034
TCG8181	3036
TCG8212	1860
TCG8213	3000
TCG8216	1863
TCG8223	5951
TCG8224	1709
TCG8226	3013
TCG8228	1869
TCG8242	3021
TCG8255	0051
TCG8301	2196
TCG8308	1070
TCG8309	5528
TCG8314	2488
TCG8316	5957
TCG8318	4145
TCG8321	5958
TCG8328	5959
TCG8368	5963
TCG8370	5964
TCG8374	5965
TCG8520	5975
TCG8542	5976
TCG8546	5977
TCG8556	5978
TCG9093	0358
TCG9094	1632
TCG9097	1481
TCG9099	0333
TCG9135	1619
TCG9157	1433
TCG9158	0466
TCG9200	4158

If replacement code is OEM, contact original manufacturer for replacement.

DEVICE TYPE	REPL CODE
TCG9370	1295
TCG9402	4162
TCG9403	1599
TCG9601	2267
TCG9602	2286
TCG9660	0618
TCG9661	2597
TCG9663	2992
TCG9664	3032
TCG9666	4142
TCG9668	2434
TCG9669	3015
TCG9670	2350
TCG9671	4149
TCG9672	0078
TCG9673	0800
TCG9675	4164
TCG9676	4170
TCG9678	4173
TCG9679	4175
TCG9680	4178
TCG9681	3045
TCG9682	3039
TCG9689	4201
TCG9691	4207
TCG9800	0850
TCG9801	0874
TCG9802	0888
TCG9803	0900
TCG9804	0919
TCG9805	0929
TCG9806	0945
TCG9807	0962
TCG9808	0982
TCG9809	0998
TCG9810	1011
TCG9811	1025
TCG9812	1039
TCG9813	4538
TCG9814	4541
TCG9904	1632
TCG9910	1257
TCG9917	OEM
TCG9923	OEM
TCG9924	2867
TCG9926	2247
TCG9930	1823
TCG9932	1041
TCG9933	2097
TCG9935	1176
TCG9936	1831
TCG9937	1835
TCG9944	0033
TCG9945	0081
TCG9946	0141
TCG9948	3399
TCG9949	1844
TCG9950	2619
TCG9951	2078
TCG9961	1859
TCG9962	0562
TCG9963	0341
TCG9969	OEM
TCG9989	3429
TCG32154	5211
TCG40085B	3450
TCG40097B	4441
TCG40098B	0432
TCG40100B	3952
TCG40106B	3626
TCG40160B	1359
TCG40161B	1373
TCG40162B	1388
TCG40163B	1407
TCG40174B	1551
TCG40175B	1528
TCG40182B	4660
TCG40192B	1764
TCG40193B	1776
TCG40194B	1769
TCG40195B	1784
TCG47368	OEM
TCG56004	0772
TCG56006	0744
TCG56008	0617
TCG56010	0875
TCG56022	3199
TCG56024	3207
TCG56026	3222
TCG65101	4604
TCG74100	4554
TCG74105	4555
TCG74107	0942
TCG74109	0968
TCG74110	0987
TCG74111	1002
TCG74116	4439
TCG74120	1116
TCG74121	0175
TCG74122	1139
TCG74123	1157
TCG74125	1182
TCG74126	1193
TCG74128	1219
TCG74132	1271
TCG74136	1316
TCG74141	1377
TCG74144	1418
TCG74145	0619
TCG74147	1451
TCG74148	1464
TCG74150	1492
TCG74151	1495
TCG74152	1517
TCG74153	1539
TCG74154	1555
TCG74155	1576
TCG74156	1592
TCG74157	1605
TCG74158	4446
TCG74160	1631
TCG74161	1645
TCG74162	1013
TCG74163	1666
TCG74164	0734
TCG74165	1685
TCG74166	0231
TCG74170	1770
TCG74173	1766
TCG74174	1770
TCG74175	1787
TCG74176	1795
TCG74177	1803
TCG74178	1813
TCG74179	1820
TCG74180	1829
TCG74181	1842
TCG74182	1856
TCG74185	1868
TCG74190	1912
TCG74191	1917
TCG74192	1921
TCG74193	1926
TCG74195	1943
TCG74196	1950
TCG74197	1956
TCG74198	1964
TCG74199	1971
TCG74221	2142
TCG74249	2290
TCG74251	2299
TCG74265	4611
TCG74278	4617
TCG74279	4619
TCG74290	2612
TCG74293	2644
TCG74298	4628
TCG74365	1459
TCG74366	1471
TCG74367	1488
TCG74368	1508
TCG74390	3240
TCG74393	3256
TCG74426	4697
TCG74490	3664
TCG75188	0508
TCG75189	0511
TCG75450B	1231
TCG75451B	1244
TCG75452B	1262
TCG75453B	1272
TCG75454B	1289
TCG75491B	1729
TCG75492B	1740
TCG75493	5021
TCG75494	5022
TCGHIDIV-1	2977
TCGHIDIV-2	2978
TCGHIDIV-3	2980
TCGHIDIV-4	2979
TCGHIDIV-12	5362
TCH98	0037
TCH99	0037
TCH99B	0037
TCIT00501P	0076
TCK1509	OEM
TCL0009450	0076
TCLT009450	0076
TCM1501	OEM
TCM1504	OEM
TCM1505	OEM
TCM1512A	OEM
TCM1701	OEM
TCM1702	OEM
TCM1703	OEM
TCM2101	OEM
TCM2401	OEM
TCM2910A	OEM
TCM2911A	OEM
TCM2912	OEM
TCM2913	OEM
TCM2914	OEM
TCM4110	OEM
TCM4910	OEM
TCM5087	OEM
TCM5089	5219
TCM5089N	OEM
TCM5091	OEM
TCM5092	OEM
TCN3C00	OEM
TC02P112	0015
TC0914	0079
TC0.09M22/1	0277
TC-0.2	0102
TCP4621AF-6002	OEM
TCP4633BN9503	OEM
TCP4633BN9905	OEM
TCP4633BN9907	OEM
TCP4633BN9908	OEM
TCP4633BN9912	OEM
TCP46307311	OEM
TCP46307311B	OEM
TCP46307322B	OEM
TCP46339903X	OEM
TCP46339904X	OEM
TCP46339905	OEM
TCR3	1103
TCR8	1103
TCR13	0240
TCR18	0240
TCR23	0240
TCR28	0240
TCR38	2659
TCR48	0676
TCR50	1103
TCR51	2491
TCR52	0240
TCR53	0240
TCR54	2659
TCR55	2659
TCR56	0676
TCR70	1103
TCR71	2491
TCR72	0240
TCR74	2659
TCR75	2659
TCR82	OEM
TCR85	OEM
TCR86	OEM
TCR102	OEM
TCR152	OEM
TCR202	OEM
TCR252	OEM
TCR302	OEM
TCR352	OEM
TCR402	OEM
TCR503	OEM
TCR505	1103
TCR510	1103
TCR550	OEM
TCR730	1103
TCR731	2491
TCR732	0240
TCR733	2659
TCR734	0676
TCR742	1103
TCR743	2491
TCR744	0240
TCR745	0240
TCR746	2659
TCR747	2659
TCR748	0676
TCR1005	0240
TCR1010	2491
TCR1050	OEM
TCR1503	OEM
TCR1505	0240
TCR1510	0240
TCR1520	OEM
TCR1550	OEM
TCR2003	OEM
TCR2005	0240
TCR2010	0240
TCR2050	OEM
TCR2503	OEM
TCR2505	2659
TCR2510	2659
TCR2550	OEM
TCR3003	OEM
TCR3005	2659
TCR3010	2659
TCR3050	OEM
TCR3503	OEM
TCR3505	0676
TCR3510	0676
TCR3550	OEM
TCR4003	OEM
TCR4005	0676
TCR4010	0676
TCR5275	OEM
TCR5276	OEM
TCR5277	OEM
TCR5278	OEM
TCR5279	OEM
TCR5280	OEM
TCR5281	OEM
TCR5282	OEM
TCR5283	OEM
TCR5284	OEM
TCR5285	OEM
TCR5286	OEM
TCR5287	OEM
TCR5288	OEM
TCR5289	OEM
TCR5290	OEM
TCR5291	OEM
TCR5292	OEM
TCR5293	OEM
TCR5294	OEM
TCR5295	OEM
TCR5296	OEM
TCR5297	OEM
TCR5298	OEM
TCR5300	OEM
TCR5301	OEM
TCR5302	OEM
TCR5303	OEM
TCR5304	OEM
TCR5305	OEM
TCR5306	OEM
TCR5307	OEM
TCR5308	OEM
TCR5309	OEM
TCR5310	OEM
TCR5311	OEM
TCR5312	OEM
TCR5313	OEM
TCR5314	OEM
TCR5315	OEM
TCS100	0595
TCS101	0477
TCS102	0595
TCS103	0477
TCT009450	0076
TCXB003H	OEM
TCXQ88535-119S	OEM
TD2SC2334	2268
TD3F012070	3938
TD3F013970	1541
TD3F013980	1541
TD3G0612KE	1790
TD3H018250	3136
TD3T00863E	1386
TD3T007340	1513
TD3T011110	OEM
TD3T012070	3938
TD4G	1875
TD5	3349
TD5F014260	1541
TD5F014270	1541
TD5F015550	0556
TD5G015540	1541
TD5Q01555S	0556
TD5Q015460	3360
TD5Q015470	3360
TD5Q015560	1541
TD5U015460	3360
TD5.B	1875
TD5.F	1875
TD5.G	1875
TD5.K	1875
TD5.L	1875
TD6	3349
TD6F	1875
TD6.B	1875
TD6.C	1875
TD7	OEM
TD7.C	1875
TD7.J	1875
TD8	OEM
TD8.A	1875
TD8.B	1875
TD8.C	1875
TD8.E	1875
TD8.I	1875
TD8.K	1875
TD8.L	1875
TD-9	OEM
TD9	OEM
TD9.C	1875
TD-13	0182
TD-15	0182
TD15	0182
TD-15-BL	0133
TD-15H	0182
TD15H	0102
TD15M	0102
TD-20W	0139
TD-25	OEM
TD-25-0	OEM
TD52	OEM
TD55	OEM
TD64	OEM
TD-80	OEM
TD80	OEM
TD-100	5018
TD100	0079
TD-101	5018
TD101	3929
TD-102	5018
TD102	0079
TD110	OEM
TD-200	5018
TD200	0079
TD-201	5018
TD201	0079
TD201A	OEM
TD-202	5018
TD202	0079
TD202A	OEM
TD203	OEM
TD203A	OEM
TD204	OEM
TD204A	OEM
TD205	OEM
TD205A	OEM
TD206	OEM
TD-250	5018
TD250	0079
TD251	OEM
TD251A	OEM
TD252	OEM
TD252A	OEM
TD253	OEM
TD253A	OEM
TD253B	OEM
TD254	OEM
TD254A	OEM
TD255	OEM
TD255A	OEM
TD256	OEM
TD256A	OEM
TD-261	OEM
TD261	OEM
TD-261A	OEM
TD261A	OEM
TD-262	OEM
TD262	OEM
TD-262A	OEM
TD262A	OEM
TD-263	OEM
TD263	OEM
TD-263A	OEM
TD263A	OEM
TD-263B	OEM
TD263B	OEM
TD-264	OEM
TD264	OEM
TD-264A	OEM
TD264A	OEM
TD-265	OEM
TD265	OEM
TD-265A	OEM
TD265A	OEM
TD-266	OEM
TD266	OEM
TD-266A	OEM
TD266A	OEM
TD271	OEM
TD271A	OEM
TD272	OEM
TD272A	OEM
TD273	OEM
TD273A	OEM
TD273B	OEM
TD274	OEM
TD274A	OEM
TD275A	OEM
TD276	OEM
TD276A	OEM
TD-400	5064
TD400	0037
TD-401	5064
TD401	0037
TD-402	5064
TD402	0037
TD-500	5064
TD500	0037
TD-501	5064
TD501	0037
TD501(SCR)	OEM
TD501F	OEM
TD501J	OEM
TD501S	OEM
TD-502	5064
TD502	0037
TD503	0179
TD510	OEM
TD-550	5064
TD550	0037
TD554	OEM
TD-600	3919
TD-601	3919
TD-602	3919
TD-700	3919
TD-701	3919
TD-702	3919
TD710	OEM
TD712	OEM
TD713	OEM
TD714	OEM
TD715	OEM
TD716	OEM
TD717	OEM
TD718	OEM
TD719	OEM
TD1001F	OEM
TD1001J	OEM
TD1001S	OEM
TD1003	0179
TD1053	OEM
TD1060	1823
TD1060P	1823
TD1061	1176
TD1061P	1176
TD1062	1041
TD1062P	1041
TD1063	2097
TD1063P	2097
TD1064	0033
TD1064P	0033
TD1065	0141
TD1065P	0141
TD1066	0562
TD1066P	0562
TD1067	3399
TD1067P	3399
TD1070	0081
TD1070P	0081
TD1072	1831
TD1072P	1831
TD1073	0358
TD1073P	0358
TD1074	0333
TD1074P	0333
TD1080	1859
TD1080P	1859
TD1083A	OEM
TD1085	1844
TD1085P	1844
TD1086	0341
TD1086P	0341
TD1401	0232
TD1401P	0232
TD1402	0512
TD1402P	0512
TD1403	0697
TD1403P	0697
TD1404	0873
TD1404P	0873
TD1405	1024
TD1405P	1024
TD1406	0743
TD1406P	0743
TD1407	1275
TD1407P	1275
TD1408	1426
TD1408P	1426
TD1409	1172
TD1409P	1172
TD1419	1168
TD1419P	1168
TD2001F	OEM
TD2001J	OEM
TD2001P	2597
TD2001S	OEM
TD2002P	2350
TD2003	0179
TD2003P	2434
TD2004P	3032
TD2005P	2992
TD2006P	3015
TD2008P	0618
TD2009P	4149
TD2010P	0078
TD2011P	2563
TD2012P	4178
TD2013P	4171
TD2015P	3066
TD-2219	5018
TD2219	0079
TD2400	OEM
TD-2905	5064
TD2905	0037
TD3001F	OEM
TD3001J	OEM
TD3001S	OEM
TD3003	0346
TD3168	OEM
TD3169	OEM
TD3171	OEM
TD3172	OEM
TD3173	OEM
TD3174	OEM
TD3400A	0232
TD3400AP	0232
TD3400P	0232
TD3401A	0268
TD3401AP	0268
TD3402A	0314
TD3402AP	0314
TD3404A	0361
TD3404AP	0361
TD3405A	0385
TD3405AP	0385
TD3409A	0492
TD3409AP	0492
TD3410A	0512
TD3410AP	0512
TD3410P	0512
TD3420A	0697
TD3420AP	0697
TD3420P	0697
TD3430A	0873
TD3430AP	0873
TD3430P	0873
TD3440A	1024
TD3440AP	1024
TD3440P	1024
TD3441A	1038
TD3441AP	1038
TD3442	OEM
TD3442A	1052
TD3442AP	1052
TD3447A	1108
TD3447AP	1108
TD3450A	0743
TD3450AP	0743
TD3450P	0743
TD3451A	1168
TD3451AP	1168
TD3451P	1168
TD3460A	1275
TD3460AP	1275
TD3460P	1275
TD3472A	1426
TD3472AP	1426
TD3473A	1172
TD3473AP	1172
TD3474A	1313
TD3474AP	1313
TD3474P	1313
TD3475A	1432
TD3475AP	1432
TD3480A	1535
TD3480AP	1535
TD3482P	1574
TD3483P	0117
TD3490	OEM
TD3490A	1208
TD3490AP	1208
TD3490P	1208
TD3491A	0980
TD3491AP	0980
TD3492A	0834
TD3492AP	0834
TD3492P	0834
TD3493A	0569
TD3493AP	0569
TD3493BP	0569
TD3493P	0569
TD3495A	1486
TD3495AP	1486
TD3503A	0734
TD3503AP	0734
TD4001F	OEM
TD4001J	OEM
TD4001S	OEM
TD4003	0346
TD5001J	OEM
TD5003	3349
TD6001J	OEM
TD-6012P	OEM
TD6103P(FA-1)	OEM
TD6104P	OEM
TD6107P	OEM
TD6110P	OEM
TD6133F	OEM
TD6301AN	OEM
TD6301AP	OEM
TD6301P	OEM
TD6306FA-1	OEM
TD6306P	5216
TD6306PFA-1	5216
TD6312P	5011
TD6314P	OEM
TD6350P	5011
TD6358P	OEM
TD6359N	OEM
TD6359P	OEM
TD6360N-02	OEM
TD6360N-04	OEM
TD6360P	OEM
TD6361N-C1	OEM
TD6365N-A1	OEM
TD6530	OEM
TD7001	OEM
TD7001J	OEM
TD7001S	OEM
TD8001	OEM
TD8001J	OEM
TD8001S	OEM
TD34121A	0175
TD34121AP	0175
TD34192A	1921
TD34192AP	1921
TD62003P	1134
TD62006P	4963
TD62007P	7142
TD62083AP	OEM
TD62103P	OEM
TD62104P	0969
TD62107P	OEM
TD62307P	OEM
TD62308AP	OEM
TD62308BP	OEM
TD62501P	OEM
TD62504	OEM
TD62504P	OEM
TD62507P	OEM
TD62553S	OEM
TD62555S	OEM
TD81515	0133
TD81518	0133
TD200401A0	1284
TD200401AL	1284
TD501052A0	2069
TD960016-1M	0133
TD3214738	3100
TD3214739	0012
TD3214740	OEM
TD3214741	OEM
TD3214742	0137
TD3214743	OEM
TD3214744	0002
TD3214745	OEM
TD3214746	OEM
TD3214747	OEM
TD3214748	OEM
TD3214749	OEM
TD3214750	OEM
TD3214751	OEM
TD3214752	OEM
TD3907562	3100
TD3907572	0057
TD3907582	OEM
TD3907592	0052
TD3909622	OEM
TD3909632	0052
TD3909642	OEM
TD3909652	0686
TD3909662	OEM
TD3909672	OEM

DEVICE TYPE	REPL CODE
TD3909682	OEM
TD3909692	OEM
TD3909702	OEM
TD3909712	OEM
TD3909722	OEM
TD3909732	OEM
TD20004710	0018
TD20008820	0161
TD30006120	1790
TD30011590	3014
TD30012070	3938
TD30016660	0060
TD30018250	3136
TD50014260	1541
TDA0470	OEM
TDA0470-D	OEM
TDA152AN	3161
TDA407	0418
TDA412	OEM
TDA420	0418
TDA1013A	3495
TDA1013AU	3495
TDA1013B	3495
TDA1032	OEM
TDA1034B	OEM
TDA1035	OEM
TDA1043	OEM
TDA1044	OEM
TDA1044F	OEM
TDA1057	OEM
TDA1060	5212
TDA1083	5195
TDA-1083/2	5076
TDA1083A	5195
TDA1151	OEM
TDA1151M	OEM
TDA-1170	4661
TDA1170	4661
TDA1170A	OEM
TDA1170D	OEM
TDA1170N	OEM
TDA1170S	4661
TDA1177	OEM
TDA1180P	7146
TDA1190P	1057
TDA1190Z	1170
TDA1200	2755
TDA1220B	OEM
TDA1330P	1808
TDA1412	0334
TDA1415	1321
TDA1424A	OEM
TDA1458D	0360
TDA1504P	OEM
TDA1506	OEM
TDA1510	OEM
TDA1512	OEM
TDA1520A	OEM
TDA1521	3156
TDA1521U	3156
TDA1524	3161
TDA1524A	3161
TDA1524AN	3161
TDA1524PN	3161
TDA1526	OEM
TDA1533	OEM
TDA1540	OEM
TDA1540D	OEM
TDA1540P	OEM
TDA1541	3237
TDA1541A	OEM
TDA1541AN	3237
TDA1541N	3237
TDA1543	OEM
TDA1559	OEM
TDA1576	OEM
TDA1578	OEM
TDA1594(A)	OEM
TDA1670A	3773
TDA1675	6227
TDA1770A	OEM
TDA1872A	OEM
TDA1905	6159
TDA1908	OEM
TDA1908A	OEM
TDA1910	OEM
TDA1940	OEM
TDA1940F	OEM
TDA1950	OEM
TDA2002	1048
TDA2003	4573
TDA2003V	4573
TDA2004	0496
TDA2005M	0496
TDA2006	4725
TDA2007	OEM
TDA2008	6237
TDA2009	OEM
TDA2009A	4375
TDA2030	4727
TDA2052	OEM
TDA2170	OEM
TDA2190	OEM
TDA2270	OEM
TDA2518Q	OEM
TDA2540	2826
TDA-2541	3738
TDA2541	2826
TDA2541Q	2826

DEVICE TYPE	REPL CODE
TDA2544	0912
TDA2545A	OEM
TDA2546A	3523
TDA2548	1780
TDA2575AQ	OEM
TDA2577	5198
TDA2577A	5198
TDA2577AS2	5198
TDA2579	OEM
TDA2579A	OEM
TDA2581Q	OEM
TDA2593	5745
TDA2593N	5745
TDA2595	OEM
TDA2595/V9	OEM
TDA2611	4601
TDA2611A	4601
TDA2611AQ	4601
TDA2653A	1254
TDA2653AN	1254
TDA2653AU	1254
TDA2730	OEM
TDA2758A	OEM
TDA2791	5785
TDA2822	OEM
TDA2822M	OEM
TDA2824	OEM
TDA3047	5246
TDA3048	5247
TDA3048D	OEM
TDA3048N	5247
TDA3190	1057
TDA3190P	OEM
TDA3333	OEM
TDA3333P	OEM
TDA3500	OEM
TDA3505	OEM
TDA3507	OEM
TDA3563	OEM
TDA3564	5391
TDA3566	OEM
TDA3568	OEM
TDA3569	OEM
TDA3569B	OEM
TDA3570	0490
TDA3651A	3713
TDA3651A2	OEM
TDA3651AQ	3713
TDA3652	2178
TDA3653B	OEM
TDA3654	2178
TDA3654AQ	2178
TDA3654Q	2178
TDA3654U	2178
TDA3769	OEM
TDA3810	OEM
TDA3830/V1	OEM
TDA4504	OEM
TDA4504A	OEM
TDA4505	3044
TDA4505A	3044
TDA4505B	3044
TDA4510	OEM
TDA4555	OEM
TDA4600-2	4189
TDA4601	4189
TDA4601D	OEM
TDA4605	OEM
TDA4605-3	OEM
TDA5030A	OEM
TDA5208T	OEM
TDA5582	OEM
TDA5708	OEM
TDA5709T	OEM
TDA6103Q	OEM
TDA6306P	OEM
TDA6350P	OEM
TDA7000	OEM
TDA7052	3802
TDA7052/N1	3802
TDA7052N1	3802
TDA7053	OEM
TDA7056	6721
TDA7231	OEM
TDA7262	OEM
TDA7263	OEM
TDA7267	OEM
TDA7318	OEM
TDA8137	OEM
TDA8170	OEM
TDA8172	4385
TDA8173	OEM
TDA8174	5824
TDA8175	6164
TDA8176	OEM
TDA8178FS	OEM
TDA8178S	OEM
TDA8179	OEM
TDA8179FS	OEM
TDA8179S	OEM
TDA8189	OEM
TDA8196	OEM
TDA8198	OEM
TDA8199	OEM
TDA8302	OEM
TDA8305	OEM
TDA8305A	OEM
TDA8368A	OEM
TDA8369	OEM

DEVICE TYPE	REPL CODE
TDA8425	OEM
TDA8426	OEM
TDA8444	OEM
TDA8702	OEM
TDA8808T	OEM
TDA8809T	OEM
TDA9400	OEM
TDA9500	OEM
TDA9852	OEM
TDA9855V2	OEM
TDA11907	1170
TDAL110	OEM
TDAL113	OEM
TDAL116	OEM
TDAL220	OEM
TDAL223	OEM
TDAL226	OEM
TDAL381S	OEM
TDAL383S	OEM
TDAL601A	OEM
TDAL601B	OEM
TDAL603A	OEM
TDAL603B	OEM
TDAL603S	OEM
TDAL606	OEM
TDAV190Z	1170
TDB0117	6941
TDB0124DP	0625
TDB0193DP	0631
TDB0555B	0973
TDB0555DP	0973
TDB0556A	3285
TDB0723	1192
TDB0723A	0026
TDB2033	OEM
TDB2905SP	1285
TDB2912SP	1838
TDB2915SP	3829
TDB7805	1922
TDB7805T	0624
TDB7806T	0923
TDB7808T	1196
TDB7812	3393
TDB7812T	0334
TDB7815T	1321
TDB7824T	2239
TDD1605S	3458
TDD1606S	3463
TDD1608S	3469
TDD1610S	3479
TDD1612S	3485
TDD1615S	3490
TDD1618S	3505
TDD1624S	3532
TDF014270	1541
TDH05	OEM
TDH1	OEM
TDH2	OEM
TDH3	OEM
TDH4	OEM
TDR05	OEM
TDR1	OEM
TDR2	OEM
TDR3	OEM
TDR4	OEM
TDRA05	OEM
TDRA1	OEM
TDRA2	OEM
TDRA3	OEM
TDRA4	OEM
T-E0137	1030
T-E0317	1030
TE35	1137
TE55	0344
TE105	0901
TE155	OEM
TE205	0901
TE305	0407
TE405	0407
TE500	0325
TE-500-E	0325
TE697	0016
TE706	0144
TE1010	0015
T-E1011	0015
TE-1011	0015
TE1011	0015
T-E1014	0143
TE-1014	0143
TE1014	0143
T-E1024	0015
TE1024	0015
T-E1024C	0015
TE1024C	0015
T-E1024D	0023
TE1024D	0015
T-E1029	0920
TE-1029	0133
TE1029	0015
T-E1031	0143
TE-1031	0143
TE1031	0143
T-E1042	0277
TE1042	0015
T-E1050	0015
TE-1050	0015
TE1050	0015
T-E1061	0196
TE-1061	0196

DEVICE TYPE	REPL CODE
TE1061	0196
T-E1061A	0196
TE-1061A	0196
TE1061A	0196
T-E1064	0015
TE1064	0102
T-E1068	0137
TE1068	0137
T-E1077	0137
TE1077	0137
T-E1078	0015
TE1078	0015
T-E1078A	0015
T-E1080	0015
TE1080	0015
T-E1086	0196
TE1086	0196
T-E1086A	0196
T-E1088	0015
TE1088	0015
T-E1089	0015
TE1089	0015
T-E1090	0015
TE1090	0087
T-E1095	0774
TE1095	0774
T-E1097	0015
TE1097	0015
T-E1098	0133
TE1098	0143
T-E1102	0015
T-E1102A	0015
T-E1105	0123
TE1105	0143
T-E1106	0137
T-E1107	1303
T-E1108	0920
TE1108	0015
T-E1114	0774
TE1114	0774
T-E1116	0143
T-E1118	0133
T-E1119	0133
T-E1121	0133
T-E1122	0087
T-E1124	0015
T-E1127	0087
T-E1133	0015
T-E1138	0015
T-E1140	0137
T-E1144	0015
T-E1145	0102
T-E1146	0102
T-E1148	0015
T-E1153	0102
T-E1155	0015
T-E1157	0015
T-E1171	0015
T-E1176	0015
T-E1177	0123
T-E1184	OEM
TE1420	0016
TE1990	0086
TE2484	0144
TE2711	0111
TE2712	0111
TE2713	0400
TE2714	0400
TE2715	0144
TE2716	0144
TE2921	0111
TE2922	0111
TE2923	0111
TE2924	0111
TE2925	0111
TE2926	0111
TE3390	0111
TE3391	0111
TE3391A	0111
TE3392	0111
TE3393	0111
TE3394	0111
TE3395	0111
TE3396	0111
TE3397	0111
TE3398	0111
TE3414	0016
TE3415	0016
TE3416	0086
TE3417	0595
TE3605	0016
TE3605A	0016
TE3606	0079
TE3606A	0079
TE3607	0016
TE3662	0007
TE3663	0007
TE3702	0791
TE3703	0791
TE3704	0016
TE3705	0016
TE3707	0144
TE3708	0144
TE3709	0144
TE3710	0144
TE3711	0144
TE3843	0111
TE3844	0111
TE3845	0111

DEVICE TYPE	REPL CODE
TE3854	0111
TE3854A	0111
TE3855	0111
TE3855A	0111
TE3859	0111
TE3859A	0595
TE3860	0111
TE3900	0111
TE3900A	0111
TE3901	0111
TE3903	0016
TE3904	0016
TE3905	0037
TE3906	0016
TE4123	0016
TE4124	0016
TE4125	0037
TE4126	0037
TE4256	0111
TE4409	0079
TE4424	0016
TE4425	0595
TE4951	0016
TE4952	0016
TE4953	0016
TE4954	0016
TE5086	0791
TE5087	0791
TE5088	0111
TE5089	0111
TE5249	0111
TE5309	0111
TE5310	0111
TE5311	0111
TE5365	0037
TE5366	0037
TE5367	0791
TE5368	0016
TE5369	0016
TE5370	0016
TE5371	0016
TE5376	0016
TE5377	0016
TE5378	0037
TE5379	0037
TE5447	0037
TE5448	0037
TE5449	0016
TE5450	0016
TE5451	0016
TE13002	OEM
TE13003	OEM
TE13004	0037
TE13005	0037
TE13006	0728
TE13007	0728
TEA0657	OEM
TEA1009	5804
TEA2014A	OEM
TEA2025	OEM
TEA2025A	OEM
TEA2261	OEM
TEA3718	OEM
TEA5120	OEM
TEA5570	OEM
TEA5582	OEM
TEA5591-4527	OEM
TEA5591A	OEM
TEA37185	OEM
TECEC-0913	OEM
TECEC-0933	OEM
TEG0129	0007
TEH0143	0233
TEH0147	OEM
TEH0149	0330
TER51	0319
TER51R	0267
TER52	0319
TER52A	0319
TER52R	0267
TER52RA	0267
TER53	0319
TER53R	0267
TER101	0319
TER101R	0267
TER102	0319
TER102A	0319
TER102R	0267
TER102RA	0267
TER103	0319
TER103R	0267
TER151	0319
TER151R	0267
TER152	0319
TER152A	0319
TER152R	0267
TER152RA	0267
TER153	0319
TER153R	0267
TER201	0319
TER201R	0267
TER202	0319
TER202A	0319
TER202R	0267
TER202RA	0267
TER203	0319
TER203R	0267
TER251	0319
TER251R	0267

DEVICE TYPE	REPL CODE
TER252	0319
TER252A	0319
TER252R	0267
TER252RA	0267
TER253	0319
TER301	0319
TER301R	0267
TER302	0319
TER302A	0267
TER302R	0267
TER302RA	0267
TER303	0319
TER303R	0267
TER401	0319
TER401R	0267
TER402	0319
TER402A	0319
TER402R	0267
TER402RA	0267
TER403	0319
TER403R	0267
TER501	0319
TER501R	0267
TER502	0319
TER502A	0319
TER502R	0267
TER502RA	0267
TER503	0319
TER503R	0267
TER601	0319
TER601R	0267
TER602	0319
TER602A	0319
TER602R	0267
TER602RA	0267
TER603	0319
TER603R	0267
TF5QS6565G	1992
TF7	0142
TF20	0015
TF21	0015
TF22	0015
TF23	0015
TF-30	0004
TF30	0004
TF44	0133
TF49	0004
TF50S6344G	2739
TF50S6565G	1992
TF65	0004
TF65/30	0004
TF65/M	0004
TF65/S/30	0004
TF65M	0164
TF66	0004
TF66/30	0004
TF66/60	0004
TF70	0600
TF71	0600
TF72	0600
TF75	0004
TF77	0004
TF77/30	OEM
TF-78	0016
TF78	0085
TF78/30	0085
TF78/30Z	0160
TF78/60	0160
TF80	OEM
TF-80/30	0160
TF80/30	0160
TF80/30Z	0160
TF80/60	OEM
TF80/80	OEM
TF80/302	0004
TF90/30	OEM
TF90/60	OEM
TF251	OEM
TF252	OEM
TF260	OEM
TF320M-A	OEM
TF320MA	0710
TF320M-AZ	0710
TF861S	3908
TFH340M	0710
TFH340MF	0710
TFH340MFLC6	0755
TFK162-12	OEM
TFK613	0017
TFR105	0071
TFR110	0071
TFR-120	0015
TFR120	0015
TFR140	0071
TFR305	OEM
TFR310	OEM
TFR320	OEM
TFR340	OEM
TFR605	OEM
TFR610	OEM
TFR620	OEM
TFR640	OEM
TFR1205	OEM
TFR1210	OEM
TFR1220	OEM
TFR1240	OEM
TFS340M	0755
TFS340MX	0755
TG1A	OEM

DEVICE TYPE	REPL CODE
TG2A	OEM
TG2SA201(C)	0050
TG2SA201(0)	0136
TG2SA201-0	0136
TG2SA201-N	0136
TG2SA608	0006
TG2SA608(C)	0006
TG2SA608-D	0006
TG2SA608-E	0006
TG2SA608C	0037
TG2SA608F	0006
TG2SA608K	0006
TG2SA608NP	0006
TG2SA608NPF	0006
TG2SA608NPG	0006
TG2SA608SP	0006
TG2SA608SPF	0006
TG2SA659	0006
TG2SA953	1348
TG2SA984	0006
TG2SA984E	0006
TG2SA984K	0006
TG2SA984KE	0006
TG2SA1016	0693
TG2SA1345	4130
TG2SA1346	3143
TG2SB375A	1675
TG2SB511-1	0925
TG2SB511-12SD325-1	OEM
TG2SB544E	0532
TG2SB560E	0477
TG2SB598	3812
TG2SB698	0006
TG2SB698F	0006
TG2SB764	0477
TG2SC2SC1175-E	OEM
TG2SC65(Y)	0233
TG2SC536	0537
TG2SC536(C)	0537
TG2SC536(E)	0537
TG2SC536-D-A	0537
TG2SC536-E	0537
TG2SC536-E-A	0537
TG2SC536-E-B	0537
TG2SC536-F	0537
TG2SC536-F-A	0537
TG2SC536E	0537
TG2SC536KN	0537
TG2SC536NP	0537
TG2SC536NPF	0537
TG2SC536SP	0537
TG2SC536SPF	0537
TG2SC927(C)	0007
TG2SC927-C-A	0007
TG2SC927-D-A	0007
TG2SC927-E-A	0007
TG2SC983-0	0066
TG2SC983-Y	0066
TG2SC983R	0066
TG2SC1025(D)	0934
TG2SC1175	0191
TG2SC1175(C)	0191
TG2SC1175-	0191
TG2SC1175-C	0191
TG2SC1175-D	0191
TG2SC1175-E	0191
TG2SC1293	1066
TG2SC1293(A)	1066
TG2SC1293-	0286
TG2SC1293-A-A	1066
TG2SC1293-B-A	1066
TG2SC1293-C-A	1066
TG2SC1293-D-A	1066
TG2SC1293A	1066
TG2SC1295(O)	1150
TG2SC1295-0	1150
TG2SC1295-O	1150
TG2SC1755-C-A	1085
TG2SC1756	0643
TG2SC1756-	0168
TG2SC1756-2SC1756	OEM
TG2SC1756-C	0643
TG2SC1756-C-A	0643
TG2SC1756-C-B	0643
TG2SC1756-D	0643
TG2SC1756-D-A	0643
TG2SC1756-D-B	0643
TG2SC1756-E	0643
TG2SC1756-E-A	0643
TG2SC1756-E-B	0643
TG2SC2057-C	0224
TG2SC2057-D	0224
TG2SC2057-E1	0224
TG2SC2057C	0224
TG2SC2057D	0224
TG2SC2228	1327
TG2SC2274	0155
TG2SC2274E	0155
TG2SC2274F	0155
TG2SC2999	0127
TG2SC3399	0898
TG2SC3400	2323
TG2SC3402	0832
TG2SD24(Y)	0142
TG2SD313	0042
TG2SD313E	0042
TG2SD325-1	0560
TG2SD386Y-D-A	0392
TG2SD386Y-E-A	0392

If replacement code is OEM, contact original manufacturer for replacement.

DEVICE TYPE	REPL CODE
TG2SD400E	2910
TG2SD438E	0191
TG2SD545	0018
TG2SD612E	1790
TG2SD613D	0060
TG2SD613E	0060
TG2SD734	1513
TG2SD894	0558
TG3A	OEM
TG4A	OEM
TG4.C	1875
TG5A	OEM
TG6A	OEM
TG7A	OEM
TG8A	OEM
TG11	0847
TG12	0015
TG21	0015
TG22	0015
TG31	0015
TG32	0015
TG41	0015
TG42	0015
TG-48	0143
TG48	0164
TG51	0015
TG52	0015
TG61	0015
TG62	0015
T-G1138	0143
T-G5033	0086
T-G5055	0086
TGB0331	0079
TGL-102	OEM
T-GMP5	OEM
T-GMP5A	OEM
T-GMP5B	OEM
TGZSA608(C)	0150
T-H1S557	0023
TH1S557	0015
T-H1S750	0917
TH-1S750	0917
T-H2SC313	3390
T-H2SC387	0144
TH2SC535	0127
T-H2SC536	0537
TH2SC536	0079
T-H2SC693	0552
TH2SC693	0079
T-H2SC715	0286
TH2SC715	0079
TH3C10	OEM
TH3J10	OEM
TH3L10	OEM
TH3L10Z	OEM
TH35	OEM
TH35S	OEM
TH40A	OEM
TH50	0015
TH55	OEM
TH55S	OEM
TH75	OEM
TH105	OEM
TH105S	OEM
TH155S	OEM
TH205S	OEM
TH252B	OEM
TH305S	OEM
TH400	0015
TH405S	OEM
TH600	0015
TH800	0071
TH801	0015
TH802	0015
TH803	0015
TH804	0015
TH805	0015
TH806	0015
TH808	0071
TH810	0071
TH1000	0071
TH1003	OEM
TH1004	OEM
TH1005	OEM
TH1006	OEM
TH1007	OEM
TH1008	OEM
TH2369	OEM
TH3638	OEM
TH4148	OEM
TH4258	OEM
TH9008P	OEM
TH-9014P	OEM
TH9015P	OEM
THD3600	0080
THF05002	OEM
TH-H2SC313	0007
THP35	OEM
THP36	OEM
THP45	OEM
THP46	OEM
THP47	OEM
THP61	OEM
THP62	OEM
THP106	OEM
THP169	OEM
THP170	OEM
THP171	OEM
THP172	OEM

DEVICE TYPE	REPL CODE
THP501	0050
THP502	0050
THS103A	OEM
T-HS6105	0015
T-HSG105	0015
THSG105	0015
T-HU60U	0150
THU60U	0004
THZ821	0471
THZ823	0471
THZ825	0471
THZ827	0471
THZ829	0471
TI010	OEM
TI025	OEM
TI031	0133
TI050	OEM
TI-1A6	0222
TI1A6	0016
TI2	OEM
TI6	OEM
TI-7A	0222
TI24A	0016
TI24B	0016
TI25A	0144
TI25B	0144
TI40A0	1103
TI-40A1	0240
TI40A1	2491
TI-40A2	0240
TI40A2	0240
TI40A3	2659
TI40A4	0676
TI40A5	OEM
TI40A6	OEM
TI42	3331
TI42A	2730
TI43	OEM
TI43A	2730
TI-51	0133
TI51	0133
TI52	0015
TI53	1335
TI54	1335
TI54A	0016
TI54D	0111
TI54E	0016
TI-55	0015
TI55	0015
TI-56	0110
TI56	0015
TI57	0609
TI58	0795
TI59	0015
TI60	0795
TI-71	0015
TI71	0015
TI-92	0016
TI92	0016
TI116	OEM
TI117	OEM
TI118	OEM
TI-132	0240
TI-136	0240
TI136	0240
TI137	2659
TI138	0676
TI140A3	0676
TI140A4	0676
TI140A5	OEM
TI140A6	OEM
TI145A5	OEM
TI145A6	OEM
TI150	OEM
TI152	0015
TI154	OEM
TI155	OEM
TI158	OEM
TI158A	OEM
TI158AL	OEM
TI158L	OEM
TI159	0609
TI-266A	0160
TI266A	0160
TI-269	0160
TI269	0160
TI320	OEM
TI321	OEM
TI338	0050
TI-363	0280
TI363	0050
TI-364	0280
TI364	0050
TI-365	0050
TI365	0050
TI-365A	0277
TI-366	0160
TI366	0160
TI-366A	0160
TI366A	0160
TI-367	0160
TI367	0160
TI-367A	0160
TI367A	0160
TI-368	0160
TI368	0160
TI-368A	0160
TI368A	0160
TI-369	0160

DEVICE TYPE	REPL CODE
TI369	0160
TI-369A	0160
TI369A	0160
TI-370	0160
TI370	0085
TI-370A	0160
TI370A	0085
TI376	OEM
TI377	OEM
TI378	OEM
TI379	OEM
TI380	OEM
TI381	OEM
TI382	OEM
TI383	OEM
TI384	OEM
TI385	0050
TI386	0050
TI387	0050
TI-388	0050
TI388	0050
TI-389	0050
TI389	0050
TI-390	0050
TI390	0050
TI-391	0050
TI391	0050
TI-393	0050
TI393	0050
TI-395	0050
TI395	0050
TI-396	0050
TI396	0050
TI-397	0050
TI397	0050
TI-398	0050
TI398	0050
TI-399	0050
TI399	0050
TI-400	0050
TI400	0050
TI-401	0050
TI401	0050
TI-402	0050
TI402	0050
TI-403	0050
TI403	0050
TI-407	0007
TI407	0144
TI-408	0007
TI408	0144
TI-409	0007
TI409	0144
TI410	0144
TI-411	0079
TI411	0016
TI-412	0016
TI412	0016
TI-413	0016
TI413	0016
TI-414	0016
TI414	0016
TI-415	0111
TI415	0016
TI-416	0111
TI416	0016
TI-417	0144
TI417	0016
TI-418	0111
TI418	0016
TI-419	0111
TI419	0016
TI-420	0144
TI420	0016
TI-421	0111
TI421	0111
TI-422	0016
TI422	0016
TI-423	0016
TI423	0016
TI-424	0086
TI424	0016
TI-425	0086
TI425	0086
TI426	OEM
TI427	OEM
TI-428	0037
TI428	0037
TI-429	0037
TI429	0037
TI-430	0144
TI430	0016
TI-431	0144
TI431	0144
TI-432	0016
TI432	0016
TI-433	0016
TI433	0016
TI440	OEM
TI442	OEM
TI443	OEM
TI444	OEM
TI445	OEM
TI446	OEM
TI447	OEM
TI448	OEM
TI450	OEM
TI451	OEM
TI457	OEM

DEVICE TYPE	REPL CODE
TI458	OEM
TI459	OEM
TI460	OEM
TI461	OEM
TI462	OEM
TI-474	0144
TI474	0144
TI-475	0086
TI475	0086
TI-480	0086
TI480	0016
TI-481	0086
TI481	0086
TI-482	0086
TI482	0016
TI-483	0086
TI483	0016
TI-484	0086
TI484	0016
TI-485	0595
TI485	0016
TI-486	0168
TI486	0042
TI487	0042
TI-490	0144
TI490	0144
TI-492	0007
TI492	0016
TI-493	0007
TI493	0016
TI-494	0007
TI494	0016
TI-495	0144
TI495	0016
TI-496	0086
TI496	0016
TI-503	0037
TI503	0037
TI539	OEM
TI540	OEM
TI550	OEM
TI551	OEM
TI602	OEM
TI605	OEM
TI607	OEM
TI607A	OEM
TI608	OEM
TI609	OEM
TI610	OEM
TI611	OEM
TI612	OEM
TI613	OEM
TI614	OEM
TI615	OEM
TI616	OEM
TI617	OEM
TI618	OEM
TI619	OEM
TI620	OEM
TI621	OEM
TI622	OEM
TI623	OEM
TI624	OEM
TI625	OEM
TI642B	0079
TI693	OEM
TI694	OEM
TI695	OEM
TI696	OEM
TI697	OEM
TI711A	OEM
TI712	OEM
TI-714	0016
TI714	0016
TI-714A	0016
TI-722	0233
TI722	0285
TI741	0325
TI-743	0037
TI743	0037
TI-744	0037
TI744	0037
TI-751	0016
TI751	0016
TI-752	0037
TI752	0037
TI801	OEM
TI802	OEM
TI802B	0016
TI803	OEM
TI803B	0016
TI804	OEM
TI805	OEM
TI806	OEM
TI-806G	0016
TI807	OEM
TI808	OEM
TI808E	0126
TI809	OEM
TI810	OEM
TI810B	0016
TI811	OEM
TI811G	0086
TI812	OEM
TI813	OEM
TI814	OEM
TI815	OEM
TI874	OEM
TI876	OEM

DEVICE TYPE	REPL CODE
TI884	OEM
TI885	OEM
TI886	OEM
TI887	OEM
TI888	OEM
TI-890	0037
TI890	0037
TI-891	0218
TI891	0791
TI896	OEM
TI897	OEM
TI898	OEM
TI899	OEM
TI903	OEM
TI904	0016
TI-905	0037
TI905	0037
TI-907	0016
TI907	0016
TI-908	0016
TI908	0016
TI910	OEM
TI951	OEM
TI952	OEM
TI953	OEM
TI1121	OEM
TI1122	OEM
TI1123	OEM
TI1124	OEM
TI1125	OEM
TI1126	OEM
TI1141	OEM
TI1142	OEM
TI1143	OEM
TI1144	OEM
TI1145	OEM
TI1146	OEM
TI1151	2662
TI1152	2662
TI1153	2662
TI1154	2662
TI1155	2662
TI1156	2662
TI1392	OEM
TI1599-5	OEM
TI1599-6	OEM
TI1722A	OEM
TI1724A	OEM
TI1779	OEM
TI1935	OEM
TI1936	OEM
TI1937	OEM
TI2150	OEM
TI2151	OEM
TI3000	OEM
TI3001	OEM
TI3010	0050
TI3011	0050
TI3012	0160
TI3013	OEM
TI3014	OEM
TI3015	0086
TI-3016	0127
TI3016	0144
TI3018	OEM
TI3027	0160
TI3028	0160
TI-3029	0160
TI3029	0160
TI-3030	0604
TI3030	0160
TI-3031	0604
TI3033	OEM
TI3034	OEM
TI3035	OEM
TI3036	OEM
TI3037	OEM
TI3038	OEM
TI3039	OEM
TI3040	OEM
TI3041	OEM
TI3042	OEM
TI8003B	0079
TI8419	0286
TI8423	OEM
TI64213	0086
TIA01	0004
TIA02	0050
TIA03	0280
TIA04	0211
TIA05	0280
TIA06	0016
TIA042	0240
TIA102	0016
TIA7630	3772
TIC01	OEM
TIC02	OEM
TIC03	OEM
TIC04	OEM
TIC05	OEM
TIC11	OEM
TIC12	OEM
TIC13	OEM
TIC14	OEM
TIC15	OEM
TIC26	OEM
TIC27	OEM
TIC29	OEM
TIC30	OEM

DEVICE TYPE	REPL CODE
TIC31	OEM
TIC44	1137
TIC45	0344
TIC46	0901
TIC47	0058
TIC54	3331
TIC55	3648
TIC56	3331
TIC57	3648
TIC106A	0717
TIC106B	0717
TIC106C	0717
TIC106D	0717
TIC106E	0308
TIC106F	0308
TIC106M	0308
TIC106Y	1396
TIC116A	2089
TIC116B	0505
TIC116C	0710
TIC116D	0710
TIC116E	0863
TIC116F	2095
TIC116M	0863
TIC122B	OEM
TIC122D	OEM
TIC122E	OEM
TIC122F	OEM
TIC126A	0397
TIC126B	0397
TIC126C	0611
TIC126D	0611
TIC126E	0952
TIC126F	2521
TIC126M	0952
TIC126N	OEM
TIC126S	OEM
TIC201	OEM
TIC201A	0572
TIC201B	0572
TIC201C	0572
TIC201D	0572
TIC201E	0935
TIC201M	0935
TIC201N	0059
TIC201S	0059
TIC205A	2352
TIC205B	2353
TIC205D	2356
TIC206A	0572
TIC206B	0572
TIC206C	0572
TIC206D	0572
TIC206E	0935
TIC206M	0935
TIC206N	0059
TIC206S	0059
TIC215A	OEM
TIC215B	OEM
TIC215D	OEM
TIC216A	0572
TIC216B	0572
TIC216C	0572
TIC216D	0572
TIC216E	0935
TIC216M	0935
TIC216N	0059
TIC216S	0059
TIC220B	OEM
TIC220D	OEM
TIC220E	OEM
TIC221B	OEM
TIC221D	OEM
TIC221E	OEM
TIC222B	0154
TIC222D	0147
TIC222E	0279
TIC225A	0572
TIC225B	0572
TIC225C	0572
TIC225D	0572
TIC225E	0935
TIC225M	0935
TIC225N	0059
TIC225S	0059
TIC226A	0572
TIC226B	0572
TIC226C	0572
TIC226D	0572
TIC226E	0935
TIC226M	0935
TIC226N	0059
TIC226S	0059
TIC230B	OEM
TIC230D	OEM
TIC230E	OEM
TIC231B	OEM
TIC231D	OEM
TIC231E	OEM
TIC232B	0154
TIC232D	0147
TIC232E	0279
TIC235M	OEM
TIC236A	0772
TIC236B	0772
TIC236C	0744
TIC236D	0744
TIC236E	0617
TIC236M	0617

DEVICE TYPE	REPL CODE
TIC236N	0875
TIC236S	0875
TIC240B	OEM
TIC240D	OEM
TIC240E	OEM
TIC241B	OEM
TIC241D	OEM
TIC241E	OEM
TIC242B	0154
TIC242D	0147
TIC242E	0279
TIC246A	5323
TIC246B	5323
TIC246C	5326
TIC246D	5326
TIC246E	1659
TIC246M	1659
TIC246N	1967
TIC246S	1967
TIC250B	OEM
TIC250D	OEM
TIC250E	OEM
TIC250M	OEM
TIC252B	3150
TIC252D	3152
TIC252E	3153
TIC252M	4100
TIC253A	0772
TIC253B	0772
TIC253C	0744
TIC253D	0744
TIC253E	0617
TIC253M	0617
TIC253N	0875
TIC253S	0875
TIC260B	OEM
TIC260D	OEM
TIC260E	OEM
TIC260M	OEM
TIC262B	3150
TIC262D	3152
TIC262E	3153
TIC262M	4100
TIC263A	5323
TIC263B	5323
TIC263C	5326
TIC263D	5326
TIC263E	1659
TIC263M	1659
TIC263N	1967
TIC263S	1967
TIC270B	OEM
TIC270D	OEM
TIC270E	OEM
TIC270M	OEM
TIC272B	2015
TIC272D	2017
TIC272E	2018
TIC272M	2018
TIC536F	0537
TIC2420	0279
TIC3010	1651
TIC3011	1651
TIC3012	1651
TIC3013	1584
TIC3014	1584
TICP100D	OEM
TICP106A	OEM
TICP106B	OEM
TICP106C	OEM
TICP106D	OEM
TICP106E	OEM
TICP106M	OEM
TICP108A	OEM
TICP108B	OEM
TICP108C	OEM
TICP108D	OEM
TICP108E	OEM
TICP108M	OEM
TICP206A	OEM
TICP206B	OEM
TICP206C	OEM
TICP206D	OEM
TICP206E	OEM
TICP206M	OEM
TID21A	OEM
TID22A	OEM
TID23A	OEM
TID24A	OEM
TID25A	OEM
TID26A	OEM
TID29A	OEM
TID30A	OEM
TID31	OEM
TID32	OEM
TID33	OEM
TID34	OEM
TID35	OEM
TID36	OEM
TID37	OEM
TID38	OEM
TID39	OEM
TID40	OEM
TID41	OEM
TID42	OEM
TID43	OEM
TID44	OEM
TID45	OEM
TID121	OEM

Original Device Types

DEVICE TYPE	REPL CODE
TID122	OEM
TID123	OEM
TID124	OEM
TID125	OEM
TID126	OEM
TID129	OEM
TID130	OEM
TID131	OEM
TID132	OEM
TID133	OEM
TID134	OEM
TID135N	OEM
TID136N	OEM
TID139N	OEM
TID140F	OEM
TID140N	OEM
TID141F	OEM
TID141N	OEM
TID142F	OEM
TID142N	OEM
TID143F	OEM
TID143N	OEM
TID144F	OEM
TID144N	OEM
TID381	1335
TID382	OEM
TID383	OEM
TID384	OEM
TID385	OEM
TID777	OEM
TID778	OEM
TID1111	OEM
TIDM155F	OEM
TIDM155J	OEM
TIDM166F	OEM
TIDM166J	OEM
TIDM168F	OEM
TIDM168J	OEM
TIDM185F	OEM
TIDM185J	OEM
TIDM186F	OEM
TIDM186J	OEM
TIDM255F	OEM
TIDM255J	OEM
TIDM266F	OEM
TIDM266J	OEM
TIDM268F	OEM
TIDM268J	OEM
TIDM285F	OEM
TIDM285J	OEM
TIDM286F	OEM
TIDM286J	OEM
TIE	0654
TIED55	OEM
TIED56	OEM
TIED59	OEM
TIED69	OEM
TIED80	OEM
TIED82	OEM
TIED83	OEM
TIED84	OEM
TIED85	OEM
TIED86	OEM
TIED87	OEM
TIED88	OEM
TIED89	OEM
TIED90	OEM
TIED91	OEM
TIED92	OEM
TIED93	OEM
TIED94	OEM
TIED95	OEM
TIED96	OEM
TIED97	OEM
TIED98	OEM
TIED451	OEM
TIED452	OEM
TIEF150	OEM
TIEF151	OEM
TIEF152	OEM
TIES06	OEM
TIES12	OEM
TIES13	OEM
TIES13A	OEM
TIES14	OEM
TIES15	OEM
TIES16A	OEM
TIES16B	OEM
TIES16C	OEM
TIES27	OEM
TIES35	OEM
TIES36	OEM
TIES471	OEM
TIES472	OEM
TIJ	0654
TIJ132-5C	OEM
TIJ132-5D	OEM
TIJ132-6A	OEM
TIJ132-6C	OEM
TIJ132-6D	OEM
TIJ132-7A	OEM
TIJ132-7C	OEM
TIJ132-7D	OEM
TIJ132-8A	OEM
TIJ132-8C	OEM
TIJ132-8D	OEM
TIJ132-9A	OEM
TIJ132-9C	OEM
TIJ132-9D	OEM
TIJ132-10A	OEM
TIJ132-10C	OEM
TIJ132-10D	OEM
TIL01	OEM
TIL09	OEM
TIL23	OEM
TIL24	OEM
TIL25	OEM
TIL26	OEM
TIL31	0591
TIL31A	OEM
TIL32	OEM
TIL33	0591
TIL33A	OEM
TIL34	0591
TIL34A	OEM
TIL38	OEM
TIL39	OEM
TIL-40	OEM
TIL40	OEM
TIL58	OEM
TIL63	OEM
TIL64	2313
TIL65	OEM
TIL66	2313
TIL67	2313
TIL77	OEM
TIL78	OEM
TIL81	5516
TIL99	OEM
TIL100	OEM
TIL102	OEM
TIL103	OEM
TIL107	OEM
TIL108	OEM
TIL111	0273
TIL112	0315
TIL113	0315
TIL114	0273
TIL115	0315
TIL116	4431
TIL117	0315
TIL118	0315
TIL119	4932
TIL120	OEM
TIL121	OEM
TIL124	0315
TIL125	0315
TIL126	0315
TIL127	1053
TIL128	1053
TIL131	OEM
TIL132	OEM
TIL133	OEM
TIL134	OEM
TIL135	OEM
TIL136	OEM
TIL137	OEM
TIL138	OEM
TIL139	OEM
TIL141	OEM
TIL142	OEM
TIL143	2636
TIL144	2636
TIL145	1417
TIL146	1417
TIL147	OEM
TIL148	OEM
TIL149	OEM
TIL153	0315
TIL154	0315
TIL155	0315
TIL156	1053
TIL157	1053
TIL158	OEM
TIL159	OEM
TIL160	OEM
TIL161	OEM
TIL201	OEM
TIL202	OEM
TIL209	OEM
TIL209A	3019
TIL210	OEM
TIL212-1	5115
TIL212-2	OEM
TIL216-1	3096
TIL216-2	OEM
TIL220	1615
TIL221	OEM
TIL224-1	3157
TIL224-2	OEM
TIL227-1	OEM
TIL227-2	OEM
TIL228-1	OEM
TIL228-2	6216
TIL231-1	OEM
TIL231-2	OEM
TIL232-1	OEM
TIL232-2	OEM
TIL234-1	1778
TIL234-2	OEM
TIL236-1	OEM
TIL236-2	OEM
TIL282	OEM
TIL302	4987
TIL303	OEM
TIL304	4988
TIL305	OEM
TIL306	OEM
TIL307	OEM
TIL308	OEM
TIL309	OEM
TIL311	OEM
TIL312	4989
TIL313	OEM
TIL321A	OEM
TIL322A	OEM
TIL327	OEM
TIL330A	OEM
TIL339	OEM
TIL340	OEM
TIL341	OEM
TIL345	OEM
TIL346	OEM
TIL347	OEM
TIL393-6	OEM
TIL393-8	OEM
TIL393-9	OEM
TIL401	OEM
TIL402	OEM
TIL403	OEM
TIL404	OEM
TIL405	OEM
TIL406	OEM
TIL411	OEM
TIL412	OEM
TIL413	OEM
TIL413S	OEM
TIL414	5006
TIL504	OEM
TIL505	OEM
TIL506	OEM
TIL507	OEM
TIL560	OEM
TIL601	OEM
TIL602	OEM
TIL603	OEM
TIL604	OEM
TIL605	OEM
TIL606	OEM
TIL607	OEM
TIL608	OEM
TIL804	OEM
TIL804-8	OEM
TIL804-10	OEM
TIL804-12	OEM
TIL804-B	OEM
TIL807	OEM
TIL808	OEM
TIL829	OEM
TIL830	OEM
TIL831	OEM
TIL832	OEM
TIL833	OEM
TIL834	OEM
TIL835	OEM
TIL836	OEM
TIL837	OEM
TIL838	OEM
TIL839	OEM
TIL840	OEM
TIL841	OEM
TIL842	OEM
TIL844	OEM
TIL906-1	6872
TIM-01	0050
TIM-1/MX	OEM
TIM-10	0050
TIM-11	0050
TIM9904ANL	OEM
TIM9905	OEM
TIM9906	OEM
TIM9907	OEM
TIM9908	OEM
TIN	0654
TIP04	0363
TIP11	OEM
TIP-14	0042
TIP14	0161
TIP24	0042
TIP27	0276
TIP-29	0042
TIP29	0042
TIP29A	0236
TIP29B	0061
TIP29C	3014
TIP29D	3014
TIP29E	3014
TIP29F	3014
TIP-29T	0168
TIP-30	0925
TIP30	0681
TIP30A	0681
TIP30B	0681
TIP30C	0681
TIP30D	5624
TIP30E	5624
TIP30F	5624
TIP31	0042
TIP31A	0042
TIP31B	0236
TIP31C	0236
TIP31D	3014
TIP31E	3014
TIP31F	3014
TIP32	0681
TIP32A	0681
TIP32B	0681
TIP32C	0681
TIP32D	OEM
TIP32E	OEM
TIP32F	OEM
TIP-33	3081
TIP33	3081
TIP33A	3081
TIP33B	3081
TIP33C	3435
TIP33D	3435
TIP33E	1506
TIP33F	1506
TIP34	0859
TIP34A	0859
TIP34B	3673
TIP34C	3673
TIP34D	3673
TIP34E	OEM
TIP34F	OEM
TIP35	3602
TIP35A	3602
TIP35B	3602
TIP35C	3602
TIP35D	OEM
TIP35E	OEM
TIP35F	OEM
TIP36	3603
TIP36A	3603
TIP36B	3603
TIP36C	3603
TIP36D	OEM
TIP36E	OEM
TIP36F	OEM
TIP37B	OEM
TIP41	0482
TIP41(B)	0728
TIP41A	0482
TIP41B	1284
TIP41C	0482
TIP41D	5637
TIP41E	5637
TIP41F	5637
TIP42	1369
TIP42(B)	1369
TIP42/A	1369
TIP42/B	1369
TIP42/C	1369
TIP42A	1369
TIP42B	1369
TIP42C	1369
TIP42D	OEM
TIP42E	OEM
TIP42F	OEM
TIP42M	OEM
TIP47	0168
TIP48	0168
TIP49	0168
TIP50	0168
TIP51	2613
TIP52	2613
TIP52(A)	OEM
TIP53	2613
TIP54	2613
TIP54(A)	OEM
TIP55A	1506
TIP56A	1506
TIP57A	1506
TIP58A	1506
TIP59	OEM
TIP60	OEM
TIP62	3906
TIP62A	3906
TIP62B	4157
TIP62C	4157
TIP63	OEM
TIP64	OEM
TIP65	OEM
TIP66	OEM
TIP73	0482
TIP73A	0482
TIP73B	0482
TIP73C	0482
TIP74	1369
TIP74A	1369
TIP74B	1369
TIP74C	1369
TIP75	3014
TIP75A	3014
TIP75B	3014
TIP75C	3014
TIP100	1212
TIP101	1212
TIP102	1212
TIP105	0602
TIP106	0602
TIP107	0602
TIP110	4330
TIP111	1212
TIP112	1212
TIP115	0602
TIP116	0602
TIP117	0602
TIP120	2235
TIP121	2235
TIP122	1212
TIP125	0602
TIP126	0602
TIP127	0602
TIP130	1212
TIP131	1212
TIP132	1212
TIP135	0602
TIP136	0602
TIP137	0602
TIP140	0134
TIP140T	1959
TIP141	0134
TIP141T	1959
TIP142	0134
TIP142T	1959
TIP145	0073
TIP145T	1968
TIP146	0073
TIP146T	1968
TIP147	0073
TIP147T	1968
TIP150	OEM
TIP151	OEM
TIP152	OEM
TIP160	5342
TIP161	5342
TIP162	5342
TIP501	0622
TIP502	1598
TIP503	0178
TIP504	0178
TIP505	4252
TIP506	4252
TIP509	0620
TIP510	0620
TIP515	0620
TIP516	0620
TIP525	1341
TIP526	2662
TIP527	1598
TIP530	0429
TIP531	1852
TIP532	1852
TIP535	1966
TIP536	1966
TIP537	1852
TIP538	1852
TIP539	1852
TIP540	1852
TIP541	0622
TIP544	1681
TIP545	1598
TIP546	1598
TIP550	5648
TIP551	0065
TIP552	0065
TIP553	0313
TIP554	1966
TIP555	1966
TIP556	4397
TIP558	0363
TIP559	1966
TIP560	1966
TIP561	1852
TIP562	0363
TIP563	0363
TIP564	OEM
TIP565	OEM
TIP575	OEM
TIP575A	OEM
TIP575B	OEM
TIP575C	OEM
TIP600	3373
TIP601	3373
TIP602	2439
TIP605	3374
TIP606	3374
TIP607	2432
TIP620	2427
TIP621	2427
TIP622	2439
TIP625	2277
TIP626	2277
TIP627	2432
TIP640	0134
TIP641	0134
TIP642	0134
TIP645	0073
TIP646	0073
TIP647	0073
TIP660	1991
TIP661	1991
TIP662	1991
TIP663	2626
TIP664	2626
TIP665	2626
TIP666	1991
TIP667	1991
TIP668	2626
TIP773B	OEM
TIP2955	3603
TIP3055	0482
TIPC47	OEM
TIPC100	OEM
TIPC102	OEM
TIPC105	OEM
TIPC106	OEM
TIPC107	OEM
TIPL751	1852
TIPL751A	4450
TIPL752	1852
TIPL752A	4450
TIPL753	3038
TIPL753A	1506
TIPL755	1852
TIPL755A	OEM
TIPL757	OEM
TIPL757A	OEM
TIPL760	4450
TIPL760A	4450
TIPL761	4450
TIPL761A	4450
TIPL762	1506
TIPL762A	1506
TIPL763	1506
TIPL763A	1506
TIPL765	OEM
TIPL765A	OEM
TIPL773	OEM
TIPL773A	OEM
TIPL774	4013
TIPL775	4013
TIPL775A	4013
TIPL777	OEM
TIPL777A	OEM
TIPL777B	OEM
TIPL780	OEM
TIPL780A	OEM
TIPL781	OEM
TIPL781A	OEM
TIPL784	OEM
TIPL785	3765
TIPL785A	3765
TIPL790	OEM
TIPL790A	OEM
TIPL790A(A)	OEM
TIR01	0015
TIR02	0015
TIR03	0015
TIR04	0015
TIR05	0015
TIR06	0015
TIR07	0071
TIR08	0071
TIR09	0071
TIR10	0071
TIR101A	OEM
TIR101B	OEM
TIR101C	OEM
TIR101D	OEM
TIR102A	OEM
TIR102B	OEM
TIR102C	OEM
TIR102D	OEM
TIR105C	OEM
TIR105D	OEM
TIR106D	OEM
TIR201A	OEM
TIR201B	OEM
TIR201C	OEM
TIR201D	OEM
TIR202A	OEM
TIR202B	OEM
TIR202C	OEM
TIR202D	OEM
TIR206A	OEM
TIR206B	OEM
TIR206C	OEM
TIR206D	OEM
TIS01	OEM
TIS02	OEM
TIS-03	0037
TIS03	0037
TIS04	0037
TIS05	OEM
TIS06	OEM
TIS07	OEM
TIS06701D20G	OEM
TIS06702D20G	OEM
TIS06704D20G	OEM
TIS06706D20G	OEM
TIS06708D20G	OEM
TIS06710D20G	OEM
TIS06712D20G	OEM
TIS06714D20G	OEM
TIS06716D20G	OEM
TIS06718D20G	OEM
TIS06720D20G	OEM
TIS06722D20G	OEM
TIS08001D20	OEM
TIS08001D20A	OEM
TIS08001D20B	OEM
TIS08001D20Y	OEM
TIS08001D20Z	OEM
TIS08002D20A	OEM
TIS08002D20B	OEM
TIS08002D20Y	OEM
TIS08002D20Z	OEM
TIS08004D20A	OEM
TIS08004D20B	OEM
TIS08004D20Y	OEM
TIS08004D20Z	OEM
TIS08006D20A	OEM
TIS08006D20B	OEM
TIS08006D20Y	OEM
TIS08006D20Z	OEM
TIS08008D20A	OEM
TIS08008D20B	OEM
TIS08008D20Y	OEM
TIS08008D20Z	OEM
TIS08010D20B	OEM
TIS08010D20Y	OEM
TIS08010D20Z	OEM
TIS08012D20A	OEM
TIS08012D20B	OEM
TIS08012D20Y	OEM
TIS08012D20Z	OEM
TIS-1B	0127
TIS11	OEM
TIS14	0325
TIS15	OEM
TIS16	OEM
TIS17	OEM
TIS-18	0007
TIS18	0127
TIS22	0016
TIS23	0016
TIS24	0144
TIS25	OEM
TIS26	OEM
TIS27	OEM
TIS34	0325
TIS37	0037
TIS38	0037
TIS39	0493
TIS41	OEM
TIS42	0325
TIS43	1175
TIS44	0016
TIS45	0016
TIS46	0016
TIS47	0016
TIS48	0016
TIS49	0016
TIS50	0037
TIS51	0016
TIS52	0016
TIS53	0037
TIS54	0037
TIS55	0016
TIS56	0007
TIS57	0007
TIS58	0325
TIS59	0325
TIS60	0155
TIS60A	0086
TIS60B	0086
TIS60C	0086
TIS60D	0086
TIS60E	0086
TIS60M	0086
TIS61	0037
TIS61A	0126
TIS61B	0126
TIS61C	0126
TIS61D	0126
TIS61E	0126
TIS61M	0126
TIS-62	0079
TIS62	0144
TIS62A	0144
TIS63	0144
TIS63A	0144
TIS64	0144
TIS64A	0144
TIS71	0016
TIS72	0016
TIS73	OEM
TIS74	OEM
TIS75	OEM
TIS78	0325
TIS79	0325
TIS82	0560
TIS83	0079
TIS84	0224
TIS85	0144
TIS86	0224
TIS87	0224
TIS-88	0325
TIS88	0325
TIS88A	OEM
TIS89	0079
TIS90	0324
TIS90-BLU	0861
TIS90-GRN	0155
TIS90-GRY	0016
TIS90-VIO	0861
TIS90-YEL	0155
TIS90M	1978
TIS90M-BLU	0861
TIS90M-GRN	0155
TIS90M-GRY	0016
TIS90M-VIO	0016
TIS90M-YEL	0155
TIS91	0006
TIS91-BLU	0037
TIS91-GRN	0037
TIS91-GRY	5605
TIS91-VIO	0037
TIS91-YEL	0037
TIS91M	0037
TIS91M-BLU	0037
TIS91M-GRN	0037
TIS91M-GRY	0016
TIS91M-VIO	0037
TIS91M-YEL	0037
TIS92	0079
TIS92-B	0016
TIS92-BLU	0079
TIS92-G	0016

If replacement code is OEM, contact original manufacturer for replacement.

DEVICE TYPE	REPL CODE
TIS92-GRN	0079
TIS92-GRY	0079
TIS92-V	0016
TIS92-VIO	0079
TIS92-Y	0016
TIS92-YEL	0079
TIS92M	0086
TIS92M-BLU	0861
TIS92M-GRN	0155
TIS92M-GRY	0016
TIS92M-VIO	0861
TIS92M-YEL	0155
TIS93	0037
TIS93-B	0037
TIS93-BLU	0037
TIS93-G	0037
TIS93-GRN	0037
TIS93-GRY	0037
TIS93-V	0037
TIS93-VIO	0037
TIS93-YEL	0037
TIS93M	0126
TIS93M-BLU	0037
TIS93M-GRN	0037
TIS93M-GRY	0602
TIS93M-VIO	0037
TIS93M-YEL	0037
TIS-94	0111
TIS94	0079
TIS94(AFAMP)	0079
TIS94(AFC)	0030
TIS94(DIO)	0030
TIS94(XSTR)	0079
TIS95	0018
TIS96	0018
TIS-97	0111
TIS97	0079
TIS-98	0079
TIS98	0127
TIS99	0127
TIS100	0715
TIS101	0715
TIS102	0233
TIS103	0233
TIS104	0037
TIS105	0224
TIS106	0861
TIS107	0086
TIS108	0224
TIS109	0698
TIS110	0086
TIS111	0086
TIS112	0037
TIS113	0016
TIS114	0016
TIS-125	0079
TIS125	0224
TIS126	0224
TIS128	0150
TIS129	0224
TIS130	OEM
TIS132	0285
TIS133	0079
TIS134	0079
TIS135	0086
TIS136	0086
TIS137	0037
TIS138	0150
TIS150	OEM
TIS151	2797
TIS175	OEM
TIS176	OEM
TIS180	OEM
TIS412	0127
TIS991	0037
TIS992	0079
TIS993	0037
TISA02	0280
TISM74LS00N	1527
TIUC226N	OEM
TI-UG-1888	0133
TI-UG1888	0133
TIV21	1029
TIV22	OEM
TIV23	OEM
TIV24	OEM
TIV25	OEM
TIV305	OEM
TIV306	0554
TIV307	OEM
TIV308	OEM
TIX09	OEM
TIX10	OEM
TIX90	0004
TIX91	0050
TIX92	0050
TIX93	OEM
TIX94	OEM
TIX95	OEM
TIX120A0	OEM
TIX120A1	OEM
TIX120A2	OEM
TIX155	OEM
TIX316	0050
TIX440	2002
TIX441	1250
TIX442	1250
TIX608	OEM
TIX609	OEM
TIX610	OEM
TIX611	OEM
TIX612	OEM
TIX613	OEM
TIX614	OEM
TIX615	OEM
TIX616	OEM
TIX617	OEM
TIX618	OEM
TIX619	OEM
TIX620	OEM
TIX621	OEM
TIX622	OEM
TIX623	OEM
TIX624	OEM
TIX690	OEM
TIX712	0079
TIX802	OEM
TIX803	OEM
TIX804	0037
TIX805	0037
TIX806	OEM
TIX807	OEM
TIX808	OEM
TIX809	OEM
TIX810	OEM
TIX811	OEM
TIX812	OEM
TIX813	OEM
TIX814	OEM
TIX815	OEM
TIX876	0144
TIX880	0144
TIX881	OEM
TIX882	OEM
TIX883	OEM
TIX888	0698
TIX890	0037
TIX891	0037
TIX895	0280
TIX896	0600
TIX1392	0698
TIX1393	0698
TIX2000	OEM
TIX2150	OEM
TIX2151	OEM
TIX3011	OEM
TIX3012	OEM
TIX3015	OEM
TIX3016	0050
TIX3016A	0050
TIX3023	OEM
TIX3032	0050
TIX3033	0161
TIX3035	0161
TIXA01	0280
TIXA02	0280
TIXA03	0280
TIXA04	0280
TIXA05	0280
TIXD747	0188
TIXD753	0636
TIXD758	0170
TIXL02	OEM
TIXL03	OEM
TIXL05	OEM
TIXL08	OEM
TIXL09	OEM
TIXL10	OEM
TIXL17	OEM
TIXL18	OEM
TIXL19	OEM
TIXL20	OEM
TIXL21	OEM
TIXL22	OEM
TIXL26	OEM
TIXL28	OEM
TIXL29	OEM
TIXL30	OEM
TIXL51	OEM
TIXL52	OEM
TIXL53	OEM
TIXL54	OEM
TIXL58	OEM
TIXL101	OEM
TIXL113	OEM
TIXL146	OEM
TIX-M01	0050
TIXM01	0050
TIX-M02	0050
TIXM02	0050
TIX-M03	0050
TIXM03	0050
TIX-M04	0050
TIXM04	0050
TIX-M05	0050
TIXM05	0050
TIX-M06	0050
TIXM06	0050
TIX-M07	0050
TIXM07	0050
TIX-M08	0050
TIXM08	0050
TIXM10	0050
TIX-M11	0050
TIXM11	0050
TIXM12	OEM
TIXM13	0050
TIX-M14	0007
TIXM14	0050
TIX-M15	0007
TIXM15	0050
TIX-M16	0007
TIXM16	0050
TIX-M17	0050
TIXM17	0050
TIXM18	0050
TIXM19	0050
TIX-M101	0050
TIXM101	0050
TIXM103	0050
TIXM104	0050
TIXM105	0050
TIXM106	0050
TIXM107	0050
TIXM108	0050
TIX-M201	0050
TIXM-201	0050
TIXM201	0050
TIX-M202	0050
TIXM202	0050
TIX-M203	0050
TIXM203	0050
TIX-M204	0050
TIXM204	0050
TIX-M205	0050
TIXM205	0050
TIX-M206	0050
TIXM206	0050
TIX-M207	0050
TIXM207	0050
TIXM301	OEM
TIXM07	OEM
TIXP39	OEM
TIXP40	OEM
TIXS09	0144
TIXS10	0144
TIXS11	OEM
TIXS12	0079
TIXS13	0079
TIXS28	0144
TIXS29	0144
TIXS30	0144
TIXS31	0144
TIXS33	OEM
TIXS37	OEM
TIXS41	OEM
TIXS42	OEM
TIXS67	OEM
TIXS78	OEM
TIXS79	OEM
TIXV01	OEM
TIXV02	OEM
TIXV03	OEM
TIXV04	OEM
TIXV06	OEM
TIXV07	OEM
TIXV08	OEM
TIXV09	OEM
TIXV10	OEM
TIXV11	OEM
TIXV12	OEM
TIXV13	OEM
TIXV14	OEM
TIXV15	OEM
TIXV16	OEM
TIXV17	OEM
TIXV18	OEM
TIXV19	OEM
TIXV304	OEM
TJ1	OEM
TJ2	OEM
TJ3	OEM
TJ5A	0015
TJ10A	0015
TJ15A	0015
TJ20A	0015
TJ25A	0015
TJ30A	0015
TJ35A	0015
TJ40A	0015
TJ60A	0015
TJ132-0A	OEM
TJ132-0C	OEM
TJ132-0D	OEM
TJ132-1A	OEM
TJ132-1C	OEM
TJ132-1D	OEM
TJ132-2A	OEM
TJ132-2C	OEM
TJ132-2D	OEM
TJ132-3A	OEM
TJ132-3C	OEM
TJ132-3D	OEM
TJ132-4A	OEM
TJ132-4C	OEM
TJ132-4D	OEM
TJ132-5A	OEM
TJ366	OEM
TJ405	OEM
TJ703DFA	OEM
TJ705DFA	OEM
TJ707DFA	OEM
TJ709DFA	OEM
TJ711DFA	OEM
TK	OEM
TK05	OEM
TK1	0468
TK1FA	OEM
TK1FB	OEM
TK2	0468
TK2FA	OEM
TK2FB	OEM
TK3FA	OEM
TK3FB	OEM
TK3L10	OEM
TK3V	OEM
TK4	0468
TK4FA	OEM
TK4FB	OEM
TK5	0609
TK5FA	OEM
TK5FB	OEM
TK6	0468
TK6FA	OEM
TK6FB	OEM
TK7	OEM
TK7FA	OEM
TK7FB	OEM
TK8	0468
TK8FA	OEM
TK8FB	OEM
TK9	OEM
TK9FA	OEM
TK9FB	OEM
TK10	0015
TK10FA	OEM
TK10FB	OEM
TK11	0015
TK12	0468
TK13	OEM
TK15	OEM
TK20	0015
TK20A	OEM
TK20B	OEM
TK20C	OEM
TK21	0015
TK21A	OEM
TK21B	OEM
TK21C	OEM
TK23	OEM
TK23A	OEM
TK23C	0004
TK24	OEM
TK24A	OEM
TK24B	OEM
TK24C	OEM
TK25	OEM
TK25A	0004
TK25B	0004
TK25C	OEM
TK26	OEM
TK26A	OEM
TK26B	OEM
TK27	OEM
TK27A	OEM
TK27B	OEM
TK28	0004
TK28C	OEM
TK30	0015
TK30C	OEM
TK31	OEM
TK31C	OEM
TK33C	0600
TK34C	0004
TK35	OEM
TK35C	0004
TK36	OEM
TK36C	0004
TK37	OEM
TK37C	0004
TK38	OEM
TK38C	OEM
TK-40	0015
TK40	0211
TK40A	0004
TK40C	0211
TK-41	0015
TK41	0211
TK41C	0050
TK42	0211
TK42C	0050
TK44	OEM
TK44C	OEM
TK45	OEM
TK45C	0211
TK46	OEM
TK46C	OEM
TK47C	OEM
TK48C	OEM
TK49C	0004
TK50	1355
TK60	1355
TK61	0015
TK70	OEM
TK71	OEM
TK72	OEM
TK76	OEM
TK-80	OEM
TK82	0861
TK105	OEM
TK110	0468
TK110FA	OEM
TK110FB	OEM
TK110FX	OEM
TK120	0468
TK120FA	OEM
TK120FB	OEM
TK120FX	OEM
TK130	OEM
TK130FA	OEM
TK130FB	OEM
TK130FX	OEM
TK140	0468
TK140FA	OEM
TK140FB	OEM
TK140FX	OEM
TK150	OEM
TK150FA	OEM
TK150FB	OEM
TK150FX	OEM
TK160	0468
TK160FA	OEM
TK160FB	OEM
TK160FX	OEM
TK170	OEM
TK170FA	OEM
TK170FB	OEM
TK170FX	OEM
TK180	0468
TK180FA	OEM
TK180FB	OEM
TK180FX	OEM
TK190	OEM
TK190FA	OEM
TK190FB	OEM
TK190FX	OEM
TK200A	0236
TK201A	0236
TK202A	OEM
TK203A	OEM
TK210	0738
TK220	0738
TK240	0738
TK250A	0861
TK251A	0861
TK252A	OEM
TK253A	OEM
TK254A	0016
TK255A	0016
TK256A	0016
TK257A	0016
TK258A	0016
TK259A	0016
TK260	0738
TK264A	0016
TK280	0738
TK310	1900
TK320	1900
TK340	0738
TK360	0738
TK380	0738
TK400	0015
TK400A	0160
TK401A	0160
TK402A	0160
TK403A	0160
TK582	0532
TK600	0015
TK-705M	0196
TK705M	0196
TK705M12/1	0196
TK800	0071
TK1000	0071
TK1100FA	OEM
TK1100FB	OEM
TK1100FX	OEM
TK1100V	0468
TK1110	OEM
TK1120V	0468
TK1130	OEM
TK1150	OEM
TK1201	OEM
TK1202	OEM
TK1204	OEM
TK1206	OEM
TK1208	OEM
TK1210	OEM
TK1212	OEM
TK1214	OEM
TK1216	OEM
TK1228-001	0050
TK1228-1001	0136
TK1228-1002	0004
TK1228-1003	0004
TK1228-1004	0004
TK1228-1005	0004
TK1228-1006	0004
TK1228-1007	0004
TK1228-1008	0198
TK1228-1009	0198
TK1228-1010	0016
TK1228-1011	0016
TK1228-1012	0016
TK1401	0272
TK1401FA	OEM
TK1401FB	OEM
TK1402	0272
TK1402FA	OEM
TK1402FB	OEM
TK1403	OEM
TK1403FA	OEM
TK1403FB	OEM
TK1404	0284
TK1404FA	OEM
TK1404FB	OEM
TK1405	OEM
TK1405FA	OEM
TK1405FB	OEM
TK1406	0284
TK1406FA	OEM
TK1406FB	OEM
TK1407	OEM
TK1407FA	OEM
TK1407FB	OEM
TK1408	0307
TK1408FA	OEM
TK1408FB	OEM
TK1409	OEM
TK1409FA	OEM
TK1409FB	OEM
TK1410	0307
TK1410FA	OEM
TK1410FB	OEM
TK1411	OEM
TK1412	0307
TK1413	OEM
TK1414	OEM
TK1415	OEM
TK1416	OEM
TK1701	0738
TK1702	0738
TK1703	OEM
TK1704	0738
TK1705	OEM
TK1706	0738
TK1707	OEM
TK1708	0738
TK1709	OEM
TK1710	0738
TK1712	0738
TK1802	OEM
TK1804	OEM
TK1806	OEM
TK1808	OEM
TK1810	OEM
TK1812	OEM
TK1814	OEM
TK2100	0738
TK2120	0738
TK2601	OEM
TK2602	OEM
TK2604	OEM
TK2606	OEM
TK2608	OEM
TK2610	OEM
TK2612	OEM
TK2614	OEM
TK2616	OEM
TK3001	OEM
TK3002	OEM
TK3004	OEM
TK3006	OEM
TK3008	OEM
TK3010	OEM
TK3012	OEM
TK3014	OEM
TK3016	OEM
TK3055	OEM
TK3100	0738
TK3120	0738
TK3601	OEM
TK3602	OEM
TK3604	OEM
TK3606	OEM
TK3608	OEM
TK3610	OEM
TK3612	OEM
TK9201	0130
TK10170	OEM
TK10280	OEM
TK10311	OEM
TK10320	OEM
TK10321	OEM
TK10360	OEM
TK10466-10A	OEM
TK10500M	OEM
TK10580M	OEM
TK15021Z	OEM
TK15050	OEM
TK19080L	OEM
TK30551	0130
TK30552	0130
TK30553	0130
TK30555	0130
TK30556	0130
TK30557	0130
TK30558	0130
TK30560	0130
TK55303	OEM
TKAL110	OEM
TKAL120	OEM
TKAL130	OEM
TKAL140	OEM
TKAL150	OEM
TKAL160	OEM
TKAL170	OEM
TKAL180	OEM
TKAL190	OEM
TKAL210	OEM
TKAL220	OEM
TKAL240	OEM
TKAL260	OEM
TKAL280	OEM
TKAL1100	OEM
TKAL1110	OEM
TKAL1120	OEM
TKAL2100	OEM
TKAL2120	OEM
TKE801	0608
TKE802	0608
TKE804	0610
TKE806	0610
TKE808	0468
TKE810	0468
TKE1201	0608
TKE1202	0608
TKE1204	0610
TKE1206	0610
TKE1208	0468
TKE1210	0468
TKF5	0015
TKF10	1355
TKF20	1355
TKF40	1355
TKF60	1355
TKF80	0072
TKF100	2637
TKSA733CY	0807
TKSA1013Y	OEM
TKSA1266Y	OEM
TKSC945C-Y	0079
TKSC3229	OEM
TKTA562-O	0006
TKTA1015Y	0006
TKTC1815Y	0286
TKTC2120Y	0155
TKTC2236AY	5044
TL010	OEM
TL010A	OEM
TL011C	OEM
TL012C	OEM
TL014C	OEM
TL021C	OEM
TL022C	OEM
TL022CGJ	OEM
TL022CP	OEM
TL022M	OEM
TL022MJG	OEM
TL022MP	OEM
TL044CJ	OEM
TL044CN	OEM
TL044M	OEM
TL044MJ	OEM
TL060AC	OEM
TL060ACJG	OEM
TL060ACP	OEM
TL060AGJG	OEM
TL060BC	OEM
TL060C	OEM
TL060CJG	OEM
TL060CP	OEM
TL060I	OEM
TL060IJG	OEM
TL060IP	OEM
TL060MJG	OEM
TL061AC	OEM
TL061ACJG	OEM
TL061ACP	3181
TL061AMJG	OEM
TL061BC	OEM
TL061BCJG	OEM
TL061BCP	OEM
TL061C	OEM
TL061CJG	OEM
TL061CP	3181
TL061I	OEM
TL061IJG	OEM
TL061IP	OEM
TL061M	OEM
TL061MJG	OEM
TL061MU	OEM
TL062AC	3320
TL062ACJG	OEM
TL062ACP	3320
TL062AMJG	OEM
TL062BC	OEM
TL062BCJG	OEM
TL062BCP	OEM
TL062C	3320
TL062CJG	OEM
TL062CP	OEM
TL062I	OEM
TL062IJG	OEM
TL062IP	OEM
TL062M	OEM
TL062MJG	OEM
TL062MU	OEM
TL064AC	OEM
TL064ACJ	OEM
TL064ACN	OEM
TL064AMJ	OEM
TL064BC	OEM
TL064BCJ	OEM
TL064BCN	OEM
TL064C	OEM
TL064CJ	OEM
TL064CN	OEM
TL064I	OEM
TL064IJ	OEM
TL064IN	OEM
TL064M	OEM
TL064MJ	OEM
TL064MW	OEM

DEVICE TYPE	REPL CODE
TL066AC	OEM
TL066ACJG	OEM
TL066ACP	OEM
TL066AMJG	OEM
TL066BC	OEM
TL066BCJG	OEM
TL066BCP	OEM
TL066C	OEM
TL066CJG	OEM
TL066CP	OEM
TL066I	OEM
TL066IJG	OEM
TL066IP	OEM
TL066MJG	OEM
TL068C	OEM
TL068CLP	OEM
TL068ILP	OEM
TL070AC	OEM
TL070ACJG	OEM
TL070ACP	OEM
TL070BC	OEM
TL070CJG	OEM
TL070CP	OEM
TL070IJG	OEM
TL070IP	OEM
TL070MJG	OEM
TL071ACJG	OEM
TL071ACP	3672
TL071AMJG	OEM
TL071BCJG	OEM
TL071BCP	OEM
TL071C	OEM
TL071CJG	OEM
TL071CP	3672
TL071I	OEM
TL071IJG	OEM
TL071IP	OEM
TL071M	OEM
TL071MJG	OEM
TL072AC	OEM
TL072ACJG	OEM
TL072ACP	OEM
TL072AMJG	OEM
TL072BC	OEM
TL072BCJG	OEM
TL072C	OEM
TL072CJG	OEM
TL072CP	3741
TL072I	OEM
TL072IJG	OEM
TL072IP	OEM
TL072M	OEM
TL072MJG	OEM
TL072ML	OEM
TL074AC	OEM
TL074ACJ	OEM
TL074ACN	OEM
TL074AMJ	OEM
TL074BC	OEM
TL074BCJ	OEM
TL074BCN	OEM
TL074C	OEM
TL074CJ	OEM
TL074CN	3391
TL074I	OEM
TL074IJ	OEM
TL074IN	OEM
TL074M	OEM
TL074MJ	OEM
TL074MW	OEM
TL075ACJ	OEM
TL075ACN	OEM
TL075BCJ	OEM
TL075BCN	OEM
TL075C	OEM
TL075CJ	OEM
TL075CN	OEM
TL075IJ	OEM
TL075IN	OEM
TL075MJ	OEM
TL080AC	OEM
TL080ACJG	OEM
TL080ACP	OEM
TL080C	OEM
TL080CJG	OEM
TL080CP	OEM
TL080I	OEM
TL080IJG	OEM
TL080IP	OEM
TL080MJG	OEM
TL081	3672
TL081AC	OEM
TL081ACJG	OEM
TL081ACP	5877
TL081AMJG	OEM
TL081BC	OEM
TL081BCDP	OEM
TL081BCJG	OEM
TL081BCP	OEM
TL081C	OEM
TL081CDP	OEM
TL081CFP	6365
TL081CJG	OEM
TL081CP	3672
TL081I	OEM
TL081IDP	OEM
TL081IJG	OEM
TL081IP	OEM
TL081M	OEM
TL081MGC	OEM
TL081MH	OEM
TL081MJG	OEM
TL082AC	OEM
TL082ACJG	OEM
TL082ACP	OEM
TL082AMJG	OEM
TL082BC	OEM
TL082BCJG	OEM
TL082BCP	OEM
TL082C	OEM
TL082CD	4105
TL082CJG	OEM
TL082CN	OEM
TL082CP	4107
TL082I	OEM
TL082IJG	OEM
TL082IP	OEM
TL082M	OEM
TL082MJG	OEM
TL083AC	OEM
TL083ACN	OEM
TL083C	OEM
TL083CJ	OEM
TL083CN	OEM
TL083I	OEM
TL083IJ	OEM
TL083IN	OEM
TL083MJ	OEM
TL084	3391
TL084AC	OEM
TL084ACDP	OEM
TL084ACJ	OEM
TL084ACN	OEM
TL084AMJ	OEM
TL084BC	OEM
TL084BCDP	OEM
TL084BCJ	OEM
TL084BCN	OEM
TL084C	OEM
TL084CD	4136
TL084CDP	3391
TL084CFP	OEM
TL084CJ	OEM
TL084CN	3391
TL084I	OEM
TL084IDP	OEM
TL084IJ	OEM
TL084IN	OEM
TL084M	OEM
TL084MDG	OEM
TL084MGC	OEM
TL084MJ	OEM
TL084MW	OEM
TL085C	OEM
TL085CJ	OEM
TL085CN	OEM
TL085MJ	OEM
TL087C	OEM
TL087CJG	OEM
TL087CP	OEM
TL087I	OEM
TL087IJG	OEM
TL087IP	OEM
TL087MJG	OEM
TL088C	OEM
TL088CJG	OEM
TL088CP	OEM
TL088CU	OEM
TL088I	OEM
TL088IJG	OEM
TL088IP	OEM
TL088IU	OEM
TL088M	OEM
TL088MJG	OEM
TL091C	OEM
TL091CJG	OEM
TL091CP	OEM
TL091MJG	OEM
TL092C	OEM
TL092CJG	OEM
TL092CP	OEM
TL092MJG	OEM
TL094C	OEM
TL094CJ	OEM
TL094CN	OEM
TL094IN	OEM
TL094M	OEM
TL094MJ	OEM
TL094MW	OEM
TL0250	OEM
TL1	0015
TL2	0015
TL2SA881Q	0006
TL2SA881R	0006
TL2SC2021R	1140
TL4	OEM
TL6	OEM
TL11	0015
TL12	0015
TL21	0015
TL22	0015
TL31	0015
TL32	0015
TL41	0015
TL42	0015
TL51	0015
TL61	0015
TL74H87N	2579
TL74H183N	4399
TL106-05	1259
TL106-1	0447
TL106-2	0940
TL106-4	0095
TL106-6	OEM
TL107-05	1259
TL107-1	0447
TL107-2	0940
TL107-4	0095
TL107-6	OEM
TL111	OEM
TL111J	OEM
TL111JG	OEM
TL170	OEM
TL170CLP	OEM
TL172	OEM
TL172CLP	OEM
TL173CLP	OEM
TL182C	OEM
TL182CL	OEM
TL182CN	OEM
TL182I	OEM
TL182IL	OEM
TL182IN	OEM
TL182MJ	OEM
TL185C	OEM
TL185CJ	OEM
TL185CN	OEM
TL185I	OEM
TL185IJ	OEM
TL185IN	OEM
TL185MJ	OEM
TL188C	OEM
TL188CL	OEM
TL188CN	OEM
TL188I	OEM
TL188IL	OEM
TL188IN	OEM
TL188MJ	OEM
TL188ML	OEM
TL191C	OEM
TL191CJ	OEM
TL191CN	OEM
TL191I	OEM
TL191IJ	OEM
TL191IN	OEM
TL191MJ	OEM
TL195C	OEM
TL287C	OEM
TL287CJG	OEM
TL287CP	OEM
TL287I	OEM
TL287IJG	OEM
TL287IP	OEM
TL287MJG	OEM
TL288C	OEM
TL288CJG	OEM
TL288CP	OEM
TL288CU	OEM
TL288I	OEM
TL288IJG	OEM
TL288IP	OEM
TL288IU	OEM
TL288MJG	OEM
TL311	OEM
TL311AJ	OEM
TL311AJG	OEM
TL311AN	OEM
TL311AP	OEM
TL311J	OEM
TL311JG	OEM
TL311N	OEM
TL311P	OEM
TL317	5280
TL317JG	OEM
TL317LP	5280
TL317P	OEM
TL321C	OEM
TL321CJG	OEM
TL321CP	OEM
TL321IJG	OEM
TL321IP	OEM
TL321MJG	OEM
TL321MP	OEM
TL322C	OEM
TL322I	OEM
TL331C	OEM
TL331CJG	OEM
TL331CP	OEM
TL331I	OEM
TL331IJG	OEM
TL331IP	OEM
TL331M	OEM
TL331MJG	OEM
TL336	OEM
TL368A	OEM
TL369	OEM
TL371	OEM
TL375	OEM
TL376	OEM
TL376C	OEM
TL376CNE	OEM
TL378	OEM
TL379	OEM
TL382	OEM
TL413	5006
TL430C	OEM
TL430CJG	OEM
TL431	OEM
TL431ACD	5294
TL431ACLP	OEM
TL431ACP	5298
TL431AILP	OEM
TL431AIP	OEM
TL431C	OEM
TL431CD	5294
TL431CJG	OEM
TL431CLP	OEM
TL431CLPB	OEM
TL431CP	5298
TL431I	OEM
TL431IJG	OEM
TL431ILP	OEM
TL431IP	OEM
TL431MJG	OEM
TL440C	OEM
TL440CJ	OEM
TL440CN	OEM
TL441C	OEM
TL441CJ	OEM
TL441CN	OEM
TL441MJ	OEM
TL441MN	OEM
TL442C	OEM
TL480CJ	OEM
TL480CN	OEM
TL481CNG	OEM
TL487C	4821
TL487CJG	OEM
TL487CP	4821
TL489	OEM
TL489C	4823
TL489CP	4823
TL490C	5720
TL490CJ	OEM
TL490CN	5720
TL491C	4824
TL491CJ	OEM
TL491CN	4824
TL493AC	OEM
TL494	1824
TL494AC	OEM
TL494C	4422
TL494CJ	1824
TL494CN	4422
TL494IJ	OEM
TL494IN	OEM
TL494MJ	OEM
TL495AC	OEM
TL495CJ	OEM
TL495CN	OEM
TL495IJ	OEM
TL495IN	OEM
TL496	OEM
TL496C	OEM
TL496CJG	OEM
TL496CP	OEM
TL497AC	OEM
TL497ACJ	OEM
TL497ACN	OEM
TL497AI	OEM
TL497AMJ	OEM
TL497MN	OEM
TL500C	OEM
TL500CN	OEM
TL501C	OEM
TL501CN	OEM
TL502C	OEM
TL502CN	OEM
TL503C	OEM
TL503CN	OEM
TL505C	OEM
TL505CN	OEM
TL506C	OEM
TL506CJ	OEM
TL506CN	OEM
TL506CW	OEM
TL506MJ	OEM
TL506MN	OEM
TL507C	OEM
TL507CP	OEM
TL507P	OEM
TL510CJ	OEM
TL510CJG	OEM
TL510CN	OEM
TL510CP	OEM
TL510MJ	OEM
TL510MJG	OEM
TL510MN	OEM
TL510MP	OEM
TL510MU	OEM
TL514C	OEM
TL514CJ	OEM
TL514CN	OEM
TL514M	OEM
TL514MJ	OEM
TL514MN	OEM
TL520C	OEM
TL520N(A)	OEM
TL521N(A)	OEM
TL531N(A)	OEM
TL532N(A)	OEM
TL533N(A)	OEM
TL560C	OEM
TL560CJG	OEM
TL560CL	OEM
TL560CP	OEM
TL601C	OEM
TL601CJG	OEM
TL601CP	OEM
TL601I	OEM
TL601IJG	OEM
TL601IP	OEM
TL601MJG	OEM
TL604C	OEM
TL604CJG	OEM
TL604CP	OEM
TL604I	OEM
TL604IJG	OEM
TL604IP	OEM
TL604MJG	OEM
TL607C	OEM
TL607CJG	OEM
TL607CP	OEM
TL607I	OEM
TL607IJG	OEM
TL607IP	OEM
TL607MJG	OEM
TL610C	OEM
TL610CJG	OEM
TL610CP	OEM
TL610I	OEM
TL610IJG	OEM
TL610IP	OEM
TL610MJG	OEM
TL702CJ	OEM
TL702CN	OEM
TL702CU	OEM
TL702MJ	OEM
TL702MU	OEM
TL710CJ	OEM
TL710CJG	OEM
TL710CN	OEM
TL710CP	OEM
TL710CU	OEM
TL710MJ	OEM
TL710MJG	OEM
TL710MN	OEM
TL710MP	OEM
TL710MU	OEM
TL711CJ	OEM
TL711CN	OEM
TL720CJ	OEM
TL720CN	OEM
TL780-00C	OEM
TL780-05C	OEM
TL780-12C	OEM
TL780-15C	OEM
TL783	OEM
TL783C	OEM
TL810C	OEM
TL810CJG	OEM
TL810CN	OEM
TL810CP	OEM
TL810CU	OEM
TL810MJ	OEM
TL810MJG	OEM
TL810MN	OEM
TL810MP	OEM
TL810MU	OEM
TL811C	OEM
TL811CJ	OEM
TL811CN	OEM
TL811CU	OEM
TL811MJ	OEM
TL811MN	OEM
TL811MU	OEM
TL820CJ	OEM
TL820CN	OEM
TL820MJ	OEM
TL820MN	OEM
TL1003	OEM
TL1006	OEM
TL1451ACNS	OEM
TL1555P	0973
TL2003	OEM
TL2006	OEM
TL3400	OEM
TL4003	OEM
TL4558P	0360
TL4558PA	OEM
TL7400N	0232
TL7401N	0268
TL7402N	0314
TL7403N	0335
TL7404N	0361
TL7405N	0385
TL7406N	1206
TL7407N	1339
TL7409N	0492
TL7410N	0512
TL7412N	2242
TL7413N	1441
TL7416N	1349
TL7420N	0697
TL7423N	3466
TL7425N	3477
TL7426N	0803
TL7430N	0873
TL7437N	3520
TL7438N	0996
TL7440N	1024
TL7442N	1052
TL7443N	1060
TL7444N	1072
TL7445N	1082
TL7446AN	1098
TL7447AN	1108
TL7448N	1125
TL7450N	0743
TL7451N	1168
TL7453N	1185
TL7454N	1202
TL7460N	1275
TL7470N	1404
TL7472N	1426
TL7473N	1172
TL7474N	1313
TL7475N	1432
TL7476N	1158
TL7480N	1535
TL7481N	3121
TL7482N	1574
TL7483N	0117
TL7485N	0374
TL7486N	1368
TL7489N	5455
TL7490N	1208
TL7491N	0980
TL7492N	0834
TL7493N	0569
TL7494N	1702
TL7495AN	1486
TL7496N	1715
TL7497N	2940
TL7601G	OEM
TL7705CP	OEM
TL7705CP-B	OEM
TL8505P	OEM
TL8506P	OEM
TL8608AP	OEM
TL8608P	OEM
TL8608P(FV)	OEM
TL8703P	OEM
TL8707P	OEM
TL8708P	OEM
TL8803P	OEM
TL74107N	0942
TL74121N	0175
TL74122N	1139
TL74123N	1157
TL74141N	1377
TL74145N	0619
TL74150N	1492
TL74151N	1495
TL74153N	1539
TL74154N	1555
TL74155N	1576
TL74156N	1592
TL74162N	1013
TL74163N	1666
TL74164N	0734
TL74165N	1685
TL74166N	0231
TL74180N	1829
TL74181N	1842
TL74182N	1856
TL74190N	1912
TL74191N	1917
TL74192N	1921
TL74193N	1926
TL74196N	1950
TL74197N	1956
TL74198N	1964
TL74199N	1971
TLA7601G	1988
TLA78601G	1988
TLB20C	OEM
TLB20M	OEM
TLBG5400	OEM
TLBG5410	5016
TLBO5160	OEM
TLBO5400	OEM
TLBO5410	OEM
TLBR5400	OEM
TLBR5410	4023
TLBV5380	OEM
TLBY5400	OEM
TLBY5410	5438
TLC27M2	OEM
TLC27M4	OEM
TLC271	OEM
TLC374	OEM
TLC548	OEM
TLC555	0585
TLC555CP	0585
TLC556	0588
TLC556CP	2869
TLFG2100	OEM
TLFG3100	OEM
TLFO2100	OEM
TLFO3100	OEM
TLFR2100	OEM
TLFR3100	OEM
TLFY2100	OEM
TLFY3100	OEM
TLG-102	OEM
TLG102	OEM
TLG-103	OEM
TLG103	3455
TLG105	3456
TLG113	OEM
TLG113A	OEM
TLG113AE	OEM
TLG113A-F	OEM
TLG-114	1778
TLG121	OEM
TLG122	1983
TLG-123A	OEM
TLG123A	OEM
TLG124	0841
TLG124A	0841
TLG124A-E	OEM
TLG143	OEM
TLG153	OEM
TLG153FA	OEM
TLG200	OEM
TLG205	OEM
TLG206	OEM
TLG208	3491
TLG211	OEM
TLG221	OEM
TLG250	OEM
TLG251	OEM
TLG253	OEM
TLG257	OEM
TLG258	OEM
TLG259	OEM
TLG321	OEM
TLG325	OEM
TLG325(TCL)	OEM
TLG332	OEM
TLG333	OEM
TLG334	OEM
TLG335	OEM
TLG367	OEM
TLG2167	OEM
TLG2168	OEM
TLG2259	OEM
TLH340MF	0710
TLN103	OEM
TLN104-B	OEM
TLN105A	OEM
TLN105B	OEM
TLN110	OEM
TLN111	OEM
TLO72BCP	OEM
TLO82CN	OEM
TLO113A	OEM
TLO153	OEM
TLO153FA	OEM
TLO200	OEM
TLO208	OEM
TLO221	OEM
TLO251	OEM
TLO253	OEM
TLO257	OEM
TLO258	OEM
TLO259	OEM
TLOG116	OEM
TLOG205	OEM
TLOG208	OEM
TLP521	3367
TLP521(YG)	OEM
TLP521-1	6889
TLP521-1(A)	OEM
TLP521-1(FA)	3367
TLP521-1(G)	OEM
TLP521-1(Y)	OEM
TLP521-1(YG)	OEM
TLP521-1-BL	1166
TLP521-1-YG	0112
TLP521-1FA	0321
TLP521-2	OEM
TLP521-YG	OEM
TLP521YG	OEM
TLP531	0315
TLP531-AUDIO	0315
TLP541G	0235
TLP550	5642
TLP551	1291
TLP551(FA)	OEM
TLP551-2	OEM
TLP551FA-1	1291
TLP551FA-2	1291
TLP560G	OEM
TLP621	0112
TLP621(GB)	6168
TLP621(GRL)	0112
TLP621-1	1166
TLP-621-1-GB	1166
TLP621-1-GB	1166
TLP621-2	OEM
TLP621-4	OEM
TLP621G	OEM
TLP621GB	0112
TLP621GR	0112
TLP631	0315
TLP634	OEM
TLP636	OEM
TLP651	OEM
TLP802	OEM
TLP-809	OEM
TLP850	OEM
TLP907-0	OEM
TLP5211YG	OEM
TLPS604	5123
TLR102	OEM
TLR102C	OEM
TLR-103	OEM
TLR103	OEM
TLR-104	OEM
TLR104	5116

If replacement code is OEM, contact original manufacturer for replacement.

DEVICE TYPE	REPL CODE
TLR105	5117
TLR-106	OEM
TLR-107	5116
TLR111	OEM
TLR112	OEM
TLR113	3182
TLR113A	OEM
TLR-113AD	OEM
TLR113DFA	OEM
TLR-114	1615
TLR114	1615
TLR114A	OEM
TLR-121	1981
TLR121	1981
TLR122	5112
TLR123	3019
TLR124	3019
TLR140	OEM
TLR143	OEM
TLR144	3163
TLR145	3169
TLR146	OEM
TLR-147	5116
TLR147	OEM
TLR200	OEM
TLR205	OEM
TLR206	OEM
TLR208	4767
TLR210	OEM
TLR211	OEM
TLR218RD	OEM
TLR221	OEM
TLR250	OEM
TLR302	OEM
TLR312	4873
TLR313	4870
TLR314	4871
TLR321	OEM
TLR324	OEM
TLR325	OEM
TLR325(US)	OEM
TLR332	4190
TLR-333	OEM
TLR-334	OEM
TLR-335	OEM
TLR2047	OEM
TLR2098	OEM
TLR2098A	OEM
TLR-2168	OEM
TLR2168	OEM
TLR2177	OEM
TLR2718	OEM
TLR-3168	OEM
TLRC221	OEM
TLRG101	3739
TLS106-4	5683
TLS113	OEM
TLS134A	OEM
TLS200	OEM
TLS221	OEM
TLS226	OEM
TLS251	OEM
TLS253	OEM
TLS257	OEM
TLS258	OEM
TLS259	OEM
TLSG116	OEM
TLSG205	OEM
TLSG208	OEM
TLSG222	OEM
TLUG163	OEM
TLUR121	OEM
TLUR122	OEM
TLUR153	OEM
TLUR163	OEM
TLY102	OEM
TLY113A	OEM
TLY124	3124
TLY153	OEM
TLY153FA	OEM
TLY200	OEM
TLY206	OEM
TLY221	OEM
TLY250	OEM
TLY251	OEM
TLY253	OEM
TLY257	OEM
TLY258	OEM
TLY259	OEM
TLYG116	OEM
TLYG205	OEM
TLYG208	OEM
TM1	0084
TM1R	2291
TM2	0708
TM2R	2291
TM2SC1685	0286
TM2SC1686	0286
TM3	0097
TM3R	2291
TM4	0097
TM4R	2291
TM5	0708
TM5R	2291
TM6	0097
TM6R	2291
TM7	2002
TM7R	2291
TM8	2002
TM8R	2291
TM9	1600
TM11	0090
TM11R	2291
TM12	0090
TM12R	2291
TM13	0580
TM13R	2291
TM14	0097
TM15	0097
TM16	0097
TM17	1250
TM17R	2291
TM18	1250
TM18R	2291
TM19	0871
TM19R	2291
TM21	0097
TM21R	2291
TM22	0097
TM22R	2291
TM23	0580
TM23R	2291
TM24	1250
TM24R	2291
TM25	0947
TM26	1250
TM27	1250
TM27R	2291
TM28	1250
TM28R	2291
TM29	0871
TM29R	2291
TM31	1000
TM31R	0476
TM32	0097
TM32R	0476
TM33	0015
TM33R	0476
TM34	0097
TM34R	0476
TM35	0097
TM36	0097
TM37	0097
TM37R	0476
TM38	0097
TM38R	0476
TM39	0097
TM39R	0476
TM41	0109
TM41R	0476
TM42	0109
TM42R	0476
TM43	0015
TM43R	0476
TM44	0097
TM44R	0476
TM45	0097
TM46	0097
TM47	0097
TM47R	0476
TM48	0097
TM48R	0476
TM49	0097
TM49R	0476
TM51	0122
TM51R	0476
TM52	0122
TM52R	0476
TM53	1112
TM53R	0476
TM54	1112
TM54R	0476
TM55	0122
TM55R	0476
TM56	0122
TM56R	0476
TM57	1112
TM57R	0476
TM58	1112
TM58R	0476
TM59	1112
TM59R	0476
TM61	0087
TM61R	0476
TM62	0015
TM62R	0476
TM63	0015
TM64	1112
TM64R	0476
TM65	0015
TM65R	0476
TM66	0015
TM66R	0476
TM67	1112
TM67R	0476
TM68	1112
TM68R	0476
TM69	1112
TM69R	0476
TM74	3011
TM74R	0449
TM75	3011
TM75R	0449
TM76	3011
TM76R	0449
TM78	3011
TM78R	0449
TM79	3011
TM79R	0449
TM84	0087
TM84R	0449
TM85	0087
TM85R	0449
TM86	0071
TM86R	0449
TM88	3011
TM88R	0449
TM89	3011
TM89R	0449
TM92	2291
TM100	0280
TM101	0600
TM102	0211
TM102A	0004
TM103	0038
TM103A	0208
TM104	0087
TM104MP	3033
TM104R	0449
TM105	0087
TM105R	0449
TM106	0150
TM106R	0449
TM107	0127
TM108	0144
TM109	0143
TM110	1114
TM110MP	0123
TM112	0917
TM113	0196
TM113A	0196
TM114	0484
TM115	1080
TM116	0015
TM117	0229
TM118	0774
TM119	1303
TM120	0474
TM121	0160
TM121MP	0265
TM122	0240
TM123	0198
TM123A	0016
TM123AP	0079
TM124	0142
TM125	0071
TM126	0136
TM127	0975
TM128	0086
TM129	0126
TM130	0103
TM130MP	1514
TM131	0222
TM131MP	0822
TM132	3712
TM133	3483
TM134	0188
TM134A	0188
TM135	0162
TM135A	0162
TM136	0157
TM136A	0157
TM137	0636
TM137A	0636
TM138	0649
TM138A	0649
TM139	0012
TM139A	0012
TM140	0170
TM140A	0170
TM141	0794
TM141A	0794
TM142	0137
TM142A	0137
TM143	0365
TM143A	0365
TM144	0100
TM144A	0100
TM145	0002
TM145A	0002
TM146	0441
TM146A	0441
TM147	0039
TM147A	0039
TM148	1834
TM148A	1834
TM149	0783
TM149A	0783
TM150A	0331
TM151	0149
TM151A	0149
TM152	0042
TM153	0925
TM154	0233
TM155	2763
TM156	0087
TM157	0276
TM158	0164
TM159	0037
TM160	0050
TM161	0007
TM162	0074
TM163A	0642
TM164	0003
TM165	0065
TM166	0277
TM167	0289
TM168	0296
TM169	0302
TM170	0250
TM171	0285
TM172A	0400
TM173BP	0293
TM174	0924
TM175	0178
TM176	0847
TM177	0133
TM178MP	0913
TM179	0604
TM180	1681
TM181	0130
TM182	0561
TM183	1199
TM184	0161
TM185	0460
TM186	0560
TM186A	0219
TM187	1267
TM187A	1051
TM188	0551
TM189	0382
TM190	0264
TM191	0338
TM192	0595
TM193	0791
TM194	0861
TM195A	0698
TM196	0423
TM197	0854
TM198	0168
TM199	0111
TM209	OEM
TM210	0566
TM211	1367
TM213	0437
TM218	0905
TM219	0491
TM220	0849
TM221	0353
TM222	0212
TM223	0543
TM224	0630
TM225	3280
TM226	1010
TM226MP	1862
TM228	1708
TM229	0224
TM230	0239
TM231	0061
TM232	3519
TM233	0330
TM234	0693
TM235	0936
TM236	0836
TM237	1411
TM238	0313
TM239	2336
TM241	2997
TM242	0313
TM243	2427
TM244	2277
TM245	3373
TM246	3374
TM247	2439
TM248	2432
TM249	1394
TM250	2448
TM251	3525
TM252	3526
TM253	0558
TM254	2897
TM257	3377
TM258	3528
TM259	3529
TM260	3530
TM261	1212
TM262	0602
TM263	2235
TM264	2237
TM265	2258
TM266	3533
TM267	3534
TM268	3535
TM269	3536
TM270	0134
TM271	0073
TM272	3538
TM273	3539
TM274	3370
TM275	3540
TM276	5822
TM277	2214
TM278	0418
TM279	2337
TM280	0177
TM280MP	2140
TM281	2013
TM282	0622
TM283	0363
TM284	0543
TM284MP	2371
TM285	1598
TM286	1027
TM287	0715
TM288	0342
TM289	0155
TM289MP	2767
TM290	0006
TM291	0236
TM292	0681
TM293	0018
TM293MP	3049
TM294	0532
TM295	1591
TM297	0324
TM297MP	3247
TM298	0436
TM299	2050
TM300	2046
TM300MP	3376
TM302	1173
TM306	1946
TM307	1430
TM308	1712
TM309K	1922
TM310	2329
TM311	0493
TM312	0325
TM313	0475
TM314	1825
TM315	1978
TM316	0259
TM317	2946
TM318	2312
TM319	0008
TM320	1974
TM321	1751
TM322	1984
TM323	0892
TM324	1480
TM325	1198
TM326	2987
TM327	2485
TM328	0620
TM329	2700
TM330	3561
TM330W	6003
TM331	0482
TM332	1369
TM333	2835
TM334	3563
TM336	3567
TM337	3568
TM338	3569
TM339	3570
TM340	2725
TM341	3572
TM342	2719
TM343	2720
TM344	2526
TM345	3041
TM346	2041
TM347	3576
TM348	3578
TM349	2170
TM350	1233
TM351	1977
TM353	3580
TM354	3582
TM355	3583
TM356	3584
TM357	3586
TM358	1030
TM358A	4465
TM358B	5036
TM359	3587
TM360	2506
TM361	2039
TM362	3560
TM363	2091
TM364	1420
TM365	2093
TM366	3589
TM367	2868
TM368	3591
TM369	2096
TM370	3810
TM373	0563
TM374	0525
TM375	0392
TM376	1085
TM377	0060
TM378	1308
TM379	0728
TM380	2939
TM381	3571
TM382	1386
TM383	0477
TM384	0429
TM385	1966
TM386	1852
TM387	2433
TM388	2414
TM389	0223
TM390	3081
TM391	0859
TM392	0859
TM393	3603
TM394	2613
TM395	3607
TM396	0187
TM397	0439
TM401	OEM
TM402	OEM
TM403	OEM
TM404	OEM
TM405	OEM
TM409	OEM
TM410	OEM
TM412	OEM
TM413	OEM
TM415	OEM
TM416	OEM
TM417	OEM
TM418	OEM
TM419	OEM
TM420	OEM
TM421	OEM
TM422	OEM
TM423	OEM
TM500A	0190
TM501A	0128
TM501B	0128
TM502	0201
TM503	0288
TM504	0378
TM505	0757
TM506	0102
TM507	0920
TM507(SCR)	1103
TM507F	OEM
TM507S	OEM
TM510	2084
TM513	1323
TM515	1217
TM516	OEM
TM518	1791
TM519	0124
TM520	2984
TM521	3858
TM522	1501
TM523	1067
TM524	3652
TM525	0348
TM526	1706
TM526A	1706
TM527	3860
TM528	2109
TM529	2546
TM530	3655
TM531	1054
TM532	0409
TM533	1710
TM534	2982
TM535	2985
TM536A	1997
TM537	1197
TM538	2984
TM539	1054
TM551	0182
TM552	0023
TM605	0120
TM703A	5986
TM704	2624
TM705A	2161
TM706	2571
TM707	0753
TM708	0664
TM709	0678
TM710	0829
TM711	2715
TM712	0167
TM713	0666
TM714	0352
TM716	3722
TM717	3724
TM718	0654
TM719	1940
TM720	0443
TM721	2279
TM722	0701
TM723	1345
TM724	0823
TM725	0692
TM726	2617
TM727	2529
TM729	0328
TM730	2743
TM731	2457
TM735	3008
TM736	1759
TM737	1443
TM738	0856
TM739	0751
TM740A	0379
TM741	3734
TM742	3735
TM743	1395
TM744	1421
TM745	2930
TM747	1808
TM748	0789
TM749	0395
TM750	3219
TM752	0517
TM753	3779
TM754	0723
TM756	1186
TM757	1186
TM758	3931
TM759	1671
TM760	2100
TM761	1876
TM763	2282
TM764	3935
TM765	3935
TM767	0623
TM768	3940
TM770	2906
TM772A	1980
TM773	1684
TM776	3944
TM778	0360
TM778A	0360
TM779-1	0372
TM780	0364
TM781	2747
TM782	0377
TM783	0802
TM784	2699
TM785	2707
TM786	2714
TM787	2257
TM788	2755
TM789	1843
TM790	0349
TM791	1337
TM793	2043
TM795	3195
TM796	1980
TM797	0521
TM799	1611
TM801	0520
TM802	3761
TM803	3763
TM804	2557
TM805	3410
TM806	3422
TM807	3766
TM808	2608
TM809	2688
TM810	2480
TM812	3770
TM813	3954
TM814	3006
TM815	0848
TM816	1077
TM817	1913
TM818	2501
TM819	1878
TM820	2817
TM821	2831
TM822	2812
TM823	3063
TM824	3079
TM825	3099
TM826	3962
TM830	1328
TM832	3781
TM833	1543
TM834	0176
TM900	2687
TM901	2689
TM903	2537
TM904	1854
TM905	2698
TM906	2701
TM907	2711
TM908	2705
TM909	1301
TM909D	1705
TM910	1797
TM910D	1800
TM911	1890
TM912	1696
TM913	3808
TM914	2727
TM915	1554
TM916	2749
TM917	2722
TM923	1192
TM923D	0026
TM925	2890
TM940	1919
TM941	0410
TM941D	1975
TM941M	0312
TM946	3663
TM947	2368
TM947D	2358
TM949	2552
TM955M	0973
TM960	0624
TM961	1285
TM962	0923
TM963	2648
TM964	1196
TM966	0334
TM967	1838
TM968	1321
TM969	3829
TM971	3575
TM972	2239
TM973	0417
TM973D	3801
TM974	0011
TM975	1300
TM976	2283
TM977	1292
TM978	3285
TM980	3831
TM981	0083
TM982	2830

If replacement code is OEM, contact original manufacturer for replacement.

DEVICE TYPE	REPL CODE
TM983	3425
TM984	3270
TM985	3832
TM986	3089
TM987	0625
TM988	2301
TM989	3381
TM990	1491
TM990/4XX	OEM
TM990/56X	OEM
TM990/100MA	OEM
TM990/101MA	OEM
TM990/102	OEM
TM990/201	OEM
TM990/202-X	OEM
TM990/203	OEM
TM990/203A-XX	OEM
TM990/204	OEM
TM990/303A	OEM
TM990/306	OEM
TM990/307	OEM
TM990/308	OEM
TM990/309	OEM
TM990/310	OEM
TM990/311	OEM
TM990/314	OEM
TM990/315	OEM
TM990/317	OEM
TM990/601	OEM
TM990/602	OEM
TM990/1241	OEM
TM990/1481	OEM
TM990/C101MA	OEM
TM990/C201	OEM
TM990/C307	OEM
TM990/C308	OEM
TM990/C310	OEM
TM991	3594
TM992	2247
TM1000	1235
TM1001	4437
TM1002	0387
TM1003	0579
TM1004	0878
TM1005	2568
TM1006	1215
TM1007	2491
TM1007(SCR)	2491
TM1007F	OEM
TM1007S	OEM
TM1008	3020
TM1009	0843
TM1010	0864
TM1011	0882
TM1012	0021
TM1013	0138
TM1014	0355
TM1015	2404
TM1016	2406
TM1017	2440
TM1018	2476
TM1019	2509
TM1020	OEM
TM1021	2155
TM1022	0240
TM1023	4557
TM1024	3550
TM1025	1187
TM1026	2464
TM1027	2876
TM1028	4559
TM1029	2262
TM1030	4561
TM1031	4562
TM1032	4563
TM1033	4564
TM1034	2889
TM1035	4565
TM1036	4566
TM1037	4093
TM1039	1621
TM1040	4569
TM1041	4570
TM1042	4571
TM1043	4390
TM1045	2393
TM1046	0528
TM1047	2882
TM1048	0673
TM1049	0653
TM1050	2891
TM1051	3785
TM1052	0433
TM1053	2573
TM1054	2576
TM1055	2578
TM1056	1837
TM1057	2577
TM1058	2124
TM1059	4025
TM1060	2581
TM1061	2582
TM1062	0403
TM1063	2584
TM1064	2585
TM1065	2586
TM1066	2587
TM1067	2588
TM1068	2590
TM1069	2599
TM1070	4076
TM1071	2603
TM1072	2591
TM1073	2596
TM1074	2592
TM1075A	0470
TM1076	0946
TM1078	3352
TM1079	4575
TM1080	0855
TM1081A	1160
TM1082	2261
TM1083	2854
TM1084	2886
TM1085	2631
TM1086	2887
TM1087	1018
TM1088	5682
TM1089	0638
TM1090	2861
TM1091	0964
TM1092	2316
TM1093	4578
TM1094	0043
TM1095	4485
TM1096	1634
TM1097	3263
TM1098	3958
TM1099	4260
TM1100	0911
TM1101	2634
TM1102	2632
TM1103	1994
TM1104	2057
TM1105	4582
TM1106	2630
TM1107	3670
TM1108	1393
TM1109	2639
TM1110	2305
TM1111	2472
TM1112	2482
TM1113	2158
TM1114	4586
TM1115	1248
TM1115A	3922
TM1116	4147
TM1117	4003
TM1118	1920
TM1119	4589
TM1120	2031
TM1121	0389
TM1122	0607
TM1123	4034
TM1124	4543
TM1125	4028
TM1126	4593
TM1127	4420
TM1128	3915
TM1130	1326
TM1131	4594
TM1132	4595
TM1133	2231
TM1134	2949
TM1135	2062
TM1136	3135
TM1137	2594
TM1139	3936
TM1140	2872
TM1141	3840
TM1142	1478
TM1149	4597
TM1150	4598
TM1152	2505
TM1153	3366
TM1154	4600
TM1155	1050
TM1156	3017
TM1158	0881
TM1159	0884
TM1160	3695
TM1161	2942
TM1162	4038
TM1163	2604
TM1164	3269
TM1165	2773
TM1166	2781
TM1167	1714
TM1168	2602
TM1169	4401
TM1170	2534
TM1171	0093
TM1172	1379
TM1173	3273
TM1174	2049
TM1175	2786
TM1176	2789
TM1177	2798
TM1178	3705
TM1179	2572
TM1180	2614
TM1181	2593
TM1182	4055
TM1183	4196
TM1184	4612
TM1185	1479
TM1186	0852
TM1187	0857
TM1188	2618
TM1189	4395
TM1190	4616
TM1191	4087
TM1192	1540
TM1193	3318
TM1194	1816
TM1195	4209
TM1196	1055
TM1197	4620
TM1200	4238
TM1201	4622
TM1202	2380
TM1203	2306
TM1204	4484
TM1205	2931
TM1206	4623
TM1207	4700
TM1208	4624
TM1209	4625
TM1210	2633
TM1211	2193
TM1212	4626
TM1213	4627
TM1214	0651
TM1215	2532
TM1216	0916
TM1217	0416
TM1218	1571
TM1219	1612
TM1221	4631
TM1222	5698
TM1223	1524
TM1224	4086
TM1225	2670
TM1226	0706
TM1227	0621
TM1228	3190
TM1229	4632
TM1230	2669
TM1231	1170
TM1232	1048
TM1233	4224
TM1234	2019
TM1235	4633
TM1236	2284
TM1237	3651
TM1238	4018
TM1239	3262
TM1240	4069
TM1241	2638
TM1242	1509
TM1243	4634
TM1244	2764
TM1245	0904
TM1246	4635
TM1247	4636
TM1248	1260
TM1249	4042
TM1250	4072
TM1251	4637
TM1252	4638
TM1253	4639
TM1254	3815
TM1255	2616
TM1256	4640
TM1257	4641
TM1258	4642
TM1259	5618
TM1260	4644
TM1261	1128
TM1262	4645
TM1263	2960
TM1264	4064
TM1265	4065
TM1266	4067
TM1267	4070
TM1268	4071
TM1269	4079
TM1270	5562
TM1271	4648
TM1272	3559
TM1273	4649
TM1274	5705
TM1300	3933
TM1301	4669
TM1302	3992
TM1303	4670
TM1304	3996
TM1305	3998
TM1306	4000
TM1307	4004
TM1308	3948
TM1309	4671
TM1310	3988
TM1311	3932
TM1312	3984
TM1313	3985
TM1613	0016
TM1614	0037
TM1711	0016
TM1712	0037
TM1814	0861
TM2004	4106
TM2007F	OEM
TM2007S	OEM
TM2010	OEM
TM2011	0845
TM2012	1007
TM2013	1134
TM2014	1261
TM2015	4935
TM2016	4936
TM2017	4937
TM2018	4939
TM2019	4940
TM2020	4941
TM2021	3893
TM2022	3920
TM2023	2750
TM2024	2827
TM2040	4358
TM2613	0079
TM2614	0037
TM2711	0079
TM2712	0037
TM3004	5112
TM3007F	OEM
TM3007S	OEM
TM3010	0841
TM4000	2024
TM4001B	0478
TM4002B	2055
TM4004	OEM
TM4006B	0646
TM4007B	2846
TM4007F	OEM
TM4007S	OEM
TM4008B	0988
TM4010	OEM
TM4011B	0215
TM4012B	0498
TM4013B	0413
TM4014B	0860
TM4015B	1014
TM4016B	1143
TM4017B	0513
TM4018B	1391
TM4019B	1525
TM4020B	1661
TM4021B	1749
TM4022B	1256
TM4023B	0519
TM4024B	1957
TM4025B	2072
TM4027B	1949
TM4028B	2228
TM4029B	2233
TM4030B	0500
TM4034B	3615
TM4035B	2777
TM4040B	0056
TM4042B	0121
TM4043B	1553
TM4044B	2308
TM4049B	0001
TM4050B	0398
TM4051B	0366
TM4052B	0024
TM4053B	0035
TM4055B	3304
TM4060B	0146
TM4063B	3727
TM4066B	0101
TM4068B	2503
TM4069	0119
TM4071B	0129
TM4072B	2524
TM4073B	1536
TM4075B	2540
TM4076B	3496
TM4077B	2558
TM4078B	0921
TM4081B	0626
TM4082B	0300
TM4085B	0303
TM4086B	0535
TM4093B	2384
TM4098B	3611
TM4099B	3330
TM4216P	OEM
TM4217P	OEM
TM4500BP	OEM
TM4510B	1963
TM4511B	1544
TM4512B	2120
TM4514B	1830
TM4515B	3600
TM4516B	2347
TM4518B	1043
TM4520B	2675
TM4522B	2837
TM4527B	3145
TM4539B	3656
TM4555B	2938
TM4556B	3431
TM4801P	OEM
TM5000A	1276
TM5001A	2874
TM5002A	0760
TM5003A	0677
TM5004	0118
TM5004A	0118
TM5005A	0299
TM5006A	0376
TM5007A	0036
TM5008A	0275
TM5009A	0140
TM5010	0041
TM5010A	0041
TM5011A	0253
TM5012A	0091
TM5013A	0471
TM5014A	0062
TM5015A	0077
TM5016	0165
TM5016A	0165
TM5017A	0322
TM5018A	0057
TM5019A	0064
TM5020A	0181
TM5021A	0052
TM5022	0053
TM5022A	0053
TM5023	0879
TM5023A	0879
TM5024A	0686
TM5025A	0445
TM5026A	0210
TM5027A	0375
TM5028A	0671
TM5029A	0700
TM5030A	0705
TM5031A	0494
TM5032A	0714
TM5033A	0455
TM5034A	0257
TM5035A	0195
TM5036A	0166
TM5037A	0010
TM5038A	0032
TM5039A	0054
TM5040A	0068
TM5041A	0092
TM5042A	0125
TM5043A	2317
TM5044A	2317
TM5045A	0173
TM5046A	0094
TM5047A	0049
TM5048A	0104
TM5049A	0156
TM5050	OEM
TM5050A	0189
TM5051A	0099
TM5052A	0089
TM5053A	0287
TM5054A	0252
TM5055A	0340
TM5056A	0370
TM5057A	0394
TM5058A	0425
TM5059A	0453
TM5060A	1473
TM5061	3003
TM5061A	3003
TM5062A	OEM
TM5063A	1312
TM5064A	3010
TM5065A	1713
TM5066A	0292
TM5067A	0456
TM5068	0533
TM5068A	0533
TM5069A	0451
TM5070	0301
TM5070A	0301
TM5071	0025
TM5071A	0025
TM5072	0244
TM5072A	0244
TM5073	1083
TM5073A	1083
TM5074	0317
TM5074A	0317
TM5075	0420
TM5075A	0420
TM5076	1649
TM5076A	1649
TM5077	0495
TM5077A	0495
TM5078A	0949
TM5079	0531
TM5079A	0531
TM5080	0565
TM5080A	0565
TM5081	0402
TM5081A	0402
TM5082	1606
TM5082A	1606
TM5083	1674
TM5083A	1674
TM5084	0726
TM5084A	0726
TM5085A	0820
TM5086	0350
TM5086A	0350
TM5087A	0931
TM5088A	0999
TM5089A	0502
TM5090A	0869
TM5091A	1156
TM5092A	2157
TM5093A	1189
TM5094A	3026
TM5095	1311
TM5095A	1311
TM5096	0098
TM5096A	0098
TM5097	0186
TM5097A	0186
TM5098	0213
TM5098A	0213
TM5099	0245
TM5099A	0245
TM5100	0028
TM5101	0255
TM5101A	0255
TM5102A	0877
TM5103A	0367
TM5104A	2858
TM5105	0421
TM5105A	0421
TM5111A	0782
TM5112A	0796
TM5113A	0806
TM5114A	0821
TM5115A	0833
TM5116	0442
TM5116A	0442
TM5117A	0876
TM5118A	3127
TM5119A	0185
TM5120A	0205
TM5121A	0480
TM5122A	0504
TM5123A	3317
TM5124A	0684
TM5125A	0225
TM5126A	0230
TM5127A	0234
TM5128A	0237
TM5129A	1397
TM5130	0247
TM5130A	0247
TM5131A	0251
TM5132A	1178
TM5133A	0256
TM5134A	2395
TM5135	0262
TM5135A	0262
TM5136	0269
TM5136A	0269
TM5137A	0274
TM5138	2399
TM5138A	2399
TM5139A	0294
TM5140A	1177
TM5141A	0309
TM5142A	0318
TM5143A	0320
TM5144A	0326
TM5145A	0337
TM5146A	0347
TM5147A	0027
TM5148A	0266
TM5149A	2856
TM5150A	0386
TM5151A	0405
TM5152A	0426
TM5153A	0444
TM5154A	3028
TM5155A	0238
TM5156A	1180
TM5157A	1190
TM5158A	1207
TM5159A	1218
TM5160	1881
TM5160A	1881
TM5161A	0647
TM5162A	1255
TM5163A	2102
TM5164A	1279
TM5165A	2225
TM5166A	0605
TM5172A	4231
TM5172AK	1006
TM5173A	2397
TM5173AK	1380
TM5174A	2035
TM5174AK	0547
TM5175A	2401
TM5175AK	2403
TM5176A	1438
TM5176AK	2113
TM5177A	2407
TM5177AK	2410
TM5178A	1445
TM5178AK	1901
TM5179A	2415
TM5179AK	2416
TM5180A	2221
TM5180AK	0696
TM5181A	1458
TM5181AK	1601
TM5182A	0221
TM5182AK	1616
TM5183A	1490
TM5183AK	1622
TM5184A	2422
TM5184AK	2424
TM5185A	1618
TM5185AK	0627
TM5186A	0510
TM5186AK	0992
TM5187A	0691
TM5187AK	0995
TM5188A	0870
TM5188AK	1263
TM5189A	1020
TM5189AK	1249
TM5190A	1153
TM5190AK	1636
TM5191A	1274
TM5191AK	1639
TM5192A	1402
TM5192AK	1703
TM5193A	1532
TM5193AK	1640
TM5194A	1078
TM5194AK	1716
TM5195A	1711
TM5195AK	1720
TM5196A	1717
TM5196AK	1731
TM5197A	1723
TM5197AK	0727
TM5198A	1736
TM5198AK	1756
TM5199A	1748
TM5199AK	1768
TM5200A	1761
TM5200AK	1782
TM5201A	2450
TM5201AK	2453
TM5202A	1772
TM5202AK	1794
TM5203A	1788
TM5203AK	1799
TM5204A	1796
TM5204AK	1809
TM5205A	1804
TM5205AK	1817
TM5206A	1194
TM5206AK	1826
TM5207A	1821
TM5207AK	1840
TM5208A	0022
TM5208AK	1853
TM5209A	0070
TM5209AK	1861
TM5210A	0132
TM5210AK	1866
TM5211A	0172
TM5211AK	1874
TM5212A	0207
TM5212AK	1884
TM5213A	0227
TM5213AK	2474
TM5214A	0263
TM5214AK	1895
TM5215A	0310
TM5215AK	1902
TM5216A	0329
TM5216AK	0736
TM5217A	0356
TM5217AK	1909
TM5218A	0381
TM5218AK	2479
TM5219A	0412
TM5219AK	1914
TM5220A	0438
TM5220AK	1163
TM5221A	0464
TM5221AK	1924
TM5222A	0488
TM5222AK	1933
TM5223A	0509
TM5223AK	1941
TM5224A	0524
TM5224AK	1947
TM5225A	0542
TM5225AK	1953
TM5226A	0063
TM5226AK	1961
TM5227A	0401
TM5227AK	0357
TM5228A	0598
TM5228AK	2500
TM5229A	0616
TM5229AK	0670
TM5230A	0634
TM5230AK	0776
TM5231A	0650
TM5231AK	2507
TM5232A	0668
TM5232AK	1071
TM5240A	0815
TM5240AK	2721
TM5241A	0827
TM5241AK	2724
TM5242A	0846
TM5242AK	2726
TM5243A	0868
TM5243AK	2729
TM5244A	0885
TM5244AK	2732
TM5245A	0897
TM5245AK	2734
TM5246A	0914
TM5246AK	2738
TM5247A	0926
TM5247AK	2740
TM5248A	0944
TM5248AK	2744

DEVICE TYPE	REPL CODE
TM5249A	0958
TM5249AK	1225
TM5250A	0978
TM5250AK	2746
TM5251A	0994
TM5251AK	1237
TM5252A	1009
TM5252AK	1252
TM5253A	1019
TM5253AK	1269
TM5254A	0889
TM5254AK	1277
TM5255A	1049
TM5255AK	1294
TM5256A	1058
TM5256AK	2752
TM5257A	0932
TM5257AK	1302
TM5258A	0932
TM5258AK	1310
TM5259A	1096
TM5259AK	2756
TM5260A	1106
TM5260AK	1324
TM5261A	1123
TM5261AK	2758
TM5262A	1135
TM5262AK	1333
TM5263A	1152
TM5263AK	1344
TM5264A	1164
TM5264AK	1356
TM5265A	1174
TM5265AK	2760
TM5266A	1184
TM5266AK	1371
TM5267A	1200
TM5267AK	2762
TM5268A	1210
TM5268AK	1387
TM5269A	1223
TM5269AK	1406
TM5270A	1232
TM5270AK	1415
TM5271A	1246
TM5271AK	1428
TM5272A	1265
TM5272AK	1440
TM5273A	1273
TM5273AK	2772
TM5274A	1290
TM5274AK	1447
TM5275A	1299
TM5275AK	2776
TM5276A	1307
TM5276AK	1461
TM5277A	1322
TM5277AK	2779
TM5278A	1331
TM5278AK	1474
TM5279A	1340
TM5279AK	2783
TM5280A	1353
TM5280AK	0613
TM5281A	1365
TM5281AK	1510
TM5282A	1384
TM5282AK	1523
TM5283A	1401
TM5283AK	1537
TM5284A	1412
TM5284AK	1550
TM5285A	1423
TM5285AK	1575
TM5286A	1436
TM5286AK	2802
TM5287A	1444
TM5287AK	2803
TM5288A	1457
TM5288AK	1507
TM5289A	1470
TM5289AK	2805
TM5290A	1484
TM5290AK	2807
TM5291A	1505
TM5291AK	2810
TM5292A	1521
TM5292AK	2815
TM5293A	1531
TM5293AK	2818
TM5294A	1548
TM5294AK	2820
TM5295A	1568
TM5295AK	2822
TM5296A	1587
TM5296AK	2824
TM5404	0058
TM5414	3620
TM5444	2270
TM5448	3404
TM5452	1396
TM5454	0447
TM5455	0940
TM5800	0110
TM5802	0242
TM5804	0540
TM5806	0965
TM5982	2901
TM5984	OEM

DEVICE TYPE	REPL CODE
TM5986	1124
TM5987	1107
TM5990	0805
TM5994	0319
TM6004	OEM
TM6010	0601
TM6402	0316
TM6407	3331
TM7004	OEM
TM7007	OEM
TM7010	OEM
TM7492	0834
TM7493A	0569
TM8004	OEM
TM8010	OEM
TM59090	OEM
TMAL110	OEM
TMAL113	OEM
TMAL116	OEM
TMAL220	OEM
TMAL223	OEM
TMAL226	OEM
TMAM4004-6	OEM
TMAM4004-10	OEM
TMAM4008-6	OEM
TMAM4008-10	OEM
TMAM4011	OEM
TMAM4014-21	OEM
TMAM4015-21	OEM
TMAM4016-21	OEM
TMAM4017-21	OEM
TMAM6001	OEM
TMAM6002	OEM
TMAM6003	OEM
TMAM6004	OEM
TMAM6058	OEM
TMAM6059	OEM
TMAM6070	OEM
TMAM6071	OEM
TMAM6073	OEM
TMAM6075	OEM
TMAM6081	OEM
TMAM6095	OEM
TMAM7001	OEM
TMAM7002	OEM
TMAM7003	OEM
TMAM7005	OEM
TMAM7010	OEM
TMAM9000	OEM
TMAM9010	OEM
TMAM9020	OEM
TMAM9040	OEM
TMC73C247-10	OEM
TMD01	0162
TMD01A	0162
TMD02	0157
TMD02A	0157
TMD03	0636
TMD03A	0636
TMD04	0062
TMD04A	0062
TMD05	OEM
TMD05A	OEM
TMD06	0165
TMD06A	0165
TMD07	0057
TMD07A	0057
TMD08	0170
TMD08A	0170
TMD09	OEM
TMD09A	OEM
TMD10	0052
TMD10A	0052
TMD20	OEM
TMD20A	OEM
TMD24	OEM
TMD25	OEM
TMD27	OEM
TMD40	OEM
TMD40A	OEM
TMD41	0133
TMD42	0609
TMD45	0609
TMD50	OEM
TMD914	OEM
TMD916	OEM
TMM314AP	2048
TMM314APL-3	2048
TMM315D	1693
TMM315D-1	1693
TMM323D	OEM
TMM416D-3	0522
TMM416D-4	0522
TMM416P-3	0522
TMM841P	OEM
TMM2016	1898
TMM2016AP	1898
TMM2016AP-10	OEM
TMM2016AP-12	1898
TMM2016BP-10	OEM
TMM2016BP-12	OEM
TMM2016BP-15	1898
TMM2016BP-90	OEM
TMM2016P	1898
TMM2016P-1	1898
TMM2016P-2	1898
TMM2016P-7	1898
TMM2063AP-10	OEM
TMM2063P-10	OEM

DEVICE TYPE	REPL CODE
TMM2063P-15	OEM
TMM2063P-150	OEM
TMM2064P-15	OEM
TMM2114AP-15	5300
TMM2116AP-15	OEM
TMM2764D-2	0812
TMM4164AP-15	OEM
TMM4164P-3	2357
TMM10010-01	OEM
TMM10010-02	OEM
TMM10010-03	OEM
TMM10010-04	OEM
TMM10010-05	OEM
TMM10010-06	OEM
TMM20000-01	OEM
TMM20000-02	OEM
TMM20000-03	OEM
TMM20000-04	OEM
TMM20000-05	OEM
TMM23256P	OEM
TMM23256P-5879	OEM
TMM24512AP-25	OEM
TMM27128AD-20	1638
TMM27256AD	OEM
TMM27256AD-20	OEM
TMM27256AD-XX	OEM
TMM27512AD-15	OEM
TMM30000-01	OEM
TMM30000-02	OEM
TMM30000-03	OEM
TMM40010-01	OEM
TMM40010-02	OEM
TMM40010-04	OEM
TMM40010-07	OEM
TMM41256CP-12	1472
TMM41257P-15	OEM
TMM41464AP-12	OEM
TMMS27256JL	OEM
TMP42C40P-10	OEM
TMP42C40P1007	OEM
TMP42C40P-1367	OEM
TMP47C00AF-68572	OEM
TMP47C25N-3055	OEM
TMP47C40P	OEM
TMP47C40P-SA6201	OEM
TMP47C40P-SB6204	OEM
TMP47C40P-SC	OEM
TMP47C200AN-2567	OEM
TMP47C200AN-2574	OEM
TMP47C200AN-2575	OEM
TMP47C200N-2572	OEM
TMP47C231AN4942	OEM
TMP47C231AN-4942Z	OEM
TMP47C231AN4943	OEM
TMP47C231AN-4943Z	OEM
TMP47C231AN4951	OEM
TMP47C231AN-4951Z	OEM
TMP47C232	OEM
TMP47C232-4981	OEM
TMP47C232-4989	OEM
TMP47C232AN-4989	OEM
TMP47C232AN-4989Z	OEM
TMP47C232N-4981	OEM
TMP47C232N4981	OEM
TMP47C232N-4989	OEM
TMP47C232N8893	OEM
TMP47C236AN-R702	OEM
TMP47C236ANR702	OEM
TMP47C337AN	OEM
TMP47C400-6764	OEM
TMP47C400AF-6857	OEM
TMP47C400AN-6265Z	OEM
TMP47C400AN-6403	OEM
TMP47C400AN-6404	OEM
TMP47C400AN-6405	OEM
TMP47C400CN-6452	OEM
TMP47C400F-6442	OEM
TMP47C400N	OEM
TMP47C400N-6402	OEM
TMP47C400N-6403	OEM
TMP47C400N-6452	OEM
TMP47C400N-6454	OEM
TMP47C400N-6455	OEM
TMP47C420AF8352	OEM
TMP47C430N-U803	OEM
TMP47C432-8857	OEM
TMP47C432AN	OEM
TMP47C432AN-8088	OEM
TMP47C432AN-8088Z	OEM
TMP47C432AN-8098	OEM
TMP47C432AN-8382	OEM
TMP47C432AN-8497	OEM
TMP47C432AN8699	OEM
TMP47C432AN-8734	OEM
TMP47C432AN-8939Z	OEM
TMP47C432AN8939Z	OEM
TMP47C432N(8083)	OEM
TMP47C432N(8091)	OEM
TMP47C432N88	OEM
TMP47C432N-8083	OEM
TMP47C432N-8091	OEM
TMP47C432N8784	OEM
TMP47C432N-8857	OEM
TMP47C432N-8858	OEM
TMP47C432N-8859	OEM
TMP47C432N-8861	OEM
TMP47C432N8861	OEM
TMP47C432N-8890	OEM
TMP47C432N8890	OEM

DEVICE TYPE	REPL CODE
TMP47C432N-8989	OEM
TMP47C434N-3119	OEM
TMP47C434N3119	OEM
TMP47C434N-3404	OEM
TMP47C434N-3464	OEM
TMP47C434N-3509Z	OEM
TMP47C434N3594	OEM
TMP47C441AN-1662	OEM
TMP47C460-9242	OEM
TMP47C634N	OEM
TMP47C634N(GS8108-01	OEM
TMP47C634N-2402	OEM
TMP47C634N-2409	OEM
TMP47C634N-2417	OEM
TMP47C634N-2418	OEM
TMP47C634N-2427Z	OEM
TMP47C634N2427Z	OEM
TMP47C634N-2438	OEM
TMP47C634N2438	OEM
TMP47C634N2453	OEM
TMP47C634N-2453Z	OEM
TMP47C634N2454	OEM
TMP47C634N2454Z	OEM
TMP47C634N2651	OEM
TMP47C634N-2654Z	OEM
TMP47C634N-2663	OEM
TMP47C634N-2679	OEM
TMP47C634N-2687	OEM
TMP47C634N-2689	OEM
TMP47C634N-R075	OEM
TMP47C634N-R301	OEM
TMP47C634NR357	OEM
TMP47C634NR357(Z)	OEM
TMP47C634N-R357Z	OEM
TMP47C634NR357Z	OEM
TMP47C634N-R401	OEM
TMP47C634N-R402	OEM
TMP47C634N-R417	OEM
TMP47C660N-1454	OEM
TMP47C670N-1264	OEM
TMP47C670N-1282	OEM
TMP47C670N-1283	OEM
TMP47C834N-R075	OEM
TMP47C870N-4627	OEM
TMP47C1237NU132	OEM
TMP47C1237N-U132(Z)	OEM
TMP47C1638NU311(F)Z	OEM
TMP47C1638NU312(F)Z	OEM
TMP47C2314951	OEM
TMP47C2324981	OEM
TMP47C2324989	OEM
TMP47C4328784	OEM
TMP47C4328857	OEM
TMP47C4328858	OEM
TMP47C4328859	OEM
TMP47C4328861	OEM
TMP47C4328890	OEM
TMP47C4328989	OEM
TMP47C6342427	OEM
TMP47P860N	OEM
TMP87C34AN-3178	OEM
TMP87C833N-3051	OEM
TMP87CH33N-3014	OEM
TMP87CH34AN-3102	OEM
TMP87CH34AN-3117	OEM
TMP87CM34AN	OEM
TMP90C041N	OEM
TMP4240P-1603	OEM
TMP4315BN0422	OEM
TMP4315BN0433	OEM
TMP4320-6214A	OEM
TMP4320-6422A	OEM
TMP4320-6432	OEM
TMP4320AN-6462	OEM
TMP4320AN6462	OEM
TMP4320AP-6201	OEM
TMP4320AP-6226A	OEM
TMP4320AP-6401A	OEM
TMP4320AP-6422	OEM
TMP4320AP-6433	OEM
TMP4320P-6202	OEM
TMP4320P-6203	OEM
TMP4320P-6204	OEM
TMP4320P-6205	OEM
TMP4321AP-FA	OEM
TMP4321AP-SA9002	OEM
TMP4321AP-SC	OEM
TMP4321AP-SC9007	OEM
TMP4323AN-8124	OEM
TMP4323AP-8605	OEM
TMP4501	OEM
TMP4720P-1802	OEM
TMP4740P	OEM
TMP4740P-5403	OEM
TMP4740P-5404	OEM
TMP4740P-5411	OEM
TMP4740P-5541	OEM
TMP4740P5541	OEM
TMP4740P-5543	OEM
TMP4740P5543	OEM
TMP4740P5659	OEM
TMP4740P-5702	OEM
TMP4746N	OEM
TMP8039P	OEM
TMP8048P-1877	OEM
TMP8049AP-6	OEM
TMP8085AP	OEM
TMP8155P	OEM

DEVICE TYPE	REPL CODE
TMP8155P-2	OEM
TMP8155P2	OEM
TMP43103312A	OEM
TMP43103323A	OEM
TMP43150414B	OEM
TMP43206432A	OEM
TMP43206462AN	OEM
TMP47201802	OEM
TMP47405541	OEM
TMP47405543	OEM
TMP47405702	OEM
TMP47465759	OEM
TMP47465768	OEM
TMP47465775	OEM
TMPD459	3895
TMPD914	OEM
TMPD2835	4917
TMPD2836	4917
TMPD2837	0907
TMPD2838	0907
TMPD4148	OEM
TMPD4150	3749
TMPD4153	OEM
TMPD4154	3749
TMPD4448	OEM
TMPD5711	OEM
TMPD6050	0080
TMPD6100	0080
TMPD6914	OEM
TMPD6916	OEM
TMPD6919	OEM
TMPD7000	OEM
TMS0117	OEM
TMS25L32-45	OEM
TMS27C128JL	OEM
TMS27C256-20JL	OEM
TMS27L08-45	OEM
TMS27PC512NL	OEM
TMS40L44-12	OEM
TMS40L44-20	OEM
TMS40L44-25	OEM
TMS40L44-45	OEM
TMS70E40	OEM
TMS70L22	OEM
TMS1000	OEM
TMS1000C	OEM
TMS1000N	OEM
TMS1000NLL	OEM
TMS1004C	OEM
TMS1018	OEM
TMS1020	OEM
TMS1023NL	OEM
TMS1024	OEM
TMS1024NLL	OEM
TMS1025	OEM
TMS1025N	OEM
TMS1025N2LC	OEM
TMS1025N2LL	OEM
TMS1050	OEM
TMS1070	OEM
TMS1070C	OEM
TMS1100	OEM
TMS1100C	OEM
TMS1100N2LL	OEM
TMS1100NLL	OEM
TMS1117	OEM
TMS1121	OEM
TMS1152NL	OEM
TMS1170	OEM
TMS1170NLHL	OEM
TMS1200	OEM
TMS1200C	OEM
TMS1270	OEM
TMS1270C	OEM
TMS1300	OEM
TMS1300C	OEM
TMS1370	OEM
TMS1400	OEM
TMS1470	OEM
TMS1600	OEM
TMS1670	OEM
TMS1670MP7572A	OEM
TMS1700	OEM
TMS1943	2295
TMS1943A2	2295
TMS1943N2	2295
TMS1943N2L	2295
TMS1943N21	2295
TMS1943NL	2295
TMS1944N2	OEM
TMS1944N2L	OEM
TMS-1952	2728
TMS1952NL	2728
TMS2016	OEM
TMS2100	OEM
TMS2114-15	5300
TMS2114-20	2048
TMS2114-25	2048
TMS2114-45	2048
TMS2114L-15	5300
TMS2114L-20	2048
TMS2114L-25	2048
TMS2114L-45	2048
TMS2132	OEM
TMS2147H-3	OEM
TMS2147H-4	OEM
TMS2147H-5	OEM
TMS2147H-7	OEM
TMS2149-3	OEM

DEVICE TYPE	REPL CODE
TMS2149-4	OEM
TMS2149-5	OEM
TMS2149-7	OEM
TMS2170	OEM
TMS2220	OEM
TMS2240	OEM
TMS2300	OEM
TMS2370	OEM
TMS2372	OEM
TMS2400	OEM
TMS2470	OEM
TMS2516-30	OEM
TMS2516-35	OEM
TMS2516-45	OEM
TMS2516JL-45	5585
TMS2532-30	5098
TMS2532-35	5098
TMS2532-45	5098
TMS2532JL-45	2607
TMS2564-35	0812
TMS2564-45	0812
TMS2564JL	OEM
TMS2600	OEM
TMS2670	OEM
TMS2708-35	OEM
TMS2708-45	OEM
TMS2716-30	OEM
TMS2716-45	OEM
TMS2732AJL-30	2697
TMS2732AJL-35	2697
TMS2764JL-25	0812
TMS3450N1	OEM
TMS3450NL	3982
TMS3451NL	OEM
TMS3452N2L	OEM
TMS3453N2L	OEM
TMS3454NR	OEM
TMS3614NC	OEM
TMS3763CNL28	OEM
TMS3763CNL8805	OEM
TMS4016-12	OEM
TMS4016-15	1898
TMS4016-20	1898
TMS4016-25	1898
TMS4044-12	OEM
TMS4044-20	OEM
TMS4044-25	OEM
TMS4044-45	OEM
TMS4116-15	OEM
TMS4116-20	0522
TMS4116-25	0522
TMS4164-15	2357
TMS4164-15NLJ	2357
TMS4164-20	2357
TMS4164-20NL	2357
TMS4164-25	2357
TMS4164NLJ	2357
TMS4256	OEM
TMS4256-15	OEM
TMS4256-15NL	1472
TMS4256FML	OEM
TMS4416	OEM
TMS4416-12NL	OEM
TMS4416-15	OEM
TMS4416-15NL	OEM
TMS4416-20	OEM
TMS4416-20NL	OEM
TMS4416-25	OEM
TMS4464-12NL	OEM
TMS4732	OEM
TMS4764	OEM
TMS4764-30NL	OEM
TMS5027	OEM
TMS5037	OEM
TMS5100	OEM
TMS5100N	OEM
TMS5100NL	OEM
TMS5110	OEM
TMS5220	OEM
TMS6100	OEM
TMS6125	OEM
TMS7000	OEM
TMS7020	OEM
TMS7040	OEM
TMS7041	OEM
TMS8080AN	4547
TMS9900-40	OEM
TMS9900NL	OEM
TMS9901	OEM
TMS9901NL	OEM
TMS9902A	OEM
TMS9903	OEM
TMS9909	OEM
TMS9914A	OEM
TMS9918A	OEM
TMS9919	OEM
TMS9927	OEM
TMS9928A	OEM
TMS9929A	OEM
TMS9937	OEM
TMS9940M	OEM
TMS9980A	OEM
TMS9981	OEM
TMS9995	OEM
TMS27128-20JL	OEM
TMS27128AD-200	OEM
TMS27128D-15	OEM
TMS27256JL	OEM
TMS27256JL-XX	OEM

DEVICE TYPE	REPL CODE
TMS99105	OEM
TMS99110	OEM
TMS99531	OEM
TMS99532	OEM
TMS99541	OEM
TMS99650	OEM
TMSK101	OEM
TMSK101A	OEM
TMSK202	OEM
TMSW330R-6	OEM
TMSW330R-10	OEM
TMSW330R-21	OEM
TMSW334H-21	OEM
TMSW340F-21	OEM
TMSW340R-6	OEM
TMSW340R-10	OEM
TMSW348E-21	OEM
TMSW352M-21	OEM
TMSW356I-21	OEM
TMSW511F-21	OEM
TMSW600P-3	OEM
TMSW754P-6	OEM
TMSW754P-10	OEM
TMSW755P-21	OEM
TMT696	0144
TMT697	0144
TMT839	0144
TMT840	0144
TMT841	0144
TMT842	0144
TMT843	0144
TMT1131	OEM
TMT1132	OEM
TMT1543	0016
TMT2427	0144
TN0102N2	OEM
TN0102N3	OEM
TN0102ND	OEM
TN0104N2	OEM
TN0104N3	OEM
TN0104ND	OEM
TN0106N2	OEM
TN0106N3	OEM
TN0106ND	OEM
TN0110N2	OEM
TN0110N3	OEM
TN0110ND	OEM
TN0202N2	OEM
TN0202N3	OEM
TN0202ND	OEM
TN0204N2	OEM
TN0204N3	OEM
TN0204ND	OEM
TN0520N2	OEM
TN0520N3	OEM
TN0520ND	OEM
TN0524N2	OEM
TN0524N3	OEM
TN0524ND	OEM
TN0602N2	OEM
TN0602N3	OEM
TN0602ND	OEM
TN0604N2	OEM
TN0604N3	OEM
TN0604ND	OEM
TN0606N2	OEM
TN0606N3	OEM
TN0606N5	OEM
TN0606N6	OEM
TN0606N7	OEM
TN0606ND	OEM
TN0610N2	OEM
TN0610N3	OEM
TN0610N5	OEM
TN0610ND	OEM
TN0620N2	OEM
TN0620N3	OEM
TN0620N5	OEM
TN0620ND	OEM
TN0624N2	OEM
TN0624N3	OEM
TN0624N5	OEM
TN0624ND	OEM
TN2SA643	1242
TN2SA733	0006
TN2SA733-4	0006
TN2SA733-Q	0006
TN-2SA733-Q-B	0006
TN2SA733-Q-B	0006
TN2SA733-R	0006
TN2SA733-R-A	0006
TN2SA733K	0006
TN2SA733P	0006
TN2SA733Q	0006
TN2SA733R	0006
TN2SA817Y	0436
TN2SA952L	0006
TN2SA953	1348
TN2SA953K	1348
TN2SA1175	3625
TN2SB810E	0532
TN2SB810H	0532
TN2SC495-R	0786
TN2SC945	0076
TN2SC945-Q	0076
TN2SC945-R	0076
TN2SC945K	0076
TN2SC945R	0076
TN2SC1449K	0161

Original Device Types

DEVICE TYPE	REPL CODE
TN2SC1507	0955
TN2SC1507-K	0955
TN2SC1507-K-A	0955
TN2SC1507-L	0955
TN2SC1507-L-A	0955
TN2SC1507-M	0955
TN2SC1507-M-A	0955
TN2SC1507L	0955
TN2SC1520-1	0168
TN2SC1520-1-1A	0168
TN2SC1520-K-1	0168
TN2SC1520-K-1A	0168
TN2SC1520-K-3A	0168
TN2SC1520-L-1	0168
TN2SC1520-L-1A	0168
TN2SC1520-L-3A	0168
TN2SC1520-M-1	0168
TN2SC1520-M-1A	0168
TN2SC1520-M-3A	0168
TN2SC1845	0530
TN2SC1941	1327
TN2SC1941-	0715
TN2SC1941-2SC1941	OEM
TN2SC2002	0951
TN2SC2002K	0951
TN2SC2002L	0951
TN2SC2785	0249
TN2SC2785E	0249
TN2SC2785H	0249
TN2SD471K	0018
TN2SD568K	0423
TN2SK68A	3342
TN3TB03001	0832
TN3TC03001	2323
TN3TC05001	0981
TN3TD03001	0898
TN7TB03001	0832
TN7TC03001	0887
TN7TC05001	3482
TN7TD03001	0898
TN10	0373
TN12	0348
TN15	OEM
TN41A	0316
TN51	OEM
TN52	OEM
TN-53	0016
TN53	0016
TN-54	0016
TN54	0324
TN55	0016
TN56	0016
TN-59	0016
TN59	0016
TN-60	0016
TN60	0016
TN-61	0016
TN61	0016
TN-62	0016
TN62	0016
TN-63	0016
TN63	0016
TN-64	0016
TN64	0016
TN71	2548
TN72	2548
TN79	0595
TN80	0016
TN81	0595
TN237	0016
TN238	0016
TN301	OEM
TN302	OEM
TN303	OEM
TN304	OEM
TN421	0086
TN591	0004
TN2001	OEM
TN2001A	OEM
TN-3200	0007
TN3200	0007
TN3866	1354
TN-3903	0016
TN3903	0016
TN-3904	0016
TN3904	0016
TN-3905	2468
TN3905	0861
TN3906	0861
TN4338	OEM
TN4339	OEM
TNC61688	0400
TNC61689	0016
TNC61690	0037
TNC61702	0016
TNC61703	0037
TNF10	OEM
TNF12	OEM
TNF15	OEM
TNH11304	5961
TNH11304EZ	OEM
TNH11306	OEM
TNH11309	OEM
TNJ6I672(RECT)	0015
TNJ1034	0111
TNJ1036	0079
TNJ60063	0050
TNJ60064	0050
TNJ60065	0050
TNJ60066	1342
TNJ60067	0050
TNJ60068	0050
TNJ60069	0050
TNJ60069(2SC74)	0127
TNJ60070	0004
TNJ60071	0050
TNJ60072	0233
TNJ60073	0136
TNJ60074	0004
TNJ60075	0160
TNJ60076	0155
TNJ60077	0136
TNJ60078	0168
TNJ60079	0004
TNJ60080	0604
TNJ60279	0050
TNJ60280	0050
TNJ60281	0050
TNJ-60282	0004
TNJ60282	0004
TNJ60283	0004
TNJ-60362	0050
TNJ60362	0136
TNJ-60363	0050
TNJ60363	0136
TNJ-60364	0050
TNJ60364	0136
TNJ-60365	0050
TNJ60365	0004
TNJ60447	0127
TNJ60448	0127
TNJ60449	0127
TNJ60450	0050
TNJ60451	0178
TNJ60453	0178
TNJ60454	0160
TNJ60455	0074
TNJ60456	0050
TNJ60457	0924
TNJ-60604	0079
TNJ60604	0127
TNJ-60605	0127
TNJ60605	0127
TNJ-60606	0079
TNJ60606	0127
TNJ-60607	0079
TNJ60607	0127
TNJ-60608	0050
TNJ60608	0280
TNJ-60610	0280
TNJ60610	0004
TNJ-60611	0280
TNJ60611	0004
TNJ-60612	0280
TNJ60612	0164
TNJ60728	0004
TNJ61217	0127
TNJ61218	0127
TNJ61219	0079
TNJ61220	0016
TNJ61221	0211
TNJ61222	0004
TNJ61223	0847
TNJ61282	0004
TNJ61671	0600
TNJ61671(2SC688)	0127
TNJ61671(2SD72)	0208
TNJ61672(2SK25)	0325
TNJ61673	0325
TNJ61673(2SB186)	0004
TNJ61673(2SK24)	0325
TNJ61674	0164
TNJ61679	0127
TNJ61730	0127
TNJ61731	0127
TNJ61734	0208
TNJ70450	0560
TNJ70478	0144
TNJ70478-1	0144
TNJ70479	0144
TNJ70479-1	0079
TNJ70480	0144
TNJ70481	0791
TNJ70482	0086
TNJ70483	0222
TNJ70484	0144
TNJ70537	0016
TNJ70539	0016
TNJ70540	0161
TNJ70541	0222
TNJ70634	0004
TNJ70635	0004
TNJ70637	0079
TNJ70638	1144
TNJ70639	0155
TNJ70640	0286
TNJ70641	0136
TNJ70688	0164
TNJ70691	0111
TNJ71034	0111
TNJ71035	0086
TNJ71036	0079
TNJ71037	0037
TNJ71143	0007
TNJ71173	0144
TNJ71234	0086
TNJ71248	0136
TNJ71252	0595
TNJ71271	0111
TNJ71277	0111
TNJ71498	0113
TNJ71629	0127
TNJ71773	0037
TNJ71774	0037
TNJ71937	0127
TNJ71963	0127
TNJ71964	0007
TNJ71965	0111
TNJ72146	0003
TNJ72147	0142
TNJ72148	0103
TNJ72149	0065
TNJ72150	0007
TNJ72151	0007
TNJ72152	0791
TNJ72153	0142
TNJ72154	0037
TNJ72275	0007
TNJ72276	0330
TNJ72277	0127
TNJ72278	0004
TNJ72279	0127
TNJ72280	0079
TNJ72281	0016
TNJ72282	0233
TNJ72283	0004
TNJ72284	0208
TNJ72285	0004
TNJ72286	0178
TNJ72287	0004
TNJ72288	0086
TNJ72289	0004
TNJ72318	0160
TNJ72319	0642
TNJ72320	0074
TNJ72368	0007
TNJ72701	0007
TNJ72774	1267
TNJ72775	0560
TNJ72783	0016
TNJ72784	0016
TNP47C231AN4943	OEM
TNP47C634N2427Z	OEM
TNQ2621	OEM
TNQ2682	OEM
TNQ2682B	OEM
TNQ2683	OEM
TNT839	0144
TNT840	0144
TNT841	0144
TNT842	0016
TNT843	0144
TNT1131	OEM
TNT1132	OEM
TNV76601G2	OEM
TNYTD03001	0898
TO-003	0050
TO-004	0050
TO-005	0004
TO-012	0160
TO-014	0004
TO-015	0160
TO-033	0016
TO-038	0016
TO-039	0016
TO-040	0016
TO-041	0004
TO1-101	0016
TO1-104	0016
TO1-105	0016
TO-36	OEM
TO-101	0280
TO101	0004
TO-102	0280
TO102	0004
TO-103	0004
TO103	0004
TO-104	0004
TO104	0004
TO507	OEM
TO1007	OEM
TO2007	OEM
TO3007	OEM
TO4007	OEM
TO5007	OEM
TO6007	OEM
TOAL113	OEM
TOAL116	OEM
TOAL223	OEM
TOAL226	OEM
TOTX172	OEM
TP0102N2	OEM
TP0102N3	OEM
TP0102ND	OEM
TP0104N2	OEM
TP0104N3	OEM
TP0104ND	OEM
TP0202N2	OEM
TP0202N3	OEM
TP0202ND	OEM
TP0204N2	OEM
TP0204N3	OEM
TP0204ND	OEM
TP0602N2	OEM
TP0602N3	OEM
TP0602ND	OEM
TP0604N2	OEM
TP0604N3	OEM
TP0604ND	OEM
TP0616N2	OEM
TP0616N3	OEM
TP0616N5	OEM
TP0616ND	OEM
TP0620N2	OEM
TP0620N3	OEM
TP0620N5	OEM
TP0620ND	OEM
TP1	OEM
TP2	OEM
TP3TC03001	3143
TP3TD06001	4130
TP5	OEM
TP7TC03001	1032
TP7TC05001	0703
TP7TD03002	4130
TP10	OEM
TP20	OEM
TP20-0284	0636
TP30	OEM
TP34	0143
TP34A	0143
TP40	OEM
TP100	0097
TP100R	2291
TP101	0015
TP107	0079
TP107A	0079
TP107B	0079
TP108	0079
TP108A	0079
TP108B	0079
TP108C	0079
TP109	0079
TP109A	0079
TP109B	0079
TP109C	0079
TP200	0097
TP200R	2291
TP201	0015
TP251A	0037
TP252A	0037
TP253A	0037
TP300	0097
TP300R	0476
TP302	0015
TP400	0097
TP400R	0476
TP402	0015
TP504	OEM
TP506	OEM
TP551	OEM
TP1004	OEM
TP1006	OEM
TP2004	OEM
TP2006	OEM
TP3004	OEM
TP3006	OEM
TP3566	0079
TP3638	0037
TP3638A	0791
TP3644	0037
TP3645	0037
TP3702	0037
TP3703	0037
TP3704	0111
TP3705	0111
TP3706	0111
TP3707	0155
TP3708	0155
TP3709	0155
TP3710	0155
TP3711	0155
TP4000	2024
TP4001	0478
TP4001AN	0478
TP4001B	OEM
TP4001BN	0478
TP4001BP	0478
TP4002	2055
TP4002AN	2055
TP4004	OEM
TP4006	OEM
TP4007	2846
TP4007UBN	2846
TP4008AN	0988
TP4008BN	0988
TP4009	1999
TP4009UBN	0001
TP4010BN	0398
TP4011	0215
TP4011AN	0215
TP4011B	0215
TP4011BN	0215
TP4011BP	0215
TP4012	0498
TP4012AN	0498
TP4013	0413
TP4013AN	0413
TP4013B	OEM
TP4013BE	0413
TP4013BN	0413
TP4014AN	0860
TP4015	1014
TP4015AN	1014
TP4016	4150
TP4016A	0101
TP4016AN	1143
TP4016BN	1143
TP4016UBN	1143
TP4017	0513
TP4017AN	0513
TP4018AN	1391
TP4018BN	1391
TP4019	1525
TP4019AN	1525
TP4019BN	1525
TP4020	1661
TP4020AN	1661
TP4020BN	1661
TP4021	1749
TP4021AN	1749
TP4022AN	1256
TP4023	0519
TP4023AN	0519
TP4024	1957
TP4024AN	1957
TP4024BN	1957
TP4025	2072
TP4025AN	2072
TP4027	1949
TP4027AN	1949
TP4027BN	1949
TP4028AN	2228
TP4028BN	2228
TP4029AN	2233
TP4030AN	0500
TP4040	0056
TP4040AN	0056
TP4040B	OEM
TP4040BN	0056
TP4042	0121
TP4042AN	0121
TP4042BN	0121
TP4043AN	1553
TP4043BN	1553
TP4044AN	2308
TP4044BN	2308
TP4049	0001
TP4049AN	0001
TP4049UBN	0001
TP4050	0398
TP4050AN	0398
TP4051	0366
TP4051BN	0366
TP4052	0024
TP4052BN	0024
TP4053BN	0035
TP4058	3519
TP4059	1411
TP4060	3519
TP4061	3519
TP4062	3519
TP4067-409	0015
TP4067-410	0111
TP4067-411	0111
TP4068BN	2503
TP4069UBN	0119
TP4069UPN	0119
TP4071BN	0129
TP4072BN	2524
TP4073BN	1536
TP4075BN	2540
TP4078BN	0921
TP4081	0626
TP4081BN	0626
TP4082BN	0300
TP4123	0016
TP4124	0016
TP4125	0037
TP4126	0037
TP4257	0037
TP4258	0037
TP4274	0600
TP4275	0144
TP4512	2120
TP4512AN	2120
TP4518	1043
TP4518AN	1043
TP4520AN	2675
TP4522AN	2837
TP5004	OEM
TP5006	OEM
TP5135	0079
TP5136	0079
TP5142	0037
TP6004	OEM
TP6006	OEM
TP6359P	OEM
TP8004	OEM
TP8006	OEM
TPA2	0023
TPA2006	OEM
TPA5000B	OEM
TPA5001A	OEM
TPA5004C	OEM
TPA5004F	OEM
TPD3004K	OEM
TPL1	OEM
TPL2	OEM
TPL3	OEM
TPL521-YG	OEM
TPLD3	OEM
TPLD3E	OEM
TPLD4	OEM
TPLD4E	OEM
TPM1	OEM
TPM2	OEM
TPM3	OEM
TPM4	OEM
TPMD4	OEM
TPMD5	OEM
TPMD6	OEM
TPND1	OEM
TPND2	OEM
TPND3	OEM
TPND4	OEM
TPQ2221	0544
TPQ2222	0544
TPQ2483	0544
TPQ2484	0544
TPQ2906	0290
TPQ2907	0290
TPQ2907A	OEM
TPQ3724	OEM
TPQ3725	OEM
TPQ3725A	OEM
TPQ3798	0282
TPQ3799	0290
TPQ3904	0544
TPQ3906	0290
TPQ4258	OEM
TPQ4354	0290
TPQ5400	2219
TPQ5401	2219
TPQ5550	2219
TPQ5551	2219
TPQ6001	OEM
TPQ6002	OEM
TPQ6100	OEM
TPQ6100A	OEM
TPQ6501	3004
TPQ6502	3005
TPQ6600	OEM
TPQ6600A	OEM
TPQ6700	OEM
TPQA05	2219
TPQA06	2219
TPQA55	OEM
TPQA56	OEM
TPS20	OEM
TPS50	OEM
TPS98	0861
TPS603	OEM
TPS604	5123
TPS605	5124
TPS606	OEM
TPS606-C	OEM
TPS703	2628
TPS703A	OEM
TPS6512	0016
TPS6513	0016
TPS6514	0016
TPS6515	0016
TPS6516	0037
TPS6517	0037
TPS6518	0037
TPS6519	0037
TPS6520	0016
TPS6521	0016
TPS6522	0037
TPS6523	0037
TPYTD03002	4130
TQ1	0144
TQ2	0144
TQ3	0144
TQ4	0016
TQ5	0144
TQ6	0144
TQ6A	OEM
TQ7	0144
TQ8	0144
TQ9	0144
TQ53	0791
TQ53A	0037
TQ54	0791
TQ54A	0037
TQ55	0791
TQ56	0532
TQ57	0791
TQ58	0791
TQ-59	0037
TQ59	0791
TQ59A	0791
TQ-60	0037
TQ60	0791
TQ60A	0791
TQ-61	0037
TQ61	0037
TQ61A	0037
TQ-62	0037
TQ62	0037
TQ62A	0037
TQ-63	0037
TQ63	0126
TQ63A	0126
TQ-64	0037
TQ64	0126
TQ64A	0126
T-Q5019	0642
T-Q5020	0136
TQ5020	0280
TQ5020(SANYO)	0196
T-Q5021	0136
TQ5021	0050
T-Q5022	0136
TQ5022	0050
T-Q5023	0633
TQ5023	0004
T-Q5025	0164
TQ5025	0004
T-Q5026	0004
TQ5026	0004
T-Q5027	0004
TQ5027	0211
T-Q5028	0160
TQ5028	0004
T-Q5030	1675
TQ5030	0604
T-Q5031	0208
TQ5031	0600
T-Q5032	0233
TQ-5032	2866
TQ5032	0600
T-Q5034	0050
TQ-5034	0050
TQ5034	0050
T-Q5035	0050
TQ5035	0050
T-Q5036	0160
TQ5036	0160
T-Q5038	0050
TQ5038	0050
T-Q5039	0208
TQ5039	0600
TQ-5044	2866
TQ5044	0208
T-Q5049	0007
TQ5049	0144
T-Q5050	0208
TQ-5050	2866
TQ5050	0208
TQ-5051	0164
TQ5051	0004
TQ-5052	0079
TQ5052	0079
T-Q5053	0079
TQ-5053	0079
TQ5053	0079
T-Q5053C	0079
TQ-5054	0079
TQ5054	0079
T-Q5055	0144
TQ-5055	0086
TQ5055	0086
T-Q5057	0142
TQ-5060	0079
TQ5060	0079
TQ-5061	0164
TQ5061	0004
TQ-5062	2866
TQ5062	0208
TQ-5062(NPN	0038
T-Q5063	0086
TQ-5063	0285
TQ5063	0233
TQ-5064	0160
TQ5064	0160
T-Q5071	0007
T-Q5073	0016
T-Q5075	0142
T-Q5077	0037
T-Q5078	0330
T-Q5079	0127
T-Q5080	0178
T-Q5081	0086
TQ5081	0086
T-Q5082	0233
T-Q5083	0003
T-Q5084	0065
T-Q5086	0007
T-Q5087	0037
T-Q5093	0111
T-Q5099	0086
T-Q5104	0142
T-Q5105	0103
T-Q5106	0127
T-Q7037	0878
TQ7037	0878
T-Q7038	0878
TQ7038	0878
TR0055	0037
TR0-1064-5	2439
TR0-1064-6	2439
TR0-1064-7	2439
TR0-1064-8	2439
TR0-2012	0050
TR0-2028-5	1199
TR0-2064-5	2432
TR0-2064-6	2432
TR0-2064-7	2432
TR0-2064-8	2432
TR0-9022	2861
TR-01	0160
TR01	0085
TR-01(PENNCREST)	0086
TR-01B	0144
TR-01B(PENNCREST)	0016
TR-01C	0160
TR01C	0160
TR-01C(PENNCREST)	0016
TR-01E	0086
TR-01E(PENNCREST)	0086
TR-01MP	0085

If replacement code is OEM, contact original manufacturer for replacement.

DEVICE TYPE	REPL CODE
TR01MP	0085
TR-02	0085
TR02	0160
TR02C	0160
TR-02E	0015
TR-03	0440
TR03	0440
TR03C	0440
TR-04	0004
TR04	0211
TR04C	0004
TR-04C(PENNCREST)	0126
TR-05	0280
TR05	0211
TR05C	0280
TR-06	0280
TR06	0050
TR-06(PENNCREST)	0016
TR06C	0050
TR-07	0050
TR07	0600
TR-07(PENNCREST)	0086
TR07C	0280
TR-08	0600
TR08	0600
TR-08(PENNCREST)	0126
TR-08C	0600
TR-09	0600
TR09	0038
TR-09C	0600
TR09C	0600
TR0140A	OEM
TR0140B	OEM
TR0140D	OEM
TR0140M	OEM
TR0141A	OEM
TR0141B	OEM
TR0141D	OEM
TR0141M	OEM
TR0142A	OEM
TR0142B	OEM
TR0142D	OEM
TR0142M	OEM
TR0146A	OEM
TR0146B	OEM
TR0146D	OEM
TR0146M	OEM
TR0911	0443
TR01014	0111
TR01015	0079
TR01026	0144
TR01027	6983
TR01037	0086
TR01040	0111
TR01042	0007
TR01045	0161
TR01051-5	6909
TR01053-1	0037
TR01054-1	6939
TR01054-5	1655
TR01054-7	0086
TR01056-5	0161
TR01057-1	6961
TR01057-3	0161
TR01059-5	5465
TR01062-1	0086
TR01062-7	0018
TR01065	0208
TR01073	0079
TR01074	0079
TR02012	0136
TR02020-2	0037
TR02051-1	0037
TR02051-5	0037
TR02051-6	0037
TR02053-5	0382
TR02053-7	0382
TR02054-1	4357
TR02054-7	0126
TR02057-1	4418
TR02057-3	0460
TR02059-5	4466
TR02062-1	0037
TR02062-6	0037
TR02063-1	0136
TR02063-8	0037
TR06011	0325
TR06014	0325
TR08004	0212
TR09004	0654
TR09005	0632
TR09006	0632
TR09007	3219
TR09008	2100
TR09009	2100
TR09010	0785
TR09011	0443
TR09018	2755
TR09019	0520
TR09022	2861
TR09023	2861
TR09024	2861
TR010602-1	0144
TR0573486	0151
TR0573491	0076
TR0573507	0151
TR0575002	0123
TR0575005	0123
TR1	OEM

DEVICE TYPE	REPL CODE
TR1AC	0593
TR1N4002	0080
TR-1R26	0050
TR1R26	0050
TR-1R31	0144
TR-1R33	0079
TR-1R35	0144
TR2	OEM
TR2A	0015
TR2N2614C	0004
TR-2R26	0050
TR2R26	0050
TR-2R31	0144
TR-2R33	0144
TR-2R35	0144
TR2S1570LH	0111
TR2SA763	0693
TR-2SC367	0286
TR-2SC371	0191
TR-2SC372	0076
TR-2SC373	0076
TR2SC373	0076
TR-2SC384	0127
TR-2SC482	1593
TR2SC535	0127
TR-2SC735	0191
TR2SC1342	0127
TR2SC1570LH	0111
TR2SD330E	1607
TR-2SK55	6979
TR2SK55	0325
TR-3	0160
TR3	OEM
TR-3R26	0050
TR3R26	0050
TR-3R31	0144
TR-3R33	0144
TR-3R35	0144
TR-3R38	0016
TR4	OEM
TR-4R26	0050
TR4R26	0050
TR-4R31	0086
TR-4R33	0016
TR-4R35	0144
TR-4R38	0037
TR4S	0036
TR5	0160
TR-5R26	0004
TR5R26	0211
TR-5R31	0086
TR-5R33	0016
TR-5R35	0016
TR-5R38	0016
TR5S	0140
TR6	OEM
TR-6R26	0004
TR6R26	0211
TR-6R33	0016
TR-6R35	0037
TR6S	0091
TR7	OEM
TR-7GS	0649
TR-7R31	0086
TR-7R35	0016
TR7S	0062
TR-7SA	0649
TR7SA	0062
TR-7SB	0137
TR7SB	0025
TR8	OEM
TR-8R35	0016
TR9	OEM
TR-9GS	0012
T-R9S	0012
TR-9S	0012
TR9S	0057
TR-9SA	0012
TR-9SB	0012
TR9SB	0012
TR-10	0600
TR10	0600
TR-10C	0600
TR-11	0136
TR11	0280
TR-11C	0050
TR11E	OEM
TR-12	0050
TR12	0050
TR-12C	0050
TR12EBM	0052
TR12S	0053
TR12SA	0052
TR12SB	0053
TR12SC	0879
TR12.5	OEM
TR12.75	OEM
TR-13	0050
TR13	0050
TR-13C	0050
TR-14	0004
TR14	0004
TR-14C	0004
TR-15	0004
TR15	0004
TR-15C	0004
TR-16	0085
TR16	0975
TR-16C	0085

DEVICE TYPE	REPL CODE
TR16C	0160
TR16H	OEM
TR16X2	3033
TR-17	0050
TR17	0050
TR-17A	0050
TR-17C	0050
TR-18	0050
TR18	0050
TR-18C	0050
TR18E	OEM
TR-19	0037
TR19	0037
TR-20	0037
TR20	0037
TR-20A	0037
TR-21	0016
TR21	0016
TR-21C	0079
TR21R	0181
TR-22	0079
TR22	0126
TR-22C	0079
TR22E	OEM
TR-23	0142
TR23	0086
TR23A	0126
TR-24	0144
TR24	0144
TR-24(PHILCO)	0079
TR-25	0086
TR25	0086
TR-26	0103
TR26	0103
TR-27	0975
TR27	0975
TR-28	0126
TR28	0037
TR-29	0491
TR-30	0037
TR30	0037
TR-30A	0037
TR-31	0037
TR31	0037
TR-31(IR)	0861
TR-32	0861
TR32	0861
TR-33	0018
TR33	0111
TR-34	0975
TR34	0975
TR-35	0604
TR35	0604
TR-36	0130
TR36	0130
TR36(IR)	0130
TR-36MP	0130
TR-36MP(IR)	0130
TR-37	0905
TR38	0144
TR43	0004
TR43A	0164
TR-43B	0160
TR44	0004
TR45	0004
TR-50	0222
TR50	0222
TR50R	0267
TR-51	0079
TR51	0050
TR-52	0532
TR52	0050
TR52(IR)	0050
TR-53	0086
TR53	0280
TR53R	0267
TR-54	0532
TR54	0004
TR-55	0560
TR55	0280
TR-55(IR)	0219
TR-56	1267
TR56	0085
TR-56(IR)	1430
TR56(IR)	1430
TR-57	0178
TR57	0440
TR-58	0854
TR58	0905
TR-59	0103
TR59	0103
TR-60	0187
TR60	0187
TR60-2	0319
TR-61	0074
TR61	0074
TR-61MP	0074
TR-62	0079
TR62	0050
TR-63	0630
TR63	0136
TR-64	0630
TR64	0280
TR-65	0698
TR65	0280
TR-66	2170
TR-67	0642
TR67	0065
TR67(IR)	0642

DEVICE TYPE	REPL CODE
TR-68	0065
TR68	0074
TR68(IR)	0003
TR-69	0400
TR69	0400
TR-69(IR)	0400
TR69(IR)	0400
TR-70	0224
TR70	0224
TR71	0004
TR71(IR)	0004
TR-72	0264
TR72	0004
TR-73	0382
TR-73(IR)	0382
TR73(IR)	0382
TR-74	0264
TR-75	0137
TR75	0137
TR-76	0161
TR76	0161
TR-76(IR)	0042
TR76(IR)	0042
TR-77	0015
TR77	0050
TR-77(IR)	0925
TR77(IR)	0925
TR-77(XSTR)	0460
TR-78	0285
TR-78(IR)	0285
TR78(IR)	0285
TR-79	0285
TR79	0285
TR-79(IR)	0285
TR79(IR)	0285
TR-80	0144
TR80	0144
TR-81	0178
TR81	0004
TR-81MP	0178
TR-82	0847
TR82	0847
TR-83	0144
TR83	0144
TR-84	0164
TR84	0164
TR-84(IR)	0164
TR84(IR)	0164
TR-85	0164
TR85	0164
TR-86	0079
TR86	0079
TR-87	0086
TR87	0050
TR-88	0126
TR88	0050
TR-91	2763
TR91(IR)	2763
TR-92	0836
TR-92(IR)	0836
TR92(IR)	0836
TR92-1	4455
TR92-2	4455
TR92-3	4455
TR92-4	4455
TR92-5	4456
TR92-6	4456
TR-93	0065
TR93	0065
TR93-1	4455
TR93-2	4455
TR93-3	4455
TR93-4	4455
TR93-5	4456
TR93-6	4456
TR94-1	4455
TR94-2	4455
TR94-3	4455
TR94-4	4455
TR94-5	4456
TR94-6	4456
TR-94MP	0822
TR94MP	0822
TR-95	0012
TR95	0007
TR-95(B)	0012
TR-95(IR)	0007
TR-95(XSTR)	0007
TR-95B	0012
TR100	2901
TR100A	OEM
TR100R	0267
TR103	2840
TR103R	0267
TR104	0280
TR105	0050
TR106A41#	OEM
TR109	0280
TR123	0050
TR125	0002
TR125B	0002
TR136A	0555
TR136B	0555
TR136D	0555
TR136M	0555
TR139	0050
TR147A	0555
TR147B	0555
TR147D	0555

DEVICE TYPE	REPL CODE
TR147M	0555
TR150	2900
TR150A	OEM
TR150R	0267
TR151	1124
TR151R	0267
TR152	0871
TR152R	0267
TR153	2840
TR153R	0267
TR-157(OLSON)	0164
TR-158(OLSON)	0164
TR-159(OLSON)	0600
TR-160(OLSON)	0600
TR-161(OLSON)	0050
TR-162(OLSON)	0016
TR-163(OLSON)	0144
TR-164(OLSON)	0086
TR-165(OLSON)	0126
TR-166(OLSON)	0050
TR167	0600
TR-167(OLSON)	0037
TR-168(OLSON)	0136
TR-169(OLSON)	0004
TR-170(OLSON)	0164
TR-171	0007
TR-172(OLSON)	0085
TR-174(OLSON)	0440
TR-175(OLSON)	0142
TR-176(OLSON)	0130
TR-178(OLSON)	0160
TR-180(OLSON)	0178
TR182	0600
TR-182(OLSON)	0460
TR183	0600
TR-183(OLSON)	0975
TR184	0600
TR-184(OLSON)	0222
TR-185(OLSON)	0331
TR-186(OLSON)	0074
TR-187(OLSON)	0642
TR-188(OLSON)	0178
TR193	0600
TR194	0600
TR200	1124
TR200R	0267
TR203	2840
TR203R	0267
TR209T2	1083
TR211	0600
TR212	0600
TR213	0600
TR214	0600
TR215	0280
TR216	0600
TR217	0280
TR218	0050
TR251	0805
TR251R	0267
TR252	0805
TR252R	0267
TR253	0805
TR253R	0267
TR262-2	0142
TR266-2	0142
TR269	0050
TR271TR26	0103
TR300	0805
TR300R	0267
TR301	0233
TR301R	0267
TR302	0016
TR302R	0267
TR303	2850
TR303R	0267
TR320	0211
TR320A	0211
TR321	0211
TR321(HFGH1)	0280
TR321A	0211
TR322	OEM
TR323	0211
TR323A	0211
TR324	OEM
TR325	OEM
TR326	OEM
TR331	0050
TR332	0211
TR333	0160
TR334	0440
TR335	0600
TR336	0600
TR337	0600
TR338	0038
TR351	0805
TR351R	0267
TR352	0805
TR352R	0267
TR353	2850
TR353R	0267
TR381	0164
TR382	0164
TR383	0211
TR383(HGFH-2)	0211
TR400	0805
TR400R	0267
TR401	0805
TR401R	0267
TR402	0805

DEVICE TYPE	REPL CODE
TR402R	0267
TR403	2850
TR403R	0267
TR460	0164
TR461	0164
TR482	0280
TR482A	0280
TR500	0319
TR500R	0267
TR501	1586
TR501R	0267
TR502	1195
TR502R	0267
TR503	0319
TR503R	0267
TR506	OEM
TR508	0211
TR508A	0211
TR515	OEM
TR515S	OEM
TR560	OEM
TR563	OEM
TR600	0319
TR600R	0267
TR601	0016
TR601R	0267
TR602	0319
TR602R	0267
TR603	0319
TR603R	0267
TR650	0211
TR650A	0004
TR653	0211
TR701	0044
TR701R	0281
TR702	0044
TR702R	0281
TR721	0211
TR722	0211
TR758A	0280
TR759	0280
TR760	0280
TR761	0136
TR762	0136
TR763	0211
TR763A	0004
TR764	0280
TR792	0280
TR800	0044
TR800R	0281
TR801	0280
TR801R	0281
TR802	0280
TR802R	0281
TR803	0050
TR1000	0126
TR-1000-2	0037
TR-1000-3	0086
TR-1000-7	0103
TR-1001	OEM
TR1001	0086
TR1001-2	0861
TR1002	0126
TR1003	0086
TR1004	0126
TR1005	0086
TR1007	0103
TR1009	0074
TR1009A	0130
TR1010	4855
TR1011	0079
TR1012	0126
TR1015	OEM
TR1015S	OEM
TR1025	0103
TR1030	0037
TR-1030-1	0037
TR1030-1	0037
TR-1030-2	0037
TR1030-2	0037
TR1031	0079
TR-1031-1	0155
TR1031-1	0861
TR-1031-2	0155
TR1031-2	0861
TR1032	0037
TR-1032-1	0150
TR1032-1	0037
TR-1032-2	0037
TR1033	0016
TR-1033-1	0016
TR1033-1	0264
TR-1033-2	0016
TR-1033-3	0016
TR1033-3	0016
TR1033-4	0264
TR1033-5	0264
TR1033-6	0264
TR1034	0037
TR-1035-3	OEM
TR1036	0604
TR-1036-1	1199
TR-1036-2	1199
TR1036-2	1199
TR-1036-3	1267
TR1036-3	1199
TR-1037-1	0561
TR1037-1	1199
TR-1037-2	0561

DEVICE TYPE	REPL CODE
TR1037-2	0561
TR-1037-3	0560
TR1037-3	0561
TR1038	0604
TR-1038-4	1681
TR1038-4	0491
TR1038-5	2013
TR1038-6	1199
TR-1039-4	0103
TR1039-4	0103
TR1039-5	0177
TR1039-6	0103
TR1077	0103
TR1110	OEM
TR1120	0084
TR1120R	2291
TR1121	0090
TR1121R	2291
TR1122	0097
TR1122R	2291
TR1123	0109
TR1123R	0476
TR1124	0109
TR1124R	0476
TR1125	0122
TR1125R	0476
TR1126	0122
TR1126R	0476
TR1128	0131
TR1128R	0449
TR1130	0145
TR1130R	0449
TR1210	OEM
TR1402A	OEM
TR1402B	OEM
TR1490	0103
TR1491	0103
TR1492	0103
TR1493	0103
TR1512-80	0144
TR1591	0178
TR1593	0178
TR1602A	4435
TR1602B	OEM
TR1605LP	0142
TR1863A	OEM
TR1863B	OEM
TR1865-PL	OEM
TR1865A	OEM
TR1865B	OEM
TR1983E	OEM
TR1983F	OEM
TR1993-2	0016
TR2007	OEM
TR2015	OEM
TR2015S	OEM
TR2083-40	0030
TR2083-41	0143
TR2083-42	0133
TR2083-44	0133
TR2083-70	0325
TR2083-71	0079
TR2083-72	0079
TR2083-73	0079
TR2083-74	0007
TR2083-75	0211
TR2880	0015
TR3015	OEM
TR3015S	OEM
TR4010-2	0016
TR4015	OEM
TR4015S	OEM
TR4083-23273	1621
TR4083-2327311	1621
TR4083-2327312	1621
TR5015	OEM
TR5528	0325
TR6010	0725
TR6015	OEM
TR7010	0679
TR7015	OEM
TR-8001	0050
TR8001	0050
TR-8002	0050
TR8002	0050
TR-8003	0050
TR8003	0050
TR-8004	0079
TR8004	0144
TR-8004-4	0144
TR8004-4	0079
TR-8004-5	0144
TR-8005	0142
TR8005	0142
TR-8006	0160
TR8006	0160
TR-8007	0037
TR8007	0211
TR-8007(FISHER)	0037
TR-8010	0127
TR8010	0079
TR-8014	0016
TR8014	0016
TR8015	2525
TR-8018	0130
TR8018	0130
TR-8019	0925
TR8019	0925
TR-8020	0126

DEVICE TYPE	REPL CODE
TR8020	0037
TR-8021	0086
TR8021	0016
TR-8022	0264
TR8022	0264
TR-8023	0086
TR8023	0264
TR-8024	0086
TR8024	0264
TR-8025	0079
TR8025	0079
TR-8026	0037
TR-8027	0325
TR8027	0325
TR-8028	0144
TR8028	0016
TR-8029	0144
TR8029	0016
TR-8030	0144
TR8030	0016
TR-8031	0144
TR8031	0016
TR-8032	0144
TR-8034	0111
TR8034	0079
TR-8035	0079
TR8035	0079
TR-8036	0086
TR8036	0086
TR-8037	0037
TR8037	0126
TR-8038	0144
TR8038	0079
TR-8039	0079
TR8039	0079
TR-8040	0111
TR8040	0079
TR-8042	0016
TR8042	0144
TR-8043	0127
TR8043	0016
TR8053	0016
TR8330	0079
TR9010	0679
TR9100	0016
TR-9100-18	0016
TR10076B	OEM
TR10076C	OEM
TR12001-4	0143
TR14001-1	0402
TR-14001-2	0247
TR-14001-4	0262
TR14002-6	0137
TR-14002-10	0039
TR14002-12	0002
TR14002-13	0531
TR19001	OEM
TR35144	0160
TR35524	0160
TR36643	0086
TR38117	0004
TR39453	0353
TR40603	0353
TR310011	0004
TR310012	0004
TR310015	0280
TR310017	0004
TR310018	0004
TR310019	0050
TR310025	0050
TR310026	0004
TR310065	0050
TR310068	0050
TR310069	0050
TR310075	0004
TR310107	0004
TR310123	0050
TR310125	0004
TR310136	0004
TR310139	0050
TR310147	0050
TR310150	0050
TR310153	0004
TR310155	0050
TR310156	0050
TR310157	0050
TR310158	0050
TR310159	0004
TR310160	0208
TR310161	0280
TR310164	0004
TR310224	0050
TR310225	0211
TR310230	0144
TR310231	0016
TR310232	0050
TR310235	0004
TR310236	0208
TR310243	0016
TR310244	0144
TR310245	0016
TR310249	0144
TR310250	0144
TR310251	0004
TR310252	0004
TR310255	0004
TR320007	0123
TR320008	0143
TR320020	0015
TR320022	0015
TR320039	0143
TR320041	0143
TR320048	0143
TR330027	0015
TR330028	0170
TR2320063	0076
TR2327031	0023
TR2327041	0023
TR2327203	0042
TR2327293	0191
TR2327312	1621
TR2327333	0191
TR2327363	0552
TR2327393	0016
TR2327411	4570
TR2327422	0443
TR2327431	0414
TR2327443	0552
TR2327444	0552
TR2327574	2793
TR2327603	0563
TR2327607	0563
TR2327723	0925
TR2327733	0062
TR2327743	0693
TR2327841	3500
TR2327852	1147
TR2337011	0133
TR2337063	0565
TR2337101	0052
TR2337103	0052
TR2337123	0091
TR2347041	0139
TR2367151	4560
TR2367161	4543
TR2367171	0520
TR3100227	0211
TR5320326	0127
TR10000102	OEM
TR20000014	OEM
TR20000040	OEM
TR30000467	OEM
TR40000021	OEM
TR40000094	OEM
TR22873700104	0443
TR228723002003	OEM
TR228735045311	0144
TR228735046011	0127
TR228735048617	0144
TR228735048618	0144
TR228735120325	0325
TR228736002003	0143
TR228736002004	0123
TR228736003026	0170
TR228779905696	0579
TRA05	OEM
TRA05D	OEM
TRA1	OEM
TRA1D	OEM
TRA-2	0136
TRA2	0633
TRA2D	OEM
TRA3	OEM
TRA3D	OEM
TRA-4	0016
TRA4	0016
TRA-4A	0016
TRA4A	0016
TRA-4B	0076
TRA4B	0016
TRA4D	OEM
TRA-7R	0160
TRA7R	0160
TRA-7RM	0160
TRA-8R	0160
TRA8R	0160
TRA9LC1013	OEM
TRA-9R	0076
TRA9R	0016
TRA-10R	0050
TRA10R	0050
TRA11	1651
TRA11D	1651
TRA-11R	0136
TRA11R	0050
TRA12	1651
TRA12D	0745
TRA-12R	0136
TRA12R	0050
TRA13	OEM
TRA13D	OEM
TRA14	1651
TRA14D	1584
TRA21	OEM
TRA21D	OEM
TRA-22	0136
TRA22	0633
TRA-22A	0136
TRA22A	0050
TRA-22B	0050
TRA22B	0050
TRA22D	OEM
TRA-23	0136
TRA23	0050
TRA-23A	0136
TRA23A	0050
TRA-23B	0136
TRA23B	0050
TRA23D	OEM
TRA-24	0136
TRA24	0050
TRA-24A	0136
TRA24A	0050
TRA-24B	0136
TRA24D	OEM
TRA32	0004
TRA33	0004
TRA34	0016
TRA36	0016
TRA71	1651
TRA71D	2449
TRA72	1651
TRA72D	0435
TRA73	OEM
TRA73D	OEM
TRA74	1584
TRA74D	1487
TRA75	OEM
TRA75D	OEM
TRA76	OEM
TRA76D	OEM
TRA105	1651
TRA105D	2519
TRA205	OEM
TRA205D	OEM
TRA516DFX	OEM
TRA516DFZ	OEM
TRA705	1651
TRA705D	2449
TRA1016DFX	OEM
TRA1016DFZ	OEM
TRA2016DFX	OEM
TRA2016DFZ	OEM
TRA3016DFX	OEM
TRA3016DFZ	OEM
TRA4016DFX	OEM
TRA4016DFZ	OEM
TRA5016DFX	OEM
TRA5016DFZ	OEM
TRA6016DFX	OEM
TRA6016DFZ	OEM
TRAL110	OEM
TRAL110D	OEM
TRAL116	OEM
TRAL116D	OEM
TRAL220	OEM
TRAL220D	OEM
TRAL226	OEM
TRAL226D	OEM
TRAL610D	OEM
TRAL615D	OEM
TRAL625D	OEM
TRAL630D	OEM
TRAL640D	OEM
TRAL1025D	OEM
TRAL1110	OEM
TRAL1110D	OEM
TRAL1115	OEM
TRAL1115D	0154
TRAL1125D	2015
TRAL1130D	OEM
TRAL1135D	2015
TRAL1225D	OEM
TRAL2210	OEM
TRAL2210D	OEM
TRAL2215	OEM
TRAL2215D	0147
TRAL2225D	2017
TRAL2230D	OEM
TRAL2235D	2017
TRAL3810D	OEM
TRAL3815D	OEM
TRAL3825D	OEM
TRAL3830D	OEM
TRAL3835D	OEM
TRANSMITTER	OEM
TRAPLC711	0016
TRAPLC871	0016
TRAPLC1013	0560
TR-BC147B	0016
TRBC147B	0016
TR-BC149C	0111
TRC15-050A	OEM
TRC15-050B	OEM
TR-C44	0280
TRC44	0280
TR-C44A	0280
TRC44A	0280
TR-C45	0211
TRC45	0280
TR-C45A	0280
TRC45A	0280
TR-C70	0211
TRC70	0211
TR-C71	0211
TRC71	0211
TR-C72	0211
TRC72	0211
TRC-P4	0196
TRDV125	OEM
TRDV225	OEM
TRDV425	OEM
TRDV625	OEM
TRDV825	OEM
TRDV1025	OEM
TRDV1226	OEM
TRF401	OEM
TRF450	OEM
TRF450A	OEM
TRF454	OEM
TRF454A	OEM
TRF641	OEM
TRF644	OEM
TRF646	OEM
TRF648	OEM
TR-FE100	0325
TRG-103	OEM
TRHA1151	3651
TR-I1835E	OEM
TRL104LED	5116
TRL2255	OEM
TRLA1230	0637
TRLA3350	0416
TRLP2004	0439
TRLP3034S	OEM
TRM13	0050
TRM14	0050
TRM15	0280
TRM16	0136
TRM17	0136
TRM21	0280
TRM34	0164
TRM81	0050
TRM1201	OEM
TRM-1202	OEM
TRM-1203	OEM
TRM-1204	OEM
TRM-1205	OEM
TRM1205	OEM
TRM-1206	OEM
TRM1206	OEM
TRM-1207	OEM
TRM-1208	OEM
TRM-1209	OEM
TRM1212	OEM
TRO2E	OEM
TRO5E	OEM
TR01015	OEM
TR01026	0079
TR01027	OEM
TR01037	0111
TR01042	0007
TR01045	OEM
TR01047	OEM
TR01048	OEM
TR01051-5	0079
TR01053-1	0150
TR01053-6	OEM
TR01054-1	0086
TR01054-7	OEM
TR01056-5	OEM
TR01064-9	OEM
TR01065	OEM
TR01073	OEM
TR01074	OEM
TR02012	OEM
TR02020-2	OEM
TR02047	OEM
TR02051-1	0150
TR02054-1	0126
TR02062-1	0150
TR02063-8	OEM
TR02064-9	OEM
TR06011	0325
TR06014	OEM
TR06015	OEM
TR08004	OEM
TRO-9004	0654
TR09004	0654
TRO-9005	0632
TR09005	0632
TRO-9006	0632
TR09006	0632
TRO-9011	0443
TR09018	2494
TR09023	OEM
TR09024	2861
TR09027	OEM
TR09032	4066
TR09033	OEM
TR09034	OEM
TR09035	OEM
TR010602-1	0079
TRPLC711	0111
TRPW5B0M050A	OEM
TRR6	0636
TR-RR38	0016
TRS02101D09A	OEM
TRS02101D09B	OEM
TRS02101D09W	OEM
TRS02101D09X	OEM
TRS02101D09Y	OEM
TRS02101D09Z	OEM
TRS02102D09A	OEM
TRS02102D09B	OEM
TRS02102D09Q	OEM
TRS02102D09W	OEM
TRS02102D09X	OEM
TRS02102D09Y	OEM
TRS02102D09Z	OEM
TRS02104D09A	OEM
TRS02104D09Q	OEM
TRS02104D09W	OEM
TRS02104D09X	OEM
TRS02104D09Z	OEM
TRS02106D09A	OEM
TRS02106D09Q	OEM
TRS02106D09W	OEM
TRS02106D09X	OEM
TRS02106D09Z	OEM
TRS02108D09A	OEM
TRS02108D09W	OEM
TRS02108D09X	OEM
TRS02108D09Z	OEM
TRS02110D09A	OEM
TRS02110D09W	OEM
TRS02110D09X	OEM
TRS02110D09Z	OEM
TRS02112D09A	OEM
TRS02112D09X	OEM
TRS02112D09Z	OEM
TRS02114D09Z	OEM
TRS02301D09Y	OEM
TRS02302D09A	OEM
TRS02302D09Q	OEM
TRS02302D09W	OEM
TRS02302D09X	OEM
TRS02302D09Z	OEM
TRS02304D09A	OEM
TRS02304D09Q	OEM
TRS02304D09W	OEM
TRS02304D09X	OEM
TRS02304D09Z	OEM
TRS02306D09A	OEM
TRS02306D09Q	OEM
TRS02306D09W	OEM
TRS02306D09X	OEM
TRS02306D09Z	OEM
TRS02308D09A	OEM
TRS02308D09W	OEM
TRS02308D09X	OEM
TRS02308D09Z	OEM
TRS02312D09Z	OEM
TRS02602D09A	OEM
TRS02602D09Q	OEM
TRS02602D09W	OEM
TRS02602D09X	OEM
TRS02602D09Z	OEM
TRS02604D09A	OEM
TRS02604D09Q	OEM
TRS02604D09W	OEM
TRS02604D09X	OEM
TRS02604D09Z	OEM
TRS02606D09A	OEM
TRS02606D09Q	OEM
TRS02606D09W	OEM
TRS02606D09X	OEM
TRS02606D09Y	OEM
TRS02606D09Z	OEM
TRS02608D09A	OEM
TRS02608D09W	OEM
TRS02608D09Z	OEM
TRS02610D09A	OEM
TRS02610D09X	OEM
TRS02610D09Z	OEM
TRS02612D09Z	OEM
TRS03802D13A	OEM
TRS03802D13B	OEM
TRS03802D13W	OEM
TRS03802D13X	OEM
TRS03802D13Y	OEM
TRS03804D13A	OEM
TRS03804D13B	OEM
TRS03804D13Q	OEM
TRS03804D13W	OEM
TRS03804D13X	OEM
TRS03804D13Y	OEM
TRS03806D13A	OEM
TRS03806D13W	OEM
TRS03806D13X	OEM
TRS03806D13Y	OEM
TRS03808D13A	OEM
TRS03808D13W	OEM
TRS03808D13X	OEM
TRS03808D13Y	OEM
TRS03810D13A	OEM
TRS03810D13B	OEM
TRS03810D13X	OEM
TRS03810D13Y	OEM
TRS03812D13B	OEM
TRS04002D13A	OEM
TRS04002D13B	OEM
TRS04002D13Q	OEM
TRS04002D13W	OEM
TRS04002D13X	OEM
TRS04002D13Y	OEM
TRS04004D13A	OEM
TRS04004D13B	OEM
TRS04004D13Q	OEM
TRS04004D13W	OEM
TRS04004D13X	OEM
TRS04006D13A	OEM
TRS04006D13B	OEM
TRS04006D13Q	OEM
TRS04006D13W	OEM
TRS04006D13X	OEM
TRS04006D13Y	OEM
TRS04008D13A	OEM
TRS04008D13B	OEM
TRS04008D13W	OEM
TRS04008D13X	OEM
TRS04008D13Y	OEM
TRS04010D13A	OEM
TRS04010D13B	OEM
TRS04010D13X	OEM
TRS04010D13Y	OEM
TRS04012D13B	OEM
TRS04302D13A	OEM
TRS04302D13Q	OEM
TRS04306D13Y	OEM
TRS04308D13X	OEM
TRS04310D13Z	OEM
TRS06602D17B	OEM
TRS06610D17B	OEM
TRS06610D17Y	OEM
TRS06610D17Z	OEM
TRS06612D17B	OEM
TRS06612D17Y	OEM
TRS06612D17Z	OEM
TRS06614D17B	OEM
TRS06614D17Y	OEM
TRS06614D17Z	OEM
TRS15X5	OEM
TRS25X	0168
TRS30X	0168
TRS35X	0168
TRS100	0233
TRS100A	0233
TRS100HC	0622
TRS101	0233
TRS120	0233
TRS125HC	0622
TRS140	0233
TRS140HP	0285
TRS140MP	0168
TRS160	0233
TRS160HP	0285
TRS160MP	0168
TRS180	0233
TRS180HP	0285
TRS180MP	0168
TRS200	0233
TRS200HP	0285
TRS200MP	0168
TRS225	0233
TRS225HP	0285
TRS225MP	0168
TRS250	0233
TRS250HP	0285
TRS250MP	0168
TRS275	0233
TRS275HP	0285
TRS275MP	0168
TRS301	0233
TRS301HP	0285
TRS301LC	0168
TRS301MP	0168
TRS325	0715
TRS325HP	0285
TRS325MP	0168
TRS350	0187
TRS350HP	0187
TRS350MP	0187
TRS375	0187
TRS375HP	0187
TRS375MP	0187
TRS401	0187
TRS401HP	0187
TRS401LC	0187
TRS401MP	0187
TRS425	0187
TRS425HP	0187
TRS425MP	0187
TRS450	0187
TRS451	0187
TRS451MP	0168
TRS475	0168
TRS475MP	0168
TRS501	0168
TRS501C	OEM
TRS501LC	0168
TRS501MP	0168
TRS525	0168
TRS525MP	0168
TRS550	0168
TRS550MP	0168
TRS1004	0264
TRS1004LP	0264
TRS1005	0142
TRS1005LP	0142
TRS1204	0168
TRS1204LP	0264
TRS1205	0142
TRS1205LP	0142
TRS1404	0168
TRS1404LP	0264
TRS1405	0142
TRS1405LP	0142
TRS1604	0168
TRS1604LP	0264
TRS1605	0142
TRS1605LP	0264
TRS1804	0168
TRS1804LP	0264
TRS1805	0142
TRS1805LP	0142
TRS2004	0168
TRS2004LP	0168
TRS2005	0142
TRS2005LP	0142
TRS2006	0168
TRS2254	0168
TRS2254LP	0168
TRS2255	0142
TRS2255LP	0142
TRS2504	0168
TRS2504LP	0168
TRS2505	0142
TRS2505LP	0142
TRS2754	0168
TRS2754LP	0168
TRS2755	0142
TRS2755LP	0142
TRS2804S	0168
TRS2805S	0168
TRS3006	0168
TRS3011	0233
TRS3012	0233
TRS3012C#	3014
TRS3014	0285
TRS3014LP	0168
TRS3015	0168
TRS3015LP	0142
TRS3016LC	0142
TRS3204S	0168
TRS3205S	0168
TRS3254	0168
TRS3254LP	0168
TRS3255	0168
TRS3501	0187
TRS3502	0187
TRS3504	0168
TRS3504LP	0168
TRS3604S	0168
TRS3742	0168
TRS3754	0168
TRS3754LP	0168
TRS4001	0187
TRS4002	0187
TRS4004	0168
TRS4006S	OEM
TRS4014	0168
TRS4014LP	0168
TRS4014S	0168
TRS4016LC	0142
TRS4254	0168
TRS4254LP	0168
TRS4296	0168
TRS4297	0168
TRS4404S	0168
TRS4405S	OEM
TRS4501	0187
TRS4504	0168
TRS4504LP	0168
TRS4754	0168
TRS4754LP	0168
TRS4926	0168
TRS4927	0168
TRS5006A	OEM
TRS5014	0168
TRS5016LC	0142
TRS5254	0168
TRS5504	0168
TRS6016LC	0142
TRS11002D25B	OEM
TRS11004D25B	OEM
TRS11006D25B	OEM
TRS11008D25B	OEM
TRS11010D25B	OEM
TRS11012D25B	OEM
TRSP2254	0439
TRSP2254S	0439
TRSP3738	OEM
TRSP3739	OEM
TRSP4000	0126
TRSP4001	0439
TRSP4002	0439
TRSR3AM	0133
TRT	OEM
TRT-1301	OEM
TR-TR38	0016
TR-U1650E	0325
TR-U1650E-1	0325
TR-U1835E	0325
TRW2307	OEM
TRW6259	OEM
TRW54001	OEM
TRW54101	OEM
TRW54201	OEM
TRW54401	OEM
TRW54501	OEM
TRW54601	OEM
TRW62601	OEM
TRW62602	OEM
TRW63601	OEM
TRW63602	OEM
TRW64601	OEM
TRW64602	OEM
TRY-1002	OEM
TS05	0015
TS035	OEM
TS035FA	0804
TS035FB	OEM
TS04700	OEM
TS05100	OEM
TS05600	OEM
TS06000	OEM
TS06200	0471
TS06800	OEM
TS07100	OEM
TS07500	OEM
TS08200	OEM
TS08700	OEM

If replacement code is OEM, contact original manufacturer for replacement.

DEVICE TYPE	REPL CODE
TS09100	0057
TS-1	0004
TS1	0015
TS-2	0004
TS2	0015
TS2A	0015
TS-3	0004
TS3	0071
TS4	0540
TS5	0015
TS6	0015
TS6.8	OEM
TS7	OEM
TS7.5	OEM
TS8	0071
TS8.2	OEM
TS9	OEM
TS9.1	OEM
TS10	OEM
TS10A	OEM
TS11	OEM
TS12	OEM
TS-13	0004
TS13	0211
TS-14	0004
TS14	0211
TS-15	0004
TS15	0211
TS16	OEM
TS17	OEM
TS18	OEM
TS20	OEM
TS22	OEM
TS24	OEM
TS27	OEM
TS30	OEM
TS30A	OEM
TS33	OEM
TS36	OEM
TS39	OEM
TS39A	OEM
TS40	OEM
TS43	OEM
TS47	OEM
TS50	OEM
TS51	OEM
TS51A	OEM
TS56	OEM
TS60	OEM
TS60A	OEM
TS62	OEM
TS68	OEM
TS75	OEM
TS80	OEM
TS82	OEM
TS82A	OEM
TS91	OEM
TS97-1	0037
TS100	OEM
TS100A	OEM
TS110	OEM
TS110A	OEM
TS120	OEM
TS135	0469
TS135FA	0804
TS135FB	OEM
TS-162	0004
TS162	0164
TS-163	0004
TS163	0164
TS-164	0211
TS164	0164
TS-165	0004
TS165	0164
TS-166	0004
TS166	0164
TS-173	0160
TS173	0160
TS-176	0160
TS176	0160
TS187	OEM
TS235	0469
TS235FA	0804
TS235FB	OEM
TS322-3	OEM
TS322-3-0	OEM
TS322-3IR	OEM
TS322-3LED	OEM
TS335	OEM
TS335FA	OEM
TS335FB	OEM
TS-337	0137
TS435	0725
TS435FA	0804
TS435FB	OEM
TS511-3	OEM
TS511-4	OEM
TS511-6	OEM
TS535	OEM
TS535FA	OEM
TS535FB	OEM
TS555A	OEM
TS555B	OEM
TS555C	OEM
TS555CD	2771
TS555CN	0585
TS555D	OEM
TS555E	OEM
TS555F	OEM

DEVICE TYPE	REPL CODE
TS555G	OEM
TS555H	OEM
TS555M	OEM
TS555N	OEM
TS555P	OEM
TS555PA	OEM
TS555PB	OEM
TS555PC	OEM
TS555S	OEM
TS555T	OEM
TS556A	OEM
TS556B	OEM
TS556C	OEM
TS556CD	OEM
TS556CJ	OEM
TS556CN	2869
TS556D	OEM
TS556E	OEM
TS556F	OEM
TS556G	OEM
TS556H	OEM
TS556ID	OEM
TS556IJ	OEM
TS556IN	OEM
TS556M	OEM
TS556MD	OEM
TS556MJ	OEM
TS556MN	OEM
TS556N	OEM
TS556P	OEM
TS556PA	OEM
TS556PB	OEM
TS556PC	OEM
TS556S	OEM
TS556T	OEM
TS560	OEM
TS-601	0280
TS601	0164
TS-602	0280
TS602	0164
TS-603	0004
TS603	0164
TS-604	0004
TS604	0164
TS605	0004
TS606	0004
TS-609	0440
TS609	0440
TS-610	0085
TS610	0160
TS-612	0085
TS612	0160
TS-613	0085
TS613	0160
TS-614	0085
TS614	0160
TS-615	0050
TS615	0280
TS-616	0004
TS616	0164
TS-617	0004
TS617	0164
TS-618	0004
TS618	0164
TS619	0164
TS-620	0050
TS620	0164
TS-621	0050
TS621	0164
TS-627	0004
TS627	0164
TS-627A	0050
TS627A	0211
TS-627B	0050
TS627B	0211
TS-629	0004
TS629	0164
TS-630	0050
TS630	0050
TS635	0725
TS635FA	0804
TS635FB	OEM
TS669A	0280
TS669B	0280
TS669C	0050
TS669D	0280
TS669E	0280
TS669F	0280
TS-672A	0050
TS672A	0211
TS-672B	0050
TS672B	0050
TS-673A	0050
TS673A	0050
TS-673B	0050
TS673B	0050
TS735	OEM
TS735FA	OEM
TS735FB	OEM
TS-739	0004
TS739	0164
TS-739B	0004
TS739B	0164
TS-740	0004
TS740	0164
TS748	OEM
TS-765	0004
TS765	0164
TS835	0679

DEVICE TYPE	REPL CODE
TS835FA	OEM
TS835FB	OEM
TS935	OEM
TS1000	OEM
TS-1007	0004
TS1007	0164
TS1035	0679
TS1060	OEM
TS-1134	OEM
TS1135	OEM
TS1235	0679
TS-1266	0004
TS1266	0164
TS-1657	0160
TS1657	0160
TS-1727	0004
TS1727	0164
TS-1728	0004
TS1728	0164
TS1792	0004
TS2000	OEM
TS2000/883B	OEM
TS2000PT	OEM
TS2060	OEM
TS2218	0086
TS2219	0086
TS2221	0016
TS2222	0016
TS2427	OEM
TS2427-1	OEM
TS2776	0233
TS2779	0330
TS2904	0037
TS2905	0037
TS2906	0037
TS2907	0037
TS3060	OEM
TS4060	OEM
TS4078BP	0921
TS9013	0144
TS10000	OEM
TSA551T/C2B	OEM
TSA5510T/C2B	OEM
TSB-245	0015
TSB245	0015
TSB-1000	0015
TSB1204L	OEM
TSB3055	3081
TSC36M	OEM
TSC36N	OEM
TSC36S	OEM
TSC40A	OEM
TSC40B	OEM
TSC40C	OEM
TSC40D	OEM
TSC40E	OEM
TSC40F	OEM
TSC40G	OEM
TSC40H	OEM
TSC40U	OEM
TSC45A	OEM
TSC45B	OEM
TSC45C	OEM
TSC45D	OEM
TSC45E	OEM
TSC45F	OEM
TSC45G	OEM
TSC45H	OEM
TSC45M	OEM
TSC45N	OEM
TSC45S	OEM
TSC45T	OEM
TSC45U	OEM
TSC46A	OEM
TSC46B	OEM
TSC46C	OEM
TSC46D	OEM
TSC46E	OEM
TSC46F	OEM
TSC46G	OEM
TSC46H	OEM
TSC46M	OEM
TSC46N	OEM
TSC46S	OEM
TSC46T	OEM
TSC46U	OEM
TSC50N	OEM
TSC50T	OEM
TSC52U	OEM
TSC136	0133
TSC150E	OEM
TSC150M	OEM
TSC150N	OEM
TSC150P	OEM
TSC150PA	OEM
TSC150PB	OEM
TSC150PC	OEM
TSC150S	OEM
TSC150T	OEM
TSC152E	OEM
TSC152M	OEM
TSC152N	OEM
TSC152P	OEM
TSC152PA	OEM
TSC152PB	OEM
TSC152PC	OEM
TSC152S	OEM
TSC152T	OEM
TSC159	0015

DEVICE TYPE	REPL CODE
TSC350AL	OEM
TSC350CL	OEM
TSC351AL	OEM
TSC351CJ	OEM
TSC351CL	OEM
TSC361AJ	OEM
TSC361CJ	OEM
TSC361CL	OEM
TSC362AJ	OEM
TSC362AL	OEM
TSC362CJ	OEM
TSC362CL	OEM
TSC363AL	OEM
TSC363CJ	OEM
TSC363CL	OEM
TSC367AJ	OEM
TSC367AL	OEM
TSC367CJ	OEM
TSC367CL	OEM
TSC368AJ	OEM
TSC368AL	OEM
TSC368CJ	OEM
TSC368CL	OEM
TSC380AJ	OEM
TSC380AL	OEM
TSC380CJ	OEM
TSC380CL	OEM
TSC381AJ	OEM
TSC381AL	OEM
TSC381CJ	OEM
TSC381CL	OEM
TSC382AJ	OEM
TSC382AL	OEM
TSC382CJ	OEM
TSC382CL	OEM
TSC383AJ	OEM
TSC383AL	OEM
TSC383BL	OEM
TSC383CJ	OEM
TSC383CL	OEM
TSC383ML	OEM
TSC390AL	OEM
TSC390CL	OEM
TSC391AL	OEM
TSC391CL	OEM
TSC392AL	OEM
TSC392CL	OEM
TSC393AL	OEM
TSC393CL	OEM
TSC394AL	OEM
TSC394CL	OEM
TSC395AL	OEM
TSC395CL	OEM
TSC396AJ	OEM
TSC396AL	OEM
TSC396CJ	OEM
TSC396CL	OEM
TSC450ACPE(M)	OEM
TSC450AIJE(M)	OEM
TSC450AMJE(M)	OEM
TSC450BCPE(M)	OEM
TSC450BIJE(M)	OEM
TSC450BMJE(M)	OEM
TSC-499	0111
TSC499	0016
TSC614	0144
TSC695	0016
TSC700A/Y(A)	OEM
TSC700AIJL(A)	OEM
TSC700AMJL(MA)	OEM
TSC-722	0086
TSC722	0086
TSC767	0111
TSC4728CHIP	OEM
TSC4729CHIP	OEM
TSC4730CHIP	OEM
TSC4731CHIP	OEM
TSC4732CHIP	OEM
TSC4733CHIP	OEM
TSC4734CHIP	OEM
TSC4735CHIP	OEM
TSC4736CHIP	OEM
TSC4737CHIP	OEM
TSC4738CHIP	OEM
TSC4739CHIP	OEM
TSC4740CHIP	OEM
TSC6082CHIP	OEM
TSC6083CHIP	OEM
TSC6084CHIP	OEM
TSC6085CHIP	OEM
TSC6086CHIP	OEM
TSC6087CHIP	OEM
TSC6088CHIP	OEM
TSC6089CHIP	OEM
TSC6090CHIP	OEM
TSC6091CHIP	OEM
TSC7106CDL	OEM
TSC7106CPL	4952
TSC7106RCPL	OEM
TSC7107CDL	OEM
TSC7107CPL	4951
TSC7107RCPL	OEM
TSC7109CPL	OEM
TSC7109IPL	OEM
TSC7109MDL	OEM
TSC7116CDL	OEM
TSC7116CPL	2589
TSC7117CDL	OEM
TSC7117CPL	OEM

DEVICE TYPE	REPL CODE
TSC7126CDL	OEM
TSC7126CPL	5526
TSC7135CJI(A)	OEM
TSC7135CPI(A)	OEM
TSC7541AD	OEM
TSC7541BD	OEM
TSC7541JN	OEM
TSC7541KN	OEM
TSC7541SD	OEM
TSC7541TD	OEM
TSC8640BN	OEM
TSC8640CJ	OEM
TSC8640CN	OEM
TSC8641BN	OEM
TSC8641CJ	OEM
TSC8641CN	OEM
TSC8700CJ	OEM
TSC8700CN	OEM
TSC8701CN	OEM
TSC8702CN	OEM
TSC8703BH	OEM
TSC8703BN	OEM
TSC8703CJ	OEM
TSC8703CN	OEM
TSC8704BH	OEM
TSC8704BN	OEM
TSC8704CJ	OEM
TSC8704CN	OEM
TSC8705BH	OEM
TSC8705BN	OEM
TSC8705CJ	OEM
TSC8705CN	OEM
TSC8750BH	OEM
TSC8750BN	OEM
TSC8750CJ	OEM
TSC8750CN	OEM
TSC9012F-001	OEM
TSC9403CJ(A)	OEM
TSC9403IL(A)	OEM
TSC9404CJ(A)	OEM
TSC9404IL(A)	OEM
TSC14433ACJ	OEM
TSC14433ACN	OEM
TSC14433CJ	OEM
TSC14433CN	OEM
TSD035	OEM
TSD135	OEM
TSD235	OEM
TSD335	OEM
TSD435	OEM
TSD536	OEM
TSD635	OEM
TSD835	OEM
TSD1035	OEM
TSD6100	OEM
TSD7000	OEM
TSF1202	OEM
TSF1202C	OEM
TSF1203	OEM
TSF1203M	OEM
TSF1204B	OEM
TSF1212M	OEM
TSF1213A	OEM
TSF1216A	OEM
TSF1218	OEM
TSF1220	OEM
TSF1220M	OEM
TSF1227L	OEM
TSF1228L	OEM
TSF1230L	OEM
TSF2202A	OEM
TSF2203	OEM
TSF3201	OEM
TSH1202C	OEM
TSI1/50	OEM
TSI1/100	OEM
TSI1/150	OEM
TSI1/200	OEM
TSI5/50	OEM
TSI5/100	OEM
TSI5/200	OEM
TSI5/400	OEM
TSK10/06	OEM
TSK10-01	OEM
TSK10-02	OEM
TSK10-04	OEM
TSK10-08	OEM
TSK10-09	OEM
TSK10-10	OEM
TSK25-01	OEM
TSK25-02	OEM
TSK25-04	OEM
TSK25-06	OEM
TSK25-08	OEM
TSK25-09	OEM
TSK25-10	OEM
TSK35-01	OEM
TSK35-02	OEM
TSK35-04	OEM
TSK35-06	OEM
TSK35-08	OEM
TSK35-09	OEM
TSK35-10	OEM
TSK50-01	OEM
TSK50-02	OEM
TSK50-04	OEM
TSK50-06	OEM
TSK50-08	OEM
TSK50-10	OEM

DEVICE TYPE	REPL CODE
TSK65-01	OEM
TSK65-02	OEM
TSK65-04	OEM
TSK65-06	OEM
TSK65-08	OEM
TSK65-09	OEM
TSK65-10	OEM
TS05400	OEM
TS05500	OEM
TSP150	OEM
TSP400	OEM
TSP1000A	OEM
TSP1000B	OEM
TSQ-2222F/C	OEM
TSQ-2907D/C	OEM
TSQ-2907F/C	OEM
TSQ-3467D/C	OEM
TSQ-3467F/C	OEM
TSQ-3725D/C	OEM
TSQ-3725F/C	OEM
TSQ-3762D/C	OEM
TSQ-3762F/C	OEM
TSS-616-1	1030
TSTD-15	0102
TSV	OEM
TSV-1000	0015
TSW61A	OEM
TSW101A	OEM
TSW201A	OEM
TSY65YN	OEM
TSZ5.1	0162
TSZ5.6	0157
TSZ6.2	0636
TSZ6.8	0025
TSZ7.5	0480
TSZ8.2	0244
TSZ9.1	0057
TSZ10	0170
TSZ11	0317
TSZ12	0137
TT05	OEM
TT1	OEM
TT1FA	OEM
TT1FB	OEM
TT2	OEM
TT2FA	OEM
TT2FB	OEM
TT2SA495-0	0006
TT2SA495-0-A	0006
TT2SA495-O	0006
TT2SA495-O-A	0006
TT2SA495-Y	0006
TT2SA495-Y-A	0006
TT2SA817Y	0436
TT2SA950	1348
TT2SA950O	1348
TT2SA965Y	5329
TT2SA1015	0148
TT2SA1015G	0148
TT2SC496Y	0786
TT2SC983	0066
TT2SC983-0-A	0066
TT2SC983-O-A	0066
TT2SC983-R-A	0066
TT2SC983-Y-A	0066
TT2SC1627(Y)	0733
TT2SC1627Y	0733
TT2SC1815	0076
TT2SC1815G	0076
TT2SC2236O	2910
TT2SC2458G	0076
TT2SD843Y	0060
TT3	OEM
TT3FA	OEM
TT3FB	OEM
TT4	OEM
TT4FA	OEM
TT4FB	OEM
TT5	OEM
TT5FA	OEM
TT5FB	OEM
TT6	OEM
TT6FA	OEM
TT6FB	OEM
TT7	OEM
TT7FA	OEM
TT7FB	OEM
TT8	OEM
TT8FA	OEM
TT8FB	OEM
TT9	OEM
TT9FA	OEM
TT9FB	OEM
TT10	OEM
TT10FA	OEM
TT10FB	OEM
TT11	OEM
TT12	OEM
TT13	OEM
TT14	OEM
TT15	OEM
TT16	OEM
TT18F400KCC	OEM
TT18F600KCC	OEM
TT18F800KCC	OEM
TT18F1000KCC	OEM
TT18F1200KCC	OEM
TT18N400	OEM
TT18N600	OEM

DEVICE TYPE	REPL CODE
TT18N800	OEM
TT18N1000	OEM
TT18N1100	OEM
TT18N1200	OEM
TT18N1300	OEM
TT18N1400	OEM
TT25F400KCC	OEM
TT25F600KCC	OEM
TT25F800KCC	OEM
TT25F1000KCC	OEM
TT25F1200KCC	OEM
TT25N400	OEM
TT25N600	OEM
TT25N800	OEM
TT25N1000	OEM
TT25N1100	OEM
TT25N1200	OEM
TT25N1300	OEM
TT25N1400	OEM
TT32N400KOF	OEM
TT32N600KOF	OEM
TT32N800KOF	OEM
TT32N1000KOF	OEM
TT32N1200KOF	OEM
TT32N1600KOF	OEM
TT45N400KOF	OEM
TT45N600KOF	OEM
TT45N800KOF	OEM
TT45N1000KOF	OEM
TT45N1200KOF	OEM
TT45N1400KOF	OEM
TT60N60CK	0652
TT60N600KOF	OEM
TT60N800KOF	OEM
TT60N1000KOF	OEM
TT60N1200KOF	OEM
TT60N1400KOF	OEM
TT66X26	0015
TT204	0144
TT204A	0144
TT204AB	0144
TT204B	0144
TT204C	0144
TT205	OEM
TT210	OEM
TT210FA	OEM
TT210FB	OEM
TT210FX	OEM
TT220	OEM
TT220FA	OEM
TT220FB	OEM
TT220FX	OEM
TT230	OEM
TT230FA	OEM
TT230FB	OEM
TT230FX	OEM
TT240	OEM
TT240FA	OEM
TT240FB	OEM
TT240FX	OEM
TT250	OEM
TT250FA	OEM
TT250FB	OEM
TT250FX	OEM
TT260	OEM
TT260FA	OEM
TT260FB	OEM
TT260FX	OEM
TT270	OEM
TT270FA	OEM
TT270FB	OEM
TT270FX	OEM
TT280	OEM
TT280FA	OEM
TT280FB	OEM
TT280FX	OEM
TT290	OEM
TT290FA	OEM
TT290FB	OEM
TT290FX	OEM
TT310FA	OEM
TT310FB	OEM
TT310V	OEM
TT320FA	OEM
TT320FB	OEM
TT320V	OEM
TT330	OEM
TT330FA	OEM
TT330FB	OEM
TT340FA	OEM
TT340FB	OEM
TT340V	OEM
TT350	OEM
TT350FA	OEM
TT350FB	OEM
TT360FA	OEM
TT360FB	OEM
TT360V	OEM
TT370	OEM
TT370FA	OEM
TT370FB	OEM
TT380FA	OEM
TT380FB	OEM
TT380V	OEM
TT390	OEM
TT390FA	OEM
TT390FB	OEM
TT500	OEM
TT501	OEM

DEVICE TYPE	REPL CODE
TT502	OEM
TT590I	OEM
TT590J	OEM
TT590K	OEM
TT590L	OEM
TT590M	OEM
TT-1083	0160
TT1083	0160
TT1097	0016
TT2100	OEM
TT2100FA	OEM
TT2100FB	OEM
TT2100FX	OEM
TT2110	OEM
TT2120	OEM
TT2130	OEM
TT2140	OEM
TT2150	OEM
TT2160	OEM
TT3100FA	OEM
TT3100FB	OEM
TT3100V	OEM
TT3110	OEM
TT3120V	OEM
TT3130	OEM
TT3140V	OEM
TT3150	OEM
TT3160V	OEM
TT3180V	OEM
TT3200V	OEM
TTAL210	OEM
TTAL220	OEM
TTAL230	OEM
TTAL240	OEM
TTAL250	OEM
TTAL260	OEM
TTAL270	OEM
TTAL280	OEM
TTAL290	OEM
TTAL2100	OEM
TTAL2110	OEM
TTAL2120	OEM
TTS2C983-Y-A	OEM
TTT-2SC1000GB	0111
TTT-2SC1000GBL	0111
TTT-2SK30A--O	3046
TTT-2SK30A-OO	3046
TU000	0164
TU1B	OEM
TU3F700H	0239
TU7F	OEM
TU10/1	OEM
TU10/2	OEM
TU11/1	OEM
TU11/2	OEM
TU12/1	OEM
TU12/2	OEM
TU13/1	OEM
TU13/2	OEM
TU14/1	OEM
TU14/2	OEM
TU25F	OEM
TU30	OEM
TU36HZ	OEM
TU101	OEM
TU102	OEM
TU105	OEM
TU110	OEM
TU120	OEM
TU220	OEM
TU420	OEM
TU620	OEM
TU820	OEM
TU920	OEM
TU1020	OEM
TU3002	OEM
TU3004	OEM
TUA008	OEM
TUA108	OEM
TUA208	OEM
TUA308	OEM
TUA408	OEM
TUA600C	OEM
TUA608	OEM
TUI1	OEM
TUI1R	OEM
TUM835	OEM
TUM1035	OEM
TUP3	OEM
TUP3S	OEM
TUP4	OEM
TUP13	OEM
TV2RD16EB2	OEM
TV2SB126	0160
TV2SB126F	0160
TV2SB126V	0160
TV2SB448	0222
TV2SC208	0007
TV4	0015
TV-6	0079
TV6	0079
TV6.5	0774
TV-7	0144
TV7	0079
TV8	0071
TV9SI	OEM
TV10SI	OEM
TV-11	0201
TV11-S	0201
TV13-11K60	0201
TV13-12K60	0288
TV13-S	0288
TV13SI	OEM
TV15	0007
TV-15A	0007
TV15A	0007
TV-15B	0144
TV15B	0007
TV15C	0007
TV-16	0007
TV16	0144
TV16SI	OEM
TV-17	0079
TV17	0016
TV-18	0016
TV18	0144
TV18-S	0378
TV18B	0079
TV18SI	OEM
TV-19	0233
TV19	0233
TV-20	0330
TV20	0007
TV20-S	0757
TV-21	0016
TV21	0595
TV-22	0144
TV22	0338
TV-23	0016
TV23	0086
TV-25	0338
TV25	0086
TV-26	0086
TV26	0595
TV-27	0208
TV27	0038
TV-28	0086
TV28	0595
TV-29	0126
TV29	0126
TV-32	0079
TV32	0144
TV-33	0008
TV33	0007
TV-34	0008
TV34	0102
TV-35	0008
TV35	0007
TV-36	0007
TV36	0079
TV-37	0007
TV37	0008
TV-38	0007
TV38	0016
TV-39	0007
TV39	0016
TV-40	0016
TV40	0016
TV-41	0086
TV41	0086
TV-42	0079
TV42	0086
TV-43	0086
TV43	0079
TV-44	0037
TV44	0037
TV-45	0086
TV45	0086
TV-46	0016
TV46	0016
TV-47	0037
TV47	0164
TV-48	0224
TV48	0016
TV-49	0233
TV49	0338
TV-50	0008
TV50	0007
TV-51	0079
TV51	0086
TV-52	0079
TV52	0086
TV-53	0016
TV53	0086
TV-54	0330
TV54	0007
TV-55	0330
TV55	0144
TV55E	OEM
TV55F	OEM
TV-56	0016
TV56	0016
TV-57	0037
TV57	0127
TV-58	0079
TV58	0127
TV59	0086
TV-60	0079
TV60	0127
TV-61	0211
TV61	0164
TV-62	0079
TV62	0400
TV-65	0079
TV65	0079
TV-66	0079
TV66	0079
TV-67	0086
TV67	0079
TV-68	0079
TV68	0079
TV-70	0233
TV70B	0233
TV-71	0150
TV-73	0560
TV-74	1267
TV74	0560
TV-75	0560
TV-77	0330
TV80	0325
TV81	0330
TV-82	0560
TV82	0161
TV-83	0325
TV83	0325
TV-84	0079
TV84	0079
TV85	0285
TV-87	0037
TV87	0037
TV-92	0079
TV92	0079
TV-93	0037
TV93	0037
TV105	0491
TV106	0975
TV108	0642
TV-109	0178
TV109	0178
TV110	2013
TV111	0975
TV-112	0423
TV112	0423
TV113	0142
TV114	0975
TV-115	0042
TV115	0144
TV-116	0925
TV116	0854
TV-117	0042
TV117	0423
TV118	0642
TV119	1966
TV-120	0276
TV120	0161
TV121	0074
TV122	0142
TV-124	0065
TV124	0065
TV-125	0065
TV125	0642
TV210B	OEM
TV215	1323
TV217C	OEM
TV217CH	OEM
TV220A	OEM
TV220B	OEM
TV220C	OEM
TV220D	OEM
TV220E	OEM
TV220F	OEM
TV220G	OEM
TV220J	OEM
TV221A	OEM
TV221B	OEM
TV221C	OEM
TV221D	OEM
TV221E	OEM
TV221F	OEM
TV221G	OEM
TV221H	OEM
TV221K	OEM
TV222A	OEM
TV222B	OEM
TV222C	OEM
TV222D	OEM
TV222E	OEM
TV222F	OEM
TV222G	OEM
TV222Z	OEM
TV225A	OEM
TV232	OEM
TV242	OEM
TV244	OEM
TV259	OEM
TV262	OEM
TV313	OEM
TV514	OEM
TV695	0774
TV848	OEM
TV851	OEM
TV1000	0144
TV1255B	OEM
TV1505	OEM
TV1546	OEM
TV1547	OEM
TV2001W	OEM
TV2002	OEM
TV2011	OEM
TV2013	OEM
TV2015	OEM
TV2017	OEM
TV2018	OEM
TV2019W	OEM
TV2022	OEM
TV2028	OEM
TV2030	OEM
TV2032	OEM
TV2042	OEM
TV2053	OEM
TV2091	OEM
TV2092	OEM
TV2125CBA	OEM
TV2125CBB	OEM
TV2202F	OEM
TV2204F	OEM
TV2206F	OEM
TV2208F	OEM
TV2210F	OEM
TV2211A	OEM
TV2211B	OEM
TV2211C	OEM
TV2211D	OEM
TV2214F	OEM
TV2215A	OEM
TV2215F	OEM
TV2216A	OEM
TV2216B	OEM
TV2216C	OEM
TV2217H	OEM
TV2226	OEM
TV2246A	OEM
TV2246B	OEM
TV2246X	OEM
TV2350	OEM
TV2403	0144
TV2404	0144
TV2411	0004
TV2412M	OEM
TV2419	0015
TV2428	0004
TV2429	0004
TV2434	0004
TV2455	0050
TV2479	0050
TV2496	0015
TV3002	1023
TV3004	1023
TV3006	1036
TV3008	1046
TV3010	1046
TV3012	6493
TV3014	6493
TV3016	OEM
TV3018	OEM
TV3020	OEM
TV3301	OEM
TV3304	OEM
TV3308	OEM
TV3309	OEM
TV3340	OEM
TV3340A	OEM
TV3340B	OEM
TV9140	OEM
TV19111	OEM
TV19112C	OEM
TV19114	OEM
TV24100	0015
TV24102	0144
TV24103	0917
TV24104	0015
TV24115	0004
TV24122	0019
TV24124	0196
TV24125	0015
TV24130	0917
TV24136	0015
TV24137	0136
TV24142	0975
TV24143	0208
TV24148	0050
TV24152	0136
TV24154	0004
TV24155	0015
TV24156	0004
TV24157	0211
TV24158	0050
TV24159	0917
TV24160	0007
TV24161	0007
TV24162	1010
TV24163	0222
TV24164	0233
TV24165	0774
TV24166	0136
TV24167	0015
TV24168	0102
TV24169	0102
TV24172	0136
TV24176	1030
TV24180	0015
TV24181	0182
TV24182	0917
TV24183	0015
TV24189	0004
TV24190	0196
TV-24191	0015
TV24191	0015
TV24193	0015
TV24194	0004
TV24200	0015
TV24203	0144
TV24204	0144
TV24209	0007
TV24210	0127
TV24211	0178
TV24214	0037
TV24215	0079
TV24216	0079
TV24217	0102
TV24218	0975
TV24219	0182
TV24220	OEM
TV24221	0015
TV24222	0102
TV24224	1083
TV24225	0087
TV24226	0196
TV24229	0050
TV24230	0050
TV-24232	0015
TV24232	0015
TV24234	0015
TV-24237	0474
TV24237	0474
TV24239	0050
TV-24266	0015
TV24266	0015
TV-24272	0649
TV24272	0649
TV24273	0123
TV24278	0015
TV24281	0016
TV24282	0015
TV24283	0015
TV24285	0277
TV24292	0015
TV24298	0015
TV24313	0144
TV24337	0160
TV24341	0975
TV24351	0050
TV24363	0037
TV24370	0004
TV24372	0016
TV24378	0540
TV24380	0127
TV24382	0127
TV24383	0127
TV24385	0127
TV24387	0144
TV-24399	0007
TV24399	0007
TV24435	0233
TV24436	0007
TV24437	0007
TV24438	0127
TV24453	0016
TV24454	0016
TV24458	0016
TV24468	0975
TV24487	0178
TV24495	0037
TV24499	0233
TV24540	0288
TV24554	0133
TV24558	OEM
TV24568	0975
TV24571	0007
TV24573	0144
TV24574	0144
TV24576	0016
TV24582	0015
TV24589	0144
TV24599	0004
TV24617	0102
TV24648	0102
TV24649	0924
TV24650	0288
TV24651	0110
TV24655	0016
TV24678	0160
TV24681	0924
TV24684	0144
TV24803	0015
TV24806	0127
TV24848	0150
TV24941	0015
TV24942	0071
TV24945	0004
TV24979	0015
TV24981	0025
TV24983	0208
TV24984	0004
TV24985	1080
TV24987	0196
TV34232	0015
TV241013	0123
TV241072	0878
TV241073	0015
TV241074	0015
TV241077	0079
TV241078	0079
TV242248	0015
TVA12W1100K	OEM
TVA12W1120K	OEM
TVA12W1140K	OEM
TVA12W1160K	OEM
TVA11393	OEM
TVAQA108SH	0165
TVC-3	0015
TVC3	0474
TVC-6	0474
TVC6	0474
TVCM-1	2161
TVCM-2	0666
TVCM-3	0753
TVCM-4	0664
TVCM-5	0678
TVCM-6	0654
TVCM-7	0443
TVCM-8	0352
TVCM-9	0354
TVCM-10	0701
TVCM-11	0167
TVCM-12	1940
TVCM-13	2279
TVCM-14	1759
TVCM-15	2457
TVCM-16	1345
TVCM-17	3008
TVCM-18	1443
TVCM-19	1421
TVCM-20	1759
TVCM-21	0751
TVCM21	0403
TVCM-22	0349
TVCM-23	2557
TVCM-24	OEM
TVCM-25	2557
TVCM-26	2557
TVCM-27	0856
TVCM-28	3735
TVCM-29	1395
TVCM-30	0802
TVCM-31	0364
TVCM-32	2549
TVCM-33	0328
TVCM-34	1337
TVCM-35	0379
TVCM-39	2755
TVCM-41	0789
TVCM-42	0395
TVCM-43	0692
TVCM-44	0632
TVCM-45	2617
TVCM-46	2624
TVCM-47	2699
TVCM-48	0823
TVCM-49	2707
TVCM-50	2571
TVCM-51	0829
TVCM-52	2714
TVCM-53	2715
TVCM-54	2529
TVCM-55	2743
TVCM-56	2747
TVCM-57	2257
TVCM-58	1843
TVCM-59	0785
TVCM-60	1478
TVCM60	1808
TVCM-61	0360
TVCM-62	0372
TVCM-63	OEM
TVCM-64	OEM
TVCM-65	2557
TVCM-66	1696
TVCM-67	2722
TVCM-68	3410
TVCM-69	OEM
TVCM69	0417
TVCM-70	OEM
TVCM-71	OEM
TVCM-72	OEM
TVCM-73	0520
TVCM-74	2124
TVCM-75	0579
TVCM-76	OEM
TVCM-77	2057
TVCM-78	0911
TVCM-79	1994
TVCM-80	OEM
TVCM-81	1050
TVCM-82	0433
TVCM-500	0232
TVCM-501	4596
TVCM-502	1313
TVCM-503	1108
TVCM-504	0569
TVCM-505	1208
TVCM-506	1645
TVCM-551	0630
TVCM-553	0086
TVC-MK3800	3135
TVCMK3800	3135
TVDFV0212	0120
TVH05	OEM
TVH08	OEM
TVH3W250M	OEM
TVH3W265M	OEM
TVH3W280M	OEM
TVH3W2100M	OEM
TVH3W2120M	OEM
TVH3W2150L	OEM
TVH3W2180L	OEM
TVH3W2220L	OEM
TVH3W2270L	OEM
TVH15	OEM
TVH30	OEM
TVH101	OEM
TVH102	OEM
TVH103	OEM
TVH104	OEM
TVH201	OEM
TVH202	OEM
TVH210	OEM
TVH211	OEM
TVH212	OEM
TVH213	OEM
TVH214	OEM
TVH-526	0484
TVH526	0484
TVH538	0196
TVH700	OEM
TVHD3/2D2100L	OEM
TVHD3/D2140L	OEM
TVHD3/D2120L	OEM
TVHD10	OEM
TVHD12	OEM
TVHD14	OEM
TVHS1R31100K	OEM
TVHS1R31300K	OEM
TVHS150	OEM
TVHS200	OEM
TVHS260	OEM
TVHS330	OEM
TVHS400	OEM
TVHS480	OEM
TVHS580	OEM
TVHS700	OEM
TVHS840	OEM
TVK1D24M	OEM
TVK1D25M	OEM
TVK1D26M	OEM
TVK1D28M	OEM
TVK-33	0128
TVK-55	0128
TVL0	OEM
TVL1	OEM
TVL2	OEM
TVL3	OEM
TVL4	OEM
TVM0	OEM
TVM-026	1997
TVM1	OEM
TVM1D246L	OEM
TVM1D256L	OEM
TVM1D267L	OEM
TVM1D279L	OEM
TVM1D292L	OEM
TVM1D2108L	OEM
TVM2	OEM
TVM3	OEM
TVM4	OEM
TVM5	OEM
TVM7K705M	0196
TVM35	0015
TVM56	0015
TVM-153	1501
TVM153	1501
TVM511	0015
TVM-526	0196
TVM526	0484
TVM529	0277
TVM-530	0913
TVM530	0196
TVM-531	0201
TVM531	0288
TVM533	0378
TVM-535	0015
TVM535	0102
TVM-537	1303
TVM537	0102
TVM-538	0196
TVM538	0196
TVM540	0757
TVM540A	0378
TVM546	0920
TVM550	0015
TVM553	0201
TVM-554	0484
TVM554	0196
TVM554A	2903
TVM563	0015
TVM567	0201
TVM569	0288
TVM-778	0190
TVM-EH2C	0015
TVM-EH2C11	0015
TVM-EH2C11/1	0015
TVM-EH2C11/1+12/1	0015
TVM-EH2C12/1	0015
TVM-HS15/1B	0201
TVM-HS20/1B	0757
TVM-HS25/1G	0196
TVM-K112C	0196
TVMK112C	0196
TVML00.09D1115	0015
TVM-L00.09M1115	1303
TVML00.09M1115	0015
TVM-M204B	0015
TVMM204B	0015
TVM026	1997
TVM-PH9D22/1	0015
TVMPH9DZ2/1	0015
TVM-PT6D22/1	0277
TVMTC00921-3	0484
TVM-TC0.2P	0484
TVM-TC0.2P11/2	0015
TVM-TC0.2P	0484
TVM-TC0.2P1	0920
TVM-TK-705M	0196

If replacement code is OEM, contact original manufacturer for replacement.

DEVICE TYPE	REPL CODE	DEVICE TYPE	REPL CODE	DEVICE TYPE	REPL CODE	DEVICE TYPE	REPL CODE	DEVICE TYPE	REPL CODE	DEVICE TYPE	REPL CODE	DEVICE TYPE	REPL CODE
TVM-TK705M	0196	TVP1014	OEM	TVP1529	OEM	TVPC505A	OEM	TVPC1022A	OEM	TVPC1539A	OEM	TVR10D	OEM
TVN5201KND	OEM	TVP1014A	OEM	TVP1529A	OEM	TVPC506	OEM	TVPC1023	OEM	TVPC5002	OEM	TVR10G	0023
TVN6000KNS	OEM	TVP1015	OEM	TVP1530	OEM	TVPC506A	OEM	TVPC1023A	OEM	TVPC5002A	OEM	TVRAK002	OEM
TVP500	OEM	TVP1015A	OEM	TVP1530A	OEM	TVPC507	OEM	TVPC1024	OEM	TVPC5003	OEM	TVS001-06SB	0301
TVP500A	OEM	TVP1016	OEM	TVP1531	OEM	TVPC507A	OEM	TVPC1024A	OEM	TVPC5003A	OEM	TVS001-06SD	OEM
TVP501	OEM	TVP1016A	OEM	TVP1531A	OEM	TVPC508	OEM	TVPC1025	OEM	TVPC5004	OEM	TVS00410	0071
TVP501A	OEM	TVP1017	OEM	TVP1532	OEM	TVPC508A	OEM	TVPC1025A	OEM	TVPC5004A	OEM	TVS03P2M	0058
TVP502	OEM	TVP1017A	OEM	TVP1532A	OEM	TVPC509	OEM	TVPC1026	OEM	TVPC5005	OEM	TVS010D1	0023
TVP502A	OEM	TVP1018	OEM	TVP1533	OEM	TVPC509A	OEM	TVPC1026A	OEM	TVPC5005A	OEM	TVS-0A70	0143
TVP503	OEM	TVP1018A	OEM	TVP1533A	OEM	TVPC510	OEM	TVPC1027	OEM	TVPC5006	OEM	TVS0A70	0143
TVP503A	OEM	TVP1019	OEM	TVP1534	OEM	TVPC510A	OEM	TVPC1027A	OEM	TVPC5006A	OEM	TVS-0A71	0143
TVP504	OEM	TVP1019A	OEM	TVP1534A	OEM	TVPC511	OEM	TVPC1028	OEM	TVPC5007	OEM	TVS0A71	0143
TVP504A	OEM	TVP1020	OEM	TVP1535	OEM	TVPC511A	OEM	TVPC1028A	OEM	TVPC5007A	OEM	TVS-0A82G	0143
TVP505	OEM	TVP1020A	OEM	TVP1535A	OEM	TVPC512	OEM	TVPC1029	OEM	TVPC5008	OEM	TVS-0A90	0143
TVP505A	OEM	TVP1021	OEM	TVP1536	OEM	TVPC512A	OEM	TVPC1029A	OEM	TVPC5008A	OEM	TVS0A90	0019
TVP506	OEM	TVP1021A	OEM	TVP1536A	OEM	TVPC513	OEM	TVPC1030	OEM	TVPC5009	OEM	TVS-0A91	0019
TVP506A	OEM	TVP1022	OEM	TVP1537	OEM	TVPC513A	OEM	TVPC1030A	OEM	TVPC5009A	OEM	TVS0A91	0019
TVP507	OEM	TVP1022A	OEM	TVP1537A	OEM	TVPC514	OEM	TVPC1031	OEM	TVPC5010	OEM	TVS-0A95	0143
TVP507A	OEM	TVP1023	OEM	TVP1538	OEM	TVPC514A	OEM	TVPC1031A	OEM	TVPC5010A	OEM	TVS-0V-02	0102
TVP508	OEM	TVP1023A	OEM	TVP1538A	OEM	TVPC515	OEM	TVPC1032	OEM	TVPC5011	OEM	TVS1/2D233M	OEM
TVP508A	OEM	TVP1024	OEM	TVP1539	OEM	TVPC515A	OEM	TVPC1032A	OEM	TVPC5011A	OEM	TVS1CE2	0015
TVP509	OEM	TVP1024A	OEM	TVP1539A	OEM	TVPC516	OEM	TVPC1033	OEM	TVPC5012	OEM	TVS1FSD-1	0015
TVP509A	OEM	TVP1025	OEM	TVP1728	OEM	TVPC516A	OEM	TVPC1033A	OEM	TVPC5012A	OEM	TVS-1N82G	0917
TVP510	OEM	TVP1025A	OEM	TVP5000	OEM	TVPC517	OEM	TVPC1034	OEM	TVPC5013	OEM	TVS1N741A	0287
TVP510A	OEM	TVP1026	OEM	TVP5000A	OEM	TVPC517A	OEM	TVPC1034A	OEM	TVPC5013A	OEM	TVS1N741H	0287
TVP511	OEM	TVP1026A	OEM	TVP5001	OEM	TVPC518	OEM	TVPC1035	OEM	TVPC5014	OEM	TVS1N4148	0124
TVP511A	OEM	TVP1027	OEM	TVP5001A	OEM	TVPC518A	OEM	TVPC1035A	OEM	TVPC5014A	OEM	TVS1N4148TE	0124
TVP512	OEM	TVP1027A	OEM	TVP5002	OEM	TVPC519	OEM	TVPC1036	OEM	TVPC5015	OEM	TVS1N4148V	0124
TVP512A	OEM	TVP1028	OEM	TVP5002A	OEM	TVPC519A	OEM	TVPC1036A	OEM	TVPC5015A	OEM	TVS1N4150V	0133
TVP513	OEM	TVP1028A	OEM	TVP5003	OEM	TVPC520A	OEM	TVPC1037	OEM	TVPC5016	OEM	TVS1N4741	0317
TVP513A	OEM	TVP1029	OEM	TVP5003A	OEM	TVPC521	OEM	TVPC1037A	OEM	TVPC5016A	OEM	TVS1N4741A	0317
TVP514	OEM	TVP1029A	OEM	TVP5004	OEM	TVPC521A	OEM	TVPC1038	OEM	TVPC5017	OEM	TVS1P20	0015
TVP514A	OEM	TVP1030	OEM	TVP5004A	OEM	TVPC522	OEM	TVPC1038A	OEM	TVPC5017A	OEM	TVS1P80	0071
TVP515	OEM	TVP1030A	OEM	TVP5005	OEM	TVPC522A	OEM	TVPC1039	OEM	TVPC5018	OEM	TVS1R80	0017
TVP515A	OEM	TVP1031	OEM	TVP5005A	OEM	TVPC523	OEM	TVPC1039A	OEM	TVPC5018A	OEM	TVS-1S18	0143
TVP516	OEM	TVP1031A	OEM	TVP5006	OEM	TVPC523A	OEM	TVPC1502	OEM	TVPC5019	OEM	TVS-1S750	0917
TVP516A	OEM	TVP1032	OEM	TVP5006A	OEM	TVPC524	OEM	TVPC1502A	OEM	TVPC5019A	OEM	TVS1S750	0917
TVP517	OEM	TVP1032A	OEM	TVP5007	OEM	TVPC524A	OEM	TVPC1503	OEM	TVPC5020	OEM	TVS1S954	0133
TVP517A	OEM	TVP1033	OEM	TVP5007A	OEM	TVPC525	OEM	TVPC1503A	OEM	TVPC5020A	OEM	TVS-1S1211	0139
TVP518	OEM	TVP1033A	OEM	TVP5008	OEM	TVPC525A	OEM	TVPC1504	OEM	TVPC5021	OEM	TVS1S1850	0023
TVP518A	OEM	TVP1034	OEM	TVP5008A	OEM	TVPC526	OEM	TVPC1504A	OEM	TVPC5021A	OEM	TVS-1S1893	0554
TVP519	OEM	TVP1034A	OEM	TVP5009	OEM	TVPC526A	OEM	TVPC1505	OEM	TVPC5022	OEM	TVS-1S1906	0015
TVP519A	OEM	TVP1035	OEM	TVP5009A	OEM	TVPC527	OEM	TVPC1505A	OEM	TVPC5022A	OEM	TVS1S1906	0015
TVP520	OEM	TVP1035A	OEM	TVP5010	OEM	TVPC527A	OEM	TVPC1506	OEM	TVPC5023	OEM	TVS-1S1922	0554
TVP520A	OEM	TVP1036	OEM	TVP5010A	OEM	TVPC528	OEM	TVPC1506A	OEM	TVPC5023A	OEM	TVS1S1922G	0917
TVP521	OEM	TVP1036A	OEM	TVP5011	OEM	TVPC528A	OEM	TVPC1507	OEM	TVPC5024	OEM	TVS1S1926K	0917
TVP521A	OEM	TVP1037	OEM	TVP5011A	OEM	TVPC529	OEM	TVPC1507A	OEM	TVPC5024A	OEM	TVS1S1950	0015
TVP522	OEM	TVP1037A	OEM	TVP5012	OEM	TVPC529A	OEM	TVPC1508	OEM	TVPC5025	OEM	TVS1S2076	0133
TVP522A	OEM	TVP1038	OEM	TVP5012A	OEM	TVPC530	OEM	TVPC1508A	OEM	TVPC5025A	OEM	TVS1S2076MC	0124
TVP523	OEM	TVP1038A	OEM	TVP5013	OEM	TVPC530A	OEM	TVPC1509	OEM	TVPC5026	OEM	TVS1S2339	0554
TVP523A	OEM	TVP1039	OEM	TVP5013A	OEM	TVPC531	OEM	TVPC1509A	OEM	TVPC5026A	OEM	TVS1S2339K	0030
TVP524	OEM	TVP1039A	OEM	TVP5014	OEM	TVPC531A	OEM	TVPC1510	OEM	TVPC5027	OEM	TVS1S2370	0017
TVP525	OEM	TVP1500	OEM	TVP5014A	OEM	TVPC532	OEM	TVPC1510A	OEM	TVPC5027A	OEM	TVS-2B-2C	0102
TVP525A	OEM	TVP1500A	OEM	TVP5015	OEM	TVPC532A	OEM	TVPC1511	OEM	TVPC5028	OEM	TVS-2B126F	0160
TVP526	OEM	TVP1501	OEM	TVP5015A	OEM	TVPC533	OEM	TVPC1511A	OEM	TVPC5028A	OEM	TVS2CS1256HG	0791
TVP526A	OEM	TVP1501A	OEM	TVP5016	OEM	TVPC533A	OEM	TVPC1512	OEM	TVPC5029	OEM	TVS2P5M	1243
TVP527	OEM	TVP1502	OEM	TVP5016A	OEM	TVPC534	OEM	TVPC1512A	OEM	TVPC5029A	OEM	TVS2PD56	OEM
TVP527A	OEM	TVP1502A	OEM	TVP5017	OEM	TVPC534A	OEM	TVPC1513	OEM	TVPC5030A	OEM	TVS2PD57	OEM
TVP528	OEM	TVP1503	OEM	TVP5017A	OEM	TVPC535	OEM	TVPC1513A	OEM	TVPC5031	OEM	TVS-2S172F	0004
TVP528A	OEM	TVP1503A	OEM	TVP5018	OEM	TVPC535A	OEM	TVPC1514	OEM	TVPC5031A	OEM	TVS-2S288A	0127
TVP529	OEM	TVP1504	OEM	TVP5018A	OEM	TVPC536	OEM	TVPC1514A	OEM	TVPC5032	OEM	TVS2S1664	3370
TVP529A	OEM	TVP1504A	OEM	TVP5019A	OEM	TVPC536A	OEM	TVPC1515	OEM	TVPC5032A	OEM	TVS-2SA71B	0050
TVP530	OEM	TVP1505	OEM	TVP5020	OEM	TVPC537	OEM	TVPC1515A	OEM	TVPC5033	OEM	TVS2SA71B	0050
TVP530A	OEM	TVP1505A	OEM	TVP5020A	OEM	TVPC537A	OEM	TVPC1516	OEM	TVPC5033A	OEM	TVS-2SA103	0136
TVP531	OEM	TVP1506	OEM	TVP5021	OEM	TVPC538	OEM	TVPC1516A	OEM	TVPC5034	OEM	TVS-2SA171	0280
TVP531A	OEM	TVP1506A	OEM	TVP5021A	OEM	TVPC538A	OEM	TVPC1517	OEM	TVPC5034A	OEM	TVS2SA171	0280
TVP532	OEM	TVP1507	OEM	TVP5022	OEM	TVPC539	OEM	TVPC1517A	OEM	TVPC5035	OEM	TVS-2SA385	0136
TVP532A	OEM	TVP1507A	OEM	TVP5022A	OEM	TVPC539A	OEM	TVPC1518	OEM	TVPC5035A	OEM	TVS-2SA385A	0136
TVP533	OEM	TVP1508	OEM	TVP5023	OEM	TVPC1002	OEM	TVPC1518A	OEM	TVPC5036	OEM	TVS-2SA385L	0136
TVP533A	OEM	TVP1508A	OEM	TVP5023A	OEM	TVPC1002A	OEM	TVPC1519	OEM	TVPC5036A	OEM	TVS2SA483	0905
TVP534	OEM	TVP1509	OEM	TVP5024	OEM	TVPC1003	OEM	TVPC1519A	OEM	TVPC5037	OEM	TVS2SA543	0693
TVP534A	OEM	TVP1509A	OEM	TVP5024A	OEM	TVPC1003A	OEM	TVPC1520	OEM	TVPC5037A	OEM	TVS2SA546	0126
TVP535	OEM	TVP1510	OEM	TVP5025	OEM	TVPC1004	OEM	TVPC1520A	OEM	TVPC5038	OEM	TVS2SA546B	0126
TVP535A	OEM	TVP1510A	OEM	TVP5025A	OEM	TVPC1004A	OEM	TVPC1521	OEM	TVPC5038A	OEM	TVS2SA546E	0126
TVP536	OEM	TVP1511	OEM	TVP5026	OEM	TVPC1005	OEM	TVPC1521A	OEM	TVPC5039	OEM	TVS2SA546H	0126
TVP536A	OEM	TVP1511A	OEM	TVP5026A	OEM	TVPC1005A	OEM	TVPC1522	OEM	TVPC5039A	OEM	TVS-2SA564	0203
TVP537	OEM	TVP1512	OEM	TVP5027	OEM	TVPC1006	OEM	TVPC1522A	OEM	TVR-06B	0015	TVS2SA564	0203
TVP537A	OEM	TVP1512A	OEM	TVP5027A	OEM	TVPC1006A	OEM	TVPC1523	OEM	TVR-06D	0023	TVS-2SA564A	0203
TVP538	OEM	TVP1513	OEM	TVP5028	OEM	TVPC1007	OEM	TVPC1523A	OEM	TVR06D	0023	TVS2SA564A	0203
TVP538A	OEM	TVP1513A	OEM	TVP5028A	OEM	TVPC1007A	OEM	TVPC1524	OEM	TVR-06G	0015	TVS2SA564C	0203
TVP539	OEM	TVP1514	OEM	TVP5029	OEM	TVPC1008	OEM	TVPC1524A	OEM	TVR06G	0015	TVS-2SA564P	0203
TVP539A	OEM	TVP1514A	OEM	TVP5029A	OEM	TVPC1008A	OEM	TVPC1525	OEM	TVR-06G23	0015	TVS2SA564P	0203
TVP1000	OEM	TVP1515	OEM	TVP5030A	OEM	TVPC1009	OEM	TVPC1525A	OEM	TVR-06GG23	0015	TVS2SA564Q	0203
TVP1000A	OEM	TVP1515A	OEM	TVP5031	OEM	TVPC1009A	OEM	TVPC1526	OEM	TVR06J	OEM	TVS-2SB126	0160
TVP1001	OEM	TVP1516	OEM	TVP5031A	OEM	TVPC1010	OEM	TVPC1527	OEM	TVR-06K	0015	TVS2SB126	0160
TVP1001A	OEM	TVP1516A	OEM	TVP5032	OEM	TVPC1010A	OEM	TVPC1527A	OEM	TVR06K	0023	TVS-2SB126F	0160
TVP1002	OEM	TVP1517	OEM	TVP5032A	OEM	TVPC1011	OEM	TVPC1528	OEM	TVR-060	OEM	TVS-2SB126V	0160
TVP1002A	OEM	TVP1517A	OEM	TVP5033	OEM	TVPC1011A	OEM	TVPC1528A	OEM	TVR-1	0102	TVS2SB126V	0160
TVP1003	OEM	TVP1518	OEM	TVP5033A	OEM	TVPC1012	OEM	TVPC1529	OEM	TVR-1B	0023	TVS-2SB171	0004
TVP1003A	OEM	TVP1518A	OEM	TVP5034	OEM	TVPC1012A	OEM	TVPC1529A	OEM	TVR1B	0023	TVS2SB171	0004
TVP1004	OEM	TVP1519	OEM	TVP5034A	OEM	TVPC1013	OEM	TVPC1530	OEM	TVR1D	0023	TVS-2SB171A	0004
TVP1004A	OEM	TVP1519A	OEM	TVP5035	OEM	TVPC1013A	OEM	TVPC1530A	OEM	TVR1D(TP)	0023	TVS2SB171A	0004
TVP1005	OEM	TVP1520	OEM	TVP5035A	OEM	TVPC1014	OEM	TVPC1531	OEM	TVR-1G	0023	TVS2SB171B	0004
TVP1005A	OEM	TVP1520A	OEM	TVP5036	OEM	TVPC1014A	OEM	TVPC1531A	OEM	TVR1G	0023	TVS-2SB171F	0004
TVP1006	OEM	TVP1521	OEM	TVP5036A	OEM	TVPC1015	OEM	TVPC1532	OEM	TVR1G(TP)	0023	TVS-2SB172	0004
TVP1006A	OEM	TVP1521A	OEM	TVP5037	OEM	TVPC1015A	OEM	TVPC1532A	OEM	TVR-1J	0023	TVS2SB172	0004
TVP1007	OEM	TVP1522	OEM	TVP5037A	OEM	TVPC1016	OEM	TVPC1533	OEM	TVR1J	0023	TVS2SB172A	0004
TVP1007A	OEM	TVP1522A	OEM	TVP5038	OEM	TVPC1016A	OEM	TVPC1533A	OEM	TVR1J(TP)	0023	TVS-2SB172F	0004
TVP1008	OEM	TVP1523	OEM	TVP5038A	OEM	TVPC1017	OEM	TVPC1534	OEM	TVR-2	0023	TVS-2SB176	0004
TVP1008A	OEM	TVP1523A	OEM	TVP5039	OEM	TVPC1017A	OEM	TVPC1534A	OEM	TVR2B	0023	TVS-2SB234	0975
TVP1009	OEM	TVP1524	OEM	TVP5039A	OEM	TVPC1018	OEM	TVPC1535	OEM	TVR-2D	0023	TVS-2SB324	1062
TVP1009A	OEM	TVP1524A	OEM	TVPC520	OEM	TVPC1018A	OEM	TVPC1535A	OEM	TVR2D	0023	TVS-2SB448	0222
TVP1010	OEM	TVP1525	OEM	TVPC502	OEM	TVPC1019	OEM	TVPC1536	OEM	TVR-2G	0023	TVS2SB449	0160
TVP1010A	OEM	TVP1525A	OEM	TVPC502A	OEM	TVPC1019A	OEM	TVPC1536A	OEM	TVR2G	0023	TVS2SB449(F)	0160
TVP1011A	OEM	TVP1526	OEM	TVPC503	OEM	TVPC1020	OEM	TVPC1537	OEM	TVR2J	0023	TVS-2SB449F	0160
TVP1012	OEM	TVP1526A	OEM	TVPC503A	OEM	TVPC1020A	OEM	TVPC1537A	OEM	TVR4J	0102	TVS2SB449F	0160
TVP1012A	OEM	TVP1527	OEM	TVPC504	OEM	TVPC1021	OEM	TVPC1538	OEM	TVR4J-TPA2	0102	TVS2SB546	1648
TVP1013	OEM	TVP1527A	OEM	TVPC504A	OEM	TVPC1021A	OEM	TVPC1538A	OEM	TVR4L	0102	TVS-2SC58	0233
TVP1013A	OEM	TVP1528A	OEM	TVPC505	OEM	TVPC1022	OEM	TVPC1539	OEM	TVR4N	0102	TVS-2SC58A	0233

DEVICE TYPE	REPL CODE
TVS2SC58A	0233
TVS-2SC183P	0475
TVS-2SC183Q	0475
TVS-2SC185A	0475
TVS-2SC206	0216
TVS-2SC208	0007
TVS-2SC208A	0007
TVS-2SC287A	0475
TVS2SC288A	1342
TVS-2SC313	3390
TVS-2SC429A	0475
TVS-2SC446	5568
TVS-2SC466	0259
TVS-2SC526	0233
TVS-2SC538	0079
TVS-2SC538A	0079
TVS2SC538A	0079
TVS-2SC562	0216
TVS2SC562	0216
TVS-2SC563	0835
TVS2SC563	0835
TVS-2SC563A	0835
TVS-2SC564	0016
TVS2SC564	0016
TVS2SC564-O	0016
TVS2SC564O	0016
TVS2SC564R	0016
TVS-2SC582	2658
TVS-2SC582A	2658
TVS-2SC605	0475
TVS-2SC606	0475
TVS-2SC644	0111
TVS2SC644	0111
TVS-2SC645	0674
TVS2SC645	0674
TVS-2SC645A	0674
TVS2SC645A	0674
TVS-2SC645B	0674
TVS2SC645B	0674
TVS-2SC645C	0674
TVS2SC645C	0674
TVS-2SC646	0620
TVS2SC647	4387
TVS-2SC683	0047
TVS2SC683	0047
TVS-2SC683V	0047
TVS-2SC684	0127
TVS2SC684	0127
TVS-2SC687	0177
TVS2SC696	0644
TVS-2SC762	1154
TVS2SC762	1154
TVS2SC783	1027
TVS-2SC828	1220
TVS2SC828	1220
TVS2SC828(Q)	1220
TVS-2SC828A	1220
TVS2SC828A	1220
TVS2SC828P	1220
TVS-2SC828Q	1220
TVS2SC828Q	1220
TVS2SC828R	1220
TVS2SC829(B)	0151
TVS2SC829B	0151
TVS-2SC840	2028
TVS2SC840	2028
TVS-2SC840A	2028
TVS2SC840A	2028
TVS2SC840C	2028
TVS-2SC901	0642
TVS2SC901	0642
TVS-2SC948	0216
TVS2SC968	2929
TVS2SC1129A	0224
TVS2SC1502	0955
TVS2SC1505	0955
TVS2SC1505(1)	0955
TVS2SC1505-1	0955
TVS2SC1505I	0955
TVS2SC1507	0955
TVS2SC1520	0168
TVS2SC1629A	0130
TVS2SC1664R	0629
TVS2SC1828	2096
TVS2SC1829	2620
TVS2SC1875Q	2660
TVS2SC1875R	2660
TVS2SC2199	0620
TVS2SC5640	0016
TVS-2SD226A	0178
TVS2SD226A	0178
TVS2SD226P	0178
TVS2SD401	1284
TVS2SD794	0161
TVS2SD794(2)	0161
TVS2SD794(2)P	0161
TVS2SD794(2)Q	0161
TVS2SD794-2	0161
TVS2SD794-2P	0161
TVS2SD794-2Q	0161
TVS2SD794-P	0161
TVS2SD794-Q	0161
TVS2SD794P	0161
TVS2SD794Q	0161
TVS2SD794R	0161
TVS2SD950	2668
TVS2SD950-Q	2668
TVS2SD950Q	2668
TVS3DL2	0023
TVS-5B-2-H5W	0015
TVS5B-2-H5W	0015
TVS-7C0.0911/10	1303
TVS8B2A	0071
TVS10D	0015
TVS10D1	0023
TVS10D2	0023
TVS-10D8	0023
TVS10D8	0023
TVS10DC4	0015
TVS10DC4R	0015
TVS10E1	0015
TVS10E2	0015
TVS10E4	0015
TVS15SB20R	0748
TVS25A103	0050
TVS25B172	0211
TVS25B448	0975
TVS25C208	0016
TVS25C562	0016
TVS25C645	0016
TVS30D1	0087
TVS45	OEM
TVS-82G	0917
TVS120Z	OEM
TVS-182G	0917
TVS-185D	0015
TVS185D	0015
TVS305	OEM
TVS310	OEM
TVS312	OEM
TVS315	OEM
TVS318	OEM
TVS324	OEM
TVS328	OEM
TVS340M	0755
TVS348	OEM
TVS360	OEM
TVS410	OEM
TVS420	OEM
TVS430	OEM
TVS505	OEM
TVS510	OEM
TVS512	OEM
TVS515	OEM
TVS518	OEM
TVS524	OEM
TVS528	OEM
TVS550	0015
TVS-828A	0016
TVS828A	0016
TVS-1303	0037
TVS-1850	0540
TVS2406M	0023
TVS3153GM1	3270
TVS25126F	0160
TVS81202	0015
TVSA81004	0735
TVSAFS01	0102
TVSAFSD1	0102
TVSAN155	0039
TVSAN225	0403
TVSAN227	2584
TVSAN230	2586
TVSAN241	4038
TVSAN5210	3787
TVSAW01-11	0317
TVSB01-02	0015
TVSB012-2	0133
TVSB0102	0609
TVSB0102V	0609
TVSB1-29	0441
TVSB24-06C	0023
TVSB24-06D	0102
TVSB1201	0015
TVSB1202	0609
TVSB1202V	0609
TVSB1204	0071
TVSB2404C	0023
TVSB2404D	0023
TVSB2406D	0102
TVSB2804D	0023
TVSB3015M	0017
TVSB4302	0023
TVSB4304	0023
TVSB4304V	0023
TVSB81004	0110
TVSB84009	2077
TVSBA236	OEM
TVSBA526	2787
TVSBAX-13	0124
TVSBAX13	0124
TVSBB2	0102
TVSBB2A	0023
TVSBB10	0182
TVSBB21	0023
TVSBN5040A	OEM
TVSBN5111	3738
TVSBN5115	OEM
TVSBN5210	3787
TVSBN5416	3853
TVSBN5416A	3853
TVSC0102FL7	0071
TVSC0102L7	0071
TVSC0410	0071
TVSC0508	0071
TVSC0510	0071
TVSC1255H	0086
TVSC2406M	0023
TVSC2615A	0017
TVSC2715M	0102
TVSC4702	0031
TVSC65010042	OEM
TVSCA3126	0521
TVSCA3137	2789
TVSCA3139GM1	2049
TVSCA3139GMI	2049
TVSCA3151EM1	OEM
TVSCA3153GM1	3270
TVSCA3153GMI	3270
TVSCN5310	2623
TVSCN5310L	2623
TVSCN5310M	2623
TVSCN5311	2623
TVSCN5311CL	2623
TVSCN5311KL	2623
TVSCN5411	5095
TVSC0510	0071
TVS-CS1255H	0086
TVSCS1255H	0086
TVS-CS1255HF	0791
TVSCS1255HF	0791
TVS-CS1256HG	0595
TVSCS1256HG	0791
TVS-CS1303	0037
TVSCSC1255H	0854
TVSCSC1255HF	0595
TVSCSC1256HG	0126
TVSD1M	0157
TVS-DG1NR	0182
TVS-DS-1K	0023
TVS-DS1K	0023
TVSDS1K	0023
TVS-DS-1M	0023
TVS-DS1M	0023
TVSDS-1M	0015
TVSDS1M	0023
TVS-DS2K	0015
TVSDS-2K	0015
TVSDS2K	0015
TVSDS2M	0023
TVSEA01-07R	0002
TVSEGA010-ST	0162
TVSEM1	0015
TVSEM1Z	0023
TVSEM1ZV	0023
TVSEN1364	1170
TVSEN11235	1429
TVSEN11238	3282
TVSEN11301	1170
TVSEN11401	3351
TVSEN11414	3359
TVSEN11436	2122
TVSEN11438	3089
TVSEQA01-05T	0041
TVSEQA01-06S	0091
TVSEQB01-12	0137
TVSEQB01-15	0002
TVSEQB01-15Z	0002
TVSERA81-004	0023
TVSERB-04D	0023
TVSERB-06B	0015
TVSERB12-02	0023
TVSERB24-04D	0023
TVSERB24-06	0023
TVSERB24-06A	0015
TVSERB24-06B	0023
TVSERC05-08	0071
TVSERC0508	0071
TVSERC0510	0071
TVSERC0510V	0071
TVSERZC10ZK241U	1993
TVSES1	0023
TVSES1Z	0023
TVSESA06	0296
TVS-ET1P	0015
TVSEU1Z	0023
TVSEU2	0023
TVSEU2A	0023
TVSEU2V	0023
TVSFG2PC	0102
TVS-FR1MD	5554
TVS-FR1N	0015
TVS-FR-1P	0023
TVS-FR1P	0023
TVSFR1P	0023
TVS-FR-1P(FR1P)	0015
TVS-FR1PC	0015
TVSFR1PC	0015
TVSFR2-005	0102
TVSFR2-02	0023
TVSFR2-02C	0015
TVSFR2-02G	0023
TVSFR2-04	0102
TVSFR2-04(SD-1)	0102
TVSFR2-06	0023
TVSFR2-10	0102
TVS-FR-2M	0015
TVS-FR2M	0015
TVS-FR2P	5405
TVS-FR2PC	0540
TVSFR2PC	0540
TVS-FR10	0015
TVSFR202	2079
TVSFT1M	0102
TVS-FT1N	0015
TVS-FT-1P	0102
TVS-FT1P	0102
TVSFT-1P	0102
TVSFT1P	0102
TVS-FT1PC	0102
TVS-FT10	0071
TVS-FU1N	0102
TVSFU1N	0102
TVSFUIN	0102
TVSG4011BCM	0215
TVSG65020054	OEM
TVSGH3F	0158
TVSGH3FLF	0158
TVSGN2050	OEM
TVSGN3002U	OEM
TVSGN3100E	OEM
TVSGN3210	OEM
TVSGN3500	OEM
TVSGN3600	OEM
TVSGP2-354	0133
TVSGRU2A	0102
TVSH20/1D	OEM
TVS-H-339W	0757
TVS-H-399W	0757
TVSHA-1	0102
TVSHA1151	3651
TVSHF-1	0102
TVSHF1	0102
TVSHFD1Z	0102
TVSHFSD-1	0102
TVSHFSD-1A	0102
TVSHFSD-1C	0102
TVS-HFSD1Z	0102
TVSHFSD-1Z	0102
TVSHFSD1Z	0102
TVS-HF-SD-12	0102
TVS-HS7/1	0774
TVSHS7/1	0774
TVS-HS25/1B	0378
TVS-HS25/16	0378
TVSHZ5C1	0248
TVSHZ5C1TD	0041
TVSHZ6A1	0883
TVSHZ7B2	0062
TVSHZ9A1	0165
TVSHZ11B1	0064
TVSHZ11C2	0181
TVSHZ11C3TE	0181
TVSHZ33S1	0166
TVSHZT33	3452
TVSHZT33S1	1329
TVSJA1200	0079
TVSJL41A	0015
TVSJL41AM	0015
TVSJN76664V	0167
TVS-K112C	0196
TVSK112C	0196
TVSKA1200	0007
TVSKB162D	0139
TVSKB265A	0133
TVSKB462F	3442
TVS-KC2-LP	0015
TVS-KC2CP12/1	0015
TVS-KC2DP12/2	0015
TVS-KC20P12/1	0015
TVSL78M12RL	0334
TVSL78N05	1292
TVSLA7530N	3817
TVSM5218L	1699
TVSM5218N	OEM
TVSM51247	3273
TVSM51247P	3273
TVSM51320P	OEM
TVSM51320P(M51320P)	OEM
TVSM51321P	OEM
TVSM51376SP	4336
TVSM53216P	1349
TVSM58479P	OEM
TVSM58725P	1898
TVSMA26	0139
TVS-MA26A	0139
TVS-MA242C	0015
TVSMB-1F	0182
TVSMB1F	0348
TVSMI-15R	1097
TVSMI15RC	0015
TVSMI-15S	1802
TVSMI15SC	0015
TVSMP0574J	1329
TVS-MPC23C	0528
TVSMPC574J	1329
TVSMPC595C	0852
TVSMPC596C	0857
TVSMPC596C2	0857
TVSMPC596C2B	0857
TVSMPC596CE	0857
TVSMPC596CZ	0857
TVSMPC1355C	0395
TVSMPS596C2	0857
TVSMR1C	0071
TVSMR-1M	0071
TVSN13T1	0316
TVSN76665	0167
TVS-OA70	0143
TVSOA71	0143
TVS-OA81	0143
TVS-OA90	0143
TVSOA90	0143
TVS-OA91	0019
TVS-OA95	0143
TVS-OV-02	0102
TVS-PC02P11/2	0015
TVSPC-713U	0315
TVS-PCD2P11/2	0015
TVSPCD2P11/2	0015
TVSPH302	2628
TVSPH302B	OEM
TVSPH309	OEM
TVSQ8112ZE	OEM
TVSQA01-06SB	0301
TVSQA01-07BE	0025
TVSQA01-07E	0002
TVSQA01-07RE	0025
TVSQA01-11SE	0317
TVSQA01-11.5E	0317
TVSQA01-12S	0137
TVSQA01-14RD	0100
TVSQA01-15S	0002
TVSQA01-25BA	0402
TVSQA01-25R	0402
TVSQA01-25RA	0402
TVSQA01-25RB	OEM
TVSQA105R	0041
TVSQA105RA	0589
TVSQA107RE	0025
TVSQA107SP	0062
TVSQA108SH	0165
TVSQA112R1	0137
TVSQA112RI	0137
TVSQA112RN-2	0052
TVSQA116R2	0445
TVSQA135R2	0032
TVSQA205AB	0140
TVSQA205C	0041
TVSQA205D	0041
TVSQA205E	0041
TVSQA205EV3	0041
TVSQA205F	0041
TVSQA205G	0041
TVSQA205GV3	0041
TVSQA206	0091
TVSQA206A	0091
TVSQA206AB	0253
TVSQA206B	0091
TVSQA206CD	0091
TVSQA206G	0295
TVSQA206M	0091
TVSQA206MV	0091
TVSQA206MV3	0091
TVSQA207D	0077
TVSQA207M	0062
TVSQA207M3	0062
TVSQA208	0165
TVSQA208C	0165
TVSQA208G	0165
TVSQA208GV3	0165
TVSQA208M	OEM
TVSQA209B	0057
TVSQA209C	0057
TVSQA209M	0057
TVSQA209M9	0057
TVSQA210D	0248
TVSQA211B	0181
TVSQA211N1	0181
TVSQA212B	0053
TVSQA213A	0053
TVSQA213B	0053
TVSQA213M2	0053
TVSQA225A	0714
TVSQA228B	0257
TVSQA230B	0195
TVSQA232M2	OEM
TVSQA233C	0166
TVSQA233CV	0166
TVSQA233CV3	0166
TVSQA233G	0010
TVSQA235AC	0714
TVSQA235C	0010
TVSQA235G	0010
TVSQB01-15ZB	0002
TVSQB01-18	0495
TVSQB105M	OEM
TVSQB105N	0162
TVSQB106P	0301
TVSQB107D	0025
TVSQB107S	0025
TVSQB107SV	0025
TVSQB109SA	6724
TVSQB110D	0317
TVSQB112	0137
TVSQB112ZE	0137
TVSQB115ZB	0002
TVSQB115ZBV	0002
TVSQB118	0495
TVSQB110D	0317
TVSRA1A	0133
TVSRA-1Z	0023
TVSRA1Z	0023
TVSRA-26	1834
TVSRA26	0502
TVSRC2	0348
TVSRD2R4E	2874
TVSRD2.7EB2	0760
TVSRD3R3EB	0299
TVSRD3R6EB2	0376
TVSRD3.0EB1	0118
TVSRD3.0EB1Z	0118
TVSRD3.0EB2	0118
TVSRD3.0EL1	0118
TVSRD3.0EL1Z	0118
TVSRD3.2EB1	OEM
TVSRD3.6EL	0376
TVSRD3.6EL1Z	0376
TVSRD4AM	0036
TVSRD4R7	0140
TVSRD4R7EB	0140
TVSRD4R7EB2	0140
TVSRD4.7EB2	0140
TVS-RD5A	0041
TVSRD5A	0041
TVSRD5AK	0041
TVSRD5R1EB2	0041
TVSRD5R1JB3	0041
TVSRD5.1EB	OEM
TVSRD5.1EB1	0041
TVSRD5.1EB1Z	0041
TVSRD5.1EB2	0041
TVSRD5.1EL2Z	0589
TVSRD5.6EB1	0041
TVSRD5.6EN3	0253
TVSRD5.6EN3Z	0253
TVSRD6A	0091
TVSRD6R8E	0062
TVSRD6R8EB	0062
TVSRD6.2EB	0471
TVSRD6.2EB1	0091
TVSRD6.2EB1Z	0091
TVSRD6.2EB3	0471
TVSRD6.2FB1	0636
TVSRD6.2FB2	0471
TVSRD6.8MB2X	0062
TVS-RD7A	0062
TVSRD7.5EB	0077
TVSRD7.5EBMV	0077
TVSRD8.2EB1	0077
TVSRD8.2EB2	0165
TVSRD9A1	OEM
TVSRD9AL	1083
TVSRD10FC	0170
TVS-RD11A	0181
TVSRD11A	0181
TVSRD11EB1	0064
TVSRD11EB3	0181
TVSRD12EB	1005
TVSRD12EB1	0052
TVSRD12EB1Z	0052
TVSRD12EB2M	0052
TVSRD12EBH	1005
TVSRD12EBM	0052
TVSRD12FB2	0137
TVSRD12FB3	0137
TVS-RD-13A	0053
TVS-RD13A	0053
TVSRD13AL	0052
TVSRD13EB2	0053
TVSRD13EB2Z	0053
TVSRD13EB3	0053
TVSRD13EL1	0053
TVSRD15EB2	0879
TVSRD15EB3	0686
TVSRD15FM	0002
TVSRD15FMZ	0002
TVSRD16EB2	0445
TVSRD22EB1	0705
TVSRD24EB1	0494
TVSRD27E	0455
TVSRD27EB	0441
TVSRD27EB4	0455
TVSRD27EB4TN	0455
TVSRD27EB4Z	0455
TVSRD27EBS	0441
TVSRD27RB4TN	OEM
TVS-RD29AN	0195
TVSRD33EB	0166
TVSRD36EB3	0010
TVSRD36EB3V	0010
TVSRD36EB3Z	0010
TVSRD47EB2	0140
TVSRF1	0023
TVSRF-1A	0023
TVSRF1A	0023
TVSRG2	0031
TVSRG4	0031
TVSRG4J5	0031
TVSRGP10J	0023
TVSRH-1	0023
TVSRH1	0023
TVSRH1M	0023
TVSRH1S	0023
TVSRH1SV	0023
TVSRH2F	0017
TVSRH2FM	0348
TVSRM1	0023
TVSRM1Z	0015
TVSRM1ZM	0015
TVSRM1ZMV	0023
TVSRM2C	0023
TVSRM10	0023
TVSRM10B	0071
TVSRM10M	0071
TVSRM10MV	0071
TVSRM11B	0071
TVSRM25	0502
TVSRM-26	2157
TVSRM26V	1268
TVSRMP5020	0015
TVSRP1A	0102
TVSRU-2	0023
TVSRU2	0023
TVSRU2AM	0023
TVSRU2AN	0023
TVSRU2M	0023
TVSRU3AM	0283
TVSRU3AN	0023
TVSRU3N	0031
TVSRVDFV211	0139
TVSRVDKB167	1925
TVS-S1B02-03C	0087
TVS-S1B02-03CR	0087
TVSS1P20	0102
TVSS1R8D	0015
TVSS1R20	0133
TVSS1R80	2528
TVSS1WB10	0241
TVSS1WBS10	0241
TVSS3-2	0015
TVSS3G4	0071
TVSS4C	0015
TVSS4LS221N	1239
TVSS4LS374N	0713
TVSS15	0102
TVSS-34	0102
TVSS34	0023
TVS-S82	0015
TVSS1854	OEM
TVSS6321FLC6	1992
TVSS6394GLC6	0755
TVSSA-2	0087
TVSSA-2B	0087
TVSSA2B	0015
TVSSA-2H	0071
TVSSA2H	0071
TVSSB-2	0102
TVSSB2-C	5313
TVS-SB-2C	0102
TVSSB-2C	5196
TVSSB2C	7041
TVSSB-2G	0102
TVSSB-2T	0102
TVS-SB-20	0071
TVS-SC15	0030
TVSSC15	0030
TVS-SD-1	0102
TVSSD1A	0102
TVS-SD-1B	0102
TVS-SD1B	0102
TVS-SD-1B(BOOST)	0015
TVSSD-1Y	0023
TVSSD-14	0143
TVS-SD82	0917
TVSSE303A	0516
TVSSEL1410G	OEM
TVS-SF-1	0348
TVSSF-1	0102
TVSSID30-15	5743
TVSSIR80	2528
TVSSJE5472-1	0276
TVSSN74LS09N	1642
TVSSN7403N	0335
TVSSN7409N	0492
TVSSN7666N	0167
TVSSN76002N	OEM
TVSSN76642N	0678
TVSSN76664N	0167
TVSSN76665	0167
TVS-SPN01	0015
TVSSR2K	0023
TVSSR2KL	OEM
TVSSR2KN	OEM
TVSSR2KN(SR2KN)	OEM
TVSSTK436TV	OEM
TVSSTK4362TV	OEM
TVSSTR371	OEM
TVSSTR380	3502
TVSSTR381	3503
TVSSTR381A	3503
TVSSTR3130	1641
TVSSTR3230	1641
TVSSTR30130	2364
TVSSV02	0015
TVSSV03	0120
TVSSV-04	0120
TVSSV-05	0015
TVSTA7676AP	OEM
TVSTA7717AP	OEM
TVS-TC009M11/10	0015
TVS-TC009M21/3	0484
TVS-TC0-09M21/3	0484
TVS-TC0.09M11/10	0015
TVS-TC-0.09M21/3	0196
TVS-TC0.09M21/3	0484
TVS-TC0.0911/10	1303
TVSTC4011BP	0215
TVSTC4052BP	0024
TVSTC4053BP	0035
TVSTC4066BP	0101
TVSTC4069UBP	0119
TVSTD15	0182
TVSTD-15H	0102
TVSTD15M	0102
TVSTDA1190Z	1170
TVSTFH340M	0755
TVSTFS340M	0755
TVSTFS340MX	0755
TVSTFS340MY	0755
TVSTLN105A	OEM
TVSTLR321	OEM
TVSTMS1943A2	2295

If replacement code is OEM, contact original manufacturer for replacement.

DEVICE TYPE	REPL CODE
TVSTMS1943N2	2295
TVSTMSI943A2	OEM
TVSUA67	0701
TVSUC78M12H2	0334
TVSUF2	0023
TVSUF-2B	0017
TVS-UFSD-1	0102
TVSUFSD-1	0015
TVSUFSD1	0102
TVSUFSD-1F	0015
TVSUFSD1F	0015
TVSUP574J	1329
TVSUPC23	0528
TVS-UPC23C	0528
TVSUPC23C	0528
TVSUPC78M-12H2	0334
TVSUPC78M12H2	0334
TVSUPC339C	0176
TVSUPC574J	1329
TVSUPC595C	0852
TVSUPC596C2	0857
TVSUPC1363CA	1945
TVSUPC1380C	1055
TVSUPD4011BC	0215
TVSUPD4013BC	0413
TVSUPD4017BC	0513
TVSUPD4027BC	1949
TVSUPD4052BC	0024
TVSUPD4053BC	0035
TVSUPD4066BC	0101
TVSUZ6R8B	0025
TVSUZ18BS	0375
TVSVD1221L	0120
TVSVFSD1	0102
TVSW04	0289
TVSW04-M	0106
TVSW04M	0106
TVSW270	0455
TVSWF2	0325
TVSWZ061	0091
TVSWZ270	0455
TVSYZ065F	0025
TVSYZ-080	0244
TVSYZ080	0244
TVSZB	0794
TVSZB1-6	0301
TVS-ZB1-11	0794
TVSZB1-11	0794
TVS-ZB1-15	0002
TVSZB1-15	0002
TVSZB1-27	0441
TVSZB1-29	0441
TVSZB-29	0441
TW3	0609
TW5	1355
TW6N100CZ	0154
TW6N100HZ	OEM
TW6N200CZ	0154
TW6N200HZ	OEM
TW6N300CZ	0147
TW6N300HZ	OEM
TW6N400CZ	0147
TW6N400HZ	OEM
TW6N500CZ	0279
TW6N500HZ	OEM
TW6N600CZ	0279
TW6N600HZ	OEM
TW6N700CZ	OEM
TW6N700HZ	OEM
TW7N400FZ	OEM
TW7N600FZ	OEM
TW7N700FZ	OEM
TW7N800FZ	OEM
TW8N100CZ	3148
TW8N100HZ	OEM
TW8N200CZ	3150
TW8N200HZ	OEM
TW8N300CZ	3152
TW8N300HZ	OEM
TW8N400CZ	3152
TW8N400HZ	OEM
TW8N500CZ	2018
TW8N500HZ	OEM
TW8N600CZ	6230
TW8N600HZ	OEM
TW8N700CZ	OEM
TW8N700HZ	OEM
TW9N400FZ	OEM
TW9N600FZ	OEM
TW9N700FZ	OEM
TW9N800FZ	OEM
TW10	0015
TW10N100CZ	3148
TW10N100HZ	OEM
TW10N200CZ	3150
TW10N200HZ	OEM
TW10N300CZ	3152
TW10N300HZ	OEM
TW10N400CZ	3152
TW10N400HZ	OEM
TW10N500CZ	4100
TW10N500HZ	OEM
TW10N600CZ	2018
TW10N600HZ	OEM
TW10N700CZ	OEM
TW10N700HZ	OEM
TW11N400FZ	OEM
TW11N600FZ	OEM
TW11N700FZ	OEM
TW11N800FZ	OEM
TW12N400CX	2017
TW12N400HX	OEM
TW12N600CX	2018
TW12N600HX	OEM
TW12N800CX	OEM
TW12N800HX	OEM
TW12N1000CX	OEM
TW12N1000HX	OEM
TW12N1100CX	OEM
TW12N1100HX	OEM
TW18N400CT	OEM
TW18N500CT	OEM
TW18N600CT	OEM
TW18N700CT	OEM
TW18N800CT	OEM
TW18N900CT	OEM
TW18N1000CT	OEM
TW18N1100CT	OEM
TW20	1355
TW25N400CT	OEM
TW25N500CT	OEM
TW25N600CT	OEM
TW25N700CT	OEM
TW25N800CT	OEM
TW25N900CT	OEM
TW25N1000CT	OEM
TW25N1100CT	OEM
TW30	1355
TW40	1355
TW50	1355
TW60	1355
TW-80	0071
TW80	1355
TW-100	0071
TW100	0071
TW120	OEM
TW135	0150
TWC5	OEM
TWC5A	OEM
TWC5B	OEM
TWC5C	OEM
TWC14	OEM
TWC14A	OEM
TWC14B	OEM
TWC14C	OEM
TWC18	OEM
TWJ30	OEM
TWLC200	OEM
TWS6	OEM
TWS7	OEM
TWS10/7642	OEM
TWS12	OEM
TWS17	OEM
TWS36	OEM
TWV	0609
TWX8	OEM
TWX16	OEM
TWX19	OEM
TWX22	OEM
TWX34	OEM
TX-0018	OEM
TX1N645	0015
TX1N647	0015
TX1N3190	0015
TX1N3191	0015
TX27A	0441
TX75	2312
TX-100-1	0086
TX100-1	0016
TX-100-2	0086
TX100-2	0016
TX-100-3	0142
TX100-3	0016
TX-100-4	0233
TX100-4	0086
TX-100-5	0142
TX100-5	0142
TX-101-8	0086
TX101-8	0142
TX-101-9	0086
TX101-9	0086
TX-101-11	0142
TX101-11	0142
TX-101-12	0016
TX101-12	0016
TX-102-1	0079
TX102-1	0016
TX102-2	0016
TX-102-4	0142
TX102-4	0142
TX-103-1	0160
TX103-1	0085
TX-104-3	0164
TX104-3	0164
TX-106-1	0164
TX106-1	0004
TX-107-1	0016
TX107-1	0016
TX-107-3	0127
TX107-3	0016
TX-107-4	0016
TX107-4	0016
TX-107-5	0016
TX107-5	0016
TX-107-6	0016
TX107-6	0016
TX-107-10	0016
TX107-10	0016
TX-107-12	0127
TX107-12	0016
TX-107-13	0142
TX107-13	0142
TX-107-16	0016
TX107-16	0016
TX-108-1	0079
TX108-1	0016
TX-111-1	0142
TX111-1	0142
TX-112-1	0016
TX112-1	0016
TX116-1	OEM
TX116-2	OEM
TX116-3	OEM
TX-119-1	0016
TX120	0127
TX121	0037
TX122	0037
TX-122-1	0037
TX-123	0400
TX123-1	0400
TX-124	0086
TX124	0086
TX124-1	0016
TX-125	0126
TX125	0037
TX125-1	0126
TX-126-1	0208
TX-128-1	0016
TX128-1	0086
TX-134-1	0037
TX134-1	0126
TX-135	0086
TX135	0086
TX-136	0126
TX136	0126
TX-138	0016
TX138	0086
TX-139	0126
TX139	0126
TX-140	0086
TX140	0086
TX-141	0016
TX141	0127
TX-145	1396
TX-183S	0074
TX287	OEM
TX-429D	OEM
TX827	OEM
TX-1005	0178
TX1005	0142
TXAL116B	0555
TXAL116C	0555
TXAL116G	OEM
TXAL116M	0555
TXAL118B	0555
TXAL118C	0555
TXAL118G	OEM
TXAL118M	0555
TXAL226B	0555
TXAL226C	0555
TXAL226G	OEM
TXAL226M	0555
TXAL228B	0555
TXAL228C	0555
TXAL228G	OEM
TXAL228M	0555
TXAL386B	OEM
TXAL386C	OEM
TXAL386M	OEM
TXAL388B	OEM
TXAL388C	OEM
TXAL388M	OEM
TXAL606B	OEM
TXAL606C	OEM
TXAL606G	OEM
TXAL606M	OEM
TXAL608B	OEM
TXAL608C	OEM
TXAL608G	OEM
TXAL608M	OEM
TXAL610B	OEM
TXAL610C	OEM
TXAL610G	OEM
TXAL610M	OEM
TXAL615B	OEM
TXAL615C	OEM
TXAL615G	OEM
TXAL615M	OEM
TXAL1110B	0555
TXAL1110C	0555
TXAL1110G	OEM
TXAL1110M	0555
TXAL1115B	OEM
TXAL1115C	OEM
TXAL1115G	OEM
TXAL1115M	OEM
TXAL2210B	0555
TXAL2210C	0555
TXAL2210G	OEM
TXAL2210M	0555
TXAL2215B	0763
TXAL2215C	OEM
TXAL2215G	OEM
TXAL2215M	0758
TXAL3810B	OEM
TXAL3810C	OEM
TXAL3810M	OEM
TXAL3815B	OEM
TXAL3815M	OEM
TXB3518MA	0037
TXC0063	OEM
TXC01A10	OEM
TXC01A20	OEM
TXC01A40	OEM
TXC01A50	OEM
TXC01A60	OEM
TXC01C60	OEM
TXC02A10	OEM
TXC02A20	OEM
TXC02A40	OEM
TXC02A50	OEM
TXC02A60	OEM
TXC03A10	OEM
TXC03A20	OEM
TXC03A40	OEM
TXC03A50	OEM
TXC03A60	OEM
TXC10K80	0875
TXC30L70	OEM
TXC38D60	OEM
TXC38E40	OEM
TXC38G70	OEM
TXC38G80	OEM
TXC38H40	OEM
TXC38H60	OEM
TXC39D60	OEM
TXC39E40	OEM
TXC39E60	OEM
TXC39G80	OEM
TXC39H40	OEM
TXC39H60	OEM
TXD10K50M	0617
TXD98A20	OEM
TXD98A40	OEM
TXD98A50	OEM
TXD98A60	OEM
TXD99A20	OEM
TXD99A40	OEM
TXD99A50	OEM
TXD99A60	OEM
TXDV208	OEM
TXDV408	OEM
TXDV608	OEM
TXDV808	OEM
TXE99A20	OEM
TXE99A40	OEM
TXE99A50	OEM
TXE99A60	OEM
TXED453C025	OEM
TXED453C050	OEM
TXED453C100	OEM
TXED455	OEM
TXED456	OEM
TXED457	OEM
TXEF402	OEM
TXEF402M001	OEM
TXEF402M003	OEM
TXEF402M006	OEM
TXEF402M010	OEM
TXEF402M020	OEM
TXEF402M030	OEM
TXEF402M040	OEM
TXEF402M050	OEM
TXES37	OEM
TXES475C025	OEM
TXES475C050	OEM
TXES475C100	OEM
TXES476C025	OEM
TXES476C050	OEM
TXES476C100	OEM
TXES478	OEM
TXES479	OEM
TXES480	OEM
TXES481	OEM
TXES482	OEM
TXES483	OEM
TXES488	OEM
TXES489	OEM
TXES490	OEM
TXES491	OEM
TXES492	OEM
TXES493	OEM
TXL-80	2312
TXL-100	2312
TXN0510	1004
TXN108(A)	OEM
TXN110	1004
TXN210	1004
TXN410	1034
TXN610	1148
TXN810	OEM
TXN1010	OEM
TXS607-8(A)	OEM
TXZ4206	0330
TY2SA933	0148
TY2SA1198E	0693
TY2SA1198S	0693
TY2SB822Q	4104
TY2SC1652(Q)	2471
TY2SC1652Q	2471
TY2SC1652R	2471
TY2SC1740	0151
TY105F	OEM
TY-107-4	0016
TY-107-12	0016
TY120SP8K2	OEM
TY120SP10K2	OEM
TY120SP12K2	OEM
TY120SP14K2	OEM
TY120SP16K2	OEM
TY120SP18K2	OEM
TY180NP2K2	OEM
TY180NP3K2	OEM
TY180NP4K2	OEM
TY180NP5K2	OEM
TY180NP6K2	OEM
TY180NP7K2	OEM
TY180NP8K2	OEM
TY180QP2K2	OEM
TY180QP3K2	OEM
TY180QP4K2	OEM
TY180QP5K2	OEM
TY180QP6K2	OEM
TY180QP7K2	OEM
TY180QP8K2	OEM
TY290SP8K2	OEM
TY290SP10K2	OEM
TY290SP12K2	OEM
TY290SP14K2	OEM
TY290SP16K2	OEM
TY290SP18K2	OEM
TY490SP8K2	OEM
TY490SP10K2	OEM
TY490SP12K2	OEM
TY490SP14K2	OEM
TY490SP16K2	OEM
TY490SP18K2	OEM
TY500SK2P8	OEM
TY500SK2P10	OEM
TY500SK2P12	OEM
TY500SK2P14	OEM
TY500SK2P16	OEM
TY500SK2P18	OEM
TY500SP8K2	OEM
TY500SP10K2	OEM
TY500SP12K2	OEM
TY500SP14K2	OEM
TY500SP16K2	OEM
TY500SP18K2	OEM
TY504	2095
TY505F	OEM
TY507	OEM
TY508	2095
TY508FA	OEM
TY510	2521
TY718	OEM
TY1004	2089
TY1007	OEM
TY1008	2089
TY1010	0397
TY2004	0505
TY2005F	OEM
TY2007	OEM
TY2008	0505
TY2008FA	OEM
TY2010	0397
TY2010GH	OEM
TY3004	0710
TY3005F	OEM
TY3007	OEM
TY3008	0710
TY3010	0611
TY4004	0710
TY4005F	OEM
TY4007	OEM
TY4008	0710
TY4008FA	OEM
TY4010	0611
TY5004	0863
TY5005F	OEM
TY5007	OEM
TY5008	0863
TY5010	0952
TY6004	0863
TY6005F	OEM
TY6007	OEM
TY6008	0863
TY6008FA	OEM
TY6010	0952
TY8008	OEM
TY8010	OEM
TYA290SP8K2	OEM
TYA290SP10K2	OEM
TYA290SP12K2	OEM
TYA290SP14K2	OEM
TYA290SP16K2	OEM
TYA290SP18K2	OEM
TYA490SP8K2	OEM
TYA490SP10K2	OEM
TYA490SP12K2	OEM
TYA490SP14K2	OEM
TYA490SP16K2	OEM
TYA490SP18K2	OEM
TYAL113B	OEM
TYAL113C	OEM
TYAL113G	OEM
TYAL113M	OEM
TYAL114B	OEM
TYAL114C	OEM
TYAL114M	OEM
TYAL116C	0555
TYAL116G	OEM
TYAL116M	0555
TYAL118B	0555
TYAL118C	0555
TYAL118G	OEM
TYAL118M	0555
TYAL223B	OEM
TYAL223C	OEM
TYAL223G	OEM
TYAL223M	OEM
TYAL226B	0555
TYAL226C	0555
TYAL226G	OEM
TYAL226M	0555
TYAL228B	0555
TYAL228C	0555
TYAL228G	OEM
TYAL228M	0555
TYAL383B	OEM
TYAL383C	OEM
TYAL383M	OEM
TYAL386B	OEM
TYAL386C	OEM
TYAL386M	OEM
TYAL388B	OEM
TYAL388C	OEM
TYAL388M	OEM
TYAL603B	OEM
TYAL603C	OEM
TYAL603G	OEM
TYAL603M	OEM
TYAL606B	OEM
TYAL606C	OEM
TYAL606G	OEM
TYAL606M	OEM
TYAL608B	OEM
TYAL608C	OEM
TYAL608G	OEM
TYAL608M	OEM
TYAL610B	OEM
TYAL610C	OEM
TYAL610G	OEM
TYAL610M	OEM
TYAL615B	OEM
TYAL615C	OEM
TYAL615G	OEM
TYAL615M	OEM
TYAL1110B	0555
TYAL1110C	0555
TYAL1110G	OEM
TYAL1110M	0555
TYAL1115B	OEM
TYAL1115C	OEM
TYAL1115G	OEM
TYAL1115M	OEM
TYAL2210B	0555
TYAL2210C	0555
TYAL2210G	OEM
TYAL2210M	0555
TYAL2215B	OEM
TYAL2215C	OEM
TYAL2215G	OEM
TYAL2215M	OEM
TYAL3810B	OEM
TYAL3810C	OEM
TYAL3810M	OEM
TYAL3815B	OEM
TYAL3815M	OEM
TYC300SK2P8	OEM
TYC300SK2P10	OEM
TYC300SK2P12	OEM
TYC300SK2P14	OEM
TYC300SK2P16	OEM
TYC300SK2P18	OEM
TYDTA124	1032
TYDTA124N	1032
TYDTA124S	1032
TYDTA144	4130
TYDTA144S	4130
TYDTC114ES	0832
TYDTC114Y	OEM
TYDTC124	0887
TYDTC124N	0887
TYDTC124S	0887
TYDTC144N	0898
TYDTC144S	0898
TYK8820	OEM
TYN058	2095
TYN0510	6900
TYN104(A)	OEM
TYN110	6413
TYN204(A)	OEM
TYN208	0505
TYN210	6413
TYN404(A)	OEM
TYN408(A)	OEM
TYN410	0710
TYN604(A)	OEM
TYN608(A)	OEM
TYN610	6444
TYN804(A)	OEM
TYN808(A)	OEM
TYN810	6453
TYN1004(A)	OEM
TYN1008(A)	OEM
TYN1010	OEM
TYP212	OEM
TYP512	OEM
TYP1012	OEM
TYP2012	OEM
TYPD0312UAR	OEM
TYS406-05(A)	OEM
TYS407-05(A)	OEM
TZ4A326	0023
TZ4C254	0015
TZ4C284	0023
TZ4E254	0015
TZ4F339-4	OEM
TZ4.7	0140
TZ5.1	0041
TZ5.6	0253
TZ6	0160
TZ6.2	0471
TZ6.8	0062
TZ6.8A	0025
TZ6.8B	0025
TZ6.8C	0025
TZ6.8D	OEM
TZ7	0160
TZ7.5	0077
TZ7.5A	0649
TZ7.5B	0649
TZ7.5C	0649
TZ7.5D	OEM
TZ-8	0160
TZ8	0160
TZ8.2	0165
TZ8.2A	0244
TZ8.2B	0244
TZ8.2C	0244
TZ8.2D	3100
TZ9	0012
TZ9.1	0057
TZ9.1A	0012
TZ9.1B	0012
TZ9.1C	0012
TZ9.1D	0012
TZ10	0064
TZ10A	0170
TZ10B	0170
TZ10C	0170
TZ10D	OEM
TZ11	0181
TZ11A	0317
TZ11B	0317
TZ11C	0317
TZ11D	OEM
TZ12	0052
TZ12A	0137
TZ12B	0137
TZ12C	0137
TZ12D	OEM
TZ13	0053
TZ13A	0365
TZ13B	0365
TZ13C	0365
TZ13D	OEM
TZ14	0100
TZ14A	0100
TZ14B	0100
TZ14C	0100
TZ14D	OEM
TZ15	0686
TZ15A	0002
TZ15B	0002
TZ15C	0002
TZ15D	OEM
TZ16	0445
TZ16A	0420
TZ16B	0420
TZ16C	0420
TZ16D	OEM
TZ17	1649
TZ17A	1649
TZ17B	1649
TZ17C	1649
TZ17D	OEM
TZ18	0375
TZ18A	0495
TZ18B	0495
TZ18C	0495
TZ18D	OEM
TZ19	0949
TZ19A	0949
TZ19B	0949
TZ19C	0949
TZ19D	OEM
TZ20	0700
TZ20A	0531
TZ20B	0531
TZ20C	0531
TZ20D	OEM
TZ22	0705
TZ22A	0565
TZ22B	0565
TZ22C	0565
TZ22D	OEM
TZ24	0494
TZ24A	0402
TZ24B	0402
TZ24C	0402
TZ24D	OEM
TZ25	1606
TZ25A	1606
TZ25B	1606
TZ25C	1606
TZ25D	OEM
TZ26	0441
TZ26A	0441
TZ27	0441
TZ27A	0441
TZ27AA	0441

DEVICE TYPE	REPL CODE
TZ27B	0441
TZ27C	0441
TZ27D	OEM
TZ28	1674
TZ28A	1674
TZ30	0726
TZ30A	0726
TZ30B	0726
TZ30C	0726
TZ30D	OEM
TZ33	0039
TZ33A	0039
TZ33B	0039
TZ33C	0039
TZ33D	OEM
TZ36	0820
TZ36A	0820
TZ36B	0820
TZ36C	0820
TZ36D	OEM
TZ39	0350
TZ39A	0350
TZ39B	0350
TZ39C	0350
TZ39D	OEM
TZ40	OEM
TZ40A	0350
TZ43	0931
TZ43A	0931
TZ43B	0931
TZ43C	0931
TZ43D	OEM
TZ45	0999
TZ45A	0999
TZ45B	0999
TZ45C	0999
TZ45D	OEM
TZ47	0999
TZ47A	0999
TZ47B	0999
TZ47C	0999
TZ47D	OEM
TZ48	0999
TZ48A	0999
TZ50	0502
TZ50A	0502
TZ50B	0502
TZ50C	0502
TZ50D	OEM
TZ51	0502
TZ51A	0502
TZ51B	0502
TZ51C	0502
TZ51D	OEM
TZ52	0502
TZ52A	0502
TZ52B	0502
TZ52C	0502
TZ52D	OEM
TZ54	0869
TZ54A	0869
TZ56	0869
TZ56A	0869
TZ56B	0869
TZ56C	0869
TZ56D	0869
TZ58	1156
TZ58A	1156
TZ60	1156
TZ60A	1156
TZ62	0783
TZ62A	0783
TZ62B	0783
TZ62C	0783
TZ62D	OEM
TZ64	0783
TZ64A	0783
TZ68	2157
TZ68A	2157
TZ68B	2157
TZ68C	2157
TZ68D	OEM
TZ75	1189
TZ75A	1189
TZ75B	1189
TZ75C	1189
TZ75D	OEM
TZ81	0111
TZ82	0111
TZ82A	0331
TZ82B	0444
TZ82C	0331
TZ82D	OEM
TZ85	3026
TZ85A	3026
TZ90	1311
TZ90A	1311
TZ91	1311
TZ91A	1311
TZ91B	1311
TZ91C	1311
TZ91D	OEM
TZ100	0098
TZ100A	0098
TZ100B	0098
TZ100C	0098
TZ100D	OEM
TZ105	0149
TZ105A	0149
TZ105B	0149
TZ105C	0149
TZ105D	OEM
TZ110	0149
TZ110A	0149
TZ110B	0149
TZ110C	0149
TZ110D	OEM
TZ120	0186
TZ120A	0186
TZ120B	0186
TZ120C	0186
TZ120D	OEM
TZ130	0213
TZ130A	0213
TZ130B	0213
TZ130C	0213
TZ130D	OEM
TZ140	0245
TZ140A	0245
TZ140B	0245
TZ140C	0245
TZ140D	OEM
TZ150	0028
TZ150A	0028
TZ150B	0028
TZ150C	0028
TZ150D	OEM
TZ160	0255
TZ160A	0255
TZ160B	0255
TZ160C	0255
TZ160D	OEM
TZ170	0877
TZ170A	0877
TZ175	0367
TZ175A	0367
TZ175B	0367
TZ175C	0367
TZ175D	OEM
TZ180	0367
TZ180A	0367
TZ180B	0367
TZ180C	0367
TZ180D	OEM
TZ200	0421
TZ200A	0421
TZ200B	0421
TZ200C	OEM
TZ200D	OEM
TZ220	OEM
TZ220A	OEM
TZ220B	OEM
TZ220C	OEM
TZ220D	OEM
TZ240	OEM
TZ240A	OEM
TZ260	OEM
TZ260A	OEM
TZ270	OEM
TZ270A	OEM
TZ275	OEM
TZ275A	OEM
TZ278	OEM
TZ278A	OEM
TZ280	OEM
TZ280A	OEM
TZ543	0037
TZ-551	0037
TZ551	0037
TZ-552	0791
TZ552	0037
TZ-553	0791
TZ553	0791
TZ554	0791
TZ581	1367
TZ582	1367
TZ966-03	0042
TZ990	OEM
TZ992-01	OEM
TZ1029-01	0065
TZ1034	0285
TZ1037	0285
TZ1088	OEM
TZ1151	0111
TZ1152	0111
TZ1153	0133
TZ1160	0016
TZ1180	0016
TZ1182	0016
TZ2001	OEM
TZ2002	OEM
TZ2003	OEM
TZ7000	0086
TZ7001	0086
TZ7002	0079
TZ7003	0079
TZ7500	0037
TZ7501	0037
TZ7502	OEM
TZ7503	0037
TZB6.8A	0025
TZB6.8B	0025
TZB7.5A	0649
TZB7.5B	0649
TZB8.2A	0244
TZB8.2B	0244
TZB9.1A	0012
TZB9.1B	0012
TZB10A	0170
TZB10B	0170
TZB11A	0317
TZB11B	0317
TZB12A	0137
TZB12B	0137
TZB13A	0365
TZB13B	0365
TZB15A	0002
TZB15B	0002
TZB16A	0420
TZB16B	0420
TZB18A	0495
TZB18B	0495
TZB20A	0531
TZB20B	0531
TZB22A	0565
TZB22B	0565
TZB24A	0402
TZB24B	0402
TZB27A	0441
TZB27B	0441
TZB30A	0726
TZB30B	0726
TZB33A	0039
TZB33B	0039
TZB36A	0820
TZB36B	0820
TZB39A	0350
TZB39B	0350
TZB43A	0931
TZB43B	0931
TZB47A	0999
TZB47B	0999
TZB51A	0502
TZB51B	0502
TZB56A	0869
TZB56B	0869
TZB62A	0783
TZB62B	0783
TZB68A	2157
TZB68B	2157
TZB75A	1189
TZB75B	1189
TZB82A	0331
TZB82B	0331
TZB91A	1311
TZB91B	1311
TZB100A	0098
TZB100B	0098
TZB110A	0149
TZB110B	0149
TZB120A	0186
TZB120B	0186
TZB130A	0213
TZB130B	0213
TZB150A	0028
TZB150B	0028
TZB160A	0255
TZB160B	0255
TZB170A	0877
TZB170B	0877
TZB180A	0367
TZB180B	0367
TZB200A	0421
TZB200B	0421
TZB220A	OEM
TZB220B	OEM
TZB250A	OEM
TZB250B	OEM
TZB300A	OEM
TZB300B	OEM
TZB350A	OEM
TZB350B	OEM
TZB400A	OEM
TZB400B	OEM
TZC6.8	0205
TZC6.8A	0205
TZC6.8B	0205
TZC6.8C	OEM
TZC6.8D	OEM
TZC6.8E	OEM
TZC7.5	0480
TZC7.5A	0480
TZC7.5B	0480
TZC7.5C	OEM
TZC7.5D	OEM
TZC7.5E	OEM
TZC8.2	0504
TZC8.2A	0504
TZC8.2B	0504
TZC8.2C	OEM
TZC8.2D	OEM
TZC8.2E	OEM
TZC9.1	0684
TZC9.1A	0684
TZC9.1B	0684
TZC9.1C	OEM
TZC9.1D	OEM
TZC9.1E	OEM
TZC10	0225
TZC10A	0225
TZC10B	0225
TZC10C	OEM
TZC10D	OEM
TZC10E	OEM
TZC11	0230
TZC11A	0230
TZC11B	0230
TZC11C	OEM
TZC11D	OEM
TZC11E	OEM
TZC12	0234
TZC12A	0234
TZC12B	0234
TZC12C	OEM
TZC12D	OEM
TZC12E	OEM
TZC13	0237
TZC13A	0237
TZC13B	0237
TZC13C	OEM
TZC13D	OEM
TZC13E	OEM
TZC14	1397
TZC14A	1397
TZC14B	1397
TZC14C	OEM
TZC14D	OEM
TZC14E	OEM
TZC15	0247
TZC15A	0247
TZC15B	0247
TZC15C	OEM
TZC15D	OEM
TZC15E	OEM
TZC17	1178
TZC17A	1178
TZC17B	1255
TZC17C	OEM
TZC17D	OEM
TZC17E	OEM
TZC18	0256
TZC18A	0256
TZC18B	0256
TZC18C	OEM
TZC18D	OEM
TZC18E	OEM
TZC19	2395
TZC19A	2395
TZC19B	2395
TZC19C	OEM
TZC19D	OEM
TZC19E	OEM
TZC20	0262
TZC20A	0262
TZC20B	0262
TZC20C	OEM
TZC20D	OEM
TZC20E	OEM
TZC22	0269
TZC22A	0269
TZC22B	0269
TZC22C	OEM
TZC22D	OEM
TZC22E	OEM
TZC24	0274
TZC24A	0274
TZC24B	0274
TZC24C	OEM
TZC24D	OEM
TZC24E	OEM
TZC25	2399
TZC25A	2399
TZC25B	2399
TZC25C	OEM
TZC25D	OEM
TZC25E	OEM
TZC27	0294
TZC27A	0294
TZC27B	0294
TZC27C	OEM
TZC27D	OEM
TZC27E	OEM
TZC30	0309
TZC30A	0309
TZC30B	0309
TZC30C	OEM
TZC30D	OEM
TZC30E	OEM
TZC33	0318
TZC33A	0318
TZC33B	0318
TZC33C	OEM
TZC33D	OEM
TZC33E	OEM
TZC36	0320
TZC36A	0320
TZC36B	0320
TZC36C	OEM
TZC36D	OEM
TZC36E	OEM
TZC39	0326
TZC39A	0326
TZC39B	0326
TZC39C	OEM
TZC39D	OEM
TZC39E	OEM
TZC43	0337
TZC43A	0337
TZC43B	0337
TZC43C	OEM
TZC43D	0337
TZC43E	OEM
TZC45	0347
TZC45A	0347
TZC45B	0347
TZC45C	OEM
TZC45D	OEM
TZC45E	OEM
TZC47	0347
TZC47A	0347
TZC47B	0347
TZC47C	OEM
TZC47D	OEM
TZC47E	OEM
TZC50	0027
TZC50A	0027
TZC50B	0027
TZC50C	OEM
TZC50D	OEM
TZC50E	OEM
TZC51	0027
TZC51A	0027
TZC51B	0027
TZC51C	OEM
TZC51D	OEM
TZC51E	OEM
TZC52	0027
TZC52A	0027
TZC52B	0027
TZC52C	OEM
TZC52D	OEM
TZC52E	OEM
TZC56	0266
TZC56A	0266
TZC56B	0266
TZC56C	OEM
TZC56D	OEM
TZC56E	OEM
TZC62	0386
TZC62A	0386
TZC62B	0386
TZC62C	OEM
TZC62D	OEM
TZC62E	OEM
TZC63C	OEM
TZC68	0405
TZC68A	0405
TZC68B	0405
TZC68C	OEM
TZC68D	OEM
TZC68E	OEM
TZC75	0426
TZC75A	0426
TZC75B	0426
TZC75C	OEM
TZC75D	OEM
TZC75E	OEM
TZC82	0444
TZC82A	0444
TZC82B	0444
TZC82C	OEM
TZC82D	OEM
TZC82E	OEM
TZC91	0238
TZC91A	0238
TZC91B	0238
TZC91C	OEM
TZC91D	OEM
TZC91E	OEM
TZC100	1180
TZC100A	1180
TZC100B	1180
TZC100C	OEM
TZC100D	OEM
TZC100E	OEM
TZC105	1190
TZC105A	1190
TZC105B	1190
TZC105C	OEM
TZC105D	OEM
TZC105E	OEM
TZC110	1190
TZC110A	1190
TZC110B	1190
TZC110C	OEM
TZC110D	OEM
TZC110E	OEM
TZC120	1207
TZC120A	1207
TZC120B	1207
TZC120C	OEM
TZC120D	OEM
TZC120E	OEM
TZC130	1218
TZC130A	1218
TZC130B	1218
TZC130C	OEM
TZC130D	OEM
TZC130E	OEM
TZC140	1881
TZC140A	1881
TZC140B	1881
TZC140C	OEM
TZC140D	OEM
TZC140E	OEM
TZC150	0647
TZC150A	0647
TZC150B	0647
TZC150C	OEM
TZC150D	OEM
TZC150E	OEM
TZC160	1255
TZC160A	1255
TZC160B	1255
TZC160C	OEM
TZC160D	OEM
TZC160E	OEM
TZC180	1279
TZC180A	1279
TZC180B	1279
TZC180C	OEM
TZC180D	OEM
TZC180E	OEM
TZC200	0605
TZC200A	0605
TZC200B	0605
TZC200C	OEM
TZC200D	OEM
TZC200E	OEM
TZP12A	OEM
TZPD3.9	0036
TZPD4.3	0275
TZPD4.7	0140
TZPD5.1	0041
TZPY3.9	0036
TZPY4.3	0275
TZPY4.7	0140
TZPY5.1	0041
TZV6.2	OEM
TZV6.2A	OEM
TZV6.8	OEM
TZV6.8A	OEM
TZV7.5	OEM
TZV7.5A	OEM
TZV8.2	OEM
TZV8.2A	OEM
TZV9.1	OEM
TZV9.1A	OEM
TZV10	OEM
TZV10A	OEM
TZV10.A	OEM
TZV11	OEM
TZV11A	OEM
TZV11.A	OEM
TZV12	OEM
TZV12A	OEM
TZV12.A	OEM
TZV13	OEM
TZV13A	OEM
TZV13.A	OEM
TZV15	OEM
TZV15A	OEM
TZV15.A	OEM
TZV16	OEM
TZV16A	OEM
TZV16.A	OEM
TZV18	OEM
TZV18A	OEM
TZV18.A	OEM
TZV20	OEM
TZV20A	OEM
TZV20.A	OEM
TZV22	OEM
TZV22A	OEM
TZV22.A	OEM
TZV24	OEM
TZV24A	OEM
TZV24.A	OEM
TZV27	OEM
TZV27A	OEM
TZV27.A	OEM
TZV30	OEM
TZV30A	OEM
TZV30.A	OEM
TZV33	OEM
TZV33A	OEM
TZV33.A	OEM
TZV36	OEM
TZV36A	OEM
TZV36.A	OEM
TZV39	OEM
TZV39A	OEM
TZV39.A	OEM
TZV43	OEM
TZV43A	OEM
TZV43.A	OEM
TZV47	OEM
TZV47A	OEM
TZV47.A	OEM
TZV51	OEM
TZV51A	OEM
TZV51.A	OEM
TZV56	OEM
TZV56A	OEM
TZV56.A	OEM
TZV62	OEM
TZV62A	OEM
TZV62.A	OEM
TZV68	OEM
TZV68A	OEM
TZV68.A	OEM
TZV75	OEM
TZV75A	OEM
TZV75.A	OEM
TZV82	OEM
TZV82A	OEM
TZV82.A	OEM
TZV91	OEM
TZV91A	OEM
TZV91.A	OEM
TZV100	OEM
TZV100A	OEM
TZV100.A	OEM
TZV110	OEM
TZV110A	OEM
TZV110.A	OEM
TZV120	OEM
TZV120A	OEM
TZV120.A	OEM
TZV130	OEM
TZV130A	OEM
TZV130.A	OEM
TZV150	OEM
TZV150A	OEM
TZV150.A	OEM
TZV160	OEM
TZV160A	OEM
TZV160.A	OEM
TZV170	OEM
TZV170A	OEM
TZV170.A	OEM
TZV180	OEM
TZV180A	OEM
TZV180.A	OEM
TZV200	OEM
TZV200A	OEM
TZV200.A	OEM
T.4HP521	OEM
T.75F50	OEM
T.75F100	OEM
T.75F200	OEM
T.75F400	OEM
U	0424
U0-5E	0540
U01R-24	OEM
U01R-30	OEM
U03C	0871
U03E	0853
U03R-2	0017
U03R-4	0017
U03R-8	0017
U03R-12	0017
U03R-16	0017
U03R-20	0017
U04C	0871
U04E	0853
U05-E	0072
U05B	0564
U05C	0087
U05E	0540
U05ES	0087
U05G	0965
U05J	0817
U06B	0031
U06C	0023
U06CS	4920
U06E	0015
U06G	0965
U06J	0817
U07J	0017
U07L	0017
U07M	0017
U07N	0017
U0A	0023
U1	0023
U1B4B42	OEM
U1BZ41	OEM
U1D4B42	OEM
U1DZ41	OEM
U1G4B42	OEM
U1GZ41	OEM
U1J4B42	OEM
U1JZ41	OEM
U1S102	0015
U1Z	0023
U2	0023
U2A	0023
U2N34	0211
U2N96	0211
U2N474A	0198
U2S93	0136
U2SB267	0004
U2T85	0208
U2T101	OEM
U2T105	OEM
U2T151	OEM
U2T155	OEM
U2T201	OEM
U2T205	OEM
U2T251	OEM
U2T255	OEM
U2T301	OEM
U2T305	OEM
U2T351	OEM
U2T355	OEM
U2T401	OEM
U2T405	OEM
U2T451	OEM
U2T455	OEM
U2T501	OEM
U2T505	OEM
U2T601	OEM
U2T605	OEM
U2TA406	1212
U2TA408	1212
U2TA410	OEM
U2TA506	OEM
U2TA508	OEM
U2TA510	OEM
U2TB406	1212
U2TB408	1212
U2TB410	OEM
U2TD410	1212

If replacement code is OEM, contact original manufacturer for replacement.

DEVICE TYPE	REPL CODE
U2TD420	OEM
U2TD430	OEM
U2TG406	1212
U2TG408	1212
U2TG410	OEM
U5B771739	OEM
U5B992328	OEM
U5D770331X	0632
U5D770339X	0632
U5D7703312	0632
U5D7703393	0632
U5D7703394	0632
U5E7746394	0753
U5L	0382
U5.6BSA	OEM
U6A7065354	0167
U6A7729394	0443
U6A7732394	0654
U6A7746394	0666
U6A7767394	0701
U6A7781394	0354
U6B7780394	0352
U6E7729394	0443
U7F7065354	0167
U7F7729394	0443
U7F7732394	0654
U7F7746394	0666
U7F7767394	0701
U7F7780394	0352
U7F7781394	0354
U8A992329	OEM
U8B770339	0632
U8B770339-825	0632
U8B770339X	0632
U8B7703394	0632
U8F7737394	2161
U8F7746394	2161
U9A7746394	0349
U9A7781394	0354
U9B7780394	0352
U9C7065354	0167
U13T1	OEM
U13T2	OEM
U13T3	OEM
U13T4	OEM
U15B	0087
U15C	0087
U15E	0087
U15G	0087
U15J	0087
U17B	0031
U17C	0031
U17D	0031
U17E	0031
U19-E	0031
U19B	0031
U19C	0031
U19E	0031
U19E-F	0031
U19E-FK	0031
U22	OEM
U33N	0023
U47ZA1	OEM
U89	OEM
U90A	OEM
U106BS	OEM
U111B	OEM
U112BA	OEM
U114	OEM
U117	0325
U119	0071
U120	0071
U139	OEM
U139D	OEM
U182	OEM
U197	0325
U199	1155
U201	1155
U203	OEM
U204	OEM
U205	OEM
U206	OEM
U207	OEM
U212	0015
U212-25	0015
U213	1228
U214	0015
U222	1155
U235	0325
U244	OEM
U248	OEM
U248A	OEM
U249	OEM
U249A	OEM
U250	OEM
U250A	OEM
U251	OEM
U251A	0652
U252	OEM
U253	OEM
U254	OEM
U255	OEM
U256	OEM
U266	OEM
U273	OEM
U273A	OEM
U274	OEM
U274A	OEM
U275	OEM
U275A	OEM
U280	OEM
U281	OEM
U282	OEM
U283	OEM
U284	OEM
U285	OEM
U308	2888
U309	2888
U310	OEM
U311	2888
U312	2888
U318	OEM
U319	OEM
U328	OEM
U329	OEM
U330	OEM
U331	OEM
U361	0071
U401B	0652
U410	2945
U410B	OEM
U411	2945
U412	2945
U417B	5195
U418B	OEM
U420B	OEM
U-422	0015
U460A	0127
U460A,B	0127
U460B	0127
U535A	0224
U535A,B	0127
U535B	0224
U535C	0016
U535M	0016
U552D	4364
U574	1329
U574(IC)	1329
U574K	1329
U821B	OEM
U1001	OEM
U1002	OEM
U-1003	OEM
U1003	OEM
U-1004	OEM
U1004	OEM
U1005	OEM
U1006	OEM
U1007	OEM
U1008	OEM
U1009	OEM
U1010	OEM
U1178	0325
U1180	3622
U1181	0325
U1285	0325
U1321	OEM
U1322	3622
U1323	0325
U1324	0325
U1420	0325
U1421	0325
U1422	0325
U1585E	0016
U1585F	0016
U1585F,H	0127
U1585G	0016
U1585H	0016
U1715	0325
U1837E	0325
U1897E	3130
U1916	0325
U1994E	0325
U2400-03	0015
U2848-1	0016
U8423	OEM
U8508E	OEM
U8514A	OEM
U8514B	OEM
U8516C	OEM
U8522A	OEM
UA0801ADM	OEM
UA0801AFM	OEM
UA0801CDC	OEM
UA0801CPC	OEM
UA0801DM	OEM
UA0801EDC	OEM
UA0801EPC	OEM
UA0801FM	OEM
UA0801HDC	OEM
UA0801HPC	OEM
UA0802ADC	OEM
UA0802APC	OEM
UA0802BDC	OEM
UA0802BPC	4316
UA0802CDC	OEM
UA0802CPC	4316
UA0802DM	OEM
UA0802FM	OEM
UA0802HDC	OEM
UA0802HPC	OEM
UA8T24DC	OEM
UA8T26	4516
UA8T26ADC	OEM
UA8T26ADM	OEM
UA8T26APC	OEM
UA8T26D	4516
UA8T26DC	4516
UA8T26P	4516
UA8T26PC	4516
UA8T28	0581
UA8T28D	0581
UA8T28DC	0581
UA8T28DM	0581
UA8T28P	0581
UA8T28PC	0581
UA67	0701
UA74H53A	OEM
UA78GC-U1	OEM
UA78GKC	5291
UA78GKM	OEM
UA78GU1C	3823
UA78H05KC	2863
UA78H12KC	3393
UA78H15KC	6043
UA78HGKC	2737
UA78L	0083
UA78L00AC	OEM
UA78L00C	OEM
UA78L02AC	OEM
UA78L02ACJG	OEM
UA78L02ACLP	OEM
UA78L02AS	OEM
UA78L02C	OEM
UA78L02CJG	OEM
UA78L02CLP	OEM
UA78L02S	OEM
UA78L05	1292
UA78L05AC	1292
UA78L05ACJG	OEM
UA78L05ACLP	1292
UA78L05ADB	OEM
UA78L05AHC	OEM
UA78L05AS	OEM
UA78L05AWC	1292
UA78L05AWV	OEM
UA78L05C	1292
UA78L05CJG	OEM
UA78L05CLP	1292
UA78L05DB	OEM
UA78L05HC	OEM
UA78L05HM	OEM
UA78L05S	1292
UA78L05WC	1292
UA78L06	4080
UA78L06AC	OEM
UA78L06ACJG	OEM
UA78L06ACLP	2301
UA78L06AS	2301
UA78L06C	OEM
UA78L06CJG	OEM
UA78L06CLP	2301
UA78L06S	2301
UA78L08	0083
UA78L08AC	0083
UA78L08ACJG	OEM
UA78L08ACLP	0083
UA78L08AWC	0083
UA78L08C	0083
UA78L08CJG	OEM
UA78L08CLP	0083
UA78L08S	0083
UA78L08WC	0083
UA78L09AC	OEM
UA78L09ACJG	OEM
UA78L09ACLP	1786
UA78L09AWC	1786
UA78L09C	OEM
UA78L09CJG	OEM
UA78L09CLP	OEM
UA78L062WV	2301
UA78L6.2AHC	2301
UA78L-8.2AWC	0083
UA78L10	OEM
UA78L10AC	OEM
UA78L10ACJG	OEM
UA78L10ACLP	OEM
UA78L10C	OEM
UA78L10CJG	OEM
UA78L10CLP	OEM
UA78L12AC	1828
UA78L12ACJG	OEM
UA78L12ACLP	1828
UA78L12ADB	OEM
UA78L12AHC	OEM
UA78L12AS	OEM
UA78L12AWC	1828
UA78L12AWV	OEM
UA78L12BD	OEM
UA78L12C	1828
UA78L12CJG	OEM
UA78L12CLP	1828
UA78L12HM	OEM
UA78L12S	OEM
UA78L15AC	3395
UA78L15ACJG	OEM
UA78L15ACLP	3395
UA78L15ADB	OEM
UA78L15AHC	OEM
UA78L15AS	OEM
UA78L15AWC	5522
UA78L15AWV	OEM
UA78L15C	3395
UA78L15CJG	OEM
UA78L15CLP	3395
UA78L15DM	OEM
UA78L15HM	OEM
UA78L15S	OEM
UA78L62AHC	2301
UA78L62AWC	2301
UA78L62AWV	2301
UA78L62HC	OEM
UA78L82	0083
UA78L82AHC	OEM
UA78L82AWC	0083
UA78L82AWV	0083
UA78L82W	0083
UA78M00C	OEM
UA78M05C	0624
UA78M05CKC	0624
UA78M05CKD	OEM
UA78M05CLA	OEM
UA78M05HC	OEM
UA78M05HM	OEM
UA78M05MLA	OEM
UA78M05U1C	OEM
UA78M05UC	0624
UA78M05UV	OEM
UA78M06C	0923
UA78M06CKC	0923
UA78M06CKD	OEM
UA78M06CLA	OEM
UA78M06HC	OEM
UA78M06HM	OEM
UA78M06MLA	OEM
UA78M06U1C	OEM
UA78M06UC	0923
UA78M06UV	OEM
UA78M08C	1196
UA78M08CKC	1196
UA78M08CKD	1196
UA78M08CLA	OEM
UA78M08HC	OEM
UA78M08HM	OEM
UA78M08MLA	OEM
UA78M08U1C	OEM
UA78M08UC	1196
UA78M08UV	OEM
UA78M10C	OEM
UA78M10CKC	OEM
UA78M10CKD	OEM
UA78M10CLA	OEM
UA78M12	1351
UA78M12C	0334
UA78M12CKC	0334
UA78M12CKD	0334
UA78M12CLA	OEM
UA78M12HC	OEM
UA78M12HM	OEM
UA78M12RM	OEM
UA78M12U1C	OEM
UA78M12UC	0334
UA78M12UV	OEM
UA78M15C	1321
UA78M15CKC	1321
UA78M15CKD	1321
UA78M15CLA	OEM
UA78M15HC	OEM
UA78M15HM	OEM
UA78M15MLA	OEM
UA78M15U1C	OEM
UA78M15UC	1321
UA78M20C	OEM
UA78M20CKC	OEM
UA78M20CKD	OEM
UA78M20CLA	OEM
UA78M20MLA	OEM
UA78M22C	OEM
UA78M22CKC	OEM
UA78M22CKD	OEM
UA78M22CLA	OEM
UA78M24C	2239
UA78M24CKC	2239
UA78M24CKD	2239
UA78M24CLA	OEM
UA78M24HC	OEM
UA78M24HM	OEM
UA78M24MLA	OEM
UA78M24U1C	OEM
UA78M24UC	2239
UA78M85C	OEM
UA78MGCH	OEM
UA78MGCU1	OEM
UA78MGH	OEM
UA78MGHC	OEM
UA78MGHM	OEM
UA78MGU1C	3823
UA78S40DC	OEM
UA78S40DM	OEM
UA78S40PC	OEM
UA79GC-U1	OEM
UA79M00C	OEM
UA79M05AHC	OEM
UA79M05AUC	1285
UA79M05C	1285
UA79M05CKC	1285
UA79M05CKD	OEM
UA79M05CLA	OEM
UA79M05HM	OEM
UA79M05MLA	OEM
UA79M05U1C	OEM
UA79M06AUC	2648
UA79M06C	2648
UA79M06CKC	2648
UA79M06CKD	OEM
UA79M06CLA	OEM
UA79M08AHC	OEM
UA79M08AUC	2791
UA79M08C	2791
UA79M08CKC	OEM
UA79M08CKD	OEM
UA79M08CLA	OEM
UA79M08HM	OEM
UA79M08MLA	OEM
UA79M08U1C	OEM
UA79M6MLA	OEM
UA79M12AHC	OEM
UA79M12AUC	1838
UA79M12AUV	OEM
UA79M12C	1838
UA79M12CKC	1838
UA79M12CKD	1838
UA79M12CLA	OEM
UA79M12HM	OEM
UA79M12MLA	OEM
UA79M12U1C	OEM
UA79M15AHC	OEM
UA79M15AUC	3829
UA79M15C	3829
UA79M15CKC	3829
UA79M15CKD	3829
UA79M15CLA	OEM
UA79M15HM	OEM
UA79M15MLA	OEM
UA79M15U1C	OEM
UA79M20C	OEM
UA79M20CKC	OEM
UA79M20CKD	OEM
UA79M20CLA	OEM
UA79M20MLA	OEM
UA79M24AUC	3575
UA79M24C	3575
UA79M24CKC	3575
UA79M24CKD	3575
UA79M24CLA	OEM
UA79M24MLA	OEM
UA79MGCH	OEM
UA79MGCU1	OEM
UA79MGH	OEM
UA79MGHM	OEM
UA79MGU1C	6026
UA101AFM	OEM
UA101AHM	OEM
UA101FM	OEM
UA101HM	OEM
UA105HM	OEM
UA107HM	OEM
UA108AFM	OEM
UA108AHM	OEM
UA108FM	OEM
UA108HM	OEM
UA109HM	OEM
UA109KM	OEM
UA110FM	OEM
UA111FM	OEM
UA111HM	OEM
UA111RM	OEM
UA117KM	OEM
UA124DM	OEM
UA139ADM	OEM
UA139DM	OEM
UA148DM	OEM
UA148FM	OEM
UA193ARM	OEM
UA193RM	OEM
UA201AFM	OEM
UA201AH	0093
UA201AHM	OEM
UA201H	0093
UA201HC	0093
UA201TC	OEM
UA207HM	OEM
UA208AFM	OEM
UA208AHM	OEM
UA208FM	OEM
UA208HM	OEM
UA209KM	1922
UA217UV	OEM
UA224DM	0625
UA224DV	OEM
UA224PV	OEM
UA239ADC	OEM
UA239APC	OEM
UA239DC	OEM
UA239PC	OEM
UA248DC	OEM
UA248PC	OEM
UA293ARC	OEM
UA293ATC	OEM
UA293RC	OEM
UA293TC	OEM
UA301AH	0093
UA301AHC	0093
UA301AT	1300
UA301ATC	1300
UA305AHC	OEM
UA305HC	6176
UA307HC	OEM
UA307T	2283
UA307TC	2283
UA308AHC	0896
UA308AHM	OEM
UA308ATC	OEM
UA308HC	0896
UA308HM	OEM
UA308TC	2246
UA309K	1922
UA309KC	1922
UA311HC	1815
UA311RC	OEM
UA311TC	2104
UA317KC	OEM
UA317UC	6251
UA318HC	OEM
UA324DC	0625
UA324PC	0625
UA339ADC	0176
UA339APC	OEM
UA339DC	OEM
UA339PC	OEM
UA348DC	2823
UA348PC	2823
UA376TC	OEM
UA393ARC	OEM
UA393ATC	OEM
UA393RC	OEM
UA393TC	OEM
UA431AWC	OEM
UA431HC	OEM
UA431HM	OEM
UA431WC	OEM
UA431WV	OEM
UA438JC	OEM
UA494DC	OEM
UA494DM	OEM
UA494PC	OEM
UA555	0973
UA555HC	3637
UA555HM	3637
UA555IC	0973
UA555TC	0973
UA556DBC20	OEM
UA556DC	3285
UA556DM	3285
UA556PC	3285
UA565JJC	OEM
UA565KJC	OEM
UA565SJM	OEM
UA565TJM	OEM
UA702MJ	OEM
UA702ML	OEM
UA703	0632
UA703A	0632
UA703C	0632
UA703CT	0632
UA-703E	0632
UA703E	0632
UA703HC	0632
UA703L	0632
UA705	2557
UA706	4586
UA706BPC	OEM
UA709A	1705
UA709AF	OEM
UA709AFM	OEM
UA709AHM	OEM
UA709AMJ	OEM
UA709AMJG	OEM
UA709AML	OEM
UA709AMU	OEM
UA709AN	OEM
UA709AN-14	OEM
UA709AT	OEM
UA709C	1705
UA709CJ	1705
UA709CJG	OEM
UA709CL	1301
UA709CN	1705
UA709CN-14	OEM
UA709CP	OEM
UA709CT	1301
UA709CU	OEM
UA709CV	OEM
UA709DC	1705
UA709F	OEM
UA709FM	OEM
UA709HC	1301
UA709HM	1301
UA709M	OEM
UA709MJ	OEM
UA709MJG	OEM
UA709ML	OEM
UA709MU	OEM
UA709N	1705
UA709N-14	OEM
UA709PC	1705
UA709T	1301
UA709TC	OEM
UA710	1797
UA710A	1800
UA710C	OEM
UA710CA	1800
UA710CF	OEM
UA710CJ	1800
UA710CL	1797
UA710CN	1800
UA710CN-14	OEM
UA710CT	1797
UA710DC	1800
UA710DM	1800
UA710F	OEM
UA710FM	OEM
UA710HC	1797
UA710HM	OEM
UA710MJ	OEM
UA710MJG	OEM
UA710ML	OEM
UA710MU	OEM
UA710N	OEM
UA710N-14	OEM
UA710PC	1800
UA710T	1797
UA711A	1890
UA711C	OEM
UA711CA	1897
UA711CF	OEM
UA711CJ	1897
UA711CK	1890
UA711CL	1890
UA711CN	1897
UA711CU	OEM
UA711DC	1897
UA711DM	1897
UA711F	OEM
UA711FM	OEM
UA711HC	1890
UA711HM	1890
UA711K	1890
UA711M	OEM
UA711MJ	OEM
UA711ML	OEM
UA711MU	OEM
UA711N	OEM
UA711PC	1897
UA-714C	OEM
UA-714E	OEM
UA714EHC	OEM
UA714HC	OEM
UA714HM	OEM
UA714LHC	OEM
UA715(METAL-CAN)	1554
UA715DC	OEM
UA715DM	OEM
UA715HC	7100
UA715HM	7100
UA-719	OEM
UA720DC	1421
UA723	1192
UA723A	OEM
UA723C	OEM
UA723CA	0026
UA723CD	OEM
UA723CF	OEM
UA723CH	OEM
UA723CJ	0026
UA723CL	1192
UA723CN	0026
UA723CP	OEM
UA723CT	1192
UA723CU	OEM
UA723DC	0026
UA723DM	0026
UA723F	OEM
UA723H	OEM
UA723HC	1192
UA723HM	0026
UA723J	OEM
UA723L	1192
UA723MJ	OEM
UA723ML	0026
UA723MU	OEM
UA723N	OEM
UA723P	OEM
UA723PC	0026
UA723T	OEM
UA725AFM	OEM
UA725AHM	OEM
UA725ARM	OEM
UA725BHC	OEM
UA725EHC	OEM
UA725ERC	OEM
UA725HC	2890
UA725HM	OEM
UA725RC	OEM
UA725RM	OEM
UA726HC	OEM
UA726HM	OEM
UA729	0443
UA729CA	0443
UA732	0654
UA732PC	0654
UA733A	OEM
UA733C	1121
UA733CA	OEM
UA733CF	1118
UA733CH	1121
UA733CJ	1118
UA733CK	OEM
UA733CL	OEM
UA733CN	1118
UA733CU	OEM
UA733DC	1118
UA733DM	OEM
UA733F	OEM
UA733FC	OEM
UA733FM	OEM
UA733H	OEM
UA733HC	1121
UA733HM	OEM
UA733K	OEM
UA733MJ	OEM
UA733MN	OEM
UA733N	OEM

If replacement code is OEM, contact original manufacturer for replacement.

DEVICE TYPE	REPL CODE
UA733PC	1118
UA734DC	OEM
UA734HC	OEM
UA734HM	OEM
UA737	0753
UA737E	2161
UA739DC	0692
UA739PC	0692
UA740AHM	OEM
UA740CT	1919
UA740EHC	1919
UA740T	OEM
UA741	0410
UA741A	0312
UA741AFM	OEM
UA741AHM	OEM
UA741ARM	OEM
UA741C	1975
UA741CA	1975
UA741CD	1976
UA741CF	OEM
UA741CFE	OEM
UA741CH	OEM
UA741CJ	1982
UA741CJG	0312
UA741CL	0410
UA741CN	1975
UA741CN-14	OEM
UA741CP	0312
UA741CT	0410
UA741CU	OEM
UA741CV	0312
UA741DC	0312
UA741DM	OEM
UA741EHC	4262
UA741ERC	OEM
UA741ETC	OEM
UA741F	OEM
UA741FC	OEM
UA741FE	OEM
UA741FM	OEM
UA741H	OEM
UA741HC	0410
UA741HM	0312
UA741M	OEM
UA741MJ	OEM
UA741MJG	OEM
UA741ML	0410
UA741MU	OEM
UA741N	OEM
UA741N-14	OEM
UA741PC	1975
UA741RC	OEM
UA741RM	OEM
UA741T	OEM
UA741TC	0312
UA741V	OEM
UA746	0753
UA746(DIP)	0666
UA746(METAL-CAN)	0753
UA746C	0666
UA746DC	0666
UA746E	2161
UA746HC	0753
UA746PC	0666
UA747ADM	OEM
UA747AFM	OEM
UA747AHM	OEM
UA747C	2368
UA747CA	2358
UA747CD	OEM
UA747CF	2358
UA747CH	OEM
UA747CJ	2358
UA747CK	2368
UA747CL	2368
UA747CN	2358
UA747CW	OEM
UA747DC	2358
UA747DM	OEM
UA747EDC	OEM
UA747EHC	OEM
UA747F	OEM
UA747FC	OEM
UA747FM	OEM
UA747H(M)	OEM
UA747HC	2368
UA747HM	2368
UA747K	2368
UA747M	OEM
UA747MJ	OEM
UA747ML	2368
UA747MW	OEM
UA747N	2358
UA747PC	2358
UA748A	OEM
UA748ADM	OEM
UA748AHM	OEM
UA748C	OEM
UA748CA	OEM
UA748CD	2444
UA748CF	OEM
UA748CFE	OEM
UA748CJ	2452
UA748CJG	2452
UA748CL	2454
UA748CN	2452
UA748CN-14	OEM
UA748CP	2452
UA748CT	2454
UA748CU	OEM
UA748CV	OEM
UA748DC	2452
UA748DM	2452
UA748F	OEM
UA748FE	OEM
UA748FM	OEM
UA748HC	0093
UA748HM	2454
UA748M	OEM
UA748MJ	2452
UA748MJG	2452
UA748ML	2454
UA748MU	OEM
UA748N	2452
UA748N-14	OEM
UA748T	2454
UA748TC	1300
UA748V	1300
UA749	2552
UA749D	2552
UA749DC	OEM
UA749DHC	6462
UA749DM	OEM
UA749PC	OEM
UA753TC	1759
UA757DC	OEM
UA757DM	OEM
UA757DMQB	OEM
UA758	1395
UA758A	OEM
UA758AP	OEM
UA758B	1395
UA758BA	1395
UA758N	1395
UA758PC	1395
UA759HC	OEM
UA759HM	OEM
UA759U1C	OEM
UA760	OEM
UA760DC	OEM
UA760DM	OEM
UA760HC	3261
UA760HM	3261
UA760RC	OEM
UA760RM	OEM
UA767	0701
UA767PC	0701
UA771ARC	3672
UA771ARM	OEM
UA771ATC	3672
UA771BRC	3672
UA771BRM	OEM
UA771BTC	3672
UA771LRC	5877
UA771LTC	3672
UA771RC	3672
UA771TC	3672
UA772ARC	3741
UA772ARM	OEM
UA772ATC	3741
UA772BRC	3741
UA772BRM	OEM
UA772BTC	3741
UA772RC	3741
UA772TC	3741
UA774LDC	3391
UA774LPC	3391
UA776HC	OEM
UA776HM	OEM
UA776TC	3942
UA777C	OEM
UA777CJ	OEM
UA777CJG	OEM
UA777CL	OEM
UA777CN	OEM
UA777CP	OEM
UA777CU	OEM
UA777DC	OEM
UA777HC	OEM
UA777MJ	OEM
UA777MJG	OEM
UA777ML	OEM
UA777MU	OEM
UA777TC	OEM
UA780	0352
UA780C	0352
UA780DC	0352
UA780PC	0352
UA781	0354
UA781-8.2AWC	OEM
UA781C	0354
UA781DC	0354
UA781PC	0354
UA783P4C	OEM
UA-785PC	OEM
UA786CKA	OEM
UA787	0521
UA787PC	0521
UA788PC	OEM
UA791KC	OEM
UA791KM	OEM
UA796	0417
UA796DC	3801
UA796HC	0417
UA796PC	3801
UA798HM	1677
UA798TC	4429
UA824PC	OEM
UA915	0247
UA1312PC	1611
UA1314PC	3761
UA1315PC	3763
UA1391TC	0848
UA1458CHC	OEM
UA1458CP	0360
UA1458CRC	OEM
UA1458CTC	0360
UA1458CTE	OEM
UA1458HC	3137
UA1458P	0360
UA1458RC	OEM
UA1458TC	0360
UA1488	0508
UA1488D	0508
UA1488DC	0508
UA1488P	0508
UA1488PC	0508
UA1489	0511
UA1489ADC	6439
UA1489APC	0511
UA1489D	0511
UA1489DC	0511
UA1489P	0511
UA1489PC	0511
UA1489PCQR	0511
UA1558HM	3137
UA1558RM	OEM
UA1748G	0093
UA2002	1048
UA2136P	1443
UA2136PC	1443
UA2240	OEM
UA2240C	2082
UA2240CJ	OEM
UA2240CN	OEM
UA2240DC	2082
UA2240DM	OEM
UA2240MJ	OEM
UA2240PC	2082
UA2525	OEM
UA2548	OEM
UA2572	OEM
UA2596	OEM
UA2656	OEM
UA2901DC	OEM
UA2901PC	0176
UA2902	0625
UA2902DV	OEM
UA2902PC	OEM
UA2902PV	OEM
UA2903RC	OEM
UA2903TC	OEM
UA3018AHM	1854
UA3018H	1854
UA3019HC	2698
UA3026HC	2701
UA3039HC	2711
UA3045DC	1696
UA3045PC	1696
UA3046DC	1696
UA3046PC	1696
UA3054DC	2722
UA3064	OEM
UA3064HC	0364
UA3064PC	0802
UA3064TC	0364
UA3065	0167
UA3065P	0167
UA3065PC	0167
UA3066DC	2549
UA3066PC	2549
UA3067DC	0328
UA3075PC	1345
UA3086DC	1696
UA3086DM	1696
UA3086PC	1696
UA3089E	2755
UA3301P	2247
UA3302DC	OEM
UA3302PC	OEM
UA3303DC	OEM
UA3303PC	0625
UA3401PC	2247
UA3403DC	0625
UA3403PC	0625
UA3503DC	OEM
UA3525	OEM
UA3533	OEM
UA3548	OEM
UA3572	OEM
UA3596	OEM
UA3656	OEM
UA3680DC	OEM
UA3680PC	OEM
UA4031P	OEM
UA4136	3024
UA4136DC	OEM
UA4136DM	OEM
UA4136PC	3024
UA4188DC	0508
UA4558T	0360
UA5116DC	OEM
UA5116JC	OEM
UA5151DC	OEM
UA5151JC	OEM
UA5156DC	OEM
UA5156JC	OEM
UA7094HC	1301
UA7300	4219
UA7351DC	OEM
UA7392DC	OEM
UA7392DM	OEM
UA7392PC	OEM
UA7508A	0624
UA7703E	OEM
UA7800	OEM
UA7800C	OEM
UA7805	0624
UA7805A	0624
UA7805C	0624
UA7805CDA	1922
UA7805CKA	1922
UA7805CKC	0624
UA7805CU	0624
UA7805DA	1922
UA7805KC	1922
UA7805KM	1922
UA7805MKA	1922
UA7805UC	0624
UA7805UV	0624
UA7806C	0923
UA7806CDA	OEM
UA7806CKC	0923
UA7806CU	0923
UA7806DA	OEM
UA7806KC	0923
UA7806KM	OEM
UA7806MKA	OEM
UA7806UC	0923
UA7806UV	0923
UA7808	1196
UA7808C	1196
UA7808CKA	OEM
UA7808CKC	1196
UA7808CU	1196
UA7808DA	OEM
UA7808KC	1196
UA7808KM	OEM
UA7808MKA	OEM
UA7808UC	1196
UA7808UV	1196
UA7810CKA	OEM
UA7810CKC	OEM
UA7810MKA	OEM
UA7812	1351
UA7812C	0334
UA7812CDA	5502
UA7812CKA	3963
UA7812CKC	0334
UA7812CU	0334
UA7812DA	3963
UA7812KC	3963
UA7812KM	3963
UA7812MKA	3963
UA7812UC	0334
UA7812UV	0334
UA7814CDA	OEM
UA7814CU	OEM
UA7814DA	OEM
UA7815	1321
UA7815C	1321
UA7815CDA	2000
UA7815CKA	2000
UA7815CKC	1321
UA7815CU	1321
UA7815DA	2000
UA7815KC	2000
UA7815KM	2000
UA7815MKA	2000
UA7815UC	1321
UA7815UV	4341
UA7818C	2259
UA7818CDA	OEM
UA7818CKA	3650
UA7818CKC	2259
UA7818CKG	2259
UA7818CU	2259
UA7818DA	OEM
UA7818KC	3650
UA7818KM	OEM
UA7818MKA	OEM
UA7818UC	2259
UA7818UV	2259
UA7822C	OEM
UA7822CKA	OEM
UA7822CKC	OEM
UA7822MKA	OEM
UA7824C	2239
UA7824CDA	OEM
UA7824CKA	3881
UA7824CKC	2239
UA7824CU	2239
UA7824DA	OEM
UA7824KC	3881
UA7824KM	OEM
UA7824MKA	OEM
UA7824UC	2239
UA7824UV	2239
UA7885C	OEM
UA7885CKA	OEM
UA7885CKC	OEM
UA7885MKA	OEM
UA7885UC	OEM
UA7900C	OEM
UA7905	1285
UA7905C	1285
UA7905CKA	2004
UA7905CKC	1285
UA7905KC	1285
UA7905KM	2004
UA7905MKA	2004
UA7905UC	1285
UA7906C	2648
UA7906CKA	OEM
UA7906CKC	2648
UA7906MKA	OEM
UA7906U	2648
UA7906UC	2648
UA7908C	2791
UA7908CKA	OEM
UA7908CKC	2791
UA7908KC	2791
UA7908KM	OEM
UA7908MKA	OEM
UA7908UC	2791
UA7912	1838
UA7912C	1838
UA7912CKA	5287
UA7912CKC	1838
UA7912KC	5287
UA7912KM	OEM
UA7912MKA	OEM
UA7912U	1838
UA7912UC	1838
UA7915C	3829
UA7915CKA	5289
UA7915CKC	3829
UA7915KC	5289
UA7915KM	5289
UA7915U	3829
UA7915UC	3829
UA7918C	3826
UA7918CKA	5290
UA7918CKC	3826
UA7918CKG	3826
UA7918MKA	OEM
UA7918UC	3826
UA7924C	3575
UA7924CKA	3573
UA7924CKC	3575
UA7924U	3575
UA7924UC	3575
UA7952C	OEM
UA8728DC	OEM
UA9636A	OEM
UA9637A	OEM
UA9637ACP	OEM
UA9638C	OEM
UA9638CP	OEM
UA9643	OEM
UA9644	OEM
UA78606	4080
UAA110	OEM
UAA170	OEM
UAA171	OEM
UAA180	OEM
UAA210	OEM
UAA1000	OEM
UAA1001	OEM
UAA1030	OEM
UAA4000	1094
UAB1010	OEM
UB-151	2785
UB-152	2785
UB-154	2785
UBB770339X	0632
UBD203	0031
UBD204	0031
UBFY11	0144
UC	OEM
UC20	0325
UC22	OEM
UC23	OEM
UC40	OEM
UC41	OEM
UC42	OEM
UC43	OEM
UC78M12H2	0334
UC100	0325
UC105	0325
UC110	0325
UC115	0325
UC120	0325
UC125	0325
UC135	OEM
UC140	OEM
UC150	2888
UC155	0325
UC201	0325
UC210	0325
UC220	0325
UC241	0325
UC258	OEM
UC300	OEM
UC305	OEM
UC310	OEM
UC315	OEM
UC320	OEM
UC325	OEM
UC330	OEM
UC335	OEM
UC337K	2723
UC340	OEM
UC450	OEM
UC451	OEM
UC547C	0016
UC588	0325
UC701	0325
UC703	0325
UC704	3622
UC705	3622
UC707	1155
UC714	0325
UC714E	OEM
UC734	0325
UC734E	0325
UC750	0325
UC751	0325
UC752	0325
UC753	0325
UC756	0325
UC801	OEM
UC803	OEM
UC804	OEM
UC805	1141
UC850	OEM
UC852	OEM
UC900	OEM
UC-1001B	OEM
UC-1002B	OEM
UC-1003B	OEM
UC-1004B	OEM
UC-1005B	OEM
UC-1006B	OEM
UC1100	0037
UC1406HA	3254
UC1496A	OEM
UC1496N	OEM
UC1700	OEM
UC2136	0325
UC2138	0325
UC2139	0325
UC2147	2945
UC2148	0325
UC2149	0325
UC3525AN	2829
UC3842	OEM
UC3842AN	OEM
UC3843AN	OEM
UC4741	0410
UC4741C	0410
UC6550	OEM
UC7550	OEM
UC7805UC	0624
UC7933	0334
UCD125	0124
UCD227	0124
UCD329	OEM
UCR10	OEM
UCR10L	OEM
UCR20	OEM
UCR20L	OEM
UCR30	OEM
UCR30L	OEM
UCR40	OEM
UCR40L	OEM
UCR50	OEM
UCR50L	OEM
UCR60	OEM
UCR60L	OEM
UCX1702	OEM
UCX2700	OEM
UD5B	0087
UD6C	OEM
UD1000	OEM
UD1001	OEM
UD-2000	3049
UD2000	OEM
UD-3005	OEM
UD3005	OEM
UD-3006	OEM
UD3006	OEM
UD-3007	OEM
UD3007	OEM
UD3008	OEM
UD-4001	OEM
UD-4024	OEM
UD4174	OEM
UD4175	OEM
UDC40013	OEM
UDGE	OEM
UDN2841	OEM
UDN2845	OEM
UDN-3611M	1244
UDN-3612M	1262
UDN3612N	OEM
UDN-3613M	1272
UDN-3614M	1289
UDN-5703A	OEM
UDN-5706A	OEM
UDN-5707A	OEM
UDN-5711M	OEM
UDN-5712M	OEM
UDN-5713M	OEM
UDN5713M	OEM
UDN-5714M	OEM
UDN5722M	OEM
UDN-5733A	OEM
UDN-6116A	OEM
UDN-6116A-1	OEM
UDN6116A-1	4081
UDN-6116A-2	OEM
UDN-6116R	OEM
UDN-6116R-2	OEM
UDN-6118A	5912
UDN6118A	3893
UDN-6118A-1	OEM
UDN-6118A-2	OEM
UDN-6118R	5914
UDN-6118R-2	OEM
UDN-6128A	5931
UDN6128A	3920
UDN-6128A-1	OEM
UDN-6128A-2	OEM
UDN-6128R	5937
UDN6128R	OEM
UDN-6128R-2	OEM
UDN6128R-2	OEM
UDN-6138A	OEM
UDN6138A	OEM
UDN-6138A-2	OEM
UDN6138A2	OEM
UDN-6144A	4943
UDN6148A	OEM
UDN6148A2	OEM
UDN-6164A	4081
UDN6164A	4081
UDN-6184A	3893
UDN6184A	OEM
UDN6510A	OEM
UDN6510R	OEM
UDN6514A	OEM
UDN6514R	OEM
UDN7180A	OEM
UDN-7183A	3893
UDN7183A	OEM
UDN7184A	OEM
UDN-7186A	3920
UDN7186A	OEM
UDQ2956R	OEM
UDQ2957R	OEM
UDS2981H	OEM
UDS2982H	OEM
UDS2983H	OEM
UDS2984H	OEM
UDS3611H	OEM
UDS3612H	OEM
UDS3613H	OEM
UDS3614H	OEM
UDS5703H	OEM
UDS5706H	OEM
UDS5707H	OEM
UDS5711H	OEM
UDS5712H	OEM
UDS5713H	OEM
UDS5714H	OEM
UDS5733H	OEM
UDS5790H	OEM
UDS5791H	OEM
UDT450	OEM
UDT451	OEM
UDT452	OEM
UDT455	OEM
UDT500D	OEM
UDT600	OEM
UDZ110	OEM
UDZ122	OEM
UDZ130	OEM
UDZ210	OEM
UDZ222	OEM
UDZ230	OEM
UDZ707	0480
UDZ708	0504
UDZ709	0684
UDZ710	0225
UDZ712	0234
UDZ715	0247
UDZ718	0256
UDZ720	0262
UDZ724	0274
UDZ727	0294
UDZ730	0309
UDZ733	0318
UDZ736	0320
UDZ740	OEM
UDZ745	OEM
UDZ760	2856
UDZ807	OEM
UDZ808	OEM
UDZ809	OEM
UDZ810	OEM
UDZ812	OEM
UDZ815	OEM
UDZ818	OEM
UDZ820	OEM
UDZ824	OEM
UDZ827	OEM
UDZ830	OEM
UDZ833	OEM
UDZ836	OEM
UDZ840	OEM
UDZ845	OEM
UDZ860	OEM
UDZ5110	OEM
UDZ5122	OEM
UDZ5130	OEM
UDZ5210	OEM
UDZ5222	OEM
UDZ5230	OEM
UDZ5707	0480
UDZ5708	0504
UDZ5709	0684

If replacement code is OEM, contact original manufacturer for replacement.

DEVICE TYPE	REPL CODE
UDZ5710	0225
UDZ5712	0234
UDZ5715	0247
UDZ5718	0256
UDZ5720	0262
UDZ5724	0274
UDZ5727	OEM
UDZ5730	OEM
UDZ5733	OEM
UDZ5736	OEM
UDZ5740	OEM
UDZ5745	OEM
UDZ5760	OEM
UDZ5807	OEM
UDZ5808	OEM
UDZ5809	OEM
UDZ5810	OEM
UDZ5812	OEM
UDZ5815	OEM
UDZ5818	OEM
UDZ5820	OEM
UDZ5824	0274
UDZ5827	OEM
UDZ5830	OEM
UDZ5833	OEM
UDZ5836	OEM
UDZ5840	OEM
UDZ5845	OEM
UDZ5860	OEM
UDZ7110	OEM
UDZ7210	OEM
UDZ7707	OEM
UDZ7708	OEM
UDZ7709	OEM
UDZ7710	OEM
UDZ7712	OEM
UDZ7715	OEM
UDZ7718	OEM
UDZ7720	OEM
UDZ7724	OEM
UDZ7727	OEM
UDZ7730	OEM
UDZ7733	OEM
UDZ7736	OEM
UDZ7740	OEM
UDZ7745	OEM
UDZ7760	OEM
UDZ7807	OEM
UDZ7808	OEM
UDZ7809	OEM
UDZ7810	OEM
UDZ7812	OEM
UDZ7815	OEM
UDZ7818	OEM
UDZ7820	OEM
UDZ7824	OEM
UDZ7827	OEM
UDZ7830	OEM
UDZ7833	OEM
UDZ7836	OEM
UDZ7840	OEM
UDZ7845	OEM
UDZ7860	OEM
UDZ8707	0649
UDZ8708	0244
UDZ8709	0012
UDZ8710	0170
UDZ8712	0137
UDZ8715	0002
UDZ8718	0495
UDZ8720	0531
UDZ8724	0402
UDZ8727	0441
UDZ8730	OEM
UDZ8733	OEM
UDZ8736	OEM
UDZ8740	OEM
UDZ8745	OEM
UDZ8760	OEM
UDZ8807	OEM
UDZ8808	OEM
UDZ8809	OEM
UDZ8810	OEM
UDZ8812	OEM
UDZ8815	OEM
UDZ8818	OEM
UDZ8820	OEM
UDZ8824	OEM
UDZ8827	OEM
UDZ8830	OEM
UDZ8833	OEM
UDZ8836	OEM
UDZ8840	OEM
UDZ8845	OEM
UDZ8860	OEM
UES101	0031
UES102	0031
UES103	0031
UES104	0031
UES201	OEM
UES202	OEM
UES203	OEM
UES204	OEM
UES301	OEM
UES302	OEM
UES303	OEM
UES304	OEM
UES501	OEM
UES502	OEM
UES503	OEM
UES504	OEM
UES505	OEM
UES601	OEM
UES602	OEM
UES603	OEM
UES604	OEM
UES605	OEM
UES606	OEM
UES701	OEM
UES702	OEM
UES703	OEM
UES704	OEM
UES705	OEM
UES706	OEM
UES801	OEM
UES802	OEM
UES803	OEM
UES804	OEM
UES805	OEM
UES806	OEM
UES1001	1335
UES1002	OEM
UES1003	OEM
UES1101	0546
UES1102	0546
UES1103	0546
UES1104	2513
UES1105	1362
UES1106	1362
UES1301	1362
UES1302	1362
UES1303	1362
UES1304	0546
UES1305	1362
UES1306	1362
UES1401	1127
UES1402	1127
UES1403	1127
UES2401	1236
UES2402	1236
UES2403	1236
UES2601	OEM
UES2602	OEM
UES2603	OEM
UES2604	OEM
UES2605	OEM
UES2606	OEM
UF-01	0050
UF01	0023
UF-1	0023
UF1	0546
UF-1A	0023
UF1A	0023
UF-1B	0017
UF1B	0546
UF-1C	0071
UF1C	0546
UF-1D	1771
UF-1E	1771
UF-1F	1771
UF-1V	0023
UF-2	0023
UF2	0023
UF-2B	0017
UF-3	0031
UF40A	OEM
UF40B	OEM
UF40C	OEM
UF55A	OEM
UF55B	OEM
UF55C	OEM
UF75A	OEM
UF75B	OEM
UF75C	OEM
UF100A	OEM
UF100B	OEM
UF100C	OEM
UF130A	OEM
UF130B	OEM
UF130C	OEM
UF160A	OEM
UF160B	OEM
UF160C	OEM
UF200A	OEM
UF200B	OEM
UF200C	OEM
UF250A	OEM
UF250B	OEM
UF250C	OEM
UF4001	1335
UF4002	0080
UF4003	0609
UF4004	0795
UF4005	0015
UF4006	0072
UF4007	0071
UF5400	0031
UF5401	0031
UF5402	0031
UF5403	0031
UF5404	0031
UF5404J	1191
UF5405	0031
UF5406	0031
UFB2.5	OEM
UFB5	OEM
UFB7	OEM
UFS5	OEM
UFS7.5	OEM
UFS10	OEM
UF-SD1	0143
UFSD-1	0015
UFSD1	0102
UFSD-1A	0102
UFSD1A	0102
UFSD-1B	0071
UFSD1B	OEM
UFSD-1C	0102
UFSD1C	0102
UFSD1F	0015
UFSD-18	0071
UG3C	0023
UG-1002	2010
UG-1003	0277
UG-1004	0106
UG1888	0133
UGB5	OEM
UGB7.5	OEM
UGB10	OEM
UGB3132AD	OEM
UGB6124ADA	OEM
UGB6124AG	OEM
UGD5	OEM
UGD7.5	OEM
UGD10	OEM
UGE0221AY4	OEM
UGE0421AY4	OEM
UGE0501NY4B	OEM
UGE1112AY4	OEM
UGE3126AY4	OEM
UGF2.5	OEM
UGF5	OEM
UGF7.5	OEM
UGH7812	0334
UGJ7109393	1922
UGN3013T	OEM
UGN3019T	OEM
UGN3020T	OEM
UGN3030T	OEM
UGN3040T	OEM
UGN3201M	OEM
UGN3203M	OEM
UGN3220S	OEM
UGN3501C	OEM
UGN3501H	OEM
UGN3501M	OEM
UGN3501T	OEM
UGN3600M	OEM
UGN3601M	OEM
UGN30501M	OEM
UGS3019T	OEM
UGS3020T	OEM
UGS3030T	OEM
UHC-001	OEM
UHC-005	OEM
UHC-006	OEM
UHC-007	OEM
UHC-023	OEM
UHC-024	OEM
UHC-036	OEM
UHC-037	OEM
UHC-400	OEM
UHC400	OEM
UHC-400-1	OEM
UHC400-1	OEM
UHC-402	OEM
UHC402	OEM
UHC-402-1	OEM
UHC402-1	OEM
UHC-403	OEM
UHC403	OEM
UHC-403-1	OEM
UHC403-1	OEM
UHC-406	OEM
UHC406	OEM
UHC-406-1	OEM
UHC406-1	OEM
UHC-407	OEM
UHC407	OEM
UHC-407-1	OEM
UHC407-1	OEM
UHC-408	OEM
UHC408	OEM
UHC-408-1	OEM
UHC408-1	OEM
UHC-420	OEM
UHC-432	OEM
UHC432	OEM
UHC-432-1	OEM
UHC432-1	OEM
UHC-433	OEM
UHC433	OEM
UHC-433-1	OEM
UHC433-1	OEM
UHC-451	OEM
UHC-459	OEM
UHC-500	OEM
UHC500	OEM
UHC-502	OEM
UHC502	OEM
UHC-503	OEM
UHC503	OEM
UHC-506	OEM
UHC506	OEM
UHC-507	OEM
UHC507	OEM
UHC-508	OEM
UHC508	OEM
UHC-520	OEM
UHC-532	OEM
UHC532	OEM
UHC-533	OEM
UHC533	OEM
UHC-551	OEM
UHC-559	OEM
UHD-400	OEM
UHD400	OEM
UHD-400-1	OEM
UHD400-1	OEM
UHD-402	OEM
UHD402	OEM
UHD-402-1	OEM
UHD402-1	OEM
UHD-403	OEM
UHD403	OEM
UHD-403-1	OEM
UHD403-1	OEM
UHD-406	OEM
UHD406	OEM
UHD-406-1	OEM
UHD406-1	OEM
UHD-407	OEM
UHD407	OEM
UHD-407-1	OEM
UHD407-1	OEM
UHD-408	OEM
UHD408	OEM
UHD-408-1	OEM
UHD408-1	OEM
UHD-420	OEM
UHD-432	OEM
UHD432	OEM
UHD-432-1	OEM
UHD432-1	OEM
UHD-433	OEM
UHD433	OEM
UHD-433-1	OEM
UHD433-1	OEM
UHD-451	OEM
UHD-459	OEM
UHD-480	OEM
UHD-481	OEM
UHD-490	OEM
UHD490	OEM
UHD-491	OEM
UHD491	OEM
UHD-500	OEM
UHD500	OEM
UHD-502	OEM
UHD502	OEM
UHD-503	OEM
UHD503	OEM
UHD-506	OEM
UHD506	OEM
UHD-507	OEM
UHD507	OEM
UHD-508	OEM
UHD508	OEM
UHD-520	OEM
UHD-532	OEM
UHD532	OEM
UHD-533	OEM
UHD533	OEM
UHD-551	OEM
UHD-559	OEM
UHIC-001	4622
UHIC001	4622
UHIC-003	2380
UHIC003	2380
UHIC-004	2306
UHIC004	2306
UHIC-004E	2306
UHIC-004F	OEM
UHIC-005	4484
UHIC005	4484
UHIC006	4636
UHIC-007	4645
UHIC007	4645
UHIC-131	OEM
UHM-412	OEM
UHM-500	OEM
UHM-505	OEM
UHP-004	0544
UHP-181	OEM
UHP-400	OEM
UHP400	OEM
UHP-400-1	OEM
UHP400-1	OEM
UHP-402	OEM
UHP402	OEM
UHP-402-1	OEM
UHP402-1	OEM
UHP-403	OEM
UHP403	OEM
UHP-403-1	OEM
UHP403-1	OEM
UHP-406	OEM
UHP406	OEM
UHP-406-1	OEM
UHP406-1	OEM
UHP-407	OEM
UHP407	OEM
UHP-407-1	OEM
UHP407-1	OEM
UHP-408	OEM
UHP408	OEM
UHP-408-1	OEM
UHP408-1	OEM
UHP-420	OEM
UHP-432	OEM
UHP432	OEM
UHP-432-1	OEM
UHP432-1	OEM
UHP-433	OEM
UHP433	OEM
UHP-433-1	OEM
UHP433-1	OEM
UHP-451	OEM
UHP-459	OEM
UHP-480	OEM
UHP480	OEM
UHP-481	OEM
UHP481	OEM
UHP-482	OEM
UHP-490	OEM
UHP490	OEM
UHP-491	OEM
UHP491	OEM
UHP-495	OEM
UHP495	OEM
UHP-500	OEM
UHP500	OEM
UHP-502	OEM
UHP502	OEM
UHP-503	OEM
UHP503	OEM
UHP-506	OEM
UHP506	OEM
UHP-507	OEM
UHP507	OEM
UHP-508	OEM
UHP508	OEM
UHP-520	OEM
UHP-532	OEM
UHP532	OEM
UHP-533	OEM
UHP533	OEM
UHP-551	OEM
UHP-559	OEM
UHV100	OEM
UHV125	OEM
UHV150	OEM
UHV175	OEM
UHV200	OEM
UHV225	OEM
UHV250	OEM
UHV275	OEM
UHV300	OEM
UIU-.KO	0212
UL-914	1069
UL-923	OEM
UL923	OEM
UL2070B	OEM
ULA2C184E	OEM
ULA9RB015E	OEM
ULA1045E	OEM
ULC3037	0664
ULM2111A	0678
ULM2114A	0349
ULM2274P	2557
ULN-2001A	0845
ULN2001A	0845
ULN2001AJ	0845
ULN2001AN	0845
ULN2001R	OEM
ULN-2002A	1007
ULN2002A	1007
ULN2002AJ	1007
ULN2002AN	1007
ULN2002R	OEM
ULN2003	1134
ULN-2003A	1134
ULN2003A	1134
ULN2003AJ	1134
ULN2003AN	1134
ULN2003N	OEM
ULN2003R	OEM
ULN-2004A	1261
ULN2004A	1261
ULN2004AJ	1261
ULN2004AN	1261
ULN2004N	OEM
ULN2004R	OEM
ULN2005A	4935
ULN2005R	OEM
ULN2011A	0845
ULN2011R	OEM
ULN2012A	1007
ULN2013A	1134
ULN2013R	OEM
ULN2014A	1261
ULN2014R	OEM
ULN2015A	4935
ULN2015R	OEM
ULN2021A	OEM
ULN2022A	OEM
ULN2023A	OEM
ULN2024A	OEM
ULN2025A	OEM
ULN-2031A	OEM
ULN2031A	OEM
ULN-2032A	OEM
ULN2032A	OEM
ULN-2033A	OEM
ULN2033A	OEM
ULN-2046A	1696
ULN2046A	1696
ULN-2054A	4839
ULN2054A	OEM
ULN2061A	OEM
ULN2061M	OEM
ULN2062M	OEM
ULN2064	OEM
ULN-2064A	4868
ULN2064A	OEM
ULN2064B	4868
ULN2064NE	4868
ULN2065	OEM
ULN2065B	4868
ULN2065NE	4868
ULN2066	OEM
ULN2066B	5314
ULN2066NE	5314
ULN2067	OEM
ULN2067B	5314
ULN2067NE	5314
ULN2068	OEM
ULN2068B	4970
ULN2068NE	4970
ULN2069	OEM
ULN2069B	4970
ULN2071B	5315
ULN2074	OEM
ULN-2074A	OEM
ULN2074B	OEM
ULN2074NE	OEM
ULN2075	OEM
ULN2075B	OEM
ULN2075NE	OEM
ULN2076B	OEM
ULN2077B	OEM
ULN-2081A	2749
ULN2081A	2749
ULN-2082A	2750
ULN2082A	2750
ULN2083A	2751
ULN-2086A	1696
ULN2086A	1696
ULN-2103M	OEM
ULN2110(A)	0520
ULN2111	0664
ULN-2111A	0664
ULN2111A	0664
ULN-2111N	0664
ULN2111N	0664
ULN2113	0678
ULN-2113A	0678
ULN2113A	0678
ULN-2113N	0678
ULN2113N	0678
ULN2114	0349
ULN-2114A	0349
ULN2114A	0666
ULN-2114K	0753
ULN2114K	0753
ULN-2114N	0349
ULN2114N	0349
ULN-2114W	2161
ULN2114W	0753
ULN-2120A	0654
ULN2120A	0654
ULN-2120N	0654
ULN-2121A	1940
ULN2121A	1940
ULN-2121N	1940
ULN2121N	1940
ULN-2122A	0443
ULN2122A	0443
ULN-2122N	0443
ULN2122N	0443
ULN2124	0352
ULN-2124A	0352
ULN2124A	0352
ULN2124N	0352
ULN-2125A	2457
ULN2125A	2457
ULN-2126A	2279
ULN2126A	2279
ULN-2126N	2279
ULN-2127A	0354
ULN2127A	0354
ULN-2127N	0354
ULN2127N	0354
ULN-2128	2279
ULN2128	0701
ULN-2128A	0701
ULN2128A	0701
ULN-2128N	0701
ULN2128N	0701
ULN-2129A	1345
ULN2129A	1345
ULN-2129N	1345
ULN-2136A	0664
ULN2136A	1443
ULN-2137A	1421
ULN2137A	1421
ULN-2139D	OEM
ULN-2139G	OEM
ULN-2139H	OEM
ULN-2139M	OEM
ULN-2140A	OEM
ULN2140A	OEM
ULN-2140H	OEM
ULN-2141A	OEM
ULN-2141H	OEM
ULN-2142A	OEM
ULN-2142H	OEM
ULN-2151D	0410
ULN-2151G	OEM
ULN-2151H	OEM
ULN-2151M	0312
ULN-2156D	0410
ULN-2156G	OEM
ULN-2156H	OEM
ULN-2156M	0312
ULN-2157A	2358
ULN-2157H	2358
ULN-2157K	2368
ULN-2158D	2562
ULN-2158G	OEM
ULN-2158H	OEM
ULN-2158M	2567
ULN-2159D	0410
ULN-2159G	OEM
ULN-2159H	1982
ULN-2159M	0312
ULN2165	6785
ULN-2165A	0167
ULN2165A	0167
ULN-2165N	0167
ULN2165N	0167
ULN-2171D	0410
ULN-2171G	OEM
ULN-2171H	OEM
ULN-2171M	0312
ULN-2172D	2562
ULN-2172G	OEM
ULN-2172H	OEM
ULN-2172M	2567
ULN-2173D	0410
ULN-2173G	OEM
ULN-2173H	OEM
ULN-2173M	0312
ULN-2174D	2562
ULN-2174G	OEM
ULN-2174H	OEM
ULN-2174M	2567
ULN-2175D	OEM
ULN-2175G	OEM
ULN-2175H	OEM
ULN-2175M	OEM
ULN-2176D	2567
ULN-2176G	OEM
ULN-2176H	OEM
ULN-2176M	2567
ULN-2177D	0410
ULN-2177G	OEM
ULN-2177H	1982
ULN-2177M	0312
ULN-2178D	2562
ULN-2178G	OEM
ULN-2178H	OEM
ULN-2178M	2567
ULN2204A	5195
ULN-2208M	3410
ULN-2209M	1759
ULN2209M	1759
ULN2209V	OEM
ULN2210	0520
ULN-2210A	0520
ULN2211	3735
ULN2211A	3735
ULN2211B	3735
ULN2211N	5602
ULN-2211P	3735
ULN-2212A	3766
ULN2212B	3766
ULN2212N	OEM
ULN-2212P	3766
ULN-2216A	OEM
ULN2224	7065
ULN-2224A	0751
ULN2224A	3705
ULN-2228A	0349
ULN2228A	0349
ULN2231A	3425
ULN2240A	OEM
ULN2241A	OEM
ULN2242A	OEM
ULN2242N	0403
ULN2243A	3423
ULN2244	1395
ULN-2244A	1395
ULN2244A	1395
ULN2244K	1395
ULN2244R	1395
ULN2245A	OEM
ULN2249A	4089
ULN2261	2501
ULN2261A	2501
ULN2262	0521
ULN2264	0364
ULN-2264A	0802
ULN2264A	0802
ULN2264AN	0349
ULN2264K	0364
ULN2266A	2549
ULN2266N	2549
ULN2267	0328
ULN2267A	0328
ULN-2267N	0328
ULN2267N	0328
ULN2268A	2830

Original Device Types

DEVICE TYPE	REPL CODE
ULN-2269A	1337
ULN2269A	1337
ULN2270B	OEM
ULN2270Q	OEM
ULN2274B	2557
ULN-2274P	2557
ULN2274P	2557
ULN2275	2557
ULN-2275A	2557
ULN2275B	2557
ULN-2275P	2557
ULN2275P	2557
ULN2276B	2557
ULN2276P	2557
ULN2276Q	2557
ULN2277	2557
ULN2277B	2557
ULN-2277P	2557
ULN2277P	2557
ULN2278	2557
ULN2278B	2557
ULN-2278P	2557
ULN2278P	2557
ULN2278Q	2557
ULN-2280A	0379
ULN2280A	0379
ULN-2280B	0379
ULN2280B	0379
ULN2280N	0379
ULN-2280P	0379
ULN2280P	0379
ULN2280Q	0379
ULN-2281B	3056
ULN2281B	3734
ULN2283B	OEM
ULN2283B1	OEM
ULN2289	2755
ULN2289(A)	2755
ULN2289A	2755
ULN2290B	1057
ULN2290Q	1170
ULN2291M	0848
ULN2293A	OEM
ULN2294M	2493
ULN2296	3788
ULN2297A	3270
ULN2298	0856
ULN-2298A	2043
ULN2298A	0856
ULN2298P	0856
ULN-2300M	OEM
ULN-2301M	OEM
ULN2401A	OEM
ULN2429A	OEM
ULN2430M	OEM
ULN-2709CM	0312
ULN-2709M	0312
ULN-2723A	0026
ULN-2723K	1192
ULN-2741D	0410
ULN2741D	0410
ULN-2747A	2358
ULN2801A	4936
ULN2801R	OEM
ULN2802A	4937
ULN2802R	OEM
ULN2803A	4939
ULN2803R	OEM
ULN2804A	4940
ULN2804R	OEM
ULN2805A	4941
ULN2805R	OEM
ULN2811A	4936
ULN2811R	OEM
ULN2812A	4937
ULN2813A	4939
ULN2813R	OEM
ULN2814A	4940
ULN2814R	OEM
ULN2815A	4941
ULN2815R	OEM
ULN2821A	OEM
ULN2822A	OEM
ULN2823A	OEM
ULN2824A	OEM
ULN2825A	OEM
ULN-3000M	OEM
ULN-3000R	OEM
ULN-3000S	OEM
ULN-3004M	OEM
ULN-3004S	OEM
ULN-3006M	OEM
ULN-3006S	OEM
ULN-3006T	OEM
ULN-3008M	OEM
ULN-3100M	OEM
ULN-3100R	OEM
ULN-3100S	OEM
ULN3262A	0521
ULN-3303M	OEM
ULN-3304M	OEM
ULN3304M	OEM
ULN-3305M	OEM
ULN3305M	OEM
ULN-3306M	OEM
ULN3306M	OEM
ULN-3330Y	5125
ULN3330Y	5125
ULN3701Z	1048

DEVICE TYPE	REPL CODE
ULN3702Z	OEM
ULN3809A	7101
ULN3810A	OEM
ULN3810A1	OEM
ULN3838A1	OEM
ULN3838A2	OEM
ULN3838A3	OEM
ULN3859A	7131
ULN3889A	OEM
ULN3941A	OEM
ULN4436A	OEM
ULN8126A	2841
ULN8126R	OEM
ULN8160A	5212
ULN8160R	5212
ULPR73B-50	OEM
ULQ8126A	OEM
ULQ8126R	OEM
ULS2001H	OEM
ULS2001R	OEM
ULS2002H	OEM
ULS2002R	OEM
ULS2003H	OEM
ULS2003R	OEM
ULS2004H	OEM
ULS2004R	OEM
ULS2005H	OEM
ULS2005R	OEM
ULS2011H	OEM
ULS2011R	OEM
ULS2012H	OEM
ULS2012R	OEM
ULS2013H	OEM
ULS2013R	OEM
ULS2014H	OEM
ULS2014R	OEM
ULS2015H	OEM
ULS2015R	OEM
ULS2021H	OEM
ULS2022H	OEM
ULS2023H	OEM
ULS2024H	OEM
ULS2025H	OEM
ULS-2045H	1696
ULS2045H	1696
ULS2083H	OEM
ULS-2139D	OEM
ULS-2139G	OEM
ULS-2139H	OEM
ULS-2139M	OEM
ULS-2140A	OEM
ULS-2140H	OEM
ULS2140H	OEM
ULS-2141A	OEM
ULS-2141H	OEM
ULS-2142A	OEM
ULS-2142H	OEM
ULS-2151D	0410
ULS-2151G	OEM
ULS-2151H	OEM
ULS-2151M	0312
ULS-2156D	0410
ULS-2156G	OEM
ULS-2156H	OEM
ULS-2156M	0312
ULS-2157A	2358
ULS-2157H	2358
ULS-2157K	2368
ULS-2158D	2562
ULS-2158G	OEM
ULS-2158H	OEM
ULS-2158M	2567
ULS-2159D	0410
ULS-2159G	OEM
ULS-2159H	1982
ULS-2159M	0312
ULS-2171D	0410
ULS-2171G	OEM
ULS-2171H	OEM
ULS-2171M	0312
ULS-2172D	2562
ULS-2172G	OEM
ULS-2172H	OEM
ULS-2172M	2567
ULS-2173D	0410
ULS-2173G	OEM
ULS-2173H	OEM
ULS-2173M	0312
ULS-2174D	2562
ULS-2174G	OEM
ULS-2174H	OEM
ULS-2174M	2567
ULS-2175D	OEM
ULS-2175G	OEM
ULS-2175H	OEM
ULS-2175M	OEM
ULS-2176D	2562
ULS-2176G	OEM
ULS-2176H	OEM
ULS-2176M	2567
ULS-2177D	0410
ULS-2177G	OEM
ULS-2177H	1982
ULS-2177M	0312
ULS-2178D	2567
ULS-2178G	OEM
ULS-2178H	OEM
ULS-2178M	2567
ULS-2723A	0026

DEVICE TYPE	REPL CODE
ULS-2723K	1192
ULS-2741D	0410
ULS2741D	0410
ULS2801H	OEM
ULS2802H	OEM
ULS2802R	OEM
ULS2803H	OEM
ULS2803R	OEM
ULS2804H	OEM
ULS2804R	OEM
ULS2805H	OEM
ULS2805R	OEM
ULS2811H	OEM
ULS2811R	OEM
ULS2812H	OEM
ULS2812R	OEM
ULS2813H	OEM
ULS2813R	OEM
ULS2814H	OEM
ULS2814R	OEM
ULS2815H	OEM
ULS2815R	OEM
ULS2821H	OEM
ULS2822H	OEM
ULS2823H	OEM
ULS2824H	OEM
ULS2825H	OEM
ULS-3006M	OEM
ULS-3006S	OEM
ULS-3006T	OEM
ULS8126R	OEM
ULS8160R	OEM
ULT2260	3720
ULT3730	0975
ULT3731	0975
ULX2210	0520
ULX2210D	0520
ULX-2231A	3425
ULX2244	1395
ULX-2267A	0328
ULX2277P	OEM
ULX-2289A	2755
ULX2298	0856
ULX3701H	OEM
ULX3701Z	1048
ULX3701ZV	OEM
ULX3788W	OEM
ULX3804A	OEM
ULX3840A	OEM
ULX8161M(A)	OEM
ULXX3777W	OEM
UM31AWC	OEM
UM1285-8	OEM
UM2010A	OEM
UM2010B	OEM
UM2010C	OEM
UM2012A	OEM
UM2012B	OEM
UM2012C	OEM
UM2014A	OEM
UM2014B	OEM
UM2016A	OEM
UM2016B	OEM
UM2016C	OEM
UM2018A	OEM
UM2018B	OEM
UM2018C	OEM
UM2020A	OEM
UM2020B	OEM
UM2020C	OEM
UM2301	OEM
UM2333-D201	OEM
UM3482	OEM
UM3482A	OEM
UM4000	OEM
UM4001A	OEM
UM4001B	OEM
UM4001C	OEM
UM4001CR	OEM
UM4001D	OEM
UM4001DR	OEM
UM4001E	OEM
UM4002A	OEM
UM4002B	OEM
UM4002C	OEM
UM4002CR	OEM
UM4002D	OEM
UM4002DR	OEM
UM4002E	OEM
UM4004A	OEM
UM4004B	OEM
UM4004C	OEM
UM4004CR	OEM
UM4004D	OEM
UM4004DR	OEM
UM4004E	OEM
UM4006A	OEM
UM4006B	OEM
UM4006C	OEM
UM4006CR	OEM
UM4006D	OEM
UM4006DR	OEM
UM4006E	OEM
UM4008A	OEM
UM4008B	OEM
UM4008C	OEM
UM4008CR	OEM
UM4008D	OEM
UM4008DR	OEM

DEVICE TYPE	REPL CODE
UM4008E	OEM
UM4010A	OEM
UM4010B	OEM
UM4010C	OEM
UM4010CR	OEM
UM4010D	OEM
UM4010DR	OEM
UM4010E	OEM
UM4012A	OEM
UM4012B	OEM
UM4012C	OEM
UM4012CR	OEM
UM4012D	OEM
UM4012DR	OEM
UM4012E	OEM
UM4300	OEM
UM4301A	OEM
UM4301B	OEM
UM4301C	OEM
UM4301CR	OEM
UM4301D	OEM
UM4301DR	OEM
UM4301E	OEM
UM4302A	OEM
UM4302B	OEM
UM4302C	OEM
UM4302D	OEM
UM4302E	OEM
UM4304A	OEM
UM4304B	OEM
UM4304C	OEM
UM4304CR	OEM
UM4304D	OEM
UM4304DR	OEM
UM4304E	OEM
UM4306A	OEM
UM4306B	OEM
UM4306C	OEM
UM4306D	OEM
UM4306E	OEM
UM4310A	OEM
UM4310B	OEM
UM4310C	OEM
UM4310CR	OEM
UM4310D	OEM
UM4310DR	OEM
UM4310E	OEM
UM4900	OEM
UM4901A	OEM
UM4901B	OEM
UM4901C	OEM
UM4901CR	OEM
UM4901D	OEM
UM4901DR	OEM
UM4901E	OEM
UM4902A	OEM
UM4902B	OEM
UM4902C	OEM
UM4904A	OEM
UM4904B	OEM
UM4904C	OEM
UM4904CR	OEM
UM4904D	OEM
UM4904DR	OEM
UM4904E	OEM
UM4906A	OEM
UM4906B	OEM
UM4906C	OEM
UM4906CR	OEM
UM4906D	OEM
UM4906DR	OEM
UM4906E	OEM
UM4910A	OEM
UM4910B	OEM
UM4910C	OEM
UM4910CR	OEM
UM4910D	OEM
UM4910DR	OEM
UM4910E	OEM
UM6000	OEM
UM6001A	OEM
UM6001B	OEM
UM6001C	OEM
UM6001CR	OEM
UM6001E	OEM
UM6001H	OEM
UM6002A	OEM
UM6002B	OEM
UM6002C	OEM
UM6002E	OEM
UM6002H	OEM
UM6004A	OEM
UM6004B	OEM
UM6004C	OEM
UM6004CR	OEM
UM6004E	OEM
UM6006A	OEM
UM6006B	OEM
UM6006C	OEM
UM6006CR	OEM
UM6006E	OEM
UM6010A	OEM
UM6010B	OEM
UM6010C	OEM
UM6010CR	OEM
UM6010E	OEM
UM6101A	OEM
UM6101B	OEM
UM6101C	OEM

DEVICE TYPE	REPL CODE
UM6101CR	OEM
UM6101E	OEM
UM6104A	OEM
UM6104B	OEM
UM6104C	OEM
UM6104CR	OEM
UM6104E	OEM
UM6108A	OEM
UM6108B	OEM
UM6108C	OEM
UM6108CR	OEM
UM6108E	OEM
UM6200	OEM
UM6201A	OEM
UM6201B	OEM
UM6201C	OEM
UM6201CR	OEM
UM6201E	OEM
UM6202A	OEM
UM6202B	OEM
UM6202C	OEM
UM6202E	OEM
UM6204A	OEM
UM6204B	OEM
UM6204C	OEM
UM6204CR	OEM
UM6204E	OEM
UM6206A	OEM
UM6206B	OEM
UM6206C	OEM
UM6206CR	OEM
UM6206E	OEM
UM6600	OEM
UM6601A	OEM
UM6601B	OEM
UM6601C	OEM
UM6601CR	OEM
UM6601E	OEM
UM6601H	OEM
UM6602A	OEM
UM6602B	OEM
UM6602C	OEM
UM6602E	OEM
UM6602H	OEM
UM6604A	OEM
UM6604B	OEM
UM6604C	OEM
UM6604CR	OEM
UM6604E	OEM
UM6606A	OEM
UM6606B	OEM
UM6606C	OEM
UM6606CR	OEM
UM6606E	OEM
UM6606H	OEM
UM6610A	OEM
UM6610B	OEM
UM6610C	OEM
UM6610D	OEM
UM6610E	OEM
UM6845	OEM
UM6845E	OEM
UM7000	OEM
UM7001A	OEM
UM7001B	OEM
UM7001C	OEM
UM7001CR	OEM
UM7001D	OEM
UM7001DR	OEM
UM7001E	OEM
UM7002A	OEM
UM7002B	OEM
UM7002C	OEM
UM7004A	OEM
UM7004B	OEM
UM7004C	OEM
UM7004CR	OEM
UM7004D	OEM
UM7004DR	OEM
UM7004E	OEM
UM7006A	OEM
UM7006B	OEM
UM7006C	OEM
UM7006D	OEM
UM7006E	OEM
UM7010A	OEM
UM7010B	OEM
UM7010C	OEM
UM7010CR	OEM
UM7010D	OEM
UM7010DR	OEM
UM7010E	OEM
UM7016A	OEM
UM7016B	OEM
UM7016C	OEM
UM7016D	OEM
UM7016E	OEM
UM7020A	OEM
UM7020B	OEM
UM7020C	OEM
UM7020CR	OEM
UM7020D	OEM
UM7020DR	OEM
UM7020E	OEM
UM7100	OEM
UM7101A	OEM
UM7101B	OEM
UM7101C	OEM
UM7101CR	OEM

DEVICE TYPE	REPL CODE
UM7101D	OEM
UM7101DR	OEM
UM7101E	OEM
UM7102A	OEM
UM7102B	OEM
UM7102C	OEM
UM7102D	OEM
UM7102E	OEM
UM7104A	OEM
UM7104B	OEM
UM7104C	OEM
UM7104CR	OEM
UM7104D	OEM
UM7104DR	OEM
UM7104E	OEM
UM7108A	OEM
UM7108B	OEM
UM7108C	OEM
UM7108CR	OEM
UM7108D	OEM
UM7108DR	OEM
UM7108E	OEM
UM7200	OEM
UM7201A	OEM
UM7201B	OEM
UM7201C	OEM
UM7201CR	OEM
UM7201D	OEM
UM7201DR	OEM
UM7201E	OEM
UM7202A	OEM
UM7202B	OEM
UM7202C	OEM
UM7202D	OEM
UM7202E	OEM
UM7204A	OEM
UM7204B	OEM
UM7204C	OEM
UM7204CR	OEM
UM7204D	OEM
UM7204DR	OEM
UM7204E	OEM
UM7206A	OEM
UM7206B	OEM
UM7206C	OEM
UM7206CR	OEM
UM7206D	OEM
UM7206DR	OEM
UM7206E	OEM
UM7300	OEM
UM7301A	OEM
UM7301B	OEM
UM7301C	OEM
UM7301CR	OEM
UM7301D	OEM
UM7301DR	OEM
UM7301E	OEM
UM7302A	OEM
UM7302B	OEM
UM7302C	OEM
UM7302D	OEM
UM7302E	OEM
UM7304A	OEM
UM7304B	OEM
UM7304C	OEM
UM7304CR	OEM
UM7304D	OEM
UM7304DR	OEM
UM7304E	OEM
UM7306A	OEM
UM7306B	OEM
UM7306C	OEM
UM7306D	OEM
UM7306E	OEM
UM7310A	OEM
UM7310B	OEM
UM7310C	OEM
UM7310CR	OEM
UM7310D	OEM
UM7310DR	OEM
UM7310E	OEM
UM8048-8-038	OEM
UM8326	OEM
UM9301	OEM
UM9302	OEM
UM9401	OEM
UM9402	OEM
UM9415	OEM
UM9441	OEM
UM9601	OEM
UM9602	OEM
UM9603	OEM
UM9604	OEM
UM9605	OEM
UM9606	OEM
UM9607	OEM
UM9608	OEM
UM20414C	OEM
UMT1006	1966
UMT1007	4397
UMT1008	1966
UMT1009	1852
UMT1011	1852
UMT1012	1852
UMT1203	OEM
UMT1204	OEM
UMT3202	0142
UMT3584	OEM
UMT3585	OEM

DEVICE TYPE	REPL CODE
UMT13004	0728
UMT13005	0728
UMT13006	OEM
UMT13007	OEM
UMT13008	OEM
UMT13009	OEM
UMX520	OEM
UMX570	OEM
UMX2020	OEM
UN11A	OEM
UN11B	OEM
UN11C	OEM
UN11D	OEM
UN12A	OEM
UN12B	OEM
UN12C	OEM
UN12D	OEM
UN23	OEM
UN211F	OEM
UN411-D	OEM
UN421-D	OEM
UN521E	OEM
UN1111	4172
UN1112	1032
UN1113	4130
UN1210	OEM
UN1211	0832
UN1212	0887
UN1213	0898
UN1214	OEM
UN1216	OEM
UN1222	OEM
UN1231	OEM
UN2069NE	OEM
UN2111	1892
UN2113	5354
UN2117	OEM
UN2119	OEM
UN2124	OEM
UN2211	5375
UN2212	5376
UN2213	4049
UN2214	OEM
UN2215	OEM
UN2216	OEM
UN4066B	0101
UN4111	4172
UN4111TA	OEM
UN4112	1032
UN4112TA	OEM
UN4113	5586
UN4113TA	OEM
UN4114	OEM
UN4114TA	OEM
UN4115	OEM
UN4115TA	OEM
UN4211	2099
UN4211TA	OEM
UN4212	0887
UN4212TA	OEM
UN4213	0898
UN4213TA	OEM
UN4214	OEM
UN4214TA	OEM
UN4215	OEM
UN4215TA	OEM
UN4216	OEM
UN5112	OEM
UN5113	OEM
UN5115	OEM
UN5212	OEM
UN5213	OEM
UN5217	OEM
UN10015	OEM
UN05	OEM
UN06	OEM
U05E	0087
U05ES	0087
U05G	0015
U06C	4920
UP11A	OEM
UP11B	OEM
UP11C	OEM
UP11D	OEM
UP12A	OEM
UP12B	OEM
UP12C	OEM
UP12D	OEM
UP32C	2891
UP1706	0007
UP1706A	0007
UP1706B	0007
UP11303	0280
UP11305	0280
UP11307	0280
UP11309	0280
UP11345	0004
UP11347	0004
UP11352	0004
UP11353	0280
UP12217	0560
UP12218	0560
UP12222	0079
UP12222B	0079
UP14046	0560
UP14046-46	0079
UP14047	0560
UP14047-46	0079
UPA16	OEM

If replacement code is OEM, contact original manufacturer for replacement.

DEVICE TYPE	REPL CODE
UPA33A	3839
UPA34A	OEM
UPA44D	OEM
UPA50A	OEM
UPA51B	OEM
UPA53C	4138
UPA54H	OEM
UPA56C	OEM
UPA60A	OEM
UPA61A	OEM
UPA62C	OEM
UPA63H	OEM
UPA64H	3126
UPA64HA	3126
UPA68H	OEM
UPA70A	OEM
UPA71A	OEM
UPA72H	OEM
UPA79C	5870
UPA80C	OEM
UPA81C	0969
UPA1428H	OEM
UPA1436H	OEM
UPA1437H	OEM
UPA1456H	OEM
UPA2003	1134
UPA2003C	1134
UPA2004C	1261
UPB74H04C	1907
UPB74H40C	0559
UPB201C	0232
UPB201D	0232
UPB202C	0512
UPB202D	0512
UPB203C	0697
UPB203D	0697
UPB204C	0873
UPB204D	0873
UPB205C	1024
UPB205D	1024
UPB206C	0743
UPB206D	0743
UPB207C	1168
UPB207D	1168
UPB208D	1185
UPB209D	1202
UPB210C	1275
UPB210D	1275
UPB211C	1426
UPB211D	1404
UPB212C	1172
UPB212D	1426
UPB213C	1313
UPB214C	1313
UPB214D	1313
UPB215C	0268
UPB215D	0268
UPB216C	2398
UPB216D	2398
UPB217C	1432
UPB217D	1432
UPB218C	1038
UPB218D	1038
UPB219C	1208
UPB219D	1208
UPB222C	0834
UPB222D	0834
UPB223C	0569
UPB223D	0569
UPB224C	1158
UPB224D	1158
UPB225C	1172
UPB225D	1172
UPB226C	1486
UPB226D	1486
UPB227D	1052
UPB230D	0117
UPB232C	0314
UPB232D	0314
UPB233C	4856
UPB233D	4856
UPB234C	0467
UPB234D	0467
UPB235D	0361
UPB236D	0385
UPB237D	3520
UPB238D	0996
UPB551C	OEM
UPB552C	OEM
UPB553	OEM
UPB553AC	OEM
UPB562	OEM
UPB562AC	OEM
UPB562C	OEM
UPB568	OEM
UPB568HA	OEM
UPB2047D	1108
UPB2080D	1535
UPB2085D	0374
UPB2086D	1368
UPB2091D	0980
UPB2150D	1492
UPB2151D	1495
UPB2154D	1555
UPB2161D	1645
UPB2170D	1722
UPB2180D	1829
UPB2181P	1842
UPB2182D	1856

DEVICE TYPE	REPL CODE
UPB2192D	1921
UPB2193D	1926
UPB2195D	1943
UPB2198D	1964
UPB6101-009	OEM
UPB7400C	0232
UPB7402C	0314
UPB7404C	0361
UPB7405C	0385
UPB7410C	0512
UPB7413C	1441
UPB7420C	0697
UPB7430C	0873
UPB7437C	3520
UPB7438C	0996
UPB7440C	1024
UPB7442C	1052
UPB7445C	1082
UPB7447C	1108
UPB7450C	0743
UPB7451C	1168
UPB7453C	1185
UPB7454C	1202
UPB7460C	1275
UPB7473C	1172
UPB7474C	1313
UPB7476C	1158
UPB7480C	1535
UPB7485C	0374
UPB7486	1368
UPB7486C	1368
UPB7491C	0980
UPB8216C	1863
UPB8284AD	OEM
UPB8288D	OEM
UPB74107C	0942
UPB74123C	1157
UPB74141D	1377
UPB74150C	1492
UPB74151C	1495
UPB74153C	1539
UPB74154C	1555
UPB74155C	1576
UPB74156	1592
UPB74156C	1592
UPB74157C	1605
UPB74161C	1645
UPB74164C	0734
UPB74170D	1722
UPB74175C	1787
UPB74180C	1829
UPB74181D	1842
UPB74182C	1856
UPB74192C	1921
UPB74193C	1926
UPB74195C	1943
UPB74198D	1964
UPC15MA	0093
UPC-16	2393
UPC16	2393
UPC-16A	2393
UPC16A	2393
UPC-16C	2393
UPC16C	2393
UPC17C	2882
UPC20	OEM
UPC-20C	0470
UPC20C	0470
UPC20C1	0470
UPC20C2	0470
UPC22C	3352
UPC23C	0528
UPC24M05HF	OEM
UPC27C	4543
UPC27CI	4543
UPC29C	0638
UPC29C2	0638
UPC29C3	0638
UPC30C	0653
UPC31C	0673
UPC32C	2891
UPC35C	OEM
UPC41C	4578
UPC46C	0930
UPC47C	0946
UPC48C1	0964
UPC49C	2887
UPC-50C	2886
UPC55D	OEM
UPC56C	OEM
UPC71A	1797
UPC78L05	1292
UPC78L05A	1292
UPC-78L-05AN	1292
UPC78L05J	1292
UPC78L05J-TP	1292
UPC78L05T	1292
UPC78L05T-E1	OEM
UPC78L08	0083
UPC78L12	1828
UPC78L15	3395
UPC78L24J	4509
UPC78M05	0624
UPC78M05H	0624
UPC78M08H	1196
UPC78M10H	2806
UPC78M12	0334
UPC78M12H	3485
UPC78M12H2	0334

DEVICE TYPE	REPL CODE
UPC78M15H	3490
UPC78M18	2259
UPC78M18H	2259
UPC78N05	OEM
UPC78N05H	OEM
UPC78N12H	OEM
UPC81	0470
UPC81C	0470
UPC143-05	0624
UPC143C08	1196
UPC151A	0410
UPC151C	0312
UPC154A	2890
UPC157A	0093
UPC157C	1300
UPC159A	OEM
UPC177C	0176
UPC206	0470
UPC213C	1313
UPC215C	0268
UPC251A	2368
UPC251C	0360
UPC262C	2891
UPC271C	2104
UPC276	4543
UPC301AC	1300
UPC301AN	1300
UPC305(C)	2169
UPC305C	2169
UPC311	OEM
UPC311C	2104
UPC311G	4169
UPC311G2	4169
UPC317	OEM
UPC317H	6251
UPC324	OEM
UPC324C	0625
UPC324G2	3202
UPC339C	0176
UPC339G2	2742
UPC358	0770
UPC358C	0770
UPC358G	6331
UPC358G2	OEM
UPC358P	0770
UPC373C	0631
UPC380C	OEM
UPC393	0631
UPC393C	0631
UPC393G2	OEM
UPC451C	0625
UPC494C	1824
UPC539C	OEM
UPC544C	1215
UPC554	1478
UPC-554C	1478
UPC554C	1478
UPC554H	1478
UPC-555A	0632
UPC555A	0632
UPC555H	2316
UPC558	3785
UPC558C	3785
UPC561	3263
UPC561C	2592
UPC562C	2891
UPC563C	OEM
UPC563H	1160
UPC563H2	1160
UPC566	0433
UPC566H	0433
UPC566H2	0433
UPC566H3	0433
UPC566H-B	0433
UPC566H-L	0433
UPC566H-M	0433
UPC566H-N	0433
UPC569C2	4395
UPC570C	2887
UPC570G	2887
UPC571	3352
UPC-571C	3352
UPC571C	3352
UPC-573C	4612
UPC573C	4612
UPC-574	1329
UPC574	1329
UPC574-J	1329
UPC574I	1329
UPC-574J	1329
UPC574J	1329
UPC574J(L)	1329
UPC574J(M)	1329
UPC574J(V)	1329
UPC574JA	1329
UPC574JAG	1329
UPC574JC	1329
UPC574J-G	1329
UPC574JK	1329
UPC574J-KL	1329
UPC574JL	1329
UPC574JM	1329
UPC574J-T	1329
UPC574JT	1329
UPC574K	1329
UPC574K02L	1329
UPC574V	1329
UPC575	2872
UPC575C	2872

DEVICE TYPE	REPL CODE
UPC575C2	2872
UPC-576H	1479
UPC576H	1479
UPC577	2261
UPC577A	2261
UPC577H	2261
UPC577N	2261
UPC580C	1055
UPC587C	0706
UPC587C2	0706
UPC592A2	4209
UPC592H	4209
UPC592H2	4209
UPC592HZ	4209
UPC592N	4209
UPC595C	0852
UPC595C2	0857
UPC596	0857
UPC-596C	0857
UPC596C	0857
UPC596C2	0857
UPC661CA	OEM
UPC661G	OEM
UPC741C	0312
UPC767PC	0701
UPC-781-05AN	1292
UPC858	2253
UPC858C	2253
UPC861C	3815
UPC861CE	3815
UPC1001	4147
UPC1001-H2	4420
UPC1001H	4420
UPC1001H2	4420
UPC1004C	OEM
UPC1008C	1379
UPC1018	2926
UPC1018C	2926
UPC1018CE	2926
UPC1018C-F	2926
UPC1020	3695
UPC1020H	3695
UPC1023H	2631
UPC1024H	OEM
UPC1025	1160
UPC1025H	6351
UPC1026	0706
UPC1026C	0706
UPC1028H	2019
UPC1031B2	OEM
UPC1031H	0904
UPC1031H2	0904
UPC1032H	2534
UPC1032HA	OEM
UPC1037H	OEM
UPC1037HA	OEM
UPC1043C	OEM
UPC1057C	5191
UPC1093	OEM
UPC1154H	1816
UPC1155H	1816
UPC1156	1816
UPC1156H	1816
UPC1156H2	1816
UPC1158H2	4812
UPC1161(C)	4803
UPC1161C	4803
UPC1161C3	OEM
UPC1170H	OEM
UPC1171C	4807
UPC1173C	4779
UPC1177H	OEM
UPC1181	0754
UPC1181H	0754
UPC1182	0756
UPC1182A	0756
UPC1182H	0756
UPC1185H	4665
UPC1185H2	4665
UPC1186H	4778
UPC1187V	4809
UPC1188H	4796
UPC1191V	4811
UPC1197	6115
UPC1197C	1260
UPC1212C	6524
UPC1213	OEM
UPC1213C	4831
UPC1216V	4835
UPC1217G	OEM
UPC1222C	5023
UPC1225H	OEM
UPC1228H	OEM
UPC1230H	4735
UPC1230H2	1081
UPC1235C	OEM
UPC1237H	OEM
UPC1237HA	OEM
UPC1238	6121
UPC1238V	1048
UPC1241H	OEM
UPC1252H2	2269
UPC1252HA	2269
UPC1253H2	1604
UPC1253HA2	1604
UPC1263C2	OEM
UPC1270H	OEM
UPC1273H	OEM
UPC1277H	4840

DEVICE TYPE	REPL CODE
UPC1278H	OEM
UPC1290C	OEM
UPC1298V	OEM
UPC1303G	OEM
UPC1342V	OEM
UPC1350C	4844
UPC1352	0490
UPC1352C	0490
UPC1352S	2623
UPC1353	2623
UPC1353C	4635
UPC1355C	0395
UPC1356C	0865
UPC1356C2	0865
UPC1358A	0553
UPC1358C	0553
UPC1358H	0553
UPC1358H2	0553
UPC1358HZ	0553
UPC1360	1627
UPC1360C	1627
UPC1362C	OEM
UPC1363C	0683
UPC1363CA	1945
UPC1365C	OEM
UPC1366C	2533
UPC1366C2	2533
UPC-1367C	2916
UPC1367C	2625
UPC1368H	0675
UPC1368H2	0675
UPC1368HR	0675
UPC1368HZ	0675
UPC1371	2623
UPC1371C	2623
UPC1372C	2623
UPC1372E	0490
UPC1373	2026
UPC1373-H	2026
UPC1373H	2026
UPC1373HA	2026
UPC1377C	5209
UPC1378	2042
UPC1378-H-L	2042
UPC1378H	2042
UPC1378H-L	2042
UPC1378H-P	2042
UPC1378JE	OEM
UPC-1379C	4350
UPC1379C	4350
UPC1380	1055
UPC1380C	6681
UPC1382	OEM
UPC1382C	2896
UPC1391H	3776
UPC1391H-1	3776
UPC1391HA	3776
UPC1393C	OEM
UPC1393CA	OEM
UPC1394C	5048
UPC1394G	OEM
UPC1394G-T2	OEM
UPC1397C	OEM
UPC1401C	OEM
UPC1401CA	0503
UPC1402CA	0583
UPC1406HA	3254
UPC1458	0360
UPC1458C	0360
UPC1470H	3250
UPC1473HA	OEM
UPC1474-HA	OEM
UPC1474HA	5256
UPC1475HA	OEM
UPC1480CA	3365
UPC1481CA	3884
UPC1483CA	OEM
UPC1484CA	OEM
UPC1486	4930
UPC1486C	4930
UPC1487C	OEM
UPC1488H	OEM
UPC1490HA	4550
UPC1491HA	OEM
UPC1498H	6796
UPC1503C	OEM
UPC1504C	OEM
UPC1505C	OEM
UPC1513HA	3159
UPC1514CA	OEM
UPC1517CA	OEM
UPC1519HA	OEM
UPC1520CA	6165
UPC1524C	OEM
UPC1530CA	1065
UPC1531HA	OEM
UPC1536C	OEM
UPC1571C	OEM
UPC1651G	OEM
UPC1675G	OEM
UPC1705C-016	OEM
UPC1712G	OEM
UPC1852	OEM
UPC1852ACT	OEM
UPC1852BCT	OEM
UPC1853CT	OEM
UPC1870CA	OEM
UPC1870CA-001	OEM
UPC1870CA-001-L	OEM

DEVICE TYPE	REPL CODE
UPC1870CA-002	OEM
UPC1870CA-003	OEM
UPC1870GH	OEM
UPC1870GH006	OEM
UPC1870GH007	OEM
UPC1871CU	OEM
UPC1873CT	OEM
UPC1875CT	OEM
UPC1880C	1055
UPC1892	OEM
UPC1982	OEM
UPC1982BCT	OEM
UPC2002	1048
UPC2002V	1048
UPC2300ACA	OEM
UPC2300CA	OEM
UPC2300G	OEM
UPC2304CA	OEM
UPC2305GH	OEM
UPC2306AGH	OEM
UPC2308ACA	OEM
UPC2308BCA	OEM
UPC2308CA	OEM
UPC2308G	OEM
UPC2309GS	OEM
UPC2310C	OEM
UPC2311C	OEM
UPC2816C	2328
UPC4001C	0478
UPC4062	3320
UPC4066	0101
UPC4512C	OEM
UPC4539BC	OEM
UPC4556(C)	OEM
UPC4556C	OEM
UPC4557C	OEM
UPC4558	0360
UPC4558/RC45	OEM
UPC4558C	0360
UPC4558D	0360
UPC4570C	OEM
UPC4570G	OEM
UPC4570G2	OEM
UPC4570G2-E1	OEM
UPC4570G2-T1	OEM
UPC4570HA	3250
UPC4570HA-1	OEM
UPC4572C	OEM
UPC4572HA	OEM
UPC4741C	OEM
UPC5742	3452
UPC6105C	OEM
UPC7805	0624
UPC7805AHF	OEM
UPC7805H	0624
UPC7812	0334
UPC7812H	0334
UPC7812HA	0334
UPC7812HF	0334
UPC7815H	3829
UPC7818HF	2259
UPC7893A	OEM
UPC7893AHF	OEM
UPC7893H	OEM
UPC7893HF	OEM
UPC7912	OEM
UPC7912H	1838
UPC11564	OEM
UPC14305	1292
UPC14305H	0624
UPC14308	1196
UPC14308H	1196
UPC14312	0334
UPC14312H	0334
UPC14315	4341
UPC14315D	1321
UPC14315H	1321
UPC143150	OEM
UPD23C512EC-143	OEM
UPD23C512EC-185	OEM
UPD23C1000-1-539	OEM
UPD23C1000-1-580	OEM
UPD23C1000-1-629	OEM
UPD23C1000C-1-580	OEM
UPD23C1000C-135	OEM
UPD74HC02G	OEM
UPD74HC04	OEM
UPD74HC14	OEM
UPD74HC367G	OEM
UPD74HC373G	OEM
UPD74HCU04C	OEM
UPD78C10G	OEM
UPD82C43C	OEM
UPD236C-1	OEM
UPD416	0522
UPD416-3	0522
UPD416C	0522
UPD416C-1	0522
UPD416C-2	0522
UPD416D	0522
UPD416D-1	0522
UPD416D-2	0522
UPD444C	OEM
UPD446C-1	OEM
UPD446G-15	OEM
UPD446G-20	OEM
UPD449	OEM
UPD471C	0129
UPD546-037	OEM

DEVICE TYPE	REPL CODE
UPD546C	OEM
UPD546C-037	OEM
UPD546C037	OEM
UPD546C-107	OEM
UPD547-060	OEM
UPD547C-049	OEM
UPD547C-060	OEM
UPD552C-083	OEM
UPD552C083	OEM
UPD553-020	OEM
UPD553C	OEM
UPD553C-020	OEM
UPD553C-042	OEM
UPD553C-149	OEM
UPD553C149	OEM
UPD553C-167	OEM
UPD554C062	OEM
UPD554C-118	OEM
UPD610BC003	OEM
UPD650C-024	OEM
UPD650C109	OEM
UPD651G-517	OEM
UPD651G-518	OEM
UPD652C-066	OEM
UPD751G-604	OEM
UPD761C	OEM
UPD765	OEM
UPD765A	OEM
UPD765AC	OEM
UPD780C	3481
UPD833G	OEM
UPD857C	3805
UPD858	2253
UPD-858C	2253
UPD858C	2253
UPD861	3815
UPD861C	3815
UPD861CE	3815
UPD861E	3815
UPD1511C-044	OEM
UPD1511C-072	OEM
UPD1511C-073	OEM
UPD1511C-074	OEM
UPD1511C-079	OEM
UPD1511C-133	OEM
UPD1511C-137	OEM
UPD1511C-144	OEM
UPD1511C-161	OEM
UPD1511C-179	OEM
UPD1512C-045	OEM
UPD1512C-061	OEM
UPD1514C-031	OEM
UPD1521	OEM
UPD1703C-011	OEM
UPD1704C-562	OEM
UPD1705C	OEM
UPD1705C-012	OEM
UPD1705C-016	OEM
UPD1705C-518	OEM
UPD1705C-520	OEM
UPD1708-G	OEM
UPD1709-113	OEM
UPD1709A-738	OEM
UPD1709A-740	OEM
UPD1709AC530	OEM
UPD1709ACT-113	OEM
UPD1709ACT-128	OEM
UPD1709ACT-733	OEM
UPD1709ACT-736	OEM
UPD1709ACT-738	OEM
UPD1709ACT-740	OEM
UPD1709ACT-741	OEM
UPD1709ACT-746	OEM
UPD1709ACT-753	OEM
UPD1709ACT-755	OEM
UPD1709C-011	OEM
UPD1709C-014	OEM
UPD1709C-113	OEM
UPD1709C-511	OEM
UPD1709C-515	OEM
UPD1709C-518	OEM
UPD1709C-520	OEM
UPD1709C-521	OEM
UPD1709C-530	OEM
UPD1709C530	OEM
UPD1709C-535	OEM
UPD1709C-536	OEM
UPD1709C-538	OEM
UPD-1709CT	OEM
UPD1709CT-113	OEM
UPD1709CT-114	OEM
UPD1709CT-118	OEM
UPD1709CT-121	OEM
UPD1709CT-127	OEM
UPD1709CT-128	OEM
UPD1709CT-713	OEM
UPD-1709CT-715	OEM
UPD-1709CT715	OEM
UPD1709CT-715	OEM
UPD1709CT-722	OEM
UPD1710G-015	OEM
UPD1711CU-517	OEM
UPD1713AGE550	OEM
UPD1913C	OEM
UPD1913C-14	OEM
UPD1937C	2090
UPD1943	OEM
UPD1943G	OEM
UPD1986C	4788

Original Device Types

DEVICE TYPE	REPL CODE
UPD-1987C	OEM
UPD1987C	2090
UPD2102AL-4	4973
UPD2102ALC-4	4973
UPD2114LC	2048
UPD2114LC-1	2048
UPD2114LC-5	2048
UPD2147A-25	OEM
UPD2147A-35	OEM
UPD2147A-45	1693
UPD2364C-1	OEM
UPD2364EC-163	OEM
UPD2732	2697
UPD2732A	2697
UPD2732D	2697
UPD2732D-4	2697
UPD2764D-FC4-A2	0812
UPD2764D-FC5-A3	0812
UPD2801C	OEM
UPD2803C	OEM
UPD2810C	OEM
UPD2812C	OEM
UPD2814C	4639
UPD2816C	2328
UPD2861C	OEM
UPD3551D	OEM
UPD3551D-02	OEM
UPD4001BC	0478
UPD4001BG	OEM
UPD4001C	0478
UPD4002BC	2055
UPD4002C	2055
UPD4011	0215
UPD4011B	0215
UPD4011BC	0215
UPD4011BCG	OEM
UPD4011BCM	0215
UPD4011C	0215
UPD4012BC	0498
UPD4012C	0498
UPD4013BC	0413
UPD4013BG	0413
UPD4013C	0413
UPD4013G	OEM
UPD4015C	1014
UPD4016	1898
UPD4016C	1898
UPD4016C-2	1898
UPD4016CX	1898
UPD4016CX-20	OEM
UPD4016D-1	1898
UPD4017BC	0513
UPD4017C	0513
UPD4020C	1661
UPD4021C	1749
UPD4023BC	0519
UPD4023C	0519
UPD4024BC	OEM
UPD4025C	2072
UPD4027	1949
UPD4027BC	1949
UPD4027C	1949
UPD4028BC	OEM
UPD4029	2233
UPD4029C	2233
UPD4030BC	0500
UPD4030C	0500
UPD4035C	2777
UPD4039C	OEM
UPD4040BC	0056
UPD4040C	0056
UPD4042C	0121
UPD4044BC	OEM
UPD4049	0001
UPD4049BC	OEM
UPD4049C	0001
UPD4049UBC	0001
UPD4050C	0398
UPD4051BC	0366
UPD4052	0024
UPD4052BC	0024
UPD4052BG	OEM
UPD4053	0035
UPD4053B	0035
UPD4053BC	0035
UPD4053BG	0035
UPD4053BP	0035
UPD4066	0101
UPD4066B	0101
UPD4066B-1	0101
UPD4066BC	0101
UPD4066BG	OEM
UPD4066BP	0101
UPD4066C	0101
UPD4066G	0101
UPD4069	0119
UPD4069BG	OEM
UPD4069C	0119
UPD4069UBC	0119
UPD4069UBP	OEM
UPD4071	0129
UPD4071BC	0129
UPD4078C	0921
UPD4081	OEM
UPD4081BC	0626
UPD4081C	0332
UPD4093BC	2384
UPD4094BC	OEM
UPD4099B	OEM
UPD4099BC	OEM
UPD4160BC	1359
UPD4164-3	2357
UPD4168C-20	OEM
UPD4364C	OEM
UPD4364C-15LL	OEM
UPD4364C-20L	OEM
UPD4364C-20LL	OEM
UPD4364C-LL	OEM
UPD4464C-15L	OEM
UPD4468-20	OEM
UPD4503BC	2053
UPD4503C	2053
UPD4512BC	OEM
UPD4518	1043
UPD4520C	2675
UPD4528B	3198
UPD4528BC	5120
UPD4528C	3198
UPD4538	1063
UPD4538BC	1063
UPD4538BG	OEM
UPD4538G	OEM
UPD4539C	3656
UPD4555C	2938
UPD4556C	3431
UPD4584BC	3626
UPD6102G	OEM
UPD6104	OEM
UPD6104C	OEM
UPD6104C-001	OEM
UPD6104C-021	OEM
UPD6105C	OEM
UPD6105C-002	OEM
UPD6105P	OEM
UPD6108C-003	OEM
UPD6108C003	OEM
UPD6108C-006	OEM
UPD6110CA	OEM
UPD6117C	OEM
UPD6121G-001	OEM
UPD6122	OEM
UPD6122G-001	OEM
UPD6122G001	OEM
UPD6122G-501	OEM
UPD6122G-504	OEM
UPD6124CA601	OEM
UPD6124CA-612	OEM
UPD6124G	OEM
UPD6125ACA-630	OEM
UPD6125AG-575	OEM
UPD6125G-531	OEM
UPD6128C-001	OEM
UPD6128C001	OEM
UPD6140-001	OEM
UPD6140C	OEM
UPD6140C-001	OEM
UPD6140C001	OEM
UPD6141C-001	OEM
UPD6143	OEM
UPD6143C	OEM
UPD6143G-602	OEM
UPD6145C-001	OEM
UPD6145G-601	OEM
UPD6147G	OEM
UPD6163ACA	OEM
UPD6250C	OEM
UPD6251C	OEM
UPD6325C	1880
UPD6325G	OEM
UPD6326C	OEM
UPD6335C	OEM
UPD6335G	OEM
UPD6336C	OEM
UPD6376GSE1	OEM
UPD6460G-601	OEM
UPD6460GT-601	OEM
UPD6510-517	OEM
UPD6510-518	OEM
UPD6900C	OEM
UPD6901C	OEM
UPD6901G	OEM
UPD7051CU-538	OEM
UPD7225G	OEM
UPD7502G229	OEM
UPD7502G245	OEM
UPD7507C-053	OEM
UPD7507C107	OEM
UPD7507CU	OEM
UPD7507HC-033	OEM
UPD7507HC-035	OEM
UPD7507SC1	OEM
UPD7507SC2	OEM
UPD7508AC-018	OEM
UPD7508BGB-505	OEM
UPD7508C-A02	OEM
UPD7508C-A37	OEM
UPD7508C-B13	OEM
UPD7508H-519	OEM
UPD7508HC035	OEM
UPD7508HG-592	OEM
UPD7508HG51922	OEM
UPD7514G	OEM
UPD7514G-064	OEM
UPD7516HCW-279	OEM
UPD7516HCW-287	OEM
UPD7519G	OEM
UPD7519G-53	OEM
UPD7519G-109	OEM
UPD7519G-124	OEM
UPD7519G-530	OEM
UPD7519G-598	OEM
UPD7519G-59812	OEM
UPD7519G-60412	OEM
UPD7520C-055	OEM
UPD7520C-087	OEM
UPD7537C	OEM
UPD7538C-021	OEM
UPD7538C021	OEM
UPD7538C-028	OEM
UPD7538C-049	OEM
UPD7538L-021	OEM
UPD7564G511	OEM
UPD7564G513	OEM
UPD7566CS-041	OEM
UPD7800	OEM
UPD7800G	OEM
UPD7801	OEM
UPD7801G-141	OEM
UPD7801G-176	OEM
UPD7801G-176-36	OEM
UPD7801G-215	OEM
UPD7805	0624
UPD7809G-029	OEM
UPD7809G-050	OEM
UPD7810	OEM
UPD7810G	OEM
UPD7810H	OEM
UPD7810HCW	OEM
UPD7810HG	OEM
UPD7811	OEM
UPD7811G	OEM
UPD7811HG	OEM
UPD8080AFC	4547
UPD8085AC	OEM
UPD8155C	OEM
UPD8155HC-2	OEM
UPD8237AC-5	OEM
UPD8243HC	OEM
UPD8251AC	OEM
UPD8253	OEM
UPD8253-5	OEM
UPD8253C	OEM
UPD8253C-2	OEM
UPD8253C-5	OEM
UPD8255AC-2	OEM
UPD8255AC-5	0051
UPD8255AP-5	OEM
UPD8284A	OEM
UPD8585C	2253
UPD17008	OEM
UPD17008CW-514	OEM
UPD17051CU-538	OEM
UPD23128EC-088	1559
UPD27128D-2	OEM
UPD27256AD	OEM
UPD27256AD-X	OEM
UPD40666BC	OEM
UPD41221C	OEM
UPD41264C	OEM
UPD41264C-15	OEM
UPD41416C-20	OEM
UPD41464L	OEM
UPD41647H-45	2357
UPD42272AGF	OEM
UPD42274LE10	OEM
UPD42832C-150	OEM
UPD65006-015	OEM
UPD65006-LC	OEM
UPD65006CW-230	OEM
UPD65006G-124	OEM
UPD65022	OEM
UPD65025GC-025	OEM
UPD75104G	OEM
UPD75108-709	OEM
UPD75108GF-718	OEM
UPD75108GF-720	OEM
UPD75206CW-025	OEM
UPD75206G-516	OEM
UPD75206G516	OEM
UPD75208CW-058	OEM
UPD75208CW-069	OEM
UPD75208CW257	OEM
UPD75216ACW-029	OEM
UPD75216ACW-095	OEM
UPD75216ACW-254	OEM
UPD75308GF4033B9	OEM
UPD75512GF-013	OEM
UPD75512GF-013-3B9	OEM
UPD75516GF-019	OEM
UPD75516GF-025	OEM
UPD75516GF-045	OEM
UPD75516GF-064-3B9	OEM
UPI404	0050
UPI706	0016
UPI706A	0016
UPI706B	0016
UPI718A	0016
UPI956	0086
UPI1301	0050
UPI1303	0211
UPI1305	0211
UPI1307	0211
UPI1309	0211
UPI1345	0164
UPI1347	0280
UPI1352	0004
UPI1353	0004
UPI1613	0086
UPI1711	0086
UPI2069	OEM
UPI2069A	OEM
UPI2070	OEM
UPI2070A	OEM
UPI2071	OEM
UPI2071A	OEM
UPI2217	0086
UPI2218	0086
UPI2222	0016
UPI2222B	0016
UPI2222P	0079
UPI4046	0016
UPI4046-46	0016
UPI4047	0016
UPI4047-46	0016
UPM5-250	OEM
UPM5-250E	OEM
UPM5-250J	OEM
UPM5-500	OEM
UPM5-500D5	OEM
UPM5-500D12	OEM
UPM5-500D28	OEM
UPM5-500E	OEM
UPM5-500J	OEM
UPM5-1000	OEM
UPM5-1000B	OEM
UPM5-1000BE	OEM
UPM5-1000BJ	OEM
UPM5-1000D12	OEM
UPM5-1000D28	OEM
UPM5-1000E	OEM
UPM5-1000J	OEM
UPM5-2000	OEM
UPM5-2000D12	OEM
UPM5-2000D28	OEM
UPM5-2000E	OEM
UPM5-2000J	OEM
UPM6-150A	OEM
UPM6-150AE	OEM
UPM6-150AJ	OEM
UPM9-100A	OEM
UPM9-100AE	OEM
UPM9-100AJ	OEM
UPM12-100A	OEM
UPM12-100AE	OEM
UPM12-100AJ	OEM
UPM15-100A	OEM
UPM15-100AE	OEM
UPM15-100AJ	OEM
UPO7500	OEM
UPS577H	OEM
UPS2003C	OEM
UPT011	0233
UPT012	0233
UPT013	0233
UPT014	0187
UPT015	0187
UPT021	0168
UPT022	0168
UPT023	0168
UPT024	0168
UPT025	0168
UPT111	0086
UPT112	0086
UPT113	0086
UPT114	0187
UPT115	0187
UPT121	0429
UPT122	0429
UPT123	0429
UPT124	0429
UPT125	0429
UPT211	0622
UPT212	0622
UPT213	0622
UPT214	0622
UPT215	0622
UPT221	0178
UPT222	0178
UPT223	2096
UPT224	2096
UPT225	2096
UPT321	2096
UPT322	2096
UPT323	2096
UPT324	2096
UPT325	2096
UPT410	0543
UPT411	0363
UPT413	0363
UPT423	0363
UPT430	0363
UPT431	0363
UPT521	0429
UPT522	0429
UPT523	0429
UPT524	0429
UPT531	1966
UPT532	1966
UPT533	1966
UPT534	1966
UPT535	1852
UPT611	0622
UPT612	0622
UPT613	0622
UPT614	0622
UPT615	0622
UPT621	1966
UPT622	1966
UPT623	1966
UPT624	1966
UPT625	1966
UPT721	0429
UPT722	0429
UPT723	0429
UPT724	0429
UPT725	0429
UPT731	1966
UPT732	1966
UPT733	1966
UPT734	1966
UPT735	1966
UPT821	0429
UPT822	0429
UPT823	0429
UPT824	0429
UPT825	0429
UPT831	1966
UPT832	1966
UPT833	1966
UPT834	1966
UPT835	1966
UPT931	2485
UPT1021	0429
UPT1022	0429
UPT1023	0429
UPT1024	0429
UPT1025	0429
UPT1031	2485
UPT1032	2485
UPT1033	2485
UPT1034	2485
UPT1035	2485
UPT1131	0620
UPT1132	0620
UPT1133	0620
UPT1134	0620
UPT1135	0620
UPT1215	0622
UPTA510	OEM
UPTA520	OEM
UPTA530	OEM
UPTB520	OEM
UPTB530	OEM
UPTB540	OEM
UPTB550	OEM
UPTO21	0142
UPTO22	0142
UPTO23	0142
UPTO24	0142
UR105	0015
UR110	0015
UR115	0015
UR120	0015
UR123AA	0084
UR123BA	0090
UR123CA	0097
UR123DA	0097
UR123EA	0109
UR123FA	0109
UR123GA	0109
UR123HA	0109
UR123KA	0109
UR123MA	0122
UR123PA	0131
UR123RA	0131
UR123VA	0145
UR125	0015
UR205	0110
UR210	0953
UR215	0242
UR220	0242
UR225	0540
UR710	OEM
URS-126-2	0074
US-0100	OEM
US-0101	OEM
US-0102	OEM
US-0103	OEM
US-0104	OEM
US-0105	OEM
US-0106	OEM
US-0107	OEM
US-0108	OEM
US-0109	OEM
US-0110	OEM
US-0111	OEM
US-0112	OEM
US-0113	OEM
US-0114	OEM
US-0115	OEM
US-0908D	3845
US-0908E	OEM
US-0909D	3418
US-0909E	OEM
US-0910D	1257
US-0910E	OEM
US-0911D	1364
US-0911E	OEM
US-0912D	3897
US-0912E	OEM
US-0913D	3901
US-0913E	OEM
US-0921D	3337
US-0921E	OEM
US-0940D	OEM
US-0960D	1069
US12	OEM
US15	OEM
US15/1	0201
US15/1B	0201
US15/1S	0201
US15/16	0288
US15-1	0288
US15-1B	0201
US18	OEM
US20	OEM
US20/1	0201
US-20/1A	OEM
US20/1A	0201
US20/1B	0201
US20/1B-S	0288
US20/1BS	0288
US20/1D	OEM
US20/1S	0201
US20-113	0201
US25	OEM
US25/1A	0288
US25/1AS	0288
US25/1CS	0378
US25/1D	OEM
US25-1B	0378
US30	OEM
US35	OEM
US40	OEM
US45	OEM
US45A	OEM
US50	OEM
US50A	OEM
US54H00A	OEM
US54H00J	OEM
US54H01A	OEM
US54H01J	OEM
US54H04A	OEM
US54H04J	OEM
US54H05A	OEM
US54H05J	OEM
US54H08A	OEM
US54H08J	OEM
US54H10A	OEM
US54H10J	OEM
US54H11A	OEM
US54H11J	OEM
US54H20A	OEM
US54H20J	OEM
US54H21A	OEM
US54H21J	OEM
US54H22A	OEM
US54H22J	OEM
US54H30A	OEM
US54H30J	OEM
US54H37A	OEM
US54H37J	OEM
US54H40A	OEM
US54H40J	OEM
US54H50A	OEM
US54H50J	OEM
US54H51A	OEM
US54H51J	OEM
US54H52A	OEM
US54H52J	OEM
US54H53A	OEM
US54H53J	OEM
US54H54A	OEM
US54H54J	OEM
US54H55A	OEM
US54H55J	OEM
US54H60A	OEM
US54H60J	OEM
US54H61A	OEM
US54H61J	OEM
US54H62A	OEM
US54H62J	OEM
US54H71A	OEM
US54H71J	OEM
US54H72A	OEM
US54H72J	OEM
US54H73A	OEM
US54H73J	OEM
US54H76A	OEM
US54H76J	OEM
US54H78A	OEM
US54H78J	OEM
US60	OEM
US60A	OEM
US70	OEM
US70A	OEM
US74H00A	0682
US74H00J	0682
US74H01A	5338
US74H01J	5338
US74H04A	1907
US74H04J	1907
US74H05A	3251
US74H05J	3251
US74H08A	5355
US74H08J	5355
US74H10A	0685
US74H10J	0685
US74H11A	2398
US74H11J	2398
US74H20A	3715
US74H20J	3715
US74H21A	4856
US74H21J	4856
US74H22A	4596
US74H22J	4596
US74H30A	5383
US74H30J	5383
US74H37A	OEM
US74H37J	OEM
US74H40A	0559
US74H40J	0559
US74H50A	1792
US74H50J	1792
US74H51A	1944
US74H51J	1944
US74H52A	2020
US74H52J	2020
US74H53A	2101
US74H53J	2101
US74H54A	2172
US74H54J	2172
US74H55A	3158
US74H55J	3158
US74H60A	5410
US74H60J	5410
US74H61A	2663
US74H61J	2663
US74H62A	2731
US74H62J	2731
US74H71A	3264
US74H71J	3264
US74H72A	3313
US74H72J	3313
US74H73A	2463
US74H73J	2463
US74H74A	2492
US74H74J	2492
US74H76A	5302
US74H76J	1715
US74H78A	5418
US74H78J	5418
US78M12MLA	OEM
US80	OEM
US80A	OEM
US100	OEM
US100A	OEM
US120	OEM
US120A	OEM
US123AA	0084
US123BA	0097
US123CA	0097
US123DA	0097
US123EA	0109
US123FA	0109
US123GA	0109
US123HA	0109
US123KA	0122
US123MA	0122
US123PA	0131
US123RA	0131
US123TA	0145
US123VA	0145
US150	OEM
US150A	OEM
US180	OEM
US180A	OEM
US183AA	1600
US183BA	0871
US183CA	0871
US183DA	0871
US183EA	0853
US183FA	0853
US183GA	0853
US183HA	0853
US183KA	1609
US183MA	1609
US183PA	1610
US183RA	1610
US183TA	1614
US183VA	1614
US200	OEM
US200A	OEM
US284HF	0757
US315HF	0757
US322-3	OEM
US322-3-0	OEM
US322-3IR	OEM
US322-3LED	OEM
US720J	OEM
US721J	OEM
US727J	OEM
US730J	OEM
US731J	OEM
US732J	OEM
US734DM	OEM
US824HFP	0757
US1020	0124
US1020M	0124
US1040	0133
US1040M	0133
US1040MT	0133
US1060	0124
US1060M	0124
US1090	0124
US1090M	0124
US1511S	0201
US5400A	OEM
US5400J	OEM
US5401A	OEM
US5401J	OEM
US5402A	OEM
US5402J	OEM

If replacement code is OEM, contact original manufacturer for replacement.

DEVICE TYPE	REPL CODE
US5403A	OEM
US5403J	OEM
US5404A	OEM
US5404J	OEM
US5405A	OEM
US5405J	OEM
US5408A	OEM
US5408J	OEM
US5409A	OEM
US5409J	OEM
US5410A	OEM
US5410J	OEM
US5411A	OEM
US5411J	OEM
US5418A	OEM
US5418J	OEM
US5420A	OEM
US5420J	OEM
US5426A	OEM
US5430A	OEM
US5430J	OEM
US5432A	OEM
US5432J	OEM
US5440A	OEM
US5440J	OEM
US5441A	OEM
US5442A	OEM
US5443A	OEM
US5444A	OEM
US5445A	OEM
US5446A	OEM
US5447A	OEM
US5448A	OEM
US5450A	OEM
US5450J	OEM
US5451A	OEM
US5451J	OEM
US5453A	OEM
US5453J	OEM
US5454A	OEM
US5454J	OEM
US5459A	OEM
US5459J	OEM
US5460A	OEM
US5460J	OEM
US5470A	OEM
US5470J	OEM
US5472A	OEM
US5472J	OEM
US5473A	OEM
US5473J	OEM
US5474A	OEM
US5474J	OEM
US5475A	OEM
US5476A	OEM
US5476J	OEM
US5477J	OEM
US5480A	OEM
US5480J	OEM
US5482A	OEM
US5482J	OEM
US5483A	OEM
US5486A	OEM
US5486J	OEM
US5490A	OEM
US5490J	OEM
US5491A	OEM
US5491J	OEM
US5492A	OEM
US5492J	OEM
US5493A	OEM
US5493J	OEM
US5494A	OEM
US5495A	OEM
US5495J	OEM
US5496A	OEM
US7051A	0041
US7082A	3003
US7091	0057
US7120B	0052
US7150	0686
US7400A	0232
US7400J	0232
US7401A	0268
US7401J	0268
US7402A	0314
US7402J	0314
US7403A	0335
US7403J	0335
US7404A	0361
US7404J	0361
US7405A	0385
US7405J	0385
US7408A	0467
US7408J	0467
US7409A	0492
US7409J	0492
US7410A	0512
US7410J	0512
US7411A	0527
US7411J	0527
US7418A	OEM
US7418J	OEM
US7420A	0697
US7420J	0697
US7426A	0803
US7427A	0818
US7430A	0873
US7430J	0873

DEVICE TYPE	REPL CODE
US7432A	0899
US7432J	0899
US7438A	0996
US7438J	0996
US7440A	1024
US7440J	1024
US7441A	1038
US7442A	1052
US7443A	1060
US7444A	1072
US7445A	1082
US7446A	1098
US7447A	1108
US7448A	1125
US7450A	0743
US7450J	0743
US7451A	1168
US7451J	1168
US7453A	1185
US7453J	1185
US7454A	1202
US7454J	1202
US7459A	OEM
US7459J	OEM
US7460A	1275
US7460J	1275
US7470A	1404
US7470J	1404
US7472A	1426
US7472J	1426
US7473A	1172
US7473J	1172
US7474A	1313
US7474J	1313
US7475A	1432
US7475J	1432
US7476A	1158
US7476J	1158
US7477J	OEM
US7480A	1535
US7480J	OEM
US7482A	1574
US7482J	OEM
US7483A	0117
US7486A	1368
US7486J	1368
US7490A	1208
US7490J	1208
US7491A	0980
US7491J	OEM
US7492A	0834
US7492J	0834
US7493A	0569
US7493J	0569
US7494A	1702
US7495A	1486
US7495J	OEM
US7496A	1715
US54107A	OEM
US54107J	OEM
US54121A	OEM
US54121J	OEM
US54145A	OEM
US54153A	OEM
US54154A	OEM
US54180A	OEM
US54180J	OEM
US74107A	0942
US74107J	0942
US74121A	0175
US74121J	0175
US74145A	0619
US74153A	1539
US74154A	1555
US74154AN	1555
US74180A	1829
US74180J	1829
US75324A	OEM
US75324G	OEM
US75324H	OEM
US75324J	OEM
USAF116ES571PML	OEM
USAF117ES552MML	OEM
USAF304	OEM
USAF377A	OEM
USAF501ES001M	1480
USAF505ES105	0004
USAF508ES020P	0861
USAF508ES021P	0861
USAF510ES030M	1480
USAF510ES031M	1480
USAF511ES035P	0018
USAF511ES036P	0018
USAF512ES040	0280
USAF515ES045M	0037
USAF515ES046M	0037
USAF516ES047M	0037
USAF516ES048M	0037
USAF518ES065M	0342
USAF519ES067M	0715
USAF519ES068M	0715
USB2.5	OEM
USB5	OEM
USB7.5	OEM
USB10	OEM
USC1420-2	OEM
USC1420-4	OEM
USC1420-6	OEM
USC1420-8	OEM

DEVICE TYPE	REPL CODE
USC1440-2	OEM
USC1440-4	OEM
USC1440-6	OEM
USC1440-8	OEM
USC1470-2	OEM
USC1470-4	OEM
USC1470-6	OEM
USC1470-8	OEM
USC1490-2	OEM
USC1490-4	OEM
USC1490-6	OEM
USC1490-8	OEM
USC2120-2	OEM
USC2120-4	OEM
USC2120-6	OEM
USC2120-8	OEM
USC2140-2	OEM
USC2140-4	OEM
USC2140-6	OEM
USC2140-8	OEM
USC2250-2	OEM
USC2250-4	OEM
USC2250-6	OEM
USC2250-8	OEM
USC2270-2	OEM
USC2270-4	OEM
USC2270-6	OEM
USC2270-8	OEM
USD320C	OEM
USD335C	OEM
USD345C	OEM
USD420	OEM
USD435	OEM
USD445	OEM
USD520	OEM
USD535	OEM
USD545	OEM
USD545HR2	OEM
USD820	OEM
USD835	OEM
USD840	OEM
USD845	OEM
USD920	OEM
USD935	OEM
USD940	OEM
USD945	OEM
USD7703394	0632
USN2N1910W	OEM
USN2N1915W	OEM
USN2N1916W	OEM
USN2N2031W	OEM
USN-7441B	1038
USN-7480A	1535
USN-7480J	OEM
USN-7482A	1574
USN-7482J	OEM
USN-7483B	0117
USN-7490A	1208
USN-7490J	OEM
USN-7491A	0980
USN-7491J	OEM
USN-7492A	0834
USN-7492J	OEM
USN-7493A	0569
USN-7493J	OEM
USR12	OEM
USR15	OEM
USR20	OEM
USR25	OEM
USR30	OEM
USR35	OEM
USR40	OEM
USR40A	OEM
USR45	OEM
USR45A	OEM
USR50	OEM
USR50A	OEM
USR60	OEM
USR60A	OEM
USR70	OEM
USR70A	OEM
USR80	OEM
USR80A	OEM
USR100	OEM
USR100A	OEM
USR120	OEM
USR120A	OEM
USR150	OEM
USR150A	OEM
USR180	OEM
USR180A	OEM
USR931	OEM
USR932	OEM
USR933	OEM
USR934	OEM
USR1171	0317
USR1172	0317
USR1173	0317
USR1174	0317
USS5	OEM
USS7.5	OEM
USS10	OEM
USS15	OEM
USS729J	OEM
UST10	OEM
UST19	OEM
UST81	OEM
UST87	OEM
UST88	OEM

DEVICE TYPE	REPL CODE
UST722	OEM
UST760	OEM
UST761	OEM
UST762	OEM
UST763	OEM
UST764	OEM
UT7-242	OEM
UT11	0015
UT12	0242
UT12.8B	0365
UT13	0242
UT14	0015
UT15	0015
UT16	0015
UT17	0015
UT18	0015
UT21	0015
UT22	0015
UT23	0015
UT24	0015
UT25	0015
UT26	0015
UT26TRIAC	0772
UT27	0015
UT100	OEM
UT101	OEM
UT111	0015
UT112	0015
UT113	0015
UT114	0015
UT115	0015
UT116	0015
UT117	0015
UT118	0071
UT119	0817
UT120	0071
UT123AA	0534
UT123BA	0765
UT123CA	0765
UT123DA	0765
UT123EA	0538
UT123FA	0538
UT123GA	0538
UT123HA	0538
UT123KA	0816
UT123MA	0816
UT123PA	0545
UT123RA	0545
UT123TA	0550
UT123VA	0550
UT183AA	5005
UT183BA	1635
UT183CA	1635
UT183DA	1635
UT183EA	1251
UT183FA	1251
UT183GA	1251
UT183HA	1251
UT183KA	1205
UT183MA	1205
UT183PA	2137
UT183RA	2137
UT183TA	2251
UT183VA	2251
UT211	1747
UT212	1747
UT213	0540
UT214	1771
UT215	0965
UT221	0015
UT222	0015
UT223	0015
UT224	0015
UT225	0015
UT226	0015
UT228	0015
UT229	0015
UT231	0015
UT232	0015
UT233	0015
UT234	0242
UT235	0540
UT236	0953
UT237	1771
UT238	0965
UT242	0242
UT244	0540
UT245	1771
UT247	0965
UT249	0953
UT251	0953
UT252	0242
UT254	0540
UT255	1771
UT257	0965
UT258	0817
UT261	0953
UT262	0242
UT264	0540
UT265	1771
UT267	0965
UT268	0817
UT345	0071
UT347	0922
UT361	0817
UT362	0817
UT363	0922
UT364	0922
UT905	1335

DEVICE TYPE	REPL CODE
UT905A	0031
UT910	0080
UT910A	0031
UT920	0609
UT920A	0031
UT940	0795
UT940A	0031
UT950A	0031
UT-1000	OEM
UT1005	0031
UT1005A	0031
UT1010	0031
UT1010A	0031
UT1020	0031
UT1020A	0031
UT1040	0031
UT1040A	0031
UT1050A	0031
UT2005	0110
UT2010	0953
UT2020	0242
UT2040	0540
UT2060	0965
UT3005	0110
UT3010	0953
UT3020	0242
UT3040	0540
UT3060	0965
UT3080	0817
UT4005	OEM
UT4010	OEM
UT4020	OEM
UT4040	OEM
UT4060	OEM
UT5105	OEM
UT5110	OEM
UT5120	OEM
UT5140	OEM
UT5160	OEM
UT6105	OEM
UT6110	OEM
UT6120	OEM
UT6140	OEM
UT6160	OEM
UT8105	OEM
UT8110	OEM
UT8120	OEM
UT8140	OEM
UT8160	OEM
UTF015	OEM
UTF025	OEM
UTF040	OEM
UTL502	OEM
UTL1001	OEM
UTL1002	OEM
UT0161	OEM
UT0250	OEM
UT0310	OEM
UT0311	OEM
UT0410	OEM
UT0416	OEM
UT0421	OEM
UT0501	OEM
UT0501R	OEM
UT0502	OEM
UT0502R	OEM
UT0503	OEM
UT0503R	OEM
UT0504	OEM
UT0504R	OEM
UT0505	OEM
UT0505R	OEM
UT0507	OEM
UT0508	OEM
UT0510	OEM
UT0510R	OEM
UT0511	OEM
UT0511R	OEM
UT0512	OEM
UT0512R	OEM
UT0513	OEM
UT0513R	OEM
UT0514	OEM
UT0515	OEM
UT0515R	OEM
UT0516	OEM
UT0516R	OEM
UT0517	OEM
UT0521	OEM
UT0521R	OEM
UT0523	OEM
UT0523R	OEM
UT0524	OEM
UT0533	OEM
UT0543	OEM
UT0544	OEM
UT0545	OEM
UT0546	OEM
UT0551	OEM
UT0551R	OEM
UT0561	OEM
UT01001	OEM
UT01001R	OEM
UT01002	OEM
UT01002R	OEM
UT01003	OEM
UT01003R	OEM
UT01004	OEM
UT01011	OEM

DEVICE TYPE	REPL CODE
UT01012	OEM
UT01013	OEM
UT01021	OEM
UT01033	OEM
UT01043	OEM
UT01044	OEM
UT01045	OEM
UT01051	OEM
UT01051R	OEM
UT01501	OEM
UT01501R	OEM
UT01502	OEM
UT01502R	OEM
UT01503	OEM
UT01503R	OEM
UT01504	OEM
UT01511	OEM
UT01511R	OEM
UT01521	OEM
UT01522	OEM
UT02001	OEM
UT02001R	OEM
UT02002	OEM
UT02002R	OEM
UT02003	OEM
UT02003R	OEM
UT02011	OEM
UT02011R	OEM
UT02012	OEM
UT02013	OEM
UT02021	OEM
UT02022	OEM
UT02023	OEM
UT02031	OEM
UT02032	OEM
UT02033	OEM
UT02302	OEM
UT02302R	OEM
UT02303	OEM
UT02303R	OEM
UT02311	OEM
UT02311R	OEM
UT02321	OEM
UTR01	0271
UTR02	1191
UTR10	0271
UTR11	0271
UTR12	1191
UTR20	0271
UTR21	0271
UTR22	1191
UTR30	0271
UTR31	0271
UTR32	1191
UTR40	0271
UTR41	0271
UTR42	1191
UTR50	0271
UTR51	0271
UTR52	1191
UTR60	0271
UTR61	0271
UTR62	1191
UTR511	OEM
UTR512	OEM
UTR905A	1145
UTR910A	1145
UTR920A	1145
UTR940A	1145
UTR950A	OEM
UTR1005A	1145
UTR1010A	1145
UTR1020A	1145
UTR1040A	1145
UTR1050A	OEM
UTR2305	OEM
UTR2310	1191
UTR2320	1191
UTR2340	1191
UTR2350	1191
UTR2360	1191
UTR3305	1191
UTR3310	1191
UTR3320	1191
UTR3340	1191
UTR3350	1191
UTR3360	1191
UTR4305	OEM
UTR4310	OEM
UTR4320	OEM
UTR4340	OEM
UTR4350	OEM
UTR4360	OEM
UTR4405	OEM
UTR4410	OEM
UTR4420	OEM
UTR4440	OEM
UTR5405	OEM
UTR5410	OEM
UTR5420	OEM
UTR5440	OEM
UTR6405	OEM
UTR6410	OEM
UTR6420	OEM
UTR6440	OEM
UTS322-3	OEM
UTS322-3-0	OEM
UTS322-3IR	OEM
UTS322-3LED	OEM

DEVICE TYPE	REPL CODE
UTX105	2513
UTX110	2513
UTX115	2513
UTX120	2513
UTX125	OEM
UTX205	OEM
UTX210	OEM
UTX215	OEM
UTX220	OEM
UTX225	OEM
UTX3105	OEM
UTX3110	OEM
UTX3115	OEM
UTX3120	OEM
UTX4105	OEM
UTX4110	OEM
UTX4115	OEM
UTX4120	OEM
UV040BQ	OEM
UV5	OEM
UV7-1	OEM
UV13	OEM
UV13-1	OEM
UV14	OEM
UV78	OEM
UV79	OEM
UV95M	OEM
UV100BQ	OEM
UV204	OEM
UV205	OEM
UV215BQ	OEM
UV229-1	OEM
UV250BQ	OEM
UV444BQ	OEM
UV1001	OEM
UV1002	OEM
UV1003	OEM
UV1004	OEM
UV1005	OEM
UV1010	OEM
UV1011	OEM
UZ1-3.3	0292
UZ1-3.4	0292
UZ1-3.6	0188
UZ1-3.9	0456
UZ1-4.3	0533
UZ1-4.7	0451
UZ1-5.1	0162
UZ1-5.6	0157
UZ1-6.2	0636
UZ1-6.8	0025
UZ1-7.5	0649
UZ1-8.2	0244
UZ1-9.1	0012
UZ1-10	0170
UZ1-11	0317
UZ1-11.5	0794
UZ1-12	0137
UZ1-14	0100
UZ1-15	0002
UZ1-20	0531
UZ1-22	0565
UZ1-24	0632
UZ1-26	0441
UZ1-27	0441
UZ1-29	0726
UZ1-33	0039
UZ1-36	0820
UZ1-39	0350
UZ1-43	0931
UZ1-43G	0931
UZ1-47	0999
UZ1-47G	0999
UZ1-51	0502
UZ1-55	1834
UZ1-56	0869
UZ1-62	0783
UZ1-68	1268
UZ1-82	0331
UZ1-92	1311
UZ1-110	0149
UZ1-120	0186
UZ1-128	0365
UZ1-130	0213
UZ1-140	0245
UZ1-150	0028
UZ1-160	0255
UZ1-164	0420
UZ1-170	0877
UZ1-176	0495
UZ1-180	0367
UZ1-190	2858
UZ1-200	0421
UZ-2.0A	OEM
UZ-2.0B	OEM
UZ-2.2A	OEM
UZ-2.2B	OEM
UZ2.2BSB	OEM
UZ-2.4A	1276
UZ-2.4B	1276
UZ-2.7A	0760
UZ-2.7B	0760
UZ2.7BSA	OEM
UZ-3.0A	0118
UZ-3.0B	0118
UZ3.0B	OEM
UZ3.3	0292
UZ-3.3A	0299
UZ-3.3B	0299

DEVICE TYPE	REPL CODE
UZ3.3BSA	OEM
UZ3.6	0188
UZ-3.6A	0376
UZ-3.6B	0376
UZ3.6B	OEM
UZ-3.9	0036
UZ3.9	0036
UZ-3.9A	0036
UZ-3.9B	0036
UZ3.9B	OEM
UZ-4.0C	0275
UZ-4.3A	0275
UZ-4.3B	0275
UZ4.3BSB	OEM
UZ4.7	0140
UZ-4.7A	0140
UZ-4.7B	0140
UZ4.7BCA	0140
UZ4.7BSB	0140
UZ5.1	0162
UZ-5.1A	0041
UZ-5.1B	0041
UZ5.1B	0162
UZ-5.1BCB	OEM
UZ-5.1BCC	OEM
UZ5.1BSA	OEM
UZ5.1BSB	OEM
UZ-5.6B	0253
UZ5.6B	OEM
UZ-5.6BCC	2121
UZ5.6BM	OEM
UZ5.6BSA	OEM
UZ5.6BSB	0253
UZ6R8B	0025
UZ6.2	0471
UZ-6.2B	0471
UZ6.2B	0471
UZ6.2BSA	OEM
UZ6.2BSB	OEM
UZ6.2BSC	0295
UZ6.2ESA	OEM
UZ-6.8B	0062
UZ6.8B	0062
UZ6.8BH	OEM
UZ6.8BM	OEM
UZ6.8BSB	0062
UZ6.8V	0062
UZ7.5	0649
UZ-7.5B	0077
UZ7.5B	OEM
UZ7.5BL	OEM
UZ7.5BM	OEM
UZ7.5BS	OEM
UZ8.2	0244
UZ-8.2B	0165
UZ8.2B	0244
UZ8.2BM	OEM
UZ8.2V	0165
UZ9B	0012
UZ9.1	0012
UZ-9.1B	0057
UZ9.1B	0012
UZ9.1BSA	OEM
UZ9.1BSB	OEM
UZ-9.1BSC	OEM
UZ9.1BSC	0057
UZ-10B	0064
UZ10B	OEM
UZ10BM	OEM
UZ10BS	OEM
UZ-11B	0064
UZ11BSA	OEM
UZ-12B	0052
UZ12B	0052
UZ-12B(AU)	0052
UZ-12BCB	4115
UZ12BSA	OEM
UZ12BSB	OEM
UZ-13B	0053
UZ13B	0137
UZ13BH	OEM
UZ13BL	OEM
UZ13BSB	OEM
UZ15	0002
UZ-15B	0686
UZ15BL	OEM
UZ15BSB	OEM
UZ-16B	0445
UZ16B	OEM
UZ16.2BH	OEM
UZ-18B	0375
UZ18B	OEM
UZ-20B	0700
UZ20B	OEM
UZ20BSB	OEM
UZ-22B	0705
UZ-24B	0494
UZ-27B	0455
UZ27BSC	OEM
UZ-30B	0195
UZ30BSB	OEM
UZ30BSC	0195
UZ-33BSB	OEM
UZ-35B	0010
UZ-36BCA	OEM
UZ36BSA	OEM
UZ-40B	0032
UZ91E	OEM
UZ91F	0012
UZ110	1180
UZ111	1190
UZ112	1207
UZ113	1218
UZ114	1881
UZ115	0647
UZ116	1255
UZ117	2102
UZ118	1279
UZ119	2225
UZ120	0605
UZ122	OEM
UZ124	OEM
UZ126	OEM
UZ128	OEM
UZ130	OEM
UZ132	OEM
UZ134	OEM
UZ136	OEM
UZ138	OEM
UZ140	OEM
UZ210	1180
UZ211	1190
UZ212	1207
UZ213	1218
UZ214	1881
UZ215	0647
UZ216	1255
UZ217	2102
UZ218	1279
UZ219	2225
UZ220	0605
UZ222	OEM
UZ224	OEM
UZ226	OEM
UZ228	OEM
UZ230	OEM
UZ232	OEM
UZ234	OEM
UZ236	OEM
UZ238	OEM
UZ240	OEM
UZ310	1180
UZ311	1190
UZ312	1207
UZ313	1218
UZ314	1881
UZ315	0647
UZ316	1255
UZ317	2102
UZ318	1279
UZ319	2225
UZ320	0605
UZ322	OEM
UZ324	OEM
UZ326	OEM
UZ328	OEM
UZ330	OEM
UZ332	OEM
UZ334	OEM
UZ336	OEM
UZ338	OEM
UZ340	OEM
UZ706	0205
UZ707	0480
UZ708	0504
UZ709	0684
UZ710	0225
UZ712	0234
UZ713	0237
UZ714	1397
UZ715	0247
UZ716	0251
UZ718	0256
UZ720	0262
UZ722	0269
UZ724	0274
UZ727	0294
UZ730	0309
UZ733	0318
UZ736	0320
UZ740	0326
UZ745	0999
UZ750	0027
UZ756	0266
UZ760	2856
UZ770	2157
UZ775	0426
UZ780	0444
UZ790	0238
UZ806	0205
UZ807	0480
UZ808	0504
UZ809	0684
UZ810	0225
UZ812	0234
UZ813	0237
UZ814	1397
UZ815	0247
UZ816	0251
UZ818	0256
UZ820	0262
UZ822	0269
UZ824	0274
UZ827	0294
UZ830	0309
UZ833	0318
UZ836	0320
UZ840	0326
UZ845	0999
UZ850	0027
UZ856	0266
UZ860	2856
UZ870	2157
UZ875	0426
UZ880	0444
UZ890	0238
UZ906	0205
UZ907	0480
UZ908	0504
UZ909	0684
UZ910	0225
UZ912	0234
UZ913	0237
UZ914	1397
UZ915	0247
UZ916	0251
UZ918	0256
UZ920	0262
UZ922	0269
UZ924	0274
UZ927	0294
UZ930	0309
UZ933	0318
UZ936	0320
UZ940	0326
UZ945	OEM
UZ950	0027
UZ956	0266
UZ960	2856
UZ970	OEM
UZ975	0426
UZ980	0444
UZ990	0238
UZ3016	0025
UZ3016A	0025
UZ3016B	0025
UZ3017	0077
UZ3017A	0649
UZ3017B	0649
UZ3018	0244
UZ3018A	0244
UZ3018B	0244
UZ3019	0012
UZ3019A	0012
UZ3019B	0012
UZ3020	0170
UZ3020A	0170
UZ3020B	0170
UZ3021	0317
UZ3021A	0317
UZ3021B	0317
UZ3022	0137
UZ3022A	0137
UZ3022B	0137
UZ3023	0365
UZ3023A	0365
UZ3023B	0365
UZ3024	0002
UZ3024A	0002
UZ3024B	0002
UZ3025	0420
UZ3025A	0420
UZ3025B	0420
UZ3026	0495
UZ3026A	0495
UZ3026B	0495
UZ3027	0531
UZ3027A	0531
UZ3027B	0531
UZ3028	0565
UZ3028A	0565
UZ3028B	0565
UZ3029	0402
UZ3029A	0402
UZ3029B	0402
UZ3030	0441
UZ3030A	0441
UZ3030B	0441
UZ3031	0726
UZ3031A	0726
UZ3031B	0726
UZ3032	0039
UZ3032A	0039
UZ3032B	0039
UZ3033	0820
UZ3033A	0820
UZ3033B	0820
UZ3034	0350
UZ3034A	0350
UZ3034B	0350
UZ3035	0931
UZ3035A	0931
UZ3035B	0931
UZ3036	0999
UZ3036A	0999
UZ3036B	0999
UZ3037	0502
UZ3037A	0502
UZ3037B	0502
UZ3038	0869
UZ3038A	0869
UZ3038B	0869
UZ3039	0783
UZ3039A	0783
UZ3039B	0783
UZ3040	2157
UZ3040A	2157
UZ3040B	2157
UZ3041	1189
UZ3041A	1189
UZ3041B	1189
UZ3042	0331
UZ3042A	0331
UZ3042B	0331
UZ3043	1311
UZ3043A	1311
UZ3043B	1311
UZ3044	0098
UZ3044A	0098
UZ3044B	0098
UZ3045	0149
UZ3045A	0149
UZ3045B	0149
UZ3046	0186
UZ3046A	0186
UZ3046B	0186
UZ3047	0213
UZ3047A	0213
UZ3047B	0213
UZ3048	0028
UZ3048A	0213
UZ3048B	0028
UZ3049	0255
UZ3049A	0255
UZ3049B	0255
UZ3050	0367
UZ3050A	0367
UZ3050B	0367
UZ3051	0421
UZ3051A	0421
UZ3051B	0421
UZ4110	1180
UZ4111	1190
UZ4112	1207
UZ4113	1218
UZ4115	0647
UZ4116	1255
UZ4118	1279
UZ4120	0605
UZ4210	1180
UZ4211	1190
UZ4212	1207
UZ4213	1218
UZ4215	0647
UZ4216	1255
UZ4218	1279
UZ4220	0605
UZ4706	0205
UZ4707	0480
UZ4708	0504
UZ4709	0684
UZ4710	0225
UZ4712	0234
UZ4713	0237
UZ4715	0247
UZ4716	0251
UZ4718	0256
UZ4720	0262
UZ4722	0269
UZ4724	0274
UZ4727	0294
UZ4730	0309
UZ4733	0318
UZ4736	0320
UZ4739	0326
UZ4743	0337
UZ4747	0347
UZ4751	0027
UZ4756	0266
UZ4762	0386
UZ4768	0405
UZ4775	0426
UZ4782	0444
UZ4791	0238
UZ4806	0205
UZ4807	0480
UZ4808	0504
UZ4809	0684
UZ4810	0225
UZ4812	0234
UZ4813	0237
UZ4815	0247
UZ4816	0251
UZ4818	0256
UZ4820	0262
UZ4822	0269
UZ4824	0274
UZ4827	0294
UZ4830	0309
UZ4833	0318
UZ4836	0320
UZ4839	0326
UZ4843	0337
UZ4847	0347
UZ4851	0027
UZ4856	0266
UZ4862	0386
UZ4868	0405
UZ4875	0426
UZ4882	0444
UZ4891	0238
UZ5110	1180
UZ5111	1190
UZ5112	1207
UZ5113	1218
UZ5114	1881
UZ5115	0647
UZ5116	1255
UZ5117	2102
UZ5118	1279
UZ5119	2225
UZ5120	0605
UZ5122	OEM
UZ5124	OEM
UZ5126	OEM
UZ5128	OEM
UZ5130	OEM
UZ5132	OEM
UZ5134	OEM
UZ5136	OEM
UZ5138	OEM
UZ5140	OEM
UZ5210	1180
UZ5211	1190
UZ5212	1207
UZ5213	1218
UZ5214	1881
UZ5215	0647
UZ5216	1255
UZ5217	2102
UZ5218	1279
UZ5219	2225
UZ5220	0605
UZ5222	OEM
UZ5224	OEM
UZ5226	OEM
UZ5228	OEM
UZ5230	OEM
UZ5232	OEM
UZ5234	OEM
UZ5236	OEM
UZ5238	OEM
UZ5240	OEM
UZ5310	1180
UZ5311	1190
UZ5312	1207
UZ5313	1218
UZ5314	1881
UZ5315	0647
UZ5316	1255
UZ5317	2102
UZ5318	1279
UZ5319	2225
UZ5320	0605
UZ5322	OEM
UZ5324	OEM
UZ5326	OEM
UZ5328	OEM
UZ5330	OEM
UZ5332	OEM
UZ5334	OEM
UZ5336	OEM
UZ5338	OEM
UZ5340	OEM
UZ5706	0205
UZ5707	0480
UZ5708	0504
UZ5709	0684
UZ5710	0225
UZ5712	0234
UZ5713	0237
UZ5714	1397
UZ5715	0247
UZ5716	0251
UZ5718	0256
UZ5720	0262
UZ5722	0269
UZ5724	0274
UZ5727	0294
UZ5730	0309
UZ5733	0318
UZ5736	0320
UZ5740	0326
UZ5745	0347
UZ5750	0027
UZ5756	0266
UZ5760	2856
UZ5770	0405
UZ5775	0426
UZ5780	0444
UZ5790	0238
UZ5806	0205
UZ5807	0480
UZ5808	0504
UZ5809	0684
UZ5810	0225
UZ5812	0234
UZ5813	0237
UZ5814	1397
UZ5815	0247
UZ5816	0251
UZ5818	0256
UZ5820	0262
UZ5822	0269
UZ5824	0274
UZ5827	0294
UZ5830	0309
UZ5833	0318
UZ5836	0320
UZ5840	0326
UZ5845	0347
UZ5850	0027
UZ5856	0266
UZ5860	2856
UZ5870	0405
UZ5875	0426
UZ5880	0444
UZ5890	0238
UZ5906	0205
UZ5907	0480
UZ5908	0504
UZ5909	0684
UZ5910	0225
UZ5912	0234
UZ5913	0237
UZ5914	1397
UZ5915	0247
UZ5916	0251
UZ5918	0256
UZ5920	0262
UZ5922	0269
UZ5924	0274
UZ5927	0294
UZ5930	0309
UZ5933	0318
UZ5936	0320
UZ5940	0326
UZ5945	0347
UZ5950	0027
UZ5956	0266
UZ5960	2856
UZ5970	0405
UZ5975	0426
UZ5980	0444
UZ5990	0238
UZ7110	OEM
UZ7110L	OEM
UZ7210	OEM
UZ7210L	OEM
UZ7310	OEM
UZ7310L	OEM
UZ7706	OEM
UZ7706L	OEM
UZ7707	OEM
UZ7707L	OEM
UZ7708	OEM
UZ7708L	OEM
UZ7709	OEM
UZ7709L	OEM
UZ7710	OEM
UZ7710L	OEM
UZ7712	OEM
UZ7712L	OEM
UZ7713	OEM
UZ7713L	OEM
UZ7714	OEM
UZ7714L	OEM
UZ7715	OEM
UZ7715L	OEM
UZ7716	OEM
UZ7716L	OEM
UZ7718	OEM
UZ7718L	OEM
UZ7720	OEM
UZ7720L	OEM
UZ7722	OEM
UZ7722L	OEM
UZ7724	OEM
UZ7724L	OEM
UZ7727	OEM
UZ7727L	OEM
UZ7730	OEM
UZ7730L	OEM
UZ7733	OEM
UZ7733L	OEM
UZ7736	OEM
UZ7736L	OEM
UZ7740	OEM
UZ7740L	OEM
UZ7745	OEM
UZ7745L	OEM
UZ7750	OEM
UZ7750L	OEM
UZ7756	OEM
UZ7756L	OEM
UZ7760	OEM
UZ7760L	OEM
UZ7770	OEM
UZ7770L	OEM
UZ7775	OEM
UZ7775L	OEM
UZ7780	OEM
UZ7780L	OEM
UZ7790	OEM
UZ7790L	OEM
UZ7806	OEM
UZ7806L	OEM
UZ7807	OEM
UZ7807L	OEM
UZ7808	OEM
UZ7808L	OEM
UZ7809	OEM
UZ7809L	OEM
UZ7810	OEM
UZ7810L	OEM
UZ7812	OEM
UZ7812L	OEM
UZ7813	OEM
UZ7813L	OEM
UZ7814	OEM
UZ7814L	OEM
UZ7815	OEM
UZ7815L	OEM
UZ7816	OEM
UZ7816L	OEM
UZ7818	OEM
UZ7818L	OEM
UZ7820	OEM
UZ7820L	OEM
UZ7822	OEM
UZ7822L	OEM
UZ7824	OEM
UZ7824L	OEM
UZ7827	OEM
UZ7827L	OEM
UZ7830	OEM
UZ7830L	OEM
UZ7833	OEM
UZ7833L	OEM
UZ7836	OEM
UZ7836L	OEM
UZ7840	OEM
UZ7840L	OEM
UZ7845	OEM
UZ7845L	OEM
UZ7850	OEM
UZ7850L	OEM
UZ7856	OEM
UZ7856L	OEM
UZ7860	OEM
UZ7860L	OEM
UZ7870	OEM
UZ7870L	OEM
UZ7875	OEM
UZ7875L	OEM
UZ7880	OEM
UZ7880L	OEM
UZ7890	OEM
UZ7890L	OEM
UZ7906	OEM
UZ7906L	OEM
UZ7907	OEM
UZ7907L	OEM
UZ7908	OEM
UZ7908L	OEM
UZ7909	OEM
UZ7909L	OEM
UZ7910	OEM
UZ7910L	OEM
UZ7912	OEM
UZ7912L	OEM
UZ7913	OEM
UZ7913L	OEM
UZ7914	OEM
UZ7914L	OEM
UZ7915	OEM
UZ7915L	OEM
UZ7916	OEM
UZ7916L	OEM
UZ7918	OEM
UZ7918L	OEM
UZ7920	OEM
UZ7920L	OEM
UZ7922	OEM
UZ7922L	OEM
UZ7924	OEM
UZ7924L	OEM
UZ7927	OEM
UZ7927L	OEM
UZ7930	OEM
UZ7930L	OEM
UZ7933	OEM
UZ7933L	OEM
UZ7936	OEM
UZ7936L	OEM
UZ7940	OEM
UZ7940L	OEM
UZ7945	OEM
UZ7945L	OEM
UZ7950	OEM
UZ7950L	OEM
UZ7956	OEM
UZ7956L	OEM
UZ7960	OEM
UZ7960L	OEM
UZ7970	OEM
UZ7970L	OEM
UZ7975	OEM
UZ7975L	OEM
UZ7980	OEM
UZ7980L	OEM
UZ7990	OEM
UZ7990L	OEM
UZ8110	0098
UZ8111	0149
UZ8112	0186
UZ8113	0213
UZ8114	0245
UZ8115	0028
UZ8116	0255
UZ8117	0877
UZ8118	0367
UZ8119	2858
UZ8120	0421
UZ8210	0098
UZ8211	0149
UZ8212	0186
UZ8213	0213
UZ8214	0245
UZ8215	0028
UZ8216	0255
UZ8217	0877
UZ8218	0367

If replacement code is OEM, contact original manufacturer for replacement.

DEVICE TYPE	REPL CODE
UZ8219	2858
UZ8220	0421
UZ8706	0025
UZ8707	0649
UZ8708	0244
UZ8709	0012
UZ8710	0170
UZ8712	0137
UZ8713	0365
UZ8714	0100
UZ8715	0002
UZ8716	0420
UZ8718	0495
UZ8720	0531
UZ8722	0565
UZ8724	0402
UZ8727	0441
UZ8730	0726
UZ8733	0039
UZ8736	0820
UZ8740	0350
UZ8745	0999
UZ8750	0502
UZ8756	0869
UZ8760	1156
UZ8770	2157
UZ8775	1189
UZ8780	0331
UZ8790	1311
UZ8806	0025
UZ8807	0649
UZ8808	0244
UZ8809	0012
UZ8810	0170
UZ8812	0137
UZ8813	0365
UZ8814	0012
UZ8815	0002
UZ8816	0420
UZ8818	0495
UZ8820	0531
UZ8822	0565
UZ8824	0402
UZ8827	0441
UZ8830	0726
UZ8833	0039
UZ8836	0820
UZ8840	0350
UZ8845	0999
UZ8850	0502
UZ8856	0869
UZ8860	1156
UZ8870	2157
UZ8875	1189
UZ8880	0331
UZ8890	1311
UZL-3.3	0292
UZL-3.4	0292
UZL-3.6	0188
UZL-3.9	0456
UZL-4.3	0533
UZL-4.7	0451
UZL-5.1	0162
UZL-5.6	0157
UZL-6H2	OEM
UZL-6.2	0636
UZL-6.8	0025
UZL7M3	OEM
UZL-7.5	0649
UZL-8.2	0244
UZL-9.1	0012
UZL-10	0170
UZL-11	0317
UZL-12	0137
UZL-14	0100
UZL-15	0002
UZL-22	0565
UZL-24	0402
UZL-26	0441
UZL-27	0441
UZL-29	0726
UZL-33	0039
UZL-36	0820
UZL-39	0350
UZL-51	0502
UZL-55	1834
UZL-56	0869
UZL-62	0783
UZL-68	1268
UZL-82	0331
UZL-91	1311
UZL-110	0149
UZL-115	0794
UZL-120	0186
UZL-128	0365
UZL-130	0213
UZL-140	0245
UZL-150	0028
UZL-160	0255
UZL-163	0420
UZL-170	0877
UZL-176	0495
UZL-180	0367
UZL-190	2858
UZL-200	0421
UZP3.3BB	OEM
UZP5.1B	OEM
UZP-5.6B	0157
UZP-6.2B	0636
UZP-6.8B	0025
UZP-7.5B	0649
UZP-8.2B	0244
UZP-9.1B	0012
UZP-10B	0170
UZP-11B	0317
UZP11B	0317
UZP-12B	0137
UZP-13B	0365
UZP-15B	0002
UZP-16B	0420
UZP-18B	0495
UZP-20B	0531
UZP-22B	0565
UZP-24B	0402
UZP-27B	0441
UZP-30B	0726
UZP-33B	0039
UZT33	OEM
V0-3C	0071
V0-6B	0015
V0-6C	0015
V0-6C-401	0015
V0-9C	0071
V01C	0242
V01E	0540
V01G	0071
V01J	0817
V01L	0922
V03-C	0071
V03C	0015
V03D	OEM
V03E	0071
V03G	0071
V03J	0071
V05B	0087
V05E	0015
V06	0015
V06(DIO)	0015
V06(RECT)	0015
V06-C	0015
V06-E	0071
V06-G	0071
V06A	0023
V06B	0023
V06BX4	0015
V06C	0023
V06C(BOOST)	0102
V06C(LAFAYETTE)	0649
V06C(POWER-RECTIFIER	0015
V06C(S)	0023
V06C5	0015
V06CS	0023
V06E	0023
V06G	0023
V06J	0102
V07E	0071
V07G	0071
V07J	0071
V08E	0071
V08G	0071
V08J	0071
V09-E	0102
V09-E(CENT.)	0920
V09-G	0023
V09C	0023
V09E	0023
V09E-TA	OEM
V09ETA	0023
V09E-Z	0023
V09G	0023
V09G-F2	0023
V09G-TA	0023
V09GTA	0023
V0BX4	0554
V0G	0015
V-1	0293
V2I	OEM
V2SC1257-/11	OEM
V2SD669C	OEM
V5R5	OEM
V5R9	OEM
V5R12	OEM
V5R12-5	OEM
V5R12-12	OEM
V5R15	OEM
V5R15-15	OEM
V6/2R	0136
V6/2RC	0004
V6/2RJ	0004
V6/4R	0136
V6/4RC	0004
V6/4RJ	0004
V6/8R	0136
V6/8RJ	0004
V6/RC	0004
V7	OEM
V7A	OEM
V7B	OEM
V7C	OEM
V7D	OEM
V7E	OEM
V7EA	OEM
V7EB	OEM
V7EC	OEM
V7ED	OEM
V8ZA05	OEM
V8ZA1	OEM
V8ZA2	OEM
V10	0005
V10/1S	0136
V10/1SJ	0136
V10/2S	0280
V10/2SJ	0280
V10/15A	0211
V10/25	0280
V10/25J	0280
V10/30A	0211
V10/50A	0211
V10/50B	OEM
V10A	0005
V10B	OEM
V10C	OEM
V10D	OEM
V10E	OEM
V10EA	OEM
V10EB	OEM
V10EC	OEM
V10ED	OEM
V11-CA	0071
V11J	0102
V11L	0102
V11M	0102
V-11N	0102
V11N	0023
V12	0720
V12A	0720
V12A12-12	OEM
V12A15-15	OEM
V12B	OEM
V12C	OEM
V12D	OEM
V12E	OEM
V12EA	OEM
V12EB	OEM
V12EC	OEM
V12ED	OEM
V12P5	OEM
V12P12/12	OEM
V12P15/15	OEM
V12R5	OEM
V12R9	OEM
V12R12	OEM
V12R12-5	OEM
V12R12-12	OEM
V12R15	OEM
V12R15-15	OEM
V12ZA05	OEM
V12ZA1	OEM
V12ZA2	OEM
V13	OEM
V13/11	0280
V15	OEM
V15/10DP	0160
V15/10P	OEM
V15/15NP	0160
V15/20DP	0160
V15/20IP	OEM
V15/20P	OEM
V15/20R	0136
V15/30DP	0160
V15/30NP	0160
V15/30P	OEM
V15A	OEM
V15B	OEM
V15C	OEM
V15C200/80-V&	0015
V15C200/80-VF	0015
V15D	OEM
V15E	OEM
V15EA	OEM
V15EB	OEM
V15EC	OEM
V15ED	OEM
V17A	0031
V17B	0031
V17C	0031
V17D	0031
V17E	0031
V17L	0015
V18MA1A	OEM
V18Z-1	OEM
V18Z-3	OEM
V18ZA05	OEM
V18ZA1	1691
V18ZA2	OEM
V18ZA3	1694
V18ZA40	OEM
V19	0023
V19B	0023
V19C	0023
V19CS	0023
V19CSS	0023
V19C-T52	0023
V19E	0023
V19E-T52	0023
V19E-Z	0023
V19G	0017
V20	OEM
V20A	OEM
V20B	OEM
V20C	OEM
V20D	OEM
V20E	OEM
V20EA	OEM
V20EB	OEM
V20EC	OEM
V20ED	OEM
V20EG	OEM
V20G	OEM
V22	OEM
V22MA1A	OEM
V22Z-1	OEM
V22Z-3	OEM
V22ZA05	OEM
V22ZA1	2008
V22ZA2	OEM
V22ZA3	2009
V23	OEM
V24E	OEM
V24Z-1	OEM
V24Z-4	OEM
V24ZA1	1314
V24ZA4	1319
V24ZA50	OEM
V27	OEM
V27A	OEM
V27A-1	OEM
V27B	OEM
V27C	OEM
V27D	OEM
V27E	OEM
V27EA	OEM
V27EB	OEM
V27EC	OEM
V27ED	OEM
V27EG	OEM
V27G	OEM
V27MA1A	OEM
V27Z-1	OEM
V27Z-4	OEM
V27ZA05	OEM
V27ZA1	2325
V27ZA2	OEM
V27ZA3	OEM
V27ZA4	2326
V27ZA60	OEM
V30/10DP	0160
V30/10P	OEM
V30/15NP	0160
V30/20DP	0160
V30/20IP	OEM
V30/20P	0160
V30/30DP	0085
V30/30NP	0160
V30/30P	OEM
V30J	0071
V30L	0348
V30M	0348
V30N	0023
V33	5541
V33A	0030
V33A-2	OEM
V33B	OEM
V33C	OEM
V33D	OEM
V33EG	0030
V33G	0030
V33MA1A	2600
V33MA1B	OEM
V33Z-1	OEM
V33Z-5	OEM
V33ZA05	OEM
V33ZA1	2790
V33ZA2	OEM
V33ZA3	OEM
V33ZA5	2796
V33ZA70	OEM
V36ZA80	OEM
V39	OEM
V39A	OEM
V39A-2	OEM
V39B	OEM
V39B/C	OEM
V39C	OEM
V39D	OEM
V39E	OEM
V39EA	OEM
V39EB	OEM
V39EC	OEM
V39EG	OEM
V39G	OEM
V39MA2A	OEM
V39MA2B	OEM
V39Z-2	OEM
V39Z-6	OEM
V39ZA05	4367
V39ZA1	4367
V39ZA3	3073
V39ZA6	3073
V40B/C	OEM
V40LA2A	OEM
V40LA2B	OEM
V47	OEM
V47A	OEM
V47A-3	OEM
V47B	OEM
V47C	OEM
V47D	OEM
V47E	OEM
V47EA	OEM
V47EB	OEM
V47EC	OEM
V47ED	OEM
V47EG	OEM
V47G	OEM
V47MA2A	OEM
V47MA2B	OEM
V47Z-2	OEM
V47Z-7	OEM
V47ZA05	5744
V47ZA1	5744
V47ZA3	OEM
V47ZA7	3075
V50	OEM
V50A260-36	0143
V50A260-36(GE-DIO)	0143
V50A260-36(SI-DIO)	0133
V51	0004
V53B/C	OEM
V55B	OEM
V55D	OEM
V55E	OEM
V55F	OEM
V56	OEM
V56A	OEM
V56A-3	OEM
V56B	OEM
V56C	OEM
V56D	OEM
V56E	OEM
V56EA	OEM
V56EB	OEM
V56EC	OEM
V56ED	OEM
V56EG	OEM
V56G	OEM
V56MA2A	OEM
V56MA2B	OEM
V56Z-2	OEM
V56Z-8	OEM
V56ZA05	OEM
V56ZA2	2935
V56ZA3	OEM
V56ZA8	2941
V58	0050
V58C	OEM
V60/10DP	0160
V60/10P	0160
V60/20DP	0160
V60/20IP	OEM
V60/20P	0160
V60/30DP	0160
V60/30P	0160
V60LA3A	OEM
V60LA3B	OEM
V64	OEM
V68	OEM
V68A	OEM
V68A-3	OEM
V68B	OEM
V68C	OEM
V68D	OEM
V68E	OEM
V68EA	OEM
V68EB	OEM
V68EC	OEM
V68ED	OEM
V68G	OEM
V68MA3A	OEM
V68MA3B	OEM
V68Z-2	OEM
V68Z-10	OEM
V68ZA05	OEM
V68ZA2	4263
V68ZA3	OEM
V68ZA10	3078
V72	OEM
V74	0143
V75	0050
V77B	OEM
V77C	OEM
V78	0229
V82	OEM
V82A	OEM
V82A-5	OEM
V82B	OEM
V82C	OEM
V82D	OEM
V82E	OEM
V82EA	OEM
V82EB	OEM
V82EC	OEM
V82ED	OEM
V82G	OEM
V82MA3A	OEM
V82MA3B	OEM
V82Z-2	OEM
V82Z-12	OEM
V82ZA05	OEM
V82ZA2	4263
V82ZA4	OEM
V82ZA12	3085
V95LA7A	OEM
V95LA7B	OEM
V100	OEM
V100A	OEM
V100A-2	OEM
V100B	OEM
V100C	OEM
V100D	OEM
V100E	OEM
V100EA	OEM
V100EB	OEM
V100EC	OEM
V100ED	OEM
V100G	OEM
V100MA4A	OEM
V100MA4B	OEM
V100Z-3	OEM
V100Z-15	OEM
V100ZA05	4222
V100ZA3	4222
V100ZA4	3090
V100ZA15	3090
V115	0143
V117	0143
V117R-2	OEM
V117R-5	OEM
V117R-6	OEM
V117R-15	OEM
V118	0127
V119	0016
V120	0050
V120A-2	OEM
V120MA1A	OEM
V120MA2B	OEM
V120PH	0144
V120RH	0086
V120Z-3	OEM
V120Z-18	OEM
V120ZA05	4921
V120ZA1	4921
V120ZA4	3104
V120ZA6	3104
V126	0002
V129	0127
V130HE150	OEM
V130LA1	0840
V130LA2	1374
V130LA5	OEM
V130LA10	0838
V130LA10A	0838
V130LA20	0830
V130LA20A	0830
V130LA20B	4268
V130PA10A	OEM
V130PA20A	2543
V130PA20B	OEM
V130PA20C	OEM
V131P	OEM
V135	0143
V139	0086
V140	0649
V140LA2	OEM
V140LA5	OEM
V140LA10A	OEM
V143	0127
V144	0142
V145	0160
V146	0086
V148	0229
V150A-3	OEM
V150HE150	OEM
V150LA1	6790
V150LA2	0246
V150LA5	OEM
V150LA10	3415
V150LA10A	1993
V150LA20	3415
V150LA20A	3415
V150LA20B	3416
V150MA1A	OEM
V150MA2B	OEM
V150PA10A	OEM
V150PA20A	OEM
V150PA20B	OEM
V150PA20C	OEM
V150R-2	OEM
V150R-6	OEM
V150R-8	OEM
V150R-17	OEM
V150ZA05	OEM
V150ZA1	5157
V150ZA4	3116
V150ZA8	3116
V151(BENCO)	1069
V152	0037
V154	0086
V157	OEM
V159	0325
V160	0137
V162	0037
V166	0086
V167	0188
V168P	OEM
V169	0016
V170P	OEM
V171	0229
V172	0086
V174	0649
V175	0441
V175LA2	OEM
V175LA10A	OEM
V176	0161
V177	0086
V180	0126
V180A-3	OEM
V180MA1A	OEM
V180MA3B	OEM
V180ZA05	OEM
V180ZA1	4135
V180ZA5	OEM
V180ZA10	3123
V181(BENCO)	3429
V183	0325
V183R-2	OEM
V183R-7	OEM
V183R-9	OEM
V183R-20	OEM
V188	0162
V188P	OEM
V191	0087
V200	OEM
V205	0050
V-210C	0143
V210C	0015
V213P	OEM
V216R-3	OEM
V216R-7	OEM
V216R-10	OEM
V216R-20	OEM
V220	0144
V220A-4	OEM
V220MA2A	2627
V220MA4B	OEM
V220SM16	OEM
V220ZA05	OEM
V221	0144
V222	0144
V230LA4	OEM
V230LA10	OEM
V230LA20A	OEM
V237C/2K	OEM
V238A/1K	OEM
V239C/1K	OEM
V239C/2K	OEM
V241C/1K	OEM
V241C/2K	OEM
V250HE250	OEM
V250LA2	3786
V250LA4	1332
V250LA10	OEM
V250LA15	1334
V250LA15A	1334
V250LA20	1334
V250LA20A	1334
V250LA40	1336
V250LA40A	1336
V250LA40B	5264
V250PA10A	OEM
V250PA20A	OEM
V250PA40A	OEM
V250PA40B	OEM
V250PA40C	OEM
V250R-3	OEM
V250R-10	OEM
V250R-12	OEM
V250R-25	OEM
V262	OEM
V270	OEM
V270/6312	OEM
V270-D1	0015
V270A-4	OEM
V270MA2A	OEM
V270MA4B	OEM
V270ZA05	OEM
V275C/3M	OEM
V275HE250	OEM
V275L2	OEM
V275LA2	1658
V275LA4	1658
V275LA10	6816
V275LA15A	1660
V275LA20A	1660
V275LA40A	4867
V275LA40B	4867
V275PA10A	OEM
V275PA20A	OEM
V275PA40A	OEM
V275PA40B	OEM
V275PA40C	OEM
V283R-3	OEM
V283R-10	OEM
V283R-12	OEM
V283R-30	OEM
V290/6314	OEM
V290P	4183
V297	0016
V300	OEM
V300LA2	4471
V300LA4	4471
V310P	OEM
V311P	OEM
V312P	OEM
V313P	OEM
V320HE300	OEM
V320LA15A	OEM
V320LA20A	2034
V320LA40A	OEM
V320LA40B	OEM
V320P	OEM
V320PA40A	OEM
V320PA40B	OEM
V320PA40C	OEM
V321P	OEM
V322	OEM
V322P	OEM
V323P	OEM
V324	OEM
V326	OEM
V327	OEM
V328	OEM
V330	0110

DEVICE TYPE	REPL CODE
V330A-5	OEM
V330MA2A	OEM
V330MA5B	OEM
V330P	OEM
V330R-4	OEM
V330R-11	OEM
V330R-13	OEM
V330R-34	OEM
V330X	0031
V330ZA05	OEM
V331	0953
V331P	OEM
V331X	0023
V332	0242
V332P	OEM
V332X	0031
V333P	OEM
V334	0540
V334X	0031
V336	0965
V336X	0031
V338	0817
V340P	OEM
V341P	OEM
V342	OEM
V342P	OEM
V343P	OEM
V344	OEM
V346	OEM
V348	OEM
V350	0087
V350R-4	OEM
V350R-11	OEM
V350R-13	OEM
V350R-36	OEM
V350X	OEM
V351	OEM
V351X	OEM
V352	OEM
V352X	OEM
V354	OEM
V354X	OEM
V356	OEM
V356X	OEM
V358	OEM
V370A-5	OEM
V383R-5	OEM
V383R-13	OEM
V383R-15	OEM
V383R-40	OEM
V390A-6	OEM
V390MA3A	OEM
V390MA6B	OEM
V390P	OEM
V390ZA05	OEM
V400	OEM
V405	0144
V405A	3607
V409	0042
V410	0037
V410A	0037
V412A	OEM
V412B	OEM
V412C	OEM
V413	2730
V413K	2730
V413L	3331
V413M	3331
V413N	2730
V415	0144
V416R-5	OEM
V416R-15	OEM
V416R-20	OEM
V416R-45	OEM
V417	0144
V420HE400	OEM
V420LA10	OEM
V420LA20A	3257
V420LA40A	3859
V420LA40B	3859
V420LB20	3257
V420LB20A	3257
V420LB40	3859
V420LB40A	3859
V420LB40B	OEM
V420PA20A	OEM
V420PA40A	OEM
V420PA40B	OEM
V420PA40C	OEM
V430A-7	OEM
V430MA3A	OEM
V430MA7B	2360
V430ZA05	1658
V433EJ	OEM
V433GZ	OEM
V433N	OEM
V433R	OEM
V433T	OEM
V433X	OEM
V433Y	OEM
V435	0144
V435A	0037
V442	0015
V450R-5	OEM
V450R-15	OEM
V450R-20	OEM
V450R-45	OEM
V460LA20A	OEM
V460LA40A	OEM
V460LA40B	OEM
V460LB20	OEM
V460LB20A	OEM
V460LB40	OEM
V460LB40A	OEM
V460LB40B	OEM
V460PA20A	OEM
V460PA40A	OEM
V460PA40B	OEM
V460PA40C	OEM
V470ZA05	OEM
V480HE450	OEM
V480LA20A	1101
V480LA40A	1101
V480LA80A	0781
V480LA80B	OEM
V480LB20	1101
V480LB20A	1101
V480LB40	1101
V480LB40A	1101
V480LB80	0781
V480LB80A	0781
V480LB80B	OEM
V480PA20A	OEM
V480PA40A	OEM
V480PA80A	OEM
V480PA80B	OEM
V480PA80C	OEM
V500	3472
V500R-6	OEM
V500R-16	OEM
V500R-22	OEM
V500R-50	OEM
V510HE500	OEM
V510LA20A	OEM
V510LA40A	OEM
V510LA80A	OEM
V510LA80B	OEM
V510LB20	OEM
V510LB20A	OEM
V510LB40	OEM
V510LB40A	OEM
V510LB80	OEM
V510LB80A	OEM
V510LB80B	OEM
V510P	OEM
V510PA20A	OEM
V510PA40A	OEM
V510PA80A	OEM
V510PA80B	OEM
V510PA80C	OEM
V511P	OEM
V512P	OEM
V513P	OEM
V518P	OEM
V520P	OEM
V521P	OEM
V522P	OEM
V523P	OEM
V530P	4775
V531P	OEM
V532P	4790
V533P	4801
V540P	OEM
V541P	4859
V542P	1970
V543P	4861
V550LA20A	OEM
V550LA40A	OEM
V550LA80A	OEM
V550LA80B	OEM
V550LB20A	OEM
V550LB40A	OEM
V550LB80A	OEM
V550LB80B	OEM
V550P	OEM
V550PA20A	OEM
V550PA40A	OEM
V550PA80A	OEM
V550PA80B	OEM
V550PA80C	OEM
V550R-6	OEM
V550R-18	OEM
V550R-24	OEM
V550R-55	OEM
V551P	OEM
V552P	OEM
V553P	OEM
V575HE550	OEM
V575LA20A	OEM
V575LA40A	OEM
V575LA80A	OEM
V575LA80B	OEM
V575LB20A	OEM
V575LB40A	OEM
V575LB80A	OEM
V575LB80B	OEM
V575PA20A	OEM
V575PA40A	OEM
V575PA80A	OEM
V575PA80B	OEM
V575PA80C	OEM
V600	0086
V601	OEM
V602	OEM
V610	OEM
V611	OEM
V612	OEM
V622P	5016
V643	0264
V653	0861
V654	0791
V655	0791
V660HE600	OEM
V660LA50A	OEM
V660LA100B	OEM
V660PA100A	OEM
V660PA100B	OEM
V660PA100C	OEM
V680ZA05	OEM
V700R-70	OEM
V721	0037
V730R-70	OEM
V741	0791
V745	0126
V750HE700	OEM
V750ZA05	OEM
V751DA40	OEM
V761	0037
V763	0037
V765	0126
V765R-70	OEM
V800	3472
V800R-80	OEM
V850R-80	OEM
V900	OEM
V900B	OEM
V900C	OEM
V900E	OEM
V900EA	OEM
V900EB	OEM
V900EC	OEM
V900ED	OEM
V907	OEM
V907A	OEM
V907B	OEM
V907C	OEM
V907D	OEM
V907E	OEM
V907EA	OEM
V907EB	OEM
V907EC	OEM
V907ED	OEM
V910	0005
V910A	0005
V910B	OEM
V910C	OEM
V910D	OEM
V910E	OEM
V910EA	OEM
V910EB	OEM
V910EC	OEM
V910ED	OEM
V912	0720
V912A	0720
V912D	OEM
V912E	OEM
V912EA	OEM
V912EB	OEM
V912EC	OEM
V912ED	OEM
V915	OEM
V915A	OEM
V915B	OEM
V915C	OEM
V915D	OEM
V915E	OEM
V915EA	OEM
V915EB	OEM
V915EC	OEM
V915ED	OEM
V918	1974
V920	OEM
V920A	OEM
V920B	OEM
V920C	OEM
V920D	OEM
V920E	OEM
V920EA	OEM
V920EB	OEM
V920EC	OEM
V920ED	OEM
V927	OEM
V927A	OEM
V927B	OEM
V927C	OEM
V927D	OEM
V927E	OEM
V927EA	OEM
V927EB	OEM
V927EC	OEM
V927ED	OEM
V933	0030
V933A	0030
V933B	OEM
V933C	OEM
V933D	OEM
V933E	OEM
V933EA	OEM
V933EB	OEM
V933EC	OEM
V933ED	OEM
V939	OEM
V939A	OEM
V939B	OEM
V939C	OEM
V939D	OEM
V939E	OEM
V939EA	OEM
V939EB	OEM
V939EC	OEM
V939ED	OEM
V947	OEM
V947A	OEM
V947B	OEM
V947C	OEM
V947D	OEM
V947E	OEM
V947EA	OEM
V947EB	OEM
V947EC	OEM
V947ED	OEM
V953A/B/G/H	OEM
V956	OEM
V956A	OEM
V956B	OEM
V956C	OEM
V956D	OEM
V956E	OEM
V956EA	OEM
V956EB	OEM
V956EC	OEM
V956ED	OEM
V968	OEM
V968A	OEM
V968B	OEM
V968C	OEM
V968D	OEM
V968EA	OEM
V968EB	OEM
V968EC	OEM
V968ED	OEM
V982	OEM
V982A	OEM
V982B	OEM
V982C	OEM
V982D	OEM
V982E	OEM
V982EA	OEM
V982EB	OEM
V982EC	OEM
V982ED	OEM
V1000	OEM
V1000LA80A	OEM
V1000LA160A	OEM
V1000LA160B	OEM
V1000LB80	OEM
V1000LB80A	OEM
V1000LB160	OEM
V1000LB160A	OEM
V1000LB160B	OEM
V1000R-100	OEM
V1112	0133
V1400R-100	OEM
V1650E-1	0325
V1650E-4	0325
V1833E	0325
V2051E	OEM
V2051H	OEM
V2051K	OEM
V2059EJ	OEM
V2059GZ	OEM
V2059N	OEM
V2059R	OEM
V2059X	OEM
V2059Z	OEM
V2060EJ	OEM
V2060M	OEM
V2060P	OEM
V2060X	OEM
V3074A20	0015
V3074A21	0015
V3310	0922
V3510	OEM
V8634-3	0917
V9446-4	0015
V9646-4	0015
V10158	0015
V10916-1	0143
V-10916-3	0143
V10916-3	0143
V11189-1	0015
V15920	0015
V27001	0015
V50260-10	0133
V50260-16	0143
V50260-36	0133
VA2SC1815GRRW-1	0076
VA10	OEM
VA10X	OEM
VA12-12	0015
VA15	0348
VA15-15	OEM
VA15X	OEM
VA20	OEM
VA20X	OEM
VA21	OEM
VA22	OEM
VA23	OEM
VA24	OEM
VA25	OEM
VA25X	OEM
VA30	OEM
VA30X	OEM
VA35	OEM
VA87E	OEM
VA92	OEM
VA92B	OEM
VA92H	OEM
VA94	OEM
VA94B	OEM
VA97	OEM
VA97B	OEM
VA98	OEM
VA98J	OEM
VA98M	OEM
VA107	OEM
VA112	OEM
VA113	OEM
VA114	OEM
VA115	OEM
VA116	OEM
VA117	OEM
VA118A	OEM
VA122	OEM
VA123	OEM
VA124	OEM
VA125	OEM
VA126	OEM
VA127	0030
VA128	0030
VA129	OEM
VA130	OEM
VA132	0005
VA133	OEM
VA134	OEM
VA135	0005
VA136	OEM
VA137	OEM
VA138	OEM
VA139	OEM
VA140	OEM
VA141	OEM
VA145B	OEM
VA145D	OEM
VA145E	OEM
VA145F	OEM
VA145H	OEM
VA145J	OEM
VA145UH	OEM
VA145UL	OEM
VA146	OEM
VA146C	OEM
VA146M	OEM
VA146N	OEM
VA150	OEM
VA151	OEM
VA152	OEM
VA153	OEM
VA154	OEM
VA155	OEM
VA156	OEM
VA157	0005
VA158	OEM
VA159	OEM
VA160	0030
VA160E	OEM
VA160M	OEM
VA160N	OEM
VA161	OEM
VA161F	OEM
VA161M	OEM
VA162	OEM
VA162F	OEM
VA162M	OEM
VA163	OEM
VA163M	OEM
VA164M	OEM
VA165	OEM
VA166	OEM
VA167	OEM
VA168	OEM
VA169	OEM
VA170	OEM
VA171	OEM
VA172	OEM
VA173	OEM
VA173M	OEM
VA173T	OEM
VA173TA	OEM
VA175M	OEM
VA175T	OEM
VA177M	OEM
VA178M	OEM
VA183G	OEM
VA183GA	OEM
VA184F	OEM
VA184M	OEM
VA184T	OEM
VA185F	OEM
VA185M	OEM
VA187M	OEM
VA200	OEM
VA201	OEM
VA201B	OEM
VA202	OEM
VA203	OEM
VA203B	OEM
VA203B/6975	OEM
VA203C	OEM
VA203G	OEM
VA203H	OEM
VA203J	OEM
VA203K	OEM
VA203L	OEM
VA204	OEM
VA205	OEM
VA206	OEM
VA207	OEM
VA208	OEM
VA209	OEM
VA210	OEM
VA210B	OEM
VA210M	OEM
VA211	OEM
VA212	OEM
VA213	OEM
VA215H	OEM
VA217C	OEM
VA217HL	OEM
VA218	OEM
VA218B	OEM
VA220A	OEM
VA220B	OEM
VA220C	OEM
VA220D	OEM
VA220E	OEM
VA220F	OEM
VA220G	OEM
VA220J	OEM
VA220Z	OEM
VA221B	OEM
VA221C	OEM
VA221D	OEM
VA221E	OEM
VA221F	OEM
VA221G	OEM
VA221H	OEM
VA222A	OEM
VA222B8821	OEM
VA222C	OEM
VA222D	OEM
VA222E	OEM
VA222F	OEM
VA222G	OEM
VA222Z	OEM
VA225	OEM
VA232	OEM
VA237A	OEM
VA237B	OEM
VA237C	OEM
VA237D	OEM
VA237E	OEM
VA237F	OEM
VA240	OEM
VA240C	OEM
VA241	OEM
VA242	OEM
VA244	OEM
VA244K	OEM
VA246	OEM
VA249	OEM
VA253	OEM
VA259	OEM
VA259A	OEM
VA259AT	OEM
VA259B	OEM
VA259BT	OEM
VA259C	OEM
VA259CT	OEM
VA259DT	OEM
VA259E	OEM
VA259ET	OEM
VA259F	OEM
VA259FT	OEM
VA259J	OEM
VA259N	OEM
VA259P	OEM
VA259R	OEM
VA259T	OEM
VA264	OEM
VA265	OEM
VA266A	OEM
VA270VP6915	OEM
VA272	OEM
VA273B	OEM
VA273D	OEM
VA273G	OEM
VA273H	OEM
VA275A	OEM
VA275B/C	OEM
VA275D	OEM
VA282	OEM
VA282AD	OEM
VA282D	OEM
VA282E	OEM
VA283	OEM
VA286	OEM
VA287	OEM
VA287A	OEM
VA287B	OEM
VA287C	OEM
VA287D	OEM
VA288A	OEM
VA297	OEM
VA297T	OEM
VA297U	OEM
VA297V	OEM
VA297W	OEM
VA297X	OEM
VA297Y	OEM
VA297Z	OEM
VA300	OEM
VA301	OEM
VA302	OEM
VA303	OEM
VA304	OEM
VA305	OEM
VA306	OEM
VA307	OEM
VA308	OEM
VA309	OEM
VA310	OEM
VA311	OEM
VA312	OEM
VA313	OEM
VA318	OEM
VA322	OEM
VA326	OEM
VA327	OEM
VA329	OEM
VA435M	OEM
VA436M	OEM
VA440M	OEM
VA440N	OEM
VA441F	OEM
VA441M	OEM
VA450C	OEM
VA450M	OEM
VA450N	OEM
VA451M	OEM
VA451N	OEM
VA460M	OEM
VA460N	OEM
VA461M	OEM
VA461N	OEM
VA462N	OEM
VA470M	OEM
VA470N	OEM
VA472M	OEM
VA480M	OEM
VA482M	OEM
VA490M	OEM
VA490N	OEM
VA503B	OEM
VA503C	OEM
VA504B	OEM
VA506	OEM
VA508	OEM
VA510B	OEM
VA510D	OEM
VA513	OEM
VA514	OEM
VA515B	OEM
VA516	OEM
VA517	OEM
VA518	OEM
VA519	OEM
VA520	OEM
VA521	OEM
VA522	OEM
VA523	OEM
VA524	OEM
VA531	OEM
VA531A	OEM
VA532	OEM
VA533	OEM
VA534	OEM
VA535	OEM
VA536	OEM
VA537	OEM
VA620A	OEM
VA621	OEM
VA622	OEM
VA623	OEM
VA624	OEM
VA625L	OEM
VA645G	OEM
VA651	OEM
VA664A	OEM
VA723	OEM
VA724	OEM
VA800E	OEM
VA800H	OEM
VA802B	OEM
VA804	OEM
VA808	OEM
VA811	OEM
VA811B	OEM
VA811C	OEM
VA812D	OEM
VA812E	OEM
VA816J	OEM
VA820	OEM
VA820D	OEM
VA820E	OEM
VA820F	OEM
VA823	OEM
VA824	OEM
VA834B	OEM
VA838	OEM
VA838A	OEM
VA838B	OEM
VA839	OEM
VA842	OEM
VA845	OEM
VA846	OEM
VA853	OEM
VA853M	OEM
VA861	OEM
VA862A	OEM
VA862B	OEM
VA862C	OEM
VA864M	OEM

If replacement code is OEM, contact original manufacturer for replacement.

DEVICE TYPE	REPL CODE
VA866	OEM
VA866S	OEM
VA869A	OEM
VA870	OEM
VA874B	OEM
VA874E	OEM
VA876	OEM
VA876A	OEM
VA879	OEM
VA879G	OEM
VA882	OEM
VA883	OEM
VA884B	OEM
VA884C	OEM
VA884D	OEM
VA888	OEM
VA888B	OEM
VA888D	OEM
VA888E	OEM
VA888H	OEM
VA890A	OEM
VA890H	OEM
VA891A	OEM
VA891H	OEM
VA892A	OEM
VA892H	OEM
VA896	OEM
VA896F	OEM
VA903	OEM
VA908	OEM
VA909	OEM
VA911	OEM
VA913A	OEM
VA914	OEM
VA914B	OEM
VA915A	OEM
VA917B	OEM
VA920A	OEM
VA920B	OEM
VA921	OEM
VA922	OEM
VA925	OEM
VA925B	OEM
VA925C	OEM
VA925E	OEM
VA926	OEM
VA927	OEM
VA928A	OEM
VA930	OEM
VA932	OEM
VA932B	OEM
VA934	OEM
VA935	OEM
VA936A	OEM
VA936B	OEM
VA936E	OEM
VA936G	OEM
VA936J	OEM
VA936K	OEM
VA936L	OEM
VA936N	OEM
VA938A	OEM
VA938C	OEM
VA938D	OEM
VA943A	OEM
VA943B	OEM
VA944A	OEM
VA944B	OEM
VA945A	OEM
VA945B	OEM
VA946A	OEM
VA946A/B	OEM
VA946C	OEM
VA946H	OEM
VA947/B	OEM
VA947A	OEM
VA947C	OEM
VA947H	OEM
VA948A	OEM
VA948A/B	OEM
VA948C	OEM
VA949SERIES	OEM
VA950A	OEM
VA950G	OEM
VA951A	OEM
VA951G	OEM
VA952A	OEM
VA952G	OEM
VA953A/B	OEM
VA954A/B	OEM
VA954A/B/G/H	OEM
VA955A/B	OEM
VA955A/B/G	OEM
VA956	OEM
VA958A	OEM
VA961	OEM
VA962	OEM
VA963A	OEM
VA963D	OEM
VA988	OEM
VA988A	OEM
VA988B	OEM
VA988C	OEM
VA5139	0554
VA5139A	0554
VA5140	OEM
VA5140A	OEM
VA5141	OEM
VA5141A	OEM

DEVICE TYPE	REPL CODE
VA5142	OEM
VA5142A	OEM
VA5143	OEM
VA5143A	OEM
VAAD009	0139
VAANZ2SC1096	0390
VAB11BN20	OEM
VAB803N20	OEM
VAB803N25	OEM
VAB814A3N44	OEM
VAB824N26	OEM
VAB825N20	OEM
VAB825N26	OEM
VAB1610AN28	OEM
VAC824N20	OEM
VAMV-1	0015
VA011A	OEM
VA011CN20	OEM
VA011DN20	OEM
VA012AN20	OEM
VA012BN20	OEM
VA012FN20	OEM
VA021CN19	OEM
VA021DN19	OEM
VAP81AEP	OEM
VAP81AM	OEM
VAP81AP	OEM
VAP81BEP	OEM
VAP81BM	OEM
VAP81BN19	OEM
VAP81BP	OEM
VAP81CEP	OEM
VAP81CM	OEM
VAP81CN9	OEM
VAP81CN15	OEM
VAP81CN20	OEM
VAP81CP	OEM
VAP81DM	OEM
VAP81DP	OEM
VAP81EEP	OEM
VAP81EM	OEM
VAP81EP	OEM
VAP81FEP	OEM
VAP81FM	OEM
VAP81FN19	OEM
VAP81FP	OEM
VAP81GEP	OEM
VAP81GM	OEM
VAP81GP	OEM
VAP81HEP	OEM
VAP81HM	OEM
VAP81HP	OEM
VAP81IEP	OEM
VAP81IM	OEM
VAP81IP	OEM
VAP82AEP	OEM
VAP82AM	OEM
VAP82AP	OEM
VAP82BEP	OEM
VAP82BM	OEM
VAP82BP	OEM
VAP82CEP	OEM
VAP82CM	OEM
VAP82CP	OEM
VAP82DEP	OEM
VAP82DM	OEM
VAP82DP	OEM
VAP82EEP	OEM
VAP82EM	OEM
VAP82EP	OEM
VAP83AEP	OEM
VAP83AM	OEM
VAP83AN9	OEM
VAP83AP	OEM
VAP83BEP	OEM
VAP83BM	OEM
VAP83BP	OEM
VAP83CEP	OEM
VAP83CM	OEM
VAP83CP	OEM
VAP83DEP	OEM
VAR	0133
VAR-1R2	0133
VARIST-5	0015
VAS13	OEM
VAS23	OEM
VAS24	OEM
VAS33	OEM
VAS34	OEM
VAS44	OEM
VAS45	OEM
VAS54	OEM
VAS55	OEM
VAS62N10	OEM
VAS62N16	OEM
VAS62N18	OEM
VAS62N20	OEM
VAS64	OEM
VAS65	OEM
VAT11N16	OEM
VAT11N18	OEM
VAT11N20	OEM
VAT11N32	OEM
VAT12N32	OEM
VAT13AN9	OEM
VAT13AN11	OEM
VAT13N32	OEM
VAT14N32	OEM
VAT15N32	OEM

DEVICE TYPE	REPL CODE
VAT16N32	OEM
VAT17N32	OEM
VAT18N32	OEM
VAT19N32	OEM
VAT51N32	OEM
VAT52N16	OEM
VAT52N18	OEM
VAT52N20	OEM
VAT52N32	OEM
VAT53N32	OEM
VAT54N32	OEM
VAT55N32	OEM
VAT56N32	OEM
VAT57N32	OEM
VAT58N32	OEM
VAT59N32	OEM
VAT71N32	OEM
VAT72N32	OEM
VAT73N32	OEM
VAT74N32	OEM
VAT75N32	OEM
VAT76N32	OEM
VAT77N32	OEM
VAT78N32	OEM
VAT79N32	OEM
VAT1002N9	OEM
VAV091	OEM
VAV092	OEM
VAV093	OEM
VAV094	OEM
VAV121	OEM
VAV122	OEM
VAV123	OEM
VAV124	OEM
VAV151	OEM
VAV152	OEM
VAV153	OEM
VAV154	OEM
VAV201	OEM
VAV202	OEM
VAV203	OEM
VAV204	OEM
VB	OEM
VB011	OEM
VB013-10N01	OEM
VB013-12N01	OEM
VB013-14N01	OEM
VB013-16N01	OEM
VB020-04N01	OEM
VB020-08N01	OEM
VB020-10N01	OEM
VB020-12N01	OEM
VB020-14N01	OEM
VB020-16N01	OEM
VB1B	OEM
VB2	OEM
VB3-80	OEM
VB10	OEM
VB10X	OEM
VB-11	0015
VB11	0004
VB12	OEM
VB13	0280
VB14	OEM
VB20	OEM
VB20X	OEM
VB30	OEM
VB30X	OEM
VB40	OEM
VB40X	OEM
VB50	OEM
VB50X	OEM
VB60	OEM
VB100	0015
VB100-1	OEM
VB100-2	OEM
VB100-3	OEM
VB100-4	OEM
VB100-5	OEM
VB100-8	OEM
VB100-18	OEM
VB100X	OEM
VB150X	OEM
VB200	OEM
VB300	0015
VB400	0015
VB500	0015
VB600	0015
VB600A	0015
VB701	OEM
VB704	OEM
VB709	OEM
VBC89C	OEM
VBC89J	OEM
VBC99J	OEM
VBC119J	OEM
VBC7506M	OEM
VBE77M	OEM
VBH600	0015
VBO2.2-14N02	OEM
VBO2.2-14N04	OEM
VBO2.2-14N05	OEM
VBO2.2-14N06	OEM
VBO13-04N01	OEM
VBO13-08N01	OEM
VBX7504M	OEM
VBX7505M	OEM
VBX7521B	OEM
VC4S0006	OEM

DEVICE TYPE	REPL CODE
VC6E	0015
VC7	OEM
VC7A	OEM
VC7B	OEM
VC10	0005
VC10A	0005
VC10B	OEM
VC12	0720
VC12A	0720
VC12B	OEM
VC15	OEM
VC15A	OEM
VC15B	OEM
VC20	3658
VC20A	OEM
VC20B	OEM
VC20X	OEM
VC27	OEM
VC27A	OEM
VC27B	OEM
VC30	3658
VC30X	OEM
VC33	0030
VC33A	0030
VC33B	OEM
VC39	OEM
VC39A	OEM
VC39B	OEM
VC40	3658
VC40X	OEM
VC47	OEM
VC47A	OEM
VC47B	OEM
VC50	4875
VC50A	OEM
VC50B	OEM
VC60	3658
VC60X	OEM
VC68	OEM
VC68A	OEM
VC68BB	OEM
VC70	3658
VC70X	OEM
VC80	3658
VC80X	OEM
VC82	OEM
VC82A	OEM
VC82B	OEM
VC99	OEM
VC99A	OEM
VC99B	OEM
VC102	OEM
VC103A	OEM
VC103B	OEM
VC103C	OEM
VC103D	OEM
VC103E	OEM
VC112	OEM
VC112A	OEM
VC112B	OEM
VC115	OEM
VC115A	OEM
VC115B	OEM
VC118	OEM
VC118A	OEM
VC118B	OEM
VC122	OEM
VC122A	OEM
VC122B	OEM
VC127	OEM
VC127A	OEM
VC127B	OEM
VC133	OEM
VC133A	OEM
VC133B	OEM
VC207	OEM
VC207A	OEM
VC207B	OEM
VC210	0005
VC210A	0005
VC210B	OEM
VC212	0720
VC212A	0720
VC212B	OEM
VC215	OEM
VC215A	OEM
VC215B	OEM
VC220	OEM
VC220A	OEM
VC220B	OEM
VC227	OEM
VC227A	OEM
VC227B	OEM
VC233	0030
VC233A	0030
VC233B	OEM
VC239	OEM
VC239A	OEM
VC239B	OEM
VC247	OEM
VC247A	OEM
VC247B	OEM
VC256	OEM
VC256A	OEM
VC256B	OEM
VC268	OEM
VC268A	OEM
VC268B	OEM
VC282	OEM

DEVICE TYPE	REPL CODE
VC282A	OEM
VC282B	OEM
VC299	OEM
VC299A	OEM
VC299B	OEM
VC306	OEM
VC306A	OEM
VC306B	OEM
VC307	OEM
VC307A	OEM
VC307B	OEM
VC309	0005
VC309A	0005
VC309B	OEM
VC310	OEM
VC310A	OEM
VC310B	OEM
VC314	OEM
VC314A	OEM
VC314B	OEM
VC315	OEM
VC315A	OEM
VC315B	OEM
VC321	0628
VC321A	0628
VC321B	OEM
VC322	OEM
VC322A	0030
VC322B	OEM
VC332	0030
VC332A	0030
VC332B	OEM
VC333	OEM
VC333A	OEM
VC333B	OEM
VC346	OEM
VC346A	OEM
VC346B	OEM
VC347	OEM
VC347A	OEM
VC347B	OEM
VC367	OEM
VC367A	OEM
VC367B	OEM
VC399	OEM
VC399A	OEM
VC399B	OEM
VC406	OEM
VC406A	OEM
VC406B	OEM
VC407	OEM
VC407A	OEM
VC407B	OEM
VC408	0554
VC408A	0554
VC408B	OEM
VC409	0005
VC409A	0005
VC409B	OEM
VC410	0005
VC410A	0005
VC410B	OEM
VC411	OEM
VC411A	OEM
VC411B	OEM
VC414	OEM
VC414A	OEM
VC414B	OEM
VC415	OEM
VC415A	OEM
VC415B	OEM
VC416	OEM
VC416A	OEM
VC416B	OEM
VC421	0628
VC421A	0628
VC421B	OEM
VC422	0628
VC422A	0628
VC422B	OEM
VC423	OEM
VC423A	OEM
VC423B	OEM
VC432	0030
VC432A	0030
VC432B	OEM
VC433	0030
VC433A	0030
VC433B	OEM
VC434	OEM
VC434A	OEM
VC434B	OEM
VC446	OEM
VC446A	OEM
VC446B	OEM
VC447	OEM
VC447A	OEM
VC447B	OEM
VC448	OEM
VC448A	OEM
VC448B	OEM
VC467	OEM
VC467A	OEM
VC467B	OEM
VC533	OEM
VC533B	OEM
VC533C	OEM
VC533D	OEM
VC539	OEM

DEVICE TYPE	REPL CODE
VC539B	OEM
VC539C	OEM
VC539D	OEM
VC547	OEM
VC547B	OEM
VC547C	OEM
VC547D	OEM
VC556	OEM
VC556B	OEM
VC556C	OEM
VC556D	OEM
VC568	OEM
VC568A	OEM
VC568B	OEM
VC568C	OEM
VC568D	OEM
VC582	OEM
VC582B	OEM
VC582C	OEM
VC582D	OEM
VC599	OEM
VC599B	OEM
VC599C	OEM
VC599D	OEM
VC605	OEM
VC605A	OEM
VC605B	OEM
VC606	OEM
VC606A	OEM
VC606B	OEM
VC607	OEM
VC607A	OEM
VC607B	OEM
VC608	OEM
VC608A	OEM
VC608B	OEM
VC610	0005
VC610A	0005
VC610B	OEM
VC611	0005
VC611A	0005
VC611B	OEM
VC612	0720
VC612A	0720
VC612B	OEM
VC613	0720
VC613A	0628
VC613B	OEM
VC615	OEM
VC615A	OEM
VC615B	OEM
VC616	OEM
VC616A	OEM
VC616B	OEM
VC618	OEM
VC618A	OEM
VC618B	OEM
VC619	OEM
VC619A	OEM
VC619B	OEM
VC622	0628
VC622A	0628
VC622B	OEM
VC623	0628
VC623A	0628
VC623B	OEM
VC627	OEM
VC627A	OEM
VC627B	OEM
VC628	OEM
VC628A	OEM
VC628B	OEM
VC633	0030
VC633A	0030
VC633B	OEM
VC634	0030
VC634A	0030
VC634B	OEM
VC639	OEM
VC639A	OEM
VC639B	OEM
VC641	OEM
VC641A	OEM
VC641B	OEM
VC647	OEM
VC647A	OEM
VC647B	OEM
VC648	OEM
VC648A	OEM
VC648B	OEM
VC720	OEM
VC722	OEM
VC722A	OEM
VC724	OEM
VC726	OEM
VC728	OEM
VC730	OEM
VC732	OEM
VC734	OEM
VC736	OEM
VC740	OEM
VC742	OEM
VC744	OEM
VC746	OEM
VC748	OEM
VC752	OEM
VC821	OEM
VC821A	OEM
VC821B	OEM

DEVICE TYPE	REPL CODE
VC822	0005
VC822A	0005
VC822B	OEM
VC823	0720
VC823A	0720
VC823B	OEM
VC824	OEM
VC824A	OEM
VC824B	OEM
VC825	OEM
VC825A	OEM
VC825B	OEM
VC826	OEM
VC826A	OEM
VC826B	OEM
VC827	0030
VC827A	0030
VC827B	OEM
VC828	OEM
VC828A	OEM
VC828B	OEM
VC829	OEM
VC829A	OEM
VC829B	OEM
VC831	OEM
VC831A	OEM
VC831B	OEM
VC832	OEM
VC832A	OEM
VC832B	OEM
VC833	OEM
VC833A	OEM
VC833B	OEM
VC834	OEM
VC834A	OEM
VC834B	OEM
VC841	0554
VC841A	0554
VC841B	OEM
VC842	OEM
VC842A	OEM
VC842B	OEM
VC843	OEM
VC843A	OEM
VC843B	OEM
VC844	OEM
VC844A	OEM
VC844B	OEM
VC845	OEM
VC845A	OEM
VC845B	OEM
VC846	OEM
VC846A	OEM
VC846B	OEM
VC847	OEM
VC847A	OEM
VC847B	OEM
VC848	OEM
VC848A	OEM
VC848B	OEM
VC849	OEM
VC849A	OEM
VC849B	OEM
VC851	OEM
VC851A	OEM
VC851B	OEM
VC852	OEM
VC852A	OEM
VC852B	OEM
VC853	OEM
VC853A	OEM
VC853B	OEM
VC854	OEM
VC854A	OEM
VC854B	OEM
VC2007	OEM
VC2008	OEM
VC2011	0230
VC2022A	OEM
VC2022C	OEM
VC2024Z	OEM
VC2070	OEM
VC2071	OEM
VC3001	OEM
VC6204	OEM
VC6233	OEM
VC6256	OEM
VC6299	OEM
VC6607	OEM
VC6615	OEM
VC6633	OEM
VCB102-1	OEM
VCB102-2	OEM
VCB104-1	OEM
VCB104-2	OEM
VCB207	OEM
VCB252-1	OEM
VCB252-2	OEM
VCB252-3	OEM
VCB252-4	OEM
VCB252-7	OEM
VCB252-9	OEM
VCB252-10	OEM
VCB252-12	OEM
VCB254-1	OEM
VCB254-2	OEM
VCB254-3	OEM
VCB254-4	OEM
VCB254-7	OEM

DEVICE TYPE	REPL CODE
VCB254-9	OEM
VCB254-10	OEM
VCB254-12	OEM
VCB256-1	OEM
VCB256-2	OEM
VCB256-3	OEM
VCB256-4	OEM
VCB256-7	OEM
VCB256-9	OEM
VCB256-10	OEM
VCB256-12	OEM
VCB502-7	OEM
VCB502-8	OEM
VCB502-10	OEM
VCB502-11	OEM
VCB504-7	OEM
VCB504-8	OEM
VCB504-10	OEM
VCB504-11	OEM
VCB506-7	OEM
VCB506-8	OEM
VCB506-10	OEM
VCB506-11	OEM
VCP193	OEM
VCP200	OEM
VCR0006	OEM
VCR0014	OEM
VCR0018	OEM
VCR0019	OEM
VCR0019-1	OEM
VCR0069	OEM
VCR0087-1	OEM
VCR00871	OEM
VCR2N	OEM
VCR3P	1141
VCR4N	1155
VCR5P	OEM
VCR6P	OEM
VCR7N	OEM
VCR10N	OEM
VCR11N	OEM
VCR12N	OEM
VCR13N	OEM
VCR20N	OEM
VCRB307	OEM
VCRB407	OEM
VCRB607	OEM
VCRC201	OEM
VCRC301	OEM
VCRC401	OEM
VCRC601	OEM
VCRE210	OEM
VCRE235	OEM
VCRE310	OEM
VCRE335	OEM
VCRE410	OEM
VCRE435	OEM
VCRE610	OEM
VCRE635	OEM
VCRH208	OEM
VCRH216	OEM
VCRH308	OEM
VCRH316	OEM
VCRH408	OEM
VCRH416	OEM
VCRH608	OEM
VCRH616	OEM
VCRS0004	OEM
VCRS0006	OEM
VCRS0008	OEM
VCRS-0009	OEM
VCRS0009	OEM
VCRS0011	2553
VCRS0012	OEM
VCRS0014	OEM
VCRS0019	OEM
VCRS0020	OEM
VCRS0027	OEM
VCRS0030	OEM
VCRS0037	6355
VCRS0038	6356
VCRS0045	OEM
VCRS0046	OEM
VCRS0047	OEM
VCRS0048	OEM
VCRS0050	OEM
VCRS0051	OEM
VCRS0052	OEM
VCRS0053	OEM
VCRS0072	OEM
VCRS0076	OEM
VCRS0083	OEM
VCRS0084	OEM
VCRS0085	OEM
VCRS0093	OEM
VCRS0101-C	OEM
VCRS0101-T-2	OEM
VCRS0102	OEM
VCRS0104	OEM
VCRS0107	OEM
VCRS0107-T	OEM
VCRS0109	OEM
VCRS0109-T	OEM
VCRS0109A	OEM
VCRS0116-TA	OEM
VCRS0117	OEM
VCRS0131	OEM
VCRS0133	OEM
VCRS0137-A	OEM

DEVICE TYPE	REPL CODE
VCRW0029	OEM
VCRW0030	OEM
VCS40	OEM
VCS50	OEM
VCS60	OEM
VCS70	OEM
VCS80	OEM
VCS90	OEM
VCV1500	OEM
VCZ80B	OEM
VD1E1////-1	0015
VD1R2JA	OEM
VD-1R4JA/B	OEM
VD-1R41B	OEM
VD3C	OEM
VD6	0015
VD6C	OEM
VD10E1////-1	0015
VD10E1LF	0015
VD10E11F	0015
VD11	0143
VD12	2232
VD13	0143
VD1150M	0133
VD-121C	0015
VD121C	0015
VD1120	0139
VD1121	0139
VD-1122	0015
VD1122	0139
VD-1123	0133
VD1123	0139
VD-1124	0139
VD1124	0139
VD1150L	0139
VD1150M	0139
VD1210	0120
VD1211	0120
VD1212	0120
VD1213	0120
VD1220	0120
VD1220-M	0120
VD-1221	0120
VD1221	0120
VD1221L	0120
VD1222	0120
VD1223	0120
VD1251L	0139
VD1260L	0139
VD1262MF	0139
VDB-8024	OEM
VDC11	OEM
VDC1006	OEM
VDC1008	OEM
VDC1020	OEM
VDC1025	OEM
VDC1026	OEM
VDC1029	OEM
VDC1032	OEM
VDI	OEM
VDI-K	OEM
VDK1000	OEM
VDS1006	OEM
VDS1007	OEM
VDS1011	OEM
VDS1013	OEM
VDS1014	OEM
VDS1015	OEM
VDS1018	OEM
VDS1019	OEM
VDS1020	OEM
VDS1021	OEM
VDS1022	OEM
VDS1023	OEM
VDS1028	OEM
VDS1032	OEM
VDU1001	OEM
VDU1002	OEM
VDU1014	OEM
VDU1015	OEM
VDU1046	OEM
VDX1001A	OEM
VDX1006	OEM
VDX1025	OEM
VDX1036	OEM
VDX1047	OEM
VDX1051	OEM
VDX1052	OEM
VDX1071	OEM
VDX1085	OEM
VDX1086	OEM
VDX1099	OEM
VDZ80B	OEM
VE08	0106
VE08X	OEM
VE18	0106
VE18X	OEM
VE27	0106
VE28	0106
VE28X	OEM
VE47	0106
VE48	0106
VE48X	OEM
VE67	2010
VE68	2010
VE68X	OEM
VE87	OEM
VE88	2400
VE108	0787

DEVICE TYPE	REPL CODE
VE79092	1844
VE79093	0033
VE79141	1835
VEC87M	OEM
VEE77M	OEM
VEKW1212	OEM
VEPS2108A1	OEM
VEPS2108B1	OEM
VEPW0721	OEM
VEPW0743	OEM
VEPW0746	OEM
VEPW0747	OEM
VERSAFLOPPYII	OEM
VERSAFLOPY	OEM
VF01	OEM
VF1	OEM
VF1A	OEM
VF1B	OEM
VF1C	OEM
VF5	OEM
VF5-5	1791
VF5-5X	1791
VF5-7	1791
VF5-7X	1791
VF5-10	1791
VF5-10X	1791
VF5-12	OEM
VF5-12X	OEM
VF5-15	OEM
VF5-15X	OEM
VF5-20	OEM
VF5-20X	OEM
VF5-25	OEM
VF5-25X	OEM
VF5-30	OEM
VF5-30X	OEM
VF5-40	OEM
VF5-40X	OEM
VF5X	OEM
VF7	OEM
VF7X	OEM
VF10	OEM
VF10-5	1791
VF10-5X	1791
VF10-7	1791
VF10-7X	1791
VF10-10	1791
VF10-10X	1791
VF10-12	OEM
VF10-12X	OEM
VF10-15	OEM
VF10-15X	OEM
VF10-20	OEM
VF10-20X	OEM
VF10-25	OEM
VF10-25X	OEM
VF10-30	OEM
VF10-30X	OEM
VF10-40	OEM
VF10-40X	OEM
VF10X	OEM
VF12	OEM
VF12X	OEM
VF15	OEM
VF15X	6956
VF20	OEM
VF20X	OEM
VF25	OEM
VF25-5	1791
VF25-5X	1791
VF25-7	1791
VF25-7X	1791
VF25-10	1791
VF25-10X	1791
VF25-12	OEM
VF25-12X	OEM
VF25-15	OEM
VF25-15X	OEM
VF25-20	OEM
VF25-20X	OEM
VF25-25	OEM
VF25-25X	OEM
VF25-30	OEM
VF25-30X	OEM
VF25-40	OEM
VF25-40X	OEM
VF25X	OEM
VF28	OEM
VF30	OEM
VF30X	OEM
VF40	OEM
VF40X	OEM
VF50	OEM
VF50X	OEM
VFA2745C	0015
VFC12	OEM
VFC12LD	OEM
VFC15	OEM
VFC15LD	OEM
VFC32BM	OEM
VFC32KP	OEM
VFC32SM	OEM
VFC42BM	OEM
VFC42BP	OEM
VFC42SM	OEM
VFC52BM	OEM
VFC52BP	OEM
VFC52SM	OEM
VFG2745B	0085

DEVICE TYPE	REPL CODE
VFG-274513	0085
VFL-2744K	0136
VFL2744K	0050
VFP-2746C	0085
VFP2746C	0160
VFP6357C	0160
VFP-6537C	0160
VFP6537C	0160
VFQ-1C	OEM
VFQ-1R	OEM
VFQ2C	OEM
VFQ3C	OEM
VFQ2745F	0004
VFS5K	0136
VFS-2745	0004
VFS2745	0164
VFS-2745J	0004
VFS2745J	0164
VFSD1	0102
VF-SD1A	0102
VFSD-1A	1303
VFT-2745H	0004
VFT2745H	0211
VFU-2746B	0160
VFU2746B	0160
VFV100K	OEM
VFW2745D	0050
VFW-27450	0136
VFX9500	OEM
VFX9507	OEM
VFY-2745E	0280
VFY2745E	0280
VG1	OEM
VG1X	OEM
VG2	OEM
VG2X	4971
VG3	OEM
VG3X	OEM
VG4	OEM
VG4X	OEM
VG5	OEM
VG5X	OEM
VG7	OEM
VG7X	OEM
VG10	OEM
VG10X	OEM
VG12	OEM
VG12X	OEM
VG15	OEM
VG15X	OEM
VG20	OEM
VG20X	OEM
VGA919S	OEM
VGB0101CAA	OEM
VGB0102CAA	OEM
VGB0103AAA	OEM
VGB0103CAA	OEM
VGB0104AAA	OEM
VGB0104CAA	OEM
VGB0124AY7A	OEM
VGB0491MY7	OEM
VGB0492MY7	OEM
VGB0493MY7	OEM
VGB0494MY7	OEM
VGC7205-0409	OEM
VGT7702-4003	OEM
VH048	0323
VH107	OEM
VH138	OEM
VH139	OEM
VH140	OEM
VH141	OEM
VH148	0323
VH160	OEM
VH167	OEM
VH175	OEM
VH183	OEM
VH191	OEM
VH200	OEM
VH201	OEM
VH225	OEM
VH226	OEM
VH227	OEM
VH228	OEM
VH229	OEM
VH230	OEM
VH231	OEM
VH247	1414
VH248	1414
VH447	0473
VH448	0473
VH647	0446
VH648	0446
VH848	1422
VH1048	2443
VHC1S2790-W-1	0628
VHC1S2790W-1	0628
VHCBB109G//-1	0030
VHCBB109G-1	0030
VHCBB1096//-1	0030
VHD02BZ2R7/1A	0760
VHD1A3-F///-1	OEM
VHD1N34///-1	0123
VHD1N34A///-1	0123
VHD1N34A///1	0123
VHD1N34A///1E	0123
VHD1N34A//-1	0123
VHD1N60////-1	0019
VHD1S34///-1	0143

DEVICE TYPE	REPL CODE
VHD1S1553//-1	0133
VHD1S1555//-1	0133
VHD1S1555//1A	0133
VHD1S1555-R-1	0133
VHD1S1834	0023
VHD1S1834//-1	0023
VHD1S1885//-1	0023
VHD1S1885-1	0023
VHD1S1887//-1	0023
VHD1S2076//-1	0133
VHD1S2076//-U	0133
VHD1S2076-//-1	0133
VHD1S2076-1	0133
VHD1S2230//1E	0015
VHD1S2835//1E	0133
VHD1S2837//-1E	0907
VHD1S2837//1E	0907
VHD1SS82///1A	0023
VHD1SS119	0396
VHD1SS119//-1	0124
VHD1SS119//1E	0124
VHD1SS198//-1	0339
VHD1SS270//-1	0133
VHD1SS270//-F	OEM
VHD10E1////-1	0015
VHD10E-4////-1	0015
VHD151885-1	0015
VHDIS1555-R-1	0133
VHDMA151WK/-1	0907
VHEHZ6B3///1A	0157
VHERD6R8EE/-1	0025
VHERD27ED//-1	0441
VHEWZ-100//1F	0064
VHEWZ-100-1F	0064
VHEWZ-1001F	0064
VHEXZ-090-1	0012
VHF1S2076-1	0133
VHFRD6R8EE/-1	OEM
VHI51494L/-1	OEM
VHIAN3211K/-1	4379
VHIAN3321K/-1	OEM
VHIAN3990//-1	OEM
VHIAN6291//-1	3437
VHIBA6209//1E	3828
VHIBA6238AU1E	4117
VHIBA6405//-1	OEM
VHIBA7021//-1	OEM
VHIBA7244S/-1	OEM
VHIBA7252S/-1	OEM
VHIBA7280AS-1	OEM
VHIBA7280S/-1	OEM
VHIBA7765AS-1	OEM
VHIBU4011B/-1	0215
VHICXA1011P-1	3083
VHID74LS136P	1628
VHIHA11845NT1	OEM
VHIHA17812W-1	1351
VHIHD74LS00P/	1527
VHIHD74LS14-1	1698
VHIHD74LS136P	1628
VHIHD74LS221/	1239
VHIHD74S04/-1	2263
VHIHD7406//-1	1206
VHIHD7417//-1	1352
VHIHD7426//-1	0803
VHIHD14042P-1	0121
VHIIR2C32//-1	OEM
VHIIR2E01//-1	OEM
VHIIR3P42//-1	OEM
VHIIR3393//-1	0631
VHIIR3702//-1	OEM
VHIIR9358//-1	6322
VHIIR94558/-1	3440
VHiKA7809PI-1	OEM
VHILA7016//-1	1028
VHiLA7838//-1	OEM
VHiLA7945//-1	OEM
VHILC7458B/-1	OEM
VHILH5047//-1	OEM
VHILH5116N-2	OEM
VHILVA516S2-1	OEM
VHILVA519S2-1	OEM
VHILVA51952-1	OEM
VHIM5223P//-1	OEM
VHIM5224P//-1	0625
VHIM5228P//-1	OEM
VHIM51321P/-1	OEM
VHIM51329P/-1	OEM
VHIM51494L/-1	OEM
VHIM54641L/-	OEM
VHIM58659P/-1	OEM
VHIMC4558S/-	OEM
VHIMN3007//-1	5201
VHIMN3101//-1	4278
VHINJM2220S-1	OEM
VHINJM4558D-1	0360
VHINS74LS32-1	0088
VHIPA0030//-1	OEM
VHiPST529C2-1	OEM
VHIPST529H2-1	OEM
VHISN74LS08-1	1633
VHISN74LS32-1	0088
VHISN74S04/-1	2263
VHISN74S11/-1	2447
VHISTK4048V-	OEM
VHISTRM6513-1	OEM
VHISTRS6301-1	OEM
VHiT8889A//-1	OEM
VHITA789S/-1	OEM

DEVICE TYPE	REPL CODE
VHITA6703S/-1	OEM
VHITA7130P2/F	2019
VHITA7133P/-1	OEM
VHITA7343P/-1	4410
VHITA7347/-1	4496
VHITA7347P/-1	4496
VHITA7348P/-1	4383
VHITA7378P/-	OEM
VHITA7764P/-1	OEM
VHiTA7809S/-1	6817
VHITA8110AP-	OEM
VHITA8216H/-1	OEM
VHITA8644N/-1	OEM
VHITA8647S/-1	OEM
VHITA8703A/-1	OEM
VHITA8703S/-1	3623
VHITA75558S-1	0770
VHITA78010AP1	2806
VHITA78055/-1	OEM
VHITC4001BP-1	0478
VHITC4013BP-1	0413
VHITC4028BP-1	2228
VHITC4030BP-1	0500
VHiTC4052BP-1	0024
VHITC4066BP-1	0101
VHITC4077BP-1	OEM
VHITC4081BP-1	0332
VHITD620304/-1	OEM
VHITDA3569/-1	OEM
VHiTDA7052/-1	3802
VHITDA7056/-1	6721
VHITDA8305A-1	OEM
VHITEA5582/-1	OEM
VHITL8708P/-1	OEM
VHIUPC574JT-1	1329
VHIUPC1373H-1	2026
VHIUPC1373HA/	2026
VHIUPC1373HA1	2026
VHIUPC1391H-1	3776
VHiUPC1406HA1	3254
VHIUPC1486C	4930
VHIUPC1486C//	OEM
VHIUPC1486C-1	4930
VHIUPC1486C1	OEM
VHIUPC1513H-1	3159
VHIUPD4066B-1	0101
VHIUPD6104C-1	OEM
VHS2SF656//1E	0901
VHS2SF1422//1E	0239
VHS3S4M//LB1E	0095
VHS3S4M/LB1E	0095
VHS326	0133
VHSCR3AM28LB1	0095
VHSCR3AMZ8L81	0095
VHSCR3AMZ8LB1	0095
VHSCRAMZ8LB1	0095
VHSS6089///1B	0239
VHSS6089///1E	0239
VHSS6142GLB1E	0095
VHSS6304G//1E	OEM
VHSS6340G	0095
VHSS6340G//1E	0095
VHSS6340GLB1E	0095
VHSS6344GLB1E	2739
VHSS6758GLBIE	OEM
VHSS6785GLB1E	4352
VHSS6785GLB2E	1992
VHSS6785GLBIE	OEM
VHSSFOR1A411E	0901
VHSSFORIA411E	0901
VHSSH3D42//-1	OEM
VHV1S1209-1	0139
VHV3K121///1E	OEM
VHVTD80////1E	OEM
VI1010	OEM
VI1023	OEM
VIDIODE	0293
VJ148	0323
VJ248	1414
VJ448	0473
VJ648	0446
VJ848	1422
VJ1048	2443
VL/8RJ	0280
VL16C450-PC	OEM
VL18RJ	0136
VL1772-02PC	OEM
VL6765-08PC	OEM
VLC9007A	OEM
VLFS0018	OEM
VLFS0031	OEM
VM08	1875
VM-0.09M	0196
VM-0.09M12/3	0196
VM18	1875
VM25	1875
VM-28	1875
VM28	1875
VM48	1875
VM68	1875
VM88	3546
VM108	3546
VM204	0554
VM204A	0554
VM205	0554
VM205A	0554
VM210	0554
VM210A	0554
VM211	0554

DEVICE TYPE	REPL CODE
VM211A	0554
VM216	0554
VM216A	0554
VM217	0554
VM217A	0554
VM222	0554
VM222A	0554
VM223	0554
VM223A	0554
VM228	0554
VM228A	0554
VM229	0554
VM229A	0554
VM234	0554
VM234A	0554
VM235	0554
VM235A	0554
VM304	0005
VM304A	0005
VM305	0005
VM305A	0005
VM310	0005
VM310A	0005
VM311	0005
VM311A	0005
VM316	0005
VM316A	0005
VM317	0005
VM317A	0005
VM328	0005
VM328A	0005
VM329	0005
VM329A	0005
VM334	0005
VM334A	0005
VM335	0005
VM335A	0005
VM504	0628
VM504A	0628
VM505	0628
VM505A	0628
VM510	0628
VM510A	0628
VM511	0628
VM511A	0628
VM516	0628
VM516A	0628
VM517	0628
VM517A	0628
VM522	0628
VM522A	0628
VM523	0628
VM523A	0628
VM528	0628
VM528A	0628
VM529	0628
VM529A	0628
VM534	0628
VM534A	0628
VM535	0628
VM535A	0628
VM604	0030
VM604A	0030
VM605	0030
VM605A	0030
VM610	0030
VM610A	0030
VM611	0030
VM611A	0030
VM616	0030
VM616A	0030
VM617	0030
VM617A	0030
VM622	0030
VM622A	0030
VM623	0030
VM623A	0030
VM628	0030
VM628A	0030
VM629	0030
VM629A	0030
VM634	0030
VM634A	0030
VM635	0030
VM20233	0142
VM-30203	0160
VM30203	0160
VM30209	0016
VM30233	0142
VM30234	0142
VM30241	0016
VM30242	0016
VM30244	0004
VM30245	0142
VM61001	OEM
VM61002	OEM
VM61003	OEM
VM61004	OEM
VM61005	OEM
VM71001	OEM
VM71002	OEM
VM71003	OEM
VM-HS9/1	0196
VMP1	OEM
VMP2	OEM
VMP4	OEM
VMP11	OEM
VMP12	OEM
VMP21	OEM

DEVICE TYPE	REPL CODE	DEVICE TYPE	REPL CODE	DEVICE TYPE	REPL CODE	DEVICE TYPE	REPL CODE	DEVICE TYPE	REPL CODE	DEVICE TYPE	REPL CODE	DEVICE TYPE	REPL CODE
VMP22	OEM	VN0350N5	OEM	VN1106ND	OEM	VN3500A	OEM	VP0206N7	OEM	VP1216N5	OEM	VR27	0441
VM-PH9D522/1	0015	VN0350ND	OEM	VN1110N1	OEM	VN3500D	OEM	VP0206ND	OEM	VP1216ND	OEM	VR27A	0441
VM-PH11D522/1	0015	VN0355N1	OEM	VN1110N2	OEM	VN3501A	OEM	VP0210N2	OEM	VP1220N1	OEM	VR27B	0441
VMPH11D522-1	0015	VN0355N5	OEM	VN1110N5	OEM	VN3501D	OEM	VP0210N3	OEM	VP1220N2	OEM	VR27E	1761
VM-PH90522/1	0015	VN0355ND	OEM	VN1110ND	OEM	VN4000A	OEM	VP0210N5	OEM	VP1220N5	OEM	VR27F	0294
VMT-01	0409	VN0360D	OEM	VN1116N1	OEM	VN4000D	OEM	VP0210ND	OEM	VP1220ND	OEM	VR30E	OEM
VMT-02	2804	VN0400A	OEM	VN1116N2	OEM	VN4001A	1668	VP0216N2	OEM	VP1304N2	OEM	VR33	0039
VMT-03	0190	VN0400D	OEM	VN1116N5	OEM	VN4001D	OEM	VP0216N3	OEM	VP1304N3	OEM	VR33A	0039
VMT-03-01	0190	VN0401A	OEM	VN1116ND	OEM	VN4501A	OEM	VP0216N5	OEM	VP1304N6	OEM	VR33B	0039
VMT-03-02	0190	VN0401D	OEM	VN1120N1	OEM	VN4501D	OEM	VP0216ND	OEM	VP1304N7	OEM	VR35E	OEM
VMT-04	3858	VN0435N1	OEM	VN1120N2	OEM	VN4502A	OEM	VP0300B	OEM	VP1304ND	OEM	VR56	1834
VMT-05	1501	VN0440N1	OEM	VN1120N5	OEM	VN4502D	OEM	VP0300L	OEM	VP1306N2	OEM	VR56A	1834
VMT-05-01	1501	VN0535N2	OEM	VN1120ND	OEM	VN5001A	OEM	VP0300M	OEM	VP1306N3	OEM	VR56B	0869
VMT-05-02	1501	VN0535N3	OEM	VN1200A	OEM	VN5001D	OEM	VP0300P	OEM	VP1306N6	OEM	VR-60	OEM
VMT-05-03	1501	VN0535ND	OEM	VN1200D	OEM	VN5002A	OEM	VP0335N1	OEM	VP1306N7	OEM	VR-60B	OEM
VMT-06	1197	VN0540N2	OEM	VN1201A	OEM	VN5002D	OEM	VP0335N2	OEM	VP1306ND	OEM	VR-60BT	OEM
VMT-07	1710	VN0540N3	OEM	VN1201D	OEM	VNC003A	OEM	VP0335N5	OEM	VP1310N2	OEM	VR-60S	OEM
VMT-08	1067	VN0540ND	OEM	VN1204N1	OEM	VNC010B	OEM	VP0335ND	OEM	VP1310N3	OEM	VR60SS	OEM
VMT-09	2546	VN0555N2	OEM	VN1204N2	OEM	VNC010D	OEM	VP0340N1	OEM	VP1310ND	OEM	VR-60T	OEM
VMT-09-01	2546	VN0555N3	OEM	VN1204N5	OEM	VNC011B	OEM	VP0340N2	OEM	VP1316N2	OEM	VR-60U	OEM
VMT-11	2982	VN0555ND	OEM	VN1204ND	OEM	VNC011D	OEM	VP0340N5	OEM	VP1316N3	OEM	VR-61	OEM
VMT-12	2984	VN0600A	OEM	VN1206B	OEM	VND010B	OEM	VP0340ND	OEM	VP1316ND	OEM	VR-61B	OEM
VMT-13	1054	VN0600D	OEM	VN1206D	OEM	VND010D	OEM	VP0345N1	OEM	VP1320D	OEM	VR-61S	OEM
VMT-14	1054	VN0606M	OEM	VN1206L	OEM	VND011B	OEM	VP0345N2	OEM	VP13202	OEM	VR61SS	OEM
VMT-15	3655	VN0610L	OEM	VN1206M	OEM	VND011D	OEM	VP0345N5	OEM	VP13203	OEM	VR-61T	OEM
VM-TC02P11/2	0015	VN0610LL	OEM	VN1206N1	OEM	VNE003A	OEM	VP0345ND	OEM	VPC1228H	OEM	VR-61U	OEM
VM-TC0.2P11/2	0015	VN0635N2	OEM	VN1206N2	OEM	VNE010B	OEM	VP0350N1	OEM	VQ1000CJ	OEM	VR62	0783
VN011BB	OEM	VN0635N3	OEM	VN1206N5	OEM	VNE010D	OEM	VP0350N2	OEM	VQ1000J	OEM	VR62A	0783
VN011BD	OEM	VN0635N5	OEM	VN1206ND	OEM	VNG004A	OEM	VP0350N5	OEM	VQ1000N6	OEM	VR62B	0783
VN0104	OEM	VN0635ND	OEM	VN1208N1	OEM	VNJ004A	OEM	VP0350ND	OEM	VQ1000N7	OEM	VR82	0331
VN0104A	OEM	VN0640N2	OEM	VN1209N1	OEM	VNL001A	OEM	VP0535N2	OEM	VQ1000P	OEM	VR82A	0331
VN0104N2	OEM	VN0640N3	OEM	VN1210L	OEM	VNL005A	OEM	VP0535N3	OEM	VQ1001J	OEM	VR82B	0331
VN0104N3	OEM	VN0640N5	OEM	VN1210M	OEM	VNM001A	OEM	VP0535ND	OEM	VQ1001P	OEM	VR110	0149
VN0104N5	OEM	VN0640ND	OEM	VN1210N1	OEM	VNM005A	OEM	VP0540N2	OEM	VQ1004J	OEM	VR110A	0149
VN0104N6	OEM	VN0645N2	OEM	VN1210N2	OEM	VNN002A	OEM	VP0540N3	OEM	VQ1004P	OEM	VR110B	0149
VN0104N7	OEM	VN0645N3	OEM	VN1210N5	OEM	VNN006A	OEM	VP0540ND	OEM	VQ1006J	OEM	VR425E	OEM
VN0104N9	OEM	VN0645N5	OEM	VN1210ND	OEM	VNP002A	OEM	VP0545N2	OEM	VQ1006P	OEM	VR475E	OEM
VN0104ND	OEM	VN0645ND	OEM	VN1216N1	OEM	VNP006A	OEM	VP0545N3	OEM	VQ2001J	OEM	VR525E	OEM
VN0106N2	OEM	VN0650N2	OEM	VN1216N2	OEM	VNS008A	OEM	VP0545ND	OEM	VQ2001P	OEM	VR575E	OEM
VN0106N3	OEM	VN0650N3	OEM	VN1216N5	OEM	VNS008D	OEM	VP0550N2	OEM	VQ2004J	OEM	VR625E	OEM
VN0106N5	OEM	VN0650N5	OEM	VN1216ND	OEM	VNS009A	OEM	VP0550N3	OEM	VQ2004P	OEM	VRBG5315S-B3	OEM
VN0106N6	OEM	VN0650ND	OEM	VN1220N1	OEM	VNS009D	OEM	VP0550ND	OEM	VQ2006J	OEM	VS-0A70	0143
VN0106N7	OEM	VN0655N2	OEM	VN1220N2	OEM	VNSS012A	OEM	VP0635N2	OEM	VQ2006P	OEM	VS-1	0015
VN0106N9	OEM	VN0655N3	OEM	VN1220N5	OEM	VNSS013A	OEM	VP0635N3	OEM	VQ3001J	OEM	VS1	0015
VN0106ND	OEM	VN0655N5	OEM	VN1220ND	OEM	VNT008A	OEM	VP0635N5	OEM	VQ3001N6	OEM	VS-1N82AG	0917
VN0109N2	OEM	VN0655ND	OEM	VN1304N2	OEM	VNT008D	OEM	VP0635ND	OEM	VQ3001N7	OEM	VS2A71BS	0050
VN0109N3	OEM	VN0660N2	OEM	VN1304N3	OEM	VNT009A	OEM	VP0640N2	OEM	VQ3001P	OEM	VS2C1740SQR1E	0151
VN0109N5	OEM	VN0660N3	OEM	VN1304N6	OEM	VNT009D	OEM	VP0640N3	OEM	VQ7254J	OEM	VS2C1741-1	3426
VN0109N9	OEM	VN0660N5	OEM	VN1304N7	OEM	VNT012A	OEM	VP0640N5	OEM	VQ7254N6	OEM	VS2S669D//1E	0563
VN0109ND	OEM	VN0660ND	OEM	VN1304ND	OEM	VNT013A	OEM	VP0640ND	OEM	VQ7254N7	OEM	VS-2SA71	0050
VN0116N2	OEM	VN0800A	OEM	VN1306N2	OEM	VO0220N2	OEM	VP0645N2	OEM	VQ7254P	OEM	VS-2SA71B	0050
VN0116N3	OEM	VN0800D	OEM	VN1306N3	OEM	VO0220N3	OEM	VP0645N3	OEM	VR	OEM	VS2SA71B	0050
VN0116N5	OEM	VN0801A	OEM	VN1306N6	OEM	VO0220N5	OEM	VP0645N5	OEM	VR1B	0023	VS-2SA71BS	0050
VN0116ND	OEM	VN0801D	OEM	VN1306N7	OEM	VO0220ND	OEM	VP0645ND	OEM	VR1D	0023	VS2SA71BS	0050
VN0120N2	OEM	VN0808M	OEM	VN1306ND	OEM	V01G	1080	VP0650N2	OEM	VR1G	0023	VS-2SA103	0136
VN0120N3	OEM	VN03601	OEM	VN1310N2	OEM	V03-G	0071	VP0650N3	OEM	VR1J	0023	VS2SA103	0136
VN0120N5	OEM	VN03605	OEM	VN1310N3	OEM	V03C	0953	VP0650N5	OEM	VR5.6	0157	VS2SA128V	0164
VN0120ND	OEM	VN2SF1422	0239	VN1310ND	OEM	V03E	OEM	VP0650ND	OEM	VR5.6A	0157	VS-2SA288A	0136
VN0204N2	OEM	VN10KE	6374	VN1316N2	OEM	V03G	0071	VP0808B	OEM	VR5.6B	0157	VS-2SA358	0136
VN0204N5	OEM	VN10KM	6376	VN1316N3	OEM	V06-C	0015	VP0808L	OEM	VR6.2	0636	VS-2SA378	0050
VN0204N6	OEM	VN10KN3	OEM	VN1316ND	OEM	V06-G	0015	VP0808M	OEM	VR6.2A	0636	VS2SA378	0050
VN0204N7	OEM	VN10KN9	OEM	VN1320N2	OEM	VO-6A	0015	VP16RP8MPC	OEM	VR6.2B	0636	VS-2SA379	0050
VN0204ND	OEM	VN10LE	OEM	VN1320N3	OEM	V06A	0540	VP130A10	1374	VR7E	OEM	VS2SA379	0050
VN0206N2	OEM	VN10LM	OEM	VN1320ND	OEM	VO-6B	0540	VP130A20	0838	VR8E	OEM	VS-2SA385	0136
VN0206N3	OEM	VN10LP	OEM	VN1706B	OEM	VO-6C	0015	VP150A10	0246	VR9B	0012	VS2SA385	0136
VN0206N5	OEM	VN30AA	OEM	VN1706D	OEM	V06C	0015	VP150A20	1993	VR9E	OEM	VS2SA448	0050
VN0206N6	OEM	VN30AB	OEM	VN1706L	OEM	VO-6C-401	0540	VP250A20	1334	VR-9.1	0012	VS2SA495-0/1E	0006
VN0206N7	OEM	VN33AJ	OEM	VN1706M	OEM	V06E	0015	VP250A40	1334	VR9.1	0012	VS2SA495-O/-1	0006
VN0206ND	OEM	VN33AK	OEM	VN1710B	OEM	V09-E	0540	VP420B40	3257	VR9.1A	0012	VS2SA495-O/1E	0006
VN0210N2	OEM	VN35AA	OEM	VN1710L	OEM	V09C	0023	VP460B40	OEM	VR9.1B	0012	VS2SA495-Y/1E	0006
VN0210N3	OEM	VN35AB	OEM	VN1710M	OEM	V09E	0102	VP480B40	1101	VR10	0170	VS2SA562-0/1E	0006
VN0210N5	OEM	VN35AJ	OEM	VN1720M	OEM	V09G	0023	VP480B80	1101	VR10A	0170	VS2SA562-O/1E	0006
VN0210ND	OEM	VN35AK	OEM	VN2020L	OEM	VOG	0015	VP510B40	OEM	VR10B	0170	VS2SA562-Q/1E	0006
VN0216N2	OEM	VN40AF	OEM	VN2222KM	OEM	VP0104N2	OEM	VP510B80	OEM	VR10E	0510	VS2SA562-Y/1E	0006
VN0216N3	OEM	VN46AF	OEM	VN2222L	OEM	VP0104N3	OEM	VP1000B80	OEM	VR10F	0225	VS2SA562TO/-1	0006
VN0216N5	OEM	VN64GA	OEM	VN2222LL	OEM	VP0104N5	OEM	VP1000B160	OEM	VR11	0794	VS2SA562T-O/1E	0006
VN0216ND	OEM	VN66AD	OEM	VN2222LM	OEM	VP0104N6	OEM	VP1008B	OEM	VR11A	0794	VS2SA562T-Q/1E	0006
VN0220N2	OEM	VN66AF	OEM	VN2306N1	OEM	VP0104N7	OEM	VP1008L	OEM	VR11B	0317	VS2SA673-B/1E	0148
VN0220N3	OEM	VN66AJ	OEM	VN2306N5	OEM	VP0104N9	OEM	VP1008M	OEM	VR11E	OEM	VS2SA673-C/1A	0148
VN0220N5	OEM	VN66AK	OEM	VN2306ND	OEM	VP0104ND	OEM	VP1106N1	OEM	VR11F	0230	VS2SA673-C/1E	0148
VN0220ND	OEM	VN67AA	OEM	VN2310N1	OEM	VP0106N2	OEM	VP1106N2	OEM	VR12	0137	VS2SA673A-B/1E	0148
VN0300B	OEM	VN67AB	OEM	VN2310N5	OEM	VP0106N3	OEM	VP1106N5	OEM	VR12A	0137	VS2SA673A-C/1E	0148
VN0300D	OEM	VN67AD	OEM	VN2310ND	OEM	VP0106N5	OEM	VP1106ND	OEM	VR12B	0137	VS2SA733AQ/-1	0006
VN0300L	OEM	VN67AF	OEM	VN2316N1	OEM	VP0106N6	OEM	VP1110N1	OEM	VR12E	0870	VS2SA738-C/-1	0460
VN0300M	OEM	VN67AJ	OEM	VN2316N5	OEM	VP0106N7	OEM	VP1110N2	OEM	VR12F	0234	VS2SA740///1E	1648
VN0330N1	1668	VN67AK	OEM	VN2316ND	OEM	VP0106N9	OEM	VP1110N5	OEM	VR13	0365	VS2SA812-M51E	1742
VN0330N5	1465	VN80AF	OEM	VN2320N1	OEM	VP0106ND	OEM	VP1110ND	OEM	VR13A	0365	VS2SA844-D/-1	0693
VN0335B1	1668	VN88AD	OEM	VN2320N5	OEM	VP0109N2	OEM	VP1116N1	OEM	VR13B	0365	VS2SA844-D/-2	0693
VN0335C1	1668	VN88AF	OEM	VN2320ND	OEM	VP0109N3	OEM	VP1116N2	OEM	VR13E	OEM	VS2SA854-1/1E	OEM
VN0335N1	1668	VN89AA	OEM	VN2335N1	OEM	VP0109N5	OEM	VP1116N5	OEM	VR13F	0237	VS2SA854-Q/1E	1348
VN0335N2	OEM	VN89AB	OEM	VN2335N5	OEM	VP0109N9	OEM	VP1116ND	OEM	VR-14	0002	VS2SA854TO/-1	1348
VN0335N5	1465	VN89AD	OEM	VN2335ND	OEM	VP0109ND	OEM	VP1120N1	OEM	VR14	0100	VS2SA933QR1E	0148
VN0335ND	OEM	VN89AF	OEM	VN2340N1	OEM	VP0116N2	OEM	VP1120N2	OEM	VR14A	0100	VS2SA933SQR1E	0148
VN0340B1	1668	VN90AA	OEM	VN2340N5	OEM	VP0116N3	OEM	VP1120N5	OEM	VR14B	0100	VS2SA937-Q/-1	2484
VN0340C1	1668	VN90AB	OEM	VN2340ND	OEM	VP0116N5	OEM	VP1120ND	OEM	VR15	0002	VS2SA950-Y/1E	1348
VN0340N1	1668	VN98AJ	OEM	VN2345N1	OEM	VP0116ND	OEM	VP1204N1	OEM	VR15A	0002	VS2SA952LK/-1	0006
VN0340N2	OEM	VN98AK	OEM	VN2345N5	OEM	VP0120N2	OEM	VP1204N2	OEM	VR15B	0002	VS2SA966-Y/-1	0532
VN0340N5	1465	VN99AA	OEM	VN2345ND	OEM	VP0120N3	OEM	VP1204N5	OEM	VR15E	1274	VS2SA968Y//-1	1648
VN0340ND	OEM	VN99AB	OEM	VN2350N1	OEM	VP0120N5	OEM	VP1204ND	OEM	VR15F	0247	VS2SA988///1E	0648
VN0345B1	OEM	VN99AJ	OEM	VN2350N5	OEM	VP0120ND	OEM	VP1206N1	OEM	VR16E	OEM	VS2SA1013//1E	1522
VN0345C1	OEM	VN99AK	OEM	VN2350ND	OEM	VP0204N2	OEM	VP1206N2	OEM	VR18	0495	VS2SA1015/1E	0148
VN0345N1	OEM	VN401D	OEM	VN2406B	OEM	VP0204N5	OEM	VP1206N5	OEM	VR18A	0495	VS2SA1015G/1E	0148
VN0345N2	OEM	VN1000A	OEM	VN2406D	OEM	VP0204N6	OEM	VP1206ND	OEM	VR18B	0495	VS2SA1015Y/-1E	0148
VN0345N5	OEM	VN1000D	OEM	VN2406L	OEM	VP0204N7	OEM	VP1210N1	OEM	VR18E	1078	VS2SA1015Y/1E	0148
VN0345ND	OEM	VN1001A	OEM	VN2406M	OEM	VP0204ND	OEM	VP1210N2	OEM	VR18F	0256	VS2SA1015Y/2E	0148
VN0350B1	OEM	VN1001D	OEM	VN2410B	OEM	VP0206N2	OEM	VP1210N5	OEM	VR20E	OEM	VS2SA1015YW-1	0148
VN0350C1	OEM	VN1106N1	OEM	VN2410L	OEM	VP0206N3	OEM	VP1210ND	OEM	VR22E	1723	VS2SA1020Y/-1	0434
VN0350N1	OEM	VN1106N2	OEM	VN2410M	OEM	VP0206N5	OEM	VP1216N1	OEM	VR22F	0269	VS2SA1037KQ-1	1742
VN0350N2	OEM	VN1106N5	OEM	VN2420L	OEM	VP0206N6	OEM	VP1216N2	OEM	VR24E	OEM	VS2SA1271-Y-1	5472

DEVICE TYPE	REPL CODE
VS-2SB126	0160
VS2SB126	0160
VS-2SB126F	0160
VS2SB126F	0160
VS-2SB126FV	0160
VS-2SB126V	0160
VS2SB126V	0160
VS-2SB128	0975
VS-2SB128V	0975
VS-2SB171	0004
VS2SB171	0004
VS-2SB172	0004
VS2SB172	0004
VS2SB172F	0004
VS-2SB172FN	0004
VS2SB172FN	0004
VS-2SB176	0004
VS2SB176	0004
VS-2SB178	0004
VS2SB178	0004
VS-2SB178A	0004
VS2SB178A	0004
VS-2SB324	1062
VS2SB324	1062
VS-2SB448	0222
VS2SB561-C/-1	0436
VS2SB561-C-1	0436
VS2SB568///-1	1911
VS2SB1212//1E	OEM
VS-2SC41	0620
VS-2SC58	0233
VS-2SC58A	0233
VS-2SC58B	0233
VS-2SC58C	0233
VS-2SC206	0216
VS2SC206	0216
VS-2SC208	0007
VS2SC208	0007
VS-2SC288A	1342
VS2SC288A	1342
VS-2SC324	0016
VS-2SC324H	0016
VS2SC324H	0016
VS-2SC371-R-1	0191
VS2SC371-R-1	0191
VS2SC372-0/1E	0076
VS2SC372-Y/1E	0076
VS2SC373//-1E	0076
VS2SC373-//1E	0076
VS2SC373-G/-1	0076
VS2SC373G-1	0076
VS2SC374-B-1	0552
VS2SC383-W/1E	0076
VS2SC383-WT-1	0076
VS-2SC385	0549
VS-2SC385L	0549
VS2SC385L	0549
VS2SC394-0-1	0155
VS2SC394-O-1	0155
VS2SC394-Y-1	0155
VS-2SC446	0659
VS2SC454-B/1E	0076
VS2SC454-C/1A	0076
VS2SC454-C/1E	0076
VS2SC454-C/3A	0076
VS-2SC458	0076
VS2SC458	0076
VS2SC458-C/1E	0076
VS2SC458-D/-1	0076
VS2SC460B-1	0151
VS-2SC466	0259
VS2SC481-1	0698
VS2SC495T/-1	0786
VS2SC495T/1	0786
VS2SC495T-1	0786
VS-2SC526	0233
VS-2SC538	0079
VS2SC538	0079
VS-2SC563	0835
VS-2SC645	0674
VS-2SC645A	0674
VS2SC645A	0674
VS-2SC645B	0674
VS-2SC645C	0674
VS2SC669C//1E	0622
VS2SC669D/1E	0622
VS2SC681A-Y2E	2437
VS-2SC683	0047
VS-2SC683Y	0047
VS-2SC684	0127
VS2SC684	0127
VS-2SC687	0177
VS2SC717///-1	0127
VS2SC732-VIF	0076
VS2SC733B-1	0111
VS2SC735Y-1	0191
VS-2SC762	1154
VS2SC784R1F	1144
VS2SC900U-1	1221
VS2SC945A(Q)/-1	0076
VS2SC945AP/-1	0076
VS2SC945AQ/-1	0076
VS2SC945LK-1	0076
VS2SC945LP-1	0076
VS2SC983-0/2E	0066
VS2SC983-O/2E	0066
VS2SC1013//1E	OEM
VS2SC1166-0-1	0733
VS2SC1166-O-1	0733

DEVICE TYPE	REPL CODE
VS2SC1166Y-1	0733
VS2SC1172-/1E	0313
VS2SC1174-/1E	0313
VS2SC1174-1E	0313
VS2SC1213-C/1A	0191
VS2SC1213AC/1A	0191
VS2SC1213AC1A	0191
VS2SC1237-/1F	0936
VS2SC1237-1F	0936
VS2SC1335D/-1	1221
VS2SC1335D/1	1221
VS2SC1335D/-1E	1221
VS2SC1447-/1E	0643
VS2SC1447LB1E	0643
VS2SC1509R/-1	0324
VS2SC1514//-1	1085
VS2SC1514/-1	1085
VS2SC1514BK-1E	1085
VS2SC1514BK1E	1085
VS2SC1569K/1E	0643
VS2SC1569K/1K	0643
VS2SC1623L51E	0724
VS2SC1627A	0733
VS2SC1627Y	0733
VS2SC1664//-1	0629
VS2SC1674L/-1	1066
VS2SC1675M-1	0076
VS2SC1681G/1E	1757
VS2SC1723//-1	1085
VS2SC1723-1	1085
VS2SC1723S/-1	1085
VS2SC1740SQR1E	0151
VS2SC1741-1	1513
VS2SC1815GR-1	0076
VS2SC1815GW-1	0076
VS2SC1815OW-1	0076
VS2SC1815Y/-1	0076
VS2SC1815YW-1	0076
VS2SC1815YW1E	0076
VS2SC1826-Y1A	0423
VS2SC1827//1E	0236
VS2SC1827Y/1E	0236
VS2SC1829/1E	2620
VS2SC1829-/1E	2620
VS2SC1841-E-1	0530
VS2SC1841E1	0530
VS2SC1855//-1	0224
VS2SC1890A/1E	0530
VS2SC1890AD-1	0530
VS2SC1890AE-1	0530
VS2SC1906//1E	0127
VS2SC1906//E	1092
VS2SC1921//-1E	0261
VS2SC1921//1E	0261
VS2SC1923-O1E	2691
VS2SC1942//1E	2660
VS2SC1959Y/1E	0286
VS2SC1983//-1	2058
VS2SC1983//1E	2058
VS2SC1983//-2	2058
VS2SC2001-M-1	1513
VS2SC2001LK-1	1513
VS2SC2001LX-1	1513
VS2SC2002-K/A	0951
VS2SC2002-K1A	0951
VS2SC2021-Q-1	1140
VS2SC2023-/1E	2613
VS2SC2059KN1E	0724
VS2SC2068LB1E	0643
VS2SC2073	1284
VS2SC2073//1E	1284
VS2SC2073//3E	1284
VS2SC2073/3E	1284
VS2SC2073LB2E	1284
VS2SC2073LBE	1284
VS2SC2120Y/-1A	0866
VS2SC2120Y/1A	0866
VS2SC2120Y/-A	0866
VS2SC22290/-1	0066
VS2SC22290/1E	0066
VS2SC2231Y/1E	0168
VS2SC2236/-Y	2910
VS2SC22360/-1	2910
VS2SC2236Y/-1	2910
VS2SC2238Y/-1	3113
VS2SC2371K	0276
VS2SC2371K/1E	0276
VS2SC2371LK	0276
VS2SC2373L	0728
VS2SC2383//-1	1562
VS2SC2383//-1E	1562
VS2SC2383//1E	1562
VS2SC2412KQ-1	0724
VS2SC2458-C/1E	0076
VS2SC2458Y/-1	0076
VS2SC2462-C-1	0724
VS2SC2481-0	0563
VS2SC24810/1E	0563
VS2SC2481Y	0563
VS2SC2481Y/1E	0563
VS2SC2482//-1	0261
VS2SC2482//-1J	0261
VS2SC2555//-1	2185
VS2SC2610/-1E	0261
VS2SC2610BK/1E	0261
VS2SC2610K/1E	0261
VS2SC2610Y/-1	0261
VS2SC2611//-1	0276
VS2SC2611/-1	0276

DEVICE TYPE	REPL CODE
VS2SC2611/-1E	0276
VS2SC2655(Y)	0018
VS2SC2655Y/-1	0018
VS2SC2712Y/-1	0724
VS2SC2735//-1	2919
VS2SC2794LB1E	0563
VS2SC2827//1E	0728
VS2SC3299Y/-1	0060
VS2SC3299Y/1E	0060
VS2SC3377-Q-1	OEM
VS2SC3619LB1E	0276
VS2SC3886A/1E	5621
VS2SC3939QR-1	OEM
VS2SC4544LB1E	OEM
VS2SC4544LB2E	OEM
VS2SC4600B-1	OEM
VS2SC14471B1E	0168
VS2SC16742/-1	0224
VS2SC17235/-1	0168
VS2SC20731B2E	OEM
VS2SC22290/1E	0261
VS2SC24810/1E	0563
VS2SD227V-1	1419
VS2SD467-C/-1	0951
VS2SD468-C/-1	0018
VS2SD471-KL1E	0018
VS2SD471-S/-1	0018
VS2SD476-C/-1	0423
VS2SD478-YL-2E	0392
VS2SD478-YL2E	0392
VS2SD478-YL-ZE	0392
VS2SD478SYL2E	0392
VS2SD478YL2E	0392
VS2SD604-//1E	0074
VS2SD655-DE1E	1978
VS2SD666-C/-1	1386
VS2SD666A-C1A	1386
VS2SD669-C/1E	0563
VS2SD669-D/1E	0563
VS2SD669C	0563
VS2SD669C//1E	0563
VS2SD669D	0563
VS2SD669D//1E	0563
VS2SD669D/1E	0563
VS2SD699C	0563
VS2SD724///1A	0392
VS2SD758-C/1E	1085
VS2SD764///-1	0223
VS2SD794A//-1	0161
VS2SD794A//1E	0161
VS2SD794ALB1E	0161
VS2SD869//1E	0055
VS2SD869-//1E	0055
VS2SD870	0055
VS2SD870///-1	0055
VS2SD871///-1	0055
VS2SD898-/1E	0055
VS2SD898A-/1E	0055
VS2SD898A/1E	0055
VS2SD898B//1E	0055
VS2SD898B-/1E	0055
VS2SD898B-13//1E	0055
VS2SD906///-1	0055
VS2SD906///1E	0055
VS2SD975	0563
VS2SD975///-1	0563
VS2SD1190//1E	5457
VS2SD1264//1E	0392
VS2SD1267-P1E	0060
VS2SD1426//1E	1541
VS2SD1554//1E	1541
VS2SD1650//-1	1541
VS2SD1650CA1E	1541
VS2SD1812Q/-1	OEM
VS2SD2045//-1	5557
VS2SD2095//1E	2129
VS2SD2125//1E	2129
VS2SD2271-1	0016
VS2SF1422/1E	0239
VS2SK30AG//1E	3046
VS2SK30AG//2E	3046
VS2SK49F-1	1280
VS2SK532///-1	OEM
VS120	0015
VS202	0015
VS247	1414
VS248	1414
VS447	0473
VS449	OEM
VS500	1586
VS600	1586
VS848	1422
VS1048	2443
VS2687	OEM
VSC2SC383-W/1E	OEM
VSCS0065	OEM
VSCS0075	OEM
VS-CS1255H	0086
VSCS1255H	0086
VS-CS1255HF	0086
VS-CS1256HG	0126
VSCS1256HG	0037
VSD101AN20	OEM
VSD101BN20	OEM
VSD101N19	OEM
VSD101N20	OEM
VSD102AN20	OEM
VSD102BN20	OEM
VSD102N10	OEM

DEVICE TYPE	REPL CODE
VSD102N19	OEM
VSD102N20	OEM
VSD102N26	OEM
VSD103AN20	OEM
VSD103BN20	OEM
VSD103N10	OEM
VSD103N19	OEM
VSD103N20	OEM
VSD103N26	OEM
VSD104AN20	OEM
VSD104BN20	OEM
VSD111N10	OEM
VSD111N20	OEM
VSD111N26	OEM
VSD112N10	OEM
VSD112N20	OEM
VSD112N26	OEM
VSD121N10	OEM
VSD121N19	OEM
VSD121N20	OEM
VSD121N26	OEM
VSD122N19	OEM
VSD122N20	OEM
VSD201N10	OEM
VSD201N20	OEM
VSD201N26	OEM
VSD203N10	OEM
VSD203N20	OEM
VSD203N26	OEM
VSD204N20	OEM
VSD211N10	OEM
VSD501AN20	OEM
VSD501BN20	OEM
VSD501N10	OEM
VSD501N20	OEM
VSD501N26	OEM
VSD502AN20	OEM
VSD502BN20	OEM
VSD502N10	OEM
VSD502N20	OEM
VSD502N26	OEM
VSD503BN20	OEM
VSD503N10	OEM
VSD503N20	OEM
VSD503N26	OEM
VSD504AN20	OEM
VSD504BN20	OEM
VSD505AN20	OEM
VSD505BN20	OEM
VSD506AN20	OEM
VSD506BN20	OEM
VSD507AN20	OEM
VSD507BN20	OEM
VSD508AN20	OEM
VSD508BN20	OEM
VSD521N10	OEM
VSD521N20	OEM
VSD521N26	OEM
VSD522N10	OEM
VSD522N20	OEM
VSD522N26	OEM
VSD701N10	OEM
VSD701N19	OEM
VSD701N20	OEM
VSD701N26	OEM
VSD702N10	OEM
VSD702N19	OEM
VSD702N20	OEM
VSD702N26	OEM
VSD711N10	OEM
VSD711N20	OEM
VSD711N26	OEM
VSD712N10	OEM
VSD712N20	OEM
VSD712N26	OEM
VSD721N10	OEM
VSD721N19	OEM
VSD721N20	OEM
VSD721N26	OEM
VSD722N10	OEM
VSD722N19	OEM
VSD722N20	OEM
VS-DG1N	0102
VS-DG1NR	0182
VSDTA114EK/-1	1892
VSDTA124EK/-1	0703
VSDTA124ES/-1	1032
VSDTA144EK/-1	3272
VSDTA144ES/-1	4130
VSDTC114EK/-1	3482
VSDTC114ES/-1	0832
VSDTC124ES/-1	0887
VSDTC124F//-1	0887
VSDTC144EK/-1	3478
VSDTC144ES/-1	0898
VSDTC144F//-1	0898
VSDTC363EK/-1	OEM
VSES0011	OEM
VSF204BN20	OEM
VSF2745	0280
VS-FR1	0015
VSFR1	0015
VS-FR1P	0015
VSFR1P	0015
VS-FR-1U	0015
VS-FT-1N	0015
VSFT1N	0015
VSG-20024	0015
VSGM380	0136

DEVICE TYPE	REPL CODE
VSH200A	1335
VSH202A	0133
VSH212A	0133
VSH222A	0133
VSH242A	0133
VSH252A	0133
VSK120	1335
VSK130	1335
VSK140	1335
VSK320	0087
VSK330	0087
VSK340	0087
VSK1520	0615
VSK1530	0615
VSK1540	0615
VSK3020S	0615
VSK3030S	0615
VSK3040S	0615
VSK4020	1545
VSK4030	1545
VSK4040	1545
VS-MA102	0015
VS-MA103	OEM
VSOA70	0143
VS-PC02P11/2	0015
VSPH9D522/1	0015
VSQS0730	OEM
VS-RD11AM	0064
VSS6142GLB1E	0095
VS-SD1	0023
VS-SD-1B	0102
VS-SD1B	0102
VSSD1B	0015
VS-SD-1Z	0023
VSSD1Z	0015
VS-SD-13	0071
VS-SD-82A	0917
VS-SD82A	0917
VS-TC0-2P11/2	0015
VS-TC02P11/2	0015
VSTC02P11/2	0015
VS-TC0.2P11/2	0920
VSZS0065	OEM
VSZS0066	OEM
VSZS0075	OEM
VSZS0076	OEM
VT005-8	OEM
VT030-18A	OEM
VT1.5040	OEM
VT14LE	OEM
VT101	OEM
VT101E	OEM
VT101H	OEM
VT101HE	OEM
VT102	OEM
VT102E	OEM
VT102L	OEM
VT102L3	OEM
VT103	OEM
VT103L	OEM
VT104	OEM
VT111	OEM
VT111E	OEM
VT111H	OEM
VT111HE	OEM
VT112	OEM
VT112E	OEM
VT112L	OEM
VT112LE	OEM
VT113	OEM
VT113L	OEM
VT121	OEM
VT121E	OEM
VT121H	OEM
VT121HE	OEM
VT122	OEM
VT122E	OEM
VT122L	OEM
VT122LE	OEM
VT123	OEM
VT123L	OEM
VT124	OEM
VT132	OEM
VT133	OEM
VT133E	OEM
VT133L	OEM
VT133LE	OEM
VT141	OEM
VT141E	OEM
VT141H	OEM
VT141HE	OEM
VT142	OEM
VT142L	OEM
VT142LE	OEM
VT143	OEM
VT143L	OEM
VT6005	OEM
VT10030	OEM
VT10150	OEM
VT120020N	OEM
VT120100N	OEM
VTA1112	3552
VTCL27-0602-01	OEM
VTCL27-0603-00	OEM
VTF200	OEM
VTF400	OEM
VTF600	OEM
VTM-01	OEM
VTM-02	OEM

DEVICE TYPE	REPL CODE
VTM-03	OEM
VTM-03-01	OEM
VTM-04	OEM
VTM-05	OEM
VTM-05-01	OEM
VTM-05-02	OEM
VTM-05-03	OEM
VTM-06	OEM
VTM-07	OEM
VTM-08	OEM
VTM-09	OEM
VTM-10	OEM
VTM-11	OEM
VTM-12	OEM
VTM-13	OEM
VTM-14	OEM
VTM-15	OEM
VTR866	OEM
VTR867	OEM
VUAS0042	OEM
VVC201	0554
VVC203	0005
VVC205	0720
VVC205A	0720
VVC205B	OEM
VVC213	0030
VVC409	0030
VVC409A	0030
VVC409B	OEM
VVC410	0030
VVC410A	0030
VVC410B	OEM
VVC412	0005
VVC412A	0005
VVC413	0628
VVC415	0628
VVC415A	0628
VVC700	0554
VVC702	0554
VVC702A	0554
VVC704	0554
VVC704A	0554
VVC705	0005
VVC705A	0005
VVC706	0005
VVC706A	0005
VVC710	0628
VVC710A	0628
VVC712	0030
VVC712A	0030
VVC713	0030
VVC713A	0030
VVC726	0554
VVC729	0005
VVC730	0005
VVC731	0720
VVC732	0720
VVC741	0628
VVC742	0628
VVC743	0628
VVC747	0030
VVC748	0030
VVC749	0030
VVC789	0554
VVC790	0554
VVC791	0554
VVC792	0554
VVC795	0005
VVC796	0005
VVC797	0005
VVC798	0005
VVC799	0005
VVC810	0720
VVC811	0720
VVC822	0030
VVC823	0030
VVC824	0030
VVC825	0030
VVC847	0554
VVC848	0554
VVC849	0554
VVC850	0005
VVC851	0005
VVC852	0005
VVC862	0628
VVC863	0628
VVC864	0628
VVC869	0030
VVC870	0030
VVC871	0030
VVC910	0005
VVC911	0005
VVC924	0628
VVC925	0628
VVC932	0030
VVC933	0030
VVC934	0030
VVC1003	0554
VVC1004	0005
VVC1005	0720
VVC1009	0628
VVC1011	0030
VVC1620	0554
VVC1624	0005
VVC1624A	0005
VVC1624B	OEM
VVC1624C	OEM
VVC1626	0720
VVC1634	0628

DEVICE TYPE	REPL CODE
VVC1638	0030
VX0006CE	OEM
VX3375	0560
VX3733	0042
VX3866	OEM
VXEW0009	OEM
VZ-050A	0162
VZ-050B	0162
VZ-050C	0162
VZ-050D	0162
VZ-050E	0162
VZ-052A	0041
VZ-052B	0041
VZ-052C	0041
VZ-052D	0041
VZ-052E	0041
VZ-056A	0157
VZ-056B	0157
VZ-056C	0157
VZ-056D	0157
VZ-056E	0157
VZ061	0091
VZ-061A	0636
VZ-061B	0636
VZ-061C	0636
VZ-061D	0636
VZ-061E	0636
VZ-063A	0471
VZ-063B	0471
VZ-063C	0471
VZ-063D	0471
VZ-063E	0471
VZ-073A	0077
VZ-073B	0077
VZ-073C	0077
VZ-073D	0077
VZ-073E	0077
VZ-075A	0077
VZ-075B	0077
VZ-075C	0077
VZ-075D	0077
VZ-075E	0077
VZ-077A	0077
VZ-077B	0077
VZ-077C	0077
VZ-077D	0077
VZ-077E	0077
VZ-079A	0165
VZ-079B	0165
VZ-079C	0165
VZ-079D	0165
VZ-079E	0165
VZ-081A	0165
VZ-081B	0165
VZ-081C	0165
VZ-081D	0165
VZ-081E	0165
VZ-083A	0165
VZ-083B	0165
VZ-083C	0165
VZ-083D	0165
VZ-083E	0165
VZ-085A	0322
VZ-085B	0322
VZ-085C	0322
VZ-085D	0322
VZ-085E	0322
VZ-090A	0057
VZ-090B	0057
VZ-090C	0057
VZ-090D	0057
VZ-090E	0057
VZ-092A	0057
VZ-092B	0057
VZ-092C	0057
VZ-092D	0057
VZ-092E	0057
VZ-094A	0057
VZ-094B	0057
VZ-094C	0057
VZ-094D	0057
VZ-096A	0064
VZ-096B	0064
VZ-096C	0064
VZ-096D	0064
VZ-096E	0064
VZ-098A	0064
VZ-098B	0064
VZ-098C	0064
VZ-098D	0064
VZ-098E	0064
VZ33CH	0292
VZ33DF	0782
VZ33EF	0782
VZ36CH	0188
VZ36DF	0796
VZ36EF	0796
VZ39CH	0456
VZ39DF	0806
VZ39EF	0806
VZ43CH	0533
VZ43DF	0821
VZ43EF	0821
VZ43F	0821
VZ47CH	0451
VZ47DF	0833
VZ47EF	0833
VZ47F	0833
VZ51CH	0162

If replacement code is OEM, contact original manufacturer for replacement.

DEVICE TYPE	REPL CODE
VZ51D	0442
VZ51EF	0442
VZ51F	0442
VZ56CH	0157
VZ56DF	0876
VZ56EF	0876
VZ56F	0876
VZ62CH	0636
VZ62DF	0185
VZ62EF	0185
VZ62F	0185
VZ68CH	0025
VZ68DF	0205
VZ68EF	0205
VZ68F	0205
VZ75CH	0649
VZ75DF	0480
VZ75EF	0480
VZ75F	0480
VZ82CH	0244
VZ82DF	0504
VZ82EF	0504
VZ82F	0504
VZ91CH	0012
VZ91DF	0684
VZ91EF	0684
VZ91F	0684
VZ-100A	0064
VZ-100B	0064
VZ-100C	0064
VZ100CH	0170
VZ-100D	0064
VZ100DF	0225
VZ-100E	0064
VZ100EF	0225
VZ100F	0225
VZ-105A	0181
VZ-105B	0181
VZ-105C	0181
VZ-105D	0181
VZ-105E	0181
VZ110	OEM
VZ-110A	0181
VZ-110B	0181
VZ-110C	0181
VZ110CH	0317
VZ-110D	0181
VZ110DF	0230
VZ-110E	0181
VZ110EF	0230
VZ-110F	0181
VZ110F	0230
VZ-115A	0794
VZ-115B	0794
VZ-115C	0794
VZ-115D	0794
VZ-115E	0794
VZ-120A	0052
VZ-120B	0052
VZ-120C	0052
VZ120CH	0137
VZ-120D	0052
VZ120DF	0234
VZ-120E	0052
VZ120EF	0234
VZ120F	0234
VZ-125A	0365
VZ-125B	0365
VZ-125C	0365
VZ-125D	0365
VZ-125E	0365
VZ-130A	0053
VZ-130B	0053
VZ-130C	0053
VZ130CH	0365
VZ-130D	0053
VZ130DF	0237
VZ-130E	0053
VZ130EF	0237
VZ130F	0237
VZ-135A	0879
VZ-135B	0879
VZ-135E	0879
VZ-140A	0879
VZ-140B	0879
VZ-140C	0879
VZ-140D	0879
VZ-140E	0879
VZ-145A	0686
VZ-145B	0686
VZ-145C	0686
VZ-145D	0686
VZ-145E	0686
VZ-150A	0686
VZ-150B	0686
VZ-150C	0686
VZ150CH	0002
VZ-150D	0686
VZ150DF	0247
VZ-150E	0686
VZ150EF	0247
VZ150F	0247
VZ-157	0445
VZ-157A	0445
VZ-157B	0445
VZ-157C	0445
VZ-157D	0445
VZ-157E	0445
VZ160CH	0420

DEVICE TYPE	REPL CODE
VZ160DF	0251
VZ160EF	0251
VZ160F	0251
VZ-162	0445
VZ-162A	0445
VZ-162B	0445
VZ-162C	0445
VZ-162D	0445
VZ-162E	0445
VZ-167A	0210
VZ-167B	0210
VZ-167C	0210
VZ-167D	0210
VZ-167E	0210
VZ-172A	0210
VZ-172B	0210
VZ-172C	0210
VZ-172D	0210
VZ-172E	0210
VZ-177A	0495
VZ-177B	0495
VZ-177C	0495
VZ-177D	0495
VZ-177E	0495
VZ180CH	0495
VZ180DF	0256
VZ180EF	0256
VZ180F	0256
VZ-182A	0375
VZ-182B	0375
VZ-182C	0375
VZ-182D	0375
VZ-182E	0375
VZ-192A	0671
VZ-192B	0671
VZ-192C	0671
VZ-192D	0671
VZ-192E	0671
VZ200CH	0531
VZ200DF	0262
VZ200EF	0262
VZ200F	0262
VZ220CH	0565
VZ220DF	0269
VZ220EF	0269
VZ220F	0269
VZ-230	0402
VZ-230A	0402
VZ-230B	0402
VZ-230C	0402
VZ-230D	0402
VZ-230E	0402
VZ-240	0402
VZ-240A	0402
VZ-240B	0402
VZ-240C	0402
VZ240CH	0402
VZ-240D	0402
VZ240DF	0274
VZ-240E	0402
VZ240EF	0274
VZ240F	0274
VZ-250	0402
VZ-250A	0402
VZ-250B	0402
VZ-250C	0402
VZ-250D	0402
VZ-250E	0402
VZ-260	0441
VZ-260A	0441
VZ-260B	0441
VZ-260C	0441
VZ-260D	0441
VZ-260E	0441
VZ-270	0441
VZ-270A	0441
VZ-270B	0441
VZ-270C	0441
VZ270CH	0441
VZ-270D	0441
VZ270DF	0294
VZ-270E	0441
VZ270EF	0294
VZ270F	0294
VZ-280	0726
VZ-280A	0726
VZ-280B	0726
VZ-280C	0726
VZ-280D	0726
VZ-280E	0726
VZ-290	0726
VZ-290A	0726
VZ-290B	0726
VZ-290C	0726
VZ-290D	0726
VZ 290E	0726
VZ-300	0726
VZ-300A	0726
VZ-300B	0726
VZ-300C	0726
VZ300CH	0726
VZ-300D	0726
VZ300DF	0309
VZ-300E	0726
VZ300EF	0309
VZ300F	0309
VZ-320	0039
VZ-320A	0039
VZ-320B	0039

DEVICE TYPE	REPL CODE
VZ-320C	0039
VZ-320D	0039
VZ-320E	0039
VZ-330	0039
VZ-330A	0039
VZ-330B	0039
VZ-330C	0039
VZ330CH	0039
VZ-330D	0039
VZ330DF	0318
VZ-330E	0039
VZ330EF	0318
VZ330F	0318
VZ-340	0039
VZ-340A	0039
VZ-340B	0039
VZ-340C	0039
VZ-340D	0039
VZ-340E	0039
W/6A	0015
W0005M	0106
W-005	0277
W005	0106
W005L	0106
W005M	0106
W0-6A	0133
W0-6B	0609
W0-61	0636
W01	0106
W01L	0106
W01M	0106
W02	0106
W02L	0106
W02M	0106
W03A	0015
W03B	0023
W03BTA	4403
W03C	0023
W04	0289
W04L	0106
W04M	0106
W06	2010
W06-A	0015
W06-B	0015
W-06A	0015
W06A	0023
W06A-T3	0023
W06A-Z	0015
W06B	0023
W06C	0023
W06F	2010
W06L	2010
W06M	2010
W06MA	OEM
W08	2400
W08F	3736
W08L	2400
W08M	2400
W08MA	OEM
W09A	0080
W09B	0609
W09C	0023
W010L	0787
W010M	0787
W075	0077
W081	OEM
W005	OEM
W1	0280
W1-050	0162
W1-052	0162
W1-054	0157
W1-056	0157
W1-058	0157
W1-061	0636
W1-063	0636
W1-065	0636
W1-071	0649
W1-073	0649
W1-075	0649
W1-077	0649
W1-079	0244
W1-081	0244
W1-083	0244
W1-085	0244
W1-088	0012
W1-090	0012
W1-092	0012
W1-094	0012
W1-098	0170
W1-100	0170
W1-110	0317
W1-115	0794
W1-120	0137
W1-125	0365
W1-130	0365
W1-140	0100
W1-150	0002
W1-157	0420
W1-162	0420
W1-167	0420
W1-172	0495
W1-177	0495
W1-182	0495
W1-230	0402
W1-240	0402
W1-250	0402
W1-260	0441
W1-270	0441
W1-290	0726

DEVICE TYPE	REPL CODE
W1-300	0726
W1-330	0039
W1-340	0039
W1A	0325
W1E	0849
W1L	0058
W1P	0325
W1R	0058
W1R(SCR)	0058
W1U	0212
W1U-1	0212
W1W	2270
W2	0600
W2-050	0162
W2-052	0157
W2-054	0157
W2-056	0157
W2-058	0157
W2-061	0636
W2-063	0636
W2-065	0636
W2-071	0649
W2-073	0649
W2-075	0649
W2-077	0649
W2-079	0244
W2-081	0244
W2-083	0244
W2-085	0244
W2-088	0012
W2-090	0012
W2-092	0012
W2-094	0012
W2-098	0170
W2-100	0170
W2-110	0317
W2-115	0794
W2-125	0365
W2-130	0365
W2-140	0100
W2-150	0002
W2-157	0420
W2-167	0420
W2-172	0495
W2-177	0495
W2-182	0495
W2-230	0402
W2-240	0402
W2-250	0402
W2-260	0441
W2-270	0441
W2-290	0726
W2-300	0726
W2-330	0039
W2-340	0441
W2AA50C	2298
W2AA50E	2298
W2B325C	OEM
W2BA25C	1434
W2BA25E	1434
W2BC25C	1434
W2BC25E	1434
W2BE25C	1434
W2BE25E	1434
W2BH25C	2294
W2BH25E	2294
W2BJ25C	1680
W2BJ25E	1680
W2BK25C	1680
W2BK25E	1680
W2CA25C	OEM
W2CA25E	OEM
W2DA25C	OEM
W2DA25E	OEM
W2DC25E	OEM
W2VB10	OEM
W3	0004
W3D	0593
W4	0208
W4-050	0162
W4-052	0162
W4-054	0157
W4-056	0157
W4-058	0157
W4-061	0007
W4-063	0636
W4-065	0636
W4-071	0649
W4-073	0649
W4-075	0649
W4-077	0649
W4-079	0244
W4-081	0244
W4-083	0244
W4-085	0244
W4-088	0012
W4-090	0012
W4-092	0012
W4-094	0012
W4-098	0170
W4-100	0170
W4-110	0317
W4-115	0794
W4D45C	1798
W4D45D	1798
W4D45M	1798
W4D45N	1798
W4D45P	1798
W4D45PB	1798

DEVICE TYPE	REPL CODE
W4D45PD	OEM
W4D45PM	OEM
W4D80C	2192
W4D80D	2192
W4D80M	2192
W4D80N	2192
W4D80P	2192
W4D80PB	2192
W4D80PD	OEM
W4D80PM	OEM
W4DA40C	4364
W4DA40D	4364
W4DA40M	4364
W4DA40N	4364
W4DA40P	4364
W4DA40PB	4364
W4DA40PD	OEM
W4DA55C	4364
W4DA55D	4364
W4DA55M	4364
W4DA55N	4364
W4DA55P	4364
W4DA55PB	4364
W4DA55PD	OEM
W4DB40C	4574
W4DB40D	4574
W4DB40M	4574
W4DB40N	4574
W4DB40P	4574
W4DB40PB	4574
W4DB40PD	OEM
W4DB55C	4364
W4DB55D	4364
W4DB55M	4364
W4DB55N	4364
W4DB55P	4364
W4DB55PB	4364
W4DB55PD	4364
W4DC40C	0652
W4DC40D	0652
W4DC40M	0652
W4DC40N	0652
W4DC40P	0652
W4DC40PB	0652
W4DC40PD	OEM
W4DC55C	0652
W4DC55D	0652
W4DC55M	0652
W4DC55N	0652
W4DC55P	0652
W4DC55PB	0652
W4DC55PD	OEM
W4H45C	OEM
W4H45D	OEM
W4H45M	OEM
W4H45N	OEM
W4H45P	OEM
W4H45PB	OEM
W4H45PD	OEM
W4H45PM	OEM
W4H80C	OEM
W4H80D	OEM
W4H80M	OEM
W4H80N	OEM
W4H80P	OEM
W4H80PB	OEM
W4H80PD	OEM
W4H80PM	OEM
W5	0085
W5-22	OEM
W5-23	OEM
W5MC/15D	OEM
W5MC/16B	OEM
W5MC/17D	OEM
W5MC/18M	OEM
W5MW1M	OEM
W5PA1M	OEM
W5R-12	OEM
W6	0440
W6-120	0137
W6-125	0365
W6-130	0365
W6-140	0100
W6-150	0002
W6-157	0420
W6-162	0420
W6-167	0420
W6-172	0495
W6-177	0495
W6-182	0495
W6-230	0402
W6-240	0402
W6-250	0402
W6-260	0441
W6-270	0441
W6-290	0726
W6-300	0726
W6-330	0039
W6-340	0039
W6MT/2A	OEM
W7	0150
W7/3G	OEM
W7/4G	OEM
W7/5GA	OEM
W7/5GC	OEM
W7/6GA	OEM
W7/6GC	OEM
W7/6GZ	OEM
W7-120	0137

DEVICE TYPE	REPL CODE
W7-125	0365
W7-130	0365
W7-140	0100
W7-150	0002
W7-157	0420
W7-162	0420
W7-167	0420
W7-172	0495
W7-177	0495
W7-182	0495
W7-230	0402
W7-240	0402
W7-250	0402
W7-260	0441
W7-270	0441
W7-290	0726
W7-300	0726
W7-330	0039
W7-340	0039
W7MC/2R	OEM
W7MC/11C	OEM
W7MC/13B	OEM
W8	0144
W8-050	0162
W8-052	0162
W8-054	0157
W8-056	0157
W8-058	0157
W8-061	0636
W8-063	0636
W8-065	0636
W8-071	0649
W8-073	0649
W8-075	0649
W8-077	0649
W8-079	0244
W8-081	0244
W8-083	0244
W8-085	0244
W8-088	0012
W8-090	0012
W8-092	0012
W8-094	0012
W8-098	0170
W8-100	0170
W8-110	0317
W8-115	0794
W8-120	0137
W8-125	0365
W8-130	0365
W8-140	0100
W8-150	0002
W8-157	0420
W8-162	0420
W8-167	0420
W8-172	0495
W8-177	0495
W8-182	0495
W8-230	0402
W8-240	0402
W8-250	0402
W8-260	0441
W8-270	0441
W8-300	0726
W8-330	0039
W8-340	0039
W8-390	0726
W9	0160
W9/2E	OEM
W9/3E	OEM
W9/340	0039
W9-050	0162
W9-054	0157
W9-056	0157
W9-058	0157
W9-061	0636
W9-063	0636
W9-065	0636
W9-071	0649
W9-073	0649
W9-075	0649
W9-077	0649
W9-079	0244
W9-081	0244
W9-083	0244
W9-085	0244
W9-088	0012
W9-090	0012
W9-092	0012
W9-094	0012
W9-098	0170
W9-100	0170
W9-110	0317
W9-115	0794
W9-120	0137
W9-125	0365
W9-130	0365
W9-140	0100
W9-150	0002
W9-157	0495
W9-162	0495
W9-167	0495
W9-172	0495
W9-177	0495
W9-182	0495
W9-230	0402
W9-240	0402
W9-250	0402
W9-260	0441

DEVICE TYPE	REPL CODE
W9-270	0441
W9-290	0726
W9-300	0039
W9ACW	0325
W10	0198
W10/3E	OEM
W10F	0787
W10M	0787
W11	0142
W12	0050
W13	0975
W14	0086
W15	0126
W16	0103
W16MP	1514
W17	0222
W17MP	0822
W18	0042
W19	0925
W20	0016
W21	0037
W22	0050
W23	0007
W24	0016
W25	0164
W26	0276
W27	2763
W28	0233
W29	0127
W45B/5E	OEM
W46D/2T	OEM
W48D/1T	OEM
W53-046-0	OEM
W55	1834
W103-254	0015
W106A1	0447
W106B1	0940
W106C1	1222
W106Y1	1396
W110	0106
W122A	2089
W122B	0505
W122C	0717
W122D	0710
W122F	2095
W208FS	OEM
W208FS-1	OEM
W208FSM	OEM
W208FSM-1	OEM
W208FSMZS	OEM
W208FSMZS-1	OEM
W208FSQ	OEM
W208FSQ-1	OEM
W208FSQZS	OEM
W208FSQZS-1	OEM
W208FST	OEM
W208FST-1	OEM
W208FSTZS	OEM
W208FSTZS-1	OEM
W208FSZS	OEM
W208FSZS-1	OEM
W264FS	OEM
W264FSQ	OEM
W264FSQZS	OEM
W264FST	OEM
W264FSTZS	OEM
W264FSZS	OEM
W265	1198
W463A	OEM
W463B	OEM
W463D	OEM
W463E	OEM
W463F	OEM
W464A	OEM
W464B	OEM
W464C	OEM
W464D	OEM
W465A	OEM
W465B	OEM
W465C	OEM
W465D	OEM
W466A	OEM
W466B	OEM
W466C	OEM
W466D	OEM
W467	OEM
W468	OEM
W469A	OEM
W469B	OEM
W469C	OEM
W470A	OEM
W470B	OEM
W470C	OEM
W470D	OEM
W496A	OEM
W496B	OEM
W496C	OEM
W496D	OEM
W496E	OEM
W497A	OEM
W497B	OEM
W497C	OEM
W497D	OEM
W497E	OEM
W498A	OEM
W498C	OEM
W498D	OEM
W498E	OEM
W498F	OEM

DEVICE TYPE	REPL CODE
W498G	OEM
W498H	OEM
W498I	OEM
W498J	OEM
W498K	OEM
W498L	OEM
W806FS	OEM
W806FS-1	OEM
W806FSM	OEM
W806FSM-1	OEM
W806FSMZS	OEM
W806FSMZS-1	OEM
W806FSQ	OEM
W806FSQ-1	OEM
W806FSQZS	OEM
W806FSQZS-1	OEM
W806FST	OEM
W806FST-1	OEM
W806FSTZS	OEM
W806FSTZS-1	OEM
W806FSZS	OEM
W806FSZS-1	OEM
W812BS	OEM
W812BS-1	OEM
W812BSM	OEM
W812BSM-1	OEM
W812BSMZS	OEM
W812BSMZS-1	OEM
W812BSQ	OEM
W812BSQ-1	OEM
W812BSQZS	OEM
W812BSQZS-1	OEM
W812BST	OEM
W812BST-1	OEM
W812BSTZS	OEM
W812BSTZS-1	OEM
W812BSZS	OEM
W812BSZS-1	OEM
W4891	OEM
W4892	OEM
W4893	OEM
W4894	OEM
W5696A	0554
WA-050	0162
WA-050A	0162
WA-050B	0162
WA-050D	0162
WA-052	0162
WA-052A	0162
WA-052B	0162
WA-052C	0162
WA-052D	0162
WA-054	0157
WA-054A	0157
WA-054B	0157
WA-054C	0157
WA-054D	0157
WA-056	0157
WA-056A	0157
WA-056B	0157
WA-056C	0157
WA-056D	0157
WA-058	0157
WA-058A	0157
WA-058B	0157
WA-058C	0157
WA-058D	0157
WA-061	0636
WA-061A	0636
WA-061B	0636
WA-061C	0636
WA-063	0636
WA-063A	0636
WA-063B	0636
WA-063C	0636
WA-063D	0636
WA-065	0636
WA-065A	0636
WA-065B	0636
WA-065C	0636
WA-065D	0636
WA-071	0649
WA-071A	0649
WA-071B	0649
WA-071C	0649
WA-071D	0649
WA-073	0649
WA-073A	0649
WA-073B	0649
WA-073C	0649
WA-073D	0649
WA-075	0649
WA-075A	0649
WA-075B	0649
WA-075C	0649
WA-075D	0649
WA-077	0649
WA-077A	0649
WA-077B	0649
WA-077C	0649
WA-077D	0649
WA-079	0244
WA-079A	0244
WA-079B	0244
WA-079C	0244
WA-079D	0244
WA-081	0244
WA-081A	0244
WA-081B	0244

DEVICE TYPE	REPL CODE
WA-081C	0244
WA-081D	0244
WA-083	0244
WA-083A	0244
WA-083B	0244
WA-083C	0244
WA-083D	0244
WA-085	0244
WA-085A	0244
WA-085B	0244
WA-085C	0244
WA-085D	0244
WA-088	0012
WA-088A	0012
WA-088B	0012
WA-088C	0012
WA-088D	0012
WA-090	0012
WA-090A	0012
WA-090B	0012
WA-090C	0012
WA-090D	0012
WA-092	0012
WA-092A	0012
WA-092B	0012
WA-092C	0012
WA-092D	0012
WA-094	0012
WA-094A	0012
WA-094B	0012
WA-094C	0012
WA-094D	0012
WA-098	0012
WA-098A	0012
WA-098B	0170
WA-098C	0170
WA-098D	0170
WA0A-90	0133
WA-26	0133
WA-100	0170
WA100	2379
WA-100A	0170
WA-100B	0170
WA-100C	0170
WA-100D	0170
WA-110	0317
WA-110A	0317
WA-110B	0317
WA-110C	0317
WA-110D	0317
WA-115	0794
WA-115A	0794
WA-115B	0794
WA-115C	0794
WA-115D	0794
WA-120	0137
WA-120A	0137
WA-120B	0137
WA-120C	0137
WA-120D	0137
WA-125	0365
WA-125A	0365
WA-125B	0365
WA-125C	0365
WA-125D	0365
WA-130	0365
WA-130A	0365
WA-130B	0365
WA-130C	0365
WA-130D	0365
WA-140	0100
WA-140A	0100
WA-140B	0100
WA-140C	0100
WA-140D	0100
WA-150	0002
WA-150A	0002
WA-150B	0002
WA-150C	0002
WA-150D	0002
WA-157	0420
WA-157A	0420
WA-157B	0420
WA-157C	0420
WA-157D	0420
WA-161D	0636
WA-162	0420
WA-162A	0420
WA-162B	0420
WA-162C	0420
WA-162D	0420
WA-167	0420
WA-167A	0420
WA-167B	0420
WA-167C	0420
WA-167D	0420
WA-172	0495
WA-172A	0495
WA-172B	0495
WA-172C	0495
WA-172D	0495
WA-177	0495
WA-177A	0495
WA-177B	0495
WA-177C	0495
WA-177D	0495
WA-182	0495
WA-182A	0495
WA-182B	0495

DEVICE TYPE	REPL CODE
WA-182C	0495
WA-182D	0495
WA200	OEM
WA-230	0402
WA-230A	0402
WA-230B	0402
WA-230C	0402
WA-230D	0402
WA-240	0402
WA-240A	0402
WA-240B	0402
WA-240C	0402
WA-240D	0402
WA-250	0402
WA-250A	0402
WA-250B	0402
WA-250C	0402
WA-250D	0402
WA-260	0441
WA-260A	0441
WA-260B	0441
WA-260C	0441
WA-260D	0441
WA-270	0441
WA-270A	0441
WA-270B	0441
WA-270C	0441
WA-270D	0441
WA-290	0726
WA-290A	0726
WA-290B	0726
WA-290C	0726
WA-290D	0726
WA-300	0726
WA300	OEM
WA-300A	0726
WA-300B	0726
WA-300C	0726
WA-300D	0726
WA-330A	0039
WA-330B	0039
WA-330C	0039
WA-330D	0039
WA-340	0039
WA-340A	0039
WA-340B	0039
WA-340C	0039
WA-340D	0039
WA400	OEM
WA500	OEM
WA600	OEM
WAC1	OEM
WAC2	OEM
WAC3	OEM
WB-050	0162
WB-050A	0162
WB-050B	0162
WB-050C	0162
WB-050D	0162
WB-052	0162
WB-052A	0162
WB-052B	0162
WB-052C	0162
WB-052D	0162
WB-054	0157
WB-054A	0157
WB-054B	0157
WB-054C	0157
WB-054D	0157
WB-056	0157
WB-056A	0157
WB-056B	0157
WB-056C	0157
WB-056D	0157
WB-058	0157
WB-058A	0157
WB-058B	0157
WB-058C	0157
WB-058D	0157
WB-061	0636
WB-061A	0636
WB-061B	0636
WB-061C	0636
WB-061D	0636
WB-063	0636
WB-063A	0636
WB-063B	0636
WB-063C	0636
WB-063D	0636
WB-065	0636
WB-065A	0636
WB-065B	0636
WB-065C	0636
WB-065D	0636
WB-071	0649
WB-071A	0649
WB-071B	0649
WB-071C	0649
WB-071D	0649
WB-073	0649
WB-073A	0649
WB-073B	0649
WB-073C	0649
WB-073D	0649
WB-075	0649
WB-075A	0649
WB-075B	0649
WB-075C	0649
WB-075D	0649

DEVICE TYPE	REPL CODE
WB-077	0649
WB-077A	0649
WB-077B	0649
WB-077C	0649
WB-077D	0649
WB-079	0244
WB-079A	0244
WB-079B	0244
WB-079C	0244
WB-081	0244
WB-081A	0244
WB-081B	0244
WB-081C	0244
WB-081D	0244
WB-083	0244
WB-083A	0244
WB-083B	0244
WB-083C	0244
WB-083D	0244
WB-085	0244
WB-085A	0244
WB-085B	0244
WB-085C	0244
WB-085D	0244
WB-088	0012
WB-088A	0012
WB-088B	0012
WB-088C	0012
WB-088D	0012
WB-090	0012
WB-090A	0012
WB-090B	0012
WB-090C	0012
WB-090D	0012
WB-092	0012
WB-092A	0012
WB-092B	0012
WB-092C	0012
WB-092D	0012
WB-094	0012
WB-094A	0012
WB-094B	0012
WB-094C	0012
WB-094D	0012
WB-098	0170
WB-098B	0170
WB-098C	0170
WB-098D	0170
WB9TUI	0325
WB23	OEM
WB-100	0170
WB-100A	0170
WB-100B	0170
WB-100C	0170
WB-100D	0170
WB-110	0317
WB-110A	0317
WB-110B	0317
WB-110C	0317
WB-110D	0317
WB-115	0794
WB-115A	0794
WB-115B	0794
WB-115C	0794
WB-115D	0794
WB-120	0137
WB-120A	0137
WB-120B	0137
WB-120C	0137
WB-120D	0137
WB-125	0365
WB-125A	0365
WB-125B	0365
WB-125C	0365
WB-125D	0365
WB-130	0365
WB-130A	0365
WB-130B	0365
WB-130C	0365
WB-130D	0365
WB-140	0100
WB-140A	0100
WB-140B	0100
WB-140C	0100
WB-140D	0100
WB-150	0002
WB-150A	0002
WB-150B	0002
WB-150C	0002
WB-150D	0002
WB-157	0420
WB-157A	0420
WB-157B	0420
WB-157C	0420
WB-157D	0420
WB-162	0420
WB-162A	0420
WB-162B	0420
WB-162C	0420
WB-162D	0420
WB-167	0420
WB-167A	0420
WB-167B	0420
WB-167C	0420
WB-167D	0420
WB-172	0495
WB-172A	0495
WB-172B	0495
WB-172C	0495

DEVICE TYPE	REPL CODE
WB-172D	0495
WB-177	0495
WB-177A	0495
WB-177B	0495
WB-177C	0495
WB-177D	0495
WB-182	0495
WB-182A	0495
WB-182B	0495
WB-182C	0495
WB-182D	0495
WB-230	0402
WB-230A	0402
WB-230B	0402
WB-230C	0402
WB-230D	0402
WB-240	0402
WB-240A	0402
WB-240B	0402
WB-240C	0402
WB-240D	0402
WB-250	0402
WB-250A	0402
WB-250B	0402
WB-250C	0402
WB-250D	0402
WB-260	0441
WB-260A	0441
WB-260B	0441
WB-260C	0441
WB-260D	0441
WB-270	0441
WB-270A	0441
WB-270B	0441
WB-270C	0441
WB-270D	0441
WB-290	0726
WB-290A	0726
WB-290B	0726
WB-290C	0726
WB-290D	0726
WB-300	0726
WB-300A	0726
WB-300B	0726
WB-300C	0726
WB-300D	0726
WB-330	0039
WB-330A	0039
WB-330B	0039
WB-330C	0039
WB-330D	0039
WB-340	0039
WB-340A	0039
WB-340B	0039
WB-340C	0039
WB-340D	0039
WBC1	OEM
WBC2	OEM
WBC3	OEM
WBC4	OEM
WBC5	OEM
WBC6	OEM
WBC7	OEM
WBC8	OEM
WBC9	OEM
WC-050	0162
WC-050A	0162
WC-050B	0162
WC-050C	0162
WC-050D	0162
WC-052	0162
WC-052A	0162
WC-052B	0162
WC-052C	0162
WC-052D	0162
WC-054	0157
WC-054A	0157
WC-054B	0157
WC-054C	0157
WC-054D	0157
WC-056	0157
WC-056A	0157
WC-056B	0157
WC-056C	0157
WC-056D	0157
WC-058	0157
WC-058A	0157
WC-058B	0157
WC-058C	0157
WC-058D	0157
WC-061	0636
WC-061A	0636
WC-061B	0636
WC-061C	0636
WC-061D	0636
WC-063	0636
WC-063A	0636
WC-063B	0636
WC-063C	0636
WC-063D	0636
WC-065	0636
WC-065A	0636
WC-065B	0636
WC-065C	0636
WC-065D	0636
WC-071	0649
WC-071A	0649
WC-071B	0649
WC-071C	0649

DEVICE TYPE	REPL CODE
WC-071D	0649
WC-073	0649
WC-073A	0649
WC-073B	0649
WC-073C	0649
WC-073D	0649
WC-075	0649
WC-075A	0649
WC-075B	0649
WC-075C	0649
WC-075D	0649
WC-077	0649
WC-077A	0649
WC-077B	0649
WC-077C	0649
WC-077D	0649
WC-079	0244
WC-079A	0244
WC-079B	0244
WC-079C	0244
WC-079D	0244
WC-081	0244
WC-081A	0244
WC-081B	0244
WC-081C	0244
WC-081D	0244
WC-083	0244
WC-083A	0244
WC-083B	0244
WC-083C	0244
WC-083D	0244
WC-085	0244
WC-085A	0244
WC-085B	0244
WC-085C	0244
WC-085D	0244
WC-088	0012
WC-088A	0012
WC-088B	0012
WC-088C	0012
WC-090	0012
WC-090A	0012
WC-090B	0012
WC-090C	0012
WC-090D	0012
WC-092	0012
WC-092A	0012
WC-092B	0012
WC-092C	0012
WC-092D	0012
WC-094	0012
WC-094A	0012
WC-094B	0012
WC-094C	0012
WC-094D	0012
WC-098	0170
WC-098A	0170
WC-098B	0170
WC-098C	0170
WC-098D	0170
WC6B	OEM
WC-88D	0012
WC-100	0170
WC-100A	0170
WC-100B	0170
WC-100C	0170
WC-100D	0170
WC-110	0317
WC-110A	0317
WC-110B	0317
WC-110C	0317
WC-110D	0317
WC-115	0794
WC-115A	0794
WC-115B	0794
WC-115C	0794
WC-115D	0794
WC-120	0137
WC120	0015
WC-120A	0137
WC-120B	0137
WC-120C	0137
WC-120D	0137
WC-125	0365
WC-125A	0365
WC-125B	0365
WC-125C	0365
WC-125D	0365
WC-130	0365
WC-130A	0365
WC-130B	0365
WC-130C	0365
WC-130D	0365
WC-140	0100
WC-140A	0100
WC-140B	0100
WC-140C	0100
WC-140D	0170
WC-150	0002
WC-150A	0002
WC-150B	0002
WC-150C	0002
WC-150D	0002
WC-157	0420
WC-157A	0420
WC-157B	0420
WC-157C	0420
WC-157D	0420
WC-162	0420

DEVICE TYPE	REPL CODE
WC-162A	0420
WC-162B	0420
WC-162C	0420
WC-162D	0420
WC-167	0420
WC-167A	0420
WC-167B	0420
WC-167C	0420
WC-167D	0420
WC-172	0495
WC-172A	0495
WC-172B	0495
WC-172C	0495
WC-172D	0495
WC-177	0495
WC-177A	0495
WC-177B	0495
WC-177C	0495
WC-177D	0495
WC-182	0495
WC-182A	0495
WC-182B	0495
WC-182C	0495
WC-182D	0402
WC-230	0402
WC-230A	0402
WC-230B	0402
WC-230C	0402
WC-230D	0402
WC-240	0402
WC-240A	0402
WC-240B	0402
WC-240C	0402
WC-240D	0402
WC-250	0402
WC-250A	0402
WC-250B	0402
WC-250C	0402
WC-250D	0402
WC-260	0441
WC-260A	0441
WC-260B	0441
WC-260C	0441
WC-260D	0441
WC-270	0441
WC-270A	0441
WC-270B	0441
WC-270C	0441
WC-270D	0441
WC-290	0726
WC-290A	0726
WC-290B	0726
WC-290C	0726
WC-290D	0726
WC-300	0726
WC-300A	0726
WC-300B	0726
WC-300C	0726
WC-300D	0726
WC-330	0039
WC-330A	0039
WC-330B	0039
WC-330C	0039
WC-330D	0039
WC-340	0039
WC-340A	0039
WC-340B	0039
WC-340C	0039
WC-340D	0039
WC501G	OEM
WC501P	OEM
WC535P	OEM
WC14020	0015
WC14027	0015
WC19862	0208
WC19862A	0164
WC19863	0004
WC19864	0004
WC19865	0015
WCN005	OEM
WCN02	OEM
WCN04	OEM
WCN06	OEM
WCN08	OEM
WCN10	OEM
WD001	0609
WD002	0609
WD003	0015
WD004	0609
WD005	0609
WD006	0609
WD007	0609
WD008	0609
WD009	0609
WD010	0609
WD011	0609
WD012	0609
WD013	3911
WD014	0015
WD015	3911
WD-050	0162
WD-050A	0162
WD-050C	0162
WD-052	0162
WD-052A	0162
WD-052B	0162
WD-052C	0162
WD-052D	0162
WD-054	0157

If replacement code is OEM, contact original manufacturer for replacement.

DEVICE TYPE	REPL CODE
WD-054A	0649
WD-054B	0157
WD-054C	0157
WD-054D	0157
WD-056	0157
WD-056A	0157
WD-056B	0157
WD-056C	0157
WD-056D	0157
WD-058	0157
WD-058A	0157
WD-058B	0157
WD-058C	0157
WD-058D	0157
WD-061	0636
WD-061A	0636
WD-061B	0636
WD-061C	0636
WD-061D	0636
WD-063	0636
WD-063A	0636
WD-063B	0636
WD-063C	0636
WD-063D	0636
WD-065	0636
WD-065A	0636
WD-065B	0636
WD-065C	0636
WD-065D	0636
WD-071	0649
WD-071A	0649
WD-071B	0649
WD-071C	0649
WD-071D	0649
WD-073	0649
WD-073A	0649
WD-073B	0649
WD-073C	0649
WD-073D	0649
WD-075	0649
WD-075A	0649
WD-075B	0649
WD-075C	0649
WD-075D	0649
WD-077	0649
WD-077A	0649
WD-077B	0649
WD-077C	0649
WD-077D	0649
WD-079	0244
WD-079A	0244
WD-079B	0244
WD-079C	0244
WD-079D	0244
WD-081	0244
WD-081A	0244
WD-081B	0244
WD-081C	0244
WD-081D	0244
WD-083	0244
WD-083A	0244
WD-083B	0244
WD-083C	0244
WD-083D	0244
WD-085	0244
WD-085A	0244
WD-085B	0244
WD-085C	0244
WD-085D	0244
WD-088	0012
WD-088A	0012
WD-088B	0012
WD-088C	0012
WD-088D	0012
WD-090	0012
WD-090A	0012
WD-090B	0012
WD-090C	0012
WD-090D	0012
WD-092	0012
WD-092A	0012
WD-092B	0012
WD-092C	0012
WD-092D	0012
WD-094	0012
WD-094A	0012
WD-094B	0012
WD-094C	0012
WD-094D	0012
WD-098	0170
WD-098A	0170
WD-098B	0170
WD-098C	0170
WD-098D	0170
WD1	0143
WD2	0917
WD3	0123
WD4	0133
WD10	0196
WD10C20-PH	OEM
WD10C20A-PH	OEM
WD11	0484
WD11C00-JT	OEM
WD11C00-JU	OEM
WD12	1080
WD16C92-PL	OEM
WD22	0474
WD37C65	OEM
WD51A	OEM
WD51B	OEM
WD55A	OEM
WD55B	OEM
WD61	0015
WD74HC200E	OEM
WD74HC200F	OEM
WD-100	0170
WD-100A	0170
WD-100B	0170
WD-100C	0170
WD-100D	0170
WD-110	0317
WD-110A	0317
WD-110B	0317
WD-110C	0317
WD-110D	0317
WD-115	0794
WD-115A	0794
WD-115B	0794
WD-115C	0794
WD-115D	0794
WD-120	0137
WD-120A	0137
WD-120B	0137
WD-120C	0137
WD-120D	0137
WD-125	0365
WD-125A	0365
WD-125B	0365
WD-125C	0365
WD-125D	0365
WD-130	0365
WD-130A	0365
WD-130B	0365
WD-130C	0365
WD-130D	0365
WD-140	0100
WD-140A	0100
WD-140B	0100
WD-140C	0100
WD-140D	0100
WD-150	0002
WD-150A	0002
WD-150B	0002
WD-150C	0002
WD-150D	0002
WD-157	0420
WD-157A	0420
WD-157B	0420
WD-157C	0420
WD-157D	0420
WD-162	0420
WD-162A	0420
WD-162B	0420
WD-162C	0420
WD-162D	0420
WD-167	0420
WD-167A	0420
WD-167B	0420
WD-167C	0420
WD-167D	0420
WD-172	0495
WD-172A	0495
WD-172B	0495
WD-172C	0495
WD-172D	0495
WD-177	0495
WD-177A	0495
WD-177B	0495
WD-177C	0495
WD-177D	0495
WD-182	0495
WD-182A	0495
WD-182B	0495
WD-182C	0495
WD-182D	0495
WD-230	0402
WD-230A	0402
WD-230B	0402
WD-230C	0402
WD-230D	0402
WD-240	0402
WD-240A	0402
WD-240B	0402
WD-240C	0402
WD-240D	0402
WD-250	0402
WD-250A	0402
WD-250B	0402
WD-250C	0402
WD-250D	0402
WD-260	0441
WD-260A	0441
WD-260B	0441
WD-260C	0441
WD-260D	0441
WD-270	0441
WD-270A	0441
WD-270B	0441
WD-270C	0441
WD-270D	0441
WD-290	0726
WD-290A	0726
WD-290B	0726
WD-290C	0726
WD-290D	0726
WD-300	0726
WD-300B	0726
WD-300C	0726
WD-300D	0726
WD-330	0039
WD-330A	0039
WD-330B	0039
WD-330C	0039
WD-330D	0039
WD-340	0039
WD-340A	0039
WD-340B	0039
WD-340C	0039
WD-340D	0039
WD900	OEM
WD1000	OEM
WD1010-PL	OEM
WD1010A-PL	OEM
WD1012	0143
WD1014-CL	OEM
WD1015-PL	OEM
WD1100-01U	OEM
WD1100-01V	OEM
WD1100-02U	OEM
WD1100-02V	OEM
WD1100-05U	OEM
WD1100-05V	OEM
WD1510E00	OEM
WD1510E01	OEM
WD1510F00	OEM
WD1510F01	OEM
WD1511E	OEM
WD1511F	OEM
WD1691PE-00-02	OEM
WD1691U	OEM
WD1691V	OEM
WD1770-PH	OEM
WD1772-PH	OEM
WD1773	OEM
WD1773-PH	OEM
WD1773PH-00	OEM
WD1793	OEM
WD1931A	OEM
WD1931B	OEM
WD1933A	OEM
WD1933B	OEM
WD1983E	OEM
WD1983F	OEM
WD1984E	OEM
WD1984F	OEM
WD1993	OEM
WD1993E	OEM
WD1993F	OEM
WD2001E	OEM
WD2001F	OEM
WD2002A	OEM
WD2002B	OEM
WD2010B-PL	OEM
WD2123	OEM
WD2123A	OEM
WD2123B	OEM
WD2143-01	OEM
WD2143L01	OEM
WD2143M	OEM
WD2143M01	OEM
WD2293-PE	OEM
WD2293A-PE	OEM
WD2501	OEM
WD2501-11	OEM
WD2511	OEM
WD2793A-PL	OEM
WD2793A-PL02-04	OEM
WD4020A	OEM
WD4020AE	OEM
WD4020B	OEM
WD4020BE	OEM
WD4200E	OEM
WD4200F	OEM
WD4210E	OEM
WD4210F	OEM
WD5869J	OEM
WD5869K	OEM
WD8250	OEM
WD8250-PL	OEM
WD8250A	OEM
WD8250B	OEM
WD8250PL-00	OEM
WD9216	OEM
WE-050	0162
WE-050A	0162
WE-050B	0162
WE-050C	0162
WE-050D	0162
WE-052	0162
WE-052A	0162
WE-052B	0162
WE-052C	0162
WE-052D	0162
WE-054	0157
WE-054A	0157
WE-054B	0157
WE-054C	0157
WE-054D	0157
WE-056	0157
WE-056A	0157
WE-056B	0157
WE-056C	0157
WE-056D	0157
WE-058	0157
WE-058A	0157
WE-058B	0157
WE-058C	0157
WE-058D	0157
WE-061	0157
WE-061A	0636
WE-061B	0636
WE-061C	0636
WE-061D	0636
WE-063	0636
WE-063A	0636
WE-063B	0636
WE-063C	0636
WE-063D	0636
WE-065	0636
WE-065A	0636
WE-065B	0636
WE-065C	0636
WE-065D	0636
WE-071	0649
WE-071A	0649
WE-071B	0649
WE-071C	0649
WE-071D	0649
WE-073	0649
WE-073A	0649
WE-073B	0649
WE-073C	0649
WE-073D	0649
WE-075	0649
WE-075A	0649
WE-075B	0649
WE-075C	0649
WE-075D	0649
WE-077	0649
WE-077A	0649
WE-077B	0649
WE-077C	0649
WE-077D	0649
WE-079	0244
WE-079A	0244
WE-079B	0244
WE-079C	0244
WE-079D	0244
WE-081	0244
WE-081A	0244
WE-081B	0244
WE-081C	0244
WE-081D	0244
WE-083	0244
WE-083A	0244
WE-083B	0244
WE-083C	0244
WE-083D	0244
WE-085	0244
WE-085A	0244
WE-085B	0244
WE-085C	0244
WE-085D	0244
WE-088	0012
WE-088A	0012
WE-088B	0012
WE-088C	0012
WE-088D	0012
WE-090	0012
WE-090A	0012
WE-090B	0012
WE-090C	0012
WE-090D	0012
WE-092	0012
WE-092A	0012
WE-092B	0012
WE-092C	0012
WE-092D	0012
WE-094	0012
WE-094A	0012
WE-094B	0012
WE-094C	0012
WE-094D	0012
WE-098	0170
WE-098A	0170
WE-098B	0170
WE-098C	0170
WE-098D	0170
WE-100	0170
WE-100A	0170
WE-100B	0170
WE-100C	0170
WE-100D	0170
WE-110	0317
WE-110A	0317
WE-110B	0317
WE-110C	0317
WE-110D	0170
WE-115	0794
WE-115A	0794
WE-115B	0794
WE-115C	0794
WE-115D	0794
WE-120	0137
WE-120A	0137
WE-120B	0137
WE-120C	0137
WE-120D	0137
WE-125	0365
WE-125A	0365
WE-125B	0365
WE-125C	0365
WE-125D	0365
WE-130	0365
WE-130A	0365
WE-130B	0365
WE-130C	0365
WE-130D	0365
WE-140	0100
WE-140A	0100
WE-140B	0100
WE-140C	0100
WE-140D	0100
WE-150	0002
WE-150A	0002
WE-150B	0002
WE-150C	0002
WE-150D	0002
WE-157	0420
WE-157A	0420
WE-157B	0420
WE-157C	0420
WE-157D	0420
WE-162	0420
WE-162A	0420
WE-162B	0420
WE-162C	0420
WE-162D	0420
WE-167	0420
WE-167A	0420
WE-167B	0420
WE-167C	0420
WE-167D	0420
WE-172	0495
WE-172A	0495
WE-172B	0495
WE-172C	0495
WE-172D	0495
WE-177	0495
WE-177A	0495
WE-177B	0495
WE-177C	0495
WE-177D	0495
WE-182	0495
WE-182A	0495
WE-182B	0495
WE-182C	0495
WE-182D	0495
WE183P	OEM
WE-230	0402
WE-230A	0402
WE-230B	0402
WE-230C	0402
WE-230D	0402
WE-240	0402
WE-240A	0402
WE-240B	0402
WE-240C	0402
WE-240D	0402
WE-250	0402
WE-250A	0402
WE-250B	0402
WE-250C	0402
WE-250D	0402
WE-260	0441
WE-260A	0441
WE-260B	0441
WE-260C	0441
WE-260D	0441
WE-270	0441
WE-270A	0441
WE-270B	0441
WE-270C	0441
WE-270D	0441
WE-290	0726
WE-290A	0726
WE-290B	0726
WE-290C	0726
WE-290D	0726
WE-300	0726
WE-300A	0726
WE-300B	0726
WE-300C	0726
WE-300D	0726
WE-330	0039
WE-330A	0039
WE-330C	0039
WE-330D	0039
WE-340	0039
WE-340A	0039
WE-340B	0039
WE-340C	0039
WE-340D	0039
WE530P	OEM
WE4110	0517
WE13400	OEM
WE0704/130	0103
WEP2	0280
WEP3	0050
WEP10	0633
WEP14	OEM
WEP15/512	2085
WEP16/511	1449
WEP17	1306
WEP17/508	1306
WEP18	2083
WEP18/509	2083
WEP19	2084
WEP19/510	2084
WEP50	0079
WEP51	0126
WEP52	0037
WEP53	0018
WEP54	2860
WEP55	2550
WEP56	0144
WEP56/108	0144
WEP57	0037
WEP58	0198
WEP58/123	0198
WEP59	0086
WEP59/128	0086
WEP60	0126
WEP60/129	0126
WEP61	0276
WEP61/157	0276
WEP62	0037
WEP62/159	0037
WEP63	0007
WEP63/161	0007
WEP64	0861
WEP64/194	0861
WEP65	0280
WEP66	0111
WEP66/199	0111
WEP67	3519
WEP67/232	3519
WEP68	0715
WEP68/287	0715
WEP89/89	0055
WEP93	OEM
WEP93/502	0201
WEP94	0288
WEP94/503	0288
WEP95	OEM
WEP95/504	0378
WEP100	0280
WEP100/100	0280
WEP101	0170
WEP102	0188
WEP103	0636
WEP103A	0649
WEP104	0012
WEP105	0137
WEP106	0150
WEP106/106	0150
WEP118	0774
WEP118/118	0774
WEP119	0102
WEP119/119	1303
WEP120	OEM
WEP125	OEM
WEP125/358	1030
WEP125/358B	5036
WEP126	OEM
WEP126/358A	4465
WEP134	0143
WEP134/109	0143
WEP138	0143
WEP139	0917
WEP139/112	0917
WEP152	0023
WEP152/552	0023
WEP154	1335
WEP155	0080
WEP156	0609
WEP157	0795
WEP158	0015
WEP158/116	0015
WEP158A	0071
WEP159	0072
WEP160	0071
WEP164	2962
WEP164/114	0484
WEP165	0196
WEP165/113	0196
WEP165A	0484
WEP166	1080
WEP166/115	1080
WEP167	2965
WEP168	0474
WEP168/120	0474
WEP169	2966
WEP170	0071
WEP170/125	0071
WEP172	0102
WEP172/506	0102
WEP173	0023
WEP175	0023
WEP177	0133
WEP177/525	0348
WEP178	1791
WEP178/518	1791
WEP179	OEM
WEP180	0293
WEP181	0190
WEP182	0293
WEP182/173AP	0293
WEP183	0288
WEP184	0378
WEP185	0757
WEP186	0102
WEP187	0004
WEP188	1501
WEP189	2977
WEP189/HI-DIV-1	2977
WEP190	2978
WEP190/HI-DIV-2	2978
WEP191	2979
WEP192	1067
WEP193	2980
WEP195	2982
WEP195/534	2982
WEP196	2983
WEP197	2984
WEP197/538	2984
WEP198	2985
WEP199	1501
WEP199/522	1501
WEP200	0030
WEP203	0409
WEP215	1323
WEP215/513	1323
WEP221	1681
WEP221/180	1681
WEP221MP	1681
WEP222	0130
WEP222/181	0130
WEP224	0630
WEP224/224	0630
WEP227	0164
WEP227/158	0164
WEP228	2763
WEP228/155	2763
WEP229	0222
WEP229/131	0222
WEP229MP	0822
WEP229MP/131MP	0822
WEP230	0085
WEP230MP	3033
WEP231	0440
WEP232	0160
WEP232MP	0265
WEP233	0440
WEP233/105	0440
WEP234	0160
WEP234/121	0160
WEP234MP	0265
WEP234MP/121MP	0265
WEP235	0975
WEP235/127	0975
WEP236	0437
WEP236/213	0437
WEP238	1424
WEP239	0847
WEP239/176	0847
WEP240	0142
WEP240/124	0142
WEP241	0178
WEP241/175	0178
WEP242	0126
WEP243	0086
WEP244	0276
WEP245	0161
WEP246	0460
WEP247	0543
WEP247MP	2371
WEP248	1598
WEP250	0211
WEP250/102	0211
WEP253	0164
WEP254	0208
WEP255	0792
WEP261/261	1212
WEP273	0563
WEP273/373	0563
WEP308	1669
WEP308/6400	1669
WEP309	2136
WEP310	1175
WEP310/6410	1175
WEP350	0922
WEP367	0222
WEP371	0191
WEP372	0076
WEP373	0076
WEP374/374	0525
WEP379/379	0728
WEP380	0286
WEP391/391	0859
WEP394	0155
WEP403	0155
WEP405	0164
WEP454	0076
WEP458	0076
WEP460	0151
WEP474	1010
WEP474/226	1010
WEP474MP	1862
WEP474MP/226MP	1862
WEP481	0222
WEP495	0006
WEP500	0632
WEP500/703A	0632
WEP501	2624
WEP501/704	2624
WEP502	2161
WEP502/705A	2161
WEP503	0753
WEP503/707	0753
WEP504	0664
WEP504/708	0664
WEP505	0678
WEP505/709	0678
WEP506	0829
WEP506/710	0829
WEP507	0167
WEP507/712	0167
WEP508	0666
WEP508/713	0666
WEP509	0352
WEP509/714	0352
WEP510	0354

DEVICE TYPE	REPL CODE
WEP510/715	0354
WEP511	0654
WEP511/718	0654
WEP512	0443
WEP512/720	0443
WEP513	0701
WEP513/722	0701
WEP514	0692
WEP515	OEM
WEP515/725	0692
WEP517	1411
WEP535	0127
WEP535/107	0127
WEP536	0537
WEP537	0191
WEP538	0079
WEP564	0203
WEP568	0111
WEP601/601	0139
WEP602	0162
WEP603	0157
WEP604	0794
WEP605	0365
WEP605/605	0120
WEP606	0100
WEP607	0002
WEP608	0441
WEP609	0039
WEP610	1834
WEP611	0783
WEP612	0331
WEP613	0149
WEP615/615	1329
WEP620	0004
WEP620/102A	0004
WEP621	0050
WEP621/160	0050
WEP624	0085
WEP624/104	0085
WEP624MP	3033
WEP624MP/104MP	3033
WEP628	0085
WEP628MP	0085
WEP630	0164
WEP631	0211
WEP632	0004
WEP633	2075
WEP634	2075
WEP635	0136
WEP635/126	0136
WEP637	0050
WEP641	0600
WEP641/101	0600
WEP641A	0038
WEP641A/103	0038
WEP641B	0208
WEP641B/103A	0208
WEP642	0222
WEP642MP	0222
WEP643	0222
WEP643MP	OEM
WEP644	0111
WEP645	0604
WEP645/179	0604
WEP700	0460
WEP701	0161
WEP702	0285
WEP702/171	0285
WEP703	0178
WEP704	0103
WEP704/130	0103
WEP705	2446
WEP706	2446
WEP707	0074
WEP707/162	0074
WEP707MP	0074
WEP708	0224
WEP709	1144
WEP710	0368
WEP711	0418
WEP712	0493
WEP713	0233
WEP713/154	0233
WEP714	OEM
WEP714/321	0130
WEP715	0532
WEP716	0037
WEP717	0037
WEP719	0224
WEP720	0224
WEP723	0016
WEP724	0016
WEP728	0016
WEP729	0016
WEP735	0016
WEP735A	0191
WEP736	0079
WEP736/123A	0016
WEP740	0642
WEP740/163	0642
WEP740A	0003
WEP740A/164	0003
WEP740B	0065
WEP740B/165	0065
WEP741	0905
WEP741/218	0905
WEP744	1027
WEP744/286	1027
WEP744MP	OEM

DEVICE TYPE	REPL CODE
WEP745	0042
WEP745/152	0042
WEP746	0925
WEP746/153	0925
WEP747	0561
WEP747/182	0561
WEP748	1199
WEP748/183	1199
WEP749	0264
WEP749/190	0264
WEP750	0164
WEP751	0560
WEP751/186	0560
WEP752	1267
WEP752/187	1267
WEP753	0595
WEP753/192	0595
WEP754	0791
WEP754/193	0791
WEP755	0698
WEP755/195A	0698
WEP756	0423
WEP756/196	0423
WEP757	0854
WEP757/197	0854
WEP758	0566
WEP758/210	0566
WEP759	1367
WEP759/211	1367
WEP760	0491
WEP760/219	0491
WEP761	2096
WEP761/369	2096
WEP762	3280
WEP762/225	3280
WEP763	0392
WEP763/375	0392
WEP764	0313
WEP764/238	0313
WEP765	3165
WEP765/242	3165
WEP766	2214
WEP766/277	2214
WEP767	0418
WEP767/278	0418
WEP768	0620
WEP768/280MP	2140
WEP768/328	0620
WEP768MP	2140
WEP769	2013
WEP769/281	2013
WEP770	0363
WEP770/283	0363
WEP771	1946
WEP771/306	1946
WEP772	1144
WEP773	0788
WEP774	3289
WEP775	0543
WEP775/284	0543
WEP775MP	2371
WEP775MP/284MP	2371
WEP776	1598
WEP776/285	1598
WEP777	1411
WEP778	1411
WEP779	0168
WEP779/198	0168
WEP780	0236
WEP780/291	0236
WEP781	0681
WEP781/292	0681
WEP782	2946
WEP782/317	2946
WEP783	2312
WEP783/318	2312
WEP784	1144
WEP785	0839
WEP785/235	0839
WEP799	1411
WEP799/237	1411
WEP801	3046
WEP801/133	3046
WEP802	0325
WEP802/132	3712
WEP803	OEM
WEP810	2170
WEP811	1198
WEP811/325	1198
WEP812	1974
WEP812/320	1974
WEP813	0493
WEP814	OEM
WEP821/821	2831
WEP822/822	2812
WEP828	1220
WEP829	0151
WEP838	0191
WEP840	0836
WEP840/236	0836
WEP850	2700
WEP850/329	2700
WEP851	0622
WEP852	0892
WEP853	1984
WEP853/322	1984
WEP854	0338
WEP854/191	0338
WEP855	0824

DEVICE TYPE	REPL CODE
WEP856	3327
WEP857	0382
WEP857/189	0382
WEP858	0551
WEP858/188	0551
WEP880	0161
WEP880/184	0161
WEP881	0482
WEP882	1369
WEP883	0460
WEP883/185	0460
WEP900	0219
WEP900/186A	0219
WEP901	1430
WEP901/307	1430
WEP903	0849
WEP903/220	0849
WEP904	0353
WEP904/221	0353
WEP905	0212
WEP905/222	0212
WEP906	0330
WEP906/233	0330
WEP907	0693
WEP907/234	0693
WEP908	2997
WEP908/241	2997
WEP909	0622
WEP909/282	0622
WEP910	0155
WEP910/289	0155
WEP911	0006
WEP911/290	0006
WEP911/290A	0006
WEP912	0018
WEP912/293	0018
WEP913	1591
WEP913/295	1591
WEP914	0324
WEP914/297	0324
WEP915	0436
WEP915/298	0436
WEP916	0532
WEP916/294	0532
WEP917	2046
WEP917/300	2046
WEP918	1173
WEP918/302	1173
WEP920	0325
WEP920/312	0325
WEP921	0475
WEP921/313	0475
WEP922	1978
WEP922/315	1978
WEP923	0259
WEP923/316	0259
WEP924	2902
WEP924/319	2902
WEP925	0124
WEP925/519	0124
WEP926	0823
WEP926/724	0823
WEP927	2617
WEP927/726	2617
WEP928	0517
WEP929	0723
WEP930	1876
WEP931	1077
WEP933	0312
WEP933/941M	0312
WEP934	1235
WEP934/1000	1235
WEP935	0021
WEP935/1012	0021
WEP936	2406
WEP937	2124
WEP937/1058	2124
WEP938	0470
WEP938/1075A	0470
WEP939	2261
WEP939/1082	2261
WEP941	0911
WEP941/1100	0911
WEP942	2632
WEP942/1102	2632
WEP943	1994
WEP943/1103	1994
WEP944	2057
WEP944/1104	2057
WEP945	2158
WEP945/1113	2158
WEP946	1920
WEP946/1118	1920
WEP948	3366
WEP948/1153	3366
WEP949	1050
WEP949/1155	1050
WEP950	0513
WEP950/4017	0513
WEP953	0087
WEP954	0087
WEP955	3331
WEP955/6407	3331
WEP956	0224
WEP956/229	0224
WEP959	3370
WEP959/274	3370
WEP962	1282
WEP962/5812	1282

DEVICE TYPE	REPL CODE
WEP963	1287
WEP963/5814	1287
WEP964	1293
WEP964/5815	1293
WEP965	1296
WEP966	1296
WEP966/5817	1296
WEP966L/966	0334
WEP968	0342
WEP968/288	0342
WEP969	0493
WEP969/311	0493
WEP970	OEM
WEP971	0400
WEP971/172	0400
WEP972	2427
WEP972/243	2427
WEP973	2277
WEP973/244	2277
WEP974	3373
WEP974/245	3373
WEP975	3374
WEP975/246	3374
WEP976	2439
WEP976/247	2439
WEP977	2432
WEP977/248	2432
WEP984	3377
WEP984/257	3377
WEP986/986	3089
WEP989	3381
WEP989/989	3381
WEP992	2258
WEP992/265	2258
WEP1008	0025
WEP1010	0244
WEP1014	2046
WEP1018	2050
WEP1018/299	2050
WEP1043	0323
WEP1044	1414
WEP1045	0473
WEP1046	0446
WEP1050	0277
WEP1051	0277
WEP1051/166	0277
WEP1052	0289
WEP1052/167	0289
WEP1053	0296
WEP1053/168	0296
WEP1054	0302
WEP1054/169	0302
WEP1056	0250
WEP1056/170	0250
WEP1060	0133
WEP1061	0913
WEP1061/178MP	0913
WEP1062	0133
WEP1062/177	0133
WEP1065	5483
WEP1066	2363
WEP1066/5322	1414
WEP1067	2369
WEP1067/5324	0473
WEP1075	0917
WEP1076	0316
WEP1076/6402	0316
WEP1077	4233
WEP1077/6403	4233
WEP1078	3407
WEP1078/6404	3407
WEP1079	4405
WEP1079/6405	4405
WEP1080	2730
WEP1080/6408	2730
WEP1082	0015
WEP1083	0015
WEP1096	0390
WEP1100	0456
WEP1100/5067	0456
WEP1102	0451
WEP1102/5069	0451
WEP1103	0162
WEP1103/135	0162
WEP1104	0157
WEP1104/136	0157
WEP1106	0025
WEP1106/5071	0025
WEP1107	0649
WEP1107/138	0649
WEP1108	0244
WEP1108/5072	0244
WEP1109	0012
WEP1109/139	0012
WEP1110	0170
WEP1110/140	0170
WEP1112	0137
WEP1112/142	0137
WEP1114	0002
WEP1114/145	0002
WEP1116	0495
WEP1116/5077	0495
WEP1118	0565
WEP1118/5080	0565
WEP1120	0441
WEP1120/146	0441
WEP1122	0039
WEP1122/147	0039
WEP1124	0350

DEVICE TYPE	REPL CODE
WEP1124/5086	0350
WEP1126	0999
WEP1126/5088	0999
WEP1128	0869
WEP1128/5090	0869
WEP1130	2157
WEP1130/5092	1268
WEP1132	0331
WEP1132/150	0331
WEP1134	0098
WEP1134/5096	0098
WEP1136	0186
WEP1136/5097	0186
WEP1138	0028
WEP1138/5100	0028
WEP1140	0367
WEP1140/5103	0367
WEP1145/5061	3003
WEP1147	1312
WEP1147/5063	1312
WEP1148	3010
WEP1148/5064	3010
WEP1149	1713
WEP1149/5065	1713
WEP1150	0292
WEP1150/5066	0292
WEP1151	0188
WEP1151/134	0188
WEP1152	0533
WEP1152/5068	0533
WEP1153	0301
WEP1153/5070	0301
WEP1154	0636
WEP1154/137	0636
WEP1155	1083
WEP1155/5073	1083
WEP1156	0317
WEP1156/5074	0317
WEP1157	0794
WEP1157/141	0794
WEP1158	0365
WEP1158/143	0365
WEP1159	0100
WEP1159/144	0100
WEP1160	0420
WEP1160/5075	0420
WEP1161	1649
WEP1161/5076	1649
WEP1162	0949
WEP1162/5078	0949
WEP1163	0531
WEP1163/5079	0531
WEP1164	0402
WEP1164/5081	0402
WEP1165	1606
WEP1165/5082	1606
WEP1166	1674
WEP1166/5083	1674
WEP1167	0726
WEP1167/5084	0726
WEP1168	0820
WEP1168/5085	0820
WEP1169	0931
WEP1169/5087	0931
WEP1172	0502
WEP1172/5089	0502
WEP1174	1834
WEP1174/148	1834
WEP1175	1156
WEP1175/5091	1156
WEP1176	0783
WEP1176/149	0783
WEP1177	1189
WEP1177/5093	1189
WEP1178	3026
WEP1178/5094	3026
WEP1179	1311
WEP1179/5095	1311
WEP1181	0149
WEP1181/151	0149
WEP1182	0213
WEP1182/5098	0213
WEP1183	0245
WEP1183/5099	0245
WEP1184	0255
WEP1184/5101	0255
WEP1185	0877
WEP1185/5102	0877
WEP1187	2858
WEP1187/5104	2858
WEP1188	0421
WEP1188/5105	0421
WEP1200	0806
WEP1204	0876
WEP1206	0205
WEP1208	0504
WEP1212	0234
WEP1214	0251
WEP1231/1231	1170
WEP1237	0936
WEP1239	1908
WEP1246/1246	4635
WEP1300	2035
WEP1300/5174	2035
WEP1300R	0547
WEP1300R/5174K	0547
WEP1302	1438
WEP1302/5176	1438
WEP1302R	2113

DEVICE TYPE	REPL CODE
WEP1302R/5176K	2113
WEP1304	1445
WEP1304/5178	1445
WEP1304R	1901
WEP1304R/5178K	1901
WEP1306	1458
WEP1306/5181	1458
WEP1306R	1601
WEP1306R/5181K	1601
WEP1307	0836
WEP1308	1490
WEP1308/5183	1490
WEP1308R	1622
WEP1308R/5183R	1622
WEP1310	0510
WEP1310/5186	0510
WEP1310R	0992
WEP1310R/5186K	0992
WEP1312	0870
WEP1312/5188	0870
WEP1312R	1263
WEP1312R/5188K	1263
WEP1314	1274
WEP1314/5191	1274
WEP1314R	1639
WEP1314R/5191K	1639
WEP1316	1078
WEP1316/5194	1078
WEP1316R	1716
WEP1316R/5194K	1716
WEP1318	1723
WEP1318/5197	1723
WEP1318R	0727
WEP1318R/5197K	0727
WEP1320	1761
WEP1320/5200	1761
WEP1320R	1782
WEP1320R/5200K	1782
WEP1322	1761
WEP1322/5203	1788
WEP1322R	1799
WEP1322R/5203K	1799
WEP1324	1804
WEP1324/5205	1804
WEP1324R	1817
WEP1324R/5205K	1817
WEP1326	0022
WEP1326/5208	0022
WEP1326R	1853
WEP1326R/5208K	1853
WEP1328	0207
WEP1328/5212	0207
WEP1328R	1884
WEP1328R/5212K	1884
WEP1330	0310
WEP1330/5215	0310
WEP1330R	1902
WEP1330R/5215K	1902
WEP1332	0356
WEP1332/5217	0356
WEP1332R	1909
WEP1332R/5217K	1909
WEP1334	0438
WEP1334/5220	0438
WEP1334R	1163
WEP1334R/5220K	1163
WEP1336	0509
WEP1336/5223	0509
WEP1336R	1941
WEP1336R/5223K	1941
WEP1338	0063
WEP1338/5226	0063
WEP1338R	1961
WEP1338R/5226K	1961
WEP1340	0616
WEP1340/5230	0616
WEP1340R	0776
WEP1340R/5230K	0776
WEP1400	1276
WEP1400/5000	1276
WEP1401	2874
WEP1401/5001	2874
WEP1402	0760
WEP1402/5002	0760
WEP1403	0677
WEP1403/5003	0677
WEP1404	0118
WEP1404/5004	0118
WEP1405	0299
WEP1405/5005	0299
WEP1406	0376
WEP1406/5006	0376
WEP1407	0036
WEP1407/5007	0036
WEP1408	0275
WEP1408/5008	0275
WEP1409	0140
WEP1409/5009	0140
WEP1410	2623
WEP1410/1410	2623
WEP1411	0041
WEP1411/5010	0041
WEP1412	0253
WEP1412/5011	0253
WEP1413	0091
WEP1413/5012	0091
WEP1414	0471
WEP1414/5013	0471
WEP1415	0062

DEVICE TYPE	REPL CODE
WEP1415/5014	0062
WEP1416	0077
WEP1416/5015	0077
WEP1416L/1416	0069
WEP1417	0165
WEP1417/5016	0165
WEP1417L/1417	2625
WEP1418	0322
WEP1418/5017	0322
WEP1419	0057
WEP1419/5018	0057
WEP1420	0064
WEP1420/5019	0064
WEP1421	0181
WEP1421/5020	0181
WEP1423	0052
WEP1423/5021	0052
WEP1424	0053
WEP1424/5022	0053
WEP1425	0879
WEP1425/5023	0879
WEP1426	0686
WEP1426/5024	0686
WEP1427	0445
WEP1427/5025	0445
WEP1428	6953
WEP1428/5026	0210
WEP1429	0375
WEP1429/5027	0375
WEP1430	0671
WEP1430/5028	0671
WEP1431	0700
WEP1431/5029	0700
WEP1432	0705
WEP1432/5030	0705
WEP1433	0494
WEP1433/5031	0494
WEP1434	0714
WEP1434/5032	0714
WEP1435	0455
WEP1435/5033	0455
WEP1436	0257
WEP1436/5034	0257
WEP1437	0195
WEP1437/5035	0195
WEP1438	0166
WEP1438/5036	0166
WEP1439	0010
WEP1439/5037	0010
WEP1440	0032
WEP1440/5038	0032
WEP1440L/1440	3738
WEP1441	0054
WEP1441/5039	0054
WEP1443	0068
WEP1443/5040	0068
WEP1445	0092
WEP1445/5041	0092
WEP1448	0125
WEP1448/5042	0125
WEP1449	2317
WEP1449/5043	2317
WEP1450	0152
WEP1450/5044	0152
WEP1451	0173
WEP1451/5045	0173
WEP1452	0094
WEP1452/5046	0094
WEP1453	0049
WEP1453/5047	0049
WEP1454	0104
WEP1454/5048	0104
WEP1455	0156
WEP1455/5049	0156
WEP1456	0189
WEP1456/5050	0189
WEP1458	0099
WEP1458/5051	0099
WEP1459	0089
WEP1459/5052	0089
WEP1460	0287
WEP1460/5053	0287
WEP1461	0252
WEP1461/5054	0252
WEP1462	0340
WEP1462/5055	0340
WEP1463	0370
WEP1463/5056	0370
WEP1464	0394
WEP1464/5057	0394
WEP1466	0425
WEP1466/5058	0425
WEP1467	0453
WEP1467/5059	0453
WEP1468	1473
WEP1468/50060	1473
WEP1504	0442
WEP1514	0225
WEP1518	0251
WEP1605	0782
WEP1605/5111	0782
WEP1606	0796
WEP1606/5112	0796
WEP1607	0806
WEP1607/5113	0806
WEP1607/5113A	0806
WEP1608	0821
WEP1608/5114	0821
WEP1609	0833

If replacement code is OEM, contact original manufacturer for replacement.

DEVICE TYPE	REPL CODE
WEP1609/5115	0833
WEP1610	0442
WEP1610/5116	0442
WEP1611	0876
WEP1611/5117	0876
WEP1612	3127
WEP1612/5118	3127
WEP1613	0185
WEP1613/5119	0185
WEP1614	0205
WEP1614/5120	0205
WEP1615	0480
WEP1615/5121	0480
WEP1616	0504
WEP1616/5122	0504
WEP1617	3317
WEP1617/5123	3317
WEP1618	0684
WEP1618/5124	0684
WEP1619	0225
WEP1619/5125	0225
WEP1620	0230
WEP1620/5126	0230
WEP1621	0234
WEP1621/5127	0234
WEP1622	0237
WEP1622/5128	0237
WEP1623	1397
WEP1623/5129	1397
WEP1624	0247
WEP1624/5130	0247
WEP1625	0251
WEP1625/5131	0251
WEP1626	1178
WEP1626/5132	1178
WEP1627	0256
WEP1627/5133	0256
WEP1628	2395
WEP1628/5134	2395
WEP1629	0262
WEP1629/5135	0262
WEP1630	0269
WEP1630/5136	0269
WEP1631	0274
WEP1631/5137	0274
WEP1632	2399
WEP1632/5138	2399
WEP1633	0294
WEP1633/5139	0294
WEP1634	1177
WEP1634/5140	1177
WEP1635	0309
WEP1635/5141	0309
WEP1636	0318
WEP1636/5142	0318
WEP1637	0320
WEP1637/5143	0320
WEP1638	0326
WEP1638/5144	0326
WEP1639	0337
WEP1639/5145	0337
WEP1641	0347
WEP1641/5146	0347
WEP1643	0027
WEP1643/5147	0027
WEP1646	0266
WEP1646/5148	0266
WEP1647	2856
WEP1647/5149	2856
WEP1648	0386
WEP1648/5150	0386
WEP1649	0405
WEP1649/5151	0405
WEP1650	0426
WEP1650/5152	0426
WEP1651	0444
WEP1651/5153	0444
WEP1652	3028
WEP1652/5154	3028
WEP1653	0238
WEP1653/5155	0238
WEP1654	1180
WEP1654/5156	1180
WEP1656	1190
WEP1656/5157	1190
WEP1657	1207
WEP1657/5158	1207
WEP1658	1218
WEP1658/5159	1218
WEP1658/5169	OEM
WEP1659	1881
WEP1659/5160	1881
WEP1660	0647
WEP1660/5161	0647
WEP1661	1255
WEP1661/5162	1255
WEP1661/6162	3309
WEP1662	2102
WEP1662/5163	2102
WEP1664	1279
WEP1664/5164	1279
WEP1665	2225
WEP1665/5165	2225
WEP1666	0605
WEP1666/5166	0605
WEP1717	0127
WEP1860	0004
WEP1945	0076
WEP2000	2571

DEVICE TYPE	REPL CODE
WEP2000/706	2571
WEP2001	2773
WEP2001/1165	2773
WEP2002	3722
WEP2002/716	3722
WEP2003	2781
WEP2003/717	6117
WEP2003/1166	2781
WEP2004	1940
WEP2004/719	1940
WEP2005	2279
WEP2005/721	2279
WEP2006	1345
WEP2006/723	1345
WEP2007	4224
WEP2007/1233	1714
WEP2008	2549
WEP2008/728	2549
WEP2009	0328
WEP2009/729	0328
WEP2010	2764
WEP2010/1244	2764
WEP2011	2457
WEP2011/731	2457
WEP2014	1759
WEP2014/736	1759
WEP2015	1443
WEP2015/737	1443
WEP2017	0751
WEP2017/739	0751
WEP2018	0379
WEP2018/740A	0379
WEP2020	3735
WEP2020/742	3735
WEP2021	1395
WEP2021/743	1395
WEP2022	1421
WEP2022/744	1421
WEP2024	0785
WEP2024/746	0785
WEP2026	0789
WEP2026/748	0789
WEP2027	0395
WEP2027/749	0395
WEP2028/370	3810
WEP2030	2715
WEP2030/711	2715
WEP2032	2529
WEP2032/727	2529
WEP2033	2743
WEP2033/730	2743
WEP2035	0856
WEP2035/738	0856
WEP2036	2930
WEP2036/745	2930
WEP2037	1808
WEP2037/747	1808
WEP2039	2707
WEP2039/785	2707
WEP2040	2755
WEP2040/788	2755
WEP2041	3195
WEP2041/795	3195
WEP2043	0521
WEP2043/797	0521
WEP2044	1753
WEP2044/798	1753
WEP2046	1753
WEP2046/804	2557
WEP2053	0360
WEP2053/778	0360
WEP2054	1301
WEP2054/909	1301
WEP2055	0364
WEP2055/780	0364
WEP2056	1705
WEP2056/909D	1705
WEP2057/782	0377
WEP2058	0802
WEP2058/783	0802
WEP2059	2699
WEP2059/784	2699
WEP2060	1797
WEP2060/910	1797
WEP2061	2714
WEP2061/786	2714
WEP2063	1982
WEP2063/941D	1982
WEP2064	1843
WEP2064/789	1843
WEP2065	0349
WEP2065/790	0349
WEP2066	1337
WEP2066/791	1337
WEP2067	0417
WEP2067/973	0417
WEP2072	3550
WEP2072/1024	3550
WEP2073	2876
WEP2073/1027	2876
WEP2074	4559
WEP2074/1028	4559
WEP2075	0520
WEP2075/801	0520
WEP2076	2262
WEP2076/1029	2262
WEP2077	4561
WEP2077/1030	4561
WEP2079	4093

DEVICE TYPE	REPL CODE
WEP2079/1037	4093
WEP2080	2393
WEP2080/1045	2393
WEP2081	2891
WEP2081/1050	2891
WEP2082	0433
WEP2082/1052	0433
WEP2086	0403
WEP2086/1062	0403
WEP2087	3352
WEP2087/1078	3352
WEP2089	0855
WEP2089/1080	0855
WEP2090	1160
WEP2090/1081A	1160
WEP2093	3263
WEP2093/1097	3263
WEP2094	2639
WEP2094/1109	2639
WEP2095	2305
WEP2095/1110	2305
WEP2097	4586
WEP2097/1114	4586
WEP2098	1248
WEP2098/1115	1248
WEP2099	3922
WEP2099/1115A	3922
WEP2100	4420
WEP2100/1127	4420
WEP2102	2231
WEP2102/1133	2231
WEP2103	3135
WEP2103/1136	3135
WEP2104	1696
WEP2104/912	1696
WEP2105	2872
WEP2105/1140	2872
WEP2106	4598
WEP2106/1150	4598
WEP2107	4600
WEP2107/1154	4600
WEP2108	3017
WEP2108/1156	3017
WEP2109	0881
WEP2109/1158	0881
WEP2110	1192
WEP2110/923	1192
WEP2111	2942
WEP2111/1161	2942
WEP2112	3269
WEP2112/1164	3269
WEP2113	1379
WEP2113/1172	1379
WEP2114	0410
WEP2114/941	0410
WEP2115	3273
WEP2115/1173	3273
WEP2117	2368
WEP2117/947	2368
WEP2118	2552
WEP2118/949	2552
WEP2119	0973
WEP2119/955M	0973
WEP2120	1540
WEP2120/1192	1540
WEP2121	3318
WEP2121/1193	3318
WEP2122	0579
WEP2122/1003	0579
WEP2123	0878
WEP2123/1004	0878
WEP2124	2568
WEP2124/1005	2568
WEP2125	1215
WEP2125/1006	1215
WEP2126	1816
WEP2126/1194	1816
WEP2127	2253
WEP2127/1198	2253
WEP2129	0864
WEP2129/1010	0864
WEP2130	0882
WEP2130/1011	0882
WEP2131	4622
WEP2131/1201	4622
WEP2132	2380
WEP2132/1202	2380
WEP2133	2404
WEP2133/1015	2404
WEP2134	2440
WEP2134/1017	2440
WEP2135	2306
WEP2135/1203	2306
WEP2136	2509
WEP2136/1019	2509
WEP2138	2155
WEP2138/1021	2155
WEP2139	4484
WEP2139/1204	4484
WEP2141	2193
WEP2141/1211	2193
WEP2142	4626
WEP2142/1212	4626
WEP2143	0416
WEP2143/1217	0416
WEP2144	1571
WEP2144/1218	1571
WEP2145	1612
WEP2145/1219	1612

DEVICE TYPE	REPL CODE
WEP2148	3190
WEP2148/1228	3190
WEP2150	1048
WEP2150/1232	1048
WEP2156	1621
WEP2156/1039	1621
WEP2160	4390
WEP2160/1043	4390
WEP2175	2786
WEP2175/1175	2786
WEP2176	2582
WEP2176/1061	2582
WEP2184	2599
WEP2184/1069	2599
WEP2185	4076
WEP2185/1070	4076
WEP2186	2603
WEP2186/1071	2603
WEP2195	OEM
WEP2207	0043
WEP2207/1094	0043
WEP2214	4582
WEP2214/1105	4582
WEP2215	2630
WEP2215/1106	2630
WEP2223	4034
WEP2223/1123	4034
WEP2228	0389
WEP2228/1121	0389
WEP2229	0607
WEP2229/1122	0607
WEP2242	2062
WEP2242/1135	2062
WEP2245	0904
WEP2245/1245	0904
WEP2248	1478
WEP2248/1142	1478
WEP2258	3695
WEP2258/1160	3695
WEP2259	0857
WEP2259/1187	0857
WEP2260	2581
WEP2260/1060	2581
WEP2261	1055
WEP2261/1196	1055
WEP2262	4069
WEP2262/1240	4069
WEP2265	1128
WEP2266	2960
WEP2266/1263	2960
WEP2268	4079
WEP2268/1269	4079
WEP2272	0478
WEP2272/4001B	0478
WEP2275	1143
WEP2275/4016	1143
WEP2279	0569
WEP2279/7493A	0569
WEP2281	2043
WEP2281/793	2043
WEP2285	2501
WEP2285/818	2501
WEP2286	0852
WEP2286/1186	0852
WEP2289	1196
WEP2289/964	1196
WEP2293	1018
WEP2293/1087	1018
WEP2294	4038
WEP2294/1162	4038
WEP2296	3705
WEP2296/1178	3705
WEP2297	4196
WEP2297/1183	4196
WEP2308	3948
WEP2308/1308	3948
WEP2319	OEM
WEP2319/309K	1922
WEP2331	0026
WEP2331/923D	0026
WEP3021	0338
WEP3023	0482
WEP3305	0926
WEP3305/5247	0926
WEP3305R	2740
WEP3305R/5247K	2740
WEP3307	0958
WEP3307/5249	0958
WEP3307R	1225
WEP3307R/5249K	1225
WEP3309	1009
WEP3309/5252	1009
WEP3309R	1252
WEP3309R/5252K	1252
WEP3311	0889
WEP3311/5254	0889
WEP3311R	1277
WEP3311R/5254K	1277
WEP3314	0932
WEP3314/5257	0932
WEP3314R	1302
WEP3314R/5257K	1302
WEP3317	1106
WEP3317/5260	1106
WEP3317R	1324
WEP3317R/5260K	1324
WEP3320	1152
WEP3320/5263	1152
WEP3320R	1344

DEVICE TYPE	REPL CODE
WEP3320R/5263K	1344
WEP3323	1184
WEP3323/5266	1184
WEP3323R	1371
WEP3323R/5266K	1371
WEP3325	1223
WEP3325/5269	1223
WEP3325R	1406
WEP3325R/5269K	1406
WEP3327	1246
WEP3327/5271	1246
WEP3327R	1428
WEP3327R/5271K	1428
WEP3334	1331
WEP3334/5278	1331
WEP3334R	1474
WEP3334R/5278K	1474
WEP3340	1436
WEP3340/5285	1436
WEP3340R	1575
WEP3340R/5285K	1575
WEP4000	0110
WEP4000/5800	0110
WEP4001	0953
WEP4001/5801	0953
WEP4002	0242
WEP4002/5802	0242
WEP4004	0540
WEP4004/5804	0540
WEP4006	0965
WEP4006/5806	0965
WEP4007	0817
WEP4007/5808	0817
WEP4008	0922
WEP4008/5809	0087
WEP4011B	0215
WEP4011B/4011B	0215
WEP4013B	0413
WEP4013B/4013B	0413
WEP4066B/4066B	0101
WEP4069	0119
WEP4069/4069	0119
WEP4081	0708
WEP4081/5912	0708
WEP4085	0580
WEP4085/5916	0580
WEP4089	1000
WEP4089/5920	1000
WEP5457	0095
WEP5457/5457	0095
WEP6097	1755
WEP6097/5633	1755
WEP6240	1396
WEP6240/5452	1396
WEP6242	1259
WEP6242/5454	1259
WEP6243	0940
WEP6243/5455	0940
WEP6244	1222
WEP6244/5456	1222
WEP6251	2095
WEP6251/5461	2095
WEP6252	2089
WEP6252/5462	2089
WEP6253	0505
WEP6253/5463	0505
WEP6254	0717
WEP6255	0710
WEP6255/5465	0710
WEP6271	1110
WEP6271/5481	1110
WEP6273	OEM
WEP6273/5483	2932
WEP6300	0061
WEP6300/231	0061
WEP6301	2327
WEP6301/230	0239
WEP6302	1712
WEP6302/308	1712
WEP6303	2329
WEP6303/310	2329
WEP6304	0920
WEP6304/507	0920
WEP6305	OEM
WEP6305/516	OEM
WEP6307	OEM
WEP6310	2336
WEP6310/239	2336
WEP6311	1652
WEP6311/276	1652
WEP6312	2337
WEP6312/279	2337
WEP6313	1825
WEP6313/314	1825
WEP6320	1137
WEP6320/5400	1137
WEP6321	0344
WEP6321/5401	0344
WEP6322	0901
WEP6322/5402	0901
WEP6323	2342
WEP6323/5403	2342
WEP6324	0058
WEP6324/5404	0058
WEP6900	OEM
WEP6901	OEM
WEP6902	OEM
WEP7400	0232
WEP7400/7400	0232

DEVICE TYPE	REPL CODE
WEP7401	0268
WEP7401/7401	0268
WEP7402	0314
WEP7402/7402	0314
WEP7408	0467
WEP7408/7408	0467
WEP7426	0803
WEP7426/7426	0803
WEP7473	1172
WEP7473/7473	1172
WEP7474	1313
WEP7474/7474	1313
WEP7476	1158
WEP7476/7476	1158
WEP7490	1208
WEP7490/7490	1208
WEP9158	0466
WEP9158/9158	0466
WEP9914	1069
WEP9917	OEM
WEP9917/9917	OEM
WEP9923	OEM
WEP9923/9923	OEM
WEP9924	2867
WEP9924/9924	2867
WEP9934	OEM
WEP9946	0141
WEP9946/9946	0141
WEP74145	0619
WEP74145/74145	0619
WEP74162	1013
WEP74162/74162	1013
WEPG6001	0604
WEPS13KV	0201
WEPS18KV	0378
WEPS20KV	0378
WEPS1000-1	0071
WEPS1200-1	1303
WEPS1400-1	0102
WEPS3000	0293
WEPS3001	0630
WEPS3002	0630
WEPS3003	0126
WEPS3020	0630
WEPS3021	0338
WEPS3023	0566
WEPS3027	1367
WEPS3031	0382
WEPS4000	0293
WEPS5000	0293
WEPS5003	0161
WEPS5004	0561
WEPS5005	1199
WEPS5007	0460
WEPS6000	0774
WEPS6500	0774
WEPS7000	0130
WEPS7001	1681
WEPS8000	0190
WEPS9100	0400
WEPSD4	0196
WEPSD5	0484
WEPSH3	2965
WEPSH4	0474
WEPSH5	2966
WF-050	0162
WF-050A	0162
WF-050B	0162
WF-050C	0162
WF-050D	0162
WF-052	0162
WF-052A	0162
WF-052B	0162
WF-052C	0162
WF-052D	0162
WF-054	0157
WF-054A	0157
WF-054B	0157
WF-054D	0157
WF-056	0157
WF-056A	0157
WF-056B	0157
WF-056C	0157
WF-056D	0157
WF-058	0157
WF-058A	0157
WF-058B	0157
WF-058C	0157
WF-058D	0157
WF-061	0002
WF-061A	0002
WF-061B	0002
WF-061C	0002
WF-061D	0002
WF-063	0002
WF-063A	0002
WF-063B	0002
WF-063C	0002
WF-063D	0002
WF-065	0002
WF-065A	0002
WF-065B	0002
WF-065C	0002
WF-065D	0002
WF-071	0649
WF-071A	0649
WF-071B	0649
WF-071C	0649
WF-071D	0649

DEVICE TYPE	REPL CODE
WF-073	0649
WF-073A	0649
WF-073B	0649
WF-073C	0649
WF-073D	0649
WF-075	0649
WF-075A	0649
WF-075B	0649
WF-075C	0649
WF-075D	0649
WF-077	0649
WF-077A	0649
WF-077B	0649
WF-077C	0649
WF-077D	0649
WF-079	0244
WF-079A	0244
WF-079B	0244
WF-079C	0244
WF-079D	0244
WF-081	0244
WF-081A	0244
WF-081B	0244
WF-081C	0244
WF-081D	0244
WF-083	0244
WF-083A	0244
WF-083B	0244
WF-083C	0244
WF-083D	0244
WF-085	0244
WF-085B	0244
WF-085C	0244
WF-085D	0244
WF-088	0012
WF-088A	0012
WF-088B	0012
WF-088C	0012
WF-088D	0012
WF-090	0012
WF-090A	0012
WF-090B	0012
WF-090C	0012
WF-090D	0012
WF-092	0012
WF-092A	0012
WF-092B	0012
WF-092C	0012
WF-092D	0012
WF-094	0012
WF-094A	0012
WF-094B	0012
WF-094C	0012
WF-094D	0012
WF-098	0170
WF-098A	0170
WF-098B	0170
WF-098C	0170
WF-098D	0170
WF1	0325
WF2	0325
WF5AM	OEM
WF5M	OEM
WF10AM	OEM
WF10M	OEM
WF15AM	OEM
WF15M	OEM
WF20	OEM
WF20A	OEM
WF20AM	OEM
WF20B	OEM
WF20M	OEM
WF21H	OEM
WF22A	OEM
WF22B	OEM
WF23A	OEM
WF23B	OEM
WF30	OEM
WF30A	OEM
WF30AM	OEM
WF30B	OEM
WF30M	OEM
WF34E	OEM
WF40	OEM
WF40A	OEM
WF40AM	OEM
WF40B	OEM
WF40M	OEM
WF42	OEM
WF43	OEM
WF49A	OEM
WF50	OEM
WF50A	OEM
WF50AM	OEM
WF50B	OEM
WF50M	OEM
WF60	OEM
WF63L	OEM
WF70	OEM
WF71	OEM
WF-100	0170
WF100	OEM
WF-100A	0170
WF100A	OEM
WF100AM	OEM
WF-100B	0137
WF100B	OEM
WF-100C	0137
WF-100D	0137

DEVICE TYPE	REPL CODE	DEVICE TYPE	REPL CODE	DEVICE TYPE	REPL CODE	DEVICE TYPE	REPL CODE	DEVICE TYPE	REPL CODE	DEVICE TYPE	REPL CODE	DEVICE TYPE	REPL CODE
WF100M	OEM	WF-340	0039	WG-092A	0012	WG-270B	0441	WH-079D	0244	WH-182C	0495	WI-079A	0244
WF-110	0244	WF-340A	0039	WG-092B	0012	WG-270C	0441	WH-081	0244	WH-182D	0495	WI-079B	0244
WF-110A	0317	WF-340B	0039	WG-092C	0012	WG-270D	0441	WH-081A	0244	WH-230	0402	WI-079C	0244
WF-110B	0317	WF-340C	0039	WG-092D	0012	WG-290	0726	WH-081B	0244	WH-230A	0402	WI-079D	0244
WF-110C	0317	WF-340D	0039	WG-094	0012	WG-290A	0726	WH-081C	0244	WH-230B	0402	WI-081	0244
WF-110D	0317	WF402	OEM	WG-094A	0012	WG-290B	0726	WH-081D	0244	WH-230C	0402	WI-081A	0244
WF-115	0794	WF402L	OEM	WG-094B	0012	WG-290C	0726	WH-083	0244	WH-230D	0402	WI-081B	0244
WF-115A	0794	WF403	OEM	WG-094C	0012	WG-290D	0726	WH-083A	0244	WH-240	0402	WI-081C	0244
WF-115B	0794	WF404L	OEM	WG-094D	0012	WG-300	0726	WH-083B	0244	WH-240A	0402	WI-081D	0244
WF-115C	0794	WF405L	OEM	WG-098	0170	WG-300A	0726	WH-083C	0244	WH-240B	0402	WI-083A	0244
WF-115D	0794	WF406L	OEM	WG-098A	0170	WG-300B	0726	WH-083D	0244	WH-240C	0402	WI-083B	0244
WF-120	0137	WF409LC	OEM	WG-098B	0170	WG-300C	0726	WH-085	0244	WH-240D	0402	WI-083C	0244
WF-120A	0137	WF412L	OEM	WG-098C	0170	WG-300D	0726	WH-085A	0244	WH-250	0402	WI-083D	0244
WF-120B	0137	WF413L	OEM	WG-098D	0170	WG-330	0039	WH-085B	0244	WH-250A	0402	WI-085	0244
WF-120C	0137	WF414L	OEM	WG0A-90	0133	WG-330A	0039	WH-085C	0244	WH-250B	0402	WI-085A	0244
WF-120D	0137	WF416	OEM	WG-10AS	0133	WG-330B	0039	WH-085D	0244	WH-250C	0402	WI-085B	0244
WF-125	0365	WF424	OEM	WG91	0012	WG-330C	0039	WH-088	0012	WH-250D	0402	WI-085C	0244
WF-125A	0365	WF461	OEM	WG-100	0170	WG-330D	0039	WH-088A	0012	WH-260	0441	WI-085D	0244
WF-125B	0365	WF471	OEM	WG-100A	0170	WG-340	0039	WH-088B	0012	WH-260A	0441	WI-088	0012
WF-125C	0365	WG-050A	0162	WG-100B	0170	WG-340A	0039	WH-088C	0012	WH-260B	0441	WI-088A	0012
WF-125D	0365	WG-050B	0162	WG-100C	0170	WG-340B	0039	WH-088D	0012	WH-260C	0441	WI-088B	0012
WF-130	0365	WG-050C	0162	WG-100D	0170	WG-340C	0039	WH-090	0012	WH-260D	0441	WI-088C	0012
WF-130A	0365	WG-050D	0162	WG-110	0317	WG-340D	0039	WH-090A	0012	WH-270	0441	WI-088D	0012
WF-130B	0365	WG-052	0162	WG-110A	0317	WG-599	0133	WH-090B	0012	WH-270A	0441	WI-090	0012
WF-130C	0365	WG-052A	0162	WG-110B	0317	WG-713	0133	WH-090C	0012	WH-270B	0441	WI-090A	0012
WF-130D	0365	WG-052B	0162	WG-110C	0317	WG713	0133	WH-090D	0012	WH-270C	0441	WI-090B	0012
WF-140	0100	WG-052C	0162	WG-110D	0317	WG713A	0124	WH-092	0012	WH-270D	0441	WI-090C	0012
WF-140A	0100	WG-052D	0162	WG-115	0794	WG-713C	0133	WH-092A	0012	WH-290	0726	WI-090D	0012
WF-140B	0100	WG-054	0157	WG-115A	0794	WG714	0133	WH-092B	0012	WH-290A	0726	WI-092	0012
WF-140C	0100	WG-054A	0157	WG-115B	0794	WG759-42	OEM	WH-092C	0012	WH-290B	0726	WI-092A	0012
WF-140D	0100	WG-054B	0157	WG-115C	0794	WG759-43	OEM	WH-092D	0012	WH-290C	0726	WI-092B	0012
WF-150	0002	WG-054C	0157	WG-115D	0794	WG759-44	OEM	WH-094	0012	WH-290D	0726	WI-092C	0012
WF-150A	0002	WG-054D	0157	WG-120	0137	WG759-51	OEM	WH-094A	0012	WH-300	0726	WI-092D	0012
WF-150B	0002	WG-056	0157	WG-120A	0137	WG1010	0124	WH-094B	0012	WH-300A	0726	WI-094	0012
WF-150C	0002	WG-056A	0157	WG-120B	0137	WG-1010A	0133	WH-094C	0012	WH-300B	0726	WI-094A	0012
WF-150D	0002	WG-056B	0157	WG-120C	0137	WG1010A	0133	WH-094D	0012	WH-300C	0726	WI-094B	0012
WF-157	0420	WG-056C	0157	WG-120D	0137	WG1010B	0133	WH-098	0170	WH-300D	0726	WI-094C	0012
WF-157A	0420	WG-056D	0157	WG-125	0365	WG-1012	0133	WH-098A	0170	WH-330	0039	WI-094D	0012
WF-157B	0420	WG-058	0157	WG-125A	0365	WG1012	0133	WH-098B	0170	WH-330A	0039	WI-098	0170
WF-157C	0420	WG-058A	0157	WG-125B	0365	WG1014A	0133	WH-098C	0170	WH-330B	0039	WI-098A	0170
WF-157D	0420	WG-058B	0157	WG-125C	0365	WG1021	0133	WH-098D	0170	WH-330C	0039	WI-098B	0170
WF-162	0420	WG-058C	0157	WG-125D	0365	WG8038BC	OEM	WH-100	0170	WH-330D	0039	WI-098C	0170
WF-162A	0420	WG-058D	0157	WG-130	0365	WG8038BM	OEM	WH-100A	0170	WH-340	0039	WI-098D	0170
WF-162B	0420	WG-061	0636	WG-130A	0365	WG8038CC	OEM	WH-100B	0170	WH-340A	0039	WI-100	0170
WF-162C	0420	WG-061A	0636	WG-130B	0365	WG0A-90	0124	WH-100C	0170	WH-340B	0039	WI-100A	0170
WF-162D	0420	WG-061B	0636	WG-130C	0365	WH-050	0402	WH-100D	0170	WH-340C	0039	WI-100B	0170
WF-167	0420	WG-061C	0636	WG-130D	0365	WH-050A	0402	WH-110	0317	WH-340D	0039	WI-100C	0170
WF-167A	0420	WG-061D	0636	WG-140	0100	WH-050B	0402	WH-110A	0317	WH1012	0133	WI-100D	0170
WF-167B	0420	WG-063	0636	WG-140A	0100	WH-050C	0402	WH-110B	0317	WI-050	0162	WI-110	0317
WF-167C	0420	WG-063A	0636	WG-140B	0100	WH-050D	0402	WH-110C	0317	WI-050A	0162	WI-110A	0317
WF-167D	0420	WG-063B	0636	WG-140C	0100	WH-052	0402	WH-110D	0317	WI-050B	0162	WI-110B	0317
WF-172	0495	WG-063C	0636	WG-140D	0100	WH-052A	0402	WH-115	0794	WI-050C	0162	WI-110C	0317
WF-172A	0495	WG-063D	0636	WG-150	0002	WH-052B	0402	WH-115A	0794	WI-050D	0162	WI-110D	0317
WF-172B	0402	WG-065	0636	WG-150A	0002	WH-052C	0402	WH-115B	0794	WI-052	0162	WI-115	0794
WF-172C	0402	WG-065A	0636	WG-150B	0002	WH-052D	0402	WH-115C	0794	WI-052A	0162	WI-115A	0794
WF-172D	0402	WG-065B	0636	WG-150C	0002	WH-054	0157	WH-115D	0794	WI-052B	0162	WI-115B	0794
WF-177	0402	WG-065C	0636	WG-150D	0002	WH-054A	0157	WH-120	0137	WI-052C	0162	WI-115C	0794
WF-177A	0402	WG-065D	0636	WG-157	0420	WH-054B	0157	WH-120A	0137	WI-052D	0162	WI-115D	0794
WF-177B	0402	WG-071	0649	WG-157A	0420	WH-054C	0157	WH-120B	0137	WI-054	0157	WI-120	0137
WF-177C	0402	WG-071A	0649	WG-157B	0420	WH-054D	0157	WH-120C	0137	WI-054A	0157	WI-120A	0137
WF-177D	0402	WG-071B	0649	WG-157C	0420	WH-056	0157	WH-120D	0137	WI-054B	0157	WI-120B	0137
WF-182	0402	WG-071C	0649	WG-157D	0420	WH-056A	0157	WH-125	0365	WI-054C	0157	WI-120C	0137
WF-182A	0402	WG-071D	0649	WG-162	0420	WH-056B	0157	WH-125A	0365	WI-054D	0157	WI-120D	0137
WF-182B	0402	WG-073	0649	WG-162A	0420	WH-056C	0157	WH-125B	0365	WI-056	0157	WI-125	0365
WF-182C	0402	WG-073A	0649	WG-162B	0420	WH-056D	0157	WH-125C	0365	WI-056A	0157	WI-125A	0365
WF-182D	0402	WG-073B	0649	WG-162C	0420	WH-058	0157	WH-125D	0365	WI-056B	0157	WI-125B	0365
WF200	OEM	WG-073C	0649	WG-162D	0420	WH-058A	0157	WH-130	0365	WI-056C	0157	WI-125C	0365
WF200A	OEM	WG-073D	0649	WG-167	0420	WH-058B	0157	WH-130A	0365	WI-056D	0157	WI-125D	0365
WF200AM	OEM	WG-075	0649	WG-167A	0420	WH-058C	0157	WH-130B	0365	WI-058	0157	WI-130	0365
WF200B	OEM	WG-075A	0649	WG-167B	0420	WH-058D	0157	WH-130C	0365	WI-058A	0157	WI-130A	0365
WF200M	OEM	WG-075B	0649	WG-167C	0420	WH-061	0636	WH-130D	0365	WI-058B	0157	WI-130B	0365
WF-230	0402	WG-075C	0649	WG-167D	0420	WH-061A	0636	WH-140	0100	WI-058C	0157	WI-130C	0365
WF-230A	0402	WG-075D	0649	WG-172	0495	WH-061B	0636	WH-140A	0100	WI-058D	0157	WI-130D	0365
WF-230B	0402	WG-077	0649	WG-172A	0495	WH-061C	0636	WH-140B	0100	WI-061	0636	WI-140	0100
WF-230C	0402	WG-077A	0649	WG-172B	0495	WH-061D	0636	WH-140C	0100	WI-061A	0636	WI-140A	0100
WF-230D	0402	WG-077B	0649	WG-172C	0495	WH-063	0636	WH-140D	0100	WI-061B	0636	WI-140B	0100
WF-240	0402	WG-077C	0649	WG-172D	0495	WH-063A	0636	WH-150	0002	WI-061C	0636	WI-140C	0100
WF-240A	0402	WG-077D	0649	WG-177	0495	WH-063B	0636	WH-150A	0002	WI-061D	0636	WI-140D	0100
WF-240B	0402	WG-079	0244	WG-177A	0495	WH-063C	0636	WH-150B	0002	WI-063	0636	WI-150	0002
WF-240C	0402	WG-079A	0244	WG-177B	0495	WH-063D	0636	WH-150C	0002	WI-063A	0636	WI-150A	0002
WF-240D	0402	WG-079B	0244	WG-177C	0495	WH-065	0636	WH-150D	0002	WI-063B	0636	WI-150B	0002
WF-250A	0402	WG-079C	0244	WG-177D	0495	WH-065A	0636	WH-157	0420	WI-063C	0636	WI-150C	0002
WF-250B	0402	WG-079D	0244	WG-182	0495	WH-065B	0636	WH-157A	0420	WI-063D	0636	WI-150D	0002
WF-250C	0402	WG-081	0244	WG-182A	0495	WH-065C	0636	WH-157B	0420	WI-065	0636	WI-157	0420
WF-250D	0402	WG-081A	0244	WG-182B	0495	WH-065D	0636	WH-157C	0420	WI-065A	0636	WI-157A	0420
WF-260	0012	WG-081B	0244	WG-182C	0495	WH-071	0649	WH-157D	0420	WI-065B	0636	WI-157B	0420
WF-260A	0441	WG-081C	0244	WG-182D	0495	WH-071A	0649	WH-162	0420	WI-065C	0636	WI-157C	0420
WF-260B	0441	WG-081D	0244	WG-230	0402	WH-071B	0649	WH-162A	0420	WI-065D	0636	WI-157D	0420
WF-260D	0441	WG-083	0244	WG-230A	0402	WH-071C	0649	WH-162B	0420	WI-071	0649	WI-162	0420
WF-270	0441	WG-083A	0244	WG-230B	0402	WH-071D	0649	WH-162C	0420	WI-071A	0649	WI-162A	0420
WF-270A	0441	WG-083B	0244	WG-230C	0402	WH-073	0649	WH-162D	0420	WI-071B	0649	WI-162B	0420
WF-270B	0441	WG-083C	0244	WG-230D	0402	WH-073A	0649	WH-167	0420	WI-071C	0649	WI-162C	0420
WF-270C	0441	WG-083D	0244	WG-240	0402	WH-073B	0649	WH-167A	0420	WI-071D	0649	WI-162D	0420
WF-270D	0441	WG-085	0244	WG-240A	0402	WH-073C	0649	WH-167B	0420	WI-073	0649	WI-167	0420
WF-290	0441	WG-085A	0244	WG-240B	0402	WH-073D	0649	WH-167C	0420	WI-073A	0649	WI-167A	0420
WF-290A	0726	WG-085B	0244	WG-240C	0402	WH-075	0649	WH-167D	0495	WI-073B	0649	WI-167B	0420
WF-290B	0726	WG-085C	0244	WG-240D	0402	WH-075A	0649	WH-172	0495	WI-073C	0649	WI-167C	0420
WF-290C	0726	WG-085D	0244	WG-250	0402	WH-075B	0649	WH-172A	0495	WI-073D	0649	WI-167D	0420
WF-290D	0726	WG-088	0012	WG-250A	0402	WH-075C	0649	WH-172B	0495	WI-075	0649	WI-172	0495
WF-300	0726	WG-088A	0012	WG-250B	0402	WH-075D	0649	WH-172C	0495	WI-075A	0649	WI-172A	0495
WF-300A	0726	WG-088B	0012	WG-250C	0402	WH-077	0649	WH-172D	0495	WI-075B	0649	WI-172B	0495
WF-300B	0726	WG-088C	0012	WG-250D	0402	WH-077A	0649	WH-177	0495	WI-075C	0649	WI-172C	0495
WF-300C	0726	WG-088D	0012	WG-260	0441	WH-077B	0649	WH-177A	0495	WI-075D	0649	WI-172D	0495
WF-300D	0726	WG-090	0012	WG-260A	0441	WH-077C	0649	WH-177B	0495	WI-077	0649	WI-177	0495
WF-330	0039	WG-090A	0012	WG-260B	0441	WH-077D	0649	WH-177C	0495	WI-077A	0649	WI-177A	0495
WF-330A	0039	WG-090B	0012	WG-260C	0441	WH-079	0244	WH-177D	0495	WI-077B	0649	WI-177B	0495
WF-330B	0039	WG-090C	0012	WG-260D	0441	WH-079A	0244	WH-182	0495	WI-077C	0649	WI-177C	0495
WF-330C	0039	WG-090D	0012	WG-270	0441	WH-079B	0244	WH-182A	0495	WI-077D	0649	WI-177D	0495
WF-330D	0039	WG-092	0012	WG-270A	0441	WH-079C	0244	WH-182B	0495	WI-079	0244	WI-182	0495

DEVICE TYPE	REPL CODE
WI-182A	0495
WI-182B	0495
WI-182C	0495
WI-182D	0495
WI-230	0402
WI-230A	0402
WI-230B	0402
WI-230C	0402
WI-230D	0402
WI-240	0402
WI-240A	0402
WI-240B	0402
WI-240C	0402
WI-240D	0402
WI-250A	0402
WI-250B	0402
WI-250C	0402
WI-250D	0402
WI-260	0441
WI-260A	0441
WI-260B	0441
WI-260C	0441
WI-260D	0441
WI-270	0441
WI-270A	0441
WI-270B	0441
WI-270C	0441
WI-270D	0441
WI-290	0726
WI-290A	0726
WI-290B	0726
WI-290C	0726
WI-290D	0726
WI-300	0726
WI-300B	0726
WI-300C	0726
WI-300D	0726
WI-330	0726
WI-330A	0039
WI-330B	0039
WI-330C	0039
WI-340	0039
WI-340A	0039
WI-340B	0039
WI-340C	0039
WI-340D	0039
WIE	0849
WJ-050	0162
WJ-050A	0162
WJ-050B	0162
WJ-050C	0162
WJ-050D	0162
WJ-052	0162
WJ-052A	0162
WJ-052B	0162
WJ-052C	0162
WJ-052D	0162
WJ-054	0157
WJ-054A	0157
WJ-054B	0157
WJ-054C	0157
WJ-054D	0157
WJ-056	0157
WJ-056A	0157
WJ-056B	0157
WJ-056C	0157
WJ-056D	0157
WJ-058	0157
WJ-058A	0157
WJ-058B	0157
WJ-058C	0157
WJ-058D	0157
WJ-061	0636
WJ-061A	0636
WJ-061B	0636
WJ-061C	0636
WJ-061D	0636
WJ-063	0636
WJ-063A	0636
WJ-063B	0636
WJ-063C	0636
WJ-063D	0636
WJ-065	0636
WJ-065A	0636
WJ-065B	0636
WJ-065C	0636
WJ-065D	0636
WJ-071	0649
WJ-071A	0649
WJ-071B	0649
WJ-071C	0649
WJ-071D	0649
WJ-073	0649
WJ-073A	0649
WJ-073B	0649
WJ-073C	0649
WJ-073D	0649
WJ-075	0649
WJ-075A	0649
WJ-075B	0649
WJ-075C	0649
WJ-075D	0649
WJ-077	0649
WJ-077A	0649
WJ-077B	0649
WJ-077C	0649
WJ-077D	0649
WJ-079	0244
WJ-079A	0244

DEVICE TYPE	REPL CODE
WJ-079B	0244
WJ-079C	0244
WJ-079D	0244
WJ-081	0244
WJ-081A	0244
WJ-081B	0244
WJ-081C	0244
WJ-081D	0244
WJ-083	0244
WJ-083A	0244
WJ-083B	0244
WJ-083C	0244
WJ-083D	0244
WJ-085	0244
WJ-085B	0244
WJ-085C	0244
WJ-085D	0244
WJ-088	0012
WJ-088A	0012
WJ-088B	0012
WJ-088C	0012
WJ-088D	0012
WJ-090	0012
WJ-090A	0012
WJ-090B	0012
WJ-090C	0012
WJ-090D	0012
WJ-092	0012
WJ-092A	0012
WJ-092B	0012
WJ-092C	0012
WJ-092D	0012
WJ-094	0012
WJ-094A	0012
WJ-094B	0012
WJ-094C	0012
WJ-094D	0012
WJ-098	0170
WJ-098A	0170
WJ-098B	0170
WJ-098C	0170
WJ-098D	0170
WJ-100	0170
WJ-100A	0170
WJ-100B	0170
WJ-100C	0170
WJ-100D	0170
WJ-110	0317
WJ-110A	0317
WJ-110B	0317
WJ-110C	0317
WJ-110D	0317
WJ-115	0794
WJ-115A	0794
WJ-115B	0794
WJ-115C	0794
WJ-115D	0794
WJ-120	0137
WJ-120A	0137
WJ-120B	0137
WJ-120C	0137
WJ-120D	0137
WJ-125	0365
WJ-125A	0365
WJ-125B	0365
WJ-125C	0365
WJ-125D	0365
WJ-130	0365
WJ-130A	0365
WJ-130B	0365
WJ-130C	0365
WJ-130D	0365
WJ-140	0100
WJ-140A	0100
WJ-140B	0100
WJ-140C	0100
WJ-140D	0100
WJ-150	0002
WJ-150A	0002
WJ-150B	0002
WJ-150C	0002
WJ-150D	0002
WJ-157	0420
WJ-157A	0420
WJ-157B	0420
WJ-157C	0420
WJ-157D	0420
WJ-162	0420
WJ-162A	0420
WJ-162B	0420
WJ-162C	0420
WJ-162D	0420
WJ-167	0420
WJ-167A	0420
WJ-167B	0420
WJ-167C	0420
WJ-167D	0420
WJ-172	0495
WJ-172A	0495
WJ-172B	0495
WJ-172C	0495
WJ-172D	0495
WJ-177	0495
WJ-177A	0495
WJ-177B	0495
WJ-177C	0495
WJ-177D	0495
WJ-182	0495
WJ-182A	0495

DEVICE TYPE	REPL CODE
WJ-182B	0495
WJ-182C	0495
WJ-182D	0495
WJ-230	0402
WJ-230A	0402
WJ-230B	0402
WJ-230C	0402
WJ-230D	0402
WJ-240	0402
WJ-240A	0402
WJ-240B	0402
WJ-240C	0402
WJ-240D	0402
WJ-250	0402
WJ-250A	0402
WJ-250B	0402
WJ-250C	0402
WJ-250D	0402
WJ-260	0441
WJ-260A	0441
WJ-260B	0441
WJ-260C	0441
WJ-260D	0441
WJ-270	0441
WJ-270A	0441
WJ-270B	0441
WJ-270C	0441
WJ-270D	0441
WJ-290	0726
WJ-290A	0726
WJ-290B	0726
WJ-290C	0726
WJ-290D	0726
WJ-300	0726
WJ-300A	0726
WJ-300B	0726
WJ-300C	0726
WJ-300D	0726
WJ-330	0039
WJ-330A	0039
WJ-330B	0039
WJ-330C	0039
WJ-330D	0039
WJ-340	0039
WJ-340A	0039
WJ-340B	0039
WJ-340C	0039
WJ-340D	0039
WJ759-33	OEM
WJ759-34	OEM
WJ759-40	OEM
WJ759-41	OEM
WK-050	0162
WK-050A	0162
WK-050B	0162
WK-050C	0162
WK-050D	0162
WK-052	0162
WK-052A	0162
WK-052B	0162
WK-052C	0162
WK-052D	0162
WK-054	0157
WK-054A	0157
WK-054B	0157
WK-054C	0157
WK-054D	0157
WK-056	0157
WK-056A	0157
WK-056B	0157
WK-056C	0157
WK-056D	0157
WK-058	0157
WK-058A	0157
WK-058B	0157
WK-058C	0157
WK-058D	0157
WK-061	0636
WK-061A	0636
WK-061B	0636
WK-061C	0636
WK-061D	0636
WK-063	0636
WK-063A	0636
WK-063B	0636
WK-063C	0636
WK-063D	0636
WK-065	0636
WK-065A	0636
WK-065B	0636
WK-065C	0636
WK-065D	0636
WK-071	0649
WK-071A	0649
WK-071B	0649
WK-071C	0649
WK-071D	0649
WK-073	0649
WK-073A	0649
WK-073B	0649
WK-073C	0649
WK-073D	0649
WK-075	0649
WK-075A	0649
WK-075B	0649
WK-075C	0649
WK-075D	0649
WK-077	0649
WK-077A	0649

DEVICE TYPE	REPL CODE
WK-077C	0649
WK-077D	0649
WK-079	0244
WK-079A	0244
WK-079B	0244
WK-079C	0244
WK-079D	0244
WK-081	0244
WK-081A	0244
WK-081B	0244
WK-081C	0244
WK-081D	0244
WK-083	0244
WK-083A	0244
WK-083B	0244
WK-083C	0244
WK-083D	0244
WK-085	0244
WK-085A	0244
WK-085B	0244
WK-085C	0244
WK-085D	0244
WK-088	0012
WK-088A	0012
WK-088B	0012
WK-088C	0012
WK-088D	0012
WK-090	0012
WK-090A	0012
WK-090B	0012
WK-090C	0012
WK-090D	0012
WK-092	0012
WK-092A	0012
WK-092B	0012
WK-092C	0012
WK-092D	0012
WK-094	0012
WK-094A	0012
WK-094B	0012
WK-094C	0012
WK-094D	0012
WK-098	0170
WK-098A	0170
WK-098B	0170
WK-098C	0170
WK-098D	0170
WK-100	0170
WK-100B	0170
WK-100C	0170
WK-100D	0170
WK-110	0317
WK-110A	0317
WK-110B	0317
WK-110C	0317
WK-110D	0317
WK-115	0794
WK-115A	0794
WK-115B	0794
WK-115C	0794
WK-115D	0794
WK-120	0137
WK-120A	0137
WK-120B	0137
WK-120C	0137
WK-120D	0137
WK-125	0365
WK-125A	0365
WK-125B	0365
WK-125C	0365
WK-125D	0365
WK-130	0365
WK-130A	0365
WK-130B	0365
WK-130C	0365
WK-130D	0365
WK-140	0100
WK-140A	0100
WK-140B	0100
WK-140C	0100
WK-140D	0100
WK-150	0002
WK-150A	0002
WK-150B	0002
WK-150C	0002
WK-150D	0002
WK-157	0420
WK-157A	0420
WK-157B	0420
WK-157C	0420
WK-157D	0420
WK-162	0420
WK-162A	0420
WK-162B	0420
WK-162C	0420
WK-162D	0420
WK-167	0420
WK-167A	0420
WK-167B	0420
WK-167C	0420
WK-167D	0420
WK-172	0495
WK-172A	0495
WK-172B	0495
WK-172C	0495
WK-172D	0495
WK-177	0495
WK-177A	0495
WK-177B	0495

DEVICE TYPE	REPL CODE
WK-177C	0495
WK-177D	0495
WK-182	0495
WK-182A	0495
WK-182B	0495
WK-182C	0495
WK-182D	0495
WK-230	0402
WK-230A	0402
WK-230B	0402
WK-230C	0402
WK-230D	0402
WK-240	0402
WK-240A	0402
WK-240B	0402
WK-240C	0402
WK-240D	0402
WK-250	0402
WK-250A	0402
WK-250B	0402
WK-250C	0402
WK-250D	0402
WK-260	0441
WK-260A	0441
WK-260B	0441
WK-260C	0441
WK-260D	0441
WK-270	0441
WK-270A	0441
WK-270B	0441
WK-270C	0441
WK-270D	0441
WK-290	0726
WK-290A	0726
WK-290B	0726
WK-290C	0726
WK-290D	0726
WK-300	0726
WK-300A	0726
WK-300B	0726
WK-300C	0726
WK-300D	0726
WK-330	0039
WK-330A	0039
WK-330B	0039
WK-330C	0039
WK-330D	0039
WK-340	0039
WK-340A	0039
WK-340B	0039
WK-340C	0039
WK-340D	0039
WK5457	OEM
WK5458	0440
WK5459	0440
WL005	OEM
WL005F	2785
WL005M	0106
WL-01	0106
WL01	0106
WL01F	0106
WL02-5004-L	OEM
WL02-5004L	OEM
WL02F	0106
WL02M	0106
WL04F	0106
WL04M	0106
WL06F	2010
WL06M	2010
WL08F	2400
WL08M	2400
WL-050	0162
WL-050A	0162
WL-050B	0162
WL-050C	0162
WL-050D	0162
WL-052	0162
WL-052A	0162
WL-052B	0162
WL-052C	0162
WL-052D	0162
WL-054	0157
WL-054A	0157
WL-054B	0157
WL-054C	0157
WL-054D	0157
WL-056	0157
WL-056A	0157
WL-056B	0157
WL-056C	0157
WL-056D	0157
WL-058	0157
WL-058A	0157
WL-058B	0157
WL-058C	0157
WL-058D	0157
WL-061	0636
WL-061A	0636
WL-061B	0636
WL-061C	0636
WL-061D	0636
WL-063	0636
WL-063A	0636
WL-063B	0636
WL-063C	0636
WL-063D	0636
WL-065	0636
WL-065A	0636
WL-065B	0636

DEVICE TYPE	REPL CODE
WL-065C	0636
WL-065D	0636
WL-071	0649
WL-071A	0649
WL-071B	0649
WL-071C	0649
WL-071D	0649
WL-073	0649
WL-073A	0649
WL-073B	0649
WL-073C	0649
WL-073D	0649
WL-075	0649
WL-075A	0649
WL-075B	0649
WL-075C	0649
WL-075D	0649
WL-077	0649
WL-077A	0649
WL-077B	0649
WL-077C	0649
WL-077D	0649
WL-079	0649
WL-079A	0244
WL-079B	0244
WL-079C	0244
WL-079D	0244
WL-081	0244
WL-081A	0244
WL-081B	0244
WL-081C	0244
WL-081D	0244
WL-083	0244
WL-083A	0244
WL-083B	0244
WL-083C	0244
WL-083D	0244
WL-085	0244
WL-085A	0244
WL-085B	0244
WL-085C	0244
WL-085D	0244
WL-088	0012
WL-088A	0012
WL-088B	0012
WL-088C	0012
WL-088D	0012
WL-090	0012
WL-090A	0012
WL-090B	0012
WL-090C	0012
WL-090D	0012
WL-092	0012
WL-092A	0012
WL-092B	0012
WL-092C	0012
WL-092D	0012
WL-094	0012
WL-094A	0012
WL-094B	0012
WL-094C	0012
WL-094D	0012
WL-098	0170
WL-098A	0170
WL-098B	0170
WL-098C	0170
WL-098D	0170
WL10F	0787
WL10M	0787
WL-100	0170
WL100	OEM
WL-100A	0170
WL-100B	0170
WL-100C	0170
WL-100D	0170
WL-110	0317
WL-110A	0317
WL-110B	0317
WL-110C	0317
WL-110D	0317
WL-115	0794
WL-115A	0794
WL-115B	0794
WL-115C	0794
WL-115D	0794
WL-120	0137
WL-120A	0137
WL-120B	0137
WL-120C	0137
WL-120D	0137
WL500	OEM
WL530P	OEM
WL900	OEM
WN-170D	0420
WO-005	0106
WO-01	2785
WO-02	2785
WO-04	2785
WO-06	2189
WO-08	3736
WO-6A	0015
WO-6B	0023
WO6C	0015
WO68	OEM
WPC23C	0528
WPD858	2253
WR-006	0110
WR006	0015
WR011	0277

DEVICE TYPE	REPL CODE
WR-013	0015
WR013	0015
WR-030	0110
WR030	0277
WR-040	0110
WR040	0277
WR1	0087
WR2	0087
WR3	0922
WR10	0435
WR30	0970
WR31	3733
WR32	0989
WR33	0197
WR34	0204
WR40	0084
WR41	0090
WR42	0097
WR43	0109
WR44	0122
WR50	3189
WR51	2901
WR52	1124
WR53	0805
WR54	0319
WR60	3189
WR61	2901
WR62	1124
WR63	0805
WR64	0319
WR100	0015
WR200	0015
WR300	0015
WR400	1355
WRE981	0015
WRR1952	0016
WRR1953	0016
WRR1954	0016
WRR1955	0540
WRR1956	0540
WRT1114	0211
WS20-1	0288
WS20-1B	0288
WS20-1S	0288
WS100	0124
WS200	0124
WS300	0124
WS531P	OEM
WS6945-2	0081
WSC2260	OEM
WSD002C	0133
WT-10C	0365
WT16X7	0196
WT16X8	0484
WT16X9	0484
WT635-01D	OEM
WT635-01M	OEM
WT635-02D	OEM
WT635-02M	OEM
WT4311	OEM
WT4312	OEM
WT4313	OEM
WT4314	OEM
WT4315	OEM
WT4316	OEM
WT4321	OEM
WT4322	OEM
WT4323	OEM
WT4324	OEM
WT4325	OEM
WT4331	OEM
WT4332	OEM
WT4333	OEM
WT4334	OEM
WT4411	OEM
WT4412	OEM
WT4413	OEM
WT4414	OEM
WT4415	OEM
WT4416	OEM
WT4421	OEM
WT4422	OEM
WT4423	OEM
WT4424	OEM
WT4425	OEM
WT4431	OEM
WT4432	OEM
WT4433	OEM
WT4434	OEM
WT5301	OEM
WT5303	OEM
WT5304	OEM
WT5305	OEM
WT5306	OEM
WT5308	OEM
WT5310	OEM
WT5503	OEM
WT5504	OEM
WT5505	OEM
WT5506	OEM
WT5510	OEM
WT5511	OEM
WT5603	OEM
WT5604	OEM
WT5605	OEM
WT5606	OEM
WT5611	OEM
WT5650	OEM
WT5651	OEM

If replacement code is OEM, contact original manufacturer for replacement.

DEVICE TYPE	REPL CODE
WT5652	OEM
WT5703	OEM
WT5704	OEM
WT5705	OEM
WT5706	OEM
WT5750(A)	OEM
WT5751(A)	OEM
WT5752(A)	OEM
WTL1010	OEM
WTL1016	OEM
WTV008CL	OEM
WTV008CP	OEM
WTV008L	OEM
WTV008P	OEM
WTV3MC	0050
WTV6MC	0050
WTV6PWR	0160
WTV12MC	0050
WTV12PWR	0160
WTV15MG	0004
WTV15VMG	0004
WTV20MC	0050
WTV20MG	0004
WTV20VH6	0004
WTV20VHG	0004
WTV20VMG	0004
WTV25PWR	0160
WTV30VHG	0004
WTV30VLG	0004
WTV30VMG	0004
WTV40PWR	0160
WTV99PWR	0160
WTV129PWR	0440
WTV199PWR	0160
WTV299PWR	0440
WTVAT6	0004
WTVAT6A	0050
WTVB5	0050
WTVB5A	0004
WTVB6	0280
WTVB6A	0211
WTVBA6	0050
WTVBA6A	0004
WTVBE6	0050
WTVBE6A	0004
WTVBMC	0164
WTVL6	0038
WTVSA7	0600
WTVSK7	0600
WTVSQ7	0600
WU-054	0157
WU-056D	0636
WU-065D	0636
WV-1	0139
WV2AA50C	2298
WV2AA50E	2298
WV2B325E	OEM
WV2BA25C	OEM
WV2BA25E	OEM
WV2BC25C	1434
WV2BC25E	1434
WV2BE25C	1434
WV2BE25E	1434
WV2BH25C	2294
WV2BH25E	2294
WV2BJ25C	1680
WV2BJ25E	1680
WV2BK25C	1680
WV2BK25E	1680
WV2CA25C	OEM
WV2CA25E	OEM
WV2DA25C	OEM
WV2DA25E	OEM
WV2DC25C	OEM
WV2DC25E	OEM
WV530P	OEM
WV535P	OEM
WX-090	0012
WX090	0012
WX6	0143
WX115UA	OEM
WX115UB	OEM
WX115UC	OEM
WX115UD	OEM
WX115VA	OEM
WX115VB	OEM
WX115VC	OEM
WX115VD	OEM
WX115WA	OEM
WX115WB	OEM
WX115WC	OEM
WX115WD	OEM
WX115XA	OEM
WX115XB	OEM
WX115XC	OEM
WX115XD	OEM
WX1015	OEM
WX1015A	OEM
WX1015B	OEM
WX1015C	OEM
WX1015D	OEM
WX1015E	OEM
WX1015F	OEM
WX1016	OEM
WX1016A	OEM
WX1016B	OEM
WX1016C	OEM
WX1016D	OEM
WX1016E	OEM
WX1016F	OEM
WZ0-90	OEM
WZ09	0057
WZ030	0299
WZ-032	0299
WZ032	0299
WZ-034	0299
WZ034	0299
WZ-036	0376
WZ036	0376
WZ-038	0036
WZ038	0036
WZ-040	0036
WZ040	0036
WZ-042	0275
WZ042	0275
WZ-044	0275
WZ044	0275
WZ-046	0140
WZ046	0140
WZ-048	0140
WZ048	0140
WZ-049	0451
WZ-050	0041
WZ050	0041
WZ-052	0041
WZ052	0041
WZ-054	0253
WZ054	0253
WZ-056	0253
WZ056	0253
WZ-058	0157
WZ058	0157
WZ060	0157
WZ-061	0471
WZ061	0091
WZ-063	0471
WZ063	0471
WZ-065	0062
WZ065	0636
WZ-067	0062
WZ067	0062
WZ-069	0062
WZ069	0025
WZ-071	0062
WZ071	0062
WZ-073	0077
WZ073	0077
WZ-075	0077
WZ075	0077
WZ-077	0077
WZ077	0077
WZ-079	0165
WZ079	0165
WZ-081	0244
WZ081	0244
WZ-083	0165
WZ083	0165
WZ-085	0012
WZ085	1083
WZ-088	0322
WZ088	0322
WZ-090	0057
WZ090	0057
WZ-092	0012
WZ092	0012
WZ-094	0012
WZ094	0057
WZ-096	0064
WZ096	0170
WZ-098	0064
WZ098	0064
WZ5.6	0157
WZ7.5	0649
WZ9.1	0012
WZ10	0170
WZ12	0137
WZ13B	0365
WZ15	0002
WZ27	0441
WZ33	0039
WZ82	0331
WZ-90	0012
WZ-100	0170
WZ100	0064
WZ-105	0181
WZ105	0181
WZ-110	0181
WZ110	0181
WZ-115	0794
WZ115	0052
WZ-120	0052
WZ120	0052
WZ-120(AU)	0052
WZ-125	0053
WZ125	0365
WZ-130	0053
WZ130	0053
WZ-135	0879
WZ135	0879
WZ-140	0879
WZ140	0100
WZ-145	0686
WZ145	0686
WZ-150	0002
WZ150	0686
WZ-157	0445
WZ157	0445
WZ-162	0445
WZ162	0445
WZ-167	0210
WZ167	0210
WZ-172	0210
WZ172	0210
WZ-177	0495
WZ177	1649
WZ-182	0375
WZ182	0375
WZ-187	0671
WZ187	0671
WZ-192	0949
WZ192	0949
WZ-197	0700
WZ197	0700
WZ-210	0565
WZ210	0705
WZ-220	0705
WZ220	0705
WZ-230	0494
WZ230	0494
WZ-240	0402
WZ240	0402
WZ-250	0714
WZ250	0714
WZ-260	0455
WZ260	0455
WZ-270	0455
WZ270	0455
WZ-280	0257
WZ280	0257
WZ-290	0195
WZ290	0195
WZ-300	0195
WZ300	0195
WZ-310	0195
WZ310	0195
WZ-320	0166
WZ320	0166
WZ-330	0166
WZ330	0166
WZ-340	0166
WZ340	0166
WZ-350	0010
WZ350	0010
WZ360	0010
WZ522	0162
WZ523	0157
WZ524	0091
WZ525	0062
WZ526	0062
WZ527	0077
WZ528	0165
WZ529	0322
WZ531	0057
WZ532	0057
WZ533	0170
WZ534	0181
WZ535	0052
WZ536	0053
WZ537	0100
WZ538	0686
WZ539	0445
WZ541	0210
WZ542	0375
WZ543	0671
WZ544	0700
WZ904	0451
WZ905	0162
WZ906	0157
WZ907	0301
WZ908	0025
WZ909	0025
WZ910	0649
WZ911	0244
WZ912	1083
WZ913	0012
WZ914	0012
WZ915	0170
WZ916	0794
WZ917	0137
WZ918	0365
WZ919	0100
WZ920	0002
WZ921	0420
WZ922	1649
WZ923	0495
WZ924	0949
WZ925	0531
WZ926	0565
WZ927	1674
WZ928	1606
WZ3000	OEM
WZ3000A	OEM
WZ3000B	OEM
WZ3002	OEM
WZ3002A	OEM
WZ3002B	OEM
WZ3003	OEM
WZ3003A	OEM
WZ3003B	OEM
WZ3004	OEM
WZ3004A	OEM
WZ3004B	OEM
WZ3005	OEM
WZ3005A	OEM
WZ3005B	OEM
X0027CE	OEM
X0043CE	0167
X0054CE	2872
X0058GE	OEM
X0065CE	1429
X0069CE	0478
X0085TA	0603
X0087TA	4830
X0088TA	OEM
X0092CE	0069
X0093CE	2054
X0094CE	2054
X0094PA	OEM
X026BK	OEM
X030AK	OEM
X0100AA	0901
X0100AB	0901
X0100BA	0058
X0100BB	0058
X0100CA	0407
X0100CB	0407
X0100DA	0407
X0100DB	0407
X0101BA	0058
X0101BB	0058
X0101BD	0179
X0101DA	0407
X0101DB	0407
X0101DD	0346
X0101MA	1683
X0101MB	1683
X0101MD	3349
X0101NA	OEM
X0101NB	OEM
X0101ND	OEM
X0102AA	0901
X0102AB	0901
X0102BA	0058
X0102BB	0058
X0102CA	0407
X0102CB	0407
X0102DA	0407
X0102DB	0407
X0103BA	0058
X0103BB	0058
X0103BD	0179
X0103DA	0407
X0103DB	0407
X0103DD	0346
X0103MA	1683
X0103MB	1683
X0103MD	3349
X0103NA	OEM
X0103NB	OEM
X0103ND	OEM
X0137CE	2194
X0137E	OEM
X0178CE	1590
X0186GE	OEM
X0212CE	3685
X0213CE	2896
X0225C	3333
X0232CE	1846
X0233CE	OEM
X0238CE	2042
X0243CE	0167
X0249CE	1329
X0252CE	0069
X0260C	2666
X0260CE	OEM
X0268GE	OEM
X0275CE	4109
X0280C	OEM
X0301BG	0179
X0301DG	0346
X0301MG	3349
X0303BG	0179
X0303DG	0346
X0303MG	3349
X0304BG	0179
X0304DG	0346
X0304MG	3349
X0308CE	OEM
X0314CE	OEM
X0359CE	OEM
X0403BE	0940
X0403BF	0940
X0403DE	0095
X0403DF	0095
X0403ME	1243
X0403MF	1243
X0403NE	OEM
X0403NF	OEM
X0404BE	0940
X0404BF	0940
X0404DE	0095
X0404DF	0940
X0404ME	OEM
X0404MF	OEM
X0404NE	OEM
X0404NF	OEM
X0424CE	OEM
X0424CE-1	OEM
X0539CE	OEM
X0564CE	OEM
X0579CE	OEM
X0580CE	OEM
X0581CE	OEM
X0600CE	3236
X0609BE	0940
X0609BF	0940
X0609DE	0095
X0609DF	0095
X0609ME	OEM
X0609MF	OEM
X0609NE	OEM
X0609NF	OEM
X0669CE	OEM
X0692CE	OEM
X0709BG	0179
X0709DG	0346
X0709MG	3349
X0735CE	OEM
X0735CE-542	OEM
X0800CE	OEM
X0820CE	OEM
X0943CE	OEM
X0950CE	OEM
X0D971	OEM
X0E861	OEM
X0E961	OEM
X0F751	OEM
X0F851	OEM
X0F951	OEM
X0G751	OEM
X0G851	OEM
X0H121	OEM
X0H131	OEM
X0H741	OEM
X1	OEM
X1C1644	0004
X1D971	OEM
X1E861	OEM
X1E961	OEM
X1F751	OEM
X1F851	OEM
X1F951	OEM
X1G751	OEM
X1G851	OEM
X1H121	OEM
X1H131	OEM
X1H741	OEM
X1RC2	OEM
X1RC3	OEM
X1RC5	OEM
X1RC7	OEM
X1RC10	OEM
X1RC20	OEM
X2	OEM
X2D971	OEM
X2E861	OEM
X2E961	OEM
X2F751	OEM
X2F851	OEM
X2F951	OEM
X2G751	OEM
X2G851	OEM
X2H121	OEM
X2H131	OEM
X2H741	OEM
X2I111	OEM
X3D971	OEM
X3E861	OEM
X3E961	OEM
X3F751	OEM
X3F851	OEM
X3F951	OEM
X3G751	OEM
X3G851	OEM
X3H741	OEM
X3I111	OEM
X4	OEM
X4B7.5	OEM
X4D971	OEM
X4E861	OEM
X4E961	OEM
X4F751	OEM
X4F851	OEM
X4F951	OEM
X4G751	OEM
X4G851	OEM
X4H741	OEM
X5	OEM
X5D971	OEM
X5E861	OEM
X5E961	OEM
X5F751	OEM
X5F851	OEM
X5F951	OEM
X5G221	OEM
X5G551	OEM
X5G751	OEM
X5G851	OEM
X5H341	OEM
X5H541	OEM
X5H741	OEM
X5I111	OEM
X5I211	OEM
X5I331	OEM
X5I531	OEM
X5J201	OEM
X5J321	OEM
X5M6	0015
X6	OEM
X6A71	OEM
X6A72	OEM
X6A73	OEM
X6A74	OEM
X6A100	OEM
X6A101	OEM
X6A102	OEM
X6A103	OEM
X6A104	OEM
X6A105	OEM
X6A106	OEM
X6A107	OEM
X6D971	OEM
X6E861	OEM
X6E961	OEM
X6F751	OEM
X6F851	OEM
X6F951	OEM
X6G551	OEM
X6G751	OEM
X6G851	OEM
X6H341	OEM
X6H541	OEM
X6H741	OEM
X6I331	OEM
X6I531	OEM
X6J321	OEM
X6P1	OEM
X6P3	OEM
X6P10	OEM
X6P11	OEM
X8	OEM
X10-3	OEM
X10-4	OEM
X10-5	OEM
X10-6	OEM
X10-7	OEM
X10-8	OEM
X10-9	OEM
X10G1829	0016
X12	OEM
X13	OEM
X13B	OEM
X13C	OEM
X16	0143
X16A545-7	0016
X16A1938	0016
X16E3860	0016
X16E3890	0595
X16E3960	0016
X16N1485	0016
X18	0143
X-24C00P	OEM
X-24C01P	OEM
X24C01P	OEM
X24C02P	OEM
X24C04	OEM
X24C04P	OEM
X24C08P	OEM
X25-3	OEM
X25-4	OEM
X25-5	OEM
X25-6	OEM
X25-7	OEM
X25-8	OEM
X25-9	OEM
X26	OEM
X30	OEM
X30A	OEM
X31	OEM
X31A	OEM
X32	OEM
X32A	OEM
X32A1389	0037
X32C4211	0079
X32C4293	0050
X32C4296	0079
X32C5099	0086
X32C5111	0086
X32C5198	0079
X32C6105	0079
X32D5422	0079
X32L4296	0079
X32M5026	0079
X32P5660	0144
X34E1226	0037
X34E2111	0079
X37C	1329
X40N121	0285
X40Z113	0285
X42	0050
X43C248	0382
X44C358	0042
X44C758	0060
X45C359	0925
X50-3	OEM
X50-4	OEM
X50-5	OEM
X50-6	OEM
X50-7	OEM
X50-8	OEM
X50-9	OEM
X61P2	OEM
X-78	0004
X78	0280
X100KS	OEM
X110	0160
X113	OEM
X125KS	OEM
X133	OEM
X134	OEM
X137	0440
X150KS	OEM
X240P	OEM
X244P	OEM
X351A	OEM
X351B	OEM
X602K	OEM
X780	OEM
X841D	OEM
X881D	OEM
X1004	OEM
X1005	0160
X-1018	1030
X1018	1030
X1030CE	OEM
X1030CE-752	OEM
X1077CE	OEM
X1132CE	OEM
X1190CE	3515
X1231CE	OEM
X1247CE	OEM
X1535CE	OEM
X1712CE	OEM
X2402P	OEM
X2404P	OEM
X2444	OEM
X2444P	OEM
X2816BP	OEM
X3002	OEM
X3002A	OEM
X3030	OEM
X3033	OEM
X3060	OEM
X3066	OEM
X3068	OEM
X3070	OEM
X3070A	OEM
X3075	OEM
X3076	OEM
X4604	0015
X5004CE	OEM
X7070	0025
X-23305-3	0540
X-23305-4	0540
X24441	OEM
X53527AC(A)	OEM
X330302	0137
X-2408473	1069
X-2408784	2078
X300071592	0525
X300073360	0006
X300101502	0148
X300101509	0148
X300102009	0434
X300106930	3577
X300115080	0006
X300117500	3625
X300117593	3625
X300145009	0477
X301076530	1968
X301077251	0525
X301079400	2897
X301096530	1968
X301131809	OEM
X302056430	0532
X302094530	0076
X302116292	0563
X302138400	0018
X302164709	0278
X302181502	0076
X302181509	0076
X302181589	0076
X302183209	0042
X302183309	0042
X302251600	1165
X302278500	0249
X302278530	0249
X302278593	0249
X302278594	0249
X302329300	0261
X302369100	0060
X302374609	0060
X302374809	0060
X302398709	3723
X303056009	1212
X303088000	0461
X303098600	2970
X303121800	2366
X303139200	2366
X303139500	2366
X303163000	1061
X303184309	7129
X310090100	OEM
X320010070	0133
X320010079	0133
X320010080	0133
X320010190	2527
X320010240	0023
X320010380	0023
X320010390	0133
X320010430	0102
X320010450	0023
X320010452	0124
X320010460	0124
X320010470	0124
X320010480	0133
X320010512	0015
X320010939	0124
X320011201	0124
X330000059	0140
X330000068	0140
X330000230	0714
X330000320	1276

DEVICE TYPE	REPL CODE
X330000342	0248
X330000422	0053
X330000442	0705
X330000492	0700
X330000522	0299
X330000542	0376
X330000550	0248
X330000560	0012
X330000570	0010
X330000582	0041
X330013009	0376
X330020020	0274
X330030288	0068
X330160109	0140
X330160209	0686
X330160309	0700
X330180069	0376
X340040030	0106
X340200010	0909
X340200040	0584
X340300010	0473
X340330017	0729
X340400030	0106
X340400040	0277
X340400060	0323
X350030030	0744
X350040090	OEM
X400004491	OEM
X400007650	OEM
X400007801	3481
X400014680	OEM
X400040161	1898
X400072010	OEM
X400072200	OEM
X400078010	OEM
X400078100	OEM
X400078101	OEM
X400081550	OEM
X400082371	OEM
X400082530	OEM
X400082550	0051
X400082591	OEM
X400104162	0522
X400120640	OEM
X400141640	2357
X400143644	OEM
X400244010	OEM
X400250300	OEM
X400662647	OEM
X420100030	0335
X420100060	1206
X420100070	1339
X420100160	1349
X420100170	1352
X420100380	0996
X420101230	1157
X420200020	2238
X420200040	2263
X420201610	0858
X420202410	OEM
X420202570	2204
X420300000	1527
X420300020	1559
X420300040	1595
X420300050	1608
X420300080	1633
X420300100	1662
X420300140	1698
X420300200	0034
X420300270	0183
X420300300	0828
X420300320	0088
X420300510	1033
X420300540	1857
X420300730	1867
X420300740	0243
X420300750	1870
X420300860	0291
X420300930	1888
X420301230	0979
X420301390	0153
X420301450	1564
X420301540	5045
X420301560	1654
X420301610	0858
X420301630	0893
X420301640	4339
X420301750	1672
X420301950	1315
X420302210	1239
X420302440	0458
X420302450	0463
X420302730	0894
X420303730	0709
X420303740	0713
X420303930	0819
X420305410	2547
X420500000	2409
X420500020	2430
X420500040	2442
X420500080	2477
X420500100	2498
X420500110	2508
X420500320	2672
X420500370	2706
X420500400	2733
X420500740	0243
X420501120	3197
X420505730	OEM
X420505740	OEM
X440029020	0625
X440029040	0770
X440032060	1206
X440040030	OEM
X440041040	0969
X440045040	OEM
X440055560	3285
X440057330	1118
X440058120	0334
X440063050	0624
X440064940	1824
X440069050	1285
X440070490	0001
X440072350	OEM
X440072920	0631
X440073860	3063
X440075631	OEM
X440078051	1292
X440078054	1292
X440079120	1838
X440137140	OEM
X440140050	0624
X440150640	3126
X440150790	5870
X440160660	0101
X440167610	OEM
X440170070	OEM
X440193050	2169
X440200650	3933
X440212000	OEM
X440223110	2104
X440672200	2765
X440672210	2765
X440751880	0508
X440751890	0511
X440755400	1596
X440756330	1596
X440757110	OEM
X440757520	OEM
X440757600	OEM
X440759820	3361
X440762120	4329
X460405000	0398
X460406901	0119
X460409300	2384
X460458400	3626
X4700078101	OEM
XA/RA-1000	OEM
XA101	0050
XA102	0050
XA103	0050
XA104	0050
XA111	0050
XA112	0050
XA121	0015
XA122	0004
XA123	0050
XA124	0050
XA126	0050
XA131	0050
XA141	0050
XA142	0050
XA143	0050
XA151	0164
XA152	0164
XA161	0050
XA162	OEM
XA494	0037
XA-495	0037
XA495	0037
XA495(C)	0037
XA495AC	0037
XA495C	0037
XA495D	0006
XA701	0600
XA702	0600
XA703	0600
XAA104	0232
XAA105	0873
XAA106	1024
XAA107	1172
XAA108	1158
XAA109	1208
XAN3061	4870
XAN3062	4871
XAN3063	OEM
XAN3064	4873
XB087AO	OEM
XB1	0211
XB2	0211
XB2/XB4	2866
XB2.XB4	2866
XB3	0211
XB3B	0211
XB3BN	0004
XB3C	0211
XB4	0038
XB-5	0160
XB5	0085
XB-7	0160
XB7	0085
XB8	0050
XB8C	OEM
XB8E	OEM
XB9	0280
XB10	0050
XB12	0086
XB13	0004
XB14	0160
XB102	0004
XB103	0004
XB104	0004
XB112	0004
XB113	0004
XB114	0004
XB121	OEM
XB152	0002
XB401	0560
XB404	0042
XB408	0042
XB413	1974
XB476	0042
XC55FA	OEM
XC55FB	OEM
XC55FC	OEM
XC55FD	OEM
XC55FDQ	OEM
XC55PA	OEM
XC55PC	OEM
XC55PD	OEM
XC55PDK	OEM
XC66-10	OEM
XC66-10K	OEM
XC66-25	OEM
XC66-25K	OEM
XC70	0025
XC88FA	OEM
XC88FAQ	OEM
XC88FB	OEM
XC88FBQ	OEM
XC88FC	OEM
XC88FCQ	OEM
XC88FD	OEM
XC88PA	OEM
XC88PAK	OEM
XC88PB	OEM
XC88PBK	OEM
XC88PC	OEM
XC88PCK	OEM
XC88PD	OEM
XC99-30	OEM
XC99-50K	OEM
XC101	0004
XC121	0004
XC131	0004
XC141	0160
XC142	0160
XC155	0085
XC156	0085
XC161	0280
XC163	OEM
XC171	0004
XC209-5VG	OEM
XC209-5VR	OEM
XC209-5VY	OEM
XC209G	0841
XC209R	OEM
XC209Y	3124
XC210G	OEM
XC210R	OEM
XC210Y	OEM
XC220G	OEM
XC220R	OEM
XC220Y	OEM
XC309	OEM
XC371	0127
XC372	0079
XC373	0079
XC374	0079
XC520R	OEM
XC521R	OEM
XC522G	OEM
XC522R	OEM
XC526G	OEM
XC526G2	OEM
XC526R	OEM
XC526R2	OEM
XC526Y	OEM
XC526Y2	OEM
XC554G	OEM
XC554G6	OEM
XC554G15	OEM
XC554G24	OEM
XC554R	OEM
XC554R9	OEM
XC554R12	OEM
XC554Y	OEM
XC554Y12	OEM
XC556G	OEM
XC556G2	OEM
XC556G3	OEM
XC556R	OEM
XC556R2	OEM
XC556R3	OEM
XC556Y	OEM
XC556Y2	OEM
XC556Y3	OEM
XC703	OEM
XC713	OEM
XC723	0103
XC880A	OEM
XC880AE	OEM
XC880B	OEM
XC880BE	OEM
XC880C	OEM
XC880CE	OEM
XC880D	OEM
XC881A	OEM
XC881AE	OEM
XC881B	OEM
XC881BE	OEM
XC881C	OEM
XC881CE	OEM
XC940A	OEM
XC940AE	OEM
XC940B	OEM
XC940BE	OEM
XC940C	OEM
XC940CE	OEM
XC940D	OEM
XC940DE	OEM
XC941A	OEM
XC941AE	OEM
XC941B	OEM
XC941BE	OEM
XC941C	OEM
XC941CE	OEM
XC941D	OEM
XC941DE	OEM
XC944	OEM
XC945	OEM
XC1011	OEM
XC1017	OEM
XC1100G	OEM
XC1100R	OEM
XC1100Y	OEM
XC1101G	OEM
XC1101R	OEM
XC1101Y	OEM
XC1209A	OEM
XC1209AB	OEM
XC1209B	OEM
XC1209BB	OEM
XC1209C	OEM
XC1209CB	OEM
XC1209D	OEM
XC1209DB	OEM
XC1288A	OEM
XC1288AB	OEM
XC1288BB	OEM
XC1288C	OEM
XC1288CB	OEM
XC1288D	OEM
XC1312	1611
XC-1312A	1611
XC1510	OEM
XC2090	OEM
XC2201	OEM
XC4550	OEM
XC4555	OEM
XC4560	OEM
XC4655	OEM
XC4850-5VG	OEM
XC4850-5VR	OEM
XC4850-5VY	OEM
XC4850-12VG	OEM
XC4850-12VR	OEM
XC4850-12VY	OEM
XC4850G	OEM
XC4850R	OEM
XC4850Y	OEM
XC4950	OEM
XC4955	OEM
XC5053G	1778
XC5053R	OEM
XC5053Y	3157
XC5057R	OEM
XC5057R3	OEM
XC5059G	OEM
XC5059R	OEM
XC5059Y	OEM
XC-5491	5386
XC5491	OEM
XC5535	OEM
XC5547R	OEM
XC5547R9	OEM
XC5547R15	OEM
XC5549G	OEM
XC5549G6	OEM
XC5549G15	OEM
XC5549G24	OEM
XC5549R	OEM
XC5549R9	OEM
XC5549R12	OEM
XC5549Y	OEM
XC5549Y6	OEM
XC5549Y12	OEM
XC5565	OEM
XC5567R	OEM
XC5567R2	OEM
XC5567R3	OEM
XC5569G	OEM
XC5569G2	OEM
XC5569G3	OEM
XC5569R	OEM
XC5569R2	OEM
XC5569R3	OEM
XC5569Y	OEM
XC5569Y2	OEM
XC5569Y3	OEM
XC89507P	OEM
XC99659P	OEM
XD/RD-1000	OEM
XD2A	0123
XD-258-42	0576
XD500	OEM
XD501	OEM
XD502	OEM
XD503	OEM
XD2201A	OEM
XD2210	OEM
XD2212D	OEM
XD-2581-42	0576
XD2804A	OEM
XD2816A	OEM
XD5081	OEM
XD5082	OEM
XDC14528-304	OEM
XDS2724P	OEM
XDS2724S	OEM
XE8340ENM	OEM
XE1503723	OEM
XF3-0.5	OEM
XF3-1.0	OEM
XF3-1.5	OEM
XF3-2.0	OEM
XF3-2.5	OEM
XF10-3	OEM
XF10-4	OEM
XF10-5	OEM
XF10-6	OEM
XF10-7	OEM
XF10-8	OEM
XF10-9	OEM
XF25-3	OEM
XF25-4	OEM
XF25-5	OEM
XF25-6	OEM
XF25-7	OEM
XF25-8	OEM
XF25-9	OEM
XF50-3	OEM
XF50-4	OEM
XF50-5	OEM
XF50-6	OEM
XF50-7	OEM
XF50-8	OEM
XF50-9	OEM
XFT2	OEM
XG1	0050
XG2	0050
XG3	0050
XG5	0050
XG8	0004
XG10	0050
XG11	0050
XG12	0050
XG24	0050
XG28	0208
XG29	0208
XG30	0016
XG32	0164
XG33	0208
XGS3025	OEM
XGS3030	OEM
XGS3035	OEM
XGS5025	OEM
XGS5030	OEM
XGS5035	OEM
XGS7001	OEM
XGS7002	0622
XGS410030	1852
XGS410035	1852
XGSA1530	OEM
XGSA1535	OEM
XGSA1540	OEM
XGSA3030	OEM
XGSA3035	OEM
XGSA3040	OEM
XGSA5030	OEM
XGSA5035	OEM
XGSA5040	OEM
XGSQ1530	OEM
XGSQ1535	OEM
XGSQ1540	OEM
XGSQ3030	OEM
XGSQ3035	OEM
XGSQ3040	OEM
XGSQ5030	OEM
XGSQ5035	OEM
XGSQ5040	OEM
XGSQ7530	OEM
XGSQ7535	OEM
XGSQ7540	OEM
XGSR3030	1966
XGSR3035	1966
XGSR3040	OEM
XGSR5030	1966
XGSR5035	1852
XGSR5040	1966
XGSR7530	OEM
XGSR7535	OEM
XGSR7540	OEM
XGSR10025	OEM
XGSR10030	OEM
XGSR10035	OEM
XGSR10040	1852
XGSR50020	OEM
XH10	OEM
XH20	OEM
XH20A	OEM
XI-548	0103
XI-549	0178
XJ13	0211
XJ71	0050
XJ72	0050
XJ73	0050
XLS24C01AP	OEM
XM203	OEM
XM541	OEM
XM542	OEM
XM543	OEM
XM544	OEM
XMP1	OEM
XMP2	OEM
XMP3	OEM
XN1B301	OEM
XN12A	0085
XN12B	0085
XN12C	0085
XN12E	0085
XN1113	OEM
XN1212	OEM
XN1213	OEM
XN1215	OEM
XN1401	OEM
XN1501	OEM
XN4113	OEM
XN4213	OEM
XN4401	OEM
XN4501	OEM
XN4601	OEM
XNC101	0038
XO238CE	OEM
XP-258I-82	0576
XP-2581-82	0576
XR082CN	OEM
XR082CP	3741
XR082D	OEM
XR082DN	OEM
XR082DP	OEM
XR082N	OEM
XR082P	OEM
XR083CN	OEM
XR083CP	OEM
XR083D	OEM
XR083DN	OEM
XR083DP	OEM
XR083M	OEM
XR083N	OEM
XR083P	OEM
XR084CN	OEM
XR084CP	5788
XR084D	OEM
XR084DN	OEM
XR084DP	OEM
XR084M	OEM
XR084N	OEM
XR084P	OEM
XR094CN	OEM
XR094CP	OEM
XR094M	OEM
XR094N	OEM
XR094P	OEM
XR095CN	OEM
XR095CP	OEM
XR095M	OEM
XR095N	OEM
XR095P	OEM
XR096CN	OEM
XR096CP	OEM
XR096N	OEM
XR096P	OEM
XR146M	OEM
XR205	OEM
XR210CN	OEM
XR210M	OEM
XR246N	OEM
XR246P	OEM
XR320P	OEM
XR340M	OEM
XR346-2CN	OEM
XR346-2CP	OEM
XR346CN	OEM
XR346CP	OEM
XR350M	OEM
XR430M	OEM
XR494CN	OEM
XR494CP	OEM
XR494M	OEM
XR495CN	OEM
XR495CP	OEM
XR495M	OEM
XR555CM	OEM
XR555CN	OEM
XR555CP	OEM
XR555M	OEM
XR556CN	OEM
XR556CP	OEM
XR556M	OEM
XR558CN	OEM
XR558CP	OEM
XR558M	OEM
XR559CN	OEM
XR559CP	OEM
XR559M	OEM
XR567CN	OEM
XR567CP	OEM
XR567M	6759
XR-1071CP	OEM
XR1310	6345
XR1458CP	0360
XR1468CN	OEM
XR1488N	0508
XR1488P	0508
XR1489AN	0511
XR1489AP	OEM
XR1524M	OEM
XR1525AN(A)	OEM
XR1527AN(A)	OEM
XR1543N(A)	OEM
XR1568M	OEM
XR1568N	OEM
XR2200	OEM
XR2200CP	OEM
XR2201CP	0845
XR2202CP	1007
XR2203CP	1134
XR2204CP	1261
XR-2206	OEM
XR2206	OEM
XR2206CN	OEM
XR2206CP	OEM
XR2206M	OEM
XR2206N	OEM
XR2206P	OEM
XR-2207CN	OEM
XR2207CN	OEM
XR2207CP	OEM
XR-2207M	OEM
XR2207M	OEM
XR-2207M/883B	OEM
XR-2207N	OEM
XR2207N	OEM
XR-2207P	OEM
XR2207P	OEM
XR2208CN	OEM
XR2208CP	OEM
XR2208M	OEM
XR2208N	OEM
XR2208P	OEM
XR2209CN	OEM
XR2209CP	OEM
XR2209M	OEM
XR-2211	OEM
XR2211	OEM
XR-2211CN	OEM
XR-2211CP	OEM
XR-2211M	OEM
XR-2211M/883B	OEM
XR-2211N	OEM
XR-2211P	OEM
XR2213CN	OEM
XR2213CP	OEM
XR2213M	OEM
XR2213P	OEM
XR2216CN	OEM
XR2216CP	OEM
XR2225AN(A)	OEM
XR2228CN	OEM
XR2228CP	OEM
XR2228M	OEM
XR2228N	OEM
XR2228P	OEM
XR2230CP	OEM
XR2235(A)	OEM
XR2240CN	OEM
XR2240CP	OEM
XR2240M	OEM
XR2240N	OEM
XR2240P	OEM
XR2242	OEM
XR2242CN	OEM
XR2242CP	OEM
XR2242M	OEM
XR2243CN	OEM
XR2243CP	OEM
XR2264CP	OEM
XR2265CP	OEM
XR2266	OEM
XR2271CN	OEM
XR2271CP	OEM
XR2272CN	OEM
XR2272CP	OEM
XR2276CP	OEM
XR2277P	OEM
XR2278P	OEM
XR2279CP	OEM
XR2284CP	OEM
XR2284P	OEM
XR2288CP	OEM
XR2288P	OEM
XR2524N	OEM
XR2524P	OEM
XR2525AC(A)	OEM
XR2527AC(A)	OEM
XR2527AN(A)	OEM
XR2556CN	OEM
XR2556CP	OEM
XR2556M	OEM
XR2567CN	OEM
XR2567CP	OEM
XR2567M	OEM
XR3403CN	1074
XR3403CP	1074
XR3503M	OEM
XR3524	5233
XR3524CN	OEM
XR3524CP	5233
XR3525AC(A)	OEM
XR3525AN(A)	OEM
XR3527AN(A)	OEM
XR3543N(A)	OEM

DEVICE TYPE	REPL CODE
XR4136CN	OEM
XR4136CP	3024
XR4136M	OEM
XR4151CP	OEM
XR4151P	OEM
XR4194CN	OEM
XR4194M	OEM
XR4194MK	OEM
XR4195CK	OEM
XR4195CP	5837
XR4195MT	OEM
XR4202M	OEM
XR4202N	OEM
XR4202P	OEM
XR4212CN	OEM
XR4212CP	OEM
XR4212M	OEM
XR4558CP	0360
XR4739CN	OEM
XR4739CP	OEM
XR4741CN	OEM
XR4741CP	OEM
XR4741M	OEM
XR5532AN	OEM
XR5532N	OEM
XR5533AN	OEM
XR5533AP	OEM
XR5533N	OEM
XR5533P	OEM
XR5534ACN	OEM
XR5534ACP	OEM
XR5534AM	OEM
XR5534CN	OEM
XR5534CP	OEM
XR5534M	OEM
XR5534N	OEM
XR5534P	OEM
XR6118AP	OEM
XR6118P	OEM
XR6128AP	OEM
XR6128P	OEM
XR8038ACN(A)	OEM
XR8038ACP(A)	OEM
XR8038AM(A)	OEM
XR8038AN(A)	OEM
XR8038AP(A)	OEM
XR-8038CN	0304
XR8038CN	0304
XR-8038CP	0304
XR8038CP	0304
XR-8038M	OEM
XR8038M	OEM
XR-8038N	0304
XR8038N	OEM
XR-8038P	0304
XR8038P	0304
XR9201CP	OEM
XR13600AP	OEM
XR13600CP	OEM
XRA5114L	OEM
XRA6459P1	OEM
XRA7305AS	OEM
XRA7305BS	OEM
XRA15218N	OEM
XRL555CN	0585
XRL555CP	OEM
XRL555M	OEM
XRL556CN	2869
XRL556CP	2869
XRL556M	OEM
XR-S200	OEM
XRU2744S	OEM
XRU27745-1	OEM
XS1	0144
XS2	0144
XS3	0144
XS4	0144
XS6	0144
XS9.1B	0012
XS10	0015
XS12	OEM
XS14	0144
XS15	0144
XS16	0015
XS16A	0015
XS17	0015
XS17A	0015
XS18	0015
XS19	0037
XS21	0016
XS22	1335
XS22(RECT.)	0015
XS23	0015
XS23A	0015
XS26	0208
XS30	0086
XS30L	OEM
XS31	0015
XS31L	OEM
XS31P	OEM
XS32L	OEM
XS32P	OEM
XS36	0007
XS37	0007
XS38	0007
XS39	0007
XS40	0127
XS40A	0015
XS40L	OEM
XS40P	OEM
XS41L	OEM
XS41P	OEM
XS42L	OEM
XS42P	OEM
XS101	0164
XS104	0164
XS121	0164
XT1-24	OEM
XT1A	0338
XT1B	0338
XT1C	OEM
XT1D	OEM
XT1.24	OEM
XT2A	OEM
XT2B	OEM
XT2C	OEM
XT2D	OEM
XT15X3	0144
XT100	0164
XT200	0164
XT200A	OEM
XT300	0136
XT400	0136
XT515	0264
XT516	0264
XT517	0264
XT518	0264
XT519	0264
XT520	0264
XT2104-401	OEM
XT2104-501	OEM
XT2104-601	OEM
XT2104-701	OEM
XT2104-801	OEM
XT2105-601	OEM
XT2105-801	OEM
XT2105-1001	OEM
XT2105-1201	OEM
XT2105-1401	OEM
XTP10	OEM
XTP11	OEM
XU604	0015
XV541	OEM
XV542	OEM
XV543	OEM
XV544	OEM
XX	OEM
XX8	OEM
XX10	OEM
XX12	OEM
XX15	OEM
XX20	OEM
XX25	OEM
XX30	OEM
XX40	OEM
XXC	OEM
XXF8	OEM
XXF10	OEM
XXF12	OEM
XXF15	OEM
XXF20	OEM
XXF25	OEM
XXF30	OEM
XXF40	OEM
XZ045	OEM
XZ047	OEM
XZ-049	0162
XZ049	0041
XZ-051	0162
XZ051	4373
XZ055	0253
XZ057	0253
XZ060	0091
XZ062	0471
XZ-064	0244
XZ064	0244
XZ-070	0301
XZ-072	0649
XZ-076	0649
XZ076	0649
XZ084	0244
XZ-086	1083
XZ-090	0012
XZ090	0012
XZ-092	0012
XZ092	0012
XZ094	0012
XZ-096	0170
XZ098	0170
XZ-100	0170
XZ100	0170
XZ102	0170
XZ107	OEM
XZ-122	0137
XZ122	0137
XZ132	1399
XZ-152	0002
XZ152	0002
XZ495C	OEM
XZH24	OEM
XZ070	0025
Y	OEM
Y0-6A	0015
Y065	0025
Y068	0062
Y10EA	OEM
Y10GA	OEM
Y23	OEM
Y25	OEM
Y100	0484
Y363	0004
Y3015	OEM
Y29979	0144
Y56601-08	0016
Y56601-50	0037
Y1343773-1	0111
Y13175962	0124
Y13189245	1137
Y13238241	0330
Y13290333	0111
Y13297762	0037
Y13297763	0037
Y13298671	0133
Y13310137	0224
Y13310138	0144
Y13310139	0037
Y13310146	0102
Y13310147	0015
Y13331721	0102
Y13331741	0285
Y13331781	0791
Y13331799	OEM
Y13331813	0313
Y13331873	0700
Y13358072	0016
Y13390991	0042
Y13391001	0925
Y13391143	0595
Y13391153	0791
Y13398601	0015
Y13399701	0037
Y13411222	0015
Y13417381	0103
Y13430051	0236
Y13432501	0133
Y13437771	0023
Y13450181	1708
Y15310157	0167
Y15310158	1329
Y15310159	1248
Y15377001	2167
Y15390751	0856
Y15415451	3380
Y15416272	2831
Y15418561	3195
Y15432511	1808
Y15432512	1808
Y15437031	0856
Y15453001	0167
Y32390915	1067
Y130201002	0522
Y130213002	0904
Y130213003	0076
Y130213004	0276
Y130213005	0191
Y130213006	2437
Y130213007	0148
Y130213041	0133
Y130213042	4920
Y130213043	0023
Y130213044	0348
Y130213045	0023
Y130800501	OEM
Y130800701	2697
Y130801001	2278
Y130801301	OEM
Y130801401	2278
Y131759610	0124
Y132929871	0133
Y133101149	0077
Y133101310	0076
Y133101317	1051
Y133101318	1165
Y133101319	0037
Y133101321	0144
Y133101322	0079
Y133101326	0595
Y133101327	0715
Y133101328	0079
Y133101332	0532
Y133101333	0018
Y133101363	0155
Y133101410	0005
Y133101411	0133
Y133101412	0143
Y133101413	0133
Y133101414	0041
Y133101415	0005
Y133101417	0133
Y133101419	0015
Y133101440	0015
Y133318728	OEM
Y310502139	OEM
Y384456911	3415
Y435800501	OEM
Y435800701	OEM
Y435800702	OEM
Y435800801	OEM
Y440800001	OEM
Y440800101	2175
Y440800102	OEM
Y440800301	2278
Y440800601	0812
Y440800701	OEM
Y440800702	OEM
Y440801001	OEM
Y440801101	0812
Y440801301	OEM
Y440801501	2278
Y440802501	OEM
Y440803801	OEM
Y440804101	OEM
Y441800101	0812
Y441800102	0812
Y441800103	OEM
Y446800001	OEM
Y446800102	OEM
Y448800001	OEM
Y448800404	OEM
Y461800002	OEM
Y461800003	OEM
Y490801601	OEM
Y560800000	OEM
YAAD001	0244
YAAD004	0015
YAAD007	0139
YAAD009	0143
YAAD010	0133
YAAD017	0244
YAAD018	0133
YAAD019	0015
YAAD020	0015
YAAD021	0244
YAAD022	0015
YAAN2SC1096K	0560
YAAN2SC1096L	0560
YAAN2SC1096N	0560
YAAN2SD141	0178
YAANL2SC1096K	0390
YAANL2SC1096L	0390
YAANL2SC1096M	0390
YAANL2SC1096N	0390
YAANL2SC1096Q	0390
YAANL2SC1096R	0390
YAANZ2S1096	0560
YAANZ2SC1096	0390
YAANZ2SC1096K	0390
YAANZ2SC1096L	0390
YAG100	OEM
YAG444	OEM
YAG444-4	OEM
YANZ2SC1096	0390
YANZ2SC1096K	0390
YANZ2SC1096L	0390
YANZ2SC1096M	0390
YAT015	OEM
YBAD009	0015
YBG1000	5155
YC02P11/2	0087
YD1121	0133
YEAD005	0005
YEAD009	OEM
YEAD010	0244
YEAD015	0012
YEAD022	OEM
YEAD024	0244
YEAD025	0012
YEAD029	0030
YEAD030	0015
YEAD032	0143
YEAD1N60P	0123
YEAD1SV53	0030
YEADAW01-06	0301
YEADISV53	0030
YEADUPC577H	2261
YEAM004	3958
YEAM53274P	1313
YEAMJM78L05	1292
YEAMLM2900N	2247
YEAMM5152L	2534
YEAMNJ58L05	1292
YEAMNJM58L05	3024
YEAMNJM78L05	1292
YEAMSCL4416	OEM
YEAMSM5118	6188
YEAMSN16889	OEM
YEAMUPC577H	2261
YEAN2SC941	0155
YEAN3SK39Q	0353
YH1000	OEM
YH1010	OEM
YH1012	OEM
YH1014	OEM
YH1020	OEM
YH1041	OEM
YH1042	OEM
YH1043	OEM
YH1045	OEM
YH1047A1	OEM
YH1047A2	OEM
YH1048	OEM
YH1049	OEM
YH1050	OEM
YH1120	OEM
YH1150	OEM
YH1160	OEM
YH1162	OEM
YH1181	OEM
YH1190	OEM
YH1191	OEM
YH1193	OEM
YH1194	OEM
YH1196	OEM
YH1197	OEM
YH1202	OEM
YH1203	OEM
YH1204	OEM
YH1205	OEM
YH1206	OEM
YH1207	OEM
YH1208	OEM
YH1209	OEM
YH1300	OEM
YH1301	OEM
YH1420	OEM
YH1421	OEM
YH1422	OEM
YH1750	OEM
YJ1050	OEM
YJ1060	OEM
YJ1071	OEM
YJ1110	OEM
YJ1112	OEM
YJ1120	OEM
YJ1121	OEM
YJ1123	OEM
YJ1124	OEM
YJ1150	OEM
YJ1160	OEM
YJ1162	OEM
YJ1164	OEM
YJ1180	OEM
YJ1180H	OEM
YJ1180L	OEM
YJ1181	OEM
YJ1181H	OEM
YJ1181L	OEM
YJ1185	OEM
YJ1185H	OEM
YJ1185L	OEM
YJ1191	OEM
YJ1193	OEM
YJ1194	OEM
YJ1200	OEM
YJ1201	OEM
YJ1210	OEM
YJ1230	OEM
YJ1280	OEM
YJ1300	OEM
YJ1301	OEM
YJ1320	OEM
YJ1321	OEM
YJ1380	OEM
YJ1400	OEM
YJ1410	OEM
YJ1420	OEM
YJ1440	OEM
YJ1441	OEM
YJ1442	OEM
YJ1443	OEM
YJ1463	OEM
YJ1464	OEM
YJ1470	OEM
YJ1480	OEM
YJ1481	OEM
YJ1490	OEM
YJ1500	OEM
YK1000	OEM
YK1001	OEM
YK1002	OEM
YK1003	OEM
YK1004	OEM
YK1005	OEM
YK1010	OEM
YK1020	OEM
YK1021	OEM
YK1023	OEM
YK1030	OEM
YK1050	OEM
YK1081	OEM
YK1082	OEM
YK1100	OEM
YK1110	OEM
YK1120	OEM
YK1151	OEM
YK1190	OEM
YK1191	OEM
YK1192	OEM
YK1195	OEM
YK1196	OEM
YK1197	OEM
YK1200	OEM
YK1210	OEM
YK1220	OEM
YK1230	OEM
YK1300	OEM
YL56	1985
YL212	3124
YL4484	3124
YL4550	3157
YL4850	3157
YM2149F	OEM
YM2201K	OEM
YM2601	OEM
YM3015	OEM
YM3015N000	OEM
YM3411	OEM
YM3428	OEM
YM3531	OEM
YM3805	OEM
YM3805N	OEM
YO	OEM
YOD4	0229
YR	OEM
YR-011	0015
YR011	0015
YS6-V139-2-2	0143
YSD2-11P4K32	OEM
YSD2-11P4K35	OEM
YSD2-11P6K32	OEM
YSD2-11P6K35	OEM
YSD2-11P8K38	OEM
YSD2-11P8K41	OEM
YSD2-11P10K38	OEM
YSD2-11P10K41	OEM
YSD2-11P12K38	OEM
YSD2-11P12K41	OEM
YSD2-11P14K38	OEM
YSD2-11P14K41	OEM
YSD2-12P4K32	OEM
YSD2-12P4K35	OEM
YSD2-12P6K32	OEM
YSD2-12P6K35	OEM
YSD2-12P8K38	OEM
YSD2-12P8K41	OEM
YSD2-12P10K38	OEM
YSD2-12P10K41	OEM
YSD2-12P12K38	OEM
YSD2-12P12K41	OEM
YSD2-12P14K38	OEM
YSD2-12P14K41	OEM
YSD5-41P10K44	OEM
YSD5-41P12K44	OEM
YSD5-41P14K44	OEM
YSD5-41P16K44	OEM
YSD5-41P18K44	OEM
YSD9-01P16K29	OEM
YSD9-01P16K32	OEM
YSD9-01P16K35	OEM
YSD9-01P18K29	OEM
YSD9-01P18K32	OEM
YSD9-01P18K35	OEM
YSD9-01P20K29	OEM
YSD9-01P20K32	OEM
YSD9-01P20K35	OEM
YSD9-02P24K38	OEM
YSD9-02P24K41	OEM
YSD9-02P24K44	OEM
YSD9-02P26K38	OEM
YSD9-02P26K41	OEM
YSD9-02P26K44	OEM
YSD9-02P28K38	OEM
YSD9-02P28K41	OEM
YSD9-02P28K44	OEM
YSD9-03P30K50	OEM
YSD9-03P30K53	OEM
YSD9-03P32K50	OEM
YSD9-03P32K53	OEM
YSD9-03P34K50	OEM
YSD9-03P34K53	OEM
YSD9-03P36K50	OEM
YSD9-03P36K53	OEM
YSD211P4K38	OEM
YSD211P6K38	OEM
YSD212P4K38	OEM
YSD902P22K38	OEM
YSG-20024	0015
YSG-V47-1-3	0102
YSG-V47-7-51	0484
YSG-V47-7-51-1	0015
YSG-V47-7-51-2	0015
YSG-V81-2-3	0913
YSG-V139-2-2	0143
YSG-V139-22	0143
YST2-01K2P8	OEM
YST2-01K2P10	OEM
YST2-01K2P12	OEM
YST2-01K2P14	OEM
YST2-01K2P16	OEM
YST2-02K1P2	OEM
YST2-02K1P4	OEM
YST2-02K1P6	OEM
YST2-02K1P8	OEM
YST3-41K3P2	OEM
YST3-41K3P4	OEM
YST3-41K3P6	OEM
YST3-41K3P8	OEM
YST3-42K3P2	OEM
YST3-42K3P4	OEM
YST3-42K3P6	OEM
YST3-42K3P8	OEM
YST3-42K3P10	OEM
YST3-42K3P12	OEM
YST3-42K3P14	OEM
YST5-01K2P8	OEM
YST5-01K2P10	OEM
YST5-01K2P12	OEM
YST5-01K2P14	OEM
YST5-01K2P16	OEM
YST5-01K2P18	OEM
YST5-21K4P18	OEM
YST5-21K4P20	OEM
YST5-21K4P22	OEM
YST5-21K4P24	OEM
YST5-21K4P26	OEM
YST5-21K4P28	OEM
YST5-21K4P30	OEM
YST8-01K2P8	OEM
YST8-01K2P10	OEM
YST8-01K2P12	OEM
YST8-01K2P14	OEM
YST8-01K2P16	OEM
YST8-01K2P18	OEM
YST8-02K1P2	OEM
YST8-02K1P4	OEM
YST8-21K3P18	OEM
YST8-21K3P20	OEM
YST8-21K3P22	OEM
YST8-21K3P24	OEM
YST8-21K3P26	OEM
YST8-21K3P28	OEM
YST8-21K3P30	OEM
YST301QP2K3	OEM
YST301QP4K3	OEM
YST301QP6K3	OEM
YST301QP8K3	OEM
YST301QP10K3	OEM
YST302QP2K3	OEM
YST302QP4K3	OEM
YST302QP6K3	OEM
YST302QP8K3	OEM
YST302QP10K3	OEM
YST302QP12K3	OEM
YST302QP14K3	OEM
YST801SP8K2	OEM
YST801SP10K2	OEM
YST801SP12K2	OEM
YST801SP14K2	OEM
YST801SP16K2	OEM
YST801SP18K2K2	OEM
YV1	0004
YV1A	0004
YV2	0004
YYAAD007	0139
YZ030	0299
YZ-037	0188
YZ037	0456
YZ-047	0451
YZ047A	0451
YZ058	0157
YZ060	3465
YZ-063	0471
YZ063	0471
YZ065F	0025
YZ-080	0244
YZ080	0244
YZ088A	OEM
YZ-284	1674
YZ284	0726
Z/D27	0441
Z0008CE	OEM
Z-00	0920
Z0021-UA	OEM
Z0027CE	OEM
Z0035-UDC	OEM
Z0039-BCA	OEM
Z0047-UBA	OEM
Z0047-VBA	OEM
Z0047UBA	OEM
Z0063-UAA	2671
Z0063-UAB	2671
Z0064-UAA	2671
Z0065-UAA	OEM
Z0065-UAB	OEM
Z0069UAA	2671
Z0069UAB	2671
Z0073CE	OEM
Z0-12	0137
Z0-12A	0137
Z0-12B	0137
Z0-27	0441
Z0-27A	0441
Z0-27B	0441
Z01-13	0365
Z-08	0649
Z0101BA	4455
Z0101BB	4456
Z0101DA	4456
Z0101DB	4456
Z0101MA	4482
Z0101MB	4482
Z0101NA	OEM
Z0101NB	OEM
Z0102BA	4481
Z0102DA	4483
Z0102MA	4499
Z0102NA	OEM
Z0104BA	4455
Z0104BB	4455
Z0104DA	4456
Z0104DB	4456
Z0104MA	4482
Z0104MB	4482
Z0104NA	OEM
Z0104NB	OEM
Z0105BA	4481
Z0105DA	4483
Z0105MA	4499
Z0105NA	OEM
Z0106BA	4455
Z0106BB	4455
Z0106DA	4456
Z0106DB	4456
Z0106MA	4482
Z0106MB	4482
Z0106NA	OEM
Z0106NB	OEM
Z0109BA	4481
Z0109DA	4483
Z0109MA	4499
Z0109NA	OEM
Z0110BA	4481
Z0110DA	4483

DEVICE TYPE	REPL CODE	DEVICE TYPE	REPL CODE	DEVICE TYPE	REPL CODE	DEVICE TYPE	REPL CODE	DEVICE TYPE	REPL CODE	DEVICE TYPE	REPL CODE	DEVICE TYPE	REPL CODE
Z0110MA	4499	Z0A1.4	OEM	Z1B9.1	0012	Z2C9.1	0684	Z3B68CF	0205	Z4B18	0495	Z5B30	0195
Z0110NA	OEM	Z0A2.0	OEM	Z1B10	0170	Z2C10	0225	Z3B75	0736	Z4B20	0531	Z5B33	0166
Z0206	0188	Z0A2.2	OEM	Z1B12	0137	Z2C12	0234	Z3B75CF	0480	Z4B22	0565	Z5C3.3	0188
Z0206A	OEM	Z0A2.4	1276	Z1B15	0002	Z2C15	0247	Z3B82CF	0504	Z4B24	0402	Z5C3.9	0188
Z0208A	OEM	Z0A2.7	OEM	Z1B18	0495	Z2C18	0256	Z3B91	1914	Z4B27	0441	Z5C4.7	0162
Z0210A	OEM	Z0A3.0	OEM	Z1B20	0531	Z2C22	0269	Z3B91CF	0684	Z4B30	0726	Z5C5.1	0162
Z0211A	OEM	Z0A3.3	0188	Z1B22	0565	Z2C33	0318	Z3B100	1163	Z4B33	0039	Z5C5.6	0157
Z0212A	OEM	Z0A3.6	0376	Z1B27	0441	Z2D3.9	0806	Z3B100CF	0225	Z4B36	0820	Z5C6.2	0636
Z0214A	OEM	Z0A3.9	0188	Z1C3.3	0188	Z2D5.6	0876	Z3B110	1933	Z4B43	0931	Z5C6.8	0062
Z0215A	OEM	Z0A4.3	OEM	Z1C3.9	0188	Z2D6.2	0185	Z3B110CF	0230	Z4B47	0999	Z5C8.2	0165
Z0216A	OEM	Z0A4.7	OEM	Z1C4.7	0162	Z2D7.5	0480	Z3B120	1941	Z4B51	0502	Z5C9.1	0012
Z0217A	OEM	Z0A5.1	0589	Z1C5.1	0162	Z2D9.1	0684	Z3B120CF	0234	Z4B56	0869	Z5C10	0170
Z0219A	OEM	Z0A5.6	0883	Z1C5.6	0162	Z2D10	0225	Z3B130	1947	Z4B62	0783	Z5C12	0137
Z0220A	OEM	Z0A6.2	0295	Z1C6.2	0636	Z2D12	0234	Z3B130CF	OEM	Z4B68	2157	Z5C15	0002
Z0222A	OEM	Z0A6.8	OEM	Z1C8.2	0165	Z2D15	0247	Z3B150	1961	Z4B75	1189	Z5C18	0375
Z0225A	OEM	Z0A7.5	OEM	Z1C9.1	0057	Z2D18	0256	Z3B150CF	0247	Z4B82	0331	Z5C20	0700
Z0228A	OEM	Z0A8.2	3100	Z1C10	0170	Z2D22	0269	Z3B160	0357	Z4B91	1311	Z5C22	0705
Z0230A	OEM	Z0A9.1	OEM	Z1C12	0137	Z2D27	0294	Z3B160CF	0251	Z4B100	0098	Z5D3.3	0188
Z0231A	OEM	Z0A10	0248	Z1C15	0002	Z2D33	0318	Z3B180	0776	Z4B110	0149	Z5D3.9	0188
Z0234A	OEM	Z0A11	OEM	Z1C18	0375	Z2.7	1312	Z3B180CF	0256	Z4B120	0186	Z5D4.7	0162
Z0254A	OEM	Z0A12	1005	Z1C20	0700	Z3	0376	Z3B200	1071	Z4B130	0213	Z5D5.1	0162
Z0255A	OEM	Z0A13	OEM	Z1C22	0705	Z3A3.0	OEM	Z3B200CF	0262	Z4B150	0028	Z5D5.6	0157
Z0302BG	2353	Z0A15	OEM	Z1C27	0441	Z3A3.3	OEM	Z3B220CF	0269	Z4B180	0367	Z5D6.2	0636
Z0302DG	2356	Z0A16	OEM	Z1D3.3	0188	Z3A3.6	OEM	Z3B240CF	0274	Z4B200	0421	Z5D6.8	0062
Z0302MG	5555	Z0A18	OEM	Z1D3.9	0036	Z3A3.9	OEM	Z3B270CF	0294	Z4K	0140	Z5D8.2	0165
Z0302NG	OEM	Z0A20	OEM	Z1D4.7	0162	Z3A4.3	OEM	Z3B300CF	0309	Z4MW333	0016	Z5D9.1	0012
Z0305BG	2353	Z0A22	OEM	Z1D5.1	0162	Z3A4.7	OEM	Z3B330CF	0318	Z4X5.1A	0162	Z5D10	0170
Z0305DG	2356	Z0A24	OEM	Z1D5.6	0157	Z3A5.1	OEM	Z3B360CF	0320	Z4X5.1B	0162	Z5D12	0137
Z0305MG	5555	Z0A27	OEM	Z1D6.2	0636	Z3A5.6	OEM	Z3B390CF	0326	Z4X5.6	0157	Z5D15	0002
Z0305NG	OEM	Z0A30	OEM	Z1D7.5	0649	Z3A6.2	OEM	Z3B430CF	0337	Z4X9.1	0012	Z5D18	0375
Z0309BG	2353	Z0A33	OEM	Z1D8.2	0165	Z3A6.8	OEM	Z3B470CF	0347	Z4X11	0794	Z5D20	0700
Z0309DG	2356	Z0B0.7	OEM	Z1D9.1	0057	Z3A7.5	OEM	Z3B510CF	0027	Z4X12	0137	Z5D22	0705
Z0309MG	5555	Z0B1.3	OEM	Z1D10	0170	Z3A8.2	OEM	Z3B560CF	0266	Z4X13	0365	Z5D82CF	1490
Z0309NG	OEM	Z0B1.4	OEM	Z1D12	0137	Z3A9.1	OEM	Z3B620CF	0386	Z4X14	0100	Z5D91CF	1618
Z0310BG	2394	Z0B2.0	OEM	Z1D15	0002	Z3A10	OEM	Z3B680CF	0405	Z4X14A	0100	Z5D100CF	0510
Z0310DG	1413	Z0B2.2	OEM	Z1D18	0375	Z3A11	OEM	Z3B750CF	0426	Z4X14B	0100	Z5D110CF	0691
Z0310MG	2387	Z0B2.4	1276	Z1D20	0700	Z3A12	OEM	Z3B820CF	0444	Z4X15	0002	Z5D120CF	0870
Z0310NG	OEM	Z0B2.7	0760	Z1D22	0705	Z3A13	OEM	Z3B910CF	0238	Z4XL6.2	0636	Z5D130CF	1020
Z0401	5659	Z0B3.0	0118	Z1D27	0455	Z3A15	OEM	Z3B1000CF	1180	Z4XL6.2B	0636	Z5D150CF	1274
Z0402	0188	Z0B3.3	0299	Z1K	OEM	Z3A16	OEM	Z3K	OEM	Z4XL7.5	0649	Z5D160CF	1402
Z0402BE	2300	Z0B3.6	0376	Z2A33F	0292	Z3A18	OEM	Z3.0	1713	Z4XL7.5B	0649	Z5D180CF	1078
Z0402BF	2300	Z0B3.9	0036	Z2A36F	0188	Z3A20	OEM	Z3.3	0292	Z4XL9.1	0012	Z5D200CF	1717
Z0402DE	2300	Z0B4.3	0275	Z2A39F	0456	Z3A22	OEM	Z3.6	0188	Z4XL9.1B	0012	Z5D220CF	1723
Z0402DF	2300	Z0B4.7	0140	Z2A43F	0533	Z3A24	OEM	Z3.9	0188	Z4XL12	0137	Z5D240CF	1736
Z0402ME	OEM	Z0B5.1	0041	Z2A47F	0451	Z3A27	OEM	Z4	0275	Z4XL12B	0137	Z5D270CF	1761
Z0402MF	OEM	Z0B5.6	0157	Z2A51F	0162	Z3A30	OEM	Z4A3.0	OEM	Z4XL14	0100	Z5D300CF	1772
Z0402NE	OEM	Z0B6.2	0471	Z2A56CF	0157	Z3A33	OEM	Z4A3.3	OEM	Z4XL16	0420	Z5D330CF	1788
Z0402NF	OEM	Z0B6.8	0062	Z2A56F	0157	Z3A36	OEM	Z4A3.6	OEM	Z4XL16B	0420	Z5D360CF	1796
Z0403	5660	Z0B7.5	0077	Z2A62F	0636	Z3A39	OEM	Z4A3.9	OEM	Z4XL18	0495	Z5D390CF	1804
Z0405	5661	Z0B8.2	0165	Z2A68F	0025	Z3A43	OEM	Z4A4.3	OEM	Z4XL18B	0495	Z5D430CF	1194
Z0405BE	2300	Z0B9.1	0057	Z2A75F	0649	Z3A47	OEM	Z4A4.7	OEM	Z4XL20	0531	Z5D470CF	0022
Z0405BF	2300	Z0B10	0170	Z2A82F	0244	Z3A51	OEM	Z4A5.1	OEM	Z4XL20B	0531	Z5D560CF	0207
Z0405DE	2300	Z0B11	0181	Z2A91F	0012	Z3A56	OEM	Z4A5.6	OEM	Z4XL22	0565	Z5D620CF	0263
Z0405DF	2300	Z0B12	0052	Z2A100F	0170	Z3A62	OEM	Z4A6.2	0636	Z4XL22B	0565	Z5D680CF	0310
Z0405ME	OEM	Z0B13	0053	Z2A110F	0317	Z3A68	OEM	Z4A6.8	OEM	Z4XL75	0649	Z5D750CF	0329
Z0405MF	OEM	Z0B15	0686	Z2A120F	0137	Z3A75	OEM	Z4A7.5	OEM	Z4.3	0533	Z5D820CF	0356
Z0405NE	OEM	Z0B16	0445	Z2A130F	0365	Z3A82	OEM	Z4A8.2	0244	Z4.7	0451	Z5D910CF	0412
Z0405NF	OEM	Z0B18	0375	Z2A150F	0002	Z3A91	OEM	Z4A9.1	OEM	Z5	0253	Z5D1000CF	0438
Z0406	0162	Z0B20	0700	Z2B3.0	OEM	Z3A100	OEM	Z4A10	OEM	Z5A3.0	OEM	Z5K	0253
Z0407	0157	Z0B22	0705	Z2B3.3	0782	Z3A110	OEM	Z4A11	OEM	Z5A3.3	0299	Z5.1	0162
Z0408	0636	Z0B24	0494	Z2B3.6	0796	Z3A120	OEM	Z4A12	OEM	Z5A3.6	OEM	Z5.1(500MW)	0041
Z0409	3465	Z0B27	0441	Z2B3.9	0806	Z3A130	OEM	Z4A13	OEM	Z5A3.9	0188	Z5.1B	0041
Z0409BE	2300	Z0B30	0195	Z2B4.3	0821	Z3A150	OEM	Z4A15	OEM	Z5A4.3	OEM	Z5.1BTA	0041
Z0409BF	2300	Z0B33	0166	Z2B4.7	0833	Z3A160	OEM	Z4A16	OEM	Z5A4.7	OEM	Z5.6	0157
Z0409DE	2300	Z0C3.3	0188	Z2B5.1	0442	Z3A180	OEM	Z4A18	OEM	Z5A5.1	0589	Z5.6(500MW)	0253
Z0409DF	2300	Z0C4.7	0162	Z2B5.6	0876	Z3A200	OEM	Z4A20	OEM	Z5A5.6	0883	Z5.6A	0157
Z0409ME	OEM	Z0C5.1	0162	Z2B6.2	0185	Z3B3.0	OEM	Z4A22	OEM	Z5A6.2	0471	Z5.6B	0157
Z0409MF	OEM	Z0C5.6	0157	Z2B6.8	0205	Z3B3.3	1006	Z4A24	OEM	Z5A6.8	OEM	Z5.6BM	0253
Z0409NE	OEM	Z0C6.2	0636	Z2B7.5	0480	Z3B3.6	1380	Z4A27	OEM	Z5A7.5	OEM	Z5.6BMTA	0253
Z0409NF	OEM	Z0C6.8	0062	Z2B8.2	0504	Z3B3.9	0547	Z4A30	OEM	Z5A8.2	OEM	Z5.6C	OEM
Z0410	0649	Z0C8.2	0165	Z2B9.1	0684	Z3B4.3	2403	Z4A33	OEM	Z5A9.1	OEM	Z5.6D	OEM
Z0410BE	2300	Z0C9.1	0012	Z2B10	0225	Z3B4.7	2113	Z4A36	OEM	Z5A10	0248	Z6	0301
Z0410BF	2300	Z0C10	0170	Z2B11	0230	Z3B5.1	2410	Z4A39	OEM	Z5A11	OEM	Z6A2	0091
Z0410DE	2300	Z0C12	0137	Z2B12	0234	Z3B5.6	1901	Z4A43	OEM	Z5A12	1005	Z6A3.6	OEM
Z0410DF	2300	Z0C15	0002	Z2B13	0237	Z3B6.2	1616	Z4A47	OEM	Z5A13	OEM	Z6A3.9	OEM
Z0410ME	OEM	Z0C18	0375	Z2B15	0247	Z3B6.8	1601	Z4A51	OEM	Z5A15	OEM	Z6A4.3	OEM
Z0410MF	OEM	Z0C20	0700	Z2B16	0251	Z3B7.5	1616	Z4A56	OEM	Z5A16	OEM	Z6A4.7	OEM
Z0410NE	OEM	Z0C22	0705	Z2B18	0256	Z3B8.2	1622	Z4A62	OEM	Z5A18	OEM	Z6A5.1	OEM
Z0410NF	OEM	Z0C27	0441	Z2B20	0262	Z3B9.1	0627	Z4A68	OEM	Z5A20	0531	Z6A5.6	OEM
Z0411	0244	Z0D3.3	0188	Z2B22	0269	Z3B10	0992	Z4A75	OEM	Z5A22	OEM	Z6A6.2	OEM
Z0412	0012	Z0D3.9	0188	Z2B24	0274	Z3B11	0995	Z4A82	OEM	Z5A24	OEM	Z6A6.8	OEM
Z0413	0170	Z0D4.7	0162	Z2B27	0294	Z3B12	1263	Z4A91	OEM	Z5A27	OEM	Z6A7.5	OEM
Z0414	0317	Z0D5.1	0162	Z2B30	0309	Z3B13	1249	Z4A100	OEM	Z5A30	OEM	Z6A8.2	OEM
Z0415	0137	Z0D5.6	0157	Z2B33	0318	Z3B15	1639	Z4A110	OEM	Z5A33	OEM	Z6A9.1	OEM
Z0416	0365	Z0D6.2	0636	Z2B36	0320	Z3B16	1703	Z4A120	OEM	Z5B3.0	0118	Z6A10	OEM
Z0417	0100	Z0D8.2	0165	Z2B39	0326	Z3B18	1716	Z4A130	OEM	Z5B3.3	0299	Z6A11	OEM
Z0418	0002	Z0D9.1	0012	Z2B43	0337	Z3B20	1731	Z4A150	OEM	Z5B3.6	0188	Z6A12	OEM
Z0419	0420	Z0D10	0170	Z2B47	0347	Z3B22	0727	Z4A160	OEM	Z5B3.9	0036	Z6A13	OEM
Z0420	0495	Z0D12	0137	Z2B51	0027	Z3B24	1756	Z4A180	OEM	Z5B4.3	0275	Z6A15	OEM
Z0421	5664	Z0D15	0002	Z2B56	0266	Z3B27	1782	Z4A200	OEM	Z5B4.7	0140	Z6A16	OEM
Z0422	5665	Z0D20	0700	Z2B62	0386	Z3B30	1794	Z4B	0036	Z5B5.1	0162	Z6A18	OEM
Z0423	0402	Z0D22	0705	Z2B68	0405	Z3B33	1799	Z4B3.0	1713	Z5B5.6	0157	Z6A20	OEM
Z0424	0441	Z0D27	0455	Z2B75	0426	Z3B33CF	0782	Z4B3.3	0292	Z5B6.2	0471	Z6A22	OEM
Z0425	OEM	Z0D27B	0441	Z2B82	0444	Z3B36	6431	Z4B3.6	0188	Z5B6.8	0062	Z6A24	OEM
Z0427	0820	Z1	OEM	Z2B91	0238	Z3B36CF	0796	Z4B3.9	0456	Z5B7.5	0077	Z6A27	OEM
Z0428	5667	Z1A2.4	1276	Z2B100	1180	Z3B39	1817	Z4B4.3	0533	Z5B8.2	0165	Z6A30	OEM
Z0430	3595	Z1A2.7	0760	Z2B110	1190	Z3B39CE	0806	Z4B4.7	0451	Z5B9.1	0057	Z6A33	OEM
Z0432	1834	Z1A3.3	0188	Z2B120	1207	Z3B39CF	0806	Z4B5.1	0162	Z5B10	0170	Z6A36	OEM
Z0433	0783	Z1A3.9	0188	Z2B130	1218	Z3B43	1826	Z4B5.6	0157	Z5B11	0181	Z6A39	OEM
Z0436	0331	Z1A4.7	0140	Z2B150	0782	Z3B43CF	0821	Z4B6.2	0636	Z5B12	0052	Z6A43	OEM
Z0438	OEM	Z1A8.2	0244	Z2B160	1255	Z3B47	1853	Z4B6.8	0025	Z5B13	0053	Z6A47	OEM
Z0439	0149	Z1B3.9	0456	Z2B180	0605	Z3B47CF	0833	Z4B8.2	0244	Z5B-15	0002	Z6A51	OEM
Z0440	5668	Z1B4.7	0451	Z2B200	0605	Z3B51	1866	Z4B9.1	0012	Z5B15	0686	Z6A56	OEM
Z0442	5669	Z1B5.1	0162	Z2B400	OEM	Z3B51CF	0442	Z4B10	0170	Z5B16	0445	Z6A62	OEM
Z0444	5671	Z1B5.6	0157	Z2C3.9	0806	Z3B56	1406	Z4B11	0317	Z5B18	0375	Z6A68	OEM
Z0445	5672	Z1B6.2	0636	Z2C4.7	0833	Z3B56CF	0876	Z4B12	0137	Z5B20	0700	Z6A75	OEM
Z0765A08PSC	OEM	Z1B6.8	0025	Z2C5.6	0876	Z3B62	1895	Z4B13	0365	Z5B22	0705	Z6A82	OEM
Z0A0.7	OEM	Z1B7.5	0649	Z2C6.2	0185	Z3B62CF	0185	Z4B15	0002	Z5B24	0494	Z6A91	OEM
Z0A1.3	OEM	Z1B8.2	0244	Z2C7.5	0480	Z3B68	1902	Z4B16	0420	Z5B27	0455	Z6A100	OEM

DEVICE TYPE	REPL CODE
Z6A110	OEM
Z6A120	OEM
Z6A130	OEM
Z6A150	OEM
Z6A160	OEM
Z6A180	OEM
Z6A200	OEM
Z6B3.6	OEM
Z6B3.9	OEM
Z6B4.3	OEM
Z6B4.7	OEM
Z6B5.1	OEM
Z6B5.6	OEM
Z6B6.2	OEM
Z6B6.8	OEM
Z6B7.5	OEM
Z6B8.2	OEM
Z6B9.1	OEM
Z6B10	OEM
Z6B11	OEM
Z6B12	OEM
Z6B13	OEM
Z6B15	OEM
Z6B16	OEM
Z6B18	OEM
Z6B20	OEM
Z6B22	OEM
Z6B24	OEM
Z6B27	OEM
Z6B30	OEM
Z6B33	OEM
Z6B36	OEM
Z6B39	OEM
Z6B43	OEM
Z6B47	OEM
Z6B51	OEM
Z6B56	OEM
Z6B62	OEM
Z6B68	OEM
Z6B75	OEM
Z6B82	OEM
Z6B91	OEM
Z6B100	OEM
Z6B110	OEM
Z6B120	OEM
Z6B130	OEM
Z6B150	OEM
Z6B160	OEM
Z6B180	OEM
Z6B200	OEM
Z6K	0062
Z6P12	OEM
Z6.2	0636
Z6.2A	0636
Z6.2B	0636
Z6.2C	0636
Z6.2D	OEM
Z6.8	0025
Z6.8A	0025
Z6.8B	0025
Z6.8BHTA	0062
Z6.8C	OEM
Z6.8D	OEM
Z7	0077
Z7A1	0062
Z7A3.0	OEM
Z7A3.3	OEM
Z7A3.6	OEM
Z7A3.9	OEM
Z7A4.3	OEM
Z7A4.7	OEM
Z7A5.1	OEM
Z7A5.6	OEM
Z7A6.2	OEM
Z7A6.8	OEM
Z7A7.5	OEM
Z7A8.2	OEM
Z7A9.1	OEM
Z7A10	OEM
Z7A11	OEM
Z7A12	OEM
Z7A13	OEM
Z7A15	OEM
Z7A16	OEM
Z7A18	OEM
Z7A20	OEM
Z7A22	OEM
Z7A24	OEM
Z7A27	OEM
Z7A30	OEM
Z7A33	OEM
Z7A36	OEM
Z7A39	OEM
Z7A43	OEM
Z7A51	OEM
Z7A56	OEM
Z7A62	OEM
Z7A68	OEM
Z7A75	OEM
Z7A82	OEM
Z7A91	OEM
Z7A100	OEM
Z7A110	OEM
Z7A120	OEM
Z7A130	OEM
Z7A150	OEM
Z7A160	OEM
Z7A180	OEM
Z7A200	OEM
Z7A400	OEM
Z7B3.0	OEM
Z7B3.3	OEM
Z7B3.7	OEM
Z7B3.9	OEM
Z7B4.3	OEM
Z7B4.7	OEM
Z7B5.1	OEM
Z7B5.6	OEM
Z7B6.2	OEM
Z7B6.8	OEM
Z7B7.5	OEM
Z7B8.2	OEM
Z7B9.1	OEM
Z7B10	OEM
Z7B11	OEM
Z7B12	OEM
Z7B13	OEM
Z7B15	OEM
Z7B16	OEM
Z7B18	OEM
Z7B20	OEM
Z7B22	OEM
Z7B24	OEM
Z7B27	OEM
Z7B30	OEM
Z7B33	OEM
Z7B36	OEM
Z7B39	OEM
Z7B43	OEM
Z7B47	OEM
Z7B51	OEM
Z7B56	OEM
Z7B62	OEM
Z7B68	OEM
Z7B75	OEM
Z7B82	OEM
Z7B91	OEM
Z7B100	OEM
Z7B110	OEM
Z7B120	OEM
Z7B130	OEM
Z7B150	OEM
Z7B160	OEM
Z7B180	OEM
Z7B200	OEM
Z7B400	OEM
Z7K	0077
Z7L101	OEM
Z7L102	OEM
Z7L121	OEM
Z7L151	OEM
Z7L181	OEM
Z7L201	OEM
Z7L220	OEM
Z7L221	OEM
Z7L270	OEM
Z7L271	OEM
Z7L330	OEM
Z7L331	OEM
Z7L390	OEM
Z7L391	OEM
Z7L441	OEM
Z7L470	OEM
Z7L471	OEM
Z7L560	OEM
Z7L561	OEM
Z7L680	OEM
Z7L681	OEM
Z7L820	OEM
Z7L821	OEM
Z7.5	0649
Z7.5A	0649
Z7.5B	0649
Z7.5BIS	0077
Z7.5BM	0077
Z7.5C	OEM
Z7.5D	OEM
Z8	0322
Z8-01CS	OEM
Z8-01PS	OEM
Z8-02QS	OEM
Z8-03RS	OEM
Z8.2	0244
Z8.2A	0244
Z8.2B	0244
Z8.2C	0244
Z8.2D	0244
Z9.1	0012
Z9.1A	0012
Z9.1B	0012
Z9.1B1	0012
Z9.1BL	0057
Z9.1BLTA	0012
Z9.1C	0012
Z9.1D	OEM
Z10	0170
Z10A	0064
Z10B	0064
Z10C	OEM
Z10D	OEM
Z10K	0170
Z10L101	OEM
Z10L102	OEM
Z10L121	OEM
Z10L151	OEM
Z10L181	OEM
Z10L201	OEM
Z10L271	OEM
Z10L330	OEM
Z10L331	OEM
Z10L390	OEM
Z10L391	OEM
Z10L441	OEM
Z10L470	OEM
Z10L471	OEM
Z10L560	OEM
Z10L561	OEM
Z10L680	OEM
Z10L681	OEM
Z10L820	OEM
Z10L821	OEM
Z10.0	0170
Z-11	0137
Z11	0317
Z11A	0317
Z11B	0181
Z11B1	0181
Z11BA	OEM
Z11BL	0181
Z11BLTA	0181
Z11BM	0181
Z11BMTA	4019
Z11C	OEM
Z11D	OEM
Z11.0	0317
Z-12	0137
Z12	0137
Z12(500MW)	0052
Z12A	0052
Z12B	0052
Z12BM	0052
Z12BMTA	0052
Z12C	0137
Z12D	OEM
Z12K	0052
Z12.0	0137
Z13	0365
Z13A	0365
Z13B	0365
Z13C	OEM
Z13D	OEM
Z13.0	0365
Z14	0100
Z14A	0100
Z14B	0100
Z14C	OEM
Z14D	OEM
Z15	0002
Z15A	0002
Z15B	0002
Z15C	0002
Z15D	OEM
Z15K	0686
Z15L101	OEM
Z15L102	OEM
Z15L121	OEM
Z15L151	OEM
Z15L181	OEM
Z15L201	OEM
Z15L221	OEM
Z15L271	OEM
Z15L331	OEM
Z15L390	OEM
Z15L391	OEM
Z15L441	OEM
Z15L470	OEM
Z15L471	OEM
Z15L560	OEM
Z15L561	OEM
Z15L680	OEM
Z15L681	OEM
Z15L820	OEM
Z15L821	OEM
Z15.0	0002
Z16	0420
Z16A	0420
Z16B	0420
Z16C	OEM
Z16D	OEM
Z16.0	0420
Z17	1649
Z17A	1649
Z17B	1649
Z17C	OEM
Z17D	OEM
Z18	0495
Z18A	0495
Z18B	1649
Z18C	OEM
Z18D	OEM
Z18K	OEM
Z18.0	0495
Z19	0949
Z19A	0949
Z19B	0949
Z19C	OEM
Z19D	OEM
Z20	0531
Z20A	0531
Z20B	0531
Z20C	OEM
Z20D	OEM
Z20.0	0531
Z21L101	OEM
Z21L102	OEM
Z21L121	OEM
Z21L151	OEM
Z21L181	OEM
Z21L201	OEM
Z21L221	OEM
Z21L271	OEM
Z21L331	OEM
Z21L391	OEM
Z21L441	OEM
Z21L470	OEM
Z21L471	OEM
Z21L560	OEM
Z21L561	OEM
Z21L680	OEM
Z21L681	OEM
Z21L820	OEM
Z21L821	OEM
Z22	0565
Z22A	0565
Z22B	0565
Z22C	OEM
Z22D	OEM
Z22.0	0565
Z24	0494
Z24A	0402
Z24B	0402
Z24BL	6014
Z24BLTA	6153
Z24C	OEM
Z24D	OEM
Z24.0	0402
Z25	1606
Z25A	1606
Z25B	1606
Z25C	OEM
Z25D	OEM
Z27	0441
Z27A	0441
Z27B	0441
Z27C	0441
Z27D	OEM
Z27.0	0441
Z30	0726
Z30(500MW)	0166
Z30A	0726
Z30B	0726
Z30BM	0195
Z30BMTA	0195
Z30C	OEM
Z30D	OEM
Z30.0	0726
Z33	0039
Z33A	0039
Z33B	0039
Z33C	OEM
Z33D	OEM
Z33M102	OEM
Z33M221	OEM
Z33M271	OEM
Z33M331	OEM
Z33M391	OEM
Z33M441	OEM
Z33M471	OEM
Z33M561	OEM
Z33M681	OEM
Z33M821	OEM
Z33.0	0039
Z36	0820
Z36A	0820
Z36B	0820
Z36C	OEM
Z36D	OEM
Z39	0350
Z39A	0350
Z39B	0350
Z39C	OEM
Z39D	OEM
Z43	0931
Z43A	0931
Z43B	0931
Z43C	OEM
Z43D	OEM
Z45	0931
Z45A	0252
Z45B	0931
Z45C	OEM
Z45D	OEM
Z47	0999
Z47A	0999
Z47B	0999
Z47C	OEM
Z47D	OEM
Z50	0502
Z50A	0502
Z50B	0502
Z50C	OEM
Z50D	OEM
Z51	0502
Z51A	0502
Z51B	0502
Z51C	OEM
Z51D	OEM
Z52	0502
Z52A	0502
Z52B	0502
Z52C	OEM
Z52D	OEM
Z56	0869
Z56A	0869
Z56B	0869
Z56C	OEM
Z56D	OEM
Z62	0783
Z62A	0783
Z62B	0783
Z62C	OEM
Z62D	OEM
Z-64	OEM
Z68	2157
Z68A	2157
Z68B	2157
Z68C	OEM
Z68D	OEM
Z75	1189
Z75A	1189
Z75B	1189
Z75C	OEM
Z75D	OEM
Z80	3481
Z80-8420	OEM
Z80-AIB	OEM
Z80-AIBN	OEM
Z80-AIO	OEM
Z80-AION	OEM
Z80-IOB	OEM
Z80-MCB4	OEM
Z80-MCB16	OEM
Z80-MDC	OEM
Z80-PPB	OEM
Z80-PPB/16	OEM
Z80-RMB	OEM
Z80-SCE4	OEM
Z80-SIB	OEM
Z80-VDB	OEM
Z80A	3481
Z80A/4MHZ	OEM
Z80ACTCCE	OEM
Z80ACTCCM	OEM
Z80ACTCCS	OEM
Z80ACTCPS	OEM
Z80ADMACE	OEM
Z80ADMACM	OEM
Z80ADMACS	OEM
Z80ADMAPS	OEM
Z80APS	OEM
Z80B	OEM
Z80B/6MHZ	OEM
Z80CTCCE	OEM
Z80CTCCM	OEM
Z80CTCCS	OEM
Z80CTCPS	OEM
Z82	0331
Z82A	0331
Z82B	0331
Z82C	OEM
Z82D	OEM
Z86C2704PSC	OEM
Z86C2704PSCR390	OEM
Z86C2704PSC-R425	OEM
Z86C2704PSCR425	OEM
Z86C2704PSC-R681	OEM
Z86C2704PSC-R1028	OEM
Z86C2704PS-R390	OEM
Z91	1311
Z91A	1311
Z91B	1311
Z91C	OEM
Z91D	OEM
Z95LA7A	OEM
Z100	0098
Z100A	0098
Z100B	0098
Z100C	OEM
Z100D	OEM
Z101	0244
Z105	0149
Z105A	0149
Z105B	0149
Z105C	OEM
Z105D	OEM
Z110	0149
Z110A	0149
Z110B	0149
Z110C	OEM
Z110D	OEM
Z111Z	0137
Z120	0186
Z120A	0186
Z120B	0186
Z120C	OEM
Z120D	OEM
Z130	0213
Z130A	0213
Z130B	0213
Z130C	OEM
Z130D	OEM
Z140	0245
Z140A	0245
Z140B	0245
Z140C	OEM
Z140D	OEM
Z150	0028
Z150A	0028
Z150B	0028
Z150C	OEM
Z150D	OEM
Z160	0255
Z160A	0255
Z160B	0255
Z160C	OEM
Z160D	OEM
Z175	0367
Z175A	0367
Z175B	0367
Z175C	OEM
Z175D	OEM
Z180	0367
Z180A	0367
Z180B	0367
Z180C	OEM
Z180D	OEM
Z200	0421
Z200A	0421
Z200B	0421
Z200C	OEM
Z200D	OEM
Z211/1G	OEM
Z220	OEM
Z220A	OEM
Z220B	OEM
Z220C	OEM
Z220D	OEM
Z237/1KW	OEM
Z501	OEM
Z501-4M	OEM
Z537	0039
Z694	0012
Z714	0012
Z765A08PSC	OEM
Z765APS	OEM
Z801	0636
Z-963B	0137
Z1000	0299
Z1002	0036
Z-1004	0451
Z1004	0140
Z-1006	0157
Z1006	0253
Z1008	0062
Z1010	0165
Z-1012	0064
Z1012	0064
Z-1014	0137
Z1014	0052
Z1015	0686
Z-1016	0002
Z1016	0686
Z1018	0375
Z1020	0705
Z-1022	0441
Z1022	0705
Z-1024	0039
Z1024	0166
Z1027	0455
Z1033	0166
Z1039	0032
Z1047	0068
Z1056	0125
Z-1068	1987
Z1068	0173
Z1082	0049
Z-1100	0188
Z1100	0189
Z1100-C	0456
Z-1102	0451
Z1102	0451
Z1102-C	0451
Z-1104	0157
Z1104	0157
Z1104-C	0157
Z-1106	0636
Z1106	0025
Z1106-C	0025
Z1108	0244
Z-1110	0170
Z1110	0170
Z1110-C	0170
Z-1112	0137
Z1112	0137
Z1112-C	0137
Z-1112C	0137
Z1112C	0137
Z-1114	0002
Z1114	0002
Z1114-C	0002
Z-1116	0495
Z1116	0495
Z1116-C	0495
Z1118	0565
Z1118-C	0565
Z-1120	0441
Z1120	0089
Z1120-C	0441
Z-1122	0039
Z1122	0039
Z1124	0350
Z-1126	0999
Z1126	0999
Z1128	0869
Z1130	0783
Z-1132	0331
Z1132	0331
Z-1134	0149
Z1134	0149
Z1136	0186
Z1138	0028
Z-1140	0157
Z1140	0253
Z-1145	0157
Z1145	0253
Z-1150	0157
Z1150	0340
Z-1155	0157
Z1155	0253
Z-1160	0157
Z1160	0253
Z-1165	0157
Z1165	0253
Z-1170	0157
Z1170	0253
Z1200	0806
Z1202	0451
Z1203	0442
Z1204	0876
Z1206	0205
Z1207	0480
Z1208	0504
Z1209	0684
Z1212	0137
Z1218	0269
Z1238	0647
Z-1240	0157
Z1240	0253
Z-1245	0157
Z1245	0253
Z-1250	0157
Z1250	0253
Z-1255	0157
Z1255	0253
Z-1260	0157
Z1260	0253
Z-1265	0157
Z1265	0253
Z-1270	0157
Z1270	0253
Z1300	0547
Z1302	2113
Z1304	1901
Z1306	1601
Z1308	1622
Z1310	0510
Z1312	0870
Z1314	1274
Z1316	1078
Z1318	1723
Z1320	1761
Z1322	1788
Z1324	1804
Z1326	0022
Z1328	0207
Z1330	0310
Z1332	0356
Z1334	0438
Z1336	0509
Z1338	0063
Z1340	0634
Z1528	0347
Z1530	0027
Z1532	1180
Z-1540	0157
Z1540	0253
Z-1545	0157
Z1545	0253
Z-1550	0157
Z1550	0253
Z-1555	0157
Z1555	0253
Z-1560	0157
Z1560	0253
Z-1565	0157
Z1565	0253
Z-1570	0157
Z1570	0253
Z2005	0442
Z2005A	0442
Z2005B	0442
Z2006	0876
Z2006A	0876
Z2006B	0876
Z2007	3127
Z2007A	3127
Z2007B	3127
Z2007U	0649
Z2008	0244
Z2008A	0185
Z2008B	0185
Z2008U	0244
Z2009	0205
Z2009A	0205
Z2009B	0205
Z2010	0170
Z2010A	0480
Z2010B	0480
Z2010U	0170
Z2011	0504
Z2011A	0504
Z2011B	0504
Z2012	0137
Z2012A	3317
Z2012B	3317
Z2012U	0137
Z2013	0684
Z2013A	0684
Z2013B	0684
Z2014	0225
Z2014A	0225
Z2014B	0225
Z2015	0002

If replacement code is OEM, contact original manufacturer for replacement.

DEVICE TYPE	REPL CODE
Z2015A	0230
Z2015B	0230
Z2015U	0002
Z2016	0234
Z2016A	0234
Z2016B	0234
Z2017	0237
Z2017A	0237
Z2017B	0237
Z2018	0100
Z2018A	1397
Z2018B	1397
Z2018U	0495
Z2019	0247
Z2019A	0247
Z2019B	0247
Z2020	0531
Z2020A	0251
Z2020B	0251
Z2021	1178
Z2021A	1178
Z2021B	1178
Z2022	0256
Z2022A	0256
Z2022B	0256
Z2022U	0565
Z2023	2395
Z2023A	2395
Z2023B	2395
Z2024	0262
Z2024A	0262
Z2024B	0262
Z2025	0269
Z2025A	0269
Z2025B	0269
Z2026	0274
Z2026A	0274
Z2026B	0274
Z2027	0441
Z2027A	2399
Z2027B	2399
Z2027U	0441
Z2028	0294
Z2028A	0294
Z2028B	0294
Z2029	1177
Z2029A	1177
Z2029B	1177
Z2030	0309
Z2030A	0309
Z2030B	0309
Z2031	0318
Z2031A	0318
Z2031B	0318
Z2033	0039
Z2033U	0039
Z2039	0350
Z2039U	0032
Z2047	0999
Z2047U	0999
Z2056	0869
Z2056U	0869
Z2068	2157
Z2068U	2157
Z2082	0331
Z2082U	0331
Z2100	0098
Z2100U	0098
Z2120	0186
Z2120U	0186
Z2150	0028
Z2150U	0028
Z2180	0367
Z2180U	0367
Z2200	0421
Z2200U	0421
Z2220	OEM
Z2220U	OEM
Z2250	OEM
Z2250U	OEM
Z2360	0102
Z2500	OEM
Z2501-6L	OEM
Z2502	OEM
Z2504	OEM
Z2506	OEM
Z2508	OEM
Z2510	OEM
Z2513	OEM
Z2514	OEM
Z2516	OEM
Z2519	OEM
Z2522	OEM
Z2525	0269
Z2526	OEM
Z2528	OEM
Z2530	0309
Z2531	OEM
Z2537	OEM
Z2542	OEM
Z2545	OEM
Z2547	OEM
Z2548	1218
Z2551	OEM
Z3305	0926
Z3307	0958
Z3309	1009
Z3311	0889
Z3314	0932

DEVICE TYPE	REPL CODE
Z3317	1106
Z3320	1152
Z3323	1184
Z3325	1223
Z3327	1246
Z3334	1331
Z3340	1423
Z3433	OEM
Z3434	0036
Z3500	OEM
Z3500(HEP)	0547
Z3500R(HEP)	2035
Z3502	OEM
Z3502(HEP)	2113
Z3502R(HEP)	1438
Z3504	OEM
Z3504(HEP)	1901
Z3504R(HEP)	1445
Z3505	OEM
Z3507	OEM
Z3511	OEM
Z3512	OEM
Z3512R(HEP)	0510
Z3514	OEM
Z3514R(HEP)	1263
Z3516	OEM
Z3516R(HEP)	1639
Z3518	OEM
Z3518R(HEP)	1716
Z3520	OEM
Z3520R(HEP)	0727
Z3522	OEM
Z3522R(HEP)	1782
Z3524	OEM
Z3524R(HEP)	1799
Z3530	OEM
Z3530R(HEP)	1884
Z3534	OEM
Z3534R(HEP)	1909
Z3536	OEM
Z3536R(HEP)	1924
Z4026	OEM
Z4729	0376
Z4729A	0376
Z-5140	0157
Z5140	0253
Z-5145	0157
Z5145	0253
Z-5150	0157
Z5150	0253
Z-5155	0157
Z5155	0253
Z-5160	0157
Z5160	0253
Z-5165	0157
Z5165	0253
Z-5170	0157
Z5170	0253
Z-5240	0157
Z5240	0253
Z-5245	0157
Z-5250	0157
Z5250	0253
Z-5255	0157
Z5255	0253
Z-5260	0157
Z5260	0253
Z-5265	0157
Z5265	0253
Z-5270	0157
Z5270	0253
Z-5540	0157
Z5540	0253
Z-5545	0157
Z5545	0253
Z-5550	0157
Z5550	0253
Z-5555	0157
Z5555	0253
Z-5560	0157
Z5560	0253
Z-5565	0157
Z5565	0253
Z-5570	0157
Z5570	0253
Z6007U	0480
Z6008	0504
Z6008U	0504
Z6010	0225
Z6010U	0225
Z6012	0234
Z6012U	0234
Z6015	0247
Z6015U	0247
Z6018	0256
Z6018U	0256
Z6022	0269
Z6022U	0269
Z6027	0294
Z6027U	0294
Z6033	0318
Z6033U	0318
Z6039	0326
Z6039U	0326
Z6047	0347
Z6047U	0347
Z6056	0266
Z6056U	0266
Z6068	0405

DEVICE TYPE	REPL CODE
Z6068U	0405
Z6082	0444
Z6082U	0444
Z6100	1180
Z6100U	1180
Z6120	1207
Z6120U	1207
Z6132-3CS	OEM
Z6132-3DS	OEM
Z6132-3PS	OEM
Z6132-4CS	OEM
Z6132-4DS	OEM
Z6132-4PS	OEM
Z6132-5CS	OEM
Z6132-5DS	OEM
Z6132-5PS	OEM
Z6132-6CS	OEM
Z6132-6DS	OEM
Z6132-6PS	OEM
Z6150	0647
Z6150U	0647
Z6180	1279
Z6180U	1279
Z6200	0605
Z6200U	0605
Z6220	OEM
Z6220U	OEM
Z6250	OEM
Z6250U	OEM
Z6300	OEM
Z6300U	OEM
Z6350	OEM
Z6350U	OEM
Z6400	OEM
Z6400U	OEM
Z6440	OEM
Z6440U	OEM
Z8000	OEM
Z8001ACE	OEM
Z8001ACM	OEM
Z8001ACMB	OEM
Z8001ACS	OEM
Z8001ADE	OEM
Z8001ADS	OEM
Z8001APE	OEM
Z8001APS	OEM
Z8001CE	OEM
Z8001CM	OEM
Z8001CMB	OEM
Z8001CS	OEM
Z8001DE	OEM
Z8001DS	OEM
Z8001PE	OEM
Z8001PS	OEM
Z8002ACE	OEM
Z8002ACM	OEM
Z8002ACMB	OEM
Z8002ACS	OEM
Z8002ADE	OEM
Z8002ADS	OEM
Z8002APE	OEM
Z8002APS	OEM
Z8002CE	OEM
Z8002CM	OEM
Z8002CMB	OEM
Z8002CS	OEM
Z8002DE	OEM
Z8002DS	OEM
Z8002PE	OEM
Z8002PS	OEM
Z8003	OEM
Z8004	OEM
Z8010ACE	OEM
Z8010ACM	OEM
Z8010ACMB	OEM
Z8010ACS	OEM
Z8010ADE	OEM
Z8010ADS	OEM
Z8010APE	OEM
Z8010APS	OEM
Z8010CE	OEM
Z8010CM	OEM
Z8010CMB	OEM
Z8010CS	OEM
Z8010DE	OEM
Z8010DS	OEM
Z8010PE	OEM
Z8010PS	OEM
Z8015	OEM
Z8030ACE	OEM
Z8030ACS	OEM
Z8030ADE	OEM
Z8030ADS	OEM
Z8030APE	OEM
Z8030APS	OEM
Z8030CS	OEM
Z8030DE	OEM
Z8030DS	OEM
Z8030PE	OEM
Z8030PS	OEM
Z8036ACE	OEM
Z8036ACS	OEM
Z8036ADE	OEM
Z8036ADS	OEM
Z8036APE	OEM
Z8036APS	OEM
Z8036CE	OEM
Z8036CS	OEM
Z8036DE	OEM

DEVICE TYPE	REPL CODE
Z8036DS	OEM
Z8036PE	OEM
Z8036PS	OEM
Z8038ACE	OEM
Z8038ACS	OEM
Z8038ADE	OEM
Z8038ADS	OEM
Z8038APE	OEM
Z8038APS	OEM
Z8038CE	OEM
Z8038CS	OEM
Z8038DE	OEM
Z8038DS	OEM
Z8038PE	OEM
Z8038PS	OEM
Z8052	OEM
Z8060	OEM
Z8068	OEM
Z8070	OEM
Z8090CE	OEM
Z8090CS	OEM
Z8090DE	OEM
Z8090DS	OEM
Z8090PE	OEM
Z8090PS	OEM
Z8091QS	OEM
Z8092QS	OEM
Z8093RS	OEM
Z8094RS	OEM
Z8400ACE	OEM
Z8400ACM	OEM
Z8400ACMB	OEM
Z8400ACS	OEM
Z8400ADE	OEM
Z8400ADS	OEM
Z8400APE	OEM
Z8400APS	3481
Z8400BCE	OEM
Z8400BCM	OEM
Z8400BCS	OEM
Z8400BDE	OEM
Z8400BDS	OEM
Z8400BPE	OEM
Z8400BPS	OEM
Z8400CE	OEM
Z8400CM	OEM
Z8400CMB	OEM
Z8400CS	OEM
Z8400DE	OEM
Z8400DS	OEM
Z8400PE	OEM
Z8400PS	OEM
Z8410ACE	OEM
Z8410ACM	OEM
Z8410ACMB	OEM
Z8410ACS	OEM
Z8410ADS	OEM
Z8410APE	OEM
Z8410APS	OEM
Z8410CE	OEM
Z8410CM	OEM
Z8410CMB	OEM
Z8410CS	OEM
Z8410DE	OEM
Z8410DS	OEM
Z8410PE	OEM
Z8410PS	OEM
Z8420ACE	OEM
Z8420ACM	OEM
Z8420ACMB	OEM
Z8420ACS	OEM
Z8420ADE	OEM
Z8420ADS	OEM
Z8420APE	OEM
Z8420APS	6253
Z8420BCE	OEM
Z8420BCM	OEM
Z8420BCMB	OEM
Z8420BCS	OEM
Z8420BDE	OEM
Z8420BDS	OEM
Z8420BPE	OEM
Z8420BPS	OEM
Z8420CE	OEM
Z8420CM	OEM
Z8420CMB	OEM
Z8420CS	OEM
Z8420DE	OEM
Z8420DS	OEM
Z8420PE	OEM
Z8420PS	OEM
Z8430AB1	7110
Z8430ACE	OEM
Z8430ACM	OEM
Z8430ACMB	OEM
Z8430ACS	OEM
Z8430ADE	OEM
Z8430ADS	OEM
Z8430APE	OEM
Z8430APS	7110
Z8430BCE	OEM
Z8430BCM	OEM
Z8430BCS	OEM
Z8430BDS	OEM
Z8430BPE	OEM
Z8430BPS	OEM
Z8430CE	OEM
Z8430CM	OEM

DEVICE TYPE	REPL CODE
Z8430CMB	OEM
Z8430CS	OEM
Z8430DE	OEM
Z8430DS	OEM
Z8430PE	OEM
Z8430PS	OEM
Z8440ACE	OEM
Z8440ACM	OEM
Z8440ACMB	OEM
Z8440ACS	OEM
Z8440ADE	OEM
Z8440ADS	OEM
Z8440APE	OEM
Z8440APS	OEM
Z8440BCE	OEM
Z8440BCM	OEM
Z8440BCMB	OEM
Z8440BCS	OEM
Z8440BDE	OEM
Z8440BDS	OEM
Z8440BPE	OEM
Z8440BPS	OEM
Z8440CE	OEM
Z8440CM	OEM
Z8440CMB	OEM
Z8440CS	OEM
Z8440DE	OEM
Z8440DS	OEM
Z8440PE	OEM
Z8440PS	OEM
Z8441ACE	OEM
Z8441ACM	OEM
Z8441ACMB	OEM
Z8441ACS	OEM
Z8441ADE	OEM
Z8441ADS	OEM
Z8441APE	OEM
Z8441APS	OEM
Z8441BCE	OEM
Z8441CE	OEM
Z8441CM	OEM
Z8441CS	OEM
Z8441DE	OEM
Z8441DS	OEM
Z8441PE	OEM
Z8441PS	OEM
Z8442ACE	OEM
Z8442ACM	OEM
Z8442ACS	OEM
Z8442ADE	OEM
Z8442ADS	OEM
Z8442APE	OEM
Z8442APS	OEM
Z8442BCE	OEM
Z8442BCM	OEM
Z8442BCMB	OEM
Z8442BCS	OEM
Z8442BDE	OEM
Z8442BDS	OEM
Z8442BPE	OEM
Z8442BPS	OEM
Z8442CE	OEM
Z8442CM	OEM
Z8442CMB	OEM
Z8442DE	OEM
Z8442DS	OEM
Z8442PE	OEM
Z8442PS	OEM
Z8449ACE	OEM
Z8449ACM	OEM
Z8449ACMB	OEM
Z8449ACS	OEM
Z8449ADE	OEM
Z8449ADS	OEM
Z8449APE	OEM
Z8449APS	OEM
Z8449BCE	OEM
Z8449BCM	OEM
Z8449BCMB	OEM
Z8449BCS	OEM
Z8449BDE	OEM
Z8449BDS	OEM
Z8449BPE	OEM
Z8449BPS	OEM
Z8449CE	OEM
Z8449CM	OEM
Z8449CMB	OEM
Z8449DE	OEM
Z8449DS	OEM
Z8449PE	OEM
Z8449PS	OEM
Z8470ACE	OEM
Z8470ACM	OEM
Z8470ACMB	OEM
Z8470ACS	OEM
Z8470ADE	OEM
Z8470ADS	OEM
Z8470APE	OEM
Z8470APS	OEM
Z8470BCE	OEM
Z8470BCM	OEM
Z8470BCS	OEM
Z8470BDE	OEM
Z8470BDS	OEM
Z8470BPE	OEM
Z8470BPS	OEM
Z8470CE	OEM
Z8470CM	OEM
Z8470CMB	OEM

DEVICE TYPE	REPL CODE
Z8470CS	OEM
Z8470DE	OEM
Z8470DS	OEM
Z8470PE	OEM
Z8470PS	OEM
Z8530ACE	OEM
Z8530ACS	OEM
Z8530ADE	OEM
Z8530ADS	OEM
Z8530APE	OEM
Z8530APS	OEM
Z8530CE	OEM
Z8530CS	OEM
Z8530PE	OEM
Z8530PS	OEM
Z8530SCCC	OEM
Z8530SCCP	OEM
Z8536ACE	OEM
Z8536ACS	OEM
Z8536ADE	OEM
Z8536ADS	OEM
Z8536APE	OEM
Z8536APS	OEM
Z8536CE	OEM
Z8536CS	OEM
Z8536DE	OEM
Z8536DS	OEM
Z8536PE	OEM
Z8536PS	OEM
Z8538ACE	OEM
Z8538ACS	OEM
Z8538ADE	OEM
Z8538ADS	OEM
Z8538APE	OEM
Z8538APS	OEM
Z8538CE	OEM
Z8538CS	OEM
Z8538DE	OEM
Z8538DS	OEM
Z8538PE	OEM
Z8538PS	OEM
Z8590CE	OEM
Z8590CS	OEM
Z8590DE	OEM
Z8590DS	OEM
Z8590PE	OEM
Z8590PS	OEM
Z8591QS	OEM
Z8592QS	OEM
Z8593RS	OEM
Z8594RS	OEM
Z8601CE	OEM
Z8601CS	OEM
Z8601DE	OEM
Z8601DS	OEM
Z8601PE	OEM
Z8601PS	OEM
Z8602QE	OEM
Z8602QS	OEM
Z8603RS	OEM
Z8611CE	OEM
Z8611CS	OEM
Z8611DE	OEM
Z8611DS	OEM
Z8611PE	OEM
Z8611PS	OEM
Z8611QE	OEM
Z8612QE	OEM
Z8612QS	OEM
Z8613RS	OEM
Z84008BCMB	OEM
Z86128	OEM
Z8612812PSC	OEM
ZA6.8	0025
ZA6.8A	0025
ZA6.8B	0025
ZA7.5	0649
ZA7.5A	0649
ZA7.5B	0649
ZA8.2	0244
ZA8.2A	0244
ZA8.2B	0244
ZA9.1	0012
ZA9.1A	0012
ZA9.1B	0012
ZA10	0170
ZA10A	0170
ZA10B	0170
ZA11	0794
ZA11A	0317
ZA11B	0317
ZA12	0137
ZA12A	0137
ZA12B	0137
ZA13	0365
ZA13A	0365
ZA13B	0365
ZA15	0002
ZA15A	0002
ZA-15B	0002
ZA15B	0002
ZA15V	0002
ZA16	0420
ZA16A	0420
ZA16B	0420
ZA18	0495
ZA18A	0495
ZA18B	0495
ZA20	0531

DEVICE TYPE	REPL CODE
ZA20A	0531
ZA20B	0531
ZA21A	0565
ZA21B	0565
ZA22	0565
ZA22A	0565
ZA22B	0565
ZA24	0402
ZA24A	0402
ZA24B	0565
ZA27	0441
ZA27A	0441
ZA27B	0441
ZA30	0726
ZA30A	0726
ZA30B	0726
ZA31A	0726
ZA33	0039
ZA33A	0039
ZA33B	0039
ZA36	0820
ZA36A	0820
ZA36B	0820
ZA39	0350
ZA39A	0350
ZA39B	0350
ZA43	0931
ZA43A	0931
ZA43B	0931
ZA47	0999
ZA47A	0999
ZA47B	0999
ZA51	0502
ZA51A	0502
ZA51B	0502
ZA56	1834
ZA56A	1834
ZA56B	0869
ZA62	0783
ZA62A	0783
ZA62B	0783
ZA68	2157
ZA68A	2157
ZA68B	1268
ZA75	1189
ZA75A	1189
ZA75B	1189
ZA82	0331
ZA82A	0331
ZA82B	0331
ZA91	1311
ZA91A	1311
ZA91B	1311
ZA100	0098
ZA100A	0098
ZA100B	0098
ZA110	0149
ZA110A	0149
ZA110B	0149
ZA120	0186
ZA120A	0186
ZA120B	0186
ZA130	0213
ZA130A	0213
ZA130B	0213
ZA150	0028
ZA150A	0028
ZA150B	0028
ZA160	0255
ZA160A	0255
ZA160B	0255
ZA180	0367
ZA180A	0367
ZA180B	0367
ZA200	0421
ZA200A	0421
ZA200B	0421
ZA1250NL	OEM
ZAC6.8	0205
ZAC6.8A	0205
ZAC6.8B	0205
ZAC7.5	0480
ZAC7.5A	0480
ZAC7.5B	0480
ZAC9.1	0684
ZAC9.1A	0684
ZAC9.1B	0684
ZAC10	0225
ZAC10A	0225
ZAC10B	0225
ZAC11	0230
ZAC11A	0230
ZAC11B	0230
ZAC12	0234
ZAC12A	0234
ZAC12B	0234
ZAC13	0237
ZAC13A	0237
ZAC13B	0237
ZAC15	0247
ZAC15A	0247
ZAC15B	0247
ZAC16	0251
ZAC16A	0251
ZAC16B	0251
ZAC18	0256
ZAC18A	0256
ZAC18B	0256
ZAC20	0262

DEVICE TYPE	REPL CODE
ZAC20A	0262
ZAC20B	0262
ZAC22	0269
ZAC22A	0269
ZAC22B	0269
ZAC24	0274
ZAC24A	0274
ZAC24B	0274
ZAC27	0294
ZAC27A	0294
ZAC27B	0294
ZAC30	0309
ZAC30A	0309
ZAC30B	0309
ZAC33	0318
ZAC33A	0318
ZAC33B	0318
ZAC36	0320
ZAC36A	0320
ZAC36B	0320
ZAC39	0326
ZAC39A	0326
ZAC39B	0326
ZAC43	0337
ZAC43A	0337
ZAC43B	0337
ZAC47	0347
ZAC47A	0347
ZAC47B	0347
ZAC51	0027
ZAC51A	0027
ZAC51B	0027
ZAC56	0266
ZAC56A	0266
ZAC56B	0266
ZAC62	0386
ZAC62A	0386
ZAC62B	0386
ZAC68	0405
ZAC68A	0405
ZAC68B	0405
ZAC75	0426
ZAC75A	0426
ZAC75B	0426
ZAC82	0444
ZAC82A	0444
ZAC82B	0444
ZAC91	0238
ZAC91A	0238
ZAC91B	0238
ZAC100	1180
ZAC100A	1180
ZAC100B	1180
ZAC110	1190
ZAC110A	1190
ZAC110B	1190
ZAC120	1207
ZAC120A	1207
ZAC120B	1207
ZAC130	1218
ZAC130A	1218
ZAC130B	1218
ZAC150	0647
ZAC150A	0647
ZAC150B	0647
ZAC160	1255
ZAC160A	1255
ZAC160B	1255
ZAC180	1279
ZAC180A	1279
ZAC180B	1279
ZAC200	0605
ZAC200A	0605
ZAC200B	0605
ZAM627-600	OEM
ZAR110	0922
ZAR210	0145
ZAR610	0922
ZAR710	0922
ZB-1	0794
ZB1	0531
ZB1-6	0301
ZB1-7	0025
ZB1-8	0244
ZB1-9	0012
ZB-1-9.5	0012
ZB1-10	0170
ZB1-10A	0170
ZB1-11	0794
ZB1-12	0137
ZB1-13	0365
ZB1-14	0100
ZB1-15	0002
ZB1-16	0420
ZB1-18	0495
ZB1-19	0949
ZB1-20	0531
ZB1-23	0402
ZB1-27	0441
ZB1-29	0441
ZB1-31	0726
ZB1-35	0039
ZB1-50	0502
ZB1-100	0098
ZB1-100-2	0098
ZB1-110	0149
ZB1-125	0213
ZB1-150	0028
ZB3.3	0188

DEVICE TYPE	REPL CODE
ZB3.3A	0292
ZB3.3B	0292
ZB3.6	0188
ZB3.6A	0188
ZB3.6B	0188
ZB3.9	0456
ZB3.9A	0456
ZB3.9B	0456
ZB4.3	0533
ZB4.3A	0533
ZB4.3B	0533
ZB4.7	0162
ZB4.7A	0451
ZB4.7B	0451
ZB5.1	0162
ZB5.1A	0162
ZB5.1B	0162
ZB5.6	0157
ZB5.6A	0157
ZB5.6B	0157
ZB6-18	OEM
ZB6C2704PSC-R425	OEM
ZB6.2	0636
ZB6.2A	0636
ZB6.2B	0636
ZB6.8	0025
ZB6.8A	0025
ZB6.8B	0025
ZB7.5	0649
ZB7.5A	0649
ZB7.5B	0649
ZB8.2	0244
ZB8.2A	0244
ZB8.2B	0244
ZB9.1	0012
ZB9.1A	0012
ZB9.1B	0012
ZB10	0170
ZB10A	0170
ZB10B	0170
ZB10X	0170
ZB11	0317
ZB11A	0317
ZB11B	0317
ZB12	0137
ZB12A	0137
ZB12B	0137
ZB13	0365
ZB13A	0002
ZB13B	0365
ZB15	0002
ZB15A	0002
ZB15B	0002
ZB16	0420
ZB16A	0420
ZB16B	0420
ZB18	0495
ZB18A	0495
ZB18B	0495
ZB20	0531
ZB20A	0531
ZB20B	0531
ZB22	0565
ZB22A	0565
ZB22B	0565
ZB24	0402
ZB24A	0402
ZB24B	0402
ZB26B	OEM
ZB27	0441
ZB27A	0441
ZB27B	0441
ZB30	0726
ZB30A	0726
ZB30B	0726
ZB33	0039
ZB33A	0039
ZB33B	0039
ZB36	0820
ZB36A	0820
ZB36B	0820
ZB39	0350
ZB39A	0350
ZB39B	0350
ZB43	0931
ZB43A	0931
ZB43B	0931
ZB47	0999
ZB47A	0999
ZB47B	0999
ZB51	0502
ZB51A	0502
ZB51B	0502
ZB56	0869
ZB56A	0869
ZB56B	0869
ZB62	0783
ZB62A	0783
ZB62B	0783
ZB68	2157
ZB68A	1268
ZB68B	2157
ZB75	1189
ZB75A	1189
ZB75B	1189
ZB82	0331
ZB82A	0331
ZB82B	0331
ZB91	1311

DEVICE TYPE	REPL CODE
ZB91A	1311
ZB91B	1311
ZB100	0098
ZB100A	0098
ZB100B	0098
ZB110	0149
ZB110A	0149
ZB110B	0149
ZB120	0186
ZB120A	0186
ZB120B	0186
ZB130	0213
ZB130A	0213
ZB130B	0213
ZB150	0028
ZB150A	0028
ZB150B	0028
ZB160	0255
ZB160A	0255
ZB160B	0255
ZB180	0367
ZB180A	0367
ZB180B	0367
ZB200	0421
ZB200A	0421
ZB200B	0421
ZB202	0760
ZB203	0188
ZB204	0533
ZB205	0157
ZB206	0636
ZB207	0649
ZB209	1083
ZB210	0170
ZB212	0137
ZB215	0244
ZB220	0531
ZB225	1606
ZB230	0726
ZB235	0820
ZB240	0931
ZB250	0502
ZB442ACMB	OEM
ZB442CS	OEM
ZBC-0128	OEM
ZBC-0128A	OEM
ZBC-0256	OEM
ZBC-0256A	OEM
ZBC-0384	OEM
ZBC-0384A	OEM
ZBC-0512	OEM
ZBC-0512A	OEM
ZBC3.3	0782
ZBC3.3A	0782
ZBC3.3B	0782
ZBC3.6	0796
ZBC3.6A	0796
ZBC3.6B	0796
ZBC3.9	0806
ZBC3.9A	0806
ZBC3.9B	0806
ZBC4.3	0821
ZBC4.3A	0821
ZBC4.3B	0821
ZBC4.7	0833
ZBC4.7A	0833
ZBC4.7B	0833
ZBC5.1	0442
ZBC5.1A	0442
ZBC5.1B	0442
ZBC5.6	0876
ZBC5.6A	0876
ZBC5.6B	0876
ZBC6.2	0185
ZBC6.2A	0185
ZBC6.2B	0185
ZBC6.8	0205
ZBC6.8A	0205
ZBC6.8B	0205
ZBC7.5	0480
ZBC7.5A	0480
ZBC7.5B	0480
ZBC8.2	0504
ZBC8.2A	0504
ZBC8.2B	0504
ZBC9.1	0684
ZBC9.1A	0684
ZBC9.1B	0684
ZBC10	0225
ZBC10A	0225
ZBC10B	0225
ZBC11	0230
ZBC11A	0230
ZBC11B	0230
ZBC12	0234
ZBC12A	0234
ZBC12B	0234
ZBC13	0237
ZBC13A	0237
ZBC13B	0237
ZBC14	1397
ZBC14A	1397
ZBC14B	1397
ZBC15	0247
ZBC15A	0247
ZBC15B	0247
ZBC17	1255
ZBC17A	1178
ZBC17B	1178

DEVICE TYPE	REPL CODE
ZBC18	0256
ZBC18A	0256
ZBC18B	0256
ZBC19	2395
ZBC19A	2395
ZBC19B	2395
ZBC20	0262
ZBC20A	0262
ZBC20B	0262
ZBC22	0269
ZBC22A	0269
ZBC22B	0269
ZBC24	0274
ZBC24A	0274
ZBC24B	0274
ZBC25	2399
ZBC25A	2399
ZBC25B	2399
ZBC27	0294
ZBC27A	0294
ZBC27B	0294
ZBC30	0309
ZBC30A	0309
ZBC30B	0309
ZBC33	0318
ZBC33A	0318
ZBC33B	0318
ZBC36	0320
ZBC36A	0320
ZBC36B	0320
ZBC39	0326
ZBC39A	0326
ZBC39B	0326
ZBC43	0337
ZBC43A	0337
ZBC43B	0337
ZBC45	0337
ZBC45A	0337
ZBC45B	0337
ZBC47	0347
ZBC47A	0347
ZBC47B	0347
ZBC50	0027
ZBC50A	0027
ZBC50B	0027
ZBC51	0027
ZBC51A	0027
ZBC51B	0027
ZBC52	0027
ZBC52A	0027
ZBC52B	0027
ZBC56	0266
ZBC56A	0266
ZBC56B	0266
ZBC62	0386
ZBC62A	0386
ZBC62B	0386
ZBC68	0405
ZBC68A	0405
ZBC68B	0405
ZBC75	0426
ZBC75A	0426
ZBC75B	0426
ZBC-80	OEM
ZBC82	0444
ZBC82A	0444
ZBC82B	0444
ZBC91	0238
ZBC91A	0238
ZBC91B	0238
ZBC100	1180
ZBC100A	1180
ZBC100B	1180
ZBC105	1190
ZBC105A	1190
ZBC105B	1190
ZBC110	1190
ZBC110A	1190
ZBC110B	1190
ZBC120	1207
ZBC120A	1207
ZBC120B	1207
ZBC130	1218
ZBC130A	1218
ZBC130B	1218
ZBC140	1881
ZBC140A	1881
ZBC140B	1881
ZBC150	0647
ZBC150A	0647
ZBC150B	0647
ZBC160	1255
ZBC160A	1255
ZBC160B	1255
ZBC180	1279
ZBC180A	1279
ZBC180B	1279
ZBC200	0605
ZBC200A	0605
ZBC200B	0605
ZBI-09	0012
ZBI-13	0157
ZBI-14	0002
ZBX218A	OEM
ZBX324	OEM
ZBX349	OEM
ZBX350	OEM
ZC008	0504
ZC009	0684

DEVICE TYPE	REPL CODE
ZC01	OEM
ZC02	OEM
ZC02-3	0441
ZC06	OEM
ZC010	0170
ZC011	0317
ZC012	0137
ZC013	0365
ZC015	0002
ZC018	0495
ZC020	0531
ZC020A	OEM
ZC020B	OEM
ZC020C	OEM
ZC021A-25A	OEM
ZC021B-25B	OEM
ZC021C-25C	OEM
ZC022	0565
ZC024	0402
ZC027	0441
ZC030	0726
ZC033	0039
ZC0181	OEM
ZC0226	OEM
ZC-0226A	OEM
ZC0226A	OEM
ZC0227	OEM
ZC0227A	OEM
ZC0229X	OEM
ZC-0231X	OEM
ZC0231X	OEM
ZC0233X	OEM
ZC-0235BX	OEM
ZC0235BX	OEM
ZC0239X	OEM
ZC0240X	OEM
ZC0241	OEM
ZC-0247A	OEM
ZC-0247A(NTSC)	OEM
ZC0247A-N	OEM
ZC-0248	OEM
ZC0248	OEM
ZC-0250	OEM
ZC0250	OEM
ZC-0253BT	OEM
ZC0256SE	OEM
ZC0261A	OEM
ZC0262	OEM
ZC0263	OEM
ZC0266X	OEM
ZC0300	OEM
ZC0310	OEM
ZC1N34A	0143
ZC1S352M	0030
ZC1S358S	0133
ZC2SA70	0050
ZC2SA71	0050
ZC2SA71A	0050
ZC2SA101	0050
ZC2SA101BA	0050
ZC2SA102	0050
ZC2SA102CA	0050
ZC2SA103	0050
ZC2SA103CA	0050
ZC2SA377	0050
ZC2SA700A	0050
ZC2SA700B	0050
ZC2SB172	0164
ZC2SB172A	0164
ZC3.3	0782
ZC3.3A	0782
ZC3.3B	0782
ZC3.6	0796
ZC3.6A	0796
ZC3.6B	0796
ZC3.9	0806
ZC3.9A	0806
ZC3.9B	0806
ZC4.3	0821
ZC4.3A	0821
ZC4.3B	0821
ZC4.7	0833
ZC4.7A	0833
ZC4.7B	0833
ZC5.1	0442
ZC5.1A	0442
ZC5.1B	0442
ZC5.6	0876
ZC5.6A	0876
ZC5.6B	0876
ZC6.2	0185
ZC6.2A	0185
ZC6.2B	0185
ZC6.8	0205
ZC6.8A	0205
ZC6.8B	0205
ZC7.5	0480
ZC7.5A	0480
ZC7.5B	0480
ZC8.2	0504
ZC8.2A	0504
ZC8.2B	0504
ZC9.1	0684
ZC9.1A	0684
ZC9.1B	0684
ZC10	0225
ZC10A	0098
ZC10B	0225
ZC11	0230

DEVICE TYPE	REPL CODE
ZC11A	0230
ZC11B	0230
ZC12	0234
ZC12A	0234
ZC12B	0234
ZC13	0237
ZC13A	0237
ZC13B	0237
ZC14	1397
ZC14A	1881
ZC14B	1397
ZC15	0247
ZC15A	0247
ZC15B	0247
ZC17	1178
ZC17A	1178
ZC17B	1178
ZC18	0256
ZC18A	0256
ZC18B	0256
ZC19	2395
ZC19A	2395
ZC19B	2395
ZC20	0262
ZC20A	0262
ZC20B	0531
ZC21	0269
ZC22	0269
ZC22A	0269
ZC22B	0269
ZC24	0274
ZC24A	0274
ZC24B	0274
ZC25	2399
ZC25A	2399
ZC25B	2399
ZC27	0294
ZC27A	0294
ZC27B	0294
ZC29BF	OEM
ZC29BG-BI	OEM
ZC29BJ	OEM
ZC29BK	OEM
ZC29BL	OEM
ZC29CG-CI	OEM
ZC29CJ	OEM
ZC29CK	OEM
ZC29CL	OEM
ZC29EF	OEM
ZC29EG-EI	OEM
ZC29EJ	OEM
ZC29EK	OEM
ZC29EL	OEM
ZC30	0309
ZC30A	0309
ZC30B	0309
ZC30BF	OEM
ZC30BG-BI	OEM
ZC30BJ	OEM
ZC30BK	OEM
ZC30BL	OEM
ZC30CF	OEM
ZC30CG-CI	OEM
ZC30CJ	OEM
ZC30CK	OEM
ZC30CL	OEM
ZC30EF	OEM
ZC30EG-EI	OEM
ZC30EJ	OEM
ZC30EK	OEM
ZC30EL	OEM
ZC31BE-BH	OEM
ZC31CE-CH	OEM
ZC31EE-EH	OEM
ZC32B	OEM
ZC32BD-BG	OEM
ZC32C	OEM
ZC32CD-CG	OEM
ZC32ED-EG	OEM
ZC33	0318
ZC33A	0318
ZC33B	0318
ZC36	0320
ZC36A	0320
ZC36B	0320
ZC39	0326
ZC39A	0326
ZC39B	0326
ZC43	0337
ZC43A	0337
ZC43B	0337
ZC45	0347
ZC45A	0347
ZC45B	0347
ZC47	0347
ZC47A	0347
ZC47B	0347
ZC50	0027
ZC50A	0027
ZC50B	0027
ZC50BB	OEM
ZC50BC-BE	OEM
ZC50BF	OEM
ZC50BG	OEM
ZC50BH	OEM
ZC50CB	OEM
ZC50CC-CE	OEM
ZC50CF	OEM
ZC50CG	OEM

DEVICE TYPE	REPL CODE
ZC50CH	OEM
ZC50EC-EE	OEM
ZC50EF	OEM
ZC50EG	OEM
ZC50EH	OEM
ZC51	0027
ZC51A	0027
ZC51B	0027
ZC51BB	OEM
ZC51BC-BG	OEM
ZC51C	OEM
ZC51CB	OEM
ZC51CC-CG	OEM
ZC51EB	OEM
ZC51EC-EG	OEM
ZC52	0027
ZC52A	0027
ZC52B	0027
ZC52BB	OEM
ZC52BC-BF	OEM
ZC52C	OEM
ZC52CB	OEM
ZC52CC-CF	OEM
ZC52EB	OEM
ZC52EC-EF	OEM
ZC53B	OEM
ZC53BB	OEM
ZC53BC-BE	OEM
ZC53C	OEM
ZC53CB	OEM
ZC53CC-CE	OEM
ZC53EB	OEM
ZC53EC-EE	OEM
ZC54B	OEM
ZC54BB	OEM
ZC54BC	OEM
ZC54BD	OEM
ZC54C	OEM
ZC54CB	OEM
ZC54CC	OEM
ZC54CD	OEM
ZC54EB	OEM
ZC54EC	OEM
ZC54ED	OEM
ZC55PB	OEM
ZC56	0266
ZC56A	0266
ZC56B	0266
ZC62	0386
ZC62A	0386
ZC62B	0386
ZC68	0405
ZC68A	0405
ZC68B	0405
ZC70BB	OEM
ZC70BC-BE	OEM
ZC70BF	OEM
ZC70BG	OEM
ZC70BH	OEM
ZC70CB	OEM
ZC70CC-CE	OEM
ZC70CF	OEM
ZC70CG	OEM
ZC70CH	OEM
ZC70EB	OEM
ZC70EC-EE	OEM
ZC70EF	OEM
ZC70EG	OEM
ZC70EH	OEM
ZC71BB	OEM
ZC71BC-BG	OEM
ZC71CB	OEM
ZC71CC-CG	OEM
ZC71EB	OEM
ZC71EC-EG	OEM
ZC72B	OEM
ZC72BB	OEM
ZC72BC-BF	OEM
ZC72C	OEM
ZC72CB	OEM
ZC72CC-CF	OEM
ZC72EB	OEM
ZC72EC-EF	OEM
ZC73B	OEM
ZC73BB	OEM
ZC73BC-BE	OEM
ZC73C	OEM
ZC73CB	OEM
ZC73CC-CE	OEM
ZC73EB	OEM
ZC73EC-EE	OEM
ZC74B	OEM
ZC74BB	OEM
ZC74BC	OEM
ZC74BD	OEM
ZC74C	OEM
ZC74CB	OEM
ZC74CC	OEM
ZC74CD	OEM
ZC74EB	OEM
ZC74EC	OEM
ZC74ED	OEM
ZC75	0426
ZC75A	0426
ZC75B	0426
ZC82	0444
ZC82A	0444
ZC82B	0444
ZC91	0238

DEVICE TYPE	REPL CODE
ZC91A	0238
ZC91B	0238
ZC99-50	OEM
ZC100	1180
ZC100A	1180
ZC100B	1180
ZC101	OEM
ZC102	1312
ZC103	0188
ZC104	0533
ZC105	0157
ZC105A	1190
ZC105B	1190
ZC106	0636
ZC107	0649
ZC109	1083
ZC110	0170
ZC110A	1190
ZC110B	1190
ZC111	OEM
ZC112	0137
ZC112BI	OEM
ZC112BJ	OEM
ZC112BK	OEM
ZC112EI	OEM
ZC112EJ	OEM
ZC112EK	OEM
ZC115	0002
ZC120	1207
ZC120A	1207
ZC120B	1207
ZC125	1606
ZC130	0726
ZC130A	1218
ZC130B	1218
ZC130T	0726
ZC135	0820
ZC-140	0010
ZC140	0931
ZC-140(TRANS)	0042
ZC140(ZENER)	0931
ZC140A	1881
ZC140B	1881
ZC150	0502
ZC150A	0647
ZC150B	0782
ZC151	OEM
ZC160	0783
ZC160A	1255
ZC160B	0504
ZC170	2157
ZC180	0331
ZC180A	1279
ZC180B	1279
ZC200	0605
ZC200A	0605
ZC200B	0605
ZC201	OEM
ZC202	OEM
ZC209	OEM
ZC240	OEM
ZC620	OEM
ZC700	0554
ZC700B	OEM
ZC700C	OEM
ZC701	OEM
ZC701B	OEM
ZC701C	OEM
ZC702	0005
ZC702B	OEM
ZC702C	OEM
ZC703	0720
ZC703B	OEM
ZC703C	OEM
ZC704	OEM
ZC704B	OEM
ZC704C	OEM
ZC705	OEM
ZC705B	OEM
ZC705C	OEM
ZC706	OEM
ZC706B	OEM
ZC706C	OEM
ZC707	OEM
ZC707B	OEM
ZC707C	OEM
ZC708	0165
ZC708B	OEM
ZC708C	OEM
ZC709	0057
ZC709B	OEM
ZC709C	OEM
ZC710	0157
ZC710B	OEM
ZC710C	OEM
ZC711	0181
ZC711B	OEM
ZC711C	OEM
ZC712	0052
ZC712B	OEM
ZC712C	OEM
ZC713	0053
ZC713B	OEM
ZC713C	OEM
ZC714	OEM
ZC714B	OEM
ZC714C	OEM
ZC715	0686
ZC716	0445

DEVICE TYPE	REPL CODE
ZC718	0375
ZC720	0700
ZC721	OEM
ZC722	0705
ZC723	OEM
ZC724	0494
ZC725	0714
ZC726	OEM
ZC727	0455
ZC728	0244
ZC729	OEM
ZC730	0195
ZC731	OEM
ZC732	OEM
ZC733	0166
ZC736	0010
ZC739	0032
ZC740	OEM
ZC741	OEM
ZC742	OEM
ZC743	0054
ZC744	OEM
ZC745	OEM
ZC746	OEM
ZC747	0068
ZC748	OEM
ZC749	OEM
ZC751	0092
ZC753	OEM
ZC754	OEM
ZC756	0125
ZC762	0152
ZC768	0173
ZC775	OEM
ZC782	0049
ZC791	0156
ZC799	0189
ZC800	0005
ZC800A	0005
ZC800B	OEM
ZC801	OEM
ZC801A	OEM
ZC801B	OEM
ZC802	0628
ZC802A	0628
ZC802B	OEM
ZC803	OEM
ZC803A	OEM
ZC803B	OEM
ZC804	OEM
ZC804A	OEM
ZC804B	OEM
ZC805	OEM
ZC805A	OEM
ZC805B	OEM
ZC806	OEM
ZC806A	OEM
ZC806B	OEM
ZC807	OEM
ZC807A	OEM
ZC807B	OEM
ZC808	OEM
ZC808A	OEM
ZC808B	OEM
ZC809	OEM
ZC809A	OEM
ZC809B	OEM
ZC810	OEM
ZC811	OEM
ZC812	OEM
ZC813	OEM
ZC814	OEM
ZC815	OEM
ZC816	OEM
ZC820	OEM
ZC821	OEM
ZC822	OEM
ZC823	OEM
ZC824	OEM
ZC825	OEM
ZC826	OEM
ZC830A	OEM
ZC830B	OEM
ZC831A	OEM
ZC831B	OEM
ZC833A	OEM
ZC833B	OEM
ZC834A	OEM
ZC834B	OEM
ZC835A	OEM
ZC835B	OEM
ZC836A	OEM
ZC836B	OEM
ZC898	OEM
ZC899	OEM
ZC2008	OEM
ZC2009	OEM
ZC2010	0225
ZC2011	0230
ZC2012	0234
ZC2013	0237
ZC2015	0247
ZC2016	0251
ZC2018	0256
ZC2020	0262
ZC2022	0269
ZC2024	0274
ZC2027	0294
ZC2030	0309

DEVICE TYPE	REPL CODE
ZC2033	0318
ZC2800	OEM
ZC2800A	0080
ZC2800B	OEM
ZC2800E	OEM
ZC2800H	OEM
ZC2810	OEM
ZC2810B	OEM
ZC2810E	OEM
ZC2810H	OEM
ZC2811	OEM
ZC2811B	OEM
ZC2811E	OEM
ZC2811H	OEM
ZC2833B	OEM
ZC5008	1490
ZC5009	1618
ZC5010	0510
ZC5011	0691
ZC5012	0870
ZC5013	1020
ZC5015	1274
ZC5016	1402
ZC5018	1078
ZC5020	1717
ZC5022	1723
ZC5024	1736
ZC5027	1761
ZC5030	1772
ZC5033	1788
ZC5036	1796
ZC5039	1804
ZC5043	1194
ZC5047	0022
ZC5051	0132
ZC5056	0207
ZC5062	0263
ZC5068	0310
ZC5075	0329
ZC5082	0356
ZC5091	0412
ZC5099	0438
ZC5800	OEM
ZC5800E	OEM
ZC5800H	OEM
ZCC3.3	0782
ZCC3.3A	0782
ZCC3.3B	0782
ZCC3.6	0796
ZCC3.6A	0796
ZCC3.6B	0796
ZCC3.9	0806
ZCC3.9A	0806
ZCC3.9B	0806
ZCC4.3	0821
ZCC4.3A	0821
ZCC4.3B	0821
ZCC4.7	0833
ZCC4.7A	0833
ZCC4.7B	0833
ZCC5.1	0442
ZCC5.1A	0442
ZCC5.1B	0442
ZCC5.6	0876
ZCC5.6A	0876
ZCC5.6B	0876
ZCC6.2	0185
ZCC6.2A	0185
ZCC6.2B	0185
ZCC6.8	0205
ZCC6.8A	0205
ZCC6.8B	0205
ZCC7.5	0480
ZCC7.5A	0480
ZCC7.5B	0480
ZCC8.2	0504
ZCC8.2A	0504
ZCC8.2B	0504
ZCC9.1	0684
ZCC9.1A	0684
ZCC9.1B	0684
ZCC10	0225
ZCC10A	0225
ZCC10B	0225
ZCC11	0230
ZCC11A	0230
ZCC11B	0230
ZCC12	0234
ZCC12A	0234
ZCC12B	0234
ZCC13	0237
ZCC13A	0237
ZCC13B	0237
ZCC14	1397
ZCC14A	1397
ZCC14B	1397
ZCC15	0247
ZCC15A	0247
ZCC15B	1881
ZCC17	1178
ZCC17A	1178
ZCC17B	1178
ZCC18	0256
ZCC18A	0256
ZCC18B	0256
ZCC19A	2395
ZCC19B	2395
ZCC20	0262
ZCC20A	0262

DEVICE TYPE	REPL CODE
ZCC20B	0262
ZCC22	0269
ZCC22A	0269
ZCC22B	0269
ZCC24	0274
ZCC24A	0274
ZCC24B	0274
ZCC25	2399
ZCC25A	2399
ZCC25B	2399
ZCC27	0294
ZCC27A	0294
ZCC27B	0294
ZCC30	0309
ZCC30A	0309
ZCC30B	0309
ZCC33	0318
ZCC33A	0318
ZCC33B	0318
ZCC36	0320
ZCC36A	0320
ZCC36B	0320
ZCC39	0326
ZCC39A	0326
ZCC39B	0326
ZCC43	0337
ZCC43A	0337
ZCC43B	0337
ZCC45	0347
ZCC45A	0347
ZCC45B	0347
ZCC47	0347
ZCC47A	0347
ZCC47B	0347
ZCC50	0027
ZCC50A	0027
ZCC50B	0027
ZCC51	0027
ZCC51A	0027
ZCC51B	0027
ZCC52	0027
ZCC52A	0027
ZCC52B	0027
ZCC56	0266
ZCC56A	0266
ZCC56B	0266
ZCC62	0386
ZCC62A	0386
ZCC62B	0386
ZCC68	0405
ZCC68A	0405
ZCC68B	0405
ZCC75	0426
ZCC75A	0426
ZCC75B	0426
ZCC82	0444
ZCC82A	0444
ZCC82B	0444
ZCC91	0238
ZCC91A	0238
ZCC91B	0238
ZCC100	1180
ZCC100A	1180
ZCC100B	1180
ZCC105	1190
ZCC105A	1190
ZCC105B	1190
ZCC110	1190
ZCC110A	1190
ZCC110B	1190
ZCC120	1207
ZCC120A	1207
ZCC120B	1207
ZCC130	1218
ZCC130A	1218
ZCC130B	6857
ZCC140	1881
ZCC140A	1881
ZCC140B	1881
ZCC150	0647
ZCC150A	0647
ZCC150B	0647
ZCC160	1255
ZCC160A	1255
ZCC160B	1255
ZCC180	1279
ZCC180A	1279
ZCC180B	1279
ZCC200	0605
ZCC200A	0605
ZCC200B	0605
ZCO12	0137
ZCOM3679	0015
ZCOM-5683-0	0015
ZD008	0244
ZD009	1083
ZD010	0064
ZD011	0317
ZD012	0052
ZD013	0365
ZD-015	0002
ZD015	0686
ZD016	0420
ZD018	0375
ZD020	0531
ZD022	0705
ZD024	0402
ZD027	0455
ZD030	0726

DEVICE TYPE	REPL CODE
ZD033	0166
ZD039	0032
ZD047	0068
ZD056	0125
ZD068	0173
ZD2.7EB1	0760
ZD3.6	OEM
ZD3.6A	0376
ZD3.6B	0188
ZD3.9	0456
ZD4.3	0533
ZD4.7	0833
ZD-5V	0162
ZD5.1	0162
ZD5.1A	0162
ZD5.1B	0442
ZD5.6	0157
ZD5.6A	0157
ZD5.6B	0157
ZD6.2	0636
ZD6.2A	0636
ZD6.2B	0636
ZD-6.2V	0471
ZD6.8	0205
ZD6.8(STUD)	2740
ZD6.8A	2740
ZD6.8B	2740
ZD6.8R	0926
ZD6.8RA	0926
ZD6.8RB	0926
ZD7.5	0480
ZD7.5(STUD)	2744
ZD7.5A	2744
ZD7.5B	2744
ZD7.5R	0944
ZD7.5RA	0944
ZD7.5RB	0944
ZD8.2	0504
ZD8.2(STUD)	1225
ZD8.2A	1225
ZD8.2B	1225
ZD8.2R	0958
ZD8.2RA	0958
ZD8.2RB	0958
ZD9.1	0684
ZD9.1(STUD)	1237
ZD9.1A	1237
ZD9.1B	1237
ZD9.1R	0994
ZD9.1RA	0994
ZD9.1RB	0994
ZD10	0225
ZD10(STUD)	1252
ZD10A	1252
ZD10B	1252
ZD10R	1009
ZD10RA	1009
ZD10RB	1009
ZD11	0230
ZD11(STUD)	1269
ZD11A	1269
ZD11B	1269
ZD11R	1019
ZD11RA	1019
ZD11RB	1019
ZD12	0234
ZD12(C)	0052
ZD12(STUD)	1277
ZD12A	1277
ZD12B	1277
ZD12C	0052
ZD12R	1772
ZD12RA	0889
ZD12RB	0889
ZD13	0237
ZD13(STUD)	1294
ZD13A	1294
ZD13B	1294
ZD13R	1049
ZD13RA	1049
ZD13RB	1049
ZD14(STUD)	2752
ZD14A	2752
ZD14B	2752
ZD14R	1058
ZD14RA	1058
ZD14RB	1058
ZD-14V	0100
ZD15	0247
ZD15(STUD)	1302
ZD15A	1302
ZD15B	1302
ZD15R	0932
ZD15RA	0932
ZD15RB	0932
ZD16	0251
ZD16(STUD)	1310
ZD16A	1310
ZD16B	1310
ZD16R	1079
ZD16RA	1079
ZD16RB	1079
ZD17(STUD)	2756
ZD17A	2756
ZD17B	2756
ZD17R	1096
ZD17RA	1096
ZD17RB	1096
ZD18	0256

DEVICE TYPE	REPL CODE
ZD18(STUD)	1324
ZD18A	1324
ZD18B	1324
ZD18R	1106
ZD18RA	1106
ZD18RB	1106
ZD19(STUD)	2758
ZD19A	2758
ZD19B	2758
ZD19R	1123
ZD19RA	1123
ZD19RB	1123
ZD20	0262
ZD20(STUD)	1333
ZD20A	1333
ZD20B	1333
ZD20R	1135
ZD20RA	1135
ZD20RB	1135
ZD22	0269
ZD22(STUD)	1344
ZD22A	1344
ZD22B	1344
ZD22R	1152
ZD22RA	1152
ZD22RB	1152
ZD24	0274
ZD24(STUD)	1356
ZD24-1L	0402
ZD24A	1356
ZD24B	1356
ZD24R	1164
ZD24RA	1164
ZD24RB	1164
ZD25(STUD)	2760
ZD25A	2760
ZD25B	2760
ZD25R	1174
ZD25RA	1174
ZD25RB	1174
ZD27	0294
ZD27(STUD)	1371
ZD27A	1371
ZD27B	1371
ZD27R	1184
ZD27RA	1184
ZD27RB	1200
ZD30	0309
ZD30(STUD)	1387
ZD30A	1387
ZD30B	1387
ZD30R	1210
ZD30RA	1210
ZD30RB	1210
ZD33	0318
ZD33(STUD)	1406
ZD33A	1406
ZD33B	1406
ZD33R	1223
ZD33RA	1223
ZD33RB	1223
ZD36	0320
ZD36(STUD)	1415
ZD36A	1415
ZD36B	1415
ZD36R	1232
ZD36RA	1232
ZD36RB	1232
ZD39	0326
ZD39(STUD)	1428
ZD39A	1428
ZD39B	1428
ZD39R	1246
ZD39RA	1246
ZD39RB	1246
ZD43	0337
ZD43(STUD)	1440
ZD43A	1440
ZD43B	1440
ZD43R	1265
ZD43RA	1265
ZD43RB	1265
ZD45(STUD)	2772
ZD45A	2772
ZD45B	2772
ZD45R	1273
ZD45RA	1273
ZD45RB	1273
ZD47	0347
ZD47(STUD)	1447
ZD47A	1447
ZD47B	1447
ZD47R	1290
ZD47RA	1290
ZD47RB	1290
ZD50(STUD)	2776
ZD50A	2776
ZD50B	2776
ZD50R	1299
ZD50RA	1299
ZD50RB	1299
ZD51	0027
ZD51(STUD)	1461
ZD51A	1461
ZD51B	1461
ZD51R	1307
ZD51RA	1307
ZD51RB	1307
ZD52(STUD)	2779

DEVICE TYPE	REPL CODE
ZD52A	2779
ZD52B	2779
ZD52R	1322
ZD52RA	1322
ZD52RB	1322
ZD56	0266
ZD56(STUD)	1474
ZD56A	1474
ZD56B	1474
ZD56R	1331
ZD56RA	1331
ZD56RB	1273
ZD62	0386
ZD62(STUD)	0613
ZD62A	0613
ZD62B	0613
ZD62R	1353
ZD62RA	1353
ZD62RB	1353
ZD68	0405
ZD68(STUD)	1510
ZD68A	1510
ZD68B	1510
ZD68R	1365
ZD68RA	1365
ZD68RB	1365
ZD75	0426
ZD75(STUD)	1523
ZD75A	1523
ZD75B	1523
ZD75R	1384
ZD75RA	1384
ZD75RB	1384
ZD82	0444
ZD82(STUD)	1537
ZD82A	1537
ZD82B	1537
ZD82R	1401
ZD82RA	1401
ZD82RB	1401
ZD91	0238
ZD91(STUD)	1550
ZD91A	1550
ZD91B	1550
ZD91R	1412
ZD91RA	1412
ZD91RB	1412
ZD100	1180
ZD100(STUD)	1575
ZD100A	1575
ZD100B	1575
ZD100R	1423
ZD100RA	1423
ZD100RB	1423
ZD105(STUD)	2802
ZD105A	2802
ZD105B	2802
ZD105R	1436
ZD105RA	1436
ZD105RB	1436
ZD110	1190
ZD110(STUD)	2803
ZD110A	2803
ZD110B	2803
ZD110R	1444
ZD110RA	1444
ZD110RB	1444
ZD120	1207
ZD120(STUD)	1507
ZD120A	1507
ZD120B	1507
ZD120R	1457
ZD120RA	1470
ZD120RB	1457
ZD130	1218
ZD130(STUD)	2805
ZD130A	2805
ZD130B	2805
ZD130R	1470
ZD130RA	1470
ZD130RB	1470
ZD140(STUD)	2807
ZD140A	2807
ZD140B	2807
ZD140R	1484
ZD140RA	1484
ZD140RB	1484
ZD150	0647
ZD150(STUD)	2810
ZD150A	2810
ZD150B	2810
ZD150R	1505
ZD150RA	1505
ZD150RB	1505
ZD160	1255
ZD160(STUD)	2815
ZD160A	2815
ZD160B	2815
ZD160R	1521
ZD160RA	1521
ZD160RB	1521
ZD175(STUD)	2818
ZD175A	2818
ZD175B	2818
ZD175R	1531
ZD175RA	1531
ZD175RB	1531
ZD180	1279
ZD180(STUD)	2820

If replacement code is OEM, contact original manufacturer for replacement.

DEVICE TYPE	REPL CODE
ZD180A	2820
ZD180B	2820
ZD180R	1548
ZD180RA	1548
ZD180RB	1548
ZD200	0605
ZD200(STUD)	2824
ZD200A	2824
ZD200B	2824
ZD200R	1587
ZD200RA	1587
ZD200RB	1587
ZD220(STUD)	OEM
ZD220A	OEM
ZD220B	OEM
ZD220R	OEM
ZD220RA	OEM
ZD220RB	OEM
ZD2008	0504
ZD2009	0684
ZD2010	0225
ZD2011	0230
ZD2012	0234
ZD2013	0237
ZD2015	0247
ZD2016	0251
ZD2018	0256
ZD2020	0262
ZD2022	0269
ZD2024	0274
ZD2027	0294
ZD2028	0269
ZD2030	0309
ZD2033	0318
ZD5008	OEM
ZD5009	OEM
ZD5010	OEM
ZD5011	OEM
ZD5012	OEM
ZD5013	OEM
ZD5015	OEM
ZD5016	OEM
ZD5018	OEM
ZD5020	OEM
ZD5022	OEM
ZD5024	OEM
ZD5027	OEM
ZD5030	OEM
ZD5033	OEM
ZD5036	OEM
ZD5039	OEM
ZD5043	OEM
ZD5047	OEM
ZD5051	OEM
ZD5056	OEM
ZD5062	OEM
ZD5068	OEM
ZD5075	OEM
ZD5082	OEM
ZD5091	OEM
ZD5099	OEM
ZDC7	OEM
ZDDGGZA10Y	0064
ZDP12	OEM
ZDRD6.8ENZ	0062
ZDT10	0144
ZDT11	0144
ZDT20	0144
ZDT21	0144
ZDT30	0144
ZDT31	0144
ZDT40	OEM
ZDT41	OEM
ZDX1F	OEM
ZDX1R	OEM
ZDX2F	OEM
ZDX2R	OEM
ZDX3F	OEM
ZDX3R	OEM
ZDX4F	OEM
ZDX4R	OEM
ZDX5	OEM
ZDX6	OEM
ZE1.5	0143
ZE4.7	0140
ZE5.6	0157
ZE6V9	0062
ZE6.8	0025
ZE6.8A	0025
ZE6.8B	0025
ZE7.5	0649
ZE7.5A	0649
ZE7.5B	0649
ZE8.2	0244
ZE8.2A	0244
ZE8.2B	0244
ZE9V4	0057
ZE9.1	0012
ZE9.1A	0012
ZE9.1B	0012
ZE10	0170
ZE10A	0170
ZE10B	0170
ZE11	0317
ZE11A	0317
ZE11B	0317
ZE12	0137
ZE12A	0137
ZE12B	0137
ZE12V7	0053
ZE13	0365
ZE13A	0137
ZE13B	0365
ZE15	0002
ZE15A	0002
ZE15B	0002
ZE16	0420
ZE16A	0420
ZE16B	0420
ZE17V2	0210
ZE18	0495
ZE18A	0495
ZE18B	0495
ZE20	0531
ZE20A	0531
ZE20B	0531
ZE22	0565
ZE22A	0565
ZE22B	0565
ZE23V2	0494
ZE24	0402
ZE24A	0402
ZE24B	0402
ZE27	0441
ZE27A	0441
ZE27B	0441
ZE30	0726
ZE30A	0726
ZE30B	0726
ZE31V	0195
ZE33	0039
ZE33A	0039
ZE33B	0039
ZE36	0820
ZE36A	0820
ZE36B	0820
ZE39	0350
ZE39A	0350
ZE39B	0350
ZE43	0931
ZE43A	0931
ZE43B	0931
ZE47	0999
ZE47A	0999
ZE47B	0999
ZE51	0502
ZE51A	0502
ZE51B	0502
ZE56	0869
ZE56A	0869
ZE56B	0869
ZE62	0783
ZE62A	0783
ZE62B	0783
ZE68	2157
ZE68A	2157
ZE68B	2157
ZE75	1189
ZE75A	1189
ZE75B	1189
ZE82	0331
ZE82A	0331
ZE82B	0331
ZE91	1311
ZE91A	1311
ZE91B	1311
ZE100	0098
ZE100A	0098
ZE100B	0098
ZE110	0149
ZE110A	0149
ZE110B	0149
ZE120	0186
ZE120A	0186
ZE120B	0186
ZE130	0213
ZE130A	0213
ZE130B	0213
ZE150	0028
ZE150A	0028
ZE150B	0028
ZE160	0255
ZE160A	0255
ZE160B	0255
ZE180	0367
ZE180A	0367
ZE180B	0367
ZE200	0421
ZE200A	0421
ZE200B	0421
ZEC3.9	0456
ZEC4.7	0451
ZEC5.6	0157
ZEC6.8	0025
ZEC8.2	0244
ZEC10	0170
ZEC12	0137
ZEC15	0002
ZEC18	0495
ZEC22	0565
ZEC27	0441
ZEC30	0726
ZEN-100	0079
ZEN100	0079
ZEN-101	0126
ZEN101	0126
ZEN-102	0079
ZEN102	0079
ZEN103	0016
ZEN-104	0144
ZEN104	0144
ZEN105	1144
ZEN-106	0532
ZEN106	0532
ZEN107	0037
ZEN108	0224
ZEN-109	0224
ZEN109	0224
ZEN-110	0079
ZEN110	0079
ZEN-111	0016
ZEN111	0016
ZEN-112	0016
ZEN112	0016
ZEN-113	0111
ZEN113	0111
ZEN-114	0016
ZEN114	0016
ZEN-115	0016
ZEN115	0016
ZEN-116	0111
ZEN116	0111
ZEN-117	0127
ZEN117	0127
ZEN-118	0127
ZEN118	0127
ZEN-119	0079
ZEN119	0079
ZEN-120	0079
ZEN120	0079
ZEN121	0079
ZEN-122	0037
ZEN122	0037
ZEN123	0325
ZEN124	0212
ZEN125	0630
ZEN-126	0079
ZEN126	0079
ZEN-127	0079
ZEN127	0079
ZEN-128	0400
ZEN128	0400
ZEN129	1175
ZEN200	0142
ZEN201	0276
ZEN202	0161
ZEN203	0460
ZEN-204	0074
ZEN204	0074
ZEN-205	0493
ZEN205	0493
ZEN206	0642
ZEN207	0630
ZEN208	0338
ZEN-209	0561
ZEN209	0561
ZEN210	0161
ZEN-211	0460
ZEN211	0460
ZEN-300	0086
ZEN300	0280
ZEN-301	0050
ZEN301	0050
ZEN302	0211
ZEN-303	0004
ZEN303	0004
ZEN-304	0164
ZEN304	0164
ZEN-305	0208
ZEN305	0208
ZEN306	0211
ZEN-307	0164
ZEN307	0164
ZEN308	0004
ZEN-309	0211
ZEN309	0211
ZEN310	0004
ZEN-311	0136
ZEN311	0136
ZEN-312	0136
ZEN312	0136
ZEN-313	0136
ZEN313	0136
ZEN-314	0004
ZEN314	0004
ZEN-315	0600
ZEN315	0600
ZEN325	0085
ZEN-326	0160
ZEN326	0160
ZEN327	0440
ZEN328	0975
ZEN-329	1424
ZEN329	1424
ZEN330	0160
ZEN331	0085
ZEN401	0087
ZEN403	0109
ZEN430	0143
ZEN431	0917
ZEN432	0196
ZEN433	0289
ZEN434	0289
ZEN450	0554
ZEN451	0005
ZEN452	0720
ZEN453	0030
ZEN500	0471
ZEN501	0649
ZEN502	0052
ZEN503	0375
ZEN504	0455
ZEN505	0649
ZEN506	0365
ZEN507	0100
ZEN508	0002
ZEN512	0039
ZEN513	0213
ZEN514	0245
ZEN515	0820
ZEN517	0309
ZEN600	OEM
ZEN601	0632
ZEN602	0701
ZEN603	0666
ZEN604	0664
ZEN605	1750
ZEN606	0753
ZEN607	0352
ZEN608	0354
ZEN609	1753
ZEN610	1754
ZENER7-1V	0649
ZENER-132	1606
ZENER-136	0949
ZF2.7	0760
ZF2.7P	0760
ZF3.3	0299
ZF3.6	0188
ZF3.6P	0376
ZF3.9	0188
ZF3.9P	0036
ZF4.3	0533
ZF4.3P	0275
ZF4.7	0140
ZF4.7P	0140
ZF5A	0157
ZF5AR	OEM
ZF5.1	0162
ZF5.6	0157
ZF6A	0301
ZF6.2	0636
ZF6.8	0062
ZF6.8A	0062
ZF6.8B	0062
ZF7.5	0077
ZF7.5A	0077
ZF7.5B	0077
ZF8A	0244
ZF-8.2	0244
ZF8.2	0165
ZF8.2A	0165
ZF8.2B	0165
ZF8.2P	0244
ZF9.1	0057
ZF9.1A	0057
ZF9.1B	0057
ZF10	0064
ZF10A	0064
ZF10B	0064
ZF11	0181
ZF11A	0181
ZF11B	0181
ZF12	0052
ZF12A	0052
ZF12B	0052
ZF13	0053
ZF13A	0053
ZF13B	0053
ZF13.3R	0292
ZF13.6R	0188
ZF15	0686
ZF15A	0686
ZF15B	0686
ZF15.1R	0162
ZF15.6R	0157
ZF-16	0002
ZF16	0445
ZF16A	0445
ZF16B	0445
ZF16.2	0636
ZF17.5	0649
ZF18	0375
ZF18A	0375
ZF18B	0375
ZF19.1	0012
ZF20	0700
ZF20A	0700
ZF20B	0700
ZF22	0705
ZF22A	0705
ZF22B	0705
ZF23.3R	0292
ZF23.6R	0376
ZF23.9R	0456
ZF24	0494
ZF24A	0494
ZF24B	0494
ZF24.3R	0533
ZF24.7R	0451
ZF25.1R	0162
ZF25.6R	0157
ZF26.2	0636
ZF26.8	0025
ZF27	0455
ZF27A	0455
ZF27B	0455
ZF27.8	0649
ZF29.1	0012
ZF30	0195
ZF30A	0195
ZF30B	0195
ZF33	0166
ZF33A	0166
ZF33B	0166
ZF36	0010
ZF36A	0010
ZF36B	0010
ZF39	0032
ZF39A	0032
ZF39B	0032
ZF43	0054
ZF43A	0054
ZF43B	0054
ZF47	0068
ZF47A	0068
ZF47B	0068
ZF51	0092
ZF51A	0092
ZF51B	0092
ZF56	0125
ZF56A	0125
ZF56B	0125
ZF62	0152
ZF62A	0152
ZF62B	0152
ZF68	0173
ZF68A	0173
ZF68B	0173
ZF75	0094
ZF75A	0094
ZF75B	0094
ZF82	0049
ZF82A	0049
ZF82B	0049
ZF91	0156
ZF91A	0156
ZF91B	0156
ZF100	0189
ZF100A	0189
ZF100B	0189
ZF110	0099
ZF110A	0099
ZF110B	0099
ZF112	0137
ZF113	0365
ZF115	0002
ZF116	0420
ZF120	0089
ZF120A	0089
ZF120B	0089
ZF122	0565
ZF124	0402
ZF127	0441
ZF130	0287
ZF130A	0287
ZF130B	0287
ZF133	0039
ZF136	0820
ZF139	0350
ZF150	0340
ZF150A	0340
ZF150B	0340
ZF156	0869
ZF160	0370
ZF160A	0370
ZF160B	0370
ZF162	0783
ZF180	0425
ZF180A	0425
ZF180B	0425
ZF182	0331
ZF200	1473
ZF200A	1473
ZF200B	1473
ZF210	0170
ZF211	0317
ZF212	0137
ZF213	0365
ZF215	0002
ZF216	0420
ZF222	0565
ZF224	0402
ZF227	0441
ZF230	0726
ZF233	0039
ZF236	0820
ZF239	0350
ZF262	0783
ZF282	0331
ZFT12	OEM
ZFT12A	OEM
ZFT14	OEM
ZFT14A	OEM
ZFT16	OEM
ZFT18	OEM
ZG2.7	0760
ZG3.3	0299
ZG3.9	0547
ZG3.9A	0547
ZG3.9B	0547
ZG3.9R	2035
ZG3.9RA	2035
ZG3.9RB	2035
ZG4.3	2403
ZG4.3A	2403
ZG4.3B	2403
ZG4.3R	2401
ZG4.3RA	2401
ZG4.3RB	2401
ZG4.7	2113
ZG4.7A	2113
ZG4.7B	2113
ZG4.7R	1438
ZG4.7RA	1438
ZG4.7RB	1438
ZG5.1	2410
ZG5.1A	2410
ZG5.1B	2410
ZG5.1R	2407
ZG5.1RA	2407
ZG5.1RB	2407
ZG5.6	1901
ZG5.6A	1901
ZG5.6B	1901
ZG5.6R	1445
ZG5.6RA	1445
ZG5.6RB	1445
ZG6.2	0696
ZG6.2A	0696
ZG6.2B	0696
ZG6.2R	2221
ZG6.2RA	2221
ZG6.2RB	2221
ZG6.8	1601
ZG6.8A	1601
ZG6.8B	1601
ZG6.8R	1458
ZG6.8RA	1458
ZG6.8RB	1458
ZG7.5	1616
ZG7.5A	1616
ZG7.5B	1616
ZG7.5R	0221
ZG7.5RA	0221
ZG7.5RB	0221
ZG8.2	1622
ZG8.2A	1622
ZG8.2B	1622
ZG8.2R	1490
ZG8.2RA	1490
ZG8.2RB	1490
ZG9.1	0627
ZG9.1A	0627
ZG9.1B	0627
ZG9.1R	1618
ZG9.1RA	1618
ZG9.1RB	1618
ZG10	0992
ZG10A	0992
ZG10B	0992
ZG10R	0510
ZG10RA	0510
ZG10RB	0510
ZG11	0995
ZG11A	0995
ZG11B	0995
ZG11R	0691
ZG11RA	0691
ZG11RB	0691
ZG12	1263
ZG12A	1263
ZG12B	1263
ZG12R	0870
ZG12RA	0870
ZG12RB	0870
ZG13	1249
ZG13A	1249
ZG13B	1249
ZG13R	1020
ZG13RA	1020
ZG13RB	1020
ZG14	1636
ZG14A	1636
ZG14B	1636
ZG14R	1153
ZG14RA	1153
ZG14RB	1153
ZG15	1639
ZG15A	1639
ZG15B	1639
ZG15R	1274
ZG15RA	1274
ZG15RB	1274
ZG16	1703
ZG16A	1703
ZG16B	1703
ZG16R	1402
ZG16RA	1402
ZG16RB	1402
ZG17	1640
ZG17A	1640
ZG17B	1640
ZG17R	1532
ZG17RA	1532
ZG17RB	1532
ZG18	1716
ZG18A	1716
ZG18B	1716
ZG18R	1078
ZG18RA	1078
ZG18RB	1078
ZG19	1720
ZG19A	1720
ZG19B	1720
ZG19R	1711
ZG19RA	1711
ZG19RB	1711
ZG20	1731
ZG20A	1731
ZG20B	1731
ZG20R	1717
ZG20RA	1717
ZG20RB	1717
ZG22	0727
ZG22A	0727
ZG22B	0727
ZG22R	1723
ZG22RA	1723
ZG22RB	1723
ZG24	1756
ZG24A	1756
ZG24B	1756
ZG24R	1736
ZG24RA	1736
ZG24RB	1736
ZG25	1768
ZG25A	1768
ZG25B	1768
ZG25R	1748
ZG25RA	1748
ZG25RB	1748
ZG27	1782
ZG27A	1782
ZG27B	1782
ZG27R	1761
ZG27RA	1761
ZG27RB	1761
ZG30	1794
ZG30A	1794
ZG30B	1794
ZG30R	1772
ZG30RA	1772
ZG30RB	1772
ZG33	1799
ZG33A	1799
ZG33B	1799
ZG33R	1788
ZG33RA	1788
ZG33RB	1788
ZG36	1809
ZG36A	1809
ZG36B	1809
ZG36R	1796
ZG36RA	1796
ZG36RB	1796
ZG39	1817
ZG39A	1817
ZG39B	1817
ZG39R	1804
ZG39RA	1804
ZG39RB	1804
ZG43	1826
ZG43A	1826
ZG43B	1826
ZG43R	1194
ZG43RA	1194
ZG43RB	1194
ZG45	1840
ZG45A	1840
ZG45B	1840
ZG45R	1821
ZG45RA	1821
ZG45RB	1821
ZG47	1853
ZG47A	1853
ZG47B	1853
ZG47R	0022
ZG47RA	0022
ZG47RB	0022
ZG50	1861
ZG50A	1861
ZG50B	1861
ZG50R	0070
ZG50RA	0070
ZG50RB	0070
ZG51	1866
ZG51A	1866
ZG51B	1866
ZG51R	0132
ZG51RA	0132
ZG51RB	0132
ZG52	1874
ZG52A	1874
ZG52B	1874
ZG52R	0172
ZG52RA	0172
ZG52RB	0172
ZG56	1884
ZG56A	1884
ZG56B	1884
ZG56R	0207
ZG56RA	0207
ZG56RB	0207
ZG62	1895
ZG62A	1895
ZG62B	1895
ZG62R	0263
ZG62RA	0263
ZG62RB	0263
ZG68	1902
ZG68A	1902
ZG68B	1902

DEVICE TYPE	REPL CODE
ZG68R	0310
ZG68RA	0310
ZG68RB	0310
ZG75	0736
ZG75A	0736
ZG75B	0736
ZG75R	0329
ZG75RA	0329
ZG75RB	0329
ZG82	1909
ZG82A	1909
ZG82B	1909
ZG82R	0356
ZG82RA	0356
ZG82RB	0356
ZG91	1914
ZG91A	1914
ZG91B	1914
ZG91R	0412
ZG91RA	0412
ZG91RB	0412
ZG100	1163
ZG100A	1163
ZG100B	1163
ZG100R	0438
ZG100RA	0438
ZG100RB	0438
ZG105	1924
ZG105A	1924
ZG105B	1924
ZG105R	0464
ZG105RA	0464
ZG105RB	0464
ZG110	1933
ZG110A	1933
ZG110B	1933
ZG110R	0488
ZG110RA	0488
ZG110RB	0488
ZG120	1941
ZG120A	1941
ZG120B	1941
ZG120R	0509
ZG120RA	0509
ZG120RB	0509
ZG130	1947
ZG130A	1947
ZG130B	1947
ZG130R	0524
ZG130RA	0524
ZG130RB	0524
ZG140	1953
ZG140A	1953
ZG140B	1953
ZG140R	0542
ZG140RA	0542
ZG140RB	0542
ZG150	1961
ZG150A	1961
ZG150B	1961
ZG150R	0063
ZG150RA	0063
ZG150RB	0063
ZG160	0357
ZG160A	0357
ZG160B	0357
ZG160R	0401
ZG160RA	0401
ZG160RB	0401
ZG175	0670
ZG175A	0670
ZG175B	0670
ZG175R	0616
ZG175RA	0616
ZG175RB	0616
ZG180	0776
ZG180A	0776
ZG180B	0776
ZG180R	0634
ZG180RA	0634
ZG180RB	0634
ZG200	1071
ZG200A	1071
ZG200B	1071
ZG200R	0668
ZG200RA	0668
ZG200RB	0668
ZG220	OEM
ZG220A	OEM
ZG220B	OEM
ZG220R	OEM
ZG220RA	OEM
ZG220RB	OEM
ZG100118	0820
ZG100119	0002
ZG100120	0565
ZGL27-6.8	OEM
ZGL27-6.8A	OEM
ZGL27-6.8B	OEM
ZGL27-7.5	OEM
ZGL27-7.5A	OEM
ZGL27-7.5B	OEM
ZGL27-8.2	OEM
ZGL27-8.2A	OEM
ZGL27-8.2B	OEM
ZGL27-8.7	OEM
ZGL27-8.7A	OEM
ZGL27-8.7B	OEM
ZGL27-9.1	OEM
ZGL27-9.1A	OEM
ZGL27-9.1B	OEM
ZGL27-10	OEM
ZGL27-10A	OEM
ZGL27-10B	OEM
ZGL27-11	OEM
ZGL27-11A	OEM
ZGL27-11B	OEM
ZGL27-12	OEM
ZGL27-12A	OEM
ZGL27-12B	OEM
ZGL27-13	OEM
ZGL27-13A	OEM
ZGL27-13B	OEM
ZGL27-14	OEM
ZGL27-14A	OEM
ZGL27-14B	OEM
ZGL27-15	OEM
ZGL27-15A	OEM
ZGL27-15B	OEM
ZGL27-16	OEM
ZGL27-16A	OEM
ZGL27-16B	OEM
ZGL27-17	OEM
ZGL27-17A	OEM
ZGL27-17B	OEM
ZGL27-18	OEM
ZGL27-18A	OEM
ZGL27-18B	OEM
ZGL27-19	OEM
ZGL27-19A	OEM
ZGL27-19B	OEM
ZGL27-20	OEM
ZGL27-20A	OEM
ZGL27-20B	OEM
ZGL27-22	OEM
ZGL27-22A	OEM
ZGL27-22B	OEM
ZGL27-24	OEM
ZGL27-24A	OEM
ZGL27-24B	OEM
ZGL27-25	OEM
ZGL27-25A	OEM
ZGL27-25B	OEM
ZGL27-27	OEM
ZGL27-27A	OEM
ZGL27-27B	OEM
ZGL27-28	OEM
ZGL27-28A	OEM
ZGL27-28B	OEM
ZGL27-30	OEM
ZGL27-30A	OEM
ZGL27-30B	OEM
ZGL27-33	OEM
ZGL27-33A	OEM
ZGL27-33B	OEM
ZGL27-36	OEM
ZGL27-36A	OEM
ZGL27-36B	OEM
ZGL27-39	OEM
ZGL27-39A	OEM
ZGL27-39B	OEM
ZGL27-43	OEM
ZGL27-43A	OEM
ZGL27-43B	OEM
ZGL27-47	OEM
ZGL27-47A	OEM
ZGL27-47B	OEM
ZGL27-51	OEM
ZGL27-51A	OEM
ZGL27-51B	OEM
ZGL27-56	OEM
ZGL27-56A	OEM
ZGL27-56B	OEM
ZGL27-60	OEM
ZGL27-60A	OEM
ZGL27-60B	OEM
ZGL27-62	OEM
ZGL27-62A	OEM
ZGL27-62B	OEM
ZGL27-68	OEM
ZGL27-68A	OEM
ZGL27-68B	OEM
ZGL27-75	OEM
ZGL27-75A	OEM
ZGL27-75B	OEM
ZGL27-82	OEM
ZGL27-82A	OEM
ZGL27-82B	OEM
ZGL27-87	OEM
ZGL27-87A	OEM
ZGL27-87B	OEM
ZGL27-91	OEM
ZGL27-91A	OEM
ZGL27-91B	OEM
ZGL27-100	OEM
ZGL27-100A	OEM
ZGL27-100B	OEM
ZGL27-110	OEM
ZGL27-110A	OEM
ZGL27-110B	OEM
ZGL27-120	OEM
ZGL27-120A	OEM
ZGL27-120B	OEM
ZGL27-130	OEM
ZGL27-130A	OEM
ZGL27-130B	OEM
ZGL27-140	OEM
ZGL27-140A	OEM
ZGL27-140B	OEM
ZGL27-150	OEM
ZGL27-150A	OEM
ZGL27-150B	OEM
ZGL27-160	OEM
ZGL27-160A	OEM
ZGL27-160B	OEM
ZGL27-170	OEM
ZGL27-170A	OEM
ZGL27-170B	OEM
ZGL27-180	OEM
ZGL27-180A	OEM
ZGL27-180B	OEM
ZGL27-190	OEM
ZGL27-190A	OEM
ZGL27-190B	OEM
ZGL27-200	OEM
ZGL27-200A	OEM
ZGL27-200B	OEM
ZGL41-6.8	OEM
ZGL41-6.8A	OEM
ZGL41-6.8B	OEM
ZGL41-7.5	OEM
ZGL41-7.5A	OEM
ZGL41-7.5B	OEM
ZGL41-8.2	OEM
ZGL41-8.2A	OEM
ZGL41-8.2B	OEM
ZGL41-9.1	OEM
ZGL41-9.1A	OEM
ZGL41-9.1B	OEM
ZGL41-10	OEM
ZGL41-10A	OEM
ZGL41-10B	OEM
ZGL41-11	OEM
ZGL41-11A	OEM
ZGL41-11B	OEM
ZGL41-12	OEM
ZGL41-12A	OEM
ZGL41-12B	OEM
ZGL41-13	OEM
ZGL41-13A	OEM
ZGL41-13B	OEM
ZGL41-15	OEM
ZGL41-15A	OEM
ZGL41-15B	OEM
ZGL41-16	OEM
ZGL41-16A	OEM
ZGL41-16B	OEM
ZGL41-18	OEM
ZGL41-18A	OEM
ZGL41-18B	OEM
ZGL41-20	OEM
ZGL41-20A	OEM
ZGL41-20B	OEM
ZGL41-22	OEM
ZGL41-22A	OEM
ZGL41-22B	OEM
ZGL41-24	OEM
ZGL41-24A	OEM
ZGL41-24B	OEM
ZGL41-27	OEM
ZGL41-27A	OEM
ZGL41-27B	OEM
ZGL41-30	OEM
ZGL41-30A	OEM
ZGL41-30B	OEM
ZGL41-33	OEM
ZGL41-33A	OEM
ZGL41-33B	OEM
ZGL41-36	OEM
ZGL41-36A	OEM
ZGL41-36B	OEM
ZGL41-39	OEM
ZGL41-39A	OEM
ZGL41-39B	OEM
ZGL41-43	OEM
ZGL41-43A	OEM
ZGL41-43B	OEM
ZGL41-47	OEM
ZGL41-47A	OEM
ZGL41-47B	OEM
ZGL41-51	OEM
ZGL41-51A	OEM
ZGL41-51B	OEM
ZGL41-56	OEM
ZGL41-56A	OEM
ZGL41-56B	OEM
ZGL41-62	OEM
ZGL41-62A	OEM
ZGL41-62B	OEM
ZGL41-68	OEM
ZGL41-68A	OEM
ZGL41-68B	OEM
ZGL41-75	OEM
ZGL41-75A	OEM
ZGL41-75B	OEM
ZGL41-82	OEM
ZGL41-82A	OEM
ZGL41-82B	OEM
ZGL41-91	OEM
ZGL41-91A	OEM
ZGL41-91B	OEM
ZGL41-100	OEM
ZGL41-100A	OEM
ZGL41-100B	OEM
ZGL41-110	OEM
ZGL41-110A	OEM
ZGL41-110B	OEM
ZGL41-120	OEM
ZGL41-120A	OEM
ZGL41-120B	OEM
ZGL41-130	OEM
ZGL41-130A	OEM
ZGL41-130B	OEM
ZGL41-140	OEM
ZGL41-140A	OEM
ZGL41-140B	OEM
ZGL41-150	OEM
ZGL41-150A	OEM
ZGL41-150B	OEM
ZGL41-160	OEM
ZGL41-160A	OEM
ZGL41-160B	OEM
ZGL41-170	OEM
ZGL41-170A	OEM
ZGL41-170B	OEM
ZGL41-180	OEM
ZGL41-180A	OEM
ZGL41-180B	OEM
ZGL41-190	OEM
ZGL41-190A	OEM
ZGL41-190B	OEM
ZGL41-200	OEM
ZGL41-200A	OEM
ZGL41-200B	OEM
ZGP10-100	1180
ZGP10-110	1190
ZGP10-120	1207
ZGP10-130	1218
ZGP10-140	1881
ZGP10-150	0647
ZGP10-160	1255
ZGP10-160A	OEM
ZGP10-160B	OEM
ZGP10-170	2102
ZGP10-170A	OEM
ZGP10-170B	OEM
ZGP10-180	1279
ZGP10-180A	OEM
ZGP10-180B	OEM
ZGP10-190	2225
ZGP10-190A	OEM
ZGP10-190B	OEM
ZGP10-200	0605
ZGP10-200A	OEM
ZGP10-200B	OEM
ZH011TA	OEM
ZH1	OEM
ZH6.8	0025
ZH6.8A	0025
ZH6.8B	0025
ZH7.5	0649
ZH7.5A	0649
ZH7.5B	0649
ZH8.2	0244
ZH8.2A	0244
ZH8.2B	0244
ZH9.1	0012
ZH9.1A	0012
ZH9.1B	0012
ZH10	0170
ZH10A	0170
ZH10B	0170
ZH11	0794
ZH11A	0794
ZH11B	0794
ZH12	0137
ZH12A	0137
ZH12B	0137
ZH13	0365
ZH13A	0365
ZH13B	0365
ZH15	0002
ZH15A	0002
ZH15B	0002
ZH16	0420
ZH16A	0420
ZH16B	0420
ZH18	0495
ZH18A	0495
ZH18AR	0367
ZH18B	0495
ZH20	0531
ZH20A	0531
ZH20B	0531
ZH22	0565
ZH22A	0565
ZH22B	0565
ZH24	0402
ZH24A	0402
ZH24B	0402
ZH27	0441
ZH27A	0441
ZH27B	0441
ZH30	0726
ZH30A	0726
ZH30B	0726
ZH33	0039
ZH33A	0039
ZH33B	0039
ZH36	0820
ZH36A	0820
ZH36B	0820
ZH39	0350
ZH39A	0350
ZH39B	0350
ZH43	0931
ZH43A	0931
ZH43B	0931
ZH47	0999
ZH47A	0999
ZH47B	0999
ZH51	0502
ZH51A	0502
ZH51B	0502
ZH56	1834
ZH56A	1834
ZH56B	1834
ZH62	0783
ZH62A	0783
ZH62B	0783
ZH68	2157
ZH68A	2157
ZH68B	2157
ZH75	1189
ZH75A	1189
ZH75B	1189
ZH82	0331
ZH82A	0331
ZH82B	0331
ZH91	1311
ZH91A	1311
ZH91B	1311
ZH100	0098
ZH100A	0098
ZH100B	0098
ZH102	1312
ZH103	0188
ZH104	0451
ZH105	0157
ZH106	0636
ZH107	0649
ZH109	1083
ZH110	0149
ZH110A	0149
ZH110B	0149
ZH112	0137
ZH115	0420
ZH120	0186
ZH120A	0186
ZH120B	0186
ZH125	1606
ZH130	0213
ZH130A	0213
ZH130B	0213
ZH135	0820
ZH140	0931
ZH150	0028
ZH150A	0028
ZH150B	0028
ZH160	0255
ZH160A	0255
ZH160B	0255
ZH170	2157
ZH180	0367
ZH180A	0367
ZH180B	0367
ZH200	0421
ZH200A	0421
ZH200B	0421
ZH203	2035
ZH204	1438
ZH205	1445
ZH206	1458
ZH209	1490
ZH210	0510
ZH212	0870
ZH225	1748
ZH304	1438
ZH305	1445
ZH306	1458
ZH309	1490
ZH310	0510
ZH312	0870
ZJ4B41	OEM
ZJ6.8	0025
ZJ6.8A	0025
ZJ6.8B	0025
ZJ7.5	0649
ZJ7.5A	0649
ZJ7.5B	0649
ZJ8.2	0244
ZJ8.2A	0244
ZJ8.2B	0244
ZJ9.1	0012
ZJ9.1A	0012
ZJ9.1B	0012
ZJ10	0170
ZJ10A	0170
ZJ10B	0170
ZJ11	0317
ZJ11A	0317
ZJ11B	0317
ZJ12	0137
ZJ12A	0137
ZJ12B	0137
ZJ13	0137
ZJ13A	0137
ZJ13B	0137
ZJ15	0002
ZJ15A	0002
ZJ15B	0002
ZJ16	0420
ZJ16A	0420
ZJ16B	0420
ZJ18	0495
ZJ18A	0495
ZJ18B	0495
ZJ20	0531
ZJ20A	0531
ZJ20B	0531
ZJ22	0565
ZJ22A	0565
ZJ22B	0565
ZJ24	0402
ZJ24A	0402
ZJ24B	0402
ZJ27	0441
ZJ27A	0441
ZJ27B	0441
ZJ30	0726
ZJ30A	0726
ZJ30B	0726
ZJ33	0039
ZJ33A	0039
ZJ33B	0039
ZJ36	0820
ZJ36A	0820
ZJ36B	0820
ZJ39	0350
ZJ39A	0350
ZJ39B	0350
ZJ40	0144
ZJ43	0931
ZJ43A	0931
ZJ43B	0931
ZJ47	0999
ZJ47A	0999
ZJ47B	0999
ZJ51	0502
ZJ51A	0502
ZJ51B	0502
ZJ56	0869
ZJ56A	0869
ZJ56B	0869
ZJ62	0783
ZJ62A	0783
ZJ62B	0783
ZJ68	2157
ZJ68A	2157
ZJ68B	2157
ZJ72	0050
ZJ73	0050
ZJ75	1189
ZJ75A	1189
ZJ75B	1189
ZJ82	0331
ZJ82A	0331
ZJ82B	0331
ZJ91	1311
ZJ91A	1311
ZJ91B	1311
ZJ100	0098
ZJ100A	0098
ZJ100B	0098
ZJ110	0149
ZJ110A	0149
ZJ110B	0149
ZJ120	0186
ZJ120A	0186
ZJ120B	0186
ZJ130	0213
ZJ130A	0213
ZJ130B	0213
ZJ150	0028
ZJ150A	0028
ZJ150B	0028
ZJ160	0255
ZJ160A	0028
ZJ160B	0255
ZJ180	0367
ZJ180A	0367
ZJ180B	0367
ZJ200	0421
ZJ200A	0421
ZJ200B	0421
ZJ224HB	OEM
ZJ224HC	OEM
ZJ224HF	OEM
ZJ224HU	OEM
ZJ224LA	OEM
ZJ224LB	OEM
ZJ224LC	OEM
ZJ224LD	OEM
ZJ224LF	OEM
ZJ224LU	OEM
ZJ227A	OEM
ZJ227B	OEM
ZJ227F	OEM
ZJ227G	OEM
ZJ227U	OEM
ZJ244A	OEM
ZJ244B	OEM
ZJ244C	OEM
ZJ244D	OEM
ZJ244E	OEM
ZJ244F	OEM
ZJ244G	OEM
ZJ244H	OEM
ZJ244U	OEM
ZJ252B	0015
ZJ253A	OEM
ZJ253B	OEM
ZJ253C	OEM
ZJ253D	OEM
ZJ253F	OEM
ZJ255A	OEM
ZJ255B	OEM
ZJ255C	OEM
ZJ255D	OEM
ZJ255F	OEM
ZJ260E	OEM
ZJ260M	OEM
ZJ260N	OEM
ZJ260P	OEM
ZJ260PA	OEM
ZJ260S	OEM
ZJ260T	OEM
ZK3.9	0547
ZK3.9A	0547
ZK3.9B	0547
ZK3.9R	2035
ZK3.9RA	2035
ZK3.9RB	2035
ZK4.3	2403
ZK4.3A	2403
ZK4.3B	2403
ZK4.3R	2401
ZK4.3RA	2401
ZK4.3RB	2401
ZK4.7	2113
ZK4.7A	2113
ZK4.7B	2113
ZK4.7R	1438
ZK4.7RA	1438
ZK4.7RB	1438
ZK5.1	2410
ZK5.1A	2410
ZK5.1B	2410
ZK5.1R	2407
ZK5.1RA	2407
ZK5.1RB	2407
ZK5.6	1901
ZK5.6A	1901
ZK5.6B	1901
ZK5.6R	1445
ZK5.6RA	1445
ZK5.6RB	1445
ZK6.2	0696
ZK6.2A	0696
ZK6.2B	0696
ZK6.2R	2221
ZK6.2RA	2221
ZK6.2RB	2221
ZK6.8	1601
ZK6.8A	1601
ZK6.8B	1601
ZK6.8R	1458
ZK6.8RA	1458
ZK6.8RB	1458
ZK7.5	1616
ZK7.5A	1616
ZK7.5B	1616
ZK7.5R	0221
ZK7.5RA	0221
ZK7.5RB	0221
ZK8.2	1622
ZK8.2A	1622
ZK8.2B	1622
ZK8.2R	1490
ZK8.2RA	1490
ZK8.2RB	1490
ZK9.1	0627
ZK9.1A	0627
ZK9.1B	0627
ZK9.1R	1618
ZK9.1RA	1618
ZK9.1RB	1618
ZK10	0992
ZK10A	0992
ZK10B	0992
ZK10R	0510
ZK10RA	0510
ZK10RB	0510
ZK11	0995
ZK11A	0995
ZK11B	0995
ZK11R	0691
ZK11RA	0691
ZK11RB	0691
ZK12	1263
ZK12A	1263
ZK12B	1263
ZK12R	0870
ZK12RA	0870
ZK12RB	0870
ZK13	1249
ZK13A	1249
ZK13B	1249
ZK13R	1020
ZK13RA	1020
ZK13RB	1020
ZK14	1636
ZK14A	1636
ZK14B	1636
ZK14R	1153
ZK14RA	1153
ZK14RB	1153
ZK15	1639
ZK15A	1639
ZK15B	1639
ZK15R	1274
ZK15RA	1274
ZK15RB	1274

If replacement code is OEM, contact original manufacturer for replacement.

DEVICE TYPE	REPL CODE
ZK16	1703
ZK16A	1703
ZK16B	1703
ZK16R	1402
ZK16RA	1402
ZK16RB	1402
ZK17	1640
ZK17A	1640
ZK17B	1640
ZK17R	1532
ZK17RA	1532
ZK17RB	1532
ZK18	1716
ZK18A	1716
ZK18B	1716
ZK18R	1078
ZK18RA	1078
ZK18RB	1078
ZK19	1720
ZK19A	1720
ZK19B	1720
ZK19R	1711
ZK19RA	1711
ZK19RB	1711
ZK20	1731
ZK20A	1731
ZK20B	1731
ZK20R	1717
ZK20RA	1717
ZK20RB	1717
ZK22	0727
ZK22A	0727
ZK22B	0727
ZK22R	1723
ZK22RA	1723
ZK22RB	1723
ZK24	1756
ZK24A	1756
ZK24B	1756
ZK24R	1736
ZK24RA	1736
ZK24RB	1736
ZK25	1768
ZK25A	1768
ZK25B	1768
ZK25R	1748
ZK25RA	1748
ZK25RB	1748
ZK27	1782
ZK27A	1782
ZK27B	1782
ZK27R	1761
ZK27RA	1761
ZK27RB	1761
ZK30	1794
ZK30A	1794
ZK30B	1794
ZK30R	1772
ZK30RA	1772
ZK30RB	1772
ZK33	1799
ZK33A	1799
ZK33B	1799
ZK33R	1788
ZK33RA	1788
ZK33RB	1788
ZK36	1809
ZK36A	1809
ZK36B	1809
ZK36R	1796
ZK36RA	1796
ZK36RB	1796
ZK39	1817
ZK39A	1817
ZK39B	1817
ZK39R	1804
ZK39RA	1804
ZK39RB	1804
ZK43	1826
ZK43A	1826
ZK43B	1826
ZK43R	1194
ZK43RA	1194
ZK43RB	1194
ZK45	1840
ZK45A	1817
ZK45B	1840
ZK45R	1821
ZK45RA	1821
ZK45RB	1821
ZK47	1853
ZK47A	1853
ZK47B	1853
ZK47R	0022
ZK47RA	0022
ZK47RB	0022
ZK50	1861
ZK50A	1861
ZK50B	1861
ZK50R	0070
ZK50RA	0070
ZK50RB	0070
ZK51	1866
ZK51A	1866
ZK51B	1866
ZK51R	0132
ZK51RA	0132
ZK51RB	0132
ZK52	1874

DEVICE TYPE	REPL CODE
ZK52A	1874
ZK52B	1874
ZK52R	0172
ZK52RA	0172
ZK52RB	0172
ZK56	1884
ZK56A	1884
ZK56B	1884
ZK56R	0207
ZK56RA	0207
ZK56RB	0207
ZK62	1895
ZK62A	1895
ZK62B	1895
ZK62R	0263
ZK62RA	0263
ZK62RB	0263
ZK68	1902
ZK68A	1902
ZK68B	1902
ZK68R	0310
ZK68RA	0310
ZK68RB	0310
ZK75	0736
ZK75A	0736
ZK75B	0736
ZK75R	0329
ZK75RA	0329
ZK75RB	0329
ZK82	1909
ZK82A	1909
ZK82B	1909
ZK82R	0356
ZK82RA	0356
ZK82RB	0356
ZK91	1914
ZK91A	1914
ZK91B	1914
ZK91R	0412
ZK91RA	0412
ZK91RB	0412
ZK100	1163
ZK100A	1163
ZK100B	1163
ZK100R	0438
ZK100RA	0438
ZK100RB	0438
ZK105	1924
ZK105A	1924
ZK105B	1924
ZK105R	0464
ZK105RA	0464
ZK105RB	0464
ZK110	1933
ZK110A	1933
ZK110B	1933
ZK110R	0488
ZK110RA	0488
ZK110RB	0488
ZK120	1941
ZK120A	1941
ZK120B	1941
ZK120R	0509
ZK120RA	0509
ZK120RB	0509
ZK130	1947
ZK130A	1947
ZK130B	1947
ZK130R	0524
ZK130RA	0524
ZK130RB	0524
ZK140	1953
ZK140A	1953
ZK140B	1953
ZK140R	0542
ZK140RA	0542
ZK140RB	0542
ZK150	1961
ZK150A	1961
ZK150B	1961
ZK150R	0063
ZK150RA	0063
ZK150RB	0063
ZK160	0357
ZK160A	0357
ZK160B	0357
ZK160R	0401
ZK160RA	0401
ZK160RB	0401
ZK175	0670
ZK175A	0670
ZK175B	0670
ZK175R	0616
ZK175RA	0616
ZK175RB	0616
ZK180	0776
ZK180A	0776
ZK180B	0776
ZK180R	0634
ZK180RA	0634
ZK180RB	0634
ZK200	1071
ZK200A	1071
ZK200B	1071
ZK200R	0668
ZK200RA	0668
ZK200RB	0668
ZK220	OEM
ZK220A	OEM

DEVICE TYPE	REPL CODE
ZK220B	OEM
ZK220R	OEM
ZK220RA	OEM
ZK220RB	OEM
ZL5	OEM
ZL6	0091
ZL7	OEM
ZL8	OEM
ZLD709	OEM
ZLD709C	OEM
ZLD709CE	OEM
ZLD709CF	OEM
ZLD709CG	OEM
ZLD709F	OEM
ZLD741	OEM
ZLD741C	OEM
ZLD741CE	OEM
ZM3.3	0292
ZM3.3A	0292
ZM3.3B	0292
ZM3.6	0188
ZM3.6A	0188
ZM3.6B	0188
ZM3.9	0456
ZM3.9A	0456
ZM3.9B	0456
ZM4.3	0533
ZM4.3A	0533
ZM4.3B	0533
ZM4.7	0451
ZM4.7A	0451
ZM4.7B	0451
ZM5.1	0162
ZM5.1A	0162
ZM5.1B	0162
ZM5.6	0157
ZM5.6A	0157
ZM5.6B	0157
ZM5.6C	OEM
ZM5.6D	OEM
ZM6	0091
ZM6A	0091
ZM6B	0091
ZM6.2	0636
ZM6.2A	0636
ZM6.2B	0636
ZM6.2C	OEM
ZM6.2D	OEM
ZM6.8	0100
ZM6.8A	0100
ZM6.8B	0025
ZM6.8C	OEM
ZM6.8D	OEM
ZM7.5	0649
ZM7.5A	0649
ZM7.5B	0649
ZM7.5C	OEM
ZM7.5D	OEM
ZM8.2	0244
ZM8.2A	0244
ZM8.2B	0244
ZM8.2C	0244
ZM8.2D	0244
ZM8.7	1083
ZM8.7A	1083
ZM8.7B	1083
ZM9.1	0012
ZM9.1A	0012
ZM9.1B	0012
ZM9.1C	0012
ZM9.1D	OEM
ZM10	0170
ZM10A	0170
ZM10B	0170
ZM10C	OEM
ZM10D	OEM
ZM11	0317
ZM11A	0317
ZM11B	0317
ZM11C	OEM
ZM11D	OEM
ZM12	0137
ZM12A	0137
ZM12B	0137
ZM12C	0137
ZM12D	OEM
ZM13	0365
ZM13A	0365
ZM13B	0365
ZM13C	OEM
ZM13D	OEM
ZM14	0100
ZM14A	0100
ZM14B	0879
ZM14C	OEM
ZM14D	OEM
ZM15	0002
ZM15A	0002
ZM15B	0002
ZM15C	0002
ZM15D	OEM
ZM16	0420
ZM16A	0420
ZM16B	0445
ZM16C	OEM
ZM16D	OEM
ZM17	1649
ZM17A	1649
ZM17B	1649

DEVICE TYPE	REPL CODE
ZM17C	OEM
ZM17D	OEM
ZM18	0495
ZM18A	0495
ZM18B	0495
ZM18C	OEM
ZM18D	OEM
ZM19	0949
ZM19A	0949
ZM19B	0949
ZM19C	OEM
ZM19D	OEM
ZM20	0531
ZM20A	0531
ZM20B	0531
ZM20C	OEM
ZM20D	OEM
ZM22	0565
ZM22A	0565
ZM22B	0565
ZM22C	OEM
ZM22D	OEM
ZM24	0402
ZM24A	0402
ZM24B	0402
ZM24C	OEM
ZM24D	OEM
ZM25	1606
ZM25A	1606
ZM25B	1606
ZM25C	OEM
ZM25D	OEM
ZM27	0441
ZM27A	0441
ZM27B	0441
ZM27C	OEM
ZM27D	OEM
ZM28	0455
ZM28A	0455
ZM28B	0455
ZM30	0726
ZM30A	0726
ZM30B	0726
ZM30C	OEM
ZM30D	OEM
ZM33	0039
ZM33A	0039
ZM33B	0039
ZM33C	OEM
ZM33D	OEM
ZM36	0820
ZM36A	0820
ZM36B	0820
ZM36C	OEM
ZM36D	OEM
ZM39	0350
ZM39A	0350
ZM39B	0350
ZM39C	OEM
ZM39D	OEM
ZM43	0931
ZM43A	0931
ZM43B	0931
ZM43C	OEM
ZM43D	OEM
ZM45	0999
ZM45A	0999
ZM45B	0999
ZM45C	OEM
ZM45D	OEM
ZM47	0999
ZM47A	0999
ZM47B	0999
ZM47C	OEM
ZM47D	OEM
ZM50	0502
ZM50A	0502
ZM50B	0502
ZM50C	OEM
ZM50D	OEM
ZM51	0502
ZM51A	0502
ZM51B	0502
ZM51C	OEM
ZM51D	OEM
ZM52	0502
ZM52A	0502
ZM52B	0502
ZM52C	OEM
ZM52D	OEM
ZM56	0869
ZM56A	0869
ZM56B	0869
ZM56C	OEM
ZM56D	OEM
ZM62	0783
ZM62A	0783
ZM62B	0783
ZM62C	OEM
ZM62D	OEM
ZM68	2157
ZM68A	2157
ZM68B	1268
ZM68C	OEM
ZM68D	OEM
ZM75	1189
ZM75A	1189
ZM75B	1189
ZM75C	OEM

DEVICE TYPE	REPL CODE
ZM75D	OEM
ZM82	0331
ZM82A	0331
ZM82B	0331
ZM82C	0331
ZM82D	OEM
ZM91	1311
ZM91A	1311
ZM91B	1311
ZM91C	OEM
ZM91D	OEM
ZM100	0098
ZM100A	0098
ZM100B	0098
ZM100C	OEM
ZM100D	OEM
ZM105	0149
ZM105A	0149
ZM105B	0149
ZM105C	OEM
ZM105D	OEM
ZM110	0149
ZM110A	0149
ZM110B	0149
ZM110C	OEM
ZM110D	OEM
ZM120	0186
ZM120A	0186
ZM120B	0186
ZM120C	OEM
ZM120D	OEM
ZM130	0213
ZM130A	0213
ZM130B	0213
ZM130C	OEM
ZM130D	OEM
ZM140	0245
ZM140A	0245
ZM140B	0245
ZM140C	OEM
ZM140D	OEM
ZM150	0028
ZM150A	0028
ZM150B	0028
ZM150C	OEM
ZM150D	OEM
ZM160	0255
ZM160A	0255
ZM160B	0255
ZM160C	OEM
ZM160D	OEM
ZM175	0367
ZM175A	0367
ZM175B	0367
ZM175C	OEM
ZM175D	OEM
ZM180	0367
ZM180A	0367
ZM180B	0367
ZM180C	OEM
ZM180D	OEM
ZM200	0421
ZM200A	0421
ZM200B	0421
ZM200C	OEM
ZM200D	OEM
ZM220	OEM
ZM220A	OEM
ZM220B	OEM
ZM220C	OEM
ZM220D	OEM
ZME60	OEM
ZMIN4935811	OEM
ZMM6.2	0471
ZMM9.1	0057
ZMM12	0052
ZMM15	0686
ZMQ9B	OEM
ZMZ205	0162
ZN6.8	OEM
ZN6.8A	OEM
ZN6.8B	OEM
ZN7.5	OEM
ZN7.5A	OEM
ZN7.5B	OEM
ZN8.2	OEM
ZN8.2A	OEM
ZN8.2B	OEM
ZN9.1	OEM
ZN9.1A	OEM
ZN9.1B	OEM
ZN10	OEM
ZN10A	OEM
ZN10B	OEM
ZN11	OEM
ZN11A	OEM
ZN11B	OEM
ZN12	OEM
ZN12A	OEM
ZN12B	OEM
ZN13	OEM
ZN13A	OEM
ZN13B	OEM
ZN15	OEM
ZN15A	OEM
ZN15B	OEM
ZN16	OEM
ZN16A	OEM
ZN16B	OEM

DEVICE TYPE	REPL CODE
ZN18	OEM
ZN18A	OEM
ZN18B	OEM
ZN20	OEM
ZN20A	OEM
ZN20B	OEM
ZN22	OEM
ZN22A	OEM
ZN22B	OEM
ZN24	OEM
ZN24A	OEM
ZN24B	OEM
ZN27	OEM
ZN27A	OEM
ZN27B	OEM
ZN30	OEM
ZN30A	OEM
ZN30B	OEM
ZN33	OEM
ZN33A	OEM
ZN33B	OEM
ZN36	OEM
ZN36A	OEM
ZN36B	OEM
ZN39	OEM
ZN39A	OEM
ZN39B	OEM
ZN43	OEM
ZN43A	OEM
ZN43B	OEM
ZN47	OEM
ZN47A	OEM
ZN47B	OEM
ZN51	OEM
ZN51A	OEM
ZN51B	OEM
ZN54L91E	OEM
ZN54L91J	OEM
ZN54L95E	OEM
ZN54L95J	OEM
ZN54L96E	OEM
ZN54L96J	OEM
ZN54L164E	OEM
ZN54L164J	OEM
ZN56	OEM
ZN56A	OEM
ZN56B	OEM
ZN62	OEM
ZN62A	OEM
ZN62B	OEM
ZN68	OEM
ZN68A	OEM
ZN68B	OEM
ZN74L91E	OEM
ZN74L91J	OEM
ZN74L95E	OEM
ZN74L95J	OEM
ZN74L96E	OEM
ZN74L96J	OEM
ZN74L164E	OEM
ZN74L164J	OEM
ZN75	OEM
ZN75A	OEM
ZN75B	OEM
ZN82	OEM
ZN82A	OEM
ZN82B	OEM
ZN91	OEM
ZN91A	OEM
ZN91B	OEM
ZN100	OEM
ZN100A	OEM
ZN100B	OEM
ZN110	OEM
ZN110A	OEM
ZN110B	OEM
ZN120	OEM
ZN120A	OEM
ZN120B	OEM
ZN130	OEM
ZN130A	OEM
ZN130B	OEM
ZN150	OEM
ZN150A	OEM
ZN150B	OEM
ZN160	OEM
ZN160A	OEM
ZN160B	OEM
ZN180	OEM
ZN180A	OEM
ZN180B	OEM
ZN200	OEM
ZN200A	OEM
ZN200B	OEM
ZN219	OEM
ZN219E	OEM
ZN220	OEM
ZN220E	OEM
ZN221	OEM
ZN221E	OEM
ZN222	OEM
ZN222E	OEM
ZN224	OEM
ZN224E	OEM
ZN229	OEM
ZN229E	OEM
ZN230	OEM
ZN230E	OEM
ZN232	OEM

DEVICE TYPE	REPL CODE
ZN232E	OEM
ZN233	OEM
ZN233E	OEM
ZN244	OEM
ZN244E	OEM
ZN246	OEM
ZN246E	OEM
ZN248	OEM
ZN248E	OEM
ZN262	OEM
ZN262E	OEM
ZN294E	OEM
ZN297E	OEM
ZN319	OEM
ZN319E	OEM
ZN320	OEM
ZN320E	OEM
ZN321	OEM
ZN321E	OEM
ZN322	OEM
ZN322E	OEM
ZN324	OEM
ZN324E	OEM
ZN329	OEM
ZN329E	OEM
ZN330	OEM
ZN330E	OEM
ZN332	OEM
ZN332E	OEM
ZN333	OEM
ZN333E	OEM
ZN342CJ-8	OEM
ZN344	OEM
ZN344E	OEM
ZN346	OEM
ZN346E	OEM
ZN348	OEM
ZN348E	OEM
ZN362	OEM
ZN362E	OEM
ZN394E	OEM
ZN397E	OEM
ZN409CE	OEM
ZN414	OEM
ZN419CE	OEM
ZN423T	OEM
ZN424E	OEM
ZN424P	OEM
ZN424T	OEM
ZN425E-6	OEM
ZN425E-7	OEM
ZN425E-8	OEM
ZN425J-8	OEM
ZN426E-6	OEM
ZN426E-7	OEM
ZN426E-8	OEM
ZN426J-8	OEM
ZN427E-8	OEM
ZN427J-8	OEM
ZN428E-8	OEM
ZN428J-8	OEM
ZN429E-6	OEM
ZN429E-7	OEM
ZN429E-8	OEM
ZN429J-8	OEM
ZN432BJ-8	OEM
ZN432BJ-9	OEM
ZN432CJ-9	OEM
ZN432CJ-10	OEM
ZN432E	OEM
ZN432J-8	OEM
ZN432J-9	OEM
ZN432J-10	OEM
ZN433BJ-8	OEM
ZN433BJ-9	OEM
ZN433BJ-10	OEM
ZN433CJ-8	OEM
ZN433CJ-9	OEM
ZN433J-8	OEM
ZN433J-9	OEM
ZN433J-10	OEM
ZN435E	OEM
ZN435J(A)	OEM
ZN436E(A)	OEM
ZN436J(A)	OEM
ZN458	OEM
ZN458A	OEM
ZN458B	OEM
ZN459	OEM
ZN459C	OEM
ZN459CP	OEM
ZN459CT	OEM
ZN459T	OEM
ZN460	OEM
ZN460C	OEM
ZN460CP	OEM
ZN1004E	OEM
ZN1005E	OEM
ZN1010E	OEM
ZN1010F	OEM
ZN1030E	OEM
ZN1034E	OEM
ZN1034P	OEM
ZN1034T	OEM
ZN1040AE	OEM
ZN1040E	OEM
ZN1060	OEM
ZN2010E	OEM
ZN2010F	OEM

If replacement code is OEM, contact original manufacturer for replacement.

DEVICE TYPE	REPL CODE
ZN5400E	OEM
ZN5400F	OEM
ZN5401E	OEM
ZN5401F	OEM
ZN5402E	OEM
ZN5402F	OEM
ZN5403J	OEM
ZN5404E	OEM
ZN5404F	OEM
ZN5405E	OEM
ZN5405F	OEM
ZN5408E	OEM
ZN5408J	OEM
ZN5409E	OEM
ZN5409J	OEM
ZN5410E	OEM
ZN5410F	OEM
ZN5412E	OEM
ZN5412J	OEM
ZN5420E	OEM
ZN5420F	OEM
ZN5425E	OEM
ZN5425J	OEM
ZN5427E	OEM
ZN5427J	OEM
ZN5428E	OEM
ZN5428J	OEM
ZN5430E	OEM
ZN5430F	OEM
ZN5432E	OEM
ZN5432J	OEM
ZN5437E	OEM
ZN5437J	OEM
ZN5438E	OEM
ZN5438J	OEM
ZN5440E	OEM
ZN5440F	OEM
ZN5442E	OEM
ZN5442J	OEM
ZN5450E	OEM
ZN5450F	OEM
ZN5451E	OEM
ZN5451F	OEM
ZN5453E	OEM
ZN5453F	OEM
ZN5454E	OEM
ZN5454F	OEM
ZN5460E	OEM
ZN5460F	OEM
ZN5470E	OEM
ZN5470F	OEM
ZN5472E	OEM
ZN5472F	OEM
ZN5473E	OEM
ZN5473F	OEM
ZN5474E	OEM
ZN5474F	OEM
ZN5475E	OEM
ZN5476E	OEM
ZN5486E	OEM
ZN5486J	OEM
ZN5491AE	OEM
ZN5492AE	OEM
ZN5492AJ	OEM
ZN5493AE	OEM
ZN5493AJ	OEM
ZN5494E	OEM
ZN5494J	OEM
ZN5495AE	OEM
ZN5495AJ	OEM
ZN5496E	OEM
ZN5496J	OEM
ZN7400E	0232
ZN7400F	0232
ZN7401E	0268
ZN7401F	0268
ZN7402E	0314
ZN7402F	0314
ZN7403E	0335
ZN7403J	0335
ZN7404E	0361
ZN7404F	OEM
ZN7405E	0385
ZN7405F	OEM
ZN7408E	0467
ZN7408J	0467
ZN7409E	0492
ZN7409J	0492
ZN7410E	0512
ZN7410F	0512
ZN7412E	2242
ZN7412J	2242
ZN7420E	0697
ZN7420F	0697
ZN7425E	3477
ZN7425J	3477
ZN7427E	0818
ZN7427J	0818
ZN7428E	4180
ZN7428J	4180
ZN7430E	0873
ZN7430F	0873
ZN7432E	0899
ZN7432J	0899
ZN7437E	3520
ZN7437J	3520
ZN7438E	0996
ZN7438J	0996
ZN7440E	1024

DEVICE TYPE	REPL CODE
ZN7440F	1024
ZN7441AE	1038
ZN7442E	1052
ZN7442J	1052
ZN7450E	0743
ZN7450F	0743
ZN7451E	1168
ZN7451F	1168
ZN7453E	1185
ZN7453F	1185
ZN7454E	1202
ZN7454F	1202
ZN7460E	1275
ZN7460F	1275
ZN7470E	1404
ZN7470F	1404
ZN7472E	1426
ZN7472F	1426
ZN7473E	1172
ZN7473F	1172
ZN7474E	1313
ZN7474F	1313
ZN7475E	1432
ZN7476E	1158
ZN7486E	1368
ZN7486J	1368
ZN7490E	1208
ZN7490F	1208
ZN7491AE	OEM
ZN7491AJ	OEM
ZN7492AE	0834
ZN7492AJ	0834
ZN7492E	0834
ZN7492F	0834
ZN7493AE	0569
ZN7493AJ	0569
ZN7493E	0569
ZN7493F	0569
ZN7494E	1702
ZN7494J	OEM
ZN7495AE	1486
ZN7495AJ	OEM
ZN7496E	1715
ZN7496J	OEM
ZN54107E	OEM
ZN54107F	OEM
ZN54121E	OEM
ZN54121J	OEM
ZN54122E	OEM
ZN54122J	OEM
ZN54123E	OEM
ZN54123J	OEM
ZN54154J	OEM
ZN54155E	OEM
ZN54155J	OEM
ZN54161E	OEM
ZN54161J	OEM
ZN54163E	OEM
ZN54163J	OEM
ZN54164E	OEM
ZN54164J	OEM
ZN54165E	OEM
ZN54165J	OEM
ZN54166J	OEM
ZN54174E	OEM
ZN54174J	OEM
ZN54175E	OEM
ZN54175J	OEM
ZN54191E	OEM
ZN54191J	OEM
ZN54192E	OEM
ZN54192J	OEM
ZN54193E	OEM
ZN54193J	OEM
ZN54194E	OEM
ZN54194J	OEM
ZN54197E	OEM
ZN54197J	OEM
ZN74107E	0942
ZN74107F	OEM
ZN74121E	0175
ZN74121J	OEM
ZN74122E	1139
ZN74122J	OEM
ZN74123E	1157
ZN74123J	OEM
ZN74154E	5970
ZN74154J	OEM
ZN74155E	1576
ZN74155J	OEM
ZN74161E	1645
ZN74161J	OEM
ZN74163E	1666
ZN74163J	OEM
ZN74164E	0734
ZN74164J	OEM
ZN74165E	1685
ZN74165J	OEM
ZN74166E	0231
ZN74166J	OEM
ZN74174E	1770
ZN74174J	OEM
ZN74175J	OEM
ZN74191E	1917
ZN74191J	OEM
ZN74192E	1921
ZN74192J	OEM
ZN74193E	1926
ZN74193J	OEM

DEVICE TYPE	REPL CODE
ZN74194E	OEM
ZN74194J	OEM
ZN74197E	OEM
ZN74197J	OEM
ZN35024712	0925
ZNA234E	OEM
ZNM442E	1052
ZNP100	OEM
ZNP102	OEM
ZNP103	OEM
ZNPCM1CE	OEM
ZNR-3K101	0139
ZNR10K241U	1993
ZNREF025A1	OEM
ZNREF025A2	OEM
ZNREF025A3	OEM
ZNREF025B1	OEM
ZNREF025B2	OEM
ZNREF025B3	OEM
ZNREF025C1	OEM
ZNREF025C2	OEM
ZNREF025C3	OEM
ZNREF040A1	OEM
ZNREF040A2	OEM
ZNREF040A3	OEM
ZNREF040B1	OEM
ZNREF040B2	OEM
ZNREF040B3	OEM
ZNREF040C1	OEM
ZNREF040C2	OEM
ZNREF040C3	OEM
ZNREF050A1	OEM
ZNREF050A2	OEM
ZNREF050A3	OEM
ZNREF050B1	OEM
ZNREF050B2	OEM
ZNREF050B3	OEM
ZNREF050C1	OEM
ZNREF050C2	OEM
ZNREF050C3	OEM
ZNREF062A1	OEM
ZNREF062A2	OEM
ZNREF062A3	OEM
ZNREF062B1	OEM
ZNREF062B2	OEM
ZNREF062B3	OEM
ZNREF062C1	OEM
ZNREF062C2	OEM
ZNREF062C3	OEM
ZNREF100A1	OEM
ZNREF100A2	OEM
ZNREF100A3	OEM
ZNREF100B1	OEM
ZNREF100B2	OEM
ZNREF100B3	OEM
ZNREF100C1	OEM
ZNREF100C2	OEM
ZNREF100C3	OEM
ZNR-K510Z	OEM
ZNV0545L	OEM
Z06.8	0025
Z06.8A	0025
Z06.8B	0025
Z07.5	0649
Z07.5A	0649
Z07.5B	0649
Z08.2	0244
Z08.2A	0244
Z08.2B	0244
Z09.1	0012
Z09.1A	0012
Z09.1B	0012
Z010	0170
Z010A	0170
Z010B	0170
Z011	0317
Z011A	0317
Z011B	0317
Z0-12	0137
Z012	0137
Z0-12A	0137
Z012A	0137
Z0-12B	0137
Z012B	0137
Z013	0365
Z013A	0365
Z013B	0365
Z0-15	0002
Z015	0002
Z0-15A	0002
Z015A	0002
Z0-15B	0002
Z015B	0420
Z016	0420
Z016A	0420
Z016B	0420
Z018	0495
Z018A	0495
Z018B	0495
Z020	0531
Z020A	0531
Z020B	0531
Z022	0565
Z022A	0565
Z022B	0565
Z024	0402
Z024A	0402
Z024B	0402
Z0-27	0441

DEVICE TYPE	REPL CODE
Z027	0441
Z0-27A	0441
Z027A	1674
Z0-27B	0441
Z027B	1674
Z030	0726
Z030A	0726
Z030B	0726
Z0-33	0039
Z033	0039
Z0-33A	0039
Z033A	0039
Z0-33B	0039
Z033B	0039
Z036	0820
Z036A	0820
Z036B	0820
Z039	0350
Z039A	0350
Z039B	0350
Z043	0931
Z043A	0931
Z043B	0931
Z047	0999
Z047A	0999
Z047B	0999
Z051	0502
Z051A	0502
Z051B	0502
Z056	0869
Z056A	0869
Z056B	0869
Z0-62	0783
Z062	0783
Z0-62A	0783
Z062A	2157
Z0-62B	0783
Z062B	2157
Z068	2157
Z068A	2157
Z068B	2157
Z075	1189
Z075A	1189
Z075B	1189
Z0-82	0331
Z082	0331
Z0-82A	0331
Z082A	0331
Z0-82B	0331
Z082B	0331
Z091	1311
Z091A	1311
Z091B	1311
Z0100	0098
Z0100A	0098
Z0100B	0098
Z0-110	0149
Z0110	0149
Z0-110A	0149
Z0110A	0149
Z0-110B	0149
Z0110B	0149
Z0120	0186
Z0120A	0186
Z0120B	0186
Z0130	0213
Z0130A	0213
Z0130B	0213
Z0150	0028
Z0150A	0028
Z0150B	0028
Z0160	0255
Z0160A	0255
Z0160B	0255
Z0180	0367
Z0180A	0367
Z0180B	0367
Z0200	0421
Z0200A	0421
Z0200B	0421
ZOA2.4	1276
ZOA2.7	0760
ZOB3.6	0188
ZOB5.1	0162
ZOB5.6	0157
ZOB6.2	0636
ZOB12	0137
ZOB15	0002
ZOB27	0441
ZOB33	0039
ZOB62	0783
ZOB82	0331
ZOB110	0149
ZOD6.2	0636
ZOD15	0002
ZOD27	0441
ZOD33	0039
ZOD62	0783
ZOD82	0331
ZOD110	0149
ZP2.7	0760
ZP3	OEM
ZP3.3	0299
ZP3.6	0188
ZP3.9	0036
ZP4.3	0533
ZP4.7	0140
ZP5.1	0162
ZP5.6	0157

DEVICE TYPE	REPL CODE
ZP6.2	0471
ZP6.8	0025
ZP6.8A	0025
ZP6.8B	0025
ZP7.5	0649
ZP7.5A	0649
ZP7.5B	0649
ZP8.2	0244
ZP8.2A	0244
ZP8.2B	0244
ZP9.1	0012
ZP9.1A	0012
ZP9.1B	0012
ZP10	0170
ZP10A	0170
ZP10B	OEM
ZP11	0317
ZP11A	0317
ZP11B	0317
ZP12	0137
ZP12A	0137
ZP12B	0137
ZP13	0365
ZP13A	0365
ZP13B	0365
ZP15	0002
ZP15A	0002
ZP15B	0002
ZP16	0420
ZP16A	0420
ZP16B	0420
ZP18	0495
ZP18A	0495
ZP18B	0170
ZP20	0531
ZP20-0120	0080
ZP20A	0531
ZP20B	0531
ZP22	0565
ZP22A	0565
ZP22B	0565
ZP24	0402
ZP24A	0402
ZP24B	0402
ZP27	0441
ZP27A	0441
ZP27B	0441
ZP30	0726
ZP30A	0726
ZP30B	0726
ZP33	0039
ZP33A	0039
ZP33B	0039
ZP36	0820
ZP36A	0820
ZP36B	0820
ZP39	0350
ZP39A	0350
ZP39B	0350
ZP43	0931
ZP43A	0931
ZP43B	0931
ZP47	0999
ZP47A	0999
ZP47B	0999
ZP51	0502
ZP51A	0502
ZP51B	0502
ZP56	0869
ZP56A	0869
ZP56B	0869
ZP62	0783
ZP62A	0783
ZP62B	0783
ZP68	2157
ZP68A	2157
ZP68B	2157
ZP75	1189
ZP75A	1189
ZP75B	1189
ZP82	0331
ZP82A	0331
ZP82B	0331
ZP91	1311
ZP91A	1311
ZP91B	1311
ZP100	0098
ZP100A	0098
ZP100B	0098
ZP110	0149
ZP110A	0149
ZP110B	0149
ZP120	0186
ZP120A	0186
ZP120B	0186
ZP130	0213
ZP130A	0213
ZP130B	0213
ZP150	0028
ZP150A	0028
ZP150B	0028
ZP160	0255
ZP160A	0255
ZP160B	0255
ZP180	0367
ZP180A	0367
ZP180B	0367
ZP200	0421
ZP200A	0421

DEVICE TYPE	REPL CODE
ZP200B	0421
ZPB-A	OEM
ZPB-PROM	OEM
ZPD1	OEM
ZPD2.7	0760
ZPD3	1713
ZPD3.3	0299
ZPD3.6	0376
ZPD3.9	0036
ZPD-4.3	OEM
ZPD4.3	0275
ZPD-4.3V	0275
ZPD4.7	0140
ZPD5V1	0162
ZPD5V6	0157
ZPD5.1	0041
ZPD5.6	0253
ZPD6V2	0471
ZPD6V8	0025
ZPD6.2	0471
ZPD6.8	0062
ZPD7V5	0649
ZPD7.1	0649
ZPD7.5	0077
ZPD8V2	0244
ZPD8.2	0244
ZPD9V1	0057
ZPD9.1	0057
ZPD10	0064
ZPD10V	0170
ZPD11	0181
ZPD11V	0317
ZPD12	0052
ZPD12V	0052
ZPD13	0053
ZPD13V	0365
ZPD15	0686
ZPD15V	0686
ZPD16	0445
ZPD16V	0420
ZPD18	0375
ZPD18V	0495
ZPD20	0700
ZPD20V	0531
ZPD-22	5248
ZPD22	0705
ZPD22V	0565
ZPD24	0494
ZPD24V	0402
ZPD27	0455
ZPD30	0195
ZPD30V	0726
ZPD33	0166
ZPD33V	0039
ZPD36	0010
ZPD39	0032
ZPD43	0054
ZPD47	0068
ZPD51	0092
ZPE1.5	OEM
ZPFAK-50701	2697
ZPFAK-50801	2697
ZPFAK-50901	2697
ZPFAK-51001	2697
ZPU	OEM
ZPU100	0098
ZPU120	0186
ZPU150	0028
ZPU180	0367
ZPU200	OEM
ZPV0545L	OEM
ZPY1	OEM
ZPY3.9	0012
ZPY4.3	0533
ZPY4.7	0451
ZPY5.1	0162
ZPY5.6	0157
ZPY-6.2	OEM
ZPY6.2	0636
ZPY6.8	0025
ZPY7.5	0649
ZPY8.2	0244
ZPY9.1	0012
ZPY10	0170
ZPY11	0317
ZPY12	0137
ZPY13	0365
ZPY15	0002
ZPY16	0420
ZPY18	0495
ZPY20	0531
ZPY22	0565
ZPY24	0402
ZPY27	0441
ZPY30	0726
ZPY33	0039
ZPY36	0820
ZPY39	0350
ZPY43	0931
ZPY47	0999
ZPY51	0502
ZPY56	0869
ZPY62	0783
ZPY68	1268
ZPY75	1189
ZPY82	0170
ZPY91	1311
ZPY100	0098
ZQ-6	0301

DEVICE TYPE	REPL CODE
ZQ6.8	0025
ZQ6.8A	0025
ZQ6.8B	0025
ZQ7.5	0649
ZQ7.5A	0649
ZQ7.5B	0649
ZQ8.2	0244
ZQ8.2A	0244
ZQ8.2B	0244
ZQ9.1	0012
ZQ9.1A	0012
ZQ9.1B	0012
ZQ10	0170
ZQ10A	0317
ZQ10B	0170
ZQ11	0317
ZQ11A	0317
ZQ11B	0317
ZQ12	0137
ZQ12A	0137
ZQ12B	0137
ZQ13	0365
ZQ13A	0365
ZQ13B	0365
ZQ15	0002
ZQ15A	0002
ZQ15B	0002
ZQ16	0420
ZQ16A	0420
ZQ16B	0420
ZQ18	0495
ZQ18A	0495
ZQ18B	0495
ZQ20	0531
ZQ20A	0531
ZQ20B	0531
ZQ22	0565
ZQ22A	0565
ZQ22B	0565
ZQ24	0402
ZQ24A	0402
ZQ24B	0402
ZQ27	0441
ZQ27A	0441
ZQ27B	0441
ZQ30	0726
ZQ30A	0726
ZQ30B	0726
ZQ33	0039
ZQ33A	0039
ZQ33B	0039
ZQ36	0820
ZQ36A	0820
ZQ36B	0820
ZQ39	0350
ZQ39A	0350
ZQ39B	0350
ZQ43	0931
ZQ43A	0931
ZQ43B	0931
ZQ47	0999
ZQ47A	0999
ZQ47B	0999
ZQ51	0502
ZQ51A	0502
ZQ51B	0502
ZQ56	0869
ZQ56A	0869
ZQ56B	0869
ZQ62	0783
ZQ62A	0783
ZQ62B	0783
ZQ68	2157
ZQ68A	2157
ZQ68B	2157
ZQ75	1189
ZQ75A	1189
ZQ75B	1189
ZQ82	0331
ZQ82A	0331
ZQ82B	0331
ZQ91	1311
ZQ91A	1311
ZQ91B	1311
ZQ100	0098
ZQ100A	0098
ZQ100B	0098
ZQ110	0149
ZQ110A	0149
ZQ110B	0149
ZQ120	0186
ZQ120A	0186
ZQ120B	0186
ZQ130	0213
ZQ130A	0213
ZQ130B	0213
ZQ150	0028
ZQ150A	0028
ZQ150B	0028
ZQ160	0255
ZQ160A	0255
ZQ160B	0255
ZQ180	0367
ZQ180A	0367
ZQ180B	0367
ZQ200	0421
ZQ200A	0421
ZQ200B	0421
ZR10	0110

If replacement code is OEM, contact original manufacturer for replacement.

DEVICE TYPE	REPL CODE
ZR10T	OEM
ZR11	0953
ZR11T	OEM
ZR12	0242
ZR12T	OEM
ZR13	0540
ZR13T	OEM
ZR14	0540
ZR14T	OEM
ZR15	1355
ZR15T	OEM
ZR20	2002
ZR21	1250
ZR22	1250
ZR23	OEM
ZR24	OEM
ZR30C	OEM
ZR31C	OEM
ZR32C	OEM
ZR33C	OEM
ZR34C	OEM
ZR35C	OEM
ZR50B793-1	0137
ZR50B793-3	0441
ZR50B849-1	0441
ZR50B921-1	0441
ZR50B921-2	0649
ZR50B921-3	0002
ZR51B155-2	0247
ZR60	0015
ZR61	1355
ZR62	1355
ZR63	1355
ZR64	1355
ZR66	1355
ZR68	0087
ZR200	0084
ZR201	0090
ZR202	0097
ZR203	0188
ZR204	0109
ZR205	0157
ZR206	0122
ZR207	0649
ZR208	0131
ZR209	1083
ZR209T1	1083
ZR209T3	1083
ZR210	1083
ZR212	0137
ZR215	0420
ZR220	0531
ZR225	1606
ZR230	0726
ZR235	0820
ZR240	0931
ZR500	0015
ZR501	OEM
ZR502	OEM
ZR503	OEM
ZR504	OEM
ZR505	OEM
ZR590	0015
ZR590A	0015
ZR601	0953
ZR602	0242
ZR604	0540
ZR606	0965
ZR608	0087
ZR821-5	0471
ZR821-10	0471
ZR821-20	0471
ZR821-50	0471
ZR821-75	0471
ZR821-100	0471
ZR821-120	0471
ZR821-150	0471
ZR823-5	0471
ZR823-10	0471
ZR823-20	0471
ZR823-50	0471
ZR823-75	0471
ZR823-100	0471
ZR823-120	0471
ZR823-150	0471
ZR825-5	0471
ZR825-10	0471
ZR825-20	0471
ZR825-50	0471
ZR825-75	0471
ZR825-100	0471
ZR825-120	0471
ZR825-150	0471
ZR827-5	0471
ZR827-10	0471
ZR827-20	0471
ZR827-50	0471
ZR827-75	0471
ZR827-100	0471
ZR827-120	0471
ZR827-150	0471
ZR829-50	0471
ZR829-75	0471
ZR829-100	0471
ZR829-120	0471
ZR935-20	OEM
ZR935-50	OEM
ZR935-75	OEM
ZR935-100	OEM

DEVICE TYPE	REPL CODE
ZR935-120	OEM
ZR935-150	OEM
ZR936-20	OEM
ZR936-50	OEM
ZR936-75	OEM
ZR936-100	OEM
ZR936-120	OEM
ZR936-150	OEM
ZR937-20	OEM
ZR937-50	OEM
ZR937-75	OEM
ZR937-100	OEM
ZR937-120	OEM
ZR937-150	OEM
ZR938-50	OEM
ZR938-75	OEM
ZR938-100	OEM
ZR938-120	OEM
ZR939-50	OEM
ZR939-75	OEM
ZR939-100	OEM
ZR939-120	OEM
ZR1025	0015
ZR1031	0015
ZR1035	0015
ZR1076	0015
ZS	OEM
ZS5.6	0157
ZS5.6A	0157
ZS5.6B	0157
ZS6	0471
ZS7	1335
ZS8	0609
ZS8LB	OEM
ZS8.5	OEM
ZS9	OEM
ZS9.1	0012
ZS9.1A	0012
ZS9.1B	0012
ZS10A	0015
ZS-10B	0015
ZS10B	0015
ZS10LB	OEM
ZS11	OEM
ZS11LB	OEM
ZS11.2	OEM
ZS11.4	OEM
ZS12	0137
ZS12A	0137
ZS12B	0137
ZS12.4	OEM
ZS13LB	OEM
ZS14	OEM
ZS14LB	OEM
ZS15	0002
ZS15A	0002
ZS15B	0002
ZS17	OEM
ZS17LB	OEM
ZS18.6	OEM
ZS20A	0015
ZS20B	0015
ZS21	0015
ZS22	0015
ZS23	0015
ZS24	0015
ZS24.8	OEM
ZS25	0015
ZS27	0441
ZS27A	0441
ZS27B	0441
ZS30	0015
ZS30A	0015
ZS30B	0015
ZS31	0015
ZS31A	0015
ZS31B	0015
ZS32	0015
ZS32A	0015
ZS32B	0015
ZS33	0039
ZS33A	0039
ZS33B	0039
ZS34	0015
ZS34A	0015
ZS34B	0015
ZS37	OEM
ZS38	0164
ZS40	0133
ZS40A	1335
ZS40F	OEM
ZS41	OEM
ZS41F	OEM
ZS42	OEM
ZS42F	OEM
ZS43	OEM
ZS50	0133
ZS51	0609
ZS52	0609
ZS53	3911
ZS56	0164
ZS60	1156
ZS61	0471
ZS62	0783
ZS62A	0783
ZS62B	0783
ZS70	1355
ZS71	1355
ZS72	1355

DEVICE TYPE	REPL CODE
ZS73	1355
ZS74	1355
ZS74B	1355
ZS75	OEM
ZS76	1355
ZS78	0072
ZS78A	0071
ZS78B	0072
ZS82	0331
ZS82A	0331
ZS82B	0331
ZS90	0609
ZS91	0080
ZS92	0609
ZS94	3911
ZS100	0015
ZS101	0080
ZS102	0609
ZS103	0795
ZS104	0795
ZS106	0015
ZS108	0072
ZS110	0149
ZS110A	0149
ZS110B	0149
ZS120	1335
ZS121	0080
ZS122	0609
ZS123	0795
ZS124	0795
ZS142	0133
ZS143	OEM
ZS150	OEM
ZS151	OEM
ZS152	OEM
ZS153	OEM
ZS154	OEM
ZS155	OEM
ZS170	OEM
ZS171	OEM
ZS172	OEM
ZS174	0015
ZS176	OEM
ZS178	OEM
ZS270	0110
ZS271	0953
ZS272	0242
ZS274	0540
ZS605	0471
ZS701	0242
ZS702	0242
ZS704	OEM
ZS706	OEM
ZS708	OEM
ZS901	OEM
ZS902	OEM
ZS905	OEM
ZS6010	0471
ZS6011	0471
ZS6020	0471
ZS6021	0471
ZSA7	OEM
ZSA11	OEM
ZSA49	0050
ZSB364	0004
ZSB439	0004
ZSD51A	OEM
ZSD81A	OEM
ZSF51B	OEM
ZSF51CT	OEM
ZSF81B	OEM
ZSF81CT	OEM
ZSS51A	OEM
ZSS51B	OEM
ZSS53A	OEM
ZSS53B	OEM
ZSS55A	OEM
ZSS55B	OEM
ZSS56B	OEM
ZSS57B	OEM
ZSS59A	OEM
ZSS59B	OEM
ZSS81A	OEM
ZSS81B	OEM
ZSS83A	OEM
ZSS83B	OEM
ZSS85A	OEM
ZSS85B	OEM
ZSS86B	OEM
ZSS87B	OEM
ZSS89A	OEM
ZSS89B	OEM
ZST51A	OEM
ZST52A	OEM
ZST52B	OEM
ZST53A	OEM
ZST53B	OEM
ZST54A	OEM
ZST81A	OEM
ZST82A	OEM
ZST82B	OEM
ZST83A	OEM
ZST83B	OEM
ZST84A	OEM
ZSY0-7	OEM
ZSY0.7	OEM
ZT06	0007
ZT5.6	0157
ZT5.6A	0157

DEVICE TYPE	REPL CODE
ZT5.6B	0157
ZT6.2	0636
ZT6.2A	0636
ZT6.2B	0636
ZT6.8	0025
ZT6.8A	0025
ZT6.8B	0025
ZT7.5	0649
ZT7.5A	0649
ZT7.5B	0244
ZT8.2	0244
ZT8.2A	0244
ZT8.2B	0244
ZT9.1	0012
ZT9.1A	0012
ZT9.1B	0012
ZT10	0170
ZT10A	0170
ZT10B	0170
ZT11	0317
ZT11A	0317
ZT11B	0317
ZT12	0137
ZT12A	0137
ZT12B	0137
ZT13	0365
ZT13A	0365
ZT13B	0365
ZT15	0002
ZT15A	0002
ZT15B	0002
ZT16	0420
ZT16A	0420
ZT16B	0420
ZT18	0495
ZT18A	0495
ZT18B	0495
ZT20	0016
ZT20(ZENER)	0531
ZT20A	0531
ZT20B	0531
ZT21	0016
ZT22	0016
ZT22(ZENER)	0565
ZT22A	0565
ZT22B	0565
ZT23	0016
ZT24	0016
ZT24(ZENER)	0402
ZT24A	0402
ZT24B	0402
ZT27	0079
ZT27(ZENER)	0441
ZT27A	0441
ZT27B	0441
ZT30	0726
ZT30A	0726
ZT30B	0726
ZT33	0039
ZT33A	0039
ZT33B	0039
ZT36	0820
ZT36A	0820
ZT36B	0820
ZT39	0350
ZT39A	0350
ZT39B	0350
ZT40	0016
ZT41	0016
ZT42	0016
ZT43	0016
ZT43(ZENER)	0931
ZT43A	0931
ZT43B	0931
ZT44	0016
ZT47	0999
ZT47A	0999
ZT47B	0999
ZT50	0016
ZT51	0502
ZT51A	0502
ZT51B	0502
ZT56	0869
ZT56A	0869
ZT56B	0869
ZT60	0016
ZT61	0016
ZT62	0016
ZT62(ZENER)	0783
ZT62A	0783
ZT62B	0783
ZT63	0016
ZT64	0016
ZT66	0016
ZT68	0016
ZT68(ZENER)	2157
ZT68A	2157
ZT68B	2157
ZT75	1189
ZT75A	1189
ZT75B	1189
ZT80	0016
ZT81	0016
ZT82	0016
ZT82(ZENER)	0331
ZT82A	0783
ZT82B	0783
ZT83	0016
ZT84	0016

DEVICE TYPE	REPL CODE
ZT85/38	OEM
ZT86	0086
ZT87	0016
ZT88	0086
ZT89	0016
ZT90	0086
ZT91	1480
ZT91(ZENER)	1311
ZT91A	1311
ZT91B	1311
ZT93	0086
ZT94	0086
ZT100	0098
ZT100A	0098
ZT100B	0098
ZT110	0016
ZT110(ZENER)	0149
ZT110A	0149
ZT110B	0149
ZT111	0016
ZT112	0016
ZT113	0016
ZT114	0016
ZT116	0016
ZT117	0016
ZT118	0016
ZT119	0016
ZT120	0186
ZT120A	0186
ZT120B	0186
ZT130	0213
ZT130A	0213
ZT130B	0213
ZT131	0037
ZT132	OEM
ZT150	0028
ZT150A	0028
ZT150B	0028
ZT152	0037
ZT153	0037
ZT154	0037
ZT160	0255
ZT160A	0255
ZT160B	0255
ZT180	0037
ZT180(ZENER)	0367
ZT180A	0367
ZT180B	0367
ZT181	0037
ZT182	0037
ZT183	0037
ZT184	0037
ZT187	0037
ZT189	0037
ZT190	0086
ZT191	0086
ZT192	0086
ZT193	0086
ZT200	0421
ZT200A	0421
ZT200B	0421
ZT202	0016
ZT202P	0016
ZT203	0016
ZT203P	0016
ZT204	0016
ZT204P	0016
ZT210	0126
ZT211	0126
ZT280	0037
ZT281	0037
ZT282	0037
ZT283	0037
ZT284	0037
ZT286	0126
ZT287	0037
ZT402	0016
ZT402P	0016
ZT403	0016
ZT403P	0016
ZT404	0016
ZT404P	0016
ZT406	0016
ZT502-10	OEM
ZT502-40	OEM
ZT502-60	OEM
ZT600	0560
ZT696	0016
ZT697	0016
ZT706	0016
ZT706A	0016
ZT708	0016
ZT709	0144
ZT805	OEM
ZT917	0144
ZT918	0144
ZT929	OEM
ZT930	0111
ZT1420	0016
ZT1479	0086
ZT1481	0086
ZT1483	0042
ZT1484	0042
ZT1485	0042
ZT1486	0042
ZT1487	0103
ZT1488	0103
ZT1489	0103
ZT1490	0103

DEVICE TYPE	REPL CODE
ZT1511	0561
ZT1512	0561
ZT1513	0561
ZT1514	OEM
ZT1613	0086
ZT1700	0086
ZT1701	0042
ZT1702	0103
ZT1703	OEM
ZT1708	0016
ZT1711	0086
ZT2102	0086
ZT2205	0016
ZT2206	0016
ZT2270	0086
ZT2368	0016
ZT2369	0016
ZT2369A	0016
ZT2475	0144
ZT2476	0016
ZT2477	0016
ZT2631	0622
ZT2708	0007
ZT2857	0144
ZT2876	0042
ZT2938	0016
ZT3053	0595
ZT3229	OEM
ZT3262	OEM
ZT3269A	0144
ZT3375	0560
ZT3439	OEM
ZT3440	0168
ZT3512	0086
ZT3600	0144
ZT3866	0086
ZT7399	OEM
ZT7488/08	OEM
ZT7488/18	OEM
ZT7488/28	OEM
ZTB3-12	OEM
ZTB3-28	OEM
ZTB3-28FL	OEM
ZTB3-28ST	OEM
ZTB12-12	OEM
ZTB12-28	OEM
ZTB12-28FL	OEM
ZTB12-28ST	OEM
ZTB25-12	OEM
ZTB25-28	OEM
ZTB25-28FL	OEM
ZTB25-28ST	OEM
ZTB70-28FL	OEM
ZTC3.12FL	OEM
ZTC238CK	OEM
ZTE1,5	OEM
ZTE1.5	OEM
ZTE2	OEM
ZTE2.4	1276
ZTE2.7	0760
ZTE3	0118
ZTE3.3	0299
ZTE3.6	0376
ZTE3.9	0036
ZTE4.3	0275
ZTE4.7	0140
ZTE5.1	0041
ZTK6,8	OEM
ZTK6.8	OEM
ZTK9	OEM
ZTK11	OEM
ZTK-11W	OEM
ZTK18	OEM
ZTK22	OEM
ZTK22Z	0565
ZTK23	OEM
ZTK27	OEM
ZTK-33	1329
ZTK33	1329
ZTK33A	1329
ZTK33BDPD	0039
ZTK33V	OEM
ZTP12	OEM
ZTR-1N60	0143
ZTR-B54	0004
ZTR-B56	0004
ZTR-F0R2B	0901
ZTR-FOR2B	0901
ZTR-SFOR2G	OEM
ZTR-W06B	0023
ZTR-W06C	0023
ZTR-W06B	0015
ZTW6,8-2	OEM
ZTW6,8-3	OEM
ZTW6.8-1	OEM
ZTW6.8-5	OEM
ZTX3IU	0144
ZTX41	OEM
ZTX42	OEM
ZTX60	OEM
ZTX107	0079
ZTX107A	OEM
ZTX107A-(RED)	0595
ZTX107B	OEM
ZTX108	0079
ZTX108A	OEM
ZTX108A-(RED)	0191
ZTX108B	OEM
ZTX108C	OEM

DEVICE TYPE	REPL CODE
ZTX109	0079
ZTX109B	OEM
ZTX109C	OEM
ZTX114	0079
ZTX114K	OEM
ZTX114L	OEM
ZTX114M	OEM
ZTX212	0037
ZTX212A-(RED)	0037
ZTX212AK	OEM
ZTX212AL	OEM
ZTX212AM	OEM
ZTX212BK	OEM
ZTX212BL	OEM
ZTX212BM	OEM
ZTX212K	OEM
ZTX212L	OEM
ZTX212M	OEM
ZTX213	0037
ZTX213A-(RED)	0037
ZTX213AK	OEM
ZTX213AL	OEM
ZTX213AM	OEM
ZTX213BK	OEM
ZTX213BL	OEM
ZTX213BM	OEM
ZTX213CK	OEM
ZTX213CL	OEM
ZTX213CM	OEM
ZTX213K	OEM
ZTX213L	OEM
ZTX213M	OEM
ZTX214	0037
ZTX214BK	OEM
ZTX214BL	OEM
ZTX214BM	OEM
ZTX214CK	OEM
ZTX214CL	OEM
ZTX214CM	OEM
ZTX214K	OEM
ZTX214L	OEM
ZTX214M	OEM
ZTX223	OEM
ZTX223AK	OEM
ZTX223AL	OEM
ZTX223AM	OEM
ZTX223BK	OEM
ZTX223BL	OEM
ZTX223BM	OEM
ZTX223K	OEM
ZTX223L	OEM
ZTX223M	OEM
ZTX237AK	OEM
ZTX237AL	OEM
ZTX237AM	OEM
ZTX237BK	OEM
ZTX237BL	OEM
ZTX237BM	OEM
ZTX237K	OEM
ZTX237L	OEM
ZTX237M	OEM
ZTX238AK	OEM
ZTX238AL	OEM
ZTX238AM	OEM
ZTX238BK	OEM
ZTX238BL	OEM
ZTX238BM	OEM
ZTX238CL	OEM
ZTX238CM	OEM
ZTX238K	OEM
ZTX238L	OEM
ZTX238M	OEM
ZTX239BL	OEM
ZTX239BM	OEM
ZTX239CK	OEM
ZTX239CL	OEM
ZTX239CM	OEM
ZTX239K	OEM
ZTX239L	OEM
ZTX239M	OEM
ZTX300	0079
ZTX300K	OEM
ZTX300L	OEM
ZTX300M	OEM
ZTX301	0079
ZTX301K	OEM
ZTX301L	OEM
ZTX301M	OEM
ZTX302	0079
ZTX302K	OEM
ZTX302L	OEM
ZTX302M	OEM
ZTX303	0079
ZTX303K	OEM
ZTX303L	OEM
ZTX303M	OEM
ZTX304	0079
ZTX304K	OEM
ZTX304L	OEM
ZTX304M	OEM
ZTX310	0079
ZTX310K	OEM
ZTX310L	OEM
ZTX310M	OEM
ZTX311	0079
ZTX311K	OEM
ZTX311L	OEM
ZTX311M	OEM
ZTX312	0079

If replacement code is OEM, contact original manufacturer for replacement.

DEVICE TYPE	REPL CODE
ZTX312K	OEM
ZTX312L	OEM
ZTX312M	OEM
ZTX313K	OEM
ZTX313L	OEM
ZTX313M	OEM
ZTX314K	OEM
ZTX314L	OEM
ZTX314M	OEM
ZTX320	0144
ZTX320K	OEM
ZTX320L	OEM
ZTX320M	OEM
ZTX321	0144
ZTX321K	OEM
ZTX321L	OEM
ZTX321M	OEM
ZTX322	OEM
ZTX322M	OEM
ZTX323	OEM
ZTX323K	OEM
ZTX323L	OEM
ZTX323M	OEM
ZTX325	0259
ZTX325K	OEM
ZTX325L	OEM
ZTX325M	OEM
ZTX326	0259
ZTX326A	OEM
ZTX326AK	OEM
ZTX326AL	OEM
ZTX326AM	OEM
ZTX326K	OEM
ZTX326L	OEM
ZTX326M	OEM
ZTX327K	OEM
ZTX327L	OEM
ZTX327M	OEM
ZTX330	0079
ZTX330K	OEM
ZTX330L	OEM
ZTX330M	OEM
ZTX331	0079
ZTX331L	OEM
ZTX331M	OEM
ZTX337	OEM
ZTX337A	OEM
ZTX337AK	OEM
ZTX337AL	OEM
ZTX337AM	OEM
ZTX337B	OEM
ZTX337BK	OEM
ZTX337BL	OEM
ZTX337BM	OEM
ZTX337C	OEM
ZTX337CK	OEM
ZTX337CL	OEM
ZTX337CM	OEM
ZTX338	OEM
ZTX338A	OEM
ZTX338AK	OEM
ZTX338AL	OEM
ZTX338AM	OEM
ZTX338B	OEM
ZTX338BK	OEM
ZTX338BL	OEM
ZTX338BM	OEM
ZTX338C	OEM
ZTX338CK	OEM
ZTX338CL	OEM
ZTX338CM	OEM
ZTX341	0285
ZTX342	0285
ZTX350	OEM
ZTX360	0079
ZTX382	OEM
ZTX382B	OEM
ZTX382BK	OEM
ZTX382BL	OEM
ZTX382BM	OEM
ZTX382C	OEM
ZTX382CK	OEM
ZTX382CL	OEM
ZTX382CM	OEM
ZTX383	OEM
ZTX383B	OEM
ZTX383BK	OEM
ZTX383BL	OEM
ZTX383BM	OEM
ZTX383C	OEM
ZTX383CK	OEM
ZTX383CL	OEM
ZTX383CM	OEM
ZTX384	OEM
ZTX384B	OEM
ZTX384BK	OEM
ZTX384BL	OEM
ZTX384BM	OEM
ZTX384C	OEM
ZTX384CK	OEM
ZTX384CL	OEM
ZTX384CM	OEM
ZTX449	OEM
ZTX449K	OEM
ZTX449L	OEM
ZTX449M	OEM
ZTX450	OEM
ZTX451	OEM
ZTX452	OEM

DEVICE TYPE	REPL CODE
ZTX453	OEM
ZTX454	OEM
ZTX455	OEM
ZTX500	0126
ZTX500K	OEM
ZTX500L	OEM
ZTX500M	OEM
ZTX501	0037
ZTX501K	OEM
ZTX501L	OEM
ZTX501M	OEM
ZTX502	0037
ZTX502K	OEM
ZTX502L	OEM
ZTX502M	OEM
ZTX503	0037
ZTX503K	OEM
ZTX503L	OEM
ZTX503M	OEM
ZTX504	0037
ZTX504K	OEM
ZTX504L	OEM
ZTX504M	OEM
ZTX510	0037
ZTX510K	OEM
ZTX510L	OEM
ZTX510M	OEM
ZTX530	0037
ZTX531	0037
ZTX537	OEM
ZTX537A	OEM
ZTX537AK	OEM
ZTX537AL	OEM
ZTX537AM	OEM
ZTX537B	OEM
ZTX537BK	OEM
ZTX537BL	OEM
ZTX537BM	OEM
ZTX537C	OEM
ZTX537CL	OEM
ZTX537CM	OEM
ZTX538	OEM
ZTX538A	OEM
ZTX538AK	OEM
ZTX538AL	OEM
ZTX538AM	OEM
ZTX538B	OEM
ZTX538BK	OEM
ZTX538BL	OEM
ZTX538BM	OEM
ZTX538C	OEM
ZTX538CK	OEM
ZTX538CL	OEM
ZTX538CM	OEM
ZTX541	0342
ZTX541K	OEM
ZTX541L	OEM
ZTX541M	OEM
ZTX542	0342
ZTX542K	OEM
ZTX542L	OEM
ZTX542M	OEM
ZTX549	OEM
ZTX550	OEM
ZTX550K	OEM
ZTX550L	OEM
ZTX550M	OEM
ZTX551	OEM
ZTX551K	OEM
ZTX551L	OEM
ZTX551M	OEM
ZTX552	OEM
ZTX600	OEM
ZTX600K	OEM
ZTX600L	OEM
ZTX600M	OEM
ZTX601	OEM
ZTX601K	OEM
ZTX601L	OEM
ZTX649	OEM
ZTX650	0079
ZTX650K	OEM
ZTX650L	OEM
ZTX650M	OEM
ZTX650S	OEM
ZTX651	0079
ZTX651K	OEM
ZTX651L	OEM
ZTX651M	OEM
ZTX651S	OEM
ZTX652	OEM
ZTX652K	OEM
ZTX652L	OEM
ZTX652M	OEM
ZTX652S	OEM
ZTX653	OEM
ZTX653K	OEM
ZTX653L	OEM
ZTX653M	OEM
ZTX653S	OEM
ZTX749	OEM
ZTX750	2260
ZTX750K	OEM
ZTX750L	OEM
ZTX750M	OEM
ZTX750S	OEM
ZTX751	2260
ZTX751K	OEM
ZTX751L	OEM

DEVICE TYPE	REPL CODE
ZTX751M	OEM
ZTX751S	OEM
ZTX752	OEM
ZTX752K	OEM
ZTX752L	OEM
ZTX752M	OEM
ZTX752S	OEM
ZTX753	OEM
ZTX753K	OEM
ZTX753L	OEM
ZTX753M	OEM
ZTX753S	OEM
ZTX3702	0037
ZTX3703	0037
ZTX3866	OEM
ZTX3905	0037
ZTX4402	0037
ZTX4403	OEM
ZU6.8	0062
ZU6.8A	0062
ZU6.8B	0062
ZU7.5	0077
ZU7.5A	0077
ZU7.5B	0077
ZU8.2	0165
ZU8.2A	0165
ZU8.2B	0165
ZU9.1	0057
ZU9.1A	0057
ZU9.1B	0057
ZU10	0064
ZU10A	0064
ZU10B	0064
ZU11	0181
ZU11A	0181
ZU11B	0181
ZU12	0052
ZU12A	0052
ZU12B	0052
ZU13	0053
ZU13A	0053
ZU13B	0053
ZU15	0879
ZU15A	0686
ZU15B	0686
ZU16	0445
ZU16A	0445
ZU16B	0445
ZU18	0375
ZU18A	0375
ZU18B	0375
ZU20	0700
ZU20A	0700
ZU20B	0700
ZU22	0705
ZU22A	0705
ZU22B	0705
ZU24	0494
ZU24A	0494
ZU24B	0494
ZU27	0455
ZU27A	0455
ZU27B	0455
ZU30	0195
ZU30A	0195
ZU30B	0195
ZU33	0166
ZU33A	0166
ZU33B	0166
ZU36	0010
ZU36A	0010
ZU36B	0010
ZU39	0032
ZU39A	0032
ZU39B	0032
ZU43	0054
ZU43A	0054
ZU43B	0054
ZU47	0068
ZU47A	0068
ZU47B	0068
ZU51	0092
ZU51A	0092
ZU51B	0092
ZU56	0125
ZU56A	0125
ZU56B	0125
ZU62	0152
ZU62A	0152
ZU62B	0152
ZU68	0173
ZU68A	0173
ZU68B	0173
ZU75	0094
ZU75A	0094
ZU75B	0094
ZU82	0049
ZU82A	0049
ZU82B	0049
ZU91	0156
ZU91A	0156
ZU91B	0156
ZU100	0189
ZU100A	0189
ZU100B	0189
ZU110	0099
ZU110A	0099
ZU110B	0099
ZU120	0089

DEVICE TYPE	REPL CODE
ZU120A	0089
ZU120B	0089
ZU130	0287
ZU130A	0287
ZU130B	0287
ZU150	0340
ZU150A	0340
ZU150B	0340
ZU160	0370
ZU160A	0370
ZU160B	0370
ZU180	0425
ZU180A	0425
ZU180B	0425
ZU200	1473
ZU200A	1473
ZU200B	1473
ZV6.8	0025
ZV6.8A	0025
ZV6.8B	0025
ZV7.5	0649
ZV7.5A	0244
ZV7.5B	0244
ZV8.2	0244
ZV8.2A	0244
ZV8.2B	0244
ZV9.1	0012
ZV9.1A	0012
ZV9.1B	0012
ZV10	0170
ZV10A	0170
ZV10B	0170
ZV11	0317
ZV11A	0317
ZV11B	0317
ZV12	0137
ZV12A	0137
ZV12B	0137
ZV13	0365
ZV13A	0365
ZV13B	0365
ZV15	0002
ZV15A	0002
ZV15B	0002
ZV16	0420
ZV16A	0420
ZV16B	0420
ZV18	0495
ZV18A	0495
ZV18B	0495
ZV20	0531
ZV20A	0531
ZV20B	0531
ZV22	0565
ZV22A	0565
ZV22B	0565
ZV24	0402
ZV24A	0402
ZV24B	0402
ZV27	0441
ZV27A	0441
ZV27B	0441
ZV30	0726
ZV30A	0726
ZV30B	0726
ZV33	0039
ZV33A	0039
ZV33B	0039
ZV36	0820
ZV36A	0820
ZV36B	0820
ZV39	0350
ZV39A	0350
ZV39B	0350
ZV43	0931
ZV43A	0931
ZV43B	0931
ZV47	0999
ZV47A	0999
ZV47B	0999
ZV51	0502
ZV51A	0502
ZV51B	0502
ZV56	0869
ZV56A	0869
ZV56B	0869
ZV62	0783
ZV62A	0783
ZV62B	0783
ZV68	2157
ZV68A	2157
ZV68B	2157
ZV75	1189
ZV75A	1189
ZV75B	1189
ZV82	0331
ZV82A	0331
ZV82B	0331
ZV91	1311
ZV91A	1311
ZV91B	1311
ZV100	0098
ZV100A	0098
ZV100B	0098
ZV110	0149
ZV110A	0149
ZV110B	0149
ZV120	0186
ZV120A	0186

DEVICE TYPE	REPL CODE
ZV120B	0186
ZV130	0213
ZV130A	0213
ZV130B	0213
ZV150	0028
ZV150A	0028
ZV150B	0028
ZV160	0255
ZV160A	0255
ZV160B	0255
ZV180	0367
ZV180A	0367
ZV180B	0367
ZV200	0421
ZV200A	0421
ZV200B	0421
ZVN0102A	OEM
ZVN0102B	OEM
ZVN0102L	OEM
ZVN0106A	OEM
ZVN0106B	OEM
ZVN0106L	OEM
ZVN0108A	OEM
ZVN0108B	OEM
ZVN0108L	OEM
ZVN0117TA	OEM
ZVN0120A	OEM
ZVN0120B	OEM
ZVN0120L	OEM
ZVN0530A	OEM
ZVN0530B	OEM
ZVN0530L	OEM
ZVN0535A	OEM
ZVN0535B	OEM
ZVN0535L	OEM
ZVN0540A	OEM
ZVN0540B	OEM
ZVN0540L	OEM
ZVN0545A	OEM
ZVN0545B	OEM
ZVN1306A	OEM
ZVN1306B	OEM
ZVN1308A	OEM
ZVN1308B	OEM
ZVN1320A	OEM
ZVN1320B	OEM
ZVN1404A	OEM
ZVN1408A	OEM
ZVN2104A	OEM
ZVN2104B	OEM
ZVN2104L	OEM
ZVN2106A	OEM
ZVN2106B	OEM
ZVN2106L	OEM
ZVN2110A	OEM
ZVN2110B	OEM
ZVN2110L	OEM
ZVN2115A	OEM
ZVN2115B	OEM
ZVN2115L	OEM
ZVN2202B	OEM
ZVN2202L	OEM
ZVN2202M	OEM
ZVN2204B	OEM
ZVN2204L	OEM
ZVN2204M	OEM
ZVN2206B	OEM
ZVN2206L	OEM
ZVN2206M	OEM
ZVN2208B	OEM
ZVN2208L	OEM
ZVN2208M	OEM
ZVN2210B	OEM
ZVN2210L	OEM
ZVN2210M	OEM
ZVN2215B	OEM
ZVN2215L	OEM
ZVN2215M	OEM
ZVN2220B	OEM
ZVN2220L	OEM
ZVN2220M	OEM
ZVN3302A	OEM
ZVN3302B	OEM
ZVN3304A	OEM
ZVN3304B	OEM
ZVN3306A	OEM
ZVN3306B	OEM
ZVN3310A	OEM
ZVN3310B	OEM
ZVN3315A	OEM
ZVN3315B	OEM
ZVP0102A	OEM
ZVP0102B	OEM
ZVP0102L	OEM
ZVP0106A	OEM
ZVP0106B	OEM
ZVP0106L	OEM
ZVP0108A	OEM
ZVP0108B	OEM
ZVP0108L	OEM
ZVP0120A	OEM
ZVP0120B	OEM
ZVP0120L	OEM
ZVP0530A	OEM
ZVP0530B	OEM
ZVP0530L	OEM
ZVP0535A	OEM
ZVP0535B	OEM
ZVP0535L	OEM

DEVICE TYPE	REPL CODE
ZVP0540A	OEM
ZVP0540B	OEM
ZVP0540L	OEM
ZVP0545A	OEM
ZVP0545B	OEM
ZVP1306A	OEM
ZVP1306B	OEM
ZVP1308A	OEM
ZVP1308B	OEM
ZVP1320A	OEM
ZVP1320B	OEM
ZVP2104A	OEM
ZVP2104B	OEM
ZVP2104L	OEM
ZVP2106A	OEM
ZVP2106B	OEM
ZVP2106L	OEM
ZVP2110A	OEM
ZVP2110B	OEM
ZVP2110L	OEM
ZVP2115A	OEM
ZVP2115B	OEM
ZVP2115L	OEM
ZVP2202B	OEM
ZVP2202L	OEM
ZVP2202M	OEM
ZVP2204B	OEM
ZVP2204L	OEM
ZVP2204M	OEM
ZVP2206B	OEM
ZVP2206L	OEM
ZVP2206M	OEM
ZVP2208B	OEM
ZVP2208L	OEM
ZVP2208M	OEM
ZVP2210B	OEM
ZVP2210L	OEM
ZVP2210M	OEM
ZVP2215B	OEM
ZVP2215L	OEM
ZVP2215M	OEM
ZVP2220B	OEM
ZVP2220L	OEM
ZVP2220M	OEM
ZVP3302A	OEM
ZVP3302B	OEM
ZVP3304A	OEM
ZVP3304B	OEM
ZVP3306A	OEM
ZVP3306B	OEM
ZVP3310A	OEM
ZVP3310B	OEM
ZVP3315A	OEM
ZVP3315B	OEM
ZW0-9.1	0012
ZW09.1	OEM
ZW2	0609
ZW2.7	1312
ZW3	5688
ZW3.3	0292
ZW3.6	0188
ZW3.9	0456
ZW4.3	0533
ZW4.7	0451
ZW5.1	0162
ZW5.6	0157
ZW6.2	0636
ZW6.8	0025
ZW7.5	0649
ZW8.2	0244
ZW9.1	0012
ZW10	0170
ZW11	0317
ZW12	0137
ZW13	0365
ZW15	0002
ZW16	0420
ZW18	0495
ZW20	0531
ZW22	0565
ZW24	0402
ZW27	0441
ZW30	5879
ZW33	5880
ZW400	OEM
ZW0-9.1	0012
ZX012	OEM
ZX028B	OEM
ZX3.9	OEM
ZX4.3	OEM
ZX4.7	OEM
ZX5.1	OEM
ZX5.6	OEM
ZX6.2	OEM
ZX6.8	OEM
ZX7.5	OEM
ZX8.2	OEM
ZX9.1	6849
ZX10	OEM
ZX11	OEM
ZX12	OEM
ZX13	OEM
ZX15	OEM
ZX16	OEM
ZX18	OEM
ZX20	OEM
ZX22	OEM
ZX24	OEM
ZX27	OEM

DEVICE TYPE	REPL CODE
ZX30	OEM
ZX33	OEM
ZX36	OEM
ZX39	OEM
ZX43	OEM
ZX47	OEM
ZX51	OEM
ZX56	OEM
ZX62	OEM
ZX68	OEM
ZX74HCT240-2N	OEM
ZX74HCT245-2N	OEM
ZX74HCT273-2N	OEM
ZX74HCT374-2N	OEM
ZX74HCTLS240N	4050
ZX74HCTLS245N	4075
ZX74HCTLS273N	4126
ZX74HCTLS374N	4311
ZX75	OEM
ZX80/05	OEM
ZX80/15	OEM
ZX82	OEM
ZX85	OEM
ZX86	OEM
ZX88	OEM
ZX91	OEM
ZX100	OEM
ZX110	OEM
ZX120	OEM
ZX130	OEM
ZX150	OEM
ZX160	OEM
ZX180	OEM
ZX200	OEM
ZX200A	OEM
ZX204	OEM
ZX208A	OEM
ZX432BJ-10	OEM
ZX609B	OEM
ZX655A	OEM
ZX660A	OEM
ZX906A	OEM
ZX907	OEM
ZX908A	OEM
ZX9700	OEM
ZY1	OEM
ZY3.9	0456
ZY4.3	0533
ZY4.7	0451
ZY5.1	0162
ZY5.6	0157
ZY6.2	0636
ZY6.8	0025
ZY7.5	0649
ZY8.2	0244
ZY9.1	0012
ZY10	0170
ZY11	0317
ZY12	0234
ZY12A	0137
ZY12B	0137
ZY13	0365
ZY15	0247
ZY15A	0002
ZY15B	0002
ZY16	0420
ZY18	0495
ZY20	0531
ZY22	0565
ZY24	0402
ZY27	0294
ZY27A	0441
ZY27B	0441
ZY30	0309
ZY33	0318
ZY33A	0039
ZY33B	0039
ZY36	0820
ZY39	0350
ZY43	0931
ZY47	0999
ZY51	0502
ZY56	0869
ZY62	0783
ZY62A	0783
ZY62B	0783
ZY68	1268
ZY75	1189
ZY82	0331
ZY82A	0331
ZY82B	0331
ZY91	1311
ZY100	0098
ZY110	1190
ZY110A	0149
ZY110B	0149
ZY120	0186
ZY130	0213
ZY150	0028
ZY160	0255
ZY180	0367
ZY200	0421
ZZ3.3	0292
ZZ3.6	0188
ZZ3.9	0456
ZZ4.1	0162
ZZ4.7	0451
ZZ5.1	0162
ZZ5.6	0157

If replacement code is OEM, contact original manufacturer for replacement.

DEVICE TYPE	REPL CODE	DEVICE TYPE	REPL CODE	DEVICE TYPE	REPL CODE	DEVICE TYPE	REPL CODE	DEVICE TYPE	REPL CODE	DEVICE TYPE	REPL CODE	DEVICE TYPE	REPL CODE
ZZ6.2	0636	ZZ9.1	0012	ZZ15	0002	ZZ22	0705	ZZ36	OEM	ZZ110	0149	ZZY36	OEM
ZZ6.8	0025	ZZ10	0170	ZZ16	OEM	ZZ27	0441	ZZ62	5906	ZZ160	OEM	ZZY62	OEM
ZZ8.2	0244	ZZ12	0137	ZZ18	0375	ZZ33	0039	ZZ82	0331	ZZY16	OEM	ZZY160	OEM

If replacement code is OEM, contact original manufacturer for replacement.

SECTION 2

Replacements

from NTE, ECG, Radio Shack, TCE

REPL. CODE	NTE PART NO.	ECG PART NO.	RADIO SHACK PART NO.	RCA PART NO.
0001	NTE4049	ECG4049	276-2449	SK4049UB
0002	NTE145A	ECG145A	276-564	SK15V
0003	NTE164	ECG164	276-2055	SK3133
0004	NTE102A	ECG102A	. . . . .	SK3004
0005	NTE611	ECG611	. . . . .	SK3324
0006	NTE290A	ECG290A	276-2023	SK3114A
0007	NTE161	ECG161	276-2016	SK3716A
0008	NTE319P	ECG319P	276-2009	SK9432
0009	NTE20	ECG20	. . . . .	SK9668
0010	NTE5037A	ECG5037A	. . . . .	SK36A
0011	NTE974	ECG974	. . . . .	SK3965
0012	NTE139A	ECG139A	276-562	SK9V1
0013	NTE2362	ECG2362	. . . . .	SK10094
0014	NTE13	ECG13	. . . . .	SK9661
0015	NTE116	ECG116	276-1104	SK3313
0016	NTE123A	ECG123A	276-1617	SK3444
0017	NTE558	ECG558	. . . . .	SK3998
0018	NTE293	ECG293	. . . . .	SK3849
0019	NTE109	ECG109	276-1123	SK3088
0020	NTE5369	ECG5369	. . . . .	SK6769
0021	NTE1012	ECG1012	. . . . .	. . . . .
0022	NTE5208A	ECG5208A	. . . . .	SK182
0023	NTE552	ECG552	. . . . .	SK9000
0024	NTE4052B	ECG4052B	. . . . .	SK4052B
0025	NTE5071A	ECG5071A	. . . . .	SK6V8
0026	NTE923D	ECG923D	276-1740	SK3165
0027	NTE5147A	ECG5147A	. . . . .	SK51X
0028	NTE5100A	ECG5100A	. . . . .	SK150V
0029	NTE6008	ECG6008	. . . . .	SK7208
0030	NTE614	ECG614	. . . . .	SK3327
0031	NTE580	ECG580	. . . . .	SK5036
0032	NTE5038A	ECG5038A	. . . . .	SK39A
0033	NTE9944	ECG9944	. . . . .	. . . . .
0034	NTE74LS20	ECG74LS20	. . . . .	SK74LS20
0035	NTE4053B	ECG4053B	. . . . .	SK4053B
0036	NTE5007A	ECG5007A	. . . . .	SK3A9
0037	NTE159	ECG159	276-2023	SK3466
0038	NTE103	ECG103	. . . . .	SK3862
0039	NTE147A	ECG147A	. . . . .	SK33V
0040	NTE213	ECG213	. . . . .	SK3012
0041	NTE5010A	ECG5010A	276-565	SK5A1
0042	NTE152	ECG152	276-2020	SK3893
0043	NTE1094	ECG1094	. . . . .	SK7769
0044	NTE6002	ECG6002	. . . . .	SK7202
0045	NTE1N415E	1N415E	. . . . .	. . . . .
0046	NTE33	ECG33	. . . . .	SK9429
0047	NTE161	ECG161	276-2016	SK3117
0048	. . . . .	. . . . .	. . . . .	SK9411
0049	NTE5047A	ECG5047A	. . . . .	SK82A
0050	NTE160	ECG160	. . . . .	SK3006
0051	NTE8255	ECG8255	. . . . .	. . . . .
0052	NTE5021A	ECG5021A	276-563	SK12A
0053	NTE5022A	ECG5022A	. . . . .	SK13A
0054	NTE5039A	ECG5039A	. . . . .	SK43A
0055	NTE89	ECG89	. . . . .	SK9119
0056	NTE4040B	ECG4040B	. . . . .	SK4040B
0057	NTE5018A	ECG5018A	276-562	SK9A1
0058	NTE5404	ECG5404	276-1067	SK3627
0059	NTE5610	ECG5610	. . . . .	SK10413
0060	NTE377	ECG377	276-2020	SK9112
0061	NTE231	ECG231	. . . . .	SK3857
0062	NTE5014A	ECG5014A	. . . . .	SK6A8
0063	NTE5226A	ECG5226A	. . . . .	SK218
0064	NTE5019A	ECG5019A	. . . . .	SK10A
0065	NTE165	ECG165	276-2055	SK3115
0066	NTE399	ECG399	. . . . .	SK3244
0067	NTE1296	ECG1296	. . . . .	SK9311
0068	NTE5040A	ECG5040A	. . . . .	SK47A
0069	NTE1413	ECG1413	. . . . .	SK9380
0070	NTE5209A	ECG5209A	. . . . .	SK184
0071	NTE125	ECG125	276-1114	SK3081
0072	NTE125	ECG125	276-1114	SK3032A
0073	NTE271	ECG271	. . . . .	SK3936
0074	NTE162	ECG162	. . . . .	SK3559
0075	NTE74LS125A	ECG74LS125A	. . . . .	SK74LS125A
0076	NTE85	ECG85	276-2009	SK3124A
0077	NTE5015A	ECG5015A	. . . . .	SK7A5
0078	NTE9672	ECG9672	. . . . .	. . . . .
0079	NTE123AP	ECG123AP	276-2009	SK3854
0080	NTE116	ECG116	276-1104	SK3311
0081	NTE9945	ECG9945	. . . . .	. . . . .
0082	NTE74LS293	ECG74LS293	. . . . .	SK74LS293
0083	NTE981	ECG981	. . . . .	SK3724
0084	NTE5870	ECG5870	. . . . .	SK3604
0085	NTE104	ECG104	. . . . .	SK3719
0086	NTE128	ECG128	276-2030	SK3024
0087	NTE156	ECG156	276-1114	SK3051
0088	NTE74LS32	ECG74LS32	. . . . .	SK74LS32
0089	NTE5052A	ECG5052A	. . . . .	SK120A
0090	NTE5872	ECG5874	. . . . .	SK3603
0091	NTE5012A	ECG5012A	. . . . .	SK6A0
0092	NTE5041A	ECG5041A	. . . . .	SK51A
0093	NTE1171	ECG1171	. . . . .	SK3565
0094	NTE5046A	ECG5046A	. . . . .	SK75A
0095	NTE5457	ECG5457	276-1020	SK3598
0096	NTE5592	ECG5592	. . . . .	SK6762
0097	NTE5874	ECG5874	. . . . .	SK3603
0098	NTE5096A	ECG5096A	. . . . .	SK100V
0099	NTE5051A	ECG5051A	. . . . .	SK110A
0100	NTE144A	ECG144A	. . . . .	SK14V
0101	NTE4066B	ECG4066B	. . . . .	SK4066B
0102	NTE506	ECG506	. . . . .	SK3925
0103	NTE130	ECG130	276-2041	SK3027
0104	NTE5048A	ECG5048A	. . . . .	SK87A
0105	NTE5876	ECG5878	. . . . .	SK3602
0106	NTE5304	ECG5304	276-1173	SK3106
0107	NTE1569	ECG1569	. . . . .	SK10081
0108	NTE5372	ECG5372	. . . . .	SK6772
0109	NTE5878	ECG5878	. . . . .	SK3602
0110	NTE5800	ECG5800	276-1141	SK9003
0111	NTE199	ECG199	276-2016	SK3245
0112	NTE3098	ECG3098	. . . . .	SK10178
0113	NTE107	ECG107	276-2016	SK3122
0114	. . . . .	ECG5552	. . . . .	SK6651
0115	. . . . .	ECG74HC109	. . . . .	SK7C109
0116	NTE5880	ECG5882	. . . . .	SK3500
0117	NTE7483	ECG7483	. . . . .	SK7483
0118	NTE5004A	ECG5004A	. . . . .	SK3A0
0119	NTE4069	ECG4069	. . . . .	SK4069UB
0120	NTE605A	ECG605A	. . . . .	SK3864
0121	NTE4042B	ECG4042B	. . . . .	SK4042B
0122	NTE5882	ECG5882	. . . . .	SK3500
0123	NTE110MP	ECG110MP	276-1123	SK3709
0124	NTE519	ECG519	276-1122	SK3100
0125	NTE5042A	ECG5042A	. . . . .	SK56A
0126	NTE129	ECG129	. . . . .	SK3025
0127	NTE107	ECG107	276-2016	SK3293
0128	. . . . .	ECG501B	. . . . .	SK3069
0129	NTE4071B	ECG4071B	. . . . .	SK4071B
0130	NTE181	ECG181	. . . . .	SK9134
0131	NTE5886	ECG5886	. . . . .	SK7086
0132	NTE5210A	ECG5210A	. . . . .	SK186
0133	NTE177	ECG177	276-1122	SK9091
0134	NTE270	ECG270	. . . . .	SK3935
0135	NTE74LS40	ECG74LS40	. . . . .	SK74LS40
0136	NTE126	ECG126A	. . . . .	SK3003A
0137	NTE142A	ECG142A	276-563	SK12V
0138	NTE1013	ECG1013	. . . . .	. . . . .
0139	NTE601	ECG601	. . . . .	SK3463
0140	NTE5009A	ECG5009A	276-565	SK4A7
0141	NTE9946	ECG9946	. . . . .	. . . . .
0142	NTE124	ECG124	. . . . .	SK3021
0143	NTE109	ECG109	276-1123	SK3090
0144	NTE108	ECG108	276-2016	SK3452
0145	NTE5890	ECG5890	. . . . .	SK7090
0146	NTE4060B	ECG4060B	. . . . .	SK4060B
0147	NTE5675	ECG5675	. . . . .	SK3508
0148	NTE290A	ECG290A	276-2023	SK9132
0149	NTE151A	ECG151A	. . . . .	SK110V
0150	NTE106	ECG106	. . . . .	SK3984
0151	NTE85	ECG85	276-2009	SK3122
0152	NTE5044A	ECG5044A	. . . . .	SK62A
0153	NTE74LS139	ECG74LS139	. . . . .	SK74LS139
0154	NTE5673	ECG5673	. . . . .	SK5173
0155	NTE289A	ECG289A	276-2009	SK3124A
0156	NTE5049A	ECG5049A	. . . . .	SK91A
0157	NTE136A	ECG136A	. . . . .	SK5V6
0158	NTE506	ECG506	. . . . .	SK3125A
0159	NTE5591	ECG5591	. . . . .	SK6761
0160	NTE121	ECG121	. . . . .	SK3717
0161	NTE184	ECG184	. . . . .	SK3190
0162	NTE135A	ECG135A	276-565	SK5V1
0163	NTE553	ECG553	. . . . .	SK9975
0164	NTE158	ECG158	. . . . .	SK3004
0165	NTE5016A	ECG5016A	. . . . .	SK8A2
0166	NTE5036A	ECG5036A	. . . . .	SK33A
0167	NTE712	ECG712	. . . . .	SK3072
0168	NTE198	ECG198	. . . . .	SK3220
0169	NTE195A	ECG195A	. . . . .	SK3195
0170	NTE140A	ECG140A	. . . . .	SK10V
0171	NTE280	ECG280	276-2041	SK3619
0172	NTE5211A	ECG5211A	. . . . .	SK188
0173	NTE5045A	ECG5045A	. . . . .	SK68A
0174	NTE5412	ECG5412	276-1067	SK3953
0175	NTE74121	ECG74121	. . . . .	SK74121
0176	NTE834	ECG834	276-1712	SK3569
0177	NTE280	ECG280	276-2041	SK3297
0178	NTE175	ECG175	. . . . .	SK3261
0179	NTE5408	ECG5408	276-1067	SK3577
0180	NTE74H108	ECG74H108	. . . . .	. . . . .
0181	NTE5020A	ECG5020A	. . . . .	SK11A
0182	NTE551	ECG551	. . . . .	SK3125A
0183	NTE74LS27	ECG74LS27	. . . . .	SK74LS27
0184	NTE229	ECG229	276-2009	SK3132
0185	NTE5119A	ECG5119A	. . . . .	SK6X2
0186	NTE5097A	ECG5097A	. . . . .	SK120V
0187	NTE396	ECG396	. . . . .	SK3103A
0188	NTE134A	ECG134A	. . . . .	SK3V6
0189	NTE5050A	ECG5050A	. . . . .	SK100A
0190	NTE500A	ECG500A	. . . . .	SK3304
0191	NTE289A	ECG289A	276-2009	SK3122
0192	NTE5590	ECG5590	. . . . .	SK6760
0193	NTE6159	ECG6159	. . . . .	SK7359
0194	NTE36	ECG36	. . . . .	SK9413
0195	NTE5035A	ECG5035A	. . . . .	SK30A
0196	NTE113A	ECG113A	. . . . .	SK9001
0197	NTE5858	ECG5858	. . . . .	SK3599
0198	NTE123	ECG123	276-1617	SK3020
0199	NTE125 N2	ECG125 N2	276-1114 N2	SK3081 N2
0200	NTE5860	ECG5862	. . . . .	SK3584
0201	NTE502	ECG502	. . . . .	SK3067
0202	NTE6163	ECG6163	. . . . .	SK10220
0203	NTE290A	ECG290A	276-2023	SK3932
0204	NTE5862	ECG5862	. . . . .	SK3584
0205	NTE5120A	ECG5120A	. . . . .	SK6X8
0206	NTE5866	ECG5866	. . . . .	SK7066
0207	NTE5212A	ECG5212A	. . . . .	SK190
0208	NTE103A	ECG103A	. . . . .	SK3835
0209	NTE74LS155	ECG74LS155	. . . . .	SK74LS155
0210	NTE5026A	ECG5026A	. . . . .	SK17A
0211	NTE102	ECG102	. . . . .	SK3722
0212	NTE222	ECG222	. . . . .	SK3065
0213	NTE5098A	ECG5098A	. . . . .	SK130V
0214	. . . . .	. . . . .	. . . . .	SK7C112
0215	NTE4011B	ECG4011B	276-2411	SK4011B
0216	NTE161	ECG161	276-2016	SK3018
0217	NTE5577	ECG5577	. . . . .	SK6727
0218	NTE192A	ECG192A	. . . . .	SK3137
0219	NTE186A	ECG186A	. . . . .	SK3357
0220	NTE94	ECG94	. . . . .	SK3559
0221	NTE5182A	ECG5182A	. . . . .	SK130
0222	NTE131	ECG131	. . . . .	SK3198
0223	NTE389	ECG389	. . . . .	SK9411
0224	NTE229	ECG229	276-2058	SK3246A
0225	NTE5125A	ECG5125A	. . . . .	SK10X
0226	NTE5534	ECG5534	. . . . .	. . . . .
0227	NTE5213A	ECG5213A	. . . . .	SK192
0228	NTE152	ECG152	276-2020	SK3054
0229	NTE117	ECG117A	. . . . .	SK5010A
0230	NTE5126A	ECG5126A	. . . . .	SK11X
0231	NTE74166	ECG74166	. . . . .	SK74166
0232	NTE7400	ECG7400	276-1801	SK7400
0233	NTE154	ECG154	. . . . .	SK3044
0234	NTE5127A	ECG5127A	. . . . .	SK12X
0235	NTE3091	ECG3091	. . . . .	SK4929
0236	NTE291	ECG291	276-2020	SK3440
0237	NTE5128A	ECG5128A	. . . . .	SK13X
0238	NTE5155A	ECG5155A	. . . . .	SK91X
0239	NTE230	ECG230	. . . . .	. . . . .
0240	NTE5483	ECG5483	. . . . .	SK3942
0241	NTE5332	ECG5332	. . . . .	SK9230
0242	NTE5802	ECG5802	276-1143	SK9005
0243	NTE74LS74A	ECG74LS74A	. . . . .	SK74LS74A
0244	NTE5072A	ECG5072A	. . . . .	SK8V2
0245	NTE5099A	ECG5099A	. . . . .	SK140V
0246	NTE1V150	ECG1V150	. . . . .	SKMV150H
0247	NTE5130A	ECG5130A	. . . . .	SK15X
0248	NTE5019T1	ECG5019T1	. . . . .	SK9970
0249	NTE2361	ECG2361	276-2009	SK3124A
0250	NTE170	ECG170	. . . . .	SK3649
0251	NTE5131A	ECG5131A	. . . . .	SK16X
0252	NTE5054A	ECG5054A	. . . . .	SK140A
0253	NTE5011A	ECG5011A	. . . . .	SK5A6
0254	NTE5983	ECG5983	. . . . .	SK7183
0255	NTE5101A	ECG5101A	. . . . .	SK160V
0256	NTE5133A	ECG5133A	. . . . .	SK18X
0257	NTE5034A	ECG5034A	. . . . .	SK28A
0258	NTE5991	ECG5991	. . . . .	SK7191
0259	NTE316	ECG316	. . . . .	SK3039
0260	NTE74LS174	ECG74LS174	. . . . .	SK74LS174
0261	NTE399	ECG399	. . . . .	SK9352
0262	NTE5135A	ECG5135A	. . . . .	SK20X
0263	NTE5214A	ECG5214A	. . . . .	SK194
0264	NTE190	ECG190	. . . . .	SK9435
0265	NTE121MP	ECG121MP	. . . . .	SK3718
0266	NTE5148A	ECG5148A	. . . . .	SK56X
0267	NTE5995	ECG5995	. . . . .	SK3518
0268	NTE7401	ECG7401	276-1801	SK7401
0269	NTE5136A	ECG5136A	. . . . .	SK22X
0270	NTE94	ECG94	. . . . .	SK9141
0271	. . . . .	ECG552	. . . . .	. . . . .
0272	. . . . .	ECG5580	. . . . .	. . . . .
0273	NTE3042	ECG3042	. . . . .	SK2042
0274	NTE5137A	ECG5137A	. . . . .	SK24X
0275	NTE5008A	ECG5008A	. . . . .	SK4A3
0276	NTE157	ECG157	. . . . .	SK3747
0277	NTE166	ECG166	276-1171	SK9075
0278	NTE2332	ECG2332	. . . . .	SK10302
0279	NTE5677	ECG5677	. . . . .	SK3520A
0280	NTE100	ECG100	. . . . .	SK3721
0281	NTE6003	ECG6003	. . . . .	SK7203
0282	NTE2322	ECG2322	. . . . .	SK10134
0283	NTE580	ECG580	. . . . .	SK3318A
0284	. . . . .	ECG5582	. . . . .	. . . . .
0285	NTE171	ECG171	. . . . .	SK3201
0286	NTE85	ECG85	276-2009	SK9229
0287	NTE5053A	ECG5053A	. . . . .	SK130A
0288	NTE503	ECG503	. . . . .	SK3068
0289	NTE167	ECG167	. . . . .	SK3647
0290	. . . . .	ECG2322	. . . . .	. . . . .
0291	NTE74LS86	ECG74LS86	. . . . .	SK74LS86
0292	NTE5066A	ECG5066A	. . . . .	SK3V3
0293	NTE173BP	. . . . .	. . . . .	. . . . .
0294	NTE5139A	ECG5139A	. . . . .	SK27X
0295	NTE5013T1	ECG5013T1	. . . . .	SK9969
0296	NTE168	ECG168	276-1173	SK3648
0297	. . . . .	ECG3086	. . . . .	SK2086
0298	. . . . .	ECG56015	. . . . .	. . . . .
0299	NTE5005A	ECG5005A	. . . . .	SK3A3
0300	NTE4082B	ECG4082B	. . . . .	SK4082B
0301	NTE5070A	ECG5070A	276-561	SK6V0
0302	NTE169	ECG169	. . . . .	SK3678
0303	NTE4085B	ECG4085B	. . . . .	SK4085B
0304	NTE864	ECG864	. . . . .	SK9851
0305	. . . . .	ECG3100	. . . . .	SK4930
0306	. . . . .	ECG56016	. . . . .	. . . . .
0307	. . . . .	ECG5584	. . . . .	. . . . .
0308	NTE5438	ECG5438	. . . . .	SK5238
0309	NTE5141A	ECG5141A	. . . . .	SK30X

REPL. CODE	NTE PART NO.	ECG PART NO.	RADIO SHACK PART NO.	RCA PART NO.
0310	NTE5215A	ECG5215A	.....	SK196
0311	.....	ECG56017	.....	.....
0312	NTE941M	ECG941M	276-007	SK3552
0313	NTE238	ECG238	276-2055	SK3710
0314	NTE7402	ECG7402	.....	SK7402
0315	NTE3041	ECG3041	.....	SK2041
0316	NTE6402	ECG6402	.....	SK3628
0317	NTE5074A	ECG5074A	.....	SK11V
0318	NTE5142A	ECG5142A	.....	SK33X
0319	NTE5994	ECG5994	.....	SK3501
0320	NTE5143A	ECG5143A	.....	SK36X
0321	.....	.....	.....	SK9965
0322	NTE5017A	ECG5017A	276-562	SK8A7
0323	NTE5312	ECG5312	.....	SK3985
0324	NTE297	ECG297	.....	SK3449
0325	NTE312	ECG312	276-2055	SK9157
0326	NTE5144A	ECG5144A	.....	SK39X
0327	NTE5468	ECG5468	.....	SK9495
0328	.....	ECG729	.....	SK3074
0329	NTE5216A	ECG5216A	.....	SK198
0330	NTE233	ECG233	276-2009	SK9387
0331	NTE150A	ECG150A	.....	SK82V
0332	NTE1538	ECG1538	.....	SK9403
0333	NTE9099	ECG9099	.....	.....
0334	NTE966	ECG966	276-1771	SK3592
0335	NTE7403	ECG7403	.....	SK7403
0336	NTE5487	ECG5487	.....	SK3944
0337	NTE5145A	ECG5145A	.....	SK43X
0338	NTE191	ECG191	.....	SK3232
0339	NTE584	ECG584	.....	.....
0340	NTE5055A	ECG5055A	.....	SK150A
0341	NTE9963	ECG9963	.....	.....
0342	NTE288	ECG288	.....	SK3434
0343	NTE605A	ECG605A	.....	SK7952
0344	NTE5401	ECG5401	276-1067	SK3638
0345	NTE5378	ECG5378	.....	.....
0346	NTE5409	ECG5409	276-1020	SK3578
0347	NTE5146A	ECG5146A	.....	SK47X
0348	NTE525	ECG525	.....	SK3925
0349	NTE790	ECG790	.....	SK3077
0350	NTE5086A	ECG5086A	.....	SK39V
0351	NTE1580	ECG1580	.....	SK7743
0352	NTE714	ECG714	.....	SK3075
0353	NTE221	ECG221	.....	SK3050
0354	NTE715	ECG715	.....	SK3076
0355	NTE1014	ECG1014	.....	.....
0356	NTE5217A	ECG5217A	.....	SK200
0357	NTE5227AK	ECG5227AK	.....	SK221
0358	NTE9093	ECG9093	.....	.....
0359	NTE19	ECG19	.....	SK9667
0360	NTE778A	ECG778A	276-038	SK3465
0361	NTE7404	ECG7404	276-1802	SK7404
0362	NTE1465	ECG1465	.....	SK3914
0363	NTE283	ECG283	.....	SK3467
0364	NTE780	ECG780	.....	.....
0365	NTE143A	ECG143A	.....	SK13V
0366	NTE4051B	ECG4051B	.....	SK4051B
0367	NTE5103A	ECG5103A	.....	SK180V
0368	NTE85	ECG85	276-2009	SK3356
0369	.....	.....	.....	SK4915
0370	NTE5056A	ECG5056A	.....	SK160A
0371	NTE455	ECG455	.....	SK4915
0372	.....	ECG779A	.....	.....
0373	NTE125	ECG125	276-1114	SK3080
0374	NTE7485	ECG7485	.....	SK7485
0375	NTE5027A	ECG5027A	.....	SK18A
0376	NTE5006A	ECG5006A	.....	SK3A6
0377	.....	ECG782	.....	.....
0378	NTE504	ECG504	.....	SK3108
0379	NTE740A	ECG740A	.....	SK3328
0380	NTE85	ECG85	276-2009	SK3899
0381	NTE5218A	ECG5218A	.....	SK202
0382	NTE189	ECG189	.....	SK3200
0383	NTE85	ECG85	.....	SK9229
0384	.....	ECG454	.....	.....
0385	NTE7405	ECG7405	.....	SK7405
0386	NTE5150A	ECG5150A	.....	SK62X
0387	NTE1002	ECG1002	.....	.....
0388	NTE6800	ECG6800	.....	SK1901
0389	NTE1121	ECG1121	.....	SK7773
0390	NTE186A	ECG186A	.....	SK3248
0391	NTE1771	ECG1771	.....	SK9749
0392	NTE375	ECG375	.....	SK9118
0393	NTE6802	ECG6802	.....	SK9797
0394	NTE5057A	ECG5057A	.....	SK170A
0395	NTE749	ECG749	.....	SK3168
0396	.....	.....	.....	SK3100
0397	NTE5552	ECG5552	.....	SK6652
0398	NTE4050B	ECG4050B	.....	SK4050B
0399	NTE5660	ECG5667A	.....	SK10459
0400	NTE172A	ECG172A	.....	SK9455A
0401	NTE5227A	ECG5227A	.....	SK220
0402	NTE5081A	ECG5081A	.....	SK24V
0403	NTE1062	ECG1062	.....	.....
0404	NTE5661	ECG5667A	.....	SK10459
0405	NTE5151A	ECG5151A	.....	SK68X
0406	NTE291	ECG291	.....	SK3440
0407	NTE5405	ECG5405	276-1020	SK3951
0408	NTE5662	ECG5667A	.....	SK10459
0409	NTE532	ECG532	.....	SK3301
0410	NTE941	ECG941	.....	SK3514
0411	.....	ECG5624	.....	SK3632
0412	NTE5219A	ECG5219A	.....	SK204

REPL. CODE	NTE PART NO.	ECG PART NO.	RADIO SHACK PART NO.	RCA PART NO.
0413	NTE4013B	ECG4013B	276-2413	SK4013B
0414	NTE222	ECG454	.....	SK3991
0415	NTE5626	ECG5626	.....	SK3633
0416	NTE1217	ECG1217	.....	SK4801
0417	NTE973	ECG973	.....	SK3233
0418	NTE278	ECG278	.....	SK3218
0419	.....	.....	.....	SK3567A
0420	NTE5075A	ECG5075A	.....	SK16V
0421	NTE5105A	ECG5105A	.....	SK200V
0422	.....	ECG5667A	.....	SK10459
0423	NTE196	ECG196	276-2020	SK3054
0424	.....	ECG116	.....	.....
0425	NTE5058A	ECG5058A	.....	SK180A
0426	NTE5152A	ECG5152A	.....	SK75X
0427	NTE74LS138	ECG74LS138	.....	SK74LS138
0428	NTE5374	ECG5374	.....	SK6774
0429	NTE384	ECG384	.....	SK3894
0430	NTE951	ECG941	.....	SK3514
0431	NTE74LS85	ECG74LS85	.....	SK74LS85
0432	NTE40098B	ECG40098B	.....	SK40098B
0433	NTE1052	ECG1052	.....	SK3249
0434	NTE25	ECG25	.....	SK3841
0435	NTE5492	ECG5492	.....	SK6792
0436	NTE298	ECG298	.....	SK3450
0437	NTE213	ECG213	.....	.....
0438	NTE5220A	ECG5220A	.....	SK206
0439	NTE397	ECG397	.....	SK3528
0440	NTE105	ECG105	.....	SK3012
0441	NTE146A	ECG146A	.....	SK27V
0442	NTE5116A	ECG5116A	.....	SK5X1
0443	.....	ECG720	.....	SK9014
0444	NTE5153A	ECG5153A	.....	SK82X
0445	NTE5025A	ECG5025A	.....	SK16A
0446	NTE5315	ECG5315	.....	SK3988
0447	NTE5454	ECG5454	276-1067	SK6754
0448	NTE6821	ECG6821	.....	SK1900
0449	NTE5911	ECG5911	.....	SK7111
0450	NTE1589	ECG1589	.....	SK9911
0451	NTE5069A	ECG5069A	276-565	SK4V7
0452	NTE74LS240	ECG74LS240	.....	SK74LS240
0453	NTE5059A	ECG5059A	.....	SK190A
0454	NTE128	ECG123	276-2009	SK3020
0455	NTE5033A	ECG5033A	.....	SK27A
0456	NTE5067A	ECG5067A	.....	SK3V9
0457	NTE74LS55	ECG74LS55	.....	SK74LS55
0458	NTE74LS244	ECG74LS244	.....	SK74LS244
0459	NTE5380	ECG5380	.....	SK6780
0460	NTE185	ECG185	.....	SK3191
0461	NTE152	ECG152	276-2020	SK3440
0462	NTE1838	ECG1838	.....	SK9750
0463	NTE74LS245	ECG74LS245	.....	SK74LS245
0464	NTE5221A	ECG5221A	.....	SK208
0465	NTE1604	ECG1604	.....	SK4835
0466	NTE9158	ECG9158	.....	.....
0467	NTE7408	ECG7408	.....	SK7408
0468	NTE5574	ECG5574	.....	SK6724
0469	NTE5543	ECG5543	.....	SK3581
0470	NTE1075A	ECG1075A	.....	SK3877
0471	NTE5013A	ECG5013A	276-561	SK6A2
0472	NTE74LS28	ECG74LS28	.....	SK74LS28
0473	NTE5314	ECG5314	.....	SK3987
0474	NTE120	ECG120	276-1101	SK3110
0475	NTE313	ECG313	.....	SK4939
0476	NTE5905	ECG5905	.....	SK7111
0477	NTE383	ECG383	.....	SK9138
0478	NTE4001B	ECG4001B	276-2401	SK4001B
0479	NTE6115	ECG6115	.....	SK10216
0480	NTE5121A	ECG5121A	.....	SK7X5
0481	NTE5651	ECG5651	276-1000	SK3506
0482	NTE331	ECG331	276-2020	SK9234
0483	NTE5387	ECG5387	.....	SK6787
0484	NTE114	ECG114	276-1101	.....
0485	NTE5605	ECG5605	276-1000	SK3666
0486	NTE1856	ECG1856	.....	SK10155
0487	NTE2055	ECG2055	.....	SK10282
0488	NTE5222A	ECG5222A	.....	SK210
0489	NTE1705	ECG1705	.....	SK9950
0490	NTE1416	ECG1416	.....	SK9016
0491	NTE219	ECG219	276-2043	SK3173
0492	NTE7409	ECG7409	.....	SK7409
0493	NTE311	ECG311	.....	SK3195
0494	NTE5031A	ECG5031A	.....	SK24A
0495	NTE5077A	ECG5077A	.....	SK18V
0496	NTE1396	ECG1396	.....	SK9225
0497	.....	ECG1793	.....	.....
0498	NTE4012B	ECG4012B	.....	SK4012B
0499	NTE552	ECG552	.....	SK5002
0500	NTE4030B	ECG4030B	.....	SK4030B
0501	NTE6355	ECG6355	.....	SK6555
0502	NTE5089A	ECG5089A	.....	SK51V
0503	.....	ECG1793	.....	SK9716
0504	NTE5122A	ECG5122A	.....	SK8X2
0505	NTE5463	ECG5463	.....	SK9292
0506	.....	ECG2049	.....	SK10278
0507	.....	.....	.....	SK3245
0508	NTE75188	ECG75188	276-2520	SK5188
0509	NTE5223A	ECG5223A	.....	SK212
0510	NTE5186A	ECG5186A	.....	SK138
0511	NTE75189	ECG75189	276-2521	SK5189
0512	NTE7410	ECG7410	.....	SK7410
0513	NTE4017B	ECG4017B	276-2417	SK4017B
0514	NTE6850	ECG6850	.....	SK9702
0515	NTE5863	ECG5863	.....	SK7063

REPL. CODE	NTE PART NO.	ECG PART NO.	RADIO SHACK PART NO.	RCA PART NO.
0516	.....	ECG3017	.....	SK10321
0517	.....	ECG752	.....	.....
0518	NTE289A	ECG289A	276-2009	SK9137
0519	NTE4023B	ECG4023B	.....	SK4023B
0520	NTE801	ECG801	.....	SK3160
0521	NTE797	ECG797	.....	SK3158
0522	NTE2117	ECG2117	.....	SK9832
0523	NTE7008	ECG7008	.....	SK10477
0524	NTE5224A	ECG5224A	.....	SK214
0525	NTE374	ECG374	.....	SK9042
0526	NTE5371	ECG5371	.....	SK6771
0527	NTE7411	ECG7411	.....	SK7411
0528	NTE1046	ECG1046	.....	SK3471
0529	.....	ECG558	.....	.....
0530	NTE90	ECG90	.....	SK3931
0531	NTE5079A	ECG5079A	.....	SK20V
0532	NTE294	ECG294	.....	SK3841
0533	NTE5068A	ECG5068A	.....	SK4V3
0534	NTE5871	ECG5871	.....	SK7171
0535	NTE4086B	ECG4086B	.....	SK4086B
0536	NTE123A	ECG123A	276-2009	SK3444
0537	NTE85	ECG85	276-2009	SK3245
0538	NTE5879	ECG5879	.....	SK7179
0539	NTE1728	ECG1728	.....	SK9748
0540	NTE5804	ECG5804	276-1144	SK9007
0541	NTE3040	ECG3040	.....	SK2040
0542	NTE5225A	ECG5225A	.....	SK216
0543	NTE284	ECG284	.....	SK3836
0544	NTE2321	ECG2321	.....	SK10119
0545	NTE5887	ECG5887	.....	SK7087
0546	NTE588	ECG588	.....	SK9938
0547	NTE5174AK	ECG5174AK	.....	SK115
0548	.....	ECG3031A	.....	SK24A
0549	NTE108	ECG107	276-2016	SK3293
0550	NTE5891	ECG5891	.....	SK7091
0551	NTE188	ECG188	.....	SK3199
0552	NTE199	ECG199	276-2016	SK3124A
0553	.....	ECG1420	.....	SK9180
0554	NTE610	ECG610	.....	SK3323
0555	NTE5645	ECG5645	.....	SK5545
0556	NTE2331	ECG2331	.....	SK9422
0557	NTE1845	ECG1845	.....	SK10082
0558	NTE253	ECG253	.....	SK3996
0559	NTE74H40	ECG74H40	.....	.....
0560	NTE186	ECG186	276-2017	SK3192
0561	NTE182	ECG182	.....	SK3188A
0562	NTE9962	ECG9962	.....	.....
0563	NTE373	ECG373	.....	SK9041
0564	NTE580	ECG580	276-1114	SK5012
0565	NTE5080A	ECG5080A	.....	SK22V
0566	NTE210	ECG210	.....	SK3202
0567	NTE5522	ECG5522	.....	SK6622
0568	NTE4950	ECG4950	.....	SK10203
0569	NTE7493A	ECG7493A	.....	SK7493
0570	.....	ECG1690	.....	.....
0571	NTE5632	ECG5632	.....	SK3937
0572	NTE5608	ECG5608	276-1000	SK6708
0573	.....	ECG5402	.....	SK3638
0574	NTE5634	ECG5635	.....	SK3533
0575	.....	.....	.....	SK3232
0576	NTE5635	ECG5635	.....	SK3533
0577	NTE5636	ECG5637	.....	SK5137
0578	NTE5631	ECG5631	.....	SK5131
0579	NTE1003	ECG1003	.....	SK3288
0580	NTE5834	ECG5834	.....	SK7042
0581	NTE6889	ECG6889	.....	SK9809
0582	NTE16	ECG16	276-2016	SK9664
0583	NTE15016	ECG1793	.....	SK9716
0584	.....	ECG6090	.....	.....
0585	NTE955MC	ECG955MC	276-1718	SK10449
0586	NTE5868	ECG5868	.....	SK7068
0587	NTE5381	ECG5381	.....	SK6781
0588	.....	ECG978C	.....	.....
0589	NTE5010T1	ECG5010T1	.....	SK9967
0590	NTE5942	ECG5944	.....	SK3606
0591	NTE3028	ECG3028	.....	SK4937
0592	NTE74LS266	ECG74LS266	.....	SK74LS266
0593	NTE5633	ECG5633	276-1000	SK3938
0594	NTE5637	ECG5637	.....	SK5137
0595	NTE192	ECG192	.....	SK9142
0596	.....	ECG56010	.....	.....
0597	NTE229	ECG229	276-2009	SK3246A
0598	NTE5228A	ECG5228A	.....	SK222
0599	NTE6154	ECG6154	.....	SK7354
0600	NTE101	ECG101	.....	SK3861
0601	NTE6010	ECG6010	.....	SK7210
0602	NTE262	ECG262	.....	SK3897
0603	.....	.....	.....	SK9405
0604	NTE179	ECG179	.....	SK3642
0605	NTE5166A	ECG5166A	.....	SK200X
0606	NTE1N416E	1N416E	.....	.....
0607	NTE1122	ECG1122	.....	.....
0608	NTE5570	ECG5570	.....	SK6720
0609	NTE116	ECG116	276-1102	SK3311
0610	NTE5572	ECG5572	.....	SK6722
0611	NTE5554	ECG5554	.....	SK6654
0612	NTE5848	ECG5848	.....	SK7048
0613	NTE5280AK	ECG5280AK	.....	SK591
0614	.....	ECG56004	.....	.....
0615	NTE6084	ECG6084	.....	SK9709
0616	NTE5229A	ECG5229A	.....	SK224
0617	NTE56008	ECG56008	.....	SK3660
0618	NTE9660	ECG9660	.....	.....

REPL. CODE	NTE PART NO.	ECG PART NO.	RADIO SHACK PART NO.	RCA PART NO.
0619	NTE74145	ECG74145	.....	SK74145
0620	NTE328	ECG328	.....	SK3895
0621	NTE1227	ECG1227	.....	SK3762
0622	NTE282	ECG282	.....	.....
0623	.....	ECG767	.....	.....
0624	NTE960	ECG960	276-1770	SK3591
0625	NTE987	ECG987	276-1711	SK3643
0626	NTE4081B	ECG4081B	.....	SK4081B
0627	NTE5185AK	ECG5185AK	.....	SK137
0628	NTE613	ECG613	.....	SK3326
0629	NTE57	ECG57	.....	SK3913
0630	NTE224	ECG224	.....	SK3049
0631	NTE943M	ECG943M	.....	SK9993
0632	NTE703	ECG703	.....	.....
0633	NTE126A	ECG126A	.....	SK3003A
0634	NTE5230A	ECG5230A	.....	SK226
0635	.....	.....	.....	SK9456A
0636	NTE137A	ECG137A	276-561	SK6V2
0637	.....	ECG788	.....	SK3829
0638	NTE1089	ECG1089	.....	SK7767
0639	.....	.....	.....	SK7795
0640	NTE552	ECG552	.....	SK3318A
0641	NTE5575	ECG5575	.....	SK6725
0642	NTE163A	ECG163A	.....	SK3439A
0643	NTE376	ECG376	.....	SK3219
0644	NTE282	ECG282	.....	SK3024
0645	NTE16006	.....	.....	SK9720
0646	NTE4006B	ECG4006B	.....	SK4006B
0647	NTE5161A	ECG5161A	.....	SK150X
0648	NTE91	ECG91	.....	SK3932
0649	NTE138A	ECG138A	.....	SK7V5
0650	NTE5231A	ECG5231A	.....	SK228
0651	NTE1214	ECG1214	.....	SK3736
0652	NTE5710	ECG5710	.....	SK10416
0653	NTE1049	ECG1049	.....	SK3470
0654	NTE718	ECG718	.....	SK3159
0655	NTE5368	ECG5368	.....	SK6768
0656	NTE74L93	ECG74L93	.....	.....
0657	NTE6158	ECG6158	.....	SK7358
0658	NTE5579	ECG5579	.....	SK6729
0659	NTE195A	ECG195A	.....	SK3039
0660	.....	.....	.....	SK9067
0661	.....	.....	.....	SK3643
0662	NTE1608	ECG1608	.....	SK9707
0663	NTE1547	ECG1547	.....	SK7676
0664	NTE708	ECG708	.....	SK3135
0665	NTE74LS348	ECG74LS348	.....	SK74LS348
0666	NTE713	ECG713	.....	SK3077
0667	NTE2428	ECG2428	.....	SK9963
0668	NTE5232A	ECG5232A	.....	SK230
0669	NTE74LS640	ECG74LS640	.....	SK7CT640
0670	NTE5229AK	ECG5229AK	.....	SK225
0671	NTE5028A	ECG5028A	.....	SK19A
0672	NTE18	ECG18	276-2030	SK3911
0673	NTE1048	ECG1048	.....	.....
0674	NTE107	ECG107	276-2016	SK3444
0675	.....	ECG1426	.....	.....
0676	NTE5485	ECG5485	.....	SK3943
0677	NTE5003A	ECG5003A	.....	SK2A8
0678	NTE709	ECG709	.....	SK3135
0679	NTE5569	ECG5569	.....	SK6719
0680	.....	ECG74HC132	.....	SK7C132
0681	NTE292	ECG292	.....	SK3441
0682	NTE74H00	ECG74H00	.....	.....
0683	NTE1409	ECG1409	.....	SK9198
0684	NTE5124A	ECG5124A	.....	SK9X1
0685	NTE74H10	ECG74H10	.....	.....
0686	NTE5024A	ECG5024A	276-564	SK15A
0687	NTE5496	ECG5496	.....	SK6796
0688	NTE453	ECG453	276-2055	SK9853
0689	NTE486	ECG486	.....	SK9658
0690	NTE74LS641	ECG74LS641	.....	SK74LS641
0691	NTE5187A	ECG5187A	.....	SK140
0692	NTE725	ECG725	.....	.....
0693	NTE234	ECG234	276-2023	SK3247
0694	NTE519 N2	ECG519 N2	276-1122 N2	SK3100 N2
0695	NTE152	ECG152	276-2020	SK9366
0696	NTE5180AK	ECG5180AK	.....	SK127
0697	NTE7420	ECG7420	.....	SK7420
0698	NTE195A	ECG195A	.....	.....
0699	NTE1689	ECG1689	.....	SK9949
0700	NTE5029A	ECG5029A	.....	SK20A
0701	NTE722	ECG722	.....	SK7812
0702	.....	.....	.....	SK120A
0703	NTE2417	ECG2417	.....	SK10109
0704	NTE74S00	ECG74S00	.....	SK74S00
0705	NTE5030A	ECG5030A	.....	SK22A
0706	NTE1226	ECG1226	.....	SK3763
0707	.....	.....	.....	SK5020
0708	NTE5830	ECG5830	.....	SK7042
0709	NTE74LS373	ECG74LS373	.....	SK74LS373
0710	NTE5465	ECG5465	.....	SK9293
0711	NTE5822	ECG5822	.....	SK3587A
0712	NTE5542	ECG5542	.....	SK6642
0713	NTE74LS374	ECG74LS374	.....	SK1918
0714	NTE5032A	ECG5032A	.....	SK25A
0715	NTE287	ECG287	.....	SK3433
0716	NTE154	ECG154	.....	SK3024
0717	NTE5437	ECG5437	.....	SK5237
0718	NTE55	ECG55	.....	SK9367
0719	NTE74LS642	ECG74LS642	.....	SK74LS642
0720	NTE612	ECG612	.....	SK3325
0721	NTE5544	ECG5545	.....	SK3582
0722	NTE5545	ECG5545	.....	SK3582
0723	NTE754	.....	.....	.....
0724	NTE2408	ECG2408	.....	SK10099
0725	NTE5547	ECG5547	.....	SK3505
0726	NTE5084A	ECG5084A	.....	SK30V
0727	NTE5197AK	ECG5197AK	.....	SK161
0728	NTE379	ECG379	.....	SK9085
0729	NTE5309	ECG5309	.....	SK5028
0730	NTE154	ECG154	.....	SK3040
0731	NTE5541	ECG5541	.....	SK6641
0732	NTE1773	ECG1773	.....	SK9752
0733	NTE289A	ECG289A	276-2009	SK3449
0734	NTE74164	ECG74164	.....	SK74164
0735	NTE585	ECG585	.....	SK9934
0736	NTE5216AK	ECG5216AK	.....	SK199
0737	NTE5310	ECG5310	.....	SK5030
0738	NTE5589	ECG5589	.....	SK6778
0739	.....	ECG123AP	.....	.....
0740	NTE5527	ECG5527	.....	SK6627
0741	NTE5502	ECG5502	.....	SK6502
0742	NTE5311	ECG5311	.....	SK5031
0743	NTE7450	ECG7450	.....	SK7450
0744	NTE56006	ECG56006	.....	SK3659
0745	NTE5504	ECG5504	.....	SK3579
0746	NTE3041	ECG3041	.....	.....
0747	NTE5507	ECG5507	.....	SK3580
0748	NTE5873	ECG5875	.....	SK3517
0749	NTE6104	ECG6104	.....	SK10211
0750	NTE5548	ECG5548	.....	SK6648
0751	NTE739	ECG739	.....	SK3235
0752	NTE5509	ECG5509	.....	SK3504
0753	.....	ECG707	.....	SK3134
0754	NTE1285	ECG1285	.....	SK3922
0755	NTE5424	ECG5424	.....	SK3575
0756	NTE1286	ECG1286	.....	SK3923
0757	NTE505	ECG505	.....	SK3108
0758	.....	ECG56019	.....	.....
0759	NTE1630	ECG1630	.....	SK7827
0760	NTE5002A	ECG5002A	.....	SK2A7
0761	NTE74LS352	ECG74LS352	.....	SK74LS352
0762	NTE5524	ECG5524	.....	SK6624
0763	.....	ECG56020	.....	.....
0764	NTE5529	ECG5529	.....	SK6629
0765	NTE5875	ECG5875	.....	SK7175
0766	NTE5531	ECG5531	.....	SK6631
0767	NTE1411	ECG1411	.....	SK9246
0768	.....	.....	.....	SK9366
0769	NTE5386	ECG5386	.....	SK6786
0770	NTE928M	ECG928M	.....	SK3692
0771	NTE74LS95B	ECG74LS95B	.....	.....
0772	NTE56004	ECG56004	.....	SK3658
0773	NTE74LS353	ECG74LS353	.....	SK74LS353
0774	NTE118	ECG118	.....	SK3066
0775	NTE74LS645	ECG74LS645	.....	SK74LS645
0776	NTE5230AK	ECG5230AK	.....	SK227
0777	NTE5877	ECG5879	.....	SK3517
0778	NTE5546	ECG5547	.....	SK3505
0779	.....	ECG779A	.....	SK3134
0780	NTE74S38	.....	.....	.....
0781	NTE524V48	ECG524V48	.....	SKMV480H
0782	NTE5111A	ECG5111A	.....	SK3X3
0783	NTE149A	ECG149A	.....	SK62V
0784	.....	.....	.....	SK7C137
0785	NTE746	ECG746	.....	SK3234
0786	NTE295	ECG295	.....	SK9041
0787	NTE5307	ECG5307	.....	SK3107
0788	NTE289A	ECG289A	276-2009	SK3132
0789	NTE748	ECG748A	.....	SK3236
0790	.....	.....	.....	SK7CT354
0791	NTE193	ECG193	.....	SK9143
0792	NTE38	ECG38	.....	SK3623
0793	NTE5699	ECG5699	.....	SK10468
0794	NTE141A	ECG141A	276-563	SK11V5
0795	NTE116	ECG116	276-1103	SK3312
0796	NTE5112A	ECG5112A	.....	SK3X6
0797	.....	ECG74HC138	.....	SK7C138
0798	NTE85	ECG85	.....	SK3849
0799	.....	.....	.....	SK7666
0800	NTE9673	ECG9673	.....	.....
0801	NTE5881	ECG5883	.....	SK3517
0802	NTE783	ECG783	.....	.....
0803	NTE7426	ECG7426	.....	SK7426
0804	NTE5360	ECG5360	.....	.....
0805	NTE5990	ECG5990	.....	SK3608
0806	NTE5113A	ECG5113A	.....	SK3X9
0807	NTE159	ECG159	.....	SK3466
0808	NTE395	ECG395	.....	SK9434
0809	.....	ECG74HC139	.....	SK7C139
0810	NTE1510	ECG1510	.....	.....
0811	.....	.....	.....	SK9464
0812	NTE2764	ECG2764	.....	.....
0813	.....	.....	.....	SK7CT356
0814	NTE855	ECG855	.....	SK7616
0815	NTE5240A	ECG5240A	.....	SK510
0816	NTE5883	ECG5883	.....	SK3517
0817	NTE5808	ECG5808	.....	SK9009
0818	NTE7427	ECG7427	.....	SK7427
0819	NTE74LS393	ECG74LS393	.....	SK74LS393
0820	NTE5085A	ECG5085A	.....	SK36V
0821	NTE5114A	ECG5114A	.....	SK4X3
0822	NTE131MP	ECG131MP	.....	SK3840
0823	NTE724	ECG724	.....	SK3525
0824	NTE49	ECG49	.....	SK3178B
0825	NTE19	ECG19	276-2023	SK3912
0826	NTE289A	ECG289A	.....	SK3124A
0827	NTE5241A	ECG5241A	.....	SK512
0828	NTE74LS30	ECG74LS30	.....	SK74LS30
0829	.....	ECG710	.....	SK3102
0830	NTE524V13	ECG524V13	276-568	SKMV130K
0831	NTE4958	ECG4958	.....	SK10205
0832	NTE2355	ECG2355	.....	SK9957
0833	NTE5115A	ECG5115A	.....	SK4X7
0834	NTE7492	ECG7492	.....	SK7492
0835	NTE161	ECG161	276-2016	SK3246A
0836	NTE236	ECG236	.....	SK3239
0837	NTE74LS160A	ECG74LS160A	.....	SK7CT160
0838	NTE2V130	ECG2V130	276-570	SKMV130J
0839	NTE235	ECG235	.....	SK3239
0840	NTE1V130	ECG1V130	.....	SKMV130G
0841	NTE3010	ECG3010	.....	SK2010A
0842	NTE85	ECG85	276-2009	SK3132
0843	NTE1009	ECG1009	.....	SK3499
0844	NTE464	ECG464	.....	SK9159
0845	NTE2011	ECG2011	.....	SK3975
0846	NTE5242A	ECG5242A	.....	SK514
0847	NTE176	ECG176	.....	SK3845
0848	NTE815	ECG815	.....	SK3255
0849	NTE220	ECG220	.....	SK3990
0850	NTE9800	ECG9800	.....	.....
0851	.....	ECG5610	.....	.....
0852	NTE1186	ECG1186	.....	SK3168
0853	NTE5920	ECG5920	.....	SK10426
0854	NTE197	ECG197	.....	SK3083
0855	NTE1080	ECG1080	.....	SK3284
0856	NTE738	ECG738	.....	SK3167
0857	NTE1187	ECG1187	.....	SK3742
0858	NTE74LS161A	ECG74LS161A	.....	SK74LS161
0859	NTE391	ECG391	.....	SK3959
0860	NTE4014B	ECG4014B	.....	SK4014B
0861	NTE194	ECG194	.....	SK3275
0862	.....	.....	276-021	.....
0863	NTE5466	ECG5466	.....	SK9294
0864	NTE1010	ECG1010	.....	SK3376
0865	NTE1483	ECG1483	.....	SK9211
0866	NTE289A	ECG289A	276-2009	SK3849
0867	NTE87	ECG87	.....	SK9237
0868	NTE5243A	ECG5243A	.....	SK516
0869	NTE5090A	ECG5090A	.....	SK56V
0870	NTE5188A	ECG5188A	.....	SK142
0871	NTE5916	ECG5916	.....	SK10424
0872	NTE1434	ECG1434	.....	SK7634
0873	NTE7430	ECG7430	.....	SK7430
0874	NTE9801	ECG9801	.....	.....
0875	NTE56010	ECG56010	.....	SK10265
0876	NTE5117A	ECG5117A	.....	SK5X6
0877	NTE5102A	ECG5102A	.....	SK170V
0878	NTE1004	ECG1004	.....	SK3365
0879	NTE5023A	ECG5023A	.....	SK14A
0880	NTE74LS162A	ECG74LS162A	.....	SK7CT162
0881	NTE1158	ECG1158	.....	.....
0882	NTE1011	ECG1011	.....	.....
0883	NTE5011T1	ECG5011T1	.....	SK9968
0884	.....	ECG1159	.....	SK3290
0885	NTE5244A	ECG5244A	.....	SK518
0886	NTE1183	ECG1183	.....	SK3480
0887	NTE2357	ECG2357	.....	SK9742
0888	NTE9802	ECG9802	.....	.....
0889	NTE5254A	ECG5254A	.....	SK538
0890	NTE373	ECG373	.....	SK3253
0891	NTE165	ECG238	276-2055	SK3710
0892	NTE323	ECG323	.....	SK9369
0893	NTE74LS163A	ECG74LS163A	.....	SK74LS163
0894	NTE74LS273	ECG74LS273	.....	SK74LS273
0895	NTE108	ECG69	276-2016	SK3132
0896	NTE938	ECG938	.....	SK9166
0897	NTE5245A	ECG5245A	.....	SK520
0898	NTE2359	ECG2359	.....	SK9959
0899	NTE7432	ECG7432	.....	SK7432
0900	NTE9803	ECG9803	.....	.....
0901	NTE5402	ECG5402	276-1067	SK3638
0902	.....	ECG6244	.....	.....
0903	NTE596	ECG596	.....	.....
0904	NTE1245	ECG1245	.....	SK3878
0905	NTE218	ECG218	.....	SK3625
0906	NTE74LS243	ECG74LS243	.....	SK74LS243
0907	NTE595	ECG595	.....	SK9944
0908	NTE5679	ECG5679	.....	SK5579
0909	.....	ECG6240	.....	.....
0910	NTE5853	ECG5855	.....	SK7069
0911	NTE1100	ECG1100	.....	SK3223
0912	NTE1545	ECG1545	.....	SK9379
0913	NTE178MP	ECG178MP	276-1122	.....
0914	NTE5246A	ECG5246A	.....	SK522
0915	.....	ECG5679	.....	.....
0916	NTE1216	ECG1216	.....	SK3735
0917	NTE112	ECG112	.....	SK3089
0918	NTE1430	ECG1430	.....	SK9247
0919	NTE9804	ECG9804	.....	.....
0920	NTE507	ECG507	276-1101	SK5014
0921	NTE4078B	ECG4078B	.....	SK4078B
0922	NTE5809	ECG5809	.....	SK9010
0923	NTE962	ECG962	.....	SK3669
0924	NTE174	.....	.....	.....
0925	NTE153	ECG153	276-2027	SK3274
0926	NTE5247A	ECG5247A	.....	SK524
0927	.....	.....	.....	SK9010

REPL. CODE	NTE PART NO.	ECG PART NO.	RADIO SHACK PART NO.	RCA PART NO.
0928	NTE56	ECG56	.....	SK3896
0929	NTE9805	ECG9805	.....	.....
0930	.....	ECG1077	.....	.....
0931	NTE5087A	ECG5087A	.....	SK43V
0932	NTE5257A	ECG5257A	.....	SK544
0933	NTE5831	ECG5831	.....	SK7049
0934	NTE175	ECG175	.....	SK3626
0935	NTE5609	ECG5609	.....	SK6709
0936	NTE235	ECG235	.....	SK3197
0937	NTE107	ECG107	276-2009	SK3293
0938	.....	.....	.....	SK7C147
0939	NTE72	ECG72	.....	SK9272
0940	NTE5455	ECG5455	276-1067	SK3597
0941	.....	.....	.....	SK3565
0942	NTE74107	ECG74107	.....	SK74107
0943	NTE74LS365A	ECG74LS365A	.....	SK7CT365
0944	NTE5248A	ECG5248A	.....	SK526
0945	NTE9806	ECG9806	.....	.....
0946	.....	ECG1076	.....	.....
0947	NTE5835	ECG5835	.....	SK7049
0948	NTE5869	ECG5869	.....	SK7069
0949	NTE5078A	ECG5078A	.....	SK19V
0950	.....	ECG1760	.....	SK10042
0951	NTE85	ECG85	276-2009	SK3449
0952	NTE5556	ECG5556	.....	SK6656
0953	NTE5801	ECG5801	276-1143	SK9004
0954	NTE5613	ECG5613	.....	SK3631
0955	NTE198	ECG198	.....	SK3219
0956	NTE74LS366A	ECG74LS366A	.....	SK7CT366
0957	NTE5615	ECG5616	.....	SK3633
0958	NTE5249A	ECG5249A	.....	SK528
0959	NTE74LS153	ECG74LS153	.....	SK74LS153
0960	NTE5616	ECG5616	.....	SK3633
0961	NTE5617	ECG5618	.....	SK6635
0962	NTE9807	ECG9807	.....	.....
0963	NTE5472	ECG5472	.....	SK10260
0964	NTE1091	ECG1091	.....	SK7768
0965	NTE5806	ECG5806	.....	SK3848
0966	NTE5618	ECG5618	.....	SK6635
0967	NTE74LS168A	ECG74LS168A	.....	SK74LS168A
0968	NTE74109	ECG74109	.....	SK74109
0969	NTE2079	ECG2079	.....	SK9357
0970	NTE5850	ECG5850	.....	SK3601
0971	.....	.....	.....	SK7658
0972	.....	ECG6241	.....	.....
0973	NTE955M	ECG955M	276-1723	SK3564
0974	NTE553	ECG553	.....	SK3322
0975	NTE127	ECG127	.....	SK3764
0976	NTE1674	ECG1674	.....	SK9325
0977	NTE74LS367	ECG74LS367	.....	SK74LS367
0978	NTE5250A	ECG5250A	.....	SK530
0979	NTE74LS123	ECG74LS123	.....	SK74LS123
0980	NTE7491	ECG7491	.....	.....
0981	NTE2416	ECG2416	.....	SK10108
0982	NTE9808	ECG9808	.....	.....
0983	NTE465	ECG465	.....	SK9158
0984	.....	ECG116	.....	SK3313
0985	NTE5851	ECG5851	.....	SK7051
0986	NTE74LS169A	ECG74LS169A	.....	SK74LS169A
0987	NTE74110	ECG74110	.....	SK74110
0988	NTE4008B	ECG4008B	.....	SK4008B
0989	NTE5854	ECG5854	.....	SK3600
0990	NTE5855	ECG5855	.....	SK7055
0991	NTE74LS368	ECG74LS368	.....	SK7CT368
0992	NTE5186AK	ECG5186AK	.....	SK139
0993	NTE5857	ECG5859	.....	SK7069
0994	NTE5251A	ECG5251A	.....	SK532
0995	NTE5187AK	ECG5187AK	.....	SK141
0996	NTE7438	ECG7438	.....	SK7438
0997	NTE5859	ECG5859	.....	SK7059
0998	NTE9809	ECG9809	.....	.....
0999	NTE5088A	ECG5088A	.....	SK47V
1000	NTE5838	ECG5838	.....	SK7038
1001	NTE5861	ECG5863	.....	SK7069
1002	NTE74111	ECG74111	.....	SK74111
1003	.....	ECG74HC151	.....	SK7C151
1004	NTE5417	ECG5417	.....	SK10255
1005	NTE5021T1	ECG5021T1	.....	SK9971
1006	NTE5172AK	ECG5172AK	.....	.....
1007	NTE2012	ECG2012	.....	SK9092
1008	NTE5867	ECG5867	.....	SK7067
1009	NTE5252A	ECG5252A	.....	SK534
1010	NTE226	ECG226	.....	SK3082
1011	NTE9810	ECG9810	.....	.....
1012	NTE5839	ECG5839	.....	SK7039
1013	NTE74162	ECG74162	.....	SK74162
1014	NTE4015B	ECG4015B	.....	SK4015B
1015	NTE125 N1	ECG125 N1	276-1114 N1	SK3081 N1
1016	NTE4532B	ECG4532B	.....	SK4532B
1017	NTE74S20	ECG74S20	.....	SK74S20
1018	NTE1087	ECG1087	.....	SK3477
1019	NTE5253A	ECG5253A	.....	SK536
1020	NTE5189A	ECG5189A	.....	SK144
1021	NTE6085	ECG6085	.....	SK9798
1022	NTE1445	ECG1445	.....	SK9187
1023	NTE6354	ECG6354	.....	SK6554
1024	NTE7440	ECG7440	.....	SK7440
1025	NTE9811	ECG9811	.....	.....
1026	.....	ECG1445	.....	.....
1027	.....	ECG286	.....	SK3194
1028	NTE1781	ECG1781	.....	SK9746
1029	NTE616	ECG616	.....	SK3320
1030	NTE358	ECG358	.....	SK9328
1031	NTE222	ECG221	.....	SK3050
1032	NTE2358	ECG2358	.....	SK9741
1033	NTE74LS51	ECG74LS51	.....	SK74LS51
1034	NTE5418	ECG5418	.....	SK10256
1035	.....	ECG4562B	.....	SK4562B
1036	NTE6356	ECG6356	.....	SK6556
1037	NTE4502B	ECG4502B	.....	SK4502B
1038	NTE7441	ECG7441	.....	SK7441
1039	NTE9812	ECG9812	.....	.....
1040	.....	ECG5329	.....	.....
1041	NTE9932	ECG9932	.....	.....
1042	NTE2329	ECG5329	.....	SK10254
1043	NTE4518B	ECG4518B	.....	SK4518B
1044	NTE5426	ECG5426	.....	.....
1045	NTE5330	ECG5330	.....	SK9972
1046	NTE6358	ECG6358	.....	SK6558
1047	.....	ECG5331	.....	SK10400
1048	NTE1232	ECG1232	.....	SK3852
1049	NTE5255A	ECG5255A	.....	SK540
1050	NTE1155	ECG1155	.....	SK3231
1051	NTE187A	ECG187A	.....	SK9076
1052	NTE7442	ECG7442	.....	SK7442
1053	NTE3084	ECG3084	.....	SK2084
1054	NTE539	ECG539	.....	SK3309
1055	.....	ECG1196	.....	.....
1056	.....	ECG5667A	.....	.....
1057	NTE1231A	ECG1231A	.....	SK9384
1058	NTE5256A	ECG5256A	.....	SK542
1059	.....	.....	.....	SK3043B
1060	NTE7443	ECG7443	.....	SK7443
1061	NTE2338	ECG2338	.....	SK9480
1062	NTE158	ECG158	.....	SK3007A
1063	NTE4538B	ECG4538B	.....	SK4538B
1064	NTE5688	ECG5688	.....	SK5588
1065	.....	.....	.....	SK9766
1066	NTE107	ECG107	276-2016	SK3132
1067	NTE523	ECG523/3306	.....	SK3306
1068	NTE171	ECG171	.....	SK3865A
1069	.....	ECG9914	.....	.....
1070	NTE8308	ECG8308	.....	.....
1071	NTE5232AK	ECG5232AK	.....	SK231
1072	NTE7444	ECG7444	.....	SK7444
1073	NTE5843	ECG5843	.....	SK7043
1074	.....	ECG987	.....	.....
1075	.....	.....	.....	SK7C157
1076	NTE3161	ECG3161	.....	SK2161
1077	.....	ECG816	.....	.....
1078	NTE5194A	ECG5194A	.....	SK154
1079	NTE5258A	ECG5258A	.....	SK546
1080	NTE115	ECG115	276-1101	SK9002
1081	.....	ECG1389	.....	.....
1082	NTE7445	ECG7445	.....	SK7445
1083	NTE5073A	ECG5073A	276-562	SK8V7
1084	NTE5580	ECG5580	.....	SK6730
1085	NTE376	ECG376	.....	SK9362A
1086	NTE5582	ECG5582	.....	SK6732
1087	NTE173BP	.....	.....	SK3710
1088	.....	.....	.....	SK7C158
1089	NTE5622	ECG5622	.....	SK6630
1090	NTE587	ECG587	.....	SK9937
1091	NTE519 N3	ECG519 N3	276-1122 N3	SK3100 N3
1092	NTE107	ECG107	.....	SK3293
1093	NTE5623	ECG5623	.....	SK3631
1094	NTE1755	ECG1755	.....	SK10038
1095	NTE5625	ECG5626	.....	SK3633
1096	NTE5259A	ECG5259A	.....	SK548
1097	NTE116 N1	ECG116 N1	276-1104 N1	SK9002
1098	NTE7446	ECG7446	.....	SK7446
1099	NTE5627	ECG5628	.....	SK6635
1100	NTE5628	ECG5628	.....	SK6635
1101	NTE2V480	ECG2V480	.....	SKMV480G
1102	NTE5584	ECG5584	.....	SK6734
1103	NTE5481	ECG5481	.....	SK3940
1104	NTE875	ECG875	.....	SK7725
1105	.....	.....	.....	SK15A
1106	NTE5260A	ECG5260A	.....	SK550
1107	NTE5987	ECG5987	.....	SK3698
1108	NTE7447	ECG7447	276-1805	SK7447
1109	NTE3083	ECG3083	.....	SK2083
1110	NTE5470	ECG5470	.....	SK10257
1111	NTE5989	ECG5989	.....	SK7191
1112	NTE5904	ECG5904	.....	SK7104
1113	NTE1N416C	1N416C	.....	.....
1114	NTE109	ECG110A	276-1123	SK3090
1115	NTE1N415C	1N415C	.....	.....
1116	NTE74120	.....	.....	.....
1117	.....	.....	.....	SK7C160
1118	NTE927D	ECG927D	.....	SK7617
1119	NTE5999	ECG5999	.....	SK7199
1120	NTE74LS377	ECG74LS377	.....	SK74LS377
1121	NTE927	ECG927	.....	SK9017
1122	.....	ECG1319	.....	SK4815
1123	NTE5261A	ECG5261A	.....	SK552
1124	NTE5986	ECG5986	.....	SK3609
1125	NTE7448	ECG7448	.....	SK7448
1126	NTE5988	ECG5988	.....	SK3608
1127	NTE597	ECG597	.....	SK9858
1128	NTE1261	ECG1261	.....	SK3872
1129	.....	.....	.....	SK7C161
1130	NTE74LS670	ECG74LS670	.....	SK7CT670
1131	NTE3167	ECG3167	.....	SK2167
1132	NTE5998	ECG5998	.....	SK7198
1133	NTE74LS378	ECG74LS378	.....	SK74LS378
1134	NTE2013	ECG2013	.....	SK9093
1135	NTE5262A	ECG5262A	.....	SK554
1136	.....	ECG56028	.....	.....
1137	NTE5400	ECG5400	276-1067	SK3950
1138	NTE5847	ECG5847	.....	SK7047
1139	NTE74122	ECG74122	.....	SK74122
1140	NTE16	ECG16	276-2016	SK3911
1141	NTE460	ECG460	.....	SK9842
1142	.....	.....	.....	SK7C162
1143	NTE4016B	ECG4016B	.....	SK4016B
1144	NTE229	ECG229	276-2058	SK3132
1145	NTE581	ECG581	.....	SK5001
1146	NTE165	ECG165	276-2055	SK3111
1147	NTE175	ECG175	.....	SK3562
1148	NTE5419	ECG5419	.....	SK10258
1149	NTE178MP	ECG178MP	276-1122	SK3100
1150	NTE389	ECG389	276-2055	SK3115
1151	NTE74LS379	ECG74LS379	.....	SK74LS379
1152	NTE5263A	ECG5263A	.....	SK556
1153	NTE5190A	ECG5190A	.....	SK146
1154	NTE161	ECG161	276-2016	SK3039
1155	NTE466	ECG466	.....	SK9163
1156	NTE5091A	ECG5091A	.....	SK60V
1157	NTE74123	ECG74123	.....	SK74123
1158	NTE7476	ECG7476	.....	SK7476
1159	.....	ECG74HC163	.....	SK7C163
1160	NTE1081A	ECG1081A	.....	SK3474
1161	NTE74LS157	ECG74LS157	.....	SK74LS157
1162	NTE377	ECG377	.....	SK9366
1163	NTE5220AK	ECG5220AK	.....	SK207
1164	NTE5264A	ECG5264A	.....	SK558
1165	NTE54	ECG54	.....	SK9366
1166	NTE3098	ECG3098	.....	SK9763
1167	.....	ECG1867	.....	.....
1168	NTE7451	ECG7451	.....	SK7451
1169	NTE4919	ECG4919	.....	SK10196
1170	NTE1231	ECG1231	.....	SK3832
1171	.....	ECG74HC164	.....	SK7C164
1172	NTE7473	ECG7473	.....	SK7473
1173	NTE302	ECG302	.....	SK3252
1174	NTE5265A	ECG5265A	.....	SK560
1175	NTE6409	ECG6409	.....	SK9124
1176	NTE9935	ECG9935	.....	.....
1177	NTE5140A	ECG5140A	.....	SK28X
1178	NTE5132A	ECG5132A	.....	SK17X
1179	NTE598	ECG598	.....	.....
1180	NTE5156A	ECG5156A	.....	SK100X
1181	.....	.....	.....	SK3569
1182	NTE74125	ECG74125	.....	SK74125
1183	.....	ECG74HC165	.....	SK7C165
1184	NTE5266A	ECG5266A	.....	SK562
1185	NTE7453	ECG7453	.....	SK7453
1186	.....	ECG757	.....	.....
1187	NTE1025	ECG1025	.....	SK3155
1188	NTE5849	ECG5849	.....	SK7049
1189	NTE5093A	ECG5093A	.....	SK75V
1190	NTE5157A	ECG5157A	.....	SK110X
1191	.....	ECG580	.....	.....
1192	NTE923	ECG923	.....	SK3164
1193	NTE74126	ECG74126	.....	SK74126
1194	NTE5206A	ECG5206A	.....	SK178
1195	NTE5992	ECG5992	.....	SK3501
1196	NTE964	ECG964	.....	SK3630
1197	NTE537	ECG537	.....	SK3903
1198	NTE325	ECG325	.....	SK9607
1199	NTE183	ECG183	.....	SK3189A
1200	NTE5267A	ECG5267A	.....	SK564
1201	NTE1471	ECG1471	.....	SK9194
1202	NTE7454	ECG7454	.....	SK7454
1203	.....	ECG5313	.....	.....
1204	NTE6112	ECG6112	.....	SK10214
1205	NTE5925	ECG5925	.....	SK10429
1206	NTE7406	ECG7406	.....	SK7406
1207	NTE5158A	ECG5158A	.....	SK120X
1208	NTE7490	ECG7490	276-1808	SK7490
1209	.....	ECG286	.....	SK3261
1210	NTE5268A	ECG5268A	.....	SK566
1211	NTE312	ECG451	276-2055	SK9164
1212	NTE261	ECG261	276-2068	SK3896
1213	NTE229	ECG229	276-2058	SK3018
1214	.....	ECG2343	.....	.....
1215	NTE1006	ECG1006	.....	SK3358
1216	NTE5021A	ECG5021A	.....	SK12A
1217	NTE515	ECG515	.....	SK9098
1218	NTE5159A	ECG5159A	.....	SK130X
1219	NTE74128	ECG74128	.....	SK74128
1220	NTE85	ECG85	276-2009	SK3931
1221	NTE199	ECG199	276-2016	SK3899
1222	NTE5456	ECG5457	276-1020	SK3598
1223	NTE5269A	ECG5269A	.....	SK568
1224	NTE289AMP	ECG289AMP	.....	SK10311
1225	NTE5249AK	ECG5249AK	.....	SK529
1226	NTE6110	ECG6110	.....	SK10213
1227	NTE199	ECG199	276-2016	SK3866A
1228	.....	.....	.....	SK3017B
1229	.....	.....	.....	SK16A
1230	NTE74LS386	ECG74LS386	.....	SK74LS386
1231	NTE75450B	ECG75450B	.....	SK9828
1232	NTE5270A	ECG5270A	.....	SK570
1233	NTE350	ECG350	.....	SK3176
1234	.....	.....	.....	SK3460
1235	NTE1000	ECG1000	.....	.....
1236	NTE6240	ECG6240	.....	SK5060

REPL. CODE	NTE PART NO.	ECG PART NO.	RADIO SHACK PART NO.	RCA PART NO.
1237	NTE5251AK	ECG5251AK	.....	SK533
1238	NTE5893	ECG5893	.....	SK7093
1239	NTE74LS221	ECG74LS221	.....	SK74LS221
1240	NTE5895	ECG5897	.....	SK7111
1241	NTE5897	ECG5897	.....	SK7097
1242	NTE193A	ECG193A	.....	SK3138
1243	NTE5458	ECG5458	.....	SK9861
1244	NTE75451B	ECG75451B	.....	SK9807
1245	NTE5899	ECG5901	.....	SK7111
1246	NTE5271A	ECG5271A	.....	SK572
1247	.....	.....	.....	SK7655
1248	NTE1115	ECG1115	.....	SK3184
1249	NTE5189AK	ECG5189AK	.....	SK145
1250	NTE5944	ECG5944	.....	SK3606
1251	NTE5921	ECG5921	.....	SK10427
1252	NTE5252AK	ECG5252AK	.....	SK535
1253	NTE5903	ECG5905	.....	SK7111
1254	NTE1804	ECG1804	.....	SK9894
1255	NTE5162A	ECG5162A	.....	SK160X
1256	NTE4022B	ECG4022B	.....	SK4022B
1257	NTE9910	ECG9910	.....	.....
1258	.....	.....	.....	SK9970
1259	NTE5453	ECG5453	276-1067	SK6753
1260	NTE1248	ECG1248	.....	SK3497
1261	NTE2014	ECG2014	.....	SK9094
1262	NTE75452B	ECG75452B	.....	SK9834
1263	NTE5188AK	ECG5188AK	.....	SK143
1264	NTE5909	ECG5909	.....	SK7109
1265	NTE5272A	ECG5272A	.....	SK574
1266	NTE7085	ECG7085	.....	.....
1267	NTE187	ECG187	.....	SK3193
1268	NTE5092A	ECG5092A	.....	SK68V
1269	NTE5253AK	ECG5253AK	.....	SK537
1270	NTE26	ECG26	.....	SK9477
1271	NTE74132	ECG74132	.....	SK74132
1272	NTE75453B	ECG75453B	.....	SK9827
1273	NTE5273A	ECG5273A	.....	SK576
1274	NTE5191A	ECG5191A	.....	SK148
1275	NTE7460	ECG7460	.....	SK7460
1276	NTE5000A	ECG5000A	.....	SK2A4
1277	NTE5254AK	ECG5254AK	.....	SK539
1278	.....	.....	.....	SK2A4
1279	NTE5164A	ECG5164A	.....	SK180X
1280	NTE312	ECG451	276-2055	SK9157
1281	.....	ECG74HC173	.....	SK7C173
1282	NTE5812	ECG5812	.....	SK3639
1283	.....	.....	.....	SK3006
1284	NTE375	ECG375	.....	SK3929
1285	NTE961	ECG961	.....	SK3671
1286	.....	ECG2117	.....	SK9832
1287	NTE5814	ECG5814	.....	SK9096
1288	NTE74LS390	ECG74LS390	.....	SK7CT390
1289	NTE75454B	ECG75454B	.....	SK75454B
1290	NTE5274A	ECG5274A	.....	SK578
1291	NTE3092	ECG3092	.....	SK9770
1292	NTE977	ECG977	.....	SK3462
1293	NTE5815	ECG5815	.....	SK3640
1294	NTE5255AK	ECG5255AK	.....	SK541
1295	NTE9370	ECG9370	.....	.....
1296	NTE5817	ECG5817	.....	SK9097
1297	NTE4583B	ECG4583B	.....	SK4583B
1298	.....	ECG74HC174	.....	SK7C174
1299	NTE5275A	ECG5275A	.....	SK580
1300	NTE975	ECG975	.....	SK3641
1301	NTE909	ECG909	.....	SK9177
1302	NTE5257AK	ECG5257AK	.....	SK545
1303	.....	ECG119	.....	SK3175A
1304	NTE74LS194	ECG74LS194A	.....	SK74LS194
1305	.....	ECG74HC175	.....	SK7C175
1306	NTE508	ECG508/R-3A3	.....	SK3756
1307	NTE5276A	ECG5276A	.....	SK582
1308	NTE378	ECG378	.....	SK9113
1309	.....	.....	.....	SK9332
1310	NTE5258AK	ECG5258AK	.....	SK547
1311	NTE5095A	ECG5095A	.....	SK91V
1312	NTE5063A	ECG5063A	.....	SK2V7
1313	NTE7474	ECG7474	.....	SK7474
1314	NTE1V015	ECG1V015	.....	SKMV15B
1315	NTE74LS195A	ECG74LS195A	.....	SK74LS195
1316	NTE74136	ECG74136	.....	SK9833
1317	NTE5689	ECG5689	.....	SK5589
1318	NTE1560	ECG1560	.....	SK7620
1319	NTE2V015	ECG2V015	.....	SKMV15C
1320	.....	ECG74LS33	.....	.....
1321	NTE968	ECG968	.....	SK3593
1322	NTE5277A	ECG5277A	.....	SK584
1323	NTE513	ECG513	.....	SK3443
1324	NTE5260AK	ECG5260AK	.....	SK551
1325	.....	ECG389	.....	SK9374
1326	NTE1130	ECG1130	.....	SK3478
1327	NTE399	ECG399	.....	SK3866A
1328	.....	ECG830	.....	.....
1329	NTE615P	ECG615A	.....	SK9976
1330	NTE74LS395A	ECG74LS395A	.....	.....
1331	NTE5278A	ECG5278A	.....	SK586
1332	NTE1V250	ECG1V250	.....	SKMV250H
1333	NTE5262AK	ECG5262AK	.....	SK555
1334	NTE2V250	ECG2V250	.....	SKMV250J
1335	NTE116	ECG116	276-1101	SK3311
1336	NTE524V25	ECG524V25	.....	SKMV250K
1337	NTE791	ECG791	.....	SK3149
1338	.....	.....	.....	SK6V2
1339	NTE7407	ECG7407	.....	SK7407
1340	NTE5279A	ECG5279A	.....	SK588
1341	NTE53	ECG53	.....	SK9140
1342	NTE65	ECG65	.....	SK9373
1343	.....	.....	.....	SK3669
1344	NTE5263AK	ECG5263AK	.....	SK557
1345	NTE723	ECG723	.....	SK3144
1346	NTE1910	ECG1910	.....	.....
1347	NTE6155	ECG6155	.....	SK7355
1348	NTE290A	ECG290A	276-2023	SK3841
1349	NTE7416	ECG7416	.....	SK7416
1350	.....	ECG9109	.....	.....
1351	.....	.....	.....	SK3592
1352	NTE7417	ECG7417	.....	SK7417
1353	NTE5280A	ECG5280A	.....	SK590
1354	.....	ECG479	.....	SK9651
1355	NTE116	ECG116	276-1104	SK3017B
1356	NTE5264AK	ECG5264AK	.....	SK559
1357	NTE7421	ECG7421	.....	SK9837
1358	NTE3020	ECG3020	.....	SK2020
1359	NTE40160B	ECG40160B	.....	SK40160B
1360	NTE277	ECG277	.....	SK3220
1361	NTE331	ECG331	276-2020	SK3620
1362	.....	ECG576	.....	.....
1363	.....	.....	.....	SK7425
1364	.....	ECG9911	.....	.....
1365	NTE5281A	ECG5281A	.....	SK592
1366	NTE576	ECG576	.....	SK10510
1367	NTE211	ECG211	.....	SK3203
1368	NTE7486	ECG7486	.....	SK7486
1369	NTE332	ECG332	.....	SK9236
1370	.....	.....	.....	SK7430
1371	NTE5266AK	ECG5266AK	.....	SK563
1372	.....	ECG577	.....	.....
1373	NTE40161B	ECG40161B	.....	SK40161B
1374	NTE1V130	ECG1V130	.....	SKMV130H
1375	NTE4585B	ECG4585B	.....	SK4585B
1376	.....	.....	.....	SK7432
1377	NTE74141	ECG74141	.....	SK74141
1378	.....	.....	.....	SK9971
1379	NTE1172	ECG1172	.....	.....
1380	NTE5173AK	ECG5173AK	.....	.....
1381	NTE159MCP	ECG159MCP	276-2009&23	SK10273
1382	NTE74LS26	ECG74LS26	.....	SK74LS26
1383	NTE74LS398	ECG74LS398	.....	SK74LS398
1384	NTE5282A	ECG5282A	.....	SK594
1385	NTE5424	ECG5424	.....	SK3857
1386	NTE382	ECG382	.....	SK9137
1387	NTE5268AK	ECG5268AK	.....	SK567
1388	NTE40162B	ECG40162B	.....	SK40162B
1389	NTE1739	ECG1739	.....	SK10031
1390	NTE74142	ECG74142	.....	SK74142
1391	NTE4018B	ECG4018B	.....	SK4018B
1392	NTE451	ECG451	276-2062	SK9164
1393	NTE1108	ECG1108	.....	.....
1394	NTE249	ECG249	.....	SK9440
1395	NTE743	ECG743	.....	SK3172
1396	NTE5452	ECG5452	276-1067	SK6752
1397	NTE5129A	ECG5129A	.....	SK14X
1398	NTE74LS399	ECG74LS399	.....	SK74LS399
1399	.....	.....	.....	SK13A
1400	NTE195A	ECG195A	.....	SK3024
1401	NTE5283A	ECG5283A	.....	SK596
1402	NTE5192A	ECG5192A	.....	SK150
1403	NTE3086	ECG3086	.....	SK2086
1404	NTE7470	ECG7470	.....	SK7470
1405	NTE4988	ECG4988	.....	SK10207
1406	NTE5269AK	ECG5269AK	.....	SK569
1407	NTE40163B	ECG40163B	.....	SK40163B
1408	NTE4989	ECG4989	.....	SK10208
1409	NTE74143	ECG74143	.....	SK74143
1410	.....	.....	.....	SK9852
1411	NTE237	ECG237	.....	SK3299
1412	NTE5284A	ECG5284A	.....	SK598
1413	NTE5642	ECG5642	276-1000	SK3506
1414	NTE5313	ECG5313	.....	SK3986
1415	NTE5270AK	ECG5270AK	.....	SK571
1416	.....	ECG9371	.....	.....
1417	NTE3100	ECG3100	.....	SK4930
1418	NTE74144	ECG74144	.....	SK74144
1419	NTE85	ECG85	.....	SK3124A
1420	NTE364	ECG364	.....	SK9637
1421	NTE744	ECG744	.....	SK3171
1422	NTE5316	ECG5316	.....	SK3989
1423	NTE5285A	ECG5285A	.....	SK600
1424	NTE176	ECG176	.....	SK3123
1425	NTE878	ECG878	.....	SK4814
1426	NTE7472	ECG7472	.....	SK7472
1427	.....	.....	.....	SK7470
1428	NTE5271AK	ECG5271AK	.....	SK573
1429	NTE1550	ECG1550	.....	SK9249
1430	NTE307	ECG307	.....	SK3203
1431	NTE108	ECG108	276-2016	SK3293
1432	NTE7475	ECG7475	.....	SK7475
1433	NTE9157	ECG9157	.....	.....
1434	NTE5700	ECG5700	.....	SK9811
1435	.....	ECG2406	.....	.....
1436	NTE5286A	ECG5286A	.....	SK602
1437	NTE1338	ECG1338	.....	SK9757
1438	NTE5176A	ECG5176A	.....	SK118
1439	NTE1879	ECG1879	.....	SK10169
1440	NTE5272AK	ECG5272AK	.....	SK575
1441	NTE7413	ECG7413	.....	SK7413
1442	NTE1656	ECG1656	.....	SK7734
1443	NTE737	ECG737	.....	SK3375
1444	NTE5287A	ECG5287A	.....	SK604
1445	NTE5178A	ECG5178A	.....	SK122
1446	NTE1479	ECG1479	.....	SK9188
1447	NTE5274AK	ECG5274AK	.....	SK579
1448	.....	ECG74HC00	.....	SK7C00
1449	NTE511	ECG511/R-2AV2	.....	SK3759
1450	.....	.....	.....	SK3065
1451	NTE74147	ECG74147	.....	SK74147
1452	.....	ECG74HC02	.....	SK7C02
1453	.....	.....	.....	SK7C03
1454	.....	.....	.....	SK7495
1455	.....	ECG74HC04	.....	SK7C04
1456	NTE5646	ECG5646	.....	SK9711
1457	NTE5288A	ECG5288A	.....	SK606
1458	NTE5181A	ECG5181A	.....	SK128
1459	NTE74365	ECG74365	.....	SK74365
1460	.....	ECG74HC08	.....	SK7C08
1461	NTE5276AK	ECG5276AK	.....	SK583
1462	.....	ECG74HC10	.....	SK7C10
1463	.....	ECG74HC11	.....	SK7C11
1464	NTE74148	.....	.....	.....
1465	NTE67	ECG67	.....	SK9509
1466	NTE2380	ECG2380	.....	SK10126
1467	.....	ECG74HC14	.....	SK7C14
1468	NTE96L02	ECG96L02	.....	.....
1469	.....	ECG286	.....	SK3131A
1470	NTE5289A	ECG5289A	.....	SK608
1471	NTE74366	ECG74366	.....	SK74366
1472	NTE21256	ECG4256	.....	SK10191
1473	NTE5060A	ECG5060A	.....	SK200A
1474	NTE5278AK	ECG5278AK	.....	SK587
1475	.....	.....	.....	SK7C20
1476	.....	ECG6157	.....	.....
1477	.....	.....	.....	SK7C21
1478	NTE1142	ECG1142	.....	SK3485
1479	NTE1185	ECG1185	.....	SK3468
1480	NTE324	ECG324	.....	SK9368
1481	NTE9097	ECG9097	.....	SK9788
1482	.....	ECG1864	.....	SK9069
1483	NTE1877	ECG1877	.....	SK10168
1484	NTE5290A	ECG5290A	.....	SK610
1485	.....	.....	.....	SK7C27
1486	NTE7495	ECG7495	.....	SK7495
1487	NTE5494	ECG5494	.....	SK6794
1488	NTE74367	ECG74367	.....	SK74367
1489	.....	.....	.....	SK7C30
1490	NTE5183A	ECG5183A	.....	SK132
1491	NTE990	ECG990	.....	SK9012
1492	NTE74150	ECG74150	.....	SK74150
1493	NTE165	ECG165	276-2055	SK9374
1494	.....	ECG74HC32	.....	SK7C32
1495	NTE74151	ECG74151	.....	SK74151
1496	NTE152	ECG152	276-2020	SK3929
1497	.....	.....	.....	SK3132
1498	NTE153	ECG153	276-2027	SK3930
1499	.....	ECG2407	.....	.....
1500	NTE11	ECG11	.....	SK9918
1501	NTE522	ECG522	.....	SK3303
1502	NTE5416	ECG5416	.....	SK3956
1503	.....	ECG2411	.....	.....
1504	NTE175	ECG286	.....	SK3194
1505	NTE5291A	ECG5291A	.....	SK612
1506	NTE2311	ECG2311	.....	SK9131
1507	NTE5288AK	ECG5288AK	.....	SK607
1508	NTE74368	ECG74368	.....	SK74368
1509	NTE1242	ECG1242	.....	SK3483
1510	NTE5281AK	ECG5281AK	.....	SK593
1511	NTE2315	ECG2315	.....	SK4909
1512	NTE5558	ECG5558	.....	SK6658
1513	NTE85	ECG85	276-2009	SK3849
1514	NTE130MP	ECG130MP	.....	SK3029
1515	.....	.....	.....	SK7C42
1516	NTE846	ECG846	.....	SK9178
1517	NTE74152	ECG74152	.....	.....
1518	.....	.....	.....	SK3122
1519	.....	.....	.....	SK74LS195
1520	NTE6007	ECG6007	.....	SK7207
1521	NTE5292A	ECG5292A	.....	SK614
1522	NTE32	ECG32	.....	SK3867A
1523	NTE5282AK	ECG5282AK	.....	SK595
1524	NTE1223	ECG1223	.....	SK3493
1525	NTE4019B	ECG4019B	.....	SK4019B
1526	.....	ECG216	.....	SK9470
1527	NTE74LS00	ECG74LS00	.....	SK74LS00
1528	NTE40175B	ECG40175B	.....	SK40175B
1529	NTE1780	ECG1780	.....	SK9731
1530	NTE6006	ECG6006	.....	SK7206
1531	NTE5293A	ECG5293A	.....	SK616
1532	NTE5193A	ECG5193A	.....	SK152
1533	.....	.....	.....	SK9760
1534	.....	ECG5534A	.....	.....
1535	NTE7480	ECG7480	.....	SK7480
1536	NTE4073B	ECG4073B	.....	SK4073B
1537	NTE5283AK	ECG5283AK	.....	SK597
1538	NTE51	ECG51	.....	SK9118
1539	NTE74153	ECG74153	.....	SK74153
1540	NTE1192	ECG1192	.....	SK3445
1541	NTE2302	ECG2302	.....	SK9422
1542	.....	ECG5548	.....	.....
1543	.....	ECG833	.....	SK7751
1544	NTE4511B	ECG4511B	.....	SK4511B
1545	NTE6094	ECG6094	.....	SK9774

REPL. CODE	NTE PART NO.	ECG PART NO.	RADIO SHACK PART NO.	RCA PART NO.
1546	NTE74LS01	ECG74LS01	.....	SK74LS01
1547	NTE5819	ECG5819	.....	SK9286
1548	NTE5294A	ECG5294A	.....	SK618
1549	.....	.....	.....	SK20X
1550	NTE5284AK	ECG5284AK	.....	SK599
1551	NTE40174B	ECG40174B	.....	SK40174B
1552	.....	ECG129&427	.....	.....
1553	NTE4043B	ECG4043B	.....	SK4043B
1554	NTE915	ECG915	.....	SK9090
1555	NTE74154	ECG74154	.....	SK74154
1556	.....	ECG9372	.....	.....
1557	.....	.....	.....	SK7C194
1558	.....	.....	.....	SK7C73
1559	NTE74LS02	ECG74LS02	.....	SK74LS02
1560	NTE6050	ECG6050	.....	SK7250
1561	.....	.....	.....	SK7C74
1562	NTE31	ECG31	.....	SK3866A
1563	.....	ECG5545	.....	.....
1564	NTE74LS145	ECG74LS145	.....	SK74LS145
1565	.....	.....	.....	SK7C75
1566	.....	.....	.....	SK9668
1567	NTE5818	ECG5818	.....	SK3586A
1568	NTE5295A	ECG5295A	.....	SK620
1569	.....	.....	.....	SK3849
1570	.....	ECG506	.....	.....
1571	NTE1218	ECG1218	.....	SK3740
1572	NTE1732	ECG1732	.....	SK7703
1573	NTE270	ECG270	.....	.....
1574	NTE7482	ECG7482	.....	.....
1575	NTE5285AK	ECG5285AK	.....	SK601
1576	NTE74155	ECG74155	.....	SK74155
1577	NTE5945	ECG5945	.....	SK7145
1578	.....	.....	.....	SK7C195
1579	NTE74LS03	ECG74LS03	.....	SK74LS03
1580	NTE1428	ECG1428	.....	SK9048
1581	NTE5948	ECG5948	.....	SK3605
1582	NTE1400	ECG1400	.....	SK7627
1583	.....	.....	.....	SK7C85
1584	NTE5515	ECG5515	.....	SK6615
1585	NTE1422	ECG1422	.....	.....
1586	NTE5952	ECG5952	.....	SK3585
1587	NTE5296A	ECG5296A	.....	SK622
1588	.....	ECG74HC86	.....	SK7C86
1589	NTE5344	ECG5344	.....	SK9105
1590	NTE1539	ECG1539	.....	SK9404
1591	NTE295	ECG295	.....	SK3253
1592	NTE74156	ECG74156	.....	SK74156
1593	NTE195A	ECG195A	.....	SK3512
1594	.....	.....	.....	SK7C93
1595	NTE74LS04	ECG74LS04	.....	SK74LS04
1596	NTE1737	ECG1737	.....	SK10030
1597	NTE129P	ECG129P	.....	SK9469
1598	NTE285	ECG285	.....	SK3846
1599	NTE9403	ECG9403	.....	SK9063
1600	NTE5912	ECG5912	.....	SK10422
1601	NTE5181AK	ECG5181AK	.....	SK129
1602	NTE74LS107	ECG74LS107	.....	SK74LS107
1603	NTE290A	ECG290A	276-2023	SK3450
1604	NTE1795	ECG1795	.....	SK9879
1605	NTE74157	ECG74157	.....	SK74157
1606	NTE5082A	ECG5082A	.....	SK25V
1607	NTE152	ECG152	276-2020	SK3239
1608	NTE74LS05	ECG74LS05	.....	SK74LS05
1609	NTE5924	ECG5924	.....	SK10428
1610	NTE5928	ECG5928	.....	SK10430
1611	NTE799	ECG799	.....	SK3238
1612	NTE1219	ECG1219	.....	SK3741
1613	NTE290A	ECG290A	276-2023	SK3247
1614	NTE5932	ECG5932	.....	SK10432
1615	NTE3018	ECG3018	.....	SK2018
1616	NTE5182AK	ECG5182AK	.....	SK131
1617	NTE74S112	ECG74S112	.....	SK74S112
1618	NTE5185A	ECG5185A	.....	SK136
1619	NTE9135	ECG9135	.....	.....
1620	NTE74LS122	ECG74LS122	.....	SK74LS122
1621	NTE1039	ECG1039	.....	SK3366
1622	NTE5183AK	ECG5183AK	.....	SK133
1623	NTE74S113	ECG74S113	.....	SK9704
1624	NTE24	ECG24	.....	SK3849
1625	NTE74LS132	ECG74LS132	.....	SK7CT132
1626	.....	.....	.....	SK3505
1627	NTE1520	ECG1520	.....	SK9043
1628	NTE74LS136	ECG74LS136	.....	SK74LS136
1629	NTE74S114	ECG74S114	.....	.....
1630	NTE5981	ECG5981	.....	SK7181
1631	NTE74160	ECG74160	.....	SK74160
1632	NTE9094	ECG9094	.....	SK9789
1633	NTE74LS08	ECG74LS08	.....	SK74LS08
1634	NTE1096	ECG1096	.....	SK3703
1635	NTE5917	ECG5917	.....	SK10425
1636	NTE5190AK	ECG5190AK	.....	SK147
1637	.....	.....	.....	SK3500
1638	NTE21128	.....	.....	.....
1639	NTE5191AK	ECG5191AK	.....	SK149
1640	NTE5193AK	ECG5193AK	.....	SK153
1641	NTE1742	ECG1742	.....	SK9995
1642	NTE74LS09	ECG74LS09	.....	SK1924
1643	NTE5340	ECG5340	.....	SK9101
1644	NTE5993	ECG5993	.....	SK3518
1645	NTE74161	ECG74161	.....	SK74161
1646	NTE74LS151	ECG74LS151	.....	SK74LS151
1647	.....	ECG991	.....	.....
1648	NTE398	ECG398	.....	SK9363

REPL. CODE	NTE PART NO.	ECG PART NO.	RADIO SHACK PART NO.	RCA PART NO.
1649	NTE5076A	ECG5076A	.....	SK17V
1650	NTE5517	ECG5517	.....	SK3615
1651	NTE5514	ECG5514	.....	SK3613
1652	NTE276	ECG276	.....	SK3296
1653	.....	ECG2385	.....	.....
1654	NTE74LS156	ECG74LS156	.....	SK74LS156
1655	.....	ECG128	.....	.....
1656	NTE74LS158	ECG74LS158	.....	SK1923
1657	NTE184&185	ECG184&185	.....	SK3190&3191
1658	NTE1V275	ECG1V275	.....	SKMV275A
1659	NTE56017	ECG56017	.....	SK9830
1660	NTE2V275	ECG2V275	.....	SKMV275B
1661	NTE4020B	ECG4020B	.....	SK4020B
1662	NTE74LS10	ECG74LS10	.....	SK74LS10
1663	NTE199	ECG199	276-2016	SK3132
1664	NTE598	ECG598	.....	SK9859
1665	NTE5516	ECG5516	.....	SK6616
1666	NTE74163	ECG74163	.....	SK74163
1667	NTE74LS11	ECG74LS11	.....	SK74LS11
1668	NTE2386	ECG2386	.....	SK10130
1669	NTE6400A	ECG6400B	.....	SK9122
1670	.....	.....	.....	SK4511B
1671	.....	ECG759	.....	.....
1672	NTE74LS175	ECG74LS175	.....	SK74LS175
1673	NTE5342	ECG5342	.....	SK9103
1674	NTE5083A	ECG5083A	.....	SK28V
1675	NTE127	ECG127	.....	SK3035
1676	NTE1573	ECG1573	.....	SK7678
1677	NTE928	ECG928	.....	SK3691
1678	NTE74LS181	ECG74LS181	.....	SK7CT181
1679	NTE74LS12	ECG74LS12	.....	SK74LS12
1680	NTE5701	ECG5701	.....	SK10414
1681	NTE180	ECG180	.....	SK9136
1682	NTE4094B	ECG4094B	.....	SK4094B
1683	NTE5406	ECG5406	.....	SK3952
1684	.....	ECG773	.....	.....
1685	NTE74165	ECG74165	.....	SK74165
1686	NTE74LS190	ECG74LS190	.....	SK7CT190
1687	NTE74LS191	ECG74LS191	.....	SK74LS191
1688	NTE74LS13	ECG74LS13	.....	SK74LS13
1689	NTE74LS192	ECG74LS192	.....	SK7CT192
1690	NTE186	ECG186	.....	SK3192
1691	NTE1V010	ECG1V010	.....	SKMV10B
1692	NTE74LS193	ECG74LS193	.....	SK74LS193
1693	NTE2147	ECG2147	.....	SK10296
1694	NTE2V010	ECG2V010	.....	SKMV10C
1695	NTE74S157	ECG74S157	.....	SK1914
1696	NTE912	ECG912	.....	SK3543
1697	.....	ECG2027	.....	.....
1698	NTE74LS14	ECG74LS14	.....	SK74LS14
1699	NTE778S	ECG778S	.....	SK10139
1700	NTE5377	ECG5377	.....	.....
1701	NTE2027	ECG2027	.....	SK10277
1702	NTE7494	ECG7494	.....	SK7494
1703	NTE5192AK	ECG5192AK	.....	SK151
1704	NTE5567	ECG5567	.....	SK6717
1705	NTE909D	ECG909D	.....	SK3590
1706	NTE526A	ECG526A	.....	SK3306
1707	NTE74LS15	ECG74LS15	.....	SK74LS15
1708	NTE228A	ECG228A	.....	SK9462
1709	NTE8224	ECG8224	.....	.....
1710	NTE533	ECG533	.....	SK3302
1711	NTE5195A	ECG5195A	.....	SK156
1712	NTE308	ECG308	.....	SK3855
1713	NTE5065A	ECG5065A	.....	SK3V0
1714	NTE1167	ECG1167	.....	SK3732
1715	NTE7496	ECG7496	.....	SK7496
1716	NTE5194AK	ECG5194AK	.....	SK155
1717	NTE5196A	ECG5196A	.....	SK158
1718	.....	ECG14	.....	.....
1719	NTE2407	ECG2407	.....	.....
1720	NTE5195AK	ECG5195AK	.....	SK157
1721	NTE1909	ECG1909	.....	SK10051
1722	NTE74170	ECG74170	.....	SK74170
1723	NTE5197A	ECG5197A	.....	SK160
1724	.....	ECG29	.....	.....
1725	NTE5599	ECG5599	.....	SK6749
1726	NTE74LS241	ECG74LS241	.....	SK74LS241
1727	NTE6244	ECG6244	.....	SK9777
1728	NTE74LS242	ECG74LS242	.....	SK74LS242
1729	NTE75491B	ECG75491B	.....	SK9786
1730	NTE74LS37	ECG74LS37	.....	SK74LS37
1731	NTE5196AK	ECG5196AK	.....	SK159
1732	NTE74LS247	ECG74LS247	.....	SK74LS247
1733	NTE2406	ECG2406	.....	SK10097
1734	NTE74LS248	ECG74LS248	.....	SK74LS248
1735	NTE74LS249	ECG74LS249	.....	SK74LS249
1736	NTE5198A	ECG5198A	.....	SK162
1737	NTE74LS251	ECG74LS251	.....	SK1919
1738	.....	ECG931	.....	.....
1739	NTE74LS253	ECG74LS253	.....	SK74LS253
1740	NTE75492B	ECG75492B	.....	.....
1741	.....	ECG8853	.....	.....
1742	NTE2409	ECG2409	.....	SK10100
1743	NTE116	ECG116	276-1102	SK5002
1744	NTE74LS257	ECG74LS257	.....	SK74LS257
1745	NTE6075	ECG6075	.....	SK7075
1746	NTE74LS258	ECG74LS258	.....	SK74LS258
1747	NTE5803	ECG5804	276-1144	SK9007
1748	NTE5199A	ECG5199A	.....	SK164
1749	NTE4021B	ECG4021B	.....	SK4021B
1750	NTE1045	ECG1045	.....	SK3072
1751	NTE321	ECG321	.....	SK3844

REPL. CODE	NTE PART NO.	ECG PART NO.	RADIO SHACK PART NO.	RCA PART NO.
1752	NTE2407	ECG2407	.....	SK10098
1753	.....	ECG798	.....	SK3216
1754	.....	ECG779A	.....	SK3170
1755	NTE5601	ECG5601	276-1000	SK6601
1756	NTE5198AK	ECG5198AK	.....	SK163
1757	NTE199	ECG199	276-2016	SK3931
1758	NTE458	ECG458	.....	SK9460
1759	NTE736	ECG736	.....	.....
1760	.....	.....	.....	SK5012
1761	NTE5200A	ECG5200A	.....	SK166
1762	.....	ECG74173	.....	.....
1763	NTE74LS21	ECG74LS21	.....	SK1920
1764	NTE40192B	ECG40192B	.....	SK40192B
1765	NTE5595	ECG5595	.....	SK6745
1766	NTE74173	ECG74173	.....	SK74173
1767	NTE4927	ECG4927	.....	SK10198
1768	NTE5199AK	ECG5199AK	.....	SK165
1769	NTE40194B	ECG40194B	.....	SK40194B
1770	NTE74174	ECG74174	.....	SK74174
1771	NTE5805	ECG5806	.....	SK3848
1772	NTE5202A	ECG5202A	.....	SK170
1773	NTE74LS280	ECG74LS280	.....	SK74LS280
1774	.....	ECG5330	.....	.....
1775	NTE74LS22	ECG74LS22	.....	SK74LS22
1776	NTE40193B	ECG40193B	.....	SK40193B
1777	NTE6357	ECG6357	.....	SK6557
1778	NTE3024	ECG3024	.....	SK2024
1779	NTE74LS283	ECG74LS283	.....	.....
1780	.....	ECG1545	.....	.....
1781	.....	.....	.....	SK3690
1782	NTE5200AK	ECG5200AK	.....	SK167
1783	NTE6077	ECG6077	.....	SK7077
1784	NTE40195B	ECG40195B	.....	SK40195B
1785	NTE126	ECG126A	.....	SK3006
1786	NTE1902	ECG1902	.....	SK3962
1787	NTE74175	ECG74175	.....	SK74175
1788	NTE5203A	ECG5203A	.....	SK172
1789	NTE6359	ECG6359	.....	SK6559
1790	NTE184	ECG184	.....	SK9041
1791	NTE518	ECG518	.....	SK9086
1792	NTE74H50	ECG74H50	.....	.....
1793	NTE2028	ECG2028	.....	SK9814
1794	NTE5202AK	ECG5202AK	.....	SK171
1795	NTE74176	ECG74176	.....	SK74176
1796	NTE5204A	ECG5204A	.....	SK174
1797	NTE910	ECG910	.....	SK9175
1798	NTE6220	ECG6220	.....	SK10434
1799	NTE5203AK	ECG5203AK	.....	SK173
1800	NTE910D	ECG910D	.....	SK9176
1801	.....	.....	.....	SK3629
1802	NTE116 N2	ECG116 N2	276-1104 N2	SK9001
1803	NTE74177	ECG74177	.....	SK74177
1804	NTE5205A	ECG5205A	.....	SK176
1805	.....	.....	.....	SK74LS322
1806	NTE23	ECG23	.....	SK9671
1807	NTE15019E	ECG15019	.....	SK10185
1808	NTE747	ECG747	.....	SK3279
1809	NTE5204AK	ECG5204AK	.....	SK175
1810	.....	ECG9974	.....	.....
1811	NTE4508B	ECG4508B	.....	SK4508B
1812	NTE960	ECG960	.....	SK3591
1813	NTE74178	ECG74178	.....	SK74178
1814	NTE5596	ECG5596	.....	SK6746
1815	NTE922	ECG922	.....	SK3567A
1816	NTE1194	ECG1194	.....	SK3484
1817	NTE5205AK	ECG5205AK	.....	SK177
1818	NTE6079	ECG6079	.....	SK7079
1819	NTE74S133	ECG74S133	.....	.....
1820	NTE74179	ECG74179	.....	SK74179
1821	NTE5207A	ECG5207A	.....	SK180
1822	.....	.....	.....	SK40061
1823	NTE9930	ECG9930	.....	.....
1824	NTE1729	ECG1729	.....	SK9912
1825	NTE314	ECG314	.....	SK3898
1826	NTE5206AK	ECG5206AK	.....	SK179
1827	NTE74S134	ECG74S134	.....	.....
1828	NTE950	ECG950	.....	SK9169
1829	NTE74180	ECG74180	.....	SK74180
1830	NTE4514B	ECG4514B	.....	SK4514B
1831	NTE9936	ECG9936	.....	.....
1832	NTE74LS33	ECG74LS33	.....	SK74LS33
1833	.....	.....	.....	SK3027
1834	NTE148A	ECG148A	.....	SK55V
1835	NTE9937	ECG9937	.....	.....
1836	NTE1903	ECG1903	.....	SK9221
1837	NTE1056	ECG1056	.....	SK3458
1838	NTE967	ECG967	.....	SK3673
1839	NTE74LS38	ECG74LS38	.....	SK74LS38
1840	NTE5207AK	ECG5207AK	.....	SK181
1841	NTE74LS42	ECG74LS42	.....	SK74LS42
1842	NTE74181	ECG74181	.....	SK74181
1843	NTE789	ECG789	.....	SK3078
1844	NTE9949	ECG9949	.....	.....
1845	NTE74LS47	ECG74LS47	.....	SK74LS47
1846	NTE2800	ECG2800	.....	.....
1847	NTE6009	ECG6009	.....	SK7209
1848	NTE5562	ECG5562	.....	SK3655
1849	NTE74LS48	ECG74LS48	.....	SK74LS48
1850	NTE74LS49	ECG74LS49	.....	SK74LS49
1851	NTE6011	ECG6011	.....	SK7211
1852	NTE386	ECG386	.....	SK9039
1853	NTE5208AK	ECG5208AK	.....	SK183
1854	NTE904	ECG904	.....	SK3542

REPL. CODE	NTE PART NO.	ECG PART NO.	RADIO SHACK PART NO.	RCA PART NO.
1855	NTE5564	ECG5564	.....	SK3656
1856	NTE74182	ECG74182	.....	SK74182
1857	NTE74LS54	ECG74LS54	.....	SK74LS54
1858	.....	ECG5404	.....	.....
1859	NTE9961	ECG9961	.....	.....
1860	NTE8212	ECG8212	.....	.....
1861	NTE5209AK	ECG5209AK	.....	SK185
1862	NTE226MP	ECG226MP	.....	SK3086
1863	NTE8216	ECG8216	.....	SK9790
1864	NTE74LS63	.....	.....	.....
1865	NTE346	ECG346	.....	SK3529
1866	NTE5210AK	ECG5210AK	.....	SK187
1867	NTE74LS73	ECG74LS73	.....	SK74LS73
1868	NTE74185	.....	.....	.....
1869	NTE8228	ECG8228	.....	.....
1870	NTE74LS75	ECG74LS75	.....	SK74LS75
1871	NTE5375	ECG5375	.....	SK6775
1872	NTE74LS77	ECG74LS77	.....	SK74LS77
1873	NTE74LS78	ECG74LS78	.....	SK74LS78
1874	NTE5211AK	ECG5211AK	.....	SK189
1875	NTE5332	ECG5332	.....	SK9232
1876	.....	ECG761	.....	.....
1877	NTE2030	ECG2030	.....	.....
1878	.....	ECG819	.....	.....
1879	.....	.....	.....	SK7683
1880	.....	.....	.....	SK9718
1881	NTE5160A	ECG5160A	.....	SK140X
1882	NTE74LS90	ECG74LS90	.....	SK74LS90
1883	.....	.....	.....	SK7698
1884	NTE5212AK	ECG5212AK	.....	SK191
1885	NTE74LS91	ECG74LS91	.....	.....
1886	NTE74S140	ECG74S140	.....	.....
1887	NTE74LS92	ECG74LS92	.....	SK74LS92
1888	NTE74LS93	ECG74LS93	.....	SK74LS93
1889	.....	.....	.....	SK6736
1890	NTE911	ECG911	.....	SK9070
1891	NTE5690	ECG5690	.....	SK5590
1892	NTE2415	ECG2415	.....	SK9962
1893	NTE6410	ECG6410	.....	SK9121
1894	NTE966	ECG966	276-1771	SK9592
1895	NTE5214AK	ECG5214AK	.....	SK195
1896	.....	.....	.....	SK7682
1897	NTE911D	ECG911D	.....	SK9071
1898	NTE2128	ECG2128	.....	SK10519
1899	NTE829	ECG829	.....	SK3891
1900	NTE5587	ECG5587	.....	SK6777
1901	NTE5178AK	ECG5178AK	.....	SK123
1902	NTE5215AK	ECG5215AK	.....	SK197
1903	NTE1345	ECG1345	.....	SK9998
1904	NTE3881	ECG3881	.....	SK2881
1905	NTE6157	ECG6157	.....	SK7357
1906	NTE74LS109A	ECG74LS109A	.....	SK74LS109
1907	NTE74H04	ECG74H04	.....	SK74H04
1908	NTE237	ECG282&427	.....	SK3299
1909	NTE5217AK	ECG5217AK	.....	SK201
1910	.....	.....	.....	SK9073
1911	NTE398	ECG398	.....	SK3930
1912	NTE74190	ECG74190	.....	SK74190
1913	.....	ECG817	.....	.....
1914	NTE5219AK	ECG5219AK	.....	SK205
1915	NTE4934	ECG4934	.....	SK10201
1916	NTE931	ECG931	.....	SK9067
1917	NTE74191	ECG74191	.....	SK74191
1918	NTE74S387	ECG74S387	.....	.....
1919	NTE940	ECG940	.....	SK9171
1920	.....	ECG1118	.....	.....
1921	NTE74192	ECG74192	.....	SK74192
1922	NTE309K	ECG309K	.....	SK3629
1923	.....	ECG5423	276-1067	SK3570
1924	NTE5221AK	ECG5221AK	.....	SK209
1925	NTE600	ECG600	.....	SK3863
1926	NTE74193	ECG74193	.....	SK74193
1927	.....	.....	.....	SK9337
1928	.....	.....	.....	SK5040
1929	.....	ECG9112	.....	.....
1930	.....	ECG1918	.....	SK9338
1931	NTE74S194	ECG74S194	.....	.....
1932	NTE4929	ECG4929	.....	SK10200
1933	NTE5222AK	ECG5222AK	.....	SK211
1934	.....	ECG5655	.....	SK5555
1935	.....	ECG5656	.....	SK5556
1936	.....	ECG109	.....	.....
1937	.....	ECG5657	.....	SK5557
1938	.....	ECG1891	.....	.....
1939	.....	ECG5720	.....	.....
1940	.....	ECG719	.....	.....
1941	NTE5223AK	ECG5223AK	.....	SK213
1942	NTE6090	ECG6090	.....	SK4911
1943	NTE74195	ECG74195	.....	SK74195
1944	NTE74H51	ECG74H51	.....	.....
1945	NTE1409N	ECG1409	.....	SK9198
1946	NTE306	ECG306	.....	SK3251
1947	NTE5224AK	ECG5224AK	.....	SK215
1948	.....	ECG5710	.....	.....
1949	NTE4027B	ECG4027B	.....	SK4027B
1950	NTE74196	ECG74196	.....	SK74196
1951	NTE1382	ECG1382	.....	SK7658
1952	NTE4935	ECG4935	.....	SK10202
1953	NTE5225AK	ECG5225AK	.....	SK217
1954	.....	.....	.....	SK3854
1955	NTE74S151	ECG74S151	.....	.....
1956	NTE74197	ECG74197	.....	SK74197
1957	NTE4024B	ECG4024B	.....	SK4024B

REPL. CODE	NTE PART NO.	ECG PART NO.	RADIO SHACK PART NO.	RCA PART NO.
1958	.....	.....	.....	SK7C237
1959	.....	ECG2343	.....	SK10091
1960	.....	ECG1382	.....	SK7658
1961	NTE5226AK	ECG5226AK	.....	SK219
1962	NTE3022	ECG3022	276-041	SK2022
1963	NTE4510B	ECG4510B	.....	SK4510B
1964	NTE74198	ECG74198	.....	SK74198
1965	.....	ECG222	.....	.....
1966	NTE385	ECG385	.....	SK3946
1967	NTE56018	ECG56018	.....	SK10268
1968	NTE2344	ECG2344	.....	SK10092
1969	.....	ECG598	.....	.....
1970	.....	ECG3173	.....	.....
1971	NTE74199	ECG74199	.....	SK74199
1972	NTE4951	ECG4951	.....	SK10204
1973	NTE6532	ECG6532	.....	.....
1974	NTE320	ECG320	.....	SK9605
1975	NTE941D	ECG941D	.....	SK3552
1976	NTE941SM	ECG941SM	.....	SK10074
1977	NTE351	ECG351	.....	SK3177
1978	NTE315	ECG315	.....	SK3250
1979	.....	ECG74HC240	.....	SK7C240
1980	.....	ECG772A	.....	.....
1981	NTE3000	ECG3000	.....	SK2000
1982	NTE941D	ECG941D	.....	.....
1983	NTE3003	ECG3003	.....	SK2003
1984	NTE322	ECG322	.....	SK3252
1985	NTE3002	ECG3002	.....	SK2002
1986	NTE6156	ECG6156	.....	SK7356
1987	NTE4959	ECG4959	.....	SK10206
1988	NTE1673	ECG1673	.....	SK9402
1989	NTE3001	ECG3001	.....	SK2001
1990	.....	.....	.....	SK7C241
1991	NTE97	ECG97	.....	SK9109
1992	NTE5424	ECG5424	.....	.....
1993	NTE2V150	ECG2V150	.....	SKMV150J
1994	NTE1103	ECG1103	.....	SK3281
1995	.....	.....	.....	SK2162
1996	.....	.....	.....	SK7C242
1997	NTE536A	ECG536A	.....	SK3900
1998	.....	ECG1914	.....	SK9331
1999	NTE4049	ECG4049	.....	SK4009UB
2000	NTE1916	ECG1916	.....	SK9332
2001	.....	.....	.....	SK7C243
2002	NTE5940	ECG5940	.....	SK3607
2003	NTE5941	ECG5941	.....	SK7145
2004	NTE1913	ECG1913	.....	SK9334
2005	.....	ECG74HC244	.....	SK7C244
2006	NTE6034	ECG6034	.....	SK7234
2007	NTE178MP	ECG178MP	276-1122	SK9091
2008	NTE1V014	ECG1V014	.....	SKMV14B
2009	NTE2V014	ECG2V014	.....	SKMV14C
2010	NTE5305	ECG5305	.....	SK3676
2011	NTE61MCP	ECG61MCP	.....	SK9925
2012	.....	.....	.....	SK3244
2013	NTE281	ECG281	276-2043	SK3359
2014	.....	.....	.....	SK7C245
2015	NTE5693	ECG5693	.....	SK3652
2016	NTE300	ECG300	.....	SK3178B
2017	NTE5695	ECG5695	.....	SK3509
2018	NTE5697	ECG5697	.....	SK3522
2019	NTE1234	ECG1234	.....	SK3487
2020	NTE74H52	ECG74H52	.....	.....
2021	NTE187	ECG187	.....	SK9076
2022	NTE852	ECG852	.....	SK9248
2023	.....	.....	.....	SK2168
2024	NTE4000	ECG4000	.....	SK4000UB
2025	NTE1715	ECG1715	.....	SK10021
2026	NTE1714S	ECG1714S	.....	SK9715
2027	.....	ECG755	.....	.....
2028	NTE175	ECG175	.....	SK3538
2029	NTE1045	ECG1045	.....	SK3101
2030	NTE382	ECG382	.....	SK3250
2031	NTE1120	ECG1120	.....	.....
2032	.....	.....	.....	SK9200
2033	.....	.....	.....	SK3466
2034	NTE2V300	ECG2V300	.....	SKMV320B
2035	NTE5174A	ECG5174A	.....	SK114
2036	NTE5569 N4	ECG5569 N4	.....	SK6719 N4
2037	.....	ECG5637	.....	.....
2038	NTE383	ECG383	.....	SK3841
2039	NTE361	ECG361	.....	SK9634
2040	.....	.....	.....	SK7715
2041	NTE346	ECG346	.....	SK9038
2042	NTE1676	ECG1676	.....	SK7653
2043	NTE793	ECG793	.....	.....
2044	.....	ECG56008	.....	SK3660
2045	NTE81	ECG81	.....	SK9114
2046	NTE300	ECG300	.....	SK3464
2047	.....	ECG3157	.....	.....
2048	NTE2114	ECG2114	.....	SK2214
2049	NTE1174	ECG1174	.....	SK3186
2050	NTE299	ECG299	.....	SK3298
2051	NTE165	ECG165	276-2055	SK3710
2052	NTE16001	ECG15	.....	SK3132
2053	NTE4503B	ECG40097B	.....	SK4503B
2054	NTE1532	ECG1532	.....	SK9320
2055	NTE4002B	ECG4002B	.....	SK4002B
2056	NTE74LS643	ECG74LS643	.....	SK7CT643
2057	NTE1104	ECG1104	.....	SK3225
2058	NTE56	ECG56	.....	SK3929
2059	NTE280	ECG87	276-2041	SK3619
2060	NTE5836	ECG5838	.....	SK7038

REPL. CODE	NTE PART NO.	ECG PART NO.	RADIO SHACK PART NO.	RCA PART NO.
2061	NTE195A	ECG123	276-2009	SK3024
2062	NTE1135	ECG1135	.....	SK3876
2063	NTE1466	ECG1466	.....	SK9212
2064	.....	ECG35	.....	SK3111
2065	.....	.....	.....	SK7C251
2066	NTE1710	ECG1710	.....	SK9951
2067	NTE1712	ECG1712	.....	SK9953
2068	NTE1713	ECG1713	.....	SK9954
2069	NTE56	ECG56	.....	SK9364
2070	NTE76	ECG76	.....	SK9600
2071	.....	ECG617	.....	.....
2072	NTE4025B	ECG4025B	.....	SK4025B
2073	NTE1628	ECG1628	.....	SK4834
2074	.....	ECG8233	.....	.....
2075	NTE85	ECG85	276-2009	SK3250
2076	NTE5840	ECG5842	.....	SK7042
2077	.....	ECG579	.....	.....
2078	NTE9951	ECG9951	.....	.....
2079	NTE588	ECG588	.....	SK5014
2080	NTE8234	ECG8234	.....	.....
2081	NTE5842	ECG5842	.....	SK7042
2082	.....	.....	.....	SK7712
2083	NTE509	ECG509/R-3AT2	.....	SK3757
2084	NTE510	ECG510/R-3DB3	.....	SK3758
2085	NTE512	ECG512/R-6DW4	.....	SK3760
2086	NTE6810	ECG6810	.....	SK6810
2087	NTE8235	ECG8235	.....	.....
2088	NTE5846	ECG5846	.....	SK7046
2089	NTE5462	ECG5462	.....	SK9291
2090	NTE1759	ECG1759	.....	SK10003
2091	NTE363	ECG363	.....	SK9636
2092	.....	ECG87	.....	.....
2093	NTE365	ECG365	.....	SK9638
2094	NTE3120	ECG3120	.....	SK10518
2095	NTE5461	ECG5461	.....	SK9290
2096	NTE369	ECG369	.....	SK3131A
2097	NTE9933	ECG9933	.....	.....
2098	.....	ECG80C97	.....	SK4503B
2099	.....	.....	.....	SK9957
2100	NTE760	ECG760	.....	.....
2101	NTE74H53	ECG74H53	.....	.....
2102	NTE5163A	ECG5163A	.....	SK170X
2103	NTE1719	ECG1719	.....	SK7719
2104	NTE922M	ECG922M	.....	SK3668
2105	NTE1558	ECG1558	.....	SK9244
2106	NTE147A	ECG147A	.....	.....
2107	NTE873	ECG873	.....	SK7718
2108	NTE3045	ECG3045	.....	SK10327
2109	NTE528	ECG528	.....	SK3906
2110	NTE374	ECG374	.....	.....
2111	NTE1847	ECG1847	.....	SK10084
2112	NTE74C00	ECG74C00	.....	SK9806
2113	NTE5176AK	ECG5176AK	.....	SK119
2114	NTE74C02	ECG74C02	.....	SK7C02
2115	.....	ECG74HC257	.....	SK7C257
2116	NTE1725	ECG1725	.....	SK7705
2117	NTE6809E	ECG6809E	.....	SK10456
2118	NTE74C08	ECG74C08	.....	SK9785
2119	NTE74C10	ECG74C10	.....	SK74C10
2120	NTE4512B	ECG4512B	.....	SK4512B
2121	.....	ECG5011A	.....	.....
2122	NTE1650	ECG1650	.....	SK7606
2123	NTE74C20	ECG74C20	.....	SK74C20
2124	NTE1058	ECG1058	.....	SK3459
2125	NTE283	ECG94	.....	SK3438
2126	NTE74S124	ECG74S124	.....	.....
2127	NTE93L08	ECG93L08	.....	.....
2128	NTE74LS112A	ECG74LS112A	.....	.....
2129	NTE2331	ECG2331	.....	SK10088
2130	NTE178MP	ECG178MP	276-1122	SK3311
2131	NTE90	ECG90	.....	SK3244
2132	NTE74S174	ECG74S174	.....	SK74S174
2133	NTE4903	ECG4903	.....	SK10194
2134	NTE74S132	.....	.....	.....
2135	.....	ECG1492	.....	.....
2136	NTE6401	ECG6401	.....	SK9123
2137	NTE5929	ECG5929	.....	SK10431
2138	NTE74S138	ECG74S138	.....	.....
2139	NTE6118	ECG6118	.....	SK10253
2140	NTE280MP	ECG280MP	.....	SK3360
2141	NTE74S175	ECG74S175	.....	.....
2142	NTE74221	ECG74221	.....	SK74221
2143	NTE74C42	ECG74C42	.....	SK9784
2144	.....	ECG840	.....	.....
2145	.....	ECG56008	.....	.....
2146	.....	.....	.....	SK9118
2147	.....	ECG5589	.....	.....
2148	NTE74C48	ECG74C48	.....	SK74C48
2149	.....	.....	.....	SK74LS266
2150	NTE6408	ECG6408	.....	SK3523
2151	NTE74S153	ECG74S153	.....	SK1915
2152	NTE4026B	ECG4026B	.....	SK4026B
2153	NTE5820	ECG5820	.....	SK9284
2154	NTE74S158	ECG74S158	.....	.....
2155	NTE1021	ECG1021	.....	.....
2156	NTE74S163	.....	.....	.....
2157	NTE5092A	ECG5092A	.....	.....
2158	.....	ECG1113	.....	SK9190
2159	NTE6113	ECG6113	.....	SK10215
2160	NTE5006A	ECG5006A	.....	.....
2161	NTE705A	ECG705A	.....	SK3134
2162	NTE74C73	ECG74C73	.....	SK74C73
2163	NTE74C74	ECG74C74	.....	SK9783

REPL. CODE	NTE PART NO.	ECG PART NO.	RADIO SHACK PART NO.	RCA PART NO.
2164	NTE74C76	ECG74C76	.....	SK74C76
2165	NTE74S181	ECG74S181	.....	.....
2166	NTE74S182	.....	.....	.....
2167	.....	ECG792	.....	.....
2168	.....	ECG870	.....	SK9992
2169	NTE1930	ECG1930	.....	SK7643
2170	NTE349	ECG349	.....	SK9624
2171	NTE74C85	ECG74C85	.....	SK9782
2172	NTE74H54	ECG74H54	.....	.....
2173	.....	ECG74176	.....	.....
2174	NTE74C90	ECG74C90	.....	SK9781
2175	NTE74S288	ECG74S288	.....	.....
2176	.....	ECG2388	.....	SK10131
2177	NTE74C93	ECG74C93	.....	SK74C93
2178	NTE1754	ECG1754	.....	SK10002
2179	NTE6051	ECG6051	.....	SK7251
2180	NTE74LS76A	ECG74LS76A	.....	SK74LS76A
2181	NTE103&411	ECG103&411	.....	.....
2182	NTE6055	ECG6055	.....	SK7255
2183	NTE289A	ECG289A	276-2009	SK3479
2184	NTE1575	ECG1575	.....	SK9195
2185	NTE2308	ECG2308	.....	SK4908
2186	NTE1893	ECG1893	.....	.....
2187	.....	ECG8250	.....	.....
2188	NTE5521	ECG5521	.....	SK6621
2189	.....	ECG5305	.....	.....
2190	NTE199	ECG199	276-2016	SK3122
2191	NTE6061	ECG6061	.....	SK7261
2192	NTE6230	ECG6230	.....	SK10435
2193	NTE1211	ECG1211	.....	SK3739
2194	NTE1751	ECG1751	.....	.....
2195	NTE3026	ECG3026	276-025	SK2026
2196	NTE8301	ECG8301	.....	.....
2197	NTE6065	ECG6065	.....	SK7265
2198	NTE74S251	ECG74S251	.....	.....
2199	NTE3079	ECG3079	.....	SK2079
2200	NTE296	ECG296	.....	.....
2201	NTE181	ECG181	.....	SK3036
2202	.....	ECG5671	.....	SK10467
2203	NTE591	ECG591	276-1122 N1	SK9940
2204	NTE74S257	.....	.....	.....
2205	NTE74S258	ECG74S258	.....	.....
2206	NTE3078	ECG3078	.....	SK2078
2207	NTE74S260	.....	.....	.....
2208	NTE77	ECG77	.....	SK9602
2209	NTE107	ECG107	276-2016	SK3356
2210	.....	ECG74HC273	.....	SK7C273
2211	.....	.....	.....	SK3672
2212	NTE329	ECG329	.....	SK3195
2213	NTE74LS490	ECG74LS490	.....	SK74LS490
2214	NTE277	ECG277	.....	.....
2215	.....	ECG3020	.....	SK2020
2216	NTE6069	ECG6069	.....	SK7269
2217	NTE74S03	ECG74S03	.....	.....
2218	NTE74LS83A	ECG74LS83A	.....	SK74LS83
2219	.....	ECG2323	.....	.....
2220	NTE74S280	.....	.....	.....
2221	NTE5180A	ECG5180A	.....	SK126
2222	NTE924	ECG924	.....	SK9128
2223	NTE15	ECG15	.....	SK9663
2224	NTE74S287	ECG74S287	.....	.....
2225	NTE5165A	ECG5165A	.....	SK190X
2226	.....	ECG6206	.....	SK9780
2227	NTE74LS295A	ECG74LS295A	.....	.....
2228	NTE4028B	ECG4028B	.....	SK4028B
2229	.....	.....	.....	SK3265
2230	.....	ECG167	.....	.....
2231	NTE1133	ECG1133	.....	SK3490
2232	NTE109	ECG109	276-1123	SK3087
2233	NTE4029B	ECG4029B	.....	SK4029B
2234	NTE6246	ECG6246	.....	SK9776
2235	NTE263	ECG263	.....	SK3180
2236	NTE3046	ECG3046	.....	SK2046
2237	NTE264	ECG264	.....	SK3181A
2238	NTE74S02	ECG74S02	.....	SK1917
2239	NTE972	ECG972	.....	SK3670
2240	.....	ECG2344	.....	.....
2241	.....	ECG5721	.....	.....
2242	NTE7412	ECG7412	.....	SK7412
2243	NTE7414	ECG7414	.....	SK7414
2244	.....	.....	.....	SK25A
2245	NTE277	ECG277	.....	SK3111
2246	NTE938M	ECG938M	.....	SK9167
2247	NTE992	ECG992	.....	SK3688
2248	.....	ECG5711	.....	.....
2249	.....	.....	.....	SK7C280
2250	.....	.....	.....	SK3444
2251	NTE5933	ECG5933	.....	SK10433
2252	NTE1488	ECG1488	.....	SK9818
2253	NTE1198	ECG1198	.....	.....
2254	NTE1912	ECG1912	.....	SK9337
2255	.....	ECG6220	.....	.....
2256	NTE74LS113	ECG74LS113	.....	SK74LS113
2257	NTE787	ECG787	.....	SK3146
2258	NTE265	ECG265	.....	SK3860
2259	NTE958	ECG958	.....	SK3699
2260	.....	ECG159	.....	.....
2261	NTE1082	ECG1082	.....	SK3461
2262	NTE1029	ECG1029	.....	SK3368
2263	NTE74S04	ECG74S04	.....	SK74S04
2264	NTE74S373	.....	.....	.....
2265	NTE5024A	ECG5024A	.....	.....
2266	NTE74S374	.....	.....	.....
2267	NTE9601	ECG9601	.....	.....
2268	NTE54	ECG54	.....	SK9391
2269	NTE1794	ECG1794	.....	SK9878
2270	NTE5444	ECG5444	.....	SK3634
2271	NTE2383	ECG2383	.....	SK9507
2272	.....	.....	.....	SK3479
2273	.....	.....	.....	SK3079
2274	.....	ECG6246	.....	.....
2275	NTE8266	ECG8266	.....	.....
2276	NTE92	ECG92	.....	SK9446
2277	NTE244	ECG244	.....	SK3183A
2278	NTE2716	ECG2716	.....	SK2716
2279	NTE721	ECG721	.....	SK7747
2280	NTE580	ECG580	.....	SK5016
2281	NTE186A&187A	ECG186A&187A	.....	SK3357&9076
2282	.....	ECG763	.....	.....
2283	NTE976	ECG976	.....	SK3596
2284	NTE1236	ECG1236	.....	SK7784
2285	.....	.....	.....	SK4515B
2286	NTE9602	ECG9602	.....	.....
2287	NTE192	ECG192	.....	SK3137
2288	.....	.....	.....	SK7474
2289	NTE1895	ECG1895	.....	.....
2290	NTE74249	ECG74249	.....	.....
2291	NTE5897	ECG5897	.....	SK7111
2292	NTE152	ECG152	276-2020	SK3197
2293	.....	ECG74249	.....	.....
2294	NTE5702	ECG5702	.....	SK10436
2295	NTE2060	ECG2060	.....	SK3966
2296	NTE282	ECG282	.....	SK3049
2297	.....	ECG1318	.....	.....
2298	NTE5703	ECG5703	.....	SK10437
2299	NTE74251	ECG74251	.....	SK74251
2300	NTE5629	ECG5629	276-1000	SK4918
2301	NTE988	ECG988	.....	SK3973
2302	NTE74LS114	ECG74LS114	.....	SK74LS114
2303	NTE1880	ECG1880	.....	SK10170
2304	NTE1687	ECG1687	.....	SK7711
2305	NTE1110	ECG1110	.....	SK3229
2306	NTE1203	ECG1203	.....	.....
2307	.....	ECG2117	.....	.....
2308	NTE4044B	ECG4044B	.....	SK4044B
2309	NTE69	ECG69	.....	.....
2310	.....	.....	.....	SK9011
2311	NTE5720	ECG5720	.....	SK10417
2312	NTE318	ECG318	.....	SK9604
2313	NTE3032	ECG3032	.....	SK2032
2314	.....	ECG282&427	.....	.....
2315	NTE6103	ECG6103	.....	SK10210
2316	NTE1092	ECG1092	.....	SK3472
2317	NTE5043A	ECG5043A	.....	SK60A
2318	NTE995	ECG995	.....	SK9209
2319	.....	.....	.....	SK9364
2320	NTE74S472	ECG74S472	.....	.....
2321	NTE74S05	ECG74S05	.....	.....
2322	NTE74S474	ECG74S474	.....	.....
2323	NTE2357	ECG2357	.....	SK10124
2324	NTE74S475	ECG74S475	.....	.....
2325	NTE1V017	ECG1V017	.....	SKMV17B
2326	NTE2V017	ECG2V017	.....	SKMV17C
2327	NTE230	ECG230	.....	SK3857
2328	NTE1528	ECG1528	.....	.....
2329	NTE310	ECG310	.....	SK3856
2330	.....	.....	.....	SK9258
2331	NTE4047B	ECG4047B	.....	SK4047B
2332	.....	ECG2410	.....	.....
2333	.....	.....	.....	SK9036
2334	NTE1883	ECG1883	.....	SK10173
2335	.....	.....	.....	SK9352
2336	NTE239	ECG239	.....	SK4914
2337	.....	ECG279A	.....	.....
2338	.....	ECG2404	.....	.....
2339	NTE1659	ECG1659	.....	SK3874
2340	NTE6073	ECG6073	.....	SK7273
2341	NTE66	ECG66	276-2072	SK9165
2342	NTE5403	ECG5404	276-1067	SK3627
2343	NTE5987	ECG5987	.....	SK7111
2344	.....	.....	.....	SK7C00
2345	.....	ECG6246	.....	SK9776
2346	NTE1855	ECG1855	.....	SK10085
2347	NTE4516B	ECG4516B	.....	SK4516B
2348	NTE56014	ECG5679	.....	SK3993
2349	NTE2406	ECG2406	.....	.....
2350	NTE9670	ECG9670	.....	.....
2351	NTE151A N3	ECG151A N3	.....	SK110V N3
2352	NTE5650	ECG5650	276-1000	SK5550
2353	NTE5651	ECG5651	276-1000	SK5551
2354	NTE329	ECG329	.....	SK3024
2355	NTE975SM	ECG975SM	.....	SK10455
2356	NTE5652	ECG5652	276-1000	SK5552
2357	NTE2164	ECG2164	.....	SK10297
2358	NTE947D	ECG947D	.....	SK3556
2359	NTE198	ECG198	.....	SK3103A
2360	.....	.....	.....	SKMV373A
2361	.....	.....	.....	SK3261
2362	.....	.....	.....	SK3883
2363	NTE5322	ECG5322	.....	SK3680
2364	NTE1777	ECG1777	.....	SK9870
2365	NTE5901	ECG5901	.....	SK7101
2366	NTE2334	ECG2334	.....	SK9475
2367	NTE2304	ECG2304	.....	SK9985
2368	NTE947	ECG947	.....	SK3526
2369	NTE5324	ECG5324	.....	SK3681
2370	NTE5326	ECG5326	.....	SK3682
2371	NTE284MP	ECG284MP	.....	SK10309
2372	NTE5327	ECG5327	.....	SK9087
2373	NTE5328	ECG5328	.....	SK9088
2374	NTE1733	ECG1733	.....	SK7809
2375	NTE289A	ECG289A	276-2009	SK3244
2376	NTE5905	ECG5905	.....	SK7105
2377	NTE128	ECG123	276-2030	SK3024
2378	NTE3080	ECG3080	.....	SK4832
2379	NTE8098	ECG8098	.....	SK849
2380	NTE1202	ECG1202	.....	.....
2381	NTE107	ECG107	276-2016	SK3018
2382	.....	.....	.....	SK9444
2383	NTE5640	ECG5640	276-1000	SK6740
2384	NTE4093B	ECG4093B	.....	SK4093B
2385	NTE1735	ECG1735	.....	SK7811
2386	NTE1667	ECG1667	.....	SK4831
2387	NTE5643	ECG5643	.....	SK3519
2388	NTE74S570	ECG74S570	.....	.....
2389	NTE74S571	ECG74S571	.....	.....
2390	NTE74S572	ECG74S572	.....	.....
2391	.....	ECG289	.....	SK3450
2392	NTE74S573	ECG74S573	.....	.....
2393	NTE1045	ECG1045	.....	.....
2394	NTE5641	ECG5641	276-1000	SK3583
2395	NTE5134A	ECG5134A	.....	SK19X
2396	NTE4900	ECG4900	.....	SK9812
2397	NTE5173A	ECG5173A	.....	.....
2398	NTE74H11	ECG74H11	.....	.....
2399	NTE5138A	ECG5138A	.....	SK25X
2400	NTE5306	ECG5306	.....	SK3677
2401	NTE5175A	ECG5175A	.....	SK116
2402	NTE123A	ECG123A	276-1617	SK3122
2403	NTE5175AK	ECG5175AK	.....	SK117
2404	NTE1015	ECG1015	.....	SK7754
2405	NTE1822	ECG1822	.....	SK9902
2406	NTE1016	ECG1016	.....	SK7755
2407	NTE5177A	ECG5177A	.....	SK120
2408	.....	.....	.....	SK7C08
2409	.....	ECG74LS00	.....	SK74LS00
2410	NTE5177AK	ECG5177AK	.....	SK121
2411	NTE42	ECG42	.....	SK9425
2412	.....	.....	.....	SK3577
2413	NTE1881	ECG1881	.....	SK10171
2414	NTE388	ECG388	.....	SK3947
2415	NTE5179A	ECG5179A	.....	SK124
2416	NTE5179AK	ECG5179AK	.....	SK125
2417	NTE1884	ECG1884	.....	SK10174
2418	NTE1823	ECG1823	.....	.....
2419	NTE1872	ECG1872	.....	SK10164
2420	NTE2350	ECG2350	.....	SK10123
2421	NTE5022A	ECG5022A	.....	.....
2422	NTE5184A	ECG5184A	.....	SK134
2423	.....	.....	.....	SK3441
2424	NTE5184AK	ECG5184AK	.....	SK135
2425	NTE1679	ECG1679	.....	SK10244
2426	.....	ECG1823	.....	.....
2427	NTE243	ECG243	.....	SK3182
2428	NTE2349	ECG2349	.....	SK10122
2429	.....	.....	.....	SK7C10
2430	.....	ECG74LS02	.....	.....
2431	NTE5023A	ECG5023A	.....	.....
2432	NTE248	ECG248	.....	SK3949
2433	NTE387	ECG387	.....	SK9040
2434	NTE9668	ECG9668	.....	.....
2435	.....	.....	.....	SK7C11
2436	.....	.....	.....	SK30A
2437	NTE283	ECG283	.....	SK3439A
2438	NTE130	ECG87	.....	SK9237
2439	NTE247	ECG247	.....	SK3948
2440	NTE1017	ECG1017	.....	.....
2441	NTE389	ECG389	.....	SK9261
2442	.....	ECG74LS04	.....	.....
2443	NTE5317	ECG5317	.....	SK9037
2444	.....	ECG975SM	.....	.....
2445	NTE74S10	ECG74S10	.....	.....
2446	NTE123A	ECG123A	276-1617	SK3132
2447	NTE74S11	ECG74S11	.....	SK74S11
2448	NTE250	ECG250	.....	SK9441
2449	NTE5491	ECG5491	.....	SK6791
2450	NTE5201A	ECG5201A	.....	SK168
2451	NTE74S15	ECG74S15	.....	.....
2452	NTE975	ECG975	.....	SK3644
2453	NTE5201AK	ECG5201AK	.....	SK169
2454	NTE1171	ECG1171	.....	SK3645
2455	.....	ECG489	.....	SK10465
2456	NTE199	ECG85	276-2016	SK3124A
2457	NTE731	ECG731	.....	SK3170
2458	NTE454	ECG454	.....	SK3991
2459	NTE123AP	ECG85	276-2016	SK3124A
2460	NTE289	ECG108	276-2016	SK3039
2461	NTE74S22	ECG74S22	.....	SK74S22
2462	NTE1918	ECG1918	.....	SK9338
2463	NTE74H73	ECG74H73	.....	.....
2464	.....	ECG1026	.....	.....
2465	.....	.....	.....	SK3899
2466	.....	.....	.....	SK7C14
2467	.....	.....	.....	SK24A
2468	NTE82	ECG82	.....	SK9115
2469	NTE74LS197	ECG74LS197	.....	SK1921
2470	NTE1928	ECG1928	.....	SK7642
2471	NTE18	ECG18	276-2009	SK9666
2472	.....	ECG1111	.....	.....

REPL. CODE	NTE PART NO.	ECG PART NO.	RADIO SHACK PART NO.	RCA PART NO.
2473	NTE6072	ECG6072	.....	SK7272
2474	NTE5213AK	ECG5213AK	.....	SK193
2475	NTE74S40	ECG74S40	.....	.....
2476	.....	ECG1018	.....	.....
2477	.....	ECG74LS08	.....	.....
2478	.....	ECG5018A	.....	.....
2479	NTE5218AK	ECG5218AK	.....	SK203
2480	NTE810A	ECG810A	.....	SK7730
2481	NTE159	ECG159	276-2023	SK3114A
2482	NTE1112	ECG1112	.....	.....
2483	NTE382	ECG382	.....	SK9352
2484	NTE17	ECG17	276-2023	SK9665
2485	NTE327	ECG327	.....	SK3945
2486	.....	ECG937	.....	SK3690
2487	NTE604	.....	.....	SK3087
2488	NTE8314	ECG8314	.....	.....
2489	.....	ECG1706	.....	.....
2490	.....	ECG2354	.....	.....
2491	NTE5482	ECG5482	.....	SK3941
2492	NTE74H74	ECG74H74	.....	.....
2493	NTE836	ECG836	.....	SK3206
2494	.....	.....	.....	SK3829
2495	NTE236	ECG236	.....	SK3197
2496	NTE74S64	ECG74S64	.....	.....
2497	NTE74S65	ECG74S65	.....	.....
2498	.....	ECG74LS10	.....	.....
2499	.....	.....	.....	SK3697
2500	NTE5228AK	ECG5228AK	.....	SK223
2501	NTE818	ECG818	.....	SK3207
2502	NTE843	ECG843	.....	SK3208
2503	NTE4068B	ECG4068B	.....	SK4068B
2504	NTE74S74	ECG74S74	.....	SK74S74
2505	NTE1152	ECG1152	.....	.....
2506	NTE360	ECG360	.....	SK9633
2507	NTE5231AK	ECG5231AK	.....	SK229
2508	.....	ECG74LS11	.....	.....
2509	NTE1019	ECG1019	.....	SK7756
2510	NTE178MP	ECG178MP	276-1122	SK3031A
2511	NTE74S86	ECG74S86	.....	SK74S86
2512	NTE1815	ECG1815	.....	SK9891
2513	.....	ECG587	.....	.....
2514	NTE1453	ECG1453	.....	SK3702
2515	NTE6087	ECG6087	.....	SK9775
2516	NTE4070B	ECG4070B	.....	SK4070B
2517	NTE924M	ECG924M	.....	SK9129
2518	NTE555	ECG555	.....	SK9150A
2519	NTE5501	ECG5501	.....	SK3579
2520	.....	.....	.....	SK3893
2521	NTE5550	ECG5550	.....	SK6650
2522	.....	ECG5013A	.....	.....
2523	NTE506	ECG506	.....	SK5014
2524	NTE4072B	ECG4072B	.....	SK4072B
2525	NTE316	ECG316	.....	SK3018
2526	NTE344	ECG344	.....	SK9620
2527	NTE506	ECG506	.....	SK3318A
2528	NTE506	ECG506	.....	SK3175A
2529	NTE727	ECG727	.....	SK3071
2530	NTE1459	ECG1459	.....	SK7797
2531	NTE4032B	ECG4032B	.....	SK4032B
2532	NTE1215	ECG1215	.....	SK3738
2533	NTE1522	ECG1522	.....	SK9197
2534	NTE1170	ECG1170	.....	SK3874
2535	NTE1609	ECG1609	.....	SK10226
2536	.....	.....	.....	SK3124A
2537	NTE903	ECG903	.....	SK3540
2538	.....	ECG1170	.....	SK3874
2539	NTE1934X	ECG1934X	.....	SK10087
2540	NTE4075B	ECG4075B	.....	SK4075B
2541	NTE74LS540	ECG74LS540	.....	SK7CT540
2542	NTE586	ECG586	.....	SK9935
2543	.....	.....	.....	SKMV130P
2544	.....	ECG471	.....	.....
2545	NTE471	ECG471	.....	SK9643
2546	NTE529	ECG529	.....	SK3307
2547	NTE74LS541	ECG74LS541	.....	SK7CT541
2548	NTE86	ECG86	.....	SK9297
2549	.....	ECG728	.....	.....
2550	NTE47	ECG47	276-2016	SK3245
2551	.....	.....	.....	SK9041
2552	NTE949	ECG949	.....	SK3166
2553	NTE1711	ECG1711	.....	SK9952
2554	.....	ECG1732	.....	SK9760
2555	.....	.....	.....	SK9042
2556	NTE130	ECG87	276-2041	SK3297
2557	NTE804	ECG804	.....	SK3455
2558	NTE4077B	ECG4077B	.....	SK4077B
2559	NTE5963	ECG5963	.....	SK7363
2560	NTE791	ECG791	.....	SK3209
2561	NTE315	ECG315	.....	SK9137
2562	NTE918	ECG918	.....	SK9144
2563	.....	ECG9662	.....	.....
2564	NTE956	ECG956	276-1778	SK9215
2565	NTE918SM	ECG918SM	.....	SK10069
2566	NTE5967	ECG5967	.....	SK7367
2567	NTE918M	ECG918M	.....	SK9145
2568	NTE1005	ECG1005	.....	SK3723
2569	NTE74S08	ECG74S08	.....	.....
2570	NTE919D	ECG919D	.....	SK9218
2571	NTE706	ECG706	.....	SK3101
2572	NTE1179	ECG1179	.....	SK3737
2573	NTE1053	ECG1053	.....	SK7761
2574	NTE919	ECG919	.....	SK9217
2575	.....	.....	.....	SK26V
2576	NTE1054	ECG1054	.....	SK3457
2577	NTE1057	ECG1057	.....	SK7762
2578	NTE1055	ECG1055	.....	SK3494
2579	NTE74H87	ECG74H87	.....	.....
2580	.....	.....	.....	SK3083
2581	NTE1060	ECG1060	.....	SK3460
2582	NTE1061	ECG1061	.....	SK3228
2583	NTE1062	ECG1062	.....	SK3235
2584	.....	ECG1063	.....	SK7763
2585	NTE1064	ECG1064	.....	.....
2586	.....	ECG1065	.....	.....
2587	.....	ECG1066	.....	SK7764
2588	.....	ECG1067	.....	SK7765
2589	.....	ECG2063	.....	.....
2590	NTE1068	ECG1068	.....	.....
2591	NTE1072	ECG1072	.....	SK3295
2592	NTE1074	ECG1074	.....	SK3495
2593	NTE1181	ECG1181	.....	SK3706
2594	NTE1137	ECG1137	.....	SK7776
2595	NTE1634	ECG1634	.....	SK10234
2596	NTE1073	ECG1073	.....	SK3496
2597	NTE9661	ECG9661	.....	.....
2598	.....	ECG80	.....	.....
2599	.....	ECG1069	.....	SK3227
2600	.....	.....	.....	SKMV18A
2601	.....	ECG1070	.....	SK3874
2602	NTE1168	ECG1168	.....	.....
2603	.....	ECG1071	.....	SK3226
2604	NTE1163	ECG1163	.....	.....
2605	NTE1002	ECG1002	.....	SK3481
2606	NTE978SM	ECG978SM	.....	SK10453
2607	.....	ECG2532	.....	.....
2608	.....	ECG808	.....	.....
2609	.....	.....	.....	SK29V
2610	.....	.....	.....	SK7C32
2611	.....	ECG5460	.....	.....
2612	NTE74290	ECG74290	.....	SK74290
2613	NTE394	ECG394	.....	SK3983
2614	NTE1180	ECG1180	.....	SK3888
2615	NTE5966	ECG5966	.....	SK7366
2616	NTE1255	ECG1255	.....	.....
2617	NTE726	ECG726	.....	SK3129
2618	NTE1188	ECG1188	.....	SK3022
2619	NTE9950	ECG9950	.....	.....
2620	NTE94	ECG86	.....	SK9297
2621	.....	ECG5550	.....	.....
2622	.....	ECG9941	.....	.....
2623	NTE1410	ECG1410	.....	SK9016
2624	NTE704	ECG704	.....	SK3023
2625	NTE1417	ECG1417	.....	SK9315
2626	NTE98	ECG98	.....	SK9111
2627	.....	.....	.....	SKMV132A
2628	NTE3033	ECG3033	.....	SK10324
2629	NTE74LS170	ECG74LS170	.....	.....
2630	NTE1106	ECG1106	.....	SK7771
2631	NTE1085	ECG1085	.....	SK3476
2632	NTE1102	ECG1102	.....	SK3224
2633	NTE1210	ECG1210	.....	SK9074
2634	NTE1101	ECG1101	.....	SK3283
2635	NTE4033B	ECG4033B	.....	SK4033B
2636	NTE3101	ECG3101	.....	SK4931
2637	NTE125	ECG125	276-1114	SK3033A
2638	NTE1241	ECG1241	.....	SK3828
2639	NTE1109	ECG1109	.....	SK3711
2640	NTE3102	ECG3102	.....	SK4932
2641	.....	ECG3103	.....	SK4916
2642	.....	ECG3539	.....	.....
2643	NTE6078	ECG6078	.....	SK9810
2644	NTE74293	ECG74293	.....	SK74293
2645	.....	.....	.....	SK3051
2646	NTE173BP	.....	.....	SK7332
2647	NTE5518	ECG5518	.....	SK3653
2648	NTE963	ECG963	.....	SK3672
2649	NTE5519	ECG5519	.....	SK3654
2650	NTE396	ECG396	.....	SK3104A
2651	.....	.....	.....	SK401B
2652	.....	.....	.....	SK403B
2653	NTE6022	ECG6022	.....	SK7222
2654	.....	.....	.....	SK7609
2655	NTE6044	ECG6044	.....	SK7244
2656	NTE1660	ECG1660	.....	SK3364
2657	NTE6026	ECG6026	.....	SK7226
2658	NTE124	ECG124	.....	SK3261
2659	NTE5484	ECG5485	.....	SK3943
2660	NTE389	ECG389	276-2055	SK9411
2661	NTE5013T	ECG5013	.....	.....
2662	NTE73	ECG73	.....	SK9273
2663	NTE74H61	ECG74H61	.....	.....
2664	NTE6030	ECG6034	.....	SK7234
2665	NTE6021	ECG6021	.....	SK7221
2666	NTE1658	ECG1658	.....	SK7736
2667	NTE74S09	ECG74S09	.....	.....
2668	NTE389	ECG389	.....	SK3710
2669	NTE1230	ECG1230	.....	.....
2670	.....	ECG1225	.....	.....
2671	NTE568	ECG568	.....	SK3909
2672	.....	ECG74LS32	.....	SK74LS32
2673	.....	.....	.....	SK7645
2674	.....	.....	.....	SK7400
2675	NTE4520B	ECG4520B	.....	SK4520B
2676	.....	.....	.....	SK7410
2677	NTE6038	ECG6040	.....	SK7240
2678	.....	.....	.....	SK7420
2679	NTE229	ECG229	276-2058	SK3124A
2680	NTE74S08	ECG74S08	.....	SK74LS08
2681	.....	ECG760	.....	.....
2682	NTE6040	ECG6040	.....	SK7240
2683	.....	ECG1693	.....	.....
2684	.....	.....	.....	SK33A
2685	NTE93L16	ECG93L16	.....	.....
2686	NTE1714M	ECG1714M	.....	SK10019
2687	NTE900	ECG900	.....	SK3547
2688	NTE809	ECG809	.....	SK7727
2689	NTE901	ECG901	.....	SK3549
2690	.....	ECG5538	.....	.....
2691	NTE107	ECG107	276-2058	SK3132
2692	NTE6860	ECG6860	.....	SK9795
2693	NTE1541	ECG1541	.....	.....
2694	NTE1525	ECG1525	.....	SK7604
2695	NTE6023	ECG6023	.....	SK7223
2696	.....	ECG5539	.....	.....
2697	NTE2732	ECG2732	.....	SK1912
2698	NTE905	ECG905	.....	SK3546
2699	NTE784	ECG784	.....	SK3524
2700	NTE329	ECG329	.....	.....
2701	NTE906	ECG906	.....	SK3548
2702	NTE488	ECG488	.....	SK9659
2703	NTE30001	ECG3099	.....	SK10177
2704	.....	.....	.....	SK3594
2705	NTE908	ECG908	.....	SK3539
2706	.....	ECG74LS37	.....	.....
2707	NTE785	ECG785	.....	.....
2708	.....	ECG3201	.....	.....
2709	.....	ECG6210	.....	.....
2710	NTE295	ECG295	.....	SK3197
2711	NTE907	ECG907	.....	SK3545
2712	NTE1461	ECG1461	.....	SK9989
2713	NTE121	ECG121	.....	SK3009
2714	.....	ECG786	.....	SK3140
2715	NTE711	ECG711	.....	.....
2716	.....	ECG5340A	.....	.....
2717	NTE121	ECG121	.....	SK3015
2718	.....	.....	.....	SK7710
2719	NTE342	ECG342	.....	SK9618
2720	NTE343	ECG343	.....	SK9619
2721	NTE5240AK	ECG5240AK	.....	SK511
2722	NTE917	ECG917	.....	SK3544
2723	NTE1911	ECG1911	.....	SK10052
2724	NTE5241AK	ECG5241AK	.....	SK513
2725	NTE340	ECG340	276-2009	.....
2726	NTE5242AK	ECG5242AK	.....	SK515
2727	NTE914	ECG914	.....	SK3541
2728	NTE2061	ECG2061	.....	SK7612
2729	NTE5243AK	ECG5243AK	.....	SK517
2730	NTE6408	ECG6408	.....	.....
2731	NTE74H62	ECG74H62	.....	.....
2732	NTE5244AK	ECG5244AK	.....	SK519
2733	.....	ECG74LS40	.....	.....
2734	NTE5245AK	ECG5245AK	.....	SK521
2735	.....	ECG380MP	.....	.....
2736	NTE957	ECG957	.....	SK9216
2737	NTE935	ECG935	.....	SK9344
2738	NTE5246AK	ECG5246AK	.....	SK523
2739	NTE5424	ECG5424	.....	SK3598
2740	NTE5247AK	ECG5247AK	.....	SK525
2741	NTE288	ECG288	.....	SK3715
2742	NTE834SM	ECG834SM	.....	SK10063
2743	NTE730	ECG730	.....	SK3143
2744	NTE5248AK	ECG5248AK	.....	SK527
2745	NTE1732	ECG1732	.....	SK9761
2746	NTE5250AK	ECG5250AK	.....	SK531
2747	.....	ECG781	.....	.....
2748	NTE902	ECG902	.....	SK9200
2749	NTE916	ECG916	.....	SK3550
2750	NTE2023	ECG2023	.....	SK3694
2751	NTE929	ECG929	.....	SK3695
2752	NTE5256AK	ECG5256AK	.....	SK543
2753	.....	ECG3202	.....	.....
2754	.....	ECG112	.....	.....
2755	NTE788	ECG788	.....	SK3829
2756	NTE5259AK	ECG5259AK	.....	SK549
2757	NTE107	ECG107	276-2016	SK3124A
2758	NTE5261AK	ECG5261AK	.....	SK553
2759	NTE3470	ECG3470	.....	SK7732
2760	NTE5265AK	ECG5265AK	.....	SK561
2761	.....	ECG1691	.....	.....
2762	NTE5267AK	ECG5267AK	.....	SK565
2763	NTE155	ECG155	.....	SK3839
2764	.....	ECG1244	.....	.....
2765	NTE1885	ECG1885	.....	SK10175
2766	NTE2303	ECG2303	.....	SK9485
2767	NTE289MP	ECG289AMP	.....	.....
2768	NTE6027	ECG6027	.....	SK7227
2769	NTE282	ECG282	.....	SK3847
2770	NTE278	ECG278	.....	SK3293
2771	NTE955SM	ECG955SM	.....	SK10450
2772	NTE5273AK	ECG5273AK	.....	SK577
2773	NTE1165	ECG1165	.....	SK3827
2774	.....	ECG76	.....	.....
2775	.....	.....	.....	SK9242
2776	NTE5275AK	ECG5275AK	.....	SK581
2777	NTE4035B	ECG4035B	.....	SK4035B
2778	NTE6074	ECG6074	.....	SK7074
2779	NTE5277AK	ECG5277AK	.....	SK585
2780	.....	ECG834	.....	.....
2781	NTE1166	ECG1166	.....	SK3827

REPL. CODE	NTE PART NO.	ECG PART NO.	RADIO SHACK PART NO.	RCA PART NO.
2782	NTE930	ECG930	.....	SK3568
2783	NTE5279AK	ECG5279AK	.....	SK589
2784	NTE2309	ECG2309	.....	SK9488
2785	.....	ECG5304	.....	.....
2786	NTE1175	ECG1175	.....	SK3212
2787	NTE1612	ECG1612	.....	SK9817
2788	.....	.....	.....	SK9243
2789	NTE1176	ECG1176	.....	SK3210
2790	NTE1V020	ECG1V020	.....	SKMV20B
2791	NTE965	ECG965	.....	SK9168
2792	NTE6105	ECG6105	.....	SK10212
2793	NTE87	ECG87	276-2041	SK3297
2794	NTE1375	ECG1375	.....	SK7608
2795	NTE2300	ECG2300	.....	SK9131
2796	NTE2V020	ECG2V020	.....	SKMV20C
2797	NTE46	ECG46	.....	SK9461
2798	NTE1177	ECG1177	.....	SK3213
2799	NTE1619	ECG1619	.....	.....
2800	.....	ECG5340	.....	.....
2801	NTE702	ECG702	.....	SK9025
2802	NTE5286AK	ECG5286AK	.....	SK603
2803	NTE5287AK	ECG5287AK	.....	SK605
2804	NTE531	ECG531	.....	SK3300
2805	NTE5289AK	ECG5289AK	.....	SK609
2806	NTE1932	ECG1932	.....	.....
2807	NTE5290AK	ECG5290AK	.....	SK611
2808	NTE1627	ECG1627	.....	SK9816
2809	NTE5486	ECG5487	.....	SK3944
2810	NTE5291AK	ECG5291AK	.....	SK613
2811	NTE506	ECG506	.....	SK3130
2812	NTE822	ECG822	.....	SK3919
2813	NTE6076	ECG6076	.....	SK7076
2814	NTE85	ECG85	.....	SK3122
2815	NTE5292AK	ECG5292AK	.....	SK615
2816	NTE948SM	ECG948SM	.....	SK10079
2817	NTE820	ECG820	.....	.....
2818	NTE5293AK	ECG5293AK	.....	SK617
2819	.....	.....	.....	SK9085
2820	NTE5294AK	ECG5294AK	.....	SK619
2821	NTE293	ECG293	.....	SK3250
2822	NTE5295AK	ECG5295AK	.....	SK621
2823	NTE948	ECG948	.....	SK9173
2824	NTE5296AK	ECG5296AK	.....	SK623
2825	NTE340	ECG340	276-2009	SK3842
2826	NTE1413	ECG1413	.....	SK7635
2827	NTE2024	ECG2024	.....	SK3667
2828	.....	.....	.....	SK3830
2829	NTE1721	ECG1721	.....	SK10024
2830	NTE982	ECG982	.....	SK3205
2831	NTE821	ECG821	.....	SK3882
2832	NTE6362	ECG6362	.....	.....
2833	NTE6068	ECG6068	.....	SK7268
2834	NTE74LS196	ECG74LS196	.....	SK74LS196
2835	NTE333	ECG333	.....	SK9609
2836	NTE6206	ECG6206	.....	SK9780
2837	NTE4522B	ECG4522B	.....	SK4522B
2838	NTE970	ECG970	.....	SK9339
2839	NTE1723	ECG1723	.....	SK10026
2840	NTE6054	ECG6054	.....	SK7254
2841	NTE1722	ECG1722	.....	SK10025
2842	NTE1655	ECG1655	.....	SK7687
2843	NTE5594	ECG5594	.....	SK6744
2844	NTE63	ECG63	.....	SK9371
2845	NTE6078	ECG6078	.....	SK7078
2846	NTE4007	ECG4007	.....	SK4007UB
2847	NTE62	ECG62	.....	.....
2848	.....	.....	.....	SK3293
2849	.....	ECG2338	.....	SK9480
2850	NTE6060	ECG6060	.....	SK7260
2851	NTE10	ECG10	.....	SK9139
2852	.....	ECG366	.....	.....
2853	.....	ECG988	.....	.....
2854	NTE1083	ECG1083	.....	.....
2855	NTE6031	ECG6035	.....	SK7245
2856	NTE5149A	ECG5149A	.....	SK60X
2857	.....	.....	.....	SK3362
2858	NTE5104A	ECG5104A	.....	SK190V
2859	.....	.....	.....	SK3249
2860	NTE47	ECG47	.....	SK9459
2861	NTE1090	ECG1090	.....	SK3291
2862	.....	ECG74S479	.....	.....
2863	NTE932	ECG932	.....	SK9340
2864	NTE1148	ECG1148	.....	SK3292
2865	.....	.....	.....	SK9241
2866	NTE102A&103A	ECG102A&103A	.....	SK3004&3835
2867	NTE9924	ECG9924	.....	.....
2868	NTE367	ECG367	.....	SK9640
2869	NTE978C	ECG978C	.....	SK10454
2870	NTE15007	ECG1860	.....	SK9945
2871	NTE6064	ECG6064	.....	SK7264
2872	NTE1140	ECG1140	.....	SK3473
2873	NTE74	ECG74	.....	SK9274
2874	NTE5001A	ECG5001A	.....	SK2A5
2875	NTE5530	ECG5531	.....	SK6631
2876	NTE1027	ECG1027	.....	.....
2877	NTE74LS126A	ECG74LS126	.....	SK7CT126
2878	NTE5722	ECG5722	.....	SK10421
2879	NTE1738	ECG1738	.....	SK9359
2880	.....	.....	.....	SK7C365
2881	NTE928SM	ECG928SM	.....	SK10073
2882	NTE1047	ECG1047	.....	SK7760
2883	.....	ECG346	.....	.....
2884	NTE320F	ECG320F	.....	SK9606

REPL. CODE	NTE PART NO.	ECG PART NO.	RADIO SHACK PART NO.	RCA PART NO.
2885	.....	.....	.....	SK7C366
2886	.....	ECG1084	.....	.....
2887	NTE1086	ECG1086	.....	.....
2888	NTE452	ECG452	.....	SK9072
2889	NTE1034	ECG1034	.....	.....
2890	NTE925	ECG925	.....	SK9174
2891	NTE1050	ECG1050	.....	.....
2892	.....	ECG928S	.....	.....
2893	NTE869	ECG869	.....	SK10066
2894	.....	.....	.....	SK7C367
2895	NTE5046A	ECG5046A	.....	SK75
2896	NTE1616	ECG1616	.....	SK7735
2897	NTE254	ECG254	.....	SK3997
2898	.....	.....	.....	SK7C368
2899	.....	ECG9110	.....	.....
2900	NTE5896	ECG5896	.....	SK7096
2901	NTE5982	ECG5982	.....	SK7182
2902	NTE319P	ECG319P	276-2009	SK3246A
2903	NTE178MP	ECG178MP	276-1122	SK9001
2904	.....	.....	.....	SK36A
2905	.....	ECG2321	.....	.....
2906	.....	ECG770	.....	.....
2907	NTE6035	ECG6035	.....	SK7235
2908	NTE2312	ECG2312	.....	SK10000
2909	.....	ECG74S478	.....	SK9703
2910	NTE382	ECG382	.....	SK3849
2911	.....	ECG375	.....	SK3929
2912	NTE52	ECG52	.....	SK3995
2913	NTE1529	ECG1529	.....	SK7804
2914	NTE1160	ECG1160	.....	SK7775
2915	.....	ECG74LS95	.....	.....
2916	NTE1552	ECG1552	.....	SK9314
2917	NTE5506	ECG5507	.....	SK3580
2918	.....	.....	.....	SK9948
2919	NTE2402	ECG2402	.....	SK10095
2920	.....	ECG74HC373	.....	SK7C373
2921	NTE1848	ECG1848	.....	SK10148
2922	.....	ECG928M	.....	.....
2923	NTE55	ECG55	.....	SK3441
2924	NTE40	ECG40	.....	SK9423
2925	.....	ECG5342	.....	.....
2926	NTE1563	ECG1563	.....	SK7648
2927	.....	ECG74HC374	.....	SK7C374
2928	NTE386	ECG386	.....	SK3995
2929	NTE123AP	ECG123A	276-2009	SK3444
2930	NTE745	ECG745	.....	.....
2931	NTE1205	ECG1205	.....	SK3482
2932	NTE5471	ECG5471	.....	SK10259
2933	NTE5473	ECG5474	276-1020	SK10261
2934	NTE96LS02	ECG96LS02	.....	.....
2935	NTE1V035	ECG1V035	.....	SKMV35B
2936	NTE5474	ECG5474	.....	SK10261
2937	.....	ECG8063	.....	SK814
2938	NTE4555B	ECG4555B	.....	SK4555B
2939	.....	ECG380	.....	.....
2940	NTE7497	ECG7497	.....	SK7497
2941	NTE2V035	ECG2V035	.....	SKMV35C
2942	NTE1161	ECG1161	.....	.....
2943	NTE8070	ECG8070	.....	SK821
2944	NTE8076	ECG8076	.....	SK827
2945	NTE461	ECG461	.....	SK9148
2946	NTE317	ECG317	.....	SK9603
2947	NTE44	ECG44	.....	SK9427
2948	NTE8081	ECG8081	.....	SK832
2949	NTE1134	ECG1134	.....	SK3489
2950	NTE457	ECG457	276-2062	SK9161
2951	NTE8085	ECG8085	.....	SK836
2952	NTE195A	ECG195A	.....	SK3479
2953	NTE5012A	ECG5021A	276-563	SK12A
2954	NTE2361	ECG2361	.....	SK10093
2955	NTE8090	ECG8090	.....	SK841
2956	NTE74C240	ECG74C240	.....	SK7C240
2957	NTE1797	ECG1797	.....	SK9753
2958	NTE74C04	ECG74C04	.....	SK74C04
2959	.....	ECG9111	.....	.....
2960	NTE1263	ECG1263	.....	SK3920
2961	NTE8096	ECG8096	.....	SK847
2962	NTE114	ECG114	276-1101	SK3296
2963	NTE8103	ECG8103	.....	SK854
2964	NTE75	ECG75	.....	SK9275
2965	NTE116 N5	ECG116 N5	276-1104 N5	SK3017B N5
2966	NTE116 N7	ECG116 N7	276-1104 N7	SK3313 N7
2967	NTE5823	ECG5823	.....	SK9287
2968	NTE8108	ECG8108	.....	SK859
2969	NTE74H103	ECG74H103	.....	.....
2970	NTE253	ECG253	.....	SK9370
2971	NTE4031B	ECG4031B	.....	SK4031B
2972	NTE8115	ECG8115	.....	SK866
2973	NTE8118	ECG8118	.....	SK869
2974	NTE6039	ECG6041	.....	SK7245
2975	.....	ECG800	.....	.....
2976	NTE8125	ECG8125	.....	SK876
2977	HVD-1	ECG-HIDIV-1	.....	SK3868
2978	HVD-2	ECG-HIDIV-2	.....	SK3869
2979	HVD-4	ECG-HIDIV-4	.....	SK3871
2980	HVD-3	ECG-HIDIV-3	.....	SK3870
2981	NTE4038B	ECG4038B	.....	SK4038B
2982	NTE534	ECG534	.....	SK3305
2983	NTE521	ECG521	.....	.....
2984	NTE538	ECG538	.....	SK3310
2985	NTE535	ECG535	.....	SK3902
2986	.....	ECG802	.....	SK3277
2987	NTE326	ECG326	.....	SK3746

REPL. CODE	NTE PART NO.	ECG PART NO.	RADIO SHACK PART NO.	RCA PART NO.
2988	NTE282	ECG282	.....	SK3444
2989	NTE2335	ECG2335	.....	SK10305
2990	NTE8139	ECG8139	.....	SK890
2991	.....	ECG64	.....	.....
2992	.....	ECG9663	.....	.....
2993	.....	.....	.....	SK9374
2994	NTE5721	ECG5721	.....	SK10419
2995	.....	ECG803	.....	SK3278
2996	NTE282	ECG282	.....	SK3265
2997	NTE241	ECG241	276-2020	SK9450
2998	NTE8149	ECG8149	.....	SK900
2999	.....	.....	.....	SK7C86
3000	NTE8213	ECG8213	.....	SK901
3001	NTE2307	ECG2307	.....	SK9483
3002	NTE5013T1	ECG5011T1	.....	.....
3003	NTE5061A	.....	.....	SK2V4
3004	.....	ECG2320	.....	.....
3005	NTE2320	ECG2320	.....	SK10118
3006	.....	ECG814A	.....	.....
3007	NTE64	ECG64	.....	SK9372
3008	.....	ECG735	.....	.....
3009	NTE282	ECG282	.....	SK9368
3010	NTE5064A	.....	.....	SK2V8
3011	NTE5910	ECG5910	.....	SK7110
3012	NTE1636	ECG1636	.....	SK9245
3013	NTE8226	ECG8226	.....	SK914
3014	NTE51	ECG51	.....	SK9452
3015	NTE9669	ECG9669	.....	.....
3016	.....	.....	.....	SK3201
3017	NTE1156	ECG1156	.....	.....
3018	.....	ECG804	.....	SK9068
3019	NTE3007	ECG3007	276-026	SK2007
3020	.....	ECG1008	.....	.....
3021	NTE8242	ECG8242	.....	SK930
3022	.....	.....	.....	SK3617
3023	NTE5318	ECG5318	.....	SK10165
3024	NTE997	ECG997	.....	SK9172
3025	NTE46	ECG46	.....	SK9442
3026	NTE5094A	ECG5094A	.....	SK87V
3027	NTE1799	ECG1799	.....	SK9881
3028	NTE5154A	ECG5154A	.....	SK87X
3029	NTE942	ECG942	.....	SK3924
3030	.....	.....	.....	SK10140
3031	NTE74C14	ECG74C14	.....	SK9805
3032	NTE9664	ECG9664	.....	.....
3033	NTE104MP	ECG104MP	.....	SK3720
3034	NTE8167	ECG8167	.....	SK948
3035	NTE5319	ECG5319	.....	SK10271
3036	NTE8181	ECG8181	.....	SK952
3037	NTE52	ECG52	.....	SK9039
3038	NTE53	ECG53	.....	SK9484
3039	NTE9682	ECG9682	.....	.....
3040	.....	ECG56031	.....	.....
3041	NTE345	ECG345	.....	SK9621
3042	.....	.....	.....	SK22100
3043	.....	ECG74HC390	.....	SK7C390
3044	.....	ECG7018	.....	.....
3045	NTE9681	ECG9681	.....	.....
3046	NTE459	ECG459	.....	SK3112
3047	NTE74C221	ECG74C221	.....	SK74C221
3048	.....	.....	.....	SK22101
3049	NTE293MP	ECG293MP	.....	SK10314
3050	NTE5320	ECG5320	.....	SK10243
3051	NTE6041	ECG6041	.....	SK7241
3052	.....	.....	.....	SK22102
3053	NTE5485	ECG5485	.....	SK10263
3054	.....	ECG9665	.....	.....
3055	NTE5487	ECG5487	.....	SK10264
3056	NTE862	ECG862	.....	SK9822
3057	NTE1565	ECG1565	.....	SK9192
3058	.....	.....	.....	SK9133
3059	.....	.....	.....	SK9335
3060	.....	.....	.....	SK9261
3061	.....	ECG9931	.....	.....
3062	NTE876	ECG876	276-1705	SK10067
3063	NTE823	ECG823	276-1731	SK9210
3064	.....	ECG74HC393	.....	SK7C393
3065	NTE6502	ECG6502	.....	.....
3066	.....	ECG9667	.....	.....
3067	.....	.....	.....	SK3673
3068	NTE1508	ECG1508	.....	SK7638
3069	NTE1509	ECG1509	.....	SK7637
3070	NTE6507	ECG6507	.....	.....
3071	NTE1549	ECG1549	.....	.....
3072	NTE453	ECG453	.....	SK9853
3073	NTE2V025	ECG2V025	.....	SKMV25B
3074	NTE1618	ECG1618	.....	.....
3075	NTE2V030	ECG2V030	.....	SKMV30C
3076	NTE194	ECG194	.....	SK3479
3077	NTE552	ECG552	.....	SK5014
3078	NTE2V040	ECG2V040	.....	SKMV40C
3079	NTE824	ECG824	.....	SK9013
3080	.....	ECG1892	.....	SK10502
3081	NTE390	ECG390	.....	SK3958
3082	NTE36	ECG36	.....	SK9389
3083	NTE15023	ECG1859	.....	SK9946
3084	NTE1610	ECG1610	.....	.....
3085	NTE2V050	ECG2V050	.....	SKMV50C
3086	NTE828	ECG828	.....	SK9127
3087	NTE576	ECG576	.....	.....
3088	NTE2382	ECG2382	.....	SK10128
3089	NTE986	ECG986	.....	SK3918
3090	NTE2V060	ECG2V060	.....	SKMV60B

Replacements

REPL. CODE	NTE PART NO.	ECG PART NO.	RADIO SHACK PART NO.	RCA PART NO.
3091	.....	ECG352	.....	.....
3092	.....	.....	.....	SK9501
3093	.....	.....	.....	SK9453
3094	NTE85	ECG85	.....	SK9465
3095	.....	.....	.....	SK3118
3096	NTE3008	ECG3008	.....	SK2008
3097	NTE1798	ECG1798	.....	SK9745
3098	NTE2311	ECG2311	.....	.....
3099	NTE825	ECG825	.....	SK9126
3100	.....	.....	.....	SK8V2
3101	.....	ECG236	.....	.....
3102	NTE5500	ECG5501	.....	SK3579
3103	NTE1439	ECG1439	.....	SK3964
3104	NTE2V075	ECG2V075	.....	SKMV75B
3105	NTE5503	ECG5504	.....	SK3579
3106	.....	.....	.....	SK40116
3107	.....	.....	.....	SK3247
3108	.....	.....	.....	SK3114A
3109	NTE5505	ECG5507	.....	SK3580
3110	.....	.....	.....	SK9317
3111	NTE37	ECG37	.....	SK9415
3112	NTE54	ECG54	.....	SK3929
3113	.....	.....	.....	SK3929
3114	NTE4902	ECG4902	.....	SK10193
3115	.....	.....	.....	SK9391
3116	NTE2V095	ECG2V095	.....	SKMV95B
3117	NTE6043	ECG6043	.....	SK7243
3118	NTE943SM	ECG943SM	.....	SK10075
3119	NTE477	ECG477	.....	SK9649
3120	.....	ECG199	.....	.....
3121	NTE7481	ECG7481	.....	SK7481
3122	NTE180MCP	ECG180MCP	.....	.....
3123	NTE2V115	ECG2V115	.....	SKMV115C
3124	NTE3011	ECG3011	.....	SK2011
3125	NTE377	ECG377	276-2020	SK9391
3126	NTE1680	ECG1680	.....	.....
3127	NTE5118A	ECG5118A	.....	SK6X0
3128	NTE74C30	ECG74C30	.....	SK74C30
3129	.....	.....	.....	SK9205
3130	NTE467	ECG467	.....	SK9162
3131	NTE289A	ECG289A	.....	SK3122
3132	.....	.....	.....	SK3054
3133	NTE459	ECG459	.....	SK9149
3134	NTE74C32	ECG74C32	.....	SK74C32
3135	.....	ECG1136	.....	.....
3136	.....	ECG2326	.....	SK10137
3137	.....	.....	.....	SK3555
3138	.....	ECG6354	.....	.....
3139	NTE5821	ECG5821	.....	SK9285
3140	.....	ECG6356	.....	.....
3141	NTE74LS624	ECG74LS624	.....	SK74LS624
3142	.....	ECG6358	.....	.....
3143	NTE2358	ECG2358	.....	SK10125
3144	.....	.....	.....	SK3714
3145	NTE4527B	ECG4527B	.....	SK4527B
3146	NTE5680	ECG5681	.....	SK6681
3147	NTE5681	ECG5681	.....	SK6681
3148	NTE5682	ECG5682	.....	SK6682
3149	NTE74LS625	ECG74LS625	.....	SK74LS625
3150	NTE5683	ECG5683	.....	SK6683
3151	NTE5684	ECG5685	.....	SK6685
3152	NTE5685	ECG5685	.....	SK6685
3153	NTE5686	ECG5687	.....	SK6686
3154	NTE74LS626	ECG74LS626	.....	SK74LS626
3155	NTE312	ECG459	.....	SK3448
3156	.....	ECG1850	.....	SK10150
3157	NTE3021	ECG3021	.....	SK2021
3158	NTE74H55	ECG74H55	.....	.....
3159	NTE1787	ECG1787	.....	SK9766
3160	.....	ECG348	.....	.....
3161	NTE1803	ECG1803	.....	SK9884
3162	NTE74LS627	ECG74LS627	.....	SK74LS627
3163	.....	ECG3012A	.....	SK10317
3164	.....	ECG345	.....	.....
3165	NTE242	ECG242	.....	SK9451
3166	.....	ECG74S454	.....	.....
3167	NTE6363	ECG6363	.....	.....
3168	NTE74S188	ECG74S188	.....	.....
3169	.....	ECG3013A	.....	SK10318
3170	.....	ECG344	.....	.....
3171	.....	ECG618	.....	.....
3172	NTE4918	ECG4918	.....	SK10195
3173	NTE584	ECG584	.....	SK3100
3174	NTE4041	ECG4041	.....	SK4041UB
3175	NTE74LS629	ECG74LS629	.....	SK74LS629
3176	.....	ECG580	.....	SK5020
3177	NTE80C95	ECG80C95	.....	.....
3178	.....	ECG377	.....	SK9391
3179	NTE80C96	ECG80C96	.....	.....
3180	NTE80C97	ECG80C97	.....	.....
3181	NTE887M	ECG887M	.....	SK9862
3182	NTE3025	ECG3025	.....	SK2025
3183	.....	.....	.....	SK9454
3184	.....	ECG5344	.....	.....
3185	NTE3035A	ECG3035A	.....	SK2035A
3186	.....	ECG5547	.....	.....
3187	NTE387MP	ECG387MP	.....	.....
3188	.....	ECG577	.....	SK5040
3189	NTE5980	ECG5980	.....	SK3610
3190	NTE1228	ECG1228	.....	SK3889
3191	NTE4926	ECG4926	.....	SK10197
3192	NTE590	ECG590	.....	SK9939
3193	NTE6508	ECG6508	.....	.....
3194	NTE352	ECG352	.....	SK9626
3195	NTE795	ECG795	.....	SK3237
3196	NTE1463	ECG1463	.....	SK9208
3197	.....	ECG74LS112A	.....	.....
3198	NTE4528B	ECG4528B	.....	SK4098B
3199	NTE56022	ECG56022	.....	SK3661
3200	.....	.....	.....	SK3945
3201	NTE4928	ECG4928	.....	SK10199
3202	NTE987SM	ECG987SM	.....	SK10452
3203	NTE123AP	ECG123AP	276-2009	SK9433
3204	NTE5092A	ECG5092A	.....	SK68A
3205	NTE74LS259	ECG74LS259	.....	SK74LS259
3206	.....	ECG461	.....	.....
3207	NTE56024	ECG56024	.....	SK3662
3208	NTE6045	ECG6045	.....	SK7245
3209	NTE51	ECG51	.....	SK9085
3210	.....	.....	.....	SK4014B
3211	NTE2308	ECG2308	.....	SK9131
3212	.....	.....	.....	SK3656
3213	.....	ECG2065	.....	.....
3214	.....	.....	.....	SK9336
3215	NTE5566	ECG5566	.....	SK3657
3216	NTE213	ECG213	.....	SK3003A
3217	NTE2327	ECG2327	.....	SK10138
3218	.....	.....	.....	SK7358
3219	NTE750	ECG750	.....	SK3280
3220	.....	.....	.....	SK3674
3221	.....	ECG1477	.....	.....
3222	NTE56026	ECG56026	.....	SK3663
3223	.....	.....	.....	SK1802
3224	.....	ECG1667	.....	.....
3225	.....	.....	.....	SK9417
3226	NTE58	ECG58	.....	SK9448
3227	NTE87	ECG87	.....	SK3559
3228	NTE1470	ECG1470	.....	SK7799
3229	.....	ECG389	.....	SK9261
3230	.....	ECG601	.....	.....
3231	NTE2309	ECG2309	.....	SK9261
3232	NTE3029A	ECG3029A	.....	SK10322
3233	NTE928S	ECG928S	.....	SK10072
3234	.....	.....	.....	SK3652
3235	.....	ECG56006	.....	.....
3236	NTE1790	ECG1790	.....	SK9850
3237	NTE2057	ECG2057	.....	.....
3238	.....	.....	.....	SK3522
3239	.....	ECG1386	.....	SK7654
3240	NTE74390	ECG74390	.....	SK74390
3241	NTE506	ECG506	.....	SK5012
3242	.....	ECG2307	.....	.....
3243	NTE5508	ECG5509	.....	SK3504
3244	.....	.....	.....	SK4001B
3245	.....	ECG5695	.....	.....
3246	NTE1454	ECG1454	.....	SK9184
3247	NTE297MP	ECG297MP	.....	SK10315
3248	.....	ECG2300	.....	SK9131
3249	NTE312	ECG312	.....	SK3116
3250	NTE1844	ECG1844	.....	SK10147
3251	NTE74H05	ECG74H05	.....	.....
3252	.....	.....	.....	SK4010B
3253	NTE1685	ECG1685	.....	SK9948
3254	NTE1792	ECG1792	.....	SK9877
3255	NTE292MCP	ECG292MCP	.....	SK10313
3256	NTE74393	ECG74393	.....	SK9835
3257	NTE2V420	ECG2V420	.....	SKMV420
3258	NTE5523	ECG5524	.....	SK6624
3259	.....	ECG328	.....	.....
3260	.....	.....	.....	SK4009UB
3261	.....	.....	.....	SK3971
3262	NTE1239	ECG1239	.....	SK3708
3263	NTE1097	ECG1097	.....	SK3446
3264	NTE74H71	ECG74H71	.....	.....
3265	NTE5943	ECG5945	.....	SK7153
3266	.....	ECG6085	.....	.....
3267	NTE53	ECG53	.....	SK9261
3268	NTE5525	ECG5527	.....	SK6627
3269	NTE1164	ECG1164	.....	.....
3270	NTE984	ECG984	.....	SK9276
3271	NTE5526	ECG5527	.....	SK6627
3272	NTE2419	ECG2419	.....	SK10111
3273	NTE1173	ECG1173	.....	.....
3274	NTE123AP	ECG216	276-2009	SK9433
3275	NTE5947	ECG5949	.....	SK7153
3276	NTE3023	ECG3023	.....	SK2023
3277	NTE5540	ECG5541	.....	SK10408
3278	.....	ECG128&427	.....	.....
3279	.....	.....	.....	SK9389
3280	NTE225	ECG225	.....	SK3045
3281	NTE5949	ECG5949	.....	SK7153
3282	NTE1669	ECG1669	.....	.....
3283	.....	ECG5431	276-1067	.....
3284	.....	ECG5422	276-1067	SK3570
3285	NTE978	ECG978	276-1728	SK3689
3286	NTE33	ECG33	.....	SK9466
3287	NTE5951	ECG5953	.....	SK7153
3288	.....	ECG74S473	.....	.....
3289	NTE195A	ECG195A	.....	SK3529
3290	NTE74LS279	ECG74LS279	.....	SK74LS279
3291	NTE5528	ECG5529	.....	SK6629
3292	NTE238	ECG238	276-2055	SK9411
3293	.....	.....	.....	SK3712
3294	NTE5953	ECG5953	.....	SK7153
3295	.....	ECG74HCT00	.....	.....
3296	.....	ECG74HCT04	.....	SK7CT04
3297	.....	ECG856	.....	SK7708
3298	.....	.....	.....	SK3123
3299	.....	ECG74HCT08	.....	SK7CT08
3300	.....	ECG74HCT14	.....	SK7CT14
3301	NTE552	ECG552	.....	.....
3302	NTE85	ECG85	276-2009	SK9407
3303	NTE93MCP	ECG93MCP	.....	SK9931
3304	NTE4055B	ECG4055B	.....	SK4055B
3305	.....	ECG1841	.....	SK10145
3306	.....	ECG74HCT32	.....	SK7CT32
3307	NTE5520	ECG5521	.....	SK6621
3308	.....	.....	.....	SK4059A
3309	NTE6162	ECG6162	.....	SK10219
3310	.....	ECG1741	.....	.....
3311	.....	.....	.....	SK4069UB
3312	.....	ECG879	.....	SK10514
3313	NTE74H72	ECG74H72	.....	.....
3314	NTE290A	ECG290A	276-2023	SK9138
3315	.....	.....	.....	SK4077B
3316	.....	.....	.....	SK3087
3317	NTE5123A	ECG5123A	.....	SK8X7
3318	.....	ECG1193	.....	SK3701
3319	.....	.....	.....	SK3613
3320	NTE889M	ECG889M	.....	SK9864
3321	NTE475	ECG475	.....	SK9647
3322	NTE476	ECG476	.....	SK9648
3323	NTE290A	ECG290A	276-2023	.....
3324	NTE5415	ECG5415	276-1020	SK3955
3325	NTE4531B	ECG4531B	.....	SK4531B
3326	NTE473	ECG473	.....	SK9645
3327	NTE50	ECG50	.....	SK3179B
3328	.....	.....	.....	SK12A
3329	NTE282	ECG282	.....	SK3512
3330	NTE4099B	ECG4099B	.....	SK4099B
3331	NTE6407	ECG6407	.....	SK3523
3332	NTE1966	.....	.....	.....
3333	NTE1377	ECG1377	.....	SK7808
3334	NTE3017	ECG3017	.....	SK10321
3335	NTE1707	ECG1707	.....	SK10018
3336	.....	ECG282&427	.....	SK3045
3337	.....	ECG9921	.....	.....
3338	NTE53	ECG53	.....	SK9133
3339	.....	ECG2337	.....	.....
3340	NTE1824	ECG1824	.....	SK9899
3341	.....	ECG1824	.....	.....
3342	NTE459	ECG459	276-2055	SK3448
3343	NTE1846	ECG1846	.....	SK10083
3344	NTE1894	ECG1894	.....	.....
3345	.....	ECG256	.....	.....
3346	.....	.....	.....	SK9283
3347	.....	ECG1748	.....	.....
3348	NTE1748	ECG1748	.....	SK9762
3349	NTE5410	ECG5410	.....	SK3503
3350	.....	.....	.....	SK9408
3351	NTE1401	ECG1401	.....	SK9354
3352	.....	ECG1078	.....	SK3498
3353	.....	ECG1825	.....	SK9898
3354	NTE451	ECG451	276-2055	SK9164
3355	NTE1825	ECG1825	.....	SK9898
3356	.....	.....	.....	SK1805
3357	NTE2325	ECG2325	.....	SK9855
3358	.....	ECG560	.....	.....
3359	NTE1403	ECG1403	.....	SK9353
3360	NTE2300	ECG2300	.....	SK9476
3361	NTE1736	ECG1736	.....	SK10029
3362	.....	.....	.....	SK3562
3363	NTE1675	ECG1675	.....	SK9815
3364	.....	.....	.....	SK7CT4002
3365	NTE1800	ECG1800	.....	SK9882
3366	NTE1153	ECG1153	.....	SK3282
3367	NTE3081	ECG3081	.....	SK9965
3368	.....	.....	.....	SK4532B
3369	.....	.....	.....	SK6A8
3370	NTE274	ECG274	.....	SK9107
3371	NTE74LS298	ECG74LS298	.....	SK74LS298
3372	.....	ECG2402	.....	.....
3373	NTE245	ECG245	.....	SK9438
3374	NTE246	ECG246	.....	SK9439
3375	.....	.....	.....	SK9032
3376	NTE300MP	ECG300MP	.....	SK10616
3377	NTE257	ECG257	.....	SK4907
3378	.....	ECG7012	.....	SK10481
3379	NTE2412	ECG2412	.....	SK10107
3380	.....	ECG794	.....	.....
3381	NTE989	ECG989	.....	SK3595
3382	.....	.....	.....	SK5014
3383	NTE5512	ECG5512	276-1020	SK3684
3384	NTE312	ECG312	276-2055	SK3112
3385	NTE238	ECG238	276-2055	SK9261
3386	NTE1786	ECG1786	.....	SK9874
3387	.....	ECG2311	.....	.....
3388	NTE2319	ECG2319	.....	SK10015
3389	.....	ECG1873	.....	SK9764
3390	NTE161	ECG161	276-2016	SK3444
3391	NTE859	ECG859	.....	SK4826
3392	NTE287	ECG287	.....	SK9352
3393	NTE933	ECG933	.....	SK9341
3394	NTE871	ECG871	.....	SK3646
3395	NTE951	ECG951	.....	SK9170
3396	.....	ECG107	.....	SK3132
3397	NTE835	ECG835	.....	.....
3398	.....	ECG382	.....	SK9453
3399	NTE9948	ECG9948	.....	.....

REPL. CODE	NTE PART NO.	ECG PART NO.	RADIO SHACK PART NO.	RCA PART NO.
3400	NTE74LS133	ECG74LS133	.....	SK74LS133
3401	NTE74C95	ECG74C95	.....	SK74C95
3402	NTE5446	ECG5446	.....	SK3635
3403	.....	.....	.....	SK9211
3404	NTE5448	ECG5448	.....	SK3636
3405	.....	.....	.....	SK7CT4015
3406	NTE20	ECG20	.....	SK9418
3407	NTE6404	ECG6404	.....	SK7900
3408	.....	ECG5360	.....	.....
3409	.....	.....	.....	SK7CT4016
3410	.....	ECG805	.....	.....
3411	.....	.....	.....	SK9116
3412	.....	.....	.....	SK7CT688
3413	.....	.....	.....	SK43A
3414	.....	ECG5315	.....	.....
3415	NTE524V15	ECG524V15	.....	SKMV150K
3416	.....	.....	.....	SKMV150L
3417	.....	.....	.....	SK7CT4017
3418	.....	ECG9900	.....	.....
3419	NTE5480	ECG5481	.....	SK3940
3420	.....	.....	.....	SK43V
3421	NTE472	ECG472	.....	SK9644
3422	NTE806	ECG806	.....	.....
3423	NTE1578	ECG1578	.....	SK9130
3424	.....	.....	.....	SK9168
3425	NTE983	ECG983	.....	SK3887
3426	.....	.....	.....	SK3024
3427	.....	ECG9976	.....	.....
3428	NTE4046B	ECG4046B	.....	SK4046B
3429	NTE9989	ECG9989	.....	.....
3430	.....	.....	.....	SK7CT4020
3431	NTE4556B	ECG4556B	.....	SK4556B
3432	.....	ECG1277	.....	.....
3433	.....	ECG5826	.....	.....
3434	NTE944M	ECG944M	.....	SK10077
3435	NTE2305	ECG2305	.....	SK4904
3436	NTE2328	ECG2328	.....	SK9991
3437	NTE1744	ECG1744	.....	SK9732
3438	NTE2301	ECG2301	.....	.....
3439	NTE5476	ECG5476	.....	SK10262
3440	.....	ECG778A	.....	.....
3441	NTE5019A	ECG5019A	.....	.....
3442	NTE607	ECG607	.....	SK9848
3443	NTE4045B	ECG4045B	.....	SK4045B
3444	.....	.....	.....	SK3926
3445	NTE1743	ECG1743	.....	SK9996
3446	.....	ECG377	.....	.....
3447	.....	ECG2361	.....	.....
3448	.....	.....	.....	SK7CT4024
3449	.....	ECG51	.....	.....
3450	NTE40085B	ECG40085B	.....	.....
3451	NTE1292	ECG1292	.....	SK9181
3452	NTE615	ECG615A	.....	SK9976
3453	.....	.....	.....	SK4052B
3454	.....	ECG10	.....	.....
3455	NTE3014	ECG3014	.....	.....
3456	NTE3015	ECG3015	.....	.....
3457	NTE4048B	ECG4048B	.....	.....
3458	.....	ECG960	.....	.....
3459	NTE4048B	ECG4048B	.....	SK4048B
3460	.....	ECG2339	.....	SK10301
3461	.....	ECG5669A	.....	SK10461
3462	NTE2337	ECG2337	.....	SK10300
3463	.....	ECG962	.....	.....
3464	.....	.....	.....	SK7CT7030
3465	.....	.....	.....	SK6V8
3466	NTE7423	ECG7423	.....	SK7423
3467	.....	ECG5699	.....	.....
3468	NTE2412	ECG2412	.....	SK10106
3469	.....	ECG964	.....	.....
3470	.....	.....	.....	SK4724B
3471	NTE2430	ECG2430	.....	SK10114
3472	.....	ECG5966	.....	.....
3473	.....	.....	.....	SK3506
3474	.....	.....	.....	SK3938
3475	.....	.....	.....	SK3736
3476	NTE2301	ECG2301	.....	SK9486
3477	NTE7425	ECG7425	.....	SK7425
3478	NTE2418	ECG2418	.....	SK10110
3479	.....	ECG1932	.....	.....
3480	.....	.....	.....	SK4572UB
3481	NTE3880	ECG3880	.....	SK2880
3482	NTE2414	ECG2414	.....	SK9961
3483	NTE133	ECG312	276-2055	SK9157
3484	.....	ECG1757	.....	.....
3485	.....	ECG966	.....	.....
3486	.....	ECG2324	.....	.....
3487	.....	ECG287	.....	.....
3488	NTE1870	ECG1870	.....	SK10161
3489	NTE70	ECG70	.....	SK9270
3490	.....	ECG968	.....	.....
3491	NTE3164	ECG3164	.....	SK2164
3492	.....	.....	.....	SK3632
3493	.....	ECG2309	.....	.....
3494	NTE2324	ECG2324	.....	SK10136
3495	NTE1852	ECG1852	.....	SK10151
3496	NTE4076B	ECG4076B	.....	SK4076B
3497	.....	ECG2339	.....	.....
3498	.....	.....	.....	SK9409
3499	NTE5603	ECG5603	276-1000	SK3665
3500	NTE88	ECG88	.....	SK9239
3501	NTE590	ECG590	276-1122 N2	SK9939
3502	NTE1548	ECG1548	.....	SK7646

REPL. CODE	NTE PART NO.	ECG PART NO.	RADIO SHACK PART NO.	RCA PART NO.
3503	NTE1546	ECG1546	.....	SK7647
3504	NTE4566B	ECG4566B	.....	SK4566B
3505	.....	ECG958	.....	.....
3506	.....	.....	.....	SK7CT4040
3507	.....	ECG2302	.....	.....
3508	NTE213	ECG213	.....	SK3035
3509	NTE209	.....	.....	.....
3510	.....	ECG2311	.....	.....
3511	.....	ECG2301	.....	.....
3512	NTE71	ECG71	.....	SK9271
3513	.....	ECG2301	.....	SK9486
3514	.....	ECG2300	.....	.....
3515	NTE7010	ECG7010	.....	SK10479
3516	NTE2330	ECG2330	.....	SK9478
3517	NTE5826	ECG5826	.....	SK7376
3518	NTE1553	ECG1553	.....	SK7720
3519	NTE232	ECG232	.....	SK9443
3520	NTE7437	ECG7437	.....	SK7437
3521	NTE15044	.....	.....	SK9854
3522	.....	.....	.....	SK9994
3523	NTE1752	ECG1752	.....	SK10036
3524	.....	ECG2305	.....	.....
3525	NTE251	ECG251	.....	SK3858
3526	NTE252	ECG252	.....	SK3859
3527	.....	.....	.....	SK7CT7046A
3528	NTE258	ECG258	.....	SK9823
3529	NTE259	ECG259	.....	SK3978
3530	NTE260	ECG260	.....	SK3979
3531	NTE6363	ECG6363	.....	SK6561
3532	.....	ECG972	.....	.....
3533	NTE266	ECG266	.....	SK9253
3534	NTE267	ECG267	.....	SK9254
3535	NTE268	ECG268	.....	SK9255
3536	NTE269	ECG269	.....	SK9256
3537	NTE125	ECG125	276-1114	SK5010A
3538	NTE272	ECG272	.....	SK9436
3539	NTE273	ECG273	.....	SK9437
3540	NTE275	ECG275	.....	SK9108
3541	NTE5828	ECG5828	.....	.....
3542	NTE2348	ECG2348	.....	.....
3543	.....	.....	.....	SK3930
3544	NTE292	ECG292	.....	SK3930
3545	NTE131&155	ECG131&155	.....	SK3198&3839
3546	NTE5334	ECG5334	.....	SK9233
3547	NTE2313	ECG2313	.....	SK10004
3548	NTE153	ECG153	276-2027	SK3441
3549	.....	ECG588	.....	.....
3550	NTE1024	ECG1024	.....	SK3152
3551	NTE599	ECG599	.....	SK9860
3552	NTE3036	ECG3036	.....	SK4936
3553	.....	.....	.....	SK3538
3554	.....	.....	.....	SK3619
3555	.....	.....	.....	SK3004
3556	.....	.....	.....	SK7CT4051
3557	.....	ECG2060	.....	.....
3558	.....	.....	.....	SK7CT4052
3559	NTE1272	ECG1272	.....	.....
3560	NTE362	ECG362	.....	SK9635
3561	NTE330	ECG330	.....	SK3012
3562	.....	.....	.....	SK3747
3563	NTE334	ECG334	.....	SK9610
3564	.....	.....	.....	SK7CT4053
3565	NTE335	ECG335	.....	SK9611
3566	.....	.....	.....	SK9016
3567	NTE336	ECG336	.....	SK9612
3568	NTE337	ECG337	.....	SK9613
3569	NTE338	ECG338	.....	SK9614
3570	NTE339	ECG339	.....	SK9616
3571	.....	ECG381	.....	SK3934
3572	NTE341	ECG341	.....	SK9617
3573	NTE1925	ECG1925	.....	SK10056
3574	NTE1840	ECG1840	.....	SK10144
3575	NTE971	ECG971	.....	SK3675
3576	NTE347	ECG347	.....	SK9660
3577	NTE378	ECG378	.....	SK9367
3578	NTE348	ECG348	.....	SK9623
3579	.....	.....	.....	SK4081B
3580	NTE353	ECG353	.....	SK9627
3581	NTE93	ECG93	.....	SK9447
3582	NTE354	ECG354	.....	SK9628
3583	NTE355	ECG355	.....	SK9629
3584	NTE356	ECG356	.....	SK9630
3585	NTE55	ECG55	.....	SK3930
3586	NTE357	ECG357	.....	SK9631
3587	NTE359	ECG359	.....	SK9632
3588	NTE2311	ECG2311	.....	SK9374
3589	NTE366	ECG366	.....	SK9639
3590	.....	ECG2202	.....	.....
3591	NTE368	ECG368	.....	SK9641
3592	.....	.....	.....	SK47A
3593	NTE5620	ECG5620	.....	SK10466
3594	NTE991	ECG991	.....	.....
3595	.....	.....	.....	SK47V
3596	NTE2331	ECG2331	.....	SK1088
3597	.....	.....	.....	SK3841
3598	.....	.....	.....	SK9001
3599	.....	.....	.....	SK3675
3600	NTE4515B	ECG4515B	.....	SK4515B
3601	NTE6059	ECG6061	.....	SK7261
3602	NTE392	ECG392	.....	SK3960
3603	NTE393	ECG393	.....	SK3961
3604	NTE579	ECG579	.....	SK10163
3605	.....	.....	.....	SK7CT4059

REPL. CODE	NTE PART NO.	ECG PART NO.	RADIO SHACK PART NO.	RCA PART NO.
3606	NTE68	ECG68	.....	SK9365
3607	NTE395	ECG395	.....	.....
3608	NTE22	ECG22	.....	SK9670
3609	NTE1485	ECG1485	.....	SK9061
3610	NTE4526B	ECG4526B	.....	SK4526B
3611	NTE4098B	ECG4098B	.....	SK4098B
3612	.....	ECG712	.....	.....
3613	.....	.....	.....	SK3039
3614	.....	.....	.....	SK7CT4060
3615	NTE4034B	ECG4034B	.....	SK4034B
3616	.....	ECG1882	.....	.....
3617	NTE159	ECG159	276-2023	SK9434
3618	.....	ECG4568B	.....	SK4568B
3619	NTE89	ECG89	.....	SK9411
3620	NTE5414	ECG5414	276-1067	SK3954
3621	NTE937	ECG937	.....	SK9146
3622	NTE456	ECG456	.....	SK3977
3623	NTE7057	ECG7057	.....	.....
3624	NTE5513	ECG5513	.....	SK3502
3625	NTE2362	ECG2362	276-2023	SK3114A
3626	NTE40106B	ECG40106B	.....	SK40106B
3627	.....	.....	.....	SK9421
3628	NTE462	ECG462	.....	SK9160
3629	.....	.....	.....	SK7CT4066
3630	NTE34	ECG34	.....	SK9430
3631	NTE470	ECG470	.....	SK9642
3632	NTE474	ECG474	.....	.....
3633	.....	.....	.....	SK7CT4067
3634	.....	ECG955M	.....	.....
3635	NTE7015	ECG7015	.....	SK10484
3636	NTE478	ECG478	.....	SK9650
3637	.....	.....	.....	SK3693
3638	NTE480	ECG480	.....	SK9652
3639	.....	ECG3751	.....	.....
3640	NTE481	ECG481	.....	SK9653
3641	NTE482	ECG482	.....	SK9654
3642	NTE483	ECG483	.....	SK9655
3643	NTE484	ECG484	.....	SK9656
3644	NTE485	ECG485	.....	SK9657
3645	NTE2429	ECG2429	.....	SK9964
3646	NTE59	ECG59	.....	SK9449
3647	.....	ECG4066B	.....	.....
3648	NTE6406	ECG6406	.....	.....
3649	NTE514	ECG514/R-3DS3	.....	SK3761
3650	NTE1920	ECG1920	.....	SK10053
3651	NTE1237	ECG1237	.....	SK3707
3652	NTE524V15	ECG524V15	.....	SK3329
3653	NTE2309	ECG2337	.....	SK10300
3654	NTE290A	ECG290A	276-2023	SK3867A
3655	NTE530	ECG530	.....	SK3308
3656	NTE4539B	ECG4539B	.....	.....
3657	.....	ECG5812	.....	.....
3658	NTE541	ECG541	.....	SK9306
3659	NTE542	ECG542	.....	SK9307
3660	NTE544	ECG544	.....	.....
3661	NTE546	ECG546	.....	SK9300
3662	NTE548	ECG548	.....	SK9305
3663	.....	ECG946	.....	.....
3664	NTE74490	ECG74490	.....	.....
3665	.....	.....	.....	SK7CT4075
3666	NTE556	ECG556	.....	SK3905
3667	NTE557	ECG557	.....	SK3904
3668	.....	ECG3830	.....	.....
3669	NTE565	ECG565	.....	.....
3670	NTE1107	ECG1107	.....	SK7772
3671	NTE5475	ECG5476	.....	SK10262
3672	NTE857M	ECG857M	.....	SK9800
3673	NTE2306	ECG2306	.....	SK4903
3674	NTE56	ECG56	.....	.....
3675	.....	ECG390	.....	.....
3676	NTE6200	ECG6200	.....	SK9773
3677	.....	.....	.....	SK7C533
3678	.....	ECG2328	.....	.....
3679	NTE6202	ECG6202	.....	SK9772
3680	NTE6204	ECG6204	.....	SK9771
3681	.....	.....	.....	SK7C534
3682	NTE379	ECG379	.....	.....
3683	NTE604	.....	.....	.....
3684	NTE606	ECG606	.....	SK9847
3685	NTE1671	ECG1671	.....	SK7841
3686	NTE921	ECG921	.....	SK7615
3687	NTE617	ECG617	.....	.....
3688	NTE618	ECG618	.....	.....
3689	NTE290AMCP	ECG290AMCP	.....	SK10312
3690	.....	ECG7009	.....	SK10478
3691	NTE56	ECG56	.....	SK9431
3692	NTE377	ECG377	276-2020	SK9445
3693	NTE54	ECG54	.....	SK9445
3694	NTE241	ECG241	276-2020	SK9445
3695	NTE1160	ECG1160	.....	.....
3696	NTE5511	ECG5511	276-1067	SK3683
3697	.....	ECG1599	.....	.....
3698	.....	.....	.....	SK3025
3699	.....	ECG293	.....	.....
3700	.....	ECG294	.....	.....
3701	NTE60	ECG60	.....	SK9033
3702	.....	.....	.....	SK9991
3703	.....	.....	.....	SK9370
3704	NTE4536B	ECG4536B	.....	SK4536B
3705	.....	ECG1178	.....	SK3480
3706	NTE4056B	ECG4056B	.....	SK4056B
3707	.....	ECG1598	.....	.....
3708	NTE56031	ECG56031	.....	SK10410

REPL. CODE	NTE PART NO.	ECG PART NO.	RADIO SHACK PART NO.	RCA PART NO.
3709	NTE376	ECG376	.....	SK9262A
3710	NTE1374	ECG1374	.....	SK3853
3711	NTE1376	ECG1376	.....	SK7707
3712	NTE132	ECG312	276-2055	SK3834
3713	NTE1567	ECG1567	.....	SK7805
3714	.....	ECG9905	.....	.....
3715	NTE74H20	ECG74H20	.....	.....
3716	NTE1486	ECG1486	.....	SK7607
3717	NTE312	ECG312	.....	SK3448
3718	.....	ECG2387	.....	.....
3719	NTE700	ECG700	.....	SK9028
3720	NTE701	ECG701	.....	SK9026
3721	.....	ECG824	.....	.....
3722	NTE716	ECG716	.....	.....
3723	NTE2336	ECG2336	.....	SK9481
3724	NTE717	ECG717	.....	.....
3725	NTE1842	ECG1842	.....	SK7674
3726	NTE74S30	ECG74S30	.....	.....
3727	NTE4063B	ECG4063B	.....	SK4063B
3728	NTE375	ECG375	.....	SK9452
3729	NTE2001	ECG2001	.....	.....
3730	.....	ECG1718	.....	.....
3731	NTE1703	ECG1703	.....	SK10016
3732	NTE153MCP	ECG153MCP	276-2020&27	SK3274&3893
3733	NTE5852	ECG5854	.....	SK3600
3734	NTE862	ECG862	.....	SK3456
3735	NTE742	ECG742	.....	SK3453
3736	.....	ECG5306	.....	.....
3737	.....	.....	.....	SK7CT4094
3738	NTE1440	ECG1440	.....	SK9056
3739	NTE3016	ECG3016	.....	SK10320
3740	.....	ECG5854	.....	SK3600
3741	NTE858M	ECG858M	276-1715	SK7641
3742	NTE4067B	ECG4067B	.....	SK4067B
3743	NTE5856	ECG5858	.....	SK3599
3744	NTE1551	ECG1551	.....	SK9298
3745	.....	ECG5335	.....	SK7313
3746	NTE214	ECG214	.....	SK9916
3747	NTE1727	ECG1727	.....	SK7618
3748	NTE215	ECG215	.....	SK9421
3749	NTE593	ECG593	.....	SK9942
3750	NTE217	ECG217	.....	SK9471
3751	NTE5900	ECG5900	.....	SK7100
3752	.....	.....	.....	SK3274
3753	NTE1625	ECG1625	.....	SK10232
3754	NTE227	ECG227	.....	SK9472
3755	.....	ECG56033	.....	.....
3756	.....	.....	.....	SK51V
3757	NTE5902	ECG5904	.....	SK7104
3758	NTE240	ECG240	.....	SK9351
3759	NTE1629	ECG1629	.....	.....
3760	.....	.....	.....	SK4536B
3761	NTE802	ECG802	.....	SK3277
3762	NTE6020	ECG6020	.....	SK7220
3763	NTE803	ECG803	.....	SK3278
3764	NTE255	ECG255	.....	SK9412
3765	NTE256	ECG256	.....	.....
3766	NTE807	ECG807	.....	SK3451
3767	NTE4506B	ECG4506B	.....	SK4056UB
3768	NTE5908	ECG5908	.....	SK7108
3769	.....	ECG1276	.....	.....
3770	NTE812	ECG812	.....	.....
3771	NTE6049	ECG6049	.....	SK7249
3772	NTE1576	ECG1576	.....	SK7672
3773	.....	ECG1862	.....	SK10159
3774	.....	.....	.....	SK74LS00
3775	.....	.....	.....	SK74LS32
3776	NTE1668	ECG1668	.....	SK9978
3777	NTE108-1	ECG108	276-2016	SK3452
3778	.....	.....	.....	SK10153
3779	.....	ECG753	.....	.....
3780	.....	.....	.....	SK3564
3781	NTE832	ECG832	.....	SK9089
3782	.....	ECG5338	.....	SK7315
3783	.....	ECG74S478	.....	.....
3784	NTE841	ECG841	.....	SK9202
3785	NTE1051	ECG1051	.....	.....
3786	NTE1V250	ECG1V250	.....	SKMV250G
3787	.....	.....	.....	SK9047
3788	NTE849	ECG849	.....	.....
3789	NTE55MCP	ECG55MCP	.....	SK9923
3790	NTE851	ECG851	.....	SK9206
3791	NTE853	ECG853	.....	SK7645
3792	NTE2053	ECG2053	.....	SK1911
3793	NTE856	ECG856	.....	SK7708
3794	NTE2312	ECG2312	.....	.....
3795	NTE860	ECG860	.....	SK7731
3796	.....	ECG2369	.....	.....
3797	NTE79	ECG79	.....	.....
3798	NTE926	ECG926	.....	SK7721
3799	NTE48	ECG48	.....	SK4906
3800	.....	.....	.....	SK74LS38
3801	NTE973D	ECG973D	.....	SK3892
3802	NTE7051	ECG7051	.....	.....
3803	.....	ECG918M	.....	.....
3804	.....	.....	.....	SK3970
3805	NTE1199	ECG1199	.....	.....
3806	.....	ECG74HC4053	.....	SK7C4053
3807	NTE107	ECG107	276-2016	SK3854
3808	.....	ECG913	.....	.....
3809	NTE4501	ECG4501	.....	SK4501UB
3810	.....	ECG370A	.....	.....
3811	NTE859	ECG859	.....	.....
3812	NTE159	ECG159	276-2023	SK3841
3813	NTE934	ECG934	.....	SK9342
3814	.....	ECG936	.....	.....
3815	NTE1254	ECG1254	.....	SK3880
3816	NTE1557	ECG1557	.....	SK7605
3817	NTE1827	ECG1827	.....	SK9885
3818	NTE939	ECG939	.....	.....
3819	NTE943	ECG943	.....	.....
3820	NTE944	ECG944	.....	SK10076
3821	NTE952	ECG952	.....	SK7752
3822	.....	.....	.....	SK3260
3823	NTE953	ECG953	.....	SK7806
3824	NTE954	ECG954	.....	SK7807
3825	.....	ECG994	.....	.....
3826	NTE959	ECG959	.....	SK9283
3827	NTE994M	ECG994M	.....	SK9224
3828	NTE1716	ECG1716	.....	SK10012
3829	NTE969	ECG969	.....	SK3674
3830	NTE4089B	ECG4089B	.....	SK4089B
3831	NTE980	ECG980	.....	SK4046B
3832	NTE985	ECG985	.....	SK3214
3833	.....	ECG9904	.....	.....
3834	NTE74C107	ECG74C107	.....	SK74C107
3835	.....	ECG993	.....	.....
3836	NTE996	ECG996	.....	SK9201
3837	.....	ECG6013	.....	SK10176
3838	.....	.....	.....	SK9919
3839	.....	ECG463	.....	.....
3840	NTE1141	ECG1141	.....	.....
3841	NTE5023A N9	ECG5023A N9	.....	SK14A N9
3842	.....	ECG468	.....	SK10462
3843	.....	ECG469	.....	SK10463
3844	.....	.....	.....	SK3021
3845	.....	ECG9908	.....	.....
3846	.....	ECG487	.....	SK10464
3847	.....	.....	.....	SK74S32
3848	NTE4095B	ECG4095B	.....	SK4095B
3849	.....	ECG490	.....	.....
3850	.....	ECG9909	.....	.....
3851	NTE4096B	ECG4096B	.....	SK4096B
3852	NTE5023A N8	ECG5023A N8	.....	SK14A N8
3853	NTE1683	ECG1683	.....	SK7626
3854	NTE5413	ECG5413	276-1067	SK3954
3855	NTE4097B	ECG4097B	.....	SK4097B
3856	.....	.....	.....	SK3764
3857	.....	ECG314	.....	SK3527
3858	NTE521	ECG521	.....	SK3304
3859	NTE524V42	ECG524V42	.....	SK10059
3860	.....	ECG527	.....	SK9035
3861	.....	ECG9903	.....	.....
3862	NTE1402	ECG1402	.....	SK9043
3863	.....	ECG540	.....	.....
3864	.....	ECG543	.....	SK9308
3865	.....	ECG545	.....	SK9303
3866	.....	ECG547	.....	SK9304
3867	NTE549	ECG549	.....	SK3901
3868	.....	ECG550	.....	.....
3869	.....	ECG5462	.....	SK9291
3870	.....	.....	.....	SK7C04
3871	NTE559	ECG559	.....	.....
3872	.....	ECG561	.....	.....
3873	NTE284	ECG284	.....	SK9134
3874	.....	.....	.....	SK9480
3875	.....	ECG562	.....	.....
3876	.....	ECG5463	.....	SK9292
3877	.....	ECG563	.....	.....
3878	NTE3034A	ECG3034	.....	.....
3879	.....	ECG564	.....	.....
3880	.....	.....	.....	SK9119
3881	NTE1924	ECG1924	.....	SK10055
3882	.....	ECG569	.....	.....
3883	.....	ECG570	.....	.....
3884	NTE1801	ECG1801	.....	SK9883
3885	.....	.....	.....	SK40192B
3886	.....	ECG2409	.....	.....
3887	NTE577	ECG577	.....	SK10511
3888	NTE578	ECG578	.....	SK10162
3889	NTE582	ECG582	.....	SK7332
3890	.....	ECG2342	.....	SK10090
3891	NTE583	ECG583	.....	.....
3892	.....	ECG2341	.....	SK10089
3893	NTE2021	ECG2021	.....	SK9020
3894	NTE591	ECG591	.....	SK9940
3895	NTE592	ECG592	.....	SK9941
3896	NTE594	ECG594	.....	SK9943
3897	.....	ECG9912	.....	.....
3898	NTE519 N8	ECG519 N8	276-1122 N8	SK3100 N8
3899	.....	ECG9915	.....	.....
3900	NTE6042	ECG6042	.....	SK7242
3901	.....	ECG9913	.....	.....
3902	NTE1779	ECG1779	.....	SK9726
3903	NTE74HC123	ECG74HC123	.....	.....
3904	NTE182&183	ECG182&183	.....	SK3188A&3189A
3905	.....	ECG74HC123	.....	.....
3906	.....	ECG153	.....	.....
3907	.....	.....	.....	SK7C123
3908	.....	ECG5440	.....	SK10401
3909	NTE285	ECG285	.....	SK9136
3910	.....	ECG5011T1	.....	.....
3911	NTE116	ECG116	276-1103	SK3031A
3912	NTE74LS148	ECG74LS148	.....	SK74LS148
3913	NTE1778	ECG1778	.....	SK9871
3914	NTE15020E	ECG15020	.....	SK10186
3915	NTE1128	ECG1128	.....	SK3488
3916	NTE1373	ECG1373	.....	SK4822
3917	.....	.....	.....	SK3644
3918	.....	.....	.....	SK3606
3919	NTE293&294	ECG293&294	.....	SK3841&3849
3920	NTE2022	ECG2022	.....	SK9019
3921	.....	ECG1574	.....	.....
3922	NTE1115A	ECG1115A	.....	SK3917
3923	.....	ECG4018B	.....	SK4018B
3924	NTE9926	ECG9926	.....	.....
3925	NTE1810	ECG1810	.....	SK9727
3926	NTE376	ECG376	.....	.....
3927	NTE15046	ECG1861	.....	SK10158
3928	NTE6362	ECG6362	.....	SK6560
3929	NTE83	ECG83	.....	SK3444
3930	.....	.....	.....	SK11V5
3931	.....	ECG758	.....	.....
3932	NTE1311	ECG1311	.....	.....
3933	NTE1300	ECG1300	.....	.....
3934	.....	ECG610	.....	.....
3935	.....	ECG765	.....	.....
3936	NTE1139	ECG1139	.....	.....
3937	NTE69	ECG69	276-2058	SK3246A
3938	NTE24	ECG24	.....	SK9840
3939	.....	ECG5562	.....	.....
3940	.....	ECG768	.....	.....
3941	NTE2354	ECG2354	.....	.....
3942	NTE888M	ECG888M	.....	SK9863
3943	.....	ECG775	.....	.....
3944	.....	ECG776	.....	.....
3945	NTE1521	ECG1521	.....	.....
3946	.....	ECG5107T2	.....	.....
3947	.....	.....	.....	SK3V6
3948	NTE1308	ECG1308	.....	SK3833
3949	NTE332MCP	ECG332MCP	.....	SK10005
3950	.....	.....	.....	SK7C139
3951	NTE1896	ECG1896	.....	SK10505
3952	NTE40100B	ECG40100B	.....	SK40100B
3953	NTE1839	ECG1839	.....	SK10143
3954	.....	ECG813	.....	.....
3955	NTE1776	ECG1776	.....	SK9743
3956	NTE955S	ECG955S	.....	SK9397
3957	NTE15041	ECG1897	.....	.....
3958	NTE1098	ECG1098	.....	SK7770
3959	NTE1778	ECG1778	.....	SK9744
3960	NTE230	ECG230	.....	SK3502
3961	.....	ECG1603	.....	.....
3962	NTE826	ECG826	.....	SK7813
3963	NTE1914	ECG1914	.....	SK9331
3964	NTE1596	ECG1596	.....	.....
3965	.....	ECG831	.....	.....
3966	.....	ECG1594	.....	.....
3967	.....	.....	.....	SK7733
3968	.....	ECG1590	.....	.....
3969	.....	ECG253	.....	SK9370
3970	.....	ECG1586	.....	.....
3971	.....	ECG837	.....	.....
3972	.....	ECG838	.....	.....
3973	NTE1582	ECG1582	.....	.....
3974	.....	ECG839	.....	SK7614
3975	.....	ECG1584	.....	.....
3976	NTE842	ECG842	.....	SK7686
3977	NTE844	ECG844	.....	SK9381
3978	NTE845	ECG845	.....	SK9382
3979	.....	ECG847	.....	.....
3980	NTE1468	ECG1468	.....	SK9322
3981	.....	ECG848	.....	.....
3982	NTE2062	ECG2062	.....	SK10295
3983	NTE850	ECG850	.....	SK9383
3984	NTE1312	ECG1312	.....	.....
3985	NTE1313	ECG1313	.....	.....
3986	.....	ECG854	.....	.....
3987	NTE3121	ECG3121	.....	SK4928
3988	NTE1310	ECG1310	.....	.....
3989	NTE861	ECG861	.....	SK4843
3990	NTE863	ECG863	.....	SK9207
3991	.....	ECG865	.....	.....
3992	.....	ECG1302	.....	.....
3993	.....	ECG74LS77	.....	.....
3994	.....	ECG866	.....	.....
3995	NTE867	ECG867	.....	SK9386
3996	NTE1304	ECG1304	.....	.....
3997	NTE868	ECG868	.....	SK7692
3998	.....	ECG1305	.....	SK7791
3999	.....	ECG872	.....	.....
4000	NTE1306	ECG1306	.....	.....
4001	NTE74C151	ECG74C151	.....	SK7C151
4002	NTE874	ECG874	.....	SK7717
4003	NTE1117	ECG1117	.....	SK9226
4004	.....	ECG1307	.....	.....
4005	NTE877	ECG877	.....	SK9257
4006	NTE1294	ECG1294	.....	SK9394
4007	NTE54	ECG54	.....	.....
4008	NTE880	ECG880	.....	.....
4009	.....	ECG1052	.....	SK3249
4010	.....	ECG5558	.....	.....
4011	NTE12	ECG12	.....	SK9919
4012	NTE28	ECG28	.....	.....
4013	NTE99	ECG99	.....	SK9110
4014	NTE74C154	ECG74C154	.....	SK74C154
4015	.....	ECG5400	.....	SK3950
4016	.....	ECG5401	.....	SK3638
4017	.....	.....	.....	SK40101B

REPL. CODE	NTE PART NO.	ECG PART NO.	RADIO SHACK PART NO.	RCA PART NO.
4018	.....	ECG1238	.....	SK3373
4019	NTE5020A	ECG5020A	.....	.....
4020	.....	ECG5404	.....	SK3627
4021	NTE225	ECG396&427	.....	SK3045
4022	NTE116	ECG116	.....	.....
4023	NTE3030	ECG3030	276-036	SK2169
4024	NTE125	ECG125	.....	.....
4025	.....	ECG1059	.....	.....
4026	.....	ECG920	.....	.....
4027	.....	ECG113A	.....	.....
4028	.....	ECG1125	.....	.....
4029	NTE96	ECG96	.....	SK9259
4030	NTE2324	ECG2324	.....	.....
4031	NTE5568	ECG5568	.....	SK6718
4032	NTE395	ECG395	.....	SK3118
4033	NTE74C157	ECG74C157	.....	SK74C157
4034	NTE1123	ECG1123	.....	SK3743
4035	.....	.....	.....	SK3465
4036	.....	.....	.....	SK3168
4037	NTE945	ECG945	.....	.....
4038	NTE1162	ECG1162	.....	SK3072
4039	NTE1068	ECG1068	.....	SK3226
4040	.....	.....	.....	SK3527
4041	NTE1843	ECG1843	.....	SK10146
4042	NTE1249	ECG1249	.....	SK3963
4043	.....	.....	.....	SK1822
4044	NTE1487	ECG1487	.....	.....
4045	NTE74C160	.....	.....	SK7C160
4046	NTE74C161	ECG74C161	.....	SK74C161
4047	NTE1450	ECG1450	.....	.....
4048	.....	ECG979	.....	.....
4049	.....	ECG2418	.....	SK10110
4050	.....	.....	.....	SK7CT240
4051	NTE1448	ECG1448	.....	.....
4052	NTE3166	ECG3166	.....	SK2166
4053	.....	ECG5554	.....	.....
4054	.....	ECG5556	.....	.....
4055	.....	ECG1182	.....	SK7779
4056	.....	.....	.....	SK3843
4057	.....	.....	.....	SK7C163
4058	NTE998	ECG998	.....	.....
4059	.....	ECG999	.....	.....
4060	NTE1406	ECG1406	.....	SK9059
4061	.....	.....	.....	SK40102B
4062	NTE74C164	ECG74C164	.....	SK74C164N
4063	NTE1446	ECG1446	.....	.....
4064	NTE1264	ECG1264	.....	.....
4065	NTE1265	ECG1265	.....	SK7787
4066	.....	.....	.....	SK3238
4067	NTE1266	ECG1266	.....	.....
4068	NTE1444	ECG1444	.....	.....
4069	NTE1240	ECG1240	.....	SK7785
4070	NTE1267	ECG1267	.....	.....
4071	NTE1268	ECG1268	.....	SK3921
4072	NTE1250	ECG1250	.....	.....
4073	.....	ECG5312	.....	.....
4074	.....	ECG363	.....	.....
4075	.....	.....	.....	SK7CT245
4076	.....	ECG1070	.....	.....
4077	.....	ECG2345	.....	SK10120
4078	.....	ECG2346	.....	SK10121
4079	NTE1269	ECG1269	.....	SK9054
4080	.....	.....	.....	SK3831
4081	NTE2026	ECG2026	.....	SK10276
4082	NTE297&298	ECG297&298	.....	SK3449&3450
4083	.....	.....	.....	SK1823
4084	.....	ECG6880	.....	.....
4085	.....	ECG389	.....	.....
4086	NTE1224	ECG1224	.....	SK7782
4087	NTE1191	ECG1191	.....	.....
4088	NTE74C173	ECG74C173	.....	SK74C173
4089	NTE1290	ECG1290	.....	SK9183
4090	.....	ECG383	.....	.....
4091	.....	.....	.....	SK40103B
4092	NTE74C174	ECG74C174	.....	SK74C174
4093	NTE1037	ECG1037	.....	SK3371
4094	NTE1863	ECG1863	.....	.....
4095	NTE186&187	ECG186&187	.....	SK9076&3357
4096	NTE74C175	ECG74C175	.....	SK74C175
4097	NTE1395	ECG1395	.....	SK9313
4098	NTE175	ECG175	.....	SK3026
4099	NTE27	ECG27	.....	SK4900
4100	NTE5687	ECG5687	.....	SK3521
4101	NTE703	ECG703	.....	SK3087
4102	.....	.....	.....	SK9713
4103	.....	.....	.....	SK1824
4104	NTE21	ECG21	.....	SK9669
4105	NTE858SM	ECG858SM	.....	SK10064
4106	NTE2004	ECG2004	.....	.....
4107	.....	ECG858M	.....	.....
4108	.....	.....	.....	SK40104B
4109	NTE1572	ECG1572	.....	SK4825
4110	NTE4597B	ECG4597B	.....	SK4597B
4111	NTE5010A	ECG5010A	.....	SK5A1
4112	NTE1581	ECG1581	.....	SK9360
4113	NTE178MP	ECG178MP	276-1122	SK9009
4114	NTE5806	ECG5806	276-1114	SK3848
4115	.....	ECG5021A	.....	.....
4116	.....	ECG87	.....	SK9031
4117	NTE1834	ECG1834	.....	SK9740
4118	.....	ECG1869SM	.....	SK10499
4119	NTE74C192	ECG74C192	.....	SK74C192
4120	.....	.....	.....	SK9748
4121	NTE5604	ECG5605	276-1000	SK3666
4122	NTE74C193	ECG74C193	.....	SK74C193
4123	.....	.....	.....	SK40105B
4124	NTE153	ECG153	.....	.....
4125	NTE153	ECG153	.....	.....
4126	.....	.....	.....	SK7CT273
4127	.....	.....	.....	SK3917
4128	.....	.....	.....	SK62V
4129	NTE1782	ECG1782	.....	SK9730
4130	NTE2360	ECG2360	.....	SK9960
4131	.....	.....	.....	SK3233
4132	.....	ECG990	.....	.....
4133	NTE282	ECG282	.....	SK3299
4134	NTE1496	ECG1496	.....	SK9321
4135	NTE1V115	ECG1V115	.....	SKMV115B
4136	NTE859SM	ECG859SM	.....	SK10065
4137	NTE1775	ECG1775	.....	SK9868
4138	NTE2084	ECG2084	.....	SK4910
4139	NTE234	ECG234	276-2023	SK3841
4140	NTE5832	ECG5834	.....	SK7042
4141	NTE15044	.....	.....	SK9738
4142	NTE9666	ECG9666	.....	.....
4143	.....	ECG1882	.....	SK10172
4144	NTE61	ECG61	.....	SK9034
4145	NTE8318	ECG8318	.....	SK9831
4146	NTE1740	ECG1740	.....	SK7817
4147	NTE1116	ECG1116	.....	SK3969
4148	NTE5598	ECG5598	.....	SK6748
4149	NTE9671	ECG9671	.....	.....
4150	.....	.....	.....	SK4016B
4151	NTE1358	ECG1358	.....	SK9987
4152	NTE5340	ECG5340	.....	SK3650
4153	.....	.....	.....	SK40107B
4154	.....	ECG9400	.....	SK9051
4155	.....	ECG399	.....	SK9352
4156	.....	ECG331	.....	.....
4157	.....	ECG332	.....	.....
4158	NTE9200	ECG9200	.....	SK10447
4159	NTE5342	ECG5342	.....	SK3651
4160	NTE385	ECG385	.....	SK3559
4161	.....	ECG9674	.....	.....
4162	NTE9402	ECG9402	.....	SK9062
4163	NTE129MCP	ECG129MCP	.....	SK9932
4164	NTE9675	ECG9675	.....	.....
4165	NTE1418	ECG1418	.....	SK9214
4166	.....	ECG1665	.....	.....
4167	.....	ECG598	.....	SK9859
4168	.....	ECG9401	.....	SK10448
4169	NTE922SM	ECG922SM	.....	SK10070
4170	NTE9676	ECG9676	.....	.....
4171	.....	ECG9677	.....	.....
4172	NTE2356	ECG2356	.....	SK9958
4173	NTE9678	ECG9678	.....	.....
4174	.....	.....	.....	SK9751
4175	NTE9679	ECG9679	.....	.....
4176	.....	ECG2114	.....	.....
4177	.....	ECG254	.....	.....
4178	NTE9680	ECG9680	.....	.....
4179	.....	.....	.....	SK9091
4180	NTE7428	ECG7428	.....	SK7428
4181	NTE6875	ECG6875	.....	SK9794
4182	.....	.....	.....	SK40108B
4183	NTE3027	ECG3027	.....	SK2027A
4184	NTE331MP	ECG331MP	.....	SK10006
4185	.....	ECG9683	.....	.....
4186	.....	.....	.....	SK4028B
4187	.....	ECG9684	.....	.....
4188	NTE6208	ECG6208	.....	SK9779
4189	NTE7002	ECG7002	.....	SK10471
4190	NTE3057	ECG3057	.....	SK10329
4191	.....	ECG9685	.....	.....
4192	NTE5332	ECG5332	276-1161	SK9230
4193	NTE7433	ECG7433	.....	SK7433
4194	NTE1783	ECG1783	.....	SK9873
4195	.....	ECG9686	.....	.....
4196	NTE1183	ECG1183	.....	.....
4197	NTE6210	ECG6210	.....	SK9778
4198	.....	.....	.....	SK3176
4199	.....	ECG9688	.....	.....
4200	NTE587	ECG587	276-1122	SK9937
4201	NTE9689	ECG9689	.....	.....
4202	.....	.....	.....	SK3512
4203	.....	.....	.....	SK3199
4204	.....	ECG9690	.....	.....
4205	.....	.....	.....	SK3200
4206	.....	.....	.....	SK74LS155
4207	NTE9691	ECG9691	.....	.....
4208	NTE237	ECG237	.....	SK3049
4209	NTE1195	ECG1195	.....	SK3469
4210	.....	.....	.....	SK40109B
4211	NTE116	ECG116	.....	SK3312
4212	.....	.....	.....	SK3588A
4213	NTE711	ECG711	.....	SK3017B
4214	.....	.....	.....	SK3112
4215	NTE74H51	ECG74H51	.....	SK74107
4216	NTE74H53	ECG74H53	.....	SK74LS03
4217	.....	.....	.....	SK4512B
4218	.....	ECG9927	.....	.....
4219	.....	ECG2002	.....	.....
4220	NTE153MCP	ECG153MCP	.....	SK9933
4221	.....	ECG9696	.....	.....
4222	NTE1V060	ECG1V060	.....	SKMV60A
4223	.....	.....	.....	SK3507
4224	NTE1233	ECG1233	.....	SK3732
4225	NTE282	ECG282	.....	SK3045
4226	.....	.....	.....	SK3921
4227	.....	.....	.....	SK7C240
4228	.....	ECG5591	.....	.....
4229	.....	.....	.....	SK40110B
4230	NTE74C244	ECG74C244	.....	SK74C244
4231	NTE5172A	ECG5172A	.....	.....
4232	NTE5079A	ECG5079A	.....	.....
4233	NTE6403	ECG6403	.....	SK9120
4234	.....	ECG9906	.....	.....
4235	.....	ECG139A	.....	.....
4236	.....	ECG9907	.....	.....
4237	.....	ECG455	.....	.....
4238	NTE1200	ECG1200	.....	SK3714
4239	.....	.....	.....	SK9177
4240	.....	.....	.....	SK3311
4241	.....	ECG7003	.....	SK10472
4242	.....	ECG1588	.....	.....
4243	NTE373	.....	.....	SK373
4244	.....	.....	.....	SK3611
4245	.....	.....	.....	SK9070
4246	NTE224	ECG224	.....	SK3529
4247	NTE54	ECG54	276-2020	SK3440
4248	.....	.....	.....	SK7CT7266
4249	.....	.....	.....	SK6810
4250	.....	ECG31	.....	.....
4251	NTE175	ECG152	276-2020	SK3026
4252	NTE95	ECG95	.....	SK9295
4253	.....	.....	.....	SK68V
4254	.....	ECG834SM	.....	SK10063
4255	.....	ECG987SM	.....	.....
4256	.....	.....	.....	SK3835
4257	.....	ECG125	.....	.....
4258	NTE1718	ECG1718	.....	SK10023
4259	.....	ECG928SM	.....	.....
4260	NTE1099	ECG1099	.....	.....
4261	.....	ECG1601	.....	.....
4262	.....	.....	.....	SK3514
4263	NTE1V040	ECG1V040	.....	SKMV40B
4264	NTE2361	ECG2361	.....	SK3124A
4265	NTE7066	ECG7066	.....	.....
4266	.....	.....	.....	SK9747
4267	.....	.....	.....	SK74195
4268	.....	.....	.....	SKMV130L
4269	.....	.....	.....	SK3165
4270	.....	ECG116	.....	SK3080
4271	NTE282	ECG282	.....	SK3529
4272	.....	.....	.....	SK3526
4273	NTE927SM	ECG927SM	.....	SK10071
4274	NTE1633	ECG1633	.....	SK7729
4275	.....	.....	.....	SK9384
4276	.....	.....	.....	SK10041
4277	NTE15038	.....	.....	.....
4278	NTE1639	ECG1639	.....	.....
4279	.....	.....	.....	SK4067B
4280	NTE1512	ECG1512	.....	.....
4281	.....	.....	.....	SK4078B
4282	.....	.....	.....	SK9460
4283	.....	.....	.....	SK7402
4284	NTE4517B	ECG4517B	.....	SK4517B
4285	NTE6107	ECG6107	.....	.....
4286	NTE6106	ECG6106	.....	.....
4287	NTE6362	ECG6362	.....	SK6558
4288	.....	ECG451	.....	SK3116
4289	.....	ECG74HC08	.....	.....
4290	.....	.....	.....	SK7623
4291	NTE1423	ECG1423	.....	SK7600
4292	NTE96S02	ECG96S02	.....	.....
4293	NTE196&197	ECG196&197	.....	SK3054&3083
4294	.....	ECG1592	.....	.....
4295	.....	.....	.....	SK4000UB
4296	.....	ECG404	.....	.....
4297	.....	ECG127	.....	SK3764
4298	.....	.....	.....	SK40114B
4299	.....	ECG586	.....	.....
4300	.....	.....	276-065	.....
4301	NTE4598B	ECG4598B	.....	SK4598B
4302	.....	.....	.....	SK4538B
4303	.....	ECG5033A	.....	.....
4304	.....	.....	276-066A	.....
4305	NTE74S51	ECG74S51	.....	.....
4306	NTE6048	ECG6048	.....	SK7248
4307	NTE1542	ECG1542	.....	SK10274
4308	NTE6058	ECG6060	.....	SK7260
4309	.....	.....	.....	SK8A2
4310	NTE1835	ECG1835	.....	SK9728
4311	.....	.....	.....	SK7CT374
4312	.....	.....	.....	SK4099B
4313	.....	ECG6087	.....	SK9775
4314	.....	ECG6094	.....	.....
4315	.....	ECG6664	.....	.....
4316	.....	ECG2056	.....	.....
4317	.....	ECG326	.....	SK3746
4318	.....	ECG1019	.....	.....
4319	.....	ECG1584	.....	SK3365
4320	.....	.....	.....	SK2A8
4321	.....	ECG7111	.....	.....
4322	.....	ECG5882	.....	.....
4323	.....	ECG5886	.....	.....
4324	.....	.....	.....	SK7CT4316
4325	.....	ECG8212	.....	.....
4326	NTE7054	ECG7054	.....	.....

REPL. CODE	NTE PART NO.	ECG PART NO.	RADIO SHACK PART NO.	RCA PART NO.
4327	NTE3074	ECG3074	.....	SK2074
4328	.....	ECG8228	.....	.....
4329	NTE1936	ECG1936	.....	SK7740
4330	NTE261	ECG261	276-2068	SK3180
4331	NTE3076	ECG3076	.....	SK2076
4332	NTE3075	ECG3075	.....	SK2075
4333	NTE3077	ECG3077	.....	SK2077
4334	NTE2317	ECG2317	.....	SK10117
4335	.....	ECG56024	.....	.....
4336	.....	ECG7014	.....	.....
4337	NTE5351	ECG5351	.....	.....
4338	.....	.....	.....	SK3591
4339	NTE74LS164	ECG74LS164	.....	SK74LS164
4340	.....	ECG74S573	.....	.....
4341	.....	.....	.....	SK3593
4342	.....	.....	.....	SK3699
4343	.....	ECG4049	.....	.....
4344	NTE30	ECG30	.....	SK9915
4345	NTE29	ECG29	.....	SK9914
4346	NTE1747	ECG1747	.....	SK7619
4347	NTE5097A	ECG5097A	.....	.....
4348	NTE21	ECG21	.....	SK3841
4349	.....	.....	.....	SK10247
4350	NTE1661	ECG1661	.....	SK4816
4351	NTE1540	ECG1540	.....	SK4837
4352	NTE5424	ECG5424	.....	.....
4353	NTE1772	ECG1772	.....	SK9866
4354	.....	ECG6809E	.....	.....
4355	.....	ECG1647	.....	SK10493
4356	NTE74LS165	ECG74LS165	.....	SK7CT165
4357	.....	ECG159	.....	SK3114A
4358	.....	ECG2040	.....	.....
4359	NTE7104	ECG7104	.....	.....
4360	NTE7039	ECG7039	.....	.....
4361	.....	ECG5046A	.....	SK75A
4362	.....	.....	.....	SK3212
4363	.....	.....	.....	SK75V
4364	NTE5711	ECG5711	.....	SK10418
4365	.....	.....	.....	SK3259
4366	NTE85	ECG85	.....	.....
4367	NTE1V025	ECG1V025	.....	SKMV25A
4368	.....	.....	276-088	.....
4369	.....	.....	.....	SK3645
4370	.....	ECG4019B	.....	.....
4371	NTE74LS166	ECG74LS166	.....	SK7CT166
4372	NTE1387	ECG1387	.....	SK7665
4373	.....	.....	.....	SK5V1
4374	NTE6400B	ECG6400B	.....	SK9122
4375	NTE7105	ECG7105	.....	.....
4376	NTE123AP	ECG123AP	276-2016	SK3854
4377	.....	.....	.....	SK9754
4378	.....	.....	.....	SK3966
4379	NTE1805	ECG1805	.....	SK9838
4380	NTE1837	ECG1837	.....	SK9755
4381	NTE1836	ECG1836	.....	SK9756
4382	NTE7004	ECG7004	.....	SK10473
4383	NTE1826	ECG1826	.....	SK9765
4384	NTE1708	ECG1708	.....	SK10017
4385	NTE1788	ECG1788	.....	SK9875
4386	.....	ECG386	.....	.....
4387	NTE130	ECG130	276-2041	SK3619
4388	.....	ECG21	.....	.....
4389	NTE1297	ECG1297	.....	SK9265
4390	NTE1043	ECG1043	.....	SK3363
4391	.....	ECG612	.....	.....
4392	.....	.....	.....	SK7CT4351
4393	.....	ECG598	.....	.....
4394	.....	.....	.....	SK7CT4352
4395	.....	ECG1189	.....	SK3705
4396	.....	.....	.....	SK7CT4353
4397	.....	.....	.....	SK9140
4398	.....	.....	.....	SK7CT137
4399	NTE74H183	ECG74H183	.....	.....
4400	NTE74LS147	ECG74LS147	.....	SK7CT147
4401	NTE1169	ECG1169	.....	SK3708
4402	.....	.....	.....	SK6621
4403	NTE552	ECG552	.....	SK3311
4404	.....	.....	.....	SK9422
4405	NTE6405	ECG6405	.....	SK9125
4406	.....	.....	.....	SK7669
4407	NTE139A	ECG139A	.....	.....
4408	.....	ECG524V15	.....	SKMV150K
4409	NTE1038	ECG1038	.....	.....
4410	NTE1657	ECG1657	.....	SK7673
4411	.....	ECG4053B	.....	.....
4412	NTE5600	ECG5601	276-1000	SK6601
4413	NTE968	ECG968	276-1771	SK3592
4414	NTE5602	ECG5602	276-1000	SK3664
4415	NTE125	ECG125	.....	SK3081
4416	NTE5606	ECG5607	.....	SK6607
4417	NTE5607	ECG5607	.....	SK6607
4418	.....	ECG183	.....	.....
4419	.....	.....	.....	SK2041
4420	NTE1127	ECG1127	.....	SK7775
4421	NTE290A	ECG290A	.....	.....
4422	NTE1789	ECG1789	.....	SK9876
4423	NTE116 N6	ECG116 N6	276-1104 N6	SK3017B N6
4424	NTE2708	ECG2708	.....	.....
4425	NTE74LS290	ECG74LS290	.....	SK74LS290
4426	NTE74LS299	ECG74LS299	.....	SK7CT299
4427	.....	.....	.....	SK3598
4428	NTE1436	ECG1436	.....	.....
4429	.....	.....	.....	SK9399
4430	NTE378	ECG378	.....	SK3274
4431	NTE3043	ECG3043	.....	SK2043
4432	NTE290A	ECG290A	.....	SK9132
4433	NTE159	ECG159	276-2023	SK3118
4434	.....	ECG1605	.....	.....
4435	.....	.....	.....	SK1854
4436	NTE101	ECG101	.....	SK3835
4437	.....	ECG1001	.....	.....
4438	.....	.....	.....	SK74107
4439	NTE74116	.....	.....	.....
4440	NTE5614	ECG5614	.....	SK3632
4441	NTE40097B	ECG40097B	.....	SK9787
4442	.....	.....	.....	SK74125
4443	.....	ECG3156	.....	.....
4444	.....	.....	.....	SK74153
4445	.....	.....	.....	SK74157
4446	NTE74158	ECG74158	.....	SK74158
4447	.....	.....	.....	SK74161
4448	.....	.....	.....	SK74162
4449	.....	.....	.....	SK74163
4450	NTE2310	ECG2310	.....	SK4919
4451	.....	.....	.....	SK74166
4452	.....	ECG107	.....	.....
4453	.....	ECG738	.....	SK9184
4454	.....	ECG738	.....	.....
4455	NTE5655	ECG5655	.....	SK5555
4456	NTE5656	ECG5656	.....	SK5556
4457	.....	.....	.....	SK74197
4458	NTE5411	ECG5411	276-1067	SK3953
4459	.....	.....	.....	SK7414
4460	.....	.....	.....	SK7CT423
4461	.....	ECG8224	.....	.....
4462	NTE74C373	ECG74C373	.....	SK74C373
4463	NTE195A	ECG195A	.....	SK3854
4464	NTE195A	ECG195A	.....	SK9645
4465	NTE358A	ECG358A	.....	SK9329
4466	.....	ECG219	.....	.....
4467	NTE74C374	ECG74C374	.....	SK74C374
4468	NTE519	ECG519	.....	SK3100
4469	NTE5093A	ECG5093A	.....	.....
4470	.....	ECG157	.....	.....
4471	NTE1V300	ECG1V300	.....	SKMV300
4472	NTE1V050	ECG1V050	.....	SKMV50B
4473	.....	.....	.....	SK9378
4474	NTE277	ECG277	.....	SK3929
4475	NTE37MCP	ECG37MCP	.....	SK9921
4476	NTE1424	ECG1424	.....	SK7601
4477	.....	ECG188	.....	.....
4478	.....	.....	.....	SK9377
4479	NTE39	ECG39	.....	SK9260
4480	.....	.....	.....	SK3962
4481	.....	ECG5655	.....	.....
4482	NTE5657	ECG5657	.....	SK5557
4483	.....	ECG5656	.....	.....
4484	NTE1204	ECG1204	.....	.....
4485	.....	ECG1095	.....	.....
4486	.....	ECG2348	.....	.....
4487	NTE4569B	ECG4569B	.....	SK4569B
4488	.....	ECG6247	.....	.....
4489	NTE2401	ECG2401	.....	SK10103
4490	NTE5552	ECG5552	.....	SK3574
4491	.....	ECG5552	.....	SK3574
4492	.....	ECG5554	.....	SK3575
4493	.....	ECG5556	.....	SK3576
4494	.....	ECG5550	.....	SK3574
4495	NTE1415	ECG1415	.....	SK9058
4496	NTE15045	ECG1873	.....	SK9764
4497	.....	ECG74LS164	.....	SK74LS164
4498	NTE2403	ECG2403	.....	SK10096
4499	.....	ECG5657	.....	.....
4500	NTE88MCP	ECG88MCP	.....	SK9929
4501	.....	ECG988	.....	SK3831
4502	.....	ECG981	.....	SK3724
4503	NTE588	ECG588	.....	SK3043B
4504	.....	ECG951	.....	SK9170
4505	.....	ECG2164	.....	.....
4506	NTE1906	ECG1906	.....	SK10048
4507	.....	.....	.....	SK7CT648
4508	NTE1917	ECG1917	.....	SK9219
4509	NTE1908	ECG1908	.....	SK10050
4510	.....	.....	.....	SK82A
4511	NTE222	ECG222	.....	SK3187
4512	.....	ECG2430	.....	.....
4513	.....	.....	.....	SK82V
4514	.....	ECG2431	.....	.....
4515	.....	.....	.....	SK7207
4516	NTE6880	ECG6880	.....	SK9701
4517	.....	.....	.....	SK3356
4518	NTE152	ECG152	.....	SK3893
4519	NTE290A	ECG290A	276-2009	SK3114A
4520	NTE1651	ECG1651	.....	SK7679
4521	.....	ECG5319	.....	.....
4522	NTE1902	ECG1910	.....	SK3962
4523	NTE5837	ECG5839	.....	SK7039
4524	.....	.....	.....	SK9946
4525	.....	ECG75452B	.....	.....
4526	NTE5024A	ECG5024A	.....	SK15A
4527	.....	ECG283	.....	.....
4528	NTE5015A	ECG5015A	.....	.....
4529	.....	ECG3041	.....	.....
4530	.....	.....	.....	SK4054B
4531	NTE4529B	ECG4529B	.....	SK4529B
4532	NTE7007	ECG7007	.....	SK10476
4533	.....	.....	.....	SK7451
4534	NTE285	ECG285	.....	SK9032
4535	.....	.....	.....	SK6632
4536	NTE2074	ECG2074	.....	SK10290
4537	.....	.....	.....	SK9961
4538	NTE9813	ECG9813	.....	.....
4539	.....	ECG331	.....	SK3188A
4540	NTE386	ECG386	.....	SK9262
4541	NTE9814	ECG9814	.....	.....
4542	.....	ECG2412	.....	.....
4543	NTE1124	ECG1124	.....	.....
4544	.....	ECG2413	.....	.....
4545	NTE5676	ECG5677	.....	SK3520A
4546	.....	ECG2401	.....	.....
4547	NTE8080A	ECG8080A	.....	.....
4548	NTE25	ECG25	.....	SK9841
4549	NTE1437	ECG1437	.....	SK3885
4550	.....	ECG7013	.....	SK10482
4551	NTE995M	ECG995M	.....	SK10451
4552	.....	.....	.....	SK7636
4553	NTE1670	ECG1670	.....	SK10241
4554	NTE74100	.....	.....	.....
4555	NTE74105	.....	.....	.....
4556	NTE1022	ECG1022	.....	.....
4557	NTE1023	ECG1023	.....	.....
4558	.....	.....	.....	SK6560
4559	NTE1028	ECG1028	.....	SK3436
4560	NTE1480	ECG1480	.....	SK7801
4561	NTE1030	ECG1030	.....	SK3372
4562	NTE1031	ECG1031	.....	.....
4563	NTE1032	ECG1032	.....	SK7758
4564	NTE1033	ECG1033	.....	SK7759
4565	NTE1035	ECG1035	.....	SK3369
4566	NTE1036	ECG1036	.....	SK3370
4567	NTE2319	ECG2319	.....	SK9261
4568	.....	.....	.....	SK3208
4569	NTE1040	ECG1040	.....	SK3362
4570	NTE1041	ECG1041	.....	SK3361
4571	NTE1042	ECG1042	.....	SK3367
4572	.....	ECG2392	.....	.....
4573	NTE1288	ECG1288	.....	SK9252
4574	NTE5712	ECG5712	.....	SK10420
4575	NTE1079	ECG1079	.....	SK7766
4576	.....	ECG5460	.....	SK10402
4577	NTE5021A	ECG5021A	.....	.....
4578	NTE1093	ECG1093	.....	.....
4579	NTE188&189	ECG188&189	.....	SK3199&3200
4580	NTE173BP	.....	.....	SK3190
4581	NTE5332	ECG5332	.....	SK9231
4582	.....	ECG1105	.....	.....
4583	.....	.....	.....	SK3189A
4584	.....	ECG189	.....	.....
4585	NTE6406	ECG6406	.....	SK9082A
4586	NTE1114	.....	.....	.....
4587	.....	.....	.....	SK3513
4588	NTE5442	ECG5442	.....	SK3634
4589	NTE1119	ECG1119	.....	.....
4590	.....	ECG2333	.....	SK10299
4591	NTE52	ECG52	.....	.....
4592	NTE6407	ECG6407	.....	SK9083A
4593	NTE1126	ECG1126	.....	SK7774
4594	NTE1131	ECG1131	.....	SK3286
4595	NTE1132	ECG1132	.....	SK3287
4596	NTE74H22	ECG74H22	.....	.....
4597	NTE1149	ECG1149	.....	.....
4598	NTE1150	ECG1150	.....	SK3890
4599	.....	ECG2042	.....	.....
4600	NTE1154	ECG1154	.....	SK3230
4601	NTE1566	ECG1566	.....	SK7726
4602	.....	.....	.....	SK7CU04
4603	NTE7422	ECG7422	.....	SK7422
4604	NTE65101	ECG65101	.....	SK9829
4605	NTE185MCP	ECG185MCP	.....	.....
4606	NTE310P	ECG310P	.....	.....
4607	NTE74C901	ECG74C901	.....	SK7C901
4608	NTE74C902	ECG74C902	.....	SK74C902
4609	NTE74C903	ECG74C903	.....	SK74C903
4610	NTE74C904	ECG74C904	.....	SK74C904
4611	NTE74265	.....	.....	.....
4612	NTE1184	ECG1184	.....	SK3486
4613	.....	ECG1768	.....	SK4802
4614	NTE31	ECG31	.....	.....
4615	NTE308P	ECG308P	.....	.....
4616	NTE1190	.....	.....	.....
4617	NTE74278	.....	.....	.....
4618	.....	ECG1769	.....	SK4804
4619	NTE74279	.....	.....	.....
4620	NTE1197	ECG1197	.....	SK3733
4621	NTE74C925	ECG74C925	.....	SK74C925
4622	NTE1201	ECG1201	.....	.....
4623	NTE1206	ECG1206	.....	SK3160
4624	NTE1208	ECG1208	.....	SK3712
4625	NTE1209	ECG1209	.....	.....
4626	NTE1212	ECG1212	.....	.....
4627	NTE1213	ECG1213	.....	SK3704
4628	NTE74298	.....	.....	.....
4629	NTE15008	ECG15008	.....	SK10183
4630	NTE117	ECG117A	276-1114	SK5010A
4631	NTE1221	ECG1221	.....	SK7781
4632	NTE1229	ECG1229	.....	SK7783
4633	NTE1235	ECG1235	.....	SK3637
4634	NTE1243	ECG1243	.....	SK3731
4635	NTE1246	ECG1246	.....	SK3879

REPL. CODE	NTE PART NO.	ECG PART NO.	RADIO SHACK PART NO.	RCA PART NO.
4636	NTE1247	ECG1247	.....	.....
4637	NTE1251	ECG1251	.....	SK7723
4638	NTE1252	ECG1252	.....	.....
4639	.....	ECG1253	.....	.....
4640	.....	ECG1256	.....	SK3873
4641	NTE1257	ECG1257	.....	SK7724
4642	NTE1258	ECG1258	.....	SK7722
4643	.....	.....	.....	SK3365
4644	NTE1260	ECG1260	.....	SK3744
4645	NTE1262	ECG1262	.....	.....
4646	.....	ECG6885	.....	.....
4647	NTE6885	ECG6885	.....	SK9793
4648	NTE1271	ECG1271	.....	SK7789
4649	NTE1273	ECG1273	.....	SK3916
4650	NTE1741	ECG1741	.....	SK9994
4651	NTE6886	ECG6886	.....	SK9792
4652	NTE1278	ECG1278	.....	SK3726
4653	NTE1281	ECG1281	.....	SK7661
4654	NTE1282	ECG1282	.....	.....
4655	.....	ECG6887	.....	.....
4656	NTE1283	ECG1283	.....	SK9267
4657	NTE1284	ECG1284	.....	SK9268
4658	NTE6887	ECG6887	.....	SK9700
4659	NTE1287	ECG1287	.....	SK9228
4660	NTE40182B	ECG40182B	.....	SK40182B
4661	NTE1289	ECG1289	.....	SK9182
4662	NTE74374	.....	.....	.....
4663	NTE1291	ECG1291	.....	SK3875
4664	NTE74376	.....	.....	.....
4665	NTE1293	ECG1293	.....	SK7649
4666	.....	ECG1770	.....	SK4803
4667	NTE6888	ECG6888	.....	SK9791
4668	.....	.....	.....	SK1851
4669	NTE1301	ECG1301	.....	.....
4670	NTE1303	ECG1303	.....	.....
4671	NTE1309	ECG1309	.....	.....
4672	NTE1298	ECG1298	.....	SK9398
4673	NTE1320	ECG1320	.....	.....
4674	.....	ECG1321	.....	.....
4675	NTE1322	ECG1322	.....	SK10007
4676	NTE1323	ECG1323	.....	SK3154
4677	.....	ECG1324	.....	.....
4678	NTE1325	ECG1325	.....	.....
4679	.....	.....	276-022	.....
4680	NTE1326	ECG1326	.....	SK7659
4681	NTE1327	ECG1327	.....	SK7660
4682	NTE1328	ECG1328	.....	.....
4683	NTE1329	ECG1329	.....	.....
4684	NTE1330	ECG1330	.....	SK9406
4685	NTE1331	ECG1331	.....	SK9189
4686	.....	ECG1332	.....	.....
4687	NTE1333	ECG1333	.....	.....
4688	.....	ECG1334	.....	.....
4689	NTE1335	ECG1335	.....	.....
4690	NTE1336	ECG1336	.....	.....
4691	NTE1337	ECG1337	.....	.....
4692	.....	.....	.....	SK7C107
4693	NTE1339	ECG1339	.....	SK9997
4694	NTE1340	ECG1340	.....	.....
4695	NTE1341	ECG1341	.....	.....
4696	NTE1342	ECG1342	.....	.....
4697	NTE74426	.....	.....	.....
4698	NTE1343	ECG1343	.....	.....
4699	NTE1344	ECG1344	.....	.....
4700	.....	ECG1207	.....	.....
4701	NTE1346	ECG1346	.....	.....
4702	NTE1347	ECG1347	.....	.....
4703	NTE1348	ECG1348	.....	.....
4704	.....	ECG1349	.....	.....
4705	NTE1350	ECG1350	.....	.....
4706	.....	ECG1351	.....	.....
4707	.....	ECG1352	.....	.....
4708	NTE1354	ECG1354	.....	.....
4709	.....	ECG74HC123	.....	SK7C123
4710	NTE1355	ECG1355	.....	.....
4711	.....	ECG74HC125	.....	SK7C125
4712	.....	ECG74HC126	.....	SK7C126
4713	NTE1361	ECG1361	.....	SK9361
4714	NTE1363	ECG1363	.....	SK7630
4715	NTE1364	ECG1364	.....	SK9395
4716	NTE1365	ECG1365	.....	.....
4717	NTE1366	ECG1366	.....	.....
4718	NTE1367	ECG1367	.....	.....
4719	NTE1368	ECG1368	.....	.....
4720	NTE1369	ECG1369	.....	.....
4721	NTE552 N2	ECG552 N2	276-1102 N2	SK9000 N2
4722	NTE1370	ECG1370	.....	SK4821
4723	NTE1371	ECG1371	.....	.....
4724	NTE1372	ECG1372	.....	.....
4725	NTE1378	ECG1378	.....	SK7706
4726	NTE1379	ECG1379	.....	SK7663
4727	NTE1380	ECG1380	.....	SK9251
4728	NTE1381	ECG1381	.....	SK7622
4729	NTE1383	ECG1383	.....	.....
4730	NTE1384	ECG1384	.....	.....
4731	.....	ECG74HC153	.....	SK7C153
4732	NTE1385	ECG1385	.....	.....
4733	.....	ECG74HC154	.....	SK7C154
4734	NTE1388	ECG1388	.....	SK9186
4735	NTE1389	ECG1389	.....	.....
4736	NTE1391	ECG1391	.....	SK7814
4737	NTE1392	ECG1392	.....	.....
4738	.....	ECG74HC161	.....	SK7C161

REPL. CODE	NTE PART NO.	ECG PART NO.	RADIO SHACK PART NO.	RCA PART NO.
4739	NTE1393	ECG1393	.....	.....
4740	NTE1394	ECG1394	.....	.....
4741	.....	.....	.....	SK7C166
4742	.....	.....	.....	SK1852
4743	NTE1404	ECG1404	.....	.....
4744	NTE1405	ECG1405	.....	SK9018
4745	NTE1407	ECG1407	.....	.....
4746	NTE1408	ECG1408	.....	SK9064
4747	NTE1412	ECG1412	.....	SK9045
4748	.....	.....	.....	SK7C181
4749	.....	.....	.....	SK7C182
4750	NTE1414	ECG1414	.....	SK9326
4751	NTE1419	ECG1419	.....	SK9046
4752	NTE1421	ECG1421	.....	SK9051
4753	.....	.....	.....	SK7C190
4754	.....	.....	.....	SK7C191
4755	.....	.....	.....	SK7C192
4756	.....	.....	.....	SK7C193
4757	NTE1467	ECG1467	.....	SK4839
4758	NTE1425	ECG1425	.....	SK9049
4759	NTE1427	ECG1427	.....	SK9050
4760	NTE1429	ECG1429	.....	.....
4761	NTE1431	ECG1431	.....	SK7793
4762	NTE1433	ECG1433	.....	.....
4763	NTE1435	ECG1435	.....	SK7794
4764	NTE1438	ECG1438	.....	.....
4765	NTE1441	ECG1441	.....	SK7795
4766	NTE1442	ECG1442	.....	.....
4767	NTE3163	ECG3163	.....	SK2163
4768	NTE1447	ECG1447	.....	.....
4769	.....	.....	.....	SK9976
4770	NTE1449	ECG1449	.....	.....
4771	NTE1451	ECG1451	.....	.....
4772	.....	ECG2043	.....	.....
4773	NTE1452	ECG1452	.....	SK7796
4774	.....	.....	.....	SK7C221
4775	.....	ECG3169	.....	.....
4776	NTE1455	ECG1455	.....	.....
4777	NTE1457	ECG1457	.....	.....
4778	NTE1458	ECG1458	.....	.....
4779	NTE1460	ECG1460	.....	SK7798
4780	NTE1462	ECG1462	.....	.....
4781	NTE1464	ECG1464	.....	.....
4782	NTE1469	ECG1469	.....	SK9356
4783	NTE552 N1	ECG552 N1	276-1104 N1	SK9000 N1
4784	.....	.....	.....	SK7C238
4785	NTE1472	ECG1472	.....	SK7800
4786	NTE284	ECG284	.....	SK3260
4787	NTE1473	ECG1473	.....	SK9355
4788	NTE1758	ECG1758	.....	SK10041
4789	NTE1474	ECG1474	.....	.....
4790	.....	ECG3170	.....	.....
4791	NTE1475	ECG1475	.....	SK4838
4792	.....	.....	.....	SK7C244
4793	NTE1476	ECG1476	.....	.....
4794	NTE1477	ECG1477	.....	.....
4795	NTE1478	ECG1478	.....	.....
4796	NTE1724	ECG1724	.....	SK7650
4797	NTE1481	ECG1481	.....	.....
4798	NTE1482	ECG1482	.....	.....
4799	NTE1484	ECG1484	.....	.....
4800	.....	.....	.....	SK7C253
4801	.....	ECG3171	.....	.....
4802	.....	.....	.....	SK7C257
4803	NTE1489	ECG1489	.....	.....
4804	.....	.....	.....	SK7C258
4805	NTE1490	ECG1490	.....	SK9312
4806	.....	.....	.....	SK7C259
4807	NTE1491	ECG1491	.....	.....
4808	NTE1492	ECG1492	.....	SK9196
4809	NTE1493	ECG1493	.....	.....
4810	.....	ECG1494	.....	.....
4811	NTE1495	ECG1495	.....	.....
4812	NTE1497	ECG1497	.....	.....
4813	NTE1498	ECG1498	.....	SK7667
4814	NTE1499	ECG1499	.....	.....
4815	NTE1500	ECG1500	.....	.....
4816	.....	.....	.....	SK1853
4817	NTE1501	ECG1501	.....	.....
4818	.....	.....	.....	SK7668
4819	NTE1502	ECG1502	.....	.....
4820	NTE1503	ECG1503	.....	SK7802
4821	NTE1504	ECG1504	.....	.....
4822	.....	.....	.....	SK7C273
4823	NTE1505	ECG1505	.....	.....
4824	.....	ECG1507	.....	.....
4825	NTE1511	ECG1511	.....	.....
4826	NTE1513	ECG1513	.....	SK9250
4827	NTE1514	ECG1514	.....	.....
4828	.....	.....	.....	SK7C283
4829	.....	ECG5309	.....	.....
4830	NTE1515	ECG1515	.....	SK7656
4831	NTE1516	ECG1516	.....	.....
4832	.....	.....	.....	SK74123
4833	.....	ECG5310	.....	.....
4834	NTE1517	ECG1517	.....	SK7704
4835	NTE1518	ECG1518	.....	.....
4836	NTE1519	ECG1519	.....	SK7803
4837	.....	ECG5311	.....	.....
4838	NTE1523	ECG1523	.....	.....
4839	.....	.....	.....	SK3544
4840	NTE1626	ECG1626	.....	SK7652
4841	.....	ECG1526	.....	SK4539B[4]539B

REPL. CODE	NTE PART NO.	ECG PART NO.	RADIO SHACK PART NO.	RCA PART NO.
4842	NTE1527	ECG1527	.....	.....
4843	.....	.....	.....	SK7C297
4844	NTE1530	ECG1530	.....	.....
4845	.....	ECG74HC299	.....	SK7C299
4846	NTE1531	ECG1531	.....	.....
4847	NTE1533	ECG1533	.....	.....
4848	.....	.....	.....	SK18A
4849	NTE1534	ECG1534	.....	.....
4850	NTE1536	ECG1536	.....	SK3884
4851	NTE1537	ECG1537	.....	.....
4852	.....	.....	.....	SK9741
4853	NTE1543	ECG1543	.....	.....
4854	NTE1544	ECG1544	.....	SK9193
4855	.....	.....	.....	SK6649
4856	NTE74H21	ECG74H21	.....	.....
4857	NTE1615	ECG1615	.....	SK9977
4858	.....	ECG123A	.....	.....
4859	.....	ECG3172	.....	.....
4860	.....	ECG5015A	.....	.....
4861	.....	ECG3174	.....	.....
4862	.....	.....	.....	SK7C354
4863	.....	.....	.....	SK7C356
4864	NTE68MCP	ECG68MCP	.....	SK9927
4865	.....	ECG74HC377	.....	SK7C377
4866	NTE377	ECG377	.....	SK9112
4867	NTE524V27	ECG524V27	.....	SK10060
4868	NTE2085	ECG2085	.....	SK9913
4869	.....	ECG1886	.....	SK10086
4870	NTE3052	ECG3052	.....	SK2052
4871	NTE3065	ECG3065	.....	SK2065
4872	NTE524V30	ECG524V30	.....	.....
4873	NTE3056	ECG3056	276-075	SK2056
4874	.....	ECG166	.....	.....
4875	.....	.....	.....	SK9306
4876	.....	.....	.....	SK7C423
4877	.....	.....	.....	SK3638
4878	.....	ECG3105	.....	.....
4879	.....	ECG4510B	.....	SK4510B
4880	.....	.....	.....	SK1855
4881	.....	ECG1910	.....	.....
4882	.....	ECG1868	.....	.....
4883	.....	ECG90	.....	SK9370
4884	.....	ECG1917	.....	.....
4885	.....	ECG1905	.....	.....
4886	NTE1907	ECG1907	.....	SK10049
4887	NTE54	ECG54	.....	SK9112
4888	.....	.....	.....	SK7C540
4889	.....	.....	.....	SK7C541
4890	.....	.....	.....	SK9401
4891	.....	.....	.....	SK7CT4510
4892	.....	.....	.....	SK7C563
4893	.....	.....	.....	SK7C564
4894	.....	.....	.....	SK7CT4511
4895	.....	.....	.....	SK7206
4896	.....	ECG74HC573	.....	SK7C573
4897	.....	ECG74HC574	.....	SK7C574
4898	NTE123AP	ECG123AP	276-2009	SK3433
4899	.....	.....	.....	SK7C583
4900	.....	ECG5328	.....	SK9088
4901	.....	.....	.....	SK7C597
4902	NTE2107	ECG2107	.....	.....
4903	.....	.....	.....	SK7CT4514
4904	NTE7488A	.....	.....	.....
4905	.....	ECG961	.....	SK3671
4906	.....	ECG963	.....	SK3672
4907	NTE5021T1	ECG5021T1	.....	SK9771
4908	.....	.....	.....	SK7CT4515
4909	.....	.....	.....	SK7CT4516
4910	.....	.....	.....	SK7CT02
4911	.....	.....	.....	SK4060B
4912	.....	.....	.....	SK7C640
4913	.....	.....	.....	SK7CT4518
4914	.....	.....	.....	SK7C643
4915	.....	.....	.....	SK7C646
4916	.....	.....	.....	SK7C648
4917	.....	ECG596	.....	.....
4918	.....	.....	.....	SK7CT4520
4919	.....	.....	.....	SK7C670
4920	NTE580	ECG580	.....	SK3925
4921	NTE1V075	ECG1V075	.....	SKMV75A
4922	NTE6116	ECG6116	.....	SK10217
4923	.....	.....	.....	SK7C688
4924	.....	.....	.....	SK10178
4925	NTE1828	ECG1828	.....	SK9736
4926	.....	ECG753	.....	SK7191
4927	NTE1684	ECG1684	.....	SK9729
4928	NTE389	ECG389	.....	SK3111
4929	NTE15039	.....	.....	.....
4930	NTE15042	ECG7019	.....	SK10488
4931	NTE15043	.....	.....	.....
4932	NTE3044	ECG3044	.....	SK4935
4933	NTE2003	ECG2003	.....	.....
4934	.....	.....	.....	SK4332
4935	NTE2015	ECG2015	.....	SK9095
4936	NTE2016	ECG2016	.....	SK9077
4937	NTE2017	ECG2017	.....	SK9078
4938	.....	.....	.....	SK16V
4939	NTE2018	ECG2018	.....	SK9079
4940	NTE2019	ECG2019	.....	SK9080
4941	NTE2020	ECG2020	.....	SK9081
4942	.....	.....	.....	SK22A
4943	NTE2025	ECG2025	.....	.....
4944	NTE280	ECG284	276-2041	SK3836

REPL. CODE	NTE PART NO.	ECG PART NO.	RADIO SHACK PART NO.	RCA PART NO.
4945	.....	ECG585	.....	.....
4946	.....	ECG578	.....	.....
4947	.....	ECG2029	.....	.....
4948	NTE580	ECG580	276-1114	SK3081
4949	NTE2031	ECG2031	.....	.....
4950	NTE2032	ECG2032	.....	SK9203
4951	NTE2050	ECG2050	.....	SK10279
4952	NTE2051	ECG2051	.....	SK10280
4953	NTE7005	ECG7005	.....	SK10474
4954	NTE2054	ECG2054	.....	SK9204
4955	NTE2070	ECG2070	.....	SK10286
4956	NTE2071	ECG2071	.....	SK10287
4957	NTE2072	ECG2072	.....	SK10288
4958	.....	.....	.....	SK7CT4538
4959	NTE2073	ECG2073	.....	SK10289
4960	.....	.....	.....	SK3115
4961	NTE2075	ECG2075	.....	SK10291
4962	NTE2076	ECG2076	.....	SK10292
4963	NTE2077	ECG2077	.....	SK10293
4964	.....	ECG5022A	.....	.....
4965	NTE2078	ECG2078	.....	SK10294
4966	NTE2080	ECG2080	.....	.....
4967	NTE2081	ECG2081	.....	.....
4968	NTE2082	ECG2082	.....	SK9903
4969	NTE2083	ECG2083	.....	SK9906
4970	NTE2087	ECG2087	.....	SK9813
4971	.....	.....	.....	SK7331
4972	.....	.....	.....	SK74LS109
4973	NTE2102	ECG2102	.....	.....
4974	NTE2104	ECG2104	.....	.....
4975	NTE129	ECG129	.....	SK3179B
4976	NTE4513B	ECG4513B	.....	SK4513B
4977	.....	.....	.....	SK9002
4978	.....	ECG6155	.....	SK7355
4979	.....	.....	.....	SK7CT4543
4980	.....	.....	.....	SK3533
4981	.....	.....	.....	SK5137
4982	.....	.....	.....	SK3937
4983	.....	ECG61	.....	.....
4984	.....	.....	.....	SK4007UB
4985	NTE213	ECG213	.....	SK3500
4986	NTE1570	ECG1570	.....	SK7681
4987	NTE3050	ECG3050	.....	SK10403
4988	NTE3051	ECG3051	.....	SK10328
4989	NTE3061	ECG3061	.....	SK2061
4990	NTE4521B	ECG4521B	.....	SK4521B
4991	.....	ECG5983	.....	.....
4992	.....	ECG5987	.....	.....
4993	NTE289A	ECG289A	.....	SK3849
4994	.....	ECG5991	.....	.....
4995	.....	ECG5995	.....	.....
4996	.....	ECG5999	.....	.....
4997	.....	ECG6003	.....	.....
4998	.....	ECG5818	.....	.....
4999	.....	.....	.....	SK3912
5000	NTE15007	ECG15007	.....	SK10182
5001	NTE75322	.....	.....	.....
5002	.....	ECG5994	.....	.....
5003	.....	ECG5998	.....	.....
5004	.....	ECG6002	.....	.....
5005	NTE5913	ECG5913	.....	SK10423
5006	.....	.....	276-145	.....
5007	.....	ECG519	.....	.....
5008	.....	ECG129&427	.....	SK3025
5009	.....	.....	.....	SK1861
5010	NTE2410	ECG2410	.....	.....
5011	NTE1677	ECG1677	.....	SK9769
5012	NTE5827	ECG5827	.....	SK7377
5013	.....	.....	.....	SK5555
5014	NTE5534	ECG5534	.....	SK3505
5015	.....	ECG517	.....	.....
5016	NTE3131	ECG3131	276-030	SK2170
5017	NTE4541B	ECG4541B	.....	SK4541B
5018	NTE83	ECG83	.....	.....
5019	.....	.....	.....	SK1862
5020	NTE4543B	ECG4543B	.....	SK4543B
5021	NTE75493	ECG75493	.....	SK75493
5022	NTE75494	ECG75494	.....	SK9825
5023	.....	.....	.....	SK7651
5024	NTE75498	ECG75498	.....	.....
5025	NTE5426	ECG5426	.....	SK3954
5026	NTE5962	ECG5962	.....	SK7362
5027	.....	ECG216	.....	.....
5028	.....	ECG9383	.....	.....
5029	NTE5534	ECG5534	.....	SK6622
5030	NTE5018A	ECG5018A	.....	SK9A1
5031	.....	ECG5552	.....	.....
5032	NTE4547B	ECG4547B	.....	SK4547B
5033	NTE2923	ECG2923	.....	.....
5034	NTE128P	ECG128P	.....	SK9468
5035	.....	ECG351	.....	SK3177
5036	NTE358B	ECG358B	.....	SK9330
5037	.....	ECG3158	.....	.....
5038	NTE5875	ECG5875	.....	SK3517
5039	NTE4551B	ECG4551B	.....	SK4551B
5040	NTE4553B	ECG4553B	.....	SK4553B
5041	NTE1905	ECG1905	.....	SK9222
5042	.....	ECG519	.....	SK3100
5043	.....	.....	.....	SK3710
5044	NTE382	ECG292	.....	.....
5045	.....	.....	.....	SK7CT154
5046	NTE6102	ECG6102	.....	SK10209
5047	.....	ECG1726	.....	.....
5048	NTE1678	ECG1678	.....	SK9947
5049	NTE4558B	ECG4558B	.....	SK4558B
5050	.....	ECG1624	.....	.....
5051	NTE3047	ECG3047	276-134	SK9708
5052	.....	.....	.....	SK7691
5053	.....	.....	.....	SK1864
5054	.....	ECG1767	.....	.....
5055	.....	.....	.....	SK7690
5056	.....	ECG35	.....	.....
5057	NTE41	ECG41	.....	SK9424
5058	NTE43	ECG43	.....	SK9426
5059	NTE45	ECG45	.....	SK9428
5060	NTE66	ECG66	.....	SK9165
5061	NTE14	ECG14	.....	SK9662
5062	NTE78	ECG78	.....	.....
5063	NTE3171	ECG3171	.....	SK10440
5064	NTE84	ECG84	.....	.....
5065	NTE3048	ECG3048	.....	SK2048
5066	NTE52	ECG52	.....	SK9261
5067	.....	.....	.....	SK7CT40102
5068	.....	.....	.....	SK7CT40103
5069	.....	.....	.....	SK7CT40104
5070	.....	.....	.....	SK7CT40105
5071	.....	ECG1622	.....	SK9044
5072	.....	.....	.....	SK7C40102
5073	.....	.....	.....	SK7C40103
5074	.....	.....	.....	SK7C40104
5075	.....	ECG74HC40105	.....	SK7C40105
5076	.....	.....	.....	SK7611
5077	.....	ECG5536	.....	.....
5078	NTE5539	ECG5539	.....	SK10407
5079	NTE5536	ECG5536	.....	SK10405
5080	NTE1151	.....	.....	.....
5081	.....	ECG5668A	.....	.....
5082	NTE1164	ECG1164	.....	SK3072
5083	.....	ECG116	.....	SK3032A
5084	.....	.....	.....	SK74LS122
5085	.....	ECG81	.....	.....
5086	.....	ECG74LS571	.....	.....
5087	.....	.....	.....	SK1866
5088	NTE152MP	ECG152MP	.....	SK9936
5089	.....	ECG5025A	.....	.....
5090	NTE2501	ECG2501	.....	SK3747
5091	NTE4584B	.....	.....	.....
5092	.....	ECG755	.....	SK3170
5093	.....	ECG48	.....	.....
5094	NTE1295	ECG1295	.....	SK9299
5095	NTE1299	ECG1299	.....	SK9314
5096	NTE2387	ECG2387	.....	SK10304
5097	.....	ECG2047	.....	.....
5098	NTE2532	ECG2532	.....	.....
5099	NTE1356	ECG1356	.....	.....
5100	NTE1357	ECG1357	.....	.....
5101	NTE1359	ECG1359	.....	.....
5102	.....	.....	.....	SK40147B
5103	NTE16007	.....	.....	.....
5104	NTE1362	ECG1362	.....	SK9396
5105	NTE977	ECG977	.....	.....
5106	NTE5426	ECG5426	.....	SK3953
5107	NTE1390	ECG1390	.....	SK7815
5108	NTE1397	ECG1397	.....	SK7603
5109	.....	ECG74S474	.....	.....
5110	NTE1398	ECG1398	.....	SK9191
5111	NTE1399	ECG1399	.....	SK9269
5112	NTE3004	ECG3004	.....	SK2004
5113	NTE3005	ECG3005	.....	SK2005
5114	.....	ECG3006	.....	SK2006A
5115	NTE3009	ECG3009	.....	SK2009A
5116	NTE3012	ECG3012	276-033	.....
5117	NTE3013	ECG3013	.....	.....
5118	NTE1432	ECG1432	.....	.....
5119	NTE3019	ECG3019	.....	SK2019
5120	.....	.....	.....	SK4098B
5121	NTE1443	ECG1443	.....	SK3967
5122	NTE3031	ECG3031	.....	SK2031
5123	NTE3037	ECG3037	.....	SK10325
5124	NTE3038	ECG3038	.....	SK10326
5125	NTE3039	ECG3039	.....	SK9705
5126	NTE3049	ECG3049	.....	SK2049
5127	NTE3053	ECG3053	.....	SK2053
5128	NTE3054	ECG3054	.....	SK2054
5129	NTE3055	ECG3055	.....	SK2055
5130	NTE3058	ECG3058	.....	SK2058
5131	NTE3059	ECG3059	.....	SK2059
5132	NTE3060	ECG3060	.....	SK2060
5133	NTE128P	ECG128P	276-2009	SK3854
5134	NTE3062	ECG3062	.....	SK2062
5135	NTE3063	ECG3063	.....	SK2063
5136	.....	.....	.....	SK74LS151
5137	NTE3064	ECG3064	.....	SK2064
5138	NTE3068	ECG3068	.....	SK2068
5139	NTE3069	ECG3069	.....	SK2069
5140	NTE3070	ECG3070	.....	SK2070
5141	NTE3071	ECG3071	.....	SK2071
5142	NTE3082	ECG3082	.....	SK4934
5143	.....	ECG5675	.....	.....
5144	NTE159	ECG159	276-2023	SK10272
5145	.....	.....	.....	SK1869
5146	.....	.....	.....	SK7CT4046A
5147	.....	.....	.....	SK3036
5148	.....	.....	.....	SK7C7046A
5149	NTE580	ECG580	.....	.....
5150	.....	ECG5331	.....	.....
5151	NTE3114	ECG3114	.....	SK2114
5152	.....	.....	.....	SK7C4046A
5153	NTE3115	ECG3115	276-081	SK2115
5154	NTE3116	ECG3116	.....	SK2116
5155	NTE3117	ECG3117	.....	SK2117
5156	NTE1535	ECG1535	.....	.....
5157	NTE1V095	ECG1V095	.....	SKMV95A
5158	NTE1554	ECG1554	.....	.....
5159	NTE1556	ECG1556	.....	SK7657
5160	NTE1559	ECG1559	.....	.....
5161	NTE2367	ECG2367	.....	.....
5162	NTE1561	ECG1561	.....	SK7610
5163	NTE1562	ECG1562	.....	.....
5164	NTE174	.....	.....	SK3130
5165	NTE3150	ECG3150	.....	SK2150
5166	.....	.....	.....	SK74LS161
5167	NTE3151	ECG3151	.....	SK2151
5168	NTE3152	ECG3152	.....	SK2152
5169	NTE1568	ECG1568	.....	.....
5170	NTE3153	ECG3153	.....	SK2153
5171	NTE3155	ECG3155	.....	.....
5172	NTE1571	ECG1571	.....	SK7677
5173	.....	ECG5002A	.....	.....
5174	NTE3160	ECG3160	.....	SK2160
5175	NTE1577	ECG1577	.....	.....
5176	NTE3162	ECG3162	.....	SK2162
5177	NTE1579	ECG1579	.....	.....
5178	NTE3165	ECG3165	.....	SK2165
5179	NTE3168	ECG3168	.....	SK2168
5180	NTE1585	ECG1585	.....	.....
5181	NTE1591	ECG1591	.....	SK7602
5182	NTE1593	ECG1593	.....	.....
5183	NTE1597	ECG1597	.....	.....
5184	.....	.....	.....	SK74LS163
5185	NTE1600	ECG1600	.....	SK10221
5186	.....	.....	.....	SK11A
5187	NTE1602	ECG1602	.....	SK10223
5188	NTE1606	ECG1606	.....	SK4828
5189	NTE1607	ECG1607	.....	.....
5190	NTE156A	.....	.....	.....
5191	NTE1611	ECG1611	.....	SK9990
5192	NTE1613	ECG1613	.....	SK10010
5193	NTE1620	ECG1620	.....	.....
5194	NTE1621	ECG1621	.....	SK10230
5195	NTE1624	ECG1624	.....	SK7611
5196	.....	.....	.....	SK3125A
5197	NTE1631	ECG1631	.....	SK10233
5198	NTE1632	ECG1632	.....	SK4812
5199	NTE1635	ECG1635	.....	SK9706
5200	NTE1637	ECG1637	.....	.....
5201	NTE1641	ECG1641	.....	SK10239
5202	NTE338F	ECG338F	.....	SK9615
5203	NTE1648	ECG1648	.....	SK9053
5204	NTE1649	ECG1649	.....	.....
5205	NTE1654	ECG1654	.....	SK7670
5206	NTE1659	ECG1659	.....	.....
5207	NTE1662	ECG1662	.....	.....
5208	NTE1663	ECG1663	.....	SK9324
5209	NTE1664	ECG1664	.....	SK4833
5210	NTE1666	ECG1666	.....	.....
5211	NTE3154	ECG3154	.....	SK2154
5212	NTE1672	ECG1672	.....	SK10242
5213	.....	.....	.....	SK3008
5214	NTE1681	ECG1681	.....	SK10246
5215	NTE1682	ECG1682	.....	SK7713
5216	.....	ECG1746	.....	.....
5217	NTE74LS173A	ECG74LS173	.....	SK1922
5218	NTE1688	ECG1688	.....	SK9725
5219	NTE1690	ECG1690	.....	.....
5220	NTE1691	ECG1691	.....	.....
5221	NTE1692	ECG1692	.....	.....
5222	NTE1693	ECG1693	.....	.....
5223	NTE139A	ECG139A	.....	SK9V1
5224	NTE5092A	ECG5092A	.....	SK62V
5225	NTE1700	ECG1700	.....	SK9980
5226	NTE1701	ECG1701	.....	SK10008
5227	NTE1702	ECG1702	.....	SK10013
5228	.....	ECG82	.....	.....
5229	NTE1704	ECG1704	.....	SK9981
5230	NTE1709	ECG1709	.....	SK10080
5231	.....	.....	.....	SK1871
5232	NTE1717	ECG1717	.....	SK10022
5233	NTE1720	ECG1720	.....	SK9983
5234	NTE1726	ECG1726	.....	SK10028
5235	.....	.....	.....	SK4A3
5236	NTE1730	ECG1730	.....	SK9955
5237	NTE1731	ECG1731	.....	.....
5238	NTE1734	ECG1734	.....	SK7810
5239	NTE1745	ECG1745	.....	SK10033
5240	.....	ECG5455	.....	SK3597
5241	NTE1749	ECG1749	.....	SK10001
5242	NTE1750	ECG1750	.....	SK10035
5243	NTE1753	ECG1753	.....	SK10037
5244	NTE1757	ECG1757	.....	.....
5245	.....	ECG4070B	.....	SK4070B
5246	NTE1761	ECG1761	.....	.....
5247	NTE1762	ECG1762	.....	SK10044
5248	.....	ECG5030A	.....	SK22A
5249	.....	.....	.....	SK10108
5250	NTE1764	ECG1764	.....	.....
5251	NTE1774	ECG1774	.....	SK9867
5252	.....	.....	.....	SK19A
5253	NTE74H106	ECG74H106	.....	.....

REPL. CODE	NTE PART NO.	ECG PART NO.	RADIO SHACK PART NO.	RCA PART NO.
5254	NTE1785	ECG1785	.....	.....
5255	.....	ECG5633	.....	.....
5256	NTE1791	ECG1791	.....	SK9717
5257	NTE1796	ECG1796	.....	SK9880
5258	NTE1808	ECG1808	.....	SK9722
5259	NTE1809	ECG1809	.....	SK9724
5260	NTE1811	ECG1811	.....	SK9733
5261	NTE1812	ECG1812	.....	SK9734
5262	NTE1813	ECG1813	.....	SK9735
5263	NTE1814	ECG1814	.....	SK9739
5264	.....	.....	.....	SKMV250L
5265	NTE1816	ECG1816	.....	SK9897
5266	.....	.....	.....	SK1872
5267	NTE1817	ECG1817	.....	SK9890
5268	NTE1818	ECG1818	.....	SK9893
5269	NTE1819	ECG1819	.....	SK9882
5270	NTE1820	ECG1820	.....	SK9889
5271	NTE1829	ECG1829	.....	SK9737
5272	NTE1830	ECG1830	.....	SK9896
5273	NTE1831	ECG1831	.....	SK9895
5274	.....	ECG595	.....	.....
5275	NTE1849	ECG1849	.....	SK10149
5276	NTE1871	ECG1871	.....	SK10140
5277	NTE1874	ECG1874	.....	SK10141
5278	NTE1875	ECG1875	.....	SK10166
5279	NTE1876	ECG1876	.....	.....
5280	NTE1900	ECG1900	.....	SK7644
5281	NTE1901	ECG1901	.....	SK10047
5282	NTE5914	ECG5916	.....	SK10424
5283	NTE5918	ECG5920	.....	SK10426
5284	.....	ECG2084	.....	SK4910
5285	NTE157	.....	.....	.....
5286	NTE5922	ECG5924	.....	SK10428
5287	NTE1915	ECG1915	.....	SK9335
5288	.....	.....	.....	SK1873
5289	NTE1919	ECG1919	.....	SK9336
5290	NTE1923	ECG1923	.....	SK10054
5291	NTE1926	ECG1926	.....	.....
5292	NTE1927	ECG1927	.....	.....
5293	NTE1934	ECG1934	.....	SK7739
5294	NTE999SM	ECG999SM	.....	SK10516
5295	.....	.....	.....	SK9037
5296	NTE1938	ECG1938	.....	SK7741
5297	NTE1940	ECG1940	.....	SK7742
5298	NTE999M	ECG999M	.....	SK10515
5299	.....	ECG193A	.....	.....
5300	.....	.....	.....	SK2214
5301	.....	.....	.....	SK10124
5302	NTE74H76	ECG74H76	.....	.....
5303	NTE350F	ECG350F	.....	SK9625
5304	NTE943M	ECG943M	.....	SK9721
5305	NTE2331	ECG2331	.....	.....
5306	NTE2000	ECG2000	.....	.....
5307	.....	.....	.....	SK9431
5308	.....	.....	.....	SK1874
5309	.....	.....	.....	SK3130
5310	.....	.....	.....	SK22859
5311	NTE2047	ECG2047	.....	.....
5312	NTE2056	ECG2056	.....	.....
5313	.....	ECG506	.....	SK3043B
5314	NTE2086	ECG2086	.....	SK9907
5315	NTE2088	ECG2088	.....	SK9908
5316	.....	ECG9615	.....	.....
5317	.....	.....	.....	SK1875
5318	.....	ECG5596	.....	SK6746
5319	.....	.....	.....	SK6747
5320	NTE1183	ECG1183	.....	SK3328
5321	.....	.....	.....	SK3023
5322	NTE5638	ECG5638	.....	SK5538
5323	NTE56015	ECG56015	.....	SK10266
5324	NTE15043	ECG1807	.....	SK9839
5325	NTE6079	ECG6079	.....	SK9710
5326	NTE56016	ECG56016	.....	SK10267
5327	.....	.....	.....	SK9284
5328	.....	.....	.....	SK9723
5329	NTE383	ECG383	.....	SK3715
5330	NTE2353	ECG2353	.....	.....
5331	.....	.....	.....	SK1876
5332	NTE178MP	ECG178MP	276-1122	SK3088
5333	NTE178MP	ECG178MP	276-1122	SK9002
5334	.....	.....	.....	SK9328
5335	.....	.....	.....	SK9719
5336	.....	ECG85	.....	SK3245
5337	.....	.....	.....	SK7664
5338	NTE74H01	ECG74H01	.....	.....
5339	NTE3882	ECG3882	.....	SK2882
5340	.....	ECG1170	.....	SK3745
5341	NTE2314	ECG2314	.....	SK9482
5342	NTE2316	ECG2316	.....	SK10116
5343	.....	.....	.....	SK1877
5344	.....	ECG1439	.....	.....
5345	NTE2318	ECG2318	.....	SK9986
5346	NTE340	ECG340	276-2009	SK3452
5347	NTE213	ECG213	.....	SK3006
5348	NTE2323	ECG2323	.....	.....
5349	NTE2329	ECG2329	.....	SK9489
5350	NTE2340	ECG2340	.....	SK9479
5351	.....	.....	.....	SK3925
5352	NTE2351	ECG2351	.....	SK9856
5353	NTE2352	ECG2352	.....	SK9857
5354	.....	ECG2419	.....	SK10111
5355	NTE74H08	ECG74H08	.....	.....
5356	NTE15006	ECG15006	.....	SK10181
5357	.....	ECG1563	.....	SK7621
5358	.....	.....	.....	SK9935
5359	.....	.....	.....	SK3108
5360	NTE2381	ECG2381	.....	SK10127
5361	NTE2390	ECG2390	.....	SK10132
5362	HVD-12	ECG-HIDIV-12	.....	SK9106
5363	NTE2392	ECG2392	.....	SK10133
5364	NTE2404	ECG2404	.....	SK10104
5365	NTE2405	ECG2405	.....	SK10105
5366	NTE2411	ECG2411	.....	SK10102
5367	.....	.....	.....	SK3715
5368	NTE2413	ECG2413	.....	SK10109
5369	NTE2426	ECG2426	.....	SK10112
5370	NTE2427	ECG2427	.....	SK10113
5371	NTE2431	ECG2431	.....	SK10115
5372	NTE85	ECG85	276-2009	SK9137
5373	.....	ECG19	.....	SK3912
5374	.....	ECG5008A	.....	.....
5375	.....	ECG2414	.....	SK9961
5376	.....	ECG2416	.....	SK10108
5377	NTE289AMP	ECG289AMP	276-2009	SK3124A
5378	NTE159	ECG159	276-2023	SK9132
5379	.....	.....	.....	SK3583
5380	.....	ECG5543	.....	.....
5381	.....	ECG3024	.....	SK2024
5382	.....	ECG374	.....	SK9370
5383	NTE74H30	ECG74H30	.....	.....
5384	NTE74HC02	ECG74HC02	.....	SK7C02
5385	.....	ECG2408	.....	.....
5386	NTE3026	ECG3026	.....	SK2026
5387	NTE398	ECG398	.....	SK9370
5388	.....	ECG6408	.....	.....
5389	NTE6411	ECG6411	.....	SK4922
5390	.....	ECG6412	.....	.....
5391	NTE1890	ECG1890	.....	SK10520
5392	.....	ECG2050	.....	.....
5393	NTE36MP	ECG36MP	.....	SK9920
5394	.....	ECG5635	.....	.....
5395	.....	ECG5646	.....	SK9711
5396	.....	.....	.....	SK3857
5397	.....	.....	.....	SK10444
5398	.....	ECG1810	.....	.....
5399	.....	.....	.....	SK10226
5400	.....	ECG5322	.....	.....
5401	.....	.....	.....	SK74LS279
5402	.....	ECG5324	.....	.....
5403	.....	ECG5326	.....	.....
5404	.....	ECG290A	.....	SK3932
5405	.....	ECG506	.....	SK3031A
5406	.....	ECG2429	.....	.....
5407	NTE238	ECG238	276-2055	2SD822
5408	.....	ECG2359	.....	SK9959
5409	.....	ECG124	.....	.....
5410	NTE74H60	ECG74H60	.....	.....
5411	NTE74HC32	ECG74HC32	.....	.....
5412	.....	ECG5671	.....	.....
5413	.....	ECG74HC161	.....	.....
5414	.....	ECG2319	.....	.....
5415	.....	ECG127	.....	.....
5416	.....	ECG199	.....	SK3018
5417	.....	ECG261	.....	.....
5418	NTE74H78	ECG74H78	.....	.....
5419	.....	.....	.....	SK9264
5420	.....	.....	.....	SK10094
5421	NTE3085	ECG3085	.....	SK10330
5422	NTE52	ECG52	.....	SK3559
5423	NTE3087	ECG3087	.....	SK2087
5424	NTE3088	ECG3088	.....	SK10331
5425	NTE3089	ECG3089	.....	SK10332
5426	NTE3090	ECG3090	.....	SK4933
5427	NTE3093	ECG3093	.....	SK10334
5428	NTE3095	ECG3095	.....	SK10335
5429	NTE3096	ECG3096	.....	SK10337
5430	NTE3097	ECG3097	.....	SK10336
5431	NTE2324	ECG2324	.....	.....
5432	NTE213	ECG213	.....	SK3717
5433	NTE3103	ECG3103	.....	SK4916
5434	NTE3111	ECG3111	.....	SK2111
5435	NTE3112	ECG3112	.....	SK2112
5436	NTE3113	ECG3113	.....	SK2113
5437	.....	ECG3054	.....	SK2054
5438	NTE3130	ECG3130	.....	SK2171
5439	.....	ECG3063	.....	SK2063
5440	.....	ECG5679	.....	SK3993
5441	.....	ECG3059	.....	SK2059
5442	NTE290A	ECG290A	.....	SK3932
5443	NTE3169	ECG3169	.....	SK10445
5444	NTE3170	ECG3170	.....	SK10439
5445	.....	.....	.....	SK9968
5446	NTE3180	ECG3180	.....	SK10446
5447	NTE3181	ECG3181	.....	SK10441
5448	NTE3182	ECG3182	.....	SK10442
5449	.....	.....	.....	SK10093
5450	.....	.....	.....	SK9457
5451	NTE390	ECG390	.....	SK9389
5452	.....	ECG2326	.....	.....
5453	.....	ECG270	.....	SK9417
5454	NTE2411	ECG2411	.....	.....
5455	NTE7489	ECG7489	.....	SK7489
5456	.....	ECG36	.....	SK9389
5457	NTE263	ECG263	.....	SK3896
5458	.....	.....	.....	SK3105
5459	.....	ECG4901	.....	SK10192
5460	.....	ECG56020	.....	SK9808
5461	.....	.....	.....	SK9972
5462	NTE217	ECG217	.....	SK3466
5463	NTE2351	ECG2351	.....	SK9417
5464	.....	.....	.....	SK9458
5465	.....	ECG130	.....	.....
5466	.....	ECG282&427	.....	SK9841
5467	.....	ECG398	.....	SK3930
5468	NTE378	ECG378	.....	SK3934
5469	NTE5429	ECG5429	.....	SK3503
5470	.....	ECG199	.....	SK3245
5471	.....	.....	.....	SK3111
5472	.....	ECG383	.....	SK9454
5473	NTE116	ECG116	.....	SK3313
5474	NTE181MP	ECG181MP	.....	SK9984
5475	.....	ECG378	.....	.....
5476	.....	.....	.....	SK4049UB
5477	.....	.....	.....	SK9390
5478	NTE37	ECG37	.....	SK9390
5479	.....	.....	.....	SK9467
5480	.....	ECG2306	.....	.....
5481	NTE2300	ECG2300	.....	SK9487
5482	NTE2302	ECG2302	.....	.....
5483	NTE5322	ECG5322	276-1185	SK3679
5484	.....	.....	.....	SK9944
5485	NTE55	ECG55	.....	.....
5486	.....	.....	.....	SK3861
5487	.....	.....	.....	SK3452
5488	.....	.....	.....	SK7C4002
5489	.....	ECG288	.....	.....
5490	.....	ECG1087	.....	SK3477
5491	NTE5915	ECG5917	.....	SK10425
5492	NTE55	ECG55	.....	SK9117
5493	.....	ECG593	.....	.....
5494	.....	ECG2329	.....	.....
5495	NTE5919	ECG5921	.....	SK10427
5496	NTE5923	ECG5925	.....	SK10429
5497	.....	ECG186A	.....	.....
5498	NTE219MCP	ECG219MCP	.....	.....
5499	NTE156 N6	ECG156 N6	276-1114 N6	SK3051 N6
5500	NTE55	ECG55	.....	.....
5501	.....	ECG1218	.....	.....
5502	.....	.....	.....	SK9331
5503	NTE358C	ECG358C	.....	.....
5504	.....	ECG263	.....	.....
5505	NTE570	ECG570	.....	.....
5506	.....	.....	.....	SK7C4015
5507	.....	.....	.....	SK7C4016
5508	.....	ECG11	.....	.....
5509	.....	ECG1714S	.....	.....
5510	.....	ECG2428	.....	.....
5511	.....	.....	.....	SK7C4017
5512	.....	.....	.....	SK7C4020
5513	.....	ECG977	.....	SK3462
5514	NTE5892	ECG5892	.....	SK7092
5515	.....	.....	.....	SK7C4024
5516	NTE3034A	ECG3034	.....	SK2032
5517	NTE5833	ECG5835	.....	SK7049
5518	.....	.....	.....	SK9169
5519	.....	.....	.....	SK7C7030
5520	.....	ECG6362	.....	SK6560
5521	NTE74H101	ECG74H101	.....	.....
5522	.....	.....	.....	SK9170
5523	NTE74H102	ECG74H102	.....	.....
5524	.....	.....	.....	SK9508
5525	.....	.....	.....	SK3313
5526	.....	ECG2049	.....	.....
5527	NTE5425	ECG5425	.....	.....
5528	NTE8309	ECG8309	.....	.....
5529	.....	.....	.....	SK9762
5530	.....	ECG74HC11	.....	.....
5531	NTE778SM	ECG778SM	.....	SK10061
5532	.....	.....	.....	SK7C4040
5533	NTE282	ECG282	.....	SK3054
5534	.....	ECG5608	.....	.....
5535	.....	ECG74HC86	.....	.....
5536	NTE179MP	ECG179MP	.....	.....
5537	.....	ECG184	.....	.....
5538	.....	.....	.....	SK7C4049
5539	.....	ECG4001B	.....	.....
5540	.....	.....	.....	SK7C4050
5541	.....	ECG614	.....	SK3327
5542	.....	ECG54	.....	.....
5543	.....	.....	.....	SK7C4051
5544	.....	.....	.....	SK7629
5545	.....	.....	.....	SK7C4052
5546	.....	ECG1701	.....	.....
5547	NTE1314	ECG1314	.....	.....
5548	.....	.....	.....	SK7C4053
5549	NTE5611	ECG5612	.....	SK6630
5550	NTE5612	ECG5612	.....	SK6630
5551	NTE5621	ECG5622	.....	SK3631
5552	.....	ECG2338	.....	.....
5553	NTE5649	ECG5649	.....	.....
5554	.....	ECG116	.....	SK3031A
5555	NTE5653	ECG5653	.....	.....
5556	.....	ECG373	.....	.....
5557	NTE2559	ECG2559	.....	.....
5558	.....	.....	.....	SK7C4059
5559	.....	ECG4002B	.....	.....
5560	NTE5672	ECG5672	.....	.....
5561	.....	.....	.....	SK7C4060
5562	.....	ECG1270	.....	SK9055

REPL. CODE	NTE PART NO.	ECG PART NO.	RADIO SHACK PART NO.	RCA PART NO.
5563	.....	ECG152	.....	.....
5564	.....	.....	.....	SK7714
5565	.....	ECG74LS259	.....	.....
5566	.....	ECG3007	.....	SK2007
5567	NTE5841	ECG5843	.....	SK7043
5568	NTE195A	ECG195A	.....	SK3444
5569	NTE5340	ECG5340	.....	SK9100
5570	.....	.....	.....	SK9007
5571	.....	.....	.....	SK7C4066
5572	.....	.....	.....	SK7C4067
5573	NTE4164	ECG2164	.....	SK10297
5574	NTE178MP	ECG178MP	276-1122	SK3033A
5575	.....	ECG5605	.....	.....
5576	.....	.....	.....	SK7C4075
5577	NTE5829	ECG5829	.....	.....
5578	.....	.....	.....	SK3517
5579	NTE376	ECG376	.....	.....
5580	.....	.....	.....	SK9635
5581	.....	ECG4034B	.....	SK4034B
5582	NTE5894	ECG5896	.....	SK7096
5583	NTE5898	ECG5900	.....	SK7100
5584	.....	ECG316	.....	.....
5585	NTE2764	ECG2764	.....	SK2716
5586	.....	.....	.....	SK9960
5587	NTE5946	ECG5948	.....	SK3605
5588	.....	ECG188	.....	SK3199
5589	NTE5950	ECG5952	.....	SK3585
5590	.....	.....	.....	SK1912
5591	NTE615P	.....	.....	.....
5592	.....	ECG189	.....	SK3200
5593	.....	ECG80	.....	SK9351
5594	NTE178MP	ECG178MP	276-1122	SK3087
5595	NTE184MP	ECG184MP	.....	.....
5596	.....	.....	.....	SK7C4094
5597	NTE74C922	ECG74C922	.....	SK74C922
5598	.....	ECG1169	.....	.....
5599	NTE74C923	ECG74C923	.....	SK74C923
5600	.....	ECG273	.....	SK9437
5601	NTE7073	ECG7073	.....	.....
5602	.....	.....	.....	SK3453
5603	NTE54MP	ECG54MP	.....	SK9922
5604	NTE5342	ECG5342	.....	SK9102
5605	.....	.....	.....	SK3450
5606	.....	ECG290A	.....	.....
5607	.....	ECG56018	.....	.....
5608	.....	ECG6084	.....	SK9709
5609	.....	.....	.....	SK9383
5610	.....	ECG56030	.....	.....
5611	.....	ECG125	.....	SK3043B
5612	NTE56020	ECG56020	.....	SK9808
5613	.....	ECG74HC4020	.....	.....
5614	.....	ECG477	.....	.....
5615	.....	ECG74HC4040	.....	.....
5616	.....	.....	.....	SK7744
5617	.....	ECG74HC4053	.....	.....
5618	.....	ECG1259	.....	.....
5619	.....	ECG74HC4060	.....	.....
5620	.....	ECG74HC4067	.....	.....
5621	NTE2324	ECG2324	.....	.....
5622	.....	.....	.....	SK3589A
5623	.....	ECG128&427	.....	SK3024
5624	.....	ECG398	.....	.....
5625	.....	ECG5693	.....	.....
5626	.....	.....	.....	SK6402
5627	.....	ECG5697	.....	SK3521
5628	.....	.....	.....	SK3509
5629	.....	ECG47	.....	.....
5630	.....	.....	.....	SK3521
5631	.....	.....	.....	SK6681
5632	NTE605	ECG605A	.....	SK3864
5633	.....	.....	.....	SK3319
5634	.....	ECG106	.....	.....
5635	.....	ECG568A	.....	SK3910
5636	.....	ECG126A	.....	.....
5637	.....	ECG379	.....	.....
5638	.....	.....	.....	SK40181B
5639	.....	ECG285	.....	.....
5640	.....	ECG284	.....	.....
5641	.....	.....	.....	SK3667
5642	.....	ECG3092	.....	.....
5643	NTE6412	ECG6412	.....	.....
5644	.....	.....	.....	SK3462
5645	NTE6415	ECG6415	.....	SK9846
5646	NTE6416	ECG6416	.....	SK9844
5647	.....	ECG468	.....	.....
5648	.....	ECG165	.....	SK3111
5649	NTE6417	ECG6417	.....	SK9843
5650	NTE6418	ECG6418	.....	SK4921
5651	NTE6419	ECG6419	.....	SK4920
5652	.....	ECG457	.....	SK9161
5653	.....	.....	.....	SK10V
5654	.....	ECG129P	.....	SK3715
5655	NTE1595	ECG1595	.....	.....
5656	.....	ECG2386	.....	SK10130
5657	.....	ECG5690	.....	.....
5658	.....	.....	.....	SK74LS03
5659	.....	.....	.....	SK3V3
5660	.....	.....	.....	SK3V9
5661	.....	.....	.....	SK4V7
5662	.....	ECG3104	.....	SK10179
5663	NTE5037A	ECG5037A	.....	.....
5664	.....	.....	.....	SK20V
5665	.....	.....	.....	SK22V

REPL. CODE	NTE PART NO.	ECG PART NO.	RADIO SHACK PART NO.	RCA PART NO.
5666	.....	.....	276-018	.....
5667	.....	.....	.....	SK39V
5668	.....	.....	.....	SK120V
5669	.....	.....	.....	SK150V
5670	.....	.....	.....	SK3146
5671	.....	.....	.....	SK180V
5672	.....	.....	.....	SK200V
5673	.....	.....	.....	SK5579
5674	NTE3094	ECG3094	.....	SK10404
5675	NTE1595	ECG1595	.....	SK3986
5676	.....	ECG612	.....	SK3325
5677	NTE213	ECG213	.....	SK3444
5678	NTE962	ECG962	.....	.....
5679	NTE760	ECG760	.....	SK3158
5680	.....	ECG55	.....	.....
5681	NTE15009	ECG15009	.....	SK10184
5682	.....	ECG1088	.....	.....
5683	.....	ECG5457	.....	.....
5684	.....	ECG1018	.....	SK3284
5685	.....	ECG3166	.....	.....
5686	.....	ECG85	.....	.....
5687	.....	ECG2357	.....	SK10124
5688	.....	ECG5065A	.....	.....
5689	.....	ECG830	.....	SK74151
5690	.....	.....	.....	SK4555B
5691	NTE1183	ECG1183	.....	SK3704
5692	.....	ECG4012B	.....	.....
5693	NTE60MP	ECG60MP	.....	SK9924
5694	NTE5314	ECG5314	276-1181	SK3987
5695	NTE3035A	ECG3035A	.....	.....
5696	.....	ECG5096A	.....	.....
5697	NTE1218	ECG1220	.....	SK3740
5698	.....	ECG1222	.....	SK3730
5699	NTE5069A	ECG5069A	.....	.....
5700	NTE5011T1	ECG5011T1	.....	.....
5701	.....	.....	.....	SK3433
5702	.....	ECG80C96	.....	.....
5703	.....	ECG1870	.....	SK10161
5704	NTE61MP	ECG61MP	.....	SK9926
5705	.....	ECG1274	.....	.....
5706	.....	ECG1275	.....	SK7790
5707	.....	ECG1279	.....	.....
5708	.....	ECG1280	.....	.....
5709	.....	.....	.....	SK3438
5710	.....	ECG3017	.....	.....
5711	NTE397	ECG397	.....	SK3053
5712	NTE5013T1	ECG5013T1	.....	.....
5713	.....	ECG116	.....	SK3017B
5714	NTE1315	ECG1315	.....	.....
5715	NTE1316	ECG1316	.....	.....
5716	NTE1317	ECG1317	.....	.....
5717	.....	.....	.....	SK9187
5718	NTE5092A	ECG5092A	.....	SK3998
5719	.....	.....	.....	SK7CT533
5720	.....	ECG1506	.....	.....
5721	.....	ECG1332	.....	SK10007
5722	.....	ECG1353	.....	.....
5723	.....	ECG2388	.....	.....
5724	.....	ECG1360	.....	.....
5725	.....	ECG5307	.....	.....
5726	.....	ECG2388	.....	SK9501
5727	NTE228A	ECG228A	.....	SK3103A
5728	.....	.....	.....	SK9165
5729	.....	ECG5196A	.....	SK159
5730	NTE5033A	ECG5033A	.....	.....
5731	NTE527A	ECG527A	.....	SK7333
5732	.....	.....	.....	SK3571
5733	NTE116	ECG116	.....	SK3113
5734	NTE5335	ECG5335	.....	SK7353
5735	.....	.....	.....	SK3188A
5736	NTE5338	ECG5338	.....	SK7313
5737	.....	.....	.....	SK7CT534
5738	NTE5346	ECG5346	.....	SK7352
5739	NTE5348	ECG5348	.....	SK7303
5740	.....	.....	.....	SK10191
5741	.....	ECG1219	.....	.....
5742	NTE5019T1	ECG5019T1	.....	.....
5743	.....	ECG125	.....	SK3125A
5744	NTE1V030	ECG1V030	.....	SKMV30A
5745	NTE7001	ECG7001	.....	SK10470
5746	.....	ECG1456	.....	.....
5747	NTE5072A	ECG5072A	.....	.....
5748	NTE261	ECG261	276-2068	SK9431
5749	.....	.....	.....	SK7315
5750	NTE5427	ECG5427	.....	SK3577
5751	NTE5428	ECG5428	.....	SK3578
5752	.....	.....	.....	SK9893
5753	NTE5018A	ECG5018A	.....	.....
5754	.....	.....	.....	SK74S20
5755	.....	ECG1555	.....	.....
5756	NTE74S37	.....	.....	.....
5757	.....	.....	.....	SK9502
5758	NTE85	ECG85	276-2009	SK9261
5759	.....	ECG1564	.....	.....
5760	.....	.....	.....	SK3679
5761	.....	.....	.....	SK3680
5762	.....	.....	.....	SK3681
5763	.....	.....	.....	SK74LS138
5764	.....	.....	.....	SK3682
5765	.....	.....	.....	SK9509
5766	.....	.....	.....	SK9087
5767	.....	ECG5314	.....	.....
5768	.....	.....	.....	SK9088

REPL. CODE	NTE PART NO.	ECG PART NO.	RADIO SHACK PART NO.	RCA PART NO.
5769	NTE2385	ECG2385	.....	SK10303
5770	.....	ECG1583	.....	.....
5771	.....	ECG1587	.....	.....
5772	NTE586	ECG586	.....	.....
5773	NTE74S85	.....	.....	.....
5774	.....	ECG1614	.....	.....
5775	.....	ECG1617	.....	.....
5776	.....	ECG1623	.....	.....
5777	NTE1761	ECG1761	.....	SK10044
5778	NTE281MCP	ECG281MCP	.....	.....
5779	.....	ECG1638	.....	.....
5780	.....	ECG1640	.....	SK10238
5781	.....	ECG1642	.....	.....
5782	.....	ECG1644	.....	.....
5783	.....	ECG1645	.....	.....
5784	.....	ECG1646	.....	.....
5785	.....	ECG1652	.....	SK10252
5786	.....	ECG1653	.....	SK7632
5787	NTE142A	ECG142A	.....	.....
5788	.....	ECG859	.....	.....
5789	.....	ECG4017B	.....	.....
5790	.....	ECG1686	.....	SK10247
5791	.....	ECG74H86	.....	.....
5792	.....	ECG54	.....	.....
5793	.....	ECG1706	.....	SK9982
5794	.....	.....	.....	SK74LS153
5795	NTE5075A	ECG5075A	.....	.....
5796	NTE6241	ECG6241	.....	SK10512
5797	NTE5016A	ECG5016A	.....	.....
5798	NTE5665	.....	.....	.....
5799	.....	ECG136A	.....	.....
5800	.....	ECG1756	.....	.....
5801	.....	.....	.....	SK74LS157
5802	.....	.....	.....	SK3214
5803	NTE587	ECG587	.....	.....
5804	NTE1763	ECG1763	.....	SK10045
5805	NTE1765	ECG1765	.....	.....
5806	.....	ECG1766	.....	SK10494
5807	.....	ECG5322	.....	SK3679
5808	NTE580	ECG580	.....	SK3036
5809	.....	ECG1784	.....	SK9227
5810	NTE1785	ECG1785	.....	SK7635
5811	NTE5013A	ECG5013A	.....	SK6A2
5812	NTE5704	ECG5704	.....	SK10438
5813	NTE5705	ECG5705	.....	SK10415
5814	.....	ECG1802	.....	.....
5815	.....	ECG1806	.....	SK9888
5816	NTE568A	.....	.....	.....
5817	.....	ECG1821	.....	.....
5818	NTE1832	ECG1832	.....	SK10142
5819	NTE1833	ECG1833	.....	SK9712
5820	NTE1851	ECG1851	.....	.....
5821	.....	ECG1853	.....	SK10152
5822	NTE230	ECG230	.....	SK3296
5823	NTE1857	ECG1857	.....	SK10156
5824	NTE1858	ECG1858	.....	.....
5825	.....	ECG1865	.....	SK10509
5826	NTE1866	ECG1866	.....	SK10496
5827	.....	ECG1869	.....	SK10500
5828	.....	ECG1878	.....	SK9757
5829	.....	ECG1887	.....	.....
5830	NTE1888	ECG1888	.....	.....
5831	.....	ECG1889	.....	SK10501
5832	NTE145A	ECG145A	.....	SK15V
5833	NTE1898	ECG1898	.....	SK10507
5834	NTE1899	ECG1899	.....	SK10508
5835	.....	ECG32	.....	.....
5836	NTE7439	ECG7439	.....	SK7439
5837	.....	ECG1941	.....	SK10058
5838	NTE1942	ECG1942	.....	.....
5839	.....	ECG1960	.....	.....
5840	.....	ECG1961	.....	.....
5841	.....	ECG1962	.....	.....
5842	.....	ECG1963	.....	.....
5843	.....	ECG1964	.....	.....
5844	.....	ECG4020B	.....	.....
5845	.....	ECG1965	.....	.....
5846	.....	ECG1966	.....	.....
5847	.....	ECG1967	.....	.....
5848	.....	ECG1968	.....	.....
5849	.....	ECG1970	.....	.....
5850	.....	ECG1971	.....	.....
5851	.....	ECG1972	.....	.....
5852	.....	ECG1973	.....	.....
5853	.....	ECG1974	.....	.....
5854	.....	ECG1975	.....	.....
5855	.....	ECG1976	.....	.....
5856	.....	ECG1977	.....	.....
5857	.....	ECG75497	.....	.....
5858	.....	ECG2200	.....	.....
5859	.....	ECG2033	.....	.....
5860	.....	ECG2041	.....	.....
5861	.....	ECG397	.....	.....
5862	.....	ECG2045	.....	.....
5863	.....	ECG2046	.....	.....
5864	.....	ECG2052	.....	SK10281
5865	NTE2057	ECG2057	.....	SK10190
5866	.....	ECG2063	.....	SK10283
5867	.....	ECG2064	.....	SK10284
5868	.....	ECG2065	.....	SK10285
5869	NTE194	ECG194	.....	SK3433
5870	.....	ECG2090	.....	.....
5871	.....	ECG1616	.....	.....

REPL. CODE	NTE PART NO.	ECG PART NO.	RADIO SHACK PART NO.	RCA PART NO.
5872	NTE5062A	.....	.....	.....
5873	.....	ECG1413	.....	.....
5874	NTE1656	ECG1656	.....	SK7633
5875	.....	ECG2203	.....	.....
5876	.....	.....	.....	SK74LS193
5877	.....	ECG857M	.....	.....
5878	NTE56030	ECG56030	.....	SK10409
5879	.....	ECG5084A	.....	.....
5880	.....	ECG147A	.....	.....
5881	.....	.....	.....	SK10234
5882	.....	ECG4022B	.....	.....
5883	NTE192&193	ECG192&193	.....	SK9142&9143
5884	.....	ECG2201	.....	.....
5885	.....	ECG2204	.....	.....
5886	.....	ECG2205	.....	.....
5887	.....	ECG2206	.....	.....
5888	.....	ECG2207	.....	SK10298
5889	.....	.....	.....	SK7C7266
5890	.....	.....	.....	SK7816
5891	.....	ECG4011B	.....	SK4011B
5892	.....	ECG4023B	.....	.....
5893	NTE329	ECG329	.....	SK3847
5894	.....	ECG5371	.....	.....
5895	.....	ECG5542	.....	.....
5896	.....	ECG2354	.....	.....
5897	.....	ECG4024B	.....	.....
5898	.....	ECG1048	.....	.....
5899	NTE2384	ECG2384	.....	SK10129
5900	.....	ECG2400	.....	.....
5901	NTE374	ECG374	.....	.....
5902	.....	.....	.....	SK4520B
5903	NTE5013A N5	ECG5013A N5	276-561 N5	SK6A2 N5
5904	.....	ECG135A	.....	SK5V1
5905	.....	.....	.....	SK7330
5906	.....	ECG149A	.....	SK62V
5907	.....	ECG1672	.....	.....
5908	.....	ECG4025B	.....	.....
5909	NTE6400	.....	.....	SK9122
5910	NTE2362	ECG2362	.....	SK3114A
5911	NTE6408	ECG6408	.....	SK7901
5912	.....	ECG2021	.....	SK9020
5913	.....	.....	.....	SK9504
5914	.....	.....	.....	SK9020
5915	.....	ECG581	.....	.....
5916	.....	ECG74HCT00	.....	SK7CT00
5917	.....	.....	.....	SK7CT03
5918	.....	.....	.....	SK7CT10
5919	.....	.....	.....	SK7CT11
5920	.....	.....	.....	SK9137
5921	.....	.....	.....	SK7CT20
5922	.....	.....	.....	SK7CT21
5923	.....	.....	.....	SK7CT27
5924	.....	.....	.....	SK3918
5925	.....	.....	.....	SK7CT30
5926	.....	.....	.....	SK3650
5927	.....	.....	.....	SK3651
5928	.....	.....	.....	SK7CT42
5929	.....	.....	.....	SK9388
5930	.....	.....	.....	SK74LS240
5931	.....	ECG2022	.....	SK9019
5932	NTE199	ECG199	276-2009	SK3245
5933	.....	.....	.....	SK7CT73
5934	.....	.....	.....	SK7CT74
5935	.....	.....	.....	SK7CT75
5936	.....	.....	.....	SK7CT85
5937	.....	.....	.....	SK9019
5938	.....	.....	.....	SK7CT86
5939	NTE8123	ECG8123	.....	.....
5940	.....	.....	.....	SK7CT93
5941	.....	.....	.....	SK74LS244
5942	NTE5653	ECG5653	.....	SK3519
5943	.....	ECG5875	.....	.....
5944	.....	ECG5879	.....	.....
5945	.....	ECG5883	.....	.....
5946	.....	ECG5887	.....	.....
5947	.....	ECG5891	.....	.....
5948	.....	.....	.....	SK9503
5949	.....	.....	.....	SK158
5950	.....	.....	.....	SK160
5951	NTE8223	ECG8223	.....	.....
5952	NTE6240	ECG6240	.....	.....
5953	.....	.....	.....	SK9967
5954	NTE290A	ECG290A	.....	SK3114A
5955	.....	.....	.....	SK7C4316
5956	NTE76MP	ECG76MP	.....	SK9601
5957	NTE8316	ECG8316	.....	.....
5958	NTE8321	ECG8321	.....	.....
5959	NTE8328	ECG8328	.....	.....
5960	NTE3017	ECG3017	.....	.....
5961	.....	.....	.....	SK7626
5962	.....	ECG311	.....	.....
5963	NTE8368	ECG8368	.....	SK4368
5964	NTE8370	ECG8370	.....	.....
5965	NTE8374	ECG8374	.....	.....
5966	.....	ECG237	.....	.....
5967	NTE6809	ECG6809	.....	SK9796
5968	.....	ECG469	.....	.....
5969	.....	ECG4021B	.....	SK4021B
5970	NTE74S151	ECG74S151	.....	SK74154
5971	.....	.....	.....	SK2A7
5972	.....	ECG4030B	.....	.....
5973	.....	ECG5638	.....	.....
5974	NTE5538	ECG5538	.....	SK10406
5975	NTE8520	ECG8520	.....	.....
5976	NTE8542	ECG8542	.....	.....
5977	NTE8546	ECG8546	.....	.....
5978	NTE8556	ECG8556	.....	.....
5979	.....	ECG46	.....	.....
5980	.....	ECG2064	.....	.....
5981	.....	.....	.....	SK7C4351
5982	.....	.....	.....	SK7C4352
5983	.....	.....	.....	SK3052
5984	.....	.....	.....	SK7C4353
5985	.....	ECG6208	.....	.....
5986	.....	ECG703	.....	.....
5987	.....	ECG580	.....	SK9000
5988	NTE74LS124	.....	.....	.....
5989	.....	ECG5020A	.....	.....
5990	.....	ECG3200	.....	.....
5991	.....	ECG291	.....	.....
5992	NTE6247	ECG6247	.....	SK10513
5993	.....	ECG5080A	.....	.....
5994	NTE7214	ECG7214	.....	.....
5995	NTE74LS260	ECG74LS260	.....	SK74LS260
5996	.....	ECG5669A	.....	.....
5997	NTE74LS283	ECG74LS283	.....	SK7CT283
5998	NTE74LS295	ECG74LS295A	.....	.....
5999	.....	.....	.....	SK4097B
6000	NTE74LS324	.....	.....	.....
6001	.....	.....	.....	SK3186
6002	NTE74LS327	.....	.....	.....
6003	NTE330	ECG330W	.....	SK3012
6004	.....	.....	.....	SK7440
6005	NTE74LS363	ECG74LS363	.....	.....
6006	NTE74LS364	ECG74LS364	.....	SK74LS364
6007	.....	ECG40097B	.....	.....
6008	NTE74LS396	.....	.....	.....
6009	NTE191	ECG191	.....	SK3865A
6010	.....	ECG1058	.....	.....
6011	NTE74LS445	.....	.....	.....
6012	.....	.....	.....	SK7V5
6013	.....	ECG156	.....	.....
6014	NTE5023A	ECG5023A	276-563	SK24A
6015	.....	.....	.....	SK40208B
6016	.....	ECG937	.....	.....
6017	.....	ECG937M	.....	.....
6018	.....	.....	.....	SK3692
6019	.....	ECG135A	.....	.....
6020	.....	.....	.....	SK3135
6021	NTE966	ECG966	.....	SK3592
6022	NTE552	ECG552	276-1104	SK3313
6023	.....	ECG270	.....	.....
6024	.....	ECG271	.....	.....
6025	.....	ECG74LS138	.....	.....
6026	.....	ECG954	.....	.....
6027	.....	.....	.....	SK9000
6028	.....	ECG7004	.....	SK10473
6029	NTE1970	ECG1970	.....	.....
6030	.....	.....	.....	SK9213
6031	.....	.....	.....	SK9369
6032	.....	.....	.....	SK9368
6033	.....	.....	.....	SK74LS367
6034	NTE87MP	ECG87MP	.....	SK9928
6035	.....	.....	.....	SK3275
6036	.....	ECG5071A	.....	.....
6037	.....	.....	.....	SK3A3
6038	NTE5021T1	ECG5021T1	.....	.....
6039	NTE937M	ECG937M	.....	SK9147
6040	NTE1747	ECG1747	.....	.....
6041	NTE88MP	ECG88MP	.....	SK9930
6042	.....	.....	.....	SK3A6
6043	.....	.....	.....	SK9343
6044	.....	ECG5065A	.....	SK3A0
6045	.....	ECG138A	.....	.....
6046	.....	ECG960	.....	SK3591
6047	NTE56	ECG56	.....	.....
6048	.....	ECG168	.....	.....
6049	.....	ECG5072A	.....	SK8V2
6050	NTE195A	ECG195A	.....	SK3049
6051	.....	.....	.....	SK3044
6052	.....	.....	.....	SK3104A
6053	.....	.....	.....	SK3894
6054	.....	.....	.....	SK10177
6055	.....	ECG74HC40105	.....	.....
6056	.....	ECG5322	.....	SK3680
6057	.....	ECG5324	.....	SK3681
6058	.....	.....	.....	SK3197
6059	.....	.....	.....	SK3847
6060	.....	ECG5326	.....	SK3682
6061	.....	ECG5327	.....	SK9087
6062	.....	ECG169	.....	.....
6063	.....	.....	.....	SK7674
6064	NTE8214	ECG8214	.....	.....
6065	.....	.....	.....	SK3596
6066	.....	.....	.....	SK9432
6067	NTE8219	ECG8219	.....	.....
6068	.....	.....	.....	SK9236
6069	.....	.....	.....	SK7407
6070	.....	ECG6090	.....	SK4911
6071	.....	ECG964	.....	SK3630
6072	.....	ECG2062	.....	.....
6073	.....	.....	.....	SK9769
6074	.....	.....	.....	SK7473
6075	.....	.....	.....	SK9097
6076	.....	ECG6156	.....	.....
6077	.....	ECG170	.....	.....
6078	.....	ECG4506	.....	.....
6079	.....	ECG1316	.....	.....
6080	NTE832SM	ECG832SM	.....	SK10062
6081	.....	ECG75451B	.....	.....
6082	.....	ECG5569	.....	.....
6083	NTE308&310	ECG308&310	.....	SK3855&3856
6084	.....	.....	.....	SK7C40105
6085	.....	ECG399	.....	.....
6086	.....	ECG5328	.....	.....
6087	.....	.....	.....	SK7C4510
6088	.....	.....	.....	SK7C4511
6089	NTE2302	ECG2303	.....	SK9422
6090	.....	.....	.....	SK7CT138
6091	NTE8613	ECG8613	.....	.....
6092	.....	.....	.....	SK10033
6093	.....	.....	.....	SK7C4514
6094	.....	.....	.....	SK7C4515
6095	.....	ECG74HCT138	.....	SK7CT158
6096	.....	ECG1409C	.....	.....
6097	.....	.....	.....	SK7C4516
6098	NTE1854M	ECG1854M	.....	SK10154
6099	.....	ECG74HCT161	.....	.....
6100	.....	.....	.....	SK7CT195
6101	.....	ECG74HCT163	.....	.....
6102	.....	.....	.....	SK7C4518
6103	.....	ECG5482	.....	SK3941
6104	.....	.....	.....	SK9959
6105	.....	ECG74HCT174	.....	.....
6106	.....	.....	.....	SK7C4520
6107	.....	ECG145A	.....	.....
6108	.....	ECG74HCT244	.....	SK7CT244
6109	NTE4000	ECG4000	.....	SK4000B
6110	.....	.....	.....	SK7675
6111	.....	ECG74HCT240	.....	.....
6112	.....	ECG5483	.....	SK3942
6113	.....	ECG74HCT273	.....	.....
6114	.....	.....	.....	SK3998
6115	.....	.....	.....	SK3497
6116	.....	ECG2410	.....	SK10101
6117	NTE717	ECG717	.....	SK3827
6118	.....	ECG4901	.....	.....
6119	.....	ECG597	.....	.....
6120	.....	.....	.....	SK7CT182
6121	.....	.....	.....	SK7706
6122	.....	ECG1854D	.....	SK10153
6123	.....	ECG74HCT373	.....	SK7CT373
6124	NTE859	ECG859	.....	SK4856
6125	.....	.....	.....	SK4A7
6126	.....	.....	.....	SK7C4538
6127	.....	ECG5485	.....	SK3943
6128	NTE15005	ECG15005	.....	SK10180
6129	NTE15021	ECG15021	.....	SK10187
6130	NTE15022	ECG15022	.....	SK10188
6131	.....	ECG74HCT374	.....	.....
6132	NTE15023	ECG15023	.....	SK10189
6133	.....	.....	.....	SK10192
6134	.....	ECG128P	.....	.....
6135	.....	.....	.....	SK7693
6136	.....	ECG5203A	.....	SK172
6137	.....	.....	.....	SK7C4543
6138	.....	ECG5609	.....	.....
6139	NTE56019	ECG56019	.....	SK10269
6140	NTE130	ECG130	276-2041	SK3111
6141	NTE285MCP	ECG285MCP	.....	SK10310
6142	.....	.....	.....	SK7697
6143	.....	.....	.....	SK7694
6144	.....	ECG1409	.....	.....
6145	.....	ECG5424	.....	SK9293
6146	.....	ECG2333	.....	.....
6147	.....	ECG5487	.....	SK3944
6148	.....	ECG74HCT573	.....	.....
6149	.....	ECG74HCT574	.....	.....
6150	.....	.....	.....	SK9046
6151	NTE56033	ECG56033	.....	SK10411
6152	NTE56028	ECG56028	.....	SK10412
6153	.....	ECG5031A	.....	.....
6154	.....	.....	.....	SK10443
6155	.....	.....	.....	SK10457
6156	.....	.....	.....	SK10458
6157	.....	ECG5668A	.....	SK10460
6158	.....	.....	.....	SK3602
6159	NTE7000	ECG7000	.....	SK10469
6160	NTE7006	ECG7006	.....	SK10475
6161	.....	ECG7011	.....	SK10480
6162	NTE7014	ECG7014	.....	SK10483
6163	NTE7016	ECG7016	.....	SK10485
6164	.....	ECG7017	.....	SK10486
6165	NTE7020	ECG7020	.....	SK10489
6166	.....	ECG2004	.....	.....
6167	.....	.....	.....	SK6749
6168	NTE3098	ECG3098	.....	.....
6169	.....	ECG5421	276-1067	SK3570
6170	.....	.....	.....	SK3440
6171	.....	ECG5432	276-1067	.....
6172	.....	ECG5433	276-1067	.....
6173	.....	ECG4069	.....	SK4069UB
6174	.....	ECG129P	.....	.....
6175	NTE5014A	ECG5014A	.....	.....
6176	.....	.....	.....	SK7642
6177	.....	.....	.....	SK9181
6178	.....	.....	.....	SK7CT574
6179	.....	.....	.....	SK74LS04
6180	.....	.....	.....	SK74LS05

REPL. CODE	NTE PART NO.	ECG PART NO.	RADIO SHACK PART NO.	RCA PART NO.
6181	.....	ECG2128	.....	.....
6182	.....	.....	.....	SK74LS08
6183	.....	ECG4011B	.....	.....
6184	.....	ECG329	.....	.....
6185	NTE941S	ECG941S	.....	.....
6186	NTE55	ECG55	.....	SK3274
6187	.....	ECG1870	.....	.....
6188	.....	.....	.....	SK3447
6189	.....	ECG556A	.....	.....
6190	.....	.....	.....	SK74LS74A
6191	.....	ECG177	.....	SK3319
6192	.....	.....	.....	SK3501
6193	NTE382	ECG293	.....	SK3849
6194	.....	ECG568A	.....	.....
6195	.....	.....	.....	SK9314
6196	.....	.....	.....	SK5101
6197	.....	.....	.....	SK3931
6198	.....	ECG3881	.....	SK2881
6199	NTE287	ECG287	.....	SK3244
6200	NTE238	ECG389	276-2055	SK3710
6201	.....	.....	.....	SK74S04
6202	.....	ECG1006	.....	.....
6203	.....	.....	.....	SK3443
6204	.....	.....	.....	SK3724
6205	.....	ECG1580	.....	.....
6206	.....	.....	.....	SK74S11
6207	.....	.....	.....	SK1924
6208	.....	.....	.....	SK5A6
6209	.....	ECG4050B	.....	.....
6210	.....	ECG74C02	.....	.....
6211	.....	ECG74C04	.....	.....
6212	.....	ECG74C08	.....	SK7C08
6213	.....	ECG152	.....	SK3893
6214	.....	ECG74C10	.....	.....
6215	.....	ECG950	.....	SK9169
6216	.....	.....	.....	SK2019
6217	.....	ECG74C20	.....	.....
6218	.....	ECG74C42	.....	SK7C42
6219	.....	ECG74C74	.....	SK7C74
6220	.....	.....	.....	SK7354
6221	.....	ECG74C85	.....	SK7C85
6222	.....	ECG4070B	.....	.....
6223	.....	ECG1462	.....	.....
6224	.....	.....	.....	SK9902
6225	.....	.....	.....	SK5X6
6226	.....	.....	.....	SK9426
6227	.....	ECG1862	.....	.....
6228	.....	ECG5387	.....	.....
6229	.....	.....	.....	SK2880
6230	.....	ECG5697	.....	SK3522
6231	.....	.....	.....	SK7210
6232	.....	ECG5111A	.....	SK3X3
6233	.....	.....	.....	SK3294
6234	.....	.....	.....	SK3552
6235	.....	ECG000	.....	.....
6236	NTE287&288	ECG287&288	.....	SK3433&3434
6237	.....	.....	.....	SK3853
6238	.....	ECG75453B	.....	.....
6239	.....	.....	.....	SK9132
6240	.....	.....	.....	SK3490
6241	NTE382	ECG382	.....	.....
6242	NTE293	.....	.....	.....
6243	.....	.....	.....	SK10A
6244	.....	.....	.....	SK3503
6245	.....	.....	.....	SK3502
6246	.....	ECG975	.....	.....
6247	NTE1110R	.....	.....	.....
6248	.....	ECG834	.....	SK3569
6249	.....	ECG1153	.....	.....
6250	.....	ECG525	.....	SK3925
6251	.....	ECG956	.....	.....
6252	.....	.....	.....	SK4560B
6253	.....	.....	.....	SK2881
6254	.....	.....	.....	SK4019B
6255	NTE5351	ECG5351	.....	SK3857
6256	.....	.....	.....	SK3219
6257	NTE1535	ECG1535	.....	SK9725
6258	.....	.....	.....	SK6A2
6259	.....	ECG1787	.....	SK9766
6260	.....	.....	.....	SK6A0
6261	.....	ECG966	.....	SK3592
6262	.....	ECG7002	.....	SK10471
6263	.....	ECG7021	.....	.....
6264	.....	ECG7022	.....	.....
6265	.....	ECG5804	.....	.....
6266	.....	ECG7023	.....	.....
6267	.....	ECG7024	.....	.....
6268	.....	ECG7025	.....	.....
6269	.....	ECG7026	.....	.....
6270	.....	ECG7027	.....	.....
6271	.....	ECG7028	.....	.....
6272	.....	ECG7029	.....	.....
6273	.....	ECG7030	.....	.....
6274	.....	ECG7031	.....	.....
6275	.....	ECG7032	.....	.....
6276	.....	ECG7033	.....	.....
6277	.....	ECG7034	.....	.....
6278	.....	ECG7035	.....	.....
6279	.....	ECG7036	.....	.....
6280	.....	ECG7037	.....	.....
6281	.....	ECG7038	.....	.....
6282	.....	ECG7039	.....	.....
6283	.....	ECG7040	.....	.....

REPL. CODE	NTE PART NO.	ECG PART NO.	RADIO SHACK PART NO.	RCA PART NO.
6284	.....	ECG7041	.....	.....
6285	.....	ECG7042	.....	.....
6286	.....	ECG7043	.....	.....
6287	.....	ECG7044	.....	.....
6288	.....	ECG7045	.....	.....
6289	.....	ECG7046	.....	.....
6290	.....	ECG7047	.....	.....
6291	.....	ECG7048	.....	.....
6292	.....	ECG7049	.....	.....
6293	.....	ECG7050	.....	.....
6294	.....	ECG7051	.....	.....
6295	.....	.....	.....	SK9768
6296	.....	ECG7052	.....	.....
6297	.....	ECG7053	.....	.....
6298	.....	ECG7054	.....	.....
6299	.....	ECG7055	.....	.....
6300	.....	ECG7056	.....	.....
6301	.....	ECG7057	.....	.....
6302	.....	ECG7058	.....	.....
6303	.....	ECG7059	.....	.....
6304	.....	ECG7060	.....	.....
6305	.....	ECG7061	.....	.....
6306	.....	ECG7062	.....	.....
6307	.....	ECG7063	.....	.....
6308	.....	ECG7064	.....	.....
6309	.....	ECG7065	.....	.....
6310	.....	ECG7066	.....	.....
6311	.....	ECG7067	.....	.....
6312	.....	ECG7068	.....	.....
6313	.....	ECG7069	.....	.....
6314	.....	ECG7070	.....	.....
6315	.....	ECG7071	.....	.....
6316	.....	ECG7072	.....	.....
6317	.....	ECG7073	.....	.....
6318	.....	ECG7074	.....	.....
6319	.....	.....	.....	SK7696
6320	.....	ECG7105	.....	.....
6321	.....	.....	.....	SK7695
6322	NTE928M	ECG928M	.....	.....
6323	NTE174	.....	.....	SK3110
6324	.....	ECG1873	.....	.....
6325	.....	ECG5540	.....	.....
6326	.....	ECG123A	.....	SK3444
6327	.....	ECG5174AK	.....	SK115
6328	.....	.....	.....	SK9735
6329	.....	.....	.....	SK4015B
6330	.....	.....	.....	SK6748
6331	.....	.....	.....	SK10073
6332	NTE74H08	ECG74H08	.....	SK7408
6333	.....	.....	.....	SK7C109
6334	.....	.....	.....	SK3691
6335	.....	.....	.....	SK9724
6336	.....	ECG1807	.....	SK9886
6337	NTE7401	ECG7401	.....	SK7401
6338	.....	ECG262	.....	.....
6339	.....	ECG1529	.....	.....
6340	NTE56014	.....	.....	SK3993
6341	.....	.....	.....	SK9297
6342	.....	ECG21	.....	SK9720
6343	.....	.....	.....	SK9505
6344	.....	.....	.....	SK7C125
6345	.....	.....	.....	SK3160
6346	.....	.....	.....	SK7C126
6347	.....	ECG74HC4020	.....	SK7C4020
6348	NTE128&401	ECG128&401	.....	.....
6349	.....	.....	.....	SK7C132
6350	.....	ECG1720	.....	.....
6351	NTE1160	ECG1160	.....	SK3474
6352	NTE1970	ECG1970	276-1771	SK3592
6353	.....	.....	.....	SK7C138
6354	NTE552	ECG515	.....	SK9000
6355	NTE519 N7	ECG519 N7	276-1122 N7	SK3100 N7
6356	NTE519 N9	ECG519 N9	276-1122 N9	SK3100 N9
6357	.....	ECG74HC4040	.....	SK7C4040
6358	.....	.....	.....	SK7C151
6359	.....	ECG912	.....	SK3543
6360	.....	.....	.....	SK7C153
6361	.....	.....	.....	SK7C154
6362	.....	ECG5318	.....	.....
6363	.....	.....	.....	SK4013B
6364	.....	.....	.....	SK9206
6365	NTE857SM	ECG857SM	.....	SK9799
6366	NTE5022A	ECG5022A	.....	SK5022A
6367	.....	.....	.....	SK4050B
6368	.....	.....	.....	SK7C164
6369	NTE955M	ECG955M	.....	SK3564
6370	.....	.....	.....	SK7C165
6371	.....	ECG177	.....	.....
6372	.....	ECG74HC4060	.....	SK7C4060
6373	.....	.....	.....	SK3993
6374	.....	.....	.....	SK9156
6375	.....	ECG2313	.....	.....
6376	.....	.....	.....	SK9155
6377	.....	.....	.....	SK7A5
6378	.....	.....	.....	SK9424
6379	.....	ECG75491B	.....	.....
6380	.....	.....	.....	SK7C173
6381	.....	ECG917	.....	SK3544
6382	.....	ECG75492P	.....	.....
6383	.....	.....	.....	SK7C174
6384	.....	.....	.....	SK7C175
6385	.....	ECG66	.....	.....
6386	.....	.....	.....	SK7700

REPL. CODE	NTE PART NO.	ECG PART NO.	RADIO SHACK PART NO.	RCA PART NO.
6387	.....	ECG1214	.....	SK3736
6388	.....	ECG5332	.....	.....
6389	.....	ECG1692	.....	.....
6390	.....	.....	.....	SK3578
6391	.....	ECG8092	.....	.....
6392	.....	.....	.....	SK9201
6393	.....	.....	.....	SK3312
6394	.....	ECG1782	.....	.....
6395	.....	ECG4988	.....	.....
6396	NTE282	ECG282	.....	SK3192
6397	NTE5018A N6	ECG5018A N6	276-562 N6	SK9A1 N6
6398	.....	.....	.....	SK7671
6399	.....	.....	.....	SK1806
6400	.....	ECG159	.....	SK3466
6401	.....	ECG1397	.....	SK7603
6402	.....	.....	.....	SK1826
6403	.....	ECG2716	.....	SK2716
6404	NTE282	ECG282	.....	SK3513
6405	.....	.....	.....	SK1878
6406	.....	.....	.....	SK1879
6407	.....	.....	.....	SK1881
6408	.....	.....	.....	SK1882
6409	.....	.....	.....	SK1883
6410	.....	ECG4072B	.....	SK4072B
6411	NTE180MCP	ECG180MCP	.....	SK9134&9136
6412	.....	.....	.....	SK24V
6413	.....	ECG5463	.....	.....
6414	.....	.....	.....	SK4075B
6415	.....	.....	.....	SK40257B
6416	.....	.....	.....	SK9759
6417	.....	ECG74290	.....	.....
6418	.....	.....	.....	SK9758
6419	NTE1773	ECG1773	.....	SK7952
6420	.....	.....	.....	SK9849
6421	.....	ECG2318	.....	SK9986
6422	.....	.....	.....	SK2131
6423	.....	ECG8552	.....	.....
6424	.....	.....	.....	SK2132
6425	.....	ECG8553	.....	.....
6426	.....	ECG8554	.....	.....
6427	.....	ECG8555	.....	.....
6428	.....	.....	.....	SK9028
6429	NTE374	.....	.....	SK374
6430	NTE7068	ECG7068	.....	.....
6431	.....	ECG5204AK	.....	SK175
6432	NTE5075A	ECG5075A	.....	SK16V2
6433	NTE966	ECG966	276-1771	.....
6434	.....	ECG230	.....	.....
6435	.....	ECG5620	.....	.....
6436	.....	.....	.....	SK9320
6437	NTE2732A	.....	.....	.....
6438	.....	.....	.....	SK3985
6439	.....	.....	.....	SK5189
6440	NTE289A	ECG289A	.....	.....
6441	.....	ECG3047	.....	.....
6442	.....	.....	.....	SK3103A
6443	NTE177 N5	ECG177 N5	276-1122 N5	SK9091 N5
6444	.....	ECG5466	.....	.....
6445	.....	.....	.....	SK9394
6446	.....	ECG534	.....	.....
6447	.....	ECG522	.....	SK3303
6448	.....	.....	.....	SK3668
6449	.....	.....	.....	SK3018
6450	.....	.....	.....	SK9031
6451	NTE506	ECG506	.....	SK5003
6452	.....	.....	.....	SK3032A
6453	.....	ECG5468	.....	.....
6454	.....	ECG327	.....	.....
6455	.....	ECG327	.....	SK9040
6456	NTE383	ECG383	.....	SK9464
6457	NTE125 N3	ECG125 N3	276-1114 N3	SK3081 N3
6458	.....	.....	.....	SK7CT22106
6459	.....	ECG7016	.....	.....
6460	.....	.....	.....	SK3090
6461	.....	.....	.....	SK7C22106
6462	NTE945	ECG945	.....	SK3166
6463	.....	.....	.....	SK9379
6464	.....	.....	.....	SK7C299
6465	.....	ECG6230	.....	.....
6466	.....	ECG489	.....	SK3201
6467	.....	ECG9301	.....	.....
6468	.....	ECG9302	.....	.....
6469	.....	ECG9303	.....	.....
6470	.....	ECG9304	.....	.....
6471	.....	ECG9306	.....	.....
6472	.....	ECG9307	.....	.....
6473	.....	.....	.....	SK2035A
6474	.....	ECG9311	.....	.....
6475	.....	ECG9312	.....	.....
6476	.....	ECG9600	.....	.....
6477	.....	.....	.....	SK74C245
6478	.....	ECG9321	.....	.....
6479	.....	ECG9322	.....	.....
6480	.....	ECG9323	.....	.....
6481	.....	ECG9324	.....	.....
6482	.....	ECG9325	.....	.....
6483	.....	ECG9326	.....	.....
6484	.....	ECG9331	.....	.....
6485	.....	ECG9332	.....	.....
6486	.....	ECG9333	.....	.....
6487	.....	ECG9334	.....	.....
6488	.....	ECG9335	.....	.....
6489	.....	ECG9342	.....	.....

REPL. CODE	NTE PART NO.	ECG PART NO.	RADIO SHACK PART NO.	RCA PART NO.
6490	.....	ECG9343	.....	.....
6491	.....	.....	.....	SK2716
6492	.....	ECG9347	.....	.....
6493	.....	ECG6362	.....	.....
6494	.....	.....	.....	SK74LS90
6495	.....	ECG9361	.....	.....
6496	.....	ECG9362	.....	.....
6497	.....	ECG9363	.....	.....
6498	.....	.....	.....	SK9947
6499	.....	ECG9367	.....	.....
6500	.....	ECG9368	.....	.....
6501	.....	ECG9375	.....	.....
6502	.....	ECG9380	.....	.....
6503	.....	ECG9381	.....	.....
6504	.....	ECG9382	.....	.....
6505	.....	ECG9390	.....	.....
6506	.....	ECG9391	.....	.....
6507	.....	ECG9392	.....	.....
6508	.....	ECG9393	.....	.....
6509	.....	ECG9394	.....	.....
6510	.....	ECG9400	.....	.....
6511	.....	ECG126A	.....	SK3008
6512	NTE121	ECG121	.....	.....
6513	NTE121MP	ECG121MP	.....	SK3015
6514	.....	.....	.....	SK40117B
6515	.....	.....	.....	SK3026
6516	.....	.....	.....	SK3028
6517	.....	ECG386	.....	SK9140
6518	NTE181MP	ECG181MP	.....	SK3037
6519	NTE123A	ECG123A	276-2030	SK3024
6520	NTE123	ECG123A	276-2009	SK3024
6521	.....	ECG131	.....	SK3052
6522	NTE941S	ECG941S	.....	SK9323
6523	NTE760	ECG760	.....	SK3077
6524	.....	ECG1516	.....	.....
6525	.....	ECG162	.....	SK3079
6526	NTE197	ECG197	.....	SK3085
6527	NTE2362	ECG2362	.....	.....
6528	.....	ECG9602	.....	.....
6529	NTE5304	ECG5304	.....	SK3105
6530	.....	ECG176	.....	SK3123
6531	.....	ECG1125	.....	SK3827
6532	.....	ECG233	.....	SK3132
6533	.....	ECG146A	.....	SK26V
6534	.....	ECG5084A	.....	SK29V
6535	NTE291	ECG291	.....	.....
6536	NTE388	ECG388	.....	SK3559
6537	.....	ECG128	.....	SK3265
6538	.....	ECG3097	.....	.....
6539	NTE129	ECG129	.....	SK3513
6540	NTE283	ECG283	.....	SK3559
6541	NTE284	ECG284	.....	SK3270
6542	NTE152	ECG152	276-2020	SK3626
6543	NTE291	ECG241	.....	SK3440
6544	.....	ECG5029A	.....	.....
6545	NTE74116	ECG8308	.....	.....
6546	.....	.....	.....	SK9222
6547	NTE7060	ECG7060	.....	.....
6548	NTE6041	ECG6041	.....	SK7245
6549	.....	.....	.....	SK3322
6550	.....	ECG53	.....	.....
6551	.....	ECG6206	.....	.....
6552	.....	ECG107	.....	SK3356
6553	.....	ECG1934	.....	SK7739
6554	.....	ECG264	.....	.....
6555	NTE604	ECG109	.....	SK3087
6556	.....	.....	.....	SK3374
6557	NTE552	ECG552	276-1104	SK9000
6558	NTE1028	ECG1028	.....	SK3435
6559	.....	ECG489	.....	.....
6560	.....	.....	.....	SK40174B
6561	.....	.....	.....	SK3481
6562	.....	.....	.....	SK10173
6563	NTE178MP	ECG178MP	276-1122	SK3709
6564	.....	.....	.....	SK9A1
6565	.....	ECG5607	.....	SK3507
6566	NTE605	ECG605A	.....	SK7951
6567	.....	.....	.....	SK9218
6568	.....	ECG129	.....	SK3513
6569	.....	.....	.....	SK3325
6570	.....	ECG286	.....	SK3538
6571	.....	.....	.....	SK3557
6572	.....	ECG9982	.....	.....
6573	.....	ECG175	.....	SK3562
6574	.....	ECG284	.....	SK9134
6575	.....	ECG9990	.....	.....
6576	.....	ECG5643	.....	SK3570
6577	NTE5554	ECG5554	.....	SK3575
6578	NTE5556	ECG5556	.....	SK3576
6579	.....	.....	.....	SK7C374
6580	.....	.....	.....	SK6746
6581	.....	.....	.....	SK7C377
6582	.....	.....	.....	SK3616
6583	.....	.....	.....	SK3361
6584	NTE328	ECG328	.....	SK3621
6585	NTE54	ECG54	.....	SK3562
6586	.....	.....	.....	SK7406
6587	.....	.....	.....	SK9986
6588	NTE549	ECG549	.....	.....
6589	.....	.....	.....	SK9277
6590	.....	ECG930	.....	SK3696
6591	.....	ECG2345	.....	SK3996
6592	.....	ECG2345	.....	.....

REPL. CODE	NTE PART NO.	ECG PART NO.	RADIO SHACK PART NO.	RCA PART NO.
6593	.....	ECG2346	.....	.....
6594	.....	ECG217	.....	.....
6595	.....	.....	.....	SK7C390
6596	.....	.....	.....	SK3745
6597	NTE40106B	ECG40106B	.....	SK4584B
6598	.....	.....	.....	SK7C393
6599	.....	ECG6204	.....	.....
6600	.....	ECG2356	.....	SK9958
6601	.....	ECG3020	.....	.....
6602	.....	.....	.....	SK3842
6603	NTE558	ECG558	.....	SK3843
6604	.....	.....	.....	SK9639
6605	NTE734	.....	.....	.....
6606	NTE568	ECG568	.....	SK3910
6607	.....	ECG823	.....	.....
6608	.....	ECG375	.....	.....
6609	.....	ECG40194B	.....	SK40194B
6610	.....	.....	.....	SK7CT107
6611	.....	.....	.....	SK7CT109
6612	NTE290A	.....	.....	.....
6613	.....	.....	.....	SK7CT112
6614	.....	.....	.....	SK3116
6615	NTE615P	ECG615A	.....	SK9776
6616	.....	ECG6006	.....	.....
6617	.....	.....	.....	SK7CT123
6618	.....	.....	.....	SK7CT125
6619	.....	.....	.....	SK7CT126
6620	.....	.....	.....	SK9393
6621	.....	.....	.....	SK9053
6622	.....	.....	.....	SK6631
6623	.....	.....	.....	SK7CT132
6624	NTE125	ECG125	276-1114	SK3311
6625	.....	ECG74HC00	.....	.....
6626	.....	ECG3181	.....	.....
6627	.....	.....	.....	SK4311
6628	.....	.....	.....	SK7CT139
6629	.....	ECG4918	.....	.....
6630	.....	ECG1V175	.....	.....
6631	.....	.....	.....	SK4320
6632	.....	.....	.....	SK4325
6633	.....	.....	.....	SK4326
6634	.....	ECG74HC02	.....	.....
6635	.....	ECG6154	.....	.....
6636	.....	ECG6158	.....	.....
6637	.....	ECG74HC04	.....	.....
6638	.....	ECG6008	.....	.....
6639	.....	ECG56017	.....	SK9830
6640	.....	ECG1938	.....	SK7741
6641	.....	ECG4926	.....	.....
6642	.....	ECG74HC109	.....	.....
6643	.....	ECG74LS132	.....	.....
6644	.....	ECG4928	.....	.....
6645	.....	.....	.....	SK7CT147
6646	.....	.....	.....	SK6V0
6647	.....	.....	.....	SK9621
6648	.....	.....	.....	SK4053B
6649	.....	.....	.....	SK7CT151
6650	.....	ECG74HC14	.....	.....
6651	.....	.....	.....	SK7CT153
6652	.....	ECG4934	.....	.....
6653	.....	.....	.....	SK9358
6654	.....	ECG4013B	.....	.....
6655	.....	ECG467	.....	.....
6656	.....	.....	.....	SK7CT157
6657	NTE85	ECG85	276-86535-3	SK31327
6658	.....	.....	.....	SK9982
6659	.....	ECG8316	.....	.....
6660	.....	.....	.....	SK7CT161
6661	.....	.....	.....	SK7CT160
6662	.....	.....	.....	SK9114
6663	.....	ECG6010	.....	.....
6664	.....	.....	.....	SK7CT163
6665	.....	ECG116	.....	SK3311
6666	.....	.....	.....	SK7CT164
6667	.....	.....	.....	SK7CT165
6668	.....	ECG967	.....	.....
6669	.....	.....	.....	SK7CT166
6670	NTE8316	ECG8316	.....	SK74161
6671	NTE1022	ECG1022	.....	SK3481
6672	.....	ECG74HC32	.....	.....
6673	.....	.....	.....	SK7CT173
6674	.....	.....	.....	SK7CT174
6675	.....	.....	.....	SK7CT175
6676	.....	ECG4950	.....	.....
6677	.....	.....	.....	SK4503B
6678	.....	ECG5587	.....	.....
6679	.....	.....	.....	SK7CT181
6680	.....	ECG5327	.....	.....
6681	.....	ECG1191	.....	.....
6682	.....	ECG298	.....	.....
6683	.....	ECG159	.....	SK3118
6684	.....	ECG2358	.....	SK9741
6685	.....	ECG1290	.....	SK9183
6686	.....	.....	.....	SK7CT190
6687	.....	.....	.....	SK7CT191
6688	NTE1139	ECG1139	.....	SK3155
6689	.....	.....	.....	SK7676
6690	.....	.....	.....	SK7CT192
6691	.....	.....	.....	SK7CT193
6692	NTE3092	ECG3092	.....	.....
6693	.....	.....	.....	SK7CT194
6694	.....	ECG4958	.....	.....
6695	NTE198	ECG198	.....	.....

REPL. CODE	NTE PART NO.	ECG PART NO.	RADIO SHACK PART NO.	RCA PART NO.
6696	.....	ECG992	.....	SK3688
6697	NTE331	ECG331	.....	SK9234
6698	.....	ECG987	.....	SK3594
6699	.....	ECG5377	.....	.....
6700	NTE159	ECG159	.....	.....
6701	.....	.....	.....	SK3630
6702	.....	.....	.....	SK3152
6703	NTE123AP	ECG123AP	.....	SK3854
6704	NTE156	ECG156	.....	SK5012
6705	.....	.....	.....	SK3246A
6706	.....	ECG580	.....	SK5032
6707	.....	.....	.....	SK5042
6708	.....	.....	.....	SK5044
6709	.....	ECG199	.....	SK3124A
6710	.....	.....	.....	SK5046
6711	.....	.....	.....	SK5048
6712	.....	.....	.....	SK5050
6713	.....	.....	.....	SK5052
6714	.....	.....	.....	SK5053
6715	.....	.....	.....	SK5054
6716	.....	ECG74S287	.....	.....
6717	.....	.....	.....	SK7679
6718	.....	ECG9301	.....	SK9301
6719	.....	.....	.....	SK5114
6720	.....	.....	.....	SK3636
6721	NTE7052	ECG7052	.....	.....
6722	.....	.....	.....	SK7CT221
6723	NTE199	ECG199	.....	SK3132
6724	NTE5073A	ECG5073A	.....	SK8V7
6725	.....	ECG327	.....	SK3945
6726	.....	ECG957	.....	.....
6727	.....	.....	.....	SK5236
6728	.....	ECG943M	.....	SK9278
6729	NTE6362	ECG6362	.....	SK6561
6730	.....	ECG1940	.....	SK7742
6731	.....	.....	.....	SK3529
6732	.....	.....	.....	SK3089
6733	.....	.....	.....	SK7CT237
6734	.....	.....	.....	SK7CT238
6735	.....	ECG4006B	.....	.....
6736	.....	ECG527	.....	.....
6737	.....	.....	.....	SK7CT241
6738	.....	.....	.....	SK7CT242
6739	.....	.....	.....	SK9110
6740	.....	ECG2380	.....	.....
6741	.....	.....	.....	SK7CT243
6742	.....	ECG379	.....	SK9085
6743	.....	ECG5004A	.....	.....
6744	.....	.....	.....	SK2068
6745	.....	ECG933	.....	SK9341
6746	.....	ECG427	.....	.....
6747	NTE383	ECG383	.....	.....
6748	.....	ECG2312	.....	.....
6749	.....	.....	.....	SK7CT251
6750	.....	.....	.....	SK2069
6751	NTE5024A	ECG5024A	276-564	SK5024A
6752	.....	.....	.....	SK7CT253
6753	.....	ECG923	.....	SK9036
6754	.....	.....	.....	SK7CT257
6755	.....	.....	.....	SK7CT258
6756	NTE5022A	ECG5022A	.....	SK14A
6757	.....	.....	.....	SK7CT259
6758	.....	.....	.....	SK9455A
6759	.....	.....	.....	SK9089
6760	.....	ECG3088	.....	.....
6761	NTE5160A N3	ECG5160A N3	.....	SK140X N3
6762	.....	ECG74HC4067	.....	SK7C4067
6763	.....	.....	.....	SK9012
6764	NTE253MCP	.....	.....	.....
6765	.....	.....	.....	SK7C573
6766	.....	.....	.....	SK3884
6767	.....	.....	.....	SK7C574
6768	.....	ECG128	.....	SK3024
6769	.....	ECG5667A	.....	SK3508
6770	.....	ECG4902	.....	.....
6771	.....	.....	.....	SK3867A
6772	.....	ECG955M	.....	SK3693
6773	.....	.....	.....	SK3434
6774	.....	ECG465	.....	SK9158
6775	.....	.....	.....	SK5704
6776	NTE2393	ECG2393	.....	.....
6777	.....	.....	.....	SK9410
6778	NTE552	ECG552	.....	SK3130
6779	.....	.....	.....	SK7CT280
6780	.....	ECG5675	.....	SK3508
6781	.....	.....	.....	SK10098
6782	.....	.....	.....	SK7CT283
6783	NTE783	ECG783	.....	SK5A6
6784	.....	.....	.....	SK7628
6785	.....	.....	.....	SK3072
6786	.....	.....	.....	.....
6787	.....	ECG5681	.....	SK6681
6788	NTE2369	.....	.....	.....
6789	NTE1230	ECG1230	.....	SK3444
6790	NTE1V150	ECG1V150	.....	SKMV150G
6791	.....	ECG832	.....	.....
6792	.....	.....	.....	SK9163
6793	.....	.....	.....	SK7CT297
6794	.....	.....	.....	SK7CT299
6795	NTE930#	ECG930#	.....	SK3568
6796	NTE1767	ECG1767	.....	.....
6797	.....	ECG4046B	.....	SK4046B
6798	NTE7214	ECG7214	.....	SK74LS253

REPL. CODE	NTE PART NO.	ECG PART NO.	RADIO SHACK PART NO.	RCA PART NO.
6799	.....	.....	.....	SK4020B
6800	.....	ECG1619	.....	.....
6801	.....	ECG5146A	.....	.....
6802	.....	ECG611	.....	.....
6803	.....	.....	.....	SK9025
6804	.....	.....	.....	SK6117
6805	.....	.....	.....	SK6118
6806	.....	ECG544	.....	.....
6807	.....	.....	.....	SK5556
6808	.....	.....	.....	SK9934
6809	.....	ECG7015	.....	.....
6810	.....	ECG1874	.....	.....
6811	.....	.....	.....	SK3066
6812	.....	.....	.....	SK68X
6813	.....	ECG5019A	.....	.....
6814	NTE198	.....	.....	.....
6815	NTE1960	ECG1960	.....	.....
6816	.....	.....	.....	SKMV275B
6817	NTE1966	ECG1966	.....	.....
6818	NTE85	ECG85	276-2009	SK3132
6819	.....	.....	.....	SK3195
6820	.....	.....	.....	SK7616
6821	.....	.....	.....	SK7CT365
6822	.....	ECG525	.....	.....
6823	.....	.....	.....	SK7701
6824	.....	.....	.....	SK7CT366
6825	.....	.....	.....	SK7CT367
6826	.....	.....	.....	SK7CT368
6827	NTE5552	ECG5552	.....	SK6651
6828	NTE290A	ECG290A	276-2026	SK3114A
6829	.....	ECG282&427	.....	SK9368
6830	.....	.....	.....	SK9121
6831	.....	.....	.....	SK7CT377
6832	.....	ECG2418	.....	SK10108
6833	NTE5608	ECG5608	.....	SK6707
6834	.....	.....	.....	SK6788
6835	.....	ECG2360	.....	SK9960
6836	NTE778A	ECG778A	.....	SK3465
6837	.....	.....	.....	SK7CT390
6838	NTE4001B	ECG4001B	276-2041	SK4001B
6839	.....	ECG5010T1	.....	.....
6840	.....	.....	.....	SK7CT393
6841	.....	ECG1766	.....	.....
6842	.....	.....	.....	SK1922
6843	.....	.....	.....	SK9500
6844	NTE289A	ECG289A	.....	SK3214A
6845	.....	ECG753	.....	SK7226
6846	.....	ECG5567	.....	.....
6847	.....	ECG5568	.....	.....
6848	.....	ECG503	.....	SK3068
6849	.....	.....	.....	SK9V1
6850	NTE293	ECG293	.....	.....
6851	.....	.....	.....	SK3487
6852	NTE153	ECG152	276-2020	SK3893
6853	NTE382	ECG382	276-2009	SK9137
6854	.....	.....	.....	SK7621
6855	.....	ECG1569	.....	.....
6856	.....	ECG2316	.....	.....
6857	.....	ECG5159A	.....	SK130X
6858	.....	ECG2308	.....	.....
6859	.....	ECG96L02	.....	.....
6860	.....	ECG5317	.....	.....
6861	.....	ECG5427	.....	SK3577
6862	NTE5348	ECG5348	.....	SK7301
6863	.....	.....	.....	SK7305
6864	.....	.....	.....	SK7307
6865	.....	ECG1620	.....	.....
6866	.....	ECG5019T1	.....	.....
6867	NTE177	ECG177	.....	SK9091
6868	.....	ECG1234	.....	.....
6869	.....	.....	.....	SK7317
6870	.....	ECG5428	.....	SK3578
6871	.....	ECG5429	.....	.....
6872	.....	.....	276-143	.....
6873	.....	.....	.....	SK9138
6874	.....	.....	.....	SK2027A
6875	.....	.....	.....	SK3574
6876	.....	ECG55	.....	SK9367
6877	.....	ECG5013T1	.....	.....
6878	.....	ECG963	.....	.....
6879	NTE174	.....	.....	SK3017B
6880	NTE5159A N3	ECG5159A N3	.....	SK130X N3
6881	NTE5162A N3	ECG5162A N3	.....	SK160X N3
6882	NTE5163A N3	ECG5163A N3	.....	SK170X N3
6883	NTE5165A N3	ECG5165A N3	.....	SK190X N3
6884	NTE5166A N3	ECG5166A N3	.....	SK200X N3
6885	.....	ECG284	.....	SK3079
6886	.....	ECG2418	.....	.....
6887	.....	.....	.....	SK9867
6888	.....	.....	.....	SK7613
6889	NTE3098	ECG3081	.....	SK9965
6890	.....	ECG6087	.....	.....
6891	.....	ECG159	.....	SK7664
6892	.....	ECG1521	.....	SK7671
6893	.....	.....	.....	SK7680
6894	.....	.....	.....	SK7699
6895	.....	.....	.....	SK7702
6896	.....	.....	.....	SK7709
6897	.....	.....	.....	SK3689
6898	.....	.....	.....	SK7738
6899	.....	ECG4900	.....	.....
6900	.....	ECG5461	.....	.....
6901	NTE1421	ECG1421	.....	SK7792
6902	.....	.....	.....	SK74LS164
6903	.....	.....	.....	SK3610
6904	.....	.....	.....	SK7CT00
6905	.....	.....	.....	SK3609
6906	.....	ECG5009A	.....	.....
6907	.....	.....	.....	SK4V3
6908	NTE1281	ECG1281	.....	.....
6909	.....	ECG81	.....	SK9114
6910	.....	.....	.....	SK9884
6911	.....	ECG2341	.....	.....
6912	.....	.....	.....	SK9918
6913	.....	.....	.....	SK7951
6914	.....	.....	.....	SK7794
6915	.....	ECG74HC125	.....	.....
6916	.....	ECG74HC126	.....	.....
6917	.....	ECG74HC132	.....	.....
6918	.....	ECG74HC138	.....	.....
6919	.....	ECG74HC139	.....	.....
6920	.....	ECG74HC151	.....	.....
6921	.....	ECG74HC153	.....	.....
6922	.....	ECG74HC154	.....	.....
6923	.....	ECG74HC163	.....	.....
6924	.....	ECG74HC164	.....	.....
6925	.....	ECG74HC165	.....	.....
6926	.....	ECG74HC173	.....	.....
6927	.....	ECG977	.....	.....
6928	.....	ECG74HC174	.....	.....
6929	.....	ECG74HC175	.....	.....
6930	.....	.....	.....	SK7C02
6931	.....	.....	.....	SK4040B
6932	.....	ECG74HC244	.....	.....
6933	.....	ECG74HC257	.....	.....
6934	.....	ECG74HC259	.....	.....
6935	.....	ECG74HC273	.....	.....
6936	.....	.....	.....	SK9157
6937	NTE584	ECG584	.....	SK9091
6938	.....	ECG74HC299	.....	.....
6939	.....	ECG123AP	.....	SK3024
6940	.....	.....	.....	SK3896
6941	NTE1199	ECG1199	.....	SK9339
6942	.....	ECG5474	.....	SK3943
6943	.....	ECG74HC373	.....	.....
6944	.....	ECG74HC374	.....	.....
6945	.....	ECG74HC377	.....	.....
6946	.....	ECG74HC390	.....	.....
6947	.....	ECG74HC393	.....	.....
6948	.....	.....	.....	SK9153
6949	NTE5534	ECG5534	.....	SK6502
6950	.....	.....	.....	SK9294
6951	.....	ECG74HC10	.....	.....
6952	.....	ECG524V17	.....	.....
6953	.....	.....	.....	SK17A
6954	.....	.....	.....	SK7CT540
6955	.....	.....	.....	SK7CT541
6956	.....	ECG518	.....	.....
6957	NTE282	ECG282	.....	SK3104A
6958	.....	ECG74HC573	.....	.....
6959	.....	ECG74HC574	.....	.....
6960	NTE5010T1	ECG5010T1	.....	.....
6961	.....	ECG182	.....	.....
6962	.....	ECG968	.....	SK3593
6963	.....	ECG2003	.....	.....
6964	.....	.....	.....	SK2V8
6965	.....	.....	.....	SK7CT563
6966	.....	.....	.....	SK7CT564
6967	NTE195A	ECG195A	.....	SK3124A
6968	NTE195A	ECG195A	.....	SK3004
6969	.....	.....	.....	SK7704
6970	.....	.....	.....	SK4046B
6971	.....	.....	.....	SK7CT573
6972	.....	.....	.....	SK9009
6973	.....	.....	.....	SK4068B
6974	NTE1398	ECG1398	.....	.....
6975	.....	.....	.....	SK7789
6976	.....	.....	.....	SK9075
6977	.....	ECG5702	.....	.....
6978	.....	ECG5703	.....	.....
6979	.....	ECG312	.....	SK3116
6980	.....	ECG5704	.....	.....
6981	.....	.....	.....	SK7CT583
6982	.....	ECG5705	.....	.....
6983	.....	ECG229	.....	.....
6984	NTE46	ECG46	.....	SK9455A
6985	NTE583	ECG583	.....	SK3311
6986	.....	.....	.....	SK3279
6987	.....	.....	.....	SK7CT597
6988	.....	ECG2303	.....	.....
6989	.....	.....	.....	SK179
6990	.....	ECG74HCT138	.....	SK7CT138
6991	.....	.....	.....	SK10244
6992	.....	ECG6809	.....	.....
6993	.....	.....	.....	SK3A0
6994	.....	ECG284	.....	SK9031
6995	.....	ECG285	.....	SK9032
6996	NTE615A	ECG615A	.....	SK9976
6997	.....	.....	.....	SK9057
6998	.....	.....	.....	SK9060
6999	.....	.....	.....	SK9065
7000	.....	.....	.....	SK3866A
7001	NTE804	ECG804	.....	SK9068
7002	.....	ECG744	.....	.....
7003	NTE5344	ECG5344	.....	SK9104
7004	.....	ECG241	.....	SK9116
7005	.....	ECG242	.....	SK9117
7006	.....	ECG5814	.....	.....
7007	.....	ECG5817	.....	.....
7008	.....	.....	.....	SK9152
7009	.....	.....	.....	SK9154
7010	.....	.....	.....	SK3717
7011	.....	ECG109	.....	SK3087
7012	.....	.....	.....	SK110V
7013	.....	.....	.....	SK74LS107
7014	.....	ECG51	.....	SK9263
7015	.....	ECG8546	.....	.....
7016	.....	.....	.....	SK9278
7017	.....	.....	.....	SK9279
7018	.....	.....	.....	SK9280
7019	.....	ECG1289	.....	SK9282
7020	.....	.....	.....	SK9288
7021	.....	.....	.....	SK9289
7022	.....	.....	.....	SK9301
7023	.....	.....	.....	SK9302
7024	.....	ECG376	.....	.....
7025	.....	.....	.....	SK9316
7026	.....	.....	.....	SK74LS191
7027	.....	.....	.....	SK9318
7028	.....	.....	.....	SK9319
7029	.....	.....	.....	SK7CT640
7030	.....	ECG941S	.....	SK9323
7031	.....	.....	.....	SK9345
7032	.....	.....	.....	SK9346
7033	.....	.....	.....	SK9347
7034	.....	.....	.....	SK9348
7035	.....	.....	.....	SK9350
7036	.....	.....	.....	SK7CT643
7037	.....	.....	.....	SK9958
7038	.....	.....	.....	SK74LS243
7039	.....	.....	.....	SK7CT646
7040	NTE5344	ECG5344	.....	SK9388
7041	.....	ECG506	.....	SK3130
7042	NTE1471	ECG1471	.....	SK9393
7043	.....	ECG85	.....	SK9407
7044	.....	ECG75492B	.....	.....
7045	.....	ECG943M	.....	SK9993
7046	.....	ECG6002	.....	SK7202
7047	.....	.....	.....	SK9445
7048	.....	ECG24	.....	SK9453
7049	.....	ECG25	.....	SK9454
7050	.....	.....	.....	SK74LS27
7051	.....	.....	.....	SK9463
7052	.....	.....	.....	SK9465
7053	.....	.....	.....	SK9466
7054	NTE227	ECG227	.....	SK9473
7055	.....	.....	.....	SK9474
7056	.....	.....	.....	SK9486
7057	.....	.....	.....	SK74LS30
7058	.....	ECG2382	.....	SK9502
7059	.....	ECG2383	.....	SK9506
7060	NTE703	ECG703	.....	SK3708
7061	.....	ECG853	.....	.....
7062	.....	.....	.....	SK100A
7063	.....	ECG1903	.....	.....
7064	NTE195A	ECG195A	.....	SK3122
7065	.....	.....	.....	SK3235
7066	.....	.....	.....	SK8A7
7067	NTE335	ECG335	.....	SK9608
7068	.....	.....	.....	SK74LS123
7069	.....	.....	.....	SK7CT670
7070	NTE346	ECG346	.....	SK9622
7071	.....	ECG74HCT161	.....	SK7CT161
7072	.....	.....	.....	SK7CT162
7073	.....	ECG74HCT163	.....	SK7CT163
7074	.....	ECG74HCT174	.....	SK7CT174
7075	.....	.....	.....	SK2150
7076	.....	.....	.....	SK9672
7077	NTE102&103	ECG102&103	.....	SK3722&3862
7078	.....	.....	.....	SK2163
7079	.....	.....	.....	SK2166
7080	.....	ECG74HCT240	.....	SK7CT240
7081	.....	ECG74HCT273	.....	SK7CT273
7082	NTE5405	ECG5405	.....	SK3951
7083	.....	ECG797	.....	SK3158
7084	.....	ECG5989	.....	.....
7085	.....	ECG5988	.....	.....
7086	.....	ECG5590	.....	.....
7087	.....	ECG74S475	.....	.....
7088	.....	ECG74HC7374	.....	SK7CT374
7089	.....	.....	.....	SK7208
7090	.....	.....	.....	SK3986
7091	.....	.....	.....	SK3129
7092	NTE53	ECG53	.....	SK9374
7093	.....	.....	.....	SK74LS75
7094	.....	.....	.....	SK7357
7095	NTE5635	ECG5635	276-1000	SK3533
7096	.....	.....	.....	SK9482
7097	.....	ECG74HCT573	.....	SK7CT573
7098	.....	ECG74HCT574	.....	SK7CT574
7099	NTE5465	ECG5465	276-1020	SK9293
7100	.....	.....	.....	SK9090
7101	.....	.....	.....	SK3763
7102	NTE5463	ECG5463	276-1067	SK9292
7103	.....	ECG382	.....	.....
7104	.....	ECG1401	.....	.....
7105	.....	.....	.....	SK68A
7106	NTE5312	ECG5312	276-1146	SK3985
7107	NTE5312	ECG5312	276-1171	SK3985

REPL. CODE	NTE PART NO.	ECG PART NO.	RADIO SHACK PART NO.	RCA PART NO.
7108	NTE5314	ECG5314	276-1173	SK9387
7109	NTE5312	ECG5312	276-1181	SK3985
7110	.....	.....	.....	SK2882
7111	.....	ECG2385	.....	SK9509
7112	.....	ECG74LS366A	.....	.....
7113	.....	.....	.....	SK3543
7114	.....	ECG2319	.....	SK9374
7115	.....	.....	.....	SK3081
7116	.....	.....	.....	SK3131A
7117	.....	.....	.....	SK3143
7118	NTE32	ECG32	.....	.....
7119	.....	ECG108	.....	SK3452
7120	NTE291	ECG291	.....	.....
7121	.....	ECG961	.....	.....
7122	.....	.....	.....	SK7CT244
7123	NTE604	.....	.....	SK3004
7124	NTE5534	ECG5534	.....	SK3581
7125	.....	.....	.....	SK3463
7126	.....	.....	.....	SK110A

REPL. CODE	NTE PART NO.	ECG PART NO.	RADIO SHACK PART NO.	RCA PART NO.
7127	.....	.....	.....	SK3581
7128	.....	.....	.....	SK3582
7129	.....	ECG2340	.....	.....
7130	.....	.....	.....	SK74LS221
7131	.....	ECG860	.....	SK7731
7132	.....	ECG5176AK	.....	SK119
7133	.....	.....	.....	SK2161
7134	.....	.....	.....	SK2151
7135	.....	ECG2386	.....	.....
7136	.....	.....	.....	SK2167
7137	.....	ECG613	.....	SK3326
7138	.....	ECG5481	.....	SK3940
7139	.....	.....	.....	SK9381
7140	.....	ECG782	.....	SK3077
7141	NTE1045	ECG1045	.....	SK3102
7142	.....	ECG2076	.....	.....
7143	.....	ECG6502	.....	.....
7144	.....	ECG943M	.....	.....
7145	.....	ECG2428	.....	SK9963

REPL. CODE	NTE PART NO.	ECG PART NO.	RADIO SHACK PART NO.	RCA PART NO.
7146	.....	.....	.....	SK9227
7147	.....	ECG89	.....	SK9119
7148	.....	.....	.....	SK7481
7149	NTE2331	ECG2331	.....	SK10098
7150	.....	.....	.....	SK3031A
7151	NTE5009A	ECG5009A	.....	SK4A7
7152	.....	.....	.....	SK9247
7153	NTE552	ECG506	.....	SK3925
7154	.....	.....	.....	SK3329
7155	.....	.....	276-068	.....
7156	.....	.....	276-069	.....
7157	NTE74LS112A	ECG74LS112A	.....	SK7CT112
7158	.....	.....	276-113	.....
7159	.....	.....	276-142	.....
7160	.....	ECG5550	.....	SK6650
7161	.....	.....	.....	SK3575
7162	.....	.....	.....	SK3576
7163	.....	ECG5043A	.....	SK60A

PROMPT® Publications

Surface-Mount Technology for PC Boards

by James K. Hollomon, Jr.

Manufacturers, managers, engineers, and others who work with printed-circuit boards will find a wealth of information about surface-mount technology (SMT) and fine-pitch technology (FPT) in this book. Practical data and clear illustrations plainly present the details of design-for-manufacturability, environmental compliance, design-for-test, and quality/reliability for today's miniaturized electronics packaging.

$26.95
Paper/309 pp./7 x 10"/Illustrated
ISBN#: 0-7906-1060-4
Pub. Date 7/95

Semiconductor Essentials

by Stephen Kamichik

Gain hands-on knowledge of semiconductor diodes and transistors with help from the information in this book. *Semiconductor Essentials* is a first course in electronics at the technical and engineering levels. Each chapter is a lesson in electronics, with problems presented at the end of the chapter to test your understanding of the material presented. This generously illustrated manual is a useful instructional tool for the student and hobbyist, as well as a practical review for professional technicians and engineers.

$16.95
Paper/112 pp./6 x 9"/Illustrated
ISBN#: 0-7906-1071-X
Pub. Date 9/95

Call for the complete list of new releases from PROMPT®. 800-428-7267